RECENT HANDBOOK PUBLICATIONS

HANDBOOK OF ANTIBIOTIC COMPOUNDS
János Bérdy
Senior Research Fellow
Institute of Drug Research
Budapest, Hungary

HANDBOOK OF CHROMATOGRAPHY: AMINO ACIDS AND AMINES
Stanley Blackburn
Research Scientist
Wool Industries Research Association
Leeds, England

HANDBOOK OF CLINICAL CHEMISTRY
Mario Werner
Professor of Pathology
(Laboratory Medicine)
The George Washington University Medical Center
Washington, D.C.

HANDBOOK OF ENVIRONMENTAL RADIATION
Alfred W. Klement, Jr.
Environmental Scientist (deceased)
Private Consultant
Kensington, Maryland

HANDBOOK OF GEOPHYSICAL EXPLORATION AT SEA
Richard A. Geyer
Professor Emeritus
Department of Oceanography
Texas A & M University
College Station, Texas

HANDBOOK OF IRRIGATION TECHNOLOGY
Herman J. Finkel
Director
Finkel & Finkel, Ltd.
Consulting Engineers
Haifa, Israel

HANDBOOK OF LASER SCIENCE AND TECHNOLOGY
Marvin J. Weber
Laser Program
Lawrence Livermore National Laboratory
University of California
Livermore, California

HANDBOOK OF LETHALITY GUIDES FOR LOW-ACID CANNED FOODS
C. R. Stumbo **D. A. Evans**
K. S. Purohit **F. J. Francis**
T. V. Ramakrishnan
Department of Food Science and Nutrition
University of Massachusetts
Amherst, Massachusetts

HANDBOOK OF LUBRICATION
E. R. Booser
Manager
Systems Engineering Subsection
Turbine Technology Laboratory
General Electric Company
Schenectady, New York

HANDBOOK OF PROCESSING AND UTILIZATION IN AGRICULTURE
Ivan A. Wolff
Director (retired)
Eastern Regional Research Center
Agricultural Research Service
U.S. Department of Agriculture
Philadelphia, Pennsylvania

HANDBOOK OF TERPENOIDS: MONOTERPENOIDS
Sukh Dev
Multi-Chem Research Centre
Nandesari, Vadodara, India
J. S. Yadav
National Chemical Laboratory
Division of Organic Chemistry
Poona, India
Anubhav Narula
Lederle Laboratory
American Cyanamid Co.
Medical Research Div.
Pearl River, New York

CRC Handbook of Chemistry and Physics

A Ready-Reference Book of Chemical and Physical Data

HANDBOOK OF CHEMISTRY AND PHYSICS
1986—1987
67th
EDITION
CRC PRESS, INC.

Editor-in-Chief

Robert C. Weast, Ph.D.

Associate Editors

Melvin J. Astle, Ph.D.
William H. Beyer, Ph.D.

In collaboration with a large number of professional chemists and physicists whose assistance is acknowledged in the list of general collaborators and in connection with the particular tables or sections involved.

CRC

CRC Press, Inc.
Boca Raton, Florida

Library of Congress Card No. 13-11056
PRINTED IN U.S.A.
ISBN-0-8493-0467-9

PREFACE

Seventy-two years ago the *Handbook of Chemistry and Physics* was first published. Its editor was Professor William R. Veazey of Case School of Applied Science. Later he was an outstanding scientist, administrator, and officer of the Dow Chemical Company. The publisher and originator of the concept of a single-volume reference book to contain data for chemists and physicists as well as containing certain frequently used mathematical tables for these scientific disciplines was Arthur Friedman, a 1907 graduate from Case. His company, the Chemical Rubber Company, sold rubber tubing, stoppers, aprons and laboratory supplies primarily to colleges and high schools; there were very few industrial laboratories in America prior to World War I. A reference book as the *Handbook of Chemistry and Physics* was a natural complement to those products being sold by the Chemical Rubber Company.

Over the past 72 years the book has been published continuously and has been revised 66 times. For several reasons during World Wars I and II the book was not revised annually. From 1915 through 1951 the book was edited by Professor Charles D. Hodgman of the Physics Department of Case School of Applied Science, a school which was later called Case Institute of Technology, and still later after the affiliation with Western Reserve University as Case Western Reserve University. Subsequent to Professor Hodgman's retirement the *Handbook of Chemistry and Physics* has been edited by the present editor who was a professor at Case Institute of Technology when he became editor.

The existance and continuing growth of the book since 1914 are the result of many factors. Certainly the most important one is the regular input from many scientists who provide data, information, advice, guidance and constructive criticism. It is these people who assure the timeliness and quality of data in this reference book. The need for such help is never-ending. Thus, the editor continues to welcome and appreciate this help. Through such help the *Handbook of Chemistry and Physics* will continue its policy of the past 72 years of being revised at frequent intervals, usually annually.

The widespread use, great power, and relatively low cost of small computers and hand-held calculators in recent years has justified removal of many of the mathematical tables which were in the earlier editions of the *Handbook of Chemistry and Physics*. Nevertheless, in other sections of the book, the general features and scheme of arrangement, which have received extensive endorsement in previous editions have been retained in this present edition, the 67th Edition.

Throughout the years there have been a number of goals regarding the preparation and publishing of the *Handbook of Chemistry and Physics*. Two of these were to present in a condensed form a large number of accurate up-to-date data in the fields of chemistry and physics consistent in content and form, and to try to assure wide utility of the data. From time to time, particularly in the more recent editions, there have been exceptions regarding the meeting of the goal to keep all tables condensed and small by application of the usual procedure of abridging extensively all the data available to the editor on a specific area of chemistry or physics. There were instances when it was not possible to select which data could be deleted without impairing the table significantly. Among these were the data in the Table of Isotopes, a table which was prepared by Mr. Russell Heath for the 1985 edition. Data in this table occupy 227 pages in the book. The table of the Line Spectra of the Elements, a table which was originally prepared under the leadership of Joseph Reader and Charles Corliss, the latter of whom is now retired, of the National Bureau of Standards requires 127 pages to present all of the data. Although this latter table has been in the *Handbook of Chemistry and Physics* for several years is frequently updated by Dr. Reader and his colleagues. The extensive section, The Elements, which is prepared and updated annually by Mr. C. R. Hammond is one containing information which is apparently used by many people. This is evidenced by the number of complimentary comments and letters received by the editor regarding this section of the *Handbook*.

The editor attempts to include material which has a high probability to find extended use in many branches of chemistry and physics and the closely allied sciences. On the other hand, in order to retain the convenience of moderate dimensions and at the same time to allow for growth of the *Handbook* due to the extension of knowledge in these sciences, and logical additions along lines already developed, it has seemed necessary to exclude types of material of use only in highly specialized lines of work. It is the selection of data and information to be excluded which presents one of the major decisions to the editor.

Chemistry and physics, always two closely related sciences, have been brought into more intimate relations by recent developments in research and our increasing understanding of matter and energy. To a growing extent people in either science are learning more of the other. One of the goals of the editor and the publisher is to provide a reference book which will assist in providing certain information to further this understanding.

Among the new and revised information appearing in this 67th edition are the following: (1) the molecular weights of all of the approximate 15,000 compounds presently in the Table of Physical Constants of Organic Compounds were recalculated when using the most recently approved IUPAC atomic weights. While performing these recalculations certain errors became apparent in some of the empirical and line formulas of certain of the compounds listed in the table. These were corrected; (2) the table, Physical Constants of Inorganic Compounds, has been completely reset to improve its legibility; (3) Physical and Photometric Data for Planets and Satellites. These tables include data on Opposition Dates, Magnitudes and Oscillating Elements for Epoch 1986 June 19·0 TDT, Ecliptic and Equinox J2000·0 for the minor planets, as well a physical and photometric data for the major planets; (3) revision and update of data on Dissociation Constants of Inorganic Bases; (3) addition of a new table representing data on the Binding Energies of Atomic Negative Ions; (4) limits of Superheat of Pure Liquids. This is a table which contains data representing the deepest penetration of a liquid into the domain of metastable states. The practical significance of this limit resides in the consequences of the phase transition that eventually occurs when this limit is reached; (5) Fine-Structure Separations in Atomic Negative Ions. This table presents data on electron affinities and is an update and revision of information in the previous edition of the *Handbook;* (6) Recommended Daily Dietary Allowances for Humans; and (7) Nutritive Value of the Edible Parts of 908 Foods. This table contains information on food energy, protein, fatty acid, mineral, vitamin, etc. contents of each of the foods listed in the table. This table and the one listed immediately above were in response to the present emphasis on foods and diets and should be of use to those many persons who are making efforts to improve their diets.

All errors called to our attention have, to the best of the editor's knowledge, been corrected. The editor and the publisher are most appreciative to all who have contacted us to help in the continuing improvement of the *Handbook of Chemistry and Physics*. We are confident that such help and cooperation are also appreciated by all users of this book. In keeping with a long established practice, the editor replies to all who write to either the publisher or to the editor.

Robert C. Weast
March 10, 1986

PHYSICS EDITORIAL BOARD

COLLABORATORS AND CONTRIBUTORS

William E. Acree, Jr., Ph.D.
Department of Chemistry
Kent State University
Kent, Ohio 44242

L. Alberts, Ph.D.
Director-General
National Institute for Metallurgy
Randburg, South Africa

C. J. Allen
Illuminating Engineer
General Electric Company
Nela Park
Cleveland, Ohio 44112

W. A. Anderson, Ph.D.
Analytical Instrument Research
Varian Associates
611 Hansen Way
Palo Alto, California

W. J. Armento
Oak Ridge National Laboratory
P.O. Box X
Oak Ridge, Tennessee

J. Askill, Ph.D.
Chairman, Physics Department
Millikin University
Decatur, Illinois

C. Thomas Avedisian, Ph.D.
Sibley School of Mechanical and Aerospace Engineering
Upson and Grumman Halls
Cornell University
Ithaca, New York 14853-7501

J. C. Bailar, Jr., Ph.D.
Professor of Inorganic Chemistry
University of Illinois
Urbana, Illinois

J. A. Bearden, Ph.D.
Department of Physics
Johns Hopkins University
Baltimore, Maryland

A. H. Benade, Ph.D.
Department of Physics
Case Western Reserve University
Cleveland, Ohio

F. F. Bentley
Air Force Materials Laboratory
Wright-Patterson Air Force Base, Ohio

William H. Beyer, Ph.D.
Professor of Mathematics and Statistics
Head of the Department of Mathematics and Statistics
University of Akron
Akron, Ohio

J. H. Billman, Ph.D.
Professor of Chemistry
Indiana University
Bloomington, Indiana

A. L. Bloom, Ph.D.
Spectra-Physics, Inc.
1255 Terra Bella Avenue
Mountain View, California

J. A. Bradley, Ph.D.
Dean, Newark College of Engineering
323 High Street
Newark, New Jersey

B. H. Brown, Ph.D.
Dartmouth College
Hanover, New Hampshire

J. E. Brown
Eastman Kodak Company
Rochester, New York

A. B. Burg, Ph.D.
Department of Chemistry
University of Southern California
Los Angeles, Californai

G. Preston Burns, Ph.D.
Staff Mathematician
Sperry Corporation
Fredericksburg, Virginia 22401

COLLABORATORS AND CONTRIBUTORS (continued)

A. F. Burr, Ph.D.
Department of Physics
New Mexico State University
Las Cruces, New Mexico

L. P. Buseth
Grevinnev. 14H
N-3100 Tønsberg, Norway

J. P. Catchpole, Ph.D.
Admiralty Materials Laboratory
Holton Heath
Dorset, England

James C. Chang, Ph.D.
Department of Chemistry
University of Northern Iowa
Cedar Falls, Iowa

Roger Clay
Department of Physics
The University of Adelaide
Adelaide, South Australia 5001
Australia

E. Richard Cohen, Ph.D.
Associate Director
Science Center/Aerospace and Systems Group
North American Rockwell Corporation
Thousand Oaks, California 91360

Charles H. Corliss
Spectroscopy Section
Optical Physics Division
National Bureau of Standards
Washington, D.C. 20234

M. Davies, Ph.D.
Edward Davies Chemical Laboratories
University of Wales
Aberystwyth, Wales

J. DeMent, D.Sc.
Dement Laboratories
4847 S.E. Division Street
Portland, Oregon

H. G. Deming, Ph.D.
2316 Tuttle Terrace
Sarasota, Florida

E. DiCyan, Ph.D.
Consulting Chemist
420 Lexington Ave.
New York, New York

Hans Dolezalek
1812 Drury Lane
Alexandria, Virginia 22307

A. P. Dunlop, Ph.D.
Director of Chemical Research and Development
John Stuart Research Laboratories
The Quaker Oats Company
Barrington, Illinois

Robert W. Dykstra, Ph.D.
Research and Development Department
Union Carbide Corporation
South Charleston, West Virginia

L. M. Foster, Ph.D.
Thomas J. Watson Research Center
International Business Machines Corp.
Yorktown Heights, New York

J. L. Franklin, Ph.D.
Welch Professor of Chemistry
William Marsh Rice University
Houston, Texas

G. Fulford, Ph.D.
Assistant Professor of Chemical Engineering
University of Waterloo
Waterloo, Ontario,
Canada

Gladys H. Fuller
NBS Institute of Science and Technology
National Bureau of Standards
Washington, D.C.

COLLABORATORS AND CONTRIBUTORS (continued)

E. F. Furtsch, Ph.D.
Department of Chemistry
Virginia Polytechnic Institute
Blacksburg, Virginia

L. A. Gillette, Ph.D.
Manager, Product Development
Pennsalt Chemical Corporation
Philadelphia, Pennsylvania

H. Gilman, Ph.D.
Department of Chemistry
Iowa State University
Ames, Iowa

B. Girling, M.Sc., F.I.M.A.
Department of Mathematics
The City University
London E.C. 1, England

A. G. Gregory
Department of Physics
The University of Adelaide
Adelaide, South Australia 5001
Australia

R. R. Gupta, M.Sc., Ph.D.
Department of Chemistry
University of Rajasthan
Jaipur-4, India

C. R. Hammond
17 Greystone
West Hartford, Connecticut 06107

W. E. Harris
Professor of Chemistry
University of Alberta
Edmonton, Alberta, Canada

H. J. Harwood, Ph.D.
Head, Durkee Famous Foods
Organic Chemistry Division
Glidden Company
Chicago, Illinois

R. L. Heath
Idaho National Engineering Lab
P.O. Box 1625
Idaho Falls, Idaho 83415

R. W. Hoffman, Ph.D.
Department of Physics
Case Western Reserve University
Cleveland, Ohio

Jesse F. Hunsberger
RD #1
East Cedarville Road
Pottstown, Pennsylvania

C. D. Hurd, Ph.D.
Chemical Laboratory
Northwestern University
Evanston, Illinois

H. D. B. Jenkins, Ph.D.
Department of Molecular Sciences
University of Warwick
Coventry CV4 7AL
England

J. L. Kassner, Ph.D.
Department of Physics
The University of Missouri-Rolla
Rolla, Missouri

Sidney O. Kastner, Ph.D.
National Aeronautics and Space
Administration
Goddard Space Flight Center
Greenbelt, Maryland

Olga Kennard, Ph.D.
University Chemical Laboratory
Cambridge, England

T. G. Kennard, Ph.D.
20747 E. Palm Drive
Glendora, California

J. A. Kerr, Ph.D.
Chemistry Department
The University of Birmingham
Haworth Building
P.O. Box 363
Birmingham B15 2TT, England

A. L. King, Ph.D.
Department of Physics
Dartmouth College
Hanover, New Hampshire

COLLABORATORS AND CONTRIBUTORS (continued)

C. R. Kinney, Ph.D.
1318 27th Street
Des Moines, Iowa

R. Kretz, Ph.D.
Department of Science and Engineering
University of Ottawa
Ottawa, Ontario, Canada K1N 6N5

G. Lang, Dipl. Ing.
A-4963
St. Peter am Hart
Austria

D. F. Lawden, Sc.D.
Department of Mathematics
University of Aston
Birmingham, England

K. Lee, Ph.D.
IBM Research Laboratory
San Jose, California

A. P. Levitt
75 Lovett Road
Newton Center, Massachusetts

M. M. MacMasters, Ph.D.
Department of Grain Science and Industry
Kansas State University
Manhattan, Kansas

George G. Maher, Ph.D.
Research Chemist
United States Department of Agriculture
Northern Regional Research Center
1815 North University Street
Peoria, Illinois 61604

W. Mahlig
Assistant Sales Manager
Laboratory Equipment Division
The W. S. Tyler Company
Cleveland, Ohio

C. J. Major, Ph.D.
Department of Chemical Engineering
University of Akron
Akron, Ohio

Georgia A. Martin, Ph.D.
Atomic and Plasma Radiation Division
National Bureau of Standards
Washington, D.C. 20234

J. C. McGowan, Ph.D., D.Sc.
''Quantock''
13 Moreton Avenue
Harpenden, Herts Al5 2EU, England

L. Meites, Ph.D.
Department of Chemistry
George Mason University
4400 University Drive
Fairfax, Virginia 22030

M. G. Mellon, Ph.D.
Professor Emeritus, Analytical Chemistry
Purdue University
West Lafayette, Indiana

H. B. Michaelson
IBM Journal of Research and Development
Armonk, New York

Thomas M. Miller
Department of Physics and Astronomy
44 West Brooks, Room 131
University of Oklahoma
Norman, Oklahoma 73019

Karl Z. Morgan, Ph.D.
School of Nuclear Engineering
Georgia Institute of Technology
Atlanta, Georgia

R. R. Nimmo, Ph.D.
Professor of Physics
University of Otago
Dunedin, New Zealand

F. J. Norton, Ph.D.
General Electric Company
1133 Eastern Avenue
Schenectady, New York

David P. Onwood, Ph.D.
Indiana-Purdue University Fort Wayne
Fort Wayne, Indiana 46805

COLLABORATORS AND CONTRIBUTORS (continued)

F. M. Page
Department of Chemistry
The University of Aston
Birmingham B4 7ET, England

B. R. Pamplin, Ph.D.
School of Physics
University of Bath
Claverton Down
Bath BA2 7AY, England

W. Parker, Ph.D.
Department of Physics
University of California
Irvine, California

M. J. Parsonage
Chemistry Department
The University of Birmingham
Birmingham 15, England

Saul Patai, Ph.D.
Department of Organic Chemistry
Hebrew University of Jerusalem
Jerusalem, Israel

I. A. Pearl, Ph.D.
The Institute of Paper Chemistry
Appleton, Wisconsin

F. R. Peart, Ph.D.
The Plessey Co. Ltd.
Allen Clark Research Centre
Northampton, England

A. C. Peed, Jr.
Eastman Kodak Company
243 State Street
Rochester, New York

R. Pepinsky, Ph.D.
Department of Physics and Astronomy
University of Florida
Gainesville, Florida

Daniel D. Pollock
Department of Mechanical and Aerospace Engineering
State University of New York at Buffalo
Buffalo, New York

H. A. Poul, Ph.D.
Professor of Physics
Oklahoma State University
Stillwater, Oklahoma

Richard L. Pratt
Staff Analyst
Data Corporation
Dayton, Ohio

Harry J. Prebluda, Ph.D.
4101 Pine Tree Dr.
Suite 803
Miami Beach, Florida 33140

I. B. Prettyman, M.S.
The Firestone Tire and Rubber Co.
1200 Firestone Parkway
Akron, Ohio

Zvi Rappoport, Ph.D.
Department of Organic Chemistry
Hebrew University of Jerusalem
Jerusalem 91904 Israel

E. H. Ratcliffe
Institution of Electrical Engineers
Stenvenage
Herts SG1 1HQ
England

Joseph Reader, Ph.D.
Atomic and Plasma Radiation Division
National Bureau of Standards
Washington, D.C. 20234

M. C. Reed, Ph.D.
1368 Wood Valley Road
Mountainside, New Jersey

B. W. Roberts, Ph.D.
General Electric Research Laboratory
Schenectady, New York

R. C. Roberts, Ph.D.
Professor Emeritus of Chemistry
Colgate University
Hamilton, New York

COLLABORATORS AND CONTRIBUTORS (continued)

R. A. Robinson
School of Chemistry
The University
Newcastle-upon-Tyne NE1 7RU, England

R. J. Rosen
Consulting Chemist
9301 Parkhill Drive
Los Angeles, California

A. H. Rosenfeld, Ph.D.
Lawrence Radiation Laboratory
University of California
Berkeley, California

Gordon D. Rowe
Specialist Lighting of GE Properties
General Electric Company
Cleveland, Ohio 44112

A. L. Rozek
Velsicol Chemical Corporation
Chicago, Illinois

S. I. Salem, Ph.D.
Professor, Department of Physics and Astronomy
California State University
Long Beach, California

Calvin E. Scheldknecht
R.D. 8
Box 398
Gettysburg, Pennsylvania

G. T. Seaborg, Ph.D.
Lawrence Berkeley Laboratory
University of California
Berkeley, California

R. Shaw
Chemical Physicist
Physical Organic Program
Stanford Research Institute
Menlo Park, California

J. R. Shelton, Ph.D.
Department of Chemistry
Case Western Reserve University
Cleveland, Ohio

G. W. Smith, Ph.D.
General Motors Corporation
Research Laboratories
Warren, Michigan

J. M. Smith, B.S.E.E.
Product Planning Large Lamp Department
General Electric Company
Cleveland, Ohio

L. D. Smithson
Air Force Materials Laboratory
Wright-Patterson Air Force Base
Dayton, Ohio

F. H. Spedding
Director, Ames Laboratory
Iowa State University
Ames, Iowa

R. H. Stokes, Ph.D.
Department of Chemistry
The University of New England
Armidale, N.S.W., Australia

Donald F. Swinehart, Ph.D.
Department of Chemistry
University of Oregon
Eugene, Oregon

A. Tarpinian
Army Materials and Mechanics Research Center
Arsenal Street
Watertown, Massachusetts

B. N. Taylor, Ph.D.
A-247-Metrology
National Bureau of Standards
Washington, D.C.

D. H. Tomlin, Ph.D.
Department of Physics
University of Reading
Reading, Berkshire, England

A. F. Trotman-Dickenson, Ph.D.
Institute of Technology
Cardiff, England

COLLABORATORS AND CONTRIBUTORS (continued)

Petr Vanýsek, Ph.D.
Department of Chemistry
Northern Illinois University
DeKalb, Illinois 60115

H. B. Vickery, Ph.D.
Connecticut Agricultural Experimental Station
112 Huntington Street
New Haven, Connecticut

John Weaver
Department of Chemical Engineering and Materials Science
University of Minnesota
Minneapolis, Minnesota

W. W. Wendlandt, Ph.D.
Professor of Chemistry
University of Houston
Houston, Texas

N. R. Whetten, Ph.D.
Research and Development Center
General Electric Company
P.O. Box 8
Schenectady, New York

Wolfgang L. Wiese, Ph.D.
Atomic Plasma and Radiation Division
National Bureau of Standards
Washington, D.C. 20234

J. H. Yoe, Ph.D.
Professor Emeritus of Chemistry
University of Virginia
Charlottesville, Virginia

G. R. Yohe, Ph.D.
Illinois State Geological Survey
University of Illinois Campus
Urbana, Illinois

T. F. Young, Ph.D.
Division of Chemical Engineering
Argonne National Laboratory
Argonne, Illinois

J. Zabicky, Ph.D.
Department of Biophysics
The Weizmann Institute of Science
Rehovoth, Israel

S. Zuffanti, A.M.
Professor of Chemistry
Northeastern University
Boston, Massachusetts

The Publishers and Editors will be grateful to readers of this Handbook who will call their attention to errors that may be discovered. Suggestions for improvement are also welcome

TABLE OF CONTENTS

MISCELLANEOUS MATHEMATICAL CONSTANTS

π CONSTANTS

π =	3.14159 26535 89793 23846 26433 83279 50288 41971 69399 37511
$1/\pi$ =	0.31830 98861 83790 67153 77675 26745 02872 40689 19291 48091
π^2 =	9.86960 44010 89358 61883 44909 99876 15113 53136 99407 24079
$\log_e \pi$ =	1.14472 98858 49400 17414 34273 51353 05871 16472 94812 91531
$\log_{10} \pi$ =	0.49714 98726 94133 85435 12682 88290 89887 36516 78324 38044
$\log_{10} \sqrt{2\pi}$ =	0.39908 99341 79057 52478 25035 91507 69595 02099 34102 92128

CONSTANTS INVOLVING e

e =	2.71828 18284 59045 23536 02874 71352 66249 77572 47093 69996
$1/e$ =	0.36787 94411 71442 32159 55237 70161 46086 74458 11131 03177
e^2 =	7.38905 60989 30650 22723 04274 60575 00781 31803 15570 55185
$M = \log_{10} e$ =	0.43429 44819 03251 82765 11289 18916 60508 22943 97005 80367
$1/M = \log_e 10$ =	2.30258 50929 94045 68401 79914 54684 36420 76011 01488 62877
$\log_{10} M$ =	9.63778 43113 00536 78912 29674 98645 −10

π^e AND e^π CONSTANTS

π^e =	22.45915 77183 61045 47342 71522
e^π =	23.14069 26327 79269 00572 90864
$e^{-\pi}$ =	0.04321 39182 63772 24977 44177
$e^{\frac{1}{2}\pi}$ =	4.81047 73809 65351 65547 30357
$i^i = e^{-\frac{1}{2}\pi}$ =	0.20787 95763 50761 90854 69556

NUMERICAL CONSTANTS

$\sqrt{2}$ =	1.41421 35623 73095 04880 16887 24209 69807 85696 71875 37695
$\sqrt[3]{2}$ =	1.25992 10498 94873 16476 72106 07278 22835 05702 51464 70151
$\log_e 2$ =	0.69314 71805 59945 30941 72321 21458 17656 80755 00134 36026
$\log_{10} 2$ =	0.30102 99956 63981 19521 37388 94724 49302 67881 89881 46211
$\sqrt{3}$ =	1.73205 08075 68877 29352 74463 41505 87236 69428 05253 81039
$\sqrt[3]{3}$ =	1.44224 95703 07408 38232 16383 10780 10958 83918 69253 49935
$\log_e 3$ =	1.09861 22886 68109 69139 52452 36922 52570 46474 90557 82275
$\log_{10} 3$ =	0.47712 12547 19662 43729 50279 03255 11530 92001 28864 19070

OTHER CONSTANTS

Euler's Constant γ =	0.57721 56649 01532 86061
$\log_e \gamma$ =	−0.54953 93129 81644 82234
Golden Ratio ϕ =	1.61803 39887 49894 84820 45868 34365 63811 77203 09180

EXPONENTIAL AND HYPERBOLIC FUNCTIONS AND THEIR COMMON LOGARITHMS

	e^x		e^{-x}	sinh x		cosh x		tanh x
x	Value	$\log^{10}$	(value)	Value	$\log^{10}$	Value	$\log^{10}$	(value)
0.00	1.0000	0.00000	1.00000	0.0000	$-\infty$	1.0000	0.00000	0.00000
0.01	1.0101	.00434	0.99005	.0100	$\bar{2}$.00001	1.0001	.00002	.01000
0.02	1.0202	.00869	.98020	.0200	$\bar{2}$.30106	1.0002	.00009	.02000
0.03	1.0305	.01303	.97045	.0300	$\bar{2}$.47719	1.0005	.00020	.02999
0.04	1.0408	.01737	.96079	.0400	$\bar{2}$.60218	1.0008	.00035	.03998
0.05	1.0513	.02171	.95123	.0500	$\bar{2}$.69915	1.0013	.00054	.04996
0.06	1.0618	.02606	.94176	.0600	$\bar{2}$.77841	1.0018	.00078	.05993
0.07	1.0725	.03040	.93239	.0701	$\bar{2}$.84545	1.0025	.00106	.06989
0.08	1.0833	.03474	.92312	.0801	$\bar{2}$.90355	1.0032	.00139	.07983
0.09	1.0942	.03909	.91393	.0901	$\bar{2}$.95483	1.0041	.00176	.08976
0.10	1.1052	.04343	.90484	.1002	$\bar{1}$.00072	1.0050	.00217	.09967
0.11	1.1163	.04777	.89583	1102	$\bar{1}$.04227	1.0061	.00262	.10956
0.12	1.1275	.05212	.88692	.1203	$\bar{1}$.08022	1.0072	.00312	.11943
0.13	1.1388	.05646	.87809	.1304	$\bar{1}$.11517	1.0085	.00366	.12927
0.14	1.1503	.06080	.86936	.1405	$\bar{1}$.14755	1.0098	.00424	.13909
0.15	1.1618	.06514	.86071	.1506	$\bar{1}$.17772	1.0113	.00487	.14889
0.16	1.1735	.06949	.85214	.1607	$\bar{1}$.20597	1.0128	.00554	.15865
0.17	1.1853	.07383	.84366	.1708	$\bar{1}$.23254	1.0145	.00625	.16838
0.18	1.1972	.07817	.83527	.1810	$\bar{1}$.25762	1.0162	.00700	.17808
0.19	1.2092	.08252	.82696	.1911	$\bar{1}$.28136	1.0181	.00779	.18775
0.20	1.2214	.08686	.81873	.2013	$\bar{1}$.30392	1.0201	.00863	.19738
0.21	1.2337	.09120	.81058	.2115	$\bar{1}$.32541	1.0221	.00951	.20697
0.22	1.2461	.09554	.80252	.2218	$\bar{1}$.34592	1.0243	.01043	.21652
0.23	1.2586	.09989	.79453	.2320	$\bar{1}$.36555	1.0266	.01139	.22603
0.24	1.2712	.10423	.78663	.2423	$\bar{1}$.38437	1.0289	.01239	.23550
0.25	1.2840	.10857	.77880	.2526	$\bar{1}$.40245	1.0314	.01343	.24492
0.26	1.2969	.11292	.77105	.2629	$\bar{1}$.41986	1.0340	.01452	.25430
0.27	1.3100	.11726	.76338	.2733	$\bar{1}$.43663	1.0367	.01564	.26362
0.28	1.3231	12160	.75578	.2837	$\bar{1}$.45282	1.0395	.01681	.27291
0.29	1.3364	.12595	.74826	.2941	$\bar{1}$.46847	1.0423	.01801	.28213
0.30	1.3499	.13029	.74082	.3045	$\bar{1}$.48362	1.0453	.01926	.29131
0.31	1.3634	.13463	.73345	.3150	$\bar{1}$.49830	1.0484	.02054	.30044
0.32	1.3771	.13897	.72615	.3255	$\bar{1}$.51254	1.0516	.02187	.30951
0.33	1.3910	.14332	.71892	.3360	$\bar{1}$.52637	1.0549	.02323	.31852
0.34	1.4049	.14766	.71177	.3466	$\bar{1}$.53981	1.0584	.02463	.32748
0.35	1.4191	.15200	.70469	.3572	$\bar{1}$.55290	1.0619	.02607	.33638
0.36	1.4333	.15635	.69768	.3678	$\bar{1}$.56564	1.0655	.02755	.34521
0.37	1.4477	.16069	.69073	.3785	$\bar{1}$.57807	1.0692	.02907	.35399
0.38	1.4623	.16503	.68386	.3892	$\bar{1}$.59019	1.0731	.03063	.36271
0.39	1.4770	.16937	.67706	.4000	$\bar{1}$.60202	1.0770	.03222	.37136
0.40	1.4918	.17372	.67032	.4108	$\bar{1}$.61358	1.0811	.03385	.37995
0.41	1.5063	.17806	.66365	.4216	$\bar{1}$.62488	1.0852	.03552	.33847
0.42	1.5220	.18240	.65705	.4325	$\bar{1}$.63594	1.0895	.03723	.39693
0.43	1.5373	.18675	.65051	.4434	$\bar{1}$.64677	1.0939	.03897	.40532
0.44	1.5527	.19109	.64404	.4543	$\bar{1}$.65738	1.0984	.04075	.41364
0.45	1.5683	.19543	.63763	.4653	$\bar{1}$.66777	1.1030	.04256	.42190
0.46	1.5841	.19978	.63128	.4764	$\bar{1}$.67797	1.1077	.04441	.43008
0.47	1.6000	.20412	.62500	.4875	$\bar{1}$.68797	1.1125	.04630	.43820
0.48	1.6161	.20846	.61878	.4986	$\bar{1}$.69779	1.1174	.04822	.44624
0.49	1.6323	.21280	.61263	.5098	$\bar{1}$.70744	1.1225	.05018	.45422
0.50	1.6487	.21715	.60653	.5211	$\bar{1}$.71692	1.1276	.05217	.46212
0.51	1.6653	.22149	.60050	.5324	$\bar{1}$.72624	1.1329	.05419	.46995
0.52	1.6820	.22583	.59452	.5438	$\bar{1}$.73540	1.1383	.05625	.47770
0.53	1.6989	.23018	.58860	.5552	$\bar{1}$.74442	1.1438	.05834	.48538
0.54	1.7160	.23452	.58275	.5666	$\bar{1}$.75330	1.1494	.06046	.49299
0.55	1.7333	.23886	.57695	.5782	$\bar{1}$.76204	1.1551	.06262	.50052
0.56	1.7507	.24320	.57121	.5897	$\bar{1}$.77065	1.1609	.06481	.50798
0.57	1.7683	.24755	.56553	.6014	$\bar{1}$.77914	1.1669	.06703	.51536
0.58	1.7860	.25189	.55990	.6131	$\bar{1}$.78751	1.1730	.06929	.52267
0.59	1.8040	.25623	.55433	.6248	$\bar{1}$.79576	1.1792	.07157	.52990
0.60	1.8221	.26058	.54881	.6367	$\bar{1}$.80390	1.1855	.07389	.53705
0.61	1.8404	.26492	.54335	.6485	$\bar{1}$.81194	1.1919	.07624	.54413
0.62	1.8589	.26926	.53794	.6605	$\bar{1}$.81987	1.1984	.07861	.55113
0.63	1.8776	.27361	.53259	.6725	$\bar{1}$.82770	1.2051	.08102	.55805
0.64	1.8965	.27795	.52729	.6846	$\bar{1}$.83543	1.2119	.08346	.56490
0.65	1.9155	.28229	.52205	.6967	$\bar{1}$.84308	1.2188	.08593	.57167
0.66	1.9348	.28664	.51685	.7090	$\bar{1}$.85063	1.2258	.08843	.57836
0.67	1.9542	.29098	.51171	.7213	$\bar{1}$.85809	1.2330	.09095	.58498
0.68	1.9739	.29532	.50662	.7336	$\bar{1}$.86548	1.2402	.09351	.59152
0.69	1.9937	.29966	.50158	.7461	$\bar{1}$.87278	1.2476	.09609	.59798
0.70	2.0138	.30401	.49659	.7586	$\bar{1}$.88000	1.2552	.09870	.60437
0.71	2.0340	.30835	.49164	.7712	$\bar{1}$.88715	1.2628	.10134	.61068
0.72	2.0544	.31269	.48675	.7838	$\bar{1}$.89423	1.2706	.10401	.61691

EXPONENTIAL AND HYPERBOLIC FUNCTIONS AND THEIR COMMON LOGARITHMS (continued)

	e^x		e^{-x}	sinh x		cosh x		tanh x
x	Value	$\log_{10}$	(value)	Value	$\log_{10}$	Value	$\log_{10}$	(value)
0.73	2.0751	.31703	.48191	.7966	$\bar{1}$.90123	1.2785	.10670	.62307
0.74	2.0959	.32138	.47711	.8094	$\bar{1}$.90817	1.2865	.10942	.62915
0.75	2.1170	.32572	.47237	.8223	$\bar{1}$.91504	1.2947	.11216	.63515
0.76	2.1383	.33006	.46767	.8353	$\bar{1}$.92185	1.3030	.11493	.64108
0.77	2.1598	.33441	.46301	.8484	$\bar{1}$.92859	1.3114	.11773	.64693
0.78	2.1815	.33875	.45841	.8615	$\bar{1}$.93527	1.3199	.12055	.65721
0.79	2.2034	.34309	.45384	.8748	$\bar{1}$.94190	1.3286	.12340	.65841
0.80	2.2255	.34744	.44933	.8881	$\bar{1}$.94846	1.3374	.12627	.66404
0.81	2.2479	.35178	.44486	.9015	$\bar{1}$.95498	1.3464	.12917	.66959
0.82	2.2705	.35612	.44043	.9150	$\bar{1}$.96144	1.3555	.13209	.67507
0.83	2.2933	.36046	.43605	.9286	$\bar{1}$.96784	1.3647	.13503	.68048
0.84	2.3164	.36481	.43171	.9423	$\bar{1}$.97420	1.3740	.13800	.68581
0.85	2.3396	.36915	.42741	.9561	$\bar{1}$.98051	1.3835	.14099	.69107
0.86	2.3632	.37349	.42316	.9700	$\bar{1}$.98677	1.3932	.14400	.69626
0.87	2.3869	.37784	.41895	.9840	$\bar{1}$.99299	1.4029	.14704	.70137
0.88	2.4100	.38218	.41478	.9981	1.99916	1.4128	.15009	.70642
0.89	2.4351	.38652	.41066	1.0122	0.00528	1.4229	.15317	.71139
0.90	2.4596	.39087	.40657	1.0265	.01137	1.4331	.15627	.21630
0.91	2.4843	39521	.40242	1.0409	.01741	1.4434	.15939	.72113
0.92	2.5093	.39955	.39852	1.0554	.02341	1.4539	.16254	.72590
0.93	2.5345	.40389	.39455	1.0700	.02937	1.4645	.16570	.73059
0.94	2.5600	.40824	.39063	1.0847	.03530	1.4753	.16888	.73522
0.95	2.5857	.41258	.38674	1.0995	.04119	1.4862	.17208	.73978
0.96	2.6117	.41692	.38289	1.1144	.04704	1.4973	.17531	.74428
0.97	2.6379	.42127	.37908	1.1294	.05286	1.5085	.17855	.74870
0.98	2.6645	.42561	.37531	1.1446	.05864	1.5199	.18181	.75307
0.99	2.6912	.42995	.37158	1.1598	.06439	1.5314	.18509	.75736
1.00	2.7183	.43429	.36788	1.1752	.07011	1.5431	.18839	.76159
1.01	2.7456	.43864	.36422	1.1907	.07580	1.5549	.19171	.76576
1.02	2.7732	.44298	.36060	1.2063	.06146	1.5669	.19504	.76987
1.03	2.8011	.44732	.35701	1.2220	.08708	1.5790	.19839	.77391
1.04	2.8292	.45167	.35345	1.2379	.09268	1.5913	.20176	.77789
1.05	2.8577	.45601	.34994	1.2539	.09825	1.6038	.20515	.78181
1.06	2.8864	.46035	.34646	1.2700	.10379	1.6164	.20855	.78566
1.07	2.9154	.46470	.34301	1.2862	.10930	1.6292	.21197	.78946
1.08	2.9447	.46904	.33960	1.3025	.11479	1.6421	.21541	.79320
1.09	2.9743	.47338	.33622	1.3190	.12025	1.6552	.21886	.79688
1.10	3.0042	.47772	.33287	1.3356	.12569	1.6685	.22233	.80050
1.11	3.0344	.48207	.32956	1.3524	.13111	1.6820	.22582	.80406
1.12	3.0659	.48641	.32628	1.3693	.13649	1.6956	.22931	.80737
1.13	3.0957	.49075	.32303	1.3863	.14186	1.7083	.23283	.81102
1.14	3.1268	.49510	.31982	1.4035	.14720	1.7233	.23636	.81441
1.15	3.1582	.49944	.31644	1.4208	.15253	1.7374	.23990	.81775
1.16	3.1899	.50378	.31349	1.4382	.15783	1.7517	.24346	.82104
1.17	3.2220	.50812	.31037	1.4558	.16311	1.7662	.24703	.82427
1.18	3.2544	.51247	.30728	1.4735	.16836	1.7808	.25062	.82745
1.19	3.2871	.51681	.30422	1.4914	.17360	1.7957	.25422	.83058
1.20	3.3201	.52115	.30119	1.5095	.17882	1.8107	.25784	.83365
1.21	3.3535	.52550	.29820	1.5276	.18402	1.8258	.26146	.83668
1.22	3.3872	.52984	.29523	1.5460	.18920	1.8412	.26510	.83965
1.23	3.4212	.53418	.29229	1.5645	.19437	1.8568	.26876	.84258
1.24	3.4556	.53853	.28938	1.5831	.19951	1.8725	.27242	.83546
1.25	3.4903	.54287	.28650	1.6019	.20464	1.8884	.27610	.84828
1.26	3.5254	.54721	.28365	1.6209	.20975	1.9045	.27979	.85106
1.27	3.5609	.55155	.28083	1.6400	.21485	1.9208	.28349	.85380
1.28	3.5996	.55590	.27804	1.6593	.21993	1.9373	.28721	.85648
1.29	3.6328	.56024	.27527	1.6788	.22499	1.9540	.29093	.85913
1.30	3.6693	.56458	.27253	1.6984	.23004	1.9709	.29467	.86172
1.31	3.7062	.56893	.26982	1.7182	.23507	1.9880	.29842	.86428
1.32	3.7434	.57327	.26714	1.7381	.24009	2.0053	.30217	.86678
1.33	3.7810	.57761	.26448	1.7583	.24509	2.0228	.30594	.86925
1.34	3.8190	.58195	.26185	1.7786	.25008	2.0404	.30972	.87167
1.35	3.8574	.58630	.25924	1.7991	.25505	2.0583	.31352	.87405
1.36	3.8962	.59064	.25666	1.8198	.26002	2.0764	.31732	.87639
1.37	3.9354	.59498	.25411	1.8406	.26496	2.0947	.32113	.87869
1.38	3.9749	.59933	.25158	1.8617	.26990	2.1132	.32495	.88095
1.39	4.0149	.60367	.24908	1.8829	.27482	2.1320	.32878	.88317
1.40	4.0552	.60801	.24660	1.9043	.27974	2.1509	.33262	.88535
1.41	4.0960	.61236	.24414	1.9259	.28464	2.1700	.33647	.88749
1.42	4.1371	.61670	.24171	1.9477	.28952	2.1894	.34033	.88960
1.43	4.1787	.62104	.23931	1.9697	.29440	2.2090	.34420	.89167

EXPONENTIAL AND HYPERBOLIC FUNCTIONS AND THEIR COMMON LOGARITHMS (continued)

x	e^x Value	e^x $\log^{10}$	e^{-x} (value)	sinh x Value	sinh x $\log^{10}$	cosh x Value	cosh x $\log^{10}$	tanh x (value)
1.44	4.2207	.62538	.23693	1.9919	.29926	2.2288	.34807	.89370
1.45	4.2631	.62973	.23457	2.0143	.30412	2.2488	.35196	.89569
1.46	4.3060	.63407	.23224	2.0369	.30896	2.2691	.35585	.89765
1.47	4.3492	.63841	.22993	2.0597	.31379	2.2896	.35976	.89958
1.48	4.3929	.64276	.22764	2.0827	.31862	2.3103	.36367	.90147
1.49	4.4371	.64710	.22537	2.1059	.32343	2.3312	.36759	.90332
1.50	4.4817	.65144	.22313	2.1293	.32823	2.3524	.37151	.90515
1.51	4.5267	.65578	.22091	2.1529	.33303	2.3738	.37545	.90694
1.52	4.5722	.66013	.21871	2.1768	.33781	2.3955	.37939	.90870
1.53	4.6182	.66447	.21654	2.2008	.34258	2.4174	.38334	.91042
1.54	4.6646	.66881	.21438	2.2251	.34735	2.4395	.38730	.91212
1.55	4.7115	.67316	.21225	2.2496	.35211	2.4619	.39126	.91379
1.56	4.7588	.67750	.21014	2.2743	.35686	2.4845	.39524	.91542
1.57	4.8066	.68184	.20805	2.2993	.36160	2.5073	.39921	.91703
1.58	4.8550	.68619	.20598	2.3245	.36633	2.5305	.40320	.91860
1.59	4.9037	.69053	.20393	2.3499	.37105	2.5538	.40719	.92015
1.60	4.9530	.69487	.20190	2.3756	.37577	2.5775	.41119	.92167
1.61	5.0028	.69921	.19989	2.4015	.38048	2.6013	.41520	.92316
1.62	5.0531	.70356	.19790	2.4276	.38518	2.6255	.41921	.92462
1.63	5.1039	.70790	.19593	2.4540	.38987	2.6499	.42323	.92606
1.64	5.1552	.71224	.19398	2.4806	.39456	2.6746	.42725	.92747
1.65	5.2070	.71659	.19205	2.5075	.39923	2.6995	.43129	.92886
1.66	5.2593	.72093	.19014	2.5346	.40391	2.7247	.43532	.93022
1.67	5.3122	.72527	.18825	2.5620	.40857	2.7502	.43937	.93155
1.68	5.3656	.72961	.18637	2.5896	.41323	2.7760	.44341	.93286
1.69	5.4195	.73396	.18452	2.6175	.41788	2.8020	.44747	.93415
1.70	5.4739	.73830	.18268	2.6456	.42253	2.8283	.45153	.93541
1.71	5.5290	.74264	.18087	2.6740	.42717	2.8549	.45559	.93665
1.72	5.5845	.74699	.17907	2.7027	.43180	2.8818	.45966	.93786
1.73	5.6407	.75133	.17728	2.7317	.43643	2.9090	.46374	.93906
1.74	5.6973	.75567	.17552	2.7609	.44105	2.9364	.46782	.94023
1.75	5.7546	.76002	.17377	2.7904	.44567	2.9642	.47191	.94138
1.76	5.8124	.76436	.17204	2.8202	.45028	2.9922	.47600	.94250
1.77	5.8709	.76870	.17033	2.8503	.45488	3.0206	.48009	.94361
1.78	5.9299	.77304	.16864	2.8806	.45948	3.0492	.48419	.94470
1.79	5.9895	.77739	.16696	2.9112	.46408	3.0782	.48830	.94576
1.80	6.0496	.78173	.16530	2.9422	.46867	3.1075	.49241	.94681
1.81	6.1104	.78607	.16365	2.9734	.47325	3.1371	.49652	.94783
1.82	6.1719	.79042	.16203	3.0049	.47783	3.1669	.50064	.94884
1.83	6.2339	.79476	.16041	3.0367	.48241	3.1972	.50476	.94983
1.84	6.2965	.79910	.15882	3.0689	.48698	3.2277	.50889	.95080
1.85	6.3598	.80344	.15724	3.1013	.49154	3.2585	.51302	.95175
1.86	6.4237	.80779	.15567	3.1340	.49610	3.2897	.51716	.95268
1.87	6.4383	.81213	.15412	3.1671	.50066	3.3212	.52130	.95359
1.88	6.5535	.81647	.15259	3.2005	.50521	3.3530	.52544	.95449
1.89	6.6194	.82082	.15107	3.2341	.50976	3.3852	.52959	.95537
1.90	6.6859	.82516	.14957	3.2682	.51430	3.4177	.53374	.95624
1.91	6.7531	.82950	.14808	3.3025	.51884	3.4506	.53789	.95709
1.92	6.8210	.83385	.14661	3.3372	.52338	3.4838	.54205	.95792
1.93	6.8895	.83819	.14515	3.3722	.52791	3.5173	.54621	.95873
1.94	6.9588	.84253	.14370	3.4075	.53244	3.5512	.55038	.95953
1.95	7.0287	.84687	.14227	3.4432	.53696	3.5855	.55455	.96032
1.96	7.0993	.85122	.14086	3.4792	.54148	3.6201	.55872	.96109
1.97	7.1707	.85556	.13946	3.5156	.54600	3.6551	.56290	96185
1.98	7.2427	.85990	.13807	3.5923	.55051	3.6904	.56707	.96259
1.99	7.3155	.86425	.13670	3.5894	.55502	3.7261	.57126	.96331
2.00	7.3891	.86859	.13534	3.6269	.55953	3.7622	.57544	.96403
2.01	7.4633	.87293	.13399	3.6647	.56403	3.7987	.57963	.96473
2.02	7.5383	.87727	.13266	3.7028	.56853	3.8335	.58382	.96541
2.03	7.6141	.88162	.13134	3.7414	.57303	3.8727	.58802	.96609
2.04	7.6906	.88596	.13003	3.7803	.57753	3.9103	.59221	.96675
2.05	7.7679	.89030	.12873	3.8196	.58202	3.9483	.59641	.96740
2.06	7.8460	.89465	.12745	3.8593	.58650	3.9867	.60061	.96803
2.07	7.9248	.89899	.12619	3.8993	.59099	4.0255	.60482	.96865
2.08	8.0045	.90333	.12493	3.9398	.59547	4.0647	.60903	.96926
2.09	8.0849	.90768	.12369	3.9806	.59995	4.1043	.61324	.96986
2.10	8.1662	.91202	.12246	4.0219	.60443	4.1443	.61745	.97045
2.11	8.2482	.91636	.12124	4.0635	.60890	4.1847	.62167	.97103
2.12	8.3311	.92070	.12003	4.1056	.61337	4.2256	.62589	.97159
2.13	8.4149	.92505	.11884	4.1480	.61784	4.2669	.63011	.97215
2.14	8.4994	.92939	.11765	4.1909	.62231	4.3085	.63433	.97269
2.15	8.5849	.93373	.11648	4.2342	62677	4.3507	.63856	.97323

EXPONENTIAL AND HYPERBOLIC FUNCTIONS AND THEIR COMMON LOGARITHMS (continued)

x	e^x Value	e^x log^{10}	e^{-x} (value)	sinh x Value	sinh x log^{10}	cosh x Value	cosh x log^{10}	tanh x (value)
2.16	8.6711	.93808	.11533	4.2779	.63123	4.3932	.64278	.97375
2.17	8.7583	.94242	.11418	4.3221	.63569	4.4362	.64701	.97426
2.18	8.8463	.94676	.11304	4.3666	.64015	4.4797	.65125	.97477
2.19	8.9352	.95110	.11192	4.4116	.64460	4.5236	.65548	.97526
2.20	9.0250	.95545	.11080	4.4571	.64905	4.5679	.65972	.97574
2.21	9.1157	.95979	.10970	4.5030	.65350	4.6127	.66396	.97622
2.22	9.2073	.96413	.10861	4.5494	.65795	4.6580	.66820	.97668
2.23	9.2999	.96848	.10753	4.5962	.66240	4.7037	.67244	.97714
2.24	9.3933	.97282	.10646	4.6434	.66684	4.7499	.67668	.97759
2.25	9.4877	.97716	.10540	4.6912	.67128	4.7966	.68093	.97803
2.26	9.5831	.98151	.10435	4.7394	.67572	4.8437	.68518	.97846
2.27	9.6794	.98585	.10331	4.7880	.68016	4.8914	.68943	.97888
2.28	9.7767	.99019	.10228	4.8372	.68459	4.9395	.69368	.97929
7.29	9.8749	.99453	.10127	4.8868	.68903	4.9881	.69794	97970
2.30	9.9742	.99888	.10026	4.9370	.69346	5.0372	.70219	.98010
2.31	10.074	1.00322	.09926	4.9876	.69789	5.0868	.70645	.98049
2.32	10.176	1.00756	.09827	5.0387	.70232	5.1370	.71071	.98087
2.33	10.278	1.01191	.09730	5.0903	.70675	5.1876	.71497	.98124
2.34	10.381	1.01625	.09633	5.1425	.71117	5.2388	.71923	.98161
2.35	10.486	1.02059	.09537	5.1951	.71559	5.2905	.72349	.98197
2.36	10.591	1.02493	.09442	5.2483	.72002	5.3427	.72776	.98233
2.37	10.697	1.02928	.09348	5.3020	.72444	5.3954	.73203	.98267
2.38	10.805	1.03362	.09255	5.3562	.72885	5.4487	.73630	.98301
2.39	10.913	1.03796	.09163	5.4109	.73327	5.5026	.74056	.98335
2.40	11.023	1.04231	.09072	5.4662	.73769	5.5569	.74484	.98367
2.41	11.134	1.04665	.08982	5.5221	.74210	5.6119	.74911	.98400
2.42	11.246	1.05099	.08892	5.5785	.74652	5.6674	.75338	.98431
2.43	11.359	1.05534	08804	5.6354	.75093	5.7235	.75766	.98462
2.44	11.473	1.05968	.08716	5.6929	.75534	5.7801	.76194	.98492
2.45	11.588	1.06402	.08629	5.7510	.75975	5.8373	.76621	.98522
2.46	11.705	1.06836	.08543	5.8097	.76415	5.8951	.77049	.98551
2.47	11.822	1.07271	.08458	5.8689	.76856	5.9535	.77477	.98579
2.48	11.941	1.07705	.08374	5.9288	.77296	6.0125	.77906	.98607
2.49	12.061	1.08139	.08291	5.9892	.77737	6.0721	.78334	.98635
2.50	12.182	1.08574	.08208	6.0502	.78177	6.1323	.78762	.98661
2.51	12.305	1.09008	.08127	6.1118	.78617	6.1931	.79191	.98688
2.52	12.429	1.09442	.08046	6.1741	.79057	6.2545	.79619	.98714
2.53	12.554	1.09877	.07966	6.2369	.79497	6.3166	.80048	.98739
2.54	12.680	1.10311	.07887	6.3004	.79937	6.3793	.80477	.98764
2.55	12.807	1.10745	.07808	6.3645	.80377	6.4426	.80906	.98788
2.56	12.936	1.11179	.07730	6.4293	.80816	6.5066	.81335	.98812
2.57	13.066	1.11614	.07654	6.4946	.81256	6.5712	.81764	.98835
2.58	13.197	1.12048	.07577	6.5607	.81695	6.6365	.82194	.98858
2.59	13.330	1.12482	.07502	6.6274	.82134	6.7024	.82623	.98881
2.60	13.464	1.12917	.07427	6.6947	.82573	6.7690	.83052	.98903
2.61	13.599	1.13351	.07353	6.7628	.83012	6.8363	.83482	.98924
2.62	13.736	1.13785	.07280	6.8315	.83451	6.9043	.83912	.98946
2.63	13.874	1.14219	.07208	6.9008	.83890	6.9729	.84341	.98966
2.64	14.013	1.14654	.07136	6.9709	.84329	7.0423	.84771	98987
2.65	14.154	1.15008	.07065	7.0417	.84768	7.1123	.85201	.99007
2.66	14.296	1.15522	.06995	7.1132	.85206	7.1831	.85631	.99026
2.67	14.440	1.15957	.06925	7.1854	.85645	7.2546	.86061	.99045
2.68	14.585	1.16391	.06856	7.2583	.86083	7.3268	.86492	.99064
2.69	14.732	1.16825	.06788	7.3319	.86522	7.3998	.86922	.99083
2.70	14.880	1.17260	.06721	7.4063	.86960	7.4735	.87352	.99101
2.71	15.029	1.17694	.06654	7.4814	.87398	7.5479	.87783	.99118
2.72	15.180	1.18128	.06587	7.5572	.87836	7.6231	.88213	.99136
2.73	15.333	1.18562	.06522	7.6338	.88274	7.6991	.89644	.99153
2.74	15.487	1.18997	.06457	7.7112	.88712	7.7758	.89074	.99170
2.75	15.643	1.19431	06393	7.7894	.89150	7.8533	.89505	.99186
2.76	15.800	1.19865	.06329	7.8683	.89588	7.9316	.89936	.99202
2.77	15.959	1.20300	.06266	7.9480	.90026	8.0106	.90367	.99218
2.78	16.119	1.20734	.06204	8.0285	.90463	8.0905	.90798	.99233
2.79	16.281	1.21168	.06142	8.1098	.90901	8.1712	.91229	.99248
2.80	16.445	1.21602	.06081	8.1919	.91339	8.2527	.91660	.99263
2.81	16.610	1.22037	.06020	8.2749	.91776	8.3351	92091	.99278
2.82	16.777	1.22471	.05961	8.3586	.92213	8.4182	.92522	.99292
2.83	16.945	1.22905	.05901	8.4432	.92651	8.5022	.92953	.99306
2.84	17.116	1.23340	.05843	8.5287	.93088	8.5871	.93385	.99320
2.85	17.288	1.23774	.05784	8.6150	.93525	8.6728	.93816	.99333
2.86	17.462	1.24208	.05727	8.7021	.93963	8.7594	.94247	.99346

EXPONENTIAL AND HYPERBOLIC FUNCTIONS AND THEIR COMMON LOGARITHMS (continued)

x	e^x Value	e^x log^{10}	e^{-x} (value)	sinh x Value	sinh x log^{10}	cosh x Value	cosh x log^{10}	tanh x (value)
2.87	17.637	1.24643	.05670	8.7902	.94400	8.8469	.94679	.99359
2.88	17.814	1.25077	.05613	8.8791	.94837	8.9352	95110	.99372
2.89	17.993	1.25511	.05558	8.9689	.95274	9.0244	.95542	.99384
2.90	18.174	1.25945	.05502	9.0596	.95711	9.1146	.95974	.99396
2.91	18.357	1.26380	.05448	9.1512	.96148	9.2056	.96405	.99408
2.92	18.541	1.26814	.05393	9.2437	.96584	9.2976	.96837	.99420
2.93	18.728	1.27248	.05340	9.3371	.97021	9.3905	.97269	.99431
2.94	18.916	1.27683	.05287	9.4315	.97458	9.4844	.97701	.99443
2.95	19.106	1.28117	.05234	9.5268	.97895	9.5791	.98133	.99454
2.96	19.298	1.28551	.05182	9.6231	.98331	9.6749	.98565	.99464
2.97	19.492	1.28985	.05130	9.7203	.98768	9.7716	.98997	.99475
2.98	19.688	1.29420	.05079	9.8185	.99205	9.8693	.99429	.99485
2.99	19.886	1.29854	.05029	9.9177	.99641	9.9680	.99861	.99496
3.00	20.086	1.30288	.04979	10.018	1.00078	10.068	1.00293	0.99505
3.05	21.115	1.32460	.04736	10.534	1.02259	10.581	1.02454	0.99552
3.10	22.198	1.34631	.04505	11.076	1.04440	11.122	1.04616	0.99595
3.15	23.336	1.36803	.04285	11.647	1.06620	11.690	1.06779	0.99633
3.20	24.533	1.38974	.04076	12.246	1.08799	12.287	1.08943	0.99668
3.25	25.790	1.41146	.03877	12.876	1.10977	12.915	1.11108	0.99700
3.30	27.113	1.43317	.03688	13.538	1.13155	13.575	1.13273	0.99728
3.35	28.503	1.45489	.03508	14.234	1.15332	14.269	1.15439	0.99754
3.40	29.964	1.47660	.03337	14.965	1.17509	14.999	1.17605	0.99777
3.45	31.500	1.49832	.03175	15.734	1.19685	15.766	1.19772	0.99799
3.50	33.115	1.52003	.03020	16.543	1.21860	16.573	1.21940	0.99818
3.55	34.813	1.54175	.02872	17.392	1.24036	17.421	1.24107	0.99835
3.60	36.598	1.56346	.02732	18.286	1.26211	18.313	1.26275	0.99851
3.65	38.475	1.58517	02599	19.224	1.28385	19.250	1.28444	9.99865
3.70	40.447	1.60689	.02472	20.211	1.30559	20.236	1.30612	0.99878
3.75	42.521	1.62860	.02352	21.249	1.32733	21.272	1.32781	0.99889
3.80	44.701	1.65032	.02237	22.339	1.34907	22.362	1.34951	0.99900
3.85	46.993	1.67203	.02128	23.486	1.37081	23.507	1.37120	0.99909
3.90	49.402	1.69375	.02024	24.691	1.39254	24.711	1.39290	0.99918
3.95	51.935	1.71546	.01925	25.958	1.41427	25.977	1.41459	0.09926
4.00	54.598	1.73718	.01832	27.290	1.43600	27.308	1.43629	0.99933
4.10	60.340	1.78061	.01657	30.162	1.47946	30.178	1.47970	0.99945
4.20	66.686	1.82404	.01500	33.336	1.52291	33.351	1.52310	0.99955
4.30	73.700	1.86747	.01357	36.843	1.56636	36.857	1.56652	0.99963
4.40	81.451	1.91090	.01227	40.719	1.60980	40.732	1.60993	0.99970
4.50	90.017	1.95433	.01111	45.003	1.65324	45.014	1.65335	0.99975
4.60	99.484	1.99775	.01005	49.737	1.69668	49.747	1.69677	0.99980
4.70	109.95	2.04118	.00910	54.969	1.74012	54.978	1.74019	0.99983
4.80	121.51	2.08461	.00823	60.751	1.78355	60.759	1.78361	0.99986
4.90	134.29	2.12804	.00745	67.141	1.82699	67.149	1.82704	0.99989
5.00	148.41	2.17147	.00674	74.203	1.87042	74.210	1.87046	0.99991
5.10	164.02	2.21490	.00610	82.008	1.91389	82.014	1.91389	0.99993
5.20	181.27	2.25833	.00552	90.633	1.95729	90.639	1.95731	0.99994
5.30	200.34	2.30176	.00499	100.17	2.00074	100.17	2.00074	0.99995
5.40	221.41	2.34519	.00452	110.70	2.04415	110.71	2.04417	0.99996
5.50	244.69	2.38862	.00409	122.34	2.08758	122.35	2.08760	0.99997
5.60	270.43	2.43205	.00370	135.21	2.13101	135.22	2.13103	0.99997
5.70	298.87	2.47548	.00335	149.43	2.17444	149.44	2.17445	0.99998
5.80	330.30	2.51891	.00303	165.15	2.21787	165.15	2.21788	0.99998
5.90	365.04	2.56234	.00274	182.52	2.26130	182.52	2.26131	0.99998
6.00	403.43	2.60577	.00248	201.71	2.30473	201.72	2.30474	0.99999
6.25	518.01	2.71434	.00193	259.01	2.41331	259.01	2.41331	0.99999
6.50	665.14	2.82291	.00150	332.57	2.52188	332.57	2.52189	1.00000
6.75	854.06	2.93149	.00117	427.03	2.63046	427.03	2.63046	1.00000
7.00	1096.6	3.04006	.00091	548.32	2.73904	548.32	2.73903	1.00000
7.50	1808.0	3.25721	.00055	904.02	2.95618	904.02	2.95618	1.00000
8.00	2981.0	3.47436	.00034	1490.5	3.17333	1490.5	3.17333	1.00000
8.50	4914.8	3.69150	.00020	2457.4	3.39047	2457.4	3.39047	1.00000
9.00	8103.1	3.90865	.00012	4051.5	3.60762	4051.5	3.60762	1.00000
9.50	13360.	4,12580	.00007	6679.9	3.82477	6679.9	3.82477	1.00000
10.00	22026.	4.34294	.00005	11013.	4.04191	11013.	4.04191	1.00000

NATURAL TRIGONOMETRIC FUNCTIONS TO FOUR PLACES

X radians	X degrees	sin x	cos x	tan x	cot x	sec x	csc x		
.0000	**0° 00′**	.0000	1.0000	.0000	—	1.000	—	**90° 00′**	1.5708
.0029	10	.0029	1.0000	.0029	343.8	1.000	343.8	50	1.5679
.0058	20	.0058	1.0000	.0058	171.9	1.000	171.9	40	1.5650
.0087	30	.0087	1.0000	.0087	114.6	1.000	114.6	30	1.5621
.0116	40	.0116	.9999	.0116	85.94	1.000	85.95	20	1.5592
.0145	50	.0145	.9999	.0145	68.75	1.000	68.76	10	1.5563
.0175	**1° 00′**	.0175	.9998	.0175	57.29	1.000	57.30	**89° 00′**	1.5533
.0204	10	.0204	.9998	.0204	49.10	1.000	49.11	50	1.5504
.0233	20	.0233	.9997	.0233	42.96	1.000	42.98	40	1.5475
.0262	30	.0262	.9997	.0262	38.19	1.000	38.20	30	1.5446
.0291	40	.0291	.9996	.0291	34.37	1.000	34.38	20	1.5417
.0320	50	.0320	.9995	.0320	31.24	1.001	31.26	10	1.5388
.0349	**2° 00′**	.0349	.9994	.0349	28.64	1.001	28.65	**88° 00′**	1.5359
.0378	10	.0378	.9993	.0378	26.43	1.001	26.45	50	1.5330
.0407	20	.0407	.9992	.0407	24.54	1.001	24.56	40	1.5301
.0436	30	.0436	.9990	.0437	22.90	1.001	22.93	30	1.5272
.0465	40	.0465	.9989	.0466	21.47	1.001	21.49	20	1.5243
.0495	50	.0494	.9988	.0495	20.21	1.001	20.23	10	1.5213
.0524	**3° 00′**	.0523	.9986	.0524	19.08	1.001	19.11	**87° 00′**	1.5184
.0553	10	.0552	.9985	.0553	18.07	1.002	18.10	50	1.5155
.0582	20	.0581	.9983	.0582	17.17	1.002	17.20	40	1.5126
.0611	30	.0610	.9981	.0612	16.35	1.002	16.38	30	1.5097
.0640	40	.0640	.9980	.0641	15.60	1.002	15.64	20	1.5068
.0669	50	.0669	.9978	.0670	14.92	1.002	14.96	10	1.5039
.0698	**4° 00′**	.0698	.9976	.0699	14.30	1.002	14.34	**86° 00′**	1.5010
.0727	10	.0727	.9974	.0729	13.73	1.003	13.76	50	1.4981
.0756	20	.0756	.9971	.0758	13.20	1.003	13.23	40	1.4952
.0785	30	.0785	.9969	.0787	12.71	1.003	12.75	30	1.4923
.0814	40	.0814	.9967	.0816	12.25	1.003	12.29	20	1.4893
.0844	50	.0843	.9964	.0846	11.83	1.004	11.87	10	1.4864
.0873	**5° 00′**	.0872	.9962	.0875	11.43	1.004	11.47	**85° 00′**	1.4835
.0902	10	.0901	.9959	.0904	11.06	1.004	11.10	50	1.4806
.0931	20	.0929	.9957	.0934	10.71	1.004	10.76	40	1.4777
.0960	30	.0958	.9954	.0963	10.39	1.005	10.43	30	1.4748
.0989	40	.0987	.9951	.0992	10.08	1.005	10.13	20	1.4719
.1018	50	.1016	.9948	.1022	9.788	1.005	9.839	10	1.4690
.1047	**6° 00′**	.1045	.9945	.1051	9.514	1.006	9.567	**84° 00′**	1.4661
.1076	10	.1074	.9942	.1080	9.255	1.006	9.309	50	1.4632
.1105	20	.1103	.9939	.1110	9.010	1.006	9.065	40	1.4603
.1134	30	.1132	.9936	.1139	8.777	1.006	8.834	30	1.4573
.1164	40	.1161	.9932	.1169	8.556	1.007	8.614	20	1.4544
.1193	50	.1190	.9929	.1198	8.345	1.007	8.405	10	1.4515
.1222	**7° 00′**	.1219	.9925	.1228	8.144	1.008	8.206	**83° 00′**	1.4486
.1251	10	.1248	.9922	.1257	7.953	1.008	8.016	50	1.4457
.1280	20	.1276	.9918	.1287	7.770	1.008	7.834	40	1.4428
.1309	30	.1305	.9914	.1317	7.596	1.009	7.661	30	1.4399
.1338	40	.1334	.9911	.1346	7.429	1.009	7.496	20	1.4370
.1367	50	.1363	.9907	.1376	7.269	1.009	7.337	10	1.4341
.1396	**8° 00′**	.1392	.9903	.1405	7.115	1.010	7.185	**82° 00′**	1.4312
.1425	10	.1421	.9899	.1435	6.968	1.010	7.040	50	1.4283
.1454	20	.1449	.9894	.1465	6.827	1.011	6.900	40	1.4254
.1484	30	.1478	.9890	.1495	6.691	1.011	6.765	30	1.4224
.1513	40	.1507	.9886	.1524	6.561	1.012	6.636	20	1.4195
.1542	50	.1536	.9881	.1554	6.435	1.012	6.512	10	1.4166
.1571	**9° 00′**	.1564	.9877	.1584	6.314	1.012	6.392	**81° 00′**	1.4137
.1600	10	.1593	.9872	.1614	6.197	1.013	6.277	50	1.4108
.1629	20	.1622	.9868	.1644	6.084	1.013	6.166	40	1.4079
.1658	30	.1650	.9863	.1673	5.976	1.014	6.059	30	1.4050
.1687	40	.1679	.9858	.1703	5.871	1.014	5.955	20	1.4021
.1716	50	.1708	.9853	.1733	5.769	1.015	5.855	10	1.3992
.1745	**10° 00′**	.1736	.9848	.1763	5.671	1.015	5.759	**80° 00′**	1.3963
.1774	10	.1765	.9843	.1793	5.576	1.016	5.665	50	1.3934
.1804	20	.1794	.9838	.1823	5.485	1.016	5.575	40	1.3904
.1833	30	.1822	.9833	.1853	5.396	1.017	5.487	30	1.3875
.1862	40	.1851	.9827	.1883	5.309	1.018	5.403	20	1.3846
.1891	50	.1880	.9822	.1914	5.226	1.018	5.320	10	1.3817
.1920	**11° 00′**	.1908	.9816	.1944	5.145	1.019	5.241	**79° 00′**	1.3788
.1949	10	.1937	.9811	.1974	5.066	1.019	5.164	50	1.3759
.1978	20	.1965	.9805	.2004	4.989	1.020	5.089	40	1.3730
.2007	30	.1994	.9799	.2035	4.915	1.020	5.016	30	1.3701
		cos x	**sin x**	**cot x**	**tan x**	**csc x**	**sec x**	**x degrees**	**x radians**

NATURAL TRIGONOMETRIC FUNCTIONS TO FOUR PLACES (continued)

x radians	x degrees	sin x	cos x	tan x	cot x	sec x	csc x		
.2036	40	.2022	.9793	.2065	3.843	1.021	4.945	20	1.3672
.2065	50	.2051	.9787	.2095	4.773	1.022	4.876	10	1.3643
.2094	**12° 00′**	.2079	.9781	.2126	4.705	1.022	4.810	**78° 00′**	1.3614
.2123	10	.2108	.9775	.2156	4.638	1.023	4.745	50	1.3584
.2153	20	.2136	.9769	.2186	4.574	1.024	4.682	40	1.3555
.2182	30	.2164	.9763	.2217	4.511	1.025	4.620	30	1.3526
.2211	40	.2193	.9757	.2247	4.449	1.025	4.560	20	1.3497
.2240	50	.2221	.9750	.2278	4.390	1.026	4.502	10	1.3468
.2269	**13° 00′**	.2250	.9744	.2309	4.331	1.026	4.445	**77° 00′**	1.3439
.2298	10	.2278	.9737	.2339	4.275	1.027	4.390	50	1.3410
.2327	20	.2306	.9730	.2370	4.219	1.028	4.336	40	1.3381
.2356	30	.2334	.9724	.2401	4.165	1.028	4.284	30	1.3352
.2385	40	.2363	.9717	.2432	4.113	1.029	4.232	20	1.3323
.2414	50	.2391	.9710	.2462	4.061	1.030	4.182	10	1.3294
.2443	**14° 00′**	.2419	.9703	.2493	4.011	1.031	4.134	**76° 00′**	1.3265
.2473	10	.2447	.9696	.2524	3.962	1.031	4.086	50	1.3235
.2502	20	.2476	.9689	.2555	3.914	1.032	4.039	40	1.3206
.2531	30	.2404	.9681	.2586	3.867	1.033	3.994	30	1.3177
.2560	40	.2532	.9674	.2617	3.821	1.034	3.950	20	1.3148
.2589	50	.2560	.9667	.2648	3.776	1.034	3.906	10	1.3119
.2618	**15° 00′**	.2588	.9659	.2679	3.732	1.035	3.864	**75° 00′**	1.3090
.2647	10	.2616	.9652	.2711	3.689	1.036	3.822	50	1.3061
.2676	20	.2644	.9644	.2732	3.647	1.037	3.782	40	1.3032
.2705	30	.2672	.9636	.2773	3.606	1.038	3.742	30	1.3003
.2734	40	.2700	.9628	.2805	3.566	1.039	3.703	20	1.2974
.2763	50	.2728	.9621	2836	3.526	1.039	3.665	10	1.2945
.2793	**16° 00′**	.2756	.9613	.2867	3.487	1.040	3.628	**74° 00′**	1.2915
.2822	10	.2784	.9605	.2899	3.450	1.041	3.592	50	1.2886
.2851	20	.2812	.9596	.2931	3.412	1.042	3.556	40	1.2857
.2880	30	.2840	.9588	.2962	3.376	1.043	3.521	30	1.2828
.2909	40	.2868	.9580	.2994	3.340	1.044	3.487	20	1.2799
.2938	50	.2896	.9572	.3026	3.305	1.045	3.453	10	1.2770
.2967	**17° 00′**	.2924	.9563	.3057	3.271	1.046	3.420	**73° 00′**	1.2741
.2996	10	.2952	.9555	.3089	3.237	1.047	3.388	50	1.2712
.3025	20	.2979	.9546	.3121	4.204	1.048	3.356	40	1.2683
.3054	30	.3007	.9537	.3153	3.172	1.049	3.326	30	1.2654
.3083	40	.3035	.9528	.3185	3.140	1.049	3.295	20	1.2625
.3113	50	.3062	.9520	.3217	3.108	1.050	3.265	10	1.2595
.3142	**18° 00′**	.3090	.9511	.3249	3.078	1.051	3.236	**72° 00′**	1.2566
.3171	10	.3118	.9502	.3281	3.047	1.052	3.207	50	1.2537
.3200	20	.3145	.9492	.3314	3.018	1.053	3.179	40	1.2508
.3229	30	.3173	.9483	.3346	2.989	1.054	3.152	30	1.2479
.3258	40	.3201	.9474	.3378	2.960	1.056	3.124	20	1.2450
.3287	50	.3228	.9465	.3411	2.932	1.057	3.098	10	1.2421
.3316	**19° 00′**	.3256	.9455	.3443	2.904	1.058	3.072	**71° 00′**	1.2392
.3345	10	.3283	.9446	.3476	2.877	1.059	3.046	50	1.2363
.3374	20	.3311	.9436	.3508	2.850	1.060	3.021	40	1.2334
.3403	30	.3338	.9426	.3541	2.824	1.061	2.996	30	1.2305
.3432	40	.3365	.9417	.3574	2.798	1.062	2.971	20	1.2275
.3462	50	.3393	.9407	.3607	2.773	1.063	2.947	10	1.2246
.3491	**20° 00′**	.3420	.9397	.3640	2.747	1.064	2.924	**70° 00′**	1.2217
.3520	10	.3448	.9387	.3673	2.723	1.065	2.901	50	1.2188
.3599	20	.3475	.9377	.3706	2.699	1.066	2.878	40	1.2159
.3578	30	.3502	.9367	.3739	2.675	1.068	2.855	30	1.2130
.3607	40	.3529	.9356	.3772	2.651	1.069	2.833	20	1.2101
.3636	50	.3557	.9346	.3805	2.628	1.070	2.812	10	1.2̇072
.3665	**21° 00′**	.3584	.9336	.3839	2.605	1.071	2.790	**69° 00′**	1.2043
.3694	10	.3611	.9325	.3872	2.583	1.072	2.769	50	1.2014
.3723	20	.3638	.9315	.3906	2.560	1.074	2.749	40	1.1985
.3752	30	.3665	.9304	.3939	2.539	1.075	2.729	30	1.1956
.3782	40	.3692	.9293	.3973	2.517	1.076	2.709	20	1.1926
.3811	50	.3719	.9283	.4006	2.496	1.077	2.689	10	1.1897
.3840	**22° 00′**	.3746	.9272	.4040	2.475	1.079	2.669	**68° 00′**	1.1868
.3869	10	.3773	.9261	.4074	2.455	1.080	2.650	50	1.1839
.3898	20	.3800	.9250	.4108	2.434	1.081	2.632	40	1.1810
.3927	30	.3827	.9239	.4142	2.414	1.082	2.613	30	1.1781
.3956	40	.3854	.9228	.4176	2.394	1.084	2.595	20	1.1752
.3985	50	.3881	.9216	.4210	2.375	1.085	2.577	10	1.1723
.4014	**23° 00′**	.3907	.9205	.4245	2.356	1.086	2.559	**67° 00′**	1.1694
.4043	10	.3934	.9194	.4279	2.337	1.088	2.542	50	1.1665
		cos x	**sin x**	**cot x**	**tan x**	**csc x**	**sec x**	**x degrees**	**x radians**

NATURAL TRIGONOMETRIC FUNCTIONS TO FOUR PLACES (continued)

x radians	x degrees	sin x	cos x	tan x	cot x	sec x	csc x		
.4072	20	.3961	.9182	.4314	2.318	1.089	2.525	40	1.1636
.4102	30	.3987	.9171	.4348	2.300	1.090	2.508	30	1.1606
4131	40	.4014	.9159	.4383	2.282	1.092	2.491	20	1.1577
.4160	50	4041	.9147	.4417	2.264	1.093	2.475	10	1.1548
.4189	**24° 00′**	.4067	.9135	.4452	2.246	1.095	2.459	**66° 00′**	1.1519
.4218	10	.4094	.9124	.4487	2.229	1.096	2.443	50	1.1490
.4247	20	.4120	.9112	.4522	2.211	1.097	2.427	40	1.1461
.4276	30	.4147	.9100	.4557	2.194	1.099	2.411	30	1.1432
.4305	40	.4173	.9088	.4592	2.177	1.100	2.396	20	1.1403
.4334	50	.4200	.9075	.4628	2.161	1.102	2.381	10	1.1374
.4363	**25° 00′**	.4226	.9063	.4663	2.145	1.103	2.366	**65° 00′**	1.1345
.4392	10	.4253	.9051	.4699	2.128	1.105	2.352	50	1.1316
.4422	20	.4279	.9038	.4734	2.112	1.106	2.337	40	1.1286
.4451	30	.4305	.9026	.4770	2.097	1.108	2.323	30	1.1257
.4480	40	.4331	.9013	.4806	2.081	1.109	2.309	20	1.1228
.4509	50	.4358	.9001	.4841	2.066	1.111	2.295	10	1.1199
.4538	**26° 00′**	.4384	.8988	.4877	2.050	1.113	2.281	**64° 00′**	1.1170
.4567	10	.4410	.8975	.4913	2.035	1.114	2.268	50	1.1141
.4596	20	.4436	.8962	.4950	2.020	1.116	2.254	40	1.1112
.4625	30	.4462	.8949	.4986	2.006	1.117	2.241	30	1.1083
.4654	40	.4488	.8936	.5022	1.991	1.119	2.228	20	1.1054
.4683	50	.4514	.8923	.5059	1.977	1.121	2.215	10	1.1025
.4712	**27° 00′**	.4540	.8910	.5095	1.963	1.122	2.203	**63° 00′**	1.0996
.4741	10	.4566	.8897	.5132	1.949	1.124	2.190	50	1.0966
.4771	20	.4592	.8884	.5169	1.935	1.126	2.178	40	1.0937
.4800	30	.4617	.8870	.5206	1.921	1.127	2.166	30	1.0908
.4829	40	.4643	.8857	.5243	1.907	1.129	2.154	20	1.0879
.4858	50	.4669	.8843	.5280	1.894	1.131	2.142	10	1.0850
.4887	**28° 00′**	.4695	.8829	.5317	1.881	1.133	2.130	**62° 00′**	1.0821
.4916	10	.4720	.8816	.5354	1.868	1.134	2.118	50	1.0792
.4945	20	.4746	.8802	.5392	1.855	1.136	2.107	40	1.0763
.4974	30	.4772	.8788	.5430	1.842	1.138	2.096	30	1.0734
.5003	40	.4797	.8774	.5467	1.829	1.140	2.085	20	1.0705
.5032	50	.4823	.8760	.5505	1.816	1.142	2.074	10	1.0676
.5061	**29° 00′**	.4848	.8746	.5543	1.804	1.143	2.063	**61° 00′**	1.0647
.5091	10	.4874	.8732	.5581	1.792	1.145	2.052	50	1.0617
.5120	20	.4899	.8718	.5619	1.780	1.147	2.041	40	1.0588
.5149	30	.4924	.8704	.5658	1.767	1.149	2.031	30	1.0559
.5178	40	.4950	.8689	.5696	1.756	1.151	2.020	20	1.0530
.5207	50	.4975	.8675	.5735	1.744	1.153	2.010	10	1.0501
.5236	**30° 00′**	.5000	.8660	.5774	1.732	1.155	2.000	**60° 00′**	1.0472
.5265	10	.5025	.8646	.5812	1.720	1.157	1.990	50	1.0443
.5294	20	.5050	.8631	.5851	1.709	1.159	1.980	40	1.0414
.5323	30	.5075	.8616	.5890	1.698	1.161	1.970	30	1.0385
.5352	40	.5100	.8601	.5930	1.686	1.163	1.961	20	1.0356
.5381	50	.5125	.8587	.5969	1.675	1.165	1.951	10	1.0327
.5411	**31° 00′**	.5150	.8572	.6009	1.664	1.167	1.942	**59° 00′**	1.0297
.5440	10	.5175	.8557	.6048	1.653	1.169	1.932	50	1.0268
.5469	20	.5200	.8542	.6088	1.643	1.171	1.923	40	1.0239
.5498	30	.5225	.8526	.6128	1.632	1.173	1.914	30	1.0210
.5527	40	.5250	.8511	.6168	1.621	1.175	1.905	20	1.0181
.5556	50	.5275	.8496	.6208	1.611	1.177	1.896	10	1.0152
.5585	**32° 00′**	.5299	8480	.6249	1.600	1.179	1.887	**58° 00′**	1.0123
.5614	10	.5324	.8465	.6289	1.590	1.181	1.878	50	1.0094
.5643	20	.5348	.8450	.6330	1.580	1.184	1.870	40	1.0065
.5672	30	.5373	.8434	.6371	1.570	1.186	1.861	30	1.0036
.5701	40	.5398	.8418	.6412	1.560	1.188	1.853	20	1.0007
.5730	50	.5422	.8403	.6453	1.550	1.190	1.844	10	.9977
.5760	**33° 00′**	.5446	.8397	.6494	1.540	1.192	1.836	**57° 00′**	.9948
.5789	10	.5471	.8371	.6536	1.530	1.195	1.828	50	.9919
.5818	20	.5495	.8355	.6577	1.520	1.197	1.820	40	.9890
.5847	30	.5519	.8339	.6619	1.511	1.199	1.812	30	.9861
.5876	40	.5544	.8323	.6661	1.501	1.202	1.804	20	.9832
.5905	50	.5568	.8307	.6703	1.492	1.204	1.796	10	.9803
.5934	**34° 00′**	.5592	.8290	.6745	1.483	1.206	1.788	**56° 00′**	.9774
.5963	10	.5616	.8274	.6787	1.473	1.209	1.781	50	.9745
.5992	20	.5640	.8258	.6830	1.464	1.211	1.773	40	.9716
.6021	30	.5664	.8241	.6873	1.455	1.213	1.766	30	.9687
.6050	40	.5688	.8225	.6916	1.446	1.216	1.758	20	.9657
.6080	50	.5712	.8208	.6959	1.437	1.218	1.751	10	.9628
		cos x	sin x	cot x	tan x	csc x	sec x	x degrees	x radians

NATURAL TRIGONOMETRIC FUNCTIONS TO FOUR PLACES (continued)

x radians	x degrees	sin x	cos x	tan x	cot x	sec x	csc x		
.6109	**35° 00′**	.5736	.8192	.7002	1.428	1.221	1.743	**55° 00′**	.9599
.6138	10	.5760	.8175	.7046	1.419	1.223	1.736	50	.9570
.6167	20	.5783	.8158	.7089	1.411	1.226	1.729	40	.9541
.6196	30	.5807	.8141	.7133	1.402	1.228	1.722	30	.9512
.6225	40	.5831	.8124	.7177	1.393	1.231	1.715	20	.9483
.6254	50	.5854	.8107	.7221	1.385	1.233	1.708	10	.9454
.6283	**36° 00′**	.5878	.8090	.7265	1.376	1.236	1.701	**54° 00′**	.9425
.6312	10	.5901	.8073	.7310	1.368	1.239	1.695	50	.9396
.6341	20	.5925	.8056	.7355	1.360	1.241	1.688	40	.9367
.6370	30	.5948	.8039	.7400	1.351	1.244	1.681	30	.9338
.6400	40	.5972	.8021	.7445	1.343	1.247	1.675	20	.9308
.6429	50	.5995	.8004	.7490	1.335	1.249	1.668	10	.9279
.6458	**37° 00′**	.6018	.7986	.7536	1.327	1.252	1.662	**53° 00′**	.9250
.6487	10	.6041	.7969	.7581	1.319	1.255	1.655	50	.9221
.6516	20	.6065	.7951	.7627	1.311	1.258	1.649	40	.9192
.6545	30	.6088	.7934	.7673	1.303	1.260	1.643	30	.9163
.6574	40	.6111	.7916	.7720	1.295	1.263	1.636	20	.9134
.6603	50	.6134	.7898	.7766	1.288	1.266	1.630	10	.9105
.6632	**38° 00′**	.6157	.7880	.7813	1.280	1.269	1.624	**52° 00′**	.9076
.6661	10	.6180	.7862	.7860	1.272	1.272	1.618	50	.9047
.6690	20	.6202	.7844	.7907	1.265	1.275	1.612	40	.9018
.6720	30	.6225	.7826	.7954	1.257	1.278	1.606	30	.8988
.6749	40	.6248	.7808	.8002	1.250	1.281	1.601	20	.8959
.6778	50	.6271	.7790	.8050	1.242	1.284	1.595	10	.8930
.6807	**39° 00′**	.6293	.7771	.8098	1.235	1.287	1.589	**51° 00′**	.8901
.6836	10	.6316	.7753	.8146	1.228	1.290	1.583	50	.8872
.6865	20	.6338	.7735	.8195	1.220	1.293	1.578	40	.8843
.6894	30	.6361	.7716	.8243	1.213	1.296	1.572	30	.8814
.6923	40	.6383	.7698	.8292	1.206	1.299	1.567	20	.8785
.6952	50	.6406	.7679	.8342	1.199	1.302	1.561	10	.8756
.6981	**40° 00′**	.6428	.7660	.8391	1.192	1.305	1.556	**50° 00′**	.8727
.7010	10	.6450	.7642	.8441	1.185	1.309	1.550	50	.8698
.7039	20	.6472	.7623	.8491	1.178	1.312	1.545	40	.8668
.7069	30	.6494	.7604	.8541	1.171	1.315	1.540	30	.8639
.7098	40	.6517	.7585	.8591	1.164	1.318	1.535	20	.8610
.7127	50	.6539	.7566	.8642	1.157	1.322	1.529	10	.8581
.7156	**41° 00′**	.6561	.7547	.8693	1.150	1.325	1.524	**49° 00′**	.8552
.7185	10	.6583	.7528	.8744	1.144	1.328	1.519	50	.8523
.7214	20	.6604	.7509	.8796	1.137	1.332	1.514	40	.8494
.7243	30	.6626	.7490	.8847	1.130	1.335	1.509	30	.8465
.7272	40	.6648	.7470	.8899	1.124	1.339	1.504	20	.8436
.7301	50	.6670	.7451	.8952	1.117	1.342	1.499	10	.8407
.7330	**42° 00′**	.6691	.7431	.9004	1.111	1.346	1.494	**48° 00′**	.8378
.7359	10	.6713	.7412	.9057	1.104	1.349	1.490	50	.8348
7389	20	.6734	.7392	.9110	1.098	1.353	1.485	40	.8319
.7418	30	.6756	.7373	.9163	1.091	1.356	1.480	30	.8290
.7447	40	.6777	.7353	.9217	1.085	1.360	1.476	20	.8261
.7476	50	.6799	.7333	.9271	1.079	1.364	1.471	10	.8232
.7505	**43° 00′**	.6820	.7314	.9325	1.072	1.367	1.466	**47° 00′**	.8203
.7534	10	.6841	.7294	.9380	1.066	1.371	1.462	50	.8174
.7563	20	.6862	.7274	.9435	1.060	1.375	1.457	40	.8145
.7592	30	.6884	.7254	.9490	1.054	1.379	1.453	30	.8116
.7621	40	.6905	.7234	.9545	1.048	1.382	1.448	20	.8087
.7650	50	.6926	.7214	.9601	1.042	1.386	1.444	10	.8058
.7679	**44° 00′**	.6947	.7193	.9657	1.036	1.390	1.440	**46° 00′**	.8029
.7709	10	.6967	.7173	.9713	1.030	1.394	1.435	50	.7999
.7738	20	.6988	.7153	.9770	1.024	1.398	1.431	40	.7970
.7767	30	.7009	.7133	.9827	1.018	1.402	1.427	30	.7941
.7796	40	.7030	.7112	.9884	1.012	1.406	1.423	20	.7912
.7825	50	.7050	.7092	.9942	1.006	1.410	1.418	10	.7883
.7854	**45° 00′**	.7071	.7071	1.0000	1.0000	1.414	1.414	**45° 00′**	.7854
		cos x	**sin x**	**cot x**	**tan x**	**csc x**	**sec x**	**x degrees**	**x radians**

RELATION OF ANGULAR FUNCTIONS IN TERMS OF ONE ANOTHER

TRIGONOMETRIC FUNCTIONS

Function	sin α	cos α	tan α	cot α	sec α	csc α
$\sin\alpha$	$\sin\alpha$	$\pm\sqrt{1-\cos^2\alpha}$	$\frac{\tan\alpha}{\pm\sqrt{1+\tan^2\alpha}}$	$\frac{1}{\pm\sqrt{1+\cot^2\alpha}}$	$\frac{\pm\sqrt{\sec^2\alpha-1}}{\sec\alpha}$	$\frac{1}{\csc\alpha}$
$\cos\alpha$	$\pm\sqrt{1-\sin^2\alpha}$	$\cos\alpha$	$\frac{1}{\pm\sqrt{1+\tan^2\alpha}}$	$\frac{\cot\alpha}{\pm\sqrt{1+\cot^2\alpha}}$	$\frac{1}{\sec\alpha}$	$\frac{\pm\sqrt{\csc^2\alpha-1}}{\csc\alpha}$
$\tan\alpha$	$\frac{\sin\alpha}{\pm\sqrt{1-\sin^2\alpha}}$	$\frac{\pm\sqrt{1-\cos^2\alpha}}{\cos\alpha}$	$\tan\alpha$	$\frac{1}{\cot\alpha}$	$\pm\sqrt{\sec^2\alpha-1}$	$\frac{1}{\pm\sqrt{\csc^2\alpha-1}}$
$\cot\alpha$	$\frac{\pm\sqrt{1-\sin^2\alpha}}{\sin\alpha}$	$\frac{\cos\alpha}{\pm\sqrt{1-\cos^2\alpha}}$	$\frac{1}{\tan\alpha}$	$\cot\alpha$	$\frac{1}{\pm\sqrt{\sec^2\alpha-1}}$	$\pm\sqrt{\csc^2\alpha-1}$
$\sec\alpha$	$\frac{1}{\pm\sqrt{1-\sin^2\alpha}}$	$\frac{1}{\cos\alpha}$	$\pm\sqrt{1+\tan^2\alpha}$	$\frac{\pm\sqrt{1+\cot^2\alpha}}{\cot\alpha}$	$\sec\alpha$	$\frac{\csc\alpha}{\pm\sqrt{\csc^2\alpha-1}}$
$\csc\alpha$	$\frac{1}{\sin\alpha}$	$\frac{1}{\pm\sqrt{1-\cos^2\alpha}}$	$\frac{\pm\sqrt{1+\tan^2\alpha}}{\tan\alpha}$	$\pm\sqrt{1+\cot^2\alpha}$	$\frac{\sec\alpha}{\pm\sqrt{\sec^2\alpha-1}}$	$\csc\alpha$

Note: The choice of sign depends upon the quadrant in which the angle terminates.

HYPERBOLIC FUNCTIONS

Function	sinh x	cosh x	tanh x
$\sinh x =$	$\sinh x$	$+\sqrt{\cosh^2 x-1}$	$\frac{\tanh x}{\sqrt{1-\tanh^2 x}}$
$\cosh x =$	$\sqrt{1+\sinh^2 x}$	$\cosh x$	$\frac{1}{\sqrt{1-\tanh^2 x}}$
$\tanh x =$	$\frac{\sinh x}{\sqrt{1+\sinh^2 x}}$	$\pm\frac{\sqrt{\cosh^2 x-1}}{\cosh x}$	$\tanh x$
$\text{cosech } x =$	$\frac{1}{\sinh x}$	$\pm\frac{1}{\sqrt{\cosh^2 x-1}}$	$\frac{\sqrt{1-\tanh^2 x}}{\tanh x}$
$\text{sech } x =$	$\frac{1}{\sqrt{1+\sinh^2 x}}$	$\frac{1}{\cosh x}$	$\sqrt{1-\tanh^2 x}$
$\coth x =$	$\frac{\sqrt{1+\sinh^2 x}}{\sinh x}$	$\frac{\pm\cosh x}{\sqrt{\cosh^2 x-1}}$	$\frac{1}{\tanh x}$

Function	cosech x	sech x	coth x
$\sinh x =$	$\frac{1}{\text{cosech } x}$	$\pm\frac{\sqrt{1-\text{sech}^2 x}}{\text{sech } x}$	$\frac{\pm 1}{\sqrt{\coth^2 x-1}}$
$\cosh x =$	$\pm\frac{\sqrt{\text{cosech}^2 x+1}}{\text{cosech } x}$	$\frac{1}{\text{sech } x}$	$\pm\frac{\coth x}{\sqrt{\coth^2 x-1}}$
$\tanh x =$	$\frac{1}{\sqrt{\text{cosech}^1 x+1}}$	$\pm\sqrt{1-\text{sech}^2 x}$	$\frac{1}{\coth x}$
$\text{cosech } x =$	$\text{cosech } x$	$\pm\frac{\text{sech } x}{\sqrt{1-\text{sech}^1 \text{zsx}}}$	$\pm\sqrt{\coth^2 x-1}$
$\text{sech } x =$	$\pm\frac{\text{cosec } x}{\sqrt{\text{cosech}^2 x+1}}$	$\text{sech } x$	$\pm\frac{\sqrt{\coth^2 x-1}}{\coth x}$
$\coth x =$	$\sqrt{\text{cosech}^2 x+1}$	$\pm\frac{1}{\sqrt{1-\text{sech}^2 x}}$	$\coth x$

Whenever two signs are shown, choose + sign if x is positive, − sign if x is negative.

*Derivatives

In the following formulas u, v, w represent functions of x, while a, c, n represent fixed real numbers. All arguments in the trigonometric functions are measured in radians, and all inverse trigonometric and hyperbolic functions represent principal values.

1. $\frac{d}{dx}(a) = 0$

2. $\frac{d}{dx}(x) = 1$

3. $\frac{d}{dx}(au) = a\frac{du}{dx}$

4. $\frac{d}{dx}(u + v - w) = \frac{du}{dx} + \frac{dv}{dx} - \frac{dw}{dx}$

5. $\frac{d}{dx}(uv) = u\frac{dv}{dx} + v\frac{du}{dx}$

6. $\frac{d}{dx}(uvw) = uv\frac{dw}{dx} + vw\frac{du}{dx} + uw\frac{dv}{dx}$

7. $\frac{d}{dx}\left(\frac{u}{v}\right) = \frac{v\frac{du}{dx} - u\frac{dv}{dx}}{v^2} = \frac{1}{v}\frac{du}{dx} - \frac{u}{v^2}\frac{dv}{dx}$

8. $\frac{d}{dx}(u^n) = nu^{n-1}\frac{du}{dx}$

9. $\frac{d}{dx}(\sqrt{u}) = \frac{1}{2\sqrt{u}}\frac{du}{dx}$

10. $\frac{d}{dx}\left(\frac{1}{u}\right) = -\frac{1}{u^2}\frac{du}{dx}$

11. $\frac{d}{dx}\left(\frac{1}{u^n}\right) = -\frac{n}{u^{n+1}}\frac{du}{dx}$

12. $\frac{d}{dx}\left(\frac{u^n}{v^m}\right) = \frac{u^{n-1}}{v^{m+1}}\left(nv\frac{du}{dx} - mu\frac{dv}{dx}\right)$

13. $\frac{d}{dx}(u^n v^m) = u^{n-1}v^{m-1}\left(nv\frac{du}{dx} + mu\frac{dv}{dx}\right)$

14. $\frac{d}{dx}[f(u)] = \frac{d}{du}[f(u)] \cdot \frac{du}{dx}$

* Let $y = f(x)$ and $\frac{dy}{dx} = \frac{d[f(x)]}{dx} = f'(x)$ define respectively a function and its derivative for any value x in their common domain. The differential for the function at such a value x is accordingly defined as

$$dy = d[f(x)] = \frac{dy}{dx}dx = \frac{d[f(x)]}{dx}dx = f'(x)\,dx$$

Each derivative formula has an associated differential formula. For example, formula 6 above has the differential formula

$$d(uvw) = uv\,dw + vw\,du + uw\,dv$$

15. $$\frac{d^2}{dx^2}[f(u)] = \frac{df(u)}{du}\cdot\frac{d^2u}{dx^2} + \frac{d^2f(u)}{du^2}\cdot\left(\frac{du}{dx}\right)^2$$

16. $$\frac{d^n}{dx^n}[uv] = \binom{n}{0}v\frac{d^nu}{dx^n} + \binom{n}{1}\frac{dv}{dx}\frac{d^{n-1}u}{dx^{n-1}} + \binom{n}{2}\frac{d^2v}{dx^2}\frac{d^{n-2}u}{dx^{n-2}} + \cdots + \binom{n}{k}\frac{d^kv}{dx^k}\frac{d^{n-k}u}{dx^{n-k}} + \cdots + \binom{n}{n}u\frac{d^nv}{dx^n}$$

where $\binom{n}{r} = \frac{n!}{r!(n-r)!}$ the binomial coefficient, n non-negative integer and $\binom{n}{0} = 1$.

17. $$\frac{du}{dx} = \frac{1}{\frac{dx}{du}} \quad \text{if } \frac{dx}{du} \neq 0$$

18. $$\frac{d}{dx}(\log_a u) = (\log_a e)\frac{1}{u}\frac{du}{dx}$$

19. $$\frac{d}{dx}(\log_e u) = \frac{1}{u}\frac{du}{dx}$$

20. $$\frac{d}{dx}(a^u) = a^u(\log_e a)\frac{du}{dx}$$

21. $$\frac{d}{dx}(e^u) = e^u\frac{du}{dx}$$

22. $$\frac{d}{dx}(u^v) = vu^{v-1}\frac{du}{dx} + (\log_e u)u^v\frac{dv}{dx}$$

23. $$\frac{d}{dx}(\sin u) = \frac{du}{dx}(\cos u)$$

24. $$\frac{d}{dx}(\cos u) = -\frac{du}{dx}(\sin u)$$

25. $$\frac{d}{dx}(\tan u) = \frac{du}{dx}(\sec^2 u)$$

26. $$\frac{d}{dx}(\cot u) = -\frac{du}{dx}(\csc^2 u)$$

27. $$\frac{d}{dx}(\sec u) = \frac{du}{dx}\sec u \cdot \tan u$$

28. $$\frac{d}{dx}(\csc u) = -\frac{du}{dx}\csc u \cdot \cot u$$

29. $$\frac{d}{dx}(\text{vers } u) = \frac{du}{dx}\sin u$$

30. $$\frac{d}{dx}(\arcsin u) = \frac{1}{\sqrt{1-u^2}}\frac{du}{dx}, \quad \left(-\frac{\pi}{2} \leq \arcsin u \leq \frac{\pi}{2}\right)$$

31. $\frac{d}{dx}(\text{arc cos } u) = -\frac{1}{\sqrt{1-u^2}}\frac{du}{dx}, \qquad (0 \leq \text{arc cos } u \leq \pi)$

32. $\frac{d}{dx}(\text{arc tan } u) = \frac{1}{1+u^2}\frac{du}{dx}, \qquad \left(-\frac{\pi}{2} < \text{arc tan } u < \frac{\pi}{2}\right)$

33. $\frac{d}{dx}(\text{arc cot } u) = -\frac{1}{1+u^2}\frac{du}{dx}, \qquad (0 \leq \text{arc cot } u \leq \pi)$

34. $\frac{d}{dx}(\text{arc sec } u) = \frac{1}{u\sqrt{u^2-1}}\frac{du}{dx}, \qquad \left(0 \leq \text{arc sec } u < \frac{\pi}{2}, -\pi \leq \text{arc sec } u < -\frac{\pi}{2}\right)$

35. $\frac{d}{dx}(\text{arc csc } u) = -\frac{1}{u\sqrt{u^2-1}}\frac{du}{dx}, \qquad \left(0 < \text{arc csc } u \leq \frac{\pi}{2}, -\pi < \text{arc csc } u \leq -\frac{\pi}{2}\right)$

36. $\frac{d}{dx}(\text{arc vers } u) = \frac{1}{\sqrt{2u-u^2}}\frac{du}{dx}, \qquad (0 \leq \text{arc vers } u \leq \pi)$

37. $\frac{d}{dx}(\sinh u) = \frac{du}{dx}(\cosh u)$

38. $\frac{d}{dx}(\cosh u) = \frac{du}{dx}(\sinh u)$

39. $\frac{d}{dx}(\tanh u) = \frac{du}{dx}(\text{sech}^2\, u)$

40. $\frac{d}{dx}(\coth u) = -\frac{du}{dx}(\text{csch}^2\, u)$

41. $\frac{d}{dx}(\text{sech } u) = -\frac{du}{dx}(\text{sech } u \cdot \tanh u)$

42. $\frac{d}{dx}(\text{csch } u) = -\frac{du}{dx}(\text{csch } u \cdot \coth u)$

43. $\frac{d}{dx}(\sinh^{-1} u) = \frac{d}{dx}[\log(u + \sqrt{u^2+1})] = \frac{1}{\sqrt{u^2+1}}\frac{du}{dx}$

44. $\frac{d}{dx}(\cosh^{-1} u) = \frac{d}{dx}[\log(u + \sqrt{u^2-1})] = \frac{1}{\sqrt{u^2-1}}\frac{du}{dx}, \qquad (u > 1, \cosh^{-1} u > 0)$

45. $\frac{d}{dx}(\tanh^{-1} u) = \frac{d}{dx}\left[\frac{1}{2}\log\frac{1+u}{1-u}\right] = \frac{1}{1-u^2}\frac{du}{dx}, \qquad (u^2 < 1)$

46. $\frac{d}{dx}(\coth^{-1} u) = \frac{d}{dx}\left[\frac{1}{2}\log\frac{u+1}{u-1}\right] = \frac{1}{1-u^2}\frac{du}{dx}, \qquad (u^2 > 1)$

47. $\frac{d}{dx}(\text{sech}^{-1} u) = \frac{d}{dx}\left[\log\frac{1+\sqrt{1-u^2}}{u}\right] = -\frac{1}{u\sqrt{1-u^2}}\frac{du}{dx}, \qquad (0 < u < 1, \text{sech}^{-1} u > 0)$

48. $\frac{d}{dx}(\text{csch}^{-1} u) = \frac{d}{dx}\left[\log\frac{1+\sqrt{1+u^2}}{u}\right] = -\frac{1}{|u|\sqrt{1+u^2}}\frac{du}{dx}$

49. $\dfrac{d}{dq}\displaystyle\int_p^q f(x)\,dx = f(q),$ $\qquad$ [p constant]

50. $\dfrac{d}{dp}\displaystyle\int_p^q f(x)\,dx = -f(p),$ $\qquad$ [q constant]

51. $\dfrac{d}{da}\displaystyle\int_p^q f(x,a)\,dx = \int_p^q \frac{\partial}{\partial a}[f(x,a)]\,dx + f(q,a)\frac{dq}{da} - f(p,a)\frac{dp}{da}$

INTEGRATION

The following is a brief discussion of some integration techniques. A more complete discussion can be found in a number of good text books. However, the purpose of this introduction is simply to discuss a few of the important techniques which may be used, in conjunction with the integral table which follows, to integrate particular functions.

No matter how extensive the integral table, it is a fairly uncommon occurrence to find in the table the exact integral desired. Usually some form of transformation will have to be made. The simplest type of transformation, and yet the most general, is substitution. Simple forms of substitution, such as $y = ax$, are employed almost unconsciously by experienced users of integral tables. Other substitutions may require more thought. In some sections of the tables, appropriate substitutions are suggested for integrals which are similar to, but not exactly like, integrals in the table. Finding the right substitution is largely a matter of intuition and experience.

Several precautions must be observed when using substitutions:

1. Be sure to make the substitution in the dx term, as well as everywhere else in the integral.
2. Be sure that the function substituted is one-to-one and continuous. If this is not the case, the integral must be restricted in such a way as to make it true. See the example following.
3. With definite integrals, the limits should also be expressed in terms of the new dependent variable. With indefinite integrals, it is necessary to perform the reverse substitution to obtain the answer in terms of the original independent variable. This may also be done for definite integrals, but it is usually easier to change the limits.

Example:

$$\int \frac{x^4}{\sqrt{a^2 - x^2}}\,dx$$

Here we make the substitution $x = |a| \sin\theta$. Then $dx = |a| \cos\theta\, d\theta$, and

$$\sqrt{a^2 - x^2} = \sqrt{a^2 - a^2 \sin^2\theta} = |a|\sqrt{1 - \sin^2\theta} = |a \cos\theta|$$

Notice the absolute value signs. It is very important to keep in mind that a square root radical always denotes the positive square root, and to assure the sign is always kept positive. Thus $\sqrt{x^2} = |x|$. Failure to observe this is a common cause of errors in integration.

Notice also that the indicated substitution is not a one-to-one function, that is, it does not have a unique inverse. Thus we must restrict the range of θ in such a way as to make the function one-to-one. Fortunately, this is easily done by solving for θ

$$\theta = \sin^{-1}\frac{x}{|a|}$$

and restricting the inverse sine to the principal values, $-\dfrac{\pi}{2} \le \theta \le \dfrac{\pi}{2}$.

Thus the integral becomes

$$\int \frac{a^4 \sin^4 \theta |a| \cos \theta \, d\theta}{|a|\,|\cos \theta|}$$

Now, however, in the range of values chosen for θ, $\cos \theta$ is always positive. Thus we may remove the absolute value signs from $\cos \theta$ in the denominator. (This is one of the reasons that the principal values of the inverse trigonometric functions are defined as they are.)

Then the $\cos \theta$ terms cancel, and the integral becomes

$$a^4 \int \sin^4 \theta \, d\theta$$

By application of integral formulas 299 and 296, we integrate this to

$$-a^4 \frac{\sin^3 \theta \cos \theta}{4} - \frac{3a^4}{8} \cos \theta \sin \theta + \frac{3a^4}{8} \theta + C$$

We now must perform the inverse substitution to get the result in terms of x. We have

$$\theta = \sin^{-1} \frac{x}{|a|}$$

$$\sin \theta = \frac{x}{|a|}$$

Then

$$\cos \theta = \pm\sqrt{1 - \sin^2 \theta} = \pm\sqrt{1 - \frac{x^2}{a^2}} = \pm\frac{\sqrt{a^2 - x^2}}{|a|}.$$

Because of the previously mentioned fact that $\cos \theta$ is positive, we may omit the $\pm$ sign. The reverse substitution then produces the final answer

$$\int \frac{x^4}{\sqrt{a^2 - x^2}} dx = -\tfrac{1}{4}x^3\sqrt{a^2 - x^2} - \tfrac{3}{8}a^2 x\sqrt{a^2 - x^2} + \frac{3a^4}{8} \sin^{-1} \frac{x}{|a|} + C.$$

Any rational function of x may be integrated, if the denominator is factored into linear and irreducible quadratic factors. The function may then be broken into partial fractions, and the individual partial fractions integrated by use of the appropriate formula from the integral table. See the section on partial fractions for further information.

Many integrals may be reduced to rational functions by proper substitutions. For example,

$$z = \tan \frac{x}{2}$$

will reduce any rational function of the six trigonometric functions of x to a rational function of z. (Frequently there are other substitutions which are simpler to use, but this one will always work. See integral formula number 484.)

Any rational function of x and $\sqrt{ax + b}$ may be reduced to a rational function of z by making the substitution

$$z = \sqrt{ax + b}.$$

Other likely substitutions will be suggested by looking at the form of the integrand.

The other main method of transforming integrals is integration by parts. This involves applying formula number 5 or 6 in the accompanying integral table. The critical factor in this method is the choice of the functions u and v. In order for the method to be successful, $v = \int dv$ and $\int v \, du$ must be easier to integrate than the original integral. Again, this choice is largely a matter of intuition and experience.

Example:

$$\int x \sin x \, dx$$

Two obvious choices are $u = x$, $dv = \sin x \, dx$, or $u = \sin x$, $dv = x \, dx$. Since a preliminary mental calculation indicates that $\int v \, du$ in the second choice would be more, rather than less, complicated than the original integral (it would contain x^2), we use the first choice.

$$u = x \qquad du = dx$$

$$dv = \sin x \, dx \qquad v = -\cos x$$

$$\int x \sin x \, dx = \int u \, dv = uv - \int v \, du = -x \cos x + \int \cos x \, dx$$

$$= \sin x - x \cos x$$

Of course, this result could have been obtained directly from the integral table, but it provides a simple example of the method. In more complicated examples the choice of u and v may not be so obvious, and several different choices may have to be tried. Of course, there is no guarantee that any of them will work.

Integration by parts may be applied more than once, or combined with substitution. A fairly common case is illustrated by the following example.

Example:

$$\int e^x \sin x \, dx$$

Let

$$u = e^x \qquad \text{Then} \quad du = e^x \, dx$$

$$dv = \sin x \, dx \qquad v = -\cos x$$

$$\int e^x \sin x \, dx = \int u \, dv = uv - \int v \, du = -e^x \cos x + \int e^x \cos x \, dx$$

In this latter integral,

$$\text{let} \quad u = e^x \qquad \text{Then} \quad du = e^x \, dx$$

$$dv = \cos x \, dx \qquad v = \sin x$$

$$\int e^x \sin x \, dx = -e^x \cos x + \int e^x \cos x \, dx = -e^x \cos x + \int u \, dv$$

$$= -e^x \cos x + uv - \int v \, du$$

$$= -e^x \cos x + e^x \sin x - \int e^x \sin x \, dx$$

This looks as if a circular transformation has taken place, since we are back at the same integral we started from. However, the above equation can be solved algebraically for the required integral:

$$\int e^x \sin x \, dx = \tfrac{1}{2}(e^x \sin x - e^x \cos x)$$

In the second integration by parts, if the parts had been chosen as $u = \cos x$, $dv = e^x \, dx$, we would indeed have made a circular transformation, and returned to the starting place.

In general, when doing repeated integration by parts, one should never choose the function u at any stage to be the same as the function v at the previous stage, or a constant times the previous v.

The following rule is called the extended rule for integration by parts. It is the result of $n + 1$ successive applications of integration by parts.

If

$$g_1(x) = \int g(x)\,dx, \qquad g_2(x) = \int g_1(x)\,dx,$$

$$g_3(x) = \int g_2(x)\,dx, \ldots, g_m(x) = \int g_{m-1}(x)\,dx, \ldots,$$

then

$$\int f(x)\cdot g(x)\,dx = f(x)\cdot g_1(x) - f'(x)\cdot g_2(x) + f''(x)\cdot g_3(x) - + \cdots$$
$$+ (-1)^n f^{(n)}(x) g_{n+1}(x) + (-1)^{n+1}\int f^{(n+1)}(x) g_{n+1}(x)\,dx.$$

A useful special case of the above rule is when $f(x)$ is a polynomial of degree n. Then $f^{(n+1)}(x) = 0$, and

$$\int f(x)\cdot g(x)\,dx = f(x)\cdot g_1(x) - f'(x)\cdot g_2(x) + f''(x)\cdot g_3(x) - + \cdots + (-1)^n f^{(n)}(x) g_{n+1}(x) + C$$

Example:

If $f(x) = x^2$, $g(x) = \sin x$

$$\int x^2 \sin x\,dx = -x^2\cos x + 2x\sin x + 2\cos x + C$$

Another application of this formula occurs if

$$f''(x) = af(x) \quad \text{and} \quad g''(x) = bg(x),$$

where a and b are unequal constants. In this case, by a process similar to that used in the above example for $\int e^x \sin x\,dx$, we get the formula

$$\int f(x)g(x)\,dx = \frac{f(x)\cdot g'(x) - f'(x)\cdot g(x)}{b - a} + C$$

This formula could have been used in the example mentioned. Here is another example.

Example:

If $f(x) = e^{2x}$, $g(x) = \sin 3x$, then $a = 4$, $b = -9$, and

$$\int e^{2x}\sin 3x\,dx = \frac{3\,e^{2x}\cos 3x - 2\,e^{2x}\sin 3x}{-9 - 4} + C = \frac{e^{2x}}{13}(2\sin 3x - 3\cos 3x) + C$$

The following additional points should be observed when using this table.

1. A constant of integration is to be supplied with the answers for indefinite integrals.
2. Logarithmic expressions are to base $e = 2.71828\ldots$, unless otherwise specified, and are to be evaluated for the absolute value of the arguments involved therein.
3. All angles are measured in radians, and inverse trigonometric and hyperbolic functions represent principal values, unless otherwise indicated.
4. If the application of a formula produces either a zero denominator or the square root of a negative number in the result, there is usually available another form of the answer which

avoids this difficulty. In many of the results, the excluded values are specified, but when such are omitted it is presumed that one can tell what these should be, especially when difficulties of the type herein mentioned are obtained.

5. When inverse trigonometric functions occur in the integrals, be sure that any replacements made for them are strictly in accordance with the rules for such functions. This causes little difficulty when the argument of the inverse trigonometric function is positive, since then all angles involved are in the first quadrant. However, if the argument is negative, special care must be used. Thus if $u > 0$,

$$\sin^{-1} u = \cos^{-1}\sqrt{1-u^2} = \csc^{-1}\frac{1}{u}, \text{ etc.}$$

However, if $u < 0$,

$$\sin^{-1} u = -\cos^{-1}\sqrt{1-u^2} = -\pi - \csc^{-1}\frac{1}{u}, \text{ etc.}$$

See the section on inverse trigonometric functions for a full treatment of the allowable substitutions.

6. In integrals 340–345 and some others, the right side includes expressions of the form

$$A \tan^{-1}[B + C \tan f(x)].$$

In these formulas, the $\tan^{-1}$ does not necessarily represent the principal value. Instead of always employing the principal branch of the inverse tangent function, one must instead use that branch of the inverse tangent function upon which $f(x)$ lies for any particular choice of x.

Example:

$$\int_0^{4\pi} \frac{dx}{2+\sin x} = \frac{2}{\sqrt{3}} \tan^{-1} \left. \frac{2\tan\frac{x}{2}+1}{\sqrt{3}} \right]_0^{4\pi}$$

$$= \frac{2}{\sqrt{3}}\left[\tan^{-1}\frac{2\tan 2\pi + 1}{\sqrt{3}} - \tan^{-1}\frac{2\tan 0 + 1}{\sqrt{3}}\right]$$

$$= \frac{2}{\sqrt{3}}\left[\frac{13\pi}{6} - \frac{\pi}{6}\right] = \frac{4\pi}{\sqrt{3}} = \frac{4\sqrt{3}\pi}{3}$$

Here

$$\tan^{-1}\frac{2\tan 2\pi + 1}{\sqrt{3}} = \tan^{-1}\frac{1}{\sqrt{3}} = \frac{13\pi}{6},$$

since $f(x) = 2\pi$; and

$$\tan^{-1}\frac{2\tan 0 + 1}{\sqrt{3}} = \tan^{-1}\frac{1}{\sqrt{3}} = \frac{\pi}{6},$$

since $f(x) = 0$.

7. B_n and E_n where used in Integrals represents the Bernoulli and Euler numbers as defined in tables of Bernoulli and Euler polynomials contained in certain mathematics reference and handbooks, as for example, Beyer, W. H., *Handbook of Mathematical Sciences,* 5th ed., CRC Press, Inc., West Palm Beach 1978, 577–583.

ELEMENTARY FORMS

1. $\int a\,dx = ax$

2. $\int a \cdot f(x)\,dx = a \int f(x)\,dx$

3. $\int \phi(y)\,dx = \int \frac{\phi(y)}{y'}\,dy, \qquad \text{where } y' = \frac{dy}{dx}$

4. $\int (u + v)\,dx = \int u\,dx + \int v\,dx, \qquad$ where u and v are any functions of x

5. $\int u\,dv = u \int dv - \int v\,du = uv - \int v\,du$

6. $\int u \frac{dv}{dx}\,dx = uv - \int v \frac{du}{dx}\,dx$

7. $\int x^n\,dx = \frac{x^{n+1}}{n+1}, \qquad \text{except } n = -1$

8. $\int \frac{f'(x)\,dx}{f(x)} = \log f(x), \qquad (df(x) = f'(x)\,dx)$

9. $\int \frac{dx}{x} = \log x$

10. $\int \frac{f'(x)\,dx}{2\sqrt{f(x)}} = \sqrt{f(x)}, \qquad (df(x) = f'(x)\,dx)$

11. $\int e^x\,dx = e^x$

12. $\int e^{ax}\,dx = e^{ax}/a$

13. $\int b^{ax}\,dx = \frac{b^{ax}}{a \log b}, \qquad (b > 0)$

14. $\int \log x\,dx = x \log x - x$

15. $\int a^x \log a\,dx = a^x, \qquad (a > 0)$

16. $\int \frac{dx}{a^2 + x^2} = \frac{1}{a} \tan^{-1} \frac{x}{a}$

17. $\int \frac{dx}{a^2 - x^2} = \begin{cases} \frac{1}{a} \tanh^{-1} \frac{x}{a} \\ \text{or} \\ \frac{1}{2a} \log \frac{a + x}{a - x}, \qquad (a^2 > x^2) \end{cases}$

18. $\int \frac{dx}{x^2 - a^2} = \begin{cases} -\frac{1}{a} \coth^{-1} \frac{x}{a} \\ \text{or} \\ \frac{1}{2a} \log \frac{x - a}{x + a}, \qquad (x^2 > a^2) \end{cases}$

19. $$\int \frac{dx}{\sqrt{a^2 - x^2}} = \begin{cases} \sin^{-1} \dfrac{x}{|a|} \\ \text{or} \\ -\cos^{-1} \dfrac{x}{|a|}, \quad (a^2 > x^2) \end{cases}$$

20. $$\int \frac{dx}{\sqrt{x^2 \pm a^2}} = \log (x + \sqrt{x^2 \pm a^2})$$

21. $$\int \frac{dx}{x\sqrt{x^2 - a^2}} = \frac{1}{|a|} \sec^{-1} \frac{x}{a}$$

22. $$\int \frac{dx}{x\sqrt{a^2 \pm x^2}} = -\frac{1}{a} \log \left(\frac{a + \sqrt{a^2 \pm x^2}}{x} \right)$$

FORMS CONTAINING $(a + bx)$

For forms containing $a + bx$, but not listed in the table, the substitution $u = \dfrac{a + bx}{x}$ may prove helpful.

23. $$\int (a + bx)^n \, dx = \frac{(a + bx)^{n+1}}{(n + 1)b}, \quad (n \neq -1)$$

24. $$\int x(a + bx)^n \, dx$$
$$= \frac{1}{b^2(n + 2)}(a + bx)^{n+2} - \frac{a}{b^2(n + 1)}(a + bx)^{n+1}, \quad (n \neq -1, -2)$$

25. $$\int x^2(a + bx)^n \, dx = \frac{1}{b^3}\left[\frac{(a + bx)^{n+3}}{n + 3} - 2a\frac{(a + bx)^{n+2}}{n + 2} + a^2\frac{(a + bx)^{n+1}}{n + 1}\right]$$

26. $$\int x^m(a + bx)^n \, dx = \begin{cases} \dfrac{x^{m+1}(a + bx)^n}{m + n + 1} + \dfrac{an}{m + n + 1}\displaystyle\int x^m(a + bx)^{n-1} \, dx \\ \text{or} \\ \dfrac{1}{a(n + 1)}\Big[-x^{m+1}(a + bx)^{n+1} \\ \qquad + (m + n + 2)\displaystyle\int x^m(a + bx)^{n+1} \, dx\Big] \\ \text{or} \\ \dfrac{1}{b(m + n + 1)}\left[x^m(a + bx)^{n+1} - ma\displaystyle\int x^{m-1}(a + bx)^n \, dx\right] \end{cases}$$

27. $$\int \frac{dx}{a + bx} = \frac{1}{b} \log (a + bx)$$

28. $$\int \frac{dx}{(a + bx)^2} = -\frac{1}{b(a + bx)}$$

29. $$\int \frac{dx}{(a + bx)^3} = -\frac{1}{2b(a + bx)^2}$$

30. $$\int \frac{x \, dx}{a + bx} = \begin{cases} \dfrac{1}{b^2}[a + bx - a \log (a + bx)] \\ \text{or} \\ \dfrac{x}{b} - \dfrac{a}{b^2} \log (a + bx) \end{cases}$$

31. $\int \frac{x\,dx}{(a+bx)^2} = \frac{1}{b^2}\left[\log(a+bx) + \frac{a}{a+bx}\right]$

32. $\int \frac{x\,dx}{(a+bx)^n} = \frac{1}{b^2}\left[\frac{-1}{(n-2)(a+bx)^{n-2}} + \frac{a}{(n-1)(a+bx)^{n-1}}\right], \quad n \neq 1, 2$

33. $\int \frac{x^2\,dx}{a+bx} = \frac{1}{b^3}\left[\frac{1}{2}(a+bx)^2 - 2a(a+bx) + a^2\log(a+bx)\right]$

34. $\int \frac{x^2\,dx}{(a+bx)^2} = \frac{1}{b^3}\left[a+bx - 2a\log(a+bx) - \frac{a^2}{a+bx}\right]$

35. $\int \frac{x^2\,dx}{(a+bx)^3} = \frac{1}{b^3}\left[\log(a+bx) + \frac{2a}{a+bx} - \frac{a^2}{2(a+bx)^2}\right]$

36. $\int \frac{x^2\,dx}{(a+bx)^n} = \frac{1}{b^3}\left[\frac{-1}{(n-3)(a+bx)^{n-3}} + \frac{2a}{(n-2)(a+bx)^{n-2}} - \frac{a^2}{(n-1)(a+bx)^{n-1}}\right], \quad n \neq 1, 2, 3$

37. $\int \frac{dx}{x(a+bx)} = -\frac{1}{a}\log\frac{a+bx}{x}$

38. $\int \frac{dx}{x(a+bx)^2} = \frac{1}{a(a+bx)} - \frac{1}{a^2}\log\frac{a+bx}{x}$

39. $\int \frac{dx}{x(a+bx)^3} = \frac{1}{a^3}\left[\frac{1}{2}\left(\frac{2a+bx}{a+bx}\right)^2 + \log\frac{x}{a+bx}\right]$

40. $\int \frac{dx}{x^2(a+bx)} = -\frac{1}{ax} + \frac{b}{a^2}\log\frac{a+bx}{x}$

41. $\int \frac{dx}{x^3(a+bx)} = \frac{2bx-a}{2a^2x^2} + \frac{b^2}{a^3}\log\frac{x}{a+bx}$

42. $\int \frac{dx}{x^2(a+bx)^2} = -\frac{a+2bx}{a^2x(a+bx)} + \frac{2b}{a^3}\log\frac{a+bx}{x}$

FORMS CONTAINING $c^2 \pm x^2$, $x^2 - c^2$

43. $\int \frac{dx}{c^2+x^2} = \frac{1}{c}\tan^{-1}\frac{x}{c}$

44. $\int \frac{dx}{c^2-x^2} = \frac{1}{2c}\log\frac{c+x}{c-x}, \quad (c^2 > x^2)$

45. $\int \frac{dx}{x^2-c^2} = \frac{1}{2c}\log\frac{x-c}{x+c}, \quad (x^2 > c^2)$

46. $\int \frac{x\,dx}{c^2 \pm x^2} = \pm\frac{1}{2}\log(c^2 \pm x^2)$

47. $\int \frac{x\,dx}{(c^2 \pm x^2)^{n+1}} = \mp\frac{1}{2n(c^2 \pm x^2)^n}$

48. $\int \frac{dx}{(c^2 \pm x^2)^n} = \frac{1}{2c^2(n-1)}\left[\frac{x}{(c^2 \pm x^2)^{n-1}} + (2n-3)\int \frac{dx}{(c^2 \pm x^2)^{n-1}}\right]$

49. $\int \frac{dx}{(x^2-c^2)^n} = \frac{1}{2c^2(n-1)}\left[-\frac{x}{(x^2-c^2)^{n-1}} - (2n-3)\int \frac{dx}{(x^2-c^2)^{n-1}}\right]$

50. $\int \frac{x\,dx}{x^2-c^2} = \frac{1}{2}\log(x^2-c^2)$

51. $$\int \frac{x\,dx}{(x^2 - c^2)^{n+1}} = -\frac{1}{2n(x^2 - c^2)^n}$$

FORMS CONTAINING $a + bx$ and $c + dx$

$$u = a + bx, \qquad v = c + dx, \qquad k = ad - bc$$

If $k = 0$, then $v = \frac{c}{a}u$

52. $$\int \frac{dx}{u \cdot v} = \frac{1}{k} \cdot \log\left(\frac{v}{u}\right)$$

53. $$\int \frac{x\,dx}{u \cdot v} = \frac{1}{k}\left[\frac{a}{b}\log(u) - \frac{c}{d}\log(v)\right]$$

54. $$\int \frac{dx}{u^2 \cdot v} = \frac{1}{k}\left(\frac{1}{u} + \frac{d}{k}\log\frac{v}{u}\right)$$

55. $$\int \frac{x\,dx}{u^2 \cdot v} = \frac{-a}{bku} - \frac{c}{k^2}\log\frac{v}{u}$$

56. $$\int \frac{x^2\,dx}{u^2 \cdot v} = \frac{a^2}{b^2ku} + \frac{1}{k^2}\left[\frac{c^2}{d}\log(v) + \frac{a(k - bc)}{b^2}\log(u)\right]$$

57. $$\int \frac{dx}{u^n \cdot v^m} = \frac{1}{k(m-1)}\left[\frac{-1}{u^{n-1} \cdot v^{m-1}} - (m + n - 2)b\int \frac{dx}{u^n \cdot v^{m-1}}\right]$$

58. $$\int \frac{u}{v}\,dx = \frac{bx}{d} + \frac{k}{d^2}\log(v)$$

59. $$\int \frac{u^m\,dx}{v^n} = \begin{cases} \dfrac{-1}{k(n-1)}\left[\dfrac{u^{m+1}}{v^{n-1}} + b(n - m - 2)\displaystyle\int \dfrac{u^m}{v^{n-1}}\,dx\right] \\ \text{or} \\ \dfrac{-1}{d(n-m-1)}\left[\dfrac{u^m}{v^{n-1}} + mk\displaystyle\int \dfrac{u^{m-1}}{v^n}\,dx\right] \\ \text{or} \\ \dfrac{-1}{d(n-1)}\left[\dfrac{u^m}{v^{n-1}} - mb\displaystyle\int \dfrac{u^{m-1}}{v^{n-1}}\,dx\right] \end{cases}$$

FORMS CONTAINING $(a + bx^n)$

60. $$\int \frac{dx}{a + bx^2} = \frac{1}{\sqrt{ab}}\tan^{-1}\frac{x\sqrt{ab}}{a}, \qquad (ab > 0)$$

61. $$\int \frac{dx}{a + bx^2} = \begin{cases} \dfrac{1}{2\sqrt{-ab}}\log\dfrac{a + x\sqrt{-ab}}{a - x\sqrt{-ab}}, & (ab < 0) \\ \text{or} & \\ \dfrac{1}{\sqrt{-ab}}\tanh^{-1}\dfrac{x\sqrt{-ab}}{a}, & (ab < 0) \end{cases}$$

62. $$\int \frac{dx}{a^2 + b^2x^2} = \frac{1}{ab}\tan^{-1}\frac{bx}{a}$$

63. $$\int \frac{x\,dx}{a + bx^2} = \frac{1}{2b}\log(a + bx^2)$$

64. $$\int \frac{x^2\,dx}{a + bx^2} = \frac{x}{b} - \frac{a}{b}\int \frac{dx}{a + bx^2}$$

65. $$\int \frac{dx}{(a + bx^2)^2} = \frac{x}{2a(a + bx^2)} + \frac{1}{2a}\int \frac{dx}{a + bx^2}$$

66. $\int \frac{dx}{a^2 - b^2x^2} = \frac{1}{2ab} \log \frac{a + bx}{a - bx}$

67. $\int \frac{dx}{(a + bx^2)^{m+1}} = \begin{cases} \frac{1}{2ma} \frac{x}{(a + bx^2)^m} + \frac{2m - 1}{2ma} \int \frac{dx}{(a + bx^2)^m} \\ \text{or} \\ \frac{(2m)!}{(m!)^2}\left[\frac{x}{2a} \sum_{r=1}^{m} \frac{r!(r-1)!}{(4a)^{m-r}(2r)!(a + bx^2)^r} + \frac{1}{(4a)^m} \int \frac{dx}{a + bx^2}\right] \end{cases}$

68. $\int \frac{x\,dx}{(a + bx^2)^{m+1}} = -\frac{1}{2bm(a + bx^2)^m}$

69. $\int \frac{x^2\,dx}{(a + bx^2)^{m+1}} = \frac{-x}{2mb(a + bx^2)^m} + \frac{1}{2mb} \int \frac{dx}{(a + bx^2)^m}$

70. $\int \frac{dx}{x(a + bx^2)} = \frac{1}{2a} \log \frac{x^2}{a + bx^2}$

71. $\int \frac{dx}{x^2(a + bx^2)} = -\frac{1}{ax} - \frac{b}{a} \int \frac{dx}{a + bx^2}$

72. $\int \frac{dx}{x(a + bx^2)^{m+1}} = \begin{cases} \frac{1}{2am(a + bx^2)^m} + \frac{1}{a} \int \frac{dx}{x(a + bx^2)^m} \\ \text{or} \\ \frac{1}{2a^{m+1}}\left[\sum_{r=1}^{m} \frac{a^r}{r(a + bx^2)^r} + \log \frac{x^2}{a + bx^2}\right] \end{cases}$

73. $\int \frac{dx}{x^2(a + bx^2)^{m+1}} = \frac{1}{a} \int \frac{dx}{x^2(a + bx^2)^m} - \frac{b}{a} \int \frac{dx}{(a + bx^2)^{m+1}}$

74. $\int \frac{dx}{a + bx^3} = \frac{k}{3a}\left[\frac{1}{2} \log \frac{(k + x)^3}{a + bx^3} + \sqrt{3} \tan^{-1} \frac{2x - k}{k\sqrt{3}}\right], \qquad \left(k = \sqrt[3]{\frac{a}{b}}\right)$

75. $\int \frac{x\,dx}{a + bx^3} = \frac{1}{3bk}\left[\frac{1}{2} \log \frac{a + bx^3}{(k + x)^3} + \sqrt{3} \tan^{-1} \frac{2x - k}{k\sqrt{3}}\right], \qquad \left(k = \sqrt[3]{\frac{a}{b}}\right)$

76. $\int \frac{x^2\,dx}{a + bx^3} = \frac{1}{3b} \log (a + bx^3)$

77. $\int \frac{dx}{a + bx^4} = \frac{k}{2a}\left[\frac{1}{2} \log \frac{x^2 + 2kx + 2k^2}{x^2 - 2kx + 2k^2} + \tan^{-1} \frac{2kx}{2k^2 - x^2}\right],$

$$\left(ab > 0, k = \sqrt[4]{\frac{a}{4b}}\right)$$

78. $\int \frac{dx}{a + bx^4} = \frac{k}{2a}\left[\frac{1}{2} \log \frac{x + k}{x - k} + \tan^{-1} \frac{x}{k}\right], \qquad \left(ab < 0, k = \sqrt[4]{-\frac{a}{b}}\right)$

79. $\int \frac{x\,dx}{a + bx^4} = \frac{1}{2bk} \tan^{-1} \frac{x^2}{k}, \qquad \left(ab > 0, k = \sqrt{\frac{a}{b}}\right)$

80. $\int \frac{x\,dx}{a + bx^4} = \frac{1}{4bk} \log \frac{x^2 - k}{x^2 + k}, \qquad \left(ab < 0, k = \sqrt{-\frac{a}{b}}\right)$

81. $\int \frac{x^2\,dx}{a + bx^4} = \frac{1}{4bk}\left[\frac{1}{2} \log \frac{x^2 - 2kx + 2k^2}{x^2 + 2kx + 2k^2} + \tan^{-1} \frac{2kx}{2k^2 - x^2}\right]$

$$\left(ab > 0, k = \sqrt[4]{\frac{a}{4b}}\right)$$

82. $\int \frac{x^2\,dx}{a+bx^4} = \frac{1}{4bk}\left[\log\frac{x-k}{x+k} + 2\tan^{-1}\frac{x}{k}\right], \qquad \left(ab<0,\ k=\sqrt[4]{-\frac{a}{b}}\right)$

83. $\int \frac{x^3\,dx}{a+bx^4} = \frac{1}{4b}\log(a+bx^4)$

84. $\int \frac{dx}{x(a+bx^n)} = \frac{1}{an}\log\frac{x^n}{a+bx^n}$

85. $\int \frac{dx}{(a+bx^n)^{m+1}} = \frac{1}{a}\int\frac{dx}{(a+bx^n)^m} - \frac{b}{a}\int\frac{x^n\,dx}{(a+bx^n)^{m+1}}$

86. $\int \frac{x^m\,dx}{(a+bx^n)^{p+1}} = \frac{1}{b}\int\frac{x^{m-n}\,dx}{(a+bx^n)^p} - \frac{a}{b}\int\frac{x^{m-n}\,dx}{(a+bx^n)^{p+1}}$

87. $\int \frac{dx}{x^m(a+bx^n)^{p+1}} = \frac{1}{a}\int\frac{dx}{x^m(a+bx^n)^p} - \frac{b}{a}\int\frac{dx}{x^{m-n}(a+bx^n)^{p+1}}$

88. $$\int x^m(a+bx^n)^p\,dx = \begin{cases} \frac{1}{b(np+m+1)}\left[x^{m-n+1}(a+bx^n)^{p+1} - a(m-n+1)\int x^{m-n}(a+bx^n)^p\,dx\right] \\ \text{or} \\ \frac{1}{np+m+1}\left[x^{m+1}(a+bx^n)^p + anp\int x^m(a+bx^n)^{p-1}\,dx\right] \\ \text{or} \\ \frac{1}{a(m+1)}\left[x^{m+1}(a+bx^n)^{p+1} - (m+1+np+n)b\int x^{m+n}(a+bx^n)^p\,dx\right] \\ \text{or} \\ \frac{1}{an(p+1)}\left[-x^{m+1}(a+bx^n)^{p+1} + (m+1+np+n)\int x^m(a+bx^n)^{p+1}\,dx\right] \end{cases}$$

FORMS CONTAINING $c^3 \pm x^3$

89. $\int \frac{dx}{c^3 \pm x^3} = \pm\frac{1}{6c^2}\log\frac{(c\pm x)^3}{c^3\pm x^3} + \frac{1}{c^2\sqrt{3}}\tan^{-1}\frac{2x\mp c}{c\sqrt{3}}$

90. $\int \frac{dx}{(c^3\pm x^3)^2} = \frac{x}{3c^3(c^3\pm x^3)} + \frac{2}{3c^3}\int\frac{dx}{c^3\pm x^3}$

91. $\int \frac{dx}{(c^3\pm x^3)^{n+1}} = \frac{1}{3nc^3}\left[\frac{x}{(c^3\pm x^3)^n} + (3n-1)\int\frac{dx}{(c^3\pm x^3)^n}\right]$

92. $\int \frac{x\,dx}{c^3\pm x^3} = \frac{1}{6c}\log\frac{c^3\pm x^3}{(c\pm x)^3} \pm \frac{1}{c\sqrt{3}}\tan^{-1}\frac{2x\mp c}{c\sqrt{3}}$

93. $\int \frac{x\,dx}{(c^3\pm x^3)^2} = \frac{x^2}{3c^3(c^3\pm x^3)} + \frac{1}{3c^3}\int\frac{x\,dx}{c^3\pm x^3}$

94. $\int \frac{x\,dx}{(c^3 \pm x^3)^{n+1}} = \frac{1}{3nc^3}\left[\frac{x^2}{(c^3 \pm x^3)^n} + (3n-2)\int \frac{x\,dx}{(c^3 \pm x^3)^n}\right]$

95. $\int \frac{x^2\,dx}{c^3 \pm x^3} = \pm\frac{1}{3}\log(c^3 \pm x^3)$

96. $\int \frac{x^2\,dx}{(c^3 \pm x^3)^{n+1}} = \mp\frac{1}{3n(c^3 \pm x^3)^n}$

97. $\int \frac{dx}{x(c^3 \pm x^3)} = \frac{1}{3c^3}\log\frac{x^3}{c^3 \pm x^3}$

98. $\int \frac{dx}{x(c^3 \pm x^3)^2} = \frac{1}{3c^3(c^3 \pm x^3)} + \frac{1}{3c^6}\log\frac{x^3}{c^3 \pm x^3}$

99. $\int \frac{dx}{x(c^3 \pm x^3)^{n+1}} = \frac{1}{3nc^3(c^3 \pm x^3)^n} + \frac{1}{c^3}\int \frac{dx}{x(c^3 \pm x^3)^n}$

100. $\int \frac{dx}{x^2(c^3 \pm x^3)} = -\frac{1}{c^3 x} \mp \frac{1}{c^3}\int \frac{x\,dx}{c^3 \pm x^3}$

101. $\int \frac{dx}{x^2(c^3 \pm x^3)^{n+1}} = \frac{1}{c^3}\int \frac{dx}{x^2(c^3 \pm x^3)^n} \mp \frac{1}{c^3}\int \frac{x\,dx}{(c^3 \pm x^3)^{n+1}}$

FORMS CONTAINING $c^4 \pm x^4$

102. $\int \frac{dx}{c^4 + x^4} = \frac{1}{2c^3\sqrt{2}}\left[\frac{1}{2}\log\frac{x^2 + cx\sqrt{2} + c^2}{x^2 - cx\sqrt{2} + c^2} + \tan^{-1}\frac{cx\sqrt{2}}{c^2 - x^2}\right]$

103. $\int \frac{dx}{c^4 - x^4} = \frac{1}{2c^3}\left[\frac{1}{2}\log\frac{c + x}{c - x} + \tan^{-1}\frac{x}{c}\right]$

104. $\int \frac{x\,dx}{c^4 + x^4} = \frac{1}{2c^2}\tan^{-1}\frac{x^2}{c^2}$

105. $\int \frac{x\,dx}{c^4 - x^4} = \frac{1}{4c^2}\log\frac{c^2 + x^2}{c^2 - x^2}$

106. $\int \frac{x^2\,dx}{c^4 + x^4} = \frac{1}{2c\sqrt{2}}\left[\frac{1}{2}\log\frac{x^2 - cx\sqrt{2} + c^2}{x^2 + cx\sqrt{2} + c^2} + \tan^{-1}\frac{cx\sqrt{2}}{c^2 - x^2}\right]$

107. $\int \frac{x^2\,dx}{c^4 - x^4} = \frac{1}{2c}\left[\frac{1}{2}\log\frac{c + x}{c - x} - \tan^{-1}\frac{x}{c}\right]$

108. $\int \frac{x^3\,dx}{c^4 \pm x^4} = \pm\frac{1}{4}\log(c^4 \pm x^4)$

FORMS CONTAINING $(a + bx + cx^2)$

$X = a + bx + cx^2$ and $q = 4ac - b^2$

If $q = 0$, then $X = c\left(x + \frac{b}{2c}\right)^2$, and formulas starting with 23 should be used in place of these.

109. $\int \frac{dx}{X} = \frac{2}{\sqrt{q}}\tan^{-1}\frac{2cx + b}{\sqrt{q}}, \qquad (q > 0)$

110. $$\int \frac{dx}{X} = \begin{cases} \dfrac{-2}{\sqrt{-q}}\tanh^{-1}\dfrac{2cx + b}{\sqrt{-q}} \\ \text{or} \\ \dfrac{1}{\sqrt{-q}}\log\dfrac{2cx + b - \sqrt{-q}}{2cx + b + \sqrt{-q}}, \qquad (q < 0) \end{cases}$$

111. $\int \frac{dx}{X^2} = \frac{2cx + b}{qX} + \frac{2c}{q}\int \frac{dx}{X}$

112. $$\int \frac{dx}{X^3} = \frac{2cx+b}{q}\left(\frac{1}{2X^2} + \frac{3c}{qX}\right) + \frac{6c^2}{q^2}\int \frac{dx}{X}$$

113. $$\int \frac{dx}{X^{n+1}} = \begin{cases} \dfrac{2cx+b}{nqX^n} + \dfrac{2(2n-1)c}{qn}\displaystyle\int \frac{dx}{X^n} \\ \text{or} \\ \dfrac{(2n)!}{(n!)^2}\left(\dfrac{c}{q}\right)^n\left[\dfrac{2cx+b}{q}\displaystyle\sum_{r=1}^{n}\left(\dfrac{q}{cX}\right)^r\left(\dfrac{(r-1)!r!}{(2r)!}\right) + \displaystyle\int \frac{dx}{X}\right] \end{cases}$$

114. $$\int \frac{x\,dx}{X} = \frac{1}{2c}\log X - \frac{b}{2c}\int \frac{dx}{X}$$

115. $$\int \frac{x\,dx}{X^2} = -\frac{bx+2a}{qX} - \frac{b}{q}\int \frac{dx}{X}$$

116. $$\int \frac{x\,dx}{X^{n+1}} = -\frac{2a+bx}{nqX^n} - \frac{b(2n-1)}{nq}\int \frac{dx}{X^n}$$

117. $$\int \frac{x^2}{X}dx = \frac{x}{c} - \frac{b}{2c^2}\log X + \frac{b^2-2ac}{2c^2}\int \frac{dx}{X}$$

118. $$\int \frac{x^2}{X^2}dx = \frac{(b^2-2ac)x+ab}{cqX} + \frac{2a}{q}\int \frac{dx}{X}$$

119. $$\int \frac{x^m\,dx}{X^{n+1}} = -\frac{x^{m-1}}{(2n-m+1)cX^n} - \frac{n-m+1}{2n-m+1}\cdot\frac{b}{c}\int \frac{x^{m-1}\,dx}{X^{n+1}} + \frac{m-1}{2n-m+1}\cdot\frac{a}{c}\int \frac{x^{m-2}\,dx}{X^{n+1}}$$

120. $$\int \frac{dx}{xX} = \frac{1}{2a}\log\frac{x^2}{X} - \frac{b}{2a}\int \frac{dx}{X}$$

121. $$\int \frac{dx}{x^2X} = \frac{b}{2a^2}\log\frac{X}{x^2} - \frac{1}{ax} + \left(\frac{b^2}{2a^2} - \frac{c}{a}\right)\int \frac{dx}{X}$$

122. $$\int \frac{dx}{xX^n} = \frac{1}{2a(n-1)X^{n-1}} - \frac{b}{2a}\int \frac{dx}{X^n} + \frac{1}{a}\int \frac{dx}{xX^{n-1}}$$

123. $$\int \frac{dx}{x^mX^{n+1}} = -\frac{1}{(m-1)ax^{m-1}X^n} - \frac{n+m-1}{m-1}\cdot\frac{b}{a}\int \frac{dx}{x^{m-1}X^{n+1}} - \frac{2n+m-1}{m-1}\cdot\frac{c}{a}\int \frac{dx}{x^{m-2}X^{n+1}}$$

FORMS CONTAINING $\sqrt{a+bx}$

124. $$\int \sqrt{a+bx}\,dx = \frac{2}{3b}\sqrt{(a+bx)^3}$$

125. $$\int x\sqrt{a+bx}\,dx = -\frac{2(2a-3bx)\sqrt{(a+bx)^3}}{15b^2}$$

126. $$\int x^2\sqrt{a+bx}\,dx = \frac{2(8a^2-12abx+15b^2x^2)\sqrt{(a+bx)^3}}{105b^3}$$

127. $$\int x^m\sqrt{a+bx}\,dx = \begin{cases} \dfrac{2}{b(2m+3)}\left[x^m\sqrt{(a+bx)^3} - ma\displaystyle\int x^{m-1}\sqrt{a+bx}\,dx\right] \\ \text{or} \\ \dfrac{2}{b^{m+1}}\sqrt{a+bx}\displaystyle\sum_{r=0}^{m}\dfrac{m!(-a)^{m-r}}{r!(m-r)!(2r+3)}(a+bx)^{r+1} \end{cases}$$

128. $$\int \frac{\sqrt{a+bx}}{x}\,dx = 2\sqrt{a+bx} + a\int \frac{dx}{x\sqrt{a+bx}}$$

129. $$\int \frac{\sqrt{a+bx}}{x^2}\,dx = -\frac{\sqrt{a+bx}}{x} + \frac{b}{2}\int \frac{dx}{x\sqrt{a+bx}}$$

130. $$\int \frac{\sqrt{a+bx}}{x^m}\,dx = -\frac{1}{(m-1)a}\left[\frac{\sqrt{(a+bx)^3}}{x^{m-1}} + \frac{(2m-5)b}{2}\int \frac{\sqrt{a+bx}}{x^{m-1}}\,dx\right]$$

131. $$\int \frac{dx}{\sqrt{a+bx}} = \frac{2\sqrt{a+bx}}{b}$$

132. $$\int \frac{x\,dx}{\sqrt{a+bx}} = -\frac{2(2a-bx)}{3b^2}\sqrt{a+bx}$$

133. $$\int \frac{x^2\,dx}{\sqrt{a+bx}} = \frac{2(8a^2-4abx+3b^2x^2)}{15b^3}\sqrt{a+bx}$$

134. $$\int \frac{x^m\,dx}{\sqrt{a+bx}} = \begin{cases} \dfrac{2}{(2m+1)b}\left[x^m\sqrt{a+bx} - ma\displaystyle\int \frac{x^{m-1}\,dx}{\sqrt{a+bx}}\right] \\ \text{or} \\ \dfrac{2(-a)^m\sqrt{a+bx}}{b^{m+1}}\displaystyle\sum_{r=0}^{m} \frac{(-1)^r m!(a+bx)^r}{(2r+1)r!(m-r)!a^r} \end{cases}$$

135. $$\int \frac{dx}{x\sqrt{a+bx}} = \frac{1}{\sqrt{a}}\log\left(\frac{\sqrt{a+bx}-\sqrt{a}}{\sqrt{a+bx}+\sqrt{a}}\right), \qquad (a > 0)$$

136. $$\int \frac{dx}{x\sqrt{a+bx}} = \frac{2}{\sqrt{-a}}\tan^{-1}\sqrt{\frac{a+bx}{-a}}, \qquad (a < 0)$$

137. $$\int \frac{dx}{x^2\sqrt{a+bx}} = -\frac{\sqrt{a+bx}}{ax} - \frac{b}{2a}\int \frac{dx}{x\sqrt{a+bx}}$$

138. $$\int \frac{dx}{x^n\sqrt{a+bx}} = \begin{cases} -\dfrac{\sqrt{a+bx}}{(n-1)ax^{n-1}} - \dfrac{(2n-3)b}{(2n-2)a}\displaystyle\int \frac{dx}{x^{n-1}\sqrt{a+bx}} \\ \text{or} \\ \dfrac{(2n-2)!}{[(n-1)!]^2}\left[-\dfrac{\sqrt{a+bx}}{a}\displaystyle\sum_{r=1}^{n-1} \frac{r!(r-1)!}{x^r(2r)!}\left(-\frac{b}{4a}\right)^{n-r-1} \right. \\ \qquad \left. + \left(-\dfrac{b}{4a}\right)^{n-1}\displaystyle\int \frac{dx}{x\sqrt{a+bx}}\right] \end{cases}$$

139. $$\int (a+bx)^{\pm\frac{n}{2}}\,dx = \frac{2(a+bx)^{\frac{2\pm n}{2}}}{b(2\pm n)}$$

140. $$\int x(a+bx)^{\pm\frac{n}{2}}\,dx = \frac{2}{b^2}\left[\frac{(a+bx)^{\frac{4\pm n}{2}}}{4\pm n} - \frac{a(a+bx)^{\frac{2\pm n}{2}}}{2\pm n}\right]$$

141. $$\int \frac{dx}{x(a+bx)^{\frac{m}{2}}} = \frac{1}{a}\int \frac{dx}{x(a+bx)^{\frac{m-2}{2}}} - \frac{b}{a}\int \frac{dx}{(a+bx)^{\frac{m}{2}}}$$

142. $\int \frac{(a+bx)^{\frac{n}{2}}\,dx}{x} = b\int (a+bx)^{\frac{n-2}{2}}\,dx + a\int \frac{(a+bx)^{\frac{n-2}{2}}}{x}\,dx$

143. $\int f(x, \sqrt{a+bx})\,dx = \frac{2}{b}\int f\left(\frac{z^2-a}{b}, z\right) z\,dz, \qquad (z = \sqrt{a+bx})$

FORMS CONTAINING $\sqrt{a+bx}$ and $\sqrt{c+dx}$

$u = a + bx \qquad v = c + dx \qquad k = ad - bc$

If $k = 0$, then $v = \frac{c}{a}u$, and formulas starting with 124 should be used in place of these.

144. $\int \frac{dx}{\sqrt{uv}} = \begin{cases} \frac{2}{\sqrt{bd}} \tanh^{-1} \frac{\sqrt{bduv}}{bv}, \; bd > o, k < o \\ \text{or} \\ \frac{2}{\sqrt{bd}} \tanh^{-1} \frac{\sqrt{bduv}}{du}, \; bd > o, k > o. \\ \text{or} \\ \frac{1}{\sqrt{bd}} \log \frac{(bv + \sqrt{bduv})^2}{v}, \qquad (bd > 0) \end{cases}$

145. $\int \frac{dx}{\sqrt{uv}} = \begin{cases} \frac{2}{\sqrt{-bd}} \tan^{-1} \frac{\sqrt{-bduv}}{bv} \\ \text{or} \\ -\frac{1}{\sqrt{-bd}} \sin^{-1}\left(\frac{2bdx + ad + bc}{|k|}\right) \qquad (bd < 0) \end{cases}$

146. $\int \sqrt{uv}\,dx = \frac{k + 2bv}{4bd}\sqrt{uv} - \frac{k^2}{8bd}\int \frac{dx}{\sqrt{uv}}$

147. $\int \frac{dx}{v\sqrt{u}} = \begin{cases} \frac{1}{\sqrt{kd}} \log \frac{d\sqrt{u} - \sqrt{kd}}{d\sqrt{u} + \sqrt{kd}} \\ \text{or} \\ \frac{1}{\sqrt{kd}} \log \frac{(d\sqrt{u} - \sqrt{kd})^2}{v}, \qquad (kd > 0) \end{cases}$

148. $\int \frac{dx}{v\sqrt{u}} = \frac{2}{\sqrt{-kd}} \tan^{-1} \frac{d\sqrt{u}}{\sqrt{-kd}}, \qquad (kd < 0)$

149. $\int \frac{x\,dx}{\sqrt{uv}} = \frac{\sqrt{uv}}{bd} - \frac{ad + bc}{2bd}\int \frac{dx}{\sqrt{uv}}$

150. $\int \frac{dx}{v\sqrt{uv}} = \frac{-2\sqrt{uv}}{kv}$

151. $\int \frac{v\,dx}{\sqrt{uv}} = \frac{\sqrt{uv}}{b} - \frac{k}{2b}\int \frac{dx}{\sqrt{uv}}$

152. $\int \sqrt{\frac{v}{u}}\,dx = \frac{v}{|v|}\int \frac{v\,dx}{\sqrt{uv}}$

153. $\int v^m \sqrt{u}\,dx = \frac{1}{(2m+3)d}\left(2v^{m+1}\sqrt{u} + k\int \frac{v^m\,dx}{\sqrt{u}}\right)$

154. $$\int \frac{dx}{v^m\sqrt{u}} = -\frac{1}{(m-1)k}\left(\frac{\sqrt{u}}{v^{m-1}} + \left(m - \frac{3}{2}\right)b\int \frac{dx}{v^{m-1}\sqrt{u}}\right)$$

155. $$\int \frac{v^m\,dx}{\sqrt{u}} = \begin{cases} \dfrac{2}{b(2m+1)}\left[v^m\sqrt{u} - mk\displaystyle\int \frac{v^{m-1}}{\sqrt{u}}dx\right] \\ \text{or} \\ \dfrac{2(m!)^2\sqrt{u}}{b(2m+1)!}\displaystyle\sum_{r=0}^{m}\left(-\frac{4k}{b}\right)^{m-r}\frac{(2r)!}{(r!)^2}v^r \end{cases}$$

FORMS CONTAINING $\sqrt{x^2 \pm a^2}$

156. $$\int \sqrt{x^2 \pm a^2}\,dx = \tfrac{1}{2}[x\sqrt{x^2 \pm a^2} \pm a^2 \log(x + \sqrt{x^2 \pm a^2})]$$

157. $$\int \frac{dx}{\sqrt{x^2 \pm a^2}} = \log(x + \sqrt{x^2 \pm a^2})$$

158. $$\int \frac{dx}{x\sqrt{x^2 - a^2}} = \frac{1}{|a|}\sec^{-1}\frac{x}{a}$$

159. $$\int \frac{dx}{x\sqrt{x^2 + a^2}} = -\frac{1}{a}\log\left(\frac{a + \sqrt{x^2 + a^2}}{x}\right)$$

160. $$\int \frac{\sqrt{x^2 + a^2}}{x}dx = \sqrt{x^2 + a^2} - a\log\left(\frac{a + \sqrt{x^2 + a^2}}{x}\right)$$

161. $$\int \frac{\sqrt{x^2 - a^2}}{x}dx = \sqrt{x^2 - a^2} - |a|\sec^{-1}\frac{x}{a}$$

162. $$\int \frac{x\,dx}{\sqrt{x^2 \pm a^2}} = \sqrt{x^2 \pm a^2}$$

163. $$\int x\sqrt{x^2 \pm a^2}\,dx = \tfrac{1}{3}\sqrt{(x^2 \pm a^2)^3}$$

164. $$\int \sqrt{(x^2 \pm a^2)^3}\,dx = \frac{1}{4}\left[x\sqrt{(x^2 \pm a^2)^3} \pm \frac{3a^2x}{2}\sqrt{x^2 \pm a^2} + \frac{3a^4}{2}\log(x + \sqrt{x^2 \pm a^2})\right]$$

165. $$\int \frac{dx}{\sqrt{(x^2 \pm a^2)^3}} = \frac{\pm x}{a^2\sqrt{x^2 \pm a^2}}$$

166. $$\int \frac{x\,dx}{\sqrt{(x^2 \pm a^2)^3}} = \frac{-1}{\sqrt{x^2 \pm a^2}}$$

167. $$\int x\sqrt{(x^2 \pm a^2)^3}\,dx = \tfrac{1}{5}\sqrt{(x^2 \pm a^2)^5}$$

168. $$\int x^2\sqrt{x^2 \pm a^2}\,dx = \frac{x}{4}\sqrt{(x^2 \pm a^2)^3} \mp \frac{a^2}{8}x\sqrt{x^2 \pm a^2} - \frac{a^4}{8}\log(x + \sqrt{x^2 \pm a^2})$$

169. $$\int x^3\sqrt{x^2 + a^2}\,dx = (\tfrac{1}{5}x^2 - \tfrac{2}{15}a^2)\sqrt{(a^2 + x^2)^3}$$

170. $$\int x^3\sqrt{x^2 - a^2}\,dx = \frac{1}{5}\sqrt{(x^2 - a^2)^5} + \frac{a^2}{3}\sqrt{(x^2 - a^2)^3}$$

171. $$\int \frac{x^2\,dx}{\sqrt{x^2 \pm a^2}} = \frac{x}{2}\sqrt{x^2 \pm a^2} \mp \frac{a^2}{2}\log(x + \sqrt{x^2 \pm a^2})$$

172. $\int \frac{x^3\,dx}{\sqrt{x^2 \pm a^2}} = \frac{1}{3}\sqrt{(x^2 \pm a^2)^3} \mp a^2\sqrt{x^2 \pm a^2}$

173. $\int \frac{dx}{x^2\sqrt{x^2 \pm a^2}} = \mp\frac{\sqrt{x^2 \pm a^2}}{a^2 x}$

174. $\int \frac{dx}{x^3\sqrt{x^2 + a^2}} = -\frac{\sqrt{x^2 + a^2}}{2a^2x^2} + \frac{1}{2a^3}\log\frac{a + \sqrt{x^2 + a^2}}{x}$

175. $\int \frac{dx}{x^3\sqrt{x^2 - a^2}} = \frac{\sqrt{x^2 - a^2}}{2a^2x^2} + \frac{1}{2|a^3|}\sec^{-1}\frac{x}{a}$

176. $\int x^2\sqrt{(x^2 \pm a^2)^3}\,dx = \frac{x}{6}\sqrt{(x^2 \pm a^2)^5} \mp \frac{a^2x}{24}\sqrt{(x^2 \pm a^2)^3} - \frac{a^4x}{16}\sqrt{x^2 \pm a^2}$

$$\mp \frac{a^6}{16}\log(x + \sqrt{x^2 \pm a^2})$$

177. $\int x^3\sqrt{(x^2 \pm a^2)^3}\,dx = \frac{1}{7}\sqrt{(x^2 \pm a^2)^7} \mp \frac{a^2}{5}\sqrt{(x^2 \pm a^2)^5}$

178. $\int \frac{\sqrt{x^2 \pm a^2}\,dx}{x^2} = -\frac{\sqrt{x^2 \pm a^2}}{x} + \log(x + \sqrt{x^2 \pm a^2})$

179. $\int \frac{\sqrt{x^2 + a^2}}{x^3}\,dx = -\frac{\sqrt{x^2 + a^2}}{2x^2} - \frac{1}{2a}\log\frac{a + \sqrt{x^2 + a^2}}{x}$

180. $\int \frac{\sqrt{x^2 - a^2}}{x^3}\,dx = -\frac{\sqrt{x^2 - a^2}}{2x^2} + \frac{1}{2|a|}\sec^{-1}\frac{x}{a}$

181. $\int \frac{\sqrt{x^2 \pm a^2}}{x^4}\,dx = \mp\frac{\sqrt{(x^2 \pm a^2)^3}}{3a^2x^3}$

182. $\int \frac{x^2\,dx}{\sqrt{(x^2 \pm a^2)^3}} = \frac{-x}{\sqrt{x^2 \pm a^2}} + \log(x + \sqrt{x^2 \pm a^2})$

183. $\int \frac{x^3\,dx}{\sqrt{(x^2 \pm a^2)^3}} = \sqrt{x^2 \pm a^2} \pm \frac{a^2}{\sqrt{x^2 \pm a^2}}$

184. $\int \frac{dx}{x\sqrt{(x^2 + a^2)^3}} = \frac{1}{a^2\sqrt{x^2 + a^2}} - \frac{1}{a^3}\log\frac{a + \sqrt{x^2 + a^2}}{x}$

185. $\int \frac{dx}{x\sqrt{(x^2 - a^2)^3}} = -\frac{1}{a^2\sqrt{x^2 - a^2}} - \frac{1}{|a^3|}\sec^{-1}\frac{x}{a}$

186. $\int \frac{dx}{x^2\sqrt{(x^2 \pm a^2)^3}} = -\frac{1}{a^4}\left[\frac{\sqrt{x^2 \pm a^2}}{x} + \frac{x}{\sqrt{x^2 \pm a^2}}\right]$

187. $\int \frac{dx}{x^3\sqrt{(x^2 + a^2)^3}} = -\frac{1}{2a^2x^2\sqrt{x^2 + a^2}} - \frac{3}{2a^4\sqrt{x^2 + a^2}}$

$$+ \frac{3}{2a^5}\log\frac{a + \sqrt{x^2 + a^2}}{x}$$

188. $\int \frac{dx}{x^3\sqrt{(x^2 - a^2)^3}} = \frac{1}{2a^2x^2\sqrt{x^2 - a^2}} - \frac{3}{2a^4\sqrt{x^2 - a^2}} - \frac{3}{2|a^5|}\sec^{-1}\frac{x}{a}$

189. $\int \frac{x^m}{\sqrt{x^2 \pm a^2}}\,dx = \frac{1}{m}x^{m-1}\sqrt{x^2 \pm a^2} \mp \frac{m-1}{m}a^2\int \frac{x^{m-2}}{\sqrt{x^2 \pm a^2}}\,dx$

190. $$\int \frac{x^{2m}}{\sqrt{x^2 \pm a^2}}\,dx = \frac{(2m)!}{2^{2m}(m!)^2}\left[\sqrt{x^2 \pm a^2}\sum_{r=1}^{m}\frac{r!(r-1)!}{(2r)!}(\mp a^2)^{m-r}(2x)^{2r-1} + (\mp a^2)^m \log\left(x + \sqrt{x^2 \pm a^2}\right)\right]$$

191. $$\int \frac{x^{2m+1}}{\sqrt{x^2 \pm a^2}}\,dx = \sqrt{x^2 \pm a^2}\sum_{r=0}^{m}\frac{(2r)!(m!)^2}{(2m+1)!(r!)^2}(\mp 4a^2)^{m-r}x^{2r}$$

192. $$\int \frac{dx}{x^m\sqrt{x^2 \pm a^2}} = \mp\frac{\sqrt{x^2 \pm a^2}}{(m-1)a^2x^{m-1}} \mp \frac{(m-2)}{(m-1)a^2}\int \frac{dx}{x^{m-2}\sqrt{x^2 \pm a^2}}$$

193. $$\int \frac{dx}{x^{2m}\sqrt{x^2 \pm a^2}} = \sqrt{x^2 \pm a^2}\sum_{r=0}^{m-1}\frac{(m-1)!m!(2r)!2^{2m-2r-1}}{(r!)^2(2m)!(\mp a^2)^{m-r}x^{2r+1}}$$

194. $$\int \frac{dx}{x^{2m+1}\sqrt{x^2 + a^2}} = \frac{(2m)!}{(m!)^2}\left[\frac{\sqrt{x^2 + a^2}}{a^2}\sum_{r=1}^{m}(-1)^{m-r+1}\frac{r!(r-1)!}{2(2r)!(4a^2)^{m-r}x^{2r}} + \frac{(-1)^{m+1}}{2^{2m}a^{2m+1}}\log\frac{\sqrt{x^2 + a^2} + a}{x}\right]$$

195. $$\int \frac{dx}{x^{2m+1}\sqrt{x^2 - a^2}} = \frac{(2m)!}{(m!)^2}\left[\frac{\sqrt{x^2 - a^2}}{a^2}\sum_{r=1}^{m}\frac{r!(r-1)!}{2(2r)!(4a^2)^{m-r}x^{2r}} + \frac{1}{2^{2m}|a|^{2m+1}}\sec^{-1}\frac{x}{a}\right]$$

196. $$\int \frac{dx}{(x-a)\sqrt{x^2 - a^2}} = -\frac{\sqrt{x^2 - a^2}}{a(x-a)}$$

197. $$\int \frac{dx}{(x+a)\sqrt{x^2 - a^2}} = \frac{\sqrt{x^2 - a^2}}{a(x+a)}$$

198. $$\int f(x, \sqrt{x^2 + a^2})\,dx = a\int f(a\tan u, a\sec u)\sec^2 u\,du, \qquad \left(u = \tan^{-1}\frac{x}{a},\ a > 0\right)$$

199. $$\int f(x, \sqrt{x^2 - a^2})\,dx = a\int f(a\sec u, a\tan u)\sec u\tan u\,du, \qquad \left(u = \sec^{-1}\frac{x}{a},\ a > 0\right)$$

FORMS CONTAINING $\sqrt{a^2 - x^2}$

200. $$\int \sqrt{a^2 - x^2}\,dx = \frac{1}{2}\left[x\sqrt{a^2 - x^2} + a^2\sin^{-1}\frac{x}{|a|}\right]$$

201. $$\int \frac{dx}{\sqrt{a^2 - x^2}} = \begin{cases} \sin^{-1}\frac{x}{|a|} \\ \text{or} \\ -\cos^{-1}\frac{x}{|a|} \end{cases}$$

202. $$\int \frac{dx}{x\sqrt{a^2 - x^2}} = -\frac{1}{a}\log\left(\frac{a + \sqrt{a^2 - x^2}}{x}\right)$$

203. $$\int \frac{\sqrt{a^2 - x^2}}{x}\,dx = \sqrt{a^2 - x^2} - a\log\left(\frac{a + \sqrt{a^2 - x^2}}{x}\right)$$

204. $$\int \frac{x\,dx}{\sqrt{a^2 - x^2}} = -\sqrt{a^2 - x^2}$$

205. $$\int x\sqrt{a^2 - x^2}\,dx = -\tfrac{1}{3}\sqrt{(a^2 - x^2)^3}$$

206. $\int \sqrt{(a^2 - x^2)^3}\, dx = \frac{1}{4}\left[x\sqrt{(a^2 - x^2)^3} + \frac{3a^2x}{2}\sqrt{a^2 - x^2} + \frac{3a^4}{2}\sin^{-1}\frac{x}{|a|}\right]$

207. $\int \frac{dx}{\sqrt{(a^2 - x^2)^3}} = \frac{x}{a^2\sqrt{a^2 - x^2}}$

208. $\int \frac{x\, dx}{\sqrt{(a^2 - x^2)^3}} = \frac{1}{\sqrt{a^2 - x^2}}$

209. $\int x\sqrt{(a^2 - x^2)^3}\, dx = -\tfrac{1}{5}\sqrt{(a^2 - x^2)^5}$

210. $\int x^2\sqrt{a^2 - x^2}\, dx = -\frac{x}{4}\sqrt{(a^2 - x^2)^3} + \frac{a^2}{8}\left(x\sqrt{a^2 - x^2} + a^2\sin^{-1}\frac{x}{|a|}\right)$

211. $\int x^3\sqrt{a^2 - x^2}\, dx = (-\tfrac{1}{5}x^2 - \tfrac{2}{15}a^2)\sqrt{(a^2 - x^2)^3}$

212. $\int x^2\sqrt{(a^2 - x^2)^3}\, dx = -\frac{1}{6}x\sqrt{(a^2 - x^2)^5} + \frac{a^2x}{24}\sqrt{(a^2 - x^2)^3}$

$+ \frac{a^4x}{16}\sqrt{a^2 - x^2} + \frac{a^6}{16}\sin^{-1}\frac{x}{|a|}$

213. $\int x^3\sqrt{(a^2 - x^2)^3}\, dx = \frac{1}{7}\sqrt{(a^2 - x^2)^7} - \frac{a^2}{5}\sqrt{(a^2 - x^2)^5}$

214. $\int \frac{x^2\, dx}{\sqrt{a^2 - x^2}} = -\frac{x}{2}\sqrt{a^2 - x^2} + \frac{a^2}{2}\sin^{-1}\frac{x}{|a|}$

215. $\int \frac{dx}{x^2\sqrt{a^2 - x^2}} = -\frac{\sqrt{a^2 - x^2}}{a^2x}$

216. $\int \frac{\sqrt{a^2 - x^2}}{x^2}\, dx = -\frac{\sqrt{a^2 - x^2}}{x} - \sin^{-1}\frac{x}{|a|}$

217. $\int \frac{\sqrt{a^2 - x^2}}{x^3}\, dx = -\frac{\sqrt{a^2 - x^2}}{2x^2} + \frac{1}{2a}\log\frac{a + \sqrt{a^2 - x^2}}{x}$

218. $\int \frac{\sqrt{a^2 - x^2}}{x^4}\, dx = -\frac{\sqrt{(a^2 - x^2)^3}}{3a^2x^3}$

219. $\int \frac{x^2\, dx}{\sqrt{(a^2 - x^2)^3}} = \frac{x}{\sqrt{a^2 - x^2}} - \sin^{-1}\frac{x}{|a|}$

220. $\int \frac{x^3\, dx}{\sqrt{a^2 - x^2}} = -\frac{2}{3}(a^2 - x^2)^{\frac{3}{2}} - x^2(a^2 - x^2)^{\frac{1}{2}} = -\frac{1}{3}\sqrt{a^2 - x^2}(x^2 + 2a^2)$

221. $\int \frac{x^3\, dx}{\sqrt{(a^2 - x^2)^3}} = 2(a^2 - x^2)^{\frac{1}{2}} + \frac{x^2}{(a^2 - x^2)^{\frac{1}{2}}} = -\frac{a^2}{\sqrt{a^2 - x^2}} + \sqrt{a^2 - x^2}$

222. $\int \frac{dx}{x^3\sqrt{a^2 - x^2}} = -\frac{\sqrt{a^2 - x^2}}{2a^2x^2} - \frac{1}{2a^3}\log\frac{a + \sqrt{a^2 - x^2}}{x}$

223. $\int \frac{dx}{x\sqrt{(a^2 - x^2)^3}} = \frac{1}{a^2\sqrt{a^2 - x^2}} - \frac{1}{a^3}\log\frac{a + \sqrt{a^2 - x^2}}{x}$

224. $\int \frac{dx}{x^2\sqrt{(a^2 - x^2)^3}} = \frac{1}{a^4}\left[-\frac{\sqrt{a^2 - x^2}}{x} + \frac{x}{\sqrt{a^2 - x^2}}\right]$

225. $\int \frac{dx}{x^3\sqrt{(a^2 - x^2)^3}} = -\frac{1}{2a^2x^2\sqrt{a^2 - x^2}} + \frac{3}{2a^4\sqrt{a^2 - x^2}}$

$- \frac{3}{2a^5}\log\frac{a + \sqrt{a^2 - x^2}}{x}$

226. $$\int \frac{x^m}{\sqrt{a^2 - x^2}}\,dx = -\frac{x^{m-1}\sqrt{a^2 - x^2}}{m} + \frac{(m-1)a^2}{m}\int \frac{x^{m-2}}{\sqrt{a^2 - x^2}}\,dx$$

227. $$\int \frac{x^{2m}}{\sqrt{a^2 - x^2}}\,dx = \frac{(2m)!}{(m!)^2}\left[-\sqrt{a^2 - x^2}\sum_{r=1}^{m} \frac{r!(r-1)!}{2^{2m-2r+1}(2r)!}a^{2m-2r}x^{2r-1} + \frac{a^{2m}}{2^{2m}}\sin^{-1}\frac{x}{|a|}\right]$$

228. $$\int \frac{x^{2m+1}}{\sqrt{a^2 - x^2}}\,dx = -\sqrt{a^2 - x^2}\sum_{r=0}^{m} \frac{(2r)!(m!)^2}{(2m+1)!(r!)^2}(4a^2)^{m-r}x^{2r}$$

229. $$\int \frac{dx}{x^m\sqrt{a^2 - x^2}} = -\frac{\sqrt{a^2 - x^2}}{(m-1)a^2x^{m-1}} + \frac{m-2}{(m-1)a^2}\int \frac{dx}{x^{m-2}\sqrt{a^2 - x^2}}$$

230. $$\int \frac{dx}{x^{2m}\sqrt{a^2 - x^2}} = -\sqrt{a^2 - x^2}\sum_{r=0}^{m-1} \frac{(m-1)!m!(2r)!2^{2m-2r-1}}{(r!)^2(2m)!a^{2m-2r}x^{2r+1}}$$

231. $$\int \frac{dx}{x^{2m+1}\sqrt{a^2 - x^2}} = \frac{(2m)!}{(m!)^2}\left[-\frac{\sqrt{a^2 - x^2}}{a^2}\sum_{r=1}^{m} \frac{r!(r-1)!}{2(2r)!(4a^2)^{m-r}x^{2r}} + \frac{1}{2^{2m}a^{2m+1}}\log\frac{a - \sqrt{a^2 - x^2}}{x}\right]$$

232. $$\int \frac{dx}{(b^2 - x^2)\sqrt{a^2 - x^2}} = \frac{1}{2b\sqrt{a^2 - b^2}}\log\frac{(b\sqrt{a^2 - x^2} + x\sqrt{a^2 - b^2})^2}{b^2 - x^2}, \qquad (a^2 > b^2)$$

233. $$\int \frac{dx}{(b^2 - x^2)\sqrt{a^2 - x^2}} = \frac{1}{b\sqrt{b^2 - a^2}}\tan^{-1}\frac{x\sqrt{b^2 - a^2}}{b\sqrt{a^2 - x^2}}, \qquad (b^2 > a^2)$$

234. $$\int \frac{dx}{(b^2 + x^2)\sqrt{a^2 - x^2}} = \frac{1}{b\sqrt{a^2 + b^2}}\tan^{-1}\frac{x\sqrt{a^2 + b^2}}{b\sqrt{a^2 - x^2}}$$

235. $$\int \frac{\sqrt{a^2 - x^2}}{b^2 + x^2}\,dx = \frac{\sqrt{a^2 + b^2}}{|b|}\sin^{-1}\frac{x\sqrt{a^2 + b^2}}{|a|\sqrt{x^2 + b^2}} - \sin^{-1}\frac{x}{|a|}$$

236. $$\int f(x, \sqrt{a^2 - x^2})\,dx = a\int f(a\sin u, a\cos u)\cos u\,du, \qquad \left(u = \sin^{-1}\frac{x}{a},\ a > 0\right)$$

FORMS CONTAINING $\sqrt{a + bx + cx^2}$

$$X = a + bx + cx^2,\ q = 4ac - b^2, \text{ and } k = \frac{4c}{q}$$

If $q = 0$, then $\sqrt{X} = \sqrt{c}\left|x + \frac{b}{2c}\right|$

237. $$\int \frac{dx}{\sqrt{X}} = \begin{cases} \frac{1}{\sqrt{c}}\log(2\sqrt{cX} + 2cx + b) \\ \text{or} \\ \frac{1}{\sqrt{c}}\sinh^{-1}\frac{2cx + b}{\sqrt{q}}, \qquad (c > 0) \end{cases}$$

238. $$\int \frac{dx}{\sqrt{X}} = -\frac{1}{\sqrt{-c}}\sin^{-1}\frac{2cx + b}{\sqrt{-q}}, \qquad (c < 0)$$

239. $$\int \frac{dx}{X\sqrt{X}} = \frac{2(2cx + b)}{q\sqrt{X}}$$

240. $\displaystyle\int \frac{dx}{X^2\sqrt{X}} = \frac{2(2cx+b)}{3q\sqrt{X}}\left(\frac{1}{X} + 2k\right)$

241. $\displaystyle\int \frac{dx}{X^n\sqrt{X}} = \begin{cases} \dfrac{2(2cx+b)\sqrt{X}}{(2n-1)qX^n} + \dfrac{2k(n-1)}{2n-1}\displaystyle\int \frac{dx}{X^{n-1}\sqrt{X}} \\ \text{or} \\ \dfrac{(2cx+b)(n!)(n-1)!4^n k^{n-1}}{q[(2n)!]\sqrt{X}} \displaystyle\sum_{r=0}^{n-1} \frac{(2r)!}{(4kX)^r (r!)^2} \end{cases}$

242. $\displaystyle\int \sqrt{X}\,dx = \frac{(2cx+b)\sqrt{X}}{4c} + \frac{1}{2k}\int \frac{dx}{\sqrt{X}}$

243. $\displaystyle\int X\sqrt{X}\,dx = \frac{(2cx+b)\sqrt{X}}{8c}\left(X + \frac{3}{2k}\right) + \frac{3}{8k^2}\int \frac{dx}{\sqrt{X}}$

244. $\displaystyle\int X^2\sqrt{X}\,dx = \frac{(2cx+b)\sqrt{X}}{12c}\left(X^2 + \frac{5X}{4k} + \frac{15}{8k^2}\right) + \frac{5}{16k^3}\int \frac{dx}{\sqrt{X}}$

245. $\displaystyle\int X^n\sqrt{X}\,dx = \begin{cases} \dfrac{(2cx+b)X^n\sqrt{X}}{4(n+1)c} + \dfrac{2n+1}{2(n+1)k}\displaystyle\int X^{n-1}\sqrt{X}\,dx \\ \text{or} \\ \dfrac{(2n+2)!}{[(n+1)!]^2(4k)^{n+1}}\left[\dfrac{k(2cx+b)\sqrt{X}}{c}\displaystyle\sum_{r=0}^{n} \frac{r!(r+1)!(4kX)^r}{(2r+2)!} + \int \frac{dx}{\sqrt{X}}\right] \end{cases}$

246. $\displaystyle\int \frac{x\,dx}{\sqrt{X}} = \frac{\sqrt{X}}{c} - \frac{b}{2c}\int \frac{dx}{\sqrt{X}}$

247. $\displaystyle\int \frac{x\,dx}{X\sqrt{X}} = -\frac{2(bx+2a)}{q\sqrt{X}}$

248. $\displaystyle\int \frac{x\,dx}{X^n\sqrt{X}} = -\frac{\sqrt{X}}{(2n-1)cX^n} - \frac{b}{2c}\int \frac{dx}{X^n\sqrt{X}}$

249. $\displaystyle\int \frac{x^2\,dx}{\sqrt{X}} = \left(\frac{x}{2c} - \frac{3b}{4c^2}\right)\sqrt{X} + \frac{3b^2-4ac}{8c^2}\int \frac{dx}{\sqrt{X}}$

250. $\displaystyle\int \frac{x^2\,dx}{X\sqrt{X}} = \frac{(2b^2-4ac)x+2ab}{cq\sqrt{X}} + \frac{1}{c}\int \frac{dx}{\sqrt{X}}$

251. $\displaystyle\int \frac{x^2\,dx}{X^n\sqrt{X}} = \frac{(2b^2-4ac)x+2ab}{(2n-1)cqX^{n-1}\sqrt{X}} + \frac{4ac+(2n-3)b^2}{(2n-1)cq}\int \frac{dx}{X^{n-1}\sqrt{X}}$

252. $\displaystyle\int \frac{x^3\,dx}{\sqrt{X}} = \left(\frac{x^2}{3c} - \frac{5bx}{12c^2} + \frac{5b^2}{8c^3} - \frac{2a}{3c^2}\right)\sqrt{X} + \left(\frac{3ab}{4c^2} - \frac{5b^3}{16c^3}\right)\int \frac{dx}{\sqrt{X}}$

253. $\displaystyle\int \frac{x^n\,dx}{\sqrt{X}} = \frac{1}{nc}x^{n-1}\sqrt{X} - \frac{(2n-1)b}{2nc}\int \frac{x^{n-1}\,dx}{\sqrt{X}} - \frac{(n-1)a}{nc}\int \frac{x^{n-2}\,dx}{\sqrt{X}}$

254. $\displaystyle\int x\sqrt{X}\,dx = \frac{X\sqrt{X}}{3c} - \frac{b(2cx+b)}{8c^2}\sqrt{X} - \frac{b}{4ck}\int \frac{dx}{\sqrt{X}}$

255. $\displaystyle\int xX\sqrt{X}\,dx = \frac{X^2\sqrt{X}}{5c} - \frac{b}{2c}\int X\sqrt{X}\,dx$

256. $\int xX^n\sqrt{X}\,dx = \frac{X^{n+1}\sqrt{X}}{(2n+3)c} - \frac{b}{2c}\int X^n\sqrt{X}\,dx$

257. $\int x^2\sqrt{X}\,dx = \left(x - \frac{5b}{6c}\right)\frac{X\sqrt{X}}{4c} + \frac{5b^2 - 4ac}{16c^2}\int \sqrt{X}\,dx$

258. $\int \frac{dx}{x\sqrt{X}} = -\frac{1}{\sqrt{a}}\log\frac{2\sqrt{aX} + bx + 2a}{x}, \quad (a > 0)$

259. $\int \frac{dx}{x\sqrt{X}} = \frac{1}{\sqrt{-a}}\sin^{-1}\left(\frac{bx + 2a}{|x|\sqrt{-q}}\right), \quad (a < 0)$

260. $\int \frac{dx}{x\sqrt{X}} = -\frac{2\sqrt{X}}{bx}, \quad (a = 0)$

261. $\int \frac{dx}{x^2\sqrt{X}} = -\frac{\sqrt{X}}{ax} - \frac{b}{2a}\int \frac{dx}{x\sqrt{X}}$

262. $\int \frac{\sqrt{X}\,dx}{x} = \sqrt{X} + \frac{b}{2}\int \frac{dx}{\sqrt{X}} + a\int \frac{dx}{x\sqrt{X}}$

263. $\int \frac{\sqrt{X}\,dx}{x^2} = -\frac{\sqrt{X}}{x} + \frac{b}{2}\int \frac{dx}{x\sqrt{X}} + c\int \frac{dx}{\sqrt{X}}$

FORMS INVOLVING $\sqrt{2ax - x^2}$

264. $\int \sqrt{2ax - x^2}\,dx = \frac{1}{2}\left[(x - a)\sqrt{2ax - x^2} + a^2\sin^{-1}\frac{x - a}{|a|}\right]$

265. $\int \frac{dx}{\sqrt{2ax - x^2}} = \begin{cases} \cos^{-1}\frac{a - x}{|a|} \\ \text{or} \\ \sin^{-1}\frac{x - a}{|a|} \end{cases}$

266. $\int x^n\sqrt{2ax - x^2}\,dx = \begin{cases} -\frac{x^{n-1}(2ax - x^2)^{\frac{3}{2}}}{n + 2} + \frac{(2n + 1)a}{n + 2}\int x^{n-1}\sqrt{2ax - x^2}\,dx \\ \text{or} \\ \sqrt{2ax - x^2}\left[\frac{x^{n+1}}{n + 2} - \sum_{r=0}^{n}\frac{(2n + 1)!(r!)^2a^{n-r+1}}{2^{n-r}(2r + 1)!(n + 2)!n!}x^r\right] \\ \qquad + \frac{(2n + 1)!a^{n+2}}{2^n n!(n + 2)!}\sin^{-1}\frac{x - a}{|a|} \end{cases}$

267. $\int \frac{\sqrt{2ax - x^2}}{x^n}\,dx = \frac{(2ax - x^2)^{\frac{3}{2}}}{(3 - 2n)ax^n} + \frac{n - 3}{(2n - 3)a}\int \frac{\sqrt{2ax - x^2}}{x^{n-1}}\,dx$

268. $\int \frac{x^n\,dx}{\sqrt{2ax - x^2}} = \begin{cases} \frac{-x^{n-1}\sqrt{2ax - x^2}}{n} + \frac{a(2n - 1)}{n}\int \frac{x^{n-1}}{\sqrt{2ax - x^2}}\,dx \\ \text{or} \\ -\sqrt{2ax - x^2}\sum_{r=1}^{n}\frac{(2n)!r!(r - 1)!a^{n-r}}{2^{n-r}(2r)!(n!)^2}x^{r-1} \\ \qquad + \frac{(2n)!a^n}{2^n(n!)^2}\sin^{-1}\frac{x - a}{|a|} \end{cases}$

269. $$\int \frac{dx}{x^n\sqrt{2ax-x^2}} = \begin{cases} \dfrac{\sqrt{2ax-x^2}}{a(1-2n)x^n} + \dfrac{n-1}{(2n-1)a}\displaystyle\int \frac{dx}{x^{n-1}\sqrt{2ax-x^2}} \\ \text{or} \\ -\sqrt{2ax-x^2}\displaystyle\sum_{r=0}^{n-1} \frac{2^{n-r}(n-1)!n!(2r)!}{(2n)!(r!)^2 a^{n-r}x^{r+1}} \end{cases}$$

270. $$\int \frac{dx}{(2ax-x^2)^{\frac{3}{2}}} = \frac{x-a}{a^2\sqrt{2ax-x^2}}$$

271. $$\int \frac{x\,dx}{(2ax-x^2)^{\frac{3}{2}}} = \frac{x}{a\sqrt{2ax-x^2}}$$

MISCELLANEOUS ALGEBRAIC FORMS

272. $$\int \frac{dx}{\sqrt{2ax+x^2}} = \log(x+a+\sqrt{2ax+x^2})$$

273. $$\int \sqrt{ax^2+c}\,dx = \frac{x}{2}\sqrt{ax^2+c} + \frac{c}{2\sqrt{a}}\log(x\sqrt{a}+\sqrt{ax^2+c}), \qquad (a>0)$$

274. $$\int \sqrt{ax^2+c}\,dx = \frac{x}{2}\sqrt{ax^2+c} + \frac{c}{2\sqrt{-a}}\sin^{-1}\left(x\sqrt{-\frac{a}{c}}\right), \qquad (a<0)$$

275. $$\int \sqrt{\frac{1+x}{1-x}}\,dx = \sin^{-1}x - \sqrt{1-x^2}$$

276. $$\int \frac{dx}{x\sqrt{ax^n+c}} = \begin{cases} \dfrac{1}{n\sqrt{c}}\log\dfrac{\sqrt{ax^n+c}-\sqrt{c}}{\sqrt{ax^n+c}+\sqrt{c}} \\ \text{or} \\ \dfrac{2}{n\sqrt{c}}\log\dfrac{\sqrt{ax^n+c}-\sqrt{c}}{\sqrt{x^n}}, \qquad (c>0) \end{cases}$$

277. $$\int \frac{dx}{x\sqrt{ax^n+c}} = \frac{2}{n\sqrt{-c}}\sec^{-1}\sqrt{-\frac{ax^n}{c}}, \qquad (c<0)$$

278. $$\int \frac{dx}{\sqrt{ax^2+c}} = \frac{1}{\sqrt{a}}\log(x\sqrt{a}+\sqrt{ax^2+c}), \qquad (a>0)$$

279. $$\int \frac{dx}{\sqrt{ax^2+c}} = \frac{1}{\sqrt{-a}}\sin^{-1}\left(x\sqrt{-\frac{a}{c}}\right), \qquad (a<0)$$

280. $$\int (ax^2+c)^{m+\frac{1}{2}}\,dx = \begin{cases} \dfrac{x(ax^2+c)^{m+\frac{1}{2}}}{2(m+1)} + \dfrac{(2m+1)c}{2(m+1)}\displaystyle\int (ax^2+c)^{m-\frac{1}{2}}\,dx \\ \text{or} \\ x\sqrt{ax^2+c}\displaystyle\sum_{r=0}^{m} \frac{(2m+1)!(r!)^2c^{m-r}}{2^{2m-2r+1}m!(m+1)!(2r+1)!}(ax^2+c)^r \\ \qquad + \dfrac{(2m+1)!c^{m+1}}{2^{2m+1}m!(m+1)!}\displaystyle\int \frac{dx}{\sqrt{ax^2+c}} \end{cases}$$

281. $$\int x(ax^2+c)^{m+\frac{1}{2}}\,dx = \frac{(ax^2+c)^{m+\frac{3}{2}}}{(2m+3)a}$$

282. $$\int \frac{(ax^2+c)^{m+\frac{1}{2}}}{x}\,dx = \begin{cases} \dfrac{(ax^2+c)^{m+\frac{1}{2}}}{2m+1} + c\displaystyle\int \frac{(ax^2+c)^{m-\frac{1}{2}}}{x}\,dx \\ \text{or} \\ \sqrt{ax^2+c}\displaystyle\sum_{r=0}^{m} \frac{c^{m-r}(ax^2+c)^r}{2r+1} + c^{m+1}\int \frac{dx}{x\sqrt{ax^2+c}} \end{cases}$$

283. $$\int \frac{dx}{(ax^2+c)^{m+\frac{1}{2}}} = \begin{cases} \dfrac{x}{(2m-1)c(ax^2+c)^{m-\frac{1}{2}}} + \dfrac{2m-2}{(2m-1)c}\displaystyle\int \frac{dx}{(ax^2+c)^{m-\frac{1}{2}}} \\ \text{or} \\ \dfrac{x}{\sqrt{ax^2+c}}\displaystyle\sum_{r=0}^{m-1} \frac{2^{2m-2r-1}(m-1)!m!(2r)!}{(2m)!(r!)^2c^{m-r}(ax^2+c)^r} \end{cases}$$

284. $$\int \frac{dx}{x^m\sqrt{ax^2+c}} = -\frac{\sqrt{ax^2+c}}{(m-1)cx^{m-1}} - \frac{(m-2)a}{(m-1)c}\int \frac{dx}{x^{m-2}\sqrt{ax^2+c}}$$

285. $$\int \frac{1+x^2}{(1-x^2)\sqrt{1+x^4}}\,dx = \frac{1}{\sqrt{2}}\log\frac{x\sqrt{2}+\sqrt{1+x^4}}{1-x^2}$$

286. $$\int \frac{1-x^2}{(1+x^2)\sqrt{1+x^4}}\,dx = \frac{1}{\sqrt{2}}\tan^{-1}\frac{x\sqrt{2}}{\sqrt{1+x^4}}$$

287. $$\int \frac{dx}{x\sqrt{x^n+a^2}} = -\frac{2}{na}\log\frac{a+\sqrt{x^n+a^2}}{\sqrt{x^n}}$$

288. $$\int \frac{dx}{x\sqrt{x^n-a^2}} = -\frac{2}{na}\sin^{-1}\frac{a}{\sqrt{x^n}}$$

289. $$\int \sqrt{\frac{x}{a^3-x^3}}\,dx = \frac{2}{3}\sin^{-1}\left(\frac{x}{a}\right)^{\frac{3}{2}}$$

FORMS INVOLVING TRIGONOMETRIC FUNCTIONS

290. $$\int (\sin ax)\,dx = -\frac{1}{a}\cos ax$$

291. $$\int (\cos ax)\,dx = \frac{1}{a}\sin ax$$

292. $$\int (\tan ax)\,dx = -\frac{1}{a}\log\cos ax = \frac{1}{a}\log\sec ax$$

293. $$\int (\cot ax)\,dx = \frac{1}{a}\log\sin ax = -\frac{1}{a}\log\csc ax$$

294. $$\int (\sec ax)\,dx = \frac{1}{a}\log(\sec ax + \tan ax) = \frac{1}{a}\log\tan\left(\frac{\pi}{4}+\frac{ax}{2}\right)$$

295. $$\int (\csc ax)\,dx = \frac{1}{a}\log(\csc ax - \cot ax) = \frac{1}{a}\log\tan\frac{ax}{2}$$

296. $$\int (\sin^2 ax)\,dx = -\frac{1}{2a}\cos ax\sin ax + \frac{1}{2}x = \frac{1}{2}x - \frac{1}{4a}\sin 2ax$$

297. $$\int (\sin^3 ax)\,dx = -\frac{1}{3a}(\cos ax)(\sin^2 ax + 2)$$

298. $$\int (\sin^4 ax)\,dx = \frac{3x}{8} - \frac{\sin 2ax}{4a} + \frac{\sin 4ax}{32a}$$

299. $$\int (\sin^n ax)\,dx = -\frac{\sin^{n-1} ax\cos ax}{na} + \frac{n-1}{n}\int(\sin^{n-2} ax)\,dx$$

300. $\int (\sin^{2m} ax)\,dx = -\frac{\cos ax}{a}\sum_{r=0}^{m-1}\frac{(2m)!(r!)^2}{2^{2m-2r}(2r+1)!(m!)^2}\sin^{2r+1} ax + \frac{(2m)!}{2^{2m}(m!)^2}x$

301. $\int (\sin^{2m+1} ax)\,dx = -\frac{\cos ax}{a}\sum_{r=0}^{m}\frac{2^{2m-2r}(m!)^2(2r)!}{(2m+1)!(r!)^2}\sin^{2r} ax$

302. $\int (\cos^2 ax)\,dx = \frac{1}{2a}\sin ax\cos ax + \frac{1}{2}x = \frac{1}{2}x + \frac{1}{4a}\sin 2ax$

303. $\int (\cos^3 ax)\,dx = \frac{1}{3a}(\sin ax)(\cos^2 ax + 2)$

304. $\int (\cos^4 ax)\,dx = \frac{3x}{8} + \frac{\sin 2ax}{4a} + \frac{\sin 4ax}{32a}$

305. $\int (\cos^n ax)\,dx = \frac{1}{na}\cos^{n-1} ax\sin ax + \frac{n-1}{n}\int (\cos^{n-2} ax)\,dx$

306. $\int (\cos^{2m} ax)\,dx = \frac{\sin ax}{a}\sum_{r=0}^{m-1}\frac{(2m)!(r!)^2}{2^{2m-2r}(2r+1)!(m!)^2}\cos^{2r+1} ax + \frac{(2m)!}{2^{2m}(m!)^2}x$

307. $\int (\cos^{2m+1} ax)\,dx = \frac{\sin ax}{a}\sum_{r=0}^{m}\frac{2^{2m-2r}(m!)^2(2r)!}{(2m+1)!(r!)^2}\cos^{2r} ax$

308. $\int \frac{dx}{\sin^2 ax} = \int (\csc^2 ax)\,dx = -\frac{1}{a}\cot ax$

309. $\int \frac{dx}{\sin^m ax} = \int (\csc^m ax)\,dx = -\frac{1}{(m-1)a}\cdot\frac{\cos ax}{\sin^{m-1} ax} + \frac{m-2}{m-1}\int\frac{dx}{\sin^{m-2} ax}$

310. $\int \frac{dx}{\sin^{2m} ax} = \int (\csc^{2m} ax)\,dx = -\frac{1}{a}\cos ax\sum_{r=0}^{m-1}\frac{2^{2m-2r-1}(m-1)!m!(2r)!}{(2m)!(r!)^2\sin^{2r+1} ax}$

311. $\int \frac{dx}{\sin^{2m+1} ax} = \int (\csc^{2m+1} ax)\,dx =$

$$-\frac{1}{a}\cos ax\sum_{r=0}^{m-1}\frac{(2m)!(r!)^2}{2^{2m-2r}(m!)^2(2r+1)!\sin^{2r+2} ax} + \frac{1}{a}\cdot\frac{(2m)!}{2^{2m}(m!)^2}\log\tan\frac{ax}{2}$$

312. $\int \frac{dx}{\cos^2 ax} = \int (\sec^2 ax)\,dx = \frac{1}{a}\tan ax$

313. $\int \frac{dx}{\cos^n ax} = \int (\sec^n ax)\,dx = \frac{1}{(n-1)a}\cdot\frac{\sin ax}{\cos^{n-1} ax} + \frac{n-2}{n-1}\int\frac{dx}{\cos^{n-2} ax}$

314. $\int \frac{dx}{\cos^{2m} ax} = \int (\sec^{2m} ax)\,dx = \frac{1}{a}\sin ax\sum_{r=0}^{m-1}\frac{2^{2m-2r-1}(m-1)!m!(2r)!}{(2m)!(r!)^2\cos^{2r+1} ax}$

315. $\int \frac{dx}{\cos^{2m+1} ax} = \int (\sec^{2m+1} ax)\,dx =$

$$\frac{1}{a}\sin ax\sum_{r=0}^{m-1}\frac{(2m)!(r!)^2}{2^{2m-2r}(m!)^2(2r+1)!\cos^{2r+2} ax}$$

$$+\frac{1}{a}\cdot\frac{(2m)!}{2^{2m}(m!)^2}\log(\sec ax + \tan ax)$$

316. $\int (\sin mx)(\sin nx)\,dx = \frac{\sin(m-n)x}{2(m-n)} - \frac{\sin(m+n)x}{2(m+n)}, \quad (m^2 \neq n^2)$

317. $\int (\cos mx)(\cos nx)\,dx = \frac{\sin(m-n)x}{2(m-n)} + \frac{\sin(m+n)x}{2(m+n)}, \quad (m^2 \neq n^2)$

318. $\int (\sin ax)(\cos ax)\,dx = \frac{1}{2a}\sin^2 ax$

319. $$\int (\sin mx)(\cos nx)\,dx = -\frac{\cos(m-n)x}{2(m-n)} - \frac{\cos(m+n)x}{2(m+n)}, \qquad (m^2 \neq n^2)$$

320. $$\int (\sin^2 ax)(\cos^2 ax)\,dx = -\frac{1}{32a}\sin 4ax + \frac{x}{8}$$

321. $$\int (\sin ax)(\cos^m ax)\,dx = -\frac{\cos^{m+1} ax}{(m+1)a}$$

322. $$\int (\sin^m ax)(\cos ax)\,dx = \frac{\sin^{m+1} ax}{(m+1)a}$$

323. $$\int (\cos^m ax)(\sin^n ax)\,dx = \begin{cases} \dfrac{\cos^{m-1} ax \sin^{n+1} ax}{(m+n)a} \\ \qquad + \dfrac{m-1}{m+n}\displaystyle\int (\cos^{m-2} ax)(\sin^n ax)\,dx \\ \text{or} \\ -\dfrac{\sin^{n-1} ax \cos^{m+1} ax}{(m+n)a} \\ \qquad + \dfrac{n-1}{m+n}\displaystyle\int (\cos^m ax)(\sin^{n-2} ax)\,dx \end{cases}$$

324. $$\int \frac{\cos^m ax}{\sin^n ax}\,dx = \begin{cases} -\dfrac{\cos^{m+1} ax}{(n-1)a\sin^{n-1} ax} - \dfrac{m-n+2}{n-1}\displaystyle\int \frac{\cos^m ax}{\sin^{n-2} ax}\,dx \\ \text{or} \\ \dfrac{\cos^{m-1} ax}{a(m-n)\sin^{n-1} ax} + \dfrac{m-1}{m-n}\displaystyle\int \frac{\cos^{m-2} ax}{\sin^n ax}\,dx \end{cases}$$

325. $$\int \frac{\sin^m ax}{\cos^n ax}\,dx = \begin{cases} \dfrac{\sin^{m+1} ax}{a(n-1)\cos^{n-1} ax} - \dfrac{m-n+2}{n-1}\displaystyle\int \frac{\sin^m ax}{\cos^{n-2} ax}\,dx \\ \text{or} \\ -\dfrac{\sin^{m-1} ax}{a(m-n)\cos^{n-1} ax} + \dfrac{m-1}{m-n}\displaystyle\int \frac{\sin^{m-2} ax}{\cos^n ax}\,dx \end{cases}$$

326. $$\int \frac{\sin ax}{\cos^2 ax}\,dx = \frac{1}{a\cos ax} = \frac{\sec ax}{a}$$

327. $$\int \frac{\sin^2 ax}{\cos ax}\,dx = -\frac{1}{a}\sin ax + \frac{1}{a}\log\tan\left(\frac{\pi}{4} + \frac{ax}{2}\right)$$

328. $$\int \frac{\cos ax}{\sin^2 ax}\,dx = -\frac{1}{a\sin ax} = -\frac{\csc ax}{a}$$

329. $$\int \frac{dx}{(\sin ax)(\cos ax)} = \frac{1}{a}\log\tan ax$$

330. $$\int \frac{dx}{(\sin ax)(\cos^2 ax)} = \frac{1}{a}\left(\sec ax + \log\tan\frac{ax}{2}\right)$$

331. $$\int \frac{dx}{(\sin ax)(\cos^n ax)} = \frac{1}{a(n-1)\cos^{n-1} ax} + \int \frac{dx}{(\sin ax)(\cos^{n-2} ax)}$$

332. $$\int \frac{dx}{(\sin^2 ax)(\cos ax)} = -\frac{1}{a}\csc ax + \frac{1}{a}\log\tan\left(\frac{\pi}{4} + \frac{ax}{2}\right)$$

333. $$\int \frac{dx}{(\sin^2 ax)(\cos^2 ax)} = -\frac{2}{a}\cot 2ax$$

334. $$\int \frac{dx}{\sin^m ax \cos^n ax} = \begin{cases} -\dfrac{1}{a(m-1)(\sin^{m-1} ax)(\cos^{n-1} ax)} + \dfrac{m+n-2}{m-1}\displaystyle\int \frac{dx}{(\sin^{m-2} ax)(\cos^n ax)} \\ \text{or} \\ \dfrac{1}{a(n-1)\sin^{m-1} ax \cos^{n-1} ax} - \dfrac{m+n-2}{n-1}\displaystyle\int \frac{dx}{\sin^m ax \cos^{n-2} ax} \end{cases}$$

335. $$\int \sin(a+bx)\,dx = -\frac{1}{b}\cos(a+bx)$$

336. $$\int \cos(a+bx)\,dx = \frac{1}{b}\sin(a+bx)$$

337. $$\int \frac{dx}{1 \pm \sin ax} = \mp \frac{1}{a}\tan\left(\frac{\pi}{4} \mp \frac{ax}{2}\right)$$

338. $$\int \frac{dx}{1+\cos ax} = \frac{1}{a}\tan\frac{ax}{2}$$

339. $$\int \frac{dx}{1-\cos ax} = -\frac{1}{a}\cot\frac{ax}{2}$$

*340. $$\int \frac{dx}{a+b\sin x} = \begin{cases} \dfrac{2}{\sqrt{a^2-b^2}}\tan^{-1}\dfrac{a\tan\frac{x}{2}+b}{\sqrt{a^2-b^2}} \\ \text{or} \\ \dfrac{1}{\sqrt{b^2-a^2}}\log\dfrac{a\tan\frac{x}{2}+b-\sqrt{b^2-a^2}}{a\tan\frac{x}{2}+b+\sqrt{b^2-a^2}} \end{cases}$$

*341. $$\int \frac{dx}{a+b\cos x} = \begin{cases} \dfrac{2}{\sqrt{a^2-b^2}}\tan^{-1}\dfrac{\sqrt{a^2-b^2}\tan\frac{x}{2}}{a+b} \\ \text{or} \\ \dfrac{1}{\sqrt{b^2-a^2}}\log\left(\dfrac{\sqrt{b^2-a^2}\tan\frac{x}{2}+a+b}{\sqrt{b^2-a^2}\tan\frac{x}{2}-a-b}\right) \end{cases}$$

*342. $$\int \frac{dx}{a+b\sin x+c\cos x}$$

$$= \begin{cases} \dfrac{1}{\sqrt{b^2+c^2-a^2}}\log\dfrac{b-\sqrt{b^2+c^2-a^2}+(a-c)\tan\frac{x}{2}}{b+\sqrt{b^2+c^2-a^2}+(a-c)\tan\frac{x}{2}}, & \text{if } a^2 < b^2+c^2, a \neq c \\ \text{or} \\ \dfrac{2}{\sqrt{a^2-b^2-c^2}}\tan^{-1}\dfrac{b+(a-c)\tan\frac{x}{2}}{\sqrt{a^2-b^2-c^2}}, & \text{if } a^2 > b^2+c^2 \\ \text{or} \\ \dfrac{1}{a}\left[\dfrac{a-(b+c)\cos x-(b-c)\sin x}{a-(b-c)\cos x+(b+c)\sin x}\right], & \text{if } a^2 = b^2+c^2, a \neq c. \end{cases}$$

*See note 6 on page A-19.

***343.** $\int \frac{\sin^2 x\, dx}{a + b\cos^2 x} = \frac{1}{b}\sqrt{\frac{a+b}{a}}\tan^{-1}\left(\sqrt{\frac{a}{a+b}}\tan x\right) - \frac{x}{b}, \qquad (ab > 0, \text{ or } |a| > |b|)$

***344.** $\int \frac{dx}{a^2\cos^2 x + b^2 \sin^2 x} = \frac{1}{ab}\tan^{-1}\left(\frac{b\tan x}{a}\right)$

***345.** $\int \frac{\cos^2 cx}{a^2 + b^2\sin^2 cx}\, dx = \frac{\sqrt{a^2+b^2}}{ab^2c}\tan^{-1}\frac{\sqrt{a^2+b^2}\tan cx}{a} - \frac{x}{b^2}$

346. $\int \frac{\sin cx \cos cx}{a\cos^2 cx + b\sin^2 cx}\, dx = \frac{1}{2c(b-a)}\log(a\cos^2 cx + b\sin^2 cx)$

347. $\int \frac{\cos cx}{a\cos cx + b\sin cx}\, dx = \int \frac{dx}{a + b\tan cx} =$

$\frac{1}{c(a^2+b^2)}[acx + b\log(a\cos cx + b\,\text{si}$

348. $\int \frac{\sin cx}{a\sin cx + b\cos cx}\, dx = \int \frac{dx}{a + b\cot cx} =$

$\frac{1}{c(a^2+b^2)}[acx - b\log(a\sin cx + b\,\text{cc}$

***349.** $\int \frac{dx}{a\cos^2 x + 2b\cos x \sin x + c\sin^2 x} =$

$\frac{1}{2\sqrt{b^2 - ac}}\log\frac{c\tan x + b - \sqrt{b^2}}{c\tan x + b + \sqrt{b^2}} \quad (b^2$

or

$\frac{1}{\sqrt{ac - b^2}}\tan^{-1}\frac{c\tan x + b}{\sqrt{ac - b^2}}, \quad (b^2$

or

$-\frac{1}{c\tan x + b}, \qquad (b^2 = ac)$

350. $\int \frac{\sin ax}{1 \pm \sin ax}\, dx = \pm x + \frac{1}{a}\tan\left(\frac{\pi}{4} \mp \frac{ax}{2}\right)$

351. $\int \frac{dx}{(\sin ax)(1 \pm \sin ax)} = \frac{1}{a}\tan\left(\frac{\pi}{4} \mp \frac{ax}{2}\right) + \frac{1}{a}\log\tan\frac{ax}{2}$

352. $\int \frac{dx}{(1 + \sin ax)^2} = -\frac{1}{2a}\tan\left(\frac{\pi}{4} - \frac{ax}{2}\right) - \frac{1}{6a}\tan^3\left(\frac{\pi}{4} - \frac{ax}{2}\right)$

353. $\int \frac{dx}{(1 - \sin ax)^2} = \frac{1}{2a}\cot\left(\frac{\pi}{4} - \frac{ax}{2}\right) + \frac{1}{6a}\cot^3\left(\frac{\pi}{4} - \frac{ax}{2}\right)$

354. $\int \frac{\sin ax}{(1 + \sin ax)^2}\, dx = -\frac{1}{2a}\tan\left(\frac{\pi}{4} - \frac{ax}{2}\right) + \frac{1}{6a}\tan^3\left(\frac{\pi}{4} - \frac{ax}{2}\right)$

355. $\int \frac{\sin ax}{(1 - \sin ax)^2}\, dx = -\frac{1}{2a}\cot\left(\frac{\pi}{4} - \frac{ax}{2}\right) + \frac{1}{6a}\cot^3\left(\frac{\pi}{4} - \frac{ax}{2}\right)$

356. $\int \frac{\sin x\, dx}{a + b\sin x} = \frac{x}{b} - \frac{a}{b}\int \frac{dx}{a + b\sin x}$

357. $\int \frac{dx}{(\sin x)(a + b\sin x)} = \frac{1}{a}\log\tan\frac{x}{2} - \frac{b}{a}\int \frac{dx}{a + b\sin x}$

358. $\int \frac{dx}{(a + b\sin x)^2} = \frac{b\cos x}{(a^2 - b^2)(a + b\sin x)} + \frac{a}{a^2 - b^2}\int \frac{dx}{a + b\sin x}$

*See note 6 on page A-19.

359. $\int \frac{\sin x\, dx}{(a + b\sin x)^2} = \frac{a\cos x}{(b^2 - a^2)(a + b\sin x)} + \frac{b}{b^2 - a^2}\int \frac{dx}{a + b\sin x}$

*360. $\int \frac{dx}{a^2 + b^2\sin^2 cx} = \frac{1}{ac\sqrt{a^2 + b^2}}\tan^{-1}\frac{\sqrt{a^2 + b^2}\tan cx}{a}$

*361. $\int \frac{dx}{a^2 - b^2\sin^2 cx} = \begin{cases} \frac{1}{ac\sqrt{a^2 - b^2}}\tan^{-1}\frac{\sqrt{a^2 - b^2}\tan cx}{a}, & (a^2 > b^2) \\ \text{or} & \\ \frac{1}{2ac\sqrt{b^2 - a^2}}\log\frac{\sqrt{b^2 - a^2}\tan cx + a}{\sqrt{b^2 - a^2}\tan cx - a}, & (a^2 < b^2) \end{cases}$

362. $\int \frac{\cos ax}{1 + \cos ax}dx = x - \frac{1}{a}\tan\frac{ax}{2}$

363. $\int \frac{\cos ax}{1 - \cos ax}dx = -x - \frac{1}{a}\cot\frac{ax}{2}$

364. $\int \frac{dx}{(\cos ax)(1 + \cos ax)} = \frac{1}{a}\log\tan\left(\frac{\pi}{4} + \frac{ax}{2}\right) - \frac{1}{a}\tan\frac{ax}{2}$

365. $\int \frac{dx}{(\cos ax)(1 - \cos ax)} = \frac{1}{a}\log\tan\left(\frac{\pi}{4} + \frac{ax}{2}\right) - \frac{1}{a}\cot\frac{ax}{2}$

366. $\int \frac{dx}{(1 + \cos ax)^2} = \frac{1}{2a}\tan\frac{ax}{2} + \frac{1}{6a}\tan^3\frac{ax}{2}$

367. $\int \frac{dx}{(1 - \cos ax)^2} = -\frac{1}{2a}\cot\frac{ax}{2} - \frac{1}{6a}\cot^3\frac{ax}{2}$

368. $\int \frac{\cos ax}{(1 + \cos ax)^2}dx = \frac{1}{2a}\tan\frac{ax}{2} - \frac{1}{6a}\tan^3\frac{ax}{2}$

369. $\int \frac{\cos ax}{(1 - \cos ax)^2}dx = \frac{1}{2a}\cot\frac{ax}{2} - \frac{1}{6a}\cot^3\frac{ax}{2}$

370. $\int \frac{\cos x\, dx}{a + b\cos x} = \frac{x}{b} - \frac{a}{b}\int \frac{dx}{a + b\cos x}$

371. $\int \frac{dx}{(\cos x)(a + b\cos x)} = \frac{1}{a}\log\tan\left(\frac{x}{2} + \frac{\pi}{4}\right) - \frac{b}{a}\int \frac{dx}{a + b\cos x}$

372. $\int \frac{dx}{(a + b\cos x)^2} = \frac{b\sin x}{(b^2 - a^2)(a + b\cos x)} - \frac{a}{b^2 - a^2}\int \frac{dx}{a + b\cos x}$

373. $\int \frac{\cos x}{(a + b\cos x)^2}dx = \frac{a\sin x}{(a^2 - b^2)(a + b\cos x)} - \frac{b}{a^2 - b^2}\int \frac{dx}{a + b\cos x}$

*374. $\int \frac{dx}{a^2 + b^2 - 2ab\cos cx} = \frac{2}{c(a^2 - b^2)}\tan^{-1}\left(\frac{a + b}{a - b}\tan\frac{cx}{2}\right)$

*375. $\int \frac{dx}{a^2 + b^2\cos^2 cx} = \frac{1}{ac\sqrt{a^2 + b^2}}\tan^{-1}\frac{a\tan cx}{\sqrt{a^2 + b^2}}$

*376. $\int \frac{dx}{a^2 - b^2\cos^2 cx} = \begin{cases} \frac{1}{ac\sqrt{a^2 - b^2}}\tan^{-1}\frac{a\tan cx}{\sqrt{a^2 - b^2}}, & (a^2 > b^2) \\ \text{or} & \\ \frac{1}{2ac\sqrt{b^2 - a^2}}\log\frac{a\tan cx - \sqrt{b^2 - a^2}}{a\tan cx + \sqrt{b^2 - a^2}}, & (b^2 > a^2) \end{cases}$

377. $\int \frac{\sin ax}{1 \pm \cos ax}dx = \mp\frac{1}{a}\log(1 \pm \cos ax)$

*See note 6 on page A-19.

378. $\int \frac{\cos ax}{1 \pm \sin ax}\, dx = \pm \frac{1}{a} \log (1 \pm \sin ax)$

379. $\int \frac{dx}{(\sin ax)(1 \pm \cos ax)} = \pm \frac{1}{2a(1 \pm \cos ax)} + \frac{1}{2a} \log \tan \frac{ax}{2}$

380. $\int \frac{dx}{(\cos ax)(1 \pm \sin ax)} = \mp \frac{1}{2a(1 \pm \sin ax)} + \frac{1}{2a} \log \tan \left(\frac{\pi}{4} + \frac{ax}{2}\right)$

381. $\int \frac{\sin ax}{(\cos ax)(1 \pm \cos ax)}\, dx = \frac{1}{a} \log (\sec ax \pm 1)$

382. $\int \frac{\cos ax}{(\sin ax)(1 \pm \sin ax)}\, dx = -\frac{1}{a} \log (\csc ax \pm 1)$

383. $\int \frac{\sin ax}{(\cos ax)(1 \pm \sin ax)}\, dx = \frac{1}{2a(1 \pm \sin ax)} \pm \frac{1}{2a} \log \tan \left(\frac{\pi}{4} + \frac{ax}{2}\right)$

384. $\int \frac{\cos ax}{(\sin ax)(1 \pm \cos ax)}\, dx = -\frac{1}{2a(1 \pm \cos ax)} \pm \frac{1}{2a} \log \tan \frac{ax}{2}$

385. $\int \frac{dx}{\sin ax \pm \cos ax} = \frac{1}{a\sqrt{2}} \log \tan \left(\frac{ax}{2} \pm \frac{\pi}{8}\right)$

386. $\int \frac{dx}{(\sin ax \pm \cos ax)^2} = \frac{1}{2a} \tan \left(ax \mp \frac{\pi}{4}\right)$

387. $\int \frac{dx}{1 + \cos ax \pm \sin ax} = \pm \frac{1}{a} \log \left(1 \pm \tan \frac{ax}{2}\right)$

388. $\int \frac{dx}{a^2 \cos^2 cx - b^2 \sin^2 cx} = \frac{1}{2abc} \log \frac{b \tan cx + a}{b \tan cx - a}$

389. $\int x(\sin ax)\, dx = \frac{1}{a^2} \sin ax - \frac{x}{a} \cos ax$

390. $\int x^2(\sin ax)\, dx = \frac{2x}{a^2} \sin ax - \frac{a^2x^2 - 2}{a^3} \cos ax$

391. $\int x^3(\sin ax)\, dx = \frac{3a^2x^2 - 6}{a^4} \sin ax - \frac{a^2x^3 - 6x}{a^3} \cos ax$

392. $$\int x^m \sin ax\, dx = \begin{cases} -\frac{1}{a} x^m \cos ax + \frac{m}{a} \int x^{m-1} \cos ax\, dx \\ \text{or} \\ \cos ax \sum_{r=0}^{\left[\frac{m}{2}\right]} (-1)^{r+1} \frac{m!}{(m-2r)!} \cdot \frac{x^{m-2r}}{a^{2r+1}} \\ + \sin ax \sum_{r=0}^{\left[\frac{m-1}{2}\right]} (-1)^r \frac{m!}{(m-2r-1)!} \cdot \frac{x^{m-2r-1}}{a^{2r+2}} \end{cases}$$

Note: $[s]$ means greatest integer $\leq s$; $[3\frac{1}{2}] = 3$, $[\frac{1}{2}] = 0$, etc.

393. $\int x(\cos ax)\, dx = \frac{1}{a^2} \cos ax + \frac{x}{a} \sin ax$

394. $\int x^2(\cos ax)\, dx = \frac{2x \cos ax}{a^2} + \frac{a^2x^2 - 2}{a^3} \sin ax$

395. $\int x^3(\cos ax)\, dx = \frac{3a^2x^2 - 6}{a^4} \cos ax + \frac{a^2x^3 - 6x}{a^3} \sin ax$

396. $$\int x^m(\cos ax)\,dx = \begin{cases} \dfrac{x^m \sin ax}{a} - \dfrac{m}{a}\displaystyle\int x^{m-1}\sin ax\,dx \\ \text{or} \\ \sin ax \displaystyle\sum_{r=0}^{\left[\frac{m}{2}\right]} (-1)^r \frac{m!}{(m-2r)!}\cdot\frac{x^{m-2r}}{a^{2r+1}} \\ \quad + \cos ax \displaystyle\sum_{r=0}^{\left[\frac{m-1}{2}\right]} (-1)^r \frac{m!}{(m-2r-1)!}\cdot\frac{x^{m-2r-1}}{a^{2r+2}} \end{cases}$$

See note integral 392.

397. $$\int \frac{\sin ax}{x}\,dx = \sum_{n=0}^{\infty} (-1)^n \frac{(ax)^{2n+1}}{(2n+1)(2n+1)!}$$

398. $$\int \frac{\cos ax}{x}\,dx = \log x + \sum_{n=1}^{\infty} (-1)^n \frac{(ax)^{2n}}{2n(2n)!}$$

399. $$\int x(\sin^2 ax)\,dx = \frac{x^2}{4} - \frac{x\sin 2ax}{4a} - \frac{\cos 2ax}{8a^2}$$

400. $$\int x^2(\sin^2 ax)\,dx = \frac{x^3}{6} - \left(\frac{x^2}{4a} - \frac{1}{8a^3}\right)\sin 2ax - \frac{x\cos 2ax}{4a^2}$$

401. $$\int x(\sin^3 ax)\,dx = \frac{x\cos 3ax}{12a} - \frac{\sin 3ax}{36a^2} - \frac{3x\cos ax}{4a} + \frac{3\sin ax}{4a^2}$$

402. $$\int x(\cos^2 ax)\,dx = \frac{x^2}{4} + \frac{x\sin 2ax}{4a} + \frac{\cos 2ax}{8a^2}$$

403. $$\int x^2(\cos^2 ax)\,dx = \frac{x^3}{6} + \left(\frac{x^2}{4a} - \frac{1}{8a^3}\right)\sin 2ax + \frac{x\cos 2ax}{4a^2}$$

404. $$\int x(\cos^3 ax)\,dx = \frac{x\sin 3ax}{12a} + \frac{\cos 3ax}{36a^2} + \frac{3x\sin ax}{4a} + \frac{3\cos ax}{4a^2}$$

405. $$\int \frac{\sin ax}{x^m}\,dx = -\frac{\sin ax}{(m-1)x^{m-1}} + \frac{a}{m-1}\int \frac{\cos ax}{x^{m-1}}\,dx$$

406. $$\int \frac{\cos ax}{x^m}\,dx = -\frac{\cos ax}{(m-1)x^{m-1}} - \frac{a}{m-1}\int \frac{\sin ax}{x^{m-1}}\,dx$$

407. $$\int \frac{x}{1 \pm \sin ax}\,dx = \mp \frac{x\cos ax}{a(1 \pm \sin ax)} + \frac{1}{a^2}\log(1 \pm \sin ax)$$

408. $$\int \frac{x}{1 + \cos ax}\,dx = \frac{x}{a}\tan\frac{ax}{2} + \frac{2}{a^2}\log\cos\frac{ax}{2}$$

409. $$\int \frac{x}{1 - \cos ax}\,dx = -\frac{x}{a}\cot\frac{ax}{2} + \frac{2}{a^2}\log\sin\frac{ax}{2}$$

410. $$\int \frac{x + \sin x}{1 + \cos x}\,dx = x\tan\frac{x}{2}$$

411. $$\int \frac{x - \sin x}{1 - \cos x}\,dx = -x\cot\frac{x}{2}$$

412. $$\int \sqrt{1 - \cos ax}\,dx = -\frac{2\sin ax}{a\sqrt{1 - \cos ax}} = -\frac{2\sqrt{2}\cos\left(\frac{ax}{2}\right)}{a}$$

413. $$\int \sqrt{1 + \cos ax}\,dx = \frac{2\sin ax}{a\sqrt{1 + \cos ax}} = \frac{2\sqrt{2}\sin\left(\frac{ax}{2}\right)}{a}$$

414. $$\int \sqrt{1 + \sin x}\,dx = \pm 2\left(\sin\frac{x}{2} - \cos\frac{x}{2}\right),$$

[use + if $(8k-1)\frac{\pi}{2} < x \le (8k+3)\frac{\pi}{2}$, otherwise $-$; k an integer]

415. $\int \sqrt{1 - \sin x}\, dx = \pm 2\left(\sin\frac{x}{2} + \cos\frac{x}{2}\right),$

[use + if $(8k - 3)\frac{\pi}{2} < x \leq (8k + 1)\frac{\pi}{2}$, otherwise $-$; k an integer]

416. $\int \frac{dx}{\sqrt{1 - \cos x}} = \pm \sqrt{2} \log \tan \frac{x}{4},$

[use + if $4k\pi < x < (4k + 2)\pi$, otherwise $-$; k an integer]

417. $\int \frac{dx}{\sqrt{1 + \cos x}} = \pm \sqrt{2} \log \tan \left(\frac{x + \pi}{4}\right),$

[use + if $(4k - 1)\pi < x < (4k + 1)\pi$, otherwise $-$; k an integer]

418. $\int \frac{dx}{\sqrt{1 - \sin x}} = \pm \sqrt{2} \log \tan \left(\frac{x}{4} - \frac{\pi}{8}\right),$

[use + if $(8k + 1)\frac{\pi}{2} < x < (8k + 5)\frac{\pi}{2}$, otherwise $-$; k an integer]

419. $\int \frac{dx}{\sqrt{1 + \sin x}} = \pm \sqrt{2} \log \tan \left(\frac{x}{4} + \frac{\pi}{8}\right),$

[use + if $(8k - 1)\frac{\pi}{2} < x < (8k + 3)\frac{\pi}{2}$, otherwise $-$; k an integer]

420. $\int (\tan^2 ax)\, dx = \frac{1}{a} \tan ax - x$

421. $\int (\tan^3 ax)\, dx = \frac{1}{2a} \tan^2 ax + \frac{1}{a} \log \cos ax$

422. $\int (\tan^4 ax)\, dx = \frac{\tan^3 ax}{3a} - \frac{1}{a} \tan x + x$

423. $\int (\tan^n ax)\, dx = \frac{\tan^{n-1} ax}{a(n - 1)} - \int (\tan^{n-2} ax)\, dx$

424. $\int (\cot^2 ax)\, dx = -\frac{1}{a} \cot ax - x$

425. $\int (\cot^3 ax)\, dx = -\frac{1}{2a} \cot^2 ax - \frac{1}{a} \log \sin ax$

426. $\int (\cot^4 ax)\, dx = -\frac{1}{3a} \cot^3 ax + \frac{1}{a} \cot ax + x$

427. $\int (\cot^n ax)\, dx = -\frac{\cot^{n-1} ax}{a(n - 1)} - \int (\cot^{n-2} ax)\, dx$

428. $\int \frac{x}{\sin^2 ax}\, dx = \int x(\csc^2 ax)\, dx = -\frac{x \cot ax}{a} + \frac{1}{a^2} \log \sin ax$

429. $\int \frac{x}{\sin^n ax}\, dx = \int x(\csc^n ax)\, dx = -\frac{x \cos ax}{a(n - 1) \sin^{n-1} ax}$

$$- \frac{1}{a^2(n - 1)(n - 2) \sin^{n-2} ax} + \frac{(n - 2)}{(n - 1)} \int \frac{x}{\sin^{n-2} ax}\, dx$$

430. $\int \frac{x}{\cos^2 ax}\, dx = \int x(\sec^2 ax)\, dx = \frac{1}{a} x \tan ax + \frac{1}{a^2} \log \cos ax$

431. $\int \frac{x}{\cos^n ax}\, dx = \int x(\sec^n ax)\, dx = \frac{x \sin ax}{a(n - 1) \cos^{n-1} ax}$

$$- \frac{1}{a^2(n - 1)(n - 2) \cos^{n-2} ax} + \frac{n - 2}{n - 1} \int \frac{x}{\cos^{n-2} ax}\, dx$$

432. $\int \frac{\sin ax}{\sqrt{1+b^2\sin^2 ax}}\,dx = -\frac{1}{ab}\sin^{-1}\frac{b\cos ax}{\sqrt{1+b^2}}$

433. $\int \frac{\sin ax}{\sqrt{1-b^2\sin^2 ax}}\,dx = -\frac{1}{ab}\log\left(b\cos ax+\sqrt{1-b^2\sin^2 ax}\right)$

434. $\int (\sin ax)\sqrt{1+b^2\sin^2 ax}\,dx = -\frac{\cos ax}{2a}\sqrt{1+b^2\sin^2 ax} - \frac{1+b^2}{2ab}\sin^{-1}\frac{b\cos ax}{\sqrt{1+b^2}}$

435. $\int (\sin ax)\sqrt{1-b^2\sin^2 ax}\,dx = -\frac{\cos ax}{2a}\sqrt{1-b^2\sin^2 ax} - \frac{1-b^2}{2ab}\log\left(b\cos ax+\sqrt{1-b^2\sin^2 ax}\right)$

436. $\int \frac{\cos ax}{\sqrt{1+b^2\sin^2 ax}}\,dx = \frac{1}{ab}\log\left(b\sin ax+\sqrt{1+b^2\sin^2 ax}\right)$

437. $\int \frac{\cos ax}{\sqrt{1-b^2\sin^2 ax}}\,dx = \frac{1}{ab}\sin^{-1}(b\sin ax)$

438. $\int (\cos ax)\sqrt{1+b^2\sin^2 ax}\,dx = \frac{\sin ax}{2a}\sqrt{1+b^2\sin^2 ax} + \frac{1}{2ab}\log\left(b\sin ax+\sqrt{1+b^2\sin^2 ax}\right)$

439. $\int (\cos ax)\sqrt{1-b^2\sin^2 ax}\,dx = \frac{\sin ax}{2a}\sqrt{1-b^2\sin^2 ax} + \frac{1}{2ab}\sin^{-1}(b\sin ax)$

440. $\int \frac{dx}{\sqrt{a+b\tan^2 cx}} = \frac{\pm 1}{c\sqrt{a-b}}\sin^{-1}\left(\sqrt{\frac{a-b}{a}}\sin cx\right), \qquad (a > |b|)$

[use + if $(2k-1)\frac{\pi}{2} < x \le (2k+1)\frac{\pi}{2}$, otherwise −; k an integer]

FORMS INVOLVING INVERSE TRIGONOMETRIC FUNCTIONS

441. $\int (\sin^{-1} ax)\,dx = x\sin^{-1} ax + \frac{\sqrt{1-a^2x^2}}{a}$

442. $\int (\cos^{-1} ax)\,dx = x\cos^{-1} ax - \frac{\sqrt{1-a^2x^2}}{a}$

443. $\int (\tan^{-1} ax)\,dx = x\tan^{-1} ax - \frac{1}{2a}\log(1+a^2x^2)$

444. $\int (\cot^{-1} ax)\,dx = x\cot^{-1} ax + \frac{1}{2a}\log(1+a^2x^2)$

445. $\int (\sec^{-1} ax)\,dx = x\sec^{-1} ax - \frac{1}{a}\log\left(ax+\sqrt{a^2x^2-1}\right)$

446. $\int (\csc^{-1} ax)\,dx = x\csc^{-1} ax + \frac{1}{a}\log\left(ax+\sqrt{a^2x^2-1}\right)$

447. $\int \left(\sin^{-1}\frac{x}{a}\right)dx = x\sin^{-1}\frac{x}{a} + \sqrt{a^2-x^2}, \qquad (a > 0)$

448. $\int \left(\cos^{-1}\frac{x}{a}\right)dx = x\cos^{-1}\frac{x}{a} - \sqrt{a^2-x^2}, \qquad (a > 0)$

449. $\int \left(\tan^{-1}\frac{x}{a}\right) dx = x\tan^{-1}\frac{x}{a} - \frac{a}{2}\log(a^2 + x^2)$

450. $\int \left(\cot^{-1}\frac{x}{a}\right) dx = x\cot^{-1}\frac{x}{a} + \frac{a}{2}\log(a^2 + x^2)$

451. $\int x[\sin^{-1}(ax)]\, dx = \frac{1}{4a^2}[(2a^2x^2 - 1)\sin^{-1}(ax) + ax\sqrt{1 - a^2x^2}]$

452. $\int x[\cos^{-1}(ax)]\, dx = \frac{1}{4a^2}[(2a^2x^2 - 1)\cos^{-1}(ax) - ax\sqrt{1 - a^2x^2}]$

453. $\int x^n[\sin^{-1}(ax)]\, dx = \frac{x^{n+1}}{n+1}\sin^{-1}(ax) - \frac{a}{n+1}\int \frac{x^{n+1}\, dx}{\sqrt{1 - a^2x^2}}, \quad (n \neq -1)$

454. $\int x^n[\cos^{-1}(ax)]\, dx = \frac{x^{n+1}}{n+1}\cos^{-1}(ax) + \frac{a}{n+1}\int \frac{x^{n+1}\, dx}{\sqrt{1 - a^2x^2}}, \quad (n \neq -1)$

455. $\int x(\tan^{-1} ax)\, dx = \frac{1 + a^2x^2}{2a^2}\tan^{-1} ax - \frac{x}{2a}$

456. $\int x^n(\tan^{-1} ax)\, dx = \frac{x^{n+1}}{n+1}\tan^{-1} ax - \frac{a}{n+1}\int \frac{x^{n+1}}{1 + a^2x^2}\, dx$

457. $\int x(\cot^{-1} ax)\, dx = \frac{1 + a^2x^2}{2a^2}\cot^{-1} ax + \frac{x}{2a}$

458. $\int x^n(\cot^{-1} ax)\, dx = \frac{x^{n+1}}{n+1}\cot^{-1} ax + \frac{a}{n+1}\int \frac{x^{n+1}}{1 + a^2x^2}\, dx$

459. $\int \frac{\sin^{-1}(ax)}{x^2}\, dx = a\log\left(\frac{1 - \sqrt{1 - a^2x^2}}{x}\right) - \frac{\sin^{-1}(ax)}{x}$

460. $\int \frac{\cos^{-1}(ax)\, dx}{x^2} = -\frac{1}{x}\cos^{-1}(ax) + a\log\frac{1 + \sqrt{1 - a^2x^2}}{x}$

461. $\int \frac{\tan^{-1}(ax)\, dx}{x^2} = -\frac{1}{x}\tan^{-1}(ax) - \frac{a}{2}\log\frac{1 + a^2x^2}{x^2}$

462. $\int \frac{\cot^{-1} ax}{x^2}\, dx = -\frac{1}{x}\cot^{-1} ax - \frac{a}{2}\log\frac{x^2}{a^2x^2 + 1}$

463. $\int (\sin^{-1} ax)^2\, dx = x(\sin^{-1} ax)^2 - 2x + \frac{2\sqrt{1 - a^2x^2}}{a}\sin^{-1} ax$

464. $\int (\cos^{-1} ax)^2\, dx = x(\cos^{-1} ax)^2 - 2x - \frac{2\sqrt{1 - a^2x^2}}{a}\cos^{-1} ax$

465. $$\int (\sin^{-1} ax)^n\, dx = \begin{cases} x(\sin^{-1} ax)^n + \frac{n\sqrt{1 - a^2x^2}}{a}(\sin^{-1} ax)^{n-1} \\ \qquad - n(n-1)\int (\sin^{-1} ax)^{n-2}\, dx \\ \text{or} \\ \sum_{r=0}^{[\frac{n}{2}]} (-1)^r \frac{n!}{(n-2r)!} x(\sin^{-1} ax)^{n-2r} \\ \qquad + \sum_{r=0}^{[\frac{n-1}{2}]} (-1)^r \frac{n!\sqrt{1 - a^2x^2}}{(n - 2r - 1)!\,a}(\sin^{-1} ax)^{n-2r-1} \end{cases}$$

Note: $[s]$ means greatest integer $\leq s$. Thus $[3.5]$ means 3; $[5] = 5$, $[\frac{1}{2}] = 0$.

466. $$\int (\cos^{-1} ax)^n\,dx = \begin{cases} x(\cos^{-1} ax)^n - \dfrac{n\sqrt{1-a^2x^2}}{a}(\cos^{-1} ax)^{n-1} \\ \qquad\qquad - n(n-1)\displaystyle\int (\cos^{-1} ax)^{n-2}\,dx \\ \qquad\text{or} \\ \displaystyle\sum_{r=0}^{\left[\frac{n}{2}\right]} (-1)^r \frac{n!}{(n-2r)!} x(\cos^{-1} ax)^{n-2r} \\ \qquad - \displaystyle\sum_{r=0}^{\left[\frac{n-1}{2}\right]} (-1)^r \frac{n!\sqrt{1-a^2x^2}}{(n-2r-1)!\,a}(\cos^{-1} ax)^{n-2r-1} \end{cases}$$

467. $$\int \frac{1}{\sqrt{1-a^2x^2}}(\sin^{-1} ax)\,dx = \frac{1}{2a}(\sin^{-1} ax)^2$$

468. $$\int \frac{x^n}{\sqrt{1-a^2x^2}}(\sin^{-1} ax)\,dx = -\frac{x^{n-1}}{na^2}\sqrt{1-a^2x^2}\sin^{-1} ax + \frac{x^n}{n^2a} + \frac{n-1}{na^2}\int \frac{x^{n-2}}{\sqrt{1-a^2x^2}}\sin^{-1} ax\,dx$$

469. $$\int \frac{1}{\sqrt{1-a^2x^2}}(\cos^{-1} ax)\,dx = -\frac{1}{2a}(\cos^{-1} ax)^2$$

470. $$\int \frac{x^n}{\sqrt{1-a^2x^2}}(\cos^{-1} ax)\,dx = -\frac{x^{n-1}}{na^2}\sqrt{1-a^2x^2}\cos^{-1} ax - \frac{x^n}{n^2a} + \frac{n-1}{na^2}\int \frac{x^{n-2}}{\sqrt{1-a^2x^2}}\cos^{-1} ax\,dx$$

471. $$\int \frac{\tan^{-1} ax}{a^2x^2+1}\,dx = \frac{1}{2a}(\tan^{-1} ax)^2$$

472. $$\int \frac{\cot^{-1} ax}{a^2x^2+1}\,dx = -\frac{1}{2a}(\cot^{-1} ax)^2$$

473. $$\int x\sec^{-1} ax\,dx = \frac{x^2}{2}\sec^{-1} ax - \frac{1}{2a^2}\sqrt{a^2x^2-1}$$

474. $$\int x^n\sec^{-1} ax\,dx = \frac{x^{n+1}}{n+1}\sec^{-1} ax - \frac{1}{n+1}\int \frac{x^n\,dx}{\sqrt{a^2x^2-1}}$$

475. $$\int \frac{\sec^{-1} ax}{x^2}\,dx = -\frac{\sec^{-1} ax}{x} + \frac{\sqrt{a^2x^2-1}}{x}$$

476. $$\int x\csc^{-1} ax\,dx = \frac{x^2}{2}\csc^{-1} ax + \frac{1}{2a^2}\sqrt{a^2x^2-1}$$

477. $$\int x^n\csc^{-1} ax\,dx = \frac{x^{n+1}}{n+1}\csc^{-1} ax + \frac{1}{n+1}\int \frac{x^n\,dx}{\sqrt{a^2x^2-1}}$$

478. $$\int \frac{\csc^{-1} ax}{x^2}\,dx = -\frac{\csc^{-1} ax}{x} - \frac{\sqrt{a^2x^2-1}}{x}$$

FORMS INVOLVING TRIGONOMETRIC SUBSTITUTIONS

479. $$\int f(\sin x)\,dx = 2\int f\left(\frac{2z}{1+z^2}\right)\frac{dz}{1+z^2}, \qquad \left(z = \tan\frac{x}{2}\right)$$

480. $$\int f(\cos x)\,dx = 2\int f\left(\frac{1-z^2}{1+z^2}\right)\frac{dz}{1+z^2}, \qquad \left(z = \tan\frac{x}{2}\right)$$

*481. $\int f(\sin x)\,dx = \int f(u)\frac{du}{\sqrt{1-u^2}}, \quad (u = \sin x)$

*482. $\int f(\cos x)\,dx = -\int f(u)\frac{du}{\sqrt{1-u^2}}, \quad (u = \cos x)$

*483. $\int f(\sin x, \cos x)\,dx = \int f(u, \sqrt{1-u^2})\frac{du}{\sqrt{1-u^2}}, \quad (u = \sin x)$

484. $\int f(\sin x, \cos x)\,dx = 2\int f\left(\frac{2z}{1+z^2}, \frac{1-z^2}{1+z^2}\right)\frac{dz}{1+z^2}, \quad \left(z = \tan\frac{x}{2}\right)$

LOGARITHMIC FORMS

485. $\int (\log x)\,dx = x\log x - x$

486. $\int x(\log x)\,dx = \frac{x^2}{2}\log x - \frac{x^2}{4}$

487. $\int x^2(\log x)\,dx = \frac{x^3}{3}\log x - \frac{x^3}{9}$

488. $\int x^n(\log ax)\,dx = \frac{x^{n+1}}{n+1}\log ax - \frac{x^{n+1}}{(n+1)^2}$

489. $\int (\log x)^2\,dx = x(\log x)^2 - 2x\log x + 2x$

490. $\int (\log x)^n\,dx = \begin{cases} x(\log x)^n - n\int (\log x)^{n-1}\,dx, & (n \neq -1) \\ \text{or} & \\ (-1)^n n!\, x \sum_{r=0}^{n} \frac{(-\log x)^r}{r!} & \end{cases}$

491. $\int \frac{(\log x)^n}{x}\,dx = \frac{1}{n+1}(\log x)^{n+1}$

492. $\int \frac{dx}{\log x} = \log(\log x) + \log x + \frac{(\log x)^2}{2 \cdot 2!} + \frac{(\log x)^3}{3 \cdot 3!} + \cdots$

493. $\int \frac{dx}{x\log x} = \log(\log x)$

494. $\int \frac{dx}{x(\log x)^n} = -\frac{1}{(n-1)(\log x)^{n-1}}$

495. $\int \frac{x^m\,dx}{(\log x)^n} = -\frac{x^{m+1}}{(n-1)(\log x)^{n-1}} + \frac{m+1}{n-1}\int \frac{x^m\,dx}{(\log x)^{n-1}}$

496. $\int x^m(\log x)^n\,dx = \begin{cases} \frac{x^{m+1}(\log x)^n}{m+1} - \frac{n}{m+1}\int x^m(\log x)^{n-1}\,dx \\ \text{or} \\ (-1)^n \frac{n!}{m+1} x^{m+1} \sum_{r=0}^{n} \frac{(-\log x)^r}{r!(m+1)^{n-r}} \end{cases}$

497. $\int x^p \cos(b\ln x)\,dx = \frac{x^{p+1}}{(p+1)^2 + b^2} \cdot [b\sin(b\ln x) + (p+1)\cos(b\ln x)] + c$

498. $\int x^p \sin(b\ln x)\,dx = \frac{x^{p+1}}{(p+1)^2 + b^2} \cdot [(p+1)\sin(b\ln x) - b\cos(b\ln x)] + c$

499. $\int [\log(ax+b)]\,dx = \frac{ax+b}{a}\log(ax+b) - x$

*The square roots appearing in these formulas may be plus or minus, depending on the quadrant of x. Care must be used to give them the proper sign.

500. $\int \frac{\log(ax+b)}{x^2}\,dx = \frac{a}{b}\log x - \frac{ax+b}{bx}\log(ax+b)$

501. $\int x^m[\log(ax+b)]\,dx = \frac{1}{m+1}\left[x^{m+1} - \left(-\frac{b}{a}\right)^{m+1}\right]\log(ax+b)$

$$- \frac{1}{m+1}\left(-\frac{b}{a}\right)^{m+1}\sum_{r=1}^{m+1}\frac{1}{r}\left(-\frac{ax}{b}\right)^r$$

502. $\int \frac{\log(ax+b)}{x^m}\,dx = -\frac{1}{m-1}\frac{\log(ax+b)}{x^{m-1}} + \frac{1}{m-1}\left(-\frac{a}{b}\right)^{m-1}\log\frac{ax+b}{x}$

$$+ \frac{1}{m-1}\left(-\frac{a}{b}\right)^{m-1}\sum_{r=1}^{m-2}\frac{1}{r}\left(-\frac{b}{ax}\right)^r, \quad (m > 2)$$

503. $\int \left[\log\frac{x+a}{x-a}\right]dx = (x+a)\log(x+a) - (x-a)\log(x-a)$

504. $\int x^m\left[\log\frac{x+a}{x-a}\right]dx = \frac{x^{m+1} - (-a)^{m+1}}{m+1}\log(x+a) - \frac{x^{m+1} - a^{m+1}}{m+1}\log(x-a)$

$$+ \frac{2a^{m+1}}{m+1}\sum_{r=1}^{\left[\frac{m+1}{2}\right]}\frac{1}{m-2r+2}\left(\frac{x}{a}\right)^{m-2r+2}$$

See note integral 392.

505. $\int \frac{1}{x^2}\left[\log\frac{x+a}{x-a}\right]dx = \frac{1}{x}\log\frac{x-a}{x+a} - \frac{1}{a}\log\frac{x^2-a^2}{x^2}$

506. $\int (\log X)\,dx = \begin{cases} \left(x + \frac{b}{2c}\right)\log X - 2x + \frac{\sqrt{4ac-b^2}}{c}\tan^{-1}\frac{2cx+b}{\sqrt{4ac-b^2}}, & (b^2 - 4ac < 0) \\ \text{or} & \\ \left(x + \frac{b}{2c}\right)\log X - 2x + \frac{\sqrt{b^2-4ac}}{c}\tanh^{-1}\frac{2cx+b}{\sqrt{b^2-4ac}}, & (b^2 - 4ac > 0) \\ \text{where} & \\ X = a + bx + cx^2 & \end{cases}$

507. $\int x^n(\log X)\,dx = \frac{x^{n+1}}{n+1}\log X - \frac{2c}{n+1}\int\frac{x^{n+2}}{X}\,dx - \frac{b}{n+1}\int\frac{x^{n+1}}{X}\,dx$

where $X = a + bx + cx^2$

508. $\int [\log(x^2+a^2)]\,dx = x\log(x^2+a^2) - 2x + 2a\tan^{-1}\frac{x}{a}$

509. $\int [\log(x^2-a^2)]\,dx = x\log(x^2-a^2) - 2x + a\log\frac{x+a}{x-a}$

510. $\int x[\log(x^2 \pm a^2)]\,dx = \frac{1}{2}(x^2 \pm a^2)\log(x^2 \pm a^2) - \frac{1}{2}x^2$

511. $\int [\log(x + \sqrt{x^2 \pm a^2})]\,dx = x\log(x + \sqrt{x^2 \pm a^2}) - \sqrt{x^2 \pm a^2}$

512. $\int x[\log(x + \sqrt{x^2 \pm a^2})]\,dx = \left(\frac{x^2}{2} \pm \frac{a^2}{4}\right)\log(x + \sqrt{x^2 \pm a^2}) - \frac{x\sqrt{x^2 \pm a^2}}{4}$

513. $\int x^m[\log(x + \sqrt{x^2 \pm a^2})]\,dx = \frac{x^{m+1}}{m+1}\log(x + \sqrt{x^2 \pm a^2})$

$$- \frac{1}{m+1}\int\frac{x^{m+1}}{\sqrt{x^2 \pm a^2}}\,dx$$

514. $\displaystyle\int \frac{\log(x + \sqrt{x^2 + a^2})}{x^2}\,dx = -\frac{\log(x + \sqrt{x^2 + a^2})}{x} - \frac{1}{a}\log\frac{a + \sqrt{x^2 + a^2}}{x}$

515. $\displaystyle\int \frac{\log(x + \sqrt{x^2 - a^2})}{x^2}\,dx = -\frac{\log(x + \sqrt{x^2 - a^2})}{x} + \frac{1}{|a|}\sec^{-1}\frac{x}{a}$

516. $\displaystyle\int x^n \log(x^2 - a^2)\,dx = \frac{1}{n+1}\left[x^{n+1}\log(x^2 - a^2) - a^{n+1}\log(x - a) - (-a)^{n+1}\log(x + a) - 2\sum_{r=0}^{\left[\frac{n}{2}\right]}\frac{a^{2r}x^{n-2r+1}}{n - 2r + 1}\right]$

See note integral 392.

EXPONENTIAL FORMS

517. $\displaystyle\int e^x\,dx = e^x$

518. $\displaystyle\int e^{-x}\,dx = -e^{-x}$

519. $\displaystyle\int e^{ax}\,dx = \frac{e^{ax}}{a}$

520. $\displaystyle\int x\,e^{ax}\,dx = \frac{e^{ax}}{a^2}(ax - 1)$

521. $\displaystyle\int x^m e^{ax}\,dx = \begin{cases} \dfrac{x^m e^{ax}}{a} - \dfrac{m}{a}\displaystyle\int x^{m-1} e^{ax}\,dx \\ \text{or} \\ e^{ax}\displaystyle\sum_{r=0}^{m}(-1)^r \frac{m!x^{m-r}}{(m - r)!a^{r+1}} \end{cases}$

522. $\displaystyle\int \frac{e^{ax}\,dx}{x} = \log x + \frac{ax}{1!} + \frac{a^2x^2}{2\cdot 2!} + \frac{a^3x^3}{3\cdot 3!} + \cdots$

523. $\displaystyle\int \frac{e^{ax}}{x^m}\,dx = -\frac{1}{m-1}\frac{e^{ax}}{x^{m-1}} + \frac{a}{m-1}\int \frac{e^{ax}}{x^{m-1}}\,dx$

524. $\displaystyle\int e^{ax}\log x\,dx = \frac{e^{ax}\log x}{a} - \frac{1}{a}\int \frac{e^{ax}}{x}\,dx$

525. $\displaystyle\int \frac{dx}{1 + e^x} = x - \log(1 + e^x) = \log\frac{e^x}{1 + e^x}$

526. $\displaystyle\int \frac{dx}{a + be^{px}} = \frac{x}{a} - \frac{1}{ap}\log(a + be^{px})$

527. $\displaystyle\int \frac{dx}{ae^{mx} + be^{-mx}} = \frac{1}{m\sqrt{ab}}\tan^{-1}\left(e^{mx}\sqrt{\frac{a}{b}}\right), \qquad (a > 0, b > 0)$

528. $\displaystyle\int \frac{dx}{ae^{mx} - be^{-mx}} = \begin{cases} \dfrac{1}{2m\sqrt{ab}}\log\dfrac{\sqrt{a}\,e^{mx} - \sqrt{b}}{\sqrt{a}\,e^{mx} + \sqrt{b}} \\ \text{or} \\ \dfrac{-1}{m\sqrt{ab}}\tanh^{-1}\left(\sqrt{\dfrac{a}{b}}\,e^{mx}\right), \qquad (a > 0, b > 0) \end{cases}$

529. $\displaystyle\int (a^x - a^{-x})\,dx = \frac{a^x + a^{-x}}{\log a}$

530. $\displaystyle\int \frac{e^{ax}}{b + ce^{ax}}\,dx = \frac{1}{ac}\log(b + ce^{ax})$

531. $\displaystyle\int \frac{x\,e^{ax}}{(1 + ax)^2}\,dx = \frac{e^{ax}}{a^2(1 + ax)}$

532. $\int x e^{-x^2}\,dx = -\frac{1}{2} e^{-x^2}$

533. $\int e^{ax}[\sin(bx)]\,dx = \frac{e^{ax}[a\sin(bx) - b\cos(bx)]}{a^2 + b^2}$

534. $\int e^{ax}[\sin(bx)][\sin(cx)]\,dx = \frac{e^{ax}[(b-c)\sin(b-c)x + a\cos(b-c)x]}{2[a^2 + (b-c)^2]} - \frac{e^{ax}[(b+c)\sin(b+c)x + a\cos(b+c)x]}{2[a^2 + (b+c)^2]}$

535. $\int e^{ax}[\sin(bx)][\cos(cx)]\,dx = \begin{cases} \frac{e^{ax}[a\sin(b-c)x - (b-c)\cos(b-c)x]}{2[a^2 + (b-c)^2]} + \frac{e^{ax}[a\sin(b+c)x - (b+c)\cos(b+c)x]}{2[a^2 + (b+c)^2]} \\ \text{or} \\ \frac{e^{ax}}{\rho}[(a\sin bx - b\cos bx)[\cos(cx - \alpha)] - c(\sin bx)\sin(cx - \alpha)] \\ \text{where} \\ \rho = \sqrt{(a^2 + b^2 - c^2)^2 + 4a^2c^2}, \\ \rho\cos\alpha = a^2 + b^2 - c^2, \qquad \rho\sin\alpha = 2ac \end{cases}$

536. $\int e^{ax}[\sin(bx)][\sin(bx + c)]\,dx$

$$= \frac{e^{ax}\cos c}{2a} - \frac{e^{ax}[a\cos(2bx + c) + 2b\sin(2bx + c)]}{2(a^2 + 4b^2)}$$

537. $\int e^{ax}[\sin(bx)][\cos(bx + c)]\,dx$

$$= \frac{-e^{ax}\sin c}{2a} + \frac{e^{ax}[a\sin(2bx + c) - 2b\cos(2bx + c)]}{2(a^2 + 4b^2)}$$

538. $\int e^{ax}[\cos(bx)]\,dx = \frac{e^{ax}}{a^2 + b^2}[a\cos(bx) + b\sin(bx)]$

539. $\int e^{ax}[\cos(bx)][\cos(cx)]\,dx = \frac{e^{ax}[(b-c)\sin(b-c)x + a\cos(b-c)x]}{2[a^2 + (b-c)^2]} + \frac{e^{ax}[(b+c)\sin(b+c)x + a\cos(b+c)x]}{2[a^2 + (b+c)^2]}$

540. $\int e^{ax}[\cos(bx)][\cos(bx + c)]\,dx$

$$= \frac{e^{ax}\cos c}{2a} + \frac{e^{ax}[a\cos(2bx + c) + 2b\sin(2bx + c)]}{2(a^2 + 4b^2)}$$

541. $\int e^{ax}[\cos(bx)][\sin(bx + c)]\,dx$

$$= \frac{e^{ax}\sin c}{2a} + \frac{e^{ax}[a\sin(2bx + c) - 2b\cos(2bx + c)]}{2(a^2 + 4b^2)}$$

542. $\int e^{ax}[\sin^n bx]\,dx = \frac{1}{a^2 + n^2b^2}\left[(a\sin bx - nb\cos bx)e^{ax}\sin^{n-1} bx + n(n-1)b^2 \int e^{ax}[\sin^{n-2} bx]\,dx\right]$

543. $\int e^{ax}[\cos^n bx]\,dx = \frac{1}{a^2 + n^2b^2}\left[(a\cos bx + nb\sin bx)e^{ax}\cos^{n-1} bx + n(n-1)b^2 \int e^{ax}[\cos^{n-2} bx]\,dx\right]$

544. $$\int x^m e^x \sin x\, dx = \frac{1}{2}x^m e^x(\sin x - \cos x) - \frac{m}{2}\int x^{m-1} e^x \sin x\, dx + \frac{m}{2}\int x^{m-1} e^x \cos x\, dx$$

545. $$\int x^m e^{ax}[\sin bx]\, dx = \begin{cases} x^m e^{ax}\dfrac{a \sin bx - b \cos bx}{a^2 + b^2} \\ \qquad - \dfrac{m}{a^2 + b^2}\displaystyle\int x^{m-1} e^{ax}(a \sin bx - b \cos bx)\, dx \\ \qquad\qquad \text{or} \\ e^{ax}\displaystyle\sum_{r=0}^{m} \dfrac{(-1)^r m!x^{m-r}}{\rho^{r+1}(m-r)!}\sin[bx - (r+1)\alpha] \\ \qquad\qquad \text{where} \\ \rho = \sqrt{a^2 + b^2}, \qquad \rho\cos\alpha = a, \qquad \rho\sin\alpha = b \end{cases}$$

546. $$\int x^m e^x \cos x\, dx = \frac{1}{2}x^m e^x(\sin x + \cos x) - \frac{m}{2}\int x^{m-1} e^x \sin x\, dx - \frac{m}{2}\int x^{m-1} e^x \cos x\, dx$$

547. $$\int x^m e^{ax} \cos bx\, dx = \begin{cases} x^m e^{ax}\dfrac{a \cos bx + b \sin bx}{a^2 + b^2} \\ \qquad - \dfrac{m}{a^2 + b^2}\displaystyle\int x^{m-1} e^{ax}(a \cos bx + b \sin bx)\, dx \\ \qquad\qquad \text{or} \\ e^{ax}\displaystyle\sum_{r=0}^{m} \dfrac{(-1)^r m!x^{m-r}}{\rho^{r+1}(m-r)!}\cos[bx - (r+1)\alpha] \\ \qquad\qquad \text{where} \\ \rho = \sqrt{a^2 + b^2}, \qquad \rho\cos\alpha = a, \qquad \rho\sin\alpha = b \end{cases}$$

548. $$\int e^{ax}(\cos^m x)(\sin^n x)\,dx = \begin{cases} \dfrac{e^{ax}\cos^{m-1}x\sin^n x[a\cos x+(m+n)\sin x]}{(m+n)^2+a^2} \\ \quad -\dfrac{na}{(m+n)^2+a^2}\displaystyle\int e^{ax}(\cos^{m-1}x)(\sin^{n-1}x)\,dx \\ \quad +\dfrac{(m-1)(m+n)}{(m+n)^2+a^2}\displaystyle\int e^{ax}(\cos^{m-2}x)(\sin^n x)\,dx \\ \text{or} \\ \dfrac{e^{ax}\cos^m x\sin^{n-1}x[a\sin x-(m+n)\cos x]}{(m+n)^2+a^2} \\ \quad +\dfrac{ma}{(m+n)^2+a^2}\displaystyle\int e^{ax}(\cos^{m-1}x)(\sin^{n-1}x)\,dx \\ \quad +\dfrac{(n-1)(m+n)}{(m+n)^2+a^2}\displaystyle\int e^{ax}(\cos^m x)(\sin^{n-2}x)\,dx \\ \text{or} \\ \dfrac{e^{ax}(\cos^{m-1}x)(\sin^{n-1}x)(a\sin x\cos x+m\sin^2 x-n\cos^2 x)}{(m+n)^2+a^2} \\ \quad +\dfrac{m(m-1)}{(m+n)^2+a^2}\displaystyle\int e^{ax}(\cos^{m-2}x)(\sin^n x)\,dx \\ \quad +\dfrac{n(n-1)}{(m+n)^2+a^2}\displaystyle\int e^{ax}(\cos^m x)(\sin^{n-2}x)\,dx \\ \text{or} \\ \dfrac{e^{ax}(\cos^{m-1}x)(\sin^{n-1}x)(a\cos x\sin x+m\sin^2 x-n\cos^2 x)}{(m+n)^2+a^2} \\ \quad +\dfrac{m(m-1)}{(m+n)^2+a^2}\displaystyle\int e^{ax}(\cos^{m-2}x)(\sin^{n-2}x)\,dx \\ \quad +\dfrac{(n-m)(n+m-1)}{(m+n)^2+a^2}\displaystyle\int e^{ax}(\cos^m x)(\sin^{n-2}x)\,dx \end{cases}$$

549. $$\int x\,e^{ax}(\sin bx)\,dx = \frac{x\,e^{ax}}{a^2+b^2}(a\sin bx - b\cos bx) - \frac{e^{ax}}{(a^2+b^2)^2}[(a^2-b^2)\sin bx - 2ab\cos bx]$$

550. $$\int x\,e^{ax}(\cos bx)\,dx = \frac{x\,e^{ax}}{a^2+b^2}(a\cos bx + b\sin bx) - \frac{e^{ax}}{(a^2+b^2)^2}[(a^2-b^2)\cos bx + 2ab\sin bx]$$

551. $$\int \frac{e^{ax}}{\sin^n x}\,dx = -\frac{e^{ax}[a\sin x+(n-2)\cos x]}{(n-1)(n-2)\sin^{n-1}x} + \frac{a^2+(n-2)^2}{(n-1)(n-2)}\int \frac{e^{ax}}{\sin^{n-2}x}\,dx$$

552. $$\int \frac{e^{ax}}{\cos^n x}\,dx = -\frac{e^{ax}[a\cos x-(n-2)\sin x]}{(n-1)(n-2)\cos^{n-1}x} + \frac{a^2+(n-2)^2}{(n-1)(n-2)}\int \frac{e^{ax}}{\cos^{n-2}x}\,dx$$

553. $$\int e^{ax}\tan^n x\,dx = e^{ax}\frac{\tan^{n-1}x}{n-1} - \frac{a}{n-1}\int e^{ax}\tan^{n-1}x\,dx - \int e^{ax}\tan^{n-2}x\,dx$$

HYPERBOLIC FORMS

554. $$\int (\sinh x)\,dx = \cosh x$$

555. $$\int (\cosh x)\,dx = \sinh x$$

556. $\int (\tanh x)\,dx = \log \cosh x$

557. $\int (\coth x)\,dx = \log \sinh x$

558. $\int (\operatorname{sech} x)\,dx = \tan^{-1}(\sinh x)$

559. $\int \operatorname{csch} x\,dx = \log \tanh \left(\frac{x}{2}\right)$

560. $\int x(\sinh x)\,dx = x \cosh x - \sinh x$

561. $\int x^n(\sinh x)\,dx = x^n \cosh x - n \int x^{n-1}(\cosh x)\,dx$

562. $\int x(\cosh x)\,dx = x \sinh x - \cosh x$

563. $\int x^n(\cosh x)\,dx = x^n \sinh x - n \int x^{n-1}(\sinh x)\,dx$

564. $\int (\operatorname{sech} x)(\tanh x)\,dx = -\operatorname{sech} x$

565. $\int (\operatorname{csch} x)(\coth x)\,dx = -\operatorname{csch} x$

566. $\int (\sinh^2 x)\,dx = \frac{\sinh 2x}{4} - \frac{x}{2}$

567. $$\int (\sinh^m x)(\cosh^n x)\,dx = \begin{cases} \frac{1}{m+n}(\sinh^{m+1} x)(\cosh^{n-1} x) \\ \qquad + \frac{n-1}{m+n}\int (\sinh^m x)(\cosh^{n-2} x)\,dx \\ \text{or} \\ \frac{1}{m+n}\sinh^{m-1} x \cosh^{n+1} x \\ \qquad - \frac{m-1}{m+n}\int (\sinh^{m-2} x)(\cosh^n x)\,dx, \quad (m+n \neq 0) \end{cases}$$

568. $$\int \frac{dx}{(\sinh^m x)(\cosh^n x)} = \begin{cases} -\frac{1}{(m-1)(\sinh^{m-1} x)(\cosh^{n-1} x)} \\ \qquad - \frac{m+n-2}{m-1}\int \frac{dx}{(\sinh^{m-2} x)(\cosh^n x)}, \quad (m \neq 1) \\ \text{or} \\ \frac{1}{(n-1)\sinh^{m-1} x \cosh^{n-1} x} \\ \qquad + \frac{m+n-2}{n-1}\int \frac{dx}{(\sinh^m x)(\cosh^{n-2} x)}, \quad (n \neq 1) \end{cases}$$

569. $\int (\tanh^2 x)\,dx = x - \tanh x$

570. $\int (\tanh^n x)\,dx = -\frac{\tanh^{n-1} x}{n-1} + \int (\tanh^{n-2} x)\,dx, \quad (n \neq 1)$

571. $\int (\operatorname{sech}^2 x)\,dx = \tanh x$

572. $\int (\cosh^2 x)\,dx = \frac{\sinh 2x}{4} + \frac{x}{2}$

573. $\int (\coth^2 x)\,dx = x - \coth x$

574. $\int (\coth^n x)\,dx = -\frac{\coth^{n-1} x}{n-1} + \int \coth^{n-2} x\,dx, \qquad (n \neq 1)$

575. $\int (\operatorname{csch}^2 x)\,dx = -\operatorname{ctnh} x$

576. $\int (\sinh mx)(\sinh nx)\,dx = \frac{\sinh (m+n)x}{2(m+n)} - \frac{\sinh (m-n)x}{2(m-n)}, \qquad (m^2 \neq n^2)$

577. $\int (\cosh mx)(\cosh nx)\,dx = \frac{\sinh (m+n)x}{2(m+n)} + \frac{\sinh (m-n)x}{2(m-n)}, \qquad (m^2 \neq n^2)$

578. $\int (\sinh mx)(\cosh nx)\,dx = \frac{\cosh (m+n)x}{2(m+n)} + \frac{\cosh (m-n)x}{2(m-n)}, \qquad (m^2 \neq n^2)$

579. $\int \left(\sinh^{-1}\frac{x}{a}\right)dx = x\sinh^{-1}\frac{x}{a} - \sqrt{x^2+a^2}, \qquad (a > 0)$

580. $\int x\left(\sinh^{-1}\frac{x}{a}\right)dx = \left(\frac{x^2}{2}+\frac{a^2}{4}\right)\sinh^{-1}\frac{x}{a} - \frac{x}{4}\sqrt{x^2+a^2}, \qquad (a > 0)$

581. $\int x^n(\sinh^{-1} x)\,dx = \frac{x^{n+1}}{n+1}\sinh^{-1} x - \frac{1}{n+1}\int \frac{x^{n+1}}{(1+x^2)^{\frac{1}{2}}}\,dx, \qquad (n \neq -1)$

582. $\int \left(\cosh^{-1}\frac{x}{a}\right)dx = \begin{cases} x\cosh^{-1}\frac{x}{a} - \sqrt{x^2-a^2}, & \left(\cosh^{-1}\frac{x}{a} > 0\right) \\ \text{or} & \\ x\cosh^{-1}\frac{x}{a} + \sqrt{x^2-a^2}, & \left(\cosh^{-1}\frac{x}{a} < 0\right), \qquad (a > 0) \end{cases}$

583. $\int x\left(\cosh^{-1}\frac{x}{a}\right)dx = \frac{2x^2-a^2}{4}\cosh^{-1}\frac{x}{a} - \frac{x}{4}(x^2-a^2)^{\frac{1}{2}}$

584. $\int x^n(\cosh^{-1} x)\,dx = \frac{x^{n+1}}{n+1}\cosh^{-1} x - \frac{1}{n+1}\int \frac{x^{n+1}}{(x^2-1)^{\frac{1}{2}}}\,dx, \qquad (n \neq -1)$

585. $\int \left(\tanh^{-1}\frac{x}{a}\right)dx = x\tanh^{-1}\frac{x}{a} + \frac{a}{2}\log(a^2-x^2), \qquad \left(\left|\frac{x}{a}\right| < 1\right)$

586. $\int \left(\coth^{-1}\frac{x}{a}\right)dx = x\coth^{-1}\frac{x}{a} + \frac{a}{2}\log(x^2-a^2), \qquad \left(\left|\frac{x}{a}\right| > 1\right)$

587. $\int x\left(\tanh^{-1}\frac{x}{a}\right)dx = \frac{x^2-a^2}{2}\tanh^{-1}\frac{x}{a} + \frac{ax}{2}, \qquad \left(\left|\frac{x}{a}\right| < 1\right)$

588. $\int x^n\left(\tanh^{-1} x\right)dx = \frac{x^{n+1}}{n+1}\tanh^{-1} x - \frac{1}{n+1}\int \frac{x^{n+1}}{1-x^2}\,dx, \qquad (n \neq -1)$

589. $\int x\left(\coth^{-1}\frac{x}{a}\right)dx = \frac{x^2-a^2}{2}\coth^{-1}\frac{x}{a} + \frac{ax}{2}, \qquad \left(\left|\frac{x}{a}\right| > 1\right)$

590. $\int x^n(\coth^{-1} x)\,dx = \frac{x^{n+1}}{n+1}\coth^{-1} x + \frac{1}{n+1}\int \frac{x^{n+1}}{x^2-1}\,dx, \qquad (n \neq -1)$

591. $\int (\operatorname{sech}^{-1} x)\,dx = x\operatorname{sech}^{-1} x + \sin^{-1} x$

592. $\int x\operatorname{sech}^{-1} x\,dx = \frac{x^2}{2}\operatorname{sech}^{-1} x - \frac{1}{2}\sqrt{1-x^2}$

593. $\int x^n\operatorname{sech}^{-1} x\,dx = \frac{x^{n+1}}{n+1}\operatorname{sech}^{-1} x + \frac{1}{n+1}\int \frac{x^n}{(1-x^2)^{\frac{1}{2}}}\,dx, \qquad (n \neq -1)$

594. $\int \operatorname{csch}^{-1} x\,dx = x\operatorname{csch}^{-1} x + \frac{x}{|x|}\sinh^{-1} x$

595. $\int x \operatorname{csch}^{-1} x\,dx = \frac{x^2}{2}\operatorname{csch}^{-1} x + \frac{1}{2}\frac{x}{|x|}\sqrt{1+x^2}$

596. $\int x^n \operatorname{csch}^{-1} x\,dx = \frac{x^{n+1}}{n+1}\operatorname{csch}^{-1} x + \frac{1}{n+1}\frac{x}{|x|}\int \frac{x^n}{(x^2+1)^{\frac{1}{2}}}\,dx, \qquad (n \neq -1)$

DEFINITE INTEGRALS

597. $\int_0^\infty x^{n-1} e^{-x}\,dx = \int_0^1 \left(\log \frac{1}{x}\right)^{n-1} dx = \frac{1}{n}\prod_{m=1}^{\infty} \frac{\left(1+\frac{1}{m}\right)^n}{1+\frac{n}{m}}$

$= \Gamma(n),\ n \neq 0, -1, -2, -3, \ldots$ (Gamma Function)

598. $\int_0^\infty t^n p^{-t}\,dt = \frac{n!}{(\log p)^{n+1}}, \qquad (n = 0, 1, 2, 3, \ldots \text{ and } p > 0)$

599. $\int_0^\infty t^{n-1} e^{-(a+1)t}\,dt = \frac{\Gamma(n)}{(a+1)^n}, \qquad (n > 0, a > -1)$

600. $\int_0^1 x^m \left(\log \frac{1}{x}\right)^n dx = \frac{\Gamma(n+1)}{(m+1)^{n+1}}, \qquad (m > -1, n > -1)$

601. $\Gamma(n)$ is finite if $n > 0$, $\Gamma(n+1) = n\Gamma(n)$

602. $\Gamma(n)\cdot\Gamma(1-n) = \frac{\pi}{\sin n\pi}$

603. $\Gamma(n) = (n-1)!$ if n = integer > 0

604. $\Gamma(\frac{1}{2}) = 2\int_0^\infty e^{-t^2}\,dt = \sqrt{\pi} = 1.7724538509\cdots = (-\frac{1}{2})!$

605. $\Gamma(n+\frac{1}{2}) = \frac{1\cdot 3\cdot 5\ldots(2n-1)}{2^n}\sqrt{\pi} \qquad n = 1, 2, 3, \ldots$

606. $\Gamma(-n+\frac{1}{2}) = \frac{(-1)^n 2^n\sqrt{\pi}}{1\cdot 3\cdot 5\ldots(2n-1)} \qquad n = 1, 2, 3, \ldots$

607. $\int_0^1 x^{m-1}(1-x)^{n-1}\,dx = \int_0^\infty \frac{x^{m-1}}{(1+x)^{m+n}}\,dx = \frac{\Gamma(m)\Gamma(n)}{\Gamma(m+n)} = B(m, n)$

(Beta function)

608. $B(m, n) = B(n, m) = \frac{\Gamma(m)\Gamma(n)}{\Gamma(m+n)}$, where m and n are any positive real numbers.

609. $\int_a^b (x-a)^m(b-x)^n\,dx = (b-a)^{m+n+1}\frac{\Gamma(m+1)\cdot\Gamma(n+1)}{\Gamma(m+n+2)},$

$(m > -1, n > -1, b > a)$

610. $\int_1^\infty \frac{dx}{x^m} = \frac{1}{m-1}, \qquad [m > 1]$

611. $\int_0^\infty \frac{dx}{(1+x)x^p} = \pi \csc p\pi, \qquad [p < 1]$

612. $\int_0^\infty \frac{dx}{(1-x)x^p} = -\pi \cot p\pi, \qquad [p < 1]$

613. $\int_0^\infty \frac{x^{p-1}\,dx}{1+x} = \frac{\pi}{\sin p\pi}$

$= B(p, 1-p) = \Gamma(p)\Gamma(1-p), \qquad [0 < p < 1]$

614. $\int_0^\infty \frac{x^{m-1}\,dx}{1+x^n} = \frac{\pi}{n \sin \frac{m\pi}{n}}, \qquad [0 < m < n]$

615. $\displaystyle\int_0^\infty \frac{x^a\,dx}{(m+x^b)^c} = \frac{m^{\frac{a+1-bc}{b}}}{b}\left[\frac{\Gamma\left(\frac{a+1}{b}\right)\Gamma\left(c-\frac{a+1}{b}\right)}{\Gamma(c)}\right]$ $\left(a>-1,\, b>0,\, m>0,\, c>\frac{a+1}{b}\right)$

616. $\displaystyle\int_0^\infty \frac{dx}{(1+x)\sqrt{x}} = \pi$

617. $\displaystyle\int_0^\infty \frac{a\,dx}{a^2+x^2} = \frac{\pi}{2}$, if $a>0$; 0, if $a=0$; $-\frac{\pi}{2}$, if $a<0$

618. $\displaystyle\int_0^a (a^2-x^2)^{\frac{n}{2}}\,dx = \frac{1}{2}\int_{-a}^a (a^2-x^2)^{\frac{n}{2}}\,dx = \frac{1\cdot 3\cdot 5\ldots n}{2\cdot 4\cdot 6\ldots(n+1)}\cdot\frac{\pi}{2}\cdot a^{n+1}$ (n odd)

619. $\displaystyle\int_0^a x^m(a^2-x^2)^{\frac{n}{2}}\,dx = \begin{cases} \frac{1}{2}a^{m+n+1}B\left(\frac{m+1}{2}, \frac{n+2}{2}\right) \\ \text{or} \\ \frac{1}{2}a^{m+n+1}\dfrac{\Gamma\left(\frac{m+1}{2}\right)\Gamma\left(\frac{n+2}{2}\right)}{\Gamma\left(\frac{m+n+3}{2}\right)} \end{cases}$

620. $\displaystyle\int_0^{\pi/2} (\sin^n x)\,dx = \begin{cases} \displaystyle\int_0^{\pi/2} (\cos^n x)\,dx \\ \text{or} \\ \dfrac{1\cdot 3\cdot 5\cdot 7\ldots(n-1)}{2\cdot 4\cdot 6\cdot 8\ldots(n)}\dfrac{\pi}{2}, & (n \text{ an even integer, } n\neq 0) \\ \text{or} \\ \dfrac{2\cdot 4\cdot 6\cdot 8\ldots(n-1)}{1\cdot 3\cdot 5\cdot 7\ldots(n)}, & (n \text{ an odd integer, } n\neq 1) \\ \text{or} \\ \dfrac{\sqrt{\pi}}{2}\dfrac{\Gamma\left(\frac{n+1}{2}\right)}{\Gamma\left(\frac{n}{2}+1\right)}, & (n>-1) \end{cases}$

621. $\displaystyle\int_0^\infty \frac{\sin mx\,dx}{x} = \frac{\pi}{2}$, if $m>0$; 0, if $m=0$; $-\frac{\pi}{2}$, if $m<0$

622. $\displaystyle\int_0^\infty \frac{\cos x\,dx}{x} = \infty$

623. $\displaystyle\int_0^\infty \frac{\tan x\,dx}{x} = \frac{\pi}{2}$

624. $\displaystyle\int_0^\pi \sin ax\cdot\sin bx\,dx = \int_0^\pi \cos ax\cdot\cos bx\,dx = 0,$ ($a\neq b$; a, b integers)

625. $\displaystyle\int_0^{\pi/a} [\sin(ax)][\cos(ax)]\,dx = \int_0^\pi [\sin(ax)][\cos(ax)]\,dx = 0$

626. $\displaystyle\int_0^\pi [\sin(ax)][\cos(bx)]\,dx = \frac{2a}{a^2-b^2}$, if $a-b$ is odd, or 0 if $a-b$ is even

627. $\displaystyle\int_0^\infty \frac{\sin x\cos mx\,dx}{x}$

$= 0$, if $m<-1$ or $m>1$; $\frac{\pi}{4}$, if $m=\pm 1$; $\frac{\pi}{2}$, if $m^2<1$

628. $\int_0^\infty \frac{\sin ax \sin bx}{x^2} dx = \frac{\pi a}{2}, \qquad (a \le b)$

629. $\int_0^\pi \sin^2 mx\, dx = \int_0^\pi \cos^2 mx\, dx = \frac{\pi}{2}$

630. $\int_0^\infty \frac{\sin^2 (px)}{x^2} dx = \frac{\pi p}{2}$

631. $\int_0^\infty \frac{\sin x}{x^p} dx = \frac{\pi}{2\Gamma(p) \sin (p\pi/2)}, \qquad 0 < p < 1$

632. $\int_0^\infty \frac{\cos x}{x^p} dx = \frac{\pi}{2\Gamma(p) \cos (p\pi/2)}, \qquad 0 < p < 1$

633. $\int_0^\infty \frac{1 - \cos px}{x^2} dx = \frac{\pi p}{2}$

634. $\int_0^\infty \frac{\sin px \cos qx}{x} dx = \left\{0, \quad q > p > 0; \quad \frac{\pi}{2}, \quad p > q > 0; \quad \frac{\pi}{4}, \quad p = q > 0\right\}$

635. $\int_0^\infty \frac{\cos (mx)}{x^2 + a^2} dx = \frac{\pi}{2|a|} e^{-|ma|}$

636. $\int_0^\infty \cos (x^2)\, dx = \int_0^\infty \sin (x^2)\, dx = \frac{1}{2}\sqrt{\frac{\pi}{2}}$

637. $\int_0^\infty \sin ax^n\, dx = \frac{1}{na^{1/n}} \Gamma(1/n) \sin \frac{\pi}{2n}, \qquad n > 1$

638. $\int_0^\infty \cos ax^n\, dx = \frac{1}{na^{1/n}} \Gamma(1/n) \cos \frac{\pi}{2n}, \qquad n > 1$

639. $\int_0^\infty \frac{\sin x}{\sqrt{x}} dx = \int_0^\infty \frac{\cos x}{\sqrt{x}} dx = \sqrt{\frac{\pi}{2}}$

640. (a) $\int_0^\infty \frac{\sin^3 x}{x} dx = \frac{\pi}{4}$ (b) $\int_0^\infty \frac{\sin^3 x}{x^2} dx \frac{3}{4} \log 3$

641. $\int_0^\infty \frac{\sin^3 x}{x^3} dx = \frac{3\pi}{8}$

642. $\int_0^\infty \frac{\sin^4 x}{x^4} dx = \frac{\pi}{3}$

643. $\int_0^{\pi/2} \frac{dx}{1 + a \cos x} = \frac{\cos^{-1} a}{\sqrt{1 - a^2}}, \qquad (a < 1)$

644. $\int_0^\pi \frac{dx}{a + b \cos x} = \frac{\pi}{\sqrt{a^2 - b^2}}, \qquad (a > b \ge 0)$

645. $\int_0^{2\pi} \frac{dx}{1 + a \cos x} = \frac{2\pi}{\sqrt{1 - a^2}}, \qquad (a^2 < 1)$

646. $\int_0^\infty \frac{\cos ax - \cos bx}{x} dx = \log \frac{b}{a}$

647. $\int_0^{\pi/2} \frac{dx}{a^2 \sin^2 x + b^2 \cos^2 x} = \frac{\pi}{2ab}$

648. $\int_0^{\pi/2} \frac{dx}{(a^2 \sin^2 x + b^2 \cos^2 x)^2} = \frac{\pi(a^2 + b^2)}{4a^3b^3}, \qquad (a, b > 0)$

649. $\int_0^{\pi/2} \sin^{n-1} x \cos^{m-1} x\, dx = \frac{1}{2} \mathrm{B}\left(\frac{n}{2}, \frac{m}{2}\right), \qquad m \text{ and } n \text{ positive integers}$

650. $\displaystyle\int_0^{\pi/2} (\sin^{2n+1}\theta)\,d\theta = \frac{2\cdot 4\cdot 6\ldots(2n)}{1\cdot 3\cdot 5\ldots(2n+1)}, \qquad (n = 1, 2, 3\ldots)$

651. $\displaystyle\int_0^{\pi/2} (\sin^{2n}\theta)\,d\theta = \frac{1\cdot 3\cdot 5\ldots(2n-1)}{2\cdot 4\ldots(2n)}\left(\frac{\pi}{2}\right), \qquad (n = 1, 2, 3\ldots)$

652. $\displaystyle\int_0^{\pi/2} \frac{x}{\sin x}\,dx = 2\left\{\frac{1}{1^2} - \frac{1}{3^2} + \frac{1}{5^2} - \frac{1}{7^2} + \ldots\right\}$

653. $\displaystyle\int_0^{\pi/2} \frac{dx}{1+\tan^m x} = \frac{\pi}{4}$

654. $\displaystyle\int_0^{\pi/2} \sqrt{\cos\theta}\,d\theta = \frac{(2\pi)^{3/2}}{[\Gamma(\frac{1}{4})]^2}$

655. $\displaystyle\int_0^{\pi/2} (\tan^h\theta)\,d\theta = \frac{\pi}{2\cos\left(\frac{h\pi}{2}\right)}, \qquad (0 < h < 1)$

656. $\displaystyle\int_0^{\infty} \frac{\tan^{-1}(ax) - \tan^{-1}(bx)}{x}\,dx = \frac{\pi}{2}\log\frac{a}{b}, \qquad (a, b > 0)$

657. The area enclosed by a curve defined through the equation $x^{\frac{b}{c}} + y^{\frac{b}{c}} = a^{\frac{b}{c}}$ where $a > 0$, c a positive odd integer and b a positive even integer is given by

$$\frac{\left[\Gamma\left(\frac{c}{b}\right)\right]^2}{\Gamma\left(\frac{2c}{b}\right)}\left(\frac{2ca^2}{b}\right)$$

658. $\displaystyle I = \iiint\limits_R x^{h-1}y^{m-1}z^{n-1}\,dv$, where R denotes the region of space bounded by the co-ordinate planes and that portion of the surface $\left(\frac{x}{a}\right)^p + \left(\frac{y}{b}\right)^q + \left(\frac{z}{c}\right)^k = 1$, which lies in the first octant and where $h, m, n, p, q, k, a, b, c$, denote positive real numbers is given by

$$\int_0^a x^{h-1}\,dx \int_0^{b\left[1-\left(\frac{x}{a}\right)^p\right]^{\frac{1}{q}}} y^m\,dy \int_0^{c\left[1-\left(\frac{x}{a}\right)^p-\left(\frac{y}{b}\right)^q\right]^{\frac{1}{k}}} z^{n-1}\,dz$$

$$= \frac{a^h b^m c^n}{pqk}\,\frac{\Gamma\left(\frac{h}{p}\right)\Gamma\left(\frac{m}{q}\right)\Gamma\left(\frac{n}{k}\right)}{\Gamma\left(\frac{h}{p}+\frac{m}{q}+\frac{n}{k}+1\right)}$$

659. $\displaystyle\int_0^{\infty} e^{-ax}\,dx = \frac{1}{a}, \qquad (a > 0)$

660. $\displaystyle\int_0^{\infty} \frac{e^{-ax} - e^{-bx}}{x}\,dx = \log\frac{b}{a}, \qquad (a, b > 0)$

661. $$\int_0^{\infty} x^n e^{-ax}\,dx = \begin{cases} \dfrac{\Gamma(n+1)}{a^{n+1}}, & (n > -1, a > 0) \\ \quad \text{or} & \\ \dfrac{n!}{a^{n+1}}, & (a > 0, n \text{ positive integer}) \end{cases}$$

662. $\displaystyle\int_0^{\infty} x^n \exp(-ax^p)\,dx = \frac{\Gamma(k)}{pa^k}, \qquad \left(n > -1, p > 0, a > 0, k = \frac{n+1}{p}\right)$

663. $\displaystyle\int_0^{\infty} e^{-a^2x^2}\,dx = \frac{1}{2a}\sqrt{\pi} = \frac{1}{2a}\Gamma\left(\frac{1}{2}\right), \qquad (a > 0)$

664. $\displaystyle\int_0^{\infty} x\,e^{-x^2}\,dx = \frac{1}{2}$

665. $\int_0^\infty x^2 e^{-x^2}\,dx = \frac{\sqrt{\pi}}{4}$

666. $\int_0^\infty x^{2n} e^{-ax^2}\,dx = \frac{1\cdot 3\cdot 5\ldots(2n-1)}{2^{n+1}a^n}\sqrt{\frac{\pi}{a}}$

667. $\int_0^\infty x^{2n+1} e^{-ax^2}\,dx = \frac{n!}{2a^{n+1}}, \qquad (a > 0)$

668. $\int_0^1 x^m e^{-ax}\,dx = \frac{m!}{a^{m+1}}\left[1 - e^{-a}\sum_{r=0}^{m}\frac{a^r}{r!}\right]$

669. $\int_0^\infty e^{\left(-x^2 - \frac{a^2}{x^2}\right)}\,dx = \frac{e^{-2a}\sqrt{\pi}}{2}, \qquad (a \geq 0)$

670. $\int_0^\infty e^{-nx}\sqrt{x}\,dx = \frac{1}{2n}\sqrt{\frac{\pi}{n}}$

671. $\int_0^\infty \frac{e^{-nx}}{\sqrt{x}}\,dx = \sqrt{\frac{\pi}{n}}$

672. $\int_0^\infty e^{-ax}(\cos mx)\,dx = \frac{a}{a^2 + m^2}, \qquad (a > 0)$

673. $\int_0^\infty e^{-ax}(\sin mx)\,dx = \frac{m}{a^2 + m^2}, \qquad (a > 0)$

674. $\int_0^\infty x\,e^{-ax}[\sin(bx)]\,dx = \frac{2ab}{(a^2 + b^2)^2}, \qquad (a > 0)$

675. $\int_0^\infty x\,e^{-ax}[\cos(bx)]\,dx = \frac{a^2 - b^2}{(a^2 + b^2)^2}, \qquad (a > 0)$

676. $\int_0^\infty x^n e^{-ax}[\sin(bx)]\,dx = \frac{n![(a + ib)^{n+1} - (a - ib)^{n+1}]}{2i(a^2 + b^2)^{n+1}}, \qquad (i^2 = -1, a > 0)$

677. $\int_0^\infty x^n e^{-ax}[\cos(bx)]\,dx = \frac{n![(a - ib)^{n+1} + (a + ib)^{n+1}]}{2(a^2 + b^2)^{n+1}}, \qquad (i^2 = -1, a > 0)$

678. $\int_0^\infty \frac{e^{-ax}\sin x}{x}\,dx = \cot^{-1} a, \qquad (a > 0)$

679. $\int_0^\infty e^{-a^2x^2}\cos bx\,dx = \frac{\sqrt{\pi}}{2a}\exp\left(-\frac{b^2}{4a^2}\right), \qquad (ab \neq 0)$

680. $\int_0^\infty e^{-t\cos\phi}\,t^{b-1}\sin(t\sin\phi)\,dt = [\Gamma(b)]\sin(b\phi), \qquad \left(b > 0, -\frac{\pi}{2} < \phi < \frac{\pi}{2}\right)$

681. $\int_0^\infty e^{-t\cos\phi}\,t^{b-1}[\cos(t\sin\phi)]\,dt = [\Gamma(b)]\cos(b\phi), \qquad \left(b > 0, -\frac{\pi}{2} < \phi < \frac{\pi}{2}\right)$

682. $\int_0^\infty t^{b-1}\cos t\,dt = [\Gamma(b)]\cos\left(\frac{b\pi}{2}\right), \qquad (0 < b < 1)$

683. $\int_0^\infty t^{b-1}(\sin t)\,dt = [\Gamma(b)]\sin\left(\frac{b\pi}{2}\right), \qquad (0 < b < 1)$

684. $\int_0^1 (\log x)^n\,dx = (-1)^n\cdot n!$

685. $\int_0^1 \left(\log\frac{1}{x}\right)^{\frac{1}{2}} dx = \frac{\sqrt{\pi}}{2}$

686. $\int_0^1 \left(\log\frac{1}{x}\right)^{-\frac{1}{2}} dx = \sqrt{\pi}$

687. $\int_0^1 \left(\log \frac{1}{x}\right)^n dx = n!$

688. $\int_0^1 x \log(1-x)\, dx = -\frac{3}{4}$

689. $\int_0^1 x \log(1+x)\, dx = \frac{1}{4}$

690. $\int_0^1 x^m (\log x)^n\, dx = \frac{(-1)^n n!}{(m+1)^{n+1}}, \qquad m > -1, n = 0, 1, 2, \ldots$

If $n \neq 0, 1, 2, \ldots$ replace $n!$ by $\Gamma(n+1)$.

691. $\int_0^1 \frac{\log x}{1+x}\, dx = -\frac{\pi^2}{12}$

692. $\int_0^1 \frac{\log x}{1-x}\, dx = -\frac{\pi^2}{6}$

693. $\int_0^1 \frac{\log(1+x)}{x}\, dx = \frac{\pi^2}{12}$

694. $\int_0^1 \frac{\log(1-x)}{x}\, dx = -\frac{\pi^2}{6}$

695. $\int_0^1 (\log x)[\log(1+x)]\, dx = 2 - 2\log 2 - \frac{\pi^2}{12}$

696. $\int_0^1 (\log x)[\log(1-x)]\, dx = 2 - \frac{\pi^2}{6}$

697. $\int_0^1 \frac{\log x}{1-x^2}\, dx = -\frac{\pi^2}{8}$

698. $\int_0^1 \log\left(\frac{1+x}{1-x}\right) \cdot \frac{dx}{x} = \frac{\pi^2}{4}$

699. $\int_0^1 \frac{\log x\, dx}{\sqrt{1-x^2}} = -\frac{\pi}{2} \log 2$

700. $\int_0^1 x^m \left[\log\left(\frac{1}{x}\right)\right]^n dx = \frac{\Gamma(n+1)}{(m+1)^{n+1}}, \qquad \text{if } m+1 > 0, n+1 > 0$

701. $\int_0^1 \frac{(x^p - x^q)\, dx}{\log x} = \log\left(\frac{p+1}{q+1}\right), \qquad (p+1 > 0, q+1 > 0)$

702. $\int_0^1 \frac{dx}{\sqrt{\log\left(\frac{1}{x}\right)}} = \sqrt{\pi}$, (same as integral 686)

703. $\int_0^\infty \log\left(\frac{e^x+1}{e^x-1}\right) dx = \frac{\pi^2}{4}$

704. $\int_0^{\pi/2} (\log \sin x)\, dx = \int_0^{\pi/2} \log \cos x\, dx = -\frac{\pi}{2} \log 2$

705. $\int_0^{\pi/2} (\log \sec x)\, dx = \int_0^{\pi/2} \log \csc x\, dx = \frac{\pi}{2} \log 2$

706. $\int_0^\pi x(\log \sin x)\, dx = -\frac{\pi^2}{2} \log 2$

707. $\int_0^{\pi/2} (\sin x)(\log \sin x)\, dx = \log 2 - 1$

708. $\int_0^{\pi/2} (\log \tan x)\,dx = 0$

709. $\int_0^{\pi} \log(a \pm b\cos x)\,dx = \pi \log\left(\frac{a + \sqrt{a^2 - b^2}}{2}\right), \qquad (a \geq b)$

710. $\int_0^{\pi} \log(a^2 - 2ab\cos x + b^2)\,dx = \begin{cases} 2\pi \log a, & a \geq b > 0 \\ 2\pi \log b, & b \geq a > 0 \end{cases}$

711. $\int_0^{\infty} \frac{\sin ax}{\sinh bx}\,dx = \frac{\pi}{2b} \tanh \frac{a\pi}{2b}$

712. $\int_0^{\infty} \frac{\cos ax}{\cosh bx}\,dx = \frac{\pi}{2b} \operatorname{sech} \frac{a\pi}{2b}$

713. $\int_0^{\infty} \frac{dx}{\cosh ax} = \frac{\pi}{2a}$

714. $\int_0^{\infty} \frac{x\,dx}{\sinh ax} = \frac{\pi^2}{4a^2}$

715. $\int_0^{\infty} e^{-ax}(\cosh bx)\,dx = \frac{a}{a^2 - b^2}, \qquad (0 \leq |b| < a)$

716. $\int_0^{\infty} e^{-ax}(\sinh bx)\,dx = \frac{b}{a^2 - b^2}, \qquad (0 \leq |b| < a)$

717. $\int_0^{\infty} \frac{\sinh ax}{e^{bx} + 1}\,dx = \frac{\pi}{2b} \csc \frac{a\pi}{b} - \frac{1}{2a}$

718. $\int_0^{\infty} \frac{\sinh ax}{e^{bx} - 1}\,dx = \frac{1}{2a} - \frac{\pi}{2b} \cot \frac{a\pi}{b}$

719. $\int_0^{\pi/2} \frac{dx}{\sqrt{1 - k^2 \sin^2 x}} = \frac{\pi}{2}\left[1 + \left(\frac{1}{2}\right)^2 k^2 + \left(\frac{1 \cdot 3}{2 \cdot 4}\right)^2 k^4 + \left(\frac{1 \cdot 3 \cdot 5}{2 \cdot 4 \cdot 6}\right)^2 k^6 + \cdots\right], \text{ if } k^2 < 1$

720. $\int_0^{\pi/2} \sqrt{1 - k^2 \sin^2 x}\,dx = \frac{\pi}{2}\left[1 - \left(\frac{1}{2}\right)^2 k^2 - \left(\frac{1 \cdot 3}{2 \cdot 4}\right)^2 \frac{k^4}{3} - \left(\frac{1 \cdot 3 \cdot 5}{2 \cdot 4 \cdot 6}\right)^2 \frac{k^6}{5} - \cdots\right], \text{ if } k^2 < 1$

721. $\int_0^{\infty} e^{-x} \log x\,dx = -\gamma = -0.5772157\ldots$

722. $\int_0^{\infty} e^{-x^2} \log x\,dx = -\frac{\sqrt{\pi}}{4}(\gamma + 2\log 2)$

723. $\int_0^{\infty} \left(\frac{1}{1 - e^{-x}} - \frac{1}{x}\right) e^{-x}\,dx = \gamma = 0.5772157\ldots$ [Euler's Constant]

724. $\int_0^{\infty} \frac{1}{x}\left(\frac{1}{1 + x} - e^{-x}\right) dx = \gamma = 0.5772157\ldots$

For n even:

725. $\int \cos^n x\,dx = \frac{1}{2^{n-1}} \sum_{k=0}^{\frac{n}{2}-1} \binom{n}{k} \frac{\sin(n-2k)x}{(n-2k)} + \frac{1}{2^n}\binom{n}{\frac{n}{2}} x$

726. $\int \sin^n x\,dx = \frac{1}{2^{n-1}} \sum_{k=0}^{\frac{n}{2}-1} \binom{n}{k} \frac{\sin[(n-2k)(\frac{\pi}{2}-x)]}{2k-n} + \frac{1}{2^n}\binom{n}{\frac{n}{2}} x$

For n odd:

727. $\int \cos^n x\,dx = \frac{1}{2^{n-1}} \sum_{k=0}^{\frac{n-1}{2}} \binom{n}{k} \frac{\sin(n-2k)x}{(n-2k)}$

728. $\int \sin^n x\,dx = \frac{1}{2^{n-1}} \sum_{k=0}^{\frac{n-1}{2}} \binom{n}{k} \frac{\sin[(n-2k)(\frac{\pi}{2}-x)]}{2k-n}$

DIFFERENTIAL EQUATIONS

SPECIAL FORMULAS

Certain types of differential equations occur sufficiently often to justify the use of formulas for the corresponding particular solutions. The following set of tables 1 to XIV covers all first, second, and *n*th order ordinary linear differential equations with constant coefficients for which the right members are of the form $P(x)e^{rx}\sin sx$ or $P(x)e^{rx}\cos sx$, where r and s are constants and $P(x)$, is a polynomial of degree n.

When the right member of a reducible linear partial differential equation with constant coefficients is not zero, particular solutions for certain types of right members are contained in tables XV to XXI. In these tables both F and P are used to denote polynomials, and it is assumed that no denominator is zero. In any formula the roles of x and y may be reversed throughout, changing a formula in which x dominates to one in which y dominates. Tables XIX, XX, XXI are applicable whether the equations are reducible or not. The symbol $\binom{m}{m}$ stands for $\frac{m!}{(m-n)!n!}$ and is the $n+1$ st coefficient in the expansion of $(a+b)^m$. Also $0! - 1$ by definition.

The tables as herewith given are those contained in the text *Differential Equations* by Ginn and Company (1955) and are published with their kind permission and that of the author, Professor Frederick H. Steen.

Solution of Linear Differential Equations with Constant Coefficients

Any linear differential equation with constant coefficients may be written in the form

$$p(D)y = R(x)$$

where D is the differential operation

$$Dy = \frac{dy}{dx}$$

$p(D)$ is a polynomial in D,
y is the dependent variable,
x is the independent variable,
$R(x)$ is an arbitrary function of x.

A power of D represents repeated differentiation, that is

$$D^n y = \frac{d^n y}{dx^n}$$

For such an equation, the general solution may be written in the form

$$y = y_c + y_p$$

where y_p is any particular solution, and y_c is called the *complementary function*. This complementary function is defined as the general solution of the *homogeneous equation*, which is the original differential equation with the right side replaced by zero, i.e.

$$p(D)y = 0$$

The complementary function y_c may be determined as follows:

1. Factor the polynomial $p(D)$ into real and complex linear factors, just as if D were a variable instead of an operator.
2. For each nonrepeated linear factor of the form $(D - a)$, where a is real, write down a term of the form

$$ce^{ax}$$

where c is an arbitrary constant.
3. For each repeated real linear factor of the form $(D - a)^n$, write down n terms of the form

$$c_1e^{ax} + c_2xe^{ax} + c_3x^2e^{ax} + \cdots + c_nx^{n-1}e^{ax}$$

where the c_i's are arbitrary constants.

4. For each non-repeated conjugate complex pair of factors of the form $(D - a + ib)(D - a - ib)$, write down 2 terms of the form

$$c_1e^{ax}\cos bx + c_2e^{ax}\sin bx$$

5. For each repeated conjugate complex pair of factors of the form $(D - a + ib)^n(D - a - ib)^n$, write down $2n$ terms of the form

$$c_1e^{ax}\cos bx + c_2e^{ax}\sin bx + c_3xe^{ax}\cos bx + c_4xe^{ax}\sin bx + \cdots + c_{2n-1}x^{n-1}e^{ax}\cos bx + c_{2n}x^{n-1}e^{ax}\sin bx$$

6. The sum of all the terms thus written down is the complementary function y_c.

To find the particular solution y_p, use the following tables, as shown in the examples. For cases not shown in the tables, there are various methods of finding y_p. The most general method is called *variation of parameters.* The following example illustrates the method:

Find y_p for $(D^2 - 4)y = e^x$.

This example can be solved most easily by use of equation 63 in the tables following. However it is given here as an example of the method of variation of parameters.

The complementary function is

$$y_c = c_1e^{2x} + c_2e^{-2x}$$

To find y_p, replace the constants in the complementary function with unknown functions,

$$y_p = ue^{2x} + ve^{-2x}$$

We now prepare to substitute this assumed solution into the original equation. We begin by taking all the necessary dervatives:

$$y_p = ue^{2x} + ve^{-2x}$$

$$y_p' = 2ue^{2x} - 2ve^{-2x} + u'e^{2x} + v'e^{-2x}$$

For each derivative of y_p except the highest, we set the sum of all the terms containing u' and v' to 0. Thus the above equation becomes

$$u'e^{2x} + v'e^{-2x} = 0 \quad \text{and} \quad y_p' = 2ue^{2x} - 2ve^{-2x}$$

Continuing to differentiate, we have

$$y_p'' = 4ue^{2x} + 4ve^{-2x} + 2u'e^{2x} - 2v'e^{-2x}$$

When we substitute into the original equation, all the terms not containing u' or v' cancel out. This is a consequence of the method by which y_p was set up.

Thus all that is necessary is to write down the terms containing u' or v' in the highest order derivative of y_p, multiply by the constant coefficient of the highest power of D in $p(D)$, and set it equal to $R(x)$. Together with the previous terms in u' and v' which were set equal to 0, this gives us as many linear equations in the first derivatives of the unknown functions as there are unknown functions. The first derivatives may then be solved for by algebra, and the unknown functions found by integration. In the present example, this becomes

$$u'e^{2x} + v'e^{-2x} = 0$$

$$2u'e^{2x} - 2v'e^{-2x} = e^x$$

We eliminate v' and u' separately, getting

$$4u'e^{2x} = e^x$$

$$4v'e^{-2x} = -e^x$$

Thus

$$u' = \tfrac{1}{4}e^{-x}$$

$$v' = -\tfrac{1}{4}e^{3x}$$

Therefore, by integrating

$$u = -\tfrac{1}{4}e^{-x}$$

$$v = -\tfrac{1}{12}e^{3x}$$

A constant of integration is not needed, since we need only one particular solution. Thus

$$y_p = ue^{2x} + ve^{-2x} = -\tfrac{1}{4}e^{-x}e^{2x} - \tfrac{1}{12}e^{3x}e^{-2x} = -\tfrac{1}{4}e^x - \tfrac{1}{12}e^x = -\tfrac{1}{3}e^x$$

and the general solution is

$$y = y_c + y_p = c_1e^{2x} + c_2e^{-2x} - \tfrac{1}{3}e^x$$

The following examples illustrate the use of the tables.

Example 1. Solve $(D^2 - 4)y = \sin 3x$.

Substitution of $q = -4$, $s = 3$ in formula 24 gives

$$y_p = \frac{\sin 3x}{-9 - 4}$$

wherefore the general solution is

$$y = c_1 e^{2x} + c_2 e^{-2x} - \frac{\sin 3x}{13}$$

Example 2. Obtain a particular solution of $(D^2 - 4D + 5)y = x^2 e^{3x} \sin x$.

Applying formula 40 with $a = 2, b = 1, r = 3, s = 1$, $P(x) = x^2$, $s + b = 2$, $s - b = 0$, $a - r = -1$, $(a - r)^2 + (s + b)^2 = 5$, $(a - r)^2 + (s - b)^2 = 1$, we have

$$y_p = \frac{e^{3x}\sin x}{2}\left[\left(\frac{2}{5} - \frac{0}{1}\right)x^2 + \left(\frac{2(-1)2}{25} - \frac{2(-1)0}{1}\right)2x + \left(\frac{3\cdot 1\cdot 2 - 2^3}{125} - \frac{3\cdot 1\cdot 0 - 0}{1}\right)2\right]$$

$$- \frac{e^{3x}\cos x}{2}\left[\left(\frac{-1}{5} - \frac{-1}{1}\right)x^2 + \left(\frac{1 - 4}{25} - \frac{1 - 0}{1}\right)2x + \left(\frac{-1 - 3(-1)4}{125} - \frac{-1 - 3(-1)0}{1}\right)2\right]$$

$$= \left(\tfrac{1}{5}x^2 - \tfrac{4}{25}x - \tfrac{2}{125}\right)e^{3x}\sin x + \left(-\tfrac{2}{5}x^2 + \tfrac{28}{25}x - \tfrac{136}{125}\right)e^{3x}\cos x$$

The special formulas effect a very considerable saving of time in problems of this type.

Example 3. Obtain a particular solution of $(D^2 - 4D + 5)y = x^2 e^{2x} \cos x$. (Compare with Example 2.)

Formula 40 is not applicable here since for this equation $r = a$, $s = b$, wherefore the denominator $(a - r)^2 + (s - b)^2 = 0$. We turn instead to formula 44. Substituting $a = 2$, $b = 1$, $P(x) = x^2$ and replacing sin by cos, cos by $-\sin$, we obtain

$$y_p = \frac{e^{2x}\cos x}{4}\left(x^2 - \tfrac{2}{4}\right) + \frac{e^{2x}\sin x}{2}\int \left(x^2 - \tfrac{1}{2}\right)dx$$

$$= \left(\frac{x^2}{4} - \frac{1}{8}\right)e^{2x}\cos x + \left(\frac{x^3}{6} - \frac{x}{4}\right)e^{2x}\sin x$$

which is the required solution.

Example 4. Find z_p for $(D_x - 3D_y)z = \ln(y + 3x)$.

Referring to Table XV we note that formula 69 (not 68) is applicable. This gives

$$z_p = x \ln(y + 3x)$$

It is easily seen that $-y/3 \ln(y + 3x)$ would serve equally well.

Example 5. Solve $(D_x + 2D_y - 4)z = y\cos(y - 2x)$.

Since R in formula 76 contains a polynomial in x, not y, we rewrite the given equation in the form

$$\left(D_y + \tfrac{1}{2}D_x - 2\right)z = \tfrac{1}{2}y\cos(y - 2x)$$

Then

$$z_c = e^{2y}F\left(x - \tfrac{1}{2}y\right) = e^{2y}f(2x - y)$$

and by the formula

$$z_p = -\tfrac{1}{2}\cos(y - 2x)\cdot\left(\frac{y}{2} + \frac{\frac{1}{2}}{2}\right)$$

$$= -\tfrac{1}{8}(2y + 1)\cos(y - 2x)$$

Example 6. Find z_p for $(D_x + 4D_y)^3 z = (2x - y)^2$.

Using formula 79, we obtain

$$z_p = \frac{\int\int\int u^2 du^3}{[2 + 4(-1)]^3} = \frac{u^5}{5\cdot 4\cdot 3\cdot(-8)} = -\frac{(2x - y)^5}{480}$$

Example 7. Find z_p for $(D_x^3 + 5D_x^2 D_y - 7D_x + 4)z = e^{2x+3y}$.

By formula 87

$$z_p = \frac{e^{2x+3y}}{2^3 + 5\cdot 2^2\cdot 3 - 7\cdot 2 + 4} = \frac{e^{2x+3y}}{58}$$

Example 8. Find z_p for

$$(D_x^4 + 6D_x^3 D_y + D_x D_y + D_y^2 + 9)z = \sin(3x + 4y)$$

Since every term in the left member is of even degree in the two operators D_x and D_y, formula 90 is applicable. It gives

$$z_p = \frac{\sin(3x + 4y)}{(-9)^2 + 6(-9)(-12) + (-12) + (-16) + 9}$$

$$= \frac{\sin(3x + 4y)}{710}$$

TABLE I: $(D - a)y = R$

R	y_p
1. e^{rx}	$\frac{e^{rx}}{r - a}$
2. $\sin sx$*	$-\frac{a \sin sx + s \cos sx}{a^2 + s^2} = -\frac{1}{\sqrt{a^2 + s^2}} \sin\left(sx + \tan^{-1}\frac{s}{a}\right)$
3. $P(x)$	$-\frac{1}{a}\left[P(x) + \frac{P'(x)}{a} + \frac{P''(x)}{a^2} + \cdots + \frac{P^{(n)}(x)}{a^n}\right]$
4. $e^{rx} \sin sx$*	Replace a by $a - r$ in formula 2 and multiply by e^{rx}.
5. $P(x)e^{rx}$	Replace a by $a - r$ in formula 3 and multiply by e^{rx}.
6. $P(x) \sin sx$*	$-\sin sx\left[\frac{a}{a^2 + s^2}P(x) + \frac{a^2 - s^2}{(a^2 + s^2)^2}P'(x) + \frac{a^3 - 3as^2}{(a^2 + s^2)^3}P''(x) + \cdots + \frac{a^k - \binom{k}{2}a^{k-2}s^2 + \binom{k}{4}a^{k-4}s^4 - \cdots}{(a^2 + s^2)^k}P^{(k-1)}(x) + \cdots\right]$ $-\cos sx\left[\frac{s}{a^2 + s^2}P(x) + \frac{2as}{(a^2 + s^2)^2}P'(x) + \frac{3a^2s - s^3}{(a^2 + s^2)^3}P''(x) + \cdots + \frac{\binom{k}{1}a^{k-1}s - \binom{k}{3}a^{k-3}s^3 + \cdots}{(a^2 + s^2)^k}P^{(k-1)}(x) + \cdots\right]$
7. $P(x)e^{rx} \sin sx$*	Replace a by $a - r$ in formula 6 and multiply by e^{rx}.
8. e^{ax}	xe^{ax}
9. $e^{ax} \sin sx$*	$-\frac{e^{ax} \cos sx}{s}$
10. $P(x)e^{ax}$	$e^{ax}\int P(x)\,dx$
11. $P(x)e^{ax} \sin sx$*	$\frac{e^{ax} \sin sx}{s}\left[\frac{P'(x)}{s^3} - \frac{P'''(x)}{s^3} + \frac{P^{v}(x)}{s^5} - \cdots\right] - \frac{e^{ax} \cos sx}{s}\left[P(x) - \frac{P''(x)}{s^2} + \frac{P^{iv}(x)}{s^4} - \cdots\right]$

* For cos sx in R replace "sin" by "cos" and "cos" by "− sin" in y_p.

$$D^n = \frac{d^n}{dx^n} \qquad \binom{m}{n} = \frac{m!}{(m - n)!n!} \qquad 0! = 1$$

TABLE II: $(D - a)^2y = R$

R	y_p
12. e^{rx}	$\frac{e^{rx}}{(r - a)^2}$
13. $\sin sx$*	$\frac{1}{(a^2 + s^2)^2}[(a^2 - s^2)\sin sx + 2as \cos sx] = \frac{1}{a^2 + s^2}\sin\left(sx + \tan^{-1}\frac{2as}{a^2 - s^2}\right)$
14. $P(x)$	$\frac{1}{a^2}\left[P(x) + \frac{2P'(x)}{a} + \frac{3P''(x)}{a^2} + \cdots + \frac{(n + 1)P^{(n)}(x)}{a^n}\right]$
15. $e^{rx} \sin sx$*	Replace a by $a - r$ in formula 13 and multiply by e^{rx}.
16. $P(x)e^{rx}$	Replace a by $a - r$ in formula 14 and multiply by e^{rx}.
17. $P(x) \sin sx$*	$\sin sx\left[\frac{a^2 - s^2}{(a^2 + s^2)^2}P(x) + 2\frac{a^3 - 3as^2}{(a^2 + s^2)^3}P'(x) + 3\frac{a^4 - 6a^2s^2 + s^4}{(a^2 + s^2)^4}P''(x) + \cdots + (k - 1)\frac{a^k - \binom{k}{2}a^{k-2}s^2 + \binom{k}{4}a^{k-4}s^4 - \cdots}{(a^2 + s^2)^k}P^{(k-2)}(x) + \cdots\right]$ $+\cos sx\left[\frac{2as}{(a^2 + s^2)^2}P(x) + 2\frac{3a^2s - s^3}{(a^2 + s^2)^3}P'(x) + 3\frac{4a^3s - 4as^3}{(a^2 + s^2)^4}P''(x) + \cdots + (k - 1)\frac{\binom{k}{1}a^{k-1}s - \binom{k}{3}a^{k-3}s^3 + \cdots}{(a^2 + s^2)^k}P^{(k-2)}(x) + \cdots\right]$
18. $P(x)e^{rx} \sin sx$*	Replace a by $a - r$ in formula 17 and multiply by e^{rx}.
19. e^{ax}	$\frac{1}{2}x^2e^{ax}$
20. $e^{ax} \sin sx$*	$-\frac{e^{ax} \sin sx}{s^2}$
21. $P(x)e^{ax}$	$e^{ax}\iint P(x)\,dx\,dx$
22. $P(x)e^{ax} \sin sx$*	$-\frac{e^{ax} \sin sx}{s^2}\left[P(x) - \frac{3P''(x)}{s^2} + \frac{5P^{iv}(x)}{s^4} - \frac{7P^{vi}(x)}{s^6} + \cdots\right] - \frac{e^{ax} \cos sx}{s^2}\left[\frac{2P'(x)}{s} - \frac{4P'''(x)}{s^3} + \frac{6P^{v}(x)}{s^5} - \cdots\right]$

* For cos sx in R replace "sin" by "cos" and "cos" by "− sin" in y_p.

TABLE III: $(D^2 + q)y = R$

R	y_p
23. e^{rx}	$\dfrac{e^{rx}}{r^2 + q}$
24. $\sin sx$*	$\dfrac{\sin sx}{-s^2 + q}$
25. $P(x)$	$\dfrac{1}{q}\left[P(x) - \dfrac{P''(x)}{q} + \dfrac{P^{\text{iv}}(x)}{q^2} - \cdots + (-1)^k \dfrac{P^{(2k)}(x)}{q^k} \cdots\right]$
26. $e^{rx}\sin sx$*	$\dfrac{(r^2 - s^2 + q)e^{rx}\sin sx - 2\,rse^{rx}\cos sx}{(r^2 - s^2 + q)^2 + (2rs)^2} = \dfrac{e^{rx}}{\sqrt{(r^2 - s^2 + q)^2 + (2rs)^2}}\sin\left[sx - \tan^{-1}\dfrac{2rs}{r^2 - s^2 + q}\right]$
27. $P(x)e^{rx}$	$\dfrac{e^{rx}}{r^2 + q}\left[P(x) - \dfrac{2r}{r^2 + q}P'(x) + \dfrac{3r^2 - q}{(r^2 + q)^2}P''(x) - \dfrac{4r^3 - 4qr}{(r^2 + q)^3}P'''(x) + \cdots + (-1)^{k-1}\dfrac{\binom{k}{1}r^{k-1} - \binom{k}{3}r^{k-3}q + \binom{k}{5}r^{k-5}q^2 - \cdots}{(r^2 + q)^{k-1}}P^{(k-1)}(x) + \cdots\right]$
28. $P(x)\sin sx$*	$\dfrac{\sin sx}{(-s^2 + q)}\left[P(x) - \dfrac{3s^2 + q}{(-s^2 + q)^2}P''(x) + \dfrac{5s^4 + 10s^2q + q^2}{(-s^2 + q)^4}P^{\text{iv}}(x) + \cdots + (-1)^k \dfrac{\binom{2k+1}{1}s^{2k} + \binom{2k+1}{3}s^{2k-2}q + \binom{2k+1}{5}s^{2k-4}q^2 + \cdots}{(-s^2 + q)^{2k}}P^{(2k)}(x) + \cdots\right]$ $- \dfrac{s\cos sx}{(-s^2 + q)}\left[\dfrac{2P'(x)}{(-s^2 + q)} - \dfrac{4s^2 + 4q}{(-s^2 + q)^3}P'''(x) + \cdots + (-1)^{k+1}\dfrac{\binom{2k}{1}s^{2k-2} + \binom{2k}{3}s^{2k-4}q + \cdots}{(-s^2 + q)^{2k-1}}P^{(2k-1)}(x) + \cdots\right]$

TABLE IV: $(D^2 + b^2)y = R$

29. $\sin bx$*	$-\dfrac{x\cos bx}{2b}$
30. $P(x)\sin bx$*	$\dfrac{\sin bx}{(2b)^2}\left[P(x) - \dfrac{P''(x)}{(2b)^2} + \dfrac{P^{\text{iv}}(x)}{(2b)^4} - \cdots\right] - \dfrac{\cos bx}{2b}\int\left[P(x) - \dfrac{P''(x)}{(2b)^2} + \cdots\right]dx$

* For cos sx in R replace "sin" by "cos" and "cos" by "− sin" in y_p.

TABLE V: $(D^2 + pD + q)y = R$

R	y_p
31. e^{rx}	$\dfrac{e^{rx}}{r^2 + pr + q}$
32. $\sin sx$*	$\dfrac{(q - s^2)\sin sx - ps\cos sx}{(q - s^2)^2 + (ps)^2} = \dfrac{1}{\sqrt{(q - s^2)^2 + (ps)^2}}\sin\left(sx - \tan^{-1}\dfrac{ps}{q - s^2}\right)$
33. $P(x)$	$\dfrac{1}{q}\left[P(x) - \dfrac{p}{q}P'(x) + \dfrac{p^2 - q}{q^2}P''(x) - \dfrac{p^3 - 2pq}{q^3}P'''(x) + \cdots + (-1)^n \dfrac{p^n - \binom{n-1}{1}p^{n-2}q + \binom{n-2}{2}p^{n-4}q^2 - \cdots}{q^n}P^{(n)}(x)\right]$
34. $e^{rx}\sin sx$*	Replace p by $p + 2r$, q by $q + pr + r^2$ in formula 32 and multiply by e^{rx}.
35. $P(x)e^{rx}$	Replace p by $p + 2r$, q by $q + pr + r^2$ in formula 33 and multiply by e^{rx}.

TABLE VI: $(D - b)(D - a)y = R$

36. $P(x)\sin sx$*	$\dfrac{\sin sx}{b - a}\left[\left(\dfrac{a}{a^2 + s^2} - \dfrac{b}{b^2 + s^2}\right)P(x) + \left(\dfrac{a^2 - s^2}{(a^2 + s^2)^2} - \dfrac{b^2 - s^2}{(b^2 + s^2)^2}\right)P'(x) + \left(\dfrac{a^3 - 3as^2}{(a^2 + s^2)^3} - \dfrac{b^3 - 3bs^2}{(b^2 + s^2)^3}\right)P''(x) + \cdots\right]$ $+ \dfrac{\cos sx}{b - a}\left[\left(\dfrac{s}{a^2 + s^2} - \dfrac{s}{b^2 + s^2}\right)P(x) + \left(\dfrac{2as}{(a^2 + s^2)^2} - \dfrac{2bs}{(b^2 + s^2)^2}\right)P'(x) + \left(\dfrac{3a^2s - s^3}{(a^2 + s^2)^3} - \dfrac{3b^2s - s^3}{(b^2 + s^2)^3}\right)P''(x) + \cdots\right]$†
37. $P(x)e^{rx}\sin sx$*	Replace a by $a - r$, b by $b - r$ in formula 36 and multiply by e^{rx}.
38. $P(x)e^{ax}$	$\dfrac{e^{ax}}{a - b}\left[\int P(x)dx + \dfrac{P(x)}{(b - a)} + \dfrac{P'(x)}{(b - a)^2} + \dfrac{P''(x)}{(b - a)^3} + \cdots + \dfrac{P^{(n)}(x)}{(b - a)^{n+1}}\right]$

* For cos sx in R replace "sin" by "cos" and "cos" by "− sin" in y_p.
† For additional terms, compare with formula 6.

TABLE VII: $(D^2 - 2aD + a^2 + b^2)y = R$

R	y_p
39. $P(x)\sin sx$*	$\frac{\sin sx}{2b}\left[\left(\frac{s+b}{a^2+(s+b)^2} - \frac{s-b}{a^2+(s-b)^2}\right)P(x) + \left(\frac{2a(s+b)}{[a^2+(s+b)^2]^2} - \frac{2a(s-b)}{[a^2+(s-b)^2]^2}\right)P'(x) + \left(\frac{3a^2(s+b)-(s+b)^3}{[a^2+(s+b)^2]^3} - \frac{3a^2(s-b)-(s-b)^3}{[a^2+(s-b)^2]^3}\right)P''(x) + \cdots\right] - \frac{\cos sx}{2b}\left[\left(\frac{a}{a^2+(s+b)^2} - \frac{a}{a^2+(s-b)^2}\right)P(x) + \left(\frac{a^2-(s+b)^2}{[a^2+(s+b)^2]^2} - \frac{a^2-(s-b)^2}{[a^2+(s-b)^2]^2}\right)P'(x) + \left(\frac{a^3-3a(s+b)^2}{[a^2+(s+b)^2]^3} - \frac{a^3-3a(s-b)^2}{[a^2+(s-b)^2]^3}\right)P''(x) + \cdots\right]^\dagger$
40. $P(x)e^{rx}\sin sx$*	Replace a by $a - r$ in formula 39 and multiply by e^{rx}.
41. $P(x)e^{ax}$	$\frac{e^{ax}}{b^2}\left[P(x) - \frac{P''(x)}{b^2} + \frac{P^{iv}(x)}{b^4} - \cdots\right]$
42. $e^{ax}\sin sx$*	$\frac{e^{ax}\sin sx}{-s^2+b^2}$
43. $e^{ax}\sin bx$*	$-\frac{xe^{ax}\cos bx}{2b}$
44. $P(x)e^{ax}\sin bx$*	$\frac{e^{ax}\sin bx}{(2b)^2}\left[P(x) - \frac{P''(x)}{(2b)^2} + \frac{P^{iv}(x)}{(2b)^4} - \cdots\right] - \frac{e^{ax}\cos bx}{2b}\int\left[P(x) - \frac{P''(x)}{(2b)^2} + \frac{P^{iv}(x)}{(2b)^4} - \cdots\right]dx$

* For $\cos sx$ in R replace "sin" by "cos" and "cos" by "− sin" in y_p.
† For additional terms, compare with formula 6.

TABLE VIII: $f(D)y = [D^n + a_{n-1}D^{n-1} + \cdots + a_1D + a_0]y = R$

R	y_p
45. e^{rx}	$\frac{e^{rx}}{f(r)}$
46. $\sin sx$*	$\frac{[a_0 - a_2s^2 + a_4s^4 - \cdots]\sin sx - [a_1s - a_3s^3 + a_5s^5 + \cdots]\cos sx}{[a_0 - a_2s^2 + a_4s^4 - \cdots]^2 + [a_1s - a_3s^3 + a_5s^5 - \cdots]^2}$

TABLE IX: $f(D^2)y = R$

47. $\sin sx$*	$\frac{\sin sx}{f(-s^2)} = \frac{\sin sx}{a_0 - a_2s^2 + \cdots \pm s^{2n}}$

TABLE X: $(D - a)^n y = R$

48. e^{rx}	$\frac{e^{rx}}{(r-a)^n}$
49. $\sin sx$*	$\frac{(-1)^n}{(a^2+s^2)^n}\{[a^n - \binom{n}{2}a^{n-2}s^2 + \binom{n}{4}a^{n-4}s^4 - \cdots]\sin sx + [\binom{n}{1}a^{n-1}s - \binom{n}{3}a^{n-3}s^3 + \cdots]\cos sx\}$
50. $P(x)$	$\frac{(-1)^n}{a^n}\left[P(x) + \binom{n}{1}\frac{P'(x)}{a} + \binom{n+1}{2}\frac{P''(x)}{a^2} + \binom{n+2}{3}\frac{P'''(x)}{a^3} + \cdots\right]$
51. $e^{rx}\sin sx$*	Replace a by $a - r$ in formula 49 and multiply by e^{rx}.
52. $e^{rx}P(x)$	Replace a by $a - r$ in formula 50 and multiply by e^{rx}.
53. $P(x)\sin sx$*	$(-1)^n\sin sx[A_nP(x) + \binom{n}{1}A_{n+1}P'(x) + \binom{n+1}{2}A_{n+2}P''(x) + \binom{n+2}{3}A_{n+3}P'''(x) + \cdots] + (-1)^n\cos sx[B_nP(x) + \binom{n}{1}B_{n+1}P'(x) + \binom{n+1}{2}B_{n+2}P''(x) + \binom{n+2}{3}B_{n+3}P'''(x) + \cdots]$ $A_1 = \frac{a}{a^2+s^2},\ A_2 = \frac{a^2-s^2}{(a^2+s^2)^2},\ \cdots,\ A_k = \frac{a^k - \binom{k}{2}a^{k-2}s^2 + \binom{k}{4}a^{k-4}s^4 - \cdots}{(a^2+s^2)^k}.$ $B_1 = \frac{s}{a^2+s^2},\ B_2 = \frac{2as}{(a^2+s^2)^2},\ \cdots,\ B_k = \frac{\binom{k}{1}a^{k-1}s - \binom{k}{3}a^{k-3}s^3 + \cdots}{(a^2+s^2)^k}.$
54. $P(x)e^{rx}\sin sx$*	Replace a by $a - r$ in formula 53 and multiply by e^{rx}.

55. $e^{ax}P(x)$ $\quad e^{ax}\int\int\cdots\int P(x)dx^n$

56. $P(x)e^{ax}\sin sx$* $\quad \dfrac{(-1)^{\frac{n-1}{2}}e^{ax}\sin sx}{s^n}\left[\binom{n}{n-1}\dfrac{P'(x)}{s}-\binom{n+2}{n-1}\dfrac{P'''(x)}{s^3}+\binom{n+4}{n-1}\dfrac{P^{v}(x)}{s^5}-\cdots\right]$

$$+\frac{(-1)^{\frac{n+1}{2}}e^{ax}\cos sx}{s^n}\left[\binom{n-1}{n-1}P(x)-\binom{n+1}{n-1}\frac{P''(x)}{s^2}+\binom{n+3}{n-1}\frac{P^{iv}(x)}{s^4}-\cdots\right]\ (n\text{ odd})$$

$$\frac{(-1)^{\frac{n}{2}}e^{ax}\sin sx}{s^n}\left[\binom{n-1}{n-1}P(x)-\binom{n+1}{n-1}\frac{P''(x)}{s^2}+\binom{n+3}{n-1}\frac{P^{iv}(x)}{s^4}-\cdots\right]$$

$$+\frac{(-1)^{\frac{n}{2}}e^{ax}\cos sx}{s^n}\left[\binom{n}{n-1}\frac{P'(x)}{s}-\binom{n+2}{n-1}\frac{P'''(x)}{s^3}+\binom{n+4}{n-1}\frac{P^{v}(x)}{s^5}-\cdots\right]\ (n\text{ even})$$

* For cos sx in R replace "sin" by "cos" and "cos" by "− sin" in y_p.

TABLE XI: $(D-a)^n f(D)y = R$

57. e^{ax} $\quad \dfrac{x^n}{n!}\cdot\dfrac{e^{ax}}{f(a)}$

* For cos sx in R replace "sin" by "cos" and "cos" by "− sin" in y_p.

TABLE XII: $(D^2+q)^n y = R$

R	y_p
58. e^{rx}	$e^{rx}/(r^2+q)^n$
59. $\sin sx$*	$\sin sx/(q-s^2)^n$
60. $P(x)$	$\dfrac{1}{q^n}\left[P(x)-\binom{n}{1}\dfrac{P''(x)}{q}+\binom{n+1}{2}\dfrac{P^{iv}(x)}{q^2}-\binom{n+2}{3}\dfrac{P^{vi}(x)}{q^3}+\cdots\right]$
61. $e^{rx}\sin sx$*	$\dfrac{e^{rx}}{(A^2+B^2)^n}\{[A^n-\binom{n}{2}A^{n-2}B^2+\binom{n}{4}A^{n-4}B^4-\cdots]\sin sx-[\binom{n}{1}A^{n-1}B-\binom{n}{3}A^{n-3}B^3+\cdots]\cos sx\}$ $A=r^2-s^2+q,\quad B=2rs$

TABLE XIII: $(D^2+b^2)^n y = R$

62. $\sin bx$* $\quad (-1)^{\frac{n+1}{2}}\dfrac{x^n\cos bx}{n!(2b)^n}$ (n odd), $\quad (-1)^{\frac{n}{2}}\dfrac{x^n\sin bx}{n!(2b)^n}$ (n even)

TABLE XIV: $(D^n-q)y = R$

63. e^{rx} $\quad e^{rx}/(r^n-q)$

64. $P(x)$ $\quad -\dfrac{1}{q}\left[P(x)+\dfrac{P^{(n)}(x)}{q}+\dfrac{P^{(2n)}(x)}{q^2}+\cdots\right]$

65. $\sin sx$* $\quad -\dfrac{q\sin sx+(-1)^{\frac{n-1}{2}}s^n\cos sx}{q^2+s^{2n}}$ (n odd), $\quad \dfrac{\sin sx}{(-s^2)^{n/2}-q}$ (n even)

66. $e^{rx}\sin sx$* $\quad \dfrac{Ae^{rx}\sin sr-Be^{rx}\cos sx}{A^2+B^2}=\dfrac{e^{rx}}{\sqrt{A^2+B^2}}\sin\left(sx-\tan^{-1}\dfrac{B}{A}\right)$

$$A=[r^n-\binom{n}{2}r^{n-2}s^2+\binom{n}{4}r^{n-4}s^4-\cdots]-q,\qquad B=[\binom{n}{1}r^{n-1}s-\binom{n}{3}r^{n-3}s^3+\cdots]$$

* For cos sx in R replace "sin" by "cos" and "cos" by "− sin" in y_p.

TABLE XV: $(D_x+mD_y)z = R$

R	z_p
67 e^{ax+by}	$\dfrac{e^{ax+by}}{a+mb}$
68. $f(ax+by)$	$\dfrac{\int f(u)du}{a+mb},\ u=ax+by$
69. $f(y-mx)$	$xf(y-mx)$
70. $\phi(x,y)f(y-mx)$	$f(y-mx)\int\phi(x,a+mx)dx$ ($a=y-mx$ after integration)

TABLE XVI: $(D_x + mD_y - k)z = R$

71. e^{ax+by}	$\dfrac{e^{ax+by}}{a + mb - k}$
72. $\sin(ax + by)$*	$-\dfrac{(a + bm)\cos(ax + by) + k\sin(ax + by)}{(a + bm)^2 + k^2}$
73. $e^{\alpha x+\beta y}\sin(ax + by)$*	Replace k in 72 by $k - \alpha - m\beta$ and multiply by $e^{\alpha x+\beta y}$
74. $e^{xk}f(ax + by)$	$\dfrac{e^{kx}\int f(u)du}{a + mb}$, $u = ax + by$
75. $f(y - mx)$	$-\dfrac{f(y - mx)}{k}$
76. $P(x)f(y - mx)$	$-\dfrac{1}{k}f(y - mx)\left[P(x) + \dfrac{P'(x)}{k} + \dfrac{P''(x)}{k^2} + \cdots + \dfrac{P^{(n)}(x)}{k^n}\right]$
77. $e^{kx}f(y - mx)$	$xe^{kx}f(y - mx)$

* For cos $(ax + by)$ replace "sin" by "cos," and "cos" by "− sin" in z_p.

$$D_x = \frac{\partial}{\partial x}; \qquad D_y = \frac{\partial}{\partial y}; \qquad D_x{}^k D_y{}^r = \frac{\partial^{k+r}}{\partial_x{}^k \partial_y{}^r}$$

TABLE XVII: $(D_x + mD_y)^n z = R$

R	z_p
78. e^{ax+by}	$\dfrac{e^{ax+by}}{(a + mb)^n}$
79. $f(ax + by)$	$\dfrac{\int\int\cdots\int f(u)du^n}{(a + mb)^n}$, $u = ax + by$
80. $f(y - mx)$	$\dfrac{x^n}{n!}f(y - mx)$
81. $\phi(x, y)f(y + mx)$	$f(y - mx)\int\int\cdots\int\phi(x, a + mx)dx^n$ ($a = y - mx$ after integration)

TABLE XVIII: $(D_x + mD_y - k)^n z = R$

82. e^{ax+by}	$\dfrac{e^{ax+by}}{(a + mb - k)^n}$
83. $f(y - mx)$	$\dfrac{(-1)^n f(y - mx)}{k^n}$
84. $P(x)f(y - mx)$	$\dfrac{(-1)^n}{k^n}f(y - mx)\left[P(x) + \binom{n}{1}\dfrac{P'(x)}{k} + \binom{n+1}{2}\dfrac{P''(x)}{k^2} + \binom{n+2}{3}\dfrac{P'''(x)}{k^3} + \cdots\right]$
85. $e^{kx}f(ax + by)$	$\dfrac{e^{kx}\int\int\cdots\int f(u)du^n}{(a + mb)^n}$, $u = ax + by$
86. $e^{kx}f(y - mx)$	$\dfrac{x^n}{n!}e^{kx}f(y - mx)$

TABLE XIX: $[D_x{}^n + a_1 D_x{}^{n-1}D_y + a_2 D_x{}^{n-2}D_y{}^2 + \cdots + a^n D_y{}^n]z = R$

87. e^{ax+by}	$\dfrac{e^{ax+by}}{a + a_1 a^{n-1}b + a_2 a^{n-2}b^2 + \cdots + a_n b^n}$
88. $f(ax + by)$	$\dfrac{\int\int\cdots\int f(u)du^n}{a^n + a_1 a^{n-1}b + a_2 a^{n-2}b^2 + \cdots + a^n b^n}$, $(u = ax + by)$

TABLE XX: $F(D_x, D_y)z = R$

89. e^{ax+by}	$\dfrac{e^{ax+by}}{F(a, b)}$

TABLE XXI: $F(D_x{}^2, D_xD_y, D_y{}^2)z = R$

90. $\sin(ax + by)$*	$\dfrac{\sin(ax + by)}{F(-a^2, -ab, -b^2)}$

* For cos $(ax + by)$ replace "sin" by "cos", and "cos" by "−sin" in z_p.

DIFFERENTIAL EQUATIONS

Differential equation	Method of solution
Separation of variables $f_1(x)\, g_1(y)\, dx + f_2(x)\, g_2(y)\, dy = 0$	$\int \frac{f_1(x)}{f_2(x)} dx + \int \frac{g_2(y)}{g_1(y)} dy = c$
Exact equation $M(x,y)dx + N(x,y)dy = 0$ where $\partial M/\partial y = \partial N/\partial x$	$\int M \partial x + \int \left(N - \frac{\partial}{\partial y} \int M \partial x \right) dy = c$ where ∂x indicates that the integration is to be performed with respect to x keeping y constant.
Linear first order equation $\frac{dy}{dx} + P(x)y = Q(x)$	$ye^{\int P dx} = \int Qe^{\int P dx} dx + c$
Bernoulli's equation $\frac{dy}{dx} + P(x)y = Q(x)y^n$	$ve^{(1-n)\int P dx} = (1-n) \int Qe^{(1-n)\int P dx} dx + c$ where $v = y^{1-n}$. If $n = 1$, the solution is $\ln y = \int (Q - P)\, dx + c$
Homogeneous equation $\frac{dy}{dx} = F\left(\frac{y}{x}\right)$	$\ln x = \int \frac{dv}{F(v) - v} + c$ where $v = y/x$. If $F(v) = v$, the solution is $y = cx$
Reducible to homogeneous $(a_1 x + b_1 y + c_1)\, dx + (a_2 x + b_2 y + c_2)\, dy = 0$ $\frac{a_1}{a_2} \neq \frac{b_1}{b_2}$	Set $u = a_1 x + b_1 y + c_1$ $v = a_2 x + b_2 y + c_2$ Eliminate x and y and the equation becomes homogenous
Reducible to separable $(a_1 x + b_1 y + c_1)\, dx + (a_2 x + b_2 y + c_2)\, dy = 0$ $\frac{a_1}{a_2} = \frac{b_1}{b_2}$	Set $u = a_1 x + b_1 y$ Eliminate x or y and equation becomes separable

$y F(xy)\,dx + x G(xy)dy = 0$	$\ln x = \int \frac{G(v)\,dv}{v\{G(v) - F(v)\}} + c$ where $v = xy$. If $G(v) = F(v)$, the solution is $xy = c$.
Linear, homogeneous second order equation $\frac{d^2y}{dx^2} + b\frac{dy}{dx} + cy = 0$ b, c are real constants	Let m_1, m_2 be the roots of $m^2 + bm + c = 0$. Then there are 3 cases: Case 1. m_1, m_2 real and distinct: $y = c_1 e^{m_1 x} + c_2 e^{m_2 x}$ Case 2. m_1, m_2 real and equal: $y = c_1 e^{m_1 x} + c_2 x e^{m_1 x}$ Case 3. $m_1 = p + qi,\ m_2 = p - qi$: $y = e^{px}(c_1 \cos qx + c_2 \sin qx)$ where $p = -b/2,\ q = \sqrt{4c - b^2}/2$
Linear, nonhomogeneous second order equation $\frac{d^2y}{dx^2} + b\frac{dy}{dx} + cy = R(x)$ b, c are real constants	There are 3 cases corresponding to those immediately above: Case 1. $y = c_1 e^{m_1 x} + c_2 e^{m_2 x} + \frac{e^{m_1 x}}{m_1 - m_2}\int e^{-m_1 x} R(x)\,dx + \frac{e^{m_2 x}}{m_2 - m_1}\int e^{-m_2 x} R(x)\,dx$ Case 2. $y = c_1 e^{m_1 x} + c_2 x e^{m_1 x} + x e^{m_1 x}\int e^{-m_1 x} R(x)\,dx - e^{m_1 x}\int x e^{-m_1 x} R(x)\,dx$ Case 3. $y = e^{px}(c_1 \cos qx + c_2 \sin qx) + \frac{e^{px}\sin qx}{q}\int e^{-px} R(x)\cos qx\,dx - \frac{e^{px}\cos qx}{q}\int e^{-px} R(x)\sin qx\,dx$

DIFFERENTIAL EQUATIONS (Continued)

Euler or Cauchy equation $x^2 \dfrac{d^2 y}{dx^2} + bx \dfrac{dy}{dx} + cy = S(x)$	Putting $x = e^t$, the equation becomes $\dfrac{d^2 y}{dt^2} + (b-1)\dfrac{dy}{dt} + cy = S(e^t)$ and can then be solved as a linear second order equation.
Bessel's equation $x^2 \dfrac{d^2 y}{dx^2} + x \dfrac{dy}{dx} + (\lambda^2 x^2 - n^2)y = 0$	$y = c_1 J_n(\lambda x) + c_2 Y_n(\lambda x)$
Transformed Bessel's equation $x^2 \dfrac{d^2 y}{dx^2} + (2p+1)x \dfrac{dy}{dx} + (\alpha^2 x^{2r} + \beta^2)y = 0$	$y = x^{-p}\left\{c_1 J_{q/r}\left(\frac{\alpha}{r}x^r\right) + c_2 Y_{q/r}\left(\frac{\alpha}{r}x^r\right)\right\}$ where $q = \sqrt{p^2 - \beta^2}$.
Legendre's equation $(1-x^2)\dfrac{d^2 y}{dx^2} - 2x\dfrac{dy}{dx} + n(n+1)y = 0$	$y = c_1 P_n(x) + c_2 Q_n(x)$

FOURIER SERIES

If $f(x)$ is a bounded periodic function of period 2L (i.e. $f x + 2L) = f(x)$, and satisfies the *Dirichlet conditions:*

A. In any period $f(x)$ is continuous, except possibly for a finite number of jump discontinuities.
B. In any period $f(x)$ has only a finite number of maxima and minima.

Then $f(x)$ may be represented by the *Fourier series*

$$\frac{a_0}{2} + \sum_{n=1}^{\infty}\left(a_n \cos\frac{n\pi x}{L} + b_n \sin\frac{n\pi x}{L}\right)$$

where a_n and b_n are as determined below. This series will converge to $f(x)$ at every point where $f(x)$ is continuous, and to

$$\frac{f(x^+) + f(x^-)}{2}$$

(i.e. the average of the left-hand and right-hand limits) at every point where $f(x)$ has a jump discontinuity.

$$a_n = \frac{1}{L}\int_{-L}^{L} f(x)\cos\frac{n\pi x}{L}\,dx, \quad n = 0, 1, 2, 3, \ldots;$$

$$b_n = \frac{1}{L}\int_{-L}^{L} f(x)\sin\frac{n\pi x}{L}\,dx, \quad n = 1, 2, 3, \ldots$$

We may also write

$$a_n = \frac{1}{L}\int_{\alpha}^{\alpha+2L} f(x)\cos\frac{n\pi x}{L}\,dx \text{ and } b_n = \frac{1}{L}\int_{\alpha}^{\alpha+2L} f(x)\sin\frac{n\pi x}{L}\,dx$$

where α is any real number. Thus if $\alpha = 0$,

$$a_n = \frac{1}{L}\int_{0}^{2L} f(x)\cos\frac{n\pi x}{L}\,dx, \quad n = 0, 1, 2, 3, \ldots;$$

$$b_n = \frac{1}{L}\int_{0}^{2L} f(x)\sin\frac{n\pi x}{L}\,dx, \quad n = 1, 2, 3, \ldots$$

2. If in addition to the above restrictions, $f(x)$), is even (i.e., $f(-x) = f(x)$ the Fourier series reduces to

$$\frac{a_0}{2} + \sum_{n=1}^{\infty} a_n \cos \frac{n\pi x}{L}$$

That is, $b_n = 0$. In this case, a simpler formula for a_n is

$$a_n = \frac{2}{L}\int_0^L f(x)\cos\frac{n\pi x}{L}\,dx, \quad n = 0, 1, 2, 3, \ldots$$

3. If in addition to the restrictions in (1), $f(x)$ is an odd function (i.e., $f(-x) = -f(x)$), then the Fourier series reduces to

$$\sum_{n=1}^{\infty} b_n \sin \frac{n\pi x}{L}$$

That is, $a_n = 0$. In this case, a simpler formula for the b_n is

$$b_n = \frac{2}{L}\int_0^L f(x)\sin\frac{n\pi x}{L}\,dx, \quad n = 1, 2, 3, \ldots$$

4. If in addition to the restrictions in (2) above, $f(x) = -f(L - x)$, then a_n will be 0 for all even values of n, including $n = 0$. Thus in this case, the expansion reduces to

$$\sum_{m=1}^{\infty} a_{2m-1} \cos \frac{(2m-1)\pi x}{L}$$

5. If in addition to the restrictions in (3) above, $f(x) = f(L - x)$, then b_n will be 0 for all even values of n. Thus in this case, the expansion reduces to

$$\sum_{m=1}^{\infty} b_{2m-1} \sin \frac{(2m-1)\pi x}{L}$$

(The series in (4) and (5) are known as *odd-harmonic series,* since only the odd harmonics appear. Similar rules may be stated for even-harmonic series, but when a series appears in the even-harmonic form, it means that $2L$ has not been taken as the smallest period of $f(x)$. Since any integral multiple of a period is also a period, series obtained in this way will also work, but in general computation is simplified if $2L$ is taken to be the smallest period.)

6. If we write the Euler definitions for $\cos\theta$ and $\sin\theta$, we obtain the complex form of the Fourier Series known either as the "Complex Fourier Series" or the "Exponential Fourier Series" of $f(x)$. It is represented as

$$f(x) = \tfrac{1}{2}\sum_{n=-\infty}^{n=+\infty} c_n e^{i\omega_n x}$$

where

$$c_n = \frac{1}{L}\int_{-L}^{L} f(x)e^{-i\omega_n x}\,dx, \quad n = 0, \pm 1, \pm 2, \pm 3, \ldots$$

with $\omega_n = \frac{n\pi}{L}$, $n = 0, \pm 1, \pm 2, \ldots$

The set of coefficients $\{c_n\}$ is often referred to as the Fourier spectrum.

7. If both sine and cosine terms are present and if $f(x)$ is of period $2L$ and expandable by a Fourier series, it can be represented as

$$f(x) = \frac{a_0}{2} + \sum_{n=1}^{\infty} c_n \sin\left(\frac{n\pi x}{L} + \phi_n\right), \text{ where } a_n = c_n \sin\phi_n,$$

$$b_n = c_n\cos\phi_n, \quad c_n = \sqrt{a_n^2 + b_n^2}, \quad \phi_n = \arctan\left(\frac{a_n}{b_n}\right)$$

It can also be represented as

$$f(x) = \frac{a_0}{2} + \sum_{n=1}^{\infty} c_n \cos\left(\frac{n\pi x}{L} + \phi_n\right), \text{ where } a_n = c_n \cos\phi_n,$$

$$b_n = -c_n\sin\phi_n, \quad c_n = \sqrt{a_n^2 + b_n^2}, \quad \phi_n = \arctan\left(-\frac{b_n}{a_n}\right)$$

where ϕ_n is chosen so as to make a_n, b_n, and c_n hold.

8. The following table of trigonometric identities should be helpful for developing Fourier Series.

	n	n even	n odd	$n/2$ odd	$n/2$ even
$\sin n\pi$	0	0	0	0	0
$\cos n\pi$	$(-1)^n$	$+1$	-1	$+1$	$+1$
$*\sin \frac{n\pi}{2}$		0	$(-1)^{(n-1)/2}$	0	0
$*\cos \frac{n\pi}{2}$		$(-1)^{n/2}$	0	-1	$+1$
$\sin \frac{n\pi}{4}$			$\frac{\sqrt{2}}{2}(-1)^{(n^2+4n+11)/8}$	$(-1)^{(n-2)/4}$	0

*A useful formula for $\sin \frac{n\pi}{2}$ and $\cos \frac{n\pi}{2}$ is given by

$$\sin \frac{n\pi}{2} = \frac{(i)^{n+1}}{2}[(-1)^n - 1] \text{ and } \cos \frac{n\pi}{2} = \frac{(i)^n}{2}[(-1)^n + 1], \text{ where } i^2 = -1.$$

AUXILIARY FORMULAS FOR FOURIER SERIES

$$1 = \frac{4}{\pi}\left[\sin \frac{\pi x}{k} + \frac{1}{3}\sin \frac{3\pi x}{k} + \frac{1}{5}\sin \frac{5\pi x}{k} + \cdots\right] \qquad [0 < x < k]$$

$$x = \frac{2k}{\pi}\left[\sin \frac{\pi x}{k} - \frac{1}{2}\sin \frac{2\pi x}{k} + \frac{1}{3}\sin \frac{3\pi x}{k} - \cdots\right] \qquad [-k < x < k]$$

$$x = \frac{k}{2} - \frac{4k}{\pi^2}\left[\cos \frac{\pi x}{k} + \frac{1}{3^2}\cos \frac{3\pi x}{k} + \frac{1}{5^2}\cos \frac{5\pi x}{k} + \cdots\right] \qquad [0 < x < k]$$

$$x^2 = \frac{2k^2}{\pi^3}\left[\left(\frac{\pi^2}{1} - \frac{4}{1}\right)\sin \frac{\pi x}{k} - \frac{\pi^2}{2}\sin \frac{2\pi x}{k} + \left(\frac{\pi^2}{3} - \frac{4}{3^3}\right)\sin \frac{3\pi x}{k} - \frac{\pi^2}{4}\sin \frac{4\pi x}{k} + \left(\frac{\pi^2}{5} - \frac{4}{5^3}\right)\sin \frac{5\pi x}{k} + \cdots\right] \qquad [0 < x < k]$$

$$x^2 = \frac{k^2}{3} - \frac{4k^2}{\pi^2}\left[\cos \frac{\pi x}{k} - \frac{1}{2^2}\cos \frac{2\pi x}{k} + \frac{1}{3^2}\cos \frac{3\pi x}{k} - \frac{1}{4^2}\cos \frac{4\pi x}{k} + \cdots\right] \qquad [-k < x < k]$$

$$1 - \frac{1}{3} + \frac{1}{5} - \frac{1}{7} + \cdots = \frac{\pi}{4}$$

$$1 + \frac{1}{2^2} + \frac{1}{3^2} + \frac{1}{4^2} + \cdots = \frac{\pi^2}{6}$$

$$1 - \frac{1}{2^2} + \frac{1}{3^2} - \frac{1}{4^2} + \cdots = \frac{\pi^2}{12}$$

$$1 + \frac{1}{3^2} + \frac{1}{5^2} + \frac{1}{7^2} + \cdots = \frac{\pi^2}{8}$$

$$\frac{1}{2^2} + \frac{1}{4^2} + \frac{1}{6^2} + \frac{1}{8^2} + \cdots = \frac{\pi^2}{24}$$

FOURIER EXPANSIONS FOR BASIC PERIODIC FUNCTIONS

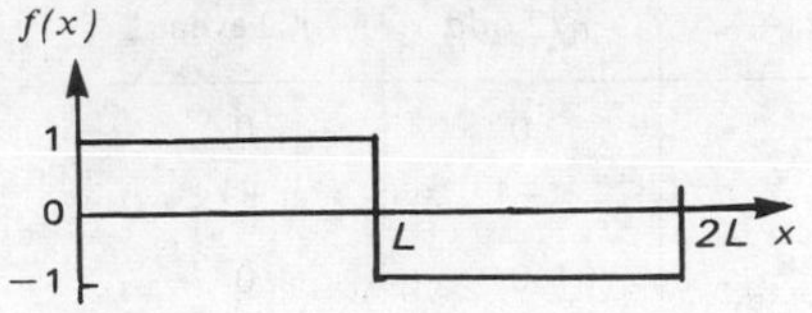

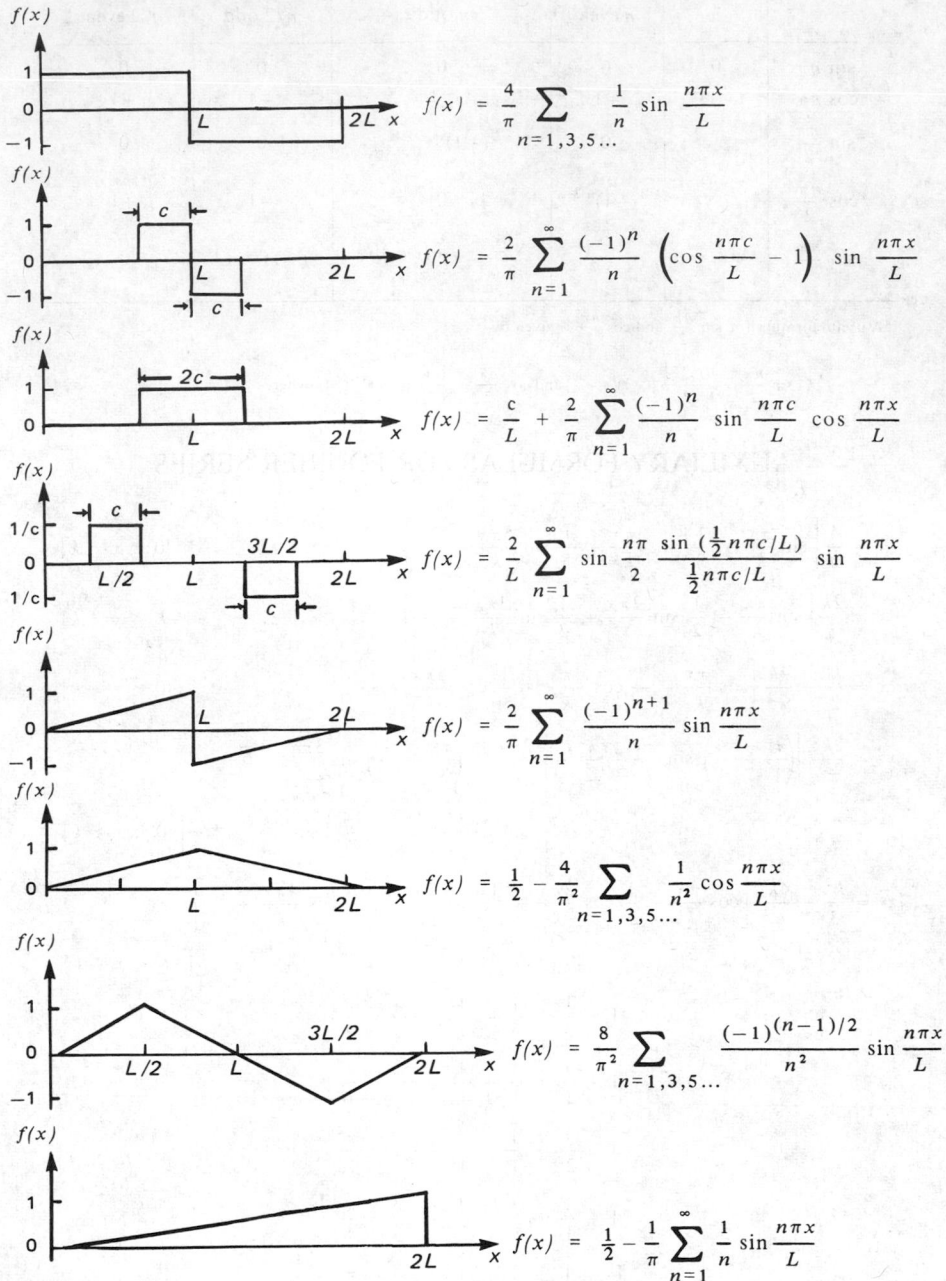

$$f(x) = \frac{4}{\pi} \sum_{n=1,3,5\ldots} \frac{1}{n} \sin \frac{n\pi x}{L}$$

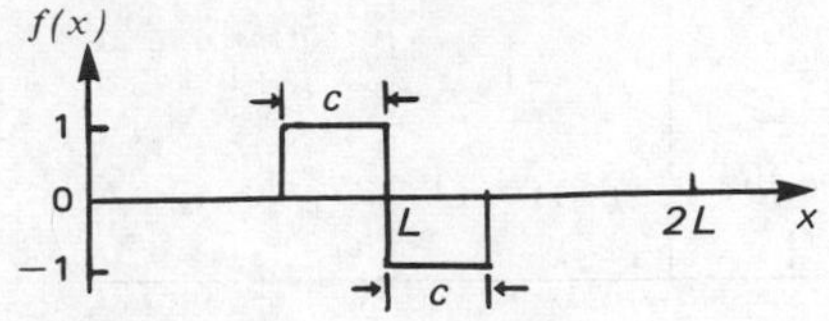

$$f(x) = \frac{2}{\pi} \sum_{n=1}^{\infty} \frac{(-1)^n}{n} \left(\cos \frac{n\pi c}{L} - 1\right) \sin \frac{n\pi x}{L}$$

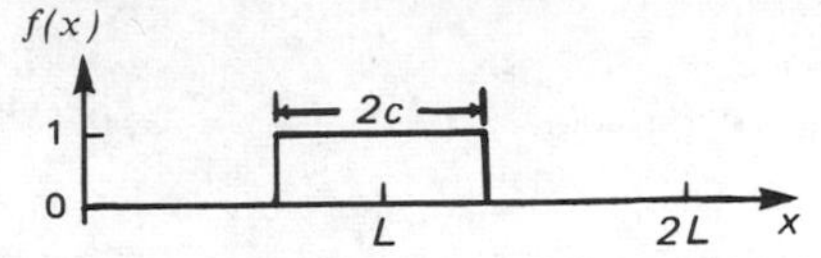

$$f(x) = \frac{c}{L} + \frac{2}{\pi} \sum_{n=1}^{\infty} \frac{(-1)^n}{n} \sin \frac{n\pi c}{L} \cos \frac{n\pi x}{L}$$

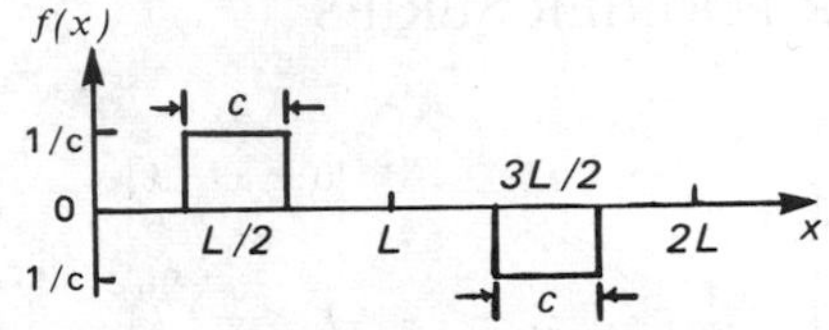

$$f(x) = \frac{2}{L} \sum_{n=1}^{\infty} \sin \frac{n\pi}{2} \, \frac{\sin\left(\frac{1}{2} n\pi c/L\right)}{\frac{1}{2} n\pi c/L} \sin \frac{n\pi x}{L}$$

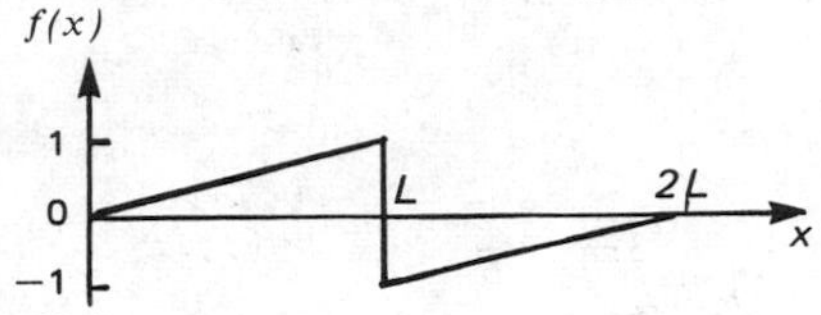

$$f(x) = \frac{2}{\pi} \sum_{n=1}^{\infty} \frac{(-1)^{n+1}}{n} \sin \frac{n\pi x}{L}$$

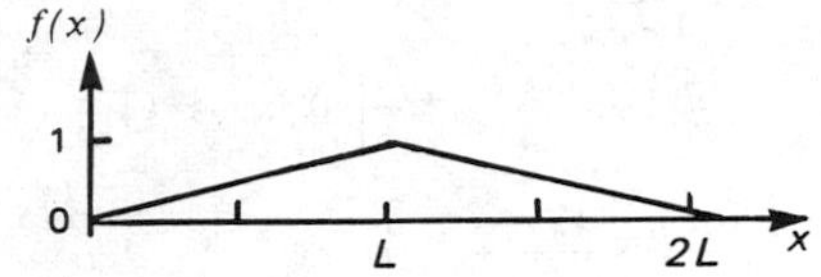

$$f(x) = \frac{1}{2} - \frac{4}{\pi^2} \sum_{n=1,3,5\ldots} \frac{1}{n^2} \cos \frac{n\pi x}{L}$$

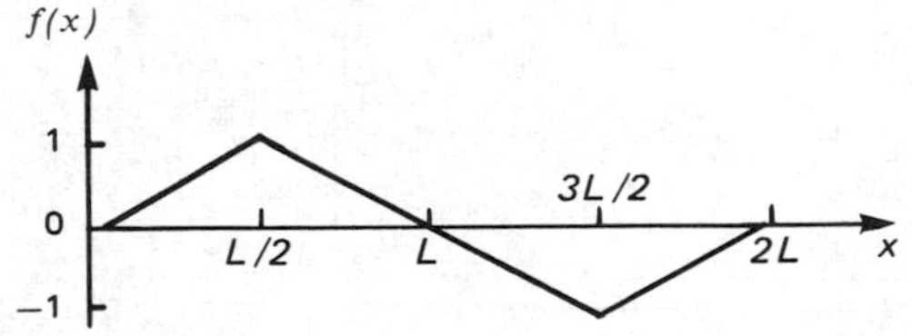

$$f(x) = \frac{8}{\pi^2} \sum_{n=1,3,5\ldots} \frac{(-1)^{(n-1)/2}}{n^2} \sin \frac{n\pi x}{L}$$

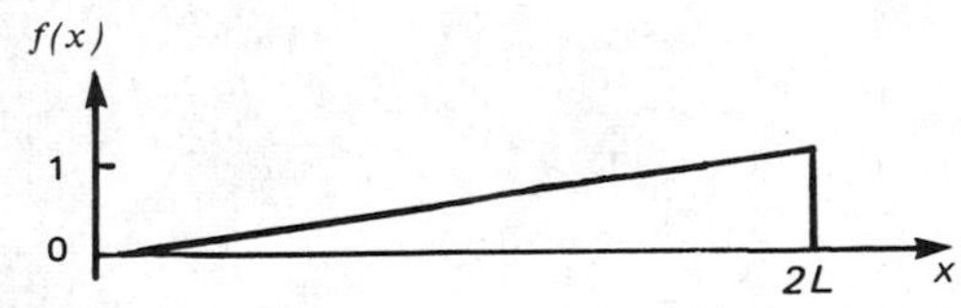

$$f(x) = \frac{1}{2} - \frac{1}{\pi} \sum_{n=1}^{\infty} \frac{1}{n} \sin \frac{n\pi x}{L}$$

FOURIER EXPANSIONS FOR BASIC PERIODIC FUNCTIONS (Continued)

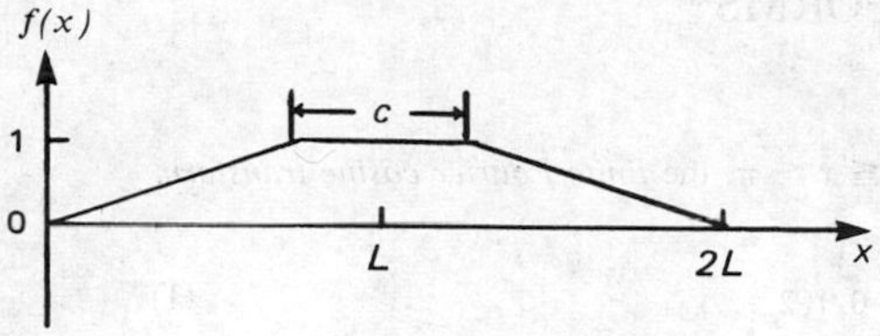

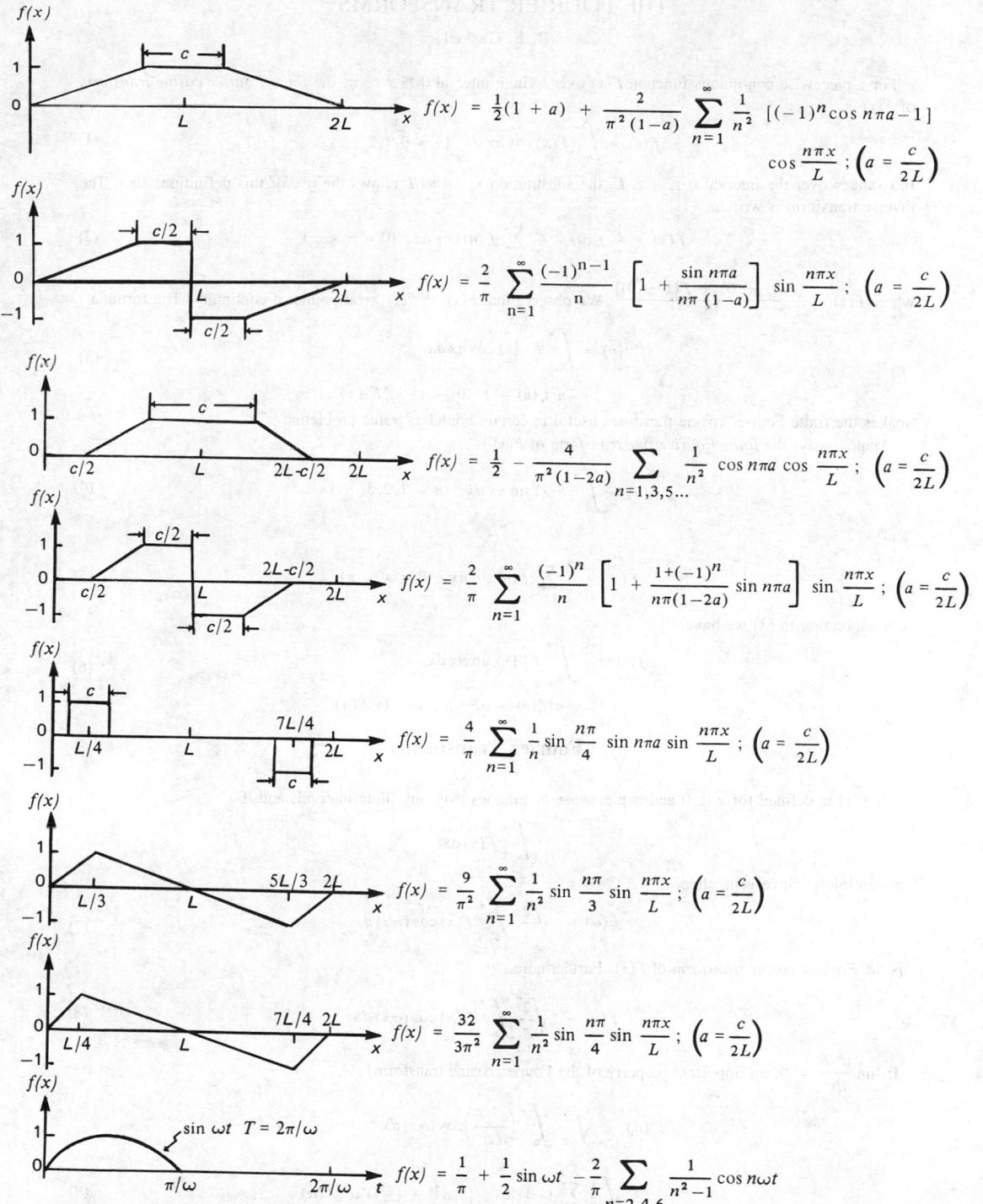

$$f(x) = \frac{1}{2}(1 + a) + \frac{2}{\pi^2(1-a)} \sum_{n=1}^{\infty} \frac{1}{n^2} \left[(-1)^n \cos n\pi a - 1\right] \cos \frac{n\pi x}{L}; \left(a = \frac{c}{2L}\right)$$

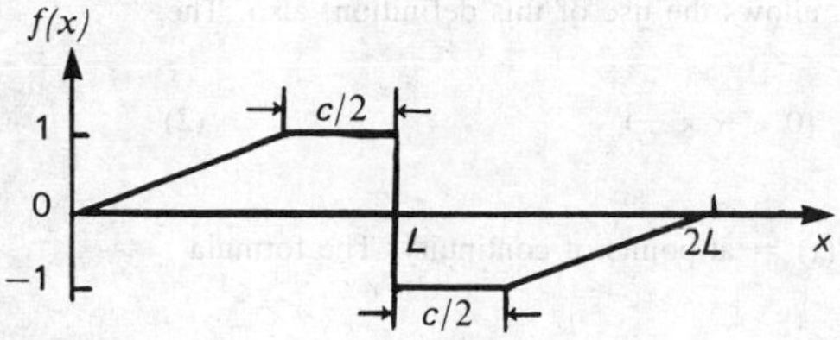

$$f(x) = \frac{2}{\pi} \sum_{n=1}^{\infty} \frac{(-1)^{n-1}}{n} \left[1 + \frac{\sin n\pi a}{n\pi(1-a)}\right] \sin \frac{n\pi x}{L}; \left(a = \frac{c}{2L}\right)$$

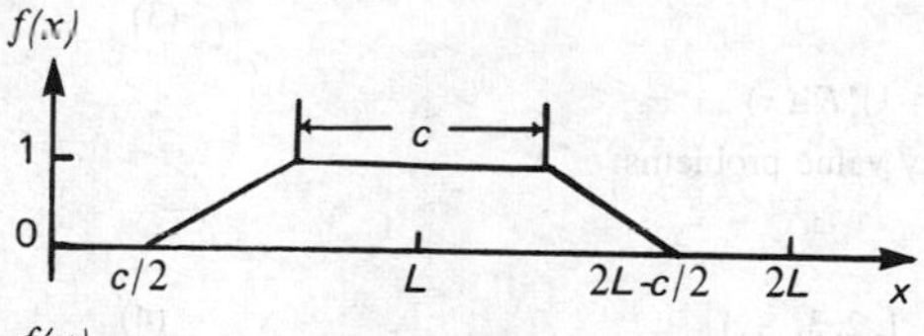

$$f(x) = \frac{1}{2} - \frac{4}{\pi^2(1-2a)} \sum_{n=1,3,5\ldots} \frac{1}{n^2} \cos n\pi a \cos \frac{n\pi x}{L}; \left(a = \frac{c}{2L}\right)$$

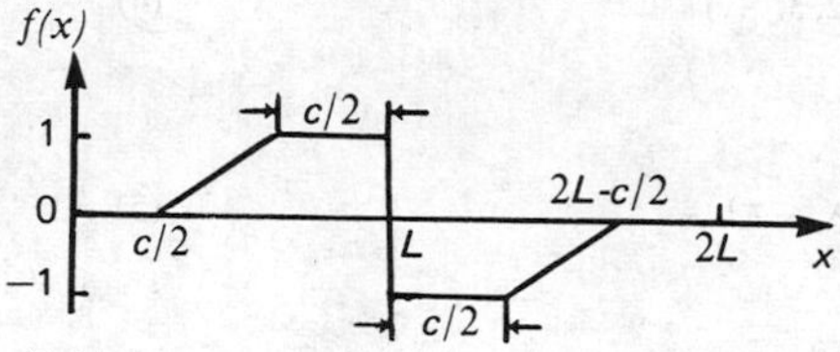

$$f(x) = \frac{2}{\pi} \sum_{n=1}^{\infty} \frac{(-1)^n}{n} \left[1 + \frac{1+(-1)^n}{n\pi(1-2a)} \sin n\pi a\right] \sin \frac{n\pi x}{L}; \left(a = \frac{c}{2L}\right)$$

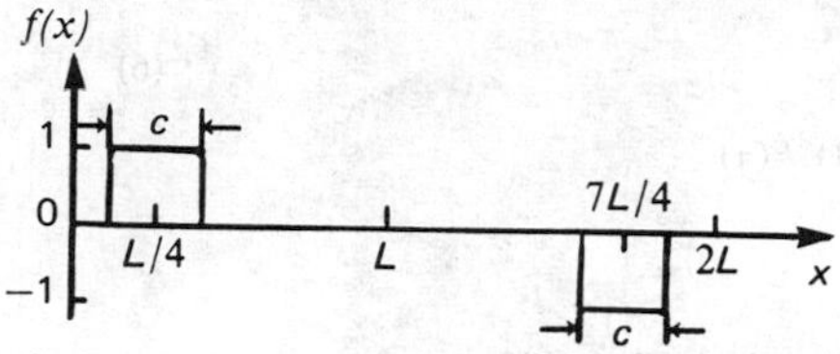

$$f(x) = \frac{4}{\pi} \sum_{n=1}^{\infty} \frac{1}{n} \sin \frac{n\pi}{4} \sin n\pi a \sin \frac{n\pi x}{L}; \left(a = \frac{c}{2L}\right)$$

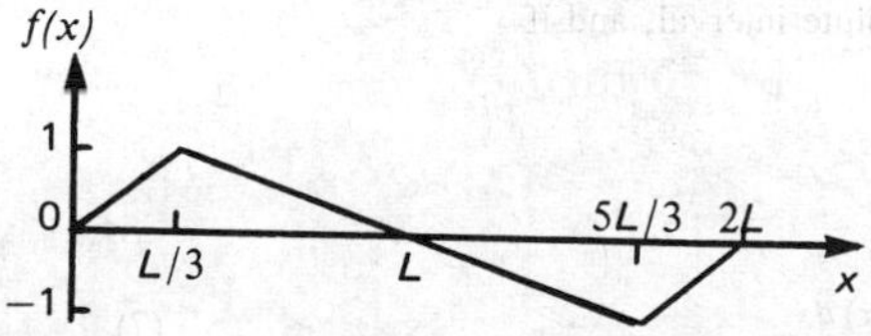

$$f(x) = \frac{9}{\pi^2} \sum_{n=1}^{\infty} \frac{1}{n^2} \sin \frac{n\pi}{3} \sin \frac{n\pi x}{L}; \left(a = \frac{c}{2L}\right)$$

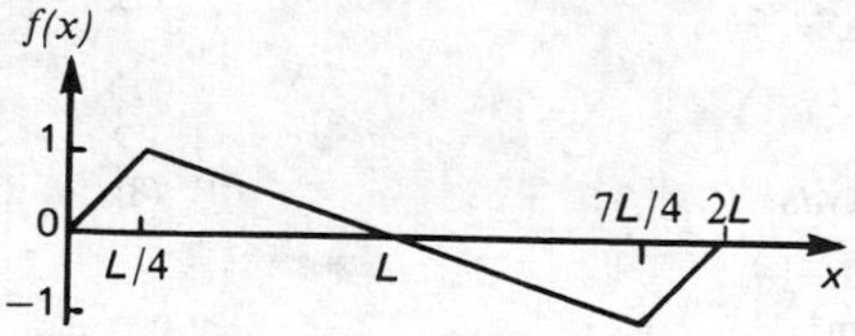

$$f(x) = \frac{32}{3\pi^2} \sum_{n=1}^{\infty} \frac{1}{n^2} \sin \frac{n\pi}{4} \sin \frac{n\pi x}{L}; \left(a = \frac{c}{2L}\right)$$

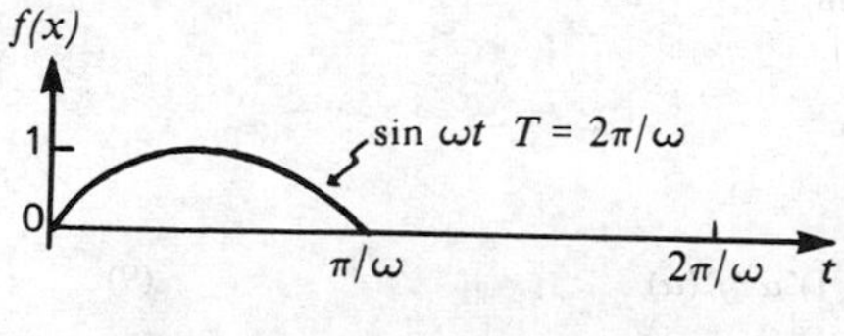

$$f(x) = \frac{1}{\pi} + \frac{1}{2} \sin \omega t - \frac{2}{\pi} \sum_{n=2,4,6\ldots} \frac{1}{n^2-1} \cos n\omega t$$

Extracted from graphs and formulas, pages 372, 373, Differential Equations in Engineering Problems, Salvadori and Schwarz, published by Prentice-Hall, Inc., 1954.

THE FOURIER TRANSFORMS*

R. E. Gaskell

For a piecewise continuous function $F(x)$ over a finite interval $0 \leqq x \leqq \pi$, the *finite Fourier cosine transform* of $F(x)$ is

$$f_c(n) = \int_0^{\pi} F(x) \cos nx \, dx \quad (n = 0, 1, 2, \ldots) \tag{1}$$

If x ranges over the interval $0 \leqq x \leqq L$, the substitution $x' = \pi x/L$ allows the use of this definition, also. The inverse transform is written.

$$\bar{F}(x) = \frac{1}{\pi} f_c(0) - \frac{2}{\pi} \sum_{n=1}^{\infty} f_c(n) \cos nx \quad (0 < x < \pi) \tag{2}$$

where $\bar{F}(x) = \dfrac{[F(x) + 0) + F(x - 0)]}{2}$. We observe that $\bar{F}(x) = F(x) =$ at points of continuity. The formula

$$\begin{aligned} f_c^{(2)}(n) &= \int_0^{\pi} F''(x) \cos nx \, dx \\ &= -n^2 f_c(n) - F'(0) + (-1)^n F'(\pi) \end{aligned} \tag{3}$$

makes the finite Fourier cosine transform useful in certain boundary value problems.

Analogously, the *finite Fourier sine transform* of $F(x)$ is

$$f_s(n) = \int_0^{\pi} F(x) \sin nx \, dx \quad (n = 1, 2, 3, \ldots) \tag{4}$$

and

$$\bar{F}(x) = \frac{2}{\pi} \sum_{n=1}^{\infty} f_s(n) \sin nx \quad (0 < x < \pi) \tag{5}$$

Corresponding to (3) we have

$$\begin{aligned} f_s^{(2)}(n) &= \int_0^{\pi} F''(x) \sin nx \, dx \\ &= -n^2 f_s(n) - nF(0) - n(-1)^n F(\pi) \end{aligned} \tag{6}$$

Fourier Transforms

If $F(x)$ is defined for $x \geqq 0$ and is piecewise continuous over any finite interval, and if

$$\int_0^{\infty} F(x) \, dx$$

is absolutely convergent, then

$$f_c(\alpha) = \sqrt{\frac{2}{\pi}} \int_0^{\infty} F(x) \cos (\alpha x) \, dx \tag{7}$$

is the *Fourier cosine transform* of $F(x)$. Furthermore.

$$\bar{F}(x) = \sqrt{\frac{2}{\pi}} \int_0^{\infty} f_c(\alpha) \cos (\alpha x) \, d\alpha \tag{8}$$

If $\lim\limits_{x \to \infty} \dfrac{d^n F}{dx^n} = 0$, an important property of the Fourier cosine transform

$$\begin{aligned} f_c^{(2r)}(\alpha) &= \sqrt{\frac{2}{\pi}} \int_0^{\infty} \left(\frac{d^{2r} F}{dx^{2r}} \right) \cos (\alpha x) \, dx \\ &= -\sqrt{\frac{2}{\pi}} \sum_{n=0}^{r-1} (-1)^n a_{2r-2n-1} \alpha^{2n} + (-1)^r \alpha^{2r} f_c(\alpha) \end{aligned} \tag{9}$$

where $\lim\limits_{x \to \infty} \dfrac{d^r F}{dx^r} = a_r$, makes it useful in the solution of many problems.

Under the same conditions.

$$f_s(\alpha) = \sqrt{\frac{2}{\pi}} \int_0^{\infty} F(x) \sin (\alpha x) \, dx \tag{10}$$

* From Beyer, W. H., Ed., *CRC Handbook of Mathematical Sciences,* 5th ed., CRC Press, Boca Raton, 1978, 592—598. With permission.

defines the *Fourier sine transform* of $F(x)$, and

$$\bar{F}(x) = \sqrt{\frac{2}{\pi}} \int_0^\infty f_s(\alpha) \sin(\alpha x)\, d\alpha \qquad (11)$$

Corresponding to (9) we have

$$f_s^{(2r)}(\alpha) = \sqrt{\frac{2}{\pi}} \int_0^\infty \frac{d^{2r}F}{dx^{2r}} \sin(\alpha x)\, dx$$

$$= -\sqrt{\frac{2}{\pi}} \sum_{n=1}^{r} (-1)^n \alpha^{2n-1} a_{2r-2n} + (-1)^{r-1} \alpha^{2r} f_s(\alpha) \qquad (12)$$

Similarly, if $F(x)$ is defined for $-\infty < x < \infty$, and if $\int_{-\infty}^{\infty} F(x)\, dx$ is absolutely convergent, then

$$f(\alpha) = \frac{1}{\sqrt{2\pi}} \int_{-\infty}^{\infty} F(x) e^{i\alpha x}\, dx \qquad (13)$$

is the *Fourier transform* of $F(x)$, and

$$\bar{F}(x) = \frac{1}{\sqrt{2\pi}} \int_{-\infty}^{\infty} f(\alpha) e^{-i\alpha x}\, d\alpha \qquad (14)$$

Also, if

$$\lim_{|x|\to\infty} \left| \frac{d^n F}{dx^n} \right| = 0 \quad (n = 1, 2, \ldots, r-1)$$

then

$$f^{(r)}(\alpha) = \frac{1}{\sqrt{2\pi}} \int_{-\infty}^{\infty} F^{(r)}(x) e^{i\alpha x}\, dx = (-i\alpha)^r f(\alpha) \qquad (15)$$

Finite Sine Transforms

	$f_s(n)$	$F(x)$
1	$f_s(n) = \int_0^\pi F(x) \sin nx\, dx \; (n = 1, 2, \cdots)$	$F(x)$
2	$(-1)^{n+1} f_s(n)$	$F(\pi - x)$
3	$\frac{1}{n}$	$\frac{\pi - x}{\pi}$
4	$\frac{(-1)^{n+1}}{n}$	$\frac{x}{\pi}$
5	$\frac{1 - (-1)^n}{n}$	1
6	$\frac{2}{n^2} \sin \frac{n\pi}{2}$	$\begin{cases} x & \text{when } 0 < x < \pi/2 \\ \pi - x & \text{when } \pi/2 < x < \pi \end{cases}$
7	$\frac{(-1)^{n+1}}{n^3}$	$\frac{x(\pi^2 - x^2)}{6\pi}$
8	$\frac{1 - (-1)^n}{n^3}$	$\frac{x(\pi - x)}{2}$
9	$\frac{\pi^2(-1)^{n-1}}{n} - \frac{2[1 - (-1)^n]}{n^3}$	x^2
10	$\pi(-1)^n \left(\frac{6}{n^3} - \frac{\pi^2}{n} \right)$	x^3
11	$\frac{n}{n^2 + c^2} [1 - (-1)^n e^{c\pi}]$	e^{cx}
12	$\frac{n}{n^2 + c^2}$	$\frac{\sinh c(\pi - x)}{\sinh c\pi}$
13	$\frac{n}{n^2 - k^2} \; (k \neq 0, 1, 2, \cdots)$	$\frac{\sin k(\pi - x)}{\sin k\pi}$
14	$\begin{cases} \frac{\pi}{2} & \text{when } n = m \\ 0 & \text{when } n \neq m \end{cases} \; (m = 1, 2, \cdots)$	$\sin mx$

	$f_s(n)$	$F(x)$
15	$\frac{n}{n^2-k^2}[1-(-1)^n \cos k\pi]$ $(k \neq 1, 2, \cdots)$	$\cos kx$
16	$\begin{cases} \frac{n}{n^2-m^2}[1-(-1)^{n+m}] & \text{when } n \neq m = 1, 2, \cdots \\ 0 & \text{when } n = m \end{cases}$	$\cos mx$
17	$\frac{n}{(n^2-k^2)^2}$ $(k \neq 0, 1, 2, \cdots)$	$\frac{\pi \sin kx}{2k \sin^2 k\pi} - \frac{x \cos k(\pi - x)}{2k \sin k\pi}$
18	$\frac{b^n}{n}$ $(\lvert b \rvert \leq 1)$	$\frac{2}{\pi} \arctan \frac{b \sin x}{1 - b \cos x}$
19	$\frac{1-(-1)^n}{n} b^n$ $(\lvert b \rvert \leq 1)$	$\frac{2}{\pi} \arctan \frac{2b \sin x}{1 - b^2}$

Finite Cosine Transforms

	$f_c(n)$	$F(x)$
1	$f_c(n) = \int_0^{\pi} F(x) \cos nx \, dx$ $(n = 0, 1, 2, \cdots)$	$F(x)$
2	$(-1)^n f_c(n)$	$F(\pi - x)$
3	0 when $n = 1, 2, \cdots$; $f_c(0) = \pi$	1
4	$\frac{2}{n} \sin \frac{n\pi}{2}$; $f_c(0) = 0$	$\begin{cases} 1 & \text{when } 0 < x < \pi/2 \\ -1 & \text{when } \pi/2 < x < \pi \end{cases}$
5	$-\frac{1-(-1)^n}{n^2}$; $f_c(0) = \frac{\pi^2}{2}$	x
6	$\frac{(-1)^n}{n^2}$; $f_c(0) = \frac{\pi^2}{6}$	$\frac{x^2}{2\pi}$
7	$\frac{1}{n^2}$; $f_c(0) = 0$	$\frac{(\pi - x)^2}{2\pi} - \frac{\pi}{6}$
8	$3\pi^2 \frac{(-1)^n}{n^2} - 6 \frac{1-(-1)^n}{n^4}$; $f_c(0) = \frac{\pi^4}{4}$	x^3
9	$\frac{(-1)^n e^c \pi - 1}{n^2 + c^2}$	$\frac{1}{c} e^{cx}$
10	$\frac{1}{n^2 + c^2}$	$\frac{\cosh c(\pi - x)}{c \sinh c\pi}$
11	$\frac{k}{n^2 - k^2}[(-1)^n \cos \pi k - 1]$ $(k \neq 0, 1, 2, \cdots)$	$\sin kx$
12	$\frac{(-1)^{n+m} - 1}{n^2 - m^2}$; $f_c(m) = 0$ $(m = 1, 2, \cdots)$	$\frac{1}{m} \sin mx$
13	$\frac{1}{n^2 - k^2}$ $(k \neq 0, 1, 2, \cdots)$	$-\frac{\cos k(\pi - x)}{k \sin k\pi}$
14	0 when $n = 1, 2, \cdots$; $f_c(m) = \frac{\pi}{2}$ $(m = 1, 2, \cdots)$	$\cos mx$

Fourier Sine Transforms*

	$F(x)$	$f_s(\alpha)$
1	$\begin{cases} 1 & (0 < x < a) \\ 0 & (x > a) \end{cases}$	$\sqrt{\frac{2}{\pi}} \left[\frac{1 - \cos \alpha}{\alpha}\right]$
2	x^{p-1} $(0 < p < 1)$	$\sqrt{\frac{2}{\pi}} \frac{\Gamma(p)}{\alpha^p} \sin \frac{p\pi}{2}$
3	$\begin{cases} \sin x & (0 < x < a) \\ 0 & (x > a) \end{cases}$	$\frac{1}{\sqrt{2\pi}} \left[\frac{\sin[a(1-\alpha)]}{1-\alpha} - \frac{\sin[a(1+\alpha)]}{1+\alpha}\right]$
4	e^{-x}	$\sqrt{\frac{2}{\pi}} \left[\frac{\alpha}{1+\alpha^2}\right]$
5	$xe^{-x^2/2}$	$\alpha e^{-\alpha^2/2}$
6	$\cos \frac{x^2}{2}$	$\sqrt{2}\left[\sin \frac{\alpha^2}{2} C\left(\frac{\alpha^2}{2}\right) - \cos \frac{\alpha^2}{2} S\left(\frac{\alpha^2}{2}\right)\right]^*$

	$F(x)$	$f_s(\alpha)$
7	$\sin \frac{x^2}{2}$	$\sqrt{2}\left[\cos \frac{\alpha^2}{2}\, C\left(\frac{\alpha^2}{2}\right) + \sin \frac{\alpha^2}{2}\, S\left(\frac{\alpha^2}{2}\right)\right]$*

* $C(y)$ and $S(y)$ are the Fresnel integrals

$$C(y) = \frac{1}{\sqrt{2\pi}} \int_0^y \frac{1}{\sqrt{t}} \cos t \, dt,$$

$$S(y) = \frac{1}{\sqrt{2\pi}} \int_0^y \frac{1}{\sqrt{t}} \sin t \, dt$$

* More extensive tables of the Fourier sine and cosine transforms can be found in Fritz Oberhettinger, *Tabellen zur-Fourier Transformation*, Springer, 1957.

Fourier Cosine Transforms

	$F(x)$	$f_c(\alpha)$
1	$\begin{cases} 1 & (0 < x < a) \\ 0 & (x > a) \end{cases}$	$\sqrt{\frac{2}{\pi}} \frac{\sin a\alpha}{\alpha}$
2	$x^{p-1} \quad (0 < p < 1)$	$\sqrt{\frac{2}{\pi}} \frac{\Gamma(p)}{\alpha^p} \cos \frac{p\pi}{2}$
3	$\begin{cases} \cos x & (0 < x < a) \\ 0 & (x > a) \end{cases}$	$\frac{1}{\sqrt{2\pi}}\left[\frac{\sin[a(1-\alpha)]}{1-\alpha} + \frac{\sin[a(1+\alpha)]}{1+\alpha}\right]$
4	e^{-x}	$\sqrt{\frac{2}{\pi}}\left(\frac{1}{1+\alpha^2}\right)$
5	$e^{-x^2/2}$	$e^{-\alpha^2/2}$
6	$\cos \frac{x^2}{2}$	$\cos\left(\frac{\alpha^2}{2} - \frac{\pi}{4}\right)$
7	$\sin \frac{x^2}{2}$	$\cos\left(\frac{\alpha^2}{2} + \frac{\pi}{4}\right)$

Fourier Transforms*

	$F(x)$	$f(\alpha)$
1	$\frac{\sin ax}{x}$	$\begin{cases} \sqrt{\frac{\pi}{2}} & \lvert\alpha\rvert < a \\ 0 & \lvert\alpha\rvert > a \end{cases}$
2	$\begin{cases} e^{iwx} & (p < x < q) \\ 0 & (x < p, x > q) \end{cases}$	$\frac{i}{\sqrt{2\pi}} \frac{e^{ip(w+\alpha)} - e^{iq(w+\alpha)}}{(w+\alpha)}$
3	$\begin{cases} e^{-cx+iwx} & (x > 0) \\ 0 & (x < 0) \end{cases} \quad (c > 0)$	$\frac{i}{\sqrt{2\pi}(w + \alpha + ic)}$
4	$e^{-px^2} \quad R(p) > 0$	$\frac{1}{\sqrt{2p}} e^{-\alpha^2/4p}$
5	$\cos px^2$	$\frac{1}{\sqrt{2p}} \cos\left[\frac{\alpha^2}{4p} - \frac{\pi}{4}\right]$
6	$\sin px^2$	$\frac{1}{\sqrt{2p}} \cos\left[\frac{\alpha^2}{4p} + \frac{\pi}{4}\right]$
7	$\lvert x\rvert^{-p} \quad (0 < p < 1)$	$\sqrt{\frac{2}{\pi}} \frac{\Gamma(1-p)\sin\frac{p\pi}{2}}{\lvert\alpha\rvert^{(1-p)}}$
8	$\frac{e^{-a\lvert x\rvert}}{\sqrt{\lvert x\rvert}}$	$\frac{\sqrt{\sqrt{(a^2+\alpha^2)} + a}}{\sqrt{a^2+\alpha^2}}$
9	$\frac{\cosh ax}{\cosh \pi x} \quad (-\pi < a < \pi)$	$\sqrt{\frac{2}{\pi}} \frac{\cos\frac{a}{2}\cosh\frac{\alpha}{2}}{\cosh\alpha + \cos a}$
10	$\frac{\sinh ax}{\sinh \pi x} \quad (-\pi < a < \pi)$	$\frac{1}{\sqrt{2\pi}} \frac{\sin a}{\cosh\alpha + \cos a}$
11	$\begin{cases} \frac{1}{\sqrt{a^2 - x^2}} & (\lvert x\rvert < a) \\ 0 & (\lvert x\rvert > a) \end{cases}$	$\sqrt{\frac{\pi}{2}} J_0(a\alpha)$

	$F(x)$	$f(\alpha)$
12	$\dfrac{\sin[b\sqrt{a^2+x^2}]}{\sqrt{a^2+x^2}}$	$\begin{cases} 0 & (\lvert\alpha\rvert > b) \\ \sqrt{\frac{\pi}{2}}\,J_0(a\sqrt{b^2-\alpha^2}) & (\lvert\alpha\rvert < b)\end{cases}$
13	$\begin{cases} P_n(x) & (\lvert x\rvert < 1) \\ 0 & (\lvert x\rvert > 1)\end{cases}$	$\dfrac{i^n}{\sqrt{\alpha}}\,J_{n+\frac{1}{2}}(\alpha)$
14	$\begin{cases} \dfrac{\cos[b\sqrt{a^2-x^2}]}{\sqrt{a^2-x^2}} & (\lvert x\rvert < a) \\ 0 & (\lvert x\rvert > a)\end{cases}$	$\sqrt{\frac{\pi}{2}}\,J_0(a\sqrt{a^2+b^2})$
15	$\begin{cases} \dfrac{\cosh[b\sqrt{a^2-x^2}]}{\sqrt{a^2-x^2}} & (\lvert x\rvert < a) \\ 0 & (\lvert x\rvert > a)\end{cases}$	$\sqrt{\frac{\pi}{2}}\,J_0(a\sqrt{\alpha^2-b^2})$

* More extensive tables of Fourier transforms can be found in W. Magnus and F. Oberhettinger, *Formulas and Theorems of the Special Functions of Mathematical Physics.* Chelsea, 1949, 116—120.

The following functions appear among the entries of the tables on transforms.

Function	Definition	Name
$Ei(x)$	$\int_{-\infty}^{x} \frac{e^v}{v}\,dv$; or sometimes defined as $-Ei(-x) = \int_x^{\infty} \frac{e^{-v}}{v}\,dv$	Sine, Cosine, and Exponential Integral tables pages 548–556
$Si(x)$	$\int_0^x \frac{\sin v}{v}\,dv$	Sine, Cosine, and Exponential Integral tables pages 548–556
$Ci(x)$	$\int_{\infty}^{x} \frac{\cos v}{v}\,dv$; or sometimes defined as negative of this integral	Sine, Cosine, and Exponential Integral tables pages 548–546
$erf(x)$	$\frac{2}{\sqrt{\pi}}\int_0^x e^{-v^2}\,dv$	Error function
$erfc(x)$	$1 - erf(x) = \frac{2}{\sqrt{\pi}}\int_x^{\infty} e^{-v^2}\,dv$	Complementary function to error function
$L_n(x)$	$\frac{e^x}{n!}\frac{d^n}{dx^n}(x^n e^{-x}),\ n = 0, 1, \cdots$	Laguerre polynomial of degree n

SERIES EXPANSION

The expression in parentheses following certain of the series indicates the region of convergence. If not otherwise indicated it is to be understood that the series converges for all finite values of x.

BINOMIAL

$$(x+y)^n = x^n + nx^{n-1}y + \frac{n(n-1)}{2!}x^{n-2}y^2 + \frac{n(n-1)(n-2)}{3!}x^{n-3}y^3 + \cdots \quad (y^2 < x^2)$$

$$(1 \pm x)^n = 1 \pm nx + \frac{n(n-1)x^2}{2!} \pm \frac{n(n-1)(n-2)x^3}{3!} + \cdots \text{ etc.} \quad (x^2 < 1)$$

$$(1 \pm x)^{-n} = 1 \mp nx + \frac{n(n+1)x^2}{2!} \mp \frac{n(n+1)(n+2)x^3}{3!} + \cdots \text{ etc.} \quad (x^2 < 1)$$

$$(1 \pm x)^{-1} = 1 \mp x + x^2 \mp x^3 + x^4 \mp x^5 + \cdots \quad (x^2 < 1)$$

$$(1 \pm x)^{-2} = 1 \mp 2x + 3x^2 \mp 4x^3 + 5x^4 \mp 6x^5 + \cdots \quad (x^2 < 1)$$

REVERSION OF SERIES

Let a series be represented by

$$y = a_1x + a_2x^2 + a_3x^3 + a_4x^4 + a_5x^5 + a_6x^6 + \cdots \quad (a_1 \neq 0)$$

to find the coefficients of the series

$$x = A_1y + A_2y^2 + A_3y^3 + A_4y^4 + \cdots$$

$$A_1 = \frac{1}{a_1} \qquad A_2 = -\frac{a_2}{a_1^3} \qquad A_3 = \frac{1}{a_1^5}(2a_2^2 - a_1a_3)$$

$$A_4 = \frac{1}{a_1^7}(5a_1a_2a_3 - a_1^2a_4 - 5a_2^3)$$

$$A_5 = \frac{1}{a_1^9}(6a_1^2a_2a_4 + 3a_1^2a_3^2 + 14a_2^4 - a_1^3a_5 - 21a_1a_2^2a_3)$$

$$A_6 = \frac{1}{a_1^{11}}(7a_1^3a_2a_5 + 7a_1^3a_3a_4 + 84a_1a_2^3a_3 - a_1^4a_6 - 28a_1^2a_2^2a_4 - 28a_1^2a_2a_3^2 - 42a_2^5)$$

$$A_7 = \frac{1}{a_1^{13}}(8a_1^4a_2a_6 + 8a_1^4a_3a_5 + 4a_1^4a_4^2 + 120a_1^2a_2^3a_4 + 180a_1^2a_2^2a_3^2 + 132a_2^6 - a_1^5a_7 - 36a_1^3a_2^2a_5 - 72a_1^3a_2a_3a_4 - 12a_1^3a_3^3 - 330a_1a_2^4a_3)$$

TAYLOR

1. $$f(x) = f(a) + (x-a)f'(a) + \frac{(x-a)^2}{2!}f''(a) + \frac{(x-a)^3}{3!}f'''(a) + \cdots + \frac{(x-a)^n}{n!}f^{(n)}(a) + \cdots \text{ (Taylor's Series)}$$

(Increment form)

2. $$f(x+h) = f(x) + hf'(x) + \frac{h^2}{2!}f''(x) + \frac{h^3}{3!}f'''(x) + \cdots$$
$$= f(h) + xf'(h) + \frac{x^2}{2!}f''(h) + \frac{x^3}{3!}f'''(h) + \cdots$$

3. If $f(x)$ is a function possessing derivatives of all orders throughout the interval $a \leqq x \leqq b$, then there is a value X, with $a < X < b$, such that

$$f(b) = f(a) + (b-a)f'(a) + \frac{(b-a)^2}{2!}f''(a) + \cdots + \frac{(b-a)^{n-1}}{(n-1)!}f^{(n-1)}(a) + \frac{(b-a)^n}{n!}f^{(n)}(X)$$

$$f(a+h) = f(a) + hf'(a) + \frac{h^2}{2!}f''(a) + \cdots + \frac{h^{n-1}}{(n-1)!}f^{(n-1)}(a) + \frac{h^n}{n!}f^{(n)}(a+\theta h), \quad b = a+h,\ 0 < \theta < 1.$$

or

$$f(x) = f(a) + (x-a)f'(a) + \frac{(x-a)^2}{2!}f''(a) + \cdots + (x-a)^{n-1}\frac{f^{(n-1)}(a)}{(n-1)!} + R_n,$$

where

$$R_n = \frac{f^{(n)}[a + \theta\cdot(x-a)]}{n!}(x-a)^n, \quad 0 < \theta < 1.$$

The above forms are known as Taylor's series with the remainder term.

4. *Taylor's series for a function of two variables*

If $\left(h\frac{\partial}{\partial x} + k\frac{\partial}{\partial y}\right)f(x,y) = h\frac{\partial f(x,y)}{\partial x} + k\frac{\partial f(x,y)}{\partial y}$;

$$\left(h\frac{\partial}{\partial x} + k\frac{\partial}{\partial y}\right)^2 f(x,y) = h^2\frac{\partial^2 f(x,y)}{\partial x^2} + 2hk\frac{\partial^2 f(x,y)}{\partial x\,\partial y} + k^2\frac{\partial^2 f(x,y)}{\partial y^2}$$

etc., and if $\left(h\frac{\partial}{\partial x} + k\frac{\partial}{\partial y}\right)^n f(x,y)\Big|_{\substack{x=a\\y=b}}$ with the bar and subscripts means that after differentiation we are to replace x by a and y by b,

$$f(a+h, b+k) = f(a,b) + \left(h\frac{\partial}{\partial x} + k\frac{\partial}{\partial y}\right)f(x,y)\Big|_{\substack{x=a\\y=b}} + \cdots + \frac{1}{n!}\left(h\frac{\partial}{\partial x} + k\frac{\partial}{\partial y}\right)^n f(x,y)\Big|_{\substack{x=a\\y=b}} + \cdots$$

MACLAURIN

$$f(x) = f(0) + xf'(0) + \frac{x^2}{2!}f''(0) + \frac{x^3}{3!}f'''(0) + \cdots + x^{n-1}\frac{f^{(n-1)}(0)}{(n-1)!} + R_n,$$

where

$$R_n = \frac{x^n f^{(n)}(\theta x)}{n!}, \quad 0 < \theta < 1.$$

EXPONENTIAL

$$e = 1 + \frac{1}{1!} + \frac{1}{2!} + \frac{1}{3!} + \frac{1}{4!} + \cdots$$

$$e^x = 1 + x + \frac{x^2}{2!} + \frac{x^3}{3!} + \frac{x^4}{4!} + \cdots \qquad \text{(all real values of } x\text{)}$$

$$a^x = 1 + x\log_e a + \frac{(x\log_e a)^2}{2!} + \frac{(x\log_e a)^3}{3!} + \cdots$$

$$e^x = e^a\left[1 + (x-a) + \frac{(x-a)^2}{2!} + \frac{(x-a)^3}{3!} + \cdots\right]$$

LOGARITHMIC

$$\log_e x = \frac{x-1}{x} + \frac{1}{2}\left(\frac{x-1}{x}\right)^2 + \frac{1}{3}\left(\frac{x-1}{x}\right)^3 + \cdots \qquad (x > \tfrac{1}{2})$$

$$\log_e x = (x-1) - \tfrac{1}{2}(x-1)^2 + \tfrac{1}{3}(x-1)^3 - \cdots \qquad (2 \geq x > 0)$$

$$\log_e x = 2\left[\frac{x-1}{x+1} + \frac{1}{3}\left(\frac{x-1}{x+1}\right)^3 + \frac{1}{5}\left(\frac{x-1}{x+1}\right)^5 + \cdots\right] \qquad (x > 0)$$

$$\log_e(1+x) = x - \tfrac{1}{2}x^2 + \tfrac{1}{3}x^3 - \tfrac{1}{4}x^4 + \cdots \qquad (-1 < x \leq 1)$$

$$\log_e(n+1) - \log_e(n-1) = 2\left[\frac{1}{n} + \frac{1}{3n^3} + \frac{1}{5n^5} + \cdots\right]$$

$$\log_e(a+x) = \log_e a + 2\left[\frac{x}{2a+x} + \frac{1}{3}\left(\frac{x}{2a+x}\right)^3 + \frac{1}{5}\left(\frac{x}{2a+x}\right)^5 + \cdots\right]$$

$$(a > 0,\ -a < x < +\infty)$$

$$\log_e \frac{1+x}{1-x} = 2\left[x + \frac{x^3}{3} + \frac{x^5}{5} + \cdots + \frac{x^{2n-1}}{2n-1} + \cdots\right], \qquad -1 < x < 1$$

$$\log_e x = \log_e a + \frac{(x-a)}{a} - \frac{(x-a)^2}{2a^2} + \frac{(x-a)^3}{3a^3} - + \cdots, \qquad 0 < x \leqq 2a$$

TRIGONOMETRIC

$$\sin x = x - \frac{x^3}{3!} + \frac{x^5}{5!} - \frac{x^7}{7!} + \cdots \qquad \text{(all real values of } x)$$

$$\cos x = 1 - \frac{x^2}{2!} + \frac{x^4}{4!} - \frac{x^6}{6!} + \cdots \qquad \text{(all real values of } x)$$

$$\tan x = x + \frac{x^3}{3} + \frac{2x^5}{15} + \frac{17x^7}{315} + \frac{62x^9}{2835} + \cdots + \frac{(-1)^{n-1}2^{2n}(2^{2n}-1)B_{2n}}{(2n)!}x^{2n-1} + \cdots,$$

$$\left[x^2 < \frac{\pi^2}{4}, \text{ and } B_n \text{ represents the } n\text{'th Bernoulli number.}\right]$$

$$\cot x = \frac{1}{x} - \frac{x}{3} - \frac{x^3}{45} - \frac{2x^5}{945} - \frac{x^7}{4725} - \cdots - \frac{(-1)^{n+1}2^{2n}}{(2n)!}B_{2n}x^{2n-1} - \cdots,$$

$$[x^2 < \pi^2, \text{ and } B_n \text{ represents the } n\text{'th Bernoulli number.}]$$

$$\sec x = 1 + \frac{x^2}{2} + \frac{5}{24}x^4 + \frac{61}{720}x^6 + \frac{277}{8064}x^8 + \cdots + \frac{(-1)^n}{(2n)!}E_{2n}x^{2n} + \cdots,$$

$$\left[x^2 < \frac{\pi^2}{4}, \text{ and } E_n \text{ represents the } n\text{'th Euler number.}\right]$$

$$\csc x = \frac{1}{x} + \frac{x}{6} + \frac{7}{360}x^3 + \frac{31}{15{,}120}x^5 + \frac{127}{604{,}800}x^7 + \cdots$$

$$+ \frac{(-1)^{n+1}2(2^{2n-1}-1)}{(2n)!}B_{2n}x^{2n-1} + \cdots,$$

$$[x^2 < \pi^2, \text{ and } B_n \text{ represents } n\text{'th Bernoulli number.}]$$

$$\sin x = x\left(1 - \frac{x^2}{\pi^2}\right)\left(1 - \frac{x^2}{2^2\pi^2}\right)\left(1 - \frac{x^2}{3^2\pi^2}\right)\cdots \qquad (x^2 < \infty)$$

$$\cos x = \left(1 - \frac{4x^2}{\pi^2}\right)\left(1 - \frac{4x^2}{3^2\pi^2}\right)\left(1 - \frac{4x^2}{5^2\pi^2}\right)\cdots \qquad (x^2 < \infty)$$

$$\sin^{-1} x = x + \frac{x^3}{2\cdot 3} + \frac{1\cdot 3}{2\cdot 4\cdot 5}x^5 + \frac{1\cdot 3\cdot 5}{2\cdot 4\cdot 6\cdot 7}x^7 + \cdots \qquad \left(x^2 < 1,\ -\frac{\pi}{2} < \sin^{-1} x < \frac{\pi}{2}\right)$$

$$\cos^{-1} x = \frac{\pi}{2} - \left(x + \frac{x^3}{2\cdot 3} + \frac{1\cdot 3}{2\cdot 4\cdot 5}x^5 + \frac{1\cdot 3\cdot 5x^7}{2\cdot 4\cdot 6\cdot 7} + \cdots\right) \qquad (x^2 < 1, 0 < \cos^{-1} x < \pi)$$

$$\tan^{-1} x = x - \frac{x^3}{3} + \frac{x^5}{5} - \frac{x^7}{7} + \cdots \qquad (x^2 < 1)$$

$$\tan^{-1} x = \frac{\pi}{2} - \frac{1}{x} + \frac{1}{3x^3} - \frac{1}{5x^5} + \frac{1}{7x^7} - \cdots \qquad (x > 1)$$

$$\tan^{-1} x = -\frac{\pi}{2} - \frac{1}{x} + \frac{1}{3x^3} - \frac{1}{5x^5} + \frac{1}{7x^7} - \cdots \qquad (x < -1)$$

$$\cot^{-1} x = \frac{\pi}{2} - x + \frac{x^3}{3} - \frac{x^5}{5} + \frac{x^7}{7} - \cdots \qquad (x^2 < 1)$$

$$\log_e \sin x = \log_e x - \frac{x^2}{6} - \frac{x^4}{180} - \frac{x^6}{2835} - \cdots \qquad (x^2 < \pi^2)$$

$$\log_e \cos x = -\frac{x^2}{2} - \frac{x^4}{12} - \frac{x^6}{45} - \frac{17x^8}{2520} - \cdots \qquad \left(x^2 < \frac{\pi^2}{4}\right)$$

$$\log_e \tan x = \log_e x + \frac{x^2}{3} + \frac{7x^4}{90} + \frac{62x^6}{2835} + \cdots \qquad \left(x^2 < \frac{\pi^2}{4}\right)$$

$$e^{\sin x} = 1 + x + \frac{x^2}{2!} - \frac{3x^4}{4!} - \frac{8x^5}{5!} - \frac{3x^6}{6!} + \frac{56x^7}{7!} + \cdots$$

$$e^{\cos x} = e\left(1 - \frac{x^2}{2!} + \frac{4x^4}{4!} - \frac{31x^6}{6!} + \cdots\right)$$

$$e^{\tan x} = 1 + x + \frac{x^2}{2!} + \frac{3x^3}{3!} + \frac{9x^4}{4!} + \frac{37x^5}{5!} + \cdots \qquad \left(x^2 < \frac{\pi^2}{4}\right)$$

$$\sin x = \sin a + (x - a)\cos a - \frac{(x-a)^2}{2!}\sin a - \frac{(x-a)^3}{3!}\cos a + \frac{(x-a)^4}{4!}\sin a + \cdots$$

VECTOR ANALYSIS

Definitions

Any quantity which is completely determined by its magnitude is called a *scalar*. Examples of such are mass, density, temperature, etc. Any quantity which is completely determined by its magnitude and direction is called a *vector*. Examples of such are velocity, acceleration, force, etc. A vector quantity is represented by a directed line segment, the length of which represents the magnitude of the vector. A vector quantity is usually represented by a boldfaced letter such as $\mathbf{V}$. Two vectors $\mathbf{V}_1$ and $\mathbf{V}_2$ are equal to one another if they have equal magnitudes and are acting in the same directions. A negative vector, written as $-\mathbf{V}$, is one which acts in the opposite direction to $\mathbf{V}$, but is of equal magnitude to it. If we represent the magnitude of $\mathbf{V}$ by v, we write $|\mathbf{V}| = v$. A vector parallel to $\mathbf{V}$, but equal to the reciprocal of its magnitude is written as $\mathbf{V}^{-1}$ or as $\frac{1}{\mathbf{V}}$.

The *unit vector* $\frac{\mathbf{V}}{v}$ $(v \neq 0)$ is that vector which has the same direction as $\mathbf{V}$, but has a magnitude of unity (sometimes represented as $\mathbf{V}_0$ or $\hat{\mathbf{v}}$).

Vector Algebra

The vector sum of $\mathbf{V}_1$ and $\mathbf{V}_2$ is represented by $\mathbf{V}_1 + \mathbf{V}_2$. The vector sum of $\mathbf{V}_1$ and $-\mathbf{V}_2$, or the difference of the vector $\mathbf{V}_2$ from $\mathbf{V}_1$ is represented by $\mathbf{V}_1 - \mathbf{V}_2$.

If r is a scalar, then $r\mathbf{V} = \mathbf{V}r$, and represents a vector r times the magnitude of $\mathbf{V}$, in the same direction as $\mathbf{V}$ if r is positive, and in the opposite direction if r is negative. If r and s are scalars, $\mathbf{V}_1$, $\mathbf{V}_2$, $\mathbf{V}_3$, vectors, then the following rules of scalars and vectors hold:

$$\mathbf{V}_1 + \mathbf{V}_2 = \mathbf{V}_2 + \mathbf{V}_1$$

$$(r + s)\mathbf{V}_1 = r\mathbf{V}_1 + s\mathbf{V}_1; \qquad r(\mathbf{V}_1 + \mathbf{V}_2) = r\mathbf{V}_1 + r\mathbf{V}_2$$

$$\mathbf{V}_1 + (\mathbf{V}_2 + \mathbf{V}_3) = (\mathbf{V}_1 + \mathbf{V}_2) + \mathbf{V}_3 = \mathbf{V}_1 + \mathbf{V}_2 + \mathbf{V}_3$$

Vectors in Space

A plane is described by two distinct vectors $\mathbf{V}_1$ and $\mathbf{V}_2$. Should these vectors not intersect each other, then one is displaced parallel to itself until they do (fig. 1.) Any other vector $\mathbf{V}$ lying in this plane is given by

$$\mathbf{V} = r\mathbf{V}_1 + s\mathbf{V}_2$$

A *position vector* specifies the position in space of a point relative to a fixed origin. If therefore $\mathbf{V}_1$ and $\mathbf{V}_2$ are the position vectors of the points A and B, relative to the origin O, then any point P on the line AB has a position vector $\mathbf{V}$ given by

$$\mathbf{V} = r\mathbf{V}_1 + (1 - r)\mathbf{V}_2$$

The scalar "r" can be taken as the parametric representation of P since $r = 0$ implies $P = B$ and $r = 1$ implies $P = A$. (fig. 2). If P divides the line AB in the ratio $r:s$ then

$$\mathbf{V} = \left(\frac{r}{r+s}\right)\mathbf{V}_1 + \left(\frac{s}{r+s}\right)\mathbf{V}_2$$

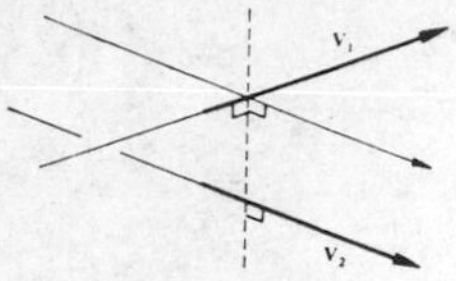

Fig. 1.

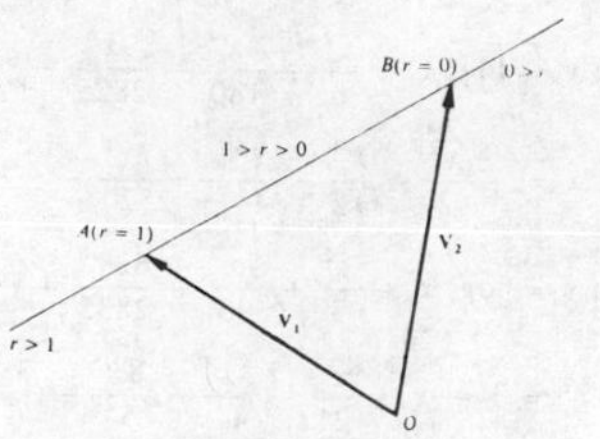

Fig. 2.

The vectors $\mathbf{V}_1$, $\mathbf{V}_2$, $\mathbf{V}_3$, . . . , $\mathbf{V}_n$ are said to be *linearly dependent* if there exist scalars r_1, r_2, r_3, . . . , r_n, not all zero, such that

$$r_1\mathbf{V}_1 + r_2\mathbf{V}_2 + \cdots + r_n\mathbf{V}_n = 0$$

A vector $\mathbf{V}$ is linearly dependent upon the set of vectors $\mathbf{V}_1$, $\mathbf{V}_2$, $\mathbf{V}_3$, . . . , $\mathbf{V}_n$ if

$$\mathbf{V} = r_1\mathbf{V}_1 + r_2\mathbf{V}_2 + r_3\mathbf{V}_3 + \cdots + r_n\mathbf{V}_n$$

Three vectors are linearly dependent if and only if they are co-planar.

All points in space can be uniquely determined by linear dependence upon three *base vectors* i.e. three vectors any one of which is linearly independent of the other two. The simplest set of base vectors are the unit vectors along the coordinate Ox, Oy and Oz axes. These are usually designated by $\mathbf{i}$, $\mathbf{j}$ and $\mathbf{k}$ respectively.

If $\mathbf{V}$ is a vector in space, and a, b and c are the respective magnitudes of the projections of the vector along the axes then

$$\mathbf{V} = a\mathbf{i} + b\mathbf{j} + c\mathbf{k}$$

and

$$v = \sqrt{a^2 + b^2 + c^2}$$

and the direction cosines of $\mathbf{V}$ are

$$\cos\alpha = a/v, \quad \cos\beta = b/v, \quad \cos\gamma = c/v.$$

The law of addition yields

$$\mathbf{V}_1 + \mathbf{V}_2 = (a_1 + a_2)\mathbf{i} + (b_1 + b_2)\mathbf{j} + (c_1 + c_2)\mathbf{k}$$

The Scalar, Dot, or Inner Product of Two Vectors $\mathbf{V}_1$ and $\mathbf{V}_2$

This product is represented as $\mathbf{V}_1 \cdot \mathbf{V}_2$ and is defined to be equal to $v_1v_2\cos\theta$, where θ is the angle from $\mathbf{V}_1$ to $\mathbf{V}_2$, i.e.,

$$\mathbf{V}_1 \cdot \mathbf{V}_2 = v_1v_2\cos\theta$$

The following rules apply for this product:

$$\mathbf{V}_1 \cdot \mathbf{V}_2 = a_1a_2 + b_1b_2 + c_1c_2 = \mathbf{V}_2 \cdot \mathbf{V}_1$$

It should be noted that this verifies that scalar multiplication is commutative.

$$(\mathbf{V}_1 + \mathbf{V}_2) \cdot \mathbf{V}_3 = \mathbf{V}_1 \cdot \mathbf{V}_3 + \mathbf{V}_2 \cdot \mathbf{V}_3$$
$$\mathbf{V}_1 \cdot (\mathbf{V}_2 + \mathbf{V}_3) = \mathbf{V}_1 \cdot \mathbf{V}_2 + \mathbf{V}_1 \cdot \mathbf{V}_3$$

If $\mathbf{V}_1$ is perpendicular to $\mathbf{V}_2$ then $\mathbf{V}_1 \cdot \mathbf{V}_2 = 0$, and if $\mathbf{V}_1$ is parallel to $\mathbf{V}_2$ then $\mathbf{V}_1 \cdot \mathbf{V}_2 = v_1v_2 = rw_1^2$

In particular

$$\mathbf{i} \cdot \mathbf{i} = \mathbf{j} \cdot \mathbf{j} = \mathbf{k} \cdot \mathbf{k} = 1,$$

and

$$\mathbf{i} \cdot \mathbf{j} = \mathbf{j} \cdot \mathbf{k} = \mathbf{k} \cdot \mathbf{i} = 0$$

The Vector or Cross Product of Vectors $\mathbf{V}_1$ and $\mathbf{V}_2$

This product is represented as $\mathbf{V}_1 \times \mathbf{V}_2$ and is defined to be equal to $v_1v_2(\sin\theta)\mathbf{1}$, where θ is the angle from $\mathbf{V}_1$ to $\mathbf{V}_2$ and $\mathbf{1}$ is a unit vector perpendicular to the plane of $\mathbf{V}_1$ and $\mathbf{V}_2$ and so directed that a right-handed screw driven in the direction of $\mathbf{1}$ would carry $\mathbf{V}_1$ into $\mathbf{V}_2$, i.e.,

$$\mathbf{V}_1 \times \mathbf{V}_2 = v_1v_2(\sin\theta)\mathbf{1}$$

and

$$\tan\theta = \frac{|\mathbf{V}_1 \times \mathbf{V}_2|}{\mathbf{V}_1 \cdot \mathbf{V}_2}$$

The following rules apply for vector products:

$$\mathbf{V}_1 \times \mathbf{V}_2 = -\mathbf{V}_2 \times \mathbf{V}_1$$
$$\mathbf{V}_1 \times (\mathbf{V}_2 + \mathbf{V}_3) = \mathbf{V}_1 \times \mathbf{V}_2 + \mathbf{V}_1 \times \mathbf{V}_3$$
$$(\mathbf{V}_1 + \mathbf{V}_2) \times \mathbf{V}_3 = \mathbf{V}_1 \times \mathbf{V}_3 + \mathbf{V}_2 \times \mathbf{V}_3$$
$$\mathbf{V}_1 \times (\mathbf{V}_2 \times \mathbf{V}_3) = \mathbf{V}_2(\mathbf{V}_3 \cdot \mathbf{V}_1 - \mathbf{V}_3(\mathbf{V}_1 \cdot \mathbf{V}_2)$$
$$\mathbf{i} \times \mathbf{i} = \mathbf{j} \times \mathbf{j} = \mathbf{k} \times \mathbf{k} = 0.\mathbf{1} \text{ (zero vector)}$$
$$= \mathbf{0}$$

$$\mathbf{i} \times \mathbf{j} = \mathbf{k}, \qquad \mathbf{j} \times \mathbf{k} = \mathbf{i}, \qquad \mathbf{k} \times \mathbf{i} = \mathbf{j}$$

If $\mathbf{V}_1 = a_1\mathbf{i} + b_1\mathbf{j} + c_1\mathbf{k}$, $\quad \mathbf{V}_2 = a_2\mathbf{i} + b_2\mathbf{j} + c_2\mathbf{k}$, $\quad \mathbf{V}_3 = a_3\mathbf{i} + b_3\mathbf{j} + c_3\mathbf{k}$,

then

$$\mathbf{V}_1 \times \mathbf{V}_2 = \begin{vmatrix} \mathbf{i} & \mathbf{j} & \mathbf{k} \\ a_1 & b_1 & c_1 \\ a_2 & b_2 & c_2 \end{vmatrix} = (b_1c_2 - b_2c_1)\mathbf{i} + (c_1a_2 - c_2a_1)\mathbf{j} + (a_1b_2 - a_2b_1)\mathbf{k}$$

It should be noted that, since $\mathbf{V}_1 \times \mathbf{V}_2 = -\mathbf{V}_2 \times \mathbf{V}_1$, the vector product is not commutative.

Scalar Triple Product

There is only one possible interpretation of the expression $\mathbf{V}_1 \cdot \mathbf{V}_2 \times \mathbf{V}_3$ and that is $\mathbf{V}_1 \cdot (\mathbf{V}_2 \times \mathbf{V}_3)$ which is obviously a scalar.

Further $\mathbf{V}_1 \cdot (\mathbf{V}_2 \times \mathbf{V}_3) = (\mathbf{V}_1 \times \mathbf{V}_2) \cdot \mathbf{V}_3 = \mathbf{V}_2 \cdot (\mathbf{V}_3 \times \mathbf{V}_1)$

$$= \begin{vmatrix} a_1 & b_1 & c_1 \\ a_2 & b_2 & c_2 \\ a_3 & b_3 & c_3 \end{vmatrix}$$

$$= v_1v_2v_3 \cos\phi \sin\theta,$$

Where θ is the angle between $\mathbf{V}_2$ and $\mathbf{V}_3$ and ϕ is the angle between $\mathbf{V}_1$ and the normal to the plane of $\mathbf{V}_2$ and $\mathbf{V}_3$.

This product is called the *scalar triple product* and is written as $[\mathbf{V}_1\mathbf{V}_2\mathbf{V}_3]$.

The determinant indicates that it can be considered as the volume of the parallelepiped whose three determining edges are $\mathbf{V}_1$, $\mathbf{V}_2$ and $\mathbf{V}_3$.

It also follows that cyclic permutation of the subscripts does not change the value of the scalar triple product so that

$$[\mathbf{V}_1\mathbf{V}_2\mathbf{V}_3] = [\mathbf{V}_2\mathbf{V}_3\mathbf{V}_1] = [\mathbf{V}_3\mathbf{V}_1\mathbf{V}_2]$$

$$\text{but}\quad [\mathbf{V}_1\mathbf{V}_2\mathbf{V}_3] = -[\mathbf{V}_2\mathbf{V}_1\mathbf{V}_3] \quad \text{etc.} \qquad \text{and}\quad [\mathbf{V}_1\mathbf{V}_1\mathbf{V}_2] \equiv 0 \text{ etc.}$$

Given three non-coplanar reference vectors $\mathbf{V}_1$, $\mathbf{V}_2$ and $\mathbf{V}_3$, the *reciprocal system is* given by $\mathbf{V}_1^*$, $\mathbf{V}_2^*$ and $\mathbf{V}_3^*$, where

$$1 = v_1v_1^* = v_2v_2^* = v_3v_3^*$$

$$0 = v_1v_2^* = v_1v_3^* = v_2v_1^* \quad \text{etc.}$$

$$\mathbf{V}_1^* = \frac{\mathbf{V}_2 \times \mathbf{V}_3}{[\mathbf{V}_1\mathbf{V}_2\mathbf{V}_3]}, \qquad \mathbf{V}_2^* = \frac{\mathbf{V}_3 \times \mathbf{V}_1}{[\mathbf{V}_1\mathbf{V}_2\mathbf{V}_3]}, \qquad \mathbf{V}_3^* = \frac{\mathbf{V}_1 \times \mathbf{V}_2}{[\mathbf{V}_1\mathbf{V}_2\mathbf{V}_3]}$$

The system $\mathbf{i}$, $\mathbf{j}$, $\mathbf{k}$ is its own reciprocal.

Vector Triple Product

The product $\mathbf{V}_1 \times (\mathbf{V}_2 \times \mathbf{V}_3)$ defines the *vector triple product*. Obviously, in this case, the brackets are vital to the definition.

$$\mathbf{V}_1 \times (\mathbf{V}_2 \times \mathbf{V}_3) = (\mathbf{V}_1 \cdot \mathbf{V}_3)\mathbf{V}_2 - (\mathbf{V}_1 \cdot \mathbf{V}_2)\mathbf{V}_3$$

$$= \begin{vmatrix} \mathbf{i} & \mathbf{j} & \mathbf{k} \\ a_1 & b_1 & c_1 \\ \begin{vmatrix} b_2 & c_2 \\ b_3 & c_3 \end{vmatrix} & \begin{vmatrix} c_2 & a_2 \\ c_3 & a_3 \end{vmatrix} & \begin{vmatrix} a_2 & b_2 \\ a_3 & b_3 \end{vmatrix} \end{vmatrix}$$

i.e. it is a vector, perpendicular to $\mathbf{V}_1$, lying in the plane of $\mathbf{V}_2$, $\mathbf{V}_3$.

Similarly

$$(\mathbf{V}_1 \times \mathbf{V}_2) \times \mathbf{V}_3 = \begin{vmatrix} \mathbf{i} & \mathbf{j} & \mathbf{k} \\ \begin{vmatrix} b_1 & c_1 \\ b_2 & c_2 \end{vmatrix} & \begin{vmatrix} c_1 & a_1 \\ c_2 & a_2 \end{vmatrix} & \begin{vmatrix} a_1 & b_1 \\ a_2 & b_2 \end{vmatrix} \\ a_3 & b_3 & c_3 \end{vmatrix}$$

$$\mathbf{V}_1 \times (\mathbf{V}_2 \times \mathbf{V}_3) + \mathbf{V}_2 \times (\mathbf{V}_3 \times \mathbf{V}_1) + \mathbf{V}_3 \times (\mathbf{V}_1 + \mathbf{V}_2) \equiv 0$$

If $\mathbf{V}_1 \times (\mathbf{V}_2 \times \mathbf{V}_3) = (\mathbf{V}_1 \times \mathbf{V}_2) \times \mathbf{V}_3$ then $\mathbf{V}_1$, $\mathbf{V}_2$, $\mathbf{V}_3$ form an *orthogonal set*. Thus $\mathbf{i}$, $\mathbf{j}$, $\mathbf{k}$ form an orthogonal set.

Geometry of the Plane, Straight Line and Sphere

The position vectors of the fixed points A, B, C, D relative to O are $\mathbf{V}_1$, $\mathbf{V}_2$, $\mathbf{V}_3$, $\mathbf{V}_4$ and the position vector of the variable point P is $\mathbf{V}$.

The vector form of the equation of the straight line through A parallel to $\mathbf{V}_2$ is

$$\mathbf{V} = \mathbf{V}_1 + r\mathbf{V}_2$$

$$\text{or} \quad (\mathbf{V} - \mathbf{V}_1) = r\mathbf{V}_2$$

$$\text{or} \quad (\mathbf{V} - \mathbf{V}_1) \times \mathbf{V}_2 = 0$$

while that of the plane through A perpendicular to $\mathbf{V}_2$ is

$$(\mathbf{V} - \mathbf{V}_1) \cdot \mathbf{V}_2 = 0$$

The equation of the line AB is

$$\mathbf{V} = r\mathbf{V}_1 + (1 - r)\mathbf{V}_2$$

and those of the bisectors of the angles between $\mathbf{V}_1$ and $\mathbf{V}_2$ are

$$\mathbf{V} = r\left(\frac{\mathbf{V}_1}{v} \pm \frac{\mathbf{V}_2}{v_2}\right)$$

$$\text{or} \qquad \mathbf{V} = r(\hat{\mathbf{v}}_1 \pm \hat{\mathbf{v}}_2)$$

The perpendicular from C to the line through A parallel to $\mathbf{V}_2$ has as its equation

$$\mathbf{V} = \mathbf{V}_1 - \mathbf{V}_3 - \hat{\mathbf{v}}_2 \cdot (\mathbf{V}_1 - \mathbf{V}_3)\hat{\mathbf{v}}_2.$$

The condition for the intersection of the two lines,

$$\mathbf{V} = \mathbf{V}_1 + r\mathbf{V}_3$$

$$\text{and} \qquad \mathbf{V} = \mathbf{V}_2 + s\mathbf{V}_4$$

$$\text{is} \qquad [(\mathbf{V}_1 - \mathbf{V}_2)\mathbf{V}_3\mathbf{V}_4] = 0.$$

The common perpendicular to the above two lines is the line of intersection of the two planes

$$[(\mathbf{V} - \mathbf{V}_1)\mathbf{V}_3(\mathbf{V}_3 \times \mathbf{V}_4)] = 0$$

$$\text{and} \qquad [(\mathbf{V} - \mathbf{V}_2)\mathbf{V}_4(\mathbf{V}_3 \times \mathbf{V}_4)] = 0$$

and the length of this perpendicular is

$$\frac{[(\mathbf{V}_1 - \mathbf{V}_2)\mathbf{V}_3\mathbf{V}_4]}{|\mathbf{V}_3 \times \mathbf{V}_4|}.$$

The equation of the line perpendicular to the plane ABC is

$$\mathbf{V} = \mathbf{V}_1 \times \mathbf{V}_2 + \mathbf{V}_2 \times \mathbf{V}_3 + \mathbf{V}_3 \times \mathbf{V}_1$$

and the distance of the plane from the origin is

$$\frac{[\mathbf{V}_1\mathbf{V}_2\mathbf{V}_3]}{|(\mathbf{V}_2 - \mathbf{V}_1) \times (\mathbf{V}_3 - \mathbf{V}_1)|}.$$

In general the vector equation

$$\mathbf{V} \cdot \mathbf{V}_2 = r$$

defines the plane which is perpendicular to $\mathbf{V}_2$, and the perpendicular distance from A to this plane is

$$\frac{r - \mathbf{V}_1 \cdot \mathbf{V}_2}{v_2}.$$

The distance from A, measured along a line parallel to $\mathbf{V}_3$, is

$$\frac{r - \mathbf{V}_1 \cdot \mathbf{V}_2}{\mathbf{V}_2 \cdot \hat{\mathbf{v}}_3} \quad \text{or} \quad \frac{r - \mathbf{V}_1 \cdot \mathbf{V}_2}{v_2 \cos \theta}$$

where θ is the angle beween $\mathbf{V}_2$ and $\mathbf{V}_3$.
(If this plane contains the point C then $r = \mathbf{V}_3 \cdot \mathbf{V}_2$ and if it passes through the origin then $r = 0$.)

Given two planes

$$\mathbf{V} \cdot \mathbf{V}_1 = r$$

$$\mathbf{V} \cdot \mathbf{V}_2 = s$$

then any plane through the line of intersection of these two planes is given by

$$\mathbf{V} \cdot (\mathbf{V}_1 + \lambda\mathbf{V}_2) = r + \lambda s$$

where λ is a scalar parameter. In particular $\lambda = \pm v_1/v_2$ yields the equation of the two planes bisecting the angle between the given planes.

The plane through A parallel to the plane of $\mathbf{V}_2$, $\mathbf{V}_3$ is

$$\mathbf{V} = \mathbf{V}_1 + r\mathbf{V}_2 + s\mathbf{V}_3$$

$$\text{or} \quad (\mathbf{V} - \mathbf{V}_1) \cdot \mathbf{V}_2 \times \mathbf{V}_3 = 0$$

$$\text{or} \quad [\mathbf{V}\mathbf{V}_2\mathbf{V}_3] - [\mathbf{V}_1\mathbf{V}_2\mathbf{V}_3] = 0$$

so that the expansion in rectangular Cartesian coordinates yields

$$\begin{vmatrix} (x - a_1) & (y - b_1) & (z - c_1) \\ a_2 & b_2 & c_2 \\ a_3 & b_3 & c_3 \end{vmatrix} = 0 \qquad (\mathbf{V} \equiv x\mathbf{i} + y\mathbf{j} + z\mathbf{k})$$

which is obviously the usual linear equation in x, y and z.

The plane through AB parallel to $\mathbf{V}_3$ is given by

$$[(\mathbf{V} - \mathbf{V}_1)(\mathbf{V}_1 - \mathbf{V}_2)\mathbf{V}_3] = 0$$

$$\text{or} \quad [\mathbf{V}\mathbf{V}_2\mathbf{V}_3] - [\mathbf{V}\mathbf{V}_1\mathbf{V}_3] - [\mathbf{V}_1\mathbf{V}_2\mathbf{V}_3] = 0$$

The plane through the three points A, B and C is

$$\mathbf{V} = \mathbf{V}_1 + s(\mathbf{V}_2 - \mathbf{V}_1) + t(\mathbf{V}_3 - \mathbf{V}_1)$$

or $\mathbf{V} = r\mathbf{V}_1 + s\mathbf{V}_2 + t\mathbf{V}_3 \quad (r + s + t \equiv 1)$

or $[(\mathbf{V} - \mathbf{V}_1)(\mathbf{V}_1 - \mathbf{V}_2)(\mathbf{V}_2 - \mathbf{V}_3)] = 0$

or $[\mathbf{V}\mathbf{V}_1\mathbf{V}_2] + [\mathbf{V}\mathbf{V}_2\mathbf{V}_3] + [\mathbf{V}\mathbf{V}_3\mathbf{V}_1] - [\mathbf{V}_1\mathbf{V}_2\mathbf{V}_3] = 0$

For four points A, B, C, D to be coplanar, then

$$r\mathbf{V}_1 + s\mathbf{V}_2 + t\mathbf{V}_3 + u\mathbf{V}_4 \equiv 0 \equiv r + s + t + u$$

The following formulae relate to a sphere when the vectors are taken to lie in three dimensional space and to a circle when the space is two dimensional. For a circle in three dimensions take the intersection of the sphere with a plane.

The equation of a sphere with center O and radius OA is

$$\mathbf{V} \cdot \mathbf{V} = v_1^2 \qquad (\text{not } \mathbf{V} = \mathbf{V}_1)$$

or $(\mathbf{V} - \mathbf{V}_1) \cdot (\mathbf{V} + \mathbf{V}_1) = 0$

while that of a sphere with center B radius v_1 is

$$(\mathbf{V} - \mathbf{V}_2) \cdot (\mathbf{V} - \mathbf{V}_2) = v_1^2$$

or $\mathbf{V} \cdot (\mathbf{V} - 2\mathbf{V}_2) = v_1^2 - v_2^2$

If the above sphere passes through the origin then

$$\mathbf{V} \cdot (\mathbf{V} - 2\mathbf{V}_2) = 0$$

(note that in two dimensional polar coordinates this is simply)

$$r = 2a \cdot \cos\theta$$

while in three dimensional Cartesian coordinates it is

$$x^2 + y^2 + z^2 - 2(a_2x + b_2y + c_2x) = 0.$$

The equation of a sphere having the points A and B as the extremities of a diameter is

$$(\mathbf{V} - \mathbf{V}_1) \cdot (\mathbf{V} - \mathbf{V}_2) = 0.$$

The square of the length of the tangent from C to the sphere with center B and radius v_1 is given by

$$(\mathbf{V}_3 - \mathbf{V}_2) \cdot (\mathbf{V}_3 - \mathbf{V}_2) = v_1^2$$

The condition that the plane $\mathbf{V} \cdot \mathbf{V}_3 = s$ is tangential to the sphere $(\mathbf{V} - \mathbf{V}_2) \cdot (\mathbf{V} - \mathbf{V}_2) = v_1^2$ is

$$(s - \mathbf{V}_3 \cdot \mathbf{V}_2) \cdot (s - \mathbf{V}_3 \cdot \mathbf{V}_2) = v_1^2 v_3^2.$$

The equation of the tangent plane at D, on the surface of sphere $(\mathbf{V} - \mathbf{V}_2) \cdot (\mathbf{V} - \mathbf{V}_2) = v_1^2$, is

$$(\mathbf{V} - \mathbf{V}_4) \cdot (\mathbf{V}_4 - \mathbf{V}_2) = 0$$

or $\mathbf{V} \cdot \mathbf{V}_4 - \mathbf{V}_2 \cdot (\mathbf{V} + \mathbf{V}_4) = v_1^2 - v_2^2$

The condition that the two circles $(\mathbf{V} - \mathbf{V}_2) \cdot (\mathbf{V} - \mathbf{V}_2) = v_1^2$ and $(\mathbf{V} - \mathbf{V}_4) \cdot (\mathbf{V} - \mathbf{V}_4) = v_3^2$ intersect orthogonally is clearly

$$(\mathbf{V}_2 - \mathbf{V}_4) \cdot (\mathbf{V}_2 - \mathbf{V}_4) = v_1^2 + v_3^2$$

The polar plane of D with respect to the circle

$$(\mathbf{V} - \mathbf{V}_2) \cdot (\mathbf{V} - \mathbf{V}_2) = v_1^2 \text{ is}$$

$$\mathbf{V} \cdot \mathbf{V}_4 - \mathbf{V}_2 \cdot (\mathbf{V} + \mathbf{V}_4) = v_1^2 - v_2^2$$

Any sphere through the intersection of the two spheres $(\mathbf{V} - \mathbf{V}_2) \cdot (\mathbf{V} - \mathbf{V}_2) = v_1^2$ and $(\mathbf{V} - \mathbf{V}_4) \cdot (\mathbf{V} - \mathbf{V}_4) = v_3^2$ is given by

$$(\mathbf{V} - \mathbf{V}_2) \cdot (\mathbf{V} - \mathbf{V}_2) + \lambda(\mathbf{V} - \mathbf{V}_4) \cdot (\mathbf{V} - \mathbf{V}_4) = v_1^2 + \lambda v_3^2$$

while the radical plane of two such spheres is

$$\mathbf{V} \cdot (\mathbf{V}_2 - \mathbf{V}_4) = -\tfrac{1}{2}(v_1^2 - v_2^2 - v_3^2 + v_4^2)$$

Differentiation of Vectors

If $\mathbf{V}_1 = a_1\mathbf{i} + b_1\mathbf{j} + c_1\mathbf{k}$, and $\mathbf{V}_2 = a_2\mathbf{i} + b_2\mathbf{j} + c_2\mathbf{k}$, and if $\mathbf{V}_1$ and $\mathbf{V}_2$ are functions of the scalar t, then

$$\frac{d}{dt}(\mathbf{V}_1 + \mathbf{V}_2 + \cdots) = \frac{d\mathbf{V}_1}{dt} + \frac{d\mathbf{V}_2}{dt} + \cdots$$

$$\text{where } \frac{d\mathbf{V}_1}{dt} = \frac{da_1}{dt}\mathbf{i} + \frac{db_1}{dt}\mathbf{j} + \frac{dc_1}{dt}\mathbf{k}, \text{ etc.}$$

$$\frac{d}{dt}(\mathbf{V}_1 \cdot \mathbf{V}_2) = \frac{d\mathbf{V}_1}{dt} \cdot \mathbf{V}_2 + \mathbf{V}_1 \cdot \frac{d\mathbf{V}_2}{dt}$$

$$\frac{d}{dt}(\mathbf{V}_1 \times \mathbf{V}_2) = \frac{d\mathbf{V}_1}{dt} \times \mathbf{V}_2 + \mathbf{V}_1 \times \frac{d\mathbf{V}_2}{dt}$$

$$\mathbf{V} \cdot \frac{d\mathbf{V}}{dt} = v \cdot \frac{dv}{dt}$$

In particular, if $\mathbf{V}$ is a vector of constant length then the right hand side of the last equation is identically zero showing that $\mathbf{V}$ is perpendicular to its derivative.

The derivatives of the triple products are

$$\frac{d}{dt}[\mathbf{V}_1\mathbf{V}_2\mathbf{V}_3] = \left[\left(\frac{d\mathbf{V}_1}{dt}\right)\mathbf{V}_2\mathbf{V}_3\right] + \left[\mathbf{V}_1\left(\frac{d\mathbf{V}_2}{dt}\right)\mathbf{V}_3\right] + \left[\mathbf{V}_1\mathbf{V}_2\left(\frac{d\mathbf{V}_3}{dt}\right)\right]$$

and
$$\frac{d}{dt}\left\{\mathbf{V}_1 \times (\mathbf{V}_2 \times \mathbf{V}_3)\right\} = \left(\frac{d\mathbf{V}_1}{dt}\right) \times (\mathbf{V}_2 \times \mathbf{V}_3) + \mathbf{V}_1$$
$$\times\left(\left(\frac{d\mathbf{V}_2}{dt}\right) \times \mathbf{V}_3\right) + \mathbf{V}_1 \times \left(\mathbf{V}_2 \times \left(\frac{d\mathbf{V}_3}{dt}\right)\right)$$

Geometry of Curves in Space

s = the *length of arc,* measured from some fixed point on the curve (fig. 3).

$\mathbf{V}_1$ = the position vector of the point A on the curve

$\mathbf{V}_1 + \delta\mathbf{V}_1$ = the position vector of the point P in the neighborhood of A

$\hat{\mathbf{t}}$ = the *unit tangent* to the curve at the point A, measured in the direction of s increasing.

The *normal plane* is that plane which is perpendicular to the unit tangent. The principal normal is defined as the intersection of the normal plane with the plane defined by $\mathbf{V}_1$ and $\mathbf{V}_1 + \delta\mathbf{V}_1$ in the limit as $\delta\mathbf{V}_1 - 0$.

$\hat{\mathbf{n}}$ = the *unit normal* (principal) at the point A. The plane defined by $\hat{\mathbf{t}}$ and $\hat{\mathbf{n}}$ is called the *osculating plane* (alternatively plane of curvature or local plane).

ρ = the radius of curvature at A

$\delta\theta$ = the angle subtended at the origin by $\delta\mathbf{V}_1$.

$$\kappa = \frac{d\theta}{ds} = \frac{1}{\rho}$$

$\hat{\mathbf{b}}$ = the *unit binormal* i.e. the unit vector which is parallel to $\hat{\mathbf{t}} \times \hat{\mathbf{n}}$ at the point A:

λ = the *torsion* of the curve at A

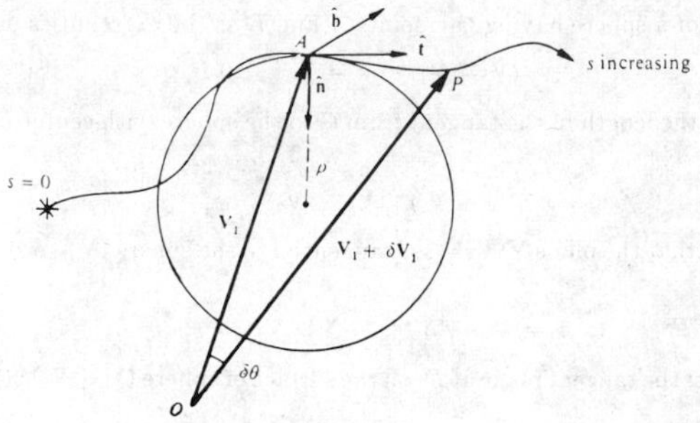

Figure 3.

Frenet's Formulae:

$$\frac{d\hat{\mathbf{t}}}{ds} = \kappa\hat{\mathbf{n}}$$
$$\frac{d\hat{\mathbf{n}}}{ds} = -\kappa\hat{\mathbf{t}} + \lambda\hat{\mathbf{b}}$$
$$\frac{d\hat{\mathbf{b}}}{ds} = -\lambda\hat{\mathbf{n}}$$

The following formulae are also applicable:

Unit tangent $\quad \hat{\mathbf{t}} = \dfrac{d\mathbf{V}_1}{ds}$

Equation of the tangent $\quad (\mathbf{V} - \mathbf{V}_1) \times \hat{\mathbf{t}} = 0$

or $\quad \mathbf{V} = \mathbf{V}_1 + q\hat{\mathbf{t}}$

Unit normal $\quad \hat{\mathbf{n}} = \dfrac{1 d^2\mathbf{V}_1}{\kappa ds^2}$

Equation of the normal plane $\quad (\mathbf{V} - \mathbf{V}_1) \cdot \hat{\mathbf{t}} = 0$

Equation of the normal $\quad (\mathbf{V} - \mathbf{V}_1) \times \hat{\mathbf{n}} = 0$

or $\quad \mathbf{V} = \mathbf{V}_1 + r\hat{\mathbf{n}}$

Unit binormal $\quad \hat{\mathbf{b}} = \hat{\mathbf{t}} \times \hat{\mathbf{n}}$

Equation of the binormal $\quad (\mathbf{V} - \mathbf{V}_1) \times \hat{\mathbf{b}} = 0$

or $\quad \mathbf{V} = \mathbf{V}_1 + u\hat{\mathbf{b}}$

or $\quad \mathbf{V} = \mathbf{V}_1 + w\dfrac{d\mathbf{V}_1}{ds} \times \dfrac{d^2\mathbf{V}_1}{ds^2}$

Equation of the osculating plane:

$$[(\mathbf{V} - \mathbf{V}_1)\hat{\mathbf{t}}\hat{\mathbf{n}}] = 0$$

or
$$\left[(\mathbf{V} - \mathbf{V}_1)\left(\frac{d\mathbf{V}_1}{ds}\right)\left(\frac{d^2\mathbf{V}_1}{ds^2}\right)\right] = 0$$

A *geodetic line* on a surface is a curve, the osculating plane of which is everywhere normal to the surface.

The differential equation of the geodetic is

$$[\hat{\mathbf{n}} d\mathbf{V}_1 \, d^2\mathbf{V}_1] = 0$$

Differential Operators—Rectangular Coordinates

$$dS = \frac{\partial S}{\partial x} \cdot dx + \frac{\partial S}{\partial y} \cdot dy + \frac{\partial S}{\partial z} \cdot dz$$

By definition

$$\nabla \equiv \text{del} \equiv \mathbf{i}\frac{\partial}{\partial x} + \mathbf{j}\frac{\partial}{\partial y} + \mathbf{k}\frac{\partial}{\partial z}$$

$$\nabla^2 \equiv \text{Laplacian} \equiv \frac{\partial^2}{\partial x^2} + \frac{\partial^2}{\partial y^2} + \frac{\partial^2}{\partial z^2}$$

If S is a scalar function, then

$$\nabla S \equiv \text{grad } S \equiv \frac{\partial S}{dx}\mathbf{i} + \frac{\partial S}{dy}\mathbf{j} + \frac{\partial S}{dz}\mathbf{k}$$

Grad S defines both the direction and magnitude of the maximum rate of increase of S at any point. Hence the name *gradient* and also its vectorial nature. ∇S is independent of the choice of rectangular coordinates.

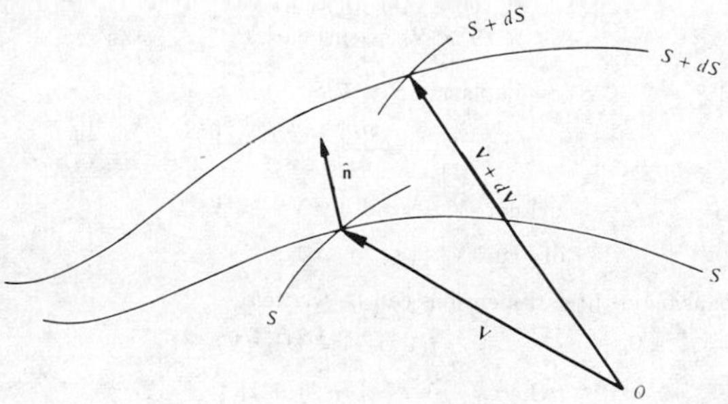

Figure 4

$$\nabla S = \frac{\partial S}{\partial n}\hat{\mathbf{n}}$$

where $\hat{\mathbf{n}}$ is the unit normal to the surface S = constant, in the direction of S increasing. The total derivative of S at a point having the position vector $\mathbf{V}$ is given by (fig. 4)

$$dS = \frac{\partial S}{\partial n}\hat{\mathbf{n}} \cdot d\mathbf{V}$$

$$= d\mathbf{V} \cdot \nabla S$$

and the directional derivative of S in the direction of $\mathbf{U}$ is

$$\mathbf{U} \cdot \nabla S = \mathbf{U} \cdot (\nabla S) = (\mathbf{U} \cdot \nabla)S$$

Similarly the directional derivative of the vector $\mathbf{V}$ in the direction of $\mathbf{U}$ is

$$(\mathbf{U} \cdot \nabla)\mathbf{V}$$

The *distributive* law holds for finding a gradient. Thus if S and T are scalar functions

$$\nabla(S + T) = \nabla S + \nabla T$$

The *associative* law becomes the rule for differentiating a product:

$$\nabla(ST) = S\nabla T + T\nabla S$$

If $\mathbf{V}$ is a vector function with the magnitudes of the components parallel to the three coordinate axes V_x, V_y, V_z, then

$$\nabla \cdot \mathbf{V} \equiv \text{div } \mathbf{V} \equiv \frac{\partial V_x}{\partial x} + \frac{\partial V_y}{\partial y} + \frac{\partial V_z}{\partial z}$$

The divergence obeys the distributive law. Thus, if $\mathbf{V}$ and $\mathbf{U}$ are vectors functions, then

$$\nabla \cdot (\mathbf{V} + \mathbf{U}) = \nabla \cdot \mathbf{V} + \nabla \cdot \mathbf{U}$$

$$\nabla \cdot (S\mathbf{V}) = (\nabla S) \cdot \mathbf{V} + S(\nabla \cdot \mathbf{V})$$

$$\nabla \cdot (\mathbf{U} \times \mathbf{V}) = \mathbf{V} \cdot (\nabla \times \mathbf{U}) - \mathbf{U} \cdot (\nabla \times \mathbf{V})$$

As with the gradient of a scalar, the divergence of a vector is invariant under a transformation from one set of rectangular coordinates to another.

$$\nabla \times \mathbf{V} \equiv \text{curl } \mathbf{V} \text{ (sometimes } \nabla \wedge \mathbf{V} \text{ or rot } \mathbf{V})$$

$$\equiv \left(\frac{\partial \mathbf{V}_z}{\partial y} - \frac{\partial \mathbf{V}_y}{\partial z}\right)\mathbf{i} + \left(\frac{\partial \mathbf{V}_x}{\partial z} - \frac{\partial \mathbf{V}_z}{\partial x}\right)\mathbf{j} + \left(\frac{\partial \mathbf{V}_y}{\partial x} - \frac{\partial \mathbf{V}_x}{\partial y}\right)\mathbf{k}$$

$$= \begin{vmatrix} \mathbf{i} & \mathbf{j} & \mathbf{k} \\ \frac{\partial}{\partial x} & \frac{\partial}{\partial y} & \frac{\partial}{\partial z} \\ V_x & V_y & V_z \end{vmatrix}$$

The *curl* (or *rotation*) of a vector is a vector which is invariant under a transformation from one set of rectangular coordinates to another.

$$\nabla \times (\mathbf{U} + \mathbf{V}) = \nabla \times \mathbf{U} + \nabla \times \mathbf{V}$$
$$\nabla \times (S\mathbf{V}) = (\nabla S) \times \mathbf{V} + S(\nabla \times \mathbf{V})$$
$$\nabla \times (\mathbf{U} \times \mathbf{V}) = (\mathbf{V} \cdot \nabla)\mathbf{U} - (\mathbf{U} \cdot \nabla)\mathbf{V} + \mathbf{U}(\nabla \cdot \mathbf{V}) - \mathbf{V}(\nabla \cdot \mathbf{U})$$

$$\text{grad } (\mathbf{U} \cdot \mathbf{V}) = \nabla(\mathbf{U} \cdot \mathbf{V})$$
$$= (\mathbf{V} \cdot \nabla)\mathbf{U} + (\mathbf{U} \cdot \nabla)\mathbf{V} + \mathbf{V} \times (\nabla \times \mathbf{U}) + \mathbf{U} \times (\nabla \times \mathbf{V})$$

If $\mathbf{V} = V_x\mathbf{i} + V_y\mathbf{j} + V_z\mathbf{k}$

$$\nabla \cdot \mathbf{V} = \nabla V_x \cdot \mathbf{i} + \nabla V_y \cdot \mathbf{j} + \nabla V_z \cdot \mathbf{k}$$

and $\nabla \times \mathbf{V} = \nabla V_x \times \mathbf{i} + \nabla V_y \times \mathbf{j} + \nabla V_z \times \mathbf{k}$

The operator ∇ can be used more than once. The number of possibilities where ∇ is used twice are

$$\nabla \cdot (\nabla \theta) \equiv \text{div grad } \theta$$
$$\nabla \times (\nabla \theta) \equiv \text{curl grad } \theta$$
$$\nabla(\nabla \cdot \mathbf{V}) \equiv \text{grad div } \mathbf{V}$$
$$\nabla \cdot (\nabla \times \mathbf{V}) \equiv \text{div curl } \mathbf{V}$$
$$\nabla \times (\nabla \times \mathbf{V}) \equiv \text{curl curl } \mathbf{V}$$

Thus: div grad $S \equiv \nabla \cdot (\nabla S) \equiv$ Laplacian $S \equiv \nabla^2 S$

$$\equiv \frac{\partial^2 S}{\partial x^2} + \frac{\partial^2 S}{\partial y^2} + \frac{\partial^2 S}{\partial z^2}$$

curl grad $S \equiv 0$; curl curl $\mathbf{V} \equiv$ grad div $\mathbf{V} - \nabla^2\mathbf{V}$;

div curl $\mathbf{V} \equiv 0$

Taylor's expansion in three dimensions can be written

$$f(\mathbf{V} + \boldsymbol{\varepsilon}) = e^{\boldsymbol{\varepsilon} \cdot \nabla} f(\mathbf{V})$$

where $\mathbf{V} = x\mathbf{i} + y\mathbf{j} + z\mathbf{k}$

and $\boldsymbol{\varepsilon} = h\mathbf{i} + l\mathbf{j} + m\mathbf{k}$

(note the analogy with $f_p = e^{phD} f_0$ in finite difference methods).

Orthogonal Curvilinear Coordinates

If at a point P there exist three uniform point functions u, v and w so that the surfaces u = const., v = const., and w = const., intersect in three distinct curves through P then the surfaces are called the *coordinate surfaces* through P. The three lines of intersection are referred to as the *coordinate lines* and their tangents a, b, and c as the *coordinate axes*. When the coordinate axes form an orthogonal set the system is said to define *orthogonal curvilinear coordinates* at P.

Consider an infinitesimal volume enclosed by the surfaces u, v, w, $u + du$, $v + dv$, and $w + dw$ (fig. 5).

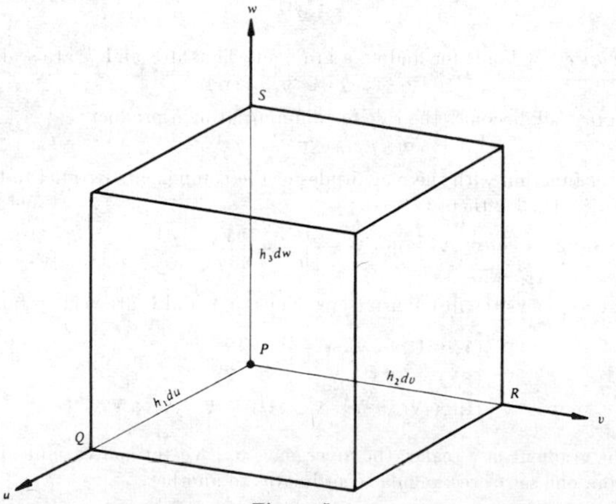

Figure 5

The surface $PRS \equiv u$ = const., and the face of the curvilinear figure immediately opposite this is $u + du$ = const. etc.

In terms of these surface constants

$$P = P(u,v,w)$$
$$Q = Q(u + du,v,w) \quad \text{and} \quad PQ = h_1 du$$

$$R = R(u,v + dv,w) \qquad PR = h_2 dv$$
$$S = S(u,v,w + dw) \qquad PS = h_3 dw$$

where h_1, h_2, and h_3 are functions of u, v, and w.

In rectangular Cartesians **i**, **j**, **k**

$$h_1 = 1, \qquad h_2 = 1, \qquad h_3 = 1.$$

$$\frac{\hat{\mathbf{a}}}{h_1}\frac{\partial}{\partial u} = \mathbf{i}\frac{\partial}{\partial x}, \qquad \frac{\hat{\mathbf{b}}}{h_2}\frac{\partial}{\partial v} = \mathbf{j}\frac{\partial}{\partial y}, \qquad \frac{\hat{\mathbf{c}}}{h_3}\frac{\partial}{\partial w} = \mathbf{k}\frac{\partial}{\partial z}.$$

In cylindrical coordinates $\hat{\mathbf{r}}$, $\hat{\boldsymbol{\phi}}$, $\hat{\mathbf{k}}$

$$h_1 = 1, \qquad h_2 = r \qquad h_3 = 1.$$

$$\frac{\hat{\mathbf{a}}}{h_1}\frac{\partial}{\partial u} = \hat{\mathbf{r}}\frac{\partial}{\partial r}, \qquad \frac{\hat{\mathbf{b}}}{h_2}\frac{\partial}{\partial v} = \frac{\hat{\boldsymbol{\phi}}}{r}\frac{\partial}{\partial \phi} \qquad \frac{\hat{\mathbf{c}}}{h_3}\frac{\partial}{\partial w} = \hat{\mathbf{k}}\frac{\partial}{\partial z}$$

In spherical coordinates $\hat{\mathbf{r}}$, $\hat{\boldsymbol{\theta}}$, $\hat{\boldsymbol{\phi}}$

$$h_1 = 1, \qquad h_2 = r, \qquad h_3 = r \sin\theta$$

$$\frac{\hat{\mathbf{a}}}{h_1}\frac{\partial}{\partial u} = \hat{\mathbf{r}}\frac{\partial}{\partial r}, \qquad \frac{\mathbf{b}}{h_2}\frac{\partial}{\partial v} = \frac{\hat{\boldsymbol{\phi}}}{r}\frac{\partial}{\partial \theta}, \qquad \frac{\hat{\mathbf{c}}}{h_3}\frac{\partial}{\partial w} = \frac{\hat{\boldsymbol{\phi}}}{r\sin\theta}\frac{\partial}{\partial \phi}$$

The general expressions for grad, div and curl together with those for ∇^2 and the directional derivative are, in orthogonal curvilinear coordinates, given by

$$\nabla S = \frac{\hat{\mathbf{a}}}{h_1}\frac{\partial S}{\partial u} + \frac{\hat{\mathbf{b}}}{h_2}\frac{\partial S}{\partial v} + \frac{\hat{\mathbf{c}}}{h_3}\frac{\partial S}{\partial w}$$

$$(\mathbf{V}\cdot\nabla)S = \frac{V_1}{h_1}\frac{\partial S}{\partial u} + \frac{V_2}{h_2}\frac{\partial S}{\partial v} + \frac{V_3}{h_3}\frac{\partial S}{\partial w}$$

$$\nabla\cdot\mathbf{V} = \frac{1}{h_1h_2h_3}\left\{\frac{\partial}{\partial u}(h_2h_3V_1) + \frac{\partial}{\partial v}(h_3h_1V_2) + \frac{\partial}{\partial w}(h_1h_2V_3)\right\}$$

$$\nabla\times\mathbf{V} = \frac{\hat{\mathbf{a}}}{h_2h_3}\left\{\frac{\partial}{\partial v}(h_3V_3) - \frac{\partial}{\partial w}(h_2V_2)\right\} + \frac{\hat{\mathbf{b}}}{h_3h_1}\left\{\frac{\partial}{\partial w}(h_1V_1) - \frac{\partial}{\partial u}(h_3V_3)\right\}$$
$$+ \frac{\hat{\mathbf{c}}}{h_1h_2}\left\{\frac{\partial}{\partial u}(h_2V_2) - \frac{\partial}{\partial v}(h_1V_1)\right\}$$

$$\nabla^2 S = \frac{1}{h_1h_2h_3}\left\{\frac{\partial}{\partial u}\left(\frac{h_2h_3}{h_1}\frac{\partial S}{\partial u}\right) + \frac{\partial}{\partial v}\left(\frac{h_3h_1}{h_2}\frac{\partial S}{\partial v}\right) + \frac{\partial}{\partial w}\left(\frac{h_1h_2}{h_3}\frac{\partial S}{\partial w}\right)\right\}$$

FORMULAS OF VECTOR ANALYSIS

	Rectangular coordinates	Cylindrical coordinates	Spherical coordinates
Conversion to rectangular coordinates		$x = r\cos\varphi \quad y = r\sin\varphi \quad z = z$	$x = r\cos\varphi\sin\theta \quad y = r\sin\varphi\sin\theta \quad z = r\cos\theta$
Gradient	$\nabla\phi = \frac{\partial\phi}{\partial x}\mathbf{i} + \frac{\partial\phi}{\partial y}\mathbf{j} + \frac{\partial\phi}{\partial z}\mathbf{k}$	$\nabla\phi = \frac{\partial\phi}{\partial r}\mathbf{r} + \frac{1}{r}\frac{\partial\phi}{\partial\varphi}\boldsymbol{\phi} + \frac{\partial\phi}{\partial z}\mathbf{k}$	$\nabla\phi = \frac{\partial\phi}{\partial r}\mathbf{r} + \frac{1}{r}\frac{\partial\phi}{\partial\theta}\boldsymbol{\theta} + \frac{1}{r\sin\theta}\frac{\partial\phi}{\partial\varphi}\boldsymbol{\phi}$
Divergence	$\nabla\cdot\mathbf{A} = \frac{\partial A_x}{\partial x} + \frac{\partial A_y}{\partial y} + \frac{\partial A_z}{\partial z}$	$\nabla\cdot\mathbf{A} = \frac{1}{r}\frac{\partial(rA_r)}{\partial r} + \frac{1}{r}\frac{\partial A_\varphi}{\partial\varphi} + \frac{\partial A_z}{\partial z}$	$\nabla\cdot\mathbf{A} = \frac{1}{r^2}\frac{\partial(r^2A_r)}{\partial r} + \frac{1}{r\sin\theta}\frac{\partial(A_\theta\sin\theta)}{\partial\theta} + \frac{1}{r\sin\theta}\frac{\partial A_\varphi}{\partial\varphi}$
Curl	$\nabla\times\mathbf{A} = \begin{vmatrix} \mathbf{i} & \mathbf{j} & \mathbf{k} \\ \frac{\partial}{\partial x} & \frac{\partial}{\partial y} & \frac{\partial}{\partial z} \\ A_x & A_y & A_z \end{vmatrix}$	$\nabla\times\mathbf{A} = \begin{vmatrix} \frac{1}{r}\mathbf{r} & \boldsymbol{\phi} & \frac{1}{r}\mathbf{k} \\ \frac{\partial}{\partial r} & \frac{\partial}{\partial\varphi} & \frac{\partial}{\partial z} \\ A_r & rA_\varphi & A_z \end{vmatrix}$	$\nabla\times\mathbf{A} = \begin{vmatrix} \frac{\mathbf{r}}{r^2\sin\theta} & \frac{\boldsymbol{\theta}}{r\sin\theta} & \frac{\boldsymbol{\phi}}{r} \\ \frac{\partial}{\partial r} & \frac{\partial}{\partial\theta} & \frac{\partial}{\partial\varphi} \\ A_r & rA_\theta & rA_\varphi\sin\theta \end{vmatrix}$
Laplacian	$\nabla^2\phi = \frac{\partial^2\phi}{\partial x^2} + \frac{\partial^2\phi}{\partial y^2} + \frac{\partial^2\phi}{\partial z^2}$	$\nabla^2\phi = \frac{1}{r}\frac{\partial}{\partial r}\left(r\frac{\partial\phi}{\partial r}\right) + \frac{1}{r^2}\frac{\partial^2\phi}{\partial\varphi^2} + \frac{\partial^2\phi}{\partial z^2}$	$\nabla^2\phi = \frac{1}{r^2}\frac{\partial}{\partial r}\left(r^2\frac{\partial\phi}{\partial r}\right) + \frac{1}{r^2\sin\theta}\frac{\partial}{\partial\theta}\left(\sin\theta\frac{\partial\phi}{\partial\theta}\right) + \frac{1}{r^2\sin^2\theta}\frac{\partial^2\phi}{\partial\varphi^2}$

Transformation of Integrals

s = the distance along some curve "C" in space and is measured from some fixed point.
S = a surface area
V = a volume contained by a specified surface
$\hat{\mathbf{t}}$ = the unit tangent to C at the point P
$\hat{\mathbf{n}}$ = the unit outward pointing normal
F = some vector function
ds = the vector element of curve ($= \hat{\mathbf{t}}\, ds$)
dS = the vector element of surface ($= \hat{\mathbf{n}}\, dS$)

Then

$$\int_{(c)} \mathbf{F} \cdot \hat{\mathbf{t}}\, ds = \int_{(c)} \mathbf{F} \cdot ds$$

and when

$$\mathbf{F} = \nabla \phi$$

$$\int_{(c)} (\nabla \phi) \cdot \hat{\mathbf{t}}\, ds = \int_{(c)} d\phi$$

Gauss' Theorem (Green's Theorem)

When S defines a closed region having a volume V

$$\iiint_{(v)} (\nabla \cdot \mathbf{F})\, dV = \iint_{(s)} (\mathbf{F} \cdot \hat{\mathbf{n}})\, dS = \iint_{(s)} \mathbf{F} \cdot dS$$

also

$$\iiint_{(v)} (\nabla \phi)\, dV = \iint_{(s)} \phi \hat{\mathbf{n}}\, dS$$

and

$$\iiint_{(v)} (\nabla \times \mathbf{F})\, dV = \iint_{(s)} (\hat{\mathbf{n}} \times \mathbf{F})\, dS$$

Stokes' Theorem

When C is closed and bounds the open surface S.

$$\iint_{(s)} \hat{\mathbf{n}} \cdot (\nabla \times \mathbf{F})\, dS = \int_{(c)} \mathbf{F} \cdot \mathrm{ds}$$

also

$$\iint_{(s)} (\hat{\mathbf{n}} \times \nabla \phi)\, dS = \int_{(c)} \phi\, \mathrm{ds}$$

Green's Theorem

$$\iint_{(s)} (\nabla \phi \cdot \nabla \theta)\, dS = \iint_{(s)} \phi \hat{\mathbf{n}} \cdot (\nabla \theta)\, dS = \iiint_{(v)} \phi(\nabla^2 \theta)\, dV$$

$$= \iint_{(s)} \theta \cdot \hat{\mathbf{n}}(\nabla \phi)\, dS = \iiint_{(v)} \theta(\nabla^2 \phi)\, dV$$

MOMENT OF INERTIA FOR VARIOUS BODIES OF MASS

The mass of the body is indicated by m

Body	Axis	Moment of inertia
Uniform thin rod	Normal to the length, at one end	$m\frac{l^2}{3}$
Uniform thin rod	Normal to the length, at the center	$m\frac{l^2}{12}$
Thin rectangular sheet, sides a and b	Through the center parallel to b	$m\frac{a^2}{12}$
Thin rectangular sheet, sides a and b	Through the center perpendicular to the sheet	$m\frac{a^2+b^2}{12}$
Thin circular sheet of radius r	Normal to the plate through the center	$m\frac{r^2}{2}$
Thin circular sheet of radius r	Along any diameter	$m\frac{r^2}{4}$
Thin circular ring. Radii r_1 and r_2	Through center normal to plane of ring	$m\frac{r_1^2+r_2^2}{2}$
Thin circular ring. Radii r_1 and r_2	Any diameter	$m\frac{r_1^2+r_2^2}{4}$
Rectangular parallelopiped, edges a, b, and c	Through center perpendicular to face ab, (parallel to edge c)	$m\frac{a^2+b^2}{12}$
Sphere, radius r	Any diameter	$m\frac{2}{5}r^2$
Spherical shell, external radius, r_1, internal radius r_2	Any diameter	$m\frac{2}{5}\frac{(r_1^5-r_2^5)}{(r_1^3-r_2^3)}$
Spherical shell, very thin, mean radius, r	Any diameter	$m\frac{2}{3}r^2$
Right circular cylinder of radius r, length l	The longitudinal axis of the solid	$m\frac{r^2}{2}$
Right circular cylinder of radius r, length l	Transverse diameter	$m\left(\frac{r^2}{4}+\frac{l^2}{12}\right)$
Hollow circular cylinder, length l, radii r_1 and r_2	The longitudinal axis of the figure	$m\frac{(r_1^2+r_2^2)}{2}$
Thin cylindrical shell, length l, mean radius, r	The longitudinal axis of the figure	mr^2
Hollow circular cylinder, length l, radii r_1 and r_2	Transverse diameter	$m\left[\frac{r_1^2+r_2^2}{4}+\frac{l^2}{12}\right]$
Hollow circular cylinder, length l, very thin, mean radius	Transverse diameter	$m\left(\frac{r^2}{2}+\frac{l^2}{12}\right)$
Elliptic cylinder, length l, transverse semiaxes a and b	Longitudinal axis	$m\left(\frac{a^2+b^2}{4}\right)$
Right cone, altitude h, radius of base r	Axis of the figure	$m\frac{3}{10}r^2$
Spheroid of revolution, equatorial radius r	Polar axis	$m\frac{2r^2}{5}$
Ellipsoid, axes $2a$, $2b$, $2c$	Axis $2a$	$m\frac{(b^2+c^2)}{5}$

Bessel Functions*

1. Bessel's differential equation for a real variable x is

$$x^2 \frac{d^2y}{dx^2} + x\frac{dy}{dx} + (x^2 - n^2)y = 0$$

* From Beyer, W. H., Ed., *CRC Handbook of Mathematical Sciences*, 5th ed., CRC Press, Boca Raton, 1978, 500—503. With permission.

2. When n is not an integer, two independent solutions of the equation are $J_n(x)$, $J_{-n}(x)$, where

$$J_n(x) = \sum_{k=0}^{\infty} \frac{(-1)^k}{k!\Gamma(n+k+1)} \left(\frac{x}{2}\right)^{n+2k}$$

3. If n is an integer $J_n(x) = (-1)^n J_{-n}(x)$, where

$$J_n(x) = \frac{x^n}{2^n n!}\left\{1 - \frac{x^2}{2^2 \cdot 1!(n+1)} + \frac{x^4}{2^4 \cdot 2!(n+1)(n+2)} - \frac{x^6}{2^6 \cdot 3!(n+1)(n+2)(n+3)} + \cdots\right\}$$

4. For $n = 0$ and $n = 1$, this formula becomes

$$J_0(x) = 1 - \frac{x^2}{2^2(1!)^2} + \frac{x^4}{2^4(2!)^2} - \frac{x^6}{2^6(3!)^2} + \frac{x^8}{2^8(4!)^2} - \cdots$$

$$J_1(x) = \frac{x}{2} - \frac{x^3}{2^3 \cdot 1!2!} + \frac{x^5}{2^5 \cdot 2!3!} - \frac{x^7}{2^7 \cdot 3!4!} + \frac{x^9}{2^9 \cdot 4!5!} - \cdots$$

5. When x is large and positive, the following asymptotic series may be used

$$J_0(x) = \left(\frac{2}{\pi x}\right)^{\frac{1}{2}}\left\{P_0(x)\cos\left(x - \frac{\pi}{4}\right) - Q_0(x)\sin\left(x - \frac{\pi}{4}\right)\right\}$$

$$J_1(x) = \left(\frac{2}{\pi x}\right)^{\frac{1}{2}}\left\{P_1(x)\cos\left(x - \frac{3\pi}{4}\right) - Q_1(x)\sin\left(x - \frac{3\pi}{4}\right)\right\}$$

where

$$P_0(x) \sim 1 - \frac{1^2 \cdot 3^2}{2!(8x)^2} + \frac{1^2 \cdot 3^2 \cdot 5^2 \cdot 7^2}{4!(8x)^4} - \frac{1^2 \cdot 3^2 \cdot 5^2 \cdot 7^2 \cdot 9^2 \cdot 11^2}{6!(8x)^6} + \cdots$$

$$Q_0(x) \sim -\frac{1^2}{1!8x} + \frac{1^2 \cdot 3^2 \cdot 5^2}{3!(8x)^3} - \frac{1^2 \cdot 3^2 \cdot 5^2 \cdot 7^2 \cdot 9^2}{5!(8x)^5} + - \cdots$$

$$P_1(x) \sim 1 + \frac{1^2 \cdot 3 \cdot 5}{2!(8x)^2} - \frac{1^2 \cdot 3^2 \cdot 5^2 \cdot 7 \cdot 9}{4!(8x)^4} + \frac{1^2 \cdot 3^2 \cdot 5^2 \cdot 7^2 \cdot 9^2 \cdot 11 \cdot 13}{6!(8x)^6} - + \cdots$$

$$Q_1(x) \sim \frac{1 \cdot 3}{1!8x} - \frac{1^2 \cdot 3^2 \cdot 5 \cdot 7}{3!(8x)^3} + \frac{1^2 \cdot 3^2 \cdot 5^2 \cdot 7^2 \cdot 9 \cdot 11}{5!(8x)^5} - \cdots$$

[In $P_1(x)$ the signs alternate from + to − after the first term]

6. If $x > 25$, it is convenient to use the formulas

$$J_0(x) = A_0(x)\sin x + B_0(x)\cos x$$

$$J_1(x) = B_1(x)\sin x - A_1(x)\cos x$$

where

$$A_0(x) = \frac{P_0(x) - Q_0(x)}{(\pi x)^{\frac{1}{2}}} \quad \text{and} \quad A_1(x) = \frac{P_1(x) - Q_1(x)}{(\pi x)^{\frac{1}{2}}}$$

$$B_0(x) = \frac{P_0(x) + Q_0(x)}{(\pi x)^{\frac{1}{2}}} \quad \text{and} \quad B_1(x) = \frac{P_1(x) + Q_1(x)}{(\pi x)^{\frac{1}{2}}}$$

7. The zeros of $J_0(x)$ and $J_1(x)$

If j_{0s} and j_{1s} are the sth zeros of $J_0(x)$ and $J_1(x)$ respectively, and if $a = 4_s - 1$, $b = 4_s + 1$

$$j_{0,s} \sim \frac{1}{4}\pi a\left\{1 + \frac{2}{\pi^2 a^2} - \frac{62}{3\pi^4 a^4} + \frac{15{,}116}{15\pi^6 a^6} - \frac{12{,}554{,}474}{105\pi^8 a^8} + \frac{8{,}368{,}654{,}292}{315\pi^{10} a^{10}} - + \cdots\right\}$$

$$j_{1,s} \sim \frac{1}{4}\pi b\left\{1 - \frac{6}{\pi^2 b^2} + \frac{6}{\pi^4 b^4} - \frac{4716}{5\pi^6 b^6} + \frac{3{,}902{,}418}{35\pi^8 b^8} - \frac{895{,}167{,}324}{35\pi^{10} b^{10}} + \cdots\right\}$$

$$J_1(j_{0,s}) \sim \frac{(-1)^{s+1} 2^{\frac{1}{2}}}{\pi a^{\frac{1}{2}}}\left\{1 - \frac{56}{3\pi^4 a^4} + \frac{9664}{5\pi^6 a^6} - \frac{7{,}381{,}280}{21\pi^8 a^8} + \cdots\right\}$$

$$J_0(j_{1,s}) \sim \frac{(-1)^s 2^{\frac{1}{2}}}{\pi b^{\frac{1}{2}}}\left\{1 + \frac{24}{\pi^4 b^4} - \frac{19{,}584}{10\pi^6 b^6} + \frac{2{,}466{,}720}{7\pi^8 b^8} - \cdots\right\}$$

8. Table of zeros for $J_0(x)$ and $J_1(x)$

$J_1(\alpha_n) = 0 \qquad J_0(\beta_n) = 0$

Roots α_n	$J_1(\alpha_n)$	Roots β_n	$J_0(\beta_n)$
2.4048	0.5191	0.0000	1.0000
5.5201	−0.3403	3.8317	−0.4028
8.6537	0.2715	7.0156	0.3001
11.7915	−0.2325	10.1735	−0.2497
14.9309	0.2065	13.3237	0.2184
18.0711	−0.1877	16.4706	−0.1965
21.2116	0.1733	19.6159	0.1801

9. Recurrence formulas

$$J_{n-1}(x) + J_{n+1}(x) = \frac{2n}{x} J_n(x) \qquad nJ_n(x) + xJ_n'(x) = xJ_{n-1}(x)$$

$$J_{n-1}(x) - J_{n+1}(x) = 2J_n'(x) \qquad nJ_n(x) - xJ_n'(x) = xJ_{n+1}(x)$$

10. If J_n is written for $J_n(x)$ and $J_n^{(k)}$ is written for $\frac{d^k}{dx^k}\{J_n(x)\}$, then the following derivative relationships are important

$$J_0^{(r)} = -J_1^{(r-1)}$$

$$J_0^{(2)} = -J_0 + \frac{1}{x}J_1 = \frac{1}{2}(J_2 - J_0)$$

$$J_0^{(3)} = \frac{1}{x}J_0 + \left(1 - \frac{2}{x^2}\right)J_1 = \frac{1}{4}(-J_3 + 3J_1)$$

$$J_0^{(4)} = \left(1 - \frac{3}{x^2}\right)J_0 - \left(\frac{2}{x} - \frac{6}{x^3}\right)J_1 = \frac{1}{8}(J_4 - 4J_2 + 3J_0), \text{ etc.}$$

11. Half order Bessel functions

$$J_{\frac{1}{2}}(x) = \sqrt{\frac{2}{\pi x}}\sin x$$

$$J_{-\frac{1}{2}}(x) = \sqrt{\frac{2}{\pi x}}\cos x$$

$$J_{n+\frac{1}{2}}(x) = -x^{n+\frac{1}{2}}\frac{d}{dx}\{x^{-(n+\frac{1}{2})}J_{n+\frac{1}{2}}(x)\}$$

$$J_{n-\frac{1}{2}}(x) = x^{-(n+\frac{1}{2})}\frac{d}{dx}\{x^{n+\frac{1}{2}}J_{n+\frac{1}{2}}(x)\}$$

n	$\left(\frac{\pi x}{2}\right)^{\frac{1}{2}} J_{n+\frac{1}{2}}(x)$	$\left(\frac{\pi x}{2}\right)^{\frac{1}{2}} J_{-(n+\frac{1}{2})}(x)$
0	$\sin x$	$\cos x$
1	$\frac{\sin x}{x} - \cos x$	$-\frac{\cos x}{x} - \sin x$
2	$\left(\frac{3}{x^2} - 1\right)\sin x - \frac{3}{x}\cos x$	$\left(\frac{3}{x^2} - 1\right)\cos x + \frac{3}{x}\sin x$
3	$\left(\frac{15}{x^3} - \frac{6}{x}\right)\sin x - \left(\frac{15}{x^2} - 1\right)\cos x$	$-\left(\frac{15}{x^3} - \frac{6}{x}\right)\cos x - \left(\frac{15}{x^2} - 1\right)\sin x$
	etc.	

12. Additional solutions to Bessel's equation are

$Y_n(x)$ (also called Weber's function, and sometimes denoted by $N_n(x)$)

$H_n^{(1)}(x)$ and $H_n^{(2)}(x)$ (also called Hankel functions)

These solutions are defined as follows

$$Y_n(x) = \begin{cases} \dfrac{J_n(x)\cos(n\pi) - J_{-n}(x)}{\sin(n\pi)} & n \text{ not an integer} \\ \lim\limits_{\nu\to n} \dfrac{J_\nu(x)\cos(\nu\pi) - J_{-\nu}(x)}{\sin(\nu\pi)} & n \text{ an integer} \end{cases}$$

$$H_n^{(1)}(x) = J_n(x) + iY_n(x)$$
$$H_n^{(2)}(x) = J_n(x) - iY_n(x)$$

The additional properties of these functions may all be derived from the above relations and the known properties of $J_n(x)$.

13. Complete solutions to Bessel's equation may be written as

$$c_1 J_n(x) + c_2 J_{-n}(x) \qquad \text{if } n \text{ is not an integer}$$

or

$$\left.\begin{array}{l} c_1 J_n(x) + c_2 Y_n(x) \\ \\ c_1 H_n^{(1)}(x) + c_2 H_n^{(2)}(x) \end{array}\right\} \text{for any value of } n$$

or

14. The modified (or hyperbolic) Bessel's differential equation is

$$x^2 \frac{d^2 y}{dx^2} + x \frac{dy}{dx} - (x^2 + n^2) y = 0$$

15. When n is not an integer, two independent solutions of the equation are $I_n(x)$ and $I_{-n}(x)$, where

$$I_n(x) = \sum_{k=0}^{\infty} \frac{1}{k!\Gamma(n+k+1)} \left(\frac{x}{2}\right)^{n+2k}$$

16. If n is an integer,

$$I_n(x) = I_{-n}(x) = \frac{x^n}{2^n n!}\left\{1 + \frac{x^2}{2^2 \cdot 1!(n+1)} + \frac{x^4}{2^4 \cdot 2!(n+1)(n+2)} + \frac{x^6}{2^6 \cdot 3!(n+1)(n+2)(n+3)} + \cdots\right\}$$

17. For $N = 0$ and $n = 1$, this formula becomes

$$I_0(x) = 1 + \frac{x^2}{2^2(1!)^2} + \frac{x^4}{2^4(2!)^2} + \frac{x^6}{2^6(3!)^2} + \frac{x^8}{2^8(4!)^2} + \cdots$$

$$I_1(x) = \frac{x}{2} + \frac{x^3}{2^3 \cdot 1!2!} + \frac{x^5}{2^5 \cdot 2!3!} + \frac{x^7}{2^7 \cdot 3!4!} + \frac{x^9}{2^9 \cdot 4!5!} + \cdots$$

18. Another solution to the modified Bessel's equation is

$$K_n(x) = \begin{cases} \dfrac{1}{2}\pi \dfrac{I_{-n}(x) - I_n(x)}{\sin(n\pi)} & n \text{ not an integer} \\ \lim\limits_{\nu \to n} \dfrac{1}{2}\pi \dfrac{I_{-\nu}(x) - I_\nu(x)}{\sin(\nu\pi)} & n \text{ an integer} \end{cases}$$

This function is linearly independent of $I_n(x)$ for all values of n. Thus the complete solution to the modified Bessel's equation may be written as

$$c_1 I_n(x) + c_2 I_{-n}(x) \qquad n \text{ not an integer}$$

or

$$c_1 I_n(x) + c_2 K_n(x) \qquad \text{any } n$$

19. The following relations hold among the various Bessel functions:

$$I_n(z) = i^{-m} J_m(iz)$$

$$Y_n(iz) = (i)^{n+1} I_n(z) - \frac{2}{\pi} i^{-n} K_n(z)$$

Most of the properties of the modified Bessel function may be deduced from the known properties of $J_n(x)$ by use of these relations and those previously given.

20. Recurrence formulas

$$I_{n-1}(x) - I_{n+1}(x) = \frac{2n}{x} I_n(x) \qquad I_{n-1}(x) + I_{n+1}(x) = 2I'_n(x)$$

$$I_{n-1}(x) - \frac{n}{x} I_n(x) = I'_n(x) \qquad I'_n(x) = I_{n+1}(x) + \frac{n}{x} I_n(z)$$

The Gamma Function*

Definition: Γ (n = $\int_0^{\infty} t^{n-1} e^{-t} dt \; n > 0$

Recursion Formula: $\Gamma(n + 1 = n\Gamma(n)$

$\Gamma(n) + 1) = n!$ if $n = 0,1,2, \ldots$ where $0! = 1$

For $n < 0$ the gamma function can be defined by using

* From Beyer, W. H., Ed., *CRC Handbook of Mathematical Sciences*, 5th ed., CRC Press, Boca Raton, 1978, 484—485. With permision.

$$\Gamma(n) = \frac{\Gamma(n+1)}{n}$$

Graph:

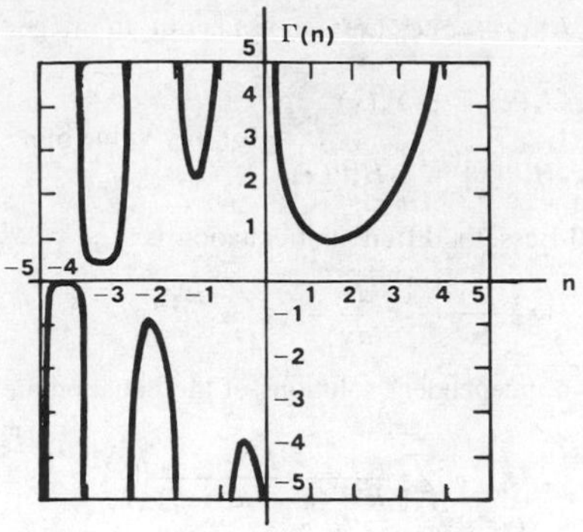

Special Values:

$$\Gamma(\tfrac{1}{2}) = \sqrt{\pi}$$

$$\Gamma(m + \tfrac{1}{2}) = \frac{1 \cdot 3 \cdot 5 \cdots (2m-1)}{2^m}\sqrt{\pi} \qquad m = 1,2,3,\ldots$$

$$\Gamma(-m + \tfrac{1}{2}) = \frac{(-1)^m 2^m \sqrt{\pi}}{1 \cdot 3 \cdot 5 \cdots (2m-1)} \qquad m = 1,2,3,\ldots$$

Definition:

$$\Gamma(x+1) = \lim_{k\to\infty} \frac{1 \cdot 2 \cdot 3 \cdots k}{(x+1)(x+2)\cdots(x+k)} k^x$$

$$\frac{1}{\Gamma(x)} = xe^{\gamma x} \prod_{m=1}^{\infty} \left\{ \left(1 + \frac{x}{m}\right) e^{-x/m} \right\}$$

This is an infinite product representation for the gamma function where γ is Euler's constant.

Properties:

$$\Gamma'(1) = \int_0^\infty e^{-x} \ln x \, dx = -\gamma$$

$$\frac{\Gamma'(x)}{\Gamma(x)} = -\gamma + \left(\frac{1}{1} - \frac{1}{x}\right) + \left(\frac{1}{2} - \frac{1}{x+1}\right) + \cdots + \left(\frac{1}{n} - \frac{1}{x+n-1}\right) + \cdots$$

$$\Gamma(x+1) = \sqrt{2\pi x}\, x^x e^{-x} \left\{ 1 + \frac{1}{12x} + \frac{1}{288x^2} - \frac{139}{51{,}840x^3} + \cdots \right\}$$

This is called *Stirling's asymptotic series.*

If we let $x = n$ a positive integer, then a useful approximation for $n!$ where n is large (e.g., $n > 10$) is given by *Stirling's formula*

$$n! \approx \sqrt{2\pi n}\, n^n e^{-n}$$

The Gamma Function*

Values of $\Gamma(n) = \int_0^\infty e^{-x} x^{n-1}\, dx;\ \Gamma(n+1) = n\Gamma(n)$

n	$\Gamma(n)$	n	$\Gamma(n)$	n	$\Gamma(n)$	n	$\Gamma(n)$
1.00	1.00000	1.25	.90640	1.50	.88623	1.75	.91906
1.01	.99433	1.26	.90440	1.51	.88659	1.76	.92137
1.02	.98884	1.27	.90250	1.52	.88704	1.77	.92376
1.03	.98355	1.28	.90072	1.53	.88757	1.78	.92623
1.04	.97844	1.29	.89904	1.54	.88818	1.79	.92877
1.05	.97350	1.30	.89747	1.55	.88887	1.80	.93138
1.06	.96874	1.31	.89600	1.56	.88964	1.81	.93408
1.07	.96415	1.32	.89464	1.57	.89049	1.82	.93685
1.08	.95973	1.33	.89338	1.58	.89142	1.83	.93969
1.09	.95546	1.34	.89222	1.59	.89243	1.84	.94261
1.10	.95135	1.35	.89115	1.60	.89352	1.85	.94561
1.11	.94740	1.36	.89018	1.61	.89468	1.86	.94869
1.12	.94359	1.37	.88931	1.62	.89592	1.87	.95184
1.13	.93993	1.38	.88854	1.63	.89724	1.88	.95507
1.14	.93642	1.39	.88785	1.64	.89864	1.89	.95838
1.15	.93304	1.40	.88726	1.65	.90012	1.90	.96177
1.16	.92980	1.41	.88676	1.66	.90167	1.91	.96523
1.17	.92670	1.42	.88636	1.67	.90330	1.92	.96877
1.18	.92373	1.43	.88604	1.68	.90500	1.93	.97240
1.19	.92089	1.44	.88581	1.69	.90678	1.94	.97610
1.20	.91817	1.45	.88566	1.70	.90864	1.95	.97988
1.21	.91558	1.46	.88560	1.71	.91057	1.96	.98374
1.22	.91311	1.47	.88563	1.72	.91258	1.97	.98768
1.23	.91075	1.48	.88575	1.73	.91466	1.98	.99171
1.24	.90852	1.49	.88595	1.74	.91683	1.99	.99581
						2.00	1.00000

* For large positive values of x, $\Gamma(x)$ approximates Stirling's asymptotic series

$$x^x e^{-x} \sqrt{\frac{2\pi}{x}} \left[1 + \frac{1}{12x} + \frac{1}{288x^2} - \frac{139}{51840x^3} - \frac{571}{2488320x^4} + \cdots\right]$$

The Beta Function*

Definition: $B(m,n) = \int_0^1 t^{m-1} (1-t)^{m-1}\, dt \qquad m > 0, n > 0$

Relationship with Gamma Function: $B(m,n) = \dfrac{\Gamma(m)\Gamma(n)}{\Gamma(m+n)}$

Properties:

$$B(m,n) = B(n,m)$$

$$B(m,n) = 2\int_0^{\pi/2} \sin^{2m-1}\theta \cos^{2n-1}\theta\, d\theta$$

$$B(m,n) = \int_0^\infty \frac{t^{m-1}}{(1+t)^{m+n}}\, dt$$

$$B(m,n) = r^n(r+1)^m \int_0^1 \frac{t^{m-1}(1-t)^{n-1}}{(r+t)^{m+n}}\, dt$$

The Error Function

Definition: $\operatorname{erf} x = \dfrac{2}{\sqrt{\pi}} \int_0^x e^{-t^2}\, dt$

Series: $\operatorname{erf} x = \dfrac{2}{\sqrt{\pi}} \left(x - \dfrac{x^3}{3} + \dfrac{1}{2!}\dfrac{x^5}{5} - \dfrac{1}{3!}\dfrac{x^7}{7} + \ldots\right)$

Property: $\operatorname{erf} x = -\operatorname{erf}(-x)$

Relationship with Normal Probability Function $f(t)$: $\int_0^x f(t)\, dt = \frac{1}{2} \operatorname{erf}\left(\frac{x}{\sqrt{2}}\right)$

To evaluate *erf* (2.3), one proceeds as follows: Since $\dfrac{x}{\sqrt{2}} = 2.3$, one finds $x = (2.3)(\sqrt{2}) = 3.25$. In the normal probability function table (page A-), one finds the entry 0.4994 opposite the value 3.25. Thus *erf* (2.3) = 2(0.4994) = 0.9988.

* From Beyer, W. H., Ed., *CRC Handbook of Mathematical Sciences*, 5th ed., CRC Press, Boca Raton, 1978, 499. With permission.

$$erfc\ z = 1 - erf\ z = \frac{2}{\sqrt{\pi}} \int_z^{\infty} e^{-t^2}\, dt$$

is known as the complimentary error function.

Orthogonal Polynomials*

I

Name: Legendre *Symbol:* $P_n(x)$ *Interval:* $[-1,1]$

Differential Equation: $(1 - x^2)\, y'' - 2\, xy' + n(n + 1)\, y = 0$

$$y = P_n(x)$$

$$\text{Explicit Expression: } P_n(x) = \frac{1}{2^n} \sum_{m=0}^{[n/2]} (-1)^m \binom{n}{m} \binom{2n - 2m}{n} x^{n-2m}$$

Recurrence Relation: $(n + 1)\, P_{n+1}(x) = (2n + 1)\, x\, P_n(x) - nP_{n-1}(x)$

Weight: 1 *Standardization:* $P_n(1) = 1$

$$\text{Norm: } \int_{-1}^{+1} [P_n(x)]^2\, dx = \frac{2}{2n + 1}$$

$$\text{Rodrigues' Formula: } P_n(x) = \frac{(-1)^n}{2^n n!} \frac{d^n}{dx^n} \{(1 - x^2)^n\}$$

$$\text{Generating Function: } R^{-1} = \sum_{n=0}^{\infty} P_n(x) z^n;\ -1 < x < 1,\ |z| < 1,$$

$$R = \sqrt{1 - 2xz + z^2}$$

Inequality: $|P_n(x)| \leq 1,\ -1 \leq x \leq 1.$

II

Name: Tschebysheff, First Kind *Symbol:* $T_n(x)$ *Interval:* $[-1,1]$

Differential Equation: $(1 - x^2)y - xy' + n^2y = 0$

$$y = T_n(x)$$

$$\text{Explicit Expression: } \frac{n}{2} \sum_{m=0}^{[n/2]} (-1)^m \frac{(n - m - 1)!}{m!(n - 2m)!} (2x)^{n-2m} = \cos(n \arccos x) = T_n(x)$$

Recurrence Relation: $T_{n+1}(x) = 2xT_n(x) - T_{n-1}(x)$

Weight: $(1 - x^2)^{-1/2}$ *Standardization:* $T_n(1) = 1$

$$\text{Norm: } \int_{-1}^{+1} (1 - x^2)^{-1/2} [T_n(x)]^2\, dx = \begin{cases} \pi/2, & n \neq 0 \\ \pi, & n = 0 \end{cases}$$

$$\text{Rodrigues' Formula: } \frac{(-1)^n (1 - x^2)^{1/2} \sqrt{\pi}}{2^{n+1} \Gamma(n + \frac{1}{2})} \frac{d^n}{dx^n} \{(1 - x^2)^{n-(1/2)}\} = T_n(x)$$

$$\text{Generating Function: } \frac{1 - xz}{1 - 2xz + z^2} = \sum_{n=0}^{\infty} T_n(x) z^n,\ -1 < x < 1,\ |z| < 1$$

Inequality: $|T_n(x)| \leq 1,\ -1 \leq x \leq 1.$

III

Name: Tschebysheff, Second Kind *Symbol:* $U_n(x)$ *Interval:* $[-1,1]$

Differential Equation: $(1 - x^2)\, y'' - 3\, xy' + n(n + 2)y = 0$

$$y = U_n(x)$$

$$\text{Explicit Expression: } U_n(x) = \sum_{m=0}^{[n/2]} (-1)^m \frac{(m - n)!}{m!(n - 2m)!} (2x)^{n-2m}$$

$$U_n(\cos\theta) = \frac{\sin[(n + 1)\theta]}{\sin\theta}$$

* From Beyer, W. H., Ed., *CRC Handbook of Mathematical Sciences,* 5th ed., CRC Press, Boca Raton, 1978, 557—560. With permission.

Recurrence Relation: $U_{n+1}(x) = 2xU_n(x) - U_{n-1}(x)$
Weight: $(1 - x^2)^{1/2}$ Standardization: $U_n(1) = n + 1$

Norm: $\int_{-1}^{+1} (1 - x^2)^{1/2}[U_n(x)]^2\,dx = \frac{\pi}{2}$

Rodrigues' Formula: $U_n(x) = \frac{(-1)^n(n+1)\sqrt{\pi}}{(1-x^2)^{1/2}2^{n+1}\Gamma(n+\frac{3}{2})}\frac{d^n}{dx^n}\{(1-x^2)^{n+(1/2)}\}$

Generating Function: $\frac{1}{1 - 2xz + z^2} = \sum_{n=0}^{\infty} U_n(x)z^n, -1 < x < 1, |z| < 1$

Inequality: $|U_n(x)| \leq n + 1, -1 \leq x \leq 1.$

IV

Name: Jacobi   Symbol: $P_n^{(\alpha,\beta)}(x)$ Interval: $[-1,1]$
Differential Equation:

$$(1 - x^2)y'' + [\beta - \alpha - (\alpha + \beta + 2)x]y' + n(n + \alpha + \beta + 1)y = 0$$
$$y = P_n^{(\alpha,\beta)}(x)$$

Explicit Expression: $P_n^{(\alpha,\beta)}(x) = \frac{1}{2^n}\sum_{m=0}^{n}\binom{n+\alpha}{m}\binom{n+\beta}{n-m}(x-1)^{n-m}(x+1)^m$

Recurrence Relation: $2(n+1)(n+\alpha+\beta+1)(2n+\alpha+\beta)P_{n+1}^{(\alpha,\beta)}(x)$
$$= (2n+\alpha+\beta+1)[(\alpha^2-\beta^2) + (2n+\alpha+\beta+2) \times (2n+\alpha+\beta)x]P_n^{(\alpha,\beta)}(x) - 2(n+\alpha)(n+\beta)(2n+\alpha+\beta+2)P_{n-1}^{(\alpha,\beta)}(x)$$

Weight: $(1 - x)^\alpha(1 + x)^\beta; \alpha,\beta > 1$ Standardization: $P_n^{(\alpha,\beta)}(x) = \binom{n+\alpha}{n}$

Norm: $\int_{-1}^{+1}(1-x)^\alpha(1+x)^\beta[P_n^{(\alpha,\beta)}(x)]^2\,dx = \frac{2^{\alpha+\beta+1}\Gamma(n+\alpha+1)\Gamma(n+\beta+1)}{(2n+\alpha+\beta+1)n!\,\Gamma(n+\alpha+\beta+1)}$

Rodrigues' Formula: $P_n^{(\alpha,\beta)}(x) = \frac{(-1)^n}{2^n n!(1-x)^\alpha(1+x)^\beta}\frac{d^n}{dx^n}\{(1-x)^{n+\alpha}(1+x)^{n+\beta}\}$

Generating Function: $R^{-1}(1 - z + R)^{-\alpha}(1 + z + R)^{-\beta} = \sum_{n=0}^{\infty} 2^{-\alpha-\beta}P_n^{(\alpha,\beta)}(x)z^n,$

$$R = \sqrt{1 - 2xz + z^2}, |z| < 1$$

Inequality: $\max_{-1\leq x\leq 1}|P_n^{(\alpha,\beta)}(x)| = \begin{cases}\binom{n+q}{n} \sim n^q \text{ if } q = \max(\alpha,\beta) \geq -\frac{1}{2} \\ |P_n^{(\alpha,\beta)}(x')| \sim n^{-1/2} \text{ if } q < -\frac{1}{2} \\ x' \text{ is one of the two maximum points nearest} \\ \frac{\beta-\alpha}{\alpha+\beta+1}\end{cases}$

V

Name: Generalized Laguerre   Symbol: $L_n^{(\alpha)}(x)$ Interval: $[0, \infty]$
Differential Equation: $xy'' + (\alpha + 1 - x)y' + ny = 0$
$$y = L_n^{(\alpha)}(x)$$

Explicit Expression: $L_n^{(\alpha)}(x) = \sum_{m=0}^{n}(-1)^m\binom{n+\alpha}{n-m}\frac{1}{m!}x^m$

Recurrence Relation: $(n + 1) L_n^{(\alpha)}+ 1(x) = [(2n + \alpha + 1) - x] L_n^{(\alpha)}(x) - (n + \alpha) L_n^{(\alpha)}- 1(x)$

Weight: $x^\alpha e^{-x}, \alpha > -1$ Standardization: $L_n^{(\alpha)}(x) = \frac{(-1)^n}{n!}x^n + \cdots$

Norm: $\int_0^\infty x^\alpha e^{-x}[L_n^{(\alpha)}(x)]^2\,dx = \frac{\Gamma(n+\alpha+1)}{n!}$

Rodrigues' Formula: $L_n^{(\alpha)}(x) = \frac{1}{n!x^\alpha e^{-x}}\frac{d^n}{dx^n}\{x^{n+\alpha}e^{-x}\}$

Generating Function: $(1 - z)^{-\alpha-1} \exp\left(\frac{xz}{z-1}\right) = \sum_{n=0}^{\infty} L_n^{(\alpha)}(x) z^n$

Inequality: $|L_n^{(\alpha)}(x)| \le \frac{\Gamma(n + \alpha + 1)}{n!\,\Gamma(\alpha + 1)} e^{x/2}; \quad x \ge 0, \; \alpha > 0$

$$|L_n^{(\alpha)}(x)| \le \left[2 - \frac{\Gamma(\alpha + n + 1)}{n!\,\Gamma(\alpha + 1)}\right] e^{x/2}; \quad x \ge 0, \; -1 < \alpha < 0$$

Orthogonal Polynomials

Name: Hermite *Symbol:* $H_n(x)$ *Interval:* $[-\infty, \infty]$
Differential Equation: $y'' - 2xy' + 2ny = 0$

Explicit Expression: $H_n(x) = \sum_{m=0}^{[n/2]} \frac{(-1)^m \, n! \, (2x)^{n-2m}}{m! \, (n-2m)!}$

Recurrence Relation: $H_{n+1}(x) = 2x\, H_n(x) - 2nH_{n-1}(x)$
Weight: e^{-x^2} *Standardization:* $H_n(1) = 2^n x^n + \ldots$

Norm: $\int_{-\infty}^{\infty} e^{-x^2} [H_n(x)]^2 \, dx = 2^n n! \sqrt{\pi}$

Rodriques' Formula: $H_n(x) = (-1)^n e^{x^2} \frac{d^n}{dx^n} (e^{-x^2})$

Generating Function: $e^{-z^2 + 2zx} = \sum_{n=0}^{\infty} H_n(x) \frac{z^n}{n!}$

Inequality: $|H_n(x)| < e^{\frac{x^2}{2}} k \, 2^{n/2} \sqrt{n!} \quad k \approx 1.086435$

NORMAL PROBABILITY FUNCTION

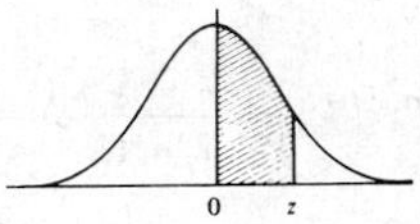

Areas under the Standard Normal Curve from 0 to z

z	0	1	2	3	4	5	6	7	8	9
0.0	.0000	.0040	.0080	.0120	.0160	.0199	.0239	.0279	.0319	.0359
0.1	.0398	.0438	.0478	.0517	.0557	.0596	.0636	.0675	.0714	.0754
0.2	.0793	.0832	.0871	.0910	.0948	.0987	.1026	.1064	.1103	.1141
0.3	.1179	.1217	.1255	.1293	.1331	.1368	.1406	.1443	.1480	.1517
0.4	.1554	.1591	.1628	.1664	.1700	.1736	.1772	.1808	.1844	.1879
0.5	.1915	.1950	.1985	.2019	.2054	.2088	.2123	.2157	.2190	.2224
0.6	.2258	.2291	.2324	.2357	.2389	.2422	.2454	.2486	.2518	.2549
0.7	.2580	.2612	.2652	.2673	.2704	.2734	.2764	.2794	.2823	.2852
0.8	.2881	.2910	.2939	.2967	.2996	.3023	.3051	.3078	.3106	.3133
0.9	.3159	.3186	.3212	.3238	.3264	.3289	.3315	.3340	.3365	.3389
1.0	.3413	.3438	.3461	.3485	.3508	.3531	.3554	.3577	.3599	.3621
1.1	.3643	.3665	.3686	.3708	.3729	.3749	.3770	.3790	.3810	.3830
1.2	.3849	.3869	.3888	.3907	.3925	.3944	.3962	.3980	.3997	.4015
1.3	.4032	.4049	.4066	.4082	.4099	.4115	.4131	.4147	.4162	.4177
1.4	.4192	.4207	.4222	.4236	.4251	.4265	.4279	.4292	.4306	.4319
1.5	.4332	.4345	.4357	.4370	.4382	.4394	.4406	.4418	.4429	.4441
1.6	.4452	.4463	.4474	.4484	.4495	.4505	.4515	.4525	.4535	.4545
1.7	.4554	.4564	.4573	.4582	.4591	.4599	.4608	.4616	.4625	.4633
1.8	.4641	.4649	.4656	.4664	.4671	.4678	.4686	.4693	.4699	.4706
1.9	.4713	.4719	.4726	.4732	.4738	.4744	.4750	.4756	.4761	.4767
2.0	.4772	.4778	.4783	.4788	.4793	.4798	.4803	.4808	.4812	.4817
2.1	.4821	.4826	.4830	.4834	.4838	.4842	.4846	.4850	.4854	.4857
2.2	.4861	.4864	.4868	.4871	.4875	.4878	.4881	.4884	.4887	.4890
2.3	.4893	.4896	.4898	.4901	.4904	.4906	.4909	.4911	.4913	.4916
2.4	.4918	.4920	.4922	.4925	.4927	.4929	.4931	.4932	.4934	.4936
2.5	.4938	.4940	.4941	.4943	.4945	.4946	.4948	.4949	.4951	.4952
2.6	.4953	.4955	.4956	.4957	.4959	.4960	.4961	.4962	.4963	.4964
2.7	.4965	.4966	.4967	.4968	.4969	.4970	.4971	.4972	.4973	.4974
2.8	.4974	.4975	.4976	.4977	.4977	.4978	.4979	.4979	.4980	.4981
2.9	.4981	.4982	.4982	.4983	.4984	.4984	.4985	.4985	.4986	.4986
3.0	.4987	.4987	.4987	.4988	.4988	.4989	.4989	.4989	.4990	.4990
3.1	.4990	.4991	.4991	.4991	.4992	.4992	.4992	.4992	.4993	.4993
3.2	.4993	.4993	.4994	.4994	.4994	.4994	.4994	.4995	.4995	.4995
3.3	.4995	.4995	.4995	.4996	.4996	.4996	.4996	.4996	.4996	.4997
3.4	.4997	.4997	.4997	.4997	.4997	.4997	.4997	.4997	.4997	.4998
3.5	.4998	.4998	.4998	.4998	.4998	.4998	.4998	.4998	.4998	.4998
3.6	.4998	.4998	.4999	.4999	.4999	.4999	.4999	.4999	.4999	.4999
3.7	.4999	.4999	.4999	.4999	.4999	.4999	.4999	.4999	.4999	.4999
3.8	.4999	.4999	.4999	.4999	.4999	.4999	.4999	.4999	.4999	.4999
3.9	.5000	.5000	.5000	.5000	.5000	.5000	.5000	.5000	.5000	.5000

F(z) below refers to area under Standard Normal Curve from $-\infty$ to z

z	1.282	1.645	1.960	2.326	2.576	3.090
F(z)	.90	.95	.975	.99	.995	.999
2[1 − F(z)]	.20	.10	.05	.02	.01	.002

PERCENTAGE POINTS, STUDENT'S *t*-DISTRIBUTION

This table gives values of t such that

$$F(t) = \int_{-\infty}^{t} \frac{\Gamma\left(\frac{n+1}{2}\right)}{\sqrt{n\pi}\,\Gamma\left(\frac{n}{2}\right)} \left(1 + \frac{x^2}{n}\right)^{-\frac{n+1}{2}} dx$$

for n, the number of degrees of freedom, equal to 1, 2, . . ., 30, 40, 60, 120, ∞; and for $F(t) = 0.60, 0.75, 0.90, 0.95, 0.975, 0.99, 0.995$, and 0.9995. The *t*-distribution is symmetrical, so that $F(-t) = 1 - F(t)$

n \ F	.60	.75	.90	.95	.975	.99	.995	.9995
1	.325	1.000	3.078	6.314	12.706	31.821	63.657	636.619
2	.289	.816	1.886	2.920	4.303	6.965	9.925	31.598
3	.277	.765	1.638	2.353	3.182	4.541	5.841	12.924
4	.271	.741	1.533	2.132	2.776	3.747	4.604	8.610
5	.267	.727	1.476	2.015	2.571	3.365	4.032	6.869
6	.265	.718	1.440	1.943	2.447	3.143	3.707	5.959
7	.263	.711	1.415	1.895	2.365	2.998	3.499	5.408
8	.262	.706	1.397	1.860	2.306	2.896	3.355	5.041
9	.261	.703	1.383	1.833	2.262	2.821	3.250	4.781
10	.260	.700	1.372	1.812	2.228	2.764	3.169	4.587
11	.260	.697	1.363	1.796	2.201	2.718	3.106	4.437
12	.259	.695	1.356	1.782	2.179	2.681	3.055	4.318
13	.259	.694	1.350	1.771	2.160	2.650	3.012	4.221
14	.258	.692	1.345	1.761	2.145	2.624	2.977	4.140
15	.258	.691	1.341	1.753	2.131	2.602	2.947	4.073
16	.258	.690	1.337	1.746	2.120	2.583	2.921	4.015
17	.257	.689	1.333	1.740	2.110	2.567	2.898	3.965
18	.257	.688	1.330	1.734	2.101	2.552	2.878	3.922
19	.257	.688	1.328	1.729	2.093	2.539	2.861	3.883
20	.257	.687	1.325	1.725	2.086	2.528	2.845	3.850
21	.257	.686	1.323	1.721	2.080	2.518	2.831	3.819
22	.256	.686	1.321	1.717	2.074	2.508	2.819	3.792
23	.256	.685	1.319	1.714	2.069	2.500	2.807	3.767
24	.256	.685	1.318	1.711	2.064	2.492	2.797	3.745
25	.256	.684	1.316	1.708	2.060	2.485	2.787	3.725
26	.256	.684	1.315	1.706	2.056	2.479	2.779	3.707
27	.256	.684	1.314	1.703	2.052	2.473	2.771	3.690
28	.256	.683	1.313	1.701	2.048	2.467	2.763	3.674
29	.256	.683	1.311	1.699	2.045	2.462	2.756	3.659
30	.256	.683	1.310	1.697	2.042	2.457	2.750	3.646
40	.255	.681	1.303	1.684	2.021	2.423	2.704	3.551
60	.254	.679	1.296	1.671	2.000	2.390	2.660	3.460
120	.254	.677	1.289	1.658	1.980	2.358	2.617	3.373
∞	.253	.674	1.282	1.645	1.960	2.326	2.576	3.291

* This table is abridged from the "Statistical Tables" of R. A. Fisher and Frank Yates published by Oliver & Boyd. Ltd., Edinburgh and London, 1938. It is here published with the kind permission of the authors and their publishers.

PERCENTAGE POINTS, CHI-SQUARE DISTRIBUTION

This table gives values of χ^2 such that

$$F(\chi^2) = \int_0^{\chi^2} \frac{1}{2^{\frac{n}{2}}\,\Gamma\left(\frac{n}{2}\right)} x^{\frac{n-2}{2}} e^{-\frac{x}{2}} dx$$

for n, the number of degrees of freedom, equal to 1, 2, . . . , 30. For $n > 30$, a normal approximation is quite accurate. The expression $\sqrt{2\chi^2} - \sqrt{2n-1}$ is approximately normally distributed as the standard normal distribution. Thus χ^2_α, the α-point of the distribution, may be computed by the formula

$$\chi^2_\alpha = \tfrac{1}{2}[x_\alpha + \sqrt{2n-1}]^2,$$

where x_α is the α-point of the cumulative normal distribution. For even values of n, $F(\chi^2)$ can be written as

$$1 - F(\chi^2) = \sum_{x=0}^{x'-1} \frac{e^{-\lambda}\lambda^x}{x!}$$

with $\lambda = \frac{1}{2}\chi^2$ and $x' = \frac{1}{2}n$. Thus the cumulative Chi-Square distribution is related to the cumulative Poisson distribution.

Another approximate formula for large n

$$\chi_\alpha^2 = n\left(1 - \frac{2}{9n} + z_\alpha\sqrt{\frac{2}{9n}}\right)^3$$

n = degrees of freedom
z_α = the normal deviate, (the value of x for which $F(x)$ = the desired percentile).

x	1.282	1.645	1.960	2.326	2.576	3.090
$F(x)$	.90	.95	.975	.99	.995	.999

$\chi^2_{.99} = 60[1 - 0.00370 + 2.326(0.06086)]^3 = 88.4$ is the 99th percentile for 60 degrees of freedom.

$$F(\chi^2) = \int_0^{\chi^2} \frac{1}{2^{\frac{n}{2}}\Gamma\left(\frac{n}{2}\right)} x^{\frac{n-2}{2}} e^{-\frac{x}{2}}\,dx$$

n \ F	.005	.010	.025	.050	.100	.250	.500	.750	.900	.950	.975	.990	.995
1	.0000393	.000157	.000982	.00393	.0158	.102	.455	1.32	2.71	3.84	5.02	6.63	7.88
2	.0100	.0201	.0506	.103	.211	.575	1.39	2.77	4.61	5.99	7.38	9.21	10.6
3	.0717	.115	.216	.352	.584	1.21	2.37	4.11	6.25	7.81	9.35	11.3	12.8
4	.207	.297	.484	.711	1.06	1.92	3.36	5.39	7.78	9.49	11.1	13.3	14.9
5	.412	.554	.831	1.15	1.61	2.67	4.35	6.63	9.24	11.1	12.8	15.1	16.7
6	.676	.872	1.24	1.64	2.20	3.45	5.35	7.84	10.6	12.6	14.4	16.8	18.5
7	.989	1.24	1.69	2.17	2.83	4.25	6.35	9.04	12.0	14.1	16.0	18.5	20.3
8	1.34	1.65	2.18	2.73	3.49	5.07	7.34	10.2	13.4	15.5	17.5	20.1	22.0
9	1.73	2.09	2.70	3.33	4.17	5.90	8.34	11.4	14.7	16.9	19.0	21.7	23.6
10	2.16	2.56	3.25	3.94	4.87	6.74	9.34	12.5	16.0	18.3	20.5	23.2	25.2
11	2.60	3.05	3.82	4.57	5.58	7.58	10.3	13.7	17.3	19.7	21.9	24.7	26.8
12	3.07	3.57	4.40	5.23	6.30	8.44	11.3	14.8	18.5	21.0	23.3	26.2	28.3
13	3.57	4.11	5.01	5.89	7.04	9.30	12.3	16.0	19.8	22.4	24.7	27.7	29.8
14	4.07	4.66	5.63	6.57	7.79	10.2	13.3	17.1	21.1	23.7	26.1	29.1	31.3
15	4.60	5.23	6.26	7.26	8.55	11.0	14.3	18.2	22.3	25.0	27.5	30.6	32.8
16	5.14	5.81	6.91	7.96	9.31	11.9	15.3	19.4	23.5	26.3	28.8	32.0	34.3
17	5.70	6.41	7.56	8.67	10.1	12.8	16.3	20.5	24.8	27.6	30.2	33.4	35.7
18	6.26	7.01	8.23	9.39	10.9	13.7	17.3	21.6	26.0	28.9	31.5	34.8	37.2
19	6.84	7.63	8.91	10.1	11.7	14.6	18.3	22.7	27.2	30.1	32.9	36.2	38.6
20	7.43	8.26	9.59	10.9	12.4	15.5	19.3	23.8	28.4	31.4	34.2	37.6	40.0
21	8.03	8.90	10.3	11.6	13.2	16.3	20.3	24.9	29.6	32.7	35.5	38.9	41.4
22	8.64	9.54	11.0	12.3	14.0	17.2	21.3	26.0	30.8	33.9	36.8	40.3	42.8
23	9.26	10.2	11.7	13.1	14.8	18.1	22.3	27.1	32.0	35.2	38.1	41.6	44.2
24	9.89	10.9	12.4	13.8	15.7	19.0	23.3	28.2	33.2	36.4	39.4	43.0	45.6
25	10.5	11.5	13.1	14.6	16.5	19.9	24.3	29.3	34.4	37.7	40.6	44.3	46.9
26	11.2	12.2	13.8	15.4	17.3	20.8	25.3	30.4	35.6	38.9	41.9	45.6	48.3
27	11.8	12.9	14.6	16.2	18.1	21.7	26.3	31.5	36.7	40.1	43.2	47.0	49.6
28	12.5	13.6	15.3	16.9	18.9	22.7	27.3	32.6	37.9	41.3	44.5	48.3	51.0
29	13.1	14.3	16.0	17.7	19.8	23.6	28.3	33.7	39.1	42.6	45.7	49.6	52.3
30	13.8	15.0	16.8	18.5	20.6	24.5	29.3	34.8	40.3	43.8	47.0	50.9	53.7

PERCENTAGE POINTS, *F*-DISTRIBUTION

This table gives values of F such that

$$F(F) = \int_0^F \frac{\Gamma\left(\frac{m+n}{2}\right)}{\Gamma\left(\frac{m}{2}\right)\Gamma\left(\frac{n}{2}\right)} m^{\frac{m}{2}} n^{\frac{n}{2}} x^{\frac{m-2}{2}} (n + mx)^{-\frac{m+n}{2}}\,dx$$

for selected values of m, the number of degrees of freedom of the numerator of F; and for selected values of n, the number of degrees of freedom of the denominator of F. The table also provides values corresponding to $F(F) = .10, .05, .025, .01, .005, .001$ since $F_{1-\alpha}$ for m and n degrees of freedom is the reciprocal of F_α for n and m degrees of freedom. Thus

$$F_{.05}(4, 7) = \frac{1}{F_{.95}(7, 4)} = \frac{1}{6.09} = .164\,.$$

$$F(F) = \int_0^F \frac{\Gamma\left(\frac{m+n}{2}\right)}{\Gamma\left(\frac{m}{2}\right)\Gamma\left(\frac{n}{2}\right)} m^{\frac{m}{2}} n^{\frac{n}{2}} x^{\frac{m}{2}-1} (n+mx)^{-\frac{m+n}{2}} dx = .90$$

n \ m	1	2	3	4	5	6	7	8	9	10	12	15	20	24	30	40	60	120	∞
1	39.86	49.50	53.59	55.83	57.24	58.20	58.91	59.44	59.86	60.19	60.71	61.22	61.74	62.00	62.26	62.53	62.79	63.06	63.33
2	8.53	9.00	9.16	9.24	9.29	9.33	9.35	9.37	9.38	9.39	9.41	9.42	9.44	9.45	9.46	9.47	9.47	9.48	9.49
3	5.54	5.46	5.39	5.34	5.31	5.28	5.27	5.25	5.24	5.23	5.22	5.20	5.18	5.18	5.17	5.16	5.15	5.14	5.13
4	4.54	4.32	4.19	4.11	4.05	4.01	3.98	3.95	3.94	3.92	3.90	3.87	3.84	3.83	3.82	3.80	3.79	3.78	3.76
5	4.06	3.78	3.62	3.52	3.45	3.40	3.37	3.34	3.32	3.30	3.27	3.24	3.21	3.19	3.17	3.16	3.14	3.12	3.10
6	3.78	3.46	3.29	3.18	3.11	3.05	3.01	2.98	2.96	2.94	2.90	2.87	2.84	2.82	2.80	2.78	2.76	2.74	2.72
7	3.59	3.26	3.07	2.96	2.88	2.83	2.78	2.75	2.72	2.70	2.67	2.63	2.59	2.58	2.56	2.54	2.51	2.49	2.47
8	3.46	3.11	2.92	2.81	2.73	2.67	2.62	2.59	2.56	2.54	2.50	2.46	2.42	2.40	2.38	2.36	2.34	2.32	2.29
9	3.36	3.01	2.81	2.69	2.61	2.55	2.51	2.47	2.44	2.42	2.38	2.34	2.30	2.28	2.25	2.23	2.21	2.18	2.16
10	3.29	2.92	2.73	2.61	2.52	2.46	2.41	2.38	2.35	2.32	2.28	2.24	2.20	2.18	2.16	2.13	2.11	2.08	2.06
11	3.23	2.86	2.66	2.54	2.45	2.39	2.34	2.30	2.27	2.25	2.21	2.17	2.12	2.10	2.08	2.05	2.03	2.00	1.97
12	3.18	2.81	2.61	2.48	2.39	2.33	2.28	2.24	2.21	2.19	2.15	2.10	2.06	2.04	2.01	1.99	1.96	1.93	1.90
13	3.14	2.76	2.56	2.43	2.35	2.28	2.23	2.20	2.16	2.14	2.10	2.05	2.01	1.98	1.96	1.93	1.90	1.88	1.85
14	3.10	2.73	2.52	2.39	2.31	2.24	2.19	2.15	2.12	2.10	2.05	2.01	1.96	1.94	1.91	1.89	1.86	1.83	1.80
15	3.07	2.70	2.49	2.36	2.27	2.21	2.16	2.12	2.09	2.06	2.02	1.97	1.92	1.90	1.87	1.85	1.82	1.79	1.76
16	3.05	2.67	2.46	2.33	2.24	2.18	2.13	2.09	2.06	2.03	1.99	1.94	1.89	1.87	1.84	1.81	1.78	1.75	1.72
17	3.03	2.64	2.44	2.31	2.22	2.15	2.10	2.06	2.03	2.00	1.96	1.91	1.86	1.84	1.81	1.78	1.75	1.72	1.69
18	3.01	2.62	2.42	2.29	2.20	2.13	2.08	2.04	2.00	1.98	1.93	1.89	1.84	1.81	1.78	1.75	1.72	1.69	1.66
19	2.99	2.61	2.40	2.27	2.18	2.11	2.06	2.02	1.98	1.96	1.91	1.86	1.81	1.79	1.76	1.73	1.70	1.67	1.63
20	2.97	2.59	2.38	2.25	2.16	2.09	2.04	2.00	1.96	1.94	1.89	1.84	1.79	1.77	1.74	1.71	1.68	1.64	1.61
21	2.96	2.57	2.36	2.23	2.14	2.08	2.02	1.98	1.95	1.92	1.87	1.83	1.78	1.75	1.72	1.69	1.66	1.62	1.59
22	2.95	2.56	2.35	2.22	2.13	2.06	2.01	1.97	1.93	1.90	1.86	1.81	1.76	1.73	1.70	1.67	1.64	1.60	1.57
23	2.94	2.55	2.34	2.21	2.11	2.05	1.99	1.95	1.92	1.89	1.84	1.80	1.74	1.72	1.69	1.66	1.62	1.59	1.55
24	2.93	2.54	2.33	2.19	2.10	2.04	1.98	1.94	1.91	1.88	1.83	1.78	1.73	1.70	1.67	1.64	1.61	1.57	1.53
25	2.92	2.53	2.32	2.18	2.09	2.02	1.97	1.93	1.89	1.87	1.82	1.77	1.72	1.69	1.66	1.63	1.59	1.56	1.52
26	2.91	2.52	2.31	2.17	2.08	2.01	1.96	1.92	1.88	1.86	1.81	1.76	1.71	1.68	1.65	1.61	1.58	1.54	1.50
27	2.90	2.51	2.30	2.17	2.07	2.00	1.95	1.91	1.87	1.85	1.80	1.75	1.70	1.67	1.64	1.60	1.57	1.53	1.49
28	2.89	2.50	2.29	2.16	2.06	2.00	1.94	1.90	1.87	1.84	1.79	1.74	1.69	1.66	1.63	1.59	1.56	1.52	1.48
29	2.89	2.50	2.28	2.15	2.06	1.99	1.93	1.89	1.86	1.83	1.78	1.73	1.68	1.65	1.62	1.58	1.55	1.51	1.47
30	2.88	2.49	2.28	2.14	2.05	1.98	1.93	1.88	1.85	1.82	1.77	1.72	1.67	1.64	1.61	1.57	1.54	1.50	1.46
40	2.84	2.44	2.23	2.09	2.00	1.93	1.87	1.83	1.79	1.76	1.71	1.66	1.61	1.57	1.54	1.51	1.47	1.42	1.38
60	2.79	2.39	2.18	2.04	1.95	1.87	1.82	1.77	1.74	1.71	1.66	1.60	1.54	1.51	1.48	1.44	1.40	1.35	1.29
120	2.75	2.35	2.13	1.99	1.90	1.82	1.77	1.72	1.68	1.65	1.60	1.55	1.48	1.45	1.41	1.37	1.32	1.26	1.19
∞	2.71	2.30	2.08	1.94	1.85	1.77	1.72	1.67	1.63	1.60	1.55	1.49	1.42	1.38	1.34	1.30	1.24	1.17	1.00

$F = \frac{s_1^2}{s_2^2} = \frac{S_1}{m} \Big/ \frac{S_2}{n}$, where $s_1^2 = S_1/m$ and $s_2^2 = S_2/n$ are independent mean squares estimating a common variance σ^2 and based on m and n degrees of freedom, respectively.

$$F(F) = \int_0^F \frac{\Gamma\left(\frac{m+n}{2}\right)}{\Gamma\left(\frac{m}{2}\right)\Gamma\left(\frac{n}{2}\right)} m^{\frac{m}{2}} n^{\frac{n}{2}} x^{\frac{m}{2}-1} (n+mx)^{-\frac{m+n}{2}} dx = .95$$

n \ m	1	2	3	4	5	6	7	8	9	10	12	15	20	24	30	40	60	120	∞
1	161.4	199.5	215.7	224.6	230.2	234.0	236.8	238.9	240.5	241.9	243.9	245.9	248.0	249.1	250.1	251.1	252.2	253.3	254.3
2	18.51	19.00	19.16	19.25	19.30	19.33	19.35	19.37	19.38	19.40	19.41	19.43	19.45	19.45	19.46	19.47	19.48	19.49	19.50
3	10.13	9.55	9.28	9.12	9.01	8.94	8.89	8.85	8.81	8.79	8.74	8.70	8.66	8.64	8.62	8.59	8.57	8.55	8.53
4	7.71	6.94	6.59	6.39	6.26	6.16	6.09	6.04	6.00	5.96	5.91	5.86	5.80	5.77	5.75	5.72	5.69	5.66	5.63
5	6.61	5.79	5.41	5.19	5.05	4.95	4.88	4.82	4.77	4.74	4.68	4.62	4.56	4.53	4.50	4.46	4.43	4.40	4.36
6	5.99	5.14	4.76	4.53	4.39	4.28	4.21	4.15	4.10	4.06	4.00	3.94	3.87	3.84	3.81	3.77	3.74	3.70	3.67
7	5.59	4.74	4.35	4.12	3.97	3.87	3.79	3.73	3.68	3.64	3.57	3.51	3.44	3.41	3.38	3.34	3.30	3.27	3.23
8	5.32	4.46	4.07	3.84	3.69	3.58	3.50	3.44	3.39	3.35	3.28	3.22	3.15	3.12	3.08	3.04	3.01	2.97	2.93
9	5.12	4.26	3.86	3.63	3.48	3.37	3.29	3.23	3.18	3.14	3.07	3.01	2.94	2.90	2.86	2.83	2.79	2.75	2.71
10	4.96	4.10	3.71	3.48	3.33	3.22	3.14	3.07	3.02	2.98	2.91	2.85	2.77	2.74	2.70	2.66	2.62	2.58	2.54
11	4.84	3.98	3.59	3.36	3.20	3.09	3.01	2.95	2.90	2.85	2.79	2.72	2.65	2.61	2.57	2.53	2.49	2.45	2.40
12	4.75	3.89	3.49	3.26	3.11	3.00	2.91	2.85	2.80	2.75	2.69	2.62	2.54	2.51	2.47	2.43	2.38	2.34	2.30
13	4.67	3.81	3.41	3.18	3.03	2.92	2.83	2.77	2.71	2.67	2.60	2.53	2.46	2.42	2.38	2.34	2.30	2.25	2.21
14	4.60	3.74	3.34	3.11	2.96	2.85	2.76	2.70	2.65	2.60	2.53	2.46	2.39	2.35	2.31	2.27	2.22	2.18	2.13
15	4.54	3.68	3.29	3.06	2.90	2.79	2.71	2.64	2.59	2.54	2.48	2.40	2.33	2.29	2.25	2.20	2.16	2.11	2.07
16	4.49	3.63	3.24	3.01	2.85	2.74	2.66	2.59	2.54	2.49	2.42	2.35	2.28	2.24	2.19	2.15	2.11	2.06	2.01
17	4.45	3.59	3.20	2.96	2.81	2.70	2.61	2.55	2.49	2.45	2.38	2.31	2.23	2.19	2.15	2.10	2.06	2.01	1.96
18	4.41	3.55	3.16	2.93	2.77	2.66	2.58	2.51	2.46	2.41	2.34	2.27	2.19	2.15	2.11	2.06	2.02	1.97	1.92
19	4.38	3.52	3.13	2.90	2.74	2.63	2.54	2.48	2.42	2.38	2.31	2.23	2.16	2.11	2.07	2.03	1.98	1.93	1.88
20	4.35	3.49	3.10	2.87	2.71	2.60	2.51	2.45	2.39	2.35	2.28	2.20	2.12	2.08	2.04	1.99	1.95	1.90	1.84
21	4.32	3.47	3.07	2.84	2.68	2.57	2.49	2.42	2.37	2.32	2.25	2.18	2.10	2.05	2.01	1.96	1.92	1.87	1.81
22	4.30	3.44	3.05	2.82	2.66	2.55	2.46	2.40	2.34	2.30	2.23	2.15	2.07	2.03	1.98	1.94	1.89	1.84	1.78
23	4.28	3.42	3.03	2.80	2.64	2.53	2.44	2.37	2.32	2.27	2.20	2.13	2.05	2.01	1.96	1.91	1.86	1.81	1.76
24	4.26	3.40	3.01	2.78	2.62	2.51	2.42	2.36	2.30	2.25	2.18	2.11	2.03	1.98	1.94	1.89	1.84	1.79	1.73
25	4.24	3.39	2.99	2.76	2.60	2.49	2.40	2.34	2.28	2.24	2.16	2.09	2.01	1.96	1.92	1.87	1.82	1.77	1.71
26	4.23	3.37	2.98	2.74	2.59	2.47	2.39	2.32	2.27	2.22	2.15	2.07	1.99	1.95	1.90	1.85	1.80	1.75	1.69
27	4.21	3.35	2.96	2.73	2.57	2.46	2.37	2.31	2.25	2.20	2.13	2.06	1.97	1.93	1.88	1.84	1.79	1.73	1.67
28	4.20	3.34	2.95	2.71	2.56	2.45	2.36	2.29	2.24	2.19	2.12	2.04	1.96	1.91	1.87	1.82	1.77	1.71	1.65
29	4.18	3.33	2.93	2.70	2.55	2.43	2.35	2.28	2.22	2.18	2.10	2.03	1.94	1.90	1.85	1.81	1.75	1.70	1.64
30	4.17	3.32	2.92	2.69	2.53	2.42	2.33	2.27	2.21	2.16	2.09	2.01	1.93	1.89	1.84	1.79	1.74	1.68	1.62
40	4.08	3.23	2.84	2.61	2.45	2.34	2.25	2.18	2.12	2.08	2.00	1.92	1.84	1.79	1.74	1.69	1.64	1.58	1.51
60	4.00	3.15	2.76	2.53	2.37	2.25	2.17	2.10	2.04	1.99	1.92	1.84	1.75	1.70	1.65	1.59	1.53	1.47	1.39
120	3.92	3.07	2.68	2.45	2.29	2.17	2.09	2.02	1.96	1.91	1.83	1.75	1.66	1.61	1.55	1.50	1.43	1.35	1.25
∞	3.84	3.00	2.60	2.37	2.21	2.10	2.01	1.94	1.88	1.83	1.75	1.67	1.57	1.52	1.46	1.39	1.32	1.22	1.00

$F = \frac{s_1^2}{s_2^2} = \frac{S_1}{m} \Big/ \frac{S_2}{n}$, where $s_1^2 = S_1/m$ and $s_2^2 = S_2/n$ are independent mean squares estimating a common variance σ^2 and based on m and n degrees of freedom, respectively.

$$F(F) = \int_0^F \frac{\Gamma\left(\frac{m+n}{2}\right)}{\Gamma\left(\frac{m}{2}\right)\Gamma\left(\frac{n}{2}\right)} m^{\frac{m}{2}} n^{\frac{n}{2}} x^{\frac{m}{2}-1} (n+mx)^{-\frac{m+n}{2}} dx = .975$$

n \ m	1	2	3	4	5	6	7	8	9	10	12	15	20	24	30	40	60	120	∞
1	647.8	799.5	864.2	899.6	921.8	937.1	948.2	956.7	963.3	968.6	976.7	984.9	993.1	997.2	1001	1006	1010	1014	1018
2	38.51	39.00	39.17	39.25	39.30	39.33	39.36	39.37	39.39	39.40	39.41	39.43	39.45	39.46	39.46	39.47	39.48	39.49	39.50
3	17.44	16.04	15.44	15.10	14.88	14.73	14.62	14.54	14.47	14.42	14.34	14.25	14.17	14.12	14.08	14.04	13.99	13.95	13.90
4	12.22	10.65	9.98	9.60	9.36	9.20	9.07	8.98	8.90	8.84	8.75	8.66	8.56	8.51	8.46	8.41	8.36	8.31	8.26
5	10.01	8.43	7.76	7.39	7.15	6.98	6.85	6.76	6.68	6.62	6.52	6.43	6.33	6.28	6.23	6.18	6.12	6.07	6.02
6	8.81	7.26	6.60	6.23	5.99	5.82	5.70	5.60	5.52	5.46	5.37	5.27	5.17	5.12	5.07	5.01	4.96	4.90	4.85
7	8.07	6.54	5.89	5.52	5.29	5.12	4.99	4.90	4.82	4.76	4.67	4.57	4.47	4.42	4.36	4.31	4.25	4.20	4.14
8	7.57	6.06	5.42	5.05	4.82	4.65	4.53	4.43	4.36	4.30	4.20	4.10	4.00	3.95	3.89	3.84	3.78	3.73	3.67
9	7.21	5.71	5.08	4.72	4.48	4.32	4.20	4.10	4.03	3.96	3.87	3.77	3.67	3.61	3.56	3.51	3.45	3.39	3.33
10	6.94	5.46	4.83	4.47	4.24	4.07	3.95	3.85	3.78	3.72	3.62	3.52	3.42	3.37	3.31	3.26	3.20	3.14	3.08
11	6.72	5.26	4.63	4.28	4.04	3.88	3.76	3.66	3.59	3.53	3.43	3.33	3.23	3.17	3.12	3.06	3.00	2.94	2.88
12	6.55	5.10	4.47	4.12	3.89	3.73	3.61	3.51	3.44	3.37	3.28	3.18	3.07	3.02	2.96	2.91	2.85	2.79	2.72
13	6.41	4.97	4.35	4.00	3.77	3.60	3.48	3.39	3.31	3.25	3.15	3.05	2.95	2.89	2.84	2.78	2.72	2.66	2.60
14	6.30	4.86	4.24	3.89	3.66	3.50	3.38	3.29	3.21	3.15	3.05	2.95	2.84	2.79	2.73	2.67	2.61	2.55	2.49
15	6.20	4.77	4.15	3.80	3.58	3.41	3.29	3.20	3.12	3.06	2.96	2.86	2.76	2.70	2.64	2.59	2.52	2.46	2.40
16	6.12	4.69	4.08	3.73	3.50	3.34	3.22	3.12	3.05	2.99	2.89	2.79	2.68	2.63	2.57	2.51	2.45	2.38	2.32
17	6.04	4.62	4.01	3.66	3.44	3.28	3.16	3.06	2.98	2.92	2.82	2.72	2.62	2.56	2.50	2.44	2.38	2.32	2.25
18	5.98	4.56	3.95	3.61	3.38	3.22	3.10	3.01	2.93	2.87	2.77	2.67	2.56	2.50	2.44	2.38	2.32	2.26	2.19
19	5.92	4.51	3.90	3.56	3.33	3.17	3.05	2.96	2.88	2.82	2.72	2.62	2.51	2.45	2.39	2.33	2.27	2.20	2.13
20	5.87	4.46	3.86	3.51	3.29	3.13	3.01	2.91	2.84	2.77	2.68	2.57	2.46	2.41	2.35	2.29	2.22	2.16	2.09
21	5.83	4.42	3.82	3.48	3.25	3.09	2.97	2.87	2.80	2.73	2.64	2.53	2.42	2.37	2.31	2.25	2.18	2.11	2.04
22	5.79	4.38	3.78	3.44	3.22	3.05	2.93	2.84	2.76	2.70	2.60	2.50	2.39	2.33	2.27	2.21	2.14	2.08	2.00
23	5.75	4.35	3.75	3.41	3.18	3.02	2.90	2.81	2.73	2.67	2.57	2.47	2.36	2.30	2.24	2.18	2.11	2.04	1.97
24	5.72	4.32	3.72	3.38	3.15	2.99	2.87	2.78	2.70	2.64	2.54	2.44	2.33	2.27	2.21	2.15	2.08	2.01	1.94
25	5.69	4.29	3.69	3.35	3.13	2.97	2.85	2.75	2.68	2.61	2.51	2.41	2.30	2.24	2.18	2.12	2.05	1.98	1.91
26	5.66	4.27	3.67	3.33	3.10	2.94	2.82	2.73	2.65	2.59	2.49	2.39	2.28	2.22	2.16	2.09	2.03	1.95	1.88
27	5.63	4.24	3.65	3.31	3.08	2.92	2.80	2.71	2.63	2.57	2.47	2.36	2.25	2.19	2.13	2.07	2.00	1.93	1.85
28	5.61	4.22	3.63	3.29	3.06	2.90	2.78	2.69	2.61	2.55	2.45	2.34	2.23	2.17	2.11	2.05	1.98	1.91	1.83
29	5.59	4.20	3.61	3.27	3.04	2.88	2.76	2.67	2.59	2.53	2.43	2.32	2.21	2.15	2.09	2.03	1.96	1.89	1.81
30	5.57	4.18	3.59	3.25	3.03	2.87	2.75	2.65	2.57	2.51	2.41	2.31	2.20	2.14	2.07	2.01	1.94	1.87	1.79
40	5.42	4.05	3.46	3.13	2.90	2.74	2.62	2.53	2.45	2.39	2.29	2.18	2.07	2.01	1.94	1.88	1.80	1.72	1.64
60	5.29	3.93	3.34	3.01	2.79	2.63	2.51	2.41	2.33	2.27	2.17	2.06	1.94	1.88	1.82	1.74	1.67	1.58	1.48
120	5.15	3.80	3.23	2.89	2.67	2.52	2.39	2.30	2.22	2.16	2.05	1.94	1.82	1.76	1.69	1.61	1.53	1.43	1.31
∞	5.02	3.69	3.12	2.79	2.57	2.41	2.29	2.19	2.11	2.05	1.94	1.83	1.71	1.64	1.57	1.48	1.39	1.27	1.00

$F = \frac{s_1^2}{s_2^2} = \frac{S_1}{m} \Big/ \frac{S_2}{n}$, where $s_1^2 = S_1/m$ and $s_2^2 = S_2/n$ are independent mean squares estimating a common variance σ^2 and based on m and n degrees of freedom, respectively.

$$F(F) = \int_0^F \frac{\Gamma\left(\frac{m+n}{2}\right)}{\Gamma\left(\frac{m}{2}\right)\Gamma\left(\frac{n}{2}\right)} m^{\frac{m}{2}} n^{\frac{n}{2}} x^{\frac{m}{2}-1} (n+mx)^{-\frac{m+n}{2}} dx = .99$$

n \ m	1	2	3	4	5	6	7	8	9	10	12	15	20	24	30	40	60	120	∞
1	4052	4999.5	5403	5625	5764	5859	5928	5982	6022	6056	6106	6157	6209	6235	6261	6287	6313	6339	6366
2	98.50	99.00	99.17	99.25	99.30	99.33	99.36	99.37	99.39	99.40	99.42	99.43	99.45	99.46	99.47	99.47	99.48	99.49	99.50
3	34.12	30.82	29.46	28.71	28.24	27.91	27.67	27.49	27.35	27.23	27.05	26.87	26.69	26.60	26.50	26.41	26.32	26.22	26.13
4	21.20	18.00	16.69	15.98	15.52	15.21	14.98	14.80	14.66	14.55	14.37	14.20	14.02	13.93	13.84	13.75	13.65	13.56	13.46
5	16.26	13.27	12.06	11.39	10.97	10.67	10.46	10.29	10.16	10.05	9.89	9.72	9.55	9.47	9.38	9.29	9.20	9.11	9.02
6	13.75	10.92	9.78	9.15	8.75	8.47	8.26	8.10	7.98	7.87	7.72	7.56	7.40	7.31	7.23	7.14	7.06	6.97	6.88
7	12.25	9.55	8.45	7.85	7.46	7.19	6.99	6.84	6.72	6.62	6.47	6.31	6.16	6.07	5.99	5.91	5.82	5.74	5.65
8	11.26	8.65	7.59	7.01	6.63	6.37	6.18	6.03	5.91	5.81	5.67	5.52	5.36	5.28	5.20	5.12	5.03	4.95	4.86
9	10.56	8.02	6.99	6.42	6.06	5.80	5.61	5.47	5.35	5.26	5.11	4.96	4.81	4.73	4.65	4.57	4.48	4.40	4.31
10	10.04	7.56	6.55	5.99	5.64	5.39	5.20	5.06	4.94	4.85	4.71	4.56	4.41	4.33	4.25	4.17	4.08	4.00	3.91
11	9.65	7.21	6.22	5.67	5.32	5.07	4.89	4.74	4.63	4.54	4.40	4.25	4.10	4.02	3.94	3.86	3.78	3.69	3.60
12	9.33	6.93	5.95	5.41	5.06	4.82	4.64	4.50	4.39	4.30	4.16	4.01	3.86	3.78	3.70	3.62	3.54	3.45	3.36
13	9.07	6.70	5.74	5.21	4.86	4.62	4.44	4.30	4.19	4.10	3.96	3.82	3.66	3.59	3.51	3.43	3.34	3.25	3.17
14	8.86	6.51	5.56	5.04	4.69	4.46	4.28	4.14	4.03	3.94	3.80	3.66	3.51	3.43	3.35	3.27	3.18	3.09	3.00
15	8.68	6.36	5.42	4.89	4.56	4.32	4.14	4.00	3.89	3.80	3.67	3.52	3.37	3.29	3.21	3.13	3.05	2.96	2.87
16	8.53	6.23	5.29	4.77	4.44	4.20	4.03	3.89	3.78	3.69	3.55	3.41	3.26	3.18	3.10	3.02	2.93	2.84	2.75
17	8.40	6.11	5.18	4.67	4.34	4.10	3.93	3.79	3.68	3.59	3.46	3.31	3.16	3.08	3.00	2.92	2.83	2.75	2.65
18	8.29	6.01	5.09	4.58	4.25	4.01	3.84	3.71	3.60	3.51	3.37	3.23	3.08	3.00	2.92	2.84	2.75	2.66	2.57
19	8.18	5.93	5.01	4.50	4.17	3.94	3.77	3.63	3.52	3.43	3.30	3.15	3.00	2.92	2.84	2.76	2.67	2.58	2.49
20	8.10	5.85	4.94	4.43	4.10	3.87	3.70	3.56	3.46	3.37	3.23	3.09	2.94	2.86	2.78	2.69	2.61	2.52	2.42
21	8.02	5.78	4.87	4.37	4.04	3.81	3.64	3.51	3.40	3.31	3.17	3.03	2.88	2.80	2.72	2.64	2.55	2.46	2.36
22	7.95	5.72	4.82	4.31	3.99	3.76	3.59	3.45	3.35	3.26	3.12	2.98	2.83	2.75	2.67	2.58	2.50	2.40	2.31
23	7.88	5.66	4.76	4.26	3.94	3.71	3.54	3.41	3.30	3.21	3.07	2.93	2.78	2.70	2.62	2.54	2.45	2.35	2.26
24	7.82	5.61	4.72	4.22	3.90	3.67	3.50	3.36	3.26	3.17	3.03	2.89	2.74	2.66	2.58	2.49	2.40	2.31	2.21
25	7.77	5.57	4.68	4.18	3.85	3.63	3.46	3.32	3.22	3.13	2.99	2.85	2.70	2.62	2.54	2.45	2.36	2.27	2.17
26	7.72	5.53	4.64	4.14	3.82	3.59	3.42	3.29	3.18	3.09	2.96	2.81	2.66	2.58	2.50	2.42	2.33	2.23	2.13
27	7.68	5.49	4.60	4.11	3.78	3.56	3.39	3.26	3.15	3.06	2.93	2.78	2.63	2.55	2.47	2.38	2.29	2.20	2.10
28	7.64	5.45	4.57	4.07	3.75	3.53	3.36	3.23	3.12	3.03	2.90	2.75	2.60	2.52	2.44	2.35	2.26	2.17	2.06
29	7.60	5.42	4.54	4.04	3.73	3.50	3.33	3.20	3.09	3.00	2.87	2.73	2.57	2.49	2.41	2.33	2.23	2.14	2.03
30	7.56	5.39	4.51	4.02	3.70	3.47	3.30	3.17	3.07	2.98	2.84	2.70	2.55	2.47	2.39	2.30	2.21	2.11	2.01
40	7.31	5.18	4.31	3.83	3.51	3.29	3.12	2.99	2.89	2.80	2.66	2.52	2.37	2.29	2.20	2.11	2.02	1.92	1.80
60	7.08	4.98	4.13	3.65	3.34	3.12	2.95	2.82	2.72	2.63	2.50	2.35	2.20	2.12	2.03	1.94	1.84	1.73	1.60
120	6.85	4.79	3.95	3.48	3.17	2.96	2.79	2.66	2.56	2.47	2.34	2.19	2.03	1.95	1.86	1.76	1.66	1.53	1.38
∞	6.63	4.61	3.78	3.32	3.02	2.80	2.64	2.51	2.41	2.32	2.18	2.04	1.88	1.79	1.70	1.59	1.47	1.32	1.00

$F = \frac{s_1^2}{s_2^2} = \frac{S_1}{m} \Big/ \frac{S_2}{n}$, where $s_1^2 = S_1/m$ and $s_2^2 = S_2/n$ are independent mean squares estimating a common variance σ^2 and based on m and n degrees of freedom, respectively.

$$F(F) = \int_0^F \frac{\Gamma\left(\frac{m+n}{2}\right)}{\Gamma\left(\frac{m}{2}\right)\Gamma\left(\frac{n}{2}\right)} m^{\frac{m}{2}} n^{\frac{n}{2}} x^{\frac{m}{2}-1} (n+mx)^{-\frac{m+n}{2}} dx = .995$$

n \ m	1	2	3	4	5	6	7	8	9	10	12	15	20	24	30	40	60	120	∞
1	16211	20000	21615	22500	23056	23437	23715	23925	24091	24224	24426	24630	24836	24940	25044	25148	25253	25359	25465
2	198.5	199.0	199.2	199.2	199.3	199.3	199.4	199.4	199.4	199.4	199.4	199.4	199.4	199.5	199.5	199.5	199.5	199.5	199.5
3	55.55	49.80	47.47	46.19	45.39	44.84	44.43	44.13	43.88	43.69	43.39	43.08	42.78	42.62	42.47	42.31	42.15	41.99	41.83
4	31.33	26.28	24.26	23.15	22.46	21.97	21.62	21.35	21.14	20.97	20.70	20.44	20.17	20.03	19.89	19.75	19.61	19.47	19.32
5	22.78	18.31	16.53	15.56	14.94	14.51	14.20	13.96	13.77	13.62	13.38	13.15	12.90	12.78	12.66	12.53	12.40	12.27	12.14
6	18.63	14.54	12.92	12.03	11.46	11.07	10.79	10.57	10.39	10.25	10.03	9.81	9.59	9.47	9.36	9.24	9.12	9.00	8.88
7	16.24	12.40	10.88	10.05	9.52	9.16	8.89	8.68	8.51	8.38	8.18	7.97	7.75	7.65	7.53	7.42	7.31	7.19	7.08
8	14.69	11.04	9.60	8.81	8.30	7.95	7.69	7.50	7.34	7.21	7.01	6.81	6.61	6.50	6.40	6.29	6.18	6.06	5.95
9	13.61	10.11	8.72	7.96	7.47	7.13	6.88	6.69	6.54	6.42	6.23	6.03	5.83	5.73	5.62	5.52	5.41	5.30	5.19
10	12.83	9.43	8.08	7.34	6.87	6.54	6.30	6.12	5.97	5.85	5.66	5.47	5.27	5.17	5.07	4.97	4.86	4.75	4.64
11	12.23	8.91	7.60	6.88	6.42	6.10	5.86	5.68	5.54	5.42	5.24	5.05	4.86	4.76	4.65	4.55	4.44	4.34	4.23
12	11.75	8.51	7.23	6.52	6.07	5.76	5.52	5.35	5.20	5.09	4.91	4.72	4.53	4.43	4.33	4.23	4.12	4.01	3.90
13	11.37	8.19	6.93	6.23	5.79	5.48	5.25	5.08	4.94	4.82	4.64	4.46	4.27	4.17	4.07	3.97	3.87	3.76	3.65
14	11.06	7.92	6.68	6.00	5.56	5.26	5.03	4.86	4.72	4.60	4.43	4.25	4.06	3.96	3.86	3.76	3.66	3.55	3.44
15	10.80	7.70	6.48	5.80	5.37	5.07	4.85	4.67	4.54	4.42	4.25	4.07	3.88	3.79	3.69	3.58	3.48	3.37	3.26
16	10.58	7.51	6.30	5.64	5.21	4.91	4.69	4.52	4.38	4.27	4.10	3.92	3.73	3.64	3.54	3.44	3.33	3.22	3.11
17	10.38	7.35	6.16	5.50	5.07	4.78	4.56	4.39	4.25	4.14	3.97	3.79	3.61	3.51	3.41	3.31	3.21	3.10	2.98
18	10.22	7.21	6.03	5.37	4.96	4.66	4.44	4.28	4.14	4.03	3.86	3.68	3.50	3.40	3.30	3.20	3.10	2.99	2.87
19	10.07	7.09	5.92	5.27	4.85	4.56	4.34	4.18	4.04	3.93	3.76	3.59	3.40	3.31	3.21	3.11	3.00	2.89	2.78
20	9.94	6.99	5.82	5.17	4.76	4.47	4.26	4.09	3.96	3.85	3.68	3.50	3.32	3.22	3.12	3.02	2.92	2.81	2.69
21	9.83	6.89	5.73	5.09	4.68	4.39	4.18	4.01	3.88	3.77	3.60	3.43	3.24	3.15	3.05	2.95	2.84	2.73	2.61
22	9.73	6.81	5.65	5.02	4.61	4.32	4.11	3.94	3.81	3.70	3.54	3.36	3.18	3.08	2.98	2.88	2.77	2.66	2.55
23	9.63	6.73	5.58	4.95	4.54	4.26	4.05	3.88	3.75	3.64	3.47	3.30	3.12	3.02	2.92	2.82	2.71	2.60	2.48
24	9.55	6.66	5.52	4.89	4.49	4.20	3.99	3.83	3.69	3.59	3.42	3.25	3.06	2.97	2.87	2.77	2.66	2.55	2.43
25	9.48	6.60	5.46	4.84	4.43	4.15	3.94	3.78	3.64	3.54	3.37	3.20	3.01	2.92	2.82	2.72	2.61	2.50	2.38
26	9.41	6.54	5.41	4.79	4.38	4.10	3.89	3.73	3.60	3.49	3.33	3.15	2.97	2.87	2.77	2.67	2.56	2.45	2.33
27	9.34	6.49	5.36	4.74	4.34	4.06	3.85	3.69	3.56	3.45	3.28	3.11	2.93	2.83	2.73	2.63	2.52	2.41	2.25
28	9.28	6.44	5.32	4.70	4.30	4.02	3.81	3.65	3.52	3.41	3.25	3.07	2.89	2.79	2.69	2.59	2.48	2.37	2.29
29	9.23	6.40	5.28	4.66	4.26	3.98	3.77	3.61	3.48	3.38	3.21	3.04	2.86	2.76	2.66	2.56	2.45	2.33	2.24
30	9.18	6.35	5.24	4.62	4.23	3.95	3.74	3.58	3.45	3.34	3.18	3.01	2.82	2.73	2.63	2.52	2.42	2.30	2.18
40	8.83	6.07	4.98	4.37	3.99	3.71	3.51	3.35	3.22	3.12	2.95	2.78	2.60	2.50	2.40	2.30	2.18	2.06	1.93
60	8.49	5.79	4.73	4.14	3.76	3.49	3.29	3.13	3.01	2.90	2.74	2.57	2.39	2.29	2.19	2.08	1.96	1.83	1.69
120	8.18	5.54	4.50	3.92	3.55	3.28	3.09	2.93	2.81	2.71	2.54	2.37	2.19	2.09	1.98	1.87	1.75	1.61	1.43
∞	7.88	5.30	4.28	3.72	3.35	3.09	2.90	2.74	2.62	2.52	2.36	2.19	2.00	1.90	1.79	1.67	1.53	1.36	1.00

$F = \frac{s_1^2}{s_2^2} = \frac{S_1}{m} \Big/ \frac{S_2}{n}$, where $s_1^2 = S_1/m$ and $s_2^2 = S_2/n$ are independent mean squares estimating a common variance σ^2 and based on m and n degrees of freedom, respectively.

$$F(F) = \int_0^F \frac{\Gamma\left(\frac{m+n}{2}\right)}{\Gamma\left(\frac{m}{2}\right)\Gamma\left(\frac{n}{2}\right)} m^{\frac{m}{2}} n^{\frac{n}{2}} x^{\frac{m}{2}-1} (n+mx)^{-\frac{m+n}{2}} dx = .999$$

n \ m	1	2	3	4	5	6	7	8	9	10	12	15	20	24	30	40	60	120	∞
1	4053*	5000*	5404*	5625*	5764*	5859*	5929*	5981*	6023*	6056*	6107*	6158*	6209*	6235*	6261*	6287*	6313*	6340*	6366*
2	998.5	999.0	999.2	999.2	999.3	999.3	999.4	999.4	999.4	999.4	999.4	999.4	999.4	999.5	999.5	999.5	999.5	999.5	999.5
3	167.0	148.5	141.1	137.1	134.6	132.8	131.6	130.6	129.9	129.2	128.3	127.4	126.4	125.9	125.4	125.0	124.5	124.0	123.5
4	74.14	61.25	56.18	53.44	51.71	50.53	49.66	49.00	48.47	48.05	47.41	46.76	46.10	45.77	45.43	45.09	44.75	44.40	44.05
5	47.18	37.12	33.20	31.09	29.75	28.84	28.16	27.64	27.24	26.92	26.42	25.91	25.39	25.14	24.87	24.60	24.33	24.06	23.79
6	35.51	27.00	23.70	21.92	20.81	20.03	19.46	19.03	18.69	18.41	17.99	17.56	17.12	16.89	16.67	16.44	16.21	15.99	15.75
7	29.25	21.69	18.77	17.19	16.21	15.52	15.02	14.63	14.33	14.08	13.71	13.32	12.93	12.73	12.53	12.33	12.12	11.91	11.70
8	25.42	18.49	15.83	14.39	13.49	12.86	12.40	12.04	11.77	11.54	11.19	10.84	10.48	10.30	10.11	9.92	9.73	9.53	9.33
9	22.86	16.39	13.90	12.56	11.71	11.13	10.70	10.37	10.11	9.89	9.57	9.24	8.90	8.72	8.55	8.37	8.19	8.00	7.81
10	21.04	14.91	12.55	11.28	10.48	9.92	9.52	9.20	8.96	8.75	8.45	8.13	7.80	7.64	7.47	7.30	7.12	6.94	6.76
11	19.69	13.81	11.56	10.35	9.58	9.05	8.66	8.35	8.12	7.92	7.63	7.32	7.01	6.85	6.68	6.52	6.35	6.17	6.00
12	18.64	12.97	10.80	9.63	8.89	8.38	8.00	7.71	7.48	7.29	7.00	6.71	6.40	6.25	6.09	5.93	5.76	5.59	5.42
13	17.81	12.31	10.21	9.07	8.35	7.86	7.49	7.21	6.98	6.80	6.52	6.23	5.93	5.78	5.63	5.47	5.30	5.14	4.97
14	17.14	11.78	9.73	8.62	7.92	7.43	7.08	6.80	6.58	6.40	6.13	5.85	5.56	5.41	5.25	5.10	4.94	4.77	4.60
15	16.59	11.34	9.34	8.25	7.57	7.09	6.74	6.47	6.26	6.08	5.81	5.54	5.25	5.10	4.95	4.80	4.64	4.47	4.31
16	16.12	10.97	9.00	7.94	7.27	6.81	6.46	6.19	5.98	5.81	5.55	5.27	4.99	4.85	4.70	4.54	4.39	4.23	4.06
17	15.72	10.66	8.73	7.68	7.02	6.56	6.22	5.96	5.75	5.58	5.32	5.05	4.78	4.63	4.48	4.33	4.18	4.02	3.85
18	15.38	10.39	8.49	7.46	6.81	6.35	6.02	5.76	5.56	5.39	5.13	4.87	4.59	4.45	4.30	4.15	4.00	3.84	3.67
19	15.08	10.16	8.28	7.26	6.62	6.18	5.85	5.59	5.39	5.22	4.97	4.70	4.43	4.29	4.14	3.99	3.84	3.68	3.51
20	14.82	9.95	8.10	7.10	6.46	6.02	5.69	5.44	5.24	5.08	4.82	4.56	4.29	4.15	4.00	3.86	3.70	3.54	3.38
21	14.59	9.77	7.94	6.95	6.32	5.88	5.56	5.31	5.11	4.95	4.70	4.44	4.17	4.03	3.88	3.74	3.58	3.42	3.26
22	14.38	9.61	7.80	6.81	6.19	5.76	5.44	5.19	4.99	4.83	4.58	4.33	4.06	3.92	3.78	3.63	3.48	3.32	3.15
23	14.19	9.47	7.67	6.69	6.08	5.65	5.33	5.09	4.89	4.73	4.48	4.23	3.96	3.82	3.68	3.53	3.38	3.22	3.05
24	14.03	9.34	7.55	6.59	5.98	5.55	5.23	4.99	4.80	4.64	4.39	4.14	3.87	3.74	3.59	3.45	3.29	3.14	2.97
25	13.88	9.22	7.45	6.49	5.88	5.46	5.15	4.91	4.71	4.56	4.31	4.06	3.79	3.66	3.52	3.37	3.22	3.06	2.89
26	13.74	9.12	7.36	6.41	5.80	5.38	5.07	4.83	4.64	4.48	4.24	3.99	3.72	3.59	3.44	3.30	3.15	2.99	2.82
27	13.61	9.02	7.27	6.33	5.73	5.31	5.00	4.76	4.57	4.41	4.17	3.92	3.66	3.52	3.38	3.23	3.08	2.92	2.75
28	13.50	8.93	7.19	6.25	5.66	5.24	4.93	4.69	4.50	4.35	4.11	3.86	3.60	3.46	3.32	3.18	3.02	2.86	2.69
29	13.39	8.85	7.12	6.19	5.59	5.18	4.87	4.64	4.45	4.29	4.05	3.80	3.54	3.41	3.27	3.12	2.97	2.81	2.64
30	13.29	8.77	7.05	6.12	5.53	5.12	4.82	4.58	4.39	4.24	4.00	3.75	3.49	3.36	3.22	3.07	2.92	2.76	2.59
40	12.61	8.25	6.60	5.70	5.13	4.73	4.44	4.21	4.02	3.87	3.64	3.40	3.15	3.01	2.87	2.73	2.57	2.41	2.23
60	11.97	7.76	6.17	5.31	4.76	4.37	4.09	3.87	3.69	3.54	3.31	3.08	2.83	2.69	2.55	2.41	2.25	2.08	1.89
120	11.38	7.32	5.79	4.95	4.42	4.04	3.77	3.55	3.38	3.24	3.02	2.78	2.53	2.40	2.26	2.11	1.95	1.76	1.54
∞	10.83	6.91	5.42	4.62	4.10	3.74	3.47	3.27	3.10	2.96	2.74	2.51	2.27	2.13	1.99	1.84	1.66	1.45	1.00

* Multiply these entries by 100.

ATOMIC WEIGHTS OF THE ELEMENTS 1981*

Abstract — The biennial review of atomic weight, $A_r(E)$, determinations and other cognate data have resulted in the following changes in recommended values (1979 values in parentheses): Hydrogen 1.00794 ± 7 (1.0079 ± 1); Silver 107.8682 ± 3 (107.868 ± 1); Lutetium 174.967 ± 1 (174.967 ± 3). These values are incorporated in the Table of Standard Atomic Weights of the Elements 1981. Whereas in the past, the Table indicated uncertainties as either 1 or 3 in the last place, other single-digit uncertainties will in the future be quoted when there is convincing evidence that by their use, a more precise standard atomic weight can be tabulated.

INTRODUCTION

The commission on Atomic Weights and Isotopic Abundances met under the chairmanship of Dr. N. E. Holden on August 26—29, 1981 during the XXXI IUPAC General Assembly in Leuven, Belgium. The Commission has for several years stressed the problems arising from the potential or actual variability of the atomic weights of many elements. The Commission's policy of recommending the greatest precision that can be reasonably supported by published measurements inevitably leads to a wide variation in the precision with which the atomic weights of the naturally occurring elements can be tabulated.

CHANGES IN ATOMIC WEIGHT VALUES

Hydrogen

The value of $A_r(H)$ = 1.0079 ± 0.0001 for the atomic weight of hydrogen was adopted by the Atomic Weights Commission in its 1971 Report to encompass all normal samples, whereas the previous value 1.0080 ± 0.0001 did not. Although the value 1.0079 is close to the atomic weight of laboratory hydrogen gas which may have been depleted of deuterium by electrolysis ($A_r(H)$ = 1.00787), it does not represent with the same accuracy the atomic weight of hydrogen from fresh water in temperate climates or in sea water ($A_r(H)$ = 1.00798). In order to improve precision, the Commission recommends the value $A_r(H)$ 1.00794 ± 0.00007 for the atomic weight of hydrogen. The quoted uncertainty covers the range of all terrestrial aqueous and gaseous sources of hydrogen. Only very exceptional geological samples may yield hydrogen with values outside this range.

Silver

The 1961 Report contained an extensive review of the atomic weights of silver, chlorine, and bromine because of the historical relation of these elements, i.e. the atomic weights of many elements were determined from their combining ratios with silver, chlorine, and bromine. The Commission recommended a value of $A_r(Ag)$ = 107.870 ± 0.003. In 1967, with physical measurements of the absolute abundance ratios of bromine and chlorine completed, the Commission recommended a value of 107.868 ± 0.001. In 1969 the Commission, after assessing the uncertainties given for the atomic weights of all the elements, reconfirmed a value of 107.868 ± 0.001.

At its 1981 meeting, the Commission considered a new determination of the absolute abundance ratio of silver by Powell, Murphy, and Gramlich who reported an atomic weight for a reference sample of silver of 107.86815 ± 0.00011 calculated from a $^{107}Ag/^{109}Ag$ ratio of 1.07638 ± 0.00022. These authors also reported results for a number of silver metal and mineral samples the average value of which was identical with the above number but with a slightly higher uncertainty. After careful examination of this and previous work, the Commission now recommends a value of $A_r(Ag)$ = 107.8682 ± 0.0003 for the atomic weight of silver.

Strontium

The Commission noted that Moore et al. had completed work on the absolute abundance ratios of a reference sample of strontium which gave an atomic weight of 87.61681 ± 0.00012 for this material. Because of known natural variations of one of the isotopes, ^{87}Sr, of this element the Commission did not feel justified in recommending a change but noted that a material with accurately known ratios is now available.

Lutetium

The value of $A_r(Lu)$ 174.97 for the atomic weight of lutetium was adopted by the Atomic Weights Commission in its 1961 Report. In the 1977 Report, the atomic weight was changed to $A_r(Lu)$ = 174.967 ± 0.003 on the basis of a new measurement by McCulloch et al. Hollinger and Devillers have recently redetermined the isotopic composition of lutetium although the absolute abundances of the lutetium isotopes were not determined. The good agreement in the mass spectrometric measurement together with the distribution of the abundances and the consequent probability of only a minor mass discrimination effect, has enabled the Commission to recommend $A_r(Lu)$ = 174.967 ± 0.001 as the most precise but still reliable value for lutetium.

* Abstracted with permission from *Pure and Applied Chemistry,* Vol. 55, No. 7, pp. 1101—1136, 1983 and the copyright owners, the International Union of Pure and Applied Chemistry. The original material was prepared for publication in *Pure and Applied Chemistry* by the IUPAC Inorganic Chemistry Division's Commission on Atomic Weights and Isotopic abundances. The journal article contains a list of references and discussions of the bases for selection of recommended values of the atomic weights.

THE TABLE OF STANDARD ATOMIC WEIGHTS 1981

The changes referred to above are incorporated in the 1981 Table of Standard Atomic Weights. A change in the 1981 Tables which the Commission has been debating for some years concerns a general policy regarding the annotations and wording of footnotes. The basic need for annotations to the Atomic Weights Tables and the Table of Isotopic Compositions arises from the necessity to impart to users additional information that is relevant to one or more elements but that cannot be made readable from numerical data in the columns. Any desire to maximize that additional information conveyed by these Tables is tempered by the need to preserve a compact format and a style that can alert the casual, yet possibly affected reader, who would look up neither the last full element by element review statement nor even the full text of a current Report.

The existing footnotes fail to give some details, such as the magnitudes or signs of differences between normal and abnormal atomic weight values or ranges, geological locations, abundancy of commercially available but abnormal sources, etc. Such additional information to be conveyed to users of these Tables will multiply in future years as materials from nuclear technologies, extra-terrestrial sources, and interest in trace-element compositions of isotopes and products from vacuum, vapor-path, and other fractionations become more widespread.

Standard Atomic Weights 1981
(Scaled to the relative atomic mass, $A_r(^{12}C) = 12$)

Names	Symbol	Atomic number	Atomic weight	Footnotes	Names	Symbol	Atomic number	Atomic weight	Footnotes
Actinium	Ac	89	227.0278	L	Mendelevium	Md	101	(258)	
Aluminium	Al	13	26.98154		Mercury	Hg	80	200.59 ± 3	
Americium	Am	95	(243)		Molybdenum	Mo	42	95.94	g
Antimony (Stibium)	Sb	51	121.75 ± 3		Neodymium	Nd	60	144.24 ± 3	g
Argon	Ar	18	39.948	g, r	Neon	Ne	10	20.179	g, m
Arsenic	As	33	74.9216		Neptunium	Np	93	237.0482	L
Astatine	At	85	(210)		Nickel	Ni	28	58.69	
Barium	Ba	56	137.33	g	Niobium	Nb	41	92.9064	
Berkelium	Bk	97	(247)		Nitrogen	N	7	14.0067	
Beryllium	Be	4	9.01218		Nobelium	No	102	(259)	
Bismuth	Bi	83	208.9804		Osmium	Os	76	190.2	g
Boron	B	5	10.81	m, r	Oxygen	O	8	15.9994 ± 3	g, r
Bromine	Br	35	79.904		Palladium	Pd	46	106.42	g
Cadmium	Cd	48	112.41	g	Phosphorus	P	15	30.97376	
Caesium	Cs	55	132.9054		Platinum	Pt	78	195.08 ± 3	
Calcium	Ca	20	40.08	g	Plutonium	Pu	94	(244)	
Californium	Cf	98	(251)		Polonium	Po	84	(209)	
Carbon	C	6	12.011	r	Potassium (Kalium)	K	19	39.0983	
Cerium	Ce	58	140.12	g	Praseodymium	Pr	59	140.9077	
Chlorine	Cl	17	35.453		Promethium	Pm	61	(145)	
Chromium	Cr	24	51.996		Protactinium	Pa	91	231.0359	L
Cobalt	Co	27	58.9332		Radium	Ra	88	226.0254	g, L
Copper	Cu	29	63.546 ± 3	r	Radon	Rn	86	(222)	
Curium	Cm	96	(247)		Rhenium	Re	75	186.207	
Dysprosium	Dy	66	162.50 ± 3		Rhodium	Rh	45	102.9055	
Einsteinium	Es	99	(252)		Rubidium	Rb	37	85.4678 ± 3	g
Erbium	Er	68	167.26 ± 3		Ruthenium	Ru	44	101.07 ± 3	g
Europium	Eu	63	151.96	g	Samarium	Sm	62	150.36 ± 3	g
Fermium	Fm	100	(257)		Scandium	Sc	21	44.9559	
Fluorine	F	9	18.998403		Selenium	Se	34	78.96 ± 3	
Francium	Fr	87	(223)		Silicon	Si	14	28.0855 ± 3	
Gadolinium	Gd	64	157.25 ± 3	g	Silver	Ag	47	107.8682 ± 3	g
Gallium	Ga	31	69.72		Sodium (Natrium)	Na	11	22.98977	
Germanium	Ge	32	72.59 ± 3		Strontium	Sr	38	87.62	g
Gold	Au	79	196.9665		Sulfur	S	16	32.06	r
Hafnium	Hf	72	178.49 ± 3		Tantalum	Ta	73	180.9479	
Helium	He	2	4.00260	g	Technetium	Tc	43	(98)	
Holmium	Ho	67	164.9304		Tellurium	Te	52	127.60 ± 3	g
Hydrogen	H	1	1.00794 ± 7	g, m, r	Terbium	Tb	65	158.9254	
Indium	In	49	114.82	g	Thallium	Tl	81	204.383	
Iodine	I	53	126.9045		Thorium	Th	90	232.0381	g, L
Iridium	Ir	77	192.22 ± 3		Thulium	Tm	69	168.9342	
Iron	Fe	26	55.847 ± 3		Tin	Sn	50	118.69 ± 3	
Krypton	Kr	36	83.80	g, m	Titanium	Ti	22	47.88 ± 3	
Lanthanum	La	57	138.9055 ± 3	g	Tungsten (Wolfram)	W	74	183.85 ± 3	
Lawrencium	Lr	103	(260)		(Unnilhexium)	(Unh)	106	(263)	
Lead	Pb	82	207.2	g, r	(Unnilpentium)	(Unp)	105	(262)	
Lithium	Li	3	6.941 ± 3	g, m, r	(Unnilquadium)	(Unq)	104	(261)	
Lutetium	Lu	71	174.967		Uranium	U	92	238.0289	g, m
Magnesium	Mg	12	24.305	g	Vanadium	V	23	50.9415	
Manganese	Mn	25	54.9380		Xenon	Xe	54	131.29 ± 3	g, m

Names	Symbol	Atomic number	Atomic weight	Footnotes
Ytterbium	Yb	70	173.04 ± 3	
Yttrium	Y	39	88.9059	
Zinc	Zn	30	65.38	
Zirconium	Zr	40	91.22	g

Atomic Number	Names	Symbol	Atomic weight	Footnotes
1	Hydrogen	H	1.00794 ± 7	g, m, r
2	Helium	He	4.00260	g
3	Lithium	Li	6.941 ± 3	g, m, r
4	Beryllium	Be	9.01218	
5	Boron	B	10.81	M, r
6	Carbon	C	12.011	r
7	Nitrogen	N	14.0067	
8	Oxygen	O	15.9994 ± 3	g, r
9	Fluorine	F	18.998403	
10	Neon	Ne	20.179	g, m
11	Sodium (Natrium)	Na	22.98977	
12	Magnesium	Mg	24.305	g
13	Aluminium	Al	26.98154	
14	Silicon	Si	28.0855 ± 3	
15	Phosphorus	P	30.97376	
16	Sulfur	S	32.06	r
17	Chlorine	Cl	35.453	
18	Argon	Ar	39.948	g, r
19	Potassium (Kalium)	K	39.0983	
20	Calcium	Ca	40.08	g
21	Scandium	Sc	44.9559	
22	Titanium	Ti	47.88 ± 3	
23	Vanadium	V	50.9415	
24	Chromium	Cr	51.996	
25	Manganese	Mn	54.9380	
26	Iron	Fe	55.847 ± 3	
27	Cobalt	Co	58.9332	
28	Nickel	Ni	58.69	
29	Copper	Cu	63.546 ± 3	r
30	Zinc	Zn	65.38	
31	Gallium	Ga	69.72	
32	Germanium	Ge	72.59 ± 3	
33	Arsenic	As	74.9216	
34	Selenium	Se	78.96 ± 3	
35	Bromine	Br	79.904	
36	Krypton	Kr	83.80	g, m
37	Rubidium	Rb	85.4678 ± 3	g
38	Strontium	Sr	87.62	g
39	Yttrium	Y	88.9059	
40	Zirconium	Zr	91.22	g
41	Niobium	Nb	92.9064	
42	Molybdenum	Mo	95.94	g
43	Technetium	Tc	(98)	
44	Ruthenium	Ru	101.07 ± 3	g
45	Rhodium	Rh	102.9055	
46	Palladium	Pd	106.42	g
47	Silver	Ag	107.8682 ± 3	g
48	Cadmium	Cd	112.41	g
49	Indium	In	114.82	g
50	Tin	Sn	118.69 ± 3	
51	Antimony (Stibium)	Sb	121.75 ± 3	
52	Tellurium	Te	127.60 ± 3	g
53	Iodine	I	126.9045	
54	Xenon	Xe	131.29 ± 3	g, m
55	Caesium	Cs	132.9054	
56	Barium	Ba	137.33	g
57	Lanthanum	La	138.9055 ± 3	g
58	Cerium	Ce	140.12	g
59	Praseodymium	Pr	140.9077	
60	Neodymium	Nd	144.24 ± 3	g
61	Promethium	Pm	(145)	
62	Samarium	Sm	150.36 ± 3	g
63	Europium	Eu	151.96	g
64	Gadolinium	Gd	157.25 ± 3	g
65	Terbium	Tb	158.9254	
66	Dysprosium	Dy	162.50 ± 3	
67	Holmium	Ho	164.9304	
68	Erbium	Er	167.26 ± 3	
69	Thulium	Tm	168.9342	
70	Ytterbium	Yb	173.04 ± 3	
71	Lutetium	Lu	174.967	
72	Hafnium	Hf	178.49 ± 3	
73	Tantalum	Ta	180.9479	
74	Wolfram (Tungsten)	W	183.85 ± 3	
75	Rhenium	Re	186.207	
76	Osmium	Os	190.2	g
77	Iridium	Ir	192.22 ± 3	
78	Platinum	Pt	195.08 ± 3	
79	Gold	Au	196.9665	
80	Mercury	Hg	200.59 ± 3	
81	Thallium	Tl	204.383	
82	Lead	Pb	207.2	g, r
83	Bismuth	Bi	208.9804	
84	Polonium	Po	(209)	
85	Astatine	At	(210)	
86	Radon	Rn	(222)	
87	Francium	Fr	(223)	
88	Radium	Ra	226.0254	g, L
89	Actinium	Ac	227.0278	L
90	Thorium	Th	232.0381	g, L
91	Protactinium	Pa	231.0359	L
92	Uranium	U	238.0289	g, m
93	Neptunium	Np	237.0482	L
94	Plutonium	Pu	(244)	
95	Americium	Am	(243)	
96	Curium	Cm	(247)	
97	Berkelium	Bk	(247)	
98	Californium	Cf	(251)	
99	Einsteinium	Es	(252)	
100	Fermium	Fm	(257)	
101	Mendelevium	Md	(258)	
102	Nobelium	No	(259)	
103	Lawrencium	Lr	(260)	
104	(Unnilquadium)	(Unq)	(261)	
105	(Unnilpentium)	(Unp)	(262)	
106	(Unnilhexium)	(Unh)	(263)	

Note: The atomic weights of many elements are not invariant but depend on the origin and treatment of the material. The footnotes to this Table elaborate the types of variation to be expected for individual elements. The values of $A_r(E)$ given here apply to elements as they exist naturally on earth and to certain artificial elements. When used with due regard to the footnotes they are considered reliable to ± 1 in the last digit, unless otherwise noted. Values in parentheses are used for radioactive elements whose atomic weights cannot be quoted precisely without knowledge of the origin of the elements; the value given is the atomic mass number of the isotope of that element of longest known half life.

g geologically exceptional specimens are known in which the element has an isotopic composition outside the limits for normal material. The difference between the atomic weight of the element in such specimens and that given in the Table may exceed considerably the implied uncertainty.

m modified isotopic compositions may be found in commercially available material because it has been subjected to an undisclosed or inadvertent isotopic separation. Substantial deviations in atomic weight of the element from that given in the Table can occur.

r range in isotopic composition of normal terrestrial material prevents a more precise atomic weight being given; the tabulated $A_r(E)$ value should be applicable to any normal material.

L Longest half-life isotope mass is chosen for the tabulated $A_r(E)$ value.

ELECTRONIC CONFIGURATION OF THE ELEMENTS

By Laurence S. Foster

References: F. H. Spedding and A. H. Daane, Editors, *The Rare Earths*, John Wiley and Sons, Inc. Publishers, New York, 1961. R. F. Gould, Editor, *Lanthanide–Actinide Chemistry*, Advances in Chemistry Series, No. 71, American Chemical Society, Washington, D.C., 1967: Paper No. 14, Mark Fred, *Electronic Structure of the Actinide Elements*.

Atomic No.	Element	K	L	M	N	O	P	Q
		1	2	3	4	5	6	7
		s	s p	s p d	s p d f	s p d f	s p d f	s p d f
1	H	1						
2	He	2						
3	Li	2	1					
4	Be	2	2					
5	B	2	2 1					
6	C	2	2 2					
7	N	2	2 3					
8	O	2	2 4					
9	F	2	2 5					
10	Ne	2	2 6					
11	Na	2	2 6	1				
12	Mg	2	2 6	2				
13	Al	2	2 6	2 1				
14	Si	2	2 6	2 2				
15	P	2	2 6	2 3				
16	S	2	2 6	2 4				
17	Cl	2	2 6	2 5				
18	Ar	2	2 6	2 6				
19	K	2	2 6	2 6 ..	1			
20	Ca	2	2 6	2 6 ..	2			
21	Sc	2	2 6	2 6 1	2			
22	Ti	2	2 6	2 6 2	2			
23	V	2	2 6	2 6 3	2			
24	Cr	2	2 6	2 6 5*	1			
25	Mn	2	2 6	2 6 5	2			
26	Fe	2	2 6	2 6 6	2			
27	Co	2	2 6	2 6 7	2			
28	Ni	2	2 6	2 6 8	2			
29	Cu	2	2 6	2 6 10*	1			
30	Zn	2	2 6	2 6 10	2			
31	Ga	2	2 6	2 6 10	2 1			
32	Ge	2	2 6	2 6 10	2 2			
33	As	2	2 6	2 6 10	2 3			
34	Se	2	2 6	2 6 10	2 4			
35	Br	2	2 6	2 6 10	2 5			
36	Kr	2	2 6	2 6 10	2 6			
37	Rb	2	2 6	2 6 10	2 6 ..	1		
38	Sr	2	2 6	2 6 10	2 6 ..	2		
39	Y	2	2 6	2 6 10	2 6 1	2		
40	Zr	2	2 6	2 6 10	2 6 2 ..	2		
41	Nb	2	2 6	2 6 10	2 6 4* ..	1		
42	Mo	2	2 6	2 6 10	2 6 5 ..	1		
43	Tc	2	2 6	2 6 10	2 6 6 ..	1		
44	Ru	2	2 6	2 6 10	2 6 7 ..	1		
45	Rh	2	2 6	2 6 10	2 6 8 ..	1		
46	Pd	2	2 6	2 6 10	2 6 10* ..	0		
47	Ag	2	2 6	2 6 10	2 6 10 ..	1		
48	Cd	2	2 6	2 6 10	2 6 10 ..	2		
49	In	2	2 6	2 6 10	2 6 10 ..	2 1		
50	Sn	2	2 6	2 6 10	2 6 10 ..	2 2		
51	Sb	2	2 6	2 6 10	2 6 10 ..	2 3		
52	Te	2	2 6	2 6 10	2 6 10 ..	2 4		
53	I	2	2 6	2 6 10	2 6 10 ..	2 5		
54	Xe	2	2 6	2 6 10	2 6 10 ..	2 6		
55	Cs	2	2 6	2 6 10	2 6 10 ..	2 6	1	
56	Ba	2	2 6	2 6 10	2 6 10 ..	2 6	2	
57	La	2	2 6	2 6 10	2 6 10 ..	2 6 1 ..	2	
58	Ce	2	2 6	2 6 10	2 6 10 2*	2 6	2	
59	Pr	2	2 6	2 6 10	2 6 10 3	2 6	2	
60	Nd	2	2 6	2 6 10	2 6 10 4	2 6	2	
61	Pm	2	2 6	2 6 10	2 6 10 5	2 6	2	
62	Sm	2	2 6	2 6 10	2 6 10 6	2 6	2	
63	Eu	2	2 6	2 6 10	2 6 10 7	2 6	2	
64	Gd	2	2 6	2 6 10	2 6 10 7	2 6 1 ..	2	
65	Tb	2	2 6	2 6 10	2 6 10 9*	2 6	2	
66	Dy	2	2 6	2 6 10	2 6 10 10	2 6	2	
67	Ho	2	2 6	2 6 10	2 6 10 11	2 6	2	
68	Er	2	2 6	2 6 10	2 6 10 12	2 6	2	
69	Tm	2	2 6	2 6 10	2 6 10 13	2 6	2	
70	Yb	2	2 6	2 6 10	2 6 10 14	2 6	2	
71	Lu	2	2 6	2 6 10	2 6 10 14	2 6 1 ..	2	
72	Hf	2	2 6	2 6 10	2 6 10 14	2 6 2 ..	2	
73	Ta	2	2 6	2 6 10	2 6 10 14	2 6 3 ..	2	
74	W	2	2 6	2 6 10	2 6 10 14	2 6 4 ..	2	
75	Re	2	2 6	2 6 10	2 6 10 14	2 6 5 ..	2	
76	Os	2	2 6	2 6 10	2 6 10 14	2 6 6 ..	2	
77	Ir	2	2 6	2 6 10	2 6 10 14	2 6 7 ..	2	
78	Pt	2	2 6	2 6 10	2 6 10 14	2 6 9 ..	1	
79	Au	2	2 6	2 6 10	2 6 10 14	2 6 10 ..	1	
80	Hg	2	2 6	2 6 10	2 6 10 14	2 6 10 ..	2	
81	Tl	2	2 6	2 6 10	2 6 10 14	2 6 10 ..	2 1	
82	Pb	2	2 6	2 6 10	2 6 10 14	2 6 10 ..	2 2	
83	Bi	2	2 6	2 6 10	2 6 10 14	2 6 10 ..	2 3	
84	Po	2	2 6	2 6 10	2 6 10 14	2 6 10 ..	2 4	
85	At	2	2 6	2 6 10	2 6 10 14	2 6 10 ..	2 5	
86	Rn	2	2 6	2 6 10	2 6 10 14	2 6 10 ..	2 6	
87	Fr	2	2 6	2 6 10	2 6 10 14	2 6 10 ..	2 6.. ..	1
88	Ra	2	2 6	2 6 10	2 6 10 14	2 6 10 ..	2 6.. ..	2
89	Ac	2	2 6	2 6 10	2 6 10 14	2 6 10 ..	2 6 1 ..	2
90	Th	2	2 6	2 6 10	2 6 10 14	2 6 10 ..	2 6 2 ..	2
91	Pa	2	2 6	2 6 10	2 6 10 14	2 6 10 2*	2 6 1 ..	2
92	U	2	2 6	2 6 10	2 6 10 14	2 6 10 3	2 6 1 ..	2
93	Np	2	2 6	2 6 10	2 6 10 14	2 6 10 4	2 6 1 ..	2
94	Pu	2	2 6	2 6 10	2 6 10 14	2 6 10 6	2 6.. ..	2
95	Am	2	2 6	2 6 10	2 6 10 14	2 6 10 7	2 6.. ..	2
96	Cm	2	2 6	2 6 10	2 6 10 14	2 6 10 7	2 6 1 ..	2
97	Bk	2	2 6	2 6 10	2 6 10 14	2 6 10 9*	2 6.. ..	2
98	Cf	2	2 6	2 6 10	2 6 10 14	2 6 10 10	2 6.. ..	2
99	Es	2	2 6	2 6 10	2 6 10 14	2 6 10 11	2 6.. ..	2
100	Fm	2	2 6	2 6 10	2 6 10 14	2 6 10 12	2 6.. ..	2
101	Md	2	2 6	2 6 10	2 6 10 14	2 6 10 13	2 6.. ..	2
102	No	2	2 6	2 6 10	2 6 10 14	2 6 10 14	2 6.. ..	2
103	Lr	2	2 6	2 6 10	2 6 10 14	2 6 10 14	2 6 1 ..	2
104	—	2	2 6	2 6 10	2 6 10 14	2 6 10 14	2 6 2 ..	2

* Note irregularity.

THE ELEMENTS

C. R. Hammond

One of the most striking facts about the elements is their unequal distribution and occurrence in nature. Present knowledge of the chemical composition of the universe, obtained from the study of the spectra of stars and nebulae, indicates that hydrogen is by far the most abundant element and may account for more than 90% of the atoms or about 75% of the mass of the universe. Helium atoms make up most of the remainder. All of the other elements together contribute only slightly to the total mass.

The chemical composition of the universe is undergoing continuous change. Hydrogen is being converted into helium, and helium is being changed into heavier elements. As time goes on, the ratio of heavier elements increases relative to hydrogen. Presumably, the process is not reversible.

Burbidge, Burbidge, Fowler, and Hoyle, and more recently, Peebles, Penzias, and others have studied the synthesis of elements in stars. To explain all of the features of the nuclear abundance curve — obtained by studies of the composition of the earth, meteorites, stars, etc. — it is necessary to postulate that the elements were originally formed by at least eight different processes: (1) hydrogen burning, (2) helium burning, (3) χ process, (4) *e* process, (5) *s* process, (6) *r* process, (7) *p* process, and (8) the X process. The X process is thought to account for the existence of light nuclei such as D, Li, Be, and B. Common metals such as Fe, Cr, Ni, Cu, Ti, Zn, etc. were likely produced early in the history of our galaxy. It is also probable that most of the heavy elements on earth and elsewhere in the universe were originally formed in supernovae, or in the hot interior of stars.

Studies of the solar spectrum have led to the identification of 67 elements in the sun's atmosphere; however, all elements cannot be identified with the same degree of certainty. Other elements may be present in the sun, although they have not yet been detected spectroscopically. The element helium was discovered on the sun before it was found on earth. Some elements such as scandium are relatively more plentiful in the sun and stars than here on earth.

Minerals in lunar rocks brought back from the moon on the Apollo missions consist predominantly of *plagioclase* {$(Ca,Na)(A1,Si)O_4O_8$} and *pyroxene* {$(Ca,Mg,Fe)_2 Si_2O_6$} — two minerals common in terrestrial volcanic rock. No new elements have been found on the moon that cannot be accounted for on earth; however, three minerals, *armalcolite* {$(Fe,Mg)Ti_2O_5$}, *pyroxferroite* {$CaFe_6 (SiO_3)_7$}, and *tranquillityite* {$Fe_8(Zr,Y)Ti_3Si_30_2$}, are new. The oldest known terrestrial rocks are about 3.75 billion years old. One rock, known as the "Genesis Rock," brought back from the Apollo 15 Mission, is about 4.15 billion years old. This is only about one-half billion years younger than the supposed age of the moon and solar system. Lunar rocks appear to be relatively enriched in refractory elements such as chromium, titanium, zirconium, and the rare earths, and impoverished in volatile elements such as the alkali metals, in chlorine, and in noble metals such as nickel, platinum, and gold.

Even older than the "Genesis Rock" are *carbonaceous chondrites,* a type of meteorite that has fallen to earth and has been studied. These are some of the most primitive objects of the solar system yet found. The grains making up these objects probably condensed directly out of the gaseous nebula from which the sun and planets were born. Most of the condensation of the grains probably was completed within 50,000 years of the time the disk of the nebula was first formed — about 4.6 billion years ago. It is now thought that this type of meteorite may contain a small percentage of presolar dust grains. The relative abundances of the elements in these meteorites are about the same as the abundances found in the solar chromosphere.

The X-ray fluorescent spectrometer sent with the Viking I spacecraft to Mars shows that the Martian soil contains about 12 to 16% iron, 14 to 15% silicon, 3 to 8% calcium, 2 to 7% aluminum, and one half to 2% titanium. The gas chromatograph — mass spectrometer on Viking II found no trace of organic compounds that should be present if life ever existed there.

F. W. Clarke and others have carefully studied the composition of rocks making up the crust of the earth. Oxygen accounts for about 47% of the crust, by weight, while silicon comprises about 28% and aluminum about 8%. These elements, plus iron, calcium, sodium, potassium, and magnesium, account for about 99% of the composition of the crust.

Many elements such as tin, copper, zinc, lead, mercury, silver, platinum, antimony, arsenic, and gold, which are so essential to our needs and civilization, are among some of the rarest elements in the earth's crust. These are made available to us only by the processes of concentration in ore bodies. Some of the so-called *rare-earth* elements have been found to be much more plentiful than originally thought and are about as abundant as uranium, mercury, lead, or bismuth. The least abundant rare-earth or *lanthanide* element, thulium, is now believed to be more plentiful on earth than silver, cadmium, gold, or iodine, for example. Rubidium, the 16th most abundant element, is more plentiful than chlorine while its compounds are little known in chemistry and commerce.

It is now thought that at least 24 elements are essential to living matter. The four most abundant in the human body are hydrogen, oxygen, carbon, and nitrogen. The seven next most common, in order of abundance, are calcium, phosphorus, chlorine, potassium, sulfur, sodium, and magnesium. Iron, copper, zinc, silicon, iodine, cobalt, manganese, molybdenum, fluorine, tin, chromium, selenium, and vanadium are needed and play a role in living matter. Boron is also thought essential for some plants, and it is possible that aluminum, nickel, and germanium may turn out to be necessary.

Ninety-one elements occur naturally on earth. Minute traces of plutonium-244 have been discovered in rocks mined in Southern California. This discovery supports the theory that heavy elements were produced during creation of the solar system. While technetium and promethium have not yet been found naturally on earth, they have been found to be present in stars. Technetium has been identified in the spectra of certain "late" type stars, and promethium lines have been identified in the spectra of a faintly visible star HR465 in Andromeda. Promethium must have been made very recently near the star's surface for no known isotope of this element has a half-life longer than 17.7 years.

It has been suggested that californium is present in certain stellar explosions known as supernovae; however, this has not been proved. At present no elements are found elsewhere in the universe that cannot be accounted for here on earth.

All atomic mass numbers from 1 to 238 are found naturally on earth except for masses 5 and 8. About 280 stable and 67 naturally radioactive isotopes occur on earth totalling 347. In addition, the neutron, technetium, promethium, and the transuranic elements (lying beyond uranium) up to Element 107 have been produced artificially. West German scientists in late 1982 confirmed the existence of Element 107 made by Soviet scientists and announced synthesis of Element 109. Laboratory processes have now extended the radioactive mass numbers beyond 238 to 266. Each element from atomic number 1 to 107 is known to have at least one radioactive isotope. About 1700 different nuclides (the name given to different kinds of nuclei, whether they are of the same or different elements) are now recognized. Many stable and radioactive isotopes are now produced and distributed by the Oak Ridge National Laboratory, Oak Ridge, Tenn., U.S.A., to customers licensed by the U.S. Department of Energy.

The nucleus of an atom is characterized by the number of protons it contains, usually denoted by Z, and by the number of neutrons, N. Isotopes of an element have the same value of Z, but different values of N. The *mass number* A, is the sum of Z and N. For example, Uranium-238 has a mass number of 238, and would contain 92 protons and 146 neutrons.

There is evidence that the definition of chemical elements must be broadened to include the electron. Several compounds known as *electrides,* have recently been made of alkaline metal elements and electrons. A relatively stable combination of a positron and electron, known as *positronium* has also been studied.

In addition to the proton, neutron, and electron, there are considerably more than 100 other fundamental particles which have been discovered or hypothesized. The majority of these fall into one of two classes, *leptons* or *hadrons.* The leptons comprise just four known particles, the electron, the *muon* (μ meson), and two kinds of *neutrinos.* The muon is essentially similar to the electron and has a charge of −1, but it is 200 times heavier. The neutrino is either of two stable particles of small (probably zero) rest mass, carrying no charge. Also there are four *antileptons,* identical to the corresponding leptons in some respects, such as mass, but they have properties exactly opposite those of the leptons. The *positron,* for example, is an antilepton, with a charge of + 1. Leptons cannot be broken into smaller units and are considered to be elementary. On the other hand, hadrons are complex and thought to have internal structure. Protons and neutrons, which make up atomic nuclei, are hadrons.

Elementary particle physics is not yet clearly understood, but groupings and arrangements of these particles have been made resembling the periodic table of chemical elements. This has led to the speculation that hadrons are composed of three (or possibly more simpler components called *quarks.* Quarks are presumed to be elementary particles. There is presently no evidence that quarks exist in isolation. Many physicists now hold that all the matter and energy in the universe is controlled by four fundamental natural forces: the electromagnetic force, gravity, a weak nuclear force, and a strong nuclear force. Each of these natural forces is passed back and forth among the basic particles of matter by unique force-carrying particles. The electromagnetic force is carried by the *photon,* the weak nuclear force by the intermediate vector *boson,* and gravity by the *gravitron.* There is now evidence of the existence of a new particle, known as the *gluon,* that binds quarks together by carrying the strong nuclear force.

The available evidence leads to the conclusion that elements 89 (actinium) through 108 (lawrencium) are chemically similar to the rare-earth or lanthanide elements (elements 57 to 71, inclusive). These elements therefore have been named *actinides* after the first member of this series. Those elements beyond uranium that have been produced artificially have the following names and symbols: neptunium, 93 (Np); plutonium, 94 (Pu); americium, 95 (Am); curium, 96 (Cm); berkelium, 97 (Bk); californium, 98 (Cf); einsteinium, 99 (Es); fermium, 100 (Fm); mendelevium, 101 (Md); nobelium, 102 (No); and lawrencium 103 (Lr). It is now claimed that Elements 104, 105, 106, 107, and 109 have been produced and identified. Names and chemical symbols have been suggested for Elements 104 and 105, but have not been officially adopted. Names for Elements 106, 107, and 109 have not yet been suggested.

Element 104 is expected to have chemical properties similar to those of hafnium and would not be a member of the actinide series. Element 105 probably would have chemical properties similar to those of tantalum, Element 106 similar to tungsten, and Element 107 similar to rhenium, and Element 109 similar to iridium.

There is still thought to be a possibility of producing elements beyond Element 109 either by bombardment of heavy isotopic targets with heavy ions, or by the irradiation of uranium or transuranic elements with the instantaneous high flux of neutrons produced by underground nuclear explosions. The limit will be set by the yields of the nuclear reactions and by the half-lives of radioactive decay. It has been suggested that Elements 102 and 103 have abnormally short lives only because they are in a pocket of instability, and that this region of instability might begin to ''heal'' around Element 105. If so, it may be possible to produce heavier isotopes with longer half-lives. It has also been suggested that Element 114, with a mass number of 298, and Element 126, with a mass number of 310, may be sufficiently stable to make discovery and identification possible. Calculations indicate that Element 110, a homolog of platinum, may have a half-life of as long as 100 million years. Searches have already been made by workers in a number of laboratories for Element 110 and its neighboring elements in naturally occurring platinum. Recent studies of the xenon component ($Xe^{131-136}$) of certain carbonaceous chronditic meteorites suggest that Elements 113, 114, or 115 may have been its progenitor.

There are many claims in the literature of the existence of various allotropic modifications of the elements, some of which are based on doubtful or incomplete evidence. Also, the physical properties of an element may change drastically by the presence of small amounts of impurities. With new methods of purification, which are now able to produce elements with 99.9999% purity, it has been necessary to restudy the properties of the elements. For example, the melting point of thorium changes by several hundred degrees by the presence of a small percentage of ThO_2 as an impurity. Ordinary commercial tungsten is brittle and can be worked only with difficulty. Pure tungsten, however, can be cut with a hacksaw, forged, spun, drawn, or extruded. In general, the value of physical properties given here applies to the pure element, when it is known.

Many of the chemical elements and their compounds are toxic and should be handled with due respect and care. In recent years there has been a greatly increased knowledge and awareness of the health hazards associated with chemicals, radioactive materials, and other agents. Anyone working with the elements and certain of their compounds should become thoroughly familiar with the proper safeguards to be taken. Reference should be made to publications such as the following:

1. *Code of Federal Regulations,* Title 29, Labor, chapter XVII, section 1910.93 of subpart G, redesignated as 1910.1000 at 40 FR (Federal Register) 23072. May 28, 1975; amended at 41 FR 11505, March 19, 1976; 41 FR 35184, August 20, 1976; FR 46784, October 22, 1976; 42 FR 3304, January 18, 1977 (corrections) and additional amendments and corrections as issued, U.S. Government Printing Office, Supt. of Documents, Washington, D.C.
2. *Code of Federal Regulations,* Title 10, Energy, Chapter 1, Nuclear Regulatory Commission, section 20.103 — 20.108; 20.201 — 207; 20.301 — 305; 20.401 — 409; 20.501 — 2; 20.601; appendices, corrections, and amendments.
3. *Occupational Safety and Health Reporter* (latest edition with amendments and corrections), Bureau of National Affairs, Washington, D.C.
4. *Atomic Energy Law Reporter,* Commerce Clearing House, Chicago, IL.
5. *Nuclear Regulation Reporter,* Commerce Clearing House, Chicago, IL.
6. *Maximum Permissible Body Burdens and Maximum Permissible Concentrations of Radionuclides in Air and in Water for Occupational Exposure,* with addenda, U.S. Department of Commerce, N.B.S, Handbook No. 69, (NCRP Report No. 22), latest edition, National Council on Radiation Protection and Measurements (NCRP), Bethesda, MD.; also refer to Permissible Quarterly Intakes of Radionuclides. Handbook of Chemistry and Physics, 67th Edition.
7. *TLVs® Threshold Limit Values for Chemical Substances and Physical Agents in Workroom Environment with Intended Changes,* latest edition, American Conference of Governmental Industrial Hygienists, Cincinnati, Ohio.

Actinium — (Gr. *aktis, aktinos,* beam or ray), Ac; at. wt. (227.0482); at no. 89; m.p. 1050°C; b.p. 3200 ± 300°C (est.); sp. gr. 10.07 (calc.). Discovered by Andre Debierne in 1899 and independently by F. Giesel in 1902. Occurs naturally in association with uranium minerals. Actinium-227, a decay product of uranium-235, is a beta emitter with a 21.6-year half-life. Its principal decay products are thorium-227 (18.5-day half-life), radium-223 (11.4-day half-life), and a number of short-lived products including radon, bismuth, polonium, and lead isotopes. In equilibrium with its decay products, it is a powerful source of alpha rays. Actinium metal has been prepared by the reduction of actinium fluoride with lithium vapor at about 1100 to 1300°C. The chemical behavior of actinium is similar to that of the rare earths, particularly lanthanum. Purified actinium comes into equilibrium with its decay products at the end of 185 days, and then decays according to its 21.6-year half-life. It is about 150 times as active as radium, making it of value in the production of neutrons.

Aluminum — (L. *alumen, alum*), Al; at. wt. 26.98154; at. no. 13; m.p. 660.37°C; b.p. 2467°C; sp. gr. 2.6989 (20°C); valence 3. The ancient Greeks and Romans used *alum* in medicine as an astringent, and as a mordant in dyeing. In 1761 de Morveau proposed the name *alumine* for the base in alum, and Lavoisier, in 1787, thought this to be the oxide of a still undiscovered metal. Wohler is generally credited with having isolated the metal in 1827, although an impure form was prepared by Oersted two years earlier. In 1807, Davy proposed the name *alumium* for the metal, undiscovered at that time, and later agreed to change it to *aluminum.* Shortly thereafter, the name *aluminium* was adopted to conform with the "ium" ending of most elements, and this spelling is now in use elsewhere in the world. *Aluminium* was also the accepted spelling in the U.S. until 1925, at which time the American Chemical Society officially decided to use the name *aluminum* thereafter in their publications. The method of obtaining aluminum metal by the electrolysis of alumina dissolved in *cryolite* was discovered in 1886 by Hall in the U.S. at about the same time by Heroult in France. Cryolite, a natural ore found in Greenland, is no longer widely used in commercial production, but has been replaced by an artifical mixture of sodium, aluminum, and calcium fluorides. *Bauxite,* an impure hydrated oxide ore, is found in large deposits in Jamaica, Australia, Surinam, Guyana, Arkansas, and elsewhere. The Bayer process is most commonly used today to refine bauxite so it can be accommodated in the Hall-Heroult refining process, used to produce most aluminum. Aluminum can now be produced from clay, but the process is not economically feasible at present. Aluminum is the most abundant metal to be found in the earth's crust (8.1%), but is never found free in nature. In addition to the minerals mentioned above, it is found in feldspars, granite, and in many other common minerals. Pure aluminum, a silvery-white metal, possesses many desirable characteristics. It is light, nontoxic, has a pleasing appearance, can easily be formed, machined, or cast, has a high thermal conductivity, and has excellent corrosion resistance. It is nonmagnetic and nonsparking, stands second among metals in the scale of malleability, and sixth in ductility. It is extensively used for kitchen utensils, outside building decoration, and in thousands of industrial applications where a strong, light, easily constructed material is needed. Although its electrical conductivity is only about 60% that of copper per area of cross section, it is used in electrical transmission lines because of its light weight. Pure aluminum is soft and lacks strength, but it can be alloyed with small amounts of copper, magnesium, silicon, manganese, and other elements to impart a variety of useful properties. These alloys are of vital importance in the construction of modern aircraft and rockets. Aluminum, evaporated in a vacuum, forms a highly reflective coating for both visible light and radiant heat. These coatings soon form a thin layer of the protective oxide and do not deteriorate as do silver coatings. They have found application in coatings for telescope mirrors, in making decorative paper, packages, toys, and in many other uses. The compounds of greatest importance are aluminum oxide, the sulfate, and the soluble sulfate with potassium (alum). The oxide, alumina, occurs naturally as ruby, sapphire, corundum, and emery, and is used in glassmaking and refractories. Synthetic ruby and sapphire have found application in the construction of lasers for producing coherent light. In 1852, the price of aluminum was about $545/lb, and just before Hall's discovery in 1886, about $11.00. The price rapidly dropped to 30c and has been as low as 15c/lb.

Americium — (the Americas), Am; at. wt. 243; at. no. 95; m.p. 994± 4°C; b.p. 2607°C; sp. gr. 13.67 (20°C); valence 2, 3, 4, 5, or 6. Americium was the fourth transuranium element to be discovered; the isotope Am^{241} was identified by Seaborg, James, Morgan, and Ghiorso late in 1944 at the wartime Metallurgical Laboratory (now the Argonne National Laboratory) of the University of Chicago as the result of successive neutron capture reactions by plutonium isotopes in a nuclear reactor:

$$Pu^{239}(n\gamma)Pu^{240}(n\gamma)Pu^{241} \xrightarrow{\beta} Am^{241}$$

Since the isotope Am^{241} can be prepared in relatively pure form by extraction as a decay product over a period of years from strongly neutron-bombarded plutonium Pu^{241} this isotope is used for much of the chemical investigation of this element. Better suited is the isotope Am^{243} due to its longer half-life (8.8×10^3 years as compared to 470 years for Am^{241}). A mixture of the isotopes Am^{241}, Am^{242}, and Am^{243} can be prepared by intense neutron irradiation of Am^{241} according to the reactions Am^{241} (n, γ) Am^{242} (n, γ) Am^{243}. Nearly isotopically pure Am^{243} can be prepared by a sequence of neutron bombardments and chemical separations as follows: neutron bombardment of Am^{241} yields Pu^{242} by the reactions Am^{241} (n, γ) Am^{242} → Pu^{242}, after chemical separation the Pu^{242} can be transformed to Am^{243} via the reactions Pu^{242} (n, γ) Pu^{243} → Am^{243}, and the Am^{243} can be chemically separated. Fairly pure Pu^{242} can be prepared more simply by very intense neutron irradiation of Pu^{239} as the result of successive neutron-capture reactions. Americium metal has been prepared by reducing the trifluoride with barium vapor at 1000 to 1200°C or the dioxide by lanthanum metal. The luster of freshly prepared americium metal is whiter and more silvery than plutonium or neptunium prepared in the same manner. It appears to be more malleable than uranium or neptunium and tarnishes slowly in dry air at room temperature. Americium is thought to exist in two forms: an alpha form which has a double hexagonal close-packed structure and a loose-packed cubic beta form. Americium must be handled with great care to avoid personal contamination. As little as 0.03 μg of Am^{241} is the maximum permissible total body burden. The alpha activity from Am^{241} is about three times that of radium. When gram quantities of Am^{241} are handled, the intense gamma activity makes exposure a serious problem. Americium dioxide, AmO_2, is the most important oxide. AmF_3, AmF_4, $AmCl_3$, $AmBr_3$, AmI_3, and other compounds have been prepared. The isotope Am^{241} has been used as a portable source for gamma radiography. It has also been used as a radioactive glass thickness gage for the flat glass industry, and as a source of ionization for smoke detectors. Americium-241 is available from the A.E.C. at a cost of $1600/g and Americium-243 at a cost of $100/mg.

Antimony — (Gr. anti plus monos — a metal not found alone), Sb; at. wt. 121.75 ± 3 at no. 51: m.p. 630.74°C; b.p. 1950°C; sp.gr. 6.691 (20°C); valence 0, −3, +3, or +5. Antimony was recognized in compounds by the ancients and was known as a metal at the beginning of the 17th centry and possibly much earlier. It is not abundant, but is found in over 100 mineral species. It is sometimes found native, but more frequently as the sulfide, *stibnite* (Sb_2S_3); it is also found as antimonides of the heavy metals, and as oxides. It is extracted from the sulfide by roasting to the oxide, which is reduced by salt and scrap iron; from its oxides it is also prepared by reduction with carbon. Two allotropic forms of antimony exists: the normal stable, metallic form, and the amorphous gray form. The so-called explosive antimony is an ill-defined material always containing an appreciable amount of halogen; therefore, it no longer warrants consideration as a separate allotrope. The yellow form, obtained by oxidation of *stibine*, SbH_3, is probably impure, and is not a distinct form. Metallic antimony is an extremely brittle metal of a flaky, crystalline texture. It is bluish white and has a metallic luster. It is not acted on by air at room temperature, but burns brilliantly when heated with the formation of white fumes of Sb_2O_3. It is a poor conductor of heat and electricity, and has a hardness of 3 to 3.5. Antimony, available commercially with a purity of 99.999 + %, is finding use in semiconductor technology for making infrared detectors, diodes, and Hall-effect devices. Commercial-grade antimony is widely used in alloys with percentages ranging from 1 to 20. It greatly increases the hardness and mechanical strength of lead. Batteries, antifriction alloys, type metal, small arms and tracer bullets, cable sheathing, and minor products use about half the metal produced. Compounds taking up the other half are oxides, sulfides, sodium antimonate, and antimony trichloride. These are used in manufacturing flame- proofing compounds, paints, ceramic enamels, glass, and pottery. Tartar emetic (hydrated potassium antimonyltartate) has been used in medicine. Antimony and many of its compounds are toxic. Exposure to antimony and its compounds should not exceed 0.5 Mg/M^3 (8-hr time weighted average 40-hr work week).

Argon — (Gr. *argos,* inactive), Ar; at. wt. 39.948; at. no. 18; freezing pt. −189.2°C; b.p.−185.7°C; density 1.7837 g/l. Its presence in air was suspected by Cavendish in 1785, discovered by Lord Rayleigh and Sir William Ramsay in 1894. The gas is prepared by fractionation of liquid air, the atmosphere containing 0.94% argon. The atmosphere of Mars contains 1.6% of Ar^{40} and 5 p.p.m. of Ar^{36}. Argon is two and one half times as soluble in water as nitrogen, having about the same solubility as oxygen. It is recognized by the characteristic lines in the red end of the spectrum. It is used in electric light bulbs and in fluorescent tubes at a pressure of about 3 mm, and in filling photo tubes, glow tubes, etc. Argon is also used as an inert gas shield for arc welding and cutting, as a blanket for the production of titanium and other reactive elements, and as a protective atmosphere for growing silicon and germanium crystals. Argon is colorless and odorless, both as a gas and liquid. It is available in high-purity form. Commercial argon is available at a cost of about 3c per cubic foot. Argon is considered to be a very inert gas and is not known to form true chemical compounds, as do krypton, xenon, and radon. However, it does form a hydrate having a dissociation pressure of 105 atm at 0°C. Ion molecules such as $(ArKr)^+$, $(ArXe)^+$, $(NeAr)^+$ have been observed spectroscopically. Argon also forms a clathrate with β hydroquinone. This clathrate is stable and can be stored for a considerable time, but a true chemical bond does not exist. Van der Waals' forces act to hold the argon. Naturally occurring argon is a mixture of three isotopes. Five other radioactive isotopes are now known to exist.

Arsenic — (L. *arsenicum,* Gr. *arsenikon,* yellow orpiment, identified with *arsenikos,* male, from the belief that metals were different sexes; Arab, *az-zernikh,* the orpiment from Persian *zerni-zar,* gold), As; at. wt. 74.9216; at. no. 33; valence −3, 0, +3 or +5. Elemental arsenic occurs in two solid modifications: yellow, and gray or metallic, with specific gravities of

1.97, and 5.73, respectively. Gray arsenic, the ordinary stable form, has a m.p. of 817°C (28 atm) and sublimes at 613°C. Several other allotropic forms of arsenic are reported in the literature. It is believed that Albertus Magnus obtained the element in 1250 A.D. In 1649 Schroeder published two methods of preparing the element. It is found native, in the sulfides *realgar* and *orpiment,* as arsenides and sulfarsenides of heavy metals, as the oxide, and as arsenates. *Mispickel* or arsenopyrite ($FeSAs$) is the most common mineral, from which on heating the arsenic sublimes leaving ferrous sulfide. The element is a steel gray, very brittle, crystalline, semimetallic solid; it tarnishes in air, and when heated is rapidly oxidized to arsenous oxide (As_2O_3) with the odor of garlic. Arsenic and its compounds are poisonous. Exposure to arsenic and its compounds (as/As) should not exceed 0.5 mg/M^3 (8-hr time-weighted average-40 hr work week.) These values, however, are being studied, and may be lowered. Arsenic is also used in bronzing, pyrotechny, and for hardening and improving the sphericity of shot. The most important compounds are white arsenic (As_2O_3), the sulfide, Paris green $3Cu(AsO_2)_2 \cdot Cu(C_2H_3O_2)_2$, calcium arsenate, and lead arsenate; the last three have been used as agricultural insecticides and poisons. Marsh's test makes use of the formation and ready decomposition of arsine (AsH_3). Arsenic is available in high-purity form. It is finding increasing uses as a doping agent in solid-state devices such as transistors. Gallium arsenide is used as a laser material to convert electricity directly into coherent light.

Astatine — (Gr. *astatos,* unstable), At; at. wt. (210); at. no. 85; m.p. 302°C; b.p. 337°C (est.); valance probably 1, 3, 5, or 7. Synthesized in 1940 by D.R. Corson, K. R. MacKenzie, and E. Segre at the University of California by bombarding bismuth with alpha particles. The longest-lived isotope, At^{210}, has a half-life of only 8.3 hr. Twenty isotopes are now known. Minute quantities of At^{215}, At^{218}, amd ^{219} exist in equilibrium in nature with naturally occurring uranium and thorium isotopes, and traces of At^{217} are in equilibrium with U^{233} and Np^{239} resulting from interaction of thorium and uranium with naturally produced neutrons. The total amount of astatine present in the earth's crust, however, totals less than 1 oz. Astatine can be produced by bombarding bismuth with energetic alpha particles to obtain the relatively long-lived $At^{209-211}$, which can be distilled from the target by heating it in air. Only about 0.05 μg of astatine have been prepared to date. The "time of flight" mass spectrometer has been used to confirm that this highly radioactive halogen behaves chemically very much like other halogens, particularly iodine. The interhalogen compounds AtI, AtBr, and AtCl are known to form, but it is not yet known if astatine forms diatomic astatine molecules. HAt and CH_3At (methyl astatide) have been detected. Astatine is said to be more metallic than iodine, and, like iodine, it probably accumulates in the thyroid gland. Workers at the Brookhaven National Laboratory have recently used reactive scattering in crossed molecular beams to identify and measure elementary reactions involving astatine.

Barium — (Gr. barys, heavy), Ba; at. wt. 137.33, at. no. 56; m.p. 725°C; b.p. 1640°C; sp. gr. 3.5 (20°C); valence 2. Baryta was distinguished from lime by Scheele in 1774; the element was discovered by Sir Humphrey Davy in 1808. It is found only in combination with other elements, chiefly in *barite* or *heavy spar* (sulfate) and *witherite* (carbonate) and is prepared by electrolysis of the chloride. Barium is a metallic element, soft, and when pure is silvery white like lead; it belongs to the alkaline earth group, resembling calcium chemically. The metal oxidizes very easily and should be kept under petroleum or other suitable oxygen-free liquids to exclude air. It is decomposed by water or alcohol. The metal is used as a "getter" in vacuum tubes. The most important compounds are the peroxide (BaO_2), chloride, sulfate, carbonate, nitrate, and chlorate. Lithopone, a pigment containing barium sulfate and zinc sulfide, has good covering power, and does not darken in the presence of sulfides. The sulfate, as permanent white or *blanc fixe,* is also used in paint, in X-ray diagnostic work, and in glassmaking. *Barite* is extensively used as a weighing agent in oilwell drilling fluids, and also in making rubber. The carbonate has been used as a rat poison, while the nitrate and chlorate give colors in pyrotechny. The impure sulfide phosphoresces after exposure to the light. The compounds and the metal are not expensive. Barium metal (99.5 + % pure) costs about $20.00/ lb. All barium compounds that are water or acid soluble are poisonous. Naturally occurring barium is a mixture of seven stable isotopes. Thirteen other radioactive isotopes are known to exist.

Berkelium — (*Berkeley,* home of the University of California), Bk; at. wt. (247); at no. 97; valance 3 or 4; sp. gr. 14 (est.). Berkelium, the eighth member of the actinide transition series, was discovered in December 1949 by Thompson, Ghiorso, and Seaborg, and was the fifth transuranium element synthesized. It was produced by cyclotron bombardment of milligram amounts of Am^{241} with helium ions at Berkeley, California. The first isotope produced had a mass number of 243 and decayed with a half-life of 4.6 hr. Eight isotopes are now known and have been synthesized. The existence of Bk^{249}, with a half-life of 314 days, make it feasible to isolate berkelium in weighable amounts so that its properties can be investigated with macroscopic quantities. One of the first visible amounts of a pure berkelium compound, berkelium chloride, was produced in 1962. It weighed 3 billionth of a gram. Berkelium probably has not yet been prepared in elemental form, but it is expected to be a silvery metal, easily soluble in dilute mineral acids, and readily oxidized by air oxygen at elevated temperatures to form the oxide. X-ray diffraction methods have been used to identify the following compounds: BkO_2, BkO_3, BkF_3, $BkCl_3$, and BkOCl. As with other actinide elements, berkelium tends to accumulate in the skeletal system. The maximum permissible body burden of Bk^{249} in the human skeleton is about 0.0004 μ. Because of its rarity, berkelium presently has no commercial or technological use. Berkelium-249 is available form O.R.N.L. at a cost of $100/mg.

Beryllium — (Gr. *beryllos, beryl;* also called Glucinium or Glucinum, Gr. *glykys,* sweet), Be; at. wt. 9.01218; at. no. 4; m.p. 1278±5°C; b.p. 2970°C; sp. gr. 1.848 (20° C); valence 2. Discovered as the oxide by Vauquelin in beryl and in emeralds in 1798. The metal was isolated in 1828 by Wohler and by Bussy independently by the action of potassium on beryllium chloride. Beryllium is found in some 30 mineral species, the most important of which are *bertrandite, beryl, chrysoberyl,* and *phenacite. Aquamarine* and emerald are precious forms of *beryl. Beryl* ($3BeO \cdot Al_2O_3 \cdot 6SiO_2$) and *bertrandite* ($4BeO \cdot 2SiO_2 \cdot H_2O$) are the most important commercial sources of the element and its compounds. Most of the metal is now prepared by reducing beryllium fluoride with magnesium metal. Beryllium metal did not become readily available to industry until 1957. The metal, steel gray in color, has many desirable properties. It is one of the lightest of all metals, and has one of the highest melting points of the light metals. Its modulus of elasticity is about one third greater than that of steel. It resists attack by concentrated nitric acid, has excellent thermal conductivity, and is nonmagnetic. It has a high permeability to X-rays, and when bombarded by alpha particles, as from radium or polonium, neutrons are produced in the ratio of about 30 neutrons/million alpha particles. At ordinary temperatures beryllium resists oxidation in air, although its ability to scratch glass is probably due to the formation of a thin layer of the oxide. Beryllium is used as an alloying agent in producing beryllium copper; which is extensively used for springs, electrical contacts, spot-welding electrodes, and nonsparking tools. It has found application as a structural material for high-speed aircraft, missiles, and spacecraft, and communication satellites. It is being used in the windshield frame, brake discs, support beams, and other structural components of the space shuttle. Because beryllium is relatively transparent to X-rays, ultra-thin Be-foil is finding use in X-ray lithography for reproduction of micro-miniature integrated circuits.

Beryllium is used in nuclear reactors as a reflector or moderator for it has a low thermal neutron absorption cross section. It is used in gyroscopes, computer parts, and instruments where lightness, stiffness, and dimensional stability are required. The oxide has a very high melting point and is also used in nuclear work and ceramic applications. Beryllium and its salts are toxic and should be handled with the greatest of care. Beryllium and its compounds should not be tasted to verify the sweetish nature of beryllium (as did early experimenters). The metal, its alloys, and its salts can be handled safely if certain work codes are observed, but no attempt should be made to work with beryllium before becoming familiar with proper safeguards. Exposure to beryllium dust in air should be limited to 2μ g/M^3 (8-hr time-weighted average — 40-hr week), with a ceiling concentration of 5μ g/M^3. A maximum peak above the acceptable ceiling concentration for an 8-hr shift is 25 μ g/M^3 for a maximum duration of 30 min. These values are being reviewed and studied. Beryllium metal in vacuum cast billet form is priced roughly at $150/lb. Fabricated forms are more expensive.

Bismuth — (Ger. *Weisse Masse*, white mass; later *Wisuth* and *Bisemutum*), Bi; at. wt. 208.9804; at. no. 83; m.p. 271.3°C; b.p. 1560 ± 5°C; sp. gr. 9.747 (20°C); valence 3 or 5. In early times bismuth was confused with tin and lead. Claude Geoffroy the Younger showed it to be distinct from lead in 1753. It is a white crystalline, brittle metal with a pinkish tinge. It occurs native. The most important ores are *bismuthinite* or bismuth glance (Bi_2S_3) and *bismite* (Bi_2O_3). Peru, Japan, Mexico, Bolivia, and Canada are major bismuth producers. Much of the bismuth produced in the U.S. is obtained as a by-product in refining lead, copper, tin, silver, and gold ores. Bismuth is the most diamagnetic of all metals, and the thermal conductivity is lower than any metal, except mercury. It has a high electrical resistence, and has the highest Hall effect of any metal (i.e., greatest increase in electrical resistance when placed in a magnetic field). "Bismanol" is a permanent magnet of high coercive force, made of MnBi, by the U.S. Naval Surface Weapons Center. Bismuth expands 3.32% on solidification. This property makes bismuth alloys particularly suited to the making of sharp castings of objects subject to damage by high temperatures. With other metals such as tin, cadmium, etc., bismuth forms low-melting alloys which are extensively used for safety devices used in fire detection and extinguishing systems. Bismuth is used in producing malleable irons and is finding use as a catalyst for making acrylic fibers. When bismuth is heated in air it burns with a blue flame, forming yellow fumes of the oxide. The metal is also used as a thermocouple material (has highest negativity known), and has found application as a carrier for U^{235} or U^{233} fuel in atomic reactors. Its soluble salts are characterized by forming insoluble basic salts on the addition of water, a property sometimes used in detection work. Bismuth oxychloride is used extensively in cosmetics. Bismuth subnitrate and subcarbonate are used in medicine. Bismuth metal costs about $8/lb in large quantities.

Boron — (Ar. *Buraq*, Pers. *Burah*), B; at. wt. 10.81; at. no. 5; m.p. 2079°C; b.p. sublimes 2550°C; sp. gr. of crystals 2.34, of amorphous variety 2.37; valence 3. Boron compounds have been known for thousands of years, but the element was not discovered until 1808 by Sir Humphry Davy and by Gay-Lussac and Thenard. The element is not found free in nature, but occurs as orthoboric acid usually in certain volcanic spring waters and as borates in *borax* and *colemanite*. *Ulexite*, another boron mineral, is interesting as it is nature's own version of "fiber optics." By far the most important source of boron is the mineral *rasorite*, also known as *kernite*, found in the Mojave desert of California. Extensive *borax* deposits are also found in Turkey. Boron exists naturally as 19.78% $_5B^{10}$ isotope and 80.22% $_5B^{11}$ isotope. High-purity crystalline boron may be prepared by the vapor phase reduction of boron trichloride or tribromide with hydrogen on electrically heated filaments. The impure, or amorphous, boron, a brownish-black powder, can be obtained by heating the trioxide with magnesium powder. Boron of 99.9999% purity has been produced and is available commercially. Elemental boron has an energy band gap of 1.50 to 1.56 eV, which is higher than that of either silicon or germanium. It has interesting optical characteristics, transmitting portions of the infrared, and is a poor conductor of electricity at room temperature, but a good conductor at high temperature. Amorphous boron is used in pyrotechnic flares to provide a distinctive green color, and in rockets as an igniter. The most important compounds of boron are boric, or boracic, acid, widely used as a mild antiseptic, and borax ($Na_2B_4O_7 \cdot 10H_2O$), which serves as a cleansing flux in welding and as a water softener in washing powders. Boron compounds are used in production of enamels for covering steel of refrigerators, washing machines, and like products. Boron compounds are also extensively used in the manufacture of borosilicate glasses. Other boron compounds show promise in treating arthritis. The isotope boron 10 is used as a control for nuclear reactors, as a shield for nuclear radiation, and in instruments used for detecting neutrons. Boron nitride has remarkable properties and can be used to make a material as hard as diamond. The nitride also behaves like an electrical insulator but conducts heat like a metal. It also has lubricating properties similar to graphite. The hydrides are easily oxidized with considerable energy liberation, and have been studied for use as rocket fuels. Demand is increasing for boron filaments, a high-strength, lightweight material chiefly employed for advanced aerospace structures. Boron is similar to carbon in that it has a capacity to form stable covalently bonded molecular networks. Carboranes, metalloboranes, phosphacarboranes, and other families comprise thousands of compounds. Crystalline boron (99%) costs about $5/g. Amorphous boron costs about $2/g. Elemental boron is not considered to be a poison, but assimilation of its compounds has a cumulative poisonous effect. Care must be taken in handling these.

Bromine — (Gr. *bromos*, stench), Br; at. wt. 79.904; at. no. 35; m.p. −7.2°C; b.p. 58.78°C; density of gas 7.59 g/ℓ, liquid 3.12 (20°C); valence 1, 3, 5, or 7. Discovered by Balard in 1826, but not prepared in quantity until 1860. A member of the halogen group of elements, it is obtained from natural brines from wells in Michigan and Arkansas. Little bromine is extracted today from seawater, which contains only about 85 ppm. Bromine is the only liquid nonmetallic element. It is a heavy, mobile, reddish-brown liquid, volatilizing readily at room temperature to a red vapor with a strong disagreeable odor, resembling chlorine, and having a very irritating effect on the eyes and throat; it is readily soluble in water or carbon disulfide, forming a red solution; it is less active than chlorine but more so than iodine; it unites readily with many elements and has a bleaching action; when spilled on the skin it produces painful sores. It presents a serious health hazard, and maximum safety precautions should be taken when handling it.Much of the bromine output in the U.S. is used in the production of ethylene dibromide, a lead scavenger used in making gasoline antiknock compounds. Lead in gasoline, however, is presently being drastically reduced, due to environmental considerations. This will greatly affect future production of bromine. Bromine is also used in making fumigants, flameproofing agents, water purification compounds, dyes, medicinals, sanitizers, inorganic bromides for photography, etc. Organic bromides are also important.

Cadmium — (L. *cadmia*; Gr. *kadmeia* — ancient name for calamine, zinc carbonate), Cd; at. wt. 112.41; at. no. 48 ; m.p. 320.9°C; b.p. 765°C; sp. gr. 8.65 (20°C); valence 2. Discovered by Stromeyer in 1817 from an impurity in zinc carbonate. Cadmium most often occurs in small quantities associated with zinc ores, such as *sphalerite* (ZnS). *Greenockite* (CdS) is the only mineral of any consequence bearing cadmium. Almost all cadmium is obtained as a by-product in the treatment of zinc, copper, and lead ores. It is a soft, bluish-white metal which is easily cut with a knife. It is similar in many respects to zinc. It is a component of some of the lowest melting alloys; it is used in bearing alloys with low coefficients of friction and great resistance to fatigue; it is used extensively in electroplating, which accounts for about 60% of its use. It is also used in many types of solder, for standard E.M.F. cells, for Ni-Cd batteries, and as a barrier to control atomic fission. Cadmium compounds are used in black and white television phosphors and in blue and green phosphors for color TV tubes. It forms a number of salts, of which the sulfate is the most common; the sulfide is used as a yellow pigment. Cadmium and solutions of its compounds are toxic. Failure to appreciate the toxic properties of cadmium may cause workers to be unwittingly exposed to dangerous fumes. Silver solder, for example, which contains cadmium, should be handled with care. Serious toxicity problems have been found from long-term exposure and work with cadmium plating baths. Exposure to cadmium dust should not exceed 0.05 mg/M^3 (8-hr time-weighted average, 40-hr week). The ceiling concentration (maximum), for a period of 15 min, should not exceed 0.15 mg/M^3. Cadmium oxide fume exposure (8-hr, 40-hr week) should not exceed 0.05 mg/M^3, and the maximum concentration should not exceed 0.05 mg/M^3. These values are presently being restudied and recommendations have been made to reduce the exposure. In 1927 the International Conference on Weights and Measures redefined the meter in terms of the wavelength of the red cadmium spectral line (i.e. 1m = 1,553,164.13 wavelengths). This definition has been changed (see under Krypton). The current price of cadmium is about $12/lb. It is available in high purity form.

Calcium — (L. *calx*, lime), Ca; at. wt. 40.08; at. no. 20; m.p. 839 ± 2°C; b.p. 1484°C; sp. gr. 1.55 (20°C); valence 2. Though lime was prepared by the Romans in the first century under the name calx, the metal was not discovered until 1808. After learning that Berzelius and Pontin prepared calcium amalgam by electrolyzing lime in mercury, Davy was able to isolate the impure metal. Calcium is a metallic element, fifth in abundance in the earth's crust, of which it forms more than 3%. It is an essential constituent of leaves, bones, teeth, and shells. Never found in nature uncombined, it occurs abundantly as *limestone* ($CaCO_3$), *gypsum* ($CaSO_4 \cdot 2H_2O$), and *fluorite* (CaF_2); *apatite* is the fluophosphate or chlorophosphate of calcium. The metal has a silvery color, is rather hard, and is prepared by electrolysis of the fused chloride to which calcium fluoride is added to lower the melting point. Chemically it is one of the alkaline earth elements; it readily forms a white coating of nitride in air, reacts with water, burns with a yellow-red flame, forming largely the nitride. The metal is used as a reducing agent in preparing other metals such as thorium, uranium, zirconium, etc., and is used as a deoxidizer, desulfurizer, or decarburizer for various ferrous and nonferrous alloys. It is also used as an alloying agent for aluminum, beryllium, copper, lead, and magnesium alloys, and serves as a "getter" for residual gases in vacuum tubes, etc. Its natural and prepared compounds are widely used. Quicklime (CaO), made by heating limestone and changed into slaked lime by the careful addition of water, is the great cheap base of chemical industry with countless uses. Mixed with sand it hardens as mortar and plaster by taking up carbon dioxide from the air. Calcium from limestone is an important element in Portland cement. The solubility of the carbonate in water containing carbon dioxide causes the formation of caves with stalactites and stalagmites and hardness in water. Other important compounds are the carbide (CaC_2), chloride ($CaCl_2$), cyanamide ($CaCN_2$), hypochlorite ($Ca(OCl)_2$), nitrate ($Ca(NO_3)_2$), and sulfide (CaS).

Californium — (State and University of California), Cf; at. wt. (251); at. no. 98. Californium, the sixth transuranium element to be discovered, was produced by Thompson, Street, Ghiorso, and Seaborg in 1950 by bombarding microgram quantities of Cm^{242} with 35 MeV helium ions in the Berkeley 60-in. cyclotron. Californium (III) is the only ion stable in aqueous solutions, all attempts to reduce or oxidize californium (III) having failed. The isotope Cf^{249} results from the beta decay of Bk^{249} while the heavier isotopes are produced by intense neutron irradiation by the reactions:

$$Bk^{249}(n\gamma)Bk^{250} \xrightarrow{\beta} Cf^{250} \text{ and } Cf^{249}(n\gamma)Cf^{250}$$

followed by

$$Cf^{250}(n\gamma)Cf^{251}(n\gamma)Cf^{252}$$

The existence of the isotopes Cf^{249}, Cf^{250}, Cf^{251}, and Cf^{252} makes it feasible to isolate californium in weighable amounts so that its properties can be investigated with macroscopic quantities. Californium-252 is a very strong neutron emitter. One microgram releases 170 million neutrons per minute, which presents biological hazards. Proper safeguards should be used in handling californium. In 1960 a few tenths of a microgram of californium trichloride $CfCl_3$, californium oxychloride, CfOCl, and californium oxide, Cf_2O_3, were first prepared. Reduction of californium to its metallic state has not yet been accomplished. Because californium is a very efficient source of neutrons, many new uses are expected for it. It has already found use in neutron moisture gages and in well-logging (the determination of water and oil-bearing layers.) It is also being used as a portable neutron source for discovery of metals such as gold or silver by on-the-spot activation analysis. Cf^{252} is now being offered for sale by the O.R.N.L. at a cost of \$10/μg. As of May 1975, more than 63 mg have been produced and sold. It has been suggested that californium may be produced in certain stellar explosions, called *supernovae,* for the radioactive decay of Cf^{254} (55-day half-life) agrees with the characteristics of the light curves of such explosions observed through telescopes. This suggestion, however, is questioned.

Carbon — (L. *carbo,* charcoal), C; at. wt. 12 exactly (C^{12}); at. wt. (natural carbon) 12.011; at. no. 6; m.p. ~ 3550°C, graphite sublimes at 3367 ± 25°C; triple point: (graphite-liquid-gas), 3627 ± 50° at a pressure of 10.1 MPa and (graphite-diamond-liquid), 3830—3930° at a pressure of 12—13 GPa; sp. gr. amorphous 1.8 to 2.1, graphite 1.9 to 2.3, diamond 3.15 to 3.53 (depending on variety); gem diamond 3.513 (25°C); valence 2, 3, or 4. Carbon, an element of prehistoric discovery, is very widely distributed in nature. It is found in abundance in the sun, stars, comets, and atmospheres of most planets. Carbon in the form of microscopic diamonds is found in some meteorites. Natural diamonds are found in *kimberlite* of ancient volcanic "pipes," such as found in South Africa, Arkansas, and elsewhere. Diamonds are now also being recovered from the ocean floor off the Cape of Good Hope. About 30% of all industrial diamonds used in the U.S. are now made synthetically. The energy of the sun and stars can be attributed at least in part to the well-known carbon-nitrogen cycle. Carbon is found free in nature in three allotropic forms: amorphous, graphite, and diamond. A fourth form, known as "white" carbon, is now thought to exist. Graphite is one of the softest known materials while diamond is one of the hardest. Graphite exists in two forms: alpha and beta. These have identical physical properties, except for their crystal structure. Naturally occurring graphites are reported to contain as much as 30% of the rhombohedral (beta) form, whereas synthetic materials contain only the alpha form. The hexagonal alpha type can be converted to the beta by mechanical treatment, and the beta form reverts to the alpha on heating it above 1000°C. In 1969 a new allotropic form of carbon was produced during the sublimation of pyrolytic graphite at low pressures. Under free-vaporization conditions above ~2550 K, "white" carbon forms as small transparent crystals on the edges of the basal planes of graphite. The interplanar spacings of "white" carbon are identical to those of a carbon form noted in the graphitic gneiss from the Ries (meteoritic) Crater of Germany. "White" carbon is a transparent birefringent material. Little information is presently available about this allotrope. In combination, carbon is found as carbon dioxide in the atmosphere of the earth and dissolved in all natural waters. It is a component of great rock masses in the form of carbonates of calcium (limestone), magnesium, and iron. Coal, petroleum, and natural gas are chiefly hydrocarbons. Carbon is unique among the elements in the vast number of variety of compounds it can form. With hydrogen, oxygen, and nitrogen, and other elements, it forms a very large number of compounds, carbon atom often being linked to carbon atom. There are upwards of a million or more known carbon compounds, many thousands of which are vital to organic and life processes. Without carbon, the basis for life would be impossible. While it has been thought that silicon might take the place of carbon in forming a host of similar compounds, it is now not possible to form stable compounds with very long chains of silicon atoms. The atmosphere of Mars contains 96.2% CO_2. Some of the most important compounds of carbon are carbon dioxide (CO_2), carbon monoxide (CO), carbon disulfide (CS_2), chloroform ($CHCl_3$), carbon tetrachloride (CCl_4), methane (CH_4), ethylene (C_2H_4), acetylene (C_2H_2), benzene (C_6H_6), ethyl alcohol (C_2H_5OH), acetic acid (CH_3COOH), and their derivatives. Carbon has seven isotopes. In 1961 the International Union of Pure and Applied Chemistry adopted the isotope carbon-12 as the basis for atomic weights. Carbon-14, an isotope with a half-life of 5730 years, has been widely used to date such materials as wood, archeological specimens, etc. Carbon-13 is now commercially available at a cost of \$700/g.

Cerium — (named for the asteroid *Ceres,* which was discovered in 1801 only 2 years before the element), Ce; at. wt. 140.12; at. no. 58; m.p. 799°C b.p. 3426°C; sp. gr. 6.657 (25°C); valence 3 or 4. Discovered in 1803 by Klaproth and by Berzelius and Hisinger; metal prepared by Hillebrand and Norton in 1875. Cerium is the most abundant of the metals of the so-called rare earths. It is found in a number of minerals including *allanite* (also known as *orthite*), *monazite, bastnasite, cerite,* and *samarskite.* Monazite and bastnasite are presently the two most important sources of cerium. Large deposits of monazite found on the beaches of Travancore, India, in river sands in Brazil, and deposits of *allanite* in the western United States, and *bastnasite* in Southern California will supply cerium, thorium, and the other rare-earth metals for many years to come. Metallic cerium is prepared by metallothermic reduction techniques, such as by reducing cerous fluoride with calcium, or by electrolysis of molten cerous chloride or other cerous halides. The metallothermic technique is used to produce high-purity cerium. Cerium is especially interesting because of its variable electronic structure. The energy of the inner 4f level is nearly the same as that of the outer or valence electrons, and only small amounts of energy are required to change the relative occupancy of these electronic levels. This gives rise to dual valency states. For example, a volume change of about 10% occurs when cerium is subjected to high pressures or low temperatures. It appears that the valence changes from about 3 to 4 when it is cooled or compressed. The low temperature behavior of cerium is complex. Four allotropic modifications are thought to exist: cerium at room temperature and at atmospheric pressure is known as γ cerium. Upon cooling to −23°C, γ cerium changes to β cerium. The remaining γ cerium starts to change to α cerium when cooled to −158°C, and the transformation is complete at −196°C. α Cerium has a density of 8.24; δ cerium exists above 726°C. At atmospheric pressure, liquid cerium is more dense than its solid form at the melting point. Cerium is an iron-gray lustrous metal. It is malleable, and oxidizes very readily at room temperature, especially in moist air. Except for europium, cerium is the most reactive of the "rare-earth" metals. It slowly decomposes in cold water, and rapidly in hot water. Alkali solutions and dilute and concentrated acids attack the metal rapidly. The pure metal is likely to ignite if scratched with a knife. Ceric salts are orange red or yellowish; cerous salts are usually white. Cerium is a component of misch metal, which is extensively used in the manufacture of pyrophoric alloys for cigarette lighters, etc. While cerium is not radioactive, the impure commercial grade may contain traces of thorium, which is radioactive. The oxide is an important constituent of incandescent gas mantles and it is emerging as a hydrocarbon catalyst in "self-cleaning" ovens. In this application it can be incorporated into oven walls to prevent the collection of cooking residues. As ceric sulfate it finds extensive use as a volumetric oxidizing agent in quantitative analysis. Cerium compounds are used in the manufacture of glass, both as a component and as a decolorizer. The oxide is finding increased use as a glass polishing agent instead of rouge, for it is much faster than rouge in polishing glass surfaces. Cerium, with other rare earths, is used in carbon-arc lighting, especially in the motion picture industry. It is also finding use as an important catalyst in petroleum refining and in metallurgical and nuclear applications. In small lots, 99.9% cerium costs about $2/g.

Cesium — (L. *caesius,* sky blue), Cs; at. wt. 132.9054; at. no. 55; m.p. 28.40± 0.01°C; b.p. 669.3°C; sp. gr. 1.873 (20°C); valence 1. Cesium was discovered spectroscopically by Bunsen and Kirchhoff in 1860 in mineral water from Durkheim. Cesium, an alkali metal, occurs in *lepidolite, pollucite* (a hydrated silicate of aluminum and cesium), and in other sources. One of the world's richest sources of cesium is located at Bernic Lake, Manitoba. The deposits are estimated to contain 300,000 tons of pollucite, averaging 20% cesium. It can be isolated by electrolysis of the fused cyanide and by a number of other methods. Very pure, gas-free cesium can be prepared by thermal decomposition of cesium azide. The metal is characterized by a spectrum containing two bright lines in the blue along with several others in the red, yellow, and green. It is silvery white, soft, and ductile. It is the most electropositive and most alkaline element. Cesium, gallium, and mercury are the only three metals that are liquid at room temperature. Cesium reacts explosively with cold water, and reacts with ice at temperatures above −116°C. Cesium hydroxide, the strongest base known, attacks glass. Because of its great affinity for oxygen the metal is used as a "getter" in electron tubes. It is also used in photoelectric cells, as well as a catalyst in the hydrogenation of certain organic compounds. The metal has recently found application in ion propulsion systems. Although these are not usable in the earth's atmosphere, 1 lb of cesium in outer space theoretically will propel a vehicle 140 times as far as the burning of the same amount of any known liquid or solid. Cesium is used in atomic clocks, which are accurate to 5 sec in 300 years. Its chief compounds are the chloride and the nitrate. The present price of cesium is about $25/g.

Chlorine — (Gr. *chloros,* greenish yellow), Cl; at. wt. 35.453; at. no. 17; m.p. −100.98°C; b.p.−34.6°C; density 3.214 g/l; sp. gr. 1.56(−33.6°C); valence 1, 3, 5, or 7. Discovered in 1774 by Scheele, who thought it contained oxygen; named in 1810 by Davy, who insisted it was an element. In nature it is found in the combined state only, chiefly with sodium as common salt ($NaCl$), *carnallite* ($KMgCl_3 \cdot 6H_2O$), and *sylvite* (KCl). It is a member of the halogen (salt-forming) group of elements and is obtained from chlorides by the action of oxidizing agents and more often by electrolysis; it is a greenish-yellow gas, combining directly with nearly all elements. At 10°C one volume of water dissolves 3.10 volumes of chlorine, at 30°C only 1.77 volumes. Chlorine is widely used in making many everyday products. It is used for producing safe drinking water the world over. Even the smallest water supplies are now usually chlorinated. It is also extensively used in the production of paper products, dyestuffs, textiles, petroleum products, medicines, antiseptics, insecticides, foodstuffs, solvents, paints, plastics, and many other consumer products. Most of the chlorine produced is used in the manufacture of chlorinated compounds of sanitation, pulp bleaching, disinfectants, and textile processing. Further use is in the manufacture of chlorates, chloroform, carbon tetrachloride, and in the extraction of bromine. Organic chemistry demands much from chlorine, both as an oxidizing agent and in substitution, since it often brings desired properties in an organic compound when substituted for hydrogen, as in one form of synthetic rubber. Chlorine is a respiratory irritant. The gas irritates the mucous membranes and the liquid burns the skin. As little as 3.5 ppm can be detected as an odor, and 1000 ppm is likely to be fatal after a few deep breaths. It was used as a war gas in 1915. Exposure to chlorine should not exceed 1 ppm(8-hr time-weighted average — 40 hr week.)

Chromium — (Gr. *chroma,* color), Cr; at. wt. 51.996; at. no. 24; m.p. 1857 ± 20°C; b.p. 2672°C; sp. gr. 7.18 to 7.20 (20°C); valence chiefly 2, 3, or 6. Discovered in 1797 by Vauquelin, who prepared the metal the next year. Chromium is a steel-gray, lustrous, hard metal that takes a high polish. The principal ore is *chromite*($FeCr_2O_4$), which is found in Zimbabwe, U.S.S.R., Transvaal, Turkey, Iran, Albania, Finland, Democratic Republic of Madagascar, and the Phillipines. The metal is usually produced by reducing the oxide with aluminum. Chromium is used to harden steel, to manufacture stainless steel, and to form many useful alloys. Much is used in plating to produce a hard, beautiful surface and to prevent corrosion. Chromium is used to give glass an emerald green color. It finds wide use as a catalyst. All compounds of chromium are colored; the most important are the chromates of sodium and potassium (K_2CrO_4) and the dichromates ($K_2Cr_2O_7$) and the potassium and ammonium chrome alums, as $KCr(SO_4)_2 \cdot 12H_2O$. The dichromates are used as oxidizing agents in quantitative analysis, also in tanning leather. Other compounds are of industrial value; lead chromate is chrome yellow, a valued pigment. Chromium compounds are used in the textile industry as mordants, and by the aircraft and other industries for anodizing aluminum. The refractory industry has found chromite useful for forming bricks and shapes, as it has a high melting point, moderate thermal expansion, and stability of crystalline structure. Chromium compounds are toxic and should be handled with proper safeguards.

Cobalt — (*Kobald,* from the German, goblin or evil spirit, *cobalos,* Greek, mine), Co; at. wt. 58.9332; at. no. 27; m.p. 1495°C; b.p. 2870°C; sp. gr. 8.9 (20°C); valence 2 or 3. Discovered by Brandt about 1735. Cobalt occurs in the minerals *cobaltite, smaltite,* and *erythrite,* and is often associated with nickel, silver, lead, copper, and iron ores, from which it is most frequently obtained as a by-product. It is also present in meteorites. Important ore deposits are found in Zaire, Morocco, and Canada. The U.S. Geological Survey has announced that the bottom of the north central Pacific Ocean may have cobalt-rich deposits at relatively shallow depths in waters close to the Hawaiian Islands and other U.S. Pacific territories. Cobalt is a brittle, hard metal, closely resembling iron and nickel in appearance. It has a magnetic permeability of about two thirds that of iron. Cobalt tends to exist as a mixture of two allotropes over a wide temperature range; the β-form predominates below 400°C, and the α above that temperature. The transformation is sluggish and accounts in part for the wide variation in reported data on physical properties of cobalt. It is alloyed with iron, nickel and other metals to make Alnico, an alloy of unusual magnetic strength with many important uses. Stellite ® alloys, containing cobalt, chromium, and tungsten, are used for high-speed, heavy-duty, high-temperature cutting tools, and for dies. Cobalt is also used in other magnet steels and stainless steels, and in alloys used in jet turbines and gas turbine generators. The metal is used in electroplating because of its appearance, hardness, and resistance to oxidation. The salts have been used for centuries for the production of brilliant and permanent blue colors in porcelain, glass, pottery, tiles, and enamels. It is the principal ingredient in Sevre's and Thenard's blue. A solution of the chloride ($CoCl_2 \cdot 6H_2O$) is used as sympathetic ink. The cobalt ammines are of interest; the oxide and the nitrate are important. Cobalt carefully used in the form of the chloride, sulfate, acetate, or nitrate has been found effective in correcting a certain mineral deficiency disease in animals. Soils should contain 0.13 to 0.30 ppm of cobalt for proper animal nutrition. Cobalt-60, an artifical isotope, is an important gamma ray source, and is extensively used as a tracer and a radiotherapeutic agent. Single compact sources of Cobalt-60 are readily available and have an equivalent gamma ray output equal to thousands of grams of radium. The cost of Cobalt-60 varies from about $1 to $10/curie, depending on quantity and specific activity. Exposure to cobalt (metal fumes and dust) should be limited to 0.05 mg/M^3 (8-hr time-weighted average, 40 hr week).

Columbium — see Niobium.

Copper — (L. *cuprum,* from the island of Cyprus), Cu; at. wt. 63.546 ± 3; at no. 29; m.p. 1083.4 ± 0.2°C; b.p. 2567 °C; sp. gr. 8.96 (20°C); valence 1 or 2. The discovery of copper dates from prehistoric times; it is said to have been mined for more than 5000 years. It is one of man's most important metals. Copper is reddish colored, takes on a bright metallic luster, and is malleable, ductile, and a good conductor of heat and electricity (second only to silver in electrical conductivity). The electrical industry is one of the greatest users of copper. Copper occasionally occurs native, and is found in many minerals such as *cuprite, malachite, azurite, chalcopyrite,* and *bornite.* Large copper ore deposits are found in the U.S., Chile, Zambia, Zaire, Peru, and Canada. The most important copper ores are the sulfides, oxides, and carbonates. From these, copper is obtained by smelting, leaching, and by electrolysis. Its alloys, brass and bronze, long used, are still very important; all American coins are now copper alloys; monel and gun metals also contain copper. The most important compounds are the oxide and the sulfate, blue vitriol; the latter has wide use as an agricultural poison and as an algicide in water purification. Copper compounds such as Fehling's solution are widely used in analytical chemistry in tests for sugar. High-purity copper (99.999 + %) is available commercially.

Curium — (Pierre and Marie Curie), Cm; at. wt. (247); at. no. 96; m.p. 1340 ± 40°C; sp. gr. 13.51 (calc.); valence 3 and 4. Although curium follows americium in the periodic system, it was actually known before americium and was the third transuranium element to be discovered. It was identified by Seaborg, James, and Ghiorso in 1944 at the wartime Metallurgical Laboratory in Chicago as a result of helium-ion bombardment of Pu^{239} in the Berkeley, California, 60-in. cyclotron. Visible amounts (30 μg) of Cm^{242}, in the form of the hydroxide, were first isolated by Werner and Perlman of the University of California in 1947. In 1950, Crane, Wallmann, and Cunningham found that the magnetic susceptibility of microgram samples of CmF_3 was of the same magnitude as that of GdF_3. This provided direct experimental evidence for assigning an electronic configuration to Cm^{+3}. In 1951, the same workers prepared curium in its elemental form for the first time. Thirteen isotopes of curium are now known. The most stable, Cm^{247}, with a half-life of 16 million years, is so short compared to the earth's age that any primordial curium must have disappeared long ago from the natural scene. Minute amounts of curium probably exist in natural deposits of uranium, as a result of a sequence of neutron captures and β decays sustained by the very low flux of neutrons naturally present in uranium ores. The presence of natural curium, however, has never been detected. Cm^{242} and Cm^{244} are available in multigram quantities. Cm^{248} has been produced only in milligram amounts. Curium is similar in some regards to gadolinium, its rare-earth homolog, but it has a more complex crystal structure. Curium is silver in color, is chemically reactive, and is more electropositive than aluminum. CmO_2, Cm_2O_3, CmF_3, CmF_4, $CmCl_3$, $CmBr_3$, and CmI_3 have been prepared. Most compounds of trivalent curium are faintly yellow in color. The A.E.C. is attempting to produce several kilograms of Cm^{244}, an isotope with a 17.6-year half-life, by neutron irradiation of plutonium in a nuclear reactor. Ultimately, it is possible that it may be produced in ton quantities by converting a number of plutonium production reactors to the manufacture of Cm^{244}. Cm^{242} generates about three thermal watts of energy per gram. This compares to one-half thermal watt per gram of Pu^{238}. This suggests use for curium as an isotope power source. Cm^{244} is now offered for sale by the A.E.C. at $100/ mg. Curium absorbed into the body accumulates in the bones, and is therefore very toxic as its radiation destroys the red-cell forming mechanism. The maximum permissible total body burden of Cm^{244}(soluble) in a human being is 0.3 μCi (microcurie).

Deuterium, an isotope of hydrogen — see Hydrogen.

Dysprosium — (Gr. *dysprositos,* hard to get at), Dy; at. wt. 162.50 ± 3; at. no. 66; m.p. 1412°C; b.p. 2562°C; sp. gr. 8.550 (25°C); valence 3. Dysprosium was discovered in 1886 by Lecoq de Boisbaudran, but not isolated. Neither the oxide nor the metal was available in relatively pure form until the development of ion-exchange separation and metallographic reduction techniques by Spedding and associates about 1950. Dysprosium occurs along with other so-called rare-earth or lanthanide elements in a variety of minerals such as *xenotime, fergusonite, gadolinite, euxenite, polycrase,* and *blomstrandine.* The most important sources, however, are from *monazite* and *bastnasite.* Dysprosium can be prepared by reduction of the trifluoride with calcium. The element has a metallic, bright silver luster. It is relatively stable in air at room temperature, and is readily attacked and dissolved, with the evolution of hydrogen, by dilute and concentrated mineral acids. The metal is soft enough to be cut with a knife and can be machined without sparking if overheating is avoided. Small amounts of impurities can greatly affect its physical properties. While dysprosium has not yet found many applications, its thermal neutron absorption cross-section and high melting point suggest metallurgical uses in nuclear control applications for alloying with special stainless steels. A dysprosium oxide-nickel cermet has found use in cooling nuclear reactor control rods. This cermet absorbs neutrons readily without swelling or contracting under prolonged neutron bombardment. In combination with vanadium and other rare earths, dysprosium has been used in making laser materials. Dysprosium-cadmium calcogenides, as sources of infrared radiation, have been used for studying chemical reactions. The cost of dysprosium metal has dropped in recent years since the development of ion- exchange and solvent extraction techniques, and the discovery of large ore bodies. The metal is still expensive, however, and costs about \$3/g in purities of 99 + %.

Einsteinium — (Albert Einstein), Es; at wt. (252); at no. 99. Einstenium, the seventh transuranic element of the actinide series to be discovered, was identified by Ghiorso and co-workers at Berkeley in December 1952 in debris from the first large thermonuclear or "hydrogen" bomb explosion, which took place in the Pacific in November 1952. The isotope produced was the 20-day Es^{253} isotope. In 1961, a sufficient amount of einsteinium was produced to permit separation of a macroscopic amount of Es^{253}. This sample weighed about 0.01 μg. A special magnetic-type balance was used in making this determination. Es^{253} so produced was used to produce mendelevium (Element 101). About 3 μg of einsteinium has been produced at Oak Ridge National Laboratores by irradiating for several years kilogram quantities of Pu^{239} in a reactor to produce Pu^{242}. This was then fabricated into pellets of plutonium oxide and aluminum powder, and loaded into target rods for an initial 1-year irradiation at the A.E.C.'s Savannah River Plant, followed by irradiation in a HFIR (High Flux Isotopic Reactor). After 4 months in the HFIR the targets were removed for chemical separation of the einsteinium from californium. Eleven isotopes of einsteinium are now recognized. Es^{254} has the longest half-life (276 days). Tracer studies using Es^{253} show that einsteinium has chemical properties typical of a heavy trivalent, actinide element.

Element 104 — In 1964, workers of the Joint Nuclear Research Institute at Dubna (U.S.S.R.) bombarded plutonium with accelerated 113 to 115 MeV neon ions. By measuring fission tracks in a special glass with a microscope, they detected an isotope that decays by spontaneous fission. They suggested that this isotope, which had a half-life of 0.3 ± 0.1 sec might be 104^{260}, produced by the following reaction:

$$_{94}Pu^{242} + {}_{10}Ne^{22} \longrightarrow 104^{260} + 4n$$

Element 104, the first *transactinide* element, is expected to have chemical properties similar to those of hafnium. It would, for example, form a relatively volatile compound with chlorine (a tetrachloride). The Soviet scientists have performed experiments aimed at chemical identification, and have attempted to show that the 0.3-sec activity is more volatile than that of the relatively nonvolatile actinide trichlorides. This experiment does not fulfill the test of chemically separating the new element from all others, but it provides important evidence for evaluation. New data, reportedly issued by Soviet scientists, have reduced the half-life of the isotope they worked with from 0.3 to 0.15 sec. The Dubna scientists suggest the name *kurchatovium* and symbol *Ku* for Element 104, in honor of Igor Vasilevich Kurchatov (1903-1960), late Head of Soviet Nuclear Research. In 1969, Ghiorso, Nurmia, Harris, K.A.Y. Eskola, and P.L. Eskola of the University of California at Berkeley reported they had positively identified two, and possibly three, isotopes of Element 104. The group also indicated that after repeated attempts so far they have been unable to produce isotope 104^{260} reported by the Dubna group in 1964. The discoveries at Berkeley were made by bombarding a target of Cf^{249} with C^{12} nuclei of 71 MeV, and C^{13} nuclei of 69 MeV. The combination of C^{12} with Cf^{249} followed by instant emission of four neutrons produced Element 104^{257}. This isotope has a half-life of 4 to 5 sec, decaying by emitting an alpha particle into No^{253}, with a half-life of 105 sec. The same reaction, except with the emission of three neutrons, was thought to have produced 104^{258}. with a half-life of about 1/100 sec. Element 104^{259} is formed by the merging of a C^{13} nuclei with Cf^{249}, followed by emission of three neutrons. This isotope has a half-life of 3 to 4 sec, and decays by emitting an alpha particle into No 255, which has a half-life of 185 sec. Thousands of atoms of 104^{257} and 104^{259} have been detected. The Berkeley group believe their identification of 104^{258} is correct, but they do not attach the same degree of confidence to this work as to their work on 104^{257} and 104^{259}. The Berkeley group proposes for the new element the name *rutherfordium* (symbol Rf), in honor of Ernest R. Rutherford, New Zealand physicist. The claims for discovery and the naming of Element 104 are still in question.

Element 105 — In 1967 G.N. Flerov reported that a Soviet team working at the Joint Institute for Nuclear Research at Dubna may have produced a few atoms of 105^{260} and 105^{261} by bombarding Am^{243} with Ne^{22}. Their evidence was based on time-coincidence measurements of alpha energies. More recently, it was reported that early in 1970 Dubna scientists synthesized Element 105 and that by the end of April 1970 "had investigated all the types of decay of the new element and had determined its chemical properties." The Soviet group has not proposed a name for Element 105. In late April 1970, it was announced that Ghiorso, Nurmia, Harris, K.A.Y. Eskola, and P.L.Eskola, working at the University of California at Berkeley, had positively identified Element 105. The discovery was made by bombarding a target of Cf^{249} with a beam of 84 MeV nitrogen nuclei in the Heavy Ion Linear Accelerator (HILAC). When a N^{15} nuclei is absorbed by a Cf^{249} nucleus, four neutrons are emitted and a new atom of 105^{260} with a half-life of 1.6 sec is formed. While the first atoms of Element 105 are said to have been detected conclusively on March 5, 1970, there is evidence that Element 105 had been formed in Berkeley experiments a year earlier by the method described. Ghiorso and his associates have attempted to confirm Soviet findings by more sophisticated methods without success. The Berkeley Group proposes the name *hahnium*, after the late German scientist Otto Hahn (1879-1968), and *Ha* for the chemical symbol.

More recently, in October 1971, it was announced that two new isotopes of Element 105 were synthesized with the heavy ion linear accelerator by A. Ghiorso and co-workers at Berkeley. Element 105^{261} was produced both by bombarding Cl^{250} with N^{15} and by bombarding Bk^{249} with O^{16}. The isotope emits 8.93-MeV α particles and decays to Lr^{257} with a half-life of about 1.8 sec. Element 105^{262} was produced by bombarding Bk^{249} with O^{18}. It emits 8.45 MeV α particles and decays to Lr^{258} with a half-life of about 40 sec.

Element 106 — In June 1974, members of the Joint Institute for Nuclear Research in Dubna, U.S.S.R., reported their discovery of Element 106, which they claim to have synthesized. In September 1974, workers of the Lawrence Berkeley and

Livermore Laboratories also reported creation of Element 106 "without any scientific doubt." The LBL and LLL Group used the SuperHILAC to accelerate O^{18} ions onto a Cf^{249} target. Element 106 was created by the reaction $Cf^{249}(O^{18}\ 4n)_{106}X^{263}$, which decayed by α emission to rutherfordium, and then by α emission to nobelium, which in turn further decayed by α emission. An elaborate detection system not only looked for correlations between the new element and its daughter, but also between daughter and granddaughter. The element so identified had α energies of 9.06 and 9.25 MeV with a half-life of 0.9 ± 0.2 sec. At Dubna, 280-MeV ions of Cr^{54} from the 310-cm cyclotron were used to strike targets of Pb^{206}, Pb^{207}, and Pb^{208}, in separate runs. Foils exposed to a rotating target disc were used to detect spontaneous fission activities, the foils being etched and examined microscopically to detect the number of fission tracts and the half-life of the fission activity. Other experiments were made to aid in confirmation of the discovery. Neither the Dubna team nor the Berkeley- Livermore Group has proposed a name as yet for Element 106.

Element l07 — In 1976 Soviet scientists at Dubna announced they had synthesized Element 107 by bombarding Bi^{204} with heavy nuclei of Cr^{54}. It is reported that earlier experiments in 1975 had allowed scientists "to glimpse" the new element for 2/1000 sec. A rapidly rotating cylinder, coated with a thin layer of the bismuth metal, was used as the target. This was bombarded by a stream of Cr^{54} ions fired tangentially. The existence of Element 107 was confirmed by a team of West German physicists at the Heavy Ion Research Laboratory at Darmstadt, who created and identified six nuclei of Element 107.

Element 109 — On August 29, 1982, Element 109 was made and discovered by physicists of the Heavy Ion Research Laboratory, Darmstadt, West Germany, by bombing a target of Bi-209 with accelerated nuclei of Fe-58. If the combined energy of two nuclei is sufficiently high, the repulsive forces between the nuclei can be overcome. In this experiment it took a week of target bombardment to produce a single fused nucleus. The team confirmed the existence of Element 109 by four independent measurements. The newly formed atom recoiled from the target at a predicted velocity and was separated from smaller, faster nuclei by a newly developed velocity filter. The time of flight to the detector and the striking energy were measured and found to match predicted values. The nucleus of ${}_{109}X^{266}$ started to decay 5 msec after striking the detector. A high-energy α particle was emitted, producing ${}_{107}X^{262}$. This in turn emitted an α particle, becoming ${}_{105}Ha^{258}$, which in turn captured an electron and became ${}_{104}Rf^{258}$. This in turn decayed into other nuclides. This experiment demonstrated the feasibility of using fusion techniques as a method of making new, heavy nuclei. West German scientists as well as those at Berkeley will soon attempt to use cold fusion to make Element 116 by bombarding curium-248 with calcium-48 nuclei. The fission product is expected to be Element 116, which should decay through a whole series of hitherto unknown nuclides. Names have not yet been proposed for Elements 106, 107, or 109.

Erbium — (*Ytterby*, a town in Sweden), Er; at. wt. 167.26 ± 3; at no. 68; m.p. 159°C; b.p. 2863°C; sp. gr. 9.066 (25°C); valence 3. Erbium, one of the so-called rare-earth elements of the lanthanide series, is found in the minerals mentioned under dysprosium above. In 1842 Mosander separated "yttria," found in the mineral *gadolinite,* into three fractions which he called *yttria, erbia,* and *terbia*. The names *erbia* and *terbia* became confused in this early period. After 1860, Mosander's *terbia* was known as *erbia,* and after 1877, the earlier known *erbia* became *terbia.* The *erbia* of this period was later shown to consist of five oxides, now known as *erbia, scandia, holmia, thulia* and *ytterbia.* By 1905 Urbain and James independently succeeded in isolating fairly pure Er_2O_3. Klemm and Bommer first produced reasonably pure erbium metal in 1934 by reducing the anhydrous chloride with potassium vapor. The pure metal is soft and malleable and has a bright, silvery, metallic luster. As with other rare-earth metals, its properties depend to a certain extent on the impurities present. The metal is fairly stable in air and does not oxidize as rapidly as some of the other rare-earth metals. Naturally occurring erbium is a mixture of six isotopes, all of which are stable. Nine radioactive isotopes of erbium are also recognized. Recent production techniques, using ion-exchange reactions, have resulted in much lower prices of the rare-earth metals and their compounds in recent years. The cost of 99+ % erbium metal is about $3/g in small quantities. Erbium is finding nuclear and metallurgical uses. Added to vanadium, for example, erbium lowers the hardness and improves workability. Most of the rare-earth oxides have sharp absorption bands in the visible, ultraviolet, and near infrared. This property, associated with the electronic structure, gives beautiful pastel colors to many of the rare-earth salts. Erbium oxide gives a pink color and has been used as a colorant in glasses and porcelain enamel glazes.

Europium — (Europe), Eu; at. wt. 151.96; at. no. 63; m.p. 822°C; b.p. 1597°C; sp. gr. 5.243 (25°C); valence 2 or 3. In 1890 Boisbaudran obtained basic fractions from samarium-gadolinium concentrates which had spark spectral lines not accounted for by samarium or gadolinium. These lines subsequently have been shown to belong to europium. The discovery of europium is generally credited to Demarcay, who separated the earth in reasonably pure form in 1901. The pure metal was not isolated until recent years. Europium is now prepared by mixture Eu_2O_3 with a 10%-excess of lanthanum metal and heating the mixture in a tantalum crucible under high vacuum. The element is collected as a silvery-white metallic deposit on the walls of the crucible. As with other rare-earth metals, except for lanthanum, europium ignites in air at about 150 to 180°C. Europium is about as hard as lead and is quite ductile. It is the most reactive of the rare-earth metals, quickly oxidizing in air. It resembles calcium in its reaction with water. *Bastnasite* and *monazite* are the principal ores containing europium. Europium has been identified spectroscopically in the sun and certain stars. Seventeen isotopes are now recognized. Europium isotopes are good neutron absorbers and are being studied for use in nuclear control applications. Europium oxide is now widely used as a phosphor activator and europium-activated yttrium vanadate is in commercial use as the red phosphor in color TV tubes. Europium-doped plastic has been used as a laser material. With the development of ion-exchange techniques and special processes, the cost of the metal has been greatly reduced in recent years. Europium is one of the rarest and most costly of the rare-earth metals. It is priced at about $50/g.

Fermium — (Enrico Fermi), Fm; at. wt. (257); at. no. 100. Fermium, the eighth transuranium element of the actinide series to be discovered, was identified by Ghiorso and co-workers in 1952 in the debris from a thermonuclear explosion in the

Pacific in work involving the University of California Radiation Laboratory, the Argonne National Laboratory, and the Los Alamos Scientific Laboratory. The isotope produced was the 20-hr Fm^{255}. During 1953 and early 1954, while discovery of elements 99 and 100 was withheld from publication for security reasons, a group from the Nobel Institute of Physics in Stockholm bombarded U^{238} with O^{16} ions, and isolated a 30-min α-emitter, which they ascribed to 100^{250}, without claiming discovery of the element. This isotope has since been identified positively, and the 30-min half-life confirmed. The chemical properties of fermium have been studied solely with tracer amounts, and in normal aqueous media only the (III) oxidation state appears to exist. The isotope Fm^{254} and heavier isotopes can be produced by intense neutron irradiation of lower elements such as plutonium by a process of successive neutron capture interspersed with beta decays until these mass numbers and atomic numbers are reached. Ten isotopes of fermium are known to exist. Fm^{257}, with a half-life of about 80 days, is the longest lived. Fm^{250}, with a half-life of 30 min, has been shown to be a product of decay of Element 102^{254}. It was by chemical identification of Fm^{250} that it was certain that Element 102 (nobelium) had been produced.

Flourine — (L. and F. *fluere*, flow, or flux), F; at. wt. 18.998403; at. no. 9; m.p. −219.62°C (1 atm.); b.p. −188.14°C (1 atm); density 1.696 g/l (0°C, 1 atm.); sp. gr. of liquid 1.108 at b.p.; valence 1. In 1529, Georigius Agricola described the use of fluorspar as a flux, and as early as 1670 Schwandhard found that glass was etched when exposed to fluorspar treated with acid. Scheele and many later investigators, including Davy, Gay-Lussac, Lavoisier, and Thenard, experimented with hydrofluoric acid, some experiments ending in tragedy. The element was finally isolated in 1886 by Moisson after nearly 74 years of continuous effort. Fluorine occurs chiefly in *fluorspar* (CaF_2) and *cryolite* (Na_2AlF_6), but is rather widely distributed in other minerals. It is a member of the halogen family of elements, and obtained by electrolyzing a solution of potassium hydrogen fluoride in anhydrous hydrogen fluoride in a vessel of metal or transparent fluorspar. Modern commercial production methods are essentially variations on the procedures first used by Moisson. Fluorine is the most electronegative and reactive of all elements. It is a pale yellow, corrosive gas, which reacts with practically all organic and inorganic substances. Finely divided metals, glass, ceramics, carbon, and even water burn in fluorine with a bright flame. Until World War II, there was no commercial production of elemental fluorine. The atom bomb project and nuclear energy applications, however, made it necessary to produce large quantities. Safe handling techniques have now been developed and it is possible at present to transport liquid fluorine by the ton. Fluorine and its compounds are used in producing uranium (from the hexafluoride) and more than 100 commercial fluorochemicals, including many well-known high-temperature plastics. Hydrofluoric acid is extensively used for etching the glass of light bulbs, etc. Fluorochloro hydrocarbons are extensively used in air conditioning and refrigeration. It has been suggested that fluorine can be substituted for hydrogen wherever it occurs in organic compounds, which could lead to an astronomical number of new fluorine compounds. The presence of fluorine as a soluble fluoride in drinking water to the extent of 2 ppm may cause mottled enamel in teeth, when used by children acquiring permanent teeth; in smaller amounts, however, fluorides are said to be beneficial and used in water supplies to prevent dental cavities. Elemental fluorine is being studied as a rocket propellant as it has an exceptionally high specific impulse value. Compounds of fluorine with rare gases have now been confirmed. Fluorides of xenon, radon, and krypton are among those reported. Elemental fluorine and the fluoride ion are highly toxic. The free element has a characteristic pungent odor, detectable in concentrations as low as 20 ppb, which is below the safe working level. The recommended maximum allowable concentration for a daily 8-hr time-weighted exposure is 0.1 ppm.

Francium — (France), Fr; at. no. 87; at. wt. (223); m.p. 27°C; b.p. 677°C; valence 1. Discovered in 1939 by Mlle. Marguerite Perey of the Curie Institute, Paris. Francium, the heaviest known member of the alkali metal series, occurs as a result of an alpha disintegration of actinium. It can also be made artificially by bombarding thorium with protons. While it occurs naturally in uranium minerals, there is probably less than an ounce of francium at any time in the total crust of the earth. It has the highest equivalent weight of any element, and is the most unstable of the first 101 elements of the periodic system. Twenty isotopes of francium are recognized. The longest lived, Fr^{223}(AcK), a daughter of Ac^{227}, has a half-life of 22 min. This is the only isotope of francium occuring in nature. Because all known isotopes of francium are highly unstable, knowledge of the chemical properties of this element comes from radiochemical techniques. No weighable quantity of the element has been prepared or isolated. The chemical properties of francium most closely resemble cesium.

Gadolinium — (*gadolinite*, a mineral named for Gadolin, a Finnish chemist), Gd; at. wt. 157.25 ± 3; at. no. 64; m.p. 1313 ± 1°C; b.p. 3266°C; sp. gr. 7.9004 (25°C); valence 3. Gadolinia, the oxide of gadolinium, was separated by Marignac in 1880 and Lecoq de Boisbaudran independently isolated the element from Mosander's "yttria" in 1886. The element was named for the mineral *gadolinite* from which this rare earth was originally obtained. Gadolinium is found in several other minerals, including *monazite* and *bastnasite,* which are of commercial importance. The element has been isolated only in recent years. With the development of ion-exchange and solvent extraction techniques, the availability and price of gadolinium and the other rare-earth metals have greatly improved. Seventeen isotopes of gadolinium are now recognized; seven occur naturally. The metal can be prepared by the reduction of the anhydrous fluoride with metallic calcium. As with other related rare-earth metals, it is silvery white, has a metallic luster, and is malleable and ductile. At room temperature, gadolinium crystallizes in the hexagonal, close-packed α form. Upon heating to 1262°C, α gadolinium transforms into the β form, which has a body-centered cubic structure. The metal is relatively stable in dry air, but in moist air it tarnishes with the formation of a loosely adhering oxide film which spalls off and exposes more surface to oxidation. The metal reacts slowly with water and is soluble in dilute acid. Gadolinium has the highest thermal neutron capture cross-section of any known element (49,000 barns). Natural gadolinium is a mixture of seven isotopes. Two of these, Gd^{155} and Gd^{157}, have excellent capture characteristics, but they are present naturally in low concentrations. As a result, gadolinium has a very fast burnout rate and has limited use as a nuclear control rod material. It has been used in making gadolinium yttrium garnets, which have microwave applications. Compounds of gadolinium are used in making phosphors for color TV tubes. The metal has unusual superconductive properties. As little as 1% gadolinium has been found to improve the workability and resistance of iron, chromium, and related alloys to high temperatures and oxidation. Gadolinium ethyl sulfate has extremely low noise characteristics and may find use in duplicating the performance of h.f. amplifiers, such as the maser. The metal is ferromagnetic. Gadolinium is unique for its high magnetic moment and for its special Curie temperature (above which ferromagnetism vanishes) lying just at room temperature. This suggests uses as a magnetic component that senses hot and cold. The price of the metal is $4/g.

Gallium — (L. *Gallia,* France; also from Latin, *gallus,* a translation of Lecoq, a cook); Ga; at. wt. 69.72; at. no. 31; m.p. 29.78°C; b.p. 2403°C; sp. gr. 5.904 (29.6°C) solid; sp. gr. 6.095 (29.6°C) liquid; valence 2 or 3. Predicted and described by Mendeleev as ekaaluminum, and discovered spectroscopically by Lecoq de Boisbaudran in 1875, who in the same year obtained the free metal by electrolysis of a solution of the hydroxide in KOH. Gallium is often found as a trace element in *diaspore, sphalerite, germanite, bauxite,* and *coal.* Some flue dusts from burning coal have been shown to contain as much as 1.5% gallium. It is the only metal, except for mercury, cesium, and rubidium, which can be liquid near room temperatures; this makes possible its use in high-temperature thermometers. It has one of the longest liquid ranges of any metal and has a low vapor pressure even at high temperatures. There is a strong tendency for gallium to supercool below its freezing point. Therefore, seeding may be necessary to initiate solidification. Ultra-pure gallium has a beautiful, silvery appearance, and the solid metal exhibits a conchoidal fracture similar to glass. The metal expands 3.1% on solidifying; therefore, it should not be stored in glass or metal containers, as they may break as the metal solidifies. Gallium wets glass or porcelain, and forms a brilliant mirror when it is painted on glass. It has found recent use in doping semiconductors and producing solid-state devices such as transistors. High-purity gallium is attacked only slowly by mineral acids. Magnesium gallate containing divalent impurities such as Mn^{+2} is finding use in commercial ultraviolet activated powder phosphors. Gallium arsenide is capable of converting electricity directly into coherent light. Gallium readily alloys with most metals, and has been used as a component in low-melting alloys. Its toxicity appears to be of a low order, but should be handled with care until more data are forthcoming. The metal can be supplied in ultrapure form (99.99999 + %). The cost is about $3/g.

Germanium — (L. *Germania,* Germany), Ge; at. wt. 72.59 ± 3; at. no. 32; m.p. 937.4°C; b.p. 2,830°C; sp. gr. 5.323 (25°C); valence 2 and 4. Predicted by Mendeleev in 1871 as ekasilicon, and discovered by Winkler in 1886. The metal is found in *argyrodite,* a sulfide of germanium and silver; in *germanite,* which contains 8% of the element; in zinc ores; in coal; and in other minerals. The element is frequently obtained commercially from flue dusts of smelters processing zinc ores, and has been recovered from the by-products of combustion of certain coals. Its presence in coal insures a large reserve of the element in the years to come. Germanium can be separated from other metals by fractional distillation of its volatile tetrachloride. The tetrachloride may then be hydrolyzed to give GeO_2; then the dioxide can be reduced with hydrogen to give the metal. Recently developed zone-refining techniques permit the production of germanium of ultra-high purity. The element is a gray-white metalloid, and in its pure state is crystalline and brittle, retaining its luster in air at room temperature. It is a very important semiconductor material. Zone-refining techniques have led to production of crystalline germanium for semiconductor use with an impurity of only one part in 10^{10}. Doped with arsenic, gallium, or other elements, it is used as a transistor element in thousands of electronic applications. Its application as a semiconductor element now provides the largest use for germanium. Germanium is also finding many other applications including use as an alloying agent, as a phosphor in fluorescent lamps, and as a catalyst. Germanium and germanium oxide are transparent to the infrared and are used in infrared spectroscopes and other optical equipment, including extremely sensitive infrared detectors. Germanium oxide's high index of refraction and dispersion has made it useful as a component of glasses used in wide-angle camera lenses and microscope objectives. The field of organogermanium chemistry is becoming increasingly important. Certain germanium compounds have a low mammalian toxicity, but a marked activity against certain bacteria, which makes them of interest as chemotherapeutic agents. The cost of germanium is about $300/lb.

Gold — (Sanskrit *Jval;* Anglo-Saxon *gold*), Au (L. *aurum,* shining dawn); at. wt. 196.9665; at. no. 79; m.p. 1064.43°C; b.p. 3080°C; sp. gr. $\sim$19.3(20°C); valence 1 or 3. Known and highly valued from earliest times, gold is found in nature as the free metal and in tellurides; it is very widely distributed and is almost always associated with quartz or pyrite. It occurs in veins and alluvial deposits, and is often separated from rocks and other minerals by sluicing or panning operations. About two thirds of the world's gold output now comes from South Africa, and about two thirds of the total U.S. production comes from South Dakota and Nevada. The metal is recovered from its ores by cyaniding, amalgamating, and smelting processes. Refining is also frequently done by electrolysis. Gold occurs in sea water to the extent of 0.1 to 2 mg/ton, depending on the location where the sample is taken. As yet, no method has been found for recovering gold from sea water profitably. It is estimated that all the gold in the world, so far refined, could be placed in a single cube 60 ft. on a side. Of all the elements, gold in its pure state is undoubtedly the most beautiful. It is metallic, having a yellow color when in a mass, but when finely divided it may be black, ruby, or purple. The Purple of Cassius is a delicate test for auric gold. It is the most malleable and ductile metal; 1 oz of gold can be beaten out to 300 ft^2. It is a soft metal and is usually alloyed to give it more strength. It is a good conductor of heat and electricity, and is unaffected by air and most reagents. It is used in coinage and is a standard for monetary systems in many countries. It is also extensively used for jewelry, decoration, dental work, and for plating. It is used for coating certain space satellites, as it is a good reflector of infrared and is inert. Gold, like other precious metals, is measured in troy weight; when alloyed with other metals, the term *carat* is used to express the amount of gold present, 24 carats being pure gold. For many years the value of gold was set by the U.S. at $20.67 /troy ounce; in 1934 this value was fixed by law at $35.00/ troy ounce, 9/10th fine. On March 17, 1968, because of a gold crisis, a two-tiered pricing system was established whereby gold was still used to settle international accounts at the old $35.00/troy ounce price while the price of gold on the private market would be allowed to fluctuate. Since this time, the price of gold on the free market has fluctuated widely. On March 19, 1968, President Johnson signed into law a bill removing the last statutory requirement for a gold backing against U.S. currency. On August 15, 1971, President Nixon announced an embargo on U.S. gold to settle international accounts, and on May 18, 1972, U.S. monetary gold was revalued at $38/ troy ounce. In February

1973 the U.S. in effect devalued the dollar by another 10%, decreasing the dollar in relation to gold from $38.00 to $42.22 / troy ounce. On September 21, 1973, President Nixon signed a bill ratifying the action taken earlier in February; the bill also restored to private U.S. citizens the right to own gold after December 31, 1973 (private possession, except for gold in the form of jewelry, certain coins, etc., had been prohibited since 1933 when the U.S. went off the gold standard). The final version of the bill gave the President discretion to lift the ban when he determined that private ownership would not impair the monetary position during this period of instability. President Ford signed final legislation on August 14, 1974, lifting the 41-year ban, to be effective after December 31, 1974. The price of gold on the free market reached a price of $620/tr. oz. in January 1980. The most common gold compounds are auric chloride ($AuCl_3$) and chlorauric acid ($HAuCl_4$), the latter being used in photography for toning the silver image. Gold has 18 isotopes; Au^{198}, with a half-life of 2.7 days, is used for treating cancer and other diseases. Disodium aurothiomalate is administered intramuscularly as a treatment for arthritis. A mixture of one part nitric acid with three of hydrochloric acid is called *aqua regia* (because it dissolved gold, the King of Metals). Gold is available commercially with a purity of 99.999 + %. For many years the temperature assigned to the freezing point of gold has been 1063.0°C; this has served as a calibration point for the International Temperature Scales (ITS-27 and ITS-48) and the International Practical Temperature Scale (ITPS-48). In 1968, a new International Practical Temperature Scale (ITPS-68) was adopted, which demands that the freezing point of gold be changed to 1064.43°C. Although workers in precision temperature measurement should adopt IPTS-68 immediately, many of the scale changes are of minor significance to the routine user. IPTS-68 has defined several other fixed temperature points, among which are the boiling points of hydrogen, neon, oxygen, and sulfur, and the freezing points of zinc, silver, tin, lead, antimony, and aluminium. The specific gravity of gold has been found to vary considerably depending on temperature, how the metal is precipitated, and cold-worked. As of December 1985, gold was priced at about $330/tr. oz. Density varies slightly with working of metal.

Hafnium — (*Hafinia*, Latin name for Copenhagen), Hf; at. wt. 178.49 ± 3; at. no. 72; m.p. 2227 ± 20°C; b.p. 4602°C; sp. gr. 13.31 (?) 20°C; valence 4. Hafnium was thought to be present in various minerals and concentrations many years prior to its discovery, in 1923, credited to D. Coster and G. von Hevesey. On the basis of the Bohr theory, the new element was expected to be associated with zirconium. It was finally identified in *zircon* from Norway, by means of X-ray spectroscope analysis. It was named in honor of the city in which the discovery was made. Most zirconium minerals contain 1 to 5% hafnium. It was originally separated from zirconium by repeated recrystallization of the double ammonium or potassium fluorides by von Hevesey and Jantzen. Metallic hafnium was first prepared by van Arkel and deBoer by passing the vapor of the tetraiodide over a heated tungsten filament. Almost all hafnium metal now produced is made by reducing the tetrachloride with magnesium or with sodium (Kroll Process). Hafnium is a ductile metal with a brilliant silver luster. Its properties are considerably influenced by the impurities of zirconium present. Of all the elements, zirconium and hafnium are two of the most difficult to separate. Their chemistry is almost identical, however, the density of zirconium is about half that of hafnium. Very pure hafnium has been produced, with zirconium being the major impurity. Because hafnium has a good absorption cross section for thermal neutrons (almost 600 times that of zirconium), has excellent mechanical properties, and is extremely corrosion resistant, it is used for reactor control rods. Such rods are used in nuclear submarines. Hafnium has been successfully alloyed with iron, titanium, niobium, tantalum, and other metals. Hafnium carbide is the most refractory binary composition known, and the nitride is the most refractory of all known metal nitrides (m.p. 3310° C). Hafnium is used in gas-filled and incandescent lamps, and is an efficient "getter" for scavenging oxygen and nitrogen. Finely divided hafnium is pyrophoric and can ignite spontaneously in air. Care should be taken when machining the metal or when handling hot sponge hafnium. At 700°C hafnium rapidly absorbs hydrogen to form the composition $HfH_{1.86}$. Hafnium is resistant to concentrated alkalis, but at elevated temperatures reacts with oxygen, nitrogen, carbon, boron, sulfur, and silicon. Halogens react directly to form tetrahalides. Exposure to hafnium should not exceed 0.5 mg/M^3 (8-hr time-weighted average — 40-hr week). The price of the metal is in the broad range of $100 to $500/lb, depending on purity and quantity. The yearly demand for hafnium in the U.S. is now in excess of 100,000 lb.

Hahnium — see Element 105.

Helium — (Gr. *helios*, the sun), He; at. wt. 4.00260; at. no. 2; m.p. below −272.2°C (26atm); b.p. − 268.934°C; density 0.1785 g/1 (0°C, 1 atm); liquid density 7.62 lb/ft³ at. b.p.; valence usually 0. Evidence of the existence of helium was first obtained by Janssen during the solar eclipse of 1868 when he detected a new line in the solar spectrum; Lockyer and Frankland suggested the name *helium* for the new element; in 1895, Ramsay discovered helium in the uranium mineral *clevite*, and it was independently discovered in clevite by the Swedish chemists Cleve and Langlet about the same time. Rutherford and Royds in 1907 demonstrated that α particles are helium nuclei. Except for hydrogen, helium is the most abundant element found throughout the universe. It has been detected spectroscopically in great abundance, especially in the hotter stars, and it is an important component in both the proton-proton reaction and the carbon cycle, which account for the energy of the sun and stars. The fusion of hydrogen into helium provides the energy of the hydrogen bomb. The helium content of the atmosphere is about 1 part in 200,000. While it is present in various radioactive minerals as a decay product, the bulk of the Free World's supply is obtained from wells in Texas, Oklahoma, and Kansas. The only known helium extraction plants, in 1984 were in Eastern Europe and the U.S.S.R.. The cost of helium fell from $2500/ft³ in 1915 to 1.5c/ft³ in 1940. The U.S. Bureau of Mines has set the price of Grade A helium at $35/1000 ft³. Helium has the lowest melting point of any element and has found wide use in cryogenic research, as its boiling point is close to absolute zero. Its use in the study of superconductivity is vital. Using liquid helium, Kurti and co-workers, and others, have succeeded in obtaining temperatures of a few microdegrees K by the adiabatic demagnetization of copper nuclei, starting from about 0.01 K. Five isotopes of helium are known. Liquid helium (He^4) exists in two forms: He^4 I and He^4 II, with a sharp transition point at 2.174 K (3.83 cm Hg). He^4I (above this temperature) is a normal liquid, but He^4II (below it) is unlike any other known substance. It expands on cooling; its conductivity for heat is enormous; and neither its heat conduction nor viscosity obeys normal rules. It has other peculiar properties. Helium is the only liquid that cannot be solidified by lowering the temperature. It remains liquid down to absolute zero at ordinary pressures, but it can readily be solidified by increasing the pressure. Solid He^3 and He^4 are unusual in that both can readily be changed in volume by more than 30% by application of pressure. The specific heat of helium gas is unusually high. The density of helium vapor at the normal boiling point is also very high, with the vapor expanding greatly when heated to room temperature. Containers filled with helium gas at 5 to 10° K should be treated as though they contained liquid helium due to the large increase in pressure resulting from warming the gas to room temperature. While helium normally has a 0 valence, it seems to have a weak tendency to combine with certain other elements. Means of preparing helium difluoride have been studied, and species such as HeNe and the molecular ions He^+ and He^{++} have been investigated. Helium is widely used as an inert gas shield for arc welding; as a protective gas in growing silicon and germanium crystals, and in titanium and zirconium production; as a cooling medium for nuclear reactors, and as a gas for supersonic wind tunnels. A mixture of 80% helium and 20% oxygen is used as an artificial atmosphere for divers and others working under pressure. Helium is extensively used for filling balloons as it is a much safer gas than hydrogen. While its density is almost twice that of hydrogen, it has about 98% of the lifting power of hydrogen. At sea level, 1000 ft^3 of helium lifts 68.5 lb. One of the recent largest uses for helium has been for pressuring liquid fuel rockets. A Saturn booster such as used on the Apollo lunar missions requires about 13 million cubic feet of helium for a firing, plus more for checkouts.

Holmium — (L. *Holmia,* for Stockholm), Ho; at. wt. 164.9304; at no. 67; m.p. 1474°C; b.p.2695°C; sp. gr. 8.795 (25°C); valence +3. The spectral absorption bands of holmium were noticed in 1878 by the Swiss chemists Delafontaine and Soret, who announced the existence of an ''Element X.'' Cleve, of Sweden, later independently discovered the element while working on erbia earth. The element is named after Cleve's native city. Pure holmia, the yellow oxide, was prepared by Homberg in 1911. Holmium occurs in *gadolinite, monazite,* and in other rare-earth minerals. It is commercially obtained from *monazite,* occurring in that mineral to the extent of about 0.05%. It has been isolated in pure form only in recent years. It can be separated from other rare earths by ion-exchange and solvent extraction techniques, and isolated by the reduction of its anhydrous chloride or fluoride with calcium metal. Pure holmium has a metallic to bright silver luster. It is relatively soft and malleable, and is stable in dry air at room temperature, but rapidly oxidizes in moist air and at elevated temperatures. The metal has unusual magnetic properties. Few uses have yet been found for the element. The element, as with other rare earths, seems to have a low acute toxic rating. The price of 99 + % holmium metal is about $8/g.

Hydrogen — (Gr. *hydro,* water, and *genes,* forming), H; at. wt. (natural) 1.00794 ± 7 at. wt. (H^1) 1.007822; at. no. 1; m.p. −259.14°C; b.p. −252.87°C; density 0.08988 g/l; density (liquid) 70.8 g/l (−253°C); density (solid) 70.6 g/l (−262°C); valence 1. Hydrogen was prepared many years before it was recognized as a distinct substance by Cavendish in 1766. It was named by Lavoisier. Hydrogen is the most abundant of all elements in the universe, and it is thought that the heavier elements were, and still are, being built from hydrogen and helium. It has been estimated that hydrogen makes up more than 90% of all the atoms or three quarters of the mass of the universe. It is found in the sun and most stars, and plays an important part in the proton-proton reaction and carbon-nitrogen cycle, which accounts for the energy of the sun and stars. It is thought that hydrogen is a major component of the planet Jupiter and that at some depth in the planet's interior the pressure is so great that solid molecular hydrogen is converted into solid metallic hydrogen. In 1973, it was reported that a group of Russian experimenters may have produced metallic hydrogen at a pressure of 2.8 Mbar. At the transition the density changed from 1.08 to 1.3 g/cm^3. Earlier, in 1972, a Livermore (California) group also reported on a similar experiment in which they observed a pressure-volume point centered at 2 Mbar. It has been predicted that metallic hydrogen may may be metastable; others have predicted it would be a superconductor at room temperature. On earth, hydrogen occurs chiefly in combination with oxygen in water, but it is also present in organic matter such as living plants, petroleum, coal, etc. It is present as the free element in the atmosphere, but only to the extent of less than 1 ppm, by volume. It is the lightest of all gases, and combines with other elements, sometimes explosively, to form compounds. Great quantities of hydrogen are required commercially for the fixation of nitrogen from the air in the Haber ammonia process and for the hydrogenation of fats and oils. It is also used in large quantities in methanol production, in hydrodealkylation, hydrocracking, and hydrodesulfurization. It is also used as a rocket fuel, for welding, for production of hydrochloric acid, for the reduction of metallic ores, and for filling balloons. The lifting power of 1 ft^3 of hydrogen gas is about 0.076 lb at 0°C, 760 mm pressure. Production of hydrogen in the U.S. alone now amounts to about 3 billion cubic feet per year. It is prepared by the action of steam on heated carbon, by decomposition of certain hydrocarbons with heat, by the electrolysis of water, or by the displacement from acids by certain metals. It is also produced by the action of sodium or potassium hydroxide on aluminum. Liquid hydrogen is important in cryogenics and in the study of superconductivity as its melting point is only a few degrees above absolute zero. The ordinary isotope of hydrogen, ${}_1H^1$, is known as *protium.* In 1932, Urey announced the preparation of a stable isotope, deuterium (${}_1H^2$ or D) with an atomic weight of 2. Two years later an unstable isotope, tritium (${}_1H^3$), with an atomic weight of 3 was discovered. Tritium has a half-life of about 12.5 years. One atom of deuterium is found mixed in with about 6000 ordinary hydrogen atoms. Tritium atoms are also present but in much smaller proportion. Tritium is readily produced in nuclear reactors and is used in the production of the hydrogen bomb. It is also used as a radioactive agent in making luminous paints, and as a tracer. The current price of tritium, to authorized personnel, is about $2/Ci; deuterium gas is readily available, without permit, at about $1/l. Heavy water, deuterium oxide (D_2O), which is used as a moderator to slow down neutrons, is available without permit at a cost of 6c to $1/g, depending on quantity and purity. Quite apart from isotopes, it has been shown that hydrogen gas under ordinary conditions is a mixture of two kinds of molecules, known as *ortho-* and *para*-hydrogen, which differ from one another by the spins of their electrons and nuclei. Normal hydrogen at room temperature contains 25% of the *para* form and 75% of the *ortho* form. The *ortho* form cannot be prepared in the pure state. Since the two forms differ in energy, the physical properties also differ. The melting and boiling points of *para*-hydrogen are about 0.1°C lower than those of normal hydrogen. Consideration is being given to an entire economy based on solar- and nuclear-generated hydrogen. Located in remote regions, power plants would electrolyze sea water; the hydrogen produced would travel to distant cities by pipelines. Pollution-free hydrogen could replace natural gas, gasoline, etc., and could serve as a reducing agent in metallurgy, chemical processing, refining, etc. It could also be used to hydrolyze trash into methane and ethylene. Public acceptance, high capital investment, and the high present cost of hydrogen with respect to present fuels are but a few of the problems facing establishment of such an economy.

Indium — (from the brilliant indigo line in its spectrum), In; at. wt. 114.82; at. no. 49; m.p. 156.61°C; b.p. 2080°C; sp. gr. 7.31 (20°C); valence 1, 2, or 3. Discovered by Reich and Richter, who later isolated the metal. Indium is most frequently associated with zinc minerals, and it is from these that most commercial indium is now obtained; however, it is also found in iron, lead, and copper ores. Until 1924, a gram or so constituted the world's supply of this element in isolated form. It is probably about as abundant as silver. About 4 million troy ounces of indium are now produced annually in the Free World. Canada is presently producing more than 1,000,000 troy ounces annually. The present cost of indium is about $1 to $5/g, depending on quantity and purity. It is available in ultrapure form. Indium is a very soft, silvery-white metal with a brilliant luster. The pure metal gives a high-pitched "cry" when bent. It wets glass, as does gallium. It has found application in making low-melting alloys; an alloy of 24% indium-76% gallium is liquid at room temperature. It is used in making bearing alloys, germanium transistors, rectifiers, thermistors, and photoconductors. It can be plated onto metal and evaporated onto glass, forming a mirror as good as that made with silver but with more resistance to atmospheric corrosion. There is evidence that indium has a low order of toxicity; however, care should be taken until further information is available.

Iodine — (Gr. *iodes,* violet), I; at. wt. 126.9045; at. no. 53; m.p. 113.5°C; b.p. 184.35°C; density of the gas 11.27 g/l; sp. gr. solid 4.93 (20°C); valence 1, 3, 5, or 7. Discovered by Courtois in 1811. Iodine, a halogen, occurs sparingly in the form of iodides in sea water from which it is assimilated by seaweeds, in Chilean saltpeter and nitrate-bearing earth, known as *caliche,* in brines from old sea deposits, and in brackish waters from oil and salt wells. Ultrapure iodine can be obtained from the reaction of potassium iodide with copper sulfate. Several other methods of isolating the element are known. Iodine is a bluish-black, lustrous solid, volatilizing at ordinary temperatures into a blue-violet gas with an irritating odor; it forms compounds with many elements, but is less active than the other halogens, which displace it from iodides. Iodine exhibits some metallic-like properties. It dissolves readily in chloroform, carbon tetrachloride, or carbon disulfide to form beautiful purple solutions. It is only slightly soluble in water. Iodine compounds are important in organic chemistry and very useful in medicine. Twenty-three isotopes are recognized. Only one stable isotope, I^{127}, is found in nature. The artificial radioisotope I^{131}, with a half-life of 8 days, has been used in treating the thyroid gland. The most common compounds are the iodides of sodium and potassium (KI) and the iodates (KIO_3). Lack of iodine is the cause of goiter. The iodide, and thyroxin which contains iodine, are used internally in medicine, and a solution of KI and iodine in alcohol is used for external wounds. Potassium iodide finds use in photography. The deep blue color with starch solution is characteristic of the free element. Care should be taken in handling and using iodine, as contact with the skin can cause lesions; iodine vapor is intensely irritating to the eyes and mucous membranes. The maximum allowable concentration of iodine in air should not exceed 1 mg/M^3 (8-hr time-weighted average — 40-hr).

Iridium — (L. *iris,* rainbow), Ir; at. wt. 192.22 ± 3; at. no. 77; m.p. 2410°C; b.p. 4130°C; sp. gr. 22.42 (17°C); valence 3 or 4. Discovered in 1803 by Tennant in the residue left when crude platinum is dissolved by aqua regia. The name iridium is appropriate, for its salts are highly colored. Iridium, a metal of the platinum family, is white, similar to platinum, but with a slight yellowish cast. It is very hard and brittle, making it very hard to machine, form, or work. It is the most corrosion-resistant metal known, and was used in making the standard meter bar of Paris, which is a 90% platinum 10% iridium alloy. This meter bar was replaced in 1960 as a fundamental unit of length (see under Krypton). Iridium is not attacked by any of the acids nor by aqua regia, but is attacked by molten salts, such as NaCl and NaCN. Iridium occurs uncombined in nature with platinum and other metals of this family in alluvial deposits. It is recovered as a by-product from the nickel mining industry. Iridium has found use in making crucibles and apparatus for use at high temperatures. It is also used for electrical contacts. Its principal use as a hardening agent for platinum. With osmium, it forms an alloy which is used for tipping pens and compass bearings. The specific gravity of iridium is only very slightly lower than that of osmium, which has been generally credited as being the heaviest known element. Calculations of the densities of iridium and osmium from the space lattices gives values of 22.65 and 22.61 g/cm^3, respectively. These values may be more reliable than actual physical measurements. At present, therefore, we know that either iridium or osmium is the densest known element, but the data do not yet allow selection between the two. Iridium costs about $300/troy ounce.

Iron — (Anglo-Saxon, *iron*), Fe (L. *ferrum*); at. wt. 55.847 ± 3 at. no. 26; m.p. 1535°C; b.p. 2750°C; sp. gr. 7.874 (20°C); valence 2, 3, 4, or 6. The use of iron is prehistoric. Genesis mentions that Tubal-Cain, seven generations from Adam, was "an instructor of every artificer in brass and iron." A rematkable iron pillar, dating to about A.D. 400, remains standing today in Delhi, India. This solid shaft of wrought iron is about 7¼ m high by 40 cm in diameter. Corrosion to the pillar has been minimal although it has been exposed to the weather since its erection. Iron is a relatively abundant element in the universe. It is found in the sun and many types of stars in considerable quantity. Its nuclei are very stable. Iron is found native as a principal component of a class of meteorites known as *siderites,* and is a minor constituent of the other two classes. The core of the earth, 2150 mi in radius, is thought to be largely composed of iron with about 10% occluded hydrogen. The metal is the fourth most abundant element, by weight, making up the crust of the earth. The most common ore is *hematite* (Fe_2O_3), from which the metal is obtained by reduction with carbon. Iron is found in other widely distributed minerals such as *magnetite,* which is frequently seen as *black sands* along beaches and banks of streams. *Taconite* is becoming increasingly important as a commercial ore. Common iron is a mixture of four isotopes. Six other isotopes are known to exist. Iron is a vital constituent of plant and animal life, and appears in hemoglobin. The pure metal is not often encountered in commerce, but is usually alloyed with carbon or other metals. The pure metal is very reactive chemically, and rapidly corrodes, especially in moist air or at elevated temperatures. It has four allotropic forms, or ferrites, known as α, β, γ, and δ, with transition points at 770, 928, and 1530°C. The α form is magnetic, but when transformed into the β form, the magnetism disappears although the lattice remains unchanged. The relations of these forms are peculiar. Pig iron is an alloy containing about 3% carbon with varying amounts of S, Si, Mn, and P. It is hard, brittle, fairly fusible, and is used to produce other alloys, including steel. Wrought iron contains only a few tenths of a percent of carbon, is tough, malleable, less fusible, and has usually a "fibrous" structure. Carbon steel is an alloy of iron with carbon, with small amounts of Mn, S, P, and Si. Alloy steels are carbon steels with other additives such as nickel, chromium, vanadium, etc. Iron is the cheapest and most abundant, useful, and important of all metals.

Krypton — (Gr. *kryptos,* hidden), Kr; at. wt. 83.80; at. no. 36; m.p. −156.6°C; b.p. −152.30± 0.10°C; density 3.733 g/l (O°C); valence usually 0. Discovered in 1898 by Ramsay and Travers in the residue left after liquid air had nearly boiled away. Krypton is present in the air to the extent of about 1 ppm. The atmosphere of Mars has been found to contain 0.3 ppm of krypton. It is one of the "noble" gases. It is characterized by its brilliant green and orange spectral lines. Naturally occurring krypton contains six stable isotopes. Fifteen other unstable isotopes are now recognized. The spectral lines of krypton are easily produced and some are very sharp. In 1960 it was internationally agreed that the fundamental unit of length, the meter, should be defined in terms of the orange-red spectral line of Kr^{86}, corresponding to the transition $5p[O_{1/2}]_1 - 6d[O_{1/2}]_1$, as follows: *1 m = 1,650,763.73 wavelengths (in vacuo) of the orange-red line of* Kr^{86}. This replaced the standard meter of Paris, which was defined in terms of a bar made of a platinum-iridium alloy. In October 1983 the meter, which originally was defined as being one ten millionth of a quadrant of the earth's polar circumference, was again redefined by the International Bureau of Weight and Measures as being the length of path traveled by light in a vacuum during a time interval of 1/299,792,458 of a second. Solid krypton is a white crystalline substance with a face-centered cubic structure which is common to all the "rare gases." While krypton is generally thought of as a rare gas that normally does not combine with other elements to form compounds, it now appears that the existence of some krypton compounds is established. Krypton difluoride has been prepared in gram quantities and can be made by several methods. A higher fluoride of krypton and a salt of an oxyacid of krypton also have been reported. Molecule ions of $ArKr^+$ and KrH^+ have been identified and investigated, and evidence is provided for the formation of KrXe or $KrXe^+$. Krypton clathrates have been prepared with hydroquinone and phenol. Kr^{85} has found recent application in chemical analysis. By imbedding the isotope in various solids, *kryptonates* are formed. The activity of these kryptonates is sensitive to chemical reactions at the surface. Estimates of the concentration of reactants are therefore made possible. Krypton is used commercially with argon as a low-pressure filling gas for fluorescent lights. It is used in certain photographic flash lamps for high-speed photography. Uses thus far have been limited because of its high cost. Krypton gas presently costs about $30/*l*.

Kurchatovium — see Element 104.

Lanthanum — (Gr. *lanthanein,* to lie hidden), La; at. wt. 138.9055 ± 3; at. no. 57; m.p. 921°C; b.p. 3457°C; sp. gr. 6.145 (25°C); valence 3. Mosander in 1839 extracted a new earth, *lanthana,* from impure cerium nitrate, and recognized the new element. Lanthanum is found in rare-earth minerals such as *cerite, monazite, allanite,* and *bastnasite.* Monazite and bastnasite are principal ores in which lanthanum occurs in percentages up to 25 and 38%, respectively. Misch metal, used in making lighter flints, contains about 25% lanthanum. Lanthanum was isolated in relatively pure form in 1923. Ion-exchange and solvent extraction techniques have led to much easier isolation of the so-called "rare-earth" elements. The availability of lanthanum and other rare earths has improved greatly in recent years. The metal can be produced by reducing the anhydrous fluoride with calcium. Lanthanum is silvery white, malleable, ductile, and soft enough to be cut with a knife. It is one of the most reactive of the rare-earth metals. It oxidizes rapidly when exposed to air. Cold water attacks lanthanum slowly, and hot water attacks it much more rapidly. The metal reacts directly with elemental carbon, nitrogen, boron, selenium, silicon, phosphorus, sulfur, and with halogens. At 310°C, lanthanum changes from a hexagonal to a face-centered cubic structure, and at 865°C it again transforms into a body-centered cubic structure. Natural lanthanum is a mixture of two stable isotopes, La^{138} and La^{139}. Seventeen other radioactive isotopes are recognized. Rare-earth compounds containing lanthanum are extensively used in carbon lighting applications, especially by the motion picture industry for studio lighting and projection. This application consumes about 25% of the rare-earth compounds produced. La_2O_3 improves the alkali resistance of glass, and is used in making special optical glasses. Small amounts of lanthanum, as an additive, can be used to produce nodular cast iron. There is current interest in hydrogen sponge alloys containing lanthanum. These alloys take up to 400 times their own volume of hydrogen gas, and the process is reversible. Heat energy is released every time they do so; therefore these alloys have possibilities in energy conservation systems.Lanthanum and its compounds have a low to moderate acute toxicity rating; therefore, care should be taken in handling them. The metal costs about $25/oz.

Lawrencium — (Ernest O. Lawrence, inventor of the cyclotron), Lr; at. no. 103; at. mass no. (260); valence +3(?). This member of the 5f transition elements (actinide series) was discovered in March 1961 by A. Ghiorso, T. Sikkeland, A. E. Larsh, and R.M. Latimer. A 3-μg californium target, consisting of a mixture of isotopes of mass number 249, 250, 251, and 252, was bombarded with either B^{10} or B^{11}. The electrically charged transmutation nuclei recoiled with an atmosphere of helium and were collected on a thin copper conveyor tape which was then moved to place collected atoms in front of a series of solid-state detectors. The isotope of element 103 produced in this way decayed by emitting an 8.6-MeV alpha particle with a half-life of 8 sec. In 1967, Flerov and associates of the Dubna Laboratory reported their inability to detect an alpha emitter with a half-life of 8 sec which was assigned by the Berkeley group to 103^{257}. This assignment has been changed to Lr^{258} or Lr^{259}. In 1965, the Dubna workers found a longer-lived lawrencium isotope, Lr^{256}, with a half-life of 35 sec. In 1968, Ghiorso and associates at Berkeley were able to use a few atoms of this isotope to study the oxidation behavior of lawrencium. Using solvent extraction techniques and working very rapidly, they extracted lawrencium ions from a buffered aqueous solution into an organic solvent, completing each extraction in about 30 sec. It was found that lawrencium behaves differently from dipositive nobelium and more like the tripositive elements earlier in the actinide series.

Lead — (Anglo-Saxon *lead*), Pb (L. *plumbum*); at. wt. 207.2; at. no. 82; m.p. 327.502°C; b.p. 1740°C; sp.gr. 11.35 (20°C); valence 2 or 4. Long known, mentioned in Exodus. The alchemists believed lead to be the oldest metal and associated it with the planet Saturn. Native lead occurs in nature, but it is rare. Lead is obtained chiefly from *galena* (PbS) by a roasting process. *Anglesite* ($PbSO_4$), *cerussite* ($PbCO_3$), and *minim* (Pb_3O_4) are other common lead minerals. Lead is a bluish-white metal of bright luster, is very soft, highly malleable, ductile, and a poor conductor of electricity. It is very resistant to corrosion; lead pipes bearing the insignia of Roman emperors, used as drains from the baths, are still in service. It is used in containers for corrosive liquids (such as in sulfuric acid chambers) and may be toughened by the addition of a small percentage of antimony or other metals. Natural lead is a mixture of four stable isotopes: Pb^{204} (1.48%), Pb^{206} (23.6%), Pb^{207} (22.6%), and Pb^{208} (52.3%). Lead isotopes are the end products of each of the three series of naturally occurring radioactive elements: Pb^{206} for the uranium series, Pb^{207} for the actinium series, and Pb^{208} for the thorium series. Seventeen other isotopes of lead, all of which are radioactive, are recognized. Its alloys include solder, type metal, and various antifriction metals. Great quantities of lead, both as the metal and as the dioxide, are used in storage batteries. Much metal also goes into cable covering, plumbing, ammunition, and in the manufacture of lead tetraethyl, used as an antiknock compound in gasoline. The metal is very effective as a sound absorber, is used as a radiation shield around X-ray equipment and nuclear reactors, and is used to absorb vibration. White lead, the basic carbonate, sublimed white lead ($PbSO_4$), chrome yellow ($PbCrO_4$), red lead (Pb_3O_4), and other lead compounds are used extensively in paints, although in recent years the use of lead in paints has been drastically curtailed to eliminate or reduce health hazards. Lead oxide is used in producing fine "crystal glass" and "flint glass" of a high index of refraction for achromatic lenses. The nitrate and the acetate are soluble salts. Lead salts such as lead arsenate have been used as insecticides, but their use in recent years has been practically eliminated in favor of less harmful organic compounds. Care must be used in handling lead as it is a cumulative poison. Environmental concern with lead poisoning has resulted in a national program to reduce the concentration of lead in gasoline.

Lithium — (Gr. *lithos*, stone), Li; at. wt. 6.941; at. no. 3; m.p. 180.54°C; b.p. 1342°C; sp. gr. 0.534 (20 °C); valence 1. Discovered by Arfvedson in 1817. Lithium is the lightest of all metals, with a density only about half that of water. It does not occur free in nature; combined it is found in small amounts in nearly all igneous rocks and in the waters of many mineral springs. *Lepidolite, spodumene, petalite,* and *amblygonite* are the more important minerals containing it. Lithium is presently being recovered from brines of Searles Lake, in California, and from those in Nevada. Large deposits of spodumene are found in North Carolina. The metal is produced electrolytically from the fused chloride. Lithium is silvery in appearance, much like Na and K, other members of the alkali metal series. It reacts with water, but not as vigorously as sodium. Lithium imparts a beautiful crimson color to a flame, but when the metal burns strongly the flame is a dazzling white. Since World War II, the production of lithium metal and its compounds has increased greatly. Because the metal has the highest specific heat of any solid element, it has found use in heat transfer applications; however, it is corrosive and requires special handling. The metal has been used as an alloying agent, is of interest in synthesis of organic compounds, and has nuclear applications. It ranks as a leading contender as a battery anode material as it has a high electrochemical potential. Lithium is used in special glasses and ceramics. The glass for the 200-in. telescope at Mt. Palomar contains lithium as a minor ingredient. Lithium chloride is one of the most hygroscopic materials known, and it, as well as lithium bromide, is used in air conditioning and industrial drying systems. Lithium stearate is used as an all-purpose and high-temperature lubricant. Other lithium compounds are used in dry cells and storage batteries. The metal is priced at about $20/lb.

Lutetium — (Lutetia, ancient name for Paris, sometimes called *cassiopeium* by the Germans), Lu; at. wt. 174.967; at. no. 71; m.p. 1663°C; b.p. 3395°C; sp. gr. 9.840 (25°C); valence 3. In 1907, Urbain described a process by which Marignac's ytterbium (1879) could be separated into the two elements, ytterbium (neoytterbium) and lutetium. These elements were identical with "aldebaranium" and "cassiopeium," independently discovered by von Welsbach about the same time. Charles James of the University of New Hampshire also independently prepared the very pure oxide, *lutecia,* at this time. The spelling of the element was changed from *lutecium* to *lutetium* in 1949. Lutetium occurs in very small amounts in nearly all minerals containing yttrium, and is present in *monazite* to the extent of about 0.003%, which is a commercial source. The pure metal has been isolated only in recent years and is one of the most difficult to prepare. It can be prepared by the reduction of anhydrous $LuCl_3$ or LuF_3 by an alkali or alkaline earth metal. The metal is silvery white and relatively stable in air. While new techniques, including ion-exchange reactions, have been developed to separate the various rare-earth elements, lutetium is still the most costly of all naturally occurring rare earths. It is slightly more abundant than thulium. It is now priced at about $75/g or $30,000/lb. Lu^{176} occurs naturally (2.6%) with Lu^{175} (97.4%). It is radioactive with a half-life of about 3×10^{10} years. Stable lutetium nuclides, which emit pure beta radiation after thermal neutron activation, can be used as catalysts in cracking, alkylation, hydrogenation, and polymerization. Virtually no other commercial uses have been found yet for lutetium, as it is one of the most costly of the rare earth metals. While lutetium, like other rare-earth metals, is thought to have a low toxicity rating, it should be handled with care until more information is available.

Magnesium — (*Magnesia*, district in Thessaly) Mg; at. wt. 24.305; at. no. 12; m.p. 648.8±0.5°C; b.p. 1090°C; sp. gr. 1.738 (20°C); valence 2. Compounds of magnesium have long been known. Black recognized magnesium as an element in 1755. It was isolated by Davy in 1808, and prepared in coherent form by Bussy in 1831. Magnesium is the eighth most abundant element in the earth's crust. It does not occur uncombined, but is found in large deposits in the form of *magnesite, dolomite,* and other minerals. The metal is now principally obtained in the U.S. by electrolysis of fused magnesium chloride derived from brines, wells, and sea water. Magnesium is a light, silvery-white, and fairly tough metal. It tarnishes slightly in air, and finely divided magnesium readily ignites upon heating in air and burns with a dazzling white flame. It is used in flashlight photography, flares, and pyrotechnics, including incendiary bombs. It is one third lighter than aluminum, and in alloys is essential for airplane and missile construction. The metal improves the mechanical, fabrication, and welding characteristics of aluminum when used as an alloying agent. Magnesium is used in producing nodular graphite in cast iron, and is used as an additive to conventional propellants. It is also used as a reducing agent in the production of pure uranium and other metals from their salts. The hydroxide (*milk of magnesia*), chloride, sulfate (*Epsom salts*), and citrate are used in medicine. Dead-burned magnesite is employed for refractory purposes such as brick and liners in furnaces and converters.

Organic magnesium compounds (Grignard's reaction) are important. Magnesium is an important element in both plant and animal life. Chlorophylls are magnesium-centered porphyrins. The adult daily requirement of magnesium is about 300 mg/day, but this is affected by various factors. Great care should be taken in handling magnesium metal, especially in the finely divided state, as serious fires can occur. Water should not be used on burning magnesium or on magnesium fires.

Manganese — (L. *magnes,* magnet, from magnetic properties of pyrolusite; It. *manganese,* corrupt form of *magnesia*), Mn; at. wt. 54.9380; at. no. 25; m.p. 1244±3°C; b.p. 1962°C; sp. gr. 7.21 to 7.44, depending on allotropic form; valence 1, 2, 3, 4, 6, or 7. Recognized by Scheele, Bergman, and others as an element and isolated by Gahn in 1774 by reduction of the dioxide with carbon. Manganese minerals are widely distributed; oxides, silicates, and carbonates are the most common. The discovery of large quantities of manganese nodules on the floor of the oceans holds promise as a source of manganese. These nodules contain about 24% manganese together with many other elements in lesser abundance. Most manganese today is obtained from ores found in the U.S.S.R., Brazil, Australia, Republic of So. Africa, Gabon, and India. *Pyrolusite* (MnO_2) and *rhodochrosite* ($MnCO_3$) are among the most common manganese minerals. The metal is obtained by reduction of the oxide with sodium, magnesium, aluminum, or by electrolysis. It is gray-white, resembling iron, but is harder and very brittle. The metal is reactive chemically, and decomposes cold water slowly. Manganese is used to form many important alloys. In steel, manganese improves the rolling and forging qualities, strength, toughness, stiffness, wear resistance, hardness, and hardenability. With aluminum and antimony, especially with small amounts of copper, it forms highly ferromagnetic alloys. Manganese metal is ferromagnetic only after special treatment. The pure metal exists in four allotropic forms. The alpha form is stable at ordinary temperature; gamma manganese, which changes to alpha at ordinary temperatures, is said to be flexible, soft, easily cut, and capable of being bent. The dioxide (pyrolusite) is used as a depolarizer in dry cells, and is used to "decolorize" glass that is colored green by impurities of iron. Manganese by itself colors glass an amethyst color, and is responsible for the color of true amethyst. The dioxide is also used in the preparation of oxygen, chlorine, and in drying black paints. The permanganate is a powerful oxidizing agent and is used in quantitative analysis and in medicine. Manganese is widely distributed throughout the animal kingdom. It is an important trace element and may be essential for utilization of vitamin B_1. Exposure to manganese dusts, fume, and compounds (as Mn) should not exceed the ceiling value of 5 mg/M³ for even short periods because of the toxicity of the element in larger quantity.

Mendelevium — (Dmitri Mendeleev), Md; at. wt. 257; at. no. 101; valence +2, +3. Mendelevium, the ninth transuranium element of the actinide series to be discovered, was first identified by Ghiorso, Harvey, Choppin, Thompson, and Seaborg early in 1955 as a result of the bombardment of the isotope Es^{253} with helium ions in the Berkeley 60-in. cyclotron. The isotope produced was Md^{256}, which has a half-life of 77 min. This first identification was notable in that Md^{256} was synthesized on a one-atom-at-a-time basis. Four isotopes are now recognized. Md^{258} has a half-life of 2 months. This isotope has been produced by the bombardment of an isotope of einsteinium with ions of helium. It now appears possible that eventually Md^{258} can be made so that some of its physical properties can be determined. Md^{256} has been used to elucidate some of the chemical properties of mendelevium in aqueous solution. Experiments seem to show that the element possesses a moderately stable dipositive (II) oxidation state in addition to the tripositive (III) oxidation state, which is characteristic of actinide elements.

Mercury — (Planet *Mercury*), Hg (*hydrargyrum,* liquid silver); at. wt. 200.59 ± 3; at. no. 80; m.p. −38.842°C; b.p. 356.58°C; sp.gr. 13.546 (20°C); valence 1 or 2. Known to ancient Chinese and Hindus; found in Egyptian tombs of 1500 B.C. Mercury is the only common metal liquid at ordinary temperatures. It only rarely occurs free in nature. the chief ore is *cinnabar* (HgS). Spain and Italy produce about 50% of the world's supply of the metal. The commercial unit for handling mercury is the "flask," which weighs 76 lb. The metal is obtained by heating cinnabar in a current of air and by condensing the vapor. It is a heavy, silvery-white metal; a rather poor conductor of heat, as compared with other metals, and a fair conductor of electricity. It easily forms alloys with many metals, such as gold, silver, and tin, which are called *amalgams.* Its ease in amalgamating with gold is made use of in the recovery of gold from its ores. The metal is widely used in laboratory work for making thermometers, barometers, diffusion pumps, and many other instruments. It is used in making mercury-vapor lamps and advertising signs, etc. and is used in mercury switches and other electrical apparatus. Other uses are in making pesticides, mercury cells for caustic soda and chlorine production, dental preparations, antifouling paint, batteries, and catalysts. The most important salts are mercuric chloride $HgCl_2$ (corrosive sublimate — a violent poison), mercurous chloride Hg_2Cl_2 (calomel, occasionally still used in medicine), mercury fulminate ($Hg(ONC)_2$, a detonator widely used in explosives), and mercuric sulfide (HgS, vermillion, a high-grade paint pigment). Organic mercury compounds are important. It has been found that an electrical discharge causes mercury vapor to combine with neon, argon, krypton, and xenon. These products, held together with van der Waals' forces, correspond to HgNe, HgAr, HgKr, and HgXe. Mercury is a virulent poison and is readily absorbed through the respiratory tract, the gastrointestinal tract, or through unbroken skin. It acts as a cumulative poison since only small amounts of the element can be eliminated at a time by the human organism. Since mercury is a very volatile element, dangerous levels are readily attained in air. Air saturated with mercury vapor at 20°C contains a concentration that exceeds the toxic limit many times. The danger increases at higher temperatures. *It is therefore important that mercury be handled with care.* Containers of mercury should be securely covered and spillage should be avoided. If it is necessary to heat mercury or mercury compounds, it should be done in a well-ventilated hood. Methyl mercury is a dangerous pollutant and is now widely found in water and streams. The U.S. National Bureau of Standards has recently redetermined the triple point of mercury and found it to be −38.84168°C.

Molybedenum — (Gr. *molybdos,* lead), Mo; at. wt. 95.94; at. no. 42; m.p. 2617°C; b.p. 4612°C; sp. gr. 10.22 (20°C); valence 2, 3, 4?, 5?, or 6. Before Scheele recognized molybdenite as a distinct ore of a new element in 1778, it was confused with graphite and lead ore. The metal was prepared in an impure form in 1782 by Hjelm. Molybdenum does not occur native, but is obtained principally from *molybdenite* (MoS_2). *Wulfenite* ($PbMoO_4$), and *powellite* ($Ca(MoW)O_4$) are also minor commercial ores. Molybdenum is also recovered as a by-product of copper and tungsten mining operations. The metal is prepared from the powder made by the hydrogen reduction of purified molybdic trioxide or ammonium molybdate. The metal is silvery white, very hard, but is softer and more ductile than tungsten. It has a high elastic modulus, and only tungsten and tantalum, of the more readily available metals, have higher melting points. It is a valuable alloying agent, as it contributes to the hardenability and toughness of quenched and tempered steels. It also improves the strength of steel at high temperatures. It is used in certain nickel-based alloys, such as the "Hastelloys ®" which are heat-resistant and corrosion-resistant to chemical solutions. Molybdenum oxidizes at elevated temperatures. The metal has found recent application as electrodes for electrically heated glass furnaces and forehearths. The metal is also used in nuclear energy applications and for missile and aircraft parts. Molybdenum is valuable as a catalyst in the refining of petroleum. It has found application as a filament material in electronic and electrical applications. Molybdenum is an essential trace element in plant nutrition. Some lands are barren for lack of this element in the soil. Molybdenum sulfide is useful as a lubricant, especially at high temperatures where oils would decompose. Almost all ultra-high strength steels with minimum yield points up to 300,000 psi(lb/in.^{2}) contain molybdenum in amounts from 0.25 to 8%.

Neodymium — (Gr. *neos*, new, and *didymos*, twin), Nd; at. wt. 144.24 ± 3; at. no. 60; m.p. 1021°C; b.p. 3068°C; sp. gr. 6.80 and 7.007, depending on allotropic form; valence 3. In 1841, Mosander, extracted from *cerite* a new rose-colored oxide, which he believed contained a new element. He named the element *didymium*, as it was "an inseparable twin brother of lanthanum." In 1885 von Welsbach separated didymium into two new elemental components, *neodymia* and *praseodymia*, by repeated fractionation of ammonium didymium nitrate. While the free metal is in *misch metal*, long known and used as a pyrophoric alloy for lighter flints, the element was not isolated in relatively pure form until 1925. Neodymium is present in misch metal to the extent of about 18%. It is present in the minerals *monazite* and *bastnasite*, which are principal sources of rare-earth metals. The element may be obtained by separating neodymium salts from other rare earths by ion-exchange or solvent extraction techniques, and by reducing anhydrous halides such as NdF_3 with calcium metal. Other separation techniques are possible. The metal has a bright silvery metallic luster. Neodymium is one of the more reactive rare-earth metals and quickly tarnishes in air, forming an oxide that spalls off and exposes metal to oxidation. The metal, therefore, should be kept under light mineral oil or sealed in a plastic material. Neodymium exists in two allotropic forms, with a transformation from a double hexagonal to a body-centered cubic structure taking place at 860°C. Natural neodymium is a mixture of seven stable isotopes. Seven other radioactive isotopes are recognized. Didymium, of which neodymium is a component, is used for coloring glass to make welder's goggles. By itself, neodymium colors glass delicate shades ranging from pure violet through wine-red and warm gray. Light transmitted through such glass shows unusually sharp absorption bands. The glass has been used in astronomical work to produce sharp bands by which spectral lines may be calibrated. Glass containing neodymium can be used as a laser material in place of ruby to produce coherent light. Neodymium salts are also used as a colorant for enamels. The price of the metal is about $2/g. Neodymium has a low-to-moderate acute toxic rating. As with other rare earths, neodymium should be handled with care.

Neon — (Gr. *neos*, new), Ne; at. wt. 20.179; at. no. 10; m.p. −248.67°C; b.p. −246.048°C (1 atm); density of gas 0.89990 g/l (1 atm, 0°C); density of liquid at b.p. 1.207 g/cm^3; valence 0. Discovered by Ramsay and Travers in 1898. Neon is a rare gaseous element present in the atmosphere to the extent of 1 part in 65,000 of air. It is obtained by liquefaction of air and separated from the other gases by fractional distillation. Natural neon is a mixture of three isotopes. Five other unstable isotopes are known. It is a very inert element; however, it is said to form a compound with fluorine. It is still questionable if true compounds of neon exist, but evidence is mounting in favor of their existence. The following ions are known from optical and mass spectrometric studies: Ne^+, $(NeAr)^+$, $(NeH)^+$, and $(HeNe)^+$. Neon also forms an unstable hydrate. In a vacuum discharge tube, neon glows reddish orange. Of all the rare gases, the discharge of neon is the most intense at ordinary voltages and currents. Neon is used in making the common neon advertising signs, which accounts for its largest use. It is also used to make high-voltage indicators, lightning arrestors, wave meter tubes, and TV tubes. Neon and helium are used in making gas lasers. Liquid neon is now commercially available and is finding important application as an economical cryogenic refrigerant. It has over 40 times more refrigerating capacity per unit volume than liquid helium and more than three times that of liquid hydrogen. It is compact, inert, and is less expensive than helium when it meets refrigeration requirements. Neon has costs about $2.00/1.

Neptunium — (Planet *Neptune*), Np; at. wt. 237.0482; at. no. 93; m.p. 640± 1°C; b.p. 3902°C (est.); sp. gr. 20.25 (20°C); valence 3, 4, 5, and 6. Neptunium was the first synthetic transuranium element of the actinide series discovered; the isotope Np^{239} was produced by McMillan and Abelson in 1940 at Berkeley, California, as the result of bombarding uranium with cyclotron-produced neutrons. The isotope Np^{237} (half-life of 2.14×10^6 years) is currently obtained in gram quantities as a by-product from nuclear reactors in the production of plutonium. Trace quantities of the element are actually found in nature due to transmutation reactions in uranium ores produced by the neutrons which are present. Neptunium is prepared by the reduction of NpF_3 with barium or lithium vapor at about 1200°C. Neptunium metal has a silvery appearance, is chemically reactive, and exists in at least three structural modifications: α-neptunium, orthorhombic, density 20.25 g/cm^3; β-neptunium (above 280°C), tetragonal, density (313° C) 19.36 g/cm^3; γ-neptunium (above 577°), cubic, density (600°C) 18.0 g/cm^3. Neptunium has four ionic oxidation states in solution: Np^{+3} (pale purple), analogous to the rare earth ion Pm^{+3}, Np^{+4} (yellow green); NpO^+ (green blue); and NpO^{++} (pale pink). These latter oxygenated species are in contrast to the rare earths which exhibit only simple ions of the (II), (III), and (IV) oxidation states in aqueous solution. The element forms tri- and tetrahalides such as NpF_3, NpF_4, $NpCl_4$, $NpBr_3$, NpI_3, and oxides of various compositions such as are found in the uranium-oxygen system, including Np_3O_8 and NpO_2. Thirteen isotopes of neptunium are now recognized. The O.R.N.L. has Np^{237} available for sale to its licensees and for export. This isotope can be used as a component in neutron detection instruments. It is offered at a price of $280/g.

Nickel — (Ger, *Nickel,* Satan or Old Nick's and from *kupfernickel,* Old Nick's copper), Ni; at. wt. 58.69; at no. 28; m.p. 1453°C; b.p. 2732°C; sp. gr. 8.902 (25°C); valence 0,1,2,3. Discovered by Cronstedt in 1751 in kupfernickel (*niccolite*). Nickel is found as a constituent in most meteorites and often serves as one of the criteria for distinguishing a meteorite from other minerals. Iron meteorites, or *siderites,* may contain iron alloyed with from 5 to nearly 20% nickel. Nickel is obtained commercially from *pentlandite* and *pyrrhotite* of the Sudbury region of Ontario, a district that produces about 30% of the nickel for the Free World. Other deposits are found in New Caledonia, Australia, Cuba, Indonesia, and elsewhere. Nickel is silvery white and takes on a high polish. It is hard, malleable, ductile, somewhat ferromagnetic, and a fair conductor of heat and electricity. It belongs to the iron-cobalt group of metals and is chiefly valuable for the alloys it forms. It is extensively used for making stainless steel and other corrosion-resistant alloys such as Invar®, Monel®, Inconel®, and the Hastelloys®. Tubing made of a copper-nickel alloy is extensively used in making desalination plants for converting sea water into fresh water. Nickel is also now used extensively in coinage and in making nickel steel for armor plate and burglar-proof vaults, and is a component in Nichrome®, Permalloy®, and constantan. Nickel added to glass gives a green color. Nickel plating is often used to provide a protective coating for other metals, and finely divided nickel is a catalyst for hydrogenating vegetable oils. It is also used in ceramics, in the manufacture of Alnico magnets, and in the Edison® storage battery. The sulfate and the oxides are important compounds. Natural nickel is a mixture of five stable isotopes; seven other unstable isotopes are known.Exposure to nickel metal and soluble compounds (as Ni) should not exceed 1mg/M^3 (8-hr time-weighted average — 40-hr week). Nickel carbonyl exposure, however, should not exceed 0.007mg/M^3, and is considered to be a very toxic material. Nickel sulphide fume and dust is recognized as having carcinogenic potential.

Niobium — (*Niobe,* daughter of Tantalus), Nb; or Columbium (*Columbia,* name for America); at. wt. 92.9064; at. no. 41; m.p. 2468 ± 10°C; b.p. 4742°C; sp. gr. 8.57 (20°C); valence 2, 3, 4?, 5. Discovered in 1801 by Hatchett in an ore sent to England more than a century before by John Winthrop the Younger, first governor of Connecticut. The metal was first prepared in 1864 by Blomstrand, who reduced the chloride by heating it in a hydrogen atmosphere. The name "niobium" was adopted by the International Union of Pure and Applied Chemistry in 1950 after 100 years of controversy. Many leading chemical societies and government organizations refer to it by this name. Most metallurgists, leading metal societies, and all but one of the leading U.S. commercial producers, however, still refer to the metal as "columbium." The element is found in *niobite* (or *columbite*), *niobite-tantalite, pyrochlore,* and *euxenite.* Large deposits of niobium have been found associated with *carbonatites* (carbon-silicate rocks), as a constituent of *pyrochlore.* Extensive ore reserves are found in Canada, Brazil, Nigeria, Zaire, and in the U.S.S.R. The metal can be isolated from tantalum, and prepared in several ways. It is a shiny, white, soft, and ductile metal, and takes on a bluish cast when exposed to air at room temperatures for a long time. The metal starts to oxidize in air at 200°C, and when processed at even moderate temperatures must be placed in a protective atmosphere. It is used as an alloying agent in carbon and alloy steels and in nonferrous metals. These alloys have improved strength and other desirable properties, and have been used in pipeline construction. The metal has a low capture cross-section for thermal neutrons. It is used in arc-welding rods for stabilized grades of stainless steel. Thousands of pounds of niobium have been used in advanced air frame systems such as were used in the Gemini space program. The element has superconductive properties; superconductive magnets have been made with Nb-Zr wire, which retains its superconductivity in strong magnetic fields. This type of application offers hope of direct large-scale generation of electric power. Sixteen isotopes of niobium are known. Niobium metal (99.5% pure) is priced at about $30/lb.

Nitrogen — (L. *nitrum,* Gr. *nitron,* native soda; genes, *forming*), N; at. wt. 14.0067; at. no. 7; m.p. −209.86 °C; b.p. −195.8°C; density 1.2506 g/l; sp. gr. liquid 0.808 (−195.8°C), solid 1.026 (−252°C); valence 3 or 5. Discovered by Daniel Rutherford in 1772, but Scheele, Cavendish, Priestley, and others about the same time studied "burnt or dephlogisticated air," as air without oxygen was then called. Nitrogen makes up 78% of the air, by volume. The atmosphere of Mars, by comparison, is 2.6% nitrogen. The estimated amount of this element in our atmosphere is more than 4000 trillion tons. From this inexhaustible source it can be obtained by liquefaction and fractional distillation. Nitrogen molecules give the orange-red, blue-green, blue-violet, and deep violet shades to the aurora. The element is so inert that Lavoisier named it *azote,* meaning without life, yet its compounds are so active as to be most important in foods, poisons, fertilizers, and explosives. Nitrogen can be also easily prepared by heating a water solution of ammonium nitrite. Nitrogen, as a gas, is colorless, odorless, and a generally inert element. As a liquid it is also colorless and odorless, and is similar in appearance to water. Two allotropic forms of solid nitrogen exist, with the transition from the α to the β form taking place at −237 °C. When nitrogen is heated, it combines directly with magnesium, lithium, or calcium; when mixed with oxygen and subjected to electric sparks, it forms first nitric oxide (NO) and then the dioxide (NO_2); when heated under pressure with a catalyst with hydrogen, ammonia is formed (Haber process). The ammonia thus formed is of the utmost importance as it is used in fertilizers, and it can be oxidized to nitric acid (Ostwald process). The ammonia industry is the largest consumer of nitrogen. Large amounts of the gas are also used by the electronics industry, which uses the gas as a blanketing medium during production of such components as transistors, diodes, etc. Large quantities of nitrogen are used in annealing stainless steel and other steel mill products. The drug industry also uses large quantities. Nitrogen is used as a refrigerant both for the immersion freezing of food products and for transportation of foods. Liquid nitrogen is also used in missile work as a purge for components, insulators for space chambers, etc., and by the oil industry to build up great pressures in wells to force crude oil upward. Sodium and potassium nitrates are formed by the decomposition of organic matter with compounds of the metals present. In certain dry areas of the world these saltpeters are found in quantity. Ammonia, nitric acid, the nitrates, the five oxides (N_2O, NO, N_2O_3, NO_2, and N_2O_5), TNT, the cyanides, etc. are but a few of the important compounds. Nitrogen gas prices vary from 2c to $2.75 per 100 ft^3, depending on purity, etc. Production of elemental nitrogen in the U.S. is more than 9 million short tons per year.

Nobelium — (Alfred Nobel, discoverer of dynamite), No; at. wt. (259); at. no. 102; valence + 2, + 3. Nobelium was unambiguously discovered and identified in April 1958 at Berkeley by A. Ghiorso, T. Sikkeland, J. R. Walton, and G. T. Seaborg, who used a new double-recoil technique. A heavy-ion linear accelerator (HILAC) was used to bombard a thin target of curium (95% Cm^{244} and 4.5% Cm^{246}) with C^{12} ions to produce 102^{254} according to the Cm^{246} (C^{12} 4*n*) reaction. Earlier in 1957 workers of the U.S., Britain, and Sweden announced the discovery of an isotope of element 102 with a 10-min half-life at 8.5 MeV, as a result of bombarding Cm^{244} with C^{13} nuclei. On the basis of this experiment the name *nobelium* was assigned and accepted by the Commission on Atomic Weights of the International Union of Pure and Applied Chemistry. The acceptance of the name was premature, for both Russian and American efforts now completely rule out the possibility of any isotope of element 102 having a half-life of 10 min in the vicinity of 8.5 MeV. Early work in 1957 on the search for this element, in Russia at the Kurchatov Institute, was marred by the assignment of 8.9± 0.4 MeV alpha radiation with a half-life of 2 to 40 sec, which was too indefinite to support claim to discovery. Confirmatory experiments at Berkeley in 1966 have shown the existence of 102^{254} with a 55-sec half-life, 102^{252} with a 2.3-sec half-life, and 102^{257} with a 23-sec half-life. Four other isotopes are now recognized, one of which — 102^{255} — has a half-life of 3 min. In view of the discoverer's traditional right to name an element, the Berkeley group, in 1967, suggested that the hastily given name *nobelium*, along with the symbol No, be retained.

Osmium — (Gr. *osme*, a smell), Os; at. wt. 190.2; at. no. 76; m.p. 3045± 30°C; b.p. 5027 ± 100°C; sp. gr. 22.57; valence 0 to +8, more usually +3, +4, +6, and +8. Discovered in 1803 by Tennant in the residue left when crude platinum is dissolved by *aqua regia*. Osmium occurs in *iridosmine* and in platinum-bearing river sands of the Urals, North America, and South America. It is also found in the nickel bearing ores of the Sudbury, Ontario region along with other platinum metals. While the quantity of platinum metals in these ores is very small, the large tonnages of nickel ores processed make commercial recovery possible. The metal is lustrous, bluish white, extremely hard, and brittle even at high temperatures. It has the highest melting point and lowest vapor pressure of the platinum group. The metal is very difficult to fabricate, but the powder can be sintered in a hydrogen atmosphere at a temperature of 2000°C. The solid metal is not affected by air at room temperature, but the powdered or spongy metal slowly gives off osmium tetroxide, which is a powerful oxidizing agent and has a strong smell. The tetroxide is highly toxic, and boils at 130° C (760 mm). Concentrations in air as low as $10^{-7}g/m^3$ can cause lung congestion, skin damage, or eye damage. Exposure to osmium tetroxide should not exceed 0.002 mg/M^3 (8-hr time-weighted average — 40-hr work week). The tetroxide has been used to detect fingerprints and to stain fatty tissue for microscope slides. The metal is almost entirely used to produce very hard alloys, with other metals of the platinum group, for fountain pen tips, instrument pivots, phonograph needles, and electrical contacts. The price of 99% pure osmium powder — the form usually supplied commercially — is about $20/g or $1000/tr. oz., depending on quantity and supplier. The measured densities of iridium and osmium seem to indicate that osmium is slightly more dense than iridium, so osmium has generally been credited with being the heaviest known element. Calculations of the density from the space lattice, which may be more reliable for these elements than actual measurements, however, give a density of 22.65 for iridium compared to 22.61 for osmium. At present, therefore, we know either iridium or osmium is the heaviest element, but the data do not allow selection between the two.

Oxygen — (Gr. *oxys*, sharp, acid, and *genes*, forming; acid former), O; at. wt. (natural) 15.9994 ± 3; at. no. 8; m.p. −218.4°C; b.p. −182.962°C; density 1.429 g/l(0°C); sp. gr. liquid 1.14 (−182.96°C); valence 2. For many centuries, workers occasionally realized air was composed of more than one component. The behavior of oxygen and nitrogen as components of air led to the advancement of the phlogiston theory of combustion, which captured the minds of chemists for a century. Oxygen was prepared by several workers, including Bayen and Borch, but they did not know how to collect it, did not study its properties, and did not recognize it as an elementary substance. Priestley is generally credited with its discovery, although Scheele also discovered it independently. Oxygen is the third most abundant element found in the sun, and it plays a part in the carbon-nitrogen cycle, one process thought to give the sun and stars their energy. Oxygen under excited conditions is responsible for the bright red and yellow-green colors of the aurora. Oxygen, as a gaseous element, forms 21% of the atmosphere by volume from which it can be obtained by liquefaction and fractional distillation. The atmosphere of Mars contains about 0.15% oxygen. The element and its compounds make up 49.2 %, by weight, of the earth's crust. About two thirds of the human body and nine tenths of water is oxygen. In the laboratory it can be prepared by the electrolysis of water or by heating potassium chlorate with manganese dioxide as a catalyst. The gas is colorless, odorless, and tasteless. The liquid and solid forms are a pale blue color and are strongly paramagnetic. Ozone (O_3), a highly active allotropic form of oxygen, is formed by the action of an electrical discharge or ultraviolet light on oxygen. Ozone's presence in the atmosphere (amounting to the equivalent of a layer 3 mm thick at ordinary pressures and temperatures) is of vital importance in preventing harmful ultraviolet rays of the sun from reaching the earth's surface. There has been recent concern that aerosols in the atmosphere may have a detrimental effect on this ozone layer. Ozone is toxic and exposure should not exceed 0.2mg/M^3 (8-hr time-weighted average — 40-hr work week.) Undiluted ozone has a bluish color. Liquid ozone is bluish black, and solid ozone is violet-black. Oxygen is very reactive and capable of combining with most elements. It is a component of hundreds of thousands of organic compounds. It is essential for respiration of all plants and animals and for practically all combustion. In hospitals it is frequently used to aid respiration of patients. Its atomic weight was used as a standard of comparison for each of the other elements until 1961 when the International Union of Pure and Applied Chemistry adopted carbon 12 as the new basis. Oxygen has eight isotopes. Natural oxygen is a mixture of three isotopes. Oxygen 18 occurs naturally, is stable, and is available commercially. Water (H_2O with 1.5% O^{18}) is also available. Commercial oxygen consumption in the U.S. is estimated to be 20 million short tons per year and the demand is expected to increase substantially in the next few years. Oxygen enrichment of steel blast furnaces accounts for the greatest use of the gas. Large quantities are also used in making synthesis gas for ammonia and methanol, ethylene oxide, and for oxy-acetylene welding. Air separation plants produce about 99% of the gas, electrolysis plants about 1%. The gas costs 5c/ ft^3 in small quantities, and about $15/ton.

Palladium — (named after the asteroid *Pallas*, discovered about the same time; Gr. *Pallas*, goddess of wisdom), Pd; at.

wt. 106.42 at. no. 46; m.p. 1554°C; b. p. 2970°C; sp. gr. 12.02 (20°C); valence 2, 3, or 4. Discovered in 1803 by Wollaston. Palladium is found along with platinum and other metals of the platinum group in placer deposits of the U.S.S.R., South and North America, Ethiopia, and Australia. It is also found associated with the nickel-copper deposits of South Africa and Ontario. Its separation from the platinum metals depends upon the type of ore in which it is found. It is a steel-white metal, does not tarnish in air, and is the least dense and lowest melting of the platinum group of metals. When annealed, it is soft and ductile; cold working greatly increases its strength and hardness. Palladium is attacked by nitric and sulfuric acid. At room temperatures the metal has the unusual property of absorbing up to 900 times its own volume of hydrogen, possibly forming Pd_2H. It is not yet clear if this is a true compound. Hydrogen readily diffuses through heated palladium and this provides a means of purifying the gas. Finely divided palladium is a good catalyst and is used for hydrogenation and dehydrogenation reactions. It is alloyed and used in jewelry trades. White gold is an alloy of gold decolorized by the addition of palladium. Like gold, palladium can be beaten into leaf as thin as 1/250,000 in. The metal is used in dentistry, watchmaking, and in making surgical instruments and electrical contracts. The metal sells for about $130/tr. oz.

Phosphorus — (Gr. *phosphoros,* light bearing; ancient name for the planet Venus when appearing before sunrise), P; at. wt. 30.97376; at. no. 15; m.p. (white) 44.1°C; b.p. (white) 280°C; sp. gr. (white) 1.82, (red) 2.20, (black) 2.25 to 2.69; valence 3 or 5. Discovered in 1669 by Brand, who prepared it from urine. Phosphorus exists in four or more allotropic forms: white (or yellow), red, and black (or violet). White phosphorus has two modifications: α and β with a transition temperature at −3.8°C. Never found free in nature, it is widely distributed in combination with minerals. *Phosphate* rock, which contains the mineral *apatite,* an impure tri-calcium phosphate, is an important source of the element. Large deposits are found in the U.S.S.R., in Morocco, and in Florida, Tennessee, Utah, Idaho, and elsewhere. Phosphorus is an essential ingredient of all cell protoplasm, nervous tissue, and bones. Ordinary phosphorus is a waxy white solid; when pure it is colorless and transparent. It is insoluble in water, but soluble in carbon disulfide. It takes fire spontaneously in air, burning to the pentoxide. It is very poisonous, 50mg constituting an approximate fatal dose. Exposure to white phosphorus should not exceed 0.1mg/M^3 (8-hr time-weighted average — 40-hr work week). White phosphorus should be kept under water, as it is dangerously reactive in air, and it should be handled with forceps, as contact with the skin may cause severe burns. When exposed to sunlight or when heated in its own vapor to 250°C, it is converted to the red variety, which does not phosphoresce in air as does the white variety. This form does not ignite spontaneously and it is not as dangerous as white phosphorus. It should, however, be handled with care as it does convert to the white form at some temperatures and it emits highly toxic fumes of the oxides of phosphorus when heated. The red modification is fairly stable, sublimes with a vapor pressure of 1 atm at 417°C, and is used in the manufacture of safety matches, pyrotechnics, pesticides, incendiary shells, smoke bombs, tracer bullets, etc. White phosphorus may be made by several methods. By one process, tri-calcium phosphate, the essential ingredient of phosphate rock, is heated in the presence of carbon and silica in an electric furnace or fuel-fired blast furnace. Elementary phosphorus is liberated as vapor and may be collected under water. If desired, the phosphorus vapor and carbon monoxide produced by the reaction can be oxidized at once in the presence of moisture or water to produce phosphoric acid, an important compound in making super-phosphate fertilizers. In recent years, concentrated phosphoric acids, which may contain as much as 70 to 75% P_2O_5 content, have become of great importance to agriculture and farm production. World-wide demand for fertilizers has caused record phosphate production in recent years. Phosphates are used in the production of special glasses, such as those used for sodium lamps. Bone-ash, calcium phosphate, is also used to produce fine chinaware and to produce monocalcium phosphate used in baking powder. Phosphorus is also important in the production of steels, phosphor bronze, and many other products. Trisodium phosphate is important as a cleaning agent, as a water softener, and for preventing boiler scale and corrosion of pipes and boiler tubes. Organic compounds of phosphorus are important.

Platinum — (Sp. *platina,* silver), Pt; at. wt. 195.08 ± 3; at. no. 78; m.p. 1772°C; b.p. 3827 ± 100°C; sp. gr. 21.45 (20°C); valence 1?, 2, 3, or 4. Discovered in South America by Ulloa in 1735 and by Wood in 1741. The metal was used by pre-Columbian Indians. Platinum occurs native, accompanied by small quantities of iridium, osmium, palladium, ruthenium, and rhodium, all belonging to the same group of metals. These are found in the alluvial deposits of the Ural mountains, of Columbia, and of certain western American states. *Sperrylite* ($PtAs_2$), occurring with the nickel-bearing deposits of Sudbury, Ontario, is the source of a considerable amount of the metal. The large production of nickel offsets there being only one part of the platinum metals in two million parts of ore. Platinum is a beautiful silvery-white metal, when pure, and is malleable and ductile. It has a coefficient of expansion almost equal to that of soda-lime-silica glass, and is therefore used to make sealed electrodes in glass systems. The metal does not oxidize in air at any temperature, but is corroded by halogens, cyanides, sulfur, and caustic alkalis. It is insoluble in hydrochloric and nitric acid, but dissolves when they are mixed as *aqua regia,* forming chloroplatinic acid (H_2PtCl_6), an important compound. The metal is extensively used in jewelry, wire, and vessels for laboratory use, and in many valuable instruments including thermocouple elements. It is also used for electrical contacts, corrosion-resistant apparatus, and in dentistry. Platinum-cobalt alloys have magnetic properties. One such alloy made of 76.7% Pt and 23.3% Co, by weight, is an extremely powerful magnet that offers a B-H (max) almost twice that of Alnico V. Platinum resistance wires are used for constructing high-temperature electric furnaces. The metal is used for coating missile nose cones, jet engine fuel nozzles, etc., which must perform reliably for long periods of time at high temperatures. The metal, like palladium, absorbs large volumes of hydrogen, retaining it at ordinary temperatures but giving it up at red heat. In the finely divided state platinum is an excellent catalyst, having long been used in the contact process for producing sulfuric acid. It is also used as a catalyst in cracking petroleum products. There is also much current interest in the use of platinum as a catalyst in fuel cells and in antipollution devices for automobiles. Platinum anodes are extensively used in cathodic protection systems for large ships and ocean-going vessels, pipelines, steel piers, etc. Fine platinum wire will glow red hot when placed in the vapor of methyl alcohol. It acts here as a catalyst, converting the alcohol to formaldehyde. This phenomenon has been used commercially to produce cigarette lighters and hand warmers. Hydrogen and oxygen explode in the presence of platinum. The price of platinum has varied widely; more than a century ago it was used to adulterate gold. It was nearly eight times as valuable as gold in 1920. The price in December 1985 was about \$360/tr. oz.

Plutonium — (Planet *pluto*), Pu; at. wt. (244); at. no 94; isotopic mass Pu^{239} 239.0522; sp. gr. (α modification) 19.84 (25°C); m.p. 641°C; b.p. 3232°C; valence 3, 4, 5, or 6. Plutonium was the second transuranium element of the actinide series to be discovered. The isotope Pu^{238} was produced in 1940 by Seaborg, McMillan, Kennedy, and Wahl by deuteron bombardment of uranium in the 60-in. cyclotron at Berkeley, California. Plutonium also exists in trace quantities in naturally occurring uranium ores. It is formed in much the same manner as neptunium, by irradiation of natural uranium with the neutrons which are present. By far of greatest importance is the isotope Pu^{239}, with a half-life of 24,400 years, produced in extensive quantities in nuclear reactors from natural uranium:

$$U^{238}(n\gamma)U^{239} \xrightarrow{\beta} Np^{239} \xrightarrow{\beta} Pu^{239}$$

Fifteen isotopes of plutonium are known. Plutonium has assumed the position of dominant importance among the transuranium elements because of its successful use as an explosive ingredient in nuclear weapons and the place which it holds as a key material in the development of industrial use of nuclear power. One kilogram is equivalent to about 22 million kilowatt hours of heat energy. The complete detonation of a kilogram of plutonium produces an explosion equal to about 20,000 tons of chemical explosive. Its importance depends on the nuclear property of being readily fissionable with neutrons and its availability in quantity. The world's nuclear-power reactors are now producing about 20,000 kg of plutonium/yr. By 1982 it was estimated that about 300,000 kg had accumulated. The various nuclear applications of plutonium are well known. Pu^{238} has been used in the Apollo lunar missions to power seismic and other equipment on the lunar surface. As with neptunium and uranium, plutonium metal can be prepared by reduction of the trifluoride with alkaline-earth metals. The metal has a silvery appearance and takes on a yellow tarnish when slightly oxidized. It is chemically reactive. A relatively large piece of plutonium is warm to the touch because of the energy given off in alpha decay. Larger pieces will produce enough heat to boil water. The metal readily dissolves in concentrated hydrochloric acid, hydroiodic acid, or perchloric acid with formation of the Pu^{+3} ion. The metal exhibits six allotropic modifications having various crystalline structures. The densities of these vary from 16.00 to 19.86 g/cm³. Plutonium also exhibits four ionic valence states in aqueous solutions: Pu^{+3} (blue lavender), Pu^{+4} (yellow brown), PuO^{+} (pink?), and PuO^{+2} (pink orange). The ion PuO^{+} is unstable in aqueous solutions, disproportionating into Pu^{+4} and PuO^{+2} the Pu^{+4} thus formed, however, oxidizes the PuO^{+} into PuO^{+2}, itself being reduced to Pu^{+3}, giving finally Pu^{+3} and PuO^{+2}. Plutonium forms binary compounds with oxygen: PuO, PuO_2, and intermediate oxides of variable composition; with the halides: PuF_3, PuF_4, $PuCl_3$, $PuBr_3$, PuI_3; with carbon, nitrogen, and silicon: PuC, PuN, $PuSi_2$. Oxyhalides are also well known: PuOCl, PuOBr, PuOI. Because of the high rate of emission of alpha particles and the element being specifically absorbed by bone marrow, plutonium, as well as all of the other transuranium elements except neptunium, are radiological poisons and must be handled with very special equipment and precautions. Plutonium is a very dangerous radiological hazard. Scientists at Berkeley believe that linear catechoylamide carboxylate has potential for removing plutonium from living tissue and bone as well as from nuclear-reactor wastes. Precautions must also be taken to prevent the unintentional formation of a critical mass. Plutonium in liquid solution is more likely to become critical than solid plutonium. The shape of the mass must also be considered where criticality is concerned. Plutonium-238 is available from the A.E.C. at a cost of about \$700/g (80 to 89% enriched.)

Polonium — (Poland, native country of Mme. Curie), Po; at. mass. (209); at. no. 84; m.p. 254°C; b.p. 962°C; sp. gr. (alpha modification) 9.32; valence −2, 0, +2, +3 (?), +4, and +6. Polonium was the first element discovered by Mme. Curie in 1898, while seeking the cause of radioactivity of pitchblende from Joachimsthal, Bohemia. The electroscope showed it separating with bismuth. Polonium is also called Radium F. Polonium is a very rare natural element. Uranium ores contain only about 100μg of the element per ton. Its abundance is only about 0.2% of that of radium. In 1934 it was found that when natural bismuth (Bi^{209}) was bombarded by neutrons, Bi^{210}, the parent of polonium, was obtained. Milligram amounts of polonium may now be prepared this way, by using the high neutron fluxes of nuclear reactors. Polonium-210 is a low-melting, fairly volatile metal, 50% of which is vaporized in air in 45 hr at 55°C. It is an alpha emitter with a half-life of 138.39 days. A milligram emits as many alpha particles as 5 g of radium. The energy released by its decay is so large (27.5 cal /Ci/day or 140 W/g) that a capsule containing about half a gram reaches a temperature above 500°C. The capsule also presents a contact gamma-ray dose rate of 1.2R/hr. A few curies of polonium exhibit a blue glow, caused by excitation of the surrounding gas. Because almost all alpha radiation is stopped within the solid source and its container, giving up its energy, polonium has attracted attention for uses as a lightweight heat source for thermoelectric power in space satellites. Polonium has more isotopes than any other element. Twenty-seven isotopes of polonium are known, with atomic masses ranging from 192 to 218. Polonium-210 is the most readily available. Isotopes of mass 209 (half-life 103 years) and mass 208 (half-life 2.9 years) can be prepared by alpha, proton, or deuteron bombardment of lead or bismuth in a cyclotron, but these are expensive to produce. Metallic polonium has been prepared from polonium hydroxide and some other polonium compounds in the presence of concentrated aqueous or anhydrous liquid ammonia. Two allotropic modifications are known to exist. Polonium is readily dissolved in dilute acids, but is only slightly soluble in alkalis. Polonium salts of organic acids char rapidly; halide ammines are reduced to the metal. Polonium can be mixed or alloyed with beryllium to provide a source of neutrons. It has been used in devices for eliminating static charges in textile mills, etc.; however, beta sources are more commonly used and are less dangerous. It is also used on brushes for removing dust from photographic films. The polonium for these is carefully sealed and controlled, minimizing hazards to the user. Polonium-210 is very dangerous to handle in even milligram or microgram amounts, and special equipment and strict control is necessary. Damage arises from the complete absorption of the energy of the alpha particle into tissue. The maximum permissible body burden for ingested polonium is only 0.03μCi, which represents a particle weighing only 6.8×10^{-12}g. Weight for weight it is about 2.5×10^{11} times as toxic as hydrocyanic acid. The maximum allowable concentration for soluble polonium compounds in air is about 2×10^{-11} μCi/cm^3. Polonium is available commercially on special order with an A.E.C. permit from the Oak Ridge National Laboratory.

Potassium — (English, *potash* — pot ashes; L. *kalium*, Arab. *qali*, alkali), K; at. wt. 39.0983; at. no. 19; m.p. 63.25°C; b.p. 760°C; sp. gr. 0.862 (20°C); valence 1. Discovered in 1807 by Davy, who obtained it from caustic potash (KOH); this was the first metal isolated by electrolysis. The metal is the seventh most abundant and makes up about 2.4% by weight of the earth's crust. Most potassium minerals are insoluble and the metal is obtained from them only with great difficulty. Certain minerals, however, such as *sylvite, carnallite, langbeinite,* and *polyhalite* are found in ancient lake and sea beds and form rather extensive deposits from which potassium and its salts can readily be obtained. Potash is mined in Germany, New Mexico, California, Utah, and elsewhere. Large deposits of potash, found at a depth of some 3000 ft in Saskatchewan, promise to be important in coming years. Potassium is also found in the ocean, but is present only in relatively small amounts, compared to sodium. The greatest demand for potash has been in its use for fertilizers. Potassium is an essential constituent for plant growth and it is found in most soils. Potassium is never found free in nature, but is obtained by electrolysis of the hydroxide, much in the same manner as prepared by Davy. Thermal methods also are commonly used to produce potassium (such as by reduction of potassium compounds with CaC_2, C, Si, or Na). It is one of the most reactive and electropositive of metals; except for lithium, it is the lightest known metal. It is soft, easily cut with a knife, and is silvery in appearance immediately after a fresh surface is exposed. It rapidly oxidizes in air and must be preserved in a mineral oil such as kerosene. As with other metals of the alkali group, it decomposes in water with the evolution of hydrogen. It catches fire spontaneously on water. Potassium and its salts impart a violet color to flames. Nine isotopes of potassium are known. Ordinary potassium is composed of three isotopes, one of which is K^{40} (0.0118%), a radioactive isotope with a half-life of 1.28×10^9 years. The radioactivity presents no appreciable hazard. An alloy of sodium and potassium (NaK) is used as a heat-transfer medium. Many potassium salts are of utmost importance, including the hydroxide, nitrate, carbonate, chloride, chlorate, bromide, iodide, cyanide, sulfate, chromate, and dichromate. Metallic potassium is available commercially for about $15/oz in small quantities.

Praseodymium — (Gr. *prasios,* green, and *didymos,* twin), Pr; at. wt. 140.9077; at. no. 59; m.p. 931°C; b.p. 3512 °C; sp. gr. (α) 6.773 (β) 6.64; valence 3 or 4. In 1841 Mosander extracted the rare earth *didymia* from *lanthana;* in 1879, Lecoq de Boisbaudran isolated a new earth, *samaria,* from didymia obtained from the mineral *samarskite.* Six years later, in 1885, von Welsbach separated didymia into two other earths, *praseodymia* and *neodymia,* which gave salts of different colors. As with other rare earths, compounds of these elements in solution have distinctive sharp spectral absorption bands or lines, some of which are only a few Angstroms wide. The element occurs along with other rare-earth elements in a variety of minerals. *Monazite* and *bastnasite* are the two principal commercial sources of the rare-earth metals. Ion-exchange and solvent extraction techniques have led to much easier isolation of the rare earths and the cost has dropped greatly in the past few years. Praseodymium can be prepared by several methods, such as by calcium reduction of the anhydrous chloride of fluoride. Misch metal, used in making cigarette lighters, contains about 5% praseodymium metal. Praseodymium is soft, silvery, malleable, and ductile. It was prepared in relatively pure form in 1931. It is somewhat more resistant to corrosion in air than europium, lanthanum, cerium, or neodymium, but it does develop a green oxide coating that spalls off when exposed to air. As with other rare-earth metals it should be kept under a light mineral oil or sealed in plastic. The rare-earth oxides, including Pr_2O_3, are among the most refractory substances known. Along with other rare earths, it is widely used as a core material for carbon arcs used by the motion picture industry for studio lighting and projection. Salts of praseodymium are used to color glasses and enamels; when mixed with certain other materials, praseodymium produces an intense and unusually clean yellow color in glass. Didymium glass, of which praseodymium is a component, is a colorant for welder's goggles. The metal (99 + % pure) is priced at about $1.00/g.

Promethium — (*Prometheus,* who, according to mythology, stole fire from heaven), Pm; at. no. 61; at. wt. (145); m.p. 1168± 6°C; b.p. 2460°C; sp. gr. 7.22±0.02 (25°C); valence 3. In 1902 Branner predicted the existence of an element between neodymium and samarium, and this was confirmed by Moseley in 1914. In 1941, workers at Ohio State University irradiated neodymium and praseodymium with neutrons, deuterons, and alpha particles, resp., and produced several new radioactivities, which most likely were those of element 61. Wu and Segre, and Bethe, in 1942, confirmed the formation; however, chemical proof of the production of element 61 was lacking because of the difficulty in separating the rare earths from each other at that time. In 1945, Marinsky, Glendenin, and Coryell made the first chemical identification by use of ion-exchange chromatography. Their work was done by fission of uranium and by neutron bombardment of neodymium. Searches for the element on earth have been fruitless, and it now appears that promethium is completely missing from the earth's crust. Promethium, however, has been identified in the spectrum of the star HR^{465} in Andromeda. This element is being formed recently near the star's surface, for no known isotope of promethium has a half-life longer than 17.7 years. Thirteen isotopes of promethium, with atomic masses from 141 to 154, are now known. Promethium — 147, with a half- life of 2.5 years, is the most generally useful. Promethium — 145 is the longest lived, and has a specific activity of 940 Ci /g. It is a soft beta emitter; although no gamma rays are emitted, X-radiation can be generated when beta particles impinge on elements of a high atomic number, and great care must be taken in handling it. Promethium salts luminesce in the dark with a pale blue or greenish glow, due to their high radioactivity. Ion-exchange methods led to the preparation of about 10 g of promethium from atomic reactor fuel processing wastes in early 1963. Little is yet generally known about the properties of metallic promethium. Two or more allotropic modifications are thought to exist. The element has applications as a beta source for thickness gages, and it can be absorbed by a phosphor to produce light. Light produced in this manner can be used for signs or signals that require dependable operation; it can be used as a nuclear-powered battery by capturing light in photocells which convert it into electric current. Such a battery, using Pm^{147}, would have a useful life of about 5 years. Promethium shows promise as a portable X-ray unit, and it may become useful as a heat source to provide auxiliary power for space probes and satellites. More than 30 promethium compounds have been prepared. Most are colored. Promethium—147 is available to A.E.C. licensees at a cost of about 50c/Ci.

Protactinium — (Gr. *protos,* first), Pa; at. wt. 231.0359; at. no. 91; m.p.< 1600°C; b.p. . . . ; sp. gr. 15.37 (calc.); valence 4 or 5. The first isotope of element 91 to be discovered was Pa^{234m}, also known as UX_2, a short-lived member of the naturally occurring U^{238} decay series. It was identified by K. Fajans and O. H. Gohring in 1913 and they named the new element *brevium.* When the longer-lived isotope Pa^{231} was identified by Hahn and Meitner in 1918, the name protoactinium was adopted as being more consistent with the characteristics of the most abundant isotope. Soddy, Cranston, and Fleck were also active in this work. The name *protoactinium* was shortened to *protactinium* in 1949. In 1927, Grosse prepared 2 mg of a white powder, which was shown to be Pa_2O_5. Later, in 1934, from 0.1 g of pure Pa_2O_5 he isolated the element by two methods, one of which was by converting the oxide to an iodide and "cracking" it in a high vacuum by an electrically heated filament by the reaction

$$2PaI_5 \longrightarrow 2Pa + 5I_2$$

Protactinium has a bright metallic luster which it retains for some time in air. The element occurs in *pitchblende* to the extent of about 1 part Pa^{231} to 10 million of ore. Ores from Zaire have about 3 ppm. Protactinium has 13 isotopes, the most common of which is Pa^{231} with a half-life of 32,500 years. A number of protactinium compounds are known, some of which are colored. The element is superconductive below 1.4 K. An indirect measurement indicates that protactinium has a vapor pressure of 5.1×10^{-5} at 1927°C. The element is a dangerous toxic material and requires precautions similar to those used when handling plutonium. In 1959 and 1961, it was announced that the Great Britain Atomic Energy Authority extracted by a 12-stage process 125 g of 99.9 % protactinium, the world's only stock of the metal for many years to come. The extraction was made from 60 tons of waste material at a cost of about $500,000. Protactinium is one of the rarest and most expensive naturally occurring elements. O.R.N.L. supplies promethium-231 at a cost of about $280/g. The element is an alpha emitter (5.0 MeV), and is a radiological hazard similar to polonium.

Radium — (L. *radius,* ray), Ra; at. wt. 226.0254; at. no. 88; m.p. 700°C; b.p. 1140°C; sp. gr. 5?; valence 2. Radium was discovered in 1898 by M. and Mme. Curie in the *pitchblende* or *uraninite* of North Bohemia, in which it occurs. There is about 1 g of radium in 7 tons of pitchblende. The element was isolated in 1911 by Mme. Curie and Debierne by the electrolysis of a solution of pure radium chloride, employing a mercury cathode; on distillation in an atmosphere of hydrogen this amalgam yielded the pure metal. Originally, radium was obtained from the rich pitchblende ore found at Joachimsthal, Bohemia. The *carnotite* sands of Colorado furnish some radium, but richer ores are found in the Republic of Zaire and the Great Bear Lake region of Canada. Radium is present in all uranium minerals, and could be extracted, if desired, from the extensive wastes of uranium processing. Large uranium deposits are located in Ontario, New Mexico, Utah, Australia, and elsewhere. Radium is obtained commercially as the bromide or chloride; it is doubtful if any appreciable stock of the isolated element now exists. The pure metal is brilliant white when freshly prepared, but blackens on exposure to air, probably due to formation of the nitride. It exhibits luminescence, as do its salts; it decomposes in water and is somewhat more volatile than barium. It is a member of the alkaline-earth group of metals. Radium imparts a carmine red color to a flame. Radium emits alpha, beta, and gamma rays and when mixed with beryllium produces neutrons. One gram of Ra^{226} undergoes 3.7×10^{10} disintegrations per sec. The *curie (Ci)* is defined as that amount of radioactivity which has the same disintegration rate as 1 g of Ra^{226}. Sixteen isotopes are now known; radium 226, the common isotope, has a half-life of 1620 years. One gram of radium produces about 0.0001 ml(stp) of emanation, or radon gas, per day. This is pumped from the radium and sealed in minute tubes, which are used in the treatment of cancer and other diseases. One gram of radium yields about 1000 cal per year. Radium is used in producing self-luminous paints, neutron sources, and in medicine for the treatment of disease. Some of the more recently discovered radioisotopes, such as Co^{60}, are now being used in place of radium. Some of these sources are much more powerful, and others are safer to use. Radium loses about 1% of its activity in 25 years, being transformed into elements of lower atomic weight. Lead is a final product of disintegration. The study of radium has greatly altered our ideas of the structure of the atom. Radium is a radiological hazard. (Stored radium should be ventilated to prevent build-up of radon.) Inhalation, injection, or body exposure to radium can cause cancer and other body disorders. The maximum permissible burden in the total body for Ra^{226} is 0.2μCi (microcuries).

Radon — (from *radium;* called *niton* at first, L. *nitens,* shining), Rn; at. wt. (~ 222); at. no. 86; m.p. − 71°C; b.p. −61.8°C; density of gas 9.73 g/l; sp. gr. liquid 4.4 at −62°C, solid 4; valence usually 0. The element was discovered in 1900 by Dorn, who called it *radium emanation.* In 1908 Ramsay and Gray, who named it *niton,* isolated the element and determined its density, finding it to be the heaviest known gas. It is essentially inert and occupies the last place in the zero group of gases in the Periodic Table. Since 1923, it has been called radon. Twenty isotopes are known. Radon—222, coming from radium, has a half-life of 3.823 days and is an alpha emitter; radon—220, emanating naturally from thorium and called *thoron,* has a half-life of 54.5 sec and is also an alpha emitter. Radon—219 emanates from actinium and is called *actinon.* It has a half-life of 3.92 sec and is also an alpha emitter. It is estimated that every square mile of soil to a depth of 6 in. contains about 1 g of radium, which releases radon in tiny amounts to the atmosphere. Radon is present in some spring waters, such as those at Hot Springs, Arkansas. On the average, one part of radon is present to 1 sextillion parts of air. At ordinary temperatures radon is a colorless gas; when cooled below the freezing point, radon exhibits a brilliant phosphorescence which becomes yellow as the temperature is lowered and orange-red at the temperature of liquid air. It has been reported that fluorine reacts with radon, forming radon fluoride. Radon clathrates have also been reported. Radon is still produced for therapeutic use by a few hospitals by pumping it from a radium source and sealing it in minute tubes, called seeds or needles, for application to patients. This practice has now been largely discontinued as hospitals can order the seeds directly from suppliers, who make up the seeds with the desired activity for the day of use. Radon is available at a cost of about $4/mCi. Care must be taken in handling radon, as with other radioactive materials. The main hazard is from inhalation of the element and its solid daughters, which are collected on dust in the air. The maximum permissible concentration of Rn^{222} in air has been set at 3× 10^{-8} μ Ci/cc (lung) for an 8-hr day, 40-hr work week. Good ventilation should be provided where radium, thorium, or actinium is stored to prevent build-up of this element. Radon build-up is also a health consideration in uranium mines.

Rhenium — (L. *Rhenus,* Rhine), Re; at. wt. 186.207; at. no. 75; m.p. 3180°C; b.p. 5627°C (est.); sp. gr. 21.02 (20°C); valence −1, +1, 2, 3, 4, 5, 6, 7. Discovery of rhenium is generally attributed to Noddack, Tacke, and Berg, who announced in 1925 they had detected the element in platinum ores and *columbite.* They also found the element in *gadolinite* and *molybdenite.* By working up 660 kg of molybdenite they were able in 1928 to extract 1 g of rhenium. The price in 1928 was $ 10,000/g. Rhenium does not occur free in nature or as a compound in a distinct mineral species. It is, however, widely spread throughout the earth's crust to the extent of about 0.001 ppm. Commercial rhenium in the U.S. today is obtained from molybdenite roaster-flue dusts obtained from copper-sulfide ores mined in the vicinity of Miami, Arizona, and elsewhere in Arizona and Utah. Some molybdenites contain from 0.002 to 0.2% rhenium. More than 150,000 troy ounces of rhenium are now being produced yearly in the United States. The total estimated Free World reserve of rhenium metal is 3500 tons. Natural rhenium is a mixture of two stable isotopes. Sixteen other unstable iostopes are recognized. Rhenium metal is prepared by reducing ammonium perrhenate with hydrogen at elevated temperatures. The element is silvery white with a metallic luster; its density is exceeded only by that of platinum, iridium, and osmium, and its melting point is exceeded only by that of tungsten and carbon. It has other useful properties. The usual commercial form of the element is a powder, but it can be consolidated by pressing and resistance-sintering in a vacuum or hydrogen atmosphere. This produces a compact shape in excess of 90% of the density of the metal. Annealed rhenium is very ductile, and can be bent, coiled, or rolled. Rhenium is used as an additive to tungsten and molybdenum-based alloys to impart useful properties. It is widely used for filaments for mass spectrographs and ion gages. Rhenium-molybdenum alloys are superconductive at 10 K. Rhenium is also used as an electrical contact material as it has good wear resistance and withstands arc corrosion. Thermocouples made of Re-W are used for measuring temperatures up to 2200°C, and rhenium wire is used in photoflash lamps for photography. Rhenium catalysts are exceptionally resistant to poisoning from nitrogen, sulfur, and phosphorus, and are used for hydrogenation of fine chemicals, hydrocracking, reforming, and the disproportionation of olefins. Rhenium costs about $250/troy ounce. Little is known of its toxicity; therefore, it should be handled with care until more data are available.

Rhodium — (Gr. *rhodon,* rose), Rh; at. wt. 102.9055; at. no. 45; m.p. 1966 ± 3°C; b.p. 3727 ± 100°C; sp. gr. 12.41 (20°C); valence 2, 3, 4, 5, and 6. Wollaston discovered rhodium in 1803-4 in crude platinum ore he presumably obtained from South America. Rhodium occurs native with other platinum metals in river sands of the Urals and in North and South America. It is also found with other platinum metals in the copper-nickel sulfide ores of the Sudbury, Ontario region. Although the quantity occurring here is very small, the large tonnages of nickel processed make the recovery commercially feasible. The annual world production of rhodium is only 7 or 8 tons. The metal is silvery white and at red heat slowly changes in air to the sesquioxide. At higher temperatures it converts back to the element. Rhodium has a higher melting point and lower density than platinum. Its major use is as an alloying agent to harden platinum and palladium. Such alloys are used for furnace windings, thermocouple elements, bushings for glass fiber production, electrodes for aircraft spark plugs, and laboratory crucibles. It is useful as an electrical contact material as it has a low electrical resistance, a low and stable contact resistance, and is highly resistant to corrosion. Plated rhodium, produced by electroplating or evaporation, is exceptionally hard and is used for optical instruments. It has a high reflectance and is hard and durable. Rhodium is also used for jewelry, for decoration, and as a catalyst. Exposure to rhodium (metal fume and dust, as Rh) should not exceed 0.1 mg/M^3 (8-hr. time-weighted average, 40-hr wk.). Soluble salts should not exceed 0.001 mg/M^3. Rhodium costs about $800/tr. oz.

Rubidium — (L. *rubidus,* deepest red), Rb; at. wt. 85.4678 ± 3; at. no. 37; m.p. 38.89°C; b.p. 686°C; sp. gr. (solid) 1.532(20°C), (liquid) 1.475 (39°C); valence 1,2,3,4. Discovered in 1861 by Bunsen and Kirchoff in the mineral *lepidolite* by use of the spectroscope. The element is much more abundant than was thought several years ago. It is now considered to be the 16th most abundant element in the earth's crust. Rubidium occurs in *pollucite, carnallite, leucite,* and *zinnwaldite,* which contains traces up to 1%, in the form of the oxide. It is found in lepidolite to the extent of about 1.5%, and is recovered commercially from this source. Potassium minerals, such as those found at Searles Lake, California, and potassium chloride recovered from brines in Michigan also contain the element and are commercial sources. It is also found along with cesium in the extensive deposits of *pollucite* at Bernic Lake, Manitoba. Rubidium can be liquid at room temperature. It is a soft, silvery-white metallic element of the alkali group and is the second most electropositive and alkaline element. It ignites spontaneously in air and reacts violently in water, setting fire to the liberated hydrogen. As with other alkali metals, it forms amalgams with mercury and it alloys with gold, cesium, sodium, and potassium. It colors a flame yellowish violet. Rubidium metal can be prepared by reducing rubidium chloride with calcium, and by a number of other methods. It must be kept under a dry mineral oil or in a vacuum or inert atmosphere. Seventeen isotopes of rubidium are known. Naturally occurring rubidium is made of two isotopes, Rb^{85} and Rb^{87}. Rubidium—87 is present to the extent of 27.85% in natural rubidium and is a beta emitter with a half-life of 5× 10^{11} years. Ordinary rubidium is sufficiently radioactive to expose a photographic film in about 30 to 60 days. Rubidium forms four oxides: Rb_2O, Rb_2O_2, Rb_2O_3, Rb_2O_4. Because rubidium can be easily ionized, it has been considered for use in "ion engines" for space vehicles; however, cesium is somewhat more efficient for this purpose. It is also proposed for use as a working fluid for vapor turbines and for use in a thermoelectric generator using the magnetohydrodynamic principle where rubidium ions are formed by heat at high temperature and passed through a magnetic field. These conduct electricity and act like an armature of a generator and cause electricity to be generated. Rubidium is used as a getter in vacuum tubes and as a photocell component. It has been used in making special glasses. $RbAg_4I_5$ is important, as it has the highest room conductivity of any known ionic crystal. At 20°C its conductivity is about the same as dilute sulfuric acid. This suggests use in thin film batteries and other applications. The present cost in small quanitites is about $20/g.

Ruthenium — (L. *Ruthenia,* Russia), Ru; at. wt. 101.07 ± 3; at. no. 44, m.p. 2310°C; b.p. 3900°C; sp. gr. 12.41 (20°C); valence 0,1,2,3,4,5,6,7,8. Berzelius and Osann in 1827 examined the residues left after dissolving crude platinum from the Ural mountains in *aqua regis.* While Berzelius found no unusual metals, Osann throught he found three new metals, one of which he named ruthenium. In 1844 Klaus, generally recognized as the discoverer, showed that Osann's ruthenium oxide was very impure and that it contained a new metal. Klaus obtained 6 g of ruthenium from the portion of crude platinum that is insoluble in *aqua regia.* A member of the platinum group, ruthenium occurs native with other members of the group in ores found in the Ural mountains and in North and South America. It is also found along with other platinum metals in small but commercial quantities in *pentlandite* of the Sudbury, Ontario, nickel-mining region, and in *pyroxinite* deposits of South Africa. The metal is isolated commercially by a complex chemical process, the final stage of which is the hydrogen reduction of ammonium ruthenium chloride, which yields a powder. The powder is consolidated by powder metallurgy techniques or by argon-arc welding. Ruthenium is a hard, white metal and has four crystal modifications. It does not tarnish at room temperatures, but oxidizes in air at about 800°C. The metal is not attacked by hot or cold acids or *aqua regia,* but when potassium chlorate is added to the solution, it oxidizes explosively. It is attacked by halogens, hydroxides, etc. Ruthenium can be plated by electrodeposition or by thermal decomposition methods. The metal is one of the most effective hardeners for platinum and palladium, and is alloyed with these metals to make electrical contacts for severe wear resistance. A ruthenium-molybdenum alloy is said to be superconductive at 10.6 K. The corrosion resistance of titanium is improved a hundredfold by addition of 0.1% ruthenium. It is a versatile catalyst. Hydrogen sulfide can be split catalytically by light using an aqueous suspension of CdS particles loaded with ruthenium dioxide. It is thought this may have application to removal of H_2S from oil refining and other industrial processes. Compounds in at least eight oxidation states have been found, but of these, the +2, +3, and +4 states are the most common. Ruthenium tetroxide, like osmium tetroxide, is highly toxic. In addition, it may explode. Ruthenium compounds show a marked resemblance to those of osmium. The metal is priced at about $4/g or $60/troy ounce.

Rutherfordium — See Element 104.

Samarium — (*Samarskite,* a mineral), Sm; at. wt. 150.36 ± 3; at. no. 62; m.p. 1077 ± 5°C; b.p. 1791°C; sp. gr. (α) 7.520 (β) 7.40; valence 2 or 3. Discovered spectroscopically by its sharp absorption lines in 1879 by Lecoq de Boisbaudran in the mineral *samarskite,* named in honor of a Russian mine official, Col. Samarski. Samarium is found along with other members of the rare-earth elements in many minerals, including *monazite* and *bastnasite,* which are commercial sources. It occurs in monazite to the extent of 2.8%. While *misch metal* containing about 1% of samarium metal, has long been used, samarium has not been isolated in relatively pure form until recent years. Ion-exchange and solvent extraction techniques have recently simplified separation of the rare earths from one another; more recently, electrochemical deposition, using an electrolytic solution of lithium citrate and a mercury electrode, is said to be a simple, fast, and highly specific way to separate the rare earths. Samarium metal can be produced by reducing the oxide with barium or lanthanum. Samarium has a bright silver luster and is reasonably stable in air. Two crystal modifications of the metal exist, with a transformation point at 917°C. The metal ignites in air at about 150°C. Sixteen isotopes of samarium exist. Natural samarium is a mixture of seven isotopes, three of which are unstable with long half-lives. Samarium, along with other rare earths, is used for carbon-arc lighting for the motion picture industry. The sulfide has excellent high-temperature stability and good thermoelectric efficiencies up to 1100°C. $SmCo_5$ has been used in making a new permanent magnet material with the highest resistance to demagnetization of any known material. It is said to have an intrinsic coercive force as high as 28,000 oersteds. Samarium oxide has been used in optical glass to absorb the infrared. Samarium is used to dope calcium fluoride crystals for use in optical masers or lasers. Compounds of the metal act as sensitizers for phosphors excited in the infrared; the oxide exhibits catalytic properties in the dehydration and dehydrogenation of ethyl alcohol. It is used in infrared absorbing glass and as a neutron absorber in nuclear reactors. The metal is priced at about $5/g. Little is known of the toxicity of samarium; therefore, it should be handled carefully.

Scandium — (L. *Scandia,* Scandinavia), Sc; at. wt. 44.9559; at. no. 21; m.p. 1541°C; b.p. 2831°C; sp. gr. 2.989 (25°C); valence 3. On the basis of the Periodic System, Mendeleev predicted the existence of *ekaboron,* which would have an atomic weight between 40 of calcium and 48 of titanium. The element was discovered by Nilson in 1876 in the minerals *euxenite* and *gadolinite,* which had not yet been found anywhere except in Scandinavia. By processing 10 kg of euxenite and other residues of rare-earth minerals, Nilson was able to prepare about 2 g of scandium oxide of high purity. Cleve later pointed out that Nilson's scandium was identical with Mendeleev's ekaboron. Scandium is apparently a much more abundant element in the sun and certain stars than here on earth. It is about the 23rd most abundant element in the sun, compared to the 50th most abundant on earth. It is widely distributed on earth, occurring in very minute quantities in over 800 mineral species. The blue color of beryl (aquamarine variety) is said to be due to scandium. It occurs as a principal component in the rare mineral *thortveitite,* found in Scandinavia and Malagasy. It is also found in the residues remaining after the extraction of tungsten from Zinnwald *wolframite,* and in *wiikite* and *bazzite.* Most scandium is presently being recovered from *thortveitite* or is extracted as a by-product from uranium mill tailings. Metallic scandium was first prepared in 1937 by Fischer, Brunger, and Grieneisen, who electrolyzed a eutectic melt of potassium, lithium, and scandium chlorides at 700 to 800°C. Tungsten wire and a pool of molten zinc served as the electrodes in a graphic crucible. Pure scandium is now produced by reducing scandium fluoride with calcium metal. The production of the first pound of 99% pure scandium metal was announced in 1960 as having been made under a U.S. Air Force contract. Scandium is a silvery-white metal which develops a slightly yellowish or pinkish cast upon exposure to air. It is relatively soft, and resembles yttrium and the rare-earth metals more than it resembles aluminum or titanium. It is a very light metal and has a higher melting point than aluminum, making it of interest to designers of space missiles. Scandium is not attacked by a 1:1 mixture of conc. HNO_3 and 48% HF. Scandium reacts rapidly with many acids. Eleven isotopes of scandium are recognized. The metal is still expensive, costing about $50/g with a purity of about 99.9%. Scandium oxide costs about $15/g. About 20 kg of scandium (as Sc_2O_3) are now being used yearly in the U.S. to produce high-intensity lights and the radioactive isotope Sc^{46} used as a tracing agent in refinery crackers for crude oil, etc. Scandium iodide added to mercury vapor lamps produces a highly efficient light source resembling sunlight, which is important for indoor or night-time color TV transmission. Little is yet known about the toxicity of scandium; therefore, it should be handled with care.

Selenium — (Gr. *Selene,* moon), Se; at. wt. 78.96 ± 3; at. no. 34; m.p. (gray) 217°C; b.p. (gray) 684.9 ± 1.0°C; sp. gr. (gray) 4.79, (vitreous) 4.28; valence −2, +4, or +6. Discovered by Berzelius in 1817, who found it associated with tellurium, named for the earth. Selenium is found in a few rare minerals, such as *crooksite* and *clausthalite.* In years past it has been obtained from flue dusts remaining from processing copper sulfide ores, but the anode muds from electrolytic copper refineries now provide the source of most of the world's selenium. Selenium is recovered by roasting the muds with soda or sulfuric acid, or by smelting them with soda and niter. Selenium exists in several allotropic forms. Three are generally recognized, but as many as six have been claimed. Selenium can be prepared with either an amorphous or crystalline structure. The color of amorphous selenium is either red, in powder form, or black, in vitreous form. Crystalline monoclinic selenium is a deep red; crystalline hexagonal selenium, the most stable variety, is a metallic gray. Natural selenium contains six stable isotopes. Fourteen other nuclides and isomers have been characterized. The element is a member of the sulfur family and resembles sulfur both in its various forms and in its compounds. Selenium exhibits both photovoltaic action, where light is converted directly into electricity, and photoconductive action, where the electrical resistance decreases with increased illumination. These properties make selenium useful in the production of photocells and exposure meters for photographic use, as well as solar cells. Selenium is also able to convert a.c. electricity to d.c., and is extensively used in rectifiers. Below its melting point selenium is a p-type semiconductor and is finding many uses in electronic and solid-state applications. It is used in Xerography for reproducing and copying documents, letters, etc. It is used by the glass industry to decolorize glass and to make ruby-colored glasses and enamels. It is also used as a photographic toner, and as an additive to stainless steel. Elemental selenium has been said to be practically nontoxic and is considered to be an essential trace element; however, hydrogen selenide and other selenium compounds are extremely toxic, and resemble arsenic in their physiological reactions. Hydrogen selenide in a concentration of 1.5 ppm is intolerable to man. Selenium occurs in some soils in amounts sufficient to produce serious effects on animals feeding on plants, such as locoweed, grown in such soils. Exposure to selenium compounds (as Se) in air should not exceed 0.2 mg/M^3 (8-hr time-weighted average — 40-hr week). Selenium is priced at about $20/lb. It is also available in high-purity form at a somewhat higher cost.

Silicon — (L. *silex, silicis,* flint), Si; at. wt. 28.0855 ± 3; at. no. 14; m.p. 1410°C; b.p. 2355°C; sp. gr. 2.33 (25°C); valence 4. Davy in 1800 thought silica to be a compound and not an element; later in 1811, Gay Lussac and Thenard probably prepared impure amorphous silicon by heating potassium with silicon tetrafluoride. Berzelius, generally credited with the discovery, in 1824 succeeded in preparing amorphous silicon by the same general method as used earlier, but he purified the product by removing the fluosilicates by repeated washings. Deville in 1854 first prepared crystalline silicon, the second allotropic form of the element. Silicon is present in the sun and stars and is a principal component of a class of meteorites known as *aerolites.* It is also a component of *tektites,* a natural glass of uncertain origin. Silicon makes up 25.7% of the earth's crust, by weight, and is the second most abundant element, being exceeded only by oxygen. Silicon is not found free in nature, but occurs chiefly as the oxide, and as silicates. *Sand, quartz, rock crystal, amethyst, agate, flint, jasper,* and *opal* are some of the forms in whch the oxide appears. *Granite, hornblende, asbestos, feldspar, clay, mica,* etc. are but a few of the numerous silicate minerals. Silicon is prepared commercially by heating silica and carbon in an electric furnace, using carbon electrodes. Several other methods can be used for preparing the element. Amorphous silicon can be prepared as a brown powder, which can be easily melted or vaporized. Crystalline silicon has a metallic luster and grayish color. The Czochralski process is commonly used to produce single crystals of silicon used for solid-state or semiconductor devices. Hyperpure silicon can be prepared by the thermal decomposition of ultra-pure trichlorosilane in a hydrogen atmosphere, and by a vacuum float zone process. This product can be doped with boron, gallium, phosphorus, or arsenic, etc. to produce silicon for use in transistors, solar cells, rectifiers, and other solid-state devices which are used extensively in the electronics and space-age industries. Hydrogenated amorphous silicon has shown promise in producing economical cells for converting solar energy into electricity. Silicon is a relatively inert element, but it is attacked by halogens and dilute alkali. Most acids, except hydrofluoric, do not affect it. Silicones are important products of silicon. They may be prepared by hydrolyzing a silicon organic chloride, such as dimethyl silicon chloride. Hydrolysis and condensation of various substituted chlorosilanes can be used to produce a very great number of polymeric products, or silicones, ranging from liquids to hard, glasslike solids with many useful properties. Elemental silicon transmits more than 95% of all wavelengths of infrared, from 1.3 to 6.7μm. Silicon is one of man's most useful elements. In the form of sand and clay it is used to make concrete and brick; it is a useful refractory material for high-temperature work, and in the form of silicates it is used in making enamels, pottery, etc. Silica, as sand, is a principal ingredient of glass, one of the most inexpensive of materials with excellent mechanical, optical, thermal, and electrical properties. Glass can be made in a very great variety of shapes, and is used as containers, window glass, insulators, and thousands of other uses. Silicon tetrachloride can be used to iridize glass. Silicon is important in plant and animal life. Diatoms in both fresh and salt water extract silica from the water to build up their cell walls. Silica is present in ashes of plants and in the human skeleton. Silicon is an important ingredient in steel; silicon carbide is one of the most important abrasives and has been used in lasers to produce coherent light of 4560 A. Regular grade silicon (97%) costs about

50c. Silicon 99.7% pure costs about $7/lb; hyperpure silicon may cost as much as $100/lb. Miners, stonecutters, and others engaged in work where siliceous dust is breathed in large quantities often develop a serious lung disease known as *silicosis.*

Silver — (Anglo-Saxon, *Seolfor siolfur),* Ag (L. argentum), at. wt. 107.8682 ± 3 at. no. 47; m.p. 961.93°C; b.p. 2212°C; sp. gr. 10.50 (20°C); valence 1,2, Siver has been known since ancient times. It is mentioned in Genesis. Slag dumps in Asia Minor and on islands in the Aegean Sea indicate that man learned to separate silver from lead as early as 3000 B.C. Silver occurs native and in ores such as *argentite* (Ag_2S) and *horn silver* (AgCl); lead, lead-zinc, copper, gold, and copper-nickel ores are principal sources. Mexico, Canada, Peru, and the U.S. are the principal silver producers in the western hemisphere. Silver is also recovered during electrolytic refining of copper. Commercial fine silver contains at least 99.9% silver. Purities of 99.999 + % are available commercially. Pure silver has a brilliant white metallic luster. It is a little harder than gold and is very ductile and malleable, being exceeded only by gold and perhaps palladium. Pure silver has the highest electrical and thermal conductivity of all metals, and possesses the lowest contact resistance. It is stable in pure air and water, but tarnishes when exposed to ozone, hydrogen sulfide, or air containing sulfur. The alloys of silver are important. Sterling silver is used for jewelry, silverware, etc. where appearance is paramount. This alloy contains 92.5% silver, the remainder being copper or some other metal. Silver is of utmost importance in photography, about 30% of the U.S. industrial consumption going into this application. It is used for dental alloys. Silver is used in making solder and brazing alloys, electrical contacts, and high capacity silver-zinc and silver-cadmium batteries. Silver paints are used for making printed circuits. It is used in mirror production and may be deposited on glass or metals by chemical deposition, electrodeposition, or by evaporation. When freshly deposited, it is the best reflector of visible light known, but it rapidly tarnishes and loses much of its reflectance. It is a poor reflector of ultraviolet. Silver fulminate ($Ag_2C_2N_2O_2$), a powerful explosive, is sometimes formed during the silvering process. Silver iodide is used in seeding clouds to produce rain. Silver chloride has interesting optical properties as it can be made transparent; it also is a cement for glass. Silver nitrate, or *lunar caustic,* the most important silver compound, is used extensively in photography. While silver itself is not considered to be toxic, most of its salts are poisonous due to the anions present. Exposure to silver (metal and soluble compounds, as Ag) in air should not exceed 0.01mg/ M^3, (8-hr time-weighted average — 40-hr week). Silver compounds can be absorbed in the circulatory system and reduced silver deposited in the various tissues of the body. A condition, known as *argyria,* results, with a greyish pigmentation of the skin and mucous membranes. Silver has germicidal effects and kills many lower organisms effectively without harm to higher animals. Silver for centuries has been used traditionally for coinage by many countries of the world. In recent times, however, consumption of silver has greatly exceeded the output. In 1939, the price of silver was fixed by the U.S. Treasury at 71c/troy ounce, and at 90.5c/ troy ounce in 1946. In November 1961 the U.S. Treasury suspended sales of nonmonetized silver, and the price stabilized for a time at about $1.29, the melt-down value of silver U.S. coins. The Coinage Act of 1965 authorized a change in the metallic composition of the three U.S. subsidiary denominations to clad or composite type coins. This was the first change in U.S. coinage since the monetary system was established in 1792. Clad dimes and quarters are made of an outer layer of 75% Cu and 25% Ni bonded to a central core of pure Cu. The composition of the one- and five-cent pieces remains unchanged. One-cent coins are 95% Cu and 5% Zn. Five-cent coins are 75% Cu and 25% Ni. Old silver dollars are 90% Ag and 10% Cu. Earlier subsidiary coins of 90% Ag and 10% Cu officially were to circulate alongside the clad coins; however, in practice they have largely disappeared (Gresham's Law), as the value of the silver is now greater than their exchange value. Silver coins of other countries have largely been replaced with coins made of other metals. On June 24, 1968, the U.S. Government ceased to redeem U.S. Silver Certificates with silver. Since that time, the price of silver has fluctuated widely. The U.S. Government discontinued selling silver to domestic users and foreign buyers on November 10, 1970. As of December 1985, the price of silver was about $6/tr. oz.

Sodium — (English, *soda;* Medieval Latin, *sodanum,* headache remedy), Na (L. *natrium*); at. wt. 22.98977; at. no. 11; m.p. 97.81 ± 0.03°C; b.p. 882.9°C; sp. gr. 0.971 (20°C); valence 1. Long recognized in compounds, sodium was first isolated by Davy in 1807 by electrolysis of caustic soda. Sodium is present in fair abundance in the sun and stars. The D lines of sodium are among the most prominent in the solar spectrum. Sodium is the sixth most abundant element on earth, comprising about 2.6% of the earth's crust; it is the most abundant of the alkali group of metals of which it is a member. The most common compound is sodium chloride, but it occurs in many other minerals, such as *soda niter, cryolite, amphibole, zeolite, sodalite,* etc. It is a very reactive element and is never found free in nature. It is now obtained commercially by the electrolysis of absolutely dry fused sodium chloride. This method is much cheaper than that of electrolyzing sodium hydroxide, as was used several years ago. Sodium is a soft, bright, silvery metal which floats on water, decomposing it with the evolution of hydrogen and the formation of the hydroxide. It may or may not ignite spontaneously on water, depending on the amount of oxide and metal exposed to the water. It normally does not ignite in air at temperatures below 115°C. Sodium should be handled with respect, as it can be dangerous when improperly handled. Metallic sodium is vital in the manufacture of sodamide and sodium cyanide, sodium peroxide, and sodium hydride. It is used in preparing tetraethyl lead, in the reduction of organic esters, and in the preparation of organic compounds. The metal may be used to improve the structure of certain alloys, to descale metal, to purify molten metals, and as a heat transfer agent. An alloy of sodium with potassium, NaK, is also an important heat transfer agent. Sodium compounds are important to the paper, glass, soap, textile, petroleum, chemical, and metal industries. Soap is generally a sodium salt of certain fatty acids. The importance of common salt to animal nutrition has been recognized since prehistoric times. Among the many compounds that are of the greatest industrial importance are common salt (NaCl), soda ash (Na_2CO_3), baking soda ($NaHCO_3$), caustic soda (NaOH), Chile saltpeter ($NaNO_3$), di- and tri-sodium phosphates, sodium thiosulfate (hypo, $Na_2S_2O_3 \cdot 5H_2O$), and borax ($Na_2B_4O_7 \cdot 10H_2O$). Seven isotopes of sodium are recognized. Metallic sodium is priced at about 15 to 20c/lb in quantity. On a volume basis, it is the cheapest of all metals. Sodium metal should be handled with great care. It should be maintained in an inert atmosphere and contact with water and other substances with which sodium reacts should be avoided.

Strontium — (*Strontian,* town in Scotland), Sr; at. wt. 87.62; at. no. 38; m.p. 769°C; b.p. 1384°C; sp. gr. 2.54; valence 2. Isolated by Davy by electrolysis in 1808; however, Adair Crawford in 1790 recognized a new mineral (strontianite) as differing from other barium minerals (baryta). Strontium is found chiefly as *celestite* ($SrSO_4$) and *strontianite* ($SrCO_3$). The metal can be prepared by electrolysis of the fused chloride mixed with potassium chloride, or is made by reducing strontium oxide with aluminum in a vacuum at a temperature at which strontium distills off. Three allotropic forms of the metal exist, with transition points at 235 and 540°C. Strontium is softer than calcium and decomposes water more vigorously. It does not absorb nitrogen below 380°C. It should be kept under kerosene to prevent oxidation. Freshly cut strontium has a silvery appearance, but rapidly turns a yellowish color with the formation of the oxide. The finely divided metal ignites spontaneously in air. Volatile strontium salts impart a beautiful crimson color to flames, and these salts are used in pyrotechnics and in the production of flairs. Natural strontium is a mixture of four stable isotopes. Twelve other unstable isotopes are known to exist. Of greatest importance is Sr^{90} with a half-life of 28 years. It is a product of nuclear fallout and presents a health problem. This isotope is one of the best long-lived high-energy beta emitters known, and is used in SNAP (Systems for Nuclear Auxiliary Power) devices. These devices hold promise for use in space vehicles, remote weather stations, navigational buoys, etc., where a lightweight, long-lived, nuclear-electric power source is needed. The major use for strontium at present is in producing glass for color television picture tubes. It has also found use in producing ferrite magnets and in refining zinc. Strontium titanate is an interesting optical material as it has an extremely high refractive index and an optical dispersion greater than that of diamond. It has been used as a gemstone, but it is very soft. It does not occur naturally. Strontium metal costs about $6 to $8/lb.

Sulfur — (Sanskrit, *sulvere;* L. *sulphurium*), S; at. wt. 32.06; at. no. 16; m.p. (rhombic) 112.8°C, (monoclinic) 119.0°C; b.p. 444.674°C; sp. gr. (rhombic) 2.07, (monoclinic) 1.957 (20°C); valence 2, 4, or 6. Known to the ancients; referred to in Genesis as *brimstone.* Sulfur is found in meteorites. A dark area near the crater Aristarchus on the moon has been studied by R. W. Wood with ultraviolet light. This study suggests strongly that it is a sulfur deposit. Sulfur occurs native in the vicinity of volcanoes and hot springs. It is widely distributed in nature as *iron pyrites, galena, sphalerite, cinnabar, stibnite, gypsum, Epsom salts, celestite, barite,* etc. Sulfur is commercially recovered from wells sunk into the salt domes along the Gulf Coast of the U.S. It is obtained from these wells by the Frasch process, which forces heated water into the wells to melt the sulfur that is then brought to the surface. Sulfur also occurs in natural gas and petroleum crudes and must be removed from these products. Formerly this was done chemically, which wasted the sulfur. New processes now permit recovery, and these sources promise to be very important. Large amounts of sulfur are being recovered from Alberta gas fields. Sulfur is a pale yellow, odorless, brittle solid, which is insoluble in water but soluble in carbon disulfide. In every state, whether gas, liquid or solid, elemental sulfur occurs in more than one allotropic form or modification; these present a confusing multitude of forms whose relations are not yet fully understood. Amorphous or "plastic" sulfur is obtained by fast cooling of the crystalline form. X-ray studies indicate that amorphous sulfur may have a helical structure with eight atoms per spiral. Crystalline sulfur seems to be made of rings, each containing eight sulfur atoms, which fit together to give a normal X-ray pattern. Ten isotopes of sulfur exist. Four occur in natural sulfur, none of which is radioactive. A finely divided form of sulfur, known as *flowers of sulfur,* is obtained by sublimation. Sulfur readily forms sulfides with many elements. sulfur is a component of black gunpowder, and is used in the vulcanization of natural rubber and as a fungicide. It is also used extensively in making phosphatic fertilizers. A tremendous tonnage is used to produce sulfuric acid, the most important manufactured chemical. It is used in making sulfite paper and other papers, as a fumigant, and in the bleaching of dried fruits. The element is a good electrical insulator. Organic compounds containing sulfur are very important. Calcium sulfate, ammonium sulfate, carbon disulfide, sulfur dioxide, and hydrogen sulfide are but a few of the other many important compounds of sulfur. Sulfur is essential to life. It is a minor constituent of fats, body fluids, and skeletal minerals. Carbon disulfide, hydrogen sulfide, and sulfur dioxide should be handled carefully. Hydrogen sulfide in small concentrations can be metabolized, but in higher concentrations it quickly can cause death by respiratory paralysis. It is insidious in that it quickly deadens the sense of smell. Sulfur dioxide is a dangerous component in atmospheric air pollution. In 1975, University of Pennsylvania scientists reported synthesis of polymeric sulfur nitride, which has the properties of a metal, although it contains no metal atoms. The material has unusual optical and electrical properties. High-purity sulfur is commercially available in purities of 99.999+%.

Tantalum — (Gr. *Tantalos,* mythological character, father of *Niobe*), Ta; at. wt. 180.9479; at. no. 73; m.p. 2996 °C; b.p. 5425 ± 100°C; sp. gr. 16.654; valence 2?,3,4?, or 5. Discovered in 1802 by Ekeberg, but many chemists thought niobium and tantalum were identical elements until Rose, in 1844, and Marignac, in 1866, indicated and showed that niobic and tantalic acids were two different acids. The early investigators only isolated the impure metal. The first relatively pure ductile tantalum was produced by von Bolton in 1903. Tantalum occurs principally in the mineral *columbite-tantalite* $(Fe, Mn)(Nb, Ta)_2O_6$. Tantalum ores are found in Australia, Brazil, Mozambique, Thailand, Portugal, Nigeria, Zaire, and Canada. Separation of tantalum from niobium requires several complicated steps. Several methods are used to commercially produce the element, including electrolysis of molten potassium fluotantalate, reduction of potassium fluotantalate with sodium, or reacting tantalum carbide with tantalum oxide. Sixteen isotopes of tantalum are known to exist. Natural tantalum contains two isotopes; one of these, Ta^{180}, is present in very small quantity(0.0123%)and is unstable with a very long half-life of > 10^{13} years. Tantalum is a gray, heavy, and very hard metal. When pure, it is ductile and can be drawn into fine wire, which is used as a filament for evaporating metals such as aluminum. Tantalum is almost completely immune to chemical attack at temperatures below 150°C, and is attacked only by hydrofluoric acid, acidic solutions containing the fluoride ion, and free sulfur trioxide. Alkalis attack it only slowly. At high temperatures, tantalum becomes much more reactive. The element has a melting point exceeded only by tungsten and rhenium. Tantalum is used to make a variety of alloys with desirable properties such as high melting point, high strength, good ductility, etc. Scientists at Los Alamos have produced a tantalum carbide graphite composite material, which is said to be one of the hardest materials ever made. The compound has a melting point of 6760°F. Tantalum has good "gettering" ability at high temperatures, and tantalum oxide films are stable and have good rectifying and dielectric properties. Tantalum is used to make electrolytic capacitors and vacuum furnace parts, which account for about 60% of its use. The metal is also widely used to fabricate chemical process equipment, nuclear reactors, and aircraft and missile parts. Tantalum is completely immune to body liquids and is a nonirritating metal. It has, therefore, found wide use in making surgical appliances. Tantalum oxide is used to make special glass with a high index of refraction for camera lenses. The metal has many other uses. The price of tantalum has greatly increased in recent years.

Technetium — (Gr. *technetos,* artifical), Tc; at. wt. (98); at. no. 43; m.p. 2172°C; b.p. 4877°C; sp. gr. 11.50 (calc.); valence 0, +2, +4, +5, +6, and +7. Element 43 was predicted on the basis of the periodic table, and was erroneously reported as having been discovered in 1925, at which time it was named *masurium.* The element was actually discovered by Perrier and Segre in Italy in 1937. It was found in a sample of molybdenum, which was bombarded by deuterons in the Berkeley cyclotron, and which E. Lawrence sent to these investigators. Technetium was the first element to be produced artificially. Since its discovery, searches for the element in terrestrial materials have been made without success. If it does exist, the concentration must be very small. Technetium has been found in the spectrum of S-, M-, and N-type stars, and its presence in stellar matter is leading to new theories of the production of heavy elements in the stars. Sixteen isotopes of technetium, with atomic masses ranging from 92 to 107, are known. Tc^{97} has a half-life of 2.6×10^6 years. Tc^{98} has a half-life of 1.5×10^6 years. The isomeric isotope Tc^{95m}, with a half-life of 61 days, is useful for tracer work, as it produces energetic gamma rays. Technetium metal has been produced in kilogram quantities. The metal was first prepared by passing hydrogen gas at 1100°C over Tc_2S_7. It is now conveniently prepared by the reduction of ammonium pertechnetate with hydrogen. Technetium is a silvery-gray metal that tarnishes slowly in moist air. Until 1960, technetium was available only in small amounts and the price was as high as $2800/g. It is now offered commercially to holders of O.R.N.L. permits at a price of $60/g. The chemistry of technetium is said to be similar to that of rhenium. Technetium dissolves in nitric acid, aqua regia, and conc. sulfuric acid, but is not soluble in hydrochloric acid of any strength. The element is a remarkable corrosion inhibitor for steel. It is reported that mild carbon steels may be effectively protected by as little as 5 ppm of $KTcO_4$ in aerated distilled water at temperatures up to 250°C. This corrosion protection is limited to closed systems, since technetium is radioactive and must be confined. Tc^{99} has a specific activity of 6.2×10^8 disintegrations per sec/g. Activity of this level must not be allowed to spread. Tc^{99} is a contamination hazard and should be handled in a glove box. The metal is an excellent superconductor at 11°K and below.

Tellurium — (L. *tellus,* earth), Te; at. wt. 127.60 ± 3 at. no. 52; m.p. 449.5 ± 0.3°C; b.p. 989.8 ± 3.8°C; sp. gr. 6.24 (20°C); valence 2, 4, or 6. Discovered by Muller von Reichenstein in 1782; named by Klaproth, who isolated it in 1798. Tellurium is occasionally found native, but is more often found as the telluride of gold (*calaverite*), and combined with other metals. It is recovered commercially from the anode muds produced during the electrolytic refining of blister copper. The U.S., Canada, Peru, and Japan are the largest Free World producers of the element. Crystalline tellurium has a silvery-white appearance, and when pure exhibits a metallic luster. It is brittle and easily pulverized. Amorphous tellurium is formed by precipitating tellurium from a solution of telluric or tellurous acid. Whether this form is truly amorphous, or made of minute crystals, is open to question. Tellurium is a p-type semiconductor, and shows greater conductivity in certain directions, depending on alignment of the atoms. Its conductivity increases slightly with exposure to light. It can be doped with silver, copper, gold, tin, or other elements. In air, tellurium burns with a greenish-blue flame, forming the dioxide. Molten tellurium corrodes iron, copper, and stainless steel. Tellurium and its compounds are probably toxic and should be handled with care. Workmen exposed to as little as 0.01 mg/m³ of air, or less, develop "tellurium breath," which has a garlic-like odor. Twenty-one isotopes of tellurium are known, with atomic masses ranging from 115 to 135. Natural tellurium consists of eight isotopes, one of which, Te^{127}, is unstable. It is present to the extent of 0.87% and has a half-life of 12×10^{13} years. Tellurium improves the machinability of copper and stainless steel, and its addition to lead decreases the corrosive action of sulfuric acid to lead and improves its strength and hardness. Tellurium is used as a basic ingredient in blasting caps, and is added to cast iron for chill control. Tellurium is used in ceramics. Bismuth telluride has been used in thermoelectric devices. Tellurium costs about $20/lb, with a purity of about 99.5%.

Terbium — (*Ytterby,* village in Sweden), Tb; at. wt. 158.9254; at. no. 65; m.p. 1356°C; b.p. 3123°C; sp. gr. 8.229; valence 3, 4. Discovered by Mosander in 1843. Terbium is a member of the lanthanide or "rare earth" group of elements. It is found in *cerite, gadolinite,* and other minerals along with other rare earths. It is recovered commercially from *monazite* in which it is present to the extent of 0.03%, from *xenotime,* and from *euxenite,* a complex oxide containing 1% or more of terbia. Terbium has been isolated only in recent years with the development of ion-exchange techniques for separating the rare-earth elements. As with other rare earths, it can be produced by reducing the anhydrous chloride or fluoride with calcium metal in a tantalum crucible. Calcium and tantalum impurities can be removed by vacuum remelting. Other methods of isolation are possible. Terbium is reasonably stable in air. It is a silvery-gray metal, and is malleable, ductile, and soft enough to be cut with a knife. Two crystal modifications exist, with a transformation temperature of 1315°C. Nineteen isotopes with atomic masses ranging from 147 to 164 are recognized. The oxide is a chocolate or dark maroon color. Sodium terbium borate is used as a laser material and emits coherent light at 5460A. Terbium is used to dope calcium fluoride, calcium tungstate, and strontium molybdate, used in solid-state devices. The oxide has potential application as an activator for green phosphors used in color TV tubes. It can be used with ZrO_2 as a crystal stabilizer of fuel cells which operate at elevated temperature. Few other uses have been found. The element is priced at about $3/g or $900/lb. Little is known of the toxicity of terbium. It should be handled with care as with other lanthanide elements.

Thallium — (Gr. *thallos,* a green shoot or twig), Tl; at. wt. 204.383; at. no. 81; m.p. 303.5°C; b.p. 1457 ± 10°C; sp. gr. 11.85 (20°C); valence 1, or 3. Thallium was discovered spectroscopically in 1861 by Crookes. The element was named after the beautiful green spectral line, which identified the element. The metal was isolated both by Crookes and Lamy in 1862 about the same time. Thallium occurs in *crooksite, lorandite,* and *hutchinsonite.* It is also present in *pyrites* and is recovered from the roasting of this ore in connection with the production of sulfuric acid. It is also obtained from the smelting of lead and zinc ores. Extraction is somewhat complex and depends on the source of the thallium. Manganese nodules, found on the ocean floor, contain thallium. When freshly exposed to air, thallium exhibits a metallic luster, but soon develops a bluish-gray tinge, resembling lead in appearance. A heavy oxide builds up on thallium if left in air, and in the presence of water the hydroxide is formed. The metal is very soft and malleable. It can be cut with a knife. Twenty isotopic forms of thallium, with atomic masses ranging from 191 to 210, are recognized. Natural thallium is a mixture of two isotopes. The element and its compounds are toxic and should be handled carefully. Contact of the metal with the skin is dangerous, and when melting the metal adequate ventilation should be provided. Exposure to thallium (soluble compounds) — skin, as Tl, should not exceed 0.1mg/M^3 (8-hr time-weighted average — 40-hr week). Thallium is suspect of carcinogenic potential for man. Thallium sulfate has been widely employed as a rodenticide and ant killer. It is odorless and tasteless, giving no warning of its presence. Its use, however, has been prohibited in the U.S. since 1975 as a household insecticide and rodenticide. Continued sales subjects dealers to civil and criminal penalties. The electrical conductivity of thallium sulfide changes with exposure to infrared light, and this compound is used in photocells. Thallium bromide-iodide crystals have been used as infrared detectors. Thallium has been used, with sulfur or selenium and arsenic, to produce low melting glasses which become fluid between 125 and 150°C. These glasses have properties at room temperatures similar to ordinary glasses and are said to be durable and insoluble in water. Thallium oxide has been used to produce glasses with a high index of refraction. Thallium has been used in treating ringworm and other skin infections; however, its use has been limited because of the narrow margin between toxicity and therapeutic benefits. A mercury-thallium alloy, which forms a eutectic at 8.5% thallium, is reported to freeze at −60°C, some 20° below the freezing point of mercury, Commercial thallium metal costs about $8/lb.

Thorium — (*Thor,* Scandinavian god of war), Th; at. wt. 232.0381; at. no. 90; m.p. 1750°C; b.p. ~4790°C; sp. gr. 11.72; valence +2(?), +3(?), +4. Discovered by Berzelius in 1828. Thorium occurs in *thorite* ($ThSiO_4$) and in *thorianite* (ThO_2 + UO_2). Large deposits of thorium minerals have been reported in New England and elsewhere, but these have not yet been exploited. Thorium is now thought to be about three times as abundant as uranium and about as abundant as lead or molybdenum. The metal is a source of nuclear power. There is probably more energy available for use from thorium in the minerals of the earth's crust than from both uranium and fossil fuels. Any sizable demand for thorium as a nuclear fuel is still several years in the future. Work has been done in developing thorium cycle converter-reactor systems. Several prototypes, including the HTGR (high-temperature gas-cooled reactor) and MSRE (molten salt converter reactor experiment), have operated. While the HTGR reactors are efficient, they are not expected to become important commercially for 10 or more years because of certain operating difficulties. Thorium is recovered commercially from the mineral *monazite,* which contains from 3 to 9% ThO_2 along with most rare-earth minerals. Much of the internal heat the earth has been attributed to thorium and uranium. Several methods are available for producing thorium metal: it can be obtained by reducing thorium oxide with calcium, by electrolysis of anhydrous thorium chloride in a fused mixture of sodium and potassium chlorides, by calcium reduction of thorium tetrachloride mixed with anhydrous zinc chloride, and by reduction of thorium tetrachloride with an alkali metal. Thorium was originally assigned a position in Group IV of the periodic table. Because of its atomic weight, valence, etc., it is now considered to be the second member of the *actinide* series of elements. When pure, thorium is a silvery-white metal which is air-stable and retains its luster for several months. When contaminated with the oxide, thorium slowly tarnishes in air, becoming gray and finally black. The physical properties of thorium are greatly influenced by the degree of contamination with the oxide. The purest specimens often contain several tenths of a percent of the oxide. High-purity thorium has been made. Pure thorium is soft, very ductile, and can be cold-rolled, swaged, and drawn. Thorium is dimorphic, changing at 1400°C from a cubic to a body-centered cubic structure. Thorium oxide has a melting point of 3300°C, which is the highest of all oxides. Only a few elements, such as tungsten, and a few compounds, such as tantalum carbide, have higher melting points. Thorium is slowly attacked by water, but does not dissolve readily in most common acids, except hydrochloric. Powdered thorium metal is often pyrophoric and should be carefully handled. When heated in air, thorium turnings ignite and burn brilliantly with a white light. The principal use of thorium has been in the preparation of the Welsbach mantle, used for portable gas lights. These mantles, consisting of thorium oxide with about 1% cerium oxide and other ingredients, glow with a dazzling light when heated in a gas flame. Thorium is an important alloying element in magnesium, imparting high strength and creep resistance at elevated temperatures. Because thorium has a low work-function and high electron emission, it is used to coat tungsten wire used in electronic equipment. The oxide is also used to control the grain size of tungsten used for electric lamps; it is also used for high-temperature laboratory crucibles. Glasses containing thorium oxide have a high refractive index and low dispersion. Consequently, they find application in high quality lenses for cameras and scientific instruments. Thorium oxide has also found use as a catalyst in the conversion of ammonia to nitric acid, in petroleum cracking, and in producing sulfuric acid. Twelve isotopes of thorium are known with atomic masses ranging from 223 to 234. All are unstable. Th^{232} occurs naturally and has a half-life of 1.41×10^{10} years. It is an alpha emitter. Th^{232} goes through six alpha and four beta decay steps before becoming the stable isotope Pb^{208}. Th^{232} is sufficiently radioactive to expose a photographic plate in a few hours. Thorium disintegrates with the production of thoron (radon^{220}), which is an alpha emitter and presents a radiation hazard. Good ventilation of areas where thorium is stored or handled is therefore essential. Thorium and is compounds are subject to licensing and control by the U.S. Atomic Energy Commission. Thorium metal (99.9%) costs about $10/lb.

Thulium — (*Thule,* the earliest name for Scandinavia), Tm; at. wt. 168.9342; at. no. 69; m.p. 1,545 ± 15° C; b.p. 1947°C; sp. gr. 9.321(25°); valence 2, 3. Discovered in 1879 by Cleve. Thulium occurs in small quantities along with other rare earths in a number of minerals. It is obtained commercially from *monazite,* which contains about 0.007% of the element. Thulium is the least abundant of the earth elements, but with new sources recently discovered, it is now considered to be about as rare as silver, gold, or cadmium. Ion-exchange and solvent extraction techniques have recently permitted much easier separation of the rare earths, with much lower costs. Only a few years ago, thulium metal was not obtainable at any cost; in 1950 the oxide sold for $450/g. Thulium metal now costs from $3 to $20/g, depending on the purity, quantity, and supplier. Thulium can be isolated by reduction of the oxide with lanthanum metal or by calcium reduction of the anhydrous fluoride. The pure metal has a bright, silvery luster. It is reasonably stable in air, but the metal should be protected from moisture in a closed container. The element is silver-gray, soft, malleable, and ductile, and can be cut with a knife. Sixteen isotopes are known, with atomic masses ranging from 161 to 176. Natural thulium, Tm^{169}, is stable. Because of the relatively high price of the metal, thulium has not yet found many practical applications. Tm^{169} bombarded in a nuclear reactor can be used as a radiation source in portable X-ray equipment. Tm^{171} is potentially useful as an energy source. Natural thulium also has possible use in *ferrites* (ceramic magnetic materials) used in microwave equipment. As with other lanthanides, thulium has a low-to-moderate acute toxic rating. It should be handled with care.

Tin — (anglo-Saxon, *tin*), Sn (L. *stannum*); at. wt. 118.69 ± 3; at. no. 50; m.p. 231.9681°C; b.p. 2270°C; sp. gr. (gray) 5.75, (white) 7.31; valence 2, 4. Known to the ancients. Tin is found chiefly in *cassiterite* (SnO_2). Most of the world's supply comes from Malaya, Boliva, Indonesia, Zaire, Thailand, and Nigeria. The U.S. produces almost none, although occurrences have been found in Alaska and California. Tin is obtained by reducing the ore with coal in a reverberatory furnace. Ordinary tin is composed of nine stable isotopes; 13 unstable isotopes are also known. Ordinary tin is a silvery-white metal, is malleable, somewhat ductile, and has a highly crystalline structure. Due to the breaking of these crystals, a "tin cry" is heard when a bar is bent. The element has two or perhaps three allotropic forms. On warming, gray, or α tin, with a cubic structure, changes at 13.2°C into white, or β tin, the ordinary form of the metal. White tin has a tetragonal structure. Some authorities believe a γ form exists between 161°C and the melting point; however, other authorities discount its existence. When tin is cooled below 13.2°C, it changes slowly from white to gray. This change is affected by impurities such as aluminum and zinc, and can be prevented by small additions of antimony or bismuth. This change from the α to β form is called the tin pest. There are few if any uses for gray tin. Tin takes a high polish and is used to coat other metals to prevent corrosion or other chemical action. Such tin plate over steel is used in the so-called tin can for preserving food. Alloys of tin are very important. Soft solder, type metal, fusible metal, pewter, bronze, bell metal, Babbitt metal, White metal, die casting alloy, and phosphor bronze are some of the important alloys using tin. Tin resists distilled, sea, and soft tap water, but is attacked by strong acids, alkalis, and acid salts. Oxygen in solution accelerates the attack. When heated in air, tin forms SnO_2, which is feebly acid, forming stannate salts with basic oxides. The most important salt is the chloride ($SnCl_2 \cdot H_2O$), which is used as a reducing agent and as a mordant in calico printing. Tin salts sprayed onto glass are used to produce electrically conductive coatings. These have been used for panel lighting and for frost-free wind- shields. Most window glass is now made by floating molten glass on molten tin (float glass) to produce a flat surface (Pilkington process). Of recent interest is a crystalline tin-niobium alloy that is superconductive at very low temperatures. This promises to be important in the construction of superconductive magnets that generate enormous field strengths but use practically no power. Such magnets, made of tin-niobium wire, weigh but a few pounds and produce magnetic fields that, when started with a small battery, are comparable to that of a 100 ton electromagnet operated continuously with a large power supply. The small amount of tin used in canned foods is quite harmless. The agreed limit of tin content in U.S. foods is 300 mg/kg. The trialkyl and triaryl tin compounds are used as biocides and must be handled carefully. Over the past 25 years the price of tin has varied from 50c/lb to its present price of $6/lb.

Titanium — (L. *Titans,* the first sons of the Earth, myth.), Ti; at. wt. 47.88 ± 3; at. no. 22; m.p. 1660 ± 10°C; b.p. 3287°C; sp. gr. 4.54; valence 2, 3, or 4. Discovered by Gregor in 1791; named by Klaproth in 1795. Impure titanium was prepared by Nilson and Pettersson in 1887; however, the pure metal (99.9%) was not made until 1910 by Hunter by heating $TiCl_4$ with sodium in a steel bomb. Titanium is present in meteorites and in the sun. Rocks obtained during the Apollo 17 lunar mission showed presence of 12.1% TiO_2. Analyses of rocks obtained during earlier Apollo missions show lower percentages. Titanium oxide bands are prominent in the spectra of M-type stars. The element is the ninth most abundant in the crust of the earth. Titanium is almost always present in igneous rocks and in the sediments derived from them. It occurs in the minerals *rutile, ilmenite,* and *sphene,* and is present in titanates and in many iron ores. Titanium is present in the ash of coal, in plants, and in the human body. The metal was a laboratory curiosity until Kroll, in 1946, showed that titanium could be produced commercially by reducing titanium tetrachloride with magnesium. This method is largely used for producing the metal today. The metal can be purified by decomposing the iodide. Titanium, when pure, is a lustrous, white metal. It has a low density, good strength, is easily fabricated, and has excellent corrosion resistance. It is ductile only when it is free of oxygen. The metal burns in air and is the only element that burns in nitrogen. Titanium is resistant to dilute sulfuric and hydrochloric acid, most organic acids, moist chlorine gas, and chloride solutions. Natural titanium consists of five isotopes with atomic masses from 46 to 50. All are stable. Four other unstable isotopes are known. Natural titanium is reported to become very radioactive after bombardment with deuterons. The emitted radiations are mostly positrons and hard gamma rays. The metal is dimorphic. The hexagonal α form changes to the cubic β form very slowly at about 880°C. The metal combines with oxygen at red heat, and with chlorine at 550°C. Titanium is important as an alloying agent with aluminum, molybdenum, manganese, iron, and other metals. Alloys of titanium are principally used for aircraft and missiles where lightweight strength and ability to withstand extremes of temperature are important. Titanium is a strong as steel, but 45% lighter. It is 60% heavier than aluminum, but twice as strong. Titanium has potential use in desalination plants for converting sea water into fresh water. The metal has excellent resistance to sea water and is used for propeller shafts, rigging, and other parts of ships exposed to salt water. A titanium anode coated with platinum has been used to provide cathodic protection from corrosion by salt water. Titanium metal is considered to be physiologically inert. When pure, titanium dioxide is relatively clear and has an extremely high index of refraction with an optical dispersion higher than diamond. It is produced artificially for use as a gemstone, but it is relatively soft. Star sapphires and rubies exhibit their asterism as a result of the presence of TiO_2. Titanium dioxide is extensively used for both house paint and artist's paint, as it is permanent and has good covering power. Titanium oxide pigment accounts for the largest use of the element. Titanium paint is an excellent reflector of infrared, and is extensively used in solar observatories where heat causes poor seeing conditions. Titanium tetrachloride is used to iridize glass. This compound fumes strongly in air and has been used to produce smoke screens. The price of titanium metal powder (99.7%) is about $25/lb.

Tungsten — (Swedish, *tung sten,* heavy stone); also known as *wolfram* (from *wolframite,* said to be named from *wolf rahm* or *spumi lupi,* because the ore interfered with the smelting of tin and was supposed to devour the tin), W; at. wt. 183.85 ±3; at. no. 74; m.p. 3410 ± 20°C; b.p. 5660°C; sp. gr. 19.3 (20°C); valence 2, 3, 4, 5, or 6. In 1779 Peter Woulfe examined the mineral now known as *wolframite* and concluded it must contain a new substance. Scheele, in 1781, found that a new acid could be made from *tung sten* (a name first applied about 1758 to a mineral now known as *scheelite*). Scheele and Bergman suggested the possibility of obtaining a new metal by reducing this acid. The de Elhuyar brothers found an acid in *wolframite* in 1783 that was identical to the acid of *tungsten* (tungstic acid) of Scheele, and in that year they succeeded in obtaining the element by reduction of this acid with charcoal. Tungsten occurs in *wolframite,* $(Fe, Mn)WO_4$; *scheelite,* $CaWO_4$; *huebnerite,* $MnWO_4$; and *ferberite,* $FeWO_4$. Important deposits of tungsten occur in California, Colorado, South Korea, Bolivia, U.S.S.R., and Portugal. China is reported to have about 75% of the world's tungsten resources. Natural tungsten contains five stable isotopes. Twelve other unstable isotopes are recognized. The metal is obtained commercially by reducing tungsten oxide with hydrogen or carbon. Pure tungsten is a steel-gray to tin-white metal. Very pure tungsten can be cut with a hacksaw, and can be forged, spun, drawn, and extruded. The impure metal is brittle and can be worked only with difficulty. Tungsten has the highest melting point and lowest vapor pressure of all metals, and at temperatures over 1650°C has the highest tensile strength. The metal oxidizes in air and must be protected at elevated temperatures. It has excellent corrosion resistance and is attacked only slightly by most mineral acids. The thermal expansion is about the same as borosilicate glass, which makes the metal useful for glass-to-metal seals. Tungsten and its alloys are used extensively for filaments for electric lamps, electron and television tubes, and for metal evaporation work; for electrical contact points for automobile distributors; X-ray targets; windings and heating elements for electrical furnaces; and for numerous space missile and high-temperature applications. High-speed tool steels, Hastelloy ®, Stellite ®, and many other alloys contain tungsten. Tungsten carbide is of great importance to the metal-working, mining, and petroleum industries. Calcium and magnesium tungstates are widely used in fluorescent lighting; other salts of tungsten are used in the chemical and tanning industries. Tungsten disulfide is a dry, high-temperature lubricant, stable to 500°C. Tungsten bronzes and other tungsten compounds are used in paints. Tungsten powder costs about $15/lb.

Uranium — (Planet *Uranus*), U; at. wt. 238.0289; at no. 92; m.p. 1132.3 ± 0.8°C; b.p. 3818°C; sp. gr. ~ 18.95; valence 2, 3, 4, 5, or 6. Yellow-colored glass, containing more than 1% uranium oxide and dating back to 79 A.D., has been found near Naples, Italy. Klaproth recognized an unknown element in *pitchblende* and attempted to isolate the metal in 1789. The metal apparently was first isolated in 1841 by Peligot, who reduced the anhydrous chloride with potassium. Uranium is not as rare as it was once thought. It is now considered to be more plentiful than mercury, antimony, silver, or cadmium, and is about as abundant as molybdenum or arsenic. It occurs in numerous minerals such as *pitchblende, uraninite, carnotite, autunite, uranophane, davidite,* and *tobernite.* It is also found in *phosphate rock, lignite, monazite sands,* and can be recovered commercially from these sources. The A.E.C. purchases uranium in the form of acceptable U_3O_8 concentrates. This incentive program has greatly increased the known uranium reserves. Uranium can be prepared by reducing uranium halides with alkali or alkaline earth metals or by reducing uranium oxides by calcium, aluminum, or carbon at high temperatures. The metal can also be produced by electrolysis of KUF_5 or UF_4, dissolved in a molten mixture of $CaCl_2$ NaCl. High-purity uranium can be prepared by the thermal decomposition of uranium halides on a hot filament. Uranium exhibits three crystallographic modifications as follows:

$$\alpha \xrightarrow{667^\circ C} \beta \xrightarrow{772^\circ C} \gamma$$

Uranium is a heavy, silvery-white metal which is pyrophoric when finely divided. It is a little softer than steel, and is attacked by cold water in a finely divided state. It is malleable, ductile, and slightly paramagnetic. In air, the metal becomes coated with a layer of oxide. Acids dissolve the metal, but it is unaffected by alkalis, Uranium has fourteen isotopes, all of which are radioactive. Naturally occurring uranium nominally contains 99.2830% by weight U^{238}, 0.7110% U^{235}, and 0.0054% U^{234}. Studies show that the percentage weight of U^{235} in natural uranium varies by as much as 0.1%, depending on the source. The U.S.D.O.E. has adopted the value of 0.711 as being their "official" percentage of U^{235} in natural uranium. Natural uranium is sufficiently radioactive to expose a photographic plate in an hour or so. Much of the internal heat of the earth is thought to be attributable to the presence of uranium and thorium. U^{238} with a half-life of 4.51×10^9 years, has been used to estimate the age of igneous rocks. The origin of uranium, the highest member of the naturally occurring elements — except perhaps for traces of neptunium or plutonium — is not clearly understood, although it may be presumed that uranium is a decay product of elements of higher atomic weight, which may have once been present on earth or elsewhere in the universe. These original elements may have been created as a result of a primordial "creation," known as "the big bang," in a supernovae, or in some other stellar processes. Uranium is of great importance as a nuclear fuel. U^{238} can be converted into fissionable plutonium by the following reactions:

$$U^{238}(n\gamma)U^{239} \xrightarrow{\beta-} Np^{239} \xrightarrow{\beta-} Pu^{239}$$

This nuclear conversion can be brought about in "breeder" reactors where it is possible to produce more new fissionable material than the fissionable material used in maintaining the chain reaction. U^{235} is of even greater importance, for it is the key to the utilization of uranium. U^{235}, while occurring in natural uranium to the extent of only 0.71%, is so fissionable with slow neutrons that a self-sustaining fission chain reaction can be made to occur in a reactor constructed from natural uranium and a suitable moderator, such as heavy water or graphite, alone. U^{235} can be concentrated by gaseous diffusion and other physical processes, if desired, and used directly as a nuclear fuel, instead of natural uranium, or used as an explosive. Natural uranium, slightly enriched with U^{235} by a small percentage, is used to fuel nuclear power reactors for the generation of electricity. Natural thorium can be irradiated with neutrons as follows to produce the important isotope U^{233}.

$$Th^{232}(n\gamma)Th^{233} \xrightarrow{\beta-} Pa^{233} \xrightarrow{\beta-} U^{233}$$

While thorium itself is not fissionable, U^{233} is, and in this way may be used as a nuclear fuel. One pound of completely fissioned uranium has the fuel value of over 1500 tons of coal. The uses of nuclear fuels to generate electrical power, to make isotopes for peaceful purposes, and to make explosives are well known. The estimated world-wide capacity of the 189 nuclear power reactors in operation in 1979 amounted to about 110 million kilowatts. Uranium in the U.S.A. is controlled by the U.S. Nuclear Regulatory Commission. New uses are being found for "depleted" uranium, i.e., uranium with the percentage of U^{235} lowered to about 0.2%. It has found use in inertial guidance devices, gyro compasses, counterweights for aircraft control surfaces, as ballast for missile reentry vehicles, and as a shielding material. Uranium metal is used for X-ray targets for production of high-energy X-rays; the nitrate has been used as photographic toner, and the acetate is used in analytical chemistry. Crystals of uranium nitrate are triboluminescent. Uranium salts have also been used for producing yellow "vaseline" glass and glazes. Uranium and its compounds are highly toxic, both from a chemical and radiological standpoint. Finely divided uranium metal, being pyrophoric, presents a fire hazard. The maximum recommended allowable concentration of soluble uranium compound in air (based on chemical toxicity) is 0.05 mg/M³ (8-hr time-weighted average — 40-hr week); for insoluble compounds the concentration is set at 0.25 mg/M³ of air. The maximum permissible total body burden of natural uranium (based on radiotoxicity) is 0.2μCi for soluble compounds.

Vanadium — (Scandinavian goddess, *Vanadis*), V; at. wt. 50.9415; at. no. 23; m.p. 1890 ± 10°C; b.p. 3380°C; sp. gr. 6.11 (18.7°C); valence 2, 3, 4, or 5. Vanadium was first discovered by del Rio in 1801. Unfortunately, a French chemist incorrectly declared del Rio's new element was only impure chromium; del Rio thought himself to be mistaken and accepted the French chemist's statement. The element was rediscovered in 1830 by Sefstrom, who named the element in honor of the Scandinavian goddess *Vanadis* because of its beautiful multicolored compounds. It was isolated in nearly pure form by Roscoe, in 1867, who reduced the chloride with hydrogen. Vanadium of 99.3 to 99.8% purity was not produced until 1927. Vanadium is found in about 65 different minerals among which are *carnotite, roscoelite, vanadinite,* and *patronite* — important sources of the metal. Vanadium is also found in phosphate rock and certain iron ores, and is present in some crude oils in the form of organic complexes. It is also found in small percentages in meteorites. Commercial production from petroleum ash holds promise as an important source of the element. High purity ductile vanadium can be obtained by reduction of vanadium trichloride with magnesium or with magnesium-sodium mixtures. Much of the vanadium metal being produced is now made by calcium reduction of V_2O_5 in a pressure vessel, an adaption of a process developed by McKechnie and Seybolt. Natural vanadium is a mixture of two isotopes, V^{50} (0.24%) and V^{51} (99.76%). V^{50} is slightly radioactive, having a half-life of 6×10^{15} years. Seven other unstable isotopes are recognized. Pure vanadium is a bright white metal, and is soft and ductile. It has good corrosion resistance to alkalis, sulfuric and hydrochloric acid, and salt waters, but the metal oxidizes readily above 660°C. The metal has good structural strength and a low-fission neutron cross section, making it useful in nuclear applications. Vanadium is used in producing rust-resistant, spring, and high-speed tool steels. It is an important carbide stabilizer in making steels. About 80% of the vanadium now produced is used as ferrovanadium or as a steel additive. Vanadium foil is used as a bonding agent in cladding titanium to steel. Vanadium pentoxide is used in ceramics and as a catalyst. It is also used as a mordant in dyeing and printing fabrics and in the manufacture of aniline black. Vanadium-gallium tape has been used in producing a superconductive magnet with a field of 175,000 gauss. Vanadium and its compounds are toxic and should be handled with care. Exposure to V_2O_2 dust (as V) should not exceed the ceiling value of 0.05mg/M³, and exposure to V_2O_2 fume (as V) should not exceed 0.1mg/M³ (8-hr time-weighted average — 40-hr week). Ductile vanadium is commercially available. Commercial vanadium metal, of about 95% purity, costs about $10/lb.

Wolfram — see Tungsten.

Xenon — (Gr. *xenon*, stranger), Xe; at. wt. 131.29 ± 3; at. no. 54; m.p. −111,9°C; b.p. −107.1 ± 3°C; density (gas) 5.887 ± 0.009 g/l, sp. gr (liquid) 3.52 (−109°C); valence usually 0. Discovered by Ramsay and Travers in 1898 in the residue left after evaporating liquid air components. Xenon is a member of the so-called noble or "inert" gases. It is present in the atmosphere to the extent of about one part in twenty million. Xenon is present in the Martian atmosphere to the extent of 0.08 ppm. The element is found in the gases evolved from certain mineral springs, and is commercially obtained by extraction from liquid air. Natural xenon is composed of nine stable isotopes. In addition to these, 22 unstable nuclides and isomers have been characterized. Before 1962, it had generally been assumed that xenon and other noble gases were unable to form compounds. Evidence has been mounting in the past few years that xenon, as well as other members of the zero valence elements, do form compounds. Among the "compounds" of xenon now reported are xenon hydrate, sodium perxenate, xenon deuterate, difluoride, tetrafluoride, hexafluoride, and $XePtF_6$ and $XeRhF_6$. Xenon trioxide, which is highly explosive, has been prepared. More than 80 xenon compounds have been made with xenon chemically bonded to fluorine and oxygen. Some xenon compounds are colored. Metallic xenon has been produced, using several hundred kilobars of pressure. Xenon in a vacuum tube produces a beautiful blue glow when excited by an electrical discharge. The gas is used in making electron tubes, stroboscopic lamps, bactericidal lamps, and lamps used to excite ruby lasers for generating coherent light. Xenon is used in the atomic energy field in bubble chambers, probes, and other applications where its high molecular weight is of value. It is also potentially useful as a gas for ion engines. The perxenates are used in analytical chemistry as oxidizing agents. $Xe^{133}$and Xe^{135} are produced by neutron irradiation in aircooled nuclear reactors. Xe^{133} has useful applications as a radioisotope. The element is available in sealed glass containers for about $20/l of gas at standard pressure. Xenon is not toxic, but its compounds are highly toxic because of their strong oxidizing characteristics.

Ytterbium — (Ytterby, village in Sweden), Yb; at. wt. 173.04 ± 3; at. no. 70; m.p. 819°C; b.p. 1194°C; sp.gr. (α) 6.965 (β) 6.54; valence 2, 3. Marignac in 1878 discovered a new component, which he called *ytterbia,* in the earth than known as *erbia.* In 1907, Urbain separated ytterbia into two components, which he called *neoytterbia* and *lutecia.* The elements in these earths are now known as *ytterbium* and *lutetium,* respectively. These elements are identical with *aldebaranium* and *cassiopeium,* discovered independently and at about the same time by von Welsbach. Ytterbium occurs along with other rare earths in a number of rare minerals. It is commercially recovered principally from *monazite sand,* which contains about 0.03%. Ion-exchange and solvent extraction techniques developed in recent years have greatly simplified the separation of the rare earths from one another. The element was first prepared by Klemm and Bonner in 1937 by reducing ytterbium trichloride with potassium. Their metal was mixed, however, with KCl. Daane, Dennison, and Spedding prepared a much purer form in 1953 from which the chemical and physical properties of the element could be determined. Ytterbium has a bright silvery luster, is soft, malleable, and quite ductile. While the element is fairly stable, it should be kept in closed containers to protect it from air and moisture. Ytterbium is readily attacked and dissolved by dilute and concentrated mineral acids and reacts slowly with water. Ytterbium normally has two allotropic forms with a transformation point at 798°C. The alpha form is a room-temperature, face-centered, cubic modification, while the high-temperature beta form is a body-centered cubic form. Another body-centered cubic phase has recently been found to be stable at high pressures at room temperatures. The alpha form ordinarily has metallic-type conductivity, but becomes a semiconductor when the pressure is increased above 16,000 atm. The electrical resistance increases tenfold as the pressure is increased to 39,000 atm and drops to about 80% of its standard temperature-pressure resistivity at a pressure of 40,000 atm. Natural ytterbium is a mixture of seven stable isotopes. Seven other unstable isotopes are known. Ytterbium metal has possible use in improving the grain refinement, strength, and other mechanical properties of stainless steel. One isotope is reported to have been used as a radiation source as a substitute for a portable X-ray machine where electricity is unavailable. Few other uses have been found. Ytterbium metal is commercially available with a purity of about 99 + % for about $1.00/g or $300/lb. Ytterbium has a low acute toxic rating.

Yttrium — (*Ytterby,* village in Sweden near Vauxholm), Y; at. wt. 88.9059; at. no. 39; m.p. 1522 ± 8°C; b.p. 3338°C; sp. gr. 4.469 (25°C); valence 3. *Yttria,* which is an earth containing yttrium, was discovered by Gadolin in 1794. *Ytterby* is the site of a quarry which yielded many unusual minerals containing rare earths and other elements. This small town, near Stockholm, bears the honor of giving names to *erbium, terbium,* and *ytterbium* as well as *yttrium.* In 1843 Mosander showed that yttria could be resolved into the earths of three elements. The name yttria was reserved for the most basic one; the others were named *erbia* and *terbia.* Yttrium occurs in nearly all of the rare-earth minerals. Analysis of lunar rock samples obtained during the Apollo missions show a relatively high yttrium content. It is recovered commercially from *monazite sand,* which contains about 3%, and from *bastnasite,* which contains about 0.2%. Wohler obtained the impure element in 1828 by reduction of the anhydrous chloride with potassium. The metal is now produced commercially by reduction of the fluoride with calcium metal. It can also be prepared by other techniques. Yttrium has a silvery-metallic luster and is relatively stable in air if their temperature exceeds 400°C, and finely divided yttrium is very unstable in air. Turnings of the metal, however, ignite in air. Yttrium oxide is one of the most important compounds of yttrium and accounts for the largest use. It is widely used in making YVO_4europium, and Y_2O_3europium phosphors to give the red color in color television tubes. Many hundreds of thousands of pounds are now used in this application. Yttrium oxide also is used to produce yttrium-iron-garnets, which are very effective microwave filters. Yttrium iron, aluminum, and gadolinium garnets, with formulas such as $Y_3Fe_5O_{12}$ and $Y_3Al_5O_{12}$, have interesting magnetic properties. Yttrium iron garnet is also exceptionally efficient as both a transmitter and transducer of acoustic energy. Yttrium aluminum garnet, with a hardness of 8.5, is also finding use as a gemstone (simulated diamond). Small amounts of yttrium (0.1 to 0.2%) can be used to reduce the grain size in chromium, molybdenum, zirconium, and titanium, and to increase strength of aluminum and magnesium alloys. Alloys with other useful properties can be obtained by using yttrium as an additive. The metal can be used as a deoxidizer for vanadium and other nonferrous metals. The metal has a low cross section for nuclear capture. Y^{90}, one of the isotopes of yttrium, exists in equilibrium with its parent Sr^{90}, a product of atomic explosions. Yttrium has been considered for use as a nodulizer for producing nodular cast iron, in which the graphite forms compact nodules instead of the usual flakes. Such iron has increased ductility. Yttrium is also finding application in laser systems and as a catalyst for ethylene polymerization. It has also potential use in ceramic and glass formulas as the oxide has a high melting point and imparts shock resistance and low expansion characteristics to glass. Natural yttrium contains but one isotope, Y^{89}. Twenty other unstable nuclides and isomers have been characterized. Yttrium metal of 99 + % purity is commercially available at a cost of about 60c/g or $150/lb.

Zinc — (Ger. *Zink,* of obscure origin), Zn; at. wt. 65.38; at. no. 30; m.p. 419.58°C; b.p. 907°C; sp. gr. 7.133 (25°C); valence 2. Centuries before zinc was recognized as a distinct element, zinc ores were used for making brass. Tubal-Cain, seven generations from Adam, is mentioned as being an "instructor in every artificer in brass and iron." An alloy containing 87% zinc has been found in prehistoric ruins in Transylvania. Metallic zinc was produced in the 13th century A.D. in India by reducing calamine with organic substances such as wool. The metal was rediscovered in Europe by Marggraf in 1746, who showed that it could be obtained by reducing *calamine* with charcoal. The principal ores of zinc are *sphalerite* or *blende* (sulfide), *smithsonite* (carbonate), *calamine* (silicate), and *franklinite* (zinc, manganese, iron oxide). Zinc can be obtained by roasting its ores to form the oxide and by reduction of the oxide with coal or carbon, with subsequent distillation of the metal. Other methods of extraction are possible. Naturally occurring zinc contains five stable isotopes. Ten other unstable nuclides and isomers are recognized. Zinc is a bluish- white, lustrous metal. It is brittle at ordinary temperatures but malleable at 100 to 150°C. It is a fair conductor of electricity, and burns in air at high red heat with evolution of white clouds of the oxide. The metal is employed to form numerous alloys with other metals. Brass, nickel silver, typewriter metal, commercial bronze, spring brass, German silver, soft solder, and aluminum solder are some of the more important alloys. Large quantities of zinc are used to produce die castings, used extensively by the automotive, electrical, and hardware industries. An alloy called *Prestal*®, consisting of 78% zinc and 22% aluminum is reported to be almost as strong as steel but as easy to mold as plastic. It is said to be so plastic that it can be molded into form by relatively inexpensive die casts made of ceramics and cement. It exhibits superplasticity. Zinc is also extensively used to galvanize other metals such as iron to prevent corrosion. Neither zinc nor zirconium is ferromagnetic, but $ZrZn_2$ exhibits ferromagnetism at temperatures below 35 K. Zinc oxide is a unique and very useful material to modern civilization. It is widely used in the manufacture of paints, rubber products, cosmetics, pharmaceuticals, floor coverings, plastics, printing inks, soap, storage batteries, textiles, electrical equipment, and other products. It has unusual electrical, thermal, optical, and solid-state properties that have not yet been fully investigated. Lithopone, a mixture of zinc sulfide and barium sulfate, is an important pigment. Zinc sulfide is used in making luminous dials, X-ray and TV screens, and fluorescent lights. The chloride and chromate are also important compounds. Zinc is an essential element in the growth of human beings and animals. Tests show that zinc-deficient animals require 50% more food to gain the same weight of an animal supplied with sufficient zinc. Zinc is not considered to be toxic, but when freshly formed ZnO is inhaled a disorder known as the *oxide shakes* or *zinc chills* sometimes occurs. It is recommended that where zinc oxide is encountered good ventilation be provided to avoid concentrations exceeding 5 mg/M^3, (time-weighted over an 8-hr exposure, 40-hr work week). The price of zinc was roughly 45c/lb. in December 1984.

Zirconium — (Arabic *zargun,* gold color), Zr; at. wt. 91.22; at. no. 40; m.p. 1852 ± 2°C; b.p. 4377°C; sp. gr. 6.506 (20°C); valence +2, +3, and +4. The name *zircon* probably originated from the arabic word *zargun,* which describes the color of the gemstone now known as *zircon, jargon, hyacinth, jacinth,* or *ligure.* This mineral, or its variations, is mentioned in biblical writings. The mineral was not known to contain a new element until Klaproth, in 1789, analyzed a jargon from Ceylon and found a new earth, which Werner named zircon (*silex circonius*), and Klaproth called *Zirkonerde* (*zirconia*). The impure metal was first isolated by Berzelius in 1824 by heating a mixture of potassium and potassium zirconium fluoride in a small iron tube. Pure zirconium was first prepared in 1914. Very pure zirconium was first produced in 1925 by van Arkel and de Boer by an iodide decomposition process they developed. Zirconium is found in abundance in S-type stars, and has been identified in the sun and meteorites. Analyses of lunar rock samples obtained during the various Apollo missions to the moon show a surprisingly high zirconium oxide content, compared with terrestial rocks. Naturally occurring zirconium contains five isotopes, one of which, Zr^{96} (abundant to the extent of 2.80%,) is unstable with a very long half-life of $> 3.6 \times 10^{17}$ years. Fifteen other unstable nuclides and isomers of zirconium have been characterized. Zircon, $ZrSiO_4$, the principal ore, is found in deposits in Florida, South Carolina, Australia, and Brazil. *Baddeleyite,* found in Brazil, is an important zirconium mineral. It is principally pure ZrO_2 in crystalline form having a hafnium content of about 1%. Zirconium also occurs in some 30 other recognized mineral species. Zirconium is produced commercially by reduction of the chloride with magnesium (the Kroll Process), and by other methods. It is a grayish-white lustrous metal. When finely divided, the metal may ignite spontaneously in air, especially at elevated temperatures. The solid metal is much more difficult to ignite. The inherent toxicity of zirconium compounds is low. Hafnium is invariably found in zirconium ores, and the separation is difficult. Commercial-grade zirconium contains from 1 to 3% hafnium. Zirconium has a low absorption cross section for neutrons, and is therefore used for nuclear energy applications, such as for cladding fuel elements. Zirconium has been found to be extremely resistant to the corrosive environment inside atomic reactors, and it allows neutrons to pass through the internal zirconium construction material without appreciable absorption of energy. Commercial nuclear power generation now takes more than 90% of zirconium metal production. Reactors of the size now being made may use as much as a half-million lineal feet of zirconium alloy tubing. Reactor-grade zirconium is essentially free of hafnium. *Zircaloy*® is an important alloy developed specifically for nuclear applications. Zirconium is exceptionally resistant to corrosion by many common acids and alkalis, by sea water, and by other agents. It is used extensively by the chemical industry where corrosive agents are employed. Zirconium is used as a getter in vacuum tubes, as an alloying agent in steel, making surgical appliances, photoflash bulbs, explosive primers, rayon spinnerets, lamp filaments, etc. It is used in poison ivy lotions in the form of the carbonate as it combines with *urushiol.* With niobium, zirconium is superconductive at low temperatures and is used to make superconductive magnets, which offer hope of direct large-scale generation of electric power. Alloyed with zinc, zirconium becomes magnetic at temperatures below 35 K. Zirconium oxide (zircon) has a high index of refraction and is used as a gem material. The impure oxide, zirconia, is used for laboratory crucibles that will withstand heat shock, for linings of metallurgical furnaces, and by the glass and ceramic industries as a refractory material. Its use as a refractory material accounts for a large share of all zirconium consumed. Commercial-grade zirconium metal sponge is priced at about $7/lb. Fabricated zirconium parts are higher in cost.

NOMENCLATURE OF INORGANIC CHEMISTRY

AMERICAN VERSION*

By permission of the Committee on Publications of the International Union of Pure and Applied Chemistry.

INDEX TO IUPAC INORGANIC RULES

*This is the United States presentation of these IUPAC Inorganic Rules as published in J. Am. Chem. Soc., 82, 5525 (1960)

1. ELEMENTS

1.1. Names and Symbols of the Elements

1.11.—The elements should have the symbols given in the following table (Table I). It is desirable that the names should differ as little as possible among the different languages, but as complete uniformity is hard to achieve, separate lists have been drawn up in English and in French. The English list only is reproduced here.

TABLE I

ELEMENTS

Name	Symbol	Atomic number	Name	Symbol	Atomic number	Name	Symbol	Atomic number
Actinium	Ac	89	Gold (Aurum)	Au	79	Praseodymium	Pr	59
Aluminum	Al	13	Hafnium	Hf	72	Promethium	Pm	61
Americium	Am	95	Helium	He	2	Protactinium	Pa	91
Antimony	Sb	51	Holmium	Ho	67	Radium	Ra	88
Argon	Ar	18	Hydrogen	H	1	Radon	Rn	86
Arsenic	As	33	Indium	In	49	Rhenium	Re	75
Astatine	At	85	Iodine	I	53	Rhodium	Rh	45
Barium	Ba	56	Iridium	Ir	77	Rubidium	Rb	37
Berkelium	Bk	97	Iron (Ferrum)	Fe	26	Ruthenium	Ru	44
Beryllium	Be	4	Krypton	Kr	36	Samarium	Sm	62
Bismuth	Bi	83	Lanthanum	La	57	Scandium	Sc	21
Boron	B	5	Lead (Plumbum)	Pb	82	Selenium	Se	34
Bromine	Br	35	Lithium	Li	3	Silicon	Si	14
Cadmium	Cd	48	Lutetium	Lu	71	Silver (Argentum)	Ag	47
Calcium	Ca	20	Magnesium	Mg	12	Sodium	Na	11
Californium	Cf	98	Manganese	Mn	25	Strontium	Sr	38
Carbon	C	6	Mendelevium	Md	101	Sulfur	S	16
Cerium	Ce	58	Mercury	Hg	80	Tantalum	Ta	73
Cesium	Cs	55	Molybdenum	Mo	42	Technetium	Tc	43
Chlorine	Cl	17	Neodymium	Nd	60	Tellurium	Te	52
Chromium	Cr	24	Neon	Ne	10	Terbium	Tb	65
Cobalt	Co	27	Neptunium	Np	93	Thallium	Tl	81
Copper (Cuprum)	Cu	29	Nickel	Ni	28	Thorium	Th	90
Curium	Cm	96	Niobium	Nb	41	Thulium	Tm	69
Dysprosium	Dy	66	Nitrogen	N	7	Tin (Stannum)	Sn	50
Einsteinium	Es	99	Nobelium	No	102	Titanium	Ti	22
Erbium	Er	68	Osmium	Os	76	Tungsten (Wolfram)	W	74
Europium	Eu	63	Oxygen	O	8	Uranium	U	92
Fermium	Fm	100	Palladium	Pd	46	Vanadium	V	23
Fluorine	F	9	Phosphorus	P	15	Xenon	Xe	54
Francium	Fr	87	Platinum	Pt	78	Ytterbium	Yb	70
Gadolinium	Gd	64	Plutonium	Pu	94	Yttrium	Y	39
Gallium	Ga	31	Polonium	Po	84	Zinc	Zn	30
Germanium	Ge	32	Potassium	K	19	Zirconium	Zr	40

◆**The committees reaffirm the name niobium for element 41 in spite of the fact that many in the United States, particularly outside of chemical circles, still retain the name columbium.**

1.12.—The names placed in parentheses (after the trivial names) in the list in Table I shall always be used when forming names derived from those of the elements, *e.g.*, aurate, ferrate, wolframate and not goldate, ironate, tungstate.

For some compounds of sulfur, nitrogen, and antimony, derivatives of the Greek name θεῖον, the French name *azote*, and the Latin name *stibium*, respectively, are used.

Although the name nickel agrees with the chemical symbol, it is essentially a trivial name and is spelled so differently in various languages (niquel, nikkel, *etc.*) that it is recommended that names of derivatives be formed from the Latin name *niccolum, e.g.*, niccolate instead of nickelate. The name mercury should be used as the root name also in languages where the element has another name (mercurate, *not* hydrargyrate).

In the cases in which different names have been used, the Commission has selected one based upon considerations of prevailing usage and practicability. It should be emphasized that their selection carries no implication regarding priority of discovery.

◆**Tungstate and nickelate are both so well established in American practice that the committees object to changing them to wolframate and niccolate.**

1.13.—Any new metallic elements should be given names ending in -ium. Molybdenum and a few other elements have long been spelled without an "i" in most languages, and the Commission hesitates to insert it.

1.14.—All new elements shall have two-letter symbols.

1.15.—All isotopes of an element should have the same name. For hydrogen the isotope names protium, deuterium, and tritium may be retained, but it is undesirable to assign isotopic names instead of numbers to other elements. They should be designated by mass numbers as, for example, "oxygen-18."

◆**The list in 3.21 implies that D is an acceptable symbol for deuterium, whereas ^{2}H is used in 1.32. It is recommended that D and T be allowed for deuterium and tritium, respectively. *Cf.* comments made at 1.31, 1.32.**

1.2. Names for Groups of Elements and their Subdivisions

1.21.—The use of the collective names: halogens (F, Cl, Br, I, and At), chalcogens (O, S, Se, Te, and Po), and halides and chalcogenides for their compounds, alkali metals (Li to Fr), alkaline earth metals (Ca to Ra), and inert gases may be continued. The name rare earth metals may be used for the elements Sc, Y, and La to Lu inclusive; the name lanthanum series for the elements no. 57–71 (La to Lu inclusive), and the name lanthanides for the elements 58–71 (Ce to Lu inclusive) are recommended. Elements no. 89 (Ac) to 103 form the actinium series, and the name actinides is reserved for the elements in which the 5*f* shell is being filled. The name transuranium elements is also approved for the elements following uranium.

◆The collective term halogenides used in the Rules has been replaced in this version by halides, which is almost universally used in English and is unambiguous.

The inclusion of Sc with the rare earths is questioned by some. No need is seen for the terms lanthanum series and actinium series, particularly since the latter term is used for a radioactive series. The use of a collective term for elements 58–71 is approved, although it is suggested that lanthan*oid* is preferable to lanthan*ide* because of the use of -ide for binary compounds; similarly, actin*oid* is preferable to actin*ide*. Definition by means of atomic numbers is recommended in both cases rather than on the basis of interpretation (*e.g.*, filling of 5*f* shells).

1.22.—The word metalloid should not be used to denote nonmetals.

1.3. Indication of Mass, Charge, etc., on Atomic Symbols

1.31.—The mass number, atomic number, number of atoms, and ionic charge of an element may be indicated by means of four indices placed around the symbol. The positions are to be occupied thus

left upper index	right lower index	mass number	number of atoms
left lower index	right upper index	atomic number	ionic charge

Ionic charge should be indicated by A^{n+} rather than by A^{+n}.

Example: ${}^{32}_{16}S_2^{2+}$ represents a doubly ionized molecule containing two atoms of sulfur, each of which has the atomic number 16 and mass number 32.

The following is an example of an equation for a nuclear reaction

$$ {}^{26}_{12}Mg + {}^{4}_{2}He = {}^{29}_{13}Al + {}^{1}_{1}H $$

◆Although the practice of American chemists and physicists in general has been to put the mass number at the upper right of the symbol, the committees recognize the advantage of putting it at the upper left so that the upper right is available for the ionic charge as needed.

1.32.—Isotopically labeled compounds may be described by adding to the name of the compounds the symbol of the isotope in parentheses.

Examples:

${}^{32}PCl_3$ phosphorus(${}^{32}P$) trichloride (spoken: phosphorus-32 trichloride)

$H^{36}Cl$ hydrogen chloride (${}^{35}Cl^{d}$) (spoken: hydrogen chloride-36)

${}^{15}NH_3$ ammonia(${}^{15}N$) (spoken: ammonia nitrogen-15)

The position of the labeled atom may be indicated by placing the isotope symbol immediately after the locant (name of the group concerned).

Example: ${}^{2}H_2{}^{35}SO_4$ sulfuric(${}^{35}S$) acid(${}^{2}H$)

If this method gives names which are ambiguous or difficult to pronounce, the whole group containing the labeled atom may be indicated.

Examples:

$HOSO_2{}^{35}SH$	thiosulfuric(${}^{35}SH$) acid
${}^{15}NO_2NH_2$	nitramide(${}^{15}NO_2$), not nitr(${}^{15}N$)amide
$NO_2{}^{15}NH_2$	nitramide(${}^{15}NH_2$)
$HO_3S^{18}O{-}^{18}OSO_3H$	peroxo(${}^{18}O_2$)disulfuric acid

1.4. Allotropes

If systematic names for gaseous and liquid modifications are required, they should be based on the size of the molecule, which can be indicated by Greek numerical prefixes (listed in 2.251). If the number of atoms is large and unknown, the prefix poly may be used. To indicate ring and chain structures the prefixes cyclo and catena may be used.

Examples:

Symbol	Trivial Name	Systematic Name
H	atomic hydrogen	monohydrogen
O_2	(common) oxygen	dioxygen
O_3	ozone	trioxygen
P_4	white phosphorus (yellow phosphorus)	tetraphosphorus
S_8	λ-sulfur	cycloöctasulfur or octasulfur
S_n	μ-sulfur	catenapolysulfur or polysulfur

For the nomenclature of solid allotropic forms the rules in Section 8 may be applied.

◆The use of prefixes to indicate ring and chain structures is favored by the committees, but it should be pointed out that ino (not catena) has been used by mineralogists for indicating chain structures in silicates, along with cyclo and other prefixes (neso, phyllo, tecto, soro) denoting structure, all used without italics or hyphens (*e.g.*, inosilicates). *Cf.* use of catena in 7.42 for chains of alternating, not self-linking, atoms.

2. FORMULAS AND NAMES OF COMPOUNDS IN GENERAL

Many chemical compounds are essentially binary in nature and can be regarded as combinations of ions or radicals; others may be treated as such for the purpose of nomenclature.

Some chemists have expressed the opinion that the name of a compound should indicate whether it is ionic or covalent. Such a distinction is made in some languages (*e.g.*, in German: Natriumchlorid *but* Chlorwasserstoff), but it has not been made consistently, and indeed it seems impossible to introduce this distinction into a consistent system of nomenclature, because the line of demarcation between these two categories is not sharp. In these rules a system of nomenclature has been built on the basis of the endings -ide and -ate, and it should be emphasized that these are intended to be applied both to ionic and covalent compounds. If it is desired to avoid such endings for neutral molecules, names can be given as coördination compounds in accordance with **2.24** and Section **7.**

2.1. Formulas

2.11.—Formulas provide the simplest and clearest method of designating inorganic compounds. They are of particular importance in chemical equations and in descriptions of chemical procedure. However, their general use in text is not recommended, although in some cases a formula, on account of its compactness, may be preferable to a cumbersome and awkward name.

2.12.—The *empirical formula* is formed by juxtaposition of the atomic symbols to give the *simplest possible* formula expressing the stoichiometric composition of the compound in question. The empirical formula may be supplemented by indication of the crystal structure—see Section **8.**

2.13.—For compounds consisting of discrete molecules the *molecular formula, i.e.*, a formula corresponding with the correct molecular weight of the compound, should be used, *e.g.*, S_2Cl_2 and $H_4P_2O_6$ and not SCl and H_2PO_3. When the molecular weight varies with temperature, *etc.*, the simplest possible formula generally may be chosen, *e.g.*, S, P, and NO_2 instead of S_8, P_4, and N_2O_4, unless it is desirable to indicate the molecular complexity.

2.14.—In the *structural formula* the sequence and spatial arrangement of the atoms in a molecule are indicated.

2.15.—In formulas the *electropositive constituent* (cation) should always be placed first, *e.g.*, KCl, $CaSO_4$. This also applies in Romance languages even though the electropositive constituent is placed last in the *name*, *e.g.*, KCl, chlorure de potassium.

If the compound contains more than one electropositive or more than one electronegative constituent, their sequence is determined by Rules **6.32** and **6.33.**

2.16.—In the case of binary compounds between nonmetals that constituent should be placed first which appears earlier in the sequence: B, Si, C, Sb, As, P, N, H, Te, Se, S, At, I, Br, Cl, O, F.

Examples: NH_3, H_2S, N_4S_4, S_2Cl_2, Cl_2O, OF_2

◆Because N_4S_4 is definitely a nitride, not a sulfide, the committees prefer to write the formula S_4N_4, name it sulfur nitride, and cite it as an exception rather than as an example.

2.161.—For compounds containing three or more elements, however, the sequence should in general follow the order in which the atoms are actually bound in the molecule or ion, *e.g.*, NCS^-, not CNS^-, HOCN (cyanic acid), and HONC (fulminic acid).

Although formulas such as HNO_3, $HClO_4$, H_2SO_4, do not agree with this rule and HNO_3 does not even follow the main rule in **2.16**, the Commission does not at this time wish to break the old custom of putting the central atom immediately after the hydrogen atom in such cases (*cf* Section **5**). The formula for hypochlorous acid may be written HOCl or HClO.

2.17.—In intermetallic compounds the constituents should be placed in the order

Fr, Cs, Rb, K, Na, Li
Ra, Ba, Sr, Ca, Mg, Be
103, No, Md, Fm, Es, Cf, Bk, Cm, Am, Pu,
Np, U, Pa, Th, Ac, Lu–La, Y, Sc
Hf, Zr, Ti
Ta, Nb, V
W, Mo, Cr
Re, Tc, Mn
Pt, Ir, Os, Pd, Rh, Ru, Ni, Co, Fe
Au, Ag, Cu
Hg, Cd, Zn
Tl, In, Ga, Al
Pb, Sn, Ge
Bi, Sb
Po

Nonmetals (except Sb) in the order given in **2.16.**

Deviations from this order may be allowed, *e.g.*, when compounds with analogous structures are compared (AgZn and AgMg).

2.18.—The number of identical atoms or atomic groups in a formula is indicated by means of Arabic numerals, placed below and to the right of the symbol or symbols in parentheses ( ) or brackets [ ] to which they refer. Water of crystallization and similar loosely bound molecules, however, are designated by means of Arabic numerals before their formulas.

Examples: $CaCl_2$ not $CaCl^2$
$[Co(NH_3)_6]Cl_3$ not $[Co6NH_3]Cl_3$
$[Co(NH_3)_6]_2(SO_4)_3$
$Na_2SO_4.10H_2O$

2.19.—The prefixes *cis, trans, sym, asym* may be used in their usual senses. The prefixes may be connected with the formula by a hyphen and it is recommended that they be italicized.
Example: *cis*-$[PtCl_2(NH_3)_2]$

2.2. Systematic Names

Systematic names of compounds are formed by indicating the constituents and their proportions according to the following rules. (For the order of the constituents see also the later sections.)

2.21.—The name of the *electropositive constituent* (or that treated as such according to **2.16**) will not be modified (see, however, **2.2531**).

In Germanic languages the electropositive constituent is placed first, but in Romance languages it is customary to place the electronegative constituent first.

2.22.—If the *electronegative constituent* is monatomic its name is modified to end in -ide. For binary compounds of the nonmetals the name of the element standing later in the sequence in **2.16** is modified to end in -ide.

Examples: Sodium chloride, calcium sulfide, lithium nitride, arsenic selenide, calcium phosphides, nickel arsenide, aluminum borides, iron carbides, boron hydrides, phosphorus hydrides, hydrogen chloride, hydrogen sulfide, silicon carbide, carbon disulfide, sulfur hexafluoride, chlorine dioxide, oxygen difluoride.

Certain polyatomic groups are also given the ending -ide—see **3.22**.

In the Romance languages the endings -ure, -uro, and -eto are used instead of -ide. In some languages the word *oxyde* is used, whereas the ending -ide is used in the names of other binary compounds; it is recommended that the ending -ide be universally adopted in these languages.

◆Nitrogen sulfide has been taken out of the examples. *Cf.* comment at 2.16.

2.23.—If the electronegative constituent is polyatomic it should be designated by the termination -ate.

In certain exceptional cases the terminations -ide and -ite are used—see **3.22**.

2.24.—In inorganic compounds it is generally possible in a polyatomic group to indicate a *characteristic atom* (as in ClO^-) or a *central atom* (as in ICl_4^-). Such a polyatomic group is designated a *complex*, and the atoms, radicals, or molecules bound to the characteristic or central atom are termed *ligands*.

In this case the name of a negatively charged complex should be formed from the name of the characteristic or central element (as indicated in **1.12**) modified to end in -ate.

Anionic ligands are indicated by the termination -o. Further details concerning the designation of ligands, the definition of "central atom," *etc.*, appear in Section **7**.

Although the terms sulfate, phosphate, *etc.*, were originally the names of the anions of particular oxo acids, the names sulfate, phosphate, *etc.*, should now designate quite generally a negative group containing sulfur or phosphorus, respectively, as the central atom, irrespective of its oxidation state (the designation of the oxidation state is discussed in later rules) and the number and nature of the ligands. The complex is indicated by brackets [ ], but this is not always necessary.

Examples:

$Na_2[SO_4]$	sodium tetraoxosulfate	$Na_3[PS_4]$	sodium tetrathiophosphate
$Na_2[SO_3]$	sodium trioxosulfate	$Na[PCl_6]$	sodium hexachlorophosphate
$Na_2[S_2O_3]$	sodium trioxothiosulfate	$K[PO_2F_2]$	potassium dioxodifluorophosphate
$Na[SO_3F]$	sodium trioxofluorosulfate	$K[POCl_2(NH)]$	potassium oxodichloroimidophosphate
$Na_3[PO_4]$	sodium tetraoxophosphate		

In many cases these names may be abbreviated, *e.g.*, sodium sulfate, sodium thiosulfate (see **2.26**), and in other cases trivial names may be used (*cf.* **2.3**, **3.224**, and Section **5**). It should be pointed out, however, that the principle is quite generally applicable, to compounds containing organic ligands also, and its use is recommended in all cases where trivial names do not exist.

The coördination principle applied in this rule may also be applied to complexes which are positive or neutral (*cf.* **3.1** and Section **7**). However, neutral complexes which are as a rule considered as binary compounds are given names according to **2.22**, **2.16**. Thus, SO_3, sulfur trioxide, not trioxosulfur.

◆In the examples it would seem that full coördination-type names should be given, *e.g.*, either sodium tetraoxosulfate (VI) or disodium tetraoxosulfate. *Cf.* 7.32 and comment at 7.312.

2.25.—Indication of the Proportions of the Constituents.

2.251.—The *stoichiometric proportions* may be denoted by means of Greek numerical prefixes (mono, di, tri, tetra, penta, hexa, hepta, octa, ennea, deca, hendeca, and dodeca) preceding without hyphen the names of the elements to which they refer. It may be necessary in some languages to supplement these numerals with hemi ($^1/_2$) and the Latin sesqui ($^3/_2$).

The prefix mono may generally be omitted. Beyond 12, Greek prefixes are replaced by Arabic numerals (with or without hyphen according to the custom of the language), because they are more readily understood.

This sytem is applicable to all types of compounds and is especially suitable for binary compounds of the nonmetals.

When it is required to indicate the number of entire groups of atoms, particularly when the name includes a numerical prefix with a different significance, the multiplicative numerals (Latin bis, Greek tris, tetrakis, *etc.*) are used and the whole group to which they refer may be placed in parentheses if necessary.

Examples:

N_2O	dinitrogen oxide	Fe_3O_4	triiron tetraoxide
NO_2	nitrogen dioxide	U_3O_8	triuranium octaoxide
N_2O_4	dinitrogen tetraoxide	MnO_2	manganese dioxide
N_2S_5	dinitrogen pentasulfide	$Ca_3[PO_4]_2$	tricalcium diorthophosphate
S_2Cl_2	disulfur dichloride	$Ca[PCl_6]_2$	calcium bis(hexachlorophosphate)

In indexes it may be convenient to italicize a numerical prefix at the beginning of the name and connect it to the rest of the name with a hyphen, but this is not desirable in text, *e.g.*, *tri*-Uranium octaoxide.

Since the degree of polymerization of many substances varies with temperature, state of aggregation, *etc.*, the name to be used should normally be based upon the simplest possible formula of the substance except when it is required specifically to draw attention to the degree of polymerization.

Example: The name nitrogen dioxide may be used for the equilibrium mixture of NO_2 and N_2O_4. Dinitrogen tetraoxide means specifically N_2O_4.

◆In accordance with the organic nomenclature rules and well-established practice, it is recommended that "or Latin" be inserted between "Greek" and "numerical prefixes" in the first sentence and that "nona" replace "ennea," and "undeca" replace "hendeca."

Extreme caution is advised in the omission of numerical prefixes, including mono (*cf.* second set of examples in 5.23), because of the frequent use of names such as chloroplatinate (*cf.* 2.26 and the last sentence in 5.24).

2.252.—The proportions of the constituents also may be indicated indirectly by *Stock's system*, that is, by Roman numerals representing the oxidation number or stoichiometric valence of the element, placed in parentheses immediately following the name. For zero the Arabic 0 will be used. When used in conjunction with symbols the Roman numeral may be placed above and to the right.

The Stock notation can be applied to both cations and anions, but preferably should *not* be applied to compounds between nonmetals.

In employing the Stock notation, use of the Latin names of the elements (or Latin roots) is considered advantageous.

Examples:

$FeCl_2$	iron(II) chloride or ferrum(II) chloride
$FeCl_3$	iron(III) chloride or ferrum(III) chloride
MnO_2	manganese(IV) oxide
BaO_2	barium(II) peroxide
$Pb^{II}{}_2Pb^{IV}O_4$	dilead(II) lead(IV) oxide or trilead tetraoxide
$K_4[Ni(CN)_4]$	potassium tetracyanoniccollate (0)
$K_4[Fe(CN)_6]$	potassium hexacyanoferrate(II)
$Na_2[Fe(CO)_4]$	sodium tetracarbonylferrate(—II)

◆While the committees favor the extended use of the Stock notation, they suggest that in some cases the system of Ewens and Bassett (designation of the aggregate charge of a complex ion by an Arabic numeral in parentheses following the name, similar to the use as superior notations with formulas) is advantageous and should be allowed as an alternate (*cf.* 3.17 and comment at 7.323). Mixed use of the two systems, while not desirable in any one context, does not affect indexing and should not lead to confusion. See comment at 1.12.

2.253.—The following systems are in use but are not recommended:

2.2531.—The system of indicating valence by means of the suffixes *-ous* and *-ic* added to the root of the name of the cation may be retained for elements exhibiting not more than two valences.

2.2532.—"*Functional*" nomenclature (such as "nitric anhydride" for N_2O_5) is not recommended apart from the name *acid* to designate the acid function (Section **5**).

◆Apparently there is no objection to acid anhydride as a class name (*cf.* 5.32). Other functional derivatives of acids are named as such in the Rules (5.3).

2.26.—In systematic names it is not always necessary to indicate stoichiometric proportions. In many instances it is permissible to omit the numbers of atoms, oxidation numbers, *etc.*, when they are not required in the particular circumstances. For instance, these indications are not generally necessary with elements of essentially constant valence.

Examples:

sodium sulfate instead of sodium tetraoxosulfate
aluminum sulfate instead of aluminum(III) sulfate
potassium chloroplatinate(IV) instead of potassium hexachloroplatinate(IV)
potassium cyanoferrate(III) instead of potassium hexacyanoferrate(III)
phosphorus pentaoxide instead of diphosphorus pentaoxide

2.3. Trivial Names

Certain well-established trivial names for oxo acids (Section **5**) and for hydrogen compounds (water, ammonia, hydrazine) are still acceptable. For some other hydrogen compounds these names are approved:

B_2H_6	diboran	PH_3	phosphine	SbH_3	stibine	P_2H_4	diphosphine
SiH_4	silane	AsH_3	arsine	Si_2H_6	disilane, *etc.*	As_2H_4	diarsine

In some languages names of the type "Chlorwasserstoff" are in use and may be retained if national nomenclature committees so wish.

Purely trivial names, free from false scientific implications, such as soda, Chile saltpeter, quicklime, are harmless in industrial and popular literature; but old incorrect scientific names such as sulfate of magnesia, Natronhydrat, sodium muriate, carbonate of lime, should be avoided under all circumstances, and they should be eliminated from technical and patent literature.

◆For BH_3 (omitted in the Rules) borane rather than the previously used borine has been recommended by the Advisory Committee on the Nomenclature of Organic Boron Compounds of the ACS in a report not yet published.

Because soda is an ambiguous term, it is suggested that it be replaced by soda ash.

3. NAMES FOR IONS AND RADICALS

3.1. Cations

3.11.—Monatomic cations should be named like the corresponding element, without change or suffix, except as provided by **2.2531**.

Examples:

Cu^+ the copper(I) ion Cu^{2+} the copper(II) ion I^+ the iodine cation

◆For I^+, iodide(I) cation is more consistent with recommended practice. *Cf.* 3.21.

3.12.—The preceding principle should apply also to polyatomic cations corresponding to radicals for which special names are given in **3.32**, *i.e.*, these names should be used without change or suffix.

Examples: NO^+ the nitrosyl cation NO_2^+ the nitryl cation

◆Polyatomic here and in 3.13, 3.14, 3.223, 3.32, and 5.2 seems to be limited to more than one *kind* of atom, and hence heteratomic would be a more precise term. It is agreed that nitryl and not nitronium should be used in all cases (*cf.* 3.151).

3.13.—Polyatomic cations formed from monatomic cations by the addition of other ions or neutral atoms or molecules (ligands) will be regarded as complex and will be named according to the rules given in Section **7**.

Examples:

$[Al(H_2O)_6]^{3+}$ the hexaaquoaluminum ion $[CoCl(NH_3)_5]^{2+}$ the chloropentamminecobalt ion

For some important polyatomic cations which fall in this section, radical names given in **3.32** may be used alternatively, *e.g.*, for UO_2^{2+} the name uranyl(VI) ion in place of dioxouranium(VI) ion.

3.14.—Names for polyatomic cations derived by addition of protons to monatomic anions are formed by adding the ending -onium to the root of the name of the anion element.

Examples: phosphonium, arsonium, stibonium, oxonium, sulfonium, selenonium, telluronium, and iodonium ions.

Organic ions derived by substitution in these parent cations should be named as such, whether the parent itself is a known compound or not: for example $(CH_3)_4Sb^+$, the tetramethylstibonium ion.

The ion H_3O^+, which is in fact the monohydrated proton, is to be known as the oxonium ion when it is believed to have this constitution, as for example in $H_3O^+ClO_4^-$, oxonium perchlorate. The widely used term hydronium should be kept for the cases where it is wished to denote an indefinite degree of hydration of the proton, as, for example, in aqueous solution. If, however, the hydration is of no particular importance to the matter under consideration, the simpler term hydrogen ion may be used. The latter also may be used for the indefinitely solvated proton in nonaqueous solvents; but definite ions such as $CH_3OH_2^+$ and $(CH_3)_2OH^+$ should be named as derivatives of the oxonium ion, *i.e.*, as methyl- and dimethyloxonium ions, respectively.

◆The committees concur in oxonium for the ion H_3O^+, but see little reason for encouraging retention of the term hydronium ion because hydrogen ion adequately designates an indeterminate degree of hydration.

3.15.—Ions from Nitrogen Bases.

3.151.—The name ammonium for the ion NH_4^+ does not conform to **3.14**, but should be retained. This decision does *not* release the word nitronium for other uses: this would lead to inconsistencies when the rules are applied to other elements.

3.152.—Substituted ammonium ions derived from nitrogen bases with names ending in -amine will receive names formed by changing -amine to -ammonium. For example, $HONH_3^+$, the hydroxylammonium ion.

3.153.—When the nitrogen base is known by a name ending otherwise than in -amine, the cation name is to be formed by adding the ending -ium to the name of the base (if necessary omitting a final -e or other vowel).

Examples: hydrazinium, anilinium, glycinium, pyridinium, guanidinium, imidazolium.

The names uronium and thiouronium, though inconsistent with this rule, may be retained.

3.16.—Cations formed by adding protons to nonnitrogenous bases may also be given names formed by adding -ium to the name of the compound to which the proton is added.

Examples: dioxanium, acetonium.

In the case of cations formed by adding protons to acids, however, their names are to be formed by adding the word acidium to the name of the corresponding anion, and not that of the acid itself. For example, $H_2NO_3^+$, the nitrate acidium ion; $H_2NO_2^+$, the nitrite acidium ion; and $CH_3COOH_2^+$, the acetate acidium ion. Note, however, that when the anion of the acid is monatomic **3.14** will apply; for example, FH_2^+ is the fluoronium ion.

◆In accord with present practice, nitric acidium ion, *etc.*, are preferred to nitrate acidium ion, *etc.* $CH_3COOH_2^+$ is organic.

3.17.—Where more than one ion is derived from one base, as, for example, $N_2H_5^+$ and $N_2H_6^{2+}$, their charges may be indicated in their names as the hydrazinium(1+) and the hydrazinium(2+) ion, respectively.

◆*Cf.* comment at 2.252 and the use of Stock notation with ions or radicals in 3.13 and 3.32.

3.2. Anions

3.21.—The names for monatomic anions shall consist of the name (sometimes abbreviated) of the element, with the termination -ide. Thus

H^-	hydride ion	Br^-	bromide ion	Se^{2-}	selenide ion	As^{3-}	arsenide ion
D^-	deuteride ion	I^-	iodide ion	Te^{2-}	telluride ion	Sb^{3-}	antimonide ion
F^-	fluoride ion	O^{2-}	oxide ion	N^{3-}	nitride ion	C^{4-}	carbide ion
Cl^-	chloride ion	S^{2-}	sulfide ion	P^{3-}	phosphide ion	Si^{4-}	silicide ion
		B^{3-}	boride ion				

Expressions of the type "chlorine ion" are used particularly in connection with crystal structure work and spectroscopy; the Commission recommends that whenever the charge corresponds with that indicated above, the termination -ide should be used.

◆*Cf.* comments at 1.15 and 1.32 regarding ^2H and D.

3.22. Polyatomic Anions.

3.221.—Certain polyatomic anions have names ending in -ide. These are

OH^-	hydroxide ion	I_3^-	triiodide ion	$NHOH^-$	hydroxylamide ion
O_2^{2-}	peroxide ion	HF_2^-	hydrogen difluoride ion	$N_2H_3^-$	hydrazide ion
O_2^-	hyperoxide ion	N_3^-	azide ion	CN^-	cyanide ion
O_3^-	ozonide ion	NH^{2-}	imide ion	C_2^{2-}	acetylide ion
S_2^{2-}	disulfide ion	NH_2^-	amide ion		

Names for other polysulfide, polyhalide, *etc.*, ions may be formed in analogous manner. The OH^- ion should not be called the hydroxyl ion. The name hydroxyl is reserved for the OH group when neutral or positively charged, whether free or as a substituent (*cf.* **3.12** and **3.32**).

◆Superoxide is well established in English for O_2^- and no advantage is seen in changing to hyperoxide.

3.222.—Ions such as SH^- and O_2H^- will be called the hydrogen sulfide ion and the hydrogen peroxide ion, respectively. This agrees with **6.2**, and names such as hydrosulfide are not needed.

◆*Cf.* comment at 6.2. All "fused" names (as hydrogensulfide here and methylisocyanide in 5.33) in the original version have been written as two words in this version. For a rule on the written form of the names of compounds see *J. Chem. Education*, 8, 1336–8 (1931)

3.223.—The names for other polyatomic anions shall consist of the name of the central atom with the termination -ate, which is used quite generally for complex anions. Atoms and groups attached to the central atom shall generally be treated as ligands in a complex (*cf.* **2.24** and Section **7**) as, for example, $[Sb(OH)_6]^-$, the hexahydroxoantimonate(V) ion.

This applies also when the exact composition of the anion is not known; *e.g.*, by solution of aluminum hydroxide or zinc hydroxide in sodium hydroxide, aluminate and zincate ions are formed.

3.224.—It is quite practicable to treat oxygen in the same manner as other ligands (**2.24**), but it has long been customary to ignore the name of this element altogether in anions and to indicate its presence and proportion by means of a series of prefixes (hypo-, per-, *etc.*, see Section **5**) and sometimes also by the suffix -ite in place of -ate.

The termination -ite has been used to denote a lower oxidation state and may be retained in trivial names in these cases

NO_2^-	nitrite	$PH_2O_2^-$	hypophosphite	$S_2O_2^{2-}$	thiosulfite
$N_2O_2^{2-}$	hyponitrite	AsO_3^{3-}	arsenite	SeO_3^{2-}	selenite
NOO_2^-	peroxonitrite	SO_3^{2-}	sulfite	ClO_2^-	chlorite
PHO_3^{2-}	phosphite	$S_2O_5^{2-}$	disulfite (pyrosulfite)	ClO^-	hypochlorite
$P_2H_2O_5^{2-}$	diphosphite (pyrophosphite)	$S_2O_4^{2-}$	dithionite		

(and correspondingly for the other halogens)

The Commission does not recommend the use of any such names other than those listed. A number of other names ending in -ite have been used, *e.g.*, antimonite, tellurite, stannite, plumbite, ferrite, manganite, but in many cases such compounds are known in the solid state to be double oxides and are to be treated as such (*cf.* **6.5**), *e.g.*, $Cu(CrO_2)_2$ copper(II) chromium(III) oxide, not copper chromite. Where there is reason to believe that a definite salt with a discrete anion exists, the name is formed in accordance with **2.24**. By dissolving, for example, Sb_2O_3, SnO, or PbO in sodium hydroxide an antimonate(III), a stannate(II), or a plumbate(II) is formed in the solution.

Concerning the use of prefixes hypo-, per-, *etc.*, see the list of acids (table in **5.214**). For all new compounds and even for the less common ones listed in the table in **3.224** or derived from the acids listed in the table in **5.214**, it is preferable to use the system given in **2.24** and in Sections **5** and **7**.

◆For phosphite names see comment at 6.2.

3.3. Radicals

3.31.—A radical is here regarded as a group of atoms which occurs repeatedly in a number of different compounds. Sometimes the same radical fulfils different functions in different cases, and accordingly different names often have been assigned to the same group. The Commission considers it desirable to reduce this diversity and recommends that formulas or systematic names be used to denote all new radicals, instead of introducing new trivial names. The list of names for ions and radicals on page B-169 gives an extensive selection of radical names at present in use in inorganic chemistry.

◆The list of names for ions and radicals following the Rules is very useful and can be made more so by additions (as of dithio, nitrilo, and azido in the last column) and by greater attempt at uniformity with organic usage. Some of the terms (as nitride and amide) listed as anions are used also of covalent compounds. A single atom may function like a radical as defined above and may be named similarly, as chloro and oxo.

3.32.—Certain radicals containing oxygen or other chalcogens have special names ending in -yl, and the Commission approves the provisional retention of

HO	hydroxyl	S_2O_5	pyrosulfuryl	PO	phosphoryl	ClO_2	chloryl
CO	carbonyl	SeO	seleninyl	VO	vanadyl	ClO_3	perchloryl
NO	nitrosyl	SeO_2	selenonyl	PuO_2	plutonyl (similarly for other actinide elements)		(and similarly for other halogens)
SO	sulfinyl (thionyl)	CrO_2	chromyl				
		UO_2	uranyl				
SO_2	sulfonyl (sulfuryl)	NpO_2	neptunyl				
		NO_2	nitryl[1]	ClO	chlorosyl		

[1]The name nitroxyl should not be used for this group since the name nitroxylic acid has been used for H_2NO_2. Although the word nitryl is firmly established in English, nitroyl may be a better model for many other languages.

Names such as the above should be used only to designate compounds containing these discrete groups. The use of thionyl and sulfuryl should be restricted to the halides. Names such as bismuthyl and antimonyl are not approved because the compounds do not contain BiO and SbO groups, respectively; such compounds are to be designated as oxide halides (**6.4**).

Radicals analogous to the above containing other chalcogens in place of oxygen are named by adding the prefixes thio-, seleno-, *etc.*

Examples:

PS thiophosphoryl   CSe selenocarbonyl

In cases where radicals may have different valences, the oxidation number of the characteristic element should be indicated by means of the Stock notation. For example, the uranyl group UO_2 may refer either to the ion UO_2^{2+} or to the ion UO_2^+; these can be distinguished as uranyl(VI) and uranyl(V), respectively. In like manner, VO may be vanadyl(V), vanadyl(IV), and vanadyl(III).

These polyatomic radicals always are treated as forming the positive part of the compound.

Examples:

$COCl_2$	carbonyl chloride	$NO_2HS_2O_7$	nitryl hydrogen disulfate
NOS	nitrosyl sulfide	S_2O_5ClF	pyrosulfuryl chloride fluoride
PON	phosphoryl nitride	$SO_2(N_3)_2$	sulfonyl azide
$PSCl_3$	thiophosphoryl chloride	SO_2NH	sulfonyl imide
POCl	phosphoryl(III) chloride	IO_2F	iodyl fluoride

By using the same radical names regardless of unknown or controversial polarity relationships, names can be formed without entering into any controversy. Thus, for example, the compounds NOCl and $NOClO_4$ are quite unambiguously denoted by the names nitrosyl chloride and nitrosyl perchlorate, respectively.

◆Caution is urged in the use of some of these radical names: Vanadyl, for example, has been used for VO_2 as well as for VO (*cf.* also the naming of $VOSO_4$ in 6.42). Most of these radical names (except hydroxyl and thionyl) can be regarded as derived from the names of acids which have lost all of their hydroxyls (analogous to -yl or -oyl organic acid radical names) by the use of -yl and -osyl for radicals from -ic and -ous acids, respectively; this is implied in the footnote about nitroxyl. The use of the Stock notation in only one example (phosphoryl(III) chloride) seems confusing. It might be clearer to indicate stoichiometric proportions, *e.g.*, phosphoryl (mono)chloride, thiophosphoryl trichloride (*cf.* phosphoryl triamide in 5.34) or to use phosphorosyl for phosphoryl(III).

The restriction of the use of thionyl and sulfuryl to the halides was agreed upon at a joint meeting of the inorganic and organic nomenclature commissions of the IUPAC in 1951.

3.33.—It should be noted that the same radical may have different names in inorganic and organic chemistry. To draw attention to such differences the prefix names of radicals as substituents in organic compounds have been listed together with the inorganic names in the list of names printed at the end of the Rules. Names of purely organic compounds, of which many are important in the chemistry of coördination compounds (Section **7**), should agree with the nomenclature of organic chemistry.

Organic chemical nomenclature is to a large extent based on the principle of substitution, *i.e.*, replacement of hydrogen atoms by other atoms or groups. Such "substitutive names" are extremely rare in inorganic chemistry; they are used, *e.g.*, in the following cases: NH_2Cl is called chloramine, and $NHCl_2$ dichloramine. These names may be retained in the absence of better terms. Other substitutive names (derived from "sulfonic acid" as a name for HSO_3H) are fluoro- and chlorosulfonic acid, aminosulfonic acid, iminodisulfonic acid, and nitrilotrisulfonic acid. These names should preferably be replaced by the following

FSO_3H	fluorosulfuric acid	$NH(SO_3H)_2$	imidodisulfuric acid
$ClSO_3H$	chlorosulfuric acid	$N(SO_3H)_3$	nitridotrisulfuric acid
NH_2SO_3H	amidosulfuric acid		

Names such as chlorosulfuric acid and amidosulfuric acid might be considered to be substitutive names derived by substitution of *hydroxyl* groups in sulfuric acid. From a more fundamental point of view, however (see **2.24**), such names are formed by adding hydroxyl, amide, imide, *etc.*, groups together with oxygen atoms to a sulfur atom, "sulfuric acid" in this connection standing as an abbreviation for "trioxosulfuric acid."

Another organic-chemical type of nomenclature, the formation of "conjunctive names," is also met in only a few cases in inorganic chemistry, *e.g.*, the hydrazine- and hydroxylaminesulfonic acids. According to the principles of inorganic chemical nomenclature these compounds should be called hydrazido- and hydroxylamidosulfuric acid.

◆These are not true "conjunctive names" since sulfonic acid is not a compound. For the naming of partial amides *cf.* also 5.34.

4. CRYSTALLINE PHASES OF VARIABLE COMPOSITION

Isomorphous replacement, interstitial solutions, intermetallic compounds, and other nonstoichiometric compounds (berthollides)

4.1.—If an intermediate crystalline phase occurs in a two-component (or more complex) system, it may obey the law of constant composition very closely, as in the case of sodium chloride, or it may be capable of varying in composition over an appreciable range, as occurs for example with FeS. A substance showing such a variation is called a *berthollide.*

In connection with the berthollides the concept of a characteristic or ideal composition is frequently used. A unique definition of this concept seems to be lacking. In one case it may be necessary to use a definition based upon lattice geometry and in another to base it on the ratio of valence electrons to atoms. Sometimes one can state several characteristic compositions, and at other times it is impossible to say whether a phase corresponds to a characteristic composition or not.

In spite of these difficulties it seems that the concept of a characteristic composition can be used in its present undefined form for establishing a system of notation for phases of variable composition. It also seems possible to use the concept even if the characteristic composition is not included in the known homogeneity range of the phase.

4.2.—For the present, mainly formulas should be used for berthollides and solid solutions, since strictly logical names tend to become inconveniently cumbersome. The latter should be used only when unavoidable (*e.g.,* for indexing), and may be written in the style of iron(II) sulfide (iron-deficient); molybdenum dicarbide (excess carbon), or the like. Mineralogical names should be used only to designate actual minerals and not to define chemical composition; thus the name calcite refers to a particular mineral (contrasted with other minerals of similar composition) and is not a term for the chemical compound whose composition is properly expressed by the name calcium carbonate. (The mineral name may, however, be used to indicate the structure type—see **6.52.**)

4.3.—A general notation for the berthollides, which can be used even when the mechanism of the variation in composition is unknown, is to put the sign ~ (read as *circa*) before the formula. (In special cases it may also be printed above the formula.)

Examples: ~FeS, $\widetilde{CuZn}$

The direction of the deviation may be indicated when required:

~FeS (iron-deficient); ~MoC_2 (excess carbon)

4.4.—For a phase where the variable composition is solely or partially caused by replacement, atoms or atomic groups which replace each other are separated by a comma and placed together between parentheses.

If possible the formula ought to be written so that the limits of the homogeneity range are represented when one or other of the two atoms or groups is lacking. For example the symbol (Ni,Cu) denotes the complete range from pure Ni to pure Cu; likewise K(Br,Cl) comprises the range from pure KBr to pure KCl. If only part of the homogeneity range is referred to, the major constituent should be placed first.

Substitution accompanied by the appearance of vacant positions (combination of substitutional and interstitial solution) receives an analogous notation. For example, $(Li_2,Mg)Cl_2$ denotes the homogeneous phase from LiCl to $MgCl_2$ where the anion lattice structure remains the same but one vacant cation position appears for every substitution of $2Li^+$ by Mg^{2+}.

The formula $(Mg_3,Al_2)Al_6O_{12}$ represents the homogeneous phase from the spinel $MgAl_2O_4$ (= $Mg_3Al_6O_{12}$) to the spinel form of Al_2O_3 (= $Al_2Al_6O_{12}$).

The solid solutions between CaF_2 and YF_3, where cation substitution is accompanied by interstitial addition of F^-, would be represented by the formula $(Ca,YF)F_2$. It is important to note that this formula is based purely on considerations of composition, and it does not imply that YF^{2+} takes over the actual physical position of Ca^{2+}. On the same basis a notation for the plagioclases would be $(NaSi,CaAl)Si_2AlO_8$.

4.5.—A still more complete notation, which should always be used in more complex cases, may be constructed by indicating in a formula the variables that define the composition. Thus, a phase involving simple substitution may be written A_xB_{1-x}.

Examples: Ni_xCu_{1-x} and KBr_xCl_{1-x}

This shows immediately that the total number of atoms in the lattice is constant. Combined substitutional and interstitial or subtractive solution can be shown in an analogous way. The commas and parentheses called for in **4.4** are not required in this case.

For example, the homogeneous phase between LiCl and $MgCl_2$ becomes $Li_{2x}Mg_{1-x}Cl_2$ and the phase between $MgAl_2O_4$ and Al_2O_3 can be written $Mg_{3x}Al_{2(1-x)}Al_6O_{12}$, which shows that it cannot contain more Mg than that corresponding to $MgAl_2O_4$ ($x = 1$). The other examples given in **4.4** will be given the formulas $Ca_xY_{1-x}F_{3-x}$ and $Na_xCa_{1-x}Si_{2+x}Al_{2-x}O_8$. In the case of the γ-phase of the Ag–Cd system, which has the characteristic formula Ag_5Cd_8, the Ag and Cd atoms can replace one another to some extent and the notation would be $Ag_{5\pm x}Cd_{8\mp x}$.

Further examples:

$Fe_{1-x}Sb$ $Fe_{1-x}O$ $Fe_{1-x}S$ $Cu_{2-x}O$ $Na_{1-x}WO_3$ (sodium tungsten bronzes)

For $x = 0$ each of these formulas corresponds to a characteristic composition. If it is desired to show that the variable denoted by x can attain only small values, this may be done by substituting ϵ for x.

Likewise a solid solution of hydrogen in palladium can be written as PdH_x, and a phase of the composition M which has dissolved a variable amount of water can be written $M(H_2O)_x$.

When this notation is used, a particular composition can be indicated by stating the actual value of the variable *x*. Probably the best way of doing this is to put the value in parentheses after the general formula. For example, $Li_{4-x}Fe_{3x}Ti_{2(1-x)}O_6$ ($x = 0.35$). If it is desired to introduce the value of *x* into the formula itself, the mechanism of solution is more clearly understood if one writes $Li_{4-0\cdot35}Fe_{3\times0\cdot35}Ti_{2(1-0\cdot35)}O_6$ instead of $Li_{3\cdot65}Fe_{1\cdot05}Ti_{1\cdot30}O_6$.

5. ACIDS

Many of the compounds which now according to some definitions are called acids do not fall into the classical province of acids. In other parts of inorganic chemistry functional names are disappearing and it would have been most satisfactory to abolish them also for those compounds generally called acids. Names for these acids may be derived from the names of the anions as in Section 2, *e.g.*, hydrogen sulfate instead of sulfuric acid. The nomenclature of acids has, however, a long history of established custom, and it appears impossible to systematize acid names without drastic alteration of the accepted names of many important and well-known substances.

The present rules are aimed at preserving the more useful of the older names while attempting to guide further development along directions which should allow new compounds to be named in a more systematic manner.

5.1. Binary and Pseudobinary Acids

Acids giving rise to the -ide anions defined by **3.21** and **3.221** will be named as binary and pseudobinary compounds of hydrogen, *e.g.*, hydrogen chloride, hydrogen sulfide, hydrogen cyanide.

For the compound HN_3 the name hydrogen azide is recommended in preference to hydrazoic acid.

5.2. Acids Derived from Polyatomic Anions

Acids giving rise to anions bearing names ending in -ate or in -ite may also be treated as in **5.1**, but names more in accordance with custom are formed by using the terminations -ic acid and -ous acid corresponding with the anion terminations -ate and -ite, respectively. Thus chloric acid corresponds to chlorate, sulfuric acid to sulfate, and phosphorous acid to phosphite.

This nomenclature may also be used for less common acids, *e.g.*, hexacyanoferric acids correspond to hexacyanoferrate ions. In such cases, however, systematic names of the type hydrogen hexacyanoferrate are preferable.

Most of the common acids are oxo acids, *i.e.*, they contain only oxygen atoms bound to the characteristic atom. It is a long-established custom not to indicate these oxygen atoms. It is mainly for these acids that long-established names will have to be retained. Most other acids may be considered as coördination compounds and be named as such.

◆Polyatomic means heteratomic here, as opposed to pseudobinary. *Cf.* comment at 3.12.

5.21. Oxo Acids.—For the oxo acids the ous–ic notation to distinguish between different oxidation states is applied in many cases. The -ous acid names are restricted to acids corresponding to the -ite anions listed in the table in **3.224**.

Further distinction between different acids with the same characteristic element is in some cases effected by means of prefixes. This notation should not be extended beyond the cases listed below.

5.211.—The prefix hypo- is used to denote a lower oxidation state, and may be retained in these cases

$H_4B_2O_4$	hypoboric acid	HPH_2O_2	hypophosphorous acid
$H_2N_2O_2$	hyponitrous acid	HOCl	hypochlorous acid (and similarly for the other halogens)
$H_4P_2O_6$	hypophosphoric acid		

5.212.—The prefix per- is used to designate a higher oxidation state and may be retained for $HClO_4$, perchloric acid, and similarly for the other elements in Group VII.

The prefix per- should not be confused with the prefix peroxo- (see **5.22**).

5.213.—The prefixes ortho- and meta- have been used to distinguish acids differing in "water content." These names are approved

H_3BO_3	orthoboric acid	H_5IO_6	orthoperiodic acid	$(H_2SiO_3)_n$	metasilicic acids
H_4SiO_4	orthosilicic acid	H_6TeO_6	orthotelluric acid	$(HPO_3)_n$	metaphosphoric acids
H_3PO_4	orthophosphoric acid	$(HBO_2)_n$	metaboric acids		

For the acids derived by removing water from orthoperiodic or orthotelluric acid, the systematic names should be used, *e.g.*, HIO_4 tetraoxoiodic(VII) acid.

The prefix pyro- has been used to designate an acid formed from two molecules of an ortho acid minus one molecule of water. Such acids can now generally be regarded as the simplest cases of isopoly acids (*cf.* **7.5**). The prefix pyro- may be retained for pyrosulfurous and pyrosulfuric acids and for pyrophosphorous and pyrophosphoric acids, although in these cases also the prefix di- is preferable.

◆The use of orthoperiodic acid and orthotelluric acid is approved, but the question of names for HIO_4, *etc.*, needs further study because of confusion in the literature.

5.214.—The accompanying Table II contains the accepted names of the oxo acids (whether known in the free state or not) and some of their thio and peroxo derivatives (**5.22** and **5.23**).

For the less common of these acids systematic names would seem preferable, for example

H_2MnO_4 manganic(VI) acid, to distinguish it from H_3MnO_4, manganic(V) acid

$HReO_4$ tetraoxorhenic(VII) acid, to distinguish it from H_3ReO_5, pentaoxorhenic(VII) acid

H_2ReO_4 tetraoxorhenic(VI) acid, to distinguish it from $HReO_3$, trioxorhenic(V) acid; H_3ReO_4, tetraoxorhenic(V) acid; and $H_4Re_2O_7$, heptaoxodirhenic(V) acid

H_2NO_2 dioxonitric(II) acid instead of nitroxylic acid.

Trivial names should not be given to such acids as HNO, $H_2N_2O_3$, $H_2N_2O_4$, of which salts have been described. These salts are to be designated systematically as oxonitrates(I), trioxodinitrates(II), tetraoxodinitrates(III), respectively.

The names germanic acid, stannic acid, antimonic acid, bismuthic acid, vanadic acid, niobic acid, tantalic acid, telluric acid, molybdic acid, wolframic acid, and uranic acid may be used for substances with indefinite "water content" and degree of polymerization.

◆Unless trivial names clash with good nomenclature practices or are ambiguous, retention of well-established ones or the use of formulas is urged (especially for HNO, $H_2N_2O_3$, *etc.*) until structures are known. Systematic coordination-type names in the case of these nitrogen acids, for example, imply a structure that is ruled out by our present state of knowledge.

If hexahydroxoantimonic acid is considered a trivial name, hexahydrox*y*- might be preferable (*cf.* the systematic name hexahydroxoantimonate(V) ion in 3.223). For the analogous use of peroxo and peroxy, see comment at 5.22.

TABLE II

NAMES FOR OXO ACIDS

H_3BO_3	orthoboric acid or (mono)boric acid	H_2SO_4	sulfuric acid
$(HBO_2)_n$	metaboric acids	$H_2S_2O_7$	disulfuric or pyrosulfuric acid
$(HBO_2)_3$	trimetaboric acid	H_2SO_5	peroxo(mono)sulfuric acid
$H_4B_2O_4$	hypoboric acid	$H_2S_2O_8$	peroxodisulfuric acid
H_2CO_3	carbonic acid	$H_2S_2O_3$	thiosulfuric acid
HOCN	cyanic acid	$H_2S_2O_6$	dithionic acid
HNCO	isocyanic acid	H_2SO_3	sulfurous acid
HONC	fulminic acid	$H_2S_2O_5$	disulfurous or pyrosulfurous acid
H_4SiO_4	orthosilicic acid	$H_2S_2O_2$	thiosulfurous acid
$(H_2SiO_3)_n$	metasilicic acids	$H_2S_2O_4$	dithionous acid
HNO_3	nitric acid	H_2SO_2	sulfoxylic acid
HNO_4	peroxonitric acid	$H_2S_xO_6$ (x = 3,4...)	polythionic acids
HNO_2	nitrous acid	H_2SeO_4	selenic acid
HOONO	peroxonitrous acid	H_2SeO_3	selenious acid
H_2NO_2	nitroxylic acid	H_6TeO_6	(ortho)telluric acid
$H_2N_2O_2$	hyponitrous acid	H_2CrO_4	chromic acid
H_3PO_4	(ortho)phosphoric acid	$H_2Cr_2O_7$	dichromic acid
$H_4P_2O_7$	diphosphoric or pyrophosphoric acid	$HClO_4$	perchloric acid
$H_5P_3O_{10}$	triphosphoric acid	$HClO_3$	chloric acid
$H_{n+2}P_nO_{3n+1}$	polyphosphoric acids	$HClO_2$	chlorous acid
$(HPO_3)_n$	metaphosphoric acids	HClO	hypochlorous acid
$(HPO_3)_3$	trimetaphosphoric acid	$HBrO_3$	bromic acid
$(HPO_3)_4$	tetrametaphosphoric acid	$HBrO_2$	bromous acid
H_3PO_5	peroxo(mono)phosphoric acid	HBrO	hypobromous acid
$H_4P_2O_8$	peroxodiphosphoric acid	H_5IO_6	(ortho)periodic acid
$(HO)_2OP{-}PO(OH)_2$	hypophosphoric acid	HIO_3	iodic acid
$(HO)_2P{-}O{-}PO(OH)_2$	diphosphoric(III,V) acid	HIO	hypoiodous acid
H_2PHO_3	phosphorous acid	$HMnO_4$	permanganic acid
$H_4P_2O_5$	diphosphorous or pyrophosphorous acid	H_2MnO_4	manganic acid
		$HTcO_4$	pertechnetic acid
HPH_2O_2	hypophosphorous acid	H_2TcO_4	technetic acid
H_3AsO_4	arsenic acid	$HReO_4$	perrhenic acid
H_3AsO_3	arsenious acid	H_2ReO_4	rhenic acid
$HSb(OH)_6$	hexahydroxoantimonic acid		

5.22. Peroxo Acids.—The prefix peroxo, when used in conjunction with the trivial names of acids, indicates substitution of –O– by –O–O– (*cf.* **7.312**).

Examples:

HNO_4	peroxonitric acid	H_2SO_5	peroxosulfuric acid
H_3PO_5	peroxophosphoric acid	$H_2S_2O_8$	peroxodisulfuric acid
$H_4P_2O_8$	peroxodiphosphoric acid		

◆Peroxy, as recommended in the 1940 Rules (inorganic), is more acceptable than peroxo to organic chemists. It is not necessary that the use with trivial names conform with the use of peroxo denoting a coördinated ligand; *e.g.*, peroxysulfuric acid or trioxoperoxosulfuric(VI) acid.

5.23. Thio Acids.—Acids derived from oxo acids by replacement of oxygen by sulfur are called *thio acids* (*cf.* **7.312**).

Examples: $H_2S_2O_2$ thiosulfurous acid $H_2S_2O_3$ thiosulfuric acid HSCN thiocyanic acid

When more than one oxygen atom can be replaced by sulfur the number of sulfur atoms generally should be indicated

H_3PO_3S	monothiophosphoric acid	H_3AsS_3	trithioarsenious acid
$H_3PO_2S_2$	dithiophosphoric acid	H_3AsS_4	tetrathioarsenic acid
H_2CS_3	trithiocarbonic acid		

The prefixes seleno- and telluro- may be used in a similar manner.

5.24. Chloro Acids, etc.—Acids containing ligands other than oxygen and sulfur are generally designated according to the rules in Section 7.

Examples:

$HAuCl_4$	hydrogen tetrachloraurate(III) or tetrachloroauric(III) acid
H_2PtCl_4	hydrogen tetrachloroplatinate(II) or tetrachloroplatinic(II) acid
H_2PtCl_6	hydrogen hexachloroplatinate(IV) or hexachloroplatinic(IV) acid
$H_4Fe(CN)_6$	hydrogen hexacyanoferrate(II) or hexacyanoferric(II) acid
$H[PHO_2F]$	hydrogen hydridodioxofluorophosphate or hydridodioxofluorophosphoric acid
HPF_6	hydrogen hexafluorophosphate or hexafluorophosphoric acid
H_2SiF_6	hydrogen hexafluorosilicate or hexafluorosilicic acid
H_2SnCl_6	hydrogen hexachlorostannate(IV) or hexachlorostannic(IV) acid
HBF_4	hydrogen tetrafluoroborate or tetrafluoroboric acid
$H[B(OH)_2F_2]$	hydrogen dihydroxodifluoroborate or dihydroxodifluoroboric acid
$H[B(C_6H_5)_4]$	hydrogen tetraphenylborate or tetraphenylboric acid

It is preferable to use names of the type hydrogen tetrachloroaurate(III).

For some of the more important acids of this type abbreviated names may be used, *e.g.*, chloroplatinic acid, fluorosilicic acid.

◆For the use of hydrido in the fifth example see comment at 7.312.

5.3. Functional Derivatives of Acids

Functional derivatives of acids are compounds formed from acids by substitution of OH and sometimes also O by other groups. In this borderline between organic and inorganic chemistry organic-chemical nomenclature principles prevail.

◆The intention of the statement "organic-chemical nomenclature principles prevail" is not clear, since most of the examples given in the sections immediately following are not named according to organic practice. *Cf.* 3.33.

5.31. Acid Halides.—The names of acid halides are formed from the name of the corresponding acid radical if this has a special name, *e.g.*, sulfuryl chloride, phosphoryl chloride.

In other cases these compounds are named as oxide halides according to rule **6.41**, *e.g.*, MoO_2Cl_2, molybdenum dioxide dichloride.

5.32. Acid Anhydrides.—Anhydrides of inorganic acids generally should be given names as oxides, *e.g.*, N_2O_5 dinitrogen pentaoxide, *not* nitric anhydride or nitric acid anhydride.

5.33. Esters.—Esters of inorganic acids are given names in the same way as the salts, *e.g.*, dimethyl sulfate, diethyl hydrogen phosphate.

If, however, it is desired to specify the constitution of the compound, a name based on the nomenclature for cöordination compounds should be used.

Example:

$(CH_3)_4[Fe(CN)_6]$	tetramethyl hexacyanoferrate(II)
or	or
$[Fe(CN)_2(CH_3NC)_4]$	dicyanotetrakis(methyl isocyanide)iron(II)

◆According to common organic practice for esters (ethers, sulfides, *etc.*) and to the naming of inorganic salts (*e.g.*, sodium sulfate, not disodium sulfate), methyl sulfate would be used instead of dimethyl sulfate. However, no objection is seen to the more specific name. Such names as methyl sulfate are better for alphabetic listing, as in indexes.

5.34. Amides.—The names for amides may be derived from the names of acids by replacing acid by amide, or from the names of the acid radicals.

Examples:

$SO_2(NH_2)_2$	sulfuric diamide or sulfonyl diamide
$PO(NH_2)_3$	phosphoric triamide or phosphoryl triamide

If not all hydroxyl groups of the acid have been replaced by NH_2 groups, names ending in -amidic acid may be used: this is an alternative to naming the compounds as complexes.

Examples:

NH_2SO_3H	amidosulfuric acid or sulfamidic acid
$NH_2PO(OH)_2$	amidophosphoric acid or phosphoramidic acid
$(NH_2)_2PO(OH)$	diamidophosphoric acid or phosphorodiamidic acid

Abbreviated names (sulfamide, phosphamide, sulfamic acid) are often used but are not recommended.

◆The use of adjectives from names of inorganic acids may lead to confusion because an -ic or -ous adjective (as chromic) may refer to a higher- or lower-valent form of the element as well as to the acid (a possibility that does not arise, of course, with adjectives from names of organic acids).

Names of the type phosphoramidic acid are recommended in the report of the Advisory Committee on the Nomenclature of Organic Phosphorus Compounds of the Division of Organic Chemistry of the ACS published in 1952, but are not as acceptable to the inorganic nomenclature committees as a whole as the amido- or cöordination-type names. *Cf.* 3.33.

The retention of sulfamic acid and sulfamide as trivial names is favored by the committees. An acceptable systematic name for NH_2SO_3H would be ammonia-sulfur trioxide, in keeping with its probable structure.

5.35. Nitriles.—The suffix -nitrile has been used in the names of a few inorganic compounds, *e.g.* $(PNCl_2)_3$, trimeric phosphonitrile chloride. According to **2.22** such compounds can be designated as nitrides, *e.g.*, phosphorus nitride dichloride. Accordingly there seems to be no reason for retention of the name nitrile (and nitrilo, *c.f.* **3.33**) in inorganic chemistry.

◆Nitrilo is used in organic chemistry, though not given in the list at the end of the Rules.

6. SALTS AND SALT-LIKE COMPOUNDS

Among salts particularly there persist many old names which are bad and misleading, and the Commission wishes to emphasize that any which do not conform to these Rules should be discarded.

6.1. Simple Salts

Simple salts fall under the broad definition of binary compounds given in Section **2**, and their names are formed from those of the constituent ions (given in Section **3**) in the manner set out in Section **2**.

6.2. Salts Containing Acid Hydrogen ("Acid" Salts[1])

Names are formed by adding the word hydrogen, to denote the replaceable hydrogen present, immediately in front of the name of the anion.

The nonacidic hydrogen present, *e.g.*, in the phosphite ion, is included in the name of the anion and is not explicitly cited (*e.g.*, Na_2PHO_3, sodium phosphite).

Examples:

$NaHCO_3$	sodium hydrogen carbonate
NaH_2PO_4	sodium dihydrogen phosphate
$NaH[PHO_3]$	sodium hydrogen phosphite

◆The use of "fused" hydrogen names (as hydrogencarbonate in the original version) is not acceptable in English; the present practice of running hydrogen as a separate word is preferred and has been followed throughout this version. *Cf.* 3.222, 6.324, 6.333. If necessary for clarity, parentheses can be used, as in naming ligands, *e.g.*, (hydrogen carbonato). The use of hydro (as in hydrocarbonato) is not acceptable because of conflicts with organic usage, where hydro denotes addition of hydrogen to unsaturated compounds.

It seems safer to cite even nonacidic hydrogen present in an anion like PHO_3^{2-} (unless the ion has a specific name), because of current usage. (*Cf.* triethyl phosphite in 7.412, seventh example).

6.3. Double Salts, Triple Salts, etc.

6.31.—In formulas all the cations shall precede the anions; in names the principles embodied in Section 2 shall be applied. In those languages where cation names are placed after anion names the adjectives double, triple, *etc.* (their equivalents in the language concerned) may be added immediately after the anion name. The number so implied concerns the number of *kinds* of cation present and *not* the total number of such ions.

[1]For "basic" salts see 6.4.

6.32.—Cations.

6.321.—Cations shall be arranged in order of increasing valence (except hydrogen, *cf.* **6.2** and **6.324**).

6.322. The cations of each valence group shall be arranged in order of decreasing atomic number, with the polyatomic radical ions (*e.g.*, ammonium) at the end of their appropriate group.

◆Alphabetical order would be simpler here and even for 6.321.

6.323. Hydration of Cations.—Owing to the prevalence of hydrated cations, many of which are in reality complex, it seems unnecessary to disturb the cation order in order to allow for this; but if it is necessary to draw attention specifically to the presence of a particular hydrated cation this may be done by writing, for example, "hexaaquo" or "tetraaquo" before the name of the simple ion. Apart from this exception, however, all complex ions should be placed after simple ones in the appropriate valence group.

◆*Cf.* comment at 7.322.

6.324. Acidic Hydrogen.—When hydrogen is considered to be present as a cation its name shall be cited last among the cations. Actually acidic hydrogen will in most cases be bound to an anion and shall be cited together with this (**6.2**). If the salt contains only one anion, acidic hydrogen shall be cited in the same place whichever view is taken of the function of this hydrogen. Nonacidic hydrogen shall be either not explicitly cited (*cf.* **6.2**) or designated hydrido (*cf.* **5.24** and **7.311**). For salts with more than one anion see **6.333.**

Examples:

$KMgF_3$	potassium magnesium fluoride
$TlNa(NO_3)_2$	thallium(I) sodium nitrate or thallium sodium dinitrate
$KNaCO_3$	potassium sodium carbonate
$NH_4MgPO_4.6H_2O$	ammonium magnesium phosphate hexahydrate
$NaZn(UO_2)_3(C_2H_3O_2)_9.6H_2O$	sodium zinc triuranyl acetate hexahydrate
$Na[Zn(H_2O)_6](UO_2)_3(C_2H_3O_2)_9$	sodium hexaaquozinc triuranyl acetate
$NaNH_4HPO_4.4H_2O$	sodium ammonium hydrogen phosphate tetrahydrate

◆*Cf.* comments at 6.2, 7.312. In the fifth and sixth examples, either triuranyl(VI) or nonaacetate should be specified. *Cf.* 3.32, 6.34.

6.33.—Anions.

6.331.—Anions are to be cited in this group order

1. H^-
2. O_2^- and OH^- (in that order)
3. Simple (*i.e.*, one element only) inorganic anions, other than H^- and O^{2-}
4. Inorganic anions containing two or more elements, other than OH^-
5. Anions of organic acids and organic substances exerting an acid function

◆The committees consider it preferable to cite H^- last in accordance with usage.

6.332.—Within group 3 the ions shall be cited in the order given in **2.16**, the inclusion of O in that list being taken as referring to all oxygen anions apart from O^{2-} (*i.e.*,O_2^{2-}, *etc.*).

Within group 4, anions containing the smallest number of atoms shall be cited first, and in the case of two ions containing the same number of atoms they shall be cited in order of decreasing atomic number of the central atoms. Thus CO_3^{2-} should precede CrO_4^{2-}, and the latter should precede SO_4^{2-}.

Within group 5 the anions shall be cited in alphabetical order.

◆Again alphabetical order would be simpler, within groups 3 and 4 as well as 5. *Cf.* comment at 7.251.

6.333.—Acidic hydrogen should be cited together with the anion to which it is attached. If it is not known to which anion the hydrogen is bound, it should be cited last among the cations.

◆*Cf.* comment at 6.2.

6.34.—The stoichiometric method is the most practicable for indicating the proportions of the constituents. It is not always essential to give the numbers of all the anions, provided the valences of all the cations are either known or indicated.

Examples:

$NaCl.NaF.2Na_2SO_4$ or $Na_6ClF(SO_4)_2$	(hexa)sodium chloride fluoride (bis)sulfate
$Ca_5F(PO_4)_3$	(penta)calcium fluoride (tris)phosphate

The parentheses in these cases mean that numerical prefixes may not be necessary. The multiplicative numerical prefixes bis, tris, *etc.*, should be used in connection with anions, because disulfate, triphosphate, *etc.*, designate isopoly anions.

6.4. Oxide and Hydroxide Salts ("Basic" salts formerly oxy and hydroxy salts)

6.41.—For the purposes of nomenclature, these should be regarded as double salts containing O^{2-} and OH^- anions, and Section **6.3** may be applied in its entirety.

6.42. Use of the Prefixes Oxy and Hydroxy.—In some languages the citation in full of all the separate anion names presents no trouble and is strongly recommended (*e.g.*, copper oxide chloride), to the exclusion of the oxy form wherever possible. In some other languages, however, such names as "oxyde et chlorure double de cuivre" are so far removed from current practice that the present system of using oxy- and hydroxy-, *e.g.*, oxychlorure de cuivre, may be retained in such cases.

Examples:

$Mg(OH)Cl$	magnesium hydroxide chloride
$BiOCl$	bismuth oxide chloride
$LaOF$	lanthanum oxide fluoride
$VOSO_4$	vanadium(IV) oxide sulfate
$CuCl_2.3Cu(OH)_2$ or $Cu_2(OH)_3Cl$	dicopper trihydroxide chloride
$ZrOCl_2.8H_2O$	zirconium oxide (di)chloride octahydrate

6.5. Double Oxides and Hydroxides

The terms "mixed oxides" and "mixed hydroxides" are not recommended. Such substances preferably should be named double, triple, *etc.*, oxides or hydroxides as the case may be.

Many double oxides and hydroxides belong to several distinct groups, each having its own characteristic structure type, which is sometimes named after some well-known mineral of the same group (*e.g.*, perovskite, ilmenite, spinel). Thus, $NaNbO_3$, $CaTiO_3$, $CaCrO_3$, $CuSnO_3$, $YAlO_3$, $LaAlO_3$, and $LaGaO_3$ all have the same structure as perovskite, $CaTiO_3$. Names such as calcium titanate may convey false implications and it is preferable to name such compounds as double oxides and double hydroxides unless there is clear and generally accepted evidence of cations and oxo or hydroxo anions in the structure. This does not mean that names such as titanates or aluminates should always be abandoned, because such substances may exist in solution and in the solid state (*cf.* **3.223**).

◆"Multiple" has been used in English as a class term including double, triple, etc. (oxides or the like).

6.51.—In the double oxides and hydroxides the metals shall be named in the same order as for double salts (**6.32**).

6.52.—When required the structure type may be added in parentheses and in italics after the name, except that when the type name is also the mineral name of the substance itself then the italics should not be used (*cf.* **4.2**).

Examples:

$NaNbO_3$	sodium niobium trioxide (*perovskite* type)
$MgTiO_3$	magnesium titanium trioxide (*ilmenite* type)
$FeTiO_3$	iron(II) titanium trioxide (ilmenite)
$4CaO.AL_2O_3.nH_2O$ or $Ca_2Al(OH)_7.nH_2O$	dicalcium aluminum hydroxide hydrate
but $Ca_3[Al(OH)_6]_2$	(tri)calcium (bis)-[hexahydroxoaluminate]
$LiAl(OH)_4.2MnO_2$ or $LiAlMn^{IV}{}_2O_4(OH)_4$	lithium aluminum dimanganese(IV) tetraoxide tetrahydroxide

7. COÖRDINATION COMPOUNDS

7.1. Definitions

In its oldest sense the term *coördination compound* is taken as referring to molecules or ions in which there is an atom (A) to which are attached other atoms (B) or groups (C) to a number in excess of that corresponding to the oxidation number of the atom A. However, the system of nomenclature originally evolved for the compounds within this narrow definition has proved useful for a much wider class of compounds, and for the purposes of nomenclature the restriction "in excess . . . oxidation number" is to be omitted. Any compound formed by addition of one or several ions and/or molecules to one or more ions or/and molecules may be named according to the same system as strict coördination compounds.

The effect of this is to bring many simple and well-known compounds under the same nomenclature rules as accepted coöordination compounds; the result is to reduce the diversity of names and avoid many controversial issues, because it should be understood that there is no intention of implying that any structural analogy necessarily exists between different compounds merely on account of a common system of nomenclature. The system extends also to many addition compounds.

In the rules which follow certain terms are used in the senses here indicated: the atom referred to above as (A) is known as the *central* or *nuclear* atom, and all other atoms which are directly attached to A are known as *coördinating atoms*. Atoms (B) and groups (C) are called *ligands*. A group containing more than one *potential* coördinating atom is termed a *multidentate* ligand, the number of potential coöordinating atoms being indicated by the terms *unidentate, bidentate, etc.* A *chelate* ligand is a ligand *attached* to *one* central atom through *two or more* coördinating atoms, while a *bridging group* is attached to *more than one* atom. The whole assembly of one or more central atoms with their attached ligands is referred to as a *complex*, which may be an uncharged molecule or an ion of either polarity. A *polynuclear complex* is a complex which contains *more than one* nuclear atom, their number being designated by the terms *mononuclear, dinuclear, etc.*

◆Some dissatisfaction with the definition of coördination compounds was expressed, though the broad definition (last sentence of first paragraph) was generally approved.

Central atom or *center of coördination* is to be preferred to *nuclear* atom because of other senses of nucleus, especially of an atom. Possible replacements for polynuclear, etc., are polycentric, *etc.*, or bridged complex since bridging group is in common use.

In the United States the Greek-Latin hybrid terms polydentate and monodentate seem to be used more than the all-Latin multidentate and unidentate.

7.2. Formulas and Names for Complex Compounds in General

7.21. Central Atoms.—In *formulas* the symbol for the central atom(s) should be placed *first* (except in formulas which are primarily structural), the anionic and neutral, *etc.*, ligands following as prescribed in **7.25**, and the formula for the whole complex entity (ion or molecule) should be placed in brackets [ ].

In *names* the central atom(s) should be placed *after* the ligands.

◆It is considered preferable to place the whole complex in brackets, especially when more than one central atom is present, but not essential with only one central atom or with complex ions or nonionic complexes, because brackets may be needed sometimes for indicating concentrations.

7.22. Indication of Valence and Proportion of Constituents.—The oxidation number of the central atom is indicated by means of the Stock notation (**2.252**). Alternatively the proportion of constituents may be given by means of stoichiometric prefixes (**2.251**).

7.23.—Formulas and names may be supplemented with the prefixes *cis, trans, etc.* (**2.19**).

7.24. Terminations.—Complex anions shall be given the termination -ate (*cf.* **2.23**, **2.24** and **3.223**). Complex cations and neutral molecules are given no distinguishing termination. For further details concerning the names of ligands see **7.3**.

7.25. Order of Citation of Ligands in Complexes.—

first: anionic ligands
next: neutral and cationic ligands

7.251.—The anionic ligands shall be cited in the order

1. H^-
2. O^{2-}, OH^- (in that sequence)
3. Other simple anions (*i.e.*, one element only)
4. Polyatomic anions
5. Organic anions in alphabetical order

The sequence within categories 3 and 4 should be that given in **6.332**.

◆H^- is preferably named last, not first (*cf.* comment at 6.331). Alphabetical order is strongly recommended for simplicity at least within 3 and 4, where only rare uses are involved. The intention seems to be to include under 3 monatomic ions (*i.e.*, one *atom*—rather than one *element*—only), in other words to exclude N_3, I_3, etc. Under 4 the insertion of "inorganic" between "Polyatomic" and "anions" is recommended.

7.252.—Neutral and cationic ligands shall be cited in the order given

first: water, ammonia (in that sequence)
then: other inorganic ligands in the sequence in which their coördinating elements appear in the list given in **2.16**
last: organic ligands in alphabetical order.

◆For the inorganic ligands alphabetical order again is recommended. The use of parentheses wherever there is any possibility of ambiguity should be stressed, as illustrated in examples under 7.321: potassium trichloro(ethylene)platinate(II), where parentheses are given in the Rules, and tetra(pyridine)platinum(II) tetrachloroplatinate(II), where they have been added in this version. Parentheses might also be helpful with "thiocyanato" preceded by a numerical prefix (see last example in 7.311) and are definitely required with two-word names for ligands recommended instead of the fused names of the original version (6.2), as in the seventh example in 7.412: di-μ-carbonyl-bis{carbonyl(triethyl phosphite)cobalt}. Since brackets denote complexes, braces can be used in formulas or names where needed in order to avoid the use of two sets of parentheses (braces were so used in some but not all such cases in the original version).

7.3. Names for Ligands

7.31.—Anionic Ligands.

7.311.—The names for anionic ligands, whether inorganic or organic, end in -o (see, however, **7.324**). In general, if the anion name ends in -ide, -ite, or -ate, the final -e is replaced by -o, giving -ido, -ito, and -ato, respectively.

Examples:

$Li[AlH_4]$	lithium tetrahydridoaluminate
$Na[BH_4]$	sodium tetrahydridoborate
$K_2[OsNCl_5]$	potassium nitridopentachloroösmate(VI)
$[Co(NH_2)_2(NH_3)_4]OC_2H_5$	diamidotetraamminecobalt(III) ethanolate
$[CoN_3(NH_3)_5]SO_4$	azidopentaamminecobalt(III) sulfate
$Na_3[Ag(S_2O_3)_2]$	sodium bis(thiosulfato)argentate(I)
$[Ru(HSO_3)_2(NH_3)_4]$	bis(hydrogen sulfito)tetraammineruthenium
(II) $NH_4[Cr(SCN)_4(NH_3)_2]$	ammonium tetrathiocyanatodiamminechromate (III)

◆For "ethanolate" in the fourth example "ethoxide" may be preferred.

7.312.—These anions do not follow exactly the above rule; modified forms have become established:

F^-	fluoride	fluoro (*not* fluo)	O_2^{2-}	peroxide	peroxo
Cl^-	chloride	chloro	HS^-	hydrogen sulfide	thiolo
Br^-	bromide	bromo	S^{2-}	sulfide	thio[1] (sulfido)
I^-	iodide	iodo	(but:S_2^{2-}	disulfide	disulfido)
O^{2-}	oxide	oxo	CN^-	cyanide	cyano
OH^-	hydroxide	hydroxo			

[1]The name thio has long been used to denote the ligand S^{2-} when it can be regarded as replacing O^{2-} in an oxo acid or its anion. The general use of this name will prevent confusion between the two interpretations of disulfido as S_2^{2-} or two S^{2-} ligands.

By analogy with hydroxo, CH_3O^-, *etc.*, are called methoxo, *etc.* For CH_3S^-, *etc.*, the systematic names methanethiolato, *etc.*, are used.

Examples:

$K[AgF_4]$	potassium tetrafluoroargentate(III)
$K_2[NiF_6]$	potassium hexafluoroniccolate(IV)
$Ba[BrF_4]_2$	barium tetrafluorobromate(III)
$Na[AlCl_4]$	sodium tetrachloroaluminate
$Cs[ICl_4]$	cesium tetrachloroiodate(III)
$K[Au(OH)_4]$	potassium tetrahydroxoaurate(III)
$K[CrOF_4]$	potassium oxotetrafluorochromate(V)
$K_2[Cr(O)_2O_2(CN)_2(NH_3)]$	potassium dioxoperoxodicyanoamminechromate(VI)
$Na[BH(OCH_3)_3]$	sodium hydridotrimethoxoborate
$K_2[Fe_2S_2(NO)_4]$	dipotassium dithiotetranitrosyldiferrate

◆It is strongly recommended on the basis of past and present usage, analogy with chloro, *etc.*, and euphony that hydro be added to the list of modified forms of names for anionic ligands and that hydrido be abandoned (*cf.* also examples in 7.311).

While the Subcommittee on Coördination Compounds recognizes the usefulness of the invariable -o ending for anionic ligands, some members of the general committees do not see a sharp enough distinction between such ligands and the same groups in organic compounds to justify a departure from organic usage by using hydroxo, methoxo, *etc.*, instead of hydroxy, methoxy, *etc.* These members therefore favor adding hydroxy, methoxy, *etc.*, to the hydrocarbon radicals excepted from the rule of -o endings for anions (7.324). For the use of peroxo, see comment at 5.22.

The -o of a negative ligand should not be elided before another vowel (chloroösmate, chloroiodo). This agrees with organic practice. *Cf.* comment at 7.322.

It should be pointed out here as well as in the list of names for ions and radicals at the end of the Rules that the approved organic name for unsubstituted HS is mercapto (not thiol, as also given in the list) and for the alkyl- and aryl-substituted radicals methylthio, *etc.*

7.313.—Ligands derived from organic compounds not normally called acids, but which function as such in complex formation by loss of a proton, should be treated as anionic and given the ending -ato. If, however, no proton is lost, the ligand must be treated as neutral—see **7.32.**

Examples:

$[Ni(C_4H_7N_2O_2)_2]$ bis(dimethylglyoximato)nickel(II)

$[Cu(C_5H_7O_2)_2]$ bis(acetylacetonato)copper(II)

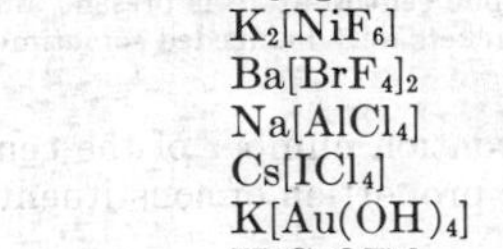

bis(8-quinolinolato)silver(II)

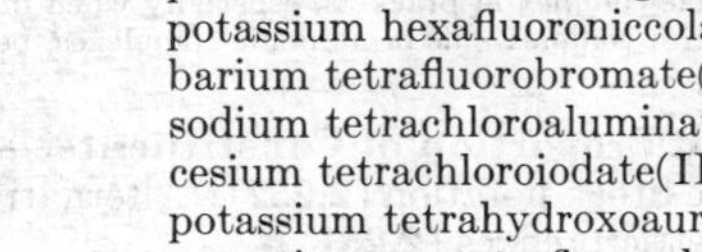

bis(4-fluorosalicylaldehydato)copper(II)

N,N'-ethylenebis(salicylideneiminato)cobalt(II)

◆Although according to 3.33 the name for the ligand in the second example should be derived from the systematic name 2,4-pentanedione instead of from acetylacetone, acetylacetonato perhaps conveys better the idea that it is the enol form that is involved. However, the coördination subcommittee questions the use of the termination -ato rather than plain -o especially in cases where the -ate terms (as dimethylglyoximate) are not accepted organic practice. Such -ato terms are especially misleading in the last two examples, where the -ato belongs with the hydroxyl part of the name, not the aldehyde or imine part to which it is attached.

7.32.—Neutral and Cationic Ligands.

7.321.—The name of the coördinated molecule or cation is to be used without change, except in the special cases provided for in **7.322.**

Examples:

$[CoCl_2(C_4H_8N_2O_2)_2]$	dichlorobis(dimethylglyoxime)cobalt(II) (*cf.* nickel derivative given in **7.313**)
cis-$[PtCl_2(Et_3P)_2]$	*cis*-dichlorobis(triethylphosphine)platinum(II)
$[CuCl_2(CH_3NH_2)_2]$	dichlorobis(methylamine)copper(II)
$[Pt\ py_4]\ [PtCl_4]$	tetra(pyridine)platinum(II) tetrachloroplatinate(II)
$[Fe(dipy)_3]Cl_2$	tris(dipyridyl)iron(II) chloride
$[Co\ en_3]_2(SO_4)_3$	tris(ethylenediamine)cobalt(III) sulfate
$[Zn\{NH_2CH_2CH(NH_2)CH_2NH_2\}_2]I_2$	bis(1,2,3-triaminopropane)zinc iodide
$K[PtCl_3(C_2H_4)]$	potassium trichloro(ethylene)platinate(II) or potassium trichloromonoethyleneplatinate(II)
$[PtCl_2\{H_2NCH_2CH(NH_2)CH_2NH_3\}]Cl$	dichloro(2,3 - diaminopropylammonium)platinum(II) chloride
$[Cr(C_6H_5NC)_6]$	hexakis(phenyl isocyanide)chromium

◆In the fifth example, bipyridine is preferred to dipyridyl in organic practice, and in the seventh example 1,2,3-propanetriamine to 1,2,3-triaminopropane.

7.322.—Water and ammonia as neutral ligands in coördination complexes are called "aquo" and "ammine," respectively.

In the tentative rules it was proposed to change the old-established "aquo" to "aqua," thus keeping the -o termination consistently for anionic ligands alone. However, as the old form is so widely used, many regarded the change as too pedantic, and the Commission has decided to retain "aquo" as an exception.

Examples:

$[Cr(H_2O)_6]Cl_3$	hexaaquochromium(III) chloride or hexaaquochromium trichloride
$[Al(OH)(H_2O)_5]^{++}$	the hydroxopentaaquoaluminum ion
$[Co(NH_3)_6]ClSO_4$	hexaamminecobalt(III) chloride sulfate
$[CoCl(NH_3)_5]Cl_2$	chloropentaamminecobalt(III) chloride
$[CoCl_3(NH_3)_2\{(CH_3)_2NH\}]$	trichlorodiammine(dimethylamine)cobalt(III)

◆Hexaquo, pentaquo, *etc.*, are used in the examples in the original version, but have been changed in this version to hexaaquo, *etc.*, for conformity with the latest approved organic practice. (*Cf.* hexaammonium, given with two a's separated by a hyphen in 7.6, second example, in the original version.)

7.323.—The groups NO, NS, CO, and CS, when linked directly to a metal atom, are to be called nitrosyl, thionitrosyl, carbonyl, and thiocarbonyl, respectively. In computing the oxidation number these radicals are treated as neutral.

Examples:

$Na_2[Fe(CN)_5NO]$	disodium pentacyanonitrosylferrate
$K_3[Fe(CN)_5CO]$	tripotassium pentacyanocarbonylferrate
$K[Co(CN)(CO)_2(NO)]$	potassium cyanodicarbonylnitrosylcobaltate(0)
$HCo(CO)_4$	hydrogen tetracarbonylcobaltate(−I)
$[Ni(CO)_2(Ph_3P)_2]$	dicarbonylbis(triphenylphosphine)nickel(0)
$[Fe\ en_3]\ [Fe(CO)_4]$	tris(ethylenediamine)iron(II) tetracarbonylferrate (−II)
$Mn_2(CO)_{10}$ or $[CO_5)Mn{-}Mn(CO)_5]$	decacarbonyldimanganese(0) or bis(pentacarbonylmanganese)

◆The necessity of arbitrarily considering these groups as always neutral can be avoided by not using the oxidation number but instead the stoichiometric proportions (as in some of these examples) or the Ewens-Bassett system (*cf.* comment at 2.252). Thus the anion in the last example in 7.312 (where NO is known to be positive) would by this system be named dithiotetranitrosyldiferrate (2−).

7.324.—Anions derived from hydrocarbons are given radical names without -o, but are counted as negative when computing the oxidation number.

The consistent introduction of the ending -o would in this case lead to unfamiliar names, *e.g.*, phenylato or phenido for $C_6H_5^-$. On the other hand, if the radicals were counted as neutral ligands, the central atom would have to be given an unusual oxidation number, *e.g.*, −I for boron in $K[B(C_6H_5)_4]$, instead of III.

Examples:

$K[B(C_6H_5)_4]$	potassium tetraphenylborate
$K[SbCl_5C_6H_5]$	potassium pentachloro(phenyl)antimonate(V)
$K_2[Cu(C_2H)_3]$	potassium triethynylcuprate(I)
$K_4[Ni(C_2C_6H_5)_4]$	potassium tetrakis(phenylethynyl)niccolate(0)
$[Fe(CO)_4(C_2C_6H_5)_2]$	tetracarbonylbis(phenylethynyl)iron(II)
$Fe(C_5H_5)_2$	bis(cyclopentadienyl)iron(II)
$[Fe(C_5H_5)_2]Cl$	bis(cyclopentadienyl)iron(III) chloride
$[Ni(NO)(C_5H_5)]$	nitrosylcyclopentadienylnickel

◆The use of radical names such as cyclopentadienyl in the examples does not seem consistent with the use of Stock notations; for the sixth example ("ferrocene"), iron(II) cyclopentadienide is preferred by some. This matter is presumably part of the whole organometallic problem to be dealt with by the new IUPAC joint organic-inorganic subcommittee.

For niccolate, see comment at 1.12.

7.33.—Alternative Modes of Linkage of Some Ligands.—Where ligands are capable of attachment by different atoms this may be denoted by adding the symbol for the atom by which attachment occurs at the end of the name of the ligand. Thus the dithioöxalato group may be attached through S or O, and these are distinguished as dithioöxalato-*S,S'* and dithioöxalato-*O,O'*, respectively.

In some cases different names are already in use for alternative modes of attachment, as, for example, thiocyanato (–SCN) and isothiocyanato (–NCS), nitro ($-NO_2$), and nitrito (–ONO). In these cases existing custom may conveniently be retained.

Examples:

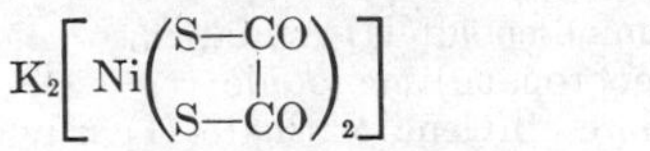

potassium bis(dithioöxalato-*S,S'*)niccolate(II)

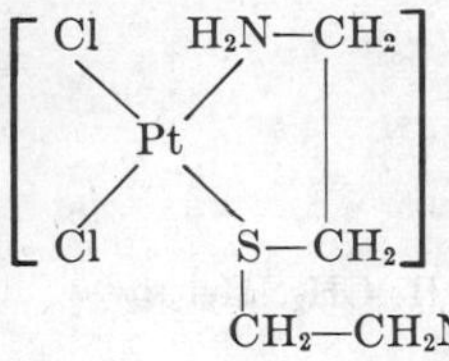

dichloro(2-*N,N*-dimethylaminoethyl 2-aminoethyl sulfide-*N',S*)platinum(II)

$K_2[Pt(NO_2)_4]$	potassium tetranitroplatinate(II)
$Na_3[Co(NO_2)_6]$	sodium hexanitrocobaltate(III)
$[Co(NO_2)_3(NH_3)_3]$	trinitrotriamminecobalt(III)
$[Co(ONO)(NH_3)_5]SO_4$	nitritopentaamminecobalt(III) sulfate
$[Co(NCS)(NH_3)_5]Cl_2$	isothiocyanatopentaamminecobalt(III) chloride

◆Thioöxalato in the original version has been changed to dithioöxalato.

7.4. Di- and Polynuclear Compounds

7.41.—Bridging Groups.

7.411.—A bridging group shall be indicated by adding the Greek letter μ immediately before its name and separating this from the rest of the complex by a hyphen. Two or more bridging groups of the same kind are indicated by di-μ-, *etc.*

7.412.—If the number of central atoms bound by one bridging group exceeds two, the number shall be indicated by adding a subscript numeral to the μ.

This system of notation allows simply of distinction between, for example, μ-disulfido (one S_2 bridge) and di-μ-sulfido (two S bridges). It is also capable of extension to much more complex and unsymmetrical structures by use of the conventional prefixes *cis, trans, asym,* and *sym* where necessary.

Examples:

$[(NH_3)_5Cr{-}OH{-}Cr(NH_3)_5]Cl_5$
μ-hydroxo-bis{pentaamminechromium(III)}chloride

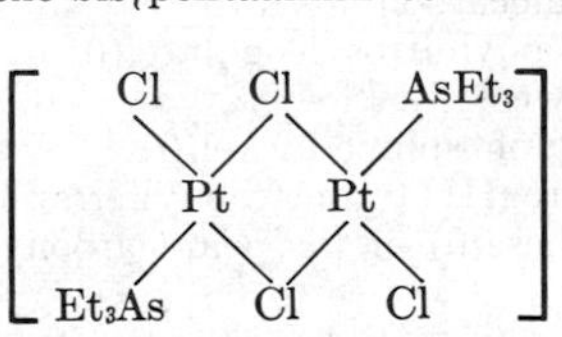

di-μ-chloro-dichlorobis(triethylarsine)diplatinum (II) (three possible isomers: *asym, sym-cis,* and *sym-trans;* the last is shown)

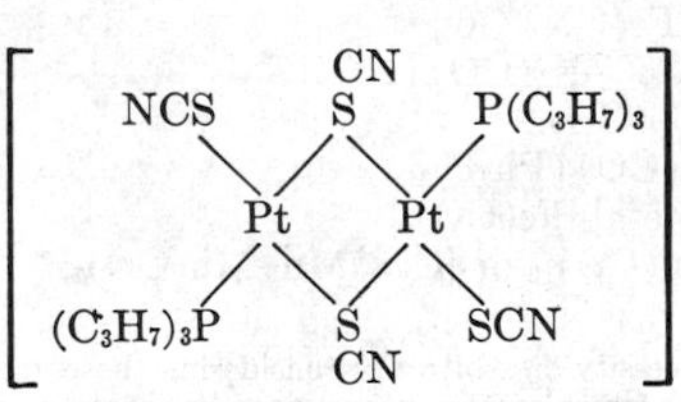

di-μ-thiocyanato-dithiocyanatobis(tripropylphosphine)diplatinum(II)

$[(CO)_3Fe(CO)_3Fe(CO)_3]$	tri-μ-carbonyl-bis(tricarbonyliron)
$[(CO)_3Fe(SEt)_2Fe(CO)_3]$	di-μ-ethanethiolato-bis(tricarbonyliron)
$[(C_5H_5)(CO)Fe(CO)_2Fe(CO)(C_5H_5)]$	di-μ-carbonyl-bis(carbonylcyclopentadienyliron)
$[(CO)\{P(OEt)_3\}Co(CO)_2Co(CO)\{P(OEt)_3\}]$	di-μ-carbonyl-bis{carbonyl(triethyl phosphite)-cobalt}
$[Au(CN)(C_3H_7)_2]_4$	cyclo-tetra-μ-cyano-tetrakis(dipropylgold)
$[CuI(Et_3As)]_4$	tetra-μ_3-iodo-tetrakis{triethylarsinecopper(I)}
$[Be_4O(CH_3COO)_6]$	μ_4-oxo-hexa-μ-acetato-tetraberyllium

7.42. Extended Structures.—Where bridging causes an indefinite extension of the structure it is best to name compounds primarily on the basis of their over-all composition; thus the compound having the composition represented by the formula $CsCuCl_3$ has an anion with the structure:

$$\left[\begin{array}{ccccccccc} & Cl & & Cl & & Cl & & Cl & \\ & | & & | & & | & & | & \\ \dots & Cl-Cu & -Cl- & Cu & -Cl- & Cu & -Cl- & Cu & \dots \\ & | & & | & & | & & | & \\ & Cl & & Cl & & Cl & & Cl & \end{array} \right]^{n-}$$

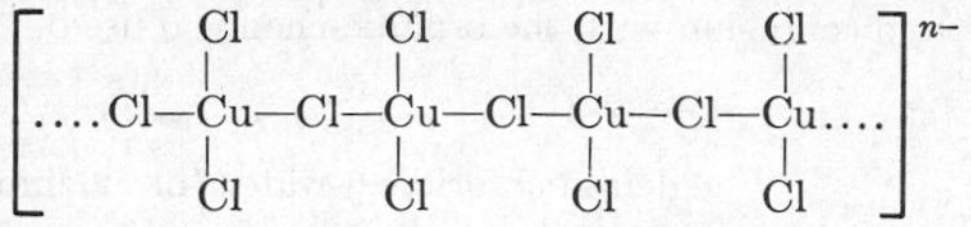

This may be expressed in the formula $(Cs^+)_n^- [(CuCl_3)_n]^{n-}$ which leads to the simple name cesium catena-μ-chloro-dichlorocuprate(II). If the structure were in doubt, however, the substance would be called cesium copper(II) chloride (as a double salt).

◆*Cf.* comment on *catena* at 1.4.

7.5. Isopoly Anions

The structure of many complicated isopoly anions has now been cleared up by X-ray work and it turns out that the indication of the several μ-oxo and oxo atoms in the name does not convey any clear picture of the structure and is therefore of little value.

For the time being it is sufficient to indicate the number of atoms by Greek prefixes, at least until isomers are found. When all atoms have their "normal" oxidation states (*e.g.*, W^{VI}), it is not necessary to give the numbers of the oxygen atoms, if all the others are indicated.

Examples:

$K_2S_2O_7$	dipotassium disulfate
$K_2S_3O_{10}$	dipotassium trisulfate
$Na_5P_3O_{10}$	pentasodium triphosphate
$K_2Cr_4O_{13}$	dipotassium tetrachromate
$Na_2B_4O_7$	disodium tetraborate
NaB_5O_8	sodium pentaborate
$Ca_3Mo_7O_{24}$	tricalcium heptamolybdate
$Na_7HNb_6O_{19}.15H_2O$	heptasodium monohydrogen hexaniobate-15-water
$K_2Mg_2V_{10}O_{28}.16H_2O$	dipotassium dimagnesium decavanadate-16-water

7.6. Heteropoly Anions

The central atom or atoms should be cited last in the name and first in the formula of the anion (*cf.* **7.21**), *e.g.*, wolframophosphate, *not* phosphowolframate.

If the oxidation number has to be given, it may be necessary to place it immediately after the atom referred to and not after the ending -ate, in order to avoid ambiguity.

The method formerly recommended for naming iso- and heteropoly anions by giving the number of atoms in parentheses is not practicable in more complicated cases.

Examples:

$(NH_4)_3PW_{12}O_{40}$	triammonium dodecawolframophosphate
$(NH_4)_6TeMo_6O_{24}.7H_2O$	hexaammonium hexamolybdotellurate heptahydrate
$Li_3HSiW_{12}O_{40}.24H_2O$	trilithium (mono)hydrogen dodecawolframosilicate-24-water
$K_6Mn^{IV}Mo_9O_{32}$	hexapotassium enneamolybdomanganate(IV)
$Na_6P^V_2Mo_{18}O_{62}$	hexasodium 18-molybdodiphosphate(V)
$Na_4P^{III}_2Mo_{12}O_{41}$	tetrasodium dodecamolybdodiphosphate(III)
$K_7Co^{II}Co^{III}W_{12}O_{42}.16H_2O$	heptapotassiumdodecawolframocobalt(II)-cobalt(III)ate-16-water
$K_3PV_2Mo_{10}O_{39}$	tripotassium decamolybdodivanadophosphate

◆The coördination subcommittee would prefer not to have sections 7.5 and 7.6 included under 7 and recommends that they should be studied by a special subcommittee. Some cyclic isomers are already known, and isopoly cations also are being investigated.

Cf. comment at 1.12 for stand on wolframate and wolframo, and 2.251 for nona instead of ennea (fourth example).

Isopoly and heteropoly are not separate words in the original version.

7.7. Addition Compounds

The ending -ate is now the accepted ending for *anions* and should generally not be used for addition compounds. Alcoholates are the *salts* of alcohols and this name should not be used to indicate alcohol of crystallization. Analogously addition compounds containing ether, ammonia, *etc.*, should *not* be termed etherates, ammoniates, *etc.*

However, one exception has to be recognized. According to the commonly accepted meaning of the ending -ate, "hydrate" would be, and was formerly regarded as, the name for a *salt* of water, *i.e.*, what is now known as a hydroxide; the name hydrate has now a very firm position as the name of a compound containing water of crystallization and is allowed also in these Rules to designate water bound in an unspecified way; it is considered to be preferable even in this case to avoid the ending -ate by using the name "water" (or its equivalent in other languages) when possible.

The names of addition compounds may be formed by connecting the names of individual compounds by hyphens (short dashes) and indicating the number of molecules by Arabic numerals. When the added molecules are organic, however, it is recommended to use multiplicative numeral prefixes (bis, tris, tetrakis, *etc.*) instead of Arabic figures to avoid confusion with the organic-chemical use of Arabic figures to indicate position of substituents.

Examples:

$CaCl_2.6H_2O$	calcium chloride–6-water (or calcium chloride hexahydrate)
$3CdSO_4.8H_2O$	3-cadmium sulfate–8-water
$Na_2CO_3.10H_2O$	sodium carbonate–10-water (or sodium carbonate decahydrate)
$AlCl_3.4C_2H_5OH$	aluminum chloride–4-ethanol or –tetrakisethanol
$BF_3.(C_2H_5)_2O$	boron trifluoride–diethyl ether
$BF_3.2CH_3OH$	boron trifluoride–bismethanol
$BF_3.H_3PO_4$	boron trifluoride–phosphoric acid
$BiCl_3.3PCl_5$	bismuth trichloride–3-(phosphorus pentachloride)
$TeCl_4.2PCl_5$	tellurium tetrachloride–2-(phosphorus pentachloride)
$(CH_3)_4NAsCl_4.2AsCl_3$	tetramethylammonium tetrachloroarsenate(III)–2-(arsenic trichloride)
$CaCl_2.8NH_3$	calcium chloride–8-ammonia
$8H_2S.46H_2O$	8-(hydrogen sulfide)–46-water
$8Kr.46H_2O$	8-krypton–46-water
$6Br_2.46H_2O$	6-dibromine–46-water
$8CHCl_3.16H_2S.136H_2O$	8-chloroform–16-(hydrogen sulfide)–136-water

These names are not very different from a pure verbal description which may in fact be used, *e.g.*, calcium chloride with 6 water, compound of aluminum chloride with 4 ethanol, *etc.*

If it needs to be shown that added molecules form part of a complex, the names are given according to **7.2** and **7.3.**

Examples:

$FeSO_4.7H_2O$ or $[Fe(H_2O)_6]SO_4.H_2O$	iron(II) sulfate heptahydrate or hexaaquoiron(II) sulfate monohydrate
$PtCl_2.2PCl_3$ or $[PtCl_2(PCl_3)_2]$	platinum(II) chloride–2-(phosphorus trichloride) or dichlorobis(phosphorus trichloride)-platinum(II)
$AlCl_3.NOCl$ or $NO[AlCl_4]$	aluminum chloride–nitrosyl chloride or nitrosyl tetrachloroaluminate
$BF_3.Et_3N$ or $[BF_3(Et_3N)]$	boron trifluoride–triethylamine or trifluoro(triethylamine)boron

◆Only some so-called "addition compounds" are known to be coordination compounds, and the formulas and names can show such structures. Those that are lattice compounds and those of unknown structure do not really belong in 7.

The committees like hydrate and ammoniate and see no particular advantage in dropping them (*cf.* the use of hydrate terms in examples in 6.324; 6-hydrate, *etc.*, as well as hexahydrate, *etc.*, are considered acceptable).

No reason can be seen for deviating here from usual organic practice by using multiplicative prefixes with simple names: tetraethanol is just as clear as tetrakisethanol, and dimethanol as bismethanol. Parentheses can always be used if there is danger of any ambiguity.

The use of short dashes ("en" dashes) instead of hyphens between the names (as in the examples here, but not in the original version) makes for greater ease in reading.

It is considered preferable by some to place the electron donor first in both the formulas and name: $(C_2H_5)_2O.BF_3$, diethyl ether–boron trifluoride (note that organic usage favors ethyl ether over diethyl ether).

8. POLYMORPHISM

Minerals occurring in nature with similar compositions have different names according to their crystal structures; thus, zinc blende, wurtzite; quartz, tridymite, and cristobalite. Chemists and metallographers have designated polymorphic modifications with Greek letters or with Roman numerals (α-iron, ice-I, *etc.*). The method is similar to the use of trivial names, and is likely to continue to be of use in the future in cases where the existence of polymorphism is established, but not the structures underlying it. Regrettably there has been no consistent system, and some investigators have designated as α the form stable at ordinary temperatures, while others have used α for the form stable immediately below the melting point, and some have even changed an already established usage and renamed α-quartz β-quartz, thereby causing confusion. If the α–β nomenclature is used for two substances A and B, difficulties are encountered when the binary system A–B is studied.

NOMENCLATURE OF INORGANIC CHEMISTRY (Continued)

LIST OF NAMES FOR IONS AND RADICALS

In inorganic chemistry substitutive names seldom are used, but the organic-chemical names are shown to draw attention to certain differences between organic and inorganic nomenclature.

	Name				
Atom or group	as neutral molecule	as cation or cationic radical[1]	as anion	as ligand	as prefix for substituent in organic compounds
H	monohydrogen	hydrogen	hydride	hydrido	
F	monofluorine		fluoride	fluoro	fluoro
Cl	monochlorine	chlorine	chloride	chloro	chloro
Br	monobromine	bromine	bromide	bromo	bromo
I	monoiodine	iodine	iodide	iodo	iodo
I_3			triiodide		
ClO		chlorosyl	hypochlorite	hypochlorito	
ClO_2	chlorine dioxide	chloryl	chlorite	chlorito	
ClO_3		perchloryl	chlorate	chlorato	
ClO_4			perchlorate		
IO		iodosyl	hypoiodite		iodoso
IO_2		iodyl			iodyl or iodoxy
O	monoöxygen		oxide	oxo	oxo or keto
O_2	dioxygen		O_2^{2-}: peroxide O_2^-: hyperoxide	peroxo	peroxy
HO	hydroxyl		hydroxide	hydroxo	hydroxy
HO_2	(perhydroxyl)		hydrogen peroxide	hydrogen peroxo	hydroperoxy
S	monosulfur		sulfide	thio (sulfido)	thio
HS	(sulfhydryl)		hydrogen sulfide	thiolo	thiol or mercapto
S_2	disulfur		disulfide	disulfido	
SO	sulfur monoxide	sulfinyl (thionyl)			sulfinyl
SO_2	sulfur dioxide	sulfonyl (sulfuryl)	sulfoxylate		sulfonyl
SO_3	sulfur trioxide		sulfite	sulfito	
HSO_3			hydrogen sulfite	hydrogen sulfito	
S_2O_3			thiosulfate	thiosulfato	
SO_4			sulfate	sulfato	
Se	selenium		selenide	seleno	seleno
SeO		seleninyl			seleninyl
SeO_2		selenonyl			selenonyl
SeO_3	selenium trioxide		selenite	selenito	
SeO_4			selenate	selenato	
Te	tellurium		telluride	telluro	telluro
CrO_2		chromyl			
UO_2		uranyl			
NpO_2		neptunyl			
PuO_2		plutonyl			
AmO_2		americyl			
N	mononitrogen		nitride	nitrido	
N_3			azide	azido	
NH			imide	imido	imino
NH_2			amide	amido	amino
NHOH			hydroxylamide	hydroxylamido	hydroxylamino
N_2H_3			hydrazide	hydrazido	hydrazino
NO	nitrogen oxide	nitrosyl		nitrosyl	nitroso
NO_2	nitrogen dioxide	nitryl		nitro	nitro
ONO			nitrite	nitrito	
NS		thionitrosyl			
$(NS)_n$		thiazyl (*e.g.*, trithiazyl)			
NO_3			nitrate	nitrato	
N_2O_3			hyponitrite	hyponitrito	
P	phosphorus		phosphide	phosphido	
PO		phosphoryl			phosphoroso
PS		thiophosphoryl			
PH_2O_3			hypophosphite	hypophosphito	
PHO_3			phosphite	phosphito	
PO_4			phosphate	phosphato	
AsO_4			arsenate	arsenato	
VO		vanadyl			
CO	carbon monoxide	carbonyl		carbonyl	carbonyl
CS		thiocarbonyl			
CH_3O	methoxyl		methanolate	methoxo	methoxy
C_2H_5O	ethoxyl		ethanolate	ethoxo	ethoxy
CH_3S			methanethiolate	methanethiolato	methylthio
C_2H_5S			ethanethiolate	ethanethiolato	ethylthio
CN		cyanogen	cyanide	cyano	cyano
OCN			cyanate	cyanato	cyanato
SCN			thiocyanate	thiocyanato and isothiocyanato	thiocyanato and isothiocyanato
SeCN			selenocyanate	selenocyanato	selenocyanato
TeCN			tellurocyanate	tellurocyanato	
CO_3			carbonate	carbonato	
HCO_3			hydrogen carbonate	hydrogen carbonato	
CH_3CO_2			acetate	acetato	acetoxy
CH_3CO	acetyl	acetyl			acetyl
C_2O_4			oxalate	oxalato	

[1]If necessary, oxidation state is to be given by Stock notation.

◆Although some additions might be made to this list, especially in the last column, and a few changes suggested, no attempt has been made to do so at this time. *Cf.* comments at such rules as 3.31, 3.32, 5.35, 6.2, and 7.312.

A rational system should be based upon crystal structure, and the designations α, β, γ, *etc.*, should be regarded as provisional, or as trivial names. The designations should be as short and understandable as possible, and convey a maximum of information to the reader. The rules suggested here have been framed as a basis for future work, and it is hoped that experience in their use may enable more specific rules to be formulated at a later date.

8.1.—For chemical purposes (*i.e.*, when particular mineral occurrences are not under consideration) polymorphs should be indicated by adding the crystal system after the name or formula. For example, zinc sulfide(cub.) or ZnS(cub.) corresponds to zinc blende or sphalerite, and ZnS(hex.) to wurtzite. The Commission considers that these abbreviations might with advantage be standardized internationally:

cub.	=	cubic	hex.	=	hexagonal
c.	=	body-centered	trig.	=	trigonal
f.	=	face-centered	mon.	=	monoclinic
tetr.	=	tetragonal	tric.	=	triclinic
o-rh.	=	orthorhombic			

Slightly distorted lattices may be indicated by use of the *circa* sign, ~. Thus, for example, a slightly distorted face-centered cubic lattice would be expressed as ~f.cub.

8.2.—Crystallographers may find it valuable to add the space-group; it is doubtful whether this system would commend itself to chemists where **8.1** is sufficient.

8.3.—Simple well-known structures may also be designated by giving the type-compound in italics in parentheses; but this system often breaks down because many structures are not referable to a type in this way. Thus, AuCd above 70° may be written as AuCd(cub.) or as AuCd(*CsCl-type*); but at low temperature only as AuCd(o-rh.), as its structure cannot be referred to a type.

◆*Cf.* 6.5 and 6.52.

ABBREVIATIONS USED IN TABLE OF PHYSICAL CONSTANTS OF INORGANIC COMPOUNDS

a	acid
abs	absolute
ac. a	acetic acid
acet	acetone
act	active
al	alcohol
alk	alkali
amm	ammonium
amor	amorphous
anh	anhydrous
appr	approximately
aq	aqua, water
aq. reg	aqua regia
asym	asymmetrical
atm	atmospheres
bipyr	bipyramidal
bl	blue
blk	black
boil	boiling
br., brn	brown
brnsh	brownish
bz	benzene
c	cold
calc	calculated
carb	carbon
caust	caustic
chl	chloroform
choc	chocolate
cit. a	citric acid
col	colorless
coll	colloidal
com'l	commercial
comp	compounds
compl	completely
conc	concentrated
const	constant
cont	contains
corros	corrosive
cr	crystalline
cub	cubic
d., dec	decomposes
deliq	deliquescent
deriv	derivative
dibas	dibasic
di-H	dihydrogen
dil	dilute
dimorph	dimorphous
disg	disagreeable
dk	dark
doubt	doubtful
duct	ductile
effl	efflorescent
em	emerald
eth	ether
ev	evolves
evln	evolution
ex	excess
exist	existence
exp	explodes
extr	extreme(ly)
f., fr	from
feath	feathery
fl	flakes
floc	flocculent
fluo, fluores	fluorescent
form	formic
fum	fuming
fus	fused
fxd	fixed
gel., gelat	gelatinous
gl	glass
glac	glacial
glit	glittering
glob	globular
glyc	glycerin
gran	granular
greas	greasy
grn	green
h	hot
hex	hexagonal
ht	heat
hyd	hydrolyzed
hydx	hydroxides
hyg	hygroscopic
i	insoluble
ign	ignites
ind	indigo
indef	indefinite
infl., inflam	inflammable
infus	infusible
irid	iridescent
leaf	leaflets
lem	lemon
lgr	ligroin
lng	long
lq., liq	liquid
lt	light
lum	luminous
lust	lustrous
me., meth	methyl
met	metal or metallic
micr	microscopic
min	mineral
misc	miscible
mixt	mixture
mod	modifications
monbas	monobasic
mon-H	monohydrogen
monocl	monoclinic
near	nearly
need	needles
nit	nitrate
oct	octahedral
odorl	odorless
offen	offensive
olv	olive
opt	optical or optically
or	orange
ord	ordinary
org	organic
oxal	oxalate or oxalic
pa	pale
pet	petroleum
pl	plates
pois	poisonous
polymorph	polymorphous
powd	powder
ppt	precipitate
pr	prisms
press	pressure
prob	probably
prop	properties
purp	purple
pyr	pyridine
quad	quadrilateral
quest	questioned
rect	rectangular
redsh	reddish
reg	regular
rhbdr	rhombohedral
rhomb	rhombic, orthorhombic
s	soluble
satd	saturated
sld	solid
sensit	sensitive
sc	scales
sec	secondary
silv	silver
sl	slightly
sly	slowly
sm	small
sod	sodium
soln	solution
solv	solvents
spont	spontaneous
st	steel
stab	stable
subl	sublimes
suffoc	suffocating
sulfd	sulfides
sulf	sulfur
sym	symmetrical
tabl	tablets
tart. a	tartaric acid
tetr	tetragonal
tetrah	tetrahedral
tol	toluene
trac	trace, traces
trans	transparent
translu	translucent
tri., trig	trigonal
tribas	tribasic
tricl	triclinic
trim	trimetric
tr	transition point
turp	turpentine
unpleas	unpleasant
unst	unstable
v	very
vac	vacuum
var	various
viol	violent, violence
visc	viscous
vitr	vitreous
vlt	violet
volt., volat	volatizes
wh	white
wh. lt	white light
yel	yellow
yelsh	yellowish
∞	soluble in all proportions
>	above
<	below

PHYSICAL CONSTANTS OF INORGANIC COMPOUNDS

No.	Name	Synonyms and Formulae	Mol. wt.	Crystalline form, properties and index of refraction	Density or spec. gravity	Melting point, °C	Boiling point, °C	Solubility, in grams per 100 cc: Cold water	Hot water	Other solvents
a1	**Actinium**	Ac	227.0278	silv wh met, cub		1050	3200 ± 300	d to $Ac(OH)_3$		
a2	bromide	$AcBr_3$	466.74	wh, hex	5.85	subl 800		s		
a3	chloride, tri-	$AcCl_3$	333.39	wh cr, hex	4.81	subl 960				
a4	fluoride, tri-	AcF_3	284.02	wh cr, hex	7.88			i	i	
a5	hydroxide	$Ac(OH)_3$	278.05	wh				i		
a6	iodide	AcI_3	607.74	wh		subl 700—800		s		
a7	oxalate	$Ac_2(C_2O_4)_3$	718.11					i		
a8	oxide, sesqui-	Ac_2O_3	502.05	wh cr, hex	9.19			i		
a9	sulfide, sesqui-	Ac_2S_3	550.24	dk, cub	6.75					
a10	**Aluminum**	Al	26.98154	silv wh duct met, cub	2.702	660.37	2467	i	i	s alk, HCl, H_2SO_4; i conc HNO_3, h ac a
a11	acetate, tri-	$Al(C_2H_3O_2)_3$	204.12	wh solid		d		v sl s	d	
a12	acetylacetonate	$Al(C_5H_7O_2)_3$	324.31	col, monocl	1.27	193, subl	314	i	i	v s al; s eth, bz
a13	*ortho*arsenate	$AlAsO_4 \cdot 8H_2O$	310.02	wh powd	3.001	vac $-H_2O$		i	i	sl s a
a14	benzoate	$Al(C_7H_5O_2)_3$	390.33	wh cr powd				v sl s		
a15	benzyloxide	Aluminum benzylate· $Al(C_7H_7O)_3$	348.38			59—60	$283—284^{0.6}$			
a16	boride	AlB_{12}	156.70	dk red-blk, monocl	2.55_4^{18}			i		s hot HNO_3; i a, alk
a17	boride, di-	AlB_2	48.60	copper red, hex.	3.19					
a18	bromate	$Al(BrO_3)_3 \cdot 9H_2O$	572.83	wh cr, hygr		62.3	d 100	s	s	sl s a
a19	bromide	$AlBr_3$ (or Al_2Br_6)	266.69	col, rhomb pl, deliq	2.64^{10} (fused)	97.5	263.3^{747}	s with viol	d	s al, acet, CS_2
a20	bromide, hexahydrate	$AlBr_3 \cdot 6H_2O$	374.79	col-yelsh need, deliq	2.54	93	d 135	s	d	s al, amyl al; sl s CS_2
a21	bromide, pentadecylhydrate	$AlBr_3 \cdot 15H_2O$	536.92	col need		−7.5	d 7	s	s	s al
a22	butoxide, tert-	$Al(C_4H_9O)_3$	246.33	wh cr.	1.0251_0^{20}	subl 180, m.p >300, sealed tube.				v s org solv
a23	carbide	Al_4C_3	143.96	yel-grn, hex.	2.36	stab to 1400	d 2200^{400}	d to CH_4		d dil a; i acet
a24	chlorate	$Al(ClO_3)_3 \cdot 6H_2O$	385.43	col, rhbdr, deliq		d		vs	vs	s dil HCl
a25	*per*chlorate	$Al(ClO_4)_3 \cdot 6H_2O$	433.43	col, hygr	2.020	82	$-6H_2O$, 178	s	s	
a26	chloride	$AlCl_3$ (or Al_2Cl_6)	133.34	wh to col, hex, odor HCl, v deliq	fus 2.44^{25} liq 1.31^{200}	$190^{2.5}$ atm	d 262 182.7^{752} subl. 177.8	69.9^{15} with viol	s d	$100^{12.5}$ abs al; 0.072^{25} chl; sCCl_4, eth sl s bz
a27	chloride, hexahydrate	$AlCl_3 \cdot 6H_2O$	241.43	col, rhomb, deliq, 1.6	2.398	d 100		s	v s ev HCl	50 abs al; s eth; sl s HCl
a28	chloride, hexammine	$AlCl_3 \cdot 6NH_3$	235.52	col cr, hygr	1.412_4^{25}	d		s		
a29	diethylmalonate deriv	$Al(C_7H_{11}O_4)_3$	504.46		1.084^{100}	98		i		s org solv
a30	ethoxide	$Al(C_2H_5O)_3$	162.16	wh cr.	1.142_0^{20}	134	205^{14}	d	s	v sl s al, eth
a31	α-ethylacetoacetate deriv	$Al(C_6H_9O_3)_3$	414.39	wh cr.	1.101^{80}	78—79	$190—200^{11}$	s		s lgr
a32	ferrocyanide	$Al_4[Fe(CN)_6]_3 \cdot 17H_2O$	1050.05	br powd				sl s	sl s	s dil a
a33	fluoride	AlF_3	83.98	col, tricl	2.882_4^{25}	1291 $subl^{760}$		0.559^{25}	s	i a, alk, al, acet
a34	fluoride	$AlF_3 \cdot 3\ 1/2\ H_2O$	147.03	wh cr powd	1.914_4^{25}	$-2H_2O$ 100	anhydr 250	i	sl s	d ac a
a35	fluoride, monohydrate	Nat. fluellite. $AlF_3 \cdot H_2O$	101.99	col, rhomb, 1.473, 1.490, 1.511	2.17			sl s	sl s	
a36	fluosilicate	Nat topaz. $2AlFO \cdot SiO_2$	184.04	rhomb, 1.619, 1.620, 1.627	3.58			i	i	
a37	hydroxide	Nat. boehmite. AlO(OH)	59.99	wh, orthorhomb microcr.	3.01	$-H_2O$, trans to γ-Al_2O_3		i	i	s h a, h alk
a38	hydroxide	Nat. diaspore. AlO(OH)	59.99	col. rhomb cr.	3.3—3.5	$-H_2O$, trans to Al_2O_3		i	i	s h a, h alk
a39	hydroxide	$Al(OH)_2$	78.00	wh, monocl.	2.42	$-H_2O$, 300		i	i	s a, alk; i al
a40	iodide	AlI_3 (or Al_2I_6)	407.70	br pl, cont free I_2, deliq	3.98^{25}	191	360	s d	s	s al, eth, CS_2, liq NH_3
a41	iodide, hexahydrate	$AlI_3 \cdot 6H_2O$	515.79	wh-yel cr, hygr	2.63	d 185	3	v s	v s	s al, CS_2
a42	isopropoxide	$Al(C_3H_7O)_3$	204.25	wh cr.	1.0346_0^{20}	118.5	140.5^8	d		s al, bz, chl
a43	lactate	$Al(C_3H_5O_3)_3$	294.19	wh-yelsh powd				v s		
a44	nitrate	$Al(NO_3)_3 \cdot 9H_2O$	375.13	col, rhomb, deliq, 1.54		73.5	d 150	63.7^{25}	v s d	100 al; s alk, acet, HNO_3. a
a45	nitride	AlN	40.99	wh cr, hex	3.26	>2200 (in N_2)	subl 2000	d (NH_3)	d	d a, alk
a46	oleate (com'l)	$Al(C_{18}H_{33}O_2)_3$(?)	871.36	wh powd, existence doubted except as basic salt				d	s	i al; v sl s bz
a47	oxalate	$Al_2(C_2O_4)_3 \cdot 4H_2O$	390.08	wh powd				i	i	i al; s a
a48	oxide	Al_2O_3	101.96	col, hex, 1.768, 1.760	3.965^{25}	2072	2980	i		v sl s a, alk
a49	oxide	α-Alumina, nat. corundum. Al_2O_3	101.96	col, rhomb cr, 1.765	3.97	2015 ± 15	2980 ± 60	0.000098^{29}	i	v sl s a, alk
a50	oxide	γ-Alumina. Al_2O_3	101.96	wh micr cr, 1.7	3.5—3.9	tr to α		i	i	sl s a, alk
a51	oxide, monohydrate	$Al_2O_3 \cdot H_2O$	119.98	col, rhomb, 1.624 ± 0.003	3.014			i	i	
a52	oxide, trihydrate	Nat. gibbsite, hydraargilite. $Al_2O_3 \cdot 3H_2O$	156.01	wh monocl cr, 1.577, 1.577, 1.595	2.42	tr to $Al_2O_3 \cdot H_2O$ (Boehmite)		i	i	s h a, alk

No.	Name	Synonyms and Formulae	Mol. wt.	Crystalline form, properties and index of refraction	Density or spec. gravity	Melting point, °C	Boiling point, °C	Solubility, in grams per 100 cc: Cold water	Hot water	Other solvents
a54	*meta*phosphate	$Al(PO_3)_3$	263.90	col, tetr	2.779			i	i	i a
a55	palmitate, mono-(com'l)	$Al(OH)_2C_{16}H_{31}O_2$	316.42	wh	1.095	200		i		s alk, hydrocarb
a56	1-phenol-4-sulfonate	$Al(C_6H_5O_4S)_3$	546.47	redsh-wh powd.				s		s al, glyc
a57	phenoxide	$Al(C_6H_5O)_3$	306.30	grayish-wh cr mass	1.23	d 265		d		s al, eth, chl
a58	*ortho*phosphate	$AlPO_4$	121.95	wh rhomb pl, 1.546, 1.556, 1.578	2.566	>1500		i	i	s a, alk, al
a59	propoxide	$Al(C_3H_7O)_3$	204.25	wh cr.	1.0578^{20}_{0}	106	248^{14}	d	d	s al
a60	salicylate	$Al(C_7H_5O_3)_3$	438.33	redsh-wh powd.				i		i al; s alk
a61	selenide	Al_2Se_3	290.84	lt brn powd, unstable in air	3.437^{15}_{4}			d	d	d a
a62	silicate	Nat. sillimanite, andalusite, cyanite. $Al_2O_3{\cdot}SiO_2$	162.05	wh, rhomb, 1.66	3.247	1545 tr to $Al_2O_3{\cdot}2SiO_2$	>1545	i	i	d HF; i HCl; s fus alk
a63	silicate	Nat. mullite. $3Al_2O_3{\cdot}2SiO_2$	426.05	col, rhomb, 1.638, 1.642, 1.653	3.156	1920		i	i	i a, HF
a64	stearate, tri-	$Al(C_{18}H_{35}O_2)_3$	877.41	wh powd	1.010	103		i		s al, bz. turp, alk
a65	sulphate	$Al_2(SO_4)_3$	342.14	wh powd, 1.47	2.71	d 770		31.3^{0}	90.1^{100}	s dil a; sl s al
a66	sulfate, hydrate	Nat. alunogenite. $Al_2(SO_4)_3{\cdot}18H_2O$	666.41	col, monocl, 1.474, 1.467, 1.483	1.69^{17}	d 86.5		86.9^{0}	1104^{100}	i al
a67	sulfide	Al_2S_3	150.14	yel, hex, odor H_2S, d moist air	2.02^{13}	1100	subl 1500 (N_2)	d		s a; i acet
a68	thallium sulfate	Aluminum thallium alum. $AlTl(SO_4)_2{\cdot}12H_2O$	639.66	col, oct, 1.50112	2.325^{20}_{4}	91		4.84^{0}	65.19^{60}	
a69	**Americium**	Am	(243)	silvery, hex		994 ± 4	2607 (extrap)			s dil a
a70	bromide	$AmBr_3$	482.71	wh, orthorhomb		subl		s		
a71	chloride	$AmCl_3$	349.49	pink, hex	5.78	subl 850		s		
a72	fluoride	AmF_3	300.00	pink, hex	9.53			i		
a73	iodide	AmI_3	623.71	yel, orthoromb	6.9			s		
a74	oxide	Am_2O_3	534.00	redsh-brn, cub or tan, or hex.						s min a
a75	oxide, di-	AmO_2	275.00	blk, cub	11.68					s min a
a76	**Ammonia**	NH_3	17.03	col gas; liq, 0.817^{-79}, $1.325^{16.5}$	0.7710 g/ℓ 760 mm	−77.7	−33.35	89.9	7.4^{100}	13.20^{20} al; s eth, org solv
a77	Ammonia-d_3	Trideuterio ammonia. ND_3	20.05			−74	−30.9			
	Ammonium									
a78	acetate	$NH_4C_2H_3O_2$	77.08	wh cr, hygr	1.17^{20}_{4}	114	d	148^{4}	d	7.89^{15} MeOH; s al; s acet
a79	acetate, hydrogen	$(NH_4)H(C_2H_3O_2)_2$	137.14	col need, deliq.		66		s		s al
a80	aluminum chloride	$NH_4Cl{\cdot}AlCl_3$	186.83	wh cr.		304		s		
a81	aluminum sulfate	$NH_4Al(SO_4)_2$	237.14	col, hex	2.45^{20}			s		s glyc; i al
a82	aluminum sulfate, hydrate	Nat. tschernigite. $NH_4Al(SO_4)_2{\cdot}12H_2O$	453.32	col, cub, 1.459.	1.64	93.5	$-10H_2O$. 120	15^{20}	v s	s dil a; i al
a83	*ortho*arsenate	$(NH_4)_3AsO_4{\cdot}3H_2O$	247.08	rhomb cr		d, $-NH_3$		sl s		
a84	*ortho*arsenate, di-H.	$NH_4H_2AsO_4$	158.97	col, tetr, 1.577, 1.522	2.311^{9}	d, $-NH_3^{300}$		33.74^{0}	122.4^{90}	
a85	*ortho*arsenate mono-H	$(NH_4)_2HAsO_4$	176.00	col, monocl, odor NH_3	1.989	d		s	d	i al
a86	*meta*arsenite	NH_4AsO_2	124.96	col, rhomb pr, hygr.				v s	d	i al, acet; sl s NH_4OH
a87	azide	NH_4N_3	60.06	col pl.	1.346	160	subl 134 expl	20.16^{30}	27.04^{40}	1.06^{20} 80% al; i eth, bz
a88	benzene sulfonate	$NH_4C_6H_5SO_3$	175.20	rhomb	1.342	d 271—275		98	320	19 c al; i eth, bz
a89	benzoate	$NH_4C_7H_5O_2$	139.15	col, rhomb	1.260	d 198	subl 160	$19.6^{14.5}$	83.3^{100}	1.63^{25}al; s glyc; i eth
a90	*penta*borate	Aluminum decaborate. $NH_4B_5O_8{\cdot}4H_2O$	272.14					7.03^{18}		
a91	*peroxy*borate	$NH_4BO_3{\cdot}1/2H_2O$	85.85	wh cr.		d		$1.55^{17.5}$		i al
a92	*tetra*borate	Ammonium biborate. $(NH_4)_2B_4O_7{\cdot}4H_2O$	263.37	col, tetr		d		7.27^{18}	52.68^{90}	sl s acet; i al
a93	bromate	NH_4BrO_3	145.94	col, hex		expl		v s	vs	
a94	bromide	NH_4Br	97.94	cub, coll hygr, 1.712^{25}	2.429	subl 452	235 vac	97^{25}	145.6^{100}	10^{78} al; s acet, eth, NH_3
a95	*di*bromoiodide	NH_4IBr_2	304.75	met-grn pr, hygr		198		v s	s	s eth
a96	bromoplatinate	$(NH_4)_2PtBr_6$	710.58	red-brn cub	4.265^{24}	d 145		0.59^{20}	0.36^{100}	
a97	bromoselenate	$(NH_4)_2SeBr_6$	594.46	red, oct cr.	3.326			d	d	sl s eth
a98	bromostannate	$(NH_4)_2SnBr_6$	634.19	col, cub	3.50	d		v s		
a99	cadmium chloride	$4NH_4Cl{\cdot}CdCl_2$	397.28	col, rhomb, 1.6038	2.01			s		
a100	calcium arsenate	$NH_4CaAsO_4{\cdot}6H_2O$	305.13	col, monocl.	1.905^{15}	d 140		0.02	s	s NH_4Cl; i NH_4OH
a101	calcium phosphate	$NH_4CaPO_4{\cdot}7H_2O$	279.20	1.561^{15}		d		i	d	s a
a102	carbamate	$NH_4NH_2CO_2$	78.07	col, rhomb		subl 60		v s	d	v s NH_4OH; sl s al; i acet
a103	carbamate acid carbonate	Sal volatile. $NH_4NH_2CO_2{\cdot}NH_4HCO_3$	157.13	wh cr.		subl		25^{15}	67^{65}	d al; s glyc; i acet
a104	carbonate	$(NH_4)_2CO_3{\cdot}H_2O$	114.10	col, cub		d 58		100^{15}	d	i al, NH_3, CS_2; s dil MeOH
a105	carbonate, hydrogen	Ammonium bicarbonate. NH_4HCO_3	79.06	col, rhomb or monocl, 1.423, 1.536, 1.555	1.58	107.5 (d 36—60)	subl	11.9^{0}	d	i al, acet

PHYSICAL CONSTANTS OF INORGANIC COMPOUNDS (continued)

No.	Name	Synonyms and Formulae	Mol. wt.	Crystalline form, properties and index of refraction	Density or spec. gravity	Melting point, °C	Boiling point, °C	Solubility, in grams per 100 cc		
								Cold water	Hot water	Other solvents
a106	cerium nitrate(ic)	$(NH_4)_2Ce(NO_3)_6$	548.23	or, monocl				141^{25}	227^{80}	s HNO_3 al
a107	cerium nitrate(ous)	$2NH_4NO_3{\cdot}Ce(NO_3)_3{\cdot}4H_2O$	558.28	monocl.		74		318.20	817.4^{65}	
a108	cerium sulfate(ous)	$(NH_4)_2SO_4{\cdot}Ce_2(SO_4)_3{\cdot}8H_2O$	844.67	monocl.	2.523	$-6H_2O$, 100	$-8H_2O$, 150	3.29^{49} (anhydr)		
a109	chlorate	NH_4ClO_3	101.49	col, monocl need.	1.80	102 expl		28.7^{0}	115^{75}	sl s al
a110	*per*chlorate	NH_4ClO_4	117.49	col, rhomb, 1.482	1.95	d		10.74^{0}	42.45^{85}	s acet; sl s al
a111	chloride	Sal ammoniac. NH_4Cl	53.49	col, cub, 1.642.	1.527	subl 340	520	29.7^{0}	75.8^{100}	0.6^{19}al; s liq NH_3; i acet, eth
a112	chloroaurate	NH_4AuCl_4	356.82	yel, monocl or rhomb			520	s		sl s al
a113	chloroaurate, hydrate	$(NH_4AuCl_4)_4{\cdot}5H_2O$	1517.34	yel, monocl		$-5H_2O$, 100		s		s al
a114	chlorogallate	NH_4GaCl_4	229.57	wh cr.		275		v s	v s	s al; i pet eth
a115	chloroiridate	$(NH_4)_2IrCl_6$	441.01	red-blk, cub.	2.856	d		0.69^{14}	4.38^{80}	i al; s HCl
a116	chloroiridite	$(NH_4)_3IrCl_6{\cdot}1\ 1/2H_2O$	486.08					s		
a117	chloroosmate	$(NH_4)_2OsCl_6$	438.99	cub	2.93					
a118	chloropalladate	$(NH_4)_2PdCl_6$	355.21	red-brn, cub	2.418	d		sl s		
a119	chloropalladite	$(NH_4)_2PdCl_4$	284.31	olive grn, tetr.	2.17	d		s		i al
a120	*hexa*chloroplatinate	$(NH_4)_2PtCl_6$	443.87	yel, cub, 1.8	3.065	d		0.7^{15}	1.25^{100}	0.005 al; i eth, conc HCl
a121	chloroplatinite	$(NH_4)_2PtCl_4$	372.97	red, rhomb (tetr)	2.936	d 140—150		s	s	i al
a122	chloroplumbate	$(NH_4)_2PbCl_6$	455.99	yel, cub	2.925	d 120		sl s	d	s a
a123	chlorostannate	$(NH_4)_2SnCl_6$	367.48	wh, cub	2.4	d		$33.^{14.5}$	v s	
a124	*tetra*chlorozincate	$ZnCl_2{\cdot}2NH_4Cl$	243.27	wh pl, rhomb, hygr.	1.879	d 150		v s		
a125	chromate	$(NH_4)_2CrO_4$	152.07	yel, monocl.	1.91^{12}	d 180		40.5^{30}	d	i al; sl s NH_3, acet
a126	*di*chromate	$(NH_4)_2Cr_2O_7$	252.06	or, monocl	2.15^{25}	d 170		30.8^{15}	89^{30}	s al; i acet
a127	*peroxy*chromate	$(NH_4)_3CrO_8$	234.11	red-brn, cub		d 40	expl 50	sl s	d	i al, eth, sl s NH_3; expl H_2SO_4
a129	chromium sulfate(ic)	$(NH_4)Cr(SO_4)_2{\cdot}12H_2O$	478.33	grn or vlt, cub; 1.4842	1.72	94, $-9H_2O$, 100		21.2^{25}	32.8^{40}	s al, dil a
a130	citrate, di(*sec.*)	$(NH_4)_2HC_6H_5O_7$	226.19	wh gran or powd	1.48			100		sl s al
a130	citrate, di(*sec.*)	$(NH_4)_2HC_6H_5O_7$	226.19	wh gran or powd	1.48			100		sl s al
a131	citrate, tri-(*tert.*	$(NH_4)_3C_6H_5O_7$	243.22	wh cr, deliq.		d		v s		i al, eth, acet
a132	cobalt *ortho*phosphate(ous)	$(NH_4)CoPO_4{\cdot}H_2O$	189.96	vlt cr powd				i	d	s a
a133	cobalt (II)-selenate	$(NH_4)_2SO_4{\cdot}CoSeSO_4{\cdot}6H_2O$	482.02	ruby-red, monocl, 1.526, 1.532, 1.541	2.228_4^{20}					
a134	cobalt sulfate(ous)	$(NH_4)_2SO_4{\cdot}CoSO_4{\cdot}6H_2O$	395.22	ruby-red, monocl, 1.490, 1,495, 1.503	1.902			20.5^{20}	45.4^{80}	i al
a135	copper chloride(ic)	$2NH_4Cl{\cdot}CuCl_2{\cdot}2H_2O$	277.47	blue, tetrag, 1.744, 1.724	1.993	d 110		33.8^{0}	99.3^{80}	s a, al; sl s NH_3
a136	copper iodide(ous)	$NH_4I{\cdot}CuI{\cdot}H_2O$	353.41	rhomb pl				d	d	s NH_4I
a137	cyanate	NH_4OCN	60.06	wh cr.	1.342_4^{20}	d 60		v s	d	sl s al; i eth
a138	cyanide	NH_4CN	44.06	col, cub	1.02^{100}	d 36	subl 40	v s	d	v s al
a139	cyanaurate	$NH_4Au(CN)_4{\cdot}H_2O$	337.09	col pl.		d 200		v s		v s al; i eth
a140	cyanaurite	$NH_4Au(CN)_2$	267.04	col, cub		d 100		v s	v s	s al: i eth
a141	cyanoplatinite	$(NH_4)_2Pt(CN)_4{\cdot}H_2O$	353.24	yel cr.				s		
a142	ethylsulfate	$NH_4C_2H_5SO_4$	143.16			99		s		
a143	ferricyanide	$(NH_4)_3Fe(CN)_6$	266.07	red cr, rhomb.		d		v s		
a144	ferrocyanide	$(NH_4)_4Fe(CN)_6{\cdot}3H_2O$	338.15	yel, monocl, turns bl in air		d		s	d	i al
a145	fluoantimonite	$(NH_4)_2SbF_5$	252.82	col, rhomb		d, subl		108		
a146	fluoborate	NH_4BF_4	104.84	wh, rhomb	1.871^{15}	subl		25^{16}	97^{100}	s NH_4OH
a147	fluogallate	$(NH_4)_3GaF_6$	237.83	wh oct cr		d >250 – GaF_3			sl s	
a148	fluogermanate	$(NH_4)_2GeF_6$	222.66	col hex pr and bi-pyram, 1.428, 1.425	2.564_{25}^{25}			s		i al, MeOH
a149	fluophosphate, di-	$NH_4PO_2F_2$	119.01	col, rhomb		213		s	s	s al, acet
a150	fluophosphate, hexa	NH_4PF_6	163.00	col, pl	2.180^{18}	d		s	s	s al, acet
a151	fluoride	NH_4F	37.04	col, hex, deliq	1.009^{25}	subl		100^{0}	d	s al; i NH_3
a152	fluoride, hydrogen	NH_4HF_2	57.04	rhomb or tetr, de-liq, 1.390	1.50	125.6		v s	v s	sl s al
a153	fluorosulfonate	NH_4FSO_3	117.10	wh need		d 245		v s	v s	sl s al; s MeOH
a154	fluosilicate	α-Nat. cryptohalite. $(NH_4)_2SiF_6$	178.15	α oct, β hex, col α 1.3696	α 2.011 β 2.152	d		18.16^{17}	55.5^{100}	sl s al; i acet
a155	fluosulfonate	NH_4SO_3F	117.10	col, need		244.7		s	s	sl s al; s MeOH; d NH_4OH
a156	fluotitanate	$(NH_4)_2TiF_6$	197.95	hex pr		d		s	s	i al, eth
a157	fluozirconate	$(NH_4)_2ZrF_6$	241.29	rhomb, hex	1.154					
a158	fluozirconate	$(NH_4)_3ZrF_7$	278.32	col, cub	1.433			sl s		
a159	formate	NH_4CHO_2	63.06	wh, monocl, deliq	1.280	116	d 180	102^{0}	531^{80}	s al, eth, NH_3
a160	gallium sulfate	Ammonium gallium alum. $Ga(NH_4)(SO_4)_2{\cdot}12H_2O$	496.06	cub, 1.468	1.777			30.84^{25}		0.00875^{25} 70% EtOH
a161	*tri*hydrogen para-*per*iodate	$(NH_4)_2H_3IO_6$	262.00	col, rhomb	2.85					
a162	hydroxide	NH_4OH	35.05	at ord temp in sol only		−77		s		
a163	iodate	NH_4IO_3	192.94	col, rhomb or monocl	3.309^{21}	d 150		2.06^{15}	14.5^{101}	

PHYSICAL CONSTANTS OF INORGANIC COMPOUNDS (continued)

No.	Name	Synonyms and Formulae	Mol. wt.	Crystalline form, properties and index of refraction	Density or spec. gravity	Melting point, °C	Boiling point, °C	Solubility, in grams per 100 cc		
								Cold water	Hot water	Other solvents
a164	iodide	NH_4I	144.94	col, cub, hygr, 1.7031	2.514^{25}	subl 551	220 vac	154.2^{0}	250.3^{100}	v s al, acet, NH_3; sl s eth
a165	*tri*iodide	NH_4I_3	398.75	dk br, rhomb	3.749	d 175		s d	d	
a166	iodoplatinate	$(NH_4)_2PtI_6$	992.58	blk, cub	4.61					
a167	iridium chloride (III)	$(NH_4)_3IrCl_6{\cdot}H_2O$	477.07	grn-br-gl, cub		d 350		10.5^{19}		i al
a168	iridium sulfate	$NH_4Ir(SO_4)_2{\cdot}12H_2O$	618.56	yel-red cr			106	s		
a169	iron (III) chloride	$2NH_4Cl{\cdot}FeCl_3{\cdot}H_2O$	287.20	ruby-red, rhomb, hygr, 1.78	1.99	234		v s	v s	
a170	iron (III) fluoride	$3NH_4F{\cdot}FeF_3$	223.95	col to lt yel, oct	1.96			sl s	sl s	
a171	iron (II) selenate	$(NH_4)_2SeO_4{\cdot}FeSeO_4{\cdot}6H_2O$	485.93	lt grn, monocl, 1.5226, 1.5260, 1.5334	2.191^{20}_{4}					
a172	iron (III) sulfate	$(NH_4)_2SO_4{\cdot}Fe_2(SO_4)_3$	532.00	wh, hex	2.49^{22}	d 420		44.15^{25}		
a173	iron sulfate(ic)	$NH_4Fe(SO_4)_2{\cdot}12H_2O$	482.18	vlt, cub oct, 1.4854	1.71	39—41	$-12H_2O$, 230	124.0^{25}	400^{100}	i al; s dil a
a174	iron sulfate(ous)	$(NH_4)_2SO_4{\cdot}FeSO_4{\cdot}6H_2O$	392.13	grn, monocl, 1.487, 1.492, 1.499	1.864^{20}_{4}	d 100—110		26.9^{20}	73.0^{80}	i al
a175	lactate	$NH_4C_3H_5O_3$	107.11	col-yelsh liq	$1.19—1.21^{15}$			∞		∞ al
a176	laurate, acid (mixt.)	$NH_4C_{12}H_{23}O_2{\cdot}C_{12}H_{24}O_2$	417.67	wh		75	d	s	s	4.8^{7} al; sl s eth, acet
a177	magnesium arsenate	$NH_4MgAsO_4{\cdot}6H_2O$	289.35	col, tetrg, 1.608	1.932^{15}	d		0.038^{20}	0.024^{80}	s a; i al
a178	magnesium carbonate	$(NH_4)_2CO_3{\cdot}MgCO_3{\cdot}4H_2O$	252.46	wh				s	v s	s a; i al
a179	magnesium chloride	$NH_4Cl{\cdot}MgCl_2{\cdot}6H_2O$	256.80	col, rhomb, deliq	1.456	$-2H_2O$, 100	d	16.7		d al
a180	magnesium chromate	$(NH_4)_2CrO_4{\cdot}MgCrO_4{\cdot}6H_2O$	400.46	yel, monocl, 1.636, 1.637, 1.653	1.84	d		v s	v s	
a181	magnesium phosphate	Guanite, struvite $NH_4MgPO_4{\cdot}6H_2O$	245.41	col, rhomb, 1.495, 1.496, 1.504	1.711	d		0.0231^{0}	0.0195^{80}	v s dil a; i al
a182	magnesium selenate	$(NH_4)_2SeO_4{\cdot}MgSeO_4{\cdot}6H_2O$	454.39	col monocl pr, 1.507, 1.509, 1.517	2.058^{20}_{4}			sl s		
a183	magnesium sulfate	Nat. boussingaulite. $(NH_4)_2SO_4{\cdot}MgSO_4{\cdot}6H_2O$	360.59	col, monocl, 1.472, 1.473, 1.479	1.723	>120	d 250	17.92^{0}	64.7^{100} (anhydr)	
a184	*l*-malate, hydrogen	$NH_4HC_4H_4O_5$	151.12	col, rhomb	1.5	161	d	$32.2^{15.7}$		
a185	*per*manganate	NH_4MnO_4	136.97	rhomb	2.208^{10}	d 110		7.9^{15}	d	
a186	manganese phosphate(ic)	$NH_4MnPO_4{\cdot}H_2O$	185.96	wh cr.				0.0031	0.05	i al, NH_4 salts
a187	manganese sulfate(ous)	$(NH_4)_2SO_4{\cdot}MnSO_4{\cdot}6H_2O$	391.22	pa red, monocl, 1.480, 1.484, 1.491	1.83			51.3^{25}	v s	
a188	molybdate	$(NH_4)_2MoO_4$	196.01	col, monocl pr	2.276^{25}_{4}	d		s, d	d	s a; i al, NH_3 acet
a189	*para*molybdate	"Molybdic acid" com'l. $(NH_4)_6Mo_7O_{24}{\cdot}4H_2O$	1235.86	col-yelsh, monocl	2.498	$-H_2O$, 90	d 190	43	d	i al; s a, alk
a190	*per*molybdate	$3(NH_4)_2{\cdot}1/2MoO_3{\cdot}2MoO_4{\cdot}6H_2O$	608.17	lt yel monocl pr	2.975	d 170		v s	v s	sl s al
a191	molybdotellurate	$(NH_4)_6TeMo_6O_{24}{\cdot}7H_2O$	1321.56	col,, rhomb	2.78	d 550	d	s	s	
a192	myristate, acid (mixt.)	$NH_4C_{14}H_{27}O_2{\cdot}C_{14}H_{28}O_2$	473.78	wh solid		75—90	d	sl s	s	s al; i eth
a193	nickel chloride	$NH_4Cl{\cdot}NiCl_2{\cdot}6H_2O$	291.18	grn, monocl, deliq	1.654			v s	v s	
a194	nickel sulfate	Double nickel salt. $(NH_4)_2SO_4{\cdot}NiSO_4{\cdot}6H_2O$	394.97	dk bl-grn, monocl, 1.495, 1.501, 1.508	1.923			10.4^{20}	30^{80}	i al; s $(NH_4)_2SO_4$
a195	nitrate	NH_4NO_3	80.04	col, rhomb, (monocl >32.1^{0})	1.725^{25}	169.6	210^{11}	118.3^{0}	871^{100}	3.8^{20} al; 17.1^{20} MeOH s acet, NH_3; i eth
a196	nitratocerate	$(NH_4)_2Ce(NO_3)_6$	548.23	yel-red, monocl				142.6^{25}	232^{90}	s al; sl s HNO_3
a197	nitrite	NH_4NO_2	64.04	wh-yelsh cr	1.69	60—70 expl	30 subl vac	v s	d	s dil al; i eth
a198	oleate, acid (mixt.)	$NH_4C_{18}H_{33}O_2{\cdot}C_{18}H_{34}O_2$	581.96	wh powd		d 78		s	s	80^{50} al; 13.3^{15} eth
a199	osmium chloride	$(NH_4)_2OsCl_6$	438.99	blk, oct	2.93^{25}	subl 170		d	d	s HCl
a200	oxalate	$(NH_4)_2C_2O_4{\cdot}H_2O$	142.11	col, rhomb, 1.439, 1.546, 1.594	1.50	d		2.54^{0}	11.8^{50}	i NH_3
a201	oxalate, acid	Ammonium binoxalate. $NH_4HC_2O_4{\cdot}H_2O$	125.08	col, rhomb	1.556	$-H_2O$, 170		s		i eth, bz
a202	oxalatoferrate (III)	$(NH_4)_3Fe(C_2O_4)_3{\cdot}3H_2O$	428.07	grn, monocl.	1.78	d 165		42.7^{0}	345^{100}	
a203	palladium (II)-chloride	$(NH_4)_2PdCl_4$	284.31	grnsh-yel, tetr	2.17	d		v s	v s	s dil al; i abs al
a204	palladium (IV)-chloride	$(NH_4)_2PdCl_6$	355.21	red oct cr	2.418	d −Cl		sl s	sl s	
a205	palmitate, (acid)- (mixt.)	$NH_4C_{16}H_{31}O_2{\cdot}C_{16}H_{32}O_2$	529.89	yelsh soapy mass or yel powd		>100	d	sl s	s	8.8^{50} al; 0.23^{13} eth
a206	*meta*per*iodate	NH_4IO_4	208.94	col, tetr	3.056^{18}	expl		2.7^{16}		
a207	*hypo*phosphate	$(NH_4)_2H_2P_2O_6$	196.04	col cr.		170		7^{25}	25^{100}	
a208	*ortho*phosphate	$(NH_4)_3PO_4{\cdot}3H_2O$	203.13	wh pr				26.1^{25}		sl s dil NH_4OH; i NH_3, acet
a209	*ortho*phosphate, di-H	$NH_4H_2PO_4$	115.03	col, tetr, 1.525, 1.479	1.803^{19}	190		22.7^{0}	173.2^{100}	i acet
a210	*ortho*phosphate-mono-H	$(NH_4)_2HPO_4$	132.06	col, monocl, 1.52	1.619	d 155	d	57.5^{10}	106.70^{70}	i al, acet, NH_3
a211	*hypo*phosphite	$NH_4H_2PO_2$	83.03	rhomb tabl	1.634	200	d 240	s	s	s al, NH_3; i acet
a212	*ortho*phosphite, di-H	$NH_4H_2PO_3$	99.03	col, monocl pr		123	d 145	171^{0}	260^{31}	i al
a213	phosphofluoride, hexa-	NH_4PF_6	163.00	col, cub	2.180^{18}_{4}	d		74.8^{20}		s al, acet; d h a
a214	phosphomolybdate	Ammonium molybdo phosphate. $(NH_4)_3P(Mo_3O_{10})_4$	1876.35	yel powd		d		sl s	sl s	s alk; i al, HNO_3

No.	Name	Synonyms and Formulae	Mol. wt.	Crystalline form, properties and index of refraction	Density or spec. gravity	Melting point, °C	Boiling point, °C	Solubility, in grams per 100 cc Cold water	Hot water	Other solvents
a215	phosphotungstate	$(NH_4)_3P(W_3O_{10})_4$	2931.27	wh				sl s	sl s	
a216	picramate	$NH_4C_6H_4N_3O_5$	216.15	redsh-brn cr powd				s		
a217	picrate	$NH_4C_6H_2N_3O_7$	246.14	red or yel, rhomb	1.719	d	expl 423	1.1^{20}	s	sl s al
a219	praseodymium sulfate	$(NH_4)_2SO_4{\cdot}Pr_2(SO_4)_3{\cdot}8H_2O$	846.24	cr	$2.531^{16.5}$	$-8H_2O$, 170		sl s		
a220	propionate	$NH_4C_3H_5O_2$	91.11	pr, deliq	1.108^{25}	45		v s		s al, ac a
a221	*per*rhenate	NH_4ReO_4	268.24	wh, hex pl	3.97	d	d	6.1^{20}	32.34^{80}	
a222	rhodanid	NH_4NCS	76.12	monocl cr, to rhomb at 92, 1.685	1.305	149.6	d 170	128^{0}	347^{60}	v s al; s MeOH, acet; i $CHCl_3$
a223	rhodium chloride	$(NH_4)_3RhCl_6{\cdot}H_2O$	387.75	dk red rhomb need		$-H_2O$, 140		s	s	sl s al; s dil NH_4Cl
a224	rhodium sulfate	Ammonium rhodium alum. $Rh(NH_4)(SO_4)_2{\cdot}12H_2O$	529.24	orange, 1.5150		102				
a225	*d*-saccharate, hydrogen	$NH_4HC_6H_8O_8$	227.17	wh need or monocl pr				1.22^{15}	24.35^{100}	i c al; s h al
a226	salicylate	$NH_4C_7H_5O_3$	155.15	col, monocl			subl	111^{25}	v s	28.8^{25} al
a227	selenate	$(NH_4)_2SeO_4$	179.03	col, monocl, 1.561, 1.563, 1.585	2.194_4^{20}	d		117^{7}	197^{100}	i al, acet, NH_3
a228	selenate, hydrogen	NH_4HSeO_4	162.00	rhomb	2.162	d				
a229	selenide	$(NH_4)_2Se$	115.04	col, or wh cr		d		s		
a230	sodium phosphate, hydrate	Microsmic salt. $NH_4NaHPO_4{\cdot}4H_2O$	209.07	col, monocl, 1.439, 1.442, 1.469	1.574	d 97				
a231	stearate, acid (mixt.)	$NH_4(C_{18}H_{35}O_2{\cdot}C_{18}H_{36}O_2$	586.00	wh cr.		d 110		v s		0.3^{25} al; 0.19^{25}eth; 0.08^{25} acet
a232	succinate	$(NH_4)_2{\cdot}C_4H_4O_4$	152.15	col cr.	1.37			s		sl s al
a233	sulfamate	$NH_4NH_2SO_3$	114.12	large, pl, deliq		125	d 160	166.6^{10}	357^{50}	i al
a234	sulfate	Nat. mascagnite. $(NH_4)_2SO_4$	132.13	col, rhomb, 1.521, 1.523, 1.533	1.769^{50}	d 235		70.6^{0}	103.8^{100}	i al, acet, NH_3
a235	sulfate, hydrogen	Ammonium bisulfate. NH_4HSO_4	115.10	col, rhomb, 1.473	1.78	146.9	d	100	v s	sl s al; i acet
a236	*peroxy*disulfate	$(NH_4)_2S_2O_8$	228.19	col, monocl, 1.498, 1.502, 1.587	1.982	d 120		58.2^{0}	v s	
a237	sulfide, hydro-	NH_4HS	51.11	wh rhomb, 1.74	1.17	118^{19atm}	88.4^{19atm}	128.1^{0}	d	s al; NH_3
a238	sulfide, mono-	$(NH_4)_2S$	68.14	col yel cr (> −18), hygr.		d		v s	d	s al; v s NH_3
a239	sulfide, penta-	$(NH_4)_2S_5$	196.39	yel pr		d 115		v s		s al, i eth, CS_2
a240	sulfite	$(NH_4)_2SO_3{\cdot}H_2O$	134.15	col, monocl, 1.515	1.41^{25}	d 60—70	subl 150	32.4^{0}	60.4^{100}	sl s al; i acet
a241	sulfite, hydrogen	Ammonium bisulfite. NH_4HSO_3	99.10	rhomb pr, deliq	2.03	subl 150 (in N_2)		71.8^{0}	84.7^{60}	
a242	*dl*-tartrate	$(NH_4)_2C_4H_4O_6$	184.15	col, monocl, d, α 1.55, β 1.581	1.601	d		58.01^{15}	81.17^{60}	sl s al
a243	*dl*-tartrate, hydrogen	$NH_4HC_4H_4O_6$	167.12	col, monocl pr, 1.561, 1.591	1.636	d		2.35^{15}	3.24^{25}	i al, s a, alk
a244	tellurate	$(NH_4)_2TeO_4$	227.67	wh powd	$3.024^{24.5}$	d		s	s	i al; s dil a
a245	thallium chloride	$3NH_4Cl{\cdot}TlCl_3{\cdot}2H_2O$	507.25	col	2.39			s		
a246	thioantimonate	$(NH_4)_3SbS_4{\cdot}4H_2O$	376.17	yel pr		d		71.2^{0}	d	i al, eth
a247	thiocarbamate	$NH_4CS_2NH_2$	110.19	yel cr.		d 50		v s		s al; sl s eth
a248	thiocarbonate, tri-	$(NH_4)_2CS_3$	144.27	yel cr, hygr		subl		v s	d	sl s al, eth
a249	thiocyanate	NH_4SCN	76.12	col, monocl, deliq	1.305	149.6	d 170	128^{0}	v s	s al, acet, NH_3
a250	*di*thionate	$(NH_4)_2S_2O_6{\cdot}1/2H_2O$	205.20	col, monocl	1.704	d 130		135^{0}	v s	i al
a251	thiosulfate	$(NH_4)_2S_2O_3$	148.20	col, monocl, hygr	1.679	d 150		v s	103.3^{100}	i al; sl s acet
a252	titanium oxalate, basic	$(NH_4)_2TiO(C_2O_4)_2{\cdot}H_2O$	294.01					v s		
a253	uranylcarbonate	$2(NH_4)_2CO_3{\cdot}UO_2CO_3{\cdot}2H_2O$	558.24	yel, monocl	2.773	d 100		5.8^{19}	d	$s(NH_4)_2CO_3$, $aq{\cdot}SO_2$
a254	uranylfluoride, penta-	$(NH_4)_3UO_2F_5$	419.14	tetr cr, 1.495	3.186	subl		s		
a255	valerate	$NH_4C_5H_9O_2$	119.16	col or wh cr		d		s		s al, eth
a256	*meta*vanadate	NH_4VO_3	116.98	wh-yelsh or col cr	2.326	d 200		0.52^{15}	6.95^{96}, d	
a257	vanadium sulfate	$NH_4V(SO_4)_2{\cdot}12H_2O$	477.28	red to bl	1.687	49		28.45^{20}		
a258	zinc sulfate	$(NH_4)_2SO_4{\cdot}ZnSO_4{\cdot}6H_2O$	401.66	wh monocl, 1.489, 1.493, 1.499	1.931	d		7^{0} (anhydr)	42^{80} (anhydr)	
a259	**Antimony**	Sb	121.75 ± 3	silv wh met, hex	6.684^{25}	630.5	1750	i	i	s hot conc H_2SO_4, aq reg
a260	bromide, tri-	$SbBr_3$	361.46	col rhomb, 1.74	4.148^{23}	96.6	280	d	d	s HCl, HBr, CS_2, NH_3, al, acet
a261	chloride, penta-	$SbCl_5$	299.02	wh liq or monocl, 1.601^{14}	liq 2.336_4^{20}	2.8	79^{22}	d	d	s HCl, tarta, $CHCl_3$
a262	chloride, tri-	Butter of antimony. $SBCl_3$	228.11	col, rhomb, deliq	3.140^{25}	73.4	283	601.6^{0}	∞^{80}	s abs al, HCl, tart a, $CHCl_3$, bz, acet
a263	fluoride, penta-	SbF_5	216.74	col oily liq	2.99^{23} liq	7	149.5	s		s KF
a264	fluoride, tri-	SbF_3	178.75	col, rhomb	$4.379^{20.9}$	292	subl 319	384.7^{0}	563.6^{20}	i NH_3
a265	hydride	SbH_3	124.77	inflamm gas	gas 4.36^{15}; liq 2.26^{-25}	−88	-17.1^{751}	0.41^{0}		1500 mℓ A; 2500 mℓ CS_2
a266	iodide, penta-	SbI_5	756.27	br		79	400.6			
a267	iodide, tri-	SbI_3	502.46	ruby-red, hex, 2.78_{Li}, 2.36_{Li}	4.917^{17}	170	401	d	d	i al; s CS_2, bz, HI, HCl
a268	iodosulfide	SbSI	280.71	dk red		392	d	i	i	d conc HCl; i CS_2

PHYSICAL CONSTANTS OF INORGANIC COMPOUNDS (continued)

No.	Name	Synonyms and Formulae	Mol. wt.	Crystalline form, properties and index of refraction	Density or spec. gravity	Melting point, °C	Boiling point, °C	Solubility, in grams per 100 cc: Cold water	Hot water	Other solvents
a269	mercaptoacetamide	Antimony thioglycolamide. $Sb(C_2H_4NOS)_3$	392.11	wh cr.		139		200		
a270	nitrate, basic	$2Sb_2O_3{\cdot}N_2O_5$(?)	691.01	wh glossy cr		d		d		d a; v sl s conc HNO_3
a271	nitride	SbN	135.76	or powd		d		d		
a272	oxide, penta-	Sb_2O_5 (or Sb_4O_{10})	323.50	yel powd	3.80 (dep on temp)	$-O$, 380 $-O_2$, 930		v sl s	v sl s	v sl s KOH, HCl, HI
a273	oxide, tetra-	Nat. cervantite. Sb_2O_4 (or $Sb_2O_3{\cdot}Sb_2O_5$)	307.50	wh powd, 2.00	5.82	$-O$, 930		v sl s	v sl s	v sl s KOH, HCl, HI
a274	oxide, tri-	Nat. senarmontite. Sb_2O_3 (or Sb_4O_6)	291.50	wh, cub, 2.087	5.2	656	1550 subl	v sl s	sl s	s KOH, HCl, tart a, ac a
a275	oxide, tri-	Nat. valentinite. Sb_2O_3 (or Sb_4O_6)	291.50	col, rhomb, 2.18, 2.35, 2.35	5.67	656	1550	v sl s	sl s	s KOH, HCl, tart, a ac a
a276	III oxychloride(ous)	SbOCl	173.20	wh monocl		d 170		i	d	s acet, HCl, CS_2; i al, eth, $CHCl_3$
a277	III oxychloride(ous)	$Sb_4O_5Cl_2$	637.90	monocl.	5.01	d 320		i		s HCl; i al, eth
a278	oxyhydrate	$H_4Sb_2O_7$	350.53	wh, amorph.		$-H_2O$, 200		sl s	sl s	s alk
a279	III oxysulfate, di-(ous)	$Sb_2O_2SO_4$	371.56	wh	4.89			d	d	s, H_2SO_4
a280	potassium tartrate	Tartar emetic. $K(SbO)C_4H_4O_6{\cdot}1/2H_2O$	333.93	col cr.	2.6	$-1/2H_2O$, 100		8.3	33.3	i al; 6.7 glyc
a281	selenide	Sb_2Se_3	480.38	gray cr.		611		v sl s		s conc HCl
a282	III sulfate	$Sb_2(SO_4)_3$	531.67	wh powd, deliq	3.625^{4}	d		i	d	s a
a283	sulfide, penta-	Sb_2S_5	403.80	yel powd, prism	4.120	d 75		i	i	i al; s HCl, alk, NH_4HS
a284	sulfide, tri-	Nat. stibnite. Sb_2S_3	339.68	blk, rhomb, 3.194, 4.064, 4.303	4.64	550	*ca* 1150	0.000175^{18}		s al, NH_4HS. K_2S, HCl; i ac a
a285	sulfide, tri-	Sb_2S_3	339.68	yel-red, amorph	4.12	550	*ca* 1150	0.000175^{18}		s al, NH_4SH, K_2S, HCl; i ac a
a286	*d*-tartrate	$Sb_2(C_4H_4O_6)_3{\cdot}6H_2O$	795.81	wh cr powd				s		
a287	telluride, tri-	Sb_2Te_3	626.30	gray	6.50^{13}	629				s HNO_3, aq reg
a288	**Argon**	Ar	39.948	col inert gas,	1.784^{0} g/ℓ cr: 1.65^{233}, liq: 1.40^{-186}	-189.2	-185.7	5.6^{0} cm³	3.01^{50} cm³	
a289	**Arsenic**	As	74.9216	gray met, hex-rhomb	5.727^{14}	817 (28 atm.)	613 subl	i	i	s HNO_3
a290	**Arsenic**	As_4	299.69	yel, cub	2.026^{18}	d 358		i	i	s CS_2, bz
a291	**Arsenic acid,** meta-	$(HO)AsO_2$	123.93	wh, hygr		d	forms ortho-arsenic acid	d		
a292	ortho-	$H_3AsO_4{\cdot}1/2H_2O$	150.95	wh translu hygr cr	2.0—2.5	35.5	$-H_2O$ 160	$302^{12.5}$	50	s al, alk, glyc
a293	pyro-	$H_4As_2O_7$	265.87	col pr		forms orthoarsenic acid 206 d				
a294	bromide, tri-	$AsBr_3$	314.63	col-yel hygr pr	3.54^{25}	32.8	221	d	d	s HCl, HBr, CS_2
a295	chloride, tri-	$AsCl_3$	181.28	oily liq or need, 1.621_F^{14}	liq 2.163^{20}	-8.5	130.2	d	d	s HBr, HCl, PCl_3, al, eth
a296	fluoride, penta-	AsF_5	169.91	col gas	7.71 g/ℓ	-80	-53	s		s alk, al, eth, bz
a297	fluoride, tri	AsF_3	131.92	oily liq.	liq 2.666	-8.5	-63^{752}	d	d	s al, eth, bz, NH_4OH
a298	hydride	Arsine. AsH_3	77.95	col gas	gas 2.695 liq $1.689^{84.9}$	-116.3	-55 (d 300)	20 mℓ		s $CHCl_3$, bz
a299	iodide, di-	AsI_2	328.73	red pr		d 136		d		s al, eth, $CHCl_3$, bz, CS_2
a300	iodide, penta-	AsI_5	709.44		3.93	76				
a301	iodide, tri-	AsI_3	455.64	red hex, *ca*. 2.59, *ca*. 2.23	4.39^{13}	146	403	6^{25}	30 d	5.2 CS_2; s al, eth, bz, $CHCl_3$
a302	oxide, pent-	As_2O_5	229.84	wh amor, deliq.	4.32	d 315		150^{16}	76.7^{100}	s al, a, alk
a303	oxide, tri-	As_2O_3	197.84	amor or vitreous	3.738	312.3		3.7^{20}	10.14^{100}	s alk, alk carb, HCl
a304	oxide, tri-	Nat. arsenolite. As_2O_3 (or As_4O_6)	197.84	col, cub or fibr, 1.755	3.865^{25}	subl 193		1.2^{2}	11.46^{100}	s al, alk, HCl
a305	oxide, tri-	Nat. claudetite. As_2O_3 (or As_4O_6)	197.84	col, monocl, 1.871, 1.92, 2.01	4.15	193, subl 312.3	457.2	1.2^{2}	11.46^{100}	s al, alk, HCl
a306	(III)oxychloride(ous)	AsOCl	126.37	brnsh		d		d	d	
a307	phosphide, mono-	AsP	105.90	br-red powd.		subl, d		d	d	sl s CS_2; s H_2SO_4, HCl; i al, eth, $CHCl_3$
a308	selenide	As_2Se_3	386.72	br cr	4.75	*ca*. 360		i	d	s alk
a309	sulfide, di-	Nat. realgar. As_2S_2	213.96	red-br monocl, 2.46, 2.59, 2.61	α 3.506^{19} β 3.254^{19}	α tr 267 β 307	565	i	i	s K_2S, $NaHCO_3$
a310	sulfide, penta-	As_2S_5	310.14	yel		subl, d 500		0.000136^{0}	i	s alk, alk sulf, HNO_3
a311	sulfide, tri-	Nat. orpiment. As_2S_3	246.03	yel or red, monocl, 2.4, 2.81, 3.02 (Li)	3.43	300	707	0.00005^{18}	sl s	s al, alk, alk carb
	Auric or **Aurous**	*See* **Gold**								
b1	**Barium**	Ba	137.33	yel-silv met	3.51^{20}	725	1640	d, ev H_2	d	s al; i bz
b2	acetate	$Ba(C_2H_3O_2)_2$	255.42	col cr.	2.468			58.8^{0}	75^{100}	
b3	acetate, hydrate	$Ba(C_2H_3O_2)_2{\cdot}H_2O$	273.43	col tricl, 1.500, 1.517, 1.525	2.19	$-H_2O$, 150		76.4^{26}	75^{100}	sl s al
b4	amide	$Ba(NH_2)_2$	169.38	gray-wh cr		280		d	d	i liq NH_3
b5	*ortho*arsenate	$Ba_3(AsO_4)_2$	689.83		5.10	1605		0.055		s a, NH_4Cl
b6	*ortho*arsenate, mono-H	$BaHAsO_4{\cdot}H_2O$	295.27	col, rhomb or monocl, 1.635	1.93^{15}	$-H_2O$, 150		sl s	d	s HCl

PHYSICAL CONSTANTS OF INORGANIC COMPOUNDS (continued)

No.	Name	Synonyms and Formulae	Mol. wt.	Crystalline form, properties and index of refraction	Density or spec. gravity	Melting point, °C	Boiling point, °C	Solubility, in grams per 100 cc		
								Cold water	Hot water	Other solvents
b7	arsenide	Ba_3As_2	561.83	br	4.1^{15}			d		d Cl_2, F_2, Br_2
b8	azide	$Ba(N_3)_2$	221.37	monocl pr	2.936	$-N_2$, 120	expl	17.3^{17}		abs al 0.017^{16}; i eth
b9	azide, hydrate	$Ba(N_3)_2 \cdot H_2O$	239.39	tricl, 1.7		expl		v s	v s	sl s al; i eth
b10	benzoate	$Ba(C_7H_5O_2)_2 \cdot 2H_2O$	415.59	col, nacreous leaf		$-2H_2O$, 100		s		sl s al
b11	boride, hexa-	BaB_6	202.19	met-blk, cub	4.36^{16}	2270		i	i	s HNO_3; i HCl
b12	bromate	$Ba(BrO_3)_2 \cdot H_2O$	411.15	col, monocl.	3.99^{18}	d 260		0.3^{0}	5.67^{100}	i al; s acet
b13	bromide	$BaBr_2$	297.14	col cr, 1.75	4.781^{24}	847	d	104.1^{20}	149^{100}	v s al, MeOH
b14	bromide, dihydrate	$BaBr_2 \cdot 2H_2O$	333.17	col, monocl, 1.713, 1.727, 1.744	3.58^{24}	880, ($-H_2O$, 75)	$-2H_2O$, 120	151^{20}	204^{100}	v s MeOH, s al
b15	bromofluoride	$BaBr_2 \cdot BaF_2$	472.46	col pl.	4.96^{18}			d	d	i al; s conc HCl, conc HNO_3
b16	bromoplatinate	$BaPtBr_6 \cdot 10H_2O$	991.99	monocl.	3.71					
b17	butyrate	$Ba(C_4H_7O_2)_2 \cdot 2H_2O$	347.56	col				37.42^{0}	42.12^{80}	
b18	carbide	BaC_2	161.35	gray, tetr	3.75			d to C_2H_2		d a
b19	carbonate(α)	$BaCO_3$	197.34	wh, hex	4.43	$1740^{90}{}_{atm}$	d	0.002^{20}	0.006^{100}	s a, NH_4Cl; i al
b20	carbonate(β)	$BaCO_3$	197.34			tr to α, 982		0.0022^{18}	0.0065^{100}	s a, NH_4Cl; i al
b21	carbonate(γ)	Nat. witherite. $BaCO_3$	197.34	wh rhomb, 1.529, 1.676, 1.677	4.43	to β, 811	d 1450	0.0022^{18}	0.0065^{100}	s a, NH_4Cl; i al
b22	chlorate	$Ba(ClO_3)_2 \cdot H_2O$	322.25	col monocl, 1.5622, 1.577, 1.635	3.18	414 ($-H_2O$, 120)	$-O$, 250	27.4^{15}	111.2^{100}	sl s al, acet, HCl
b23	*per*chlorate	$Ba(ClO_4)_2$	336.23	col, hex	3.2	505		198.5^{25}	562.3^{100}	v s al
b24	*per*chlorate, hydrate	$Ba(ClO_4)_2 \cdot 3H_2O$	390.28	col, hex, 1.533	2.74	d 400		198^{33}	s	al 124^{62}
b25	chloride α	$BaCl_2$	208.24	col, monocl; 1.7303, 1.7367, 1.7420	3.856^{24}	tr to cub 962	1560	37.5^{26}	59^{100}	sl s HCl, HNO_3; v sl s al
b26	chloride β	$BaCl_2$	208.24	col, cub	3.917	963	1560			v sl s al
b27	chloride, hydrate	$BaCl_2 \cdot 2H_2O$	244.27	col, monocl, 1.629, 1.642, 1.658 n_D^{25}	3.097_4^{24}	$-2H_2O$, 113	37.5^{20}		58.7^{100}	sl s HCl, HNO_3; v sl s al
b28	*hypo*chlorite	$Ba(ClO)_2 \cdot 2H_2O$	276.27	col cr.		d				
b29	chlorofluoride	$BaCl_2 \cdot BaF_2$	383.56	col, tetr, 16.40	4.51^{18}			d	d	i al; s conc HCl, HNO_3
b30	chlorofluoride	$BaPtCl_6H_2O$	653.22	or-yel, monocl	2.868	$-5H_2O$, 70	d	s		d a, al; i MeOH, eth
b31	chloroplatinite	$BaPtCl_4 \cdot 3H_2O$	528.27	dk red pr	2.868	$-3H_2O$, 150		v s		s al
b32	chromate	$BaCrO_4$	253.32	yel, rhomb	4.498^{15}			0.00034^{16}	0.00044^{28}	s min a
b33	*di*chromate	$BaCr_2O_7$	353.32	red, monocl.				sl s		s h conc H_2SO_4
b34	*di*chromate, hydrate	$BaCr_2O_7 \cdot 2H_2O$	389.35	bright red-yel need		$-2H_2O$, 120		d		s conc CrO_3 soln
b35	chromite	$BaO \cdot 4Cr_2O_3$	761.29	grn-blk, hex	5.4^{15}			i	i	s a, fus carb
b36	citrate	$Ba_3(C_6H_5O_7)_2 \cdot 7H_2O$	916.30	wh powd		$-7H_2O$, 150		0.0406^{18}	0.0572^{25}	sl s al; s HCl
b37	cyanide	$Ba(CN)_2$	189.37	wh cr powd.				80^{14}		18^{14} 70% al
b38	cyanoplatinite	$BaPt(CN)_4 \cdot 4H_2O$	508.54	(a) monocl, yel, α 1.6704	a) 2.076	$-2H_2O$, 100		3^{16}	25^{100}	i al
				(b) grn, rhomb	b) 2.085					
b39	dithionate	$Ba(SO_3)_2 \cdot 2H_2O$	333.48	col,, rhomb or monocl, 1.586, 1.595, 1.607	$4.536^{13.5}$	d 120		24.75^{18}	90.9^{100}	sl s al
b40	ethylsulfate	$Ba(C_2H_5SO_4)_2 \cdot 2H_2O$	423.60	wh lust leaf				s		sl s al
b41	ferrocyanide	$Ba_2Fe(CN)_6 \cdot 6H_2O$	594.70	yel, monocl.	2.666	$-H_2O$, 40		0.17^{15}	0.9^{100}	i al
b42	fluogallate	$Ba_3(GaF_6)_2 \cdot H_2O$	797.43	wh cr.	4.06	$-1/2$ H_2O, 110; $-1/2$ H_2O, 230		i		s HF
b43	fluoride	BaF_2	175.33	col, cub, 1.4741 n_D^{25}	4.89	1355	2137	0.12^{25}	sl s	s a, NH_4Cl
b44	fluoroiodide	$BaF_2 \cdot BaI_2$	566.47	pl	5.21^{18}			d	d	i al; s conc HCl, conc NHO_3
b45	fluosilicate	$BaSiF_6$	279.41	rhomb need	4.29_4^{21}	d 300		0.026^{100}	0.09^{100}	i al; sl s a, NH_4Cl
b46	formate	$Ba(CHO_2)_2$	227.37	col, rhomb, 1.573, 1.597, 1.636	3.21	d		27.76^{0}	39.71^{90}	i al, eth
b47	gluconate	$Ba(C_6H_{11}O_7)_2 \cdot 3H_2O$	581.67	pr or rhomb leaf		-3 H_2O, 100; d 120		$3.3^{15.5}$		i al
b48	hydride	BaH_2	139.35	gray cr.	4.21^{0}	d 675	1400(?)	d to $Ba(OH)_2$ $+H_2$		d a
b49	hydroxide	$Ba(OH)_2 \cdot 8H_2O$	315.47	col, monocl, 1.471, 1.502, 1.50.	2.18^{16}	78	$-8H_2O$, 78	5.6^{15}	94.7^{78}	sl s al; i acet
b50	hyponitrite	$BaN_2O_2 \cdot 4H_2O$	269.40	wh cr powd.	2.742^{25}					
b51	iodate	$Ba(IO_3)_2$	487.14	monocl.	4.998	d		0.008^{0}	197^{100}	s HNO_3, HCl
b52	iodate, hydrate	$Ba(IO_3)_2 \cdot H_2O$	505.15	col, monocl	4.657^{18}	$-H_2O$, 200		v sl s	sl s	s NHO_3, HCl; i al, acet, H_2SO_4
b53	iodide	BaI_2	391.14	col cr.	5.15_4^{25}	740		170^{0}		al 77^{20}
b54	iodide, hydrate	$BaI_2 \cdot 2H_2O$	427.17	col rhomb, deliq	5.15	$-H_2O$, 98.9; $-2H_2O$, 539; d 740	200^{15}	269^{100}		1.07^{15}al; s acet
b55	iodide, hydrate	$BaI_2 \cdot 6H_2O$	499.23	col, hex		25.7		410^{0}	v s	v s al
b56	laurate	$Ba(C_{12}H_{23}O_2)_2$	535.96	wh leaf cr		260		$0.008^{15.3}$	0.011^{50}	0.008^{25} al; 0.006^{25} eth

PHYSICAL CONSTANTS OF INORGANIC COMPOUNDS (continued)

No.	Name	Synonyms and Formulae	Mol. wt.	Crystalline form, properties and index of refraction	Density or spec. gravity	Melting point, °C	Boiling point, °C	Solubility, in grams per 100 cc		
								Cold water	Hot water	Other solvents
b57	*l*-malate	$BaC_4H_4O_5$	269.40					0.883^{20}	1.044^{80}	
b58	malonate	$BaC_3H_2O_4 \cdot H_2O$	257.39	col				0.143^{0}	0.326^{80}	
b59	manganate	$BaMnO_4$	256.27	gray-grn, hex	4.85			v sl s		s a
b60	*per*-manganate	$Ba(MNO_4)_2$	375.20	br-vlt cr	3.77	d 200		62.5^{11}	75.4^{25}	d al
b61	methylsulfate	$Ba(CH_3SO_4)_2 \cdot 2H_2O$	395.55	col effl cr				s		s al
b62	molybdate	$BaMoO_4$	297.27	wh powd	4.65	1480		0.0058^{23}		sl s a
b63	myristate	$Ba(C_{14}H_{27}O_2)_2$	592.06					0.007^{25}	0.010^{50}	0.009^{25} al; 0.003^{25} eth 0.046^{15} MeOH
b64	nitrate	Nitrobarite. $Ba(NO_3)_2$	261.34	col cub, 1.572	3.24^{23}	592	d	8.7^{20}	34.2^{100}	i al; sl s a
b65	nitride	Ba_3N_2	440.00	yel-br	4.783^{25}_{4}		1000 vac	d	d	
b66	nitrite	$Ba(NO_2)_2$	229.34	col, hex	3.23^{23}	d 217		67.5^{20}	300^{100}	sl s al
b67	nitrite, hydrate	$Ba(NO_2)_2 \cdot H_2O$	247.36	col-yelsh, hex	3.173^{20}	d 115		63^{20}	109.6^{80}	1.6 al; v s HCl; i acet
b68	oxalate	$BaCr_2O_4$	225.35	cr	2.658	d 400		0.0093^{18}	0.0228^{100}	i al; s NH_4Cl, a
b69	oxide	BaO	153.33	col, cub, wh-yelsh powd, 1.98	5.72	1918	*ca* 2000	3.48^{20}	90.8^{100}	s dil a, al; i acet, NH_3
b70	oxide, per-	BaO_2	169.33	wh-gray powd	4.96	450	−O, 800	v sl s	d	s dil a; i acet
b71	oxide, per-, hydrate	BaO_28H_2O	313.45	col, hex	2.292	$-8H_2O$, 100		0.168	d	s dil a; i al, eth, acet
b72	palmitate	$Ba(C_{16}H_{31}O_2)_2$	648.17	wh cr powd		d		0.004^{15}	0.007^{50}	$0.008^{16.5}$ al; 0.001^{15} eth
b73	*hypo*phosphate	$BaPO_3$	216.30	need				sl s		s al; v sl s ac a
b74	*ortho*phosphate di-	$BaHPO_4$	233.31	wh, rhomb, 1.635, 1.617	4.165^{15}	d 410^{710}		0.01—0.02		s a, NH_4Cl
b75	*ortho*phosphate, mono-	$Ba(H_2PO_4)_2$	331.30	tricl.	2.9^{4}			d	d	s a
b76	*ortho*phosphate, tri-	$Ba_3(PO_4)_2$	601.93	wh, cub	4.1^{16}			i	i	s a
b77	*pyro*phosphate	$Ba_2P_2O_7$	448.60	wh, rhomb	3.9^{20}			0.01	sl s	s a, NH_4 salts
b78	*hypo*phosphite	$Ba(H_2PO_2)_2 \cdot H_2O$	285.32	wh, monocl	2.90^{17}_{4}	d 100—150		30^{15}	33^{100}	i al
b79	propionate	$Ba(C_3H_5O_2)_2 \cdot H_2O$	301.49	rhomb, β 1.518		d 300		48^{0}	67.9^{90}	0.08 al
b80	salycylate	$Ba(C_7H_9O_3)_2 \cdot H_2O$	437.64	wh need				s		
b81	selenate	$BaSeO_4$	280.29	wh, rhomb	4.75	d		0.0118	0.138^{100}	s HCl; i HNO_3
b82	selenide	BaSe	216.29	wh cub disc, n_D, 2.268	5.02			d	d	d HCl
b83	*meta*silicate	$BaSiO_3$	213.41	col, rhomb, 1.673, 1.674, 1.678	4.399	1604		i	d	s HCl
b84	*meta*silicate, hydrate	$BaSiO_3 \cdot 6H_2O$	321.51	rhomb, 1.542, 1.548, 1.548	2.59			0.17	d	
b85	stearate	$Ba(C_{16}H_{35}O_2)_2$	704.28	wh powd				0.004^{15}	0.006^{50}	$0.005^{16.5}$ al, 0.008^{25} al, 0.001^{25} eth
b86	succinate	$BaC_4H_4O_4$	253.40	wh powd				0.421^{0}	0.237^{80}	sl s al
b87	sulfate	Nat. barite, prec. blanc fixe. $BaSO_4$	233.39	wh, rhomb (monocl), 1.637, 1.638, 1.649	4.50^{15}	1580	tr 1149 monocl	0.000222^{50} 0.000246^{25}	0.000336^{50} 0.000413^{100}	0.006 s 3% HCl; sl s H_2SO_4
b88	*peroxydi*sulfate	$BaS_2O_3 \cdot 4H_2O$	401.51	wh, monocl		d		52.2^{0}	d	d al
b89	sulfide, hydro-	$Ba(HS)_2 \cdot 4H_2O$	275.53	yel, rhomb		d 50		s		i al
b90	sulfide, mono-	BaS	169.39	col, cub, n_D 2.155	4.25^{13}	1200		d	d	i al
b91	sulfide, tetra-	$BaS_4 \cdot H_2O$	283.59	red or yel, rhomb	2.988	d 300		41^{15}	v s	i al, CS_2
b92	sulfide, tri-	BaS_3	233.51	yel-grn cr		d 554		s	s	
b93	sulfite	$BaSO_3$	217.39	col, cub hex		d		0.02^{20}	0.002^{80}	v s HCl
b94	tartrate	$BaC_4H_4O_6 \cdot H_2O$	303.42	wh cr.	$2.980^{20.8}$			0.026^{18}	0.056^{30}	0.032^{18} al
b95	tellurate	$BaTeO_4 \cdot 3H_2O$	382.97	volum wh	4.2^{200}	d >200		sl s	sl s	s HCl, NHO_3
b96	*pyro*tellurate, hydrogen	$Ba(HTe_2O_7)_2 \cdot H_2)$	891.75	volum ppt; hot:yel; cold: wh				s	s	s a
b97	telluride	BaTe	264.93	yel-wh, cub, disc, n_D 2.440	5.13					d a
b98	thiocarbonate	$BaCS_2$	245.42	yel, hex		d		1.08^{0}	d	i al
b99	thiocyanate	$Ba(SCN)_2 \cdot 2H_2$	289.52	need, deliq	2.286^{18}	$-H_2O$, 160		4.3^{20}	s	35^{20} al
b100	thiosulfate	BaS_2O_3	249,45	wh, rhomb		d220		0.2		
b101	thiosulfate, hydrate	$BaS_2O_3 \cdot H_2O$	267.46	wh, rhomb	3.5^{18}	d 100		0.208^{20}		i al
b102	titanate	$BaTiO_3$	233.21	tetr and hex, 2.40	tetr 6.017, hex 5.806					
b103	tungstate	$BaWO_4$	385.18	col, tetr	5.04			sl s	sl s	d a
b104	*pyro*vanadate	$Ba_2V_2O_7$	488.54	wh cr.		863				
b105	**Beryllium**	Glucinum, Be	9.01218	grey met, hex	1.85^{20}	1278 ± 5	2970 (5 mm)	i	sl s, d	s dil a, alk; i Hg
b106	acetate	$Be(C_2H_3O_2)_2$	127.10	col pl.		d 300		i		i al, eth, CCl_4
b107	acetate, basic	$Be_4O(C_2H_3O_2)_6$	406.32	oct	1.36^{4}	284	331	sl d	d	s chl, ac a; sl s al, eth
b108	acetate propionate, basic	$Be_3O(C_2H_3O_2)_3 \cdot (C_3H_5O_2)_3$	448.40			127	330			
b109	aluminate	Nat. chrysoberyl. $BeAl_2O_4$	126.97	rhomb, 1.747, 1.748, 1.757	3.76	1870				i a
b110	aluminum silicate	Nat. beryl. $Be_3Al_2(SiO_3)_6$	537.50	transp, hex, col, 1.580, 1.547	2.66	1410 ± 100				i a
b111	aluminum silicate	Nat. euclase. $Be_2Al_2(SiO_4)_2 \cdot (OH)_2$	290.17	monocl, 1.652, 1.655, 1.671	3.1					
b112	benzenesulfonate	$Be(C_6H_5O_3S)_2$	323.34	monocl.				v s	v s	v s al, acet, ac a; i eth, bz, CS_2, CCl_4
b113	*ortho*borate, basic	Nat. hamgergite, $Be_2(OH)BO_3$	93.84	rhomb, 1.560, 1,591, 1.631	2.35					
b114	bromide	$BeBr_2$	168.83	wh need, deliq	3.465^{25}	490 ± 10 subl	520	s	v s	s al, eth; i bz; 18.56 pyr

PHYSICAL CONSTANTS OF INORGANIC COMPOUNDS (continued)

No.	Name	Synonyms and Formulae	Mol. wt.	Crystalline form, properties and index of refraction	Density or spec. gravity	Melting point, °C	Boiling point, °C	Solubility, in grams per 100 cc: Cold water	Hot water	Other solvents
b115	butyrate, basic	$Be_4O(C_4H_7O_2)_6$	574.64				239^{19}			
b116	carbide	Be_2C	33.03	yel, hex	1.90^{15}	>2100 d		d	d	s a; d alk
b117	carbonate, basic	$BeCO_3 + Be(OH)_2$	112.05	wh powd				i	d	s a, alk
b118	chloride	$BeCl_2$	79.92	col need, deliq	1.899^{25}	405	520 (488)	v s	v s, d	v s al, eth, pyrl sl s bx, chl, CS_2, i NH_3, acet
b119	fluoride	BeF_2	47.01	col, amorph, <1.33	1.986^{25}	subl 800		α	α	sl s al; s H_2SO_4
b120	hydride	BeH_2	11.03	wh cr.		d 125		d	d	i eth, tol
b121	iodide	BeI_2	262.82	col need	4.325^{26}	510 ± 10	590	d	d	s al, eth, CS_2
b122	nitrate	$Be(NO_3)_2 \cdot 3H_2O$	187.97	wh-yel cr, deliq	1.557	60	142	v s	v s	v s al
b123	nitride	Be_2N_2	55.05	col, cub		2200 ± 100	d 2240	d	d	d a, conc alk; i al
b124	oxalate	$BeC_2O_4 \cdot 3H_2O$	151.08	rhomb, β 1.487		$-H_2O$, 100; $-3H_2O$, 220	d 350	38.22^{26}		
b125	oxide	Nat. bromillite, BeO	25.01	wh hex, 1.719, 1.733	3.01	2530 ± 30	*ca*3900	0.00002^{30}		a conc H_2SO_4, fus KOH
b126	oxide	$BeO \cdot xH_2O$		wh amorph powd or gel		d		i	i	s a, alk, $(NH_4)_2CO_3$
b127	2,4-epntanedione deriv	Beryllium acetylacetonate $Be(C_5H_7O_3)_2$	207.23	wh, monocl	1.168^{4}	108	270	sl s	d	s al, eth, a
b128	*ortho*phosphate	$Be_3(PO_4)_2 \cdot 3H_2O$	271.03			$-H_2O$, 100		s	s	s ac a
b129	propionate basic	$Be_4O(C_3H_6O_2)_6$	490.48			120				
b130	selenate	$BeSeO_4 \cdot 4H_2O$	224.03	col, rhomb, 1.466, 1.501, 1,503	2.03	$-2H_2O$, 100; $-4H_2O$, 300		56.7^{25}	s	
b131	*di*silicate	Nat. bertrandite. $Be_4Si_2O_7(OH)_2$	238.23	rhomb, 1.591, 1.605, 1.604	2.6					
b132	*ortho*silicate	Nat. phenazite, Be_2SiO_4	110.11	tricl, col, 1.654, 1.670	3.0					i a
b133	stearate (com'l)	$Be(C_{18}H_{35}O_2)_2$	575.96	wh, waxy		45		i	i	i al; s eth, CCl_4
b134	sulfate	$BeSO_4$	105.07		2.443	d 550—600		i	d to $BeSO_4$, $4H_2O$	
b135	sulfate, hydrate	$BeSO_4 \cdot H_2O$	177.13	col, tetr, 1,472, 1.440	$1.713^{10.5}$	$-2H_2O$, 100	$-4H_2O$, 400	42.4^{26}	100^{100}	sl a conc H_2SO_4; i al
b136	sulfide	BeS	41.07	reg	2.36			d	d	
b137	**Bismuth**	Bi	208.9804	rhomb silver-wh or redsh met	9.80	271.3	1560 ± 5^{760}	i	i	s h H_2SO_4, HNO, aq, reg; sl s h HCl
b138	acetate	$Bi(C_2H_3O_2)_3$	386.11	wh cr.		d		i	i	s ac a
b139	*ortho*arsenate	$BiAsO_4$	347.90	wh, monocl, 2.14,2.15, 2.18	7.14			i	i	sl s h conc HNO_3
b140	benzoate	$Bi(C_7H_5O_2)_3$	572.33	wh powd						s a; i eth
b141	bromide, tri-	$BiBr_3$	448,69	yel cr powd, deliq	5.72_4^{25}	218	453	d to BiOBr	d	i al; s HCl, HBr, eth; v s liq NH_3
a142	carbonate, basic	Bismuth oxycarbonate, $Bi_2O_2CO_3$	509.97	wh powd	6.86		d	i	i	s a
b143	chloride, tetra-	$BiCl_4$	350.79	col cr.		226				
b144	chloride, tri-	$BiCl_3$	315.34	wh cr, deliq.	4.75_4^{25}	230—232	447	d to BiOCl	d	s a, al, eth, acet
b145	*di*chromate, basic	$(BiO)_2 \cdot Cr_2O_7$	665.95	yel-or-red cr				i	i	s a; i alk
b146	citrate	$BiC_6H_5O_7$	398.08	wh cr.	3.458	d		sl s	sl s	sl s al; a NH_4OH
b147	fluoride, tri-	BiF_3	265.98	gray cr, cub, 1.74	5.32^{20}	727		i		s inorg a; i liq NH_3, al
b148	gallate, basic	Com'l dermatol. $Bi(OH)_2 \cdot C_7H_5O_5$	412.11	yel, amorph.		d		i		i al, eth
b149	hydroxide	$Bi(OH)_3$	260.00	wh amorph powd	4.36	$-H_2O$, 100; d 415	$-1^1/_2$ H_2O, 400	0.00014	d	s a; i or sl s conc alk
b150	iodate	$Bi(IO_3)_3$	733.69	wh				i		i HNO_3
b151	iodide, di-	BiI_2	762.79	red need, rhomb		d 400	subl vac	s		s al, MeOH
b152	iodide, tri-	BiI_3	589.69	redsh, hex.	5.778^{15}	408	*ca* 500	i	d	3.5^{20}al; i a; sl s NH_3; s HCl, HI
b153	lactate, *dl*	$Bi(C_6H_9O_6) \cdot 7H_2O$	512.22	pr, need				v sl s		i al
b154	molybdate	$Bi_2(MoO_4)_3$	485.07	col tricl, sl hygr	2.83	d 30	$-$ $5H_2O$, 80	d	d	v s HNO_3; s a, gluc; i al, 42^{19} acet
b156	nitrate, basic	$BiONO_3 \cdot H_2O$	305.00	hex leaf	4.928^{18}	$-H_2O$, 105	$-HNO_3$, 260	i	i	s a; i al
b157	oxalate	$Bi_2(C_2O_4)_3 \cdot 7H_2O$	808.13	wh powd		$-6H_2O$, *ca*130		d		s inorg a; i al, eth
b158	oxide, mono-	BiO	224.98	dk-gray powd.	7.15^{19}	d *ca*180 (Bi_2O_3)		sl d	d	d dil a; s dil KOH
b159	oxide pent-	Bi_2O_5	497.96	dk red or br.	5.10	$-O$, 150	$-2O$, 357	i	i	s KOH
b160	oxide, tetr-	$Bi_2O_4 \cdot 2H_2O$	517.99	br powd	5.6	$-H_2O$, 110	$-2H_2O$, 180	i	i	s a
b161	oxide, tri-	Bi_2O_3	465.96	yel, rhomb	8.9	825 ± 3	1890 (?)	i	i	s a
b162	oxide tri-	Bi_2O_3	465.96	gray-blk, cub	8.20	tr 704		i	i	s a
b163	oxide, tri-	Bi_2O_3	465.96	wh-lt yel, rhomb, 1.91	8.55	860		i	i	sl s a
b164	oxybromide	BiOBr	304.88	col cr or wh powd	8.082^{15}	d red ht		i	i	i al; s a
b165	oxychloride	BiOCl	260.43	wh cr or powd, 2.15	7.72^{15}	red ht.		i	i	s a; i NH_3, tart a, acet
b166	oxyfluoride	BiOF	243.98	wh cr or powd	7.5_{20}^{20}	d red ht		i		s a
b167	oxyiodide	BiOI	351.88	red cr, tetr	7.922	d red ht		i	i	s a; i al, $CHCl_3$
b168	*ortho*phosphate	$BiPO_4$	303.95	wh, monocl	6.323^{15}	d		i	i	s HCl; i al, dil HNO_3

No.	Name	Synonyms and Formulae	Mol. wt.	Crystalline form, properties and index of refraction	Density or spec. gravity	Melting point, °C	Boiling point, °C	Solubility, in grams per 100 cc		
								Cold water	Hot water	Other solvents
b169	propionate, basic	$BiOC_3H_5O_2$	298.05	wh powd				i		v s dil HCl; i al
b170	salicylate, basic	Bismuth subsalicylate, $Bi(C_7H_5O_3)_3 \cdot Bi_2O_3$ (appr)	1086.28	wh micr cr (variable comp)				i		s a, alk; i al, eth
b171	selenide, tri-	Nat. guanajuatite, Bi_2Se_3	654.84	blk, rhomb	6.82	710	d	i		i alk
b172	silicate	Nat. eulytite, $2Bi_2O_3 \cdot 3SiO_2$	1112.17	yel, cub, 2.05	6.11			i	i	s d HCl, HNO_3
b173	sulfate	$Bi_2(SO_4)_3$	706.13	wh need	5.08^{15}	d 405		d	d	s a
b174	mono-sulfide	BiS	241.04	dk gray powd.	7.6—7.8	680 (in CO_2)	d	v sl s		
b175	sulfide, tri-	Nat. bismuthinite, bismuthglance, Bi_2S_4	514.14	br-blk, rhomb, 1.340, 1.456, 1.459	7.39	d 685		0.000018^{18}		s HNO_3; i dil sl
b176	tartrate	$Bi_2(C_4H_4O_6)_3 \cdot 6H_2O$	970.27	wh powd	2.595_5^{25}	$-3H_2O$, 105		i	i	s a, alk, i al
b177	tellurate	Montanite. $Bi_2TeO_6 \cdot 2H_2O$	677.59	biaxial, β: 2.09	3.79					
b178	telluride, tri-	Nat. tetradymite. Bi_2Te_3	800.76	gray, rhbdr	7.7_4^{20}	573				d HNO_3
b179	vanadate	Nat. pucherite, $Bi_2O_3 \cdot V_2O_5$	647.84	red-grn, rhomb, 2.41, 2.50, 2,51 (Li)	6.25^{25}					
b180	**Bismuthic acid**	$HBiO_3$	257.99	red	5.75	$-H_2O$, 120	$-2O$, 357	i	i	s a, KOH
b181	**Borazole**	$B_3N_3H_6$	80.50	col liq	1^{-65}, 0.8614^{06}	−58	55	sl d	d	
b182	**Boric Acid**, meta-	HBO_2	43.82	wh cr, cub, 1.691	2.486	236 ± 1		v sl s	sl s	
b183	ortho-	Boracic acid, H_3BO_3	61.83	col, tricl, 1.337, 1.461, 1.462	1.435^{15}	169 ± 1 tr to HBO_2	$-1^1/_2H_2O$, 300	6.35^{30}	27.6^{100}	28^{20} glyc; 0.0078 eth; 5.56 al; 20. 20^{25} MeOH; 1.92^{25} liq NH_3; sl s acet
b184	tetra- (pyro-)	$H_2B_4O_7$	157.25	vitr or wh powd				s	s	s al
b185	fluo-	HBF_4	87.81	col liq			d 130	∞	∞	s al
b186	**Borinoaminoborine**	B_2H_7N	42.68	col liq		−66.5	76.2			s triborine triamine
b187	**Boron**	B.	10.81	yel monocl or br amorph powd	2.34, 2.37 amorph	2300	2550	i	i	v sl s HNO_3
b188	arsenate	$BAsO_4$	149.73	wh cr tetrag, 1.681, 1.690	3.64	subl *ca* 700		v sl s	1.4^{100}	i al; s inorg a
b189	bromide, tri-	BBr_2	250.52	col fum liq, $n_D^{16.3}$ 1.5312	$2.6431_4^{18.4}$	−46	91.3 ± 0.25	d		s al, CCl_4
b190	bromide, di-, iodide,	BBr_2I	297.52	col liq			125	d	d	
b191	bromide, mono-, diiodide	$BBrI_2$	344.52	col liq			180	d	d	
b192	(di-) bromide, mono-, pentahydride	B_2H_4Br	106.56	col gas		−104	*ca* 10	hyd to HBO_2 HBr + H_2		
b193	(tetra-) carbide	B_4C	55.26	blk rhbdr	2.52	2350	>3500	i	i	i a; s fus alk
b194	chloride, tri-	BCl_3	117.17	col fum liq, $1.419^{5.7}$ α line H_2	1.349_4^{11}	−107.3	12.5		d to HCl + H_3BO_3	d al
b196	fluoride dihydrate	$BF_3 \cdot 2H_2O$	103.84	col liq; n_D^{20}	1.6316^{20}	6		d	d	s eth, dioxan
b197	hydride	Diborane, borethane, B_2H_6	27.67	col gas	liq; 0.447^{-11}	−1.65.5	−92.5	sl s d to H_3BO_3 + H_2		d 1.6^0 al; s NH_4OH, conc H_2SO_4
b198	hydride	Dihydrotetraborane, borobutane. B_4H_{10}	53.32	col gas, pois	0.56^{-35}	−120.8	16	sl s, d		s bz; d al
b199	hydride	Pentaborane. B_5H_9	63.12	col liq	0.66^0	−46.82	58.4	d		
b200	hydride	Hexaborane. B_6H_{10}	74.94	col liq	0.69^0	−65	0^{72}	d		
b201	hydride	Decaborane. $B_{10}H_{14}$	122.21	wh cr.	0.94^{25}	99.5	213	sl s	d	v s CS_2; s al, eth bz
b202	iodide, tri-	BI_3	391.52	col pl, hygr	3.35^{50}	49.9	210	d	d	d al; v s CS_2, bz, CCl_4
b203	nitride	BN	24.82	wh, hex	2.25	subl *ca* 3000		i	sl d	sl s h a
b204	oxide	B_2O_3	69.92	rhomb cr, 1.64, 1.61	2.46 ± 0.01	45 ± 2	*ca* 1860	sl s	s	
b205	oxide glass	B_2O_3	69.62	col, vitr 1.485	1.812_4^{25}	*ca* 450		1.1^0	15.7^{100}	s al, a
b206	phosphide	BP	41.78	maroon powd.		ign 200	i	i	i all solv	
b207	*tri*selenide	B_2Se_3	258.50	yel-gray powd				d	d	
b208	(hexa-) silicide	B_6Si	92.95	blk cr	2.47			i		s HNO_3; d H_2SO_4; i KOH
b209	(tri-) silicide	B_2Si	60.52	blk rhomb	2.52			i		sl s HNO_3; d H_2SO_4, KOH
b210	sulfide, penta-	B_2S_5	181.92	col, tetrag	1.85	390		d	d	d al
b211	sulfide, tri-	B_2S_3	117.80	wh cr or vitr	1.55	310		d		sl s PCl_3, SCl_2; d al
b212	**Borotungstic acid**	$H_5BW_{12}O_{40} \cdot 30H_{20}$	3402.48	tetr, cr	3	45—51		s		s al, eth
b213	**Bromic acid**	$HBrO_3$	128.91	known in sol only, col or yelsh		d 100		v s	d	
b214	**Bromine**	Br_2	159.808	dk red liq, 1.661	2.928^{59}, 3.119^{20}	−7.2	58.78	4.17^0, 3.58^{20}	3.52^{50}	v s al, eth, chl, CS_2
b215	azide	Bromoazide, BrN_3	121.92	cr, red liq		*ca* 45	exp			s eth, KI; sl s bz, ligr
b216	chloride	BrCl.	115.36	red-col liq or gas		*ca* −66 d 10	*ca* 5	s d		s eth, CS_2
b217	fluoride, mono-	BrF	98.90	red-br gas		d −33	−20			
b218	fluoride, penta-	BrF_5	174.90	col liq	2.466^{25}	−61.3	40.5	d	d	
b219	fluoride, tri-	BrF_3	136.90	col-gray-yel liq.	2.49^{135}	(−2) 8.8	135	d viol. to O_2, HOBr, HF, $HBrO_3$		d alk

No.	Name	Synonyms and Formulae	Mol. wt.	Crystalline form, properties and index of refraction	Density or spec. gravity	Melting point, °C	Boiling point, °C	Solubility, in grams per 100 cc		
								Cold water	Hot water	Other solvents
b220	hydrate	$Br_2{\cdot}10H_2O$	339.96	red oct	1.49	d 6.8		s		
b221	oxide, di-	BrO_2	111.91	lt yel		d 0				
b222	oxide, mon-	Br_2O	175.81	dk br		−17 to −18				s, d CCl_4
b223	(tri-) oxide, oct-	Br_3O_8(or Br_3O_8)*n*	367.71	wh		stable at −40				
	Bromauric acid									
b224	**Bromauric acid**	$HAuBr_4{\cdot}5H_2O$	607.67	red-br cr		27		v s		s al
b225	**Bromous acid, hypo-**	HBrO	96.91	known only in soln, col-yel		40 (vac)		s	s d	s al, eth, chl
b226	**Bromoplatinic acid**	$H_2PtBr_6{\cdot}9H_2O$	838.66	red monocl, deliq		d −100		v s	v s	
c1	**Cadmium**	Cd	112.41	hex silv-wh malleable met	8.642	320.9	765	i	i	s a, NH_4NO_3, h H_2SO_4
c2	acetate	$Cd(C_2H_3O_2)_2$	230.50	col	2.341	256	d	v s		s MeOH
c3	acetate, hydrate	$Cd(C_2H_3O_2)_2{\cdot}2H_2O$	266.53	col monocl, odor ac a	2.01	$-H_2O$, 130		v s	v s	v s al
c4	amide	$Cd(NH_2)_2$	144.46		3.05^{25}	d 120				
c5	ammonium chloride	$CdCl_2{\cdot}NH_4Cl$	236.81	col need, rhomb	2.93	289		33.45^{16}	$43.99^{63.8}$	s al, MeOH
c6	ammonium sulfate	$Cd(NH_4)_2(SO_4)_2{\cdot}6H_2O$	448.69	col monocl pr	2.061^{20}_{4}	100 d $-H_2O$		s	s	
c7	arsenate, hydrogen	$CdHAsO_4{\cdot}H_2O$	270.35		4.164^{15}_{4}	>120				
c8	arsenide	Cd_3As_2	487.07	dk gray cub	6.21^{15}_{4}	721		i	i	sl s HCl; s HNO_3; i aq reg
c9	benzoate	$Cd(C_7H_5O_2)_2{\cdot}2H_2O$	390.67					3.34^{20}		sl s al
c10	borate	$Cd(BO_3)_2{\cdot}H_2O$	248.04	wh rhomb	3.758	d		125^{17}		i al
c11	borotungstate	$Cd_5(BW_{12}O_{40})_2{\cdot}18H_2O$	6600.30	yel tricl		75		1250^{19}	v s	
c12	bromide	$CdBr_2$	227.22	yel cr.	5.192^{25}	567	863	57^{10}	162^{104}	26.6^{15} sl; 0.4^{15} eth; s HCl; 1.6^{18} acet
c13	bromide, tetrahydrate	$CdBr_2{\cdot}4H_2O$	344.28	sm wh need, effl		tr 36		121^{10}	v s	25 al; s acet; sl s eth
c14	carbonate	$CdCO_3$	172.42	wh, trig	4.258^{4}	d <500		i	i	s a, KCN, NH_4 salts; i NH_3
c15	chlorate	$Cd(ClO_3)_2{\cdot}2H_2O$	315.34	col pr, deliq	2.28^{18}	80		298^{0}	487^{65}	s al, a, acet
c16	chloride	$CdCl_2$	183.32	col, hex	4.047^{25}	568	960	140^{20}	150^{100}	1.52^{15} al; $1.7^{15.5}$ MeOH; i acet, eth
c17	chloride	$CdCl_2{\cdot}2\,{}^1/_2\,H_2O$	228.35	col monocl, 1.6513	3.327	tr 34		168^{20}	180^{100}	2.05^{15} MeOH; sl s al
c18	chloroacetate, di-	$Cd(C_2HCl_2O_2)_2{\cdot}2H_2O$	386.29	need	2.132^{15}					
c19	chloroacetate, mono-	$Cd(C_2H_2ClO_2)_2{\cdot}6H_2O$	407.48		1.942^{25}					
c20	chloroacetate, tri-	$Cd(C_2Cl_3O_2)_2{\cdot}1{}^1/_2H_2O$	464.19	rhomb	2.093^{25}					
c21	chloroplatinate	$CdPtCl_6{\cdot}3H_2O$	574.25	yel trig need	2.882	$-H_2O$, 170, d		s	s	
c23	chromite	$CdCr_2O_4$	280.40	grn to blk, cub	5.79^{17}			i	i	i a
c24	cyanide	$Cd(CN)_2$	167.45	cr		dec >200		1.7^{15}	s	s a, KCN, NH_4OH; i al
c25	ferrocyanide	$Cd_2Fe(CN)_6{\cdot}xH_2O$						i	i	s HCl
c26	fluogallate	$[Cd(H_2O)_6]{\cdot}[GaG_5H_2O]$	403.23	col cr 1.45	2.79	$-5H_2O$, 110		v s		
c27	fluoride	CdF_2	150.41	wh cub, 1.56	6.64	1100	1758	4.35^{25}		s a, HF; i al, NH_3
c28	fluosilicate	$CdSeF_6{\cdot}6H_2O$	362.58	col hex.				s	s	s 50% al
c29	formate	$Cd(CHO_2)_2{\cdot}2H_2O$	238.47	monocl.	2.44	d		v s		
c30	fumarate	$CdC_4H_2O_4$	226.47					0.9^{30}		
c31	hydroxide	$Cd(OH)_2$	146.42	wh, trig or amorph	4.79^{15}	d 300		0.00026^{26}		s a, NH_4 salts; i alk
c32	iodate	$Cd(IO_3)_2$	462.22	wh cr.	6.43	d		s		s NHO_3, $NH_4{\cdot}OH$
c33	idodie	CdI_2	366.22	grn-yel powd	5.670^{30}_{4}	387	796	86.2^{25}	125^{100}	110.5^{20} al; 41^{25} acet; sl s NH_3; 206.7^{25} MeOH
c34	lactate	$Cd(C_3H_5O_3)_2$	290.55	need				10	12.5	i al
c35	maleate	$Cd(C_4H_2O_4){\cdot}2H_2O$	262.50					0.66^{30}		
c36	*per*manganate	$Cd(MnO_4)_2{\cdot}6H_2O$	458.37		2.81	d 95		v s	v s	
c37	molybdate	$CdMoO_4$	272.35	yel pl.	5.347			sl s		s a, NH_4OH, KCN
c38	nitrate	$Cd(NO_3)_2$	236.42	col		350		109^{0}	326^{50} 682^{100}	v s a; s et ac
c39	nitrate, tetra-hydrate	$Cd(NO_3)_2{\cdot}4H_2O$	308.48	wh pr, need, hygr	2.455^{17}_{4}	59.4	132	215		s al, NH_3; i HNO_3
c40	nitrocovaltate (III)	Cadmium cobaltinitrite. $Cd_3[Co(NO_2)_6]_2$	1007.16	yel		d 175		sl s	v s	d a, alk, org solv
c41	oxalate	CdC_2O_4	200.43	col cr.	3.32^{18}_{4}	d 340				s a; i al
c42	oxalate, trihydrate	$CdC_2O_4{\cdot}3H_2O$	254.48	col cr.		d		0.005^{18}	0.009	
c43	oxide	CdO	128.41	br, amorph	6.95	>1500	d 900-1000	i	i	s a, NH_4 salts; i alk
c44	oxide	CdO	128.41	br cub, 1.49 (Li)	8.15	>1500	subl 1559	i	i	s a, NH_4 salts; i alk
c45	*ortho*phosphate	$Cd_3(PO_4)_2$	527.17	col amorph		1500		i		s a, NH_4 salts
c46	*pyro*posophate	$Cd_2P_2O_7$	398.76	wh cr leaf	4.965^{15}	above red heat		sl s	s	s a, NH_3
c47	phosphate, dihydrogen	$Cd(H_2PO_4)_2{\cdot}2H_2O$	342.42	col, tricl	2.74^{15}_{4}	d 100				i al, eth; s HCl
c48	phosphide	Cd_3P_2	399.18	grn, tetr need	5.60	700				s d HCl; s exp conc HNO_3
c49	potassium cyanide	$Cd(CN)_2{\cdot}2KCN$	294.68	col, glossy, oct.	1.847			33.3	100	i al
c50	potassium sulfate	$CdK_2(SO_4){\cdot}2H_2O$	322.69	tricl col tab	2.922^{16}			42.89^{16}	47.40^{40}	
c51	salicylate	$Cd(C_7H_6O_3)_2{\cdot}H_2O$	404.66	wh need				sl s	s	s al, eth, glycerol a, NH_4OH
c52	selenate	$CdSeO_4{\cdot}2H_2O$	291.40	rhomb	3.63	$-1H_2O$, 100; $-2H_2O$, 170		v s		
c53	selenide	CdSe	191.37	grn-br or red powd, hex	5.81^{15}_{4}	>1350		i		d a
c54	*meta*silicate	$CdSiO_3$	188.49	col. rhomb, 1.739	4.93	1242		v sl s		

No.	Name	Synonyms and Formulae	Mol. wt.	Crystalline form, properties and index of refraction	Density or spec. gravity	Melting point, °C	Boiling point, °C	Solubility, in grams per 100 cc: Cold water	Hot water	Other solvents
c55	sulfate	$CdSO_4$	208.47	wh, rhomb	4.691^{20}_{4}	1000		75.5^{0}	60.8^{100}	i al, acet, NH_3
c56	sulfate, hydrate	$CdSO_4{\cdot}H_2O$	226.48	monocl.	3.79^{20}	tr 108		s	s	i al
c57	sulfate, hydrate	$CdSO_4{\cdot}7H_2O$	334.57	col. monocl	2.48	tr 4		s		i al
c58	sulfate hydrate	$3CdSO_4{\cdot}8H_2O$	769.53	col. monocl, 1.565	3.09	tr 41.5		113^{0}	s	
c59	sulfide	Nat. greenockite, CdS	144.47	yel-or hex, 2.506, 2.529	4.82	1750^{100atm}	subl in N_2, 980	0.00013^{18}	colloid	s a; v sl s NH_4OH
c60	sulfite	$CdSO_3$	192.47	cr		d		sl s		i al; s a, NH_4OH
c61	tartrate	$CdC_4H_4O_4$	260.48	wh cr powd				sl s		s a, NH_4OH
c62	telluride	CdTe	240.01	blk, cub	5.850^{15}	1121	1091	i		i a; d HNO_3
c63	tungstate	$CdWO_4$	360,26	yel cr.				0.05		s NH_4OH
	Cadmium complexes									
c64	tetramminecadmium *per*rhenate	$[Cd)NH_3)_4](ReO_4)_2$	680.94		3.714^{25}_{4}					0.037 conc NH_4OH
c65	tetrapyridine cadmiumfluosilicate	$[Cd(C_5H_5N)_4]SiF_6$	570.89	wh, tricl	2.282					
c66	**Calcium**	Ca	40.08	silv wh soft met, cub	1.54	839 ± 2	1484	d to H_2 + $Ca(OH)_2$	d	s a, liq NH_3, sl s al, i bz
c67	acetate	$Ca(C_2H_3O_2)_2$	158.17	col cr; 1.55, 1.56, 1.57		d		37.4^{0}	29.7^{100}	sl s al
c68	acetate, dihydrate	$Ca(C_2H_3O_2)_2{\cdot}2H_2O$	194.20	col need		$-1H_2O$, 84		34.7^{20}	33.5^{50}	
c69	acetate, monohydrate	$Ca(C_2H_3O_2)_2{\cdot}H_2O$	176.18	col need		d		43.6^{0}	34.3^{100}	sl s al
c70	aluminate	$CaAl_2O_4$ (or $CaO{\cdot}Al_2O_2$)	158.04	wh monocl, tricl or rhomb; 1.643, 1.665, 1.663	2.981^{25}	1600		d		s HCl; i HNO_3, H_2SO_4
c71	(tri-)aluminate	$Ca_3Al_2O_6$ (or $3CaO{\cdot}Al_2O_2$)	270.20	wh, cub, 1.710	3.038^{25}	d 1535		i		
c72	(tri-)aluminate hexahydrate	$3CaO{\cdot}Al_2O_3{\cdot}6H_2O$	378.29	col, oct, 1.603	2.52^{20}	d 700—800		d		
c73	aluminosilicate	$2CaO{\cdot}Al_2O_3{\cdot}SiO_3$	274.20	col, tetr, 1.669, 1.658	3.048	1590 ± 2				d a
c74	aluminosilicate	Nat. anorthite. $CaAl_3{\cdot}Si_2O_6$ (or $CaO{\cdot}Al_2O_3{\cdot}2SiO_2$)	278.21	wh, tricl, 1.5832	2.765	1551				
c75	*ortho*arsenate	$Ca_3(AsO_4)_2$	398.08	col amorph powd	3.620	1.455		0.013^{25}		
c76	arsenate, trihydrate	Nat. haidingerite, $2CaO{\cdot}As_2O_5{\cdot}3H_2O$	396.04	col, rhomb, 1.590, 1.602, 1.638	2.967					
c77	arsenide	Ca_3As_2	270.08	red cr	3.031^{25}	d		d	d	d a; s h HNO_3
c78	azide	$Ca(N_3)^2$	124.12	col, rhomb, hyg		$-3H_2O$, 110; exp 144-156		38.1^{0}	45^{15}	d a; s h HNO_3 0.211^{16} al; i eth
c79	benzoate	$Ca(C_7H_5O_2)_2{\cdot}3H_2O$	336.36	col, rhomb	1.436	$-3H_2O$, 110		2.7^{0}	8.3^{80}	
c80	*meta*borate	$Ca(BO_2)_2$	125.70	cil, flat rhomb pr, 1.550, 1.660, 1.680		1154		sl s		s a, NH_4 salts; sl s ac a
c81	*meta*borate, hexahydrate	$Ca(BO_2)_3{\cdot}6H_2O$	233.79	col, tetr, 1.520, 1.502	1.88			0.25^{20}		
c82	*tetra*borate	CaB_4O_7	195.32	readily vitrified		986				
c83	boride	CaB_6	104.94	blk, cub	2.3^{30}	2235		i	i	s HNO_3; sl s conc H_2SO_4
c84	bromide	$CaBr_2$	199.89	col, rhomb need, deliq	3.353^{25}	sl d 730	806—812	142^{30}	312^{106}	s al, acet, a; sl a NH_3, MeOH
c85	bromate	$Ca(BrO_3)_2{\cdot}H_2O$	313.90	monocl cr	3.329	$-H_2O$, 180		v s	v s	
c86	bromide, hexahydrate	$CaBr_2{\cdot}6H_2O$	307.98	col, hex cr	2.295	38.2	149	594^{0}	1360^{25}	s al, acet a
c87	butyrate	$Ca(C_4H_7O_2)_2{\cdot}3H_2O$	268.32	col cr.				s	sl s	
c88	carbide	CaC_2	64.10	col, tetr, 1.75	2.22	stab 25—447	2300	d	d	
c89	carbonate	Nat. aragonite, $CaCO_3$	100.09	col, rhomb, 1.530, 1.681, 1.685	2.930	tr to calcite 520	d 825	0.00153^{25}	0.00190^{75}	s a, NH_4Cl
c90	carbonate	Nat. calcite. $CaCO_3$	100.09	col, rhomb or hex. 1.6583, 1.4864	2.710^{18}	1339^{1025}	d 898.6	0.0014^{25}	0.0018^{75}	s a, NH_4Cl
c91	carbonate, hexahydrate	$CaCO_3{\cdot}6H_2O$	208.16	col, monocl, 1.460, 1.535, 1.545	1.771^{0}					
c92	chlorate	$Ca(ClO_3)_2$	206.98	wh cr, hyg		340 ± 10 (− some O)		s	s	s al, acet
c93	chlorate, dihydrate	$Ca(ClO_3)_2{\cdot}2H_2O$	243.01	wh-yelsh, rhomb. or monoc, deliq	2.711	$-H_2O$, 100		177.8^{8}	v s	s al, acet
c94	*per*chlorate	$(CaClO_4)_2$	238.98	col cr.	2.651	d 270		188.6^{25}	v s	166.2^{25}_{26} al; 237.4 MeOH
c95	chloride	$CaCl_2$	110.99	col, cub, deliq 1.52	2.15^{25}_{4}	782	>1600	74.5^{20}	159^{100}	s al, acet, ace a
c96	chloride aluminate	$3CaO{\cdot}Al_2O_3{\cdot}CaCl_2{\cdot}10H_2O$	561.34	col, monocl or hex, hex, 1.550, 1.535	1.892^{14}	$-H_2O$, 105	$-8H_2O$, 350	sl s	d	s a
c97	chloride, dihydrate	$CaCl_2{\cdot}2H_2O$	147.02	col cr.	0.835			97.7^{0}	326^{60}	50^{80} al
c98	chloride, hexahydrate	$CaCl_2{\cdot}6H_2O$	219.08	col, trig, deliq, 1.417, 1.393	1.71^{25}	26.92	$-4H_2O$, 30, $-6H_2O$, 200	279^{0}	536^{20}	s al
c99	chloride, monohydrate	$CaCl_2{\cdot}H_2O$	129.00	col cr, deliq.		260		76.8^{0}	249^{100}	s al; i acet
c100	chloride fluoride *ortho*phosphate	$3Ca_3(PO_4)_2{\cdot}CaClF$	1025.08	col cr, 1.634, 1.631	3.14	1270		v sl s		
c101	chlorite	$Ca(ClO_2)_2$	174.98	wh, cub	2.71			d	d	i al
c102	*hypo*chlorite	$Ca(ClO)_2$	142.98	wh powd or flat pl, 1.545, 1.69	2.35	d 100		s		i al

No.	Name	Synonyms and Formulae	Mol. wt.	Crystalline form, properties and index of refraction	Density or spec. gravity	Melting point, °C	Boiling point, °C	Solubility, in grams per 100 cc		
								Cold water	Hot water	Other solvents
c103	chlorite, basic	$Ca(ClO)_2 \cdot 2Ca(OH)_2$	291.17	wh, hex, 1.51, 1.585	2.10			sl s solns with 5—6 % avail Cl	d	d a
c104	*hypo*chlorite, basic	Bleaching powder, chlorinated lime, $Ca(ClO)_2 \cdot CaCl_2 \cdot xCa(OH)_2 \cdot xH_2O$	comp varies	wh powd strong Cl odor		d		d evin Cl		d a
c105	*hypo*chlorite, trihydrate	$Ca(ClO)_2 \cdot 3H_2O$	197.03	tetr pl, 1.535, 1.63	2.1	$-3H_2O$, 60				
c106	chromate	$CaCrO_4 \cdot 2H_2O$	192.10	yel, monocl pr		$-2H_2O$, 200		16.3^{20}	18.2^{45}	s a, al
c107	chromite	$CaCr_2O_4$	208.07	ol grn, cub need	4.8^{18}	2090		i	i	i a; s fus K_2CO_3
c108	cinnamate	$Ca(C_9H_7O_2)_2 \cdot 3H_2O$	388.43	col cr.				0.22^{2}	1.34^{100}	
c109	citrate	$Ca_3(C_6H_5O_7)_2 4H_2)$	570.50	wh need		$-4H_2O$, 120		0.085^{18}	0.096^{23}	0.0065^{18} al
c110	cyanamide	$CaCN_2$	80.10	col, hex, rhbdr.		1300 subl >1150		d evl NH_3	d	
c111	cyanide	$Ca(CN)_2$	92.12	wh powd		d>350		d	d	
c112	cyanoplatinite	$CaPt(CN)_4 \cdot 5H_2O$	429.31	yel-grn fluoresc, rhomb, 1.6226		$-5H_2O$, 100		s		
c113	ferricyanide	$Ca_3[Fe(CN)_6]_2 12H_2O$	760.33	red need, deliq				v s	v s	
c114	ferrite, mono-	$CaO \cdot Fe_2O_3$	215.77	dk redsh r, rhomb, 2.58, 2.43 (Na)	5.08	1250		i	i	v sl s a
c115	ferrocyanide	$Ca_2Fe(CN)_6 11$ or $12H_2O$	490.28 or 508.30	yel tricl, 1.570, 1.582, 1.596	1.68	d		86.8^{25}	115^{65}	i al
c116	fluosilicate	$CaSiF_6$	182.16	col, tetr	2.66^{18}			sl s		s al, HF, HCl
c117	fluoride	Nat. fluorite. CaF_2	78.08	cool, cub luminisc w heat, 1.434	3.180	1423	*ca*2500	0.0016^{18}	0.0017^{26}	s HN_4 salts; sl s a; i acet
c118	fluosilicate, dihydrate	$CaSiF_6 \cdot 2H_2O$	218.19	col, tetrag	2.254			sl s d		s HCl, HF; i al
c119	formate	$Ca(CHO_2)_2$	130.12	col, rhomb, 1.510, 1.514, 1.578	2.015	d		16.2^{0}	18.4^{100}	i al
c120	fumarate	$CaC_4H_2O_4 \cdot 3H_2O$	208.18	col, rhomb				2.11^{30}		
c121	*d*gluconate	$Ca(C_6H_{11}O_7)_2 \cdot H_2O$	448.39	wh cr powd, need		$-H_2O$, 120		3.3^{15}		v sl s al
c122	glycerophosphate	$CaC_3H_5(OH)_2PO_4$	210.14	wh cr powd, hyg		d 170		2^{25}	sl s	i al
c123	hydride	CaH_2	42.10	wh, rhomb cr.	1.9	816 (in H_2) d *ca* 600		d H_2 + $Ca(OH)_2$		d a
c124	hydroxide	$Ca(OH)_2$	74.09	col, hex, 1.574, 1.545	2.24	$-H_2O$, 580	d	0.185^{0}	0.077^{100}	s NH_4 salts, a; i al
c125	hyponitrite	$CaN_2O_2 \cdot 4H_2O$	172.15	wh cr.	1.834	d 320				d dil a
c126	iodate	Nat. lautarite, $Ca(IO_3)_2$	389.89	col, monocl.	4.519^{15}	d 540		0.20^{15}	0.67^{90}	s HNO_3; i al
c127	iodate, hexahydrate	$Ca(IO_3)_2 \cdot 6H_2O$	497.98	col, rhomb		d 35		0.13^{0}	1.22^{100}	s HNO_3
c128	iodide	CaI_2	293.89	yelsh-wh, hex, deliq	3.956^{25}	784	*ca*1100	209^{20}	426^{100}	126^{20} MeOH; s al, acet, a
c129	iodide, hexahydrate	$CaI_2 \cdot 6H_2O$	401.98	yel, hex need	2.55	d 42	160	757^{0}	1680^{30}	s a, al, acet
c130	iron (III) aluminate	Calcium (tetra-) aluminoferite, nat. celite. $4CaO \cdot Fe_2O_3 \cdot Al_2O_3$	485.97	brn, rhomb, 1.98, 2.05, 2.08 all for λ	3.77	1418				
c131	isobutyrate	$Ca(C_4H_7O_2)_2 \cdot 5H_2O$	304.35	col powd				20	sl s	
c132	lactate	$Ca(C_3H_5O_3)_2 \cdot 5H_2O$	308.30	wh need, effl		$-3H_2O$, 100		3.1°	7.9^{30}	sl s a; i al, eth
c133	laurate	$Ca(C_{12}H_{23}O_2)_2 \cdot H_2O$	456.73	wh need, effl		182—183		0.004^{15}	0.055^{100}	0.059^{15}, 1.72^{78} al
c134	linoleate	$Ca(C_{18}H_{31}O_2)_2$	598.97	wh amorph powd				i		s al, eth
c135	magnesium carbonate	Nat. dolomite. $CaCO_3 \cdot MgCO_3$	184.40	col, trig, 1.6817, 1.5026	2.872	d 730—760		0.0078^{18}		
c136	magnesium *meta*silicate	Nat. diopside. $CaO \cdot MgO \cdot 2SiO_2$	216.55	col, monocl, 1.665, 1.672, 1.695	3.275	1391		i	i	i HCl
c137	magnesium *ortho*silicate	Nat. mervinite. $Ca_3Mg(SiO_4)_2$	328.71	col to pa grn, monocl, 1.708, 1.711, 1.718	3.150					
c138	dl-malate	$CaC_4H_4O_5 \cdot 3H_2O$	226.20	col, rhomb, 1.545, 1.555, 1.575				0.321°	$0.451^{37.5}$	i al
c139	*l*-malate	$CaC_4H_4O_5 \cdot 2H_2O$	208.18	col				0.812°	$1.224^{37.5}$	s al
c140	malate, dihydrogen	$Ca(HC_4H_4O_5)_2 \cdot 6H_2O$	414.33	rhomb, or wh cr powd, 1.493, 1.507, 1.545				sl s		
c141	maleate	$CaC_4H_2O_4 \cdot H_2O$	172.15	col rhomb, 1.495, 1.575, 1.640				2.89^{25}	3.21^{40}	
c142	malonate	$CaC_3H_2O_4 \cdot 4H_2O$	214.19					0.44°	0.72^{100}	
c143	*per*manganate	$Ca(MnO_4)_2 \cdot 5H_2O$	368.03	purp cr.	2.4	d		331^{14}	338^{25}	s NH_4OH
c144	α-methylbutyrate	Calcium ethylmethylacetate. $Ca(C_5H_9O_2)_2$	242.33					24.24°	25.65^{70}	
c145	molybdate	Nat. powellite $CaMoO_4$	200.02	col, tetr, 1.967, 1.978	4.38—4.53			i	d	s a i al, eth
c146	nitrate	$Ca(NO_3)_2$	164.09	col, cub, hyg	2.504^{18}	561		121.2^{18}	376^{100}	14^{15} al; s MeOH, liq NH_3, acet; i eth
c147	nitrate, tetrahydrate	$Ca(NO_3)_2 \cdot 4H_2O$	236.15	col, monocl, deliq, 1.465, 1.498, 1.504	α 1.896, β 1.82	α 42.7, β 39.7	d 132	266°	660^{30}	s al, acet
c148	nitrate, trihydrate	$Ca(NO_3)_2 \cdot 3H_2O$	218.14	col, tricl		51.1				
c149	nitride	Ca_3N_2	148.25	brn cr, hex	2.63^{17}	1195		d	d	s dil a; d abs al
c150	nitrite	$Ca(NO_2)_2 \cdot H_2O$	150.11	col-yelsh, hex, deliq	2.23^{34}	$-H_2O$, 100		45.9^{0}	89.6^{91}	sl s al
c151	nitrite, tetrahydrate	$Ca(NO)_2)_2 \cdot 4H_2O$	204.15	col cr, tetr.	1.674^{0}_{0}	$-2H_2O$, 44		74.9°	106^{42}	s al
c152	oleate	$Ca(C_{18}H_{33}O_2)_2$	603.00	wh wax-like cr		83—84		0.04^{25}	0.03^{50}	sl s eth

PHYSICAL CONSTANTS OF INORGANIC COMPOUNDS (continued)

No.	Name	Synonyms and Formulae	Mol. wt.	Crystalline form, properties and index of refraction	Density or spec. gravity	Melting point, °C	Boiling point, °C	Solubility, in grams per 100 cc: Cold water	Hot water	Other solvents
c153	oxalate	CaC_2O_4	128.10	col, cub	2.2^4	d		0.00067^{13}	0.0014^{95}	s a; i ac a
c154	oxalate, hydrate	$CaC_2O_4 \cdot H_2O$	146.11	col	2.2	$-H_2O$, 200		i	i	s a; i ac a
c155	oxide	Lime, calcia. CaO	56.08	col, cub, 1.838	3.25—3.38	2614	2850	0.131^{10} d	0.07^{80} d	s a
c156	oxide, per-	CaO_2	72.08	wh, tetr, 1.895	2.92_4^{25}	d 275		sl s		s a
c157	oxide, peroctahydrate	$CaO_2 \cdot 8H_2O$	216.20	wh, tetr, pearly	1.70	$-8H_2O$, 200	d 275 expl	sl s	d	s a, NH_4 salts; i al, eth
c158	palmitate	$Ca(C_{16}H_{31}O_2)_2$	550.92	wh or yelsh, wh fatty powd				0.003^{25}		v sl s al; 0.008^{25} eth
c159	1-phenol-4 sulfonate(p-)	$Ca[C_6H_4(OH)SO_3]_2 \cdot H_2O$	404.42	wh to pinkish powd				s		s al
c160	phenoxide	$Ca(OC_6H_5)_2$	226.29	redsh powd				sl s		sl s al
c161	*hypo*phosphate	$Ca_2P_2O_6 \cdot 2H_2O$	274.13	gel				i		s HCl
c162	*meta*phosphate	$Ca(PO_3)_2$	198.02	col, 1.588, 1.595	2.82	975		i	i	i a
c163	*ortho*phosphate, di-(sec)	Nat. brushite. $CaHPO_4 \cdot 2H_2O$	172.09	wh, tricl, 1.5576, 1.5457, 1.5392	2.306^{16}	$-H_2O$, 109		0.0316^{38}	0.075^{100}	i al, s a
c164	*ortho*phosphate, mono-(prim.)	$Ca(H_2PO_4)_2 \cdot H_2O$	252.07	col, tricl, deliq, 1.5292, 1.5176, 1.4392	2.220^{16}	$-H_2O$, 109	d 203	1.8^{30}	d	s a
c165	*ortho*phospate, tri-(tert.)	Nat. whitlockite. $Ca_3(PO)_4)_2$	310.18	wh amorph powd, 1.629, 1.626	3.14	1670		0.002	d	i al; s a
c166	*pyro*phosphate	$Ca_2P_2O_7$	254.10	col, biax, 1.585, 1.604	3.09	1230		i		s a
c167	*pyro*phosphate, pentahydrate	$Ca_2P_2O_7 \cdot 5H_2O$	344.18	col, monocl, 1.539, 1.545, 1.551	2.25			sl s		s a; i HN_4Cl
c168	phosphide	Ca_3P_2	182.19	gray lumps	2.51	*ca* 1600		d ev PH_3		s a; i al, eth, bz
c169	*hypo*phosphite	$Ca(H_2PO_2)_2$	170.06	wh-gray, monocl		d		15.4^{25}	12.5^{100}	i al
c170	*ortho*phosphite, di-	$2CaHPO_3 \cdot 3H_2O$	294.17					sl s	d	s NH_4Cl
c171	*ortho*plumbate	Ca_2PbO_4	351.36	red-br cr	5.71	d		i	d	s a
c172	propionate	$Ca(C_3H_5O_2)_2 \cdot H_2O$	204.24	col, monocl tabl				49^0	55.8^{100}	i al
c173	*l*-quinate	$Ca(C_7H_{11}O_6)_2 \cdot 10H_2O$	602.55	rhomb leaf		50, $-10H_2O$ 120		16^{18}		i al
c174	salicylate	$Ca(C_7H_5O_3)_2 \cdot 2H_2O$	350.34	wh, oct		$-2H_2O$, 120		4^{25}	s	s al
c175	selenate	$CaSeO_4$	183.04	col	2.88			7.9^6	5.4^{67}	
c176	selenate, dihydrate	$CaSeO_4 \cdot 2H_2O$	219.07	col, monocl	2.68					
c177	selenide	CaSe	119.04	cub, 2.274	3.57					
c178	*meta*silicate(α)	Nat. pseudowollastonite. $CaSiO_3$	116.16	col, monocl, 1.610, 1.611, 1.664	2.905	1540		0.0095^{17}		s HCl
c179	*meta*silicate(β)	Nat. wollastonite. $CaSiO_3$	116.16	col, monocl, 1.616, 1.629, 1.631	2.5	tr 1200				
c180	di-*ortho*silicate (I)	Ca_2SiO_4	172.24	col, monocl, 1.717, 1.735	3.27	2130				
c181	di-*ortho*silicate (II)	Ca_2SiO_4	172.24	col, rhomb, 1.717, 1.735	3.28	tr to (I) 1420				
c182	di-*ortho*silicate (III)	Ca_2SiO_4	172.24	col, monocl, 1.642, 1.645, 1.654	2.97	tr to 675				
c183	(tri-)silicate	Nat. alite. Ca_3SiO_5 or $(3CaO \cdot SiO_2)$	228.32	col, monocl, α 1.718, β 1.724		1900 (incogr)				
c184	silicide	$CaSi_2$	96.25		2.5			i	d	s a, alk
c185	stearate	$Ca(C_{18}H_{35}O_2)_2$	607.03	cr powd		179—180		0.004^{15}		i al, eth
c186	succinate	$CaC_4H_6O_4 \cdot 3H_2O$	212.22	col, 1.460, 1.540, 1.610				0.193^{10}	0.89^{80}	
c187	sulfate	Nat. anhydrite. $CaSO_4$	136.14	col, rhomb, or monocl, 1.569, 1.575, 1.613	2.960	monocl 1450	rhomb tr to monocl 1193	0.209^{30}	0.1619^{100}	s a, NH_4 salts, $Na_2S_2O_3$, glyc
c188	sulfate	Soluble anhydrite. $CaSO_4$	136.14	col, hex or tricl, 1.505, 1.548	2.61	tr to rhomb >200				
c189	sulfate half-hydrate	Plaster of Paris. $CaSO_4 \cdot {}^1/_2H_2O$	145.15	wh powd		$-{}^1/_2H_2O$, 163		0.3^{20}	sl s	s a NH_4 salts, $Na_2S_2O_3$, glyc
c190	sulfate dihydrate	Nat. gypsum. $CaSO_4 \cdot 2H_2O$	172.17	col, monocl, 1.521, 1.523, 1.530	2.32	$-1{}^1/_2H_2O$, 128	$-2H_2O$, 163	0.241	0.222^{100}	s a, NH_4 salts, $Na_2S_2O_3$, glyc
c191	sulfide	Nat. oldhamite. CaS	72.14	col, cub, 2.137	2.5	d		0.021^{15} d	0.048^{60} d	d a
c192	sulfide, hydro-	$Ca(HS)_2 \cdot 6H_2O$	214.31	col pr		d 15—18		v s		s al
c193	sulfite	$CaSO_3 \cdot {}^1/_2H_2O$	129.15	col, hex		$-{}^1/_2H_2O$, >250		0.0043^{18}	0.0011^{100}	s H_2SO_4
c194	sulfite, dihydrogen	$Ca(HSO_3)_2$	202.21	yelsh liq, strong SO_2, odor				s		s a
c195	*d*-tartrate	$CaC_4H_4O_6 \cdot 4H_2O$	260.21	col, rhomb, 1.525, 1.535, 1.550		d		0.0266^0	$0.0689^{37.5}$	sl s al
c196	*dl*-tartrate	$CaC_4H_4O_6 \cdot 4H_2O$	260.21	tricl, powd or need		$-4H_2O$, 200		0.0032^0	$0.0078^{37.5}$	s HCl; i ac, a
c197	*meso*tartrate	$CaC_4H_4O_6 \cdot 3H_2O$	242.20	wh, monocl or tricl pr		$-3H_2O$ <170		i	0.16^{100}	0.28^{18}, 0.85^{100} ac a
c198	telluride	CaTe	167.68	cub, 251, 2.58	4.873					
c199	tellurite	$CaTeO_3$	215.68	wh fl		>960		sls	s	s a
c200	thiocarbonate, tri-	$CaCS_3$	148.27	yel cr.				s		i al
c201	thiocyanate	$Ca(SCN)_2 \cdot 3H_2O$	210.28	wh cr, deliq.				v s	v s	v s al
c202	*di*-thionate	$Ca(SO_3)_2 \cdot 4H_2O$	272.26	col, trig, 1.5496	2.176			16^0	30^{30}	

No.	Name	Synonyms and Formulae	Mol. wt.	Crystalline form, properties and index of refraction	Density or spec. gravity	Melting point, °C	Boiling point, °C	Solubility, in grams per 100 cc		
								Cold water	Hot water	Other solvents
c203	thiosulfate	$CaS_2O_3·6H_2O$	260.29	tricl.	1.872	d		100^3	d	s al
c204	*meta*titanate	Nat. perovskite. $CaTiO_3$	135.96	col, cub, rhomb, β 2.34	4.10	1975				
c205	tungstate	$CaWO_4$	287.93	wh, tetr, 1.9263, 1.9107	6.062^{20}			0.00064^{15}	0.00012^{100}	
c206	tungstate	Nat. scheelite. $CaWO_4$	287.93	col or w sc, tetr. 1.918, 1.934	6.06			0.2		i al, a; s NH_4Cl
c207	*meta*tungstate	$Ca_3H_4[H_2(W_2O_7)_6]·27H_2O$	3490.88	col, tric		$-7H_2O$, 105	$-10H_2O$, d			d a
c208	valerate	$Ca(C_5H_9O_2)_2$	242.33					8.28^0	7.39^{100}	
c209	*meta*zirconate	$CaZrO_3$	179.30	col, monocl	4.78	2550				
c210	**Carbon**	Diamond. C	12.011	col, cub, 2.4173	3.51	Diamond trans. to graphite		i	i	i a, alk
c211	carbon	Graphite C	12.011	blk, hex	2.25^{20}	subl 3652 − in vac. at 1500°C 97			i	s liq Fe; i a, alk
c212	carbon, amorphous	C.	12.011	amorph, blk.	1.8—2.1	subl 3652—97		i	i	i a, alk
c213	(di)-bromide, hexa-	Hexabromomethane. C_2Br_6	503.45	rhomb pr, 1.740, 1.847, 1.863	3.823	148—149 d	210	i		s CS_2; v sl s al, eth
c214	bromide, tetra-	Tetrabromomethane. CBr_4	331.63	col, monocl or oct	3.42	tr to oct 48.4; m.p. 90.1	189.5	0.024^{30}		s al, eth, chl
c215	(di)-bromide, tetra-	Tetrabromethylene.C_2Br_4	343.64			57.5	227			
c216	(di)-chloride, hexa-	Hexachloro ethane.C_2Cl_6	236.74	col, rhmb, tricl or cub	2.091	subl 187		i		s al, eth, oils
c217	chloride, tetra-	Tetrachloromethane. CCl_4	153.82	col liq, 1.4601	1.5867^{20}_{20}	−23	76.8		v sl s	s al, bz, chl, eth
c218	(di)-chloride, tetra,-	Tetrachloroethylene. C_2Cl_4	165.83	col liq, eth odor, 1.5055	1.6311^{15}_{4}	−22.4	120.8			s al, eth
c219	fluoride, tetra-	Tetrafluoromethane. CF_4	88.00	col gas	1.96^{-184}	−184	−128	sl s		
c220	iodide, tetra-	Tetraiodomethane. CI_4	519.63	dk red, cub	4.34^{20}	d 171		i	d	s al, CS_2, eth, MeOH, bz
c221	oxide, di-	CO_2	44.01	col gas or col liq	1.977^0 g/ℓ, liq 1.101^{-37}, solid 1.56^{-79}	$-56.6^{5.2atm}$	−78.5 subl	171.3^0 cm³ 0.348^0g 0.145^{25}g	90.1^{20} cm³ 0.097^{40} g 0.058^{60}g	31^{15} cm³ al, s acet
c222	oxide, mon-	CO	28.01	col odorl pois gas	1.250^0 g/ℓ liq 0.793	−199	−191.5	3.5^0 cm	2.32^{20} cm³	s al, bz, ac a, Cu_2Cl_2
c223	oxide, sub-	C_3O_2	68.03	col gas or liq, 1.4538	liq 1.114^0	−111.3	7	d		
c224	oxysulfide	COS	60.07	col gas, pois	gas 1.073 g/ℓ° liq 1.24^{-87}	−138.2	−50.2	54^{20} mℓ		s al; v s CS_2
c225	selenide, di-	CSe_2	169.93	golden yel liq, 1.845^{20}	2.6626^{25}_{4}	−45.5	125—126	i		d al; s CS_2, tol
c226	selenide, sulfide	CSeS	123.03	yel oily liq	1.9874	−85	84.5	i	i	sl s al; s CS_2
c227	sulfide, di-	CS_2	76.13	col liq, inflamm, 1.62950^{18}	1.261^{22}_{20}	−110.8	46.3	0.22^{22}	0.14^{50}	s al, eth
c228	sulfide, mono-	CS	44.07	red powd	1.66	d −160		i		ial; s eth, CS_2
c229	sulfide, sub-	C_3S_2	100.15	red liq	1.274	−0.5	d 90			
c230	sulfide telluride	CSTe	171.67	yel-red liq	2.9^{-50}	−54	d > −54			s CS_2, bz
c231	sulfochloride	Thiophosgene. $CSCl_2$	114.98	yel-red liq	1.509^{15}		73.5			
c232	**Carbonic acid**	H_2CO_3	62.03	exists in solution only				s		
c233	**Carbonyl bromide**	Carbon oxybromide. $COBr_2$	187.82	col liq			64.5			
c234	**Carbonyl chloride**	Phosgene, carbon oxychloride, $COCl_2$	98.92	col gas, pois	1.392	−104	8.3	d		d al, a; v s bz, tol s ac a
c235	**Carbonyl fluoride, di-**	COF_2	66.01	col gas, hyg	sol 1.388^{-190} liq 1.139^{-114}	−114	−83.1	d		
c236	**Carbonyl selenide**	COSe	106.97	col gas, very pois	liq $1.812^{4.1}$	−124.4	−21.7	d		s $COCl_2$
c237	**Cerium**	Ce	140.12	gray met, cub or hex	hex 6.657 cub 6.757	799	3426	sl d	d	s dil min a, i alk
c238	(III) acetate	$Ce(c_2H_3O_2)_3$	317.25	col		d 308		20^{15}	12^{75}	
c239	(III) acetate hydrate	$Ce(C_2H_3O_2)_3·1^1/_2H_2O$	344.28	wh-redsh cr powd		$-1^1/_2H_2O$, 115	d	26.5^{15}	16.2^{75}	
c240	boride, hexa-	CeB_6	204.98	blue met, cub.		2190	d	i	i	i HCl
c241	boride, tetra-	CeB_4	183.36	tetr	5.74					
c242	III bromate	$Ce(BrO_3)_3·9H_2O$	685.96	redsh-wh, hex		49		s		
c243	bromide	$CeBr_3·H_2O$	397.85	col need, deliq		d		v s	v s	v s al
c244	carbide	CeC_2	164.14	red, hex	5.23			d	d	s a
c245	carbonate	$Ce_2(CO_3)_3·5H_2O$	550.34	wh cr.				i		s a; sl s $(NH_4)_2CO_3$
c246	carbonate fluoride	Nat. bastnaesite. $CeFCO_3$	219.13	hex, 1.717, 1.817	5					
c247	chloride	$CeCl_3$	246.48	col cr, deliq.	3.92^0	848	1727	100	d	30 al; s acet
c248	citrate	$Ce(C_6H_5O_7)·3^1/_2H_2O$	392.27	wh powd				i		s dil min a
c249	(III) cyanoplatinite	$Ce_2[CN)_4]_3·18H_2O$	1501.97	yel-bl lust, monocl	2.657	$-13^1/_2H_2O$ 100—110	d	s		
c250	(III) fluoride	CeF_3	197.12	wh, hex	6.16	1460	2300	i		
c251	(IV) fluoride	$CeF_4·H_2O$	234.13	col microcr, 1.614	4.77	*ca* 650	d	i		s a
c252	hydride	CeH_3	143.14	dk bl amorph powd		ign		d		

PHYSICAL CONSTANTS OF INORGANIC COMPOUNDS (continued)

No.	Name	Synonyms and Formulae	Mol. wt.	Crystalline form, properties and index of refraction	Density or spec. gravity	Melting point, °C	Boiling point, °C	Solubility, in grams per 100 cc: Cold water	Hot water	Other solvents
c253	(III) hydroxide	$Ce(OH)_3$	191.14	wh gelat ppt						s a, $(NH_4)_2CO_3$; i alk
c254	(IV) iodate	$Ce(IO_3)_4$	839.73	yel cr.				0.015^{20}		
c255	(III) iodate	$Ce(IO_3)_3 \cdot 2H_2O$	700.86	cr				0.16^{25}		s HNO_3
c256	(III) iodide	$CeI_3 \cdot 9H_2O$	682.97	wh or redsh-wh cr		752	1397	v s		v s al
c257	(III) molybdate	$Ce_2(MoO_4)_3$	760.05	yel, tetr, 2.019, 2.007	4.83	973				
c258	(III) nitrate	$Ce(NO_3)_3 \cdot 6H_2O$	434.23	col or redsh cr (trac La, Di), deliq		$-3H_2O$, 150	d 200	v s	v s	50 al; s acet
c259	(IV) nitrate, basic	$Ce(OH)(NO_3)_3 \cdot 3H_2O$	397.19	long red need.				s		
c260	(III) oxalate	$Ce_2(C_2O_4)_3 \cdot 9H_2O$	706.44	yel-wh cr		d		v sl s		s H_2SO_4, HCl; i $H_2C_2O_4$, alk, eth, al
c261	(IV) oxide	$CeO_2 \cdot xH_2O$		yelsh gelat ppt						s a; sl s alk carb, i alk
c262	(III) oxide	Ce_2O_3	328.24	gray-grn, trig	6.86	1692, ign 200		i	i	s H_2SO_4, i HCl
c263	(IV) oxide(di-)	Ceria. CeO_2	172.12	brn-wh, cub.	7.132^{23}	*ca* 2600		i	i	s H_2SO_4, HNO_3; i dil a
c264	oxychloride	CeOCl	191.57	purp leaf.				i		s dil a
c265	(III) 2,4- pentanedione	Cerium acetylacetonate. $Ce(C_5H_7O_2)_3 \cdot 3H_2O$	491.49	lt yel cr ppt		131—132		d		v s al
c266	(III) *meto*phosphate	$Ce(PO_3)_3$	377.04	micr need	3.272					i a
c267	(III) *ortho*phosphate	Nat. monazite. $CePO_4$	235.09(?)	red, monocl or yel, rhomb, 1.795	5.22			i	i	s a; i al
c268	(III) salicylate	$Ce(C_7H_5O_3)_3$	551.46	wh-redsh wh powd				i		i al
c269	(III) selenate	$Ce_2(SeO_4)_3$	709.11	rhomb	4.456			39.55^{0}	2.513^{00}	
c270	silicide	$CeSi_2$	196.29		5.67^{17}			i		
c271	(IV) sulfate	$Ce(SO_4)_2$	332.24	deep yel cr	3.91^{18}	d 195		sl d, forms basic salts		
c272	(III) sulfate	$Ce_2(SO_4)_3$	568.41	col to grn, monocl or rhomb.	3.912	d 920^{746}		10.1^{0}	2.25^{100}	
c273	(IV) sulfate, dihydrate	$Ce(SO_4)_2 \cdot 4H_2O$	404.30	yel, rhomb				v s d		s dil H_2SO_4
c274	(III) sulfate, monohydrate	$Ce_2(SO_4)_3 \cdot 9H_2O$	730.55	hex need.	2.831			11.87^{15}	0.42^{30}	
c275	(III) sulfate, octahydrate	$Ce_2(SO_4)_3 \cdot 8H_2O$	712.54	pink cr, tricl	2.886^{17}	$-8H_2O$, 630		12^{20}	6^{50}	
c276	(III) sulfate pentahydarte	$Ce_2(SO_4)_3 \cdot 5H_2O$	658.49	monocl.	3.17			3.90^{50}	0.514^{100}	
c277	(III) sulfide	Ce_2S_3	376.42	red cr, br-dk powd, purp	5.020^{11}	d 2100 (vac)		i	d	s dil a
c278	(III) tungstate	$Ce_2(WO_4)_3$	1023.78	yel, tetr	6.77^{17}	1089				
	Cerium complexes									
c280	hexaantipyrinecerium perchlorate	$[Ce(C_{11}H_{12}N_2O)_6] \cdot (ClO_4)_3$	1567.85	col, hex cr		d 295—300		1.08^{20}		
c281	hexaantipyrinecerium iodide (III)	$Ce(C_{11}H_{12}N_2O)_6 \cdot I_3$	1650.21	large yel cr		268—270		15.10^{20}		
c282	**Cesium**	Cs	132.9054	silv met cr hex	1.8785^{15}	28.40 ± 0.01	669.3	d		s liq NH_3
c283	acetate	$CsC_2H_3O_2$	191.95	deliq		194		$945.1^{-2.5}$	$1345.5^{88.5}$	
c284	aluminum sulfate	$CsAl(SO_4)_2 \cdot 12H_2O$	568.19	col, cub, 1.4587	1.97	117		0.34^{0}	42.51^{100}	s dil al
c285	amide	$CsNH_2$	148.93	wh need	3.44_4^{25}	262 ± 1		d		s liq NH_3
c286	azide	CsN_3	174.93	col need, deliq		310		224.2^{0}		1.037^{16} al; i eth
c287	benzoate	$CsC_7H_5O_2$	254.02					294.5^{0}	398.5^{100}	
c288	borofluoride	$CsBF_4$	219.71	rhomb, 1.350.	3.20	550 d		1.6^{17}	*ca* 30^{100}	s dil NH_3
c289	borohydride	$CsBN_4$	147.75	wh, cub, 1.498.	2.404			v s		sl s al; i eth, bz
c290	bromate	$CsBrO_3$	260.81	hex, *ca* 2.15	4.109^{16}	*ca* 420 d		3.66^{25}	5.32^{35}	
c291	bormide, mono-	CsBr	212.81	col cub, 1.6984	4.44, liq 3.04^{700}	636	1300	124.3^{25}	v s	s a
c292	bromide, tri-	$CsBr_3$	372.62	rhomb		180				
c293	*di*bromochloride	$CsBr_2Cl$	328.17	yel-red, rhomb		191	150, $-Br_2$	s		d al, acet
c294	bromochloride iodide	CsIBrCl.	375.17	yel-red, rhomb		235	d 290	s		s al
c295	bromoiodide di-	$CsIBr_2$	419.62	rhomb	4.25	248	d 320	4.61^{20}		s al
c296	carbonate	Cs_2CO_3	325.82	col cr, deliq.		d 610		260.5^{15}	v s	11^{19} al; s eth
c297	carbonate, hydrogen	$CsHCO_3$	193.92	rhomb		$175 - {}^{1}/_{2}H_2O$		209.3^{15}	v s	s al
c298	chlorate	$CsClO_3$	216.36	sm cr.	3.57			$6.28^{19.8}$	76.5^{90}	s al
c299	*per*chlorate	$CsClO_4$	232.36	rhomb, at 219 cub, 1.4752, 1.4788, 1.4804	3.327^{4}	d 250		2.00^{25}	28.57^{99}	0.093^{25} al; 0.7878^{25} al; 0.150^{25} acet
c300	chlorobromide	$CsBrCl_2$	283.72	glossy-yel, rhomb.		205		s		d al, eth
c301	chloride	CsCl.	168.36	col, cub, deliq, 1.6418	3.988	645	1290	$162.22^{0.7}$	$259.56^{89.5}$	33.7^{25} MeOH; v s al; i acetone
c302	chloroiodide	$CsICl_2$	330.72	or, trig.	3.86	230	d 290	s		s al
c303	chloroaurate	$CsAuCl_4$	471.68	yel, monocl.				0.5^{10}	27.5^{100}	s al
c304	chlorobromide, di-	$CsBrCl_2$	283.72	glossy-yel, rhomb		205				
c305	chloro*di*bromide	$CsBr_2Cl$	328.17	yel		191				
c306	chloroiodide, di-	$CsICl_2$	330.72	or, trig.	3.86	230	d 290	s		s al
c307	chloroplatinate	Cs_2PtCl_6	673.61	yel, cub	4.197 ± 0.004	d 570		0.024^{0}	0.377^{100}	i al
c309	chlorostannate	Cs_2SnCl_6	597.22	wh, cub	3.33					
c310	chromate	Cs_2CrO_4	381.80	yel pr, rhomb.	4.237			71.4^{13}	95.5^{30}	
c311	chromium sulfate	Cesium chromium alum. $Cs[Cr(H_2O)_6](SO_4)_2 \cdot 6H_2O$	593.20	vlt cr.	2.064	116		9.4^{25}		
c312	cyanide	CsCN	158.92	very sm wh cr	2.93			v s	v s	
c313	fluoride	CsF	151.90	cub, deliq, 1.478 ± 0.005^{18}	4.115	682	1251	367^{18}		191^{15} MeOH; i Diox, Pyr

No.	Name	Synonyms and Formulae	Mol. wt.	Crystalline form, properties and index of refraction	Density or spec. gravity	Melting point, °C	Boiling point, °C	Solubility, in grams per 100 cc: Cold water	Hot water	Other solvents
c314	fluoride	$CsF \cdot 1^1/_2H_2O$	178.93			703		366.5^{18}		
c315	fluorogermanate	Cs_2GeF_6	452.39	isotrop cr, reg oct	4.10			sl s	v s	sl s a
c316	fluosilicate	Cs_2SiF_6	407.89	wh, cub	3.372^{17}			60^{17}	sl s	i al
c317	fluotellurite	$CsTeF_5$	355.50	col need				d	d	s HF soln
c318	formate	$CsCHO_2$	177.92		1.0169^{21}_{4}				$2012^{96.4}$	
c319	formate	$CsCHO_2 \cdot H_2O$	195.94			41, $-H_2O$				
c320	gallium selenate	$CaGa(SeO_4)_2 \cdot 2H_2O$	704.72	col cr.				4.14^{25}		
c321	gallium sulfate	$CsGa(SeO_4)_2 \cdot 12H_2O$	610.92	col cub, 1.46495		2.113		1.21^{25}		0.0035^{25} 75% al
c322	hydride	CsH	133.91	wh cr, cub	3.41	d		d	d	d a; i org solv
c323	hydrofluoride	$CsF \cdot HF$	171.91	need, deliq		160		v s		v s a; i al
c324	hydrogencarbide	$CsHC_2$	157.94	trsp cr		300		d		
c325	hydroxide	$CsOH$	149.90	lt yel, deliq	3.675	272.3		395.5^{15}		s al
c326	iodate	$CsIO_3$	307.81	wh, monocl	4.85			2.6^{24}		
c327	*meta*periodate	$CsIO_4$	323.81	wh rhomb pl	4.259^{15}_{4}			2.15^{15}	s	
c328	iodide	CsI	259.81	rhomb, deliq, 1.7876	4.510^{25}_{4}	626	1280	44^{0}	160^{61}	s al
c329	iodide, penta-	CsI_5	767.43	bl, tricl		73				
c330	iodide, tri-	CsI_3	513.62	blk, rhomb	4.47	207.5		sl s	sl s	s al
c331	iodotetrachloride	$CsICl_4$	401.62	pale or needles	3.374^{-10}	228	d	sl s	sl s	
c332	iron (II) sulfate	$Cs_2SO_4 \cdot FeSO_4 \cdot 6H_2O$	621.86	lt grn, monocl, 1.500, 1.504, 1.509	2.791^{20}_{4}	*ca* 70		101.1^{25} (anhyd)		
c333	iron (III) sulfate	$CsFe(SO_4)_2 \cdot 12H_2O$	597.05	pa-vlt cr, 1.484	2.061^{20}	*ca* 90		s	s	
c334	magnesium sulfate	$Cs_2SO_4 \cdot MgSO_4 \cdot 6H_2O$	590.32	col, monocl, 1.486, 1.452	2.676^{20}_{4}					
c335	*per*manganate	$CsMnO_4$	251.84		3.597	d 320		0.097^{1}	1.27^{50}	
c336	mercury bromide(ic)	$CsBr \cdot 2HgBr_2$	933.61	rhomb				0.807^{17}		sl s al
c337	mercury chloride(ic)	$CsCl \cdot HgCl_2$	439.85	col, cub or rhomb, 1.792				1.44^{17}		i abs al
c338	nitrate	$CsNO^3$	194.91	col, hex or cub, 1.55, 1.56	3.685 liq 2.71^{500}	414	d	9.16^{0}	196.8^{100}	s acet, sl s al
c339	nitrate, hydrogen	$CsNO_3 \cdot NHO_3$	257.92	oct		100				
c340	nitrate, dihydrogen	$CsNO_3 \cdot 2HNO_3$	320.94	col pl.		32—36				
c341	nitrite	$CsNO_2$	178.91	yel cr.				v s	v s	
c342	oxalate	$Cs_2C_2O_4$	353.83		3.230^{15}			282.9^{25}		
c343	oxide	Cs_2O	281.81	or need	4.25	d 400; m.p. 490 (in N_2)		v s	d	s a
c344	oxide, per	Cs_2O_2	297.81	pa yel need	4.25	400	650. $-O_2$	s	d	s a
c345	oxide, tri-	Cesium oxide, sesqui- Cs_2O_3	313.81	choc br cr, cub.	4.25	400		d		s a
c346	phthalate, hydrogen	$CsHC_8H_4O_4$	298.03	rhomb	2.178					
c347	polonium chloride	Cs_2PoCl_6	687.53	cub, 1.86	3.82					
c348	rhodium sulfate	$CsRh(SO_4)_2 \cdot 12H_2O$	644.12	yel, oct	2.238	110—111		sl s		
				or cr	2.22^{20}_{4}	111		sl s	s	
c350	salicylate	$CsC_7H_5O_3$	270.02					196.2^{0}	1522^{100}	
c351	selenate	Cs_2SeO_4	408.77	rhomb, deliq, 1.5950, 1.5060, 1.5964	4.4528^{20}_{4}			244.8^{12}		
c352	sulfate	Cs_2SO_4	361.87	col rhomb, or hex, 1.560, 1.564, 1.566	4.243	1010	tr hex 600	167^{0}	220^{100}	i al, acet
c353	sulfate, hydrogen	$CsHSO_4$	229.97	col rhomb pr	3.352^{16}	d		s		
c354	sulfide	$Cs_2S \cdot 4H_2O$	369.93	wh cr, deliq.				v s	v s	
c355	sulfide, di-	Cs_2S_2	329.93	dk red, amorph.		460	>800	hgr		
c356	sulfide, di-	$Cs_2S_2 \cdot H_2O$	347.95	tetr				s		
c357	sulfide, hexa-	Cs_2S_6	458.17	br red		186				
c358	sulfide, penta-	Cs_2S_5	426.11		2.806^{15}	210				
c359	sulfide, tetra-	Cs_2S_4	394.05	yel		d 160				
c360	sulfide, tri-	Cs_2S_3	361.99	yel leaf		217	780			
c361	tartrate, hydrogen	$CsHC_4H_4O_6$	281.99	wh, rhomb cr.				9.7^{25}	98^{100}	
c362	*l*-tartrate	$Cs_2C_4H_4O_6$	413.88	col, trig	3.03^{14}			v s d	v s	
c363	vanadium sulfate	Cesium vanadium alum·$VCs(SO_4)_2 \cdot 12H_2O$	592.15	red, cub, 1.4780	2.033^{20}_{4}	82	$-12H_2O$, 230; d 300	0.464^{10}	sl s	
c364	**Chloramine, mono-**	NH_2Cl	51.48	yel liq		−66		s		s al, eth; v sl s CCl_4, bz
c365	**Chloric acid**	$HClO_3 \cdot 7H_2O$	210.57	known only as col sol.	$1.282^{14.2}$	<−20	d 40	v s		
c366	**Chloric acid, per**	$HClO_4$	100.46	col liq unstable	1.764^{22}	−112	39^{56}	∞		
c367	**Chloric acid, per**	Hydronium perchlorate·$HClO_4 \cdot H_2O$ or $(H_3O)^+(ClO_4)^-$	118.47	need, fairly stab	1.88, liq 1.776^{50}	50	exp 100	v s	v s	
c368	per, dihydrate	$HClO_4 \cdot 2HO$	136.49	stab liq.	1.65	−17.8	200	v s	v s	s al
c369	**Chlorine**	Cl_2	70.906	grnsh-yel gas, or liq, or rhomb cr; gas 1.000768, liq 1.367	3.214° g/ℓ	−100.98	−34.6	310^{10} cm^3 1.46° g	177^{30} cm^3 0.57^{30} g	s alk
c370	azide	chlor(o)azide Cln_3	77.47	gas, expl				sl s		d alk
c371	fluoride, mono-	ClF	54.45	col gas	1.62^{-100}	−154 ± 5	−100.8	d	d	
c372	fluoride, tri-	ClF_3	92.45	col gas	1.77^{13}	−83	11.3	d	d	
c373	hydrate	$Cl_2 \cdot 8H_2O$	215.03	lt yel, rhomb	1.23	d 9.6		i		s alk

PHYSICAL CONSTANTS OF INORGANIC COMPOUNDS (continued)

No.	Name	Synonyms and Formulae	Mol. wt.	Crystalline form, properties and index of refraction	Density or spec. gravity	Melting point, °C	Boiling point, °C	Solubility, in grams per 100 cc: Cold water	Hot water	Other solvents
c374	oxide, di-	ClO_2	67.45	yel red gas, or red cr, expl.	3.09^{11} g/ℓ	−59.5	9.9^{731} exp	2000^{4} cm³	d to $HClO_3$, Cl_2, O_2	s alk, H_2SO_4
c375	oxide, hept-	Cl_2O_7	182.90	col oil		−91.5	82	s d		s bz
c376	oxide, mono-	Cl_2O	86.91	yel-red gas, or red-br liq	3.89° g/ℓ	−20	3.8^{766} exp	200 cm³	d to HOCl	s alk, H_2SO_4
c377	oxide, tetr-	ClO_4 or Cl_2O_8	99.45				d	s d		s bz
c378	chloroauric acid	$HAuCl_4{\cdot}4H_2O$	411.85	brt yel need, deliq		d		s	v s	s al, eth
c379	chloroplatinic acid	$H_2PtCl_6{\cdot}6H_2O$	517.91	red br pr,deliq	2.431	60		v s	v s	s al, eth
c380	**chlorostannic acid**	$H_2SnCl_6{\cdot}6H_2O$	441.52	col leaf	1.93	9		s		
c381	**Chlorosulfonic acid**	$ClSO_3H$	116.52	col fum liq, 1.437^{14}	1.766^{18}	−80	158	d to H_2SO_4 + HCl		d al, a; i CS_2
c382	**Chlorotetroxy fluoride**	ClO_4F	118.45	col gas, v exp		−167.3	−15.9			
c383	**Chloryl(per-)fluoride**	ClO_3F	102.45	gas	1.392^{25}	−146	−46.8			
c384	**Chromium**	Cr	51.996	steel gray, cub v hard	7.20^{28}	1857 ± 20	2672	i	i	s dil H_2SO_4, HCl; i HNO_3, aq reg
c385	(II) acetate	$Cr(C_2H_3O_2)_2$	170.09	red cr				sl s	s	sl s al
c386	(III) acetate	$Cr(C_2H_3O_2)_3{\cdot}H_2O$	247.15	gray-grn powd or blsh-grn pasty mass				s		i al
c387	arsenide, mon-	CrAs	126.92	gray, hex	6.35^{16}			i	i	i a
c388	boride, mono-	CrB	62.81	silv cr, orthorhomb	6.17	2760(?)		i	i	s fus Na_2O_2
c389	(II) bromide	$CrBr_2$	211.80	wh cr.	4.356	842		s	s	s al
c390	(III) bromide	$CrBr_3$	291.71	olv gr, hex	4.250^{25}_{4}	subl		i	s	v s al; d alk
c391	bromide, hexahydrate	$[CrBr_2(H_2O)_4]Br{\cdot}2H_2O$	399.80	grn cr, deliq				s	s tr to vlt	s al; i eth
c392	bromide, hexahydrate	$[Cr(H_2O)_6]Br_3$	399.80	blsh gray to vlt.	5.4^{17}			v s	v s	i al
c393	(tri-)carbide, di-	Cr_3C_2	180.01	gray, rhomb	6.68	1980	3800	i	i	
c394	carbonyl	$Cr(CO)_6$	220.06	col, orthorhomb	1.77	d 110	210 exp	i	i	i al, eth, ac a; sl s CHl_3, CCl_4
c395	(II) chloride	$CrCl_2$	122.90	wh need, deliq	2.878^{25}	824		v s	v s	i al, eth
c396	(III) chloride	$CrCl_3$	158.36	vlt, trig	2.76^{15}	*ca* 1150	subl 1300	i	sl s	i al, acet, MeOH, eth
c397	chloride, hexahydrate	$[Cr(H_2O)_4Cl_2]Cl{\cdot}2H_2O$	266.45	vlt, monocl	1.76	83		58.5^{25}	s	s al; i eth; sl s acet
c398	(II) fluoride	CrF_2	89.99	grn, cr, monocl	4.11	1100	>1300	sl s		i al; s h HCl
c399	(III) fluoride	CrF_3	108.99	grn, rhomb	3.8	>1000	subl 1100−1200	i		i al, NH_3; sl s a; s HF
c400	(II) hydroxide	$Cr(OH)_2$	86.01	yel-br		v		d		s a
c401	iodate, hydrate	$[Cr(H_2O)_6]I_3{\cdot}3H_2O$	594.85	dk vlt cr, hygro	4.915^{25}_{4}	41 − HI		v s	v s	s al, acet; i CHl
c402	(II) iodide	CrI_2	305.80	graysh powd	5.196	856	subl vac 800	s		
c403	(III) iodide	CrI_3	432.71	shiny blk cr.	4.915^{25}_{5}	>600	$-I_2$, vac 350			
c404	(III) nitrate	$Cr(NO_3)_3{\cdot}7{}^1/_2H_2O$	373.13	br, monocl		100	d	s	s	
c405	(III) nitrate	$Cr(NO_3)_3{\cdot}9H_2O$	400.15	purple, monocl		60	d 100	s	s	s a, slk, al, acet
c406	nitride, mono-	CrN	66.00	cub or amorph	5.9	d 1700		i	i	sl s aq reg
c407	(II) oxlate	$CrC_2O_4{\cdot}H_2O$	158.03	yel cr powd	2.468			sl s	s	s dil a
c408	(III) oxalate	$Cr_2(C_2O_4)_3{\cdot}6H_2O$	476.14	red, amorph, hyg		120, $-H_2O$ tr to grn		s	s	v s (red) al, eth; i (grn) al
c409	oxide, di-	CrO_2	83.99	br-blk powd.		300, −O		i		s HNO_3
c410	(II) oxide, mon-	CrO	68.00	blk powd				i	i	i dil HNO_3
c411	(III) oxide, sesqui-	Cr_2O_3	151.99	grn, hex, 2.551	5.21	2266 ± 25	4000	i	i	i a, alk, al
c412	(III) oxide, sesqui-	$Cr_2O_3{\cdot}xH_2O$	varies	vlt, amorph or bl-gray grn gel				i	i	s a, alk; sl s NH_4OH
c413	oxide, tri-	Chromic anhydride, "chromic acid", CrO_3	99.99	red, rhomb, deliq	2.70	196	d	61.7°	67.45^{100}	s al, eth, H_2SO_4 HNO_3
c414	oxychloride	CrO_2Cl_2	154.90	dk red liq	1.911	−96.5	117	d	d	d al; s eth, ac a
c415	2,4-pentandione	Chrominium acetylaceton-ate. $Cr(C_5H_7O_2)_3$	349.32			216	340	i		s org solv; i lgr
c416	(III) *ortho*phosphate	$CrPO_4{\cdot}2H_2O$	183.00	vlt cr	$2.42^{32.5}$			sl s		s a, alk; i ac a
c417	(III) *ortho*phosphate	$CrPO_4{\cdot}6H_2O$	255.06	vlt, tricl, 1.568, 1.591, 1.699	2.121^{14}	100				
c418	*pyro*phosphate	$Cr_4(P_2O_7)_3$	729.81	pa grn, monocl.	3.2			i	i	s alk
c419	phosphide, mono-	CrP	82.97	gray-blk cr	5.7^{15}			i		s HNO_3, HF
c420	silicide	Cr_3Si_2	212.16	gray, tetr pr.	5.5°			i	i	s HCl, HF; i H_2SO_4, HNO_3
c421	(II) sulfate	$CrSO_4{\cdot}7H_2O$	274.16	bl cr				12.35°	d	s ls al; s NH_4OH
c422	(III) sulfate	$Cr_2(S)_4)_3$	392.16	vlt or red powd	3.012			i, s*		sl s al; i a
c423	(III) sulfate	$Cr_2(SO_4)_3{\cdot}15H_2O$	622.39	vlt, amorph sc	1.867^{17}	100	$-10H_2O$, 100	s	d 67	i al
c424	(III) sulfate	$Cr_2(SO_4)_3{\cdot}18H_2O$	716.44	bl vlt, cub oct, 1.564	1.7^{22}	$-12H_2O$,100		120^{20}	s	s al
c425	(II) sulfide, mono-	CrS	84.06	blk powd, hex	4.85	1550		i		v s a
c426	(III) sulfide, sesqui-	Cr_2S_3	200.17	brn-blk powd	3.77^{19}	−S, 1350		i, d	i	s HNO_3; d al
c427	(III) sulfite	$Cr_2(SO_3)_3$	344.17	grnsh-wh	2.2	d				
c428	(II) tartrate	$CrC_4H_4O_6$	200.07	bl powd	2.33			i	i	sl s a; i ac a
c429	**Chromium complexes** *hex*ammine chromium-(III) chloride	$[Cr(NH_3)_6]Cl_3{\cdot}H_2O$	278.55	yel cr.	1.585			s		
c430	*hex*aureachromium- (III) fluosilicate	$[Cr(CON_2H_4)_6]_2\cdot[SiF_6]_3{\cdot}3H_2O$	1304.93	lt grn leaf				0.522^{20}		i al

* Several chromic salts exist in two forms, a soluble and insoluble modification.

PHYSICAL CONSTANTS OF INORGANIC COMPOUNDS (continued)

No.	Name	Synonyms and Formulae	Mol. wt.	Crystalline form, properties and index of refraction	Density or spec. gravity	Melting point, °C	Boiling point, °C	Solubility, in grams per 100 cc: Cold water	Hot water	Other solvents
c431	*hex*aureachromium- (III) *per*rhenate	$[Cr(CON_2H_4)_6]\cdot(ReO_4)_3$	1162.44	grn need	2.652^{25}_{4}			1.786		0.667 al
c432	chloropentammine chromium chloride	$[Cr(NH_3)_5Cl]Cl_2$	243.51	red, oct	1.696			0.65^{16}		i HCl
c433	**Cobalt**	Co	58.9332	silv gray met, cub	8.9	1495	2870	i	i	s a
c434	(III) acetate	$Co(C_2H_3O_2)_3$	236.07	grn, oct		d 100		hydr		s a, glac ac a
c435	(II) acetate	$Co(C_2H_3O_2)_2\cdot4H_2O$	249.08	red-vlt, monocl, deliq, 1.542	1.705^{19}	$-4H_2O$, 140		s	s	s a, al
c436	aluminate	(approx) Thenard's blue. $CoAl_2O_4$	176.89	bl, cub				i	i	
c437	(II) *ortho*arsenate	$Co_3(AsO_4)_2\cdot8H_2O$	598.76	vlt-red, monocl, 1.626, 1.661, 1.669	3.178^{15}	d		i	i	s dil a, NH_4OH
c438	arsenic sulfide	Nat. cobaltite. CoAsS	165.91	gray-redsh	6.2—6.3	d				
c439	arsenide	Co_2As	192.79	cr powd	8.28	950		i	i	i CHl, H_2SO_4; s HNO_3, aq reg
c440	(II) benzoate	$Co(C_7H_5O_2)_2\cdot4H_2O$	373.23	gray red leaf		$-4H_2O$, 115		v s		
c441	boride, mono-	CoB	69.74	pr	7.25^{18}			d	d	s HNO_3, aq reg
c442	(II) bromate	$Co(BrO_3)_2\cdot6H_2O$	422.83	red, oct				45.5^{17}		s NH_4OH
c443	(II) bromide	$CoBr_2$	218.74	grn, hex. deliq	4.909^{25}_{4}	678 (in N_2)		66.7^{59}	68.1^{97}	77.1^{20} al; 58.6^{30} MeOH; s eth, acet
c444	(II) bromide hexahydrate	$CoBr_2\cdot6H_2O$	626.84	red-vlt pr, deliq	2.46	47—48, $-4H_2O$ 100	$-6H_2O$, 130	s red color	153.2^{97}	s blk color, al, a, eth
c445	bromoplatinate	$CoPtBr_6\cdot12H_2O$	949.62	trig	2.762					
c446	carbonate	Nat. spherocobaltite. $CoCO_3$	118.94	red, trig, 1.855, 1.60.	4.13	d		i	i	s a; i NH_3
c447	(II) carbonate, basic	$2CoCO_3\cdot3Co(OH)_2\cdot H_2O$	534.74	vlt-red pr				i	d	s a, $(NH_4)_2CO_3$
c448	carbonyl tetra-	Dicobalt octacarbonyl. $[Co(CO)_4]_2$ or $Co_2(CO)_8$	341.95	or cr or dk br, microcr.	1.73^{18}	51	d 52	i	i	sl s al; s CS_2, eth
c449	carbonyl, tri-	Tetracobalt dodecacarbonyl. $[Co(CO)_3]_4$ or $Co_4(CO)_{12}$	571.86	blk cr				sl s		s bz; d Br
c450	(II) chlorate	$Co(ClO_3)_2\cdot6H_2O$	333.93	red, cub, deliq, 1.55	1.92	50	d 100	558.3^{0}	v s	s al
c451	(II) *per*chlorate	$Co(ClO_4)_2$	257.83	red need 1.510, 1.490	3.327			100^{0}	115^{45}	s al, acet
c452	(II) *per*chlorate	$Co(ClO_4)_2\cdot5H_2O$	347.91	red, hex		143		100.13^{0}	115.10^{65}	v s al, acet i $CHCl_3$
c453	*per*chlorate	$Co(ClO_4)_2\cdot6H_2O$	365.93	red pr		d 1534	d	259^{18}		s al, acet
c454	(II) *per*chlorate	$Co(ClO_4)_2\cdot6H_2O$	365.93	red, oct, deliq, 1.55		d 182		255^{18}		s al, acet
c455	(II) chloride	$CoCl_2$	129.84	bl, hex, hygr	3.356^{36}_{4}	724 (in HCl gas)	1049	45^{7}	105^{96}	54.4 al; 8.6 acet; 38.5 MeOH; sl s eth
c456	(III) chloride	$CoCl_3$	165.29	red cr or yel cr.	2.94	subl		s		
c457	(II) chloride, dihydrate	$CoCl_2\cdot2H_2O$	165.87	red-vlt, monocl or tricl, 1.625, 1.671, 1.67	2.477^{25}_{26}			s	s	v sl s eth
c458	(II) chloride, hexahydrate	$CoCl_2\cdot6H_2O$	237.93	red, monocl.	1.924^{25}_{25}	86	$-6H_2O$, 110	76.7^{0}	190.7^{100}	v s (bl col) al; s acet; 0.29 eth
c459	chloroplatinate	$CoPtCl_6\cdot6H_2O$	574.82	trig	2.699	d				
c460	chlorostannate	$CoSnCl_6\cdot6H_2O$	498.43	rhomb or trig		d 100				
c461	(II) chromate	$CoCrO_4$	174.93	gray blk cr		d		i	d	s a, NH_4OH
c462	(II) citrate	$Co_3(C_6H_5O_7)_2\cdot2H_2O$	591.03	rose-red		$-2H_2O$, 150		0.8		
c463	(II) cyanide dihydrate	$Co(CN)_2\cdot2H_2O$	147.00	buff anhydr bl-vlt powd	anhydr 1.872^{25}	$-2H_2O$, 280	d 300	0.00418^{18}		s KCN, HCl, NH_4OH
c464	(II) cyanide, trihydrate	$Co(CN)_2\cdot3H_2O$	165.01	red-gray powd. amorph		$-3H_2O$, 250		i		s KCN
c465	(II) ferricyanide	$Co_3[Fe(CN)_6]_2$	600.71	red need				i		s NH_4OH; i HCl
c466	(II) ferrocyanide	$Co_2Fe(CN)_6\cdot xH_2O$		gray-grn				i		s KCN; i HCl
c467	(II) fluogallate	$[Co(H_2O)_6][GaF_5\cdot H_2O]$	349.75	pink cr, monocl (?), 1.45	2.35	$-5H_2O$, 110		sl s		d a
c468	(II) fluoride	CoF_2	96.93	pink monocl	4.46^{25}_{4}	*ca* 1200	1400	1.5^{25}	s	sl s a; i al, eth, bz
c469	(III) fluoride	CoF_3	115.93	br, hex.	3.88			d to $Co(OH)_3$		i, al,eth, bz
c470	fluoride	$Co_2F_6\cdot7H_2O$	357.96	grn powd	2.314^{25}			d		s H_2SO_4
c471	(II) fluoride, tetrahydrate	$CoF_2\cdot4H_2O$	168.99	α: red, rhomb oct, β: rose cr powd	2.192^{25}_{4}	d 200		s	s	i al
c472	fluosilicate	$CoSiF_6\cdot6H_2O$	309.10	pink trig, 1.382, 1.387	2.113^{19}			$118.1^{21.5}$		
c473	(II) formate	$Co(CHO_2)_2\cdot2H_2O$	185.00	red cr	2.129^{22}	$-2H_2O$, 140	d 175	5.03^{20}		
c474	(III) hydroxide	$Co(OH)_2$	92.95	rose-red, rhomb	3.597^{15}	d		0.00032		s a, NH_4 salts; i alk
c475	(III) hydroxide	$Co_2O_3\cdot3H_2O$	219.91	blk-brn powd	4.46	d	$-H_2O$, 100	0.00032		s a; i al
c476	(II) iodate	$Co(IO_3)_2$	408.74	bl-vlt need	5.008^{18}	d 200		0.45^{18}	1.33^{100}	s HCl, HNO_3. h H_2SO_4
c477	(II) iodate, hexahydrate	$Co(IO_3)_2\cdot6H_2O$	516.83	red, oct	3.689^{21}	d 61	$-4H_2O$, 135	s		
c478¹	(II) iodide (α) stable	CoI_2	312.74	blk hex, hyg	5.68	515(vac)	570 (vac)	159^{0}	420^{100}	v s al, acet
c478²	(II) iodide (β)	CoI_2	312.74	yel need, unstab	5.45^{25}	d 400		s		
c479	(II) iodide, dihydrate	$CoI_2\cdot2H_2O$	348.77	grn, deliq		d 100		376.2^{45}	s	
c480	(II) iodide, hexahydrate	$CoI_2\cdot6H_2O$	420.83	br-red hex, hygr	2.90	d 27, $-6H_2O$,		s	s	s al, eth, chl
c481	iodoplatinate	$CoPtI_6\cdot9H_2O$	1177.59	trig	3.618					

No.	Name	Synonyms and Formulae	Mol. wt.	Crystalline form, properties and index of refraction	Density or spec. gravity	Melting point, °C	Boiling point, °C	Solubility, in grams per 100 cc		
								Cold water	Hot water	Other solvents
c482	linoleate	$Co(C_{18}H_{31}O_2)_2$	617.82	br, amorph				i		s al,eth,acet
c483	(II) nitrate	$Co(NO_3)_2·6H_2O$	291.03	red, monocl, 1.52	1.87_4^{25}	55—56	$-3H_2O$, 55	133.8^{0}	217^{80}	$100.0^{12.5}$ al; s acet; sl s NH_3
c484	nitrosylcarbonyl	$Co(NO)(CO)_3$	172.97	cherryred liq		−1.05	48.6; d 55	i		s al, eth, acet, bz
c485	(II) oleate	$Co(C_{18}H_{35}O_2)_2$	621.85	br, amorph						s al, eth, oils, bz
c486	(II) oxalate	CoC_2O_4	146.95	wh or redsh	3.021^{25}	d 250		i		s a, NH_4OH
c487	oxalate, dihydrate	$CoC_2O_4·2H_2O$	182.98	pink cr		$-H_2O$, *ca* 190		v sl s	sl s	v sl s a; s NH_4OH
c488	(II) oxide	CoO	74.93	pink cub	6.45	1795 ± 20		i	i	s a; i al, NH_4OH
c489	(III) oxide	Co_2O_3	165.86	blk-gray, hex, or rhomb	5.18	d 895		i	i	s a; i al
c490	(II, III) oxide	Co_3O_4	240.80	blk, cub	6.07	tr to CoO 900—950		i	i	v sl s a; i sq reg
c491	palmitate	$Co(C_{16}H_{31}O_2)_2$	569.78			70.5				s pyr, hot CS_2, CCl_4; sl s eth; i MeOH, acet
c492	(II) *ortho*phosphate	$Co_3(PO_4)_2$	366.74	redsh cr	2.587^{25}			i	i	s H_3PO_4, NH_4OH
c493	(II) *ortho*phosphate, dihydrate	$Co_3(PO_4)_2·2H_2O$	402.77	pink powd.				i		s H_3PO_4
c494	(II) *ortho*phosphate, octahydrate	$Co_3(PO_4)_2·8H_2O$	510.86	redsh powd	2.769^{25}	$-8H_2O$, 200		sl s		s min a, H_3PO_4; i al
c495	phosphide	Co_2P	148.84	gray need	6.4^{15}	1386		i	i	s HNO_3, aq reg
c496	(II) propionate	$Co(C_3H_5O_2)_2·3H_2O$	259.12	dk-red cr		*ca*250		anh 33.5^{11}		v s al
c497	(II) *per*rhenate	$Co(ReO_4)_2·5H_2O$	649.42	dk pink		d		d		
c498	(II) selenate, heptahydrate	$CoSeO_4·7H_2O$	328.00	monocl.	2.135					
c499	selenate, hexahydrate	$CoSeO_4·6H_2O$	309.98	red, monocl, 1.5225	2.25^{17}			s	s	
c500	(II) selenate, pentahydrate	$CoSeO_4·5H_2O$	291.97	ruby red, tricl-	2.512	d		v s		
c501	selenide, mono-	CoSe	137.89	yel, hex	7.65	red heat				sHNO_3,aq reg; i alk
c502	(II) *ortho*silicate	CO_2SiO_4	209.95	vlt cr, rhomb	4.63	1345		i	i	s dil HCl
c503	silicide	CoSi	87.02	rhomb		1395				s HCl; i HNO_3, H_2SO_4
c504	silicide,di-	$CoSi_2$	115.10	rhomb	5.3	1277				
c505	(di-)silicide	Co_2Si	145.95	gray cr	7.28°	1327				d a
c506	(II) *ortho*stannate	Co_2SnO_4	300.55	grnsh-bl, cub	6.30^{18}					i H_2SO_4; s h HCl
c507	(II) sulfate	$CoSo_4$	154.99	dk blsh, cub	3.71_{25}^{25}	d 735		36.2^{20}	83^{100}	1.04^{18} MeOH; i NH_3
c508	(II) sulfate, heptahydrate	Nat. bieberite. $CoSO_4·7H_2O$	281.10	red-pink, monocl, 1.477, 1.483, 1.489	1.948_{25}^{25}	96.8	$-7H_2O$, 420	60.4^{3}	67^{70}	2.5^{3} al; 54.5^{18} MeOH
c509	(II) sulfate, hexahydrate	$CoSO_4·6H_2O$	263.08	red, monocl, 1.531, 1.549, 1.552	2.019_{15}^{15}	$-2H_2O$, 95				
c510	(II) sulfate, monohydrate	$CoSO_4·H_2O$	173.01	red cr, 1.603, 1.639, 1.683	3.075^{25}	d		s	s	
c511	(III) sulfate	$Co_2 (SO_4)_3·18H_2O$	730.31	bl-grn		d 35		s d		s H_2SO_4; i pyr
c512	sulfide, di-	CoS_2	123.05	blk, cub	4.269			i		s HNO_3, aq reg
c513	sulfide, mono-	Nat. sycoporite. CoS	90.99	redsh, silv-wh, oct	5.45^{18}	>1116		0.00038^{18}		sl s a
c514	(III) sulfide, sesqui	Co_2S_3	214.05	blk cr	4.8					d a, aq reg
c515	(tri-) sulfide	Cobalt sulfide, tetra-(ous, ic) Nat. linneite. Co_3S_4	305.04	dk gray, cub	4.86	d 480				
c516	(II) sulfite	$CoSO_3·5H_2O$	229.07	red				i		s H_2SO_3
c517	tartrate	$CoC_4H_4O_6$	207.01	redsh, monocl				sl s		s dil a
c518	thiocyanate	$Co(SCN)_2·3H_2O$	229.13	vlt, rhomb.		$-3H_2O$, 105		s		s al, MeOH, eth
c519	*ortho*titanate	Co_2TiO_4	229.74	grnsh-blk, cub.	5.07—5.12					s conc HCl; sl s dil HCl
c520	(II) tungstate	$CoWO_4$	306.78	bl-grn, monocl	8.42			i		s h conc a; sl s c dil a
	Cobalt complexes									
c521	hexammine cobalt (II) bromide	$CoBr_2·6NH_3$	320.92	dk pink cr.	1.871_4^{25}	d 258		d		
c522	diamminecobalt (II) chloride [α]	$CoCl_2·2NH_3$	163.90	rose cr	2.097	273				
c523	diamminecobalt (II) chloride (β)	$CoCl_2·2NH_3$	163.90	bl-vlt	2.073	tr to α, 210 (ín NH_3)				
c524	hexamminecobalt (II) chloride	$[Co(NH_3)_6]Cl_2$	232.02	rose red, oct	1.497	d		d		s NH_4OH; i abs al
c525	hexamminecobalt (III) chloride	$Co(NH_3)_6 Cl_3$	267.48	wine-red, monocl	1.710_4^{25}	$-1NH_3$, 215		5.9^{10}	$12.74^{46.6}$	s conc HCl; i al,, NH_4OH
c526	hexamminecobalt (II) iodide	$CoI_2·6NH_3$	414.93	dk pink, cub	2.096_4^{25}	141^{100mm}				
c527	hexamminecobalt (III) nitrate	$Co(NH_3)_6·(NO_3)_3$	347.13	yel, tetr	1.804_4^{25}			1.7^{25}	v s	v sl s dil a
c528	hexamminecobalt (III) *per*rhenate	$[Co(NH_3)_6](ReO_4)_3·2H_2O$	947.76	or-yel pr.	3.329^{25}			0.0469		
c529	hexamminecobalt (II) sulfate	$CoSO_4·6NH_3$	257.17	pink powd.	1.654_4^{25}	d 116^{760}		d		v s dil NH_3
c530	hexamminecobalt (III) sulfate	$[Co(NH_3)_6]_2(SO_4)_3·5H_2O$	700.48	dk yel, monocl.	1.797^{25} anhydr	$-4H_2O$, 100	$-5H_2O$, 150	$1.4^{17.4}$		
c531	ammonium tetranitrodiammine (III) cobaltate	Erdmann's salt. $NH_4[Co(NH_3)_2(NO_2)_4]$	295.05	redsh-pa brn, rhomb, 1.78, 1.78. 1.74	1.876^{25}					

No.	Name	Synonyms and Formulae	Mol. wt.	Crystalline form, properties and index of refraction	Density or spec. gravity	Melting point, °C	Boiling point, °C	Solubility, in grams per 100 cc		
								Cold water	Hot water	Other solvents
c532	aquapentamminecobalt (III) chloride (roseo)	$[Co(NH_3)_5 \cdot H_2O]Cl_3$	268.46	brick red cr	1.7^{25}	d 100	...	24.87^{25}	...	sl s HCl; i al
c533	aquapentamminecobalt (III)-sulfate (roseo)	$[Co(NH_2)_5H_2O]_2(SO_4)_3 \cdot 2H_2O$	638.33	red, tetr	1.854^{20}	$-3H_2O$, 99	d 110	$1^{17.2}$	1.72^{27}	s H_2SO_4
c534	cis-chloroaquotetrammine-cobalt (III) chloride	$[Co(NH_3)_4H_2O)Cl]Cl_2$	251.43	vlt, rhomb	1.847	d	...	1.4^{0}	...	s a; i al
c535	chloropentamminecobalt-(III) chloride (purpureo)	$[Co(NH_3)_5Cl]Cl_2$	250.44	dk red-vlt, rhomb	1.819^{25}_{25}	d	...	0.4^{25}	$1.031^{46.6}$	s conc H_2SO_4; i al
c536	triethylenediammine-cobalt-(III) chloride	$Co[C_2H_4(NH_2)_2]_3Cl_3 \cdot 3H_2O$	399.64	br pr	1.542^{17}	256; $-3H_2O$, 100	...	v s	...	...
c537	trinitrotriamminecobalt	$CO(NH_3)_3(NO_2)_3$	248.04	yel, rhomb pl or leaf	1.992^{25}_{4}	d 158	exp 164	$0.177^{16.5}$	0.28^{25}	...
c538	trinitrotetramminecobalt-(III)nitrate	$[Co(NH_3)_4(NO_2)_2]NO_2$	265.07	hel, rhomb	1.922^{17}	...	...	3^{20}	...	...
c539	potassium tetranitrodiamminecobaltate (III)	$K[Co(NH_3)_2(NO_2)_4]$	316.11	yel, rhomb	2.076^{15}	...	...	$1.758^{16.5}$	...	...
	Columbium	see Niobium.		...						
c540	**Copper**	Cu	63.546 ± 3	redsh met, cub	8.92	1083.4±0.2	2567	i	i	s HNO_3, h H_2SO_4; v sl s HCl, NH_4OH
c541	acetate, basic	Blue verdigris. $Cu(C_2H_3O_2)_2 \cdot CuO.6H_2O$	369.27	grnsh-bl powd	...	...	...	sl s	...	s dil a, NH_4OH; sl s al
c542	(II) acetate	Neutral verdigris. $Cu(C_2H_3O_2)_2 \cdot H_2O$	199.65	dk grn powd, 1.545, 1.550	1.882, anhydr-1.93	115	d 240	7.2	20	7.14 al; s eth
c543	(II) acetate metaarsenate	Paris green. $Cu(C_2H_3O_2)_2 \cdot 3Cu(AsO_2)_2$ (approx)	1013.80	em grn powd	...	...	...	i	...	s a, NH_4OH; i al
c544	(III) acetylide	Cu_2C_2	151.11	red, amorph, expl.	...	exp	...	v sl s	...	s a, KCN
c545	amine azide	$Cu(NH_3)_2(N_3)_2$	181.65	dk grn cr, exp	...	d 100—105	exp 202	i	d	d a; i MeOH
c546	(II) *di*amminechloride, di-	$Cu(NH_3)_2Cl_2$	168.51	grn cr	2.32^{25}_{4}	260—270	d 300	i	...	s NH_4OH; i abs a
c547	(II) *hex*amminechloride, di-	$Cu(NH_3)_6Cl_2$	236.64	bl, cub	1.48^{25}_{4}	...	...	v s	...	...
c548	*tet*rammine dithionate	$[Cu(NH_3)_4]S_2O_6$	291.78	vlt-bl cr	...	d 160	...	s	d	...
c549	(II) *tet*rammine nitrate	$[Cu(NH_3)_4](NO_3)_2$	225.68	dk-bl, oct	1.91^{25}_{4}	d 210 exp	...	s	...	...
c550	(II) amine nitrate	$[Cu(NH_3)_4](NO_2)_2$	223.68	vlt-bl, tetr	...	$-2NH_3$ 97	...	v s	...	...
c551	*tet*rammine sulfate	Cuprum ammoniacale. $[Cu(NH_3)_4]SO_4 \cdot H_2O$	245.74	dk-bl, rhomb, unstab	1.79^{25}_{4}	$-NH_3.H_2O$, 30	...	$18.05^{21.5}$	...	...
c552	(tri-)antimonide	Cu_3Sb	312.39	gray	8.51	687	...	...	...	...
c553	(II) *ortho*arsenate	$Cu_3(AsO_4)_2 \cdot 4H_2O$	540.54	blsh-grn	...	...	...	i	i	s a, NH_4OH
c554	(II) *ortho*arsenate, di-H	$Cu_5H_2(AsO_4)_4 \cdot 2H_2O$	911.45	bl	...	...	...	i	...	s a, NH_4OH
c555	arsenide	Cu_5As_2	467.57	bl, oct	7.56	...	...	i	i	s a, NH_4OH
c556	tri-arsenide	Nat. domeykite, Cu_3As.	265.56	hex	8.0	830	...	...	...	...
c557	(II) *ortho*arsenite, hydrogen(?)	Scheele's green. $CuHAsO_3$(?)	187.47	grn powd	...	d	...	i	i	s a, NH_4OH; i al
c558	(I) azide	CuN_3	105.57	col cr, v exp	3.26	...	...	0.00075^{20}	...	d conc H_2SO_4; s NH_4Cl
c559	(II) azide	$Cu(N_3)_2$	147.59	brn-red or brn-yel cr, exp	2.604	exp 215	...	0.008^{20}	...	v s dil a
c560	(II) benzoate	$Cu(C_7H_5O_2)_2 \cdot 2H_2O$	341.81	lt bl cr powd	...	$-H_2O$, 110	...	sl s	...	s dil a; sl s al
c561	(II) *meta*borate	$Cu(BO_2)_2$	149.16	blsh grn cr powd	3.859	...	...	s	...	...
c562	boride	Cu_3B_2	212.26	yel	8.116	...	...	...	...	...
c563	(II) bromate	$Cu(BrO_3)_2 \cdot 6H_2O$	427.44	bl-grn, cub	2.583	d 180	$-6H_2O$, 200	v s	...	s NH_4 OH
c564	(I) bromide	CuBr (or Cu_2Br_2)	143.45	wh, cub, 2.116.	4.98	492	1345	v sl s	d	s HBr, HCl, HNO_3, NH_4OH; i acet
c565	(II) bromide	$CuBr_2$	223.35	blk, monocl, deliq	4.77^{25}_{4}	498	...	v s	...	s al, acet, NH_3, pyr; i bz
c566	*trioxy*bromide	$CuBr_2 \cdot 3Cu(OH)_2$	516.04	em grn, rhomb	4.00	$-H_2O$, 210—215	d 240—250	i	d	s dil min a, NH_4OH; v s ac a
c567	(II) butyrate	$Cu(C_4H_7O_2)_2 \cdot 2H_2O$	273.77	dk grn cr	...	...	...	v sl s	...	s al, eth, NH_4OH, dil a
c568	(I) carbonate	Cu_2CO_3	187.10	yel	4.40	d	...	i	i	s a, NH_4OH
c569	(II) carbonate, basic	Nat. malachite. $CuCO_3 \cdot Cu(OH)_2$	221.12	dk grn, monocl, 1.655, 1.875, 1.909	4.0	d 200	...	i	d	0.026 aq CO_2; s a, NH_4OH, KCN; i al
c570	(II) carbonate, basic	Nat. azurite, chessylite. $2CuCO_3 \cdot Cu(OH)_2$	344.67	bl, monocl, 1.730, 1.758, 1.838	3.88	d 220	...	i	d	s NH_4OH, h $NaHCO_3$
c571	(II) chlorate	$Cu(ClO_3)_2 \cdot 6H_2O$	338.54	grn, cub, deliq	...	65	d 100	207^{0}	v s	s al, acet
c572	(II) chlorate, basic	$Cu(ClO_3)_2 \cdot 3Cu(OH)_2$	523.13	grn cr or amorph	3.55	d	...	i	i	s dil a
c573	*per*chlorate	$Cu(ClO_4)_2$	262.45	monocl, 1.495, 1.505, 1.522	2.225^{23}	82.3	...	s	s	...
c574	*per*chlorate, hexahydrate	$Cu(ClO_4)_2 \cdot 6H_2O$	370.54	lt bl, tricl, deliq, 2.505	2.225^{25}_{4}	82	d 120	ws	...	s al, eth
c575	(I) chloride(ous)	Nat. nantokite. CuCl (or Cu_2Cl_2)	99.00	wh, cub, 1.93	4.14	430	1490	0.0062	...	s HCl, NH_4OH, eth; i al
c576	(II) chloride	$CuCl_2$	134.45	br, yel powd, hygr	3.386^{25}_{4}	620	993 d to CuCl	70.6^{0}	107.9^{100}	53^{15} al; 68^{15} MeOH; s h H_2SO_4, acet
c577	(II) chloride, basic	$CuCl_2 \cdot Cu(OH)_2$	232.01	yel-grn, hex.	3.78	$-H_2O$, 250	d red heat	d	d	...
c578	(II) chloride, dihydrate	Nat. eriochalcite. $CuCl_2 \cdot 2H_2O$	170.43	bl-grn, rhomb, deliq. 1.644, 1.683, 1.731	2.54	$-2H_2O$, 100	d	110.4^{0}	192.4^{100}	s al, NH_4OH
c579	chloride, thioureate	$CuCl \cdot 3[CS(NH_2)_2]$	327.35	col pr, 1.758, 1.17719	1.73	168	...	v s	...	...

PHYSICAL CONSTANTS OF INORGANIC COMPOUNDS (continued)

No.	Name	Synonyms and Formulae	Mol. wt.	Crystalline form, properties and index of refraction	Density or spec. gravity	Melting point, °C	Boiling point, °C	Solubility, in grams per 100 cc: Cold water	Hot water	Other solvents
c580	(II) chromate, basic	$CuCrO_4·2CuO·2H_2O$	374.66	yel br		$-2H_2O$, 260		i		s dil a, NH_4OH; i al
c581	(II) *di*chromate	$CuCr_3O_7·2H_2O$	315.56	blk cr, deliq	2.283	$-2H_2O$, 100		v s	d	s a, NH_4OH, al
c582	(I) chromite	$Cu_2Cr_2O_4$	295.08	gray blk cub pl.	5.24^{20}			i	i	s HNO_3
c583	(II) citrate	$2Cu_2C_6H_4O_7·5H_2O$	720.45	blsh grn powd		$-H_2O$, 100		i		s a, NH_4OH
c584	(I) cyanide	CuCN	89.56	wh, monocl pr	2.92	473 (in N_2)	d	i	i	s HCl, KCN, NH_4OH; sl s
c585	(II) cyanide	$Cu(CN)_2$	115.58	yel-grn powd		d		i		s a, alk, KCN, pyr,
c586	ethylacetoacetate	$Cu(C_6H_9O_3)_2$	321.82	grn need.		192—193	subl	i		v s al, eth; 10^{80} bz
c587	(I) ferricyanide	$Cu_3Fe(CN)_6$	402.59	br red				i		i HCl; s NH_4OH
c588	(II) ferricyanide	$Cu_3[Fe(CN)_6]_2·14H_2O$	866.76	yel-grn				i		i HCl; s NH_4OH
c589	(II) ferrocyanide	Hatchett's brown $Cu_2Fe(CN)_2·xH_2O$		red brn.				i	i	i a, NH_3; s NH_4OH
c590	(I) fluogallate	$[Cu(H_2O)_6][GaF_5·H_2O]$	354.36	pa bl, monocl(?), 1.45	2.20	$-5H_2O$, 110		sl s		s HF
c591	(I) fluoride	CuF (or Cu_2F_2)	82.54	red cr, (exist?)		908	subl, 1100	i		s HCl, HF; d HNO_3; i al
c592	(II) fluoride	CuF_2	101.54	wh, tricl	4.23	d 950		4.7^{20}	s	s dil min a; i al
c593	(II) fluoride dihydrate	$CuF_2·2H_2O$	137.57	bl, monocl	2.93^{25}_{4}	d		4.7^{20}	d	s HCl, HF, HNO_3, al; i acet, NH_3
c594	(I) fluosilicate	Cu_2SiF_6	269.17	red powd		d to SiF_4			d 100	
c595	(II) fluosilicate	$CuSiF_6·4H_2O$	277.68	monocl pr.	2.158			42.8		
c596	(II) fluosilicate hexahydrate	$CuSiF_6·6H_2O$	313.71	bl, rhomb, deliq, 1.409, 1.408	2.207			232^{17}		0.16^{20} 92% al
c597	(II) formate	$Cu(CHO_2)_2$	153.58	bl, monocl	1.831			12.5	d	0.25 al
c598	(II) formate tetrahydrate	$Cu(CHO_2)_2·4H_2O$	225.64	bl cr	1.81	$-H_2O$, 130		6.2		s alk; sl s al
c599	(II) glycerine deriv	$Cu(C_2H_4NO_2)_2·H_2O$	229.68	bl need.		$-H_2O$, 130		0.57^{15}		s alk
c600	hydride	CuH (or Cu_2H_2)	64.55	red-brn, (exist?)	6.38	d sl 55—60		i	d 65	d HCl
c601	(II) hydroxide	$Cu(OH)_2$	97.56	bl gel cr powd	3.368	$-H_2O$, d		i	d	s a, NH_4OH, KCN
c602	(II) *tri*hydroxychloride	γ: Paratacamite δ: atacamite $CuCl_2·3Cu(OH)_2$	427.13	γ: grn, hex, δ: grn rhomb; γ: 1.743, 1.849, δ: 1.861, 1.861, 1.880, grn lt	(γ) 3.75	$-H_2O$, 250		i	i	v s a
c603	(II) *tri*hydroxynitrate	$Cu(NO_3)_2·3Cu(OH)_2$	480.24	dk grn, rhomb or moncl.	rhomb, 3.41 monocl, 3.378	$-H_2O$ ~400		i	d	v s a
c604	(II) iodate	$Cu(IO_3)_2$	413.35	grn, moncl	5.241^{15}	d		0.1364^{15}	i	s dil HNO_3, dil H_2SO_4
c605	(II) iodate, basic	$Cu(OH)IO_3$	255.46	grn, rhomb	4.873	d 290		i	i	s dil H_2SO_4
c606	(II) iodate, monohydrate	Nat. bellingerite. $Cu_3(IO_3)_6·2H_2O$	431.37	bl, tricl	4.872	$-H_2O$, 248	d 290	0.33^{15}	0.65^{100}	s dil H_2SO_4, NH_4OH; i al, dil HNO_3
c607	*para*periodate	Cu_2HIO_6	351.00	grn cr powd.		d 110		i	i	s HNO_3, NH_4OH
c608	(I) iodide	Nat. marshite. CuI (or Cu_2I_2)	190.45	wh or brnsh-wh, cub, 2.346	5.62	605	1290	0.008^{18}		s dil HCl, KI, KCN, con c H_2SO_4, liq NH_3
c609	(II) lactate	$Cu(C_3H_5O_3)_2·2H_2O$	277.72	dk bl, monocl				16.7	45^{100}	s NH_4OH; sl s al
c610	(II) laurate	$Cu(C_{12}H_{23}O_2)_2$	462.17	lt bl powd.		111—113		sl s	sl s	
c611	mercury iodide (α)	Cu_2HgI_4	835.30	red, tetr	6.116	*tr ca* 67		i		
c612	mercury iodide (β)	Cu_2HgI_4	835.30	choc, cub	6.102			i		
c613	(II) nitrate, hexahydrate	$Cu(NO_3)_2·H_2O$	295.65	bl cr, deliq	2.074	$-3H_2O$, 26.4		243.7^{0}	∞	s al
c614	(II) nitrate, trihydrate	$Cu(NO_3)_2·3H_2O$	241.60	bl cr, deliq	2.32^{25}_{4}	114.5	$-HNO_3$, 170	137.8^{0}	1270^{100}	$100^{12.5}$ al; v s liq NH_3
c615	nitride	Cu_3N	204.64	dk grn powd	5.84^{25}_{4}	d 300		d		d a
c616	(II) nitrite, basic	$Cu(NO_2)_2·3Cu(OH)_2$	448.24	grn powd		d 120		i	d	v s dil a; sl s al; s NH_4OH
c617	(II) *hypo*nitrite, basic	$Cu(NO)_2·Cu(OH)_2$	221.12	pea grn amorph, hygr.		d<100		i		s dil a; v s NH_4OH; d NaOH
c618	(II) nitroprusside	$CuFe(CN)_5NO·2H_2O$	315.52	wh-grnsh powd.				i		s alk; i al
c619	(II) oleate	$Cu(C_{18}H_{33}O_2)_2$	626.46	br powd or grn-bl mass, pois				i	i	s eth
c620	(II) oxalate	$CuC_2O_4·{}^1/_2H_2O$	160.57	bl wh.				0.00253^{25}		s NH_4OH; i ac a
c621	(I) oxide	Nat. cuprite. Cu_2O	143.09	red, oct cub, 2.705	6.0	1235	−O, 1800	i	i	s HCl, NH_4Cl, NH_4OH; sl s HNO_3; i al
c622	(II) oxide	Nat. tenorite. CuO	79.55	blk, monocl, β 2.63	6.3—6.49	1326		i	i	s a, NH_4Cl, KCN
c623	oxide, per-	$CuO_2·H_2O$	113.56	br or brnsh-blk cr		d 60		i		i al, s d a
c624	oxide, sub-	Cu_4O	270.18	olv grn, (exist?)		d		i		d a
c625	(II) oxychloride	Nat. atacamite. $Cu_2(OH)_3Cl$ (or $CuCl_2·3Cu(OH)_2$)	213.57	grn, orthorhomb.	3.76—3.78					
c626	(II) oxychloride	Brunswick green. $CuCl_2·3CuO·4H_2O$(?)	445.15	grn powd, or em grn to grnsh-blk, rhomb		$-H_2O$, 140			d 100	s a, NH_4OH
c627	(II) palmitate	$Cu(C_{16}H_{31}O_2)_2$	574.39	grn-bl powd.		120		i		s h bz, CS_2, CCl_4; sl s al, eth; i MeOH, acet
c628	2,4-pentanedione	Copper acetylacetonate $Cu(C_5H_7O_2)_2$	261.76	bl cr		>230	subl	i		sl s al; s chl
c629	(I) phenyl	C_6H_5Cu	140.65	col powd		d 80		d	d	i al, CS_2; s pyr

PHYSICAL CONSTANTS OF INORGANIC COMPOUNDS (continued)

No.	Name	Synonyms and Formulae	Mol. wt.	Crystalline form, properties and index of refraction	Density or spec. gravity	Melting point, °C	Boiling point, °C	Solubility, in grams per 100 cc Cold water	Hot water	Other solvents
c630	(II) *ortho*phosphate	$Cu_3(PO_4)_2 \cdot 3H_2O$	434.63	bl, rhomb		d		i	sl s	s a, NH_4OH, H_3PO_4; i NH_3
c631	(tri-) phosphide	Cu_3P	221.61	gray-blk	6.4—6.8	d		i		s HNO_3; i HCl
c632	(di-) pyridine chloride(di)	$Cu(C_5H_5N)_2Cl_2$	292.65	grn-bluish, monocl, 1.60, 1.75	1.76	d 263		s	d	sl s c al, chl
c633	(II) salicylate	$Cu(C_7H_5O_3)_2 \cdot 4H_2O$	409.84	bl-grn need				v s		v s al, NH_4OH
c634	(II) selenate	$CuSeO_4 \cdot 5H_2O$	296.58	bl, tricl, 1.56	2.559	$-4H_2O$, 50—100	$-5H_2O$, 150	25.7^{15}	d	s a, NH_4OH; v sl s acet; i a
c635	(I) selenide	Cu_2Se	206.05	blk, cub	6.749^{30}_{4}	1113				d HCl
c636	(II) selenide	CuSe	142.51	grn-blk hex pl, unstab	5.99	d red heat		i	i	sl s HCl, NH_4OH; s h HNO_3
c637	selenite	$CuSeO_3 \cdot 2H_2O$	226.53	bl-grn, rhomb	3.31^{25}_{4}	$-H_2O$, 100		i	i	
c638	silicide	Cu_4Si	282.27	wh met	7.53	850				i HCl; d HNO_3
c639	(II) stearate	$Cu(C_{18}H_{35}O_2)_2$	630.50	lt grn-bl amorph powd		125		i		s eth, h bz, chl, turp; sl s pyr; i MeOH, acet
c640	(I) sulfate	Cu_2SO_4	223.15	gray powd, 1.724, 1.733, 1.739	3.605	+O, 200		d		s conc HCl, NH_3 glac ac a
c641	(II) sulfate	Nat. hydrocyanite. $CuSO_4$	159.60	grn, wh, rhomb, 1.733	3.603	sl d above 200	d 650 to CuO	14.3^{0}	75.4^{100}	1.04^{18} MeOH; i al
c642	(II) sulfate, basic	Nat. brochantite. $CuSO_4 \cdot 3Cu(OH)_2$	452.29	grn, monocl, 1.728, 1.771, 1.800	3.78	d 300		i	i	s a, NH_4OH
c643	(II) sulfate, pentahydrate	Bluevitriol, nat. chalcanthite. $CuSO_4 \cdot 5H_2O$	249.68	bl, tricl, 1.514, 1.537, 1.543	2.284	$-4H_2O$, 110	$-5H_2O$, 150	31.6^{0}	203.3^{100}	15.6^{18} MeOH; i al
c644	(I) sulfide	Nat. chalcocite. Cu_2S	159.15	blk, rhomb	5.6	1100		$X10^{-14}$		s HNO_3, NH_4OH; i acet
c645	(II) sulfide	Nat. covellite. CuS	95.61	blk, monocl or hex, 1.45	4.6	tr 103	d 220	0.000033^{18}		s HNO_3, KCN, h HCl, H_2SO_4; i al, alk
c646	(I) sulfite, monohydrate	$Cu_2SO_3 \cdot H_2O$	225.17	red pr	4.46^{15}			i		d dil a; s NH_4OH
c647	(I) sulfite, monohydrate	$Cu_2SO_3 \cdot H_2O$	225.17	wh, hex	3.83^{15}	d		sl s		s HCl, NH_4OH; i al, eth
c648	(I, II) sulfite, dihydrate	Chevreul's salt. $Cu_2SO_3 \cdot CuSO_3 \cdot 2H_2O$	386.78	red cr	3.57	d 200		i	i	s HCl, NH_4OH; sl s HNO_3
c649	(II) tartrate	$CuC_4H_4O_6$	211.62	lt bl powd				v sl s		s a, alk
c650	(II) tartrate, trihydrate	$CuC_4H_4O_6 \cdot 3H_2O$	265.66	lt gray-bl powd		d		0.02^{15}	0.14^{35}	s a, alk
c651	telluride	Cu_2Te	254.69	bl-blk, oct	7.27					s Br + H_2O; i HCl, H_2SO_4
c652	telluride	Nat. rickardite. Cu_4Te_3	636.98	purp, tetr	7.54					
c653	tellurite	$CuTeO_3$	239.14	blk glass				i	i	s conc HCl
c654	(I) thiocyanate	CuSCN	121.62	wh	2.843	1084		0.0005^{18}		s NH_4OH; sl s ac a; i al; d min a
c655	(II) thiocyanate	$Cu(SCN)_2$	179.70	blk		d 100		d	d	s a, NH_4OH
c656	(II) tungstate	$CuWO_4 \cdot 2H_2O$	3437.42	lt-grn, oct		red heat		0.1^{15}		s NH_4OH; sl s ac a; i al; d min a
c657	xanthate	Copper ethylxanthogenate. $Cu(C_3H_5OS_2)_2$	305.93	yel ppt		d		i		s NH_4OH; v sl s al; i CS_2
	Copper complexes									
c658	diamminecopper (II) acetate	$Cu(C_2H_2O_2)_2 \cdot 2NH_3$	215.70	vlt bl cr		d *ca* 175		s d		s ac a, NH_4OH; i al
c659	tetrammine copper (II) sulfate	$[Cu(NH_3)_4]SO_4 \cdot H_2O$	245.74	bl, rhomb	1.81	d 150		$18.5^{21.5}$	d	i al
c660	tetrapyridine copper- (II) fluosilicate	$[Cu(C_5H_5N)_4]SiF_6$	522.03	purp-bl, rhomb	2.108					
c661	tetrapyridine copper- (II) *per*rhenate	$[Cu(C_5H_5N)_4](ReO_4)_2$	880.36	bl cr, monocl	2.338			0.5555		
c662	**Cyanic acid** isocyanic acid	HOCN	43.03	liq	1.14^{20}_{4}			s d		s eth, bz, tol
c663	Cyanoauric acid	$HAu(CN)_4 \cdot 3H_2O$	356.09	tab		50	d	s		s al, eth
c664	**Cobalticyanic acid**	$[H_3Co(CN)_6]2 \cdot H_2O$	454.14	col need, deliq		d 100		s		s al, HCl, dil HNO_3, dil H_2SO_4
c665	**Cyanogen**	$(CN)_2$	52.04	col gas, pungent odor v pois	2.335 g/ℓ, liq: $0.9577^{-21.17}$	−27.9	−20.7	450^{20} cm^3		230 cm^3 al, 500 cm^3 eth
c666	**Cyanogen compounds**	See organic tables								
d1	**Deuterium**	Heavy hydrogen. D_2	4.032	col gas	lig $0.169^{-250.9}$	−254.6	−249.7	sl s		
d2	deuterium chloride	DCl	37.47	gas		−114.8	−81.6	24.1 cm^2 11.9^{25} cm^3	8.4^{50} cm^3 7.12^{50} cm^3	
d3	deuterium oxide	Heavy water. D_2O	20.03	col liq or hex cr, 1.33844^{20}	1.105^{20}_{4}	3.82	101.42^{760}			
d4	**Dysprosium**	Dy	162.50±3	met, hex	8.5500	1412	2562	i	i	s a
d5	acetate	$Dy(C_2H_3O_2)_3 \cdot 4H_2O$	411.69	yel need		d 120		s		v sl s al
d6	bromate	$Dy(BrO_3)_3 \cdot 9H_2O$	708.34	yel hex need		78	$-6H_2O$, 110	v s		sl s al
d7	bromide	$DyBr_3$	402.21	col cr		881	1480	s	s	
d8	carbonate	$Dy_2(CO_3)_3 \cdot 4H_2O$	577.09			$-3H_2O$, 15		i		
d9	chloride	$DyCl_3$	268.86	shining yel pl	3.67^{0}_{4}	718	1500	s	s	
d10	chromate	$Dy_2(CrO_4)_3 \cdot 10H_2O$	853.13	yel cr		$-3^1/_2H_2O$, 150	d		1.002^{25}	
d11	fluoride	DyF_3	219.50	col cr		1360	>2200	i	i	i dil a

PHYSICAL CONSTANTS OF INORGANIC COMPOUNDS (continued)

No.	Name	Synonyms and Formulae	Mol. wt.	Crystalline form, properties and index of refraction	Density or spec. gravity	Melting point, °C	Boiling point, °C	Solubility, in grams per 100 cc: Cold water	Hot water	Other solvents
d12	iodide	DyI_3	543.21	yelsh grn cr		955	1320	s	s	
d13	nitrate	$Dy(NO_3)_3 \cdot 5H_2O$	438.59	yel cr.		88.6		s		
d14	oxalate	$Dy_2(C_2O_4)_3 \cdot 10H_2O$	769.21	wh pr		$-H_2O$, 40		i	i	s dil a
d15	oxide	Dysprosia. Dy_2O_3	373.00	wh powd	7.81^{27}	2340 ± 10				grn soln a
d16	*ortho*phosphate	$DyPO_4 \cdot 5H_2O$	347.55	yelsh-wh powd		$-5H_2O$> 200		i		s dil a, ac a
d17	selenate	$Dy_2(SeO_4)_3 \cdot 8H_2O$	898.00	yel need		$-8H_2O$, 200		v s		i al
d18	sulfate	$Dy_2(SO_4)_3 \cdot 8H_2O$	757.30	bril yel cr		stab 110	$-8H_2O$, 360	5.072^{20}	3.34^{40}	
e1	**Erbium**	Er	167.26 ± 3	dk gray powd.	9.006	1529	2863	i	i	s a
e2	acetate	$Er(C_2H_3O_2)_3 \cdot 4H_2O$	416.45	wh cr, tricl	2.114					
e3	bromide	$ErBr_3 \cdot 9H_2O$	569.11	vlt-rose cr		950	1460	s	s	
e4	chloride	$ErCl_3 \cdot 6H_2O$	381.71	pink cr, deliq		774	1500	s	s	s al
e5	fluoride	ErF_3	224.26	rose cr		1350	2200	i	i	i dil a
e6	iodide	ErI_3	547.97	vlt-red cr		1020	1280	s	s	
e7	nitrate	$Er(NO_3)_3 \cdot 5H_2O$	443.35	redsh cr		$-4H_2O$, 130		s		s al, eth, acet
e8	oxalate	$Er_2(C_2O_4)_3 \cdot 10H_2O$	778.73	redsh micr powd	2.04(?)	d 575		i	i	i dil a
e9	oxide	Erbia. Er_2O_3	382.52	rose red powd, tr to cub at 1300	8.640	infus		0.00049^{24}		sl s min a
e10	sulfate	$Er_2(SO_4)_3$	622.69	wh powd, hygr.	3.678	d 630		43^{0}		
e11	sulfate, octahydrate	$Er_2(SO_4)_3 \cdot 8H_2O$	766.82	rose red, monocl	3.217	$-8H_2O$, 400		16^{20}	6.53^{40}	
e12	**Europium**	Eu	151.96	steel gray met, cub	5.2434	822	1597	i	i	
e13	(II) bromide	$EuBr_2$	311.77			677	1880	s	s	
e14	(III) bromide	$EuBr_3$	391.67			702	d	s	s	
e15	(II) chloride	$EuCl_2$	222.87	wh amorph		727	>2000	s	s	s a
e16	(III) chloride	$EuCl_3$	258.32	yel need	4.89^{20}	850				
e17	(II) fluoride	EuF_2	189.96	brt yel	6.495	1380	>2400	i	i	
e18	(III) fluoride	EuF_3	208.96	col		1390	2280	i	i	i dil a
e19	(II) iodide	EuI_2	405.77	br to olv grn cr.	5.50_4^{26}	527	1580	s	s	
e20	(III) iodide	EuI_3	532.67			877	d	s	s	
e21	(III) nitrate	$Eu(NO_3)_3 \cdot 6H_2O$	446.07	col		85 (sealed tube)		v s	v s	
e22	oxide	Eu_2O_3	351.92	pa rose powd	7.42					
e23	(II) sulfate	$EuSO_4$	248.02	col, orthorhomb	4.989^{20}			i	i	i dil a
e24	(III) sulfate	$Eu_2(SO_4)_3 \cdot 8H_2O$	736.22	pa rose cr	4.95 (anh)	$-8H_2O$, 375		2.563^{20}	1.93^{40}	
	Ferric or ferrous	See *Iron*								
f1	**Ferricyanic acid**	$H_3Fe(CN)_6$	214.98	grn-brn need, deliq		d		s	s	s al
f2	**Ferrocyanic acid**	$H_4Fe(CN)_6$	215.98	wh need, bl in moist air		d		s	s	s al; i eth
f3	**Fluoboric acid**	HBF_4	87.81	col liq		d 130		∞	s	∞ al
f4	**Fluorophosphoric acid, di-**	HPO_3F_2	101.98	col fum liq	1.583_4^{26}	−75	116	s		
f5	**Fluorophosphoric acid, hexa-**	HPF_6	145.97	col fum liq	*ca* 1.65 (65%)	31 ($6H_2O$)				
f6	**Fluorophosphonic acid, mono-**	H_2PO_3F	99.99	col visc liq	1.818_4^{26}			∞		
f7	**Fluorine**	F	18.998403	grn yel gas, pois, 1.000195.	1.69^{15}g/ℓ 1.51^{-188}	−219.62	−188.14	HF + O_3	d	
f8	(di-)oxide	F_2O	54.00	col gas or yel brn liq.	liq $1.90^{-233.8}$	−223.8	−144.8	sl s d	i	sl s alk, a
f9	oxide, di-	Dioxygen fluoride, F_2O_2	70.00	brn gas, red liq, orange solid	sol 1.912^{-165} li1 1.45^{-57}	−163.5	−57			
f10	**Fluosilicic acid**	$H_2SiF_6 \cdot xH_2O$	hydr (not known)	col fum coros liq, 1.3465^{25} (60.97% soln)	1.4634^{25} (60.97% soln)		d	s	s	sl s alk
f11	dihydrate	$H_2SiF_6 \cdot 2H_2O$	180.12	wh cr, fum, deliq		d		s	s	s alk
f12	**Fluosulfonic acid**	HSO_3F	100.06	col liq	1.743^{15}	−87.3	165.5	s		
g1	**Gadolinium**	Gd	157.25 ± 3	col or lt yel met, hex	7.9004	1313	3266	i	i	s a
g2	acetate tetrahydrate	$Gd(C_2H_3O_2)_3 \cdot 4H_2O$	406.44	col. tricl	1.611			11.6^{25}		
g3	acetylacetonate, trihydrate	$Gd[CH(COCH_3)_2]_3 \cdot 3H_2O$	508.62			143.5—145		i	i	
g4	bromide, hexahydrate	$GdBr_3 \cdot 6H_2O$	505.05	rhomb pl	2.844^{16}			s	s	s HBr
g5	chloride	$GdCl_3$	263.61	col monocl, pr	4.52^{0}	609		s	s	
g6	chloride, hexahydrate	$GdCl_3 \cdot 6H_2O$	371.70	wh pr, deliq	2.424^{0}			s	s	
g7	iodide	GdI_3	537.96	citr yel		926	1340	s	s	
g8	fluoride	GdF_3	214.25					i		sl s h HF
g9	*di*methylphosphate	$Gd[(CH_3)_2PO_4]_3$	532.37					23.0^{25}	6.7^{100}	
g10	nitrate, hexahydrate	$Gd(NO_3)_3 \cdot 6H_2O$	451.36	tricl, deliq	2.332	91		v s	v s	s al
g11	nitrate, pentahydrate	$Gd(NO_3)_3 \cdot 5H_2O$	433.34	pr	2.406^{15}	92		i	i	v sl s conc HNO_3
g12	oxalate	$Gd_2(C_2O_4)_3 \cdot 10H_2O$	758.71	monocl.		$-6H_2O$, 110		i		s HNO_3; v sl s H_2SO_4
g13	oxide	Gadolinia. Gd_2O_3	362.50	wh amorph powd, hygr.	7.407^{15}	2330 ± 20		v sl s		s a
g14	selenate	$Gd_2(SeO_4)_3 \cdot 8H_2O$	887.50	monocl, pearly	3.309	$-8H_2O$, 130		s	s	
g15	sulfate	$Gd_2(SO_4)_3$	602.67	col	$4.139^{14.5}$	d 500		3.98^{0}	$2.26^{34.4}$	
g16	sulfate, octahydrate	$Gd_2(SO_4)_3 \cdot 8H_2O$	746.80	col, monocl	$3.010^{14.6}$			3.28^{20}	2.54^{40}	
g17	sulfide	Gd_2S_3	418.68	yel mass, hyg	3.8			d		d a
g18	**Gallium**	Ga	69.72	gray-blk orthor-homb, tendency to undercool	sol $5.904^{29.6}$ liq $6.095^{29.8}$	29.78	2403	i	i	s a, i alk

PHYSICAL CONSTANTS OF INORGANIC COMPOUNDS (continued)

No.	Name	Synonyms and Formulae	Mol. wt.	Crystalline form, properties and index of refraction	Density or spec. gravity	Melting point, °C	Boiling point, °C	Solubility, in grams per 100 cc: Cold water	Hot water	Other solvents
g19	acetate, basic	$4Ga(C_2H_3O_2)_3 \cdot 2Ga_2O_3 \cdot 6H_2O$	1452.37	wh micro cr.		d 160		s d	d	i ac a
g20	acetylacetonate	2,4-Pentanedione deriv. $Ga(C_5H_7)_2)_3$	367.04	monocl or pl, rhomb or pyram, rhomb pyram	1.42, 1.41	194—95	subl 140^{10}	s	s	s acet
g21	arsenide	$GaAs$	144.64	dk-gray cub cr		1238				
g22	bromide, tri-	$GaBr_3$	309.43	wh cr.	3.69_{4}^{25}	121.5 ± 0.6	278.8	s	s	sl s NH_3
g23	bromide, tri-, hexammine	$GaBr_3 \cdot 6NH_3$	411.62	wh powd				d	d	sl s NH_2
g24	bromide, tri-, monammine	$GaBr_3 \cdot NH_3$	326.46	wh powd	3.11^{25}	124		d	d	sl s NH_3
g25	*per*chlorate	$Ga(ClO_4)_3 \cdot 6H_2O$	476.16			d 175		v s		v s al
g26	chloride, di-	$GaCl_2$	140.63	wh cr, deliq.		164	535	d	d	s bz
g27	chloride, tri-	$GaCl_3$	176.08	wh need, deliq	2.47_{4}^{26} liq 2.36_{80}^{80}	77.9 ± 02	201.3	v s	v s	s bz, CCl_4, CS_2
g28	chloride, tri-, hexammine	$GaCl_3 \cdot 6NH_3$	278.26					d	d	s NH_3
g29	chloride, tri-, monoammine	$GaCl_3 \cdot NH_3$	193.11	wh powd	2.189^{25}	124		d	d	s NH_3
g30	ferrocyanide	$Ga_4[Fe(CN)_6]_3$	914.74			d		d	d	i conc HCl
g31	fluoride, tri-	GaF_3	126.72	wh powd	4.47 ± 0.01	subl 800 (in N_2)	*ca* 1000	0.002	i	v sl s dil a; s HF
g32	fluoride, tri-, trihydrate	$GaF_3 \cdot 3H_2O$	180.76	wh powd		$-H_2O$ (vac) 140		i	sl s	sl s dil H_2F_3; v s dil HCl
g33	fluoride, tri-, triammine	$GaF_3 \cdot 3HN_3$	177.81	wh powd		$-NH_3$, 100		d	d	
g34	hydride	Digallane, galloethane, Ga_2H_6	145.49	col liq		−21.4	139, d > 130	d	d	d a, alk
g35	hydroxide	$Ga(OH)_3$	120.74	wh		d 440		i	i	a dil a
g36	hydroxyquinoline deriv.	$Ga(C_9H_6NO)_3$	502.18	grn-yel cr		>150	subl vac	0.0001	0.0012	s a, alk; sl s al
g37	iodide, tri-	GaI_3	450.43	lt yel cr	4.15_{4}^{25}	212 ± 1	subl 345		d	
g38	iodide, tri-, hexammine	$GaI_3 \cdot 6NH_3$	552.62	wh powd				d	d	
g39	iodide, tri-, monoammine	$GaI_3 \cdot NH_3$	467.46	wh powd	3.635_{4}^{25}			d	d	
g40	nitrate	$Ga(NO_3)_3 \cdot \times H_2O$		wh cr, deliq.		d 110	d to Ga_2O_3 200	v s	v s	s abs al; i eth
g41	nitride	GaN	83.73	dk gray powd.	6.1	subl 800		i	i	i dil a; sl s h conc H_2So_4, h conc NaOH
g42	oxide, sesqui-(α)	Ga_2O_3	187.44	wh, hex, rhomb, 1.92, 1.95	6.44	1900; tr to β 600		i	i	s alk; v sl s h a
g43	oxide, sesqui-(β)	Ga_2O_3	187.44	monocl, rhomb.	5.88	1795 ± 15		i	i	s alk; v sl s ha a
g44	oxide, sesqui-, monohydrate	$Ga_2O_3 \cdot H_2O$	205.45	wh micr cr, orthorhom, 1.84	5.2	$-H_2O$, 400 tr to Ga_2O_3		i	i	sl s a; s alk
g45	oxide, sub-	Ga_2O	155.44	blk brn powd	4.77_{4}^{26}	>660	subl > 500	i	i	s a, alk
g46	oxalate	$Ga_2(C_2O_4)_3 \cdot 4H_2O$	475.56	wh micro cr, hygr		$-4H_2O$, 180	d 200	0.4		
g47	oxychloride	$6GaOCl \cdot 14H_2O$	979.25	oct				i		v a KOH; i dil HNO_3; s acet
g48	selenate	$Ga_2(SeO_4)_3 \cdot 16H_2O$	856.56	col, monocl or tricl cr				57.5^{25}	v s	
g49	selenide, mono-	$GaSe$	148.68	dk red-br greasy leaf	5.03_{4}^{25}	960 ± 10				
g50	selenide, sesqui-	Ga_2Se_3	376.32	rdsh-bl brittle, hard	4.92_{4}^{26}	> 1020 ± 10				
g51	selenide, sub-	Ga_2Se	218.40	bl	5.02_{4}^{25}					
g52	sulfate	$Ga_2(SO_4)_3$	427.61	wh powd		diss 690^{760}		v s	v s	s al; i eth
g53	sulfate, hydrate	$Ga_3(SO_4)_3 \cdot 18H_2O$	751.89	oct cr.				v s	v s	s 60% al; i eth
g54	sulfide, mono-	GaS	101.78	yel cr.	3.86_{4}^{25}	965 ± 10		i	d	s a, alk
g55	sulfide, sesqui-	Ga_2S_3	235.62	yel cr, or wh amorph	3.65^{25}	1255 ± 10		d	d	a s, alk
g56	sulfide, sub-	Ga_2S	171.50	dk gray	4.18_{4}^{25}	d vac 800		d	d	s a, alk
g57	telluride, mono-	$GaTe$	197.32	blk soft cr.	5.44_{4}^{25}	824 ± 2				
g58	telluride, sesqui-	Ga_2Te_3	522.24	blk brittle cr	5.57_{4}^{25}	790 ± 2				
	Germane									
g59	bromo-	GeH_3Br	155.52	col liq	$2.34^{29.5}$	−32	52	d	d	i al; d alk
g60	chloro-	GeH_3Cl	111.07	col liq	1.75^{-25}	−52	28.0	d	d	i al; d alk
g61	chloro trifluoro-	GeF_3Cl	165.04	col gas		−66.2	−20.63	d	d	s abs al
g62	dibromo-	GeH_2Br_2	234.41		2.80^{0}	−15.0	89.0	d	d	i al; d alk
g63	dichloro-	GeH_2Cl_2	145.51	col liq	1.90^{-68}	−68.0	69.5	d	d	i al; d alk
g64	dichlorodifluoro-	$GeCl_2F_2$	181.49	col gas		−51.8	−2.8	d		s abs al
g65	tribromo-	Germanium bromoform $GeHBr_3$	313.31	col liq		−24	d	d	d	d alk
g66	trichloro-	Germanium chloroform. $GeHCl_3$	179.96	col liq	1.93^{0}	−71	75.2 d	d	d	d alk
g67	trichlorofluoro-	$GeCl_3F$	197.95	col liq		−49.8	37.5	d		s abs al
g68	**Germanium**	Ge	72.59 ± 3	gray-wh met-cub	5.35_{20}^{20}	937.4	2830	i	i	s h H_2SO_4, aq reg; i alk
g69	bromide, di-	$GeBr_2$	232.40	col need or pl		122	d	d	d	s a, $GeBr_4$, al; i bz
g70	bromide, tetra-	$GeBr_4$	392.21	gray-wh oct, 1.6269	3.132_{29}^{29}	26.1	186.5	d	d	s abs al, eth, bz; i conc H_2SO_4
g71	chloride, di-	$GeCl_2$	143.50	wh powd		d to Ge + $GeCl_4$		d	d	s $GeCl_4$; i al, chl
g72	chloride, tetra-	$GeCl_4$	214.40	col liq, 1.464.	1.8443^{30}	−49.5	84	d	d	s al, eth; v s dil HCl; i conc HCl conc H_2SO_4

No.	Name	Synonyms and Formulae	Mol. wt.	Crystalline form, properties and index of refraction	Density or spec. gravity	Melting point, °C	Boiling point, °C	Solubility, in grams per 100 cc: Cold water	Hot water	Other solvents
g73	fluoride, di-	GeF_2	110.59	wh cr, hygr		d > 350	subl	s	v s	
g74	fluoride, tetra-	GeF_4	148.58	col gas or liq, not liq at atm press	$2.46^{36.5}$	subl −37		d to GeO + H_2GeF_6		
g75	fluoride, tetra-	$GeF_4 \cdot 3H_2O$	202.63	wh cr, deliq.		d		s	s	
g76	hydride	Digermane. Ge_2H_6	151.23	liq	1.98^{-100}	−109	29; d 215	d		s liq NH_3
g77	hydride	Trigermane. Ge_3H_3	225.84	col liq	2.2^{30}	−105.6	110.5; d 195	i	i	s CCl_4
g78	hydride, tetra-	Germane. GeH_4	76.62	col gas	1.523^{-142}	−165	−88.5; d 350	i	i	s liq NH_3, NaOCl; sl s h HCl
g79	imide	$Ge(NH)_2$	102.62	wh amorph powd		d 150			d to NH_3 + GeO_2	
g80	iodide, di-	GeI_2	326.40	or hex pl	5.37	d	subl vac 240	s	s d	s conc HI, dil a; sl s CCl_4, chl; i CS_2
g81	iodide, tetra-	GeI_4	580.21	red-or, cub	4.322^{20}_{26}	144	d 440	s d		d al, acet; s CS_2, CCl_4, bz, MeOH
g82	(tri-) nitride, di-	Ge_3N_2	245.78	blk cr			subl 650			
g83	(tri-) nitride, tetra-	Ge_3N_4	273.80	wh-lt brn powd	5.25^{25}	d 450		i	i	i a, alk
g84	oxide, d-(insoluble)	GeO_2	104.59	tetr	6.239	1086 ± 5		i		sl s NaOH; i HCl
g85	oxide, di-(soluble)	GeO_2	104.59	col, hex, 1.650	4.228^{25}	1115.0 ± 4		0.447^{25}	1.07^{100}	s a, alk; i HCl, HF; one form NaOH, NH_4OH; one form
g86	oxide, mono-	GeO	88.59	blk cr powd, 1.607		subl 710		i		s Cl_2 water, H_2O_2 + NH_4OH; i a, alk
g87	oxychloride	$GeOCl_2$	159.50	col liq		−56	d > 20	d	d	i all solv
g88	selenide	$GeSe_2$	230.51	orange, rhomb(?)	4.56^{25}	707 ± 3	d	i	i	v sl s a; sl s alk
g89	sulfide, di-	GeS_2	136.71	wh powd, or wh, orthorhomb	2.94^{14}	*ca* 800	subl > 600	0.45 d	d to GeO_2 + H_2S	s alk, alk sulf; i al, eth, a; 3.112 liq NH_3
g90	sulfide, mono-	GeS	104.65	yel-red amorph, or rhomb bipyram, blk	amorph: 3.31 rhomb: 4.01	530	subl 430	0.24	i	s HCl, alk or alk sulf; sl s NH_4OH; 0.0473 liq NH_3
	Glucinum	See **Beryllium**								
g91	**Gold**	Au	196.9665	yel duct met, cub, coll blue-viol	19.31	1064.43	3080	i	i	s aq reg, KCN, h H_2SO_4; i a
g92	(I) bromide	AuBr	276.87	yel-gray mass, or cr powd	7.9	d 115		i	i	d a; s NaCN
g93	(III) bromide	$AuBr_3$	436.68	gray powd, or br cr		97.5–Br, 160		sl s		s eth, al
g94	(I) chloride	AuCl	232.42	yel cr.	7.4	170 d to $AuCl_3$	d 289.5	v sl s d	d	s HCl, HBr
g95	(III) chloride	$AuCl_3$ or Au_2Cl_6	303.33	claret-red cr pr	3.9	d 254	subl 265	68	v s	s al, eth; sl s NH_3; i CS_2
g96	(I) cyanide	AuCN	222.98	lt yel cr powd	7.12^{25}	d		v sl s	v sl s	s KCN, NH_4OH; i eth, alk
g97	(III) cyanide	Cyanoauric acid. $Au(CN)_3 \cdot 3H_2O$ or $HAu(CN)_4 \cdot 3H_2O$	329.07 or 356.09	col pl, hygr		d 50		v s	d, v s	s al, eth
g98	(I) iodide	AuI	323.87	grnsh-yel powd.	8.25	d 120		v sl s	sl s d	s KI
g99	(III) iodide	AuI_3	577.68	dk grn				i	d	s iodides
g100	(III) nitrate, hydrogen	Nitratoauric acid. $AuH(NO_3)_4 \cdot 3H_2O$ or $H[Au(NO_3)_4] \cdot 3H_2O$	500.04	yel, tricl, oct	2.84	d 72		s, d		s HNO_3
g101	(III) oxide	Au_2O_3	441.93			−O, 160	−3O, 250	i	i	s HCl, conc HNO;3, NaCN
g102	(III) oxide	$Au_2O_3 \cdot xH_2O$				$-1\frac{1}{2}H_2O$, 250		5.7×10^{-11} 25		s HCl, NaCN, conc HNO_3
g103	phosphide	Au_2P_3	486.85	gray	6.67		d			i HCl, dil HNO_3
g104	selenide	Au_2Se_2	630.81		4.65^{22}					
g105	(I) sulfide	Au_2S	425.99	br blk powd.		d 240		i fresh sol	ppt coll	s sq reg, KCN; i a
g106	(III) sulfide	Au_2S_3	490.11	br blk powd.	8.754	d 197		i		i al, eth, s Na_2S
g107	telluride, di-	Nat. krennerite. $AuTe_2$	452.17	1) rhomb, 2) mon-ocl, 3) tricl yel earthy to massive	8.2—9.3	d 472		i	i	
h1	**Hafnium**	Hf	178.49 ± 3	hex	13.31^{20}	2227 ± 20	4602	i	i	s HF
h2	bromide	$HfBr_4$	498.11	wh		subl 420				
h3	carbide	HfC	190.50		12.20	*ca* 3890		i		
h4	chloride	$HfCl_4$	320.30	wh		subl 319		d		s MeOH, acet
h5	fluoride	HfF_4	254.48	monocl, 1.56						
h6	iodide	HfI_4	686.11				400 subl vac			
h7	nitride	HfN	192.50	yel-brn, cub.		3305				
h8	oxide	Hafnia. HfO_2	210.49	wh, cub	9.68^{20}	2758 ± 25	~5400(?)	i	i	
h9	oxychloride	$HfOCl_2 \cdot 8H_2O$	409.52	col				s		
h10	**Helium**	He	4.00260	col gas, inert odorless	0.1785^{0} g/ℓ liq $0.147^{-270.3}$	-272.2^{26atm}	−268.9	0.94^{0} cm^3 0.94^{25} cm^3	1.05^{50} cm^3 1.21^{75} cm^3	i al; absorbed by Pt
h11	**Holmium**	Ho	164.9304	met, hex	8.7947	1474	2695	i	i	
h12	bromide	$HoBr_3$	404.64	lt yel		914	1470	s	s	
h13	chloride	$HoCl_3$	271.29	lt yel		718	1500	s	s	
h14	iodide	HoI_3	545.64	lt yel		989	1300	s	s	
h15	fluoride	HoF_3	221.93	lt yel		1143	>2200	i	i	i dil a

No.	Name	Synonyms and Formulae	Mol. wt.	Crystalline form, properties and index of refraction	Density or spec. gravity	Melting point, °C	Boiling point, °C	Solubility, in grams per 100 cc: Cold water	Hot water	Other solvents
h16	oxalate	$Ho_2(C_2O_4)_3 \cdot 10H_2O$	774.07	pa tan		$-H_2O$, 40	d	i	i	i dil a
h17	oxide	Holmia. Ho_2O_3	377.86	tan				i	i	s a
h18	**Hydrazine**	HN_2NH_2	32.05	col liq or wh cr, 1.470^{22}	1.004^{25}	2.0	113.5	v s		s al
h19	azide	$N_2H_4 \cdot HN_3$	75.07	wh pr, deliq, 1.53, 1.76		75.4		v s	v s	1.2^{23} al; 6.1^{23} MeOH; i CS_2 bz
h20	fluogermanate	$2N_2H_4 \cdot H_2GeF_6$	252.69	monocl pr, 1.452, 1.460, 1.464	2.406_{25}^{25}			s		
h21	fluosilicate	$N_2H_4 \cdot H_2SiF_6$	176.14	cr		d 186		v s		sl s al
h22	formate	$N_2H_4 \cdot 2CH_2O_2$	124.10			128		s		
h23	hydrate	$N_2H_4 \cdot H_2O$	50.06	col fum liq or cub cr, 1.42842	1.03^{21}	−40	118.5^{740}	∞	∞	s al; i eth, chl
h24	hydrochloride, di-	$N_2H_4 \cdot 2HCl$	104.97	col vitr, oct	1.42	198, −HCl	d 200	27.2^{32}	v s	sl s al
h25	hydrochloride, mono-	$N_2H_4 \cdot HCl$	68.51	wh need		89	d 240	v s		v sl s al; v s liq NH_3
h26	hydroiodide	Hydrazine monoiodide, $N_2H_4 \cdot HI$	159.96	col pr		124—126		s		
h27	nitrate, di-	$N_2H_4 \cdot 2HNO_3$	158.07	col cr		104 (rapid heat); d 80 (slow heat)		v s	d	
h28	nitrate, mono-	$N_2H_4 \cdot HNO_3$	95.06	col dimorph need (α, β)		α70.71; β62.09	subl 140	174.9^{10}	2127^{60}	sl s al
h29	oxalate	$2N_2H_4 \cdot H_2C_2O_4$	154.13	wh need		148		200^{35}		0.0003^{22} al; i eth
h30	*per*chlorate	$N_2H_4 \cdot HClO_4 \cdot {}^1/_2H_2O$	141.51	exp	1.939	137	d 145	d		s al; i eth, bz, chl, CS_2
h31	*hypo*phosphate	$N_2H_4 \cdot 2H_2PO_3$	194.02			152				
h32	*ortho*phosphate	$N_2H_4 \cdot H_3PO_4$	130.04	cr, hygr		82		v s		
h33	*ortho*phosphite	$N_2H_4 \cdot 2H_3PO_3$	196.04			82				
h34	*ortho*phosphite	$N_2H_4 \cdot H_3PO_3$	114.04			36		v s		
h35	picrate	$N_2H_4 \cdot HC_6H_2N_3O_7 \cdot {}^1/_2H_2O$	270.16			201.3		s	s	
h36	selenate	$N_2H_4 \cdot H_2SeO_4$	177.02	col cr powd, unstab		exp		v sl s	v s	
h37	sulfate	$N_2H_4H_2SO_4$	130.12	col, rhomb	1.37	254	d	3.415^{25}	14.39^{80}	i al
h38	sulfate	$(N_2H_4)_2 \cdot H_2SO_4$	162.16	col cr, hygr		85		202.2^{25}	554.4^{60}	i al
h39	tartrate	$(N_2H_4)C_4H_6O_6$	182.13	col cr; $[\alpha]_D^{20}$ + 22.5		182—183		6.0^0		
h40	**Hydroazoic acid**	Azoïmide. HN_3	43.03	col liq	1.09_4^{25}	−80	37	∞	∞	s al, alk, eth
h41	**Hydrogen**	H_2	2.01588 ± 14	col gas, cub sol	gas 0.0899 g/ℓ li1 0.070	−259.14	−252.8	2.14^0 cm³ 1.91^{25} cm³	0.85^{30} cm³ 1.89^{50} cm³	6.925^0 cm³ al
h42	antimonide	Stibine H_3Sb	124.77	col gas, pois	gas 5.30^0 g/ℓ liq 2.26^{-25}	−88.5	−17	20 cm³	4 cm³	1500 cm³ al; 2500 cm³ CS_2
h43	arsenide	Arsine H_3As	77.95	col gas, pois	3.484 g/ℓ	−113.5	−55, d 230	20 cm³	sl s	sl s al, alk
h44	arsenide (solid)	H_2As_2	151.86	br powd		d 200		i	d	i al, eth, CS_2, alk; s HNO_3
h45	bismuthide	Bismuthine. H_3Bi	212.00	liq, v unstab			22			
h46	bromide	Hydrobromic acid. HBr	80.91	col gas or pa yel liq, 1.325	gas 3.5^0 g/ℓ liq 2.77^{-67}	−88.5	−67.0	221^0	130^{100}	s al
h47	bromide (const. boiling)	HBr(47%) + H_2O		col liq	1.49	−11	126			
h48	bromide, dihydrate	$HBr \cdot 2H_2O$	116.94	wh cr, col liq	2.11^{-15}	−11		s	s	
h49	bromide, monohydrate	$HBr \cdot H_2O$	98.93	col liq	1.78	stab	−3.6 to −15.5 between 1—2.5 atm			
h50	chloride	Hydrochloric acid. HCl	36.46	col gas or col liq, pois	$1.187^{-84.9}$ gas 1.00045 g/ℓ	−114.8	−84.9	82.3°	56.1^{60}	327 cm³ al; s eth, bz
h51	chloride (const. boiling)	HCl(20.24%) + H_2O		col liq	1.097		110			
h52	chloride, dihydrate	$HCl \cdot 2H_2O$	72.49	col liq	$1.46^{18.3}$	−17.7	d	d	∞	s al
h53	chloride, monohydrate	$HCl \cdot H_2O$	54.48	col liq	1.48	−15.35		∞	∞	s al
h54	chloride, trihydrate	$HCl \cdot 3H_2O$	90.51	col liq		−24.4	d	∞		s al
h55	cyanide	Hydrocyanic acid. HCN	27.03	col liq or gas, pois, liq 1.2675^{10}	gas 0.901 g/ℓ liq 0.699^{22}	−14	26	∞	∞	∞ al; s eth
h56	fluoride	Hydrofluoric acid. HF	20.01	col fum cor liq, or gas; gas 1.90	$0.991^{19.54}$ liq 0.987	−83.1	19.54	∞	v s	
h57	fluoride (const. boiling)	HF(35.35%) + H_2O		col liq			120			
h58	iodide	Hydroiodic acid. HI	127.91	col gas, or pa yel liq, $n_D^{16.5}$ 1.466	gas 5.66^0 g/ℓ liq $2.85^{-4.7}$	−50.8	-35.38^4 atm	42.5^0 cm³	v s	s al
h59	iodide (const. boiling)	HI(57%) + H_2O		col or pa yel fum liq	1.70^{15}		127^{774}			
h60	iodide, dihydrate	$HI \cdot 2H_2O$	163.94	col liq		−43		∞		
h61	iodide, tetrahydrate	$HI \cdot 4H_2O$	199.97	col liq		−36.5		∞		
h62	iodide, trihydrate	$HI \cdot 3H_2O$	181.96	col liq		−48		∞		
h63	oxide	Water. H_2O	18.01528	col liq or hex cr, liq 1.333 sol 1.309, 1.313	1.000_4^4	0.000	100.000			∞ al
h64	oxide, per-	H_2O_2	34.01	col liq; 1.414^{22}	1.4067^{25}	−0.41	150.2^{760}	∞		s al, eth; i pet eth
h65	phosphide	Phosphine, H_3P	34.00	col pois inflam gas or col liq, 1.317 liq	gas 1.529 g/ℓ liq 0.746^{-90}	−133.5	−87.4	26^{17} cm³	i	s al, eth, Cu_2Cl_2

PHYSICAL CONSTANTS OF INORGANIC COMPOUNDS (continued)

No.	Name	Synonyms and Formulae	Mol. wt.	Crystalline form, properties and index of refraction	Density or spec. gravity	Melting point, °C	Boiling point, °C	Solubility, in grams per 100 cc		
								Cold water	Hot water	Other solvents
h66	phosphide	H_4P_2	65.98	col liq	1.012	−90	57.5^{735}	i	i	s al, turp
h67	phosphide	$(H_2P_4)_3$	377.73	yel solid	1.83^{19}	ign 160	d	i	i	i al; s P, P_2H_4
h68	sulfide	H_2S	34.08	col gas, infl, liq 1.374	1.539^{0} g/ℓ	−85.5	−60.7	437^{0} cm^3	186^{40} cm^3	9.54^{20} cm^3 al; s CS_2
h69	sulfide, di-	H_2S_2	66.14	yel oil, 1.885	1.334^{20}	−89.6	70.7^{760}	d		s bz, eth, CS_2; i al
h70	sulfide, penta-	H_2S_5	162.32	clear yel oil	1.67^{16}	−50	50^{4}			s bz, eth, CS_2; i al
h71	sulfide, tetra-	H_2S_4	130.26	lt yel liq	1.588^{15}	−85				
h72	sulfide, tri-	H_2S_3	98.20	brt yel liq, 1.705^{15}	1.496^{15}	−52	d 90			s bz, eth, CS_2; i al
h73	telluride	H_2Te	129.62	col gas or yel need	gas 5.81 g/ℓ liq 2.57^{-20}	−49	−2	s unstab		s al, alk
h74	**Hydroxylamine**	NH_2OH	33.03	wh need or col liq, deliq	1.204	33.05	56.5	s	d	s a, al, MeOH; v sl s eth, chl, bz, CS_2
h75	acetate	$NH_2OH{\cdot}CH_3CO_2$	92.07	col cr.		87	subl 90	v s		
h76	bromide	$NH_2OH{\cdot}HBr$	113.94	wh, monocl	2.35$_4^{22}$			v s	v s	i eth
h77	fluogermanate	$(NH_2OH)_2{\cdot}H_2GeF_6{\cdot}2H_2O$	290.69	monocl pr, 1.418, 1.438, 1.443	2.229^{25}			s		s abs al
h78	fluosilicate	$(NH_2OH)_2H_2SiF_6{\cdot}2H_2O$	246.18	scales				v s		i al
h79	formate	$NH_2OH{\cdot}HCO_2$	78.05	col need		76	d 80	v s	d	s h al; i eth
h80	hydrochloride	$NH_2OH{\cdot}HCl$	69.49	col, monocl	1.67^{17}	151	d	83^{17}	v s	4.43^{20} al; 16.4^{20} MeO H; s gluc; i eth
h81	iodide	$NH_2OH{\cdot}HI$	160.94	col need, hygr		83—84 exp		v s	d	v s MeO H; sl s eth
h82	nitrate	$NH_2OH{\cdot}HNO_3$	96.04	wh		48	d <100	v s	d	v s al
h83	*ortho*phosphate	$(NH_2OH)_3{\cdot}H_3PO_4$	197.08			148 exp		1.9^{20}	16.8^{90}	
h84	sulfate	$(NH_2OH)_2{\cdot}H_2SO_4$	164.13	col, monocl		d 170	d	32.9^{0}	68.5^{20}	sl s al; s eth
i1	**Indium**	In	114.82	soft silv wh met, tetr	7.30^{20}	156.61	2080	i	i	s a; v sl s NaOH
i2	antimonide	InSb	236.57	cr		535				
i3	arsenide	InAs	189.74	met cr		943				i a
i4	bromide, di-	$InBr_2$	274.63	pa yel solid	4.22^{25}	235	632 subl	d		s a
i5	bromide, mono-	InBr	194.72	red br solid	4.96^{25}	220	662 subl	d		s a
i6	bromide, tri-	$InBr_3$	354.53	wh to yel need, deliq	4.74^{25}	436 ± 2	subl	v s		
i7	*per*chlorate	$In(ClO_4)_3{\cdot}8H_2O$	557.29	col cr, deliq.		ca 800	d 200	v s	d	s abs al; sl s eth
i8	chloride, di-	$InCl_2$	185.73	wh rhomb, deliq	3.655^{25}	235	550—570	d	d	
i9	chloride, mono-	InCl	150.27	1) yel or 2) dk red deliq	4.19^{25} yel 4.18^{25} red	225 ± 1	608	d	d	s a
i10	chloride, tri-	$InCl_3$	221.18	wh pl, deliq.	3.46$_4^{25}$	586 subl 300	volat 600	v s	v s	sl s al, eth
i11	cyanide	$In(CN)_3$	192.87	wh ppt				unstab		i dil a; s HCN; v sl s NaOH
i12	fluoride	InF_3	171.82	col	4.39^{25} ± 1	1170 ± 10	>1200	0.040^{25}		
i13	fluoride, trihydrate	$InF_3{\cdot}3H_2O$	225.86	cr		$-3H_2O$, 100		8.49^{22}		s a; i al, eth
i14	fluoride, nonahydrate	$InF_3{\cdot}9H_2O$	333.95	wh need		d		sl s	d	s HCl, HNO_3; i al, eth
i15	hydroxide	$In(OH)_3$	165.84	wh ppt		$-H_2O$<150		i		s a; v sl s NaOH; i NH_4OH
i16	iodate	$In(IO_3)_3$	639.53	wh cr.			d	0.067^{20}		s dil HNO_3, dil H_2SO_4; s d HCl
i17	iodide, di-	InI_2	368.63		4.71^{25}	212				s a
i18	iodide, mono-	InI	241.72	br red solid	5.31	351	711—715		sl d	i al, eth, chl; s dil a
i19	iodide, tri-	InI_3	495.53	carmine red, or yel cr	4.69	210		v unstable	s	s a, chl, bz yxl
i20	methylate	$In(CH_3)_3$	159.92	col cr.	1.568$_{19}^{19}$			d		d al, MeOH; s liq NH_3, eth; v s acet, bz
i21	nitrate	$In(NO_3)_3{\cdot}3H_2O$	354.88	pl, deliq		$-2H_2O$, 100	d	v s		s al
i22	nitrate	$In(NO_3)_3{\cdot}4^1/_2H_2O$	381.90	need, deliq		$-4^1/_2H_2O$	d	v s		s al
i23	oxide, mon-	InO	130.82	wh gray				i		s a
i24	oxide, sesqui-	In_2O_3	277.64	red brn, (h) pa yel (c) amorph and trig	7.179		volat 850	i		amorph s a; cr i a
i25	oxide, sub-	In_2O	245.64	blk cr	6.99$_4^{25}$	subl vac 565—700				s HCl
i26	phosphide	InP	145.79	brittle mass met		1070				v sl s min a
i27	selenate	$In_2(SeO_4)_3{\cdot}10H_2O$	838.67	cr, deliq				v s		
i28	selenide, sesqui-	In_2Se_3	466.52	blk cr, or soft dk scales	5.67$_4^{25}$	890 ± 10				s, d conc a
i29	sulfate	$In_2(SO_4)_3$	517.81	wh gray powd, monocl pr, hygr	3.438			s	v s	
i30	sulfate	$In_2(SO_4)_3{\cdot}9H_2O$	679.95	wh powd, hygr.	3.44	d 250		v s		
i31	sulfate, dihydrate	$In_2(SO_4)_3{\cdot}H_2SO_4{\cdot}7H_2O$	741.99	rhomb cr		$-7H_2O$, $-H_2SO_4$, *ca* 250				
i32	sulfide, mono-	InS	146.88	dk.	5.18^{25}	692 ± 5	subl vac 850			s HCl, HNO_3
i33	sulfide, sesqui-	In_2S_3	325.82	red cr or yel ppt	4.90	1050	subl *ca* 850 in high vac	i		s a; sl s Na_2S
i34	sulfide, sub-	In_2S	261.70	yel or blk need	5.87^{25}	653 ± 5				
i35	sulfite, basic	$2In_2O_3{\cdot}3SO_2{\cdot}8H_2O$	891.58	wh cr.		$-3H_2O$, 100	$-8H_2O$, 260	i		
i36	telluride, sesqui-	InTe	242.42	dk met, shiny	6.29$_4^{25}$	696 ± 2				i HCl, s HNO_3
i37	telluride	In_2Te_3	612.44	bl brittle cr	5.78	667				

PHYSICAL CONSTANTS OF INORGANIC COMPOUNDS (continued)

No.	Name	Synonyms and Formulae	Mol. wt.	Crystalline form, properties and index of refraction	Density or spec. gravity	Melting point, °C	Boiling point, °C	Solubility, in grams per 100 cc		
								Cold water	Hot water	Other solvents
i38	**Iodic acid**	HIO_3	175.91	col or pa yel cr powd, rhomb	4.629^0	d 110		286^0 310^{16}	473^{80} 576^{101}	v s 87% al; sl s HNO_3; i abs al, eth, chl
i39	metaper-	HIO_4	191.91	col		subl 110	d 138	v s, d		
i40	*ortho*paraper-	H_5IO_6 or $HIO_4·2H_2O$	227.94	wh monocl, deliq			d 140	113	v s	s al, eth
i41	**Iodine**	I_2	253.809	vlt blk met lust, rhomb, 3.34	4.93	113.5	184.35^{atm}	0.029^{20} 0.030^{25}	0.078^{50}	20.5^{15} al; 16.46^{25} bz 20.6^{17} eth; s chl glyc, KI; 24^{25} eth 23^{25} MeOH 20.15^{25} CS_2; $2.91^{25}CCl_4$
i42	azide	Iod(o)azide. IN_3	168.92	yel, exp				s d		s $Na_2S_2O_3$
i43	bromide, mono-	IBr.	206.81	dk gray cr	4.4157^0	(42) subl 50	d 116	s d		s al, eth, chl, CS_2
i44	bromide, tri-	IBr_3	366.62	br liq				s		s al
i45	chloride, mono-(α)	ICl.	162.36	dk-red need, cub, red br oily liq	3.1822^0	27.2	97.4	d to HIO_3 + Cl		s al, eth, CS_2, HCl
i46	chloride, mono-(β)	ICl.	162.36	brn red, rhombic 6 sided pl	liq 3.24^{34}	13.92	97.4; d 100	d		s al, eth, HCl
i47	chloride, tri-	ICl_3	233.26	yel brn, rhomb, red liq	3.117^{15}	101^{16atm}	d 77	s d		s al, eth, CCl_4, ac a, bz
i48	cyanide	ICN	152.92	wh cr.				sl s	sl s	s al, eth
i49	fluoride, hepta-	IF_7	259.89	col cr or liq	liq 2.8^6	5.5	4.5 subl	v s, d	d	d a, alk
i50	fluoride, penta-	IF_6	221.90	col liq	3.75	9.6	98	d	d	d a, ak
i51	oxide, di- (or tetra-)	IO_2 or I_2O_4	158.90	lem-yel cr	4.2^{18}_{10}	d slow 75 rap 130		d to HIO_3 + I_2		s H_2SO_4; sl s acet; i al, eth
i52	oxide, pent-	I_2O_5	333.81	wh trim	4.799_4^{25}	d 300—350		187.4^{13}	v s	i abs al, eth, chl, CS_2; sl s dil a
i53	(tetra-) oxide non-	I_4O_9	651.61	yel powd, hygr.			d 75			
i54	**Iodo platinic acid**	$H_2PtI_6·9H_2O$	1120.66	blk, monocl, deliq		>100		v s, d	d	
i55	**Iodous acid, hypo-**	Iodine hydroxide. HOI	143.91	only in sol, yel to grayish				d	d	
i56	**Iridium**	Ir.	192.22 ± 3	silv wh met, cub	22.421	2410	4130	i	i	sl s aq reg; i a, alk
i57	bromide, tetra-	$IrBr_4$	511.84	blk, deliq		d		s d		s al
i58	bromide, tri-	$IrBr_3·4H_2O$.	503.99	olv-grn cr		$-3H_2O$,100		v s		i al
i59	carbonyl	$Ir_2(CO)_8$	608.52	yel cr.		subl 160 (in CO_2)				s CCl_4
i60	carbonyl	$Ir_4(CO)_{12}$	1105.00	yel cr.		subl 250 (in CO_2)				i CCl_4
i61	carbonyl chloride	$Ir(CO)_2Cl_2$	319.15	col need		d 140		d		d HCl, KOH
i62	chloride, di-	$IrCl_2$.	263.13	blk gray cr (exist?)		d 733		s		i a, alk
i63	chloride, tetra-	$IrCl_4$.	334.00	dk-brn, amorph, hygr.		d		s	d	s al, dil HCl
i64	chloride, tri-	$IrCl_3$	298.58	olv-grn hex, or trig	5.30	d 763		i	i	i a, alk
i65	fluoride, hexa-	IrF_6	306.21	yel glass or tetr	6.0	44.4^{30}	53	d	d	
i66	iodide, tetra-	IrI_4	699.84	blk		d 100		i	i	i al, s KI
i67	iodide, tri-	IrI_3	572.93	grn		d		sl s	s	sl s al
i68	oxalic acid	$H_3[Ir(C_2O_4)_3·xH_2O$		pa yel cr.				v s	v s	sl s al; i eth
i69	oxide, di-	IrO_2	224.22	blk tetr or bl cr.	11.665	d 1100		0.0002^{20}		i a, alk
i70	oxide, di- hydrate.	$IrO_2·2H_2O$ or $Ir(OH)_4$	260.25	indigo bl cr		$-2H_2O$, 350		i	i	s HCl
i71	oxide, sesqui-	Ir_2O_3	432.44	bl-blk (exist?)		$-O$, 400		i		s H_2SO_4, h HCl; i alk
i72	oxide, sesqui-	$Ir_2O_2·xH_2O$		olive green		d		i		s a, alk
i73	phosphor chloride.	IrP_3Cl_{12} or $IrCl_3·3PCl_3$	710.58	yel pr		d 250		sl s	d 100	sl s, d al
i74	selenide.	$IrSe_2$	350.14	dk gray cr powd		d 600—700 (in CO_2)				sl s aq reg; i a
i75	sulfate.	$Ir_2(SO_4)_3·xH_2O$		yel pr		d		s		
i76	sulfide, di	IrS_2	256.34	br-blk	8.43_4^{25}	d 300		i		s aq reg; i a
i77	sulfide, hydro-	$Ir(HS)_3·2H_2O$	327.45	choc br		d		i		s HNO_3
i78	sulfide, mono-	IrS	224.28	bl-blk			d	i		s k_2S, i a
i79	sulfide, sesqui-	Ir_2S_3.	480.62	br-blk	9.64_4^{25}	d		sl s		s K_2S, HNO_3
i80	telluride.	$IrTe_3$	575.02	dk gray cr.	9.5_4^{25}			i	i	i a; s h aq reg
	Iridium complexes									
i81	aquopentammine iridium-chloride	$[Ir(NH_3)_5(H_2O)]Cl_3$.	366.29	wh micro cr.	2.474_4^{15}	$-H_2O$, 100		ca 75^{25}	d	i al, eth
i82	chloropentammine iridium chloride	$[Ir(NH_3)_5Cl]Cl_2$	383.73	pa yel, rhomb	$2.681_4^{15.5}$	d		sl s	s	i HCl
i83	hexammine iridium chloride	$[Ir(NH_3)_6]Cl_3$	400.76	col, rhomb	$2.434_4^{15.5}$			20		
i84	hexammine iridium nitrate	$[Ir(NH_3)_6](NO_3)_3$	480.42	col, tetr micro cr	2.395_1^{15}			1.7^{14}		
i85	nitratopentammine iridium nitrate	$[Ir(NH_3)_5(NO)_3)](NO_3)_2$	463.39	wh micro cr.	2.515_4^{18}	heat exp		0.28	2.5^{100}	
i86	**Iron**	Fe	55.847 ± 3	silv met, cub	7.86	1535	2750	i	i	s a; i alk, al, eth
i87	(II) acetate	$Fe(C_2H_3O_2)_2·4H_2O$	246.00	lt grn need, monocl		d		v s		
i88	(III) acetate, basic	$FeOH(C_2H_3O_2)_2$	190.94	br-red powd.				i		s a, al
i89	(III) acetylacetonate	$Fe(C_5O_2H_7)_3$.	353.18	rubyred, rhomb	1.33	184		sl s	sl s	s al, acet, bz, chl
i90	(III) *ortho*arsenate	$Fe_3(AsO_4)_2·6H_2O$	553.47	grn amorph powd		d		i	i	s dil HCl; sl s NH_4OH
i91	(III) *ortho*arsenate	Nat. scorodite. $FeAsO_4·2H_2O$	230.80	grn, rhomb, 1.765, 1.774, 1.797	3.18	d		i	i	s HCl; i HNO_3

No.	Name	Synonyms and Formulae	Mol. wt.	Crystalline form, properties and index of refraction	Density or spec. gravity	Melting point, °C	Boiling point, °C	Solubility, in grams per 100 cc: Cold water	Hot water	Other solvents
i92	(III) *ortho*arsenite, basic	$2FeAsO_3{\cdot}Fe_2O_3{\cdot}5H_2O$	607.30	br-yel powd		d		sl s		s a, alk
i93	(II) *pyro*arsenite	$Fe_2As_2O_5$	341.53	grn-wh				i		s NH_4OH
i94	arsenide	FeAs	130.77	wh	7.83	1020		v sl s		
i95	arsenide, di-	Arsenoferrite. $FeAs_2$	205.69	silv gray, cub.	7.4	990		i		sl s HNO_3; i HCl
i96	(III) benzoate	$Fe(C_7H_5O_2)_3$	419.19	br powd				i		s h al, h eth
i97	boride	FeB	66.66	gray cr	7.15^{18}			i		s NHO_3, h conc H_2SO_4
i98	(II) bromide	$FeBr_2$	215.66	grn-yel, hex.	4.636^{25}	d 684 (?)		109^{10}	170^{96}	s al; sl s bz
i99	(III) bromide	$FeBr_3$ or Fe_2Br_6	295.56	dk red-brn, rhomb (?) deliq		subl d		s	s	s al, eth, sl s NH_3
i100	(III) bromide-, hexahydrate	$FeBr_3{\cdot}6H_2O$	403.65	dk grn		27		v s	v s	s al, eth
i101	(III) cacodylate	$Fe[(CH_3)_2AsO_2]_3$	466.82	yelsh amorph powd				6.67		sl s al
i102	carbide	Fe_3C	179.55	gray, cub	7.694	1837		i	i	s a
i103	(II) carbonate	Nat. siderite. $FeCO_3$	115.86	gray, trig, 1.875, 1.633	3.8	d		0.0067^{25}		s CO_2 sol
i104	(II) carbonate, hydrate	$FeCO_3{\cdot}H_2O$	133.87	amorph		d		sl s		s a, CO_2 sol
i105	carbonyl, ennea-	$Fe_2(CO)_9$	363.79	yel met cr, hex.	2.085^{18}	d 80		i	i	v sl s al, MeOH; d HNO_3; i a
i106	carbonyl, penta-	$Fe(CO)_5$	195.90	visc yel liq	liq 1.457^{21}	−21	102.8^{749}	i		s al, eth, bz, alk, conc H_2SO_4
i107	carbonyl, tetra-	$Fe(CO)_4$	167.89	dk grn lust cr, tetr	1.996^{18}	d 140—150		i		s org solv, conc HNO_3 h H_2SO_4
i108	(II) *per*chlorate	$Fe(ClO_4)_2$	254.75	wh or grnsh-wh need, hygr		d>100		v s		
i109	(II) *per*chlorate hexahydrate	$Fe(ClO_4)_2{\cdot}6H_2O$	362.84	grn, 1.493, 1.478		d>100		97.8^{0}	116.1^{60}	86.5^{20} al; s $HClO_4$
i110	*oxy*chloride	FeOCl	107.30	brn, rhomb	3.1	d 200				
i111	(II) chloride	Nat. lawrencite. $FeCl_2$	126.75	grn to yel, hex deliq, 1.567	3.16^{25}_{4}	670—674	subl	64.4^{10}	105.7^{100}	100 al; s acet; i eth
i112	(II) chloride, dihydrate	$FeCl_2{\cdot}2H_2O$	162.78	grn, monocl.	2.358					
i113	(II) chloride, tetrahydrate	$FeCl_2{\cdot}4H_2O$	198.81	bl-grn, monocl, deliq	1.93			160.1^{10}	415.5^{100}	s al; sl s acet
i114	(III) chloride	Nat. molysite. $FeCl_3$ or Fe_2Cl_6	162.21	blk-brn hex	2.898^{25}_{4}	306	d 315	74.4^{0}	535.7^{100}	v s al, MeOH, eth; 63^{18} acet
i115	(III) chloride, hydrate	$FeCl_3{\cdot}2^1/_2H_2O$	207.24	dk yel-red, rhomb, deliq		56		v s	v s	v s al, eth
i116	(III) chloride, hexahydrate	$FeCl_3{\cdot}6H_2O$	270.30	br yel cr mass, v deliq		37	280—285	91.9^{20}	∞	s al, eth
i117	(II) chloride, hexammine	$FeCl_2{\cdot}6NH_3$	228.94	wh powd	1.928^{25}_{4}					
i118	(II) chloroplatinate	$FePtCl_6{\cdot}6H_2O$	571.74	yel, hex	2.714	d		v s	v s	
i119	(III) *di*chromate	$Fe_2(Cr_2O_7)_3$	759.66	red-brn gran				s		s a
i120	(II) chromite	$FeCr_2O_4$	223.84	brn-blk, cub	4.97^{20}			i	i	sl s a
i121	(II) citrate	$FeC_6H_6O_7{\cdot}H_2O$	263.97	wh micr, rhomb		d 350 (in H_2)		sl s		s NH_4OH
i122	(III) citrate	$FeC_6H_5O_7{\cdot}5H_2O$	335.03	red-brn scales				sl s	s	i al
i123	hydroxide	Nat. goethite. FeO(OH)	88.85	brn, blksh, rhomb, 2.260, 2.394, 2.400	4.28	$-^1/_2H_2O$, 136				s HCl
i124	(II) ferricyanide	$Fe_3[Fe(CN)_6]_2$(?)	591.45	deep bl.		d		i		i al, dil a
i125	(III) ferricyanide	Berlin green. $Fe[Fe(CN)_6]$	267.80	cub						
i126	(II) ferrocyanide	$Fe_2[Fe(CN)_6]$	323.65	bl-wh, amorph	1.601^{25}_{4}	d 100	d 430 (vac)	i		
i127	(III) ferrocyanide	$Fe_4[Fe(CN)_6]_3$	859.25	dk bl cr		d		i	i	s HCl, H_2SO_4; i al, eth
i128	(II) fluoride	FeF_2	93.84	wh, rhomb	4.09^{25}_{4}	>1000 (?)		sl s		s a; i al, eth
i129	(II) fluoride, tetrahydrate	$FeF_2{\cdot}8H_2O$	237.97	grn-bl	4.20 (anh)	$-8H_2O$, 100		sl s	s	s HF, a; i al, eth
i130	(II) fluoride, tetrahydrate	$FeF_2{\cdot}4H_2O$	165.90	wh, rhomb	2.095	d		v sl s		s a; sl s al, eth
i131	(III) fluoride	FeF_3 or Fe_2F_6	112.84	grn, rhomb	3.52	>1000		sl s	s	s a; i al, eth
i132	(III) fluoride, tetrahydrate	$FeF_3{\cdot}4^1/_2H_2O$	193.91	yel cr.		$-3H_2O$, 100	d	sl s	s	i al
i133	(II) fluosilicate	$FeSiF_6{\cdot}H_2O$	306.01	col, trig, 1.361, 1.385	1.961			128.2		
i134	(III) fluosilicate	$Fe_2(SiF_6)_3$ (exist ?)	537.92	flesh col, gel				s	s, d	
i135	(III) formate	$Fe(CHO_2)_3{\cdot}H_2O$	208.92	red cr or powd				s		v sl s al
i136	(III) glycerophosphate	$Fe_3[C_3H_5(OH)_2{\cdot}PO_4]_3$	677.72	yelsh-grn sc or powd				50^{25}		i al
i137	(II) hydrogen cyanide	$H_4[Fe(CN)_6]$	215.98	wh, rhomb	1.536^{25}_{4}	d 190		s	s	v s al; s a; i acet
i138	(III) hydrogen cyanide	$H_3[Fe(CN)_6]$	214.98	brn-yel need		d 50—60		s	s	v s al; i eth
i139	(II) hydroxide	$Fe(OH)_2$	89.86	pa grn hex, or wh amorph	3.4	d		0.00015^{18}		s a, NH_4Cl; i alk
i140	(III) hydrosulfate	Iron(ic) tetrasulfate. Nat. rhomboclase $Fe_2O_3{\cdot}4SO_3{\cdot}9H_2O$	642.06	wh to pink powd, 1.533, 1.550, 1.635	2.172	$-6H_2O$, 80		s		sl s abs al
i141	(III) iodate	$Fe(IO_3)_3$	580.56	grn yel powd	4.80^{20}_{4}	d 130		sl s		i dil HNO_3
i142	(II) iodide	FeI_2	309.66	gray hex, hygr	5.315	red heat		s	s	s al, acet
i143	(II) iodide, tetrahydrate	$FeI_2{\cdot}4H_2O$	381.72	gray blk cr, deliq	2.873	d 90—98		v s	d	s al, eth
i144	(II) lactate	$Fe(C_3H_5O_3)_2{\cdot}3H_2O$	288.03	grn-wh cr or powd		d		2.1^{10}	8.5^{100}	s alk citrate; v sl s al; i eith
i145	(III) lactate	$Fe(C_3H_5O_3)_3$	323.06	br, amorph, deliq				s	v s	i eth
i146	(III) malate	$Fe_2(C_4H_4O_5)_3$	507.91	br scales, hygr				s		s al

PHYSICAL CONSTANTS OF INORGANIC COMPOUNDS (continued)

No.	Name	Synonyms and Formulae	Mol. wt.	Crystalline form, properties and index of refraction	Density or spec. gravity	Melting point, °C	Boiling point, °C	Solubility, in grams per 100 cc: Cold water	Hot water	Other solvents
i147	methanoarsenate	$Fe_2(CH_3AsO_3)_3$	525.56	redsh br lustr scales				50		i al, eth
i148	(II) nitrate	$Fe(NO_3)_2 \cdot 6H_2O$	287.95	grn, rhomb		60.5		83.5^{20}	166.7^{61}	
i149	(III) nitrate	$Fe(NO_3)_3 \cdot 6H_2O$	349.95	cub		35	d	150^{0}	∞	
i150	(III) nitrate	$Fe(NO_3)_3 \cdot 9H_2O$	404.00	col-pa vlt, monocl, deliq	1.684	47.2	d 125	s	s	s al, acet; sl s HNO_3
i151	nitride	Fe_4N	237.39		6.57(?)					
i152	nitride	Fe_2N or Fe_4N_2	125.70	gray	6.35	d 200		i		s HCl, H_2SO_4
i153	nitrosyl carbonyl	$Fe(NO)_2(CO)_2$	171.88	dk red cr	1.56	18.5	d 50, 110	i		s org solv
i154	(III) oleate	$Fe(C_{18}H_{33}O_2)_3$	900.22	br-red fatty lumps				i		s a, al, eth
i155	(II) oxalate	$FeC_2O_4 \cdot 2H_2O$	179.90	pa yel, rhomb	2.28	d 190		0.022	0.026	s a
i156	(III) oxalate	$Fe_2(C_2O_4)_3 \cdot 5H_2O$	465.83	yel micro cr powd		d 100		v s	v s	s a; i al
i157	(II) oxide	Nat. wuestite. FeO	71.85	blk, cub, 2.32	5.7	1369 ± 1		i	i	s a; i al, alk
i158	(III) oxide	Nat. hematitie. Fe_2O_3	159.69	red-brn to blk, trig, 3.01, 2.94(Li)	5.24	1565		i	i	s HCl, H_2SO_4; sl s HNO_3
i159	oxide	Iron ferrosoferric, nat. magnetite. Fe_3O_4	231.54	blk, cub or red-blk powd, 2.42	5.18	1594 ± 5		i	i	s conc a; i al, eth
i160	(III) oxide, hydrate	$Fe_2O_3 \cdot xH_2O$		red-brn amorph powd or gelat	2.44—3.60	all H_2O, 350—400		i	i	s a; i al, eth
i161	(II) *ortho*phosphate	Nat. vivianite. $Fe_3(PO_4)_2 \cdot 8H_2O$	501.61	wh-bl, monocl, 1.579, 1.603, 1.633	2.58			i	i	s a; i ac a
i162	(III) *ortho*phosphate	$FePO_4 \cdot 2H_2O$	186.85	pink, monocl	2.74	d		v sl s	0.67^{100}	s HCl, H_2SO_4; i HNO_3
i163	(III) *pyro*phosphate	$Fe_4(P_2O_7)_3 \cdot 9H_2O$	907.36	yel-wh powd				i		s a, alk citr
i164	phosphide, mono-	FeP	86.82	rhomb	6.07 (5.2^{20})					
i165	(di-)phosphide	Fe_2P	142.67	bl-gray cr or powd	6.56	1290		i	i	s aq reg, HNO_3 + HF; i dil a
i166	(tri-)phosphide	Fe_3P	198.51	gray	6.74	1100		i		
i167	(III) *hypo*phosphite	$Fe(H_2PO_2)_3$	250.81	wh or gray-wh powd		d		0.043^{25}	0.083^{100}	s alk citr
i168	(II) *meta*silicate	Nat. gruenerite. $FeSiO_3$	131.93	gray-grn, rhomb, 1.672, 1.697, 1.717	3.5	1146				
i169	*ortho*silicate	Nat. fayalite; Fe_2SiO_4	203.78	col, rhomb	4.34	1503 (?)		i	i	d HCl
i170	silicide	FeSi	83.93	yel-gray, oct	6.1			i	i	i aq reg
i171	(II) sulfate	Nat. szomolnokite. $FeSO_4 \cdot H_2O$	169.92	off-wh, monocl	2.970^{25}			sl s	s	
i172	(III) sulfate	$Fe_2(SO_4)_3$	399.87	yel rhomb, hygr, 1.814	3.097^{18}	d 480		sl s	d	i H_2SO_4, NH_3
i173	(II) sulfate, heptahydrate	Nat. melanterite. $FeSO_4 \cdot 7H_2O$	278.01	bl-grn, monocl, 1.471, 1.478, 1.486	1.898	64 − $6H_2O$, 90	− $7H_2O$, 300	15.65	48.6^{50}	sl s al; s abs MeOH
i174	(II) sulfate, pentahydrate	Nat. siderotil. $FeSO_4 \cdot 5H_2O$	241.98	wh, tricl, 1.526, 1.536, 1.542	2.2	− $5H_2O$, 300		s	s	i al
i175	(II) sulfate, tetrahydrate	$FeSO_4 \cdot 4H_2O$	223.97	grn monocl pr, 1.533, 1.535	2.23—2.29					
i176	(III) sulfate, enneahydrate	Nat. coquimbite. $Fe_2(SO_4)_3 \cdot 9H_2O$	562.00	rhomb, deliq, 1.552, 1.558	2.1	− $7H_2O$, 175		440	d	s abs al
i177	sulfide, di-	Nat. pyrite. FeS_2	119.97	yel, cub	5.0	1171		0.00049		d HNO_3. dil a
i178	sulfide, di-	Nat. marcasite. FeS_2	119.97	yel, rhomb	4.87	tr 450	d	0.00049		d HNO_3, i dil a
i179	(II) sulfide	Nat. troilite. FeS	87.91	blk-brn, hex	4.74	1193—1199	d	0.00062^{18}	d	s d a; i NH_3
i180	(III) sulfide	Fe_2S_3	207.87	yel-grn	4.3	d		sl d	d FeS + S	d a
i181	(II) sulfite	$FeSO_3 \cdot 3H_2O$	189.95	grnsh or wh cr		d 250		s sl s		s SO_2 sol; i al
i182	tantalate	Nat. tapiolite. $Fe(TaO_3)_2$	513.74	lt brn, tetr, 2.27, 2.42 (Li)	7.33					
i183	*d*-tartrate	$FeC_4H_4O_6$	203.92	wh cr.				0.877^{16}	v sl s	v s min a; s NH_4OH
i184	(II) thiocyanate	$Fe(SCN)_2 \cdot 3H_2O$	226.05	grn, rhomb		d		v s		s al, eth, acet
i185	(III) thiocyanate	$Fe(SCN)_3$ or $Fe_2(SCN)_6$	230.08	blk-red, cub, deliq				v s	d	s al, eth, acet
i186	(II) thiosulfate	$FeS_2O_3 \cdot 5H_2O$	258.04	grn cr, deliq				v s	d	v s al
i187	tungstate	Nat. ferberite. $FeWO_4$	303.69	tetr, 2.40 (Li)	6.64					
i188	*meta*vanadate	$Fe(VO_3)_3$	352.67	grayish-brn powd.				i		s a; i al
k1	**Krypton**	Kr	83.80	inert gas	gas 3.736 g/ℓ liq 2.155 *at* −152.9	−156.6	−152.30 ± 0.10	11.0^{0} cm^3 6.0^{25} cm^3	4.67^{50} cm^3	
l1	**Lanthanum**	La	138.9055 ± 3	wh met, tarnish in air, α: hex, β: cut above 350	α 6.1453, β 6.17	921	3457	d	d	s min a; i conc H_2SO_4
l2	acetate	$La(C_2H_3O_2)_3 \cdot 1^1/_2H_2O$	343.06					16.88^{25}		
l3	boride, hexa-	LaB_6	203.77	purp met cub	2.61	2210	d	i	i	i HCl
l4	bromate	$La(BrO_3)_3 \cdot 9H_2O$	684.75	hex pr		37.5	−7 H_2O, 100	28.5^{16}		i al
l5	bromide	$LaBr_3 \cdot 7H_2O$	504.72	col cr.	5.057^{26} anh	783 ± 3 anh	1577	v s		v s al; i eth
l6	carbide	LaC_2	162.93	yel cr.	5.02			d	d	s H_2SO_4; i conc HNO_3
l7	carbonate	$La_2(CO_3)_3 \cdot 8H_2O$	601.96	wh	2.6—2.7			i	i	s dil a; sl s aq CO_2; i acet
l8	chloride	$LaCl_3$	245.26	wh cr, deliq.	3.842^{25}	860	>1000	v s	d	v s al, pyr; i eth, bz
l9	chloride, heptahydrate	$LaCl_3 \cdot 7H_2O$	371.37	wh, tricl, hygr		d 91		v s	v s	v s al
l10	hexaantipyrin *per*chlorate	$[La(C_{11}H_{12}N_2O)]_6 \cdot (ClO_4)_3$	1566.63	col, hex cr		d 290—295		1.50^{20}		
l11	hydroxide	$La(OH)_3$	189.93	wh powd		d		i		s a

No.	Name	Synonyms and Formulae	Mol. wt.	Crystalline form, properties and index of refraction	Density or spec. gravity	Melting point, °C	Boiling point, °C	Solubility, in grams per 100 cc		
								Cold water	Hot water	Other solvents
112	iodate	$La(IO_3)_3$	663.62	col, cr				1.7^{25}		
113	iodide	LaI_3	519.62	gray-wh, rhomb, hygr.	5.63	772		v s		s acet
114	molybdate	$La_2(MoO_4)_3$	757.62	tetr	4.77^{16}	1181		0.00179^{25}	0.0033^{85}	
115	nitrate	$La(NO_3)_3 \cdot 6H_2O$	433.01	col cr, deliq, tricl		40	d 126	151.1^{25}	v s	v s al; s acet
116	oxalate	$La_2(C_2O_4)_3 \cdot 9H_2O$	704.01	wh cr.		d		0.00008^{25}		s min a
117	oxide	Lanthana. La_2O_3	325.81	wh rhomb, or amorph.	6.51^{15}	2307	4200	0.0004^{29}	d	s a, NH_4Cl; i acet
118	sulfate	$La_2(SO_4)_3$	565.98	wh powd, hygr.	3.60^{15}	d 1150		3.0	0.69^{100}	sl s al; i acet
119	sulfate, hydrate	$La_2(SO_4)_3 \cdot 9H_2O$	728.12	col hex, 1.564	2.821	d white heat		3.8^{0}	0.87^{100}	sl s HCl; i al
120	sulfide	La_2S_3	373.99	red-yel cr, hex	4.911^{11}	2100—2150 vac		d	d	s a
121	**Lead**	Pb	207.20	silv-blsh wh soft met, cub, 2.01	11.3437^{16} Ra-Pb 11.288^{20}_{20} UPb 11.2960^{16}	327.502	1740	i	i	s HNO_3, h conc H_2SO_4
122	abietate	$Pb(C_{20}H_{29}O_2)_2$	810.10	brn lumps or yelsh-wh powd				i		
123	acetate	$Pb(C_2H_3O_2)_2$	325.29	wh cr.	3.25^{20}_{4}	280		44.30^{20}	221^{50}	s glyc; v sl s al
124	acetate, basic	$Pb_2(OH)(C_2H_3O_2)_3$	608.54	wh				v s		sl s al
125	acetate, basic	$Pb(C_2H_3O_2)_2 \cdot 3PbO \cdot H_2O$	1012.90	wh powd						
126	acetate, basic	$Pb(C_2H_3O_2)_2 \cdot Pb(OH)_2 \cdot H_2O$	584.52	wh, monocl				v s		v s al
127	acetate, decahydrate	$Pb(C_2H_3O_2)_2 \cdot 10H_2O$	505.44	wh, rhomb cr.	1.69	22		s	s	i al
128	acetate, trihydrate	Sugar of lead. $Pb(C_2H_3O_2)_2 \cdot 3H_2O$	379.34	wh, monocl, β 1.567	2.55	– H_2O, 75	d 200	45.61^{15}	200^{100}	i al
129	acetate, tetra-	$Pb(C_2H_3O_2)_4$	443.38	col, monocl	2.228^{17}	175		d		d al; s chl, h ac a
130	*di*antimonate	$Pb_2Sb_2O_7$	769.90	dk yel powd	6.72			i	i	v sl s HCl
131	*ortho*antimonate	$Pb_3(SbO_4)_2$	993.10	or-yel powd.	6.58^{20}_{4}			i	i	v sl s HCl
132	*ortho*antimonate	Nat. monimolite. $Pb_3(SbO_4)_2$	993.10	orange powd	6.58^{20}_{4}			i		i dil a
133	*meta*arsenate	$Pb(AsO_3)_2$	453.04	hex tabl	6.42^{15}			d	d	s HNO_3
134	*ortho*arsenate	$Pb_3(AsO_4)_2$	899.44	wh cr, v pois	7.80	1042, sl d 1000		v sl s		s HNO_3
135	*ortho*arsenate, di-	Nat. schultenite. $PbHAsO_4$	347.13	monocl leaf, α 1.90, γ 1.97	5.79	d 720	– H_2O, 220	i	sl s	s HNO_3, caust alk
136	*ortho*arsenate, mono-	$Pb(H_2AsO_4)_2$	489.07	tricl, 1.74, 1.82	4.46^{15}	d 140		d		s HNO_3
137	*pyro*arsenate	$Pb_2As_2O_7$	676.24	rhomb, β 2.03	6.85^{15}_{5}	802		i	d	s HCl, HNO_3; i ac a
138	*meta*arsenite	$Pb(AsO_2)_2$	421.04	wh powd	5.85			i		s HNO_3
139	*ortho*arsenite	$Pb_3(AsO_3)_2 \cdot xH_2O$		wh powd	5.85			i		s alk, HNO_3
140	azide	$Pb(N_3)_2$	291.24	col need, or powd.			expl 350	0.023^{18}	0.09^{70}	v s ac a; i NH_4OH
141	*meta*borate	$Pb(BO_2)_2 \cdot H_2O$	310.83	wh cr powd	5.598, anhydr	– H_2O, 160		i	i	s a; i alk
142	borofluoride	$Pb(BF_4)_2$	380.81	cr pr				d		d al
143	bromate	$Pb(BrO_3)_2 \cdot H_2O$	481.02	col. monocl	5.53	d 180		1.38^{20}	sl s	
144	bromide	$PbBr_2$	367.01	wh, rhomb	6.66	373	916	0.4554^{0} 0.8441^{20}	4.71^{100}	s a, KBr; sl s NH_3; i al
145	buryrate	$Pb(C_4H_7O_2)_2$	381.40	col scales, pois		90		i	i	s dil HNO_3
146	caprate	$Pb(C_{10}H_{19}O_2)_2$	549.72			103—104		i	i	0.0029^{20} eth
147	caproate	$Pb(C_6H_{11}O_2)_2$	437.50			73—74				1.09^{25} eth
148	caprylate	Lead octoate. $Pb(C_3H_{15}O_2)_2$	493.61	wh leaf		83.5—84.5		i	i	s al; 0.0938 eth
149	carbonate	Nat. cerussite. $PbCO_3$	267.21	col, rhomb, 1.804, 2.076, 2.078	6.6	d 315		0.00011^{20}	d	s a, alk; i NH_3, al
150	carbonate, basic	White lead, hydrocerussite. $2PbCO_3 \cdot Pb(OH)_2$	775.63	wh powd, or hex.	6.14	d 400		i	i	sl s aq CO_2; s HNO_2; i al
151	cerotate	$Pb(C_{26}H_{51}O_2)_2$	998.58	wh need		113		i		i al, eth; s bz
152	chlorate	$Pb(ClO_3)_2$	374.10	wh monocl, deliq	3.89	d 230		v s		s al
153	chlorate, hydrate	$Pb(ClO_3)_2 \cdot H_2O$	392.12	wh, monocl, deliq.	4.037	d 110		151.3^{18}	171^{80}	s al
154	*per*chlorate	$Pb(ClO_4)_2 \cdot 3H_2O$	460.15	wh, rhomb	2.6	d 100		499.7^{26}		s al
155	chloride	Nat. cotunnite. $PbCl_2$	278.11	wh, rhomb, 2.199, 2.217, 2.260	5.85	501	950^{760}	0.99^{20}	3.34^{100}	sl s dil HCl,NH_3; i al; s NH_4 salts
156	chloride, tetra-	$PbCl_4$	349.01	yel oily liq	3.18^{0}	–15	expl 105	d (Cl_2)	d	s conc HCl
157	chloride, sulfide	$PbCl_2 \cdot 3PbS$	995.89	red				i	d	d a, alk; i dil a
158	chlorite	$Pb(ClO_2)_2$	342.10	yel, monocl		espl 126		0.095^{20}	0.42^{100}	s KOH
159	chromate	Nat. crocoite, chrome yellow. $PbCrO_4$	323.19	yel, monocl, 2.31, 2.37(Li), 2.66	6.12^{15}	844	d	0.000058^{25}	i	s a, alk; i ac a, NH_3
160	chromate, basic	Chrome red. $PbCrO_4 \cdot PbO$	546.39	red cr powd.	6.63			i	i	s a, alk
161	chromate, basic	$Pb_2(OH)_2CrO_4$	564.41	red amorph or cr	6.63	920		i	i	s KOH
162	*di*chromate	$PbCr_2O_7$	423.19	red cr				d		s a, alk
163	citrate	$Pb_3(C_6H_5O_7)_2 \cdot 3H_2O$	1053.85	wh cr powd				s		v sl s al
164	cyanate	$Pb(OCN)_2$	291.23	wh need		d		i	sl s	
165	cyanide	$Pb(CN)_2$	259.24	yelsh-wh powd, pois				sl s	s	s KCN
166	enanthate	$Pb(C_7H_{13}O_2)_2$	465.56	wh leaf		91.5		sl s		i al
167	ethylsulfate	$Pb(C_2H_5SO_4)_2 \cdot 2H_2O$	493.47	col liq, pois.				s		
168	ferricyanide	$Pb_3[Fe(CN)_6]_2 \cdot 5$ (or 6) H_2O	1135.58	blk-brn to red, monocl pr		–H_2O, 110—120 d		sl s	s, d 100	s alk, HNO_3
169	ferrite	$PbFe_2O_4$	382.89	hex		1530 d, 725				
170	ferrocyanide	$Pb_2Fe(CN)_6$ $3H_2O$	680.40	yelsh-wh powd		–H_2O, 100		i		sl s H_2SO_4
171	fluoride	PbF_2	245.20	col, rhomb, pois	8.24	855	1290	0.064^{20}		s HNO_3; i acet, NH_3

PHYSICAL CONSTANTS OF INORGANIC COMPOUNDS (continued)

No.	Name	Synonyms and Formulae	Mol. wt.	Crystalline form, properties and index of refraction	Density or spec. gravity	Melting point, °C	Boiling point, °C	Solubility, in grams per 100 cc: Cold water	Hot water	Other solvents
172	fluorochloride	Nat. matlockite. PbFCl	261.65	wh, tetr, 2.145, 2.006	7.05	601		0.037^{25}	0.1081^{100}	
173	fluosilicate	$PbSiF_6·2H_2O$	385.31	col, monocl		d		s	v s	
174	fluosilicate, tetrahydrate	$PbSiF_6·4H_2O$	421.34	col, monocl		d<100				
175	formate	$Pb(CHO_2)_2$	297.24	wh, rhomb, lust, 1.789, 1.852, 1.877	4.63	d 190		1.6^{15}	20^{100}	i al
176	hydride, di-	PbH_2	209.22	gray powd		d				
177	hydroxide	$Pb(OH)_2$	241.21	wh, amorph		d 145		0.0155^{20}	sl s	s a, alk; i ac a
178	hydroxide	$Pb_2O(OH)_2$ or $2PbO·H_2O$	464.41	wh cub, or amorph powd, pois	7.592	d 145		0.014	sl s	s alk, ac a, HNO_3
179	iodate	$Pb(IO_3)_2$	557.01	wh	6.155^{20}	d 300		0.0012^{2}	0.003^{25}	sl s HNO_3; i NH_3
180	*para*periodate	$PbHIO_5$	415.11	wh cr		d 130		i	i	s dil HNO_3
181	*para*periodate, hydrate	$PbHIO_5·H_2O$	433.12	amorph		$-H_2O$, 110		i	i	sl s dil HNO_3
182	iodide, basic	$PbI_2·PbO·H_2O$	702.22	rhomb cr	6.83^{20}	d 100				
183	iodide, di-	PbI_2	461.01	yel hex powd, pois	6.16	402	954	0.044^{0} 0.063^{20}	0.41^{100}	s alk, KI; i al
184	iodide, mono-	PbI	334.10	pa yel		d 300		0.1		
185	isobutyrate	$Pb(C_4H_7O_2)_2$	381.40	wh pr		<100		9.1^{15}		
186	lactate	$Pb(C_3H_5O_3)_2$	385.34	wh cr powd				s		s h al
187	laurate	$Pb(C_{12}H_{23}O_2)_2$	605.83	chalky wh powd		104.7		0.009^{35}		0.008^{25} al; $0.007^{14.5}$ eth
188	lignocenate	$Pb(C_{24}H_{47}O_2)_2$	942.47	wh powd		117		i		v s h bz; sl s al; i eth
189	malate	$Pb(C_4H_4O_5)·3H_2O$	393.32	wh powd				sl s		v sl s al
190	melissate	$Pb(C_{31}H_{61}O_2)_2$	1138.85	wh powd		115—116		i	i	s biol tol, ac a; sl s h bz, chl; i al, eth
191	molybdate	Nat. wulfenite. $PbMoO_4$	367.14	col-lt yel, tetr pl	6.92_4^{25}	1060—1070			i	d conc H_2SO_4; s a, KOH; i al
192	myristate	$Pb(C_{14}H_{27}O_2)_2$	661.93	wh powd		107		0.005^{35}	0.006^{50}	0.004^{25} al; $0.010^{14.5}$
193	2-naphthalenesulfonate	$Pb(C_{10}H_7SO_3)_2$	621.65	wh cr powd, pois				i		s al
194	nitrate	$Pb(NO_3)_2$	331.21	col, cub or monocl, pois, 1.782	4.53^{20}	d 470		37.65^{0} 56.5^{0}	127^{100}	8.77^{22} 43% al; s alk, NH_3
195	nitrate, basic	$Pb(OH)NO_3$	286.21	wh rhomb cr	5.93	d 180		19.4^{19}	s	s a
196	nitrite	$3PbO·N_2O_3·H_2O$	763.63	lt yel powd				v s		s dil HNO_3
197	oleate	$Pb(C_{18}H_{32}O_2)_2$	770.12					i		6.46^{20} eth; s pet eth; sl s al
198	oxalate	PbC_2O_4	295.22	wh powd	5.28	d 300		0.00016^{18}		s HNO_3; i al
199	oxide-, di.	Plattnerite. PbO_2	239.20	bz, tetr, ω 2.3(Li)	9.375	d 290		i	i	s dil HCl; sl s ac a; i al
1100	oxide, mono-	Litharge. PbO	223.20	yel, tetr	9.53	886		0.0017^{20}		s HNO_3. alk, Pb acet, NH_4Cl, $CaCl_2$, $SrCl_2$
1101	oxide, mono-	Massicot. PbO	223.20	yel, rhomb, 2.51, 2.61(Li), 2.71	8.0			0.0023^{22}	i	s alk
1102	oxide, red	Minium. Pb_3O_4	685.60	red cr sc, or amorph powd	9.1	d 500		i	i	s HCl, acet a; i al
1103	oxide, sesqui-	Pb_2O_3	462.40	or-yel powd, amorph		d 370		i	d	d a
1104	oxide, sub	Pb_2O	430.40	blk, amorph	8.342	d		i	i	s a, alk
1105	oxychloride	$PbCl_2·3PbO$	947.70	yel				0.0056^{18}	0.07^{74}	
1106	oxychloride	Cassel yellow. $PbCl_2·7PBO$	1840.50	yel cr, or powd				i		
1107	oxychloride	Fiedlerite. $2PbCl_2·PbO·H_2O$	797.43	monocl, 1.816 2.1023, 2.026	5.88^{20}	d 150				s HNO_3
1108	oxychloride	Nat. laurionite. $PbCl_2·Pb(OH)_2$	519.32	rhomb	6.24	d 142				
1109	oxychloride	Matlockite. $PbCl_2·PbO$	501.31						i	s alkalies, hot conc HCl
1110	oxychloride	Nat. matlockite. $PbCl_2·Pb(OH)_2$	519.32	wh, tetrag, 2.04 2.15, 2.15	7.21	d 524		0.0095^{18}		s alk
1111	oxychloride	Nat. mendipite. $PbCl_2·2PbO$	724.50	yel, rhomb, 2.24, 2.27, 2.31	7.08	693		i	i	s alk
1112	oxychloride	Nat. paralaurionite. $PbCl_2·PbO·H_2O$	519.32	col to wh, monocl pr, 2.146	6.05^{15}	d 150				
1113	palmitate	$Pb(C_{16}H_{31}O_2)_2$	718.04	wh powd		112.3		0.005^{35}	0.007^{50}	s al; 0.148^{20} eth
1114	phenolate	Lead phenate, lead carbolate. $Pb(OH)OC_6H_5$	317.31	yelsh-wh powd				i		
1115	phenolsulfonate	Lead sulfocarbolate. $Pb[C_6H_4(OH)SO_3]_2·5H_2O$	643.60	wh lustr need				s		s al
1116	*meta*phosphate	$Pb(PO_3)_2$	365.14	col cr		800(?)			v sl s	
1117	*ortho*phosphate	$Pb_2(PO_4)_2$	811.54	col or wh powd, hex, 1.970, 1.936	6.9—7.3	1014		0.000014^{20}	i	s HNO_3, alk; i ac a, al
1118	*ortho*phosphate, di-	$PbHPO_4$	303.18	rhomb	5.661^{15}	d				s HNO_3, alk, NH_4Cl
1119	*ortho*phosphate, mono-	$Pb(H_2PO_4)_2$	401.17	need						s alk, dil HNO_3, h conc HCl
1120	phosphide	PbP_5	362.07	blk unstable, inflam		d 400 (vac)		d	d	d dil a
1122	*ortho*phosphite	$PbHPO_3$	287.18	wh, powd		d		i		s HNO_3

No.	Name	Synonyms and Formulae	Mol. wt.	Crystalline form, properties and index of refraction	Density or spec. gravity	Melting point, °C	Boiling point, °C	Solubility, in grams per 100 cc		
								Cold water	Hot water	Other solvents
1123	picrate	$Pb(C_6H_2N_3O_7)_2{\cdot}H_2O$	681.41	yel, need	2.831^{20}	$-H_2O$, 130	expl	0.88^{15}		
1124	proprionate, tetra-	$Pb(C_3H_5O_2)_4$	499.49	solid		132				
1125	*pyro*phosphate	$Pb_2P_2O_7$	588.34	wh, rhomb	5.8^{20}	824		i	i	s HNO_3, KOH
1126	*pyro*phosphate, hydrate	$Pb_2P_2O_7{\cdot}H_2O$	606.36	wh, rhomb		806 anhydr		i	d	s HNO_3, KOH, $Na_4P_2O_7$
1127	selenate	$PbSeO_4$	350.16	wh, rhomb	6.37_4^{20}	d		i	i	s conc a
1128	selenide	Nat. clausthalite. PbSe	286.16	gray, cub	8.10^{15}	1065		i		s HNO_3
1129	metasilicate	Nat. alamosite. $PbSiO_3$	283.28	col or wh, monocl	6.49	766		i		d a
1130	*ortho*silicate, di-	Nat. barysilite. $Pb_2Si_2O_7$	582.57	wh, trig, 2.070, 2.050	6.707			i	i	
1131	stearate	$Pb(C_{18}H_{35}O_2)_2$	774.15	wh powd		115.7		0.05^{35}	0.06^{50}	$0.005^{14.5}$ eth; i al
1132	sulfate	Nat. anglesite. $PbSO_4$	303.26	wh, monocl, or rhomb, 1.877, 1.822, 1.894	6.2	1170		0.00425^{25}	0.0056^{40}	s NH_4 salts; sl s conc H_2SO_4; i a
1133	sulfate, basic	Nat. lanarkite. $PbSO_4{\cdot}PbO$	526.46	wh, monocl, 1.93, 1.99, 2.02	6.92	977		0.0044^{0}	v sl s	sl s H_2SO_4
1134	sulfate, hydrogen	$Pb(HSO_4)_2{\cdot}H_2O$	419.35	wh cr		d		0.0001^{18} d		sl s H_2SO_4
1135	*peroxydi*sulfate	$PbS_2O_3{\cdot}3H_2O$	453.36	deliq				v s		
1136	sulfide	Nat. galena. PbS	239.26	bl met cub, 3.921	7.5	1114		0.000086^{13}		s a; i al, KOH
1137	sulfite	$PbSO_3$	287.26	wh powd		d		i	i	s HNO_3
1138	tartrate, *dl*-	$PbC_4H_4O_6$	355.27	wh cr powd	2.53^{19}			0.0025^{20}	0.0074^{100}	s HNO_3, KOH; i al, ac a, NH_4 ac
1139	*di*thionate	$PbS_2O_6{\cdot}4H_2O$	439.38	trig, 1.635, 1.653	3.22	d		115.0^{20}		
1140	thiosulfate	PbS_2O_3	319.32	wr cr	5.18	d		0.03		s a, $Na_2S_2O_3$
1141	*meta*titanate	$PbTiO_3$	303.08	yel. rhomb-pyr	7.52			i	i	
1142	telluride	Nat. altaite. PbTe	334.80	Wh, cub	8.164_4^{20}	917				i a
1143	thiocyanate	$Pb(SCN)_2$	323.36	wh, monocl	3.82	d 190		0.05^{20}	0.2^{100}	s KCNS, HNO_3
1144	tungstate	Nat. stolzite. $PbWO_4$	455.05	tetr, 2.269, 2.182	8.23			i		i HNO_3, s KOH
1145	tungstate	Nat. raspite. $PbwO_4$	455.05	col, monocl, 2.27, 2.27, 2.30	1123			0.03		d a; i al
1146	*meta*vanadate	$Pb(VO_3)_2$	405.08	yel powd				sl s		d HCl; s dil HNO_3
1147	**Lithium**	Li	6.941 ± 3	silver white, soft	0.534^{20}	180.54	1342	d		
1148	acetate	$LiC_2H_3O_2{\cdot}2H_2O$	102.02	wh, rhomb, α 1.40, β 1.50		70	d	300^{15}	v s	21.5 al
1149	acetylsalicylate	$LiC_9H_7O_4$	186.09	wh powd hygr, d in moist air				100		25 al
1150	*meta*aluminate	$LiAlO_2$ (or $Li_2Al_2O_4$)	65.92	wh, rhomb, 1.604, 1.614	2.55_4^{25}	1900—2000		i		
1151	aluminum hydride	$LiAlH_4$	37.95	wh cr powd	0.917	d 125		d		ca 30 eth
1152	amide	$LiNH_2$	22.96	col need, cub	$1.178^{17.5}$	380—400	d 750—200 subl	s	d	sl s liq NH_3, al; i eth, bz
1153	antimonide	Li_3Sb	142.57		3.2^{17}	>950		d	d	d a
1154	*ortho*arsenate	Li_3AsO_4	159.74	wh powd, rhomb	3.07^{15}			v sl s		s dil ac a; i pyr
1155	azide	LiN_3	48.96	col cr, hygr		d 115—298		66.41^{16}		20.26^{16} abs al; i eth
1156	benzoate	$LiC_7H_5O_2$	128.06	wh cr or powd				33^{25}	40^{100}	7.7^{26} al, 10^{78} al
1157	*meta*borate	$LiBO_2$	49.75	wh, tricl	$1.397^{41.7}$	845		2.57^{20}	11.8^{80}	
1158	*meta*borate	$LiBO_2{\cdot}8H_2O$	193.87	col, trig	$1.38^{14.9}$	47				
1159	*penta*borate	$Li_2B_{10}O_{16}{\cdot}8H_2O$	522.09	wh	1.72	300—350 $-8H_2O$		36.3^{45}	194^{100}	3.9^{30} al; 22^{33} glycerine; i bz
1160	*tetra*borate	$Li_2B_4O_7$	169.12	2h cr		930		2.89^{20}	5.45^{100}	i org solv
1161	borohydrate	$LiBH_4$	21.78	rhomb cr	0.66	d 279		s d		s eth
1162	borohydrate	$LiBH_4$	21.78	wh, orthorhomb	0.666	275 d		v sl s		d al; 2.5 eth
1163	bromide	LiBr	86.85	wh, cub, deliq, 1.784	3.464^{25}	550	1265	145^{4}	254^{90}	73^{40} al; 8 MeOH; s al, eth; sl s pyrid
1164	bromide, dihydrate	$LiBr{\cdot}2H_2)$	122.88	wh cr		$-1H_2O$; 44		$246.^{20}$	v s	s al
1165	carbide	Li_2C_2	37.90	wh cr or powd	1.65^{18}			d	d	s a
1166	carbonate	Li_2CO_2	73.89	wh, monocl, 1.428, 1.567, 1.572	2.11	723	d 1310^{760}	1.54^{0}	0.72^{100}	i al; acet
1167	carbonate, acid	Lithium bicarbonate. $LiHCO_3$	67.96	wh				5.5^{13}		
1168	chlorate	$LiClO_3$	90.39	col, rhomb need, deliq, α 1.63, γ 1.64	1.1190_4^{18} (18%Soln)	127.6	300 d	500^{27}		v s al; 0.142^{25} acetone
1169	chlorate	$LiClO_3{\cdot}{}^1/_2H_2O$ (or ${}^1/_3H_2O$)	99.40	wh, tetr, deliq		65(?)	$-{}^1/_2H_2O$, 90 d 290	v s	v s	v s al
1170	perchlorate	$LiClO_4$	106.39	wh	2.428	236	430 d	60.0^{25}	150^{89}	152^{25} al; 182^{25} MeOH; 114^{25} eth; 137^{25} acetone
1171	perchlorate, trihydrate	$LiClO_4{\cdot}3H_2O$	160.44	wh, hex	1.841	95 deliq 236 (anhydr)	d 100 $-2H_2O$	130^{25}		72.9^{25} al; 156^{25} MeOH; 96.2^{25} acetone; 0.096^{25} eth
1172	chloride	LiCl	42.39	wh, cub, 1.662	2.068^{25}	605	1325—1360	63.7^{0}	130^{96}	25.10^{30} al; 42.36^{25} MeOH; 4.11^{25} acetone; $0.538^{33.9}$ NH_4OH
1173	chloride, monohydrate	$LiCl{\cdot}H_2O$	60.41	wh cr, hydr	1.78	$-H_2O$ >98		86.2^{30}	2	s HCl
1174	chloroplatinate	$Li_2PtCl_6{\cdot}6H_2O$	529.77	or prism		$-6H_2O$, 180		v s	v s	v s al; i eth
1175	dichromate, dihydrate	$Li_2Cr_2O_7{\cdot}2H_2O$	265.90	orange-red cr, deliq	2.34^{30}	187 d	110 $-2H_2O$	187^{30}	278^{100}	s reacts al

PHYSICAL CONSTANTS OF INORGANIC COMPOUNDS (continued)

No.	Name	Synonyms and Formulae	Mol. wt.	Crystalline form, properties and index of refraction	Density or spec. gravity	Melting point, °C	Boiling point, °C	Solubility, in grams per 100 cc: Cold water	Hot water	Other solvents
1177	citrate	$Li_3C_6H_5O_7{\cdot}4H_2O$	281.99	col cr or powd, deliq		$-4H_2O$, 105		74.5^{25}	66.7^{100}	sl s al, eth
1178	fluoride	LiF	25.94	wh, cub, 1.3915	2.635^{20}	845	1676	0.27^{18}		i al; s HF
1179	fluosilicate	$Li_2SiF_6{\cdot}2H_2O$	191.99	wh, monocl, 1.300, 1.296	2.33^{12}	$-2H_2O$, 100	d	73^{17}		s al; i eth, acet
1180	fluosulfonate	$LiSO_3F$	106.00	wh powd		360		v s	s	v s al, eth, acet; i ligorin
1181	formate, monohydrate	$H{\cdot}COOLi{\cdot}H_2O$	69.97	wh, rhomb	1.46	$-H_2O$, 94	d 230	27.85^{18}	57.05^{98}	sl s al, acet; i bz
1182	gallium hydride	$LiGaH_4$	80.69	wh cr.				d	d	s eth
1183	gallium nitride	Li_3GaN_2	118.56	lt gr powd.	3.35	d 800		d	d	s a, alk
1184	*meta*germanate	Li_2GeO_3	134.47	monocl, 1.7.	3.53^{21}	1239		0.85^{25}		s a
1185	hydride	LiH	7.95	wh cr.	0.82	680		d		v sl s a
1186	hydroxide	LiOH	23.95	wh tetr, 1.464, 1.452	1.46	450	d 924	12.8^{20}	17.5^{100}	sl s al
1187	hydroxide, monohydrate	$LiOH{\cdot}H_2O$	41.96	wh monocl, 1.460, 1.524	1.51			22.3^{10}	26.8^{80}	sl s al; i eth
1188	iodate	$LiIO_3$	181.84	wh, hex, hygr	4.502^{32}_{4}			80.3^{18}		i al
1189	iodide	LiI	133.85	wh, cub, 1.955 ± 0.003	4.076	449	1180 ± 10	165^{20}	433^{80}	250.8^{20} al; 42.6^{18} acet 343.4^{20} MeOH; v s NH_4OH
1190	iodide, trihydrate	$LiI{\cdot}3H_2O$	187.89	col-yelsh, hex, hygr.	3.48	73 $-H_2O$	$-2H_2O$, 80 $-H_2O$, 300	151^{0}	201.2^{60}	s abs al, acet
1191	laurate	$LiC_{12}H_{23}O_2$	206.25	wh powd		229.2—229.8		$0.154^{16.3}$	0.178^{25}	0.322^{25} al; $0.008^{15.3}$
1192	*per*manganate	$LiMnO_4{\cdot}3H_2O$	179.92	cub	2.06	d 190		71.43^{16}		d alk
1193	molybdate	Li_2MoO_4	173.82	wh trig, hygr	2.66	705		v s		
1194	myristate	$LiC_{14}H_{27}O_2$	234.31			223.6—224.2		$0.027^{16.3}$ 0.036^{25}	0.062^{50}	$0.010^{15.3}$ eth; 0.331^{15} acet; 0.155^{20} al
1195	nitrate	$LiNO_3$	68.95	wh, trig, 1.735, 1.735	2.38	264	d 600	$89.8^{27.55}$	234^{100}	s NH_4OH, al; 37.15 pyridine
1196	nitrate, trihydrate	$LiNO_3{\cdot}3H_2O$	122.99	col need		$-2^1/_2H_2O$, 29.9	$-3H_2O$, 61.1	34.8^{0}	$57.48^{29.6}$	s al, MeOH, acet
1197	nitride	Li_3N	34.83	red-brn amorph, or blk-gray cr, cub		tr 840—850 (in N_2)				
1198	nitrite	$LiNO_2{\cdot}H_2O$	70.96	col flat need	1.615^{0}	>100	d	125^{0}	459^{50}	v s abs al
1199	oxalate	$Li_2C_2O_4$	101.90	col, rhomb, 1.465, 1.53, 1.696	$2.121^{17.5}$	d		$8^{19.5}$		i al, eth
1200	oxalate, acid	$LiHC_2O_4{\cdot}H_2O$	113.98			d		8^{17}		
1201	oxide	Li_2O	29.88	wh cr, cub, n_D 1.644	$2.013^{25.2}$	>1700	1200^{600}	6.67^{0} d	10.02^{100}	
1202	palmitate	$LiC_{16}H_{31}O_2$	262.36	wh powd		224.5		0.01^{18}	0.015^{25}	0.347^{15} acet; 0.077^{20} al; $0.005^{15.8}$ eth
1203	*meta*phosphate	$LiPO_3$	85.91	col pl.	2.461	red heat		i	i	s a
1204	*ortho*phosphate	Li_3PO_4	115.79	col, rhomb	$2.537^{17.5}$	837		0.039^{18}		s a, NH_4OH; i acet
1205	*ortho*phosphate	$Li_3PO_4{\cdot}{}^1/_2H_2O$	124.80	wh cr powd	2.41	$-{}^1/_2H_2O$, 100		0.04^{25}		s a
1206	phosphate, di- H	LiH_2PO_4	103.93	col cr, hygr	2.461	>100				
1207	potassium sulfate	$LiKSO_4$	142.10	col, hex; n_D 1.4723, 1.4717	2.393^{20}			s	s	
1208	potassium *dl*-tartrate	$LiKC_4H_4O_6{\cdot}H_2O$	212.13	col, monocl, β 1.523 (red)	1.610			s		
1209	salicylate	$LiC_7H_5O_3$	144.06	wh, powd, deliq		d		133.3		50 al
1210	selenide	$Li_2Se{\cdot}9H_2O$	254.98	col, rhomb, deliq				d		
1211	*meta*silicate	Li_2SiO_3	89.97	col, rhomb; α 1.584, γ 1.604	2.52^{25}_{4}	1204		i	s d	s dil HCl
1212	*ortho*silicate	Li_4SiO_4	119.85	col, rhomb; α 1.594, γ 1.614	2.392^{25}_{4}	1256		i	d	d a
1213	silicide	Li_6Si_2	97.82	bl cr, hygr	*ca.* 1.12	d 600 (vac)		d	d	d a; i NH_3 turp
1214	sodium fluoaluminate	$Li_3Na_3(AlF_6)_2$	371.74	cub cr, 1.3395	2.774	710		0.074^{18}		
1215	stearate	$LiC_{18}H_{35}O_2$	290.42	wh cr.	220.5—221.5			0.010^{18}		0.010^{25} al; 0.040^{18} eth; 0.457^{15} acet
1216	sulfate	Li_2SO_4	109.94	α monocl; β hex or rhomb, γ cub 500°C; β 1.465	2.221		845	26.1^{0}	23^{108}	i abs al, acet
1217	sulfate, hydrogen	$LiHSO_4$	104.01	col pr	2.123^{13}	120		d		
1218	sulfate, monohydrate	$Li_2SO_4{\cdot}H_2O$	127.95	col cr, monocl, 1.465, 1.477, 1.488	880			34.9^{25}	29.2^{100}	11.5^{30} al + H_2O (23.9% alco); i acet, pryidine
1219	sulfide	Li_2S	45.94	wh-yel, cub, deliq	1.66	900—975		v s	v s	v s al
1220	sulfide, hydro-	LiHS	40.01	wh powd, hygr.				s	s	s al
1221	sulfite, monohydrate	$Li_2SO_3{\cdot}H_2O$	111.96	wh need, α 1.53, γ 1.59		455 d	140 $-H_2O$	24.9^{30}	22^{80}	i org solv
1222	tartrate	$Li_2C_4H_4O_6{\cdot}H_2O$	179.97	wh cr powd				s		
1223	thallium *dl*-tartrate	$LiTlC_4H_4O_6{\cdot}2H_2O$	395.43	tricl.	3.144					
1224	thiocyanate	LiSCN	65.02	wh cr, deliq, n_D 1.333				v s		s methylacet
1225	*di*thionate	$Li_2S_2O_6{\cdot}2H_2O$	210.03	col, rhomb, 1.5602	2.158	d		v s		
1226	tungstate	Li_2WO_4	261.73	col, trig	3.71	742		v s	v s	d a; i al
1227	**Lutetium**	Cassiopeium. Lu	174.967	met, hex.	9.8404	1663	3395			

PHYSICAL CONSTANTS OF INORGANIC COMPOUNDS (continued)

No.	Name	Synonyms and Formulae	Mol. wt.	Crystalline form, properties and index of refraction	Density or spec. gravity	Melting point, °C	Boiling point, °C	Solubility, in grams per 100 cc: Cold water	Hot water	Other solvents
1228	bromide	$LuBr_3$	414.68		1025	1400		s	s	
1229	chloride	$LuCl_3$	281.33	col cr.	3.98	905	subl 750	s		
1230	fluoride	LuF_3	231.96			1182	220	i	i	
1231	iodide	LuI_3	555.68			1050	1200	s	s	
1232	oxalate	$Lu_2(C_3O_4)_3 \cdot 6H_2O$	722.08	wh cr.		50 ($-H_2O$)		i	i	i dil a
1233	oxide	Lu_2O_3	397.93	cub cr	9.42					
1234	sulfate	$Lu_2(SO_4)_3 \cdot 8H_2O$	782.23	col cr.				42.27^{20}	16.93^{40}	
m1	**Magnesium**	Mg	24.305	silv wh met, hex	1.74^{5}	648.8	1107	i	d to $Mg(OH)_2$	s min a, conc. HF, NH_4 salts; i CrO_3, alk.
m2	acetate	$Mg(C_2H_3O_2)_2$	142.39	wh cr.	1.42	323 d		v s	v s	5.25^{15} MeOH
m3	acetate, tetrahydrate	$Mg(C_2H_3O_2)_2 \cdot 4H_2O$	214.46	col, monocl deliq, β 1.491	1.454	80		120^{15}	∞	v s al
m4	aluminate	Nat. spinel. $MgAl_2O_4$	142.27	col, cub, 1.723.	3.6	2135				sl s H_2SO_4; v sl s dil HCl; i HNO_3
m5	amide	$Mg(NH_2)_2$	56.35	gray powd.		d 350—400	d	d		v sl s liq NH_3; d al
m6	antimonate	$MgO \cdot Sb_2O_5 \cdot 12H_2O$	579.98	hex or tricl cr.	2.60 (hex)	$-12H_2O$, 200		v sl s		d HCl
m7	antimonide	Mg_3Sb_2	316.42	met hex pl	4.088_4^{25}	961		i		
m8	*ortho*arsenate	Nat. hoernesite. $Mg_3(AsO_4)_2 \cdot 8H_2O$	494.88	wh monocl	2.60—2.61					
m9	*ortho*arsenate	$Mg_3(AsO_4)_2 \cdot 22H_2O$	747.09	wh cr.	1.788	$-17H_2O$, 100	$-21H_2O$, 220	i	i	s a, NH_4Cl
m10	*ortho*arsenate, mono-H	Nat. roesslerite. $MgHAsO_4 \cdot 7H_2O$	290.34	monocl.	1.943^{15}	$-5H_2O$, 100		d		
m11	arsenide	Mg_3As_2	222.76	brn-red, cub	3.148_4^{25}	800		d	d	d dil a, al
m12	*ortho*arsenite	$Mg_3(AsO_3)_2$	318.75	wh				s	v s	s a, NH_4Cl; i NH_4OH
m13	benzoate	$Mg(C_7H_5O_2)_2 \cdot 3H_2O$	320.58	wh powd		$-3H_2O$, 110	d 200	6.16^{15}	19.6^{100}	s al
m14	bismuthide	Mg_3Bi_2	490.88	met, hex.	5.945_4^{25}	823				
m15	bismuth nitrate	$3\ Mg(NO_3)_2 \cdot 2Bi(NO_3)_2 \cdot 24H_2O$	1543.29	col cr, deliq.	2.32_{18}^{16}	71		d	d	s HNO_3
m16	diborate	Nat. ascharite. $Mg_2B_2O_5 \cdot H_2O$	168.24	orthorhmb, 1.54	2.60—2.70					
m17	*meta*borate	Nat. pinnoite. $Mg(BO_2)_2 \cdot 3H_2O$	163.97	yel, tetr, pyram, 1.565, 1.575	2.27—2.30					
m18	metaborate, octahydrate	$Mg(BO_2)_2 \cdot 8H_2O$	254.04	col, tetr, 1.565, 1.575	2.30			i	v sl s	s a
m19	*ortho*borate	$Mg_3(BO_3)_2$	190.53	col, rhomb, 1.6527, 1.6537, 1.6548	2.99^{21}			i	i	s min a; i ac a
m20	boride	MgB_4	89.17	bl		d 1200 (vac)		d		sl s a
m21	bromate	$Mg(BrO_3)_2 \cdot 6H_2O$	388.20	col, cub, 1.514.	2.29	$-6H_2O$, 200	d	42^{18}	v s	i al
m22	bromide	$MgBr_2$	184.11	wh hex cr, deliq	3.72_4^{25}	700		101.50^{20}	125.6^{100}	6.9 al; 21.8^{20} MeOH
m23	bromide, hexahydrate	$MgBr_2 \cdot 6H_2O$	292.20	col, hex pr or need, hygr, fluo in x-rays	2.00	172.4		316^{0}	v s	s al, acet; sl s NH_3
m24	bromoplatinate	$MgPtBr_6 \cdot 12H_2O$	914.99	trig	2.802					
m25	carbonate	Nat. magnesite. $MgCO_3$	84.31	wh, trig, 1.717, 1.515	2.958	d 350	$-CO_2$, 900	0.0106		s a, aq + CO_2; i acet, NH_3
m26	carbonate, basic artinite	Nat. Artinite. $MgCO_3 \cdot Mg(OH)_2 \cdot 3H_2O$	196.68	wh, rhomb, 1.489, 1.534, 1.557	2.02^{20}					
m27	carbonate, basic	Nat. hydromagnesite. $3MgCO_3 \cdot Mg(OH)_2 \cdot 3H_2O$	365.31	wh, rhomb, 1.527, 1.530, 1.540	2.16	d		0.04	0.011	s a, NH_4 salts
m28	carbonate, pentahydrate	Nat. lansfordite. $MgCO_3 5H_2O$	174.39	wh, monocl, 1.456, 1.476, 1.502	1.73	d in air		0.176^{7}	0.375^{20}	s HCl, $MgSO_4$ soln
m29	carbonate, trihydrate	Nat. nesquehonite. $MgCO_3 \cdot 3H_2O$	138.36	col, rhomb need, 1.495, 1.501, 1.526	1.850	165		0.179^{16}	d	s a; 1.4 aq + CO_2
m30	chlorate	$Mg(ClO_3)_2 \cdot 6H_2O$	299.30	wh, rhomb need, deliq	1.80^{25}	35	d 120	128.6^{18}	v s	s al
m31	*per*chlorate	$Mg(ClO_4)_2$	223.21	deliq	2.21^{18}	d 251		99.3^{25}	v s	23.96^{25} al
m32	*per*chlorate, hexahydrate	$Mg(ClO_4)_2 \cdot 6H_2O$	331.30	wh, rhomb cr, 1.482, 1.458	1.98	185—190		v s	v s	
m33	*per*chlorate, hexammine	$Mg(ClO_4)_2 \cdot 6NH_3$	325.39	wh, cub	1.41_4^{20}					s liq NH_3; s d al
m34	chloride	$MgCl_2$	95.21	wh, lustr hex cr, 1.675, 1.59	2.316—2.33	714	1412	54.25^{20}	72.7^{100}	7.40^{30} al
m35	chloride, hexahydrate	Nat. bischofite. $MgCl_2 \cdot 6H_2O$	203.30	col, monocl, deliq, 1.495, 1.507, 1.528	1.569	d 116—118	d	167	367	s al
m36	chloropalladate	$MgPdCl_6 \cdot 6H_2O$	451.53	hex	2.12	d				
m37	chloroplatinate	$MgPtCl_6 \cdot 6H_2O$	540.19	yel, trig	3.692	$-H_2O$, 180		s	s	
m38	chlorostannate	$MgSnCl_6 \cdot 6H_2O$	463.80	tricl.	2.08	d 100				
m39	chromate	$MgCrO_4 \cdot 7H_2O$	266.41	yel, rhomb, 1.521, 1.550, 1.568	1.695		211.5^{18}	v s		
m40	chromite	$MgCr_2O_4$	192.29	dk-grn or red, cub	4.6^{20}			i	i	s conc H_2SO_4; i dl a, dil alk
m41	citrate, nono-H	$MgHC_6H_5O_7 \cdot 5H_2O$	304.49	wh gran powd				20^{25}	s	s a; i al
m42	cyanide	$Mg(CN)_2$	76.34			d 300 to $MgCN_2$	d 600	s	d	
m43	cyanoplatinite	$MgPt(CN)_4 \cdot 7H_2O$	449.56	red, pr	2.185^{16}	$-H_2O$, 45		v s	v s	i al, eth

No.	Name	Synonyms and Formulae	Mol. wt.	Crystalline form, properties and index of refraction	Density or spec. gravity	Melting point, °C	Boiling point, °C	Solubility, in grams per 100 cc		
								Cold water	Hot water	Other solvents
m44	ferrite	$MgFe_2O_4$	200.00	blk, oct, 2.35	4.44—4.60	1750 ± 25				s conc HCl; i dil a, h HNO_3, al
m45	ferrocyanide	$Mg_2Fe(CN)_6 \cdot 12H_2O$	476.75	pa yel cr		d *ca* 200		33		i al
m46	fluoride	Nat. sellaite. MgF_2	62.30	col, tetr, faint vlt, lumin, 1.378, 1.390; 3.14		1261	2239	0076^{18}	i	s HNO_3; sl s a; i al
m47	fluosilicate	$MgSiF_6$	166.38	wh, cr or powd				65		
m48	fluosilicate, hexahydrate	$MgSiF_6 \cdot 6H_2O$	274.47	wh, trig	1.788	d 120		$64.8^{17.5}$		i al
m49	formate	$Mg(CHO_2)_2 \cdot 2H_2O$	150.37	col, rhomb		$-2H_2O$, 100		14^{0} (anh)	24^{100} (anh)	i al, eth
m50	orthogermanate	Mg_2GeO_4	185.20	wh ppt				0.0016^{25}		s a; i alk
m51	germanide	Mg_2Ge	121.20			1115				
m52	hydride	MgH_2	26.32	wh tetr cr or mass		d 280 (vac)		d viol		i eth
m53	hydroxide	Nat. brucite. $Mg(OH)_2$	58.32	col, hex pl, 1.559, 1.580	2.36	$-H_2O$, 350		0.0009^{18}	0.004^{100}	s a, NH_4 salts
m54	iodate	$Mg(IO_3)_2 \cdot 4H_2O$	446.17	wh, monocl	$3.3^{13.5}$	$-4H_2O$, 210	d	10.2^{20}	19.3^{100}	
m55	iodide	MgI_2	278.11	wh, hex, deliq	4.43_4^{25}	d<637		148^{18}	164.9^{110}	s al, eth, NH_3
m56	iodide, octahydrate	$MgI_2 \cdot 8H_2O$	422.24	wh powd. deliq		d 41		81^{20}	90.3^{80}	s al, eth
m57	lactate	$Mg(C_3H_5O_3)_2 \cdot 3H_2O$	256.49	wh cr powd, v bitter taste				3.3	16.7^{100}	i al, eth
m58	laurate	$Mg(C_{12}H_{23}O_2)_2 \cdot 2H_2O$	458.96	wh lumps		150.4		0.007^{25}	0.041^{100}	0.415^{15} al; 0.012^{25} eth
m59	*per*manganate	$Mg(MnO_4)_2 \cdot 6H_2O$	370.27	dk purp need, deliq	2.18(?)	d		v s	d	s MeOH, ac a
m60	molybdate	$MgMoO_4$	184.24	rhomb, tricl	2.208			13.7^{25}		
m61	myristate	$Mg(C_{14}H_{27}O_2)_2$	479.04	wh powd		131.6		0.006^{15}	0.014^{50}	0.189^{15} al; 0.007^{25} eth
m62	nitrate, dihydrate	$Mg(NO_3)_2 \cdot 2H_2O$	184.35	col pr	2.0256^{25}	129		s	s	s al, liq NH_3; sl s conc HNO_3
m63	nitrate, hexahydrate	$Mg(NO_3)_2 \cdot 6H_2O$	256.41	col, monocl, deliq.	1.6363_4^{25}	89	d 330	125	v s	s al, liq NH_3
m64	nitride	Mg_3N_2	100.93	grn-yel, powd or mass	2.712_4^{25}	d 800	subl 700 (vac)	d	d	s a; i al
m65	nitrite, trihydrate	$Mg(NO_2)_2 \cdot 3H_2O$	170.36	wh pr, hygr		d 100		s		s al
m66	oleate	$Mg(C_{18}H_{33}O_2)_2$	587.22	yel powd or mass				0.024^{25}		6.64^{20} al; s linseed oil; sl s eth
m67	oxalate	$MgC_2O_4 \cdot 2H_2O$	148.36	wh powd	2.45	d 150		0.07^{16}	0.08^{100}	s alk, a, oxalate
m68	oxide	Nat. periclase. MgO	40.30	col, cub, 1.736	3.58^{25}	2852	3600	0.00062	0.0086^{30}	s a, NH_4 salts, i al
m69	oxide, per-	MgO_2	56.30	wh powd				i	i	s a
m70	palmitate	$Mg(C_{16}H_{31}O_2)_2$	535.15	wh cr need or lumps		121.5		0.008^{25}	0.009^{50}	0.047^{25} al; 0.003^{25} eth
m71	*ortho*phosphate	$Mg_3(PO_4)_2$	262.86	rhomb pl, iridisc		1184		i	i	s NH_4 salts, i liq NH_3
m72	*ortho*phosphate	$Mg_3(PO_4)_2 \cdot 22H_2O$	659.19	col, monocl pr	1.640^{15}	$-18H_2O$, 100	d 200	v sl s		d a
m73	*ortho*phosphate, mono-H	Nat. newberyite. $MgHPO_4 \cdot 3H_2O$	174.33	wh, rhomb, 1.514, 1.518, 1.533	2.123^{15}	$-H_2O$, 205	d 550—650	sl s		s a
m74	*ortho*phosphate, mono-H, heptahydrate	$MgHPO_4 \cdot 7H_2O$	246.39	wh, monocl need	1.728^{15}	$-4H_2O$, 100	d 550—650	0.3	0.2	s a; i al
m75	*ortho*phosphate, octahydrate	Nat. bobierite. $Mg_3(PO_4)_2 \cdot 8H_2O$	406.98	wh, monocl pl, 1.510, 1.520, 1.543	2.195^{15}	$-5H_2O$, 150	$-8H_2O$, 400	v sl s		s NH_4 citrate
m76	*ortho*phosphate, tetrahydrate	$Mg_3(PO_4)_2 \cdot 4H_2O$	334.92	monocl.	1.64^{15}			0.0205		s a; i NH_4 salts
m77	phosphide	Mg_3P_2	134.86	yel-grn cub cr	2.055			d	d	d dil min a; sl d conc H_2SO_4
m78	*hypo*phosphite	$Mg(H_2PO_2)_2 \cdot 6H_2O$	262.37	wh, ditetrag.	$1.59_4^{12.5}$	$-5H_2O$, 100	$-6H_2O$, 180	20^{25}		i al, eth
m79	*ortho*phosphite	$MgHPO_3 \cdot 3H_2O$	158.33						0.25	s a
m80	*pyro*phosphate	$Mg_2P_2O_7$	222.55	col, tab monocl, 1.602, 1.604, 1.615	2.559, (3.06)	1383		i	i	s a; i al
m81	platinocyanide	$MgPt(CN)_4 \cdot 7H_2O$	449.56	red cr, 1.561	2.185^{16}	$-2H_2O$, 45		s	s	s al; i eth
m82	salicylate	$Mg(C_7H_5O_3)_2 \cdot 4H_2O$	370.60	col or sl redsh cr powd, effl				s		s al
m83	selenate	$MgSeO_4 \cdot 6H_2O$	275.35	col, monocl, 1.468, 1.489, 1.491	1.928			v s	v s	
m84	selenide	MgSe	103.27	lght gray powd or cr, 2.44	4.21			d	d	d a
m85	*meta*silicate	Nat. clinoenstatite. $MgSiO_3$	100.39	wh, monocl, α 1.651, γ 1.660	3.192_4^{25}	d 1557		i	i	v sl s HF
m86	*ortho*silicate	Nat. forsterite. Mg_2SiO_4	140.69	wh, orthorhmb, 1.65, 1.66, 1.67	3.21	1910		i		d h HCl
m87	(di-) silicide	Mg_2Si	76.70	blue cub	1.94	1102		i	d	s a, NH_4Cl, HCl
m88	silicofluoride	$MgSiF_6 \cdot 6H_2O$	274.47	wh hex-rhomb, 1.3439, 1.3602	1.788	d 100		60^{25}	5	s dil a, v sl s HF; i al
m89	stannide	Mg_2Sn	167.30	blsh-wh meet		778		s		s dil HCl
m90	stearate	$Mg(C_{18}H_{35}O_2)_2$	591.25	wh powd or lumps		86—88		0.003^{15} 0.004^{25}	0.008^{50}	0.020^{25} al; 0.003^{25} eth
m91	sulfate	$MgSO_4$	120.36	col, rhomb cr, 1.56.	2.66	d 1124		26^{0}	73.8^{100}	s al, glyc; 1.16^{18} eth; i acet
m92	sulfate, heptahydrate	Epsom salt, nat. epsomite. $MgSO_4 \cdot 7H_2O$	246.47	col, rhomb or monocl, 1.433, 1.455, 1.461	1.68	$-6H_2O$, 150	$-7H_2O$, 200	71^{20}	91^{40}	sl s al, glyc

No.	Name	Synonyms and Formulae	Mol. wt.	Crystalline form, properties and index of refraction	Density or spec. gravity	Melting point, °C	Boiling point, °C	Solubility, in grams per 100 cc		
								Cold water	Hot water	Other solvents
m93	sulfate, monohydrate	Nat. kieserite, $MgSO_4 \cdot H_2O$	138.38	col, monocl pr, 1.523, 1.535, 1.586	2.445				68.4^{100}	
m94	sulfide	MgS	56.37	pa red-brn, cub, phosph, 2.271	2.84	d >2000		d	d	s a, PCl_3
m95	sulfite	$MgSO_3 \cdot 6H_2O$	212.45	wh, rhomb or hex, 1.511, 1.464 (hex)	1.725	$-6H_2O$, 200	d	1.25	s	i al, NH_3
m96	*d*-tartrate	$MgC_4H_4O_6 \cdot 5H_2O$	262.45	wh, rhomb	1.67	$-4H_2O$, 100	$-5H_2O$, 200	0.8^{18}	1.44^{90}	s min a; i al, NH_3
m97	*d*-tartrate, hydrogen	$Mg(HC_4H_4O_6)_2 \cdot 4H_2O$	394.53	wh, rhomb	1.72				1.893^{100}	
m98	telluride	MgTe	151.91	wh, hex cr	3.86			d		d a
m99	thiosulfate	$MgS_2O_3 \cdot 6H_2O$	244.51	col, rhomb pr.	1.818^{24}	$-3H_2O$, 170	d	v s	v s	i al
m100	thiotellurite	Mg_3TeS_5	360.82	pa yel cr mass				s	s	s al
m101	tungstate	$MgWO_4$	272.15	col, monocl	5.66			i		d a; i al
m102	**Manganese**	Mn	54.9380	gray-pink met, cub or tetr.	7.20	1244 ± 3	1962	d	d	s dil a
m103	(II) acetate	$Mn(C_3H_3O_2)_2$	1173.03	brn cr	1.74			s, d		s al
m104	acetate, tetrahydrate	$Mn(C_2H_3O_2)_2 \cdot 4H_2O$	245.09	pa red, monocl	1.589			s		s al
m105	arsenide, mono-	MnAs	129.86	blk, hex	6.17—6.20 (5.55)	d 400		i	i	s HCl, aq reg
m106	arsenide, di-	Mn_2As	184.80			1400		i	i	s aq reg
m107	arsenide, tri-	Mn_2As_2	314.66	magnetic, (exist?)				i	i	s aq reg
m108	(II) benzoate	$Mn(C_7H_5O_2)_2 \cdot 4H_2O$	369.23	pa red pr				6.55^{15}		
m109	boride, di-	MnB_2	76.56	gray-vlt cr.	6.9			d	d	s a
m110	boride, mono-	MnB	65.75	cr powd	6.2^{15}					
m111	bromide, di-	$MnBr_2$	214.75	rose cr	4.385_4^{25}	d		127.3^{0}	228^{100}	i NH_3
m112	bromide, di-, tetrahydrate	$MnBr_2 \cdot 4H_2O$	286.81	α stable, rose monocl, deliq β labile, col, rhomb		d 64.3		296.7^{0}		
m113	carbide	Mn_3C	176.83	tetr	6.89^{17}			d	d	s a
m114	(II) carbonate	Nat rhodochrosite $MnCO_3$	114.95	rose, rhomb, lt brn in air	3.125	d		0.0065^{25}		s dil a, aq CO_2; i al, NH_3
m115	chloride, di-	Scacchite. $MnCl_2$	125.84	pink, cub cr, deliq	2.977_4^{25}	650	1190	72.3^{25}	123.8^{100}	s al; i eth, NH_3
m116	chloride, di-, tetrahydrate	$MnCl_2 \cdot 4H_2O$	197.91	rose, monocl, deliq	2.01	58	$-H_2O$, 106; $-4H_2O$, 198	151^{8}	656^{100}	s al; i eth
m117	chloride, tri-	$MnCl_3$	161.30	brn cr or grnsh-blk		d sl				s abs al
m118	chloroplatinate	$MnPtCl_6 \cdot 6H_2O$	570.83	trig	2.692	d				
m119	chromite	$MnCr_2O_4$	222.93	gray-blk, cub	4.97^{20}			i	i	i a
m120	(II) citrate	$Mn_3(C_6H_5O_7)_2$	543.02	wh-redsh powd.				v sl s		s Na-citr sol
m121	(II) ferrocyanide	$Mn_2Fe(CN)_6 \cdot 7H_2O$	447.94	grnsh-wh powd.				i		s HCl; i NH_4 salts
m122	fluogallate	$[Mn(H_2O)_6][GaF_5 \cdot H_2O]$	345.76	pink, orthorhomb, 1.45	2.22	d 230		v s		s HF
m123	fluosilicate	$MnSiF_6 \cdot 6H_2O$	305.11	rose, hex pr, 1.357, 1.374	1.903	d		140^{18}	v s	s al
m124	fluoride, di-	MnF_2	92.93	red, tetr, or redsh powd	3.98	856		0.66^{40}	0.48^{100}	s a; i al, eth
m125	fluoride, tri-	MnF_3	111.93	red cr	3.54	d		d	d	s a
m126	formate	$Mn(CHO_2)_2 \cdot 2H_2O$	181.00	rhomb	1.953	d		s	s	
m127	(II) glycerophosphate	$MnC_3H_7O_6P$	225.00	wh or sl red powd				sl s		s a, citr a; i al
m128	hydroxide	$MnO(OH)_2$	104.95	blk-brn, amorph (exist?)	2.58			v sl s		
m129	(II) hydroxide	Nat. pyrochroite. $Mn(OH)_2$	88.95	wh-pink, trig 1.723, 1.681	3.258^{13}	d		0.0002^{18}		s a, NH_4 salts; i alk
m130	(III) hydroxide	Magnanite. MnO(OH)	87.94	br-blk, rhomb. 2.24, 2.24, 2.53 (Li)	4.2—4.4	d		i	i	s HCl, h H_2SO_4
m131	iodide, di-	MnI_2	308.75	pink, hex cr, deliq, br. in air	5.0^{1}	638 (vac) d *ca*80	subl vac 500	s	s	0.02^{25} NH_3
m132	iodide, di-, tetrahydrate	$MnI_2 \cdot 4H_2O$	380.81	rose, monocl, deliq		d		s	v s	
m133	*hexa*iodoplatinate	$MnPtI_6 \cdot 9H_2O$	1173.58	trig	3.604_4^{20}	d				
m134	(II) nitrate	$Mn(NO_3)_2 \cdot 4H_2O$	251.01	col, or rose, monocl	1.82	25.8	129.4	426.4^{0}	∞	v s al
m135	(II) lactate	$Mn(C_3H_5O_3)_2 \cdot 3H_2O$	287.13	pa red, monocl		d		10	v s	s al
m136	(II) oxalate	MnC_2O_4	142.96	wh cr powd	$2.43^{21.7}$	d 150		i	i	s a, NH_4Cl
m137	(II) oxalate, dihydrate	$MnC_2O_4 \cdot 2H_2O$	178.99	redsh-wh oct cr powd		$-2H_2O$, 100	d	0.0312^{25}	0.037^{36}	
m138	(II) oxalate, trihydrate	$MnC_2O_4 \cdot 3H_2O$	197.00	pink, tricl		$-H_2O$, 25				
m139	(II, III) oxide	Nat. hausmannite. Mn_3O_4	228.81	blk, tetr (rhomb), 2.46 (Li) 2.15 (Li)	4.856	1564		i	i	s HCl
m140	oxide, di-	Nat. pyrolusite. MnO_2	86.94	blk, rhomb, or brn-blk powd	5.026	−O, 535		i	i	s HCl; i HNO_3, acet
m141	oxide, hept-	Mn_2O_7	221.87	dk red oil, hyg, exp	2.396_4^{20}	5.9	d 55, exp 95	v s	d	s H_2SO_4
m142	(II) oxide, mono-	Nat. manganosite, MnO	70.94	grn, cub, 2.16	5.43—5.46 (3.7—3.9)			i	i	s a, NH_4Cl
m143	(III) oxide, sesqui-	Nat. braunite. Mn_2O_3	157.87	blk, cub (tetr)	4.50	−O, 1080		i	i	s a; i ac a
m144	oxide, tri-	MnO_3	102.94	redsh, deliq (exist?)		d		s	d	s alk, H_2SO_4
m145	(III) *meta*phosphate	$Mn_2(PO_3)_6 \cdot 2H_2O$	619.74					sl s	s	

No.	Name	Synonyms and Formulae	Mol. wt.	Crystalline form, properties and index of refraction	Density or spec. gravity	Melting point, °C	Boiling point, °C	Solubility, in grams per 100 cc: Cold water	Hot water	Other solvents
m146	(II) *ortho*phosphate	Nat. reddingite. $Mn_3(PO_4)_2 \cdot 3H_2O$	408.80	rose or yelsh-wh rhomb, 1.651, 1.656, 1.683	3.102					
m147	(III) *ortho*phosphate	$MnP_4 \cdot H_2O$	167.92	gray cr powd		$-H_2O$, 300	d	i		s h conc H_2SO_4, conc HCl, molten H_3PO_4
m148	(II) *ortho*phosphate, di-H	$Mn(H_2PO_4)_2 \cdot 2H_2O$	284.94			$-H_2O$, >100		s		i al
m149	(II) *ortho*phosphate mono-H	$MnHPO_4 \cdot 3H_2O$	204.96	red, rhomb or pink powd, 1.656				sl s	d	s a; i al
m150	(II) *pyro*phosphate	$Mn_2P_2O_7$	283.82	br-pink, monocl, 1.695, 1.704, 1.710	3.707^{25}	1196		i		s a
m151	(II) *pyro*phosphate, trihydrate	$Mn_2P_2O_7 \cdot 3H_2O$	337.87	wh, amorph powd				i		s $K_2P_2O_7$ sol, H_2SO_3 i acet
m152	phosphide, mono-	MnP	85.91	dk gray	5.39^{21}	1190		i	i	sl s HNO_3
m153	(tri-)phosphide, di-	Mn_3P_2	226.76	dk gray	5.12^{18}	1095		i	i	sl s dil HNO_3
m154	(II) *hypo*phosphite	$Mn(H_2PO_2)_2 \cdot H_2O$	202.93	rose cr or powd		$-H_2O$, >150		12.5	16.7	i al
m155	(II) *ortho*phosphite	$MnHPO_3 \cdot H_2O$	152.93	redsh		$-H_2O$, 200		sl s		s $MnSO_4$, $MnCl_2$
m156	selenate, dihydrate	$MnSeO_4 \cdot 2H_2O$	233.93	rhomb	2.95—3.01			s	s	
m157	selenate, penta-hydrate	$MnSeO_4 \cdot 5H_2O$	287.97	pa red, trig	2.33—2.39					
m158	selenide	MnSe	133.90	gray, cub	5.55^{15}			i		d dil a
m159	selenite	$MnSeO_3 \cdot 2H_2O$	217.93	monocl cr				v sl s	v sl s	
m160	(II) *meta*slicate	Nat. rhodonite. $MnSiO_3$	131.02	red, tricl, 1.733, 1.740, 1.744	3.72^{25}	1323		i		i HCl
m161	silicide, di-	$MnSi_2$	111.11	gray, oct.	5.24^{13}			i	i	s HF, alk; i HNO_3, H_2SO_3
m162	silicide, mono-	MnSi	83.02	tetrah.	5.90^{15}	1280		i	i	s HF; v sl s a
m163	(di-)silicide,	Mn_3Si	137.96	quadr pr	6.20^{15}	1316		i	i	s HCl, NaOH: i HNO_3
m164	(II) sulfate	$MnSO_4$	151.00	redsh	3.25	700	d 850	52^{5}	70^{70}	s al; i eth
m165	(III) sulfate	$Mn_2(SO_4)_3$	398.05	grn cr, deliq, hex	3.24	d 160		d	d	s HCl, dil H_2SO_4; i conc. H_2SO_4, HNO_3
m166	(II) sulfate, dihydrate	$MnSO_4 \cdot 2H_2O$	187.03	(exist?).	2.526^{15}	stab 57—117		85.25^{35}	106.8^{55}	
m167	(II) sulfate, heptahydrate	$MnSO_4 7H_2O$	277.10	red monocl or rhomb	2.09	$-7H_2O$, 280; stab + 9		172	118^{13}	i al
m168	(II) sulfate, hexahydrate	$MnSO_4 \cdot 6H_2O$	259.09	(exist?).		stab + 5 to + 8		147.4	1.345^{38}	
m169	(II) sulfate, monohydrate	Nat. szmikite. $MnSO_4 \cdot H_2O$	169.01	pa pink monocl, 1.562, 1.595, 1.632	2.95	stab 57—117		98.47^{48}	79.8^{100}	
m170	(II) sulfate, pentahydrate	$MnSO_4 \cdot 5H_2O$	241.07	rose, tricl, 1.495, 1.508, 1.514	2.103^{15}	stab 9—26		124^{0}	142^{54}	
m171	(II) sulfate, tetrahydrate	Common form. $MnSO_4 \cdot 4H_2O$	223.06	pink, monocl or rhomb effl, 1.508, 1.522	2.107	stab 26—27		105.3^{0}	111.2^{54}	i al
m172	(II) sulfate, trihydrate	$MnSO_4 \cdot 3H_2O$	205.04	(exist?).	2.356^{15}	stab 30—40		74.22^{5}	99.31^{57}	
m173	(II) sulfide	Nat. alabandite. MnS	87.00	grn cub or pink amorph, 2.70 (Li)	3.99	d		0.00047^{18}		s dil a, al; i $(NH_4)_2S$
m174	(II) sulfide	$3MnS \cdot H_2O$	279.01	gray-pink		d		0.0006	i	s dil a; i $(NH_4)_2$ S
m175	(IV) sulfide	Nat. hauerite. MnS_2	119.06	blk cub, 2.69 (Li)	3.463	d		i	i	d HCl
m176	(II) tantalate	$Mn(TaO_3)_2$	512.83	blk, rhomb, 2.22, 2.25, 2.29	7.03					
m177	(II) tartrate	$MnC_4H_4O_6$	203.01	wh powd				v sl s		s dil a
m178	(II) thiocyanate	$Mn(SCN)_2 \cdot 3H_2O$	225.14	deliq		$-3H_2O$, 160—170		s	v s	v s al
m179	(II) dithionate	$Mn(SO_3)_2$	215.05	tricl cr	1.757			s	v s	
m180	(II) titanate	Nat. pyrophanite. $MnTiO_3$	150.82	yel, trig, 2.481, 2.210	4.54	1360				
m181	valerate	$Mn(C_5H_9O_2)_2 \cdot 2H_2O$	293.22	br powd				s		
m182	**Manganic acid, per-**	$HMnO_4$	118.84					v s	d	
m183	**Manganocyanic acid**	$H_4Mn(CN)_6$	215.08			d		i		v s al; i eth
m184	**Mercury**	Quicksilver. Hg	200.59 ± 3	silv wh met, liq.	13.5939_4^{20}	−38.87	356.58	i	i	s HNO_3; i dil HCl, HBr, HI cold H_2SO_4
m185	(I) acetate	$Hg_2(C_2H_3O_2)_2$	519.27	micaceous plates		d		0.75^{12}		s HNO_3, H_2SO_4; i al eth
m186	(II) acetate	$Hg(C_2H_3O_2)_2$	318.68	wh, sc or powd	3.270	d		25^{10}	100^{100}	s al, ac a
m187	(II) acetylide	$3HgC_2 \cdot H_2O$	691.85	wh powd	5.3	expl		i	i	i al
m188	(II) *ortho*arsenate	$Hg_3(AsO_4)_2$	879.61	yel				v sl s		s HCl, HNO_3
m189	(I) *ortho*arsenate mono-H	Hg_2HAsO_4	541.11					i		s HNO_3; i ac a, NH_4OH
m190	(I) azide	$Hg_2(N_3)_2$	485.22	wh cr.		expl, d by light		0.025		
m191	(II) benzoate	$Hg(C_7H_5O_2)_2 \cdot H_2O$	460.84	wh cr powd		165		1.2^{15}	2.5^{100}	s al, NaCl, NH_4Cl, bz
m192	(I) bromate	$Hg_2(BrO_3)_2$	656.98	cr		d		d		sl s HNO_3
m193	(II) bromate	$Hg(BrO_3)_2 \cdot 2H_2O$	492.42	cr		d 130—140		0.15	1.6	s HCl, HNO_3, $Hg(NO_3)_2$
m194	(I) bromide	Hg_2Br_2	560.99	wh, yel, tetr	7.307	subl 345		0.000004^{26}		s a; i al, acet

PHYSICAL CONSTANTS OF INORGANIC COMPOUNDS (continued)

No.	Name	Synonyms and Formulae	Mol. wt.	Crystalline form, properties and index of refraction	Density or spec. gravity	Melting point, °C	Boiling point, °C	Solubility, in grams per 100 cc		
								Cold water	Hot water	Other solvents
m195	(II) bromide	$HgBr_2$	360.40	col rhomb	6.109^{25} 5.12^{240}	236	322	0.61^{25}	4.0^{100}	15^{0} al; s MeOH; v sl s eth
m196	bromide iodide	HgBrI	407.40	yel, rhomb	...	229	360	...	...	s al, eth
m197	(I) carbonate	Hg_2CO_3	461.19	yel br cr	...	d 130	...	0.0000045	d	s NH_4Cl; i al
m198	(II) carbonate, basic	$HgCO_3 \cdot 2HgO$	693.78	br red	...	...	...	i	...	s NH_4Cl, aq CO_2
m199	(I) chlorate	$Hg_2(ClO_3)_2$	568.08	wh, rhomb	6.409	d 250	...	s	d	s al, ac a
m200	(II) chlorate	$Hg(ClO_3)_2$	367.49	wh need	4.998	d	...	25	...	...
m201	(I) chloride	Calomel. Hg_2Cl_2	472.09	wh, tetr, 1.973, 2.656	7.150	subl 400	...	0.00020^{25}	0.001^{43}	s aq reg, $Hg(NO_3)_2$; sl s HCl, h HNO_3; i al, eth
m202	(II) chloride	Corrosive sublimate. $HgCl_2$	271.50	col, rhomb or wh powd pois, 1.859	5.44^{25}, liq 4.44^{280}	276	302	6.9^{20}	48^{100}	33^{25} al; 4 eth; s ac a, pyr
m203	(I) chromate	Hg_2CrO_4	517.17	red need, or powd	...	d	...	v sl s	sl s	s HCN, HNO_3; i al, ac
m204	(II) chromate	$HgCrO_4$	316.58	red, rhomb	...	d	...	sl s, d	d	s HN_4Cl; d a; i acet
m205	(II) cyanide	$Hg(CN)_2$	252.63	col, tetr, or wh powd pois, 1.645	3.996	d	...	9.3^{14}	33^{100}	$25^{19.5}$ MeOH; 8 al; s NH_3, glyc; i bz
m206	(I) fluoride	Hg_2F_2	439.18	yel, cub	8.73^{15}_{4}	570	d	d to Hg_2O	...	...
m207	(II) fluoride	HgF_2	238.59	col, cub	8.95^{15}	d 645	650	d	...	s HF, dil HNO_3
m208	(I) fluosilicate	$Hg_2SiF_6 \cdot 2H_2O$	579.29	col pr	2.134	...	...	sl s	...	i HCl
m209	(II) fluosilicate	$HgSiF_6 \cdot 6H_2O$	450.76	col, rhomb, deliq	...	d easily	...	...	...	...
m210	(I) formate	$Hg_2(CHO_2)_2$	491.22	glist scales	...	d	...	0.4^{17}	d	i al
m211	(II) fulminate	$Hg(CNO)_2$	284.62	wh, cub	4.42	expl	...	sl s	s	s al, NH_4OH
m212	(I) iodate	$Hg_2(IO_3)_2$	750.99	yelsh	...	d 250	...	i	i	s dil HCl, conc HNO_3
m213	(II) iodate	$Hg(IO_3)_2$	550.40	wh, amorph powd						
	...			...		...	...	i	...	s HCl, NH_4Cl, NaCl, KI; i HNO_3
m214	(I) iodide	Hg_2I_2	654.99	yel, tetr or amorph powd	7.70	subl 40	d 290	v sl s	...	s KI, NH_4OH; i al, eth
m215	(II) iodide (α)	HgI_2	454.40	red, tetr	6.36^{25}_{4}	tr 127	...	0.01^{25}	...	3.18^{25} acet; 2.23^{25} al; s chl; d NH_4OH
m216	(II) iodide (β)	HgI_2	454.40	yel, rhomb cr or powd	6.094^{127}_{4}	259	354	v sl s	sl s	s eth, KI, $Na_2S_2O_3$; v sl s al
m217	(I) nitrate	$Hg_2(NO_3)_2 \cdot 2H_2O$	561.22	col, monocl, effl.	4.79^{4}	70	...	d	s, d	s dil HNO_3; i NH_4OH
m218	(II) nitrate	$Hg(NO_3)_2 \cdot {}^1/_2H_2O$	333.61	wh-yelsh cr or powd, deliq	4.39	79	d	v s	d	s acet, NH_3, NH_3; i al
m219	(II) nitrate	$Hg(NO_3)_2 \cdot H_2O$	342.62	col cr or wh powd, deliq	4.3	...	...	s	...	s HNO_3; i al
m220	(I) nitrite	$Hg_2(NO_2)_2$	493.19	yel	7.33	d 100	...	d	...	...
m221	nitride	Hg_3N_2	629.78	br powd	...	expl	...	d	...	s NH_4OH, NH_4 salts; d a
m222	(I) oxalate	$Hg_2C_2O_4$	489.20	...	...	...	...	i	i	sl s HNO_3
m223	(II) oxalate	HgC_2O_4	288.61	...	...	d	...	0.0107^{20}	...	s HCl; sl s HNO_3
m224	(I) oxide	Hg_2O	417.18	blk or brnish-blk powd	9.8	d 100	...	i	i	s HNO_3
m225	(II) oxide	Nat. montroydite. HgO	216.59	yel or red, rhomb, 2.37, 2.5, 2.65	1.1^{4}	d 500	...	0.0053^{25}	0.0395^{100}	s a; i al, eth, acet, alk, NH_3
m226	(II) oxybromide	$HgBr_2 \cdot 3HgO$	1010.17	yel cr.	...	...	...	i	sl s	v s al
m227	(II) oxychloride	$HgCl_2 \cdot 2HgO$	704.67	red hex, or blk monocl	red 8.16—8.43 blk 8.53	...	...	...	...	...
m228	(II) oxychloride	$HgCl_2 \cdot 3HgO$	921.26	yel, hex	7.93	d 260	...	i	d	...
m229	(II) oxycyanide	$Hg(CN)_2 \cdot HgO$	469.21	wh need or cr powd	4.437^{19}	expl	...	1.25	s	...
m230	(II) oxyfluoride	$HgF_2 \cdot HgO \cdot H_2O$	473.19	yel cr.	...	d 100	...	d	...	s dil HNO_3
m231	(II) oxyiodide	$HgI_2 \cdot 3HgO$	1104.17	yel br	...	...	...	...	d	s HI
m232	(II) selenide	Nat. tiemannite. HgSe	279.55	gray plates	8.266	vac subl	...	i	...	s aq reg
m233	(I) sulfate	Hg_2SO_4	497.24	col monocl, wh-yelsh powd	7.56	d	d	0.06^{25}	0.09^{100}	s HNO_3, H_2SO_4
m234	(II) sulfate	$HgSO_4$	296.65	col rhomb or wh powd	6.47	d	...	d	...	s a, NaCl; i al, acet, NH_3
m235	(II) sulfate, basic	$HgSO_4 \cdot 2HgO$	729.83	lem yel powd	6.44	...	volat	0.003^{16}	sl s	s a; i al
m236	(I) sulfide	Hg_2S	433.24	blk	...	d	...	i	...	i al, $(NH_4)_2$ S
m237	(II) sulfide (α)	Cinnabar, vermillion, HgS	232.65	red cr hex, or powd, 2.854, 3.201	8.10	subl 583.5	...	0.000001^{18}	...	s aq reg, Na_2S; i al, HNO_3
m238	(II) sulfide (β)	Metacinnabar. HgS	232.65	blk, cub or amorph powd	7.73	583.5	...	i	...	s aq reg, Na_2S, alk; i al, HNO_3
m239	(I) tartrate	$Hg_2C_4H_4O_6$	549.25	yelsh-wh cr powd	...	...	...	i	...	i a
m240	(I) *ortho*tellurate	HgH_4TeO_6	428.22	trans, orthorhomb	...	d 20	...	slow d	rapid d	...
m241	(II) *ortho*tellurate	Hg_3TeO_6	825.37	amber, cub	...	unalt at 140	...	i	i	s HCl, HNO_3
m242	(I) thiocyanate	$Hg_2(SCN)_2$	517.34	...	...	d	...	i	...	s HCl, KCNS
m243	(II) thiocyanate	$Hg(SCN)_2$	316.75	wh powd, pois.	...	d	...	0.07^{25}	s	s NH_4 salts, HCl, NH_3, KCN; sl s al, eth
m244	(I) tungstate	Hg_2WO_4	649.03	yel, amorph	...	d	...	i	i	d a; i al
m245	(II) tungstate	$HgWO_4$	448.44	yel	...	d	...	i	d	d a; i al

PHYSICAL CONSTANTS OF INORGANIC COMPOUNDS (continued)

No.	Name	Synonyms and Formulae	Mol. wt.	Crystalline form, properties and index of refraction	Density or spec. gravity	Melting point, °C	Boiling point, °C	Solubility, in grams per 100 cc: Cold water	Hot water	Other solvents
	Mercury nitrogen compounds									
m246	mercury (II) bromide, ammonobasic	$Hg(NH_2)Br$	296.52	wh powd		d		d		s NH_4OH; i al
m247	mercury (II) bromide, diammine	$Hg(NH_3)_2Br_2$	394.46	wh powd		180		d		s NH_4Cl, NH_4Br, NH_4I
m248	mercury (II) chloride ammonobasic	Infusible ppt. $Hg(NH_2)Cl$	252.07	wh powd or sm pr	5.70	infus		0.14	d 100	d a; i al
m249	mercury (II) chloride, aquobasic ammonobasic	Chloride of Millon's base. OHg_2NH_2Cl	468.65	pa yel or wh powd		d >120		sl s		s HCl, HNO_3
m250	mercury (II) chloride, diammine	Fusible white ppt. $Hg(NH_3)_2Cl_2$	305.56	rhombd		300		i	d	s a, KI
m251	mercury (II) iodide, ammonobasic	$Hg(NH_2)I$	343.52							i eth
m252	mercury (II) iodide aquobasic ammonobasic	Iodide of Millon's base. OHg_2NH_2I	560.11	yel to brn		>128	expl	i		s d, HCl, KI soln
m253	mercury (II) iodide, diammine	$Hg(NH_3)_2I_2$	488.46	col or pa yel powd or need				d		s NH_4OH
m254	**Millons's base**	$(HO)_2Hg_2NH_2OH$	468.22		4.083^{18}					
m255	**Molybdenum**	Mo	95.94	silv-wh met, or gray-blk powd, cub	10.2	2610	5560	i	i	s h conc HNO_3, h conc H_2SO_4, aq reg; sl s HCl; i HF, NH_3
m256	boride, (di-)	MoB_2	117.56	rhomb	7.12					
m257	boride, (mono-)	MoB	106.75	tetr	8.65					
m258	(di-) boride	Mo_2B	202.69	tetr	9.26					
m259	bromide, di-	$MoBr_2$ (or Mo_3Br_6)	255.75	yel red, amorph	$4.88^{17.5}$			i	i	s alk; i a, aq reg
m260	bromide, tetra-	$MoBr_4$	415.56	blk need, deliq		d	volat	v s		d alk
m261	bromide, tri-	$MoBr_3$	335.65	dk-grn need		d		i	i	d alk, NH_3; i a
m262	carbide, mono-	MoC	107.95	gray, hex	8.20^{20}	2692		i	i	sl s HNO_3, HF, h H_2SO_4, HCl; i alk hydr
m263	carbide(di-)	Mo_2C	203.89	wh, hex pr	8.9	2687		i		sl s HNO_3, HF h H_2SO_4 aq reg, HCl; i alk
m264	carbonyl	$Mo(CO)_6$	264.00	wh cr, rhomb diamagnet	1.96	d 150, without meltg	156.4^{766}	i	i	s bz; sl s eth
m265	carbonyl, tripyridine, tri-	$Mo(CO)_3(C_5H_5N)_3$	417.28	yel-brn cr		d				
m266	chloride, di-	$MoCl_2$ (or Mo_3Cl_6)	166.85	yel, amorph.	3.714^{25}	d		i	i	s HCl, H_2SO_4, alk, NH_4OH, al, acet
m267	chloride, penta-	$MoCl_5$	273.21	grn-blk cr, trig, deliq	2.928	194	268	d	d	s conc min a, liq NH_3, CCl_4, chl; s d al, eth
m268	chloride, tetra-	$MoCl_4$	237.75	brn powd or cr, deliq		d	vol	d	d	s conc min a; d al, eth
m269	chloride, tri-	$MoCl_3$	202.30	dk red need or powd	3.578^{25}_{4}	d		i	sl d	s conc H_2SO_4, conc HNO_3; v sl s al, eth; i HCl; d alk
m270	fluoride, hexa-	MoF_6	209.93	col cr.	liq $2.55^{17.5}$	17.5^{406}	37^{760}	s d	d	s NH_4OH, alk
m271	hydrotetrachlorohydroxide, di-	$[Mo_3Cl_4(H_2O_2)(OH)_2 \cdot 6H_2O$	607.77	lt yel cr		$-H_2O$ 35—300		i	i	i a, al
m272	hydroxide	$Mo(OH)_3$ (or $Mo_2O_3 \cdot 3H_2O$)	146.96	blk powd		d		0.2		sl s H_2SO_4, HCl; s 30% H_2O_2
m273	hydroxide	$MoO(OH)_3$ (or $Mo_2O_5 \cdot 3H_2O$)	162.96	br to blk powd				0.2 (coll)		s a, alk carb; i alk hydr
m274	(VI) hydroxide	$MoO_3 \cdot 2H_2O$	179.97	lt yel, monocl pr.	3.124^{15}			0.05^{15}		s dil H_2O_2, alk hydr; sl s a
m275	hydroxy*tetra*bromide, di-	$Mo_3Br_4(OH)_2$	641.45	red powd						s alk
m276	hydroxy*tetra*bromide, dioctahydrate	$Mo_3Br_4(OH)_2 \cdot 8H_2O$	785.57	golden yel cr		d	d	4lf		s HCl; d alk, HNO_3
m277	hydroxy*tetra*chloride, di-	$Mo_3Cl_4(OH)_2 \cdot 2H_2O$	499.68	pa yel, amorph.					i	s conc a; i al
m278	iodide, di-	MoI_2	349.75	brn powd	$5.278^{25.4}$			i	d	v sl s a
m279	iodide, tetra-	MoI_4	603.56	blk cr		d 100				
m280	oxide, di-	MoO_2	127.94	lead gray, tetr or monocl	6.47			i	i	sl s h conc H_2SO_4; i alk, HCl, HF
m281	oxide, pent-	Mo_2O_5	271.88	vlt-bl powd (exist?)						s h H_2SO_4, h CHl
m282	oxide, pent-	"Molybd. blue" $Mo_2O_5 \cdot xH_2O$ (variations in Mo and O)		dk blue coll or powd	3.6^{18}	d				s a, MeOH; i acet, bz, chl
m283	oxide, sesqui-	Mo_2O_3	239.88	blk, opaque (exist?)				i	i	i a, alk, NH_4OH
m284	oxide, tri-	Molybdic anhydride	143.94	col or wh-yel, rhomb	4.692^{21}	795	subl 1155^{760}	0.1066^{18}	2.055^{70}	s a, alk sulf, NH_4OH
m285	oxydibromide, di-	MoO_2Br_2	287.75	yel-red tabl, deliq				s		
m286	oxy*tetra*chloride	$MoOCl_4$	253.75	grn cr, deliq		subl		s		
m287	oxy*tri*chloride	$MoOCl_3$	218.30	grn cr		subl 100		d		
m288	oxy*di*chloride, di-	MoO_2Cl_2	198.84	yelsh wh scaly cr	3.31^{17}	subl		s	s	s al, eth
m289	oxy*di*chloride, dihydrate	$MoO_2Cl_2 \cdot H_2O$	216.86	pa yel cr.		subl		s	s	sl s al, acet, eth
m290	oxy*hexa*chloride, tri-	$Mo_2O_3Cl_6$	452.60	rubyred or dk vlt cr		d		d		s eth

PHYSICAL CONSTANTS OF INORGANIC COMPOUNDS (continued)

No.	Name	Synonyms and Formulae	Mol. wt.	Crystalline form, properties and index of refraction	Density or spec. gravity	Melting point, °C	Boiling point, °C	Solubility, in grams per 100 cc: Cold water	Hot water	Other solvents
m291	oxy*penta*chloride, tri-	$Mo_2O_3Cl_5$	417.14	dk brn-blk cr, deliq		melts easily	subl	s	s	
m292	oxychloride acid	$MoO(OH)_2Cl_2$	216.86	wh need, deliq		d 160		v s		s al, eth, acet
m293	oxy*di*fluoride, di-	MoO_2F_2	165.94	wh cr, hygr	3.49^{25}	subl 270		v s	v s	s al, MeOH; i eth, chl, tol
m294	oxy*tetra*fluoride	$MoOF_4$	187.93	col-wh, deliq	3.001^{25}	98	180	s		s al, eth, CCl_4, s d H_2So_4; v sl s bz
m295	*meta*phosphate	$Mo(PO_3)_6$	569.77	yel powd	3.28^{0}			i	i	sl s h aq reg; i HCL, HNO_3, H_2SO_4
m296	phosphide	MoP (or Mo_2P_2)	126.91	gray-grn cr powd	6.167					s h HNO_3
m297	phosphide	MoP_2	157.89	blk powd	5.35^{25}					s HNO_3, h conc H_2SO_4, aq reg; i conc HCl
m298	silicide	$MoSi_2$	152.11	gray, met, tetr	$6.31^{20.5}$					i a, aq reg; v s $HF+HNO_3$
m299	sulfide, di-	Nat. molybdenite. MoS_2	160.06	blk luster, hex	4.80^{14}	1185	subl 450, d in air	i	i	s h H_2SO_4, aq reg, HNO_3, i dil a, conc H_2SO_4
m300	sulfide, penta	$Mo_2S_5{\cdot}3H_2O$	406.23	dk br powd		$-H_2O$, 135	d	i	i	s NH_4OH, alk, sulfides
m301	sulfide, sesqui-	Mo_2S_3	288.06	steel gray need	5.91^{15}	d 1100	vol 1200			i conc HCl; d h HNO_3
m302	sulfide, tetra-	MoS_4	224.18	brn powd		d		i	i	i a; s h H_2SO_4, alk sulfide
m303	sulfide, tri-	MoS_3	192.12	blk pl		d	d	sl s	s	s alk sulf, conc KOH
m304	**Molybdic acid**	$H_2MoO_4{\cdot}H_2O$ (or $MoO_3{\cdot}2H_2O$)	179.97	yel, monocl	3.124^{15}	$-H_2O$, 70	d	0.133^{18}	2.568^{70}	s alk hydr, alk carb; sl s a
m305	anhydrous	H_2MoO_4 (or $MoO_3{\cdot}H_2O$)	161.95	wh or sl yelsh, hex	3.112	$-H_2O$, 70		sl s	sl s	s alk, NH_4OH, H_2SO_4; i NH_3
m306	**Molybdic arsenic acid**	$As_2O_5{\cdot}6MoO_3{\cdot}18H_2O$	1417.74	col, trig	$2.493^{19.8}$	$-15H_2O$, 150		v s		s abs al; i chl, CS_2
m307	**Molybdic phosphoric acid**	$H_7[P(Mo_2O_7)_6]{\cdot}28H_2O$	2365.71	yel, oct	2.53	78		d		
m308	**Molybdic silicic acid**	$H_8[Si(Mo_2O_7)_6]{\cdot}28H_2O$	2363.83	yel, tetr		45	d 100	600^{14}		s dil a; i bz, chl, CS_2
n1	**Neodymium**	Nd	144.24 ± 3	silv-wh to yelsh met, hex to m.p. 868, cub from 868	hex 7.004 cub 6.80	1024	3027	d		
n2	acetate	$Nd(C_2H_3O_2)_3{\cdot}H_2O$	339.39					26.2		
n3	acetylacetonate	$Nd[CH(COCH_3)_2]_3$	441.57	pink cr	1.618	150—152				
n4	*hexa*antipyrin *per*-chlorate	$[Nd(C_{11}H_{12}N_2O)_6]{\cdot}(ClO_4)_3$	1571.97	rose, hex cr		d 285—289		0.99^{20}		
n5	bromate	$Nd(BrO_3)_3{\cdot}9H_2O$	690.08	red, hex		66.7	$-9H_2O$, 150	151^{25}		
n6	bromide	$NdBr_3$	383.95	grn cr		684	540	sl s		
n7	carbide	NdC_2	168.26	yel, hex leaf	5.15	d		d	d	s dil a; i conc HNO_3
n8	chloride	$NdCl_3$	250.60	rose-vlt pr	4.134^{25}	784	1600	96.7^{13}	140^{100}	44.5 al; i eth, chl
n9	chloride, hexahydrate	$NdCl_3{\cdot}6H_2O$	358.69	red, rhomb		124	$-6H_2O$, 160	246^{13}	511^{100}	v s al
n10	chromate	$Nd_2(CrO_4)_3{\cdot}8H_2O$	780.58	yel cr.				0.027		
n11	fluoride	NdF_3	201.24	pa lilac		1410	2300	i	i	
n12	iodide	NdI_3	524.95	blk cr powd		775 ± 3	370	s	s	
n13	kojate	$Nd(C_6H_5O_4)_3$	567.55	lt choc		d 275		i		
n14	manganous nitrate	$2Nd(NO_3)_3{\cdot}3Mn(NO_3)_2{\cdot}24H_2O$	1629.72	vlt-red	2.114	77		77.4^{30}		
n15	magnesium nitrate	$2Nd(NO_3)_3{\cdot}3Mg(NO_3)_2{\cdot}24H_2O$	1537.82	vlt-red	2.020	109		69.5^{30}		
n16	dimethylphosphate	$Nd[(CH_3)_2PO_4]_3$	519.36	pa lilac, hex pl				56.1^{25}	22.3^{100}	
n17	molybdate	$Nd_2(MoO_4)_3$	768.29	tetr, 2.005	5.14^{18}	176				
n18	nickel nitrate	$2Nd(NO_3)_3{\cdot}3Ni(NO_3)_2{\cdot}24H_2O$	1640.98	blsh-grn	2.202	105.6		71.5^{30}		
n19	nitrate	$Nd(NO_3)_3{\cdot}6H_2O$	438.35	tricl				152.9^{25}		s al, acet
n20	nitride	NdN	158.25	blk powd				d		
n21	oxalate	$Nd_2(C_2O_4)_3{\cdot}10H_2O$	732.69	rose cr				0.000074^{25}		
n22	oxide	Neodymia. Nd_2O_3	336.48	lt bl powd, red fluores	7.24	~1900		0.00019^{29}	0.003^{75}	s a
n23	sulfate	$Nd_2(SO_4)_3{\cdot}8H_2O$	720.89	red, monocl, 1.41, 1.551, 1.562	2.85	1176		8^{20}	5.4^{40}	
n24	sulfide	Nd_2S_3	384.66	oliv grn powd	5.179^{11}	d		i	d	s dil a
n25	**Neon**	Ne	20.179	inert gas col sol, cub	gas: 0.9002^{0} g/ℓ; liq: $1.204^{-245.9}$	−248.67	−245.9	1.47^{20} cm^3		s liq O_2
n26	**Neptunium**	Np	237.0482	α: orthorhomb silvery β: tetr (above 278) υ: cub (above 500)	α; 20.45 β: 19.36^{313} γ: 18.0^{600}	630 ± 1 278 ± 5 stab 278—570				s HCl
n27	bromide, tri-	$NpBr_3$	476.76	α: hex β: grn orthorhomb	α 6.92	subl *ca* 800		s		
n28	chloride, tetra-	$NpCl_4$	378.86	red-brn tetr	4.92	538		s		
n29	chloride, tri-	$NpCl_3$	343.41	wh, hex	5.38	*ca*800		s		
n30	fluoride, hexa-	NpF_6	351.04	brn, orthorhomb		53	d	d		
n31	fluoride, tetra-	NpF_4	313.04	grn, monocl	6.8			i		i conc HNO_3
n32	fluoride, tri-	NpF_3	294.04	purple, hex	9.12			i		
n33	iodide, tri-	NpI_3	617.76	brn, orthorhomb	6.82			s		
n34	oxide, di-	NpO_2	269.05	apple grn, cub	11.11			i		s conc a

PHYSICAL CONSTANTS OF INORGANIC COMPOUNDS (continued)

No.	Name	Synonyms and Formulae	Mol. wt.	Crystalline form, properties and index of refraction	Density or spec. gravity	Melting point, °C	Boiling point, °C	Solubility, in grams per 100 cc		
								Cold water	Hot water	Other solvents
n35	(tri-) oxide, octa-	Np_3O_3	839.14	brn, cub		d 500				s HNO_3
n36	**Nickel**	Ni	58.69	silv met, cub	8.90	1455	2730	i	i	s dil HNO_3; sl s HCl, H_2SO_4; i NH_3
n37	acetate	$Ni(C_2H_3O_2)_2$	176.78	grn pr	1.798	d	16.6			i al
n38	acetate, tetrahydrate	$Ni(C_2H_3O_2)_2 \cdot 4H_2O$	248.84	grn pr	1.744	d	16			s dil al
n39	antimonide	Nat. breithauptite. NiSb	180.44	lt copper red, hex	7.54	1158	d 1400			
n40	*ortho*arsenate, octahydrate	$Ni_3(AsO_4)_2 \cdot 8H_2O$	598.03	helsh-grn powd.	4.98			i		s a
n41	arsenide	Nat. niccolite. NiAs	133.61	hex	7.57^{0}	968		i	i	s aq reg
n42	*ortho*arsenite, acid	$Ni_3H_6(AsO_3)_4 \cdot H_2O$	691.81	grn-wh		d		i		s a, alk
n43	benzenesulfonate	$Ni(C_6H_5SO_3)_2 \cdot 6H_2O$	481.11	grn, monocl.	1.628^{25}	$-H_2O$	d	14.3^{18}	51.5^{22}	5.9 al; 4.5 eth
n44	boride	NiB	69.50	pr	7.39^{18}			d	d	s aq reg, HNO_3
n45	bromate	$Ni(BrO_3)_2 \cdot 6H_2O$	422.59	monocl.	2.575	d		28		
n46	bromide	$NiBr_2$	218.50	yel brn, deliq	5.098^{27}	963		112.8°	155.1^{100}	s al, eth, NH_4OH
n47	bromide, trihydrate	$NiBr_2 \cdot 3H_2O$	272.54	yelsh-grn need, deliq		$-3H_2O$, 300		199°	315.7^{100}	s al, eth, NH_4OH
n48	bromoplatinate	$NiPtBr_6 \cdot 6H_2O$	841.29	trig	3.715					
n49	di-*N*-butyldithiocarbamate	(NBC) $Ni[(C_4H_9)_2NCSS]_2$	467.43	dk oliv grn powd	1.29	89—90		i	i	sl s bz, pet comp; i al
n50	carbide	Ni_3C	188.08	dk gray powd.	7.957^{25}					
n51	carbonate	$NiCO_3$	118.70	lt grn, rhomb		d		0.0093^{25}	i	s a
n52	carbonate, basic	$2NiCO_3 \cdot 3Ni(OH)_2 \cdot 4H_2O$	587.57	lt grn cr or brn powd		d		i	d	s a, NH_4 salts
n53	carbonate, basic	Zaratite. $NiCO_3$ $2Ni(OH)_2 \cdot 4H_2O$(?)	376.17	emerald grn, cub, 1.56—1.61	2.6			i	i	s h dil HCl, NH_4OH
n54	carbonyl	$Ni(CO)_4$	170.73	col, volat, inflamm, liq, or need	1.32^{17}	−25	43	$0.018^{9.8}$		s aq reg, al, eth, bz, HNO_3; i dil a, dil alk
n55	chlorate	$Ni(ClO_3)_2 \cdot 6H_2O$	333.68	dk red	2.07	d 80		0.9^{27}		
n56	*per*chlorate	$Ni(Clo_2)_2 \cdot 6H_2O$	365.68	grn, hex need, 1.518, 1.498		140		222.5°	273.7^{45}	s al, acet, chl
n57	chloride	$NiCl_2$	129.60	yel sc, deliq	3.55	1001	subl 973	64.2^{20}	87.6^{100}	s al, NH_4OH; i NH_3
n58	chloride, hexahydrate	$NiCl_2 \cdot 6H_2O$	237.69	grn, monocl, deliq ~1.57				254^{20}	599^{100}	v s al
n59	chloropalladate	$NiPdCl_6 \cdot 6H_2O$	485.92	hex	2.353					
n60	chloroplatinate	$NiPtCl_6 \cdot 6H_2O$	574.58	trig	2.798					
n61	cyanide	$Ni(CN)_2$	110.73	yel-brn				i	i	s KCN
n62	cyanide, tetrahydrate	$Ni(CN)_2 \cdot 4H_2O$	182.79	lt grn pl or powd pois		$-4H_2O$, 200	d	i	i	s KCN, NH_4OH, alk; sl s dil a
n63	ferrocyanide	$Ni_2Fe(CN)_6 \cdot xH_2O$		grn-wh	1.982(?)			i		s KCN, NH_4OH; i HCl
n64	fluogallate	$[Ni(H_2O)_6][GaF_5 \cdot H_2O]$	349.51	pa grn, monocl(?) 1.45	2.45	$-5H_2O$, 110		sl s		s HF
n65	fluoride	NiF_2	96.69	grn, tetr	4.63	subl 1000 (in HF)		4^{25}		s a, alk, eth, NH_3
n66	fluosilicate	$NiSiF_6 \cdot 6H_2O$	308.86	grn, trig, 1.391, 1.407	2.134	d				
n67	formate, dihydrate	$Ni(CHO_2)_2 \cdot 2H_2O$	184.76	grn cr	2.154	d		s		
n68	(II) hydroxide	$Ni(OH)_2$ (or $NiO \cdot xH_2O$)	92.70	grn cr, or amorph	4.15(3.65)	d 230		0.013		s a, NH_4OH
n69	iodate	$Ni(IO_3)_2$	408.50	yel need	5.07			1.1^{30}	1.0^{30}	
n70	iodate, tetrahydrate	$Ni(IO_3)_2 \cdot 4H_2O$	480.56	hex		d 100		1.4^{30}	1.1^{90}	
n71	iodide	NiI_2	312.50	blk cr, deliq	5.834	797		124.2°	188.2^{100}	s al
n72	dimethylglyoxime	$Ni(HC_4H_6N_2O_2)_2$	288.91	scarlet red cr		subl 250		i	i	s a, abs al; i a ac, NH_4OH
n73	nitrate, hexahydrate	$Ni(NO_3)_2 \cdot 6H_2O$	290.79	grn, monocl, deliq	2.05	56.7	136.7	238.5°	v s	s al, NH_4OH
n74	oleate	$Ni(C_{13}H_{33}O_2)_2$	621.61	grn oil		18—20				
n75	oxalate	$NiC_2O_4 \cdot 2H_2O$	182.74	lt grn powd				i		s a, NH_4 salts; v sl s oxal a
n76	oxide, mono-	Nat. bunsenite. NiO	74.69	grn-blk, cub, 2.1818(red)	6.67	1984		i	i	s a, NH_4OH
n77	*ortho*phosphate	$Ni_3(PO_4)_2 \cdot 8H_2O$	510.13	apple grn pl or emerald cr gran		d		i	i	s a, NH_4 salts; i me acet. et acet
n78	*pyro*phosphate	$Ni_2P_2O_7 \cdot xH_2O$		grn	3.93 (anhydr)					s a, NH_4OH
n79	(di-)phosphide	Ni_2P	148.35	gray cr.	6.31^{15}	1112		i		s HNO_3 + HF; i a
n80	(penta)phosphide, (di)-	Ni_5P_2	355.40	need or tabl cr		1185				
n81	(tri-)phosphide, (di-)	Ni_3P_2	238.01	dk grn-blk	5.99			i	i	s HNO_3; i HCl
n82	*hypo*phosphite	$Ni(H_2PO_2)_2 \cdot 6H_2O$	296.76	grn	$1.82^{19.8}$	d 100		s		
n83	selenate	$NiSeO_4 \cdot 6H_2O$	309.74	grn, tetr, 1.5393	2.314			s		
n84	selenide	NiSe	137.65	wh or gray, cub	8.46	red heat		i		s aq reg, HNO_3; i a, HCl
n85	silicide	Ni_2Si	145.47		7.2^{17}	1309		i		i a
n86	stearate	$Ni(C_{18}H_{35}O_2)_2$	625.64	grn powd		100		i		s CCl_4, pyr; sl s acet; i MeOH, eth
n87	sulfate	$NiSO_4$	154.75	yel, cub	3.68	d 848^{760}		29.3°	87.3^{100}	i al, eth, acet
n88	sulfate, hepatahydrate	Morenosite. $NiSO_4 \cdot 7H_2O$	280.85	grn, rhomb, 1.467, 1.489, 1.492	1.948	99; $-H_2O$ 31.5	$-6H_2O$, 103	$75.6^{15.5}$	475.8^{100}	s al
n89	sulfate, hexahydrate	Single nickel salt. $NiSO_4 \cdot 6H_2O$	262.84	α: bl, tetr β: grn, monocl, 1.511, 1.487	2.07	tr: 53.3	$-6H_2O$, 103	65.52°	340.7^{100}	12.5 MeOH; v s al, NH_4OH
n90	sulfide, mono-	Nat. millerite. NiS	90.75	blk; trig or amorph	5.3—5.65	797		0.00036^{18}		s aq reg, HNO_3, KHS; sl s a
n91	sulfide, sub-	Heazlewoodite. Ni_3S_2	240.19	pa yelsh bronze met, lust	5.82	790		i		s HNO_3

No.	Name	Synonyms and Formulae	Mol. wt.	Crystalline form, properties and index of refraction	Density or spec. gravity	Melting point, °C	Boiling point, °C	Solubility, in grams per 100 cc: Cold water	Hot water	Other solvents
n92	(II, III) sulfide	Poydomite. Ni_3S_4	304.31	gray-blk, cub	4.7			i		s HNO_3
n93	sulfite	$NiSO_3 \cdot 6H_2O$	246.84	grn, tetrah				i		s HCl, H_2SO_4
	dithionate	$NiS_2O_6 \cdot 6H_2O$	326.90	grn, tricl.	1.908	d				
n94	**Nickel Complexes**									
n95	diaquotetrammine nickel (II) nitrate	$[Ni(NH_3)_4(H_2O)_2] \cdot (NO_3)_2$	286.85	grn cr				s		i al
n96	hexamminenickel (II) bromide	$[Ni(NH_3)_6]Br_2$	320.68	vlt powd.	1.837			v s	d	
n97	hexamminenickel (II) chlorate	$[Ni(NH_3)_6](ClO_3)_2$	327.78		1.52	180		d to $Ni(NH_3)_4$		
n98	hexamminenickel (II) chloride	$[Ni(NH_3)_6]Cl_2$	231.78	blsh, cub	1.468^{25}			s	d	s NH_4OH; i al
n99	hexamminenickel (II) iodide	$[Ni(NH_3)_6]I_2$	414.68	pa bl, cub	2.101	d		d		s NH_4OH
n100	hexamminenickel(II) nitrate	$[Ni(NH_3)_6](NO_3)_2$	284.88	bl, oct or cub				4.46		
n101	tetrapyridinnickel (II) fluosilic	$[Ni(C_5H_5N)_4]SiF_6$	517.17	bl grn, rhomb	2.307					
n102	**Niobium**	Columbium. Nb.	92.9064	steel gray, lustr met cub, 1.80	8.57^{20}	2468 ± 10	5127	i	i	s fus alk; i HCl, HNO_3, aq reg
n103	boride	NbB_2	114.53	hex	6.97	2900(?)				
n104	bromide, penta-	$NbBr_5$	492.43	purp red		265.2	361.6	d		s al, ethyl bromide
n105	carbide	NbC	104.92	blk, cub or lavender-gray powd	7.6	3500		i		s HNO_3, HF
n106	chloride, penta-	$NbCl_5$	270.17	yel-wh, deliq	2.75	204.7	254	d		s al, HCl, CCl_4
n107	fluoride, penta-	NbF_5	187.90	col, monocl pr, hygr.	3.293	72—73	236	d		s al; sl s chl, CS_2, H_2SO_4
n108	hydride	NbH	93.91	gray powd.	6.6	infus				s HF, conc H_2SO_4; i HCl, alk, HNO_3
n109	nitride	NbN	106.91	blk, cub	8.4	2573		i		s HF + HNO_3; i HNO_3
n110	oxalate, hydrogen	$Nb(HC_2O_4)_5$	538.04	col, monocl				d	d	s $H_2C_2O_4$; d al
n111	oxide, di-	NbO_2	124.91	blk	5.9			i	i	sl s alk; i a
n112	oxide, mon- (or di-)	NbO (or Nb_2O_2)	108.91	blk, cub	7.30			i	i	s a, alk; i al, NHO_3
n113	oxide, pent-	Nb_2O_5	265.81	wh, rhomb	4.47	1485 ± 5		i	i	s HF, alk; i a
n114	oxide, pent-, hydrate	$Nb_2O_5 \cdot \chi H_2O$				d		i		s conc H_2SO_4, conc HCl, HF, alk; i NH_3
n115	oxide, tri-(seequi)	Nb_2O_3	233.81	bl-blk		1780				
n116	oxybromide	$NbOBr_3$	348.62	yel cr.		subl		d		s a
n117	oxychloride	$NbOCl_3$	215.26	col need		subl 400		s, d	d	s al, H_2SO_4; i HCl
n118	potassium fluoride	$NbOF_3 \cdot 2KF \cdot H_2O$	300.11	monocl leaf, lustr fatty				7.8	v s	
n119	**Nitric acid**	HNO_3	63.01	col liq, corr, pois, $1.397^{16.4}$	1.5027^{25}_{4}	−42	83	∞	∞	d al, viol; s eth
n120	const boil	68% HNO_3 + 32% H_2O		col liq	1.41		120.5	∞	∞	
n121	**Nitrogen**	N_2	28.0134	col gas, col liq, sol cub cr	gas 1.2506 g/ℓ; liq $0.8081^{-195.8}$ sol $1.026^{-252.5}$	−209.86	−195.8	2.33° cm^3	1.42^{40} cm^3	sl s al
n122	chloride, tri-	NCl_3	120.37	yel oil or rhomb cr	1.653	<−40	<71, expl 95	i	d	s chl, bz, CCl_4, CS_2, PCl_3
n123	fluoride, tri-	NF_3	71.00	col gas	liq: 1.537^{-129}	−206.60	−128.8	v sl s		
n124	iodide, tri-	NI_3	394.72	blk, expl.		expl	subl vac	i	d	s KCNS, $Na_2S_2O_3$
n125	iodide, tri-, monoammine	$NI_3 \cdot NH_3$	411.75	dk red, bhomb	3.5	d>20	expl	i	d	s HCl, KCNS, $Na_2S_2O_3$; i abs al
n126	oxide(ic)	NO	30.01	col gas, bl liq, sol liq 1.330^{-90}	gas 1.3402 g/ℓ liq; $1.269^{-150.2}$	−163.6	−151.8	7.34° cm^3	2.37^{60} cm^3	3.4 cm^3 H_2SO_4; 26.6 cm^3 al; s $FeSO_4$, CS_2
n127	oxide(ous)	N_2O	44.01	col gas or liq or cub cr, 1.005^{20}_{760}	1.977^{0}_{760} g/ℓ	−90.8	−88.5	130° cm^3	56.7^{25} cm^3	s al, eth, H_2SO_4
n128	oxide, pent-	Nitric anhydride. N_2O_5	108.01	wh, rhomb or hex	1.642^{18}	30	d 47	s	d to HNO_3	d chl
n129	oxide, tri-	NO_3	62.00	blsh gas		d at ord temp				s eth
n130	peroxide	Nitrogen oxide, di-. NO_2	46.01	col sol, yel liq or brn gas, 1.40^{20}	1.4494^{20}_{20}	−11.20	21.2	s, d		s alk, CS_2, chl
n131	(di-) oxide(tri-)	Nitrous anhydride. N_2O_3	76.01	red-bron gas, bl sol or liq	1.447^{2}	−102	d 3.5	s	d	s eth, a alk
n132	oxi (tri-) fluoride	NO_3F	81.00	col gas expl.	liq: $1.507^{-45.9}$ sol: $1.951^{-193.2}$	−175	−45.9	d		s acet; expl al, eth
n133	sulfide, penta-	N_2S_5	188.31	gray cr.		10—11	d	d	d	s CS_2, eth; i bz, al
n134	sulfide, tetra-	Tetranitrogen tetrasulfide sulfurnitride. N_4S_4	184.27	hel cr, 2.046	2.24^{18}	d 178				s al, bz, CS_2
n135	**Nitrosyl bromide**	NOBr	109.91	br gas or dk br liq	>1.0	−55.5	−2	d	d	s alk
n136	*per*chlorate	$NOClO_4H_2O$	147.47	rhomb, deliq	2.169	d 100		d		expl al, eth
n137	chloride	NOCl	65.46	yel gas or yel-red liq or cr	gas: 2.99 g/ℓ liq: 1.417^{-12}	−64.5	−5.5	d	d	s fum H_2SO_4
n138	fluoborate	$NOBF_4$	116.81	col, rhomb, cr, hygr.	2.185^{25}_{4}	subl $250^{0.01}$		d		

No.	Name	Synonyms and Formulae	Mol. wt.	Crystalline form, properties and index of refraction	Density or spec. gravity	Melting point, °C	Boiling point, °C	Solubility, in grams per 100 cc: Cold water	Hot water	Other solvents
n139	fluoride	NOF	49.00	col gas	2.176 g/ℓ	−134	−56			d to HNO_2 + HF
n140	**Nitrosylsulfuric acid**	Chamber crystals. $NOHSO_4$	127.07	col, rhomb		d 73.5		d		s H_2SO_4
n141	**Nitrosylsulfuric anhydride**	$(NOSO_3)_2O$	236.13	tetr		217	360	d		s H_2SO_4
n142	**Nitrous acid**	HNO_2	47.01	only in sol (pa bl)				d		
n143	hypo-	$H_2N_2O_2$	62.03	wh, sol		expl		s		
n144	**Nitryl** chloride	NO_2Cl	81.46	pa yel-br gas	gas: 2.57 g/ℓ liq: 1.32^{14}	<−31	5	d		
n145	fluoride	NO_2F	65.00	col gas, col sol.	2.90 g/ℓ	−139	−63.5	d		d al, eth, chl
o1	**Osmium**	Os	190.20	gray-blsh met, hex	22.48^{20}	2700	>5300	i	i	sl s aq reg, HNO_3; i NH_3
o2	carbonyl chloride	$Os(CO)_3Cl_2$	345.14	col pr	269—273	d 280		i	i	s NaOH; i a
o3	chloride, di-	$OsCl_2$	261.11	dk brn, deliq	d			i	sl d	s al, eth, HNO_3; sl s alk
o4	chloride, tetra-	$OsCl_4$	332.01	red br need		subl		sl s, d		i al
o5	chloride, tri-	$OsCl_3$	296.56	br, cub		d 500—600		v s		s alk, al, a; sl s eth
o6	chloride, tri-, trihydrate	$OsCl_3{\cdot}3H_2O$	250.60	dk gr cr		d		v s		s al
o7	fluoride, hexa-	OsF_6	304.19	brn cr		32.1	45.9	d	d	
o8	fluoride, tetra-	OsF_4	266.19	br powd				d	d	
o9	iodide	OsI_4	697.82	vlt-blk, hygr met, lust				v s	d	s al
o10	oxide, di-, brown	OsO_2	222.20	brn cr	$11.37^{21.4}$	30% tr to OsO_4, 500		i	i	i a
o11	oxide, di- black	OsO_2	222.20	blk powd	7.71^{21}	tr to br 350—400		i	i	s dil HCL
	oxide, mon-	OsO	206.20	blk				i	i	
013	oxide, sesqui-	Os_2O_3	428.40	dkg brn		d		i	i	i a
014	oxide, tetra-	OsO_4	254.20	col, monocl	4.906^{22}	40.6	130	5.70^{10}	6.23^{25}	250 ± 10^{20} CCl_4; s al, eth, NH_4OH, $POCl_3$
o15	sulfide, di-	OsS_2	254.32	blk, cub	9.47	d		i	i	s HNO_3; i alk
o16	sulfide, tetra-	OsS_4	318.44	br blk (exist?)		d		i		s dil HNO_3; i $(NH_4)_2S$
o17	sulfite	$OsSO_3$	270.26	bl blk		d		i		s dil HCL, alk
o18	telluride	$OsTe_2$	445.40	gray-blk cr		*ca*600				i a; d dil HNO_3
o19	**Oxygen**	O_2	31.9988	col gas, sol hex cr	gas: 1.429° g/ℓ, liq: 1.149^{-183} sol: $1.426^{-252.5}$	−218.4	−182.962	4.89° cm^3 3.16^{25} cm^3	2.46^{50} cm^3 2.30^{100} cm^3	2.78^{25} al
o20	Fluoride	OF_2	54.00	col gas, unst	liq: $1.90^{-223.8}$	−223.8	−144.8	sl s, d	i	sl s a, alk
o21	**Ozone**	O_3	47.9982	col gas, or dk bl liq, or bl-blk cr, liq: 1.2226.	gas: 2.144° g/ℓ liq: $1.614^{-195.4}$ g/ℓ	−192.7 ± 2 1.0	−111.9	49° cm^3		s alk sol, oils
p1	**Palladium**	Pd	106.42	silv wh, met, cub	12.02^{20} $11.40^{22.5}$	1554	2970	i	i	s aq reg, h HNO_3, H_2SO_4; sl s HCl
p2	bromide	$PdBr_2$	266.23	red br	5.173^{16}	d		i	i	s H Br
p3[1]	chloride	$PdCl_2$	177.33	dk red, cub need, deliq	4.0^{18}	d 500		s	s	s HBr; acet
p3[2]	chloride, dihydrate	$PdCl_2{\cdot}2H_2O$	213.36	br pr, deliq		d		v s	v s	s HCl, acet
p4	cyanide	$Pd(CN)_2$	158.46	yelsh-wh.		d		i	i	s KCN, NH_4OH; i dil a
p5	fluoride, di-	PdF_2	144.42	br, etr	5.80	volat	d red heat	sl s, d		s HF
p6	fluoride, tri-	PdF_3	163.42	blk, rhomb	5.06	d	d	d	d	s HF
p7	hydride	Pd_2H (or Pd_4H_2)	213.85	silv met (exist?)	10.76	d				
p8	iodide	PdI_2	360.23	blk powd	6.003^{18}	d 350		i	i	s KI; i al, eth, dil HCl
p9	nitrate	$Pd(NO_3)_2$	230.43	br yel, rhomb, deliq		d		s, d		s HNO_3
p10	oxide, di-	$PdO_2{\cdot}\chi H_2O$		dull red		d −H_2O, −O		i	i,d	s a, alk
p11	oxide, mon-	PdO	122.42	grnsh-bl or amber mass, or blk powd	9.70^{20}_{4}	870		i	i	i aq reg
p12	oxide, mon-, hydrate	$PdO{\cdot}\chi H_2O$		yel to brn		d		i	i	s a, NH_3,NH_4Cl
p13	selenate	$PdSeO_4$	249.38	dk brn-red, rhomb, deliq	6.5	d red heat		v s	v s	i al, eth, alk; s NH_3
p14	selenide	PdSe	185.38	dk gray		<960		i		s aq reg
p15	selenide, di-	$PdSe_2$	264.34	olive gray, hex		<1000		i	i	v s aq reg; v sl s HNO_3; i alk
p16	silicide	PdSi	134.51	cr	7.31^{15}			i		
p17	sulfate	$PdSO_4{\cdot}2H_2O$	238.51	red-br cr, deliq		d		v s	d	
p18	sulfide, di-	PdS_2	170.54	dk br cr	4.7—4.8^{25}_{4}	d		i	i	s aq reg, $(NH_4)_2S$
p19	sulfide, mono-	PdS	138.48	brn-blk tetr	6.6^{25}_{4}	d 950		i	i	sl s HNO_3, aq reg; i HCl, $(NH_4)_2S$
p20	sulfide, sub-	Pd_2S	244.90	grn gray (exist?)	7.303^{15}	d 800		i	i	sl s a, aq reg
p21	telluride, di-	$PdTe_2$	361.62	silvery cr, hex				i	i	v s aq reg; s HNO_3; i alk

No.	Name	Synonyms and Formulae	Mol. wt.	Crystalline form, properties and index of refraction	Density or spec. gravity	Melting point, °C	Boiling point, °C	Solubility, in grams per 100 cc: Cold water	Hot water	Other solvents
	Palladium complexes									
p22	diamminepalladium (II) hydroxide	$Pd(NH_3)_2(OH)_2$	174.50	yel micr cr		>105		v s	d	
p23	dichlorodiammine-palladium (II) *trans* (or α)	$Pd(NH_3)_2 \cdot Cl_2$	211.39	yel, tetr	2.5	d		0.304^{10}	s, d	s, d a; s NH_4OH; i chl, acet
p24	tetramminepalladium (II) chloride	$Pd(NH_3)_4Cl_2 \cdot H_2O$	263.46	col, tetr	1.91^{18}	d 120		v s		
p25	tetramminepalladium tetrachloropalladate	Vauquelin's salt. $Pd(NH_3)_4 \cdot PdCl_4$	422.77	pink powd or need	2.489^{21}	tr yel 184 d above 192		i	sl s	sl s dil HCl; s KOH
p26	**Phospham**	PN_2H	60.00	wh, amorph		infus		i	d	s conc H_2SO_4, alk; i a
p27	**Phosphomolybdic acid**	Molybdophosphoric acid $P_2O_5 \cdot 20MoO_3 \cdot 51H_2O$	3939.49	yel, tetr		78—90		s	s	
	Phosphonium									
p28	bromide	PH_4Br	114.91	col, cub	gas: 2.464 g/ℓ	subl *ca* 30	38.8^{794}	d	d	
p29	chloride	PH_4Cl	70.46	col, cub		$78^{46\ atm}$	subl	d		
p30	iodide	PH_4I	161.91	col, tetr, deliq	2.86	18.5, subl 61.8	80	d		d, s a, alk
p31	sulfate	$(PH_4)_2SO_4$	166.07					d		
p32	**Phosphoramide**	Phosphorylamide. $PO(NH_2)_3$	94.04	wh, cr		d		40.66^{15}	i	sl, me
	Phosphoramide									
p33	difluoro-	$H_2PO_2F_2$	102.99	col, fum liq	1.583_4^{25}	−96.5 ± 0.1	115.9 sl d			
p34	hypo-	$H_4P_2O_4 \cdot 2H_2O$	198.01	col, rhomb deliq		70	d 100	d to H_3PO_3 + HPO_3		
p35	meta-	HPO_3	79.98	col, vitrous deliq	2.2—2.5	subl		d to H_3PO_4d		s al; i liq CO_2
p36	monofluo-	H_2PO_3F	99.99	oily, col liq	1.818	−80				
p37	ortho-	H_3PO_4	98.00	col, liq, or rhomb cr, deliq	1.834^{18}	42.35	−½ H_2O, 213	548	v s	s al
p38	ortho-	$2H_3PO_4 \cdot H_2O$	214.01	col, hex pr deliq		29.32	d	v s		
p39	pyro-	$H_4P_2O_7$	177.98	col, need or liq, hygr.		61		709^{23}	d to H_3PO_4	v s al, eth
p40	**Phosphorus,** black	P_4	123.89504	blk, incombust	2.70					i CS_2, conc H_2SO_4
p41	red	P_4	123.89504	redsh-brn, cub, or amorph powd, (mix of col and vlt?).	2.34	$590^{43\ atm}$	ign 200 280	v sl s	i	s abs al; i CS_2, eth, NH_3
p42	violet	P_4	123.89504	vlt, monocl	2.36	590				i org solv
p43	yellow	Phosphorus, white. P_4	123.89504	yel (or wh) cub or wax like solid 2.144	1.82^{20}	44.1	280	0.0003^{15}	sl s	0.3 al; $880^{10}CS_2$; s bz, NH_3, alk, eth, chl, tol
p44	bromide, penta-	PBr_5	430.49	yel, rhomb		d < 100	d 106	d		s CS_2, CCl_4, bz
p45	bromide, tri-	PBr_3	270.69	col, fum liq, $1.697^{26.6}$	2.852^{15}	−40	172.9	d		d al; s eth, chl, CS_2, CCl_4
p46	bromide(di-) chloride, tri	PBr_2Cl_3	297.14	or cr		d 35		d		
p47	bromide(hepta-) chloride, di-	PBr_7Cl_2	661.21	pr				d		s PCl_3, PCl_5
p48	bromide(mono-)chloride, tetra-	$PBrCl_4$	252.69	yel cr.				d		
p49	bromide(octa-) chloride, tri-	PBr_8Cl_3	776.56	brn need		25		d		
p50	bromide(di-) fluoride, tri-	PBr_2F_3	247.78	pa yel liq		−20	d 15	d		d glass
p51	bromide nitride	$(PNBr_2)_3$	614.37	col, rhomb		190	subl v 150	i		s eth; sl s chl, CS_2
p52	chloride, di-	PCl_2 (or P_2Cl_4?)	101.88	col liq		−28	180	hydr		
p53	chloride, penta-	PCl_5	208.24	yelsh-wh, tetr. fum	gas: 4.65^{296} g/ℓ	d 166.8 (press)	subl 162	d		d a; s CS_2, CCl_4
p54	chloride, tri-	PCl_3	137.33	col, fum liq, 1.516^{14}	1.574^{21}	−112	75.5^{749}	d	d	s eth, bz, chl, CS_2, CCl_4
p55	chloride(di-) fluoride, tri-	PCl_2F_3	158.87	col liq	5.4 g/ℓ	−8	10			
p56	chloride(tri-)iodide, di-	PCl_3I_2	391.14	red, hex		d 259		d		s CS_2
p57	chloride(di-)nitride	$(PNCl_2)_3$	347.66	rhomb	1.98	114	256.5	i	d	s al, eth, bz, chl, a ac, CS_2
p58	chloride(di-)nitride	$(PNCl_2)_4$	463.55		2.18_{24}^{24}	123.5	328.5			
p59	chloride(di-)nitride	$(PNCl_2)_5$	579.43			41	224^{13}, polym >250			
p60	chloride(di-)nitride	$(PNCl_2)_6$	695.32			90	262^{13}, polym >250			
p61	cyanide	$P(CN)_3$	109.03	wh need		subl 130		d		v s eth; sl s h bz
p62	fluoride, penta-	PF_5	125.97	col gas	5.805 g/ℓ	−83	−75	d		
p63	fluoride, tri-	PF_3	87.97	col gas	3.907 g/ℓ	−151.5	−101.5	d		s al; d alk
p64	hydride, tri-	Phosphine. PH_3	34.00	col gas, pois		−133	−87.7	o.26 vol 20		
p65	iodide, di	P_2I_4	569.57	or, tricl		110		d	d	s CS_2
p66	iodide, tri-	PI_3	411.69	red, hex deliq	4.18	61	d	d	d	v s CS_2
p67	oxide, pent-	Phosphoric anhydride. P_2O_5 (or P_4O_{10})	141.94	wh, monocl or powd, v deliq	2.39	580—585	subl 300	d to H_3PO_4	d	s H_2SO_4; i acet, NH_3
p68	oxide, sesqui-	Phosphorus trioxide. P_4O_6 (or P_2O_3)	219.89	col, or wh powd, monocl cr, deliq	2.135^2s;1	23.8	175.4	d to H_3PO_3	d	s chl, bz, eth, CS_2
p69	oxide, tetra-	P_2O_4	125.95	col, rhomb, deliq	2.54^{23}	>100	180	v s to H_3PO_3	d	
p70	oxide, tri-	P_2O_3(or P_4O_6)	109.95	col, or wh powd, or monocl, deliq	2.135^{21}	23.8	173.8 (in N_2)	d to H_3PO_3	d	s CS_2, eth, chl, bs.

No.	Name	Synonyms and Formulae	Mol. wt.	Crystalline form, properties and index of refraction	Density or spec. gravity	Melting point, °C	Boiling point, °C	Solubility, in grams per 100 cc		
								Cold water	Hot water	Other solvents
p71	oxybromide	$POBr_3$	286.69	col pl.	2.822	56	189.5	d		s H_2SO_4 CS_2, eth, bz, chl
p72	oxy*di*bromide chloride	$POBr_2Cl$	242.23	sol or liq	liq: 2.45^{50}	30	165	d		
p73	oxybromide chloride, di	$POBrCl_2$	197.78	tabl or liq	liq:2.104^{14}	13	137.6	d		
p74	oxychloride	$POCl_3$	153.33	col, fum liq, $1.460^{25.1}$	1.675	2	105.3	d	d	d al, a
p75	oxychloride	$P_2O_3Cl_4$	251.76	col fum liq	liq: 1.58^{7}	< −50	212	d		
p76	oxyfluoride	POF_3	103.97	col gas	4.69 g/ℓ	−68	−39.8	d		d al
p77	oxynitride	PON	60.98	wh, amorph		red heat		i	i	i a, alk
p78	oxysulfide	$P_4O_6S_4$	348.13	wh, tetr, deliq		102	295	d		50 CS_2
p79	selenide, penta-	P_2Se_5	456.75	dk red-blk need		d		d		s CCl_4; i CS_2
p80	(tetra-)selenide, tri-	P_4Se_3	360.78	or red cr	1.31	242	360—400			
p81	(tetra-)sulfide, hepta-	P_4S_7	348.32	lt yel cr	2.19^{17}	310	523			sl s CS_2
p82	sulfide, penta-	P_2S_5 (or P_4S_{10})	222.25	gray-yel cr, deliq	2.03	286—90	514	i	d	0.22 CS_2; s alk
p83	sesquisulfide	Tetraphosphorus trisulfide. P_4S_3	220.08	yel, rhomb	2.03^{17}	174	408	i	d	100^{17} CS_2; 11.1^{80} bz
p84	thiobromide	$PSBr_3$	302.75	yel, oct	2.85^{17}	38	d 212	s		s eth, CS_2, PCl_3
p85	thiochloride	$PSCl_3$	169.39	col, fum liq	1.668	−35	125	sl d	d	s CS_2
p86	thiocyanate	$P(SCN)_3$	205.21	liq	1.625^{18}	*ca* −4	265	d		s al, eth, bz, CS_2
	Phosphorous acid									
p87	hypo-	$H(H_2PO_2)$	66.00	col oily liq or deliq cr	1.493^{19}	26.5	d 130	s	v s	v s al, eth
p88	meta-	HPO_2	63.98	feather like cr				d		
p89	ortho-	$H_2(HPO_3)$	82.00	col-yel cr, deliq	$1.651^{21.2}$	73.6	d 200	309°	694^{40}	s al
p90	pyro-	$H_4P_2O_5$	145.98	need		38	d 120	d		
p91	**Phosphotungstic acid**	Tungstophosphoric acid $H_3[P(W_3O_{10})_4]·14H_2O$	3132.39	yel-grn cr, tricl.		d		s		s al, eth
p92	phosphotungstic acid	Dodecatungtophosphoric acid. $H_3[P(W_3O_{10})_4]·24H_2O$	3312.54	trig		89		s		
p93	**Platinic acid,** *hexa*chloro	$H_2PtCl_6·6H_2O$	517.91	red, brn, deliq	2.431	60		v s	v s	s al, eth
p94	*tetra*cyano	$H_2Pt(CN)_4$	301.17			d 100		v s	v s	v s al, eth, chl
p95	*hexa*hydroxy-	$H_2Pt(OH)_6$	299.14	yel need		$-2H_2O$, 100	$-3H_2O$, 120	i	v sl s	s H_2SiF_6, dil a, alk
p96	*hexa*iodo-	$H_2PtI_6·9H_2O$	1120.66	blk-red, deliq				d		
p97	**Platinum**	Pt	195.08 ± 3	silv met, cub	21.45^{20}	1772	3827 ± 100			s aq reg, fus alk
p98	arsenide	Nat. sperrylite. $PtAs_2$	344.92	gray, cub	11.8	d >800		sl d	sl d	v sl s a
p99	bromic acid	$H_2PtBr_6·9H_2O$	838.66	br-red, monocl, hygr.				v s	v s	v s al, eth
p100	(II) bromide, di-	$PtBr_2$	345.89	br	6.65_4^{25}	d 250		i	i	i al; s HBr, KBr, aq Br
p101	(IV) bromide, tetra-	$PtBr_4$	514.70	dk cr	5.69_4^{25}	d 180		0.41^{20}	sl s	v s al, eth, HBr
p102	carbonyl bromide	$[Pt_2(CO)_2]Br_4$	765.80	lt red, need, hygr	5.115_4^{25}	d 180		s d		s abs al, CCl_4, bz
p103	carbonyl chloride, di-	$Pt(CO)Cl_2$	294.00	yel need	4.2326_4^{25}	195, subl 240 in CO_2	d 300	d	d	s conc HCl H_2SO_4, al
p104	*di*carbonyl chloride, di-	$Pt(CO)_2Cl_2$	322.01	lt yel need.	3.4882_4^{25}	142	−CO, 210	d	d	d HCl; s CCl_4
p105	*di*platinum dicarbonyl tetrachloride	$Pt_2(CO)_2Cl_4$	587.99	or-yel need	4.235_4^{25}	195	subl 240 (in CO_2)	d		d HCl
p106	*di*platinum tricarbonyl tetrachloride	$Pt_2(CO)_3Cl_4$	616.00	or-yel need		130	d 250	d		s h CCl_4, d al
p107	carbonyl diiodide	PtI_2CO	476.90	red cr	5.157	d 140—150		d		s d, al; s bz
p108	carbonyl sulfide	$Pt(CO)S$	255.15	br-blk		d 300—400		d		d alk, al
p109	chloric acid	$H_2PtCl_6·6H_2o$	517.91	brn-red cr, hygr	2.431	60	d >115	s	s	s abs al, acet; v s eth
p110	(II) chloride, di-	$PtCl_2$	265.99	olive grn, hex	6.05	d 581 (in Cl_2)		v sl s		i al, eth; s HCl, NH_4OH
p111	(IV) chloride, tetra-	$PtCl_4$	336.89	br-red cr	4.303_4^{25}	d 370 (in Cl_2)		58.7^{25}	v s	sl s al, NH_3; s acet; i eth
p112	(IV) chloride, tetra-, hydrage	$PtCl_4·5H_2O$	426.97	red, monocl.	2.43	$-H_2O$, 100		v s	v s	s al, eth
p113	chloride, tri-	$PtCl_3$	301.44	grnsh-blk	5.256^{25}	435		sl s	s	s h HCl; v sl s conc HCl
p114	*di*chlorocarbonyl, dichloride	$Pt(COCl_2)Cl_2$	364.90	yel cr.		d		v s		sl s al; v sl s CCl_4
p115	(II) cyanide, di-	$Pt(CN)_2$	247.12	yel-br cr				i	i	i al, a alk; s KCN
p116	(II) fluoride, di-	PtF_2	233.98	yelsh-grn				i	i	
p117	fluoride, hex-	PtF_6	309.07	dk red solid, very unstable		57.6				
p118	(IV) fluoride, tetra-	PtF_4	271.07	deep red, fused, or yel-lt brn cr, deliq		d red heat		s d	v s	s a, alk
p119	(II) hydroxide	$Pt(OH)_2$	229.09	blk		d		i	i	s HCl, HBr, alk; i H_2SO_4, dil HNO_3
p120	(II) hydroxide, hydrage	$Pt(OH)_2·2H_2O$	265.13			$-2H_2O$, 100		i	i	s conc a
p121	*mono*hydroxy chloric acid	$H_2[PtCl_5(OH)]·H_2O$	409.38	red-brn cr, hygr						
p122	(II) iodide, di-	PtI_2	448.89	blk powd	6.403_4^{25}	d 360		i	i	i eth, a; s HI; sl s Na_2SO_3
p123	(IV) iodide, tetra-	PtI_4	702.70	brn, amorph, or blk cr	6.064_4^{25}	d 130		s d		s al, acet, alk, HI, KI, liq NH_3
p124	iodide, tri-	PtI_3	575.79	blk, like graphite	7.414_4^{25}	d 270		i	i	i al, eth; s KI
p125	(II) oxide, mon-	PtO	211.08	vlt-blk	14.9^{15}	d 550		i	i	s HCl; i a, aq reg

No.	Name	Synonyms and Formulae	Mol. wt.	Crystalline form, properties and index of refraction	Density or spec. gravity	Melting point, °C	Boiling point, °C	Solubility, in grams per 100 cc: Cold water	Hot water	Other solvents
p126	(II) oxide, mon-, dihydrate	$PtO \cdot 2H_2O$	247.11			$-2H_2O$, 140—150				s conc HCl, conc H_2SO_4, conc HNO_3
p127	(IV) oxide, di-	PtO_2	227.08	blk	10.2	450		i	i	i a, aq reg
p128	(IV) oxide, di-dihydrate	Platinic hydroxide· $PtO_2 \cdot 2H_2O$ or $Pt(OH)_4$	263.11			$-2H_2O$, 100		i	i	s HCl, aq reg, KOH
p129	(IV) oxide, di-, hydrate	$PtO_2 \cdot H_2O$	245.09					i	i	i aq reg, ac a, HCl; sl s NaOH
p130	(IV) oxide, di- trihydrate	$PtO_2 \cdot 3H_2O$	281.12	ochre		d 300		i	i	i aq reg, HCl
p131	(IV) oxide, di-, tetrahydrate	Hydroxoplatinic acid. $PtO_2 \cdot 4H_2O$ or $H_2Pt(OH)_6$	299.14	yel need		$-2H_2O$, 100	$-3H_2O$, 120	i	i	s a, dil alk
p132	(II, IV) oxide,	Pt_3O_4	649.24			d		i		i a, aq reg
p133	oxide, sesqui,	$Pt_2O_3 \cdot 3H_2O$	492.20			$-H_2O$, 100		i	i	s conc H_2SO_4, caust alk
p134	oxide, tri-	PtO_3	243.09	redsh-brn powd						s HCl, H_2SO_3 sl s HNO_3, H_2SO_4
p135	*pyro*phosphate	PtP_2O_7	369.02	grn-yel	4.85	d 600		v sl s		
p136	phosphide	PtP_2	257.03	met shine	9.01_4^{25}	ca 1500		i	i	i a; v sl s aq reg
p137	selenide, di-	$PtSe_2$	353.00	blk or gray cr or amorph	7.65	d when dry				s aq reg; sl s HNO_3, H_2SO_4
p138	selenide, tri-	$PtSe_3$	431.96	bl flakes	7.15	d 140		i	i	i conc a, CS_2; s aq reg
p139	sulfate	$Pt(SO_4)_2 \cdot 4H_2O$	459.26	yel pl				s	d	s al, eth, a
p140	(IV) sulfide, di-	PtS_2	259.20	blk-brn powd	7.66_4^{25}	d 225—250		i	i	s HCl, HNO_3; i $(NH_4)_2S$
p141	(II) sulfide, mono-	PtS	227.14	blk, tetr	10.04_4^{25}	d		i	i	s $(NH_4)_2$ S; i a, alk
p142	sulfide, sesqui-	Pt_2S_3	486.34	gray (exist?)	5.52	d		i	i	sl s aq reg; i a
p143	(III) sulfuric acid	$H_2[Pt_2(SO_4)_4(H_2O)_2] \cdot 9^1/_2H_2O$	983.58	or, red, tricl		d 150		s	s	d alk
p144	telluride	$PtTe_2$	450.28	gray, hex		1200—1300				sl s Na_2S, $(NH_4)_2S$
	Platinum Complexes									
p145	tetramine platinum (II) chloride	$[Pt(NH_3)_4]Cl_2 \cdot H_2O$	352.12	col, tetr, 1.672, 1.667	2.737	250, $-H_2O$, 100				
p146	tetrammineplatinum (II) chloroplatinite	Magnus, salt. $[Pt(NH_3)_4]PtCl_4$	600.09	grn or red, tetr	<4.1	d		sl s	sl s	
p147	tetrachlorodiammine platinum (IV) *trans-*	$[Pt(NH_3)_2]Cl_4$	370.95		3.3	200—216				
p148	tetrachlorodiammine platinum (IV), *cis-*	$[Pt(NH_3)_2]Cl_4$	370.95	or-yel, rhomb or hexag pl or need		240				
p149	**Plumbous, plumbic**	see Lead								
p150	**Plutonium**α	Pu	242.059	sil wh, monocl	19.84	641	3232			s HCl; i HNO_3, conc H_2SO_4
p151	β	Pu	239.0522	monocl	17.70	stab 122 ± 2 to 206 ± 3				
p152	γ	Pu	239.0522	orthorhomb	17.14	stab 206 ± 3 to 319 ± 5				
p153	δ	Pu	239.0522	cub	15.92	stab 319 ± 5 to 451 ± 4				
p154	δ′	Pu	239.0522	tetrag	16.00	stab 451 ± 4 to 476 ± 5				
p155	ϵ	Pu	239.0522	cub	16.51	stab 476 ± 5 to 639.5 ± 2				
p156	bromide, tri-	$PuBr_3$	478.76	grn, orthorhomb	6.69	681		s		
p157	chloride, tri-	$PuCl_3$	345.41	emerald grn, hex	5.70	760		s		s dil a
p158	fluoride, hexa-	PuF_6	353.04	redsh-brn, orthorhomb		50.75	62.3	d		
p159	fluoride, tetra-	PuF_4	315.05	pa brn, monocl	7.0 ± 0.2	1037				
p160	fluoride, tri-	PuF_3	296.05	purple, hex	9.32	1425(±3)		i		
p161	iodide, tri-	PuI_3	619.77	bright grn, orthorhomb	6.92	777		s		
p162	nitride	PuN	253.06	blk, cub	14.25			hydrol		s HCl, H_2SO_2
p163	oxalate	$Pu(C_2O_4)_2 \cdot 6H_2O$	523.18	yel-grn				i		
p164	oxide, di-	PuO_2	271.05	yelsh-grn, cub	11.46					sl s h conc H_2SO_4, HNO_3, HF
p165	sulfate	$Pu(SO_4)_2$	431.17	light pink						s dil min a
p166	sulfate, tetrahydrate	$Pu(SO_4)_2 \cdot 4H_2O$	503.23	coral pink		d 280				s dil min a
p167	**Polonium**	Po	(209)	α-Po: simple cub; β-Po: rhbr	9.4 (for β-Po)	254	962	sl s		s dil min a; v sl s dil KOH
p168	ammonium chloride	$(NH_4)_2PoCl_6$	457.78		2.76					
p169	*tetra*bromide	$PoBr_4$	528.60	bright red, cub		330 (in Br atm)	360^{200}			s al, acet; i bz, CCl_4
p170	*di*chloride	$PoCl_2$	279.89	ruby red, orthorhomb	6.50	subl 190				s dil HNO_3
p171	*tetra*chloride	$PoCl_4$	350.79	yel, monocl or tric		300 (in Cl atm)	390	s, d		s HCL; sl s al, acet
p172	*tetra*iodide	PoI_4	716.60	blk cr		200 (in N atm subl)				sl s al, acet; i bz, CCl_4
p173	*di*oxide	PoO_2	240.98	red, tetr		d 500				
p174	selenate	$2PoO_2 \cdot SeO_3$	608.92	wh powd		>400				s dil HCl
p175	sulfate, basic	$2PoO_2 \cdot SO_3$	562.02	wh powd		>400, d 550				s dil HCl

No.	Name	Synonyms and Formulae	Mol. wt.	Crystalline form, properties and index of refraction	Density or spec. gravity	Melting point, °C	Boiling point, °C	Solubility, in grams per 100 cc Cold water	Hot water	Other solvents
p176	*di*sulfate	$Po(SO_4)_2$	401.10	purp		d 550				i al; v s dil HCl
p177	monosulfide	PoS	241.04	blk		d 275				i al; sl s dil HCl
p178	**Potassium**	Kalium. K	39.0983	cub silv met	0.86^{20}	63.25	760	d to KOH	d	d al; s a, Hg, NH_3
p179	acetate	$KC_2H_3O_2$	98.14	wh, lust powd, deliq	1.57^{25}	292		253^{20}	492^{62}	33 al; 24.24^{15} MeOH; s liq NH_3; i eth, acet
p180	acetate, acid	$K_2C_2H_3O_2{\cdot}HC_2H_3O_2$	158.20	col, need or pl, hygr.		148	d 200	d	d	s al, acet
p181	acetyl salicylate	$KC_9H_7O_4{\cdot}2H_2O$	254.28			65				
p182	*meta*asluminate	$K_2Al_2O_4{\cdot}3H_2O$	250.20	col cr.				v s, d	v s, d	s alk; i al
p183	aluminosilicate	Nat. orthoclase. $KAlSi_3O_8$ (or $K_2O{\cdot}Al_2O_3{\cdot}6SiO_2$)	278.33	wh, monocl, 1.518, 1.524, 1.526	2.56	*ca*1200				
p184	aluminosilicate	Nat. microcline. $KAlSi_3O_8$ (or $K_2O{\cdot}Al_2O_3{\cdot}6SiO_2$)	278.33	wh, tricl, 1.522, 1.526, 1530	2.54—2.57	1140—1300				
p185	aluminosilicate	Nat. muscovite, white mica. $KAl_3Si_3O_{10}$, $\cdot(OH)_2$ (or $K_2O{\cdot}3Al_2O_3{\cdot}6SiO_2{\cdot}2H_2O$	398.31	col, monocl, 1.551, 1.587, 1.581	2.76—2.80	d		i		
p186	aluminum *meta*silicate	Nat. leucite. $KAlSi_2O_6$	218.25	col cr, 1.508	2.47	1686 ± 5		i	i	d a
p187	aluminum *ortho*silicate	Nat. kaliophilite. $KAlSiO_4$	158.16	col, hex or rhomb (hex→rhomb 2540°) hex: 1.532, 1.572; rhomb: 1.528, 1.536	2.5	*ca*1800 (rhomb)				
p188	aluminum sulfate	Nat. kalinite. $KAl(SO_4)_2{\cdot}12H_2O$	474.38	col, cub, oct or monocl, cub: 1.454, 1.4564; hex: 1.456, 1.429	1.757_{14}^{20}	92.5 $-9H_4O$, 64.5	$-12H_2O$, 200	11.4^{20}	v s	i al, acet; s dil a
p189	amide	Potassamide. KNH_2	55.12	col-wh, or yel-grn, hygr.		335	subl 400	d	d	d al; s liq NH_3
p190	*peroxy*almmine sulfonate	$(KSO_3)_2NO$	268.32	yel cr, expl				o.62^{3}, d	6.6^{29}, d	i al
p191	ammonium tartrate	$KNH_4C_4H_4O_6$	205.21	wh, cr powd				v s		
p192	antimonate, hydroxo-	"*Pyro*"-antimonate. $KSb(OH)_6{\cdot}{}^1/_2H_2O$	271.90	wh gran or cr powed				2.82^{26}	s	
p193	antimonide	K_3Sb	239.04	yel-grn		812		d		d air
p194	antimony tartrate	$KSbC_4H_4O_7{\cdot}{}^1/_2H_2O$	333.93	col, rhomb, 1.620, 1.636, 1.638	2.607	$-H_2O$, 100		$5.26^{8.7}$	35.7^{100}	i al; 6.67^{25} glyc
p195	*ortho*arsenate	K_3AsO_4	256.21	col need, deliq		1310		18.87	v s	4 al
p196	*ortho*arsenate, di-H	KH_2AsO_4	180.03	col, tetr, 1.567, 1.518	2.867	288		19^{6}	v s	i al; s NH_3, a; 52.5 glyc
p197	*ortho*arsenate, mono H	K_2HAsO_4	218.12	col monocl pr		d 300		18.86^{6}	s	i al
p198	*meta*arsenite	$KAsO_2$	146.02	wh powd, hygr.				s	s	sl s al
p199	*ortho*arsenite	K_3AsO_3	240.21	col need				v s		s al
p200	*meta*arsenite, acid	$KH(AsO_2)_2{\cdot}H_2O$	271.96					s		sl s al
p201	aurate	$KAuO_2{\cdot}3H_2O$ (or $2H_2o$)	322.11	lt yel need.		d		s	d	s al
p202	azide	KN_3	81.12	col, tetr	2.04	350 (vac)		49.6^{17}	105.7^{100}	s al; i eth
p203	benzoate	$KC_7H_5O_2{\cdot}3H_2O$	214.26	wh cr powd		$-3H_2O$, 110	d	52^{25}	112^{100}	s al
p204	*di*borane	Diboranidex. $K_2B_2H_6$	105.86	wh, cub cr, 1.493	1.18	subl 400, vac		d		
p205	*penta*borate	Pentaboranidex. $K_2B_5H_9$	141.32	wh powd		d<180		s, d	d	
p206	*di*borane, dihydroxy	$K_2B_2H_6O_2$	137.86	col, cub cr	1.39	d→K, 400—500		s, d		s al; d a
p207	*meta*borate	KBO_2 (or $K_2B_2O_4$)	81.91	col, hex 1.526, 1.450		950		71^{30}	v s	i al, eth
p208	*penta*borate	$KB_5O_8{\cdot}4H_2O$	293.20	col, rhomb		780			0.007°	
p209	*peroxy*borate	$KBO_3{\cdot}{}^1/_2H_2O$	106.91	wh cr.		$-O_2$, 100	d 150	1.22°		i al, eth
p210	*tetra*borate	$K_2B_4O_7{\cdot}8H_2O$	377.55	col, monocl	1.74 (anhydr)	d		26.7^{30}	v s	
p211	borohydride	KBH_4	53.94	wh, cub, 1.494	1.178	d 500		19.3^{20}	v s	o.25 al; 0.56 MeOH; i eth
p213	boroxalate	$KHC_2O_4{\cdot}HBO_2{\cdot}2H_2O$	207.97			$-H_2O$, 110		v s	v s	i, d al
p214	borotartrate	Solution: cream of tartar. $KC_4H_4BO_7$	213.98	wh cr powd	1.832			v s		i al, eth, chl
p215	bromate	$KBrO_3$	167.00	col, trig	$3.27^{17.5}$	434 d 370		13.3^{40}	49.75^{100}	sl s al; i acet
p216	bromide	KBr	119.00	col, cub, sl hygr, 1.559	2.75^{25}	734	1435	53.48°	102^{100}	0.142^{25} al; sl s eth; s glyc
p217	bromoaurate	$K[AuBr_4]$	555.68	red-brn, rhomb		d 120		sl s		s al
p218	bromoaurate, dihydrate	$K[AuBr_4]{\cdot}2H_2O$	591.71	vlt, monocl cr	4.08			19.5^{15}	204^{67}	s al, KBr; d eth
p219	bromoiodide, di-	$KIBr_2$	325.81	red, rhomb		60	d 180	v s		
p220	*hexa*bromoplatinate	$K_2[PtBr_6]$	752.70	dk red-brn, cub	4.66^{24}	d>400		2.02^{20}	10^{100}	i al
p221	*tetra*bromoplatinite	$K_2[PtBr_4]$	592.89	br, rhomb				v s	v s	
p222	bromoplatinite, dihydrate	$K_2[PtBr_4]{\cdot}2H_2O$	628.92	blk, rhomb	3.747_4^{25}	$-H_2O$, vac		v s	v s	
p223	bromostannate	$K_2[SnBr_6]$	671.31	wh cr.	3.783					
p224	cacodylate	$K[(CH_3)_2AsO_2]{\cdot}H_2O$	194.10	wh cr.				s		sl s al; i eth
p225	cadmium cyanide	$K_2[Cd(CN)_4]$	294.68	col, oct	1.846	450		33.3	100^{100}	s al
p226	cadmium iodide	$2KI{\cdot}CdI_2{\cdot}2H_2O$	734.26	wh-yelsh cr powd, deliq	3.359			137^{15}		s a, al, eth
p227	calcium chloride	Chlorocalcite $KCl{\cdot}CaCl_2$	184.54	col cub, β 1.52	754			s		

No.	Name	Synonyms and Formulae	Mol. wt.	Crystalline form, properties and index of refraction	Density or spec. gravity	Melting point, °C	Boiling point, °C	Solubility, in grams per 100 cc: Cold water	Hot water	Other solvents
p228	calcium magnesium sulfate	Polyhalite. $K_2Ca_2Mg(SO_4)_4 \cdot 2H_2O$	602.92	wh, trig, 1.548, 1.562, 1.567	2.775					
p229	calcium sulfate	Kaluszite, syngenite. $K_2Ca(SO_4)_2 \cdot H_2O$	328.41	col, monocl, 1.500, 1.517, 1.518	2.60	1004		0.25	d	i al; s a
p230	*d*-camphorate	$K_2C_{10}H_{14}O_4 \cdot 5H_2O$	366.49	col, need cluster, hygr.		$-5H_2O$, 110		260^{14}		s al
p231	carbide	KHC_2	64.13	col, rhomb cr.	1.37					
p232	carbonate	K_2CO_3	138.21	col, monocl, hygr, 1.531	2.428^{19}	891	d	112^{20}	156^{100}	i al, acet
p233	*peroxy*carbonate	$K_2C_2O_6 \cdot H_2O$	216.23			200—300		s		
p234	carbonate, dihydrate	$K_2CO_3 \cdot 2H_2O$	174.24	col, monocl, hygr, 1.380, 1.432, 1.573	2.043	$-H_2O$, 130		146.9	331^{100}	
p235	carbonate, hydrogen	$KHCO_3$	100.12	col, monocl, 1.482	2.17	d 100—200		22.4	60^{60}	i al
p236	carbonate, trihydrate	$2K_2CO_3 \cdot 3H_2O$	330.46	col, monocl, 1.380, 1.482, 1.573	2.043			129.4	268.3^{100}	i al, conc HN_4OH
p237	carbonyl	$(KCO)_6$	402.65	gray-red		expl		expl		d al
p238	chlorate	$KClO_3$	122.55	col, monocl, 1.409, 1.517, 1.524	2.32	356	d 400	7.1^{20}	57^{100}	14.1^{100} 50% al; sl s glyc, liq NH_3; i acet; s alk
p239	*per*chlorate	$KClO_4$	138.55	col, rhomb, 1.4717, 1.4724, 1.476	2.52^{10}	610 ± 10	d 400	0.75°	21.8^{100}	v sl s al; i eth
p240	chloride	Nat. sylvite. KCl	74.55	cub, col 1.490	1.984	770	subl 1500	34.4	56.7^{100}	sl s al; s eth, glyc, alk
p241	*hypo*chlorite	KClO	90.55	in sol only		d		v s	v s	
p242	chloroaquoruthenate (III) penta-	$K_2[Ru(H_2O)Cl_5]$	374.55	rose pr		$-H_2O$, 200		s	s	sl s al
p243[1]	chloroaurate	$KAuCl_4$	377.88	yel, monocl.	3.75	d 357		61.8^{20}	80.2^{60}	25 al; s a
p243[2]	chloroaurate, dihydrate	$K[AuCl_4] \cdot 2H_2O$	413.91	yel, rhomb pl.				s	s	s al, eth
p244	chlorochromate	Peligot's salt. $KCrO_3Cl$.	174.55	red, monocl.	2.497	d		s, d		s a
p245	chlorohydroxoruthenate	$K_2[Ru(OH)Cl_5]$	373.54	brn-red cr		d		s, d	d	i al
p246	chloroiodate (III)	$KICl_4$	307.81	yel, rhomb	1.76^{45}	d		d		d eth
p247	chloriodide, di-	$KICl_2$	236.91	col, monocl.		60	d 215	d		
p248	chloroiridate	K_2IrCl_6	483.13	blk, cub	3.546	d		125^{19}	6.67	i al, KCl, HN_4OH
p249	chloronitrosylruthenate (III) penta-	$K_2Ru(NO)Cl_5$	386.54	dk red, rhomb		d		12^{25}	80^{60}	i al
p250	chloroösmate (III)	$K_3OsCl_6 \cdot 3H_2O$	574.26	dk red cr		$-3H_2O$, 150		v s		s a; i eth
p251	chloroösmate (IV)	K_2OsCl_6	481.11	red, cub		d		sl s	s	i al; s dil HCl
p252	chloropalladate	K_2PdCl_6	397.33	red, cub	2.738	d		sl s, d	d	i al; sl s HCl
p253	chloropalladite	K_2PdCl_4	326.43	red-brn, tetr (yel cub)	2.67	d 105		s	v s	s KCl, HN_4OH; i al
p254	*hexa*chloroplatinate	K_2PtCl_6	485.99	yel, cub, ~ 1.825	3.499^{24}	d 250		0.481^{2}	5.22^{100}	i al, eth
p255	*tetra*chloroplatinite	K_2PtCl_4	415.09	red-brn tetr, 1.64 1.67	3.38	d		0.93^{16}	3.3^{100}	i al
p256	chloroplumbate	K_2PbCl_6	498.11	lt yel, cub		d 190		d		s h HCl
p257	chlororhenate (IV)	K_2ReCl_6	477.12	yel-grn, oct	3.34			0.8	d	d alk; sl s HCl
p258	chlororhenate (V)	K_2ReOCl_5	457.67	grn hex pl				d	d	s a; i al, eth
p259	chlororhodate, hexa-	$K_3RhCl_6 \cdot 3H_2O$	486.96	red, tricl	3.291	d		d		sl s al, KCl
p260	chlororhodite, penta-	K_2RhCl_5	358.37	red, rhomb		d		sl s	d	i al
p261	chlororuthenate (IV)	K_2RuCl_6	391.98	blk, cub		d		s, d		i al
p262	chlorostannate	K_2SnCl_6	409.60	col, cub, 1.657	2.71			s	s	
p263	chlorotellurate	K_2TeCl_6	418.51	pale yel, octahedral				d	d	s HCl
p264	chromate	Nat. tarapacaite. K_2CrO_4	194.19	hel, rhomb, β 1.74	2.732^{18}	968.3		62.9^{20}	79.2^{100}	i al
p265	*di*chromate	$K_2Cr_2O_7$	294.18	red, monocl or tricl, 1.738	2.676_4^{25}	tricl→monocl 241.6 m.p. 398	d 500	4.9°	102^{100}	i al
p266	*peroxy*chromate	K_3CrO_8	297.29	brn-red, cub		d 170		sl s		i a, al, eth
p267	chromium sulfate	Potassium chromium alum. $K[Cr(SO_4)_2] \cdot 12H_2O$	499.39	vlt-ruby red, but, oct, 1.4814	1.826^{25}	89	$-10H_2O$, 100 $-12H_2O$, 400	24.39^{25}	50	i al; s dil a
p268	chromium chromate, basic	$K_2CrO_4 \cdot 2[Cr(OH) \cdot CrO_4]$	564.18	vlt brn amorph powd	2.28^{14}	300		i		i al, acet a
p269	citrate	$K_3C_6H_5O_7 \cdot H_2O$	324.41	wh cr.	1.98	d 230		167^{15}	199.7^{31}	sl s al; s glyc
p270	citrate, monobasic	$KH_2C_6H_5O_7$	230.22	wh cr powd				s		
p271	cobalt carbonate, hydrogen(ous)	$KHCO_3 \cdot CoCO_3 \cdot 4H_2O$	291.12	rose need				d		
p272	cobalt (II) cyanide	$K_4Co(CN)_6$	371.43	redsh-brn cr, deliq	2.039_4^{25}			v s	v s	d a; i al, eth, $CHCl_3$
p273	cobalt (III)cyanide	$K_3Co(CN)_6$	332.33	wh-yel, monocl pr	1.878_4^{25}			sl s	sl s	s dil HCL, dil HNO_3; sl s al
p274	cobaltinitrite	Fischer's salt. $K_3[Co(NO_2)_6]$	452.56	yel pr, cub				0.9^{17}	d	i al
p275	cobaltinitrite, hydrate	$K_3[Co(NO_2)_6] \cdot H_2O$	470.28	yel cr powd				i	s, d	s min a; sl s ac a; i al, eth
p276	cobaltinitrite, hydrate	$K_3[Co(NO_2)_6] \cdot 1\frac{1}{2}H_2O$	479.28	yel, tetr		d 200		0.089^{17}	sl s	i al, meth
p277	cobaltmalonate (II)	$K_2[Co(C_3H_2O_4)_2]$	341.22		2.234					
p278	cobalt sulfate (II)	$K_2SO_4 \cdot CoSO_4 \cdot 6H_2O$	437.34	red pr, monocl, 1.481, 1.487, 1.500	2.218			25.5°	108.4^{49}	

PHYSICAL CONSTANTS OF INORGANIC COMPOUNDS (continued)

No.	Name	Synonyms and Formulae	Mol. wt.	Crystalline form, properties and index of refraction	Density or spec. gravity	Melting point, °C	Boiling point, °C	Solubility, in grams per 100 cc		
								Cold water	Hot water	Other solvents
p279	copperchloride	$KCl \cdot CuCl_2$	209.00	red need	2.86					
p280	cyanate	KOCN	81.12	col, tetrag	2.056^{20}	d 700—900		75^{25}	s	i al
p281	cyanide	KCN	65.12	col, cub, wh gran, deliq, very pois, 1.410	1.52^{16}	634.5		50	100	$0.88^{19.5}$ al; $4.91^{19.5}$ MeOH; s glyc
p282	cyanoargentate (I)	Potassium argentocyanide. $K[Ag(CN)_2]$	199.00	cub, 1.625	2.36			25^{20}	100	4.85% al; i a
p283	cyanoaurate	$K[Au(CN_2]$	288.10	col, rhomb	3.45			14.3	200	sl s al; i eth
p284	cyanoaurate (III)	$K[Au(CN)_4 \cdot 1^1/_2H_2O$	367.16	col tabl		d 200		s	v s	s al
p285	cyanocadmate	$K_2[Cd(CN_4]$	294.68	col, cub	1.85			33	100^{100}	sl s al
p286	cyanochromate (III)	$K_3[Cr(CN)_6]$	325.40	yel, rhomb	1.71			30.9^{20}		i al
p287	cyanocobaltate (II)	$K_4[Co(CN)_6]$	371.43					s	s	i al, eth
p288	cyanocobaltate (III)	$K_3[CO(CN)_6]$	332.33	yel, monocl	1.906	d		s	s	i al; sl s NH_3
p289	cyanocuprate (I)	$K_3[Cu(CN)_4]$	284.91	col, rhbdr		d		v s		
p290	cyanomanganate (II)	$K_4[Mn(CN)_6] \cdot 3H_2O$	421.48	deep bl, tetr.				s	d	
p291	cyanomanganate (III)	$K_3[Mn(CN)_6]$	328.34	red, rhomb, 1.553, 1.555, 1.571 (Li)				s		
p292	cyanomercurate	$K_2[Hg(CN)_4]$	382.86	col, cr pois				s		s al
p293	cyanomolybdate	$K_4[Mo(CN)_8] \cdot 2H_2O$	496.51	yel, rhomb	2.337_4^{25} (anhyd)	$-H_2O$, 105-119		v s	v s	i eth; 0.0017^{20} al
p294	cyanonickelate (II)	$K_2[Ni(CN)_4]H_2O$	258.97	red-yel, monocl cr or powd	1.875^{11}	$-H_2O$, 100		s		d a
p295	cyanoosmite	$K_4[Os(CN)_6] \cdot 3H_2O$	556.75	col, hel, monocl, β 1.607		d		sl s	s	i al, eth
p296	cyanoplatinite	$K_2[Pt(CN)_4] \cdot 3H_2O$	431.39	col, yel, rhomb, blue fluor, deliq	2.455^{16}	$-3H_2O$, 100	d 400-600	sl s	s	sl s al, eth, H_2SO_4
p297	cyanotungstate (IV)	$K_4[W(CN)_8] \cdot 2H_2O$	584.42	lt yel-grn cr powd	1.989_4^{25} (anhydr)	$-2H_2O$, 115		130^{18}	s	i al, eth
p298	ethylsulfate	$KC_2H_5SO_4$	164.22	wh, monocl	1.843			s		s al
p299	ferricyanide	$K_3Fe(CN)_6$	329.25	red, monocl, 1.566, 1.569, 1,583	1.85^{25}	d		33^4	77.5^{100}	i al; s acet
p300	ferrocyanide	Yellow prussate of potash, $K_4Fe(CN)_6 \cdot 3H_2O$	422.39	lem yel, monocl, β 1.577	1.85^{17}	$-3H_2O$, 70	d	27.8^{12} anhydr 14.5^0	$90.6^{96.3}$ anhydr 74^{98}	s acet; i al, eth, NH_2
p301	fluoberyllate	K_2BeF_4	163.20	col, rhomb		red heat		2^{20}	5.26^{100}	
p302	fluoborate	Nat. avogadrite, KBF_4	125.90	col, rhomb or cub, 1.324, 1.325, 1,325	2.498^{20}	d 350	d	0.44^{20}	6.27^{100}	sl s al, eth; i alk
p303	fluogermanate	K_2GeF_6	264.78	wh, hex		730	*ca* 835	0.542^{18}	2.58^{130}	
p304	fluomanganate (IV)	K_2MnF_6	247.13	yel, hex, tabl		d		d	d	s conc HCl
p305	fluoniobate, penta-	Potassium oxyniobate, $K_2NbOF_8 \cdot H_2O$	300.11	col, monocl pl or leaf				7.69	s	s d conc HF
p306	fluorescein deriv	$K_2C_{20}H_{10}O_5$	408.49	yelsh-red powd.				s		
p307	fluoride	KF	58.10	col, cub deliq, 1.363	2.48	858	1505	92.3^{18}	v s	s HF, NH_2, i al
p308	fluoride, acid	KHF_2	78.10	col, cub, deliq	2.37	d *ca* 225	d	41^{21}	v s	S $KC_2H_3O_2$; i al
p309	fluoride, dihydrate,	$KF \cdot 2H_2)$	94.13	col, monco pr, deliq, 1.352	2.454	41	156	349.3^{18}	v s	s HF; i al
p310	*hex*fluorophosphate	KPF_6	184.06			*ca*575	d	9.3^{25}	20.6^{50}	
p311	fluorotungstate	$2KF \cdot WO_2F_2 \cdot H_2O$	388.05	monocl.		$-H_2O$, red heat		6^{17}	s	
p312	fluosilicate	Nat. hieratite, K_2SiF_6	220.27	col, bub or hex 1.3991	hex 3.08 cub 2.665^{17}	d		$0.12^{17.5}$ 6.9^{19}	0.954^{100}	s HCl; i NH_3; v sl s al
p313	fluostannate	$K_2SnF_6 \cdot H_2O$	328.89	monocl pr	3.053			3.7^{18}	33.3^{100}	i al, NH_3
p314	fluosulfonate	$KFSO_3$	138.15	short, thick pr		311	6.9^{19}			
p315	fluotantalate	K_2TaF_7	392.13	col, rhomb	4.56; 5.24			sl s, d		sl s HF
p316	fluotellurate, di-	$K_2TeO_3F_2 \cdot 3H_2O$	345.84	micros oct, monocl		d		sl s	sl s	s HF
p317	fluothorate	$K_2ThF_6 \cdot 4H_2O$	496.29	col		d		2.15	6.6	d a; i al
p318	fluotitanate	$K_2TiF_6 \cdot H_2O$	258.08	col, monocl lust leaf		$-H_2O$, 32 m.p. 780	d	0.556^0	1.27^{21}	sl s min a; i NH_3
p319	fluozirconate	K_2ZrF_6	283.41	col, monocl, 1.466, 1.455	3.48			0.781^2	25^{100}	i NH_3
p320	formate	$KCHO_2$	84.12	col, rhomb deliq	1.91	167.5	d	331^{18}	657^{80}	s al; i eth
p321	gadolinium sulfate	$D_2SO_4 \cdot Gd_2(SO_4)_3 \cdot 2H_2O$	812.96	cr	3.503^{16}			s	s	s K_2SO_4
p322	gallium sulfate	$KGa(SO_4)_2 \cdot 12H_2O$	517.12	col cr.	1.895			s		s a
p323	*di*germanate	$K_2Ge_2O_4$	303.37	wh cr.	$4.31^{21.6}$	>83		s		s a
p324	*meta*germanate	K_2GeO_2	198.78	wh cr.	$3.40^{21.5}$	823		s		s a
p325	*tetra*germanate	$K_2Ge_4O_9$	512.55	wh cr.	$4.12^{21.5}$	1083		s		s a
p326	glycerophosphate	$K_2C_3H_7PO_6$	248.26	colsl yelsh mass, hygr.				v s	v s	s al
p327	hydride	KH	40.11	wh need 1.453	1.47	d		d	d	i CS_2, eth, bz
p328	hydroxide	KOH	56.11	wh, rhomb deliq.	2.044	360.4 ±0.7	1320-1324	107^{15}	178^{100}	v s al; i eth, NH_2
p329	(tri-)hydroxylammine trisulfonate	$(KSO_3)_3 \cdot NO \cdot 1\frac{1}{2} H_2O$	414	50col, nonocl pr		$-H_2O$, 100-200		4^{18}	sl d	
p330	*hexa*hydroxyplatinate	$K_2[Pt(OH)_6]$	375.32	yel, rhomb	5.18	d 160		s	s	i al
p331	imidolsulfonate	$(KSO_3)_2NH$	253.33	col, monocl.	2.515	d 170-180	d 360-440, vac	1.3^{23}	d	i HNO_3
p332	iodate	KIO_3	214.00	col, monocl.	3.93_4^{32}	560	d>100	4.74^0	32.3^{100}	s KI; i al, NH_3
p333	iodate, acid	$KIO_3 \cdot HIO_3$	389.91	col monocl				1.33^{15}		i al

PHYSICAL CONSTANTS OF INORGANIC COMPOUNDS (continued)

No.	Name	Synonyms and Formulae	Mol. wt.	Crystalline form, properties and index of refraction	Density or spec. gravity	Melting point, °C	Boiling point, °C	Solubility, in grams per 100 cc		
								Cold water	Hot water	Other solvents
p334	iodate, acid	$KIO_3 \cdot 2HIO_3$	565.82	col, tircl				4.15		
p335	*meta*periodate	KIO_4	230.00	col, tetr, 1.6205	3.618_4^{15}	582	−0, 300	0.66^{12}	s	v sl s KOH
p336	iodide	KI	166.00	col or wh, cub or gran, 1.677	3.13	681	1330	127.5^{6}	208^{100}	1.88^{25} al; 1.31^{25} acet; sl s eth; NH_3
p337	iodide, tri-	$KI_3 \cdot \frac{1}{2}H_2O$	428.82	dk bl, monocl,deliq	3.498	31	d 225	v s		s al, KI
p338	iodoaurate	$KAuI_4$	743.68	blk lust cr		d 150		s, d		s dil KI sol
p339	iodoridite	K_3IrI_6	1070.94	gr cr		d		i	i	i al
p340	iodomercurate (II) tetra-	K_2HgI_4 (or $KI \cdot HgI_2$)	786.40	yel cr, deliq				v s		i al
p341	iodomercurate (II) tri-	Potassium mercury iodide, $KHgI_3$ (or $KI \cdot HgI_2$)	620.40	yel pr, deliq		105		v s		341^{34} al; s KI sol, ac a eth
p342	iodoplatinate	K_2PtI_6	1034.70	blk, rect	4.96_4^{25}			s	s, d	i al
p343	iridium chloride	Potassium hexachloro iridate, K_2IrCl_4	483.13	redsh-blk, cub	3.549	d		1.12^{20}	s	i al
p344	iridium oxalate	$K_3[Ir(C_2O_4)_3] \cdot 4H_2O$	645.63	orange, tricl cr	2.510^{19}	$-H_2O$, 120	d 160	s	v s	i al, eth
p345	iron chloride (III)	Nat. erythrosiderite, $2KCl \cdot FeCl_3 \cdot H_2O$	329.32	red, orthorhomb	2.372					
p346	iron (III) oxalate	$K_2Fe(C_2O_4)_3 \cdot 3H_2O$	491,25	emerald grn, monocl, 1.5019, 1.5558, 1.5960	2.133_4^{20}	$-3H_2O$, 100	d 230	4.7^{0}	118^{100}	i al
p348	iron sulfate (III)	$KFe(SO_4)_3 \cdot 12H_2O$	599.30	vlt, cub oct, 1.452	1.83	33		$20^{12.6}$	v s	i al
p349	iron sulfate (III)	Nat. krausite. $K_2SO_4Fe_2(SO_4)_3 \cdot 24H_2O$	1006.49	pa yel-grn, monocl, 1.482	1.806	28	d 33			
p350	iron sulfate (II)	$K_2SO_4FeSO_4 \cdot 6H_2O$	434.25	grn pr, monocl, 1.476, 1,482, 1.497	2.169	d		s	s	
p351	iron sulfide	$KFeS_2$	159.07	purp. hex	2.563			d		
p352	lactate	$KC_3H_5O_3 \cdot xH_2O$						s		s al; i eth
p353	laurate	$KC_{12}H_{23}O_2$	238.41	amorph						4.5^{15} al
p354	laurate, acid (mixt)	$KC_{12}H_{23}O_2 \cdot C_{12}H_{24}O_2$	438.73	wh, wax like sol		160				$0.904^{13.5}$al
p355	lead chloride	Nat. pseudocotunnite, $2HCl \cdot PbCl_2$	427.21	yel		490		s		
p356	magnesium carbonate, hydrogen	$KHCO_3 \cdot MgCO_3 \cdot 4H_2O$	256.49	col, tricl or rhomb	2.98			d		
p357	magnesium chloride	Nat. carnalite $KCL \cdot MgCl_2 \cdot 6H_2O$	277.85	col, rhomb, deliq 1.466, 1.475, 1,494	1.61	265		64.5^{19}	d	d al
p358	magnesium chloride sulfate	Nat. kainite $[KMgCl(SO_4)]_4 \cdot 11H_2O$	977.82	col, monocl	2.131			79.56^{18}		i al, eth
p359	magnesium chromate	$K_2CrO_4 \cdot MgCrO_4 \cdot 2H_2O$	370.52	tricl	2.59					
p360	magnesium phosphate, hexahydrate	$KMgPO_4 \cdot 6H_2O$	266.47	wh, rhomb cr		$-5H_2O$, 110		d		
p361	magnesium selenate	$K_2SeO_4 \cdot MgSeO_4 \cdot 6H_2O$	496.51	col, monocl, 1.497, 1.499, 1,511	2.3645_4^{20}	$-2H_2O$, 33		s	s	
p362	magnesium sulfate	Nat. langbeinite. $K_2SO_4 \cdot 2MgSO_4$	414.98	tetrah	2.829	927				
p363	magnesium sulfate	Nat. leonite, $K_2SO_4 \cdot MgSO_4 \cdot 4H_2O$	366.68	col, monocl, 1.483, 1.487, 1,490	2.201^{20}			v s		
p364	magnesium sulfate	$K_2SO_4 \cdot MgSO_4 \cdot 6H_2O$	402.71	col, monocl 1.461, 1.463, 1.476	2.15	d 72		19.26^{0} 25^{20}	59.8^{75}	
p365	malate	$K_2C_4H_4O_4$	210.27	col, viscid mass				s		
p366	manganate	K_2MnO_4	197.13	grn, rhomb		d 190		d	d	s KOH
p367	*per*manganate	$KMnO_4$	158.03	purple rhomb, 1.59	2.703	d<240		6.38^{20}	25^{65}	d al; v s MeOH, acet; s H_2SO_4
p368	manganese chloride(ous)	Chloromanganokalite, $4KCl \cdot MnCl_2$	424.05	trig. 1.50	2.31			s	s	
p369	manganese sulfate(ic)	$KMn(SO_4)_2 \cdot 12H_2O$	502.33	vlt, cub, (oct)				d		
p370	manganese sulfate(ous)	Manganolongbeinite. $K_2SO_4 \cdot 2MnSO_4$	476.25	rose, tetrah, 1.572	3.02	850				
p371	mercury tartrate(ous)	$KHgC_4H_4O_6$	387.76	wh cr powd				i		i al
p372	methionate	Potassium methane disulfonate. $K_2CH_2(SO_3)_2$	252.34	monocl, β 1.539	2.376			s		
p373	methylsulfate	$2KCH_3SO_4 \cdot H_2O$	318.40	wh cr.				s		s al
p374	molybdate	K_2MoO_4	238.13	wh powd or 4-sid pr, deliq	2.91^{18}	919	d 1400	184.6^{25}	v s	i al
p375	*per*molybdate	$K_2O \cdot 3MoO_4 \cdot 3H_2O$	628.05	lt yel cr, monocl		d 180		sl s	s	v sl s al
p376	*tri*molybdate	$K_2O \cdot 3MoO_3 \cdot 3H_2O$	580.06	wh need		571 (anhydr)		0.22^{15}	s	
p377	molybdenum cyanate	$K_4[Mo(CN)_8] \cdot 2H_2O$	496.51	yel, rhomb	2.337_4^{25} (anhydr)	$-2H_2O$ 105-110		v s	v s	0.0017^{20} abs al; i eth; d HCl, H_2SO_4
p378	myristate, acid (mixt)	$KC_{14}H_{27}O_2 \cdot C_{14}H_{28}O_2$	494.84	wh, waxlike sol		153				$0.453^{13.5}$ al
p379	naphthalene -1,5-disulfonate	$K_2C_{10}H_6(SO_3)_2 \cdot 2H_2O$	400.50	monocl, 1.485, 1.669, 1.697	1.797			s		
p380	nickelsulfate	$K_2SO_4 \cdot NiSO_4 \cdot 6H_2O$	437.09	bl, monocl, 1.484, 1.492, 1.505	2.124	d < 100		7^{0}	60.8^{75}	
p381	nitrate	Saltpeter, KNO_3	101.10	col, rhomb or trig 1.335, 1.5056, 1.5064	2.109^{16}	tr-trig 129 m.p. 334	d 400	13.3^{0}	247^{100}	i dil al, eth; s liq NH_3, glyc
p382	nitride	K_3N	131.30	grnsh-blk		d		d		

No.	Name	Synonyms and Formulae	Mol. wt.	Crystalline form, properties and index of refraction	Density or spec. gravity	Melting point, °C	Boiling point, °C	Solubility, in grams per 100 cc		
								Cold water	Hot water	Other solvents
p383	nitrite	KNO_3	85.10	wh-yelsh pr, deliq	1.915	440	d	281^{0}	413^{100}	s hot al; v s liq NH_3
p384	*m*-nitrophenoxide	$KOC_4H_4NO_2·2H_2O$	213.23	flat or need	1.691^{20}	$-H_2O$, 130	d	16.3^{15}		s al
p385	*p*-nitrophenoxide	$KOC_6H_4NO_2·2H_2O$	213.23	yel leaf	1.652^{20}	$-2H_2O$, 130	d	7.5^{15}		sl s al
p386	nitroplatinite	$K_2Pt(NO_2)_4$	457.30	col, monocl			d	3.8^{15}	s	
p387	nitroprusside	$K_2[Fe(NO)(CN)_5]2H_2O$	330.17	red, monocl, hydr				100^{16}		s al
p388	nitrososulfate	$K_2SO_3(NO)^2$	218.27	col need		d 127, espl		$12^{14.5}$, d		i al
p389	oleate	$KC_{18}H_{33}O_2$	320.56	yelsh or brnsh soft mass or cr, 1.452				25	s	$4.315^{13.5}$ al; 100^{50} al; 3.5^{35} eth
p390	oleate, acid (mixt)	$KC_{18}H_{32}O_2·C_{18}H_{34}O_2$	603.02	wh, wax-like solid		95		s	s	$5.2^{13.5}$ al
p391	osmate	$K_2OsO_4·2H_2O$	367.42	vlt, cub, hybr.		$-H_2O$, > 100		sl s	s, d	i al, eth
p392	osmiumchloride	K_2OsCl_6	481.11	blk, oct	3.42^{16}	d 600		s	s	i al; s HCl
p393	osmylchloride	$K_2OsO_2Cl_4$	442.21	red, tetr	3.42	d 200 (in H atm)		s; s; i al; s HCl		
p394	osmyloxalate	$K_2[OsO_2(C_2O_4)_2]·2H_2O$	512.47	brn need, tricl		$-H_2O$, 80	d 180	0.75^{15}	3.0^{15}	
p395	oxalate	$K_2C_2O_4·H_2O$	184.23	wh, monocl, 1.440, 1.485, 1.550	$2.127^{3.9}$	$-H_2O$, 100			33^{16}	
p396	oxalate, hydrogen	KHC_2O_4	128.13	col, monocl, 1.382, 1.553, 1.573	2.044	d		2.5	16.7^{100}	i al, eth
p396	oxalate, hydrogen, monohydrate	$KHC_2O_4·H_2O$	146.14	rhomb	$2.044^{3.9}$					
p398	oxalate, tetra-	$KHC_2O_4·H_2C_2O_4·2H_2O$	254.19	col, tricl, 1.415, 1.536, 1.560	1.836	d		1.8^{13}		d al
p399	oxaloferrate (II)	$K_2·Fe(C_2O_4)_3]·2H_2O$	346.11	gold need		d		s	s	
p400	oxaloferrate (II)	$K[Fe(C_2O_4)_2]·2\frac{1}{2}H_2O$	316.02	br cr		d		92^{21}	d	i al
p401	oxaloferrate (III)	$K_2[Fe(C_2O_4)_3]·3H_2O$	491.25	grn, monocl.		$-3H_2O$, 100	d 230	4.7^{0}	117.7^{100}	i al, NH_3; s acet
p402	oxalatoplatinate	$K_2[Pt(C_2O_4)_2]·2H_2O$	485.35	col, monocl pr	3.04^{12}	$-H_2O$, 100		sl s	s	
p403	oxalatouranate (IV)	$K_4[U(C_2O_4)_4]·5H_2O$	836.58	yel, monocl.	2.563					
p404	oxide, mon-	K_2O	94.20	col, cub, hygr	2.32^{0}	d 350		v s	v s	s al eth
p405	*per*oxide	K_2O_2	110.20	wh, amorph, deliq		490	d			
p406	oxide, super-	KO_2	71.10	yel, cub leaf	2.14	380	d	v s, d		d al
p407	oxide, tri- (sesqui-)	K_2O_3	126.20	red		430		ev O_2		d dil H_2SO_4
p408	palladium chloride	$K_2(PdCl_4)$	326.43	yel-grnsh-br cr tetr, 1.710, 1.523	2.67	524		s	s	sl s^{80} al
p409	palladium chloride	$K_2(PdCl_4)$	397.33	lt red, oct	2.738	d 170		v sl s	sl s	s HCl; i al
p410	palladium oxalate	$K_2[Pd(C_2O_4)_3]·4H_2O$	432.72	yel need		dec in air, $-4H_2O$, 80		0.833^{27}	$9.98^{32.9}$	
p411	palmitate, acid (mixt)	$KC_{16}H_{31}O_2·C_{16}H_{32}O_2$	550.95	wh, fatty sol		138				0.198^{13} al
p412	1-phenol-2-sulfonate (*o*-)	$KC_6H_4(OH)SO_3·H_2O$	230.28	rhomb, 1.527, 1.568, 1.647	1.87	400		s		s al
p413	1-phenol-4-sulfonate(*p*-)	$KC_6H_4(OH)SO_3$	212.26	rhomb, 1.571, 1.608, 1,694	1.87	>260				
p414	phenylsulfate	$KC_6H_5SO_4$	212.26	rhomb leaf		d 150-160	d	14^{15}		v sl s al
p415	*meta*phosphate, hexa-	$(KPO_3)_6$	708.42	col mass hygr	1.207	810	1320	s	s	i al
p416	*meta*phosphate, tetra-	$(KPO_3)_4·2H_2O$	508.31	col cr.		$-2H_2O$, 100		100^{15}	s	
p417	*ortho*phosphate	K_3PO_4	212.27	col, rhomb, deliq	2.564^{17}	1340		90^{20}	s	i al
p418	*ortho*phosphate, di-H	KH_2PO_4	136.09	col, tetr, deliq, 1.510, 1.4864.	2.338	252.6		33^{25}	83.5^{20}	i al
p419	*ortho*phosphate, mono-H	K_2HPO_4	174.18	wh, amorph, deliq		d		167^{20}	v s	v s al
p420	*pyro*phosphate	$K_4P_2O_7·3H_20$	384.38	col, deliq	2.33	$-2H_2O$, 180	$-3H_2O$, 300	s	v s	i al
p421	*sub*phosphate	$K_2PO_3·4H_2O$	229.23	col, rhomb		40	$1-4H_2O$, 150	v s	v s	
p422	hypophosphite	KH_2PO_2	104.09	wh, hex, deliq		d		200^{25}	330	v sl s abs al, NH_2; i eth; 11.1^{25} chl
p423	orthophosphite, di-H	KH_2PO_3	120.09	wh cr, deliq.		d		220^{20}	v s	i al
p424	phthalate, hydrogen	$KHC_8H_4O_4$	204.44	col, rhomb	1.636			10^{25}	33^{100}	
p425	picrate	$KC_6H_2N_3O_7$	267.20	yel-rdsh or grnsh, rhomb, 1.527, 1.903, 1.952	1.852	expl 310		0.5^{15}	25^{100}	0.184^{25} al
p426	Piperate	$KC_{12}H_9O_4$	256.30	lt yel cr powd				sl s	v s	
p427	platinate, hydroxo-	$K_2Pt(OH)_6$	375.32	yel, rhomb		d 160		s		i al
p428	platinorhodanide	$K_2[Pt(CNS)_6]·2H_2O$	657.77	red, rhomb				s	8^{60}	s h al
p429	platinum iodide, hexa-	K_2PtI_6	1034.70	blk, cub	4.963_4^{25}			s		sl s al
p430	plumbate, hydroxo-	$K_2Pb(OH)_6$	387.44	col, rhomb				d	d	s KOH
p431	praseodymium sulfate	$3K_2SO_4·Pr_2(SO_4)_3·H_2O$	1110.77	cr	3.275^{16}			sl s		s HCl, HNO_3
p432	propionate	$KC_3H_5O_2·H_2O$	130.19	wh cr, hygr, or wh leaf, deliq		$-H_2O$, 120		207^{16}	359	22.2^{13} 95% al
p433	propyl sulfate	$KC_3H_7SO_4$	178.24	wh cr powd				v s		
p434	*per*rhenate	$KReO_4$	289.30	wh, tetr, 1.643	4.887	550	1360-1370	1.21^{20}	14.0^{100}	v sl s al
p435	rhenium (IV) chloride	$K_2[Recl_6]$	477.12	yel-grn cr, oct	3.34	d		s	s	s a; i conc H_2SO_4; d h H_2SO_4
p437	rhenium (V) oxychloride	$K_2[ReOCl_4]$	457.67	yel-grn cr, rhomb or monocl, 1.52				s		sl s HCl; s H_2SO_4; i al, eth
p438	rhenium oxycyanide	$K_3[ReO_2(CN)_4]$	439.57	red cr, monocl	2.70_4^{25}	d 300-400, vac		s	s	v sl s al; i alk
p439	rhodium cyanide	$K_3[Rh(CH)_6]$	376.31	pa yel, monocl, 1.5498, 1.5513, 1.5634				v s	v s	

No.	Name	Synonyms and Formulae	Mol. wt.	Crystalline form, properties and index of refraction	Density or spec. gravity	Melting point, °C	Boiling point, °C	Solubility, in grams per 100 cc: Cold water	Hot water	Other solvents
p440	rhodium oxalate	$K_3[Rh(C_2O_4)_3]\cdot4^1/_2H_2O$	565.33	col, tricl	2.171^{20}_4	$-4^1/_2H_2O$, 190		v s	v s	
p441	rhodium sulfate	$KRh(SO_4)_3\cdot12H_2O$	646.36	yel, cub	2.23			s		
p442	ruthenate	$K_2RuO_6\cdot H_2O$	261.28	blk, tetr		$-H_2O$, 200	d 400 vac	v s	d	d a, al
p443	perruthenate	$KRuO_4$	204.17	blk, tetr		d 44		sl s	s, d	
p444	d-saccharate, acid	$KHC_6H_8O_8$	248.23	rhomb need				1.1^{6}	s	
p445	salicylate	$KC_7H_5O_3$	176.21	wh powd				s		s al
p446	santoninate	$KC_{15}H_{19}O_4$	302.41	wh cr powd, deliq				s		s al
p447	selenate	K_2SeO_4	221.15	col, rhomb, hygr, 1.535, 1.539, 1.545	3.066			110.5^{0}	122.2^{100}	
p448	selenide	K_2Se	157.16	wh cub, reddens in air, hygr	2.851^{15}			s d	s	
p449	selenite	K_2SeO_3	205.15	wh, deliq		d 875		s		sl s al
p450	selenocyanate	KSeCN	144.08	need, deliq	2.347	d 100		s	s	d a; s al
p451	selenocyanoplatinate	$K_2Pt(SeCN)_6$	903.14	rhomb	$3.378^{12.5}$	d 80				
p452	selenothionate	$K_2SeS_2O_4$	317.27	col, monocl pr		d 250		s, d		
p453	*di*silicate	$K_2Si_2O_5$	214.36	col, rhomb, 1.502, 1.513	2.456^{25}_4	1015 ± 10				
p454	*meta*silicate	K_2SiO_3	154.28	col, rhomb (?), 1.520, 1.528		976		s	s	i al
p455	*tetra*silicate	$K_2Si_4O_9\cdot H_2O$	352.55	wh, rhomb, α 1.495, β 1.535	2.417	d 400		s	s	i al
p456	*di*silicate, hydrogen	$KHSi_2O_5$	176.27	wh, rhomb, 1.480, 1.530	2.417^{15}_4	515				d HCl
p457	silicotungstate	$K_4SiW_{12}O_{40}\cdot18H_2O$	3354.93	col, hex		$-17H_2O$, 100		33.3^{20}	v s	v s acet; a MeOH; sl s al; eth, bz
p458	silver carbonate	$KAgCO_3$	206.98	rect pl	3.769	d		d	d	
p459	silver nitrate	$KNO_3\cdot AgNO_3$	270.98	monocl.	3.219	125		v s	v s	
p460	sodium antimony tartrate	$KNaSbC_4H_3O_7$	346.90	wh, scales or powd				s		
p461	sodium carbonate	$KNaCO_3\cdot6H_2O$	230.19	monocl, hygr. eff	$1.61\text{-}1.63^{14}$	$-6H_2O$, 100		185.2^{15}		
p462	sodium ferricyanide	$K_2Na[Fe(CN)_6]$	313.14	or-red, monocl		d		50^{25}	80^{80}	
p463	sodium nitrocobaltate (III)	$K_2Na[Co(NO_2)_6]\cdot H_2O$	454.17	yel cr.	1.633	135		0.07^{25}		i al
p464	sodium sulfate	$3K_2SO_4\cdot Na_2SO_4$	664.80	col, rhbdr	2.7			s	s	
p465	sodium tartrate	Rochelle salt, seignette salt, $KNaC_4H_4O_6\cdot4H_2O$	282.22	col, rhomb, 1.492, 1.493, 1.496	1.790	70-80	$-4H_2O$, 215	26^{0}	66^{26}	v sl s al
p466	sorbate	$KC_6H_7O_2$	150.22	col cr.	1.363^{25}_{25}	d 270		58.5^{25}		sl s MeOH
p467	stannate, hydroxo-	$K_2Sn(OH)_6$	298.93	col, trig	3.197			85^{10}	110.5^{20}	sl s KOH; i al, acet
p468	stearate	$KC_{18}H_{35}O_2$	322.57	wh cr powd					s	$0.145^{13.5}$ al; i eth, chl, CS_2
p469	stearate, acid (mixt)	$KC_{18}H_{35}O_2\cdot C_{18}H_{36}O_2$	607.06	wh powd		153		s	s	$0.091^{13.5}$ al
p470	strontium chromium oxalate(ic)	$KSrCr(C_2O_4)_3\cdot6H_2O$	550.86	grnsh blk cr.	2.155^{13}					
p471	styphnate	$KC_6H_2N_3O_8\cdot H_2O$	301.21	yel, monocl pr		$-H_2O$, 120	expl	1.54^{30}		v sl s al
p472	succinate	$K_2C_4H_4O_4\cdot3H_2O$	248.32	rhomb, hygr	1.564			s		
p473	succinate, hydrogen	$KHC_4H_4O_4$	156.18	monocl.	1.767	d 242				
p474	succinate, hydrogen, dihydrate	$KHC_4H_4O_4\cdot2H_2O$	192.21	rhomb, 1.417, 1.530, 1.533	1.616			s		s al
p475	succinate, hydrogen	$KHC_4H_4O_4\cdot C_4H_6O_4$	274.27	monocl.	1.56	162				
p476	sulfate	Nat. arcanite, K_2SO_4	174.25	col, rhomb or hex 1.494, 1.495, 1.497	2.662	tr 558 m. p. 1069	1689	12^{25}	24.1^{100}	i al, acet, CS_2
p477	*peroxydi*sulfate	$K_2S_2O_8$	270,31	col, tricl, 1.461, 1.467, 1.566	2.477	d<100		1.75^{0}	5.2^{20}	i al
p478	*pyro*sulfate	$K_2S_2O_7$	254.31	col need	2.512^{25}_4	>300	d	s	d	
p479	sulfate, hydrogen	Nat. mercallite, misenite, $KHSO_4$	136.16	col, rhomb, deliq 1.480	2.322	214	d	36.3^{0}	121.6^{100}	i al, acet
p480	sulfide, di-	K_2S_2	142.32	red-yel cr		470		s	d	s al
p481	sulfide, hydro-	KHS	72.17	yel, rhomb deliq	1.68-1.70	455		d	d	s al
p482	sulfide, mono-	K_2S	110.26	yel-br, cub, deliq	1.805^{14}	840		s	v s	s al, glyc; i eth
p483	sulfide, mono-, pentahydrate	$K_2S\cdot5H_2O$	200.33	col, rhomb		60	$-3H_2O$, 150	s		s al, glycl i eth
p484	sulfide, penta-	K_2S_5	238.50	or cr, hygr		206	d 300	v s	v s	sl s al
p485	sulfide, tetra-	K_2S_4	206.44	red-brn cr		145	d 850	s		s al
p486	sulfide, tetra-dihydrate	$K_2S_4\cdot2H_2O$	242.47	yel				s	s	sl s al
p487	sulfide, tri-	K_2S_3	174.38	br-yel cr		252		s	d	s al
p488	sulfide, di-, trihydrate	$K_2S_2\cdot3H_2O$	196.36	yel				v s	v s	s al
p489	sulfite	$K_2SO_3\cdot2H_2O$	194.29	wh-yelsh, hex		d		100	<100	sl s al; i NH_3; d dil a
p490	*pyro*sulfite	Potassium *meta*sulphite $K_2S_2O_4$	222.31	col, monocl pl	2.34	d 190		sl s		sl s al; i eth
p491	sulfite, hydrogen	$KHSO_2$	120.16	col cr.		d 190		s	s	i al
p492	*d*-tartrate	$K_2C_4H_4O_6\cdot{}^1/_2H_2O$	235.28	col, monocl, β 1.526	1.98^{20}_4	$-H_2O$, 155	d 200-220	150^{14}	278^{100}	sl s al
p493	*dl*-tartrate	$K_2C_4H_4O_6\cdot2H_2O$	262.30	col, monocl.	1.984	$-2H_2O$, 100		100^{25}		
p494	*d*-tartrate, hydrogen	$KHC_4H_4O_6$	188.18	col, rhomb, 1.511, 1.550, 1,590	1.984^{18}			0.37	6.1^{100}	s a, alk; i al, ac a
p495	*dl*-tartrate, hydrogen;.	$KHC_4H_4O_6$	188.10	col, monocl.	1.954			0.42^{15}	7.0^{100}	i al; s min a
p496	*meta*tellurate	K_2TeO_4	269.79	soft glutinous mass		d 200		d		
p497	*ortho*tellurate	$K_2H_4TeO_6\cdot3H_2O$	359.87	col, rhomb, deliq		$-H_2O$	−O, 300	sl s	s	i al; sl s KOH
p498	telluride	K_2Te	205.80	col, cub, hygr	2.51			s, d	s	

PHYSICAL CONSTANTS OF INORGANIC COMPOUNDS (continued)

No.	Name	Synonyms and Formulae	Mol. wt.	Crystalline form, properties and index of refraction	Density or spec. gravity	Melting point, °C	Boiling point, °C	Solubility, in grams per 100 cc Cold water	Hot water	Other solvents
p499	tellurite	K_2TeO_3	253.79	wh cr, deliq.		d 460-470		v s	v s	s hK_2CO_3, KOH
p500	tellurium chloride	K_2TeCl_6	418.51	yel, oct, hygr.	2.645			s d		s, d al; s dil HCl
p501	thioantimonate	$2K_3SbS_4·4\frac{1}{2}H_2O$	815.64	yel cr, deliq.				300^{0}	400^{80}	i al
p502	thioarsenate	K_2AsS_4	320.46	wh cr, deliq.			d	v s		i al
p503	thioarsenite	K_3AsS_3	288.40			d		s		i al
p504	thiocarbonate, tri-	K_2CS_3	186.39	yel-red-brn cr, deliq		d		v s	s	s NH_3; sl s al; i eth
p505	thiocyanate	KNCS	97.18	col, rhomb pr, deliq	1.886^{14}	173.2	d 500	177.2^{0}	217^{20}	s al; 20.75^{22} acet; 0.18^{13} amyl al; v s liq NH_3
p506	*di*thionate	$K_2S_2O_6$	238.31	col, trig, 1.455, 1.515	2.278	d		6	66^{100}	i al
p507	*penta*thionate	$K_2S_4O_6·1^1/_2H_2O$	361.52	col, rhomb. -1.63	2.112	d		s	d	i al
p508	*tetra*thionate	$K_2S_4O_6$	302.43	col, monocl, 1.6057	2.296			v s		i al
p509	*tri*thionate	$K_2S_3O_6$	207.37	col, rhomb, 1.475, 1.480, 1.487	2.304	d 30-40		s	d	i al
p510	thioplatinate	$K_2Pt_4S_4$	1050.88	bl gray cr	6.44^{15}	d, ign		i		d HCl
p511	thiosulfate, penta-hydrate	$3K_2S_2O_3·5H_2O$	661.02	col, rhomb		d		$150.2^{17.2}$		
p512	*meta*thiostannate	$K_2SnS_3·3H_2O$	347.11	dk brn oil	1.847^{18}	$-3H_2O$, 100		s		i al
p513	thiosulfate	$K_2S_2O_3·^1/_3H_2O$	196.32	col, monocl, deliq	2.590 anhydr 2.23	$-H_2O$, 200	d	96.1^{0}	312^{90}	i al
p514	tungstate	$K_2WO_4·2H_2O$	362.07	col, monocl, deliq	3.113	tr 388, 921		51.5	151.5	d a; i al
p515	*meta*tungstate	$K_6[H_2W_{12}O_{40}]·18H_2O$	3407.06	hex		*ca*930		s	v s	d a
p516	*meta*tungstate	K_2UO_4	380.22	or-yel, rhomb.					i	v s a
p517	uranyl acetate	$KUO_2(C_2H_3O_2)_3·H_2O$	504.28	tetr	3.296^{15} (1/2H2O)	$-H_2O$), 275		s		
p518	uranyl carbonate	$2K_2CO_3·UO_2CO_3$	606.45	yel, hex		$-CO_2$		7.4^{15}	d	s K_2CO_3 sol; i a
p519	uranyl sulfate	$K_2SO_4·UO_2SO_4·2H_2O$	576.37	yel, monocl.	$3.363^{19.1}$	$-2H_2O$, 120		s		
p520	urate, acid	$KHC_5H_2N_4O_3$	206.20	wh powd				sl s		
p521	*meta*vanadate	KVO_3	138.04	col cr.				sl s	s	sl s KOH; i al
p522	vanadium sulfate	Potassium vanadium alum $KV(SO_4)_2·12H_2O$	498.34	violet cubic	1.783^{20}	20	$-12H_2O$, 230	1984^{10}		
p523	*ethyl*xanthate	KC_2H_5OCSS	160.29	wh to pa yel cr or cr powd	$1.558^{21.5}$	d>200		v s	d	s al; i eth
p524	**Praseodymium** (α form)	Pr	140.9077	pa yel, met, hex up to 798	6.773	931	3512	d		s a
p525	(β form)	Pr	140.9077	cub	6.64	935	3127	d		s a
p526	acetate	$Pr(C_2H_3O_2)_3·3H_2)$	372.09	grn need				v s		
p527	acetylacetonate	$Pr(C_5H_7O_2)_3$	438.24	cr ppt		146				s CS_2
p528	bromate	$Pr(BrO_3)_3·9H_2O$	686.75	grn, hex		56.5	$-7H_2O$, 170	196^{25}		
p529	bromide	$PrBr_3$	380.62	grn cr powd.		691	1547	d, sl s		
p530	carbide	PrC_2	164.93	yel cr.	5.10	d		d	d	s dil a
p531	carbonate	$Pr_2(CO_3)_3·8H_2O$	605.97	grn silky pl		$-6H_2O$, 100		i		s a
p532	chloride	$PrCl_2$	247.27	br grn need	4.02^{25}	786	1700	103.9^{12}	∞^{100}	v s al; 2.4 pyr; i chl eth
p533	chloride, heptahydrate	$PrCl_3·7H_2O$	373.37	grn, tricl.	2.25^{17}	115		334^{13}	∞^{100}	s al, HCl
p534	hexantipyrine *per*chlorate	$[Pr(C_{11}H_{12}N_2O)_6]·(ClO_4)_3$	1568.63	grn hex leaf		d 286-291				
p535	iodide	PrI_3	521.62	gr cr, hygr		737		v s		
p536	molybdate	$Pr_2(MoO_4)_3$	761.63	grass-green tetr.	4.84	1030				
p537	oxalate	$Pr_2(C_2O_4)_3·10H_2O$	726.03	lt grn cr				i		s a
p538	oxide, di-	PrO_2	172.91	br-bl powd	6.82	>350 tr to Pr_6O_{10}				
p539	oxide, sesqui-	Preseodymia, Pr_2O_3	329.81	yel-grn, amorph	7.07	d		0.000020^{29}		s a
p540	selenate	$Pr_2(Se)_4)_3$	710.69		4.30^{15}			36^{0}	3^{92}	
p541	sulfate	$Pr_2(SO_4)_3$	569.99	lt grn powd	3.72^{16}			23.7^{0} 17.7^{20}	1.02^{96}	
p542	sulfate, octahydrate	$Pr_2(SO_4)_3·8H_2O$	714.11	grn, monocl, 1.540, 1.549, 1.561	$2.827^{13.3}$			17.4^{20}	sl s	
p543	sulfate, pentahydrate	$Pr_2(SO_4)_3·5H_2O$	660.06	monocl pr	3.176^{16}				1.85^{3}	
p544	sulfide	Pr_2S_3	378.00	br powd	5.042^{11}	d		o	d	s dil a
p545	**Protactinium**	Pa	231.0359	gray met, tetrag	15.37	<1600				
p546	chloride	$PaCl_4$	372.85	yel-grn, tetrag		subl 400 in vac		s		s dil HCl
p547	fluoride	PaF_4	307.03	monocl.				i		
p548	oxide, di-	PaO_2	263.03	blk, cub						
p549	oxide, pent-	Pa_2O_5	542.07	wh, cub						s dil HF
r1	**Radium**	Ra	226.0254	silver wh met.	5(?)	700	<1140	d, ev H_2		d a
r2	bromide	$RaBr_2$	385.83	col-yelsh, moncl	5.79	728	subl 900	s	s	s al
r3	bromide dihydrate	$RaBr_2·2H_2O$	421.86	wh, monocl.		$-2H_2O$, 100		s	s	
r4	carbonate	$RaCO_3$	286.03	wh, or sl brnsh, monocl				i		d a
r5	chloride	$RaCl_2$	296.93	col-yelsh, monocl	4.91	1000		s	s	s al
r6	chloride, dihydrate	$RaCl_2·2H_2O$	332.96	wh, monocl, discol		$-2H_2O$, 100		s	s	s HCl
r7	iodate	$Ra(IO_3)_2$	575.83					0.0175^{0}	0.170^{100}	
r8	nitrate	$Ra(NO_3)_2$	350.04	cr				13.9^{20}		
r9	sulfate	$RaSO_4$	322.08	col, rhomb				0.000002^{25}	0.000005^{45}	i a
r10	**Radon**	Niton, Radium emanation, Rn	(222)	col gas, opaque cr	gas 9.73 g/ℓ liq 4.4^{-62} sol 4.0	−71	−61.8	51.0^{0} cm³ 22.4^{25} cm³	13.0^{60} cm³	sl s al, org liqu

No.	Name	Synonyms and Formulae	Mol. wt.	Crystalline form, properties and index of refraction	Density or spec. gravity	Melting point, °C	Boiling point, °C	Solubility, in grams per 100 cc: Cold water	Hot water	Other solvents
r11	**Rhenium**	Re	186.207	met lust, hex	20.53	3180	5627 (est)	i	i	s dil HNO_3, H_2O_2; sl s H_2SO_4; i HCl
r12	bromide	$ReBr_3$	425.92	grn-blk cr		subl 500, vac				s dil H_2SO_4, HBr, liq NH_3
r13	carbonyl-penta-	$[Re(CO)_5]_2$	652.52	col. cub cr		d 250				v sl s org solv
r14	chloride, penta-	$ReCl_5$	363.47	dr grn to blk	4.9	d	d	d	d	S HCl, alk
r15	chloride, tetra-	$ReCl_4$	328.02	bok (exist. ?)			500	s d	s d	s HCl
r16	chloride, tri-	ReC_3	292.57	dk red, hex			>500	s	s	s a, alk, liq NH_3, al; sl s eth
r17	fluoride	ReF_4	262.20	de grn	5.383^{26}	124.5	d 500	d		s a
r18	fluoride, hexa-	ReF_6	300.20	pa yel, v hygr	liq 6.1573, sld $3.616^{18.8}$	18.8	47.6	2, d	s, d	d HNO_3, H_2SO_4
r19	iodide pentacarbonyl	ReI.5CO	453.16	yel, rhombd.		200	d 400, subl vac 90	i		s bz
r20	oxide,di-	ReO_2	218.21	blk	11.4_4^{25}	d 1000		i	i	s conc HCl, H_2O_2
r21	oxide, hept-	Re_2O_7	484.41	yel pl or hex or powd, hygr	6.103	ca 297	subl 250	v s	v s	v s al; s alk, a
r22	oxide, per-	Re_2O_3(?)	500.41	wh	8.4	145		v s	v s	s alk; sl s eth
r23	oxide, sesqui-	Dirhenium trioxide $Re_2O_3{\cdot}xH_2O$		unstable			d, ev H_2			
r24	oxide, tri-	ReO_3	234.21	red, blue, cub	6.9-7.4		d 400	i	i	s H_2O_2, HNO_3
r25	oxybromide, tri-	ReO_3Br	314.11	wh		39.5	163			
r26	oxychloride, tri-	ReO_3Cl	296.66	col liq	3.867_4^{20}	4.5	131^{760}	d	d	s CCl_4
r27	oxy*tetra*chloride	$ReOCl_4$	344.02	or need		29.3	223.00	d	d	
r28	oxy*tetra*fluoride	$ReOF_4$	278.20	wh	lq 3.717 sol 4.032	39.7	62.7			
r29	oxy*di*fluoride, di-	ReO_2F_2	256.20	col		156				
r30	sulfide, di-	ReS_2	250.33	blk tr leaf	7.506_4^{20}		d	i	i	s HNO_2; i al, alk, HCl
r31	sulfide, hepta-	Re_2S_7	596.83	blk powd	4.866		d	i	i	s HNO_3, H_2O_2, alk; i CHl
r32	**Rhodium**	Rh	102.9055	gray-wh, cub	12.4	1966 ± 3	3727 ± 100	i	i	s H_2SO_4 + HCl, conc H_2SO_4; sl s a, aq reg
r33	amminechloride, hexa-	$[Rh(NH_3)_4]Cl_3$	311.45	rhomb pl	2.008_5^{25}	$-NH_3$ 210, d		12.5^8	s	
r34	carbonylchloride, basic	$RhCl_2{\cdot}RhO{\cdot}3CO$	376.75	ruby red need.		subl 125.5		sl s	d	s CCl_4, ac a, bz
r35	chloride, tri-	$RhCl_3$	209.26	br-red powd, deliq		d 450-500	subl 800	i	i	a a, aq reg
r36	chloride, tri-	$RhCl_3{\cdot}xH_2O$		dk red, deliq		d 100		v s		s al, HCl; i eth
r37	fluoride, tri-	RhF_3	159.90	red, rhomb	5.38	>600 subl		i	i	i a, alk
r38	iodide, tri-	RhI_3	483.62	blk				i		
r39	nitrate	$Rh(NO_3)_3{\cdot}2H_2O$	324.95	red, deliq				s	s	i al
r40	oxide, di-	RhO_2	134.90	br				i	i	i a, alk
r41	oxide, di-, dihydrate	$RhO_2{\cdot}2H_2O$	170.93	olive grn.		d		i		s HCl, acet a, alk
r42	oxide, sesqui-	Rh_2O_3	253.81	gray cr or amorph	8.20	d 1100-1150		i	i	i a, aq reg, KOH
r43	oxide, sesqui-, pentahydrate	$Rh_2O_3{\cdot}5H_2O$	343.89	yel ppt		d		i	s	s a
r44	sulfate	$Rh_2(SO_4)_3{\cdot}4H_2O$	566.04	red		d		s	s	
r45	sulfate	$Rh_2(S)_4)_3{\cdot}12H_2O$	710.17	lt yel cr				v s	d	i al
r46	sulfate	$Rh_2(SO_4)_3{\cdot}15H_2O$	764.21	pa yel cr.		d		v s	d	i al, eth
r47	sulfide, hydro-	$Rh(HS)_3$	202.11	blk		d		i	d	s aq reg, aq Br; i NA_2S
r48	sulfide, mono-	RhS	134.97	gray-blk cr		d		i	i	i a, aq reg
r49	sulfide, sesqui-	Rh_2S_3	302.00	blk	6.40_4^{25}	d		i	i	i a, aq reg, aq Br
r50	sulfite	$Rh_2(SO_3)_3{\cdot}6H_2O$	554.08	yel cr.		d		s		i al
r51	**Rubidium**	Rb	85.4678 ± 3	soft, silver wh met	1.532 liq: $1.475^{36.5}$	38.89	686	d	d	d al; s a
r52	acetate	$RbC_2H_2O_2$	144.51	col, nacreous leaf, hygr.		246		$86^{44.7}$	$89.3^{99.4}$	
r53	aluminum sulfate	$RbAl(SO_4)_2{\cdot}12H_2O$	520.75	col, cub, oct, 1.457, 1.45232, 1.46618	1.867^0	99		2.59^{20}	43.25^{80}	
r54	azide	RbN_2	127.49	col need or plates	2.7876	d *ca*310		107.1^{16}		0.182^{16} al; i eth
r55	borofluoride	$RbBF_4$	172.27	very sm rhomb cr, 1.333	2.829^{30}	590	d 500	0.6^{17}	10^{100}	
r56	borohydride	$RbBH_4$	100.31	white, cubic	1.920	1.487		v s		sl s al, i ether bz
r57	bromate	$RbBrO_3$	213.37	cub	3.68	430		$.293^{25}$	5.08^{40}	
r58	bromide	RbBr	165.37	col cub, 1.5530	3.35 liq:2.79^{730}	693	1340	98^5	$205.2^{113.5}$	i al, sl s acet
r59	bromide, tri-	$RbBr_3$	325.18	red, rhomb		d 140				
r60	bromochloroiodide	RbIBrCl	327.73	rhomb		d 200				s al, d eth
r61	bromoiodide, di-	$RbIBr_2$	372.18	red, rhomb	3.84	225	d 265	s		s al, d eth
r62	carbonate	Rb_2Co_3	230.94	col cr, deliq.		837	d 740	450^{20}	s	0.7abs sl
r63	carbonate acid	$RbHCO_3$	146.48	wh, rhomb		d 175		53.73^{20}		2.0 al
r64	chlorate	$RbClO_3$	168.92	trim.	3.19			5.0^{19}	62.8^{100}	
r65	perchlorate	$RbClO_4$	184.92	rhomb, 1.4701	2.80	fus	d	0.5^0	18^{100}	0.009^{25} al, 0.06^{25} MeOH
r66	chloride	RbCl	120.92	col, cub, 1.493^{35}	2.80: liq: 2.088^{750}	718	1390	77^0	138.9^{100}	0.08^{25} al, 1.41^{25} MeOH; v sl s NH_3
r67	chlorobromide, di-	$RbBrCl_2$	236.29	rhomb		d 110				
r68	chloro*di*bromide	$RbBr_2Cl$	280.73	rhomb		76				

PHYSICAL CONSTANTS OF INORGANIC COMPOUNDS (continued)

No.	Name	Synonyms and Formulae	Mol. wt.	Crystalline form, properties and index of refraction	Density or spec. gravity	Melting point, °C	Boiling point, °C	Solubility, in grams per 100 cc		
								Cold water	Hot water	Other solvents
r69	chloroiodide, di-	$RbICl_2$	283.28	dk orange, rhomb		180-200	d 265	v s	v s	
r70	chloroplatinate	Rb_2PtCl_4	578.73	yel, cub	$3.94^{17.5}$	d		0.184^{0}	0.634^{100}	i al
r71	chloroplatinate, hexa-	$Rb_2[PtCl_6]$	578.73	yel, cub	$3.94^{17.5}$	d		0.014^{0}	0.33^{100}	i al
r72	chromate	Rb_2CrO_4	286.93	yel, rhomb, ~1.71	3.518			62^{0}	95.7^{60}	
r73	*di*chromate	$Rb_2Cr_2O_7$	386.92	tricl or monocl, >1.95, 1.70	3.02 monocl 3.125 tricl		tricl: monocl:	4.96^{16} 5.42^{18}	27.3 28.1^{60}	
r74	chromiumsulfate	$RbCr(SO_4)_2 \cdot 12H_2O$	545.76	vlt cub, 1.482	1.946	107		43.4^{25}		
r75	cobalt (II) sulfate	$RbSO_4 \cdot CoSO_4 \cdot 6H_2O$	444.61	rubyred, monocl, 1.486, 1.491, 1.501	2.56^{15}			9.3^{25}	s	
r76	coppersulfate	$Rb_2SO_4 \cdot CuSO_4 \cdot 6H_2O$	534.69	monocl, 1.489, 1.491, 1.504	2.57			10.28^{25}		
r77	cyanide	$RbCN$	111.49	col cr powder	2.32			s	s	i al, eth
r78	galliumsulfate	$RbGa(SO_4)_2 \cdot 12H_2O$	563.49	col cr, 1.46579	1.962			s		
679	fluoride	RbF	104.47	col, cubic, 1.398	3.557	795	1410	130.6^{18}	s dil HF; i al, eth, NH_2	
r80	fluorogermanate	Rb_2GeF_6	357.52	wh cr				sl s	v s	
r81	rluosilicate	Rb_2SiF_6	313.01	cub, oct	3.332^{20}			0.16^{20}	1.35^{100}	s a, i al
r82	fluosulfonate	$RbFSO_3$	184.52	need		304				
r83	iodate	$RbIO_2$	260.37	monocl or cub	$4.33^{19.5}$	d		2.1^{23}		v s HCl
r84	*meta*periodate	$RbIO_4$	276.37	tetr	3.918^{18}			0.65^{13}		
r85	iodide	RbI	212.37	col, cub, 1.6474	3.55; liq: 2.878^{25}	647	1300	$152.^{17}$	163^{25}	s liq NH_3; 0.674^{25} acet
r86	iodide, tri-	RbI_3	466.18	blk, rhomb	4.03^{22}	190		s		
r87	iodide, cmpd, with SO_2	$RbI \cdot 4SO_2$	468.61	lemon yel		13.5				
r88	iron (II) selenate	$Rb_2SeO_4 \cdot FeSeO_4 \cdot 6H_2O$	620.79	blue-grn monocl, prism, 1.513, 1.520, 1.532	2.819					
r89	iron (III) selenate	$RbFe(SeO_4)_2 \cdot 12H_2O$	643.41	cub, 1.507^{18}	2.31^{11}	45	$-12H_2O$, 100			
r90	iron (II) sulfate	$Rb_2SO_4 \cdot FeSO_4 \cdot 6H_2O$	526.99	gr monocl, prism, 1.4815, 1.4874, 1.4977	2.516	d 60		24.2^{25} (anhydr)		
r91	iron (III) sulfate	$RbFe(SO_4)_2 \cdot 12H_2O$	549.61	cub, 1.4823	1.91-1.95	48.53	$4.55^{6.6}$	52.6^{90}		
r92	hydride	RbH	86.48	col need	2.60	d 300		d	d	d a
r93	hydroxide	$RbOH$	102.48	gray-wh, deliq	3.203^{11}	301 ± 0.9		180^{15}	v s	s al
r94	neodymium nitrate	$2(?)RbNO_3 \cdot Nd(NO_3)_3 \cdot 4H_2O$	697.26	redsh-vlt pl	2.56	47	$-4H_2O$, 60			
r95	nitrate	$RbNO_3$	147.47	col, hex cub, rhomb or tricl, hygr, 1.51, 1.52, 1.524	3.11: liq: 2.395^{100}	tr cub 161.4 m.p. 310	tricl rhom 219	44.28^{16}	452^{100}	sl s acet: v s HNO_3
r96	nitrate, hydrogen	$RbNO_3 \cdot HNO_3$	210.49	tetr		62				
r97	nitrate, hydrogen	$RbNO_3 \cdot 2HNO_3$	273.50	col need		45				
r98	magnesium sulfate	$Rb_2SO_4 \cdot MgSO_4 \cdot 6H_2O$	495.45	col, monocl, 1.467, 1.469, 1.478	2.386_4^{20}			20.2^{25} (anhydr)		
r99	permanganate	$RbMnO_4$	204.40	cr	$3.235^{10.4}$	d 295		0.5^{0}	4.7^{60}	
r100	oxide, mon-	Rb_2O	186.94	col-yel, cub	3.72	d 400		s d	s d	s liq NH_3
r101	oxide, per-	Rb_2O_2	202.93	yel, cub	3.65^{0}	570	d 1011^{760} mm	dec to RbOH + H_2		
r102	oxide, super	RbO_2, unstable	117.47	yel plates	3.80	432	d 1157^{1} atm			
r103	oxide (tetr-)	Rb_2O_4	234.93	dk orange cr, deliq		dec 500 vac				
r104	oxide, tri- (sesqui)	Rb_2O_3 (or Rb_4O_6)	218.93	blk cub	3.53^{0}	489		s d		
r105	praseodymium nitrate	$2RbNO_3 \cdot Pr(NO_3)_3 \cdot 4H_2O$	693.93	grnsh, monocl, need hygr	2.50	63.5	$-4H_2O$, 60			
r106	selenate	Rb_2SeO_4	313.89	col, rhomb, 1.5515, 1.5537, 1.5582	3.90			159^{12}		
r107	silicofluoride	Rb_2SiF_6	313.01	col, oct or hex	3.3383_4^{20}			sl s	s	s a, i al
r108	sulfate	Rb_2SO_4	266.99	col, rhomb hex, 1.513, 1.513, 1.514	3.613_4^{20} liq2.53^{1100}	1060 trig 653	*ca* 1700	42.4^{10}	81.8^{100}	i acet; sl s NH_2
r109	sulfate, hydrogen	$RbHSO_4$	182.53	rhomb, 1.473	2.892^{16}	$<$red heat				
r110	sulfide, di-	Rb_2S_2	235.06	dk red		420	volat > 850			
r111	sulfide, hexa-	Rb_2S_6	363.30	brn-red		201				
r112	sulfide, mono-	Rb_2S	203.00	wh-pale yel	2.912	530 d vac		v s	v s	
r113	sulfide, mono-tetrahydrate	$Rb_2S \cdot 4H_2O$	275.06	cr, deliq				v s	v s	
r114	sulfide, penta-	RbS_4	331.24	red, rhomb, deliq	2.628^{15}	225		d		s 70· al; i eth, chl
r115	sulfide, tri-	Rb_2S_3	267.12	redsh yel		213				
r116	tartrate, d & l	$Rb_2C_4H_6$	319.01	trig	2.658_4^{20}			200^{25}		i toluol
r117	*dl*-tartrate, hydrogen,	$RbHC_4H_4O_6$	234.55	trim pr	2.282	201 d		1.18^{25}	11.7^{100}	
r118	vanadium sulfate	Rubidium vanadium alum. $RbV(SO_4) \cdot 12H_2O$	544.71	yellow, cubic, 1.4689	1.915_4^{20}	64	230-$12H_2O$, 300 dec	2.56^{10}		
r119	**Ruthenium**	Ru	101.07 ± 3	gray-wh or silv brittle met, hex	12.30	2310	3900	i	i	i sq reg,a, al; s fus alk
r120	carbonyl, penta-	$Ru(CO)_6$	241.12	col liq		−22				s al, bz
r121	chloride, tetra-	$RuCl_4 \cdot 5H_2O$	332.96	rdsh-br cr, hygr		d		s		s al

PHYSICAL CONSTANTS OF INORGANIC COMPOUNDS (continued)

No.	Name	Synonyms and Formulae	Mol. wt.	Crystalline form, properties and index of refraction	Density or spec. gravity	Melting point, °C	Boiling point, °C	Solubility, in grams per 100 cc: Cold water	Hot water	Other solvents
r122	chloride, tri-	$RuCl_3$	207.43	br cr, deliq	3.11	d > 500		i	d	sl s al; s HCl; i CS_2
r123	fluoride penta-	RuF_5	196.06	dk grn cr	$2.963^{16.5}$	101	250	d	d	
r124	hydroxide	$Ru(OH)_3$	152.09	blk powd				v sl s		s a; i alk
r125	oxide, di-	RuO_2	133.07	dk bl, tetr	6.97	d		i	i	i a; s fus alk
r126	oxide, tetr-	RuO_4	165.07	yel, rhomb need	3.29^{21}	25.5	d 108	2.033^{20}	2.249^{76}	s al, a alk, CCl_4
r127	oxychloride ammoniated	Ruthenium red. $Ru_2(OH)_2Cl_4{\cdot}7NH_3{\cdot}3H_2O$	551.23	brn-red powd				s		
r128	silicide	RuSi.	129.16	met pr	5.40^4			i	i	s HNO_3 + HF
r129	sulfide	Nat. laurite RuS_2	165.19	gray-blk, cub	6.99	d 1000		i	i	i a; s fus alk
s1	**Samarium**	Sm.	150.36 ± 3	wh-gray met, hex	7.520	1077	1791	i	i	s a
s2	acetate	$Sm(C_2H_3O_2)_3{\cdot}3H_2O$	381.54		1.94			15^{25}		
s3	acetylacetonate	$Sm(C_6H_7O_2)_3$	447.69	cr mass		146-147		i		
s4	benzylacetonate	$Sm(C_{10}H_9O_2)_3{\cdot}2H_2O$	669.93	straw color		103-105		i		s org solv
s5	bromate	$Sm(BrO_3)_3{\cdot}9H_2O$	696.20	yel, hex		75	$-9H_2O$, 150	114^{25}		v sl s al
s6	(II) bromide	$SmBr_2$	310.17	dk brn	5.1	508	1880	d		
s7	(III) bromide	$SmBr_3{\cdot}6H_2O$	498.16	yel cr, deliq.	2.971^{22}	640				
s8	carbide	SmC_2	174.38	yel, hex	5.86			d	d	s, d a
s9	(II) chloride	$SmCl_2$.	221.27	red-brn cr	4.56^{25}	740		s, d		i al, CS_2
s10	(III) chloride	SmC_3	256.72	yelsh-wh cr, hygr	4.46^{18}	678 ± 1	d	92.4^{10}	99.9^{50}	v s al; 6.4^{25} pyr
s11	(III) chloride, hexahydrate	$SmCl_3{\cdot}6H_2O$	364.81	grn-yel, tricl, hygr	2.383	$-5H_2O$, 100				
s12	chromate	$Sm_2(CrO_4)_3{\cdot}8H_2O$	792.82	yel				0.043^{25}		
s13	(II) fluoride	SmF_2	188.36			1306	>2400			
s14	(III) fluoride	SmF_3	207.36			1306	2323	i	i	
s15	hydroxide	$Sm(OH)_3$	201.38	pa yel powd				i		s a; i alk
s17	(II) iodide	SmI_2	404.17	dk brn		527	1580	d		
s18	kojate	$Sm(C_6H_5O_4)_3$	573.67			d 275		i		
s19	(III) methyl-phosphate, di-	$Sm[(CH_3)_2PO_4]_3$	525.48	cream col, hex pr				35.2^{25}	10.8^{95}	
s20	(III) molybdate	$Sm_2(MoO_4)_3$.	780.53	vlt, rhomb oct	5.36					
s21	(III) nitrate)	$Sm(NO_3)_3{\cdot}6H_2O$	444.47	pa yel, tricl	2.375	78-79		v s		
s22	(III) oxalate	$Sm_2(C_2O_4)_3{\cdot}10H_2O$.	744.93	wh cr.				0.000054		s H_2SO_4
s23	oxide, sesqui-	Samaria. Sm_2O_2	348.72	wh-yelsh powd	8.347			i		v s a
s24	(III) sulfate.	$Sm_2(SO_4)_3{\cdot}8H_2O$	733.02	lt yel, monocl, 1.543, 1.552, 1.563	2.930	$-8H_2O$, 450		2.67^{30} 4.4^{25}	1.99^{40}	
s25	(III) sulfide	Sm_2S_3.	396.90	yelsh-pink	5.729	1900			d	d dil a
s26	sulfate, basic	$Sm_2O_2SO_4$	428.78	yel powd		d 1100		i		i dil H_2SO_4
s27	**Scandium**	Sc	44.9559	silv met, cubic or hex	2.9890	1541	2831	d, ev H_2		
s28	acetylacetonate	$Sc(C_5H_7O_2)_3$.	342.28	col pl.		187.5	subl 210-215			s al, bz, chl
s29	bromide.	$ScBr_3$	284.67		3.914	subl>1000				
s30	chloride.	$ScCl_2$	151.31	col cr.	2.39_4^{25}	939	subl 800-850	v s	v s	i abs al
s31	hydroxide	$Sc(OH)_3$	95.98	col amorph				i		s dil a
s32	nitrate	'$Sc(NO_3)_3$	230.97	col, deliq		150		s		s al
s33	nitrate, tetrahydrate	$Sc(NO_3)_3{\cdot}4H_2O$	303.03	col, pr, deliq		$-4H_2O$, 100		v s		
s34	oxalate	$Sc_2(C_2O_4)_3{\cdot}5H_2O$	444.05	cr		$-4H_2O$, 140				
s35	oxide	Scandia. ScO_2.	137.91	wh powd	3.864			i	i	s h a
s36	sulfate.	$Sc_2(SO_4)_3$	378.08	col cr.	2.579	d		10.3^{25}	v s	
s37	sulfate, hexahydrate	$Sc_2(SO_4)_3{\cdot}6H_2O$.	486.18			$-4H_2O$, 100 $-6H_2O$, 250		v s		
s38	sulfate, pentahydrate	$Sc_2(SO_4)_3{\cdot}5H_2O$.	468.16		2.519	$-3H_2O$, 100		54.6^{25}		
s39	**Selenic acid**	H_2SeO_4.	144.97	wh, hex prism, hygr.	3.004_4^{15}	58 eas undercools	d 260	1300^{30}	∞^{60}	s H_2SO_4; d al; i NH_3
s40	monohydrate.	$H_2SeO_4{\cdot}H_2O$.	162.99	wh, need	2.627_4^{15} liq 2.3564_4^{15}	26 eas undercools	205	v s	v s	
s41	tetrahydrate	$H_2SeO_4{\cdot}4H_2O$	217.03	col liq		51.7 eas undercools	$-H_2O$, 172^{35}	α		s H_2SO_4; d org solv
s42	**Selenium**.	Se	78.96 ± 3	blsh-gray, met hex	4.81_4^{20}	217	684 ± 1.0	i	i	s H_2SO_4 $CHCl_3$; i al; v sl s CS_2
s43	selenium	Se	78.96 ± 3	red, monocl prism	4.50	170-180 traf to hex	684.8	i	i	$0.1^{46.6}$ CS_2; s H_2SO_4, HNO_3
s44	selenium	Se	78.96 ± 3	red amorph, blk vitr	red 4.26 blk 4.28	tr to hex, 60-80	684.8	i	i	s H_2SO_4, CS_2 bz
s45	bromide, ''mono-''	Diselenium dibromide Se_2Br_2.	317.73	dk red liq	3.604^{15}		227d	d	d	d al; s CS_2, $CHCl_3$, C_2H_5Br
s46	bromide, tetra-	$SeBr_4$	398.58	or-red-brn cr		d 75		d	d	s CS_2, chl, C_2H_5Br, HCl
s47	bromide (mono-) chloride, tri-	$SeBrCl_3$.	265.22	yel br cr		190				i Cs_2
s48	bromide (tri-) chloride	$SeBe_2Cl$	354.13	or cr, hygr		d				v sl s CS_2
s49	carbide	SeC_2	102.98	yel liq, 1.845.	2.682_4^{20}	45.5	$125—126^{760}$	i		s CS_2, eth, CCl_4, bz, al
s50	chloride ''mono''-	Diselenium dichloride, Se_2Cl_2.	228.83	br red liq, 1.596	2.77_4^{25}	−85	d 130	d	d	d al, eth; s CS_2, chl, CCl_4, bz
s51	chloride, tetra-	$SeCl_4$	220.77	wh-yel, cub, deliq, 1.807	$3.78\text{-}3.85^{360}$	205, subl 170—196	d 288	d	d	i al, eth, CS_2: s $POCl_3$; d a, alk
s52	fluoride, hexa-	SeF_6.	192.95	col gas, 1.895	3.25^{-28} g/ℓ	−39, subl −46.6	−34.5	s d		
s53	fluoride, tetra-	SeF_4.	154.95	col liq or wh cr		m.p. −13.8 frz −90	>100	d	d	

PHYSICAL CONSTANTS OF INORGANIC COMPOUNDS (continued)

No.	Name	Synonyms and Formulae	Mol. wt.	Crystalline form, properties and index of refraction	Density or spec. gravity	Melting point, °C	Boiling point, °C	Solubility, in grams per 100 cc: Cold water	Hot water	Other solvents
s54	hydride	H_2Se	80.98	col gas, pois	gas 3.664^{760} air; liq: $2.004^{-41.5}$	60.4	−41.5	3.77^{4}	$270^{22.5}$	s CS_2, $COCl_2$
s55	iodide, ''mono''-	Diselenium diiodide, Se_2I_2	411.73	steelgray cr (exist ?)	…	68-70	d 100	d	d	…
s56	nitride	Se_4N_4	371.87	amorph, or yel-brickred, hygr.	…	expl 160—200	d	u	sl d	i al, eth; v sl s acet ac, bz
s57	oxide, di-	SeO_2	110.96	wh, monocl, col, tetr, pois, >1.76	3.95^{15}_{15}	340—350 subl 315—317	…	38.4^{14}	82.5^{65}	6.67^{14} al; $4.35^{15.3}$ acet; $1.11^{13.9}$ac a; s bz
s58	oxide, tri-	SeO_3	126.96	pa yel cub or fiber, deliq	3.6	118	d 180	d, v s	d, v s	s al, conc H_2SO_4; i eth, bz, chl CCl_4
s59	oxybromide	Selenyl bromide. $SeOBr_2$	254.77	red yel cr	liq 3.38^{50}	41.6	217^{710} d	d	…	s CS_2, CCl_4, chl, H_2SO_4, bz
s60	oxychloride	$SeOCl_2$	165.87	col-yel, liq 1.651^{20}	2.42^{22}	8.5	176.4	d	…	s CS_2, CCl_4, chl, bz
s61	sulfur oxy*tetra*chloride	$SeSO_3Cl_4$	300.83	hex pr	…	165	183	d	…	…
s62	oxyfluoride	Selenyl fluoride, $SeOF_2$	132.96	col liq	2.67	4.6	124	d	…	s al, CCl_4
s63	sulfide, di-	SeS_2	143.08	br red-yel	…	<100	d	i	…	d aq reg, HNO_3; s $(NH_4)_2S$
s64	sulfide, ''mono''-	SeS	111.02	or-yel tabl or powd	3.056^{0}	d 118—119	…	i	i	s CS_2; i ethl d al
s65	sulfur oxide	$SeSO_2$	159.02	grn pr or yel powd	…	$-SO_2$, 40	…	d, 118	…	s H_2SO_4; i SO_3
s67	**Selenious acid**	H_2SeO_3	128.97	col, hex, deliq	3.004^{15}_{4}	d 70	$-H_2O$	167^{20}	v s	v s al; i NH_3
s68	**Silane,** bromo-	SiH_3Br	111.01	col gas, expl in air	1.72^{-80} 1.533^{0}	−94	1.9	…	…	…
s69	bromotrichloro-	$SiBrCl_3$	214.35	col liq	1.826	−62	80.3	d	d	…
s70	chloro-	SiH_3Cl	66.56	col gas	gas: 3.033 g/ℓ liq: 1.145^{-113}	−118.1	−30.4	…	…	…
s71	dibromo-	SiH_2Br_2	189.91	col liq, inflam	2.17^{0}	−70.1	66	d	…	d alk
s72	dibromo-	Silicobromoform, $SiHBr_2$	268.81	col liq, inflam	2.7^{17}	−73	109	d	d	d NH_3
s73	dibromodichloro-	$SiBr_2Cl_2$	258.80	col liq	2.172^{25}_{4}	−45.5	104	d	d	…
s74	dichloro-	SiH_2Cl_2	101.01	gas	gas 4.599g/ℓ liq 1.42^{-122}	−122	8.3	d	d	…
s75	dichloro difluoro-	$SiCl_2F_2$	136.99	gas	6.2784 g/ℓ	−144 ± 2	−31.7 ± 0.2	d	d	…
s76	(hexa-) hexaoxocyclo-	Siloxane $Si_6O_3H_6$	222.56	wh, pl	1.32^{20}	d 140, inflam	…	sl d	d	…
s77	monochloro trifluoro-	$SiClF_3$	120.53	gas	5.455 g/ℓ	−138.0 ± 2	−70.0 ± 0.2d	d	d	…
s78	monoiodo-	SiH_2I	158.01	col liq	$2.035^{14.8}$	−57.0	45.5	d	…	…
s79	(tri-)nitrilo-	Silicylamine, tri-$(SiH_3)_3N$	107.33	col inflam liq	0.895^{-106}	−105.6	…	…	…	…
s81	tribromochloro-	$SiBr_3Cl$	303.25	col liq	2.497^{25}_{4}	−20.8 ± 1	126-128	d	d	…
s82	trichloro-	Silicochloroform, $SiHCl_3$	135.45	col liq	1.34	−126.5	33^{768} mm	d	d	s CS_2, CCl_4, chl, bz
s83	trichloroiodo-	$SiCl_3I$	261.35	col liq	…	>−60	113.5	d	…	…
s84	trifluoro-	Silicofluoroform, $SiHF_3$	86.09	col gas	3.86^{0} g/ℓ	−131.4	*ca*-95	d	d	d al, eth, alk; s tol
s85	triiodo-	Silicoiodoform. $SiHI_2$	409.81	red liq	3.314^{20}	8	220	d	d	s CS_2, bz
s86	**Silicane cyanate**	$Si(OCN)_4$	196.15	sol or liq	1.414^{20}_{4}	34.5 ± 0.5	247.2 ± 0.5^{760}	d	…	…
s87	diimide	$Si(NH)_2$	58.11	wh powd	…	d 900	…	…	…	…
s88	isocyanate	$Si(NCO)_4$	196.15	sol or liq	1.434^{25}_{4}	26.0 ± 0.5	185.6 ± 0.3^{760}	d	…	i acet; s bz, CCl_4, CS_2
s89	**Silicic acid, di-**	$H_2Si_2O_5$	138.18	col cr.	…	d 150	…	i	i	s NH_3, HF
s90	meta	H_2SiO_3	78.10	col, amorph.	…	d room temp	…	i	i	s NH_3, HF, h alk
s91	**Silicon**	Si	28.0855 ± 3	steel gray, large to micr cr, cub	2.32—2.34	1410	2355	i	i	s HF + HNO_3; i HF
s92	acetate, tetra-	$Si(C_2H_3O_2)_4$	264.26	col cr, hygr	…	subl 110 d 160—170	148^{6} mm	d	…	d al; sl s acet, bz
s93	bromide, tetra-	Tetrabromosilane. $SiBr_4$	347.70	col fum liq, sol cub	liq: 2.7715^{25}_{4}	5.4	154	d	d	d H_2SO_4
	…	…		…	sol: 3.292^{-79}					
s94	(di-) bromide, hexa-	Si_2Br_6	535.60	wh, rhomb	…	95	240	d	d	s CS_2; d KOH
s95	bromide(di-) sulfide	$SiSBr_2$	219.95	col pl.	…	93	$150^{18.3}$ mm	d	d	s bz, CS_2
s96	carbide	SiC	40.10	col-blk, hex or cub, 2.654, 2.697	3.217	~2700, subl, d	…	i	i	i a; s fus KOH
s97	chloride, tetra-	Tetrachlorosilane. $SiCl_4$	169.90	col fum liq	liq: 1.483^{20}, sol: 1.90^{-97}gas: 7.59 g/ℓ	−70	57.57	d	d	d al
s98	(di-)chloride, hexa-	Hexachlorodisilane. Si_2Cl_6	268.89	col liq, 1.4748^{18}	1.58^{0}	−1	145^{769}	d	d	d al
s99	chloride(di-) sulfide	$SiSCl_2$	131.05	col pr	…	75	$92^{22.5}$	d	d	s CCl_4, CS_2 bz
s100	chloride(tri-) sulfide, hydro-	$SiCl_3HS$	167.51	col liq	1.45	…	96—100	d	d	d al
s101	fluoride, tetra-	Tetrafluorosilane. SiF_4	104.08	col gas	gas: 4.69 g/$ℓ^{760}$ liq: 1.66^{-95}	−90.2	−86	d	d	s abs al, HF; i eth
s102	(di-)fluoride, hexa	Hexa-fluorodisilane. Si_2F_6	170.16	gas	7.759 g/ℓ	−18.7	−18.5	d	d	…
s103	hydride	Silane, silicane. SiH_4	32.12	col gas	liq: 0.68^{-85} gas: 1.44 g/ℓ	−185	-111.8^{760} mm	i	…	d KOH
s104	hydride	Disilane, disilicane. Si_2H_6	62.22	col gas	gas: 2.865 g/ℓ liq: 0.686^{-20}	−132.5	−14.5	sl d	…	s al, bz, CS_2

No.	Name	Synonyms and Formulae	Mol. wt.	Crystalline form, properties and index of refraction	Density or spec. gravity	Melting point, °C	Boiling point, °C	Solubility, in grams per 100 cc		
								Cold water	Hot water	Other solvents
s105	hydride	Trisilane, trisilanepropane. Si_3H_8	92.32	col liq	liq: 0.743^0; gas: 4.15 g/ℓ^{760}	−117.4	52.9	d	d	d CCl_4
s106	hydride	Tetrasilane, tetrasilane butane. Si_4H_{10}	122.42	col liq	liq: 0.79^0 gas 5.48 g/$\ell^{0.760}$	−108	84.3	d	...	...
s107	iodide, tetra-	Tetraiodosilane. SiI_4	535.70	col, cub	4.198	120.5	287.5	d	...	2.2^{27} CS_2
s108	(di-)iodide, hexa-	Hexaiodosilane. Si_2I_6	817.60	col, hex	...	d 250	d	d	d	19^{19} CS_2
s109	nitride	Si_3N_4	140.28	gray-wh amorph powd	3.44	1900 press	...	i	i	s HF
s110	oxide, di-	Nat. cristobalite. SiO_2	60.08	col, cub or tetr, 1.487, 1.484	2.32	1723 ± 5	2230 (2590)	i	i	s HF; v sl s alk
s111	oxide, di-	Nat. lechatelierite. SiO_2	60.08	col, amorph, vitr, 1.4588	2.19	...	2230 (2590)	i	i	s HF; v sl s alk
s112	oxide, di-	Nat. opal. $SiO_2{\cdot}xH_2O$	...	col. amorph 1.41—1.46	2.17—2.20	>1600	...	i	i	s HF; v sl s alk
s113	oxide, di-	Nat. tridymite. SiO_2	60.08	col, rhomb, 1.469, 1.470, 1.471	2.26_4^{25}	1703	2230 (2590)	i	i	s HF; v sl s alk
s114	oxide, di-	Nat. quartz. SiO_2	60.00	col, hex, 1.544, 1.553	2.635—2.660	1610	2230 (2590)	i	i	s HF; v sl s alk
s115	oxide, mon-	SiO	44.08	wh, cub	2.13	>1702	1880	i	i	s dil HF + HNO_3
s116	oxychloride	Chlorosiloxane. Si_2OCl_6	284.89	col liq	...	28.1 ± 0.2	137	d	...	∞ CS_2, CCl_4
s117	oxyfluoride	Si_2OF_6	186.16	col gas	1.358 liq	−47.8 ± 0.5	−23.3	d	d	d alk
s118	sulfide, di-	SiS_2	92.21	wh need, rhomb	2.02	subl 1090	white heat	d	...	d al, liq NH_3; s dil alk; i bz
s119	sulfide, mono-	SiS	60.15	yel need	1.853^{15}	subl 940^{20}	...	d	d	d al, alk
s120	thiocyanate	$Si(CNS)_4$	260.40	wh, rhomb need	1.409_4^{20}	143.8	314.2	d	...	d al, a, alk; i eth, CS_2, $CHCl_3$
s121	**Silicotungstic acid**	$H_4[Si(W_3O_{10})_4]{\cdot}26H_2O$	3346.69	wh-sl yel cr, deliq	...	...	...	v s	v s	v s al
s122	**Silicyl oxide**	Disiloxane $(SiH_3)_2O$	78.22	col gas	gas: 3.491 g/ℓ liq: 0.881^{-80}	−144	−15.2	v sl s	sl d	...
s123	**Siloxane, (di-), oxide**	$[H(O)Si]_2{\cdot}O$	106.19	wh volum subst	...	expl *ca* 300	...	sl s	...	s, d HF; d al
s124	**Silver**	Ag	107.8682 ± 3	wh met, cub 0.54	10.5^{20}	961.93	2212	i	i	s HNO_3, h H_2SO_4, KCN; i alk
s125	acetate	$AgC_2H_3O_2$	166.91	wh pl.	3.259^{15}	d	...	1.02^{20}	2.52^{80}	s dil HNO_3
s126	acetylide	Ag_2C_2	239.76	wh ppt	...	expl	...	i	...	s a; sl s al
s127	*ortho*arsenate	Ag_3AsO_4	462.52	dk red, cub	6.657^{25}	d	...	0.00085^{20}	...	s NH_4OH, ac a
s128	*ortho*arsenite	Ag_3AsO_3	446.52	yel, powd	...	d 150	...	0.00115^{20}	i	s ac a, NH_4OH, HNO_3; i al
s129	azide	AgN_3	149.89	wh rhomb pr, expl	...	252	297	i	0.01^{100}	s KCN, dil HNO_3; sl s NH_4OH
s130	benzoate	$AgC_7H_5O_2$	228.98	wh powd	...	...	...	0.262^{25}	s	0.017 al
s131	*tetra*borate	$Ag_2B_4O_7{\cdot}2H_2O$	407.00	wh cr.	...	...	...	sl s	...	s a
s132	bromate	$AgBrO_3$	235.77	col, tetr, 1.874, 1.920	5.206	d	...	0.196^{25}	1.33^{90}	s NH_4OH; sl s HNO_3
s133	bromide	Bromyrite: AgBr	187.77	pa yel, 2.253	6.473^{25}	432	d>1300	8.4×10^{-6}	0.00037^{100}	s KCN, $Na_2Sr_2O_3$, NaCl sol; sl s NH_4OH; i al
s134	carbonate	Ag_2CO_3	275.75	yel powd	6.077	d 218	...	0.0032^{20}	0.05^{100}	s NH_4OH, $Na_2S_2O_3$; i al
s135	chlorate	$AgClO_3$	191.32	wh, tetr	4.430_4^{20}	230	d 270	10^{15}	50^{80}	sl s al
s136	*per*chlorate	$AgClO_4$	207.32	wh, cr, deliq	2.806^{25}	d 486	...	557^{25}	s	s al; 101 tol; 5.28 bz
s137	chloride	Nat. cerargyrite. AgCl	143.32	wh, cub, 2.071	5.56	455	1550	0.000089^{10}	0.0021^{100}	s NH_4OH, $Na_2S_2O_3$, KCN
s138	chlorite	$AgClO_2$	175.32	yel cr.	...	105 expl	...	0.45^{25}	2.13^{100}	...
s139	chromate	Ag_2CrO_4	331.73	red, monocl.	5.625	...	...	0.0014^0	0.008^{70}	s NH_4OH, KCN
s140	*di*chromate	$Ag_2Cr_2O_7$	431.72	red, tricl	4.770	d	...	0.0083^{15}	d	s a, NH_4OH, KCN
s141	citrate	$Ag_3C_6H_5O_7$	512.71	wh need	...	d	...	0.028^{18}	sl s	s a, NH_4OH, KCN, $Na_2S_2O_3$
s142	cyanate	AgOCN	149.89	col	4.00	d	...	sl s	s	s KCN, HNO_3, NH_4OH
s143	cyanide	AgCN	133.89	wh, hex	3.95	d 320	...	0.000023^{20}	...	s HNO_3, NH_4OH, KCN, $Na_2S_2O_3$
s144	ferricyanide	$Ag_3Fe(CN)_6$	535.56	...	...	...	...	0.000066^{20}	...	i a; s NH_4OH, h $(NH_4)_2CO_3$
s145	ferrocyanide	$Ag_4Fe(CN)_6{\cdot}H_2O$	661.44	wh	...	...	...	i	i	s KCN; i a, NH_4 salts, NH_4OH
s146	fluogallate	$Ag_3[GaF_6]{\cdot}10H_2O$	687.47	col, orthorhomb cr, 1.493	2.90	...	...	v s	...	i al
s147	fluoride	AgF	126.87	yel, cub, deliq	$5.852^{15.5}$	435	*ca* 1159	$182^{15.5}$	205^{108}	sl s NH_4OH
s148	fluoride, di-	AgF_2	145.87	brn, rhomb	4.57—4.58	690	d 700	d	d	...
s149	(di-)fluoride	Ag_2F	234.73	yel, hex	8.57	d 90	...	d	...	...
s150	fluosilicate	$Ag_2SiF_6{\cdot}4H_2O$	429.87	wh powd or col cr, deliq	...	>100	d	v s	...	...
s151	fulminate	$Ag_2C_2N_2O_2$	299.77	need	...	expl	...	0.075^{13}	s	i NHO_3; s NH_4OH
s152	iodate	$AgIO_3$	282.77	col, rhomb	$5.525^{16.5}$	>200	d	0.03^{10}	0.019^{60}	s HNO_3, NH_4OH, KI
s153	*per*iodate	$AgIO_4$	298.77	or yel, tetrag	5.57	d 180	...	d	...	s HNO_3
s154	iodide(α)	miersite AgI	234.77	ye tetr 2.02 ± .02	5.683_4^{30}	tr 146 to β	...	$2.8 \times 10^{-7.25}$	$2.5 \times 10^{-6.60}$	s KCN, $Na_2S_2O_3$, KI; sl s NH_4OH
s155	iodide (β)	iodyrite AgI	234.77	pa ye hex	$6.010_4^{14.6}$	558	1506	...	...	...
s156	iodomercurate (α)	Ag_2HgI_4	923.94	yel, tetrag	6.02	tr to β 50.7	...	i	...	s KI, KCN; i dil a
s157	iodomercurate (β)	Ag_2HgI_4	923.94	red, cub	5.90	d 158	...	i	...	s KI, KCN; i dil a

No.	Name	Synonyms and Formulae	Mol. wt.	Crystalline form, properties and index of refraction	Density or spec. gravity	Melting point, °C	Boiling point, °C	Solubility, in grams per 100 cc: Cold water	Hot water	Other solvents
s158	hydrogen(tri-) *para*periodate	$Ag_2H_3IO_6$	441.66	yel, rhomb	5.68^{25}	60 d		1.68^{25}		s HNO_3
s159	hyponitrite	$Ag_2N_2O_2$	275.75	yel	5.75^{30}	d 110		v sl s		d HNO_3, H_2SO_4
s160	lactate	$AgC_3H_5O_3{\cdot}H_2O$	214.95	wh or sl gray cr, powd				*ca* 7.7		
s161	laurate	$AgC_{12}H_{23}O_2$	307.18	wh, greasy powd		212.5				0.007^{25} al; 0.008^{15} eth
s162	levunilate	$AgC_5H_7O_8$	222.98	leaf				0.67^{17}	d	
s163	*per*manganate	$AgMnO_4$	226.80	dk vlt, monocl	4.27^{25}	d		0.55^{0}	$1.69^{28.5}$	d al
s164	mercury iodide (α)	Ag_2HgI_4	923.94	yel, tetrag	6.02	trst 50.7		i		
s165	mercury iodide (β)	Ag_2HgI_4	923.94	red, cub	5.90	158 d		i		
s166	myristate	$AgC_{14}H_{27}O_2$	335.24			211		0.007^{25}		0.006^{25} al; 0.007^{15} eth
s167	nitrate	$AgNO_3$	169.87	col, rhomb, 1.729, 1.744, 1.788	4.352^{19}	212	d 444	122^{0}	952^{190}	s eth, glyc; v sl s abs al
s168	nitrite	$AgNO_2$	153.87	wh, rhomb	4.453^{26}	d 140		0.155^{0}	1.363^{60}	s ac a, NH_4OH; i al
s169	nitroplatinite	$Ag_2[Pt(NOp_2)_4]$	594.84	yel-brn monocl pr		d 100		sl s	s	
s170	nitroprusside	$Ag_2[FeNO(CN)_5]$	431.68	lt pink				i		s NH_4OH; i al, HNO_3
s171	oxalate	$Ag_2C_2O_4$	303.76	col cr.	5.029^{4}	expl 140		0.00339^{18}		s KCN, NH_4OH, a
s172	oxide	Ag_2O	231.74	br-blk, cub	$7.143^{16.6}$	d 230		0.0013^{20}	0.0053^{80}	s a, KCN, NH_4OH, al
s173	oxide, per	Ag_2O_2 (or AgO)	247.74	gray-blk, cub	7.44	d>100		i		s H_2SO_4, HNO_3, NH_4OH
s174	palmitate	$AgC_{16}H_{31}O_2$	363.29	wh, greasy powd		209		0.0012^{20}	0.006^{25}	0.007^{15} eth; 0.006^{25} al
s175	*meta*phosphate	$AgPO_3$	186.84	wh, amorph	6.37	*ca*482		i		s HNO_3, NH_4OH
s176	*ortho*phosphate	Ag_3PO_4	418.58	yel, cub	6.370^{25}	849		$0.00065^{19.5}$		s a, KCN, NH_4OH, NH_3
s177	*ortho*phosphate, mono-H	Ag_2HPO_4	311.72	wh, trig	1.8036	d 110				
s178	*pyro*phosphate	$Ag_4P_2O_7$	605.42	wh	$5.306^{7.5}$	585		i	i	s a, NH_4OH, KCN, ac a
s179	propionate	$AgC_3H_5O_2$	180.94	wh leaf or need	2.687_4^{25}			0.842^{20}	2.04^{80}	
s180	*per*rhenate	$AgReO_4$	358.07	wh cr, tetrag or rhomb	7.05	430		0.32^{20}		
s181	salicylate	$AgC_7H_5O_3$	244.98	wh to redsh-wh cr				sl s		s al
s182	selenate	Ag_2SeO_4	358.69	wh, orthorhomb cr	5.72			0.118^{20}		
s183	selenide	Ag_2Se	294.70	thin gray pl, cub	8.0	880	d	i		s NH_4OH, h HNO_3
s184	stearate	$AgC_{18}H_{35}O_2$	391.34	wh powd amorph		205		0.006^{20}		0.006^{25} al; 0.006^{25} eth
s185	sulfate	Ag_2SO_4	311.79	wh, rhomb. 1.7583, 1.7748, 1.7852	$5.45^{29.2}$	652	d 1085	0.57^{0}	1.41^{100}	s a, NH_4OH; i al
s186	sulfide	Nat. acanthite. Ag_2S	247.80	gray-blk, rhomb	7.326	tr 175	d		v sl s	s KCN, conc H_2SO_4, HNO_3
s187	sulfide	Nat. argentite. Ag_2S	247.80	blk, cub	7.317	825	d	8.4×10^{-15}		s KCN, a
s188	sulfite	Ag_2SO_3	295.79	wh cr.		d 100		v sl s		s a, NH_4OH, KCN; i HNO_3
2189	*d*-tartrate	$Ag_2C_4H_4O_6$	363.81	wh, scales	3.423^{15}	d		0.2^{18}	0.203^{25}	s a, KCN, NH_4OH
s190	*ortho*tellurate, tetra-H	$Ag_2H_4TeO_6$	443.36	straw yel, rhomb bipyr		d>200		i	i	s KCN, NH_4OH
s191	telluride	Nat. hessite, Ag_2Te	343.34	gray, cub	8.5	955		i	i	s KCN, NH_4OH
s192	tellurite	Ag_2TeO_3	391.33	yel-wh ppt		250-bl 450-pa yel		i	i	s KCN, NH_3
s193[1]	thioantimonite	Nat. pyrargyrite. Ag_3SbS_3	541.53	red, trig, 3.084 2.881 (Li)	5.76	486		i	i	s HNO_3
s193[2]	thioarsenite	Nat. proustite. Ag_3AsS_3	494.71	scarlet red, trig, 3.088, 2.792	5.49	490		i	i	s HNO_3
s194	thiocyanate	AgSCN	165.95	col cr.		d		0.000021^{25}	0.00064^{100}	s NH_4OH; i a
s195	*di*-thionate	$Ag_2S_2O_6{\cdot}2H_2O$	411.88	rhomb cr, ~1.662	3.61					
s196	thiosulfate	$Ag_2S_2O_3$	327.85	wh cr.		d		sl s		s $Na_2S_2O_3$, NH_4OH
s197	tungstate	Ag_2WO_4	463.58	pa yel cr.				0.015^{15}		s KCN, NH_4OH, HNO_3
	Silver complex									
s198	diamminesilver *per*rhenate	$[Ag(NH_3)_2]ReO_4$	392.13	col monocl cr.	3.901					1.618 conc NH_4OH
s199	**Sodium**	Na	22.98977	silv, met cub, 4.22	0.97	97.81 ± 0.03	882.9	d to NaOH + H_2		d al; i eth, bz
s200	acetate	$NaC_2H_3O_2$	82.03	wh gr powd, monocl, 1.464	1.528	324		119^{0}	170.15^{100}	sl s al
s201	acetate trihydrate	$NaC_2H_3O_2{\cdot}3H_2O$	136.08	col, monocl pr, effl, β 1.464	1.45	58	123, $-3H_2O$, 120	76.2^{0}	138.8^{50}	2.1^{18} al; s eth
s202	alumina trisilicate	Nat. albite. $NaAlSi_3O_8$ (or $Na_2O{\cdot}Al_2O_3{\cdot}6SiO_2$)	262.22	col, tricl, 1.525, 1.529, 1.536	2.61	1100			sl d	s HCl; d dil al
s203	*meta*aluminate	$NaAlO_2$	81.97	wh amorph powd, hygr, 1.566, 1.595, 1.580		1800		s	v s	i al
s204	aluminum chloride	$NaCl{\cdot}AlCl_3$	191.78	wh-yelsh cr powd, hygr.		185		s	s	
s205	aluminum *meta*-silicate	Nat. jadeite. $Na_2O{\cdot}Al_2O_3{\cdot}4SiO_2$	404.28	col, monocl	3.3	1000—1060		i	i	d HCl

PHYSICAL CONSTANTS OF INORGANIC COMPOUNDS (continued)

No.	Name	Synonyms and Formulae	Mol. wt.	Crystalline form, properties and index of refraction	Density or spec. gravity	Melting point, °C	Boiling point, °C	Solubility, in grams per 100 cc Cold water	Hot water	Other solvents
s206	aluminum *ortho*-silicate	Nat. nephelite. $Na_2O·Al_2O_3·2SiO_2$	284.11	col, hex, 1.537 ± 0.002	2.619^{21}	1526		i	d	d a
s207	aluminum sulfate	$NaAl(SO_4)_2·12H_2O$	458.27	col, cub oct, 1.4388	1.6754^{20}	61		110^{15} (anhydr)	146^{30} (anhydr)	
s208	amide	Sodamide. $NaNH_2$	39.01	wh, conchoid fract		210	400	d	d	d hot al; 0.1 liq NH_3
s209	ammonium phosphate	Microcosmic salt, stercorite. $NaNH_4HPO_4·4H_2O$	209.07	col, monoc, 1.439, 1.441, 1.469	1.554	d 79		16.7	100	i al, acet
s210	ammonmium sulfate	$NaNH_4SO_4·2H_2O$	173.12	wh, rhomb	1.63^{15}	d 80		s	s	
s211	ammonium tartrate	$NaNH_4C_4H_4O_6·4H_2O$	261.16	wh, rhomb	1.590			21.09^{0}		
s212	*meta*antimonate	Leuconine. $NaSbO_3$	192.74	wh powd				i	s	s Na_2S sol
s213	antimonate, hydroxy	"Pyroantimonate". $NaSb(OH)_6$	246.78	pseudo cub				$0.03^{12.3}$	0.3^{100}	sl s al
s214	*pyro*antimonate, dihydro-	$Na_2H_2Sb_2O_7·6H_2O$(?)	511.58	wh, tetrag		d 280		i	0.28^{100}, d	
s215	antimonide	Na_3Sb	190.72	blk powd or bl cr, inflamm		856		d		sl s NH_3
s216	*meta*antimonite	$NaSbO_2·3H_2O$	230.78	col, rhomb	2.864	d		d		
s217	*meta*arsenate	$NaAsO_3$	145.91	rhomb, effl, 1.479, 1.502, 1.527	2.301	615		v s		
s218	*ortho*arsenate	$Na_3AsO_4·12H_2O$	424.07	col, trig or hex prism, 1.457, 1.466	1.752—1.804	86.3		$38.9^{15.5}$		1.67 al; 50^{15} glyc
s219	*ortho*arsenate, di-H	$NaH_2AsO_4·H_2O$	181.94	col, rhomb or monocl, 1.583, 1.553, 1.507	2.53	130, $-H_2O$, 100	d 200—280	s		
s220	*ortho*arsenate, mono-H	$Na_2HAsO_4·7H_2O$	312.01	col, monocl, pois, 1.462, 1.466, 1.478	1.88	130, $-5H_2O$, 50	d 180	5.46^{0}	100^{100}	s glyc; sl s al
s221	*ortho*arsenate, mono-H	$Na_2HAsO_4·12H_2O$	402.09	col, monocl, effl. 1.445, 1.466, 1.451	1.736	28	$-12H_2O$, 100	56^{14}	140.7^{30}	sl s al; i liq Cl
s222	*pyro*arsenate	$Na_4As_2O_7$	353.80	wh cr.	2.205	850	d 1000	v s		
s223	arsenate fluoride	$2Na_3AsO_4·NaF·19H_2O$	800.06	wh, cub, 1.4657, 1.4693, 1.4726	2.849^{25}			10^{75}		
s224	arsenite	Sodium metaarsenite (?) (com'l) $NaAsO_2$ (or mixt with Na_3AsO_3)	129.91	gray-wh powd, pois	1.87			v s	v s	sl s al
s225	arsenotartrate	$Na(AsO)C_4H_4O_6·2^1/_2H_2O$	307.02	shiny cr, pois.		$-2^1/_2H_2O$, 275	d 275	6.5^{19}		i al
s226	azide	NaN_3	65.01	col, hex	1.846^{20}	d Na + N	d in vac	41.7^{17}		0.314^{16} al; s liq NH_3, i eth
s227	barbital	$NaC_8H_{11}N_2O_3$	206.18	wh powd				20^{25}	40^{100}	sl s al; i eth
s228	benzenesulfonate	$NaC_6H_5SO_3$	180.15	wh cr.				35.8^{30}	v s	
s229	benzoate	$NaC_7H_5O_2$	144.11	col cr, or wh amorph, or gran powd				66^{20}	74.2^{100}	1.64^{25} al
s230	*meta*bismuthate	$NaBiO_3$	297.97	yel-brn powd (com'l), yel (pure)				i	d	d a
s231	*meta*borate	$NaBO_2$	65.80	col, hex pr	2.464	966	1434	26^{20}	36^{35}	
s232	*meta*borate, tetrahydrate	$NaBO_2·4H_2O$	137.86	tricl, coll		57	$-H_2O$, 120	v s	v s	
s233	*meta*borate, peroxyhydrate	Sodium perborate (com'l). $NaBO_2·H_2O_2·3H_2O$	153.86	col, monocl		63.0	$-H_2O$, 130—150	2.55^{15}	3.75^{32}	s a, al, glyc
s234	*tetra*borate	$Na_2B_4O_7$	201.22	cr, 1.5010	2.367	741	d 1575	1.06^{0}	8.79^{40}	i al
s235	*tetra*borate, decahydrate	Borax. $Na_2B_4O_7·10H_2O$	381.37	col, monocl, effl, 1.447, 1.469, 1.472	1.73	75, $-8H_2O$, 60	$-10H_2O$, 320	2.01^{0}	170^{100}	v sl s al; s glyc; i a
s236	*tetra*borate, pentahydrate	$Na_2B_4O_7·5H_2O$	291.29	col, cub or hex, deliq, 1.461	1.815	$-H_2O$, 120		22.65^{65} (anhydr)	52.3^{100}	
s237	borohydride	$NaBH_4$	37.83	white cub, 1.542	1.074	400 dec		55^{25}	v s	4 al; 16.4 MeOH; s pyr; i eth
s238	bromate	$NaBrO_3$	150.89	col, cub, 1.594	$3.339^{17.5}$	381		27.5^{0}	90.9^{100}	i al
s239	bromide	$NaBr$	102.89	col, cub, hygr, 1.6412	3.203^{25}_{4}	747	1390	116.0^{50}	121^{100}	sl s al
s240	bromide, dihydrate	$NaBr·2H_2O$	138.92	col, monocl pr	2.176	$-2H_2O$, 51		79.5^{0}	$118.6^{80.5}$	2.31^{25} al; s liq NH_3; 17.42^{15} MeOH
s241	bromoaurate	$NaAuBr_4·2H_2O$	575.60	br-blk cr				s		
s242	bromoiridite	$Na_3IrBr_6·12H_2O$	956.80	dk grn, rhomb, eff		100	$-H_2O$, 150			s NH_4OH
s243	bromoplatinate	$Na_2PtBr_6·6H_2O$	828.58	dk red, tricl	3.323	d 150		v s	v s	v s al
s244	cacodylate	$Na[(CH_3)_2AsO_2]·3H_2O$	214.03	wh		ca 60	$-H_2O$, 120	200^{15-20}		40^{25} al; 100^{15-20} 90% al
s245	calcium sulfate	$Na_2Ca(SO_4)_2·2H_2O$	314.21	col, monocl need.	2.64	$-2H_2O$, 80		d	d	
s246	*d*-camphorate	$Na_2C_{10}H_{14}O_4·3H_2O$	298.24	wh need, hygr		$-3H_2O$, 100		122^{14}		s al
s247	carbide	Na_2C_2	70.00	wh powd	1.575^{15}	*ca* 700		d	d	s a; d al
s248	carbonate	Na_2CO_3	105.99	wh powd, hygr, 1.535	2.532	851	d	7.1^{0}	45.5^{100}	sl s abs al; i acet
s249	carbonate, decahydrate	Washing soda. $Na_2CO_3·10H_2O$	286.14	wh, monocl, 1.405, 1.425, 1.440	1.44^{15}	32.5—34.5	$-H_2O$, 33.5	21.52^{0}	421^{104}	i al
s250	carbonate, heptahydrate	$Na_2CO_3·7H_2O$	232.10	rhomb bipyr, effl	1.51	$-H_2O$, 32		16.90	33.9^{35}	
s251	carbonate, monohydrate	Crystal carbonate, thermonatrite, $Na_2CO_3·H_2O$	124.00	col, rhomb, deliq, 1.506, 1.509	2.25	$-H_2O$, 100		33	52.08	14^{25} glyc; i al, eth

No.	Name	Synonyms and Formulae	Mol. wt.	Crystalline form, properties and index of refraction	Density or spec. gravity	Melting point, °C	Boiling point, °C	Solubility, in grams per 100 cc: Cold water	Hot water	Other solvents
s252	carbonate, sesqui-	$Na_2CO_3·NaHCO_3·2H_2O$	226.03	col, monocl, 1.5073	2.112	d		13^0	42^{100}	
s253	carbonate hydrogen	$NaHCO_3$	84.01	wh, monocl pr, 1.500	2.159	$-CO_2$, 270		6.9^0	16.4^{60}	sl s al
s254	chlorate	$NaClO_3$	106.44	col, cub or trig, 1.513	2.490^{15}	248—261	d	79^0	230^{100}	s al, liq NH_3, glyc
s255	*per*chlorate	$NaClO_4$	122.44	wh, rhomb, deliq, 1.4606, 1.4617, 1.4731		d 482	d	s	v s	s al
s256	*per*chlorate, hydrate	$NaClO_4·H_2O$	140.46	col rhbdr, deliq	2.02	130	d 482	209^{15}	284^{50}	s al
s257	chloride	Common salt, nat. halite. NaCl	58.44	col, cub, 1.5442	2.165^{25}_{4}	801	1413	35.7^0	39.12^{100}	sl s al, liq, NH_3; s glyc; i HCl
s258	chlorite	$NaClO_2$	90.44	wh, cr, hygr		d 180—200		39^{17}	55^{60}	
s259	hypochlorite, pentahydrate	$NaOCl·5H_2O$	164.52	col		18		29.3^0	94.2^{23}	
s260	*hypo*chlorite	NaOCl	74.44	in solution only						
s261	*hypo*chlorite, dihydrate	$NaOCl·2^1/_2H_2O$	119.48	col, hygr		57.5		v s		
s262	chloroaurate	$NaAuCl_4·2H_2O$	397.80	yel, rhomb, ω 1.545 ε> 1.75		d 100		150^{10}	990^{60}	v s al, eth
s263	chloroiridate	$Na_2IrCl_6·6H_2O$	559.01	dull red-blk, tricl		d 600		v s	v s	sl s al
s264	chloroiridite	$Na_3IrCl_6·12H_2O$	690.09	dk grn cr		$-H_2O$, 50		31.46^{15}	307.26^{85}	
s265	chloroosmate	$Na_2OsCl_6·2H_2O$	484.93	or-red, rhomb pr				v s		s al
s266	chloropalladite	$Na_2PdCl_4·3H_2O$	348.26	br-red cr, deliq				v s		s al
s267	chloroplatinate	Na_2PtCl_6	453.78	or-yel powd, hygr		tr 150—160		s	v s	s al
s268	*hexa*chloroplatinate	$Na_2PtCl_6·6H_2O$	561.87	or-red, tricl	2.500	$-6H_2O$, 100		66^{15}	v s	11.9 al, MeOH; i eth
s269	chloroplatinite	$Na_2PtCl_4·4H_2O$	454.93	red pr		100	$-H_2O$, 150	s		s al
s270	chlororhodite, hexa-	Na_3RhCl_6	384.59	red, tricl		d>550		v s		
s271	chlororhodite, hexa-, hydrate	$Na_3RhCl_6·18H_2O$	708.87	garnet red, oct, effl		d 904, effl		v s		i al
s272	chromate	Na_2CrO_4	161.97	yel, rhomb bipyram	2.710—2.736			87.3^{30}		sl s al; s MeOH
s273	chromate decahydrate	$Na_2CrO_4·10H_2O$	342.13	yel, monocl, deliq	1.483	19.92		50^{10}	126^{100}	sl s al; i ac a
s274	*di*chromate	$Na_2Cr_2O_7·2H_2O$	298.00	red, monocl pr, deliq, 1.661, 1.699, 1.751	2.52^{13}	$-2H_2O$, 100 356.7 (anhydr)	400 (anhydr)	238^0 (anhydr) 180^{20}	508^{80} (anhydr) 433^{98}	i al
s275	*peroxy*chromate	Na_3CrO_8	248.96	or pl		d 115		sl s		i al, eth
s276	cinnamate	$NaC_9H_7O_2$	170.14	wh cr powd				9.1	5^{100}	0.625 90% al; s glyc
s277	citrate, dihydrate	$Na_3C_6H_5O_7·2H_2O$	294.10	wh cr, gran or powd		$-2H_2O$, 150		72^{25}	167^{100}	0.625 90% al; s glyc
s278	citrate, pentahydrate	$Na_3C_6H_5O_7·5(\text{or } 5^1/_2)H_2O$	348.15	wh, rhomb	$1.857^{23.5}$	$-5H_2O$, 150	d	92.6^{25}	250^{100}	sl s al
s279	cobaltinitrite	$Na_3Co(NO_2)_6$	403.94	yelsh-brnsh cr powd				v s, d		sl s al; d min a; i dil ac a
s280	cyanamide, mono-	$NaHCN_2$	64.02	wh cr powd, hygr				v s		
s281	cyanate	NaOCN	65.01	col need	1.937^{20}	d 700 vac		s	s	v sl s eth, bz
s282	cyanide	NaCN	49.01	col, cub, deliq, pois, 1.452		563.7	1496	48^{10}	82^{35}	sl s al; s NH_3
s283	cyanoaurite	Sodium aurocyanide. $NaAu(CN)_2$	271.99					s		
s284	cyanocuprate (I)	$NaCu(CN)_2$	138.57		1.013^{20}	d 100		s		
s285	cyanoplatinite	$Na_2[Pt(CN)_4]·3H_2O$	399.18	col, tricl	2.646	$-3H_2O$, 120—125		s	s	s al
s286	enanthate	Sodium heptanoate. $NaC_7H_{13}O_2$	152.17	wh cr powd or leaf		240—350		s		s al
s287	ethyl acetoacetate	$NaC_6H_9O_3$	152.13	need		d		d		s eth
s288	ethyl sulfate	$NaC_2H_5SO_4·H_2O$	166.12	wh, hex pl, deliq		d		164^{17}		d alk, H_2SO_4; 142al
s289	ferrate (III)	Ferrite. $Na_23Fe_2O_4$	221.67	br, hex pl or need	4.05			d		v s dil HCl
s290	ferricyanide	$Na_3Fe(CN)_6·H_2O$	298.94	red cr, deliq				18.9^0	67^{100}	i al
s291	ferrocyanide	Yellow prussiate of soda. $Na_4Fe(CN)_6·10H_2O$	484.07	pa yel, monocl, 1.519, 1.530, 1.544	1.458			31.85^{20}	156.5^{98}	i al
		Sodium hexacvanoferrate (II).								
s292	fluoaluminate	Na_3AlF_6	209.94	col, monocl, β 1.364	2.90	1000		sl s		i HCl.; d alk
s293	fluoantimonate	$NaSbF_6$	258.73	rhomb	3.375^{18}	<1360		128.6^{20}		s al, acet
s294	fluoberyllate	Na_2BeF_4	130.99	wh, rhomb or monocl		d		1.47^{18}	2.94^{100}	
s295	fluoborate	$NaBF_4$	109.79	wh, rhomb	2.47^{20}	sl d 384	d	108^{26}	210^{100}	sl s al; d H_2SO_4
s296	fluoride	Nat. villiaumite. NaF	41.99	col, cub or tetr, 1.336	2.558^{41}	993	1695	4.22^{18}		s HF; v sl s al
s297	fluoride, hydrogen	NaF·HF	61.99	col, or wh cr powd, rhdr	2.08			s	s	
s298	fluoride *ortho*phosphate	$NaF·Na_3PO_4·12H_2O$	422.11		2.2165			12^{25}	57.5^{50}	
s299	fluoroacetate, mono-	$NaC_2H_2FO_2$	100.02	wh powd		200		111^{25}		1.4^{25} al; 5^{25} MeOH; 0.04^{25} acet; 0.0049^{25} CCl_4
s300	fluorophosphate, hexa-	$NaPF_6·H_2O$	185.97		2.369^{19}			103.2^0		
s301	fluorophosphate, mono-	Na_2PO_3F	143.95	col		ca 625		25		
s302	fluosilicate	Na_2SiF_6	188.06	col, hex, 1.312, 1.309	2.679	d		0.0652^{17}	2.46^{100}	i al
s303	fluosulfonate	$NaSO_3F$	122.05	shiny leaf, hygr		d red heat		s		s al, acet; i eth

PHYSICAL CONSTANTS OF INORGANIC COMPOUNDS (continued)

No.	Name	Synonyms and Formulae	Mol. wt.	Crystalline form, properties and index of refraction	Density or spec. gravity	Melting point, °C	Boiling point, °C	Solubility, in grams per 100 cc Cold water	Hot water	Other solvents
s304	formaldehydesulfoxylate	$NaHSO_2·CH_2O·2H_2O$	154.11	rhomb pr, hygr		64	d	v s		d a; s al, alk
s305	formate	$NaCHO_2$	68.01	col, monocl, deliq	1.92^{20}	253	d	97.2^{20}	160^{100}	sl s al; i eth
s306	2-furanacrylate	$NaC_7H_5O_3$	160.10	lt brn powd	1.919	d		s	s	sl s al; i eth
s307	*meta*germanate	Na_2GeO_3	166.57	wh, monocl, deliq, 1.59	3.31^{22}	1083			d	s a
s308	*meta*germanate, heptahydrate	$Na_2GeO_3·7H_2O$	292.67	col, rhomb		83		24.6^{0}		s a
s309	(mono-) d-glutamate	$NaC_5H_8NO_4$	169.11	wh cr.		d		v s		sl s al
s310	glycerophosphate, monohydrate	$Na_2C_3H_7O_6P·H_2O$	234.05	yelsh visc liq; wh cr or powd.				s		s al
s311	glycerophosphate, pentahydrate	$Na_2C_3H_7O_6P·5^1/_2H_2O$	315.12	wh pl, sc or powd		>130		v s		i al
s312	gold sulfide	$NaAuS·4H_2O$	324.08	col, monocl.		d		s		s al
s313	hydride	NaH	24.00	silver need, 1.470	0.92	d 800		d	d	s molten Na; i CS_2, CCl_4, NH_3, bz
s314	hydroxide	$NaOH$	40.00	wh, deliq, 1.3576	2.130	318.4	1390	42^{0}	347^{100}	v s al, glyc; i acet, eth
s315	iodate	$NaIO_3$	197.89	wh, rhomb	$4.277^{17.5}$	d		9^{20}	34^{100}	i al; s ac a
s316	*meta*periodate	$NaIO_4$	213.89	col, tetr	4.174	d 300		14.44^{25}	$38.9^{51.5}$	s H_2SO_4, HNO_3, ac a
s317	*meta*periodate, trihydrate	$NaIO_4·3H_2O$	267.94	col, rhombdsh, effl	3.219^{18}	d 175		18.78^{25}	$36.4^{34.5}$	
s318	*para*periodate	Na_5IO_6	337.85	wh		800 d		d		
s319	(tri-)*para*periodate	$Na_3H_2IO_6$	293.89	col, hexag.				sl s		s con NaOH sol
s320	iodide	NaI	149.89	col, cub, 1.7745	3.667^{25}	661	1304	184^{25}	302^{100}	42.57^{25} al; 39.9^{25} acet; s glyc
s321	iodide, dihydrate	$NaI·2H_2O$	185.92	col, monocl.	$2.448^{20.8}$	752		317.9^{0}	1550^{100}	v s NH_3
s322	iodoplatinate	$Na_2PtI_6·6H_2O$	1110.58	brn, monocl.	3.707			v s		s al
s323	iridium chloride	Sodium hexachloroiridate. $Na_3IrCl_6·12H_2O$	690.09	olive cr, rhomb or trig-rhomb		50		s	s	i al
s324	iron (III) nitrosopentacyanide	$Na_2[Fe(CN)_5NO]·2H_2O$	297.95	ruby red, rhomb, 1.605, 1.575, 1.56	1.687^{25}	$-H_2O$, 100	d 160	40^{16}	s	
s325	iron (III) oxalate	$Na_3[Fe(C_2O_4)_3]·5^1/_2H_2O$	487.96	grn, monocl.	$1.973^{17.5}$	$-4H_2O$	d 300	32^{0}	182^{100}	
s326	iron (III) sulfate	$3Na_2SO_4·Fe_2(SO_4)_3·6H_2O$	934.07	wh, trig, 1.558, 1.613	2.5	$-6H_2O$, 100		d v sl		i al
s327	lactate	$NaC_3H_5O_3$	112.06	col or yelsh liq, very hygr		17	d 140	v s		s al; i eth
s328	lithium sulfate	$Na_3Li(SO_4)_2·6H_2O$	376.12	col, ditrig	2.009	$-6H_2O$, 50		s	s	
s329	magnesium carbonate	$Na_2CO_3·MgCO_3$	190.30	wh, rhomb	2.729^{15}	677 CO_2 1240/atm		d	d	
s330	magnesium sulfate	Nat. bloedite. $Na_2SO_4·MgSO_4·4H_2O$	334.46	col, monocl, 1.486, 1.488, 1.489	2.23			s		
s331	magnesium tartrate	$Na_2Mg(C_4H_4O_6)_2·10H_2O$	546.58	wh, monocl pr or powd				s	s	
s332	manganate	$Na_2MnO_4·10H_2O$	543.07	grn, monocl.		17		s	d	
s333	*per*manganate	$NaMnO_4$	141.93	red cr. deliq		d		v s	v s	
s334	*per*manganate, trihydrate	$NaMnO_4·3H_2O$	195.97	purp cr, deliq.	2.47	d 170		v s	v s	s NH_3; d alk
s335	methanearsenate	$Na_2CH_3AsO_2·6H_2O$	292.03	wh cr powd		130—140		ca 100		sl s al; i bz, eth. oils
s336	methoxide	$CH_3ONa·2CH_3OH$	118.11	wh powd		d, $-CH_3OH$		s, d		s CH_3OH
s337	methylsulfate	$NaCH_3SO_4·H_2O$	152.10	col cr, hygr				s		s al
s338	molybdate	Na_2MoO_4	205.92	opaque wh	3.28^{18}	687		s 44.3	84^{100}	
s339	molybdate, dihydrate	$Na_2MoO_4·2H_2O$	241.95	wh, rhbdr	3.28(?)	$-2H_2O$, 100		56.2^{0}	115.5^{100}	i meth acet
s340	*deca*molybdate	$Na_2Mo_{10}O_{31}·21H_2O$	1879.68	wh, monocl pr				sl s	0.842^{100}	
s341	*di*molybdate	$Na_2Mo_2O_7$	349.86	wh need		612		sl s	sl s	
s342	*octa*molybdate	$Na_2Mo_8O_{25}17H_2O$	1519.74	monocl cr		$-H_2O$, 20		v s	v s	
s343	*para*molybdate	$Na_6Mo_7O_{24}·22H_2O$	1589.84	col, monocl, effl		700 $-H_2O$, 100—120		117.9^{30} (anhydr)	s	
s344	*tetra*molybdate	$Na_2Mo_4O_{13}·6H_2O$	745.82	yel need				39.8^{21}	v s	
s345	*tri*molybdate	$Na_2Mo_3O_{10}·7H_2O$	619.90	need acicular		528 $-6H_2O$, 100—120		3.878^{20}	13.7^{100}	
346	*meta*niobate	$Na_2Nb_2O_6·7H_2O$	453.90	pseudo-cub	4.512—4.559	$-H_2O$, 100		s		
s347	nitrate	Soda niter. $NaNO_3$	84.99	col, trig or rhbdr, 1.587, 1.336	2.261	306.8	d 380	92.1^{25}	180^{100}	s al, MeOH; v s NH_3; v sl s acet; sls glyc
s348	nitride	Na_3N	82.98	dk gray		d 300		d		
s349	nitrite	$NaNO_2$	69.00	col-yel, rhomb pr, hygr.	2.168^{0}	271	d 320	81.5^{15}	163^{100}	0.3^{20} eth; 4.4^{20} MeOH; 3 abs al; v s NH_3
s350	*hypo*nitrite	$Na_2N_2O_2$	105.99		1.728^{25}	d 300		d		i al
s351	*p*-nitrophenoxide	$NaOC_6H_4NO_2·4H_2O$	233.15	yel, monocl pr		$-2H_2O$, 36 $-4H_2O$, 120	d	5.97^{25}		sl s al
s352	nitroplatinite	$Na_2Pt(NO_2)_4$	425.08	pa yel rhomb or monocl pr, effl				s	s	
s353	nitroprusside	$Na_2[Fe(NO)(CN)_5]·2H_2O$	297.95	red, rhomb	1.72			40^{16}		s al
s354	oleate	$NaC_{18}H_{38}O_2$	304.45	wh cr, or yel amorph gran.		232—235		10^{12}		s al; sl s eth
s355	oxalate	$Na_2C_2O_4$	134.00	col cr, or wh powd	2.34	d 250—270		3.7^{20}	6.33^{100}	i al, eth
s356	oxalate, hydrogen	$NaHC_2O_4·H_2O$	130.03	wh, monocl.		$-H_2O$, 100	d 200	1.7^{15}	21^{100}	
s357	oxalatoferrate (III)	$Na_3Fe(C_2O_4)_3·xH_2O$	388.88 $+xH_2O$	grn, monocl cr	$1.973^{17.5}$	$-H_2O$, 100—120		32.5	182^{100}	

No.	Name	Synonyms and Formulae	Mol. wt.	Crystalline form, properties and index of refraction	Density or spec. gravity	Melting point, °C	Boiling point, °C	Solubility, in grams per 100 cc: Cold water	Hot water	Other solvents
s358	oxide, mon-	Na_2O	61.98	wh-gray, deliq	2.27	subl 1275		d	d	d al
s359	oxide, per-	Na_2O_2	77.98	yel-wh powd	2.805	d 460	d 657	s	d	i al
s360	oxide, per-, octahydrate	$Na_2O_2{\cdot}8H_2O$	222.10	wh, hex		d 30	d	s	d	i al
s361	palmitate	$NaC_{16}H_{31}O_2$	278.41	wh cr.		270				
s362	pentobarbital	$NaC_{11}H_{17}N_2O_3$	248.26					s	s, d	s al
s363	phenobarbital	$NaC_{12}H_{11}N_2O_3$	254.22	wh				v s		s al; i eth, chl
s364	1-phenol-4-sulfonate (*p*-)	$NaC_6H_4(OH)SO_3{\cdot}2H_2O$	232.18	col, monocl or gran, sl effl		d		23.8^{25}	125^{100}	0.75^{25} al; 20^{25}glyc
s365	phenoxide	$NaOC_6H_5$	116.09	wh cr need, deliq				v s		s al, acet; d a
s366	phenylcarbonate	$NaC_7H_5O_3$	160.10	col powd		d 120		d		d acet
s367	*hypo*phosphate	$Na_4P_2O_6{\cdot}10H_2O$	430.06	col, monocl, 1.477, 1.482, 1.504	1.823	d		1.49^{25}	5.46^{50}	
s368	*hypo*phosphate, di-H	$Na_2H_2P_2O_6{\cdot}6H_2O$	314.03	col, monocl, 1.468, 1.490, 1.504	1.849	250 (anhydr)	$-6H_2O$, 100	2.35	25	s dil H_2SO_4, NH_4OH; i al
s369	*meta*phosphate, hexa-	Graham's salt. $(NaPO_3)_6$	611.77	col glass, 1.482 ± 0.002				v s		
s370	*meta*phosphate, tri-, hexahydrate	Knorre's salt. $(NaPO_3)_3{\cdot}6H_2O$	413.98	col, tricl, effl, 1.433, 1.442, 1.446		53; $-6H_2O$, 50		s		
s371	*ortho*phosphate	$Na_3PO_4{\cdot}10H_2O$	344.09	col, oct	$2.536^{17.5}$ (anhydr)	100		8.8 (anhydr)		
s372	*ortho*phosphate	$Na_3PO_4{\cdot}12H_2O$	380.12	col, trig, 1.446, 1.452	1.62^{20}	d 73.3—76.7	$-12H_2O$, 100	1.5^{0}	157^{70}	i CS_2, al
s373	*ortho*phosphate, di-H	$NaH_2PO_4{\cdot}H_2O$	137.99	col, rhomb, 1.456, 1.458, 1.487	2.040	$-H_2O$, 100	d 204	59.9^{0}	427^{100}	v sl s eth, chl, tol; i al
s374	*ortho*phosphate, di-H	$NaH_2PO_4{\cdot}2H_2O$	156.01	col, rhomb, 1.4629	1.91	60		v s	v s	
s375	*ortho*phosphate, mono-H	Sörensen's sodium phosphate. $Na_2HPO_4{\cdot}2H_2O$	177.99	rhomb bispheroidal, 1.463	2.066^{16}	$-2H_2O$, 95		100^{50}	117^{80}	
s376	*ortho*phosphate, mono-H	$Na_2HPO_4{\cdot}7H_2O$	268.07	col, monocl pr, 1.442	1.679	$-5H_2O$, 48.1		104^{40}		i al
s377	*ortho*phosphate, mono-H	$Na_2HPO_4{\cdot}12H_2O$	358.14	col, rhomb or monocl, eff, wh powd, 1.432, 1.436, 1.437	1.52	$-5H_2O$, 35.1	$-12H_2O$, 100	4.15	87.4^{34}	i al
s378	*pyro*phosphate	$Na_4P_2O_7$	265.90	wh cr, 1.425	2,534	880		3.16^{0}	40.26^{100}	
s379	*pyro*phosphate	$Na_4P_2O_7{\cdot}10H_2O$	446.06	col, monocl, 1.450, 1.453, 1.460	1.815—1.836	$-H_2O$, 93.8 m.p. 880		5.41^{0}	93.11^{100}	i al, NH_3
s380	*pyro*phosphate, di-H	$Na_2H_2P_2O_7{\cdot}H_2O$	330.03	monocl, 1.4599, 1.4646, 1.4649	1.85	$-H_2O$,220		6.9^{0}	35^{40}	
s381	phosphide	Na_3P	99.94	red		d		d, PH_3		
s382	*hypo*phosphite	$NaH_2PO_2{\cdot}H_2O$	105.99	col, monocl, deliq		d viol		100^{25}	667^{100}	v s al; s glyc; sl s NH_3 NH_4OH
s383	*ortho*phosphite, di-H	$NaH_2PO_3{\cdot}2^1/_2H_2O$	149.02	col, monocl, 1.419, 1.431, 1.449		42	$-2^1/_2H_2O$, 100	56^{0}	193^{42}	
s384	*ortho*phosphite, mono-H	$Na_2HPO_3{\cdot}5H_2O$	216.04	wh, rhomb deliq, β, 1.443	53	d 200-250		s	v s	i al, NH_4OH
a385	*tri*phosphate	Sodium tripolyphosphate, $Na_5P_3O_{10}$	367.86	powd and gran				14.5^{25}	32.5^{100}	
s386	phthalate	$Na_2C_8H_4O_4$	210.10	wh powd or pearly pl		$-H_2O$, 150				
s387	platinate, hydroxo-	$Na_2Pt(OH)_6$	343.10	yel or red-brn. hex		$-3H_2O$, 150—170	d	s		i al; sl s HCl
s388	platinum cyanide	$Na_2[Pt(CN)_4]{\cdot}3H_2O$	399.18	col, tricl	2.646	$-H_2O$, 120—125		s		s al
s389	plumbate, hydroxo-	$Na_2Ph(OH)_6$	355.22	yel-wh lumps, hygr				d to PbO_2		d a; s alk
s390	potassium(*dl*)- tartrate	$NaKC_4H_4O_6$	210.16	col, tricl		90—100	d 200	47.4_6 (anhydr)	v s	
s391	propionate	$NaC_2H_5O_2$	96.06	wh, gran powd				s		s al
s392	*per*rhenate	$NaReO_4$	273.19	col, hex pl, hygr	5.39	300(in O_2) d 440 (vac)		100^{20}		s al
s393	*pyrohypo*rhenate	$Na_4Re_2O_7{\cdot}H_2O$	594.38	sandy yel cr.				0.004		
s394	rhodiumchloride	$Na_2RhCl_6{\cdot}12H_2O$	600.78	dk red cr, monocl pr		$-12H_2O$, 120		v s	v s	i al
s395	rhodiumnitrite	$Na_3[Rh(NO_2)_6]$	447.91	wh cr.		d 360		40^{17}	s	i al; d a
s396	*per*ruthenate	$NaRuO_4{\cdot}H_2O$	206.07	blk cr, lamellar		d 440 vac		v s	d	
s397	salicylate	$NaC_7H_5O_3$	160.10	wh cr powd				111^{15}	125^{25}	17^{15} al; 25 glyc
s398	selenate	Na_2SeO_4	188.94	col, rhomb	$3.213^{17.4}$			84^{35}	72.8^{100}	
s399	selenate, decahydrate	$Na_2SeO_4{\cdot}10H_2O$	369.09	col, monocl	1.603—1.620	*ra*32 trans		43.5^{20}	340^{100}	
s400	selenide	Na_2Se	124.94	wh to red, cr, deliq	2.625^{10}	>875		d		i NH_3
s401	selenite	$Na_2SeO_3{\cdot}5H_2O$	263.01	wh cr, tetrag				s	s	i al
s402	silicate	Waterglass. $Na_2O{\cdot}xSiO_2(r = 3 - 5)$		col, amorph, deliq				s	s	i al, K and Na salts
s403	*di*silicate	$Na_2Si_2O_5$	182.15	rhomb pearly luster 1.500, 1.510		874		s	s	
s404	*meta*silicate	Na_2SiO_3	122.06	col, monocl, α 1.518, γ 1.527	2.4	1088		s	s, d	i al, K and Na salts

PHYSICAL CONSTANTS OF INORGANIC COMPOUNDS (continued)

No.	Name	Synonyms and Formulae	Mol. wt.	Crystalline form, properties and index of refraction	Density or spec. gravity	Melting point, °C	Boiling point, °C	Solubility, in grams per 100 cc: Cold water	Hot water	Other solvents
s405	*meta*silicate	$Na_2SiO_3 \cdot 9H_2O$	284.20	col, rhomb bi-pyramid, effl	...	40—48	$-6H_2O$, 100	v s	v s	s dil NaOH; i al, a
s406	*ortho*silicate	Na_4SiO_4	184.04	col, hex, 1.530	...	1018	...	s	s	...
s407	silicotungstate, dodeca-	$Na_4[Si(W_3O_{10})_4] \cdot 20H_2O$	3326.53	col, tricl	...	$-7H_2O$, 100	d	v s	v s	s a, sl s al
s408	stannate, hydroxo-	$Na_2Sn(OH)_6$	266.71	col, hex or wh powd, or lumps	...	$-3H_2O$, 140	...	$61.3^{15.5}$	50^{100}	i al, acet
s409	stearate	$NaC_{18}H_{35}O_2$	306.46	wh fatty powd	...	...	...	s	s	s h al
s410	succinate	$Na_2C_4H_4O_4 \cdot 6H_2O$	270.14	wh, gran or powd	...	$-6H_2O$, 120	...	21.45^{0}	86.63^{75}	v sl s al
s411	succinate, tetrahydroxy-	Sodium dihydroxy tartrate. $Na_2C_4H_4O_8 \cdot 3H_2O$	280.10	...	...	d	...	0.032^{0}	d	d min a; i al, eth
s412	sulfanilate	$NaC_6H_4(NH_2)SO_3$	195.17	wh. lust cr leaf	...	...	...	s	...	...
s413	sulfate, anhydr	Na_2SO_4	142.04	monocl (between *ca*160—185), 1.480	...	884; tr to hex *ca* 241	...	s	$42—5^{100}$	s HI
s414	sulfate, anhydr	Nat. thenardite. Na_2SO_4	142.04	orthorhomb, 1.484, 1.477, 1.471	2.68	...	...	4.76^{0}	42.7^{100}	s glyc; i al
s415	sulfate, decahydrate	Glauber's salt, mirabilite. $Na_2SO_4 \cdot 10H_2O$	322.19	col, monocl, effl, 1.394, 1.396, 1.398	1.464	32.38	$-10H_2O$, 100	11^{0}	92.7^{30}	i al
s416	sulfate, heptahydrate	$Na_2SO_4 \cdot 7H_2O$	268.14	wh, rhomb or tetrag	...	tr to anhydr 24 4	...	19.5^{0}	44^{20}	i al
s417	*pyro*sulfite	Sodium metabisulfite, $Na_2S_2O_5$	190.10	wh powd or cr ($+7H_2O$)	1.4	>d 150	...	54^{20}	81.7^{100}	sl s al; s glyc
s418	*pyro*sulfate	$Na_2S_2O_7$	222.10	wh, transluc cr, deliq	2.658^{25}	400.9	d 460	s	...	s fum H_2SO_4
s419	sulfate hydrogen	$NaHSO_4$	120.06	col, triel	2.435^{13}	>315	d	28.6^{25}	100^{100}	sl s al; i NH_3
s420	sulfate hydrogen, monohydrate	$NaHSO_4 \cdot H_2O$	138.07	col, monocl, deliq ~ 1.46	$2.103^{13.5}_{4}$	58.54 ± 0.5	...	*ca* 67, d	d	d al
s421	sulfide, hydro-	NaHS	56.06	col, rhomb or wh gran cr, deliq	...	350	...	vs	...	s al
s422	sulfide, hydrodihydrate	$NaHS \cdot 2H_2O$	92.09	col need, deliq	...	d	...	s	s	d a; s al
s423	sulfide, mono-	Na_2S	78.04	wh cr, deliq	1.856^{14}	1180	...	15.4^{10}	57.2^{90}	d a; sl s al; i eth
s424	sulfide, monohydrate	$Na_2S \cdot 9H_2O$	240.18	col, tetr, deliq	1.427^{16}_{4}	d 920	...	47.5^{10}	96.7^{10}	d, sl s al
s425	sulfide, penta-	Na_2S_5	206.28	yel (exist ?)	...	251.8	...	s	s	s al
s426	sulfide, tetra-	Na_2S_4	174.22	yel, cub, hygr	...	275	d	s	...	s al
s427	sulfide, hydrotrihydrate	$NaHS \cdot 3H_2O$	110.11	col, lust rhomb cr	...	22	d	s	s	s al
s428	sulfite	Na_2SO_3	126.04	wh powd or hex, prism, 1.565, 1.515	$2.633^{15.4}$	d red heat	d	12.54^{0}	28.3^{80}	sl s al; i liq Cl_2, NH_3
s429	sulfite hydrate	$Na_2SO_3 \cdot 7H_2O$	252.14	col, monocl, effl	1.539^{15}	$-7H_2O$, 150	d	32.8^{0}	196^{40}	sl s al
s430	*hydro*sulfite	Dithionite, hyposulfite $Na_2S_2O_4 \cdot 2H_2O$	210.13	col, monocl(?) cr, or yel-wh powd	...	d 52	...	25.4^{20}	d	d a; s alk; i al
s431	sulfite, hydrogen	$NaHSO_3$	104.06	wh, monocl, yel in sol, 1.526	1.48	d	...	v s	v s	sl s al
s432	*d*(& *l*)-tartrate	$Na_2C_4H_4O_6 \cdot 2H_2O$	230.08	col, rhomb. 1.545, 1.49	1.818	$-2H_2O$, 150	...	29^{6}	66^{43}	i al
s433	*d*-tartrate, hydrogen	$NaHC_4H_4O_6 \cdot H_2O$	190.09	wh cr powd, rhomb, 1.53, 1.54, 1.60	...	$-H_2O$, 100	d 234	6.7^{18}	9.2^{30}	...
s434	*di*-tartrate hydrogen	$NaHC_4H_4O_6 \cdot H_2O$	190.09	col, monocl or tricl, 1.53, 1.54, 1.60	...	$-H_2O$, 100	d 219	8.9^{19}	...	...
s435	*ortho*tellurate, tetra-H	$Na_2H_4TeO_6$	273.61	col, hex pl	...	d	d	0.77^{18}	2^{100}	s h dil HNO_3: i NaOH
s436	telluride	Na_2Te	173.58	wh cr powd very hygr, d in air	2.90	953	...	v s, d	v s, d	...
s437	tellurite	Na_2TeO_3	221.58	wh, rhomb pr.	...	...	...	sl s	s	...
s438	thioantimonate	Schlippe's salt, $Na_3SBS_4 \cdot 9H_2O$	481.10	pa yel, cub	1.806	87	d 234	20.15^{0}	100^{100}	i al, eth
s439	thioarsenate	$Na_3AsS_4 \cdot 8H_2O$	416.25	yel, monocl, β 1.6802	...	d	...	v s	d	i al
s440	thiocarbonate, tri	$Na_2CS_3 \cdot H_2O$	172.19	yel need, deliq	...	d 75	...	s	d	s al; i eth, bz
s441	thiocyanate	NaSCN	81.07	col, rhomb deliq pois, ~1.625	...	287	...	$139.31^{21.3}$	225^{100}	v s al, acet
s442	*di*thionate	$Na_2S_2O_6 \cdot 2H_2O$	242.13	col, rhomb, 1.482, 1.495, 1.519	2.189	$-2H_2O$, 110	$-SO_2$, 267	47.6^{16}	90.9^{100}	s HCl; i al
s443	thiosulfate	$Na_2S_2O_3$	158.10	col, monocl	1,667	...	...	50	231^{100}	i al
s444	thiosulfate, pentahydrate	"Hypo", sodium hyposulfite. $Na_2S_2O_3 \cdot 5H_2O$	248.17	col, monocl, effle, 1.489, 1.508, 1.536	1.729^{17}	40—45 d 48	$-5H_2O$, 100	79.4^{0}	291.1^{45}	s NH_3; i al
s445	thiosulfosurate (I)	$Na_3[Au(S_2O_3)_2] \cdot 2H_2O$	526.20	wh cr, monocl	3.09	$-H_2O$, 150	d	50	...	i al
s446	*trit*itanate	$Na_2Ti_2O_7$	301.62	wh need, monocl	3.35—3.50	1128	...	i	...	s h HCl
s447	tungstate	Na_2WO_4	293.83	wh, rhomb	4.179	698	...	57.5^{0} 73.2^{21}	96.9^{100}	...
s448	tungstate, dihydrate	$Na_2WO_4/2H_2O$	329.86	col pl, rhomb. 1.5533	3.23—3.25	$-2H_2O$, 100 anhydr 698	...	41^{0}	123.5^{100}	sl s NH_3; i al, a
s449	*meta*tungstate	$Na_2O \cdot 4WO_2 \cdot 10H_2O$	1169.52	col, oct	...	706.6	...	s	v s	i a
s450	*para*tungstate	$Na_6W_7O_{34} \cdot 16H_2O$	2097.12	col, tricl	3.987	$-12H_2O$, 100; $-16H_2O$, 300	...	8	d	...

No.	Name	Synonyms and Formulae	Mol. wt.	Crystalline form, properties and index of refraction	Density or spec. gravity	Melting point, °C	Boiling point, °C	Solubility, in grams per 100 cc: Cold water	Hot water	Other solvents
s451	*meta*uranate	Na_2UO_4	348.01	gr yel or red pl, rhomb pr.				i	i	s dil a, alk carb
s452	uranyl acetate	$NaUO_2(C_2H_2O_2)_3$	470.15	yel, tetr pr, 1.501	2.56					
s453	uranyl carbonate	$2Na_2CO_2{\cdot}UO_2CO_3$	542.01	yel cr.		d 400		sl s		i al
s454	urate	$Na_2C_5H_2N_4O_2{\cdot}H_2O$	230.09	wh gran powd or hard cr nodules					1.3^{100}	v sl s 90% al
s455	urate, acid	$NaHC_5H_2N_4O_3$	190.09	wh gran powd				0.083	0.8^{100}	
s456	valerate	$NaC_5H_9O_2$	124.12	wh cr or mass, hygr.		140		s		s al
s457	*meta*vanadate	$NaVO_3$	121.93	col, monocl pr		630		21.1^{25}	38.8^{75}	
s458	*ortho*vanadate	Na_3VO_4	183.91	col, hex pr		850—866		s		i al
s459	*ortho*vanadate, decahydrate	$Na_2VO_4{\cdot}10H_2O$	364.06	wh, cub or hex cr, 1.5305, 1.5398, 1.5475				s	s	
s460	*ortho*vanadate, hexadecylhydrate	$Na_2VO_4{\cdot}16H_2O$	472.15	col need		866 (anhydr)		v s	d	i al
s461	*pyro*vanadate	$Na_4V_2O_7$	305.87	col, hex		632—654		s	s	i al
s462	ethylxanthate	NaC_2H_5OCSS	144.18	yelsh powd				s		s al
s463	zinc uranyl acetate	$NaZn(UO_2)_3(C_2H_3O_2)_9{\cdot}9H_2O$	1591.99	monocl cr. α 1.475, γ 1.480				i		s al
	Stannous	See under tin								
	Stannic	See under tin								
s464	**Strontium**	Sr	87.62	silv wh to pa yel met	2.6^{20}	769	1384	d	d	s a, al, liq NH_3
s465	acetate	$Sr(C_2H_3O_2)_2$	205.71	wh cr.	2.099	d		36.9	36.4^{97}	0.26^{15} MeOH
s466	acetate	$Sr(C_2H_3O_2)_2{\cdot}H_2O$	214.72	wh cr powd		$-{}^1/_2H_2O$ 150		s		sl s al
s467	*ortho*arsenate, acid	$SrHAsO_4{\cdot}H_2O$	245.56	rhomb, need	3.606^{15}; 4.035 (anhydr)	$-H_2O$, 125		$0.284^{15.5}$d	s a	
s468	*ortho*arsenite	$Sr_2(AsO_3)_2{\cdot}4H_2O$	580.76	cr, or wh powd				sl s		s a; sl s al
s469	borate, *tetra-*	$SrB_4O_7{\cdot}4H_2O$	314.92						77^{100}	s HNO_3; NH_4 salts
s470	boride, hexa-	SrB_6	152.48	blk, cub	3.39^{15}	2235		i	i	s HNO_2; i HCl
s471	bromate	$Sr(BrO_3)_2{\cdot}H_2O$	361.44	col yelsh, monocl, hygr.	3.773	$-H_2O$, 120	d 240	33^{16}		
s472	bromide	$SrBr_2$	247.43	wh, hex need, hygr, 1,575	4.216^{24}	643	d	100^{20}	222.5^{100}	s al, amyl al
s473	bromide, hexahydrate	$SrBr_2{\cdot}6H_2O$	355.52	col, hex, hygr	2.386_4^{25}	tr to $2H_2O$, 88.6	$-6H_2O$, >180	204.2^0	∞	63.9^{30} al; 113.4^{30} MeOH; 0.6^{30} acet; i eth
s474	carbide	SrC_2	111.64	blk, tetr	3.2	>1700		d	d	d a
s475	carbonate	Nat. strontianite. $SrCO_3$	147.63	col, rhomb, or wh powd trfrs to −hex at 926, 1.516, 1.664, 1.666	3.70	1497^{69atm}	$-CO_2$, 1340	0.0011^{18}	0.065^{100}0	.12 aq CO_2; s a, NH_4 salts
s476	chlorate	$Sr(ClO_3)_2$	254.52	col, rhomb. or wh powd, 1.516, 1.605, 1.626	3.152	d 120		174.9^{18}	v s	s dil al; i abs al
s477	*per*chlorate	$Sr(ClO_4)_2$	286.52	col cr, hygr.				310^{25}	v s	212 MeOH; 181 al; i eth
s478	chloride	$SrCl_2$	158.53	col, cub 1.650^{25}	3.052	875	1250	53.8^{20}	100.8^{100}	v sl s abs al, acet; i NH_3
s479	chloride, dihydrate	$SrCl_2{\cdot}2H_2O$	194.56	transp leaf, 1.594, 1.595, 1.617	2.6715^{25}					
s480	chloride, fluoride	$SrCl_2{\cdot}SrF_2$	284.14	col, tetr. 1.651, 1.627	4.18	962		d	d	s conc HNO_3, conc HCl, i al
s481	chloride, hexahydrate	$SrCl_2{\cdot}6H_2O$	266.62	col, trig, 1.536, 1.487	1.93	115, $-4H_2O$, 60	$-6H_2O$, 100	106.2^0	205.8^{40}	3.8^6 al
s482	chromate	$SrCrO_4$	203.61	yel, monocl.	3.895^{15}			0.12^{15}	3^{100}	s HCl, HNO_3, ac a, NH_4 salts
s483	cyanide	$Sr(CN)_2{\cdot}4H_2O$	211.72	wh. rhomb. deliq		d		v s		
s484	cyanoplatinite	$Sr[Pt(CN_4]{\cdot}5H_2O$	476.85	col, monocl pr, 1.696		$-5H_2O$, 150				s abs al
s485	glycerophosphate	$SrC_2H_7O_6P$	257.68	wh powd				sl s		i al
s486	ferrocyanide	$Sr_2Fe(CN)_6{\cdot}15H_2O$	657.42	yel, monocl.				50	100	
s487	fluoride	SrF_2	125.62	col, cub or wh powd, 1.442	4.24	1473	2489^{760}	0.011^0	0.012^{27}	a hot HCl; i HF, al acet
s488	fluosilicate	$SrSiF_6{\cdot}2H_2O$	265.73	monocl.	$2.99^{17.5}$d		$3\ .2^{15}$	v s	s HCl; 0.065^{16} 50% al	
s489	formate	$Sr(CHO_2)_2$	177.66	col, rhomb, 1.559, 1.547, 1.598	2.693	71.9		9.1^0	34.4^{100}	
490	formate, dihydrate	$Sr(CHO_2)_2{\cdot}2H_2O$	213.69	col, rhomb, 1.484, 1.521, 1.538	2.25	d, $-2H_2O$, 100		$11.62^{36.6}$	26.57^{100}	i al, eth
s491	hydride	SrH_2(?)	89.64	wh, rhomb. hygr	3.72	d 675	subl 1000 (in H_2)	d	d	d al
s492	hydroxide	$Sr(OH)_2$	121.63	wh, deliq	3.625	375 (in H_2)	$-H_2O$, 710	0.41^0	21.83^{100}	s a, NH_4Cl
s493	hydroxide, octahydrate	$Sr(OH)_2{\cdot}8H_2O$	265.76	col, tetr, deliq. 1.499, 1.476	1.90	$-8H_2O$, 100		0.90^0	47.71^{100}	s a, NH_4Cl; i acet
s494	iodate	$Sr(IO_2)_2$	437.43	tricl.	5.045^{16}			0.03^{15}	0.8^{100}	

No.	Name	Synonyms and Formulae	Mol. wt.	Crystalline form, properties and index of refraction	Density or spec. gravity	Melting point, °C	Boiling point, °C	Solubility, in grams per 100 cc: Cold water	Hot water	Other solvents
s495	iodide	SrI_2	341.43	col pl.	4.549^{26}_{4}	515	d	165.3^{0}	383^{100}	4.5^{30} al; 0.31^{0} NH_4OH; a MeOH
s496	iodide, hexahydrate	$SrI_2 \cdot 6H_2O$	449.52	col-yelah, hex, deliq	2.672^{25}	d 90	...	448.9^{0}	∞	s al; i eth
s497	lactate	$Sr(C_3H_5O_3)2 \cdot 3H_2O$	319.81	wh cr or gran powd	...	$-3H_2O$, 120	...	25	200^{100}	sl s al
s498	*per*manganate	$Sr(MnO_4)_2 \cdot 3H_2O$	379.54	purpl, cub	2.75	d 175	...	270^{0}	291^{18}	...
s499	molybdate	$SrMoO_4$	247.56	col, tetr ~1.91	4.54^{26}_{25}	d	...	0.0104^{17}	...	s a
s500	nitrate	$Sr(NO_3)_2$	211.63	col, cub	2.986	570	...	70.9^{18}	100^{00}	0.012 aba al; v a NH_3; al s acet
s501	nitrate, tetrahydrate	$Sr(NO_3)_2 \cdot 4H_2O$	283.69	col, monocl	2.2	$-4H_2O$, 100	1100 tr SrO	60.43^{0}	206.5^{100}	s liq NH_3; v sl s abs al, acet; i HNO_3
s502	nitride	Sr_3N_2	290.87	...	...	...	...	d	d	s HCl
s503	nitrite	$Sr(NO_2)_2 \cdot H_2O$	197.65	col, hex, 1.588	2.408^{0}_{0}	$-H_2O$>100	d 240	58.9^{0}	182^{100}	0.42^{30} 90% al
s504	*hypo*nitrite	$SrN_2O_2 \cdot 5H_2O$	237.71	wh need	2.173^{25}_{4}	...	...	v sl s	sl s	v sl s NH_3
s505	oxalate	$SrC_2O_4 \cdot H_2O$	193.65	col cr.	...	$-H_2O$, 150	...	0.0051^{18}	0.15^{100}	s HCl, HNO_3
s506	oxide	SrO	103.62	gray wh, cub, 1.810	4.7	2430	~3000	0.69^{30}	22.85^{100}	30 fus KOH; sl s al; i eth, acet
s507	oxide, per-	SrO_2	119.62	wh powd	4.56	d 215^{760}	0.018^{20}	d	v s al, NH_4Cl; i acet	
s508	oxide, per-, octahydrate	$SrO_2 \cdot 8H_2O$	263.74	col cr.	1.951	$-8H_2O$, 100	d	0.018^{20}	d	s NH_4Cl; i al, acet, NH_4OH
s509	*ortho*phosphate, di-	$SrHPO_4$	183.60	col, rhomb	3.544^{15}	1.62	...	i	i	s a, NH_4 salts
s510	salicylate	$Sr(C_7H_5O_3)_2 \cdot 2H_2O$	397.88	col cr.	...	d	...	5.6^{25}	28.6^{100}	1.5^{25}, 9.5^{78} al
s511	selenate	$SrSeO_4$	230.58	col, rhomb	4.23	...	...	i	i	s hot HCl; i HNO_3
s512	selenide	SrSe	166.58	wh, cub, 2.220	4.38	...	...	d	d	s HCl
s513	*meto*silicate	$SrSiO_2$	163.70	col, pr monocl, 1.599, 1.637	3.65	1580	...	i	i	...
s514	*ortho*silicate	$SrSiO_4$	179.70	monocl, 1.728, 1.732, 1.758	3.84	>1750	...	...	...	...
s515	sulfate	Nat. celestite, $SrSO_4$	183.68	col, rhomb, 1.622, 1.624, 1.631	3.96	1605	...	0.0113^{0}	0.014^{30}	sl s a; i al, dil H_2SO_4
s516	sulfate, hydrogen	$Sr(HSO_4)_2$	281.75	col	...	d	...	d	...	$14^{70}H_2SO_4$
s517	sulfide, hydro	$Sr(HS)_2$	153.76	col, cub need, 2.107	...	d	...	s	d	...
s518	sulfide, mono	SrS	119.68	col, lt gray, cub, 2.107	3.70^{15}	>2000	...	i	d	d a
s519	sulfide, tetra-	$SrS_4 \cdot 6H_2O$	323.95	redsh cr, hygr	...	25	$-4H_2O$, 100	s	s	s al
s520	sulfite	$SrSO_2$	167.68	col cr.	...	d	...	0.0033^{17}	...	v s H_2SO_4; s a, HCl
s521	tartrate	$SrC_4H_4O_6 \cdot 4H_2O$	307.75	wh, monocl	1.966	...	...	0.112^{0}	0.755^{85}	s dil HCl, dil HNO_3
s522	telluride	SrTe	215.22	wh, cub, 2.408	4.83	...	...	...	...	...
s523	thiocyanate	$Sr(SCN)_2 \cdot 3H_2O$	257.82	deliq	...	$-3H_2O$, 100	d 160—170	v s	...	v s al
s524	thiosulfate	$SrS_2O_2 \cdot 5H_2O$	289.81	monocl need	2.17^{17}	$-4H_2O$, 100	...	2.5^{12}	57^{100}	i al
s525	*di*thionate	$SrS_2O_4 \cdot 4H_2O$	319.80	trig. 1.530, 1.525	2.373	$-4H_2O$, 78	...	22^{16}	67^{100}	i al
s526	tungstate	$SrWO_4$	335.47	col, tetr	6.187	d		0.14^{15}	...	id a; i al
s527	**Sulfamic acid**	Amidosulfuric, amino-sulfonic acid, NH_2SO_3H	97.09	col, rhomb	2.126^{25}	200 d	d	14.68	47.08^{60}	v sl s al, eth, acet; i CS_2, CCl_4
s528	**Sulfamide**	Sulfuryl amide, $SO_2(NH_2)_2$	96.10	rhomb pl	1.611	91.5	d 250	s	...	s al
s529	**Sulfur**(α)	S_8	256.48	yel, rhomb, 1.957	2.07^{20}	112.8 95.5 (revers.) 444.6	444.674	i	i	23^{0} CS_2; sl s tol, al, bz, eth, liq NH_3; s CCl_4
s530	(β)	S_8	256.48	pa yel, monocl	1.96	119.0	444.674	i	i	70 CS_2; s al, bz
s531	(γ)	S_8	256.48	pa yel, amorph	1.92	*ca* 120	444.6	i	i	i CS_2
s532	bromide, mono-	S_2Br_2	223.93	red liq, 1.730	2.63	−40	$54^{0.2}$	d	d	s CS_2
s533	chloride, di-	SCl_2	102.97	dk red liq. 1.557^{11}	1.621^{15}_{15}	−78	d 59	...	...	s CCl_4, bz; d al, eth
s534	chloride, mono-	S_2Cl_2	135.03	yel-red liq, 1.666^{14}	1.678	−80	135.6	d	d	s bz, eth, CS_2
s535	chloride, tetra-	SCl_4	173.87	yel-br liq	...	−30	d −15	d	d	...
s536	fluoride, hexa-	SF_6	146.05	col gas	gas 6.602 g/ℓ liq 1.88^{-50}	−50.5	−63.8 (subl)	sl s	sl s	s al, KOH
s537	fluoride, mono-	S_2F_2	102.12	col gas	liq 1.5^{-100}	−120.5	−38.4	d	d	d KOH
s538	fluoride, tetra-	SF_4	108.05	gas (exist ?)	...	−124	−40	d	d	...
s539	(di) fluoride, deca-	S_2F_{10}	254.10	col liq	2.08^{0}_{4}	−92	29	...	...	d fus caust
s540	(tetra-) nitride, di-	S_4N_2	156.25	red liq or gray solid	1.901^{18}	23	d 100 expl	i	...	s eth; sl s al, CS_2
s541	(tetra-) nitride, tetra-	S_4N_4	184.27	or red, monocl	2.22^{15}	subl 179	expl 160	d	...	s CS_2, chl, bz, NH_3; sl s al, eth
s542	(tri)-*di*nitrogen dioxide	$S_2N_2O_2$	156.19	pa yel cr.	...	100.7	d	i	...	s al, bz
s543	oxide, di-	SO_2	64.06	col gas or liq suffoc odor	gas 2.927 g/ℓ liq 1.434	−72.7	−10	22.8^{0}	0.58^{90}	s al, ac a, H_2SO_4
s544	oxide, hept-	Sulfur oxide, per-S_2O_7	176.12	visc liq, or need	...	0	subl 10	d	d	s H_2SO_4
s545	oxide, mono-	SO (or S_2O_2)	48.06	col gas	...	d	d	d	...	...
s546	oxide, sesqui-	S_2O_2	112.12	bl-grn cr.	...	d 70—95	...	d	d	s al, eth, fum H_2SO_4
s547	oxide, tetra-	Sulfurperoxide, SO_4	96.06	wh	...	d 0—3	...	s, d	...	d dil H_2SO_4
s548	oxide, tri-(α)	SO_2	80.06	silky fibr need, stable modific.	1.97^{20}	16.83	44.8	d	d	forms fum H_2SO_4
s549	oxide, tri-(β)	$(SO_3)_2$	160.12	asbestos like fiber metastable	...	62.4	50 (subl)	d	d	forms fum H_2SO_4

PHYSICAL CONSTANTS OF INORGANIC COMPOUNDS (continued)

No.	Name	Synonyms and Formulae	Mol. wt.	Crystalline form, properties and index of refraction	Density or spec. gravity	Melting point, °C	Boiling point, °C	Solubility, in grams per 100 cc		
								Cold water	Hot water	Other solvents
s550	oxide, tri-(γ)	SO_2	80.06	vitreous, orthorhomb, metastable	liq 1.920_4^{20} sld 2.29^{-10}	16.8	44.8	d	d	forms fum H_2SO_4
s551	oxy*tetra*chloride, mono-	S_2OCl_4	221.93	dk red liq	1.656^0		60	d	d	d al
s552	oxy*tetra*chloride, tri-	$S_2O_2Cl_4$	253.93	wh, rhomb need or pl		d 57		d	d	d al
s553	*tri*thiazyl chloride	S_4N_2Cl	205.71	pa yel cr		d 170 (vac)		d	d	
s554	**Sulfuric acid**	H_2SO_4	98.07	col liq	1.841 (96—98%)	10.36 (100%) 3.0 (98%)	330±0.5 (100%)	∞ ev heat	∞	d al
s555	dihydrate	$H_2SO_4 \cdot 2H_2O$	134.10	col liq, 1.405	1.650^0	−38.9	167	∞	∞	d al, eth
s556	hexahydrate	$H_2SO_4 \cdot 6H_2o$	206.17	liq		−54		v s	v s	
s557	monohydrate	$H_2SO_4 \cdot H_2O$	116.09	col liq or monocl cr, 1.438	1.788	8.62	290	∞	∞	d al
s558	octahydrate	$H_2SO_4 \cdot 8H_2O$	242.20	liq		−62		v s	v s	
s559	peroxidi-	Per(di-)sulfuric acid, $H_2S_2O_8$	194.13	hygr cr		d 65	d	d	d	s al, eth, H_2SO_4
s560	peroximono-	Permonosulfuric acid, Caro's acid H_2SO_5	114.07			d 45		d	d	s H_2PO_4
s561	pyro-	$H_2S_2O_7$	178.13	col cr, hygr	1.9^{20}	35	d	d	d	d al
s562	tetrahydrate	$H_2SO_4 \cdot 4H_2O$	170.13			−27		∞	∞	d al,eth
s563	**Sulfurous acid**	H_2SO_2	82.07	in sol only	*ca* 1.03			s		s al, eth, ac a
s564	**Sulfuryl** chloride	SO_2Cl_2	134.96	col liq, 1.444	1.6674_4^{20}	−54.1	69.1	d	d	s bz, ac a
s565	chloride fluoride	SO_2ClF	118.51	col gas	1.623^0 g/ℓ	−124.7	7.1	d		
s566	fluoride	SO_2F_2	102.06	col gas	gas 3.72 g/ℓ liq 1.7	−136.7	−55.4	10^9		s al, CC_4; sl s alk
s567	pyro-, chloride	$S_2O_5Cl_2$	215.02	col liq, 1.937^{20}	gas 9.6 g/ℓ .iq 1.818_4^{11}	−39 to −37	152.5	d	d	d a
t1	**Tantalum**	Ta	180.9479	gray black hard metal, cub or powd	met 16.6^{20} powd 14.401	2996	5425 ± 100	i	i	s HF, fus alk; i a
t2	boride, di-	TaB_2	202.57		11.15	3000(?)				
t3	bromide	$TaBr_5$	580.47	yel cr	4.67	265	348.8	d	d	s abs al, eth
t4	carbide	TaC	192.96	blk, cub	13.9	3880	5500	i	i	sl s H_2SO_4. HF
t5	chloride, penta-	$TaCl_5$	358.21	1t yel, vitr cr powd	3.68^{27}	216	242	d		s abs al, H_2SO_4
t6	fluoride	TaF_8	275.94	col, tetrag, deliq	4.74	96.8	229.5	s		s HF, eth
t7	nitride	TaN	194.95	br bronze or blk, hex	16.30	3360 ± 50		i	i	sl s aq reg, HF, HNO_3
t8	oxide, pent-	Ta_2O_5	441.89	col, rhomb	8.2	1872 ± 10		i	i	s fus $KHSO_4$, HF; i a
t9	oxide, pent- hydrate	Tantalic acid $Ta_2O_5xH_2O$		col gel				s		s alk, exc conc HNO_3; i a
t10	oxide, tetr-	Ta_2O_4 (or TaO_2)	425.89	dk gray powd.		oxidizes		i	i	i a
t11	sulfide	Ta_2S_4 (or TaS_2)	490.14	blk powd or cr		>1300		i	i	al s HF + HNO_3; i HCl
t12	**Telluric acid, ortho-**	$Te(OH)_4$ or $H_2TeO_4 \cdot 2H_2O$	229.64	wh, monocl pr	3.071	136		s	s	sl a dil a, HNO_3; i abs al, acet, eth
t13	**Telluric acid**	$Te(OH)_4$ or H_6TeO_6	229.64	wh cub	3.158	136		s	s	sl s dil a, HNO_3; i abs al, acet, eth
t14	**Tellurium**	Te	127.60 ± 3	br blk, amorph, 1.0025	6.00	449.5 ± 0.3	989.8 ± 3.8	i	i	s H_2SO_4; HNO_3, aq reg, KCN, KOH; i HCl, CS_2
t15	**Tellurium**	Te	127.60 ± 3	rhomb silv wh met, 1.0025	6.25	452	1390	i	i	s H_2SO_4; HNO_3, aq keg, KCN KOH; i HCl, CS_2
t16	bromide, di-	$TeBr_2$	287.41	brn to gray grn, need, unstable		210	339	d		s eth; al s a; d NaOH
t17	bromide, tetra-	$TeBr_4$	447.22	or cr	4.31_4^{15}	380 ± 6	d 421	sl s	d	s eth, a tart a, NaOH
t18	chloride, di-	$TeCl_2$	198.51	blk cr or amorph, unstable	7.05	209 ± 5	327	d	d	s min a, tart a; d NaOH
t19	chloride, tetra	$TeCl_4$	269.41	wh to yel cr, deliq	3.26^{18} 2.559^{232}	224	380^{760}	s d	s d	s HCl, bz, al, chl, CCl_4; i CS_2
t20	ethoxide	$Te(OC_2H_5)_4$	307.84			20	107—$107.5^{5.5}$			
t21	fluoride, hexa-	TeF_4	241.59	col gas unpleas odor	sol 4.006^{-191} liq 2.56^{-36}	−36	+35.5	d	d	d a, alk
t22	fluoride, tetra-	TeF_4	203.59	wh cr hygr		subl	>97	d	d	
t23	hydride	H_2Te	129.62	col gas pois	4.49	−48.9	-2.2^{760}	v s	s	d al
t24	iodide, di-	TeI_2	381.41	blk cr (exist ?)		subl		i	i	
t25	iodide, tetra,-	TeI_4	635.22	blk cr	5.403_4^{15}	280	d	sl s	d	s alk, aq NH_3, HI
t26	methoxide	$Te(OCH_3)_4$	251.74	solid			123—124			
t27	oxide, di-	Tellurite, TeO_2	159.60	wh, tetr or rhomb, 2.00, 2.18(Li), 2.35	tetr 5.67^{15} rhomb 5.91^0	733	1245	i	i	s HCl, hot HNO_3, alk; i NH_4OH
t28	oxide, mon-	TeO	143.60	blk, amorph (exist ?)	5.682	d 370 (in CO_2)	d	i	i	s dil a, H_2SO_4, KOH
t29	oxide, tri-	TeO_3	175.60	α yel amorph β gray cr	α 5.0752_4^{105} β 6.21	d 395		i	i	d conc HCl; s hot KOH; i a, al
t30	sulfide	TeS_2	191.72	red-blk powd amorph (exist ?)				i		i a; s alk sulf
t31	sulfoxide	$TeSO_3$	207.66	deep red amorph		d 30	d	d	d	s H_2SO_4
t32	**Tellurous acid**	H_2TeO_3(?)	177.61	wh flocks, indef not isolated	3.05	d 40		0.00067	d	s a, NaOH; sl s NH_4OH; i al

PHYSICAL CONSTANTS OF INORGANIC COMPOUNDS (continued)

No.	Name	Synonyms and Formulae	Mol. wt.	Crystalline form, properties and index of refraction	Density or spec. gravity	Melting point, °C	Boiling point, °C	Solubility, in grams per 100 cc: Cold water	Hot water	Other solvents
t33	**Terbium**	Tb	158.9254	silv-gray met, hex	8.2294	1360 ± 4	3123	i	i	s a
t34	bromide	$TbBr_3$	398.64			827	1490	s	s	
t35	chloride hexahydrate	$TbCl_2{\cdot}6H_2O$	373.38	col pr cr, deliq	4.35 (anhydr)	588 (anhydr)	$-H_2O$, 180-200 (in HCl gas)	v s		
t36	fluoride	TbF_3	215.92			1172	2280(?)	i	i	i dil a
t37	iodide	TbI_3	539.64			946	>1300	s	s	
t38	*di*methylphosphate	$Tb[(CH_3)_2PO_4]_3$	534.05					12.6^{25}	8.07^{40}	
t39	nitrate	$Tb(NO_3)_3{\cdot}6H_2O$	453.03	col, monocl cr		893		s		
t40	oxalate	$Tb_2(C_2O_4)_3{\cdot}10H_2O$	762.06	wh cr.	2.60	$-H_2O$, 40		i	i	i dil a
t41	oxide	Terbia. Tb_2O_3	365.85	wh solid						s dil a
t42	oxide, per-	Tb_4O_7	747.70	dk-brn or blk solid		−O2		i	i	s hot conc a
t43	sulfate	$Tb_2(SO_4)_3{\cdot}8H_2O$	750.15	wh cr.		$-8H1_2O$, 360		3.561^{20}	2.51^{40}	
t44	**Thallium**	Tl	204.383	bl-wh met, tetr	11.85	303.5	1457 ± 10	i	i	s HNO_3. H_2SO_4; sl s HCl
t45	acetate	$TlC_2H_2O_2$	263.43	silk wh cr, deliq	3.765^{137}	131		v s		v s al, $CHCl_3$; i acet
t46	aluminum sulfate	$TlAl(SO_4)_2{\cdot}12H_2O$	639.66	oct, 1.488	2.306^{20}	91		11.78^{23}		
t47	azide	TlN_3	246.40	yel, tetr		330 (vac)		0.1712^{0}	0.3^{16}	i al, eth
t48	bromate	$TlBrO_2$	332.29	col, need				0.35^{20}	s	s dil al
t49	bromide, di-	Bromothallate(ous). Tl_2Br_4 or $Tl^{I}_3[Tl^{III}Br_4]$	728.38	yel need				d	d	
t50	bromide, mono-	TlBr	284.29	yel-wh, cub, 2.4—2.8	$7.557^{17.3}$	480	815	0.05^{25}	0.25^{68}	s al; i HB_2, acet
t51	bromide, tri-	$TlBr_2$	444.10	yel, deliq. unstable		d		s	v s	v s al
t52	carbonate	Tl_2CO_2	468.78	col, monocl.	7.11	273		$4.03^{15.5}$	27.2^{100}	i abs al, eth, acet
t53	chlorate	$TlClO_2$	287.83	need (rhomb ?)	5.047^{9}			2^{0}	57.31^{100}	
t54	*per*chlorate	$TlClO_4$	303.83	col, rhomb	4.89	501	d	20.5^{30}	167^{100}	sl s al
t55	chloride	TlCl	239.84	wh reg discol in air, 2.247	7.004^{30}_{4}	430	720	$0.29^{15.6}$	$2.41^{99.35}$	i al, acet; d a
t56	chloride, tri-	$TlCl_3$	310.74	hex pl, hygr		25	d	v s		s al, eth
t57	chloride, tri-	$TlCl_3{\cdot}H_2O$	328.76	col, need		$-H_2O$, 60	d 100	v s	d	v s al, eth
t58	chloride, tri-	$TlCl_3{\cdot}4H_2O$	382.80	col, need		37	$-4H_2O$, 100	86.2^{17}	d	s al, eth
t59	chloroplatinate	Tl_2PtCl_6	816.56	pale or cr	5.76^{17}			0.0064^{15}	0.05^{100}	
t60	chromate	Tl_2CrO_4	524.76	yel				0.03^{60}	0.2^{100}	sl s a, alk; i ac a
t61	*di*chromate	$Tl_2Cr_2O_7$	624.75	red				i		d a
t62	chromium sulfate	Thallium chromium alum. $Tl[Cr(H_2O)_6](SO_4)_2{\cdot}6H_2O$	664.68	vlt cr	2.394	92		163.8^{25}		
t63	cyanate	TlCNO	246.40	col, need	5.487^{20}_{4}			s	v s	sl s al
t64	cyanide	TlCN	230.40	tabl	6.523	d		$16.8^{28.5}$	s a	
t65	ethoxide	$(TlOC_2H_5)_4$	997.78	col liq	3.522	−3	d 80	s d		9.11^{2}ls5 al; s bz; i liq NH_3
t66	ethylate	$TlOC_2H_5$	249.44	liq, 1.6714^{30}	3.493^{20}_{4}	−3	d 130			sl s al; s eth
t67	ferrocyanide	$Tl_4Fe(CN)_6{\cdot}2H_2O$	1065.52	yel, tricl	4.641			0.37^{18}	3.93^{101}	
t68	fluogallate	$Tl_2(GaF_5H_2O)$	591.49	col, orthorhomb.	6.44			s		
t69	fluoride, mono-	TlF	223.38	col, cub, oct	8.23^{4}	327	655	78.6^{15}	d	sl s al
t70	fluoride, tri-	TlF_3	261.38	olive grn.	8.36^{25}	d 550		d		i conc HCl
t71	fluosilicate	$Tl_2SiF_6{\cdot}2H_2O$	586.87	hex pl	5.72			v s		
t72	formate	$TlHCO_2$	249.40	col, need, hygr.	4.967^{104}	101		500^{10}		v a MeOH, sl s al; i $ChCl_3$
t73	(I) hydroxide	TlOH	221.39	pa yel, need		d139		25.9^{0}	52^{40}	s al
t74	iodate	$TlIO_3$	379.29	wh need				0.058^{20}	sl s	sl s HNO_3
t75	iodide (α)	TlI	331.29	yel, rhomb	7.29	tr to (β) 170		0.0006^{20}	0.12^{100}	s liq NH_3
t76	iodide (β)	TlI	331.29	red, cub	$7.098^{14.7}$	440	823	i	i	i al
t77	iodide, tri-	TlI_3	585.10	blk, lust rhomb.		d		s		
t78	iron (III) sulfate	$TlFe(SO_4)_2{\cdot}12H_2O$	668.53	pink, oct, n^{17}_{D} 1.524	2.351^{15}	$-H_2O$ *ca* 100		36.15^{25} (anhydr)		
t79	magnesium sulfate	$Tl_2SO_4{\cdot}MgSO_4{\cdot}6H_2O$	733.28	wh dull cr, 1.5660, 1.5836, 1.5900	3.573^{20}_{4}	$-6H_2O$, 40		d 0		
t80	methoxide	$TlOCH_3$	235.42	wh cr powd		d>120		s d		1.70^{25} CH_3OH; 3.16^{24} bz
t81	molybdate	Tl_2MoO_4	568.70	wh powd or cr		vol red heat		i	v sl s	i al; s alk carb, conc NH_4OH, HF
t82	myristate	$TlC_{14}H_{27}O_2$	431.75	wh powd		120—3				0.52^{25} 50% al
t83	(I) nitrate (α)	$TlNO_3$	266.39	cubic		206	430	9.55^{20}	4.13^{100}	i al, s acet
t84	(I) nitrate (β)	$TlNO_3$	266.39	trig		tr 145 to (α)				
t85	(I) nitrate (γ)	$TlNO_3$	266.39	rhomb, α 1.817	$5.556^{21.4}$	tr 75 to (β)		3.91^{0}	414^{100}	i al; s acet
t86	(III) nitrate	$Tl(NO_3)_3$	390.40	cr				s		
t87	(III) nitrate	$Tl(NO_3)_3{\cdot}3H_2O$	444.44	col, rhomb, deliq			s 100	d	d	
t88	nitrite	$TlNO_2$	250.39	yel micro cr.		182		32.10^{25}	95.78^{96}	i a
t89	oleate	$TlC_{18}H_{33}O_2$	485.84	wh cr clusters		131—2		0.05^{15}	0.3^{80}	3.0^{25} al
t90	oxalate	$Tl_2C_2O_4$	496.79	monocl pr	6.31			1.48^{15}	9.02^{100}	
t91	oxalate, tetra-	$TlH_3(C_2O_4)_2{\cdot}2H_2O$	419.48	tricl, leaf, 1.5097, 1.6319, 1.6538	2.992^{17}	d 100		76.9^{23}	v s	i cold al; s hot al
t92	(I) oxide	Tl_2O	424.77	blk, deliq	9.52^{16}	300	1080^{760} −O, 1865	v s d to TlOH		s a, al
t93	(III) oxide	Tl_2O_2	456.76	col, amorph pr, hex	hex 10.19^{22} am 9.65^{21}	717 ± 5	−2O, 875	i	i	s a; i alk
t94	palmitate	$TlC_{16}H_{31}O_2$	459.80	crn need		115—117		0.01^{15}	0.07^{60}	1.04^{45} al

No.	Name	Synonyms and Formulae	Mol. wt.	Crystalline form, properties and index of refraction	Density or spec. gravity	Melting point, °C	Boiling point, °C	Solubility, in grams per 100 cc		
								Cold water	Hot water	Other solvents
t95	phenoxide	$TlOC_6H_5$	297.49	wh cr.		233—5		d		s hot bz; sl s lgr
t96	*ortho*phosphate	Tl_3PO_4	708.12	col, need	6.89^{10}			0.5^{15}	0.67^{100}	i al, s NH_4 salts
t97	*ortho*phosphate, (di)-β	TlH_2PO_4	301.37	monocl.	4.726	*ca*190		sl s	sl s	i al
t98	*pyro*phosphate	Tl_4Pli2O_7	991.48	monocl pr	6.786^{20}	>120		40		
t99	picrate	$TlC_4H_2N_3O_7$	432.48	red, monocl or yel tricl	red 3.164^{17} yel 2.993^{17}	expl 723—725		0.135^0	2.43^{70}	0.40 CH_2OH
t100	rhodanide	TlCNS	262.46	glossy leaflets, rhomb, tetr cr	4.954^{30}_4	d low temp		0.393^{25}		0.024^0 liq SO_2; i acet; a MeOH
t101	selenate	Tl_2SeO_4	551.72	rhomb need, 1.949, 1.959, 1.964		>400		2.13^{10}	8.5^{50}	i al, eth
t102	selenide	Tl_2Se	487.73	gray leaf.	9.05^{25}_4	340		i		s a; i acet a
t103	silver nitrate	$TlNO_2·AgNO_3$	436.26	wh cr powd		75		s		
t104	stearate	$TlC_{18}H_{35}O_2$	487.86	need		119		0.005^{15}	0.095^{75}	0.18^{15} al, 0.060^{50} al
t105	(I) sulfate	Tl_2SO_4	504.82	col, rhomb. 1.860, 1.867, 1.885	6.77	632	d	4.87^{30}	19.14^{100}	
t106	(I) sulfate hydrogen	$TlHSO_4$	301.45	pr need		120 d				v sl s dil H_2SO_4
t107	(III) sulfate	$Tl_2(SO_4)_3·7H_2O$	823.05	col leaf		$-6H_2O$, 220		d	d	s dil H_2SO_4
t108	(I) sulfide	Tl_2S	440.83	bl-blk tetr	8.46	448.5	d	0.02^{20}	sl s	s a; i alk, acet
t109	(III) sulfide	Tl_2S_3	504.95	blk, amorph.		260 (in N_2)	d	i	i	s hot H_2SO_4
t110	sulfite	Tl_2SO_3	488.82	wh cr.	6.427			3.34^{15}	v s	i al
t111	tartrate(dl)	$Tl_2C_4H_4O_6$	556.84	monocl.	4.659	d 165		13.3^{15}		
t112	*meta*tellurate	Tl_2TeO_4	600.36	heavy wh ppt	$6.760^{17.6}$	red heat		sl s	sl s	
t113	thiocyanate	TlSCN	262.46	col, tetr	4.956^{30}			0.315^{30}	0.727^{40}	i al
t114	*di*thionate	$Tl_2S_2O_4$	568.88	monocl.	5.57^{20}_4	d		41.8^{19}		
t115	thiosulfate	$Tl_2S_2O_3$	520.88	wh rhomb cr		d 130		sl s	v s	
t116	*meta*vanadate	$TlVO_3$	303.32	gray cr.	6.09^{17}	424		0.87^{11}	0.21^{100}	
t117	*pyro*vanadate	$Tl_4V_2O_7$	1031.41	light yel	8.21^{19}	454		0.2^{14}	0.26^{100}	
t118	**Thiocarbonyl chloride**	Thiophoagene. $CSCl_2$	114.98	red yel liq, 1.5442	1.509^{15}		73.5	d		d al; s eth
t119	**Thiocarbonyl chloride, tetra-**	$CSCl_4$	185.88	yel	1.712^{13}		146—147		d	
t120	**Thiocyanic acid(iso)**	HSCN(HNCS)	59.09	col mass or gas		> −110	polym to solid −90	v s		v s al, eth, bz
t121	**Thiocyanogen**	$(SCN)_2$	116.16	liq. or yel solid		−2 to −3		d		s al, eth, CS_2, CCl_4
t122	**Thionyl** bromide	$SOBr_2$	207.87	or, yel liq	2.68^{18}	−52	138^{773}, 68^{40}	d	d	s bz, chl, CS_2, CCl_4
t123	chloride	$SOCl_2$	118.97	col, or yel liq; 1.527^{10}	1.655^{10}_4	−105	78.8^{746} d 140	d	d	d a, al, alk; s bz, chl
s124	chloride fluoride	SOClF	102.51	gas		−139.5	12.2			
t125	fluoride	SOF_2	86.06	col, gas	gas 2.93 g/ℓ liq 1.780^{-100}	−110.5	−43.8	d	d	s eth, bz, chl, acet, $AsCl_3$
t126	**Thiophosphoramide**	Thiophosphorylamide. $PS(NH_2)_3$	111.10	yel wh cr	1.7^{13}	d 200		14.3^{25}	d	sl, me
t127	**Thiophosphoryl** bromide	$PSBr_3$	302.75	yel, cub	2.85^{17}	37.8	125^{25}	d		s eth, CS_2, PCl_3
t128	thiophosphoryl bromide, hydrate	$PSBr_3·H_2O$	320.76	yel cr.	2.794^{18}	35				
t129	thiophosphoryl bromide (mono-) chloride, di-	$PSBrCl_2$	213.84	yel liq	2.12^0	−30	d 150	d		
t130	thiophosphoryl bromide (di-) chloride	$PSBr_2Cl$	258.29	pa grn fum liq	2.48^0	−60	95^{60}	d		
t131	thiophosphoryl chloride	$PSCl_3$	169.39	col liq, 1.563 (c)	1.635	−35	125	d		s bz, CS_2, CCl_4
t132	thiophosphoryl fluoride	PSF_3	120.03	gas		$3.8^{7.6atm}$	d	sl s d		s eth; i bz, CS_2
t133	**Thiosulfuric acid**	$H_2S_2O_3$	114.13	in sol only				s		
t134	**Thorium**	Th	232.0381	gray, cub radioactive	11.7			i	i	s HCl, H_2SO_4, aq reg; sl s HNO_3
t135	boride, hexa-	ThB_6	296.90	dk viol-blk met, cub	6.4^{15}	2195		i	i	s HNO_3; i H_2SO_4, HCl, HF, aq alk
t136	boride, tetra	ThB_4	275.28	tetr pr	7.5^{15}			i	i	s HNO_3 HCl, hot H_2SO_4
t137	bromide	$ThBr_4$	551.65	col cr, hygr	5.67	subl 610	725	s	s	
t138	carbide	ThC_2	256.06	yel, tetr	8.96^{18}	2655 ± 25	*ca* 5000 (?)	d		v sl s conc a
t139	carbonate	$Th(CO_3)_2$	352.06	exist?.				i	d	s conc Na_2CO_3
t140	chloride	$ThCl_4$	373.85	wh, rhomb, deliq	4.59	770 ± 2 subl 820	d 928	v s	v s	s al, a, KCl; sl s eth
t142	*tetra*cyanoplatinate	$Th[Pt(CN)_4]_2·16H_2O$	1118.58	yel-grn, rhomb	2.460			sl s	s	
t143	fluoride	ThF_4	308.03	wh cub powd	6.32^{24}	>900				sl d dil H_2SO_4, HCl; i conc H_2SO_4
t144	fluoride	$ThF_4·4H_2O$	380.09	cr		$-H_2O$, 100	$-2H_2O$, 140 — 200	0.017^{25}		i HF
t145	hydroxide	$Th(OH)_4$	300.07	wh gelat		d		i	i	s a; i alk, HF
t146	iodate	$Th(IO_3)_4$	931.65					i	id	s dil H_2SO_4; i dil HNO_3
t147	iodide, tetra-	ThI_4	739.66	yel		566	839	s		
t148	nitrate	$Th(NO_3)_4$	480.06	plates, deliq.		d 500		v s		s al
t149	nitrate	$Th(NO_3)_4·4H_2O$	552.12	col cr.		swells		v s		v s al; sl s acet; 36.9 eth
t150	nitrate	$Th(NO_3)_4·12H_2O$	696.24	col leaf, deliq		d		v s		v s al, a
t151	nitride	Th_3N_4	752.14	dk brn powd, or blk cr.				sl d	d	s HCl
t152	oxalate	$Th(C_2O_4)_2$	408.08	wh cr.	4.637^{16}	d		0.0017^{17}	0.0017^{50}	s h aq $(NH_4)_2C_2O_4$; sl s a

PHYSICAL CONSTANTS OF INORGANIC COMPOUNDS (continued)

No.	Name	Synonyms and Formulae	Mol. wt.	Crystalline form, properties and index of refraction	Density or spec. gravity	Melting point, °C	Boiling point, °C	Solubility, in grams per 100 cc		
								Cold water	Hot water	Other solvents
t153	oxalate	$Th(C_2O_4)_2·6H_2O$	516.17	wh amorph powd				i		s Na_2CO_3, $(NH_4)_2C_2O_4$ sol; i HNO_3
t154	oxide, di-	Thorianite. ThO_2	264.04	wh cub, 2.20 (liq)	9.86	3220 ± 50	4400	i	i	s hot H_2SO_4; i dil a, alk
t155	oxysulfide	ThOS	280.10	yel cr.	6.44^{0}	d		i		s aq reg; sl s HNO_3
t156	2,4-pentanedione	Thorium acetylacetonate. $Th(C_5H_7O_2)_4$	628.48	col cr.		171 subl 160^{10}	260 — 270^{10}	sl s		v s al, chl; s eth
t157	*hypo*phosphate	$ThP_2O_6·11H_2O$	588.15	wh amorph ppt		$-11H_2O$, 160		i	i	i a, alk
t158	*meta*phosphate	$Th(PO_3)_4$	547.93	col rhomb pr	$4.08^{16.4}$					
t159	*ortho*phosphate	$Th_3(PO_4)_4·4H_2O$	1148.06	wh gelat				i	i	s 30^0 HCl; i a
t160	picrate	$Th(C_6H_2N_3O_7)_4·10H_2O$	1324.58					0.305^{25}		
t161	selenate	$Th(SeO_4)_2·9H_2O$	680.09	col, monocl	3.026	$-8H_2O$, 200	d 1500	0.5^0	2.0^{100}	
t162	*ortho*silicate	Thorite $ThSiO_4$	324.12	col, tetr, 1.80, 1.81	6.82^{16}			v sl s		i a
t163	silicide	$ThSi_2$	288.21	blk, tetr	7.96^{16}					s hot HCl; sl s H_2SO_4
t164	sulfate	$Th(SO_4)_2$	424.15	wh cr, hygr	4.225^{17}			s	s	i a; v s $NH_4C_3H_3O_2$
t165	sulfate	$Th(SO_4)_2·4H_2O$	496.21	wh need, or cr powd		$-4H_2O$, 400		9.41^{17} (anhydr)	2.54^{50} (anhydr)	i a
t166	sulfate	$Th(SO_4)_2·6H_2O$	532.24					1.63^{15}	6.64^{60}	i a
t167	sulfate	$Th(SO_4)_2·8H_2O$	568.28	monocl, prism, 1.5168		$-4H_2O$, 42		1.88^{25}	3.71^{44}	i a
t168	sulfate	$Th(SO_4)_2·9H_2O$	586.29	wh monocl	2.77	$-9H_2O$, 400		1.57^{20}	6.67^{55}	i a
t169	sulfide	ThS_2	296.16	dk brn-blk cr	7.30^{25}_{4}	1925 ± 50 (vac)		i	d 200	s hot aq reg; sl s a
t170	*pyro*vanadate	$ThV_2O_7·6H_2O$	554.01	yel				i	i	s conc a
t171	**Thulium**	Tm	168.9342	silv wh met, hex	9.3208	1545	1947	i	i	
t172	bromide	$TmBr_3$	408.65			952	1440	s	s	
t173	chloride	$TmCl_3·7H_2O$	401.40	grn cr, deliq		824	1440	v s		v s al
t174	fluoride	TmF_3	225.93			1158	>2200	i	i	s dil a
t175	iodide	TmI_3	549.65	brt yel cr		1015	1260	s	s	
t176	oxalate	$Tm_2(C_2O_4)_3·6H_2O$	710.02	grn-wh ppt		$-H_2O$, 50		i		s alk oxal sol; i dil a
t177	oxide	Thulia. Tm_2O_3	385.87	grn-wh powd						sl s min a
t178	**Tin** gray	Sn	118.69 ± 3	gray, cub	5.75	231.9681	2270	i	i	s HCl, H_2SO_4, aq reg, alk; sl s dil HNO_3
t179	**Tin** white	Sn	118.69 ± 3	wh met, tetr	7.28	231.88 stable 13.2 — 161	2260	i	i	s HCl, H_2SO_4, aq reg, alk; sl s dil HNO_3
t180	**Tin** brittle	Sn	118.69 ± 3	wh, rhomb	6.52 — 56	231.89 stable > 161	2260	i	i	s HCl, H_2SO_4, aq reg, alk; sl s dil HNO_3
t181	(II) acetate	$Sn(C_2H_3O_2)_2$	236.78	yelsh powd		182	240	d		s dil HCl
t182	*pyro*arsenate	$Sn_2As_2O_7$	499.22	flocculent ppt		d As_2O_3 + SnO_2		i	i	i conc ac a
t183	(II) bromide	$SnBr_2$	278.50	pa yel, rhomb	5.117^{17}	215.5	620	85.2^0	222.5^{100}	s al, eth, acet
t184	(IV) bromide	$SnBr_4$	438.31	col, rhomb pyr, deliq	liq 3.34^{35}	31	202^{734}	s d	d	s acet, PCl_3, $AsBr_3$
t185	bromide chloride (tri-)	$SnBrCl_3$	304.95	col liq	2.51^{13}	− 31	50^{30}			
t186	bromide (di-) chloride (di-)	$SnBr_2Cl_2$	349.40		2.82^{13}	− 20	65^{30} d 191	d	d	
t187	bromide (tri-) chloride	$SnBr_2Cl$	393.86	liq	3.12^{13}	1	73^{30}			
t188	bromide (di-) iodide (di-)	$SnBr_2I_2$	532.31	or-red, hex pl.	3.631^{15}	50	225	s	d < 80	
t189	(II) chloride	$SnCl_2$	189.60	wh, rhomb	3.95^{25}_{4}	246	652	83.9^0	259.8^{15} d	s al, eth, acet, et acet, me acet, pyr
t190	(II) chloride dihydrate	$SnCl_2·2H_2O$	225.63	wh, monocl	$2.710^{15.5}$	37.7	d	d	d	s al, eth, acet, glac ac a
t191	(IV) chloride	$SnCl_4$	260.50	col liq, solid cub, 1.512	liq 2.226	− 33	114.1	s	d	s eth
t192	(IV) chloride pentahydrate	$SnCl_4·5H_2O$	350.58	monocl cr		stable 19 — 56		s		
t193	(IV) chloride tetrahydrate	$SnCl_4·4H_2O$	332.56	opaque			stable 56 — 83	s		
t194	(IV) chloride trihydrate	$SnCl_4·3H_2O$	314.55	col, monocl cr		80	stable 64 — 83	s		
t195	(IV) chloride *di*ammine	$SnCl_4·2NH_3$	294.56	cr				s		d HCl
t196	chloride (tri-) bromide	$SnCl_3Br$	304.95	col liq	2.51^{13}	− 31	50^{30}			
t197	chloride (di-) iodide (di-)	$SnCl_2I_2$	443.40	red mobile liq	3.287^{15}		297	s	d	s chl, bz CS_2
t198	(IV) chloride nitrosyl-chloride	$SnCl_4·2NOCl$	391.42	pa yel, oct cr	2.60	180		d	d	
t199	(IV) chromate	$Sn(CrO_4)_2$	350.68	br yel cr powd		d		s		
t200	(II) ferricyanide	$Sn_3[Fe(CN)_6]_2$	779.98	wh		d		i	i	s HCl
t201	(II) ferrocyanide	$Sn_2Fe(CN)_6$	449.33	wh gel				i	i	d HCl
t202	(IV) ferrocyanide	$SnFe(CN)_6$	330.64					i	i	d h HCl
t203	(II) fluoride	Fluoristan. SnF_2	156.69	wh, monocl cr				s		
t204	(IV) fluoride	SnF_4	194.68	wh, monocl cr, hygr.	4.780^{19}	705 subl		v s	d	
205	hydride	Stannane. SnH_4	122.72	gas		d − 150	− 52			s $AgNO_3$, $HgCl_2$, conc alk, conc H_2SO_4

PHYSICAL CONSTANTS OF INORGANIC COMPOUNDS (continued)

No.	Name	Synonyms and Formulae	Mol. wt.	Crystalline form, properties and index of refraction	Density or spec. gravity	Melting point, °C	Boiling point, °C	Solubility, in grams per 100 cc: Cold water	Hot water	Other solvents
t206	iodide	SnI_2	372.50	yelsh-red to red monocl need	5.285	320	717	0.98^{30}	4.03^{100}	v s NH_4OH, HI soln
t207	(IV) iodide	SnI_4	626.31	or red cub, 2.106	4.473^{0}	144.5	364.5	s	d	141.1^{25} CS_2; 6.03^{15} CCl_4; 17.88^{25} bz
t208	(II) nitrate	$Sn(NO_3)_2{\cdot}20H_2O$	603.01	col leaf		−20		d	d	d HNO_3
t209	(II) nitrate, basic	$SnO{\cdot}Sn(NO_3)_2$	377.39	wh cr mass		d > 100 expl		d	d	
t210	(IV) nitrate	$Sn(NO_3)_4$	366.71	silky need		d 50		d		
t211	(II) oxide, mon-	SnO	134.69	blk, cub (tetr)	6.446^{0}	d 1080^{600}		i	i	s a, alk; sl s NH_4Cl
t212	oxide, mon-hydrate	$SnO{\cdot}xH_2O$		wh powd or yellow-brn cr					d to SnO	d a; alk; s alk carb; i NH_4OH
t213	(IV) oxide, di-	Nat. cassiterite. SnO_2	150.69	wh, tetr, (also hex or rhomb), 1.997, 2.093	6.95	1630	subl 1800 — 1900	i	i	d KOH, NaOH; i aq reg
t214	oxide, di-hydrate	α-Stannic acid or "ordinary" stannic acid. $SnO_2{\cdot}xH_2O$		amorph or gel				i	i	s a, alk, K_2CO_3
t215	oxide, di-hydrate	β-Stannic acid or "meta" stannic acid. $SnO_2{\cdot}xH_2O$		wh, amorph or gel				i	i	i a, K_2Co_3; sol alk
t216	(II) *meta*phosphate	$Sn(PO_3)_2$	276.63	amorph mass	$3.380^{22.8}$					
t217	(II) *ortho*phosphate	$Sn_3(PO_4)_2$	546.01	wh, amorph	3.823^{17}			i	i	d a, alk
t218	(II) *ortho*phosphate, di-H	$Sn(H_2PO_4)_2$	312.66	wh, rhomb cr	$3.167^{22.8}$	d	d		d	
t219	(II) *ortho*phosphate, mono-H	$SnHPO_4$	214.67	cr	$3.476^{15.5}$	stabl > 100	d	i	i	s dil min a
t220	(II) *pyro*phosphate	$Sn_2P_2O_7$	411.32	amorph powd	$4.009^{16.4}$					s conc a
t221	phosphide, mono-	SnP	149.66	silv wh	6.56	d	d	i	i	s HCl; i HNO_3
t222	phosphide, tri-	SnP_3	211.61	cr	4.10^{0}	<415 d to Sn_4P_3		i	i	d HNO_3; i HCl
t223	*tetra*phosphide, tri-	Sn_4P_3	567.68	wh cr	5.181	d < 480		i	i	d fixed alk hydr, HCl
t224	phosphorus chloride	$SnCl_4{\cdot}PCl_5$	468.74	col cr		subl 200		d	d	
t225	(II) selenide	SnSe	197.65	steelgray cr	6.179^{0}	861		i	i	d HCl, HNO_3, aq reg, alk sulf
t226	(II) sulfate	$SnSO_4$	214.75	wh-yelsh cr powd		>360 (SO_2)		33^{25}		s H_2SO_4
t227	(IV) sulfate	$Sn(SO_4)_2{\cdot}2H_2O$	346.84	wh, hex pr, deliq				v s	d	s eth, dil H_2SO_4, HCl
t228	(II) sulfide	SnS	150.75	gray-blk cub, monocl	5.22^{25}	882	1230	0.000002^{18}		d HCl, alk, $(NH_4)_2S$
t229	(IV) sulfide	Mosaic gold. SnS_2	182.81	gold yel, hex	4.5	d 600		0.0002^{18}		d alk sulf, aq reg, alk hydr, PCl_5, $SnCl_2$; i a
t230	(IV) sulfur chloride	$SnCl_4{\cdot}2SCl_4$	608.25	yel cr		37	d < 40	d	d	s eth, bz, CS_2, ethyl acet; d HNO_3
t231	tartrate	$SnC_4H_4O_6$	266.76	heavy wh powd				s		v s dil HCl
t232	(II) telluride	SnTe	246.29	gray cr	6.48	780	d	i	i	d alk sulf
t233	(IV) telluride	$SnTe_2$	373.89	blk, flocc ppt				i	i	d dil a, alk
t234	**Titanic acid, ortho-**	α-Titanic acid. H_2TiO_4	113.89	wh		d		v sl s d		s dil HCl, dil H_2SO_4, conc alk
t235	**Titanium**	Ti	47.88 ± 3	α hex, tr β cub 838, silv gray	4.5^{20}	1660 ± 10	3287	i	i	s dil a
t236	boride, di-	TiB_2	69.50	hex	4.50	2900				
t237	bromide, di-	$TiBr_2$	207.69	blk powd	4.31	d > 500		s ev H_2		
t238	bromide, tetra-	$TiBr_4$	367.50	or yel, deliq	2.6	39	230	d		s abs al, abs eth
t239	bromide, tri-	$TiBr_3{\cdot}6H_2O$	395.68	redsh-viol or dk blue cr, deliq		115	d 400	v s		v s al, acet
t240	carbide	TiC	59.89	gr met, cub	4.93	3140 ± 90	4820	i	i	s aq reg, HNO_3
t241	chloride, di-	$TiCl_2$	118.79	lt br-blk, hex, deliq	3.13	subl H_2	d 475 vac	d		s al, i eth, chl, CS_2
t242	chloride, tetra-	$TiCl_4$	189.69	lt yel liq, $1.61^{10.5}$	liq 1.726 sol 2.06^{-79}	−25	136.4	s	d	s dil HCl, al
t243	chloride, tri-	$TiCl_3$	154.24	dk viol, deliq	2.64	d 440	660^{108}	s	s	v s al; s HCl; i eth
t244	fluoride, tetra	TiF_4	123.87	wh powd, hygr	$2.798^{20.5}$	>400 (pressure)	284 (subl)	s d		s H_2SO_4, al, C_6H_5N; i eth
t245	fluoride, tri-	TiF_3	104.88	purp-red or vlt	3.40	1200	1400	red s vlt i		
t246	hydride	TiH_2	49.90	gray powd	3.9^{12}	d 400				
t247	iodide, di-	TiI_2	301.69	blk, hygr	4.99	600	1000	d		d alk; s conc HF, conc HCl
t248	iodide, tetra-	TiI_4	555.50	red, cub	4.3	150	377.1	v s	d	
t249	nitride	TiN	61.89	yel-bronze, cub	5.22	2930		i	i	sl s hot aq reg + HF
t250	oxalate	$Ti_2(C_2O_4)_3{\cdot}10H_2O$	539.97	yel pr				s	s	i al, eth
t251	oxide, di-	Nat. brookite. TiO_2	79.88	wh, rhomb, 2.583, 2.586, 2.741	4.17	1825		i	i	s H_2SO_4, alk; i a
t252	oxide, di-	Nat. octahedrite, anatase. TiO_2	79.88	br-blk, tetr, 2.554, 2.493	3.84			i	i	s H_2SO_4, alk; i a
t253	oxide, di-	Nat. rutile. TiO_2	79.88	col, tetr, 2.616, 2.903	4.26	1830 — 1850	2500 — 3000	i	i	s H_2SO_4, alk; i a
t254	oxide, mon-	TiO	63.85	yel blk, pr.	4.93	1750	>3000			s dil H_2SO_4; i HNO_3
t255	oxide, sesqui-	Ti_2O_3	143.76	vlt blk, trig	4.6	2130 d		i	i	s H_2SO_4; i HCl, HNO_3
t256	phosphide	TiP	78.85	gray, met	3.95^{25}			i	i	i a
t257	sulfate	$Ti_2(SO_4)_3$	383.93	green powd				i	i	s dil a; i al, eth, conc H_2SO_4

No.	Name	Synonyms and Formulae	Mol. wt.	Crystalline form, properties and index of refraction	Density or spec. gravity	Melting point, °C	Boiling point, °C	Solubility, in grams per 100 cc Cold water	Hot water	Other solvents
t258	sulfate, basic	$TiOSO_4$	159.94	wh or sl yelsh powd, 1.80 — 1.89				d		
t259	sulfide, di-	TiS_2	112.00	yel sc	3.22^{20}			hyd sl	d in steam	d HCl; s dil HNO_3, H_2SO_4
t260	sulfide, mono-	TiS	79.94	redsh solid	4.12			i		s conc H_2SO_4; i HCl, HF, dil H_2SO_4
t261	sulfide, sesqui	Ti_2S_3	191.94	grayish-blk cr.	3.584			i	i	s conc H_2SO_4, conc HNO_3; i dil H_2SO_4, dil HCl
t262	**Tungsten**	Wolfram. W	183.85 ± 3	gray-blk, cub	19.35^{20}_{4}	3410 ± 20	5660	i	i	v sl s HNO_3, H_2SO_4, aq reg; s HNO_3 + HF, fus NaOH + $NaNO_3$; i HF, KOH
t263	arsenide	WAs_2	333.69	blk cr	6.9^{18}	d red heat		i		d hot HNO_3, hot H_2SO_4
t264	boride, di-	WB_2	205.47	silvery, oct	10.77	*ca* 2900		i	i	s aq reg
t265	bromide, di-	WBr_2	343.66	bl-blk need		d 400		d		
t266	bromide, penta-	WBr_5	583.37	vlt-brn need, hygr		276	333	d		s abs al, chl, eth, alk
t267	bromide, hexa-	WBr_6	663.27	bl-blk, need.	6.9	232		i	d	s abs a, eth, CS_2, NH_4OH
t268	carbide	WC	195.86	blk, hex	15.63^{18}	2870 ± 50	6000	i		s HNO_3 + HF, aq reg
t269	(di-)carbide	W_2C	379.71	blk, hex	17.15	2860	6000	i		s HNO_3 + HCl
t270	carbonyl	$W(CO)_6$	351.91	col, rhomb cr.	2.65	d ~ 150	175^{766}	i	i	s fum HNO_3; v sl
t271	chloride, di-	WCl_2	254.76	gray, amorph	5.436			d		s al, eth, bz
t272	chloride, hexa-	WCl_6	396.57	dk bl, cub	3.52^{25}_{4}	275	346.7		d^{60}	s al, eth, bz, CCl_4; v s CS_2, POCl
t273	chloride, penta-	WCl_5	361.12	blk, deliq	3.875^{25}_{4}	248	275.6		d to W_2O_5	v sl s CS_2
t274	chloride, tetra-	WCl_4	325.66	gray, deliq	4.624^{25}_{4}	d		d		
t275	fluoride, hexa-	WF_6	297.84	col gas, or lt yel liq.	liq 3.44 gas 12.9 g/ℓ	2.5^{420}	17.5	d	d	s alk
t276	iodide, di-	WI_2	437.66	br-gr, amorph	6.799^{25}_{4}	d		i	d	s alk; i al CS_2
t277	iodide, tetra-	WI_4	691.47	blk, cr	5.2^{18}	d		i	d	s abs al; i eth, chl, turp
t278	nitride, di-	WN_2	211.86	brn, cub		above 400 (vac)		d	d	
t279	oxide, di-	WO_2	215.85	br, cub.	12.11	1500 — 1600 (in N_2)	*ca* 1430 subl 800	i	i	s a, KOH
t280	oxide, pent-	Mineral blue. W_2O_5 or W_4O_{11}	447.70 or 911.39	blue-vlt, tricl		subl 800 — 900	*ca* 1530 d 2000	i	i	i a
t281	oxide, tri-	WO_3	231.85	yel, rhomb, or yel-or powd	7.16	1473		i	i	s hot alk; sl s HF; i a
t282	oxy*di*bromide, di	WO_2Br_2	375.66	red, prism		d				
t283	oxy*tetra*bromide	$WOBr_4$	519.47	blk, deliq		277	327	d	d	
t284	oxy*tetra*chloride	$WOCl_4$	341.66	red, need		211	227.5	d	d	s CS_2, S_2Cl_2, bz
t285	oxy*di*chloride, di-	WO_2Cl_2	286.75	lt yel tabl		266		s	d	i al; s NH_4OH, alk
t286	oxy*tetra*fluoride-	WOF_4	275.84	col pl, hygr		110	187.5	d		sl s CS_2; i CCl_4
t287	phosphide	WP	214.82	gray, prism	8.5			i		s HNO_3 + HF; i alk, HCl
t288	phosphide	WP_2	245.80	blk cr	5.8	d		i	i	s HNO_3 + HF, aq reg; i al, eth
t289	phosphide	W_2P	398.67	dk gray prism	5.21	d				s fus Na_2CO_3 + $NaNO_3$; i a, aq reg
t290	silicide	WSi_2	240.02	blue, gray, tetrag	9.4	above 900		i	i	s HNO_3 + HF; i aq reg
t291	sulfide, di-	Nat. tungstenite. WS_2	247.97	dk gray, hexag	7.5^{10}	d 1250		i		s HNO_3 + HF, fus alk; i al
t292	sulfide, tri-	WS_3	280.03	choc brn powd				sl s	s	s alk
t293	**Tungstic acid,** meta-	$H_2W_4O_{13}·9H_2O$	1107.55	col, tetrag	3.93	d 50		88.57^{22}	$111.87^{43.5}$	$110.76^{24.3}$ eth; s al
t294	**Tungstic acid,** ortho-	H_2WO_4	249.86	yel powd, 2.24	5.5	$-H_2O$, 100	1473	i	sl s	s alk, HF, NH_3; i most a
t295	**Tungstic acid,** ortho-	$H_2WO_4·H_2O$	267.88	wh		$H_2W_2O_7$ *at* 100		sl s		s alk
u1	**Uranic acid** meta-	Uranyl hydroxide. H_2UO_4 (or $UO_2(OH)_2$)	304.04	yel, rhomb, or powd	5.926	$-H_2O$ 250 — 300		i	i	s a, alk carb
u2	**Uranium**	U	238.0289	silvery, cubic, radioactive	19.05 ± 0.02^{25}	1132.3 ± 0.8	3818	i	i	s a; i alk, al
u3	boride, di-	UB_2	259.65	hex	12.70	2365				
u4	bromide tetra-	UBr_4	557.64	br leaf, deliq	5.35	516	792^{760}	v s	v s	s d al, MeOH; i bz; s liq HN_3
u5	bromide tri-	UBr_3	477.74	dk brn need, hygr	6.53	730	volat	s		d al
u6	*di*carbide	UC_2	262.05	met cr	11.28^{16}	2350 — 2400	4370^{760}	d	d	i al; d dil inorg a
u7	chloride, penta-	UCl_5	415.29	dk green, gray need, red by trans light, hydr	3.81 (?)	d 300		d		s abs als, a acet, NH_4Cl; d ac a; i bz, eth
u8	chloride, tetra-	UCl_4	379.84	dk grn met, cub oct, hygr	4.87	590 ± 1	792^{760}	v s	s	s al, acet, ac a; i eth, $CHCl_3$
u9	chloride, tri-	UCl_3	344.39	dk red need, hygr	5.44^{25}_{4}	842 ± 5		s	s	s MeOH, acet, glac acet a; i eth

PHYSICAL CONSTANTS OF INORGANIC COMPOUNDS (continued)

No.	Name	Synonyms and Formulae	Mol. wt.	Crystalline form, properties and index of refraction	Density or spec. gravity	Melting point, °C	Boiling point, °C	Solubility, in grams per 100 cc Cold water	Hot water	Other solvents
u10	fluoride, hexa-	UF_6	352.02	col cr, deliq, monocl	4.68^{21}	64.5 — 64.8	56.2^{765}	d		d al, eth; s CCl_4, chl; i CS_2
u11	fluoride, tetra-	UF_4	314.02	green, tricl need	6.70 ± 0.10	960 ± 5		v sl s		i dil a, alk; s conc a, conc alk
u12	fluoride, tri-	UF_3	295.02	blk cr or fused		d above 1000		sl d		v sl s dil inorg a
u13	hydride	UH_3	241.05	blk-brn powd	10.95			i	i	i al, acet, liq NH_3; sl s dil HCl; d HNO_3
u14	hydride	UH_3	241.05	blk powd, cub	11.4					
u15	iodide, tetra-	UI_4	745.65	blk, need	5.6^{15}	506	759	s	s d	
u16	nitride, mono-	UN	252.04	br powd	14.31	*ca* 2630 ± 50				i HCl, H_2SO_4
u17	oxide, di-	UO_2	270.03	br-blk rhomb, or cub	10.96	2878 ± 20		i	i	s HNO_3, conc H_2SO_4
u18	oxide, per-	$UO_4 \cdot 2H_2O$	338.06	pa yel cr, hygr		d 115		0.0006^{20}	0.008^{90}	d HCl
u19	oxide, tri-	Uranyl oxide. UO_3	286.03	yel-red powd	7.29	d		i		s HNO_3, HCl
u20	*tri*-oxide, oct-	U_2O_3	842.08	olive green-blk	8.30	d 1300 to UO_2		i	i	s HNO_3, H_2SO_4
u21	(IV) sulfate	$U(SO_4)_2 \cdot 4H_2O$	502.21	grn, rhomb		$-4H_2O$, 300		23^{11}	9^{63} (anhydr)	s dil a
u22	(IV) sulfate	$U(SO_4)_2 \cdot 8H_2O$	574.27			d 90		11.3^{18}	58.2^{62}	i al; s dil a
u23	(IV) sulfate	$U(SO_4)_2 \cdot 9H_2O$	592.28	grnsh, monocl		$-7H_2O$, 230	$-9H_2O$ red heat			s dil H_2SO_4
u24	sulfide, di-	US_2	302.15	gray-blk, tetr	7.96^{25}	>1100	oxidizes	sl d		s conc HCl; d HNO_3 v s al
u25	sulfide, mono-	US	270.09	blk amorph powd	10.87	above 2000				i HCl, HNO_3
u26	sulfide, sesqui-	U_2S_3	572.24	gray blk, rhomb need		ign				s + O aq reg conc HNO_3; i dil a
u27	**Uranyl acetate**	$UO_2(C_2H_3O_2)_2 \cdot 2H_2O$	424.15	yel, rhomb	2.893^{15}	$-2H_2O$, 110	d 275	7.694^{15}	d	v s al
u28	benzoate	$UO_2(C_7H_5O_2)_2$	512.26	yel powd				sl s		sl s al
u29	bromide	UO_2Br_2	429.84	grn-yel need, hygr				s d		s al, eth
u30	*per*chlorate	$UO_2(ClO_4)_2 \cdot 6H_2O$	577.02	yel cr, deliq, rhomb		90 d 110				
u31	chloride	UO_2Cl_2	340.93	yel, deliq		578	d	320^{18}	v s	s al, amyl al, eth
u32	formate	$UO_2(CHO_2)_2 \cdot H_2O$	378.08	yel, oct	3.695^{19}	$-H_2O$, 110		7.2^{15}		sl s form a; 0.74^{15} MeOH, 2.37^{15} acet
u33	iodate	$UO_2(IO_3)_2$	619.83	yel, rhomb	5.2	d 250		s	s	i HNO_3
u34	iodate	$UO_2(IO_3)_2 \cdot H_2O$	637.85	α prismatic, stable, β pyramidal	α 5.220^{18} β 5.052^{18}			α 0.1049^{18} β 0.1214^{18}		
u35	iodide	UO_2I_2	523.84	red, deliq		d in air				s al, eth, bz
u36	nitrate	$UO_2(NO_3)_2 \cdot 6H_2O$	502.13	yel, rhomb, deliq, 1.4967	2.807^{13}	60.2 d 100	118		∞ 60	v s al, eth, ac a, acet, MeOH
u37	oxalate	$UO_2C_2O_4 \cdot 3H_2O$	412.09	yel cr.		$-H_2O$, 110		0.8^{14}	3.3^{100}	s inorg a, alk, oxal a
u38	phosphate, mono-H	$UO_2HPO_4 \cdot 4H_2O$	438.07	yel pl, tetr.				i	i	s HNO_3, aq Na_2CO_3; i ac a
u39	potassium carbonate	$UO_2CO_3 \cdot 2K_2CO_3$	606.45	yel cr.		$-CO_2$, 300		7.4^{15}	d	i al
u40	sodium carbonate	$UO_2CO_3 \cdot 2Na_2CO_3$	542.01	yel cr.				sl s		i al
u41	sulfate	$UO_2SO_4 \cdot 3H_2O$	420.13	yel-grn cr	$3.28^{16.5}$	d 100		$20.5^{15.5}$	22.2^{100}	24.3^{13} conc H_2SO_4; 30^{13} conc HCl
u42	sulfate	$2(UO_2SO_4) \cdot 7H_2O$	858.28	yel		anh 300		sl s		s H_2SO_3
u43	sulfide	UO_2S	302.09	brn-blk, tetr.		d 40-50		sl s	s d	s dil a, dil al, $(NH_4)_2CO_3$; i abs al
u44	sulfite	$UO_2SO_3 \cdot 4H_2O$	422.15	pa-gr cr						s H_2SO_3
v1	**Vanadic acid,** meta	HVO_3	99.95	yel sc				i		s a, alk,; i NH_4OH
v2	tetra-	$H_2V_4O_{11}$	381.78	br amorph				i		s a, alk, NH_4OH
v3	**Vanadium**	V	50.9415	lt gray met, cub, 3.03	5.96	1890 ± 10	3380	i	i	s aq reg, HNO_3, H_2SO_4, HF; i HCl, alk
v4	boride, di-	VB_2	72.56	hex	5.10					
v5[1]	bromide, tri-	VBr_3	290.65	grn-blk, deliq	4.00^{18}	d		s		s al, eth; i HBr
v5[2]	carbide	VC	62.95	blk. cub	5.77	2810	3900	i		s HNO_3, fus KNO_3; i HCl, H_2SO_4
v6	chloride, di-	VCl_2	121.85	grn, hex, deliq	3.23^{18}			s d	s d	s al, eth
v7	chloride, tetra-	VCl_4	192.75	red-br liq	1.816^{30}	−28 ± 2	148.5^{755}	s d		s abs al, eth, chl, acet a
v8	chloride, tri-	VCl_3	157.30	pink cr, deliq	3.00^{18}	d		s d	s d	s abs al, eth
v9	fluoride, penta.	VF_5	145.93		2.177^{19}		111.2^{758}			s al
v10	fluoride, tetra-	VF_4	126.94	br yel	2.975^{23}	d 325		s		s acet; sl s al, chl
v11	fluoride, tri-	VF_3	107.94	grn, rhomb	3.363^{19}	>800	subl	i		s al, chl, CS_2
v12	fluoride, tri-	$VF_3 \cdot 3H_2O$	161.98	dk gr, rhomb		$-3H_2O$, 100		s	v s d	i abs al
v13	iodide, di-	VI_2	304.75	vlt-rose, hex	5.44	750-800 subl vac		s		i al, CCl_4, CS_2, bz
v14	iodide, tri-	$VI_3 \cdot 6H_2O$	539.75	gr cr, deliq		d		v s		s al
v15	nitride	VN	64.95	blk, cub	6.13	2320		i		sl s aq reg
v16	oxide	VO (or V_2O_2)	66.94	lt gray cr	5.758^{14}	ign		i	i	s a
v17	oxide, di- (or tetr)-	VO_2 (or V_2O_4)	82.94	bl cr	4.339	1967		i	i	s a, alk
v18	oxide, pent-	V_2O_5	181.88	yel-red, rhomb, 1.46, 1.52, 1.76	3.357^{18}	690	d 1750	0.8^{20}		s a, alk, i abs al
v19	oxide, sesqui	Vanadium trioxide. V_2O_3	149.88	blk cr	4.87^{18}_{4}	1970		sl s	s	s HNO_3, HF, alk
v20	oxybromide	VOBr	146.84	vlt, oct.	4.00^{18}	d 480		v sl s		s acet, anhydr eth, acet
v21	oxy di-bromide	$VOBr_2$	226.75	br powd, deliq		d 180		s		

PHYSICAL CONSTANTS OF INORGANIC COMPOUNDS (continued)

No.	Name	Synonyms and Formulae	Mol. wt.	Crystalline form, properties and index of refraction	Density or spec. gravity	Melting point, °C	Boiling point, °C	Solubility, in grams per 100 cc		
								Cold water	Hot water	Other solvents
v22	oxy*tri*bromide	$VOBr_3$	306.65	red liq	$2.933^{14.5}$	d 180	130^{100}	s		
v23	oxychloride	$VOCl$	102.39	yel brn powd	$2.824, 3.64^{20}$		127	i		v s HNO_3
v24	oxy*di*chloride	$VOCl_2$	137.85	grn, deliq	2.88^{13}			d		s dil HNO_3
v25	oxy*tri*chloride	$VOCl_3$	173.30	yel liq	1.829	−77 ± 2	126.7	s d		s al, eth, ac a
v26	oxy*di*fluoride	VOF_2	104.94	yel	3.396^{19}	d				sl s acet
v27	oxy*tri*fluoride	VOF_3	123.94	yel-wh, hygr	2.459^{19}	300	480			
v28	silicide, di-	VSi_2	107.11	met pr	4.42			i	i	s HF; i al, eth, a
v29	(*dl*-)silicide	V_2Si	129.97	silv wh pr	5.48^{17}			i	i	s HF; i al, eth, a
v30	sulfate (hypovanadous)	$VSO_4 \cdot 7H_2O$	273.11	vit, monocl		d in air				
v31	sulfide, mono- or (di-)	VS (or V_2S_2)	83.00	blk pl (exist ?)	4.20	d				s hot H_2SO_4, HNO_3; sl s KSH; i HCl, alk
v32	sulfide, penta-	V_2S_5	262.18	blk-grn powd	3.0	d		i		s HNO_3, alk sulf, alk
v33	sulfide, sesqui- or (tri-)	V_2S_3	198.06	grn-blk pl, or powd	4.72^{21}	d>600		i		s alk sulf; sl s, alk, HCl, HNO_3. H_2SO_4
v34	**Vanadyl sulfate**	$VOSO_4$	163.00	bl				v s		
w1	**Water**	H_2O	18.01528	col liq, or col hex cr	liq 1.000_4 sld 0.9168^0	0.00	100.00			s al
w2	**Water heavy**	Deuterium oxide. D_2O	20.0312	col liq or hex cr, 1.33844^{20}	1.105_4^{20}	3.82	101.42	∞	∞	∞ al; sl s eth
w3	**Wolfram**	See tungsten								
x1	**Xenon**	Xe	131.29 ± 3	col inert gas	gas 5.887 g/ℓ ± 0.009 liq 3.52^{-100} solid 2.7^{-140}	−111.9	−107.1 ± 3	24.1° cm^3, 8.4^{50}, 11.9^{25} cm^3 7.12^{00}		
y1	**Ytterbium**	Yb	173.04 ± 3	cub	6.9654 up to 789 6.54 above 789	819 ± 5	1194	i		s a
y2	(III) acetate	$Yb(C_2H_3O_2)_3 \cdot 4H_2O$	422.23	hex pl	2.09	$-4H_2O$, 100		v s	v s	
y3	(II) bromide	$YbBr_2$	332.85		5.91_4^{25}	677	1800	s	s	s dil a
y4	(III) bormide	$YbBr_3$	412.75	col cr.		956	d	s	s	
y5	(II) chloride	$YbCl_2$	243.95	grn-yel cr	5.08	702	1900	s	s	s dil a
y6	(III) chloride	$YbCl_3 \cdot 6H_2O$	387.49	grn, rhomb cr, deliq	2.575	865 $-6H_2O$, 180		v s	v s	s abs al
y7	(II) fluoride	YbF_2	211.04			1052	2380	i	i	
y8	fluoride	YbF_3	230.04			1157	2200	i	i	i dil a
y9	(II) iodide	YbI_2	426.85	lt yel, hex cr	5.40_4^{25}	780 ± 4	1300 d(700) vac	s	s	s dil a
y10	(III) iodide	YbI_3	553.75	gold yel cr		d 700	d	s	s	s dil a
y11	(III) oxalate	$Yb_3(C_2O_4)_3 \cdot 10H_2O$	790.29	col cr.	2.644			0.00033^{25}		sl s dil a
y12	(III) oxide	Ytterbia. Yb_2O_3	394.08	col	9.17			i	i	s h dil a
y13	(III) selenate	$Yb_2(SeO_4)_3 \cdot 8H_2O$	919.08	hex pl	3.30			s d	s	
y14	(III) selenite	$Yb_2(SeO_3)_3$	726.95					i		
y15	(III) sulfate	$Yb_2(SO_4)_3$	634.25	col cr.	3.793	d 900		44.2^0	4.7^{100}	
y16	(III) sulfate, octohydrate	$Yb_2(SO_4)_3 \cdot 8H_2O$	778.38	prism.	3.286			35.9^{25}	21.1^{40}	
y17	**Yttrium**	Y	88.9059	gray-blk met, hex	4.4689	1522	3338	sl d	d	v s dil a; s h KOH
y18	acetate	$Y(C_2H_3O_2)_3 \cdot 4H_2O$	338.10	col, tricl					9.03^{25}	
y19	bromate	$Y(BrO_3)_3 \cdot 9H_2O$	634.75	hex pr		74	$-6H_2O$, 100	168^{25}		sl s al; i eth
y20	bromide	YBr_3	328.62	deliq		904		v s		s al; i eth
y21	bromide hydrate	$YBr_3 \cdot 9H_2O$	490.76	col tabl, deliq				v s		sl s al; i eth
y22	carbide	YC_2	112.93	yel., microcr	4.13^{18}			d		
y23	carbonate	$Y_2(CO_3)_3 \cdot 3H_2O$	411.89	wh-redsh powd.						s dil min a, $(NH_4)_2CO_3$; sl s aq CO_2; i al, eth
y24	chloride	YCl_3	195.26	shiny wh leaf	2.67	721	1507	78^{10}	82^{50}	60.1^{15} al; 60.6^{15} pyr
y25	chloride, hexahydrate	$YCl_3 \cdot 6H_2O$	303.36	redsh-wh, rhomb, deliq	2.18^{18}	$-5H_2O$, 100		217^{20}	235^{50}	s al; i eth
y26	chloride, monohydrate	$YCl_3 \cdot H_2O$	213.28	col cr.		$-H_2O$, 160		v s		
y27	fluoride	YF_3	145.90	gelat	4.01	1387		i		v sl s dil a
y28	hydroxide	$Y(OH)_3$	139.93	wh-yel gelat or powd		d		i	i	s a, NH_4Cl; i alk
y29	iodide	YI_3	469.62	wh, cr, deliq		1004	$650\text{-}700^{0.02}$	v s		s al, acet; sl s eth
y30	molybdate	$Y_2(MoO_4)_3 \cdot 4H_2O$	729.69	grayish or yelsh, tetr pl, 2.03	$4.79_{|8}^{|6}$	1347				
y31	nitrate, hexahydrate	$Y(NO_3)_3 \cdot 6H_2O$	383.01	col, redsh cr, deliq	2.68	$-3H_2O$, 100		$134.7^{22.5}$		v s al, eth, HNO_3
y32	nitrate, tetrahydrate	$Y(NO_3)_3 \cdot 4H_2O$	346.98	redsh-wh pr	2.682			s		s al, HNO_3
y33	oxalate	$Y_2(C_2O_4)_3 \cdot 9H_2O$	604.01	wh cr powd		d		0.0001		sl s HCl
y34	oxide	Yttria. Y_2O_3	225.81	col-yelsh, cub or powd	5.01	2410		0.00018^{29}		s a; i alk
y35	sulfate	$Y_2(SO_4)_3$	465.98	wh powd	2.52	d 1000		5.38^{25}	s	s sat K_2SO_4 sol
y36	sulfate, octahydrate	$Y_2(SO_4)_3 \cdot 8H_2O$	610.11	col-redsh, monocl, 1.543, 1.549, 1.576	2.558	$-8H_2O$, 120	d 700	7.47^{16} (anhydr)	1.99^{95} (anhydr)	i al, alk; s conc H_2SO_4
y37	sulfide	Y_2S_3	273.99	yel-gr powd.						d a
y38	**Yttrium hexaantipyrine** *per*chlorate	$[Y(C_{11}H_{12}N_2O)_6](ClO_4)_3$	1516.63	col, nex cr		d 293-296		0.55^{20}		
y39	hexaantipyrine iodide	$[Y(C_{11}H_{12}N_2O)_6]I_3$	1598.99	col cr.		280-282		4.65^{20}		
z1	**Zinc**	Zn	65.38	bluish-wh met, hex	7.14	419.58	907	i	i	s a, alk, ac a
z2	acetate	$Zn(C_2H_3O_2)_2$	183.47	col, monocl	1.84	d 200	subl vac	30^{20}	44.6^{100}	2.8^{25} al; 166.79^{79}
z3	acetate, dihydrate	$Zn(C_2H_3O_2)_2 \cdot 2H_2O$	219.50	col. monocl, β 1.494	1.735	237	$-2H_2O$, 100	31.1^{20}	66.6^{100}	2 al

PHYSICAL CONSTANTS OF INORGANIC COMPOUNDS (continued)

No.	Name	Synonyms and Formulae	Mol. wt.	Crystalline form, properties and index of refraction	Density or spec. gravity	Melting point, °C	Boiling point, °C	Solubility, in grams per 100 cc		
								Cold water	Hot water	Other solvents
z4	acetylacetonate	$Zn(C_5H_7O_2)_2$	263.60	need		138	subl	v s d		v s bz, acet; s al
z5	aluminate	Nat. gahnite. $ZnAl_2O_4$	183.34	cub, grn 1.78	4.58			i	i	i a; sl s alk
z6	amide	$Zn(NH_2)_2$	97.43	wh powd, amorph	2.13^{25}	d 200 vac		d	d	i al, eth
z7	antimonide	Zn_3Sb_2	439.64	silv wh, rhomb pr	6.33	570		d		
z8	*ortho*arsenate	Nat. koettigite. $Zn_3(AsO_4)_2 \cdot 8H_2O$	618.10	monocl, 1.662, 1.683, 1.717	3.309^{15}	$-1H_2O$, 100		i	i	s HNO_3, H_3PO_4, alk
z9	*ortho*arsenate, basic	Nat. adamite. $Zn_3(AsO_4)_2 \cdot Zn(OH)_2$	573.37	col, rhomb	4.475^{15}	d 250				
z10	*ortho*arsenate, hydrogen	$ZnHAsO_4 \cdot H_2O$	277.37	wh, rhomb		$-H_2O$, 327		d	d	
z11	arsenide	Zn_3As_2	345.98	met-gray, tetr	5.528	1015		i		d a
z12	benzoate	$Zn(C_7H_5O_2)_2$	307.61	wh powd				2.46^{20}	1.44^{20}	
z13	borate	$3ZnO \cdot 2B_2O_3$	383.37	wh tricl cr, or amorph powd	cr 4.22 powd 3.64	980		s		cr i HCl; amorph; s HCl
z14	bromate	$Zn(BrO_3)_2 \cdot 6H_2O$	429.28	wh, cub, 1.5452	2.566	100	$-6H_2O$, 200	v s	∞	
z15	bromide	$ZnBr_2$	225.19	col, rhomb, hygr n_D^{18} 1.5452	4.201_4^{25}	394	650	447^{20}	675^{100}	v s al, eth, acet; s NH_4OH
z16	butyrate	$Zn(C_4H_7O_2)_2 \cdot 2H_2O$	275.61	wh pr				10.7^{16}	d	
z17	caproate	$Zn(C_6H_{11}O_2)_2$	295.68					$1.03^{24.5}$		
z18	carbonate	Nat. smithsonite. $ZnCO_3$	125.39	col, trig, 1.818, 1.618	4.398	$-CO_2$, 300		0.001^{15}		s a, alk, NH_4 salts; i NH_3, acet, pyr
z19	chlorate	$Zn(ClO_3)_2 \cdot 4H_2O$	304.34	col yelsh, cub, deliq	2.15	d 60	d	262^{20}	v s	167 al; s acet, eth, glyc
z20	chlorate, per-	$Zn(ClO_4)_2 \cdot 6H_2O$	372.37	wh, rhomb, deliq, 1.508, 1.480	2.252 ± 0.01	105-107	d 200	s		s al
z21	chlordie	$ZnCl_2$	136.29	wh, hex, deliq, 1.681, 1.713	2.91^{25}	283	732	432^{25}	615^{100}	$100^{12.5}$ al; v s eth; i NH_3
z22	chloroplatinate	$ZnPtCl_6.6H_2O$	581.27	yel, trig, hygr	2.717^{12}	d 160		v s	v s	v s al; d H_2SO_4
z23	chromate	$ZnCrO_4$	181.37	lem-yel pr	3.40			i	d	s a, liq NH_3; i acet
z24	chromate	$ZnCr_2O_4$	233.37	dk grn to black, cub	5.30^{15}					
z25	*di*chromate	$ZnCr_2O_7 \cdot 3H_2O$	335.41	redsh-brn cr, or or-yel powd, hygr				v s	d	i al, eth; s a
z26	citrate	$Zn_3(C_6H_5O_7)_2 \cdot 2H_2O$	610.37					sl s		
z27	cyanide	$Zn(CN)_2$	117.42	col, rhomb	1.852	d 800		0.0005^{20}		s alk, KCN, NH_3, i al
z28	ferrate (III)	Ferrite. $ZnFe_2O_4$	241.07	blk, oct	5.33^{20}	1590				s conc HCl; i dil a, alk
z29	ferrocyanide	$Zn_2Fe(CN)_6$	342.71	wh powd	1.85_4^{25}			i		s excess alk; i dil a
z30	ferrocyanide, trihydrate	$Zn_2Fe(CN)_6 \cdot 3H_2O$	396.76	wh powd		d		i	i	i al HCl; d NaOH; s NH_4OH; v sl s NH_3
z31	fluoride	ZnF_2	103.38	col, monocl or tricl	4.95_4^{25}	872	ca 1500	1.62^{20}	s	s hot a, NH_4OH; i al, NH_3
z32	fluoride, tetrahydrate	$ZnF_2 \cdot 4H_2O$	175.44	col, rhomb	2.255	$-4H_2O$, 100	tr to ZNO, 3000	1.6^{18}	s	s a, alk, NH_4OH
z33	fluosilicate	$ZnSiF_6 \cdot 6H_2O$	315.55	col, hex pr, 1.3824, 1.3956	2.104	d 100		v s		
z34	formaldehydesulfoxylate	$Zn(HSO_2 \cdot CH_2O)_2$	255.57	rhomb pr		d		v s	v s	d a; i al
z35	formaldehydesulfoxylate, basic	$Zn(OH)HSO_2 \cdot CH_2O$	177.48	rhomb pr		d		i	i	d a; i al
z36	formate	$Zn(CHO_2)_2$	155.42	col, cr	2.368	d		3.80	62^{100}	
z37	formate	$Zn(CHO_2)_2 \cdot 2H_2O$	191.45	wh, monocl, 1.513, 1.526, 1.566	2.207^{20}	$-2H_2O$, 140	d	5.2^{20}	38^{100}	i al
z38	gallate	$ZnGa_2O_4$	268.82	wh fine cr, 1.74	6.15 calc	<800		i	i	i org solv; s dil a, NH_4OH
z39	glycerophosphate	$ZnC_3H_7O_6P$	235.44	wh amorph powd				s		i al, eth
z40	hydroxide(ε)	$Zn(OH)_2$	99.39	col, rhomb	3.053	d 125		v sl s		s a, alk
z41	iodate	$Zn(IO_3)_2$	415.19	wh, need	5.0632^5	d		0.87	1.31	s alk, HNO_3
z42	iodate, dihydrate	$Zn(IO_3)_2 \cdot 2H_2O$	451.22	wh, cr powd	4.223_4^{25}	$-H_2O$, 200		0.877	1.32	s HNO_3, NH_4OH
z43	iodide	ZnI_2	319.19	col, hexag	4.7364_4^{25}	446	d 624	432^{18}	511^{100}	s a, al, eth, NH_3, $(NH_4)_2CO_3$
z44	*d*-lactate	$Zn(C_3H_5O_3)_2 \cdot 2H_2O$	279.55					5.7^{15}	9^{33}	0.104 h 98% al
z45	*di*-lactate	$Zn(C_3H_5O_3)_2 \cdot 3H_2O$	297.57	wh, rhomb cr				1.67^{106}	16.7^{100}	v sl s al
z46	laurate	$Zn(C_{12}H_{23}O_2)_2$	464.01	wh powd		128		0.01^{15}	0.019^{100}	0.010^{15} al
z47	*per*manganate	$Zn(MnO_4)_2 \cdot 6H_2O$	411.34	vlt-br or bl, deliq	2.47	$-5H_2O$, 100		33.3	v s	d al, a
z48	nitrate, trihydrate	$Zn(NO_3)_2 \cdot 3H_2O$	243.44	col, need		45.5		327.3^{40}		
z49	nitrate, hexahydrate	$Zn(NO_3)_2 \cdot 6H_2O$	297.48	col, tetrag	2.065^{14}	36.4	$-6H_2O$, 105-131	184.3^{20}	∞	v s al
z50	nitride	Zn_3N_2	224.15	gray	6.22_4^{25}			d		s HCl
z51	oleate	$Zn(C_{13}H_{33})_2)_2$	628.30	wax-like solid		70		i		s al, eth, bz, CS_2; al s acet
z52	oxalate	$ZnC_2O_4 \cdot 2H_2O$	189.43	wh powd	3.28_4^{25}	d 100		0.00079^{18}		s a, alk
z53	oxide	Nat. zincite. ZnO	81.38	wh, hex, 2.008, 2.029	5.606	1975		0.00016^{29}		s a, alk, NH_4Cl; i al, NH_3
z54	oxide, per-	$ZnO_2 \cdot {}^1/_2H_2O$	106.39	yelsh, powd.	3.00 ± 0.08	$-O_2$, vac		sl d	d	d al, eth, acet
z55	1-phenol-4-sulfonate(p)	$Zn(C_6H_5SO_4)_2 \cdot 8H_2O$	555.83	col cr or fine wh powd, effl		$-8H_2O$, 125		62.5	250^{100}	55.6^{25} al
z56	*ortho*phosphate	$Zn_3(PO_4)_2$	386.08	col, rhomb	3.998^{15}	900		i	i	s a, NH_4OH; i al·
z57	*ortho*phosphate, dihydrogen	$Zn(H_2PO_4)_2 \cdot 2H_2O$	295.39	tricl.		d 100		d		
z58	*ortho*phosphate, octahydrate	$Zn_3(PO_4)_2 \cdot 8H_2O$	530.20	rhomb pl	3.109^{15}			i		s alk

No.	Name	Synonyms and Formulae	Mol. wt.	Crystalline form, properties and index of refraction	Density or spec. gravity	Melting point, °C	Boiling point, °C	Solubility, in grams per 100 cc: Cold water	Hot water	Other solvents
z59	*ortho*phosphate, tetrahydrate	α-Hopeite. $Zn_3(PO_4)_2{\cdot}4H_2O$	458.14	col, rhomb, 1.572, 1.591, 1.59	3.04	tr > 105		i	i	v s a, NH_4OH, HN_4 salts
z60	*ortho*phosphate tetrahydrate	β-Hopeite. $Zn_3(PO_4)_2{\cdot}4H_2O$	458.14	col, rhomb, 1.574, 1.582, 1.582	3.03	tr > 140		i	i	v s a, NH_4OH, NH_4 salts
z61	*ortho*phosphate tetrahydrate	Parahopeite. $Zn_3(PO_4)_2{\cdot}4H_2O$	458.14	col, tricl, 1.614, 1.625, 1.665	3.75	tr > 163		i	i	v s a, NH_4OH, NH_4 salts
z62	*pyro*phosphate	$Zn_2P_2O_7$	304.70	wh powd	3.75^{23}			i	i	s a, alk, NH_4OH
z63	phosphide	Zn_3P_2	258.09	dk gray, tetrag, pois	4.55^{13}	>420	1100; subl in H_2	d		d H_2SO_4 ev H_3P s HNO_3; s (viol) dil a; i al
z64	hypophosphite	$Zn(H_2PO_2)_2{\cdot}H_2O$	213.37	col, cr powd, hygr				s		s alk
z65	picrate	$Zn(C_6H_2N_3O_7)_2{\cdot}8H_2O$	665.70	yel cr powd, expl		expl		s		
z66	salicylate	$Zn(C_7H_5O_3)_2{\cdot}3H_2O$	393.66	need				5^{20}		s al
z67	selenate	$ZnSeO_4{\cdot}5H_2O$	298.41	wh, tricl	2.591_{4}^{20}	d > 50		s		
z68	selenide	ZnSe	144.34	yelsh to redsh, cub, 2.89	5.42_{4}^{15}	>1100		i		s a; d NHO_3
z69	silicate	Nat. hemimorphite. $2ZnO{\cdot}SiO_2{\cdot}H_2O$	240.86	rhomb, or trigon 1.614, 1.617, 1.636	3.43			i	i	
z70	*meta*silicate	$ZnSiO_3$	141.46	col, rhomb	3.42	1437		i		i a
z71	*ortho*silicate	Nat. willemite. Zn_2SiO_4	222.84	trig, 1.694, 1.723	4.103	1509		i	i	s acet a
z72	stearate	$Zn(C_{18}H_{35}O_2)_2$	632.33	light powd		130		i		i al, eth
z73	sulfate	Nat. zinkosite. $ZnSO_4$	161.44	sol, rhomb, 1.658, 1.669, 1.670	3.54_{4}^{25}	d 600		s	s	sl s al; s MeOH, glyc
z74	sulfate, heptahydrate	Nat. goslarite. $ZnSO_4{\cdot}7H_2O$	287.54	col, rhomb, effl, 1.457, 1.480, 1.484	1.957_{4}^{25}	100	$-7H_2O$, 280	96.5^{20}	663.6^{100}	sl s al, glyc
z75	sulfate, hexahydrate	$ZnSO_4{\cdot}6H_2O$	269.53	col, monocl or tetrag	2.072^{15}	$-5H_2O$, 70		s	117.5^{40}	
z76	sulfide,(α)	Nat. wurtzite. ZnS	97.44	col, hex, 2.356, 2.378	3.98	1700 ± 20^{50} atm	1185	0.00069^{18}		v s a; i ac a
z77	sulfide(β)	Nat. sphalerite. ZnS	97.44	col, cub, 2.368	4.102^{25}	tr 1020		0.000065^{18}		v s a
z78	sulfide, monohydrate	$ZnS{\cdot}H_2O$	115.46	yelsh-wh powd	3.98	1049		i		s a
z79	sulfite	$ZnSO_3{\cdot}2H_2O$	181.47	wh, cr powd		$-2H_2O$, 100	d 200	0.16	d	i al; s H_2SO_3
z80	tartrate	$ZnC_4H_4O_6{\cdot}H_2O$ (or $2H_2O$)	231.47	wh powd				0.055^{30}		s KOH, NaOH
z81	tellurate	Zn_3TeO_6	419.74	wh, gran ppt				i	i	s a
z82	telluride	ZnTe	192.98	red, cub, 3.56	6.34^{15}	1238.5		d		s d a
z83	thiocyanate	$Zn(SCN)_2$	181.54	wh powd, deliq				s		s al, NH_4OH
z84	valerate	$Zn(C_5H_9O_2)_2{\cdot}2H_2O$	303.66	wh glist sc or powd				$2.6^{24\text{-}25}$	s	*ca*2.5 al; v sl s eth
	Zinc Complexes									
z85	diamminezinc chloride	$[Zn(NH_3)_2]Cl$	170.35	col, rhomb, 1.625, 1.590	2.10	210.8	d 271	d		
z86	tetrammine perrhenate	$[Zn(NH_3)_4](ReO_4)_2$	633.91	wh, cub cr	3.608_{4}^{25}					0.1852 conc NH_4OH
z87	tetrapyridine fluosilicate	$[Zn(C_5H_5N)_4]SiF_6$	523.86	wh, rhomb	2.197					
z88	**Zirconium**	Zr	91.22	silver gray, met	6.49	1852 ± 2	4377	i	i	s HF, aq reg; sl s a
z89	boride, di	ZrB_2	112.84	hex	6.085	*ca*3200				
z90	bromide, di	$ZrBr_2$	251.03	blk powd, ign in air		d 350		d ev H_2		
z93	bromide, tetra-	$ZrBr_4$	410.84	wh cr powd, deliq		450 ± 1^{15atm}	357 subl	i d		s liq NH_3, acetone; i bz, CCl_4
z94	bromide, tri-	$ZrBr_3$	330.93	bl-blk powd		d 350		d ev H_2		
z95	carbide	ZrC	103.23	gray met, cub	6.73	3540	5100	i		sl s conc H_2SO_4
z96	carbonate, basic	$3ZrO_2{\cdot}CO_2{\cdot}H_2O$	431.68	wh, amorph powd				i		s a
z97	chloride, di-	$ZrCl_2$	162.13	blk	3.6^{18}	d 350		d ev H_2		
z98	chloride, tetra-	$ZrCl_4$	233.03	wh cr	2.803^{15}	437^{25atm}	subl 331	s	d	s al, eth, conc HCl
z99	chloride, tri-	$ZrCl_3$	197.58	br cr	3.00^{18}	d 350		d ev H_2		s $-H_2$ conc al; i org cpd
z100	fluoride	ZrF_4	167.21	wh hex, 1.59	4.43	subl ~ 600		1.388^{25}	d	sl s HF
z101	hydride	ZrH_2	93.24	gray-blk powd						s di HF, conc a
z102	hydroxide	$Zr(OH)_4$	159.25	wh amorp powd	3.25	$-2H_2O$, 500		0.02	i	s min a
z103	iodide	ZrI_4	598.84	wh need, hygr		499 ± 2 6.3^{atm}	d ~ 600	s d	s	d al; s eth; v sl s CS_2, bz; i liq NH_3
z104	nitrate	$Zr(NO_3)_4{\cdot}5H_2O$	429.32	col cr, deliq, 1.60, 1.61				v s		s al
z105	nitride	ZrN	105.23	yel-brn cr	7.09	2980 ± 50		i	i	sl s inorg ac; s conc H_2SO_4 HF, aq reg
z106	oxide	Nat. baddeleyite. ZrO_2	123.22	col-yel-brn, monocl, 2.13, 2.19, 2.20	5.89	*ca* 2700	*ca* 5000	i	i	s H_2SO_4, HF
z107	oxide	Zirconia. ZrO_2 HfO_2 < 2%	123.22	wh, monocl below 1000°, cub, above	5.6	2715		i	i	s H_2SO_4, HF
z108	oxide	Zirconium hydroxide, zirconic acid. $ZrO_2{\cdot}xH_2O$		gel or wh amorph powd	3.25	$-2H_2O$, 550		0.02		s acids; i al, alk
z109	phosphide	ZrP_2	153.17	gray, brittle	4.77_{4}^{25}			i		v s conc hot H_2SO_4
z110	selenate	$Zr(SeO_4)_2{\cdot}4H_2O$	449.20	hex trsp cr				s		sl s al, conc a
z111	selenite	$Zr(SeO_3)_2$	345.14	wh sm cr	4.3	d ~ 400		i	i	sl s H_2SO_4
z112	*ortho*silicate	Zircon, hyacinth. $ZrSiO_4$	183.30	tetr, var colors, 1.92-96; 1.97-2.02	4.56	2550		i	i	i a, aq reg, alk

No.	Name	Synonyms and Formulae	Mol. wt.	Crystalline form, properties and index of refraction	Density or spec. gravity	Melting point, °C	Boiling point, °C	Solubility, in grams per 100 cc		
								Cold water	Hot water	Other solvents
z113	silicide	$ZrSi_2$	147.39	steel gray rhomb, lust met	4.88^{22}			i	i	s HF; i inorg a, aq reg
z114	sulfate	$Zr(SO_4)_2$	283.34	microcr powd, hygr.	3.22^{16}	410 d		s		
z115	sulfate	$Zr(SO_4)_2 \cdot 4H_2O$	355.40	wh cr powd, rhomb	3.22^{16}	$-3H_2O$, 135-150		v s		i al
z116	sulfide	ZrS_2	155.34	steelgray cr, hexag	3.87	~ 1550		i	i	i a
z117	**Zirconyl bromide**	$ZrOBr_2 \cdot xH_2O$		brill need, deliq		$-H_2O$, 120		s		s hot conc HBr
z118	chloride	$ZrOCl_2 \cdot 8H_2O$	322.25	wh, need, tetr, effl, 1.552, 1.563		$-6H_2O$, 150	$-8H_2O$, 210	s	d	s al, eth; sl s HCl
z119	iodide	$ZrOI_2 \cdot 8H_2O$	505.15	col, need, hygr.		d		v s	v s	s al; v s eth
z120	sulfide	Zirconium sulfoxide, ZrOS	139.28	yel powd	4.87	ign in air		i	i	

GRAVIMETRIC FACTORS AND THEIR LOGARITHMS

Rudolf Loebel

Compiled from International atomic weights of 1964. To facilitate use of this table the group of substances weighed under each element as well as the substance sought under each substance weighed are arranged in alphabetical order of their formulas.

Weighed	Sought	Factor	Log of Factor +10	Reciprocal of Factor	Log of Reciprocal of Factor +10
Aluminum					
Al	Al_2O_3	1.88946	10.27634	0.52925	9.72366
	$AlPO_4$	4.51987	10.65513	0.22125	9.34488
Al_4C_3	Al_2O_3	1.41653	10.15122	0.70595	9.84877
$Al(C_9H_6ON)_3$	Al	0.05873	8.76886	17.02811	11.23114
	Al_2O_3	0.11096	9.04516	9.01226	10.95483
$AlCl_3$	Al_2O_3	0.38233	9.58244	2.61554	10.41756
AlF_3	CaF_2	1.39464	10.14446	0.71703	9.85554
Al_2O_3	Al	0.52925	9.72366	1.88946	10.27634
	Al_4C_3	0.70595	9.84877	1.41653	10.15122
	$AlCl_3$	2.61552	10.41756	0.38233	9.58244
	$AlPO_4$	2.39214	10.37879	0.41804	9.62122
	$Al_2(SO_4)_3$	3.35567	10.52578	0.29800	9.47422
	$Al_2(SO_4)_3.18H_2O$	6.53605	10.81531	0.15300	9.18469
	$K_2SO_4.Al_2(SO_4)_3.24H_2O$	9.30532	10.96873	0.10747	9.03129
	$(NH_4)_2SO_4.Al_2(SO_4)_3.24H_2O$	8.89216	10.94901	0.11246	9.05100
$AlPO_4$	Al	0.22125	9.34488	4.51977	10.65512
	Al_2O_3	0.41804	9.62122	2.39211	10.37878
	P_2O_5	0.58196	9.76489	1.71833	10.23511
$Al_2(SO_4)_3$	Al_2O_3	0.29800	9.47422	3.35570	10.52578
$Al_2(SO_4)_3.18H_2O$	Al_2O_3	0.15300	9.18469	6.53595	10.81531
CaF_2	AlF_3	0.71703	9.85554	1.39464	10.14446
$K_2SO_4.Al_2(SO_4)_3.24H_2O$	Al_2O_3	0.10747	9.03129	9.30492	10.96872
$(NH_4)_2SO_4.Al_2(SO_4)_3.24H_2O$	Al_2O_3	0.11246	9.05100	8.89205	10.94900
P_2O_5	$AlPO_4$	1.71831	10.23510	0.58196	9.76489
Ammonium					
Ag	NH_4Br	0.90802	9.95810	1.10130	10.04191
	NH_4Cl	0.49589	9.69538	2.01657	10.30462
	NH_4I	1.34366	10.12829	0.74424	9.87171
AgBr	NH_4Br	0.52161	9.71735	1.91714	10.28265
	NH_4Cl	0.37323	9.57198	2.67931	10.42802
AgI	NH_4I	0.61737	9.79055	1.61977	10.20945
$BaSO_4$	$(NH_4)_2SO_4$	0.56615	9.75293	1.76632	10.24707
Br	NH_4Br	1.22574	10.08840	0.81583	9.91160
Cl	NH_4	0.50880	9.70655	1.96541	10.29345
	NH_4Cl	1.50881	10.17864	0.66277	9.82136
HCl	NH_4Cl	1.46710	10.16646	0.68162	9.83354
I	NH_4I	1.14214	10.05772	0.87555	9.94229
$MgNH_4PO_4.6H_2O$	NH_3	0.06941	8.84142	14.40714	11.15857
	NH_4	0.07352	8.86641	13.60174	11.13360
	$(NH_4)_2O$	0.10613	9.02584	9.42241	10.97416
N	NH_3	1.21589	10.08489	0.82244	9.91510
	NH_4	1.28785	10.10987	0.77649	9.89014
	NH_4Cl	3.81903	10.58195	0.26184	9.41804
	NH_4NO_3	5.71466	10.75699	0.17499	9.24302
	$(NH_4)_2O$	1.85899	10.26928	0.53793	9.73072
	$(NH_4)_2SO_4$	4.71699	10.67367	0.21200	9.32634
NH_3	$MgNH_4PO_4.6H_2O$	14.40648	11.15855	0.06941	8.84142
	N	0.82244	9.91510	1.21589	10.08489
	NH_4	1.05919	10.02498	0.94412	9.97503
	NH_4Cl	3.14093	10.49706	0.31837	9.50293
	$(NH_4)_2CO_3$	2.82099	10.45040	0.35448	9.54959
	NH_4HCO_3	4.64199	10.66671	0.21542	9.33329
	NH_4NO_3	4.69998	10.67210	0.21277	9.32791
	$(NH_4)_2O$	1.52891	10.18438	0.65406	9.81562
	NH_4OH	2.05783	10.31341	0.48595	9.68659
	$(NH_4)_2PtCl_6$	13.03213	11.11501	0.07673	8.88497
	$(NH_4)_2SO_4$	3.87945	10.58877	0.25777	9.41123
	N_2O_5	3.17106	10.50121	0.31535	9.49879
	Pt	5.72763	10.75797	0.17459	9.24202
	SO_3	2.35053	10.37117	0.42544	9.62884
NH_4	Cl	1.96540	10.29345	0.50880	9.70655
	$MgNH_4PO_4.6H_2O$	13.60144	11.13359	0.07352	8.86641
	N	0.77648	9.89013	1.28786	10.10986
	NH_3	0.94412	9.97503	1.05919	10.02498

Weighed	Sought	Factor	Log of Factor +10	Reciprocal of Factor	Log of Reciprocal of Factor +10
Ammonium (contd.)					
NH_3	NH_4Cl	2.96542	10.47208	0.33722	9.52792
	$(NH_4)_2PtCl_6$	12.30389	11.09005	0.08128	8.90998
	Pt	5.40757	10.73301	0.18493	9.26701
NH_4Br	Ag	1.10130	10.04191	0.90802	9.95810
	AgBr	1.91713	10.28265	0.52161	9.71735
	Br	0.81583	9.91160	1.22575	10.08840
NH_4Cl	Ag	2.01656	10.30461	0.49589	9.69538
	AgCl	2.67934	10.42802	0.37323	9.57198
	Cl	0.66277	9.82136	1.50882	10.17864
	HCl	0.68162	9.83354	1.46709	10.16646
	N	0.26185	9.41805	3.81898	10.58194
	NH_3	0.31838	9.50294	3.14090	10.49706
	NH_4	0.33722	9.52791	2.96542	10.47208
	$(NH_4)_2O$	0.48677	9.68732	2.05436	10.31268
	NH_4OH	0.65516	9.81635	1.52634	10.18365
	$(NH_4)_2PtCl_6$	4.14913	10.61795	0.24101	9.38204
	Pt	1.82354	10.26091	0.54838	9.73908
$(NH_4)_2CO_3$	NH_3	0.35448	9.54959	2.82103	10.45040
NH_4HCO_3	NH_3	0.21542	9.33329	4.64209	10.66672
NH_4I	Ag	0.74422	9.87170	1.34369	10.12830
	AgI	1.61977	10.20946	0.61737	9.79055
	I	0.87555	9.94228	1.14214	10.05772
NH_4NO_3	NH_3	0.21277	9.32791	4.69991	10.67209
	$(NH_4)_2PtCl_6$	2.77280	10.44292	0.36064	9.55709
	N_2O_5	0.67470	9.82911	1.48214	10.17089
	Pt	1.21865	10.08588	0.82058	9.91412
$(NH_4)_2O$	$MgNH_4PO_4.6H_2O$	9.42281	10.97418	0.10613	9.02584
	N	0.53733	9.73069	1.86105	10.26931
	NH_3	0.65407	9.81562	1.52889	10.18438
	NH_4Cl	2.05437	10.31268	0.48677	9.68732
	$(NH_4)_2PtCl_6$	8.52378	10.93063	0.11732	9.06937
	N_2O_5	2.07406	10.31682	0.48215	9.68319
	Pt	3.74621	10.57359	0.26694	9.42641
NH_4OH	N	0.39967	9.60170	2.50206	10.39830
	NH_3	0.48595	9.68660	2.05782	10.31341
	NH_4	0.51471	9.71156	1.94284	10.28843
	NH_4Cl	1.52633	10.18365	0.65517	9.81635
	$(NH_4)_2PtCl_6$	6.33297	10.80158	0.15790	9.19841
	Pt	2.78334	10.44456	0.35928	9.55544
$(NH_4)_2PtCl_6$	NH_3	0.07674	8.88502	13.03101	11.11497
	NH_4	0.08128	8.90998	12.30315	11.09002
	NH_4Cl	0.24102	9.38206	4.14903	10.61794
	NH_4NO_3	0.36065	9.55709	2.77277	10.44291
	$(NH_4)_2O$	0.11732	9.06937	8.52370	10.93063
	NH_4OH	0.15791	9.19841	6.33272	10.80159
	$(NH_4)_2SO_4$	0.29768	9.47375	3.35931	10.52625
$(NH_4)_2SO_4$	$BaSO_4$	1.76632	10.24709	0.56615	9.75291
	H_2SO_4	0.74223	9.87054	1.34729	10.12947
	N	0.21200	9.32634	4.71698	10.67367
	NH_3	0.25777	9.41123	3.87943	10.58877
	$(NH_4)_2PtCl_6$	3.35927	10.52625	0.29768	9.47375
	$(NH_4)_2SO_4$	1.47640	10.16921	0.67732	9.83079
	SO_3	0.60589	9.78239	1.65046	10.21761
N_2O_5	NH_3	0.31535	9.49879	3.17108	10.50121
	NH_4NO_3	1.48214	10.17089	0.67470	9.82911
	$(NH_4)_2O$	0.48214	9.68318	2.07409	10.31682
Pt	NH_3	0.17459	9.24202	5.72770	10.75798
	NH_4	0.18493	9.26701	5.40745	10.73300
	NH_4Cl	0.54838	9.73908	1.82355	10.26092
	NH_4NO_3	0.82058	9.91412	1.21865	10.08588
	$(NH_4)_2O$	0.26694	9.42641	3.74616	10.57364
	NH_4OH	0.35928	9.55544	2.78334	10.44456
	$(NH_4)_2SO_4$	0.67732	9.83079	1.47641	10.16921
SO_3	NH_3	0.42543	9.62883	2.35056	10.37118
	$(NH_4)_2SO_4$	1.65046	10.21760	0.60589	9.78239
Antimony					
$K(SbO)C_4H_4O_6.\frac{1}{2}H_2O$	Sb	0.36460	9.56182	2.74273	10.43819
	Sb_2O_3	0.43647	9.63995	2.29111	10.36005

Weighed	Sought	Factor	Log of Factor +10	Reciprocal of Factor	Log of Reciprocal of Factor +10
Antimony (contd.)					
$K(SbO)C_4H_4O_6.\frac{1}{2}H_2O$	Sb_2O_4	0.46043	9.66317	2.17188	10.33684
	Sb_2S_3	0.50862	9.70640	1.96610	10.29360
Sb	$K(SbO)C_4H_4O_6.\frac{1}{2}H_2O$	2.74275	10.43819	0.36460	9.56182
	Sb_2O_3	1.19713	10.07814	0.83533	9.92186
	Sb_2O_4	1.26283	10.10137	0.79187	9.89866
	Sb_2O_5	1.32854	10.12340	0.75271	9.87659
	Sb_2S_3	1.39503	10.14458	0.71683	9.85540
	Sb_2S_5	1.65840	10.21969	0.60299	9.78031
Sb_2O_3	$K(SbO)C_4H_4O_6.\frac{1}{2}H_2O$	2.29111	10.36005	0.43647	9.63995
	Sb	0.83533	9.92186	1.19713	10.07814
	Sb_2O_4	1.05489	10.02320	0.94797	9.97680
	Sb_2O_5	1.10978	10.04523	0.90108	9.95476
	Sb_2S_3	1.16532	10.06645	0.85813	9.93356
	Sb_2S_5	1.38532	10.14155	0.72185	9.85845
Sb_2O_4	$K(SbO)C_4H_4O_6.\frac{1}{2}H_2O$	2.17190	10.33684	0.46043	9.66317
	Sb	0.79187	9.89866	1.26283	10.10137
	Sb_2O_3	0.94797	9.97680	1.05489	10.02300
	Sb_2O_5	1.05203	10.02203	0.95054	9.97797
	Sb_2S_3	1.10468	10.04324	0.90523	9.95671
	Sb_2S_5	1.31324	10.11834	0.76148	9.88166
Sb_2O_5	Sb	0.75270	9.87663	1.32853	10.12337
	Sb_2O_3	0.90108	9.95477	1.10977	10.04523
	Sb_2O_4	0.95054	9.97798	1.05202	10.02202
	Sb_2S_5	1.24828	10.09632	0.80109	9.90368
Sb_2S_3	$K(SbO)C_4H_4O_6.\frac{1}{2}H_2O)$	1.96603	10.29360	0.50863	9.70640
	Sb	0.71681	9.85542	1.39503	10.14458
	Sb_2O_3	0.85814	9.93356	1.16532	10.06645
	Sb_2O_4	0.90524	9.95676	1.10469	10.04332
	Sb_2O_5	0.95234	9.97879	1.05006	10.02121
Sb_2S_5	Sb	0.60299	9.78031	1.65840	10.21969
	Sb_2O_3	0.72186	9.85834	1.38570	10.14167
	Sb_2O_4	0.76148	9.88166	1.31323	10.11834
	Sb_2O_5	0.80110	9.90369	1.24828	10.09631
Arsenic					
As	As_2O_3	1.32032	10.12068	0.75739	9.87932
	As_2O_5	1.53387	10.18579	0.65195	9.81422
	As_2S_3	1.64194	10.21535	0.60904	9.78405
	As_2S_5	2.06991	10.31597	0.48311	9.68405
	$BaSO_4$	4.67291	10.66959	0.21400	9.33041
	$Mg_2As_2O_7$	2.07191	10.31637	0.48265	9.68364
	$MgNH_4AsO_4.\frac{1}{2}H_2O$	2.53968	10.40478	0.39375	9.59523
AsO_3	$BaSO_4$	2.84821	10.45457	0.35110	9.54543
	$Mg_2As_2O_7$	1.26286	10.10135	0.79185	9.89865
	$MgNH_4AsO_4.\frac{1}{2}H_2O$	1.54797	10.18976	0.64601	9.81024
AsO_4	$BaSO_4$	2.52018	10.40143	0.39680	9.59857
	$Mg_2As_2O_7$	1.11742	10.04821	0.89492	9.95178
	$MgNH_4AsO_4.\frac{1}{2}H_2O$	1.36969	10.13662	0.73009	9.86337
As_2O_3	As	0.75739	9.87932	1.32032	10.12068
	As_2O_5	1.16173	10.06511	0.86079	9.93490
	As_2S_3	1.24359	10.09468	0.80412	9.90532
	As_2S_5	1.56773	10.19527	0.63781	9.80472
	$BaSO_4$	3.53922	10.54891	0.28255	9.45110
	$Mg_2As_2O_7$	1.56925	10.19569	0.63725	9.80431
	$MgNH_4AsO_4.\frac{1}{2}H_2O$	1.92353	10.28410	0.51988	9.71590
As_2O_5	As	0.65195	9.81422	1.53386	10.18579
	As_2O_3	0.86077	9.93489	1.16175	10.06511
	As_2S_3	1.07046	10.02957	0.93418	9.97043
	As_2S_5	1.34947	10.13016	0.74103	9.86984
	$BaSO_4$	3.04648	10.48380	0.32825	9.51621
	$Mg_2As_2O_7$	1.35077	10.13058	0.74032	9.86942
	$MgNH_4AsO_4.\frac{1}{2}H_2O$	1.65573	10.21899	0.60396	9.78101
As_2S_3	As	0.60903	9.78464	1.64196	10.21536
	As_2O_3	0.80412	9.90532	1.24360	10.09468
	As_2O_5	0.93418	9.97043	1.07046	10.02957
	As_2S_5	1.26065	10.10060	0.79324	9.89939
	$Mg_2As_2O_7$	1.26187	10.10102	0.79247	9.89898
As_2S_5	As	0.48311	9.68405	2.06992	10.31595
	As_2O_3	0.63786	9.80472	1.56774	10.19527

Weighed	Sought	Factor	Log of Factor +10	Reciprocal of Factor	Log of Reciprocal of Factor +10
Arsenic (contd.)					
As_2S_5	As_2O_5	0.74103	9.86984	1.34947	10.13016
	As_2S_3	0.79324	9.89940	1.26065	10.10060
$BaSO_4$	As	0.21399	9.33039	4.67312	10.66961
	AsO_3	0.35110	9.54543	2.84819	10.45457
	AsO_4	0.39679	9.59856	2.52022	10.40143
	As_2O_3	0.28255	9.45110	3.53920	10.54890
	As_2O_5	0.32825	9.51621	3.04646	10.48379
$Ca_2As_2O_7$	As	0.43814	9.64161	2.28238	10.35839
	As_2O_3	0.57849	9.76230	1.72864	10.23770
$Mg_2As_2O_7$	As	0.48264	9.68363	2.07194	10.31638
	AsO_3	0.79184	9.89864	1.26288	10.10136
	AsO_4	0.89491	9.95178	1.11743	10.04822
	As_2O_3	0.63725	9.80431	1.56924	10.19569
	As_2O_5	0.74031	9.86942	1.35079	10.13059
	As_2S_3	0.79248	9.89899	1.26186	10.10102
$MgNH_4AsO_4.\frac{1}{2}H_2O$	As	0.39375	9.59523	2.53968	10.40478
	AsO_3	0.64600	9.81023	1.54799	10.18977
	AsO_4	0.73009	9.86337	1.36969	10.13662
	As_2O_3	0.51988	9.71590	1.92352	10.28410
	As_2O_5	0.60396	9.78101	1.65574	10.21899
Barium					
Ba	$BaCO_3$	1.43694	10.15744	0.69592	9.84256
	$BaCrO_4$	1.84455	10.26589	0.54214	9.73411
	$BaSiF_6$	2.03444	10.30844	0.49154	9.69156
	$BaSO_4$	1.69943	10.23030	0.58843	9.76969
$BaCl_2$	$BaCO_3$	0.94766	9.97665	1.05523	10.02334
	$BaCrO_4$	1.21647	10.08510	0.82205	9.91490
	$BaSO_4$	1.12077	10.04952	0.89224	9.95048
$BaCl_2.2H_2O$	$BaSO_4$	0.95546	9.98021	1.04662	10.01979
$BaCO_3$	Ba	0.69592	9.84256	1.43695	10.15747
	$BaCl_2$	1.05523	10.02334	0.94766	9.97665
	$BaCrO_4$	1.28366	10.10845	0.77902	9.89155
	$Ba(HCO_3)_2$	1.31426	10.11869	0.76088	9.88132
	BaO	0.77700	9.89042	1.28700	10.10958
	$BaSO_4$	1.18267	10.07286	0.84554	9.92713
	CO_2	0.22300	9.34830	4.48430	10.65170
$BaCrO_4$	Ba	0.54214	9.73411	1.84455	10.26589
	$BaCl_2$	0.82205	9.91490	1.21647	10.08510
	$BaCO_3$	0.77902	9.89155	1.28366	10.10846
	BaO	0.60530	9.78197	1.65207	10.21803
BaF_2	$BaSiF_6$	1.59359	10.20238	0.62751	9.79762
$Ba(HCO_3)_2$	$BaCO_3$	0.76088	9.88132	1.31427	10.11869
$Ba(NO_3)_2$	$BaSO_4$	0.89306	9.95088	1.11975	10.04912
BaO	$BaCO_3$	1.28701	10.10958	0.77699	9.89042
	$BaCrO_4$	1.65208	10.21803	0.60530	9.78197
	$BaSiF_6$	1.82233	10.26061	0.54878	9.73934
	$BaSO_4$	1.52211	10.18244	0.65698	9.81756
	CO_2	0.28701	9.45790	3.48420	10.54210
BaO_2	$BaSO_4$	1.37829	10.13934	0.72554	9.86066
BaS	$BaSO_4$	1.37780	10.13919	0.72579	9.86081
$BaSiF_6$	Ba	0.49152	9.69154	2.03451	10.30846
	BaF_2	0.62751	9.79762	1.59360	10.20238
	BaO	0.54878	9.73939	1.82222	10.26060
$BaSO_4$	Ba	0.58843	9.76969	1.69944	10.23030
	$BaCl_2$	0.89225	9.95049	1.12076	10.04952
	$BaCl_2.2H_2O$	1.04662	10.01979	0.95546	9.98021
	$BaCO_3$	0.84554	9.92713	1.18268	10.07286
	$Ba(NO_3)_2$	1.11975	10.04912	0.89306	9.95088
	BaO	0.65698	9.81756	1.52212	10.18244
	BaO_2	0.72554	9.86066	1.37828	10.13934
	BaS	0.72579	9.86081	1.37781	10.13911
CO_2	$BaCO_3$	4.48421	10.65169	0.22300	9.34830
	BaO	3.48421	10.54211	0.28701	9.45790
Beryllium					
Be	BeO	2.77530	10.44331	0.36032	9.55668
$BeCl_2$	BeO	0.31297	9.49551	3.19519	10.50450
BeO	Be	0.36032	9.55668	2.77531	10.44331
	$BeCl_2$	3.19525	10.50451	0.31296	9.49549

Weighed	Sought	Factor	Log of Factor +10	Reciprocal of Factor	Log of Reciprocal of Factor +10
Beryllium (contd.)					
BeO	$BeSO_4.4H_2O$	7.08211	10.85017	0.14120	9.14983
$Be_2P_2O_7$	Be	0.09389	8.97262	10.65076	11.02738
$BeSO_4.4H_2O$	BeO	0.14120	9.14983	7.08215	10.85017
Bismuth					
Bi	$BiAsO_4$	1.66475	10.22135	0.60069	9.77865
	Bi_2O_3	1.11484	10.04721	0.89699	9.95279
	BiOCl	1.24620	10.09559	0.80244	9.90441
	Bi_2S_3	1.23014	10.08996	0.81292	9.91005
$BiAsO_4$	Bi	0.60069	9.77865	1.66475	10.22135
	Bi_2O_3	0.66968	9.82587	1.49325	10.17413
$Bi(NO_3)_3.5H_2O$	Bi_2O_3	0.48030	9.68151	2.08203	10.31849
	BiOCl	0.53689	9.72988	1.86258	10.27011
Bi_2O_3	Bi	0.89699	9.95279	1.11484	10.04722
	$BiAsO_4$	1.49326	10.17414	0.66968	9.82587
	$Bi(NO_3)_3.5H_2O$	2.08204	10.31849	0.48030	9.68151
	BiOCl	1.11783	10.04837	0.89459	9.95163
	$BiONO_3$	1.23180	10.09054	0.81182	9.90946
	Bi_2S_3	1.10343	10.04275	0.90627	9.95726
BiOCl	Bi	0.80244	9.90441	1.24619	10.09559
	$Bi(NO_3)_3.5H_2O$	1.86256	10.27011	0.53690	9.72989
	Bi_2O_3	0.89458	9.95162	1.11784	10.04838
	$BiONO_3$	1.10195	10.04215	0.90748	9.95784
$BiONO_3$	Bi_2O_3	0.81182	9.90946	1.23180	10.09054
	BiOCl	0.90748	9.95784	1.10195	10.04216
$BiPO_4$	Bi	0.68754	9.83730	1.45446	10.16270
	Bi_2O_3	0.76651	9.88452	1.30461	10.11548
Bi_2S_3	Bi	0.81291	9.91005	1.23015	10.08996
	Bi_2O_3	0.90627	9.95726	1.10342	10.04274
Boron					
B	B_2O_3	3.21987	10.50784	0.31057	9.49216
	KBF_4	11.64619	11.06618	0.08586	8.93379
BO_2	B_2O_3	0.81314	9.91016	1.22981	10.08984
BO_3	B_2O_3	0.59192	9.77226	1.68942	10.22774
B_2O_3	B	0.31057	9.49216	3.21990	10.50785
	BO_2	1.22982	10.08985	0.81313	9.91016
	BO_3	1.68943	10.22774	0.59192	9.77226
	B_4O_7	1.11491	10.04724	0.89693	9.95276
	H_3BO_3	1.77630	10.24952	0.56297	9.75049
	KBF_4	3.61678	10.55832	0.27647	9.44165
	$Na_2B_4O_7.10H_2O$	2.73896	10.43758	0.36510	9.56241
B_4O_7	B_2O_3	0.89693	9.95276	1.11491	10.04724
$C_{20}H_{16}N_4.HBF_4$	B	0.02702	8.43169	37.00962	11.56832
	B_2O_3	0.08698	8.93942	11.49690	11.06058
H_3BO_3	B_2O_3	0.56297	9.75049	1.77629	10.24951
	KBF_4	2.03624	10.30883	0.49110	9.69117
KBF_4	B	0.08587	8.93384	11.64551	11.06616
	B_2O_3	0.27648	9.44167	3.61690	10.55834
	H_3BO_3	0.49110	9.69117	2.03625	10.30885
	$Na_2B_4O_7.10H_2O$	0.75725	9.87924	1.32057	10.12076
$Na_2B_4O_7.10H_2O$	B_2O_3	0.36510	9.56241	2.73898	10.43759
	KBF_4	1.32056	10.12075	0.75725	9.87924
Bromine					
Ag	Br	0.74079	9.86969	1.34991	10.13030
	BrO_3	1.18575	10.07399	0.84335	9.92601
	HBr	0.75013	9.87514	1.33310	10.12487
AgBr	Br	0.42555	9.62895	2.34990	10.37105
	BrO_3	0.68116	9.83325	1.46808	10.16675
	HBr	0.43092	9.63440	2.32062	10.36561
AgCl	Br	0.55754	9.74627	1.79359	10.25372
Br	Ag	1.34991	10.13030	0.74079	9.86969
	AgBr	2.34991	10.37105	0.42555	9.62895
	AgCl	1.79358	10.25372	0.55754	9.74627
	O	0.10011	9.00047	9.98901	10.99952
BrO_3	Ag	0.84335	9.92601	1.18575	10.07399
	AgBr	1.46809	10.16676	0.68116	9.83325
O	Br	9.98899	10.99952	0.10011	9.00047
HBr	Ag	1.33309	10.12486	0.75014	9.87514
	AgBr	2.32064	10.36561	0.43092	9.63440

Weighed	Sought	Factor	Log of Factor +10	Reciprocal of Factor	Log of Reciprocal of Factor +10
Cadmium					
Cd	$CdCl_2$	1.63087	10.21242	0.61317	9.78758
	$Cd(NO_3)_2$	2.10329	10.32290	0.47545	9.67711
	CdO	1.14235	10.05780	0.87539	9.94221
	CdS	1.28523	10.10898	0.77807	9.89102
	$CdSO_4$	1.85463	10.26825	0.53919	9.73174
$Cd(C_{13}H_8O_2N)_2$	Cd	0.21095	9.32418	4.74046	10.67582
	CdO	0.24098	9.38198	4.14972	10.61802
$CdCl_2$	Cd	0.61317	9.78758	1.63087	10.21242
	CdO	0.70045	9.84538	1.42765	10.15462
	CdS	0.78807	9.89657	1.26892	10.10343
	$CdSO_4$	1.13720	10.05584	0.87935	9.94417
$Cd(C_9H_6NO)_2$	Cd	0.28050	9.44793	3.56506	10.55207
	CdO	0.32043	9.50573	3.12081	10.49426
CdMoO4	Cd	0.41272	9.61566	2.42295	10.38434
	CdO	0.47147	9.67345	2.12103	10.32655
$Cd(NO_3)_2$	Cd	0.47545	9.67711	2.10327	10.32290
	CdO	0.54312	9.73490	1.84121	10.26510
	CdS	0.61106	9.78608	1.63650	10.21392
	$CdSO_4$	0.88177	9.94536	1.13408	10.05464
CdO	Cd	0.87539	9.94221	1.14235	10.05781
	$CdCl_2$	1.42765	10.15462	0.70045	9.85438
	$Cd(NO_3)_2$	1.84120	10.26510	0.54312	9.73490
	CdS	1.12508	10.05118	0.88883	9.94882
	$CdSO_4$	1.62352	10.21046	0.61595	9.78955
CdS	Cd	0.77807	9.89102	1.28523	10.10898
	$CdCl_2$	1.26893	10.10343	0.78807	9.89657
	$Cd(NO_3)_2$	1.63651	10.21392	0.61106	9.78608
	CdO	0.88883	9.94882	1.12507	10.05118
	$CdSO_4$	1.44303	10.15928	0.69299	9.84072
$CdSO_4$	Cd	0.53919	9.73174	1.85463	10.26825
	$CdCl_2$	0.87935	9.94417	1.13720	10.05584
	$Cd(NO_3)_2$	1.13408	10.05464	0.88177	9.94536
	CdO	0.61595	9.78955	1.62351	10.21046
	CdS	0.69299	9.84072	1.44302	10.15928
Calcium					
$BaSO_4$	CaS	0.30908	9.49007	3.23541	10.50993
	$CaSO_4$	0.58329	9.76588	1.71441	10.23411
	$CaSO_4.2H_2O$	0.73766	9.86786	1.35564	10.13214
Ca	$CaCl_2$	2.76921	10.44235	0.36111	9.55764
	$CaCO_3$	2.49726	10.39746	0.40044	9.60253
	CaF_2	1.94810	10.28961	0.51332	9.71039
	CaO	1.39920	10.14588	0.71469	9.85411
	$CaSO_4$	3.39671	10.53106	0.29440	9.46894
	Cl	1.76911	10.24776	0.56526	9.75225
$Ca_3(AsO_4)_2$	$Mg_2As_2O_7$	0.77990	9.89204	1.28222	10.10798
$Ca_2As_2O_7$	Ca	0.23439	9.36994	4.26639	10.63006
	CaO	0.32795	9.51581	3.04925	10.48420
$CaC_2O_4.H_2O$	Ca	0.27430	9.43823	3.64564	10.56178
	CaO	0.38379	9.58410	2.60550	10.41589
$CaCl_2$	Ca	0.36111	9.55764	2.76924	10.44236
	$CaCO_3$	0.90179	9.95511	1.10891	10.04489
	CaO	0.50527	9.70352	1.97914	10.29648
	$CaSO_4$	1.22660	10.08870	0.81526	9.91130
	Cl	0.63885	9.80540	1.56531	10.19460
$CaCO_3$	Ca	0.40044	9.60253	2.49725	10.39751
	$CaCl_2$	1.10890	10.04489	0.90179	9.95511
	$Ca(HCO_3)_2$	1.61964	10.20942	0.61742	9.79058
	CaO	0.56030	9.74842	1.78476	10.25158
	$CaSO_4$	1.36018	10.13360	0.73520	9.86641
	$CaSO_4.2H_2O$	1.72015	10.23557	0.58134	9.76442
	CO_2	0.43970	9.64316	2.27428	10.35684
	HCl	0.72854	9.86245	1.37261	10.13755
CaF_2	Ca	0.51332	9.71039	1.94810	10.28961
	$CaSO_4$	1.74360	10.24145	0.57353	9.75855
$Ca(HCO_3)$	$CaCO_3$	0.61742	9.79058	1.61964	10.20942
	CaO	0.34594	9.53900	2.89067	10.46100
$Ca(IO_3)_2$	Ca	0.10280	9.01199	9.72763	10.98801
	CaO	0.14384	9.15788	6.95217	10.84212

GRAVIMETRIC FACTORS AND THEIR LOGARITHMS (Continued)

Weighed	Sought	Factor	Log of Factor +10	Reciprocal of Factor	Log of Reciprocal of Factor +10
Calcium (contd.)					
$Ca(NO_3)_2$	N_2O_5	0.65824	9.81838	1.51920	10.18162
CaO	Ca	0.71470	9.85412	1.39919	10.14588
	$CaCl_2$	1.97914	10.29648	0.50527	9.70352
	$CaCO_3$	1.78477	10.25158	0.56030	9.74842
	CaF_2	1.39230	10.14373	0.71824	9.85627
	$Ca(HCO_3)_2$	2.89069	10.46100	0.34594	9.53900
	$Ca_3(PO_4)_2$	1.84368	10.26569	0.54239	9.73431
	$CaSO_4$	2.42760	10.38518	0.41193	9.61482
	$CaSO_4.2H_2O$	3.07008	10.48715	0.32572	9.51285
	Cl	1.26437	10.10187	0.79091	9.89813
	CO_2	0.78477	9.89474	1.27426	10.10527
	MgO	0.71879	9.85660	1.39123	10.14340
	SO_3	1.42760	10.15461	0.70048	9.84540
$Ca_3(PO_4)_2$	CaO	0.54239	9.73431	1.84369	10.26569
	$CaSO_4$	1.31672	10.11950	0.75946	9.88051
	$Mg_2P_2O_7$	0.71755	9.85585	1.39363	10.14415
	$(NH_4)_3PO_4.12MoO_3$	12.09843	11.08273	0.08266	8.91730
	P_2O_5	0.45761	9.66050	2.18527	10.33950
CaS	$BaSO_4$	3.23538	10.50992	0.30908	9.49007
$CaSO_4$	$BaSO_4$	1.71441	10.23411	0.58329	9.76581
	Ca	0.29440	9.46894	3.39674	10.53107
	$CaCl_2$	0.81526	9.91130	1.22660	10.08870
	$CaCO_3$	0.73520	9.86641	1.36017	10.13360
	CaF_2	0.57353	9.75855	1.74359	10.24145
	CaO	0.41193	9.61482	2.42760	10.38518
	$Ca_3(PO_4)_2$	0.75946	9.88051	1.31673	10.11950
	SO_3	0.58809	9.76944	1.70042	10.23056
$CaSO_4.2H_2O$	$BaSO_4$	1.35564	10.13213	0.73766	9.86786
	$CaCO_3$	0.58134	9.76443	1.72016	10.23557
	CaO	0.32572	9.51285	3.07012	10.48716
	SO_3	0.46502	9.66747	2.15045	10.33252
$CaWO_4$	WO_3	0.80523	9.90592	1.24188	10.09408
Cl	Ca	0.56526	9.75225	1.76910	10.24775
	$CaCl_2$	1.56531	10.19460	0.63885	9.80540
	CaO	0.79091	9.89813	1.26437	10.10188
CO_2	$CaCO_3$	2.27428	10.35684	0.43970	9.64316
	CaO	1.27427	10.10526	0.78476	9.89474
HCl	$CaCO_3$	1.37256	10.13753	0.72857	9.86247
$Mg_2As_2O_7$	$Ca_3(AsO_4)_2$	1.28221	10.10796	0.77990	9.89204
MgO	CaO	1.39118	10.14339	0.71881	9.85661
$Mg_2P_2O_7$	$Ca_3(PO_4)_2$	1.39365	10.14415	0.71754	9.85585
$(NH_4)_3PO_4.12MoO_3$	$Ca_3(PO_4)_2$	0.08265	8.91725	12.09892	11.08275
N_2O_5	$Ca(NO_3)_2$	1.51920	10.18162	0.65824	9.81838
P_2O_5	$Ca_3(PO_4)_2$	2.18521	10.33949	0.45762	9.66051
SO_3	CaO	0.70046	9.84539	1.42763	10.15462
	$CaSO_4$	1.70043	10.23056	0.58809	9.76944
	$CaSO_4.2H_2O$	2.15046	10.33253	0.46502	9.66747
WO_3	$CaWO_4$	1.24188	10.09408	0.80523	9.90592
Carbon					
Ag	CN	0.24120	9.38238	4.14594	10.61762
	HCN	0.25054	9.39888	3.99138	10.60112
	KCN	0.60369	9.78081	1.65648	10.21918
$AgCN$	CN	0.19433	9.28854	5.14589	10.71146
	HCN	0.20185	9.30503	4.95417	10.69497
	KCN	0.48638	9.68698	2.05601	10.31302
$AgCNS$	CNS	0.34999	9.54406	2.85722	10.45594
$Ag_4Fe(CN)_6$	C	0.11200	9.04922	8.92857	10.95078
$AgOCN$	OCN	0.28033	9.44767	3.56722	10.55233
$BaCO_3$	C	0.06086	8.78433	16.43115	11.21567
	CO_2	0.22300	9.34830	4.48431	10.65170
	CO_3	0.30408	9.48299	3.28861	10.51701
BaO	CO_2	0.28701	9.45790	3.48420	10.54210
	CO_2 (bicarbonate)	0.57402	9.75893	1.74210	10.24107
$BaSO_4$	CNS	0.24885	9.39594	4.01849	10.60406
C	$BaCO_3$	16.4305	11.21565	0.06086	8.78433
	CO_2	3.66409	10.56397	0.27292	9.43604
$CaCO_3$	CO_2	0.43970	9.64316	2.27428	10.35684
$Ca(HCO_3)_2$	CO_2	0.54295	9.73476	1.84179	10.26524

Weighed	Sought	Factor	Log of Factor +10	Reciprocal of Factor	Log of Reciprocal of Factor +10
Carbon (contd.)					
CaO	CO_2	0.78477	9.89474	1.27426	10.10526
	CO_2 (bicarbonate)	1.56954	10.19577	0.63713	9.80423
CN	Ag	4.14599	10.61762	0.24120	9.38238
	AgCN	5.14599	10.71147	0.19433	9.28854
CNS	AgCNS	2.85720	10.45594	0.34999	9.54406
	$BaSO_4$	4.01846	10.60406	0.24885	9.39594
	CuCNS	2.09394	10.32097	0.47757	9.67904
CO_2	$BaCO_3$	4.48420	10.65169	0.22301	9.34832
	$Ba(HCO_3)_2$	2.94672	10.46934	0.33963	9.53066
	BaO	3.48421	10.54211	0.28701	9.45790
	C	0.27292	9.43603	3.66408	10.56397
	$CaCO_3$	2.27426	10.35684	0.43970	9.64316
	$Ca(HCO_3)_2$	1.84174	10.26523	0.54296	9.73477
	CaO	1.27426	10.10526	0.78477	9.89474
	CO_3	1.36354	10.13467	0.73339	9.86534
	Cs_2CO_3	7.40329	10.86943	0.13508	9.13059
	$CsHCO_3$	4.40632	10.64407	0.22695	9.35593
	$FeCO_3$	2.63249	10.42037	0.37987	9.57964
	$Fe(HCO_3)_2$	2.02092	10.30555	0.49482	9.69445
	K_2CO_3	3.14049	10.49700	0.31842	9.50300
	$KHCO_3$	2.27491	10.35696	0.43958	9.64304
	K_2O	2.14049	10.33050	0.46718	9.66948
	Li_2CO_3	1.67887	10.22502	0.59564	9.77498
	$LiHCO_3$	1.54410	10.18868	0.64763	9.81133
	Li_2O	0.67887	9.83179	1.47304	10.16821
	$MgCO_3$	1.91595	10.28239	0.52193	9.71761
	$Mg(HCO_3)_2$	1.66265	10.22080	0.60145	9.77920
	MgO	0.91595	9.96188	1.09176	10.03812
	$MnCO_3$	2.61185	10.41695	0.38287	9.58305
	$Mn(HCO_3)_2$	2.01059	10.30332	0.49737	9.69668
	MnO	1.61185	10.20733	0.62041	9.79268
	Na_2CO_3	2.40829	10.38171	0.41523	9.61829
	$NaHCO_3$	1.90882	10.28077	0.52388	9.71923
	Na_2O	1.40829	10.14869	0.71008	9.85131
	$(NH_4)_2CO_3$	2.18329	10.33911	0.45802	9.66088
	NH_4HCO_3	1.79632	10.25439	0.55669	9.74561
	$PbCO_3$	6.07135	10.78328	0.16471	9.21672
	Rb_2CO_3	5.24767	10.71996	0.19056	9.28003
	$RbHCO_3$	3.32856	10.52225	0.30043	9.47774
	Rb_2O	4.24767	10.62815	0.23542	9.37184
	$SrCO_3$	3.35446	10.52562	0.29811	9.47438
	$Sr(HCO_3)_2$	2.3818	10.37690	0.41985	9.62309
	SrO	2.35446	10.37189	0.42473	9.62811
CO_3	$BaCO_3$	3.2886	10.51701	0.30408	9.48299
	CO_2	0.73339	9.86533	1.36353	10.13466
Cs_2CO_3	CO_2	0.13507	9.13056	7.40357	10.86944
$CsHCO_3$	CO_2	0.22695	9.35593	4.40626	10.64407
CuCNS	CNS	0.47757	9.67904	2.09393	10.32097
$FeCO_3$	CO_2	0.37987	9.57964	2.63248	10.42037
$Fe(HCO_3)_2$	CO_2	0.49482	9.69445	2.02094	10.30556
HCN	Ag	3.99137	10.60112	0.25054	9.39888
	AgCN	4.95408	10.69497	0.20185	9.30503
KCN	Ag	1.65648	10.21918	0.60369	9.78081
	AgCN	2.05602	10.31302	0.48638	9.68698
K_2CO_3	CO_2	0.31842	9.50300	3.14051	10.49700
$KHCO_3$	CO_2	0.43958	9.64304	2.27490	10.35696
K_2O	CO_2	0.46718	9.66948	2.14050	10.33052
Li_2CO_3	CO_2	0.59564	9.77498	1.67887	10.22502
Li_2O	CO_2	1.47304	10.16821	0.67887	9.83179
$MgCO_3$	CO_2	0.52193	9.71761	1.91597	10.28239
$Mg(HCO_3)_2$	CO_2	0.60145	9.77920	1.66265	10.22080
MgO	CO_2	1.09176	10.03812	0.91595	9.96188
$MnCO_3$	CO_2	0.38287	9.58305	2.61185	10.41695
MnO	CO_2	0.62041	9.79268	1.61184	10.20733
Na_2CO_3	CO_2	0.41523	9.61829	2.40830	10.38171
$NaHCO_3$	CO_2	0.52388	9.71923	1.90883	10.28077
Na_2O	CO_2	0.71008	9.85131	1.40829	10.14869
$(NH_4)_2CO_3$	CO_2	0.45802	9.66088	2.18331	10.33911

Weighed	Sought	Factor	Log of Factor +10	Reciprocal of Factor	Log of Reciprocal of Factor +10
Carbon (contd.)					
NH_4HCO_3	CO_2	0.55669	9.74561	1.79633	10.25439
$PbCO_3$	CO_2	0.16471	9.21672	6.07128	10.78328
Rb_2CO_3	CO_2	0.19056	9.28003	5.24769	10.71996
$RbHCO_3$	CO_2	0.30043	9.47774	3.32856	10.52225
Rb_2O	CO_2	0.23542	9.37184	4.24773	10.62815
$SrCO_3$	CO_2	0.29811	9.47438	3.35447	10.52562
$Sr(HCO_3)_2$	CO_2	0.41984	9.62308	2.38186	10.37691
SrO	CO_2	0.42472	9.62810	2.35449	10.37190
Cerium					
Ce	$Ce_2(C_2O_4)_3.3H_2O$	2.13513	10.32943	0.46836	9.67058
	$Ce(NO_3)_4$	2.77005	10.44249	0.36100	9.55751
	$Ce(NO_3)_4(NH_4NO_3)_2.H_2O$	4.04111	10.60650	0.24746	9.39351
	CeO_2	1.22838	10.08933	0.81408	9.91067
	Ce_2O_3	1.17128	10.06866	0.85377	9.93134
	$Ce_2(SO_4)_3$	2.02833	10.30714	0.49302	9.69286
$Ce_2(C_2O_4)_3.3H_2O$	Ce	0.46835	9.67057	2.13516	10.32943
	$Ce_2(SO_4)_3$	0.95000	9.97772	1.05263	10.02228
$Ce(NO_3)_4$	Ce	0.36100	9.55751	2.77008	10.44249
	CeO_2	0.44345	9.64684	2.25505	10.35316
	Ce_2O_3	0.42284	9.62618	2.36496	10.37382
$Ce(NO_3)_4(NH_4NO_3)_2.H_2O$	Ce	0.24745	9.39349	4.04122	10.60651
	CeO_2	0.30396	9.48282	3.28991	10.51719
	Ce_2O_3	0.28984	9.46216	3.45018	10.53784
CeO_2	Ce	0.81408	9.91067	1.22838	10.08933
	$Ce(NO_3)_4$	2.25505	10.35316	0.44345	9.64685
	$Ce(NO_3)_4(NH_4NO_3)_2.H_2O$	3.28986	10.51718	0.30396	9.48281
	Ce_2O_3	0.95352	9.97933	1.04875	10.02068
Ce_2O_3	Ce	0.85377	9.93134	1.17128	10.06866
	$Ce(NO_3)_4$	2.36498	10.37383	0.42284	9.62618
	$Ce(NO_3)_4(NH_4NO_3)_2.H_2O$	3.45016	10.53784	0.28984	9.46216
	CeO_2	1.04874	10.02067	0.95353	9.97933
	$Ce_2(SO_4)_3$	1.73172	10.23848	0.57746	9.76152
$Ce_2(SO_4)_3$	Ce	0.49302	9.69286	2.02832	10.30714
	$Ce_2(C_2O_4)_3.3H_2O$	1.05265	10.02220	0.94998	9.97771
	Ce_2O_3	0.57746	9.76152	1.73172	10.23848
Cesium					
AgCl	CsCl	1.17468	10.06992	0.85130	9.93008
Cl	Cs	3.74877	10.57389	0.26675	9.42610
	CsCl	4.74877	10.67658	0.21058	9.32342
Cs	Cl	0.26675	9.42610	3.74883	10.57390
	CsCl	1.26675	10.10269	0.78942	9.89731
	Cs_2O_3	1.22576	10.08840	0.81582	9.91159
	Cs_2O	1.06019	10.02539	0.94323	9.97462
	Cs_2PtCl_6	2.53422	10.40385	0.39460	9.59616
	Cs_2SO_4	1.36139	10.13398	0.73454	9.86602
$CsB(C_6H_5)_4$	Cs	0.29394	9.46826	3.40205	10.53174
	Cs_2O	0.31164	9.49365	3.20883	10.50635
CsCl	AgCl	0.85130	9.93008	1.17467	10.06993
	Cl	0.21058	9.32342	4.74879	10.67658
	Cs	0.78942	9.89731	1.26675	10.10269
	Cs_2O	0.83693	9.92269	1.19484	10.07731
	Cs_2PtCl_6	2.00055	10.30114	0.49986	9.69885
	Cs_2SO_4	1.07470	10.03129	0.93049	9.96871
Cs_2CO_3	Cs	0.81582	9.91159	1.22576	10.08841
	Cs_2PtCl_6	2.06746	10.31544	0.48369	9.68457
	Cs_2SO_4	1.11065	10.04557	0.90037	9.95442
$CsClO_4$	Cs	0.57199	9.75739	1.74828	10.24261
	CsCl	0.72457	9.86008	1.38013	10.13992
	Cs_2O_3	0.70110	9.84578	1.42631	10.15421
	Cs_2O	0.60642	9.78277	1.64902	10.21723
	Cs_2SO_4	0.77870	9.89137	1.28419	10.10863
Cs_2O	Cs	0.94323	9.97462	1.06019	10.02539
	CsCl	1.19483	10.07731	0.83694	9.92269
	Cs_2PtCl_6	2.39034	10.37846	0.41835	9.62154
	Cs_2SO_4	1.28410	10.10860	0.77876	9.89140
	SO_3	0.28410	9.45347	3.51989	10.54653
Cs_2PtCl_6	Cs	0.39460	9.59616	2.53421	10.40385
	CsCl	0.49986	9.69885	2.00056	10.30115

Weighed	Sought	Factor	Log of Factor +10	Reciprocal of Factor	Log of Reciprocal of Factor +10
Cesium (contd.)					
Cs_2PtCl_6	Cs_2CO_3	0.48368	9.68456	2.06748	10.31544
	Cs_2O	0.41835	9.62154	2.39034	10.37846
Cs_2SO_4	Cs	0.73454	9.86602	1.36140	10.13399
	CsCl	0.93048	9.96871	1.07471	10.03129
	Cs_2CO_3	0.90037	9.95442	1.11065	10.04557
	Cs_2O	0.77876	9.89140	1.28409	10.10860
SO_3	Cs_2O	3.51987	10.54652	0.28410	9.45347
	Cs_2SO_4	4.51988	10.65513	0.22124	9.34486
Chlorine					
Ag	Cl	0.32866	9.51675	3.04266	10.48325
	HCl	0.33801	9.52893	2.95850	10.47107
AgCl	Cl	0.24736	9.39333	4.04269	10.60667
	ClO_3	0.58226	9.76512	1.71745	10.23488
	ClO_4	0.69389	9.84129	1.44115	10.15871
	HCl	0.25440	9.40552	3.93082	10.59448
$BaCrO_4$	Cl	0.27990	9.44700	3.57270	10.55299
Ca	Cl	1.76911	10.24776	0.56526	9.75225
Cl	Ag	3.04262	10.48325	0.32866	9.51675
	AgCl	4.04262	10.60666	0.24736	9.39333
	$BaCrO_4$	3.57276	10.55300	0.27990	9.44700
	Ca	0.56526	9.75225	1.76910	10.24775
	HCl	1.02843	10.01217	0.97236	9.98783
	K	1.10292	10.04255	0.90668	9.95745
	KCl	2.10292	10.32282	0.47553	9.67718
	Li	0.19572	9.29164	5.10934	10.70837
	Mg	0.34288	9.53514	2.91647	10.46486
	$MgCl_2$	1.34288	10.12804	0.74467	9.87196
	MnO_2	1.22609	10.08852	0.81560	9.91148
	Na	0.64846	9.81188	1.54212	10.18811
	NaCl	1.64846	10.21708	0.60663	9.21708
Cl	NH_4	0.50880	9.70655	1.96541	10.29346
	$PbCrO_4$	4.55787	10.65876	0.21940	9.34124
ClO_3	AgCl	1.71745	10.23488	0.58226	9.76512
	KCl	0.89340	9.95105	1.11932	10.04895
	NaCl	0.70032	9.84530	1.42792	10.15471
ClO_4	AgCl	1.44114	10.15870	0.69390	9.84130
	KCl	0.74967	9.87487	1.33392	10.12513
	NaCl	0.58765	9.76912	1.70169	10.23088
HCl	Ag	2.95850	10.47107	0.33801	9.52893
	AgCl	3.93086	10.59448	0.25440	9.40552
	NH_4Cl	1.46710	10.16646	0.68162	9.83354
	$(NH_4)_2SO_4$	1.81206	10.25817	0.55186	9.74183
K	Cl	0.90668	9.95745	1.10292	10.04255
KCl	Cl	0.47553	9.67718	2.10292	10.32282
	ClO_3	1.11932	10.04895	0.89340	9.95105
	ClO_4	1.33393	10.12513	0.74966	9.87486
Li	Cl	5.10924	10.70836	0.19572	9.29164
Mg	Cl	2.91650	10.46487	0.34288	9.53514
$MgCl_2$	Cl	0.74467	9.87196	1.34288	10.12804
MnO_2	Cl	0.81560	9.91148	1.22610	10.08853
Na	Cl	1.54212	10.18811	0.64896	9.81188
NaCl	Cl	0.60663	9.78293	1.64845	10.21708
	ClO_3	1.42791	10.15470	0.70032	9.84530
	ClO_4	1.70168	10.23088	0.58765	9.76912
NH_4	Cl	1.96539	10.29345	0.50880	9.70655
NH_4Cl	HCl	0.68162	9.83354	1.46709	10.16646
$(NH_4)_2SO_4$	HCl	0.55186	9.74184	1.81205	10.25817
$PbCrO_4$	Cl	0.21940	9.34124	4.55789	10.65876
Chromium					
Ag_2CrO_4	Cr	0.15674	9.19518	6.37999	10.80482
	Cr_2O_3	0.22895	9.35974	4.36777	10.64026
	CrO_3	0.30143	9.47919	3.31752	10.52082
	CrO_4	0.34966	9.54365	2.85992	10.45635
$BaCrO_4$	Cr	0.20525	9.31228	4.87211	10.68772
	CrO_3	0.39472	9.59629	2.53344	10.40371
	CrO_4	0.45788	9.66075	2.18398	10.33924
	Cr_2O_3	0.29998	9.47709	3.33356	10.52291
	$Cr_2(SO_4)_3.18H_2O$	1.41404	10.15046	0.70719	9.84954

Weighed	Sought	Factor	Log of Factor +10	Reciprocal of Factor	Log of Reciprocal of Factor +10
Chromium (contd.)					
$BaCrO_4$	$PbCrO_4$	1.27573	10.10576	0.78386	9.89424
Cr	$BaCrO_4$	4.87210	10.68772	0.20525	9.31228
	Cr_2O_3	1.46155	10.16482	0.68421	9.83519
	$PbCrO_4$	6.21547	10.79347	0.16089	9.20653
$Cr(C_9H_6NO)_3$	Cr	0.10733	9.03072	9.31706	10.96928
	Cr_2O_3	0.15677	9.19526	6.37877	10.80473
CrO_3	$BaCrO_4$	2.53345	10.40372	0.39472	9.59629
	Cr	0.51999	9.71600	1.92311	10.28401
	Cr_2O_3	0.75999	9.88081	1.31581	10.11920
	K_2CrO_4	1.94210	10.28827	0.51491	9.71173
	$K_2Cr_2O_7$	1.47105	10.16762	0.67979	9.83237
	$PbCrO_4$	3.23199	10.50947	0.30941	9.49053
CrO_4	$BaCrO_4$	2.18399	10.33925	0.45788	9.66075
	$PbCrO_4$	2.78618	10.44501	0.35891	9.55499
Cr_2O_3	$BaCrO_4$	3.33351	10.52291	0.29998	9.47709
	Cr	0.68420	9.83518	1.46156	10.16482
	CrO_3	1.31580	10.11919	0.75999	9.88081
	CrO_4	1.52634	10.18364	0.65516	9.81635
	$PbCrO_4$	4.25265	10.62866	0.23515	9.37135
$Cr_2(SO_4)_3.18H_2O$	$BaCrO_4$	0.70718	9.84953	1.41407	10.15047
	$PbCrO_4$	0.90217	9.95529	1.10844	10.04471
K_2CrO_4	CrO_3	0.51491	9.71173	1.94209	10.28827
	$PbCrO_4$	1.66418	10.22120	0.60090	9.77880
$K_2Cr_2O_7$	CrO_3	0.67979	9.83237	1.47104	10.16762
	$PbCrO_4$	2.19707	10.34184	0.45515	9.65815
$PbCrO_4$	Cr	0.16089	9.20652	6.21543	10.79347
	CrO_3	0.30941	9.49053	3.23196	10.50946
	CrO_4	0.35891	9.55499	2.78621	10.44501
	Cr_2O_3	0.23514	9.37133	4.25279	10.62868
	$Cr_2(SO_4)_3.18H_2O$	1.10843	10.04471	0.90218	9.95529
	K_2CrO_4	0.60090	9.77880	1.66417	10.22119
	$K_2Cr_2O_7$	0.45514	9.65815	2.19713	10.34186
Cobalt					
Co	$Co(NO_3)_2.6H_2O$	4.93731	10.69349	0.20254	9.30651
	$Co(NO_2)_3.(KNO_2)_3$	7.67433	10.88504	0.13030	9.11494
	CoO	1.27148	10.10431	0.78649	9.89569
	Co_3O_4	1.36197	10.13417	0.73423	9.86583
	$CoSO_4$	2.63001	10.41996	0.38023	9.58005
	$CoSO_4.7H_2O$	4.76984	10.67851	0.20965	9.32149
	$(CoSO_4)_2.(K_2SO_4)_3$	7.06550	10.84914	0.14153	9.15085
$Co(C_{10}H_6O_2N).2H_2O$	Co	0.09638	8.98399	10.37560	11.01602
	CoO	0.12255	9.08831	8.15993	10.91169
$(Co[C_5H_5N]_4).(SCN)_2$	Co	0.11990	9.07882	8.34028	10.92118
	CoO	0.15246	9.18316	6.55910	10.81684
$Co(C_9H_6ON)_2$	Co	0.16972	9.22973	5.89206	10.77027
	Co_2O_3	0.23883	9.37809	4.18708	10.62191
$Co(NO_3)_2.6H_2O$	Co	0.20254	9.30651	4.93730	10.69349
$Co(NO_2)_3.(KNO_2)_3$	Co	0.13030	9.11494	7.67460	10.88506
	CoO	0.16568	9.21927	6.03573	10.78073
CoO	Co	0.78648	9.89569	1.27149	10.10432
	$Co(NO_2)_3.(KNO_2)_3$	6.03571	10.78073	0.16568	9.21927
	Co_3O_4	1.07117	10.02986	0.93356	9.97014
	$CoSO_4$	2.06845	10.31564	0.48345	9.68435
	$(CoSO_4)_2.(K_2SO_4)_3$	5.55690	10.74483	0.17996	9.25518
Co_3O_4	Co	0.73423	9.86583	1.36197	10.13417
	CoO	0.93356	9.97014	1.07117	10.02986
$Co_2P_2O_7$	Co	0.40392	9.60630	2.47574	10.39371
	CoO	0.51357	9.71060	1.94715	10.28940
	Co_3O_4	0.55012	9.74046	1.81779	10.25953
	$CoSO_4$	1.06230	10.02624	0.94135	9.97375
	$CoSO_4.7H_2O$	1.92661	10.28479	0.51905	9.71521
$CoSO_4$	Co	0.38023	9.58005	2.62999	10.41996
	CoO	0.48345	9.68435	2.06847	10.31565
$CoSO_4.7H_2O$	Co	0.20965	9.32150	4.76985	10.67851
	CoO	0.26657	9.42581	3.75136	10.57419
$(CoSO_4)_2.(K_2SO_4)_3$	Co	0.14153	9.15085	7.06553	10.84914
	CoO	0.17996	9.25518	5.55679	10.74482
Columbium	See Niobium				

Weighed	Sought	Factor	Log of Factor +10	Reciprocal of Factor	Log of Reciprocal of Factor +10
Copper					
Cu	$Cu_2C_2H_3O_2.(AsO_2)_3$	3.98875	10.60084	0.25071	9.39917
	CuCNS	1.91407	10.28196	0.52245	9.71804
	CuO	1.25181	10.09754	0.79884	9.90246
	Cu_2O	1.12590	10.05150	0.88818	9.94850
	Cu_2S	1.25228	10.09770	0.79854	9.90230
	$CuSO_4.5H_2O$	3.92949	10.59433	0.25449	9.40567
$Cu_2C_2H_3O_2.(AsO_2)_3$	Cu	0.25071	9.39917	3.98867	10.60083
CuCNS	Cu	0.52245	9.71804	1.91406	10.28195
	CuO	0.65400	9.81558	1.52905	10.18443
$(Cu[C_5H_5N]_2.(SCN)_2$	Cu	0.18804	9.27425	5.31802	10.72575
	CuO	0.23539	9.37179	4.24827	10.62821
$Cu(C_7H_6NO_2)_2$	Cu	0.18922	9.27697	5.28485	10.72303
	CuO	0.23687	9.37451	4.22172	10.62549
$Cu(C_9H_6NO)_2$	Cu	0.18059	9.25669	5.53741	10.74330
	CuO	0.22606	9.35422	4.42360	10.64578
$Cu(C_{12}H_{10}ONS)_2$	Cu	0.12795	9.10704	7.81555	10.89296
	CuO	0.16017	9.20458	6.24337	10.79542
$Cu(NH_2C_6H_4CO_2)_2$	Cu	0.18922	9.27697	5.28485	10.72303
	CuO	0.23687	9.37451	4.22172	10.62549
CuO	Cu	0.79884	9.90246	1.25182	10.09754
	CuCNS	1.52904	10.18442	0.65401	9.81558
	Cu_2S	1.00038	10.00016	0.99962	9.99983
	$CuSO_4.5H_2O$	3.13905	10.49680	0.31857	9.50320
Cu_2O	Cu	0.88817	9.94850	1.12591	10.05150
	CuO	1.11183	10.04603	0.89942	9.95396
	Cu_2S	1.11224	10.04620	0.89909	9.95380
Cu_2S	Cu	0.79854	9.90230	1.25229	10.09770
	CuO	0.99962	9.99983	1.00004	10.00002
	Cu_2O	0.89908	9.95380	1.11225	10.04620
	$CuSO_4.5H_2O$	3.13787	10.49663	0.31869	9.50337
$CuSO_4.5H_2O$	Cu	0.25449	9.40567	3.92943	10.59433
	CuO	0.31857	9.50320	3.13903	10.49679
	Cu_2S	0.31869	9.50337	3.13785	10.49663
$Mg_2As_2O_7$	$Cu_2C_2H_3O_2.(AsO_2)_3$	1.08844	10.03681	0.91875	9.96320
Dysprosium					
Dy_2O_3	Dy	0.87131	9.94017	1.14770	10.05983
Erbium					
Er	Er_2O_3	1.14349	10.05824	0.87452	9.94177
Er_2O_3	Er	0.87452	9.94177	1.14349	10.05824
Europium					
Eu_2O_3	Eu	0.86361	9.93632	1.15793	10.06368
Fluorine					
BaF_2	$BaSiF_6$	1.59359	10.20238	0.62751	9.79762
$BaSiF_6$	BaF_2	0.62751	9.79762	1.59360	10.20238
	F	0.40795	9.61061	2.45128	10.38939
	HF	0.42960	9.63306	2.32775	10.36694
	H_2SiF_6	0.51568	9.71238	1.93919	10.28762
	SiF_4	0.37248	9.57110	2.68471	10.42889
	SiF_6	0.50847	9.70627	1.96668	10.29373
CaF_2	F	0.48664	9.68721	2.05491	10.31279
	6HF	0.51248	9.70968	1.95130	10.29033
	H_2SiF_6	0.61517	9.78900	1.62557	10.21100
	SiF_6	0.60657	9.78288	1.64861	10.21712
$CaSO_4$	F	0.27910	9.44576	3.58295	10.55424
	HF	0.29391	9.46821	3.40240	10.53178
F	$BaSiF_6$	2.45125	10.38939	0.40796	9.61062
	CaF_2	2.05491	10.31279	0.48664	9.68721
	$CaSO_4$	3.58293	10.55433	0.27910	9.44576
	H_2SiF_6	1.26407	10.10177	0.79110	9.89823
	K_2SiF_6	1.93244	10.28611	0.51748	9.71389
2HF	H_2SiF_6	3.60115	10.55644	0.27769	9.44356
6HF	H_2SiF_6	1.20038	10.07932	0.83307	9.92068
H_2SiF_6	$BaSiF_6$	1.93917	10.28762	0.51568	9.71237
	CaF_2	1.62555	10.21100	0.61518	9.78900
	F	0.79109	9.89823	1.26408	10.10178
	2HF	0.27769	9.44356	3.60114	10.55644
	6HF	0.83307	9.92068	1.20038	10.07932
	K_2SiF_6	1.52875	10.18434	0.65413	9.81566

GRAVIMETRIC FACTORS AND THEIR LOGARITHMS (Continued)

Weighed	Sought	Factor	Log of Factor +10	Reciprocal of Factor	Log of Reciprocal of Factor +10
Fluorine (contd.)					
H_2SiF_6	SiF_4	0.72232	9.85873	1.38443	10.14129
	SiF_6	0.98601	9.99388	1.01419	10.00612
2KF	K_2SiF_6	1.89568	10.27777	0.52752	9.72224
K_2SiF_6	F	0.51748	9.71389	1.93244	10.28611
	6HF	0.54493	9.73634	1.83510	10.26366
	H_2SiF_6	0.65413	9.81566	1.52875	10.18434
	KF	0.52751	9.72223	1.89570	10.27777
	SiF_6	0.64498	9.80955	1.55044	10.19045
Na_2SiF_6	F	0.60615	9.78258	1.64976	10.21742
	6HF	0.63831	9.80503	1.56664	10.19498
	H_2SiF_6	0.70022	9.88435	1.30511	10.11505
	2NaF	0.44655	9.64987	2.23939	10.35013
	SiF_4	0.55344	9.74307	1.80688	10.25693
	SiF_6	0.75550	9.87823	1.32363	10.12177
PbClF	F	0.07261	8.86100	13.77221	11.13900
SiF_4	$BaSiF_6$	2.68467	10.42889	0.37249	9.57111
	H_2SiF_6	1.38444	10.14128	0.72231	9.85872
SiF_6	$BaSiF_6$	1.96666	10.29373	0.50848	9.70627
	CaF_2	1.64862	10.21712	0.60657	9.78288
	H_2SiF_6	1.01419	10.00612	0.98601	9.99388
	K_2SiF_6	1.55043	10.19045	0.64498	9.80955
Gallium					
Ga	Ga_2O_3	1.34423	10.12847	0.74392	9.87153
$Ga(C_9H_4NOBr_2)_3$	Ga	0.07146	8.85406	13.99384	11.14594
	Ga_2O_3	0.09606	8.98254	10.41016	11.01746
Ga_2O_3	Ga	0.74392	9.87153	1.34423	10.12847
Ga_2S_3	Ga	0.59178	9.77216	1.68982	10.22785
Germanium					
Ge	GeO_2	1.44083	10.15861	0.69404	9.84138
$(C_9H_6NOH)_4.(GeO_2.12MoO_3)$	Ge	0.03009	8.47842	33.23363	11.52158
	GeO_2	0.04335	8.63699	23.06805	11.36301
GeO_2	Ge	0.69404	9.84139	1.44084	10.15861
K_2GeF_6	Ge	0.27415	9.43799	3.64764	10.56201
Mg_2GeO_4	Ge	0.39193	9.59321	2.55148	10.40679
	GeO_2	0.56471	9.75183	1.77082	10.24817
Gold					
Au	$AuCl_3$	1.53998	10.18749	0.64936	9.81249
	$HAuCl_4.4H_2O$	2.09095	10.32034	0.47825	9.67966
	$KAu(CN)_4.H_2O$	1.81836	10.25968	0.54995	9.74032
$AuCl_3$	Au	0.64938	9.81250	1.53993	10.18750
C_6H_5SAu	Au	0.64339	9.80847	1.55427	10.19153
$HAuCl_4.4H_2O$	Au	0.47825	9.67966	2.09096	10.32034
$KAu(CN)_4.H_2O$	Au	0.54995	9.74032	1.81835	10.25967
Hafnium					
HfO_2	Hf	0.84797	9.92838	1.17929	10.07162
$Hf(C_6H_5CHOHCO_2)_4$	Hf	0.22794	9.35782	4.38712	10.64218
	HfO_2	0.26880	9.42943	3.72024	10.57057
Holmium					
Ho_2O_3	Ho	0.87297	9.94100	1.14551	10.05900
Hydrogen					
AgCNS	HCNS	0.35606	9.55152	2.80852	10.44848
$BaSO_4$	HCNS	0.25317	9.40341	3.94992	10.59659
CuCNS	HCNS	0.48586	9.68651	2.05821	10.31349
H	H_2O	8.93644	10.95116	0.11190	9.04883
	O	7.93644	10.89963	0.12600	9.10037
HCNS	AgCNS	2.80846	10.44847	0.35607	9.55154
	$BaSO_4$	3.94991	10.59659	0.25317	9.40341
	CuCNS	2.05822	10.31350	0.48586	9.68651
H_2O	H	0.11190	9.04883	8.93655	10.95117
O	H	0.12600	9.10037	7.93651	10.89963
Indium					
In	In_2O_3	1.20902	10.08244	0.92712	9.91757
	In_2S_3	1.41888	10.15194	0.70478	9.84805
$In(C_9H_6ON)_3$	In	0.20980	9.32181	4.76644	10.67819
	In_2O_3	0.25365	9.40423	3.94244	10.59577
In_2O_3	In	0.82711	9.91756	1.20903	10.08244
In_2S_3	In	0.70478	9.84805	1.41888	10.15194

GRAVIMETRIC FACTORS AND THEIR LOGARITHMS (Continued)

Weighed	Sought	Factor	Log of Factor +10	Reciprocal of Factor	Log of Reciprocal of Factor +10
Iodine					
Ag	HI	1.18580	10.07401	0.84331	9.92599
	I	1.17646	10.07058	0.85001	9.92942
AgCl	I	0.88544	9.94716	1.12938	10.05284
AgI	HI	0.54483	9.73626	1.83543	10.26374
	I	0.54054	9.73283	1.85000	10.26717
	IO_3	0.74498	9.87214	1.34232	10.12786
	IO_4	0.81313	9.91016	1.22982	10.08985
	I_2O_5	0.71091	9.85181	1.40665	10.14819
	I_2O_7	0.77906	9.89157	1.28360	10.10843
HI	Ag	0.84331	9.92599	1.18580	10.07401
	AgI	1.83542	10.26374	0.54483	9.73626
	Pd	0.41590	9.61899	2.40442	10.38101
	PdI_2	1.40799	10.14860	0.71023	9.85140
	TlI	2.58981	10.41327	0.38613	9.58673
I	Ag	0.85000	9.92942	1.17647	10.07058
	AgCl	1.12937	10.05283	0.88545	9.94716
	AgI	1.85000	10.26717	0.54054	9.73283
	Pd	0.41921	9.62243	2.38544	10.37757
	PdI_2	1.41917	10.15203	0.70464	9.84797
	TlI	2.61039	10.41671	0.38308	9.58329
IO_3	AgI	1.34231	10.12785	0.74498	9.87214
	PdI_2	1.02971	10.01272	0.97115	9.98729
	TlI	1.89402	10.27738	0.52798	9.72262
IO_4	AgI	1.22981	10.08984	0.81313	9.91016
	PdI_2	0.94341	9.97470	1.05998	10.02530
	TlI	1.73528	10.23937	0.57628	9.76063
I_2O_5	AgI	1.40665	10.14819	0.71091	9.95181
	PdI_2	1.07907	10.03305	0.92672	9.96695
	TlI	1.98481	10.29772	0.50383	9.70228
I_2O_7	AgI	1.28360	10.10843	0.77906	9.89157
	PdI_2	0.98467	9.99329	1.01557	10.00671
	TlI	1.81120	10.25797	0.55212	9.74203
Pd	HI	2.40437	10.38100	0.41591	9.61900
	I	2.38542	10.37757	0.41921	9.62243
PdI_2	HI	0.71023	9.85140	1.40799	10.14860
	I	0.70463	9.70463	1.41918	10.15204
	IO_3	0.97114	9.98728	1.02972	10.01272
	IO_4	1.05998	10.02530	0.94341	9.97470
	I_2O_5	0.92672	9.96695	1.07907	10.03305
	I_2O_7	1.01556	10.00671	0.98468	9.99330
TlI	HI	0.38613	9.58673	2.58980	10.41327
	I	0.38308	9.58329	2.61042	10.41671
	IO_3	0.52798	9.72262	1.89401	10.27738
	IO_4	0.57627	9.76063	1.73530	10.23937
	I_2O_5	0.50382	9.70228	1.98484	10.29772
	I_2O_7	0.55212	9.74203	1.81120	10.25797
Iron					
Ag	$Fe_7(CN)_{18}$ (Prussian blue)	0.44253	9.64594	2.25973	10.35406
CN	$Fe_7(CN)_{18}$	1.83474	10.26358	0.54504	9.73643
CO_2	$FeCO_3$	2.63249	10.42037	0.37987	9.57964
	$Fe(HCO_3)_2$	2.02092	10.30555	0.49482	9.69445
	FeO	1.63249	10.21287	0.61256	9.78715
Fe	$Fe(HCO_3)_2$	3.18517	10.50313	0.31395	9.49686
	FeO	1.28648	10.10940	0.77731	9.89059
	Fe_2O_3	1.42973	10.15526	0.69943	9.84474
	$FePO_4$	2.70056	10.43145	0.37029	9.56854
	FeS	1.57414	10.19704	0.63527	9.80296
	$FeSO_4$	2.72009	10.43458	0.36763	9.56541
	$FeSO_4.7H_2O$	4.97817	10.69707	0.20088	9.30294
	$FeSO_4.(NH_4)_2SO_4.6H_2O$	7.02054	10.84637	0.14244	9.15363
$FeAsO_4$	$Mg_2As_2O_7$	0.79702	9.90147	1.25467	10.09853
$Fe(C_6H_5N[NO]O)_3$	Fe	0.11953	9.07748	8.36610	10.92252
	Fe_2O_3	0.17090	9.23274	5.85138	10.76726
$Fe(C_9H_6NO)_3$	Fe	0.11437	9.05821	8.74355	10.94169
	Fe_2O_3	0.16351	9.21354	6.11583	10.78646
$FeCl_3$	Fe_2O_3	0.49225	9.69219	2.03149	10.30781
$Fe_7(CN)_{18}$	Ag	2.25971	10.35405	0.44253	9.64594
	CN	0.54503	9.73642	1.83476	10.26358

GRAVIMETRIC FACTORS AND THEIR LOGARITHMS (Continued)

Weighed	Sought	Factor	Log of Factor +10	Reciprocal of Factor	Log of Reciprocal of Factor +10
Iron (contd.)					
$FeCO_3$	CO_2	0.37986	9.57962	2.63255	10.42038
	FeO	0.62013	9.79248	1.61257	10.20752
	Fe_2O_3	0.68918	9.83833	1.45100	10.16167
$Fe(HCO_3)_2$	CO_2	0.49482	9.69445	2.02094	10.30555
	Fe	0.31396	9.49687	3.18512	10.50313
	FeO	0.40390	9.60627	2.47586	10.39373
	Fe_2O_3	0.44887	9.65212	2.22782	10.34788
FeO	CO_2	0.61256	9.78715	1.63249	10.21285
	Fe	0.77732	9.89060	1.28647	10.10940
	$FeCO_3$	1.61256	10.20751	0.62013	9.79248
	$Fe(HCO_3)_2$	2.47588	10.39373	0.40390	9.00027
	Fe_2O_3	1.11134	10.04584	0.89981	9.95415
	$FePO_4$	2.09918	10.32205	0.47638	9.67795
	FeS	1.22360	10.08764	0.81726	9.91236
	SO_3	1.11436	10.04703	0.89738	9.95298
Fe_2O_3	Fe	0.69943	9.84474	1.42974	10.15524
	$FeCl_3$	2.03149	10.30781	0.49225	9.69219
	$FeCO_3$	1.45099	10.16167	0.68918	9.83833
	$Fe(HCO_3)_2$	2.22781	10.34788	0.44887	9.65212
	$Fe(HCO_3)_3$	2.99200	10.47596	0.33422	9.52403
	FeO	0.89981	9.95415	1.11135	10.04584
	Fe_3O_4	0.96660	9.98525	1.03455	10.01475
	$FePO_4$	1.88886	10.27620	0.52942	9.72380
	FeS	1.10101	10.04113	0.90826	9.95821
	$FeSO_4$	1.90252	10.27933	0.52562	9.72067
	$FeSO_4.7H_2O$	3.48190	10.54182	0.28720	9.45818
	$Fe_2(SO_4)_3$	2.50406	10.39864	0.39935	9.60135
	$FeSO_4.(NH_4)_2SO_4.6H_2O$	4.91040	10.69112	0.20365	9.30888
Fe_3O_4	Fe_2O_3	1.03455	10.01475	0.96660	9.98525
$FePO_4$	Fe	0.37029	9.56854	2.70059	10.43145
	FeO	0.47638	9.67795	2.09916	10.32204
	Fe_2O_3	0.52942	9.72380	1.88886	10.27620
FeS	Fe	0.63527	9.80296	1.57413	10.19704
	FeO	0.81726	9.91236	1.22360	10.08764
	Fe_2O_3	0.90826	9.95821	1.10101	10.04179
$FeSO_4$	Fe	0.36763	9.56541	2.72013	10.43459
	Fe_2O_3	0.52562	9.72067	1.90252	10.27933
	SO_3	0.52704	9.72184	1.89739	10.27817
$FeSO_4.7H_2O$	Fe	0.20088	9.30294	4.97810	10.69706
	Fe_2O_3	0.28720	9.45818	3.48189	10.54182
$FeSO_4.(NH_4)_2SO_4.6H_2O$	Fe	0.14244	9.15363	7.02050	10.84637
	Fe_2O_3	0.20365	9.30888	4.91039	10.69112
$Fe_2(SO_4)_3$	Fe_2O_3	0.39935	9.60135	2.50407	10.39864
$Mg_2As_2O_7$	$FeAsO_4$	1.25468	10.09853	0.79702	9.90147
SO_3	FeO	0.89738	9.95298	1.11436	10.04703
	$FeSO_4$	1.89739	10.27816	0.52704	9.72184
Lanthanum					
La	La_2O_3	1.17277	10.06921	0.85268	9.93079
La_2O_3	La	0.85268	9.93079	1.17277	10.06921
Lead					
$BaSO_4$	$PbSO_4$	1.29927	10.11370	0.76966	9.88630
Pb	$PbCl_2$	1.34225	10.12783	0.74502	9.87217
	$PbCO_3$	1.28958	10.11044	0.77545	9.88955
	$(PbCO_3)_2.Pb(OH)_2$	1.24781	10.09615	0.80140	9.90385
	$PbCrO_4$	1.55982	10.19307	0.64110	9.80693
	PbO	1.07722	10.03231	0.92832	9.96770
	PbO_2	1.15445	10.06238	0.86621	9.93762
	$Pb(OH)_2$	1.16415	10.06601	0.85900	9.93399
	PbS	1.15474	10.06248	0.86600	9.93752
	$PbSO_4$	1.46363	10.16543	0.68323	9.83457
$Pb(C_2H_3O_2).3H_2O$	$PbCrO_4$	0.85198	9.93043	1.17374	10.06957
	$PbSO_4$	0.79944	9.90279	1.25088	10.09722
$Pb(C_7H_5NO_2)_2$	Pb	0.43396	9.63746	2.30436	10.36255
	PbO	0.46747	9.66975	2.13971	10.33024
$PbCl_2$	Pb	0.74502	9.87218	1.34225	10.12783
	PbO	0.80255	9.90448	1.24603	10.09553
$PbCO_3$	Pb	0.77541	9.88954	1.28964	10.11047
	PbO	0.83529	9.92184	1.19719	10.07816

Weighed	Sought	Factor	Log of Factor +10	Reciprocal of Factor	Log of Reciprocal of Factor +10
Lead (contd.)					
$PbCO_3$	$PbSO_4$	1.13492	10.05497	0.88112	9.94504
$(PbCO_3)_2.Pb(OH)_2$	Pb	0.80141	9.90386	1.24780	10.09614
	$PbCrO_4$	1.25007	10.09693	0.79996	9.90307
	$PbSO_4$	1.17298	10.06929	0.85253	9.93017
$PbCrO_4$	Pb	0.64110	9.80692	1.55982	10.19307
	$Pb(C_2H_3O_2)_2.3H_2O$	1.17371	10.06956	0.85200	9.93044
	$(PbCO_3)_2.Pb(OH)_2$	0.79996	9.90307	1.25006	10.09693
	PbO	0.69061	9.83922	1.44800	10.16077
	Pb_3O_4	0.70710	9.84947	1.41423	10.15052
	$PbSO_4$	0.93833	9.97235	1.06572	10.02765
$Pb(NO_3)_2$	PbO	0.67388	9.82859	1.48394	10.17141
	PbO_2	0.72219	9.85865	1.38468	10.14135
	$PbSO_4$	0.91561	9.96171	1.09217	10.03829
PbO	Pb	0.92831	9.96770	1.07723	10.03231
	$PbCl_2$	1.24602	10.09553	0.80256	9.90448
	$PbCO_3$	1.19718	10.07816	0.83530	9.92184
	$PbCrO_4$	1.44800	10.16077	0.69061	9.83923
	$Pb(NO_3)_2$	1.48393	10.17141	0.67389	9.82859
	PbO_2	1.07169	10.03007	0.93311	9.96993
	PbS	1.07196	10.03018	0.93287	9.96982
	$PbSO_4$	1.35871	10.13313	0.73599	9.86687
PbO_2	Pb	0.86622	9.93763	1.15444	10.06237
	$Pb(NO_3)_2$	1.38467	10.14135	0.72219	9.85865
	PbO	0.93311	9.96993	1.07169	10.03007
	$PbSO_4$	1.26782	10.10306	0.78876	9.89694
Pb_3O_4	$PbCrO_4$	1.41423	10.15052	0.70710	9.84948
	$PbSO_4$	1.32702	10.12288	0.75357	9.87712
$Pb(OH)_2$	Pb	0.85900	9.93399	1.16414	10.06601
$Pb_2P_2O_7$	Pb	0.70434	9.84778	1.41977	10.15222
	PbO	0.75874	9.88009	1.31797	10.11991
PbS	Pb	0.86600	9.93752	1.15473	10.06249
	PbO	0.93287	9.96982	1.07196	10.03017
	$PbSO_4$	1.26750	10.10296	0.78895	9.89705
$PbSO_4$	$BaSO_4$	0.76966	9.88630	1.29928	10.11370
	Pb	0.68323	9.83457	1.46364	10.16543
	$Pb(C_2H_3O_2)_2.3H_2O$	1.25088	10.09722	0.79944	9.90279
	$PbCO_3$	0.88112	9.94504	1.13492	10.05497
	$(PbCO_3)_2.Pb(OH)_2$	0.85253	9.93071	1.17298	10.06929
	$PbCrO_4$	1.06572	10.02765	0.93833	9.97236
	$Pb(NO_3)_2$	1.09216	10.03828	0.91562	9.96172
	PbO	0.73599	9.86687	1.35871	10.13313
	PbO_2	0.78876	9.89694	1.26781	10.10305
	Pb_3O_4	0.75357	9.87712	1.32702	10.12288
	PbS	0.78895	9.89705	1.26751	10.10295
Lithium					
CO_2	Li_2CO_3	1.67887	10.22504	0.59564	9.77498
	$LiHCO_3$	1.54410	10.18868	0.64763	9.81133
	Li_2O	0.67887	9.83179	1.47304	10.16821
Li	LiCl	6.10924	10.78599	0.16369	9.21402
	Li_2CO_3	5.32404	10.72624	0.18783	9.27377
	Li_2O	2.15283	10.33201	0.46450	9.66699
	Li_3PO_4	5.56218	10.74524	0.17979	9.25477
	Li_2SO_4	7.92189	10.89883	0.12623	9.10115
LiCl	Li	0.16369	9.21402	6.10911	10.78598
	Li_2CO_3	0.87147	9.94025	1.14749	10.05975
	Li_2O	0.35239	9.54702	2.83776	10.45297
	Li_3PO_4	0.91045	9.95926	1.09836	10.04074
	Li_2SO_4	1.29670	10.11284	0.77119	9.88716
Li_2CO_3	CO_2	0.59564	9.77498	1.67887	10.22502
	Li	0.18782	9.27374	5.32425	10.72626
	LiCl	1.14747	10.05974	0.87148	9.94026
	$LiHCO_3$	1.83963	10.26473	0.54359	9.73527
	Li_2O	0.40436	9.60677	2.47304	10.39323
	Li_3PO_4	1.04473	10.01901	0.95719	9.98100
$LiHCO_3$	CO_2	0.64762	9.81132	1.54412	10.18868
	Li_2CO_3	0.54363	9.73530	1.83949	10.26470
	Li_2O	0.21982	9.34207	4.54918	10.65794
	Li_3PO_4	0.56795	9.75431	1.76072	10.24569

Weighed	Sought	Factor	Log of Factor +10	Reciprocal of Factor	Log of Reciprocal of Factor +10
Lithium (contd.)					
Li_2O	CO_2	1.47304	10.16821	0.67887	9.83179
	Li	0.46449	9.66698	2.15290	10.33303
	LiCl	2.83767	10.45296	0.35240	9.54704
	Li_2CO_3	2.47304	10.39323	0.40436	9.60677
	$LiHCO_3$	4.54890	10.65791	0.21983	9.34209
	Li_3PO_4	2.58364	10.41223	0.38705	9.58777
	Li_2SO_4	3.67975	10.56582	0.27175	9.43417
	SO_3	2.67872	10.42809	0.37317	9.57191
Li_3PO_4	Li	0.17979	9.25477	5.56204	10.74523
	LiCl	1.09835	10.04074	0.91046	9.95926
	Li_2CO_3	0.95718	9.98099	1.04474	10.01901
	$LiHCO_3$	1.76070	10.24569	0.56796	9.75432
	Li_2O	0.38705	9.58777	2.58365	10.41223
	$Li_2.SO_4.H_2O$	1.65761	10.21949	0.60328	9.78052
Li_2SO_4	Li	0.12623	9.10116	7.92205	10.89883
	LiCl	0.77118	9.88716	1.29671	10.11284
	Li_2O	0.27175	9.43417	3.67975	10.56582
	Li_3PO_4	0.70213	9.84642	1.42424	10.15358
	SO_3	0.72823	9.86227	1.37319	10.13773
$Li_2SO_4.H_2O$	Li_3PO_4	0.60328	9.78052	1.65761	10.22948
SO_3	Li_2O	0.37317	9.57191	2.67974	10.42809
	Li_2SO_4	1.37318	10.13773	0.72824	9.86227
Lutetium					
Lu_2O_3	Lu	0.87938	9.94418	1.13716	10.05582
Magnesium					
$BaSO_4$	$MgSO_4$	0.51574	9.71243	1.93896	10.28757
	$MgSO_4.7H_2O$	1.05605	10.02368	0.94692	9.97631
Br	Mg	0.15212	9.18219	6.57376	10.81781
	$MgBr_2$	1.15212	10.06150	0.86797	9.93850
	$MgBr_2.6H_2O$	1.82807	10.26200	0.54703	9.73801
Cl	Mg	0.34288	9.53514	2.91647	10.46486
	$MgCl_2$	1.34288	10.12804	0.74467	9.87196
	$MgCl_2.6H_2O$	2.86643	10.45734	0.34887	9.54266
CO_2	$MgCO_3$	1.91595	10.28239	0.52193	9.71761
	MgO	0.91595	9.96187	1.09176	10.03812
I	Mg	0.09579	8.98132	10.43950	11.01868
	MgI_2	1.09578	10.03972	0.91259	9.96028
Mg	Br	6.57262	10.81780	0.15212	9.18219
	Cl	2.91650	10.46487	0.34288	9.53514
	I	10.43965	11.01869	0.09579	8.98132
	$MgCO_3$	3.46829	10.54011	0.28833	9.45989
	MgO	1.65807	10.21960	0.60311	9.78040
	$Mg_2P_2O_7$	4.57731	10.66061	0.21847	9.33939
	$MgSO_4$	4.95122	10.69471	0.20197	9.30529
$MgBr_2$	Br	0.86796	9.93850	1.15213	10.06150
$MgBr.6H_2O$	Br	0.54702	9.73800	1.82809	10.26200
MgC_2O_4	Mg	0.21643	9.33532	4.62043	10.66468
	MgO	0.35886	9.55493	2.78660	10.44508
$Mg(C_9H_6NO)_2$	Mg	0.07777	8.89081	12.85843	11.10919
	$MgBr_2$	0.58898	9.77010	1.69785	10.22990
	$MgBr_2.6H_2O$	0.93455	9.97060	1.07003	10.02939
	$MgCO_3$	0.26972	9.43091	3.70755	10.56909
	$MgCl_2$	0.30458	9.48370	3.28321	10.51630
	$MgCl_2.6H_2O$	0.65014	9.81301	1.53813	10.18699
	$Mg(HCO_3)_2$	0.46813	9.67037	2.13616	10.32963
	MgO	0.12895	9.11042	7.75494	10.88958
	$MgSO_4$	0.38505	9.58552	2.59707	10.41448
	$MgSO_4.7H_2O$	0.78842	9.89676	1.26834	10.10324
$MgCl_2$	Cl	0.74467	9.87196	1.34288	10.12804
	$Mg_2P_2O_7$	1.16872	10.06771	0.85564	9.93229
$MgCl_2.6H_2O$	Cl	0.34887	9.54266	2.86640	10.45723
	$Mg_2P_2O_7$	0.54753	9.73841	1.82638	10.26159
$MgCl_2.KCl.6H_2O$	$Mg_2P_2O_7$	0.40059	9.60270	2.49632	10.39730
$MgCO_3$	CO_2	0.52193	9.71761	1.91597	10.28239
	Mg	0.28833	9.45989	3.46825	10.54011
	$Mg(HCO_3)_2$	1.73559	10.23945	0.57617	9.76055
	MgO	0.47807	9.67949	2.09174	10.32051
	$Mg_2P_2O_7$	1.31977	10.12049	0.75771	9.87950

Weighed	Sought	Factor	Log of Factor +10	Reciprocal of Factor	Log of Reciprocal of Factor +10
Magnesium (contd.)					
$Mg(HCO_3)_2$	$MgCO_3$	0.57617	9.76055	1.73560	10.23945
	MgO	0.27545	9.44004	3.63042	10.55996
	$Mg_2P_2O_7$	0.76041	9.88105	1.31508	10.11895
MgI_2	I	0.91258	9.96027	1.09579	10.03973
MgO	CO_2	1.09176	10.03812	0.91595	9.96187
	Mg	0.60311	9.78040	1.65807	10.21960
	$MgCO_3$	2.09176	10.32051	0.47807	9.67949
	$Mg(HCO_3)_2$	3.63045	10.55996	0.27545	9.44004
	$Mg_2P_2O_7$	2.76064	10.44101	0.36223	9.55898
	$MgSO_4$	2.98613	10.47511	0.33488	9.52489
	SO_3	1.98611	10.29800	0.50350	9.70200
$Mg_2P_2O_7$	Mg	0.21847	9.33939	4.57729	10.66061
	$MgCl_2$	0.85563	9.93229	1.16873	10.06771
	$MgCl_2.6H_2O$	1.82638	10.26159	0.54753	9.73841
	$MgCl_2.KCl.6H_2O$	2.49633	10.39730	0.40059	9.60270
	$MgCO_3$	0.75771	9.87950	1.31977	10.12049
	$Mg(HCO_3)_2$	1.31508	10.11896	0.76041	9.88105
	MgO	0.36224	9.55900	2.76060	10.44101
	$MgSO_4$	1.08168	10.03410	0.92449	9.96590
	$MgSO_4.7H_2O$	2.21488	10.34535	0.45149	9.65465
$MgSO_4$	$BaSO_4$	1.93896	10.28757	0.51574	9.71243
	Mg	0.20197	9.30529	4.85123	10.69471
	MgO	0.33488	9.52489	2.98614	10.47511
	$Mg_2P_2O_7$	0.92449	9.96590	1.08168	10.03410
	SO_3	0.66511	9.82289	1.50351	10.17711
$MgSO_4.7H_2O$	$BaSO_4$	0.94693	9.97632	1.05604	10.02368
	$Mg_2P_2O_7$	0.45149	9.65465	2.21489	10.34535
	SO_3	0.32482	9.51164	3.07863	10.48836
SO_3	MgO	0.50350	9.70200	1.98610	10.29800
	$MgSO_4$	1.50351	10.17711	0.66511	9.82289
	$MgSO_4.7H_2O$	3.07863	10.48836	0.32482	9.51164
Manganese					
$BaSO_4$	$MnSO_4$	0.64696	9.81088	1.54569	10.18912
CO_2	$MnCO_3$	2.61184	10.41694	0.38287	9.58305
	MnO	1.61184	10.20733	0.62041	9.79268
Mn	$MnCO_3$	2.09230	10.32062	0.47794	9.67937
	MnO	1.29122	10.11100	0.77446	9.88900
	MnO_2	1.58246	10.19933	0.63193	9.80067
	Mn_2O_3	1.43684	10.15741	0.69597	9.84259
	Mn_3O_4	1.38830	10.14248	0.72031	9.85752
	$Mn_2P_2O_7$	2.58308	10.41213	0.38713	9.58786
MnC_2O_4	Mn	0.38429	9.58466	2.60220	10.41534
	Mn_3O_4	0.53352	9.72715	1.87434	10.27285
$Mn(C_9H_6NO)_2$	Mn	0.16005	9.20426	6.24805	10.79574
	Mn_3O_4	0.22220	9.34674	4.50045	10.65326
$MnCO_3$	CO_2	0.38287	9.58305	2.61185	10.41695
	Mn	0.47794	9.67937	2.09231	10.32063
	$Mn(HCO_3)_2$	1.53961	10.18741	0.64952	9.81259
	MnO	0.61713	9.79038	1.62040	10.20962
	Mn_3O_4	0.66353	9.82186	1.50709	10.17814
	$Mn_2P_2O_7$	1.23456	10.09152	0.81001	9.90849
	MnS	0.75689	9.87903	1.32120	10.12097
	$MnSO_4$	1.31365	10.11848	0.76124	9.88152
$Mn(HCO_3)_2$	$MnCO_3$	0.64952	9.81259	1.53960	10.18741
	MnO	0.40084	9.60297	2.49476	10.39703
	Mn_3O_4	0.43098	9.63446	2.32029	10.36555
MnO	CO_2	0.62041	9.79268	1.61184	10.20733
	Mn	0.77446	9.88900	1.29122	10.11099
	$MnCO_3$	1.62041	10.20963	0.61713	9.79038
	$Mn(HCO_3)_2$	2.49479	10.39703	0.40084	9.60297
	Mn_2O_3	1.11277	10.04641	0.89866	9.95360
	Mn_3O_4	1.07518	10.03148	0.93008	9.96852
	$Mn_2P_2O_7$	2.00049	10.30114	0.49988	9.69887
	MnS	1.22647	10.08865	0.81535	9.91134
	$MnSO_4$	2.12865	10.32811	0.46978	9.67189
	SO_3	1.12864	10.05255	0.88602	9.94744
MnO_2	Mn	0.63193	9.80067	1.58245	10.19933
	Mn_3O_4	0.87731	9.94315	1.13985	10.05684

Weighed	Sought	Factor	Log of Factor +10	Reciprocal of Factor	Log of Reciprocal of Factor +10
Manganese (contd.)					
MnO_2	$Mn_2P_2O_7$	1.63233	10.21281	0.61262	9.78719
Mn_2O_3	Mn	0.69597	9.84259	1.43684	10.15741
	MnO	0.89865	9.95359	1.11278	10.04641
	Mn_3O_4	0.96622	9.98508	1.03496	10.01492
Mn_3O_4	Mn	0.72030	9.85751	1.38831	10.14249
	$MnCO_3$	1.50709	10.17814	0.66353	9.82186
	$Mn(HCO_3)_2$	2.32032	10.36555	0.43098	9.63446
	MnO	0.93007	9.96852	1.07519	10.03149
	MnO_2	1.13984	10.05684	0.87732	9.94316
	Mn_2O_3	1.03496	10.01492	0.96622	9.98508
	$MnSO_4$	1.97978	10.29662	0.50511	9.70339
$Mn_2P_2O_7$	Mn	0.38713	0.58786	2.58311	10.41214
	$MnCO_3$	0.81000	9.90849	1.23457	10.09152
	MnO	0.49987	9.69886	2.00052	10.30114
	MnO_2	0.61262	9.78719	1.63233	10.21281
	$MnSO_4$	1.06405	10.02696	0.93981	9.97304
MnS	Mn	0.63146	9.80035	1.58363	10.19966
	$MnCO_3$	1.32120	10.12097	0.75689	9.87903
	MnO	0.81535	9.91134	1.22647	10.08865
	$MnSO_4$	1.73559	10.23945	0.57617	9.76055
$MnSO_4$	$BaSO_4$	1.54570	10.18913	0.64696	9.81088
	Mn	0.36383	9.56090	2.74854	10.43910
	MnO	0.46978	9.67189	2.12866	10.32811
	Mn_3O_4	0.50510	9.70338	1.97981	10.29663
	$Mn_2P_2O_7$	0.93980	9.97304	1.06406	10.02696
	MnS	0.57617	9.76055	1.73560	10.23945
	SO_3	0.53021	9.72445	1.88605	10.27555
SO_3	MnO	0.88603	9.94745	1.12863	10.05255
	$MnSO_4$	1.88604	10.27555	0.53021	9.72445
Mercury					
Hg	HgCl	1.17673	10.07068	0.84981	9.92932
	$HgCl_2$	1.35351	10.13146	0.73882	9.86854
	HgO	1.07976	10.03332	0.92613	9.96667
	HgS	1.15983	10.06440	0.86220	9.93561
$Hg_3(AsO_4)_2$	Hg	0.68413	9.83514	1.46171	10.16486
$Hg(C_{12}H_{10}ONS)_2$	Hg	0.31681	9.50080	3.15647	10.49920
HgCl	Hg	0.84981	9.92932	1.17673	10.07068
	$HgCl_2$	1.15023	10.06079	0.86939	9.93921
	$HgNO_3$	1.11248	10.04629	0.89889	9.95371
	HgO	0.91760	9.96265	1.08980	10.03735
	Hg_2O	0.88369	9.94630	1.13162	10.05370
	HgS	0.98564	9.99372	1.01457	10.00629
$HgCl_2$	Hg	0.73882	9.86854	1.35351	10.13146
	HgCl	0.86939	9.93921	1.15023	10.06079
	HgS	0.85691	9.93294	1.16698	10.06706
$Hg(CN)_2$	HgS	0.92091	9.96422	1.08588	10.03578
Hg_2CrO_4	Hg	0.77572	9.88971	1.28912	10.11029
$HgNO_3$	HgCl	0.89889	9.95371	1.11248	10.04629
	HgS	0.88598	9.94742	1.12869	10.05257
$Hg(NO_3)_2$	HgS	0.71673	9.85536	1.39523	10.14464
$Hg(NO_3)_2.H_2O$	HgS	0.67903	9.83189	1.47269	10.16812
HgO	Hg	0.92613	9.96667	1.07977	10.03333
	HgCl	1.08980	10.03735	0.91760	9.96265
	HgS	1.07415	10.03106	0.93097	9.96894
Hg_2O	HgCl	1.13160	10.05369	0.88370	9.94630
	HgS	1.11535	10.04741	0.89658	9.95259
HgS	Hg	0.86220	9.93561	1.15982	10.06440
	HgCl	1.01457	10.00629	0.98564	9.99372
	$HgCl_2$	1.16698	10.06706	0.85691	9.93294
	$Hg(CN)_2$	1.08588	10.03578	0.92091	9.96422
	$HgNO_3$	1.12869	10.05257	0.88598	9.94743
	$Hg(NO_3)_2$	1.39522	10.14464	0.71673	9.85536
	$Hg(NO_3)_2.H_2O$	1.47268	10.16811	0.67903	9.83189
	HgO	0.93097	9.96894	1.07415	10.03106
	Hg_2O	0.89656	9.95258	1.11537	10.04741
	$HgSO_4$	1.27509	10.10554	0.78426	9.89446
$HgSO_4$	HgS	0.78426	9.89446	1.27509	10.10554
Molybdenum					
Mo	MoO_3	1.50031	10.17618	0.66653	9.82382

Weighed	Sought	Factor	Log of Factor +10	Reciprocal of Factor	Log of Reciprocal of Factor +10
Molybdenum					
Mo	MoC	1.12518	10.05122	0.88875	9.94878
	MoS_3	2.00261	10.30159	0.49935	9.69841
	$PbMoO_4$	3.82666	10.58282	0.26132	9.41717
MoC	C	0.11127	9.04638	8.89715	10.95362
	Mo	0.88874	9.94877	1.12519	10.05122
$MoO_2(C_9H_6NO)_2$	Mo	0.23049	9.36265	4.33858	10.63735
	MoO_3	0.34580	9.53883	2.89184	10.46118
MoO_3	Mo	0.66653	9.82382	1.50031	10.17617
	MoS_3	1.33479	10.12541	0.74918	9.87459
	$(NH_4)_2MoO_4$	1.36175	10.13410	0.73435	9.86590
	$(NH_4)_3PO_4.12MoO_3$	1.08630	10.03595	0.92056	9.96405
	$PbMoO_4$	2.55058	10.40664	0.39207	9.59336
MoS_3	Mo	0.49935	9.69841	2.00260	10.30159
	MoO_3	0.74918	9.87459	1.33479	10.12541
	$(NH_4)_2MoO_4$	1.02019	10.00868	0.98021	9.99132
MoSi	Mo	0.77352	9.88847	1.29279	10.11152
	Si	0.22645	9.35497	4.41599	10.64503
$(NH_4)_2MoO_4$	MoO_3	0.73435	9.86590	1.36175	10.13410
	MoS_3	0.98021	9.99132	1.02019	10.00868
	$(NH_4)_3PO_4.12MoO_3$	0.79771	9.90185	1.25359	10.09816
	$PbMoO_4$	1.87302	10.27254	0.53390	9.72746
$(NH_4)_3PO_4.12MoO_3$	MoO_3	0.92054	9.96404	1.08632	10.03596
	$(NH_4)_2MoO_4$	1.25359	10.09816	0.79771	9.90185
$PbMoO_4$	Mo	0.26132	9.41717	3.82673	10.58283
	MoO_3	0.39207	9.59336	2.55056	10.40664
	$(NH_4)_2MoO_4$	0.53390	9.72746	1.87301	10.27254
Neodymium					
Nd	Nd_2O_3	1.16639	10.06684	0.85735	9.93315
Nd_2O_3	Nd	0.85735	9.93316	1.16638	10.06684
Nickel					
Ni	$Ni(C_4H_7N_2O_2)_2$	4.92148	10.69209	0.20319	9.30790
	$Ni(NO_3)_2.6H_2O$	4.95231	10.69481	0.20193	9.30520
	NiO	1.27254	10.10467	0.78584	9.89533
	$NiSO_4$	2.63618	10.42099	0.37934	9.57903
Ni	$NiSO_4.7H_2O$	4.78419	10.67981	0.20902	9.32019
$Ni(C_4H_7N_2O_2)_2$	Ni	0.20319	9.30790	4.92150	10.69210
	NiO	0.25857	9.41258	3.86742	10.58742
$(Ni[C_5H_5N]_4)(SCN)_2$	Ni	0.11950	9.07737	8.36820	10.92263
	NiO	0.15207	9.18204	6.57592	10.81776
$Ni(C_7H_6O_2N)_2$	Ni	0.17736	9.24886	5.63825	10.75114
	NiO	0.22573	9.35359	4.43007	10.64641
$Ni(C_9H_6NO)_2$	Ni	0.16918	9.22835	5.91086	10.77165
	NiO	0.21529	9.33303	4.64490	10.66698
$Ni(NO_3)_2.6H_2O$	Ni	0.20193	9.30520	4.95221	10.69480
	NiO	0.25696	9.40987	3.89166	10.59013
	$NiSO_4$	0.53231	9.72616	1.87860	10.27383
NiO	Ni	0.78584	9.89533	1.27252	10.10467
	$Ni(C_4H_7N_2O_2)_2$	3.86749	10.58743	0.25857	9.41258
	$Ni(NO_3)_2.6H_2O$	3.89171	10.59014	0.25696	9.40987
	$NiSO_4$	2.07161	10.31631	0.48272	9.68370
	$NiSO_4.7H_2O$	3.75960	10.57514	0.26599	9.42487
$NiSO_4$	Ni	0.37934	9.57903	2.63616	10.42098
	$Ni(NO_3)_2.6H_2O$	1.87859	10.27384	0.53231	9.72616
	NiO	0.48272	9.68370	2.07159	10.31630
	$NiSO_4.7H_2O$	1.81482	10.25884	0.55102	9.74117
$NiSO_4.7H_2O$	Ni	0.20902	9.32019	4.78423	10.67981
	NiO	0.26599	9.42487	3.75954	10.57514
	$NiSO_4$	0.55102	9.74117	1.81482	10.25884
Niobium					
C	Nb	7.73497	10.88846	0.12928	9.11153
	NbC	8.73493	10.94126	0.11448	9.05873
	$\frac{1}{2}Nb_2O_5$	11.06508	11.04396	0.09038	8.95607
Nb	C	0.12928	9.11153	7.73515	10.88848
	Nb_2O_5	1.43053	10.15550	0.69904	9.84450
NbC	C	0.11448	9.05873	8.73515	10.94127
	Nb	0.88552	9.94720	1.12928	10.05280
Nb_2O_5	Nb	0.69904	9.84450	1.43053	10.15550
Nitrogen					
$AgNO_2$	HNO_2	0.30552	9.48504	3.27311	10.51496

Weighed	Sought	Factor	Log of Factor +10	Reciprocal of Factor	Log of Reciprocal of Factor +10
Nitrogen (contd.)					
$AgNO_2$	N_2O_3	0.24698	9.39266	4.04891	10.60733
$C_{20}H_{16}N_4.HNO_3$	N	0.37312	9.57185	2.68010	10.42815
	NO_3	0.16517	9.21793	6.05437	10.78207
HNO_2	$AgNO_2$	3.27310	10.51496	0.30552	9.48504
HNO_3	N	0.22228	9.34690	4.49883	10.65310
	NH_3	0.27027	9.43180	3.70000	10.56820
	NH_4Cl	0.84889	9.92885	1.17801	10.07115
	$(NH_4)_2PtCl_6$	3.52221	10.54682	0.28391	9.45318
	NO	0.47619	9.67778	2.10000	10.32222
	Pt	1.54801	10.18977	0.64599	9.81023
	SO_3	0.63528	9.80297	1.57411	10.19703
	N_2O_5	0.53413	9.72765	1.87220	10.27235
N	HNO_3	4.49877	10.65310	0.22229	9.34692
	$NaNO_3$	6.06816	10.78306	0.16479	9.21693
	NH_3	1.21589	10.08489	0.82244	9.91510
	NH_4Cl	3.81902	10.58195	0.26185	9.41805
	$(NH_4)_2PtCl_6$	15.84563	11.19991	0.06311	8.80010
	$(NH_4)_2SO_4$	4.71699	10.67367	0.21200	9.32634
	NO_2	3.28454	10.51648	0.30446	9.48353
	NO_3	4.42680	10.64609	0.22590	9.35392
	N_2O_3	2.71341	10.43352	0.36854	9.56648
	N_2O_5	3.85566	10.58610	0.25936	9.41390
	Pt	6.96416	10.84287	0.14359	9.15712
	SO_3	2.85798	10.45606	0.34990	9.54394
$NaNO_3$	N	0.16479	9.21693	6.06833	10.78307
	N_2O_5	0.63539	9.80304	1.57384	10.19696
NH_3	HNO_3	3.69998	10.56820	0.27027	9.43180
	N	0.82244	9.91510	1.21589	10.08489
	NO_3	3.64079	10.56119	0.27467	9.43881
	N_2O_5	3.17107	10.50121	0.31535	9.49879
$NH_4B(C_6H_5)_4$	N	0.04153	8.61836	24.07898	11.38164
	NH_3	0.05049	8.70320	19.80590	11.29679
	NH_4	0.05348	8.72819	18.69858	11.27181
NH_4Cl	HNO_3	1.17799	10.07115	0.84890	9.92886
	N	0.26185	9.41805	3.81898	10.58195
	NO_3	1.15914	10.06413	0.86271	9.93586
	N_2O_5	1.00959	10.00415	0.99050	9.99585
$(NH_4)_2PtCl_6$	HNO_3	0.28392	9.45320	3.52212	10.54680
	N	0.06311	8.80010	15.84535	11.19990
	NO_3	0.27938	9.44620	3.57935	10.55380
	N_2O_5	0.24333	9.38620	4.10965	10.61380
$(NH_4)_2SO_4$	N	0.21200	9.32634	4.71698	10.67367
	N_2O_5	0.81740	9.91243	1.22339	10.08757
NO	HNO_3	2.10000	10.32222	0.47619	9.67778
	NO_2	1.53320	10.18560	0.65223	9.81440
	NO_3	2.06641	10.31522	0.48393	9.68478
	N_2O_3	1.26660	10.10264	0.78952	9.89736
	N_2O_5	1.79980	10.25522	0.55562	9.74478
NO_2	N	0.30446	9.48353	3.28450	10.51647
	NO	0.65223	9.81440	1.53320	10.18560
NO_3	N	0.22590	9.35392	4.42674	10.64608
	NH_3	0.27467	9.43881	3.64073	10.56119
	NH_4Cl	0.86270	9.93586	1.15915	10.06413
	NO	0.48393	9.68478	2.06641	10.31522
	Pt	1.57318	10.19678	0.63566	9.80322
N_2O_3	$AgNO_2$	4.04875	10.60732	0.24699	9.39268
	N	0.36854	9.56648	2.71341	10.43352
	NO	0.78951	9.89736	1.26661	10.10264
N_2O_5	KNO_3	1.87217	10.27235	0.53414	9.72766
	N	0.25936	9.41390	3.85564	10.58610
	$NaNO_3$	1.57382	10.19695	0.63540	9.80305
	NH_3	0.31535	9.49879	3.17108	10.50121
	NH_4Cl	0.99049	9.99585	1.00960	10.00415
	$(NH_4)_2PtCl_6$	4.10965	10.61380	0.24333	9.38620
	$(NH_4)_2SO_4$	1.22339	10.08757	0.81740	9.91243
	NO	0.55562	9.74478	1.79980	10.25522
	Pt	1.80621	10.25677	0.55365	9.74324
	SO_3	0.74124	9.86996	1.34909	10.13004

Weighed	Sought	Factor	Log of Factor +10	Reciprocal of Factor	Log of Reciprocal of Factor +10
Nitrogen (contd.)					
Pt	HNO_3	0.64595	9.81020	1.54811	10.18980
	N	0.14358	9.15709	6.96476	10.84291
	NO_3	0.63562	9.80320	1.57327	10.19680
	N_2O_5	0.55364	9.74323	1.80623	10.25678
SO_3	HNO_3	1.57410	10.19703	0.63528	9.80297
	N	0.34990	9.54394	2.85796	10.45605
	N_2O_5	1.34908	10.13004	0.74125	9.86996
Osmium					
Os	OsO_4	1.33649	10.12597	0.74823	9.87404
OsO_4	Os	0.74823	9.87404	1.33649	10.12597
Palladium					
K_2PdCl_6	Pd	0.26781	9.42783	3.73399	10.57217
	$PdCl_2.2H_2O$	0.53687	9.72987	1.86265	10.27013
Pd	K_2PdCl_6	3.73402	10.57217	0.26781	9.42783
	$PdCl_2.2H_2O$	2.00460	10.30205	0.49883	9.69795
	PdI_2	3.38534	10.52960	0.29539	9.47040
	$Pd(NO_3)_2$	2.16541	10.33554	0.46181	9.66446
$Pd(C_4H_7N_2O_2)_2$	Pd	0.31610	9.49982	3.16356	10.50018
$PdCl_2.2H_2O$	K_2PdCl_6	1.86263	10.27012	0.53688	9.72988
	Pd	0.49883	9.69795	2.00469	10.30204
$PdCl_2.C_{12}H_8N_2$	Pd	0.29762	9.47366	3.35999	10.52634
PdI_2	Pd	0.29539	9.47040	3.38536	10.52960
$Pd(NO_3)_2$	Pd	0.46181	9.66446	2.16539	10.33554
Phosphorus					
Ag_3PO_4	P	0.07400	8.86923	13.51351	11.13077
	PO_4	0.22689	9.35582	4.40742	10.64418
	P_2O_5	0.16954	9.22927	5.89831	10.77072
$Ag_4P_2O_7$	P	0.10232	9.00996	9.77326	10.99004
	PO_4	0.31374	9.49657	3.18735	10.50343
	P_2O_5	0.23446	9.37007	4.26512	10.62993
Al_2O_3	P_2O_5	1.39214	10.14368	0.71832	9.85632
$AlPO_4$	PO_4	0.77875	9.89140	1.28411	10.10860
	P_2O_5	0.58196	9.76489	1.71833	10.23511
$Ca_3(PO_4)_2$	P_2O_5	0.45762	9.66051	2.18522	10.33949
$FePO_4$	PO_4	0.62971	9.79914	1.58803	10.20086
	P_2O_5	0.47058	9.67263	2.12504	10.32737
$Mg_2P_2O_7$	H_3PO_4	0.88059	9.94471	1.13560	10.05523
	Na_2HPO_4	1.27565	10.10573	0.78391	9.89427
	$Na_2HPO_4.12H_2O$	3.21828	10.50763	0.31073	9.49238
	$NaNH_4.HPO_4.4H_2O$	1.87870	10.27386	0.53228	9.72614
	P	0.27833	9.44456	3.59286	10.55544
	PO_4	0.85341	9.93116	1.17177	10.06884
	P_2O_5	0.63776	9.80466	1.56799	10.19535
Na_2HPO_4	$Mg_2P_2O_7$	0.78392	9.89427	1.27564	10.10573
	P_2O_5	0.49995	9.69893	2.00020	10.30105
$Na_2HPO_4.12H_2O$	$Mg_2P_2O_7$	0.31072	9.49237	3.21833	10.50763
	P_2O_5	0.19816	9.29702	5.04643	10.70299
$NaNH_4HPO_4.4H_2O$	$Mg_2P_2O_7$	0.53228	9.72614	1.87871	10.27386
	P_2O_5	0.33946	9.53079	2.94586	10.46921
$(NH_4)_3PO_4.12MoO_3$	P	0.01651	8.21775	60.56935	11.78223
	PO_4	0.05061	8.70424	19.75894	11.29577
	P_2O_5	0.03782	8.57772	26.44104	11.42227
P	Ag_3PO_4	13.51403	11.13078	0.07400	8.86923
	$Ag_4P_2O_7$	9.77314	10.99004	0.10232	9.00996
	$Mg_2P_2O_7$	3.59282	10.55544	0.27833	9.44456
	$(NH_4)_3PO_4.12MoO_3$	60.5786	11.78234	0.01651	8.21775
	P_2O_5	2.29136	10.36009	0.43642	9.63990
	$P_2O_5.24MoO_3$	58.0564	11.76385	0.01723	8.23616
	$U_2P_2O_{11}$	11.52587	11.06167	0.08676	8.93832
PO_4	Ag_3PO_4	4.40744	10.64418	0.22689	9.35582
	$Ag_4P_2O_7$	3.18739	10.50343	0.31374	9.49657
	$AlPO_4$	1.28410	10.10860	0.77876	9.89140
	$FePO_4$	1.58803	10.20086	0.62971	9.79914
	$Mg_2P_2O_7$	1.17175	10.06884	0.85342	9.93116
	$(NH_4)_3PO_4.12MoO_3$	19.7570	11.29572	0.05062	8.70432
	$P_2O_5.24MoO_3$	18.9344	11.27725	0.05281	8.72272
	$U_2P_2O_{11}$	3.75902	10.57507	0.26603	9.42493
P_2O_5	Ag_3PO_4	5.89780	10.77069	0.16955	9.22930

Weighed	Sought	Factor	Log of Factor +10	Reciprocal of Factor	Log of Reciprocal of Factor +10
Phosphorus (contd.)					
P_2O_5	$Ag_4P_2O_7$	4.26520	10.62994	0.23446	9.37007
	Al_2O_3	0.71831	9.85631	1.39216	10.14369
	$AlPO_4$	1.71831	10.23510	0.58197	9.76440
	$Ca_3(PO_4)_2$	2.18521	10.33949	0.45762	9.66051
	$FePO_4$	2.12502	10.32736	0.47058	9.67263
	$Mg_2P_2O_7$	1.56798	10.19534	0.63776	9.80466
	Na_2HPO_4	2.00020	10.30107	0.49995	9.69893
	$Na_2HPO_4.12H_2O$	5.04623	10.70297	0.19817	9.29704
	$NaNH_4HPO_4.4H_2O$	2.94578	10.46920	0.33947	9.53080
	$(NH_4)_3PO_4.12MoO_3$	26.4377	11.42222	0.03783	8.57784
	P	0.43642	9.63990	2.29137	10.36010
	$P_2O_5.24MoO_3$	25.3370	11.40376	0.03947	8.59622
	$U_2P_2O_{11}$	5.03013	10.70158	0.19880	9.29842
$P_2O_5.24MoO_3$	P	0.01722	8.23603	58.07201	11.76397
	PO_4	0.05281	8.72272	18.93581	11.27728
	P_2O_5	0.03947	8.59627	25.33570	11.40373
$U_2P_2O_{11}$	P	0.08676	8.93832	11.52605	11.06176
	PO_4	0.26603	9.42493	3.75897	10.57507
	P_2O_5	0.19880	9.29842	5.03018	10.70158
Platinum					
$H_2PtCl_6.6H_2O$	K_2PtCl_6	0.93852	9.97244	1.06551	10.02756
	Pt	0.37673	9.57603	2.65442	10.42397
K_2PtCl_6	$H_2PtCl_6.6H_2O$	1.06551	10.02756	0.93852	9.97244
	Pt	0.40141	9.60359	2.49122	10.39541
	$PtCl_4$	0.69320	9.84086	1.44259	10.15915
	$PtCl_4.5H_2O$	0.87854	9.94376	1.13825	10.05624
$(NH_4)_2PtCl_6$	Pt	0.43950	9.64296	2.27531	10.35704
	$PtCl_4$	0.75897	9.88022	1.31758	10.11978
	$PtCl_6$	0.91854	9.96310	1.08868	10.03690
Pt	$H_2PtCl_6.6H_2O$	2.65442	10.42397	0.37673	9.57603
	K_2O	0.48287	9.68383	2.07095	10.31617
	K_2PtCl_6	2.49121	10.39641	0.40141	9.60359
	$(NH_4)_2PtCl_6$	2.27531	10.35704	0.43950	9.64296
	$PtCl_4$	1.72690	10.23727	0.57907	9.76273
	$PtCl_4.5H_2O$	2.18863	10.34017	0.45691	9.65983
$PtCl_4$	K_2PtCl_6	1.44259	10.15915	0.69320	9.84086
	$(NH_4)_2PtCl_6$	1.31757	10.11978	0.75897	9.88022
	Pt	0.57907	9.76273	1.72691	10.23727
$PtCl_4.5H_2O$	K_2PtCl_6	1.13825	10.05624	0.87854	9.94376
	Pt	0.45691	9.65983	2.18861	10.34017
$PtCl_6$	K_2PtCl_6	1.19199	10.07628	0.83893	9.92373
	$(NH_4)_2PtCl_6$	1.08869	10.03691	0.91854	9.96310
Potassium					
Ag	KBr	1.10328	10.04269	0.90639	9.95732
	KCl	0.69116	9.83958	1.44684	10.16042
	$KClO_3$	1.13612	10.05543	0.88019	9.94458
	$KClO_4$	1.28444	10.10872	0.77855	9.89129
	KCN	0.60369	9.78081	1.65648	10.21919
	KI	1.53895	10.18723	0.64979	9.81277
AgBr	KBr	0.63378	9.80194	1.57783	10.19806
	$KBrO_3$	0.88939	9.94909	1.12437	10.05091
AgCl	KCl	0.52019	9.71616	1.92237	10.28384
	$KClO_3$	0.85508	9.93201	1.16948	10.06799
	$KClO_4$	0.96672	9.98530	1.03443	10.01470
AgCN	KCN	0.48638	9.68698	2.05601	10.31302
AgI	KI	0.70709	9.84947	1.41425	10.15053
	KIO_3	0.91154	9.95978	1.09704	10.04022
$BaCrO_4$	K_2CrO_4	0.76658	9.88456	1.30450	10.11544
	$K_2Cr_2O_7$	0.58064	9.76391	1.72224	10.23609
$BaSO_4$	$KHSO_4$	0.58343	9.76599	1.71400	10.23401
	K_2S	0.47244	9.67435	2.11667	10.32565
	K_2SO_4	0.74664	9.87311	1.33933	10.12689
Br	K	0.48933	9.68960	2.04361	10.31040
	KBr	1.48933	10.17299	0.67144	9.82701
CaF_2	$KF.2H_2O$	2.41114	10.38223	0.41474	9.61778
$CaSO_4$	$KF.2H_2O$	1.38286	10.14078	0.72314	9.85922
Cl	K	1.10292	10.04255	0.90668	9.95745
	KCl	2.10292	10.32282	0.47553	9.67718

Weighed	Sought	Factor	Log of Factor +10	Reciprocal of Factor	Log of Reciprocal of Factor +10
Potassium (contd.)					
Cl	$KClO_3$	3.45677	10.53867	0.28929	9.46133
	$KClO_4$	3.90808	10.59196	0.25588	9.40804
	K_2O	1.32856	10.12338	0.75269	9.87662
CO_2	$KHCO_3$	2.27491	10.35696	0.43958	9.64304
	K_2CO_3	3.14049	10.49700	0.31842	9.50300
	K_2O	2.14049	10.33051	0.46718	9.66948
I	KI	1.30811	10.11665	0.76446	9.88335
	KIO_3	1.68634	10.22695	0.59300	9.77306
K	Br	2.04360	10.31040	0.48933	9.68960
	Cl	0.90668	9.95745	1.10292	10.04256
	KBr	3.04360	10.48339	0.32856	9.51661
	KCl	1.90668	10.28028	0.52447	9.71972
	$KClO_3$	3.13419	10.49614	0.31906	9.50387
	$KClO_4$	3.54337	10.54941	0.28222	9.45059
	KI	4.24546	10.62793	0.23555	9.37208
	KNO_3	2.58572	10.41259	0.38674	9.58742
	K_2O	1.20458	10.08084	0.83016	9.91916
	K_2PtCl_6	6.21467	10.79342	0.16091	9.20658
	K_2SO_4	2.22835	10.34799	0.44876	9.65201
	Pt	2.49463	10.39701	0.40086	9.60299
K_3AsO_4	$Mg_2As_2O_7$	0.60584	9.78236	1.65060	10.21764
$K(B[C_6H_5]_4)$	K	0.10912	9.03790	9.16422	10.96210
	K_2O	0.13144	9.11873	7.60803	10.88127
KBr	Ag	0.90639	9.95732	1.10328	10.04269
	AgBr	1.57783	10.19806	0.63378	9.80194
	Br	0.67144	9.82701	1.48934	10.17299
	K	0.32856	9.51662	3.04358	10.48338
	K_2O	0.39580	9.59748	2.52653	10.40253
$KBrO_3$	AgBr	1.12436	10.05091	0.88939	9.94909
$KC_{12}H_4O_{12}N_7$	K	0.08192	8.91339	12.20703	11.08661
	K_2O	0.09868	8.99423	10.13377	11.00577
KCl	Ag	1.44685	10.16043	0.69116	9.83958
	AgCl	1.92238	10.28384	0.52019	9.71616
	Cl	0.47553	9.67718	2.10292	10.32283
	K	0.52447	9.71972	1.90669	10.28028
	$KClO_3$	1.64379	10.21585	0.60835	9.78415
	$KClO_4$	1.85840	10.26914	0.53810	9.73086
	K_2CO_3	0.92692	9.96704	1.07884	10.03296
	$K_2Cr_2O_7$	1.97299	10.29512	0.50684	9.70487
	$KHCO_3$	1.34289	10.12804	0.74466	9.87196
	KNO_3	1.35614	10.13230	0.73739	9.86770
	K_2O	0.63180	9.80058	1.58278	10.19942
	K_2PtCl_6	3.25941	10.51314	0.30680	9.48686
	K_2SO_4	1.16871	10.06770	0.85564	9.93229
	Pt	1.30836	10.11673	0.76432	9.88328
$KClO_3$	Ag	0.88019	9.94458	1.13612	10.05543
	AgCl	1.16948	10.06799	0.85508	9.93201
	Cl	0.28929	9.46133	3.45674	10.53867
	KCl	0.60835	9.78415	1.64379	10.21584
$KClO_4$	Ag	0.77855	9.89129	1.28444	10.10872
	AgCl	1.03443	10.01470	0.96672	9.98530
	Cl	0.25588	9.40804	3.90808	10.59196
	K	0.28222	9.45059	3.54333	10.54941
	KCl	0.53810	9.73086	1.85839	10.26914
	K_2O	0.33995	9.53142	2.94161	10.46858
KCN	Ag	1.65648	10.21918	0.60369	9.78081
	AgCN	2.05602	10.31302	0.48638	9.68698
K_2CO_3	CO_2	0.31842	9.50300	3.14051	10.49700
	KCl	1.07883	10.03295	0.92693	9.96705
	K_2O	0.68158	9.83352	1.46718	10.16648
	KOH	0.81191	9.90951	1.23166	10.09049
	K_2PtCl_6	3.51638	10.54610	0.28438	9.45390
	K_2SO_4	1.26085	10.10067	0.79312	9.89934
K_2CrO_4	$BaCrO_4$	1.30449	10.11544	0.76658	9.88456
$K_2Cr_2O_7$	$BaCrO_4$	1.72220	10.23608	0.58065	9.76391
	KCl	0.50685	9.70488	1.97297	10.29512
	K_2O	0.32021	9.50543	3.12295	10.49456
$KF.2H_2O$	CaF_2	0.41474	9.61778	2.41115	10.38222

Weighed	Sought	Factor	Log of Factor +10	Reciprocal of Factor	Log of Reciprocal of Factor +10
Potassium (contd.)					
$KF.2H_2O$	$CaSO_4$	0.72314	9.85922	1.38286	10.14078
K_2HAsO_4	$Mg_2As_2O_7$	0.71165	9.85227	1.40519	10.14774
$KHCO_3$	KCl	0.74466	9.87196	1.34289	10.12804
	K_2O	0.47046	9.67252	2.12558	10.32748
	K_2PtCl_6	2.42716	10.38510	0.41200	9.61490
	K_2SO_4	0.87029	9.93966	1.14904	10.06034
$KHSO_4$	$BaSO_4$	1.71401	10.23401	0.58343	9.76599
	K_2SO_4	0.63988	9.80610	1.56279	10.19389
KI	Ag	0.64980	9.81278	1.53894	10.18722
	AgI	1.41425	10.15053	0.70709	9.84947
	I	0.76446	9.88335	1.30811	10.11665
	K	0.23555	9.37208	4.24538	10.62792
	K_2O	0.28374	9.45292	3.52435	10.54708
KIO_3	AgI	1.09705	10.04023	0.91154	9.95978
	I	0.59300	9.77305	1.68634	10.22695
$KMnO_4$	Mn_2O_3	0.49948	9.69852	2.00208	10.30148
	MnS	0.55051	9.74007	1.81650	10.25924
K_2MnO_4	Mn_2O_3	0.40041	9.60250	2.49744	10.39749
	MnS	0.44132	9.64475	2.26593	10.35525
KNO_2	K_2SO_4	1.02380	10.01022	0.97675	9.98978
	N_2O_3	0.44656	9.64988	2.23934	10.35012
KNO_3	K	0.38674	9.58742	2.58572	10.41260
	KCl	0.73739	9.86770	1.35613	10.13230
	K_2O	0.46586	9.66826	2.14657	10.33174
	K_2PtCl_6	2.40344	10.38083	0.41607	9.61917
	K_2SO_4	0.86179	9.93540	1.16038	10.06460
	N	0.13853	9.14154	7.21865	10.85846
	NH_3	0.16844	9.22645	5.93683	10.77356
	NO	0.29678	9.47243	3.36950	10.52757
	N_2O_5	0.53414	9.72766	1.87217	10.27235
K_2O	Cl	0.75269	9.87662	1.32857	10.12338
	CO_2	0.46718	9.66948	2.14050	10.33052
	K	0.83016	9.91916	1.20459	10.08085
	KBr	2.52667	10.40255	0.39578	9.59745
	KCl	1.58284	10.19944	0.63178	9.80057
	$KClO_3$	2.60186	10.41529	0.38434	9.58472
	$KClO_4$	2.94155	10.46858	0.33996	9.53143
	K_2CO_3	1.46718	10.16648	0.68158	9.83352
	$K_2Cr_2O_7$	3.12294	10.49456	0.32021	9.50543
	$KHCO_3$	2.12558	10.32748	0.47046	9.67252
	KI	3.52439	10.54709	0.28374	9.45292
	KNO_3	2.14655	10.33174	0.46586	9.66826
	KOH	1.19122	10.07599	0.83948	9.92401
	K_2PtCl_6	5.15918	10.71258	0.19383	9.28742
	K_2SO_4	1.84990	10.26715	0.54057	9.73285
	N_2O_5	1.14657	10.05940	0.87217	9.94060
KOH	K_2CO_3	1.23164	10.09048	0.81193	9.90952
	K_2O	0.83946	9.92400	1.19124	10.07600
K_2PtCl_6	K	0.16091	9.20658	6.21465	10.79342
	KCl	0.30680	9.48686	3.25948	10.51315
	K_2CO_3	0.28438	9.45390	3.51642	10.54610
	$KHCO_3$	0.41200	9.61490	2.42718	10.38510
	KNO_3	0.41606	9.61916	2.40350	10.38084
	K_2O	0.19383	9.28742	5.15916	10.71258
	K_2SO_4	0.35856	9.55456	2.78893	10.44544
	$K_2SO_4.Al_2(SO_4)_3.24H_2O$	1.95218	10.29052	0.51225	9.70948
	$K_2SO_4.Cr_2(SO_4)_3.24H_2O$	2.05512	10.31283	0.48659	9.68716
K_2S	$BaSO_4$	2.11666	10.32565	0.47244	9.67435
	K_2SO_4	1.58039	10.19877	0.63276	9.80124
K_2SiO_3	SiO_2	0.38943	9.59043	2.56786	10.40958
K_2SO_4	$BaSO_4$	1.33933	10.12689	0.74664	9.87311
	K	0.44876	9.65201	2.22836	10.34799
	KCl	0.85565	9.93230	1.16870	10.06770
	K_2CO_3	0.79312	9.89934	1.26084	10.10066
	$KHCO_3$	1.14904	10.06034	0.87029	9.93966
	$KHSO_4$	1.56281	10.19390	0.63987	9.80609
	KNO_2	0.97676	9.98979	1.02379	10.01021
	KNO_3	1.16038	10.06460	0.86179	9.93540

Weighed	Sought	Factor	Log of Factor +10	Reciprocal of Factor	Log of Reciprocal of Factor +10
Potassium (contd.)					
K_2SO_4	K_2O	0.54057	9.73285	1.84990	10.26715
	K_2PtCl_6	2.78889	10.44543	0.35857	9.55457
	K_2S	0.63276	9.80124	1.58038	10.19876
	SO_3	0.45942	9.66221	2.17666	10.33779
$K_2SO_4.Al_2(SO_4)_3.24H_2O$	K_2PtCl_6	0.51224	9.70947	1.95221	10.29053
$K_2SO_4.Cr_2(SO_4)_3.24H_2O$	K_2PtCl_6	0.48658	9.68715	2.05516	10.31284
$Mg_2As_2O_7$	K_3AsO_4	1.65059	10.21764	0.60584	9.78236
	K_2HAsO_4	1.40519	10.14774	0.71165	9.85227
Mn_2O_3	$KMnO_4$	2.00207	10.30148	0.49948	9.69852
	K_2MnO_4	2.49731	10.39747	0.40043	9.60253
MnS	$KMnO_4$	1.81649	10.25923	0.55051	9.74077
	K_2MnO_4	2.26592	10.35524	0.44132	9.64475
N	KNO_3	7.21847	10.85845	0.13853	9.14154
NH_3	KNO_3	5.93678	10.77355	0.16844	9.22645
NO	KNO_3	3.36954	10.52757	0.29678	9.47243
N_2O_3	KNO_2	2.23934	10.35012	0.44656	9.64988
N_2O_5	KNO_3	1.87217	10.27235	0.53414	9.72765
	K_2O	0.87217	9.94060	1.14657	10.05940
Pt	K	0.40086	9.60299	2.49464	10.39701
	KCl	0.76431	9.88327	1.30837	10.11673
SiO_2	K_2SiO_3	2.56783	10.40957	0.38943	9.59043
SO_3	K_2SO_4	2.17664	10.33779	0.45942	9.66221
Praseodymium					
Pr	Pr_2O_3	1.17032	10.06832	0.85447	9.93169
Pr_2O_3	Pr	0.85447	9.93169	1.17032	10.06831
Pr_6O_{11}	Pr	0.82770	9.91787	1.20817	10.08213
Rhenium					
$C_{20}H_{16}N_4.HReO_4$	Re	0.33038	9.51901	3.02682	10.48098
$(C_6H_5)_4AsReO_4$	Re	0.29392	9.46823	3.40229	10.53177
Rhodium					
Na_3RhCl_6	Rh	0.26757	9.42744	3.73734	10.57256
Rh	Na_3RhCl_6	3.73735	10.57256	0.26757	9.42744
$RhCl_3$	Rh	0.49175	9.69174	2.03355	10.30826
Rubidium					
AgCl	Rb	0.59635	9.77550	1.67687	10.22450
	RbCl	0.84368	9.92618	1.18528	10.07382
Cl	Rb	2.41080	10.38216	0.41480	9.61784
	RbCl	3.41071	10.53284	0.29319	9.46715
Rb	AgCl	1.67688	10.22450	0.59635	9.77550
	Cl	0.41480	9.61784	2.41080	10.38216
	RbCl	1.41477	10.15069	0.70683	9.84932
	Rb_2CO_3	1.35106	10.13068	0.74016	9.86933
	Rb_2O	1.09360	10.03886	0.91441	9.96114
	Rb_2PtCl_6	3.38569	10.52965	0.29536	9.47035
	Rb_2SO_4	1.56195	10.19367	0.64023	9.80634
$RbB(C_6H_5)_4$	Rb	0.21119	9.32467	4.73507	10.67533
	Rb_2O	0.23096	9.36354	4.32975	10.63647
RbCl	AgCl	1.18527	10.07382	0.84369	9.92618
	Cl	0.29319	9.46715	3.41076	10.53285
	Rb	0.70683	9.84932	1.41477	10.15069
	Rb_2CO_3	0.95493	9.97997	1.04720	10.02003
	Rb_2O	0.77296	9.08816	1.29373	10.11184
	Rb_2PtCl_6	2.39301	10.37896	0.41788	9.62105
	Rb_2SO_4	1.10399	10.04297	0.90581	9.95704
Rb_2CO_3	Rb	0.74019	9.86934	1.35100	10.12066
	RbCl	1.04720	10.02003	0.45493	9.97997
	$RbHCO_3$	1.26864	10.10335	0.78825	9.89666
	Rb_2PtCl_6	2.50595	10.39897	0.39905	9.60103
	Rb_2SO_4	1.15609	10.06299	0.86498	9.93701
$RbClO_4$	Rb	0.46220	9.66483	2.16357	10.33517
	Rb_2CO_3	0.62446	9.79550	1.60138	10.20449
	RbCl	0.65390	9.81551	1.52929	10.18448
	$RbHCO_3$	0.79218	9.89883	1.26234	10.10118
	Rb_2O	0.50546	9.70369	1.97840	10.29632
	Rb_2SO_4	0.72193	9.85850	1.38518	10.14151
$RbHCO_3$	Rb_2CO_3	0.78831	9.89670	1.26854	10.10330
	Rb_2PtCl_6	1.97546	10.29567	0.50621	9.70433
	Rb_2SO_4	0.91136	9.95969	1.09726	10.04031

Weighed	Sought	Factor	Log of Factor +10	Reciprocal of Factor	Log of Reciprocal of Factor +10
Rubidium (contd.)					
Rb_2O	Rb	0.91441	9.96114	1.09360	10.03887
	RbCl	1.29368	10.11182	0.77299	9.88817
	Rb_2PtCl_6	3.09591	10.49079	0.32301	9.50922
	Rb_2SO_4	1.42827	10.15481	0.70015	9.84519
Rb_2PtCl_6	Rb	0.29537	9.47037	3.38558	10.52963
	RbCl	0.41787	9.62103	2.39309	10.37896
	Rb_2CO_3	0.39905	9.60103	2.50595	10.39898
	$RbHCO_3$	0.50624	9.70436	1.97535	10.29565
	Rb_2O	0.32301	9.50922	3.09588	10.49079
Rb_2SO_4	Rb	0.64022	9.80633	1.56196	10.19367
	RbCl	0.90577	9.95701	1.10403	10.04298
	Rb_2CO_3	0.86498	9.93701	1.15610	10.06300
	$RbHCO_3$	1.09730	10.04033	0.91133	9.95968
	Rb_2O	0.70015	9.84520	1.42827	10.15481
Samarium					
Sm_2O_3	Sm	0.86235	9.93568	1.15962	10.06431
Scandium					
Sc_2O_3	Sc	0.65196	9.81422	1.53384	10.18578
$Sc(C_9H_6NO)_3.(C_9H_6NOH)$	Sc	0.07221	8.85860	13.84850	11.14140
	Sc_2O_3	0.11076	9.04438	9.02853	10.95562
Selenium					
H_2SeO_3	Se	0.61224	9.78692	1.63335	10.21308
H_2SeO_4	Se	0.54466	9.73613	1.83601	10.26389
$PbSeO_4$	Se	0.22550	9.35315	4.43459	10.64685
	SeO_2	0.31689	9.50091	3.15567	10.49909
Se	H_2SeO_3	1.63336	10.21308	0.61223	9.78691
	H_2SeO_4	1.83599	10.26387	0.54467	9.73613
	SeO_2	1.40527	10.14776	0.71161	9.85224
	SeO_3	1.60790	10.20626	0.62193	9.79374
SeO_2	Se	0.71161	9.85224	1.40526	10.14776
SeO_3	Se	0.62193	9.79374	1.60790	10.20626
Silicon					
$BaSiF_6$	SiF_4	0.37249	9.57111	2.68464	10.42888
	SiO_2	0.21503	9.33250	4.65051	10.66750
H_2SiO_3	SiO_2	0.76933	9.88611	1.29983	10.11388
K_2SiF_6	SiF_4	0.47249	9.67439	2.11645	10.32561
	SiO_2	0.27277	9.43580	3.66609	10.56420
Si	SiO_2	2.13932	10.33027	0.46744	9.66973
SiC	C	0.29955	9.47647	3.33834	10.52353
	Si	0.70045	9.84538	1.42765	10.15462
SiF_4	$BaSiF_6$	0.26847	9.42890	3.72481	10.57110
	K_2SiF_6	2.11645	10.32561	0.47249	9.67439
SiF_4	SiO_2	0.57730	9.76140	1.73220	10.23860
SiO_2	$BaSiF_6$	4.65041	10.66749	0.21503	9.33250
	H_2SiO_3	1.29983	10.11388	0.76933	9.88611
	K_2SiF_6	3.66614	10.56421	0.27277	9.43580
	Si	0.46744	9.66973	2.13931	10.33027
	SiF_4	1.73221	10.23861	0.57730	9.76140
	SiO_3	1.26627	10.10252	0.78972	9.89747
	SiO_4	1.53256	10.18542	0.65250	9.81458
	Si_2O	0.60058	9.77857	1.66506	10.22182
	$Si(OH)_4$	1.59965	10.20403	0.62514	9.79597
$SiO_2.12MoO_3$	Si	0.01571	8.19618	63.65372	11.80382
	SiO_2	0.03362	8.52660	29.74420	11.47340
SiO_3	SiO_2	0.78972	9.89747	1.26627	10.10252
SiO_4	SiO_2	0.65250	9.81458	1.53257	10.18542
Si_2O	SiO_2	1.66505	10.22142	0.60058	9.77857
$Si(OH)_4$	SiO_2	0.62514	9.79598	1.59964	10.20402
Silver					
Ag	AgBr	1.74079	10.24075	0.57445	9.75925
	AgCl	1.32866	10.12341	0.75264	9.87659
	AgCN	1.24120	10.09384	0.80567	9.90616
	AgI	2.17645	10.33775	0.45946	9.66225
	$AgNO_3$	1.57481	10.19723	0.63500	9.80277
	Ag_2O	1.07416	10.03107	0.93096	9.96893
	Ag_3PO_4	1.29347	10.11176	0.77311	9.88824

Weighed	Sought	Factor	Log of Factor +10	Reciprocal of Factor	Log of Reciprocal of Factor +10
Silver (contd.)					
Ag	$Ag_4P_2O_7$	1.40313	10.14710	0.71269	9.85290
	Br	0.74079	9.86970	1.34991	10.13030
	Cl	0.32866	9.51675	3.04266	10.48325
	I	1.17646	10.07058	0.85001	9.92942
AgBr	Ag	0.57445	9.75925	1.74080	10.24075
	Br	0.42555	9.62895	2.34990	10.37105
AgCl	Ag	0.75264	9.87659	1.32866	10.12342
	$AgNO_3$	1.18526	10.07382	0.84370	9.92619
	Ag_2O	0.80845	9.90765	1.23693	10.09235
	Br	0.55754	9.74628	1.79359	10.25372
	Cl	0.24736	9.39333	4.04269	10.60667
AgCN	Ag	0.80567	9.90616	1.24120	10.09384
Ag_2CrO_4	Ag	0.65034	9.81314	1.53766	10.18686
AgI	Ag	0.45946	9.66225	2.17647	10.33775
	I	0.54054	9.73283	1.85000	10.26717
$AgNO_3$	Ag	0.63499	9.80277	1.57483	10.19723
	AgCl	0.84370	9.92619	1.18526	10.07381
Ag_2O	Ag	0.93096	9.96893	1.07416	10.03107
	AgCl	1.23693	10.09235	0.80845	9.90765
Ag_3PO_4	Ag	0.77311	9.88824	1.29348	10.11176
$Ag_4P_2O_7$	Ag	0.71269	9.85290	1.40313	10.14710
Br	Ag	1.34991	10.13030	0.74079	9.86970
	AgBr	2.34991	10.37105	0.42555	9.62895
	AgCl	1.79358	10.25372	0.55754	9.74628
Cl	Ag	3.04261	10.48325	0.32867	9.51676
	AgCl	4.04262	10.60666	0.24736	9.39333
I	Ag	0.85000	9.92942	1.17647	10.07058
	AgI	1.85000	10.26717	0.54054	9.73283
Sodium					
Ag	NaBr	0.95392	9.97951	1.04831	10.02049
	NaCl	0.54179	9.73383	1.84573	10.26617
	NaI	1.38958	10.14288	0.71964	9.85712
AgBr	NaBr	0.54798	9.73876	1.82488	10.26123
AgCl	NaCl	0.40777	9.61042	2.45236	10.38959
	$NaClO_3$	0.74267	9.87080	1.34649	10.21921
	$NaClO_4$	0.85429	9.93161	1.17056	10.06839
AgI	NaI	0.63846	9.80513	1.56627	10.19487
$BaSO_4$	$NaHSO_4$	0.51439	9.71129	1.94405	10.28871
	$NaHSO_4.H_2O$	0.59157	9.77201	1.69042	10.22799
	Na_2S	0.33438	9.52424	2.99061	10.47577
	Na_2SO_3	0.54002	9.73241	1.85178	10.26759
	$Na_2SO_3.7H_2O$	1.08032	10.03355	0.92565	9.96645
	Na_2SO_4	0.60857	9.78431	1.64320	10.21569
	$Na_2SO_4.10H_2O$	1.38043	10.14001	0.72441	9.85998
B_2O_3	$Na_2B_4O_7$	1.44511	10.15990	0.69199	9.84010
	$Na_2B_4O_7.10H_2O$	2.73895	10.43758	0.36510	9.56241
Br	Na	0.28770	9.45894	3.47584	10.54106
	NaBr	1.28770	10.10981	0.77658	9.89019
	Na_2O	0.38781	9.58862	2.57858	10.41138
$CaCl_2$	NaCl	1.05321	10.02252	0.94948	9.97749
$CaCO_3$	Na_2CO_3	1.05894	10.02488	0.94434	9.97513
CaF_2	NaF	1.07551	10.03161	0.92979	9.96838
CaO	Na_2CO_3	1.88996	10.27645	0.52911	9.72355
$CaSO_4$	Na_2CO_3	0.77853	9.89128	1.28447	10.10873
Cl	Na	0.64846	9.81188	1.54212	10.18812
	NaCl	1.64846	10.21708	0.60663	9.78292
	Na_2O	0.87410	9.94156	1.14403	10.05844
CO_2	Na_2CO_3	2.40829	10.38171	0.41523	9.61829
	Na_2O	1.40829	10.14869	0.71008	9.85131
H_3BO_3	$Na_2B_4O_7$	0.81356	9.91039	1.22917	10.08961
	$Na_2B_4O_7.10H_2O$	1.54195	10.18807	0.64853	9.81193
I	Na	0.18116	9.25806	5.51998	10.74194
	NaI	1.18115	10.07231	0.84663	9.92769
	Na_2O	0.24419	9.38773	4.09517	10.61227
KBF_4	$Na_2B_4O_7$	0.39954	9.60156	2.50288	10.39844
	$Na_2B_4O_7.10H_2O$	0.75725	9.87924	1.32057	10.12076
$Mg_2As_2O_7$	Na_2HAsO_3	1.09454	10.03923	0.91363	9.96077
	Na_2HAsO_4	1.19761	10.07832	0.83500	9.92169

Weighed	Sought	Factor	Log of Factor +10	Reciprocal of Factor	Log of Reciprocal of Factor +10
Sodium (contd.)					
$MgCl_2$	NaCl	1.22756	10.08904	0.81462	9.91096
$Mg_2P_2O_7$	Na_2HPO_4	1.27565	10.10573	0.78391	9.89427
	$Na_2HPO_4.12H_2O$	3.21828	10.50763	0.31072	9.49237
	$NaNH_4HPO_4.4H_2O$	1.87870	10.27386	0.53228	9.72614
	Na_3PO_4	1.47318	10.16825	0.67880	9.83174
	$Na_4P_2O_7.10H_2O$	2.00414	10.30193	0.49897	9.69807
N	$NaNO_3$	6.06815	10.78306	0.16479	9.21693
Na	Br	3.47585	10.54106	0.28770	9.45894
	Cl	1.54212	10.18811	0.64846	9.81188
	I	5.52003	10.74194	0.18116	9.25806
	NaBr	4.47585	10.65088	0.22342	9.34912
	NaCl	2.54213	10.40520	0.39337	9.59480
	Na_2CO_3	2.30513	10.36269	0.43382	9.63731
	$NaHCO_3$	3.65410	10.56278	0.27367	9.43723
	NaI	6.52003	10.81425	0.15337	9.18574
	Na_2O	1.34797	10.12968	0.74186	9.87032
	Na_2SO_4	3.08921	10.48985	0.32371	9.51016
$Na_2B_4O_7$	B_2O_3	0.69198	9.84009	1.44513	10.15991
	H_3BO_3	1.22916	10.08961	0.81356	9.91039
	KBF_4	2.50287	10.39844	0.39954	9.60156
$Na_2B_4O_7.10H_2O$	B_2O_3	0.36510	9.56241	2.73898	10.43759
	H_3BO_3	0.64853	9.81193	1.54195	10.18807
	KBF_4	1.32057	10.12076	0.75725	9.87924
NaBr	Ag	1.04831	10.02048	0.95392	9.97951
	AgBr	1.82489	10.26123	0.54798	9.73876
	Br	0.77658	9.89019	1.28770	10.10981
	Na	0.22342	9.34912	4.47588	10.65088
	Na_2O	0.30116	9.47880	3.32049	10.52120
$NaC_2H_3O_2.Mg(C_2H_3O_2)_2.$ $3UO_2(C_2H_3O_2)_2.6\frac{1}{2}H_2O$	Na	0.01527	8.18375	65.50075	11.81625
	NaBr	0.06833	8.83464	14.63400	11.16536
	Na_2CO_3	0.03519	8.54646	28.41474	11.45354
	NaCl	0.03881	8.58895	25.76589	11.41104
	NaF	0.02788	8.44536	35.86286	11.55465
	$NaHCO_3$	0.05579	8.74654	17.92500	11.25346
	NaI	0.09954	8.99801	10.04590	11.00199
	$NaNO_3$	0.05644	8.75162	17.71667	11.24837
	Na_2O	0.02058	8.31345	48.59086	11.68655
	NaOH	0.02656	8.42426	37.64746	11.57574
	Na_2SO_4	0.04716	8.67361	21.20261	11.32638
$NaC_2H_3O_2.Zn(C_2H_3O_2)_2.$ $3UO_2(C_2H_3O_2)_2.6\frac{1}{2}H_2O$	Na	0.01495	8.17461	66.89410	11.85239
	NaBr	0.06691	8.82549	14.94545	11.17450
	Na_2CO_3	0.03446	8.53730	29.01999	11.46270
	NaCl	0.03800	8.57981	26.31440	11.42019
	NaF	0.02733	8.43621	36.62601	11.56379
	$NaHCO_3$	0.05466	8.73739	18.30663	11.26261
	NaI	0.09747	8.98886	10.25977	11.01114
	$NaNO_3$	0.05527	8.74247	18.09398	11.25753
	NaOH	0.02601	8.41511	38.44970	11.58489
	Na_2O	0.02015	8.30430	49.62532	11.69570
	Na_2SO_4	0.04618	8.66446	21.65392	11.33554
Na_2CO_3	$CaCO_3$	0.94434	9.97513	1.05894	10.02488
	CaO	0.52911	9.72355	1.88997	10.27645
	$CaSO_4$	1.28447	10.10873	0.77853	9.89128
	CO_2	0.41523	9.61829	2.40830	10.38171
	Na	0.43381	9.63730	2.30516	10.36270
	NaCl	1.10281	10.04250	0.90677	9.95750
	$NaHCO_3$	1.58520	10.20008	0.63084	9.79992
	NaOH	0.75474	9.87780	1.32496	10.12220
	Na_2O	0.58477	9.76699	1.71007	10.23301
	Na_2SO_4	1.34015	10.12715	0.74619	9.87285
$Na_2CO_3.10H_2O$	Na_2SO_4	0.49639	9.69582	2.01455	10.30417
NaCl	Ag	1.84573	10.26617	0.54179	9.73383
	AgCl	2.45236	10.38958	0.40777	9.61042
	Cl	0.60663	9.78292	1.64845	10.21707
	Na	0.39337	9.59480	2.54214	10.40520
	$NaClO_3$	1.82129	10.26038	0.54906	9.73962

Weighed	Sought	Factor	Log of Factor +10	Reciprocal of Factor	Log of Reciprocal of Factor +10
Sodium (contd.)					
	$NaClO_4$	2.09503	10.32118	0.47732	9.67881
	Na_2CO_3	0.90677	9.95750	1.10282	10.04251
	$NaHCO_3$	1.43742	10.15759	0.69569	9.84242
	Na_2HPO_4	1.21451	10.08440	0.82338	9.91560
	Na_2O	0.53025	9.72448	1.88590	10.27552
$NaCl$	Na_2SO_4	1.21521	10.08465	0.82290	9.91535
$NaClO_3$	$AgCl$	1.34650	10.12921	0.74267	9.87080
	$NaCl$	0.54906	9.73961	1.82129	10.26038
$NaClO_4$	$AgCl$	1.17055	10.06839	0.85430	9.93161
	$NaCl$	0.47732	9.67881	2.09503	10.32119
NaF	CaF_2	0.92979	9.96838	1.07551	10.03161
Na_2HAsO_3	$Mg_2As_2O_7$	0.91362	9.96077	1.09455	10.03923
Na_2HAsO_4	$Mg_2As_2O_7$	0.83499	9.92168	1.19762	10.07832
$NaHCO_3$	Na	0.27366	9.43721	3.65417	10.56279
	$NaCl$	0.69569	9.84242	1.43742	10.15759
	Na_2CO_3	0.63083	9.79991	1.58521	10.20009
	Na_2O	0.36889	9.56690	2.71084	10.43310
Na_2HPO_4	$Mg_2P_2O_7$	0.78392	9.89427	1.27564	10.10573
	$NaCl$	0.82338	9.91560	1.21451	10.08440
	Na_2O	0.43660	9.64008	2.29043	10.35992
	$Na_4P_2O_7$	0.93654	9.97153	1.06776	10.02847
	P_2O_5	0.49994	9.69892	2.00024	10.30108
$Na_2HPO_4.12H_2O$	$Mg_2P_2O_7$	0.31072	9.49237	3.21833	10.50763
	$Na_4P_2O_7$	0.37122	9.56963	2.69382	10.43037
	P_2O_5	0.19816	9.29702	5.04643	10.70298
$NaHSO_3$	SO_2	0.61564	9.78933	1.62433	10.21076
$NaHSO_4$	$BaSO_4$	1.94404	10.28871	0.51439	9.71129
$NaHSO_4.H_2O$	$BaSO_4$	1.69039	10.22799	0.59158	9.77201
NaI	Ag	0.71964	9.85712	1.38958	10.14288
	AgI	1.56626	10.19486	0.63846	9.80513
	I	0.84662	9.92769	1.18117	10.07231
	Na	0.15337	9.18574	6.52018	10.81426
	Na_2O	0.20674	9.31542	4.83699	10.68458
$NaNH_4HPO_4.4H_2O$	$Mg_2P_2O_7$	0.53228	9.72614	1.87871	10.27386
	NH_3	0.08146	8.91094	12.27596	11.08905
	P_2O_5	0.33946	9.53079	2.94586	10.46921
$NaNO_3$	N	0.16479	9.21693	6.06833	10.78307
	Na_2O	0.36461	9.56183	2.74266	10.43818
	NH_3	0.20037	9.30183	4.99077	10.69817
	NO	0.35303	9.54781	2.83262	10.45219
	N_2O_5	0.63539	9.80304	1.57384	10.19696
Na_2O	Br	2.57854	10.41137	0.38782	9.58863
	Cl	1.14401	10.05843	0.87412	9.94157
	CO_2	0.71008	9.85131	1.40829	10.14869
	I	4.09501	10.61225	0.24420	9.38775
	Na	0.74185	9.87032	1.34798	10.12968
	$NaBr$	3.32039	10.52119	0.30117	9.47881
	$NaCl$	1.88587	10.27551	0.53026	9.72449
	Na_2CO_3	1.71008	10.23302	0.58477	9.76699
	$NaHCO_3$	2.71078	10.43310	0.36890	9.56691
	Na_2HPO_4	2.29044	10.35992	0.43660	9.64008
	NaI	4.83686	10.68457	0.20675	9.31545
	$NaNO_3$	2.74265	10.43817	0.36461	9.56183
	$NaOH$	1.29065	10.11081	0.77480	9.88919
	Na_2SO_4	2.29176	10.36017	0.43635	9.63984
	N_2O_5	1.74269	10.24122	0.57383	9.75878
	SO_3	1.29176	10.11118	0.77414	9.88882
$NaOH$	Na_2CO_3	1.32495	10.12220	0.75475	9.87780
	Na_2O	0.77479	9.88918	1.29067	10.11082
$Na_4P_2O_7$	Na_2HPO_4	1.06775	10.02847	0.93655	9.97153
	$Na_2HPO_4.12H_2O$	0.26938	9.43037	3.71223	10.56963
$Na_4P_2O_7.10H_2O$	$Mg_2P_2O_7$	0.49897	9.69807	2.00413	10.30193
Na_2S	$BaSO_4$	2.99062	10.47576	0.33438	9.52424
Na_2SO_3	$BaSO_4$	1.85176	10.26758	0.54003	9.73242
	SO_2	0.50827	9.70609	1.96746	10.29391
$Na_2SO_3.7H_2O$	$BaSO_4$	0.92564	9.96644	1.08033	10.03356
	SO_2	0.25407	9.40495	3.93592	10.59505
Na_2SO_4	$BaSO_4$	1.64319	10.21569	0.60857	9.78431
	Na	0.32370	9.51014	3.08928	10.48986

Weighed	Sought	Factor	Log of Factor +10	Reciprocal of Factor	Log of Reciprocal of Factor +10
Sodium (contd.)					
	NaCl	0.82289	9.91534	1.21523	10.08466
	Na_2CO_3	0.74619	9.87285	1.34014	10.21715
	$Na_2CO_3.10H_2O$	2.01451	10.30417	0.49640	9.69583
	Na_2O	0.43635	9.63984	2.29174	10.36016
	SO_3	0.56365	9.75101	1.77415	10.24899
$Na_2SO_4.10H_2O$	$BaSO_4$	0.72441	9.85998	1.38043	10.14001
$Na_2U_2O_7.2ZnU_2O_7$	Na	0.02369	8.37457	42.21190	11.62544
	Na_2O	0.03193	8.50420	31.31851	11.49579
NH_3	$NaNH_4HPO_4.4H_2O$	12.27607	11.08906	0.08146	8.91094
	$NaNO_3$	4.99070	10.69816	0.20037	9.30183
NO	$NaNO_3$	2.83258	10.45218	0.35304	9.54782
N_2O_5	$NaNO_3$	1.57382	10.19695	0.63540	9.80305
	Na_2O	0.57382	9.75878	1.74271	10.24123
P_2O_5	Na_2HPO_4	2.00020	10.30107	0.49995	9.69893
	$Na_2HPO_4.12H_2O$	5.04623	10.70297	0.19817	9.29704
	$NaNH_4HPO_4.4H_2O$	2.94578	10.46920	0.33947	9.53080
SO_2	$NaHSO_3$	1.62442	10.21070	0.61560	9.78930
	Na_2SO_3	1.96756	10.29393	0.50824	9.70607
SO_3	Na_2O	0.77414	9.88882	1.29176	10.11118
	Na_2SO_4	1.77414	10.24899	0.56365	9.75101
Strontium					
CO_2	$SrCO_3$	3.35446	10.52562	0.29811	9.47438
SO_3	SrO	1.29424	10.11202	0.77265	9.88798
	$SrSO_4$	2.29422	10.36063	0.43588	9.63937
Sr	$SrCO_3$	1.68489	10.22657	0.59351	9.77343
	$Sr(NO_3)_2$	2.41532	10.38298	0.41402	9.61702
	SrO	1.18261	10.07284	0.84559	9.92716
	$SrSO_4$	2.09633	10.32146	0.47702	9.67854
$SrCl_2$	$SrCO_3$	0.93124	9.96906	1.07384	10.03094
	SrO	0.65363	9.81533	1.52992	10.18467
	$SrSO_4$	1.15865	10.06395	0.86307	9.93605
$SrCO_3$	CO_2	0.29811	9.47438	3.35447	10.52562
	Sr	0.59351	9.77343	1.68489	10.22657
	$SrCl_2$	1.07383	10.03095	0.93125	9.96907
	$Sr(HCO_3)_2$	1.42010	10.15232	0.70418	9.84768
	$Sr(NO_3)_2$	1.43352	10.15640	0.69758	9.84359
	SrO	0.70189	9.84627	1.42472	10.15373
	$SrSO_4$	1.24419	10.09489	0.80374	9.90511
SrC_2O_4	Sr	0.49886	9.69798	2.00457	10.30202
	SrO	0.58996	9.77082	1.69503	10.22918
$Sr(HCO_3)_2$	$SrCO_3$	0.70417	9.84768	1.42011	10.15232
	SrO	0.49425	9.69395	2.02327	10.30602
$Sr(IO_3)_2$	Sr	0.20031	9.30170	4.99226	10.69829
	SrO	0.23688	9.37453	4.22155	10.69829
$Sr(NO_3)_2$	Sr	0.41402	9.61702	2.41434	10.38298
	$SrCO_3$	0.69759	9.84360	1.43351	10.15640
	SrO	0.48963	9.68987	2.04236	10.31014
	$SrSO_4$	0.86793	9.93849	1.15217	10.06151
SrO	SO_3	0.77265	9.88798	1.29425	10.11202
	Sr	0.84559	9.92716	1.18261	10.07284
	$SrCl_2$	1.52992	10.18467	0.65363	9.81533
	$SrCO_3$	1.42472	10.15373	0.70189	9.84627
	$Sr(HCO_3)_2$	2.02326	10.30605	0.49425	9.69395
	$Sr(NO_3)_2$	2.04237	10.31014	0.48963	9.68987
	$SrSO_4$	1.77263	10.24862	0.56413	9.75138
$SrSO_4$	SO_3	0.43588	9.63937	2.29421	10.36063
	Sr	0.47703	9.67855	2.09630	10.32145
	$SrCl_2$	0.86308	9.93605	1.15864	10.06395
	$SrCO_3$	0.80373	9.90511	1.24420	10.09489
	$Sr(NO_3)_2$	1.15217	10.06151	0.86793	9.93848
	SrO	0.56413	9.75138	1.77264	10.24862
Sulfur					
As_2S_3	H_2S	0.41555	9.61862	2.40645	10.38138
	S	0.39097	9.59214	2.55774	10.40786
$BaSO_4$	FeS_2	0.25702	9.40997	3.89077	10.59003
	H_2S	0.14602	9.16441	6.84838	10.83559
	H_2SO_3	0.35166	9.54612	2.84366	10.45388
	H_2SO_4	0.42021	9.62347	2.37976	10.37653

Weighed	Sought	Factor	Log of Factor +10	Reciprocal of Factor	Log of Reciprocal of Factor +10
Sulfur (contd.)					
	S	0.13738	9.13792	7.27908	10.86208
	SO_2	0.27448	9.43851	3.64325	10.56149
	SO_3	0.34302	9.53532	2.91528	10.46468
	SO_4	0.41158	9.61445	2.42966	10.38555
$C_{12}H_{12}N_2.H_2SO_4$	S	0.11357	9.05526	8.80514	10.94474
	SO_4	0.34026	9.53181	2.93893	10.46819
CdS	H_2S	0.23591	9.37275	4.23890	10.62725
	S	0.22196	9.34627	4.50532	10.65372
FeS_2	$BaSO_4$	3.89081	10.59004	0.25702	9.40997
H_2S	As_2S_3	2.40646	10.38138	0.41555	9.61862
	$BaSO_4$	6.84859	10.83560	0.14602	9.16441
	CdS	4.23885	10.62725	0.23591	9.37275
	SO_3	2.34924	10.37093	0.42567	9.62907
H_2SO_3	$BaSO_4$	2.84363	10.45287	0.35166	9.54612
H_2SO_4	$BaSO_4$	2.37974	10.37653	0.42021	9.62347
	$(NH_4)_2SO_4$	1.34728	10.12946	0.74224	9.87054
	SO_3	0.81631	9.91186	1.22502	10.08814
$(NH_4)_2SO_4$	H_2SO_4	0.74223	9.87054	1.34729	10.12946
	SO_3	0.60589	9.78239	1.65046	10.21760
S	As_2S_3	2.55775	10.40786	0.39097	9.59214
	$BaSO_4$	7.27919	10.86208	0.13738	9.13792
	CdS	4.50536	10.65373	0.22196	9.34627
SO_2	$BaSO_4$	3.64329	10.56149	0.27448	9.43851
SO_2	$BaSO_4$	2.91524	10.46468	0.34302	9.53532
	H_2S	0.42567	9.62907	2.34924	10.37093
	$(NH_4)_2SO_4$	1.65047	10.21761	0.60589	9.78239
SO_4	$BaSO_4$	2.42968	10.38555	0.41158	9.61445
Tantalum					
Ta	$TaCl_5$	1.97965	10.29659	0.50514	9.70341
	Ta_2O_5	1.22105	10.08674	0.81897	9.91327
TaC	C	0.06225	8.79413	16.06451	11.20587
	Ta	0.93775	9.97209	1.06638	10.02791
$TaCl_5$	Ta	0.50514	9.70341	1.97965	10.29659
	Ta_2O_5	0.61680	9.79041	1.62127	10.20985
Ta_2O_4	Ta_2O_5	1.03757	10.01602	0.96379	9.98398
Ta_2O_5	Ta	0.81897	9.91327	1.22105	10.08673
	$TaCl_5$	1.62126	10.20985	0.61680	9.79014
	Ta_2O_4	0.96379	9.98398	1.03757	10.01602
Tellurium					
H_2TeO_4	Te	0.65906	9.81893	1.51731	10.18108
$H_2TeO_4.2H_2O$	Te	0.55565	9.74480	1.79969	10.25520
Te	H_2TeO_4	1.51732	10.18108	0.65906	9.81893
	$H_2TeO_4.2H_2O$	1.79969	10.25520	0.55565	9.74480
	TeO_2	1.25078	10.09718	0.79950	9.90282
	TeO_3	1.37618	10.13868	0.72665	9.86133
	$(TeO_2)_2SO_3$	1.56450	10.19438	0.63918	9.80562
TeO_2	Te	0.79950	9.90282	1.25078	10.09718
TeO_3	Te	0.72665	9.86133	1.37618	10.13868
$(TeO_2)_2SO_3$	Te	0.63918	9.80562	1.56450	10.19438
Terbium					
Tb_4O_7	Tb	0.85021	9.92953	1.17618	10.07048
Thallium					
$(C_6H_5)_4AsTlCl_4$	Tl	0.28014	9.44738	3.56964	10.55263
Tl	TlCl	1.17346	10.06947	0.85218	9.93053
	Tl_2CO_3	1.14682	10.05949	0.87198	9.94051
	Tl_2CrO_4	1.28377	10.10849	0.77896	9.89151
	$TlHSO_4$	1.47497	10.16878	0.67798	9.83122
	TlI	1.62093	10.20976	0.61693	9.79024
	$TlNO_3$	1.30337	10.11507	0.76724	9.88493
	Tl_2O	1.03914	10.01668	0.96233	9.98332
	Tl_2PtCl_6	1.99772	10.30055	0.50057	9.69946
	Tl_2SO_4	1.23501	10.09167	0.80971	9.90833
TlCl	Tl	0.85218	9.93053	1.17346	10.06947
	Tl_2PtCl_6	1.70240	10.23106	0.58741	9.76894
Tl_2CO_3	Tl	0.87198	9.94051	1.14682	10.05951
	Tl_2PtCl_6	1.74197	10.24104	0.57406	9.75896
Tl_2CrO_4	Tl	0.77895	9.89151	1.28378	10.10850
$TlHSO_4$	Tl	0.67798	9.83122	1.47497	10.16878

GRAVIMETRIC FACTORS AND THEIR LOGARITHMS (Continued)

Weighed	Sought	Factor	Log of Factor +10	Reciprocal of Factor	Log of Reciprocal of Factor +10
Thallium (contd.)					
TlI	Tl	0.61693	9.79024	1.62093	10.20976
	Tl_2PtCl_6	1.23244	10.09076	0.81140	9.90924
$TlNO_3$	Tl	0.76724	9.88493	1.30338	10.11507
	Tl_2PtCl_6	1.53272	10.18546	0.65243	9.81453
Tl_2O	Tl	0.96223	9.98332	1.03914	10.01668
	Tl_2PtCl_6	1.92247	10.28386	0.52016	9.71614
Tl_2PtCl_6	Tl	0.50057	9.69946	1.99772	10.30054
	TlCl	0.58741	9.76894	1.70239	10.23106
	Tl_2CO_3	0.57406	9.75896	1.74198	10.24105
	TlI	0.81140	9.90924	1.23244	10.00076
	$TlNO_3$	0.65244	9.81454	1.53271	10.18546
	Tl_2O	0.52016	9.71614	1.92249	10.28386
	Tl_2SO_4	0.61821	9.79114	1.61757	10.20886
Tl_2SO_4	Tl	0.80971	9.90833	1.23501	10.09167
	Tl_2PtCl_6	1.61757	10.20886	0.61821	9.79114
Thorium					
Th	ThO_2	1.13790	10.05610	0.87881	9.94390
$Th(C_9H_6NO)_4.(C_9H_6NOH)$	Th	0.24327	9.38609	4.11066	10.61391
	ThO_2	0.27682	9.44220	3.61246	10.55781
$ThCl_4$	ThO_2	0.70626	9.84896	1.41590	10.15103
$Th(NO_3)_4.6H_2O$	ThO_2	0.44898	9.65223	2.22727	10.34777
ThO_2	Th	0.87881	9.94390	1.13790	10.05610
	$ThCl_4$	1.41590	10.15103	0.70626	9.84896
	$Th(NO_3)_4.6H_2O$	2.22729	10.34778	0.44898	9.65223
Thulium					
Tm_2O_3	Tm	0.87561	9.94231	1.14206	10.05769
Tin					
Sn	$SnCl_2$	1.59744	10.20342	0.62600	9.79657
	$SnCl_2.2H_2O$	1.90100	10.27898	0.52604	9.72102
	$SnCl_4$	2.19479	10.34140	0.45562	9.65860
	$SnCl_4.(NH_4Cl)_2$	3.09622	10.49083	0.32297	9.50916
	SnO	1.13480	10.05492	0.88121	9.94508
	SnO_2	1.26961	10.10367	0.78764	9.89633
$SnCl_2$	Sn	0.62600	9.79657	1.59744	10.20342
	SnO_2	0.79478	9.90025	1.25821	10.09975
$SnCl_2.2H_2O$	Sn	0.52604	9.72102	1.90100	10.27898
	SnO_2	0.66786	9.82469	1.49732	10.17531
$SnCl_4$	Sn	0.45562	9.65860	2.19481	10.34139
	SnO_2	0.57846	9.76227	1.72873	10.23772
$SnCl_4.(NH_4Cl)_2$	Sn	0.32297	9.50916	3.09626	10.49084
	SnO_2	0.41005	9.61284	2.43873	10.38716
SnO	Sn	0.88121	9.94508	1.13480	10.05491
	SnO_2	1.11879	10.04875	0.89382	9.95125
SnO_2	Sn	0.78764	9.89633	1.26962	10.10367
	$SnCl_2$	1.25821	10.09976	0.79474	9.90023
	$SnCl_2.2H_2O$	1.49731	10.17531	0.66786	9.82469
	$SnCl_4$	1.72871	10.23772	0.57847	9.76228
	$SnCl_4.(NH_4Cl)_2$	2.43872	10.38716	0.41005	9.61284
	SnO	0.89382	9.95125	1.11879	10.04875
Titanium					
K_2TiF_6	F	0.47472	9.67644	2.10650	10.32356
	K	0.32573	9.51286	3.07003	10.48714
	Ti	0.19951	9.29996	5.01228	10.70004
	TiO_2	0.33279	9.52217	3.00490	10.47783
Ti	K_2TiF_6	5.01232	10.70004	0.19951	9.29996
	TiC	1.25073	10.09717	0.79953	9.90283
	TiO_2	1.66806	10.22222	0.59950	9.77779
TiC	C	0.20049	9.30210	4.98778	10.69791
	Ti	0.79953	9.90283	1.25073	10.09717
$TiO(C_9H_4NOCl_2)_2$	Ti	0.09776	8.99016	10.22913	11.00984
	TiO_2	0.16306	9.21235	6.13271	10.78765
$TiO(C_9H_6ON)_2$	Ti	0.13600	9.13354	7.35294	10.86646
	TiO_2	0.22685	9.35574	4.40820	10.64426
TiO_2	K_2TiF_6	3.00488	10.47782	0.33279	9.52217
	Ti	0.59950	9.77779	1.66806	10.22221
	TiC	0.74969	9.87488	1.33388	10.12512

Weighed	Sought	Factor	Log of Factor +10	Reciprocal of Factor	Log of Reciprocal of Factor +10
Tungsten					
$FeWO_4$	W	0.60539	9.78203	1.65183	10.21797
	WO_3	0.76344	9.88277	1.30986	10.11722
$MgWO_4$	W	0.67552	9.82964	1.48034	10.17036
	WO_3	0.85189	9.93038	1.17386	10.06962
$MnWO_4$	W	0.60719	0.78332	1.64693	10.21668
	WO_3	0.76571	9.88406	1.30598	10.11593
$PbWO_4$	W	0.40403	9.60641	2.47506	10.39359
	WO_3	0.50952	9.70716	1.96263	10.29284
W	W_2C	1.03266	10.01396	0.96837	9.98604
	WC	1.06532	10.02748	0.93869	9.97252
	WO_2	1.17405	10.06969	0.85175	9.93031
	WO_3	1.26108	10.10075	0.79297	9.89925
W_2C	C	0.03163	8.50010	31.61555	11.49990
	W	0.96837	9.98604	1.03266	10.01396
WC	C	0.06133	8.78767	16.30523	11.21233
	W	0.93868	9.97252	1.06533	10.02749
WO_2	W	0.85175	9.93031	1.17405	10.06969
$WO_2(C_9H_6ON)_2$	W	0.36467	9.56190	2.74221	10.43810
	WO_3	0.45987	9.66264	2.17453	10.33736
WO_3	W	0.79297	9.89926	1.26108	10.10074
Uranium					
U	UO_2	1.13444	10.05478	0.88149	9.94522
	U_3O_8	1.17925	10.07160	0.84800	9.92840
	$U_2P_2O_{11}$	1.49981	10.17604	0.66675	9.82396
UO_2	U	0.88149	9.94522	1.13444	10.05478
	U_3O_8	1.03950	10.01682	0.96200	9.98318
	$U_2P_2O_{11}$	1.32268	10.12145	0.75604	9.87854
$UO_2(C_9H_6ON)_2.(C_9H_7ON)$	U	0.33835	9.52937	2.95552	10.47063
	UO_2	0.38384	9.58415	2.60525	10.41585
U_3O_8	U	0.84799	9.92839	1.17926	10.07161
	UO_2	0.96199	9.98317	1.03951	10.01683
	$UO_2(NO_3)_2.6H_2O$	1.78864	10.25252	0.55908	9.74747
$UO_2(NO_3)_2.6H_2O$	U_3O_8	0.55909	9.74748	1.78862	10.25252
$U_2P_2O_{11}$	U	0.66675	9.82396	1.49981	10.17603
	UO_2	0.75639	9.87875	1.32207	10.12125
Vanadium					
V	VC	1.23578	10.09194	0.80921	9.90806
	V_2O_5	1.78518	10.25168	0.56017	9.74832
VC	C	0.19080	9.28058	5.24109	10.71942
	V	0.80921	9.90806	1.23577	10.09194
VO_4	V_2O_5	0.79120	9.89829	1.26390	10.10171
V_2O_5	V	0.56017	9.74832	1.78517	10.25168
	VC	0.69225	9.84026	1.44456	10.15974
	VO_4	1.26390	10.10171	0.79120	9.89829
Ytterbium					
Yb	YbO_3	1.13870	10.05641	0.87819	9.94359
Yb_2O_3	Yb	0.87820	9.94359	1.13869	10.05640
Yttrium					
Y	Y_2O_3	1.26994	10.10378	0.78744	9.89622
Y_2O_3	Y	0.78744	9.89622	1.26994	10.10378
Zinc					
$BaSO_4$	ZnS	0.41744	9.62059	2.39555	10.37941
	$ZnSO_4.7H_2O$	1.23196	10.09059	0.81171	9.90940
Zn	$ZnNH_4PO_4$	2.72877	10.43596	0.36647	9.56404
	ZnO	1.24476	10.09508	0.80337	9.90492
	$Zn_2P_2O_7$	2.33043	10.36744	0.42911	9.63257
	ZnS	1.49044	10.17332	0.67094	9.82668
$Zn(C_9H_6NO)_2$	Zn	0.18483	9.26677	5.41038	10.73323
	ZnO	0.23007	9.36186	4.34650	10.63814
$ZnCl_2$	ZnO	0.59708	9.77603	1.67482	10.22397
$ZnCO_3$	ZnO	0.64899	9.81224	1.54086	10.18777
$(Zn[C_5H_5N]_2).(SCN)_2$	Zn	0.19241	9.28423	5.19724	10.71577
	ZnO	0.23951	9.37932	4.17519	10.62068
$ZnNH_4PO_4$	Zn	0.36646	9.56403	2.72881	10.43597
	ZnO	0.45616	9.65912	2.19221	10.34089
ZnO	Zn	0.80337	9.90492	1.24476	10.09509
	$ZnCl_2$	1.67482	10.22397	0.59708	9.77603

Weighed	Sought	Factor	Log of Factor +10	Reciprocal of Factor	Log of Reciprocal of Factor +10
Zinc (contd.)					
	$ZnCO_3$	1.54086	10.18776	0.64899	9.81224
	$ZnNH_4PO_4$	2.19221	10.34088	0.45616	9.65912
	$Zn_2P_2O_7$	1.87219	10.27235	0.53413	9.72765
	ZnS	1.19737	10.07823	0.83516	9.92177
	$ZnSO_4.7H_2O$	3.53373	10.54823	0.28299	9.45177
$Zn_2P_2O_7$	Zn	0.42911	9.63257	2.33040	10.36743
	ZnO	0.53413	9.72765	1.87220	10.27235
ZnS	$BaSO_4$	2.39556	10.37941	0.41744	9.62059
	Zn	0.67094	9.82668	1.49045	10.17332
	ZnO	0.83516	9.92177	1.19738	10.07822
	$ZnSO_4.7H_2O$	2.95125	10.47001	0.33884	9.52999
$ZnSO_4.7H_2O$	$BaSO_4$	0.81171	9.90940	1.23197	10.09060
	ZnO	0.28299	9.45177	3.53369	10.54823
	ZnS	0.33884	9.52999	2.95125	10.47001
Zirconium					
K_2ZrF_6	Zr	0.32187	9.50768	3.10684	10.49232
	ZrF_4	0.58999	9.77084	1.69494	10.22915
	ZrF_6	0.72407	9.85978	1.38108	10.14022
	ZrO_2	0.43478	9.63827	2.30001	10.36173
Zr	ZrO_2	1.35080	10.13059	0.74030	9.86941
	ZrC	1.13166	10.05372	0.88366	9.94629
$Zr(BrC_8H_6O_3)_4$	Zr	0.09020	8.95521	11.08647	11.04479
	ZrO_2	0.12184	9.08579	8.20749	10.91421
$Zr(C_8H_7O_3)_4$	Zr	0.13110	9.11760	7.62777	10.88240
	ZrO_2	0.17709	9.24819	5.64685	10.75181
ZrC	C	0.11635	9.06577	8.59476	10.93424
	Zr	0.88366	9.94629	1.13166	10.05372
ZrO_2	Zr	0.74030	9.86941	1.35080	10.13059
	ZrC	0.83777	9.92312	1.19365	10.07687
$Zr_2P_2O_7$	Zr	0.34402	9.53658	2.90681	10.46342
	ZrO_2	0.46470	9.66717	2.15193	10.33283

PHYSICAL CONSTANTS OF MINERALS

Compiled by Ralph Kretz

The following table presents data for many of the more common minerals.

In order to avoid duplication and save space, very few cross references are given in the body of the table. If the name sought is not found in the table, consult the **synonym index** given below.

Specific gravities are given at normal atmospheric temperatures, a more precise statement being valueless considering the large variations in natural minerals.

Hardness is given in terms of Mohs' scale. (See under Hardness.)

Indices of refraction for the sodium line, $\lambda = 5893$ Å, unless otherwise indicated. Li, $\lambda = 6708$ Å. Indices will invariably be given in the order ω, ϵ or α, β, γ. Uniaxial crystals are considered positive if $\epsilon > \omega$, negative if $\omega > \epsilon$. Biaxial crystals are considered positive if β is nearer α in value than it is γ and negative if β is nearer γ than α.

ABBREVIATIONS

Abbreviation	Meaning of abbreviation	Abbreviation	Meaning of abbreviation	Abbreviation	Meaning of abbreviation
bl	blue	grn	green	rhbdr	rhombohedral
blk	black	grnsh	greenish	rhomb	rhombic
blksh	blackish	hex	hexagonal	somet	sometimes
blsh	bluish	iridesc	iridescent	tarn	tarnishes
br	brown	monocl.	monoclinic	tetr	tetragonal
brnsh	brownish	oft	often	tricl	triclinic
col	colorless	pa	pale	vlt	violet
cub	cubic	purp	purple	wh	white
dk	dark	(R)	radioactive	yel	yellow
Fe	Fe, ferrous iron	redsh	redish	yelsh	yellowish
Fe^{+3}	Fe, ferric iron				

SYNONYM INDEX

Compound sought	Listed
Acmite	Aegirine
Agate	Quartz (impure)
Aluminum hydroxide	Boehmite, Diaspore, Gibbsite
Amphibole	Actinolite, Anthophyllite, Cummingtonite, Glaucophane, Hornblende, Riebeckite, Tremolite
Antimony oxide	Senarmontite, Valentinite
Antimony sulfide	Stibnite
Arsenic oxide	Arsenolite, Claudetite
Arsenic sulfide	Orpiment, Realgar
Barium carbonate	Witherite
Barium sulfate	Barite
Barytes	Barite
Bauxite	Gibbsite, Boehmite, Diaspore
Brimstone	Sulfur
Bronzite	Orthopyroxene
Cadmium sulfide	Greenockite
Calamine	Hemimorphite
Calcium carbonate	Aragonite, Calcite, Vaterite
Calcium sulfate	Anhydrite, Gypsum
Calcium sulfide	Oldhamite
Carborundum	Moissanite
Chalcedony	Quartz (impure, fibrous)
Chinaclay	Kaolinite
Chloanthite	Skutterodite
Chromespinel	Chromite
Chrysolite	Serpentine
Clinoptolite	Heulandite
Clayminerals	Illite, Kaolinite, Montmorillonite
Clinochlore	Chlorite
Cobaltbloom	Erythrite
Copper chloride	Nantokite
Copper oxide	Cuprite
Copper sulfide	Chalcocite, Covellite, Digenite
Emerald	Beryl
Emery	Mixture of Corundum, Magnetite and other minerals
Epsom salt	Epsomite
Feldspar	Orthoclase, Microcline, Anorthoclase, Albite, Oligoclase, Andesine, Anorthite
Fibrolite	Sillimanite
Flint	Quartz (impure)
Fluorapatite	Apatite
Fluorspar	Fluorite
Garnet	Almandine, Pyrope, Spessartite, Andradite, Grossularite, Uvarovite, Hydrogrossularite
Garnierite	Serpentine (Ni-bearing)
Glauber salt	Mirabilite
Hyacinth	Zircon
Iceland spar	Calcite
Idocrase	Vesuvianite
Iron carbonate	Siderite
Iron hydroxide	Goethite, Lepidocrocite
Iron oxide	Hematite, Magnetite
Iron spinel	Hercynite
Iron sulfide	Marcasite, Pyrite, Pyrrhotite
Lapis lazuli	Lazurite
Lead carbonate	Cerussite
Lead chloride	Cotunnite
Lead chromate	Crocoite
Lead oxide	Litharge, Minium
Lead sulfate	Anglesite
Lead sulfide	Galena
Limonite	Goethite (impure)
Lithiophyllite	Triphylite
Lithium mica	Lepidolite
Lodestone	Magnetite
Magnesium carbonate	Magnesite
Magnesium hydroxide	Brucite
Magnesium oxide	Periclase
Magnesium sulfate	Kieserite
Manganese carbonate	Rhodochrosite
Manganese hydroxide	Pyrochroite
Manganese oxide	Hausmannite, Manganosite, Pyrolusite
Manganese sulfide	Alabandite
Meerschaum	Serpentine
Mica	Muscovite, Paragonite, Phlogopite, Biotite, Lepidolite
Native copper	Copper
Native gold	Gold
Nickel oxide	Bunsenite
Nickel sulfide	Millerite
Orthite	Allanite
Penninite	Chlorite
Peridote	Olivine
Pistacite	Epidote
Pitchblende	Uraninite
Plagioclase	Albite, Oligoclase, Andesine, Anorthite
Potassium chloride	Sylvite
Potassium sulfate	Arcanite
Pyroxene	Diopsite, Angite, Aegirine, Jadeite, Pigeonite, Eustatite, Orthopyroxene
Rocksalt	Halite
Ruby	Corundum
Sapphire	Corundum
Silica	Christobalite, Quartz, Tridymite
Silver chloride	Cerargyrite
Silver iodide	Jodyrite, Miersite
Silver sulfide	Acanthite, Argentite
Smalltite	Skutterotite
Soapstone	Mixture of Talc and other minerals
Sodium chloride	Halite
Sodium sulfate	Thenardite
Strontium carbonate	Strontianite
Strontium sulfate	Celestite
Thorium oxide	Thorianite
Tin oxide	Cassiterite
Titanite	Sphene
Titanium oxide	Anatase, Brookite, Rutile
Uranium oxide	Uraninite
Zeolite	Natrolite, Mesolite, Scolecite, Thomasonite, Harmatome, Eddingtonite, Heulandite, Stilbite, Phillipsite, Chabazite, Gmelinite, Levyn, Laumontite, Mordenite
Zincblende	Sphalerite
Zinc carbonate	Smithsonite
Zinc oxide	Zincite
Zinc spinel	Gahnite
Zinc sulfide	Sphalerite, Wurtzite
Zirconium oxide	Baddeleyite

PHYSICAL CONSTANTS OF MINERALS

Name	Formula	Sp. gr.	Hardness	Crystalline form and color	Index of refraction (Na) η; ω ϵ / α β γ
Acanthite	AgS	7.2–7.3	2–2.5	rhomb.(?), iron-blk.	...
Actinolite	$Ca_2((Mg,Fe)_5Si_8O_{22}(OH,F)_2$	3.02–3.44	5–6	monocl., pa. to dk. grn.	1.599–1.688, 1.612–1.697, 1.622–1.705
Aegirine	$NaFe^{+3}Si_2O_6$	3.55–3.60	6	monocl., dk. grn. to grnsh. blk.	1.750–1.776, 1.780–1.820, 1.800–1.836
Åkermanite	$Ca_2MgSi_2O_7$	2.944	5–6	tetr., col., gray-grn., br.	1.632, 1.640
Alabandite	MnS	4.050	3.5–4	cub., iron-blk., tarn., br.	...
Albite	$NaAlSi_3O_8$	2.63	6–6.5	tricl., col., wh., somet. yel., pink, grn.	1.527, 1.531, 1.538
Allanite	$(Ca,Mn,Ce,La,Y,Th)_2(Fe,Fe^{+3},Ti)(Al,Fe^{+3})_2Si_3O_{12}(OH)$	3.4–4.2	5–6.5	monocl., pa. br. to blk.	1.690–1.791, 1.700–1.815, 1.706–1.828
Allemontite	$AsSb$	5.8–6.2	3–4	hex., tin-wh. to redsh., gray, tarn. gray–brnsh. blk.	...
Almandine	$Fe_3Al_2Si_3O_{12}$	4.318	6–7.5	cub., red, dk. red, blk.	1.830
Altaite	$PbTe$	8.15	3	cub., tin-wh., yelsh., tarn. bronze-yel.	...
Aluminite	$Al_2(SO_4)(OH)_4.7H_2O$	1.66–1.82	1–2	monocl.(?), wh.	1.459, 1.464, 1.470
Alunite	$(K,Na)Al_3(SO_4)_2(OH)_6$	2.6–2.9	3.5–4	rhbdr., wh., gray, yel., redsh., br.	1.572, 1.592
Alunogen	$Al_2(SO_4)_3.18H_2O$	1.77	1.5–2	tricl., col., wh., yelsh. wh., redsh. wh.	1.459–1.475, 1.461–1.478, 1.470–1.485
Amblygonite	$(Li,Na)Al(PO_4)(F,OH)$	3.0–3.1	5.5–6	tricl., wh., yelsh. wh., grnsh. wh., blsh. wh., gray	1.591, 1.604, 1.613
Analcite	$NaAlSi_2O_6.H_2O$	2.24–2.29	5.5	cub., wh., pink, gray	1.479–1.493
Anatase	TiO_2	3.90	5.5–6	tetr., br., yelsh. br., redsh. br., bl., blk., grn., gray	2.5612, 2.4880
Andalusite	Al_2OSiO_4	3.13–3.16	6.5–7.5	rhomb., pink, wh., red	1.629–1.640, 1.633–1.644, 1.638–1.650
Andesine	$([NaSi]_{0.7-0.5}[CaAl]_{0.3-0.5})AlSi_2O_8$	2.65–2.68	6–6.5	tricl., wh., gray, grn.	1.544–1.555, 1.548–1.558, 1.551–1.563
Andorite	$PbAgSb_3S_6$	5.33–5.37	3–3.5	rhomb., dk. steel gray, somet. tarn. yel. or iridesc.	...
Andradite	$Ca_3Fe_2^{+3}Si_3O_{12}$	3.859	6–7.5	cub., brnsh. red, blk., somet. yel., grn.	1.887
Anglesite	$PbSO_4$	6.37–6.39	2.5–3	rhomb., col., wh., somet. gray, yelsh., grn. tinge	1.8771, 1.8826, 1.8937
Anhydrite	$CaSO_4$	2.98	3.5	rhomb., col., blsh. wh., vlt.	1.5698, 1.5754, 1.6136
Ankerite	$Ca(Fe,Mg,Mn)(CO_3)_2$	2.8–3.1	3.5–4	rhbdr., br., yelsh. br., grnsh. br., pink	1.690–1.750, 1.510–1.548
Anorthite	$CaAl_2Si_2O_8$	2.76	6–6.5	tricl., wh., yel., grn., blk.	1.577, 1.585, 1.590
Anorthoclase	$(Na,K)AlSi_3O_8$	2.56–2.60	6	tricl., col., wh.	1.523, 1.528, 1.529
Anthophyllite	$(Mg,Fe)_7Si_8O_{22}(OH,F)_2$	2.85–3.57	5.5–6	rhomb., wh., gray, grn., br., yelsh. br., dk. br.	1.596–1.694, 1.605–1.710, 1.615–1.722
Antimony	Sb	6.61–6.72	3–3.5	hex., tin-wh.	...
Apatite	$Ca_5(PO_4)_3(OH,F,Cl)$	3.1–3.35	5	hex., grn., wh., yel., br. red, bl.	1.629–1.667, 1.624–1.666
Apophyllite	$KFCa_4Si_8O_{20}.8H_2O$	2.33–2.37	4.5–5	tetr., col., wh., pink, pa. yel., pa. grn.	1.534–1.535, 1.535–1.537
Aragonite	$CaCO_3$	2.94–2.95	3.5–4	rhomb., col., wh.	1.530–1.531, 1.680–1.681, 1.685–1.686
Arcanite	K_2SO_4	2.663	...	rhom., col., wh.	1.4935, 1.4947, 1.4973
Argentite	Ag_2S	7.2–7.4	2–2.5	cub., blksh. lead gray	...
Arsenic	As	5.63–5.78	3.5	hex., tin-wh., tarn. dk. gray	...
Arsenolite	As_2O_3	3.86–3.88	1.5	cub., wh., somet. blsh., yelsh., redsh. tinge	1.755
Arsenopyrite	$FeAsS$	5.9–6.2	5.5–6	monocl., silver-wh., to steel gray	...
Atacamite	$Cu_2(OH)_3Cl$	3.74–3.78	3–3.5	rhomb., grn., dk. grn., blksh. grn.	1.831, 1.861, 1.880
Augelite	$Al_2(PO_4)(OH)_3$	2.696	4.5–5	monocl., col., wh., yelsh. wh., rose	1.5736, 1.5759, 1.5877
Augite	$(Ca,Mg,Fe,Fe^{+3},Ti,Al)_2(Si,Al)_2O_6$	3.23–3.52	5.5–6	monocl., pa. br., br., purp. br., grn., blk.	1.671–1.735, 1.672–1.741, 1.703–1.761
Autunite	$Ca(UO_2)_2(PO_4)_2.10–12H_2O$	3.1–3.2	2–2.5	tetr., yel., somet. grnsh. yel. to pa. grn.	1.577, 1.553
Axinite	$(Ca,Mn,Fe)_3Al_2BO_3Si_4O_{12}(OH)$	3.26–3.36	6.5–7	tricl., br., yelsh.	1.674–1.693, 1.681–1.701, 1.684–1.704
Azurite	$Cu_3(OH)_2(CO_3)_2$	3.77	3.5–4	monocl., azure bl., dk. bl., pa. bl.	1.730, 1.758, 1.838
Baddeleyite	ZrO_2	5.4–6.02	6.5	monocl., col., yel., gr., redsh. br., br., blk.	2.13, 2.19, 2.20
Barite	$BaSO_4$	4.50	3–3.5	rhomb., col., wh., somet. br., dk. br., gray	1.6362, 1.6373, 1.6482
Benitoite	$BaTi(SiO_3)_3$	3.65	6–6.5	rhbdr., bl., purp., col.	1.757, 1.804
Bertrandite	$Be_4Si_2O_7(OH)_2$	2.6	6	rhomb., col.	1.589, 1.602, 1.613
Beryl	$Be_3Al_2Si_6O_{18}$	2.66–2.83	7.5–8	hex., col., wh., blsh. grn., grnsh. yel., yel., bl.	1.565–1.590, 1.567–1.598
Beryllonite	$NaBe(PO)_4$	2.81	5.5–6	monocl., col., wh., pa. yel.	1.5520, 1.5579. 1.561
Biotite	$K(Mg,Fe)_3AlSi_3O_{10}(OH,F)_2$	2.7–3.3	2.5–3	monocl., blk., dk. br., redsh. br.	1.565–1.625–1.605–1.696, 1.605–1.696
Bismuth	Bi	9.70–9.83	2–2.5	rhbdr., silver-wh. to redsh wh.	...
Bismuthinite	Bi_2S_3	6.75–6.81	2	rhomb., lead gray to tin-wh., tarn. yel. or iridesc.	...
Bixbyite	$(Mn,Fe)_2O_3$	4.945	6–6.5	cub., blk.	...
Bloedite	$Na_2Mg(SO_4)_2.4N_2O$	2.22–2.28	2.5–3	monocl., col., somet. blsh.-grn. or redsh.	1.483, 1.486, 1.487
Boehmite	$AlO(OH)$	3.01–3.06	3.5–4	rhomb., wh.	1.64–1.65, 1.65–1.66, 1.65–1.67
Boracite	$Mg_3B_7O_{13}Cl$	2.91–2.97	7–7.5	rhomb., col., wh., gray, yel., blsh.-grn., grn.	1.66, 1.66, 1.67
Borax	$Na_2B_4O_7.10H_2O$	1.715	2–2.5	monocl., col., wh., gray, blsh. or grnsh-wh.	1.4466, 1.4687, 1.4717
Bornite	Cu_5FeS_4	5.06–5.08	3	cub., copper red to pinchbeck br., tarn. purp., iridesc.	...
Boulangerite	$Pb_5Sb_4S_{11}$	6.0–6.2	2.5–3	monocl., blsh. lead gray, oft, with yel. spots	...
Bournonite	$PbCuSbS_3$	5.80–5.86	2.5–3	rhomb., steel gray to blk.	...
Braggite	PtS	10.0	...	tetr., steel gray	...
Braunite	$(Mn,Si)_2O_3$	4.72–4.83	6–6.5	tetr., brns. blk. to steel gray	...
Bravoite	$(Ni,Fe)S_2$	4.62	5.5–6	cub., steel gray	...
Breithauptite	$NiSb$	8.23	5.5	hex., pa. copper red to vlt., tarn.	...
Brochantite	$Cu_4(SO_4)(OH)_6$	3.79	3.5–4	monocl., emerald-grn. to blksh. grn., pa. grn.	1.728, 1.771, 1.800
Bromyrite	$AgBr$	6.47	2.5	cub., col., gray, yelsh., grnsh.-br.	2.253
Brookite	TiO_2	4.08–4.20	5.5–6	rhomb., br., yelsh. br., redsh. br., blk.	2.5831, 2.5843, 2.7004
Brucite	$Mg(OH)_2$	2.38–3.40	2.5	hex., wh., pa. grn., gray, bl., yel., br.	1.560–1.590, 1.580–1.600
Bunsenite	NiO	6.898	5.5	cub., dk. pistachio-grn.	(Li) 2.37
Cacoxenite	$Fe_4(PO_4)_3(OH)_3.12H_2O$	2.2–2.4	3–4	hex., yel. to brnsh.-yel., redsh. yel., somet. grnsh.	1.575–1.585, 1.635–1.656
Calcite	$CaCO_3$	2.715–2.94	3	rhbdr., col., wh., somet. gray, yel., pink, bl.	1.658–1.740, 1.486–1.550
Caledonite	$Cu_2Pb_5(SO_4)_3(CO_3)(OH)_6$	5.75–5.77	2.5–3	rhomb., dk. grn., blsh. grn.	1.815–1.821, 1.863–1.869, 1.906–1.912
Calomel	$HgCl$	7.15	1.5	tetr., col., wh., gray, yelsh. wh., br.	1.973, 2.656
Cancrinite	$(Na,Ca)_{7-8}Al_6Si_6O_{24}(CO_3SO_4Cl)_{1.5-2}.1–5H_2O$	2.51–2.42	5–6	hex., col., wh., pa. bl., pa. grn., yel., redsh.	1.528–1.507, 1.503–1.495
Carnalite	$KMgCl_3.6H_2O$	1.602	2.5	rhomb., col., wh., oft. redsh., somet. yel., bl.	1.466, 1.475, 1.494
Carnotite	$K_2(UO_2)_2(VO_4)_2.3H_2O$	...	1–2	rhomb. or monocl., bright yel., yel., grnsh. yel.	1.75, 1.92, 1.95
Cassiterite	SnO_2	6.99	6–7	tetr., yelsh. or redsh. br., brnsh.-blk.	2.006, 2.0972
Celestite	$SrSO_4$	3.96	3–3.5	rhomb., col., wh.. pa. bl., redsh., grnsh., brnsh.	1.621–1.622, 1.623–1.624, 1.630–1.631
Celsian	$BaAl_2Si_2O_8$	3.10–3.39	6–6.5	monocl., col., wh., yel.	1.579–1.587, 1.583–1.593, 1.588–1.600
Cervantite	Sb_2O_4(?)	6.64	4–5	rhomb.(?)., yel., wh., somet. redsh.-wh.	...
Cerargyrite	$AgCl$	5.55	2.5	cub., col., gray, grnsh.-br., tarn. purp., yelsh.	2.071
Cerussite	$PbCO_3$	6.53–6.57	3–3.5	rhomb., col., wh., gray, somet. bl., blk., grn.	1.8036, 2.0765, 2.0786
Chabazite	$(Ca,Na_2)Al_2Si_4O_{12}.6H_2O$	2.05–2.10	4.5	rhbdr., redsh.-wh., wh., yelsh., grnsh.	1.470–1.494
Chalcocite	Cu_2S	5.5–5.8	2.5–3	rhomb., blksh., lead gray	...
Chalcanthite	$CuSO_4.5H_2O$	2.28	2.5	tricl., dk. bl. to sky bl., somet. grnsh.	1.514, 1.537, 1.543
Chalcopyrite	$CuFeS_2$	4.1–4.3	3.5–4	tetr., brass-yel., tarn., iridisc.	...
Chiolite	$Na_5Al_3F_{14}$	3.00	3.5–4	tetr., wh. to col.	1.349, 1.342
Chlorite	$(Mg,Al,Fe)_{12}(Si,Al)_8O_{20}(OH)_{16}$	2.6–3.3	2–3	monocl., grn., wh., yel., pink, br., red	1.57–1.66, 1.57–1.67, 1.57–1.67
Chloritoid	$(Fe,Mg,Mn)_2(AlFe^{+3})Al_3O_2SiO_4(OH)_4$	3.51–3.80	6.5	monocl., tricl., dk. grn.	1.713–1.730, 1.719–1.734, 1.723–1.740

PHYSICAL CONSTANTS OF MINERALS (Continued)

Name	Formula	Sp. gr.	Hardness	Crystalline form and color	Index of refraction (Na) η; ω ε / α β γ
Chondrodite	$Mg(OH,F)_2.2Mg_2SiO_4$	3.16–3.26	6.5	monocl., yel., br., red	1.592–1.615, 1.602–1.627, 1.621–1.646
Chromite	$FeCr_2O_4$	4.5–5.1	5.5	cub., blk.	2.16
Chrysoberyl	$BeAl_2O_4$	3.65–3.85	8.5	rhomb., grn., yel.	1.746, 1.748, 1.756
Chrysocolla	$CuSiO_3.2H_2O$	~2.4	2	rhomb., (?)., grn., bl., br., blk.	1.575, 1.597, 1.598
Cinnabar	HgS	8.090	2–2.5	hex., red, brnsh. red., gray	(Li) 2.814, 3.143
Claudetite	As_2O_3	4.15	2.5	monocl., col. to wh.	1.87, 1.92, 2.01
Clinozoisite	$Ca_2Al_3Si_3O_{12}(OH)$	3.21–3.38	6.5	monocl., col., pa. yel., gray, grn.	1.670–1.715, 1.674–1.725, 1.690–1.734
Cobaltite	$CoAsS$	6.33	5.5	cub., silver wh., redsh., steel gray, blk.	......
Colemanite	$Ca_2B_6O_{11}.5H_2O$	2.42–2.43	4.5	monocl., col., wh., yelsh. wh., gray	1.586, 1.592, 1.614
Columbite	$(Fe,Mn)(Cb,Ta)_2O_6$	5.15–5.25	6	rhomb., iron blk. to br. blk.	......
Connellite	$Cu_{19}(SO_4)Cl_4(OH)_{32}.3H_2O$(?)	3.36	3	hex., azure bl.	1.724–1.738, 1.746–1.758
Copiapite	$(Fe,Mg)Fe_4^{+3}(SO_4)_6(OH)_2.20H_2O$	2.08–2.17	2.5–3	tricl., yel., grnsh. yel.	1.51–1.53, 1.53–1.55, 1.58–1.60
Copper	Cu	8.95	2.5–3	cub., red	......
Coquimbite	$Fe_2(SO_4)_3.9H_2O$	2.10–2.12	2.5	hex., pa. vlt. to dk. amethystine, yelsh., grnsh.	1.53–1.55, 1.55–1.57
Cordierite	$Al_3(Mg,Fe)_2Si_5AlO_{18}$	2.53–2.78	7	rhomb., gray-bl., bl., dk. bl.	1.522–1.558, 1.524–1.574, 1.527–1.578
Corundum	Al_2O_3	4.022	9	hex., col., bl., yel., purp., grn., pink, red	1.767–1.772, 1.759–1.763
Cotunnite	$PbCl_2$	5.80	2.5	rhomb., col. to wh., somet. yelsh., grnsh.	2.199, 2.217, 2.260
Covellite	CuS	4.6–4.76	1.5–2	hex., indigo bl., dk. bl., iridesc. brass yel. to red	......
Cristobalite	SiO_2	2.33	6–7	tetr.(?)., col., wh., yel.	1.487, 1.484
Crocoite	$PbCrO_4$	5.96–6.02	2.5–3	monocl., red, orange red, orange yel.	2.29, 2.36, 2.66
Cryolite	Na_3AlF_6	2.96–2.98	2.5	monocl., col. to wh., brnsh., redsh., blk.	1.338, 1.338, 1.339
Cryolithionite	$Na_3Li_3Al_2F_{12}$	2.77	2.5–3	cub., col. to wh.	1.3395
Cubanite	$CuFe_2S_3$	4.03–4.18	3.5	rhomb., brass to bronze yel.	......
Cummingtonite	$(Mg,Fe)_7Si_8O_{22}(OH)_2$	3.2–3.5	5–6	monocl., dk. grn., br.	1.635–1.665, 1.644–1.675, 1.655–1.698
Cuprite	Cu_2O	6.14	3.5–4	cub., red, somet. blk.	......
Danburite	$CaSi_2B_2O_8$	3.0	7	rhomb., pa. yel., col., dk. yel., yelsh. br.	1.63, 1.63–1.64, 1.63–1.64
Datolite	$CaBSiO_4(OH)$	2.96–3.00	5–5.5	monocl. col., wh., yelsh., grnsh., pinksh.	1.622–1.626, 1.649–1.654, 1.666–1.670
Daubreelite	Cr_2FeS_4	3.80–3.82	?	cub., blk.	......
Derbylite	$Fe_6Ti_6Sb_2O_{23}$(?)	4.53	5	rhomb., pitch blk.	2.45, 2.45, 2.51
Diamond	C	3.50–3.53	10	cub., col., pa. yel. to dk. yel., pa. br. to dk. br., wh., blsh. wh.	2.4175
Diaspore	$AlO(OH)$	3.3–3.5	6.5–7	rhomb., wh., graysh. wh., col.	1.682–1.706, 1.705–1.725, 1.730–1.752
Digenite	$Cu_{2x}S$	5.546	2.5–3	cub., bl. to blk.	......
Diopside	$CaMgSi_2O_6$	3.22–3.38	5.5–6.5	monocl., wh., pa. grn., dk. grn.	1.664–1.695, 1.672–1.701, 1.695–1.721
Dioptase	$Cu_6Si_6O_{18}.6H_2O$	3.5	5	rhbdr., emerald grn.	1.64–1.66, 1.70–1.71
Dolomite	$CaMg(CO_3)_2$	2.86	3.5–4	rhbdr., wh., oft. yel. or br. tinge, col.	1.679, 1.500
Douglasite	$K_2FeCl_4.2H_2O$(?)	2.16	......	pa. grn., tarn. brnsh. red	1.485–1.491, 1.497–1.503
Dyscrasite	Ag_3Sb	9.67–9.81	3.5–4	rhomb., silver wh., tarn. gray, yelsh. or blksh.	......
Eddingtonite	$BaAl_2Si_3O_{10}.4H_2O$	2.7–2.8	......	rhomb. or monocl., col., pink, br. wh.	1.541, 1.553, 1.557
Eglestonite	Hg_4OCl_2	8.4	2.5	cub., yel., orange-yel. to dk. brnsh., tarn. bl.	2.47–2.51
Emplectite	$CuBiS_2$	6.38	2	rhomb., gray to tin wh.	......
Empressite	$AgTe$	7.510	3–3.5	pa. bronze	......
Enargite	Cu_3AsS_4	4.4–4.5	3	rhomb., gray-blk. to iron-blk.	......
Enstatite	$MgSiO_3$	3.209	5–6	rhomb., col., gray, grn., yel., brn.	1.650–1.662, 1.653–1.671, 1.658–1.680
Epidote	$Ca_2Fe^{+3}Al_2Si_3O_{12}(OH)$	3.38–3.49	6	monocl., grn., yel., gray	1.715–1.751, 1.725–1.784, 1.734–1.797
Epsomite	$MgSO_4.7H_2O$	1.675–1.679	2–2.5	rhomb., col., wh. pink, grn.	1.4325, 1.4554, 1.4609
Erythrite	$(Co,Ni)_3(AsO_4)_2.8H_2O$	3.06	1.5–2.5	monocl., crimson-red, red, pa. pink	1.626, 1.661, 1.699
Eucairite	$CuAgSe$	7.6–7.8	2.5	silver wh. to lead gray	......
Euclasite	$BeAlSiO_4(OH)$	3.0–3.1	7.5	monocl., col., pa. grn., bl.	1.651, 1.655, 1.671
Eudialite	$(Na,Ca,Fe)_6ZrSi_6O_{18}(OH,Cl)$(?)	2.8–3.1	5–6	hex., pa. pink, red, br.	1.59–1.61, 1.59–1.61
Eulytite	$Bi_4Si_3O_{12}$	6.6	4.5	cub., br., yel., gray	2.05
Euxenite	$(Y,Ca,Ce,U,Th)(Cb,Ta,Ti)_2O_6$	5.0–5.9	5.5–6.5	rhomb., blk., grnsh. or brnsh. tint.	~2.2
Fayalite	Fe_2SiO_4	4.392	6.5	rhomb., grnsh., yelsh.	1.827, 1.869, 1.879
Ferberite	$FeWO_4$	7.51	4–4.5	monocl., br. to blk.	(Li)2.37–2.43
Fergussonite	$(Y,Er,Ce,Fe)(Cb,Ta,Ti)O_4$	5.6–5.8	5.5–6.5	tetr., gray, yel., br., dk. br.	2.1
Fluorite	CaF_2	3.18	4	cub., bl., purp., wh., col., yel., grn.	1.433–1.435
Forsterite	Mg_2SiO_4	3.222	7	rhomb., wh., grnsh., yelsh.	1.635, 1.651, 1.670
Franklinite	$ZnFe_2^{+3}O_4$	5.07–5.34	5.5–6.5	Cub., blk. to br.-blk.	(Li) ~2.36
Gahnite	$ZnAl_2O_4$	4.62	7.5–8	cub., dk. bl.-grn., somet. yelsh. or brnsh.	1.79–1.81
Galena	PbS	7.57–7.59	2.5–2.75	cub. lead gray	......
Galenabismuthite	$PbBi_2S_4$	7.04	2.5–3.5	rhomb., pa. gray to tin-wh., lead gray, somet. tarn., yel. or irid.	......
Ganomalite	$(Ca,Pb)_{10}(OH,Cl)_2(Si_2O_7)_3$	5.4–5.7	3–4	hex., col., gray	1.910, 1.945
Gaylussite	$Na_2Ca(CO_3)_2.5H_2O$	1.991	2.5–3	monocl., col. to yelsh. wh., graysh. wh., wh.	1.4435, 1.5156, 1.5233
Gehlenite	$Ca_2Al_2SiO_7$	3.038	5–6	tetr., col., gray-grn., br.	1.669, 1.658
Geikielite	$MgTiO_3$	4.05	5–6	rhbdr., brnsh blk., blsh.	2.31, 1.95
Gibbsite	$Al(OH)_3$	2.38–2.42	2.5–3.5	monocl., wh., graysh., grnsh. or redsh.-wh.	1.56–1.58, 1.56–1.58, 1.58–1.60
Glauberite	$Na_2Ca(SO_4)_2$	2.75–2.85	2.5–3	monocl., gray, yelsh., somet. col., redsh.	1.515, 1.535, 1.536
Glauconite	$(K,Na,Ca)_{1.2-2}(Fe^{+3},Al,Fe,Mg)_4Si_{7-7.6}Al_{1-0.4}O_{20}(OH)_4.nH_2O$	2.4–2.95	2	monocl., col., ye.lsh. grn., grn., blsh. gray	1.592–1.610, 1.614–1.641, 1.614–1.641
Glaucophane	$Na_2Mg_3Al_2Si_8O_{22}(OH)_2$	3.08–3.30	6	monocl., gray, lavender bl.	1.606–1.661, 1.622–1.667, 1.627–1.670
Gmelinite	$(Ca,Na_2)Al_2Si_4O_{12}.6H_2O$	~2.1	4.5	rhbdr., wh., redsh.-wh., yelsh., grnsh.	1.476–1.494, 1.474–1.480
Goethite	$FeO(OH)$	3.3–4.3	5–5.5	rhomb., blksh.-br., yelsh. or redsh.-br., yel.	2.260–2.275, 2.393–2.409, 2.398–2.515
Gold	Au	19.3	2.5–3	cub., yel.	......
Goslarite	$ZnSO_4.7H_2O$	1.978	2–2.5	rhomb., col., wh., somet. br., grn., bl.	1.4568, 1.4801, 1.4844
Graphite	C	2.09–2.23	1–2	hex., iron-blk. to steel gray	......
Greenockite	CdS	4.9	3–3.5	hex., yel. to orange	2.506, 2.529
Grossularite	$Ca_3Al_2Si_3O_{12}$	3.594	6–7.5	cub., wh., yel., grn., br., red	1.734
Gummite (R)	$UO_3.H_2O$	3.9–6.4	2.5–5	yel., orange, redsh.-yel., red, br. blk.	......
Gypsum	$CaSO_4.2H_2O$	2.30–2.37	2	monocl., wh., col., somet. gray, red, yel., br.	1.519–1.521, 1.523–1.526, 1.529–1.531
Halite	$NaCl$	2.16–2.17	2.5	cub., col., wh., orange, red	1.544
Hambergite	$Be_2(OH)(BO_3)$	2.36	7.5	rhomb., col. to gray, wh., yel.	1.56, 1.59, 1.63
Hanksite	$Na_{22}K(SO_4)_9(CO_3)_2Cl$	2.562	3–3.5	hex., col., somet. pa.-yelsh. or gray	1.481, 1.461
Harmotome	$BaAl_2Si_6O_{16}.6H_2O$	2.41–2.47	4.5	monocl., or rhomb., col., wh., pink, gray, yel.	1.503–1.508, 1.505–1.509, 1.508–1.514
Hausmannite	Mn_3O_4	4.83–4.85	5.5	tetr., brnsh.-blk.	(Li) 2.46, 2.15
Haüyne	$(Na,Ca)_{4-8}Al_6Si_6O_{24}(SO_4,S)_{1-2}$	2.44–2.50	5.5–6	cub., wh., gray, grn., bl.	1.496–1.505
Hedenbergite	$CaFeSi_2O_6$	3.50–3.56	6	monocl., brnsh.-grn., dk. grn., blk.	1.716–1.726, 1.723–1.730, 1.741–1.751
Helvite	$Mn_4Be_3Si_3O_{12}S$	3.20–3.44	6	cub., yel., br., redsh.-brn.	1.728–1.749
Hematite	Fe_2O_3	5.26	5–6	rhbdr., steel gray, dull red to bright red	3.22, 2.94
Hemimorphite	$Zn_4Si_2O_7(OH)_2.H_2O$	3.45	5	rhomb., col., wh., pa. bl., pa. grn., br.	1.614, 1.617, 1.636
Hercynite	$FeAl_2O_4$	4.40	7.5–8	cub., blk.	1.835
Herderite	$CaBe(PO_4)(Fe,OH)$	2.95–3.01	5–5.5	monocl., col. to pa. yel. or grnsh.-wh.	1.592, 1.612, 1.621
Hessite	Ag_2Te	8.24–8.45	2–3	monocl., (<149.5°), cub. (>149.5°)., gray	......
Heulandite	$(Ca,Na_2)Al_2Si_7O_{18}.6H_2O$	2.1–2.2	3.5–4	pseudo-monocl., col., wh., yel., pink, red, gray, br.	1.491–1.505, 1.493–1.503, 1.500–1.512

Name	Formula	Sp. gr.	Hard-ness	Crystalline form and color	Index of refraction (Na) η; ω ε ω β γ
Hopeite	$Zn_3(PO_4)_2.4H_2O$	3.0–3.1	3.25	rhomb., col. to grayish-wh., pa. yel.	1.57–1.59, 1.58–1.60, 1.58–1.60
Hornblende	$(Ca,Na,K)_{2-3}(Mg,Fe,Fe^{+3}Al)_5Si_6(Si,Al)_2O_{22}(OH,F)_2$	3.02–3.45	5–6	monocl., grn., dk. grn., blk.	1.615–1.705, 1.618–1.714, 1.632–1.730
Huebnerite	$MnWO_4$	7.12	4–4.5	monocl., yel.-br. to red br., somet. br., blk.	2.17, 2.22, 2.32
Humite	$Mg(OH,F)_2.3Mg_2SiO_4$	3.2–3.32	6	rhomb., yel., orange	1.607–1.643, 1.619–1.653, 1.639–1.675
Huntite	$Mg_3Ca(CO_3)_4$	2.696	......	rhomb.(?), wh.	......
Hydrogrossularite	$Ca_3Al_2Si_2O_8(SiO_4)_{1-m}(OH)_{4m}$	3.594–3.13	6–7.5	cub., wh., buff, pa. grn., gray, pink	1.734–1.675
Hydromagnesite	$Mg_4(OH)_2(CO_3.3H_2O$	2.236	3.5	monocl., col. to wh.	1.520–1.526, 1.524–1.530, 1.544–1.546
Illite	$K_{1-1.5}Al_4Si_{7-6.5}Al_{1-1.5}O_{20}(OH)_4$	2.6–2.9	1–2	monocl., wh.	1.54–1.57, 1.57–1.61, 1.57–1.61
Ilmenite	$FeTiO_3$	4.68–4.76	5–6	rhbdr., iron-blk.	......
Iodyrite	AgI	5.69	1.5	hex., col. on exposure to light, yel., br.	2.21, 2.22
Jadeite	$NaAlSi_2O_6$	3.24–3.43	6	monocl., col., wh., grn., grnsh. bl.	1.640–1.658, 1.645–1.663, 1.652–1.673
Jamesonite	$Pb_4FeSb_6S_{14}$	5.63	2.5	monocl., gray-blk., somet. tarn. iridesc.	......
Jarosite	$KFe_3(SO_4)_2(OH)_6$	2.91–3.26	2.5–3.5	rhbdr., ocherous, amber yel. to dk. br.	1.820, 1.715
Kainite	$KMg(SO_4)Cl{\cdot}3H_2O$	2.15	2.5–3	monocl., col., gray bl., vlt., yelsh., redsh.	1.494, 1.505, 1.516
Kaliophyllite	$KAlSiO_4$	2.61	6	hex., col.	1.532, 1.537
Kaolinite	$Al_4Si_4O_{10}(OH)_8$	2.61–2.68	2–2.5	tricl. or monocl., wh., redsh.-wh., grnsh.-wh	1.533–1.565, 1.559–1.569, 1.560–1.570
Kernite	$Na_2B_4O_7.4H_2O$	1.908	2.5	monocl., col., wh.	1.454, 1.472, 1.488
Kieserite	$MgSO_4.H_2O$	2.571	3.5	monocl., col., gray, wh., yelsh.	1.520, 1.533, 1.584
Kyanite	Al_2OSiO_4	3.53–3.65	5.5–7	tricl., bl., wh., gray, grn., yel., pink	1.712–1.718, 1.721–1.723, 1.727–1.734
Lanarkite	$Pb_2(SO_4)O$	6.92	2–2.5	monocl., gray to grnsh. wh., pa. yel.	1.925–1.931, 2.004–2.010, 2.033–2.039
Lanthanite	$(La,Ce)_2(CO_3)_38H_2O$	2.69–2.74	2.5–3	rhomb., col. to wh., pink, yelsh.	1.51–1.53, 1.584–1.590, 1.610–1.616
Laumontite	$CaAl_2Si_4O_{12}.4\text{–}3.5H_2O$	2.2–2.3	3–3.5	monocl., col., wh., red, yel., brn.	1.502–1.514, 1.512–1.522, 1.514–1.525
Laurionite	$Pb(OH)Cl$	6.24	3–3.5	rhomb., col. to wh.	2.08, 2.12, 2.16
Lawsonite	$CaAl_2(OH)_2Si_2O_7.H_2O$	3.05–3.10	6	rhomb., col., wh.	1.655, 1.674–1.675, 1.684–1.686
Lazulite	$(Mg,Fe)Al_2(PO_4)_2(OH)_2$	3.08–3.38	5.5–6	monocl., bl., blsh. wh., dk. bl., blsh. grn.	1.604–1.626, 1.626–1.654, 1.637–1.663
Lazurite	$Na_4SSi_3Al_3O_{12}$	2.38–2.45	5–5.5	cub., berlin bl., azure bl., grnsh. bl., vlt.	1.500
Leadhillite	$Pb_4(SO_4)(CO_3)_2(OH)_2$	6.55	2.5–3	monocl., col. to wh., gray, pa. grn., pa. bl., yelsh.	1.87, 2.00, 2.01
Lepidocrocite	$FeO(OH)$	4.05–4.31	5	rhomb., ruby-red to red-br.	1.94, 2.20, 2.51
Lepidolite	$K_2(Li,Al)_{5-6}Si_{6-7}Al_{2-1}O_{20}(OH,F)_4$	2.80–2.90	2.5–4	monocl., col., pa. pink, pa. purp.	1.525–1.548, 1.551–1.585, 1.554–1.587
Leucite	$KAlSi_2O_6$	2.47–2.50	5.5–6	tetr., (pseudo-cub.) wh., gray	1.508–1.511
Levyne	$(Ca,Na_2)Al_2Si_4O_{12}.6H_2O$	~2.1	4.5	rhbdr., wh., redsh. wh., yelsh., grnsh.	1–496–1.505, 1.491–1.500
Litharge	PbO	9.14	2	tetr., red	(Li) 2.665, 2.535
Loellingite	$FeAs_2$	7.39–7.41	5–5.5	rhomb., silver wh. to steel-gray	......
Magnesite	$MgCO_3$	2.98–3.44	3.5–4.5	rhbdr., wh., col., somet. yel., br.	1.700–1.782, 1.509–1.563
Magnetite	Fe_3O_4	5.175	5.5–6.5	cub., blk. to br.-blk.	2.42
Malachite	$Cu_2(OH)_2(CO_3)$	4.03–4.07	3.5–4	monocl., bright grn. to dk. grn., blksh. grn.	1.652–1.658, 1.872–1.878, 1.906–1.912
Manganite	$MnO(OH)$	4.32–4.43	4	monocl., dk. steel-gray to iron-blk.	(Li) 2.25, 2.25, 2.53
Manganosite	MnO	5.364	5.5	cub., emerald grn., tarn. bl.	......
Marcasite	FeS_2	4.887	6–6.5	rhomb., pa. bronze-yel., tin-wh.	......
Marialite	$Na_4Al_3Si_9O_{24}Cl$	2.50–2.62	5–6	tetr., col., wh., pa. grnsh. yel., gray, br.	1.546–1.550, 1.540–1.541
Marshite	CuI	5.68	2.5	cub., col. to pa. yel., on exposure to light, red	2.346
Mascagnite	$(NH_4)_2SO_4$	1.768	2–2.5	rhomb., col., gray, yelsh.	1.5202, 1.5230, 1.5330
Matlockite	$PbFCl$	7.12	2.5–3	tetr., col. or yel. to pa. amber, grnsh.	2.145, 2.006
Meionite	$Ca_4Al_6Si_6O_{24}CO_3$	2.78	5–6	tetr., col., wh., pa. grnsh. yel., gray, br.	1.590–1.600, 1.556–1.562
Melanterite	$FeSO_4.7H_2O$	1.898	2	monocl., grn., grnsh. bl., grnsh. wh.	1.47, 1.48, 1.49
Melilite	$(Ca,Na,K)_2(Mg,Fe,Fe^{+3},Al,Si)_3O_7$	2.95–3.05	5–6	tetr., yelsh., br., grn.-br.	1.624–1.666, 1.616–1.661
Mellite	$Al_2C_{12}O_{12}.18H_2O$	1.64	2–2.5	tetr., yel., redsh., brnsh., somet. wh.	1.5393, 1.5110
Mendipite	$Pb_3O_2Cl_2$	7.24	2.5	rhomb., col. to wh., gray, oft. yel., red, bl. tinge	2.22 2.26, 2.25–2.29, 2.29–2.33
Mesolite	$Na_2Ca_2(Al_2Si_3O_{10}).8H_2O$	~2.20	5	monocl., col., wh., gray, yel., pink, red	β = 1.504–1.508
Metacinnabar	HgS	7.65	3	cub., graysh.-blk.	......
Microcline	$KAlSi_3O_8$	2.56–2.63	6–6.5	tricl., col., wh., pink, red, yel., grn.	1.514–1.529, 1.518–1.533, 1.521–1.539
Microlite	$(Na,Ca)_2Ta_2O_6(O,OH,F)$	4.2–6.4	5–5.5	cub., pa. yel. to br., somet. red, grn.	~2.0
Miersite	AgI	5.64–5.68	2.5	cub., canary-yel.	2.18–2.22
Millerite	NiS	5.3–5.7	3–3.5	hex., pa. brass-yel. to bronze-yel., gray, tarn. iridesc.	......
Mimetite	$Pb_5(AsO_4,PO_4)_3Cl$	7.24	3.5–4	hex. pa. yel. to yelsh. br., orange-yel., wh.	2.147, 2.128
Minium	Pb_3O_4	8.9–9.2	2.5	scarlet red, bl. red, somet. yel. tint.	(Li) 2.40–2.44
Mirabilite	$Na_2SO_4.10H_2O$	1.490	1.5–2	monocl., col. to wh.	1.391–1.397, 1.393–1.399, 1.395–1.401
Moissanite	SiC	3.218	9.5	hex., grn. to blk., somet. blsh., red	2.647–2.649, 2.689–2.693
Molybdenite	MoS_2	4.62–4.73	1–1.5	hex., lead-gray	......
Monazite	$(Ce,La,Th)PO_4$	5.0–5.3	5	monocl., yel., br., redsh. br.	1.774–1.800, 1.777–1–801, 1.828–1.851
Monetite	$CaH(PO_4)$	2.929	3.5	tricl., wh., pa. yelsh.-wh.	1.587, ~1.615, 1.640
Monticellite	$CaMgSiO_4$	3.08–3.27	5.5	rhomb., col.	1.639–1.654, 1.646–1.664, 1.653–1.674
Montmorillonite	$(0.5Ca,Na)_{0.7}(Al,Mh,Fe)_4(Si,Al)_8O_{20}(OH)_4.nH_2O$	2–3	1–2	monocl., wh., yel., grn.	1.48–1.61, 1.50–1.64, 1.50–1.64
Montroydite	HgO	11.23	2.5	rhomb., dk. red to brnsh. red, br.	(Li) 2.37, 2.5, 2.65
Mordenite	$(Na_2,K_2Ca)Al_2Si_{10}O_{24}.7H_2O$	2.12–2.15	3–4	rhomb., col., wh., red, yel., br.	1.472–1.483, 1.475–1.485, 1.477–1.487
Muscovite	$KAl_2Si_3AlO_{10}(OH,F)_2$	2.77–2.88	2.5–3	monocl., col., pa. grn., pa. red, pa. br.	1.552–1.574, 1.582–1.610, 1.587–1.616
Nantokite	$CuCl$	4.136	2.5	cub., col. to wh., grayish, grn.	1.925–1.935
Natrolite	$Na_2Al_2Si_3O_{10}.2H_2O$	2.20–2.26	5	rhomb., col., wh., gray, yel., pink, red	1.473–1.483, 1.476–1.486, 1.485–1.496
Nepheline	$Na_3KAl_4Si_4O_{16}$	2.56–2.665	5.5–6	hex., col., wh., gray	1.529–1.546, 1.526–1.542
Newberyite	$MgH(PO_4).3H_2O$	2.10	3.0–3.5	rhomb., col.	1.511–1.517, 1.514–1.520, 1.530–1.536
Niccolite	$NiAs$	7.784	5–5.5	hex., pa. copper-red, tarn. gray to blk.	......
Nosean	$Na_8Al_6Si_6O_{24}SO_4$	2.30–2.40	5.5	cub., gray, bl., br.	1.495
Oldhamite	CaS	2.58	4	cub., pa. chestnut-br.	2.137
Oligoclase	$([NaSi]_{0.9\text{–}0.7}[CaAl]_{0.1\text{–}0.3})AlSi_2O_8$	2.63–2.65	6–6.5	tricl., col., wh., gray, grnsh., pink	1.533–1.544, 1.537–1.548, 1.543–1.552
Olivenite	$Cu_2(AsO_4)(OH)$	3.9–4.5	3	rhomb., olive grn., grnsh.-br., br., gray	1.75–1.78, 1.79–1.82, 1.83–1.87
Olivine	$(Mg,Fe)SiO_4$	3.22–4.39	6.5–7	rhomb., olive grn., grayish grn. to yelsh. br.	1.63–1.83, 1.65–1.87, 1.67–1.88
Opal	$SiO_2.nH_2O$	1.73–2.16	~6	col., wh., yel., br., red, grn., bl., blk., amorp.	1.41–1.46
Orpiment	As_2S_3	3.49	1.5–2	monocl., yel., brnsh. yel.	(Li) 2.4, 2.81, 3.02
Orthoclase	$KAlSi_3O_8$	2.55–2.63	6–6.5	monocl., col., wh., pink, red, yel., grn.	1.518–1.529, 1.522–1.533, 1.522–1.539
Orthopyroxene	$(Mg,Fe)SiO_3$	3.209–3.96	5–6	rhomb., col., gray, grn., yel., dk. brn.	1.650–1.768, 1.653–1.770, 1.658–1.788
Paragonite	$NaAl_2Si_3AlO_{10}(OH)_2$	2.85	2.5	monocl., col., pa. yel.	1.564–1.580, 1.594–1.609, 1.600–1.609
Parisite	$(Ce,La,Na)FCO_3.CaCO_3$	4.42	4.5	hex., brnsh., yel.	1.672, 1.771
Pectolite	$Ca_2NaH(SiO_3)_3$	2.86–2.90	4.5–5	tricl., col., wh.	1.595–1.610, 1.605–1.615, 1.632–1.645
Penfieldite	$Pb_4Cl_6(OH)_2$	6.6		hex., wh.	2.13, 2.21
Pentlandite	$(Fe,Ni)_9S_8$	4.6–5.0	3.5–4	cub., pa. bronze-yel.	......
Percylite	$PbCuCl_2(OH)_2(?)$	?	2.5	cub(?), sky bl.	2.04–2.06
Periclase	MgO	3.55–3.68	5.5	cub., col. to gray-wh., yel., brnsh. yel., grn., bl.	1.7350
Pekovskite	$CaTiO_3$	3.97–4.26	5.5	pseudo cub., blk., gray-blk., brnsh. bl., redsh. br., br., yel.	2.30–2.38
Petalite	$LiAlSi_4O_{10}$	2.412–2.422	6.5	monocl., wh., gray, somet. pink, grn.	1.504–1.507, 1.510–1.513, 1.516–1.523

Name	Formula	Sp. gr.	Hardness	Crystalline form and color	Index of refraction (Na) η; ω ϵ / α β γ
Pharmacosiderite	$Fe_3(AsO_4)_2(OH)_3.5H_2O$	2.797	2.5	cub., olive-grn. to yel., br., redsh.	1.676–1.704
Phenakite	Be_2SiO_4	2.98	7.5	rhbder., col., rose, yel., br.	1.654, 1.670
Phillipsite	$(0.5Ca,Na,K)_3Al_3Si_5O_{16}.6H_2O$	2.2	4–4.5	monocl. or rhomb., col., wh., pink, gray, yel.	1.483–1.504, 1.484–1.509, 1.496–1.514
Phlogopite	$KMg_3AlSi_3O_{10}(OH,F)_2$	2.76–2.90	2–2.5	monocl., col., yelsh.-br., grn., redsh.-br., br.	1.530–1.590, 1.557–1.637, 1.558–1.637
Phosgenite	$Pb_2(CO_3)Cl_2$	6.133	2–3	tetr., yelsh. wh. to yelsh. br., br., somet. wh., rose, gray	2.1181, 2.1446
Piemontite	$Ca_2(Mn,Fe^{+3},Al)_2AlSi_3O_{12}(OH)$	3.45–3.52	6	monocl., redsh. brn., blk.	1.732–1.794, 1.750–1.807, 1.762–1.829
Pigeonite	$(Mg,Fe,Ca)(Mg,Fe)Si_2O_6$	3.30–3.46	6	monocl., br., grnsh. br., blk.	1.682–1.722, 1.684–1.722, 1.705–1.751
Platinum	Pt	14–19	4–4.5	cub., whitish, steel gray to dk. gray	
Pollucite	$CsAlSi_2O_6$	2.9	6.5	tetr., (pseudo-cub.) col.	1.507–1.527
Polybasite	$(Ag,Cu)_{16}Sb_2S_{11}$	6.0–6.2	2–3	monocl., iron-blk.	
Powellite	$Ca(Mo,W)O_4$	4.21–4.25	3.5–4	tetr., straw-yel., br., grnsh., somet. gray, bl., blk.	1.959–1.982, 1.967–1.993
Prehnite	$Ca_2Al_2Si_3O_{10}(OH)_2$	2.90–2.95	6–6.5	rhomb., pa. grn., yel., gray, wh.	1.611–1.632, 1.615–1.642, 1.632–1.665
Proustite	Ag_3AsS_3	5.57	2–2.5	rhbdr., scarlet-vermillion	3.0877, 2.7924
Pseudobrookite	Fe_2TiO_5	4.33–4.39	6	rhomb., dk. red-br. to brnsh. blk. and blk.	(Li) 2.38, 2.39, 2.42
Psilomelane	$BaMn^{+2}Mn_8{}^{+4}O_{16}(OH)_4$	4.71	5–6	rhomb., iron-blk. to steel-gray	
Pumpellyite	$Ca_4(Mg,Fe,Mn)(Al,Fe^{+3},Ti)_5(OH)_3Si_6O_{23}.2H_2O$	3.18–3.23	6	monocl., grn., blsh. grn., br.	1.674–1.702, 1.675–1.715, 1.688–1.722
Pyrargyrite	Ag_3SbS_3	5.85	2.5	rhbdr., deep red	(Li) 3.084, 2.881
Pyrite	FeS_2	5.018	6–6.5	cub., pa. brass-yel., tarn. iridesc.	
Pyrochlore	$NaCaCb_2O_6F$	4.2–6.4	5–5.5	cub., br. to blk., yelsh., redsh. or blksh. br.	
Pyrochroite	$Mn(OH)_2$	3.23–3.27	2.5	hex., col. to pa. grn. or bl., tarn. br. to blk.	1.72, 1.68
Pyrolusite	MnO_2	5.04–5.08	6–6.5	tetr., pa. steel-gray, iron-gray, blk., blsh.	
Pyromorphite	$Pb_5(PO_4,AsO_4)_3Cl$	7.00–7.08	3.5–4	hex., grn., yel., br., orange, brnsh. red., gray	2.058, 2.048
Pyrope	$Mg_3Al_2Si_3O_{12}$	3.582	6–7.5	cub., red, pink	1.714
Pyrophyllite	$Al_2Si_4O_{10}(OH)_2$	2.65–2.90	1–2	monocl., wh., yel., pa. bl., gray-grn., brnsh.-grn.	1.534–1.556, 1.568–1.589, 1.596–1.601
Pyrrhotite	$Fe_{1-0.8}S$	4.58–4.65	3.5–4.5	hex., bronze-yel. to br., tarn., somet. iridesc.	
Quartz	SiO_2	2.65	7	rhbdr., col., wh., blk., purp., grn., bl., rose	1.544, 1.553
Rammelsbergite	$NiAs_2$	7.0–7.2	5.5–6	tin. wh., redsh. tinge	
Raspite	$PbWO_4$	8.46	2.5–3	monocl., yelsh. br., pa. yel., gray	1.25–1.29, 1.25–1.29, 1.28–1.32
Realgar	AsS	3.56	1.5–2	monocl., aurora-red to orange-yel.	2.538, 2.684, 2.704
Riebeckite	$Na_2Fe_3Fe_2{}^{+3}Si_8O_{22}(OH,F)_2$	3.02–3.42	5	monocl., dk. bl., bl.	1.654–1.701, 1.662–1.711, 1.668–1.717
Rhodochrosite	$MnCO_3$	3.70	3.5–4	rhbdr., pink, red, br., brnsh.-yel.	1.816, 1.597
Rhodonite	$(Mn,Fe,Ca)SiO_3$	3.57–3.76	5.5–6.5	tricl., pink to brnsh. red	1.711–1.738, 1.716–1.741, 1.724–1.751
Rutile	TiO_2	4.23–5.5	6–6.5	tetr., redsh. brn. to red, somet. yelsh., blsh.	2.605–2.613, 2.899–2.901
Safflorite	$(Co,Fe)As_2$	7.0–7.5	4.5–5	rhomb., tin-wh., tarn. dk. gray	
Samarskite	$(Y,Er,Ce,U,Ca,Fe,Pb,Th)(Cb,Ta,Ti,Sn)_2O_6$	5.69	5–6	rhomb., velvet blk., somet. brnsh. tint	~2.20
Sapphirine	$(Mg,Fe)_2Al_4O_6SiO_4$	3.40–3.58	7.5	monocl., pa. bl., pa. grn.	1.701–1.717, 1.703–1.720, 1.705–1.724
Scapolite	$(Na,Ca)_4Al_3(Al,Si)_3Si_6O_{24}(Cl,F,OH,CO_3,SO_4)$	2.50–2.78	5–6	tetr., col., wh., pa. grnsh. yel., gray, bl.	1.546–1.600, 1.540–1.562
Scheelite	$CaWO_4$	6.08–6.12	4.5–5	tetr., yelsh. wh., pa. yel., brnsh., col., wh., gray	1.920, 1.936
Scolecite	$CaAl_2Si_3O_{10}.3H_2O$	2.25–2.29	5	monocl., col., wh., gray, yel., pink, red	1.507–1.513, 1.516–1.520, 1.517–1.521
Scorodite	$Fe^{+3}(AsO_4).2H_2O$	3.28	3.5–4	rhomb., pa. grn., gray grn., br. somet. col., blsh., yel.	1.784, 1.795, 1.814
Sellaite	MgF_2	3.15	5	tetr., col. to wh.	1.378, 1.390
Senarmontite	Sb_2O_3	5.50	2–2.5	pseudo-cub., col., gray-wh.	2.087
Serpentine	$Mg_3Si_2O_5(OH)_4$	~2.55	2.5–3.5	monocl., wh., yel., gray, grn., blsh. grn.	1.53–1.57, 1.56, 1.54–1.57
Siderite	$FeCO_3$	3.96	4–4.5	rhbdr., yelsh. br., br., dk. br.	1.875, 1.635
Sillimanite	Al_2OSiO_4	3.23–3.27	6.5–7.5	rhomb., col., wh., yelsh., br., grnsh.	1.654–1.661, 1.658–1.662, 1.637–1.683
Silver	Ag	10.1–11.1	2.5–3	cub., wh., tarn. gray or blk.	
Skutterudite	$(Co,Ni)As_3$	6.1–6.9	5.5–6	cub., between tin-wh. and silver-gray, tarn. gray or iridesc.	
Smithsonite	$ZnCO_3$	4.42–4.44	4–4.5	rhbdr., grayish wh. to dk. gray, grnsh., brnsh. wh.	1.848, 1.621
Sodalite	$Na_8Al_6Si_6O_{24}Cl_2$	2.27–2.33	5.5–6	cub., bl., grn., yel., gray, pink	1.483–1.487
Sperrylite	$PtAs_2$	10.58	6–7	cub , tin-wh.	
Spessartite	$Mn_3Al_2Si_3O_{12}$	4.190	6–7.5	cub., blk., dk. red, brnsh. red., bl., yelsh. orange	1.800
Sphalerite	ZnS	3.9–4.1	3.5–4	cub., br., blk., yel., red, wh.	2.369
Sphene	$CaTiSiO_4(O,OH,F)$	3.45–3.55	5	monocl., col., yel., grn., br., blk.	1.843–1.950, 1.870–2.034, 1.943–2.110
Spinel	$MgAl_2O_4$	3.55	7.5–8	cub., grn., red, bl., br. to col.	1.719
Spodumene	$LiAlSi_2O_6$	3.03–3.22	6.5–7	monocl., col., gray-wh., pa. bl., pa. grn., yelsh.	1.648–1.663, 1.655–1.669, 1.662–1.679
Stannite	Cu_2FeSn_4	4.3–4.5	4	tetr., steel gray to iron blk.	
Staurolite	$(Fe,Mg)_2(AlFe^{+3})_9O_6SiO_4(O,OH)_2$	3.74–3.83	7.5	monocl., brn., redsh., yelsh.	1.739–1.747, 1.745–1.753, 1.752–1.761
Stercorite	$Na(NH_4)H(PO_4).4H_2O$	1.615	2	tricl., wh., yelsh., brnsh.	1.439, 1.442, 1.469
Stibiotantalite	$Sb(Ta,Cb)O_4$	5.7–7.5	5.5	rhomb., dk. br. to pa. yel.-br., red-br., grnsh.-yel.	2.38, 2.41, 2.46
Stibnite	Sb_2S_3	4.61–4.65	2	rhomb., lead-gray to steel-gray	
Stilbite	$(Ca,Na_2K_2)Al_2Si_7O_{18}.7H_2O$	2.1–2.2	3.5–4	monocl., col., wh., yel., pink, red, gray, br.	1.484–1.500, 1.492–1.507, 1.494–1.513
Stilpnomelane	$(K,Na,Ca)_{0-1.4}(Fe^{+3}Fe,Mg,Al)_{6-8}Si_8O_{20}(OH)_4(O,OH,H_2O)_{4-8}$	2.59–2.96	3–4	monocl., br., dk. br., redsh. br., blk., dk. grn.	1.543–1.634, 1.576–1.745, 1.576–1.745
Stolzite	$PbWO_4$	7.9–8.4	2.5–3	tetr., redsh. br., yelsh. gray, straw-yel., grnsh.	2.26–2.28, 2.18–2.20
Strengite	$Fe^{+3}(PO_4).2H_2O$	2.90	3.5–4.5	rhomb., red, carmine, vlt., near col.	1.707, 1.719, 1.741
Strontianite	$SrCO_3$	3.72	3.5	rhomb., col., wh., yel., grnsh., brnsh.	1.516–1.520, 1.664–1.667, 1.666–1.669
Struvite	$Mg(NH_4)(PO_4).6H_2O$	1.71	2	rhomb., col., somet. yelsh., brnsh.	1.495, 1.496, 1.504
Sulfur	S	2.07	1.5–2.5	rhomb., yel., brnsh., grnsh., redsh., gray	1.9579, 2.0377, 2.2452
Sylvanite	$(Ag,Au)Te_2$	8.161	1.5–2	monocl., steel-gray to silver-wh.	
Sylvite	KCl	1.99	2	cub., col., wh., somet. grayish, blsh., yelsh., red	1.49031
Talc	$Mg_3Si_4O_{10}(OH)_2$	2.58–2.83	1	monocl., col., wh., pa. grn., dk. grn., br.	1.539–1.550, 1.589–1.594, 1.589–1.600
Tantalite	$(Fe,Mn)(Ta,Cb)_2O_6$	7.90–8.00	6.5	rhomb., iron-bl. to br.-blk.	2.26, 2.32, 2.43
Tapiolite	$FeTa_2O_6$	7.9	6–6.5	tetr., blk.	(Li) 2.27, 2.42
Tellurobismuthite	Bi_2Te_3	7.800–7.830	1.5–2	rhbdr., pa. lead-gray	
Terlinguaite	Hg_2OCl	8.725	2.5	monocl., yel. to grnsh.-yel., somet. br.	(Li) 2.33–2.37, 2.62–2.66, 2.64–2.68
Tetrahedrite	$(Cu,Fe)_{12}Sb_4S_{13}$	4.6–5.1	3–4.5	cub., flint-gray to iron-blk. to dull-blk.	
Thenardite	Na_2SO_4	2.664	2.5–3	rhomb., col., grayish-wh., yelsh., yelsh. br., redsh.	1.464–1.471, 1.473–1.477, 1.481–1.485
Thermonatrite	$Na_2CO_3.H_2O$	2.255	1–1.5	rhomb., col. to wh., grayish, yelsh.	1.420, 1.506, 1.524
Thomsenolite	$NaCaAlF_6.H_2O$	2.981	2	monocl., col. to wh., somet. brnsh., redsh.	1.4072, 1.4136, 1.4150
Thomsonite	$NaCa_2([Al,Si]_5O_{10})_2.6H_2O$	2.10–2.39	5–5.5	rhomb., col., wh., pink, br.	1.497–1.530, 1.513–1.533, 1.518–1.544
Thorianite (R)	ThO_2	9.7	6.5	cub., dk. gray to brnsh.-blk., blk.	~2.20
Thorite (R)	$ThSiO_4$	5.2–5.4	4.5–5	tetr., orange-yel., brnsh. to blk.	~1.8
Topaz	$Al_2SiO_3(OH,F)_2$	3.49–3.57	8	rhomb., col., wh., yel., gray, grn., red, bl.	1.606–1.629, 1.609–1.631, 1.616–1.638

PHYSICAL CONSTANTS OF MINERALS (Continued)

Name	Formula	Sp. gr.	Hardness	Crystalline form and color	Index of refraction (Na) η; ω ϵ; ω β γ
Torbernite (R)	$Cu(UO_2)_2(PO_4)_2.8\text{-}12H_2O$	3.22	2-2.5	tetr., various shades of grn.	1.592, 1.582
Tourmaline	$Na(Mg,Fe,Mn,Li,Al)_3Al_6Si_6O_{18}(BO_3)_3(OH,F)_4$	3.03-3.25	7	rhbdr., blk., bl., grn., yel., red, col., br.	1.635-1.675, 1.610-1.650
Tremolite	$Ca_2Mg_5Si_8O_{22}(OH,F)_2$	3.0	5-6	monocl., col., gray, wh.	1.599, 1.612, 1.622
Tridymite	SiO_2	2.27	7	rhomb., col., wh.	1.471-1.479, 1.472-1.480, 1.474-1.483
Triphyllite-Lithiophyllite	$Li(Fe,Mn)PO_4$	3.34-3.58	4-5	rhomb., blsh. or grnsh. gray to yelsh. br., br.	1.66-1.70, 1.67-1.70, 1.68-1.71
Troegerite (R)	$(UO_2)_3(AsO_4)_2.12H_2O$	....	2-3	tetr., lemon-yel.	1.58-1.59, 1.625-1.635
Trona	$Na_3H(CO_3)_2.2H_2O$	2.14	2.5-3	monocl.. gray or yelsh. wh., col.	1.412, 1.492, 1.540
Turquois	$Cu(Al,Fe^{+3})_6(PO_4)_4(OH)_8.4H_2O$	2.6-3.2	4.5-6	tricl., bl., grn., grnsh.-gray	1.61-1.78, 1.62-1.84, 1.65-1.84
Ullmannite	$NiSbS$	6.61-6.69	5-5.5	cub., steel-gray to silver-wh.	....
Uraninite (R)	UO_2	8.0-11	5-6	cub., steel-blk., brnsh.-blk., grayish, grn.	....
Uvarovite	$Ca_3Cr_2Si_3O_{12}$	3.90	6-7.5	cub., emerald-grn.	1.86
Valentinite	Sb_2O_3	5.76	2.5-3	rhomb., col. to wh., somet. yelsh., redsh., gray, br.	2.18, 2.35, 2.35
Vanadinite	$Pb_5(VO_4)_3Cl$	6.5-7.1	2.75-3	hex., orange-red, red, brnsh.-red, br., brnsh.-yel., yel.	2.416, 2.350
Variscite-Strengite	$(AlFe^{+3})(PO_4).2H_2O$	2.57-2.87	3.5-4.5	rhomb., pa. grn., grn., blsh.-grn., red, vlt., col.	1.563-1.707, 1.588-1.719, 1.594-1.741
Vaterite	$CaCO_3$	2.645	....	hex., col.	1.550, 1.640-1.650
Vermiculite	$(Mg,Ca)_{0.7}(Mg,Fe^{+3}Al)_6(Al,Si)_8O_{20}(OH)_4.8H_2O$	~2.3	~1.5	monocl., col., yel., grn., br.	1.525-1.564, 1.545-1.583, 1.545-1.583
Vesuvianite	$Ca_{10}(Mg,Fe)_2Al_4(Si_2O_7)_2(SiO_4)_5(OH,F)_4$	3.33-3.43	6-7	tetr., yel., grn., br.	1.700-1.746, 1.703-1.752
Villiaumite	NaF	2.79	2-2.5	cub., carmine, (nat.), col. (artif.)	1.327
Vivianite	$Fe_3(PO_4)_2.8H_2O$	2.67-2.69	1.5-2	monocl., col., tarn. pa. bl., grnsh. bl., dk. bl., blsh. blk.	1.579-1.616, 1.602-1.656, 1.629-1.675
Wagnerite	$Mg_2(PO_4)F$	3.15	5-5.5	monocl., yel., gray, somet. red, grn.	1.568, 1.572, 1.582
Wavellite	$Al_3(OH)_3(PO_4)_2.5H_2O$	2.36	3.25-4	rhomb., grnsh. wh., grn. to yel., somet. br., bl., wh.	1.520-1.535, 1.526-1.543, 1.545-1.561
Whewellite	$Ca(C_2O_4).H_2O$	2.23	2.5-3	monocl., col., somet. yelsh., brnsh.	1.491, 1.554, 1.650
Willemite	Zn_2SiO_4	3.9-4.1	5.5	rhbdr., wh., yel., grn., red, gray, br.	1.691, 1.719
Witherite	$BaCO_3$	4.29-4.30	3.5	rhomb., col., wh., gray, yelsh. br.	1.529, 1.676, 1.677
Wolframite	$(Fe,Mn)WO_4$	7.12-7.51	4-4.5	monocl., dk. gray, brnsh. blk. to iron blk.	(Li) ~2.26, 2.32, 2.42
Wollastonite	$CaSiO_3$	2.87-3.09	4.5-5	tricl., wh., col., gray, pa. grn.	1.616-1.640, 1.628-1.650, 1.631-1.653
Wulfenite	$PbMoO_4$	6.5-7.0	2.75-3	tetr., orange-yel. to yel., gray, grn., br., red	2.403, 2.283
Wurtzite	ZnS	3.98	3.5-4	hex., brnsh. blk.	2.356, 2.378
Xenotime	$Y(PO_4)$	4.4-5.1	4-5	tetr., yelsh. br. to redsh. br., somet. gray, wh., pa. yel., grnsh.	1.721, 1.816
Zeunerite (R)	$Cu(UO_2)_2(AsO_4)_2.10\text{-}16H_2O$	....	....	tetr.	1.602-1.610
Zincite	ZnO	5.64-5.68	4	hex., orange-yel. to dk. red, somet. yel.	2.013, 2.029
Zircon	$ZrSiO_4$	4.6-4.7	7.5	tetr., redsh. br., yel., gray, grn., col.	1.923-1.960, 1.968-2.015
Zoisite	$Ca_2Al_3Si_3O_{12}(OH)$	3.15-3.365	6	rhomb., gray, grnsh., brnsh.	1.685-1.705, 1.688-1.710, 1.697-1.725

X-Ray Crystallographic Data, Molar Volumes, and Densities of Minerals and Related Substances

From U.S. Geological Survey Bulletin 1248 by
Richard A. Robie, Philip M. Bethke and Keith M. Beardsley

An extensive list of references and the bases for the calculations and the selection of data are given in the above referenced Bulletin. Bulletin 1248 may be obtained from the Superintendent of Documents, U.S. Government Printing Office, Washington, D.C., 20402.

Z; The number of gram formula weights per unit cell.

r; Indicates the data were obtained at an unspecified room temperature and may be taken as 25°+5°C.

*; Indicates the measurements were made on a natural specimen which may have deviated slightly from the listed formula. Densities for these minerals were calculated using the formula weight for the stoichiometric phase.

hex-R; Rhombohedral symmetry. To distinguish from true hexagonal symmetry.

X-Ray Crystallographic Data of Minerals

	Name and formula	Crystal system	Space group	Structure type	Z	a_o	b_o	c_o	α_o	β_o	γ_o	Cell volume 10^{-24} cm^3	Molar volume cm^3	Molar volume cal bar^{-1}	X-Ray density grams cm^{-3}	Temp. °C	
	Elements																
1	Silver Ag	cubic	Fm3m(225)	face-centered cubic	4	4.0862 ±.0002						68.227 ± .010	10.272 ± .002	.24556 ±.00008	10.501 ±.002	25	1
2	Arsenic As	hex-R	R$\bar{3}$m(166)	arsenic	6	3.760 ±.001		10.555 ±.003				129.23 ± .08	12.972 ± .002	.31007 ±.00023	5.776 ±.004	26	2
3	Gold Au	cubic	Fm3m(225)	face-centered cubic	4	4.0786 ±.0002						67.847 ± .010	10.215 ± .002	.24420 ±.00008	19.282 ±.003	25	3
4	Bismuth Bi	hex-R	R$\bar{3}$m(166)	arsenic	6	4.5459 ±.0010		11.8622 ±.0030				212.29 ± .11	21.309 ± .011	.50934 ±.00030	9.8071 ±.0050	26	4
5	Diamond C*	cubic	Fd3m(227)	diamond	8	3.5670 ±.0001						45.385 ± .004	3.4166 ± .0003	.08170 ±.00005	3.5155 ±.0003	25	5
6	Graphite C*	hex.	C6/mmc(194)	graphite	4	2.4612 ±.0001		6.7079 ±.0010				35.189 ± .006	5.2982 ± .0009	.12668 ±.00007	2.2670 ±.0004	15	6
7	Copper Cu	cubic	Fm3m(225)	face-centered cubic	4	3.6150 ±.0005						47.242 ± .020	7.1128 ± .0030	.17005 ±.00012	8.9331 ±.0037	25	7
8	α-Iron Fe	cubic	Im3m(229)	body-centered cubic	2	2.8664 .0005						23.551 ± .012	7.0918 ± .0037	.16954 ±.00013	7.8748 ±.0041	25	8
9	Nickel Ni	cubic	Fm3m(225)	face-centered cubic	4	3.5238 ±.0005						43.756 ± .019	6.5880 ± .0028	.15750 ±.00011	8.9117 ±.0038	25	9
10	Lead Pb	cubic	Fm3m(225)	face-centered cubic	4	4.9505 ±.0005						121.32 ± .04	18.267 ± .006	.43663 ±.00018	11.342 ±.003	25	10
11	Platinum Pt	cubic	Fm3m(225)	face-centered cubic	4	3.9231 ±.0005						60.379 ± .023	9.0909 ± .0035	.21732 ±.00013	21.460 ±.008	25	11
12	orthorhombic Sulfur S	orth.	Fddd(70)	S_8 ring molecules	128	10.4646 ±.0020	12.8660 ±.0020	24.4860 ±.0040				3296.73 ± .97	15.511 ± .005	.37078 ±.00015	2.0671 ±.0006	25	12
13	monoclinic Sulfur S	mon.	P2$_1$/c(14)	S_8 ring molecules	48	11.04 ±.03	10.98 ±.03	10.92 ±.03		96.73 ±.50		1314.6 ± 6.4	16.49 ± .08	.3943 ±.0020	1.944 ±.009	103	13
14	rhombohedral Sulfur S	hex-R	R$\bar{3}$(148)	S_6 ring molecules	18	10.818 ±.002		4.280 ±.001				433.78 ± .19	14.514 ± .006	.34693 ±.00020	2.2092 ±.0010	r	14
15	Antimony Sb	hex-R	R$\bar{3}$m(166)	arsenic	6	4.310 ±.001		11.279 ±.003				181.45 ± .09	18.213 ± .010	.43535 ±.00028	6.685 ±.004	26	15
16	Selenium Se	hex.	P3$_1$21(152) P3$_2$21(154)		3	4.3642 ±.0008		4.9588 ±.0008				81.793 ± .033	16.420 ± .007	.39249 ±.00020	4.8088 ±.0019	26	16
17	Silicon Si	cubic	Fd3m(227)	diamond	8	5.4305 ±.0003						160.15 ± .03	12.056 ± .002	.28819 ±.00009	2.3296 ±.0004	25	17
18	β-Tin (white) Sn	tet.	I4$_1$/amd(141)		4	5.8315 ±.0008		3.1813 ±.0006				108.18 ± .04	16.289 ± .005	.38935 ±.00017	7.2867 ±.0024	26	18
19	Tellurium Te	hex.	P3$_1$21(152) P3$_2$21(154)		3	4.4570 ±.0008		5.9290 ±.0010				102.00 ± .04	20.476 ± .008	.48944 ±.00024	6.2316 ±.0025	25	19

X-Ray Crystallographic Data of Minerals

	Name and formula	Crystal system	Space group	Structure type	Z	a_o	b_o	c_o	α_o	β_o	γ_o	Cell volume 10^{-24} cm³	Molar volume cm³	Molar volume cal bar⁻¹	X-Ray density grams cm⁻³	Temp. °C	
20	Zinc Zn	hex.	$P6_3/mmc(194)$	hexagonal close packed	2	2.665 ±.001		4.947 ±.001				30.428 ± .024	9.162 ± .007	.2190 ±.0002	7.134 ±.006	25	20
	Sulfides, arsenides, tellurides, selenides, and sulfosalts																
21	Shandite β-$Ni_3Pb_2S_2$*	hex-R	$R\bar{3}m(166)$		3	5.576 ±.010		13.658 ±.010				367.76 ± 1.35	73.83 ± .27	1.765 ±.007	8.867 ±.033	r	21
22	High-Argentite Ag_2S I	cubic			4	6.269 ±.020						246.4 ± 2.4	37.09 ± .36	.8866 ±.0085	6.680 ±.064	600	22
23	Argentite Ag_2S II	cubic			2	4.870 ±.008						115.5 ± .6	34.78 ± .17	.8313 ±.0041	7.125 ±.035	189	23
24	Acanthite Ag_2S III	mon.	$P2_1/c(14)$		4	4.228 ±.002	6.928 ±.005	7.862 ±.003		99.58 ±.30		227.08 ± .29	34.19 ± .04	.8172 ±.0011	7.248 ±.009	25	24
25	High-Naumanite Ag_2Se	cubic			2	4.993 ±.016						124.48 ± 1.20	37.48 ± .36	.8959 ±.0087	7.862 ±.076	170	25
26	Ag_2Te I	cubic			2	5.29 ±.01						148.0 ± .8	44.58 ± .26	1.065 ±.006	7.702 ±.044	825	26
27	Ag_2Te II	cubic			4	6.585 ±.010						285.54 ± 1.30	42.99 ± .20	1.028 ±.005	7.986 ±.036	250	27
28	Hessite Ag_2 Te III	mon.	$P2_1/c(14)$		4	8.09 ±.02	4.48 ±.01	8.96 ±.02		123.33 ±.30		271.33 ± 1.43	40.85 ± .22	.9764 ±.0052	8.405 ±.044	r	28
29	$Ag_{1.55}Cu_{.45}S$ I	cubic			4	6.110 ±.010						228.10 ± 1.12	34.34 ± .17	.8209 ±.0041	6.635 ±.033	300	29
30	$Ag_{1.55}Cu_{.45}S$ II	cubic			2	4.825 ±.005						112.33 ± .35	33.83 ± .11	.8085 ±.0026	6.736 ±.021	116	30
31	Jalpaite $Ag_{1.55}Cu_{.45}S$ III	tet.			16	8.673 ±.004		11.756 ±.006				884.30 ± .93	33.286 ± .035	.79559 ±.00088	6.8455 ±.0072	r	31
32	$Ag_{.93}Cu_{1.07}S$ I	cubic			4	5.961 ±.009						211.82 ± .96	31.89 ± .14	.7623 ±.0035	6.283 ±.029	196	32
33	$Ag_{.93}Cu_{1.07}S$ II	hex.			2	4.138 ±.004		7.105 ±.007				105.36 ± .23	31.73 ± .07	.7583 ±.0017	6.316 ±.014	100	33
34	Stromeyerite $Ag_{.93}Cu_{1.07}S$ III	orth.	Cmcm(63)		4	4.066 ±.002	6.628 ±.003	7.972 ±.004				214.84 ± .18	32.35 ± .03	.7732 ±.0007	6.194 ±.005	r	34
35	Eucairite AgCuSe	orth.	pseudo P4/nmm(129)		10	4.105 ±.010	20.35 ±.02	6.31 ±.01				527.12 ± 1.62	31.75 ± .10	.7588 ±.0024	7.887 ±.024	r	35
36	Petzite Ag_3AuTe_2*	cubic	$I4_132(214)$		8	10.38 ±.02						1118.4 ± 5.5	84.19 ± .49	2.012 ±.012	9.214 ±.053	r	36
37	Maldonite Au_2Bi	cubic	Fd3m(227)	Cu_2Mg	8	7.958 ±.002						503.98 ± .38	37.94 ± .03	.9068 ±.0007	15.891 ±.012	r	37
38	High-Digenite Cu_2S I	cubic			4	5.725 ±.010						187.64 ± .98	28.25 ± .15	.6753 ±.0036	5.633 ±.030	465	38
39	High-Chalcocite Cu_2S II	hex.			2	3.961 ±.004		6.722 ±.007				91.34 ± .21	27.50 ± .06	.6574 ±.0015	5.786 ±.013	152	39
40	Chalcocite Cu_2S III	orth.	Ab2m(39)		96	11.881 ±.004	27.323 ±.010	13.491 ±.004				4379.5 ± 2.5	27.475 ± .016	.65671 ±.00043	5.7924 ±.0034	r	40
41	Digenite $Cu_{1.79}S$ (Cu rich side)	cubic		deformed fluorite	4	5.5695 ±.0010						172.76 ± .09	26.012 ± .014	.6217 ±.0004	5.605 ±.003	25	41
42	Digenite $Cu_{1.77}S$ (S rich side)	cubic		deformed fluorite	4	5.5542 ±.0010						171.34 ± .09	25.798 ± .014	.6166 ±.0004	5.602 ±.003	25	42
43	Berzelianite Cu_2Se	cubic			4	5.85 ±.01						200.2 ± 1.0	30.14 ± .15	.7205 ±.0037	6.835 ±.035	170	43
44	High-Bornite Cu_5FeS_4*	cubic			1	5.50 ±.01						166.4 ± .9	100.2 ± .5	2.395 ±.013	5.008 ±.027	240	44

X-Ray Crystallographic Data of Minerals

	Name and formula	Crystal system	Space group	Structure type	Z	a_o	b_o	c_o	α_o	β_o	γ_o	Cell volume 10^{-24} cm³	Molar volume cm³	Molar volume cal bar⁻¹	X-Ray density grams cm⁻³	Temp. °C	
45	Metastable Bornite Cu_5FeS_4	cubic			8	10.94 ±.02						1309.34 ± 7.18	98.57 ± .54	2.356 ±.013	5.091 ±.028	r	45
46	Low-Bornite Cu_5FeS_4*	tet.	$P\bar{4}2_1c$(144)		16	10.94 ±.02		21.88 ±.04				2618.7 ±10.7	98.57 ± .40	2.356 ±.010	5.091 ±.021	r	46
47	Umangite Cu_3Se_2	tet.	P4/mmm(123)		2	6.402 ±.010		4.276 ±.010				175.25 ± .68	52.77 ± .21	1.261 ±.005	6.604 ±.026	r	47
48	Heazelwoodite Ni_3S_2	hex-R	R32(155)		3	5.746 ±.001		7.134 ±.002				203.98 ± .09	40.95 ± .02	.9788 ±.0005	5.867 ±.003	r	48
49	Maucherite $Ni_{11}As_8$	tet.	$P4_12_12$(92)		4	6.870 ±.001		21.81 ±.01				1029.36 ± .56	154.98 ± .08	3.7043 ±.0021	8.0343 ±.0044	r	49
50	Pentlandite $Fe_{5.25}Ni_{3.75}S_8$	cubic	Fm3m(225)		4	10.196 ±.010						1059.96 ± 3.12	159.59 ± .47	3.8144 ±.0113	4.823 ±.014	r	50
51	Pentlandite $Fe_{4.75}Ni_{5.25}S_8$	cubic	Fm3m(225)		4	10.095 ±.010						1028.77 ± 3.06	154.89 ± .46	3.702 ±.011	4.998 ±.015	r	51
52	Sternbergite $AgFe_2S_3$*	orth.	Ccmm(63)		8	11.60 ±.02	12.675 ±.020	6.63 ±.01				974.81 ± 2.71	73.39 ± .20	1.754 ±.005	4.303 ±.012	r	52
53	Argentopyrite $AgFe_2S_3$*	orth.	Pmmm(47)		4	6.64 ±.01	11.47 ±.02	6.45 ±.02				491.2 ± 1.9	73.96 ± .29	1.768 ±.007	4.269 ±.017	r	53
54	Realgar AsS*	mon.	$P2_1/m$(11)		16	9.29 ±.05	13.53 ±.05	6.57 ±.03		106.55 ±.30		791.6 ± 6.4	29.80 ± .24	.7122 ±.0058	3.591 ±.029	r	54
55	Oldhamite CaS	cubic	Fm3m(225)	rock salt	4	5.689 ±.006						184.12 ± .58	27.722 ± .088	.6626 ±.0021	2.602 ±.008	r	55
56	Greenockite CdS	hex.	$P6_3mc$(186)	zincite	2	4.1354 ±.0010		6.7120 ±.0010				99.407 ± .050	29.934 ± .015	.71549 ±.00041	4.8261 ±.0024	r	56
57	Hawleyite CdS	cubic	$F\bar{4}3m$(216)	sphalerite	4	5.833 ±.002						198.46 ± .20	29.88 ± .03	.7142 ±.0008	4.835 ±.005	r	57
58	(hypothetical) CdS	cubic	Fm3m(225)	rock salt	4	5.516 ±.002						167.83 ± .18	25.27 ± .03	.6040 ±.0007	5.717 ±.006	r	58
59	Cadmoselite CdSe	hex.	$P6_3mc$(186)	zincite	2	4.2977 ±.0010		7.0021 ±.0010				112.00 ± .05	33.727 ± .016	.80614 ±.00044	5.6738 ±.0028	r	59
60	CdTe	cubic	$F\bar{4}3m$(216)	sphalerite	4	6.4805 ±.0006						272.16 ± .08	40.977 ± .012	.97943 ±.00032	5.8569 ±.0016	25	60
61	(hypothetical) CoS	cubic	$F\bar{4}3m$(216)	sphalerite	4	5.339 ±.001						152.19 ± .09	22.91 ± .02	.5477 ±.0004	3.971 ±.002	r	61
62	Chalcopyrite ($CuFeS_2$) $CuFeS_{1.90}$	tet.	I42d(122)		4	5.2988 ±.0010		10.434 ±.005				292.96 ± .18	44.109 ± .027	1.0543 ±.0007	4.0878 ±.0025	r	62
63	Cubanite $CuFe_2S_3$*	orth.	Pcmn(62)		4	6.46 ±.01	11.12 ±.01	6.23 ±.01				447.53 ± 1.08	67.38 ± .16	1.611 ±.004	4.026 ±.010	r	63
64	Covellite CuS	hex.	$P6_3/mmc$(194)		6	3.792 ±.001		16.34 ±.01				203.48 ± .16	20.42 ± .02	.4882 ±.0005	4.682 ±.001	r	64
65	Klockmannite CuSe	hex.		deformed covellite	78	14.206 ±.010		17.25 ±.05				3014.8 ± 9.7	23.28 ± .08	.5564 ±.0018	6.122 ±.020	r	65
66	Troilite FeS	hex.	$P6_3/mmc$(194)	niccolite	2	3.446 ±.003		5.877 ±.001				60.439 ± .106	18.20 ± .03	.4350 ±.0008	4.830 ±.009	28	66
67	Pyrrhotite $Fe_{.980}S$	hex.	$P6_3/mmc$(194)	defect niccolite	2	3.446 ±.001		5.848 ±.002				60.14 ± .04	18.11 ± .01	.4329 ±.0003	4.793 ±.003	28	67
68	Pyrrhotite $Fe_{.885}S$	hex.	$P6_3/mmc$(194)	defect niccolite	2	3.440 ±.001		5.709 ±.003				58.507 ± .046	17.62 ± .02	.4211 ±.0004	4.625 ±.004	28	68
69	(hypothetical) FeS	cubic	$F\bar{4}3m$(216)	sphalerite	4	5.455 ±.001						162.32 ± .09	24.44 ± .01	.5842 ±.0004	3.597 ±.002	r	69
70	(hypothetical) FeS	hex.	$P6_3mc$(186)	zincite	2	3.872 ±.001		6.345 ±.002				82.38 ± .05	24.81 ± .02	.5930 ±.0004	3.544 ±.002	r	70

X-Ray Crystallographic Data of Minerals

	Name and formula	Crystal system	Space group	Structure type	Z	a_o	b_o	c_o	α_o	β_o	γ_o	Cell volume 10^{-24} cm³	Molar volume cm³	Molar volume cal bar⁻¹	X-Ray density grams cm⁻³	Temp. °C	
71	Cinnabar HgS	hex.	$P3_121(152)$ $P3_21(154)$	cinnabar	3	4.149 ±.001		9.495 ±.002				141.55 ± .07	28.416 ± .015	.6792 ±.0004	8.187 ±.004	r	71
72	Metacinnabar HgS	cubic	$F\bar{4}3m(216)$	sphalerite	4	5.8517 ±.0010						200.38 ± .10	30.169 ± .016	.7211 ±.0004	7.712 ±.004	r	72
73	Tiemannite $HgSe$	cubic	$F\bar{4}3m(216)$	sphalerite	4	6.0853 ±.0050						225.34 ± .56	33.928 ± .084	.8110 ±.0020	8.239 ±.020	r	73
74	Coloradoite $HgTe$	cubic	$F\bar{4}3m(216)$	sphalerite	4	6.4600 ±.0006						269.59 ± .08	40.590 ± .011	.97016 ±.00032	8.0855 ±.0023	r	74
75	Alabandite MnS	cubic	$Fm3m(225)$	rock salt	4	5.2234 ±.0005						142.51 ± .04	21.457 ± .006	.51289 ±.00019	4.0546 ±.0012	r	75
76	(hypothetical) MnS	cubic	$F\bar{4}3m(216)$	sphalerite	4	5.611 ±.002						176.65 ± .19	26.60 ± .03	.6357 ±.0007	3.271 ±.004	r	76
77	(hypothetical) MnS	hex.	$P6_3mc(186)$	zincite	2	3.986 ±.001		6.465 ±.002				88.96 ± .05	26.79 ± .02	.6403 ±.0004	3.248 ±.002	r	77
78	Niccolite $NiAs$	hex.	$P6_3/mmc(194)$	niccolite	2	3.618 ±.001		5.034 ±.001				57.07 ± .03	17.18 ± .01	.4108 ±.0003	7.776 ±.005	r	78
79	Millerite NiS	hex-R	$R3m(160)$		9	9.616 ±.001		3.152 ±.001				252.41 ± .10	16.891 ± .006	.40374 ±.00020	5.3743 ±.0020	r	79
80	Breithauptite $NiSb$	hex.	$P6_3/mmc(194)$	niccolite	2	3.942 ±.001		5.155 ±.001				69.37 ± .04	20.89 ± .01	.4994 ±.0004	8.639 ±.005	r	80
81	Galena PbS	cubic	$Fm3m(225)$	rock salt	4	5.9360 ±.0005						209.16 ± .05	31.492 ± .008	.75272 ±.00024	7.5973 ±.0019	26	81
82	Clausthalite $PbSe$	cubic	$Fm3m(225)$	rock salt	4	6.1255 ±.0005						229.84 ± .06	34.605 ± .009	.82713 ±.00025	8.2690 ±.0020	r	82
83	Teallite $PbSnS_2$	orth.	$Pbnm(62)$	GeS	2	4.266 ±.003	11.419 ±.007	4.090 ±.002				199.24 ± .21	59.996 ± .063	1.4340 ±.0016	6.501 ±.007	r	83
84	Altaite $PbTe$	cubic	$Fm3m(225)$	rock salt	4	6.4606 ±.0005						269.66 ± .06	40.601 ± .009	.97043 ±.00027	8.2459 ±.0019	r	84
85	Cooperite PtS	tet.	$P4_2/mmc(131)$		2	3.4699 ±.0006		6.1098 ±.0010				73.563 ± .028	22.152 ± .008	.5295 ±.0003	10.254 ±.004	r	85
86	Herzenbergite SnS	orth.	$Pbnm(62)$	GeS	4	4.328 ±.002	11.190 ±.004	3.978 ±.001				192.66 ± .12	29.01 ± .02	.6933 ±.0005	5.197 ±.003	r	86
87	Sphalerite ZnS	cubic	$F\bar{4}3m(216)$	sphalerite	4	5.4093 ±.0005						158.28 ± .04	23.831 ± .007	.56962 ±.00020	4.0885 ±.0011	r	87
88	Wurtzite ZnS	hex.	$P6_3mc(186)$	zincite	2	3.8230 ±.0010		6.2565 ±.0010				79.190 ± .043	23.846 ± .013	.56998 ±.00036	4.0859 ±.0022	r	88
89	Stilleite $ZnSe$	cubic	$F\bar{4}3m(216)$	sphalerite	4	5.6685 ±.0005						182.14 ± .05	27.424 ± .007	.65548 ±.00022	5.2630 ±.0014	r	89
90	$ZnTe$	cubic	$F\bar{4}3m(216)$	sphalerite	4	6.1020 ±.0006						227.20 ± .07	34.209 ± .010	.81765 ±.00029	5.6410 ±.0017	r	90
91	Orpiment As_2S_3*	mon.	$P2_1/n(14)$		4	11.49 ±.02	9.59 ±.02	4.25 ±.01		90.45 ±.30		468.3 ± 1.7	70.51 ± .25	1.685 ±.006	3.490 ±.013	r	91
92	Bismuthinite Bi_2S_3	orth.	$Pbnm(62)$	stibnite	4	11.150 ±.004	11.300 ±.004	3.981 ±.001				501.59 ± .28	75.520 ± .043	1.8050 ±.0011	6.8081 ±.0038	26	92
93	Tellurobismuthite Bi_2Te_3	hex-R	$R\bar{3}m(166)$	Bi_2Te_2S	3	4.3835 ±.0020		30.487 ±.003				507.33 ± .47	101.85 ± .09	2.4342 ±.0023	7.862 ±.007	25	93
94	Stibnite Sb_2S_3	orth.	$Pbnm(62)$	stibnite	4	11.229 ±.004	11.310 ±.004	3.8389 ±.0010				487.54 ± .28	73.406 ± .042	1.7545 ±.0010	4.6276 ±.0026	25	94
95	Linnaeite Co_3S_4	cubic	$Fd3m(227)$	spinel	8	9.401 ±.001						830.85 ± .27	62.548 ± .020	1.4950 ±.0005	4.8772 ±.0016	r	95
96	Greigite Fe_3S_4	cubic	$Fd3m(227)$	spinel	8	9.876 ±.002						963.26 ± .59	72.52 ± .04	1.733 ±.001	4.079 ±.003	r	96

X-Ray Crystallographic Data of Minerals

	Name and formula	Crystal system	Space group	Structure type	Z	a_o	b_o	c_o	α_o	β_o	γ_o	Cell volume 10^{-24} cm³	Molar volume cm³	Molar volume cal bar⁻¹	X-Ray density grams cm⁻³	Temp. °C	
97	Daubreeite $FeCr_2S_4$	cubic	Fd3m(227)	spinel	8	9.966 ±.005						989.83 ± 1.49	74.52 ± .11	1.781 ±.003	3.866 ±.006	r	97
98	Violarite $FeNi_2S_4$	cubic	Fd3m(227)	spinel	8	9.464 ±.005						847.66 ± 1.34	63.81 ± .10	1.525 ±.002	4.725 ±.008	r	98
99	Polymidite Ni_3S_4	cubic	Fd3m(227)	spinel	8	9.480 ±.001						851.97 ± .27	64.138 ± .020	1.5330 ±.0005	4.7458 ±.0015	r	99
100	Co-Safflorite $CoAs_2$	mon.		deformed marcasite	2	5.049 ±.002	5.872 ±.002	3.127 ±.001		90.45 ±.20		92.706 ± .057	27.92 ± .02	.6672 ±.0005	7.479 ±.005	26	100
101	Safflorite $(CO_{.5}Fe_{.5})As_2$	orth.	Pnnm(58)	marcasite	2	5.231 ±.002	5.953 ±.002	2.962 ±.002				92.237 ± .078	27.775 ± .024	.6639 ±.0006	7.461 ±.006	26	101
102	Cobaltite CoAsS*	cubic	$P2_13$(198)	NiSbS	4	5.60 ±.05						175.62 ± 4.70	26.44 ± .71	.6320 ±.0170	6.275 ±.168	r	102
103	Glaucodot (Co,Fe)AsS*	orth.	Cmmm(65)		24	6.64 ±.05	28.39 ±.10	5.64 ±.05				1063.2 ±12.9	26.68 ± .32	.6377 ±.0078	6.161 ±.075	r	103
104	Cattierite CoS_2	cubic	Pa3(205)	pyrite	4	5.5345 ±.0005						169.53 ± .05	25.524 ± .007	.61009 ±.00021	4.8213 ±.0013	r	104
105	Trogtalite $CoSe_2$	cubic	Pa3(205)	pyrite	4	5.8588 ±.0010						201.11 ± .10	30.279 ± .016	.72374 ±.00042	7.1618 ±.0037	r	105
106	Loellingite $FeAs_2$	orth.	Pnnm(58)	marcasite	2	5.300 ±.002	5.981 ±.002	2.882 ±.001				91.357 ± .056	27.51 ± .02	.6576 ±.0005	7.477 ±.005	26	106
107	Arsenopyrite FeAsS*	tri.	P$\bar{1}$(2)		4	5.760 ±.010	5.690 ±.005	5.785 ±.005	90.00 ±.20	112.23 ±.20	90.00 ±.20	175.51 ± .44	26.42 ± .07	.6316 ±.0016	6.162 ±.015	r	107
108	Gudmundite FeSbS*	mon.	$B2_1/d$(14)		8	10.00 ±.05	5.93 ±.03	6.73 ±.03		90.00 ±.50		399.09 ± 3.35	30.04 ± .25	.7181 ±.0061	6.978 ±.059	r	108
109	Pyrite FeS_2	cubic	Pa3(205)	pyrite	4	5.4175 ±.0005						159.00 ± .04	23.940 ± .007	.57221 ±.00020	5.0116 ±.0014	r	109
110	Marcasite FeS_2*	orth.	Pnnm(58)	marcasite	2	4.443 ±.002	5.423 ±.002	3.3876 ±.0015				81.622 ± .060	24.579 ± .018	.58749 ±.00047	4.8813 ±.0036	25	110
111	Ferroselite $FeSe_2$	orth.	Pnnm(58)	marcasite	2	4.801 ±.005	5.778 ±.005	3.587 ±.004				99.50 ± .17	29.96 ± .05	.7162 ±.0013	7.134 ±.013	r	111
112	Frohbergite $FeTe_2$	orth.	Pnnm(58)	marcasite	2	5.265 ±.005	6.265 ±.005	3.869 ±.002				127.62 ± .17	38.43 ± .05	.9185 ±.0013	8.094 ±.011	r	112
113	Hauerite MnS_2	cubic	Pa3(205)	pyrite	4	6.1014 ±.0006						227.14 ± .07	34.198 ± .010	.81741 ±.00029	3.4816 ±.0010	28	113
114	Molybdenite MoS_2	hex.	$P6_3/mmc$(194)	molybdenite	2	3.1604 ±.0010		12.295 ±.002				106.35 ± .07	32.025 ± .021	.76547 ±.00055	4.9982 ±.0033	26	114
115	Rammelsbergite $NiAs_2$	orth.	Pnnm(58)	marcasite	2	4.757 ±.002	5.797 ±.004	3.542 ±.002				97.645 ± .096	29.41 ± .03	.7030 ±.0007	7.091 ±.007	26	115
116	Pararammelsbergite $NiAs_2$	orth.	Pbca(61)		8	5.75 ±.01	5.82 ±.01	11.428 ±.02				382.42 ± 1.15	28.79 ± .09	.6882 ±.0021	7.244 ±.022	r	116
117	Gersdorfite NiAsS	cubic	$P2_13$(198)		4	5.693 ±.001						184.51 ± .10	27.78 ± .01	.6640 ±.0004	5.964 ±.003	26	117
118	Vaesite NiS_2	cubic	Pa3(205)	pyrite	4	5.6873 ±.0005						183.96 ± .05	27.697 ± .007	.66203 ±.00022	4.4350 ±.0012	r	118
119	$NiSe_2$	cubic	Pa3(205)	pyrite	4	5.9604 ±.0010						211.75 ± .11	31.882 ± .016	.76204 ±.00043	6.7948 ±.0034	20	119
120	Melonite $NiTe_2$	hex.	P$\bar{3}$ml(164)	cadmium iodide	1	3.869 ±.010		5.308 ±.010				68.81 ± .38	41.44 ± .23	.9905 ±.0055	7.575 ±.042	84	120
121	Sperrylite $PtAs_2$	cubic	Pa3(205)	pyrite	4	5.968 ±.005						212.56 ± .53	32.00 ± .08	.7650 ±.0020	10.778 ±.027	r	121
122	Laurite RuS_2	cubic	Pa3(205)	pyrite	4	5.60 ±.02						175.6 ± 1.9	26.44 ± .28	.6320 ±.0068	6.248 ±.067	r	122

X-Ray Crystallographic Data of Minerals

	Name and formula	Crystal system	Space group	Structure type	Z	a_o	b_o	c_o	α_o	β_o	γ_o	Cell volume 10^{-24} cm³	Molar volume cm³	Molar volume cal bar⁻¹	X-Ray density grams cm⁻³	Temp. °C	
123	Tungstenite WS_2	hex.	P6₃/mmc(194)	molybdenite	2	3.154 ±.001		12.362 ±.004				106.50 ± .08	32.069 ± .023	.76652 ±.00059	7.7325 ±.0055	26	123
124	Co-Skutterudite $CoAs_{3-x}$ $CoAs_{2.95}$	cubic	Im3(204)		8	8.2060 ±.0010						552.58 ± .20	41.599 ± .015	.99428 ±.00041	6.7298 ±.0025	r	124
125	Fe-Skutterudite $FeAs_{3-x}$ $FeAs_{2.95}$	cubic	Im3(204)		8	8.1814 ±.0010						547.62 ± .20	41.226 ± .015	.98537 ±.00041	6.7158 ±.0025	r	125
126	Ni-Skutterudite $NiAs_{3-x}$ $NiAs_{2.95}$	cubic	Im3(204)		8	8.3300 ±.0010						578.01 ± .21	43.513 ± .016	1.0400 ±.0004	6.4286 ±.0023	r	126
127	Tennantite $Cu_{12}As_4S_{13}$	cubic	I4̄3m(217)	tetrahedrite	2	10.190 ±.004						1058.09 ± 1.25	318.62 ± .38	7.1652 ±.0090	4.642 ±.006	r	127
128	Tetrahedrite $Cu_{12}Sb_4S_{13}$	cubic	I4̄3m(217)	tetrahedrite	2	10.327 ±.004						1101.3 ± 1.3	331.64 ± .39	7.9266 ±.0094	5.024 ±.006	r	128
129	Enargite Cu_3AsS_4	orth.	Pnn2(34)		2	6.426 ±.005	7.422 ±.005	6.144 ±.005				293.03 ± .38	88.24 ± .12	2.109 ±.003	4.463 ±.006	26	129
130	Luzonite Cu_3AsS_4*	tet.	I4̄2m(121)		2	5.289 ±.005		10.440 ±.008				292.04 ± .60	87.94 ± .18	2.1019 ±.0043	4.478 ±.009	26	130
131	Famatimite Cu_3SbS_4*	tet.	I4̄m(121)		2	5.384 ±.005		10.770 ±.008				312.19 ± .62	94.01 ± .19	2.2469 ±.0045	4.687 ±.009	26	131
132	Proustite Ag_3AsS_3	hex-R	R3c(161)		6	10.816 ±.001		8.6948 ±.0013				880.89 ± .21	88.420 ± .021	2.1133 ±.0006	5.595 ±.001	26	132
133	Pyrargyrite Ag_3SbS_3	hex-R	R3c(161)		6	11.052 ±.002		8.7177 ±.0020				922.18 ± .40	92.564 ± .040	2.2124 ±.0010	5.8506 ±.0025	26	133
134	Miargyrite $AgSbS_2$*	mon.	Cc(9)		8	12.862 ±.013	4.111 ±.004	13.220 ±.010		98.63 ±.15		691.10 ± 1.14	52.027 ± .086	1.244 ±.002	5.646 ±.009	r	134

Oxides and hydroxides

	Name and formula	Crystal system	Space group	Structure type	Z	a_o	b_o	c_o	α_o	β_o	γ_o	Cell volume 10^{-24} cm³	Molar volume cm³	Molar volume cal bar⁻¹	X-Ray density grams cm⁻³	Temp. °C	
135	Corundum Al_2O_3	hex-R	R3̄c(167)	corundum	6	4.7591 ±.0004		12.9894 ±.0030				254.78 ± .07	25.575 ± .007	.61128 ±.00022	3.9869 ±.0011	25	135
136	Boehmite AlO(OH)*	orth.	Cmcm(63)	lepidocrocite	4	2.868 ±.003	12.227 ±.003	3.700 ±.003				129.75 ± .17	19.535 ± .026	.46695 ±.00067	3.071 ±.004	26	136
137	Diaspore AlO(OH)*	orth.	Pbnm(62)		4	4.401 ±.005	9.421 ±.005	2.845 ±.002				117.96 ± .17	17.760 ± .026	.4245 ±.0007	3.378 ±.005	r	137
138	Gibbsite $Al(OH)_3$	mon.	P2₁/n(14)		8	9.719 ±.002	5.0705 ±.0010	8.6412 ±.0010		94.57 ±.25		424.49 ± .20	31.956 ± .015	.7638 ±.0004	2.441 ±.001	r	138
139	Arsenolite As_2O_3	cubic	Fd3m(227)	diamond	16	11.074 ±.005						1358.0 ± 1.8	51.118 ± .069	1.2218 ±.0017	3.870 ±.005	25	139
140	Claudetite As_2O_3	mon.	P2₁/n(14)		4	5.339 ±.002	12.984 ±.005	4.5405 ±.0010		94.27 ±.10		313.88 ± .19	47.259 ± .028	1.1296 ±.0007	4.1863 ±.0025	25	140
141	Bromellite BeO	hex.	P6₃mc(186)	zincite	2	2.6979 ±.0005		4.3772 ±.0005				27.592 ± .011	8.3086 ± .0032	.19862 ±.00012	3.0104 ±.0012	26	141
142	Bismite α-Bi_2O_3	mon.	P2₁/c(14)	pseudo orthorhombic	8	8.166 ±.005	13.827 ±.010	5.850 ±.004		90.00 ±.20		660.53 ± .77	49.73 ± .06	1.1885 ±.0014	9.371 ±.011	25	142
143	Lime CaO	cubic	Fm3m(225)	rock salt	4	4.8108 ±.0005						111.34 ± .03	16.764 ± .005	.40071 ±.00017	3.3453 ±.0010	26	143
144	Portlandite $Ca(OH)_2$	hex.	P3̄ml(164)	CdI_2	1	3.5933 ±.0005		4.9086 ±.0020				54.388 ± .027	33.056 ± .016	.79011 ±.00043	2.2415 ±.0011	26	144
145	Monteponite CdO	cubic	Fm3m(225)	rock salt	4	4.6953 ±.0010						103.51 ± .07	15.585 ± .010	.37254 ±.00028	8.2386 ±.0053	27	145
146	Cerianite CeO_2	cubic	Fm3m(225)	fluorite	4	5.4110 ±.0020						158.43 ± .18	23.853 ± .026	.57016 ±.00068	7.216 ±.008	26	146
147	CoO	cubic	Fm3m(225)	rock salt	4	4.260 ±.002						77.31 ± .11	11.64 ± .02	.2782 ±.0004	6.438 ±.009	26	147

X-Ray Crystallographic Data of Minerals

No.	Name and formula	Crystal system	Space group	Structure type	Z	a_o	b_o	c_o	α_o	β_o	γ_o	Cell volume 10^{-24} cm³	Molar volume cm³	Molar volume cal bar⁻¹	X-Ray density grams cm⁻³	Temp. °C	No.
148	Eskolaite Cr_2O_3	hex-R	R$\bar{3}$c(167)	corundum	6	4.9607 ±.0020		13.599 ±.010				289.82 ± .32	29.090 ± .032	.6953 ±.0008	5.225 ±.006	r	148
149	Tenorite CuO	mon.	C2/c(15)		4	4.684 ±.005	3.425 ±.005	5.129 ±.005		99.47 ±.17		81.16 ± .17	12.22 ± .03	.2921 ±.0007	6.509 ±.014	26	149
150	Cuprite Cu_2O	cubic	Pn3m(224)		2	4.2696 ±.0010						77.833 ± .055	23.437 ± .016	.56021 ±.00044	6.1047 ±.0043	26	150
151	Wustite $Fe_{.953}O$	cubic	Fm3m(225)	defect rock salt	4	4.3088 ±.0003						79.996 ± .017	12.044 ± .003	.28791 ±.00011	5.7471 ±.0012	17	151
152	Hematite Fe_2O_3	hex-R	R$\bar{3}$c(167)	corundum	6	5.0329 ±.0010				13.7492 ±.0010		301.61 ± .12	30.274 ± .012	.72361 ±.00034	5.2749 ±.0021	25	152
153	Magnetite Fe_3O_4	cubic	Fd3m(227)	spinel	8	8.3940 ±.0005						591.43 ± .11	44.524 ± .008	1.0642 ±.0002	5.2003 ±.0009	22	153
154	Goethite α-FeO(OH)*	orth.	Pbnm(62)		4	4.596 ±.005	9.957 ±.010	3.021 ±.003				138.2 ± .2	20.82 ± .04	.4975 ±.0009	4.269 ±.008	r	154
155	Lepidocrocite γ-FeO(OH)*	orth.	Amam(63)		4	3.868 ±.010	12.525 ±.010	3.066 ±.003				148.54 ± .43	22.364 ± .064	.5346 ±.0016	3.973 ±.011	r	155
156	α-Ga_2O_3	hex-R	R$\bar{3}$c(167)	corundum	6	4.9793 ±.0010		13.429 ±.003				288.34 ± .13	28.943 ± .013	.69179 ±.00036	6.4762 ±.0030	24	156
157	Low-germania GeO_2	tet.	P4/mnm(136)	rutile	2	4.3963 ±.0010		2.8626 ±.0010				55.327 ± .032	16.660 ± .010	.39824 ±.00027	6.2777 ±.0036	25	157
158	High-germania GeO_2	hex.	$P3_121(152)$ $P3_221(154)$	α-quartz	3	4.987 ±.002		5.652 ±.002				121.73 ± .11	24.438 ± .021	.58413 ±.00056	4.2797 ±.0038	26	158
159	Ice H_2O	hex.	$P6_3$/mmc(194)		4	4.5212 ±.0010		7.3666 ±.0010				130.41 ± .06	19.635 ± .009	.46932 ±.00026	.9175 ±.0004	0	159
160	Hafnia HfO_2	mon.	$P2_1$/c(14)	baddeleyite	4	5.1156 ±.0010	5.1722 ±.0010	5.2948 ±.0010		99.18 ±.08		138.30 ± .06	20.823 ± .008	.49772 ±.00025	10.108 ±.004	r	160
161	Montroydite HgO	orth.	Pnma(62)		4	6.608 ±.003	5.518 ±.003	3.519 ±.003				128.3 ± .1	19.32 ± .02	.4618 ±.0006	11.21 ±.01	25	161
162	Periclase MgO	cubic	Fm3m(225)	rock salt	4	4.2117 ±.0005						74.709 ± .027	11.248 ± .004	.26889 ±.00014	3.5837 ±.0013	25	162
163	Brucite $Mg(OH)_2$	hex.	P$\bar{3}$ml(164)	CdI_2	1	3.147 ±.004		4.769 ±.004				40.90 ± .11	24.63 ± .07	.5888 ±.0016	2.368 ±.006	26	163
164	Manganosite MnO	cubic	Fm3m(225)	rock salt	4	4.4448 ±.0005						87.813 ± .030	13.221 ± .004	.31604 ±.00015	5.3653 ±.0018	26	164
165	Pyrolusite MnO_2	tet.	P4/mnm(136)	rutile	2	4.388 ±.003		2.865 ±.002				55.16 ± .08	16.61 ± .02	.3971 ±.0007	5.234 ±.008	r	165
166	Bixbyite Mn_2O_3	cubic	Ia3(206)	Tl_2O_3	16	9.411 ±.005						833.5 ± 1.3	31.37 ± .05	.7499 ±.0012	5.032 ±.008	25	166
167	Hausmanite Mn_3O_4	tet.	$I4_1$/amd(141)		8	8.136 ±.005		9.422 ±.005				623.68 ± .84	46.95 ± .06	1.1222 ±.0016	4.873 ±.007	20	167
168	Molybdite MoO_3	orth.	Pbnm(62)		4	3.962 ±.002	13.858 ±.005	3.697 ±.004				202.98 ± .25	30.56 ± .04	.7305 ±.0010	4.710 ±.006	26	168
169	Bunsenite NiO	cubic	Fm3m(225)	rock salt	4	4.177 ±.002						72.88 ± .10	10.97 ± .02	.2623 ±.0004	6.809 ±.010	26	169
170	Litharge PbO red	tet.	P4/nmm(129)		2	3.9759 ±.0040		5.023 ±.004				79.40 ± .17	23.91 ± .05	.5715 ±.0013	9.334 ±.020	27	170
171	Massicot PbO yellow	orth.	Pb2a(32)		4	5.489 ±.003	4.755 ±.004	5.891 ±.004				153.8 ± .2	23.15 ± .03	.5533 ±.0007	9.641 ±.012	27	171
172	Minium Pb_3O_4	tet.	$P4_2$/mbc(135)		4	8.815 ±.005		6.565 ±.003				510.13 ± .62	76.81 ± .09	1.836 ±.002	8.926 ±.009	25	172
173	Senarmontite Sb_2O_3	cubic	Fm3m(225)	arsenic trioxide	16	11.152 ±.003						1386.9 ± 1.1	52.206 ± .042	1.2478 ±.0011	5.5837 ±.0045	26	173

X-Ray Crystallographic Data of Minerals

	Name and formula	Crystal system	Space group	Structure type	Z	a_o	b_o	c_o	α_o	β_o	γ_o	Cell volume 10^{-2} cm³	Molar volume cm³	Molar volume cal bar⁻¹	X-Ray density grams cm⁻³	Temp. °C	
174	Valentinite Sb_2O_3	orth.	Pccn(56)	antimony trioxide	4	4.914 ±.002	12.468 ±.005	5.421 ±.004				332 13 ± 31	50.007 ± .047	1.1952 ±.0012	5.8292 ±.0054	25	174
175	Cervantite Sb_2O_4	cubic	Fd3m(227)		8	10.305 ±.005						1094 3 ± 1 6	82.38 ± .12	1.9690 ±.0029	3.733 ±.005	26	175
176	Selenolite SeO_2	tet.	$P4_2/mbc(135)$ $P4_2bc(106)$		8	8.35 ±.01		5.08 ±.01				354 2 ± 1 1	26.66 ± .08	.6373 ±.0020	4.161 ±.013	26	176
177	α-Quartz SiO_2*	hex.	$P3_121(152)$ $P3_221(154)$		3	4.9136 ±.0001		5.4051 ±.0001				113 01 ± 01	22.688 ± .001	.54229 ±.00007	2.6483 ±.0001	25	177
178	β-Quartz SiO_2*	hex.	$P6_422(181)$ $P6_222(180)$		3	4.999 ±.001		5.4592 ±.0020				118 15 ± 06	23.718 ± .013	.5669 ±.0004	2.533 ±.002	575	178
179	α-Cristobalite SiO_2	tet.	$P4_12_12(92)$ $P4_32_12(96)$		4	4.971 ±.003		6.918 ±.003				170 95 ± 22	25.739 ± .033	.61521 ±.00083	2.3344 ±.0030	25	179
180	β-Cristobalite SiO_2	cubic	Fd3m(227)		8	7.1382 ±.0010						363 72 ± 15	27.381 ± .012	.65447 ±.00032	2.1944 ±.0009	405	180
181	Keatite SiO_2	tet.	$P4_12_12(92)$ $P4_32_12(96)$		12	7.456 ±.003		8.604 ±.005				478 3 ± 5	24.01 ± .02	.5738 ±.0006	2.503 ±.003	r	181
182	β-Tridymite SiO_2	hex.	$P\bar{6}2c(172)$ $P6_3/mmc(194)$		4	5.0463 ±.0020		8.2563 ±.0030				182 08 ± 16	27.414 ± .024	.65527 ±.00062	2.1917 ±.0019	405	182
183	Coesite SiO_2*	mon.	B2/b(15)		16	7.152 ±.001	12.379 ±.002	7.152 ±.001		120.00 ±.17		548.37 ± .95	20.641 ± .036	.49338 ±.00090	2.9110 ±.0050	25	183
184	Stishovite SiO_2*	tet.	P4/mnm(136)	rutile	2	4.1790 ±.0010		2.6649 ±.0010				46.540 ± .028	14.014 ± .009	.33500 ±.00025	4.2874 ±.0026	r	184
185	Melanophlogite SiO_2*	cubic	Pm3n(223)	clathrate type	46	13.402 ±.004						2407.2 ± 2.2	31.516 ± .028	.75325 ±.00072	1.9065 ±.0017	r	185
186	Cassiterite SnO_2	tet.	P4/mnm(136)	rutile	2	4.738 ±.003		3.188 ±.003				71.57 ± .11	21.55 ± .03	.5151 ±.0009	6.992 ±.011	26	186
187	Tellurite TeO_2*	orth.	Pbca(61)	tellurite	8	5.607 ±.003	12.034 ±.005	5.463 ±.003				368.61 ± .32	27.750 ± .024	.66328 ±.00062	5.7514 ±.0050	25	187
188	Paratellurite TeO_2	tet.	$P4_12_12(92)$ $P4_32_12(96)$		4	4.810 ±.002		7.613 ±.002				176.14 ± .15	26.52 ± .02	.6339 ±.0006	6.018 ±.005	25	188
189	Thorianite ThO_2	cubic	Fm3m(225)	fluorite	4	5.5952 ±.0005						175.16 ± .05	26.373 ± .007	.63038 ±.00021	10.012 ±.003	25	189
190	Rutile TiO_2	tet.	P4/mnm(136)		2	4.5937 ±.0005		2.9618 ±.0010				62.500 ± .025	18.820 ± .008	.44986 ±.00023	4.2453 ±.0017	25	190
191	Anatase TiO_2	tet.	$I4_1/amd(141)$		4	3.785 ±.002		9.514 ±.006				136.30 ± .17	20.522 ± .025	.4905 ±.0007	3.893 ±.005	r	191
192	Brookite TiO_2*	orth.	Pcab(61)		8	5.456 ±.002	9.182 ±.005	5.143 ±.003				257.6 ± .2	19.40 ± .02	.4636 ±.0005	4.119 ±.004	r	192
193	Titanium sesquioxide Ti_2O_3	hex-R	$R\bar{3}c(167)$	corundum	6	5.149 ±.002		13.642 ±.010				313.2 ± .3	31.44 ± .03	.7515 ±.0009	4.574 ±.005	r	193
194	Uraninite UO_2	cubic	Fm3m(225)	fluorite	4	5.4682 ±.0010						163.51 ± .09	24.618 ± .014	.58843 ±.00037	10.969 ±.006	26	194
195	Karelianite V_2O_3	hex-R	$R\bar{3}c(167)$	corundum	6	4.952 ±.002		14.002 ±.010				297.36 ± .32	29.848 ± .032	.71342 ±.00081	5.0216 ±.0054	r	195
196	Zincite ZnO	hex.	$P6_3mc(186)$	zincite	2	3.2495 ±.0005		5.2069 ±.0005				47.615 ± .015	14.338 ± .005	.34273 ±.00016	5.6750 ±.0018	25	196
197	Baddeleyite ZrO_2	mon.	$P2_1/c(14)$	baddeleyite	4	5.1454 ±.0010	5.2075 ±.0010	5.3107 ±.0010		99.23 ±.08		140.46 ± .06	21.148 ± .009	.50548 ±.00025	5.8267 ±.0023	r	197
	Multiple oxides																
198	Spinel $MgAl_2O_4$	cubic	Fd3m(227)	spinel	8	8.080 ±.002						527.5 ± .4	39.71 ± .03	.9492 ±.0008	3.583 ±.003	26	198

X-Ray Crystallographic Data of Minerals

	Name and formula	Crystal system	Space group	Structure type	Z	a_o	b_o	c_o	α_o	β_o	γ_o	Cell volume 10^{-24} cm^3	Molar volume cm^3	Molar volume cal bar^{-1}	X-Ray density grams cm^{-3}	Temp. °C	
199	Hercynite $FeAl_2O_4$	cubic	Fd3m(227)	spinel	8	8.150 ±.004						541.3 ± .6	40.75 ± .05	.9740 ±.0011	4.265 ±.005	25	199
200	Galaxite $MnAl_2O_4$	cubic	Fd3m(227)	spinel	8	8.258 ±.002						563.2 ± .4	42.39 ± .03	1.013 ±.001	4.078 ±.003	25	200
201	Gahnite $ZnAl_2O_4$	cubic	Fd3m(227)	spinel	8	8.0848 ±.0020						528.45 ± .39	39.783 ± .030	.95088 ±.00075	4.6083 ±.0034	26	201
202	Magnetite $FeFe_2O_4$	cubic	Fd3m(227)	spinel	8	8.3940 ±.0005						591.43 ± .11	44.524 ± .008	1.0642 ±.0002	5.2003 ±.0009	22	202
203	Jacobsite $MnFe_2O_4$	cubic	Fd3m(227)	spinel	8	8.499 ±.002						613.9 ± .4	46.22 ± .03	1.105 ±.001	4.990 ±.004	25	203
204	Trevorite $NiFe_2O_4$	cubic	Fd3m(227)	spinel	8	8.339 ±.003						579.9 ± .6	43.65 ± .05	1.043 ±.001	5.370 ±.006	25	204
205	Picrochromite $MgCr_2O_4$	cubic	Fd3m(227)	spinel	8	8.333 ±.003						578.6 ± .6	43.56 ± .05	1.041 ±.001	4.415 ±.005	26	205
206	Ilmenite $FeTiO_3$	hex-R	$R\bar{3}$(148)	ilmenite	6	5.093 ±.005		14.055 ±.020				315.73 ± .75	31.69 ± .08	.7574 ±.0019	4.788 ±.012	r	206
207	Geikielite $MgTiO_3$	hex-R	$R\bar{3}$(148)	ilmenite	6	5.054 ±.005		13.898 ±.010				307.44 ± .65	30.86 ± .07	.7376 ±.0016	3.896 ±.008	26	207
208	Pyrophanite $MnTiO_3$	hex-R	$R\bar{3}$(148)	ilmenite	6	5.155 ±.005		14.18 ±.01				326.3 ± .7	32.76 ± .07	.7829 ±.0017	4.605 ±.010	r	208
209	Cobalt Titanate $CoTiO_3$	hex-R	$R\bar{3}$(148)	ilmenite	6	5.066 ±.001		13.918 ±.005				309.34 ± .17	31.05 ± .02	.7422 ±.0004	4.986 ±.003	r	209
210	Perovskite $CaTiO_3$	orth.	Pcmn(62)	perovskite	4	5.3670 ±.0010	7.6438 ±.0010	5.4439 ±.0010				223.33 ± .07	33.626 ± .010	.80371 ±.00028	4.0439 ±.0012	r	210
211	Chrysoberyl $BeAl_2O_4$	orth.	Pmnb(62)	olivine	4	5.4756 ±.0020	9.4041 ±.0030	4.4267 ±.0020				227.94 ± .15	34.320 ± .023	.82031 ±.00059	3.6997 ±.0025	25	211
	Halides																
212	Halite $NaCl$	cubic	Fm3m(225)	rock salt	4	5.6402 ±.0002						179.43 ± .02	27.015 ± .003	.64571 ±.00011	2.1634 ±.0002	26	212
213	Sylvite KCl	cubic	Fm3m(225)	rock salt	4	6.2931 ±.0002						249.23 ± .02	37.524 ± .004	.89690 ±.00013	1.9868 ±.0002	25	213
214	Villiaumite NaF	cubic	Fm3m(225)	rock salt	4	4.6342 ±.0005						99.523 ± .032	14.984 ± .005	.35818 ±.00016	2.8021 ±.0009	25	214
215	Chlorargyrite $AgCl$	cubic	Fm3m(225)	rock salt	4	5.5491 ±.0005						170.87 ± .05	25.727 ± .007	.61493 ±.00021	5.5710 ±.0015	26	215
216	Bromargyrite $AgBr$	cubic	Fm3m(225)	rock salt	4	5.7745 ±.0005						192.55 ± .05	28.991 ± .008	.69294 ±.00022	6.4772 ±.0017	26	216
217	Nantockite $CuCl$	cubic	$F\bar{4}3m$(216)	sphalerite	4	5.416 ±.003						158.87 ± .26	23.92 ± .04	.5717 ±.0010	4.139 ±.007	25	217
218	Marshite CuI	cubic	$F\bar{4}3m$(216)	sphalerite	4	6.0507 ±.0010						221.52 ± .11	33.353 ± .017	.7972 ±.0004	5.710 ±.003	26	218
219	Miersite AgI	cubic	$F\bar{4}3m$(216)	sphalerite	4	6.4963 ±.0010						274.16 ± .10	41.278 ± .020	.9866 ±.0004	5.688 ±.003	r	219
220	Iodargyrite AgI	hex.	$P6_3mc$(186)	zincite	2	4.5955 ±.0010		7.5005 ±.0033				137.18 ± .10	41.308 ± .030	.9873 ±.0009	5.683 ±.004	25	220
221	Calomel $HgCl$	tet.	14/mm(139)		4	4.478 ±.005		10.910 ±.005				218.77 ± .50	32.939 ± .075	.7873 ±.0018	7.166 ±.016	26	221
222	Fluorite CaF_2	cubic	Fm3m(225)	fluorite	4	5.4638 ±.0004						163.11 ± .04	24.558 ± .005	.58701 ±.00017	3.1792 ±.0007	25	222
223	Sellaite MgF_2	tet.	$P4_2/mnm$(136)	rutile	2	4.621 ±.001		3.050 ±.001				65.13 ± .04	19.61 ± .01	.4688 ±.0003	3.177 ±.002	18	223

X-Ray Crystallographic Data of Minerals

	Name and formula	Crystal system	Space group	Structure type	Z	a_o	b_o	c_o	α_o	β_o	γ_o	Cell volume 10^{-24} cm³	Molar volume cm³	Molar volume cal bar⁻¹	X-Ray density grams cm⁻³	Temp. °C	
224	Chloromagnesite $MgCl_2$	hex-R	R3m(166)		3	3.632 ±.004		17.795 ±.016				203.29 ± .48	40.81 ± .10	.9754 ±.0024	2.333 ±.006	r	224
225	Lawrencite $FeCl_2$	hex-R	$R\bar{3}m$(166)		3	3.593 ±.003		17.58 ±.09				196.55 ± 1.06	39.46 ± .21	.9431 ±.0051	3.212 ±.017	r	225
226	Scacchite $MnCl_2$	hex-R	$R\bar{3}m$(166)		3	3.711 ±.002		17.59 ±.07				209.79 ± .86	42.11 ± .17	1.007 ±.004	2.988 ±.012	r	226
227	Cotunnite $PbCl_2$	orth.	Pnmb(62)		4	4.535 ±.005	7.62 ±.01	9.05 ±.01				312.74 ± .64	47.09 ± .10	1.1254 ±.0023	5.906 ±.012	26	227
228	Matlockite PbFCl	tet.	P4/nmm(129)		2	4.106 ±.005		7.23 ±.01				121.89 ± .34	36.70 ± .10	.8773 ±.0025	9.853 ±.028	26	228
229	Cryolite Na_3AlF_6*	mon.	$P2_1/n$(14)		2	5.40 ±.01	5.60 ±.01	7.776 ±.010		90.18 ±.25		235.1 ± .7	70.81 ± .20	1.692 ±.005	2.965 ±.009	r	229
230	Neighborite $NaMgF_3$	orth.	Pcmn(62)	perovskite	4	5.363 ±.001	7.676 ±.001	5.503 ±.001				226.54 ± .07	34.11 ± .01	.8152 ±.0003	3.058 ±.001	18	230
	Carbonates and nitrates																
231	Calcite $CaCO_3$	hex-R	$R\bar{3}c$(167)	calcite	6	4.9899 ±.0010		17.064 ±.002				367.96 ± .15	36.934 ± .015	.88278 ±.00041	2.7100 ±.0011	26	231
232	Otavite $CdCO_3$	hex-R	$R\bar{3}c$(167)	calcite	6	4.9204 ±.0010		16.298 ±.003				341.72 ± .15	34.300 ± .015	.81983 ±.00041	5.0265 ±.0022	26	232
233	Cobalticalcite $CoCO_3$	hex-R	$R\bar{3}c$(167)	calcite	6	4.6581 ±.0010		14.958 ±.003				281.07 ± .13	28.213 ± .013	.67435 ±.00036	4.2159 ±.0020	26	233
234	Siderite $FeCO_3$	hex-R	$R\bar{3}c$(167)	calcite	6	4.6887 ±.0010		15.373 ±.003				292.68 ± .14	29.378 ± .014	.70219 ±.00037	3.9436 ±.0018	26	234
235	Magnesite $MgCO_3$	hex-R	$R\bar{3}c$(167)	calcite	6	4.6330 ±.0010		15.016 ±.003				279.13 ± .13	28.018 ± .013	.66969 ±.00036	3.0095 ±.0014	26	235
236	Rhodochrosite $MnCO_3$	hex-R	$R\bar{3}c$(167)	calcite	6	4.7771 ±.0010		15.664 ±.003				309.57 ± .14	31.073 ± .014	.74272 ±.00039	3.6992 ±.0017	26	236
237	Nickelous Carbonate $NiCO_3$	hex-R	$R\bar{3}c$(167)	calcite	6	4.5975 ±.0010		14.723 ±.002				269.51 ± .12	27.052 ± .012	.64660 ±.00034	4.3886 ±.0020	26	237
238	Smithsonite $ZnCO_3$	hex-R	$R\bar{3}$(167)	calcite	6	4.6528 ±.0010		15.025 ±.003				281.69 ± .13	28.275 ± .013	.67583 ±.00037	4.4343 ±.0021	26	238
239	Dolomite $CaMg(CO_3)_2$*	hex-R	$R\bar{3}$(148)	calcite	3	4.8079 ±.0010		16.010 ±.003				320.50 ± .15	64.341 ± .029	1.5378 ±.0008	2.8661 ±.0013	26	239
240	Huntite $Mg_3Ca(CO_3)_4$*	hex-R	R32(155)	calcite	3	9.498 ±.003		7.816 ±.004				610.63 ± .50	122.58 ± .10	2.9299 ±.0024	2.880 ±.002	26	240
241	Norsethite $BaMg(CO_3)_2$*	hex-R	R32(155)	calcite	3	5.020 ±.005		16.75 ±.02				365.6 ± .8	73.39 ± .17	1.754 ±.004	3.838 ±.009	r	241
242	Vaterite $CaCO_3$	hex.			6	7.135 ±.005		8.524 ±.007				375.80 ± .61	37.72 ± .06	.9016 ±.0015	2.653 ±.004	r	242
243	Witherite $BaCO_3$	orth.	Pnam(62)	aragonite	4	6.430 ±.005	8.904 ±.005	5.314 ±.005				304.24 ± .41	45.81 ± .06	1.095 ±.002	4.308 ±.006	26	243
244	Aragonite $CaCO_3$	orth.	Pnam(62)	aragonite	4	5.741 ±.005	7.968 ±.005	4.959 ±.005				226.85 ± .33	34.15 ± .05	.8164 ±.0012	2.930 ±.004	26	244
245	Cerussite $PbCO_3$	orth.	Pnam(62)	aragonite	4	6.152 ±.005	8.436 ±.005	5.195 ±.005				269.61 ± .38	40.59 ± .06	.9702 ±.0014	6.582 ±.009	26	245
246	Strontianite $SrCO_3$	orth.	Pnam(62)	aragonite	4	6.029 ±.005	8.414 ±.005	5.107 ±.005				259.07 ± .37	39.01 ± .06	.9323 ±.0014	3.785 ±.005	26	246
247	Shortite $Na_2Ca_2(CO_3)_3$	orth.	Amm2(38)		2	4.961 ±.005	11.03 ±.02	7.12 ±.01				389.6 ± [illegible].0	117.3 ± .3	2.804 ±.007	2.610 ±.007	r	247
248	Malachite $Cu_2(OH)_2CO_3$	mon.	$P2_1/a$(14)		4	9.502 ±.007	11.974 ±.007	3.240 ±.003		98.75 ±.25		364.35 ± .54	54.86 ± .08	1.311 ±.002	4.030 ±.006	25	248

X-Ray Crystallographic Data of Minerals

No.	Name and formula	Crystal system	Space group	Structure type	Z	a_o	b_o	c_o	α_o	β_o	γ_o	Cell volume 10^{-24} cm^3	Molar volume cm^3	Molar volume cal bar^{-1}	X-Ray density grams cm^{-3}	Temp. °C	No.
249	Azurite $Cu_3(OH)_2(CO_3)_2$	mon.	$P2_1/a$(14)		2	5.008 ±.005	5.844 ±.005	10.336 ±.005		92.45 ±.25		302.22 ± .43	91.01 ± .13	2.1752 ±.0031	3.787 ±.005	25	249
250	Niter KNO_3	orth.	Pnam(62)	aragonite	4	6.431 ±.005	9.164 ±.005	5.414 ±.005				319.07 ± .42	48.04 ± .06	1.148 ±.002	2.105 ±.003	26	250
251	Soda Niter $NaNO_3$	hex-R	$R\bar{3}c$(167)	calcite	6	5.0696 ±.0010		16.829 ±.005				374.57 ± .19	37.598 ± .019	.89866 ±.00049	2.2606 ±.0011	25	251
252	Gerhardite $Cu_2(NO_3)(OH)_3$	orth.	$P2_12_12_1$(19)		4	6.075 ±.004	13.812 ±.008	5.592 ±.004				469.21 ± .53	70.65 ± .08	1.689 ±.002	3.399 ±.004	r	252
	Sulfates and borates																
253	Barite $BaSO_4$	orth.	Pnma(62)	barite	4	8.878 ±.005	5.450 ±.005	7.152 ±.003				346.05 ± .40	52.10 ± .06	1.245 ±.002	4.480 ±.005	26	253
254	Anhydrite $CaSO_4$	orth.	Amma(63) Ccmm(63)	anhydrite	4	6.991 ±.005	6.996 ±.005	6.238 ±.005				305.09 ± .39	45.94 ± .06	1.098 ±.002	2.964 ±.004	26	254
255	Anglesite $PbSO_4$	orth.	Pnma(62)	barite	4	8.480 ±.005	5.398 ±.005	6.958 ±.003				318.50 ± .38	47.95 ± .06	1.146 ±.002	6.324 ±.008	25	255
256	Celestite $SrSO_4$	orth.	Pnma(62)	barite	4	8.359 ±.005	5.352 ±.005	6.866 ±.005				307.17 ± .41	46.25 ± .06	1.105 ±.002	3.972 ±.005	26	256
257	Zinkosite $ZnSO_4$	orth.	Pnma(62)	barite	4	8.588 ±.008	6.740 ±.006	4.770 ±.005				276.10 ± .46	41.57 ± .07	.9936 ±.0017	3.883 ±.006	25	257
258	Arcanite KS_2SO_4	orth.	Pnma(62)	arcanite	4	5.772 ±.005	10.072 ±.005	7.483 ±.004				435.03 ± .49	65.50 ± .07	1.566 ±.002	2.661 ±.003	25	258
259	Mascagnite $(NH_4)_2SO_4$	orth.	Pnma(62)	arcanite	4	7.782 ±.005	5.993 ±.005	10.636 ±.005				496.04 ± .57	74.68 ± .09	1.7851 ±.0021	1.7693 ±.0020	25	259
260	Thenardite Na_2SO_4	orth.	Fddd(70)	thenardite	8	5.863 ±.005	12.304 ±.005	9.821 ±.005				708.47 ± .76	53.33 ± .06	1.275 ±.002	2.663 ±.003	25	260
261	Gypsum $CaSO_4.2H_2O$*	mon.	C2/c(15)		4	5.68 ±.01	15.18 ±.01	6.29 ±.01		113.83 ±.22		496.1 ± 1.5	74.69 ± .22	1.785 ±.005	2.305 ±.007	r	261
262	Epsomite $MgSO_4.7H_2O$	orth.	$P2_12_12_1$(19)		4	11.86 ±.01	11.99 ±.01	6.858 ±.007				975.22 ± 1.53	146.83 ± .23	3.5094 ±.0055	1.679 ±.003	25	262
263	Goslarite $ZnSO_4.7H_2O$	orth.	$P2_12_12_1$(19)	epsomite	4	11.779 ±.005	12.050 ±.005	6.822 ±.003				968.29 ± .72	145.79 ± .11	3.4845 ±.0026	1.9723 ±.0015	25	263
264	Mirabilite $Na_2SO_4.10H_2O$	mon.	$P2_1/c$(14)		4	11.51 ±.01	10.38 ±.01	12.83 ±.01		107.75 ±.17		1459.9 ± 2.6	219.8 ± .4	5.253 ±.009	1.466 ±.003	24	264
265	Chalcanthite $CuSO_4.5H_2O$	tri.	$P\bar{1}$(2)		2	6.1045 ±.0050	10.72 ±.01	5.949 ±.007	97.57 ±.17	107.28 ±.17	77.43 ±.17	361.88 ± .72	108.97 ± .22	2.6045 ±.0052	2.2912 ±.0046	r	265
266	Brochantite $Cu_4SO_4(OH)_6$*	mon.	$P2_1/c$(14)		4	13.066 ±.010	9.85 ±.01	6.022 ±.010		103.27 ±.25		754.3 ± 1.8	113.6 ± .2	2.715 ±.006	3.982 ±.009	r	266
267	Syngenite $K_2Ca(SO_4)_2.H_2O$	mon.	$P2_1/m$(11)		2	9.775 ±.005	7.156 ±.005	6.251 ±.005		104.00 ±.25		424.27 ± .68	127.76 ± .20	3.0535 ±.0049	2.5707 ±.0041	r	267
268	Alunite $KAl_3(SO_4)_2(OH)_6$	hex-R	R3m(160)		3	6.982 ±.005		17.32 ±.01				731.2 ± 1.1	146.8 ± .2	3.508 ±.005	2.822 ±.004	r	268
269	Natroalunite $NaAl_3(SO_4)_2(OH)_6$	hex-R	R3m(160)		3	6.974 ±.005		16.69 ±.01				702.99 ± 1.09	141.1 ± .2	3.373 ±.005	2.821 ±.004	r	269
270	Hexahydrite $MgSO_4.6H_2O$	mon.	C2/c(15)		8	10.110 ±.005	7.212 ±.004	24.41 ±.01		98.30 ±.10		1761.2 ± 1.6	132.58 ± .12	3.1689 ±.0029	1.7232 ±.0015	r	270
271	Leonhardtite $MgSO_4.4H_2O$	mon.	$P2_1/n$(14)		4	5.922 ±.006	13.604 ±.004	7.905 ±.005		90.85 ±.20		636.78 ± .78	95.88 ± .12	2.2915 ±.0029	2.0071 ±.0025	r	271
272	Melanterite $FeSO_4.7H_2O$	mon.	$P2_1/c$(14)		4	14.072 ±.010	6.503 ±.007	11.041 ±.010		105.57 ±.15		973.29 ± 1.69	146.54 ± .25	3.5025 ±.0061	1.8972 ±.0033	r	272
273	Vanthoffite $MgSO_4.3Na_2SO_4$	mon.	$P2_1/c$(14)		2	9.797 ±.003	9.217 ±.003	8.199 ±.003		113.50 ±.10		678.96 ± .65	204.45 ± .20	4.8866 ±.0047	2.6730 ±.0025	r	273

X-Ray Crystallographic Data of Minerals

	Name and formula	Crystal system	Space group	Structure type	Z	a_o	b_o	c_o	α_o	β_o	γ_o	Cell volume 10^{-24} cm^3	Molar volume cm^3	Molar volume cal bar^{-1}	X-Ray density grams cm^{-3}	Temp. °C	
274	Dolerophanite $Cu_2O(SO_4)$	mon.	C2/m(15)		4	9.355 ±.010	6.312 ±.005	7.628 ±.005		122.29 ±.10		380.77 ± .70	57.33 ± .11	1.3703 ±.0026	4.171 ±.008	r	274
275	Retgersite $NiSO_4.4H_2O$	tet.	$P4_12_12(92)$ $P4_32_1(96)$		4	6.782 ±.004		18.28 ±.01				840.80 ± 1.09	126.59 ± .16	3.0257 ±.0040	2.076 ±.003	25	275
276	Colemanite $CaB_3O_4(OH)_3.H_2O$*	mon.	$P2_1/a(14)$		4	8.743 ±.004	11.264 ±.002	6.102 ±.003		110.12 ±.08		564.26 ± .49	84.957 ± .073	2.0306 ±.0018	2.4194 ±.0021	r	276
277	Borax $Na_2B_4O_7.10H_2O$	mon.	C2/c(15)		4	11.858 ±.005	10.674 ±.005	12.197 ±.005		106.68 ±.03		1478.8 ± 1.1	222.66 ± .17	5.3217 ±.0041	1.7128 ±.0013	r	277
278	Kernite $Na_2B_4O_7.4H_2O$	mon.	$P2_1/c(14)$		4	7.022 ±.003	9.151 ±.004	15.676 ±.008		108.83 ±.25		953.40 ± 1.61	143.55 ± .24	3.4309 ±.0058	1.9038 ±.0032	r	278
279	Hambergite $Be_2BO_3.(OH,F)$*	orth.	Pbca(61)		8	9.755 ±.001	12.201 ±.001	4.426 ±.001				526.79 ± .14	39.658 ± .011	.9479 ±.0003	2.3663 ±.0006	r	279
	Phosphates, molybdates, and tungstates																
280	Berlinite $AlPO_4$	hex.	$P3_121(152)$ $P3_221(154)$	α-quartz	3	4.942 ±.005		10.97 ±.007				232.03 ± .50	46.58 ± .10	1.113 ±.002	2.618 ±.006	25	280
281	Xenotime YPO_4	tet.	$I4_1/amd(141)$	zircon	4	6.885 ±.005		5.982 ±.005				283.57 ± .48	42.69 ± .07	1.020 ±.002	4.307 ±.008	26	281
282	Hydroxylapatite $Ca_5(PO_4)_3OH$	hex.	$P6_3/m(176)$	apatite	2	9.418 ±.003		6.883 ±.003				528.7 ± .5	159.2 ± .2	3.805 ±.004	3.155 ±.004	r	282
283	Fluorapatite $Ca_5(PO_4)_3F$	hex.	$P6_3/m(176)$	apatite	2	9.3684 ±.0030		6.8841 ±.0030				523.25 ± .41	157.56 ± .12	3.7659 ±.0030	3.2007 ±.0025	25	283
284	Chlorapatite $Ca_5(PO_4)_3Cl$	hex.	$P6/_3m(176)$	apatite	2	9.629 ±.005		6.777 ±.003				544.76 ± .61	163.86 ± .19	3.916 ±.004	3.178 ±.004	r	284
285	Carbonate-apatite $Ca_{10}(PO_4)_6CO_3H_2O$	hex.	$P6_3/m(176)$	apatite	1	9.436 ±.010		6.883 ±.010				530.74 ± 1.76	319.6 ± .8	7.640 ±.020	3.281 ±.008	r	285
286	Turquois $CuAl_6(PO_4)_4(OH)_8.4H_2O$*	tri.	$P\bar{1}(2)$		1	7.424 ±.008	7.629 ±.008	9.910 ±.010	68.61 ±.20	69.71 ±.20	65.08 ±.20	461.40 ± 1.12	277.9 ± .7	6.6416 ±.0162	2.927 ±.007	r	286
287	Powellite $CaMoO_4$	tet.	$I4_1/a(100)$	scheelite	4	5.226 ±.005		11.43 ±.007				312.17 ± .63	47.00 ± .09	1.1234 ±.0023	4.256 ±.009	25	287
288	Wulfenite $PbMoO_4$	tet.	$I4_1/a(100)$	scheelite	4	5.435 ±.005		12.110 ±.007				357.72 ± .69	53.859 ± .104	1.2873 ±.0025	6.816 ±.013	25	288
289	Scheelite $CaWO_4$	tet.	$I4_1/a(100)$	scheelite	4	5.242 ±.005		11.372 ±.005				312.49 ± .61	47.049 ± .092	1.1245 ±.0023	6.120 ±.012	25	289
290	Stolzite $PbWO_4$	tet.	$I4_1/a(100)$	scheelite	4	5.4616 ±.0030		12.046 ±.005				359.22 ± .42	54.100 ± .064	1.2931 ±.0016	8.4110 ±.0099	25	290
291	Ferberite $FeWO_4$	mon.	P2/c(13)	wolframite	2	4.732 ±.004	5.708 ±.003	4.965 ±.004		90.00 ±.05		134.21 ± .27	40.38 ± .05	.9652 ±.0013	7.520 ±.010	r	291
292	Huebnerite $MnWO_4$	mon.	P2/c(13)	wolframite	2	4.834 ±.004	5.758 ±.005	4.999 ±.004		91.18 ±.10		139.21 ± .20	41.89 ± .06	1.001 ±.002	7.228 ±.010	r	292
293	Wolframite $Fe_{.5}Mn_{.5}WO_4$	mon.	P2/c(13)	wolframite	2	4.782 ±.004	5.731 ±.004	4.982 ±.004		90.57 ±.10		136.53 ± .18	41.11 ± .06	.9826 ±.0014	7.376 ±.010	r	293
294	Sanmartinite $ZnWO_4$	mon.	P2/c(13)	wolframite	2	4.691 ±.003	5.720 ±.003	4.925 ±.003		89.36 ±.20		132.14 ± .14	39.79 ± .04	.9511 ±.0010	7.872 ±.008	25	294
	Ortho and ring structure silicates																
295	Forsterite Mg_2SiO_4	orth.	Pbnm(62)	olivine	4	4.758 ±.002	10.214 ±.003	5.984 ±.002				290.81 ± .18	43.786 ± .027	1.0465 ±.0007	3.2136 ±.0020	25	295
296	Fayalite Fe_2SiO_4	orth.	Pbnm(62)	olivine	4	4.817 ±.005	10.477 ±.005	6.105 ±.010				308.21 ± .62	46.389 ± .093	1.1088 ±.0023	4.3928 ±.0088	r	296
297	Tephroite Mn_2SiO_4*	orth.	Pbnm(62)	olivine	4	4.871 ±.005	10.636 ±.005	6.232 ±.005				322.87 ± .45	48.612 ± .067	1.1619 ±.0017	4.1545 ±.0058	r	297

X-Ray Crystallographic Data of Minerals

	Name and formula	Crystal system	Space group	Structure type	Z	a_o	b_o	c_o	α_o	β_o	γ_o	Cell volume 10^{-24} cm³	Molar volume cm³	Molar volume cal bar⁻¹	X-Ray density grams cm⁻³	Temp. °C	
298	Lime Olivine γCa_2SiO_4	orth.	Pbnm(62)	olivine	4	5.091 ±.010	11.371 ±.020	6.782 ±.010				392.61 ± 1.19	59.11 ± .18	1.4129 ±.0043	2.914 ±.009	r	298
299	Nickel Olivine Ni_2SiO_4	orth.	Pbnm(62)	olivine	4	4.727 ±.002	10.121 ±.005	5.915 ±.002				282.98 ± .21	42.61 ± .03	1.0184 ±.0008	4.917 ±.004	r	299
300	Cobalt Olivine Co_2SiO_4	orth.	Pbnm(62)	olivine	4	4.782 ±.002	10.301 ±.005	6.003 ±.002				295.70 ± .21	44.52 ± .03	1.0642 ±.0008	4.716 ±.003	r	300
301	Monticellite $CaMgSiO_4$	orth.	Pbnm(62)	olivine	4	4.827 ±.005	11.084 ±.005	6.376 ±.005				341.13 ± .47	51.362 ± .071	1.2276 ±.0017	3.046 ±.004	r	301
302	Kerschsteinite $CaFeSiO_4$	orth.	Pbnm(62)	olivine	4	4.886 ±.005	11.146 ±.005	6.434 ±.010				350.39 ± .67	52.756 ± .101	1.2609 ±.0025	3.564 ±.007	r	302
303	Knebelite $MnFeSiO_4$*	orth.	Pbnm(62)	olivine	4	4.854 ±.010	10.602 ±.010	6.162 ±.010				317.11 ± .88	47.74 ± .13	1.1412 ±.0032	4.249 ±.012	r	303
304	Glauchroite $CaMnSiO_4$	orth.	Pbnm(62)	olivine	4	4.944 ±.004	11.19 ±.01	6.529 ±.005				361.2 ± .9	54.38 ± .14	1.2997 ±.0032	3.441 ±.009	r	304
305	Fluor-Norbergite $Mg_2SiO_4.MgF_2$	orth.	Pnmb(62)		4	8.727 ±.005	10.271 ±.010	4.709 ±.002				422.09 ± .51	63.551 ± .077	1.5190 ±.0019	3.194 ±.004	25	305
306	Chondrodite $2Mg_2SiO_4.MgF_2$*	mon.	$P2_1/c(14)$		2	7.89 ±.03	4.743 ±.020	10.29 ±.03		109.03 ±.30		364.0 ± 2.4	109.6 ± .7	2.620 ±.017	3.136 ±.021	r	306
307	Fluor-Humite $3Mg_2SiO_4.MgF_2$	orth.	Pnma(62)		4	10.243 ±.005	20.72 ±.02	4.735 ±.002				1004.9 ± 1.2	151.31 ± .18	3.6163 ±.0042	3.2017 ±.0037	25	307
308	Clinohumite $4Mg_2SiO_4.MgF_2$*	mon.	$P2_1/c(14)$		2	13.68 ±.04	4.75 ±.02	10.27 ±.02		100.83 ±.50		655.5 ± 3.8	197.4 ± 1.1	4.717 ±.027	3.167 ±.018	r	308
309	Grossularite $Ca_3Al_2Si_3O_{12}$	cubic	Ia3d(230)	garnet	8	11.851 ±.001						1664.43 ± .42	125.30 ± .03	2.9948 ±.0008	3.595 ±.001	25	309
310	Uvarovite $Ca_3Cr_2Si_3O_{12}$	cubic	Ia3d(230)	garnet	8	11.999 ±.002						1727.57 ± .86	130.05 ± .07	3.1084 ±.0016	3.848 ±.002	26	310
311	Andradite $Ca_3Fe_2Si_3O_{12}$	cubic	Ia3d(230)	garnet	8	12.048 ±.001						1748.82 ± .44	131.65 ± .03	3.1466 ±.0008	3.860 ±.001	25	311
312	Goldmanite $Ca_3V_2Si_3O_{12}$	cubic	Ia3d(230)	garnet	8	12.070 ±.005						1758.42 2.19	132.38 ± .16	3.1639 ±.0040	3.765 ±.005	r	312
313	Almandite $Fe_3Al_2Si_3O_{12}$	cubic	Ia3d(230)	garnet	8	11.526 ±.001						1531.21 ± .40	115.27 ± .04	2.7551 ±.0008	4.318 ±.001	25	313
314	Pyrope $Mg_3Al_2Si_3O_{12}$	cubic	Ia3d(230)	garnet	8	11.459 ±.001						1504.67 ± .39	113.27 ± .03	2.7074 ±.0008	3.559 ±.001	25	314
315	Spessartite $Mn_3Al_2Si_3O_{12}$	cubic	Ia3d(230)	garnet	8	11.621 ±.001						1569.39 ± .41	118.15 ± .03	2.8238 ±.0008	4.190 ±.001	25	315
316	Zircon $ZrSiO_4$*	tet.	I4/amd(141)	zircon	4	6.604 ±.005		5.979 ±.005				260.76 ± .45	39.261 ± .068	.9384 ±.0017	4.669 ±.008	25	316
317	Thorite $ThSiO_4$	tet.	I4/amd(141)	zircon	4	7.143 ±.004		6.327 ±.003				322.82 ± .39	48.60 ± .06	1.1617 ±.0015	6.668 ±.008	r	317
318	Coffinite $USiO_4$	tet.	I4/amd(141)	zircon	4	6.995 ±.004		6.263 ±.005				306.45 ± .43	46.140 ± .064	1.103 ±.002	7.155 ±.010	r	318
319	Kyanite Al_2SiO_5*	tri.	$P\bar{1}(2)$		4	7.123 ±.001	7.848 ±.002	5.564 ±.008	89.92 ±.15	101.25 ±.08	105.97 ±.08	292.83 ± .45	44.09 ± .07	1.054 ±.002	3.675 ±.006	25	319
320	Andalusite Al_2SiO_5*	orth.	Pnnm(58)		4	7.7959 ±.0050	7.8983 ±.0020	5.5583 ±.0020				342.25 ± .27	51.530 ± .040	1.2316 ±.0010	3.145 ±.002	25	320
321	Sillimanite Al_2SiO_5*	orth.	Pbnm(62) Pnma(62)		4	7.4843 ±.0030	7.6730 ±.0030	5.7711 ±.0040				331.42 ± .30	49.899 ± .044	1.1927 ±.0011	3.248 ±.003	25	321
322	3.2 Mullite $3AL_2O_3.2SiO_2$	orth.			3/4	7.557 ±.002	7.6876 ±.0020	2.8842 ±.0010				167.56 ± .09	134.55 ± .07	3.2159 ±.0016	3.166 ±.002	r	322
323	2.1 Mullite $2Al_2O_3.SiO_2$	orth.	Pbam(55)		6/5	7.5788 ±.0020	7.6909 ±.0020	2.8883 ±.0010				168.35 ± .09	84.492 ± .043	2.0195 ±.0011	3.125 ±.002	r	323

X-Ray Crystallographic Data of Minerals

	Name and formula	Crystal system	Space group	Structure type	Z	a_o	b_o	c_o	α_o	β_o	γ_o	Cell volume 10^{-24} cm	Molar volume cm³	Molar volume cal bar⁻¹	X-Ray density grams cm⁻³	Temp. °C	
324	Staurolite $Fe_2Al_9Si_4O_{22}(OH)_2$*	mon.	C2/m(15)		2	7.90 ±.10	16.65 ±.15	5.63 ±.10		90.00 ±.25		740.5 ±17.5	223.0 ± 5.3	5.330 ±.126	3.825 ±.090	r	324
325	Topaz $Al_2(SiO_4)(OH)$*	orth.	Pmnb(62)		4	8.394 ±.005	8.792 ±.007	4.649 ±.003				343.10 ± .4[illegible]	51.66 ± .06	1.2347 ±.0015	3.563 ±.005	26	325
326	Phenacite Be_2SiO_4*	hex-R	R$\bar{3}$(148)	phenacite	18	12.472 ±.005		8.252 ±.005				1111.6 ± 1.1	37.194 ± .037	.8890 ±.0009	2.960 ±.003	25	326
327	Willemite Zn_2SiO_4	hex-R	R$\bar{3}$(148)	phenacite	18	13.94 ±.01		9.309 ±.003				1566.6 ± 2.3	52.42 ± .08	1.253 ±.002	4.251 ±.006	25	327
328	Dioptase CuH_2SiO_4*	hex-R	R$\bar{3}$(148)	phenacite	18	14.61 ±.02		7.80 ±.01				1441.9 ± 4.4	48.24 ± .15	1.153 ±.004	3.247 ±.010	r	328
329	Larnite β-Ca_2SiO_4*	mon.	$P2_1/n$(14)		4	5.48 ±.02	6.76 ±.02	9.28 ±.02		94.55 ±.33		342.7 ± 1.8	51.60 ± .27	1.233 ±.006	3.338 ±.017	r	329
330	Akermanite $Ca_2MgSi_2O_7$	tet.	P$\bar{4}2_1$m(113)	melilite	2	7.8435 ±.0030		5.010 ±.003				308.2[illegible] ± .3[illegible]	92.812 ± .090	2.2183 ±.0022	2.9375 ±.0029	r	330
331	Gehlenite $Ca_2Al_2SiO_7$	tet.	P$\bar{4}2_1$m(113)	melilite	2	7.690 ±.003		5.0675 ±.0030				299.6[illegible] ± .2[illegible]	90.239 ± .088	2.1568 ±.0022	3.0387 ±.0030	r	331
332	Fe-Gehlenite $Ca_2Fe_2SiO_7$	tet.	P$\bar{4}2_1$m(113)	melilite	2	7.54 ±.01		4.855 ±.005				276.0[illegible] ± .7[illegible]	83.12 ± .24	1.9865 ±.0057	3.994 ±.011	r	332
333	Hardystonite $Ca_2ZnSi_2O_7$*	tet.	P$\bar{4}2_1$m(113)	melilite	2	7.87 ±.03		5.01 ±.02				310.3 ± 2.7	93.44 ± .80	2.233 ±.019	3.357 ±.029	r	333
334	Sodium Melilite $NaCaAlSi_2O_7$	tet.	P$\bar{4}2_1$m(113)	melilite	2	8.511 ±.005		4.809 ±.003				348.35 ± .45	104.90 ± .14	2.507 ±.003	2.462 ±.003	r	334
335	Beryl $Be_3Al_2(Si_6O_{18})$*	hex.	P6/mmc(192)	beryl	2	9.215 ±.005		9.192 ±.005				675.98 ± .82	203.55 ± .25	4.8651 ±.0060	2.641 ±.003	25	335
336	Indialite high Cordierite $Mg_2Al_3(AlSi_5O_{18})$	hex.	P6/mmc(192)	beryl	2	9.7698 ±.0030		9.3517 ±.0030				773.02 ± .54	232.78 ± .16	5.5636 ±.0039	2.513 ±.002	25	336
337	Low Cordierite $Mg_2Al_3(AlSi_5O_{18})$	orth.	Cccm(66)		4	9.721 ±.003	17.062 ±.006	9.339 ±.003				1548.96 ± .88	233.22 ± .13	5.5741 ±.0032	2.508 ±.001	25	337
338	Fe-Indialite $Fe_2Al_3(AlSi_5O_{18})$	hex.	P6/mmc(192)	beryl	2	9.860 ±.010		9.285 ±.010				781.75 ± 1.80	235.40 ± .54	5.6264 ±.0130	2.753 ±.006	r	338
339	Fe-Cordierite $Fe_2Al_3(AlSi_5O_{18})$	orth.	Cccm(66)	cordierite	4	9.726 ±.010	17.065 ±.010	9.287 ±.010				1541.40 ± 2.47	232.08 ± .37	5.5468 ±.0089	2.792 ±.005	r	339
340	Mn-Indialite $Mn_2Al_3(AlSi_5O_{18})$	hex.	P6/mmc(192)	beryl	2	9.925 ±.010		9.297 ±.010				793.11 ± .81	238.8 ± .5	5.708 ±.013	2.706 ±.006	r	340
341	Sapphirine $Mg_2Al_4O_6SiO_4$*	mon.	$P2_1/c$(14)		8	11.26 ±.03	14.46 ±.03	9.95 ±.02		125.33 ±.50		1321.7 ± 9.7	99.50 ± .73	2.378 ±.017	3.464 ±.025	r	341
342	Elbaite $NaLiAl_{7.67}B_3Si_6O_{27}(OH)_4$*	hex-R	R3m(160)	tourmaline	3	15.842 ±.010		7.009 ±.010				1523.4 ± 2.9	305.82 ± .58	7.3093 ±.0140	3.271 ±.006	r	342
343	Schorl $NaFe_3Al_6B_3Si_6O_{27}(OH)_4$*	hex-R	R3m(160)	tourmaline	3	16.032 ±.010		7.149 ±.010				1591.3 ± 3.0	319.45 ± .60	7.635 ±.014	3.297 ±.006	r	343
344	Dravite $NaMg_3Al_6B_3Si_6O_{27}(OH)_4$	hex-R	R3m(160)	tourmaline	3	15.942 ±.010		7.224 ±.010				1589.[illegible] ± 2.[illegible]	319.19 ± .60	7.629 ±.014	3.004 ±.006	r	344
345	Uvite $CaMg_4Al_5B_3Si_6O_{27}(OH)_4$	hex-R	R3m(160)	tourmaline	3	15.86 ±.01		7.19 ±.01				1566.[illegible] ± 2.9	314.4 ± .6	7.515 ±.014	3.095 ±.006	r	345
346	Sphene $CaTiSiO_5$*	mon.	A2/a(15)		4	7.07 ±.01	8.72 ±.01	6.56 ±.01		113.95 ±.25		369.61 ± 1.13	55.65 ± .17	1.330 ±.004	3.523 ±.011	r	346
347	Datolite $CaBSiO_4(OH)$*	mon.	$P2_1/c$(14)		4	9.62 ±.03	7.60 ±.03	4.84 ±.02		90.15 ±.25		353.9 ± 2.3	53.28 ± .35	1.273 ±.008	3.003 ±.020	r	347
348	Euclase $AlBeSiO_4(OH)$*	mon.	$P2_1/a$(14)		4	4.763 ±.005	14.29 ±.02	4.618 ±.005		100.25 ±.10		309.30 ± .64	46.57 ± .10	1.113 ±.002	3.116 ±.007	r	348
349	Chloritoid $H_2FeAl_2SiO_7$*	mon.	C2/c(15)		8	9.48 ±.01	5.48 ±.01	18.18 ±.01		101.77 ±.25		924.6 ± 2.[illegible]	69.61 ± .16	1.664 ±.004	3.619 ±.008	r	349

X-Ray Crystallographic Data of Minerals

	Name and formula	Crystal system	Space group	Structure type	Z	a_o	b_o	c_o	α_o	β_o	γ_o	Cell volume 10^{-24} cm^3	Molar volume cm^3	Molar volume cal bar^{-1}	X-Ray density grams cm^{-3}	Temp. °C	
350	Hemimorphite $Zn_4(OH)_2Si_2O_7.H_2O$*	orth.	Imm2(35)		2	8.370 ±.005	10.719 ±.005	5.120 ±.005				459.36 ± .57	138.32 ± .17	3.306 ±.004	3.482 ±.004	25	350
351	Zoisite $Ca_2Al_3(SiO_4)_3OH$	orth.	Pnma(62)		4	16.15 ±.01	5.581 ±.005	10.06 ±.01				906.74 ± 1.34	136.52 ± .20	3.263 ±.005	3.328 ±.005	r	351
352	Clinozoisite $Ca_2Al_3(SiO_4)_3OH$	mon.	$P2_1/m(11)$		2	8.887 ±.007	5.581 ±.005	10.14 ±.01		115.93 ±.33		452.30 ± 1.45	136.20 ± .44	3.255 ±.010	3.336 ±.011	r	352
353	Epidote $Ca_2Al_{1.5}Fe_{1.5}(SiO_4)_3OH$*	mon.	$P2_1/m(11)$		2	8.89 ±.02	5.63 ±.01	10.19 ±.02		115.40 ±.30		460.72 ± 1.97	138.7 ± .6	3.316 ±.014	3.587 ±.015	r	353
354	Piemontite $Ca_2Al_{1.5}Mn_{1.5}(SiO_4)_3OH$*	mon.	$P2_1/m(11)$		2	8.95 ±.02	5.70 ±.01	9.41 ±.02		115.70 ±.50		432.56 ± 2.38	130.3 ± .7	3.113 ±.017	3.810 ±.021	r	354
355	Lawsonite $CaAl_2Si_2O_7(OH)_2.H_2O$	orth.	Cccm(63)		4	8.787 ±.005	5.836 ±.005	13.123 ±.008				672.96 ± .80	101.32 ± .12	2.4217 ±.0029	3.101 ±.004	r	355
	Chain and band structure silicates																
356	Enstatite $MgSiO_3$*	orth.	Pcab(61)		16	8.829 ±.010	18.22 ±.01	5.192 ±.005				835.21 ± 1.32	31.44 ± .05	.7514 ±.0012	3.194 ±.005	r	356
357	Clinoenstatite $MgSiO_3$	mon.	$P2_1/c(15)$		8	9.620 ±.005	8.825 ±.005	5.188 ±.005		108.33 ±.17		418.10 ± .66	31.47 ± .05	.7523 ±.0012	3.190 ±.005	r	357
358	Protoenstatite $MgSiO_3$	orth.	Pbcn(60)		8	9.25 ±.01	8.74 ±.01	5.32 ±.01				430.10 ± 1.05	32.38 ± .08	.7739 ±.0019	3.101 ±.008	r	358
359	High Clinoenstatite $MgSiO_3$	tri.			8	10.000 ±.005	8.934 ±.004	5.170 ±.003	88.27 ±.05	70.03 ±.04	91.01 ±.04	433.72 ± .40	32.65 ± .03	.7804 ±.0008	3.075 ±.003		359
360	Clinoferrosilite $FeSiO_3$	mon.	$P2_1/c(14)$		8	9.7085 ±.0010	9.0872 ±.0011	5.2284 ±.004		108.43 ±.05		437.60 ± .15	32.943 ± .011	.7874 ±.0003	4.005 ±.002	r	360
361	Orthoferrosilite $FeSiO_3$	orth.	Pcab(61)	enstatite	16	9.080 ±.002	18.431 ±.004	5.238 ±.001				876.6 ± .54	33.00 ± .02	.7887 ±.0008	3.998 ±.004	r	361
362	Diopside $CaMg(SiO_3)_2$	mon.	C2/c(15)	diopside	4	9.743 ±.005	8.923 ±.005	5.251 ±.003		105.93 ±.25		438.97 ± .69	66.09 ± .10	1.580 ±.003	3.277 ±.005	r	362
363	Hedenbergite $CaFe(SiO_3)_2$*	mon.	C2/c(15)	diopside	4	9.854 ±.010	9.024 ±.010	5.263 ±.010		104.23 ±.33		453.64 ± 1.28	68.30 ± .19	1.632 ±.005	3.632 ±.010	r	363
364	Johannsenite $CaMn(SiO_3)_2$*	mon.	C2/c(15)	diopside	4	9.83 ±.03	9.04 ±.03	5.27 ±.02		105.00 ±.50		452.35 ± 2.87	68.11 ± .43	1.628 ±.010	3.629 ±.023	r	364
365	Ureyite $NaCr(SiO_3)_2$	mon.	C2/c(15)	diopside	4	9.550 ±.016	8.712 ±.007	5.273 ±.008		107.44 ±.16		418.6 ± 1.1	63.02 ± .16	1.506 ±.004	3.605 ±.009	r	365
366	Jadeite $NaAl(SiO_3)_2$*	mon.	C2/c(15)	diopside	4	9.409 ±.005	8.564 ±.005	5.220 ±.005		107.50 ±.20		401.15 ± .67	60.40 ± .10	1.444 ±.002	3.347 ±.006	r	366
367	Acmite (Aegirine) $NaFe(SiO_3)_2$	mon.	C2/c(15)	diopside	4	9.658 ±.005	8.795 ±.005	5.294 ±.005		107.42 ±.20		429.06 ± .70	64.60 ± .11	1.544 ±.003	4.411 ±.007	r	367
368	Ca Tschermak Molecule $CaAl_2SiO_6$	mon.	C2/c(15)	diopside	4	9.615 ±.005	8.661 ±.005	5.272 ±.003		106.12 ±.20		421.77 ± .59	63.50 ± .09	1.518 ±.002	3.435 ±.005	r	368
369	Spodumene $LiAl(SiO_3)_2$	mon.	C2/c(15)	diopside	4	9.451 ±.002	8.387 ±.002	5.208 ±.001		110.07 ±.03		387.7 ± .1	58.37 ± .02	1.395 ±.001	3.188 ±.001	r	369
370	β-Spodumene $LiAl(SiO_3)_2$	tet.	$P4_32_12(96)$ $P4_12_12(92)$		4	7.5332 ±.0008		9.1540 ±.0008				519.48 ± .12	78.215 ± .018	1.8694 ±.0005	2.379 ±.001	r	370
371	Pectolite $Ca_2NaH(SiO_3)_3$*	tri.	$P\bar{1}(2)$		2	7.99 ±.01	7.04 ±.01	7.02 ±.01	90.05 ±.25	95.27 ±.25	102.47 ±.25	383.84 ± .99	115.58 ± .30	2.763 ±.007	2.876 ±.007	r	371
372	Wollastonite $CaSiO_3$*	tri.	$P\bar{1}(2)$		6	7.94 ±.01	7.32 ±.01	7.07 ±.01	90.03 ±.25	95.37 ±.25	103.43 ±.25	397.82 ± 1.03	39.93 ± .10	.9544 ±.0025	2.909 ±.008	r	372
373	Parawollastonite $CaSiO_3$*	mon.	$P2_1(4)$		12	15.417 ±.004	7.321 ±.002	7.066 ±.003		95.40 ±.10		793.98 ± .47	39.85 ± .02	.9524 ±.0006	2.915 ±.002	r	373
374	Pseudowollastonite $CaSiO_3$*	tri.			24	6.90 ±.02	11.78 ±.02	19.65 ±.02	90.00 ±.30	90.80 ±.30	90.00 ±.30	1597.0 ± 5.6	40.08 ± .14	.9579 ±.0034	2.899 ±.010	r	374

X-Ray Crystallographic Data of Minerals

No.	Name and formula	Crystal system	Space group	Structure type	Z	a_o	b_o	c_o	α_o	β_o	γ_o	Cell volume 10^{-24} cm^3	Molar volume cm^3	Molar volume cal bar^{-1}	X-Ray density grams cm^{-3}	Temp. °C	
375	Rhodonite $MnSiO_3$*	tri.	P$\bar{1}$(2)		10	7.682 ±.002	11.818 ±.003	6.707 ±.002	92.36 ±.05	93.95 ±.05	105.66 ±.05	583.77 ± .37	35.158 ± .019	.8403 ±.0005	3.727 ±.002	r	375
376	Bustamite $CaMn(SiO_3)_2$*	tri.	A$\bar{1}$(2)		6	7.736 ±.003	7.157 ±.003	13.824 ±.010	90.52 ±.25	94.58 ±.25	103.87 ±.25	740.38 ± 1.08	74.32 ± .11	1.776 ±.003	3.326 ±.005	r	376
377	Pyroxmangite $MnFe(SiO_3)_2$*	tri.	P$\bar{1}$(2)		7	7.56 ±.02	17.45 ±.05	6.67 ±.02	84.00 ±.30	94.30 ±.30	113.70 ±.30	800.77 ± 4.29	68.90 ± .36	1.647 ±.009	3.817 ±.020	r	377
378	Tremolite $Ca_2Mg_5[Si_8O_{22}](OH)_2$*	mon.	C2/m(12)	tremolite	2	9.840 ±.010	18.052 ±.020	5.275 ±.010		104.70 ±.25		906.34 ± 2.43	272.92 ± .73	6.523 ±.018	2.977 ±.008	r	378
379	Fluor-tremolite $Ca_2Mg_5[Si_8O_{22}]F_2$	mon.	C2/m(12)	tremolite	2	9.781 ±.007	18.01 ±.01	5.267 ±.005		104.52 ±.25		898.18 ± 1.56	270.46 ± .47	6.464 ±.011	3.018 ±.005	20	379
380	Ferrotremolite $Ca_2Fe_5[Si_8O_{22}](OH)_2$	mon.	C2/m(12)	tremolite	2	9.97 ±.01	18.34 ±.02	5.30 ±.01		104.50 ±.10		938.24 ± 2.92	282.53 ± .69	6.753 ±.017	3.434 ±.008	r	380
381	Grunerite $Fe_7[Si_8O_{22}](OH)_2$	mon.	C2/m(12)	tremolite	2	9.572 ±.005	18.44 ±.01	5.342 ±.007		101.77 ±.25		923.08 ± 1.63	277.96 ± .49	6.644 ±.012	3.603 ±.006	r	381
382	Cummingtonite (hypo.) $Mg_7[Si_8O_{22}](OH)_2$	mon.	C2/m(12)	tremolite	2	9.476 ±.010	17.935 ±.010	5.292 ±.005		102.23 ±.25		878.97 ± 1.58	264.68 ± .47	6.326 ±.011	2.950 ±.005	r	382
383	Riebeckite $Na_2Fe_3Fe_2[Si_8O_{22}](OH)_2$	mon.	C2/m(12)	tremolite	2	9.729 ±.020	18.065 ±.020	5.334 ±.010		103.31 ±.25		912.29 ± 2.89	274.71 ± .87	6.566 ±.021	3.407 ±.011	r	383
384	Magnesioriebeckite $Na_2Mg_3Fe_2[Si_8O_{22}](OH)_2$	mon.	C2/m(12)	tremolite	2	9.733 ±.010	17.946 ±.020	5.299 ±.010		103.30 ±.25		900.74 ± 2.37	271.24 ± .71	6.483 ±.017	3.102 ±.008	r	384
385	Gaucophane I $Na_2Mg_3Al_2[Si_8O_{22}](OH)_2$	mon.	C2/m(12)	tremolite	2	9.748 ±.010	17.915 ±.020	5.273 ±.010		102.78 ±.25		898.04 ± 2.35	270.42 ± .71	6.463 ±.017	2.898 ±.008	r	385
386	Glaucophane II $Na_2Mg_3Al_2[Si_8O_{22}](OH)_2$	mon.	C2/m(12)	tremolite	2	9.663 ±.010	17.696 ±.020	5.277 ±.010		103.67 ±.10		876.79 ± 2.17	264.02 ± .65	6.310 ±.016	2.968 ±.007	r	386
387	Fluor-edenite $NaCa_2Mg_5[AlSi_7O_{22}]F_2$	mon.	C2/m(12)	tremolite	2	9.847 ±.005	18.00 ±.01	5.282 ±.005		104.83 ±.25		905.03 ± 1.51	272.53 ± .46	6.514 ±.011	3.076 ±.005	r	387
388	Fluor-richterite $Na_2CaMg_5[Si_8O_{22}]F_2$	mon.	C2/m(12)	tremolite	2	9.823 ±.005	17.96 ±.01	5.268 ±.005		104.33 ±.25		900.47 ± 1.48	271.15 ± .45	6.481 ±.011	3.033 ±.005	r	388
389	Anthophyllite $Mg_7[Si_8O_{22}](OH)_2$	orth.	Pnma(62)		4	18.61 ±.02	18.01 ±.06	5.24 ±.01				1756.3 ± 7.0	264.4 ± 1.1	6.320 ±.025	2.953 ±.012	r	389
	Framework structure silicates																
390	Microcline $KAlSi_3O_8$	tri.	C$\bar{1}$(2)		4	8.582 ±.002	12.964 ±.005	7.222 ±.002	90.62 ±.10	115.92 ±.10	87.68 ±.10	722.06 ± .67	108.72 ± .10	2.5984 ±.0025	2.560 ±.002	r	390
391	High Sanidine $KAlSi_3O_8$	mon.	C2/m(12)		4	8.615 ±.002	13.031 ±.003	7.177 ±.002		115.98 ±.10		724.28 ± .69	109.05 ± .10	2.6064 ±.0025	2.552 ±.002	r	391
392	Orthoclase $KAlSi_3O_8$*	mon.	C2/m(12)		4	8.562 ±.003	12.996 ±.004	7.193 ±.003		116.02 ±.15		719.25 ± 1.02	108.29 ± .15	2.5883 ±.0037	2.570 ±.004	r	392
393	Fe-Sanidine $KFeSi_3O_8$	mon.	C2/m(12)		4	8.689 ±.008	13.12 ±.01	7.319 ±.007		116.10 ±.30		749.28 ± 2.24	112.81 ± .34	2.6964 ±.0081	2.723 ±.008	r	393
394	Fe-Microcline $KFeSi_3O_8$	tri.	C$\bar{1}$(2)		4	8.68 ±.01	13.10 ±.01	7.340 ±.007	90.75 ±.25	116.05 ±.25	86.23 ±.25	748.09 ± 1.92	112.63 ± .29	2.692 ±.007	2.727 ±.007	r	394
395	Low Albite $NaAlSi_3O_8$	tri.	C$\bar{1}$(2)		4	8.139 ±.002	12.788 ±.003	7.160 ±.002	94.27 ±.10	116.57 ±.10	87.68 ±.10	664.65 ± .60	100.07 ± .09	2.3918 ±.0022	2.620 ±.002	26	395
396	High Albite (Analbite) $NaAlSi_3O_8$	tri.	C$\bar{1}$(2)		4	8.160 ±.002	12.870 ±.003	7.106 ±.002	93.54 ±.10	116.36 ±.10	90.19 ±.10	667.00 ± .60	100.43 ± .09	2.4003 ±.0022	2.611 ±.002	r	396
397	Anorthite $CaAl_2Si_2O_8$	tri.	P$\bar{1}$(2)	primitive cell	8	8.177 ±.002	12.877 ±.003	14.169 ±.003	93.17 ±.02	115.85 ±.02	91.22 ±.02	1338.9 ± .6	100.79 ± .04	2.4090 ±.0011	2.760 ±.001	r	397
398	Synthetic $CaAl_2Si_2O_8$	hex.	P6_3/mcm(193)		2	5.10 ±.02		14.72 ±.02				331.57 ± 2.64	99.85 ± .79	2.386 ±.019	2.786 ±.022	r	398
399	Synthetic $CaAl_2Si_2O_8$	orth.	P2_12_12(18)		2	8.22 ±.02	8.60 ±.02	4.83 ±.01				341.44 ± 1.35	102.82 ± .41	2.457 ±.010	2.706 ±.011	r	399

X-Ray Crystallographic Data of Minerals

	Name and formula	Crystal system	Space group	Structure type	Z	a_o	b_o	c_o	α_o	β_o	γ_o	Cell volume 10^{-24} cm^3	Molar volume cm^3	Molar volume cal bar^{-1}	X-Ray density grams cm^{-3}	Temp. °C	
400	Celsian $BaAl_2Si_2O_8$*	mon.	$I2_1/c(15)$		8	8.627 ±.010	13.045 ±.010	14.408 ±.020		115.20 ±.25		1467.1 ± 4.2	110.45 ± .31	2.640 ±.008	3.400 ±.010	r	400
401	Paracelsian $BaAl_2Si_2O_8$*	mon.	$P2_1/a(14)$		4	8.58 ±.02	9.583 ±.020	9.08 ±.02		90.00 ±.50		746.6 ± 2.9	112.4 ± .4	2.687 ±.010	3.340 ±.013	r	401
402	Banalsite $BaNa_2Al_4Si_4O_{16}$*	orth.			4	8.50 ±.02	9.97 ±.02	16.72 ±.03				1416.9 ± 5.1	213.3 ± .8	5.099 ±.018	3.092 ±.011	r	402
403	Danburite $CaB_2Si_2O_8$*	orth.	Pnam(62)		4	8.04 ±.02	8.77 ±.02	7.74 ±.02				545.8 ± 2.3	82.17 ± .35	1.964 ±.008	2.992 ±.013	r	403
404	Low Nepheline $NaAlSiO_4$	hex.	$C6_3(178)$		8	9.986 ±.005		8.330 ±.004				719.38 ± .80	54.16 ± .06	1.294 ±.002	2.623 ±.003	r	404
405	High Carnegeite $NaAlSiO_4$	cubic			4	7.325 ±.004						393.03 ± .64	59.18 ± .10	1.414 ±.002	2.401 ±.004	750	405
406	Kaliophilite natural $KAlSiO_4$*	hex.	$P6_322(182)$		54	26.930 ±.010		8.522 ±.004				5352.4 ± 4.7	59.69 ± .05	1.427 ±.001	2.650 ±.002	r	406
407	Kaliophilite synthetic $KAlSiO_4$	hex.	$P6_3(173)$ $P6_322(182)$		2	5.180 ±.002		8.559 ±.004				198.89 ± .18	59.89 ± .05	1.431 ±.001	2.641 ±.002	r	407
408	Kalsilite $KAlSiO_4$	hex.	$P6_3(173)$		2	5.1597 ±.0020		8.7032 ±.0030				200.66 ± .17	60.424 ± .051	1.4442 ±.0031	2.618 ±.002	r	408
409	Leucite $KAlSi_2O_6$	tet.	$I4_1/a(100)$		16	13.074 ±.003		13.738 ±.003				2348.23 ± 1.19	88.389 ± .045	2.1126 ±.0011	2.469 ±.001	25	409
410	High Leucite $KAlSi_2O_6$*	cubic	Ia3d(230)		16	13.43 ±.05						2422.3 ±27.1	91.18 ± 1.02	2.179 ±.024	2.394 ±.027	625	410
411	Fe-Leucite $KFeSi_2O_6$	tet.	$I4_1/a(100)$		16	13.205 ±.002		13.970 ±.003				2435.98 ± .91	91.692 ± .034	2.1915 ±.0009	2.695 ±.001	25	411
412	Petalite $LiAlSi_4O_{10}$*	mon.	$P2_1/n(14)$		2	11.32 ±.03	5.14 ±.01	7.62 ±.01		105.90 ±.20		426.41 ± 1.57	128.4 ± .5	3.069 ±.011	2.385 ±.009	r	412
413	Marialite $Na_4Al_3Si_9O_{24}Cl$	tet.	I4/m(87) P4/m(83)		2	12.064 ±.008		7.514 ±.004				1093.6 ± 1.6	329.3 ± .5	7.871 ±.011	2.566 ±.004	r	413
414	Meionite $Ca_4Al_6Si_6O_{24}CO_3$	tet.	I4/m(87) P4/m(83)		2	12.174 ±.008		7.652 ±.015				1134.07 ± 2.68	341.5 ± .8	8.162 ±.019	2.737 ±.007	r	414
	Sheet structure silicates																
415	Muscovite $KAl_2[AlSi_3O_{10}](OH)_2$*	mon.	C2/c(15)	$2M_2$ mica	4	5.203 ±.005	8.995 ±.005	20.030 ±.010		94.47 ±.33		934.57 ± 1.21	140.71 ± .18	3.363 ±.004	2.831 ±.004	r	415
416	Paragonite $NaAl_2[AlSi_3O_{10}](OH)_2$*	mon.	C2/c(15)	$2M_1$ mica	4	5.13 ±.03	8.89 ±.05	19.32 ±.10		95.17 ±.50		877.52 ± 8.47	132.1 ± 1.3	3.158 ±.031	2.893 ±.028	r	416
417	Lepidolite $K_2Al_3Li_2[AlSi_7O_{20}](OH)_4$*	mon.	C2/c(15)	$2M_2$ mica	2	9.2 ±.1	5.3 ±.1	20.0 ±.2		98.00 ±.50		965.7 ±23.2	290.8 ± 7.0	6.950 ±.167	2.698 ±.065	r	417
418	Phlogopite $KMg_3[AlSi_3O_{10}](OH)_2$	mon.	Cm(8)	1M mica	2	5.326 ±.010	9.210 ±.010	10.311 ±.010		100.17 ±.10		497.83 ± 1.19	149.91 ± .36	3.5830 ±.0086	2.784 ±.007	r	418
419	Fluor-phlogopite $KMg_3[AlSi_3O_{10}]F_2$	mon.	Cm(8)	1M mica	2	5.299 ±.005	9.188 ±.005	10.135 ±.005		99.92 ±.10		486.07 ± .60	146.37 ± .18	3.498 ±.004	2.878 ±.004	r	419
420	Annite $KFe_3[AlSi_3O_{10}](OH)_2$	mon.	Cm(8)	1M mica	2	5.391 ±.010	9.350 ±.005	10.313 ±.020		99.70 ±.25		512.40 ± 1.45	154.30 ± .44	3.688 ±.010	3.318 ±.009	r	420
421	Ferriannite $KFe_3[FeSi_3O_{10}](OH)_2$	mon.	C2/m(12)		2	5.430 ±.002	9.404 ±.003	10.341 ±.006		100.07 ±.20		519.92 ± .51	156.56 ± .15	3.7419 ±.0037	3.454 ±.003	r	421
422	Margarite $CaAl_2[Al_2Si_2O_{10}](OH)_2$*	mon.	C2/c(15)	2M mica	4	5.13 ±.02	8.92 ±.03	19.50 ±.05		95.00 ±.50		888.9 ± 5.2	133.8 ± .8	3.199 ±.019	2.975 ±.017	r	422
423	Talc $Mg_3Si_4O_{10}(OH)_2$*	mon.	C2/c(15)	$2M_1$	4	5.287 ±.007	9.158 ±.010	18.95 ±.01		99.50 ±.20		904.94 ± 1.71	136.25 ± .26	3.2565 ±.0062	2.784 ±.005	r	423
424	Pyrophyllite $Al_2Si_4O_{10}(OH)_2$*	mon.	C2/c(15)	$2M_1$	4	5.14 ±.02	8.90 ±.02	18.55 ±.03		99.92 ±.20		835.9 ± 4.0	125.9 ± .6	3.008 ±.015	2.863 ±.014	r	424

X-Ray Crystallographic Data of Minerals

	Name and formula	Crystal system	Space group	Structure type	Z	a_o	b_o	c_o	α_o	β_o	γ_o	Cell volume 10^{-24} cm³	Molar volume cm³	Molar volume cal bar⁻¹	X-Ray density grams cm⁻³	Temp. °C	
425	Minnesotaite $Fe_3Si_4O_{10}(OH)_2$*	mon.	C2/c(15)		4	5.4 ±.1	9.42 ±.04	19.4 ±.1		100.00 ±.50		971.8 ±19.2	146.3 ± 2.9	3.497 ±.069	3.239 ±.064	r	425
426	Dickite $Al_2Si_2O_5(OH)_4$*	mon.	Cc(9)		4	5.150 ±.002	8.940 ±.003	14.736 ±.005		103.58 ±.10		659.4[illegible] ± .4[illegible]	99.30 ± .07	2.3733 ±.0018	2.600 ±.002	r	426
427	Kaolinite $Al_2Si_2O_5(OH)_4$*	tri.	P1(1)		2	5.155 ±.007	8.959 ±.010	7.407 ±.008	91.68 ±.35	104.87 ±.35	89.93 ±.35	330.43 ± .85	99.52 ± .26	2.3785 ±.0062	2.594 ±.007	r	427
428	Nacrite $Al_2Si_2O_5(OH)_4$*	mon.	Cc(9)		4	8.909 ±.010	5.146 ±.010	15.697 ±.020		113.70 ±.25		658.9 ± 2.1	99.21 ± .32	2.3713 ±.0076	2.602 ±.008	r	428
	Zeolites																
429	Analcite $NaAlSi_2O_6.H_2O$	cubic	Ia3d(230)		16	13.733 ±.005						2589.98 ± 2.83	97.49 ± .11	2.3301 ±.0026	2.258 ±.003	r	429
430	Natrolite $Na_2Al_2Si_3O_{10}.2H_2O$*	orth.	Fdd2(43)		8	18.30 ±.02	18.63 ±.02	6.60 ±.01				2250.1 ± 4.8	169.39 ± .37	4.049 ±.009	2.245 ±.005	r	430

HEAT CAPACITY OF ROCK FORMING MINERALS

The units of heat capacity at constant pressure, C_p, in this table are cal kg⁻¹ deg⁻¹. Values of these units are given for several temperatures with the temperatures being in degrees C.

Heat capacity at other temperatures may be calculated by use of the equation $C_p = a + bT - cT^{-5}$. The units of these constants are cal kg⁻¹ deg⁻¹.

Mineral	Heat capacity at various temperatures: -200	0	200	400	800	1200	Constants for heat capacity equation: a	10^3b	10^5c
Albite	–	0.1695	0.236	0.26	0.286	–	1.018	0.187	0.268
Amphibole	–	0.177	0.246	0.27	0.296	–	1.067	1.183	0.281
Apatite	–	0.24	–	–	–	–	–	–	–
Arsenopyrite	–	0.103 at 55°C	–	–	–	–	–	–	–
Asbestos	–	0.195	–	–	–	–	–	–	–
Barite	0.047	0.1076	0.12	0.1315	0.1555	–	0.382	0.253	0
Cassiterite	–	0.0814	0.103	0.115	0.132	–	0.387	0.157	0.007
Chalcopyrite	–	0.129 at 50°C	–	–	–	–	–	–	–
Diamond	–	0.104	0.253	0.328	0.445	–	0.754	1.067	0.454
Dolomite	–	0.222 at 60°C	–	–	–	–	–	–	–
Fluorite	0.0525	0.203	0.213	0.222	0.24	0.263	0.798	0.204	0
Galena	0.034	0.0496	0.0528	0.0562	–	–	0.188	0.007	0
Garnet	–	0.177 at 58°C	–	–	–	–	–	–	–
β-Graphite	–	0.152	0.282	0.348	0.45	–	0.932	0.913	0.4077
Hematite	–	0.146	0.189	0.215	0.258	–	0.64[illegible]	0.420	0.111
Ice	0.156	0.492	–	–	–	–	–	–	–
Kaolinite	–	0.222	0.244	–	–	–	0.80[illegible]	0.463	0.0
Labradorite	–	0.196 at 60°C	–	–	–	–	–	–	–
Magnesite	0.0385	0.207	–	–	–	–	–	–	–
Magnetite	–	0.1435	0.19[illegible]5	0.222	–	–	0.744	0.340	0.177
Mica (mono-crystal)	–	0.208	–	–	–	–	–	–	–
Microcline	–	0.163	0.227	0.248	0.342	–	0.988	0.166	0.263
Oligoclase	–	0.2035 at 60°C	–	–	–	–	–	–	–
Olivine	–	0.189 at 36°C	–	–	–	–	–	–	–
Orthoclase	–	0.146	0.22[illegible]	0.251	0.347	–	0.043	0.124	0.351
Pyrite	0.0179	0.1195	0.14[illegible]	0.165	–	–	0.373	0.466	0
Pyroxene	–	0.18	0.24[illegible]	0.275	–	–	0.973	0.336	0.233
α-Quartz	0.0414	0.167	0.23[illegible]	0.2695	–	–	0.7574	0.607	0.168
β-Quartz	–	–	–	–	0.28	0.3165	0.763	0.383	0
Serpentine	–	0.227	–	–	–	–	–	–	–
Siderite	0.056	0.163	–	–	–	–	–	–	–
Talc	–	0.208 at 59°C	–	–	–	–	–	–	–
Zircon	–	0.146 at 60°C	–	–	–	–	–	–	–

RESISTIVITIES OF SEMICONDUCTING MINERALS (ZERO FREQUENCY)

Native elements	ϱ (ohm-m)
Diamond (C)	2.7
Sulfides	
Argentite, Ag_2S	1.5 to 2.0×10^{-3}
Bismuthinite, Bi_2S_3	3 to 570
Bornite, $Fe_2S_3 \cdot nCu_2S$	1.6 to 6000×10^{-6}
Chalcocite, Cu_2S	80 to 100×10^{-6}
Chalcopyrite, $Fe_2S_3 \cdot Cu_2S$	150 to 9000×10^{-6}
Covellite, CuS	0.30 to 83×10^{-6}
Galena, PbS	6.8×10^{-6} to 9.0×10^{-2}
Haverite, MnS_2	10 to 20
Marcasite, FeS_2	1 to 150×10^{-3}
Metacinnabarite, 4HgS	2×10^{-6} to 1×10^{-3}
Millerite, NiS	2 to 4×10^{-7}
Molybdenite, MoS_2	0.12 to 7.5
Pentlandite, $(Fe, Ni)_9S_8$	1 to 11×10^{-6}
Pyrrhotite, Fe_7S_8	2 to 160×10^{-6}
Pyrite, FeS_2	1.2 to 600×10^{-3}
Sphalerite, ZnS	2.7×10^{-3} to 1.2×10^{4}
Antimony-sulfur compounds	
Berthierite, $FeSb_2S_4$	0.0083 to 2.0
Boulangerite, $Pb_5Sb_4S_{11}$	2×10^{3} to 4×10^{4}
Cylindrite, $Pb_3Sn_4Sb_2S_{14}$	2.5 to 60
Franckeite, $Pb_5Sn_3Sb_2S_{14}$	1.2 to 4
Hauchecornite, $Ni_9(Bi, Sb)_2S_8$	1 to 83×10^{-6}
Jamesonite, $Pb_4FeSb_6S_{14}$	0.020 to 0.15
Tetrahedrite, Cu_3SbS_3	0.30 to 30,000
Arsenic-sulfur compounds	
Arsenopyrite, FeAsS	20 to 300×10^{-6}
Cobaltite, CoAsS	6.5 to 130×10^{-3}
Enargite, Cu_3AsS_4	0.2 to 40×10^{-3}
Gersdorffite, NiAsS	1 to 160×10^{-6}
Glaucodote, (Co, Fe)AsS	5 to 100×10^{-6}
Antimonide	
Dyscrasite, Ag_3Sb	0.12 to 1.2×10^{-6}
Arsenides	
Allemonite, $SbAs_3$	70 to 60,000
Lollingite, $FeAs_2$	2 to 270×10^{-6}
Nicollite, NiAs	0.1 to 2×10^{-6}
Skutterudite, $CoAs_3$	1 to 400×10^{-6}
Smaltite, $CoAs_2$	1 to 12×10^{-6}
Tellurides	
Altaite, PbTe	20 to 200×10^{-6}
Calavarite, $AuTe_2$	6 to 12×10^{-6}
Coloradoite, HgTe	4 to 100×10^{-6}
Hessite, Ag_2Te	4 to 100×10^{-6}
Nagyagite, $Pb_6Au(S, Te)_{14}$	20 to 80×10^{-6}
Sylvanite, $AgAuTe_4$	4 to 20×10^{-6}
Oxides	
Braunite, Mn_2O_3	0.16 to 1.0
Cassiterite, SnO_2	4.5×10^{-4} to 10,000
Cuprite, Cu_2O	10 to 50
Hollandite, $(Ba, Na, K)Mn_8O_{16}$	2 to 100×10^{-3}
Ilmenite, $FeTiO_3$	0.001 to 4
Magnetite, Fe_3O_4	52×10^{-6}
Manganite, $MnO \cdot OH$	0.018 to 0.5
Melaconite, CuO	6000
Psilomelane, $KMnO \cdot MnO_2 \cdot nH_2O$	0.04 to 6000
Pyrolusite, MnO_2	0.007 to 30
Rutile, TiO_2	29 to 910
Uraninite, UO	1.5 to 200

From Carmichael, R. S., ed., *Handbook of Physical Properties of Rocks*, Vol. I, CRC Press, 1982.

MINERALS ARRANGED IN ORDER OF INCREASING VICKERS HARDNESS NUMBERS

Mineral species	Mean	Range	Remarks
Graphite	12	12	
Molybdenite	17	16—19	⊥ to cleavage
	23	21—28	∥ to cleavage
Bismuth	18	16—19	
Tellurbismuth	21	20—21	
Argentite	24	20—30	
Hessite	33	28—41	
Orpiment	38	23—52	
Electrum	40	34—44	
Stromeyerite	41	38—44	
Altaite	51	48—57	
Gold	51	50—52	
Silver	53	48—63	
Realgar	56	53—60	
Digenite	61	56—67	
Arsenic	63	57—69	
Pyrargyrite	71	50—97	⊥ to cleavage
	106	98—126	∥ to cleavage
Covellite	72	69—78	
Galena	76	71—84	
Pyrolusite	76	76	Average hardness ⊥ to fibers
	252	252	Average hardness ∥ to fibers
	279	256—346	Isotropic sections
	292	225—405	Microcrystalline
Stibnite	77	42—109	
Chalcophanite	81	71—85	⊥ to cleavage
	124	103—165	∥ to cleavage
	133	110—178	Isotropic sections
Chalcocite	84	68—98	
Antimony	89	83—99	
Jamesonite	99	96—105	Granular allotriomorphic sections
	113	105—121	Prismatic sections
Bornite	103	97—105	
Bismuthinite	107	92—119	
Miargyrite	110	104—123	
Sylvanite	110	102—125	
Kobellite	116	69—173	
Proustite	123	109—135	
Platinum	126	125—127	
Copper	134	120—143	
Naumannite	148	115—185	
Zincite	154	150—157	⊥ to cleavage
	304	295—318	∥ to cleavage
Pearceite	160	153—164	
Enargite	160	133—185	⊥ to cleavage
	272	245—346	∥ to cleavage
Boulangerite	166	157—183	
Dyscrasite	167	162—178	
Berthierite	171	155—185	
Zinkenite	178	162—207	
Emplectite	191	168—213	∥ to elongation
	222	197—238	⊥ to elongation
Bournonite	192	185—199	
Chalcopyrite	194	186—219	
Blende	198	186—209	
Cuprite	199	192—218	
Stannite	210	197—221	
Cubanite	213	199—228	
Pentlandite	215	202—230	
Tenorite	236	209—254	
Millerite	236	225—256	Isotropic sections
	254	235—280	∥ to elongation
	348	318—376	⊥ to elongation
Pyrrhotite	248	230—259	Anistropic sections
	303	280—318	Isotropic sections
Alabandite	251	240—266	
Coffinite	258	236—333	
Chalcostibite	276	264—285	
Niccolite	336	328—348	Anistropic sections
	446	433—455	Isotropic sections
Tennantite	338	320—361	
Freibergite	345	317—375	
Scheelite	348	285—429	
Tetrahedrite	351	328—367	
Famatinite	363	333—397	
Wolframite	373	357—394	
Manganite	410	367—459	
Carrollite	463	351—566	
Lollingite	486	421—556	⊥ to elongation
	825	739—920	∥ to elongation
Siegenite	524	503—533	
Ullmannite	525	498—542	
Betafite	525	503—560	
Ilmenite	536	519—553	Possible differences in composition
	681	659—703	
Goethite	554	525—620	Microcrystalline
	803	772—824	Coarsely crystalline
Magnetite	560	530—599	
Breithauptite	563	542—584	
Psilomelane	572	503—627	
Hausmannite	587	541—613	
Braunite	595	584—605	

MINERALS ARRANGED IN ORDER OF INCREASING VICKERS HARDNESS NUMBERS (Continued)

Mineral species	Mean	Range	Remarks
Pyrochlore	613	572—665	
Hollandite	620	560—724	
Skutterudite	653	589—724	
Gersdorffite	698	665—743	
Maucherite	704	685—724	
Euxenite	707	599—782	
Rammelsbergite	712	687—778	
Brannerite	720	710—730	
Pitchblende	720	673—803	Fresh specimens, oxidation produces marked decrease in hardness
Lepidocrocite	724	690—782	
Jacobsite	734	724—745	
Davidite	745	707—803	
Hematite	755	739—822	Microcrystalline
	1,000	920—1,062	Coarsely crystalline
Pararammelsbergite	772	762—803	
Coronadite	784	767—813	
Columbite-tantalite	803	724—882	
Uraninite	808	782—839	
Maghemite	946	894—988	
Bixbyite	1,018	1,003—1,033	
Thorianite	1,918	988—1,115	
Cassiterite	1,053	1,027—1,075	
Arsenopyrite	1,094	1,048—1,127	
Bravoite	1,097	1,003—1,288	
Marcasite	1,113	941—1,288	
Glaucodot	1,124	1,071—1,166	
Rutile	1,139	1,074—1,210	
Pyrite	1,165	1,027—1,240	
Cobaltite	1,200	1,176—1,226	
Chromite	1,206	1,195—1,210	

From Carmichael, R. S., ed., *Handbook of Physical Properties of Rocks,* Vol. I, CRC Press, 1982.

SOLUBILITY PRODUCT CONSTANTS

James C. Chang

The following solubility product constants are calculated from the free energies of formation of the substances as solids and those of the aqueous ions at their standard states of m = 1. Thus, for the reaction

$$M_mX_n(s) \leftrightarrows mM^{n+}(aq) + nX^{m-}(aq),$$

$$\Delta G^\circ = m\Delta G_f^\circ(M^{n+}, aq) + n\Delta G_f^\circ(X^{m-}, aq) - \Delta G^\circ(M_mX_n, s)$$

where M_mX_n is the slightly soluble substance, M^{n+} and X^{m-} are the two ions produced in solution by the dissociation of $M_m X_n$. Then the solubility product constant, K_{sp}, is calculated by using the equation

$$\ln K_{sp} = -\frac{\Delta G^\circ}{RT}$$

The values in the following table are for K_{sp} at 25°C.

Substance	Formula	Solubility Product
Aluminum phosphate	$AlPO_4$	9.83×10^{-21}
Barium carbonate	$BaCO_3$	2.58×10^{-9}
Barium chromate	$BaCrO_4$	1.17×10^{-10}
Barium fluoride	BaF_2	1.84×10^{-7}
Barium hydroxide 8-hydrate	$Ba(OH)_2{\cdot}8H_2O$	2.55×10^{-4}
Barium iodate	$Ba(IO_3)_2$	4.01×10^{-9}
Barium iodate 1-hydrate	$Ba(IO_3)_2{\cdot}H_2O$	1.67×10^{-9}
Barium sulfate	$BaSO_4$	1.07×10^{-10}
Bismuth arsenate	$BiAsO_4$	4.43×10^{-10}
Bismuth sulfide	Bi_2S_3	1.82×10^{-99}
Cadmium arsenate	$Cd_3(AsO_4)_2$	2.17×10^{-33}
Cadmium carbonate	$CdCO_3$	6.18×10^{-12}
Cadmium fluoride	CdF_2	6.44×10^{-3}
Cadmium hydroxide	$Cd(OH)_2$	5.27×10^{-15}
Cadmium iodate	$Cd(IO_3)_2$	2.49×10^{-8}
Cadmium oxalate 3-hydrate	$CdC_2O_4{\cdot}3H_2O$	1.42×10^{-8}
Cadmium phosphate	$Cd_3(PO_4)_2$	2.53×10^{-33}
Cadmium sulfide	CdS	1.40×10^{-29}
Calcium carbonate	$CaCO_3$	4.96×10^{-9}
Calcium fluoride	CaF_2	1.46×10^{-10}
Calcium hydroxide	$Ca(OH)_2$	4.68×10^{-6}
Calcium iodate	$Ca(IO_3)_2$	6.47×10^{-6}
Calcium iodate 6-hydrate	$Ca(IO_3)_2{\cdot}6H_2O$	7.54×10^{-7}
Calcium oxalate 1-hydrate	$CaC_2O_4{\cdot}H_2O$	2.34×10^{-9}
Calcium phosphate	$Ca_3(PO_4)_2$	2.07×10^{-33}
Calcium sulfate	$CaSO_4$	7.10×10^{-5}
Cobalt(II) arsenate	$Co_3(AsO_4)_2$	6.79×10^{-29}
Cobalt(II) hydroxide (pink)	$Co(OH)_2$	1.09×10^{-15}
Cobalt(II) hydroxide (blue)	$Co(OH)_2$	5.92×10^{-15}
Cobalt(II) iodate 2-hydrate	$Co(IO_3)_2{\cdot}2H_2O$	1.21×10^{-2}
Cobalt(II) phosphate	$Co_3(PO_4)_2$	2.05×10^{-35}
Copper(I) bromide	$CuBr$	6.27×10^{-9}
Copper(I) chloride	$CuCl$	1.72×10^{-7}
Copper(I) iodide	CuI	1.27×10^{-12}
Copper(I) sulfide	Cu_2S	2.26×10^{-48}
Copper(I) thiocyanate	$CuSCN$	1.77×10^{-13}
Copper(II) arsenate	$Cu_3(AsO_4)_2$	7.93×10^{-36}
Copper(II) iodate 1-hydrate	$Cu(IO_3)_2{\cdot}H_2O$	6.94×10^{-8}
Copper(II) oxalate	CuC_2O_4	4.43×10^{-10}
Copper(II) phosphate	$Cu_3(PO_4)_2$	1.39×10^{-37}
Copper(II) sulfide	CuS	1.27×10^{-36}
Iron(II) carbonate	$FeCO_3$	3.07×10^{-11}
Iron(II) fluoride	FeF_2	2.36×10^{-6}
Iron(II) hydroxide	$Fe(OH)_2$	4.87×10^{-17}
Iron(II) sulfide	FeS	1.59×10^{-19}
Iron(III) hydroxide	$Fe(OH)_3$	2.64×10^{-39}
Iron(III) phosphate 2-hydrate	$FePO_4{\cdot}2H_2O$	9.92×10^{-29}
Lead bromide	$PbBr_2$	6.60×10^{-6}
Lead carbonate	$PbCO_3$	1.46×10^{-13}
Lead chloride	$PbCl_2$	1.17×10^{-5}
Lead fluoride	PbF_2	7.12×10^{-7}
Lead hydroxide	$Pb(OH)_2$	1.42×10^{-20}
Lead iodate	$Pb(IO_3)_2$	3.68×10^{-13}
Lead iodide	PbI_2	8.49×10^{-9}
Lead oxalate	PbC_2O_4	8.51×10^{-10}
Lead sulfate	$PbSO_4$	1.82×10^{-8}
Lead sulfide	PbS	9.04×10^{-29}
Lead thiocyanate	$Pb(SCN)_2$	2.11×10^{-5}
Lithium carbonate	Li_2CO_3	8.15×10^{-4}
Magnesium carbonate	$MgCO_3$	6.82×10^{-6}
Magnesium carbonate 3-hydrate	$MgCO_3{\cdot}3H_2O$	2.38×10^{-6}
Magnesium carbonate 5-hydrate	$MgCO_3{\cdot}5H_2O$	3.79×10^{-6}
Magnesium fluoride	MgF_2	7.42×10^{-11}
Magnesium hydroxide	$Mg(OH)_2$	5.61×10^{-12}
Magnesium oxalate 2-hydrate	$MgC_2O_4{\cdot}2H_2O$	4.83×10^{-6}
Magnesium phosphate	$Mg_3(PO_4)_2$	9.86×10^{-25}
Manganese(II) carbonate	$MnCO_3$	2.24×10^{-11}
Manganese(II) hydroxide	$Mn(OH)_2$	2.06×10^{-13}
Manganese(II) iodate	$Mn(IO_3)_2$	4.37×10^{-7}
Manganese(II) oxalate 2-hydrate	$MnC_2O_4{\cdot}2H_2O$	1.70×10^{-7}
Manganese(II) sulfide	MnS	4.65×10^{-14}
Mercury(I) bromide	Hg_2Br_2	6.41×10^{-23}
Mercury(I) carbonate	Hg_2CO_3	3.67×10^{-17}
Mercury(I) chloride	Hg_2Cl_2	1.45×10^{-18}
Mercury(I) fluoride	Hg_2F_2	3.10×10^{-6}
Mercury(I) iodide	Hg_2I_2	5.33×10^{-29}
Mercury(I) oxalate	$Hg_2C_2O_4$	1.75×10^{-13}
Mercury(I) sulfate	Hg_2SO_4	7.99×10^{-7}
Mercury(I) thiocyanate	$Hg_2(SCN)_2$	3.12×10^{-20}
Mercury(II) hydroxide	$Hg(OH)_2$	3.13×10^{-26}
Mercury(II) iodide	HgI_2	2.82×10^{-29}
Mercury(II) sulfide (black)	HgS	6.44×10^{-53}
Mercury(II) sulfide (red)	HgS	2.00×10^{-53}
Nickel(II) carbonate	$NiCO_3$	1.42×10^{-7}

Substance	Formula	Solubility Product
Nickel(II) hydroxide	$Ni(OH)_2$	5.47×10^{-16}
Nickel(II) iodate	$Ni(IO_3)_2$	4.71×10^{-5}
Nickel(II) phosphate	$Ni_3(PO_4)_2$	4.73×10^{-32}
Nickel(II) sulfide	NiS	1.07×10^{-21}
Palladium(II) sulfide	PdS	2.03×10^{-58}
Palladium(II) thiocyanate	$Pd(SCN)_2$	4.38×10^{-23}
Platinum(II) sulfide	PtS	9.91×10^{-74}
Potassium hexachloroplatinate(IV)	$K_2[PtCl_6]$	7.48×10^{-6}
Potassium perchlorate	$KClO_4$	1.05×10^{-2}
Silver(I) acetate	$AgC_2H_3O_2$	1.94×10^{-3}
Silver(I) arsenate	Ag_3AsO_4	1.03×10^{-22}
Silver(I) bromate	$AgBrO_3$	5.34×10^{-5}
Silver(I) bromide	AgBr	5.35×10^{-13}
Silver(I) carbonate	Ag_2CO_3	8.45×10^{-12}
Silver(I) chloride	AgCl	1.77×10^{-10}
Silver(I) chromate	Ag_2CrO_4	1.12×10^{-12}
Silver(I) cyanide	AgCN	5.97×10^{-17}
Silver(I) iodate	$AgIO_3$	3.17×10^{-8}
Silver(I) iodide	AgI	8.51×10^{-17}
Silver(I) oxalate	$Ag_2C_2O_4$	5.40×10^{-12}
Silver(I) phosphate	Ag_3PO_4	8.88×10^{-17}
Silver(I) sulfate	Ag_2SO_4	1.20×10^{-5}
Silver(I) sulfide (α-form)	Ag_2S	6.69×10^{-50}
Silver(I) sulfide (β-form)	Ag_2S	1.09×10^{-49}
Silver(I) sulfite	Ag_2SO_3	1.49×10^{-14}
Silver(I) thiocyanate	AgSCN	1.03×10^{-12}
Strontium arsenate	$Sr_3(AsO_4)_2$	4.29×10^{-19}
Strontium carbonate	$SrCO_3$	5.60×10^{-10}
Strontium fluoride	SrF_2	4.33×10^{-9}
Strontium iodate	$Sr(IO_3)_2$	1.14×10^{-7}
Strontium iodate 1-hydrate	$Sr(IO_3)_2 \cdot H_2O$	3.58×10^{-7}
Strontium iodate 6-hydrate	$Sr(IO_3)_2 \cdot 6H_2O$	4.65×10^{-7}
Strontium sulfate	$SrSO_4$	3.44×10^{-7}
Tin(II) hydroxide	$Sn(OH)_2$	5.45×10^{-27}
Tin(II) sulfide	SnS	3.25×10^{-28}
Zinc arsenate	$Zn_3(AsO_4)_2$	3.12×10^{-28}
Zinc carbonate	$ZnCO_3$	1.19×10^{-10}
Zinc carbonate 1-hydrate	$ZnCO_3 \cdot H_2O$	5.41×10^{-11}
Zinc fluoride	ZnF_2	3.04×10^{-2}
Zinc hydroxide (γ-form)	$Zn(OH)_2$	6.86×10^{-17}
Zinc hydroxide (β-form)	$Zn(OH)_2$	7.71×10^{-17}
Zinc hydroxide (ε-form)	$Zn(OH)_2$	4.12×10^{-17}
Zinc iodate	$Zn(IO_3)_2$	4.29×10^{-6}
Zinc oxalate 2-hydrate	$ZnC_2O_4 \cdot 2H_2O$	1.37×10^{-9}
Zinc sulfide	ZnS	2.93×10^{-25}

PROPERTIES OF RARE EARTH METALS

F. H. Spedding

Symbol	M. P. °C[1]	B. P. °C[1]	Heat of Vaporization $\Delta H_{v,o}$ Kcal/g atm[2]	Density	Atomic Vol. (cm³/mole)	Metallic Radius Å	Electrical Resistivity Polycrystalline Wire 298°K (ohm-cm x 10^{-6})[3]	Residual Resistivity Wire 4.2°K (ohm-cm x 10^{-6})[3]	Compressibility cm²/kg x 10^{-6} **
Sc	1541	2831	89.9	2.989	15.041	1.640	52	3	2.26
Y	1522	3338	101.3	4.469	19.894	1.801	59	2	2.68
La	921	3457	103.1	6.145	22.603	1.879	61-80*	SC	4.04
Ce	799	3426	101.1	γ=6.767	20.400	1.820	70-80	10	4.10
				β=6.657	21.049				
Pr	931	3512	85.3	6.773	20.805	1.828	68	1	3.21
Nd	1021	3068	78.5	7.007	20.585	1.821	65	7	3.0
Pm	1168	2700 est.							
Sm	1077	1791	49.2	7.520	20.001	1.804	91	7	3.34
Eu	822	1597	41.9	5.243	28.981	1.984	91	1	8.29
			ΔH°298						
Gd	1313	3266	95.3	7.900	19.904	1.801	127	1	2.56
Tb	1356	3123	93.4	8.229	19.312	1.783	114	4	2.45
Dy	1412	2562	70.0	8.550	19.006	1.774	100	5	2.55
Ho	1474	2695	72.3	8.795	18.753	1.766	88	3	2.47
Er	1529	2863	76.1	9.066	18.450	1.757	71	3	2.39
Tm	1545	1947	55.8	9.321	18.124	1.746	74	3	2.47
Yb	819	1194	36.5	6.965	24.843	1.939	28	2	7.39
Lu	1663	3395	102.2	9.840	17.781	1.735	60	2	2.38

* Crystal usually mixture of α hcp and fcc lattice. SC - Superconductor

** Best values in author's opinion. Many numbers are taken from P. W. Bridgman's publication.

[1] Corrected for new temperature scale. Best values Ames Laboratory 2/1/72.

[2] Values from Thermodynamic Properties of Metals and Alloys (Review) R. Hultgren, R. Orr and K. Kelley. Supplements 1967.

[3] Weighted average of original papers 1/1/72.

Symbol	Crystal Structure at Room Temperature 25°C		Allotropic Forms Transition Temperatures Expressed in °C	
Sc	Hex. (to 1335°)	a=3.3088Å, c=5.2680Å	bcc(above 1335°)	
Y	Hex. (to 1478°)	a=3.6482Å, c=5.7318Å	bcc(above 1478°)	a=4.08Å
La	Hex. (to 310°) (usually contains fcc also)	a=3.7740Å, c=12.171Å	fcc(310°-865°) bcc(above 865°)	 a=4.26Å
Ce	fcc(~ 0° to 726°)	a=5.160Å	fcc(below-157° on cooling) (up to -94° on heating)	a=4.85Å
			Hex. (below-23° on cooling) (up to 168° on heating)	a=3.68Å c=11.92Å
Pr	Hex. (to 795°)	a=3.6721Å, c=11.832Å	bcc(above 795°)	a=4.12Å
Nd	Hex. (to 863°)	a=3.6583Å, c=11.7966Å	bcc(above 863°)	a=4.13Å
Sm	*Rhom (to 926°)	a=8.9834Å, α=23° 49.5′	bcc(above 926°)	a=4.07Å
Eu	bcc	a=4.5827Å		
Gd	Hex. (to 1235°)	a=3.6336Å, c=5.7810Å	bcc(above 1235°)	a=4.05Å†
Tb	Hex. (to 1289°)	a=3.6055Å, c=5.6966Å	bcc(above 1289°)	a=4.02Å †
Dy	Hex. (to 1381°)	a=3.5915Å, c=5.6501Å	bcc(above 1381°)	a=3.98Å †
Ho	Hex.	a=3.5778Å, c=5.6178Å	Not present pure metals	
Er	Hex.	a=3.5592Å, c=5.5850Å	Not present pure metals	
Tm	Hex.	a=3.5375Å, c=5.5540Å	Not present pure metals	
Yb	Hex. (to 795°)	a=5.4848Å,	bcc(above 795°)	a=4.44Å
Lu	Hex.	a=3.5052Å, c=5.5494Å	Not present pure metals	

† Extrapolated from magnesium alloy studies. Hex. refers to close packed hexagonal. La, Ce, Pr, and Nd have a stacking order ABAC, the other rare earths ABAB. fcc refers to close packed face centered cubic with a stacking order ABC, ABC, bcc refers to body centered cubic. These high temperature forms are very soft and deform very easily.

* The rhombic Sm cell can be expressed as hexagonal with a=3.6290Å, c=26.207 and possess a stacking order ABABCBCAC

DENSITY OF LIQUID ELEMENTS

Gernot Lang

Data in this table consist of those published in various places in the literature. Generally, there has been excellent agreement among values obtained by different methods of measurement. Temperatures are in degrees C and densities in grams per cubic centimeter. Grams per cc × 62.427961 = pounds per cu ft.

Element:	Purity:	ρ_{mp}	ρ_{t_1}		ρ_{t_2}		ρ_{t_3}		Atm.:	Method:	Ref.
			t	ρ	t	ρ	t	ρ			
Ag	—	9.32	1100	9.17	1200	9.07	—	—	—	IBF	74
Ag	—	9.33	1100	9.20	1200	9.10	1300	9.00	—	DBF	33
Ag	99.9	—	1100	9.170	1200	9.070	1300	8.980	—	IBF	Q
Ag	—	9.285	1100	9.15	1200	9.07	1300	8.97	—	IBF	24
Ag	—	9.345	1100	9.20	—	—	—	—	H_2	Bubble pressure	46
Ag	—	9.33 ±0.01	1100	9.18	1200	9.08	1300	8.97	Ar	Bubble pressure	47
Ag	—	9.346	1100	9.22	1200	9.13	1300	9.04	—	DBF	39
Al	99.4	2.41	—	—	—	—	—	—	—	BF	62
Al	99.4	2.38_4	—	—	—	—	—	—	—	Pycnometer	15
Al	99.996	2.368	700	2.357	800	2.332	900	2.304	—	BF	25
Al	99.99	2.39	—	—	—	—	—	—	—	Pycnometer	60
Al	99.997	2.39	677	2.380	807	2.334	912	2.292	Ar	Bubble pressure	11
Au	99.96	17.28	1100	17.24	1200	17.12	1300	16.99	H_2	IBF	24
Au	—	17.361	1100	17.221	1200	17.099	1300	16.950	—	IBF	P
B	99.8	2.08 ±0.03	—	—	—	—	—	—	vac.	Weighing and vol. determin.	80
Ba	—	3.325		$\rho_t = 3.476 - (2.14 \cdot 10^{-4})t$ (t°C)					Ar	DBF	1
Ba	—	3.320		$\rho_t = 3.847 - 5.26 \cdot 10^{-4}T$ (T°K)					—	calculated	K
Be	—	—	1500	1.42 ±0.04		—	—	—	—	calculated	22
Bi	—	10.07	300	10.03	400	9.91	500	9.78	—	Manometer pressure	31
Bi	—	10.02	400	9.87	600	9.62	800	9.40	—	Dilatometer	7
Bi	—	10.07	400	9.90	500	9.78	600	9.66	—	Dilatometer	53
Bi	—	10.04	400	9.91	600	9.67	800	9.44	—	DBF	34
Bi	—	—	400	9.86	600	9.62	700	9.49	N_2	Bubble pressure	78
Bi	99.94	10.03	400	9.88	600	9.62	700	9.51	—	DBF	66
Bi	99.90	—	800	9.43	900	9.28	1000	9.15	H_2	Bubble pressure	57
Bi	—	10.02	300	9.98	400	9.85	500	9.73	—	Dilatometer	30
Bi	99.98	10.07	500	9.78	700	9.53	900	9.30	Ar	Bubble pressure	49
Bi	—	10.057	700	9.51 ±0.03		—	—	—	—	—	L
Ca	—	1.365		$\rho_t = 1.613 - 2.21 \cdot 10^{-4}T$ (T°K)					—	calculated	K
Cd	—	8.02		$\rho_t = 8.02 - 0.00110\,(t - 320)$ (t°C)					—	Manometer pressure	31
Cd	99.97	—	340	8.009	400	7.942	500	7.821	N_2	Bubble pressure	28
Cd	—	—	385	7.94	479	7.83	560	7.74	Ar	Bubble pressure	76
Co	—	—	1600	8.08 – 8.13		—	—	—	—	Drop volume	59
Co	—	—	1500	ca. 8.05	—	—	—	—	—	BF	26
Co	99.99	7.67	1500	7.66	1600	7.54	1700	7.42	Ar	Bubble pressure	50
Co	—	—	1500	7.70	—	—	—	—	—	—	81
Co	99.53	7.76		$\rho_t = 9.51 - 9.88 \cdot 10^{-4}T$ (T°K)					vac.	Sessile drop (mp – 2200)	U
Cr	—	—	1950	6.00 ±0.13		—	—	—	vac.	Sessile drop	18
Cr	—	6.46		—	—	—	—	—	—	calculated	3
Cs	—	1.843	110	1.800	310	1.691	510	1.575	—	Pycnometer	H
Cs	—	1.851		$\rho_t = 1.853 - 5.71 \cdot 10^{-4}t$ (t°C)					—	Pycnometer (100–750°C)	F
Cu	—	7.99	1100	7.97	1200	7.81	1300	7.66	—	IBF	6
Cu	—	7.96_2	1100	7.94	1200	7.77	1300	7.62	—	Dilatometer	85
Cu	—	7.92_4	1100	7.91	1300	7.76	1500	7.61	—	DBF	84
Cu	99.99	7.940	1100	7.924	1200	7.846	1300	7.764	—	IBF	Q
Cu	—	—	1100	8.10	—	—	—	—	—	Drop volume	43
Cu	—	7.87_5	1100	7.86	1300	7.70	1500	7.55	—	DBF	45
Cu	—	—	1100	7.90	—	—	—	—	—	Pycnometer	52
Cu	—	7.99_2	1100	7.98	1300	7.81	1500	7.66	Ar	DBF	8
Cu	—	—	1100	8.07	—	—	—	—	—	Bubble pressure	Beer in 49
Cu	—	8.090		$\rho_t = 9.370 - 9.442 \cdot 10^{-4}T$ (T°K)					—	Drop volume	Mehairy in 49
Cu	—	8.03	1100	8.02	1300	7.86	1500	7.70	Ar	Bubble pressure	49
Fe	—	7.13	—	—	—	—	—	—	—	DBF	75
Fe	—	7.24	—	—	—	—	—	—	vac.	Drop volume	35
Fe	—	—	1550	7.01	—	—	—	—	air	Casting	79
Fe	—	7.15	—	—	—	—	—	—	—	DBF	42
Fe	—	6.99		$\rho_t = 8.523 - 8.358 \cdot 10^{-4}T$ (T°K)					Ar	DBF	38
Fe	Carbonyl	—	1550	7.189	—	—	—	—	—	—	27
Fe	—	7.015		$\rho_t = 8.618 - 8.83 \cdot 10^{-4}T$ (T°K)					Ar	DBF	C
Fe	—	—	1555	6.98	—	—	—	—	—	Bubble pressure	Beer in 49
Fe	99.96	7.03_5	1550	7.01	1600	6.94	1700	6.80	Ar	Bubble pressure	49
Fe	—	—	1550	7.13	—	—	—	—	—	—	70
Fe	99.9	7.02		$\rho_t = 8.50 - 8.17 \cdot 10^{-4}T$ (T°K)					vac.	Sessile drop	U

Element:	Purity:	ρ_{mp}	ρ_{t_1} t	ρ_{t_1} ρ	ρ_{t_2} t	ρ_{t_2} ρ	ρ_{t_3} t	ρ_{t_3} ρ	Atm.:	Method:	Ref.
Ga	—	6.20 ±0.01	—	—	—	—	—	—	—	—	E
Ge	—	5.52	1000	5.50	1100	5.44	1200	5.39	—	Volumetric measurement	58
Ge	—	5.57$_5$	1000	5.56	1100	5.51	1200	5.45	N_2	Pycnometer	40
Ge	99.990	5.49	1000	5.46	1200	5.36	1500	5.21	Ar	Bubble pressure	49
Hf	—	12.0	—	—	—	—	—	—	—	calculated	3
Hg	—	—	0	13.5951	20	13.5457	100	13.3514	—	—	—
In	—	—	231	6.99	302	6.93	421	6.84	Ar	Bubble pressure	76
Ir	—	20.0	—	—	—	—	—	—	—	Calculated	3
K	—	0.826	$\rho_t = 0.826 - 0.000222\ (t - 62.4)$ (t°C)						—	—	72
K	—	0.819	$\rho_t = 0.819 - 0.238 \cdot 10^{-3}\ (t - 64)$ (t°C)						—	Pycnometer (64–1400°C)	D
K	—	0.828	110	0.817	310	0.772	510	0.724	—	Pycnometer	H
Li	—	0.515	$\rho_t = 0.515 - 0.101 \cdot 10^{-3}\ (t - 200)$ (t°C)						—	Pycnometer (200–1600°C)	D
Li	—	0.518	$\rho_t = 0.5368 - 1.021 \cdot 10^{-4}\ t$ (t°C)						—	Pycnometer (400–1125°C)	F
Mg	—	—	700	1.575	800	1.555	—	—	—	DBF	67
Mg	99.5	1.585	700	1.570	800	1.550	900	1.525	—	IBF	N
Mg	—	1.57	—	—	—	—	—	—	—	Estimated	17
Mg	—	1.590	$\rho_t = 1.834 - 2.67 \cdot 10^{-4}\ T$ (T°K)						Ar	DBF	B
Mn	—	—	1440	5.84 ±2%	—	—	—	—	He	Volumetric method	M
Mn	—	6.43	—	—	—	—	—	—	—	—	71
Mo	99.7	9.33	—	—	—	—	—	—	vac.	Drop volume	64
Mo	—	9.35	—	—	—	—	—	—	—	calculated	3
Na	—	0.938$_5$	$\rho_t = 0.938_5 - 0.000260\ (t - 96.5)$ (t°C)						—	—	72
Na	—	—	$\rho_t = 0.938 - 1.9 \cdot 10^{-4}\ T$ (T°K)						—	Bubble pressure	82
Na	—	0.927	$\rho_t = 0.927 - 0.238 \cdot 10^{-3}\ (t - 100)$ (t°C)						—	Pycnometer	D
Na	—	0.928	410	0.854	610	0.806	810	0.754	—	Pycnometer	H
Nb, Cb	—	7.83	—	—	—	—	—	—	—	calculated	3
Ni	—	—	1500	ca. 8.04	—	—	—	—	—	BF	26
Ni	—	7.905	$\rho_t = 9.908 - 11.598 \cdot 10^{-4}\ T$ (T°K)						Ar	DBF	C
Ni	99.85	7.77	1500	7.71$_5$	1600	7.59$_5$	1700	7.48	Ar	Bubble pressure	49
Ni	99.99	7.78	$\rho_t = 7.78 - 0.006\ (t - 1453)$ (t°C)						vac.	Sessile drop	21
Ni	—	—	1500	7.78	—	—	—	—	—	—	81
Ni	99.95	7.91	$\rho_t = 9.51 - 10.00 \cdot 10^{-4}\ T$ (T°K)						vac.	Sessile drop	U
Os	—	20.1	—	—	—	—	—	—	—	calculated	3
Pb	—	10.71	$\rho_t = 10.71 - 0.00139\ (t - 327)$ (t°C)						—	Manometer pressure	31
Pb	99.98	—	340	10.57	400	10.49	440	10.43	N_2	Bubble pressure	28
Pb	—	—	568	10.380	640	10.294	720	10.201	—	DBF	O
Pb	—	—	400	10.56	500	10.43	700	10.17	N_2	Bubble pressure	78
Pb	—	—	437	10.52	623	10.28	705	10.16	Ar	Bubble pressure	76
Pb	99.9923	10.678	$\rho_t = 10.678 - 13.174 \cdot 10^{-4}\ T$ (T°K)						Ar	DBF	A
Pd	—	10.7	—	—	—	—	—	—	—	calculated	20
Pd	—	10.7	—	—	—	—	—	—	—	calculated	3
Pd	99.95	—	1600	10.43	1700	10.31	1800	10.18$_5$	Ar	Bubble pressure	49
Pt	99.84	19.7 ±0.25	—	—	—	—	—	—	vac.	Drop volume	19
Pt	99.999	—	1800	18.82 ±0.02	—	—	—	—	Ar	Sessile drop	44
Pt	—	18.91	—	—	—	—	—	—	Ar	Bubble pressure	48
Pt	—	—	1800	18.82	1850	18.67$_5$	1875	18.60$_5$	Ar	Bubble pressure	49
Pu	—	—	$\rho_t = 17.57 - 1.45 \cdot 10^{-3}\ t$ (t°C) ±0.21						—	(655–P60°C)	T
Pu	99.95	16.623	699	16.548	824	16.370	950	16.185	—	Pycnometer	32
Rb	—	1.48	—	—	—	—	—	—	—	estimated	56
Rb	—	1.463	$\rho_t = 1.481 - 4.51 \cdot 10^{-4}\ t$ (t°C)						—	Pycnometer	F
Rb	—	—	45.5	1.4505	47.4	1.4498	49.3	1.4469	vac.	Pycnometer	G
Re	99.4	18.9	—	—	—	—	—	—	vac.	Drop volume	64
Re	—	18.7	—	—	—	—	—	—	—	calculated	3
Rh	—	10.65	—	—	—	—	—	—	—	calculated	20
Rh	—	11.1	—	—	—	—	—	—	—	calculated	3
Ru	—	10.9	—	—	—	—	—	—	—	calculated	3
S	—	1.819	120–160°C: $\rho_t = 1.901 - 8.00 \cdot 10^{-4}\ t$ (t°C)						N_2	Bubble pressure	61
Sb	—	6.49	700	6.45	800	6.39	1000	6.27	—	IBF	7
Sb	99.52	—	650	6.530	700	6.509	800	6.424	N_2	Bubble pressure	28
Sb	—	—	700	6.41	800	6.38	900	6.35	H_2	Bubble pressure	78

DENSITY OF LIQUID ELEMENTS (Continued)

Element:	Purity:	ρ_{mp}	ρ_{t_1}		ρ_{t_2}		ρ_{t_3}		Atm.:	Method:	Ref.
			t	ρ	t	ρ	t	ρ			
Sb	—	—	650	6.52	700	6.50	—	—	H_2	Bubble pressure	23
Sb	—	—	728	6.44	827	6.38	917	6.32	Ar	Bubble pressure	76
Sb	99.9	6.50	650	6.49	700	6.45	—	—	—	DBF	66
Sb	99.992	6.483	$\rho_t = 6.596 + 2.022 \cdot 10^{-4}\,T - 3.629 \cdot 10^{-7}\,T^2$ (T°K)							DBF, Bubble pressure	37
Sb	99.6	6.46_5	700	6.42	900	6.30	1200	6.13	Ar	Bubble pressure	49
Se	—	3.987	$\rho_t = 3.987 - 16 \cdot 10^{-4}\,(T - 493)$ (T°K)						—	—	13
Se	—	3.985	$\rho_t = 3.985 - 15.5 \cdot 10^{-4}\,(T - 490)$ (T°K)						—	Pycnometer	10
Se	—	4.06	$\rho_t = 4.06 - 5 \cdot 10^{-4}\,(T - 491)$ (T°K)						—	—	4
Se	—	3.984	—	—	—	—	—	—	—	—	I
Se	—	4.01	250	3.97	300	3.91	400	3.79	—	BF	51
Si	—	—	1550	2.54	—	—	—	—	—	—	14
Si	99.9	2.52_5	1450	2.51	1500	2.49	1600	2.46	Ar	Bubble pressure	49
Sn	—	6.988	300	6.94	500	6.81	800	6.64	—	DBF	12
Sn	—	6.98	—	—	—	—	—	—	—	DBF	63
Sn	—	7.01	$\rho_t = 7.01 - 0.00074\,(t - 232)$ (t°C)						—	Manometer pressure	31
Sn	—	6.97	300	6.92	500	6.78	800	6.57	—	IBF	74
Sn	—	6.896	300	6.93	500	6.78	600	6.71	—	Dilatometer	53
Sn	—	6.966	300	6.92	600	6.72	1200	6.34	—	DBF	84
Sn	—	6.983	300	6.93	400	6.86	500	6.79	—	Pycnometer	73
Sn	—	7.00	300	6.95	500	6.80	600	6.74	N_2	Bubble pressure	65
Sn	—	—	625	6.67	—	—	—	—	N_2	Bubble pressure	78
Sn	—	6.99	300	6.94	400	6.87	—	—	H_2	Bubble pressure	23
Sn	—	6.993	$\rho_t = 6.98 - 0.00074\,(t - 250)$ (t°C)						—	Manometer pressure	69
Sn	—	6.968 ±0.005	300	6.93	400	6.87	500	6.79	—	Dilatometer	30
Sn	—	6.968	300	6.96	400	6.84	500	6.77	Ar	Bubble pressure	83
Sn	—	6.978 ±0.022	300	6.93	600	6.70	1200	6.29	Ar	Bubble pressure	49
Ta	—	15.0	—	—	—	—	—	—	—	calculated	3
Te	—	5.75	460	5.75	500	5.74	600	5.70	—	Pycnometer	40
Te	—	—	460	5.58	600	5.53	—	—	N_2	Bubble pressure	77
Te	99.9999	5.797	460	5.79	600	5.72	700	5.66	Ar	Bubble pressure	49
Ti	98.7	4.11 ±0.08	—	—	—	—	—	—	—	Capillary method	16
Ti	—	4.15	—	—	—	—	—	—	—	calculated	3
Tl	—	11.29	412	11.13	504	11.01	651	10.77	Ar	Bubble pressure	76
U	—	17.907	$\rho_t = 19.356 - 10.328 \cdot 10^{-4}\,T$ (T°K)						Ar	DBF	29
V	—	5.55	—	—	—	—	—	—	—	calculated	3
W	—	17.6	—	—	—	—	—	—	—	estimated	9
W	99.8	17.7	—	—	—	—	—	—	vac.	Drop volume	64
W	—	17.5	—	—	—	—	—	—	—	calculated	3
Zn	—	6.59	$\rho_t = 6.59 - 0.00097\,(t - 419)$ (t°C)						—	Manometer pressure	31
Zn	—	—	600	6.35	—	—	—	—	—	IBF	7
Zn	—	6.56_2	500	6.47	600	6.37	700	6.29	—	Dilatometer	53
Zn	—	6.55_1	500	6.47	600	6.38	700	6.29	—	Pycnometer	73
Zn	—	6.64_5	500	6.56	600	6.45	800	6.25	—	DBF	67
Zn	—	6.64_4	500	6.58	600	6.49	800	6.22	—	DBF	68
Zn	—	6.64	500	6.57	600	6.50	700	6.43	N_2	Bubble pressure	65
Zn	—	—	500	6.55	—	—	—	—	H_2	Bubble pressure	78
Zn	—	6.64_5	—	—	—	—	—	—	—	Pycnometer	41
Zn	99.995	6.562	500	6.417	600	6.370	700	6.259	—	IBF	25
Zn	—	6.57_5	500	6.49	600	6.37	700	6.26	Ar	Bubble pressure	83
Zn	99.999	6.57_7 ±0.012	500	6.48	600	6.37	700	6.27	Ar	Bubble pressure	49
Zr	—	5.80	—	—	—	—	—	—	—	calculated	3

DENSITY OF LIQUID ELEMENTS

SUPPLEMENTARY TABLE

Gernot Lang

Element	Purity	ρ_{mp}	ρt_1		ρt_2		ρt_3		Atm.	Method	Ref.
			t°C	ρ	t°C	ρ	t°C	ρ			
Ag		9.36	$\rho_t = 9.36 - 0.00114\,(t - t_{mp})$ (t°C)							Bubble pressure	110
Ag	99.95		$\rho_t = 9.318 - 1.08 \cdot 10^{-3}\,(t - 1000) \pm 0.006$ (1000–1200°C)						H_2	Bubble pressure	81
Ag	99.999	9.31	$\rho_t = 10.31 - 1.05 \cdot 10^{-3} \cdot t$(t°C) (960–1150)						Ar	Sessile drop	88
Al	99.99	2.39	$\rho_t = 2.60 - 3.2 \cdot 10^{-4}\,t$ (t°C) (900–1750°C)						He	Drop volume	87
Al	99.999	2.365	$\rho_t = 2.576 - 3.2 \cdot 10^{-4}\,t$ (t°C)						He	Sessile drop	108
Al	99.998	2.375	$\rho_t = 2.375 - 5.4 \cdot 10^{-4}\,(t - 660)$ (t°C)						Ar	Sessile drop	97

DENSITY OF LIQUID ELEMENTS (*Continued*)

Element	Purity	ρ_{mp}	ρ_{t_1} t°C	ρ_{t_1} ρ	ρ_{t_2} t°C	ρ_{t_2} ρ	ρ_{t_3} t°C	ρ_{t_3} ρ	Atm.	Method	Ref.
Ba	99.5	3.32		$\rho_t = 3.59 - 2.74 \cdot 10^{-4} T$				(T°K)		Bubble pressure	90
Bi	99.9999	10.05		$\rho_t = 10.05 - 1.41 \cdot 10^{-3} (t - t_{mp})$				(t°C)	N_2	Volume determin.	96
Bi	99.98	10.04		$\rho_t = 10.04 - 1.44 \cdot 10^{-3} (t - t_{mp})$				(t°C)	N_2	Volume determin.	96
Bi	99.999	10.114		$\rho_t = 10.406 - 1.078 \cdot 10^{-3} t$ (271–414)				(t°C)	N_2	DBF	89
Bi	–	10.031		$\rho_t = 10.336 - 12.367 \cdot 10^{-4} t$				(t°C)	vac.	Pycnometer	92
Ca	–	1.365		$\rho_t = 1.613 - 2.21 \cdot 10^{-4} T$				(T°K)		Bubble pressure	90
Cd	99.999	7.996		$\rho_t = 8.388 - 12.205 \cdot 10^{-4} t$				(t°C)	Ar	Pycnometer	91
Co	99.53	7.78		$\rho_t = 9.5_7 - 10.1_7 \cdot 10^{-4} T$ (1780–2200)				(T°K)	Ar	Drop volume	103
Co	99.67	7.75		$\rho_t = 9.7_1 - 1.1_1 \cdot 10^{-3} T$				(T°K)	Ar	Bubble pressure	109
Cs	–	1.86		$\rho_t = 1.84/1 + 1.1755 \cdot 10^{-4} (t - 82.4) + 7.656 \cdot 10^{-8} (t - 82.4)^2$ (t°F, 29–704°C)					Ar	Pycnometer	86
Cs	–	–	40	1.840	271	1.711	376	1.655	Ar	Dilatometer	106
Fe	–	7.06		$\rho_t = 7.05 - 7.3 \cdot 10^{-4} (t - 1550)$				(t°C)	H_2, He	Sessile drop	101
Fe	99.9	7.03		$\rho_t = 8.5_7 - 8.5_3 \cdot 10^{-4} T$ (1800–2150)				(T°K)	Ar	Drop volume	103
Fe	99.98	7.05		$\rho_t = 8.7_8 - 0.95_8 \cdot 10^{-3} T$				(T°K)	Ar	Bubble pressure	109
Fr	–		100	2.29						calculated	100
Ga	–	6.1136		$\rho_t = 6.11564 - 7.37437 \cdot 10^{-4} t + 1.37767 \cdot 10^{-7} t^2$ (t°C) (50–600°C)						Pycnometer	98
Hg	–	–	0	13.5951	20	13.5457	100	13.3514		–	105
In	99.999	7.032		$\rho_t = 7.153 - 0759 \cdot 10^{-3} t$				(t°C)	N_2	DBF (160–532)	89
In	99.999	7.016		$\rho_t = 7.146 - 8.362 \cdot 10^{-4} t$				(t°C)	Ar	Pycnometer	91
Mg	99.5		959	1.52	1053	1.47	–	–		Bubble pressure	90
Mo	–	9.1							vac.	Pendant drop and drop weight	95
Nb, Cb	–	7.6							vac.	Pendant drop and drop weight	95
Ni	–	7.82		$\rho_t = 7.81 - 8.7 \cdot 10^{-4} (t - 1460)$				(t°C)	H_2, He	Sessile drop	U
Ni	–	7.65							vac.	Pendant drop and drop weight	95
Ni	99.95	7.95		$\rho_t = 9.8_1 - 10.8_0 \cdot 10^{-4} T$ (1890–2150)				(T°K)	Ar	Drop volume	103
Pb	99.999	10.330		$\rho_t = 10.650 - 9.8 \cdot 10^{-4} t$ (1000–1600)				(t°C)	He	Sessile drop	108
Pb	99.997	10.660		$\rho_t = 11.060 - 12.220 \cdot 10^{-4} t$				(t°C)		Pycnometer	107
Pd	99.998	10.379		$\rho_t = 12.193 - 11.69 \cdot 10^{-4} t$ (MP–1700)				(t°C)	He	Sessile drop	108
Rb	99.4	1.385		$\rho_t = 1.55643 - 2.6511 \cdot 10^{-4} t - 6.26779/t$ (t°F, 66–1076°C)					Ar	Dilatometer	106
Rb	–	1.484		$\rho_t = 1.472/1 + 1.3309 \cdot 10^{-4} (t - 102) + 5.2106 \cdot 10^{-8} (t - 102)^2$ (t°F, 39–730°C)					Ar	Pycnometer	86
Sb	99.999	6.452		$\rho_t = 6.818 - 5.8 \cdot 10^{-4} t$ (1000–1600°C)				(t°C)	He	Sessile drop	108
Sb	99.999	6.535		$\rho_t = 6.962 - 0.673 \cdot 10^{-3} t$ (635–745°C)				(t°C)	N_2	D B F	89
Sb	–	6.493		$\rho_t = 6.902 - 6.486 \cdot 10^{-4} t$ (MP–746°C)				(t°C)	Ar	Pycnometer	93
Si	99.9999		1500	2.46					Ar		94
Sn	99.999		246	6.964	336	6.900	400	6.854	vac.	Pycnometer	111
Sn	99.999	7.01		$\rho_t = 7.16 - 6.3 \cdot 10^{-4} t$				(t°C)		Bubble pressure	104
Sn	99.999	6.981		$\rho_t = 7.135 - 0.663 \cdot 10^{-3} t$ (MP–438°C)				(t°C)	N_2	D B F	89
Sn	99.999	6.973		$\rho_t = 7.139 - 7.125 \cdot 10^{-4} t$				(t°C)		Pycnometer	107
Sr	99.5	2.375		$\rho_t = 2.648 - 2.62 \cdot 10^{-4} T$				(T°K)		Bubble pressure	90
Te	–	5.71		$\rho_t = 5.71 - 3.6 \cdot 10^{-4} (t - t_{mp})$				(t°C)		Bubble pressure	110
Te	99.7	5.86		$\rho_t = 5.86 - 0.73 \cdot 10^{-3} (t - t_{mp})$				(t°C)	N_2	Volume determin.	96
Ti	–	4.10							vac.	Pendant drop and drop weight	95
U		17.27		$\rho_t = 19.520 - 16.01 \cdot 10^{-4} T$				(T°K)	vac.	Pycnometer	102
V	–	5.3							vac.	Pendant drop and drop weight	95
Zr	–	5.60							vac.	Pendant drop and drop weight	95

References

1. Addison and Pulham, J. Chem. Soc., 3873 (1962).
2. Allen, Trans, AIME, **227,** 1175 (1963).
3. Allen, Trans. AIME, **230,** 1537 (1964).
4. Astachov, Penin and Dobkina, Zh. Fiz. Chim., **20,** 403 (1946).
5. Beer (in Lucas), Mém. Sci. Rev. Mét., **61,** 1, 97 (1964).
6. Bornemann and Sauerwald, Z. Metallkunde, **14,** 145 (1922).
7. Bornemann and Sauerwald, Z. Metallkunde, **14,** 254 (1922).
8. Cahill and Kirshenbaum, J. Phys. Chem., **66,** 1080 (1962).
9. Calverley, Proc. Phys. Soc., **70,** 1040 (1957).
10. Campbell and Epstein, J. Amer. Chem. Soc., **64,** 2679 (1942).
11. Coy and Mateer, Trans. ASM, **58,** 99 (1965).
12. Day, Sosman and Hostetter, Am. J. Sci., **187,** 1 (1914).
13. Dobinsky and Weselowsky, Bull. Int. Acad. Polon., **A,** 446 (1936).
14. Dshemilev, Popel and Zarevski, Fiz. Met. i Met., **18,** 83 (1964).
15. Edwards and Moorman, Chem & Met. Eng., **24,** 61 (1921).
16. Eljutin and Maurach, Izv. A. N., OTN, **4,** 129 (1956).
17. Eremenko, Ukr. Chim. Zh., **28,** 427 (1962).
18. Eremenko and Naidich, Izv. A. N., OTN, **2,** 111 (1959).
19. Eremenko and Naidich, Izv. A. N., OTN, **6,** 129 (1959).
20. Eremenko and Naidich, Izv. A. N., OTN, **6,** 100 (1961).
21. Eremenko and Nishenko, Ukr. Chim. Zh., **30,** 125 (1964).
22. Eremenko, Nishenko and Taj-Shou-Wej, Izv. A. N., OTN, **3,** 116 (1960).
23. Fisher and Phillips, J. Metals, **6,** 1060 (1954).
24. Gebhardt and Becker, Z. Metallkunde, **42,** 111 (1951).
25. Gebhardt, Becker and Dorner, Aluminium, **31,** 315 (1955).
26. Geld and Vertman, Fiz. Met i Met., **10,** 793 (1960).
27. Gogiberidse and Kekelidse, Izv. A. N., **3,** 125 (1963).
28. Greenaway, J. Inst. Met., **74,** 133 (1947).
29. Grosse, Cahill and Kirshenbaum, J. Amer. Chem. Soc., **83,** 4665 (1961).
30. Herczynska, Naturwiss., **47,** 200 (1960).
31. Hogness, J. Amer. Chem. Soc., **43,** 1621 (1921).
32. Jones, Ofte, Rohr and Wittenberg, Trans. ASM, **55,** 819 (1962).
33. Jouniaux, Bull. Soc. Chim. France, **47,** 524 (1930).
34. Jouniaux, Bull. Soc. Chim. France, **51,** 677 (1932).
35. Kingery and Humenik, J. Phys. Chem., **57,** 359 (1953).
36. Kirshenbaum and Cahill, Trans. ASM, **55,** 844 (1962).
37. Kirshenbaum and Cahill, Trans. ASM, **55,** 849 (1962).
38. Kirshenbaum and Cahill, Trans. AME, **224,** 816 (1962).
39. Kirshenbaum, Cahill and Grosse, J. Inorg. Nucl. Chem., **24,** 333 (1962).
40. Klemm et al., Monatsh. Chem., **83,** 629 (1952).
41. Knappwost and Restle, Z. Elektrochemie, **58,** 112 (1954).
42. Königer and Nagel, GieBerei TWB, **13,** 57 (1960).
43. Kozakevitch, Châtel, Urbain and Sage, Rev. de Mét., **52,** 139 (1955).
44. Kozakevitch and Urbain, C. R., Paris, **253,** 2229 (1961).
45. Lang, WADC-Techn. Rep., 57–488 (1957).
46. Lauermann and Metzger, Z. Phys. Chem., **216,** 37 (1961).
47. Lucas, C. R., Paris, **250,** 1850 (1960).
48. Lucas, C. R., Paris, **253,** 2526 (1961).
49. Lucas, Mém. Sci. Rev. Mét., **61,** 1 (1964).
50. Lucas, Mém. Sci. Rev. Mét., **61,** 97 (1964).
51. Lucas and Urbain, C. R., Paris, **258,** 6403 (1964).
52. Malmberg, J. Inst. Met., **89,** 137 (1960).
53. Matuyama, Sci. Rep. RITU, **18,** 19 (1929).
54. Mehairy and Wood (in Lucas), Mém. Sci. Rev. Mét., **61,** 1 (1964).
55. Metals Handbook: Properties and Selection of Metals, 8th edit., **1,** (1961).
56. Metals Reference Book, C. J. Smithells, 4th edit., **1,** 688 (1967).
57. Metzger, Z. Phys. Chem., **211,** 1 (1959).
58. Mokrovski and Regel, J. Phys. Techn., **22,** 1281 (1952).
59. Monma and Suto, J. Jap. Inst. Met., **1,** 69 (1960).
60. Naidich and Eremenko, Fiz. Met. i Met., **6,** 62 (1961).
61. Ono and Matsushima, Sci. Rep. RITU, **9,** 309 (1957).
62. Pascal and Jouniaux, C. R., Paris, **158,** 414 (1914).
63. Pascal and Jouniaux, Z. Elektrochemie, **22,** 72 (1916).
64. Pekarev, Izv. Vyss. Utch. Saved., Tsvetn. Met., **6,** 111 (1963).
65. Pelzel, Berg-u. Hütt. Mon. Hefte, Leoben, **93,** 248 (1948).

66. Pelzel, Z. Metallkunde, **50,** 392 (1959).
67. Pelzel and Sauerwald, Z. Metallkunde, **33,** 229 (1941).
68. Pelzel and Schneider, Z. Metallkunde, **35,** 121 (1943).
69. Pokrovski and Saidov, Zh. Fiz. Chim., **29,** 1601 (1955).
70. Popel, Smirnov, Zarevski, Dshemilev and Pastuchov, Izv. A. N., **1,** 62 (1965).
71. Popel, Zarevski and Dshemilev, Fiz. Met. i Met., **18,** 468 (1964).
72. Rink, C. R., Paris, **189,** 39 (1929).
73. Saeger and Ash, J. Res. Nat. Bur. Stand., **8,** 37 (1932).
74. Sauerwald, Z. Metallkunde, **14,** 457 (1922).
75. Sauerwald and Widawski, Z. anorg. allg. Chem., **155.** 1 (1926).
76. Schneider and Heymer, Z. anorg. allg. Chem., **286,** 118 (1956).
77. Smith and Spitzer, J. Phys. Chem., **66,** 946 (1962).
78. Stauffer, Thesis Göttingen (1953).
79. Stott and Rendall, J. Iron & Steel Inst., **175,** 374 (1953).
80. Tavadse, Bairamashvili, Chantadse and Zagareishvili, Doklady A. N., **150,** 544 (1963).
81. Tavadse, Bairamashvili and Chantadse, Doklady A. N., **162,** 67 (1965).
82. Taylor, J. Inst. Met., **83,** 143 (1954).
83. Übelacker and Lucas, C. R., Paris, **254,** 1622 (1962).
84. Widawski and Sauerwald, Z. anorg. allg. Chem., **192,** 145 (1930).
85. Zimmermann and Esser, Arch. f. Eisenh., **2,** 867 (1929).
A) Kirshenbaum, Cahill and Grosse, J. Inorg. Nucl. Chem., **22,** 33 (1961).
B) McGonigal, Kirshenbaum and Grosse, J. Phys. Chem., **66,** 737 (1962).
C) Grosse and Kirshenbaum, J. Inorg. Nucl. Chem., **25,** 331 (1963).
D) Golchova, Teplofiz. Vysok. Temp., **4,** 360 (1966).
E) Bosio, C. R., Paris, **259,** 4545 (1964).
F) Spilrajn and Jakimovich, Teplofiz. Vysok. Temp., **5,** 239 (1967).
G) Jakimovich and Saars, Teplofiz. Vysok. Temp., **5,** 532 (1967).
H) Stone, Ewing, Spann, Steinkuller, Williams and Miller, J. Chem. Engng. Data, **11,** 320 (1966).
I) Shirai, Hamada and Kobayashi, J. Chem. Soc. Japan, **84,** 968 (1963).
K) Grosse and McGonigal, J. Phys. Chem., **68,** 414 (1964).
L) Cubicciotti, J. Phys. Chem., **68,** 537 (1964).
M) Watolin and Esin, Fiz. Met. i Met., **16,** 936 (1963).
N) Gebhardt, Becker and Trägner, Z. Metallkunde, **46,** 90 (1955).
O) Kubaschewski and Hörnle, Z. Metallkunde, **42,** 129 (1951).
P) Gebhardt and Dorner, Z. Metallkunde, **42,** 353 (1951).
Q) Gebhardt and Wörwag, Z. Metallkunde, **42,** 358 (1951).
R) Been, Edwards, Teeter and Chalkins, NEPA-1585, US-AEC (1950).
S) da C. Andrade and Dobbs, Proc. Roy. Soc., **211,** 12 (1952).
T) Wilkinson: Extractive and Physical Metallurgy of Plutonium and its Alloys, New York-London, 1960.
U) Saito and Sakuma, J. Jap. Inst. Met., **31,** 1140 (1967).
86. Achener, HTLMHTTM., Vol. I, Oak Ridge, p. 5, November 1964.
87. Ayushina, Lewin and Geld, Zh. Fiz. Chim., **42,** 2799 (1968).
88. Bernard and Lupis, Met. Trans., **2,** 555 (1971).
89. Berthou and Tougas, Met. Trans., **1,** 2978 (1970).
90. Bohdansky and Schins, J. Inorg. Nucl. Chem., **30,** 2331 (1968).
91. Crawley, Trans. AIME, **242,** 2237 (1968).
92. Crawley and Kiff, Met. Trans., **2,** 609 (1971).
93. Crawley and Kiff, Met. Trans., **3,** 158 (1972).
94. Eljutin, Kostikow and Lewin, Izv. Vys. Uch. Sav., Tsvetn. Met., 2, 131 (1970).
95. Eljutin, Kostikow and Penkow, Poroshk. Met., **9,** 46 (1970).
96. Keskar and Hruska, Met. Trans., **1,** 2357 (1970).
97. Körber and Löhberg, Gießereiforschung, **23,** 173 (1971).
98. Koster, Hensel and Franck, Ber. Bunsenges., **74,** 43 (1970).
99. Nagamori, Trans. AIME, **245,** 1897 (1969).
100. Osminin, Zh. Fiz. Chim., **43,** 2610 (1969).
101. Popel, Shergin and Zarewski, Zh. Fiz. Chim., **43,** 2365 (1969).
102. Rohr and Wittenberg, J. Phys. Chem., **74,** 1151 (1970).
103. Saito and Sakuma, Sci. Rep. RITU, **22,** 57 (1970).
104. Schwaneke and Folke, Us-Bur. Min., Invest. Rep. No. **7372** (1970).
105. Stoffhütte. Taschenbuch der Werkstoffkunde, 4th ed. p. 1059, 1967.
106. Tepper, Murchison, Zelenak and Roehlich, HTLMHTTM, Vol. I, Oak Ridge, p. 26, November 1964.
107. Thresh, Crawley and White, Trans. AIME, **242**, 819 (1968).
108. Watolin, Esin, Uchow and Dubinin, Trudy Inst. Met. Sverdlovsk, **18,** 73 (1969).
109. Watanabe, Trans. Jap. Inst. Met., **12,** 17 (1971).
110. Wobst and Rentzsch, Z. Phys. Chem., **240,** 36 (1969).
111. Crawley, Trans. AIME, **245,** 1655 (1969).

HEAT OF FUSION OF SOME INORGANIC COMPOUNDS

Rudolf Loebel

Values in parentheses are of uncertain reliability

Compound	Formula	M.P., °C	H_f cal/g mole	H_f cal/g
Actinium[227]	Ac	1050 ± 50	(3400)	(11.0)
Aluminum	Al	658.5	2550	94.5
Aluminum bromide	Al_2Br_6	87.4	5420	10.1
Aluminum chloride	Al_2Cl_6	192.4	19600	63.6
Aluminum iodide	Al_2I_6	190.9	7960	9.8
Aluminum oxide	Al_2O_3	2045.0	(26000)	(256.0)
Antimony	Sb	630	4770	39.1
Antimony tribromide	$SbBr_3$	96.8	3510	9.7
Antimony trichloride	$SbCl_3$	73.3	3030	13.3
Antimony pentachloride	$SbCl_5$	4.0	2400	8.0
Antimony trioxide	Sb_4O_6	655.0	(26990)	(46.3)
Antimony trisulfide	Sb_4S_6	546.0	11200	33.0
Argon	Ar	− 190.2	290	7.25
Arsenic	As	816.8	(6620)	(22.0)
Arsenic pentafluoride	AsF_4	− 80.8	2800	16.5
Arsenic tribromide	$AsBr_3$	30.0	2810	8.9
Arsenic trichloride	$AsCl_3$	− 16.0	2420	13.3
Arsenic trifluoride	AsF_3	− 6.0	2486	18.9
Arsenic trioxide	As_4O_6	312.8	8000	22.2
Barium	Ba	725	1830	13.3
Barium bromide	$BaBr_2$	846.8	6000	21.9
Barium chloride	$BaCl_2$	959.8	5370	25.9
Barium fluoride	BaF_2	1286.8	3000	17.1
Barium iodide	BaI_2	710.8	(6800)	(17.3)
Barium nitrate	$Ba(NO_3)_2$	594.8	(5900)	(22.6)
Barium oxide	BaO	1922.8	13800	93.2
Barium phosphate	$Ba_3(PO_4)_2$	1727	18600	30.9
Barium sulfate	$BaSO_4$	1350	9700	41.6
Beryllium	Be	1278	—	260.0
Beryllium bromide	$BeBr_2$	487.8	(4500)	(26.6)
Beryllium chloride	$BeCl_2$	404.8	(3000)	(30)
Beryllium fiuoride	BeF_2	796.8	(6000)	(127.6)
Beryllium iodide	BeI_2	479.8	(4500)	(17.1)
Beryllium oxide	BeO	2550.0	17000	679.7
Bismuth	Bi	271	2505	12.0
Bismuth tribromide	$BiBr_3$	216.8	(4000)	(8.9)
Bismuth trichloride	$BiCl_3$	223.8	2600	8.2
Bismuth trifluoride	BiF_3	726.0	(6200)	(23.3)
Bismuth trioxide	Bi_2O_3	815.8	6800	14.6
Boron	B	2300	(5300)	(490)
Boron tribromide	BBr_3	− 48.8	(700)	(2.9)
Boron trichloride	BCl_3	− 107.8	(500)	(4.3)
Boron trifluoride	BF_3	− 128.0	480	7.0
Boron triiodide	BI_3	31.8	(1000)	(2.5)
Boron trioxide	B_2O_3	448.8	5500	78.9
Bromine	Br_2	− 7.2	2580	16.1
Bromine pentafluoride	BrF_5	− 61.4	1355	7.07
Cadmium	Cd	320.8	1460	12.9
Cadmium bromide	$CdBr_2$	567.8	(5000)	(18.4)
Cadmium chloride	$CdCl_2$	567.8	5300	28.8
Cadmium fluoride	CdF_2	1110	(5400)	(35.9)
Cadmium iodide	CdI_2	386.8	3660	10.0
Cadmium sulfate	$CdSO_4$	1000	4790	22.9
Calcium	Ca	851	2230	55.7
Calcium bromide	$CaBr_2$	729.8	4180	20.9
Calcium carbonate	$CaCO_3$	1282	(12700)	(126)
Calcium chloride	$CaCl_2$	782	6100	55
Calcium fluoride	CaF_2	1382	4100	52.5
Calcium nitrate	$Ca(NO_3)_2$	560.8	5120	31.2
Calcium oxide	CaO	2707	(12240)	(218.1)
Calcium metasilicate	$CaSiO_3$	1512	13400	115.4
Calcium sulfate	$CaSO_4$	1297	6700	49.2
Carbon dioxide	CO_2	− 57.6	1900	43.2
Carbon monoxide	CO	− 205	199.7	7.13
Cyanogen	C_2N_2	− 27.2	2060	39.6
Cyanogen chloride	CNCl	− 5.2	2240	36.4
Cerium	Ce	775	2120	15.1
Cerium (III) chloride	$CeCl_3$	820.8	(8000)	(32.4)
Cerium (IV) fluoride	CeF_4	976	(10000)	(46.3)
Cesium	Cs	28.3	500	3.7
Cesium chloride	CsCl	641.8	3600	21.4
Cesium fluoride	CsF	704.8	(2450)	(16.1)
Cesium nitrate	$CsNO_3$	406.8	3250	16.6
Chlorine	Cl_2	− 103 ± 5	1531	22.8
Chromium	Cr	1890	3660	62.1
Chromium (II) chloride	$CrCl_2$	814	7700	65.9
Chromium (III) sequioxide	Cr_2O_3	2279	4200	27.6
Chromium trioxide	CrO_3	197	3770	37.7
Cobalt	Co	1490	3640	62.1
Cobalt (II) chloride	$CoCl_2$	727	7390	56.9
Cobalt (II) fluoride	CoF_2	1201	(9000)	(92.9)
Copper	Cu	1083	3110	49.0
Copper (I) bromide	CuBr	487	(2300)	(16.03)
Copper (II) chloride	$CuCl_2$	430	4890	24.7
Copper (I) chloride	CuCl	429	2620	26.4
Copper (I) cyanide	$Cu_2(CN)_2$	473	(5400)	(30.1)
Copper (I) iodide	CuI	587	(2600)	(13.6)
Copper (II) oxide	CuO	1446	2820	35.4
Copper (I) oxide	Cu_2O	1230	(13400)	(93.6)
Copper (I) sulfide	Cu_2S	1129	5500	34.6
Dysprosium	Dy	1407	4100	25.2
Dysprosium chloride	$DyCl_3$	646	(7000)	(26.0)
Erbium	Er	1496	4100	24.5
Erbium trichloride	$ErCl_3$	775	(7000)	(26.0)
Europium	Eu	826	2500	16.4
Europium trichloride	$EuCl_3$	622	(8000)	(20.9)
Fluorine	F_2	− 219.6	244.0	6.4
Gadolinium	Gd	1312	3700	23.8
Gadolinium trichloride	$GdCl_3$	608	(7000)	(26.6)

HEAT OF FUSION OF SOME INORGANIC COMPOUNDS (Continued)

Compound	Formula	M.P., °C	H_f cal/g mole	H_f cal/g
Gallium	Ga	29	1336	19.1
Germanium	Ge	959	(8300)	(114.3)
Gold	Au	1063	3030	15.3
Hafnium	Hf	2214	(6000)	(34.1)
Holmium	Ho	1461	4100	24.8
Holmium trichloride	$HoCl_3$	717	(8000)	(29.5)
Hydrogen	H_2	−259.25	28	13.8
Hydrogen bromide	HBr	−86.96	575.1	7.1
Hydrogen chloride	HCl	−114.3	476.0	13.0
Hydrogen fluoride	HF	−83.11	1094	54.7
Hydrogen iodide	HI	−50.91	686.3	5.4
Hydrogen nitrate	HNO_3	−47.2	601	9.5
Hydrogen oxide (water)	H_2O	0	1436	79.72
Deuterium oxide	D_2O	3.78	1516	75.8
Hydrogen peroxide	H_2O_2	−0.7	2920	8.58
Hydrogen selenate	H_2SeO_4	57.8	3450	23.8
Hydrogen sulfate	H_2SO_4	10.4	2360	24.0
Hydrogen sulfide	H_2S	−85.6	5683	16.8
Hydrogen sulfide (di-)	H_2S_2	−89.7	1805	27.3
Hydrogen telluride	H_2Te	−49.0	1670	12.9
Indium	In	156.3	781	6.8
Iodine	I_2	112.9	3650	14.3
Iodine chloride (α)	ICl	17.1	2660	16.4
Iodine chloride (β)	ICl	13.8	2270	13.3
Iron	Fe	1530.0	3560	63.7
Iron carbide	Fe_3C	1226.8	12330	68.6
Iron (III) chloride	Fe_2Cl_6	303.8	20500	63.2
Iron (II) chloride	$FeCl_2$	677	7800	61.5
Iron (II) oxide	FeO	1380	(7700)	(107.2)
Iron oxide	Fe_3O_4	1596	33000	142.5
Iron pentacarbonyl	$Fe(CO)_5$	−21.2	3250	16.5
Iron (II) sulfide	FeS	1195	5000	56.9
Lanthanum	La	920	2400	17.4
Lanthanum chloride	$LaCl_3$	861	(9000)	(36.6)
Lead	Pb	327.3	1224	5.9
Lead bromide	$PbBr_2$	487.8	4290	11.7
Lead chloride	$PbCl_2$	497.8	5650	20.3
Lead fluoride	PbF_2	823	1860	7.6
Lead iodide	PbI_2	412	5970	17.9
Lead molybdate	$PbMoO_4$	1065	(25800)	70.8
Lead oxide	PbO	890	2820	12.6
Lead sulfide	PbS	1114	4150	17.3
Lead sulfate	$PbSO_4$	1087	9600	31.6
Lead tungstate	$PbWO_4$	1123	(15200)	(33.4)
Lithium	Li	178.8	1100	158.5
Lithium borate, meta-	$LibO_2$	845	(5570)	(111.9)
Lithium bromide	LiBr	552	2900	33.4
Lithium chloride	LiCl	614	3200	75.5
Lithium fluoride	LiF	896	(2360)	(91.1)
Lithium hydroxide	LiOH	462	2480	103.3
Lithium iodide	LiI	440	(1420)	(10.6)
Lithium metasilicate	Li_2SiO_3	1177	7210	80.2
Lithium molybdate	Li_2MoO_4	705	4200	24.1
Lithium nitrate	$LiNO_3$	250	6060	87.8

Compound	Formula	M.P., °C	H_f cal/g mole	H_f cal/g
Lithium orthosilicate	Li_4SiO_4	1249	7430	60.5
Lithium sulfate	Li_2SO_4	857	3040	27.6
Lithium tungstate	Li_2WO_4	742	(6700)	(25.6)
Lutetium	Lu	1651	4600	26.3
Lutetium chloride	$LuCl_3$	904	(9000)	(32)
Magnesium	Mg	650	2160	88.9
Magnesium bromide	$MgBr_2$	711	8300	45.0
Magnesium chloride	$MgCl_2$	712	8100	82.9
Magnesium fluoride	MgF_2	1221	5900	94.7
Magnesium oxide	MgO	2642	18500	459.0
Magnesium phosphate	$Mg_3(PO_4)_2$	1184	(11300)	(42.9)
Magnesium silicate	$MgSiO_3$	1524	14700	146.4
Magnesium sulfate	$MgSO_4$	1327	3500	28.9
Manganese	Mn	1220	3450	62.7
Manganese dichloride	$MnCl_2$	650	7340	58.4
Manganese metasilicate	$MnSiO_3$	1274	(8200)	(62.6)
Manganese (II) oxide	MnO	1784	13000	183.3
Manganese oxide	Mn_3O_4	1590	(39000)	(170.4)
Mercury	Hg	−39	557.2	2.7
Mercury bromide	$HgBr_2$	241	3960	10.9
Mercury chloride	$HgCl_2$	276.8	4150	15.3
Mercury iodide	HgI_2	250	4500	9.9
Mercury sulfate	$HgSO_4$	850	(1440)	(4.8)
Molybdenum	Mo	2622	(6600)	(68.4)
Molybdenum dichloride	$MoCl_2$	726.8	6000	3.58
Molybdenum hexafluoride	MoF_6	17	2500	11.9
Molybdenum trioxide	MoO_3	795	(2500)	(17.3)
Neodymium	Nd	1020	1700	11.8
Neodymium trichloride	$NdCl_3$	758	(8000)	(31.9)
Neon	Ne	−248.6	77.4	3.83
Nickel	Ni	1452	4200	71.5
Nickel chloride	$NiCl_2$	1030	18470	142.5
Nickel subsulfide	Ni_3S_2	790	5800	25.8
Niobium	Nb	2496	(6500)	(68.9)
Niobium pentachloride	$NbCl_5$	211	8400	30.8
Niobium pentoxide	Nb_2O_5	1511	24200	91.0
Nitrogen	N_2	−210	172.3	6.15
Nitrogen oxide (ic)	NO	−163.7	549.5	18.3
Nitrogen oxide (ous)	N_2O	−90.9	1563	35.5
Nitrogen tetroxide	N_2O_4	−13.2	5540	60.2
Osmium	Os	2700	(7000)	(36.7)
Osmium tetroxide (white)	OsO_4	41.8	2340	9.2
Osmium tetroxide (yellow)	OsO_4	55.8	4060	15.5
Oxygen	O_2	−218.8	106.3	3.3
Palladium	Pd	1555	4120	38.6
Phosphoric acid	H_3PO_4	42.3	2520	25.8
Phosphoric acid, hypo-	$H_4P_2O_6$	54.8	8300	51.2
Phosphorous acid, hypo-	H_3PO_2	17.3	2310	35.0
Phosphorous acid, ortho-	H_3PO_3	73.8	3070	37.4

HEAT OF FUSION OF SOME INORGANIC COMPOUNDS (Continued)

Compound	Formula	M.P., °C	H_f cal/g mole	H_f cal/g
Phosphorus, yellow	P_4	44.1	600	4.8
Phosphorus oxychloride	$POCl_3$	1.0	3110	20.3
Phosphorus pentoxide	P_4O_{10}	569.0	17080	60.1
Phosphorus trioxide	P_4O_6	23.7	3360	15.3
Platinum	Pt	1770	4700	24.1
Potassium	K	63.4	574	14.6
Potassium borate, meta-	KBO_2	947	(5660)	(69.1)
Potassium bromide	KBr	742	5000	42.0
Potassium carbonate	K_2CO_3	897	7800	56.4
Potassium chloride	KCl	770	6410	85.9
Potassium chromate	K_2CrO_4	984	6920	35.6
Potassium cyanide	KCN	623	(3500)	(53.7)
Potassium dichromate	$K_2Cr_2O_7$	398	8770	29.8
Potassium fluoride	KF	875	6500	111.9
Potassium hydroxide	KOH	360	(1980)	(35.3)
Potassium iodide	KI	682	4100	24.7
Potassium nitrate	KNO_3	338	2840	28.1
Potassium peroxide	K_2O_2	490	6100	55.3
Potassium phosphate	K_3PO_4	1340	8900	41.9
Potassium pyro-phosphate	$K_4P_2O_7$	1092	14000	42.4
Potassium sulfate	K_2SO_4	1074	8100	46.4
Potassium sulfocyanide	KCNS	179	2250	23.1
Potassium tungstate	K_2WO_4	929	(4400)	(13.6)
Praseodymium	Pr	931	2700	19.0
Praseodymium chloride	$PrCl_3$	786	(8000)	(32.3)
Promethium	Pm	(1027)	(3000)	(21)
Rhenium	Re	3167 ± 60	(7900)	(42.4)
Rhenium heptoxide	Re_2O_7	296	15340	30.1
Rhenium hexafluoride	ReF_6	19.0	5000	16.6
Rubidium	Rb	38.9	525	6.1
Rubidium bromide	RbBr	677	3700	22.4
Rubidium chloride	RbCl	717	4400	36.4
Rubidium fluoride	RbF	833	4130	39.5
Rubidium iodide	RbI	638	2990	14.0
Rubidium nitrate	$RbNO_3$	305	1340	9.1
Samarium	Sm	1072	2600	17.3
Samarium chloride	$SmCl_2$	562	(6000)	(27.1)
Scandium	Sc	1538	3800	84.4
Scandium trichloride	$ScCl_3$	940	(19000)	(125.4)
Selenium	Se	217	1220	15.4
Selenium oxychloride	$SeOCL_3$	9.8	1010	6.1
Silicon	Si	1427	9470	337.0
Silicondioxide (Cristobalite)	SiO_2	2100	2100	35.0
Silicondioxide (Quartz)	SiO_2	1470	3400	56.7
Disilane, hexafluoro-	Si_2F_6	−28.6	3900	22.9
Silicon tetrachloride	$SiCl_4$	−67.7	1845	10.8
Silver	Ag	961	2700	25.0
Silver bromide	AgBr	430	2180	11.6
Silver chloride	AgCl	455	3155	22.0
Silver cyanide	AgCN	350	2750	20.5
Silver iodide	AgI	557	2250	9.5
Silver nitrate	$AgNO_3$	209	2755	16.2
Silver sulfide	Ag_2S	841	3360	13.5
Silver sulfate	Ag_2SO_4	657	(4280)	(13.7)
Sodium	Na	97.8	630	27.4
Sodium borate, meta-	$NaBO_2$	966	8660	134.6
Sodium bromide	NaBr	747	6140	59.7
Sodium carbonate	Na_2CO_3	854	7000	66.0
Sodium chloride	NaCl	800	7220	123.5
Sodium chlorate	$NaClO_3$	255	5290	49.7
Sodium cyanide	NaCN	562	(4360)	(88.9)
Sodium fluoride	NaF	992	7000	166.7
Sodium hydroxide	NaOH	322	2000	50.0
Sodium iodide	NaI	662	5340	35.1
Sodium molybdate	Na_2MoO_4	687	3600	17.5
Sodium nitrate	$NaNO_3$	310	3760	44.2
Sodium phosphate, meta-	$NaPO_3$	988	(4960)	(48.6)
Sodium peroxide	Na_2O_2	460	5860	75.1
Sodium pyro-phosphate	$Na_4P_2O_7$	970	(13700)	(51.5)
Sodium silicate, meta-	Na_2SiO_3	1087	10300	84.4
Sodium silicate, di-	$Na_2Si_2O_5$	884	8460	46.4
Sodium silicate, aluminum-	$NaAlSi_3O_8$	1107	13150	50.1
Sodium sulfide	Na_2S	920	(1200)	15.4
Sodium sulfate	Na_2SO_4	884	5830	41.0
Sodium thiocyanate	NaCNS	323	4450	54.8
Sodium tungstate	Na_2WO_4	702	5800	19.6
Strontium	Sr	757	2190	25.0
Strontium bromide	$SrBr_2$	643	4780	19.3
Strontium chloride	$SrCl_2$	872	4100	26.5
Strontium fluoride	SrF_2	1400	4260	34.0
Strontium oxide	SrO	2430	16700	161.2
Sulfur (monatomic)	S	119	295	9.2
Sulfur dioxide	SO_2	−73.2	2060	32.2
Sulfur trioxide (α)	SO_3	16.8	2060	25.8
Sulfur trioxide (β)	SO_3	32.3	2890	36.1
Sulfur trioxide (γ)	SO_3	62.1	6310	79.0
Tantalum	Ta	2996 ± 50	(7500)	34.6 to 41.5
Tantalum pentachloride	$TaCl_5$	206.8	9000	25.1
Tantalum pentoxide	Ta_2O_5	1877	48000	108.6
Tellurium	Te	453	3230	25.3
Terbium	Tb	1356	3900	24.6
Thallium	Tl	302.4	1030	5.0
Thallium bromide, mone-	TlBr	460	5990	21.0
Thallium carbonate	Tl_2CO_3	273	4400	9.5

Compound	Formula	M.P.,°C	H_f cal/g mole	H_f cal/g
Thallium chloride, mono-	TlCl	427	4260	17.7
Thallium iodide, mono-	TlI	440	3125	9.4
Thallium nitrate	$TlNO_3$	207	2290	8.6
Thallium sulfide	Tl_2S	449	3000	6.8
Thallium sulfate	Tl_2SO_4	632	5500	10.9
Thorium	Th	1845	(<4600)	(<19.8)
Thorium dioxide	ThO_2	2952	291100	1102.0
Thorium chloride	$ThCl_4$	765	22500	61.6
Thulium	Tm	1545	4400	26.0
Thulium chloride	$TmCl_3$	821	(9000)	(32.6)
Tin	Sn	231.7	1720	14.4
Tin bromide, ic-	$SnBr_4$	29.8	3000	6.8
Tin bromide, ous-	$SnBr_2$	231.8	(1720)	(6.1)
Tin chloride, ic-	$SnCl_4$	−33.3	2190	8.4
Tin chloride, ous-	$SnCl_2$	247	3050	16.0
Tin iodide, ic-	SnI_4	143.4	(4330)	(6.9)
Tin oxide	SnO	1042	(6400)	(46.8)
Titanium	Ti	1800	(5000)	(104.4)
Titanium bromide, tetra-	$TiBr_4$	38	(2060)	(5.6)
Titanium chloride, tetra-	$TiCl_4$	−23.2	2240	11.9
Titanium dioxide	TiO_2	1825	(11400)	(142.7)
Titanium oxide	TiO	991	14000	219
Tungsten	W	3387	(8420)	(45.8)
Tungsten dioxide	WO_2	1270	13940	60.1
Tungsten hexafluoride	WF_6	−0.5	1800	6.0
Tungsten tetrachloride	WCl_4	327	6000	18.4
Tungsten trioxide	WO_3	1470	13940	60.1
Uranium[235]	U	~1133	3700	20
Uranium tetrachloride	UCl_4	590	10300	27.1
Vanadium	V	1917	(4200)	(70)
Vanadium dichloride	VCl_2	1027	8000	65.6
Vanadium oxide	VO	2077	15000	224.0
Vanadium pentoxide	V_2O_5	670	15560	85.5
Xenon	Xe	111.6	740	5.6
Ytterbium	Yb	823	2200	12.7
Yttrium	Y	1504	4100	46.1
Yttrium oxide	Y_2O_3	2227	25000	110.7
Yttrium trichloride	YCl_3	709	(9000)	(46)
Zinc	Zn	419.4	1595	24.4
Zinc chloride	$ZnCl_2$	283	(5540)	(40.6)
Zinc oxide	ZnO	1975	4470	54.9
Zinc sulfide	ZnS	1700 ± 20	(9100)	(93.3)
Zirconium	Zr	1857	(5500)	(60)
Zirconium dichloride	$ZrCl_2$	727	7300	45.0
Zirconium oxide	ZrO_2	2715	20800	168.8

TABLE OF THE ISOTOPES (1985 UPDATE)

Compiled by
Russell L. Heath
Idaho National Engineering Laboratory
EG&G Idaho Inc.
Idaho Falls, Idaho

The following information is provided to assist in the use of the Table of Isotopes. This compilation of nuclear data presents a selected set of currently adopted values for experimental quantities which characterize the decay of radioactive nuclides. The approach used for the presentation of gamma rays and particles emitted in the decay of radionuclides was to include the major photons and beta groups which dominate the observed energy spectrum of photons and particles emitted in the decay of a given nuclide. In general, gamma rays are listed which span the first 2 decades on a relative intensity scale. To provide the applied spectroscopist with information on the relative intensity of X-rays emitted in electron-capture decay or associated with the internal conversion process, the intensity of the most intense K ($K_{\alpha 1}$) or L (L_β) X-ray line is listed when the X-ray component represents a significant contribution to the total photon spectrum. Values used in this case represent calculated intensities derived from level-scheme information. Values adopted and presented in the table have been compiled from original published data and evaluated data sets in major data compilations. A list of the major references used in this evaluation is given below. The effective literature cutoff date for data in this edition of the Table is 1984.

TABLE LAYOUT

Column no.	Column title	Description
1	Isotope	This column lists the isotopes with atomic number and mass number. Isomers are indicated by the addition of the letter m. The elements and their corresponding atomic number are also listed.
2	A	Mass number
3	Z	Atomic number
4	Isotopic abundance	Isotopic abundance in percent
5	Atomic mass	Lists values for atomic mass in the physical scale. Values of atomic mass are all relative to the mass of the Carbon 12 isotope which has been assigned a value of 12.000000 atomic mass units.
6	Half-life	Half-life in decimal notation. The notation used is: ns = nanoseconds, μs = microseconds, ms = milliseconds, s = seconds, m = minutes, d = days, and y = years.
7	Decay mode	Observed modes of decay for all radioactive species. Symbols used are: α = alpha particle emission; β− = negative beta emission; β+ = positron emission; E.C. = orbital electron emission; I.T. = isomeric transition from upper to lower isomeric state; n = neutron emission; and S.F. = spontaneous fission. Contained in ( ) are values for the % decay by a given decay mode.
8	Decay energy	The currently adopted values for total disintegration energy (Q) in MeV. Where more than one mode of decay exists, separate values are given.
9	Particle energy	The end-point energies of beta particle transitions or discrete energies of alpha particles are given in MeV. Estimated experimental error for alpha particle energies is indicated in ( ), representing variation in the least significant figure.
10	Particle intensity	The intensity of beta groups or alpha particle transitions are given in percent of total decay rate.

Column no.	Column title	Description
11	Thermal neutron cross section	Cross sections for thermal-neutron capture is given in barns (10^{-24} cm^2) or in millibarns (mb). Where neutron capture for a given nucleus may produce two isomers, the separate cross sections are listed. Fission cross sections are listed for the heavy elements and are designated by the symbol σ_f.
12	Spin	Values for the nuclear spin or angular momentum of the ground state of the isotope is given in units h/2π.
13	μ	Values of the magnetic dipole moments of the isotopes are given in nuclear magnetons, with diamagnetic correction.
14	Gamma-ray energy	Gamma-ray energies, together with estimated experimental error are given in MeV. The abbreviation ann.rad. refers to 511.006 keV photons associated with the annihilation of positrons in matter. Where positron emission is a significant component in the decay, annihilation radiation will be present.
15	Gamma-ray intensity	Values for gamma-ray intensity are given in percent of total decay rate for a given gamma-ray transition. When complete level-scheme information is not available relative photon intensities are indicated by the symbol +.

GENERAL NUCLEAR DATA REFERENCES

The following references represent the major sources of compiled nuclear data:

1. Nuclear Data Sheets, edited by The National Nuclear Data Center for the International Network for Nuclear Structure Data Evaluation, Academic Press, New York, Vol. 1—36.
2. **Lederer, C. M. and Shirley, V., Eds.,** *Table of Isotopes,* 7th ed., Wiley Interscience, New York, 1978.
3. **Browne, E. and Firestone, R.,** *Radioactivity Handbook,* to be published by Isotopes Project, LBL.
4. Chart of the Nuclides, 1983 Ed., prepared by Walker, F. W., Miller, D. G., and Feiner, F., Knolls Atomic Power Lab. operated by General Electric Co.
5. **Wapstra, A. H. and Audi, G.,** Atomic mass table, *Nuclear Physics A,* (preprint), July 1984.
6. **Mughabghab, S. F., Divadeenam, M., and Holden, N. E.,** *Neutron Cross Sections,* Part A, Academic Press, 1981; **Mughabghab, S. F.,** *Neutron Cross Section,* Part B, Academic Press, 1984.
7. Gamma-ray Spectrum Catalogue — Ge(Li) and Si(Li) Spectrometry, DOE Report ANCR-1000, 1974.
8. **Reus, U. and Westmeier, W.,** Catalogue of gamma rays from radioactive decay, *Atomic Data and Nuclear Data Tables,* 29, No. 2, September 1983.
9. ENSDF — a computer file of evaluated nuclear data maintained by the National Nuclear Data Center at Brookhaven National Laboratory.
10. ENDF-B Data File summary documentation, Ed. by Kinsey, R., Report BNL-NCS-17541(ENDF-201), National Nuclear Data Center, BNL, 1979.

Isotope	A	Z	% Natural abundance	Atomic mass	Half-life	Decay mode	Decay energy (MeV)	Particle energy (MeV)	Particle intensity	Thermal neutron cross section	Spin (h/2π)	μ Nucl. mag. moment	Gamma-ray energy (MeV)	Gamma-ray intensity
n	1			1.008665	12m.	β-	0.7825	0.7825	100%		1/2	-1.9131		
H		1		1.00797						0.332 ± 2 mb				
$_{1}H^{1}$	1	1	99.985%	1.007825							1/2	+2.79284		
$_{1}H^{2}$	2	1	0.015%	2.0140						0.51 ± 0.01mb	1	+0.85743		
$_{1}H^{3}$	3	1		3.01605	12.26 yr	β-	0.01861	0.01861	100%	≤6μb	1/2	+2.97896		
He				4.002603										
$_{2}He^{3}$	3	2	0.00014%	3.01603						<0.1 mb	1/2	-2.12762		
$_{2}He^{4}$	4	2	99.99986%	4.00260						0	3/2-			
$_{2}He^{5}$	5	2		5.01222										
$_{2}He^{6}$	6	2		6.018886	0.808 s.	β-	3.5097	3.5097	100%		0+			
$_{2}He^{8}$	8	2		8.03392	0.122 s.	β-	14	13	88%		0+		0.99	88%
						n			12%					
Li				6.9409										
$_{3}Li^{5}$	5	3		5.01254							3/2-			
$_{3}Li^{6}$	6	3	7.5%	6.015121						38 ± 3 mb	1	+.822056		
$_{3}Li^{7}$	7	3	92.5%	7.016003						45 ± 5 mb	3/2	+3.25644		
$_{3}Li^{8}$	8	3		8.022485	0.844 s.	β-,α	16.005	12.5	100%		2+	+1.6532		
								α(1.6)						
$_{3}Li^{9}$	9	3		9.026789	0.178 s.	β-	13.6068	13.5	75%		3/2-			
						β-		11	25%					
						n2α(35%)		(n)0.3	96 †					
Be				9.012182						7.6 ± 0.8 mb				
$_{4}Be^{6}$	6	4		6.019725							0+			
$_{4}Be^{7}$	7	4		7.016928	53.29d	EC	0.862				3/2-		0.47759	10.35%
$_{4}Be^{8}$	8	4		8.005305	0.067 fs	2α		0.046			0+			
$_{4}Be^{9}$	9	4	100%	9.012182						7.6 ± 0.8 mb	3/2-	-1.1776		
$_{4}Be^{10}$	10	4		10.013534	1.6 x 10^{6}y	β-	0.556	0.555	100%		0+			
$_{4}Be^{11}$	11	4		11.021658	13.8 s.	β-,β-α	11.48	11.48	61%		1/2+		1.7722	0.28%
	11	4											2.1248	33%
	11	4											2.8931	0.09%
	11	4											4.6663	2%
	11	4											5.019	0.5%
	11	4											5.8518	2.1%
	11	4											6.7905	4.5%
	11	4											7.9747	1.7%
B				10.81003										
$_{5}B^{8}$	8	5		8.024605	0.772 s.	β+,2α		13.7(β+)	93%		2+	1.0355	ann. rad.	
$_{5}B^{9}$	9	5		9.013328	0.85 as	p2α								
$_{5}B^{10}$	10	5	19.8%	10.012937	σ_{α}3837b						3+	+1.8007		
$_{5}B^{11}$	11	5	80.2%	11.009305						5 ± 3 mb	3/2-	+2.6886		
$_{5}B^{12}$	12	5		12.014352	0.0202 s.	β-	13.369				1+	+1.0031	4.439	1.3%
						β-α(1.6%)								
$_{5}B^{13}$	13	5		13.01778	0.0173 s.	β-	13.436	13.4			3/2-	+3.17778	3.68	7.6%
	13	5				β-n(.25%)		2.43(n)	0.09%					
	13	5						3.55(n)	0.16%					

TABLE OF THE ISOTOPES (Continued)

Isotope	A	Z	% Natural abundance	Atomic mass	Half-life	Decay mode	Decay energy (MeV)	Particle energy (MeV)	Particle intensity	Thermal neutron cross section	Spin (h/2π)	μ Nucl. mag. moment	Gamma-ray energy (MeV)	Gamma-ray intensity
C				12.0111						3.5 mb.				
$_6C^9$	9	6		9.031039	127 ms.	β+,p,2α	16.497						ann.rad.	
$_6C^{10}$	10	6		10.01686	19.3 s.	β+	3.650	1.865			0+		ann. rad.	100%
	10	6											0.71829 ± 0.0001	98.5%
	10	6											1.02178 ± 0.0002	1.5%
$_6C^{11}$	11	6		11.01143	20.3 m.	β+,E.C.	1.982	0.9608	99%		3/2-	-0.964	ann.rad.	99+%
$_6C^{12}$	12	6	98.90%	12.000000						3.5 mb.	0+			
$_6C^{13}$	13	6	1.10%	13.003355						1.4 mb.	1/2-	+0.70241		
$_6C^{14}$	14	6		14.003241	5730 y.	β-	0.15648	0.1565	100%		0+			
$_6C^{15}$	15	6		15.010599	2.45 s.	β-	9.772	4.51	68%		1/2+		5.29887 ± 0.0001	68%
	15	6						9.82	32%				7.3011 ± 0.0005	0.008%
	15	6											8.3129 ± 0.001	0.032%
	15	6											9.0500 ± 0.001	0.031%
$_6C^{16}$	16	6		16.014701	0.75 s.	β-,n	8.012							
N				14.0067						1.91 b.				
$_7N^{12}$	12	7		12.018613	11.00 ms.	β+,β+α	17.338	16.38	95%		1+		ann.rad.	
	12	7											4.4389	2.1%
$_7N^{13}$	13	7		13.005738	9.97 m.	β+	2.2205	1.190	100%		1/2-	0.32224		
$_7N^{14}$	14	7	99.63%	14.003074						1.83 b.	1+	+0.40376		
$_7N^{15}$	15	7	0.37%	15.000108						0.02 mb.	1/2-	-0.28319		
$_7N^{16}$	16	7		16.006099	7.13 s.	β-,α	10.4187	4.27 β	68%		2-		6.129170 ± 0.0004	69%
	16	7				β-		10.44 β	26%				7.11515 ± 0.0001	5%
	16	7				α		1.7					2.75	1%
$_7N^{17}$	17	7		17.008450	4.17 s.	β-,β-n	8.680	3.7	100%		1/2-		0.871	3%
	17	7						0.4-1.7n	95%				2.1842 ± 0.	0.3%
$_7N^{18}$	18	7		18.014081	0.63 s.	β-	14.057	9.4	100%				0.82	60%
	18	7											1.65	60%
	18	7											1.982	100%
	18	7											2.47	40%
$_7N^{19}$	19	7		19.017040	0.42 s.	β-	12.53						2.47	
O				15.9994						0.28 mb.				
$_8O^{13}$	13	8		13.02810	8.9 ms.	β+,p	17.77	1.56 (p)					ann. rad.	
$_8O^{14}$	14	8		14.008595	70.6 s.	β+	5.1430	1.81	99%		0+		ann.rad.	
	14	8											2.31264 ± 0.0001	99%
$_8O^{15}$	15	8		15.003065	122 s.	β+	2.754	1.723	100%		1/2-	0.7189	ann.rad.	
$_8O^{16}$	16	8	99.762%	15.994915						0.19 mb.	0+			
$_8O^{17}$	17	8	0.038%	16.999131						0.4 mb.	5/2+	-1.89379		
$_8O^{18}$	18	8	0.200%	17.999160						0.16 mb.	0+			
$_8O^{19}$	19	8		19.003577	26.9 s.	β-	4.819	3.25	60%		5/2+		0.197	100%
	19	8						4.60	40%				1.3569 ± 0.001	55%
	19	8											1.4437 ± 0.001	3%
	19	8											1.5544 ± 0.001	1%
$_8O^{20}$	20	8		20.004075	13.5 s.	β-	3.82				0+		1.057	
$_8O^{21}$	21	8		21.008730	3.14 s.	β-	8.17						(0.21 - 4.2)	
F				18.998403						9.6 mb.				
$_9F^{17}$	17	9		17.002095	64.7 s.	β+	2.761	1.75			5/2+	+4.7223	ann.rad.	
$_9F^{18}$	18	9		18.000937	109.8 m.	β+,E.C.	1.655	0.635	97%				ann.rad.	
$_9F^{19}$	19	9	100%	18.998403						9.6 mb.	1+	+2.62887		
$_9F^{20}$	20	9		19.999981	11.0 s.	β-	7.029	5.398	100%		2+	+2.0935	1.6326 ± 0.0008	100%
	20	9											3.3343 ± 0.0007	0.02%
$_9F^{21}$	21	9		20.999948	4.33 s.	β-	5.686	3.7	8%		5/2+		0.3505 ± 0.0005	71%
	21	9						5.0	63%				1.3951 ± 0.0003	7%
	21	9						5.4	29%				1.746	
$_9F^{22}$	22	9		22.003030	4.23 s.	β-	10.85	3.48	15%		4+		1.2746 ± 0.0003	100%
	22	9						4.67	7%				1.9000 ± 0.0006	8.7%
	22	9						5.50	62%				2.0826 ± 0.0005	82%
	22	9											2.1661 ± 0.0005	62%
	22	9											2.2839 ± 0.0007	5.1%
	22	9											2.9877 ± 0.0009	7.0%
	22	9											3.9835 ± 0.001	1.2%
	22	9											4.2479 ± 0.001	1.0%
	22	9											4.3661 ± 0.001	11.3%
$_9F^{23}$	23	9		23.003600	2.2 s.	β-	8.51				5/2+		0.49289 ± 0.0007	6%
	23	9											0.81523 ± 0.0005	12%
	23	9											1.01672 ± 0.0005	10%
	23	9											1.70144 ± 0.0001	48%
	23	9											1.82225 ± 0.0002	24%
	23	9											1.91932 ± 0.0005	9%
	23	9											2.12877 ± 0.000	34%
	23	9											2.41434 ± 0.0004	7.5%
	23	9											2.73424 ± 0.0005	6%
	23	9											3.43139 ± 0.0004	12%
	23	9											3.83071 ± 0.0004	3.3%
Ne				20.179						40 mb.				
$_{10}Ne^{17}$	17	10		17.017690	109 ms.	β+.p	14.526	1.4-10.6			1/2-		ann.rad.	
$_{10}Ne^{18}$	18	10		18.005710	1.67 s.	β+	4.45	3.416	92%		0+		ann.rad.	
	18	10											0.659 ± 0.001	0.14%
	18	10											1.0413 ± 0.001	7.8%
$_{10}Ne^{19}$	19	10		19.001879	17.22 s.	β+	3.238	2.24	99%		1/2+	-1.887	ann.rad.	
	19	10											1.35692 ± 0.0001	0.0019%

Isotope	A	Z	% Natural abundance	Atomic mass	Half-life	Decay mode	Decay energy (MeV)	Particle energy (MeV)	Particle intensity	Thermal neutron cross section	Spin (h/2π)	μ Nucl. mag. moment	Gamma-ray energy (MeV)	Gamma-ray intensity
$_{10}Ne^{20}$	20	10	90.51%	19.992435							0+			
$_{10}Ne^{21}$	21	10	0.21%	20.993843						<1.5 b.	3/2+	-0.66179		
$_{10}Ne^{22}$	22	10	9.22%	21.991383						48 mb.	0+			
$_{10}Ne^{23}$	23	10		22.994465	37.2 s.	β-	4.376	3.95	32%		5/2+	-1.08	0.440 ± 0.001	33%
	23	10						4.39	67%				1.639 ± 0.003	1%
$_{10}Ne^{24}$	24	10		23.993613	3.38 m.	β-	2.468	1.10	8%		0+		0.4722 ± 0.0001	100%
	24	10						1.98	92%				0.87435 ± 0.0001	9%
$_{10}Ne^{25}$	25	10		24.997690	0.61 s.	β-	7.2	6.3			½+		0.0885 ± 0.0001	96%
	25	10						7.3					0.97977 ± 0.0001	20%
Na				22.98997						0.53 b.				
$_{11}Na^{19}$	19	11		19.013879	0.03 s.	β+,p	11.18							
$_{11}Na^{20}$	20	11		20.007344	0.446 s.	β+	13.89				2+	+0.3694	ann.rad.	
	20	11				α		2.15					1.633	
$_{11}Na^{21}$	21	11		20.997650	22.5 s.	β+	3.547	2.50	95%		3/2+	+2.3863	ann.rad.	
	21	11											0.351	5.1%
$_{11}Na^{22}$	22	11		21.994434	2.605 y.	β+(90%)	2.842	0.545	90%		3+	+1.746	ann.rad.	
	22	11				E.C.(10%)							1.2745 ± 0.00005	99.9%
$_{11}Na^{23}$	23	11	100%	22.989767						0.53 b.	3/2+	+2.21752		
$_{11}Na^{24m}$	24	11			20.2 ms.	I.T.,β-					1+		0.4723	100%
$_{11}Na^{24}$	24	11		23.990961	14.97 h.	β-	5.514	1.389	>99%		4+	+1.6903	1.3686 ± 0.0003	100%
	24	11											2.7541 ± 0.0003	100%
	24	11											3.8672 ± 0.0003	0.06%
$_{11}Na^{25}$	25	11		24.989953	59.3 s.	β-	3.833	2.6	7%		5/2+	+3.683	0.38966 ± 0.0001	13.5%
	25	11						3.15	25%				0.58506 ± 0.0001	13%
	25	11						4.0	65%				0.9752 ± 0.0004	15%
	25	11											1.3797 ± 0.0005	0.003%
	25	11											1.6117	0.1%
$_{11}Na^{26}$	26	11		25.992586	1.07 s.	β-	9.31							
$_{11}Na^{27}$	27	11		26.993940	0.29 s.	β-	8.96	7.95					0.98477 ± 0.0002	86%
	27	11				β-,n							1.6985 ± 0.0005	14%
$_{11}Na^{28}$	28	11		27.978780	30 ms.	β-	13.9	12.3			1+		1.475 ± 0.01	30%
	28	11				β-,n							2.380 ± 0.02	16%
$_{11}Na^{29}$	29	11		29.002830	43 ms.	β-,n	13.4	11.5					1.510 ± 0.02	7%
	29	11											2.100 ± 0.02	4%
	29	11											2.570 ± 0.03	12%
	29	11											3.16 ± 0.04	3%
$_{11}Na^{30}$	30	11		30.008800	53 ms.	β-	18.1							
$_{11}Na^{31}$	31	11		31.012680	17 ms.	15.7								
	31	11				β-,n								
Mg				24.305						63 mb.				
$_{12}Mg^{20}$	20	12		20.018864	0.1 s.	β+,p	10.73							
$_{12}Mg^{21}$	21	12		21.011716	122 ms.	β+,p	13.10				5/2+			
$_{12}Mg^{22}$	22	12		21.999574	3.86 s.	β+	4.788	3.05			0+		0.0739 ± 0.001	60%
	22	12											0.5830 ± 0.007	100%
	22	12											1.2797 ± 0.001	6%
	22	12											1.9347 ± 0.002	9%
$_{12}Mg^{23}$	23	12		22.994124	11.32 s.	β+	4.058	3.09	92%		3/2+		0.438 ± 0.001	9%
$_{12}Mg^{24}$	24	12	78.99%	23.985042						53.mb.	0+			
$_{12}Mg^{25}$	25	12	10.00%	24.985837						18 mb.	5/2+	-0.85545		
$_{12}Mg^{26}$	26	12	11.01%	25.982593						36 mb.	0+			
$_{12}Mg^{27}$	27	12		26.984341	9.45 m.	β-	2.610	1.59	41%		½+		0.17068 ± 0.00001	1%
	27	12						1.75	58%				0.84376 ± 0.00003	73%
	27	12						2.65	0.3%				1.01443 ± 0.0004	30%
$_{12}Mg^{28}$	28	12		27.983876	21.0 h.	β-	1.832	0.459	95%		0+		0.0306 ± 0.0006	95%
	28	12											0.4006 ± 0.0002	38%
	28	12											0.9417 ± 0.0004	38%
	28	12											1.3422 ± 0.0002	57%
	28	12											1.3726 ± 0.0002	5%
	28	12											1.5894 ± 0.0004	5%
$_{12}Mg^{29}$	29	12		28.98848	1.3 s.	β-	7.49	5.4			3/2+		0.9603 ± 0.0005	52 +
	29	12											1.3981 ± 0.001	64
	29	12											1.4300 ± 0.001	34
	29	12											1.7539 ± 0.0007	22
	29	12											2.2237 ± 0.0004	100
	29	12											2.865 ± 0.001	1
$_{12}Mg^{30}$	30	12		29.990230	0.33s.	β-	6.10				0+			
$_{12}Mg^{31}$	31	12		30.995930	0.25 s.	β-	11.2							
Al				26.98154						233 mb.				
$_{13}Al^{22}$	22	13		22.079370	70 ms.	β+	18.5						ann.rad.	
	22	13				β+,p								
$_{13}Al^{23}$	23	13		23.007265	0.47 s.	β+	12.24						ann.rad.	
	23	13				β+,p								
$_{13}Al^{24}$	24	13		23.999941	2.07 s.	β+	13.87	3.40	48%		4+		1.078 ± 0.002	16%
	24	13						4.42	41%				1.368 ± 0.002	96%
	24	13						6.80	3%				2.753 ± 0.002	43%
	24	13						8.74	8%				3.205 ± 0.002	4%
	24	13											3.505 ± 0.002	2%
	24	13											3.886 ± 0.002	6%
	24	13											4.200 ± 0.002	4.4%

Isotope	A	Z	% Natural abundance	Atomic mass	Half-life	Decay mode	Decay energy (MeV)	Particle energy (MeV)	Particle intensity	Thermal neutron cross section	Spin (h/2π)	μ Nucl. mag. moment	Gamma-ray energy (MeV)	Gamma-ray intensity
	24	13											4.237 ± 0.002	3.6%
	24	13											4.315 ± 0.003	15%
	24	13											4.640 ± 0.003	3.6%
	24	13											5.177 ± 0.003	1%
	24	13											5.392 ± 0.003	20%
	24	13											7.0662 ± 0.002	41%
	24	13											7.928 ± 0.003	1.4%
$_{13}Al^{25}$	25	13		24.990429	7.17 s.	β+	4.277	3.27			5/2+	3.6455	ann.rad.	
	25	13											1.6115 ± 0.0002	100 +
	25	13											0.975 ± 0.002	5
$_{13}Al^{26m}$	26	13			6.34 s.	β+		3.2			0+		ann.rad.	
$_{13}Al^{26}$	26	13		25.986892	7.2 x 10^5y	β+ (82%)	4.005	1.16			5+		ann.rad.	
	26	13				E.C. (18%)							1.80865 ± 0.00007	99.8%
	26	13											1.12967 ± 0.0001	2.5%
	26	13											2.938 ± 0.002	.24%
$_{13}Al^{27}$	27	13	100%							233 mb.	5/2+	+3.64150		
$_{13}Al^{28}$	28	13		27.981910	2.25 m.	β-	4.642	2.865	100%		3+	2.791	1.7778 ± 0.0006	100%
$_{13}Al^{29}$	29	13		28.980446	6.5 m.	β-	3.68	1.4	30%		5/2+		1.2732 ± 0.0008	89%
	29	13						2.5	70%				2.0282 ± 0.0008	4%
	29	13											2.4262 ± 0.0008	7%
$_{13}Al^{30}$	30	13		29.982940	3.69 s.	β-	8.54	5.05					1.26313 ± 0.00003	35%
	30	13											1.3115 ± 0.0006	2.5%
	30	13											1.7330 ± 0.0005	2%
	30	13											2.23525 ± 0.00005	65%
	30	13											2.5951 ± 0.0005	5%
$_{13}Al^{31}$	31	13		30.983800	0.64 s.	β-	7.9	6.25					0.6281 ± 0.0003	10 +
	31	13											0.75223 ± 0.0003	18
	31	13											1.56449 ± 0.0003	17
	31	13											1.69473 ± 0.0003	59
	31	13											2.31664 ± 0.0004	73
	31	13											2.7876 ± 0.0005	4
Si				28.0855						171 mb.				
$_{14}Si^{24}$	24	14		24.011546	0.1 ms.	β+,p	10.82						ann.rad.	
$_{14}Si^{25}$	25	14		25.004109	220 ms.	β+,p	12.74						ann.rad.	
$_{14}Si^{26}$	26	14		25.992330	2.20 s.	β+	5.065	3.282			0+		ann.rad.	
	26	14											0.8294 ± 0.0008	22%
	26	14											1.6223 ± 0.001	2.6%
	26	14											1.8442 ± 0.002	0.3%
$_{14}Si^{27}$	27	14		26.986704	4.14 s.	β+	4.812	3.85	100%		5/2+		ann.rad.	
	27	14					1.45						2.211 ± 0.005	0.2%
$_{14}Si^{28}$	28	14	92.23%	27.976927						171 mb.	0+			
$_{14}Si^{29}$	29	14	4.67%	28.976495						0.1 b.	1/2+	-0.5553		
$_{14}Si^{30}$	30	14	3.10%	29.973770						0.107 b.	0+			
$_{14}Si^{31}$	31	14		30.975362	2.62 h.	β-	1.49	1.471	99.9%		3/2+		1.2662 ± 0.0005	0.07%
$_{14}Si^{32}$	32	14		31.974148	≈100 y.	β-	0.227	0.213	100%		0+			
$_{14}Si^{33}$	33	14		32.977920	6.2 s.	5.77	3.92						1.4313 ± 0.0005	13 +
	33	14											1.84769 ± 0.0005	100
	33	14											2.538 ± 0.002	10
$_{14}Si^{34}$	34	14		33.97636	2.8 s.	β-	4.7	3.09			0+		0.42907 ± 0.0005	60 +
	34	14											1.17852 ± 0.0002	64
	34	14											1.60756 ± 0.0005	36
P		15		30.97376						0.180 b.				
$_{15}P^{26}$	26	15		26.012080	≈20 ms.	β+,p								
$_{15}P^{28}$	28	15		27.992313	270 ms.	β+	14.33	3.94	13%		3+		ann.rad.	
	28	15						5.25	13%				1.779 ± 0.002	98%
	28	15						6.96	16%				2.839 ± 0.002	2.8%
	28	15						8.8	7%				3.040 ± 0.002	3.2%
	28	15						11.49	52%				4.498 ± 0.002	12%
	28	15											6.021 ± 0.002	1.9%
	28	15											6.810 ± 0.002	2%
	28	15											7.537 ± 0.002	9%
	28	15											7.933 ± 0.002	2%
$_{15}P^{29}$	29	15		28.981803	4.14 s.	β+	4.944	3.945	98%		1/2+	1.2349	ann.rad.	
	29	15											1.28	0.8%
	29	15											2.43	0.2%
$_{15}P^{30}$	30	15		29.978307	2.50 m.	β+	4.226	3.245	99.9%		1+		ann.rad.	
	30	15											2.230 ± 0.003	0.07%
$_{15}P^{31}$	31	15	100%	30.973762						0.233 b.	1/2+	+1.13160		
$_{15}P^{32}$	32	15		31.973907	14.28 d.	β-	1.710	1.710	100%		1+	-0.2524		
$_{15}P^{33}$	33	15		32.971725	25.3 d.	β-	0.249	0.249	100%		1/2+			
$_{15}P^{34}$	34	15		33.973636	12.4 s.	β-	5.37	3.2	15%		1+		1.78 -4.1 (weak)	
	34	15						5.1	85%				2.127 ± 0.005	15%
$_{15}P^{35}$	35	15		34.973232	47.s.	β-	3.9	2.34	100%				1.5722 ± 0.001	100%
$_{15}P^{36}$	36	15		35.977570	5.9 s.	β-	9.8						0.902	
	36	15											3.291	
S		16		32.066						0.52 b.				
$_{16}S^{29}$	29	16		28.996610	0.19 s.	β+	13.79				5/2+		ann.rad.	
	29	16				β+,p								

TABLE OF THE ISOTOPES (Continued)

Isotope	A	Z	% Natural abundance	Atomic mass	Half-life	Decay mode	Decay energy (MeV)	Particle energy (MeV)	Particle intensity	Thermal neutron cross section	Spin (h/2π)	μ Nucl. mag. moment	Gamma-ray energy (MeV)	Gamma-ray intensity
$_{16}S^{30}$	30	16		29.984903	1.18 s.	β+	6.144	4.42	78%		0+		ann.rad.	
	30	16						5.08	20%				0.678	79%
$_{16}S^{31}$	31	16		30.979554	2.55 s.	β+	5.396	4.39	99%		1/2+	0.48793	ann.rad.	
	31	16											1.2662 ± 0.0005	1.2%
	31	16											3.135 ± 0.005	0.03%
	31	16											3.505 ± 0.005	0.01%
$_{16}S^{32}$		16	95.02%	31.972070						0.52 b.	0+			
$_{16}S^{33}$	33	16	0.75%	32.971456						1.4 mb.	3/2+	+0.64382		
$_{16}S^{34}$	34	16	4.21%	33.967866						0.2 b.	0+			
$_{16}S^{35}$	35	16		34.969031	87.2 d.	β-	0.167	0.1674	100%		3/2+	+1.00		
$_{16}S^{36}$	36	16	0.02%	35.967080						0.23 b.	0+			
$_{16}S^{37}$	37	16		36.971125	5.05 m.	β-	4.865	1.64	94%				0.9083 ± 0.0004	0.06%
	37	16						4.75	5.6%				3.1033 ± 0.00002	94.2%
	37	16											3.7416 ± 0.0005	0.2%
$_{16}S^{38}$	38	16		37.971162	2.84 h.	β-	2.94	1.00			0+		0.1962 ± 0.0004	0.2%
	38	16											1.8459 ± 0.0005	2.4%
	38	16											1.9421 ± 0.0003	84%
	38	16											2.7516 ± 0.0005	1.6%
$_{16}S^{39}$	39	16		38.975310	11.5 s.	β-	6.8						1.301	
	39	16											1.697	
Cl		17		35.453						33.5 b.				
$_{17}Cl^{31}$	31	17		30.992410	0.15 s.	β+	12.0						ann.rad.	
	31	17				β+,p								
$_{17}Cl^{32}$	32	17		31.985690	297 ms.	β+	12.69	4.75	25%		1+		ann.rad.	
	32	17						6.18	10%				1.548 ± 0.002	3.5%
	32	17						7.48	14%				2.2305 ± 0.001	92%
	32	17						9.47	50%				2.4638 ± 0.001	4%
	32	17						11.6	1%				2.885 ± 0.001	1%
	32	17											3.3175 ± 0.001	2.4%
	32	17											4.281 ± 0.001	2.5%
	32	17											4.433 ± 0.001	0.8%
	32	17											4.694 ± 0.001	2.7%
	32	17											4.770 ± 0.001	20%
	32	17											5.549 ± 0.002	1.6%
	32	17											7.194 ± 0.003	0.4%
$_{17}Cl^{33}$	33	17		32.977451	2.51 s.	β+	5.583	4.51	98%		3/2+		ann.rad.	
	33	17											0.8405 ± 0.001	0.56%
	33	17											1.9661 ± 0.0005	0.56%
	33	17											2.8665 ± 0.0005	0.56%
$_{17}Cl^{34m}$	34	17			32.2 m.	β+		1.35	24%		3+		ann.rad.	
	34	17						2.47	28%					
	34	17				I.T.							0.1457 ± 0.0008	42%
	34	17											1.1758 ± 0.0005	10%
	34	17											2.1276 ± 0.0005	42%
	34	17											3.3037 ± 0.001	10%
$_{17}Cl^{34}$	34	17		33.973763	1.53 s.	β+	5.492	4.50	100%		0+		ann.rad.	
$_{17}Cl^{35}$	35	17	75.77%	34.968852						43.6 b.	3/2+	+0.82187		
$_{17}Cl^{36}$	36	17		35.968306	3 x 10^5y.	β-	0.7093	0.7093	98%		2+	+1.28547		
	36	17				β+,E.C.	1.142	0.115	0.002%				ann.rad.	
$_{17}Cl^{37}$	37	17	24.23%	36.965903						0.4 b.	3/2+	+0.68412		
$_{17}Cl^{38m}$	38	17			0.70 s.	I.T.					5-		0.67138 ± 0.0002	100%
$_{17}Cl^{38}$	38	17		37.968010	37.2 m.	β-	4.917	1.11	31%		2-	2.05	1.64216 ± 0.0001	31%
	38	17						2.77	11%				2.16760 ± 0.0002	42%
	38	17						4.91	58%					
$_{17}Cl^{39}$	39	17		38.968005	55.7 m.	β-	3.44	1.91	85%		3/2+		0.25026 ± 0.0001	47%
	39	17						2.18	8%				0.98579 ± 0.0001	3.2%
	39	17						3.45	7%				1.09097 ± 0.0001	4.0%
	39	17											1.26720 ± 0.0005	54%
	39	17											1.51736 ± 0.0008	38%
$_{17}Cl^{40}$	40	17		39.970440	1.35 m.	β-	7.5				2-		0.6431 ± 0.0003	6%
	40	17											0.6591 ± 0.0002	2%
	40	17											0.8810 ± 0.0003	2.4%
	40	17											1.0629 ± 0.0002	2.4%
	40	17											1.4608 ± 0.0001	77%
	40	17											1.5889 ± 0.0002	7%
	40	17											1.7465 ± 0.0002	2.5%
	40	17											1.7978 ± 0.0002	2.4%
	40	17											2.0505 ± 0.0004	1%
	40	17											2.2201 ± 0.0002	7.5%
	40	17											2.4579 ± 0.0002	5%
	40	17											2.6219 ± 0.0002	14%
	40	17											2.8402 ± 0.0002	17%
	40	17											3.1010 ± 0.0002	11%
	40	17											3.9186 ± 0.0002	4%
	40	17											5.8799 ± 0.0008	4%
$_{17}Cl^{41}$	41	17		40.970590	34 s.	β-	5.67	3.8					(0.167 - 1.359)	
Ar		18		39.948						0.66 b.				
$_{18}Ar^{32}$	32	18		31.997660	≈0.1 s.	β+,p	11.2						ann. rad.	
$_{18}Ar^{33}$	33	18		32.989930	17 ms.	β+	11.62	3.12			1/2+		ann.rad.	
	33	18				β+,p							0.810 ± 0.002	48%
$_{18}Ar^{34}$	34	18		33.980269	0.844 s.	β+	6.061	5.0	95%		0+		ann.rad.	
	34	18											0.4608 ± 0.001	0.8%

Isotope	A	Z	% Natural abundance	Atomic mass	Half-life	Decay mode	Decay energy (MeV)	Particle energy (MeV)	Particle intensity	Thermal neutron cross section	Spin (h/2π)	μ Nucl. mag. moment	Gamma-ray energy (MeV)	Gamma-ray intensity
	34	18											0.6658 ± 0.001	2.5%
	34	18											2.5795 ± 0.001	0.8%
	34	18											3.1290 ± 0.001	1.3%
$_{18}Ar^{35}$	35	18		34.975256	1.77 s.	β+	5.965	4.94	93%		3/2+	+0.633	ann.rad.	
	35	18											1.2185 ± 0.0005	1.22%
	35	18											1.763 ± 0.001	0.25%
	35	18											2.964 ± 0.001	0.2%
	35	18											3.003 ± 0.005	0.1%
$_{18}Ar^{36}$	36	18	0.337%	35.967545						5.5 mb.	0+			
$_{18}Ar^{37}$	37	18		36.966776	34.8 d.	E.C.					3/2+	+0.95		
$_{18}Ar^{38}$	38	18	0.63%	37.962732							0+			
$_{18}Ar^{39}$	39	18		38.964314	269 y.	β-	0.565	0.565	100%	600 b.	7/2-	-1.3		
$_{18}Ar^{40}$	40	18	99.60%	39.962384						0.65 b.	0+			
$_{18}Ar^{41}$	41	18		40.964501	1.83 h.	β-	2.492	1.198		0.5 b.	7/2-		1.29364 ± 0.00005	99%
	41	18											1.6770 ± 0.0003	0.05%
$_{18}Ar^{42}$	42	18		41.963050	33 y.	β-	0.60	0.60	100%		0+			
$_{18}Ar^{43}$	43	18		42.965670	5.4 m.	β-	4.6						0.4791 ± 0.002	10 +
	43	18											0.7300 ± 0.0001	13
	43	18											0.9752 ± 0.0001	100
	43	18											1.4400 ± 0.0003	39
	43	18											2.3455 ± 0.0005	28
$_{18}Ar^{44}$	44	18		43.96365	11.9 m.	β-	3.54				0+		0.182	
	44	18											1.703	
	44	18											1.866	
$_{18}Ar^{45}$	45	18		44.968090	21 s.	β-	6.9				7/2-		0.0610	
	45	18											1.020	
	45	18											3.707	
$_{18}Ar^{46}$	46	18		45.968090	8.3 s.	β-	5.70				0+		1.944	
K		19		39.0983						2.1 b.				
$_{19}K^{35}$	35	19		34.988011	0.19 s.	β+	11.88				3/2+		ann.rad.	
	35	19				β+,p							1.751	
	35	19											2.5698	
	35	19											2.9827	
$_{19}K^{36}$	36	19		35.981293	0.342 s.	β+	12.81	5.3	42%		2+	0.548	ann.rad.	
	36	19						9.9	44%				1.97044 ± 0.0005	82%
	36	19											2.17029 ± 0.0002	3%
	36	19											2.20783 ± 0.0005	30%
	36	19											2.43343 ± 0.0002	32%
	36	19											2.47046 ± 0.0004	5%
	36	19											4.44079 ± 0.0003	8%
	36	19											6.61213 ± 0.0004	7%
$_{19}K^{37}$	37	19		36.973377	1.23 s.	β+	6.149	5.13			3/2+	+0.20321	ann.rad.	
	37	19											2.7944 ± 0.0008	2%
	37	19											3.602 ± 0.002	0.05%
$_{19}K^{38m}$	38	19			0.926 s.	β+	6.742	5.02	100%		0+		ann.rad.	
$_{19}K^{38}$	38	19		37.969080	7.63 m.	β+	5.913	2.60	99.8%		3+	+1.374	ann.rad.	
	38	19											2.1675 ± 0.0003	99.8%
	38	19											3.9356 ± 0.0005	0.2%
$_{19}K^{39}$	39	19	93.2581%	38.963707						2.10 b.	3/2+	+0.39146		
$_{19}K^{40}$	40	19	0.0117%	39.963999	1.25x10^{9}y.	β-	1.32	1.312	89%		4-	-1.298	ann.rad.	
	40	19				β+,E.C.	1.50		10.7%				1.46081 ± 0.00005	10.5%
$_{19}K^{41}$	41	19	6.7302%	40.961825						1.46 b.	3/2+	+0.21487		
$_{19}K^{42}$	42	19		41.962402	12.36 h.	β-	3.523	1.97	19%		2-	-1.1425	0.31260 ± 0.0002	0.3%
	43	19						3.523	81%				1.5246 ± 0.0003	18.9%
$_{19}K^{43}$	43	19		42.960717	22.3 h.	β-	1.82	0.465	8%		3/2+	0.163	0.2211 ± 0.0002	4%
	43	19						0.825	87%				0.3729 ± 0.0002	88%
	43	19						1.24	3.5%				0.3971 ± 0.0002	11%
	43	19						1.814	1.3%				0.6178 ± 0.0002	81%
$_{19}K^{44}$	44	19		43.96156	22.1 m.	β-	5.66	5.66	34%		2-		0.368207 ± 0.0001	2.2%
	44	19											0.65135 ± 0.00001	2.0%
	44	19											0.72649 ± 0.00001	2.4%
	44	19											0.74763 ± 0.00001	1.2%
	44	19											0.87653 ± 0.00003	1.1%
	44	19											1.01955 ± 0.00007	5.5%
	44	19											1.02474 ± 0.00002	6%
	44	19											1.10799 ± 0.00001	5%
	44	19											1.12608 ± 0.00001	7%
	44	19											1.15700 ± 0.00001	58%
	44	19											1.24475 ± 0.00005	8%
	44	19											1.75263 ± 0.00001	4%

Isotope	A	Z	% Natural abundance	Atomic mass	Half-life	Decay mode	Decay energy (MeV)	Particle energy (MeV)	Particle intensity	Thermal neutron cross section	Spin (h/2π)	μ Nucl. mag. moment	Gamma-ray energy (MeV)	Gamma-ray intensity
	44	19											1.77797 ± 0.00002	2%
	44	19											2.14423 ± 0.00008	7%
	44	19											2.15079 ± 0.00002	22%
	44	19											2.51899 ± 0.00002	9%
	44	19											2.65641 ± 0.00003	4.6%
	44	19											3.39551 ± 0.00004	1.6%
	44	19											3.66136 ± 0.00002	6%
$_{19}K^{45}$	45	19		44.960696	17.3 m.	β-	4.20	1.1	23%		3/2+	0.1734	0.1743 ± 0.0005	80%
	45	19						2.1	69%				1.2607 ± 0.0008	7%
	45	19						4.0	8%				1.7056 ± 0.0006	69%
	45	19											2.3542 ± 0.0005	14%
	45	19											2.5988 ± 0.001	3%
$_{19}K^{46}$	46	19		45.961976	107 s.	β-	7.72	6.3			2-		1.347 ± 0.001	91%
	45	19											1.439 ± 0.002	3%
	45	19											1.670 ± 0.002	3%
	45	19											1.780 ± 0.002	8%
	45	19											2.274 ± 0.002	8%
	45	19											3.015 ± 0.005	9%
	45	19											3.700 ± 0.005	28%
$_{19}K^{47}$	47	19		46.961677	17.5 s.	β-	6.65	4.1	99%		1/2+		0.56474 ± 0.0003	15%
	47	19						6.0	1%				0.58575 ± 0.0003	85%
	47	19											2.01313 ± 0.0003	100%
$_{19}K^{48}$	48	19		47.965514	69 s.	β-	12.1	5.0			(2-)		0.67122 ± 0.0001	4%
	48	19											0.6723 ± 0.00005	20%
	48	19											0.78016 ± 0.0001	32%
	48	19											0.7931 ± 0.0001	10%
	48	19											0.86275 ± 0.0001	4.5%
	48	19											1.3009 ± 0.0002	9%
	48	19											1.5378 ± 0.0001	15%
	48	19											2.2830 ± 0.0003	3%
	48	19											2.3881 ± 0.0001	11%
	48	19											2.7889 ± 0.0001	18%
	48	19											3.06229 ± 0.0003	5%
	48	19											3.83153 ± 0.00007	80%
	48	19											4.5072 ± 0.0003	4%
	48	19											6.6137 ± 0.0005	14%
	48	19											7.3009 ± 0.0005	2%
$_{19}K^{49}$	49	19		48.966940	1.3 s.	β-	11						2.025	
	49	19											2.252	
$_{19}K^{50}$	50	19			≈0.7 s.	β-	16							
$_{19}K^{51}$					0.38 s.	β-								
Ca		20		40.078						0.43 b.				
$_{20}Ca^{36}$	36	20		35.993090	0.1 s.	β+	11.0						ann.rad.	
	36	20				β+,n								
$_{20}Ca^{37}$	37	20		36.985873	175 ms.	β+	11.6				3/2+		ann.rad.	
	37	20				β+,n								
$_{20}Ca^{38}$	38	20		37.976318	0.45 s.	β+	6.74				0+		ann.rad.	
	38	20											1.5677 ± 0.0005	25%
	38	20											3.210 ± 0.002	1%
$_{20}Ca^{39}$	39	20		38.970718	0.86 s.	β+	6.531	5.49	100%		3/2+	1.02168	ann.rad.	
$_{20}Ca^{40}$	40	20	96.941%	39.962591						0.41 b.	0+			
$_{20}Ca^{41}$	41	20		40.962278	1 x 10^5 y.	E.C.	0.421				7/2-	-1.595		
$_{20}Ca^{42}$	42	20	0.647%	41.958618						0.7 b.	0+			
$_{20}Ca^{43}$	43	20	0.135%	42.958766						6 b.	7/2-	-1.3173		
$_{20}Ca^{44}$	44	20	2.086%	43.955480						0.8 b.	0+			
$_{20}Ca^{45}$	45	20		44.956185	163.8 d.	β-	0.257	0.257	100%		7/2-			
$_{20}Ca^{46}$	46	20	0.004%	45.953689						0.7 b.	0+			
$_{20}Ca^{47}$	47	20		46.954543	4.536 d.	β-	1.988	0.684	84%		7/2-		0.4889 ± 0.0003	9%
	47	20						1.98	16%				0.8079 ± 0.0003	9%
	47	20											1.29680 ± 0.0002	77%
$_{20}Ca^{48}$	48	20	0.187%	47.952533						1.1 b.	0+			
$_{20}Ca^{49}$	49	20		48.955672	8.72 m.	β-	5.263	0.89	7%		3/2-		3.0844 ± 0.0001	92%
	49	20						1.95	92%				4.0719 ± 0.0001	7%
$_{20}Ca^{50}$	50	20		49.957519	14 s.	β-	4.97	3.12			0+		0.2569	
	50	20											(0.0715 - 1.591)	
$_{20}Ca^{51}$	51	20		50.961420	10 s.	β-	0.728							
Sc		21		44.95591						27.2 b.				
$_{21}Sc^{40}$	40	21		39.977963	0.182 s.	β+	14.32	5.73	50%		4-		ann.rad.	
	40	21						7.53	15%				0.7556 ± 0.0008	41%
	40	21						8.76	15%				1.126 ± 0.003	12%
	40	21						9.58	20%				1.8778 ± 0.0007	25%
	40	21											2.0458 ± 0.0007	25%
	40	21											3.1679 ± 0.0007	12%
	40	21											3.7356 ± 0.0008	100%
	40	21											3.920 ± 0.001	13%

TABLE OF THE ISOTOPES (Continued)

Isotope	A	Z	% Natural abundance	Atomic mass	Half-life	Decay mode	Decay energy (MeV)	Particle energy (MeV)	Particle intensity	Thermal neutron cross section	Spin (h/2π)	μ Nucl. mag. moment	Gamma-ray energy (MeV)	Gamma-ray intensity
$_{21}Sc^{41}$	41	21		40.96250	0.596 s.	β+	6.495	5.61	100%		7/2-	5.43	ann.rad.	
$_{21}Sc^{42m}$	42	21			61.6 s.	β+		2.82			7+		ann.rad.	
	42	21											0.4375 ± 0.0005	100%
	42	21											1.2270 ± 0.0005	100%
	42	21											1.5245 ± 0.0005	100%
$_{21}Sc^{42}$	42	21		41.965514	0.68 s.	β+	6.424	5.32	100%		0+		ann.rad.	
$_{21}Sc^{43}$	43	21		42.961150	3.89 h.	β+,E.C.	2.221	0.82	22%		7/2-	+4.62	ann.rad.	
	43	21						1.22	78%				0.3729 ± 0.0001	22%
$_{21}Sc^{44m}$	44	21			58.6 h.	I.T.	0.27				6+	+3.88	0.27124 ± 0.0001	98.4%
	44	21				E.C.	3.926						1.0018 ± 0.00003	1.4%
	44	21											1.12606 ± 0.00003	1.4%
	44	21											1.15700 ± 0.0003	1.4%
$_{21}Sc^{44}$	44	21		43.959404	3.93 h.	β+, E.C.	3.655	1.47			0+	+2.56	ann.rad.	
	44	21											1.15700 ± 0.00001	100%
	44	21											1.49945 ± 0.00002	1%
$_{21}Sc^{45}$	45	21	100%	44.955910						(0.1 + 17)b.	7/2	+4.756		
$_{21}Sc^{46m}$	46	21			18.7 s.	I.T.	0.14253				1-		0.14253 ± 0.00002	62%
$_{21}Sc^{46}$	46	21		45.995170	83.8 d.	β-	2.367	0.357	100%		4+	+3.03	0.88925 ± 0.00003	100%
	46	21											1.12051 ± 0.00001	100%
$_{21}Sc^{47}$	47	21		46.952408	3.42 d.	β-	0.601	0.439	69%		7/2-	+5.34	0.15938 ± 0.00001	68%
	47	21						0.601	31%					
$_{21}Sc^{48}$	48	21		47.952235	43.7 h.	β-	3.99	0.655			6+		0.98350 ± 0.0001	100%
	48	21											1.03750 ± 0.00001	97%
	48	21											1.21285 ± 0.00007	2%
	48	21											1.31209 ± 0.00003	100%
$_{21}Sc^{49}$	49	21		48.950022	57.3 m.	β-	2.005	2.00	99.9%		7/2-		1.7619 ± 0.0003	0.05%
$_{21}Sc^{50}$	50	21		49.952186	1.71 m.	β-	3.05	76%			(5+)		0.5235 ± 0.0001	88%
	50	21						3.60	24%				1.1210 ± 0.0001	100%
	50	21											1.5537 ± 0.0002	100%
$_{21}Sc^{51}$	51	21		50.953602	12.4 s.	β-	6.51	4.4			7/2-		0.7177 ± 0.0004	7%
	51	21						5.0					0.9072 ± 0.0004	9%
	51	21											1.2938 ± 0.0004	6%
	51	21											1.4373 ± 0.0004	52%
	51	21											1.5675 ± 0.0004	15%
	51	21											2.0511 ± 0.0004	8%
	51	21											2.1441 ± 0.0004	31%
Ti		22		47.88						6.1 b.				
$_{22}Ti^{41}$	41	22		40.983150	80 ms.	β+,p	12.94				3/2+		ann.rad.	
$_{22}Ti^{42}$	42	22		41.973031	0.20 s.	β+	7.001	6.0					ann.rad.	
	42	22											0.6107 ± 0.0005	56%
$_{22}Ti^{43}$	43	22		42.968523	0.49 s.	β+	6.867	5.80			7/2-		ann.rad.	
$_{22}Ti^{44}$	44	22		43.959689	47 y.	E.C.	0.265				0+		0.06785 ± 0.00004	88%
	44	22											0.07838 ± 0.00004	93%
$_{22}Ti^{45}$	45	22		44.958124	3.078 h.	β+ (86%)	2.063	1.04			7/2-	0.095	ann.rad.	
	45	22				E.C. (14%)							(0.36-1.66)weak	
$_{22}Ti^{46}$	46	22	8.0%	45.952629						0.6 b.	0+			
$_{22}Ti^{47}$	47	22	7.3%	46.951764						1.7 b.	5/2-	-0.7885		
$_{22}Ti^{48}$	48	22	73.8%	47.947947						7.9 b.	0+			
$_{22}Ti^{49}$	49	22	5.5%	48.947871						2.2 b.	7/2-	-1.10417		
$_{22}Ti^{50}$	50	22	5.4%	49.944792						0.177 b.	0+			
$_{22}Ti^{51}$	51	22		50.946616	5.76 m.	β-	2.472	1.50	92%		3/2-		0.3197 ± 0.0002	93%
	51	22						2.13					0.6094 ± 0.0003	1%
	51	22											0.9291 ± 0.0003	6%
$_{22}Ti^{52}$	52	22		51.946898	1.7 m.	β-	1.97	1.8	100%		0+		0.0170 ± 0.0005	100%
	52	22											0.12445 ± 0.00005	100%
$_{22}Ti^{53}$	53	22		52.949730	33 s.	β-	5.02	(2.2-3)			3/2-		0.1008 ± 0.0001	20%
	53	22											0.1276 ± 0.0001	45%
	53	22											0.2284 ± 0.0001	39%
	53	22											0.6796 ± 0.001	4%
	53	22											1.001 ± 0.001	4%
	53	22											1.3211 ± 0.001	6%
	53	22											1.4217 ± 0.001	10%
	53	22											1.6755 ± 0.0005	45%
	53	22											(1.72-2.8)	
V		23		50.9415						5.06 b.				
$_{23}V^{44}$	44	23		43.974450	0.09 s.	β+,α	13.7						ann.rad.	
$_{23}V^{46}$	46	23		45.960198	0.422 s.	β+	7.05	6.03	100%		0+		ann.rad.	
$_{23}V^{47}$	47	23		46.954906	31.3 m.	β+,E.C.	2.927	1.90	99+%		3/2-		ann.rad.	
	47	23											1.7949 ± 0.0008	0.19%
	47	23											(0.2-2.16)weak	

TABLE OF THE ISOTOPES (Continued)

Isotope	A	Z	% Natural abundance	Atomic mass	Half-life	Decay mode	Decay energy (MeV)	Particle energy (MeV)	Particle intensity	Thermal neutron cross section	Spin (h/2π)	μ Nucl. mag. moment	Gamma-ray energy (MeV)	Gamma-ray intensity
$_{23}V^{48}$	48	23		47.952257	15.98 d.	β+	4.015	0.698	50%		4+	1.63	ann.rad.	
	48	23											0.94410 ± 0.00002	8%
	48	23											0.98350 ± 0.00002	100%
	48	23											(1.3-2.4)weak	
$_{23}V^{49}$	49	23		48.948517	331 d.	E.C.	0.601				7/2-	4.47		
$_{23}V^{50}$	50	23	0.25%	49.947161	>3.9x10^{17}y	E.C., β-				0.1 b.	6+	+3.34745		
$_{23}V^{51}$	51	23	99.75%	50.943962						4.9 b.	7/2-	+5.1514		
$_{23}V^{52}$	52	23		51.944778	3.76 m.	β-	3.976	2.47			3/2-		1.4341 ± 0.0001	100%
$_{23}V^{53}$	53	23		52.944340	1.61 m.	β-	3.436	2.52			7/2-		1.0060 ± 0.0005	90%
	53	23											1.2891 ± 0.0003	10%
$_{23}V^{54}$	54	23		53.946442	49.8 s.	β-	7.04	1.00	5%		(5+)		0.564 ± 0.002	4%
	54	23						2.00	12%				0.8351 ± 0.0001	100%
	54	23						2.95	45%				0.986 ± 0.001	82%
	54	23						5.20	11%				1.462 ± 0.002	7%
	54	23											1.784 ± 0.003	7%
	54	23											2.255 ± 0.003	50%
	54	23											2.353 ± 0.003	12%
	54	23											3.170 ± 0.005	12%
$_{23}V^{55}$	55	23		54.947240	6.5 s.	β-	6.0	6.0					0.517	
	55	23											0.8806	
Cr		24								3.1 b.				
$_{24}Cr^{45}$	45	24		44.979110	0.05 s.	β+,p	12.4				7/2-		ann.rad.	
$_{24}Cr^{46}$	46	24		45.968360	≈0.26 s.	β+	7.61						ann.rad.	
$_{24}Cr^{47}$	47	24		46.962905	460 ms.	β+	7.45						ann.rad.	
$_{24}Cr^{48}$	48	24		47.954033	21.6 h.	E.C.	1.65						ann.rad.	
	48	24											0.116 ± 0.002	95%
	48	24											0.305 ± 0.010	100%
$_{24}Cr^{49}$	49	24		48.951338	42.1 m.	β+,E.C.	2.627	1.39			5/2-	0.476	ann.rad.	
	49	24						1/45					0.06229 ± 0.00001	0.04%
	49	24						1.54					0.09064 ± 0.00001	51%
	49	24											0.15293 ± 0.00001	27%
	49	24											(0.2-1.6)weak	
$_{24}Cr^{50}$	50	24	4.35%	49.946046						15.8 b.	0+			
$_{24}Cr^{51}$	51	24		50.944768	27.70 d.	E.C.	0.751				7/2-	-0.934	0.320076 ± 0.0001	10.2%
$_{24}Cr^{52}$	52	24	83.79%	51.940509						0.8 b.	0+			
$_{24}Cr^{53}$	53	24	9.50%	52.940651						18 b.	3/2-	-0.47454		
$_{24}Cr^{54}$	54	24	2.36%	53.938882						0.36 b.	0+			
$_{24}Cr^{55}$	55	24		54.940842	3.50 m.	β-	2.603	2.5			3/2-		1.5282 ± 0.0002	0.04%
	55	24											2.2518 ± 0.0003	0.004%
$_{24}Cr^{56}$	56	24		55.940643	5.9 m.	β-	1.62	1.50	100%		0+		0.026 ± 0.002	100%
	56	24											0.083 ± 0.003	100%
$_{24}Cr^{57}$	57	24		56.943440	21 s.	β-	≈4.7	3.3			3/2-	0.0834		
	57	24						3.5					0.850	
	57	24											1.752	
Mn		25		54.9380						13.3 b.				
$_{25}Mn^{49}$	49	25		48.951338	0.38 s.	β+	7.72	6.69			5/2-		ann.rad.	
$_{25}Mn^{50m}$	50	25			1.7 m.	β+	7.887	3.54			5+		ann.rad.	
	50	25											1.0980 ± 0.0002	100%
	50	25											1.2824 ± 0.0005	33%
	50	25											1.4433 ± 0.0002	70%
	50	25											1.9445 ± 0.0005	4%
	50	25											3.1152 ± 0.001	1%
$_{25}Mn^{50}$	50	25		49.954239	0.283 s.	β+	7.632	6.61			0+		ann.rad.	
$_{25}Mn^{51}$	51	25		50.948213	46.2 m.	β+,E.C.	3.209	2.2			5/2-	3.568	ann.rad.	
	51	25											0.7491 ± 0.0001	0.26%
	51	25											1.1480 ± 0.0001	0.1%
	51	25											1.1644 ± 0.0001	0.1%
	51	25											2.00135 ± 0.001	0.05%
$_{25}Mn^{52m}$	52	25			21.1 m.	β+(98%)	5.09	2.631			2+	0.0076	ann.rad.	
	52	25				I.T.(2%)	0.378						0.3778 (I.T.)	
	52	25											1.43406 ± 0.00001	98%
	52	25											(0.7-4.8)weak	
$_{25}Mn^{52}$	52	25		51.945568	5.59 d.	β+	4.712	0.575			6+	+3.0621	ann.rad.	
	52	25				E.C.							0.74421 ± 0.00001	90%
	52	25											0.84816 ± 0.00005	3%
	52	25											1.2462 ± 0.0003	4%
	52	25											1.3336 ± 0.0001	5%
	52	25											1.43406 ± 0.00001	100%
$_{25}Mn^{53}$	53	25		52.941291	3.7x10^{6} y.	E.C.	0.596			70 b.	7/2-	5.024		
$_{25}Mn^{54}$	54	25		53.940361	312 d.	E.C.	1.377				3+	+3.2818	0.83403 ± 0.00005	100%
$_{25}Mn^{55}$	55	25	100%	54.938047						13.3 b.	5/2-	+3.4687		
$_{25}Mn^{56}$	56	25		55.938906	2.579 h.	β-	3.696	0.718	18%		3+	+3.2266	0.84675 ± 0.0002	98.9%

TABLE OF THE ISOTOPES (Continued)

Isotope	A	Z	% Natural abundance	Atomic mass	Half-life	Decay mode	Decay energy (MeV)	Particle energy (MeV)	Particle intensity	Thermal neutron cross section	Spin (h/2π)	μ Nucl. mag. moment	Gamma-ray energy (MeV)	Gamma-ray intensity
								1.028	34%				1.81072 ± 0.00004	27%
	56	25											2.11305 ± 0.00005	14.5%
	56	25											2.5229 ± 0.0005	1%
$_{25}Mn^{57}$	57	25		56.938285	1.45 m.	β-	2.691				5/2-			
$_{25}Mn^{58}$	58	25		57.940060	65 s.	β-	6.32	3.8			3+		0.45916 ± 0.0002	20%
	58	25						5.1					0.81076 ± 0.00001	82%
	58	25											0.86394 ± 0.00003	14%
	58	25											1.26574 ± 0.00005	8%
	58	25											1.32309 ± 0.00005	53%
	58	25											1.67472 ± 0.00007	10%
	58	25											2.42245 ± 0.0001	1%
	58	25											2.63815 ± 0.0001	1.2%
$_{25}Mn^{59}$	59	25		58.940440	4.6 s.	β-	5.18	4.5					0.471	
	59	25											0.531	
	59	25											0.726	
$_{25}Mn^{60}$	60	25		59.943210	1.8 s.	β-	8.5	5.7			3+		0.824	
	60	25											1.969	
$_{25}Mn^{62}$	62	25			0.9 s.	β-					(3+)		0.877	
	62	25											0.942	
	62	25											1.299	
Fe		26		55.847						2.56 b.				
$_{26}Fe^{49}$	49	26			0.08 s.	β+	13.1						ann.rad.	
$_{26}Fe^{51}$	51	26		50.956825	0.25 s.	β+	8.02						ann.rad.	
$_{26}Fe^{52m}$	52	26			46 s.	β+	4.4						ann.rad.	
	52	26											(0.622-2.286)	
$_{26}Fe^{52}$	52	26		51.948114	8.28 h.	β+(57%)	2.37	0.804			0+		ann.rad.	
	52	26				E.C.(43%)							0.16868 ± 0.00001	99%
	52	26				I.T.							0.377 (I.T.)	
$_{26}Fe^{53m}$	53	26			2.54 s.	I.T.	3.0407				19/2-		0.7011 ± 0.0001	99%
	53	26											1.0115 ± 0.0001	87%
	53	26											1.3281 ± 0.0001	87%
	53	26											2.3396 ± 0.0001	13%
$_{26}Fe^{53}$	53	26		52.945310	8.51 m.	β+	3.774	2.40	42%		7/2-		ann.rad.	
	53	26						2.80	57%				0.3779 ± 0.0001	42%
	53	26											(1.2 - 3.2)weak	
$_{26}Fe^{54}$	54	26	5.8%	53.939612						2.3 b.	0+			
$_{26}Fe^{55}$	55	26		54.938296	2.7 y.	E.C.	0.2314				3/2-			
$_{26}Fe^{56}$	56	26	91.72%	55.934939						2.6 b.	0+			
$_{26}Fe^{57}$	57	26	2.2%	56.935396						2.5 b.	½-	+0.09044		
$_{26}Fe^{58}$	58	26	0.28%	57.933277						1.26 b	0+			
$_{26}Fe^{59}$	59	26		58.934877	44.51 d.	β-	1.565	0.273	48%		3/2-	0.29	0.14265 ± 0.00002	1%
	59	26						0.475	51%				0.19234 ± 0.00006	3%
	59	26											1.09922 ± 0.00002	56%
	59	26											1.29156 ± 0.00003	43%
	59	26											1.48178 ± 0.00006	0.06%
$_{26}Fe^{60}$	60	26		59.93408	≈10⁵ y.	β-	0.243	0.184	100%		0+		0.0586 ± 0.0005	100%(IT)
$_{26}Fe^{61}$	61	26		60.936748	6.0 m.	β-	3.97	2.5	13%		3/2-		0.12034 ± 0.0001	4.4%
	61	26						2.63	54%				0.29790 ± 0.00007	22%
	61	26						2.80	31%				1.02742 ± 0.0001	43%
	61	26											1.20507 ± 0.0001	44%
	61	26											1.64595 ± 0.0001	7%
	61	26											2.0116 ± 0.0002	4%
$_{26}Fe^{62}$	62	26		61.936773	68 s.	β-	2.50	2.5	100%		0+		0.5061 ± 0.0001	100%
$_{26}Fe^{63}$	63	26		62.94075	4.9 s.	β-							0.995	
	63	26											1.365	
	63	26											1.427	
Co		27		58.9332						37.2 b.				
$_{27}Co^{53m}$	53	27			0.25 s.	β+,p					19/2-		ann.rad.	
$_{27}Co^{53}$	53	27		52.954225	0.26 s.	β+	8.30				7/2-		ann.rad.	
$_{27}Co^{54m}$	54	27			1.46 m.	β+	8.44	4.25	100%		7+		ann.rad.	
	54	27											0.411 ± 0.001	99%
	54	27											1.130 ± 0.001	100%
	54	27											1.408 ± 0.001	100%
$_{27}Co^{54}$	54	27		53.948460	0.19 s.	β+	8.242	7.34	100%		0+		ann.rad.	
$_{27}Co^{55}$	55	27		54.942001	17.5 h.	β+	3.452	0.53			7/2-	+4.822	ann.rad.	
	55	27				E.C.		1.03					0.0918 ± 0.0003	2.7%
	55	27						1.50					0.4772 ± 0.0002	20%
	55	27											0.9315 ± 0.0003	75%
	55	27											1.3167 ± 0.0003	7%

Isotope	A	Z	% Natural abundance	Atomic mass	Half-life	Decay mode	Decay energy (MeV)	Particle energy (MeV)	Particle intensity	Thermal neutron cross section	Spin (h/2π)	μ Nucl. mag. moment	Gamma-ray energy (MeV)	Gamma-ray intensity
	55	27											1.3700 ± 0.0005	3%
	55	27											1.4087 ± 0.0003	16%
$_{27}Co^{56}$	56	27		55.939841	77.7 d.	β+	4.566	1.459	18%		0+	3.830	ann.rad.	
	56	27				E.C.							0.84678 ± 0.00006	99.9%
	56	27											1.03783 ± 0.00007	14%
	56	27											1.23828 ± 0.00004	68%
	56	27											1.36022 ± 0.00002	4%
	56	27											1.77149 ± 0.00005	16%
	56	27											2.01536 ± 0.00005	3%
	56	27											2.0349 ± 0.00005	8%
	56	27											2.59858 ± 0.00008	17%
	56	27											3.20230 ± 0.0001	3%
	56	27											3.25360 ± 0.0001	7.4%
	56	27											3.27325 ± 0.0001	1.7%
	56	27											3.54805 ± 0.0002	0.18%
	56	27											3.60060 ± 0.0004	0.16%
$_{27}Co^{57}$	57	27		56.936294	271 d.	E.C.	0.836				7/2-	+4.733	0.01441 ± 0.00005	11%
	57	27											0.12206 ± 0.00002	85.6%
	57	27											0.13647 ± 0.00003	10%
$_{27}Co^{58m}$	58	27			9.1 h.	I.T.					5+		0.024889 ± 0.00002	0.035%
$_{27}Co^{58}$	58	27		57.935755	70.91 d.	β+	2.30				2+	+4.044	ann.rad.	
	58	27				E.C.							0.810755 ± 0.00003	99%
	58	27											0.86347 ± 0.0001	0.7%
	58	27											1.67473 ± 0.00006	0.52%
$_{27}Co^{59}$	59	27	100%	58.933198						(20 + 17) b.	7/2-	+4.627		
$_{27}Co^{60m}$	60	27			10.48 m.	I.T.(99.8%	0.059				2+	+4.40	0.058603 ± 0.00001	2.0%
	60	27				β-(0.02%)	1.56							
$_{27}Co^{60}$	60	27		59.933819	5.272 y.	β-	2.824	0.315	99.7%	2 b.	5+	+3.799	1.173210 ± 0.00002	100%
	60	27											1.332470 ± 0.00002	100%
$_{27}Co^{61}$	61	27		60.932478	1.65 h.	β-	1.322	1.22	95%		7/2-		0.067415 ± 0.00001	86%
	61	27											0.8417 ± 0.0005	0.7%
	61	27											0.9092 ± 0.0005	3%
$_{27}Co^{62m}$	62	27			13.9 m.	β-		0.88	25%		5+		1.1635 ± 0.0003	70%
	62	27						2.88	75%				1.1730 ± 0.0003	98%
	62	27											1.7191 ± 0.0003	7%
	62	27											2.0039 ± 0.0003	19%
	62	27											2.1049 ± 0.0003	6%
$_{27}Co^{62}$	62	27		61.934060	1.5 m.	β-	5.32	1.03	10%		2+		1.1292 ± 0.0003	13%
	62	27						1.76	5%				1.1730 ± 0.0003	83%
	62	27						2.9	20%				1.9851 ± 0.001	3%
	62	27						4.05	60%				2.3020 ± 0.001	19%
	62	27											2.3458 ± 0.001	1%
	62	27											3.159 ± 0.002	1%
$_{27}Co^{63}$	63	27		62.933614	27.5 s.	β-	3.67	3.6			7/2-		0.08713 ± 0.0001	49%
	63	27											0.1556 ± 0.0001	1.8%
	63	27											0.9817 ± 0.0003	2.6%
	63	27											1.0691 ± 0.0001	1.6%
	63	27											2.1745 ± 0.0005	1.2%
$_{27}Co^{64}$	64	27		63.935812	0.30 s.	β-	8.12	7.0			1+			
Ni		28		58.67						37.2 b.				
$_{28}Ni^{53}$	53	28		52.968430	0.05 s.	β+,p	13.2				7/2-		ann.rad.	
$_{28}Ni^{55}$	55	28		54.951336	0.19 s.	β+	8.7	7.66			7/2-		ann.rad.	
$_{28}Ni^{56}$	56	28		55.943124	6.10 d.	E.C.	2.14				0+		0.15838 ± 0.00003	98.8%
	56	28											0.26950 ± 0.00002	36%
	56	28											0.48044 ± 0.00002	32%
	56	28											0.74995 ± 0.00003	49%
	56	28											0.81185 ± 0.00003	87%
	56	28											1.56180 ± 0.00005	14%
$_{28}Ni^{57}$	57	28		56.39799	36.1 h.	β+	3.265	0.712	10%		3/2-	0.88	ann.rad.	
	57	28				E.C.		0.849	76%				0.12719 ± 0.00002	13.6%

Isotope	A	Z	% Natural abundance	Atomic mass	Half-life	Decay mode	Decay energy (MeV)	Particle energy (MeV)	Particle intensity	Thermal neutron cross section	Spin (h/2π)	μ Nucl. mag. moment	Gamma-ray energy (MeV)	Gamma-ray intensity
	57	28											1.37759 ± 0.00004	78%
	57	28											1.75748 ± 0.00008	7%
	57	28											1.91943 ± 0.00008	15%
$_{28}Ni^{58}$	58	28	68.27%	57.935346						4.6 b.	0+			
$_{28}Ni^{59}$	59	28		58.934349	≈7.6x10^{4}y.	E.C.					3/2-			
$_{28}Ni^{60}$	60	28	26.10%	59.930788						2.9 b.	0+			
$_{28}Ni^{61}$	61	28	1.13%	60.931058						2.4 b.	3/2-	-0.75002		
$_{28}Ni^{62}$	62	28	3.59%	61.928346						14.5 b.	0+			
$_{28}Ni^{63}$	63	28		62.929669	100 y.	β-	0.065	0.065			$^1/_2$-			
$_{28}Ni^{64}$	64	28	0.91%	63.927968						1.55 b.	0+			
$_{28}Ni^{65}$	65	28		64.930086	2.52 h.	β-	2.134	0.65	30%		5/2-	0.69	0.36627 ± 0.00003	5%
	65	28						1.020	11%				1.11553 ± 0.00004	16%
	65	28						2.140	58%				1.48184 ± 0.00005	23%
$_{28}Ni^{66}$	66	28		65.929116	54.8 h.	β-	0.24				0+			
$_{28}Ni^{67}$	67	28		66.9315709	20 s.	β-	3.56	3.8					0.1406 ± 0.0008	39 +
	67	28											0.2080 ± 0.0006	68
	67	28											0.5531 ± 0.0004	43
	67	28											0.7085 ± 0.0005	92
	67	28											0.7515 ± 0.002	23
	67	28											0.7791 ± 0.0004	27
	67	28											0.8741 ± 0.0004	88
	67	28											1.0722 ± 0.0005	100
	67	28											1.1004 ± 0.0007	31
	67	28											1.6539 ± 0.0004	100
	67	28											1.7602 ± 0.0005	39
	67	28											1.809 ± 0.001	30
	67	28											1.938 ± 0.001	25
	67	28											1.975 ± 0.001	65
Cu		29		63.546						3.78 b.				
$_{29}Cu^{58}$	58	29		57.944538	3.21 s.	β+	8.563	4.5	15%		1+		ann.rad.	
	58	29				E.C.		7.439	83%				0.0403 ± 0.0004	5%
	58	29											1.4483 ± 0.0002	11%
	58	29											1.4546 ± 0.0002	16%
$_{29}Cu^{59}$	59	29		58.939503	82 s.	β+	4.801	1.9			3/2-		ann.rad.	
	59	29						3.75					0.3393 ± 0.0001	8%
	59	29											0.8780 ± 0.0001	12%
	59	29											1.3015 ± 0.0001	15%
	59	29											(0.4 - 2.6)weak	
$_{29}Cu^{60}$	60	29		59.937366	23.2 m.	β+	6.127	2.00	69+		2+	+1.219	ann.rad.	
	60	29				E.C.		3.00	18				0.4673 ± 0.0002	3.6%
	60	29						3.92	6				0.9524 ± 0.0002	2.8%
	60	29											1.0352 ± 0.0002	3.8%
	60	29											1.3325 ± 0.0002	88%
	60	29											1.8618 ± 0.0003	5%
	60	29											1.9369 ± 0.0003	2.2%
	60	29											2.1589 ± 0.0002	3.6%
	60	29											2.7461 ± 0.0003	1.1%
	60	29											3.1941 ± 0.0003	2.1%
	60	29											(0.4 - 5.0)weak	
$_{29}Cu^{61}$	61	29		60.933461	3.41 h.	β+	2.239	0.56	3%		3/2-	+2.14	ann.rad.	
	61	29						0.94	5%				0.06711 ± 0.0002	6%
	61	29					1.15	2%					0.28370 ± 0.0002	13%
	61	29						1.220	51%				0.3729 ± 0.0005	2.3%
	61	29											0.65604 ± 0.0002	11%
	61	29											0.90868 ± 0.0004	1.2 %
	61	29											1.8516 ± 0.0003	4.8%
	61	29											(0.5 - 2.1)weak	
$_{29}Cu^{62}$	62	29		61.932586	9.74 m.	β+(98%)	3.95	2.93	98%		1+	-0.380	ann.rad.	
	62	29				E.C.							1.17302 ± 0.0001	0.6%
	62	29											(0.87 - 3.37)weak	
$_{29}Cu^{63}$	63	29	69.17%	62.939598						4.47 b.	3/2-	+2.2233		
$_{29}Cu^{64}$	64	29		63.929765	12.701 h.	β-(39%)	0.578	0.578			1+	-0.217	ann.rad.	
	64	29				β+(19%)	1.675	0.65					1.3459 ± 0.0003	0.6%
	64	29				E.C.(41%)								
$_{29}Cu^{65}$	65	29	30.83%	64.927793						2.17 b.	3/2-	+2.3817		
$_{29}Cu^{66}$	66	29		65.928872	5.10 m.	β-	2.642	1.65	6%		1+	-0.282	0.8330 ± 0.001	0.15%
	66	29						2.7	94%				1.0392 ± 0.0002	8%
$_{29}Cu^{67}$	67	29		66.927747	61.9 h.	β-	0.58	0.395	56%		3/2-		0.09125 ± 0.0001	7%
	67	29						0.484	23%				0.09325 ± 0.0001	17%
	67	29						0.577	20%				0.18453 ± 0.0001	47%
	67	29											0.30022 ± 0.0001	1.5%
$_{29}Cu^{68m}$	68	29			3.8 m.	I.T.(86%)					6-		0.0843 ± 0.0005	70%
	68	29				β-(14%)	1.8						0.1112 ± 0.0005	18%
	68	29											0.5259 ± 0.0005	74%
	68	29											0.6369 ± 0.0005	8%
	68	29											1.0410 ± 0.0005	8%
	68	29											1.3403 ± 0.0005	12%

TABLE OF THE ISOTOPES (Continued)

Isotope	A	Z	% Natural abundance	Atomic mass	Half-life	Decay mode	Decay energy (MeV)	Particle energy (MeV)	Particle intensity	Thermal neutron cross section	Spin (h/2π)	μ Nucl. mag. moment	Gamma-ray energy (MeV)	Gamma-ray intensity
$_{29}Cu^{68}$	68	29		67.929620	31 s.	β-	4.45	3.5	40%		1+		0.8059 ± 0.0005	0.8%
	68	29						4.6	31%				1.0774 ± 0.0005	58%
	68	29											1.2613 ± 0.0005	17%
	68	29											1.8832 ± 0.0005	1.2%
	68	29											(0.15 - 2.34)weak	
$_{29}Cu^{69}$	69	29		68.929425	3.0 m.	β-	2.67	2/48	80%		3/2-		0.5307 ± 0.0003	3%
	69	29											0.6490 ± 0.0005	1.5%
	69	29											0.8340 ± 0.0005	6%
	69	29											1.0065 ± 0.0008	10%
	69	29											1.1795 ± 0.0001	1%
$_{29}Cu^{70m}$	70	29			46 s.	β-		2.52	10%		5-		0.3865 ± 0.0004	8%
	70	29											0.8848 ± 0.0002	100%
	70	29											0.9017 ± 0.0002	90%
	70	29											1.1087 ± 0.0004	8%
	70	29											1.2517 ± 0.0005	60%
	70	29											1.6906 ± 0.0006	5%
	70	29											2.0614 ± 0.0005	4%
	70	29											3.062 ± 0.002	1.4%
$_{29}Cu^{70}$	70	29		69.932386	5 s.	β-	6.58	5.42	54%		1+		0.8848 ± 0.0002	54%
	70	29						6.09	46%					
$_{29}Cu^{71}$	71	29		70.932560	20 s.	β-					3/2-		0.490	
$_{29}Cu^{72}$	72	29			6.6 s.						(1+)		0.652	
$_{29}Cu^{73}$	73	29			3.9 s.	β-							0.450	
Zn		30		65.39						1.1 b.				
$_{30}Zn^{57}$	57	30		56.964990	0.04 s.	β+,p	15						ann.rad.	
$_{30}Zn^{59}$	59	30		58.949270	184 ms.	β+,p	9.1	8.1			3/2-		ann.rad.	
	59	30											0.491	
	59	30											0.914	
$_{30}Zn^{60}$	60	30		59.941830	2.4 m.	β+(97%) E.C.(3%)	4.16				0+		ann.rad.	
	60	30											0.0614 ± 0.0005	24%
	60	30											0.2734 ± 0.0005	10%
	60	30											0.3344 ± 0.0005	9%
	60	30											0.3646 ± 0.0003	3%
	60	30											0.6703 ± 0.0004	68%
$_{30}Zn^{61}$	61	30		60.939514	89.1 s.	β+	5.64	4.38	68%		3/2-		ann.rad.	
	61	30											0.2664 ± 0.0004	16%
	61	30											0.4752 ± 0.0003	7.4%
	61	30											0.6904 ± 0.0004	1.5%
	61	30											0.9700 ± 0.0003	2.4%
	61	30											1.1854 ± 0.0003	1.5%
	61	30											1.6605 ± 0.0004	7.4%
	61	30											1.9971 ± 0.0005	1%
$_{30}Zn^{62}$	62	30		61.934332	9.26 h.	β+(3%) E.C.(93%)	1.63	0.66	7%		0+		ann.rad.	
	62	30											0.04094 ± 0.0006	25%
	62	30											0.5075 ± 0.0004	15%
	62	30											0.5481 ± 0.0004	15%
	62	30											0.59665 ± 0.00001	24%
	62	30											(0.2 - 1.5)weak	
$_{30}Zn^{63}$	63	30		62.933214	38.1 m.	β+(93%) E.C.(7%)	3.367	1.02			3/2-	-0.28164	ann.rad.	
	63	30						1.40					0.66962 ± 0.00005	8.4%
	63	30						1.71					0.96206 ± 0.00005	6.6%
	63	30						2.36	84%				1.13067 ± 0.0002	0.01%
	63	30											1.3270 ± 0.0003	0.07%
	63	30											1.3926 ± 0.0004	0.1%
	63	30											1.41208 ± 0.00005	0.76%
	63	30											1.54704 ± 0.0005	0.13%
	63	30											(0.24 - 3.1)weak	
$_{30}Zn^{64}$	64	30	48.6%	63.929145						0.76 b.	0+			
$_{30}Zn^{65}$	65	30		64.929243	243.8 d.	β+(98%) E.C.(1.5%)	1.352	0.325			5/2-	+0.7690	ann.rad.	
	65	30											1.11552 ± 0.00002	50.8%
$_{30}Zn^{66}$	66	30	27.9%	65.926034						0.9 b.	0+			
$_{30}Zn^{67}$	67	30	4.1%	66.927129						7.25 b.	5/2-	+0.8755		
$_{30}Zn^{68}$	68	30	18.8%	67.924846						(0.072 + 0.9) b.	0+			
$_{30}Zn^{69m}$	69	30			13.8 h.	I.T. (99+%)	0.439				9/2+		0.4390 ± 0.0002	95%
$_{30}Zn^{69}$	69	30		68.926552	57 m	β-	0.905	0.905	99.9%		1/2-		0.318	0.001%
$_{30}Zn^{70}$	70	30	0.6%	69.925325						(0.008 + 0.083) b.	0+			
$_{30}Zn^{71m}$	71	30			3.97 h.	β-		1.45			9/2+		0.12148 ± 0.00005	2.9%
	71	30											0.14260 ± 0.00005	5.4%
	71	30											0.38628 ± 0.00005	92%
	71	30											0.48734 ± 0.00005	60%
	71	30											0.51155 ± 0.00005	28%

TABLE OF THE ISOTOPES (Continued)

Isotope	A	Z	% Natural abundance	Atomic mass	Half-life	Decay mode	Decay energy (MeV)	Particle energy (MeV)	Particle intensity	Thermal neutron cross section	Spin (h/2π)	μ Nucl. mag. moment	Gamma-ray energy (MeV)	Gamma-ray intensity
	71	30											0.59607 ± 0.00005	27%
	71	30											0.62019 ± 0.00005	56%
	71	30											0.9647 ± 0.0001	4%
	71	30											1.1074 ± 0.0002	2.1%
	71	30											1.7596 ± 0.0002	1%
	71	30											2.3177 ± 0.0002	0.1%
$_{30}Zn^{71}$	71	30		70.927727	2.4 m.						1/2-		0.12152 ± 0.00005	2.7%
	71	30											0.3900 ± 0.0003	3.6%
	71	30											0.5116 ± 0.0001	30%
	71	30											0.9103 ± 0.0001	7.5%
	71	30											1.1200 ± 0.0001	2.1%
	71	30											(0.39 - 2.29)weak	
$_{30}Zn^{72}$	72	30		71.926856	β-	0.457	0.25	14%			0+		0.0164 ± 0.0003	8%
	72	30						0.30	86%				0.1887 ± 0.0001	2%
	72	30											0.1447 ± 0.0001	83%
	72	30											0.1915 ± 0.0002	9.4%
$_{30}Zn^{73}$	73	30		72.929780	24 s.	β-	4.70	4.7			3/2-		0.216 ± 0.001	100 +
	73	30											0.496 ± 0.001	26
	73	30											0.911 ± 0.001	26
$_{30}Zn^{74}$	74	30		73.929461	96 s.	β-	2.4	2.1					0.0503 ± 0.0005	18%
	74	30											0.0531 ± 0.0005	10%
	74	30											0.0573 ± 0.0005	80%
	74	30											0.0861 ± 0.0005	4%
	74	30											0.1167 ± 0.0001	4%
	74	30											0.1400 ± 0.0005	37%
	74	30											0.1904 ± 0.0005	27%
	74	30											0.3473 ± 0.0005	6.3%
$_{30}Zn^{75}$	75	30		74.932690	10.2 s.	β-	6.1							
$_{30}Zn^{76}$	76	30		75.932940	5.7 s.	β-	4.0	3.6					0.56	
	76	30											1.10	
$_{30}Zn^{77}$	77	30		76.936750	1.4 s.	β-	7.5	4.8			7/2+		0.189	
	77	30											0.473	
$_{30}Zn^{78}$	78	30		77.937780	1.5 s.	β-	6.0						0.1817	
	78	30											0.2248	
Ga		31		69.723						2.9 b.				
$_{31}Ga^{62}$	62	31		61.944178	0.116 s.	β+	9.17	8.3			0+		ann.rad.	
	63	31				E.C.								
$_{31}Ga^{63}$	63	31		62.939140	32 s.	β+	5.5	4.5					ann.rad.	
	63	31				E.C.							0.1930 ± 0.0002	5%
	63	31											0.2480 ± 0.0002	3.4%
	63	31											0.6271 ± 0.0002	10%
	63	31											0.6370 ± 0.0002	11%
	63	31											0.6501 ± 0.0002	4.5%
	63	31											0.7685 ± 0.0002	2%
	63	31											1.0652 ± 0.0004	45 %
	63	31											1.6917 ± 0.0005	3%
$_{31}Ga^{64}$	64	31		63.936836	2.63 m.	β+	7.16	2.79			0+		ann.rad.	195%
	64	31						6.05					0.80785 ± 0.0001	14%
	64	31											0.91878 ± 0.0001	8%
	64	31											0.99152 ± 0.0001	43%
	64	31											1.38727 ± 0.0001	12%
	64	31											1.6175 ± 0.0002	1.6&
	64	31											1.79943 ± 0.0001	3.5%
	64	31											1.9958 ± 0.0003	1.6%
	64	31											2.2704 ± 0.0001	2.1%
	64	31											3.3659 ± 0.0001	13%
	64	31											3.4251 ± 0.0001	4.0%
	64	31											3.7951 ± 0.0001	1%
	64	31											4.454 ± 0.0001	0.7%
$_{31}Ga^{65}$	65	31		64.932738	15.2 m.	β+(86%)	3.256	0.82	10+		3/2-		ann.rad.	
	65	31				E.C.		1.39	19				0.0538 ± 0.0002	5%
	65	31						2.113	56				0.0611 ± 0.0002	12%
	65	31						2.237	15				0.1151 ± 0.0002	55%
	65	31											0.1530 ± 0.0002	96%
	65	31											0.2069 ± 0.0002	39%
	65	31											0.7518 ± 0.0002	8.2%
	65	31											0.7689 ± 0.0002	1/3%
	65	31											0.7946 ± 0.0002	0.25%
	65	31											0.9097 ± 0.0002	0.5%
	65	31											0.9322 ± 0.0002	1.8%
	65	31											1.0474 ± 0.0003	0.9%
	65	31											1.2288 ± 0.0002	0.7%
	65	31											1.3547 ± 0.0002	0.8%
	65	31											2.2121 ± 0.0003	0.13%
	65	31											(0.06 - 2.4)weak	
$_{31}Ga^{66}$	66	31		65.931590	9.4 h.	β+(56%)	5.175	0.74	1%		0+		ann.rad.	
	66	31				E.C.(43%)		1.84	54%				0.8337 ± 0.0003	6.1%
	66	31						4.153	51%				1.03935 ± 0.00008	38%
	66	31											1.2322 ± 0.0002	0.5%
	66	31											1.3334 ± 0.0001	1.2%

Isotope	A	Z	% Natural abundance	Atomic mass	Half-life	Decay mode	Decay energy (MeV)	Particle energy (MeV)	Particle intensity	Thermal neutron cross section	Spin (h/2π)	μ Nucl. mag. moment	Gamma-ray energy (MeV)	Gamma-ray intensity
	66	31											1.8992 ± 0.0001	0.43%
	66	31											1.9187 ± 0.0001	2.4%
	66	31											2.1902 ± 0.0001	5.7%
	66	31											2.4225 ± 0.0001	1.8%
	66	31											2.7523 ± 0.0001	23%
	66	31											3.3813 ± 0.0002	1.4%
	66	31											3.4225 ± 0.0002	0.8%
	66	31											4.0865 ± 0.0001	1.1%
	66	31											4.2955 ± 0.0002	3.5%
	66	31											4.8066 ± 0.0002	1.5%
	66	31											(0.29 - 4.8)weak	
$_{31}Ga^{67}$	67	31		66.928204	78.25 h.	E.C.	1.001				3/2-	+1.8507	0.09128 ± 0.00002	4%
	67	31											0.09332 ± 0.0002	38%
	67	31											0.18459 ± 0.0004	23%
	67	31											0.20896 ± 0.0006	3%
	67	31											0.30024 ± 0.0006	19%
	67	31											0.3936 ± 0.0006	5.6%
	67	31											0.8880 ± 0.0002	0.2%
$_{31}Ga^{68}$	68	31		67.927981	68.1 m.	β+(90%)	2.921	1/83			1+	0.01175	ann.rad.	
	68	31				E.C.(10%)							1.0774 ± 0.0001	3%
	68	31											(0.57 - 2.33)weak	
$_{31}Ga^{69}$	69	31	60.1%	68.925580						1.7 b.	3/2-	+2.01659		
$_{31}Ga^{70}$	70	31		69.926028	21.1 m.	E.C.(0.2%)	0.655				1+		0.1755 ± 0.0005	0.15%
	70	31				β-(99.8%)	1.653	1.65	99%				1.042 ± 0.005	0.48%
$_{31}Ga^{71}$	71	31	39.9%	70.924700						4.7 b.	3/2-	+2.56227		
$_{31}Ga^{72}$	72	31		71.926365	13.95 h.	β-	3.99	0.64	40%		3-	-0.13224	0.60005 ± 0.00005	5.8%
	72	31						1.51	9%				0.62986 ± 0.00005	24%
	72	31						2.52 8%					0.89422 ± 0.0001	10%
	72	31						3.15	11%				1.0507 ± 0.0001	7%
	72	31											1.2309 ± 0.0002	1.4%
	72	31											1.2601 ± 0.0002	1.15%
	72	31											1/2768 ± 0.0002	1.6%
	72	31											1/2768 ± 0.0002	1.6%
	72	31											1.4640 ± 0.0001	3.6%
	72	31											1.5968 ± 0.0002	4.3%
	72	31											1.8611 ± 0.0001	5.3%
	72	31											2.1095 ± 0.0002	1%
	72	31											2.2016 ± 0.0002	26%
	72	31											2.4910 ± 0.0002	7.5%
	72	31											2.5077 ± 0.0002	12.8%
	72	31											(0.11 - 3.3)weak	
$_{31}Ga^{73}$	73	31		72.925169	4.87 h.	1.59					3/2-		0.05344 ± 0.00005	10%
	73	31											0.0687 ± 0.0002	2%
	73	31											0.29732 ± 0.00005	47%
	73	31											0.32570 ± 0.00007	7%
	73	31											0.73942 ± 0.00005	2%
	73	31											0.9936 ± 0.0005	0.1%
	73	31											(0.01 - 1.00)weak	
$_{31}Ga^{74m}$	74	31			9.5 s.	I.T.					1+		0.0565 ± 0.0001	75%
$_{31}Ga^{74}$	74	31		73.926940	8.1 m.	β-	5.4	2.6			3-		0.59588 ± 0.00004	91%
	74	31											0.6042 ± 0.00001	2.9%
	74	31											0.60840 ± 0.00005	14.5%
	74	31											0.8678 ± 0.0001	9%
	74	31											1.10134 ± 0.00006	5.5%
	74	31											1.2043 ± 0.0001	7.5%
	74	31											1.3322 ± 0.0003	1.8%
	74	31											1.7448 ± 0.0001	4.8%
	74	31											1.0406 ± 0.0001	5.5%
	74	31											2.2790 ± 0.0001	2.4%
	74	31											2.3535 ± 0.0001	44%
	74	31											2.58007 ± 0.0001	1.3%
	74	31											2.97090 ± 0.0001	1%
	74	31											(0.49 - 3.99)weak	
$_{31}Ga^{75}$	75	31		74.926499	2.10 m.	β-	3.39	3.3			3/2-		0.1770	10 +
	75	31											0.2528	100
	75	31											0.5747	32
	75	31											0.8854	11
	75	31											0.9272	7
	75	31											1.2485	5
	75	31											1.5011	5
	75	31											1.5430	1
$_{31}Ga^{76}$	76	31		75.928670	29.1 s.	β-	6.8				3-		0.4310 ± 0.0005	9%
	76	31											0.54551 ± 0.00003	26%

Isotope	A	Z	% Natural abundance	Atomic mass	Half-life	Decay mode	Decay energy (MeV)	Particle energy (MeV)	Particle intensity	Thermal neutron cross section	Spin (h/2π)	μ Nucl. mag. moment	Gamma-ray energy (MeV)	Gamma-ray intensity
	76	31											0.56293 ± 0.00003	65.8%
	76	31											0.8472 ± 0.0001	3.5%
	76	31											0.97650 ± 0.00005	4.5%
	76	31											1.10841 ± 0.00008	18%
	76	31											1.2080 ± 0.0001	1.3%
	76	31											2.1295 ± 0.0001	2.2%
	76	31											2.2144 ± 0.0001	1.1%
	76	31											2.3569 ± 0.0001	2.4%
	76	31											2.5786 ± 0.0001	2.2%
	76	31											2.6192 ± 0.0001	2.1%
	76	31											2.9199 ± 0.0001	9.1%
	76	31											3.1414 ± 0.0001	4.2%
	76	31											3.3888 ± 0.0002	3%
	76	31											3.9517 ± 0.0002	4.2%
	76	31											(1.0 - 4.2)weak	
$_{31}Ga^{77}$	77	31		76.928700	13 s.	β-	5/3	5/2					0.469	
	77	31											0.459	
$_{31}Ga^{78}$	78	31		77.931760	5.09 s.	β-	7/9				3+		0.619	
	78	31											1.025	
	78	31											1.186	
$_{31}Ga^{79}$	79	31		78.932530	3.0 s.	β-	6.8	4.6					2.18	
$_{31}Ga^{80}$	80	31		79.936250	1.66 s.	β-	10	10						
$_{31}Ga^{81}$	81	31		80.937750	1.23 s.	β-	8.3	5.1					0.217	
$_{31}Ga^{82}$	82	31			1.9 s.	β-	7.4							
$_{31}Ga^{83}$	83	31			1.2 s.	β-								
Ge		32								2.2 b.				
$_{32}Ga^{64}$	64	32		63.941570	63 s.	β+ E.C.	4.4	3.0			0+		ann.rad.	
	64	32											0.0651 ± 0.0002	2%
	64	32											0.1282 ± 0.0002	11%
	64	32											0.3841 ± 0.0003	4.7%
	64	32											0.4270 ± 0.0003	37%
	64	32											0.6671 ± 0.0003	17%
	64	32											0.7745 ± 0.0003	7%
$_{32}Ge^{65}$	65	32		64.939440	31 s.	β+	6.2	0.82	10+				ann.rad.	
	65	32				E.C.		1.39	19				0.0621 ± 0.0005	27%
	65	32						2.113	56				0.1908 ± 0.0002	10%
	65	32						2.237	15				0.4591 ± 0.0005	10%
	65	32											0.5877 ± 0.0002	3%
	65	32											0.6187 ± 0.0004	1.6%
	65	32											0.6497 ± 0.0002	33%
	65	32											0.8091 ± 0.0002	21%
	65	32											1.0759 ± 0.0003	0.8%
	65	32											1.2371 ± 0.0003	1.3%
	65	32											2.0996 ± 0.0004	1.5%
	65	32											(0.4 - 3.2)weak	
$_{32}Ge^{66}$	66	32		65.933847	2.27 h.	β+(27%) E.C.(73%)	2.10				0+		ann.rad.	
	66	32											0.0224 ± 0.0002	1.6%
	66	32											0.0483 ± 0.0002	1.0%
	66	32											0.04389 ± 0.0001	29%
	66	32											0.06512 ± 0.0001	6%
	66	32											0.10885 ± 0.00003	10.5%
	66	32											0.18203 ± 0.00004	5.7%
	66	32											0.19020 ± 0.00003	5.7%
	66	32											0.27297 ± 0.00004	11%
	66	32											0.33805 ± 0.00004	10%
	66	32											0.38185 ± 0.00005	28%
	66	32											0.47062 ± 0.00006	7.4%
	66	32											0.4720 ± 0.0001	3%
	66	32											0.53674 ± 0.00007	6%
	66	32											0.70594 ± 0.00003	4.3%
$_{32}Ge^{67}$	67	32		66.932737	19.0 m.	β+(96%)	4.22	1.6			1/2-		ann.rad.	
	67	32				E.C.(4%)		2.3					0.16701 ± 0.00005	84%
	67	32						3.15					0.3595 ± 0.0002	1.5%
	67	32											0.7282 ± 0.0005	2.5%
	67	32											0.8283 ± 0.0003	3.0%
	67	32											0.9112 ± 0.0003	3.1%
	67	32											0.9148 ± 0.0003	3.0%
	67	32											1.0818 ± 0.0003	1%
	67	32											1.4728 ± 0.0003	4.9%
	67	32											1.8094 ± 0.0006	1.3%
	67	32											(0.4 - 3.7)weak	

Isotope	A	Z	% Natural abundance	Atomic mass	Half-life	Decay mode	Decay energy (MeV)	Particle energy (MeV)	Particle intensity	Thermal neutron cross section	Spin (h/2π)	μ Nucl. mag. moment	Gamma-ray energy (MeV)	Gamma-ray intensity
$_{32}Ge^{68}$	68	32		67.928096	288 d.	E.C.	0.11				0+		Ga k x-ray	39%
$_{32}Ge^{69}$	69	32		68.927969	39.1 h.	β+(36%)	2.225	0.70			5/2-	0.735	ann.rad.	72%
	69	32				E.C.(64%)		1.2					0.3184 ± 0.0002	1.25%
	69	32											0.5739 ± 0.0002	12%
	69	32											0.8717 ± 0.0002	11%
	69	32											1.1064 ± 0.0002	27%
	69	32											1.3362 ± 0.0002	3%
	69	32											(0.2 - 2.04)weak	
$_{32}Ge^{70}$	70	32	20.5%	69.924250						3.3 b.	0+			
$_{32}Ge^{71}$	71	32		70.924953	11.2 d.	E.C.	0.236				1/2-	+0.547		
$_{32}Ge^{72}$	72	32	27.4%	71.922079						1.0 b.	0+			
$_{32}Ge^{73}$	73	32	7.8%	72.923463						14 b.	9/2+	-0.87946		
$_{32}Ge^{74}$	74	32	36.5%	73.921177						(0.16 + 0.36)b.	0+			
$_{32}Ge^{75m}$	75	32			48 s.	I.T.					7/2+		0.13968 ± 0.00003	39%
$_{32}Ge^{75}$	75	32		74.922858	82.8 m.	β-	1.178	1.19			1/2-	+0.510	0.26461 ± 0.00005	11%
	75	32											0.41931 ± 0.00005	0.2%
$_{32}Ge^{76}$	76	32	7.8%	75.921401						(.09 + .06)b.	0+			
$_{32}Ge^{77m}$	77	32			53 s.	I.T.(20%)					1/2-		0.1597 ± 0.0001	10%
	77	32				β-(80%)	2.861	2.9					0.1948 ± 0.0001	0.2%
	77	32											0.21551 ± 0.0001	21%
$_{32}Ge^{77}$	77	32		76.923548	11.30 h.	β-	2.70	0.71	23%		7/2+		0.2108 ± 0.0005	29%
	77	32						1.38	35%				0.2156 ± 0.0005	27%
	77	32						2.19	42%				0.2645 ± 0.0005	53%
	77	32											0.3674 ± 0.0005	12%
	77	32											0.4163 ± 0.0005	20%
	77	32											0.5579 ± 0.0005	15%
	77	32											0.6316 ± 0.0005	7%
	77	32											0.7141 ± 0.0005	7.5%
	77	32											0.8101 ± 0.0005	2.2%
	77	32											1.0851 ± 0.0005	6.4%
	77	32											1.3684 ± 0.0005	3.1%
	77	32											1.9996 ± 0.0005	0.6%
	77	32											2.3415 ± 0.0005	0.5%
	77	32											(0.15 - 2.37)weak	
$_{32}Ge^{78}$	78	32		77.922853	1,45 h.	-	0.95	0.70			0+		0.2773 ± 0.0005	96%
	78	32											0.2939 ± 0.0005	4%
$_{32}Ge^{79m}$	79	32			19 s.	β-								
$_{32}Ge^{79}$	79	32		78.925360	42 s.	β-	4.1	4.0	20%		1/2-		0.2164 ± 0.0004	5%
	79	32						4.3	80%				0.2304 ± 0.0004	25%
	79	32											0.5427 ± 0.0004	15%
	79	32											0.6331 ± 0.0004	3%
	79	32											0.7450 ± 0.0004	5%
	79	32											0.7818 ± 0.0005	6%
$_{32}Ge^{80}$	80	32		79.925520	29 s.	β-	2/69	2.4			0+		0.1104 ± 0.0004	6%
	80	32											0.2656 ± 0.0004	25%
	80	32											0.9372 ± 0.0004	4%
	80	32											1.0140 ± 0.0004	2.5%
	80	32											1.1160 ± 0.0004	2.5%
	80	32											1.2561 ± 0.0005	3%
	80	32											1.5643 ± 0.0005	4.5%
$_{32}Ge^{81m}$	81	32			7.6 s.	β-		3.75			9/2+		0.3362 ± 0.0004	
	81	32											0.7935 ± 0.0004	
$_{32}Ge^{81}$	81	32		80.928820	7.6 s.	β-	6.2	3.44			1/2+		0.1976 ± 0.0004	21 +
	81	32											0.3362 ± 0.0004	100
$_{32}Ge^{82}$	82	32		81.929810	4.6 s.	β-	4.7				0+		1.093	
$_{32}Ge^{83}$	83	32		82.934250	1.9 s.	β-	7.4							
As		33		74.9216						4.5 b.				
$_{33}As^{67}$	67	33		66.939190	43 s.	β+	6.0	5.0			5/2-		0.121	
	67	33				E.C.							0.123	
	67	33											0.244	
$_{33}As^{68}$	68	33		67.936790	2.53 m.	β+	8.1				3+		ann.rad.	
	68	33											0.6135 ± 0.0005	6%
	68	33											0.6512 ± 0.0002	24%
	68	33											0.7626 ± 0.0002	23%
	68	33											1.0165 ± 0.0001	66%
	68	33											1.2534 ± 0.0004	1%
	68	33											1.2635 ± 0.0005	4%
	68	33											1.4125 ± 0.0003	12%
	68	33											1.622 ± 0.001	3%
	68	33											1.7787 ± 0.0002	23%
	68	33											2.008 ± 0.001	3%
	68	33											2.457 ± 0.001	3%
	68	33											3.058 ± 0.003	1%
	68	33											3.220 ± 0.003	0.6%
$_{33}As^{69}$	69	33		68.932280	15.1 m.	β+(98%)	4.02	2.95			5/2-		ann.rad.	
	69	33				E.C.(2%)							0.0868 ± 0.0005	1.5%
	69	33											0.1458 ± 0.0003	2.4%
	69	33											0.2327 ± 0.0003	5%
	69	33											0.2871 ± 0.0005	0.9%
	69	33											0.3741 ± 0.0005	0.7%
	69	33											0.3981 ± 0.0005	0.6%

Isotope	A	Z	% Natural abundance	Atomic mass	Half-life	Decay mode	Decay energy (MeV)	Particle energy (MeV)	Particle intensity	Thermal neutron cross section	Spin (h/2π)	μ Nucl. mag. moment	Gamma-ray energy (MeV)	Gamma-ray intensity
$_{33}As^{70}$	70	33		69.930929	52.6 m.	β+(84%)	6.22	1.44			4+	2.1	ann.rad.	
	70	33				E.C.(16%)		2.14					0.1753 ± 0.0005	2.6%
	70	33						2.89					0.5952 ± 0.0005	16%
	70	33											0.6684 ± 0.0005	21%
	70	33											0.7448 ± 0.0005	21%
	70	33											0.9057 ± 0.0005	12%
	70	33											1.0395 ± 0.0007	82%
	70	33											1.0993 ± 0.0005	4.4%
	70	33											1.1143 ± 0.001	21%
	70	33											1.1181 ± 0.0007	3.2%
	70	33											1.3394 ± 0.0008	8.5%
	70	33											1.4125 ± 0.0005	9%
	70	33											1.4961 ± 0.0005	1.6%
	70	33											1.7079 ± 0.0007	18%
	70	33											1.7813 ± 0.0007	3.9%
	70	33											2.0077 ± 0.001	3%
	70	33											(0.17 - 4.4)weak	
$_{33}As^{71}$	71	33		70.927111	62 h.	β+(32%)	2.013				5/2-	+1.6735	ann.rad.	
	71	33				E.C.(68%)							0.1749 ± 0.0002	81%
	71	33											0.3274 ± 0.0002	2.7%
	71	33											0.5000 ± 0.0002	2.8%
	71	33											1.0957 ± 0.0002	4.2%
	71	33											1.2127 ± 0.0003	0.3%
	71	33											1.2988 ± 0.0003	0.2%
$_{33}As^{72}$	72	33		71.926755	26.0 h.	β+(77%)	4.355	0.669	5 +		2-	-2.1578	ann.rad.	
	72	33						1.884	12				0.6299 ± 0.0001	8%
	72	33						2.498	62				0.83395 ± 0.00005	80%
	72	33						3.339	19				1.0507 ± 0.0001	9.6%
	72	33											1.4640 ± 0.0001	1.1%
	72	33											1.4758 ± 0.0002	0.5%
	72	33											2.1059 ± 0.0002	0.6%
	72	33											2.1095 ± 0.0002	0.3%
	72	33											2.2016 ± 0.0002	0.5%
	72	33											2.5077 ± 0.0002	0.3%
	72	33											(0.1 - 4.0)weak	
$_{33}As^{73}$	73	33		72.923827	80.3 d.	E.C.	0.346				3/2-		0.013263 ± 0.00001	0.1%
	73	33											0.053437 ± 0.00001	10.5%
	73	33											Se k x-ray	90%
$_{33}As^{74}$	74	33		73.923827	17.78 d.	β+(31%)	2.562	0.94	26%		2-	-1.597	ann.rad.	
	74	33				E.C.(37%)	1.53	3%					0.59588 ± 0.0001	60%
	74	33				β-	1.354	0.71	16%				0.6084 ± 0.0001	0.6%
	74	33						1.35	16%				0.6348 ± 0.0001	15%
	74	33											1.2043 ± 0.0001	0.25%
$_{33}As^{75}$	75	33	100%	74.921594						4.5 b.	3/2-	+1.43947		
$_{33}As^{76}$	76	33		75.922393	26.3 h.	β-	0.54	3%			2	-0.906	0.5591 ± 0.0001	45%
	76	33						1.184	2%				0.5632 ± 0.0001	1.2%
	76	33						1.785	8%				0.65703 ± 0.0005	5.7%
	76	33						2.410	36%				1.21272 ± 0.0001	1.3%
	76	33						2.97	51%				1.21602 ± 0.0001	3.4%
	76	33											1.2285 ± 0.0001	1.2%
	76	33											(0.3 - 2.6)weak	
$_{33}As^{77}$	77	33		76.920646	38.8 h.	β-	0.6904	0.70	98%		3/2-		0.0880 ± 0.0003	0.27%
	77	33											0.2391 ± 0.0002	1.6%
	77	33											0.2500 ± 0.0003	0.4%
	77	33											0.52078 ± 0.0001	0.43%
$_{33}As^{78}$	78	33		77.921830	1.515 h.	β-	4.21	3.00	12%		2-		0.3543 ± 0.0003	1.7%
	78	33						3.70	17%				0.5454 ± 0.0003	2.5%
	78	33						4.42	37%				0.6136 ± 0.0003	54%
	78	33											0.6862 ± 0.0003	1.8%
	78	33											0.6954 ± 0.0003	18%
	78	33											0.8276 ± 0.0003	7.2%
	78	33											1.0800 ± 0.0003	1.8%
	78	33											1.1445 ± 0.0003	2.1%
	78	33											1.2399 ± 0.0003	5.6%
	78	33											1.3088 ± 0.0003	10%
	78	33											1.3731 ± 0.0003	3.3%
	78	33											1.5300 ± 0.0003	2.5%
	78	33											1.7143 ± 0.0003	1.8%
	78	33											1.8360 ± 0.0003	1.4%
	78	33											1.9216 ± 0.0003	1.6%
	78	33											1.9955 ± 0.0003	1.1%
	78	33											2.6825 ± 0.0003	1.5%
$_{33}As^{79}$	79	33		78.920946	9.0 m.	β-	2.28	1.80	95%		3/2-		0.0955 ± 0.0005	16%
	79	33											0.3645 ± 0.0005	1.9%
	79	33											0.4320 ± 0.0005	1.5%
	79	33											0.3645 ± 0.0005	1.5%
	79	33											0.8785 ± 0.0008	1.4%
$_{33}As^{80}$	80	33		79.922528	16 s.	β-	5.58	3.38			1+		0.6662 ± 0.0002	42%
	80	33											1.2072 ± 0.0002	4%
	80	33											1.6454 ± 0.0002	7%
	80	33											2.3578 ± 0.0005	0.9%
	80	33											(2.5 - 3.0)weak	

Isotope	A	Z	% Natural abundance	Atomic mass	Half-life	Decay mode	Decay energy (MeV)	Particle energy (MeV)	Particle intensity	Thermal neutron cross section	Spin (h/2π)	μ Nucl. mag. moment	Gamma-ray energy (MeV)	Gamma-ray intensity
$_{33}As^{81}$	81	33		80.922131	33 s.	β-	3.87				3/2-		0.4676 ± 0.0002	20%
	81	33											0.4911 ± 0.0002	8%
	81	33											0.5211 ± 0.0002	1%
	81	33											1.4060 ± 0.0002	1%
	81	33											2.8324 ± 0.0002	0.3%
$_{33}As^{82m}$	83	33			14 s.	β-		3.6			5-		0.3435 ± 0.001	23%
	82	33											0.5605 ± 0.0001	14%
	82	33											0.6544 ± 0.0001	72%
	82	33											0.8151 ± 0.0004	7%
	82	33											0.8186 ± 0.0004	27%
	82	33											1.0799 ± 0.0004	23%
	82	33											1.7180 ± 0.0005	3%
	82	33											1.7313 ± 0.0002	27%
	82	33											1.8954 ± 0.0002	38%
$_{33}As^{82}$	82	33		81.924769	19 s.	β-	7.52	7.2	80%		1+		0.6544 ± 0.0001	15%
	82	33											0.7552 ± 0.0002	2%
	82	33											1.7313 ± 0.0002	3.8%
	82	33											1.9709 ± 0.0003	1.5%
	82	33											2.3462 ± 0.001	1.6%
	82	33											2.3534 ± 0.001	1.7%
	82	33											2.4412 ± 0.001	1.6%
	82	33											2.6038 ± 0.001	1.5%
	82	33											2.7227 ± 0.001	1.5%
	82	33											2.8348 ± 0.001	1.5%
	82	33											3.6688 ± 0.001	1.2%
	82	33											3.7730 ± 0.001	0.9%
$_{33}As^{83}$	83	33		82.924980	13 s.	β-	5.5						0.7345	100 +
	83	33											0.8338	19
	83	33											1.1131	34
	83	33											1.8954	18
	83	33											2.0767	28
	83	33											2.2029	22
	83	33											2.8579	16
$_{33}As^{84}$	84	33		83.929060	5.5 s.	β-,n	9.8				1-		0.6671 ± 0.0002	21%
	84	33											1.4439 ± 0.0005	49%
	84	33											1.8437 ± 0.0002	3%
	84	33											2.0866 ± 0.0003	5%
	84	33											2.4612 ± 0.0003	4%
	84	33											3.0379 ± 0.0005	1.2%
	84	33											4.9459 ± 0.001	1.1%
	84	33											5.0877 ± 0.001	0.7%
	84	33											5.1510 ± 0.001	0.8%
$_{33}As^{85}$	85	33		84.931820	2.03 s.	β-,n					3/2-		0.667 ± 0.001	42 +
	85	33											1.1115 ± 0.0005	12
	85	33											1.4551 ± 0.0002	100
	85	33											3.7494 ± 0.0007	3
$_{33}As^{86}$	86	33			0.9 s.	β-,n							0.704	
Se		34		78.96						11.7 b.				
$_{34}Se^{69}$	69	34		68.939570	27.4 s.	β+	6.79	5.006					ann.rad.	
	69	34				E.C.							0.0664 ± 0.0004	27%
	69	34											0.0982 ± 0.0004	63%
	69	34											0.69114 ± 0.0005	14%
$_{34}Se^{70}$	70	34		69.933880	41.1 m.	β+	2.8				0+		ann.rad.	
	70	34											0.03205 ± 0.00005	1.9%
	70	34											0.04951 ± 0.00005	35%
	70	34											0.13254 ± 0.00005	2.6%
	70	34											0.13563 ± 0.0005	2.6%
	70	34											0.2027 ± 0.0001	4.8%
	70	34											0.2441 ± 0.0001	2.8%
	70	34											0.3767 ± 0.0002	9.6%
	70	34											0.4262 ± 0.0002	29%
	70	34											0.49969 ± 0.0001	1.4%
$_{34}Se^{71}$	71	34		70.932270	4.7 m.	β+	4.8	3.4	36%		5/2-		ann.rad.	
	71	34				E.C.							0.1472 ± 0.0003	47%
	71	34											0.7241 ± 0.0003	6%
	71	34											0.8309 ± 0.0003	13%
	71	34											0.8711 ± 0.0003	7%
	71	34											0.9784 ± 0.0003	4.7%
	71	34											1.0960 ± 0.0003	10%
	71	34											1.2432 ± 0.0003	7%
	71	34											1.265 ± 0.0003	1.7%
$_{34}Se^{72}$	72	34		71.927110	8.4 d.	E.C.	0.33				0+		0.0460 ± 0.0002	57%
$_{34}Se^{73m}$	73	34				I.T.(73%)	0.0257	0.85			3/2-		ann.rad.	36%
	73	34				β+(27%)	2.77	1.45					0.0257 ± 0.0002	27%
	73	34						1.70					0.1807 ± 0.0001	0.5%
	73	34											0.2538 ± 0.0001	2.5%
	73	34											0.3204 ± 0.0001	0.8%
	73	34											0.3934 ± 0.0001	1.6%
	73	34											0.4016 ± 0.0001	1.2%
	73	34											0.5775 ± 0.0001	1.2%
	73	34											1.0778 ± 0.0002	0.6%

Isotope	A	Z	% Natural abundance	Atomic mass	Half-life	Decay mode	Decay energy (MeV)	Particle energy (MeV)	Particle intensity	Thermal neutron cross section	Spin (h/2π)	μ Nucl. mag. moment	Gamma-ray energy (MeV)	Gamma-ray intensity
$_{34}Se^{73}$	73	34		72.926768	7.1 h.	β+(65%)	2.74	0.80			9/2+		ann.rad.	125%
	73	34				E.C.(35%)		1.32	95%				0.0670 ± 0.0001	72%
	73	34						1.68	1%				0.3609 ± 0.0001	97%
	73	34											(0.6 - 1.5)weak	
$_{34}Se^{74}$	74	34	0.9%	73.922475						52 b.	0+			
$_{34}Se^{75}$	75	34		74.922521	118.5 d.	E.C.	0.864				5/2+	0.67	0.09673 ± 0.00001	3%
	75	34											0.121115 ± 0.00001	15.8%
	75	34											0.136000 ± 0.00001	55%
	75	34											0.198596 ± 0.00001	1.4%
	75	34											0.264651 ± 0.00001	58%
	75	34											0.279528 ± 0.00001	15.9%
	75	34											0.303913 ± 0.00001	1.3%
	75	34											0.400646 ± 0.00001	11.6%
$_{34}Se^{76}$	76	34	9.0%	75.919212						(21 + 64) b.	0+			
$_{34}Se^{77m}$	77	34			17.4 s.	I.T.					7/2+		0.1619 ± 0.0002	52%
$_{34}Se^{77}$	77	34	7.6%	76.919912						42 b.	1/2-	+0.53506		
$_{34}Se^{78}$	78	34	23.5%							(0.43 + 0.2)b.	0+			
$_{34}Se^{79m}$	79	34			3.89 m.	I.T.					1/2-		0.09573 ± 0.00003	9.5%
$_{34}Se^{79}$	79	34		78.918498	6x10⁴ y.	β-	0.149				7/2+	-1.018		
$_{34}Se^{80}$	80	34	49.6%	79.916520						(0.07 + 0.54) b.	0+			
$_{34}Se^{81m}$	81	34			57.3 m.	I.T.(99%)	0.1031				7/2+		0.1031 ± 0.0003	9.7%
	81	34											0.2602 ± 0.0002	0.06%
	81	34											0.27599 ± 0.0001	0.06%
$_{34}Se^{81}$	81	34		803917990	18.5 m.	β-	1.59	1.6	98%		1/2-		0.27594 ± 0.00005	0.85%
	81	34											0.29008 ± 0.00005	0.75%
	81	34											0.5524 ± 0.0001	0.12%
	81	34											0.56604 ± 0.00005	0.25%
	81	34											0.6498 ± 0.0001	0.25%
	81	34											0.82827 ± 0.00005	0.32%
$_{34}Se^{82}$	82	34	9.4%	81.916698						(0.039 + 0.005) b.	0+			
$_{34}Se^{83m}$	83	34			70 s.	β-	1.78				1/2-		0.35666 ± 0.00006	17%
	83	34											0.7990 ± 0.0001	11%
	83	34						2.88					0.9879 ± 0.0001	15%
	83	34						3.92					0.9976 ± 0.0001	1.1%
	83	34											1.0206 ± 0.0001	1.9%
	83	34											1.0305 ± 0.0001	20.6%
	83	34											1.0634 ± 0.0002	3.5%
	83	34											1.6600 ± 0.0001	1.7%
	83	34											2.0514 ± 0.0002	10.7%
$_{34}Se^{83}$	83	34		82.919117	22.3 m.	β-	3.67	0.93			9/2+		0.22516 ± 0.00006	33%
	83	34						1.51					0.35666 ± 0.00006	69%
	83	34											0.51004 ± 0.00008	45%
	83	34											0.7180 ± 0.0001	16%
	83	34											0.7990 ± 0.0001	16%
	83	34											0.8666 ± 0.0001	9.1%
	83	34											0.8836 ± 0.0001	7.2%
	83	34											1.0641 ± 0.0001	6%
	83	34											1.0820 ± 0.0001	2.8%
	83	34											1.1917 ± 0.0001	4.2%
	83	34											1.299 ± 0.0001	6.0%
	83	34											1.8948 ± 0.0001	7.2%
	83	34											2.2902 ± 0.0003	9.5%
	83	34											2.3374 ± 0.0003	3.5%
$_{34}Se^{84}$	84	34		83.918463	3.3 m.	β-	1.83	1.41	100%		0+		0.4088 ± 0.0005	100%
$_{34}Se^{85}$	85	34		84.922260	32 s.	β-	6.2	5.9			5/2+		0.3450 ± 0.001	22%
	85	34											0.6094 ± 0.001	41%
	85	34											0.941 ± 0.001	2.2%
	85	34											0.954 ± 0.001	2.9%
	85	34											1.207 ± 0.001	2.9%
	85	34											1.373 ± 0.001	1%
	85	34											1.428 ± 0.001	2.2%
	85	34											2.237 ± 0.001	1.3%
	85	34											2.418 ± 0.001	1%
	85	34											2.456 ± 0.001	1%
	85	34											3.376 ± 0.001	3.2%
	85	34											3.657 ± 0.001	1.5%
	85	34											3.685 ± 0.001	1%

Isotope	A	Z	% Natural abundance	Atomic mass	Half-life	Decay mode	Decay energy (MeV)	Particle energy (MeV)	Particle intensity	Thermal neutron cross section	Spin (h/2π)	μ Nucl. mag. moment	Gamma-ray energy (MeV)	Gamma-ray intensity
	85	34											3.775 ± 0.001	1%
$_{34}Se^{86}$	86	34		85.924270	15 s.	β-	5.1				5/2+		0.7881 ± 0.001	4 +
	86	34											1.1183 ± 0.001	5
	86	34											1.4003 ± 0.0006	14
	86	34											2.0124 ± 0.001	24
	86	34											2.2416 ± 0.001	17
	86	34											2.4433 ± 0.0008	100
	86	34											2.6619 ± 0.001	49
$_{34}Se^{87}$	87	34		86.928390	5.6 s.	β-	≈7						0.468 ± 0.001	100 +
	87	34				n							1.4979 ± 0.001	23
$_{34}Se^{88}$	88	34			1.5 s.	β-,n	≈7						0.5346	
$_{34}Se^{89}$	89	34			0.41 s.	β-,n								
Br		35		79.904						6.8 b.				
$_{35}Br^{72}$	72	35		71.936630	1.31 m.	β+	≈9.0				3		ann.rad.	
	72	35											0.3799 ± 0.0003	3.4%
	72	35											0.4547 ± 0.0003	14%
	72	35											0.7528 ± 0.0004	2.5%
	72	35											0.7748 ± 0.0003	7%
	72	35											0.8620 ± 0.0002	70%
	72	35											1.0547 ± 0.0003	3.5%
	72	35											1.0616 ± 0.0003	5%
	72	35											1.1364 ± 0.0004	17%
	72	35											1.3167 ± 0.0003	17%
	72	35											1.5098 ± 0.0004	2.7%
	72	35											1.5713 ± 0.0004	3.8%
	72	35											1.7240 ± 0.0005	3.4%
	72	35											2.3719 ± 0.0007	7%
$_{35}Br^{73}$	73	35		72.931680	3.4 m.	β+	4.6	3.7			3/2-		ann.rad.	
	73	35											0.0649 ± 0.0001	100 +
	73	35											0.1255 ± 0.0001	23
	73	35											0.2751 ± 0.0002	10
	73	35											0.3352 ± 0.0002	34
	73	35											0.4006 ± 0.0002	20
	73	35											0.6995 ± 0.0002	40
	73	35											0.8487 ± 0.0004	20
	73	35											0.9136 ± 0.0002	19
	73	35											0.9307 ± 0.0002	22
	73	35											0.9956 ± 0.0002	7
$_{35}Br^{74m}$	74	35			41.5 m.	β+		4.5			4-		ann.rad.	200%
	74	35						5.2					0.6152 ± 0.0001	8%
	74	35											0.6343 ± 0.0002	19%
	74	35											0.6348 ± 0.0001	98%
	74	35											0.7285 ± 0.0001	38%
	74	35											0.8389 ± 0.0002	6%
	74	35											1.2495 ± 0.0002	7.6%
	74	35											1.2691 ± 0.0002	8.8%
	74	35											1.7149 ± 0.0003	6.5%
	74	35											2.2838 ± 0.0003	3%
	74	35											2.3119 ± 0.0003	3.7%
	74	35											3.9576 ± 0.0006	3.8%
	74	35											(0.2 - 4.38)weak	
$_{35}Br^{74}$	74	35		73.929898	25.3 m.	β+	6.92						ann.rad.	
	74	35											0.6341 ± 0.0002	20%
	74	35											0.6348 ± 0.0001	68%
	74	35											1.0228 ± 0.0001	5.6%
	74	35											1.2689 ± 0.0001	7.6%
	74	35											1.8428 ± 0.0002	2.5%
	74	35											2.1306 ± 0.0002	2%
	74	35											2.3961 ± 0.0002	8%
	74	35											2.6152 ± 0.0002	8%
	74	35											2.6616 ± 0.0003	5.5%
	74	35											2.7708 ± 0.0005	2.4%
	74	35											3.2499 ± 0.0005	6.7%
	74	35											3.6246 ± 0.0003	6%
	74	35											3.6319 ± 0.0005	2.6%
	74	35											3.9727 ± 0.0002	2.5%
	74	35											4.3796 ± 0.0004	4.5%
	74	35											(0.2 - 4.7)weak	
$_{35}Br^{75}$	75	35		74.925753	98 m.	β+(76%)	3.0				3/2-		ann.rad.	150%
	75	35				E.C.(24%)							0.14119 ± 0.0001	7%
	75	35											0.28650 ± 0.0002	92%
	75	35											0.29285 ± 0.0002	2.8%
	75	35											0.37739 ± 0.0001	4.1%
	75	35											0.43175 ± 0.0001	4%
	75	35											0.57293 ± 0.0001	2%
	75	35											0.73394 ± 0.0001	1.6%
	75	35											0.91205 ± 0.0001	1.1%
	75	35											0.95210 ± 0.0001	1.7%
	75	35											(0.1 - 1.56)weak	
$_{35}Br^{76m}$	76	35			1.49 s.	I.T.	5.05				4+		0.104548 ± 0.00005	
	76	35											0.05711 ± 0.00005	
$_{35}Br^{76}$	76	35		75.924528	16.1 h.	β+(57%)	4.956	1.9			1-	0.5482	ann.rad.	130%

Isotope	A	Z	% Natural abundance	Atomic mass	Half-life	Decay mode	Decay energy (MeV)	Particle energy (MeV)	Particle intensity	Thermal neutron cross section	Spin (h/2π)	μ Nucl. mag. moment	Gamma-ray energy (MeV)	Gamma-ray intensity
	76	35				E.C.(43%)		3.15					0.47291 ± 0.00006	1.9%
	76	35						3.68					0.55911 ± 0.00005	74%
	76	35											0.56322 ± 0.00005	3%
	76	35											0.65700 ± 0.00005	5.3%
	76	35											1.12985 ± 0.00006	4.5%
	76	35											1.2160 ± 0.00005	9%
	76	35											1.22865 ± 0.00006	2.1%
	76	35											1.4752 ± 0.00006	2.4%
	76	35											1.85368 ± 0.00005	14.9%
	76	35											2.11127 ± 0.00008	2.4%
	76	35											2.79272 ± 0.00006	5.4%
	76	35											2.95055 ± 0.00005	7.6%
	76	35											(0.4 - 4.6)weak	
$_{35}Br^{77m}$	77	35			4.3 m.	I.T.	0.1059				9/2+		0.1059 ± 0.0002	13.7%
$_{35}Br^{77}$	77	35		76.921378	57.0 h.	E.C.(99%)	1.365				1+		ann.rad.	
	77	35				β+(0.74%)							0.08759 ± 0.0007	1.3%
	77	35											0.16183 ± 0.00008	1.0%
	77	35											0.20040 ± 0.00007	1.1%
	77	35											0.23898 ± 0.00007	23%
	77	35											0.24977 ± 0.00007	2/9%
	77	35											0.29723 ± 0.00008	4.0%
	77	35											0.30376 ± 0.00009	2.0%
	77	35											0.43947 ± 0.00006	1.5%
	77	35											0.52069 ± 0.00006	22%
	77	35											0.57464 ± 0.00008	1.1%
	77	35											0.57891 ± 0.00007	2.8%
	77	35											0.58548 ± 0.00007	1.5%
	77	35											0.75535 ± 0.00007	1.6%
	77	35											0.81779 ± 0.00006	2.0%
	77	35											1/00505 ± 0.00006	0.9%
	77	35											(0.08 - 1.2)weak	
$_{35}Br^{78}$	78	35		77.921144	6.46 m.	β+(92%)	3.3574	1.2			1+		ann.rad.	185%
	78	35				E.C.(8%)		2.5					0.61363 ± 0.00006	13.6%
	78	35											0.8848 ± 0.0001	0.07%
	78	35											1.3086 ± 0.0001	0.04%
	78	35											1.7210 ± 0.0002	0.04%
	78	35											1.9239 ± 0.0002	0.05%
	78	35											2.4767 ± 0.0004	0.02%
	78	35											(0.7 - 3.0)weak	
$_{35}Br^{79m}$	79	35			4.86 s.	I.T.	0.207				9/2+	+2.106	0.2072 ± 0.0004	76%
$_{35}Br^{79}$	79	35	50.69%	78.918336						(2.5 + 8.2)b.	3/2-	+2.1064		
$_{35}Br^{80m}$	80	35			4.42 h.	I.T.	0.04885				5-	+1.3177	Br k x-ray	93%
	80	35											0.03705 ± 0.00002	39%
	80	35											0.04885 ± 0.00003	0.5%
$_{35}Br^{80}$	80	35		79.918528	17.6 m.	β-(92%)	2.00	1.38 β-	7.6%		1+	0.5140	ann.rad.	
	80	35				E.C.(5.7%)	1.870	1.99 β-	82%				0.6162 ± 0.0005	6.7%
	80	35				β+(2.6%)		0.85 β+	2.8%				0.6394 ± 0.0002	0.2%
	80	35											0.6658 ± 0.0002	1.1%
	80	35											0.7038 ± 0.0002	0.13%
	80	35											1.2561 ± 0.0004	0.09%
$_{35}Br^{81}$	81	35	49.31%	80.916289						(2.4 + 0.26)b.	3/2-	+2.2706		
$_{35}Br^{82m}$	82	35				I.T.(98%)	0.046				2-		0.046	0.2%
	82	35				β-(2%)	3.139						0.6985 ± 0.0002	0.025%
	82	35											0.77645 ± 0.0002	0.2%
	82	35											1.4748 ± 0.0002	0.02%
$_{35}Br^{82}$	82	35		81.916802	35.30 h.	β-	3.093	0.444			5-	+1.6270	0.221411 ± 0.00002	2.3%

Isotope	A	Z	% Natural abundance	Atomic mass	Half-life	Decay mode	Decay energy (MeV)	Particle energy (MeV)	Particle intensity	Thermal neutron cross section	Spin (h/2π)	μ Nucl. mag. moment	Gamma-ray energy (MeV)	Gamma-ray intensity
	82	35											0.55432 ± 0.00001	71%
	82	35											0.61905 ± 0.00002	43%
	82	35											0.69832 ± 0.00002	29%
	82	35											0.77649 ± 0.00003	83%
	82	35											0.82781 ± 0.00002	24%
	82	35											1.04398 ± 0.00003	27%
	82	35											1.3747 ± 0.00005	27%
	82	35											1.47482 ± 0.00008	17%
$_{35}Br^{83}$	83	35		82.915179	2.39 h.	β-	0.98	0.395	1%		3/2-		0.52041 ± 0.00005	0.06%
	83	35						0.925	99%				0.52964 ± 0.00001	1.3%
	83	35											0.5526 ± 0.0001	0.02%
$_{35}Br^{84m}$	84	35			6.0 m.	β-		2.2	100%		(6-)		0.4240 ± 0.001	100%
	84	35											0.4472 ± 0.001	2%
	84	35											0.8816 ± 0.001	98%
	84	35											1.0160 ± 0.001	1%
	84	35											1.4628 ± 0.0005	97%
	84	35											1.8972 ± 0.0008	2%
$_{35}Br^{84}$	84	35		83.916503	31.8 m.	β-	4.65	2.70	11%		2-		0.6048 ± 0.0003	1.7%
	84	35						3.81	20%				0.7365 ± 0.0003	1.2%
	84	35						4.63	34%				0.8022 ± 0.0002	5.7%
	84	35											0.8816 ± 0.0001	42%
	84	35											1.0159 ± 0.0003	6.2%
	84	35											1.2133 ± 0.0002	2.6%
	84	35											1.4638 ± 0.0007	2.0%
	84	35											1.7412 ± 0.0004	1.6%
	84	35											1.8775 ± 0.0004	2.1%
	84	35											1.8976 ± 0.0002	14.9%
	84	35											2.0296 ± 0.0005	2.1%
	84	35											2.4841 ± 0.0003	6.7%
	84	35											2.8241 ± 0.0004	1.1%
	84	35											3.0454 ± 0.0004	2.5%
	84	35											3.2353 ± 0.0005	2.0%
	84	35											3.3658 ± 0.0004	2.9%
	84	35											3.9275 ± 0.0004	6.8%
$_{35}Br^{85}$	85	35		84.915612	2.87 m.	β-	2.5				3/2-		0.30486 ± 0.00002	14%
	85	35											0.80241 ± 0.0001	2.5%
	85	35											0.92463 ± 0.00005	1.6%
	85	35											1.7270 ± 0.0001	0.4%
	85	35											1.8325 ± 0.0001	0.1%
	85	35											(0.09 - 2.4)weak	
$_{35}Br^{86}$	86	35		85.918800	β-	7.6					(2-)		0.5012 ± 0.0005	1.6%
	86	35											0.6853 ± 0.0001	1.1%
	86	35											0.78496 ± 0.0001	3.6%
	86	35											1.21702 ± 0.0001	6.6%
	86	35											1.2861 ± 0.0001	7.7%
	86	35											1.38973 ± 0.0001	10.2%
	86	35											1.46509 ± 0.0001	7.4%
	86	35											1.53424 ± 0.0001	7.7%
	86	35											1.56460 ± 0.0001	62%
	86	35											1.9663 ± 0.0001	6%
	86	35											2.34937 ± 0.0001	10%
	86	35											2.75106 ± 0.0001	19%
	86	35											3.0090 ± 0.0003	1%
	86	35											3.7588 ± 0.0003	1%
	86	35											3.7831 ± 0.0003	1.1%
	86	35											4.4012 ± 0.0003	1%
	86	35											4.88512 ± 0.0002	1.2%
	86	35											5.40580 ± 0.0002	5.6%
	86	35											5.5176 ± 0.0003	3.6%
	86	35											6.2116 ± 0.0003	0.8%
	86	35											(0.5 - 6.8)weak	
$_{35}Br^{87}$	87	35		86.920690	56.1 s.	β-	6.8	6.1			3/2-		0.42182 ± 0.00006	5%
	87	35											0.5294 ± 0.0001	1.2%
	87	35											0.53190 ± 0.00007	8%
	87	35											0.6105 ± 0.0001	1.1%
	87	35											0.9528 ± 0.0001	1%
	87	35											1.0213 ± 0.0001	1.9%
	87	35											1.36089 ± 0.0001	5%
	87	35											1.41983 ± 0.00009	32%
	87	35											1.4494 ± 0.0001	1.8%
	87	35											1.4762 ± 0.0001	12%

Isotope	A	Z	% Natural abundance	Atomic mass	Half-life	Decay mode	Decay energy (MeV)	Particle energy (MeV)	Particle intensity	Thermal neutron cross section	Spin (h/2π)	μ Nucl. mag. moment	Gamma-ray energy (MeV)	Gamma-ray intensity
	87	35											1.5777 ± 0.0001	9%
	87	35											1.6073 ± 0.0001	1.8%
	87	35											1.9784 ± 0.0001	1%
	87	35											2.07156 ± 0.0001	3.5%
	87	35											2.4526 ± 0.0001	1%
	87	35											2.7051 ± 0.0001	2.6%
	87	35											2.8212 ± 0.0001	2.5%
	87	35											2.8366 ± 0.0001	2.4%
	87	35											3.0273 ± 0.0001	1.9%
	87	35											3.9173 ± 0.0001	2.9%
	87	35											4.1826 ± 0.0001	7%
	87	35											4.7848 ± 0.0001	27%
	87	35											4.9618 ± 0.0002	2.8%
	87	35											5.1205 ± 0.0002	0.8%
	87	35											5.1955 ± 0.0002	0.7%
	87	35											5.2013 ± 0.0003	1.0%
	87	35											(0.2 - 6.1)weak	
$_{35}Br^{88}$	88	35		87.924080	16.4 s.	β-	9.0				1-		0.7649 ± 0.0005	1.6%
	88	35				n							0.7753 ± 0.0001	77%
	88	35											0.7934 ± 0.0005	1.2%
	88	35											0.8021 ± 0.0001	16%
	88	35											1.0537 ± 0.0001	1.8%
	88	35											1.3690 ± 0.0001	1.1%
	88	35											1.4407 ± 0.0001	5%
	88	35											1.5670 ± 0.0001	2.3%
	88	35											1.6442 ± 0.0001	3.1%
	88	35											2.6245 ± 0.0001	1.8%
	88	35											2.9457 ± 0.0002	2.4%
	88	35											3.0194 ± 0.0002	1.2%
	88	35											3.9322 ± 0.0002	5%
	88	35											4.0217 ± 0.0002	2.6%
	88	35											4.1479 ± 0.0001	4%
	88	35											4.5628 ± 0.0002	3.2%
	88	35											4.7134 ± 0.0002	1.1%
	88	35											4.7212 ± 0.0005	1%
	88	35											4.985 ± 0.0005	1%
	88	35											5.0193 ± 0.0005	2.1%
	88	35											(0.1 - 6.99)weak	
$_{35}Br^{89}$	89	35		88.926550	4.4 s.	β-	8.5				3/2-		0.7753	
	89	35				n							1.0978	
$_{35}Br^{90}$	90	35		89.931010	1.9 s.	β-	10.7	8.3			2-		0.6555	18 +
	90	35				n		9.8					0.7071	100
	90	35											1.3626	27
	90	35											2.1282	17
	90	35											2.2528	16
	90	35											3.3454	10
$_{35}Br^{91}$	91	35			0.54 s.	β-(90%)							0.263	
	91	35				β-n(10%)							0.803	
$_{35}Br^{92}$	92	35			0.36 s.	β-							0.740	
	92	35				β-n								
Kr		36	83.80							25 b.				
$_{36}Kr^{72}$	72	36		71.942060	17 s.	β+	5.1				0+		ann.rad.	190%
	72	36				E.C.							0.1626 ± 0.0002	8%
	72	36											0.2522 ± 0.0002	3%
	72	36											0.3100 ± 0.0002	15%
	72	36											0.4150 ± 0.0002	19%
	72	36											0.5766 ± 0.0002	6%
$_{36}Kr^{73}$	73	36		72.938920	27 s.	β+	6.7						ann.rad.	
	73	36				E.C.							0.1511 ± 0.0004	13%
	73	36				β+,p							0.1781 ± 0.0003	66%
	73	36											0.2136 ± 0.0004	9%
	73	36											0.2413 ± 0.0004	7%
	73	36											0.3036 ± 0.0004	4%
	73	36											0.3292 ± 0.0004	4%
	73	36											0.3919 ± 0.0004	7%
	73	36											0.4736 ± 0.0004	11%
$_{36}Kr^{74}$	74	36		73.933290	11.5 m.	β+	3.3				0+		ann.rad.	74%
	74	36				E.C.							0.00985 ± 0.00002	5%
	74	36											0.0628 ± 0.0001	10%
	74	36											0.06740 ± 0.00001	1.2%
	74	36											0.08970 ± 0.0001	31%
	74	36											0.0938 ± 0.0001	3%
	74	36											0.1234 ± 0.0001	9.3%
	74	36											0.1403 ± 0.0001	3.3%
	74	36											0.14970 ± 0.0001	1%
	74	36											0.2030 ± 0.0001	19%
	74	36											0.2169 ± 0.0001	10%
	74	36											0.2339 ± 0.0001	5.3%
	74	36											0.2967 ± 0.0001	1%
	74	36											0.3065 ± 0.0001	10.5%
	74	36											0.6091 ± 0.0001	1%
	74	36											0.7013 ± 0.0001	1.7%

Isotope	A	Z	% Natural abundance	Atomic mass	Half-life	Decay mode	Decay energy (MeV)	Particle energy (MeV)	Particle intensity	Thermal neutron cross section	Spin (h/2π)	μ Nucl. mag. moment	Gamma-ray energy (MeV)	Gamma-ray intensity
	74	36											(0.02 - 0.9)weak	
$_{36}Kr^{75}$	75	36		74.931029	4.5 m.	β+	5.0	3.2					ann.rad.	190%
	75	36				E.C.							0.0884 ± 0.0001	1.6%
	75	36											0.1325 ± 0.0001	31%
	75	36											0.1533 ± 0.0001	3.7%
	75	36											0.1547 ± 0.0002	9.6%
	75	36											0.7931 ± 0.0002	0.8%
	75	36											1.3463 ± 0.0004	0.65%
	75	36											1.6018 ± 0.0004	0.84%
$_{36}Kr^{76}$	76	36		75.925959	14.8 h.	E.C.	1.35				0+		Br k x-ray	51%
	76	36											0.0455 ± 0.0001	18%
	76	36											0.1032 ± 0.0001	3%
	76	36											0.2522 ± 0.0002	6.5%
	76	36											0.2718 ± 0.0002	4%
	76	36											0.2992 ± 0.0002	1%
	76	36											0.3099 ± 0.0002	2%
	76	36											0.3158 ± 0.0002	38%
	76	36											0.3353 ± 0.0002	5%
	76	36											0.4065 ± 0.0002	11%
	76	36											0.4521 ± 0.0002	8.8%
	76	36											0.5528 ± 0.0003	1.3%
	76	36											0.5815 ± 0.0004	0.9%
	76	36											0.5825 ± 0.0004	0.9%
	76	36											(0.03 - 1.07)	
$_{36}Kr^{77}$	77	36		76.924610	1.24 h.	β+(80%)	0.90				5/2+		ann.rad.	
	77	36				E.C>(20%)		1.55					0.1062 ± 0.0002	1.2%
	77	36						1.70					0.1297 ± 0.0001	84%
	77	36						1.87					0.1465 ± 0.0001	3.9%
	77	36											0.2762 ± 0.0002	3.2%
	77	36											0.3122 ± 0.0002	3.6%
	77	36											0.6060 ± 0.0001	0.4%
	77	36											0.7346 ± 0.0002	0.4%
	77	36											(0.1 - 2.3)weak	
$_{36}Kr^{78}$	78	36	0.35%							(0.17 + 6) b.	0+			
$_{36}Kr^{79m}$	79	36			50 s.	I.T.	0.1300				7/2+		Kr x-ray	
	79	36											0.13001 ±	27%
													0.00001	
$_{36}Kr^{79}$	79	36		78.920084	35.0 h.	β+(7%)	1.63				1/2-		ann.rad.	
	79	36				E.C.(93%)							0.21702 ± 0.0001	2.4%
	79	36											0.2613 ± 0.0001	12.7%
	79	36											0.2995 ± 0.0001	1.7%
	79	36											0.3063 ± 0.0001	2.6%
	79	36											0.3890 ± 0.0001	1.6%
	79	36											0.39756 ± 0.0001	9.5%
	79	36											0.6061 ± 0.0001	8.1%
	79	36											0.8320 ± 0.0001	1.3%
	79	36											1.11514 ± 0.0003	0.4%
	79	36											1.33213 ± 0.0001	0.44%
	79	36											(0.13 - 1.3)weak	
$_{36}Kr^{80}$	80	36	2.25%	79.916380						(4.6 + 8) b.	0+			
$_{36}Kr^{81m}$	81	36			13.3 s.	I.T.	0.1903				1/2-		0.19030 ± 0.0001	67%
$_{36}Kr^{81}$	81	36		80.916590	$2.1x10^5$ y.	E.C.	0.276				7/2+		Br k x-ray	53%
	81	36											0.2760 ± 0.0001	3.6%
$_{36}Kr^{82}$	82	36	11.6%	81.913482						(16 + 20) b.	0+			
$_{36}Kr^{83m}$	83	36			1.86 h.	I.T.	0.04155				1/2-		Kr k x-ray	13.2%
	83	36											0.00940 ±	4.9%
													0.00002	
	83	36											0.03216 ±	0.04%
													0.00002	
$_{36}Kr^{83}$	83	36	11.5%	82.914135						1800 b.	9/2+	-0.9707		
$_{36}Kr^{84}$	84	36	57.0%	83.911507						(0.09 + 0.042) b.	0+			
$_{36}Kr^{85m}$	85	36			4.48 h.	β-(79%)		0.83	79%		1/2-		0.30487 ±	14%
													0.00002	
	85	36				I.T.(21%)	0.304						0.15118 ±	79%
													0.00001	
$_{36}Kr^{85}$	85	36		84.912531	10.72 y.	β-	0.687	0.15	0.4%		9/2+	1.005	0.51399 ±	0.43%
													0.00002	
$_{36}Kr^{86}$	86	36	17.3%	85.910616						0.003 b.	0+			
$_{36}Kr^{87}$	87	36		86.913360	76.3 m.	β-	3.886	0.93	4%		5/2+		0.40258 ±	50%
													0.00002	
	87	36						1.33	8%				0.6739 ± 0.0001	1.1%
	87	36					1.45	5%					0.84545 ±	8%
													0.00004	
	87	36						3.0	7%				1.1747 ± 0.0001	1.1%
	87	36					3.49	43%					1.7405 ± 0.0001	2%
	87	36						3.89	30%				2.01181 ± 0.0001	2.1%
	87	36											2.4084 ± 0.0002	0.5%
	87	36											2.5548 ± 0.0002	9%
	87	36											2.5583 ± 0.0002	4%
	87	36											3.3084 ± 0.0005	0.5%
$_{36}Kr^{88}$	88	36		87.914453	2.84 h.	β-	2.91				0+		0.0275 ± 0.00001	2%
	88	36											0.16598 ±	3.2%
													0.00004	

Isotope	A	Z	% Natural abundance	Atomic mass	Half-life	Decay mode	Decay energy (MeV)	Particle energy (MeV)	Particle intensity	Thermal neutron cross section	Spin (h/2π)	μ Nucl. mag. moment	Gamma-ray energy (MeV)	Gamma-ray intensity
	88	36											0.19632 ± 0.00002	26.6%
	88	36											0.36223 ± 0.00001	2.3%
	88	36											0.83482 ± 0.00001	13.3%
	88	36											0.98578 ± 0.00002	1.3%
	88	36											1.14133 ± 0.00006	1.3%
	88	36											1.17951 ± 0.00003	1.0%
	88	36											1.25067 ± 0.00004	1.1%
	88	36											1.3695 ± 0.0002	1.5%
	88	36											1.5184 ± 0.0001	2.2%
	88	36											1.5298 ± 0.0001	1.3%
	88	36											2.0298 ± 0.0003	4.6%
	88	36											2.0354 ± 0.0001	3.8%
	88	36											2.1958 ± 0.0001	1.3%
	88	36											2.2318 ± 0.0001	3.5%
	88	36											2.39202 ± 0.0001	3.5%
$_{36}Kr^{89}$	89	36		88.917640	3.16 m.	β-	4.99	3.8			5/2+		0.19746 ± 0.00003	1.9%
	89	36						4.6					0.2209 ± 0.0001	2%
	89	36						4.9					0.3450 ± 0.0001	1.2%
	89	36											0.3561 ± 0.0001	4.2%
	89	36											0.3693 ± 0.0001	1.4%
	89	36											0.4114 ± 0.0001	2.6%
	89	36											0.4975 ± 0.0001	6.8%
	89	36											0.4986 ± 0.0003	1.2%
	89	36											0.5769 ± 0.0001	5.8%
	89	36											0.5858 ± 0.0001	16.8%
	89	36											0.6962 ± 0.0001	1.8%
	89	36											0.7384 ± 0.0001	4.3%
	89	36											0.8671 ± 0.0001	6.0%
	89	36											0.9043 ± 0.0001	7.3%
	88	36											1.1078 ± 0.0001	3%
	89	36											1.1161 ± 0.0001	1.7%
	89	36											1.2737 ± 0.0001	1.4%
	89	36											1.3243 ± 0.0001	3.12%
	89	36											1.4728 ± 0.0001	7%
	89	36											1.5010 ± 0.0001	1.3%
	89	36											1.5300 ± 0.0001	1.3%
	89	36											1.5337 ± 0.0001	5.2%
	89	36											1.6937 ± 0.0001	4.5%
	89	36											1.9034 ± 0.0001	1%
	89	36											2.0122 ± 0.0001	1.6%
	89	36											2.8662 ± 0.0002	1.8%
	89	36											3.1403 ± 0.0002	1.1%
	89	36											3.3617 ± 0.0002	1.1%
	89	36											3.5329 ± 0.0002	1.4%
	89	36											(0.2 - 4.7)weak	
$_{36}Kr^{90}$	90	36		89.919520	32.3 s.	β-	4.39	2.6	77%		0+		0.12182 ± 0.00003	33.5%
	90	36						2.8	6%				0.49263 ± 0.00005	1.2%
	90	36											0.5395 ± 0.0001	30.7%
	90	36											0.5544 ± 0.0001	5.0%
	90	36											0.6191 ± 0.0001	1.1%
	90	36											0.7313 ± 0.0001	1.5%
	90	36											1.1187 ± 0.0001	38.9%
	90	36											1.4238 ± 0.0001	2.9%
	90	36											1.5378 ± 0.0001	9.6%
	90	36											1.5522 ± 0.0001	2.2%
	90	36											1.6582 ± 0.0001	1.3%
	90	36											1.7801 ± 0.0001	6.7%
	90	36											2.1275 ± 0.0001	1.4%
	90	36											2.7267 ± 0.0001	0.9%
	90	36											(0.1 - 4.2)weak	
$_{36}Kr^{91}$	91	36		90.923380	8.6 s.	β-	6.4	4.33			5/2+		0.10878 ± 0.00004	42%
	91	36						4.59					0.4120 ± 0.0001	2.3%
	91	36						4.98					0.50658 ± 0.0001	19%
	91	36						5.4					0.5556 ± 0.0001	2%
	91	36											0.6129 ± 0.0001	7.6%
	91	36											0.6301 ± 0.0001	2.2%
	91	36											0.6624 ± 0.0001	1.3%
	91	36											0.7610 ± 0.0001	1.0%
	91	36											0.8749 ± 0.0001	1.3%
	91	36											1.0249 ± 0.0002	2.9%
	91	36											1.1087 ± 0.0001	7.1%
	91	36											1.1368 ± 0.0001	1.0%
	91	36											1.1780 ± 0.0001	1.3%
	91	36											1.3043 ± 0.0001	1.24%

Isotope	A	Z	% Natural abundance	Atomic mass	Half-life	Decay mode	Decay energy (MeV)	Particle energy (MeV)	Particle intensity	Thermal neutron cross section	Spin (h/2π)	μ Nucl. mag. moment	Gamma-ray energy (MeV)	Gamma-ray intensity
	91	36											1.5016 ± 0.0001	4.8%
	91	36											1.6667 ± 0.0001	1.0%
	91	36											2.4843 ± 0.0001	2.8%
	91	36											2.7358 ± 0.0002	1.5%
	91	36											2.9818 ± 0.0002	1.3%
	91	36											3.0568 ± 0.0002	0.9%
	91	36											3.1135 ± 0.0002	2.1%
	91	36											(0.1 - 4.4)weak	
$_{36}Kr^{92}$	92	36		91.926270	1.84 s.	β-	6.2						0.1424 ± 0.0001	64%
	92	36				n							0.3168 ± 0.0001	5.8%
	92	36											0.3423 ± 0.0001	2.1%
	92	36											0.4847 ± 0.0001	3.2%
	92	36											0.54830 ± 0.0001	14%
	92	36											0.6237 ± 0.0001	1.3%
	92	36											0.8126 ± 0.0001	14.5%
	92	36											0.8763 ± 0.0001	4.2%
	92	36											1.0442 ± 0.0001	4.7%
	92	36											1.21860 ± 0.0001	59.8%
	92	36											1.3608 ± 0.0001	3.5%
	92	36											1.8968 ± 0.0002	0.8%
	92	36											2.0390 ± 0.0002	0.4%
	92	36											2.8328 ± 0.0002	0.3%
	92	36											3.1995 ± 0.0002	0.4%
	92	36											(0.15 - 3.7)weak	
$_{36}Kr^{93}$	93	36		92.931130	1.29 s.	β-	8.5	7.1					0.1820 ± 0.0001	5.4%
	93	36				n							0.2523 ± 0.0002	19.6%
	93	36											0.2536 ± 0.0002	41%
	93	36											0.26683 ± 0.00005	21%
	93	36											0.32309 ± 0.00002	24%
	93	36											0.4966 ± 0.0001	1.8%
	93	36											0.5702 ± 0.0001	1.2%
	93	36											0.8204 ± 0.001	3.7%
	93	36											1.0262 ± 0.0001	2.2%
	93	36											1.2150 ± 0.0001	1.8%
	93	36											1.2388 ± 0.0001	1.1%
	93	36											1.2961 ± 0.0001	1.9%
	93	36											1.3879 ± 0.0001	1.3%
	93	36											1.4353 ± 0.0001	1.0%
	93	36											1.5058 ± 0.0001	2.2%
	93	36											1.5962 ± 0.0001	1.4%
	93	36											1.6271 ± 0.0001	2.0%
	93	36											1.6411 ± 0.0001	1.4%
	93	36											1.6978 ± 0.0001	1.4%
	93	36											1.7425 ± 0.0001	1.3%
	93	36											1.9618 ± 0.0001	1.8%
	93	36											2.0189 ± 0.0001	1.4%
	93	36											2.0354 ± 0.0001	1.8%
	93	36											2.1815 ± 0.0001	1.2%
	93	36											2.3499 ± 0.0001	7.4%
	93	36											2.4960 ± 0.0001	2.3%
	93	36											2.5613 ± 0.0001	1.0%
	93	36											2.6026 ± 0.0001	4.2%
	93	36											2.8559 ± 0.0001	2.2%
	93	36											3.2267 ± 0.0002	1.0%
	93	36											3.4607 ± 0.0006	0.7%
$_{36}Kr^{94}$	94	36			0.21 s.	β-	7.5						0.1353 ± 0.001	13 +
	94	36				n							0.1874 ± 0.001	37
	94	36											0.2196 ± 0.001	67
	94	36											0.2881 ± 0.001	33
	94	36											0.3208 ± 0.001	26
	94	36											0.3546 ± 0.001	28
	94	36											0.3590 ± 0.001	39
	94	36											0.3948 ± 0.001	21
	94	36											0.4022 ± 0.001	20
	94	36											0.5933 ± 0.001	29
	94	36											0.6293 ± 0.0001	100
	94	36											0.6957 ± 0.001	25
Rb		37		85.4678						0.38 b.				
$_{37}Rb^{75}$	75	37		74.938510	17 s.	β+	6.6	2.31					ann. rad.	71%
	75	37											0.0628 ± 0.0001	10%
	75	37											0.08970 ± 0.0001	31%
	75	37											0.0938 ± 0.0001	3.3%
	75	37											0.1234 ± 0.0001	9.3%
	75	37											0.1403 ± 0.0001	9%
	75	37											0.1497 ± 0.0001	2.1%
	75	37											0.2030 ± 0.0001	19/4%
	75	37											0.2169 ± 0.0001	10%
	75	37											0.2339 ± 0.0001	5.3%
	75	37											0.2967 ± 0.0001	11%
	75	37											0.3065 ± 0.0001	10.5%
	75	37											0.6091 ± 0.0001	1%
	75	37											0.7013 ± 0.0001	1.7%

Isotope	A	Z	% Natural abundance	Atomic mass	Half-life	Decay mode	Decay energy (MeV)	Particle energy (MeV)	Particle intensity	Thermal neutron cross section	Spin (h/2π)	μ Nucl. mag. moment	Gamma-ray energy (MeV)	Gamma-ray intensity
	75	37											0.9699 ± 0.0001	0.25%
	75	37											(0.06 - 1.1)weak	
$_{37}Rb^{76}$	76	37		75.934960	17 s.	β+	8.2	5.2					ann.rad.	425 +
	76	37											0.3452	16
	76	37											0.3549	25
	76	37											0.4235	100
	76	37											0.800	7
	76	37											0.885	9
	76	37											0.919	8
	76	37											0.974	4
	76	37											1/173	4
	76	37											1.219	5
$_{37}Rb^{77}$	77	37		76.930280	3.8 m.	β+	5.10	3.86			3/2-		ann.rad.	
	77	37											0.0665 ± 0.0001	66%
	77	37											0.1785 ± 0.0003	26%
	77	37											0.3933 ± 0.0003	12%
	77	37											0.6085 ± 0.0008	3.6%
	77	37											0.6265 ± 0.0004	4.3%
	77	37											0.7798 ± 0.0008	1.2%
	77	37											0.9587 ± 0.0004	2.4%
	77	37											0.9708 ± 0.0004	4%
	77	37											0.9883 ± 0.0004	2.4%
	77	37											(0.04 - 1.8)weak	
$_{37}Rb^{78m}$	78	37			5.7 m.	I.T.	0.1034				4-		ann.rad.	247 +
	78	37				β+							0.10336 ± 0.0001	8
	78	37				E.C.							0.4553 ± 0.0001	100
	78	37											0.6647 ± 0003	44
	78	37											0.6927 ± 0.0003	13
	78	37											0.7251 ± 0.0003	7
	78	37											0.7534 ± 0.0003	8
	78	37											1.1994 ± 0.0003	9
	78	37											1.2325 ± 0.0001	2
	78	37											1.5301 ± 0.0005	5
	78	37											1.6303 ± 0.0004	6.6
	78	37											1.6446 ± 0.0004	9
	78	37											1.8530 ± 0.0005	7
	78	37											1.9440 ± 0.0005	9
	78	37											2.0136 ± 0.0005	5
	78	37											2.1180 ± 0.0007	4
	78	37											2.6273 ± 0.0007	4
	78	37											3.0830 ± 0.0007	1.7
	78	37											3.3180 ± 0.001	1.5
$_{37}Rb^{78}$	78	37		77.928090	17.66 m.	β+	7.02				0+		ann.rad.	200 +
	78	37				E.C.							0.4553 ± 0003	100
	78	37											0.5624 ± 0.0003	17
	78	37											0.6353 ± 0.0003	2.3
	78	37											0.6647 ± 0.0003	5
	78	37											0.6929 ± 0.0003	20
	78	37											0.7351 ± 0.0005	3
	78	37											0.8591 ± 0.0005	5
	78	37											1.1481 ± 0.0003	12
	78	37											1.3006 ± 0.0005	5
	78	37											1.7810 ± 0.001	11
	78	37											2.4202 ± 0.0007	7
	78	37											2.4953 ± 0.0007	4
	78	37											2.5152 ± 0.0007	6
	78	37											2.8930 ± 0.001	3
	78	37											2.9830 ± 0.001	7
	78	37											3.0830 ± 0.0007	5
	78	37											3.4380 ± 0.0007	20
	78	37											3.5410 ± 0.001	2
	78	37											3.5740 ± 0.0015	2
	78	37											3.8930 ± 0.001	4
$_{37}Rb^{79}$	79	37		78.923954	23 m.	β+(84%)	3.53				5/2-		ann.rad.	168%
	79	37				E.C.(16%)							0.01788 ± 0.00006	1.6%
	79	37											0.13601 ± 0.00002	10%
	79	37											0.14349 ± 0.00005	11%
	79	37											0.14723 ± 0.00003	8%
	79	37											0.15484 ± 0.00002	6%
	79	37											0.16068 ± 0.00002	7%
	79	37											0.18282 ± 0.0001	16%
	79	37											0.35066 ± 0.00006	7%
	79	37											0.39765 ± 0.00005	5%
	79	37											0.50530 ± 0.00004	13%
	79	37											0.62208 ± 0.0001	7%

TABLE OF THE ISOTOPES (Continued)

Isotope	A	Z	% Natural abundance	Atomic mass	Half-life	Decay mode	Decay energy (MeV)	Particle energy (MeV)	Particle intensity	Thermal neutron cross section	Spin (h/2π)	μ Nucl. mag. moment	Gamma-ray energy (MeV)	Gamma-ray intensity
	79	37											0.68812 ± 0.00004	24%
	79	37											(0.14 - 2.0)weak	
$_{37}Rb^{80}$	80	37		79.922519	34.s.	β+	5.71	3.86	22%		1+	-0.0834	ann.rad.	198%
	80	37						4/7	74%				0.6167 ± 0.0005	25%
	80	37											0.6396 ± 0.0005	1.5%
	80	37											0.7043 ± 0.0005	1.9%
	80	37											1.2571 ± 0.001	0.6%
$_{37}Rb^{81m}$	81	37			32 m.	I.T.	0.85	1.4			9/2+		ann.rad.	46%
	81	37				β+ ,E.C.							0.085	
$_{37}Rb^{81}$	81	37		80.918990	4.58 h.	β+(27%)	2.24	1.05			3/2-	+2.05	ann.rad.	46%
	81	37				E.C.(73%)							0.19030 ± 0.00005	66%(D)
	81	37											0.44614 ± 0.00002	19%
	81	37											0.45671 ± 0.00003	2.4%
	81	37											0.53760 ± 0.00004	1.5%
	81	37											(0.05 - 1.5)weak	
$_{37}Rb^{82m}$	82	37			6.47 h.	β+(26%)		0.80			5-	+1.5	ann.rad.	
	82	37				E.C.(74%)							0.5542 ± 0.0005	63%
	82	37											0.6189 ± 0.0004	37%
	82	37											0.6982 ± 0.0004	24%
	82	37											0.7768 ± 0.0004	82%
	82	37											0.8278 ± 0.0004	21%
	82	37											1/0079 ± 0.0006	6.9%
	82	37											1.0442 ± 0.0006	33%
	82	37											1.3172 ± 0.0006	26%
	82	37											1.4749 ± 0.0006	17%
	82	37											(0.1 - 2.3)weak	
$_{37}Rb^{82}$	82	37		81.918195	1.273 m.	β+(96%)	4.36	3.3			1+	+1.6434	ann.rad.	191%
	82	37				E.C.(4%)							0.7665 ± 0.0002	13.6%
	82	37											1.3952 ± 0.0003	0.5%
	82	37											(0.6 - 2.9)weak	
$_{37}Rb^{83}$	83	37		82.915144	86.2 d.	E.C.	0.96				5/2-	+1.43	Kr x-ray	60%
	83	37											0.009396 ± 0.00009	6%
	83	37											0.52039 ± 0.00003	46%
	83	37											0.52958 ± 0.00001	30%
	83	37											0.55254 ± 0.00002	16.6%
	83	37											0.6485 ± 0.0002	0.1%
	83	37											0.7892 ± 0.0002	0.8%
	83	37											0.7986 ± 0.0002	0.3%
$_{37}Rb^{84m}$	84	37			20.3 m.	I.T.	0.216				6-		0.2163 ± 0.0002	29%
	84	37											0.2482 ± 0.0002	64.5%
	84	37											0.4645 ± 0.0002	53%
$_{37}Rb^{84}$	84	37		83.91439	32.9 d.	β+(22%)	2.682	0.780	11%		2-	-1.297	ann.rad.	
	84	37				E.C.(75%)		1.658	11%				0.88160 ± 0.00002	74%
	84	37				β-(3%)	0.893	0.893					1.01614 ± 0.00002	0.35%
	84	37											1.89773 ± 0.00004	1%
$_{37}Rb^{85}$	85	37	72.17%	84.911794						(0.05 + 0.43)b.	5/2-	+1.35302		
$_{37}Rb^{86m}$	86	37			1.018 m.	I.T.	0.5560				6-		0.5558 ± 0.0002	98%
$_{37}Rb^{86}$	86	37		85.911172	18.63 d.	β-	1.774	1.774	8.8%		2-	-1.6920	1.0768 ± 0.0005	8.8%
$_{37}Rb^{87}$	87	37	27.83%	86.909187	4.9x10^{10}y.	β-	0.273	0.273	100%		3/2-	+2.7512		
$_{37}Rb^{88}$	88	37		87.911326	17.7 m.	β-	5.32	5.32		1.0 b.	2-	0.508	0.89803 ± 0.00004	14%
	88	37											1.83601 ± 0.00005	21.4%
	88	37											2.11889 ± 0.00007	0.42%
	88	37											2.57772 ± 0.00006	0.2%
	88	37											2.57772 ± 0.00006	0.2%
	88	37											2.67781 ± 0.00005	1.95 %
	88	37											3.00945 ± 0.00007	0.24%
	88	37											3.21848 ± 0.00008	0.21%
	88	37											3.21850 ± 0.0001	2.2%
	88	37											3.486 ± 0.0001	1.35%
	88	37											4.7424 ± 0.0002	1.5%
$_{37}Rb^{89}$	89	37		88.912278	β-	4.50	1.26	38%			3/2-		0.6571 ± 0.0001	1.0%
	89	37						1.9	5%				1.03188 ± 0.0002	58%
	89	37						2.2	34%				1.2481 ± 0.0007	43%
	89	37						4.49	18%				1.5381 ± 0.0001	2.6%
	89	37											2.0075 ± 0.0001	2.4%

Isotope	A	Z	% Natural abundance	Atomic mass	Half-life	Decay mode	Decay energy (MeV)	Particle energy (MeV)	Particle intensity	Thermal neutron cross section	Spin (h/2π)	μ Nucl. mag. moment	Gamma-ray energy (MeV)	Gamma-ray intensity
	89	37											2.1960 ± 0.0001	13%
	89	37											2.5701 ± 0.0001	10%
	89	37											2.7072 ± 0.0001	2%
	89	37											3.5088 ± 0.0002	1.1%
$_{37}Rb^{90m}$	90	37			4.28 m.	β-	4.50	1.7			4-		0.1069 ± 0.0001	0.22%
	90	37						6.5					0.3145 ± 0.0003	0.9%
	90	37											0.5512 ± 0.0002	0.9%
	90	37											0.7207 ± 0.0001	0.5%
	90	37											0.8242 ± 0.0001	8.9%
	90	37											0.83169 ± 0.00005	47%
	90	37											0.8720 ± 0.0001	0.5%
	90	37											0.9524 ± 0.0001	1.7%
	90	37											1.0607 ± 0.0001	7.8%
	90	37											1.1405 ± 0.0001	0.9%
	90	37											1.2428 ± 0.0001	3.1%
	90	37											1.2718 ± 0.0001	1.6%
	90	37											1.3754 ± 0.00004	17.1%
	90	37											1.3772 ± 0.0001	2.3%
	90	37											1.4890 ± 0.0004	0.4%
	90	37											1.6035 ± 0.0002	0.5%
	90	37											1.6656 ± 0.0001	4.9%
	90	37											1.6962 ± 0.0001	1.7%
	90	37											1.7389 ± 0.0001	1.9%
	90	37											1.8382 ± 0.0001	0.8%
	90	37											2.1283 ± 0.0001	5.3%
	90	37											2.2565 ± 0.0001	0.7%
	90	37											2.4973 ± 0.0001	0.7%
	90	37											2.5923 ± 0.0001	0.6%
	90	37											2.6178 ± 0.0001	0.6%
	90	37											2.7527 ± 0.0001	11.8%
	90	37											2.8344 ± 0.0001	1.9%
	90	37											3.0321 ± 0.0005	0.4%
	90	37											3.2051 ± 0.0002	1.1%
	90	37											3.3170 ± 0.0001	14.7%
	90	37											3.3708 ± 0.0004	0.4%
$_{37}Rb^{90}$	90	37		89.914811	2.6 m.	β-	6.57				1-		0.8317 ± 0001	27.8%
	90	37											0.9978 ± 0.0001	0.3%
	90	37											1.0386 ± 0.0001	0.2%
	90	37											1.0607 ± 0.0004	6.6%
	90	37											1.8041 ± 0.0001	0.4%
	90	37											1.8923 ± 0.0001	0.4%
	90	37											2.1393 ± 0.0002	0.3%
	90	37											2.2075 ± 0.0001	0.3%
	90	37											2.2163 ± 0.0002	0.34%
	90	37											2.4739 ± 0.0002	.42%
	90	37											3.2951 ± 0.0002	0.6%
	90	37											3.3170 ± 0.0001	0.6%
	90	37											3.3619 ± 0.0001	0.68%
	90	37											3.3832 ± 0.0001	4.7%
	90	37											3.5342 ± 0.0001	2.8%
	90	37											3.8144 ± 0.0001	0.4%
	90	37											4.1355 ± 0.0002	4.7%
	90	37											4.6464 ± 0.0002	1.6%
	90	37											5.1874 ± 0.0002	0.81%
	90	37											5.3330 ± 0.0002	0.30%
$_{37}Rb^{91}$	91	37		90.916485	58.4 s.	β-	5.86				3/2-		0.09363 ± 0.0002	32%
	91	37											0.3454 ± 0.0001	7.9%
	91	37											0.4391 ± 0.0001	2.0%
	91	37											0.5932 ± 0.0001	1.2%
	91	37											0.6028 ± 0.0001	2.7%
	91	37											0.9485 ± 0.0001	1.1%
	91	37											1.0420 ± 0.0001	2.1%
	91	37											1.1372 ± 0.0001	3.7%
	91	37											1.3678 ± 0.0001	0.7%
	91	37											1.4822 ± 0.0001	1.4%
	91	37											1.6158 ± 0.0001	2.3%
	91	37											1.7402 ± 0.0001	1.3%
	91	37											1.8492 ± 0.0001	3.1%
	91	37											1.9716 ± 0.0001	6.4%
	91	37											2.5059 ± 0.0002	1.3%
	91	37											2.5642 ± 0.0002	11.9%
	91	37											2.9257 ± 0.0002	1.5%
	91	37											3.4465 ± 0.0002	1.4%
	91	37											3.5997 ± 0.0002	9.9%
	91	37											3.6391 ± 0.0002	1.1%
	91	37											3.6437 ± 0.0002	0.75%
	91	37											3.8443 ± 0.0003	0.98%
	91	37											4.0782 ± 0.0002	3.9%
	91	37											4.2654 ± 0.0002	1.4%
$_{37}Rb^{92}$	92	37		91.919661	β-	8.12	8.1	94%			1-		0.5698 ± 0.0001	0.7%
	92	37											0.8147 ± 0.0001	4.0%
	92	37											1.3846 ± 0.0003	0.44%
	92	37											1.7123 ± 0.0002	0.53%
	92	37											2.8206 ± 0.0002	0.76%

Isotope	A	Z	% Natural abundance	Atomic mass	Half-life	Decay mode	Decay energy (MeV)	Particle energy (MeV)	Particle intensity	Thermal neutron cross section	Spin (h/2π)	μ Nucl. mag. moment	Gamma-ray energy (MeV)	Gamma-ray intensity
	92	37											3.1100 ± 0.0005	0.12%
	92	37											4.6377 ± 0.001	0.3%
	92	37											5.1881 ± 0.0008	0.3%
	92	37											5.55842 ± 0.001	0.2%
	92	37											5.6322 ± 0.001	0.24%
	92	37											(0.1 - 6.1)	
$_{37}Rb^{93}$	93	37			5.85 s.	β-	7.47	7.4					0.2134 ± 0.0005	4.5%
	93	37				n(1%)							0.2192 ± 0.0006	1.9%
	93	37											0.4326 ± 0.0001	11.8%
	93	37											0.7099 ± 0.0001	3.6%
	93	37											0.9861 ± 0.0001	4.6%
	93	37											1.1482 ± 0.0001	1.0%
	93	37											1.2383 ± 0.0001	1.0%
	93	37											1.3852 ± 0.0001	3.9%
	93	37											1.5629 ± 0.0001	0.7%
	93	37											1.6129 ± 0.0001	1.1%
	93	39											1.8085 ± 0.0001	1.9%
	93	37											1.8697 ± 0.0001	1.3%
	93	37											2.0541 ± 0.0001	0.9%
	93	37											2.2294 ± 0.0001	0.64%
	93	37											2.5052 ± 0.0002	0.55%
	93	37											2.7050 ± 0.0002	0.7%
	93	37											2.8613 ± 0.0002	0.75%
	93	37											3.4582 ± 0.0002	2.5%
	93	37											3.8040 ± 0.0002	1.1%
	93	37											3.8676 ± 0.0002	1.7%
	93	37											3.9343 ± 0.0002	0.7%
	93	37											4.2712 ± 0.0002	0.23%
	93	37											4.8751 ± 0.0003	0.12%
$_{37}Rb^{94}$	94	37		93.926432	2.73 s.	β-	10.2	9.3					0.6777 ± 0.0003	4 +
	93	37				n(10%)							0.8369 ± 0.0001	100
	94	37											1.0894 ± 0.0002	21
	94	37											1.3091 ± 0.0001	16
	94	37											1.5775 ± 0.0001	39
$_{37}Rb^{95}$	95	37		94.929352	0.38 s.	β-	9.3	8.6					0.2040	22 +
	95	37				n(8%)							0.3289	17
	95	37											0.3522	100
	95	37											0.6602	10
	95	37											0.6808	36
	95	37											0.7691	11
$_{37}Rb^{96}$	96	37		95.934370	0.20 s.	β-	10.3	10.8					0.412 ± 0.001	4 +
	96	37				n(13%)							0.593 ± 0.001	8
	96	37											0.606 ± 0.001	6
	96	37											0.691 ± 0.001	10
	96	37											0.813 ± 0.001	100
	96	37											0.978 ± 0.001	8
	96	37											1.036 ± 0.001	9
	96	37											1.179 ± 0.001	4
	96	37											1.335 ± 0.001	4
$_{37}Rb^{97}$	97	37		96.937440	0.17 s.	β-							0.167 ± 0.001	100 +
	97	37				n(27%)							0.418 ± 0.001	18
	97	37											0.519 ± 0.001	38
	97	37											0.585 ± 0.001	79
	97	37											0.599 ± 0.001	56
	97	37											0.652 ± 0.001	21
	97	37											0.697 ± 0.001	21
	97	37											1.258 ± 0.001	52
$_{37}Rb^{98}$	98	37		97.941960	0.13 s.	β-								
	98	37				n(13%)								
Sr		38		87.62						1.2 b.				
$_{38}Sr^{79}$	79	38		78.929860	2.1 m.	β+	5.2						ann.rad.	
	79	38											0.1889 ± 0.0003	100 +
	79	38											0.4442 ± 0.0003	37
	79	38											0.5599 ± 0.0008	63
	79	38											0.6112 ± 0.00008	26
$_{38}Sr^{80}$	80	38		79.924650	106 m.	β+	2.00				0+		ann.rad.	
	80	38											0.1750 ± 0.0005	26 +
	80	38											0.2359 ± 0.0008	11
	80	38											0.3788 ± 0.0005	11
	80	38											0.4141 ± 0.0005	8
	80	38											0.5534 ± 0.0005	18
	80	38											0.5890 ± 0.0005	100
$_{38}Sr^{81}$	81	38		80.923270	22.2 m.	β+(87%)	3.99	2.43			1/2+		ann.rad.	173%
	81	38				E.C.(13%)		2.68					0.1478 ± 0.0003	30%
	81	38											0.1423 ± 0.0003	4%
	81	38											0.1534 ± 0.0003	36%
	81	38											0.1883 ± 0.0003	21%
	81	38											0.3865 ± 0.0005	4%
	81	38											0.4435 ± 0.0003	17%
	81	38											0.5746 ± 0.0003	6%
	81	38											0.7015 ± 0.001	1.4%
	81	38											0.7213 ± 0.0003	2%
	81	38											0.9093 ± 0.0003	3.0%

Isotope	A	Z	% Natural abundance	Atomic mass	Half-life	Decay mode	Decay energy (MeV)	Particle energy (MeV)	Particle intensity	Thermal neutron cross section	Spin (h/2π)	μ Nucl. mag. moment	Gamma-ray energy (MeV)	Gamma-ray intensity
	81	38											0.9386 ± 0.0003	2.8%
	81	38											(0.08 - 1.6)	
$_{38}Sr^{82}$	82	38		81.918414	25.6 d.	E.C.	0.21						Rb x-ray	
$_{38}Sr^{83m}$	83	38			5.0 s.	I.T.	0.2591				1/2-		0.2593 ± 0.0001	87.6%
$_{38}Sr^{83}$	83	38		82.917566	32.4 h.	β+(24%)	2.25	0.465			7/2+		ann.rad.	
	83	38				E.C.(76%)		0.803					0.0423 ± 0.0005	1.6%
	83	38						1.227					0.3816 ± 0.0003	12%
	83	38											0.3894 ± 0.0003	1.6%
	83	38											0.4184 ± 0.0003	5%
	83	38											0.7627 ± 0.0003	30%
	83	38											0.7785 ± 0.0003	2.0%
	83	38											1.5623 ± 0.0004	2.0%
	83	38											(0.1 - 1.6)	
$_{38}Sr^{84}$	84	38	0.56%	83.913430						(0.53 + 0.27)b.	0+			
$_{38}Sr^{85m}$	85	38			67.6 m.	I.T.(87%)	0.2387				1/2-		0.23168 ± 0.00001	84%
	85	38											0.2386 ± 0.0001	0.4%
$_{38}Sr^{85}$	85	38		84.912937	64.8 d.	E.C.	1.08				9/2+		0.51300 ± 0.00001	99%
$_{38}Sr^{86}$	86	38	9.86%	85.909267						(0.84 + ?)	0+			
$_{38}Sr^{87m}$	87	38			2.80 h.	I.T.	0.3885				1/2-		0.38852 ± 0.00002	82%
$_{38}Sr^{87}$	87	38	7.00	86.908884						16 b.	9/2+	-1.093		
$_{38}Sr^{88}$	88	38	82.58%	87.905619						5.8 mb.	0+			
$_{38}Sr^{89}$	89	38		88.907450	50.52 d.	β-	1.492	1.492	100%		5/2+		0.9092 ± 0.0001	0.0009%
$_{38}Sr^{90}$	90	38		89.907738	29 y.	β-	0.546	0.546	100%	0.8 b.	0+			
$_{38}Sr^{91}$	91	38		90.910187	9.5 h.	β-	2.685	0.61	7%		5/2+		0.2745 ± 0.0002	1.1%
	91	38						1.09	33%				0.55562 ± 0.00005	59%(D)
	91	38						1.36	29%				0.6202 ± 0.0001	1.7%
	91	38						1.36	29%				0.6529 ± 0.0002	7.7%
	91	38						2.03	4%				0.7498 ± 0.0001	23.0%
	91	38						2.03	4%				0.9258 ± 0.0002	4.0%
	91	38						2.66	26%				1.0243 ± 0.0001	33.4%
	92	38											1.2810 ± 0.0001	0.93%
	91	38											1.4135 ± 0.0001	0.91%
	91	38											1.6511 ± 0.001	0.37%
$_{38}Sr^{92}$	92	38		91.910944	2.71 h.	β-	1.89	0.55	96%		0+		0.214162 ± 0.00003	3.4%
	92	38						1.5	3%				0.43062 ± 0.00005	3.5%
	92	38											0.95332 ± 0.00009	3.6%
	92	38											1.1423 ± 0.0001	2.9%
	92	38											1.38309 ± 0.00006	90%
$_{38}Sr^{93}$	93	38		92.913987	7.5 m.	β	4.13	2.2	10%		5/2+		0.2601 ± 0.0001	7%
	93	38						2.6	25%				0.3465 ± 0.0001	3%
	93	38						3.2	65%				0.3774 ± 0.0001	1.4%
	93	38											0.4327 ± 0.0001	1.4%
	93	38											0.4462 ± 0.0001	2.3%
	93	38											0.4820 ± 0.0001	1.1%
	93	38											0.4837 ± 0.0001	1.6%
	93	38											0.59028 ± 0.00005	30.7%
	93	38											0.5938 ± 0.0002	1.1%
	93	38											0.5961 ± 0.0002	1.3%
	93	38											0.6109 ± 0.0001	1.0%
	93	38											0.6636 ± 0.0001	1.6%
	93	38											0.7104 ± 0.0001	20.8%
	93	38											0.8349 ± 0.0001	1.6%
	93	38											0.87573 ± 0.00005	23.4%
	93	38											0.8881 ± 0.0001	2.1%
	93	38											1.0406 ± 0.0001	3.0%
	93	38											1.0940 ± 0.0001	1.7%
	93	38											1.1225 ± 0.0001	3.8%
	93	38											1.2155 ± 0.0001	2.4%
	93	38											1.2664 ± 0.0001	1.1%
	93	38											1.3212 ± 0.0001	2.5%
	93	38											1.3871 ± 0.0001	3.3%
	93	38											1.6340 ± 0.0001	1.4%
	93	38											1.6941 ± 0.0001	2.5%
	93	38											1.6941 ± 0.0001	2.5%
	93	38											1.7066 ± 0.0001	1.1%
	93	38											1.7654 ± 0.0001	1.0%
	93	38											1.8115 ± 0.0001	1.3%
	93	38											1.9288 ± 0.0001	1.1%
	93	38											2.2303 ± 0.0001	1.5%
	93	38											2.2961 ± 0.0001	0.7%
	93	38											2.3647 ± 0.0001	1.5%
	93	38											2.5438 ± 0.0001	2.9%
	93	38											2.6886 ± 0.0001	2.0%
$_{38}Sr^{94}$	94	38		93.915367	75 s.	β-	3.51	2.1			0+		0.6219	2.1%
	94	38						3.3					0.7043	2.1%

Isotope	A	Z	% Natural abundance	Atomic mass	Half-life	Decay mode	Decay energy (MeV)	Particle energy (MeV)	Particle intensity	Thermal neutron cross section	Spin (h/2π)	μ Nucl. mag. moment	Gamma-ray energy (MeV)	Gamma-ray intensity
	94	38											0.7241	2.5%
	94	38											0.8064	1.8%
	94	38											1.4283 ± 0.0005	95.4%
$_{38}Sr^{95}$	95	38		94.919380	25 s.	β-	5.5						0.6859	24%
	95	38						6.1	50%				0.8269	3%
	95	38											0.9453	2.4%
	95	38											1.2777	2.2%
	95	38											2.6835	1.3%
	95	38											2.7173	5%
	95	38											2.8908	1%
	95	38											2.8908	1%
	95	38											2.9332	4%
	95	38											3.3528	1%
	95	38											3.6159	1.8%
	95	38											3.7432	0.9%
	95	38											4.0753	1.3%
$_{38}Sr^{96}$	96	38		95.921760	1.06 s.	β-	5.4	4.2			0+		0.1222	66%
	96	38											0.5305	8%
	96	38											0.8094	75%
	96	38											0.9318	16%
$_{38}Sr^{97}$	97	38		96.926140	0.4 s.	β-	7.2						0.2164	1.1%
	97	38											0.3071	12%
	97	38											0.3658	5%
	97	38											0.4125	2.5%
	97	38											0.4741	2.4%
	97	38											0.4799	2.8%
	97	38											0.6522	12%
	97	38											0.6973	4.9
	97	38											0.8014	4.9%
	97	38											0.8920	4.6%
	97	38											0.9538	22.6%
	97	38											1.2580	10.3%
	97	38											1.5150	2.4%
	97	38											1.9050	24%
	97	38											2.1213	2.6%
	97	38											2.2875	3.6%
$_{38}Sr^{98}$	98	38		97.928620	0.65 s.	β-							0.0365	8.4%
	98	38											0.0525	2.4%
	98	38											0.1190	30%
	98	38											0.1701	1.2%
	98	38											0.4286	9.6%
	98	38											0.4447	10.5%
	98	38											0.5636	3.9%
	98	38											0.6006	1.2%
Y		39		88.9059						1.28 b.				
$_{39}Y^{80}$	80	39		79.936250	36 s.	β+	6.9	4.9					ann.rad.	
	80	39						5.6					0.3858	
	80	39											0.5951	
	80	39											1.1852	
$_{39}Y^{81}$	81	39		80.929200	72 s.	β+	5.5	3.7					ann.rad.	
	81	39						4.2					0.428	
	81	39											0.469	
$_{39}Y^{82}$	82	39		81.926810	9.5 s.	β+	7.8	6.3			1+		ann.rad.	
	82	39											0.5736	
	82	39											0.6017	
	82	39											0.7375	
$_{39}Y^{83m}$	83	39			2.85 m.	β+(95%)					9/2+		ann.rad.	
	83	39				E.C.(5%)							0.2591 ± 0.0001	90%(D)
	83	39											0.4218 ± 0.0003	32%
	83	39											0.4945 ± 0.0002	13%
$_{39}Y^{83}$	83	39		82.922300	7.1 m.	β+	4.4				1/2-		ann.rad.	190%
	83	39				E.C.							0.0355 ± 0.0001	20%
	83	39											0.2591 ± 0.0001	2%(D)
	83	39											0.3916 ± 0.0002	1.5%
	83	39											0.4544 ± 0.0002	2%
	83	39											0.4899 ± 0.0002	6%
	83	39											0.6821 ± 0.0001	1.3%
	83	39											0.7435 ± 0.0001	1.2%
	83	39											0.8587 ± 0.0001	3.0%
	83	39											0.8821 ± 0.0001	6%
	83	39											0.9273 ± 0.0001	1%
	83	39											0.9518 ± 0.0001	2%
	83	39											1.2392 ± 0.0002	0.6%
	83	39											1.3365 ± 0.0001	3%
	83	39											1.3719 ± 0.0001	1%
	83	39											1.9157 ± 0.0004	0.3%
	83	39											2.9053 ± 0.001	0.2%
	83	39											2.9440 ± 0.001	0.2%
	83	39											(0.03 - 3.4)	
$_{39}Y^{84m}$	84	39			4.6 s.	β+					1+		ann.rad.	
	84	39				E.C.							0.7930 ± 0.0003	35%
$_{39}Y^{84}$	84	39		83.920310	40 m.	β+	6.5	1.64	47 +		5-		ann.rad.	
	84	39				E.C.		2.24	25				0.4628 ± 0.0002	10%
	84	39						2.64	21				0.6022 ± 0.0002	10%

Isotope	A	Z	% Natural abundance	Atomic mass	Half-life	Decay mode	Decay energy (MeV)	Particle energy (MeV)	Particle intensity	Thermal neutron cross section	Spin (h/2π)	μ Nucl. mag. moment	Gamma-ray energy (MeV)	Gamma-ray intensity
	84	39						3.15	7				0.6583 ± 0.0002	4.5%
	84	39											0.6606 ± 0.0002	11.5%
	84	39											0.7931 ± 0.0002	89%
	84	39											0.9744 ± 0.0002	79%
	84	39											0.9942 ± 0.0002	4.2%
	84	39											1.0398 ± 0.0002	58%
	84	39											1.2550 ± 0.0002	6.7%
	84	39											1.2626 ± 0.0002	2.5%
	84	39											1.3309 ± 0.0002	3.1%
	84	39											1.4534 ± 0.0002	1.8%
	84	39											1.5028 ± 0.0002	6.8%
	84	39											1.6145 ± 0.0002	1.8%
	84	39											1.6546 ± 0.0002	2.6%
	84	39											1.7444 ± 0.0002	2.2%
	84	39											1.7636 ± 0.0002	1.9%
	84	39											1.9180 ± 0.0004	2.3%
	84	39											2.2953 ± 0.0004	1.9%
	84	39											2.3095 ± 0.0004	1.1%
	84	39											(0.2 - 3.3)	
$_{39}Y^{85m}$	85	39			4.9 h.	β+(70%)					9/2+		ann.rad.	
	85	39				E.C.(30%)							0.2317 ± 0.0001	23%(D)
	85	39											0.5044 ± 0.0002	1.5%
	85	39											0.5356 ± 0.0002	3.5%
	85	39											0.5467 ± 0.0002	1.2%
	85	39											0.5684 ± 0.0002	1.7%
	85	39											0.6119 ± 0.0002	1%
	85	39											0.6980 ± 0.0002	1.2%
	85	39											0.7673 ± 0.0001	3.7%
	85	39											0.7686 ± 0.0002	1.2%
	85	39											0.7879 ± 0.0002	1.4%
	85	39											1.0301 ± 0.0002	2.1%
	85	39											1.1232 ± 0.0002	1.8%
	85	39											2.1238 ± 0.0002	5%
	85	39											2.1721 ± 0.0002	2.3%
	85	39											2.3517 ± 0.0002	0.5%
	85	39											2.7822 ± 0.0003	0.35%
	85	39											(0.1 - 3.1)	
$_{39}Y^{85}$	85	39		84.916437	2.6 h.	β+(55%)	3.26	1.54			1/2-		ann.rad.	
	85	39				E.C.(45%)							0.2317 ± 0.0001	80%(D)
	85	39											0.5045 ± 0.0002	64%
	85	39											0.9140 ± 0.0002	6.8%
	85	39											1.2780 ± 0.0005	0.3%
	85	39											1.3208 ± 0.0005	0.35%
	85	39											(0.07 - 1.4)	
$_{39}Y^{86m}$	86	39			48 m.	I.T.(99%)					8+		ann.rad.	
	86	39				β+							0.0102 ± 0.0001	
	86	39				E.C.							0.2080 ± 0.0003	94%
	86	39											(0.09 - 1.1)	
$_{39}Y^{86}$	86	39		85.914893	14.74 h.	β+	5.24				4		ann.rad.	
	86	39				E.C.							0.3070 ± 0.0001	3%
	86	39											0.3829 ± 0.0002	3.6%
	86	39											0.4431 ± 0.0001	17%
	86	39											0.5806 ± 0.0001	4.8%
	86	39											0.6088 ± 0.0001	2%
	86	39											0.6277 ± 0.0001	33%
	86	39											0.7033 ± 0.0001	15%
	86	39											0.7774 ± 0.0001	12%
	86	39											0.8260 ± 0.0001	3%
	86	39											1.0240 ± 0.0001	3.8%
	86	39											1.0766 ± 0.0001	82%
	86	39											1.1531 ± 0.0001	31%
	86	39											1.8017 ± 0.0001	1.6%
	86	39											1.8544 ± 0.0001	16%
	86	39											1.9207 ± 0.0001	21%
	86	39											2.5679 ± 0.0002	2.3%
	86	39											2.6101 ± 0.0002	1.2%
	86	39											(0.1 - 3.8)	
$_{39}Y^{87m}$	87	39			13 h.	I.T.(98%)					9/2+		0.3807 ± 0.0005	78%(D)
	87	39				β+(0.7%)		1.15	0.7%					
	87	39				E.C.								
$_{39}Y^{87}$	87	39		86.910882	80.3 h.	E.C. (99+%)	1.861	0.78					0.3880 ± 0.0001	90%(D)
	87	39											0.4870 ± 0.0001	92%
$_{39}Y^{88}$	88	39		87.909508	106.61 d.	E.C. (99+%)	3.623	0.76			4-		ann.rad.	0.4%
	88	39				β+(0.2%)							0.89802 ± 0.00002	92%
	88	39											1.83601 ± 0.00003	99.3%
	88	39											2.73404 ± 0.00005	0.5%
	88	39											3.2190 ± 0.0002	0.007%
$_{39}Y^{89m}$	89	39			15.7 s.	I.T.	0.909				9/2+		0.9092 ± 0.0002	99.1%
$_{39}Y^{89}$	89	39	100%	88.905849						(0.001 + 1.28)b.	1/2-	-0.1373		

Isotope	A	Z	% Natural abundance	Atomic mass	Half-life	Decay mode	Decay energy (MeV)	Particle energy (MeV)	Particle intensity	Thermal neutron cross section	Spin (h/2π)	μ Nucl. mag. moment	Gamma-ray energy (MeV)	Gamma-ray intensity
$_{39}Y^{90m}$	90	39			3.19 h.	I.T. (99+%)	0.68204				7+		0.2025 ± 0.0001	97%
	90	39				β-(0.002%)							0.4794 ± 0.0001	91%
	90	39											0.6820 ± 0.0002	0.4%
$_{39}Y^{90}$	90	39		89.907152	64.0 h.	β-	2.283	2.283			2-	-1.630		
$_{39}Y^{91m}$	91	39			49.7 m.	I.T.	0.555				9/2+		0.55562 ± 0.00005	95%
$_{39}Y^{91}$	91	39		90.907303	58.5 d.	β-	1.546	1.546			1/2-	0.1641	1.208 ± 0.001	0.3%
$_{39}Y^{92}$	92	39		91.908917	3.54 h.	β-	3.62	3.64			2-		0.4485 ± 0.0001	2.4%
	92	39											0.4926 ± 0.0001	0.4%
	92	39											0.5611 ± 0.0001	2.4%
	92	39											0.8443 ± 0.0001	1.2%
	92	39											0.9128 ± 0.0001	0.6%
	92	39											0.9345 ± 0.0001	13.9%
	92	39											1.1324 ± 0.0001	0.2%
	92	39											1.4054 ± 0.0001	4.7%
	92	39											1.8473 ± 0.0001	0.35%
	92	39											(0.4 - 3.3)	
$_{39}Y^{93m}$	93	39			0.82 s.	I.T.	0.759				9/2+		0.1684 ± 0.0001	51%
	93	39											0.5902 ± 0.0001	100%
$_{39}Y^{93}$	93	39		92.909571	10.2 h.	β-	2.89	2.89	90%		1/2-		0.2669 ± 0.0002	6.8%
	93	39											0.6802 ± 0.0001	0.6%
	93	39											0.9471 ± 0.0001	2.0%
	93	39											1.4254 ± 0.0001	0.24%
	93	39											1.45050 ± 0.0001	0.34%
	93	39											1.9178 ± 0.0002	1.4%
	93	39											2.1846 ± 0.0002	0.15%
	93	39											2.1908 ± 0.0002	0.17%
$_{39}Y^{94}$	94	39		93.911597	18.7 m.	β-	4.92	4.92			2-		0.3816 ± 0.0001	2%
	94	39											0.5509 ± 0.0001	5%
	94	39											0.7526 ± 0.0001	1.4%
	94	39											0.9188 ± 0.0001	56%
	94	39											1.1389 ± 0.0001	65
	94	39											1.6714 ± 0.0001	2.5%
	94	39											2.1406 ± 0.0002	1%
	94	39											2.8463 ± 0.0003	0.4%
	94	39											2.8987 ± 0.0006	0.1%
	94	39											(0.3 - 4.1)	
$_{39}Y^{95}$	95	39		94.912814	10.3 m.	β-	4.45				1/2-		0.4324 ± 0.0002	2%
	95	39											0.9542 ± 0.0002	19%
	95	39											1.0485 ± 0.0004	1%
	95	39											1.1740 ± 0.0004	0.7%
	95	39											1.3243 ± 0.0003	5.3%
	95	39											1.6185 ± 0.0008	1.7%
	95	39											1.8050 ± 0.0005	1.5%
	95	39											1.8928 ± 0.0005	0.7%
	95	39											1.9406 ± 0.0005	2.8%
	95	39											2.1760 ± 0.0005	8.2%
	95	39											2.2529 ± 0.0006	1.4%
	95	39											2.3733 ± 0.0008	0.8%
	95	39											2.6330 ± 0.0008	5.1%
	95	39											3.1298 ± 0.001	0.7%
	95	39											3.2502 ± 0.001	1.3%
	95	39											3.5770 ± 0.001	7.6%
	95	39											3.6840 ± 0.001	0.4%
	95	39											3.8870 ± 0.002	0.3%
$_{39}Y^{96m}$	96	39			9.8 s.	β-							0.1467 ± 0.0002	36%
	96	39											0.1744 ± 0.0002	2%
	96	39											0.2268 ± 0.0002	2%
	96	39											0.2891 ± 0.0002	1%
	96	39											0.3636 ± 0.0002	22%
	96	39											0.4753 ± 0.0002	3.3%
	96	39											0.6174 ± 0.0002	55%
	96	39											0.6316 ± 0.0002	7.6%
	96	39											0.6526 ± 0.0002	1.5%
	96	39											0.9062 ± 0.0002	18%
	96	39											0.9150 ± 0.0002	59%
	96	39											0.9603 ± 0.0002	4.2%
	96	39											0.9793 ± 0.0002	3.6%
	96	39											1.1071 ± 0.0002	48%
	96	39											1.1850 ± 0.0002	3.4%
	96	39											1.2227 ± 0.0002	26%
	96	39											1.7507 ± 0.0002	89%
	96	39											1.8976 ± 0.0002	5.2%
	96	39											2.2255 ± 0.0002	5.8%
$_{39}Y^{96}$	96	39		95.915940	6.2 s.	β-	7.12	7.12			0-		1.594 ± 0.0005	25%
$_{39}Y^{97m}$	97	39			1.21 s.	β-	7.4	4.8			9/2+		0.1614	77%
	97	39						6.0					0.9700	43%
	97	39											1.1030	92%
	97	39											1.2441	9%
	97	39											1.2642	3%
	97	39											1.4000	5%
$_{39}Y^{97}$	97	39		96.918120	3.7 s.	β-					1/2-		0.2969	1.3%
	97	39											0.5448	1%
	97	39											0.7560	1.1%

Isotope	A	Z	% Natural abundance	Atomic mass	Half-life	Decay mode	Decay energy (MeV)	Particle energy (MeV)	Particle intensity	Thermal neutron cross section	Spin (h/2π)	μ Nucl. mag. moment	Gamma-ray energy (MeV)	Gamma-ray intensity
	97	39											1.1030	5%
	97	39											1.2910	5.7%
	97	39											1.4000	4.5%
	97	39											1.8870	1.8%
	97	39											1.9960	7.4%
	97	39											2.7431	6.5%
	97	39											3.2876	18%
	97	39											3.4013	14%
	97	39											3.5495	3.1%
$_{39}Y^{98m}$	98	39			2.0 s.	β-	9.8	8.7					0.2415	7%
	98	39											0.2531	4%
	98	39											0.5830	18%
	98	39											0.6205	75%
	98	39											0.6473	55%
	98	39											0.7526	7%
	98	39											1.2228	97%
	98	39											1.5907	3%
	98	39											1.7873	4.2%
	98	39											1.0016	15%
$_{39}Y^{98}$	98	39		97.922300	0.65 s.	β-	8.9	5.5			1+		0.2131	1.3%
	98	39											0.2686	2.3%
	98	39											0.5216	0.7%
	98	39											1.2228	11%
	98	39											1.5907	4.4%
	98	39											1.7441	1.3%
	98	39											2.4206	1.5%
	98	39											2.9413	5.3%
	98	39											3.2037	0.7%
	98	39											3.2283	1.3%
	98	39											3.3100	2.2%
	98	39											3.3757	0.6%
	98	39											3.4686	0.6%
	98	39											4.4501	3.1%
$_{39}Y^{99}$	99	39		98.924720	1.5 s.	β-	7.6				1/2-		0.1217	44%
	99	39											0.1307	6.2%
	99	39											0.1940	2.2%
	99	39											0.2764	2.6%
	99	39											0.4060	1.8%
	99	39											0.4538	4.8%
	99	39											0.5362	8.8%
	99	39											0.5754	11%
	99	39											0.6000	6.2%
	99	39											0.6026	4.4%
	99	39											0.6399	3.5%
	99	39											0.7242	19.8%
	99	39											0.7300	1.3%
	99	39											0.7822	4.8%
	99	39											0.9301	4.4%
	99	39											1.0130	7.9%
Zr		40		91.224						0.184 b.				
$_{40}Zr^{82}$	82	40		81.931100	2.5 m.	β+	4.0						ann.rad.	
$_{40}Zr^{83m}$	83	40			8 s.	β+							ann.rad.	
$_{40}Zr^{83}$	83	40		82.928760	44 s.	β+	6.0						ann.rad.	
	83	40											0.0556	
	83	40											0.1050	10%
	83	40											0.2560	8%
	83	40											0.303	7%
	83	40											0.474	9%
	83	40											0.791	4%
	83	40											1.525	9%
$_{40}Zr^{84}$	84	40		83.923320	28 m.	β+	2.7				0+		ann.rad.	
	84	40				E.C.							0.0449	
	84	40											0.1125	
	84	40											0.3729	
	84	40											0.667	
$_{40}Zr^{85m}$	85	40			10.9 s.	I.T.	0.292				1/2-		ann.rad.	
	85	40				β+,E.C.							0.2922 ± 0.0003	100 +
	85	40											0.4165 ± 0.0003	9
$_{40}Zr^{85}$	85	40		84.921470	7.9 m.	β+	4.4				7/2+		ann.rad.	
	85	40				E.C.							0.2663 ± 0.0002	23%
	85	40											0.4163 ± 0.0002	25%
	85	40											0.4543 ± 0.0002	41%
	85	40											1.1984 ± 0.0002	4.5%
	85	40											1.7682 ± 0.0003	1.8%
	85	40											1.8762 ± 0.0003	0.4%
	85	40											1.9557 ± 0.0003	0.4%
$_{40}Zr^{86}$	86	40		85.916290	16.5 h.	E.C.	1.3				0+		0.0280 ± 0.0005	20%
	86	40											0.243 ± 0.001	96%
	86	40											0.612 ± 0.001	5%
$_{40}Zr^{87m}$	87	40			14.0 s.	I.T.	0.3362				1/2-		0.1352 ± 0.0003	27%
	87	40											0.2010 ± 0.0003	97%
$_{40}Zr^{87}$	87	40		86.914817	1.73 h.	β+	3.67				9/2+		ann.rad.	
	87	40				E.C.							0.3811 ± 0.0002	93%(D)
	87	40											0.793 ± 0.001	0.6%

TABLE OF THE ISOTOPES (Continued)

Isotope	A	Z	% Natural abundance	Atomic mass	Half-life	Decay mode	Decay energy (MeV)	Particle energy (MeV)	Particle intensity	Thermal neutron cross section	Spin (h/2π)	μ Nucl. mag. moment	Gamma-ray energy (MeV)	Gamma-ray intensity
	87	40											1.210 ± 0.001	1.3%
	87	40											1.228 ± 0.001	4%
	87	40											1.808 ± 0.001	0.2%
	87	40											2.220 ± 0.001	0.4%
	87	40											2.616 ± 0.002	0.2%
${}_{40}Zr^{88}$	88	40		87.910225	83.4 d.	E.C.	0.67				0+		0.3929 ± 0.001	97%
${}_{40}Zr^{89m}$	89	40			4.18 m.	I.T.(94%)	0.5878				1/2-		ann.rad.	
	89	40				β+(1.5%)							0.5878 ± 0.0002	94%
	89	40				E.C.(4.7%)							0.9092 ± 0.0001	6%(D)
${}_{40}Zr^{89}$	89	40		88.908890	78.4 h.	β+(23%)	2.83	0.9			9/2+		ann.rad.	46%
	89	40				E.C.(77%)							1.7129 ± 0.0008	0.7%
${}_{40}Zr^{90m}$	90	40			0.809 s.	I.T.					5-		0.1326 ± 0.0002	5%
	90	40											2.1862 ± 0.0001	16%
	90	40											2.3189 ± 0.0001	84%
${}_{40}Zr^{90}$	90	40	51.45%	89.904703						0.05 b.	0+			
${}_{40}Zr^{91}$	91	40	11.27%	90.905644						0.9 b.	5/2+	-1.3036		
${}_{40}Zr^{92}$	92	40	17.17%	91.905039						0.2 b.	0+			
${}_{40}Zr^{93}$	93	40		92.906474	$1.5x10^6$ y.	β-	0.08			≈1 b.	5/2+		0.0304	(D)
${}_{40}Zr^{94}$	94	40	17.33%	93.906314						0.05 b.	0+			
${}_{40}Zr^{95}$	95	40		94.908042	64.03 d.	β-	1.124	0.360	55%		5/2+		0.23569 ± 0.00005	0.9%
	95	40						0.396	44%				0.72418 ± 0.00001	44%
	95	40											0.75672 ± 0.00001	55%
${}_{40}Zr^{96}$	96	40	2.78%	95.908275						0.022 b.	0+			
${}_{40}Zr^{97}$	97	40		96.910950	16.8 h.	β-	2.658	1.91			1/2-		0.2541 ± 0.0002	1.2%
	97	40											0.3554 ± 0.0001	2.3%
	97	40											0.5076 ± 0.0001	5%
	97	40											0.6024 ± 0.0002	1.4%
	97	40											0.7434 ± 0.0003	93%(D)
	97	40											1.0213 ± 0.0003	1.3%
	97	40											1.1479 ± 0.0001	2.6%
	97	40											1.2761 ± 0.0001	1.0%
	97	40											1.3627 ± 0.0001	1.3%
	97	40											1.7505 ± 0.0001	1.3%
${}_{40}Zr^{98}$	98	40		97.912735	30.7 s.	β-	2.24	2.2	100%		0+			
${}_{40}Zr^{99}$	99	40		98.916540	2.1 s.	β-	4.6	3.5			1/2+		0.0284	1.7%
	99	40											0.0558	2.2%
	99	40											0.0818	2.8%
	99	40											0.1792	5.6%
	99	40											0.3872	7.8%
	99	40											0.4150	5.0%
	99	40											0.4617	11.8%
	99	40											0.4691	56%
	99	40											0.4899	0.8%
	99	40											0.5460	45%
	99	40											0.5940	27%
	99	40											0.6279	2.2%
	99	40											0.6500	2.2%
${}_{40}Zr^{100}$	100	40		99.917750	7.1 s.	β-	3.3				0+		0.4005	
	100	40											0.5042	
${}_{40}Zr^{101}$	101	40		100.921520	2.0 s.	β-	5.8	6.2					0.1194	
	101	40											0.2057	
	101	40											0.2089	
Nb		41								1.15 b.				
${}_{41}Nb^{86}$	86	41		85.925310	1.45 m.	β+	8.0						ann.rad.	
	86	41											0.751	100 +
	86	41											1.003	57
${}_{41}Nb^{87m}$	87	41			3.7 m.	β+					1/2-		ann.rad.	
	87	41				E.C.							0.1352 ± 0.0003	100%(D)
	87	41											0.2010 ± 0.0003	100%(D)
${}_{41}Nb^{87}$	87	41		86.920370	2.6 m.	β+5.2							ann.rad.	
	87	41				E.C.							0.2010 ± 0.0003	26%(D)
	87	41											0.4706 ± 0.0002	19%
	87	41											0.6000 ± 0.0006	2%
	87	41											0.6165 ± 0.0002	9%
	87	41											0.9142 ± 0.0003	6%
	87	41											1.0665 ± 0.0004	10%
	87	41											1.6832 ± 0.0003	4%
	87	41											1.8842 ± 0.0003	9%
	87	41											2.1533 ± 0.0007	0.9%
${}_{41}Nb^{88m}$	88	41			7.8 m.	β+					4-		ann.rad.	
	88	41				E.C.							0.2625 ± 0.0005	10%
	88	41											0.3996 ± 0.0003	42%
	88	41											0.4510 ± 0.0003	24%
	88	41											0.5341 ± 0.0003	13%
	88	41											0.6382 ± 0.0003	25%
	88	41											0.7607 ± 0.0003	16%
	88	41											0.9184 ± 0.0005	11%
	88	41											1.0569 ± 0.0003	90%
	88	41											1.0825 ± 0.0003	58%
	88	41											1.3992 ± 0.0005	5.5%
	88	41											1.8179 ± 0.0006	10%

TABLE OF THE ISOTOPES (Continued)

Isotope	A	Z	% Natural abundance	Atomic mass	Half-life	Decay mode	Decay energy (MeV)	Particle energy (MeV)	Particle intensity	Thermal neutron cross section	Spin (h/2π)	μ Nucl. mag. moment	Gamma-ray energy (MeV)	Gamma-ray intensity
	88	41											1.9754 ± 0.0008	6%
$_{41}Nb^{88}$	88	41		87.917950	14.3 m.	β+	7.2	3.2			8+		ann.rad.	
	88	41				E.C.							0.0767 ± 0.0002	23%
	88	41											0.2714 ± 0.0002	53%
	88	41											0.3994 ± 0.0001	33%
	88	41											0.5029 ± 0.0002	64%
	88	41											0.6711 ± 0.0002	65%
	88	41											1.0570 ± 0.0001	100%
	88	41											1.0828 ± 0.0001	100%
	88	41											1.5430 ± 0.0006	1%
	88	41											2.3119 ± 0.0007	0.8%
	88	41											(0.07 - 2.5)	
$_{41}Nb^{89m}$	89	41			122 m.	β+		3.3			9/2+		0.5324 ± 0.0001	0.5%
	89	41				E.C.							0.5880 ± 0.0001	10%(D)
	89	41											1.1270 ± 0.0002	2%
	89	41											1.2590 ± 0.0002	1%
	89	41											1.3323 ± 0.0002	1%
	89	41											1.5114 ± 0.0003	1.8%
	89	41											1.6272 ± 0.0002	3.6%
	89	41											2.5723 ± 0.0004	3%
	89	41											3.0927 ± 0.0002	3.1%
	89	41											(0.17 - 4.0)	
$_{41}Nb^{89}$	89	41		88.913449	66 m.	β+(74%)	4.44	2.8			1/2-		ann.rad.	150%
	89	41				E.C.(26%)							0.5074 ± 0.0007	85%
	89	41											0.5880 ± 0.0002	100%
	89	41											0.7696 ± 0.0005	6%
	89	41											1.2775 ± 0.002	1.6%
$_{41}Nb^{90m}$	90	41			18.8 s.	I.T.	0.1246				4-		0.002	
	90	41											0.1225 ± 0.0001	64%
$_{41}Nb^{90}$	90	41		89.911263	14.6 h.	β+(53%)	6.111	0.86	5%		8+	4.941	ann.rad.	100%
	90	41				E.C.(47%)		1.5	92%				0.1412 ± 0.0001	64%
	90	41											0.3713 ± 0.0001	1.8%
	90	41											0.8277 ± 0.0001	1%
	90	41											0.8906 ± 0.0001	1.9%
	90	41											1.1292 ± 0.0001	91%
	90	41											1.6117 ± 0.0001	3%
	90	41											2.1862 ± 0.0001	18%
	90	41											2.3189 ± 0.0001	82%
	90	41											(0.1 - 3.3)	
$_{41}Nb^{91m}$	91	41			62 d.	I.T.(97%)					1/2-		0.1045 ± 0.0005	0.6%
	91	41				E.C.(3%)							1.2050 ± 0.0007	3.4%
$_{41}Nb^{91}$	91	41		90.906991	7 x 10^2 y.						5/2+		Mo k x-ray	54%
$_{41}Nb^{92m}$	92	41			10.13 d.	E.C. (99+%)					2+	6.114	0.9126 ± 0.0002	1.7%
	92	41											0.9345 ± 0.0001	99%
	92	41											1.8475 ± 0.0003	0.8%
$_{41}Nb^{92}$	92	41		91.907192	3x10^7 y.	E.C.	2.006				7+		0.5611 ± 0.0001	100%
	92	41											0.9345 ± 0.0001	100%
$_{41}Nb^{93m}$	93	41			13.6 y.	I.T.	0.0304				1/2-		Nb x-ray	
	93	41											0.0304	
$_{41}Nb^{93}$	93	41	100%	92.906377						1.15 b.	9/2+	+6.1705		
$_{41}Nb^{94m}$	94	41			6.26 m.	I.T. (99+%)	2.086				3+		Nb k x-ray	
	94	41				β-(0.5%)							0.0409 ± 0.0002	0.07%
	94	41											0.87109 ± 0.00002	0.5%
$_{41}Nb^{94}$	94	41		93.907280	2.4x10^4y.	β-	2.04	0.47			6+		0.70263 ± 0.00002	99%
	94	41											0.87109 ± 0.00002	100%
$_{41}Nb^{95m}$	95	41			3.61 d.	I.T.(97.5%	0.2357				1/2-		0.2040 ± 0.0001	2%
	95	41				β-(2.5%)							0.2356 ± 0.0005	26.1%
$_{41}Nb^{95}$	95	41		94.906835	34.98 d.	β-	0.160	0.926			9/2+	6.123	0.76578 ± 0.00002	99.8%
$_{41}Nb^{96}$	96	41		95.908100	23.4 h.	β-	3.187	0.5	10%		6+		0.2191 ± 0.0002	3.8%
	96	41						0.75	90%				0.2414 ± 0.0002	3.9%
	96	41											0.3501 ± 0.0002	1.1%
	96	41											0.3718 ± 0.0001	2.8%
	96	41											0.4600 ± 0.0006	28.2%
	96	41											0.4807 ± 0.0008	6.3%
	96	41											0.5689 ± 0.0006	55.7%
	96	41											0.7195 ± 0.0002	7.3%
	96	41											0.7782 ± 0.0001	96.8%
	96	41											0.8102 ± 0.0007	10%
	96	41											0.8124 ± 0.0003	3.4%
	96	41											0.8476 ± 0.0003	1.6%
	96	41											1.0913 ± 0.0006	49.3%
	96	41											1.2002 ± 0.0006	20%
	96	41											1.4977 ± 0.0008	3%
$_{41}Nb^{97m}$	97	41			54 s.	I.T.	0.7434	98%			6+		0.7434 ± 0.0001	98%
$_{41}Nb^{97}$	97	41		96.908096	73.6 m.	β-	1.934	1.2	98%		9/2+	7.3	0.4809 ± 0.0002	0.15%
	97	41											0.6579 ± 0.0001	98.3%
	97	41											1.0245 ± 0.0001	1.1%
	97	41											1.2686 ± 0.0001	0.2%
	97	41											1.5156 ± 0.0001	0.2%

Isotope	A	Z	% Natural abundance	Atomic mass	Half-life	Decay mode	Decay energy (MeV)	Particle energy (MeV)	Particle intensity	Thermal neutron cross section	Spin (h/2π)	μ Nucl. mag. moment	Gamma-ray energy (MeV)	Gamma-ray intensity
$_{41}$Nb^{98m}	98	41			51 m.	β-							0.1726 ± 0.0006	1.7%
	98	41											0.3354 ± 0.0006	9.3%
	98	41											0.4344 ± 0.0006	1.0%
	98	41											0.6448 ± 0.0006	5.3%
	98	41											0.7131 ± 0.0006	9.7%
	98	41											0.7227 ± 0.0003	69.7%
	98	41											0.7874 ± 0.0003	93%
	98	41											0.7916 ± 0.0006	6.4%
	98	41											0.8235 ± 0.0005	2.5%
	98	41											0.8336 ± 0.0005	10.9%
	98	41											0.9097 ± 0.0006	1.2%
	98	41											0.9937 ± 0.0006	1.3%
	98	41											0.9966 ± 0.0006	1.7%
	98	41											1.0243 ± 0.0006	1.0%
	98	41											1.1689 ± 0.0006	17.1%
	98	41											1.2302 ± 0.0006	1.5%
	98	41											1.3176 ± 0.0006	1.0%
	98	41											1.3355 ± 0.0006	1.4%
	98	41											1.4326 ± 0.0006	5.5%
	98	41											1.4367 ± 0.0006	1.9%
	98	41											1.4671 ± 0.0006	1.0%
	98	41											1.5121 ± 0.0006	5.1%
	98	41											1.5412 ± 0.0006	2.1%
	98	41											1.5412 ± 0.0006	2.1%
	98	41											1.5465 ± 0.0006	3.0%
	98	41											1.7019 ± 0.0006	9.1%
	98	41											1.8851 ± 0.0006	3.0%
	98	41											1.9454 ± 0.0006	1.5%
	98	41											1.8905 ± 0.0006	3.5%
$_{41}$Nb^{98}	98	41		97.910330	2.8 s.	β-	4.59	4.6			1+		0.6451 ± 0.0003	0.8%
	98	41											0.7874 ± 0.0003	3.2%
	98	41											0.9717 ± 0.0003	0.8%
	98	41											1.0243 ± 0.0003	1.6%
	98	41											1.4324 ± 0.0003	0.8%
$_{41}$Nb^{99m}	99	41			2.6 m.	β-		3.2			1/2-		0.0978 100 +	
	99	41				I.T.							0.1375	22
	99	41											0.254	57
	99	41											0.264	11
	99	41											0.352	42
	99	41											0.451	20
	99	41											0.525	13
	99	41											0.549	10
	99	41											0.554	8
	99	41											0.598	12
	99	41											0.631	19
	99	41											0.655	8
	99	41											0.673	11
	99	41											0.793	20
	99	41											0.889	5.5
	99	41											0.905	8
	99	41											0.945	9.5
	99	41											1.100	6.5
	99	41											1.259	7
	99	41											1.317	5
	99	41											1.475	11
	99	41											1.698	11
	99	41											1.735	8.5
	99	41											2.010	6.5
	99	41											2.241	8
	99	41											2.544	10
	99	41											2.642	50
	99	41											2.693	14
	99	41											2.734	15
	99	41											2.791	7.5
	99	41											2.854	48
	99	41											3.010	4
$_{41}$Nb^{99}	99	41		98.911619	15.0 s.	β-	3.62	3.5	100%		9/2+		0.0978 ± 0.0005	50%
	99	41											0.1375 ± 0.0005	90%
$_{41}$Nb^{100m}	100	41			1.5 s.	β-							Nb k x-ray	
	100	41											0.159	
	100	41											0.6364	
	100	41											1.0637	
$_{41}$Nb^{100}	100	41		99.914180	3.1 s.	β-	6.2	5.8					0.5354	100 +
	100	41											0.6001	46
	100	41											0.9661	
	100	41											1.0637	
	100	41											1.2803	
	100	41											1.5658	
$_{41}$Nb^{101}	101	41		100.915320	7.1 s.	β-	4.6	4.3					0.1105 ± 0.0005	
	101	41											0.1577 ± 0.0005	
	101	41											0.1806 ± 0.0005	
	101	41											0.2762 ± 0.0005	
	101	41											0.2897 ± 0.0005	
	101	41											0.4409 ± 0.0005	
	101	41											0.4659 ± 0.0005	

TABLE OF THE ISOTOPES (Continued)

Isotope	A	Z	% Natural abundance	Atomic mass	Half-life	Decay mode	Decay energy (MeV)	Particle energy (MeV)	Particle intensity	Thermal neutron cross section	Spin (h/2π)	μ Nucl. mag. moment	Gamma-ray energy (MeV)	Gamma-ray intensity
	101	41											0.7969 ± 0.0005	
	101	41											0.8100 ± 0.0005	
$_{41}Nb^{102m}$	102	41			1.3 s.	β-								
$_{40}Nb^{102}$	102	41		101.918040	4.3 s.	β-	7.2						0.2960 ± 0.0002	
	102	41											0.3976 ± 0.0002	
	102	41											0.4006	
	102	41											0.5514	
	102	41											0.8474	
	102	41											0.9490	
	102	41											1.2354	
	102	41											1.6330	
	102	41											1.7375	
	102	41											2.1844	
$_{41}Nb^{103}$	103	41		102.919370	1.5 s.	β-	5.5							
Mo		42		95.94						2.60 b.				
$_{42}Mo^{88}$	88	42		87.921820	8.0 m.	β+	3.6				0+		ann.rad.	
	88	42				E.C.							0.0800 ± 0.0005	80 +
	88	42											0.1399 ± 0.0005	60
	88	42											0.1707 ± 0.0005	100
$_{42}Mo^{89m}$	89	42			0.19 s.	I.T.	0.119				1/2-		0.119	
	89	42											0.268	
$_{42}Mo^{89}$	89	42		88.919480	2.2 m.	β+					9/2+		ann.rad.	
	89	42				E.C.							0.659	
	89	42											0.803	
	89	42											1.155	
	89	42											1.272	
$_{42}Mo^{90}$	90	42		89.913933	5.67 h.	β+(25%)	2.49	1.085			0+		ann.rad.	50%
	90	42				E.C>(75%)							0.04274 ± 0.00001	2%
	90	42											0.12237 ± 0.00005	64%
	90	42											0.1629 ± 0.0001	6%
	90	42											0.2031 ± 0.0001	6%
	90	42											0.25734 ± 0.00005	78%
	90	42											0.3232 ± 0.0001	6%
	90	42											0.4454 ± 0.0002	6%
	90	42											0.9415 ± 0.0004	6%
	90	42											0.9902 ± 0.0006	1%
	90	42											1.2713 ± 0.0006	4%
	90	42											1.3874 ± 0.0005	2%
	90	42											1.4546 ± 0.0007	2%
$_{42}Mo^{91m}$	91	42			65 s.	I.T.(50%)	0.653				1/2-		ann.rad.	75%
	91	42				β+, EC (50%)		2.5					0.6529 ± 0.0001	48%
	91	42						2.8					1.2081 ± 0.0001	20%
	91	42						2.8					1.5080 ± 0.0001	24%
	91	42						4.0					2.2407 ± 0.0004	0.8%
$_{42}Mo^{91}$	91	42		90.911755	15.5 m.	β+(94%)	4.44	3.44	94%		9/2-		ann.rad.	190%
	91	42				E.C.(6%)							1.6373 ± 0.0001	0.3%
	91	42											2.6321 ± 0.0002	0.1%
	91	42											3.0286 ± 0.0002	0.08%
	91	42											(0.1 - 4.2)	
$_{42}Mo^{92}$	92	42	14.84%	91.906808							0+			
$_{42}Mo^{93m}$	93	42			6.9 h.	I.T.	2.425				21/2+	+9.21	0.26306 ± 0.00001	58%
	93	42				(99+%)							0.68461 ± 0.00008	99.7%
	93	42											1.47711 ± 0.00002	99%
$_{42}Mo^{93}$	93	42		92.906813	3.5x10^3 y.	E.C.	0.406				5/2+		0.0304	(D)
$_{42}Mo^{94}$	94	42	9.25%	93.905085							0+			
$_{42}Mo^{95}$	95	42	15.92%	94.905840						14.5 b.	5/2+	-0.9133		
$_{42}Mo^{96}$	96	42	16.68%	95.904678							0+			
$_{42}Mo^{97}$	97	42	9.55%	96.906020						2 b.	5/2+	-0.9335		
$_{42}Mo^{98}$	98	42	24.13%	97.905406						0.132 b.	0+			
$_{42}Mo^{99}$	99	42		98.907711	65.94 h.	β-	1.357	0.45	14%		1/2+		0.144048 ± 0.0002	88%
	99	42						0.84	2%				0.18109 ± 0.0003	6.0%
	99	42						1.21	84%				0.36644 ± 0.0005	1.2%
	99	42											0.73947 ± 0.0005	12.6%
	99	42											0.77787 ± 0.00002	0.1%
	99	42											0.82298 ± 0.0004	0.1%
	99	42											0.9607 ± 0.0001	0.07%
$_{42}Mo^{100}$	100	42	9.63%	99.907477						0.195 b.	0+			
$_{42}Mo^{101}$	101	42		100.910345	14.6 m.	β-	2.81	2.23			1/2+		0.0063 ± 0.0001	80%
	101	42											0.0093 ± 0.0001	
	101	42											0.19193 ± 0.00004	22%
	101	42											0.5058 ± 0.0001	11.4%
	101	42											0.5908 ± 0.0001	
	101	42											0.6955 ± 0.0001	6.6%
	101	42											0.7130 ± 0.0001	

TABLE OF THE ISOTOPES (Continued)

Isotope	A	Z	% Natural abundance	Atomic mass	Half-life	Decay mode	Decay energy (MeV)	Particle energy (MeV)	Particle intensity	Thermal neutron cross section	Spin (h/2π)	μ Nucl. mag. moment	Gamma-ray energy (MeV)	Gamma-ray intensity
$_{43}Tc^{94}$	94	43		93.909654	4.88 h.	β+(11%)	4.26				7+	5.20	ann.rad.	22%
	94	43				E.C.(89%)							0.4491 ± 0.0002	2.6%
	94	43											0.5321 ± 0.0001	2.6%
	94	43											0.7026 ± 0.0001	100%
	94	43											0.8496 ± 0.0001	98%
	94	43											0.8710 ± 0.0001	100%
	94	43											0.9161 ± 0.0001	7.4%
	94	43											1.5917 ± 0.0003	2.4%
$_{43}Tc^{95m}$	95	43			61 d.	I.T.(4%))					1/2-		ann.rad.	
	95	43				β+(0.3%)		0.5					0.0389 ± 0.0001	1%
	95	43				E.C.(96%)		0.7					0.2041 ± 0.0001	66%
	95	43											0.5821 ± 0.0001	32%
	95	43											0.7658 ± 0.0001	3.7%
	95	43											0.7862 ± 0.0001	9.1%
	95	43											0.8206 ± 0.0002	5%
	95	43											0.8351 ± 0.0002	28%
	95	43											1.0392 ± 0.0002	3%
$_{43}Tc^{95}$	95	43		94.907657	20.0 h.	E.C.(100%)	1.69				9/2+	9.058	0.7657 ± 0.0001	93%
	95	43											0.8699 ± 0.0001	0.3%
	95	43											0.9478 ± 0.0001	2.2%
	95	43											1.0738 ± 0.0001	4%
$_{43}Tc^{96m}$	96	43			52 m.	I.T.(90%)	1.69				4+		0.0342 ± 0.0002	
	96	43				β+, EC							0.7782 ± 0.0001	1.9%
						(2%)								
	96	43											1.2002 ± 0.0001	1.1%
$_{43}Tc^{96}$	96	43		95.907870	4.3 d.	E.C.	2.97	2.0			7+	+5.37	Mo k x-ray	55%
	96	43											0.3143 ± 0.0001	2.4%
	96	43											0.3165 ± 0.0001	1.4%
	96	43											0.5689 ± 0.0001	1%
	96	43											0.7782 ± 0.0001	99%
	96	43											0.8125 ± 0.0001	82%
	96	43											0.8498 ± 0.0001	98%
	96	43											1.0913 ± 0.0002	1.1%
	96	43											1.12168 ± 0.0001	15%
	96	43											1.2002 ± 0.0002	0.4%
$_{43}Tc^{97m}$	97	43			90 d.	I.T.	0.0965				1/2-		Tc k x-ray	41%
	97	43											0.0965	0.32%
$_{43}Tc^{97}$	97	43		96.906364	$2.6x10^6$ y.	E.C.(100%)	0.320				9/2+		Mo k x-ray	54%
$_{43}Tc^{98}$	98	43		97.907215	$4.2x10^6$ y.	β-	1.79	0.597	100%		6+		0.65241 ± 0.00005	100%
	98	43											0.74535 ± 0.00005	100%
$_{43}Tc^{99m}$	99	43			6.01 h.	I.T.(100%)	0.142				1/2-		Tc k x-ray	6.7%
	99	43											0.14049 ± 0.00001	89%
	99	43											0.14261 ± 0.00001	0.04%
$_{43}Tc^{99}$	99	43		98.906254	$2.13x10^5$y.	β-	0.293	0.293	100%		9/2+	+5.6847		
$_{43}Tc^{100}$	100	43		99.907657	15.8 s.	β-	3.203	2.2			1+		0.5396 ± 0.0001	7%
	100	43						2.9					0.5908 ± 0.0001	5.7%
	100	43					3.3						1.5122 ± 0.0003	0.4%
	100	43											(0.3 - 2.6)weak	
$_{43}Tc^{101}$	101	43		100.907327	14.2 m.	β-	1.62	1.32			9/2+		0.1272 ± 0.0001	2.9%
	101	43											0.1841 ± 0.0001	
	101	43											0.3068 ± 0.0001	88%
	101	43											0.5314 ± 0.0001	1%
	101	43											0.5451 ± 0.0001	6%
	101	43											0.7156 ± 0.0001	0.7%
	101	43											(0.1 - 0.93)weak	
$_{43}Tc^{102m}$	102	43			4.4 m.	I.T.(2%)	4.8				5+		0.4184 ± 0.0001	4%
	102	43				β-(98%)							0.4752 ± 0.0001	85$
	102	43											0.4972 ± 0.0006	5.7%
	402	43											0.6281 ± 0.0006	25%
	402	43											0.6302 ± 0.0005	15$
	402	43											0.6969 ± 0.0009	6%
	102	43											1.0464 ± 0.0003	12%
	102	43											1.1033 ± 0.0004	12%
	102	43											1.1131 ± 0.0006	2%
	102	43											1.1976 ± 0.0005	7%
	102	43											1.2925 ± 0.0003	4%
	102	43											1.3386 ± 0.0003	4%
	102	43											1.5962 ± 0.0008	2.7%
	102	43											1.6163 ± 0.0007	15%
	102	43											1.7112 ± 0.0001	2.7%
	102	43											1.8107 ± 0.0001	5.6%
	102	43											2.2257 ± 0.001	5.5%
	102	43											2.2447 ± 0.001	11.4%
	102	43											2.4384 ± 0.001	4.4%
$_{43}Tc^{102}$	102	43		101.909208	5.3 s.	β-	4.5	3.4			1+		0.4686 ± 0.0001	0.8%
	102	43						4.2					0.4751 ± 0.0001	6%
	102	43											0.6281 ± 0.0001	0.7%
	102	43											0.6368 ± 0.0001	0.4%
	102	43											1.1032 ± 0.0001	0.35%
	102	43											1.1055 ± 0.0002	0.7%
	102	43											1.3620 ± 0.0002	0.3%

TABLE OF THE ISOTOPES (Continued)

Isotope	A	Z	% Natural abundance	Atomic mass	Half-life	Decay mode	Decay energy (MeV)	Particle energy (MeV)	Particle intensity	Thermal neutron cross section	Spin (h/2π)	μ Nucl. mag. moment	Gamma-ray energy (MeV)	Gamma-ray intensity
$_{43}Tc^{103}$	103	43		102.909172	54 s.	β-	2.64	2.0					0.1361 ± 0.0001	15%
	103	43						2.2					0.1743 ± 0.0001	2.6%
	103	43											0.2104 ± 0.0001	9%
	103	43											0.3435 ± 0.0002	3.8%
	103	43											0.3464 ± 0.0001	16%
	103	43											0.3886 ± 0.0001	2%
	103	43											0.4032 ± 0.0001	2%
	103	43											0.5012 ± 0.0001	2%
	103	43											0.5629 ± 0.0001	6%
	103	43											0.6612 ± 0.0001	0.7%
	103	43											0.9024 ± 0.0003	0.6%
	103	43											(0.13 - 1.0)weak	
$_{43}Tc^{104}$	104	43		103.911460	182 m.	β-	5.6	2.4			(3)		0.3483 ± 0.0001	15%
	104	43						3.3					0.3580 ± 0.0001	89%
	104	43						4.4					0.5305 ± 0.0001	16%
	104	43											0.5351 ± 0.0001	15%
	104	43											0.8844 ± 0.0001	11%
	104	43											0.8931 ± 0.0001	10%
	104	43											1.1574 ± 0.0001	3%
	104	43											1.2818 ± 0.0001	2%
	104	43											1.3805 ± 0.0001	1.7%
	104	43											1.3966 ± 0.0001	2.4%
	104	43											1.5413 ± 0.0001	1%
	104	43											1.5967 ± 0.0001	4%
	104	43											1.6124 ± 0.0001	6%
	104	43											1.6768 ± 0.0001	8%
	104	43											1.7369 ± 0.0001	2%
	104	43											1.9110 ± 0.0001	2%
	104	43											1.9771 ± 0.0002	1.6%
	104	43											2.0157 ± 0.0001	1.8%
	104	43											2.1238 ± 0.0001	2.2%
	104	43											2.1905 ± 0.0001	1.8%
	104	43											2.4655 ± 0.0002	1.2%
	104	43											2.6085 ± 0.0002	1.6%
	104	43											3.1492 ± 0.0002	1.2%
	104	43											(0.3 - 3.7)	
$_{43}Tc^{105}$	105	43		104.911820	7.6 m.	β-	3.7	3.4			5/2+		0.1079 ± 0.0001	9.6%
	105	43											0.1384 ± 0.0001	3%
	105	43											0.1432 ± 0.0001	11%
	105	43											0.1578 ± 0.0001	1.8%
	105	43											0.1593 ± 0.0001	7%
	105	43											0.2256 ± 0.0001	2%
	105	43											0.2520 ± 0.0001	4%
	105	43											0.2726 ± 0.0001	2.4%
	105	43											0.3215 ± 0.0001	7.6%
	105	43											0.3223 ± 0.0001	1%
	105	43											0.3583 ± 0.0003	1.7%
	105	43											0.4419 ± 0.0003	1.2%
	105	43											0.4459 ± 0.0002	1.1%
	105	43											0.4628 ± 0.0001	3%
	105	43											0.4663 ± 0.0003	1%
	105	43											0.4801 ± 0.0005	1%
	105	43											0.4906 ± 0.0003	1.6%
	105	43											0.5779 ± 0.0002	1.6%
	105	43											0.5402 ± 0.0002	1.8%
	105	43											0.7393 ± 0.0003	1.1%
	105	43											0.8960 ± 0.0005	1%
	105	43											1.0084 ± 0.0005	1.2%
	105	43											1.3663 ± 0.0003	2.2%
	105	43											1.5106 ± 0.0005	1.6%
	105	43											1.5601 ± 0.0005	1.4%
	105	43											2.0539 ± 0.0005	1.0%
	105	43											2.1554 ± 0.0005	1.5%
$_{43}Tc^{106}$	106	43		105.914510	36 s.	β-	6.7						0.2703 ± 0.0001	100%
	106	43											0.5222 ± 0.0001	7%
	106	43											0.7923 ± 0.0001	5%
	106	43											1.5043 ± 0.0001	1.2%
	106	43											1.6155 ± 0.0001	1.7%
	106	43											1.9694 ± 0.0001	9%
	106	43											2.2393 ± 0.0001	14%
	106	43											2.7014 ± 0.0001	2%
	106	43											2.7770 ± 0.0002	2%
	106	43											2.7893 ± 0.0002	8%
	106	43											2.9163 ± 0.0002	3%
	106	43											2.9459 ± 0.0002	3%
	106	43											3.0471 ± 0.0002	3%
	106	43											3.1864 ± 0.0002	5%
	106	43											3.2595 ± 0.0002	3%
	106	43											3.3642 ± 0.0003	1%
	106	43											3.5510 ± 0.0003	1%
$_{43}Tc^{107}$	107	43		106.915230	21.2 s.	β-	4.8						0.1027 ± 0.0001	21%
	107	43											0.1063 ± 0.0001	8%
	107	43											0.1421 ± 0.0001	3%
	107	43											0.1455 ± 0.0001	2%
	107	43											0.1770 ± 0.0001	9%

TABLE OF THE ISOTOPES (Continued)

Isotope	A	Z	% Natural abundance	Atomic mass	Half-life	Decay mode	Decay energy (MeV)	Particle energy (MeV)	Particle intensity	Thermal neutron cross section	Spin (h/2π)	μ Nucl. mag. moment	Gamma-ray energy (MeV)	Gamma-ray intensity
	107	43											0.1997 ± 0.0001	2%
	107	43											0.2915 ± 0.0001	4%
	107	43											0.3225 ± 0.0001	1%
	107	43											0.3354 ± 0.0001	1%
	107	43											0.3545 ± 0.0001	2%
	107	43											0.3603 ± 0.0001	3%
	107	43											0.4587 ± 0.0001	6%
	107	43											0.4898 ± 0.0001	1%
	107	43											0.5954 ± 0.0001	5%
	107	43											0.8569 ± 0.0001	1%
	107	43											1.1181 ± 0.0001	1%
	107	43											1.2187 ± 0.0001	1%
	107	43											1.5732 ± 0.0001	1%
	107	43											1.7419 ± 0.0001	0.5%
	107	43											2.3789 ± 0.0003	0.5%
	107	43											2.5023 ± 0.0003	0.6%
	107	43											2.5374 ± 0.0003	0.6%
$_{43}Tc^{108}$	108	43		107.918420	5.0 s.	β-	7.8						0.2422 ± 0.0001	82%
	108	43											0.4656 ± 0.0001	14%
	108	43											0.7078 ± 0.0001	11%
	108	43											0.7326 ± 0.0001	10%
	108	43											1.1180 ± 0.0002	5%
	108	43											1.4170 ± 0.0001	4%
	108	43											1.5835 ± 0.0001	10%
	108	43											1.7604 ± 0.0001	2%
Ru		44		101.07						2.6 b.				
$_{44}Ru^{92}$	92	44		91.920120	3.7 m.	β+ (53%)	4.5				0+		ann.rad.	106%
	92	44				E.C.(47%)							0.1346 ± 0.0001	63%
	92	44											0.2138 ± 0.0001	92%
	92	44											0.2593 ± 0.0001	89%
	92	44											0.4507 ± 0.0001	6.6%
	92	44											0.8670 ± 0.0001	11%
	92	44											0.9102 ± 0.0001	3%
	92	44											0.9450 ± 0.0003	2.6%
	92	44											0.9472 ± 0.0003	2.6%
	92	44											1.2196 ± 0.0001	6%
	92	44											1.2291 ± 0.0001	3%
	92	44											1.4036 ± 0.0002	1.6%
	92	44											1.5176 ± 0.0003	1.8%
	92	44											1.6976 ± 0.00001	3.5%
	92	44											2.0597 ± 0.0002	3%
	92	44											2.3023 ± 0.001	1%
$_{44}Ru^{93m}$	93	44			10.8 s.	I.T.(21%)					1/2-		ann.rad.	
	93	44				β+, EC (79%)							0.7344 ± 0.0001	23%
	93	44											0.9283 ± 0.0002	1.7%
	93	44											1.1112 ± 0.0001	27%
	93	44											1.3962 ± 0.0001	40%
	93	44											2.0931 ± 0.0002	20%
$_{44}Ru^{93}$	93	44		92.917050	60 s.	β+	6.3				9/2+		ann.rad.	
	93	44				E.C.							0.6807 ± 0.0001	5.9%
	93	44											1.4349 ± 0.0001	0.7%
	93	44											1.8014 ± 0.0001	0.4%
	93	44											3.2343 ± 0.0002	0.3%
	93	44											3.9147 ± 0.0002	0.3%
	93	44											4.3895 ± 0.0003	0.1%
	93	44											(0.5- 4.2)weak	
$_{44}Ru^{94}$	94	44		93.911361	52 m.	E.C.(100%)	1.59				5/2+		0.3672 ± 0.0005	80%
	94	44											0.5247 ± 0.0005	2%
	94	44											0.8922 ± 0.0005	20%
$_{44}Ru^{95}$	95	44		94.910414	1.64 h.	E.C.(85%)	2.57				5/2+		ann.rad.	
	95	44				β+(15%)							0.2904 ± 0.0001	3.5%
	95	44											0.3010 ± 0.0001	2%
	95	44											0.3364 ± 0.0001	71%
	95	44											0.6268 ± 0.0001	18%
	95	44											0.7485 ± 0.0001	1.6%
	95	44											0.8063 ± 0.0001	4.1%
	95	44											0.8422 ± 0.0001	1.3%
	95	44											0.8890 ± 0.0001	1.9%
	95	44											1.0507 ± 0.0001	2.6%
	95	44											1.0968 ± 0.0001	21%
	95	44											1.1787 ± 0.0002	5%
	95	44											1.4106 ± 0.0001	2.5%
	95	44											1.4593 ± 0.0001	2.1%
	95	44											1.7854 ± 0.0002	0.6%
	95	44											1.9881 ± 0.0002	0.7%
	95	44											2.3445 ± 0.0002	1.4%
$_{44}Ru^{96}$	96	44	5.52%	95.907599						0.2 b.	0+			
$_{44}Ru^{97}$	97	44		96.907556	2.89 d.	E.C.	1.12				5/2+	0.687	Tc k x-ray	58%
	97	44											0.1088 ± 0.0001	0.1%
	97	44											0.2157 ± 0.0001	10%
	97	44											0.3245 ± 0.0001	86%
	97	44											0.4606 ± 0.0001	10%
	97	44											0.5693 ± 0.0001	0.9%

TABLE OF THE ISOTOPES (Continued)

Isotope	A	Z	% Natural abundance	Atomic mass	Half-life	Decay mode	Decay energy (MeV)	Particle energy (MeV)	Particle intensity	Thermal neutron cross section	Spin (h/2π)	μ Nucl. mag. moment	Gamma-ray energy (MeV)	Gamma-ray intensity
$_{44}Ru^{98}$	98	44	1.88%	97.905287							0+			
$_{44}Ru^{99}$	99	44	12.7%	98.905939						5 b.	5/2+	-0.6413		
$_{44}Ru^{100}$	100	44	12.6%	99.904219						5.8 b.	0+			
$_{44}Ru^{101}$	101	44	17.0%	100.905582						5 b.	5/2+	-0.7188		
$_{44}Ru^{102}$	102	44	31.6%	101.904348						1.2 b.	0+			
$_{44}Ru^{103}$	103	44		102.906323	39.24 d.	β-	0.767	0.117	5%		5/2+	0.67	0.05329 ± 0.00001	0.36%
	103	44						0.225	91%				0.29498 ± 0.00001	0.36%
	103	44						0.725	3%				0.4438 ± 0.0001	0.31%
	103	44											0.49708 ± 0.00001	86%
	103	44											0.55704 ± 0.00004	0.8%
	103	44											0.61033 ± 0.00001	5.3%
	103	44											(0.04 - 1.6)weak	
$_{44}Ru^{104}$	104	44	18.7%	103.905424						0.35 b.	0+			
$_{44}Ru^{105}$	105	44		104.907744	4.44 h.	β-	1.917	1.109	22%	0.4 b.	3/2+		0.12968 ± 0.00007	5.7%
	105	44						1.134	13%				0.1491 ± 0.0001	1.8%
	105	44						1.187	49%				0.2629 ± 0.0001	6.6%
	105	44											0.31664 ± 0.00005	11%
	105	44											0.3502 ± 0.0001	1%
	105	44											0.3934 ± 0.0001	3.8%
	105	44											0.4135 ± 0.0001	2.2%
	105	44											0.46943 ± 0.00005	17.6%
	105	44											0.4993 ± 0.0003	2%
	105	44											0.6562 ± 0.0001	2%
	105	44											0.67634 ± 0.00006	15.7%
	105	44											0.72420 ± 0.00005	47%
	105	44											0.87585 ± 0.00005	2.5%
	105	44											0.90763 ± 0.00006	0.5%
	105	44											0.9694 ± 0.0001	2.1%
	105	44											1.0174 ± 0.0001	0.3%
	105	44											1.3214 ± 0.0001	0.2%
	105	44											1.6983 ± 0.0002	0.2%
	105	44											(0.1 - 1.8)	
$_{44}Ru^{106}$	106	44		105.907321	372.6 d.	β-	0.039	0.0394	100%	0.15 b.	0+			
$_{44}Ru^{107}$	107	44		106.910130	3.8 m.	β-	3.2	2.1					0.1939 ± 0.0002	10.9%
	107	44						3.2					0.3741 ± 0.0003	3.5%
	107	44											0.4055 ± 0.0002	2.6%
	107	44											0.4625 ± 0.0002	4.3%
	107	44											0.4891 ± 0.0002	1.4%
	107	44											0.5791 ± 0.0002	2.6%
	107	44											0.8488 ± 0.0002	5.3%
	107	44											1.0429 ± 0.0002	1.9%
	107	44											1.2724 ± 0.0002	2.2%
$_{44}Ru^{108}$	108	44		107.910140	4.6 m.	β-	1.2	1.2					0.0923 ± 0.0004	2%
	108	44											0.1651 ± 0.0004	32%
	108	44											0.4339 ± 0.0004	46%
	108	44											0.4975 ± 0.0004	7%
	108	44											0.5112 ± 0.0004	1.8%
	108	44											0.6189 ± 0.0004	15%
	108	44											0.9324 ± 0.0004	4%
$_{44}Ru^{109}$	109	44		108.913240	35 s.	β-	4.2						0.1164	
	109	44											0.3584	
$_{44}Ru^{110}$	110	44		109.913760	15 s.	β-	2.5						0.1121 ± 0.0003	100 +
	110	44											0.3737 ± 0.0003	100
	110	44											0.4397 ± 0.0003	10
	110	44											0.5729 ± 0.0003	4
	110	44											0.7967 ± 0.0005	10
	110	44											0.8134 ± 0.0005	3
	110	44											1.0962 ± 0.0005	2
Rh		45		102.9055						145 b.				
$_{45}Rh^{94m}$	94	45			71 s.	β+		6.4					ann.rad.	
	94	45											0.1264 ± 0.0002	4%
	94	45											0.3117 ± 0.0001	12%
	94	45											0.4381 ± 0.0002	7%
	94	45											0.4926 ± 0.0003	4%
	94	45											0.5529 ± 0.0003	2%
	94	45											0.7562 ± 0.0001	51%
	94	45											1.0681 ± 0.0003	5%
	94	45											1.0752 ± 0.0002	31%
	94	45											1.1107 ± 0.0002	3%
	94	45											1.4307 ± 0.0001	100%
	94	45											1.5397 ± 0.0003	4%
	94	45											1.8043 ± 0.001	2%

Isotope	A	Z	% Natural abundance	Atomic mass	Half-life	Decay mode	Decay energy (MeV)	Particle energy (MeV)	Particle intensity	Thermal neutron cross section	Spin (h/2π)	μ Nucl. mag. moment	Gamma-ray energy (MeV)	Gamma-ray intensity
	94	45											1.9025 ± 0.001	2%
	94	45											2.1245 ± 0.001	2%
	94	45											2.7786 ± 0.001	1.1%
	94	45											3.0077 ± 0.001	1.1%
	94	45											3.2103 ± 0.001	1%
	94	45											3.2560 ± 0.001	2%
$_{45}Rh^{94}$	94	45		93.921670	25.8 s.	β+	9.6				8+		ann.rad.	
	94	45											0.1461 ± 0.0001	75%
	94	45											0.3117 ± 0.0001	97%
	94	45											0.4381 ± 0.0002	3%
	94	45											0.7562 ± 0.0001	100%
	94	45											1.4307 ± 0.0001	100%
	94	45											2.0995 ± 0.001	2%
	94	45											2.1245 ± 0.001	1.1%
	94	45											3.2103 ± 0.001	0.9%
	94	45											3.2560 ± 0.001	1.4%
$_{45}Rh^{95m}$	95	45			1.96 m.	I.T.(88%)		0.5433			$^{1}/_{2}$+		ann.rad.	
	95	45				β+, EC (12%)							0.5433 ± 0.0003	80%
	95	45											0.7837 ± 0.0004	8%
	95	45											2.8210 ± 0.0008	0.8%
	95	45											3.1862 ± 0.0008	0.9%
	95	45											3.8244 ± 0.0007	1.3%
$_{45}Rh^{95}$	95	45		94.915900	5.0 m.	β+	5.1	0.7	1.3 +		9/2+		ann.rad.	
	95	45						1.04	12				0.2293 ± 0.0003	2%
	95	45						1.33	3.5				0.4103 ± 0.0003	1%
	95	45											0.6225 ± 0.0005	2.6%
	95	45											0.6610 ± 0.0003	105%
	95	45											0.6776 ± 0.0003	6%
	95	45											0.7644 ± 0.0007	2%
	95	45											0.8950 ± 0.0003	2%
	95	45											0.9416 ± 0.0003	72%
	95	45											1.3170 ± 0.0003	3%
	95	45											1.3520 ± 0.0003	21%
	95	45											1.4893 ± 0.0003	3.4%
	95	45											1.4947 ± 0.0003	5%
	95	45											1.5245 ± 0.0005	2%
	95	45											2.1210 ± 0.0003	1.6%
	95	45											2.7918 ± 0.0003	2.4%
	95	45											3.0632 ± 0.0005	1%
	95	45											(0.2 - 3.8)	
$_{45}Rh^{96m}$	96	45			1.51 m.	I.T.(60%)	0.0520				2+		ann.rad.	
	96	45				β+, EC (40%)		4.70					Tc,Ru x-rays	
	96	45											0.0520 ± 0.0001	0.2%
	96	45											0.6853 ± 0.0001	4%
	96	45											0.8087 ± 0.0002	2.5%
	96	45											0.8326 ± 0.0001	39%
	96	45											1.0985 ± 0.0002	8%
	96	45											1.4512 ± 0.0002	1.3%
	96	45											1.6921 ± 0.0002	6.4%
	96	45											1.7436 ± 0.0002	1%
	96	45											1.9073 ± 0.0002	1%
	96	45											2.1636 ± 0.0002	2.5%
	96	45											2.2575 ± 0.0003	1.7%
	96	45											2.4591 ± 0.0003	0.7%
	96	45											(0.4 - 3.3)	
$_{45}Rh^{96}$	96	45		95.914515	9.9 m.	β+	6.44	3.3			5+		ann.rad.	
	96	45				E.C.							0.4299 ± 0.0002	2%
	96	45											0.6315 ± 0.0001	80%
	96	45											0.6441 ± 0.0001	5%
	96	45											0.6853 ± 0.0001	98%
	96	45											0.7418 ± 0.0001	32%
	96	45											0.8007 ± 0.0002	3%
	96	45											0.8326 ± 0.0001	100%
	96	45											0.9155 ± 0.0002	1%
	96	45											0.9441 ± 0.0002	2%
	96	45											1.0703 ± 0.0002	1.7%
	96	45											1.2279 ± 0.0003	8.7%
	96	45											1.2306 ± 0.0003	7.5%
	96	45											1.2421 ± 0.0002	1.3%
	96	45											1.2757 ± 0.0002	3.1%
	96	45											1.5667 ± 0.0005	2%
	96	45											1.5887 ± 0.0005	1%
	96	45											1.6055 ± 0.0002	2.7%
	96	45											1.6487 ± 0.0002	2%
	96	45											1.6921 ± 0.0002	1.9%
	96	45											1.7325 ± 0.0002	5.7%
	96	45											1.7886 ± 0.0002	1.9%
	96	45											1.8596 ± 0.0002	1.6%
	96	45											1.9630 ± 0.0002	1.2%
	96	45											(0.2 - 3.4)	
$_{45}Rh^{97m}$	97	45			46 m.	I.T.(5%)		1.8			$^{1}/_{2}$-		ann.rad.	
	97	45				β+, EC (95%)		2.1					0.1886 ±	51%

Isotope	A	Z	% Natural abundance	Atomic mass	Half-life	Decay mode	Decay energy (MeV)	Particle energy (MeV)	Particle intensity	Thermal neutron cross section	Spin (h/2π)	μ Nucl. mag. moment	Gamma-ray energy (MeV)	Gamma-ray intensity
	97	45						2.5					0.4215 ± 0.0003	13%
	97	45											0.5278 ± 0.0002	9%
	97	45											0.5824 ± 0.0003	2.6%
	97	45											0.7190 ± 0.0003	3%
	97	45											0.7711 ± 0.0003	4.8%
	97	45											0.9085 ± 0.0005	2%
	97	45											0.9955 ± 0.0003	3.0%
	97	45											1.0134 ± 0.0003	3.6%
	97	45											1.1838 ± 0.0006	4.5%
	97	45											1.1870 ± 0.0006	4.5%
	97	45											1.4264 ± 0.0004	2.3%
	97	45											1.5866 ± 0.0004	8.3%
	97	45											1.7184 ± 0.0006	2.4%
	97	45											2.0074 ± 0.0005	4%
	97	45											2.0361 ± 0.0005	4%
	97	45											2.1223 ± 0.0007	3%
	97	45											2.2452 ± 0.0005	13%
	97	45											2.6088 ± 0.0006	2%
	97	45											2.6478 ± 0.0006	3%
	97	45											3.3741 ± 0.0006	1%
$_{45}Rh^{97}$	97	45		96.911320	31.0 m.	β+	3.51	1.8			9/2+		ann.rad.	
	97	45						2.1					0.1886 ± 0.0002	1%
	97	45						2.52					0.3892 ± 0.0004	1%
	97	45											0.4515 ± 0.0003	75%
	97	45											0.7772 ± 0.0004	1.6%
	97	45											0.8073 ± 0.0004	1.4%
	97	45											0.8398 ± 0.0003	12%
	97	45											0.8788 ± 0.0003	9.3%
	97	45											1.0537 ± 0.0004	1.7%
	97	45											1.2280 ± 0.0005	1.1%
	97	45											1.3101 ± 0.0006	1.1%
	97	45											1.91310 ± 0.0009	0.7%
	97	45											(0.2 - 3.5)	
$_{45}Rh^{98m}$	98	45			3.5 m.	β+		3.4			2+		ann.rad.	
	98	45											0.6154 ± 0.0001	5%
	98	45											0.6524 ± 0.0001	96%
	98	45											0.7452 ± 0.0002	78%
	98	45											1.1440 ± 0.0005	8%
	98	45											1.4149 ± 0.0005	4%
$_{45}Rh^{98}$	98	45		97.910716	8.6 m.	β+(90%)	5.06	2.8			5+		ann.rad.	180%
	98	45						3.45					0.6524 ± 0.0001	94%
	98	45											0.7453 ± 0.0001	5%
	98	45											0.7623 ± 0.0002	78%
	98	45											1.1644 ± 0.0004	1.1%
	98	45											1.4149 ± 0.0004	1.1%
	98	45											1.8170 ± 0.0004	4.8%
$_{45}Rh^{99m}$	99	45			4.7 h.	β+(8%)		.74			9/2+		ann.rad.	16%
	99	45				E.C.(92%)							0.2766 ± 0.0004	1.6%
	99	45											0.3408 ± 0.0004	69%
	99	45											0.5282 ± 0.0004	1.4%
	99	45											0.6178 ± 0.0004	12%
	99	45											0.7193 ± 0.0004	1.2%
	99	45											0.9366 ± 0.0004	2.2%
	99	45											1.2612 ± 0.0004	11%
$_{45}Rh^{99}$	99	45		98.908192	16.1 d.	β+(4%)	2.09	0.54			1/2-		ann.rad.	
	99	45				E.C.(97%)		0.68					0.0894 ± 0.0001	31%
	99	45											0.3224 ± 0.0004	4%
	99	45											0.3530 ± 0.0004	32%
	99	45											0.5277 ± 0.0004	41%
	99	45											0.6180 ± 0.0004	5%
	99	45											0.8066 ± 0.0004	1.4%
	99	45											0.9145 ± 0.0004	1.4%
	99	45											(0.1 - 2.0)	
$_{45}Rh^{100m}$	100	45			4.7 m.	I.T.(99%)					5+		ann.rad.	
	100	45				β+(0.4%)					5+		0.0748 ± 0.0005	70%
	100	45											0.2647 ± 0.0005	30%
	100	45											0.5396 ± 0.001	0.4%
	100	45											0.6869 ± 0.0005	0.2%
$_{45}Rh^{100}$	100	45		99.908116	20.8 h.	β+	3.63	1.3			1-		0.4462 ± 0.0001	11%
	100	45				E.C.		2.1					0.5396 ± 0.0001	78%
	100	45						2.6					0.5882 ± 0.0002	4%
	100	45											0.5908 ± 0.0001	1.4%
	100	45											0.8225 ± 0.0002	20%
	100	45											1.1071 ± 0.0002	13%
	100	45											1.3476 ± 0.0002	4.8%
	100	45											1.3621 ± 0.0001	15%
	100	45											1.5534 ± 0.0002	20%
	100	45											1.6275 ± 0.0003	1.6%
	100	45											1.9297 ± 0.0002	12%
	100	45											2.3761 ± 0.0003	35%
	100	45											2.5302 ± 0.0002	2.7%
$_{45}Rh^{101m}$	101	45			4.34 d.	E.C.(92%)					9/2+	+5.51	Rh k x-ray	51%
	101	45				I.T.(8%)	0.1573						0.1272 ± 0.0001	0.6%
	101	45											0.1573 ± 0.0001	0.2%
	101	45											0.3069 ± 0.0001	86%
	101	45											0.5451 ± 0.0001	4%

TABLE OF THE ISOTOPES (Continued)

Isotope	A	Z	% Natural abundance	Atomic mass	Half-life	Decay mode	Decay energy (MeV)	Particle energy (MeV)	Particle intensity	Thermal neutron cross section	Spin (h/2π)	μ Nucl. mag. moment	Gamma-ray energy (MeV)	Gamma-ray intensity
$_{45}Rh^{101}$	101	45		100.906159	3.3 y.	E.C.	0.54				1/2-		Ru k x-ray	66%
	101	45											0.1272 ± 0.0001	73%
	101	45											0.1980 ± 0.0002	71%
	101	45											0.2950 ± 0.0003	0.7%
	101	45											0.3252 ± 0.0002	13%
	101	45											0.4880 ± 0.0003	0.4%
$_{45}Rh^{102m}$	102	45			206 d.	I.T.(5%)							ann.rad.	28%
	102	45				β-(19%)							0.4686 ± 0.0001	3%
	102	45				β+(14%)							0.4751 ± 0.0001	44%
	102	45				E.C.(62%)							0.5566 ± 0.0001	2%
	102	45											0.6280 ± 0.0001	4%
	102	45											1.1032 ± 0.0001	2.8%
	102	45											(0.4 - 1.6)	
$_{45}Rh^{102}$	102	45		101.906814	2.9 y.	E.C.	2.33				(6+)	4.11	Ru k x-ray	55%
	102	45											0.4152 ± 0.0002	2%
	102	45											0.4185 ± 0.0002	9%
	102	45											0.4204 ± 0.0002	3%
	102	45											0.4751 ± 0.0001	94%
	102	45											0.6280 ± 0.0001	8%
	102	45											0.6313 ± 0.0001	55%
	102	45											0.6924 ± 0.0002	1.6%
	102	45											0.6956 ± 0.0003	2.8%
	102	45											0.6975 ± 0.0001	43%
	102	45											0.7668 ± 0.0001	34%
	102	45											1.0466 ± 0.0001	34%
	102	45											1.1032 ± 0.0001	45%
	102	45											1.1128 ± 0.0001	19%
$_{45}Rh^{103}$	103	45	100%	102.905500						(11 + 134)b.	1/2-	-0.0884		
$_{45}Rh^{104m}$	104	45			4.36 m.	I.T. (99+%)				800 b.	5+		Rh k x-ray	55%
	104	45											0.0514 ± 0.0001	48%
	104	45											0.0775 ± 0.0001	2%
	104	45											0.0971 ± 0.0001	3%
	104	45											0.5558 ± 0.0001	2.4%(D)
	104	45											0.7678 ± 0.0001	0.1%
$_{45}Rh^{104}$	104	45		103.906651	41.8 s.	β-(99+%)	2.44	1.88	2%		1+		0.3581 ± 0.0002	0.02%
	104	45				E.C.(0.4%)	1.14	2.44	98%				0.5558 ± 0.0001	2%
	104	45											1.2370 ± 0.0001	0.07%
	104	45											(0.35 - 1.8)weak	
$_{45}Rh^{105m}$	105	45			45 s.	I.T.	1.298				1/2-		Rh k x-ray	35%
	105	45											0.1296 ± 0.0001	20%
$_{45}Rh^{105}$	105	45		104.905686	35.4 h.	β-	0.57	0.247	30%	5x10³ b.	7/2+	+4.428	0.2801 ± 0.0002	0.2%
	105	45						0.567	70%				0.3061 ± 0.0002	5.1%
	105	45											0.3189 ± 0.0001	19.2%
$_{45}Rh^{106m}$	106	45			2.18 h.	β-		0.92			6+		0.2217 ± 0.0001	6.5%
	106	45											0.2286 ± 0.0001	2%
	106	45											0.3910 ± 0.0001	3.5%
	106	45											0.4062 ± 0.0001	12%
	106	45											0.4296 ± 0.0001	13%
	106	45											0.4510 ± 0.0001	24%
	106	45											0.5119 ± 0.0001	86%
	106	45											0.6012 ± 0.0001	3%
	106	45											0.6162 ± 0.0001	20%
	106	45											0.6460 ± 0.0001	2.8%
	106	45											0.6904 ± 0.0001	2%
	106	45											0.7031 ± 0.0001	4.5%
	106	45											0.7173 ± 0.0001	29%
	106	45											0.7484 ± 0.0001	19%
	106	45											0.7932 ± 0.0001	5.7%
	106	45											0.8043 ± 0.0001	13%
	106	45											0.8084 ± 0.0001	7.5%
	106	45											0.8247 ± 0.0001	14%
	106	45											0.8473 ± 0.0001	3.6%
	106	45											1.0197 ± 0.0002	2%
	106	45											1.0458 ± 0.0001	31%
	106	45											1.1280 ± 0.0001	14%
	106	45											1.1994 ± 0.0001	11%
	106	45											1.2229 ± 0.0001	8%
	106	45											1.3944 ± 0.0002	2.8%
	106	45											1.5277 ± 0.0002	18%
	106	45											1.5724 ± 0.0002	6.7%
	106	45											1.7228 ± 0.0002	2%
	106	45											1.8390 ± 0.0002	2%
$_{45}Rh^{106}$	106	45		105.907279	29.8 s.	β-	3.54	2.4	2%		1+		0.51186 ± 0.00001	21%
	106	45						3.0	12%				0.61612 ± 0.00005	0.8%
	106	45						3.54	79%				0.62187 ± 0.00005	9.8%
	106	45											1.0504 ± 0.0001	1.5%
	106	45											1.12807 ± 0.00005	0.4%
	106	45											1.5622 ± 0.0001	0.16%
	106	45											1.7664 ± 0.0001	0.03%
	106	45											1.9885 ± 0.0001	0.03%

Isotope	A	Z	% Natural abundance	Atomic mass	Half-life	Decay mode	Decay energy (MeV)	Particle energy (MeV)	Particle intensity	Thermal neutron cross section	Spin (h/2π)	μ Nucl. mag. moment	Gamma-ray energy (MeV)	Gamma-ray intensity
	106	45											2.1126 ± 0.0001	0.035%
	106	45											2.3660 ± 0.0001	0.03%
	106	45											(0.05 - 3.04)	
$_{45}Rh^{107}$	107	45		106.906751	21.7 m.	β-	1.2	65%			5/2+		0.2776 ± 0.0002	2%
	107	45						1.5	17%				0.3028 ± 0.0002	66%
	107	45											0.3128 ± 0.0002	5%
	107	45											0.3218 ± 0.0002	2%
	107	45											0.3482 ± 0.0002	2%
	107	45											0.3673 ± 0.0002	2%
	107	45											0.3925 ± 0.0002	9%
	107	45											0.5677 ± 0.0002	1%
	107	45											0.6701 ± 0.0002	2%
$_{45}Rh^{108m}$	108	45			17 s.	β-	4.5				1+		0.4339 ± 0.0004	43%
	0.8	45											0.4973 ± 0.0004	6%
	108	45											0.5112 ± 0.0004	2%
	108	45											0.6189 ± .0004	14%
	108	45											0.9324 ± 0.0004	3%
$_{45}Rh^{108}$	108	45		107.908650	6.0 m.	β-	4.4	1.57					0.4046 ± 0.0002	27%
	108	45											0.4339 ± 0.0002	92%
	108	45											0.4973 ± 0.0002	23%
	108	45											0.5811 ± 0.0002	58%
	108	45											0.6146 ± 0.0002	28%
	108	45											0.7230 ± 0.0002	7%
	108	45											0.9014 ± 0.0002	30%
	108	45											0.9313 ± 0.0002	7%
	108	45											0.9471 ± 0.0002	50%
	108	45											1.0927 ± 0.0004	3%
	108	45											1.2343 ± 0.0004	8%
	108	45											1.5280 ± 0.0004	1%
	108	45											1.8156 ± 0.0004	6%
$_{45}Rh^{109}$	109	45		108.908734	81 s.	β-	2.58				5/2+		0.1134 ± 0.0001	6%
	109	45											0.1780 ± 0.0001	8%
	109	45											0.2153 ± 0.0001	2%
	109	45											0.2450 ± 0.0001	1.3%
	109	45											0.2492 ± 0.0001	6%
	109	45											0.2763 ± 0.0001	2.2%
	109	45											0.2914 ± 0.0001	7.8%
	109	45											0.3254 ± 0.0002	1.5%
	109	45											0.3268 ± 0.0001	56%
	109	45											0.3780 ± 0.0001	1.3%
	109	45											0.4261 ± 0.0001	8%
	109	45											(0.1 - 1.6)	
$_{45}Rh^{110m}$	110	45			3.1 s.	β-		≈5					0.3737 ± 0.0003	100%
	110	45											0.4397 ± 0.0003	10%
	110	45											0.5729 ± 0.0005	4%
	110	45											0.7967 ± 0.0005	9.5%
	110	45											0.8134 ± 0.0005	3%
	110	45											1.0962 ± 0.0005	2%
$_{45}Rh^{110}$	110	45		109.910960	29 s.	β-	5.4	2.6					0.3737 ± 0.0002	91%
	110	45											0.3985 ± 0.0002	15%
	110	45											0.4400 ± 0.0002	26%
	110	45											0.4788 ± 0.0005	4%
	110	45											0.5312 ± 0.0005	2%
	110	45											0.5463 ± 0.0002	36%
	110	45											0.5849 ± 0.0002	17%
	110	45											0.6534 ± 0.0002	17%
	110	45											0.6877 ± 0.0002	28%
	110	45											0.8137 ± 0.0002	9%
	110	45											0.8381 ± 0.0002	22%
	110	45											0.8905 ± 0.0002	13%
	110	45											0.9045 ± 0.0002	27%
	110	45											0.9796 ± 0.0005	5%
	110	45											1.0483 ± 0.0005	8%
	110	45											1.0865 ± 0.0005	3%
	110	45											1.2165 ± 0.0005	7%
	110	45											1.2309 ± 0.0005	13%
	110	45											1.3921 ± 0.0005	4.6%
	110	45											1.5258 ± 0.001	2%
	110	45											1.5792 ± 0.001	2%
	110	45											1.5936 ± 0.0005	6%
	110	45											1.8717 ± 0.001	1%
	110	45											1.8851 ± 0.0005	4%
$_{45}Rh^{111}$	111	45		110.911630	11 s.	β-							0.3753	
$_{45}Rh^{112}$	112	45		111.914410	0.8 s.	β-							0.3489 ± 0.0002	
Pd		46		106.42						6.9 b.				
$_{46}Pd^{96}$	96	46		95.918010	2.0 m.	E.C.	3.3						0.1248	100 +
	96	46											0.4995	42
$_{46}Pd^{97}$	97	46		96.916480	3.1 m.	β+.E.C.	4.8				5/2+		ann.rad.	
	97	46											0.2653 ± 0.0001	50%
	97	46											0.4752 ± 0.0001	24%
	97	46											0.7927 ± 0.0001	12%
	97	46											0.9337 ± 0.0001	2%
	97	46											0.9403 ± 0.0003	3%
	97	46											1.0536 ± 0.0005	3%
	97	46											1.0554 ± 0.0005	5%

Isotope	A	Z	% Natural abundance	Atomic mass	Half-life	Decay mode	Decay energy (MeV)	Particle energy (MeV)	Particle intensity	Thermal neutron cross section	Spin (h/2π)	μ Nucl. mag. moment	Gamma-ray energy (MeV)	Gamma-ray intensity
	97	46											1.0585 ± 0.0005	2.4%
	97	46											1.1718 ± 0.0003	3%
	97	46											1.2378 ± 0.0005	2%
	97	46											1.4942 ± 0.0002	5.5%
	97	46											1.5198 ± 0.0005	2%
	97	46											1.6387 ± 0.0003	3.4%
	97	46											1.6411 ± 0.0003	3%
	97	46											1.7596 ± 0.0001	6%
	97	46											1.7972 ± 0.0005	1%
	97	46											1.8468 ± 0.0003	2.6%
	97	46											1.9939 ± 0.0005	1%
	97	46											2.0295 ± 0.0005	2.5%
	97	46											2.4284 ± 0.001	1%
	97	46											2.6844 ± 0.001	1%
	97	46											2.9746 ± 0.001	1%
	97	46											3.3421 ± 0.001	0.7%
	97	46											(0.2 - 3.4)	
$_{46}Pd^{98}$	98	46		97.912722	18 m.	β+	1.87				0+		ann.rad.	
	98	46				E.C.							0.0677 ± 0.0002	18 +
	98	46											0.1068 ± 0.0002	26
	98	46											0.1125 ± 0.0002	100
	98	46											0.1745 ± 0.0002	21
	98	46											0.6630 ± 0.0004	53
	98	46											0.7257 ± 0.0004	10
	98	46											0.8379 ± 0.0004	30
$_{46}Pd^{99}$	99	46		98.911763	21.4 m.	β+(49%)	3.36	1.58			5/2+		ann.rad.	
	99	46				E.C.(51%)		1.93					0.1360 ± 0.0001	73%
	99	46						2.18					0.2636 ± 0.0001	15%
	99	46											0.3867 ± 0.0001	2.8%
	99	46											0.3998 ± 0.0001	3.6%
	99	46											0.4103 ± 0.0002	1.3%
	99	46											0.4273 ± 0.0001	2.0%
	99	46											0.6504 ± 0.0005	1.3%
	99	46											0.6528 ± 0.0005	1.4%
	99	46											0.6531 ± 0.0003	2.6%
	99	46											0.6734 ± 0.0002	6.9%
	99	46											0.7866 ± 0.0002	3.3%
	99	46											0.8098 ± 0.0002	2.0%
	99	46											1.0134 ± 0.0003	1.4%
	99	46											1.0996 ± 0.0004	1.0%
	99	46											1.2564 ± 0.0004	1.0%
	99	46											1.3356 ± 0.0004	4.6%
	99	46											1.6805 ± 0.001	0.6%
	99	46											1.7176 ± 0.0005	0.6%
	99	46											2.2462 ± 0.0007	0.8%
	99	46											2.5363 ± 0.001	0.6%
	99	46											(0.2 - 2.85)	
$_{46}Pd^{100}$	100	46		99.908527	3.6 d.	E.C.	0.38				0+		0.03271 ± 0.0001	1.6 +
	100	46											0.0421 ± 0.0001	1.5
	100	46											0.0748 ± 0.0001	98
	100	46											0.0840 ± 0.0001	100
	100	46											0.1261 ± 0.0001	11
	100	46											0.1588 ± 0.0003	2
$_{46}Pd^{101}$	101	46		100.908287	8.4 h.	β+(5%)	1.98				5/2+		ann.rad.	
	101	46				E.C.(95%)							0.0244 ± 0.0001	4%
	101	46											0.2697 ± 0.0001	6%
	101	46											0.2963 ± 0.0001	19%
	101	46											0.5660 ± 0.00014	3.5%
	101	46											0.5904 ± 0.0001	12%
	101	46											0.7238 ± 0.0001	2%
	101	46											1.2020 ± 0.0001	1.5%
	101	46											1.2890 ± 0.0001	2.3%
$_{46}Pd^{102}$	102	46	1.02%	101.905634						3 b.	0+			
$_{46}Pd^{103}$	103	46		102.906114	16.97 d.	E.C.	0.572				5/2+		Rh k x-ray	64%
	103	46											0.03975 ± 0.0002	0.07 (D)
	103	46											0.3575 ± 0.0001	0.02
	103	46											0.4971 ± 0.0001	0.004
$_{46}Pd^{104}$	104	46	11.14%	103.904029							0+			
$_{46}Pd^{105}$	105	46	22.33%	104.905079						22 b.	5/2+	-0.642		
$_{46}Pd^{106}$	106	46	+27.33%	105.903478						(0.015 + 28)b.	0+			
$_{46}Pd^{107m}$	107	46			20.9 s.	I.T.	0.2149				11/2-		Pd k x-ray	16%
	107	46											0.2149 ± 0.0005	69%
$_{46}Pd^{107}$	107	46			$6.5x10^6$ y.	β-	0.033	0.033			5/2+			
$_{46}Pd^{108}$	108	46	26.46%	107.903895						(0.19 + 8) b.	0+			
$_{46}Pd^{109m}$	109	46			4.68 m.	I.T.	0.189				11/2-		Pd x-ray	
	109	46											0.18903 ± 0.0002	56%
$_{46}Pd^{109}$	109	46		108.905954	β-	1.116	1.028				5/2+		0.08803 ± 0.0002	3.6%
	109	46											0.31134 ± 0.0003	0.03%
	109	46											0.6024 ± 0.0001	0.08%
	109	46											0.6363 ± 0.0001	0.01%
	109	46											0.6472 ± 0.0001	0.02%
	109	46											0.7813 ± 0.0001	0.01%
	109	46											(0.08 - 1.0)weak	
$_{46}Pd^{110}$	110	46	11.72%	109.905167						0.02 b.	0+			
$_{46}Pd^{111m}$	111	46			5.5 h.	I.T.(73%)	0.172				11/2-		0.0704 ± 0.0001	8%

Isotope	A	Z	% Natural abundance	Atomic mass	Half-life	Decay mode	Decay energy (MeV)	Particle energy (MeV)	Particle intensity	Thermal neutron cross section	Spin (h/2π)	μ Nucl. mag. moment	Gamma-ray energy (MeV)	Gamma-ray intensity
	111	46				β-(27%)							0.1722 ± 0.0001	33%
	111	46											0.3913 ± 0.0003	5%
	111	46											0.4135 ± 0.0003	1.7%
	111	46											0.4155 ± 0.0003	1.5%
	111	46											0.5256 ± 0.0001	1.2%
	111	46											0.5750 ± 0.0001	3%
	111	46											0.6328 ± 0.0002	3%
	111	46											0.6942 ± 0.0001	2%
	111	46											0.7622 ± 0.0001	1.2%
	111	46											1.1159 ± 0.0002	1%
	111	46											1.2825 ± 0.0002	1%
	111	46											1.6911 ± 0.0002	1.2%
	111	46											(0.1 - 1.97)	
$_{46}Pd^{111}$	111	46		110.907660	22 m.	β-	2.20	2.12	95%		5/2+		0.0598 ± 0.0001	0.5%(D)
	111	46											0.2454 ± 0.0001	0.5%
	111	46											0.3761 ± 0.0001	0.4%
	111	46											0.5800 ± 0.0001	0.8%
	111	46											0.6504 ± 0.0001	0.55%
	111	46											1.3885 ± 0.0002	0.54%
	111	46											1.4590 ± 0.0003	0.56%
$_{46}Pd^{112}$	112	46		111.907323	21.03 h.	β-	0.29	0.28			0+		0.0185 ± 0.0005	27%
$_{46}Pd^{113m}$	113	46			89 s.	β-							0.0958 ± 0.0002	
$_{46}Pd^{113}$	113	46		112.910110	98 s.	β-	3.4				5/2+		0.0958 ± 0.0002	100 +
	113	46											0.2220 ± 0.0002	37
	113	46											0.4824 ± 0.0004	62
	113	46											0.5679 ± 0.0003	21
	113	46											0.6436 ± 0.0002	81
	113	46											0.7394 ± 0.0002	76
	113	46											0.8695 ± 0.0006	4.2
$_{46}Pd^{114}$	114	46		113.910310	2.48 m.	β-	1.5				0+		0.1266 ± 0.0002	3%
	114	46											0.2320 ± 0.0002	3.5%
	114	46											0.5582 ± 0.0002	12%(D)
	114	46											0.5760 ± 0.0002	9%(D)
$_{46}Pd^{115}$	115	46		114.913590	47 s.	β-	4.6						0.1255 ± 0.0003	64 +
	115	46											0.2554 ± 0.0003	59
	115	46											0.3040 ± 0.0002	32
	115	46											0.3428 ± 0.0002	100
	115	46											0.3606 ± 0.0003	28
	115	46											0.5944 ± 0.0003	34
$_{46}Pd^{116}$	116	46		115.914000	12.7 s.	β-	2.6						0.1015 ± 0.0002	8%
	116	46											0.1147 ± 0.0001	88%
	116	46											0.1778 ± 0.0002	12%
	116	46											0.2161 ± 0.0002	2%
	116	46											0.2795 ± 0.0003	6%
Ag		47		107.8682						63.6 b.				
$_{47}Ag^{96}$	96	47			5.1 s.	β+							ann.rad.	
	96	47				E.C.							0.1248	100 +
	96	47											0.4995	42
$_{47}Ag^{97}$	97	47		96.923890	19 s.	β+							ann.rad.	
	97	47				E.C.							0.6862 ± 0.0001	100 +
	97	47											1.2941 ± 0.0002	53
$_{47}Ag^{98}$	98	47		97.921560	46.7 s.	β+	8.6				5+		ann.rad.	
	98	47				E.C.							0.5711 ± 0.0002	59%
	98	47											0.6611 ± 0.0002	12%
	98	47											0.6786 ± 0.0002	88%
	98	47											0.8631 ± 0.0002	100%
$_{47}Ag^{99m}$	99	47			11 s.	I.T.(100%)					1/2-		Ag k x-ray	26%
	99	47											0.1636 ± 0.0003	37%
	99	47											0.3426 ± 0.0002	99%
$_{47}Ag^{99}$	99	47		98.917590	2.07 m.	β+(87%)	3.5				9/2+		ann.rad.	
	99	47				E.C.(13%)							0.2199 ± 0.0001	4%
	99	47											0.2645 ± 0.0001	63%
	99	47											0.4637 ± 0.0001	1%
	99	47											0.5682 ± 0.0001	3.8%
	99	47											0.5962 ± 0.0001	1.5%
	99	47											0.6360 ± 0.0001	1.4%
	99	47											0.6870 ± 0.0001	3.2%
	99	47											0.8056 ± 0.0001	12%
	99	47											0.8157 ± 0.0003	7%
	99	47											0.8323 ± 0.0001	13%
	99	47											0.8385 ± 0.0001	2%
	99	47											0.8640 ± 0.0001	4%
	99	47											0.9632 ± 0.0003	1%
	99	47											1.5319 ± 0.0001	4.5%
	99	47											1.5404 ± 0.0001	1.4%
	99	47											1.5853 ± 0.0005	1%
	99	47											1.8734 ± 0.0001	2.3%
	99	47											1.8810 ± 0.0002	1.3%
	99	47											1.9071 ± 0.0004	1.2%
	99	47											2.7085 ± 0.0004	0.9%
	99	47											3.1818 ± 0.0004	1.3%
	99	47											(0.2 - 3.5)	
$_{47}Ag^{100m}$	100	47			2.3 m.	β+					2+		ann.rad.	
	100	47				E.C.							0.6657 ± 0.0002	93%

TABLE OF THE ISOTOPES (Continued)

Isotope	A	Z	% Natural abundance	Atomic mass	Half-life	Decay mode	Decay energy (MeV)	Particle energy (MeV)	Particle intensity	Thermal neutron cross section	Spin (h/2π)	μ Nucl. mag. moment	Gamma-ray energy (MeV)	Gamma-ray intensity
	100	47											0.9222 ± 0.0002	10%
	100	47											1.5877 ± 0.0003	7%
	100	47											1.6941 ± 0.0003	13%
	100	47											1.8208 ± 0.0008	6%
	100	47											1.9560 ± 0.0007	5%
	100	47											2.0130 ± 0.001	1%
	100	47											2.1190 ± 0.0005	11%
$_{47}Ag^{100}$	100	47		99.916140	2.0 m.	β+	7.1	5.4			5+		ann.rad.	
	100	47				E.C.							0.2807 ± 0.0002	10%
	100	47											0.4503 ± 0.0002	19%
	100	47											0.6657 ± 0.0002	100%
	100	47											0.7309 ± 0.0002	8%
	100	47											0.7508 ± 0.0002	84%
	100	47											0.7732 ± 0.0002	23%
	100	47											0.8625 ± 0.0002	4%
	100	47											0.8905 ± 0.0002	3.6%
	100	47											0.9607 ± 0.0002	1.9%
	100	47											1.0538 ± 0.0002	13%
	100	47											1.1157 ± 0.0002	4.3%
	100	47											1.2607 ± 0.0005	5.2%
	100	47											1.2780 ± 0.0002	3.4%
	100	47											1.4053 ± 0.0002	3.6%
	100	47											1.5039 ± 0.0002	13.6%
	100	47											1.6860 ± 0.0003	8%
	100	47											1.7679 ± 0.0004	1%
	100	47											2.1190 ± 0.0005	2.9%
	100	47											2.2148 ± 0.0005	1.9%
$_{47}Ag^{101m}$	101	47			3.1 s.	I.T.	0.23				1/2-		Ag k x-ray	47%
	101	47											0.0981 ± 0.0002	62%
	101	47											0.1762 ± 0.0002	47%
$_{47}Ag^{101}$	101	47		100.912810	11.1 m.	β+(69%)	4.2	1.08			9/2+		ann.rad.	135%
	101	47				E.C.(31%)		1.56					0.2610 ± 0.0001	52%
	101	47						2.18					0.2747 ± 0.0002	2%
	101	47						2.73					0.3269 ± 0.0002	2%
	101	47						3.38					0.4392 ± 0.0003	3%
	101	47											0.5076 ± 0.0004	2%
	101	47											0.5433 ± 0.0002	2%
	101	47											0.5880 ± 0.0002	10%
	101	47											0.6673 ± 0.0001	10%
	101	47											0.6778 ± 0.0002	4%
	101	47											0.7347 ± 0.0002	3%
	101	47											0.8932 ± 0.0002	1%
	101	47											0.9383 ± 0.0002	1%
	101	47											0.9443 ± 0.0002	1%
	101	47											1.0936 ± 0.0002	2.6%
	101	47											1.1739 ± 0.0002	9%
	101	47											1.2053 ± 0.0002	2.6%
	101	47											2.0531 ± 0.0003	0.8%
	101	47											2.6991 ± 0.0003	0.4%
	101	47											(0.2 - 3.1)	
$_{47}Ag^{102m}$	102	47			7.8 m.	β+(38%)					2+	+4.14	ann.rad.	76%
	102	47				E.C.(13%)							0.5567 ± 0.0002	42%
	102	47				I.T.(49%)							0.9777 ± 0.0003	2.6%
	102	47											1.3878 ± 0.0003	2.6%
	102	47											1.4611 ± 0.0004	4.5%
	102	47											1.5348 ± 0.0004	2.6%
	102	47											1.5888 ± 0.0004	1.2%
	102	47											1.6923 ± 0.0004	2.2%
	102	47											1.8347 ± 0.0003	9.8%
	102	47											2.0178 ± 0.0004	2.8%
	102	47											2.0545 ± 0.0004	6.6%
	102	47											2.1594 ± 0.0004	5%
	102	47											2.6821 ± 0.0004	1.7%
	102	47											2.7165 ± 0.0004	1.8%
	102	47											3.2386 ± 0.0004	5%
$_{47}Ag^{102}$	102	47		101.911950	13 m.	β+(78%)	5.9	2.26			5+		ann.rad.	156%
	102	47				E.C.(22%)							0.5567 ± 0.0002	98%
	102	47											0.7194 ± 0.0002	58%
	102	47											0.8354 ± 0.0003	14%
	102	47											0.8657 ± 0.0003	3.7%
	102	47											0.8915 ± 0.0003	4%
	102	47											0.9777 ± 0.0003	2%
	102	47											1.2571 ± 0.0003	13%
	102	47											1.3057 ± 0.0004	2%
	102	47											1.4733 ± 0.0004	2.7%
	102	47											1.5227 ± 0.0004	2.7%
	102	47											1.5348 ± 0.0004	2.3%
	102	47											1.5558 ± 0.0004	2.6%
	102	47											1.5816 ± 0.0003	14%
	102	47											1.7446 ± 0.0003	17%
	102	47											1.8007 ± 0.0004	2.8%
	102	47											2.2429 ± 0.0005	1%
	102	47											2.6130 ± 0.0004	3.5%
	102	47											2.7269 ± 0.0005	1.4%
	102	47											3.3980 ± 0.0006	1.4%

Isotope	A	Z	% Natural abundance	Atomic mass	Half-life	Decay mode	Decay energy (MeV)	Particle energy (MeV)	Particle intensity	Thermal neutron cross section	Spin (h/2π)	μ Nucl. mag. moment	Gamma-ray energy (MeV)	Gamma-ray intensity
	102	47											3.4065 ± 0.0006	1.7%
$_{47}Ag^{103m}$	103	47			5.7 s.	I.T.	0.134				1/2-		Ag k x-ray	33%
	103	47											0.1344 ± 0.0001	21%
$_{47}Ag^{103}$	103	47		102.908980	66 m.	β+(28%)	2.67				7/2+	+4.47	ann.rad.	56%
	103	47				E.C.(72%)							0.1187 ± 0.0001	31%
	103	47											0.1482 ± 0.0001	28%
$_{47}Ag^{104m}$	104	47			33 m.	β+(43%)		2.70			2+	+3.7	ann.rad.	86%
	104	47				E.C.(24%)							0.5558 ± 0.0001	60%
	104	47				I.T.(33%)							0.7657 ± 0.0002	1%
	104	47											1.2388 ± 0.0003	2.6%
	104	47											1.3418 ± 0.0003	1%
	104	47											1.7208 ± 0.0004	1.4%
	104	47											2.1392 ± 0.0005	1%
	104	47											2.2767 ± 0.0004	1.6%
	104	47											2.7295 ± 0.0005	0.8%
	104	47											3.2136 ± 0.0005	1%
	104	47											3.4078 ± 0.0005	1%
	104	47											(0.5 - 3.4)	
$_{47}Ag^{104}$	104	47		103.908623	69 m.	β+(16%)	4.28	0.99			5+	+4.0	ann.rad.	32%
	104	47				E.C.(84%)							0.5558 ± 0.0001	92%
	104	47											0.6232 ± 0.0002	2.5%
	104	47											0.7405 ± 0.0002	7%
	104	47											0.7587 ± 0.0002	10%
	104	47											0.8579 ± 0.0002	10%
	104	47											0.8630 ± 0.0003	7%
	104	47											0.9080 ± 0.0003	4.5%
	104	47											0.9233 ± 0.0005	7%
	104	47											0.9259 ± 0.0005	12%
	104	47											0.9416 ± 0.0003	25%
	104	47											1.0753 ± 0.0003	2%
	104	47											1.2652 ± 0.0003	4.3%
	104	47											1.3418 ± 0.0003	7.3%
	104	47											1.5266 ± 0.0003	7.1%
	104	47											1.6258 ± 0.0003	5%
	104	47											1.7818 ± 0.0004	3%
	104	47											(0.18 - 2.27)	
$_{47}Ag^{105m}$	105	47			7.2 m.	I.T.(98%)	0.025				7/2+		Ag x-ray	1%
	105	47				E.C.(2%)							0.3063 ± 0.0001	0.2%
	105	47											0.3192 ± 0.0001	0.9%
	105	47											(0.1 - 1.0)weak	
$_{47}Ag^{105}$	105	47		104.906520	41.3 d.	E.C.	1.34				1/2-	0.1014	0.0640 ± 0.0001	11%
	105	47											0.2804 ± 0.0001	31%
	105	47											0.3192 ± 0.0001	4%
	105	47											0.3315 ± 0.0001	4%
	105	47											0.3445 ± 0.0001	42%
	105	47											0.3726 ± 0.0001	2%
	105	47											0.4434 ± 0.0001	11%
	105	47											0.6179 ± 0.0001	1%
	105	47											0.6445 ± 0.0001	10%
	105	47											0.6507 ± 0.0001	2%
	105	47											0.8075 ± 0.0001	1%
	105	47											1.0879 ± 0.0001	3.6%
$_{47}Ag^{106m}$	106	47			8.5 d.	E.C.					6+	3.71	Pd k x-ray	58%
	106	47											0.2217 ± 0.0001	6.6%
	106	47											0.2286 ± 0.0001	2%
	106	47											0.3910 ± 0.0001	4%
	106	47											0.4062 ± 0.0001	13%
	106	47											0.4296 ± 0.0001	13%
	106	47											0.4510 ± 0.0001	28%
	106	47											0.5118 ± 0.0001	88%
	106	47											0.6162 ± 0.0001	22%
	106	47											0.6802 ± 0.0001	2%
	106	47											0.7031 ± 0.0001	4%
	106	47											0.7173 ± 0.0001	29%
	106	47											0.7484 ± 0.0001	21%
	106	47											0.7932 ± 0.0001	6%
	106	47											0.8043 ± 0.0001	12%
	106	47											0.8084 ± 0.0001	4%
	106	47											0.8247 ± 0.0001	15%
	106	47											0.8478 ± 0.0001	4%
	106	47											1.0458 ± 0.0001	30%
	106	47											1.1280 ± 0.0001	12%
	106	47											1.1994 ± 0.0001	11%
	106	47											1.2229 ± 0.0001	7%
	106	47											1.5277 ± 0.0001	16%
	106	47											1.5724 ± 0.0002	7%
	106	47											1.8390 ± 0.0001	1%
$_{47}Ag^{106}$	106	47		105.906662	24.0 m.	β+(59%)	2.98		1.96		1+	+2.85	ann.rad.	120%
	106	47				E.C.(41%)							0.5119 ± 0.0001	17%
	106	47											0.6219 ± 0.0001	0.3%
	106	47											0.8735 ± 0.0001	0.2%
	106	47											1.0503 ± 0.0001	0.16%
$_{47}Ag^{107m}$	107	47			44.2 s.	I.T.	0.093				7/2+		Ag x-ray	
	107	47											0.0931 ± 0.0001	4.7%
$_{47}Ag^{107}$	107	47	51.84%	106.905092						(0.35 + 38)b.	1/2-	-0.1135		

Isotope	A	Z	% Natural abundance	Atomic mass	Half-life	Decay mode	Decay energy (MeV)	Particle energy (MeV)	Particle intensity	Thermal neutron cross section	Spin (h/2π)	μ Nucl. mag. moment	Gamma-ray energy (MeV)	Gamma-ray intensity
$_{47}Ag^{108m}$	108	47			1.3x10² y.	E.C.(92%)					6+	3.580	Ag k x-ray	11%
	108	47				I.T.(8%)	0.079						Pd k x-ray	54%
	108	47											0.0791 ± 0.0001	6.5%
	108	47											0.43392 ± 0.00004	91%
	108	47											0.61427 ± 0.00005	91%
	108	47											0.72290 ± 0.00005	91%
$_{47}Ag^{108}$	108	47		107.905952	2.42 m.	β-(97%)		1.02	1.7%		1+	+2.6884	ann.rad.	
	108	47				E.C.(2%)		1.65	96%				0.43392 ± 0.00004	0.5%
	108	47				β+(1%)		0.88	0.3%				0.61885 ± 0.00005	0.25%
	108	47											0.63298 ± 0.00003	1.75%
$_{47}Ag^{109m}$	109	47			39.8 s.	I.T.	0.088				7/2+		Ag k x-ray	28%
	109	47											0.0880 ± 0.0001	3.6%
$_{47}Ag^{109}$	109	47	48.16%	108.904757						(4.6 + 87) b.	1/2-	-0.1305		
$_{47}Ag^{110m}$	110	47			249.8 d.	β-(99%)				80 b.	6+	+3.607	0.4468 ± 0.0001	3.7%
	110	47				I.T.(1%)	0.1164						0.6203 ± 0.0001	2.8%
	110	47											0.65774 ± 0.00002	95%
	110	47											0.6776 ± 0.0001	11%
	110	47											0.6870 ± 0.0001	6.5%
	110	47											0.70667 ± 0.00002	16.7%
	110	47											0.7443 ± 0.0001	4.7%
	110	47											0.76393 ± 0.00002	22%
	110	47											0.81802 ± 0.00002	7.3%
	110	47											0.88467 ± 0.00002	73%
	110	47											0.93748 ± 0.00002	34%
	110	47											1.38427 ± 0.00003	24%
	110	47											1.47575 ± 0.00003	4%
	110	47											1.50501 ± 0.00003	13%
	110	47											1.56226 ± 0.00003	1.2%
$_{47}Ag^{110}$	110	47		109.906111	24.6 s.	β-	2.89	2.22	5%		1+	2.7271	0.65774 ± 0.00002	4.5%
	110	47						2.89	95%				0.8154 ± 0.0002	0.04%
	110	47											1.1257 ± 0.0001	0.015%
$_{47}Ag^{111m}$	111	47			65 s.	I.T.(99%)	0.059				7/2+		Ag k x-ray	16%
	111	47				β-(1%)							0.0598 ± 0.0001	0.5%
	111	47											0.1713 ± 0.0001	0.1%
	111	47											0.2454 ± 0.0001	0.5%
	111	47											0.6201 ± 0.0003	0.1%
$_{47}Ag^{111}$	111	47		110.905295	7.47 d.	β-	1.037	1.035		3 b.	1/2-	-0.146	0.2454 ± 0.0001	1.2%
	111	47											0.3421 ± 0.0001	6.7%
$_{47}Ag^{112}$	112	47		111.907010	3.14 h.	β-	3.96	3.94			2-	0.0547	0.6067 ± 0.0002	3%
	112	47											0.6174 ± 0.0002	42%
	112	47											0.6948 ± 0.0002	3%
	112	47											0.8512 ± 0.0002	1%
	112	47											1.3123 ± 0.0002	1%
	112	47											1.3877 ± 0.0002	5%
	112	47											1.6136 ± 0.0002	2.8%
	112	47											2.1062 ± 0.0002	2.4%
	112	47											2.5068 ± 0.0002	1%
	112	47											(0.4 - 2.9)	
$_{47}Ag^{113m}$	113	47			68 s.	I.T.(80%)	0.043				7/2+		0.1422 ± 0.0002	1.6%
	113	47				β-(20%)							0.2983 ± 0.0001	6%
	113	47											0.3161 ± 0.0001	10%
	113	47											0.3923 ± 0.0002	6%
	113	47											0.5838 ± 0.0003	2%
	113	47											0.7083 ± 0.0004	2%
$_{47}Ag^{113}$	113	47		112.906558	5.3 h.	β-	2.01	2.01			1/2-	0.159	0.2588 ± 0.0001	1.6%
	113	47											0.2986 ± 0.0001	10%
	113	47											0.3163 ± 0.0001	1.3%
	113	47											0.6723 ± 0.0001	0.9%
	113	47											0.6906 ± 0.0001	0.7%
$_{47}Ag^{114}$	114	47		113.908760	4.5 s.	β-	5.0	4.9			1+		0.5582 ± 0.0002	12%
	114	47											0.5760 ± 0.0002	1%
	114	47											1.3041 ± 0.0005	0.7%
	114	47											1.9946 ± 0.0005	1%
$_{47}Ag^{115m}$	115	47			18.7 s.	β-					7/2+		0.1134 ± 0.0002	11 +
	115	47											0.1315 ± 0.0002	89
	115	47											0.2288 ± 0.0002	100
	115	47											0.2753 ± 0.0002	6
	115	47											0.3887 ± 0.0002	46
	115	47											0.451 ± 0.001	2

TABLE OF THE ISOTOPES (Continued)

Isotope	A	Z	% Natural abundance	Atomic mass	Half-life	Decay mode	Decay energy (MeV)	Particle energy (MeV)	Particle intensity	Thermal neutron cross section	Spin (h/2π)	μ Nucl. mag. moment	Gamma-ray energy (MeV)	Gamma-ray intensity
	115	47											0.4734 ± 0.0005	2
$_{47}Ag^{115}$	115	47		114.908800	20.0 m.	β-	3.14				1/2-		0.1316 ± 0.0002	3%
	115	47											0.2128 ± 0.0001	4.4%
	115	47											0.2291 ± 0.0001	18%
	115	47											0.3261 ± 0.0001	2%
	115	47											0.3722 ± 0.0001	2%
	115	47											0.4727 ± 0.0001	4%
	115	47											0.6491 ± 0.0001	3%
	115	47											0.6981 ± 0.0001	2%
	115	47											1.5069 ± 0.0003	1.2%
	115	47											1.8416 ± 0.0003	1.8%
	115	47											1.9269 ± 0.0003	1.3%
	115	47											2.1132 ± 0.0003	2.8%
	115	47											(0.13 - 2.49)	
$_{47}Ag^{116m}$	116	47			10 s.	I.T.(2%)		3.2			5+		0.1027 ± 0.0002	2%
	116	47				β-(98%)							0.2549 ± 0.0003	7%
	116	47											0.2643 ± 0.0003	5%
	116	47											0.4577 ± 0.0005	2%
	116	47											0.5134 ± 0.0002	92%
	116	47											0.6670 ± 0.0003	8%
	116	47											0.6995 ± 0.0002	8%
	116	47											0.7055 ± 0.0002	61%
	116	47											0.7088 ± 0.0003	20%
	116	47											0.8068 ± 0.0001	16%
	116	47											0.9312 ± 0.0004	4%
	116	47											0.9744 ± 0.0006	2%
	116	47											1.0289 ± 0.0003	30%
	116	47											1.2130 ± 0.0003	6%
	116	47											1.4086 ± 0.0004	2%
$_{47}Ag^{116}$	116	47		115.911200	2.68 m.	β-	5.0				2-		0.5134 ± 0.0002	76%
	116	47											0.6399 ± 0.0003	3%
	116	47											0.6993 ± 0.0002	11%
	116	47											0.7058 ± 0.0002	2%
	116	47											0.8668 ± 0.0004	1.4%
	116	47											0.9935 ± 0.0005	1.7%
	116	47											1.1285 ± 0.0006	2.5%
	116	47											1.2126 ± 0.0006	7%
	116	47											1.3041 ± 0.0005	5.5%
	116	47											1.4018 ± 0.0005	1.2%
	116	47											1.4077 ± 0.0006	3%
	116	47											1.4371 ± 0.0006	1.6%
	116	47											1.5696 ± 0.0006	1%
	116	47											1.6046 ± 0.0006	2%
	116	47											1.6414 ± 0.0006	2%
	116	47											1.6910 ± 0.0006	1.7%
	116	47											2.0039 ± 0.0006	1.9%
	116	47											2.0908 ± 0.0006	1.1%
	116	47											2.1344 ± 0.0006	1.7%
	116	47											2.2463 ± 0.0006	2.4%
	116	47											2.2890 ± 0.0006	1.7%
	116	47											2.3487 ± 0.0006	1.1%
	116	47											2.4779 ± 0.0006	12%
	116	47											2.5009 ± 0.0006	1%
	116	47											2.6618 ± 0.0006	4.2%
	116	47											2.7032 ± 0.0006	1.8%
	116	47											2.8281 ± 0.0007	1.3%
	116	47											2.8341 ± 0.0007	2.5%
$_{47}Ag^{117m}$	117	47			5.3 s.	β-		3.3			7/2+		0.1354 ± 0.0001	46%
	117	47											0.2981 ± 0.0001	20%
	117	47											0.3221 ± 0.0001	7%
	117	47											0.3377 ± 0.0001	9%
	117	47											0.3868 ± 0.0001	38%
	117	47											0.5221 ± 0.0001	9%
	117	47											0.5578 ± 0.0001	2%
	117	47											0.6373 ± 0.0001	1%
	117	47											0.6846 ± 0.0001	7%
	117	47											0.7548 ± 0.0001	1%
	117	47											0.7863 ± 0.0001	2%
	117	47											0.8199 ± 0.0001	2%
	117	47											1.2204 ± 0.0002	1%
$_{47}Ag^{117}$	117	47		116.911700	73 s.	β-	4.17				1/2-		0.1354 ± 0.0001	23%
	117	47											0.1571 ± 0.0001	8%
	117	47											0.3072 ± 0.0001	2%
	117	47											0.3123 ± 0.0001	6%
	117	47											0.3377 ± 0.0001	10%
	117	47											0.4426 ± 0.0001	7%
	117	47											0.4677 ± 0.0001	2%
	117	47											0.5299 ± 0.0001	1%
	117	47											0.6651 ± 0.0001	1%
	117	47											0.7795 ± 0.0001	2%
	117	47											1.6090 ± 0.0001	4%
	117	47											1.6576 ± 0.0001	2%
	117	47											1.6962 ± 0.0001	1%
	117	47											1.7488 ± 0.0001	1%
	117	47											1.8544 ± 0.0001	2%

Isotope	A	Z	% Natural abundance	Atomic mass	Half-life	Decay mode	Decay energy (MeV)	Particle energy (MeV)	Particle intensity	Thermal neutron cross section	Spin (h/2π)	μ Nucl. mag. moment	Gamma-ray energy (MeV)	Gamma-ray intensity
	117	47											1.9955 ± 0.0001	4%
	117	47											2.0133 ± 0.0002	4%
	117	47											1.0354 ± 0.0002	1%
	117	47											2.0567 ± 0.0002	3%
	117	47											2.1921 ± 0.0002	2%
	117	47											2.2459 ± 0.0002	3%
	117	47											2.5141 ± 0.0002	1%
	117	47											2.8883 ± 0.0002	2.6%
$_{47}Ag^{118m}$	118	47			2.8 s.	β-(59%)							0.1277 ± 0.0001	7%
	118	47				I.T.(41%)	0.1277						0.4878 ± 0.0001	59%
	118	47											0.6771 ± 0.0001	58%
	118	47											0.7709 ± 0.0001	20%
	118	47											0.8083 ± 0.0001	6%
	118	47											1.0586 ± 0.0002	32%
$_{47}Ag^{118}$	118	47		117.914570	4.0 s.	β-	7.1						0.4878 ± 0.0001	81%
	110	47											0.6771 ± 0.0001	34%
	118	47											0.7815 ± 0.0001	6%
	118	47											0.7978 ± 0.0001	7%
	118	47											1.0586 ± 0.0002	2%
	118	47											1.2700 ± 0.0003	4%
	118	47											1.9390 ± 0.0003	4%
	118	47											2.1015 ± 0.0003	8%
	118	47											2.7792 ± 0.0003	6%
	118	47											2.7894 ± 0.0003	9%
	118	47											3.2259 ± 0.0003	11%
$_{47}Ag^{119}$	119	47		118.915630	2.1 s.	β-	5.4				7/2+		0.0674 ± 0.0001	6%
	119	47											0.1990 ± 0.0001	7%
	119	47											0.2134 ± 0.0001	7%
	119	47											0.3662 ± 0.0001	10%
	119	47											0.3706 ± 0.0001	3%
	119	47											0.3991 ± 0.0001	9%
	119	47											0.4071 ± 0.0001	2%
	119	47											0.4827 ± 0.0001	2%
	119	47											0.4979 ± 0.0001	3%
	119	47											0.5439 ± 0.0001	3%
	119	47											0.6264 ± 0.0002	11%
	119	47											0.6282 ± 0.0002	2%
	119	47											0.6544 ± 0.0003	3%
	119	47											0.6561 ± 0.0002	1%
	119	47											0.6604 ± 0.0001	6%
	119	47											0.7374 ± 0.0001	1%
	119	47											0.7792 ± 0.0001	4%
	119	47											0.8254 ± 0.0001	2%
	119	47											0.8514 ± 0.0001	2%
	119	47											1.0085 ± 0.0001	2%
	119	47											1.0265 ± 0.0001	6%
	119	47											1.1733 ± 0.0002	1%
	119	47											1.3748 ± 0.0002	1%
	119	47											1.5266 ± 0.0002	1%
	119	47											1.8983 ± 0.0002	2%
	119	47											2.0282 ± 0.0002	1%
	119	47											2.0607 ± 0.0005	4%
	119	47											2.4709 ± 0.0004	1%
	119	47											2.7864 ± 0.0002	2%
	119	47											2.9518 ± 0.0003	1.5%
$_{47}Ag^{120m}$	120	47			0.32 s.	β-							0.2030 ± 0.0002	
	120	47				I.T.							0.5059 ± 0.0002	
	120	47											0.6978 ± 0.0002	
	120	47											0.8300 ± 0.0002	
	120	47											0.9258 ± 0.0002	
$_{47}Ag^{120}$	120	47		119.918650	1.2 s.	β-	8.2						0.5059 ± 0.0002	
	120	47											0.6978 ± 0.0002	
	120	47											0.8171 ± 0.0002	
	120	47											1.3231 ± 0.0002	
$_{47}Ag^{121}$	121	47		120.919970	0.8 s.	β-	6.4						0.1150 ± 0.0001	21 +
	121	47											0.1464 ± 0.0004	2
	121	47											0.1785 ± 0.0001	10
	121	47											2.2030 ± 0.0003	2
	121	47											0.2736 ± 0.0001	8
	121	47											0.3148 ± 0.0001	100
	121	47											0.3537 ± 0.0001	57
	121	47											0.3622 ± 0.0001	11
	121	47											0.3696 ± 0.0001	17
	121	47											0.3720 ± 0.0001	9
	121	47											0.4306 ± 0.0001	12
	121	47											0.5007 ± 0.0001	24
	121	47											0.8020 ± 0.0002	5
	121	47											0.8172 ± 0.0002	11
	121	47											1.1570 ± 0.0002	8
	121	47											1.1705 ± 0.0003	6
	121	47											1.1959 ± 0.0002	15
	121	47											1.3711 ± 0.0002	4
	121	47											1.5105 ± 0.0002	17
	121	47											1.8120 ± 0.0003	5
	121	47											2.2052 ± 0.0005	7

Isotope	A	Z	% Natural abundance	Atomic mass	Half-life	Decay mode	Decay energy (MeV)	Particle energy (MeV)	Particle intensity	Thermal neutron cross section	Spin (h/2π)	μ Nucl. mag. moment	Gamma-ray energy (MeV)	Gamma-ray intensity
	121	47											2.5194 ± 0.0005	4
	121	47											(0.11 - 2.5)many	
$_{47}Ag^{122}$	122	47			1.5 s.	β-							0.5695 ± 0.0001	96%
	122	47				n							0.6502 ± 0.0001	20%
	122	47											0.7597 ± 0.0001	32%
	122	47											0.7884 ± 0.0003	12%
	122	47											1.3678 ± 0.0005	4%
	122	47											1.4231 ± 0.0009	3%
Cd		48		112.41						2.45 b.				
$_{48}Cd^{99}$	99	48		98.924860	16 s.	β+,E.C.							ann.rad.	
$_{48}Cd^{100}$	100	48		99.920230	1.1 m.	β+,E.C.	≈4.0						ann.rad.	
	100	48											0.0935 ± 0.0005	
	100	48											0.1238 ± 0.0005	
	100	48											0.1388 ± 0.0005	
	100	48											0.1781 ± 0.0005	
	100	48											0.2198 ± 0.0005	
	100	48											0.3676 ± 0.0005	
	100	48											0.4275 ± 0.0005	
	100	48											0.5670 ± 0.0005	
	100	48											0.9353 ± 0.0005	
$_{48}Cd^{101}$	101	48		100.918740	1.2 m.	β+(83%)	5.5				5/2+		In k x-ray	
	101	48				E.C.(17%)							0.0985 ± 0.0002	47%
	101	48											0.3089 ± 0.0002	2%
	101	48											0.5234 ± 0.0002	5%
	101	48											0.6379 ± 0.0002	2%
	101	48											0.6869 ± 0.0002	4%
	101	48											0.7058 ± 0.0002	2%
	101	48											0.9247 ± 0.0002	7%
	101	48											1.0226 ± 0.0002	2%
	101	48											1.1873 ± 0.0002	5%
	101	48											1.2030 ± 0.0002	5%
	101	48											1.2589 ± 0.0002	8%
	101	48											1.3318 ± 0.0002	2%
	101	48											1.4171 ± 0.0002	6%
	101	48											1.6314 ± 0.0002	2%
	101	48											1.6909 ± 0.0002	3%
	101	48											1.6967 ± 0.0002	4%
	101	48											1.7225 ± 0.0002	11%
	101	48											1.8597 ± 0.0002	4%
	101	48											1.9609 ± 0.0002	2%
	101	48											1.9902 ± 0.0002	1.4%
	101	48											2.1300 ± 0.0002	1%
	101	48											2.8419 ± 0.0002	2%
$_{48}Cd^{102}$	102	48		101.914440	5.5 m.	β+(27%)	2.4				0+		ann.rad.	
	102	48				E.C.(73%)							0.0974 ± 0.0002	3%
	102	48											0.1160 ± 0.0002	6%
	102	48											0.1204 ± 0.0002	2%
	102	48											0.2133 ± 0.0002	4%
	102	48											0.3603 ± 0.0002	4%
	102	48											0.4148 ± 0.0002	7%
	102	48											0.4810 ± 0.0002	61%
	102	48											0.5051 ± 0.0002	9%
	102	48											0.6757 ± 0.0002	4%
	102	48											1.0366 ± 0.0002	12%
	102	48											1.3598 ± 0.0002	5%
$_{48}Cd^{103}$	103	48		102.913451	7.7 m.	β+(33%)	4.17				5/2+		ann.rad.	
	103	48				E.C.(67%)							Ag k x-ray	
	103	48											0.1344 ± 0.0001	3%
	103	48											0.3870 ± 0.0001	3%
	103	48											0.5630 ± 0.0004	1.7%
	103	48											0.6262 ± 0.0004	1.8%
	103	48											0.9631 ± 0.0004	2%
	103	48											1.0799 ± 0.0001	5.7%
	103	48											1.0993 ± 0.0001	1.8%
	103	48											1.3117 ± 0.0001	1.9%
	103	48											1.4487 ± 0.0001	5.8%
	103	48											1.4618 ± 0.0001	12%
	103	48											1.4763 ± 0.0001	2%
	103	48											1.7485 ± 0.0001	1.5%
	103	48											1.8220 ± 0.0001	1.1%
	103	48											1.8342 ± 0.0001	1.0%
	103	48											1.8800 ± 0.0001	3.5%
	103	48											1.9302 ± 0.0001	2.0%
	103	48											2.0119 ± 0.0001	1.3%
	103	48											2.0225 ± 0.0001	1.1%
	103	48											2.1330 ± 0.0002	2%
	103	48											2.1995 ± 0.0002	1.5%
	103	48											2.2451 ± 0.0002	1.2%
	103	48											2.3737 ± 0.0002	1.6%
	103	48											2.4011 ± 0.0002	1.2%
	103	48											2.6814 ± 0.0003	1.5%
	103	48											(0.1 - 2.8)	
$_{48}Cd^{104}$	104	48		103.909851	58 m.	E.C.	1/40				0+		Ag k x-ray	
	104	48											0.0666 ± 0.0002	2.2%
	104	48											0.0835 ± 0.0002	47%

TABLE OF THE ISOTOPES (Continued)

Isotope	A	Z	% Natural abundance	Atomic mass	Half-life	Decay mode	Decay energy (MeV)	Particle energy (MeV)	Particle intensity	Thermal neutron cross section	Spin (h/2π)	μ Nucl. mag. moment	Gamma-ray energy (MeV)	Gamma-ray intensity
	104	48											0.5590 ± 0.0002	6.7%
	104	48											0.6257 ± 0.0002	2.1%
	104	48											0.7093 ± 0.0002	20%
$_{48}Cd^{105}$	105	48		104.909459	55.3 m.	β+(26%)	2.74	1.69			5/2+		Ag k x-ray	
	105	48				E.C.(74%)							0.3469 ± 0.0001	4%
	105	48											0.4332 ± 0.0001	3%
	105	48											0.6072 ± 0.0001	4%
	105	48											0.6485 ± 0.0001	1.6%
	105	48											0.9341 ± 0.0001	1.3%
	105	48											0.9618 ± 0.0001	4.7%
	105	48											1.0716 ± 0.0001	1.3%
	105	48											1.3025 ± 0.0001	4%
	105	48											1.3885 ± 0.0001	2.7%
	105	48											1.4161 ± 0.0005	1.6%
	105	48											1.5578 ± 0.0001	2%
	105	40											1.6358 ± 0.0001	1%
	105	48											1.6653 ± 0.0001	1.3%
	105	48											1.6933 ± 0.0001	3.5%
	105	48											1.8975 ± 0.0001	1.4%
	105	48											1.9331 ± 0.0001	1.6%
	105	48											2.2728 ± 0.0002	1.0%
	105	48											2.3333 ± 0.0001	2%
	105	48											(0.25 - 2.4)	
$_{48}Cd^{106}$	106	48	1.25%	105.906461						1 b.	0+			
$_{48}Cd^{107}$	107	48		106.906613	6.5 h.	E.C. (99+%)	1.42				5/2+	-0.6144	Ag k x-ray	
	107	48				β+							0.0931 ± 0.0001	5%
	107	48											0.7965 ± 0.0001	0.06%
	107	48											0.8289 ± 0.0001	0.16%
$_{48}Cd^{108}$	108	48	0.89%	107.904176						1.1 b.	0+			
$_{48}Cd^{109}$	109	48		108.904953	462.3 d.	E.C.	0.184			700 b.	5/2+	-0.8270	Ag k x-ray	
	109	48											0.08804 ± 0.00008	3.6%
$_{48}Cd^{110}$	110	48	122.49%	109.903005						(0.1 + 11) b.	0+			
$_{48}Cd^{111m}$	111	48			48.7 m.	I.T.					11/2-		Cd k x-ray	
	111	48											0.15082 ± 0.00001	29%
	111	48											0.24539 ± 0.00002	94%
$_{48}Cd^{111}$	111	48	12.80%	110.904182						24 b.	1/2+	-0.5943		
$_{48}Cd^{112}$	112	48	24.13%	111.902758						2.2 b.	0+			
$_{48}Cd^{113m}$	113	48			13.7y.	β-(99.9%)	0.59	0.59	99.9%		11/2-	-1.087	0.2637 ± 0.0003	0.02%
$_{48}Cd^{113}$	113	48	12.22%	112.904400						2x10[4]b.	1/2+	-0.6217		
$_{48}Cd^{114}$	114	48	28.73%	113.903357						(0.04 + 0.3)b.	0+			
$_{48}Cd^{115m}$	115	48			44.6 d.	β-	1.614	0.68	1.6%		11/2-	-1.042	0.48450 ± 0.0005	0.3%
	115	48						1.62	97%				0.93381 ± 0.0003	2%
	115	48											1.13261 ± 0.00004	0.1%
	115	48											1.29064 ± 0.00004	0.9%
$_{48}Cd^{115}$	115	48		114.905430	53.5 h.	β-	1.441	0.58	42%		1/2+	-0.648	0.23141 ± 0.00003	0.7%
	115	48						1.11	58%				0.26085 ± 0.00003	1.9%
	115	48											0.33624 ± 0.00003	50%(D)
	115	48											0.49227 ± 0.00003	8.5%
	115	48											0.52780 ± 0.00003	29%
$_{48}Cd^{116}$	116	48	7.49%	115.904754						(0.2 + 0.8)b.	0+			
$_{48}Cd^{117m}$	117	48			3.4 h.	β-		0.67			11/2-		0.1586 ± 0.0002	90%(D)
	117	48											0.3669 ± 0.0001	3%
	117	48											0.4609 ± 0.0001	1.6%
	117	48											0.5529 ± 0.0002	100%
	117	48											0.5644 ± 0.0001	15%
	117	48											0.6318 ± 0.0001	2.8%
	117	48											0.7127 ± 0.0001	1%
	117	48											0.7481 ± 0.0001	4%
	117	48											0.8604 ± 0.0001	8%
	117	48											0.9314 ± 0.0001	3.6%
	117	48											1.0291 ± 0.0001	12%
	117	48											1.0660 ± 0.0001	23%
	117	48											1.2346 ± 0.0001	11%
	117	48											1.3393 ± 0.0005	2%
	117	48											1.3655 ± 0.0001	1.5%
	117	48											1.4329 ± 0.0001	13%
	117	48											1.9973 ± 0.0001	26%
	117	48											2.0964 ± 0.0001	7.4%
	117	48											2.3228 ± 0.0001	7.9%
	117	48											2.4174 ± 0.0001	1%
$_{48}Cd^{117}$	117	48		116.907228	2.49 h.	β-	2.53	0.67	51%		1/2+		0.2209 ± 0.0001	1.2%
	117	48						1.29	10%				0.2733 ± 0.0001	29%
	117	48											0.3445 ± 0.0001	18%
	117	48											0.4342 ± 0.0001	10%

Isotope	A	Z	% Natural abundance	Atomic mass	Half-life	Decay mode	Decay energy (MeV)	Particle energy (MeV)	Particle intensity	Thermal neutron cross section	Spin (h/2π)	μ Nucl. mag. moment	Gamma-ray energy (MeV)	Gamma-ray intensity
	117	48											0.8318 ± 0.0001	2.3%
	117	48											0.8807 ± 0.0001	4%
	117	48											1.0517 ± 0.0001	3.8%
	117	48											1.1166 ± 0.0001	1%
	117	48											1.1424 ± 0.0001	1.7%
	117	48											1.2479 ± 0.0001	1.2%
	117	48											1.2600 ± 0.0001	1.1%
	117	48											1.3033 ± 0.0001	18%
	117	48											1.3376 ± 0.0001	1.6%
	117	48											1.4087 ± 0.0001	1.3%
	117	48											1.5622 ± 0.0001	1.4%
	117	48											1.5766 ± 0.0001	11%
	117	48											1.7069 ± 0.0001	1%
	117	48											1.7231 ± 0.0001	2%
$_{48}$Cd^{118}	118	48		117.906914	50.3 m.	β-	0.75	0.75			0+			
$_{48}$Cd^{119m}	119	48			2.20 m.	β-					11/2-		0.1056 ± 0.0001	3.3%
	119	48											0.3603 ± 0.0002	1%
	119	48											0.4115 ± 0.0003	2%
	119	48											0.4224 ± 0.0001	10%
	119	48											0.5850 ± 0.0001	4.8%
	119	48											0.6330 ± 0.0003	1.5%
	119	48											0.7090 ± 0.0005	1.2%
	119	48											0.7208 ± 0.0002	18%
	119	48											0.8177 ± 0.0001	1.3%
	119	48											0.9025 ± 0.0005	1.1%
	119	48											0.9232 ± 0.0001	6.9%
	119	48											0.9963 ± 0.0002	2%
	119	48											1.0250 ± 0.0001	25%
	119	48											1.1019 ± 0.0001	9.8%
	119	48											1.1855 ± 0.0002	2.9%
	119	48											1.2037 ± 0.0001	13%
	119	48											1.3441 ± 0.0002	6.8%
	119	48											1.3608 ± 0.0002	1.5%
	119	48											1.3641 ± 0.0002	5.2%
	119	48											1.3771 ± 0.001	1.2%
	119	48											1.4363 ± 0.0003	5.2%
	119	48											1.4742 ± 0.0004	2%
	119	48											1.6685 ± 0.0002	3%
	119	48											1.7019 ± 0.0002	2.8%
	119	48											1.7729 ± 0.0005	1.3%
	119	48											1.9607 ± 0.0006	1.1%
	119	48											2.0213 ± 0.0002	22%
	119	48											2.1043 ± 0.0002	6%
	119	48											2.4226 ± 0.0003	4.5%
	119	48											2.52033 ± 0.0003	1.1%
$_{48}$Cd^{119}	119	48		118.909890	2.69 m.	β-	3.79	3.5			1/2+		0.1340 ± 0.0001	8%
	119	48											0.2929 ± 0.0001	40%
	119	48											0.3429 ± 0.0001	19%
	119	48											0.4462 ± 0.0003	2%
	119	48											0.7730 ± 0.0002	1.8%
	119	48											0.7847 ± 0.0003	1%
	119	48											0.9413 ± 0.0001	3%
	119	48											1.0184 ± 0.0002	1%
	119	48											1.0503 ± 0.0001	7%
	119	48											1.1326 ± 0.0002	1%
	119	48											1.2878 ± 0.0001	3%
	119	48											1.3169 ± 0.0002	10%
	119	48											1.6097 ± 0.0001	12%
	119	48											1.7140 ± 0.0002	2.5%
	119	48											1.7338 ± 0.0001	9%
	119	48											1.7637 ± 0.0001	10%
	119	48											2.0266 ± 0.0004	1.5%
	119	48											2.0565 ± 0.0003	2.5%
	119	48											2.3564 ± 0.0003	7%
	119	48											2.5483 ± 0.0001	2%
$_{48}$Cd^{120}	120	48		119.909852	50.8 s.	β-	1.8	1.5			0+			
$_{48}$Cd^{121m}	121	48			8 s.	β-					11/2-		0.1008 ± 0.0001	3%
	121	48											0.4201 ± 0.0001	4%
	121	48											0.4471 ± 0.0001	3%
	121	48											0.5722 ± 0.0001	2%
	121	48											0.9525 ± 0.0001	5%
	121	48											0.9878 ± 0.0001	14%
	121	48											1.0209 ± 0.0001	19%
	121	48											1.0695 ± 0.0001	2%
	121	48											1.1393 ± 0.0001	6%
	121	48											1.1815 ± 0.0001	12%
	121	48											1.2713 ± 0.0001	3%
	121	48											1.3367 ± 0.0001	2%
	121	48											1.3820 ± 0.0001	3%
	121	48											1.4569 ± 0.0001	1.8%
	121	48											1.4675 ± 0.0001	3%
	121	48											1.4873 ± 0.0002	2%
	121	48											1.5041 ± 0.0001	4%
	121	48											2.0594 ± 0.0001	21%
	121	48											2.1148 ± 0.0001	2%

Isotope	A	Z	% Natural abundance	Atomic mass	Half-life	Decay mode	Decay energy (MeV)	Particle energy (MeV)	Particle intensity	Thermal neutron cross section	Spin (h/2π)	μ Nucl. mag. moment	Gamma-ray energy (MeV)	Gamma-ray intensity
	121	48											2.2918 ± 0.0001	2%
	121	48											2.3319 ± 0.0001	2.8%
	121	48											2.3648 ± 0.0001	8%
	121	48											2.3698 ± 0.0002	1%
	121	48											2.4550 ± 0.0001	4.8%
	121	48											2.5108 ± 0.0001	2.8%
	121	48											2.5623 ± 0.0001	3.7%
$_{48}Cd^{121}$	121	48		120.913100	13.5 s.		β-	5.0			(3/2 +		0.2102 ± 0.0001	3%
	121	48											0.3242 ± 0.0001	49%
	121	48											0.3492 ± 0.0001	13%
	121	48											0.4025 ± 0.0001	4%
	121	48											1.4411 ± 0.0002	2%
	121	48											0.5947 ± 0.0002	2%
	121	48											0.6506 ± 0.0002	4%
	121	48											0.6736 ± 0.0002	3%
	121	48											0.7661 ± 0.0001	6%
	121	48											0.9098 ± 0.0002	2%
	121	48											0.9786 ± 0.0003	2%
	121	48											0.9878 ± 0.0001	2%
	121	48											1.0403 ± 0.0001	18%
	121	48											1.0960 ± 0.0002	6%
	121	48											1.1499 ± 0.0002	2%
	121	48											1.2774 ± 0.0003	3%
	121	48											1.2969 ± 0.0001	4%
	121	48											1.3152 ± 0.0001	7%
	121	48											1.3236 ± 0.0003	1%
	121	48											1.3279 ± 0.0003	2%
	121	48											1.4512 ± 0.0002	1%
	121	48											1.5841 ± 0.0001	5%
	121	48											1.6271 ± 0.0002	2%
	121	48											1.6475 ± 0.0002	4%
	121	48											1.6618 ± 0.0002	1.6%
	121	48											1.6989 ± 0.0001	5.7%
	121	48											1.8226 ± 0.0002	1.6%
	121	48											1.8348 ± 0.0001	3%
	121	48											1.8540 ± 0.0001	3%
	121	48											1.8853 ± 0.0002	2%
$_{48}Cd^{122}$	122	48		121.913500	5.8 s.	β-	3.0				0+			
$_{48}Cd^{124}$	124	48			0.9 s.	β-					0+		0.0365 ± 0.0001	5%
	124	48											0.0628 ± 0.0001	23%
	124	48											0.1433 ± 0.0001	13%
	124	48											0.1799 ± 0.0001	50%
In		49		114.82						194 b.				
$_{49}In^{102}$	102	49		101.92440	24 s.	E.C.	9.2						0.1566	10%
	102	49											0.3965	12%
	102	49											0.5930	30%
	102	49											0.7768	100%
	102	49											0.8614	96%
	102	49											0.9237	10%
$_{49}In^{103}$	103	49		102.920110	1.1 m.	β+.E.C.	6.2				9/2+		ann.rad.	
	103	49											0.1879 ± 0.0001	100 +
	103	49											0.2020 ± 0.0001	19
	103	49											0.6995 ± 0.001	10
	103	49											0.7200 ± 0.001	32
	103	49											0.7399	19
$_{49}In^{104}$	104	49		103.918440	1.82 m.	β+.E.C.	≈8.0				(6+)		ann.rad.	
	104	49											0.3212 ± 0.0002	3.5%
	104	49											0.4739 ± 0.0002	5%
	104	49											0.5026 ± 0.0008	4%
	104	49											0.5330 ± 0.0003	3%
	104	49											0.6222 ± 0.0005	12%
	104	49											0.6580 ± 0.0002	100%
	104	49											0.8341 ± 0.0003	100%
	104	49											0.8781 ± 0.0002	28%
	104	49											0.9433 ± 0.0006	17%
	104	49											1.0002 ± 0.0006	10%
	104	49											1.1249 ± 0.0003	2%
	104	49											1.2816 ± 0.0003	2.5%
	104	49											1.7023 ± 0.0003	1.2%
	104	49											2.0062 ± 0.0003	2%
$_{49}In^{105m}$	105	49			43 s.	I.T.					1/2-		In k x-ray	
	105	49											0.6740 ± 0.0001	94%
$_{49}In^{105}$	105	49		104.914558	4.9 m.	β+.E.C.					9/2+		0.1310 ± 0.0001	43%
	105	49											0.1956 ± 0.0001	5%
	105	49											0.2282 ± 0.0003	1%
	105	49											0.2600 ± 0.0001	14%
	105	49											0.4735 ± 0.0003	1%
	105	49											0.5694 ± 0.0003	1.8%
	105	49											0.6038 ± 0.0001	10%
	105	49											0.6394 ± 0.0002	4.7%
	105	49											0.6680 ± 0.0001	7.6%
	105	49											0.7017 ± 0.0003	1%
	105	49											0.7702 ± 0.0003	1.6%
	105	49											0.8323 ± 0.0001	6.9%

Isotope	A	Z	% Natural abundance	Atomic mass	Half-life	Decay mode	Decay energy (MeV)	Particle energy (MeV)	Particle intensity	Thermal neutron cross section	Spin (h/2π)	μ Nucl. mag. moment	Gamma-ray energy (MeV)	Gamma-ray intensity
	108	49											1.2996 ± 0.0002	16%
	108	49											1.4862 ± 0.0003	4%
	108	49											1.6066 ± 0.0003	8%
$_{49}In^{109m}$	109	49			1.3 m.	I.T.	0.649				1/2-		In k x-ray	3.7%
	109	49											0.6498 ± 0.0002	94%
$_{49}In^{109}$	109	49		108.907133	4.2 h.	β+(8%)	2.03	0.79			9/2+	+5.53	ann.rad.	
	109	49				E.C.(92%)							Cd k x-ray	60%
	109	49											0.2035 ± 0.0002	73%
	109	49											0.3475 ± 0.0003	2%
	109	49											0.4262 ± 0.0003	4%
	109	49											0.6136 ± 0.0004	2.5%
	109	49											0.6235 ± 0.0004	6%
	109	49											0.6498 ± 0.0004	3%
	109	49											1.1491 ± 0.0006	4.3%
	109	49											1.6223 ± 0.0008	2.1%
$_{49}In^{110m}$	110	49			4.9 h.	E.C.					7+	10.5	Cd k x-ray	60%
	110	49											0.4611 ± 0.0001	2.2%
	110	49											0.4618 ± 0.0001	4.6%
	110	49											0.5819 ± 0.0001	8.4%
	110	49											0.5842 ± 0.0001	6.4%
	110	49											0.6417 ± 0.0001	26%
	110	49											0.6577 ± 0.0001	97%
	110	49											0.6776 ± 0.0004	4%
	110	49											0.7074 ± 0.0001	29%
	110	49											0.7599 ± 0.0001	3%
	110	49											0.8180 ± 0.0001	2%
	110	49											0.8447 ± 0.0001	3%
	110	49											0.8847 ± 0.0001	92%
	110	49											0.9375 ± 0.0001	67%
	110	49											0.9972 ± 0.0001	10%
	110	49											1.1174 ± 0.0001	4%
	110	49											1.4758 ± 0.0001	1.2%
	110	49											(0.1 - 1.98)	
$_{49}In^{110}$	110	49		109.907230	69 m.	β+(62%)	3.94	2.26			2+		ann.rad.	
	110	49				E.C.(38%)							Cd k x-ray	23%
	110	49											0.6577 ± 0.0001	98%
	110	49											1.1258 ± 0.0001	1%
	110	49											2.1295 ± 0.0001	2%
	110	49											2.2115 ± 0.0001	1.8%
	110	49											2.3175 ± 0.0001	1.3%
	110	49											(0.6 - 3.6)	
$_{49}In^{111m}$	111	49			7.7 m.	I.T.	0.357				1/2-	+5.53	In k x-ray	7.4%
	111	49											0.5372 ± 0.0001	87%
$_{49}In^{111}$	111	49		110.905109	2.806 d.	E.C.							Cd k x-ray	68%
	111	49											0.1712 ± 0.0001	94%
	111	49											0.2453 ± 0.0001	90%
$_{49}In^{112m}$	112	49			20.9 m.	I.T.	0.155				4+		In k x-ray	47%
	112	49											0.1555 ± 0.0002	13%
$_{49}In^{112}$	112	49		111.905536	14.4 m.	β+(22%)	2.588				1+		ann.rad.	
	112	49				E.C.(34%)							Cd k x-ray	20%
	112	49											0.6064 ± 0.0001	1.2%
	112	49											0.6171 ± 0.0001	5%
	112	49											0.8509 ± 0.0002	0.2%
	112	49											1.2531 ± 0.0002	0.2%
$_{49}In^{113m}$	113	49			1.657 h.	I.T.	0.3917				1/2-	-0.210	In k x-ray	20%
	113	49											0.3917 ± 0.0001	64%
$_{49}In^{113}$	113	49	4.3%	112.904061						(8 + 3.9)b.	9/2+	+5.523		
$_{49}In^{114m}$	114	49			49.51 d.	I.T.(97%)	0.190				5+	+4.7	In k x-ray	28%
	114	49				E.C.(3%)	1.6						0.19027 ± 0.00003	16%
	114	49											0.55843 ± 0.00003	3.4%
	114	49											0.72524 ± 0.00003	3.4%
	114	49											1.29983 ± 0.00007	0.16%
$_{49}In^{114}$	114	49		113.904916	71.9 s.	β-(97%)	1.986	1.986			1+	+1.7	Cd k x-ray	2%
	114	49				E.C.(3%)	1.452						0.5584 ± 0.0002	0.07%
	114	49											0.5727 ± 0.0001	0.004%
	114	49											1.2998 ± 0.0001	0.14%
$_{49}In^{115m}$	115	49			4.486 h.	I.T.(95%)	0.336	0.83			1/2-	-0.255	In k x-ray	28%
	115	49				β-(5%)							0.3362 ± 0.0001	46%
	115	49											0.4974 ± 0.0001	0.05%
$_{49}In^{115}$	115	49		114.903880	4.4x10^{14}y.	β-	0.496	1348		(87+75+41)b.	9/2+	+5.534		
$_{49}In^{116m2}$	116	49			2.16 s.	I.T.	0.162				8-		In k x-ray	28%
	116	49											0.1624 ± 0.0001	37%
$_{49}In^{116m1}$	116	49			54.1 m.	β-					5+	+4.3	0.13792 ± 0.0003	3%
	116	49											0.41688 ± 0.00002	29%
	116	49											0.81865 ± 0.00002	11%
	116	49											1.09723 ± 0.00002	56%
	116	49											1.29349 ± 0.00005	84%

Isotope	A	Z	% Natural abundance	Atomic mass	Half-life	Decay mode	Decay energy (MeV)	Particle energy (MeV)	Particle intensity	Thermal neutron cross section	Spin (h/2π)	μ Nucl. mag. moment	Gamma-ray energy (MeV)	Gamma-ray intensity
	105	49											0.8554 ± 0.0003	1%
	105	49											0.9633 ± 0.0003	1%
	105	49											1.1396 ± 0.0003	1.3%
	105	49											1.1903 ± 0.0003	1.2%
	105	49											1.2560 ± 0.0003	2.4%
	105	49											1.3870 ± 0.0002	5.1%
$_{49}In^{106m}$	106	49			6.2 m.	β+(85%)					3+		ann.rad.	
	106	49				E.C.(15%)							0.6326 ± 0.0001	92%
	106	49											0.8611 ± 0.0001	11%
	106	49											1.0838 ± 0.0002	3%
	106	49											1.1626 ± 0.0001	1.6%
	106	49											1.6209 ± 0.0001	4.9%
	106	49											1.7164 ± 0.0001	19%
	106	49											1.7379 ± 0.0003	1.5%
	106	49											1.9336 ± 0.0001	8.3%
	106	49											1.9975 ± 0.0003	2%
	106	49											2.0873 ± 0.0002	2%
	106	49											2.2569 ± 0.0002	5%
	106	49											2.8621 ± 0.0005	1.5%
	106	49											2.9182 ± 0.0003	4%
	106	49											3.4945 ± 0.0001	2%
$_{49}In^{106}$	106	49		105.913490	5.3 m.	β+(65%)					6+		ann.rad.	
	106	49				E.C.(35%)							0.2259 ± 0.0003	7%
	106	49											0.4331 ± 0.0003	2.3%
	106	49											0.5246 ± 0.0005	2%
	106	49											0.5412 ± 0.0002	13%
	106	49											0.5524 ± 0.0001	25%
	106	49											0.5586 ± 0.0005	2%
	106	49											0.5925 ± 0.0005	3%
	106	49											0.6107 ± 0.0002	4%
	106	49											0.6232 ± 0.0005	2%
	106	49											0.6327 ± 0.0001	100%
	106	49											0.7535 ± 0.0005	2%
	106	49											0.8368 ± 0.0005	2%
	106	49											0.8611 ± 0.0001	96%
	106	49											0.9978 ± 0.0001	48%
	106	49											1.0091 ± 0.0002	30%
	106	49											1.1390 ± 0.0005	2.4%
	106	49											1.4719 ± 0.0003	3%
	106	49											1.7801 ± 0.0005	1.5%
$_{49}In^{107m}$	107	49			51 s.	I.T.	0.678				1/2-		In k x-ray	3%
	107	49											0.6785 ± 0.0003	94%
$_{49}In^{107}$	107	49		106.910284	32.5 m.	β+(35%)	3.4	2.3			9/2+		ann.rad.	
	107	49				E.C(65%)							Cd k x-ray	41%
	107	49											0.2050 ± 0.0001	48%
	107	49											0.3209 ± 0.0001	10%
	107	49											0.3653 ± 0.0001	4%
	107	49											0.5055 ± 0.0001	12%
	107	49											0.7280 ± 0.0001	3.3%
	107	49											0.8090 ± 0.0001	3.3%
	107	49											1.2683 ± 0.0001	5.5%
	107	49											1.9222 ± 0.0001	1.8%
	107	49											2.0648 ± 0.0003	1.7%
	107	49											2.2848 ± 0.0002	1.1%
	107	49											2.3040 ± 0.0004	1.2%
	107	49											(0.2 - 2.99)	
$_{49}In^{108m}$	108	49			40 m.	β+(53%)					3+		ann.rad.	
	108	49				E.C.(47%)							Cd k x-ray	28%
	108	49											0.6329 ± 0.0002	76%
	108	49											0.8455 ± 0.0004	2.4%
	108	49											0.9685 ± 0.0005	4.3%
	108	49											1.5294 ± 0.0005	7.3%
	108	49											1.6012 ± 0.0003	4%
	108	49											1.7321 ± 0.0004	4%
	108	49											1.8519 ± 0.0005	3%
	108	49											1.9863 ± 0.0005	12%
	108	49											2.0483 ± 0.0004	3%
	108	49											3.0468 ± 0.0004	2.4%
	108	49											3.4522 ± 0.0005	9.1%
	108	49											3.8255 ± 0.002	2.3%
$_{49}In^{108}$	108	49		107.909678	57 m.	β+(33%)	5.13	1.3			6+		ann.rad.	
	108	49				E.C.(67%)							Cd k x-ray	41%
	108	49											0.2429 ± 0.0002	37%
	108	49											0.2667 ± 0.0003	3%
	108	49											0.3259 ± 0.0002	13%
	108	49											0.5689 ± 0.0002	5.3%
	108	49											0.6331 ± 0.0001	100%
	108	49											0.6489 ± 0.0002	5%
	108	49											0.7311 ± 0.0003	9%
	109	49											0.7547 ± 0.0005	3%
	108	49											0.8756 ± 0.0002	93%
	108	49											1.0331 ± 0.0002	24%
	108	49											1.0568 ± 0.0002	30%
	108	49											1.0935 ± 0.0003	5%
	108	49											1.1985 ± 0.0005	4%

TABLE OF THE ISOTOPES (Continued)

Isotope	A	Z	% Natural abundance	Atomic mass	Half-life	Decay mode	Decay energy (MeV)	Particle energy (MeV)	Particle intensity	Thermal neutron cross section	Spin (h/2π)	μ Nucl. mag. moment	Gamma-ray energy (MeV)	Gamma-ray intensity
	116	49											1.50752 ± 0.00005	10%
	116	49											1.7524 ± 0.0001	2.5%
	116	49											2.11221 ± 0.00006	15%
$_{49}In^{116}$	116	49		115.905264	14.1 s.	β-	3.27	3.3	99%		1+		0.46313 ± 0.00003	0.25%
	116	49											1.2526 ± 0.0005	0.03%
	116	49											1.29349 ± 0.00005	1.3%
$_{49}In^{117m}$	117	49			1.933 h.	β-(53%)	1.769	1.77			1/2-	0.25	In k x-ray	13%
	117	49				I.T.(47%)							0.15855 ± 0.00001	81%
	117	49											0.31531 ± 0.00002	19%
	117	49											0.55294 ± 0.00002	75%(D)
$_{49}In^{117}$	117	49		116.904517	43.1 m.	β-	1.453	0.74			9/2+		0.15855 ± 0.00001	86%
	117	49											0.3966 ± 0.0004	0.14%
	117	49											0.55294 ± 0.00002	99%
$_{49}In^{118m2}$	118	49			8.5 s.	I.T.(98%)					(8-)		In k x-ray	31%
	118	49				β-(2%)							0.1382 ± 0.0005	22%
	118	49											0.2537 ± 0.0001	1.4%
	118	49											1.0507 ± 0.0001	1.5%
	118	49											1.2296 ± 0.0001	1.5%
$_{49}In^{118m1}$	118	49			4.40 m.	β-					5+		0.2086 ± 0.0003	2.3%
	118	49											0.4458 ± 0.0002	6%
	118	49											0.4744 ± 0.0004	3%
	118	49											0.5602 ± 0.001	1.3%
	118	49											0.6373 ± 0.0004	3.5%
	118	49											0.6833 ± 0.0002	55%
	118	49											1.0970 ± 0.0004	3%
	118	49											1.1732 ± 0.0006	1%
	118	49											1.2295 ± 0.0002	96%
	118	49											1.2591 ± 0.0005	4%
	118	49											2.0423 ± 0.001	3%
$_{49}In^{118}$	118	49		117.906120	5.0 s.	β-	4.2	4.2			1+		0.5282 ± 0.0004	0.7%
	118	49											1.1734 ± 0.0005	0.4%
	118	49											1.2295 ± 0.0002	5%
	118	49											2.0432 ± 0.0005	0.1%
$_{49}In^{119m}$	119	49			18.0 m.	β-(97%)		2.7			1/2-		0.3114 ± 0.0001	1%
	119	49				I.T.(3%)	0.311						0.7631 ± 0.0001	2.5%(D)
$_{49}In^{119}$	119	49		118.905819	2.4 m.	β-	2.34	1.6			9/2+		0.0239 ± 0.0001	16%
	119	49											0.6495 ± 0.0001	0.5%
	119	49											0.7631 ± 0.0001	99%
	119	49											1.2149 ± 0.0001	0.4%
$_{49}In^{120m}$	120	49			3 s.	β-		536			1+		0.7042 ± 0.0006	1.4%
	120	49											1.1725 ± 0.0003	19%
	120	49											2.0398 ± 0.001	2%
	120	49											2.3902 ± 0.001	1%
$_{49}In^{120}$	120	49		119.907890	44 s.	β-	5.3	2.2			(5+)		0.4146 ± 0.0003	2%
	120	49						3.1					0.5924 ± 0.0005	2%
	120	49											0.6371 ± 0.0004	3%
	120	49											0.7029 ± 0.0004	2%
	120	49											0.7134 ± 0.0002	7%
	120	49											0.8637 ± 0.0002	32%
	120	49											0.9849 ± 0.0004	2%
	120	49											1.0232 ± 0.0002	54%
	120	49											1.1714 ± 0.0002	96%
	120	49											1.1840 ± 0.0004	2.4%
	120	49											1.2945 ± 0.0003	12%
	120	49											1.4723 ± 0.00033	4%
	120	49											1.8871 ± 0.0005	5%
	120	49											2.0081 ± 0.0005	6%
	120	49											2.1790 ± 0.0005	3%
	120	49											2.6061 ± 0.0009	2%
	120	49											(0.4 - 2.7)	
$_{49}In^{121m}$	121	49			3.9 m.	β-(99%)		5.3			1/2-		0.0601 ± 0.0002	20%
	121	49				I.T.(1%)	0.313						0.3136 ± 0.0001	0.5%
	121	49											0.9256 ± 0.0001	1%
	121	49											1.0412 ± 0.0005	1%
	121	49											1.1022 ± 0.0005	1%
	121	49											1.1204 ± 0.0002	0.5%
	122	49											2.8038 ± 0.0007	0.1%
	121	49											2.8643 ± 0.0007	0.1%
$_{49}In^{121}$	121	49		120.907847	23 s.	β-	3.36				9/2+		0.2620 ± 0.0001	8%
	121	49											0.6573 ± 0.0001	7%
	121	49											0.8693 ± 0.0001	1%
	121	49											0.9193 ± 0.0001	4%
	121	49											0.9256 ± 0.0001	87%
	121	49											1.0928 ± 0.0004	0.3%
$_{49}In^{122m}$	122	49			1.5 s.	β-		5.3			(1+)		1.0131 ± 0.0001	3%
	122	49											1.1403 ± 0.0001	29%

Isotope	A	Z	% Natural abundance	Atomic mass	Half-life	Decay mode	Decay energy (MeV)	Particle energy (MeV)	Particle intensity	Thermal neutron cross section	Spin (h/2π)	μ Nucl. mag. moment	Gamma-ray energy (MeV)	Gamma-ray intensity
	122	49											1.3897 ± 0.0001	2%
	122	49											2.0656 ± 0.0002	2%
	122	49											2.7591 ± 0.0002	3%
	122	49											2.9757 ± 0.0004	0.8%
	122	49											3.8197 ± 0.0003	0.3%
$_{49}In^{122}$	122	49		121.910280	10.1 s.	β-	6.37	4.4					0.2391 ± 0.0002	2%
	122	49											0.6435 ± 0.0002	3%
	122	49											0.8194 ± 0.0002	9%
	122	49											0.8313 ± 0.0002	8%
	122	49											0.9024 ± 0.0002	6%
	122	49											0.9745 ± 0.0001	14%
	122	49											1.0014 ± 0.0001	54%
	122	49											1.0131 ± 0.0001	12%
	122	49											1.0915 ± 0.0003	10%
	122	49											1.1403 ± 0.0001	100%
	122	49											1.1363 ± 0.0001	20%
	122	49											1.1903 ± 0.0001	28%
	122	49											1.3010 ± 0.0001	7%
	122	49											2.0931 ± 0.0002	4%
	122	49											2.4419 ± 0.0004	1%
$_{49}In^{123m}$	123	49			48 s.	β-		4.6			(1/2-)		0.1258 ± 0.0001	38%
	123	49											1.170 ± 0.001	0.1%
	123	49											3.234 ± 0.003	0.1%
$_{49}In^{123}$	123	49		122.910450	6.0 s.	β-	4.4	3.3			(9/2+)		0.6188 ± 0.0003	3%
	123	49											0.8455 ± 0.0002	1%
	123	49											1.0197 ± 0.0002	32%
	123	49											1.1305 ± 0.0002	63%
	123	49											1.3823 ± 0.0003	1%
	123	49											2.0012 ± 0.0005	0.3%
$_{49}In^{124m}$	124	49			2.4 s.		β-				8-		0.1029 ± 0.0001	45%
	124	49											0.1203 ± 0.0001	38%
	124	49											0.2431 ± 0.0001	11%
	124	49											0.2535 ± 0.0001	4%
	124	49											0.3635 ± 0.0001	17%
	124	49											0.8497 ± 0.0002	2%
	124	49											0.9154 ± 0.0002	3%
	124	49											0.9559 ± 0.0001	12%
	124	49											0.9699 ± 0.0001	52%
	124	49											1.0729 ± 0.0001	47%
	124	49											1.1168 ± 0.0001	15%
	124	49											1.1316 ± 0.0001	100%
	124	49											1.1990 ± 0.0001	9%
	124	49											1.3599 ± 0.0001	39%
	124	49											1.4401 ± 0.0001	9%
	124	49											1.1847 ± 0.0002	2%
	124	49											1.8560 ± 0.0004	1%
	124	49											2.6976 ± 0.0004	3%
$_{49}In^{124}$	124	49		123.912980	3.2 s.	β-	7.61	5			3+		0.7070 ± 0.0001	2%
	124	49											0.9699 ± 0.0001	3%
	124	49											0.9978 ± 0.0001	21%
	124	49											1.0899 ± 0.0001	3%
	124	49											1.1316 ± 0.0001	68%
	124	49											1.3147 ± 0.0001	4%
	124	49											1.4707 ± 0.0001	6%
	124	49											1.5813 ± 0.0001	2%
	124	49											1.7435 ± 0.0002	2%
	124	49											2.0825 ± 0.0002	3%
	124	49											2.4264 ± 0.0002	1%
	124	49											3.2142 ± 0.0002	21%
	124	49											3.2641 ± 0.0002	1%
	124	49											3.7615 ± 0.0003	1%
	124	49											3.9170 ± 0.0003	2%
	124	49											(0.3 - 4.6)	
$_{49}In^{125m}$	125	49			12.2 s.	β-		5.5			1/2-		0.1876 ± 0.0005	100 +
$_{49}In^{125}$	125	49		124.913670	2.33 s.	β-	5.5	4.1			9/2+		0.4260 ± 0.0001	2%
	125	49											0.6179 ± 0.0001	8%
	125	49											0.7446 ± 0.0001	6%
	125	49											0.8271 ± 0.0001	2%
	125	49											0.9365 ± 0.0001	3%
	125	49											1.0318 ± 0.0001	10%
	125	49											1.3350 ± 0.0001	76%
	125	49											1.5582 ± 0.0004	1%
$_{49}In^{126m}$	126	49			1.45 s.			4.9			(8-)		0.9086 ± 0.0001	4%
	126	49											0.9696 ± 0.0001	15%
	126	49											1.1357 ± 0.0001	2%
	126	49											1.1411 ± 0.0001	56%
	126	49											1.5710 ± 0.0001	3%
	126	49											1.6872 ± 0.0001	2%
	126	49											2.1053 ± 0.0002	2%
	126	49											2.11035 ± 0.0002	2%
	126	49											2.3704 ± 0.0002	2%
$_{49}In^{126}$	126	49		125.916470	1.53 s.	β-	8.1	4.2			3+		0.1118 ± 0.0001	88%
	126	49											0.2585 ± 0.0001	9%
	126	49											0.2693 ± 0.0001	6%
	126	49											0.3159 ± 0.0001	12%

Isotope	A	Z	% Natural abundance	Atomic mass	Half-life	Decay mode	Decay energy (MeV)	Particle energy (MeV)	Particle intensity	Thermal neutron cross section	Spin (h/2π)	μ Nucl. mag. moment	Gamma-ray energy (MeV)	Gamma-ray intensity
	126	49											0.4439 ± 0.0001	2%
	126	49											0.5014 ± 0.0001	6%
	126	49											0.5717 ± 0.0001	3%
	126	49											0.7883 ± 0.0001	8%
	126	49											0.9058 ± 0.0001	11%
	126	49											0.9086 ± 0.0001	100%
	126	49											0.9627 ± 0.0001	2%
	126	49											0.9774 ± 0.0002	3%
	126	49											1.0648 ± 0.0001	5%
	126	49											1.1411 ± 0.0001	100%
	126	49											1.1925 ± 0.0001	4%
	126	49											1.2359 ± 0.0001	2%
	126	49											1.3674 ± 0.0001	3%
	126	49											1.3780 ± 0.0001	23%
	126	49											1.4069 ± 0.0001	3%
	126	49											1.4954 ± 0.0003	2%
	126	49											1.5644 ± 0.0001	2%
	126	49											1.6117 ± 0.0001	6%
	126	49											1.6365 ± 0.0001	30%
	126	49											1.7583 ± 0.0002	5%
	126	49											2.5601 ± 0.0003	4%
	126	49											2.8286 ± 0.0003	5%
$_{49}In^{127m}$	127	49			3.76 s.	β-		6.4			(1/2-)		0.2523 ± 0.0003	77%
	127	49											0.8328 ± 0.0003	4%
	127	49											0.9484 ± 0.0003	2%
	127	49											1.0851 ± 0.0003	3%
	127	49											3.074 ± 0.001	6%
$_{49}In^{127}$	127	49		126.917320	1.12 s.	β-	6.5	4.9			(9/2+)		0.4680 ± 0.0003	1%
	127	49											0.6387 ± 0.0003	3%
	127	49											0.6461 ± 0.0003	8%
	127	49											0.7154 ± 0.0003	2%
	127	49											0.7926 ± 0.0003	2%
	127	49											0.8051 ± 0.0003	8%
	127	49											0.9563 ± 0.0003	6%
	127	49											0.9637 ± 0.0003	5%
	127	49											1.0486 ± 0.0003	7%
	127	49											1.0947 ± 0.0003	5%
	127	49											1.2141 ± 0.0003	1%
	127	49											1.5558 ± 0.0003	2%
	127	49											1.5977 ± 0.0003	67%
	127	49											20.190 ± 0.001	0.5%
$_{49}In^{128m}$	128	49			0.835 s.	β-		5.4			(8-)		0.1205 ± 0.0001	11%
	128	49											0.2572 ± 0.0001	4%
	128	49											0.3212 ± 0.0001	10%
	128	49											0.4577 ± 0.0001	2%
	128	49											0.8315 ± 0.0001	12%
	128	49											1.0549 ± 0.0001	6%
	128	49											1.1688 ± 0.0001	12%
	128	49											1.7800 ± 0.0001	3%
	128	49											1.8670 ± 0.0001	32%
	128	49											1.9739 ± 0.0001	19%
	128	49											2.1221 ± 0.0001	4%
$_{49}In^{128}$	128	49		127.920560	0.9 s.		β-				3+		0.9352 ± 0.0001	8%
	128	49											1.0895 ± 0.0001	7%
	128	49											1.1688 ± 0.0001	50%
	128	49											1.4643 ± 0.0001	2%
	128	49											1.5877 ± 0.0002	2%
	128	49											1.7393 ± 0.0001	2%
	128	49											1.8167 ± 0.0001	2%
	128	49											2.1041 ± 0.0001	6%
	128	49											2.2585 ± 0.0001	3%
	128	49											3.0511 ± 0.0002	2%
	128	49											3.5198 ± 0.0002	17%
	128	49											3.8862 ± 0.0002	4%
	128	49											3.9548 ± 0.0002	4%
	128	49											4.0380 ± 0.0002	2%
	128	49											4.2970 ± 0.0002	12%
$_{49}In^{129m}$	129	49			1.25 s.	β-(98%)		≈7.5					0.3153 ± 0.0003	68%
	129	49				n(2%)							0.9067 ± 0.0003	4%
	129	49											0.9732 ± 0.0003	1%
	129	49											1.2220 ± 0.0003	6%
	129	49											1.2885 ± 0.0003	2%
$_{49}In^{129}$	129	49		128.921600	0.59 s.	β-	7.6	5.5					0.2853 ± 0.0003	1%
	129	49											0.7288 ± 0.0003	5%
	129	49											0.7693 ± 0.0003	9%
	129	49											1.0083 ± 0.0003	6%
	129	49											1.0546 ± 0.0003	4%
	129	49											1.0745 ± 0.0003	2%
	129	49											1.0957 ± 0.0003	3%
	129	49											1.1010 ± 0.0003	1%
	129	49											1.3487 ± 0.0003	2%
	129	49											1.7814 ± 0.0003	2%
	129	49											1.8650 ± 0.0003	32%
	129	49											2.1180 ± 0.0003	44%
	129	49											2.5460 ± 0.001	2%

Isotope	A	Z	% Natural abundance	Atomic mass	Half-life	Decay mode	Decay energy (MeV)	Particle energy (MeV)	Particle intensity	Thermal neutron cross section	Spin (h/2π)	μ Nucl. mag. moment	Gamma-ray energy (MeV)	Gamma-ray intensity
$_{49}In^{130m}$	130	49			0.53 s.	β-							0.0892 ± 0.0001	17%
	130	49											0.1298 ± 0.0001	7%
	130	49											0.1380 ± 0.0001	9%
	130	49											0.4082 ± 0.0001	7%
	130	49											0.7744 ± 0.0001	39%
	130	49											0.8070 ± 0.0001	5%
	130	49											1.2212 ± 0.0001	76%
	130	49											1.3402 ± 0.0001	3%
	130	49											1.4292 ± 0.0002	2%
	130	49											2.0283 ± 0.0001	11%
	130	49											2.3171 ± 0.0001	13%
	130	49											2.4099 ± 0.0002	2%
	130	49											3.1840 ± 0.0003	8%
	130	49											3.2417 ± 0.0003	4%
	130	49											4.0415 ± 0.0003	3%
$_{49}In^{130}$	130	49		129.924870	0.31 s.	β-	≈10.2				10-		0.0892 ± 0.0001	41%
	130	49											0.1298 ± 0.0001	65%
	130	49											0.1380 ± 0.0001	20%
	130	49											0.2191 ± 0.0001	2%
	130	49											0.7744 ± 0.0001	54%
	130	49											0.8070 ± 0.0001	2%
	130	49											0.9526 ± 0.0001	17%
	130	49											1.2212 ± 0.0001	65%
	130	49											1.9052 ± 0.0001	80%
	130	49											1.9458 ± 0.0001	7%
	130	49											2.0283 ± 0.0001	4%
	130	49											2.0915 ± 0.0002	5%
	130	49											2.8985 ± 0.0003	3%
$_{49}In^{131}$	131	49		130.926410	0.28 s.	β-	8.8	6.4			(9/2+)		0.3328 ± 0.0002	
	131	49											2.433	
$_{49}In^{132}$	132	49			0.22 s.	β-		≈5					0.1320	22%
	132	49											0.2992	63%
	132	49											0.3747	74%
	132	49											0.4791	24%
	132	49											0.5764	22%
	132	49											2.2863	19%
	132	49											2.3797	31%
	132	49											4.0406	65%
	132	49											4.3513	26%
	132	49											4.4158	8%
Sn		50		118.710						0.63 b.				
$_{50}Sn^{106}$	106	50		105.917030	2.1 m.	β+(20%)	3.3						ann.rad.	
	136	50				E.C.(80%)							In k x-ray	27%
	106	50											0.1223 ± 0.0001	15%
	106	50											0.2241 ± 0.0002	14%
	106	50											0.2532 ± 0.0001	29%
	106	50											0.3262 ± 0.0003	7%
	106	50											0.3865 ± 0.0002	51%
	106	50											0.4772 ± 0.0002	32%
	106	50											0.7123 ± 0.0003	17%
	106	50											0.8639 ± 0.0003	11%
	106	50											1.097 ± 0.001	1%
	106	50											1.1896 ± 0.0003	17%
$_{50}Sn^{107}$	107	50		106.915870	2.9 m.	E.C.	5.2						0.4218	5 +
	107	50				β+							0.6105	3
	107	50											0.6785 ± 0.0003	100(D)
	107	50											1.0013 ± 0.0003	29
	107	50											1.1290 ± 0.0002	100
	107	50											1.1860	12
	107	50											1.358	6
	107	50											1.396	21
	107	50											1.424	10
	107	50											1.542	30
	107	50											1.808	25
	107	50											2.116	10
	107	50											2.216	8
	107	50											2.316	6
	107	50											2.547	10
	107	50											2.825	13
	107	50											3.060	7
$_{50}Sn^{108}$	108	50		107.911880	10.3 m.	β+(1%)	2.05				0+		In k x-ray	66%
	108	50				E.C.(99%)							0.1046 ± 0.0004	12%
	108	50											0.1678 ± 0.0003	18%
	108	50											0.2357 ± 0.0002	6%
	108	50											0.2724 ± 0.0003	41%
	108	50											0.3965 ± 0.0002	58%
	108	50											0.6692 ± 0.0004	20%
	108	50											0.8293 ± 0.0005	3%
	108	50											0.8587 ± 0.0006	2%
	108	50											0.8891 ± 0.0005	3%
	108	50											1.6544 ± 0.0005	2%
	108	50											1.6848 ± 0.0006	2%
$_{50}Sn^{109}$	109	50		108.911294	18.0 m.	β+(9%)	3.88				7/2+		ann.rad.	
	109	50				E.C.(91%)							In k x-ray	56%

Isotope	A	Z	% Natural abundance	Atomic mass	Half-life	Decay mode	Decay energy (MeV)	Particle energy (MeV)	Particle intensity	Thermal neutron cross section	Spin (h/2π)	μ Nucl. mag. moment	Gamma-ray energy (MeV)	Gamma-ray intensity
	109	50											0.3312 ± 0.0002	10%
	109	50											0.3845 ± 0.0003	3%
	109	50											0.4374 ± 0.0003	3%
	109	50											0.5219 ± 0.0002	3%
	109	50											0.6234 ± 0.0004	2%
	109	50											0.6498 ± 0.0002	32%(D)
	109	50											0.7909 ± 0.0003	2%
	109	50											1.0264 ± 0.0002	5%
	109	50											1.0390 ± 0.0002	5%
	109	50											1.0992 ± 0.0002	31%
	109	50											1.1192 ± 0.0003	3%
	109	50											1.3213 ± 0.0002	12%
	109	50											1.4620 ± 0.0006	2%
	109	50											1.4636 ± 0.0004	3%
	109	50											1.4642 ± 0.0002	4%
	109	50											1.4887 ± 0.0002	4%
	109	50											1.5744 ± 0.0002	6%
	109	50											1.8898 ± 0.0003	2%
	109	50											1.9111 ± 0.0002	6%
	109	50											2.0552 ± 0.0003	2%
	109	50											2.1956 ± 0.0002	1.5%
	109	50											2.5418 ± 0.0003	2.7%
	109	50											2.7854 ± 0.0003	1.8%
	109	50											2.8586 ± 0.0002	1%
$_{50}Sn^{110}$	110	50		109.907858	4.0 h.	E.C.	0.58				0+		In k x-ray	61%
	110	50											0.283 ± 0.001	97%
$_{50}Sn^{111}$	111	50		110.907741	35.3 m.	β+(31%)	2.45	1.5			7/2+		In k x-ray	42%
	111	50				E.C.(69%)							0.7620 ± 0.0001	1.4%
	111	50											1.1530 ± 0.0001	2.5%
	111	50											1.6105 ± 0.0002	1.2%
	111	50											1.9147 ± 0.0002	1.9%
	111	50											2.1071 ± 0.0002	0.4%
	111	50											2.1795 ± 0.0003	0.3%
	111	50											2.2121 ± 0.0002	0.2%
	111	50											2.3233 ± 0.0002	0.3%
$_{50}Sn^{112}$	112	50	1.0%	111.904826						(0.3 + 0.7)b.	0+			
$_{50}Sn^{113m}$	113	50			21.4 m.	I.T.(92%)	0.077				7/2+		Sn k x-ray	36%
	113	50				E.C.(8%)							In x-ray	5%
	113	50											0.0774 ± 0.0001	0.5%
$_{50}Sn^{113}$	113	50		112.905176	115.1 d.	E.C.					1/2+		In k x-ray	80%
	113	50											0.25511 ± 0.00001	1.9%
	113	50											0.39169 ± 0.00001	64%
	113	50											0.6380 ± 0.0001	0.001%
$_{50}Sn^{114}$	114	50	0.7%	113.902784						0.1 b.	0+			
$_{50}Sn^{115}$	115	50	0.4%	114.903348						30 b.	1/2+	-0.918		
$_{50}Sn^{116}$	116	50	14.7%	115.901747						(0.006+0.1) b.	0+			
$_{50}Sn^{117m}$	117	50			13.6 d.	I.T.	0.3146				11/2-		Sn k x-ray	54%
	117	50											0.15602 ± 0.00002	2%
	117	50											0.15856 ± 0.00002	86%
$_{50}Sn^{117}$	117	50	7.7%	116.902956						2.3 b.	1/2+	-1.000		
$_{50}Sn^{118}$	118	50	24.3%	117.901609						(0.054 + 0.2) b.	0+			
$_{50}Sn^{119m}$	119	50			293 d.	I.T.	0.0895				11/2-	+0.67	Sn k x-ray	
	119	50											0.02387 ± 0.00002	16%
	119	50											0.0657 ± 0.0001	0.02%
$_{50}Sn^{119}$	119	50	8.6%	118.903310						2 b.	1/2+	-1.046		
$_{50}Sn^{120}$	120	50	32.4%	119.902200						(0.001 + 0.16) b.	0+			
$_{50}Sn^{121m}$	121	50			≈55y.	I.T.(78%)	0.006				11/2-		Sn k x-ray	
	121	50				β-(22%)	0.394	0.354					0.03715 ± 0.0004	2%
$_{50}Sn^{121}$	121	50		120.904238	27.0 h.	β-	0.388	0.383	100%		3/2+	0.70		
$_{50}Sn^{122}$	122	50	4.6%	121.903440						(0.16 + 0.001)b.	0+			
$_{50}Sn^{123m}$	123	50			40.1 m.	β-	1.42	1.26	99%		3/2+		0.1603 ± 0.0001	86%
	123	50											0.3814 ± 0.0004	0.04%
$_{50}Sn^{123}$	123	50		122.905722	129.2 d.	β-	1.398	1.42	99.4%		11/2-		0.1603 ± 0.0001	0.002%
	123	50											1.0302 ± 0.0001	0.03%
	123	50											1.0886 ± 0.0001	0.6%
$_{50}Sn^{124}$	124	50	5.66%	123.905274						(0.13 + 0.004) b.	0+			
$_{50}Sn^{125m}$	125	50			9.52 m.	β-		2.03	98%		3/2+		0.3321 ± 0.0001	97%
	125	50											0.5896 ± 0.0005	0.2%
	125	50											1.4040 ± 0.0005	0.7%
	125	50											1.4839 ± 0.0005	0.2%
$_{50}Sn^{125}$	125	50		124.907785	β-	2.352	2.35	82%			11/2-		0.3321 ± 0.0001	1.3%
	125	50											0.4698 ± 0.0001	1.4%
	125	50											0.8008 ± 0.0001	1%
	125	50											0.8225 ± 0.0001	4%
	125	50											0.9155 ± 0.0001	4%
	125	50											1.0671 ± 0.0001	9%
	125	50											1.0877 ± 0.0001	1.1%
	125	50											1.0892 ± 0.0001	4.3%

TABLE OF THE ISOTOPES (Continued)

Isotope	A	Z	% Natural abundance	Atomic mass	Half-life	Decay mode	Decay energy (MeV)	Particle energy (MeV)	Particle intensity	Thermal neutron cross section	Spin (h/2π)	μ Nucl. mag. moment	Gamma-ray energy (MeV)	Gamma-ray intensity
	125	50											2.0018 ± 0.0001	1.8%
$_{50}Sn^{126}$	126	50		125.907654	≈10^5y.	β-	0.38	0.25	100%		0+		0.0643 ± 0.0001	9.6%
	126	50											0.0869 ± 0.0001	9%
	126	50											0.0876 ± 0.0001	37%
	126	50											0.4148 ± 0.0002	98%(D)
	126	50											0.6663 ± 0.0002	100%(D)
	126	50											0.6950 ± 0.0002	100%(D)
$_{50}Sn^{127m}$	127	50			4.15 m.	β-		2.72			1/2+		0.4909 ± 0.0004	90%
	127	50											1.3480 ± 0.001	5%
	127	50											1.5640 ± 0.002	4%
$_{50}Sn^{127}$	127	50		126.910355	2.1 h.	β-	3.2	2.42			11/2-		0.1197 ± 0.0004	2%
	127	50						3.2					0.1692 ± 0.0004	2%
	127	50											0.2625 ± 0.0004	2%
	127	50											0.2662 ± 0.0004	2%
	127	50											0.2843 ± 0.0004	3%
	127	50											0.4382 ± 0.0004	0%
	127	50											0.4909 ± 0.0004	5%
	127	50											0.4932 ± 0.0004	5%
	127	50											0.5454 ± 0.0004	2%
	127	50											0.5833 ± 0.0004	3%
	127	50											0.5923 ± 0.0004	2%
	127	50											0.8059 ± 0.0004	8%
	127	50											0.8231 ± 0.0004	11%
	127	50											0.8247 ± 0.0004	6%
	127	50											0.8595 ± 0.0004	8%
	127	50											0.9792 ± 0.0004	7%
	127	50											1.0361 ± 0.0004	2%
	127	50											1.0933 ± 0.0007	4%
	127	50											1.0956 ± 0.0004	19%
	127	50											1.1143 ± 0.0004	38%
	127	50											1.1604 ± 0.0004	2%
	127	50											1.5843 ± 0.0004	2%
	127	50											2.0034 ± 0.0005	5%
	127	50											2.3174 ± 0.0005	1%
	127	50											2.5849 ± 0.0005	1.6%
	127	50											2.6959 ± 0.0005	1.6%
	127	50											2.8464 ± 0.0005	1%
$_{50}Sn^{128}$	128	50		127.910560	59.1 m.	β-	1.29	0.48			0+		0.0321 ± 0.0002	4%
	128	50						0.63					0.0457 ± 0.0002	13%
	128	50											0.0751 ± 0.0002	27%
	128	50											0.1527 ± 0.0002	6%
	128	50											0.1604 ± 0.0002	2%
	128	50											0.4044 ± 0.0002	6%
	128	50											0.4823 ± 0.0002	58%
	128	50											0.5573 ± 0.0002	16%
	128	50											0.6805 ± 0.0002	16%
$_{50}Sn^{129m}$	129	50			6.9 m.	β-					11/2-		1.1611 ± 0.0003	
$_{50}Sn^{129}$	129	50		128.913440	2.5 m.	β-	4.0				3/2+		0.6456 ± 0.0003	
$_{50}Sn^{130m}$	130	50		1.7 m.	β-		3.0				(7-)		0.0847 ± 0.0001	14%
	130	50											0.1449 ± 0.0001	34%
	130	50											0.3113 ± 0.0001	14%
	130	50											0.5436 ± 0.0002	10%
	130	50											0.8992 ± 0.0002	17%
	130	50											0.9624 ± 0.0005	3%
$_{50}Sn^{130}$	130	50		129.913920	3.7 m.	β-	2.1	1.10			0+		0.0700 ± 0.0001	36%
	130	50											0.1925 ± 0.0002	71%
	130	50											0.2292 ± 0.0002	24%
	130	50											0.3413 ± 0.0002	2%
	130	50											0.4347 ± 0.0002	14%
	130	50											0.5505 ± 0.0002	3%
	130	50											0.7431 ± 0.0001	19%
	130	50											0.7798 ± 0.0001	59%
	130	50											0.8723 ± 0.0003	1%
$_{50}Sn^{131m}$	131	50			39 s.	β-							0.0823 ± 0.0001	21 +
	131	50											0.3043 ± 0.0001	32
	131	50											0.4500 ± 0.0001	90
	131	50											0.7985 ± 0.0001	86
	131	50											0.8851 ± 0.0001	9
	131	50											1.0734 ± 0.0001	9
	131	50											1.1415 ± 0.0003	12
	131	50											1.2026 ± 0.0002	19
	131	50											1.2260 ± 0.0001	100
	131	50											1.2292 ± 0.0001	30
	131	50											1.4811 ± 0.0001	12
	131	50											1.9311 ± 0.0001	9
	131	50											2.0293 ± 0.0002	5
	131	50											2.0825 ± 0.0002	5
	131	50											2.1864 ± 0.0002	4
	131	50											(0.08 - 3.21)	
$_{50}Sn^{131}$	131	50		130.916940	61 s.	β-	4.6						see Sn^{131m}	
$_{50}Sn^{132}$	132	50		131.917760	40 s.	β-	3.1	1.76					0.0855 ± 0.0001	49%
	132	50											0.2467 ± 0.0001	42%
	132	50											0.3402 ± 0.0001	43%
	133	50											0.5287 ± 0.0002	2%
	132	50											0.5488 ± 0.0002	2%

Isotope	A	Z	% Natural abundance	Atomic mass	Half-life	Decay mode	Decay energy (MeV)	Particle energy (MeV)	Particle intensity	Thermal neutron cross section	Spin (h/2π)	μ Nucl. mag. moment	Gamma-ray energy (MeV)	Gamma-ray intensity
	132	50											0.6519 ± 0.0002	2%
	132	50											0.8985 ± 0.0001	42%
	132	50											0.9922 ± 0.0001	38%
	132	50											1.0778 ± 0.0003	2%
	132	50											1.2388 ± 0.0002	13%
Sb		51								5.4 b.				
$_{51}Sb^{109}$	109	51		108.918143	18.3 s.	β+	6.83						ann.rad.	
	109	51				E.C.							0.2467	2 +
	109	51											0.2610	2
	109	51											0.5448	11
	109	51											0.6645	63
	109	51											0.6876	19
	109	51											0.9254	100
	109	51											0.9506	2
	109	51											1.0617	75
	109	51											1.0780	4
	109	51											1.3435	2
	109	51											1.3435	2
	109	51											1.4958	30
$_{51}Sb^{110}$	110	51		109.916770	23.5 s.	β+	8.4	6.9			3+		ann.rad.	
	110	51				E.C.							0.6365 ± 0.0004	4%
	110	51											0.7515 ± 0.0004	4%
	110	51											0.8271 ± 0.0003	9%
	110	51											0.9089 ± 0.0003	8%
	110	51											0.9847 ± 0.0001	31%
	110	51											1.0258 ± 0.0004	2%
	110	51											1.2117 ± 0.0001	92%
	110	51											1.2433 ± 0.0003	13%
	110	51											1.4825 ± 0.0004	4%
	110	51											1.6095 ± 0.0005	2%
	110	51											1.7359 ± 0.0005	7%
	110	51											1.7653 ± 0.0005	4%
	110	51											1.9709 ± 0.0006	4.9%
	110	51											2.0291 ± 0.0006	3.7%
	110	51											2.1208 ± 0.0008	7.3%
	110	51											2.2349 ± 0.0008	2%
	110	51											2.6732 ± 0.001	2%
$_{51}Sb^{111}$	111	51		110.913220	75 s.	β+(87%)	5.2	3.3			5/2+		ann.rad.	
	111	51				E.C.(13%)							0.1002 ± 0.0001	3%
	111	51											0.1545 ± 0.0001	71%
	111	51											0.3888 ± 0.0001	4%
	111	51											0.4891 ± 0.0001	42%
	111	51											0.6436 ± 0.0002	2%
	111	51											0.7554 ± 0.0002	5%
	111	51											0.7778 ± 0.0002	3%
	111	51											0.8974 ± 0.0003	2%
	111	51											1.0326 ± 0.0001	10%
	111	51											1.1475 ± 0.0002	4%
	111	51											1.8413 ± 0.0003	0.5%
$_{51}Sb^{112}$	112	51		111.912411	51 s.	β+(90%)	7.06	4.75			3+		ann.rad.	
	112	51				E.C.(10%)							0.6700 ± 0.0004	4%
	112	51											0.8946 ± 0.0002	2.6%
	112	51											0.9909 ± 0.0001	14%
	112	51											1.0980 ± 0.0002	2%
	112	51											1.2571 ± 0.0001	96%
	112	51											1.5664 ± 0.0002	1.6%
	112	51											1.7102 ± 0.0002	1.3%
	112	51											2.1609 ± 0.0002	1%
	112	51											2.24 791 ±	1%
													0.0002	
	112	51											(0.3 - 3.6)	
$_{51}Sb^{113}$	113	51		112.909372	6.7 m.	β+(65%)	3.89	2.4			5/2+		ann.rad.	
	113	51				E.C.(35%)							Sn k x-ray	20%
	113	51											0.3324 ± 0.0001	14%
	113	51											0.4980 ± 0.0001	77%
	113	51											0.9406 ± 0.0001	2.5%
	113	51											1.0133 ± 0.0001	0.9%
	113	51											1.5563 ± 0.0002	1%
$_{51}Sb^{114}$	114	51		113.909110	3.5 m.	β+(78%)	6.09	4.0			3/2+		ann.rad.	
	114	51				E.C.(22%)							Sn k x-ray	9%
	114	51											0.3272 ± 0.0001	7%
	114	51											0.7173 ± 0.0001	5%
	114	51											0.8876 ± 0.0001	18%
	114	51											0.9748 ± 0.0001	3%
	114	51											1.2999 ± 0.0001	98%
	114	51											1.5600 ± 0.0002	1%
	114	51											1.6438 ± 0.0001	1.3%
	114	51											1.9079 ± 0.0001	1%
	114	51											1.9262 ± 0.0001	1.7%
	114	51											2.2398 ± 0.0002	1%
$_{51}Sb^{115}$	115	51		114.906601	32.1 m.	β+(67%)	3.03	1.50			5/2+		ann.rad.	
	115	51				E.C.(33%)							Sn k x-ray	27%
	115	51											0.4973 ± 0.0001	98%
	115	51											1.2366 ± 0.0002	0.6%

Isotope	A	Z	% Natural abundance	Atomic mass	Half-life	Decay mode	Decay energy (MeV)	Particle energy (MeV)	Particle intensity	Thermal neutron cross section	Spin (h/2π)	μ Nucl. mag. moment	Gamma-ray energy (MeV)	Gamma-ray intensity
	115	51											1.2799 ± 0.0002	0.3%
	115	51											1.6338 ± 0.0002	0.3%
$_{51}Sb^{116m}$	116	51			60 m.	β+(78%)		1.2			8-		ann.rad.	
	116	51				E.C.(22%)							Sn k x-ray	51%
	116	51											0.09982 ± 0.00002	32%
	116	51											0.13552 ± 0.00005	29%
	116	51											0.4073 ± 0.0001	42%
	116	51											0.5429 ± 0.0002	52%
	116	51											0.8440 ± 0.0001	12%
	116	51											0.9725 ± 0.0001	72%
	116	51											1.0724 ± 0.0001	28%
	116	51											1.2935 ± 0.0001	100%
$_{51}Sb^{116}$	116	51		115.906800	16 m.	β+(50%)	4.71	1.3			3+		ann.rad.	
	116	51				E.C.(50%)		2.3					Sn k x-ray	20%
	116	51											0.93180 ± 0.00005	26%
	116	51											1.29354 ± 0.00004	85%
	116	51											2.22533 ± 0.00007	17%
$_{51}Sb^{117}$	117	51		116.904841	2.80 h.	β+(2%)	1.76	0.57			5/2+	+2.67	Sn k x-ray	44%
	117	51				E.C.(98%)							0.1586 ± 0.00001	86%
	117	51											0.8614 ± 0.0001	0.3%
	117	51											1.0045 ± 0.0002	0.2%
	117	51											1.0206 ± 0.0005	0.1%
	117	51											1.0210 ± 0.0005	0.1%
$_{51}Sb^{118m}$	118	51			5.00 h.	E.C.(99%)					8-		Sn k x-ray	57%
	118	51											0.0410 ± 0.0001	18%
	118	51											0.25368 ± 0.00001	99%
	118	51											1.05069 ± 0.00003	97%
	118	51											1.09151 ± 0.00008	4%
	118	51											1.22964 ± 0.00004	100%
$_{51}Sb^{118}$	118	51		117.905534	3.6 m.	β+(74%)	3.66	2.65			1+		ann.rad.	150%
	118	51				E.C.(26%)							Sn k x-ray	10%
	118	51											0.5282 ± 0.00004	0.4%
	118	51											0.8269 ± 0.0006	0.4%
	118	51											1.22964 ± 0.00006	2.5%
	118	51											1.2670 ± 0.0005	0.6%
$_{51}Sb^{119}$	119	51		118.903948	36.1 h.	E.C.	0.59				5/2+	+3.45	Sn k x-ray	39%
	119	51											0.0239 ± 0.0001	16%
$_{51}Sb^{120m}$	120	51			5.76 d.	E.C.					8-		Sn k x-ray	50%
	120	51											0.0898 ± 0.0002	80%
	120	51											0.19730 ± 0.00003	88%
	120	51											1.02301 ± 0.00004	99%
	120	51											1.1130 ± 0.0006	1.3%
	120	51											1.17121 ± 0.00006	100%
$_{51}Sb^{120}$	120	51		119.903821	15.9 m.	β+(41%)	2.68	1.75			1+	+2.3	ann.rad.	
	120	51				E.C.(59%)							Sn k x-ray	23%
	120	51											0.7038 ± 0.0003	0.15%
	120	51											0.9886 ± 0.0007	0.06%
	120	51											1.17121 ± 0.00005	1.7%
$_{51}Sb^{121}$	121	51	57.3%	120.903821						(0.05 + 6.2)b.	5/2+	+3.359		
$_{51}Sb^{122m}$	122	51			4.21 m.	I.T.	0.162				8-		Sb x-ray	42%
	122	51											0.0614 ± 0.0001	57%
	122	51											0.0761 ± 0.0001	20%
$_{51}Sb^{122}$	122	51		121.905179	2.71 d.	β-(98%)	1.982	1.41	65%		2-	-1.90	0.56409 ± 0.00005	71%
	122	51				β+(2%)	1.622	1.98	26%				0.69277 ± 0.00008	3.7%
	122	51											1.14050 ± 0.0001	0.6%
	122	51											1.2569 ± 0.0001	0.8%
$_{51}Sb^{123}$	123	51	42.7%	122.904216						(0.02 + 0.04 + 4.1)b.	7/2+	+2.547		
$_{51}Sb^{124m2}$	124	51			20.3 m.	I.T.	0.035							
$_{51}Sb^{124m1}$	124	51			96 s.	I.T.(80%)		1.2			5+		0.4984 ± 0.0001	20%
	124	51				β-(20%)		1.7					0.6027 ± 0.0001	20%
	124	51											0.6458 ± 0.0001	20%
	124	51											1.1010 ± 0.0002	0.3%
$_{51}Sb^{124}$	124	51		123.905038	60.20 d.	β-	2.905	0.61	52%	17 b.	3-		0.60271 ± 0.0001	98%
	124	51						2.3	23%				0.64583 ± 0.00001	7%
	124	51											0.7093 ± 0.0001	1.4%
	124	51											0.71376 ± 0.00001	2.4%

Isotope	A	Z	% Natural abundance	Atomic mass	Half-life	Decay mode	Decay energy (MeV)	Particle energy (MeV)	Particle intensity	Thermal neutron cross section	Spin (h/2π)	μ Nucl. mag. moment	Gamma-ray energy (MeV)	Gamma-ray intensity
	124	51											0.72277 ± 0.00002	11%
	124	51											0.96819 ± 0.00002	1.9%
	124	51											1.04511 ± 0.00002	1.9%
	124	51											1.3255 ± 0.0001	1.5%
	124	51											1.3681 ± 0.0001	2.5%
	124	51											1.4366 ± 0.0001	1.2%
	124	51											1.69094 ± 0.00004	49%
	124	51											2.09089 ± 0.00004	5.6%
$_{51}Sb^{125}$	125	51		124.905252	2.76 y.	β-	0.767	0.13	30%		$^1/_2$+		0.0355 ± 0.0001	6%
	125	51						0.30	45%				0.17632 ± 0.00002	6.8%
	125	51						0.62	13%				0.38044 ± 0.00002	1.5%
	125	51											0.42786 ± 0.00002	29%
	125	51											0.46336 ± 0.00001	10%
	125	51											0.60060 ± 0.00001	18%
	125	51											0.60672 ± 0.00003	5%
	125	51											0.63595 ± 0.00003	11%
	125	51											0.67144 ± 0.00004	2%
$_{51}Sb^{126m2}$	126	51			11 s.	I.T.					3-		L x-ray	
	126	51											0.0227 ± 0.0001	0.13%
$_{51}Sb^{126m1}$	126	51			19.0 m.	β-(86%)					5+		0.4148 ± 0.0002	86%
	126	51				I.T.(14%)							0.6663 ± 0.0002	86%
	126	51											0.6950 ± 0.0002	86%
	126	51											0.9282 ± 0.0003	1.3%
	126	51											1.0348 ± 0.0002	1.8%
$_{51}Sb^{126}$	126	51		125.907250	12.4 d.	β-	3.66				8-		0.2786 ± 0.0002	2%
	126	51											0.2965 ± 0.0003	4%
	126	51											0.4148 ± 0.0002	84%
	126	51											0.5738 ± 0.0002	7%
	126	51											0.5930 ± 0.0002	7%
	126	51											0.6563 ± 0.0002	2%
	126	51											0.6663 ± 0.0002	100%
	126	51											0.6950 ± 0.0002	100%
	126	51											0.6970 ± 0.0002	30%
	126	51											0.7205 ± 0.0002	54%
	126	51											0.8567 ± 0.0002	18%
	126	51											0.9893 ± 0.0002	7%
	126	51											1.0348 ± 0.0002	1%
	126	51											1.2130 ± 0.0002	2.4%
$_{51}Sb^{127}$	127	51		126.906919	3.84 d.	β-	1.58	0.89			7/2+		0.2524 ± 0.0003	8%
	127	51						1.1					0.2908 ± 0.0005	2%
	127	51						1.50					0.4121 ± 0.0005	4%
	127	51											0.4451 ± 0.0005	4%
	127	51											0.4370 ± 0.0004	25%
	127	51											0.5433 ± 0.0005	3%
	127	51											0.6035 ± 0.0005	4%
	127	51											0.6857 ± 0.0005	35%
	127	51											0.6985 ± 0.0005	3%
	127	51											0.7222 ± 0.0005	2%
	127	51											0.7837 ± 0.0005	14%
	127	51											0.9244 ± 0.0009	0.5%
$_{51}Sb^{128m}$	128	51			10 m.	β-(96%)					5+		0.3140 ± 0.0001	92%
	128	51				I.T.(4%)							0.5941 ± 0.0001	3%
	128	51											0.7432 ± 0.0001	96%
	128	51											0.7539 ± 0.0001	96%
	128	51											0.7876 ± 0.0001	7%
	128	51											0.8440 ± 0.0003	2%
	128	51											0.9083 ± 0.0002	2%
	128	51											1.0409 ± 0.0003	1%
	128	51											1.1417 ± 0.0003	0.8%
	128	51											1.1580 ± 0.0003	1.7%
$_{51}Sb^{128}$	128	51		127.909180	9.1 h.	β-	4.39	2.3			8-		0.2148 ± 0.0002	2%
	128	51											0.3141 ± 0.0001	61%
	128	51											0.3177 ± 0.0002	3%
	128	51											0.3224 ± 0.0002	3%
	128	51											0.5265 ± 0.0001	45%
	128	51											0.6287 ± 0.0001	31%
	128	51											0.6362 ± 0.0001	36%
	128	51											0.6542 ± 0.0002	17%
	128	51											0.6671 ± 0.0003	2%
	128	51											0.6839 ± 0.0003	3%
	128	51											0.6929 ± 0.0003	2%
	128	51											0.7276 ± 0.0003	4%

Isotope	A	Z	% Natural abundance	Atomic mass	Half-life	Decay mode	Decay energy (MeV)	Particle energy (MeV)	Particle intensity	Thermal neutron cross section	Spin (h/2π)	μ Nucl. mag. moment	Gamma-ray energy (MeV)	Gamma-ray intensity
	128	51											0.7433 ± 0.0001	100%
	128	51											0.7540 ± 0.0001	100%
	128	51											0.8136 ± 0.0002	13%
	128	51											0.8458 ± 0.0004	2%
	128	51											0.8780 ± 0.0004	3%
	128	51											1.0475 ± 0.0004	3%
	128	51											1.0786 ± 0.0004	2%
	128	51											1.1127 ± 0.0004	2%
	128	51											1.1816 ± 0.0004	4%
	128	51											1.3780 ± 0.0004	1.8%
$_{51}Sb^{129m}$	129	51			17.7 m.	β-							0.4338	
	129	51											0.6578	
	129	51											0.7598	
$_{51}Sb^{129}$	129	51		128.909146	4.4 h.	β-	2.38	1.4					0.0278 ± 0.0001	19%(D)
	129	51						1.5					0.1808 ± 0.0005	3%
	129	51						1.8					0.3591 ± 0.0005	[illegible]
	129	51											0.4596 ± 0.0001	8%(D)
	129	51											0.5447 ± 0.0003	18%
	129	51											0.6337 ± 0.0005	3%
	129	51											0.6543 ± 0.0005	3%
	129	51											0.6836 ± 0.0003	6%
	129	51											0.7610 ± 0.0005	4%
	129	51											0.7734 ± 0.0006	3%
	129	51											0.8128 ± 0.0005	45%
	129	51											0.8762 ± 0.0007	3%
	129	51											0.9146 ± 0.0005	21%
	129	51											0.9664 ± 0.0006	8%
	129	51											1.0301 ± 0.0006	13%
	129	51											1.2085 ± 0.0007	1%
	129	51											1.5687 ± 0.0008	0.8%
	129	51											1.6546 ± 0.001	1%
	129	51											1.7365 ± 0.001	6%
$_{51}Sb^{130m}$	130	51			6.3 m.	β-	2.6	2.12					0.1023 ± 0.0001	41%
	130	51											0.3485 ± 0.0002	5%
	130	51											0.6271 ± 0.0003	5%
	130	51											0.6477 ± 0.0003	3%
	130	51											0.6974 ± 0.0003	4%
	130	51											0.7489 ± 0.0003	4%
	130	51											0.7934 ± 0.0001	86%
	130	51											0.8163 ± 0.0002	12%
	130	51											0.8394 ± 0.0001	100%
	130	51											0.9208 ± 0.0001	4%
	130	51											0.9422 ± 0.0004	3%
	130	51											1.0175 ± 0.0002	3%
	130	51											1.0465 ± 0.0004	3%
	130	51											1.0717 ± 0.0004	2%
	130	51											1.1028 ± 0.0004	4%
	130	51											1.1420 ± 0.0004	6%
	130	51											1.1773 ± 0.0004	2%
	130	51											1.2000 ± 0.0004	4%
	130	51											1.5980 ± 0.0005	3%
	130	51											2.1166 ± 0.0008	1%
$_{51}Sb^{130}$	130	51		129.911590	38.4 m.	β-	5.0	2.9			8-		0.1823 ± 0.0001	65%
	130	51											0.2580 ± 0.0002	4%
	130	51											0.2853 ± 0.0002	3%
	130	51											0.3033 ± 0.0002	2%
	130	51											0.3309 ± 0.0001	78%
	130	51											0.4554 ± 0.0002	5%
	130	51											0.4680 ± 0.0001	18%
	130	51											0.4836 ± 0.0003	2%
	130	51											0.5067 ± 0.0003	2%
	130	51											0.6267 ± 0.0003	3%
	130	51											0.6547 ± 0.0003	2%
	130	51											0.6809 ± 0.0003	6%
	130	51											0.7320 ± 0.0001	22%
	130	51											0.7394 ± 0.0001	100%
	130	51											0.8394 ± 0.0001	100%
	130	51											0.9349 ± 0.0002	19%
	130	51											1.0002 ± 0.0004	2%
	130	51											1.0895 ± 0.0004	4%
	130	51											1.1414 ± 0.0004	2%
	130	51											1.2923 ± 0.0004	4%
	130	51											1.4437 ± 0.0005	2%
	130	51											1.5819 ± 0.0008	2%
	130	51											1.7626 ± 0.0005	2%
	130	51											1.9974 ± 0.0005	2%
$_{51}Sb^{131}$	131	51		130.911950	23.0 m.	β-	3.2	1.3			7/2 +		0.6423 ± 0.0001	22%
	131	51						3.0					0.6579 ± 0.0004	7%
	131	51											0.7263 ± 0.0001	4%
	131	51											0.8546 ± 0.0005	3%
	131	51											0.9331 ± 0.0001	25%
	131	51											0.9434 ± 0.0001	44%
	131	51											1.1236 ± 0.0002	8%
	131	51											1.2074 ± 0.0001	4%
	131	51											1.2338 ± 0.0002	2%

TABLE OF THE ISOTOPES (Continued)

Isotope	A	Z	% Natural abundance	Atomic mass	Half-life	Decay mode	Decay energy (MeV)	Particle energy (MeV)	Particle intensity	Thermal neutron cross section	Spin (h/2π)	μ Nucl. mag. moment	Gamma-ray energy (MeV)	Gamma-ray intensity
	131	51											1.2675 ± 0.0002	3%
	131	51											1.7220 ± 0.0005	2.3%
	131	51											1.8544 ± 0.0003	4%
	131	51											2.1799 ± 0.0004	2%
	131	51											2.3356 ± 0.0004	2%
	131	51											2.3989 ± 0.0007	1%
	131	51											2.6623 ± 0.0002	1%
$_{51}Sb^{132m}$	132	51			3.07 m.	β-		3.0			4+		0.1034 ± 0.0001	14%
	132	51						4.0					0.3538 ± 0.0002	3%
	132	51											0.3823 ± 0.0001	8%
	132	51											0.4368 ± 0.0002	3%
	132	51											0.4473 ± 0.0002	2%
	132	51											0.6099 ± 0.0003	2%
	132	51											0.6356 ± 0.0002	10%
	132	51											0.6968 ± 0.0001	86%
	132	51											0.8141 ± 0.0003	5%
	132	51											0.8166 ± 0.0002	11%
	132	51											0.9739 ± 0.0001	98%
	132	52											0.9896 ± 0.0001	15%
	132	51											1.0932 ± 0.0003	5%
	132	51											1.1335 ± 0.0002	6%
	132	51											1.1522 ± 0.0004	3%
	132	51											1.1965 ± 0.0004	3%
	132	51											1.2133 ± 0.0004	2%
	132	51											1.4363 ± 0.0004	2%
	132	51											1.5135 ± 0.0005	2%
	132	51											1.6445 ± 0.0008	2%
	132	51											1.7880 ± 0.0008	3%
	132	51											2.2804 ± 0.0008	1%
	132	51											2.5883 ± 0.0008	1%
$_{51}Sb^{132}$	132	51		131.914410	4.12 m.	β-	5.5				8-		0.1034 ± 0.0001	35%
	132	51											0.1506 ± 0.0001	66%
	132	51											0.2760 ± 0.0002	4%
	132	51											0.2930 ± 0.0002	4%
	132	51											0.3686 ± 0.0002	7%
	132	51											0.3823 ± 0.0001	7%
	132	51											0.4965 ± 0.0002	13%
	132	51											0.6968 ± 0.0001	100%
	132	51											0.8819 ± 0.0003	6%
	132	51											0.9739 ± 0.0001	100%
	132	51											1.0415 ± 0.0003	18%
	132	51											1.1669 ± 0.0004	10%
	132	51											1.3783 ± 0.0004	4%
	132	51											1.7637 ± 0.0008	
	132	51											1.8546 ± 0.0008	2%
	132	51											2.664 ± 0.001	4%
$_{51}Sb^{133}$	133	51		132.915150	2.5 m.	β-	4.0	1.0			7/2+		0.4235 ± 0.0005	6%
	133	51											0.6318 ± 0.0005	19%
	133	51											0.8165 ± 0.0005	15%
	133	51											0.8385 ± 0.0005	8%
	133	51											1.0250 ± 0.0005	2%
	133	51											1.0764 ± 0.0005	30%
	133	51											1.3050 ± 0.0005	2%
	133	51											1.4900 ± 0.001	1%
	133	51											1.7282 ± 0.001	5%
	133	51											2.7517 ± 0.001	9%
$_{51}Sb^{134m}$	134	51			0.85 s.	β-		8.4						
$_{51}Sb^{134}$	134	51		133.920550	10.4 s.	β-	8.4	6.1					0.1152 ± 0.0001	49%
	134	51											0.2970 ± 0.0001	97%
	134	51											0.7063 ± 0.0001	57%
	134	51											1.2791 ± 0.0001	100%
$_{51}Sb^{135}$	135	51		134.924520	1.71 s.	β-					7/2+		1.127	
	135	51											1.279	
Te		52								5.4 b.				
$_{52}Te^{108}$	108	52		107.929550	2.1 s.	α(68%)					0+			
	108	52				β+, EC (32%)	6.9							
$_{52}Te^{109}$	109	52		108.927370	4.2 s.	β+, EC (96%)	8.6							
	109	52				α(4%)								
$_{52}Te^{110}$	110	52		109.922560	18.5 s.	β+,E.C.	5.3				0+		ann.rad.	
	110	52											0.2191 ± 0.0006	
	110	52											0.6059 ± 0.0006	
$_{52}Te^{111}$	111	52		110.921130	19.3 s.	β+,E.C.	≈7				(7/2+)		ann.rad.	
	111	52											0.267	
	111	52											0.322	
	111	52											0.341	
$_{52}Te^{112}$	112	52		111.917020	2.0 m.	β+,E.C.	4.3				0+		ann.rad.	
	112	52											0.0386 ± 0.0003	16 +
	112	52											0.1042 ± 0.0003	27
	112	52											0.1327 ± 0.0002	23
	112	52											0.1674 ± 0.0002	7
	112	52											0.2364 ± 0.0004	9
	112	52											0.2962 ± 0.0002	86
	112	52											0.3509 ± 0.0003	36

Isotope	A	Z	% Natural abundance	Atomic mass	Half-life	Decay mode	Decay energy (MeV)	Particle energy (MeV)	Particle intensity	Thermal neutron cross section	Spin (h/2π)	μ Nucl. mag. moment	Gamma-ray energy (MeV)	Gamma-ray intensity
	112	52											0.3727 ± 0.0002	100
	112	52											0.4187 ± 0.0002	57
	112	52											0.4769 ± 0.0003	14
	112	52											0.4940 ± 0.0003	30
	112	52											0.6904 ± 0.0004	10
	112	52											0.7430 ± 0.0002	11
	112	52											0.7973 ± 0.0002	24
	112	52											0.8074 ± 0.0004	9
	112	52											0.8201 ± 0.0002	17
	112	52											0.8819 ± 0.0003	10
	112	52											0.9248 ± 0.0006	11
	112	52											0.9284 ± 0.0004	11
	112	52											0.9713 ± 0.0002	23
	112	52											1.2824 ± 0.0009	17
	112	52											1.2872 ± 0.0008	10
	112	52											1.5026 ± 0.0006	15
	112	52											1.6576 ± 0.0003	14
	112	52											1.9637 ± 0.0004	17
$_{52}Te^{113}$	113	52		112.915920	1.7 s.	β+(85%)	≈ 6.1	4.5			(7/2+)		ann.rad.	
	113	51				E.C.(15%)							Sb k x-ray	6%
	113	52											0.6448 ± 0.0002	6%
	113	52											0.8144 ± 0.0003	22%
	113	52											1.0181 ± 0.0004	12%
	113	52											1.1812 ± 0.0004	12%
	113	52											1.2567 ± 0.0005	5.5%
	113	52											1.5503 ± 0.0007	2.2%
	113	52											1.8681 ± 0.0009	2.4%
	113	52											2.0937 ± 0.001	3%
	113	52											2.1155 ± 0.001	2%
	113	52											2.2212 ± 0.0009	2%
	113	52											2.5352 ± 0.0005	2.6%
	113	52											2.5524 ± 0.0009	1.5%
	113	52											2.6065 ± 0.0005	1.8%
$_{52}Te^{114}$	114	52		113.912230	15 m.	β+(40%)	2.7				0+		ann.rad.	
	114	52				E.C.(60%)							Sb k x-ray	55%
	114	52											0.0838 ± 0.0001	7%
	114	52											0.0903 ± 0.0001	10%
	114	52											0.2446 ± 0.0001	3%
	114	52											0.4972 ± 0.0001	3%
	114	52											0.7266 ± 0.0002	4%
	114	52											1.4176 ± 0.0002	3%
	114	52											1.8417 ± 0.0006	2%
	114	52											1.8973 ± 0.0005	4%
$_{52}Te^{115m}$	115	52			6.7 m.	β+(45%)					(1/2+)		ann.rad.	
	115	52				E.C.(55%)							Sb k x-ray	20%
	115	52											0.5487 ± 0.0002	4%
	115	52											0.6106 ± 0.0002	4%
	115	52											0.7236 ± 0.0001	18%
	115	52											0.7704 ± 0.0001	35%
	115	52											1.0319 ± 0.0002	8%
	115	52											1.0987 ± 0.0001	9%
	115	52											1.1557 ± 0.0004	3%
	115	52											1.2793 ± 0.0002	10%
	115	52											1.3508 ± 0.0002	8%
	115	52											1.4081 ± 0.0003	3%
	115	52											1.4917 ± 0.0003	3%
	115	52											1.5041 ± 0.0002	10%
	115	52											1.5617 ± 0.0004	4%
	115	52											1.6548 ± 0.0004	6%
	115	52											1.9360 ± 0.0003	3%
	115	52											2.1044 ± 0.0002	8%
	115	52											2.2153 ± 0.0004	6%
$_{52}Te^{115}$	115	52		114.911700	5.8 m.	β+(45%)	4.8	2.7			7/2+		ann.rad.	
	115	52				E.C.(55%)							Sb k x-ray	20%
	115	52											0.3745 ± 0.0002	3%
	115	52											0.5684 ± 0.0001	3%
	115	52											0.6033 ± 0.0001	4%
	115	52											0.6570 ± 0.0001	7%
	115	52											0.7236 ± 0.0001	32%
	115	52											1.0986 ± 0.0001	17%
	115	52											1.2905 ± 0.0001	6%
	115	52											1.3004 ± 0.0003	2%
	115	52											1.3268 ± 0.0001	22%
	115	52											1.3806 ± 0.0001	24%
	115	52											1.5997 ± 0.0001	3%
	115	52											(0.22 - 2.7)	
$_{52}Te^{116}$	116	52		115.908450	2.5 h.	E.C.	≈ 1.6				0+		Sb k x-ray	60%
	116	52											0.0937 ± 0.0001	30%
	116	52											0.1030 ± 0.0001	2%
	116	52											0.6287 ± 0.0001	3%
	116	52											0.6379 ± 0.0002	0.7%
	116	52											1.0553 ± 0.0002	0.6%
$_{52}Te^{117}$	117	52		116.908630	62 m.	E.C.(75%)	3.53	1.75			1/2+		ann.rad.	
	117	52				β+(25%)							Sb k x-ray	30%
	117	52											0.9197 ± 0.0006	65%

TABLE OF THE ISOTOPES (Continued)

Isotope	A	Z	% Natural abundance	Atomic mass	Half-life	Decay mode	Decay energy (MeV)	Particle energy (MeV)	Particle intensity	Thermal neutron cross section	Spin (h/2π)	μ Nucl. mag. moment	Gamma-ray energy (MeV)	Gamma-ray intensity
	117	52											0.9239 ± 0.0007	6%
	117	52											0.9967 ± 0.0007	4%
	117	52											1.0907 ± 0.0007	7%
	117	52											1.5651 ± 002	1%
	117	52											1.7164 ± 0.0007	16%
	117	52											2.3000 ± 0.0007	11%
$_{52}Te^{118}$	118	52		117.905908	6.00 d.	E.C.					0+		Sb k x-ray	40%
$_{52}Te^{119m}$	119	52			4.7 d.	E.C.					11/2-		Sb k x-ray	40%
	119	52											0.15360 ± 0.00003	66%
	119	52											0.2705 ± 0.0001	28%
	119	52											0.9126 ± 0.0001	6%
	119	52											0.9422 ± 0.0001	5%
	119	52											0.9764 ± 0.0001	3%
	119	52											0.9793 ± 0.0001	3%
	119	52											1.0132 ± 0.0001	2.5%
	119	52											1.0484 ± 0.0001	3%
	119	52											1.0957 ± 0.0001	2%
	119	52											1.1368 ± 0.0001	8%
	119	52											1.21271 ± 00005	66%
	119	52											1.3664 ± 0.0002	1%
	119	52											2.0896 ± 0.0001	5%
$_{52}Te^{119}$	119	52		118.906411	16.05 h.	β+(2%)	2.29	0.63			1/2+		ann.rad.	
	119	52				E.C.(98%)							Sb k x-ray	40%
	119	52											0.6440 ± 0.0001	84%
	119	52											0.6998 ± 0.0001	10%
	119	52											1.4132 ± 0.0001	1%
	119	52											1.7497 ± 0.0001	4%
	119	52											1.1056 ± 0.0001	0.6%
	119	52											1.1770 ± 0.0001	0.7%
$_{52}Te^{120}$	120	52	0.096%	119.904048						0.3 b.	0+			
$_{52}Te^{121m}$	121	52			I.T.(89%)	0.29							Te k x-ray	19%
	121	52			E.C.(11%)								0.2122 ± 0.0001	83%
	121	52											1.1021 ± 0.0001	2.5%
$_{52}Te^{121}$	121	52		120.904947	16.8 d.	E.C.	1.05				1/2+		Sb k x-ray	40%
	121	52											0.4705 ± 0.0001	1%
	121	52											0.5076 ± 0.0001	18%
	121	52											0.5731 ± 0.0001	80%
$_{52}Te^{122}$	122	52	2.60%	121.903054						3 b.	0+			
$_{52}Te^{123m}$	123	52			119.7 d.	I.T.	0.247				11/2-		Te k x-ray	26%
	123	52											0.0885 ± 0.0001	0.09%
	123	52											0.1590 ± 0.0001	84%
$_{52}Te^{123}$	123	52	0.903%	122.904271	1.3x10^{13}y.	E.C.	0.052			420 b.	1/2+	-0.7359		
$_{52}Te^{124}$	124	52	4.816%	123.902823						(0.04 + 7) b.	0+			
$_{52}Te^{125m}$	125	52			58 d.	I.T.	0.145				11/2-	+0.7	Te k x-ray	63%
	125	52											0.0355 ± 0.0001	6.7%
	125	52											0.1093 ± 0.0001	0.3%
$_{52}Te^{125}$	125	52	7.14%	124.904433						1.6 b.	1/2+	-0.8871		
$_{52}Te^{126}$	126	52	18.95%	125.903314						(0.13 + 0.9) b.	0+			
$_{52}Te^{127m}$	127	52			109 d.	I.T.(98%)	0.088				11/2-		Te k x-ray	19%
	127	52				β-(2%)	0.77						0.0883 ± 0.0001	0.08%
$_{52}Te^{127}$	127	52		126.905227	9.5 h.	β-	0.697	0.697			3/2+		0.3603 ± 0.0001	0.1%
$_{52}Te^{128}$	128	52	31.69%	127.904463						(0.015 + 0.20) b.	0+			
$_{52}Te^{129m}$	129	52			33.4 d.	I.T.(63%)	0.105	0.91			11/2-		Te k x-ray	15%
	129	52				β-(37%)		1.60					0.45984 ± 0.00004	4.5%(D)
	129	52											0.6959 ± 0.0001	3%
	129	52											0.7296 ± 0.0001	0.7%
$_{52}Te^{129}$	129	52		128.906594	69.5 m.	β-	1.499	0.99	9%		3/2-		0.0278 ± 0.0001	16%
	129	52						1.45	89%				0.45984 ± 00004	7%
	129	52											0.48728 ± 0.00003	1.3%
	129	52											0.80198 ± 0.00004	0.2%
	129	52											1.08378 ± 0.00004	0.5%
	129	52											1.11157 ± 0.00004	0.2%
$_{52}Te^{130}$	130	52	33.80%	129.906229	2.4x10^{21}y.					(0.02 + 0.22) b.	0+			
$_{52}Te^{131m}$	131	52			32.4 h.	β-(78%)	2.4	0.42			11/2-		0.0811 ± 0.0001	4%
	131	52				I.T.(22%)	0.18						0.1021 ± 0.0001	8%
	131	52											0.14973 ± 0.00002	20%(D)
	131	52											0.20066 ± 0.00004	7%
	131	52											0.24095 ± 0.00004	7.6%
	131	52											0.33431 ± 0.00003	9.5%
	131	52											0.66506 ± 0.00005	4%
	131	52											0.77369 ± 0.00004	38%

Isotope	A	Z	% Natural abundance	Atomic mass	Half-life	Decay mode	Decay energy (MeV)	Particle energy (MeV)	Particle intensity	Thermal neutron cross section	Spin (h/2π)	μ Nucl. mag. moment	Gamma-ray energy (MeV)	Gamma-ray intensity
	131	52											0.78249 ± 0.00004	7.8%
	131	52											0.79375 ± 0.00003	14%
	131	52											0.82278 ± 0.00004	6%
	131	52											0.85225 ± 0.00005	21%
	131	52											0.9099 ± 0.0001	3%
	131	52											1.059 ± 0.001	1.5%
	131	52											1.12551 ± 0.00006	11%
	131	52											1.20657 ± 0.00006	9.7%
	131	52											1.6457 ± 0.0001	1.2%
	131	52											1.8876 ± 0.0001	1.3%
	131	52											2.0011 ± 0.0001	2%
$_{52}Te^{131}$	131	52		131.908528	25.0 m.	β-	2.249	1.35	12%		3/2+		0.14973 ± 0.00002	69%
	131	52						1.69	22%				0.45327 ± 0.00004	18%
	131	52						2.14	60%				0.49269 ± 0.00005	5%
	131	52											0.60205 ± 0.00005	4%
	131	52											0.65426 ± 0.00005	1.5%
	131	52											0.94857 ± 0.00004	2.3%
	131	52											0.99719 ± 0.00004	3.3%
	131	52											1.14698 ± 0.00006	5%
$_{52}Te^{132}$	132	52		131.908517	78.2 h.	β-	2.249	1.35	12%		3/2+		0.049725 ± 0.00005	14%
	132	52						1.69	22%				0.11198 ± 0.00005	1.8%
	132	52											0.11645 ± 0.00004	1.9%
	132	52											0.22830 ± 0.00002	88%
$_{52}Te^{133m}$	133	52			55.4 m.	β-(82%)		2.4	30%		11/2-		Te k x-ray	
	133	52				I.T.(18%)	0.334						0.0949 ± 0.0002	3%
	133	52											0.1689 ± 0.0001	5%
	133	52											0.2134 ± 0.0001	2%
	133	52											0.3121 ± 0.0003	17%(D)
	133	52											0.3341 ± 0.0001	7%
	133	52											0.4290 ± 0.0001	2%
$_{52}Te^{133}$	133	52		132.910910	12.5 m.	β-	3.0	2.3	25%				0.3121 ± 0.0003	73%
	133	52											0.4079 ± 0.0003	31%
	133	52											0.7201 ± 0.0004	7%
	133	52											0.7874 ± 0.0005	6%
	133	52											0.8445 ± 0.0006	3%
	133	52											0.9311 ± 0.0005	5%
	133	52											1.0003 ± 0.0008	6%
	133	52											1.0210 ± 0.0006	3%
	133	52											1.0618 ± 0.0008	1%
	133	52											1.2520 ± 0.0007	1%
	133	52											1.3334 ± 0.0005	10%
	133	52											1.7175 ± 0.0006	3%
	133	52											1.8818 ± 0.0007	1.5%
$_{52}Te^{134}$	134	52		133.911520	42 m.	β-	1.6	0.6			0+		0.0794 ± 0.0001	21%
	134	52						0.7					0.1809 ± 0.0001	18%
	134	52											0.2012 ± 0.0001	9%
	134	52											0.2105 ± 0.0001	22%
	134	52											0.2780 ± 0.0001	21%
	134	52											0.4351 ± 0.0001	19%
	134	52											0.4610 ± 0.0001	10%
	134	52											0.4646 ± 0.0001	4.7%
	134	52											0.5660 ± 0.0001	18%
	134	52											0.6363 ± 0.0002	2%
	134	52											0.6658 ± 0.0001	1%
	134	52											0.7130 ± 0.0001	4.7%
	134	52											0.7426 ± 0.0001	15%
	134	52											0.7672 ± 0.0001	29%
	134	52											0.8441 ± 0.0001	1%
	134	52											0.9255 ± 0.0001	1.5%
$_{52}Te^{135}$	135	52		134.916420	19.2 s.	β-	6.0	5.4					0.267	8 +
	135	52											0.603	100
	135	52											0.870	23
$_{52}Te^{136}$	136	52		135.920120	18 s.	β-	5.1	2.5			0+		0.0873 ± 0.0002	13%
	136	52											0.1350 ± 0.0003	3%
	136	52											0.3326 ± 0.0002	21%
	136	52											0.3561 ± 0.0004	2%
	136	52											0.4913 ± 0.0003	3%

Isotope	A	Z	% Natural abundance	Atomic mass	Half-life	Decay mode	Decay energy (MeV)	Particle energy (MeV)	Particle intensity	Thermal neutron cross section	Spin (h/2π)	μ Nucl. mag. moment	Gamma-ray energy (MeV)	Gamma-ray intensity
	136	52											0.5423 ± 0.0003	2%
	136	52											0.5786 ± 0.0002	20%
	136	52											0.6307 ± 0.0002	12%
	136	52											0.7382 ± 0.0002	6%
	136	52											1.3412 ± 0.0005	2%
	136	52											1.5669 ± 0.0005	1%
	136	52											2.0779 ± 0.0003	25%
	136	52											2.4969 ± 0.0005	5%
	136	52											2.5694 ± 0.0003	17%
	136	52											2.6048 ± 0.0006	1%
	136	52											2.6560 ± 0.0006	1%
	136	52											2.8040 ± 0.0006	2%
	136	52											3.0495 ± 0.0006	2%
	136	52											3.2351 ± 0.0004	17%
$_{52}Te^{137}$	137	52		136.925410	4 s.	β-(98%)							0.2436 ± 0.0003	
	137	52				n(2%)								
I		53								6.2 b.				
$_{53}I^{110}$	110	53		109.935060	0.65 s.	β+, EC (83%)	≈ 12						ann.rad.	
	110	53				α(17%)								
	110	53				p (11%)								
$_{53}I^{111}$	111	53		110.930250	7.5 s.	β+,E.C.	≈ 8.5						ann.rad.	
	111	53											0.2665 ± 0.0006	
	112	53											0.3215 ± 0.0006	
	112	53											0.3412 ± 0.0006	
$_{53}I^{112}$	112	53		111.927940	3.4 s.	β+,E.C.	≈ 10						ann.rad.	
	112	53											0.6889 ± 0.0006	
	112	53											0.7869 ± 0.0006	
$_{53}I^{113}$	113	53		112.923650	≈ 5.9 s.	β+,E.C.	≈ 7.2						ann.rad.	
	113	53											0.0550 ± 0.0002	32 +
	113	53											0.1600 ± 0.0002	14
	113	53											0.2165 ± 0.0002	7
	113	53											0.3204 ± 0.0002	33
	113	53											0.3515 ± 0.0002	43
	113	53											0.4061 ± 0.0002	8
	113	53											0.4625 ± 0.0002	100
	113	53											0.5230 ± 0.0005	7
	113	53											0.5674 ± 0.0002	36
	113	53											0.6086 ± 0.0005	6
	113	53											0.6224 ± 0.0002	74
	113	53											0.6280 ± 0.0002	13
	113	53											0.6514 ± 0.0005	3
	113	53											0.6902 ± 0.0005	8
	113	53											0.6962 ± 0.0005	3
	113	53											0.7740 ± 0.0005	8
	113	53											0.7982 ± 0.0002	12
	113	53											0.8021 ± 0.0005	8
	113	53											0.8960 ± 0.0005	10
	113	53											0.9291 ± 0.0003	8
	113	53											1.1610 ± 0.0005	9
	113	53											1.4224 ± 0.0003	11
$_{53}I^{114}$	114	53		113.921790	2.1 s.	β+,E.C.	≈ 8.9						ann.rad.	
	114	53											0.6826 ± 0.0006	
	114	53											0.7088 ± 0.0006	
$_{53}I^{115}$	115	53		114.918090	28 s.	β+,E.C.	≈ 6.0						ann.rad.	
	115	53											0.275	
	115	53											0.284	
	115	53											0.460	
	115	53											0.709	
$_{53}I^{116}$	116	53		115.916780	2.9 s.	β+(97%)	7.8	6.7			1+		ann.rad.	
	116	53				E.C.(3%)							0.5402 ± 0.0004	1.2%
	116	53											0.6789 ± 0.0003	8.3%
$_{53}I^{117}$	117	53		116.913460	2.3 m.	β+,E.C.	4.3	3.3			(5/2+)		ann.rad.	
	117	53											0.2744	27 +
	117	53											0.3032	2
	117	53											0.3259	100
	117	53											0.4972	1
	117	53											0.6831	2
	117	53											0.8373	2
$_{53}I^{118m}$	118	53			≈ 8.5 m.	β+,E.C.		4.9					ann.rad.	
	118	53				I.T.							0.104	
	118	53											0.5998 ± 0.0003	100 +
	118	53											0.6052 ± 0.0004	100
	118	53											0.6138 ± 0.0007	54
$_{53}I^{118}$	118	53		117.912780	14.3 m.	β+,E.C.	≈ 6.4						ann.rad.	
	118	53											0.3524 ± 0.0005	3%
	118	53											0.5448 ± 0.0004	12%
	118	53											0.5518 ± 0.0006	2%
	118	53											0.6052 ± 0.0004	95%
	118	53											0.7407 ± 0.0005	2%
	118	53											1.1499 ± 0.0005	5%
	118	53											1.2570 ± 0.0006	4%
	118	53											1.3384 ± 0.0005	12%
$_{53}I^{119}$	119	53		118.910030	19.2 m.	β+(54%)	3.4	2.1			(5/2+)		ann.rad.	

Isotope	A	Z	% Natural abundance	Atomic mass	Half-life	Decay mode	Decay energy (MeV)	Particle energy (MeV)	Particle intensity	Thermal neutron cross section	Spin (h/2π)	μ Nucl. mag. moment	Gamma-ray energy (MeV)	Gamma-ray intensity
	119	53				E.C.(46%)							Te k x-ray	
	119	53											0.2575 ± 0.0001	90%
	119	53											0.3206 ± 0.0001	2%
	119	53											0.5570 ± 0002	2%
	119	53											0.6356 ± 0.0001	3%
	119	53											0.7062 ± 0.0002	1.3%
	119	53											1.0034 ± 0.0002	0.5%
$_{53}I^{120m}$	120	53			53 m.	β+(80%)							ann.rad.	
	120	53				E.C.(20%)							Te k x-ray	
	120	53											0.4257 ± 0.0005	3%
	120	53											0.5604 ± 0.0003	100%
	120	53											0.6011 ± 0.0003	8%
	120	53											0.6147 ± 0.0003	67%
	120	53											0.6545 ± 0.0005	2%
	120	53											0.7039 ± 0.0005	2%
	120	53											0.7622 ± 0.0001	2%
	120	53											0.8818 ± 0.0005	2%
	120	53											0.9213 ± 0.0004	4%
	120	53											1.0315 ± 0.0006	1%
	120	53											1.0399 ± 0.0005	6%
	120	53											1.0592 ± 0.0005	5%
	120	53											1.1586 ± 0.0006	3%
	120	53											1.1973 ± 0.0006	2%
	120	53											1.2613 ± 0.0007	2%
	120	53											1.3346 ± 0.0007	4%
	120	53											1.3459 ± 0.0004	19%
	120	53											1.3635 ± 0.0007	4%
	120	53											1.4021 ± 0.0007	4%
	120	53											1.4050 ± 0.0005	9%
	120	53											1.7614 ± 0.001	4%
	120	53											1.7758 ± 0.001	5%
	120	53											1.8083 ± 0.001	4%
	119	53											2.4032 ± 0.001	7%
	120	53											2.4628 ± 0.0015	4%
	120	53											2.6025 ± 0.002	3%
	120	53											2.8110 ± 0.0015	4%
	120	53											2.8643 ± 0.002	2%
	120	53											2.9329 ± 0.0015	4%
	120	53											3.1051 ± 0.0015	2%
$_{53}I^{120}$	120	53		119.909840	1.35 h.	β+(81%)	5.4				2-		ann.rad.	
	120	53				E.C.(19%)							Tek x-ray	8%
	120	53											0.5427 ± 0.0003	1%
	120	53											0.5604 ± 0.0003	73%
	120	53											0.6011 ± 0.0003	6%
	120	53											0.6411 ± 0.0003	9%
	120	53											1.2016 ± 0.0005	2%
	120	53											1.5230 ± 0.0004	11%
	120	53											1.5347 ± 0.0005	2%
	120	53											2.1880 ± 0.001	1.4%
	120	53											2.4548 ± 0.0005	2%
	120	53											2.4918 ± 0.001	1%
	120	53											2.5644 ± 0.001	2%
	120	53											2.9329 ± 0.0015	0.7%
	120	53											(0.43 - 3.1)	
$_{53}I^{121}$	121	53		120.907394	2.12 h.	β+(13%)	2.28	1.2			5/2+		ann.rad.	
	121	53				E.C.(87%)							Te k x-ray	37%
	121	53											0.2122 ± 0.0001	85%
	121	53											0.5321 ± 0.0001	5.4%
	121	53											0.5988 ± 0.0001	1.4%
	121	53											(0.14 - 1.1)weak	
$_{53}I^{122}$	122	53		121.907595	3.6 m.	β+	4.23	3.1			1+		ann.rad.	
	122	53				E.C.							Te k x-ray	10%
	122	53											0.5641 ± 0.0001	18%
	122	53											0.6928 ± 0.0001	1.3%
	122	53											0.7933 ± 0.0001	1.3%
	122	53											1.7469 ± 0.0001	0.33%
	122	53											2.1923 ± 0.0001	0.26%
$_{53}I^{123}$	123	53		122.905594	13.1 h.	E.C.	1.23				5/2+		Te k x-ray	46%
	123	53											0.1590 ± 0.0001	83%
	123	53											0.4400 ± 0.0001	0.4%
	123	53											0.5290 ± 0.0001	1.4%
	123	53											0.5385 ± 0.0001	0.4%
$_{53}I^{124}$	124	53		123.906207	4.17 d.	β+(23%)	3.16	1.5			2-		ann.rad.	
	124	53				E.C.(77%)		2.1					Te k x-ray	31%
	124	53											0.6027 ± 0.0001	61%
	124	53											0.7228 ± 0.0001	10%
	124	53											1.3255 ± 0.0001	1.4%
	124	53											1.3760 ± 0.0001	1.7%
	124	53											1.5095 ± 0.0001	3%
	124	53											1.6910 ± 0.0001	10%
	124	53											2.0910 ± 0.0001	0.6%
	124	53											2.2833 ± 0.0001	0.7%
	124	53											2.7469 ± 0.0001	0.5%
$_{53}I^{125}$	125	53		124.904620	59.9 d.	E.C.	0.178			900 b.	5/2+	+3.0	Te k x-ray	74%
	125	53											0.0355 ± 0.0001	6.7%

TABLE OF THE ISOTOPES (Continued)

Isotope	A	Z	% Natural abundance	Atomic mass	Half-life	Decay mode	Decay energy (MeV)	Particle energy (MeV)	Particle intensity	Thermal neutron cross section	Spin (h/2π)	μ Nucl. mag. moment	Gamma-ray energy (MeV)	Gamma-ray intensity
$_{53}I^{126}$	126	53		125.905624	13.0 d.	E.C.				9 x 10^3 b.	2-		ann.rad.	
	126	53				β+	2.16	1.13					Te k x-ray	22%
	126	53				β-	1.25	0.86					0.3887 ± 0.0002	32%
	126	53						1,25					0.6622 ± 0.0002	31%
	126	53											0.7538 ± 0.0002	3.9%
	126	53											0.8799 ± 0.0002	0.7%
	126	53											1.4201 ± 0.0004	0.3%
$_{53}I^{127}$	127	53		126.904473						6.2 b.	5/2+	+2.808		
$_{53}I^{128}$	128	53		127.905810	25.00 m.	β-	2.128	2.12					Te k x-ray	2%
	128	53				E.C.	1.257						0.44287 ± 0.00002	16%
	128	53											0.52658 ± 0.00003	1.5%
	128	53											0.74321 ± 0.00004	0.15%
	128	53											0.96943 ± 0.00007	0.38%
$_{53}I^{129}$	129	53		128.904986	1.6x10^7y.	β-	0.193	0.15		(20 + 10) b.	7/2+	+2.617	Xe k x-ray	
	129	53											0.0396 ± 0.0001	7.5%
$_{53}I^{130m}$	130	53		129.906713	9.0 m.	I.T.(83%)	0.048				2+		I k x-ray	14%
	130	53				β-(17%)							0.5361 ± 0.0001	17%
	130	53											0.5861 ± 0.0001	1%
	130	53											1.6141 ± 0.0001	0.5%
$_{53}I^{130}$	130	53		129.906713	12.36 h.	β-	2.98	1.04		18 b.	5+		0.4180 ± 0.0001	34%
	130	53											0.5361 ± 0.0001	99%
	130	53											0.5391 ± 0.0001	1.4%
	130	53											0.5861 ± 0.0001	1.7%
	130	53											0.6685 ± 0.0001	96%
	130	53											0.7395 ± 0.0001	82%
	130	53											1.1575 ± 0.0001	11%
	130	53											1.2721 ± 0.0001	0.8%
$_{53}I^{131}$	131	53		130.906114	8.040 d.	β-	0.971	0.606		80 b.	7/2+	+2.74	0.08017 ± 0.0005	2.6%
	131	53											0.17725 ± 0.0009	0.27%
	131	53											0.27248 ± 0.00008	0.9%
	131	53											0.28431 ± 0.00003	6%
	131	53											0.32574 ± 0.00005	0.25%
	131	53											0.36446 ± 0.00003	81%
	131	53											0.50300 ± 0.00005	0.4%
	131	53											0.63699 ± 0.00005	7.3%
	131	53											0.64266 ± 0.00005	0.2%
	131	53											0.72288 ± 0.00005	1.3%
$_{53}I^{132m}$	132	53			83 m.	β-(14%)	3.58	0.80			4+	3.08	I k x-ray	13%
	132	53				I.T.(86%)		1.03					0.0980 ± 0.001	4%
	132	53						1.2					0.5059 ± 0.0001	5%
	132	53						1.6					0.52264 ± 0.00003	16%
	132	53						2.16					0.63019 ± 0.00003	14%
	132	53											0.6506 ± 0.0002	2.7%
	132	53											0.66768 ± 0.00003	99%
	132	53											0.6698 ± 0.0003	5%
	132	53											0.6716 ± 0.0004	5.2%
	132	53											0.72695 ± 0.00003	6.5%
	132	53											0.77260 ± 0.00003	76%
	132	53											0.8098 ± 0.0001	3%
	132	53											0.81228 ± 0.00001	5.6%
	132	53											0.87697 ± 0.00001	1.1%
	132	53											0.95457 ± 0.00003	18%
	132	53											1.13602 ± 0.00004	3%
	132	53											1.1436 ± 0.0001	1.4%
	132	53											1.1732 ± 0.0001	1.1%
	132	53											1.2953 ± 0.00001	2%
	132	53											1.37201 ± 0.00005	2.5%
	132	53											1.39895 ± 0.00005	7.1%
	132	53											1.44254 ± 0.00005	1.4%
	132	53											1.92096 ± 0.00008	1.2%

Isotope	A	Z	% Natural abundance	Atomic mass	Half-life	Decay mode	Decay energy (MeV)	Particle energy (MeV)	Particle intensity	Thermal neutron cross section	Spin (h/2π)	μ Nucl. mag. moment	Gamma-ray energy (MeV)	Gamma-ray intensity
	132	53											2.00225 ± 0.00007	1.1%
	132	53											2.0866 ± 0.0001	0.2%
	132	53											2.1724 ± 0.0001	0.2%
$_{53}I^{133m}$	133	53			9 s.	I.T.	1.63				19/2-		I kx-ray	35%
	133	53											0.0730 ± 0.001	4%
	133	53											0.6474 ± 0.0001	100%
	133	53											0.9126 ± 0.0001	100%
$_{53}I^{133}$	133	53		132.907780	20.8 h.	β-	1.76	1.23	85%		7/2+	+2.84	0.51056 ± 0.00002	1.8%
	133	53											0.52989 ± 0.00002	86%
	133	53											0.61794 ± 0.00003	0.5%
	133	53											0.68031 ± 0.00004	0.65%
	133	53											0.70661 ± 0.00002	1.5%
	133	53											0.76836 ± 0.00004	0.46%
	133	53											0.82061 ± 0.00003	0.15%
	133	53											0.85636 ± 0.00003	1.2%
	133	53											0.87537 ± 0.00002	45%
	133	53											0.909	0.2%
	133	53											1.05231 ± 0.00005	0.56%
	133	53											1.06017 ± 0.00005	0.14%
	133	53											1.23653 ± 0.00003	1.5%
	133	53											1.29833 ± 0.00003	2.3%
	133	53											1.35054 ± 0.00008	0.15%
$_{53}I^{134m}$	134	53			3.5 m.	I.T.(98%)	0.316				8-		I k x-ray	39%
	134	53				β-(2%)							0.0444 ± 0.0002	10%
	134	53											0.2719 ± 0.0003	79%
$_{53}I^{134}$	134	53		133.909850	52.5 m.	β-	4.2	1.25			4+		0.1354 ± 0.0001	4%
	134	53											0.2355 ± 0.0001	2%
	134	53											0.40545 ± 0.00002	7%
	134	53											0.4333 ± 0.0001	4%
	134	53											0.5144 ± 0.0001	2%
	134	53											0.5408 ± 0.0001	7.6%
	134	53											0.5954 ± 0.0001	11%
	134	53											0.6218 ± 0.0001	11%
	134	53											0.6280 ± 0.0001	2%
	134	53											0.67734 ± 0.00003	7.8%
	134	53											0.7307 ± 0.0001	1.8%
	134	53											0.76667 ± 0.00003	4.1%
	134	53											0.84702 ± 0.00003	95%
	134	53											0.8573 ± 0.0001	7%
	134	53											0.88409 ± 0.00002	65%
	134	53											0.9479 ± 0.0001	4%
	134	53											0.9747 ± 0.0001	4.8%
	134	53											1.0403 ± 0.0001	1.9%
	134	53											1.07255 ± 0.00003	15%
	134	53											1.13616 ± 0.00004	9.2%
	134	53											1.4552 ± 0.0001	2.3%
	134	53											1.61380 ± 0.00004	4.3%
	134	53											1.71419 ± 0.00005	2.7%
	134	53											1.80684 ± 0.00004	5.5%
$_{53}I^{135}$	135	53		134.910023	6.585 h.	β-	2.71	0.9			7/2+		0.2884 ± 0.0001	3%
	135	53						1.3					0.41768 ± 0.00003	3.5%
	135	53											0.52658 ± 0.00002	14%(D)
	135	53											0.54658 ± 0.00002	7.1%
	135	53											0.83686 ± 0.00002	6.7%
	135	53											0.97233 ± 0.00004	1.2%

Isotope	A	Z	% Natural abundance	Atomic mass	Half-life	Decay mode	Decay energy (MeV)	Particle energy (MeV)	Particle intensity	Thermal neutron cross section	Spin (h/2π)	μ Nucl. mag. moment	Gamma-ray energy (MeV)	Gamma-ray intensity
	135	53											1.03877 ± 0.00004	7.9%
	135	53											1.10160 ± 0.00005	1.6%
	135	53											1.12402 ± 0.00003	3.6%
	135	53											1.13156 ± 0.00003	22.5%
	135	53											1.26046 ± 0.00003	28.6%
	135	53											1.45766 ± 0.00005	8.6%
	135	53											1.67817 ± 0.00004	9.5%
	135	53											1.70658 ± 0.00004	4.1%
	135	53											1.79133 ± 0.00004	7.7%
	135	53											1.83080 ± 0.00005	0.60%
	135	53											1.9274 ± 0.0001	0.30%
	135	53											2.04610 ± 0.00006	0.90%
	135	53											2.25518 ± 0.00008	0.60%
	135	53											2.40877 ± 0.00005	0.95%
$_{53}I^{136m}$	136	53			45 s.	β-		4.7			6-		0.1973 ± 0.0001	78%
	136	53						5.2					0.3468 ± 0.0001	3%
	136	53											0.3701 ± 0.0001	17%
	136	53											0.3814 ± 0.0001	100%
	136	53											0.7500 ± 0.0001	6%
	136	53											0.8126 ± 0.0001	3%
	136	53											0.9141 ± 0.0002	3%
	136	53											1.3130 ± 0.0001	100%
	136	53											(0.16 - 2.36)	
$_{53}I^{136}$	136	53		135.914650	83.6 s.	β-	6.9	4.7			2-		0.3447 ± 0.0001	2.4%
	136	53											0.9765 ± 0.0002	2.7%
	136	53											1.2468 ± 0.0001	2.3%
	136	53											1.3130 ± 0.0001	67%
	136	53											1.3211 ± 0.0001	25%
	136	53											1.5364 ± 0.0001	1%
	136	53											1.9622 ± 0.0003	2.3%
	136	53											2.2896 ± 0.0002	10%
	136	53											2.4146 ± 0.0002	6.9%
	136	53											2.6342 ± 0.0002	6.8%
	136	53											2.8689 ± 0.0002	4%
	136	53											2.9563 ± 0.0003	0.7%
	136	53											3.1411 ± 0.0003	0.7%
	136	53											3.2118 ± 0.0003	0.5%
	136	53											4.2695 ± 0.0002	0.4%
	136	53											(0.3 - 6.1)	
$_{53}I^{137}$	137	53		136.917870	24.5 s.	β-	5.9	5.0			(7/2+)		0.6010 ± 0.0001	5%
	137	53											1.2180 ± 0.0001	13%
	137	53											1.2201 ± 0.00022	3%
	137	53											1.3026 ± 0.0001	4.4%
	137	53											1.5122 ± 0.0001	1.2%
	137	53											1.5343 ± 0.0001	3.2%
	137	53											1.7661 ± 0.0001	1.2%
	137	53											1.8730 ± 0.0001	1.5%
	137	53											2.0298 ± 0.0001	1.7%
	137	53											2.6297 ± 0.0001	0.8%
	137	53											3.1944 ± 0.0002	0.75%
	137	53											3.7956 ± 0.0002	0.85%
	137	53											3.9962 ± 0.0003	0.5%
	137	53											(0.25 - 4.4)weak	
$_{53}I^{138}$	138	53		137.922370	6.4 s.	β-	7.8	6.9					0.4836 ± 0.0001	5%
	138	53						7.4					0.5888 ± 0.0001	77%
	138	53											0.8307 ± 0.0001	2.2%
	138	53											0.8702 ± 0.0002	4%
	138	53											0.8752 ± 0.0001	13%
	138	53											1.2775 ± 0.0001	3.3%
	138	53											1.3143 ± 0.0001	1.4%
	138	53											1.4269 ± 0.0003	1.6%
	138	53											1.6733 ± 0.0001	1.7%
	138	53											1.8093 ± 0.0002	2.8%
	138	53											2.2622 ± 0.0001	5.3%
	138	53											2.5724 ± 0.0002	1.6%
	138	53											2.8356 ± 0.0002	1.7%
	138	53											3.3103 ± 0.0002	1.5%
	138	53											3.4963 ± 0.0002	1.0%
	138	53											4.1820 ± 0.0002	0.9%
	138	53											4.3189 ± 0.0002	0.6%
	138	53											(0.4 - 5.3)	
$_{53}I^{139}$	139	53		138.926050	2.3 s.	β-	6.8						0.192	

Isotope	A	Z	% Natural abundance	Atomic mass	Half-life	Decay mode	Decay energy (MeV)	Particle energy (MeV)	Particle intensity	Thermal neutron cross section	Spin (h/2π)	μ Nucl. mag. moment	Gamma-ray energy (MeV)	Gamma-ray intensity
	139	53				n							0.198	
	139	53											0.273	
	139	53											0.382	
	139	53											0.386	
	139	53											0.468	
	139	53											0.683	
	139	53											1.313	
$_{53}I^{140}$	140	53			0.86 s.	β-							0.372	
	140	53				n							0.377	
	140	53											0.457	
Xe		54		131.29						25 b.				
$_{54}Xe^{114}$	114	54		113.928170	10.3 s.	β+,E.C.	≈ 6.0				0+		ann.rad.	
	114	54											0.1031 ± 0.0002	
	114	54											0.1616 ± 0.0002	
	114	54											0.3083 ± 0.0002	
	114	54											0.6826 ± 0.0006	(D)
	114	54											0.7088 ± 0.0006	(D)
$_{54}Xe^{115}$	115	54		114.926280	18 s.	β+,E.C.	≈ 7.6						ann.rad.	
$_{54}Xe^{116}$	116	54		115.921610	57 s.	β+,E.C.	≈ 4.5	3.3			0+		ann.rad.	
	116	54											0.1042 ± 0.0002	100 +
	116	54											0.1916 ± 0.0002	38
	116	54											0.2264 ± 0.0002	29
	116	54											0.2477 ± 0.0003	40
	116	54											0.3000 ± 0.0004	12
	116	54											0.3107 ± 0.0004	42
	116	54											0.4127 ± 0.0002	36
	116	54											0.9230 ± 0.001	25
$_{54}Xe^{117}$	117	54		116.920250	61 s.	β+,E.C.	≈ 6.4						ann.rad.	
	117	54											0.0737	15 +
	117	54											0.0949	12
	117	54											0.1121	5
	117	54											0.1171	41
	117	54											0.1554	17
	117	54											0.1607	34
	117	54											0.2033	4
	117	54											0.2214	100
	117	54											0.2570	17
	117	54											0.	2947
	117	54											0.3034	23
	117	54											0.3158	26
	117	54											0.3532	20
	117	54											0.4392	20
	117	54											0.5190	55
	117	54											0.6097	4.5
	117	54											0.6389	50
	117	54											0.6613	56
	117	54											1.5232	24
$_{54}Xe^{118}$	118	54		117.916210	4 m.	β+,E.C.	≈ 3.2	2.7			0+		ann.rad.	
	118	54											0.0535 ± 0.0005	100 +
	118	54											0.0600 ± 0.0005	90
	118	54											0.1199 ± 0.0005	76
	118	54											0.1505 ± 0.0005	44
	118	54											0.2740 ± 0.0005	30
$_{54}Xe^{119}$	119	54		118.915390	5.8 m.	β+,E.C.	5.0	3.5					0.0873 ± 0.0006	43 +
	119	54											0.0910 ± 0.0006	16
	119	54											0.0960 ± 0.0006	38
	119	54											0.1000 ± 0.0006	95
	119	54											0.1417 ± 0.0006	11
	119	54											0.1466 ± 0.0006	8
	119	54											0.2082 ± 0.0006	60
	119	54											0.2318 ± 0.0006	100
	119	54											0.2357 ± 0.0006	10
	119	54											0.2948 ± 0.0006	8
	119	54											0.3080 ± 0.0006	11
	119	54											0.3205 ± 0.0006	12
	119	54											0.4377 ± 0.0006	13
	119	54											0.4615 ± 0.0006	97
	119	54											0.5365 ± 0.0006	9
	119	54											0.6930 ± 0.0006	12
$_{54}Xe^{120}$	120	54		119.911940	40 m.	β+, EC (97%)	2.0				0+		I k x-ray	55%
	120	54				β+(3%)							0.0251 ± 0.0002	30%
	120	54											0.0726 ± 0.0002	9%
	120	54											0.0772 ± 0.0002	4%
	120	54											0.1760 ± 0.0003	5%
	120	54											0.1781 ± 0.0002	7%
	120	54											0.2956 ± 0.0002	1.1%
	120	54											0.3359 ± 0.0002	1%
	120	54											0.4243 ± 0.0003	1.2%
	120	54											0.4492 ± 0.0003	1.7%
	120	54											0.5294 ± 0.0003	1.4%
	120	54											0.5556 ± 0.0003	1.5%
	120	54											0.5904 ± 0.0003	1.6%

Isotope	A	Z	% Natural abundance	Atomic mass	Half-life	Decay mode	Decay energy (MeV)	Particle energy (MeV)	Particle intensity	Thermal neutron cross section	Spin (h/2π)	μ Nucl. mag. moment	Gamma-ray energy (MeV)	Gamma-ray intensity
	120	54											0.6311 ± 0.0003	1.0%
	120	54											0.6789 ± 0.0002	1.6%
	120	54											0.7484 ± 0.0004	1.1%
	120	54											0.7533 ± 0.0003	1.4%
	120	54											0.7625 ± 0.0003	4.5%
	120	54											0.9655 ± 0.0003	1.2%
	120	54											(0.1 - 1.03)many	
$_{54}Xe^{121}$	121	54		120.911450	39 m.	β+(44%)	3.8	2.8			5/2+		ann.rad.	
	121	54				E.C.(56%)							I k x-ray	33%
	121	54											0.0801 ± 0.0002	3%
	121	54											0.0957 ± 0.0002	6%
	121	54											0.1328 ± 0.0002	15%
	121	54											0.1758 ± 0.0003	6%
	121	54											0.2527 ± 0.0002	18%
	121	54											0.3007 ± 0.0002	4%
	121	54											0.3105 ± 0.0002	8%
	121	54											0.4452 ± 0.0002	11%
	121	54											0.5291 ± 0.0004	2%
	121	54											0.6497 ± 0.0003	2%
	121	54											0.8425 ± 0.0003	1%
	121	54											0.9310 ± 0.0003	2%
	121	54											0.9580 ± 0.0003	1.5%
	121	54											1.0356 ± 0.0003	1.5%
	121	54											1.1864 ± 0.0005	0.9%
	121	54											1.5408 ± 0.0004	0.8%
	121	54											1.6315 ± 0.0004	0.8%
	121	54											2.5447 ± 0.0005	1.5%
	121	54											2.6217 ± 0.0005	1.1%
	121	54											2.6434 ± 0.0005	3.1%
	121	54											2.7977 ± 0.0004	1.2%
	121	54											(0.1 - 3.1)many	
$_{54}Xe^{122}$	122	54		121.908170	20.0 h.	E.C.	0.7				0+		I k x-ray	41%
	122	54											0.1486 ± 0.0001	2.6%
	122	54											0.3501 ± 0.0001	7.7%
	122	54											0.4166 ± 01.0001	1.9%
$_{54}Xe^{123}$	123	54		122.908469	2.0 h.	β+(23%)	2.68	1.51			1/2+		ann.rad.	
	123	54				E.C.(77%)							I k x-ray	37%
	123	54											0.1489 ± 0.0002	49%
	123	54											0.1781 ± 0.0002	15%
	123	54											0.3302 ± 0.0002	8%
	123	54											0.8996 ± 0.0004	2.4%
	123	54											1.0934 ± 0.0003	2.8%
	123	54											1.1131 ± 0.0003	1.6%
	123	54											1.8073 ± 0.0003	1.2%
	123	54											(0.1 - 2.1)weak	
$_{54}Xe^{124}$	124	54	0.10%	123.905894						(28 + 140)b.				
$_{54}Xe^{125m}$	125	54			57 s.	I.T.	0.253				(9/2-)		Xe k x-ray	33%
	125	54											0.1111 ± 0.001	62%
	125	54											0.141 ± 0.001	20%
$_{54}Xe^{125}$	125	54		124.906397	17.1 h.	E.C.	1.68	0.47			1/2+		I k x-ray	54%
	125	54											0.0550 ± 0.0001	6%
	125	54											0.1884 ± 0.0001	55%
	125	54											0.2434 ± 0.0001	29%
	125	54											0.4538 ± 0.0001	4%
	125	54											0.8465 ± 0.0004	1%
	125	54											1.1810 ± 0.0004	0.6%
$_{54}Xe^{126}$	126	54	0.09%	125.904281						(0.4 + 3) b.	0+			
$_{54}Xe^{127m}$	127	54			69 s.	I.T.	0.297				(9/2-)		Xe k x-ray	29%
	127	54											0.1246 ± 0.0003	69%
	127	54											0.1725 ± 0.0003	38%
$_{54}Xe^{127}$	127	54		126.905182	36.341 d.	E.C.	0.66				1/2+		I k x-ray	46%
	127	54											0.0576 ± 0.0001	1.2%
	127	54											0.1453 ± 0.0001	4.3%
	127	54											0.1721 ± 0.0001	25%
	127	54											0.2029 ± 0.0001	68%
	127	54											0.3750 ± 0.0001	17%
$_{54}Xe^{128}$	128	54	1.91%	127.903531						(0.5 + 4) b.	0+			
$_{54}Xe^{129m}$	129	54			8.88 d.	I.T.	0.236				11/2-		Xe k x-ray	67%
	129	54											0.0396 ± 0.0001	7.5%
	129	54											0.1966 ± 0.0001	4.6%
$_{54}Xe^{129}$	129	54	26.4%	128.904780						21 b.	1/2+	-0.7768		
$_{54}Xe^{130}$	130	54	4.1%	129.903509						(0.4 + 5) b.	0+			
$_{54}Xe^{131m}$	131	54			11.92 d.	I.T.	0.164				11/2-	+0.6908	Xe k x-ray	28%
	131	54											0.16398 ± 0.00005	2%
$_{54}Xe^{131}$	131	54	21.2%	130.905072						90 b.	3/2+			
$_{54}Xe^{132}$	132	54	26.9%	131.904144						(0.05 + 0.4) b.	0+			
$_{54}Xe^{133m}$	133	54			2.19 d.	I.T.	0.233				11/2-		Xe k x-ray	30%
	133	54											0.23325 ± 0.00005	10%
$_{54}Xe^{133}$	133	54		132.905888	5.25 d.	β-	0.427	0.346	99%		3/2+		Cs k x-ray	24%
	133	54											0.080998 ± 0.0001	36%
	133	54											0.1606 ± 0.0001	0.06%

Isotope	A	Z	% Natural abundance	Atomic mass	Half-life	Decay mode	Decay energy (MeV)	Particle energy (MeV)	Particle intensity	Thermal neutron cross section	Spin (h/2π)	μ Nucl. mag. moment	Gamma-ray energy (MeV)	Gamma-ray intensity
$_{54}Xe^{134}$	134	54	10.4%	133.905395						(0.003 + 0.26) b.	0+			
$_{54}Xe^{135m}$	135	54			15.3 m.	I.T.					11/2-		Xe k x-ray	7%
	135	54											0.52658 ± 0.00002	80%
$_{54}Xe^{135}$	135	54		134.907130	9.10 h.	β-	1.16	0.91		2.6x10⁶ b.	3/2+		0.24975 ± 0.00002	90%
	135	54											0.3584 ± 0.0001	0.2%
	135	54											0.4080 ± 0.0001	0.4%
	135	54											0.60807 ± 0.00004	2.9%
$_{54}Xe^{136}$	136	54	8.9%	135.907214						0.26 b.	0+			
$_{54}Xe^{137}$	137	54		136.911557	3.84 m.	β-	4	18	3.7		7/2-		0.45549 ± 0.00005	31%
	137	54						4.2					0.8489 ± 0.0001	0.6%
	137	54											0.9822 ± 0.0001	0.2%
	137	54											1.2732 ± 0.0001	0.2%
	137	54											1.7834 ± 0.0001	0.4%
	137	54											2.8498 ± 0.0001	0.2%
$_{54}Xe^{138}$	138	54		137.913980	14.1 m.	β-	2.7	0.8			0+		0.1538 ± 0.0001	6%
	138	54						2.4					0.2426 ± 0.0001	3.5%
	138	54											0.2583 ± 0.0001	31%
	138	54											0.3964 ± 0.0001	6%
	138	54											0.4014 ± 0.0001	2.2%
	138	54											0.4345 ± 0.0001	20%
	138	54											0.9171 ± 0.0001	0.9%
	138	54											1.1143 ± 0.0001	1.5%
	138	54											1.76826 ± 0.0001	16.7%
	138	54											1.8509 ± 0.0001	1.5%
	138	54											2.0047 ± 0.0001	5.4%
	138	54											2.0158 ± 0.0001	12.3%
	138	54											2.0792 ± 0.0001	1.4%
	138	54											2.2523 ± 0.0001	2.3%
$_{54}Xe^{139}$	139	54		138.918740	40.4 s.	β-	5.0	4.5			7/2-		0.1750 ± 0.0001	18%
	139	54						5.0					0.2186 ± 0.0001	52%
	139	54											0.2254 ± 0.0001	3%
	139	54											0.2898 ± 0.0001	8%
	139	54											0.2965 ± 0.0001	20%
	139	54											0.3935 ± 0.0001	6%
	139	54											0.4915 ± 0.0001	1.3%
	139	54											0.6128 ± 0.0001	5.1%
	139	54											0.7238 ± 0.0001	1.7%
	139	54											0.7324 ± 0.0001	1.6%
	139	54											0.7880 ± 0.0001	3.1%
	139	54											1.2065 ± 0.0001	0.6%
	139	54											1.2429 ± 0.0001	0.5%
	139	54											1.3449 ± 0.0001	1.1%
	139	54											1.5202 ± 0.0001	0.8%
	139	54											1.6703 ± 0.0001	1.0%
	139	54											1.8960 ± 0.0001	0.6%
	139	54											2.0859 ± 0.0001	0.6%
	139	54											2.3288 ± 0.0001	1.6%
	139	54											(0.1 - 3.37)weak	
$_{54}Xe^{140}$	140	54		139.921620	13.6 s.	β-	4.1				0+		0.0801 ± 0.0001	5%
	140	54											0.1125 ± 0.0001	4%
	140	54											0.1184 ± 0.0001	5%
	140	54											0.2120 ± 0.0001	2%
	140	54											0.2810 ± 0.0001	1.5%
	140	54											0.3900 ± 0.0001	1.5%
	140	54											0.4387 ± 0.0001	2.7%
	140	54											0.4618 ± 0.0001	1.6%
	140	54											0.5149 ± 0.0002	1.1%
	140	54											0.5189 ± 0.0002	1.1%
	140	54											0.5478 ± 0.0002	1.2%
	140	54											0.5573 ± 0.0001	5.3%
	140	54											0.6080 ± 0.0001	2.4%
	140	54											0.6220 ± 0.0001	8%
	140	54											0.6273 ± 0.0002	1%
	140	54											0.6392 ± 0.0002	1.4%
	140	54											0.6534 ± 0.0001	5.1%
	140	54											0.7741 ± 0.0001	4%
	140	54											0.8055 ± 0.0001	21%
	140	54											0.8798 ± 0.0001	3%
	140	54											0.9250 ± 0.0002	1.5%
	140	54											0.9890 ± 0.0001	3%
	140	54											1.1371 ± 0.0001	2.2%
	140	54											1.1767 ± 0.0002	1.2%
	140	54											1.2091 ± 0.0001	1.4%
	140	54											1.3091 ± 0.0001	6.7%
	140	54											1.4137 ± 0.0001	13%
	140	54											1.4276 ± 0.0001	1.2%
	140	54											1.8859 ± 0.0001	0.4%
	140	54											(0.04 - 2.3)weak	
$_{54}Xe^{141}$	141	54		140.926610	1.72 s.	β-	6.2	4.9					0.1187 ± 0.0001	12%
	141	54											0.1876 ± 0.0001	2.6%

TABLE OF THE ISOTOPES (Continued)

Isotope	A	Z	% Natural abundance	Atomic mass	Half-life	Decay mode	Decay energy (MeV)	Particle energy (MeV)	Particle intensity	Thermal neutron cross section	Spin (h/2π)	μ Nucl. mag. moment	Gamma-ray energy (MeV)	Gamma-ray intensity
	141	54											0.4591 ± 0.0001	4.5%
	141	54											0.4678 ± 0.0001	2.9%
	141	54											0.5399 ± 0.0001	5.2%
	141	54											0.5566 ± 0.0002	4.7%
	141	54											0.7553 ± 0.0001	1.2%
	141	54											0.9095 ± 0.0001	22%
	141	54											0.9800 ± 0.0001	1%
	141	54											1.0281 ± 0.0001	1.7%
	141	54											1.0519 ± 0.0001	1.0%
	141	54											1.5570 ± 0.0002	2.9%
	141	54											(0.05 - 2.55)weak	
$_{54}Xe^{142}$	142	54		141.929630	1.2 s.	β-	5.0	3.7			0+		0.0338 ± 0.0001	25 +
	142	54						4.2					0.0729 ± 0.0001	27
	142	54											0.1575 ± 0.0001	17
	142	54											0.1651 ± 0.0001	22
	142	54											0.1917 ± 0.0001	36
	142	54											0.1974 ± 0.0002	7
	142	54											0.2038 ± 0.0001	92
	142	54											0.2117 ± 0.0004	4
	142	54											0.2191 ± 0.0003	5
	142	54											0.2507 ± 0.0001	32
	142	54											0.2867 ± 0.0001	17
	142	54											0.2919 ± 0.0001	17
	142	54											0.3091 ± 0.0001	27
	142	54											0.3347 ± 0.0001	12
	142	54											0.3530 ± 0.0002	13
	142	54											0.3799 ± 0.0001	10
	142	54											0.3942 ± 0.0001	17
	142	54											0.4065 ± 0.0001	11
	142	54											0.4145 ± 0.0001	47
	142	54											0.4324 ± 0.0002	12
	142	54											0.4382 ± 0.0002	6
	142	54											0.4531 ± 0.0001	20
	142	54											0.4682 ± 0.0001	24
	142	54											0.5382 ± 0.0001	77
	142	54											0.5477 ± 0.0002	7
	142	54											0.5578 ± 0.0002	7
	142	54											0.5718 ± 0.0001	100
	142	54											0.6056 ± 0.0001	22
	142	54											0.6181 ± 0.0001	72
	142	54											0.6448 ± 0.0001	63
	142	54											0.6646 ± 0.0001	17
	142	54											0.6722 ± 0.0002	7
	142	54											0.7355 ± 0.0004	6
	142	54											0.7374 ± 0.0002	13
	142	54											0.8012 ± 0.0002	6
	142	54											0.8074 ± 0.0003	6
	142	54											0.8914 ± 0.0001	11
	142	54											0.9912 ± 0.0001	9
	142	54											0.1568 ± 0.0002	7
	142	54											1.1874 ± 0.0002	6
	142	54											1.2192 ± 0.0002	6
	142	54											1.2270 ± 0.0001	19
	142	54											1.2580 ± 0.0001	8
	142	54											1.3001 ± 0.0001	31
	142	54											1.3123 ± 0.0001	21
	142	54											1.4106 ± 0.0001	6
	142	54											1.9020 ± 0.0002	8
Cs		55		132.9054						29 b.				
$_{55}Cs^{114}$	114	55		113.941270	0.57 s.	β+,E.C.							ann.rad.	
	114	55											0.6826 ± 0.0006	
	114	55											0.7088 ± 0.0006	
$_{55}Cs^{115}$	115	55		114.936070	1.4 s.	β+,E.C.							ann.rad.	
$_{55}Cs^{116m}$	116	55			0.7 s.	β+,E.C.							ann.rad.	
	116	55											0.3935 ± 0.0002	
$_{55}Cs^{116}$	116	55		115.933110	3.8 s.	β+,E.C.	≈ 11.1				5		ann.rad.	
	116	55											0.3222 ± 0.0004	4%
	116	55											0.3935 ± 0.0002	93%
	116	55											0.4583 ± 0.0003	3%
	116	55											0.5243 ± 0.0002	75%
	116	55											0.5412 ± 0.0003	6%
	116	55											0.5602 ± 0.0003	7%
	116	55											0.6113 ± 0.0003	6%
	116	55											0.6151 ± 0.0003	30%
	116	55											0.6223 ± 0.0003	10%
	116	55											0.6393 ± 0.0003	7%
	116	55											0.6774 ± 0.0006	3%
	116	55											0.9037 ± 0.0008	3%
	116	55											0.9059 ± 0.0008	2%
	116	55											0.9112 ± 0.0004	3%
	116	55											0.9656 ± 0.0006	1%
	116	55											1.0158 ± 0.0004	3%
	116	55											1.0339 ± 0.0008	2%
	116	55											1.0449 ± 0.0006	1%

Isotope	A	Z	% Natural abundance	Atomic mass	Half-life	Decay mode	Decay energy (MeV)	Particle energy (MeV)	Particle intensity	Thermal neutron cross section	Spin (h/2π)	μ Nucl. mag. moment	Gamma-ray energy (MeV)	Gamma-ray intensity
	116	55											1.0615 ± 0.0004	7%
	116	55											1.0721 ± 0.0006	2%
	116	55											1.0807 ± 0.0004	7%
	116	55											1.1680 ± 0.0008	3%
	116	55											1.2470 ± 0.0008	2%
	116	55											1.3215 ± 0.0008	2%
	116	55											1.4460 ± 0.0008	1.5%
$_{55}Cs^{117}$	117	55		116.928900	6.7 s.	β+,E.C.	≈ 8.1						ann.rad.	
$_{55}Cs^{118}$	118	55		117.926740	15 s.	β+,E.C.	≈ 9.8						ann.rad.	
	118	55											0.3372 ± 0.0002	100 +
	118	55											0.4727 ± 0.0002	37
	118	55											0.4930 ± 0.0002	5
	118	55											0.5559 ± 0.0002	5
	118	55											0.5862 ± 0.0003	3
	118	55											0.5865 ± 0.0002	15
	118	55											0.5006 ± 0.0002	11
	118	55											0.6765 ± 0.0002	3
	118	55											0.9281 ± 0.0003	4
	118	55											1.0218 ± 0.0002	4
	118	55											1.0288 ± 0.0002	5
	118	55											1.1120 ± 0.0002	1.5
	118	55											1.1437 ± 0.0002	1.6
$_{55}Cs^{119}$	119	55		118.922490	38 s.	β+,E.C.	6.6				9/2+		ann.rad.	
	119	55											0.169	67 +
	119	55											0.176	81
	119	55											0.224	100
	119	55											0.257	57
	119	55											0.314	38
	119	55											0.390	16
	119	55											0.667	17
$_{55}Cs^{120}$	120	55		119.920800	64 s.	β+,E.C.	8.2						ann.rad.	
	120	55											0.3224 ± 0.0001	100 +
	120	55											0.3955 ± 0.0003	3
	120	55											0.4735 ± 0.0001	30
	120	55											0.5251 ± 0.0001	3
	120	55											0.5534 ± 0.0001	19
	120	55											0.5858 ± 0.0003	4
	120	55											0.6012 ± 0.0002	11
	120	55											0.6051 ± 0.0002	4
	120	55											0.7018 ± 0.0002	2
	120	55											0.8758 ± 0.0003	6
	120	55											0.9491 ± 0.0003	7
	120	55											1.2750 ± 0.0008	7
	120	55											1.3894 ± 0.0003	2
	120	55											1.4451 ± 0.0003	2
	120	55											1.6727 ± 0.0008	2
	120	55											1.9820 ± 0.0005	1.5
	120	55											2.0560 ± 0.0008	1
	120	55											2.3154 ± 0.0004	0.9
	120	55											2.4673 ± 0.0004	0.7
													(0.3 - 3.28)	
$_{55}Cs^{121m}$	121	55			121 s.	I.T.(60%)		4	45		(9/2+)		ann.rad.	
	121	55				β+(40%)							0.1794 ± 0.0001	10%
	121	55											0.1961 ± 0.0001	12%
	121	55											0.2345 ± 0.0001	2%
	121	55											0.4146 ± 0.0002	2%
	121	55											0.4273 ± 0.0001	4%
	121	55											0.4598 ± 0.0001	5%
	121	55											0.5540 ± 0.0002	0.8%
$_{55}Cs^{121}$	121	55		120.917250	136 s.	β+,E.C.	5.4	4.4			3/2+		ann.rad.	
	121	55											0.1537 ± 0.0001	1%
	121	55											(0.08 - 0.56)weak	
$_{55}Cs^{122m}$	122	55			4.4 m.						8		ann.rad.	
	122	55											0.3311 ± 0.0002	91%
	122	55											0.3710 ± 0.0002	3%
	122	55											0.4971 ± 0.0002	77%
	122	55											0.5120 ± 0.0004	8%
	122	55											0.5601 ± 0.0002	14%
	122	55											0.5740 ± 0.0002	5%
	122	55											0.6385 ± 0.0002	61%
	122	55											0.6541 ± 0.0003	7%
	122	55											0.6846 ± 0.0003	4%
	122	55											0.7506 ± 0.0002	11%
	122	55											0.8155 ± 0.0004	2.4%
	122	55											0.8430 ± 0.0002	3%
	122	55											0.8829 ± 0.0002	7%
	122	55											0.9459 ± 0.0003	4%
	122	55											0.9942 ± 0.0003	2%
	122	55											1.0888 ± 0.0003	4%
	122	55											1.0978 ± 0.0002	11%
	122	55											1.2981 ± 0.0004	4.5%
	122	55											1.3801 ± 0.0005	3.4%
	122	55											1.4057 ± 0.0004	8%
	122	55											1.4600 ± 0.0004	4%

Isotope	A	Z	% Natural abundance	Atomic mass	Half-life	Decay mode	Decay energy (MeV)	Particle energy (MeV)	Particle intensity	Thermal neutron cross section	Spin (h/2π)	μ Nucl. mag. moment	Gamma-ray energy (MeV)	Gamma-ray intensity
	122	55											1.4955 ± 0.0004	3%
	122	55											1.6846 ± 0.0004	2%
	122	55											2.0521 ± 0.0004	2.5%
	122	55											2.1021 ± 0.0005	2.3%
	122	55											(0.27 - 2.22)weak	
$_{55}Cs^{122}$	122	55		121.916090	21 s.	β+,E.C.	7.4	5.8			(1+)		ann.rad.	
	122	55											0.3311 ± 0.0002	100 +
	122	55											0.4971 ± 0.0002	2
	122	55											0.5120 ± 0.0004	9
	122	55											0.8179 ± 0.0002	6
	122	55											0.8430 ± 0.0002	4
	122	55											0.8829 ± 0.0002	1.7
	122	55											1.0359 ± 0.0003	1
	122	55											1.3852 ± 0.0003	1
	122	55											1.4955 ± 0.0004	1
	122	55											1.7344 ± 0.0004	1
	122	55											2.1991 ± 0.0007	1
$_{55}Cs^{123m}$	123	55			1.7 s.	I.T.					11/2-		Cs k x-ray	25%
	123	55											0.0640 ± 0.0001	4%
	123	55											0.0946 ± 0.0001	25%
$_{55}Cs^{123}$	123	55		122.912990	5.9 m.	β+(75%)	4.12	2.4			½+		ann.rad.	
	123	55				E.C.(25%)		3.0					Xe k x-ray	17%
	123	55											0.0834 ± 0.0001	2.7%
	123	55											0.0974 ± 0.0001	13%
	123	55											0.2619 ± 0.0001	1.7%
	123	55											0.3071 ± 0.0001	2.7%
	123	55											0.5964 ± 0.0002	7.4%
	123	55											0.6109 ± 0.0002	2.3%
	123	55											0.6441 ± 0.0002	2%
	123	55											0.7415 ± 0.0001	2%
	123	55											1.1762 ± 0.0004	1%
	123	55											1.2732 ± 0.0002	1.8%
$_{55}Cs^{124}$	124	55		123.912270	30.8 s.	β+(92%)	5.8				1+		ann.rad.	
	124	55				E.C.(8%)							Xe k x-ray	4%
	124	55											0.3539 ± 0.0002	41%
	124	55											0.4925 ± 0.0004	3%
	124	55											0.8462 ± 0.0002	1%
	124	55											0.9418 ± 0.0001	4%
	124	55											1.3362 ± 0.0002	0.6%
	124	55											1.6890 ± 0.0003	0.5%
	124	55											2.0199 ± 0.0002	0.8%
$_{55}Cs^{125}$	125	55		124.909725	45 m.	β+(40%)	3.06	2.05			½+	+1.41	ann.rad.	
	125	55				E.C.(60%)							Xe k x-ray	26%
	125	55											0.112	9%
	125	55											0.412	5%
	125	55											0.526	24%
	125	55											0.540	3%
	125	55											0.600	3%
	125	55											0.712	3.5%
	125	55											0.922	0.8%
	125	55											0.995	0.5%
	125	55											1.158	0.4%
	125	55											1.579	0.3%
	125	55											1.698	0.3%
	125	55											2.116	0.8%
$_{55}Cs^{126}$	126	55		125.909465	1.	64 m.	β+ (81%)	4.83	3.4		1+		ann.rad.	
	126	55				E.C.(19%)		3.7					Xe k x-ray	8%
	126	55											0.3886 ± 0.0001	42%
	126	55											0.4912 ± 0.0002	5%
	126	55											0.8798 ± 0.0003	1.5%
	126	55											0.9252 ± 0.0002	5%
	126	55											1.6786 ± 0.0005	0.7%
	126	55											2.0673 ± 0.0005	0.3%
$_{55}Cs^{127}$	127	55		126.907428	6.2 h.	β+(96%)	2.11	0.65			½+	+1.46	Xe k x-ray	8%
	127	55				E.C.(4%)							0.1247 ± 0.0002	16%
	127	55											0.2871 ± 0.0002	3.4%
	127	55											0.4119 ± 0.0002	58%
	127	55											0.4623 ± 0.0002	4%
	127	55											0.5872 ± 0.0002	3.5%
	127	55											0.8066 ± 0.0004	0.3%
	127	55											0.9314 ± 0.0003	0.3%
$_{55}Cs^{128}$	128	55		127.907755	3.62 m.	β+(68%)	3.93	2.4			1+		ann.rad.	
	128	55				E.C.(32%)		2.9					Xe k x-ray	13%
	128	55											0.4429 ± 0.0001	26%
	128	55											0.5266 ± 0.0001	2%
	128	55											0.9695 ± 0.0001	0.6%
	128	55											1.1401 ± 0.0001	1%
$_{55}Cs^{129}$	129	55		128.906027	32.3 h.	E.C.	1.17				½+	+1.479	Xe k x-ray	55%
	129	55											0.0396 ± 0.0001	3%
	129	55											0.2786 ± 0.0001	1.3%
	129	55											0.3182 ± 0.0001	2.5%
	129	55											0.3719 ± 0.0001	31%
	129	55											0.4115 ± 0.0001	23%
	129	55											0.5489 ± 0.0001	3.5%

Isotope	A	Z	% Natural abundance	Atomic mass	Half-life	Decay mode	Decay energy (MeV)	Particle energy (MeV)	Particle intensity	Thermal neutron cross section	Spin (h/2π)	μ Nucl. mag. moment	Gamma-ray energy (MeV)	Gamma-ray intensity
	129	55											0.5885 ± 0.0001	0.6%
$_{55}Cs^{130}$	130	55		129.906753	29.2 m.	β+(55%)	3.02	1.98			1+	+1.4	ann.rad.	
	130	55				E.C.(43%)							Xe k x-ray	22%
	130	55				β-(1.6%)	0.44	0.44	1.6%				0.5361 ± 0.0001	4%
	130	55											0.5861 ± 0.0001	0.5%
	130	55											0.8945 ± 0.0002	0.4%
	130	55											1.6150 ± 0.0002	0.3%
	130	55											1.6874 ± 0.0002	0.2%
	130	55											1.7070 ± 0.0002	0.1%
	130	55											1.9973 ± 0.0003	0.2%
$_{55}Cs^{131}$	131	55		130.905444	9.69 d.	E.C.	0.35				5/2+	+3.54	Xe k x-ray	39%
$_{55}Cs^{132}$	132	55		131.906431	6.47 d.	E.C.(98%)					3+	+2.22	Xe k x-ray	39%
	132	55				β+(0.3%)							0.4646 ± 0.0002	1.9%
	132	55				β-(2%)							0.6302 ± 0.0002	1%
	132	55											0.66769 ± 0.00002	97%
	132	55											1.13605 ± 0.00002	0.5%
	132	55											1.31791 ± 0.00002	0.6%
$_{55}Cs^{133}$	133	55	100%	132.905429						(2.6 + 27) b.	7/2+	+2.579		
$_{55}Cs^{134m}$	134	55			2.91 h.	I.T.	0.139				8-	+1.096	Cs k x-ray	16%
	134	55											0.0112 ± 0.0001	1%
	134	55											0.12749 ± 0.00001	13%
$_{55}Cs^{134}$	134	55		133.906696	2.065 y.	β-	2.06	0.089	27%	140 b.	4+	+2.990	0.56327 ± 0.00002	8.4%
	134	55						0.658	70%				0.56935 ± 0.00002	15.4%
	134	55											0.60473 ± 0.00003	97.6%
	134	55											0.79584 ± 0.00003	85.4%
	134	55											0.80194 ± 0.00005	8.7%
	134	55											1.03864 ± 0.00006	1.0%
	134	55											1.16798 ± 0.00006	1.8%
	134	55											1.36519 ± 0.00005	3.0%
$_{55}Cs^{135m}$	135	55			53 m.	I.T.	1.627				19/2-		0.7869 ± 0.0001	99%
	135	55											0.8402 ± 0.0002	96%
$_{55}Cs^{135}$	135	55		134.905885	3x10^6 y.	β-	0.205	0.205	100%	8.7 b.	7/2+	+2.729		
$_{55}Cs^{136m}$	136	55			19 s.	I.T.					8			
$_{55}Cs^{136}$	136	55		135.907289	13.1 d.	β-	2.55	0.341			5+	+3.70	0.06691 ± 0.00001	12%
	136	55											0.08629 ± 0.00005	6%
	136	55											0.15322 ± 0.00005	7.5%
	136	55											0.17656 ± 0.00005	14%
	136	55											0.27365 ± 0.00004	13%
	136	55											0.34057 ± 0.00005	47%
	136	55											0.81850 ± 0.00004	86%
	136	55											1.04807 ± 0.00007	65%
	136	55											1.23534 ± 0.00005	20%
$_{55}Cs^{137}$	137	55		136.907073	30.17 y.	β-	1.17	0.514	95%		7/2+	+2.838	Ba k x-ray	4%
	137	55											0.66164 ± 0.00001	85.1%(D)
$_{55}Cs^{138m}$	138	55			2.9 m.	I.T.(75%)	0.080				6-		Cs k x-ray	20%
	138	55				β-(25%)							0.0799 ± 0.0003	0.1%
	138	55											0.1125 ± 0.0003	2%
	138	55											0.1917 ± 0.0002	20%
	138	55											0.3249 ± 0.0001	1.5%
	138	55											0.4628 ± 0.0001	25%
	138	55											1.43579 ± 0.00004	25%
$_{55}Cs^{138}$	138	55		137.911004	β-	5.38	3.2				3-	+0.5	0.1381 ± 0.0001	2%
	138	55											0.2278 ± 0.0001	1.5%
	138	55											0.40844 ± 0.00005	4.7%
	138	55											0.46269 ± 0.00003	31%
	138	55											0.54685 ± 0.00003	11%
	138	55											0.87170 ± 0.00003	5.1%

Isotope	A	Z	% Natural abundance	Atomic mass	Half-life	Decay mode	Decay energy (MeV)	Particle energy (MeV)	Particle intensity	Thermal neutron cross section	Spin (h/2π)	μ Nucl. mag. moment	Gamma-ray energy (MeV)	Gamma-ray intensity
	138	55											1.00969 ± 0.00003	30%
	138	55											1.43579 ± 0.00003	76.3%
	138	55											1.5553 ± 0.0001	0.4%
	138	55											2.21788 ± 0.00006	15.2%
	138	55											2.58285 ± 0.0001	0.2%
	138	55											2.63929 ± 0.00008	7.6%
	138	55											3.3390 ± 0.0003	0.15%
	138	55											3.3669 ± 0.0003	0.23%
$_{55}Cs^{139}$	139	55		138.913349	β-	4.21	4.2				7/2+		0.6272 ± 0.0001	0.6%
	139	55											1.2832 ± 0.0001	7.7%
	139	55											1.4207 ± 0.0001	0.8%
	139	55											1.6807 ± 0.0001	0.6%
	139	55											2.1109 ± 0.0001	0.7%
	139	55											2.3499 ± 0.0001	0.6%
	139	55											2.5318 ± 0.0001	0.45%
	139	55											2.6058 ± 0.0001	0.3%
	139	55											(0.4 - 3.66)weak	
$_{55}Cs^{140}$	140	55		139.917256	63.7 s.	β-	6.22	5.7			1-		0.5283 ± 0.0001	4%
	140	55						6.2					0.6023 ± 0.0001	70%
	140	55											0.6722 ± 0.0001	1.5%
	140	55											0.9084 ± 0.0001	11%
	140	55											1.0082 ± 0.0002	1%
	140	55											1.1300 ± 0.0001	3.1%
	140	55											1.2005 ± 0.0001	6%
	140	55											1.2216 ± 0.0001	3%
	140	55											1.3914 ± 0.0002	2.2%
	140	55											1.4220 ± 0.0002	1.1%
	140	55											1.6349 ± 0.0002	3.2%
	140	55											1.7073 ± 0.0002	1.6%
	140	55											1.8533 ± 0.0002	4%
	140	55											2.1016 ± 0.0002	3.9%
	140	55											2.2373 ± 0.0002	3.9%
	140	55											2.2684 ± 0.0002	1.7%
	140	55											2.3306 ± 0.0002	4.6%
	140	55											2.4297 ± 0.0002	1.7%
	140	55											2.5220 ± 0.0002	4.1%
	140	55											3.0540 ± 0.0002	1.5%
	140	55											3.4515 ± 0.0003	0.6%
	140	55											(0.41 - 3.94)weak	
$_{55}Cs^{141}$	141	55		140.920006	24.9 s.	β-	5.26	5.2			7/2+		Ba k x-ray	33%
	141	55											0.0485 ± 0.0001	10%
	141	55											0.5551 ± 0.00001	4.7%
	141	55											0.5616 ± 0.0001	5.8%
	141	55											0.5887 ± 0.0001	5.1%
	141	55											0.6919 ± 0.0001	3.7%
	141	55											1.0618 ± 0.0001	1.3%
	141	55											1.1406 ± 0.0002	1.3%
	141	55											1.1471 ± 0.0002	3.8%
	141	55											1.1536 ± 0.0002	1.2%
	141	55											1.1715 ± 0.0002	1.1%
	141	55											1.1775 ± 0.0002	1.3%
	141	55											1.1940 ± 0.0002	5.4%
	141	55											1.8940 ± 0.0004	0.5%
	141	55											1.9407 ± 0.0003	0.65%
	141	55											2.0596 ± 0.0006	0.5%
	141	55											2.0660 ± 0.0004	0.5%
	141	55											3.0722 ± 0.0003	0.7%
	141	55											(0.05 - 3.33)many	
$_{55}Cs^{142}$	142	55		141.924220	1.8 s.	β-	7.32	6.9					0.3596 ± 0.0001	100 +
	142	55											0.9668 ± 0.0001	28
	142	55											1.1759 ± 0.0001	10
	142	55											1.3265 ± 0.0001	34
	142	55											1.4222 ± 0.0002	3
	142	55											1.9821 ± 0.0002	4.4
	142	55											2.3978 ± 0.0002	2.6
	142	55											3.2834 ± 0.0003	2.1
	142	55											3.5733 ± 0.0003	2.3
$_{55}Cs^{143}$	143	55		142.927220	1.78 s.	β-	6.1				(3/2+)		0.1955 ± 0.0001	1%
	143	55											0.2324 ± 0.0001	14%
	143	55											0.2633 ± 0.0001	5.6%
	143	55											0.2727 ± 0.0001	5.6%
	143	55											0.3064 ± 0.0001	10%
	143	55											0.4666 ± 0.0002	7%
	143	55											0.5274 ± 0.0002	4%
	143	55											0.5348 ± 0.0002	2%
	143	55											0.5707 ± 0.0002	2%
	143	55											0.6052 ± 0.0002	2.4%
	143	55											0.6267 ± 0.0002	4%
	143	55											0.6599 ± 0.0002	7%
	143	55											0.6617 ± 0.0002	7%
	143	55											0.7293 ± 0.0002	2%

Isotope	A	Z	% Natural abundance	Atomic mass	Half-life	Decay mode	Decay energy (MeV)	Particle energy (MeV)	Particle intensity	Thermal neutron cross section	Spin (h/2π)	μ Nucl. mag. moment	Gamma-ray energy (MeV)	Gamma-ray intensity
	143	55											0.7927 ± 0.0002	1%
	143	55											0.8337 ± 0.0002	2%
	143	55											0.8679 ± 0.0002	1%
	143	55											1.0213 ± 0.0002	0.4%
	143	55											1.2081 ± 0.0002	0.4%
	143	55											1.9775 ± 0.0002	2.6%
	143	55											(0.17 - 1.98)many	
$_{55}Cs^{144}$	144	55		143.931930	1.01 s.	β-	8.5	8			1		0.1993 ± 0.0001	47%
	144	55											0.3083 ± 0.0002	2.2%
	144	55											0.3309 ± 0.0002	5%
	144	55											0.5598 ± 0.0002	9.2%
	144	55											0.6392 ± 0.0002	9.6%
	144	55											0.7587 ± 0.0002	9.6%
	144	55											0.8204 ± 0.0003	2%
	144	55											1.0095 ± 0.0004	1.4%
	144	55											1.0889 ± 0.0004	1.4%
	144	55											1.1160 ± 0.0004	1.7%
	144	55											1.3190 ± 0.0005	1.5%
	144	55											2.0130 ± 0.0006	2.6%
	144	55											2.1395 ± 0.0008	2.6%
	144	55											2.1765 ± 0.0008	2.2%
	144	55											2.4095 ± 0.001	3.2%
	144	55											2.7110 ± 0.0015	2.2%
	144	55											3.057 ± 0.003	1%
$_{55}Cs^{145}$	145	55		144.935320	0.58 s.	β-	7.8	7.4			3/2 +		0.1126 ± 0.0001	10%
	145	55											0.1755 ± 0.0001	20%
	145	55											0.1990 ± 0.0001	13%
	145	55											0.2410 ± 0.0001	5.3%
	145	55											0.4357 ± 0.0001	8.5%
	145	55											0.4548 ± 0.0001	4.5%
	145	55											0.4921 ± 0.0001	2.2%
	145	55											0.5471 ± 0.0001	5.1%
	145	55											0.7532 ± 0.0001	3%
Ba		56		137.33						1.3 b.				
$_{56}Ba^{120}$	120	56		119.926490	≈ 32 s.	β+,E.C.	≈ 5.3				0+		ann.rad.	
	120	56											0.051	
	120	56											0.182	
$_{56}Ba^{121}$	121	56		120.924700	30 s.	β+,E.C.	≈ 6.9						ann.rad.	
$_{56}Ba^{122}$	122	56		121.920170	2.0 m.	β+,E.C.	≈ 3.8				0+		ann.rad.	
$_{56}Ba^{123}$	123	56		122.919210	2.7 m.	β+,E.C.	≈ 5.7						ann.rad.	
	123	56											0.0306 ± 0.0006	56 +
	123	56											0.0583 ± 0.0006	2
	123	56											0.0639 ± 0.0006	14
	123	56											0.0927 ± 0.0006	51
	123	56											0.1161 ± 0.0006	54
	123	56											0.1200 ± 0.0006	23
	123	56											0.1235 ± 0.0006	69
	123	56											0.1370 ± 0.0006	23
$_{56}Ba^{124}$	124	56		123.915380	11.4 m.	β+,E.C.	≈ 2.9						ann.rad.	
	124	56											0.1568 ± 0.0005	3%
	124	56											0.1695 ± 0.0003	21%
	124	56											0.1888 ± 0.0003	11%
	124	56											0.2116 ± 0.0004	4%
	124	56											0.2531 ± 0.0003	5%
	124	56											0.2716 ± 0.0005	8%
	124	56											0.7155 ± 0.001	4%
	124	56											0.9330 ± 0.001	3%
	124	56											1.0480 ± 0.001	3%
	124	56											1.2160 ± 0.001	13%
$_{56}Ba^{125}$	125	56		124.914640	3.5 m.	β+,E.C.							ann.rad.	
	125	56											0.0550 ± 0.0006	48 +
	125	56											0.0631 ± 0.0006	8
	125	56											0.0776 ± 0.0006	100
	125	56											0.0854 ± 0.0006	82
	125	56											0.1001 ± 0.0006	6
	125	56											0.1080 ± 0.0006	8
	125	56											0.1409 ± 0.0006	86
$_{56}Ba^{126}$	126	56		125.911260	99 m.	β+(2%) E.C.(98%)	≈ 1.8				0+		Cs k x-ray	42%
	126	56											0.2179 ± 0.0001	4%
	126	56											0.2336 ± 0.0001	20%
	126	56											0.2410 ± 0.0001	6%
	126	56											0.2576 ± 0.0001	8%
	126	56											0.2812 ± 0.0002	3%
	126	56											0.3283 ± 0.0002	2%
	126	56											0.4893 ± 0.0002	3%
	126	56											0.5389 ± 0.0002	2%
	126	56											0.6818 ± 0.0002	4.6%
	126	56											0.7098 ± 0.0003	1.6%
	126	56											0.8416 ± 0.0005	1%
	126	56											0.8639 ± 0.0002	1%
	126	56											0.9768 ± 0.0002	2%
	126	56											0.9934 ± 0.0003	2%
	126	56											1.0354 ± 0.0003	2%
													1.0520 ± 0.0003	1%

Isotope	A	Z	% Natural abundance	Atomic mass	Half-life	Decay mode	Decay energy (MeV)	Particle energy (MeV)	Particle intensity	Thermal neutron cross section	Spin (h/2π)	μ Nucl. mag. moment	Gamma-ray energy (MeV)	Gamma-ray intensity
	126	56											1.2108 ± 0.0003	2%
	126	56											1.2418 ± 0.0003	1%
	126	56											1.2930 ± 0.0003	4%
$_{56}Ba^{127}$	127	56		126.911130	12 m.	β+(54%)	3.5				1/2+		ann.rad.	
	127	56				E.C.(46%)							Cs k x-ray	26%
	127	56											0.1148 ± 0.0003	9%
	127	56											0.1808 ± 0.0003	12%
	127	56											1.2010 ± 0.0003	1.6%
	127	56											1.5001 ± 0.0003	0.4%
	127	56											1.5660 ± 0.0003	0.4%
	127	56											(0.07 - 2.5)weak	
$_{56}Ba^{128}$	128	56		127.908237	2.43 d.	E.C.	0.45				0+		Cs k x-ray	41%
	128	56											0.27344 ± 0.00005	14%
$_{56}Ba^{129m}$	129	56			2.1 h.	E.C.(98%)					7/2+		Cs k x-ray	45%
	129	56				β+(2%)							0.1769 ± 0.0003	6%
	129	56											0.1823 ± 0.0001	47%
	129	56											0.2023 ± 0.0001	16%
	129	56											0.2143 ± 0.0001	4%
	129	56											0.3924 ± 0.0004	6%
	129	56											0.4202 ± 0.0002	12%
	129	56											0.4596 ± 0.0002	3%
	129	56											0.4816 ± 0.0002	3%
	129	56											0.5432 ± 0.0002	2%
	129	56											0.5468 ± 0.0003	5%
	129	56											0.5661 ± 0.0003	3%
	129	56											0.5969 ± 0.0001	7%
	129	56											0.6789 ± 0.0001	7%
	129	56											0.7486 ± 0.0002	3%
	129	56											0.7806 ± 0.0002	3%
	129	56											0.8040 ± 0.0003	2%
	129	56											0.8203 ± 0.0003	2%
	129	56											0.8725 ± 0.0001	3%
	129	56											0.8927 ± 0.0001	10%
	129	56											0.9573 ± 0.0003	2%
	129	56											0.9997 ± 0.0001	4%
	129	56											1.0351 ± 0.0003	4%
	129	56											1.0447 ± 0.0003	7.5%
	129	56											1.0476 ± 0.0006	3%
	129	56											1.1224 ± 0.0002	3%
	129	56											1.2092 ± 0.0002	4%
	129	56											1.2218 ± 0.0001	4%
	129	56											1.4593 ± 0.0001	26%
	129	56											1.6238 ± 0.0002	5%
$_{56}Ba^{129}$	129	56		128.908642	2.5 h.	β+(20%)	2.43				1/2+		ann.rad.	
	129	56				E.C.(80%)							Cs k x-ray	35%
	129	56											0.1291 ± 0.0001	6%
	129	56											0.2143 ± 0.0001	10%
	129	56											0.2208 ± 0.0001	6%
	129	56											0.5541 ± 0.0002	1.5%
	129	56											1.1646 ± 0.0002	1.1%
	129	56											1.8304 ± 0.0002	0.5%
	129	56											1.9540 ± 0.0002	0.5%
$_{56}Ba^{130}$	130	56	0.106%	129.906282						(2.5 + 9) b.	0+			
$_{56}Ba^{131m}$	131	56			14.6 m.	I.T.	0.187				9/2-		Ba k x-ray	25%
	131	56											0.0790 ± 0.0002	1.2%
	131	56											0.1085 ± 0.0002	55%
$_{56}Ba^{131}$	131	56		130.906902	11.8 d.	E.C.	1.36				1/2+		Cs k x-ray	52%
	131	56											0.12381 ± 0.00001	29%
	131	56											0.13360 ± 0.00001	2.2%
	131	56											0.21608 ± 0.00001	20%
	131	56											0.37324 ± 0.00001	14%
	131	56											0.49636 ± 0.00001	47%
	131	56											0.58499 ± 0.00002	1.2%
	131	56											0.62016 ± 0.00002	1.4%
	131	56											0.9239 ± 0.0001	0.7%
	131	56											1.0476 ± 0.0001	1.3%
$_{56}Ba^{132}$	132	56	0.101%	131.905042						(0.6 + 7) b.	0+			
$_{56}Ba^{133m}$	133	56			38.9 h.	I.T.	0.288				11/2-		Ba k x-ray	28%
	133	56											0.2761 ± 0.0002	17%
$_{56}Ba^{133}$	133	56		132.905988	10.53 y.	E.C.	0.52				1/2+		Cs k x-ray	63%
	133	56											0.0796 ± 0.0001	2%
	133	56											0.08099 ± 0.00001	33%
	133	56											0.27639 ± 0.00001	7.3%
	133	56											0.30285 ± 0.00001	19%

Isotope	A	Z	% Natural abundance	Atomic mass	Half-life	Decay mode	Decay energy (MeV)	Particle energy (MeV)	Particle intensity	Thermal neutron cross section	Spin (h/2π)	μ Nucl. mag. moment	Gamma-ray energy (MeV)	Gamma-ray intensity
	133	56											0.35600 ± 0.00002	62%
	133	56											0.38385 ± 0.00002	8.8%
$_{56}Ba^{134}$	134	56	2.417%	133.904486						(0.16 + 2) b.	0+			
$_{56}Ba^{135m}$	135	56			28.7 h.	I.T.	0.268				11/2-		Ba k x-ray	28%
	135	56											0.2682 ± 0.0001	16%
$_{56}Ba^{135}$	135	56	6.592%	134.905665						(0.014 + 6) b.	3/2+	+0.8365		
$_{56}Ba^{136m}$	136	56			0.306 s.	I.T.	2.0305				7-		Ba k x-ray	16%
	136	56											0.1639 ± 0.0001	31%
	136	56											0.8185 ± 00.0001	100%
	136	56											1.0481 ± 0.0001	100%
$_{56}Ba^{136}$	136	56	7.854%	135.904553						(0.010 + 0.4) b.	0+			
$_{56}Ba^{137m}$	137	56			2.552 m.	I.T.	0.66165				11/2-		Ba k x-ray	4%
	137	56											0.66164 ± 0.00002	90%
$_{56}Ba^{137}$	137	56	11.23%	136.905812						5.1 b.	3/2+	+0.9357		
$_{56}Ba^{138}$	138	56	71.70%	137.905232						0.4 b.	0+			
$_{56}Ba^{139}$	139	56		138.908826	833.1 m.	β-	2.3	2.2	27%	6 b.	7/2-		0.16585 ± 0.00001	24%
	139	56						2.3	72%				1.2544 ± 0.0001	0.03%
	139	56											1.42033 ± 0.00005	0.26%
$_{56}Ba^{140}$	140	56		139.910581	12.76 d.	β-	1.03	0.48		1.6 b.	0+		0.16268 ± 0.00001	7.1%
	140	56						1.0					0.30485 ± 0.00001	4.7%
	140	56											0.42372 ± 0.00001	3.3%
	140	56											0.43757 ± 0.00001	2%
	140	56											0.53727 ± 0.00002	24.5%
$_{56}Ba^{141}$	141	56		140.914363	18.3 m.	β-	3.23	2.59			3/2-		0.1903 ± 0.0001	46%
	141	56				2.73							0.2770 ± 0.0001	24%
	141	56											0.3042 ± 0.0001	25%
	141	56											0.3437 ± 0.0001	14%
	141	56											0.4576 ± 0.0001	4%
	141	56											0.4621 ± 0.0001	4%
	141	56											0.4673 ± 0.0001	5.6%
	141	56											0.6252 ± 0.0001	3.2%
	141	56											0.6479 ± 0.0001	6%
	141	56											0.7390 ± 0.0001	4.2%
	141	56											0.8761 ± 0.0001	3.3%
	141	56											1.1975 ± 0.0002	4.3%
	141	56											1.3240 ± 0.0004	0.8%
	141	56											1.4370 ± 0.0003	0.8%
	141	56											1.6823 ± 0.0002	1.4%
	141	56											(0.1 - 2.5)many	
$_{56}Ba^{142}$	142	56		141.916360	10.7 m.	β-	2.13	1.0			0+		0.23152 ± 0.00004	11.5%
	142	56					1.7						0.25512 ± 0.00004	20%
	142	56											0.3090 ± 0.0001	25%
	142	56											0.3638 ± 0.0001	4.5%
	142	56											0.4250 ± 0.0001	5.5%
	142	56											0.5998 ± 0.0001	1.8%
	142	56											0.8402 ± 0.0001	3.5%
	142	56											0.8949 ± 0.0001	12%
	142	56											0.9488 ± 0.0001	10%
	142	56											1.0009 ± 0.0001	8.9%
	142	56											1.0785 ± 0.0001	10.5%
	142	56											1.0936 ± 0.0001	2.6%
	142	56											1.1265 ± 0.0001	1.8%
	142	56											1.2022 ± 0.0001	6.0%
	142	56											1.2040 ± 0.0001	15.5%
	142	56											1.3799 ± 0.0001	3.9%
$_{56}Ba^{143}$	143	56		142.920480	15 s.	β-	4.2				3/2-		0.1786 ± 0.0001	1.5%
	143	56											0.21148 ± 0.00003	10%
	143	56											0.2912 ± 0.0001	3.5%
	143	56											0.4315 ± 0.0001	1.2%
	143	56											0.7190 ± 0.0001	1.6%
	143	56											0.7988 ± 0.0001	5.6%
	143	56											0.8952 ± 0.0001	1.5%
	143	56											0.9250 ± 0.0001	1.8%
	143	56											0.9805 ± 0.0001	3.8%
	143	56											1.0103 ± 0.0001	3.2%
	143	56											1.1964 ± 0.0001	2.4%
	143	56											1.6492 ± 0.0002	0.35%
	143	56											(0.17 - 2.4)weak	
$_{56}Ba^{144}$	144	56		143.922840	11.5 s.	β-	3.0				0+		La k x-ray	40%
	144	56											0.0690 ± 0.0001	4%
	144	56											0.0818 ± 0.0001	7%

Isotope	A	Z	% Natural abundance	Atomic mass	Half-life	Decay mode	Decay energy (MeV)	Particle energy (MeV)	Particle intensity	Thermal neutron cross section	Spin (h/2π)	μ Nucl. mag. moment	Gamma-ray energy (MeV)	Gamma-ray intensity
	144	56											0.10386 ± 0.00005	25%
	144	56											0.1113 ± 0.0001	6%
	144	56											0.1150 ± 0.0001	3%
	144	56											0.1566 ± 0.0001	16%
	144	56											0.1728 ± 0.0001	16%
	144	56											0.2076 ± 0.0002	2.4%
	144	56											0.2282 ± 0.0002	2%
	144	56											0.2594 ± 0.0002	4%
	144	56											0.2893 ± 0.0002	2%
	144	56											0.2917 ± 0.0003	3%
	144	56											0.3734 ± 0.0002	2%
	144	56											0.3882 ± 0.0001	17%
	144	56											0.43048 ± 0.00005	21%
	144	56											0.5158 ± 0.0002	8%
	144	56											0.5412 ± 0.0002	5%
	144	56											0.5704 ± 0.0002	3.8%
	144	56											0.5834 ± 0.0002	3%
	144	56											0.7032 ± 0.0003	1.4%
	144	56											0.7851 ± 0.0003	1.4%
$_{56}Ba^{145}$	145	56		144.926960	4.0 s.	β-	4.9	4.9			(5/2-)		La k x-ray	25%
	145	56											0.0656 ± 0.0002	6%
	145	56											0.0918 ± 0.0002	8%
	145	56											0.09709 ± 0.00005	20%
	145	56											0.1618 ± 0.0002	3.6%
	145	56											0.1892 ± 0.0002	1.8%
	145	56											0.2542 ± 0.0002	1.8%
	145	56											0.2860 ± 0.0002	1.6%
	145	56											0.3032 ± 0.0002	3.6%
	145	56											0.3255 ± 0.0002	2.0%
	145	56											0.3344 ± 0.0002	1.0%
	145	56											0.3437 ± 0.0002	1.2%
	145	56											0.3521 ± 0.0002	1.0%
	145	56											0.3788 ± 0.0002	6.5%
	145	56											0.4175 ± 0.0002	6%
	145	56											0.4775 ± 0.0002	2%
	145	56											0.5330 ± 0.0002	2.6%
	145	56											0.5714 ± 0.0002	1.6%
	145	56											0.5785 ± 0.0002	2.0%
	145	56											0.5985 ± 0.0002	3.6%
	145	56											0.6838 ± 0.0002	1.4%
	145	56											0.7306 ± 0.0002	1.6%
	145	56											0.8435 ± 0.0002	1.2%
	145	56											1.1104 ± 0.0003	1.4%
$_{56}Ba^{146}$	146	56		145.930120	2.2 s.	β-	4.3	3.9			0+		0.0644 ± 0.0001	16 +
	146	56											0.2513 ± 0.0002	30
	146	56											0.3270 ± 0.0003	38
	146	56											0.3329 ± 0.0002	100
	146	56											0.3622 ± 0.0003	
$_{56}Ba^{147}$	147	56		146.934230	0.70 s.	β-	≈ 6.1							
$_{56}Ba^{148}$	148	56		147.937190	0.47 s.	β-,n	≈ 4.9							
La		57		138.9055						8.98 b.				
$_{57}La^{125}$	125	57			≈ 76 s.	β+,E.C.					11/2-		ann.rad.	
	125	57											0.0436	
	125	57											0.0676	
$_{57}La^{126}$	126	57			1.0 m.	β+,E.C.							ann.rad.	
	126	57											0.2561 ± 0.001	
	126	57											0.340 ± 0.005	
	126	57											0.4555 ± 0.001	
	126	57											0.6214 ± 0.001	
$_{57}La^{127}$	127	57		126.916280	3.8 m.	β+,E.C.	≈ 4.8						ann.rad.	
	127	57											0.025	
	127	57											0.0562	
$_{57}La^{128}$	128	57		127.915320	4.6 m.	β+(80%) E.C.(20%)	≈ 6.9				(5-)		ann.rad.	
	128	57											Ba k x-ray	10%
	128	57											0.2841 ± 0.0001	87%
	128	57											0.4399 ± 0.0003	2.1%
	128	57											0.4757 ± 0.0005	2%
	128	57											0.4793 ± 0.0001	54%
	128	57											0.4879 ± 0.0002	10%
	128	57											0.5670 ± 0.0002	3.9%
	128	57											0.6005 ± 0.0002	10%
	128	57											0.6090 ± 0.0003	8.2%
	128	57											0.6266 ± 0.0002	3.8%
	128	57											0.6325 ± 1.0002	5.6%
	128	57											0.6436 ± 0.0002	15%
	128	57											0.8845 ± 0.0002	8.1%
	128	57											0.9150 ± 0.0003	3.5%
	128	57											0.9389 ± 0.0003	2.6%
	128	57											1.0363 ± 0.0003	2%
	128	57											1.0404 ± 0.0002	10%
	128	57											1.0532 ± 0.0002	10%

TABLE OF THE ISOTOPES (Continued)

Isotope	A	Z	% Natural abundance	Atomic mass	Half-life	Decay mode	Decay energy (MeV)	Particle energy (MeV)	Particle intensity	Thermal neutron cross section	Spin (h/2π)	μ Nucl. mag. moment	Gamma-ray energy (MeV)	Gamma-ray intensity
	128	57											1.0704 ± 0.0002	4.5%
	128	57											1.0882 ± 0.0002	9%
	128	57											1.31009 ± 0.0003	4.6%
	128	57											1.2761 ± 0.0005	5%
	128	57											1.4123 ± 0.0003	3.5%
	128	57											1.5059 ± 0.0004	3.6%
	128	57											1.7555 ± 0.0004	1.2%
	128	57											1.9196 ± 0.0004	1.2%
	128	57											2.2120 ± 0.0006	0.9%
$_{57}La^{129}$	129	57		128.912640	11.60 m.	β+(58%)	3.72	2.4			3/2+		ann.rad.	
	129	57				E.C.(42%)							Ba k x-ray	23%
	129	57											0.1105 ± 0.0001	17%
	129	57											0.2538 ± 0.0001	8%
	129	57											0.2786 ± 0.0001	24%
	129	57											0.3184 ± 0.0001	2%
	129	57											0.3166 ± 0.0001	6%
	129	57											0.4486 ± 0.0001	5%
	129	57											0.4570 ± 0.0001	8%
	129	57											0.4582 ± 0.0001	2%
	129	57											0.6013 ± 0.0002	1%
	129	57											0.6178 ± 0.0002	1%
	129	57											(0.1 - 1.8)	
$_{57}La^{130}$	130	57		129.912400	8.7 m.	β+(78%)	≈ 5.7				(3-)		ann.rad.	
	130	57				E.C.(22%)							Ba k x-ray	10%
	130	57											0.3573 ± 0.0001	81%
	130	57											0.4529 ± 0.0001	3.7%
	130	57											0.5444 ± 0.0001	18%
	130	57											0.5506 ± 0.0001	27%
	130	57											0.5694 ± 0.0001	3%
	130	57											0.5758 ± 0.0003	3%
	130	57											0.6494 ± 0.0001	1.7%
	130	57											0.7180 ± 0.0001	2.9%
	130	57											0.9079 ± 0.0001	17%
	130	57											0.9748 ± 0.0001	3%
	130	57											1.0036 ± 0.0001	8%
	130	57											1.1200 ± 0.0001	2%
	130	57											1.1708 ± 0.0001	3.9%
	130	57											1.1772 ± 0.0001	2.2%
	130	57											1.2006 ± 0.0001	3%
	130	57											1.4387 ± 0.0001	2.4%
	130	57											1.5254 ± 0.0001	7%
	130	57											1.7216 ± 0.0001	2.1%
	130	57											2.7521 ± 0.0003	1.0%
	130	57											2.7967 ± 0.0004	1.3%
	130	57											2.8101 ± 0.0003	1.4%
$_{57}La^{131}$	131	57		130.910080	59 m.	β+(76%)	3.0	1.4			3/2+		ann.rad.	
	131	57				E.C.(24%)		1.9					Ba k x-ray	40%
	131	57											0.1085 ± 0.0002	23%
	131	57											0.1609 ± 0.0002	2%
	131	57											0.2097 ± 0.0004	3%
	131	57											0.2575 ± 0.0004	3%
	131	57											0.3658 ± 0.0006	16%
	131	57											0.4184 ± 0.0007	6%
	131	57											0.4542 ± 0.0007	6%
	131	57											0.5263 ± 0.0008	10%
	131	57											0.5617 ± 0.0005	1.3%
	131	57											0.5941 ± 0.0005	1.5%
	131	57											0.6111 ± 0.0005	1.0%
	131	57											0.8660 ± 0.001	1.2%
	131	57											0.9742 ± 0.001	0.7%
$_{57}La^{132m}$	132	57			24 m.	I.T.(76%)					6-		La k x-ray	19%
	132	57				β+, EC (24%)							0.1352 ± 0.0002	44%
	132	57											0.2376 ± 0.0005	4%
	132	57											0.2856 ± 0.0005	7%
	132	57											0.4645 ± 0.0001	22%
	132	57											0.5671 ± 0.0001	4%
	132	57											0.6631 ± 0.0001	5%
	132	57											0.6977 ± 0.0001	4%
	132	57											0.8993 ± 0.0001	7%
	132	57											1.0317 ± 0.0001	3%
	132	57											1.0466 ± 0.00014	6%
$_{57}La^{132}$	132	57		131.910100	4.8 h.	β+(40%)	4.71	2.6			2-		ann.rad.	
	132	57				E.C.(60%)		3.2					Ba k x-ray	24%
	132	57											0.4645 ± 0.0001	77%
	132	57											0.5158 ± 0.0001	5%
	132	57											0.5404 ± 0.0001	8%
	132	57											0.5671 ± 0.0001	16%
	132	57											0.6631 ± 0.0001	9%
	132	57											0.8993 ± 0.0001	5%
	132	57											1.0317 ± 0.0001	8%
	132	57											1.0466 ± 0.0001	3%
	132	57											1.2212 ± 0.0001	3%
	132	57											1.5337 ± 01.0001	1.5%
	132	57											1.6046 ± 0.0001	4%

Isotope	A	Z	% Natural abundance	Atomic mass	Half-life	Decay mode	Decay energy (MeV)	Particle energy (MeV)	Particle intensity	Thermal neutron cross section	Spin (h/2π)	μ Nucl. mag. moment	Gamma-ray energy (MeV)	Gamma-ray intensity
	132	57											1.9099 ± 0.0001	9%
	132	57											2.1028 ± 0.0001	6%
	132	57											2.3914 ± 0.0001	1.0%
	132	57											2.7547 ± 0.0001	1.6%
	132	57											3.1990 ± 0.0001	0.7%
$_{57}$La^{133}	133	57		132.908140	3.91 h.	β+(4%)	2.0	1.2			5/2+		Ba k x-ray	38%
	133	57				E.C.(96%)							0.2788 ± 0.0001	1.9%
	133	57											0.2901 ± 0.0001	1.1%
	133	57											0.3024 ± 0.0001	1.2%
	133	57											0.5653 ± 0.0001	0.5%
	133	57											0.6183 ± 0.0001	0.8%
	133	57											0.6328 ± 0.0004	0.9%
$_{57}$La^{134}	134	57		133.908460	6.5 m.	β+(63%)	3.70	2.67			1+		ann.rad.	
	134	57				E.C.(37%)							Ba k x-ray	15%
	134	57											0.6047 ± 0.0001	5%
	134	57											1.5549 ± 0.0001	0.4%
	134	57											(0.5 - 1.9)weak	
$_{57}$La^{135}	135	57		134.906953	19.5 h.	E.C.	1.20				5/2+		Ba k x-ray	45%
	135	57											0.4805 ± 0.0001	1.5%
	135	57											0.5878 ± 0.0001	0.11%
	135	57											0.8745 ± 0.0001	0.16%
$_{57}$La^{136}	136	57		135.907630	9.87 m.	β+(36%)	2.9	1.9			1+		ann.rad.	
	136	57				E.C.(64%)							Ba k x-ray	26%
	136	57											0.8185 ± 0.0001	2.3%
	136	57											1.3230 ± 0.0001	0.3%
$_{57}$La^{137}	147	57		136.906460	6x10^{4}y.	E.C>	0.61				7/2+			
$_{57}$La138	138	57	0.09%	137.907105	1.06x10^{11}y	β-(34%)	1.04	0.26			5+	+3.707	Ba k x-ray	13%
	138	57				E.C.(66%)	1.75						0.7887 ± 0.0001	34%
	138	57											1.4359 ± 0.0001	66%
$_{57}$La^{139}	139	57	99.91%	138.906346						8.94 b.	7/2+			
$_{57}$La^{140}	140	57		139.909471	40.28 h.	β-	3.761	1.24	19%	2.7 b.	3-		0.32876 ± 0.00005	20.5%
	140	57						1.36	41%				0.43252 ± 0.00002	2.9%
	140	57						1.68	20%				0.48701 ± 0.00003	45%
	140	57						2.16	10%				0.75165 ± 0.00003	4.5%
	140	57											0.81577 ± 0.00003	24%
	140	57											0.86784 ± 0.00003	5.6%
	140	57											0.91954 ± 0.00004	2.7%
	140	57											0.92519 ± 0.00004	2.7%
	140	57											1.59617 ± 0.00006	95%
	140	57											2.34780 ± 0.00006	0.85%
	140	57											2.52132 ± 0.00006	3.5%
	140	57											2.54714 ± 0.00006	0.1%
	140	57											2.8995 ± 0.0002	0.07%
	140	57											3.1185 ± 0.0002	0.03%
$_{57}$La^{141}	141	57		140.910896	3.93 h.	β-	2.45	2.43			(7/2+)		1.3545 ± 0.0001	1.6%
	141	57											1.6933 ± 0.0001	0.07%
	141	57											2.2670 ± 0.0002	0.04%
$_{57}$La^{142}	142	57		141.914090	92 m.	β-	4.52	1.98			2-		0.6412 ± 0.0001	47%
	142	57						2.11					0.8948 ± 0.0001	8%
	142	57											1.0114 ± 0.0001	4%
	142	57											1.0437 ± 0.0001	2.7%
	142	57											1.1602 ± 0.0001	1.7%
	142	57											1.2331 ± 0.0001	1.9%
	142	57											1.3629 ± 0.0001	2.1%
	142	57											1.5458 ± 0.0001	3%
	142	57											1.7229 ± 0.0002	1.5%
	142	57											1.7564 ± 0.0001	2.7%
	142	57											1.9013 ± 0.0001	7.2%
	142	57											2.0042 ± 0.0002	0.9%
	142	57											2.0255 ± 0.0002	1.0%
	142	57											2.0552 ± 0.0001	2.2%
	142	57											2.1004 ± 0.0002	1.0%
	142	57											2.1872 ± 0.0001	3.7%
	142	57											2.3977 ± 0.0001	13%
	142	57											2.5426 ± 0.0001	10%
	142	57											2.6668 ± 0.0002	1.8%
	142	57											2.9718 ± 0.0003	3.1%
	142	57											3.6121 ± 0.0002	0.9%
	142	57											3.6327 ± 0.0002	1.0%
	142	57											(0.17 - 3.8)many	
$_{57}$La^{143}	143	57		142.915920	14.1 m.	β-	3.30	3.3					0.6203 ± 0.0001	1.0%
	143	57											0.6214 ± 0.0001	0.65%
	143	57											0.6437 ± 0.0001	0.66%

Isotope	A	Z	% Natural abundance	Atomic mass	Half-life	Decay mode	Decay energy (MeV)	Particle energy (MeV)	Particle intensity	Thermal neutron cross section	Spin (h/2π)	μ Nucl. mag. moment	Gamma-ray energy (MeV)	Gamma-ray intensity
	143	57											0.7981 ± 0.0001	0.5%
	143	57											1.1461 ± 0.0002	0.36%
	143	57											1.1485 ± 0.0002	0.5%
	143	57											1.5564 ± 0.0001	0.43%*
	143	57											1.9614 ± 0.0001	0.43%
	143	57											2.5001 ± 0.0001	0.31%
	143	57											2.6247 ± 0.0001	0.13%
$_{57}La^{144}$	144	57		143.919650	40 s.	β-	5.6	4.3					0.3973 ± 0.0002	94%
	144	57						4.6					0.4314 ± 0.0002	3.6%
	144	57											0.5411 ± 0.0002	39%
	144	57											0.5849 ± 0.0003	7.9%
	144	57											0.7054 ± 0.0004	4.1%
	144	57											0.7352 ± 0.0003	7.0%
	144	57											0.8448 ± 0.0002	22%
	144	57											0.9522 ± 0.0003	2.9%
	144	57											0.9688 ± 0.0005	3.3%
	144	57											0.9785 ± 0.0005	1.9%
	144	57											1.0527 ± 0.0003	2%
	144	57											1.2763 ± 0.0005	1.6%
	144	57											1.2943 ± 0.0005	6.5%
	144	57											1.4314 ± 0.0004	4.2%
	144	57											1.4896 ± 0.0006	1.4%
	144	57											1.5235 ± 0.00074	3.5%
	144	57											1.6737 ± 0.0006	1.4%
	144	57											1.7555 ± 0.0008	1.0%
	144	57											1.9423 ± 0.0009	1.7%
	144	57											1.9964 ± 0.0007	2.8%
	144	57											2.0078 ± 0.0009	1.2%
	144	57											2.0505 ± 0.001	1.3%
	144	57											2.3530 ± 0.001	1.9%
	144	57											2.6627 ± 0.001	1.9%
	144	57											2.8652 ± 0.0012	1.0%
$_{57}La^{145}$	145	57		144.921650	25 s.	β-	4.1	4.1					Pr k x-ray	20%
	145	57											0.0700 ± 0.0002	12%
	145	57											0.1182 ± 0.0002	4%
	145	57											0.1641 ± 0.0001	3%
	145	57											0.1698 ± 0.0002	3.5%
	145	57											0.3558 ± 0.0002	4.2%
	145	57											0.4474 ± 0.0002	3.5%
	145	57											0.5052 ± 0.0002	1.9%
	145	57											0.6718 ± 0.0002	2.0%
	145	57											0.7435 ± 0.0002	1.6%
	145	57											0.7865 ± 0.0002	1.9%
	145	57											0.8835 ± 0.0002	0.9%
	145	57											0.8896 ± 0.0002	1.1%
	145	57											0.9320 ± 0.0002	3.1%
	145	57											1.0307 ± 0.0003	1.9%
	145	57											1.0508 ± 0.0003	1.6%
	145	57											1.2380 ± 0.0003	1.0%
	145	57											1.5965 ± 0.0003	1.3%
	145	57											1.8195 ± 0.0003	3.4%
	145	57											1.9461 ± 0.0003	1.0%
	145	57											2.0878 ± 0.0003	0.9%
	145	57											2.1552 ± 0.0003	1.0%
	145	57											2.2047 ± 0.0003	0.9%
	145	57											2.3594 ± 0.0003	1.5%
	145	57											2.3771 ± 0.0005	0.67%
	145	57											2.4792 ± 0.0003	0.8%
	145	57											2.5426 ± 0.0003	0.8%
$_{57}La^{146}$	146	57		145.925530	10 s.	β-	6.2						0.2585 ± 0.0001	100 +
	146	57											0.4099 ± 0.0001	7
	146	57											0.6661 ± 0.0001	10
	146	57											0.7023 ± 0.0001	10
	146	57											0.7847 ± 0.0001	5
	146	57											0.9246 ± 0.0001	12
	146	57											1.0159 ± 0.0001	5
	146	57											1.2744 ± 0.0002	2
	146	57											1.3821 ± 0.0002	3
	146	57											1.4982 ± 0.0002	2.2
	146	57											1.7567 ± 0.0003	1.2
$_{57}La^{147}$	147	57		146.928100	4.1 s.	β-	4.8	4.3-4.6					0.1176 ± 0.0003	15%
	147	57											0.1869 ± 0.0003	7.1%
	147	57											0.2150 ± 0.0003	8%
	147	57											0.2357 ± 0.0003	3.1%
	147	57											0.2735 ± 0.0003	2.5%
	147	57											0.2834 ± 0.0003	2.9%
	147	57											0.3532 ± 0.0003	2%
	147	57											0.3996 ± 0.0005	2%
	147	57											0.4384 ± 0.0003	6%
	147	57											0.4953 ± 0.0005	2%
	147	57											0.5074 ± 0.0005	1%
	147	57											0.5168 ± 0.0005	3%
	147	57											0.5709 ± 0.0005	1%
	147	57											0.5984 ± 0.0005	1%
	147	57											0.7098 ± 0.0005	0.6%

Isotope	A	Z	% Natural abundance	Atomic mass	Half-life	Decay mode	Decay energy (MeV)	Particle energy (MeV)	Particle intensity	Thermal neutron cross section	Spin (h/2π)	μ Nucl. mag. moment	Gamma-ray energy (MeV)	Gamma-ray intensity
$_{57}La^{148}$	148	57		147.931390	≈ 2.6 s.	β-	≈ 6.5						0.1584 ± 0.0001	56%
	148	57											0.2524 ± 0.0001	1.7%
	148	57											0.2949 ± 0.0001	6.7%
	148	57											0.3790 ± 0.0001	4%
	148	57											0.6018 ± 0.0001	7.7%
	148	57											0.6829 ± 0.0001	6.5%
	148	57											0.7603 ± 0.0001	8.6%
	148	57											0.7771 ± 0.0001	7.2%
	148	57											0.8313 ± 0.0001	5.2%
	148	57											0.9583 ± 0.0001	4.0%
	148	57											0.9899 ± 0.0001	9.4%
	148	57											1.3386 ± 0.0001	1.8%
	148	57											1.4315 ± 0.0001	1.3%
	148	57											1.8905 ± 0.0001	1.2%
	148	57											1.9854 ± 0.0001	2.5%
	148	57											1.9947 ± 0.0001	3.3%
	148	57											2.0307 ± 0.0001	1.2%
	148	57											2.0931 ± 0.0001	7.1%
	148	57											2.2191 ± 0.0002	1.5%
	148	57											2.3914 ± 0.0001	3.9%
$_{57}La^{149}$	149	57		148.934310	1.2 s.	β-	≈ 5.1							
Ce		58		140.12						0.6 b.				
$_{58}Ce^{129}$	129	58			≈ 3.5 m.	β+,E.C.							ann.rad.	
	129	58											0.0675 ± 0.0001	
$_{58}Ce^{130}$	130	58			25 m.	β+,E.C.	≈ 2.3				0+		ann.rad.	
	130	58											La k x-ray	15%
$_{58}Ce^{131m}$	131	58			5 m.	β+,E.C.							ann.rad.	
	131	58											0.2304	36 +
	131	58											0.3955	
	131	58											0.4213	54
$_{58}Ce^{131}$	131	58		130.914270	9.5 m.	β+,E.C.							ann.rad.	
	131	58											0.119	
	131	58											0.169	
	131	58											0.414	
$_{58}Ce^{132}$	132	58		131.911490	3.5 h.	E.C.	≈ 1.3				0+		La k x-ray	40%
	132	58											0.1554	11%
	132	58											0.1821	79
	132	58											0.1901	3
	132	58											0.2167	5
	132	58											0.2514	2
	132	58											0.2799	2
	132	58											0.3027	2
	132	58											0.3296	3
	132	58											0.3681	1
	132	58											0.4244	1
	132	58											0.4315	1
	132	58											0.4515	1.5
	132	58											0.5762	1
$_{58}Ce^{133m}$	133	58			97 m.	β+,E.C.					1/2+		ann.rad.	
	133	58											0.0769 ± 0.0005	35 +
	133	58											0.0973 ± 0.0001	100
	133	58											0.1740 ± 0.0005	1
	133	58											0.3767 ± 0.0003	2
	133	58											0.5577 ± 0.0003	25
$_{58}Ce^{133}$	133	58		132.911360	5.4 h.	β+(8%) E.C.(92%)	≈ 3.0	1.3			9/2-		ann.rad.	
	133	58											La k x-ray	55%
	133	58											0.0584 ± 0.0001	19%
	133	58											0.0879 ± 0.0001	5%
	133	58											0.1308 ± 0.0001	18%
	133	58											0.3464 ± 0.0001	4%
	133	58											0.4326 ± 0.0001	3.5%
	133	58											0.4442 ± 0.0001	2.3%
	133	58											0.4755 ± 0.0001	3.2%
	133	58											0.4722 ± 0.0001	39%
	133	58											0.5104 ± 0.0001	21%
	133	58											0.5238 ± 0.0001	3.1%
	133	58											0.5419 ± 0.0001	2.9%
	133	58											0.6118 ± 0.0001	2.6%
	133	58											0.6447 ± 0.0001	2.0%
	133	58											0.6895 ± 0.0001	4.1%
	133	58											0.7845 ± 0.0001	9.6%
	133	58											0.9510 ± 0.0001	1.3%
	133	58											0.9901 ± 0.0001	2.9%
	133	58											1.3772 ± 0.0001	1.7%
	133	58											1.4322 ± 0.0001	1.2%
	133	58											1.4948 ± 0.0001	3.2%
	133	58											1.5004 ± 0.0001	4.8%
	133	58											1.5266 ± 0.0001	2.5%
	133	58											1.5846 ± 0.0001	2.4%
	133	58											1.7202 ± 0.0002	1.3%
	133	58											1.7694 ± 0.0001	1.2%
	133	58											1.8873 ± 0.0003	1.0%
	133	58											2.0182 ± 0.0001	1.4%
	133	58											2.1192 ± 0.0002	1.2%
$_{58}Ce^{134}$	134	58		133.908890	76 h.	E.C.	≈ 0.4				0+		La k x-ray	40%

TABLE OF THE ISOTOPES (Continued)

Isotope	A	Z	% Natural abundance	Atomic mass	Half-life	Decay mode	Decay energy (MeV)	Particle energy (MeV)	Particle intensity	Thermal neutron cross section	Spin (h/2π)	μ Nucl. mag. moment	Gamma-ray energy (MeV)	Gamma-ray intensity
	134	58											0.1304 ± 0.0001	0.21%
	134	58											0.1623 ± 0.0001	0.23%
	134	58											0.6047 ± 0.0001	5%(D)
$_{58}Ce^{135m}$	135	58			20 s.	I.T.	0.446				11/2-		Ce k x-ray	
	135	58											0.0826 ± 0.0001	23%
	135	58											0.1497 ± 0.0001	23%
	135	58											0.2134 ± 0.0001	78%
	135	58											0.2961 ± 0.0001	17%
$_{58}Ce^{135}$	135	58		134.909117	17.8 h.	β+(1%)	2.02	0.80			1/2+		La k x-ray	43%
	135	58				E.C.(99%)							0.0345 ± 0.0001	1.9%
	135	58											0.2065 ± 0.0001	8%
	135	58											0.2656 ± 0.0001	42%
	135	58											0.3001 ± 0.0001	23%
	135	58											0.4836 ± 0.0001	1.9%
	135	58											0.5181 ± 0.0001	13.6%
	135	58											0.5725 ± 0.0001	11%
	135	58											0.5771 ± 0.0001	5.1%
	135	58											0.6046 ± 0.0001	2.9%
	135	58											0.6068 ± 0.0001	19%
	135	58											0.6658 ± 0.0001	3.3%
	135	58											0.7836 ± 0.0001	11%
	135	58											0.8284 ± 0.0001	5.2%
	135	58											0.8714 ± 0.0001	3.2%
	135	58											0.9059 ± 0.0001	1.6%
	135	58											1.1841 ± 0.0001	1.1%
$_{58}Ce^{136}$	136	58	0.19%	135.907140						(1 + 6) b.	0+			
$_{58}Ce^{137m}$	137	58			34.4 h.	I.T.(99%)	0.254				11/2-		Ce k x-ray	29%
	137	58				E.C.(0.8%)							0.1693 ± 0.0001	0.4%
	137	58											0.2543 ± 0.0001	11%
	137	58											0.8248 ± 0.0001	0.4%
$_{58}Ce^{137}$	137	58		136.907780	9.0 h.	β+	1.22				3/2+		La k x-ray	40%
	137	58											0.0106 ± 0.0001	0.6%
	137	58											0.4332 ± 0.0001	0.06%
	137	58											0.4366 ± 0.0001	0.33%
	137	58											0.4472 ± 0.0001	2.2%
$_{58}Ce^{138}$	138	58	0..25%	137.905985						(0.015 + 1.1) b.	0+			
$_{58}Ce^{139m}$	139	58			56 s.	I.T.	0.7542				11/2-		Ce k x-ray	3%
	139	58											0.7542 ± 0.0001	92.5%
$_{58}Ce^{139}$	139	58		138.906631							3/2+		La k x-ray	42%
	139	58											0.16585 ± 0.00003	80%
$_{58}Ce^{140}$	140	58	88.48%	139.905433						0.58 b.	0+			
$_{58}Ce^{141}$	141	58		140.908271	32.5 d.	β-	0.581	0.444	69%		7/2-		Pr k x-ray	9%
	141	58						0.582	31%				0.14544 ± 0.00003	48.4%
$_{58}Ce^{142}$	142	58	11.08%	141.909241						0.95 b.	0+			
$_{58}Ce^{143}$	143	58		142.912383	33.0 h.	β-	1.462	0.74	12%		3/2-		Pr k x-ray	34%
	143	58						1.110	47%				0.0574 ± 0.0001	12%
	143	58											0.2316 ± 0.0001	2%
	143	58											0.2933 ± 0.0001	43%
	143	58											0.3506 ± 0.0001	3.3%
	143	58											0.4904 ± 0.0001	2.1%
	143	58											0.6645 ± 0.0001	5.6%
	143	58											0.7220 ± 0.0001	5.3%
	143	58											0.8804 ± 0.0001	1.0%
	143	58											1.1030 ± 0.0002	0.4%
$_{58}Ce^{144}$	144	58		143.913643	284.4 d.	β-	0.318	0.185	20%		0+		Pr k x-ray	4%
	144	58						0.238	5%				0.0801 ± 0.0001	1.1%
	144	58											0.1335 ± 0.0001	11%
$_{58}Ce^{145}$	145	58		144.917230	2.9 m.	β-	2.5	1.7	24%		5/2-		Pr k x-ray	31%
	145	58						2.1	76%				0.0627 ± 0.0001	15%
	145	58											0.0320 ± 0.0001	2.6%
	145	58											0.2845 ± 0.0001	9.7%
	145	58											0.3514 ± 0.0001	6.3%
	145	58											0.4236 ± 0.0001	4.6%
	145	58											0.4397 ± 0.0001	7.3%
	145	58											0.4924 ± 0.0001	1.5%
	145	58											0.5123 ± 0.0001	1.3%
	145	58											0.6562 ± 0.0001	1.3%
	145	58											0.7245 ± 0.0001	64%
	145	58											0.7832 ± 0.0001	2.5%
	145	58											0.8599 ± 0.0002	2.1%
	145	58											1.1107 ± 0.0001	2.7%
	145	58											1.1481 ± 0.0001	10%
	145	58												
													1.2105 ± 0.0002	1%
$_{58}Ce^{146}$	146	58		145.918670	13.6 m.	β-	1.0	0.75	90%		0+		Pr k x-ray	7%
	146	58											0.0986 ± 0.0001	3%
	146	58											0.1009 ± 0.0001	2.6%
	146	58											0.1335 ± 0.0001	8.3%
	146	58											0.1413 ± 0.0001	3.3%
	146	58											0.2105 ± 0.0001	5.0%
	146	58											0.2182 ± 0.0001	21%
	146	58											0.2509 ± 0.0001	2.6%
	146	58											0.2646 ± 0.0001	9.1%
	146	58											0.3167 ± 0.0001	57%
	146	58											0.3515 ± 0.0001	2.3%

Isotope	A	Z	% Natural abundance	Atomic mass	Half-life	Decay mode	Decay energy (MeV)	Particle energy (MeV)	Particle intensity	Thermal neutron cross section	Spin (h/2π)	μ Nucl. mag. moment	Gamma-ray energy (MeV)	Gamma-ray intensity
	146	58											0.4157 ± 0.0001	1.3%
	146	58											0.5030 ± 0.0001	1.0%
$_{58}Ce^{147}$	147	58		146.922530	56 s.	β-	3.3	3.3					0.0930 ± 0.0003	4%
	147	58											0.1987 ± 0.0003	1.7%
	147	58											0.2183 ± 0.0003	1.9%
	147	58											0.2687 ± 0.0003	5.5%
	147	58											0.2891 ± 0.0003	1.0%
	147	58											0.3591 ± 0.0003	1.3%
	147	58											0.3741 ± 0.0003	3.1%
	147	58											0.4520 ± 0.0003	2.3%
	147	58											0.4671 ± 0.0003	2.3%
	147	58											0.5779 ± 0.0003	1.0%
	147	58											0.5804 ± 0.0003	1.7%
	147	58											0.7011 ± 0.0003	1.3%
	147	58											0.8223 ± 0.0003	1.3%
$_{58}Ce^{148}$	148	58		147.924410	48 s.	β-	2.0	1.66			0+		0.0904 ± 0.0005	15 +
	148	58											0.0985 ± 0.0005	76
	148	58											0.1052 ± 0.0005	30
	148	58											0.1168 ± 0.0005	18
	148	58											0.1212 ± 0.0005	72
	148	58											0.1301 ± 0.0005	3
	148	58											0.1683 ± 0.0005	4
	148	58											0.1917 ± 0.0005	8
	148	58											0.1957 ± 0.0005	40
	148	58											0.2337 ± 0.0005	6
	148	58											0.2697 ± 0.0005	29
	148	58											0.2738 ± 0.0005	33
	148	58											0.2918 ± 0.0005	100
	148	58											0.3250 ± 0.0005	45
	148	58											0.3327 ± 0.0005	5
	148	58											0.3472 ± 0.0005	9
	148	58											0.3693 ± 0.0005	14
	148	58											0.3744 ± 0.0005	4
	148	58											0.3906 ± 0.0005	3
	148	58											0.3997 ± 0.0005	6
	148	58											0.4220 ± 0.0005	22
	148	58											0.5207 ± 0.0005	2
$_{58}Ce^{149}$	149	58		148.927760	5.2 s.	β-	≈ 3.5						0.0577 ± 0.0003	100 +
	149	58											0.0864 ± 0.0003	20
	149	58											0.3800 ± 0.0003	34
	149	58											0.3900 ± 0.0003	2
	149	58											0.4600 ± 0.0003	2
	149	58											0.7028 ± 0.0003	2
	149	58											0.8645 ± 0.0003	8
	149	58											0.8927 ± 0.0003	8
$_{58}Ce^{150}$	150	58		149.929670	4.4 s.	β-	≈ 3.1						0.1099 ± 0.0003	100 +
$_{58}Ce^{151}$	151	58		150.933170	1.0 s.	β-							0.0526 ± 0.0002	
	151	58											0.0848 ± 0.0001	
	151	58											0.0968 ± 0.0002	
	151	58											0.1186 ± 0.0001	
Pr		59		140.9077						11.4 b.				
$_{59}Pr^{132}$	132	59		131.919120	1.6 m.	β+,E.C.	≈ 7.2						ann.rad.	
	132	59											0.325 ± 0.001	
	132	59											0.496 ± 0.001	
	132	59											0.533 ± 0.001	
$_{59}Pr^{133}$	133	59		132.916190	6.7 m.	β+,E.C.	≈ 4.6				5/2+		ann.rad.	
	133	59											0.074	74 +
	133	59											0.1343	100
	133	59											0.2419	40
	133	59											0.2767	6
	133	59											0.3156	85
	133	59											0.3308	43
	133	59											0.3626	12
	133	59											0.4605	17
	133	59											0.4650	50
	133	59											0.6449	10
	133	59											0.8536	3
	133	59											1.4610	3
	133	59											1.4948	8
	133	59											1.8033	3
	133	59											1.8312	3.5
	133	59											1.8641	5
	133	59											1.8755	4
$_{59}Pr^{134m}$	134	59			11 m.	β+,E.C.							ann.rad.	
	134	59											0.294	
	134	59											0.460	
	134	59											0.495	
	134	59											0.632	
$_{59}Pr^{134}$	134	59		133.915440	17 m.	β+,E.C.	≈ 6.1				2+		ann.rad.	
	134	59											0.294	100 +
	134	59											0.460	15
	134	59											0.495	60
	134	59											0.495	60
	134	59											0.632	10
$_{59}Pr^{135}$	135	59		134.913140	25 m.	β+,E.C.	3.6				3/2+		ann.rad.	

Isotope	A	Z	% Natural abundance	Atomic mass	Half-life	Decay mode	Decay energy (MeV)	Particle energy (MeV)	Particle intensity	Thermal neutron cross section	Spin (h/2π)	μ Nucl. mag. moment	Gamma-ray energy (MeV)	Gamma-ray intensity
	135	59											0.0826 ± 0.0001	50 +
	135	59											0.2135 ± 0.0001	48
	135	59											0.2961 ± 0.0001	100
	135	59											0.4843 ± 0.0003	6
	135	59											0.5832 ± 0.0002	30
	135	59											0.6138 ± 0.0002	6
	135	58											0.6209 ± 0.0002	6
	135	58											0.6975 ± 0.0002	5
	135	59											0.8069 ± 0.0002	4
	135	59											0.9341 ± 0.0002	3
	135	59											1.0169 ± 0.0002	2
	135	59											1.5389 ± 0.0004	4
	135	59											1.7519 ± 0.0006	2
$_{59}Pr^{136}$	136	59		135.912640	13.1 m.	β+(57%)	5.10	2.98			2+		ann.rad.	
	136	59				E.C.(43%)							Ce k x-ray	18%
	136	59											0.4608 ± 0.0002	7.7%
	136	59											0.5398 ± 0.0002	52%
	136	59											0.5522 ± 0.0002	76%
	136	59											1.0008 ± 0.0003	5%
	136	59											1.0925 ± 0.0003	18%
	136	59											1.3597 ± 0.0004	1%
	136	59											1.4250 ± 0.0003	1%
	136	59											1.5148 ± 0.0004	1.9%
	136	59											1.6028 ± 0.0003	3.9%
	136	59											1.8990 ± 0.0005	1%
	136	59											2.0668 ± 0.0003	3%
	136	59											2.2407 ± 0.0004	0.7%
	136	59											2.3136 ± 0.0004	0.6%
	136	59											2.4508 ± 0.0003	0.7%
$_{59}Pr^{137}$	137	59		136.910680	77 m.	β+(26%)	2.70	1.68			5/2+		ann.rad.	
	137	59				E.C.(74%)							Ce k x-ray	30%
	137	59											0.4339 ± 0.0002	1.3%
	137	59											0.5140 ± 0.0002	1.1%
	137	59											0.8367 ± 0.0001	1.8%
	137	59											1.0886 ± 0.0002	0.4%
	137	59											(0.16 - 1.8)	
$_{59}Pr^{138m}$	138	59			2.1 h.	β+(24%)		1.65			7-		ann.rad.	
	138	59				E.C.(76%)							Ce k x-ray	36%
	138	59											0.3027 ± 0.0001	80%
	138	59											0.3909 ± 0.0001	6.1%
	138	59											0.5475 ± 0.0001	5.2%
	138	59											0.6359 ± 0.0001	1.8%
	138	59											0.7887 ± 0.0001	99%
	138	59											1.0378 ± 0.0001	100%
	138	59											(0.07 - 2.0)	
$_{59}Pr^{138}$	138	59		137.910748	1.5 m.	β+(75%)	4.44	3.44			1+		ann.rad.	150%
	138	59				E.C.(25%)							Ce k x-ray	10%
	138	59											0.6882 ± 0.0001	0.8%
	138	59											0.7887 ± 0.0001	2.4%
	138	59											1.4478 ± 0.0002	0.1%
	138	59											1.5511 ± 0.0001	0.4%
$_{59}Pr^{139}$	139	59		138.908917	4.41 h.	β+(8%)	2.13	1.09			5/2+		ann.rad.	16%
	139	59				E.C.(92%)							Ce k x-ray	40%
	139	59											0.2551 ± 0.0001	0.2%
	139	59											1.3473 ± 0.0001	0.4%
	139	59											1.6307 ± 0.0001	0.3%
$_{59}Pr^{140}$	140	59		139.909071	3.39 m.	β+(51%)	3.39	2.37			1+		ann.rad.	100%
	140	59				E.C.(49%)							Ce k x-ray	20%
	140	59											0.3069 ± 0.0002	0.2%
	140	59											1.5965 ± 0.0001	0.5%
$_{59}Pr^{141}$	141	59	100%	140.907647						(3.9 + 7.5) b.	5/2+	+4.3		
$_{59}Pr^{142m}$	142	59			14.6 m.	I.T.	0.004	c.e.			5/2-			
$_{59}Pr^{142}$	142	59		141.910039	19.13 h.	β-	2.160	0.58	4%	20 b.	2-		0.5088 ± 0.0005	0.02%
	142	59						2.16	96%				1.57580 ± 0.00005	3.7%
$_{59}Pr^{143}$	143	59		142.910814	13.58 d.	β-	0.934	0.935		90 b.	7/2+		0.7420 ± 0.0001	0.00001%
$_{59}Pr^{144m}$	144	59			7.2 m.	I.T. (99+%)	0.059				3-		Pr k x-ray	16%
	144	59				β-							0.0590 ± 0.0001	0.08%
	144	59											0.6965 ± 0.0001	0.04%
	144	59											0.8142 ± 0.0002	0.04%
$_{59}Pr^{144}$	144	59		143.913301	β-	2.997	0.807	1%			0-		0.69649 ± 0.00002	1.3%
	144	59					2.30	1.1%					1.48912 ± 0.00004	0.27%
	144	59					2.997	98%					2.18562 ± 0.00005	0.7%
$_{59}Pr^{145}$	145	59		144.914501	5.98 h.	β-	1.81	1.81	97%		7/2+		0.0725 ± 0.00011	0.3%
	145	59											0.6758 ± 0.0001	0.5%
	145	59											0.7483 ± 0.0001	0.52%
	145	59											0.9790 ± 0.0001	0.2%
	145	59											1.1503 ± 0.0001	0.2%
$_{59}Pr^{146}$	146	59		145.917570	24.1 m.	β-	4.1	2.6	30%				0.4539 ± 0.0001	48%
	146	59						3.7	10%				0.6017 ± 0.0001	3%
	146	59						4.1	40%				0.7357 ± 0.0001	7.5%

TABLE OF THE ISOTOPES (Continued)

Isotope	A	Z	% Natural abundance	Atomic mass	Half-life	Decay mode	Decay energy (MeV)	Particle energy (MeV)	Particle intensity	Thermal neutron cross section	Spin (h/2π)	μ Nucl. mag. moment	Gamma-ray energy (MeV)	Gamma-ray intensity
	146	59											0.7889 ± 0.0001	6.3%
	146	59											0.9229 ± 0.0001	2.3%
	146	59											1.0168 ± 0.0001	1.2%
	146	59											1.3767 ± 0.0001	4.4%
	146	59											1.5247 ± 0.0001	45.6%
	146	59											1.6904 ± 0.0002	0.6%
	146	59											2.2275 ± 0.0003	0.5%
	146	59											2.2252 ± 0.0001	1%
	146	59											2.3566 ± 0.0002	0.8%
	146	59											2.6816 ± 0.0003	0.4%
$_{59}Pr^{147}$	147	59		146.918980	13.4 m.	β-	2.68	1.5			5/2+		0.0780 ± 0.0001	10%
	147	59						2.1					0.1279 ± 0.0001	9%
	147	59											0.3146 ± 0.0001	24%
	147	59											0.3357 ± 0.0001	6%
	147	59											0.4778 ± 0.0001	5%
	147	59											0.5548 ± 0.0001	8%
	147	59											0.5779 ± 0.0001	16%
	147	59											0.6413 ± 0.0001	19%
	147	59											0.9960 ± 0.0001	1.6%
	147	59											1.1365 ± 0.0001	1.6%
	147	59											1.2611 ± 0.0002	5.3%
	147	59											1.3004 ± 0.0001	3%
	147	59											1.3245 ± 0.0003	1%
	147	59											1.4168 ± 0.0002	0.3%
	147	59											1.5433 ± 0.0003	0.3%
	147	59											1.5433 ± 0.0003	0.3%
	147	59											1.5465 ± 0.0008	0.3%
	147	59											1.5935 ± 0.0003	0.3%
	147	59											1.6237 ± 0.0003	0.3%
	147	59											1.6734 ± 0.0003	0.3%
	147	59											1.7932 ± 0.0002	0.3%
$_{59}Pr^{148m}$	148	59			2.0 m.	β-		4.0			(4)		0.5503 ± 0.0001	22%
	148	59											0.6113 ± 0.0001	1%
	148	59											0.8964 ± 0.0001	1%
	148	59											0.9149 ± 0.0001	11%
	148	59											1.4651 ± 0.0001	22%
$_{59}Pr^{148}$	148	59		147.922210	2.28 m.	β-	5.0	4.8			(1)		0.3017 ± 0.0001	58%
	148	59											0.6150 ± 0.0003	2%
	148	59											0.6975 ± 0.0001	4%
	148	59											0.7212 ± 0.0005	4%
	148	59											0.8693 ± 0.0002	4%
	148	59											1.0230 ± 0.0002	5%
	148	59											1.2486 ± 0.0002	3%
	148	59											1.3578 ± 0.0002	5%
	148	59											1.3817 ± 0.0003	2%
	148	59											1.9080 ± 0.0003	1%
	148	59											2.1304 ± 0.0002	1.7%
	148	59											2.6297 ± 0.0004	1%
$_{59}Pr^{149}$	149	59		148.923792	2.3 m.	β-	3.3	3.3	55%		(5/2+)		0.1085 ± 0.0001	9.5%
	149	59											0.1385 ± 0.0001	11%
	149	59											0.1623 ± 0.0001	3%
	149	59											0.1651 ± 0.0001	10%
	149	59											0.2077 ± 0.0002	3%
	149	59											0.2583 ± 0.0001	5.7%
	149	59											0.3164 ± 0.0001	3%
	149	59											0.3213 ± 0.0001	2.5%
	149	59											0.3330 ± 0.0001	6%
	149	59											0.3660 ± 0.0001	3.1%
	149	59											0.4063 ± 0.0001	2.4%
	149	59											0.4330 ± 0.0001	2.4%
	149	59											0.4746 ± 0.0001	2.8%
	149	59											0.5174 ± 0.0001	4.8%
	149	59											0.6230 ± 0.0001	1.8%
	149	59											0.6625 ± 0.0001	1.8%
	149	59											0.7491 ± 0.0001	1.4%
	149	59											0.7820 ± 0.0002	1.3%
$_{59}Pr^{150}$	150	59		149.926360	6.2 s.	β-	≈ 5.0				1+		0.1302 ± 0.0003	100 +
	150	59											0.2512 ± 0.0003	13
	150	59											0.5459 ± 0.0003	16
	150	59											0.8044 ± 0.0003	63
	150	59											0.8527 ± 0.0003	25
	150	59											0.9315 ± 0.0003	18
	150	59											1.0616 ± 0.0003	12
$_{59}Pr^{151}$	151	59		150.927910	4 s.	β-	≈ 3.7						0.1640 ± 0.0001	
$_{59}Pr^{152}$	152	59		151.930700	3.2 s.	β-	≈ 5.4						0.0726	
	151	59											0.164	
	152	59											0.285	
Nd		60		144.24						49 b.				
$_{60}Nd^{133}$	133	60			1.2 m.	β+,E.C.							ann.rad.	
	133	60											0.061	
	133	60											0.106	
	133	60											0.166	
	133	60											0.227	
	133	60											0.251	

Isotope	A	Z	% Natural abundance	Atomic mass	Half-life	Decay mode	Decay energy (MeV)	Particle energy (MeV)	Particle intensity	Thermal neutron cross section	Spin (h/2π)	μ Nucl. mag. moment	Gamma-ray energy (MeV)	Gamma-ray intensity
	133	60											0.369	
$_{60}Nd^{134}$	134	60		133.918870	≈ 8.5 m.	β+(17%)	≈ 2.8				0+		ann.rad.	34%
	134	60				E.C.(83%)							Pr k x-ray	40%
	134	60											0.0901	2%
	134	60											0.1012	2%
	134	60											0.1631	58%
	134	60											0.2168	12%
	134	60											0.2889	13%
	134	60											0.4679	3%
	134	60											0.4835	2%
	134	60											0.9920	2%
	134	60											1.000	4%
$_{60}Nd^{135}$	135	60		134.918190	12 m.	β+(65%)	≈ 4.8				9/2-		ann.rad.	130%
	135	60				E.C.(35%)							Pr k x-ray	35%
	135	60											0.0415 ± 0.0001	23%
	135	60											0.1126 ± 0.0001	5%
	135	60											0.1647 ± 0.0001	4%
	135	60											0.1851 ± 0.0001	3%
	135	60											0.2041 ± 0.0001	51%
	135	60											0.2060 ± 0.0003	3%
	135	60											0.2454 ± 0.0002	3%
	135	60											0.2561 ± 0.0002	3%
	135	60											0.2719 ± 0.0002	2%
	135	60											0.3728 ± 0.0002	2%
	135	60											0.4411 ± 0.0002	15%
	135	60											0.4519 ± 0.0002	4%
	135	60											0.4758 ± 0.0002	8%
	135	60											0.5016 ± 0.0002	10%
	135	60											0.5937 ± 0.0004	4%
	135	60											0.6165 ± 0.0003	2%
	135	60											0.9666 ± 0.0007	3%
	135	60											1.1721 ± 0.0007	1%
	135	60											1.4807 ± 0.0007	1%
	135	60											1.7520 ± 0.001	2%
$_{60}Nd^{136}$	136	60		135.915010	50.7 m.	E.C.(94%)	2.21	1.04			0+		Pr kx-ray	56%
	136	60				β+(6%)							0.0401 ± 0.0002	20%
	136	60											0.1091 ± 0.0001	33%
	136	60											0.1446 ± 0.0002	2%
	136	60											0.1492 ± 0.0001	9%
	136	60											0.4766 ± 0.0001	1.6%
	136	60											0.5749 ± 0.0001	12%
	136	60											0.9724 ± 0.0002	1.1%
$_{60}Nd^{137m}$	137	60			1.6 s.	I.T.	0.5196				11/2-		Nd k x-ray	30%
	137	60											0.1084 ± 0.0005	34%
	137	60											0.1775 ± 0.0005	57%
	137	60											0.2337 ± 0.0003	64%
	137	60											0.2861 ± 0.0005	21%
$_{60}Nd^{137}$	137	60		136.914760	38 m.	β+(40%)	3.80	1.7	20%		1/2+		ann.rad.	80%
	137	60				E.C.(60%)		2.40		20%			Pr k x-ray	45%
	137	60											0.0755 ± 0.0001	17%
	137	60											0.2382 ± 0.0001	4%
	137	60											0.3066 ± 0.0002	10%
	137	60											0.5051 ± 0.0003	9%
	137	60											0.5806 ± 0.0001	13%
	137	60											0.7616 ± 0.0002	9%
	137	60											0.7816 ± 0.0001	9%
	137	60											0.9257 ± 0.0002	7%
	137	60											0.9272 ± 0.0002	3%
	137	60											1.2431 ± 0.0002	1.4%
	137	60											1.6264 ± 0.0002	0.9%
	137	60											1.8131 ± 0.0002	0.8%
	137	60											2.0573 ± 0.0002	1%
$_{60}Nd^{138}$	138	60		137.911820	5.1 h.	E.C.	≈ 1.1				0+		Pr k x-ray	40%
	138	60											0.1995 ± 0.0001	0.6%
	138	60											0.3258 ± 0.0001	3%
$_{60}Nd^{139m}$	139	60			5.5 h.	I.T.(12%)	0.231	1.7			11/2-		Nd k x-ray	3%
	139	60				β+(88%)							Pr k x-ray	53%
	139	60											0.1139 ± 0.0001	34%
	139	60											0.3624 ± 0.0001	2%
	139	60											0.4038 ± 0.0001	2%
	139	60											0.5476 ± 0.0001	2%
	139	60											0.7012 ± 0.0001	3%
	139	60											0.7081 ± 0.0001	22%
	139	60											0.7382 ± 0.0002	30%
	139	60											0.7965 ± 0.0003	4%
	139	60											0.8020 ± 0.0003	6%
	139	60											0.8096 ± 0.0003	5%
	139	60											0.8278 ± 0.0003	9%
	139	60											0.9101 ± 0.0003	6.5%
	139	60											0.9822 ± 0.0002	22%
	139	60											1.0062 ± 0.0004	2.7%
	139	60											1.0123 ± 0.0003	2.3%
	139	60											1.0752 ± 0.0003	3%
	139	60											1.1053 ± 0.0003	2.3%
	139	60											1.3223 ± 0.0004	1.6%

Isotope	A	Z	% Natural abundance	Atomic mass	Half-life	Decay mode	Decay energy (MeV)	Particle energy (MeV)	Particle intensity	Thermal neutron cross section	Spin (h/2π)	μ Nucl. mag. moment	Gamma-ray energy (MeV)	Gamma-ray intensity
	139	60											2.0609 ± 0.0003	4%
$_{60}Nd^{139}$	139	60		138.911920	30 m.	β+(25%)	2.80	1.77			3/2+		ann.rad.	50%
	139	60				E.C.(75%)							Pr k x-ray	31%
	139	60											0.4050 ± 0.0001	6%
	139	60											0.4755 ± 0.0003	1%
	139	60											0.6217 ± 0.0003	1%
	139	60											0.6690 ± 0.0003	1.3%
	139	60											0.9169 ± 0.0003	1.3%
	139	60											0.9234 ± 0.0003	1.1%
	139	60											1.0742 ± 0.0004	2.1%
	139	60											1.4055 ± 0.0005	0.5%
$_{60}Nd^{140}$	140	60		139.909306	3.37 d.	E.C.	0.22				0+		Pr k x-ray	40%
$_{60}Nd^{141m}$	141	60			61 s.	I.T. (99+%)	0.757				11/2-		Nd k x-ray	3%
	141	60											0.7565 ± 0.0003	91.5%
$_{60}Nd^{141}$	141	60		140.909594	2.5 h.	E.C.(98%)	1.81	0.79			3/2+		Pr k x-ray	39%
	141	60				β+(2%)							0.1454 ± 0.0001	0.2%
	141	60											1.1269 ± 0.0002	0.8%
	141	60											1.1473 ± 0.0002	0.3%
	141	60											1.2926 ± 0.0002	0.5%
	141	60											1.2986 ± 0.0002	0.13%
$_{60}Nd^{142}$	142	60	27.13%	141.907719						19 b.	0+			
$_{60}Nd^{143}$	143	60	12.18%	142.909810						330 b.	7/2-	-1.08		
$_{60}Nd^{144}$	144	60	23.80%	143.910083	2.1x10^{15}y.					3.6 b.	0+			
$_{60}Nd^{145}$	145	60	8.30%	144.912570						45 b.	7/2-	-0.66		
$_{60}Nd^{146}$	146	60	17.19%	145.913113						1.4 b.	0+			
$_{60}Nd^{147}$	147	60		146.916097	10.99 d.	β-	0.895	0.805		400 b.	5/2-	0.59	Pr k x-ray	23%
	147	60											0.09111 ± 0.00002	28%
	147	60											0.12048 ± 0.00005	0.4%
	147	60											0.19644 ± 0.00005	0.2%
	147	60											0.27537 ± 0.00002	0.8%
	147	60											0.31941 ± 0.00002	1.95%
	147	60											0.39816 ± 0.00002	0.9%
	147	60											0.43989 ± 0.00002	1.2%
	147	60											0.48924 ± 0.00003	0.15%
	147	60											0.53102 ± 0.00002	13%
	147	60											0.58934 ± 0.00004	0.04%
	147	60											0.59480 ± 0.00003	0.27%
	147	60											0.68052 ± 0.00002	0.03%
	147	60											0.68590 ± 0.00004	0.81%
$_{60}Nd^{148}$	148	60	5.76%	147.916889						2.5 b.	0+			
$_{60}Nd^{149}$	149	60		148.920145	1.73 h.	β-	1.688	1.03	25%		5/2-		Pr k x-ray	16%
	149	60						1.13	26%				0.0970 ± 0.0001	1.4%
	149	60											0.11432 ± 0.00002	19%
	149	60											0.15588 ± 0.00001	5.9%
	149	60											0.1886 ± 0.0001	1.8%
	149	60											0.1989 ± 0.0001	1.4%
	149	60											0.2081 ± 0.0001	2.5%
	149	60											0.21131 ± 0.00001	26%
	149	60											0.2402 ± 0.0001	3.9%
	149	60											0.2677 ± 0.0001	6.0%
	149	60											0.2702 ± 0.0001	11%
	149	60											0.32656 ± 0.00001	4.6%
	149	60											0.3492 ± 0.0001	1.4%
	149	60											0.42355 ± 0.00001	7.5%
	149	60											0.4436 ± 0.0001	1.1%
	149	60											1.54051 ± 0.00001	6.6%
	149	60											1.65483 ± 0.00002	7.9%
	149	60											1.2342 ± 0.0001	0.32%
	149	60											(0.06 - 1.6)	
$_{60}Nd^{150}$	150	60	5.64%	149.920887						2.5 b.	0+			
$_{60}Nd^{151}$	151	60		150.923825	12.4 m.	β-	2.443	1.2			(3/2+)		Pm k x-ray	14%
	151	60											0.1168 ± 0.0001	46%
	151	60											0.1389 ± 0.0001	8.7%
	151	60											0.1707 ± 0.0001	3.8%
	151	60											0.1751 ± 0.0001	7.2%
	151	60												

Isotope	A	Z	% Natural abundance	Atomic mass	Half-life	Decay mode	Decay energy (MeV)	Particle energy (MeV)	Particle intensity	Thermal neutron cross section	Spin (h/2π)	μ Nucl. mag. moment	Gamma-ray energy (MeV)	Gamma-ray intensity
	151	60											0.2557 ± 0.0001	15%
	151	60											0.3006 ± 0.0001	1.9%
	151	60											0.4023 ± 0.0001	2%
	151	60											0.4235 ± 0.0001	6.4%
	151	60											0.4606 ± 0.0001	1.1%
	151	60											0.5852 ± 0.0001	1.6%
	151	60											0.6778 ± 0.0001	7.7%
	151	60											0.7364 ± 0.0002	7.2%
	151	60											0.7394 ± 0.0007	1.5%
	151	60											0.7555 ± 0.0005	1.4%
	151	60											0.7975 ± 0.0002	5.5%
	151	60											0.8411 ± 0.0005	1.1%
	151	60											0.9141 ± 0.0008	1.2%
	151	60											1.0164 ± 0.0002	2.9%
	151	60											1.1221 ± 0.0003	4.6%
	151	60											1.1806 ± 0.0002	15%
	151	60											(0.10 - 1.9)many	
$_{60}Nd^{152}$	152	60		151.924680	11.4 m.	β	1.1				0+		0.0160 ± 0.0005	8%
	152	60											0.0746 ± 0.0002	1%
	152	60											0.2501 ± 0.0002	22%
	152	60											0.2785 ± 0.0002	32%
	152	60											0.2946 ± 0.0002	4%
$_{60}Nd^{154}$	154	60		153.929400	≈ 40 s.	β-	≈ 2.5						0.40	
	154	60											0.70	
Pm		61												
$_{61}Pm^{134}$	134	61			24 s.	β+,E.C.							ann.rad.	
	134	61											0.294	100 +
	134	61											0.460	15
	134	61											0.495	60
	134	61											0.632	10
$_{61}Pm^{135}$	135	61			0.8 m.	β+,E.C.					11/2-		0.129	
	135	61											0.1987	100 +
	135	61											0.271	
	135	61											0.3622	20
	135	61											0.465	
$_{61}Pm^{136}$	136	61		135.923980	1.8 m.	β+(89%)	≈ 7.9				(5/2+)		ann.rad.	180%
	136	61				E.C.(11%)							Nd k x-ray	7%
	136	61											0.3028 ± 0.0005	14%
	136	61											0.3700 ± 0.0005	10%
	136	61											0.3735 ± 0.0003	89%
	136	61											0.4880 ± 0.0005	9.2%
	136	61											0.6027 ± 0.0003	50%
	136	61											0.6780 ± 0.0008	7%
	136	61											0.6930 ± 0.001	3%
	136	61											0.6960 ± 0.001	10%
	136	61											0.7704 ± 0.0003	18%
	136	61											0.8150 ± 0.0003	18%
	136	61											0.8580 ± 0.0003	31%
	136	61											0.8621 ± 0.0005	7.7%
	136	61											1.0597 ± 0.0005	14%
	136	61											1.0700 ± 0.0008	2.6%
$_{61}Pm^{137}$	137	61		136.920450	2.4 m.	β+,E.C.	≈ 5.1				(11/2-		ann.rad.	
	137	61											0.0870 ± 0.0002	12%
	137	61											0.1086 ± 0.0002	73%
	137	61											0.1775 ± 0.0002	84%
	137	61											0.2687 ± 0.0003	18%
	137	61											0.3251 ± 0.0005	6%
	137	61											0.3288 ± 0.0005	7%
	137	61											0.3523 ± 0.0003	2.7%
	137	61											0.3706 ± 0.0003	6%
	137	61											0.3892 ± 0.0003	6%
	137	61											0.4106 ± 0.0005	15%
	137	61											0.4140 ± 0.0005	4%
	137	61											0.4573 ± 0.0005	9%
	137	61											0.4592 ± 0.0005	4%
	137	61											0.4707 ± 0.0003	8%
	137	61											0.5060 ± 0.001	15%
	137	61											0.5251 ± 0.0004	3%
	137	61											0.5296 ± 0.0004	5%
	137	61											0.5338 ± 0.0004	11%
	137	61											0.5488 ± 0.0003	8%
	137	61											0.5656 ± 0.0003	5%
	137	61											0.5810 ± 0.001	28%
	137	61											0.6908 ± 0.0003	6%
	137	61											0.7595 ± 0.0005	2%
	137	61											0.8368 ± 0.0006	2%
	137	61											0.9230 ± 0.0005	5%
	137	61											1.0647 ± 0.0005	2%
	137	61											1.0922 ± 0.0005	9%
	137	61											1.1899 ± 0.0005	2.4%
	137	61											1.2847 ± 0.0005	14%
$_{61}Pm^{138}$	138	61		137.919340	3.2 m.	β+(50%)	≈ 7.0	3.9			3+		ann.rad.	100%
	138	61				E.C.(50%)							Nd k x-ray	21%
	138	61											0.4372 ± 0.0002	9.5%

TABLE OF THE ISOTOPES (Continued)

Isotope	A	Z	% Natural abundance	Atomic mass	Half-life	Decay mode	Decay energy (MeV)	Particle energy (MeV)	Particle intensity	Thermal neutron cross section	Spin (h/2π)	μ Nucl. mag. moment	Gamma-ray energy (MeV)	Gamma-ray intensity
	138	61											0.4931 ± 0.0002	20%
	138	61											0.5209 ± 0.0002	91%
	138	61											0.7290 ± 0.0002	35%
	138	61											0.7406 ± 0.0003	6%
	138	61											0.8103 ± 0.0003	2.8%
	138	61											0.8290 ± 0.0003	6%
	138	61											0.9306 ± 0.0002	4.7%
	138	61											0.9721 ± 0.0003	4%
	138	61											1.0116 ± 0.0003	3.5%
	138	61											1.0140 ± 0.0003	7%
	138	61											1.1346 ± 0.0003	2.3%
	138	61											1.2791 ± 0.0003	10%
	138	61											1.4828 ± 0.0003	2%
	138	61											1.6753 ± 0.0003	3%
	138	61											1.8029 ± 0.0005	1.6%
	138	61											2.6050 ± 0.0004	2.7%
	138	61											3.4605 ± 0.0004	2.7%
	138	61											3.4799 ± 0.0004	1%
$_{61}Pm^{139}$	139	61		138.916780	4.1 m.	β+(68%)	4.4	3.0			(5/2+)		ann.rad.	
	139	61				E.C.(32%)							Nd k x-ray	14%
	139	61											0.3678 ± 0.0002	3%
	139	61											0.4028 ± 0.0002	12%
	139	61											0.4631 ± 0.0002	3.5%
	139	61											0.7565 ± 0.0003	1.7%
	139	61											1.7046 ± 0.0003	0.6%
	139	61											(0.27 - 2.4)	
$_{61}Pm^{140m}$	140	61			5.9 m.	β+(70%)					7/2-		ann.rad.	140%
	140	61				E.C.(30%)							Nd k x-ray	15%
	140	61											0.4199 ± 0.0002	92%
	140	61											0.7738 ± 0.0002	100%
	140	61											1.0283 ± 0.0002	100%
	140	61											1.1975 ± 0.0002	3.8%
	140	61											2.1457 ± 0.0005	0.75%
$_{61}Pm^{140}$	140	61		139.915820	9.2 s.	β+(89%)	6.0	5.0	74%		1+		ann.rad.	180%
	140	61				E.C.(11%)							Nd k x-ray	5%
	140	61											0.7738 ± 0.0002	5.3%
	140	61											1.4898 ± 0.0005	1.0%
$_{61}Pm^{141}$	141	61		140.913600	20.9 m.	β+(52%)	3.72				5/2+		ann.rad.	100%
	141	61				E.C.(48%)							Nd k x-ray	21%
	141	61											0.1937 ± 0.0001	1.6%
	141	61											0.8862 ± 0.0001	2.4%
	141	61											1.2233 ± 0.0001	4.6%
	141	61											1.3455 ± 0.0001	1.3%
	141	61											1.4031 ± 0.0001	0.7%
	141	61											1.5647 ± 0.0001	0.8%
	141	61											2.0738 ± 0.0001	0.6%
$_{61}Pm^{142}$	142	61		141.912970	40.5 s.	β+(86%)	4.9	3.8			1+		ann.rad.	170%
	142	61				E.C.(20%)							Nd k x-ray	9%
	142	61											0.6414 ± 0.0005	0.6%
	142	61											1.5758 ± 0.0004	3.3%
	142	61											2.3843 ± 0.0006	0.1%
$_{61}Pm^{143}$	143	61		142.910930	265 d.	E.C.	1.04				5/2+		Nd k x-ray	40%
	143	61											0.7420 ± 0.0001	38%
$_{61}Pm^{144}$	144	61		143.912588	363 d.	E.C.	2.333				5-		Nd k x-ray	41%
	144	61											0.4768 ± 0.0001	42%
	144	61											0.6180 ± 0.0001	99%
	144	61											0.6965 ± 0.0001	100%
	144	61											0.7786 ± 0.0001	1.5%
	144	61											0.8141 ± 0.0001	0.6%
$_{61}Pm^{145}$	145	61		144.912743	17.7 y.	E.C.	0.161				5/2+		Nd k x-ray	37%
	145	61											0.0672 ± 0.0001	0.55%
	145	61											0.0723 ± 0.0001	1.8%
$_{61}Pm^{146}$	146	61		145.914708	5.53 y.	E.C.(63%)	1.48			8 x 10^3 b.	3-		Nd k x-ray	25%
	146	61				β-(37%)	1.542	0.795					0.4538 ± 0.0002	62%
	146	61											0.6333 ± 0.0003	2.5%
	146	61											0.7362 ± 0.0004	22%
	146	61											0.7474 ± 0.0003	36%
$_{61}Pm^{147}$	147	61		146.915135	2.6234 y.	β-	0.225	0.224		(85 + 97) b.	7/2+	± 2.7	0.1213 ± 0.0001	0.003%
	147	61											0.1974 ± 0.0001	0.0001%
$_{61}Pm^{148m}$	148	61			41.3 d.	β-(95%)	2.6	0.4	60%	1.1 x 10^4 b.	6-		0.0985 ± 0.0001	2.5%
	148	61				I.T.(5%)	0.137	0.5	17%				0.2881 ± 0.0001	13%
	148	61						0.7	21%				0.3116 ± 0.0001	3.9%
	148	61											0.4141 ± 0.0001	19%
	148	61											0.4328 ± 0.0001	5.4%
	148	61											0.5013 ± 0.0001	6.7%
	148	61											0.5503 ± 0.0001	96%
	148	61											0.5997 ± 0.0001	13%
	148	61											0.6113 ± 0.0001	5.5%
	148	61											0.6300 ± 0.0001	89%
	148	61											0.7257 ± 0.0001	33%
	148	61											0.9153 ± 0.0001	17%
	148	61											1.0138 ± 0.0001	20%
$_{61}Pm^{148}$	148	61		147.917473	5.37 d.	β-	2.47	1.02		2 x 10^3 b.	1-	+2.0	0.5503 ± 0.0001	22%
	148	61											0.6113 ± 0.0001	1%
	148	61											0.8964 ± 0.0001	1%

Isotope	A	Z	% Natural abundance	Atomic mass	Half-life	Decay mode	Decay energy (MeV)	Particle energy (MeV)	Particle intensity	Thermal neutron cross section	Spin (h/2π)	μ Nucl. mag. moment	Gamma-ray energy (MeV)	Gamma-ray intensity
	148	61											0.9149 ± 0.0001	11%
	148	61											1.4651 ± 0.0001	22%
$_{61}Pm^{149}$	149	61		148.918332	53.1 h.	β-	1.073	0.78	9%	1400 b.	7/2+		0.2859 ± 0.0001	3.1%
	149	61						1.062	90%				0.5909 ± 0.0003	0.07%
	149	61											0.8305 ± 0.0002	0.03%
	149	61											0.8332 ± 0.0002	0.03%
	149	61											0.8594 ± 0.0003	0.1%
	149	61											0.8819 ± 0.0005	0.02%
$_{61}Pm^{150}$	150	61		149.920981	2.69 h.	β-	3.45	1.6			(1-)		0.3339 ± 0.0001	69%
	150	61						2.1					0.4065 ± 0.0001	5.6%
	150	61											0.7122 ± 0.0001	4.4%
	150	61											0.7375 ± 0.0001	2.3%
	150	61											0.8318 ± 0.0001	12%
	150	61											0.8600 ± 0.0001	3.4%
	150	61											0.8764 ± 0.0001	7.4%
	150	61											1.1658 ± 0.0001	16%
	150	61											1.1939 ± 0.0001	5%
	150	61											1.2233 ± 0.0001	2.9%
	150	61											1.3245 ± 0.0001	18%
	150	61											1.3793 ± 0.0001	3.2%
	150	61											1.7364 ± 0.0001	7%
	150	61											1.9367 ± 0.0001	1.5%
	150	61											(0.25 - 2.9)	
$_{61}Pm^{151}$	151	61		150.921203	28.4 h.	β-	1.187	0.84		700 b.	5/2+	± 1.8	0.1000 ± 0.0001	2.5%
	151	61											0.1048 ± 0.0001	3.5%
	151	61											0.1636 ± 0.0001	1.5%
	151	61											0.1677 ± 0.0001	7.8%
	151	61											0.1772 ± 0.0001	3.6%
	151	61											0.2090 ± 0.0001	1.6%
	151	61											0.2324 ± 0.0001	1.0%
	151	61											0.2401 ± 0.0001	3.6%
	151	61											0.2751 ± 0.0001	6.6%
	151	61											0.3239 ± 0.0001	1.2%
	151	61											0.3401 ± 0.0001	22%
	151	61											0.3449 ± 0.0001	2.1%
	151	61											0.4408 ± 0.0001	1.5%
	151	61											0.4457 ± 0.0001	4.0%
	151	61											0.6361 ± 0.0001	1.4%
	151	61											0.7176 ± 0.0001	4.0%
	151	61											0.7527 ± 0.0001	1.2%
	151	61											0.7726 ± 0.0001	0.8%
	151	61											0.8078 ± 0.0001	0.5%
$_{61}Pm^{152m2}$	152	61			15 m.	β-,I.T.					(>6)		0.1218 ± 0.0001	
	152	61											0.1374 ± 0.0005	
	152	61											0.2003 ± 0.0005	
	152	61											0.2299 ± 0.0003	
	152	61											0.2447 ± 0.0001	
	152	61											0.3404 ± 0.0001	
	152	61											0.3604 ± 0.0005	
	152	61											1.2140 ± 0.001	
	152	61											1.2338 ± 0.0001	
	152	61											1.4375 ± 0.0003	
$_{61}Pm^{152m1}$	152	61			7.52 m.	β-					(4-)		0.1218 ± 0.0001	45%
	152	61											0.2447 ± 0.0001	78%
	152	61											0.3404 ± 0.0001	31%
	152	61											0.6562 ± 0.0003	3.5%
	152	61											0.6883 ± 0.0002	2.3%
	152	61											0.7808 ± 0.0001	4.2%
	152	61											0.8102 ± 0.0002	5.2%
	152	61											1.0051 ± 0.0001	2.9%
	152	61											1.0971 ± 0.0001	29%
	152	61											1.1121 ± 0.0001	4%
	152	61											1.4375 ± 0.0001	23%
	152	61											2.2007 ± 0.0002	0.7%
$_{61}Pm^{152}$	152	61		151.923490	4.1 m.	β-	3.5	3.45	20%		1+		0.1218 ± 0.0001	16%
	152	61						3.50	60%				0.6959 ± 0.0001	1.3%
	152	61											0.8414 ± 0.0001	2.2%
	152	61											0.9609 ± 0.0004	1.9%
	152	61											0.9633 ± 0.0004	1.8%
	152	61											1.3212 ± 0.0002	0.6%
	152	61											(0.12 - 2.1)	
$_{61}Pm^{153}$	153	61		152.924134	5.4 m.	β-	1.90	1.65			(5/2-)		0.0910 ± 0.0003	3.5%
	153	61											0.1198 ± 0.0001	6%
	153	61											0.1273 ± 0.0001	14%
	153	61											0.1294 ± 0.0001	1.8%
	153	61											0.1754 ± 0.0001	2%
	153	61											0.1829 ± 0.0001	2.7%
$_{61}Pm^{154m}$	154	61			2.7 m.	β-		2.0					0.0820 ± 0.0001	15%
	154	61											0.1848 ± 0.0001	30%
	154	61											0.2311 ± 0.0003	10%
	154	61											0.2802 ± 0.0002	11%
	154	61											0.3589 ± 0.0003	3%
	154	61											0.5467 ± 0.0002	10%
	154	61											0.7428 ± 0.0004	2.8%
	154	61											0.7453 ± 0.0004	3.2%

Isotope	A	Z	% Natural abundance	Atomic mass	Half-life	Decay mode	Decay energy (MeV)	Particle energy (MeV)	Particle intensity	Thermal neutron cross section	Spin (h/2π)	μ Nucl. mag. moment	Gamma-ray energy (MeV)	Gamma-ray intensity
	154	61											0.8342 ± 0.0003	3.5%
	154	61											0.8396 ± 0.0002	2.1%
	154	61											0.9305 ± 0.0002	4.9%
	154	61											1.2047 ± 0.0006	1.6%
	154	61											1.2737 ± 0.0005	1.7%
	154	61											1.3586 ± 0.0002	8.8%
	154	61											1.3940 ± 0.0002	3.4%
	154	61											1.4403 ± 0.0002	20%
	154	61											1.4577 ± 0.0004	3.3%
	154	61											1.5494 ± 0.0006	2.6%
	154	61											1.5513 ± 0.0006	1.6%
	154	61											1.6256 ± 0.0002	3.7%
	154	61											0.6561 ± 0.0002	3.7%
	154	61											1.7339 ± 0.0003	2.0%
	154	61											1.7976 ± 0.0004	2.0%
	154	61											1.8409 ± 0.0002	2.9%
	154	61											2.0589 ± 0.0002	5.5%
	154	61											2.1409 ± 0.0002	3.1%
$_{61}Pm^{154}$	154	61		153.926500	1.7 m.	β-	4.0	1.9					0.0820 ± 0.0001	12%
	154	61											0.1848 ± 0.0001	4.7%
	154	61											0.7543 ± 0.0006	4.5%
	154	61											0.8396 ± 0.0002	12%
	154	61											0.8915 ± 0.0002	6.6%
	154	61											0.9111 ± 0.0003	4.5%
	154	61											0.9216 ± 0.0002	8.3%
	154	61											0.9700 ± 0.0002	5.0%
	154	61											1.0176 ± 0.0002	9.9%
	154	61											1.0962 ± 0.0003	5.6%
	154	61											1.1481 ± 0.0002	9.0%
	154	61											1.1778 ± 0.0008	3.6%
	154	61											1.3940 ± 0.0002	12%
	154	61											2.0589 ± 0.0002	19%
	154	61											2.1409 ± 0.0002	11%
	154	61											2.3477 ± 0.0003	1.7%
	154	61											2.5106 ± 0.0002	1.3%
	154	61											(0.08 - 2.8)	
$_{61}Pm^{155}$	155	61		154.927960	48 s.	β-	≈ 3.1				(5/2)		0.0531 ± 0.0005	0.94%
	155	61											0.4098 ± 0.00002	2.2%
	155	61											0.7254 ± 0.0002	5.3%
	155	61											0.7620 ± 0.0003	1.5%
	155	61											0.7786 ± 0.0002	7.8%
Sm		62		150.36						59 x 10^2 b.				
$_{62}Sm^{138}$	138	62		137.923420	3.0 m.	β+,E.C.	≈ 3.5				0+		ann.rad.	
	138	62											0.0536	
	138	62											0.0747	
$_{62}Sm^{139m}$	139	62			≈ 9.5 s.	I.T.(94%)	0.457				(11/2-		Sm k x-ray	29%
	139	62				β+(6%)							0.1118 ± 0.0003	23%
	139	62											0.1553 ± 0.0003	33%
	139	62											0.1901 ± 0.0003	37%
	139	62											0.2673 ± 0.0003	36%
$_{62}Sm^{139}$	139	62		138.922600	2.6 m.	β+(75%)	5.4	4.1			(1/2+)		Pm k x-ray	14%
	139	62				E.C.(25%)							0.3678 ± 0.0002	3%
	139	62											0.4028 ± 0.0002	12%
	139	62											0.4631 ± 0.0002	3.5%
	139	62											0.7565 ± 0.0003	1.7%
	139	62											0.8158 ± 0.0002	1.0%
	139	62											0.9816 ± 0.0004	0.9%
	139	62											1.7046 ± 0.0003	0.6%
	139	62											(0.27 - 2.4)	
$_{62}Sm^{140}$	140	62		139.919040	14.8 m.	β+,E.C.	≈ 3.0	1.9			0+		ann.rad.	
	140	62											Pm k x-ray	35%
	140	62											0.1141 ± 0.0005	1.5%
	140	62											0.1201 ± 0.0005	3%
	140	62											0.1396 ± 0.0005	8%
	140	62											0.2207 ± 0.0005	2%
	140	62											0.2255 ± 0.0005	13%
	140	62											0.3398 ± 0.0005	2%
	140	62											0.3448 ± 0.0005	1.1%
	140	62											1.13800 ± 0.0007	1.5%
	140	62											1.2745 ± 0.0007	1.3%
	140	62											1.5298 ± 0.0007	1%
	140	62											(0.07 - 1.7)	
$_{62}Sm^{141m}$	141	62			22.6 m.	β+(32%)		1.6			11/2-		ann.rad.	64%
	141	62				E.C.(68%)		2.19					Pm k x-ray	36%
	141	62				I.T.(0.3%)	0.1758						0.1966 ± 0.0003	75%
	141	62											0.4318 ± 0.0001	41%
	141	62											0.5385 ± 0.0003	8.5%
	141	62											0.6387 ± 0.0001	2.7%
	141	62											0.6768 ± 0.0003	1.4%
	141	62											0.6846 ± 0.0002	8%
	141	62											0.7257 ± 0.0005	1.5%
	141	62											0.7503 ± 0.0003	1.6%
	141	62											0.7774 ± 0.0003	21%
	141	62											0.7859 ± 0.0001	7%

Isotope	A	Z	% Natural abundance	Atomic mass	Half-life	Decay mode	Decay energy (MeV)	Particle energy (MeV)	Particle intensity	Thermal neutron cross section	Spin (h/2π)	μ Nucl. mag. moment	Gamma-ray energy (MeV)	Gamma-ray intensity
	141	62											0.8059 ± 0.0001	3.6%
	141	62											0.8371 ± 0.0002	3.6%
	141	62											0.7850 ± 0.0001	1.3%
	141	62											0.8965 ± 0.0001	1.5%
	141	62											0.9113 ± 0.0003	9.3%
	141	62											0.9247 ± 0.0001	2.3%
	141	62											0.9833 ± 0.0003	7.4%
	141	62											1.0091 ± 0.0004	3%
	141	62											1.1084 ± 0.0002	1.3%
	141	62											1.1176 ± 0.0002	3.3%
	141	62											1.1451 ± 0.0002	8.9%
	141	62											1.4634 ± 0.0006	1.8%
	141	62											1.4903 ± 0.0001	9.4%
	141	62											1.7864 ± 0.0004	11.1%
	141	62											2.0737 ± 0.0002	1.4%
$_{62}Sm^{141}$	141	62		140.918473	10.2 m.	β+(52%)	4.55	3.2			1/2+		ann.rad.	104%
	141	62				E.C.(48%)							Pm k x-ray	23%
	141	62											0.3244 ± 0.0002	2.5%
	141	62											0.4039 ± 0.0001	2.5%
	141	62											0.4382 ± 0.0001	38%
	141	62											1.0571 ± 0.0002	3.3%
	141	62											1.0919 ± 0.0002	2.6%
	141	62											1.2926 ± 0.0002	6.8%
	141	62											1.4639 ± 0.0004	1.9%
	141	62											1.4957 ± 0.0002	1.8%
	141	62											1.6007 ± 0.0003	4.0%
	141	62											2.0378 ± 0.0003	2.8%
$_{62}Sm^{142}$	142	62		141.915206	72.5 m.	β+(6%)	2.1	1.0			0+		ann.rad.	12%
	142	62				E.C.(94%)							Pm k x-ray	38%
$_{62}Sm^{143m}$	143	62			66 s.	I.T.(99%)	0.7540				11/2-		Sm k x-ray	4%
	143	62				E.C.(0.2%)							0.2718 ± 0.0004	0.2%
	143	62											0.6886 ± 0.0004	0.2%
	143	62											0.7540 ± 0.002	90%
$_{62}Sm^{143}$	143	62		142.914626	8.83 m.	β+(46%)	3.44	2.47			3/2+		ann.rad.	92%
	143	62				E.C.(54%)							Pm k x-ray	22%
	143	62											0.2718 ± 0.0003	0.3%
	143	62											1.0565 ± 0.0002	1.8%
	143	62											1.1734 ± 0.0004	0.4%
	143	62											1.5149 ± 0.0002	0.6%
$_{62}Sm^{144}$	144	62	3.1%	143.911998						0.7 b.	0+			
$_{62}Sm^{145}$	145	62		144.913409	340 d.	E.C.	0.621				7/2-		Pm k x-ray	72%
	145	62											0.0613 ± 0.0001	12%
	145	62											0.4924 ± 0.0002	0.003%
$_{62}Sm^{146}$	146	62		145.913053	1.03x10^{8}y.	α		2.50			0+			
$_{62}Sm^{147}$	147	62	15.0%	146.914895	1.08x10^{11}y.	α		2.23		57 b.	7/2-	-0.813		
$_{62}Sm^{148}$	148	62	11.3%	147.914820	7x10^{15} y.	α		1.96		3 b.	0+			
$_{62}Sm^{149}$	149	62	13.8%	148.917181	10^{16} y.	α				4 10^{4} b.	7/2-	-0.66		
$_{62}Sm^{150}$	150	62	7.4%	149.917273						103 b.	0+			
$_{62}Sm^{151}$	151	62		150.919929	90 y.	β-	0.076	0.076		1.5 x 10^{4} b.	5/2-		0.02154 ± 0.00001	0.03%
$_{62}Sm^{152}$	152	62	26.7%	151.919729						208 b.	0+			
$_{62}Sm^{153}$	153	62		152.922094	46.7 h.	β-	0.810	0.64		400 b.	3/2+		Eu k x-ray	31%
	153	62						0.71					0.069676 ± 0.00004	5.3%
	153	62											0.10318 ± 0.00001	28%
	153	62											0.17286 ± 0.00001	0.7%
$_{62}Sm^{154}$	154	62	22.7%	153.922206						8 b.	0+			
$_{62}Sm^{155}$	155	62		154.924636	22.2 m.	β1.622	1.52				3/2-		Eu k x-ray	9%
	155	62											0.10432 ± 0.00005	75%
	155	62											0.1414 ± 0.0001	2%
	155	62											0.2457 ± 0.0001	3.8%
	155	62											0.5225 ± 0.0002	0.15%
$_{62}Sm^{156}$	156	62		155.925518	9.4 h.	β-	0.71	0.43			0+		0.0381 ± 0.0001	3%
	156	62						0.71					0.0872 ± 0.0002	24%
	156	62											0.1657 ± 0.0004	15%
	156	62											0.2038 ± 0.0001	23%
	156	62											0.2464 ± 0.0003	1.3%
	156	62											0.2687 ± 0.0006	2.5%
	156	62											0.2907 ± 0.0005	3%
$_{62}Sm^{157}$	157	62		156.928210	8.1 m.	β-	2.6	2.4			3/2-		Eu k x-ray	9%
	157	62											0.0767 ± 0.0003	2%
	157	62											0.1210 ± 0.0002	6%
	157	62											0.1964 ± 0.0002	22%
	157	62											0.1978 ± 0.0002	62%
	157	62											0.2631 ± 0.0002	1.6%
	157	62											0.3175 ± 0.0003	1.6%
	157	62											0.3942 ± 0.0002	13%
	157	62											0.8440 ± 0.0002	6%
	157	62											1.3861 ± 0.0003	1.2%
	157	62											1.4630 ± 0.0003	3.4%
$_{62}Sm^{158}$	158	62		157.930000	5.5 m.	β-	≈ 1.9				0+		0.1002 ± 0.0003	4.6%
	158	62											0.1490 ± 0.0003	4.7%
	158	62											0.1777 ± 0.0003	4%

TABLE OF THE ISOTOPES (Continued)

Isotope	A	Z	% Natural abundance	Atomic mass	Half-life	Decay mode	Decay energy (MeV)	Particle energy (MeV)	Particle intensity	Thermal neutron cross section	Spin (h/2π)	μ Nucl. mag. moment	Gamma-ray energy (MeV)	Gamma-ray intensity
	158	62											0.1894 ± 0.0003	15%
	158	62											0.1907 ± 0.0003	4%
	158	62											0.2241 ± 0.0003	8.5%
	158	62											0.2266 ± 0.0003	5.2%
	158	62											0.2297 ± 0.0003	6.7%
	158	62											0.2854 ± 0.0003	1.7%
	158	62											0.2997 ± 0.0003	2.1%
	158	62											0.3213 ± 0.0003	8.3%
	158	62											0.3245 ± 0.0003	11%
	158	62											0.3268 ± 0.0003	2%
	158	62											0.3386 ± 0.0003	3.7%
	158	62											0.3617 ± 0.0003	6.6%
	158	62											0.3636 ± 0.0003	12%
	158	62											0.5512 ± 0.0003	3%
	158	62											0.7914 ± 0.0003	1.6%
	158	62											1.1269 ± 0.0003	1.2%
Eu		63								4600 b.				
$_{63}Eu^{141m}$	141	63			3.3 s.	β+(58%)					11/2-		ann.rad.	118%
	141	63				E.C.(9%)							Eu k x-ray	4%
	141	63				I.T.(33%)	0.0964						0.0964 ± 0.0002	0.7%
	141	63											0.3940 ± 0.0002	0.6%
	141	63											0.8829 ± 0.0002	0.5%
	141	63											1.5953 ± 0.0003	0.4%
	141	63											(0.09 - 1.6)	
$_{63}Eu^{141}$	141	63		140.924870	40 s.	β+(81%)	6.0				5/2+		ann.rad.	170%
	141	63				E.C.(15%)							Sm k x-ray	7%
	141	63											0.3695 ± 0.0002	2.5%
	141	63											0.3829 ± 0.0002	4.5%
	141	63											0.3845 ± 0.0002	8.5%
	141	63											0.3940 ± 0.0002	14%
	141	63											0.3956 ± 0.0002	2.5%
	141	63											0.5931 ± 0.0002	4.5%
	141	63											0.5979 ± 0.0002	1.9%
	141	63											0.6059 ± 0.0002	1.5%
	141	63											0.8829 ± 0.0002	1.1%
	141	63											0.9961 ± 0.0003	0.6%
	141	63											0.9998 ± 0.0003	0.4%
	141	63											1.2454 ± 0.0003	0.4%
	141	63											1.7662 ± 0.0005	0.6%
$_{63}Eu^{142m}$	142	63			1.22 m.	β+(83%)		4.8					ann.rad.	160%
	142	63				E.C.(17%)							Sm k x-ray	7%
	142	63											0.5400 ± 0.0002	5%
	142	63											0.5566 ± 0.0002	86%
	142	63											0.5637 ± 0.0002	8%
	142	63											0.6287 ± 0.0002	4%
	142	63											0.7680 ± 0.0002	99%
	142	63											1.0161 ± 0.0002	11%
	142	63											1.0233 ± 0.0002	91%
	142	63											1.3419 ± 0.0002	3%
	142	63											1.7001 ± 0.0003	0.8%
	142	63											2.2584 ± 0.0002	0.6%
$_{63}Eu^{142}$	142	63		141.923150	β+(94%)	≈ 7.5	7.0				1+		ann.rad.	190%
	142	63				E.C.(6%)							0.7680 ± 0.0002	11%
	142	63											0.8896 ± 0.0003	1.5%
	142	63											1.2874 ± 0.0003	1.5%
	142	63											1.6581 ± 0.0005	1.5%
	142	63											1.7541 ± 0.0004	1.5%
	142	63											2.0555 ± 0.001	0.5%
$_{63}Eu^{143}$	143	63		142.920150	2.62 m.	β+(72%)	5.1	4.1			5/2+		ann.rad.	140%
	143	63				E.C.(28%)		5.1					Sm k x-ray	12%
	143	63											0.1077 ± 0.0001	2%
	143	63											0.8053 ± 0.0002	1%
	143	63											1.4684 ± 0.0002	1.1%
	143	63											1.5368 ± 0.0003	3.3%
	143	63											1.6073 ± 0.0002	1%
	143	63											1.8049 ± 0.0002	1.6%
	143	63											1.9127 ± 0.0002	2.1%
	143	63											2.1045 ± 0.0002	0.9%
$_{63}Eu^{144}$	144	63		143.918792	10.2 s.	β+(86%)	6.32	5.2			1+		ann.rad.	4%
	144	63				E.C.(13%)							Sm k x-ray	5%
	144	63											0.8177 ± 0.0002	1.5%
	144	63											1.6601 ± 0.0002	9.6%
	144	63											2.4233 ± 0.0002	1%
$_{63}Eu^{145}$	145	63		144.916267	5.93 d.	β+(2%)	2.72	0.80			5/2+		ann.rad.	4%
	145	63				E.C.(98%)	1.70						Sm k x-ray	41%
	145	63											0.5426 ± 0.0002	4.2%
	145	63											0.6535 ± 0.0001	15%
	145	63											0.7648 ± 0.0002	1.6%
	145	63											0.8937 ± 0.0002	66%
	145	63											1.6587 ± 0.0002	16%
	145	63											1.8044 ± 0.00022	1.1%
	145	63											1.8768 ± 0.0002	1.4%
	145	63											1.9970 ± 0.0002	7%
$_{63}Eu^{146}$	146	63		145.917215	4.58 d.	β+(5%)	3.87	1.47			4-		ann.rad.	10%

TABLE OF THE ISOTOPES (Continued)

Isotope	A	Z	% Natural abundance	Atomic mass	Half-life	Decay mode	Decay energy (MeV)	Particle energy (MeV)	Particle intensity	Thermal neutron cross section	Spin (h/2π)	μ Nucl. mag. moment	Gamma-ray energy (MeV)	Gamma-ray intensity
	146	63				E.C.(95%)							Sm k x-ray	38%
	146	63											0.4305 ± 0.0001	4.8%
	146	63											0.5223 ± 0.0002	4.9%
	146	63											0.6336 ± 0.0001	43%
	146	63											0.6341 ± 0.0001	37%
	146	63											0.6655 ± 0.0001	6.9%
	146	63											0.7025 ± 0.0001	6.5%
	146	63											0.7032 ± 0.0002	8.5%
	146	63											0.7047 ± 0.0002	2.7%
	146	63											0.7470 ± 0.0001	98%
	146	63											0.9002 ± 0.0003	3.7%
	146	63											1.0583 ± 0.0002	6.7%
	146	63											1.1760 ± 0.0002	2.1%
	146	63											1.2969 ± 0.0003	5.6%
	146	63											1.4088 ± 0.0004	3.2%
	146	63											1.5396 ± 0.0002	6.1%
	146	63											1.9310 ± 0.0002	1.2%
	146	63											2.0808 ± 0.0003	2.2%
	146	63											2.0808 ± 0.0003	2.2%
	146	63											2.4365 ± 0.0003	1.0%
	146	63											(0.27 - 2.64)	
$_{63}Eu^{147}$	147	63		146.916742	24.3 d.	E.C.(99.%)					5/2 +		Sm k x-ray	52%
	147	63				β + (0.4%)							0.12113 ± 0.0008	23%
	147	63											0.19725 ± 0.0009	26%
	147	63											0.6014 ± 0.0001	6.8%
	147	63											0.6776 ± 0.0001	11%
	147	63											0.7988 ± 0.0001	5.5%
	147	63											0.8571 ± 0.0001	3.1%
	147	63											0.9331 ± 0.0001	3.6%
	147	63											0.9559 ± 0.0001	3.9%
	147	63											1.0772 ± 0.0001	6.4%
	147	63											1.2559 ± 0.0002	1.0%
$_{63}Eu^{148}$	148	63		147.918125	54.5 d.	E.C.	3.12				5-		Sm k x-ray	41%
	148	63											0.2415 ± 0.0002	1%
	148	63											0.4139 ± 0.0002	19%
	148	63											0.4327 ± 0.0002	2.8%
	148	63											0.5503 ± 0.0001	99%
	148	63											0.5532 ± 0.0001	17%
	148	63											0.5719 ± 0.0001	9.1%
	148	63											0.6113 ± 0.0001	19%
	148	63											0.6299 ± 0.0001	71%
	148	63											0.6543 ± 0.0001	2%
	148	63											0.7257 ± 0.0001	13%
	148	63											0.8700 ± 0.0001	5.5%
	148	63											0.9133 ± 0.0001	2.4%
	148	63											0.9305 ± 0.0001	2.9%
	148	63											0.9673 ± 0.0001	2.9%
	148	63											1.0341 ± 0.0001	7.9%
	148	63											1.1469 ± 0.0001	1.9%
	148	63											1.1833 ± 0.0001	1.7%
	148	63											1.3285 ± 0.0002	1.2%
	148	63											1.3446 ± 0.0001	3.6%
	148	63											1.6215 ± 0.0001	4.6%
	148	63											1.6504 ± 0.0001	3.7%
$_{63}Eu^{149}$	149	63		148.917926	93.1 d.	E.C.	0.69				5/2 +		Sm k x-ray	39%
	149	63											0.2545 ± 0.0002	0.6%
	149	63											0.2770 ± 0.0002	3.3%
	149	63											0.3275 ± 0.0002	3.9%
	149	63											0.3500 ± 0.0003	0.3%
	149	63											0.5059 ± 0.0002	0.55%
	149	63											0.5285 ± 0.0002	0.53%
	149	63											0.5359 ± 0.0003	0.05%
	149	63											0.5584 ± 0.0003	0.05%
$_{63}Eu^{150m}$	150	63			36 y.	E.C.					4,5-		Sm k x-ray	43%
	150	63											0.3340 ± 0.0001	94%
	150	63											0.4394 ± 0.0001	79%
	150	63											0.5055 ± 0.0001	4.7%
	150	63											0.5843 ± 0.0001	52%
	150	63											0.7122 ± 0.0001	1.1%
	150	63											0.7374 ± 0.0001	9.4%
	150	63											0.7480 ± 0.0001	5.0%
	150	63											0.8692 ± 0.0001	2.1%
	150	63											1.0490 ± 0.0001	5.2%
	150	63											1.1706 ± 0.0001	1.3%
	150	63											1.1971 ± 0.0001	1.1%
	150	63											1.2469 ± 0.0001	1.9%
	150	63											1.3437 ± 0.0001	2.5%
	150	63											1.4855 ± 0.0001	1.8%
	150	63											(0.25 - 1.8)	
$_{63}Eu^{150}$	150	63		149.919702	12.6 h.	β-(92%)	1.009	1.010			0-		Sm k x-ray	3%
	150	63				β + (0.4%)	2.29	1.24					0.3339 ± 0.0001	3.7%
	150	63				E.C.(8%)							0.4065 ± 0.0001	2.4%
	150	63											0.7122 ± 0.0001	0.1%
	150	63											0.8319 ± 0.0001	0.2%
	150	63											0.9214 ± 0.0001	0.2%

Isotope	A	Z	% Natural abundance	Atomic mass	Half-life	Decay mode	Decay energy (MeV)	Particle energy (MeV)	Particle intensity	Thermal neutron cross section	Spin (h/2π)	μ Nucl. mag. moment	Gamma-ray energy (MeV)	Gamma-ray intensity
	150	63											1.1657 ± 0.0001	0.2%
	150	63											1.2233 ± 0.0001	0.2%
	150	63											1.9637 ± 0.0001	0.1%
$_{63}Eu^{151}$	151	63	47.8%	150.919847						(4 + 3300 + 5900) b.	5/2 +	+ 3.464		
$_{63}Eu^{152m2}$	152	63			96 m.	I.T.	0.1478				8-		Eu k x-ray	12%
	151	63											0.0898 ± 0.0001	70%
$_{63}Eu^{152m1}$	152	63			9.3 h.	β-(72%)		1.85			0-		Sm k x-ray	13%
	152	63				E.C.(28%)		0.89					0.12178 ± 0.00001	7.2%
	152	63											0.34427 ± 0.00001	2.4%
	152	63											0.84153 ± 0.00002	14.5%
	152	63											0.96334 ± 0.00002	12%
	152	63											1.31461 ± 0.00002	0.9%
	152	63											1.38900 ± 0.00002	0.8%
$_{63}Eu^{152}$	152	63		151.921742	13.4 y.	E.C.(72%)	1.876	0.69			3-	± 1.924	Sm k x-ray	38%
	152	63				β-(28%)	1.822	1.47					Gd k x-ray	11%
	152	63											0.12178 ± 0.00001	28%
	152	63											0.24470 ± 0.00001	7.5%
	152	63											0.34427 ± 0.00001	27%
	152	63											0.44396 ± 0.00001	3.1%
	152	63											0.77887 ± 0.00001	13%
	152	63											0.86737 ± 0.00002	4.2%
	152	63											0.96404 ± 0.00001	14.6%
	152	63											1.08583 ± 0.00002	9.9%
	152	63											1.11209 ± 0.00002	13.6%
	152	63											1.2129 ± 0.0001	1.4%
	152	63											1.2991 ± 0.0001	1.6%
	152	63											1.40802 ± 0.00003	21%
$_{63}Eu^{153}$	153	63	52.2%	152.921225						350 b.	5/2 +	+ 1.530		
$_{63}Eu^{154m}$	154	63			46.1 m.	I.T.	≈ 0.16				8-		Eu k x-ray	6%
	154	63											0.0358 ± 0.0001	10%
	154	63											0.0682 ± 0.0001	36%
	154	63											0.1009 ± 0.0001	26%
$_{63}Eu^{154}$	154	63		153.922975	8.5 y.	β-(99.9%)	1.978	0.27	29%		3-	± 2.000	Gd k x-ray	13%
	154	63				EC.(0.02%)	0.728	0.58	38%				0.12299 ± 0.00001	40%
	154	63						0.84	17%				0.2477 ± 0.0001	6.6%
	154	63						0.98	4%				0.59178 ± 0.00002	4.8%
	154	63						1.87	11%				0.6924 ± 0.0001	1.7%
	154	63											0.72331 ± 0.00001	19.7%
	154	63											0.7568 ± 0.0001	4.3%
	154	63											0.8732 ± 0.0001	11.5%
	154	63											0.9963 ± 0.10.3%	
	154	63											1.2745 ± 0.0001	35.5%
	154	63											1.5965 ± 0.0002	1.8%
$_{63}Eu^{155}$	155	63		154.922889	4.73 y.	β-	0.25	0.25	13%		5/2 +		Gd k x-ray	12%
	155	63											0.0600 ± 0.0001	1.2%
	155	63											0.0865 ± 0.0001	33%
	155	63											0.1053 ± 0.0001	22%
$_{63}Eu^{156}$	156	63		155.924752	15.2 d.	β-	2.45	0.30	11%		1 +		0.08397 ± 0.0005	9%
	156	63						0.49	30%				0.5995 ± 0.0001	2%
	156	63						1.2	12%				0.64623 ± 0.00008	6.7%
	156	63						2.45	31%				0.723441 ± 0.0001	58%
	156	63											0.8118 ± 0.0001	10%
	156	63											0.8670 ± 0.0001	1.4%
	156	63											0.9444 ± 0.0001	1.4%
	156	63											0.9605 ± 0.0001	1.6%
	156	63											1.0651 ± 0.0001	5.2%
	156	63											1.0792 ± 0.0001	4.9%
	156	63											1.1535 ± 0.0001	7.2%
	156	63											1.1542 ± 0.0001	5.0%
	156	63											1.2307 ± 0.0001	8.5%
	156	63											1.2424 ± 0.0001	7%
	156	63											1.2774 ± 0.0001	3.1%
	156	63											1.3664 ± 0.0001	1.7%

TABLE OF THE ISOTOPES (Continued)

Isotope	A	Z	% Natural abundance	Atomic mass	Half-life	Decay mode	Decay energy (MeV)	Particle energy (MeV)	Particle intensity	Thermal neutron cross section	Spin (h/2π)	μ Nucl. mag. moment	Gamma-ray energy (MeV)	Gamma-ray intensity
	156	63											1.8770 ± 0.0002	1.6%
	156	63											1.9377 ± 0.0001	2.1%
	156	63											1.9660 ± 0.0001	4.1%
	156	63											2.0266 ± 0.0001	3.5%
	156	63											2.0977 ± 0.0001	4.0%
	156	63											2.1810 ± 0.0001	2.3%
	156	63											2.1868 ± 0.0001	3.7%
	156	63											2.2056 ± 0.0001	1%
	156	63											2.2699 ± 0.0001	1.1%
$_{63}Eu^{157}$	157	63		156.925418	15.15 h.	β-	1.36	0.91			(5/2+)		Gd k x-ray	32%
	157	63						1.30	41%				0.0545 ± 0.0001	4%
	157	63											0.064 ± 0.001	22%
	157	63											0.320 ± 0.001	3%
	157	63											0.373 ± 0.001	10%
	157	63											0.401 ± 0.001	1%
	157	63											0.413 ± 0.001	17%
$_{63}Eu^{158}$	158	63		157.927800	β-	3.5	2.5				(1-)		0.0795 ± 0.0001	11%
	158	63											0.1820 ± 0.0001	1.9%
	158	63											0.5280 ± 0.0001	1.3%
	158	63											0.6064 ± 0.0001	3.3%
	158	63											0.7430 ± 0.0001	3.0%
	158	63											0.8241 ± 0.0001	1.1%
	158	63											0.8976 ± 0.0001	10%
	158	63											0.9065 ± 0.0001	1.5%
	158	63											0.9225 ± 0.0002	1.6%
	158	63											0.9442 ± 0.0001	25%
	158	63											0.9530 ± 0.0001	1.6%
	158	63											0.9621 ± 0.0001	1.6%
	158	63											0.9771 ± 0.0001	13.6%
	158	63											0.9870 ± 0.0001	1.1%
	158	63											1.0054 ± 0.0003	1.0%
	158	63											1.1076 ± 0.0001	4.3%
	158	63											1.1165 ± 0.0001	1.0%
	158	63											1.2636 ± 0.0002	2.0%
	158	63											1.3479 ± 0.0001	1.4%
	158	63											1.8846 ± 0.0002	1.0%
	158	63											1.9445 ± 0.0002	1.3%
	158	63											2.3677 ± 0.0003	0.66%
	158	63											2.4474 ± 0.0004	0.63%
$_{63}Eu^{159}$	159	63		158.929084	18 m.	β-	2.51	2.4			(5/2+)		0.0678 ± 0.0001	33%
	159	63						2.6					0.0786 ± 0.0001	15%
	159	63											0.0804 ± 0.0004	2%
	159	63											0.0957 ± 0.0001	12%
	159	63											0.1464 ± 0.0001	6%
	159	63											0.1598 ± 0.0002	2%
	159	63											0.1769 ± 0.0001	2%
	159	63											0.2275 ± 0.0003	3%
	159	63											0.6022 ± 0.00022	1.5%
	159	63											0.6134 ± 0.0002	2%
	159	63											0.6595 ± 0.0001	2%
	159	63											0.6649 ± 0.0001	5%
	159	63											0.6766 ± 0.0001	3%
	159	63											0.6819 ± 0.0001	4%
	159	63											0.7265 ± 0.0003	1%
	159	63											0.7443 ± 0.0002	1.6%
	159	63											0.7539 ± 0.0002	1.6%
	159	63											0.8047 ± 0.0002	4%
	159	63											1.0948 ± 0.0002	2%
	159	63											1.1284 ± 0.0003	0.9%
	159	63											1.59202 ± 0.0002	1.1%
$_{63}Eu^{160}$	160	63		159.931880	53 s.	β-	$\approx$ 4.4	2.7			(0-)		0.0753 ± 0.0001	17 +
	160	63						4.1					0.1735 ± 0.0002	32
	160	63											0.2666 ± 0.0003	8
	160	63											0.3020 ± 0.0003	31
	160	63											0.3980 ± 0.0003	39
	160	63											0.4131 ± 0.0003	76
	160	63											0.5155 ± 0.0003	100
	160	63											0.6578 ± 0.0003	6
	160	63											0.7370 ± 0.0003	9
	160	63											0.8217 ± 0.0003	68
	160	63											0.9110 ± 0.0003	97
	160	63											0.9246 ± 0.0003	60
	160	63											0.9953 ± 0.0005	32
	160	63											1.1480 ± 0.0005	33
Gd		64		157.25						49 x 10^3 b.				
$_{64}Gd^{143m}$	143	64			1.83 m.	β+(67%)					11/2-		ann.rad.	145%
	143	64				E.C.(33%)							Eu k x-ray	42%
	143	64				I.T.							0.1176 ± 0.0001	6.4%
	143	64											0.2719 ± 0.0001	83%
	143	64											0.3895 ± 0.0001	3.4%
	143	64											0.5880 ± 0.0001	15%
	143	64											0.6681 ± 0.0001	9.5%
	143	64											0.7586 ± 0.0001	5.4%
	143	64											0.7999 ± 0.0001	10%

TABLE OF THE ISOTOPES (Continued)

Isotope	A	Z	% Natural abundance	Atomic mass	Half-life	Decay mode	Decay energy (MeV)	Particle energy (MeV)	Particle intensity	Thermal neutron cross section	Spin (h/2π)	μ Nucl. mag. moment	Gamma-ray energy (MeV)	Gamma-ray intensity
	143	64											0.8244 ± 0.0001	4.9%
	143	64											0.8905 ± 0.0001	1.7%
	143	64											0.9070 ± 0.0001	2.1%
	143	64											0.9165 ± 0.0001	4.2%
	143	64											0.9849 ± 0.0001	2%
	143	64											1.0083 ± 0.0001	1.3%
	143	64											1.0414 ± 0.0001	3%
	143	64											1.0873 ± 0.0001	1.6%
	143	64											1.2192 ± 0.0001	4.1%
	143	64											1.3736 ± 0.0001	1.1%
	143	64											1.3867 ± 0.0001	1.2%
	143	64											1.4046 ± 0.0001	2.8%
	143	64											1.6293 ± 0.0001	1.9%
	143	64											1.7932 ± 0.0001	2.6%
	143	64											1.8071 ± 0.0001	7.5%
	143	64											1.8203 ± 0.0001	3.0%
	143	64											1.8860 ± 0.0002	0.7%
$_{64}Gd^{143}$	143	64		142.926490	39 s.	β+(82%)	≈ 5.9				1/2+		ann.rad.	160%
	143	64				E.C.(18%)							Eu k x-ray	13%
	143	64											0.2048 ± 0.0001	19%
	143	64											0.2588 ± 0.0001	75%
	143	64											0.4637 ± 0.0001	9.9%
	143	64											0.8129 ± 0.0001	5.4%
	143	64											1.2842 ± 0.0004	1%
	143	64											1.4648 ± 0.0004	0.9%
$_{64}Gd^{144}$	144	64		143.922760	4.5 m.	β+(45%)	≈ 3.7	3.3			0+		ann.rad.	90%
	144	64				E.C.(55%)							Eu k x-ray	22%
	144	64											0.3332 ± 0.0005	12%
	144	64											0.3470 ± 0.0005	4%
	144	64											0.6220 ± 0.0005	2%
	144	64											0.6298 ± 0.0005	4%
	144	64											0.6419 ± 0.0005	2%
	144	64											0.8677 ± 0.0005	2%
$_{64}Gd^{145m}$	145	64			85 s.	I.T.(95%)	0.749				11/2-		0.0273 ± 0.0001	4.5%
	145	64				β+(4%)	5.7						0.3295 ± 0.0003	4.4%
	145	64											0.3866 ± 0.0003	4.1%
	145	64											0.7214 ± 0.0004	83%
$_{64}Gd^{145}$	145	64		144.921690	23 m.	β+(33%)	5.07	2.5			1/2+		ann.rad.	66%
	145	64				E.C.(67%)							Eu k x-ray	27%
	145	64											0.3299 ± 0.0001	2.7%
	145	64											0.8044 ± 0.0001	8.6%
	145	64											0.9526 ± 0.0001	1.5%
	145	64											1.0408 ± 0.0001	9.9%
	145	64											1.0723 ± 0.0001	2.8%
	145	64											1.6001 ± 0.0001	1.8%
	145	64											1.7579 ± 0.0001	34%
	145	64											1.8806 ± 0.0001	33%
	145	64											2.4948 ± 0.0001	1.3%
	145	64											2.6422 ± 0.0001	2.1%
	145	64											(0.32 - 3.69)	
$_{64}Gd^{146}$	146	64		145.918304	48.3 d.	E.C.	1.02	0.35			0+		Eu k x-ray	96%
						(99.9%)								
	146	64				β+(0.2%)							0.1147 ± 0.0001	44%
	146	64											0.1155 ± 0.0001	44%
	146	64											0.1546 ± 0.0001	46%
$_{64}Gd^{147}$	147	64		146.918943	38.1 h.	E.C.	2.08	0.93			7/2-		Eu k x-ray	48%
						(99.8%)								
	147	64				E.C.(0.2%)							0.2293 ± 0.0001	64%
	147	64											0.3099 ± 0.0001	3.7%
	147	64											0.3185 ± 0.0001	2%
	147	64											0.3463 ± 0.0001	1.9%
	147	64											0.3699 ± 0.0001	17%
	147	64											0.3960 ± 0.0001	34%
	147	64											0.4849 ± 0.0001	2.8%
	147	64											0.5591 ± 0.0001	6.2%
	147	64											0.6191 ± 0.0001	3.5%
	147	64											0.6252 ± 0.0001	4.5%
	147	64											0.7550 ± 0.0001	1.9%
	147	64											0.7658 ± 0.0001	10%
	147	64											0.7764 ± 0.0001	5%
	147	64											0.7780 ± 0.0001	4.6%
	147	64											0.8616 ± 0.0001	1.7%
	147	64											0.8934 ± 0.0001	737%
	147	64											0.9289 ± 0.0001	19%
	147	64											1.0692 ± 0.0001	6.5%
	147	64											1.1307 ± 0.0001	5.7%
	147	64											1.2357 ± 0.0001	1.0%
	147	64											(0.1 - 1.8)	
$_{64}Gd^{148}$	148	64		147.918113	75 y.	α		3.1828				0+		
$_{64}Gd^{149}$	149	64		148.919344	9.3 d.	E.C.	1.32				7/2-		Eu k x-ray	52%
	149	64											0.1496 ± 0.0002	42%
	149	64											0.2605 ± 0.0003	1%
	149	64											0.2720 ± 0.0001	2.6%
	149	64											0.2985 ± 0.0003	23%
	149	64											0.3465 ± 0.0003	18%

TABLE OF THE ISOTOPES (Continued)

Isotope	A	Z	% Natural abundance	Atomic mass	Half-life	Decay mode	Decay energy (MeV)	Particle energy (MeV)	Particle intensity	Thermal neutron cross section	Spin (h/2π)	μ Nucl. mag. moment	Gamma-ray energy (MeV)	Gamma-ray intensity
	149	64											0.4964 ± 0.0003	1.3%
	149	64											0.5164 ± 0.0003	2%
	149	64											0.5342 ± 0.0003	2.4%
	149	64											0.6452 ± 0.0003	1.1%
	149	64											0.7482 ± 0.0003	6.2%
	149	64											0.7886 ± 0.0003	5.3%
	149	64											0.9391 ± 0.0004	1.6%
$_{64}Gd^{150}$	150	64		149.918662	$1.8x10^6$ y.	α		2.73			0+			
$_{64}Gd^{151}$	151	64		150.920346	≈ 120 d.	E.C.	0.48				7/2-		Eu k x-ray	43%
	151	64											0.1536 ± 0.0001	5.1%
	151	64											0.1747 ± 0.0001	2.4%
	151	64											0.2432 ± 0.0001	4.6%
	151	64											0.3074 ± 0.0001	0.8%
$_{64}Gd^{152}$	152	64	0.20%	151.919786						10 b.	0+			
$_{64}Gd^{153}$	153	64		152.921745	241.6 d.	E.C.				3 x 10^4 b.	3/2-		Eu k x-ray	61%
	153	61											0.06968 ± 0.00002	2.3%
	153	64											0.09743 ± 0.00001	30%
	153	64											0.10318 ± 0.00001	22%
$_{64}Gd^{154}$	154	64	2.18%	153.920861						80 b.	0+			
$_{64}Gd^{155}$	155	64	14.80%	154.922618						61 x 10^3 b.	3/2-	-0.27		
$_{64}Gd^{156}$	156	64	20.47%	155.922118						2 b.	0+			
$_{64}Gd^{157}$	157	64	15.65%	156.923956						2.55 x 10^5 b.	3/2-	-0.36		
$_{64}Gd^{158}$	158	64	24.84%	157.924099						2.4 b.	0+			
$_{64}Gd^{159}$	159	64		158.926384	18.6 h.	β-	0.60	11%			3/2-		Tb k x-ray	10%
	159	64						0.89	26%				0.05845 ± 0.00005	2.3%
	159	64						0.95	63%				0.36351 ± 0.00001	10.8%
$_{64}Gd^{160}$	160	64	21.86%	159.927049						0.8 b.	0+			
$_{64}Gd^{161}$	161	64		160.929664	3.7 m.	β-	1.958	1.56	85%	3 x 10^4 b.	5/2-		Tb k x-ray	25%
	161	64											0.0563 ± 0.0001	3.8%
	161	64											0.0774 ± 0.0001	1.1%
	161	64											0.1023 ± 0.0001	14%
	161	64											0.1652 ± 0.0001	2.6%
	161	64											0.2836 ± 0.0001	6%
	161	64											0.3149 ± 0.0001	23%
	161	64											0.3381 ± 0.0001	1.7%
	161	64											0.3609 ± 0.0001	61%
	161	64											0.4801 ± 0.0001	2.74%
	161	64											0.5295 ± 0.0001	1.3%
$_{64}Gd^{162}$	162	64		161.931010	8.4 m.	β-	1.4	1.0			0+		0.4030	
	162	61											0.4421	
$_{64}Gd^{163}$	163	64			68 s.	β-							0.2868	
	163	64											0.214	
	163	64											1.685	
Tb		65		158.9254						23 b.				
$_{65}Tb^{145}$	145	65		144.928940	30 s.	β+,E.C.	≈ 6.6						ann.rad.	
	145	65											0.2003 ± 0.0003	7%
	145	65											0.2466 ± 0.0003	4%
	145	65											0.2577 ± 0.0003	39%
	145	65											0.2685 ± 0.0003	3%
	145	65											0.5240 ± 0.0003	10%
	145	65											0.5370 ± 0.0003	23%
	145	65											0.5721 ± 0.0003	14%
	145	65											0.6980 ± 0.0003	5%
	145	65											0.9085 ± 0.0003	7%
	145	65											0.9351 ± 0.0003	5%
	145	65											0.9876 ± 0.0003	37%
	145	65											1.0149 ± 0.0003	5%
	145	65											1.1093 ± 0.0003	14%
	145	65											1.3880 ± 0.0003	6%
	145	65											1.4325 ± 0.0003	10%
	145	65											1.4467 ± 0.0003	15%
$_{65}Tb^{146}$	146	65		145.927150	23 s.	β+(76%)	≈ 8.3				(4-)		ann.rad.	150%
	146	65				E.C.(24%)							Gd k x-ray	10%
	146	65											0.4410 ± 0.0001	13%
	146	65											0.6550 ± 0.0002	2%
	146	65											0.9876 ± 0.0004	1%
	146	65											1.0319 ± 0.0004	3%
	146	65											1.0789 ± 0.0002	50%
	146	65											1.5795 ± 0.0002	97%
	146	65											1.9718 ± 0.0003	3%
	146	65											3.1396 ± 0.0004	11%
$_{65}Tb^{147m}$	147	65			1.8 m.	β+(35%)					11/2-		ann.rad.	70%
	147	65				E.C.(65%)							Gd k x-ray	26%
	147	65											1.1789 ± 0.0004	2%
	147	65											1.3977 ± 0.0002	83%
	147	65											1.7978 ± 0.0003	14%

Isotope	A	Z	% Natural abundance	Atomic mass	Half-life	Decay mode	Decay energy (MeV)	Particle energy (MeV)	Particle intensity	Thermal neutron cross section	Spin (h/2π)	μ Nucl. mag. moment	Gamma-ray energy (MeV)	Gamma-ray intensity
$_{65}Tb^{147}$	147	65		146.923820	1.6 h.	β+(42%)	≈ 4.6				5/2+		ann.rad.	85%
	147	65				E.C.(58%)							Gd k x-ray	26%
	147	65											0.1197 ± 0.0004	4%
	147	65											0.1398 ± 0.0004	20%
	147	65											0.3474 ± 0.0006	1.7%
	147	65											0.4070 ± 0.0004	1.4%
	147	65											0.5472 ± 0.0004	2.0%
	147	65											0.5547 ± 0.0004	3.7%
	147	65											0.6944 ± 0.0004	31%
	147	65											1.1522 ± 0.0004	72%
	147	65											1.6281 ± 0.0004	2.7%
	147	65											1.9483 ± 0.0004	1.4%
	147	65											2.5619 ± 0.0004	1.7%
	147	65											2.6814 ± 0.0004	2.6%
$_{65}Tb^{148m}$	148	65			2.2 m.	β+(25%)					9+		ann.rad.	50%
	148	65				E.C.(75%)							Gd k x-ray	33%
	148	65											0.1295 ± 0.0002	2%
	148	65											0.1429 ± 0.0002	2%
	148	65											0.3945 ± 0.0001	86%
	148	65											0.4817 ± 0.0001	3%
	148	65											0.4888 ± 0.0001	5%
	148	65											0.6319 ± 0.0001	94%
	148	65											0.7530 ± 0.0001	2%
	148	65											0.7845 ± 0.0001	99%
	148	65											0.8081 ± 0.0006	3%
	148	65											0.8824 ± 0.0001	91%
$_{65}Tb^{148}$	148	65		147.924140	60 m.	β+,EC					2-		ann.rad.	
	148	65											Gd k x-ray	
	148	65											0.4888 ± 0.0001	22 +
	148	65											0.6319 ± 0.0001	15
	148	65											0.7845 ± 0.0001	100
	148	65											1.0781 ± 0.0002	12
	148	65											1.4025 ± 0.001	2
	148	65											1.491 ± 0.001	5
	148	65											1.8626 ± 0.0005	8
	148	65											(0.14 - 3.8)weak	
$_{65}Tb^{149m}$	149	65			4.2 m.	E.C.(88%)					11/2-		ann.rad.	24%
	149	65				β+(12%)							Gd k x-ray	36%
	149	65											0.1650 ± 0.0001	7%
	149	65											0.6307 ± 0.0003	2.6%
	149	65											0.7960 ± 0.0001	92%
$_{65}Tb^{149}$	149	65		148.923248	4.15 h.	β+(4%)	3.7	1.8			(1/2+)		Gd k x-ray	38%
	149	65				α(16%)		3.97					0.1650 ± 0.0001	27%
	149	65											0.1872 ± 0.0002	4.3%
	149	65											0.3522 ± 0.0001	30%
	149	65											0.3886 ± 0.0001	19%
	149	65											0.4648 ± 0.0001	5.7%
	149	65											0.6521 ± 0.0001	16%
	149	65											0.8171 ± 0.0001	12%
	149	65											0.8534 ± 0.0001	16%
	149	65											0.8619 ± 0.0001	8.4%
	149	65											1.1755 ± 0.0001	3.5%
	149	65											1.3412 ± 0.0001	2.3%
	149	65											1.6403 ± 0.0001	3.2%
	149	65											1.8274 ± 0.0001	1.2%
	149	65											2.0079 ± 0.0001	0.8%
	149	65											2.9613 ± 0.0001	0.8%
	149	65											(0.1 - 3.2)weak	
$_{65}Tb^{150m}$	150	65			6.0 m.	β+(17%)							ann.rad.	35%
	150	65				E.C.(83%)							Gd k x-ray	37%
	150	65											0.1620 ± 0.0002	7%
	150	65											0.3431 ± 0.0001	25%
	150	65											0.4124 ± 0.0002	10%
	150	65											0.4153 ± 0.0002	4%
	150	65											0.4384 ± 0.0001	42%
	150	65											0.4557 ± 0.0002	12%
	150	65											0.4963 ± 0.0001	23%
	150	65											0.510 ± 0.0.001	26%
	150	65											0.5665 ± 0.0001	22%
	150	65											0.6380 ± 0.0001	99%
	150	65											0.6484 ± 0.0003	18%
	150	65											0.6504 ± 0.0003	69%
	150	65											0.7899 ± 0.0004	2%
	150	65											0.8275 ± 0.0001	41%
$_{65}Tb^{150}$	150	65		149.923669	3.3 h.	β+,E.C.	≈ 4.7				2-		ann.rad.	
	150	65											0.4963 ± 0.0001	
	150	65											0.5691 ± 0.0001	2.5%
	150	65											0.6380 ± 0.0001	72%
	150	65											0.6504 ± 0.0002	4%
	150	65											0.7925 ± 0.0003	4.4%
	150	65											0.8803 ± 0.0001	3%
	150	65											1.2917 ± 0.0001	1.6%
	150	65											1.4305 ± 0.0001	2.4%
	150	65											1.4536 ± 0.0001	2.4%
	150	65											1.5185 ± 0.0002	2.3%

Isotope	A	Z	% Natural abundance	Atomic mass	Half-life	Decay mode	Decay energy (MeV)	Particle energy (MeV)	Particle intensity	Thermal neutron cross section	Spin (h/2π)	μ Nucl. mag. moment	Gamma-ray energy (MeV)	Gamma-ray intensity
	150	65											1.5927 ± 0.0001	1.6%
	150	65											1.7881 ± 0.0001	1.6%
	150	65											2.0917 ± 0.0003	1.4%
	150	65											2.1487 ± 0.0003	1.0%
	150	65											2.2078 ± 0.0003	1.0%
	150	65											2.4263 ± 0.0003	0.9%
	150	65											(0.3 - 4.29)weak	
$_{65}Tb^{151m}$	151	65			≈ 50 s.	I.T.(95%)					11/2-		0.0229 ± 0.0001	3%
	151	65				β+, E.C.							0.0495 ± 0.0001	24%
						(7%)								
	151	65											0.3797 ± 0.0001	7%
	151	65											0.5224 ± 0.0001	1.7%
	151	65											0.8305 ± 0.0001	3.7%
$_{65}Tb^{151}$	151	65		150.923100	17.6 h.	β+(1%)	≈ 4.7	3.7			1/2+		Gd k x-ray	60%
	151	65				E.C.(99%)							0.1083 ± 0.0001	25%
	151	65											0.1804 ± 0.0001	11%
	151	65											0.1921 ± 0.0002	4%
	151	65											0.2517 ± 0.0001	26%
	151	65											0.2870 ± 0.0001	25%
	151	65											0.3804 ± 0.0002	4%
	151	65											0.3953 ± 0.0002	10%
	151	65											0.4265 ± 0.0002	4%
	151	65											0.4437 ± 0.0002	10%
	151	65											0.4790 ± 0.0003	16%
	151	65											0.5873 ± 0.0001	17%
	151	65											0.6048 ± 0.0001	3%
	151	65											0.6166 ± 0.0001	10%
	151	65											0.7038 ± 0.0002	4%
	151	65											0.7311 ± 0.0002	9%
	151	65											(0.1 - 1.8)weak	
$_{65}Tb^{152m}$	152	65			4.1 m.	I.T.(79%)	0.5018				(6+)		Tb k x-ray	40%
	152	65				E.C.(21%)	4.35						Gd k x-ray	9%
	152	65											0.0480 ± 0.0002	1%
	152	65											0.0589 ± 0.0002	7%
	152	65											0.1596 ± 0.0001	16%
	152	65											0.2354 ± 0.0001	4%
	152	65											0.2772 ± 0.0001	8.5%
	152	65											0.2833 ± 0.0001	60%
	152	65											0.3443 ± 0.0001	20%
	152	65											0.3859 ± 0.0001	3%
	152	65											0.4111 ± 0.0001	18%
	152	65											0.4719 ± 0.0001	12%
	152	65											0.5194 ± 0.0001	5%
	152	65											0.5326 ± 0.0001	4%
	152	65											0.5062 ± 0.0002	1%
	152	65											0.6474 ± 0.0002	4%
	152	65											0.7260 ± 0.0002	3%
	152	65											1.1062 ± 0.0002	3%
	152	65											1.1669 ± 0.0002	4%
$_{65}Tb^{152}$	152	65		151.923919	17.6 h.	β+(20%)	3.85	2.45			2-		ann.rad.	40%
	152	65				E.C.(80%)		2.82					Gd k x-ray	33%
	152	65											0.2711 ± 0.0001	8%
	152	65											0.3443 ± 0.0001	57%
	152	65											0.4111 ± 0.0001	3.6%
	152	65											0.5863 ± 0.0001	8%
	152	65											0.7033 ± 0.0002	2.2%
	152	65											0.7649 ± 0.0001	2.6%
	152	65											0.7789 ± 0.0001	5%
	152	65											0.9741 ± 0.0002	2.8%
	152	65											1.1092 ± 0.0002	2.3%
	152	65											1.2991 ± 0.0001	1.9%
	152	65											1.3147 ± 0.0002	1.2%
	152	65											1.9042 ± 0.0002	1.7%
	152	65											2.4050 ± 0.0003	1.3%
	152	65											(0.2 - 2.88)	
$_{65}Tb^{153}$	153	65		152.923440	2.34 d.	E.C.	1.58				5/2+		Gd k x-ray	55%
	153	65											0.0829 ± 0.0001	6%
	153	65											0.0876 ± 0.0001	2%
	153	65											0.1022 ± 0.0001	6%
	153	65											0.1097 ± 0.0001	7%
	153	65											0.1704 ± 0.0001	7%
	153	65											0.2119 ± 0.0001	32%
	153	65											0.2496 ± 0.0001	2.4%
	153	65											0.3036 ± 0.0001	1.0%
	153	65											0.8354 ± 0.0002	1.2%
	153	65											0.9451 ± 0.0002	1.1%
	153	65											0.9917 ± 0.0002	1.2%
	153	65											(0.05 - 1.1)weak	
$_{65}Tb^{154m2}$	154	65			23 h.	E.C.(98%)					(7-)		Gd k x-ray	61%
	154	65				I.T.(2%)							0.1231 ± 0.0001	44%
	154	65											0.1413 ± 0.0001	7%
	154	65											0.1720 ± 0.0001	5%
	154	65											0.2259 ± 0.0001	27%
	154	65											0.2479 ± 0.0001	81%
	154	65											0.2658 ± 0.0001	4%

Isotope	A	Z	% Natural abundance	Atomic mass	Half-life	Decay mode	Decay energy (MeV)	Particle energy (MeV)	Particle intensity	Thermal neutron cross section	Spin (h/2π)	μ Nucl. mag. moment	Gamma-ray energy (MeV)	Gamma-ray intensity
	154	65											0.2675 ± 0.0003	4%
	154	65											0.3467 ± 0.0001	71%
	154	65											0.4268 ± 0.0001	18%
	154	65											0.4792 ± 0.0002	4%
	154	65											0.5064 ± 0.0002	4%
	154	65											0.5180 ± 0.0001	3.9%
	154	65											0.6423 ± 0.0002	4.2%
	154	65											0.6495 ± 0.0001	7%
	154	65											0.8732 ± 0.0001	3.4%
	154	65											0.8927 ± 0.0001	4.8%
	154	65											0.9930 ± 0.0002	17%
	154	65											0.9963 ± 0.0001	3%
	154	65											1.0047 ± 0.0001	4%
	154	65											1.0612 ± 0.0003	4%
	154	65											1.1407 ± 0.0002	2.3%
	154	65											1.1934 ± 0.0009	3%
	154	65											1.4199 ± 0.0001	47%
$_{65}Tb^{154m1}$	154	65			9 h.	β+(78%)					(3-)		Gd k x-ray	41%
	154	65				I.T.(22%)							0.1231 ± 0.0001	31%
	154	65											0.2479 ± 0.0001	22%
	154	65											0.5180 ± 0.0001	6%
	154	65											0.5401 ± 0.0001	20%
	154	65											0.6495 ± 0.0001	11%
	154	65											0.6765 ± 0.0001	3%
	154	65											0.6924 ± 0.0001	3%
	154	65											0.7567 ± 0.0001	2.7%
	154	65											0.8732 ± 0.0001	9.3%
	154	65											0.8927 ± 0.0001	3%
	154	65											0.9963 ± 0.0001	8.7%
	154	65											1.0047 ± 0.0001	11%
	154	65											1.1287 ± 0.0002	1.6%
	154	65											1.1407 ± 0.0002	1.4%
	154	65											1.1521 ± 0.0006	2.2%
	154	65											1.2581 ± 0.0002	1.6%
	154	65											1.2884 ± 0.0002	1.4%
	154	65											1.4906 ± 0.0002	1.0%
	154	65											1.9650 ± 0.0001	2%
	154	65											2.1538 ± 0.0002	1%
	154	65											(0.12 - 2.57)many	
$_{65}Tb^{154}$	153	65		153.924690	22 h.	E.C.(99%)	3.56	1.86			(0-)		Gd k x-ray	49%
	154	65				β+(1%)		2.45					0.1231 ± 0.0001	28%
	154	65											0.5576 ± 0.0001	5.8%
	154	65											0.6924 ± 0.0001	3.4%
	154	65											0.7051 ± 0.0001	5%
	154	65											0.7221 ± 0.0002	8.2%
	154	65											0.8732 ± 0.0001	5.6%
	154	65											0.8783 ± 0.0002	3%
	154	65											0.9963 ± 0.0001	5.2%
	154	65											1.1181 ± 0.0003	2.5%
	154	65											1.1232 ± 0.0002	6.1%
	154	65											1.2744 ± 0.0001	11%
	154	65											1.2913 ± 0.0002	7.4%
	154	65											1.4146 ± 0.0002	2.0%
	154	65											1.9966 ± 0.0001	8.0%
	154	65											2.0419 ± 0.0001	2.1%
	154	65											2.0641 ± 0.0001	7.6%
	154	65											2.1197 ± 0.0002	4.5%
	154	65											2.1872 ± 0.0001	11%
	154	65											2.3075 ± 0.0002	1.6%
	154	65											2.3425 ± 0.0003	1.6%
	154	65											2.3453 ± 0.0003	1.6%
	154	65											2.4305 ± 0.0001	2.3%
	154	65											2.4862 ± 0.0002	1.4%
	154	65											2.9000 ± 0.0004	1.0%
	154	65											3.0232 ± 0.0003	0.8%
	154	65											(0.12 - 3.14)many	
$_{65}Tb^{155}$	155	65		154.923499	5.3 d.	E.C.	0.82				3/2+		Gd k x-ray	55%
	155	65											0.0453 ± 0.0001	1.5%
	155	65											0.08654 ± 0.00001	29%
	155	65											0.10530 ± 0.00001	23%
	155	65											0.1486 ± 0.0001	2.4%
	155	65											0.1613 ± 0.0001	2.5%
	155	65											0.16330 ± 0.00002	4.1%
	155	65											0.18008 ± 0.00002	6.8%
	155	65											0.3407 ± 0.0001	1.1%
	155	65											0.36738 ± 0.00002	2.1%
$_{65}Tb^{156m2}$	156	65			24 h.	I.T.					(4+)		Tb k x-ray	
	156	65											0.0496 ± 0.0001	73%
$_{65}Tb^{156m1}$	156	65			5.0 h.	I.T.	0.0884				(0+)		Tb k x-ray	
	156	65											0.0884 ± 0.0001	1%

TABLE OF THE ISOTOPES (Continued)

Isotope	A	Z	% Natural abundance	Atomic mass	Half-life	Decay mode	Decay energy (MeV)	Particle energy (MeV)	Particle intensity	Thermal neutron cross section	Spin (h/2π)	μ Nucl. mag. moment	Gamma-ray energy (MeV)	Gamma-ray intensity
$_{65}Tb^{156}$	156	65		155.924742	5.3 d.	E.C.	2.438				3-	± 1.4	Gd k x-ray	54%
	156	65											0.08896 ± 0.00001	19%
	156	65											0.19921 ± 0.00002	40%
	156	65											0.2625 ± 0.0001	5.7%
	156	65											0.2965 ± 0.0001	4.5%
	156	65											0.35645 ± 0.00002	14%
	156	65											0.42244 ± 0.00003	8%
	156	65											0.53435 ± 0.00002	67%
	156	65											0.7801 ± 0.0001	2.4%
	156	65											0.9259 ± 0.0001	3.4%
	156	65											1.1541 ± 0.0001	10%
	156	65											1.1589 ± 0.0001	7.3%
	156	65											1.22245 ± 0.00007	31%
	156	65											1.3345 ± 0.0001	2.5%
	156	65											1.4217 ± 0.0001	12%
	156	65											1.6461 ± 0.0001	3.8%
	156	65											1.8454 ± 0.0001	4.1%
	156	65											2.0143 ± 0.0002	1.1%
$_{65}Tb^{157}$	157	65		156.924023	≈ 150 y.	E.C.	0.058				3/2+		Gd k x-ray	14%
	157	65											0.0545 ± 0.0001	0.02%
$_{65}Tb^{158m}$	158	65			10.5 s.	I.T.	0.11				0-		Gd k x-ray	25%
	158	65											0.0110 ± 0.0001	0.9%
$_{65}Tb^{158}$	158	65		157.925411	≈ 150 y.	E.C.(80%)	1.216				3-	± 1.74	Gd k x-ray	39%
	158	65				β-(20%)	0.936						0.0795 ± 0.0001	11%
	158	65											0.0989 ± 0.0001	4.6%
	158	65											0.1820 ± 0.0001	9.2%
	158	65											0.7801 ± 0.0002	9.3%
	158	65											0.9442 ± 0.0001	43%
	158	65											0.9621 ± 0.0001	20%
	158	65											1.1076 ± 0.0001	2.1%
	158	65											1.1871 ± 0.0002	1.6%
$_{65}$Tb159	159	65	100%	158.925342						23.0 b.	3/2+	+1.95		
$_{65}Tb^{160}$	160	65		159.927163	72.4 d.	β-	1.834	0.57	47%	600 b.	3-		Dy k x-ray	11%
	160	65						0.87	27%				0.08678 ± 0.00001	13%
	160	65											0.19703 ± 0.00001	5.2%
	160	65											0.21564 ± 0.00001	4.0%
	160	65											0.29857 ± 0.00001	27.5%
	160	65											0.30956 ± 0.00004	0.9%
	160	65											0.87936 ± 0.00002	30%
	160	65											0.9623 ± 0.0001	10%
	160	65											0.96615 ± 0.00002	25.5%
	160	65											1.17793 ± 0.00002	15.5%
	160	65											1.1999 ± 0.00005	2.4%
	160	65											1.27185 ± 0.00003	7.6%
	160	65											1.31216 ± 0.00005	3.0%
$_{65}Tb^{161}$	161	65		160.927566	6.91 d.	β-	0.591	0.46	23%		3/2+		Dy k x-ray	11%
	161	65						0.52	66%				0.02565 ± 0.00005	21%
	161	65						0.6	10%				0.04892 ± 0.00005	15%
	161	65											0.05720 ± 0.00005	1.6%
	161	65											0.07458 ± 0.00004	9.8%
	161	65											0.08793 ± 0.00004	0.2%
$_{65}Tb^{162}$	162	65		161.929510	7.6 m.	β-2.5	1.3				(1/2-)		Dy k x-ray	10%
	162	65											0.0807 ± 0.0001	8.8%
	162	65											0.1850 ± 0.0001	2.7%
	162	65											0.1853 ± 0.0001	15%
	162	65											0.2600 ± 0.0001	81%
	162	65											0.6974 ± 0.0001	2.6%
	162	65											0.8075 ± 0.0001	43%
	162	65											0.8823 ± 0.0001	14%
	162	65											0.8882 ± 0.0001	39%
$_{65}Tb^{163}$	163	65		162.930550	19.5 m.	β-	1.70	1.3			3/2+		Dy k x-ray	4%
	163	65											0.2509 ± 0.0002	6.7%
	163	65											0.3163 ± 0.0002	8%
	163	65											0.3385 ± 0.0002	4.5%

TABLE OF THE ISOTOPES (Continued)

Isotope	A	Z	% Natural abundance	Atomic mass	Half-life	Decay mode	Decay energy (MeV)	Particle energy (MeV)	Particle intensity	Thermal neutron cross section	Spin (h/2π)	μ Nucl. mag. moment	Gamma-ray energy (MeV)	Gamma-ray intensity
	163	65											0.3479 ± 0.0001	6%
	163	65											0.3511 ± 0.0001	26%
	163	65											0.3542 ± 0.0002	4.6%
	163	65											0.3865 ± 0.0002	4.5%
	163	65											0.3897 ± 0.0002	24%
	163	65											0.4019 ± 0.0002	2.5%
	163	65											0.4151 ± 0.0002	5.4%
	163	65											0.4219 ± 0.0002	11%
	163	65											0.4277 ± 0.0002	3.5%
	163	65											0.4624 ± 0.0002	2.2%
	163	65											0.4754 ± 0.0002	2.9%
	163	65											0.4945 ± 0.0002	72%
	163	65											0.5075 ± 0.0002	4.6%
	163	65											0.5331 ± 0.0002	9.5%
	163	65											0.5596 ± 0.0002	2%
	163	65											0.5840 ± 0.0002	7%
	163	65											0.6084 ± 0.0002	3.7%
	163	65											0.6301 ± 0.0002	1.1%
	163	65											0.8334 ± 0.0001	1.0%
$_{65}Tb^{164}$	164	65		163.933320	3.0 m.	β-	3.86	1.7			(5+)		Dy k x-ray	13%
	164	65											0.0734 ± 0.0001	8%
	164	65											0.1488 ± 0.0001	4%
	164	65											0.1689 ± 0.0005	24%
	164	65											0.2111 ± 0.0001	6%
	164	65											0.2157 ± 0.0001	20%
	164	65											0.2591 ± 0.0002	4%
	164	65											0.2775 ± 0.0001	8%
	164	65											0.2947 ± 0.0001	6.37%
	164	65											0.3448 ± 0.0005	5%
	164	65											0.4103 ± 0.0002	6%
	164	65											0.5485 ± 0.0002	8%
	164	65											0.5859 ± 0.0002	3.9%
	164	65											0.6110 ± 0.0002	19%
	164	65											0.6473 ± 0.0005	5.8%
	164	65											0.6737 ± 0.0002	9%
	164	65											0.6885 ± 0.0002	20%
	164	65											0.7548 ± 0.0002	22%
	164	65											0.7617 ± 0.0002	16%
	164	65											0.7826 ± 0.0002	5%
	164	65											0.8430 ± 0.001	3%
	164	65											0.845 ± 0.001	5%
	164	65											0.9660 ± 0.0005	1.5%
	164	65											1.1043 ± 0.001	1.1%
	164	65											1.1662 ± 0.001	1.9%
	164	65											1.1694 ± 0.001	2.3%
	164	65											1.2898 ± 0.0005	5.6%
	164	65											1.3775 ± 0.0005	5%
	164	65											1.4339 ± 0.0005	8.1%
	164	65											1.6567 ± 0.001	1%
	164	65											2.5110 ± 0.0015	1.3%
$_{65}Tb^{165}$	165	65			2.1 m.	β-	≈ 3.0				3/2+		0.5389	
	165	65											1.1785	
	165	65											1.2920	
	165	65											1.6648	
Dy		66		162.50						920 b.				
$_{66}Dy^{147m}$	147	66			58 s.	I.T.(40%)					(11/2-		Dy k x-ray	
	147	66				β+, EC(60%)							0.072 ± 0.001	5%
	147	66											0.6787 ± 0.0002	33%
$_{66}Dy^{147}$	147	66		146.930680	≈ 80 s.	E.C.,β+	≈ 6.4				$^1/_2$+		ann.rad.	
	147	66											0.1007	
	147	66											0.2534	
	147	66											0.3653	
$_{66}Dy^{148}$	148	66		147.927020	3.1 m.	β+(4%)	2.9	1.2			0+		ann.rad.	
	148	66				E.C.(96%)							Tb k x-ray	38%
	148	66											0.6202 ± 0.0001	100%
$_{66}Dy^{149}$	149	66		148.927110	4.2 m.	β+,E.C.	≈ 3.6				(7/2-)		ann.rad.	
	149	66											0.1008 ± 0.0001	100 +
	149	66											0.1063 ± 0.0001	51
	149	66											0.2534 ± 0.0001	50
	149	66											0.6536 ± 0.0001	60
	149	66											0.7365 ± 0.0001	19
	149	66											0.7417 ± 0.0001	17
	149	66											0.7753 ± 0.0001	35
	149	66											0.7894 ± 0.0001	65
	149	66											1.2742 ± 0.0003	18
	149	66											1.7765 ± 0.0003	79
	149	66											1.8062 ± 0.0003	64
$_{66}Dy^{150}$	150	66		149.925577	7.17 m.	β+, EC(67%)	≈ 1.8				0+		Tb k x-ray	26%
	150	66				α (33%)		4.233					0.3967 ± 0.0002	66%
$_{66}Dy^{151}$	151	66		150.926032	17 m.	β+(5%)	2.76				7/2-		Tb k x-ray	38%
	151	66				E.C.(89%)							0.1764 ± 0.0001	11%
	151	66				α (6%)		4.067					0.3030 ± 0.0001	1.4%
	151	66											0.3861 ± 0.0001	20%

TABLE OF THE ISOTOPES (Continued)

Isotope	A	Z	% Natural abundance	Atomic mass	Half-life	Decay mode	Decay energy (MeV)	Particle energy (MeV)	Particle intensity	Thermal neutron cross section	Spin (h/2π)	μ Nucl. mag. moment	Gamma-ray energy (MeV)	Gamma-ray intensity
	151	66											0.4322 ± 0.0001	4.1%
	151	66											0.4632 ± 0.0001	2.5%
	151	66											0.4766 ± 0.0001	8.2%
	151	66											0.5463 ± 0.0001	15%
	151	66											0.6892 ± 0.0001	2.6%
	151	66											0.7003 ± 0.0001	1.9%
	151	66											0.7556 ± 0.0001	2.1%
	151	66											0.8339 ± 0.0001	2.3%
	151	66											0.8455 ± 0.0001	2.1%
	151	66											0.9837 ± 0.0001	2.1%
	151	66											1.0104 ± 0.0001	3.1%
	151	66											1.1143 ± 0.0001	2.7%
	151	66											1.1298 ± 0.0001	2.3%
	151	66											1.1418 ± 0.0001	2.1%
	151	66											1.4757 ± 0.0001	2.2%
	151	66											1.5381 ± 0.0001	2.0%
	151	66											1.5931 ± 0.0001	2.7%
	151	66											1.7016 ± 0.0001	4.7%
	151	66											(0.16 - 2.09)weak	
$_{66}Dy^{152}$	152	66		151.924716	2.3 h.	E.C.	0.74				0+		Tb k x-ray	40%
	152	66				α		3.63					0.2569 ± 0.0001	97.5%
$_{66}Dy^{153}$	153	66		152.925769	6.3 h.	β+(1%)	2.170	0.89			(7/2-)		Tb k x-ray	90%
	153	66				E.C.(99%)							0.0807 ± 0.0001	11%
	153	66				α (0.01%)		3.46					0.0997 ± 0.0001	10%
	153	66											0.1476 ± 0.0001	3.7%
	153	66											0.2137 ± 0.0001	11%
	153	66											0.2442 ± 0.0001	4.2%
	153	66											0.2543 ± 0.0001	8.3%
	153	66											0.2747 ± 0.0004	6.9%
	153	66											0.2967 ± 0.0001	1.0%
	153	66											0.3237 ± 0.0001	1.1%
	153	66											0.3895 ± 0.0001	1.5%
	153	66											0.4156 ± 0.0001	1.1%
	153	66											0.4341 ± 0.0001	1.2%
	153	66											0.4487 ± 0.0001	1.0%
	153	66											0.4714 ± 0.0001	1.3%
	153	66											0.5105 ± 0.0002	1.1%
	153	66											0.5372 ± 0.0001	1.3%
	153	66											0.5937 ± 0.0001	1.1%
	153	66											0.6598 ± 0.0001	1.1%
	153	66											1.0240 ± 0.0001	1.1%
	153	66											1.0499 ± 0.0001	1.1%
	153	66											1.1043 ± 0.0001	1.0%
	153	66											(0.08 - 1.66)weak	
$_{66}Dy^{154}$	154	66		153.924429	≈ 3 x 10^6y.	α		2.87			0+			
$_{66}Dy^{155}$	155	66		154.925747	10 h.	β+(2%)	2.09	0.845			3/2-		Tb k x-ray	48%
	155	66				E.C.(98%)							0.0655 ± 0.0001	1.8%
	155	66											0.0903 ± 0.0001	1.1%
	155	66											0.1614 ± 0.0001	1.1%
	155	66											0.1846 ± 0.0001	3.4%
	155	66											0.2269 ± 0.0001	69%
	155	66											0.2711 ± 0.0001	1.2%
	155	66											0.4842 ± 0.0001	1.1%
	155	66											0.4986 ± 0.0001	1.8%
	155	66											0.5084 ± 0.0001	1.2%
	155	66											0.6411 ± 0.0001	1.3%
	155	66											0.6642 ± 0.0001	2.3%
	155	66											0.9055 ± 0.0001	2.5%
	155	66											0.9997 ± 0.0001	2.4%
	155	66											1.0899 ± 0.0001	2.8%
	155	66											1.1555 ± 0.0001	2.1%
	155	66											1.1662 ± 0.0001	1.7%
	155	66											1.2512 ± 0.0001	0.9%
	155	66											1.3678 ± 0.0001	0.8%
	155	66											1.6650 ± 0.0001	0.9%
$_{66}Dy^{156}$	156	66	0.06%	155.925277						33 b.	0+			
$_{66}Dy^{157}$	157	66		156.925460	8.1 h.	E.C.	1.34				3/2-	-0.30	Tb k x-ray	43%
	157	66											0.1822 ± 0.0001	2%
	157	66											0.3262 ± 0.0001	93%
$_{66}Dy^{158}$	158	66	0.10%	157.924403						40 b.	0+			
$_{66}Dy^{159}$	159	66		158.925735	144 d.	E.C.	0.366				3/2-		Tb k x-ray	48%
	159	66											0.0582 ± 0.0001	2%
	159	66											0.3262 ± 0.0001	93%
$_{66}Dy^{160}$	160	66	2.34%	159.925193						60 b.	0+			
$_{66}Dy^{161}$	161	66	18.9%	160.926930						580 b.	5/2+	-0.48		
$_{66}Dy^{162}$	162	66	25.5%9	161.926795						180 b.	0+			
$_{66}Dy^{163}$	163	66	24.9%	162.928728						130 b.	5/2-	+0.673		
$_{66}Dy^{164}$	164	66	28.2%	163.929171						(1.7 + 10^3) b.	0+			
$_{66}Dy^{165m}$	165	66			1.26 m.	I.T.(98%)	0.108			2 x 10^3 b.	1/2-		Dy k x-ray	4.8%
	165	66				β-(2%)							0.1082 ± 0.0001	3%
	165	66											0.1538 ± 0.0001	0.2%
	165	66											0.3617 ± 0.0001	0.5%
	165	66											0.5155 ± 0.0001	1.5%
$_{66}Dy^{165}$	165	66		164.931700	2.33 h.	β-	1.286	1.29		3.6 x 10^3 b.	7/2+	0.51	Ho k x-ray	5%
	165	66											0.09468 ± 0.00001	3.6%

Isotope	A	Z	% Natural abundance	Atomic mass	Half-life	Decay mode	Decay energy (MeV)	Particle energy (MeV)	Particle intensity	Thermal neutron cross section	Spin (h/2π)	μ Nucl. mag. moment	Gamma-ray energy (MeV)	Gamma-ray intensity
	165	66											0.27974 ± 0.00001	0.5%
	165	66											0.36166 ± 0.00003	0.8%
	165	66											0.63340 ± 0.00003	0.6%
	165	66											0.71534 ± 0.00004	0.5%
	165	66											1.07964 ± 0.00001	0.07%
$_{66}Dy^{166}$	166	66		165.932803	81.6 h.	β-	0.48	0.40			0+		Ho k x-ray	25%
	166	66											0.0282 ± 0.0001	1%
	166	66											0.0825 ± 0.0001	13%
	166	66											0.3717 ± 0.0001	0.5%
	166	66											0.4260 ± 0.0001	0.5%
$_{66}Dy^{167}$	167	66		166.935650	6.2 m.	β-	2.35	1.78			(1/2-)		Ho k x-ray	20%
	167	66											0.1332 ± 0.0001	3%
	167	66											0.2500 ± 0.0001	10%
	167	66											0.2593 ± 0.0001	28%
	167	66											0.3103 ± 0.0001	25%
	167	66											0.5697 ± 0.0002	48%
	167	66											0.7071 ± 0.0002	1%
	167	66											0.9970 ± 0.0002	0.5%
	167	66											(0.06 - 1.4)	
$_{66}Dy^{168}$	168	66			≈ 8.5 m.	β-					0+		Ho k x-ray	14%
	168	66											0.1435 ± 0.0002	8%
	168	66											0.1925 ± 0.0002	31%
	168	66											0.4430 ± 0.0005	16%
	168	66											0.4867 ± 0.0002	22%
	168	66											0.6302 ± 0.0003	15%
Ho		67		164.9304						65 b.				
$_{67}Ho^{148}$	148	67		147.937340	9 s.	β+,E.C.	9.61						ann.rad.	
	148	67											0.5043 ± 0.0003	17 +
	148	67											0.6615 ± 0.0002	69
	148	67											1.6883 ± 0.0002	100
$_{67}Ho^{149}$	149	67		148.933600	21 s.	β+,E.C.	6.04				(9+)		ann.rad.	
	149	67											1.0733 ± 0.0001	13 +
	149	67											1.0911 ± 0.0001	100
	149	67											1.5836 ± 0.0002	9
$_{67}Ho^{150m}$	150	67			26 s.	β+,E.C.					(9+)		ann.rad.	
	150	67											0.3939 ± 0.0001	93%
	150	67											0.4112 ± 0.0002	7%
	150	67											0.5511 ± 0.0001	88%
	150	67											0.6243 ± 0.0002	3%
	150	67											0.6534 ± 0.0001	100
	150	67											0.8034 ± 0.0001	100
$_{67}Ho^{150}$	150	67		149.933200	≈ 88 s.	β+,E.C.							ann.rad.	
	150	67											0.5913 ± 0.0002	31%
	150	67											0.6534 ± 0.0003	30%
	150	67											0.8034 ± 0.0001	100%
$_{67}Ho^{151m}$	151	67			48 s.	β+, EC(87%)							ann.rad.	
	151	67				α(13%)		4.605					0.2102 ± 0.0002	
	151	67											0.4889 ± 0.0004	
	151	67											0.6948 ± 0.0002	
	151	67											0.7762 ± 0.0001	
$_{67}Ho^{151}$	151	67		150.931510	β+, EC(80%)	≈ 5.10							ann.rad.	
	151	67				α(20%)		4.519					0.3522 ± 0.0004	
	151	67											0.5274 ± 0.0001	100 +
	151	67											0.9676 ± 0.0003	3
	151	67											1.0471 ± 0.0001	
$_{67}Ho^{152m}$	152	67			51 s.	β+, EC(90%)					(9+)		ann.rad.	
	152	67				α(10%)		4.453					0.4929 ± 0.0004	53%
	152	67											0.6138 ± 0.0001	90%
	152	67											0.6474 ± 0.0001	90%
	152	67											0.6835 ± 0.0001	77%
	152	67											0.75850 ± 0.0002	10%
$_{67}Ho^{152}$	152	67		151.931580	2.4 m.	β+, EC(88%)	6.4				(3+)		ann.rad.	
	152	67				α(12%)		4.387					0.6140 ± 0.0001	88%
	152	67											0.6476 ± 0.0001	14%
$_{67}Ho^{153m}$	153	67			2.0 m.	β+, EC (99+%)							ann.rad.	
	153	67				α		3.91					0.2958 ± 0.0001	100 +
	153	67											0.3346 ± 0.0001	45
	153	67											0.3661 ± 0.0001	4
	153	67											0.4381 ± 0.0001	16
	153	67											0.6383 ± 0.0001	29
	153	67											1.0872 ± 0.0002	5
	153	67											1.2770 ± 0.001	10
$_{67}Ho^{153}$	153	67		152.930195	9.3 m.	β+, EC (99+%)	4.12						ann.rad.	

Isotope	A	Z	% Natural abundance	Atomic mass	Half-life	Decay mode	Decay energy (MeV)	Particle energy (MeV)	Particle intensity	Thermal neutron cross section	Spin (h/2π)	μ Nucl. mag. moment	Gamma-ray energy (MeV)	Gamma-ray intensity
	153	67				α		4.01					0.0905 ± 0.0001	6%
	153	67											0.1089 ± 0.0001	99%
	153	67											0.1215 ± 0.0001	2%
	153	67											0.1618 ± 0.0001	95%
	153	67											0.1990 ± 0.0002	5%
	153	67											0.2302 ± 0.0001	58%
	153	67											0.2590 ± 0.0002	12%
	153	67											0.2707 ± 0.0001	78%
	153	67											0.3659 ± 0.0001	100%
	153	67											0.3917 ± 0.0002	10%
	153	67											0.4054 ± 0.0003	4%
	153	67											0.4202 ± 0.0002	17%
	153	67											0.4565 ± 0.0002	46%
	153	67											0.5510 ± 0.0002	9%
	153	67											0.5537 ± 0.0002	23%
	153	67											0.5656 ± 0.0002	22%
$_{67}Ho^{154m}$	154	67			3.2 m.	β+,E.C.					(8+)		ann.rad.	
	154	67											0.2894 ± 0.0002	5%
	154	67											0.3095 ± 0.0002	4%
	154	67											0.3346 ± 0.0001	94%
	154	67											0.3466 ± 0.0002	10%
	154	67											0.4058 ± 0.0004	3%
	154	67											0.4069 ± 0.0001	19%
	154	67											0.4124 ± 0.0001	79%
	154	67											0.4347 ± 0.0002	2%
	154	67											0.4433 ± 0.0002	5%
	154	67											0.4771 ± 0.0001	55%
	154	67											0.5047 ± 0.0002	16%
	154	67											0.5238 ± 0.0001	18%
	154	67											0.5706 ± 0.0001	10%
	154	67											0.7251 ± 0.0001	12%
	154	67											0.7328 ± 0.0002	3%
	154	67											0.7406 ± 0.0002	2%
	154	67											0.8141 ± 0.0001	14%
	154	67											0.9591 ± 0.0003	2%
	154	67											0.9683 ± 0.0003	2.5%
	154	67											0.9929 ± 0.0003	5%
	154	67											0.9997 ± 0.0003	2.3%
	154	67											1.1385 ± 0.0003	1.0%
	154	67											1.2488 ± 0.0002	18%
$_{67}Ho^{154}$	154	67		153.930610	12 m.	β+,E.C.	5.76				(3+)		ann.rad.	135%
	154	67											Dy k x-ray	15%
	154	67											0.3262 ± 0.0003	4.7%
	154	67											0.3346 ± 0.0001	83%
	154	67											0.4125 ± 0.0001	15%
	154	67											0.5049 ± 0.0003	1.5%
	154	67											0.5700 ± 0.0008	11%
	154	67											0.6925 ± 0.0002	5.1%
	154	67											0.6955 ± 0.0003	1%
	154	67											0.7297 ± 0.0003	1.2%
	154	67											0.8734 ± 0.0001	12%
	154	67											0.9052 ± 0.0002	2.2%
	154	67											0.9997 ± 0.0002	3.2%
	154	67											1.0271 ± 0.0002	5.6%
	154	67											1.0859 ± 0.0003	1.6%
	154	67											1.1732 ± 0.0002	1.8%
	154	67											1.3006 ± 0.0002	3.7%
	154	67											1.4204 ± 0.0003	2.0%
	154	67											1.4983 ± 0.0003	1.4%
	154	67											1.5103 ± 0.0003	1.8%
	154	67											1.6565 ± 0.0003	1.4%
	154	67											1.8341 ± 0.0004	1.0%
	154	67											1.8493 ± 0.0004	0.9%
	154	67											1.9378 ± 0.0005	0.6%
	154	67											2.0103 ± 0.0006	0.5%
$_{67}Ho^{155}$	155	67		154.929078	48 m.	β+(6%) E.C.(94%)	3.10				(5/2+)		ann.rad.	12%
	155	67											Dy k x-ray	47%
	155	67											0.0474 ± 0.0001	3%
	155	67											0.1039 ± 0.0001	2.7%
	155	67											0.1363 ± 0.0001	4.1%
	155	67											0.1385 ± 0.0001	1.1%
	155	67											0.1608 ± 0.0001	1.4%
	155	67											0.1630 ± 0.0001	1.2%
	155	67											0.1851 ± 0.0001	2.8%
	155	67											0.2009 ± 0.0001	1.9%
	155	67											0.2024 ± 0.0001	2.1%
	155	67											0.2084 ± 0.0001	2.0%
	155	67											0.2189 ± 0.0001	1.8%
	155	67											0.2478 ± 0.0001	1.9%
	155	67											0.2622 ± 0.0001	1.4%
	155	67											0.3097 ± 0.0001	1.2%
	155	67											0.3254 ± 0.0001	3.4%
	155	67											0.3692 ± 0.0001	1.0%
	155	67											0.3829 ± 0.0002	1.2%
	155	67											0.4086 ± 0.0001	1.5%

Isotope	A	Z	% Natural abundance	Atomic mass	Half-life	Decay mode	Decay energy (MeV)	Particle energy (MeV)	Particle intensity	Thermal neutron cross section	Spin (h/2π)	μ Nucl. mag. moment	Gamma-ray energy (MeV)	Gamma-ray intensity
	155	67											0.8971 ± 0.0001	1.2%
	155	67											1.0235 ± 0.0001	1.3%
	155	67											(0.06 - 2.24)weak	
$_{67}Ho^{156m}$	156	67			56 m.	I.T.	0.0352				1+		ann.rad.	50%
	156	67				β+(25%)		1.8			1+		Dy k x-ray	47%
	156	67				E.C.(75%)		2.9					0.1378 ± 0.0001	52%
	156	67											0.2666 ± 0.0001	54%
	156	67											0.3664 ± 0.0001	11%
	156	67											0.6844 ± 0.0002	5%
	156	67											0.6911 ± 0.0002	4.3%
	156	67											0.7644 ± 0.0002	3.6%
	156	67											0.8846 ± 0.0002	7.1%
	156	67											0.8909 ± 0.0002	2.7%
	156	67											0.9317 ± 0.0002	3.0%
	156	67											1.0310 ± 0.0002	3.2%
	156	67											1.1221 ± 0.0001	3.4%
	156	67											1.2236 ± 0.0002	2.4%
	156	67											1.4169 ± 0.0002	1.1%
	156	67											1.4540 ± 0.0002	1.1%
	156	67											1.4723 ± 0.0002	1.1%
	156	67											2.0310 ± 0.0004	1.0%
	156	67											2.0362 ± 0.0004	0.7%
	156	67											2.0534 ± 0.0005	0.7%
	156	67											2.4178 ± 0.0005	1.5%
	156	67											(0.28 - 2.9)weak	
$_{67}Ho^{156}$	156	67		155.929640	≈ 2 m.	β+,E.C.	5.00				(5+)		ann.rad.	
	156	67											0.1378	
	156	67											0.2665	
$_{67}HO^{157}$	157	67		156.928190	12.6 m.	β+(5%)	2.54	1.18			7/2-		ann.rad.	10%
	157	67				E.C.(95%)							Dy k x-ray	71%
	157	67											0.0611 ± 0.0001	5%
	157	67											0.0865 ± 0.0001	5%
	157	67											0.1477 ± 0.0001	2%
	157	67											0.1531 ± 0.0001	3%
	157	67											0.1624 ± 0.0001	1%
	157	67											0.1881 ± 0.0001	4%
	157	67											0.1934 ± 0.0001	7%
	157	67											0.2722 ± 0.0002	4%
	157	67											0.2800 ± 0.0001	21%
	157	67											0.3202 ± 0.0001	2%
	157	67											0.3411 ± 0.0001	16%
	157	67											0.5083 ± 0.0001	3%
	157	67											0.5556 ± 0.0001	3%
	157	67											0.7086 ± 0.0002	1.4%
	157	67											0.8354 ± 0.0002	1.1%
	157	67											0.8700 ± 0.0002	1.0%
	157	67											0.8966 ± 0.0002	4.2%
	157	67											1.2111 ± 0.0002	2.3%
$_{67}Ho^{158m2}$	158	67			21 m.	β+,E.C.					(9+)		ann.rad.	
	158	67											0.0981 ± 0.0001	
	158	67											0.1664 ± 0.0002	
	158	67											0.2182 ± 0.0001	
	158	67											0.3205 ± 0.0001	
	158	67											0.4062 ± 0.0001	
	158	67											0.9774 ± 0.0001	
	158	67											1.0532	
	158	67											0.4846	
$_{67}Ho^{158m1}$	158	67			27 m.	I.T.(44%)					2-		ann.rad.	
	158	67				E.C.(56%)							Dy k x-ray	67%
	158	67											0.0989 ± 0.0001	31%
	158	67											0.2182 ± 0.0001	46%
	158	67											0.9945 ± 0.0001	3%
	158	67											1.1615 ± 0.0001	1%
	158	67											1.2982 ± 0.0001	1%
	158	67											1.6237 ± 0.0001	2%
	158	67											1.7905 ± 0.0001	8%+
	158	67											2.2213 ± 0.0002	1.6%
	158	67											2.5456 ± 0.0002	1.2%
	158	67											2.6054 ± 0.0002	2.4%
$_{67}Ho^{158}$	158	67		157.928930	11.3 m.	β+(8%)	4.22	1.30			5+		ann.rad.	
	158	67				E.C.(92%)							Dy k x-ray	53%
	158	67											0.0989 ± 0.0001	53%
	158	67											0.2182 ± 0.0001	43%
	158	67											0.3205 ± 0.0001	8%
	158	67											0.7274 ± 0.0001	3%
	158	67											0.7314 ± 0.0001	4%
	158	67											0.8466 ± 0.0001	7.6%
	158	67											0.8474 ± 0.0001	18%
	158	67											0.8505 ± 0.0001	15%
	158	67											0.9457 ± 0.0001	16%
	158	67											0.9464 ± 0.0001	9%
	158	67											0.9488 ± 0.0001	24%
	158	67											0.9976 ± 0.0001	4%
	158	67											1.0648 ± 0.0001	3%
	158	67											1.1842 ± 0.0001	1%

Isotope	A	Z	% Natural abundance	Atomic mass	Half-life	Decay mode	Decay energy (MeV)	Particle energy (MeV)	Particle intensity	Thermal neutron cross section	Spin (h/2π)	μ Nucl. mag. moment	Gamma-ray energy (MeV)	Gamma-ray intensity
	158	67											1.2110 ± 0.0001	1.5%
	158	67											1.4634 ± 0.0001	2.3%
	158	67											1.5781 ± 0.0001	5.8%
	158	67											2.0651 ± 0.0001	2.3%
	158	67											2.1194 ± 0.0001	1.7%
	158	67											2.2016 ± 0.0002	3.4%
$_{67}Ho^{159m}$	159	67			8.3 s.	I.T.	0.206				$^1/_2$+		Ho k x-ray	10%
	159	67											0.1660 ± 0.0001	5%
	159	67											0.2059 ± 0.0001	40%
$_{67}Ho^{159}$	159	67		158.927706	33 m.	E.C.	1.836						Dy k x-ray	73%
	159	67											0.0567 ± 0.0002	5%
	159	67											0.1006 ± 0.0002	3.7%
	159	67											0.1210 ± 0.0002	33%
	159	67											0.1320 ± 0.0002	22%
	159	67											0.1558 ± 0.0002	2%
	159	67											0.1731 ± 0.0002	2%
	159	67											0.1776 ± 0.0002	6%
	159	67											0.1863 ± 0.0002	3%
	159	67											0.2177 ± 0.0002	3%
	159	67											0.2529 ± 0.0002	13%
	159	67											0.3096 ± 0.0002	14%
	159	67											0.8387 ± 0.0002	3%
	159	67											(0.06 - 1.2)weak	
$_{67}Ho^{160m}$	160	67			4.9 h.	I.T.(67%)	0.060				2-		0.0868 ± 0.0001	14%
	160	67				E.C.(33%)	3.35						0.1970 ± 0.0001	12%
	160	67											0.5385 ± 0.0003	4%
	160	67											0.6455 ± 0.0003	14%
	160	67											0.6464 ± 0.0003	39%
	160	67											0.7281 ± 0.0003	30%
	160	67											0.7529 ± 0.0003	2.6%
	160	67											0.7652 ± 0.0003	3.7%
	160	67											0.8715 ± 0.0004	6.6%
	160	67											0.8791 ± 0.0004	19%
	160	67											0.9619 ± 0.0004	17%
	160	67											0.9658 ± 0.0004	17%
	160	67											1.0684 ± 0.0005	2.7%
	160	67											1.1985 ± 0.0006	2.2%
	160	67											1.2715 ± 0.0006	2.8%
	160	67											1.2854 ± 0.0006	1.7%
	160	67											1.370 ± 0.001	1%
	160	67											1.4310 ± 0.0007	1%
	160	67											2.5430 ± 0.001	1.2%
	160	67											2.6127 ± 0.001	1.1%
	160	67											2.6305 ± 0.0015	1.3%
	160	67											2.6732 ± 0.0014	1.6%
$_{67}Ho^{160}$	160	67		159.928720	25.6 m.	β+,E.C.	3.29	0.57			5+		See Ho^{166m}	
	160	67											0.7282	
	160	67											0.8794	
$_{67}Ho^{161m}$	161	67			6.7 s.	I.T.	0.211						Ho k x-ray	10%
	161	67											0.2112 ± 0.0001	44%
$_{67}Ho^{161}$	161	67		160.927849	2.5 h.	E.C.	0.856				7/2-		Dy k x-ray	44%
	161	67											0.0256 ± 0.0001	27%
	161	67											0.0592 ± 0.0001	1.2%
	161	67											0.0774 ± 0.0001	2.5%
	161	67											0.1031 ± 0.0001	3.3%
$_{67}Ho^{162m}$	162	67			68 m.	I.T.(61%)					6-		Dy k x-ray	61%
	162	67				E.C.(39%)							Ho k x-ray	23%
	162	67											0.0578 ± 0.0001	4%
	162	67											0.0807 ± 0.0001	11%
	162	67											0.1850 ± 0.0001	29%
	162	67											0.2828 ± 0.0001	11%
	162	67											0.9372 ± 0.0002	11%
	162	67											1.1249 ± 0.0002	1.2%
	162	67											1.2200 ± 0.0002	22%
$_{67}Ho^{162}$	162	67		161.929092	15 m.	E.C.(96%)	0.295				1+		Dy k x-ray	48%
	162	67				β+(4%)							0.0807 ± 0.0001	8%
	162	67											1.3196 ± 0.0002	3.7%
	162	67											1.3728 ± 0.0002	0.8%
$_{67}Ho^{163m}$	163	67			1.09 s.	I.T.	0.298				($^1/_2$+)		Ho k x-ray	5.7%
	163	67											0.2798 ± 0.0001	77.5%
$_{67}Ho^{163}$	163	67		162.928731	33 ± .2 y.	E.C.	0.004				7/2-		Dy M x-rays	
$_{67}Ho^{164m}$	164	67			37.5 ± 1 m.	I.T.	0.140				(6-)		Ho k x-ray	37%
	164	67											0.0373 ± 0.0001	11%
	164	67											0.0566 ± 0.0001	6.7%
	164	67											0.0940 ± 0.0001	0.15%
$_{67}Ho^{164}$	164	67		163.930285	29 ± 0.5 m.	E.C.(58%)	1.029				1+		Dy k x-ray	25%
	164	67				β-(42%)	1.013						0.0734 ± 0.0001	1.8%
	164	67											0.0914 ± 0.0001	2.5%
$_{67}Ho^{165}$	165	67	100%	164.930319						(3.5 + 62) b.	7/2-	+4.173		
$_{67}Ho^{166m}$	166	67			1.2 ± 0.2 x 10^3y	β-					7-	4.1	Er k x-ray	20%
	166	67											0.08057 ± 0.00001	13%
	166	67											0.18407 ± 0.00002	74%

Isotope	A	Z	% Natural abundance	Atomic mass	Half-life	Decay mode	Decay energy (MeV)	Particle energy (MeV)	Particle intensity	Thermal neutron cross section	Spin (h/2π)	μ Nucl. mag. moment	Gamma-ray energy (MeV)	Gamma-ray intensity
	166	67											0.2159 ± 0.0001	2.6%
	166	67											0.28046 ± 0.00002	30%
	166	67											0.3007 ± 0.0001	3.8%
	166	67											0.3657 ± 0.0001	2.5%
	166	67											0.4109 ± 0.0001	12%
	166	67											0.4515 ± 0.0001	3.1%
	166	67											0.4648 ± 0.0001	1.2%
	166	67											0.5298 ± 0.0001	10%
	166	67											0.5710 ± 0.0001	6%
	166	67											0.6115 ± 0.0001	1.4%
	166	67											0.6705 ± 0.0001	5.8%
	166	67											0.6912 ± 0.0001	1.6%
	166	67											0.71169 ± 0.00004	59%
	166	67											0.7523 ± 0.0001	13%
	166	67											0.7788 ± 0.0001	3%
	166	67											0.81031 ± 0.00004	63%
	166	67											0.8306 ± 0.0001	11%
	166	67											0.9509 ± 0.0001	3.1%
	166	67											1.2414 ± 0.0001	1%
	166	67											1.4270 ± 0.0001	0.6%
$_{67}Ho^{166}$	166	67		165.932281	1.117 ± 0.002 d.	β-	1.854	1.776	48%		0-		Er k x-ray	5%
	166	67						1.855	51%				0.08057 ± 0.00002	6%
	166	67											1.37943 ± 0.00005	0.9%
	166	67											1.5819 ± 0.0001	0.2%
	166	67											1.6624 ± 0.0001	0.1%
$_{67}Ho^{167}$	167	67		166.933127	3.1 ± 0.1 h.	β-	0.97	0.31	43%		(7/2-)		Er k x-ray	12%
	167	67						0.61	21%				0.0793 ± 0.0002	2%
	167	67						0.96	15%				0.0835 ± 0.0002	2%
	167	67						0.97	15%				0.2379 ± 0.0002	5%
	167	67											0.3213 ± 0.0002	24%
	167	67											0.3465 ± 0.0002	57%
	167	67											0.3862 ± 0.0002	3%
	167	67											0.4030 ± 0.0002	3%
	167	67											0.4600 ± 0.0002	2%
$_{67}Ho^{168}$	168	67		167.935290	3.0 ± 0.1 m.	β-	2.7	2.0			3+		Er k x-ray	9%
	168	67											0.0798 ± 0.0001	10%
	168	67											0.1843 ± 0.0001	6%
	168	67											0.1982 ± 0.0001	2%
	168	67											0.4475 ± 0.0001	1.4%
	168	67											0.6317 ± 0.0001	3%
	168	67											0.7306 ± 0.0001	1.5%
	168	67											0.7413 ± 0.0001	36%
	168	67											0.8159 ± 0.0001	18%
	168	67											0.8211 ± 0.0001	34%
	168	67											1.3718 ± 0.0001	1.3%
	168	67											(0.08 - 2.34)weak	
$_{67}Ho^{169}$	169	67		168.936869	4.7 ± 0.1 m.	β-	2.12	1.2			(7/2-)			
	169	67						2.0					0.1496 ± 0.0003	7 +
	169	67											0.1519 ± 0.0003	26
	169	67											0.6289 ± 0.0003	13
	169	67											0.6765 ± 0.0002	20
	169	67											0.7170 ± 0.0002	15
	169	67											0.7610 ± 0.0002	48
	169	67											0.7784 ± 0.0002	47
	169	67											0.7884 ± 0.0001	100
	169	67											0.8494 ± 0.0006	5
	169	67											0.8529 ± 0.0002	536
	169	67											0.8664 ± 0.0002	21
	169	67											0.8764 ± 0.0003	10
$_{67}Ho^{170m}$	170	67			43 ± 2 s.	β-							0.0787 ± 0.0002	100 +
	170	67											0.1816 ± 0.0002	8
	170	67											0.4820 ± 0.0003	13
	170	67											0.5409 ± 0.0002	13
	170	67											0.6998 ± 0.0003	7.6
	170	67											0.8123 ± 0.0002	59
	170	67											0.8812 ± 0.0002	12
	170	67											0.9594 ± 0.0005	7
	170	67											1.0227 ± 0.0004	9
	170	67											1.1875 ± 0.0003	15
	170	67											1.2663 ± 0.0007	7.9
	170	67											1.8940 ± 0.0003	27
	170	67											1.9401 ± 0.0003	6.2
	170	67											1.9726 ± 0.0003	21
	170	67											1.9925 ± 0.0005	3
	170	67											2.6061 ± 0.0004	2
	170	67											2.6465 ± 0.0004	2
	170	67											2.6848 ± 0.0004	3

TABLE OF THE ISOTOPES (Continued)

Isotope	A	Z	% Natural abundance	Atomic mass	Half-life	Decay mode	Decay energy (MeV)	Particle energy (MeV)	Particle intensity	Thermal neutron cross section	Spin (h/2π)	μ Nucl. mag. moment	Gamma-ray energy (MeV)	Gamma-ray intensity
	170	67											2.7151 ± 0.0008	1.5
	170	67											2.7892 ± 0.0015	0.7
$_{67}Ho^{170}$	170	67		169.939620	2.8 ± 0.2 m.	β-							Er k x-ray	29%
	170	67											0.0787 ± 0.0001	13%
	170	67											0.0947 ± 0.0001	3%
	170	67											0.1035 ± 0.0001	5%
	170	67											0.1239 ± 0.0001	4%
	170	67											0.14125 ± 0.0001	2%
	170	67											0.1654 ± 0.0001	4%
	170	67											0.1816 ± 0.0001	27%
	170	67											0.2274 ± 0.0001	4%
	170	67											0.2582 ± 0.0001	43%
	170	67											0.2804 ± 0.0001	3%
	170	67											0.2834 ± 0.0001	3%
	170	67											0.4132 ± 0.0002	4%
	170	67											0.4774 ± 0.0002	4%
	170	67											0.7504 ± 0.0002	6%
	170	67											0.7863 ± 0.0005	6%
	170	67											0.8435 ± 0.0002	3%
	170	67											0.8547 ± 0.0005	14%
	170	67											0.8670 ± 0.0002	2.5%
	170	67											0.8902 ± 0.0002	25%
	170	67											0.9321 ± 0.0002	42%
	170	67											0.9346 ± 0.0005	4%
	170	67											0.9414 ± 0.0005	24%
	170	67											0.9574 ± 0.0003	4%
	170	67											0.9765 ± 0.0003	3%
	170	67											1.0247 ± 0.0004	1.7%
	170	67											1.1118 ± 0.0003	2.4%
	170	67											1.1387 ± 0.0002	24%
	170	67											1.1530 ± 0.0003	2%
	170	67											1.2260 ± 0.0003	4%
	170	67											1.3069 ± 0.0003	0.5%
Er		68		167.26						160 b.				
$_{68}Er^{150}$	150	68		149.997710	20 s.	β+(36%)	4.2				0+		ann.rad.	125%
	150	68				E.C.(64%)							Ho k x-ray	
	150	68											0.4758 ± 0.0003	99%
$_{68}Er^{151}$	151	68		150.937200	23 s.	β+,E.C.	5.3						ann.rad.	
$_{68}Er^{152}$	152	68		151.934920	10.3 s.	β+, EC (10%)	3.12				0+		ann.rad.	
	152	68				α(90%)		4.804						
$_{68}Er^{153}$	153	68		152.934870	37.1 s.	α(53%)		4.674					ann.rad.	
	153	68				β+, EC (47%)		4.35						
$_{68}Er^{154}$	154	68		153.932772	3.7 m.	β+, EC (99+%)	2.014				0+		ann.rad.	
	154	68				α(0.5%)		4.166						
$_{68}Er^{155}$	155	68		154.933060	5.3 m.	β+, EC (47%)					(7/2-)		ann.rad.	94%
	155	68				E.C.(53%)							Ho k x-ray	32%
	155	68											0.1101 ± 0.0001	7%
	155	68											0.1238 ± 0.0001	2%
	155	68											0.1851 ± 0.0001	1%
	155	68											0.2011 ± 0.0001	2%
	155	68											0.2340 ± 0.0001	3%
	155	68											0.2415 ± 0.0002	5%
	155	68											0.3287 ± 0.0002	1%
	155	68											0.3586 ± 0.0003	1%
	155	68											0.4227 ± 0.0001	1.5%
	155	68											0.4526 ± 0.0002	1.8%
	155	68											0.5122 ± 0.0002	2.8%
$_{68}Er^{156}$	156	68		155.931290	20 m.	β+,E.C.	1.53				0+		ann.rad.	
	156	68											0.0298 ± 0.0001	17 +
	156	68											0.0352 ± 0.0001	100
	156	68											0.0522 ± 0.0001	
	156	68											0.1336 ± 0.0004	44
$_{68}Er^{157}$	157	68		156.931910	24 m.	β+,E.C.	3.47				3/2-		ann.rad.	
	157	68											0.117	
	157	68											0.385	
	157	68											1.320	
	157	68											1.660	
	157	67											1.820	
	157	68											2.000	
$_{68}Er^{158}$	158	68		157.930010	2.3 h.	E.C. (99.5%)	1.00	0.74			0+		Ho k x-ray	71%
	158	68				β+(0.5%)							0.0719 ± 0.0001	10%
	158	68											0.2486 ± 0.0001	3%
	158	68											0.3108 ± 0.0001	1.6%
	158	68											0.3868 ± 0.0001	7%
	158	68											0.5161 ± 0.0003	1%
$_{68}Er^{159}$	159	68		158.930678	36 m.	β+(7%)	2.77				3/2-		ann.rad.	14%
	159	68				E.C.(93%)							Ho k x-ray	42%
	159	68											0.1660 ± 0.0001	4%

TABLE OF THE ISOTOPES (Continued)

Isotope	A	Z	% Natural abundance	Atomic mass	Half-life	Decay mode	Decay energy (MeV)	Particle energy (MeV)	Particle intensity	Thermal neutron cross section	Spin (h/2π)	μ Nucl. mag. moment	Gamma-ray energy (MeV)	Gamma-ray intensity
	159	68											0.2523 ± 0.0003	4%
	159	68											0.3146 ± 0.0001	1%
	159	68											0.5054 ± 0.0001	2%
	159	68											0.5517 ± 0.0002	2.5%
	159	68											0.5810 ± 0.0001	4.6%
	159	68											0.6245 ± 0.0001	35%
	159	68											0.6493 ± 0.0001	25%
	159	68											0.9425 ± 0.0004	1%
	159	68											1.8383 ± 0.0007	1%
	159	68											2.0016 ± 0.0007	0.9%
	159	68											(0.07 - 2.5)many	
$_{68}Er^{160}$	160	68		159.929080	28.6 h.	E.C.	0.33				0+		Ho k x-ray	38%
	160	68											(0.05 - 0.96)weak	
$_{68}Er^{161}$	161	68		160.929996	3.24 h.	E.C.	2.00				3/2-	-0.370	Ho k x-ray	46%
	161	68											0.2015 ± 0.0001	1%
	161	68											0.3148 ± 0.0001	2.6%
	161	68											0.5926 ± 0.0001	3%
	161	68											0.8265 ± 0.0001	61%
	161	68											0.8650 ± 0.0002	1.2%
	161	68											0.9317 ± 0.0001	1.9%
	161	68											1.1745 ± 0.0003	1.2%
	161	68											(0.07 - 1.74)weak	
$_{68}Er^{162}$	162	68	0.14%	161.928775						19 b.	0+			
$_{68}Er^{163}$	163	68		162.930030	75.1 m.	E.C.	1.21				5/2-	+0.57	Ho k x-ray	40%
	163	68											0.4361 ± 0.0001	0.03%
	163	68											0.4399 ± 0.0001	0.03%
	163	68											1.1135 ± 0.0003	0.05%
$_{68}Er^{164}$	164	68	1.61%	163.929198						13 b.	0+			
$_{68}Er^{165}$	165	68		164.930723	10.36 h.	E.C.	0.377				3/2-		Ho k x-ray	39%
$_{68}Er^{166}$	166	68	33.6%	165.930290						(15 + 5) b.	0+			
$_{68}Er^{167m}$	167	68			2.28 s.	I.T.	0.208				1/2-		Er k x-ray	10%
	167	68											0.2078 ± 0.0001	42%
$_{68}Er^{167}$	167	68	22.95%	166.932046						670 b.	7/2+	-0.5665		
$_{68}Er^{168}$	168	68	26.8%	167.932368						2.7 b.	0+			
$_{68}Er^{169}$	169	68		168.934588	9.4 ± .02 d.	β-	0.351	0.35	≈ 100%		1/2-	+0.515	Tm k x-ray	0.006%
	169	68											0.1098 ± 0.0001	0.0013%
	169	68											0.1182 ± 0.0001	0.0001%
$_{68}Er^{170}$	170	68	14.9%	169.935461						5.7 b.	0+			
$_{68}Er^{171}$	171	68		170.938027	7.52 ± 0.03 h.	β-					5/2-	0.70	Tm k x-ray	24%
	171	68											0.11160 ± 0.00002	20%
	171	68											0.11669 ± 0.00001	2.3%
	171	68											0.12409 ± 0.00001	9%
	171	68											0.29591 ± 0.00003	29%
	171	68											0.30832 ± 0.00001	64%
	171	68											0.79634 ± 0.00008	0.6%
	171	68											0.90795 ± 0.00008	0.6%
	171	68											(0.08 - 1.4)weak	
$_{68}Er^{172}$	172	68		171.939353	2.05 ± 0.02 d.	β-	0.889	0.28	48%				Tm k x-ray	17%
	172	68						0.36	46%				0.0597 ± 0.0001	2.9%
	172	68											0.0681 ± 0.0001	3.5%
	172	68											0.1278 ± 0.0001	2.3%
	172	68											0.2027 ± 0.0001	1.1%
	172	68											0.3835 ± 0.0001	2.5%
	172	68											0.4073 ± 0.0001	45%
	172	68											0.4460 ± 0.0001	3.1%
	172	68											0.4754 ± 0.0001	1.1%
	172	68											0.6101 ± 0.0001	47%
$_{68}Er^{173}$	173	68		172.942280	1.40 ± 0.1 m.	β-	2.50				(7/2-)		Tm k x-ray	34%
	173	68											0.0942 ± 0.0002	5%
	173	68											0.1161 ± 0.0002	19%
	173	68											0.1186 ± 0.0002	2%
	173	68											0.1224 ± 0.0001	21%
	173	68											0.1928 ± 0.0002	46%
	173	68											0.1992 ± 0.0002	48%
	173	68											0.8008 ± 0.0006	10%
	173	68											0.8952 ± 0.0004	54%
Tm		69		168.9342						105 b.				
$_{69}Tm^{152}$	152	69		151.944460	5.2 s.	β+,E.C.	8.88						ann.rad.	
$_{69}Tm^{153}$	153	69		152.941830	1.6 s.	β+, EC (10%)	6.49						ann.rad.	
	153	69				α(90%)		5.11						
$_{69}Tm^{154m}$	154	69			3.4 s.	β+, EC (15%)							ann.rad.	
	154	69				α		5.03						

Isotope	A	Z	% Natural abundance	Atomic mass	Half-life	Decay mode	Decay energy (MeV)	Particle energy (MeV)	Particle intensity	Thermal neutron cross section	Spin (h/2π)	μ Nucl. mag. moment	Gamma-ray energy (MeV)	Gamma-ray intensity
$_{69}Tm^{154}$	154	69		153.941360	8.3 s.	β+, EC (56%)	7.99						ann.rad.	
	154	69				α (44%)		4.96						
$_{69}Tm^{155}$	155	69		154.939010	25 s.	β+,E.C.	5.55						0.0315	5 +
	155	69				α		4.45					0.0638 ± 0.0001	3
	155	69											0.0881 ± 0.0002	17
	155	69											0.1520 ± 0.0001	7
	155	69											0.1716 ± 0.0001	2
	155	69											0.2268 ± 0.0002	100
	155	69											0.2476 ± 0.0002	6
	155	69											0.3153 ± 0.0003	2
	155	69											0.3235 ± 0.0003	8
	155	69											0.3790 ± 0.0004	4
	155	69											0.4334 ± 0.0003	3
	155	69											0.5187 ± 0.0004	3
	155	69											0.5320 ± 0.0005	20
	155	69											0.5333 ± 0.0005	5
	155	69											0.5757 ± 0.0003	2
	155	69											0.6067 ± 0.0002	11
$_{69}Tm^{156m}$	156	69			19 s.	α		4.46						
$_{69}Tm^{156}$	156	69		155.938840	80 s.	β+,E.C.	7.03						ann.rad.	
	156	69				α		4.23					0.3446 ± 0.0001	86%
	156	69											0.4208 ± 0.0001	1.6%
	156	69											0.4529 ± 0.0001	17%
	156	69											0.5860 ± 0.0002	14%
	156	69											0.6089 ± 0.0002	1.4%
	156	69											0.7000 ± 0.0002	1.2%
	156	69											0.8762 ± 0.0002	2.3%
	156	69											0.8985 ± 0.0002	1.3%
	156	69											0.9304 ± 0.0001	5.0%
	156	69											0.9590 ± 0.0001	8.8%
	156	69											1.0068 ± 0.0002	3.1%
	156	69											1.0171 ± 0.0003	1.1%
	156	69											1.2208 ± 0.0002	2.9%
	156	69											0.2261 ± 0.0003	1.2%
	156	69											1.2861 ± 0.0002	2.7%
	156	69											1.3661 ± 0.0003	1.5%
	156	69											1.5163 ± 0.0003	1.2%
	156	69											1.5180 ± 0.0004	1.9%
	156	69											1.5651 ± 0.0003	1.6%
	156	69											1.6640 ± 0.0004	1.2%
	156	69											0.6700 ± 0.0003	1.4%
	156	69											1.8253 ± 0.0003	0.8%
$_{69}Tm^{157}$	157	69		156.936880	3.6 m.	β+,E.C.	4.63						ann.rad.	
	157	69				α		3.97					0.1104 ± 0.0001	9%
	157	69											0.1312 ± 0.0002	4%
	157	69											0.1754 ± 0.0002	3%
	157	69											0.1960 ± 0.0001	3%
	157	69											0.2416 ± 0.0001	7%
	157	69											0.2475 ± 0.0001	3%
	157	69											0.3080 ± 0.0002	2%
	157	69											0.3484 ± 0.0002	9%
	157	69											0.3570 ± 0.0002	7%
	157	69											0.3578 ± 0.0002	5%
	157	69											0.3606 ± 0.0002	4%
	157	69											0.3674 ± 0.0002	4%
	157	69											0.3707 ± 0.0001	5%
	157	69											0.3810 ± 0.0001	2%
	157	69											0.3855 ± 0.0001	10%
	157	69											0.3873 ± 0.0002	2%
	157	69											0.4550 ± 0.0002	10%
	157	69											0.4846 ± 0.0002	3%
	157	69											0.5250 ± 0.0002	4%
	157	69											0.5353 ± 0.0002	4%
	157	69											0.5491 ± 0.0003	5%
	157	69											0.5556 ± 0.0003	3%
	157	69											0.5750 ± 0.0001	3%
	157	69											0.6855 ± 0.0002	2%
	157	69											0.9234 ± 0.0003	1%
	157	69											0.9564 ± 0.0003	1%
	157	69											1.2624 ± 0.0005	2%
	157	69											(0.1 - 1.58)weak	
$_{69}Tm^{158}$	158	69		157.936990	4.0 m.	β+, EC (74%)	6.50				(2-)		ann.rad.	150%
	158	69				E.C.(26%)							Er k x-ray	18%
	158	69											0.1921 ± 0.0001	68%
	158	69											0.3351 ± 0.0001	19%
	158	69											0.6143 ± 0.0001	1.9%
	158	69											0.6280 ± 0.0001	7.4%
	158	69											0.6566 ± 0.0001	1.9%
	158	69											0.7969 ± 0.0002	1.2%
	158	69											0.8148 ± 0.0001	1.3%
	158	69											0.8201 ± 0.0001	3.6%
	158	69											0.8512 ± 0.0001	5.2%
	158	69											0.9891 ± 0.0001	4.0%
	158	69											1.0651 ± 0.0001	1.7%

Isotope	A	Z	% Natural abundance	Atomic mass	Half-life	Decay mode	Decay energy (MeV)	Particle energy (MeV)	Particle intensity	Thermal neutron cross section	Spin (h/2π)	μ Nucl. mag. moment	Gamma-ray energy (MeV)	Gamma-ray intensity
	158	69											1.1498 ± 0.0001	8.4%
	158	69											1.2259 ± 0.0001	1.5%
	158	69											1.3340 ± 0.0001	3.6%
	158	69											1.4186 ± 0.0001	1.5%
	158	69											1.5505 ± 0.0001	1.8%
	158	69											1.5772 ± 0.0001	1.0%
	158	69											(0.18 - 2.81)weak	
$_{69}Tm^{159}$	159	69		158.934970	9.0 m.	β+(23%)	4.00				5/2+		ann.rad.	45%
	159	69				E.C.(77%)							Er k x-ray	79%
	159	69											0.0591 ± 0.0001	5%
	159	69											0.0848 ± 0.0001	7%
	159	69											0.1197 ± 0.0001	3%
	159	69											0.1277 ± 0.0001	5%
	159	69											0.1441 ± 0.0001	2%
	159	69											0.1609 ± 0.0001	4%
	159	69											0.1629 ± 0.0001	2%
	159	69											0.1965 ± 0.0001	2%
	159	69											0.2202 ± 0.0001	5%
	159	69											0.2477 ± 0.0001	2%
	159	69											0.2527 ± 0.0001	2%
	159	69											0.2713 ± 0.0001	6%
	159	69											0.2890 ± 0.0001	5%
	159	69											0.3483 ± 0.0001	4%
	159	68											0.3749 ± 0.0002	2%
	159	69											0.4085 ± 0.0002	2%
	159	69											0.4503 ± 0.0001	1%
	159	69											0.4617 ± 0.0001	1%
	159	69											0.5417 ± 0.0002	1%
	159	69											1.2701 ± 0.0003	0.8%
	159	69											(0.05 - 1.27)weak	
$_{69}Tm^{160}$	160	69		159.935090	9.2 m.	β+(15%)	5.60				1-		ann.rad.	30%
	160	69				E.C.(85%)							Er k x-ray	
	160	69											0.1264	100 +
	160	69											0.2642	26
	160	69											0.5971	5
	160	69											0.6175	5
	160	69											0.6401	5
	160	69											0.7285	36
	160	69											0.7678	8
	160	69											0.7977	7
	160	69											0.8544	23
	160	69											0.8614	20
	160	69											0.8820	6
	160	69											1.0077	7
	160	69											1.2491	8
	160	69											1.2641	4
	160	69											1.2697	8
	160	69											1.3685	24
	160	69											1.3947	10
	160	69											1.4606	12
	160	69											1.5264	11
	160	69											1.5366	3
	160	69											1.5869	3
	160	69											1.7685	2
	160	69											1.8944	2
	160	69											2.0685	3
	160	69											2.1334	3
	160	69											2.2022	4
$_{69}Tm^{161}$	161	69		160.933430	38 m.	β+,E.C.	3.20				7/2+		ann.rad.	98%
	161	69											Er k x-ray	25%
	161	69											0.0595 ± 0.0001	5%
	161	69											0.0844 ± 0.0001	9%
	161	69											0.0998 ± 0.0001	2.4%
	161	69											0.1059 ± 0.0001	3.4%
	161	69											0.1126 ± 0.0001	3%
	161	69											0.1256 ± 0.0001	1.6%
	161	69											0.1289 ± 0.0001	3%
	161	69											0.1439 ± 0.0001	4%
	161	69											0.1534 ± 0.0001	3%
	161	69											0.1578 ± 0.0001	2%
	161	69											0.1720 ± 0.0001	5%
	161	69											0.1902 ± 0.0001	3.4%
	161	69											0.2071 ± 0.0001	2%
	161	69											0.2129 ± 0.0001	3.2%
	161	69											0.2157 ± 0.0001	1.6%
	161	69											0.2181 ± 0.0001	1%
	161	69											0.2446 ± 0.0001	1%
	161	69											0.2525 ± 0.0001	1.6%
	161	69											0.2655 ± 0.0001	1%
	161	69											0.3538 ± 0.0001	1%
	161	69											0.3695 ± 0.0001	1.4%
	161	69											0.3726 ± 0.0001	1%
	161	69											0.5236 ± 0.0004	0.8%
	161	69											1.0032 ± 0.0004	0.7%
	161	69											1.6481 ± 0.0003	19%

Isotope	A	Z	% Natural abundance	Atomic mass	Half-life	Decay mode	Decay energy (MeV)	Particle energy (MeV)	Particle intensity	Thermal neutron cross section	Spin (h/2π)	μ Nucl. mag. moment	Gamma-ray energy (MeV)	Gamma-ray intensity
	161	69											1.7880 ± 0.0003	1.7%
	161	69											1.8500 ± 0.0003	1.6%
	161	69											1.8547 ± 0.0003	0.8%
	161	69											1.8941 ± 0.0004	0.7%
	161	69											(0.04 - 2.15)many	
$_{69}Tm^{162m}$	162	69			24 s.	I.T.(90%)					5+		Tm k x-ray	27%
	162	69				β+, EC							Er k x-ray	5%
						(10%)								
	162	69											0.0669 ± 0.0001	7%
	162	69											0.1020 ± 0.0001	3%
	162	69											0.2275 ± 0.0001	5%
	162	69											0.3775 ± 0.0002	1.6%
	162	69											0.7100 ± 0.0001	3.6%
	162	69											0.7987 ± 0.0002	5.6%
	162	69											0.8115 ± 0.0001	6.6%
	162	69											0.9003 ± 0.0004	6.8%
$_{69}Tm^{162}$	162	69		161.933920	21.7 m.	β+(8%)	4.99				1-		ann.rad.	16%
	162	69				E.C.(92%)							Er k x-ray	48%
	162	69											0.1020 ± 0.0001	17%
	162	69											0.2275 ± 0.0001	7%
	162	69											0.5707 ± 0.0001	2.1%
	162	69											0.6723 ± 0.0002	6.9%
	162	69											0.7987 ± 0.0001	9.1%
	162	69											0.8998 ± 0.0004	5.6%
	162	69											0.9007 ± 0.0004	6.5%
	162	69											0.9851 ± 0.0001	1.1%
	162	69											1.0690 ± 0.0001	1.1%
	162	69											1.1000 ± 0.0001	1.4%
	162	69											1.2500 ± 0.0001	4.8%
	162	69											1.2547 ± 0.0001	1.5%
	162	69											1.3184 ± 0.0001	5.6%
	162	69											1.3522 ± 0.0001	3.4%
	162	69											1.4042 ± 0.0001	2.8%
	162	69											1.5064 ± 0.0001	1.4%
	162	69											1.9747 ± 0.0001	1.2%
	162	69											2.0158 ± 0.0001	1.1%
	162	69											2.1402 ± 0.0001	1.3%
	162	69											2.2317 ± 0.0001	0.8%
	162	69											3.2979 ± 0.0002	0.65%
	162	69											3.5746 ± 0.0002	0.4%
	162	69											(0.1 - 3.75)many	
$_{69}Tm^{163}$	163	69		162.932648	1.8 h.	E.C.(98%)	2.44				1/2+	0.081	Er k x-ray	76%
	163	69				β+(1%)							0.0692 ± 0.0001	11%
	163	69											0.1043 ± 0.0001	19%
	163	69											0.1901 ± 0.0001	1.3%
	163	69											0.2396 ± 0.0001	4.1%
	163	69											0.2414 ± 0.0001	9.3%
	163	69											0.2414 ± 0.0001	9.3%
	163	69											0.2752 ± 0.0001	2.4%
	163	69											0.2997 ± 0.0001	4.1%
	163	69											0.3457 ± 0.0002	1.1%
	163	69											0.3934 ± 0.0002	1.2%
	163	69											0.4041 ± 0.0002	1.0%
	163	69											0.4712 ± 0.0002	3.8%
	163	69											0.5502 ± 0.0002	1.5%
	163	69											0.5796 ± 0.0002	1.7%
	163	69											0.6659 ± 0.0002	1.7%
	163	69											0.7798 ± 0.0002	0.7%
	163	69											0.9452 ± 0.0002	0.7%
	163	69											1.1300 ± 0.0001	1.9%
	163	69											1.2048 ± 0.0001	2.4%
	163	69											1.2240 ± 0.0001	2.0%
	163	69											1.2649 ± 0.0001	4.9%
	163	69											1.3182 ± 0.0001	1.4%
	163	69											1.3743 ± 0.0001	4.2%
	163	69											1.3869 ± 0.0001	1.0%
	163	69											1.3974 ± 0.0001	7.1%
	163	69											1.4343 ± 0.0001	7.6%
	163	69											1.4558 ± 0.0001	3.4%
	163	69											1.4656 ± 0.0002	1.8%
	163	69											1.4694 ± 0.0002	2.7%
	163	69											1.7492 ± 0.0001	0.9%
	163	69											1.8037 ± 0.0001	1.2%
$_{69}Tm^{164m}$	164	69			5.1 m.	I.T.(80%)					6-		0.0914 ± 0.0001	12%
	164	69				β+, EC							0.1394 ± 0.0001	2.9%
						(20%)								
	164	69											0.2081 ± 0.0001	18%
	164	69											0.2405 ± 0.0001	8.6%
	164	69											0.3149 ± 0.0001	11%
	164	69											0.4102 ± 0.0001	1.6%
	164	69											0.5470 ± 0.0001	5.1%
	164	69											0.7689 ± 0.0001	1.9%
	164	69											0.8207 ± 0.0001	1.5%
	164	69											0.8549 ± 0.0001	1.2%
$_{69}Tm^{164}$	164	69		163.933451	2.0 m.	β+(36%)	3.96	2.94			1+		ann.rad.	70%

TABLE OF THE ISOTOPES (Continued)

Isotope	A	Z	% Natural abundance	Atomic mass	Half-life	Decay mode	Decay energy (MeV)	Particle energy (MeV)	Particle intensity	Thermal neutron cross section	Spin (h/2π)	μ Nucl. mag. moment	Gamma-ray energy (MeV)	Gamma-ray intensity
	164	69				E.C.(64%)							Er k x-ray	31%
	164	69											0.0914 ± 0.0001	6.7%
	164	69											0.2081 ± 0.0001	1.2%
	164	69											0.7689 ± 0.0001	1.4%
	164	69											0.8603 ± 0.0001	1.1%
	164	69											1.1546 ± 0.0002	1.7%
	164	69											1.6107 ± 0.0001	1.1%
	164	69											1.6744 ± 0.0001	1.0%
	164	69											1.8625 ± 0.0001	0.5%
	164	69											2.0816 ± 0.0001	1.6%
	164	69											2.3836 ± 0.0001	0.4%
$_{69}Tm^{165}$	165	69		164.932432	30.06 h.	E.C.	1.59				1/2+	0.139	Er k x-ray	52%
	165	69											0.0472 ± 0.0001	17%
	165	69											0.0544 ± 0.0001	7.2%
	165	69											0.1136 ± 0.0001	1.6%
	165	69											0.2189 ± 0.0001	3.3%
	165	69											0.24296 ± 0.00005	3.5%
	165	69											0.2924 ± 0.0001	1.3%
	165	69											0.29728 ± 0.00005	14%
	165	69											0.3469 ± 0.0001	3.1%
	165	69											0.3565 ± 0.0001	2.7%
	165	69											0.3894 ± 0.0001	2.8%
	165	69											0.4483 ± 0.0002	2.6%
	165	69											0.46024 ± 0.00006	4.0%
	165	69											0.4873 ± 0.0002	1.2%
	165	69											0.5424 ± 0.0002	1.7%
	165	69											0.5639 ± 0.0002	2.4%
	165	69											0.5897 ± 0.0002	2.3%
	165	69											0.80636 ± 0.00008	8.4%
	165	69											1.1313 ± 0.0002	1.4%
	165	69											1.18456 ± 0.00008	2.6%
	165	69											1.4272 ± 0.0003	0.9%
$_{69}Tm^{166}$	166	69		165.933561	7.70 h.	E.C.(98%)	3.047				2+	0.092	Er k x-ray	51%
	166	69				β+(2%)							0.0806 ± 0.0001	11%
	166	69											0.1844 ± 0.0001	18%
	166	69											0.2152 ± 0.0001	6.1%
	166	69											0.4596 ± 0.0001	2.7%
	166	69											0.5499 ± 0.0001	3.7%
	166	69											0.5988 ± 0.0001	2.2%
	166	69											0.6743 ± 0.0001	6.7%
	166	69											0.6748 ± 0.0001	2.8%
	166	69											0.6912 ± 0.0001	8.0%
	166	69											0.7043 ± 0.0002	1.1%
	166	69											0.7053 ± 0.0001	12%
	166	69											0.7578 ± 0.0001	2.7%
	166	69											0.7789 ± 0.0001	21%
	166	69											0.7859 ± 0.0001	11%
	166	69											0.8103 ± 0.0001	1.2%
	166	69											0.8756 ± 0.0001	4.7%
	166	69											1.1523 ± 0.0002	1.8%
	166	69											1.1766 ± 0.0002	11%
	166	69											1.2353 ± 0.0002	2.1%
	166	69											1.2734 ± 0.0001	17%
	166	69											1.3006 ± 0.0001	1.6%
	166	69											1.3469 ± 0.0002	1.3%
	166	69											1.3741 ± 0.0001	6.7%
	166	69											1.5050 ± 0.0002	1.0%
	166	69											1.6529 ± 0.0002	1.2%
	166	69											1.8680 ± 0.0002	4.8%
	166	69											1.8953 ± 0.0002	1.5%
	166	69											2.0524 ± 0.0002	20%
	166	69											2.0796 ± 0.0002	7.5%
	166	69											2.0923 ± 0.0002	1.9%
$_{69}Tm^{167}$	167	69		166.932848	9.25 d.	E.C.	0.747				1/2+	-0.197	Er k x-ray	48%
	167	69											0.0571 ± 0.0001	3.5%
	167	69											0.20778 ± 0.00008	41%(D)
	167	69											0.5315 ± 0.0001	1.6%
$_{69}Tm^{168}$	168	69		167.934170	93.1 d.	E.C.	1.680				3+		Er k x-ray	47%
	168	69											0.0798 ± 0.0001	11%
	168	69											0.0992 ± 0.0001	4.4%
	168	69											0.1843 ± 0.0001	16.4%
	168	69											0.19825 ± 0.00002	50%
	168	69											0.4475 ± 0.0001	22%
	168	69											0.5468 ± 0.0001	2.4%
	168	69											0.6317 ± 0.0001	7.7%
	168	69											0.6457 ± 0.0001	1.4%
	168	69											0.7203 ± 0.0001	11%
	168	69											0.7306 ± 0.0001	5%

TABLE OF THE ISOTOPES (Continued)

Isotope	A	Z	% Natural abundance	Atomic mass	Half-life	Decay mode	Decay energy (MeV)	Particle energy (MeV)	Particle intensity	Thermal neutron cross section	Spin (h/2π)	μ Nucl. mag. moment	Gamma-ray energy (MeV)	Gamma-ray intensity
	168	69											0.74132 ± 0.00003	11%
	168	69											0.81595 ± 0.00003	46%
	168	69											0.8211 ± 0.0001	11%
	168	69											0.8299 ± 0.0001	6.2%
	168	69											0.91490 ± 0.00003	2.9%
	168	69											1.27741 ± 0.00005	1.6%
$_{69}Tm^{169}$	169	69	100%	168.934212						105 b.	$^1/_2$+	-0.2316		
$_{69}Tm^{170}$	170	69		169.935198	128.6 ± 0.3 d.	β-(99.8%)	0.968	0.883	24%	92 b.	1-	0.2476	Yb k x-ray	2%
	170	69				E.C.(0.2%)	0.314	0.968	76%				0.08425 ± 0.00003	3.3%
$_{69}Tm^{171}$	171	69		170.936427	1.92 ± 0.1 y.	β-	0.096	0.03	2%		$^1/_2$+	0.2303	0.06674 ± 0.00001	0.14%
	171	69						0.096	98%					
$_{69}Tm^{172}$	172	69		171.938397	2.65 ± 0.01 d.	β-	1.88	1.79	36%		2-		Yb k x-ray	5%
	172	69						1.88	29%				0.07879 ± 0.00001	6.5%
	172	69											0.18156 ± 0.00001	2.7%
	172	69											0.91211 ± 0.00001	1.4%
	172	69											1.38722 ± 0.00002	5.5%
	172	69											1.46601 ± 0.00002	4.5%
	172	69											1.4705 ± 0.0001	1.9%
	172	69											1.52982 ± 0.00002	5.1%
	172	69											1.60861 ± 0.00003	4.0%
$_{69}Tm^{173}$	173	69		172.939596	8.24 ± 0.08 h.	β-	1.29	0.80	21%		$^1/_2$+		Yb k x-ray	3%
	173	69						0.86	71%				0.3988 ± 0.0005	88%
	173	69											0.4613 ± 0.0005	6.9%
$_{69}Tm^{174}$	174	69		173.942180	5.4 ± 0.1 m.	β-	3.09	0.70	14%		(4-)		Yb k x-ray	18%
	174	69						1.20	83%				0.07664 ± 0.00004	9.1%
	174	69											0.17669 ± 0.00004	66%
	174	69											0.27332 ± 0.00008	86%
	174	69											0.3666 ± 0.0001	92%
	174	69											0.49433 ± 0.00009	11.4%
	174	69											0.62845 ± 0.00009	2.7%
	174	69											0.99205 ± 0.00008	87%
	174	69											1.2419 ± 0.0001	1.7%
	174	69											1.2654 ± 0.0001	2.2%
	174	69											(0.08 - 1.6)	
$_{69}Tm^{175}$	175	69		174.942180	15.2 ± 0.5 m.	β-	2.40	0.9	36%		($^1/_2$+)		Yb k x-ray	10%
	175	69						1.9	23%				0.36396 ± 0.00001	13%
	175	69											0.3946 ± 0.0002	3.3%
	175	69											0.51487 ± 0.00001	65%(D)
	175	69											0.63926 ± 0.00001	6.1%
	175	69											0.81143 ± 0.00001	4.3%
	175	69											0.85808 ± 0.00005	5.7%
	175	69											0.94125 ± 0.00004	14.2%
	175	69											0.98247 ± 0.00004	9.9%
	175	69											1.3770 ± 0.0002	3.0%
	175	69											1.5251 ± 0.0005	1.3%
$_{69}Tm^{176}$	176	69		175.946750	1.9 ± 0.1 m.	β-	3.90				(4+)		Yb k x-ray	18%
	176	69											0.0817 ± 0.001	12%
	176	69											00.1898 ± 0.0001	44%
	176	69											0.2344 ± 0.0001	3.2%
	176	69											0.2383 ± 0.0001	2.5%
	176	69											0.2398 ± 0.0001	7.9%
	176	69											0.2929 ± 0.0002	3.5%
	176	69											0.2996 ± 0.0001	3.2%

TABLE OF THE ISOTOPES (Continued)

Isotope	A	Z	% Natural abundance	Atomic mass	Half-life	Decay mode	Decay energy (MeV)	Particle energy (MeV)	Particle intensity	Thermal neutron cross section	Spin (h/2π)	μ Nucl. mag. moment	Gamma-ray energy (MeV)	Gamma-ray intensity
	176	69											0.3303 ± 0.0002	8.6%
	176	69											0.3435 ± 0.0001	6.9%
	176	69											0.3819 ± 0.0001	23%
	176	69											0.4106 ± 0.0002	4.6%
	176	69											0.4570 ± 0.0002	2.8%
	176	69											0.6216 ± 0.0002	3.4%
	176	69											0.9005 ± 0.0003	2.6%
	176	69											1.0499 ± 0.0002	7.0%
	176	69											1.0691 ± 0.0002	33%
	176	69											1.0881 ± 0.0002	5.7%
	176	69											1.1787 ± 0.0002	2.9%
	176	69											1.2541 ± 0.0003	2.0%
	176	69											1.2609 ± 0.0002	2.3%
	176	69											1.5893 ± 0.0002	2.8%
	176	69											1.9711 ± 0.0002	2.4%
	176	69											2.6214 ± 0.0005	2.9%
	176	69											2.8716 ± 0.0003	2.1%
	176	69											2.9142 ± 0.0005	4.3%
Yb		70		173.04						35 b.				
$_{70}$Yb^{154}	154	70		153.946190	0.40 s.	β,EC(7%)							ann.rad.	
	154	70				α(93%)		5.32						
$_{70}$Yb^{155}	155	70		154.945530	1.7 s.	β+, EC (16%)							ann.rad.	
	155	70				α(84%)		5.19						
$_{70}$Yb^{156}	156	70		155.942690	24 s.	β+, EC (21%)	3.7				0+		ann.rad.	
	156	70				α(79%)		4.69						
$_{70}$Yb^{157}	157	70		156.942430	39 s.	β+, EC (99+%)	5.2						ann.rad.	
	157	70				α(0.5%)		4.69						
$_{70}$Yb^{158}	158	70		157.939858	1.5 m.	β+,E.C.	≈ 2.7				0+		ann.rad.	
	158	70											0.0741 ± 0.0001	100 +
	158	70											0.1477 ± 0.0001	1.7
	158	70											0.1603 ± 0.0001	2.1
	158	70											0.2526 ± 0.0002	3.3
$_{70}$Yb^{159}	159	70		158.939950	12 s.	E.C.,β+	≈ 4.4						Tm k x-ray	61%
	159	70											0.0777 ± 0.0001	7
	159	70											0.1131 ± 0.0001	12
	159	70											0.1661 ± 0.0001	100
	159	70											0.1761 ± 0.0001	14
	159	70											0.1772 ± 0.0001	20
	159	70											0.1919 ± 0.0001	4
	159	70											0.1937 ± 0.0001	5
	159	70											0.1976 ± 0.0001	5
	159	70											0.2391 ± 0.0001	10
	159	70											0.3297 ± 0.0001	18
	159	70											0.3903 ± 0.0001	18
	159	70											0.4972 ± 0.0003	9
$_{70}$Yb^{160}	160	70		159.937670	4.8 m.	β+,E.C.	≈ 2.1				0+		ann.rad.	
	160	70											0.0342 ± 0.0001	3 +
	160	70											0.0420 ± 0.0001	7
	160	70											0.0982 ± 0.0001	3
	160	70											0.1322 ± 0.0001	14
	160	70											0.1404 ± 0.0001	22
	160	70											0.1737 ± 0.0001	100
	160	70											0.1744 ± 0.0001	13
	160	70											0.2158 ± 0.0001	48
	160	70											0.3200 ± 0.0002	3
	160	70											0.3276 ± 0.0002	6
	160	70											0.3730 ± 0.0001	10
	160	70											0.3863 ± 0.0002	3
	160	70											0.3894 ± 0.0002	5
	160	70											0.5821 ± 0.0002	3
$_{70}$Yb^{161}	161	70		160.937940	4.2 m.	β+,E.C.	≈ 4.1				3/2-		ann.rad.	
	161	70											Tm k x-ray	44%
	161	70											0.0782 ± 0.0001	41%
	161	70											0.1403 ± 0.0001	3%
	161	70											0.1444 ± 0.0001	5.5%
	161	70											0.1883 ± 0.0001	4.2%
	161	70											0.2985 ± 0.0001	1.6%
	161	70											0.3147 ± 0.0001	3.2%
	161	70											0.3301 ± 0.0001	3.3%
	161	70											0.3447 ± 0.0003	1.7%
	161	70											0.3810 ± 0.0003	2%
	161	70											0.4582 ± 0.0002	3.4%
	161	70											0.5555 ± 0.0002	1.7%
	161	70											0.5605 ± 0.0002	2.5%
	161	70											0.5697 ± 0.0002	6.7%
	161	70											0.5999 ± 0.0001	31%
	161	70											0.6315 ± 0.0002	16%
	161	70											0.6591 ± 0.0001	3.8%
	161	70											1.0072 ± 0.0004	1.1%
	161	70											1.0427 ± 0.0002	1.2%
	161	70											1.1456 ± 0.0005	1.1%
	161	70											1.1825 ± 0.0005	1%

Isotope	A	Z	% Natural abundance	Atomic mass	Half-life	Decay mode	Decay energy (MeV)	Particle energy (MeV)	Particle intensity	Thermal neutron cross section	Spin (h/2π)	μ Nucl. mag. moment	Gamma-ray energy (MeV)	Gamma-ray intensity
	161	70											1.3649 ± 0.0005	1.1%
	161	70											1.5178 ± 0.0005	1.1%
$_{70}Yb^{162}$	162	70		161.935860	18.9 ± 0.2 m.	β+,E.C.	1.81				0+		ann.rad.	
	162	70											Tm k x-ray	47%
	162	70											0.1188 ± 0.0001	25%
	162	70											0.1635 ± 0.0001	36%
$_{70}Yb^{163}$	163	70		162.936270	11.05 ± 0.25m.	β+(26%)	3.37	1.4			3/2-		ann.rad.	52%
	163	70											Tm k x-ray	36%
	163	70											0.0636 ± 0.0001	7%
	163	70											0.1232 ± 0.0001	2.1%
	163	70											0.1615 ± 0.0001	1.1%
	163	70											0.3262 ± 0.0001	1.7%
	163	70											0.6872 ± 0.0001	1.7%
	163	70											0.8603 ± 0.0001	10.8%
	163	70											1.3318 ± 0.0001	0.7%
	163	70											1.6891 ± 0.0001	0.8%
	163	70											1.7467 ± 0.0002	1.8%
	163	70											1.9078 ± 0.0001	1.6%
	163	70											(0.06 - 1.9)weak	
$_{70}Yb^{164}$	164	70		163.934530	1.26 ± 0.03 h.	E.C.	1.00				0+		Tm k x-ray	38%
	164	70											0.0914 ± 0.0001	6.9%(D)
	164	70											0.6752 ± 0.0001	0.3%
$_{70}Yb^{165}$	165	70		164.935398	9.9 ± 0.3 m.	β+(10%)	2.76	1.58			(5/2-)		ann.rad.	10%
	165	70				E.C.(90%)							Tm k x-ray	60%
	165	70											0.0801 ± 0.0001	33%
	165	70											0.1181 ± 0.0001	1.6%
	165	70											0.1473 ± 0.0001	0.7%
	165	70											0.9567 ± 0.0001	0.7%
	165	70											1.0903 ± 0.0001	3%
	165	70											1.5013 ± 0.0001	0.4%
$_{70}Yb^{166}$	166	70		165.933875	2.36 ± .004 d.	E.C.	0.292				0+		Tm k x-ray	67%
	166	70											0.0828 ± 0.0001	15%
	166	70											0.1844 ± 0.0001	21%
	166	70											0.7789 ± 0.0001	25%
	166	70											1.2734 ± 0.0001	20%(D)
	166	70											2.0524 ± 0.0002	24%(D)
$_{70}Yb^{167}$	167	70		166.934946	17.5 ± 0.2 m.	β+(0.5%)	1.954	0.639			5/2-		Tm k x-ray	91%
	167	70				E.C. (99.5%)							0.06296 ± 0.00008	5%
	167	70											0.10616 ± 0.00004	22%
	167	70											0.11337 ± 0.00002	55%
	167	70											0.1166 ± 0.0001	2.8%
	167	70											0.1435 ± 0.0001	2.1%
	167	70											0.17633 ± 0.00006	20%
	167	70											0.1772 ± 0.0001	2.7%
	167	70											1.0371 ± 0.0001	0.6%
$_{70}Yb^{168}$	168	70	0.13%	167.933894						2.3 x 10^3 b.	0+			
$_{70}Yb^{169m}$	169	70			46 ± 2 s.	I.T.	0.0242				1/2-		Yb L x-ray	16%
	169	70											0.0242 ± 0.0001	0.0004%
$_{70}Yb^{169}$	169	70		168.935186	32.02 ± 0.01d.	E.C.	0.908			3.6 x 10^3 b.	7/2+		Tm k x-ray	95%
	169	70											0.06306 ± 0.00003	45%
	169	70											0.09365 ± 0.00001	2.7%
	169	70											0.10977 ± 0.00001	18%
	169	70											0.1182 ± 0.0001	1.9%
	169	70											0.13051 ± 0.00001	11.5%
	169	70											0.17718 ± 0.00002	22%
	169	70											0.19795 ± 0.00003	36%
	169	70											0.26106 ± 0.00003	1.8%
	169	70											0.30772 ± 0.00003	11.1%
$_{70}Yb^{170}$	170	70	3.05%	169.934759						10 b.	0+			
$_{70}Yb^{171}$	171	70	14.3%	170.936323						50 b.	1/2-			
$_{70}Yb^{172}$	172	70	21.9%	171.936378						1 b.	0+			
$_{70}Yb^{173}$	173	70	16.12%	172.938208						17 b.	5/2-			
$_{70}Yb^{174}$	174	70	31.8%	173.938859							65 b.	0+		
$_{70}Yb^{175}$	175	70		174.941273	4.19 ± 0.01 d.	β-	0.468	0.467			7/2-		Lu k x-ray	2%
	175	70											0.11378 ± 0.00001	1.9%

Isotope	A	Z	% Natural abundance	Atomic mass	Half-life	Decay mode	Decay energy (MeV)	Particle energy (MeV)	Particle intensity	Thermal neutron cross section	Spin (h/2π)	μ Nucl. mag. moment	Gamma-ray energy (MeV)	Gamma-ray intensity
	175	70											0.28248 ± 0.00001	3.1%
	175	70											0.39629 ± 0.00002	6.5%
$_{70}Yb^{176m}$	176	70			11.4 ± 0.5 s.	I.T.	1.051				(8-)		Yb k x-ray	31%
	176	70											0.0821 ± 0.0002	12%
	176	70											0.0961 ± 0.0003	72%
	176	70											0.1901 ± 0.0002	76%
	176	70											0.2929 ± 0.0003	93%
	176	70											0.3897 ± 0.0004	97%
$_{70}Yb^{176}$	176	70	12.7%	175.942564						3 b.	0+			
$_{70}Yb^{177m}$	177	70			6.41 ± 0.02 s.	I.T.	0.3315				1/2-		Yb k x-ray	38%
	177	70											0.1131 ± 0.0001	6.6%
	177	70											0.2084 ± 0.0001	11%
	177	70											0.2497 ± 0.0001	0.2%
	177	70											0.3213 ± 0.0001	0.2%
$_{70}Yb^{177}$	177	70		176.945253	1.9 ± 0.1 h.	β-	1.398	1.40			9/2+		Lu k x-ray	7%
	177	70											0.1216 ± 0.0001	3%
	177	70											0.1504 ± 0.0001	20%
	177	70											0.9417 ± 0.0003	1%
	177	70											1.0801 ± 0.0003	5.5%
	177	70											1.2414 ± 0.0003	3.4%
$_{70}Yb^{178}$	178	70		177.946639	1.23 ± 0.05 h.	β-	0.630	0.25			0+			
$_{70}Yb^{179}$	179	70			8 m.								0.1415 ± 0.0004	6 +
	179	70											0.1473 ± 0.0003	14
	179	70											0.3246 ± 0.0004	20
	179	70											0.3516 ± 0.0003	43
	179	70											0.3815 ± 0.0003	26
	179	70											0.4111 ± 0.0003	17
	179	70											0.4265 ± 0.0006	6
	179	70											0.4312 ± 0.0004	8
	179	70											0.4711 ± 0.0005	6
	179	70											0.5001 ± 0.0004	11
	179	70											0.6125 ± 0.0003	100
	179	70											0.6430 ± 0.0004	11
	179	70											0.9942 ± 0.001	4
	179	70											1.0244 ± 0.001	7
Lu		71		174.967						84 b.				
$_{71}Lu^{154}$	154	71		153.957460	1.0 s.	β+,E.C.	10.560							
$_{71}Lu^{155}$	155	71		154.954080	0.07 s.	E.C.	7.97							
	155	71				α		5.66						
$_{71}Lu^{156m}$	156	71			0.21 s.	β+,E.C.							ann.rad.	
	156	71				α		5.57						
$_{71}Lu^{156}$	156	71		155.953070	0.5 s.	β+,E.C.	9.67						ann.rad.	
	156	71				α		5.45						
$_{71}Lu^{157}$	157	71		156.949940	5.5 s.	β+, EC (94%)	6.99						ann.rad.	
	157	71				α		5.00						
$_{71}Lu^{158}$	158	71		157.949290	10 s.	β+, EC (99%)	8.78						ann.rad.	
	158	71				α		4.67					0.3682 ± 0.0001	100 +
	158	71											0.4770 ± 0.0002	21
$_{71}Lu^{159}$	159	71		158.946480	12 s.	β+,E.C.	6.08						ann.rad.	
	159	71											0.1505 ± 0.0001	100 +
	159	71											0.1875 ± 0.0001	25
	159	71											0.3693 ± 0.0001	19
$_{71}Lu^{160}$	160	71		159.946040	35 s.	β+,E.C.	7.80						ann.rad.	
	160	71											0.2434 ± 0.0001	100 +
	160	71											0.3756 ± 0.0002	8
	160	71											0.3957 ± 0.0002	30
	160	71											0.5773 ± 0.0002	13
	160	71											0.7044 ± 0.0002	6
	160	71											0.7382 ± 0.0002	7
	160	71											0.8201 ± 0.0003	9
	160	71											0.8707 ± 0.0004	9
$_{71}Lu^{161}$	161	71		160.943630	1.2 m.	β+,E.C.	5.30						ann.rad.	
	161	71											0.0437 ± 0.0003	70 +
	161	71											0.0671 ± 0.0002	48
	161	71											0.0868 ± 0.0002	17
	161	71											0.1003 ± 0.0001	95
	161	71											0.1052 ± 0.0001	28
	161	71											0.1108 ± 0.0001	100
	161	71											0.1562 ± 0.0001	49
	161	71											0.1701 ± 0.00002	14
	161	71											0.1771 ± 0.0002	14
	161	71											0.2046 ± 0.0002	30
	161	71											0.2111 ± 0.0002	20
	161	71											0.2218 ± 0.0002	20
	161	71											0.2562 ± 0.0003	49
$_{71}Lu^{162}$	162	71		161.943470	1.4 m.	β+,E.C.	7.09						ann.rad.	
	162	71											0.1666 ± 0.0001	100 +

Isotope	A	Z	% Natural abundance	Atomic mass	Half-life	Decay mode	Decay energy (MeV)	Particle energy (MeV)	Particle intensity	Thermal neutron cross section	Spin (h/2π)	μ Nucl. mag. moment	Gamma-ray energy (MeV)	Gamma-ray intensity
	162	71											0.3209 ± 0.0001	20
	162	71											0.6314 ± 0.0001	28
	162	71											0.6564 ± 0.0002	7
	162	71											0.8253 ± 0.0002	18
	162	71											0.8398 ± 0.0003	8
$_{71}Lu^{163}$	163	71		162.941240	4.1 ± 0.2 m.	β+,E.C.	4.63						ann.rad.	
	163	71											0.0539	82
	163	71											0.0581	43
	163	71											0.0730	2
	163	71											0.0792	5
	163	71											0.0935	5
	163	71											0.1023	9
	163	71											0.1504	44
	163	71											0.1631	100
	163	71											0.1674	9
	163	71											0.2066	6
	163	71											0.2213	20
	163	71											0.2527	8
	163	71											0.3026	24
	163	71											0.3135	25
	163	71											0.3169	13
	163	71											0.3717	49
	163	71											0.3818	10
	163	71											0.3912	6
	163	71											0.3954	8
	163	71											0.4002	4
	163	71											0.4530	11
	163	71											0.4568	6
	163	71											0.4611	7
	163	71											0.4827	6
	163	71											0.4843	19
	163	71											0.5382	12
	163	71											0.5623	16
	163	71											0.5665	7
	163	71											0.6331	8
	163	71											0.6951	15
$_{71}Lu^{164}$	164	71		163.941290	3.17 ± 0.03 m	β+,E.C.	6.30	1.6					0.1238 ± 0.0002	100 +
	164	71						3.8					0.2621 ± 0.0002	32
	164	71											0.5520 ± 0.0002	12
	164	71											0.6082 ± 0.0002	6
	164	71											0.6880 ± 0.0003	6
	164	71											0.7404 ± 0.0002	38
	164	71											0.7479 ± 0.0002	16
	164	71											0.8521 ± 0.0003	9
	164	71											0.8639 ± 0.0002	29
	164	71											0.8804 ± 0.0002	21
	164	71											0.9494 ± 0.0003	7
	164	71											0.9796 ± 0.0004	3
	164	71											1.0738 ± 0.0005	10
	164	71											1.1148 ± 0.0004	4
	164	71											1.1994 ± 0.0004	8
	164	71											1.2124 ± 0.0003	14
	164	71											1.2925 ± 0.0004	6
	164	71											1.3356 ± 0.0006	11
	164	71											1.3760 ± 0.0004	6
	164	71											1.3895 ± 0.0004	6
	164	71											1.5134 ± 0.0005	6
$_{71}Lu^{165}$	165	71		164.939480	11.8 ± 0.5 m.	β+,E.C.	3.80	2.06			1/2+		ann.rad.	
	165	71											0.0393 ± 0.0001	8%
	165	71											0.1206 ± 0.0001	25%
	165	71											0.1324 ± 0.0001	23%
	165	71											0.1742 ± 0.0001	12%
	165	71											0.2036 ± 0.0001	10%
	165	71											0.2174 ± 0.0001	5%
	165	71											0.2534 ± 0.0001	4%
	165	71											0.2710 ± 0.0001	5%
	165	71											0.3565 ± 0.0001	5%
	165	71											0.3605 ± 0.0001	8%
	165	71											0.3605 ± 0.0001	8%
	165	71											0.3725 ± 0.3%	3%
	165	71											0.5523 ± 0.0002	2%
	165	71											0.6091 ± 0.0002	2%
	165	71											0.6866 ± 0.0002	2.5%
	165	71											0.7535 ± 0.0002	2.2%
	165	71											1.0734 ± 0.0003	1.9%
	165	71											1.5600 ± 0.0003	1.9%
	165	71											1.6016 ± 0.0002	4.0%
	165	71											1.6135 ± 0.0002	4.0%
	165	71											1.7344 ± 0.0003	2.2%
	165	71											1.8019 ± 0.0004	1.9%
	165	71											(0.04 - 2.0)weak	
$_{71}Lu^{166m2}$	166	71			2.1 ± 0.1 m	β+(35%)					(0-)		ann.rad.	70%

Isotope	A	Z	% Natural abundance	Atomic mass	Half-life	Decay mode	Decay energy (MeV)	Particle energy (MeV)	Particle intensity	Thermal neutron cross section	Spin (h/2π)	μ Nucl. mag. moment	Gamma-ray energy (MeV)	Gamma-ray intensity
	166	71				E.C.(65%)							Yb k x-ray	26%
	166	71											0.1024 ± 0.0001	11%
	166	71											0.2281 ± 0.0001	4%
	166	71											1.0673 ± 0.0002	15%
	166	71											1.2494 ± 0.0008	13%
	166	71											1.2566 ± 0.0001	23%
	166	71											1.4775 ± 0.0003	2.7%
	166	71											1.5297 ± 0.0001	11%
	166	71											1.9232 ± 0.0004	2.4%
	166	71											1.9963 ± 0.0002	3.3%
	166	71											2.0986 ± 0.0002	16%
	166	71											2.3246 ± 0.0003	9%
$_{71}Lu^{166m1}$	166	71			1.4 ± 0.1 m.	β+, EC (58%)					(3-)		ann.rad.	
	166	71				I.T.(42%)	0.0344						0.1024 ± 0.0001	21%
	166	71											0.2281 ± 0.0001	26%
	166	71											0.2861 ± 0.0001	19%
	166	71											0.4213 ± 0.0001	4%
	166	71											0.5260 ± 0.0001	5%
	166	71											0.5709 ± 0.0001	5.5%
	166	71											0.5810 ± 0.0006	2%
	166	71											0.6432 ± 0.0001	6%
	166	71											0.7051 ± 0.0001	7.5%
	166	71											0.7088 ± 0.0001	2.4%
	166	71											0.8119 ± 0.0001	17%
	166	71											0.8301 ± 0.0001	18%
	166	71											0.8322 ± 0.0001	4.5%
	166	71											0.8664 ± 0.0004	2.1%
	166	71											0.9324 ± 0.0001	14%
	166	71											0.9368 ± 0.0001	14%
	166	71											0.9846 ± 0.0006	4%
	166	71											1.2769 ± 0.0002	2%
	166	71											1.2835 ± 0.0002	6.6%
	166	71											1.3544 ± 0.0002	2%
	166	71											1.5049 ± 0.0006	2%
	166	71											1.6787 ± 0.0004	2.2%
	166	71											1.8013 ± 0.0006	1.7%
	166	71											1.9740 ± 0.0006	1.0%
$_{71}Lu^{166}$	166	71		165.939760	2.8 ± 0.2 m.	β+(25%)	2.2				(6-)		ann.rad.	50%
	166	71				E.C.(75%)							Yb k x-ray	51%
	166	71											0.0676 ± 0.0001	4%
	166	71											0.1024 ± 0.0001	25%
	166	71											0.2087 ± 0.0001	4%
	166	71											0.2281 ± 0.0001	77%
	166	71											0.2485 ± 0.0001	4.8%
	166	71											0.27440 ± 0.0001	9.9%
	166	71											0.2763 ± 0.0001	14%
	166	71											0.3375 ± 0.0001	41%
	166	71											0.3601 ± 0.0001	3.6%
	166	71											0.3679 ± 0.0001	31%
	166	71											0.3830 ± 0.0001	3.0%
	166	71											0.4303 ± 0.0001	5%
	166	71											0.4747 ± 0.0001	2.7%
	166	71											0.5376 ± 0.0001	8.1%
	166	71											0.5777 ± 0.0001	4.0%
	166	71											0.6293 ± 0.0001	7%
	166	71											0.6599 ± 0.0001	3.7%
	166	71											0.7944 ± 0.0001	3%
	166	71											0.8145 ± 0.0001	6.7%
	166	71											0.8322 ± 0.0001	6%
	166	71											0.8376 ± 0.0001	2.7%
	166	71											0.8606 ± 0.0001	3.3%
	166	71											0.9368 ± 0.0001	5.7%
	166	71											0.9974 ± 0.0001	18%
	166	71											1.0563 ± 0.0006	2.1%
	166	71											1.0673 ± 0.0002	2.5%
	166	71											1.1224 ± 0.0001	4.0%
	166	71											1.1748 ± 0.0002	4.4%
	166	71											1.2907 ± 0.0002	9.7%
	166	71											1.3544 ± 0.0002	1.7%
	166	71											1.4596 ± 0.0001	7.8%
	166	71											1.4873 ± 0.0004	1.1%
	166	71											1.6266 ± 0.0003	0.9%
	166	71											1.6858 ± 0.0003	0.5%
$_{71}Lu^{167}$	167	71		166.938310	51.5 ± 0.1 m.	β+(2%)	3.13	2.1			7/2+		Yb k x-ray	50%
	167	71				E.C.(98%)							0.0297 ± 0.0001	15%
	167	71											0.0339 ± 0.0001	3%
	167	71											0.0787 ± 0.0001	1.5%
	167	71											0.1450 ± 0.0001	2.2%
	167	71											0.1789 ± 0.0001	2.6%
	167	71											0.1887 ± 0.0001	2%
	167	71											0.2132 ± 0.0001	3.5%
	167	71											0.2392 ± 0.0001	8.2%

Isotope	A	Z	% Natural abundance	Atomic mass	Half-life	Decay mode	Decay energy (MeV)	Particle energy (MeV)	Particle intensity	Thermal neutron cross section	Spin (h/2π)	μ Nucl. mag. moment	Gamma-ray energy (MeV)	Gamma-ray intensity
	167	71											0.2585 ± 0.0001	1.4%
	167	71											0.2618 ± 0.0001	1.3%
	167	71											0.2787 ± 0.0001	1.9%
	167	71											0.3177 ± 0.0001	1.5%
	166	71											0.4011 ± 0.0001	2.5%
	167	71											0.4454 ± 0.0001	1.0%
	167	71											0.5733 ± 0.0008	1.1%
	166	71											0.7848 ± 0.0001	0.6%
	167	71											0.9884 ± 0.0001	0.8%
	167	71											1.1883 ± 0.0001	1.2%
	167	71											1.2272 ± 0.0002	1.2%
	167	71											1.2672 ± 0.0001	3.3%
	167	71											1.5068 ± 0.0001	2.4%
	167	71											1.6445 ± 0.0001	1.2%
	167	71											1.9414 ± 0.0001	1.3%
	167	71											1.9740 ± 0.0001	1.2%
	167	71											2.0131 ± 0.0002	1.2%
	167	71											(0.03 - 2.0)weak	
$_{71}Lu^{168m}$	168	71			6.7 ± 0.4 m.	β+(12%)					3+		ann.rad.	24%
	168	71				E.C.(88%)							Yb k x-ray	48%
	168	71											0.0877 ± 0.0001	13%
	168	71											0.1988 ± 0.0001	28%
	168	71											0.2987 ± 0.0001	2.6%
	168	71											0.7303 ± 0.0003	1.9%
	168	71											0.7525 ± 0.0008	2.0%
	168	71											0.7805 ± 0.0003	3.7%
	168	71											0.8535 ± 0.0002	4.8%
	168	71											0.8846 ± 0.0002	14%
	168	71											0.8960 ± 0.0002	16%
	168	71											0.9792 ± 0.0002	20%
	168	71											0.9838 ± 0.0002	12%
	168	71											1.0326 ± 0.0002	11%
	168	71											1.0717 ± 0.0003	3.7%
	168	71											1.0838 ± 0.0003	5.5%
	168	71											1.1368 ± 0.0002	12%
	168	71											1.2199 ± 0.0002	11%
	168	71											1.2335 ± 0.0002	3.3%
	168	71											1.2645 ± 0.00003	3.0%
	168	71											1.3377 ± 0.0002	4.2%
	168	71											1.3639 ± 0.0002	3.9%
	168	71											1.4208 ± 0.0002	10%
	168	71											1.4635 ± 0.0003	2.5%
	168	71											2.1414 ± 0.0005	2.8%
	168	71											2.3401 ± 0.0002	1.1%
$_{71}Lu^{168}$	168	71		167.938690	5.5 ± 0.1 m.	β+(6%)	4.47	1.2			(6-)		ann.rad.	12%
	168	71				E.C.(94%)							Yb k x-ray	
	168	71											0.1114 ± 0.0002	16%
	168	71											0.1124 ± 0.0002	16%
	168	71											0.1566 ± 0.0002	7.3%
	168	71											0.1796 ± 0.0002	6.0%
	168	71											0.2236 ± 0.0001	8.2%
	168	71											0.2286 ± 0.0002	17%
	168	71											0.3247 ± 0.0002	7.4%
	168	71											0.3483 ± 0.0002	18%
	168	71											0.3874 ± 0.0002	3.4%
	168	71											0.4011 ± 0.0004	6.7%
	168	71											0.4794 ± 0.0004	2.4%
	168	71											0.5398 ± 0.0002	11%
	168	71											0.8600 ± 0.0003	3.1%
	168	71											1.1850 ± 0.0002	11%
	168	71											1.2335 ± 0.0005	2.6%
	168	71											1.3875 ± 0.0002	6%
	168	71											1.4135 ± 0.0003	3.9%
	168	71											1.4836 ± 0.0002	17%
	168	71											1.6860 ± 0.0005	4.8%
$_{71}Lu^{169m}$	169	71			2.7 ± 0.2 m.	I.T.	0.0290				3+		Lu L x-ray	17%
	169	71											0.0290 ± 0.0002	0.001%
$_{71}Lu^{169}$	169	71		168.937648	1.419 ± 0.002 d.	E.C.	2.293	1.271			7/2+		Yb k x-ray	53%
	169	71											0.0874 ± 0.0001	2.1%
	169	71											0.19121 ± 0.0001	18%
	169	71											0.3786 ± 0.0001	1.8%
	169	71											0.8898 ± 0.0001	4.6%
	169	71											0.9606 ± 0.0001	20%
	169	71											1.0075 ± 0.0001	1.6%
	169	71											1.0603 ± 0.0001	1.6%
	169	71											1.1849 ± 0.0001	1.9%
	169	71											1.2833 ± 0.0001	1.8%
	169	71											1.3388 ± 0.0001	1.4%
	169	71											1.3790 ± 0.0001	2.8%
	169	71											1.4497 ± 0.0001	8.6%
	169	71											1.4634 ± 0.0001	1.3%
	169	71											1.4668 ± 0.0001	2.9%

Isotope	A	Z	% Natural abundance	Atomic mass	Half-life	Decay mode	Decay energy (MeV)	Particle energy (MeV)	Particle intensity	Thermal neutron cross section	Spin (h/2π)	μ Nucl. mag. moment	Gamma-ray energy (MeV)	Gamma-ray intensity
	169	71											(0.08 - 2.1)weak	
$_{71}Lu^{170m}$	170	71			0.7 ± 0.1 s.	I.T.	0.0929						Lu L x-ray	
	170	71											0.04449 ± 0.00006	0.85%
	170	71											0.0484 ± 0.0001	0.4%
$_{71}Lu^{170}$	170	71		169.938452	2.01 ± 0.03 d.	E.C.	3.44	2.44			0+		Yb k x-ray	46%
	170	71											0.19319 ± 0.00004	2.1%
	170	71											1.57227 ± 0.00001	1.2%
	170	71											0.58711 ± 0.00001	12.7%
	170	71											0.5908 ± 0.0001	15%
	170	71											0.93886 ± 0.00005	1.6%
	170	71											0.98512 ± 0.00005	5.4%
	170	71											0.98721 ± 0.00005	1.7%
	170	71											0.9996 ± 0.0001	1.5%
	170	71											1.00317 ± 0.00005	3.4%
	170	71											1.0543 ± 0.0001	4.6%
	170	71											1.0615 ± 0.0001	2.1%
	170	71											1.13368 ± 0.00006	1.0%
	170	71											1.13862 ± 0.00003	3.5%
	170	71											1.21841 ± 0.00006	1.4%
	170	71											1.22556 ± 0.00005	4.8%
	170	71											1.2571 ± 0.0001	1.4%
	170	71											1.28029 ± 0.00004	7.9%
	170	71											1.29476 ± 0.00006	2.8%
	170	71											1.30746 ± 0.00005	1.1%
	170	71											1.34101 ± 0.00004	3.2%
	170	71											1.36460 ± 0.00004	4.5%
	170	71											1.39565 ± 0.00006	2.2%
	170	71											1.40521 ± 0.00006	2.5%
	170	71											1.42816 ± 0.00004	3.4%
	170	71											1.45032 ± 0.00004	1.6%
	170	71											1.45532 ± 0.00005	1.1%
	170	71											1.45988 ± 0.00007	1.0%
	170	71											1.51246 ± 0.00004	2.5%
	170	71											1.9557 ± 0.0001	1.3%
	170	71											2.0400 ± 0.0001	2.5%
	170	71											2.0419 ± 0.0001	5.9%
	170	71											2.12621 ± 0.00005	5.0%
	170	71											2.1912 ± 0.0001	1.6%
	170	71											2.36417 ± 0.00004	1.4%
	170	71											2.66390 ± 0.00005	1.2%
	170	71											2.69145 ± 0.00008	2.2%
	170	71											2.74821 ± 0.00005	2.1%
	170	71											2.7832 ± 0.0001	1.0%
	170	71											2.8544 ± 0.0001	1.7%
	170	71											2.93982 ± 0.00005	1.5%
	170	71											2.9657 ± 0.0001	1.2%
	170	71											3.0309 ± 0.0001	1.3%
	170	71											(0.1 - 3.38)many	
$_{71}Lu^{171m}$	171	71			1.3 ± 0.3 m.	I.T.	0.0711				1/2-		Lu k x-ray	61%
	171	71											0.07119 ± 0.0001	0.02%
$_{71}Lu^{171}$	171	71		170.937911	8.24 ± 0.03 d.	E.C.	1.481	0.362			7/2+	2.03	Yb k x-ray	64%

Isotope	A	Z	% Natural abundance	Atomic mass	Half-life	Decay mode	Decay energy (MeV)	Particle energy (MeV)	Particle intensity	Thermal neutron cross section	Spin (h/2π)	μ Nucl. mag. moment	Gamma-ray energy (MeV)	Gamma-ray intensity
	171	71											0.01939 ± 0.00001	14%
	171	71											0.06674 ± 0.00002	2.5%
	171	71											0.072387 ± 0.00002	2.0%
	171	71											0.075899 ± 0.00002	6.1%
	171	71											0.085611 ± 0.00002	1.0%
	171	71											0.66744 ± 0.00001	11%
	171	71											0.68931 ± 0.00001	2.4%
	171	71											0.71268 ± 0.00001	1.1%
	171	71											0.73983 ± 0.00001	48%
	171	71											0.78072 ± 0.00001	4.3%
	171	71											0.84001 ± 0.00001	3.0%
	171	71											0.85311 ± 0.00001	2.5%
	171	71											1.2822 ± 0.0001	0.3%
	171	71											(0.02 - 1.3)weak	
$_{71}Lu^{172m}$	172	71			3.7 ± 0.5 m.	I.T.	0.0419				1-		Lu L x-rays	
	172	71											0.04186 ± 0.00004	0.004%
$_{71}Lu^{172}$	172	71		171.939085	6.70 ± 0.03 d.	E.C.	2.524				4-	2.25	Yb k x-ray	57%
	172	71											0.07879 ± 0.00001	11%
	172	71											0.0966 ± 0.0002	5.1%
	172	71											0.11276 ± 0.00002	1.5%
	172	71											0.18156 ± 0.00001	20%
	172	71											0.20342 ± 0.00002	4.8%
	172	71											0.26993 ± 0.00004	1.8%
	172	71											0.27974 ± 0.00003	1.1%
	172	71											0.32392 ± 0.00004	1.4%
	172	71											0.37251 ± 0.00002	2.6%
	172	71											0.37756 ± 0.00002	3.2%
	172	71											0.41033 ± 0.00002	2.0%
	172	71											0.43256 ± 0.00002	1.5%
	172	71											0.49046 ± 0.00001	1.9%
	172	71											0.52828 ± 0.00002	3.9%
	172	71											0.54020 ± 0.00004	1.3%
	172	71											0.69737 ± 0.00002	5.8%
	172	71											0.81012 ± 0.00002	16%
	172	71											0.90079 ± 0.00002	29%
	172	71											0.91211 ± 0.00001	15%
	172	71											0.92909 ± 0.00005	3.1%
	172	71											1.00278 ± 0.00002	5.3%
	172	71											1.02241 ± 0.00005	1.5%
	172	71											1.0808 ± 0.0004	1.1%
	172	71											1.09367 ± 0.00001	63%
	172	71											1.11307 ± 0.00005	1.9%
	172	71											1.48898 ± 0.00004	1.1%
	172	71											1.58416 ± 0.00002	2.5%
	172	71											1.62195 ± 0.00002	2.1%

Isotope	A	Z	% Natural abundance	Atomic mass	Half-life	Decay mode	Decay energy (MeV)	Particle energy (MeV)	Particle intensity	Thermal neutron cross section	Spin (h/2π)	μ Nucl. mag. moment	Gamma-ray energy (MeV)	Gamma-ray intensity
	172	71											(0.07 - 2.2)weak	
$_{71}Lu^{173}$	173	71		172.938929	1.37 ± 0.01 y.	E.C.	0.675				7/2+		Yb k x-ray	45%
	173	71										2.34	0.07860 ± 0.00002	7.8%
	173	71											0.10066 ± 0.00001	3.1%
	173	71											0.17132 ± 0.00003	1.8%
	173	71											0.27198 ± 0.00004	13%
	173	71											0.63586 ± 0.00002	0.9%
$_{71}Lu^{174m}$	174	71			142 ± 2 d.	I.T.(99.3%)	0.17086				6-	2.34	Lu k x-ray	33%
	174	71				E.C.(0.7%)							0.067055 ± 0.00008	6.8%
	174	71											0.1767 ± 0.0001	0.5%
	174	71											0.2733 ± 0.0001	0.6%
	174	71											0.99205 ± 0.00008	0.6%
$_{71}Lu^{174}$	174	71		173.940336	3.31 ± 0.05 y.	E.C.	1.378				1-	1.94	Yb k x-ray	42%
	174	71											0.07664 ± 0.00004	5.8%
	174	71											1.2419 ± 0.0001	6.5%
$_{71}Lu^{175}$	175	71	97.40%	174.940770						(16 + 9) b.	7/2+	+2.2327		
$_{71}Lu^{176m}$	176	71			3.63 ± 0.01 h.	β-	1.313	1.229			1-	+0.318	Hf k x-ray	5%
	176	71						1.317					0.088372 ± 0.00009	8.9%
$_{71}Lu^{176}$	176	71	2.59%	175.942679	3.6 x 10^{10}y.					(5 + 2300) b.	7-	+3.19	Hf k x-ray	16%
	176	71											0.08837 ± 0.00001	13%
	176	71											0.20187 ± 0.00003	84%
	176	71											0.30691 ± 0.00005	93%
$_{71}Lu^{177m}$	177	71			160 ± 0.3 d.	I.T.(22%)	0.9702				23/2-	2.75	Lu k x-ray	9.7%
	177	71				β-(78%)							Hf k x-ray	58%
	177	71											0.10534 ± 0.00001	11%
	177	71											0.11295 ± 0.00001	21%
	177	71											0.12164 ± 0.00002	6.3%
	177	71											0.12850 ± 0.00004	15%
	177	71											0.13670 ± 0.00001	1.4%
	177	71											0.14717 ± 0.00003	3.8%
	177	71											0.15329 ± 0.00003	17.8%
	177	71											0.17186 ± 0.00005	5.2%
	177	71											0.17440 ± 0.00001	12.7%
	177	71											0.17700 ± 0.00002	3.4%
	177	71											0.20410 ± 0.00005	14.4%
	177	71											0.20836 ± 0.00001	61%
	177	71											0.21443 ± 0.00001	6.6%
	177	71											0.21809 ± 0.00001	3.2%
	177	71											0.22847 ± 0.00005	37%
	177	71											0.23384 ± 0.00001	5.6%
	177	71											0.24965 ± 0.00002	6.1%
	177	71											0.26879 ± 0.00001	3.6%
	177	71											0.28179 ± 0.00005	14%
	177	71											0.2915 ± 0.0001	1.0%
	177	71											0.29645 ± 0.00003	5.4%
	177	71											0.29905 ± 0.00003	1.6%
	177	71											0.30550 ± 0.00002	1.7%

Isotope	A	Z	% Natural abundance	Atomic mass	Half-life	Decay mode	Decay energy (MeV)	Particle energy (MeV)	Particle intensity	Thermal neutron cross section	Spin (h/2π)	μ Nucl. mag. moment	Gamma-ray energy (MeV)	Gamma-ray intensity
	177	71											0.31371 ± 0.00001	1.2%
	177	71											0.31903 ± 0.00001	11%
	177	71											0.32769 ± 0.00001	17.4%
	177	71											0.3417 ± 0.0001	1.8%
	177	71											0.36743 ± 0.00001	3.2%
	177	71											0.37850 ± 0.00001	28%
	177	71											0.38504 ± 0.00004	2.9%
	177	71											0.41366 ± 0.00001	17%
	177	71											0.41853 ± 0.00001	20%
	177	71											0.46583 ± 0.00005	2.3%
$_{71}Lu^{177}$	177	71		176.943752	6.71 d.	β-	0.497	0.497			7/2+	+2.239	0.11295 ± 0.00001	6.4%
	177	71											0.20836 ± 0.00001	11%
$_{71}Lu^{178m}$	178	71			23 m.	β-					(9-)		0.2166 ± 0.0001	2.5%
	178	71											0.3317 ± 0.0001	11.6%
$_{71}Lu^{178}$	178	71		177.945963	28.5 m.	β-	2.11	2.03			1+		Hf k x-ray	4%
	178	71											0.0932 ± 0.0001	6.6%
	178	71											1.2692 ± 0.0001	1.0%
	178	71											1.3099 ± 0.0001	1.5%
	178	71											1.3408 ± 0.0001	4.7%
	178	71											(0.09 - 1.7)weak	
$_{71}Lu^{179}$	179	71		178.947260	4.6 h.	β-	1.35	1.35			7/2+		0.2143 ± 0.0001	12%
	179	71											0.3377 ± 0.0001	0.2%
$_{71}Lu^{180}$	180	71		179.949870	5.7 ± 0.1 m.	β-	3.1	1.49					0.09331 ± 0.00006	13%
	180	71											0.21525 ± 0.00001	21%
	180	71											0.31651 ± 0.00005	14.9%
	180	71											0.40795 ± 0.00005	50%
	180	71											0.9830 ± 0.0001	2.2%
	180	71											1.1068 ± 0.0001	23%
	180	71											1.1982 ± 0.0001	15%
	180	71											1.2001 ± 0.0001	26%
	180	71											1.2995 ± 0.0001	14%
	180	71											1.4349 ± 0.0003	2.0%
	180	71											1.5147 ± 0.0001	8.0%
	180	71											1.8885 ± 0.0009	1.2%
$_{71}Lu^{181}$	181	71			3.5 ± 0.3 m.	β-					(7/2+)		0.0458 ± 0.0002	6.5%
	181	71											0.0530 ± 0.0002	4%
	181	71											0.0988 ± 0.0002	3.5%
	181	71											0.1056 ± 0.0003	4%
	181	71											0.1250 ± 0.0004	3.2%
	181	71											0.1530 ± 0.0003	2.6%
	181	71											0.2059 ± 0.0003	16%
	181	71											0.2404 ± 0.0004	4.5%
	181	71											0.3293 ± 0.0003	5.0%
	181	71											0.3344 ± 0.0004	3.7%
	181	71											0.3418 ± 0.0004	3.2%
	181	71											0.4637 ± 0.0005	4.5%
	181	71											0.5749 ± 0.0003	15%
	181	71											0.5899 ± 0.0001	3.2%
	181	71											0.6525 ± 0.0004	2.5%
	181	71											0.6999 ± 0.0004	4.1%
	181	71											0.8054 ± 0.0003	8.6%
	181	71											0.8584 ± 0.0003	7.6%
$_{71}Lu^{182}$	182	71			2.0 m.	β-	≈ 4.1						0.0978 ± 0.0002	14%
	182	71											0.2240 ± 0.0005	4%
	182	71											0.7208 ± 0.0005	29%
	182	71											0.8081 ± 0.0005	14%
	182	71											0.8182 ± 0.0005	29%
Hf		72		178.49						104 b.				
$_{72}Hf^{158}$	158	72		157.954590	2.9 s.	E.C.(54%)	≈ 4.9				0+			
	158	72				α(46%)		5.27						
$_{72}Hf^{159}$	159	72		158.953740	5.6 s.	β+, EC (88%)	6.76						ann.rad.	
	159	72				α(12%)		5.09						
$_{72}Hf^{160}$	160	72		159.950550	12 s.	β+, EC (97%)	4.21				0+		ann.rad.	
	160	72				α	4.78							
$_{72}Hf^{161}$	161	72		160.950110	17 s.	α		4.60						
$_{72}Hf^{162}$	162	72		161.947204	37.6 s.	β+,E.C.	3.48				0+		ann.rad.	

Isotope	A	Z	% Natural abundance	Atomic mass	Half-life	Decay mode	Decay energy (MeV)	Particle energy (MeV)	Particle intensity	Thermal neutron cross section	Spin (h/2π)	μ Nucl. mag. moment	Gamma-ray energy (MeV)	Gamma-ray intensity
	162	72											0.1739 ± 0.0001	100 +
	162	72											0.1963 ± 0.0001	25
	162	72											0.4101 ± 0.0001	17
$_{72}Hf^{163}$	163	72		162.946980	40 s.	β+,E.C.	5.35						ann.rad.	
	163	72											0.0454 ± 0.0001	48 +
	163	72											0.0621 ± 0.0001	64
	163	72											0.0710 ± 0.0001	100
	163	72											0.0849 ± 0.0001	1
	163	72											0.1331 ± 0.0001	24
	163	72											0.1622 ± 0.0002	16
	163	72											0.2333 ± 0.0001	17
	163	72											0.4961 ± 0.0001	13
	163	72											0.5203 ± 0.0001	19
	163	72											0.5352 ± 0.0002	4
	163	72											0.6882 ± 0.0001	33
$_{72}Hf^{166}$	166	72		165.942250	6.8 m.	E.C.(93%)	2.32						ann.rad.	
	166	72				β+(7%)							Lu k x-ray	48%
	166	72											0.0788 ± 0.0001	42%
	166	72											0.0930 ± 0.0002	3%
	166	72											0.2446 ± 0.0004	1.6%
	166	72											0.2839 ± 0.0002	1.6%
	166	72											0.3068 ± 0.0004	1.8%
	166	72											0.3418 ± 0.0001	4.8%
	166	72											0.3776 ± 0.0005	4.1%
	166	72											0.4079 ± 0.0001	4.6%
	166	72											0.4830 ± 0.0001	4.2%
$_{72}Hf^{167}$	167	72		166.942600	2.05 m.	β+(40%)	4.00				(5/2-)		ann.rad.	80%
	167	72				E.C.(60%)							Lu k x-ray	24%
	167	72											0.1399 ± 0.0002	3.1%
	167	72											0.1754 ± 0.0002	4.9%
	167	72											0.3152 ± 0.0001	81%
$_{72}Hf^{168}$	168	72		167.940730	25.9 m.	β+,E.C.	1.90				0+		ann.rad.	
	168	72											0.1572	70 +
	168	72											0.1838	100
	168	72											0.1988 ± 0.0001	38
$_{72}Hf^{169}$	169	72		168.941240	3.25 m.	E.C.(85%)	3.35				(5/2-)		ann.rad.	30%
	169	72				β+(15%)							Lu k x-ray	38%
	169	72											0.1236 ± 0.0002	4.1%
	169	72											0.3695 ± 0.0002	10.2%
	169	72											0.4929 ± 0.0001	89%
$_{72}Hf^{170}$	170	72		169.939740	16.0 h.	E.C.	1.20				0+		Lu k x-ray	58%
	170	72											0.0985 ± 0.0001	4%
	170	72											0.0999 ± 0.0001	2.5%
	170	72											0.1202 ± 0.0001	19%
	170	72											0.1647 ± 0.0001	33%
	170	72											0.2081 ± 0.0002	3.4%
	170	72											0.2255 ± 0.0002	1.1%
	170	72											0.2914 ± 0.0002	1.3%
	170	72											0.3089 ± 0.0003	2.6%
	170	72											0.4813 ± 0.0002	4.7%
	170	72											0.5016 ± 0.0002	4.7%
	170	72											0.5402 ± 0.0002	3.1%
	170	72											0.5729 ± 0.0002	18.5%
	170	72											0.6207 ± 0.0002	22.9%
$_{72}Hf^{171}$	171	72		170.940490	12.1 ± 0.4 h.	E.C.,β+	2.40				7/2+		ann.rad.	
	171	72											Lu k x-ray	58%
	171	72											0.1221 ± 0.0001	13%
	171	72											0.1370 ± 0.0001	7%
	171	72											0.1471 ± 0.0001	2.4%
	171	72											0.2691 ± 0.0001	2.2%
	171	72											0.2958 ± 0.0001	7.9%
	171	72											0.3475 ± 0.0001	9.7%
	171	72											0.4695 ± 0.0001	5.5%
	171	72											0.5401 ± 0.0002	2.1%
	171	72											0.6620 ± 0.0001	15%
	171	72											0.6660 ± 0.0001	4.0%
	171	72											0.7883 ± 0.0002	1.8%
	171	72											0.8525 ± 0.0001	5.0%
	171	72											1.0714 ± 0.0002	12%
	171	72											1.1616 ± 0.0003	2.2%
	171	72											1.2926 ± 0.0003	1.1%
	171	72											1.3008 ± 0.0003	1.4%
	171	72											1.3085 ± 0.0004	1.1%
	171	72											1.3402 ± 0.0003	1.7%
	171	72											1.5050 ± 0.0003	1.4%
	171	72											1.5580 ± 0.0003	1.4%
	171	72											1.7473 ± 0.0003	2.1%
	171	72											1.8350 ± 0.0003	1.4%
	171	72											2.0195 ± 0.0004	2.3%
$_{72}Hf^{172}$	172	72		171.939460	1.87 ± 0.03 y.	E.C.	0.350				0+		Lu k x-ray	57%
	172	72											0.02399 ± 0.00005	20%
	172	72											0.06735 ± 0.0001	5.3%

Isotope	A	Z	% Natural abundance	Atomic mass	Half-life	Decay mode	Decay energy (MeV)	Particle energy (MeV)	Particle intensity	Thermal neutron cross section	Spin (h/2π)	μ Nucl. mag. moment	Gamma-ray energy (MeV)	Gamma-ray intensity
	172	72											0.08175 ± 0.00005	4.5%
	172	72											0.1141 ± 0.0001	2.6%
	172	72											0.1229 ± 0.0001	1.1%
	172	72											0.12582 ± 0.00005	11.3%
	172	72											0.1279 ± 0.0001	1.5%
$_{72}Hf^{173}$	173	72		172.940650	23.6 ± 0.1 h.	E.C.	1.60				1/2-		Lu k x-ray	55%
	173	72											0.12367 ± 0.00002	83%
	173	72											0.13495 ± 0.00002	4.8%
	173	72											0.13963 ± 0.00003	12%
	173	72											0.1620 ± 0.0001	6.5%
	173	72											0.29697 ± 0.00002	34%
	173	72											0.30656 ± 0.00002	6.3%
	173	72											0.31124 ± 0.00002	11%
	173	72											0.89910 ± 0.00006	1.0%
	173	72											1.0340 ± 0.0001	0.4%
	173	72											1.0387 ± 0.0001	0.3%
	173	72											1.2056 ± 0.0001	0.3%
	173	72											(0.1 - 2.1)weak	
$_{72}Hf^{174}$	174	72	0.16%	173.940044						500 b.	0+			
$_{72}Hf^{175}$	175	72		174.941507	70 ± 2 d.	E.C.	0.686				5/2-	0.70	Lu k x-ray	47%
	175	72											0.08936 ± 0.0001	2.3%
	175	72											0.34340 ± 0.00008	87%
	175	72											0.43275 ± 0.00008	1.6%
$_{72}Hf^{176}$	176	72	5.2%	175.941406						26 b.	0+			
$_{72}Hf^{177m2}$	177	72			51.4 ± 0.5 m.	I.T.	2.740				37/2-		Hf k x-ray	29%
	177	72											0.2140 ± 0.0001	40%
	177	72											0.2951 ± 0.0001	68%
	177	72											0.2951 ± 0.0001	6.8%
	177	72											0.3115 ± 0.0001	58%
	177	72											0.3267 ± 0.0001	65%
	177	72											0.5724 ± 0.0001	7%
	177	72											0.6065 ± 0.0001	11%
	177	72											0.6382 ± 0.0001	20%
$_{72}Hf^{177m1}$	177	72			1.08 ± 0.06 s.	1.315					23/2+		Hf k x-ray	75%
	177	72											0.10534 ± 0.00001	15%
	177	72											0.11295 ± 0.00001	27%
	177	72											0.12849 ± 0.00001	20%
	177	72											0.15329 ± 0.00001	23%
	177	72											0.17440 ± 0.00002	16%
	177	72											0.17700 ± 0.00002	4.3%
	177	72											0.20410 ± 0.00001	19%
	177	72											0.20836 ± 0.00006	79%
	177	72											0.21443 ± 0.00001	8%
	177	72											0.22847 ± 0.00005	48%
	177	72											0.23384 ± 0.00001	7%
	177	72											0.24965 ± 0.00002	8%
	177	72											0.28179 ± 0.00001	18%
	177	72											0.29645 ± 0.00003	7%
	177	72											0.29905 ± 0.00003	2.0%
	177	72											0.30550 ± 0.00003	2.2%
	177	72											0.31371 ± 0.00001	1.6%
	177	72											0.32769 ± 0.00001	22%

TABLE OF THE ISOTOPES (Continued)

Isotope	A	Z	% Natural abundance	Atomic mass	Half-life	Decay mode	Decay energy (MeV)	Particle energy (MeV)	Particle intensity	Thermal neutron cross section	Spin (h/2π)	μ Nucl. mag. moment	Gamma-ray energy (MeV)	Gamma-ray intensity
	177	72											0.37851 ± 0.00001	37%
	177	72											0.38504 ± 0.00004	4%
	177	72											0.41853 ± 0.00001	25%
	177	72											0.46583 ± 0.00005	2.8%
$_{72}Hf^{177}$	177	72	18.6%	176.943217						(1 + 370) b.	7/2-	+0.7935		
$_{72}Hf^{178m2}$	178	72			31 ± 1 y.	I.T.					16+		Hf k x-ray	47%
	178	72											0.08886 ± 0.00002	62%
	178	72											0.09316 ± 0.00001	17%
	178	72											0.21342 ± 0.00001	81%
	178	72											0.21665 ± 0.00001	64%
	178	72											0.23738 ± 0.00002	9%
	178	72											0.25761 ± 0.00002	174%
	178	72											0.29680 ± 0.00003	10%
	178	72											0.32555 ± 0.00002	94%
	178	72											0.42635 ± 0.00002	97%
	178	72											0.45403 ± 0.00002	16%
	178	72											0.49499 ± 0.00002	69%
	178	72											0.53499 ± 0.00003	9%
	178	72											0.57418 ± 0.00003	84%
$_{72}Hf^{178m1}$	178	72			4.0 ± 0.2 s.	I.T.					8-		Hf k x-ray	35%
	178	72											0.08886 ± 0.00002	62%
	178	72											0.09316 ± 0.00001	17%
	178	72											0.21342 ± 0.00001	81%
	178	72											0.32555 ± 0.00002	94%
	178	72											0.42635 ± 0.00002	97%
$_{72}Hf^{178}$	178	72	27.1%	177.943696						(50 + 30) b.	0+			
$_{72}Hf^{179m2}$	179	72			25.1 d.	I.T.	1.1057				25/2-		Hf k x-ray	56%
	179	72											0.1227 ± 0.0001	27%
	179	72											0.1461 ± 0.0001	26%
	179	72											0.1698 ± 0.0001	19%
	179	72											0.1928 ± 0.0002	21%
	179	73											0.2170 ± 0.0002	8.8%
	179	72											0.2366 ± 0.0002	18%
	179	72											0.2575 ± 0.0003	3.2%
	179	72											0.2689 ± 0.0002	11%
	179	72											0.3160 ± 0.0002	20%
	179	72											0.3626 ± 0.0002	38%
	179	72											0.4098 ± 0.0003	21%
	179	72											0.4537 ± 0.0003	66%
$_{72}Hf^{179m1}$	179	72			18.7 s.	I.T.	0.375				1/2-		Hf k x-ray	28%
	179	72											0.1607 ± 0.0001	2.8%
	179	72											0.2141 ± 0.0001	95%
	179	72											0.3748 ± 0.0001	0.005%
$_{72}Hf^{179}$	179	72	13.74%	178.945812						(45 + 41) b.	9/2+	-0.6409		
$_{72}Hf^{180m}$	180	72			5.519 ± 0.004 h.	I.T.	1.1416				8-	+8.7	Hf k x-ray	18%
	180	72											0.0575 ± 0.0001	48%
	180	72											0.0933 ± 0.0001	17%
	180	72											0.2152 ± 0.0001	82%
	180	72											0.3323 ± 0.0001	94%
	180	72											0.4432 ± 0.0001	85%
	180	72											0.5007 ± 0.0001	13%
$_{72}Hf^{180}$	180	72	35.2%	179.946545						13 b.	0+			
$_{72}Hf^{181}$	181	72		180.949096	42.4 ± 0.06 d.	β-	1.027	0.408		30 b.	1/2-		Ta k x-ray	13%
	181	72											0.13294 ± 0.00007	36%
	181	72											0.13617 ± 0.00007	6%
	181	72											0.34583 ± 0.00007	15%
	181	72											0.48200 ± 0.00005	81%

TABLE OF THE ISOTOPES (Continued)

Isotope	A	Z	% Natural abundance	Atomic mass	Half-life	Decay mode	Decay energy (MeV)	Particle energy (MeV)	Particle intensity	Thermal neutron cross section	Spin (h/2π)	μ Nucl. mag. moment	Gamma-ray energy (MeV)	Gamma-ray intensity
$_{72}Hf^{182m}$	182	72			62 m.	β-(54%)	1.60	0.49	43%		8-		Hf k x-ray	8%
	182	72				I.T.(46%)	1.1729	0.95	10%				0.0509 ± 0.0002	13%
	182	72											0.0978 ± 0.0002	14%
	182	72											0.1143 ± 0.0002	7%
	182	72											0.1328 ± 0.0002	3%
	182	72											0.1432 ± 0.0002	4.7%
	182	72											0.1468 ± 0.0002	4.0%
	182	72											0.1734 ± 0.0002	3.0%
	182	72											0.1787 ± 0.0002	2%
	182	72											0.2244 ± 0.0002	38%
	182	72											0.3396 ± 0.0002	6.2%
	182	72											0.3441 ± 0.0002	46%
	182	72											0.4558 ± 0.0002	20%
	182	72											0.5066 ± 0.0002	24%
	182	72											0.6032 ± 0.0002	6%
	182	72											0.6133 ± 0.0002	1.2%
	182	72											0.6276 ± 0.0002	1.1%
	182	72											0.7997 ± 0.0002	10%
	182	72											0.8231 ± 0.0002	3%
	182	72											0.9428 ± 0.0002	21%
$_{72}Hf^{182}$	182	72		181.950550	9 x 10^6 y.	β-	0.431				0+		Ta k x-ray	7%
	182	72											0.1143 ± 0.0001	3%
	182	72											0.1561 ± 0.0001	7%
	182	72											0.2704 ± 0.0001	80%
$_{72}Hf^{183}$	183	72		182.953530	64 m.	β-	2.01	1.18	68%		3/2-		Ta k x-ray	13%
	183	72						1.54	25%				0.0732 ± 0.0001	38%
	183	72											0.3159 ± 0.0001	1%
	183	72											0.3979 ± 0.0001	3%
	183	72											0.4591 ± 0.0001	27%
	183	72											0.7837 ± 0.0001	65%
	183	72											1.4702 ± 0.0001	2.7%
$_{72}Hf^{184}$	184	72		183.955440	4.1 h.	β-	1.3	0.74	38%		0+		Ta k x-ray	15%
	184	72						0.85	16%				0.0414 ± 0.0002	10%
	184	72						1.10	46%				0.0439 ± 0.0002	6%
	184	72											0.0479 ± 0.0002	1%
	184	72											0.1391 ± 0.0002	48%
	184	72											0.1810 ± 0.0002	15%
	184	72											0.3449 ± 0.0002	38%
Ta		73		180.9479						20.5 b.				
$_{73}Ta^{159}$	159	73		158.962860	0.6 s.	β+, EC (20%)	8.49						ann.rad.	
	159	73				α(80%)		5.60						
$_{73}Ta^{160}$	160	73		159.961630		β+,E.C.	10.3						ann.rad.	
	160	73				α		5.41						
$_{73}Ta^{161}$	161	73		160.958210		β+,E.C.	7.55						ann.rad.	
	161	73				α		5.15						
$_{73}Ta^{164}$	164	73		163.953370	13.6 s.	β+	8.35						ann.rad.	
	164	73				α		4.62					0.2110	
	164	73											0.3768	
$_{73}Ta^{166}$	166	73		165.950280	32 s.	β+(82%)	7.48						ann.rad.	160%
	166	73				E.C.(18%)							Hf k x-ray	16%
	166	73											0.1587 ± 0.0002	53%
	166	73											0.3117 ± 0.0003	28%
	166	73											0.5360 ± 0.0004	4.0%
	166	73											0.5524 ± 0.0004	3.0%
	166	73											0.5945 ± 0.0003	3.5%
	166	73											0.6514 ± 0.0004	8.5%
	166	73											0.7428 ± 0.0004	7.0%
	166	73											0.7500 ± 0.0005	5.5%
	166	73											0.8101 ± 0.0004	9.8%
	166	73											0.8474 ± 0.0004	7.2%
	166	73											0.8622 ± 0.0006	3.7%
	166	73											0.8641 ± 0.0005	4.9%
	166	73											0.9062 ± 0.0006	6.1%
	166	73											0.9770 ± 0.0008	2.5%
	166	73											1.0549 ± 0.001	4.4%
	166	73											1.1738 ± 0.001	5%
	166	73											1.2883 ± 0.0012	3.1%
	166	73											1.4470 ± 0.002	3%
$_{73}Ta^{167}$	167	73		166.948080	3 m.	β+,E.C.	5.10						ann.rad.	
$_{73}Ta^{168}$	168	73		167.947820	2.4 m.	β+(77%)	6.60						ann.rad.	150%
	168	73				E.C.(23%)							Hf k x-ray	21%
	168	73											0.1239 ± 0.0002	37%
	168	73											0.2615 ± 0.0002	28%
	168	73											0.3711 ± 0.0004	4%
	168	73											0.5270 ± 0.0006	2.7%
	168	73											0.6464 ± 0.0008	3%
	168	73											0.7502 ± 0.0006	10%
	168	73											0.7730 ± 0.0008	6%
	168	73											0.8156 ± 0.0008	2%
	168	73											0.8340 ± 0.0008	2%
	168	73											0.8741 ± 0.0008	6%
	168	73											0.8975 ± 0.001	2%
	168	73											0.9072 ± 0.001	6%

Isotope	A	Z	% Natural abundance	Atomic mass	Half-life	Decay mode	Decay energy (MeV)	Particle energy (MeV)	Particle intensity	Thermal neutron cross section	Spin (h/2π)	μ Nucl. mag. moment	Gamma-ray energy (MeV)	Gamma-ray intensity
	168	73											0.9342 ± 0.0012	3%
	168	73											0.9866 ± 0.001	5%
	168	73											1.0581 ± 0.001	2%
	168	73											1.2481 ± 0.002	2%
	168	73											1.2824 ± 0.002	2%
	168	73											1.4063 ± 0.002	1.6%
	168	73											1.4414 ± 0.002	1.1%
	168	73											1.6682 ± 0.003	1.1%
$_{73}Ta^{169}$	169	73		168.946020	5 m.	β+,E.C.	4.45						ann.rad.	
	169	73											0.0288 ± 0.0001	100 +
	169	73											0.0382 ± 0.0001	25
	169	73											0.0777 ± 0.0001	7
	169	73											0.1328 ± 0.0001	9
	169	73											0.1535 ± 0.0001	35
	169	73											0.1770 ± 0.0001	10
	169	73											0.1878 ± 0.0002	5
	169	73											0.1924 ± 0.0001	43
	169	73											0.2300 ± 0.0001	12
	169	73											0.3945 ± 0.0001	15
	169	73											0.4040 ± 0.0002	9
	169	73											0.4408 ± 0.0001	17
	169	73											0.5204 ± 0.0002	9
	169	73											0.5290 ± 0.0002	11
	169	73											0.5474 ± 0.0003	9
	169	73											0.5950 ± 0.0002	26
$_{73}Ta^{170}$	170	73		169.945970	6.76 ± 0.06 m.	β+(70%)	5.80				(3+)		ann.rad.	140%
	170	73				E.C.(35%)							Hf k x-ray	22%
	170	73											0.1008 ± 0.0002	21%
	170	73											0.2212 ± 0.0002	16%
	170	73											0.6650 ± 0.0003	1%
	170	73											0.7655 ± 0.0002	1%
	170	73											0.8348 ± 0.0004	1.5%
	170	73											0.8604 ± 0.0002	7.4%
	170	73											0.9870 ± 0.0003	5.8%
	170	73											1.1190 ± 0.0006	1.1%
	170	73											1.3442 ± 0.0006	1.4%
$_{73}Ta^{171}$	171	73		170.944680	23.4 m.	β+,E.C.	3.90				(5/2-)		0.0496 ± 0.0001	100 +
	171	73											0.0619 ± 0.0001	9
	171	73											0.0667 ± 0.0001	4
	171	73											0.0807 ± 0.0001	4
	171	73											0.0920 ± 0.0001	11
	171	73											0.1171 ± 0.0001	5
	171	73											0.1524 ± 0.0001	6
	171	73											0.1663 ± 0.0001	19
	171	73											0.1755 ± 0.0001	16
	171	73											0.3524 ± 0.0001	3
	171	73											0.4067 ± 0.0001	5
	171	73											0.4444 ± 0.0001	16
	171	73											0.4547 ± 0.0001	4
	171	73											0.4713 ± 0.0002	9
	171	73											0.4927 ± 0.0002	15
	171	73											0.5018 ± 0.0002	23
	171	73											0.5064 ± 0.0002	54
	171	73											0.5223 ± 0.0002	11
	171	73											0.5380 ± 0.0002	15
	171	73											0.5545 ± 0.0002	7
	171	73											0.5709 ± 0.0002	3
	171	73											0.6068 ± 0.0002	4
	171	73											0.6217 ± 0.0002	4
	171	73											0.7676 ± 0.0002	9
	171	73											0.7889 ± 0.0002	4
	171	73											0.9871 ± 0.0002	9
	171	73											1.0078 ± 0.0002	3
	171	73											(0.05 - 1.02)many	
$_{73}Ta^{172}$	172	73		171.944740	36.8 ± 0.3 m.	β+(25%)	4.92				(3-)		ann.rad.	50%
	172	73				E.C.(75%)							Hf k x-ray	40%
	172	73											0.21396 ± 0.00005	52%
	172	73											0.3187 ± 0.0002	5.0%
	172	73											0.5035 ± 0.0001	1.3%
	172	73											0.6431 ± 0.0001	2.2%
	172	73											0.7760 ± 0.0001	2.4%
	172	73											0.8203 ± 0.0001	3.0%
	172	73											0.8571 ± 0.0001	4.1%
	172	73											0.9523 ± 0.0001	1.8%
	172	73											0.9800 ± 0.0001	3.7%
	172	73											0.9954 ± 0.0001	2.1%
	172	73											1.0500 ± 0.0001	2.2%
	172	73											1.0752 ± 0.0001	3.5%
	172	73											1.0856 ± 0.0001	7.6%
	172	73											1.10923 ± 0.00006	14%
	172	73											1.18646 ± 0.00005	2.5%

Isotope	A	Z	% Natural abundance	Atomic mass	Half-life	Decay mode	Decay energy (MeV)	Particle energy (MeV)	Particle intensity	Thermal neutron cross section	Spin (h/2π)	μ Nucl. mag. moment	Gamma-ray energy (MeV)	Gamma-ray intensity
	172	73											1.2404 ± 0.0001	2.0%
	172	73											1.2656 ± 0.0002	2.5%
	172	73											1.2775 ± 0.0001	2.7%
	172	73											1.3303 ± 0.0001	7.6%
	172	73											1.3869 ± 0.0001	2.5%
	172	73											1.4796 ± 0.0002	2.2%
	172	73											1.5443 ± 0.0001	6.2%
	172	73											(0.09 - 3.8)many	
$_{73}Ta^{173}$	173	73		172.943650	3.65 ± 0.05 h	β+(24%)	2.80				(5/2-)		ann.rad.	48%
	173	73				E.C.(76%)							Hf k x-ray	56%
	173	73											0.06972 ± 0.00007	6%
	173	73											0.17219 ± 0.00006	17%
	173	73											0.18058 ± 0.00007	2.1%
	173	73											0.7011 ± 0.0001	1.2%
	173	73											1.0299 ± 0.0001	1.6%
	173	73											1.2082 ± 0.0001	2.7%
	173	73											1.4322 ± 0.0003	0.6%
	173	73											(0.06 - 2.7)weak	
$_{73}Ta^{174}$	174	73		173.944340	1.18 ± 0.05 h.	β+(27%)	4.00				(3+)		ann.rad.	54%
	174	73				E.C.(73%)							Hf k x-ray	44%
	174	73											0.09089 ± 0.00002	16%
	174	73											0.20638 ± 0.00003	58%
	174	73											0.31080 ± 0.00004	1.0%
	174	73											0.76472 ± 0.00004	1.3%
	174	73											0.97110 ± 0.00005	1.2%
	174	73											1.15135 ± 0.00005	1.1%
	174	73											1.20582 ± 0.00004	4.8%
	174	73											1.22831 ± 0.00004	1.4%
	174	73											1.35773 ± 0.00006	0.8%
	174	73											(0.09 - 3.64)many	
$_{73}Ta^{175}$	175	73		174.943650	10.5 ± 0.2 h.	E.C.	2.00				7/2+		Hf k x-ray	64%
	175	73											0.0816 ± 0.0001	5.7%
	175	73											0.1046 ± 0.0001	3.0%
	175	73											0.1261 ± 0.0001	5.5%
	175	73											0.1410 ± 0.0001	2.2%
	175	73											0.2077 ± 0.0001	13.3%
	175	73											0.2671 ± 0.0001	10%
	175	73											0.3487 ± 0.0001	11%
	175	73											0.4368 ± 0.0002	3.8%
	175	73											0.4754 ± 0.0002	1.9%
	175	73											0.8578 ± 0.0002	3.0%
	175	73											0.9987 ± 0.0002	2.4%
	175	73											1.1439 ± 0.0002	1.1%
	175	73											1.2255 ± 0.0002	2.4%
	175	73											1.7121 ± 0.0002	1.1%
	175	73											1.7218 ± 0.0003	1.1%
	175	73											1.7447 ± 0.0002	1.3%
	175	73											1.7936 ± 0.0002	4.4%
	175	73											1.8263 ± 0.0002	1.2%
$_{73}Ta^{176}$	176	73		175.944730	8.08 ± 0.07 h.	E.C.	3.10				1-		Hf k x-ray	45%
	176	73											0.08837 ± 0.00001	11%
	176	73											0.20187 ± 0.00003	5.5%
	176	73											0.46623 ± 0.00005	1.1%
	176	73											0.50775 ± 0.00008	1.4%
	176	73											0.52152 ± 0.00005	2%
	176	73											0.61121 ± 0.00005	1.2%
	176	73											0.61690 ± 0.00004	1.0%
	176	73											0.71053 ± 0.00004	5.2%
	176	73											1.02317 ± 0.00004	2.6%
	176	73											1.15735 ± 0.00003	24.6%

Isotope	A	Z	% Natural abundance	Atomic mass	Half-life	Decay mode	Decay energy (MeV)	Particle energy (MeV)	Particle intensity	Thermal neutron cross section	Spin (h/2π)	μ Nucl. mag. moment	Gamma-ray energy (MeV)	Gamma-ray intensity
	176	73											1.19023 ± 0.00006	4.4%
	176	73											1.22503 ± 0.00003	5.5%
	176	73											1.25298 ± 0.00003	3.0%
	176	73											1.26887 ± 0.00006	1.3%
	176	73											1.29106 ± 0.00004	1.28%
	176	73											1.34135 ± 0.00004	3.2%
	176	73											1.35748 ± 0.00004	1.9%
	176	73											1.55507 ± 0.00004	3.9%
	176	73											1.58402 ± 0.00004	5.1%
	176	73											1.61627 ± 0.00005	1.2%
	176	73											1.63084 ± 0.00005	1.7%
	176	73											1.63371 ± 0.00005	2.8%
	176	73											1.64344 ± 0.00004	2.3%
	176	73											1.69653 ± 0.00005	4.5%
	176	73											1.72208 ± 0.00005	3.1%
	176	73											1.82370 ± 0.00004	4.3%
	176	73											1.86287 ± 0.00004	3.8%
	176	73											2.04485 ± 0.00006	1.3%
	176	73											2.83193 ± 0.00007	4.2%
	176	73											2.92031 ± 0.00007	2.1%
$_{73}Ta^{177}$	177	73		176.944460	2.36 ± 0.01 d.	E.C.	1.158				7/2+		Hf k x-ray	42%
	177	73											0.11295 ± 0.00001	7.2%
	177	73											0.20836 ± 0.00001	1.0%
	177	73											0.42460 ± 0.00005	0.1%
	177	73											0.74591 ± 0.00005	0.2%
	177	73											1.0577 ± 0.0001	0.3%
	177	73											(0.07 - 1.06)weak	
$_{73}Ta^{178m}$	178	73			2.45 ± 0.05 h.	E.C.					(7-)		Hf k x-ray	75%
	178	73											0.08886 ± 0.00002	62%
	178	73											0.09316 ± 0.00003	17.4%
	178	73											0.21342 ± 0.00002	81%
	178	73											0.32555 ± 0.00002	94%
	178	73											0.33166 ± 0.00006	32%
	178	73											0.42635 ± 0.00002	97%
$_{73}Ta^{178}$	178	73		177.945750	9.3 ± 0.03 m.	E.C.(99%)	1.910				1+		ann.rad.	2%
	178	73				β+(1%)							Hf k x-ray	42%
	178	73											0.09316 ± 0.00003	6.7%
	178	73											1.10614 ± 0.00007	1.53%
	178	73											1.18345 ± 0.00004	0.17%
	178	73											1.3409 ± 0.0001	1.0%
	178	73											1.3506 ± 0.0001	1.2%
	178	73											1.4961 ± 0.0001	0.27%
$_{73}Ta^{179}$	179	73		178.945930	1.82 ± 0.05 y.	E.C.	0.110				7/2+		Hf k x-ray	29%
$_{73}Ta^{180m}$	180	73			8.15 ± 0.01 h.	E.C.(87%)	0.865				1+		Hf k x-ray	36%
	180	73				β-(13%)	0.710	0.61	3%				W k x-ray	0.3%
	180	73						0.71	10%				0.09333 ± 0.00006	5%

TABLE OF THE ISOTOPES (Continued)

Isotope	A	Z	% Natural abundance	Atomic mass	Half-life	Decay mode	Decay energy (MeV)	Particle energy (MeV)	Particle intensity	Thermal neutron cross section	Spin (h/2π)	μ Nucl. mag. moment	Gamma-ray energy (MeV)	Gamma-ray intensity
	180	73											0.10340 ± 0.00001	0.7%
$_{73}Ta^{180}$	180	73	0.012%	179.947462						600 b.	(9-)			
$_{73}Ta^{181}$	181	73	99.998%	180.947992						(0.011 + 20) b.	7/2+	+2.370		
$_{73}Ta^{182m}$	182	73			15.9 m.	I.T.	0.5198				10-		Ta k x-ray	46%
	182	73											0.14678 ± 0.00002	36%
	182	73											0.17157 ± 0.00002	47%
	182	73											0.18493 ± 0.00002	23%
	182	73											0.31837 ± 0.00005	6.5%
$_{73}Ta^{182}$	182	73		181.950149	114.5 d.	β-	1.814	0.25	30%		3-	2.6	W k x-ray	17%
	182	73						0.44	20%				0.06775 ± 0.00001	41%
	182	73						0.52	40%				0.10010 ± 0.00001	14%
	182	73											0.11367 ± 0.00002	1.9%
	182	73											0.15243 ± 0.00001	7.1%
	182	73											0.15639 ± 0.00001	2.7%
	182	73											0.17939 ± 0.00001	3.1%
	182	73											0.19836 ± 0.00001	1.5%
	182	73											0.22211 ± 0.00001	7.6%
	182	73											0.22932 ± 0.00001	3.6%
	182	73											0.26407 ± 0.00001	3.6%
	182	73											1.12127 ± 0.00003	35.0%
	182	73											1.18902 ± 0.00003	16.4%
	182	73											1.22138 ± 0.00003	27.4%
	182	73											1.23099 ± 0.00003	11.6%
	182	73											1.25739 ± 0.00003	1.5%
	182	73											1.27370 ± 0.00003	0.67%
	182	73											1.28913 ± 0.00003	1.4%
	182	73											1.3427 ± 0.0001	0.26%
	182	73											1.37381 ± 0.00003	0.23%
$_{73}Ta^{183}$	183	73		182.951369	5.1 d.	β-	1.07	0.45	5%		7/2+		W k x-ray	44%
	183	73						0.62	91%				0.0847 ± 0.0001	1.3%
	183	73											0.0991 ± 0.0001	6.6%(D)
	183	73											0.1079 ± 0.0001	11%(D)
	183	73											0.1441 ± 0.0001	2.5%
	183	73											0.1613 ± 0.0001	8.9%
	183	73											0.1623 ± 0.0001	4.9%
	183	73											0.2099 ± 0.0001	4.5%
	183	73											0.2443 ± 0.0001	8.6%
	183	73											0.2461 ± 0.0001	27%
	183	73											0.2917 ± 0.0001	3.8%
	183	73											0.3131 ± 0.0002	7.3%
	183	73											0.3540 ± 0.0001	11.4%
$_{73}Ta^{184}$	184	73		183.954005	8.7 h.	β-	2.86	1.11	15%		(5-)		W k x-ray	15%
	184	73						1.17	81%				0.1112 ± 0.0001	24%
	184	73											0.1613 ± 0.0001	3.3%
	184	73											0.2153 ± 0.0001	12%
	184	73											0.2267 ± 0.0001	6.8%
	184	73											0.2444 ± 0.0001	3.6%
	184	73											0.2528 ± 0.0001	49%
	184	73											0.3180 ± 0.0001	23%
	184	73											0.3843 ± 0.0001	12.8%
	184	73											0.4140 ± 0.0001	74%
	184	73											0.4611 ± 0.0001	11%
	184	73											0.5367 ± 0.0001	13%
	184	73											0.6420 ± 0.0001	1.4%
	184	73											0.7921 ± 0.0001	15%
	184	73											0.8948 ± 0.0001	11%
	184	73											0.9033 ± 0.0001	15%
	184	73											0.9209 ± 0.0001	33%
	184	73											1.1101 ± 0.0001	2.3%
	184	73											1.1738 ± 0.0001	4.9%
$_{73}Ta^{185}$	185	73		184.955553	49 m.	β-	1.994	1.21	5%		(7/2+)		W k x-ray	23%
	185	73						1.77	81%				0.0697 ± 0.0002	2%

Isotope	A	Z	% Natural abundance	Atomic mass	Half-life	Decay mode	Decay energy (MeV)	Particle energy (MeV)	Particle intensity	Thermal neutron cross section	Spin (h/2π)	μ Nucl. mag. moment	Gamma-ray energy (MeV)	Gamma-ray intensity
	185	73											0.1078 ± 0.0001	2.7%
	185	73											0.1473 ± 0.0001	1.1%
	185	73											0.1739 ± 0.0001	22%
	185	73											0.1776 ± 0.0001	26%
	185	73											0.2435 ± 0.0001	3.7%
	185	73											0.3944 ± 0.0005	0.8%
	185	73											0.5417 ± 0.0005	0.8%
	185	73											0.5887 ± 0.001	0.8%
$_{73}Ta^{186}$	186	73		185.958540	10.5 m.	β-	3.9	2.2			(3-)		W k x-ray	15%
	186	73											0.1223 ± 0.0001	23%
	186	73											0.1979 ± 0.0001	59%
	186	73											0.2149 ± 0.0001	50%
	186	73											0.2925 ± 0.0005	4%
	186	73											0.3075 ± 0.0001	11%
	186	73											0.3092 ± 0.0001	3%
	186	73											0.4177 ± 0.0002	15%
	186	73											0.4570 ± 0.001	2.5%
	186	73											0.5106 ± 0.0005	44%
	186	73											0.5672 ± 0.0003	4.0%
	186	73											0.6153 ± 0.0002	33%
	186	73											0.7375 ± 0.0003	34%
	186	73											0.7392 ± 0.0003	11.8%
	186	73											0.7594 ± 0.0005	2%
	186	73											0.7998 ± 0.0005	3%
	186	73											0.8300 ± 0.0005	2%
	186	73											0.8841 ± 0.001	2%
	186	73											0.9230 ± 0.001	1.4%
	186	73											(0.09 - 1.5)	1.4%
W		74								18.4 b.				
$_{74}W^{160}$	160	74		159.968480	0.08 s.	α		5.92			0+			
$_{74}W^{161}$	161	74		160.967140	0.41 s.	β+, EC (18%)	8.34							
	164	74				α(82%)		5.78						
$_{74}W^{162}$	162	74		161.963290	1.39 s.	β+, EC (54%)	5.72				0+			
	162	74				α(46%)		5.54						
$_{74}W^{163}$	163	74		162.962270	2.8 s.	β+, EC (59%)	7.54							
	163	74				α(41%)		5.38						
$_{74}W^{164}$	164	74		163.958820	6 s.	β+, EC (97%)	5.08				0+		ann.rad.	
	164	74				α(3%)		5.15						
$_{74}W^{165}$	165	74		164.958110	≈ 5.1 s.	β+, EC (99%)	6.86						ann.rad.	
	165	74				α(1%)		4.91						
$_{74}W^{166}$	166	74		165.955020	16 s.	β+, EC (99%)	4.42				0+		ann.rad.	
	166	74				α(1%)		4.74						
$_{74}W^{172}$	172	74		171.947430	≈ 6.7 m.	β+,E.C.	2.50						0.0359 ± 0.0003	39 +
	172	74											0.0396 ± 0.0003	10
	172	74											0.1093 ± 0.0002	6
	172	74											0.1302 ± 0.0002	27
	172	74											0.1452 ± 0.0005	5
	172	74											0.1538 ± 0.0003	10
	172	74											0.1749 ± 0.0003	21
	172	74											0.3244 ± 0.0002	7
	172	74											0.4234 ± 0.0002	6
	172	74											0.4576 ± 0.0002	100
	172	74											0.6236 ± 0.0002	28
	172	74											0.6360 ± 0.0003	5
	172	74											0.7708 ± 0.0006	4
$_{74}W^{173}$	173	74		172.947710	16.1 ± 0.5 m.	E.C.	3.78						0.0499	
	173	74											0.1057	
	173	74											0.3652	
	173	74											0.707	
$_{74}W^{174}$	174	74		173.946160	29 ± 1 m.	E.C.	1.70				0+		0.0354 ± 0.0001	148 +
	174	74											0.0619 ± 0.0004	20
	174	74											0.0964 ± 0.0001	11
	174	74											0.1252 ± 0.0001	81
	174	74											0.1365 ± 0.0001	78
	174	74											0.1437 ± 0.0001	24
	174	74											0.1627 ± 0.0001	14
	174	74											0.1740 ± 0.0001	5
	174	74											0.1930 ± 0.0001	56
	174	74											0.2020 ± 0.0001	41
	174	74											0.2164 ± 0.0002	7
	174	74											0.2334 ± 0.0001	32
	174	74											0.2395 ± 0.0001	13
	174	74											0.2898 ± 0.0002	10
	174	74											0.3287 ± 0.0001	100
	174	74											0.3398 ± 0.0001	36
	174	74											0.3549 ± 0.0001	21
	174	74											0.3645 ± 0.0001	37

TABLE OF THE ISOTOPES (Continued)

Isotope	A	Z	% Natural abundance	Atomic mass	Half-life	Decay mode	Decay energy (MeV)	Particle energy (MeV)	Particle intensity	Thermal neutron cross section	Spin (h/2π)	μ Nucl. mag. moment	Gamma-ray energy (MeV)	Gamma-ray intensity
	174	74											0.3770 ± 0.0001	57
	174	74											0.3785 ± 0.0001	84
	174	74											0.4288 ± 0.0001	123
	174	74											0.4722 ± 0.0001	4
	174	74											0.5475 ± 0.0001	4
	174	74											0.5676 ± 0.0001	4
	174	74											0.8350 ± 0.0001	6
$_{74}W^{175}$	175	74		174.946770	34 ± 1 m.	E.C.	2.90				$^1/_2$-		0.01498 ± 0.00002	
	173	74											0.03641 ± 0.00002	
	175	74											0.05138 ± 0.00002	
	175	74											0.1211 ± 0.0001	
	175	74											0.1491 ± 0.0001	
	175	74											0.1667 ± 0.0001	
	175	74											0.2703 ± 0.0001	
$_{74}W^{176}$	176	74		175.945590	2.5 ± 0.2 h.	β+,E.C.	0.800				0+		0.03358 ± 0.00004	0.08 +
	176	74											0.06129 ± 0.00004	9
	176	74											0.08414 ± 0.0004	5
	176	74											0.09487 ± 0.00004	9
	176	74											0.10020 ± 0.00005	100
$_{74}W^{177}$	177	74		176.946610	2.21 ± 0.04 h.	E.C.	2.00				($^1/_2$-)		Ta k x-ray	78%
	177	74											0.15505 ± 0.00004	59%
	177	74											0.15594 ± 0.00004	4%
	177	74											0.18569 ± 0.00007	16%
	177	74											0.42694 ± 0.00004	13%
	177	74											0.5284 ± 0.0001	2.4%
	177	74											0.6116 ± 0.0001	6%
	177	74											0.6173 ± 0.0001	2.5%
	177	74											1.0149 ± 0.0001	4.8%
	177	74											1.0364 ± 0.0001	10%
	177	74											1.0668 ± 0.0001	3%
	177	74											1.1825 ± 0.0001	3.7%
$_{74}W^{178}$	178	74		177.945840	21.5 ± 0.1 d.	E.C.	0.089				0+		Ta k x-ray	13%
$_{74}W^{179m}$	179	74			6.4 m.	I.T.(99.7%) E.C.(0.3%)	0.222				($^1/_2$-)		W k x-ray	27%
	179	74											0.2220 ± 0.0001	8.6%
	179	74											0.2387 ± 0.0003	0.2%
	179	74											0.2817 ± 0.0003	0.2%
$_{74}W^{179}$	179	74		178.947067	37.5 m.	E.C.	1.060				(7/2-)		Ta k x-ray	39%
	179	74											0.0307 ± 0.0001	28%
	179	74											0.0339 ± 0.0002	0.2%
$_{74}W^{180}$	180	74	0.13%	179.946701						30 b.	0+			
$_{74}W^{181}$	181	74		180.948192	121 ± 0.2 d.	E.C.	0.187				9/2+		Ta k x-ray	33%
	181	74											0.13617 ± 0.00007	0.032%
	181	74											0.15221 ± 0.00002	0.08%
$_{74}W^{182}$	182	74	26.3%	181.948202						21 b.	0+			
$_{74}W^{183m}$	183	74			5.15 s.	I.T.					(11/2+		W k x-ray	63%
	183	74											0.0465 ± 0.0001	6%
	183	74											0.0526 ± 0.0001	7%+
	183	74											0.0991 ± 0.0001	9%
	183	74											0.1025 ± 0.0001	2.3%
	183	74											0.1605 ± 0.0001	4.9%
$_{74}W^{183}$	183	74	14.3%	182.950220						10 b.	$^1/_2$-	+0.11778		
$_{74}W^{184}$	184	74	30.67%	183.950928						(0.002 + 1.8) b.	0+			
$_{74}W^{185m}$	185	74			1.65 m.	I.T.	0.1974				11/2+		W k x-ray	4%
	185	74											0.0659 ± 0.0001	5.8%
	185	74											0.1315 ± 0.0001	4.3%
	185	74											0.1737 ± 0.0001	3.3%
$_{74}W^{185}$	185	74		184.953416	74.8 d.	β-	0.433	0.433	99.9%		3/2-		0.12536 ± 0.00003	0.019%
$_{74}W^{186}$	186	74	28.6%	185.954357										
$_{74}W^{187}$	187	74		186.957153	23.9 h.	β-	1.312	0.624 1.315		60 b.	3/2-	0.688	Re k x-ray	14%
	187	74											0.0725 ± 0.0001	13%
	178	74											0.13424 ± 0.00003	10%
	187	74											0.47951 ± 0.00003	25%
	187	74											0.55151 ± 0.00001	5.9%
	187	74											0.61824 ± 0.00004	7.3%

TABLE OF THE ISOTOPES (Continued)

Isotope	A	Z	% Natural abundance	Atomic mass	Half-life	Decay mode	Decay energy (MeV)	Particle energy (MeV)	Particle intensity	Thermal neutron cross section	Spin (h/2π)	μ Nucl. mag. moment	Gamma-ray energy (MeV)	Gamma-ray intensity
	187	74											0.68572 ± 0.00004	32%
	187	74											0.77295 ± 0.00007	4.8%
$_{74}W^{188}$	188	74		187.958480	69.4 d.	β-	0.349	0.349	99%		0+		0.0636 ± 0.0001	0.1%
	188	74											0.2271 ± 0.0001	0.2%
	188	74											0.2907 ± 0.0001	0.4%
$_{74}W^{189}$	189	74		188.961900	11.5 m.	β-	2.50	1.4 2.5			(3/2-(		0.258 ± 0.0003	100 +
	189	74											0.417 ± 0.004	96
	189	74											0.550 ± 0.001	28
	189	74											0.855 ± 0.015	20
	189	74											0.955 ± 0.020	17
$_{74}W^{190}$	190	74		189.963210	30 m.	β-	1.3	0.95			0+		Re k x-ray	54%
	190	74											0.1576 ± 0.0001	39%
	190	74											0.1621 ± 0.0001	11%
Re		75		186.207						90 b.				
$_{75}Re^{162}$	162	75		161.976060	0.10 s.	α		6.12						
$_{75}Re^{163}$	163	75		162.971970	0.26 s.	β+,E.C.	9.04							
	163	75				α		5.92						
$_{75}Re^{164}$	164	75		163.970590	0.9 s.	β+,E.C.	10.9							
	164	75				α		5.78						
$_{75}Re^{165}$	165	75		164.961890	2.4 s.	β+, EC (87%)	8.18							
	165	75				α		5.51						
$_{75}Re^{166}$	166	75		165.965740	2.2 s.	β+,E.C.	9.98							
	166	75				α		5.50						
$_{75}Re^{167}$	167	75		166.962590	2.0 s.	β+,E.C.	7.52							
	167	75				α		5.35						
$_{75}Re^{168}$	168	75		167.961540	2.9 s.	β+,E.C.	9.08							
	168	75				α		5.14						
$_{75}Re^{170}$	170	75		169.958040	8 s.	β+,E.C.	8.15						0.1560 ± 0.0004	57%
	170	75											0.3055 ± 0.0004	85%
	170	75											0.4125 ± 0.0004	50%
$_{75}Re^{172m}$	172	75			15 s.	β+,E.C.							ann.rad.	
	172	75											0.1234 ± 0.0001	45 +
	172	75											0.2537 ± 0.0002	100
	172	75											0.3504 ± 0.0005	55
	172	75											0.4194 ± 0.0003	10
$_{75}Re^{172}$	172	75		171.953180	2.3 m.	β+,E.C.	7.29						ann.rad.	
	172	75											0.1234 ± 0.0007	100 +
	172	75											0.2537 ± 0.0002	74
	172	75											0.7430 ± 0.0002	19
$_{75}Re^{174}$	174	75		173.953180	2.3 m.	β+,E.C.	6.54						ann.rad.	
	174	75											0.1119 ± 0.0004	29%
	174	75											0.2430 ± 0.0004	23%
	174	75											0.3490 ± 0.0004	11%
$_{75}Re^{175}$	175	75		174.951430	4.6 m.	β+,E.C.	4.35						ann.rad.	
$_{75}Re^{176}$	176	75		175.951500	5.3 m.	β+,E.C.	5.50				(3+)		ann.rad.	
	176	75											0.1089 ± 0.0003	26%
	176	75											0.2406 ± 0.0003	48%
$_{75}Re^{177}$	177	75		176.950370	14 m.	E.C.(78%) β+(22%)	3.50				(5/2-)		ann.rad.	44%
	177	75											W k x-ray	62%
	177	75											0.0797 ± 0.0001	7%
	177	75											0.0843 ± 0.0002	6%
	177	75											0.0949 ± 0.0001	4%
	177	75											0.1014 ± 0.0002	3%
	177	75											0.1968 ± 0.0002	8%
	177	75											0.2098 ± 0.0003	3%
	177	75											0.7081 ± 0.0006	2%
	177	75											0.7234 ± 0.0006	2%
	177	75											1.7705 ± 0.0008	2%
	177	75											1.9112 ± 0.0008	1%
	177	75											1.9646 ± 0.0008	3%
	177	75											1.9861 ± 0.0008	1%
$_{75}Re^{178}$	178	75		177.950850	13.2 m.	β+(11%) E.C.(89%)	4.66	3.3			(3)		ann.rad.	22%
	178	75											W k x-ray	47%
	178	75											0.1059 ± 0.0003	23%
	178	75											0.2373 ± 0.0003	45%
	178	75											0.7779 ± 0.0004	4%
	178	75											0.9391 ± 0.0005	9%
	178	75											0.9766 ± 0.0005	3%
	178	75											1.1108 ± 0.0004	2.7%
	178	75											1.1306 ± 0.0004	3.3%
	178	75											1.2553 ± 0.0004	1.4%
	178	75											1.2756 ± 0.0004	1.7%
	178	75											1.3115 ± 0.0002	1.2%
	178	75											1.4500 ± 0.0005	1.1%
	178	75											1.5984 ± 0.0004	1.3%
	178	75											1.5984 ± 0.0004	1.3%
	178	75											2.9576 ± 0.0005	0.9%
	178	75											3.1686 ± 0.0005	0.9%
$_{75}Re^{179}$	179	75		178.949960	19.7 m.	E.C.(99%) β+(1%)	2.69	0.95			(5/2+)		W k x-ray	56%
	179	75											0.1199 ± 0.0001	4.7%
	179	75											0.1891 ± 0.0001	7.3%
	179	75											0.2900 ± 0.0001	26%

TABLE OF THE ISOTOPES (Continued)

Isotope	A	Z	% Natural abundance	Atomic mass	Half-life	Decay mode	Decay energy (MeV)	Particle energy (MeV)	Particle intensity	Thermal neutron cross section	Spin (h/2π)	μ Nucl. mag. moment	Gamma-ray energy (MeV)	Gamma-ray intensity
	179	75											0.2963 ± 0.0001	8.7%
	179	75											0.3089 ± 0.0002	3.2%
	179	75											0.4018 ± 0.0001	7.0%
	179	75											0.4154 ± 0.0001	10%
	179	75											0.4302 ± 0.0001	27%
	179	75											0.4648 ± 0.0001	3.7%
	179	75											0.4773 ± 0.0001	9.0%
	179	75											0.4983 ± 0.0001	5.6%
	179	75											0.8326 ± 0.0001	2.9%
	179	75											1.3713 ± 0.0001	1.0%
	179	75											1.5604 ± 0.0001	3.1%
	179	75											1.6803 ± 0.0001	13%
	179	75											1.8087 ± 0.0001	2.2%
$_{75}Re^{180}$	180	75		179.950780	2.45 m.	E.C.(92%)	3.79	1.76			($^1/_2$-)		ann.rad.	16%
	180	75				β+(8%)							W k x-ray	45%
	180	75											0.1036 ± 0.0001	22%
	180	75											0.8254 ± 0.0001	9.8%
	180	75											0.9028 ± 0.0001	89%
	180	75											(0.07 - 2.2)weak	
$_{75}Re^{181}$	181	75		180.950020	20 h.	E.C.	1.70				5/2+	3.242	W k x-ray	61%
	181	75											0.1775 ± 0.0002	1.6%
	181	75											0.3186 ± 0.0003	1.1%
	181	75											0.3319 ± 0.0003	1.3%
	181	75											0.3561 ± 0.0003	1.7%
	181	75											0.3607 ± 0.0003	12%
	181	75											0.3655 ± 0.0003	56%
	181	75											0.5578 ± 0.0004	2.1%
	181	75											0.6390 ± 0.0004	6.4%
	181	75											0.6512 ± 0.0004	1.0%
	181	75											0.6618 ± 0.0004	3.0%
	181	75											0.8052 ± 0.0004	3.1%
	181	75											0.9074 ± 0.0005	1.0%
	181	75											0.9536 ± 0.0005	3.5%
	181	75											1.0002 ± 0.0005	3.3%
	181	75											1.0094 ± 0.0005	2.4%
	181	75											1.0756 ± 0.0005	1.0%
	181	75											1.4407 ± 0.0005	1.9%
$_{75}Re^{182m}$	182	75			12.7 h.	E.C.		0.55			2+		W k x-ray	52%
	182	75						1.74					0.0677 ± 0.0001	38%
	182	75											0.1004 ± 0.0001	14%
	182	75											0.1524 ± 0.0001	6.7%
	182	75											0.2293 ± 0.0001	2.1%
	182	75											0.4703 ± 0.0002	2.0%
	182	75											0.8949 ± 0.0002	2.1%
	182	75											1.1214 ± 0.0002	32%
	182	75											1.1892 ± 0.0002	15%
	182	75											1.2215 ± 0.0002	25%
	182	75											1.2311 ± 0.0002	1.3%
	182	75											1.2573 ± 0.0002	1.4%
	182	75											1.2893 ± 0.0002	1.2%
	182	75											(0.06 - 2.2)weak	
$_{75}Re^{182}$	182	75		181.951210	64 h.	E.C.	2.80				(7+)	0.399	W k x-ray	91%
	182	75											0.0678 ± 0.0001	25%
	182	75											0.1001 ± 0.0001	16%
	182	75											0.1137 ± 0.0001	5.3%
	182	75											0.1308 ± 0.0001	8.1%
	182	75											0.1338 ± 0.0001	2.6%
	182	75											0.1489 ± 0.0001	1.9%
	182	75											0.1524 ± 0.0001	9.2%
	182	75											0.1564 ± 0.0001	7.8%
	182	75											0.1692 ± 0.0001	12%
	182	75											0.1729 ± 0.0001	3.9%
	182	75											0.1785 ± 0.0001	2.5%
	182	75											0.1794 ± 0.0001	3.3%
	182	75											0.1914 ± 0.0001	7.3%
	182	75											0.1983 ± 0.0001	4.4%
	182	75											0.2143 ± 0.0001	1.2%
	182	75											0.2175 ± 0.0001	3.5%
	182	75											0.2216 ± 0.0001	7.0%
	182	75											0.2221 ± 0.0001	9.2%
	182	75											0.2262 ± 0.0001	3.3%
	182	75											0.2293 ± 0.0001	30%
	182	75											0.2475 ± 0.0001	5.5%
	182	75											0.2564 ± 0.0001	10%
	182	75											0.2641 ± 0.0001	3.9%
	182	75											0.2763 ± 0.0001	9.5%
	182	75											0.2814 ± 0.0001	6.2%
	182	75											0.2866 ± 0.0001	7.6%
	182	75											0.3391 ± 0.0001	6.0%
	182	75											0.3511 ± 0.0001	11%
	182	75											1.0017 ± 0.0001	2.7%
	182	75											1.0762 ± 0.0002	11%
	182	75											1.1133 ± 0.0001	5.1%
	182	75											1.1213 ± 0.0001	24%
	182	75											1.1890 ± 0.0001	9.8%

TABLE OF THE ISOTOPES (Continued)

Isotope	A	Z	% Natural abundance	Atomic mass	Half-life	Decay mode	Decay energy (MeV)	Particle energy (MeV)	Particle intensity	Thermal neutron cross section	Spin (h/2π)	μ Nucl. mag. moment	Gamma-ray energy (MeV)	Gamma-ray intensity
	182	75											1.2214 ± 0.0001	19%
	182	75											1.2310 ± 0.0001	16%
	182	75											1.3427 ± 0.0002	2.8%
	182	75											1.4273 ± 0.0002	11%
$_{75}Re^{183}$	183	75		182.950817	70 d.	E.C.	0.556				(5/2+)		W k x-ray	62%
	183	75											0.09908 ± 0.00001	2.7%
	183	75											0.10793 ± 0.00001	2.2%
	183	75											0.10972 ± 0.00001	2.9%
	183	75											0.16232 ± 0.00001	23%
	183	75											0.20880 ± 0.00001	3%
	183	75											0.2461 ± 0.0001	1.3%
	183	75											0.29172 ± 0.00001	3.2%
$_{75}Re^{184m}$	184	75			165 d.	I.T.(75%)	0.188				8+		Re k x-ray	24%
	184	75				E.C.(25%)							0.1047 ± 0.0001	13%
	184	75											0.16127 ± 0.00001	6.6%
	184	75											0.2165 ± 0.0001	9.6%
	184	75											0.31800 ± 0.00001	5.9%
	184	75											0.38425 ± 0.00001	3.2%
	184	75											0.53667 ± 0.00001	3.4%
	184	75											0.92093 ± 0.00002	8.3%
	184	75											(0.10 - 1.1)weak	
$_{75}Re^{184}$	184	75		183.952530	38 d.	E.C.	1.492			9 x 10^3 b.	3-	2.499	W k x-ray	45%
	184	75											0.1112 ± 0.0001	17%
	184	75											0.25284 ± 0.00001	3.0%
	184	75											0.6419 ± 0.0001	1.9%
	184	75											0.79207 ± 0.00002	37%
	184	75											0.8948 ± 0.0001	16%
	184	75											0.90328 ± 0.00003	38%
	184	75											(0.1 - 1.4)weak	
$_{75}Re^{185}$	185	75	37.40%	184.952951						111 b.	5/2+			
$_{75}Re^{186m}$	186	75			2.0 x 10^5 y.	I.T>	0.150						Re k x-ray	
	186	75											0.0590 ± 0.0001	18.6%
	186	75											0.0993 ± 0.0001	1.1%
$_{75}Re^{186}$	186	75		185.954984	β-(92%)	1.074	0.973	21%			1-	+1.739	W k x-ray	3%
	186	75				E.C.(8%)	0.585	1.07	71%				0.1227 ± 0.0001	0.7%
	186	75											0.1372 ± 0.0001	9%
	186	75											0.7675 ± 0.0001	0.03%
$_{75}Re^{187}$	187	75	62.6%	186.955744	4.5 x 10^{10}y.	β-	0.0025	0.0025		(2.8 + 75) b.	5/2+	+3.2197		
$_{75}Re^{188m}$	188	75			18.6 m.	I.T.	0.172				(6-)		Re k x-ray	31%
	188	75											0.0925 ± 0.0001	5.1%
	188	75											0.1059 ± 0.0001	11%
	188	75											0.1560 ± 0.0001	0.6%
	186	75											0.1695 ± 0.0001	0.1%
$_{75}Re^{188}$	188	75		187.958106	16.98 h.	β-	2.210	1.962	20%		1-	+1.788	Os k x-ray	2%
	188	75						2.118	79%				0.15502 ± 0.00002	15%
	188	75											0.47798 ± 0.00005	1.0%
	188	75											0.63312 ± 0.00004	1.2%
	188	75											0.82952 ± 0.00005	0.4%
	188	75											0.93141 ± 0.00006	0.6%
$_{75}Re^{189}$	189	75		188.959219	24 h.	β-	1.01	1.01			(5/2+)		0.1471 ± 0.0001	1.4%
	189	75											0.1854 ± 0.0001	2.1%
	189	75											0.2167 ± 0.0001	6.0%
	189	75											0.2194 ± 0.0001	5.0%
	189	75											0.2451 ± 0.0001	3.7%
	189	75											0.2759 ± 0.0001	0.34%
	189	75											0.5634 ± 0.0001	0.6%
$_{75}Re^{190m}$	190	75			3.0 h.	β-(51%)					(6-)		Re k x-ray	14%
	190	75				I.T.(49%)							0.1191 ± 0.0001	11%
	190	75											0.2238 ± 0.0001	14%
	190	75											0.2238 ± 0.0001	14%
	190	75											0.2829 ± 0.0001	2.4%
	190	75											0.2948 ± 0.0001	3.0%
	190	75											0.3902 ± 0.0001	5.3%
	190	75											0.4316 ± 0.0001	8.85%
	190	75											0.4908 ± 0.0001	3.8%
	190	75											0.5026 ± 0.0001	3.7%
	190	75											0.5186 ± 0.0001	7.3%

Isotope	A	Z	% Natural abundance	Atomic mass	Half-life	Decay mode	Decay energy (MeV)	Particle energy (MeV)	Particle intensity	Thermal neutron cross section	Spin (h/2π)	μ Nucl. mag. moment	Gamma-ray energy (MeV)	Gamma-ray intensity
	190	75											0.5587 ± 0.0005	5.9%
	190	75											0.6309 ± 0.0002	9.2%
	190	75											0.6731 ± 0.0001	10%
	190	75											0.7686 ± 0.0001	3.7%
	190	75											0.9582 ± 0.0001	2.2%
	190	75											(0.1 - 1.79)weak	
$_{75}Re^{190}$	190	75		189.961850	3.0 m.	β-	3.2	1.8			(2-)		Os k x-ray	6.8%
	190	75											0.1867 ± 0.0001	49%
	190	75											0.2238 ± 0.0001	25%
	190	75											0.3611 ± 0.0001	15%
	190	75											0.3712 ± 0.0001	21%
	190	75											0.3974 ± 0.0001	8.2%
	190	75											0.4072 ± 0.0001	15%
	190	75											0.4316 ± 0.0001	18%
	190	75											0.5580 ± 0.0001	29%
	190	75											0.5693 ± 0.0001	26%
	190	75											0.6051 ± 0.0001	66%
	190	75											0.6309 ± 0.0002	19%
	190	75											0.7686 ± 0.0001	2.9%
	190	75											0.8290 ± 0.0001	23%
	190	75											0.8391 ± 0.0001	7.8%
	190	75											1.2002 ± 0.0001	3.1%
	190	75											1.3870 ± 0.0002	1.3%
	190	75											1.4375 ± 0.0003	0.7%
	190	75											1.7945 ± 0.0003	0.5%
$_{75}Re^{191}$	191	75		190.963112	≈ 9.8 m.	β-	2.042	1.8						
$_{75}Re^{192}$	192	75		191.965870	16 s.	β-	4.10	≈ 2.5					0.2058 ± 0.0001	
	192	75											0.2832	
	192	75											0.4673	
	192	75											0.4890	
	192	75											0.7505	
Os		76								15 b.				
$_{76}Os^{166}$	166	76		165.972470	0.18 s.	β+, EC (28%)		6.27			0+		ann. rad.	
	166	76				α(72%)		5.98						
$_{76}Os^{167}$	167	76		166.971290	0.7 s.	β+, EC (76%)	8.11						ann.rad.	
	167	76				α(24%)		5.81						
$_{76}Os^{168}$	168	76		167.967670	2.2 s.	β+, EC (51%)	7.57				0+		ann. rad.	
	168	76				α(49%)								
$_{76}Os^{169}$	169	76		168.966850	3.3 s.	β+, EC (89%)	5.16						ann.rad.	
	169	76				α(17%)		5.57						
$_{76}Os^{170}$	170	76		169.963571	7.1 ± 0.2 s.	β+,E.C.	6.78				0+		ann.rad.	
	170	76				α		5.40						
$_{76}Os^{171}$	171	76		170.962890	7.9 ± 0.6 s.	β+, EC (98%)	6.78						ann.rad.	
	171	76				α(2%)		5.24						
$_{76}Os^{172}$	172	76		171.960000	19 ± 2 s.	β+, EC (99%)	4.43				0+		ann.rad.	
	172	76				α(1%)		5.10					0.177	100 +
	172	76											0.187	50
	172	76											0.276	25
	172	76											0.285	30
$_{76}Os^{173}$	173	76		172.957120	≈ 16 ± 5 s.	β+,E.C.	6.01						ann.rad.	
	173	76				α(0.02%)		4.94					ann.rad.	
$_{76}Os^{174}$	174	76		173.957120	44 ± 4 s.	β+,E.C.	3.67				0+		0.118 ± 0.001	100 +
	174	76				α(0.02%)		4.76					0.138 ± 0.001	25
	174	76											0.158 ± 0.001	
	174	76											0.302	26
	174	76											0.325	43
	174	76											0.372	20
	174	76											0.387	10
$_{76}Os^{175}$	175	76		174.956980	1.4 ± 0.1 m.	β+,.E.C.	5.17						0.125	100 +
	175	76											0.170	6
	175	76											0.181	11
	175	76											0.226	4
	175	76											0.248	9
	175	76											0.3.8	3
	175	76											0.410	5
$_{76}Os^{176}$	176	76		175.954880	3.6 ± 0.5 m.	β+,E.C.	3.15				0+		0.8155 ± 0.001	36 +
	176	76											0.7758 ± 0.0001	98
	176	76											0.8573 ± 0.00001	69
	176	76											1.2093 ± 0.0001	71
	176	76											1.2909 ± 0.0001	100
$_{76}Os^{177}$	177	76		176.954980	2.8 ± 0.3 m.	β+,E.C.	4.30				(1/2-)		0.0848 ± 0.0002	100 +
	177	76											0.1572 ± 0.0002	20
	177	76											0.1958 ± 0.0002	61
	177	76											0.3002 ± 0.0002	29
	177	76											0.4110 ± 0.0002	17

Isotope	A	Z	% Natural abundance	Atomic mass	Half-life	Decay mode	Decay energy (MeV)	Particle energy (MeV)	Particle intensity	Thermal neutron cross section	Spin (h/2π)	μ Nucl. mag. moment	Gamma-ray energy (MeV)	Gamma-ray intensity
	177	76											0.4570 ± 0.0002	21
	177	76											0.5394 ± 0.0002	22
	177	76											0.5762 ± 0.0003	13
	177	76											0.64492 ± 0.0004	10
	177	76											0.6861 ± 0.0001	12
	177	76											0.7333 ± 0.0003	26
	177	76											0.7914 ± 0.0004	13
	177	76											0.9524 ± 0.0005	18
	177	76											1.2686 ± 0.0006	33
	177	76											1.3368 ± 0.0006	11
	177	76											1.7439 ± 0.0006	12
$_{76}Os^{178}$	178	76		177.953250	5.0 ± 0.4 m.	β+,E.C.	2.24				0+		ann.rad.	
	178	76											0.3200 ± 0.001	4 +
	178	76											0.3508 ± 0.001	24
	178	76											0.5331 ± 0.001	52
	178	76											0.5518 ± 0.001	29
	178	76											0.5946 ± 0.0008	72
	178	76											0.6006 ± 0.001	41
	178	76											0.613 ± 0.001	22
	178	76											0.6325 ± 0.001	40
	178	76											0.6850 ± 0.0012	65
	178	76											0.9687 ± 0.0008	100
	178	76											1.3311 ± 0.0012	94
$_{76}Os^{179}$	179	76		178.953830	7 m.	β+,E.C.	3.61						ann.rad.	
	179	76											0.0654 ± 0.0001	100 +
	179	76											0.1657 ± 0.0002	7
	179	76											0.2186 ± 0.0002	17
	179	76											0.5328 ± 0.00041	10
	179	76											0.5938 ± 0.0003	16
	179	76											0.6334 ± 0.0005	5
	179	76											0.68947 ± 0.0005	11
	179	76											0.6975 ± 0.0005	5
	179	76											0.7453 ± 0.0005	6
	179	76											0.7508 ± 0.0005	5
	179	76											0.7594 ± 0.0003	8
	179	76											0.8177 ± 0.0003	6
	179	76											0.9684 ± 0.0003	14
	179	76											1.3110 ± 0.0004	10
	179	76											1.3303 ± 0.0004	13
	179	76											1.3642 ± 0.0005	3
	179	76											1.3835 ± 0.0005	4
	179	76											1.4295 ± 0.0005	3
	179	76											1.4488 ± 0.0005	3
$_{76}Os^{180}$	180	76		179.952390	21.7 ± 0.6 m.	β+,E.C.	1.510				0+		Re k x-ray	42%
	180	76											0.0202 ± 0.0001	17 +
	180	76											0.0316 ± 0.0002	
	180	76											0.0482 ± 0.0002	
	180	76											0.0499 ± 0.0002	
	180	76											0.0544 ± 0.0002	
	180	76											0.0746 ± 0.0002	
	180	76											0.1040 ± 0.0002	
	180	76											0.1070 ± 0.0002	
	180	76											0.1137 ± 0.0002	
	180	76											0.1838 ± 0.0002	
	180	76											0.2001 ± 0.0002	
	180	76											0.2182 ± 0.0002	
	180	76											0.2500 ± 0.0002	
	180	76											0.3194 ± 0.0002	
	180	76											0.3290 ± 0.0002	
	180	76											0.3491 ± 0.0002	
	180	76											0.4013 ± 0.0002	
	180	76											0.4857 ± 0.0002	
	180	76											0.6670 ± 0.0002	
	180	76											0.7174 ± 0.0002	
$_{76}Os^{181m}$	181	76			2.7 ± 0.1 m.	β+,E.C.	1.8				(7/2-)		0.11794 ± 0.00004	28 +
	181	76											0.14493 ± 0.00006	100
	181	76											1.1187 ± 0.0001	4.2
	181	76											1.4679 ± 0.001	1.3
$_{76}Os^{181}$	181	76		180.953270	1.75 ± 0.5 h.	E.C.	3.03				(1/2-)		ann.rad.	
	181	76											0.11794 ± 0.00004	55%
	181	76											0.16712 ± 0.00005	3.0%
	181	76											0.23868 ± 0.00001	44%
	181	76											0.24277 ± 0.00006	6.1%
	181	76											0.75120 ± 0.00002	3.2%
	181	76											0.7512 ± 0.0002	3.2%
	181	76											0.7590 ± 0.0002	2.4%

TABLE OF THE ISOTOPES (Continued)

Isotope	A	Z	% Natural abundance	Atomic mass	Half-life	Decay mode	Decay energy (MeV)	Particle energy (MeV)	Particle intensity	Thermal neutron cross section	Spin (h/2π)	μ Nucl. mag. moment	Gamma-ray energy (MeV)	Gamma-ray intensity
	181	76											0.7875 ± 0.0004	5.3%
	181	76											0.8267 ± 0.0002	20%
	181	76											0.9549 ± 0.0005	5.1%
	181	76											1.0603 ± 0.0002	5.7%
	181	76											1.1107 ± 0.0005	2.1%
	181	76											1.1814 ± 0.0008	1.0%
	181	76											1.3051 ± 0.0003	1.8%
	181	76											1.3465 ± 0.0005	1.1%
	181	76											1.3852 ± 0.0009	1.2%
	181	76											1.4923 ± 0.0004	1.0%
	181	76											1.5679 ± 0.0004	1.0%
	181	76											1.5726 ± 0.0003	1.1%
	181	76											1.7053 ± 0.0003	1.4%
	181	76											1.7397 ± 0.0003	1.2%
	181	76											1.9816 ± 0.0002	1.2%
	181	76											(0.07 - 2.64)many	
$_{76}Os^{182}$	182	76		181.952120	21.5 h.	E.C.	0.850				0+		Re k x-ray	43%
	182	76											0.1308 ± 0.0001	3.5%
	182	76											0.1802 ± 0.0001	37%
	182	76											0.2633 ± 0.0001	7.0%
	182	76											0.2743 ± 0.0001	1.8%
	182	76											0.5100 ± 0.0001	55%
$_{76}Os^{183m}$	183	76			9.9 h.	E.C.(84%)					1/2-		Os k x-ray	2.5%
	183	76				I.T.(16%)							Re k x-ray	34%
	183	76											0.4845 ± 0.0001	1.6%
	183	76											0.8784 ± 0.0005	1.6%
	183	76											0.9548 ± 0.0003	1.1%
	183	76											1.0347 ± 0.0003	6.5%
	183	76											1.1020 ± 0.0003	50%
	183	76											1.1080 ± 0.0003	23%
$_{76}Os^{183}$	183	76		182.953290	13 h.	E.C.	2.30				9/2+		Re k x-ray	72%
	183	76											0.1144 ± 0.0001	21%
	173	76											0.1679 ± 0.0001	7.7%
	183	76											0.2363 ± 0.0001	2.2%
	183	76											0.3818 ± 0.0001	77%
	183	76											0.8510 ± 0.0002	3.9%
	183	76											0.8876 ± 0.0003	1.1%
	183	76											0.8875 ± 0.0003	1.1%
	183	76											1.1633 ± 0.0004	1.2%
	183	76											1.4389 ± 0.0006	0.5%
$_{76}Os^{184}$	184	76	0.02%	183.952488						30 x 10^2 b.	0+			
$_{76}Os^{185}$	185	76		184.954041	93.6 d.	E.C.	1.015				1/2-		Re k x-ray	35%
	185	76											0.5921 ± 0.0001	1.3%
	185	76											0.6461 ± 0.0001	81%
	185	76											0.7174 ± 0.0001	4.1%
	185	76											0.8748 ± 0.0001	6.6%
	185	76											0.8805 ± 0.0001	5.0%
$_{76}Os^{186}$	186	76	1.58%	185.953830	2 x 10^{15} y.	α		≈ 2.75		80 b.	0+			
$_{76}Os^{187}$	187	76	1.6%	186.955741						320 b.	1/2-	+0.0646		
$_{76}Os^{188}$	188	76	13.3%	187.955860						5 b.	0+			
$_{76}Os^{189m}$	189	76			5.8 h.	I.T.	0.0308				9/2-		Os L x-ray	13%
	189	76											0.0308 ± 0.0001	0.0003%
$_{76}Os^{189}$	189	76	16.1%	188.958137						(0.0002 + 20) b.	3/2+	0.6599		
$_{76}Os^{190m}$	190	76			9.9 m.	I.T.	1.705				10-		Os k x-ray	9.8%
	190	76											0.1867 ± 0.0001	70%
	190	76											0.3611 ± 0.0001	95%
	190	76											0.5026 ± 0.0001	98%
	190	76											0.6161 ± 0.0002	98.5%
$_{76}Os^{190}$	190	76	26.4%	189.958436						(9 + 4) b.	0+			
$_{76}Os^{191m}$	191	76			13.1 h.	I.T.	0.0744				3/2-		Os k x-ray	4.1%
	191	76											0.0744 ± 0.0001	0.07%
$_{76}Os^{191}$	191	76		190.960920	15.4 d.	β-	0.313	0.140	100%		9/2-		Ir k x-ray	28%(D)
	191	76											0.1294 ± 0.0001	26%(D)
$_{76}Os^{192m}$	192	76			6.1 s.	I.T.	2.0154				(10-)		Os k x-ray	22%
	192	76											0.2058	69%
	192	76											0.2832	6.5%
	192	76											0.2924	4.2%
	192	76											0.3024	54%
	192	76											0.3068	5.8%
	192	76											0.3745	24%
	192	76											0.4204	6.1%
	192	76											0.4522	4.4%
	192	76											0.4531	59%
	192	76											0.4845	5.9%
	192	76											0.4890	15%
	192	76											0.5083	12%
	192	76											0.5632	12%
	192	76											0.5692	72%
	192	76											0.5883	4.1%
	192	76											0.6057	11%
	192	76											0.6195	11%
	192	76											0.6240	1.7%
$_{76}Os^{192}$	192	76	41.0%	191.961467						2.0 b.	0+			
$_{76}Os^{193}$	193	76		192.964138	β-	1.14	1.04	20%			3/2-	1.30	Ir k x-ray	69.5%
	193	76											0.1389 ± 0.0001	4.3%

Isotope	A	Z	% Natural abundance	Atomic mass	Half-life	Decay mode	Decay energy (MeV)	Particle energy (MeV)	Particle intensity	Thermal neutron cross section	Spin (h/2π)	μ Nucl. mag. moment	Gamma-ray energy (MeV)	Gamma-ray intensity
	193	76											0.2804 ± 0.0001	1.2%
	193	76											0.3216 ± 0.0001	1.3%
	193	76											0.3875 ± 0.0001	1.3%
	193	76											0.4605 ± 0.0001	3.9%
	193	76											0.5574 ± 0.0001	1%
$_{76}Os^{194}$	194	76		193.965173	6.0 y.	β-	0.097	0.054	33%		0+		Ir L x-ray	11%
	194	76						0.096	67%				0.0429 ± 0.0002	5.4%
$_{76}Os^{195}$	195	76		194.968110	6.5 m.	β-	≈ 2.0	≈ 2.0						
$_{76}Os^{196}$	196	76		195.969620	34.9 m.	β-	≈ 0.84	0.84			0+	0.1262	5%	
	196	76											0.2071 ± 0.0002	2.4%
	196	76											0.2578 ± 0.0002	2.3%
	196	76											0.3154 ± 0.0002	2.5%
	196	76											0.4079 ± 0.0002	5.9%
	196	76											0.6291 ± 0.0004	1.6%
$_{77}Ir^{170}$	170	77		169.974970	1.05 ± 0.1 s.	α		6.03						
$_{77}Ir^{171}$	171	77		170.971700	1.6 ± 0.1 s.	α		5.91						
$_{77}Ir^{172}$	172	77		171.970550	2.1 ± 0.1 s.	α		5.811						
$_{77}Ir^{173}$	173	77		172.967560	3.0 ± 0.1 s.	α		5.665						
$_{77}Ir^{174}$	174	77		173.966660	4 ± 1 s.	α		5.478						
$_{77}Ir^{175}$	175	77		174.964150	4.5 ± 1.0 s.	α		5.393						
$_{77}Ir^{176}$	176	77		175.963480	8 ± 1 s.	α		5.118						
$_{77}Ir^{177}$	177	77		176.961350	21 ± 2 s.	α		5.011						
$_{77}Ir^{178}$	178	77		177.961050	12 ± 2 s.	β+,E.C.	7.26						0.1320 ± 0.0005	71 +
	178	77											0.2667 ± 0.0003	100
	178	77											0.2700 ± 0.0001	5
	178	77											0.3633 ± 0.0004	35
	178	77											0.3987 ± 0.0004	13
	178	77											0.4329 ± 0.0005	5
	178	77											0.5329 ± 0.0005	5
	178	77											0.5469 ± 0.0005	6
	178	77											0.6250 ± 0.0005	15
	178	77											0.6395 ± 0.0004	13
	178	77											0.7002 ± 0.0004	8
	178	77											0.7329 ± 0.0004	5
	178	77											0.8649 ± 0.0004	9
	178	77											0.9000 ± 0.0004	13
	178	77											1.0176 ± 0.0005	5
													1.2015 ± 0.0004	7
$_{77}Ir^{179}$	179	77		178.959190	4 m.	E.C.	4.99							
$_{77}Ir^{180}$	180	77		179.959260	1.5 ± 0.1 m.	E.C.	6.40						0.1321 ± 0.0003	40%
	180	77											0.2765 ± 0.0003	42%
	180	77											0.4928 ± 0.0003	2.9%
	180	77											0.6141 ± 0.0004	2.1%
	180	77											0.6445 ± 0.0005	7.0%
	180	77											0.6990 ± 0.0005	9.4%
	180	77											0.7883 ± 0.0004	3.8%
	180	77											0.8463 ± 0.0005	2.4%
	180	77											0.8703 ± 0.0005	8.6%
	180	77											0.8905 ± 0.0004	9.1%
	180	77											0.9689 ± 0.0005	3.1%
	180	77											1.0143 ± 0.0005	1.1%
	180	77											1.0648 ± 0.0004	6.0%
	180	77											1.3306 ± 0.0005	4.4%
$_{77}Ir^{181}$	181	77		180.957640	4.9 ± 0.1 m.	β+,E.C.	4.07				(7/2+)		ann.rad.	
	181	77											0.0196 ± 0.0002	6 +
	181	77											0.0653 ± 0.0002	20
	181	77											0.0938 ± 0.0002	29
	181	77											0.1025 ± 0.0002	25
	181	77											0.1076 ± 0.0002	100
	181	77											0.1235 ± 0.0002	28
	181	77											0.1846 ± 0.0002	28
	181	77											0.2185 ± 0.0005	14
	181	77											0.2270 ± 0.0002	58
	181	77											0.2316 ± 0.0002	30
	181	77											0.3090 ± 0.0002	14
	181	77											0.3189 ± 0.0002	46
	181	77											0.3505 ± 0.0002	7
	181	77											0.3752 ± 0.0002	16
	181	77											0.5755 ± 0.0002	9
	181	77											0.7001 ± 0.0002	9
	181	77											1.1823 ± 0.0003	9
	181	77											1.1926 ± 0.0003	11
	181	77											1.3471 ± 0.0003	13
	181	77											1.3810 ± 0.0003	13
	181	77											1.5288 ± 0.0003	29
	181	77											1.5450 ± 0.0003	6
	181	77											1.5656 ± 0.0003	13
	181	77											1.5934 ± 0.0003	9
	181	77											1.6396 ± 0.0003	52
	181	77											1.6464 ± 0.0003	27
	181	77											1.6525 ± 0.0003	17
													1.7149 ± 0.0003	6
$_{77}Ir^{182}$	182	77		181.957970	15 m.	β+(44%)	5.45						ann.rad.	

Isotope	A	Z	% Natural abundance	Atomic mass	Half-life	Decay mode	Decay energy (MeV)	Particle energy (MeV)	Particle intensity	Thermal neutron cross section	Spin (h/2π)	μ Nucl. mag. moment	Gamma-ray energy (MeV)	Gamma-ray intensity
	182	77				E.C.(56%)							Os k x-ray	33%
	182	77											0.1273 ± 0.0003	35%
	182	77											0.2363 ± 0.0003	9.1%
	182	77											0.2370 ± 0.0002	43%
	182	77											0.3931 ± 0.0002	3%
	182	77											0.4000 ± 0.0003	3.1%
	182	77											0.7643 ± 0.0002	5.6%
	182	77											0.7901 ± 0.0003	3.2%
	182	77											0.8909 ± 0.0002	5.7%
	182	77											0.9123 ± 0.0002	8.8%
	182	77											1.0633 ± 0.0003	2.2%
	182	77											1.1180 ± 0.0006	2.6%
	182	77											1.2516 ± 0.0005	1.9%
	182	77											1.6520 ± 0.0006	2.5%
$_{77}Ir^{183}$	183	77		182.956710	56 m.	β+,E.C.	3.19						ann.rad.	
	183	77											0.0877 ± 0.0002	63 +
	183	77											0.1022 ± 0.0002	18
	183	77											0.1368 ± 0.0002	17
	183	77											0.1657 ± 0.0002	14
	183	77											0.1945 ± 0.0002	23
	183	77											0.2285 ± 0.0002	100
	183	77											0.2367 ± 0.0002	29
	183	77											0.2397 ± 0.0002	26
	183	77											0.2506 ± 0.0002	10
	183	77											0.2544 ± 0.0002	23
	183	77											0.2824 ± 0.0002	70
	183	77											0.3144 ± 0.0002	11
	183	77											0.3422 ± 0.0002	29
	183	77											0.3477 ± 0.0002	29
	183	77											0.4122 ± 0.0002	19
	183	77											0.4577 ± 0.0002	6
	183	77											0.4619 ± 0.0002	5
	183	77											0.4984 ± 0.0002	15
	183	77											0.6174 ± 0.0002	8
	183	77											0.6551 ± 0.0002	20
	183	77											0.6708 ± 0.0002	10
	183	77											0.6922 ± 0.0002	33
	183	77											0.7061 ± 0.0002	5
	183	77											0.7248 ± 0.0002	5
	183	77											0.8001 ± 0.0002	33
	183	77											0.8966 ± 0.0002	18
$_{77}Ir^{184}$	184	77		183.957560	3.0 h.	β+(12%)	4.72	2.3			5		ann.rad.	24%
	184	77				E.C.(88%)		2.9					Os k x-ray	48%
	184	77						3.3					0.11968 ± 0.0001	30%
	184	77											0.2640 ± 0.0001	67%
	184	77											0.3904 ± 0.0001	26%
	184	77											0.4931 ± 0.0001	5.8%
	184	77											0.5029 ± 0.0002	2.9%
	184	77											0.5397 ± 0.0001	6.7%
	184	77											0.6012 ± 0.0001	3.2%
	184	77											0.6266 ± 0.0001	2.4%
	184	77											0.8239 ± 0.0001	3.8%
	184	77											0.8413 ± 0.0002	7.9%
	184	77											0.9429 ± 0.0002	3.6%
	184	77											0.9441 ± 0.0002	2.7%
	184	77											0.9613 ± 0.0002	12%
	184	77											1.0445 ± 0.0002	5.3%
	184	77											1.0622 ± 0.0002	2.9%
	184	77											1.1053 ± 0.0002	5.3%
	184	77											1.0622 ± 0.0003	2.9%
	184	77											1.1053 ± 0.0002	5.3%
	184	77											1.2369 ± 0.0001	2.1%
	184	77											1.2478 ± 0.0001	2.6%
	184	77											1.3343 ± 0.0003	2.3%
	184	77											1.4579 ± 0.0002	1.4%
	184	77											1.6725 ± 0.0003	3.7%
	184	77											2.0630 ± 0.0004	4.5%
	184	77											2.2430 ± 0.0006	0.9%
$_{77}Ir^{185}$	185	77		184.956730	14 h.	β+(3%)					(5/2-)		ann.rad.	6%
	185	77				E.C.(97%)							Os k x-ray	55%
	185	77											0.0974 ± 0.0002	4.1%
	185	77											0.1007 ± 0.0002	2.4%
	185	77											0.1536 ± 0.0002	2.0%
	185	77											0.1582 ± 0.0002	2.4%
	185	77											0.2238 ± 0.0002	2.1%
	185	77											0.2543 ± 0.0002	13%
	185	77											0.5392 ± 0.0002	1.3%
	185	77											0.6462 ± 0.0002	1.2%
	185	77											1.6418 ± 0.0005	1.1%
	185	77											1.6683 ± 0.0005	3.6%
	185	77											1.7322 ± 0.0005	2.7%
	185	77											1.7384 ± 0.0005	2.4%
	185	77											1.8288 ± 0.0005	9.8%
	185	77											1.8700 ± 0.0005	1.2%
$_{77}Ir^{186m}$	186	77			1.7 h.	E.C.					(2-)		Os k x-ray	45%
	186	77											0.1371 ± 0.0001	30%

TABLE OF THE ISOTOPES (Continued)

Isotope	A	Z	% Natural abundance	Atomic mass	Half-life	Decay mode	Decay energy (MeV)	Particle energy (MeV)	Particle intensity	Thermal neutron cross section	Spin (h/2π)	μ Nucl. mag. moment	Gamma-ray energy (MeV)	Gamma-ray intensity
	186	77											0.2969 ± 0.0001	11%
	186	77											0.6303 ± 0.0001	21%
	186	77											0.6363 ± 0.0001	2.1%
	186	77											0.7126 ± 0.0001	4.4%
	186	77											0.7675 ± 0.0001	24%
	186	77											0.7732 ± 0.0001	16%
	186	77											0.7832 ± 0.0001	2.3%
	186	77											0.8441 ± 0.0001	2.7%
	186	77											0.9334 ± 0.0001	2.3%
	186	77											0.9380 ± 0.0001	2.6%
	186	77											0.9870 ± 0.0001	12.6%
	186	77											1.0463 ± 0.0001	1.2%
	186	77											1.6172 ± 0.0002	4.8%
	186	77											1.7111 ± 0.0002	2.3%
	186	77											1.7544 ± 0.0003	5.3%
	186	77											2.1870 ± 0.0002	3.4%
	186	77											2.2241 ± 0.0004	1.2%
$_{77}Ir^{186}$	186	77		185.957943	15.7 h.	E.C.(98%) β+(2%)	3.83				(5+)		Os k x-ray	48%
	186	77											0.1372 ± 0.0001	41%
	186	77											0.2968 ± 0.0001	62%
	186	77											0.3577 ± 0.0001	1.9%
	186	77											0.4348 ± 0.0001	34%
	186	77											0.5844 ± 0.0001	5.4%
	186	77											0.6303 ± 0.0001	4.9%
	186	77											0.6362 ± 0.0001	6.9%
	186	77											0.7673 ± 0.0001	5.3%
	186	77											0.7731 ± 0.0001	8.8%
	186	77											0.8413 ± 0.0001	5.1%
	186	77											0.8466 ± 0.0002	6.3%
	186	77											0.9332 ± 0.0001	5.3%
	186	77											0.9436 ± 0.0001	8.6%
	186	77											1.0571 ± 0.0001	3.1%
	186	77											1.1879 ± 0.0001	2.0%
	186	77											1.3144 ± 0.0001	2.0%
	186	77											1.6474 ± 0.0001	4.7%
	186	77											1.7010 ± 0.0001	2.1%
	186	77											2.2420 ± 0.0002	1.4%
	186	77											2.8352 ± 0.0003	0.8%
	186	77											(0.13 - 3.0)weak	
$_{77}Ir^{187}$	187	77		186.957350	10.5 h.	E.C.	1.500				3/2+		Os k x-ray	65%
	187	77											0.0743 ± 0.0001	4.6%
	187	77											0.1777 ± 0.0001	2.8%
	187	77											0.1874 ± 0.0001	1.8%
	187	77											0.4009 ± 0.0003	4.1%
	187	77											0.4271 ± 0.0002	4.4%
	187	77											0.4917 ± 0.0001	1.4%
	187	77											0.5015 ± 0.0001	1.6%
	187	77											0.6109 ± 0.0001	4.0%
	187	77											0.7997 ± 0.0002	0.9%
	187	77											0.9128 ± 0.0001	5.0%
	187	77											0.9774 ± 0.0002	3.1%
	187	77											0.9873 ± 0.0002	2.8%
$_{77}Ir^{188}$	188	77		187.958830	41.4 h.	β+	2.79		1.13		(2-)		Os k x-ray	44%
	188	77				E.C. (99+%)		1.64					0.1550 ± 0.0001	30%
	188	77											0.4780 ± 0.0001	15%
	188	77											0.6330 ± 0.0001	18%
	188	77											0.6349 ± 0.0002	5.0%
	188	77											0.8294 ± 0.0001	5.2%
	188	77											1.2098 ± 0.0001	7.0%
	188	77											1.4354 ± 0.0002	1.5%
	188	77											1.4572 ± 0.0002	1.8%
	188	77											1.4652 ± 0.0002	1.3%
	188	77											1.5745 ± 0.0002	2.6%
	188	77											1.7157 ± 0.0002	6.2%
	188	77											1.9441 ± 0.0002	3.9%
	188	77											2.0498 ± 0.0002	5.0%
	188	77											2.0596 ± 0.0004	7.0%
	188	77											2.0969 ± 0.0004	5.7%
	188	77											2.0991 ± 0.0004	4.8%
	188	77											2.1937 ± 0.0004	2.4%
	188	77											2.2146 ± 0.0002	19%
$_{77}Ir^{189}$	189	77		188.958712	13.2 d.	E.C.	0.535				3/2+		Os k x-ray	38%
	189	77											0.0952 ± 0.0001	0.4%
	189	77											0.1859 ± 0.0001	0.2%
	189	77											0.1974 ± 0.0001	0.3%
	189	77											0.2167 ± 0.0001	0.5%
	189	77											0.2194 ± 0.0001	0.5%
	189	77											0.2335 ± 0.0001	0.3%
	189	77											0.2449 ± 0.0001	6.0%
	189	77											0.2758 ± 0.0001	0.5%
$_{77}Ir^{190m2}$	190	77			3.2 h.	β+, EC (95%)					(11-)			
	190	77				I.T.(5%)								
$_{77}Ir^{190m1}$	190	77			1.2 h.	I.T.	0.0263						Ir L x-ray	14%

TABLE OF THE ISOTOPES (Continued)

Isotope	A	Z	% Natural abundance	Atomic mass	Half-life	Decay mode	Decay energy (MeV)	Particle energy (MeV)	Particle intensity	Thermal neutron cross section	Spin (h/2π)	μ Nucl. mag. moment	Gamma-ray energy (MeV)	Gamma-ray intensity
$_{77}Ir^{190}$	190	77			11.8 d.	E.C.	2.0				(4+)		Os k x-ray	45%
	190	77											0.1867 ± 0.0001	48%
	190	77											0.1969 ± 0.0002	2.5%
	190	77											0.2338 ± 0.0001	3.6%
	190	77											0.2948 ± 0.0001	6.2%
	190	77											0.3611 ± 0.0001	13%
	190	77											0.3712 ± 0.0001	22%
	190	77											0.3800 ± 0.0001	2.0%
	190	77											0.3974 ± 0.0001	6.3%
	190	77											0.4072 ± 0.0001	27%
	190	77											0.4206 ± 0.0001	1.6%
	190	77											0.4316 ± 0.0001	2.6%
	190	77											0.4478 ± 0.0001	2.5%
	190	77											0.5186 ± 0.0001	33%
	190	77											0.5580 ± 0.0001	29%
	190	77											0.6051 ± 0.0001	38%
	190	77											0.6309 ± 0.0001	2.8%
	190	77											0.7262 ± 0.0001	3.6%
	190	77											0.7686 ± 0.0001	2.1%
	190	77											0.8290 ± 0.0001	3.3%
	190	77											1.0360 ± 0.0002	2.3%
	190	77											(0.2 - 1.4)weak	
$_{77}Ir^{191m}$	191	77			4.93 s.	I.T.	0.1714				11/2-		Ir k x-ray	28%
	191	77											0.0824 ± 0.0001	0.02%
	191	77											0.1294 ± 0.0001	25.7%
$_{77}Ir^{191}$	191	77	37.3%	190.960584						(0.2 + 310) b.	3/2+	+0.1461		
$_{77}Ir^{192m2}$	192	77			241 y.	I.T.	0.161				(9+)		Ir k x-ray	
$_{77}Ir^{192m1}$	192	77			1.44 m.	I.T.	0.0580				(1+)		Ir L x-ray	11%
	192	77											0.0580 ± 0.0004	0.04%
	192	77											0.2959 ± 0.0001	0.002%
	192	77											0.3165 ± 0.0001	0.01%
$_{77}Ir^{192}$	192	77		191.962580	73.83 d.	β-	1.454			14 x 10² b.	(4-)	+1.880	Pt k x-ray	5%
	192	77											0.20577 ± 0.00001	3.2%
	192	77											0.29595 ± 0.00001	29%
	192	77											0.30844 ± 0.00001	30%
	192	77											0.31649 ± 0.00001	83%
	192	77											0.46806 ± 0.00001	48%
	192	77											0.48457 ± 0.00001	3.2%
	192	77											0.58857 ± 0.00001	4.6%
	192	77											0.60440 ± 0.00001	8.4%
	192	77											0.88452 ± 0.00002	0.3%
	192	77											1.06148 ± 0.00004	0.05%
$_{77}Ir^{193m}$	193	77			10.6 d.	I.T.	0.0802				11/2-		Ir L x-ray	12%
	193	77											0.0803 ± 0.0001	0.005%
$_{77}Ir^{193}$	193	77	62.7%	192.962917						111 b.	3/2+	+0.1591		
$_{77}Ir^{194m}$	194	77			171 d.	β-					11		Pt k x-ray	8%
	194	77											0.1117 ± 0.0005	8.9%
	194	77											0.3284 ± 0.0001	93%
	194	77											0.3388 ± 0.0005	55%
	194	77											0.3908 ± 0.0005	35%
	194	77											0.4829 ± 0.0001	97%
	194	77											0.5624 ± 0.0005	70%
	194	77											0.6005 ± 0.0005	62%
	194	77											0.6878 ± 0.0005	59%
	194	77											1.0118 ± 0.0005	3.6%
$_{77}Ir^{194}$	194	77		193.965069	19.2 h.	β-	2.248	1.92	9%		1-	0.37	0.2935 ± 0.0001	2.5%
	194	77						2.25	86%				0.3284 ± 0.0001	13%
	194	77											0.6451 ± 0.0001	1.2%
	194	77											0.9387 ± 0.0001	0.6%
	194	77											1.1508 ± 0.0001	0.6%
	194	77											1.1835 ± 0.0001	0.3%
	194	77											1.4689 ± 0.0001	0.2%
	194	77											(0.1 - 2.2)weak	
$_{77}Ir^{195m}$	195	77			3.9 h.	β-		0.41			(11/2-		Pt k x-ray	36%
	195	77						0.97					0.0989 ± 0.0001	10%
	195	77											0.1297 ± 0.0001	1.7%
	195	77											0.1728 ± 0.0001	5.0%
	195	77											0.2018 ± 0.0002	1.4%
	195	77											0.2113 ± 0.0001	2.2%
	195	77											0.2392 ± 0.0001	1.8%
	195	77											0.2516 ± 0.0001	1.8%
	195	77											0.2878 ± 0.0002	1.0%
	195	77											0.2903 ± 0.0002	1.9%
	195	77											0.3065 ± 0.0001	2.2%
	195	77											0.3199 ± 0.0001	9.6%
	195	77											0.3564 ± 0.0002	1.8%

Isotope	A	Z	% Natural abundance	Atomic mass	Half-life	Decay mode	Decay energy (MeV)	Particle energy (MeV)	Particle intensity	Thermal neutron cross section	Spin (h/2π)	μ Nucl. mag. moment	Gamma-ray energy (MeV)	Gamma-ray intensity
	195	77											0.3593 ± 0.0002	4.6%
	195	77											0.3649 ± 0.0001	9.5%
	195	77											0.4090 ± 0.0001	1.4%
	195	77											0.4329 ± 0.0001	9.6%
	195	77											0.4812 ± 0.0001	2.7%
	195	77											0.5754 ± 0.0001	1.5%
	195	77											0.6849 ± 0.0001	9.6%
	195	77											0.8009 ± 0.0001	1.0%
$_{77}Ir^{195}$	195	77		194.965966	2.8 h.	β-	1.118	1.0	80%		(3/2+)		Pt k x-ray	28%
	195	77						1.11	13%				0.0989 ± 0.0001	9.7%
	195	77											0.1297 ± 0.0001	1.4%
	195	77											0.2113 ± 0.0001	1.5%
$_{77}Ir^{196m}$	196	77			1.40 h.	β-		1.16					Pt k x-ray	10%
	196	77											0.1033 ± 0.0002	16%
	196	77											0.3557 ± 0.0002	94%
	196	77											0.3935 ± 0.0002	97%
	196	77											0.4209 ± 0.0003	2.5%
	196	77											0.4471 ± 0.0002	94%
	196	77											0.5214 ± 0.0002	96%
	196	77											0.6335 ± 0.0003	1.1%
	196	77											0.6473 ± 0.0002	91%
	196	77											0.6939 ± 0.0002	4.2%
	196	77											0.7273 ± 0.0002	2.6%
	196	77											0.8356 ± 0.0002	6.3%
	196	77											1.4825 ± 0.0004	2.3%
$_{77}Ir^{196}$	196	77		195.968370	52 s.	β-	3.210	2.1	15%		0-		0.3329 ± 0.0002	4.3%
	196	77						3.2	80%				0.3557 ± 0.0002	19%
	196	77											0.4468 ± 0.0002	4.5%
	196	77											0.7796 ± 0.0002	10%
	196	77											1.0470 ± 0.0002	1%
	196	77											1.4684 ± 0.0002	0.8%
	196	77											1.5642 ± 0.0002	0.9%
$_{77}Ir^{197m}$	197	77			8.9 m.	β-					(11/2-		0.3465	(D)
	197	77				I.T.							See Ir^{197}	
$_{77}Ir^{197}$	197	77		196.969629	5.85 m.	β-	2.156	1.5			(3/2+)		0.0531 ± 0.0001	9 +
	197	77						2.0					0.1351 ± 0.0001	27
	197	77											0.2689 ± 0.0001	13
	197	77											0.2996 ± 0.0001	24
	197	77											0.3783 ± 0.0001	38
	197	77											0.4306 ± 0.0001	61
	197	77											0.4568 ± 0.0002	37
	197	77											0.4697 ± 0.0001	100
	197	77											0.4964 ± 0.0004	36
	197	77											0.5091 ± 0.0003	21
	197	77											0.5272 ± 0.0001	24
	197	77											0.5392 ± 0.0001	14
	197	77											0.5420 ± 0.0001	11
	197	77											0.7153 ± 0.0001	9
	197	77											0.8091 ± 0.0001	32
	197	77											0.8159 ± 0.0001	45
	197	77											0.8664 ± 0.0001	13
	197	77											0.9394 ± 0.0001	21
	197	77											0.9871 ± 0.0001	15
	197	77											1.3432 ± 0.0001	21
$_{77}Ir^{198}$	198	77		197.972160	8 s.	β-	4.00						0.4074 ± 0.0003	100 +
	198	77											0.5070 ± 0.0003	76
Pt		78		195.08						10 b.				
$_{78}Pt^{172}$	172	78		171.977220	0.10 s.	α		6.31			0+			
$_{78}Pt^{173}$	173	78		172.976280	0.34 s.	β+,E.C.	8.12							
	173	78				α		6.20						
$_{78}Pt^{174}$	174	78		173.972811	0.90 ± 0.01 s.	β+, EC (17%)	5.73				0+			
	174	78				α(83%)		6.040						
$_{78}Pt^{175}$	175	78		174.972130	2.52 ± 0.08 s.	β+, EC (65%)	7.43						0.0774 ± 0.0008	
	175	78				α(35%)		5.831	5%				0.1354 ± 0.0008	
	175	78						5.96	54%				0.2128 ± 0.0008	
	175	78						6.038	5%					
$_{78}Pt^{176}$	176	78		175.968930	6.3 ± 0.1 s.	β+, EC (60%)	5.08				0+		ann.rad.	
	176	78				α(40%)		5.528	0.6%				0.2277	
	176	78						5.750	41%					
$_{78}Pt^{177}$	177	78		176.968360	11 ± 2 s.	E.C.(91%)	6.	53					0.0908	
	177	78				α(9%)		5.485	3%					
	177	78						5.525	6%					
$_{78}Pt^{178}$	178	78		177.965700	21.0 ± 0.7 s.	E.C.(93%)	4.34				0+			
	178	78				α(7%)		5.286	0.2%					
	178	78						5.442	7%					
$_{78}Pt^{179}$	179	78		178.965270	≈ 43 s.	β+,E.C.	5.66							
	179	78				α		5.16						
$_{78}Pt^{180}$	180	78		179.963130	52 ± 3 s.	β+, EC (99.7%	3.61				0+			
	180	78				α(0.3%)		5.140						
$_{78}Pt^{181}$	181	78		180.963100	51 ± 5 s.	β+,E.C.	5.08							

Isotope	A	Z	% Natural abundance	Atomic mass	Half-life	Decay mode	Decay energy (MeV)	Particle energy (MeV)	Particle intensity	Thermal neutron cross section	Spin (h/2π)	μ Nucl. mag. moment	Gamma-ray energy (MeV)	Gamma-ray intensity
$_{78}Pt^{182}$	182	78		181.961160	2.7 m.	β+,E.C.	2.97				0+		ann.rad.	
	182	78											0.1360 ± 0.0002	100 +
	182	78											0.1460 ± 0.0015	15
	182	78											0.1860 ± 0.0015	7
	182	78											0.2100 ± 0.0015	12
$_{78}Pt^{183m}$	183	78		182.961630	43 s.	β+,E.C.					(7/2-)		ann.rad.	
	183	78				I.T.							0.3132 ± 0.0003	28 +
	183	78											0.3164 ± 0.0003	53
	183	78											0.3290 ± 0.0003	36
	183	78											0.6296 ± 0.0003	100
	183	78											0.6453 ± 0.0003	23
$_{78}Pt^{183}$	183	78		182.961630	7 m.	β+,E.C.	4.58						ann.rad.	
$_{78}Pt^{184}$	184	78		183.959920	17.3 m.	β+,E.C.	2.20						ann.rad.	
	184	78											0.0926 ± 0.0003	16 +
	184	78											0.1170 ± 0.0004	7
	184	78											0.1395 ± 0.0003	6
	184	78											0.1445 ± 0.0004	4
	184	78											0.1495 ± 0.0006	4
	184	78											0.1549 ± 0.0003	100
	184	78											0.1616 ± 0.0004	6
	184	78											0.1829 ± 0.0004	8
	184	78											0.1919 ± 0.0003	94
	184	78											0.2093 ± 0.0004	6
	184	78											0.2165 ± 0.0003	17
	184	78											0.3946 ± 0.0003	16
	184	78											0.5484 ± 0.0003	77
	184	78											0.6107 ± 0.0009	12
	184	78											0.7312 ± 0.0004	43
$_{78}Pt^{185m}$	185	78			33 m.	β+,E.C.					1/2-			
$_{78}Pt^{185}$	185	78		184.960700	71 m.	β+,E.C.	3.700				(9/2+)		ann.rad.	
	185	78											0.0857 ± 0.0001	4 +
	185	78											0.1056 ± 0.0001	6
	185	78											0.1198 ± 0.0001	15
	185	78											0.1353 ± 0.0001	80
	185	78											0.1974 ± 0.0001	74
	185	78											0.2068 ± 0.0002	6
	185	78											0.2126 ± 0.0001	12
	185	78											0.2296 ± 0.0001	100
	185	78											0.2430 ± 0.0002	6
	185	78											0.2512 ± 0.0003	8
	185	78											0.2551 ± 0.0002	51
	185	78											0.2644 ± 0.0002	8
	185	78											0.2943 ± 0.0001	7
	185	78											0.3001 ± 0.0002	8
	185	78											0.3354 ± 0.0002	13
	185	78											0.3845 ± 0.0002	15
	185	78											0.4188 ± 0.0002	6
	185	78											0.4598 ± 0.0002	7
	185	78											0.4650 ± 0.0002	25
	185	78											0.5849 ± 0.0002	17
	185	78											0.6408 ± 0.0002	7
	185	78											0.7205 ± 0.0002	20
	185	78											0.7353 ± 0.0002	8
	185	78											0.8376 ± 0.0003	7
	185	78											0.8952 ± 0.0003	7
	185	78											0.9625 ± 0.0004	5
	185	78											1.2928 ± 0.0004	4
	185	78											1.3958 ± 0.0004	4
$_{78}Pt^{186}$	186	78		185.959360	2.0 h.	β+,E.C.	2.320				0+		ann.rad.	
	186	78											0.1805 ± 0.0004	1.7 +
	186	78											0.2808 ± 0.0004	2.4
	186	78											0.3667 ± 0.0004	3.3
	186	78											0.6115 ± 0.0004	8.6
	186	78											0.6892 ± 0.0003	100
$_{78}Pt^{187}$	187	78		186.960470	2.35 h.	β+,E.C.		2.90			3/2		ann.rad.	
	187	78											Ir k x-ray	61%
	187	78											0.1064 ± 0.0001	8%
	187	78											0.1100 ± 0.0001	5%
	187	78											0.1220 ± 0.0001	2.8%
	187	78											0.1869 ± 0.0001	3.5%
	187	78											0.2015 ± 0.0003	6.8%
	187	78											0.2476 ± 0.0001	3.5%
	187	78											0.2849 ± 0.0001	5.2%
	187	78											0.3048 ± 0.0001	4.1%
	187	78											0.4272 ± 0.0001	2.1%
	187	78											0.6296 ± 0.0001	2.6%
	187	78											0.7092 ± 0.0001	4.9%
	187	78											0.8193 ± 0.0002	3.5%
	187	78											0.9127 ± 0.0003	1.7%
	187	78											1.1455 ± 0.0003	1.6%
	187	78											1.1570 ± 0.0007	1.1%
	187	78											1.2554 ± 0.0003	2.2%
$_{78}Pt^{188}$	188	78		187.959386	10.2 d.	E.C.	0.518				0+		Ir k x-ray	49%
	188	78											0.1876 ± 0.0001	19%
	188	78											0.1951 ± 0.0001	19%

Isotope	A	Z	% Natural abundance	Atomic mass	Half-life	Decay mode	Decay energy (MeV)	Particle energy (MeV)	Particle intensity	Thermal neutron cross section	Spin (h/2π)	μ Nucl. mag. moment	Gamma-ray energy (MeV)	Gamma-ray intensity
	188	78											0.3814 ± 0.0001	7.5%
	188	78											0.4233 ± 0.0001	4.4%
$_{78}Pt^{189}$	189	78		188.960817	10.9 h.	β+,E.C.	1.961						Ir k x-ray	55%
	189	78											0.0943 ± 0.0001	4.7%
	189	78											0.1411 ± 0.0001	2.6%
	189	78											0.1867 ± 0.0001	1.4%
	189	78											0.2435 ± 0.0001	4.4%
	189	78											0.3005 ± 0.0001	2.3%
	189	78											0.3177 ± 0.0001	2.0%
	189	78											0.5449 ± 0.0001	3.6%
	189	78											0.5688 ± 0.0001	4.4%
	189	78											0.6076 ± 0.0001	5.1%
	189	78											0.6271 ± 0.0001	1.5%
	189	78											0.7214 ± 0.0001	5.8%
	189	78											(0.09 - 1.47)weak	
$_{78}Pt^{190}$	190	78	0.01%	189.959917						800 b.	0+			
$_{78}Pt^{191}$	191	78		190.961665	2.96 d.	E.C.	1.021				(3/2-)		Ir k x-ray	66%
	191	78											0.0824 ± 0.0001	4.9%
	191	78											0.0965 ± 0.0001	3.3%
	191	78											0.1294 ± 0.0001	3.2%
	191	78											0.1722 ± 0.0001	3.5%
	191	78											0.3687 ± 0.0001	1.6%
	191	78											0.3512 ± 0.0001	3.4%
	191	78											0.3599 ± 0.0001	6.0%
	191	78											0.4094 ± 0.0001	8.0%
	191	78											0.4565 ± 0.0001	3.4%
	191	78											0.5389 ± 0.0001	13.7%
	191	78											0.6241 ± 0.0001	1.4%
$_{78}Pt^{192}$	192	78	0.79%	191.961019						(1 + 9) b.	0+			
$_{78}Pt^{193m}$	193	78			4.33 d.	I.T.	0.1498				13/2+		Pt k x-ray	7%
	193	78											0.1355 ± 0.0001	0.11%
$_{78}Pt^{193}$	193	78		192.962977	50 y.	E.C.	0.057				(1/2-)		Ir k x-rays	
$_{78}Pt^{194}$	194	78	32.9%	193.962655						(0.1 + 1.1) b.	0+			
$_{78}Pt^{195m}$	195	78			4.02 d.	I.T.	0.2952				13/2+	0.597	Pt k x-ray	39%
	195	78											0.0989 ± 0.0001	11%
	195	78											0.1297 ± 0.0001	2.8%
$_{78}Pt^{195}$	195	78	33.8%	194.964766						29 b.	1/2-	+0.6095		
$_{78}Pt^{196}$	196	78	25.3%	195.964926						(0.05 + 0.7) b.	0+			
$_{78}Pt^{197m}$	197	78			95.4 m.	I.T.(97%)					13/2+		Pt k x-ray	24%
	197	78				β-(3%)							0.0530 ± 0.0001	1.1%
	197	78											0.3465 ± 0.0002	11%
$_{78}Pt^{197}$	197	78		196.967315	18.3 h.	β-	0.719				1/2-	0.51	Au k x-ray	17%
	197	78											0.1914 ± 0.0001	3.7%
	197	78											0.2688 ± 0.0001	0.2%
$_{78}Pt^{198}$	198	78	7.2%	197.967869						(0.3 + 3.5) b.	0+			
$_{78}Pt^{199m}$	199	78			13.5 s.	I.T.	0.424				13/2+		Pt k x-ray	3.4%
	199	78											0.3919 ± 0.0001	85%
$_{78}Pt^{199}$	199	78		198.970552	30.8 m.	β-	1.688	0.90	18%		(5/2-)		0.0772 ± 0.0001	1.5%
	199	78						1.14	14%				0.18579 ± 0.00002	3.3%
	199	78											0.19169 ± 0.00003	2.4%
	199	78											0.24646 ± 0.00003	2.2%
	199	78											0.31703 ± 0.00003	4.9%
	199	78											0.4681 ± 0.0001	1.0%
	199	78											0.4747 ± 0.0001	1.1%
	199	78											0.49375 ± 0.00003	5.7%
	199	78											0.54298 ± 0.00004	15%
	199	78											0.71455 ± 0.00003	1.9%
	199	78											0.96831 ± 0.00004	1.1%
$_{78}Pt^{200}$	200	78		199.971417	12.5 ± 0.3 h.	β-	0.690				0+		Au k x-ray	10%
	200	78											0.13590 ± 0.00009	3.1%
	200	78											0.20004 ± 0.00004	0.6%
	200	78											0.22747 ± 0.00004	2.0%
	200	78											0.24371 ± 0.00003	2.4%
	200	78											0.33024 ± 0.00003	1.1%
$_{78}Pt^{201}$	201	78		200.974500	2.5 ± 0.1 m.	β-	2.660				(5/2-)		0.070	
	201	78											0.152	
	201	78											0.222	
	201	78											1.760	
Au		79		196.9665						98.7 b.				
$_{79}Au^{176}$	176	79		175.980060	1.3 ± 0.3 s.	β+,E.C.	10.37							

Isotope	A	Z	% Natural abundance	Atomic mass	Half-life	Decay mode	Decay energy (MeV)	Particle energy (MeV)	Particle intensity	Thermal neutron cross section	Spin (h/2π)	μ Nucl. mag. moment	Gamma-ray energy (MeV)	Gamma-ray intensity
	176	79				α		6.260	80 +					
	176	79						6.290	20					
$_{79}Au^{177}$	177	79		176.976920	1.3 ± 0.4 s.	α		6.115						
	176	79						6.150						
$_{79}Au^{178}$	178	79		177.975760	2.6 ± 0.5 s.	α		5.920						
$_{79}Au^{179}$	179	79		178.973170	7.5 s.	α		5.85						
$_{79}Au^{180}$	180	79		179.972310	8.1 ± 0.3 s.	E.C.	8.55						0.1522 ± 0.0003	100 +
	180	79											0.2564 ± 0.0003	30
	180	79											0.3240 ± 0.0003	18
	180	79											0.3434 ± 0.0003	14
	180	79											0.4505 ± 0.0005	7
	180	79											0.5242 ± 0.0003	44
	180	79											0.5524 ± 0.0004	7
	180	79											0.6765 ± 0.0004	20
	180	79											0.7077 ± 0.0005	4
	180	79											0.8084 ± 0.0004	30
	180	79											0.8597 ± 0.0006	35
	180	79											1.0321 ± 0.0007	23
$_{79}Au^{181}$	181	79		180.970130	11.4 ± 0.5 s.	E.C.(99%)	6.55	5.482						
$_{79}Au^{182}$	182	79		181.969580	21 s.	β+,E.C.	7.85						ann.rad.	
	182	79											0.1549 ± 0.0002	46%
	182	79											0.2649 ± 0.0003	18%
	182	79											0.3449 ± 0.0003	2.7%
	182	79											0.3561 ± 0.0003	1.2%
	182	79											0.5126 ± 0.0004	4.1%
	182	79											0.6138 ± 0.0004	2.2%
	182	79											0.6388 ± 0.0004	1.2%
	182	79											0.6672 ± 0.0003	3.1%
	182	79											0.7870 ± 0.0002	6.2%
	182	79											0.8553 ± 0.0002	6.6%
	182	79											0.8997 ± 0.0004	1.2%
	182	79											1.0264 ± 0.0003	3.4%
	182	79											1.0843 ± 0.0004	1.4%
	182	79											(0.13 - 1.4)weak	
$_{79}Au^{183}$	183	79		182.967660	42 s.	E.C.	7.09						0.1630 ± 0.0001	52%
	184	79											0.2730 ± 0.0001	42%
	184	79											0.3625 ± 0.0001	18%
	184	79											0.4327 ± 0.0004	2.0%
	184	79											0.4353 ± 0.0004	2.0%
	184	79											0.4860 ± 0.0001	6.2%
	184	79											0.5921 ± 0.0002	3.4%
	184	79											0.6487 ± 0.0002	3.2%
	184	79											0.6645 ± 0.0002	2.0%
	184	79											0.7770 ± 0.0002	6.9%
	184	79											0.8312 ± 0.0003	2.2%
	184	79											0.8440 ± 0.0002	5.6%
	184	79											0.8710 ± 0.0003	3.9%
	184	79											1.0096 ± 0.0003	2.7%
	184	79											1.0713 ± 0.0003	2.5%
	184	79											1.0903 ± 0.0003	1.9%
	184	79											1.2457 ± 0.0003	2.4%
	184	79											1.3087 ± 0.0003	1.5%
	184	79											1.3976 ± 0.0003	1.8%
	184	79											1.5256 ± 0.0003	1.2%
	184	79											1.7138 ± 0.0004	1.5%
	184	79											1.7138 ± 0.0004	1.5%
	184	79											1.7546 ± 0.0003	3.3%
	184	79											1.8142 ± 0.0003	2.8%
	184	79											2.1963 ± 0.0003	1.5%
	184	79											2.1963 ± 0.0003	1.5%
	184	79											2.4752 ± 0.0004	2.0%
	184	79											2.4909 ± 0.0003	1.0%
$_{79}Au^{185m}$	185	79			6.8 m.	β+,E.C.								
	185	79				I.T.	0.145							
$_{79}Au^{185}$	185	79		184.965800	4.3 m.	β+,E.C.	4.76				(5/2-)		ann.rad.	
$_{79}Au^{186m}$	186	79			2 m.	β+,E.C.							0.1915 ± 0.0001	
$_{79}Au^{186}$	186	79		185.966100	10.7 m.	β+,E.C.	6.28				3		ann.rad.	
	186	79											0.1915 ± 0.0001	56%
	186	79											0.2988 ± 0.0001	23%
	186	79											0.4156 ± 0.0002	7.7%
	186	79											0.6070 ± 0.0002	4.8%
	186	79											0.6765 ± 0.0003	3.0%
	186	79											0.7654 ± 0.0003	9.6%
	186	79											0.7987 ± 0.0004	4.8%
	186	79											0.8816 ± 0.0003	1.9%
	186	79											1.2162 ± 0.0003	3.4%
	186	79											1.2892 ± 0.0005	1.9%
	186	79											1.7259 ± 0.0004	1.1%
	186	79											1.7376 ± 0.0004	1.7%
	186	79											2.0246 ± 0.0005	1.8%
	186	79											2.0356 ± 0.0005	2.6%
$_{79}Au^{187}$	187	79		186.964460	8.2 m.	β+,E.C.	3.72				1/2		ann.rad.	
	187	79											0.0512 ± 0.0001	1.9%
	187	79											0.0653 ± 0.0001	1.9%

Isotope	A	Z	% Natural abundance	Atomic mass	Half-life	Decay mode	Decay energy (MeV)	Particle energy (MeV)	Particle intensity	Thermal neutron cross section	Spin (h/2π)	μ Nucl. mag. moment	Gamma-ray energy (MeV)	Gamma-ray intensity
	187	79											0.1811 ± 0.0001	1.9%
	187	79											0.1853 ± 0.0001	1.6%
	187	79											0.1903 ± 0.0001	1.8%
	187	79											0.2474 ± 0.0001	1.8%
	187	79											0.2474 ± 0.0001	1.8%
	187	79											0.2510 ± 0.0001	1.9%
	187	79											0.3516 ± 0.0001	1.7%
	187	79											0.3903 ± 0.0001	2.6%
	187	79											0.4262 ± 0.0001	3.2%
	187	79											0.5601 ± 0.0003	2.7%
	187	79											0.6210 ± 0.0003	3.1%
	187	79											0.7069 ± 0.0003	3.7%
	187	79											0.7213 ± 0.0003	2.1%
	187	79											0.8344 ± 0.0003	2.0%
	187	79											0.9152 ± 0.0003	5.2%
	187	79											1.1897 ± 0.0003	1.9%
	187	79											1.2668 ± 0.0003	4.2%
	187	79											1.3191 ± 0.0003	3.4%
	187	79											1.3321 ± 0.0003	12%
	187	79											1.4081 ± 0.0003	4.7%
	187	79											1.4521 ± 0.0003	2.8%
	187	79											1.9605 ± 0.0003	1.7%
	187	79											1.9881 ± 0.0003	2.1%
	187	79											2.0270 ± 0.0003	1.9%
	187	79											2.0528 ± 0.0003	1.3%
	187	79											2.0813 ± 0.0003	1.3%
$_{79}Au^{188}$	188	79		187.965080	8.8 m.	$\beta+$,E.C.	5.30				(1-)		ann.rad.	
	188	79											0.2660 ± 0.0001	100 +
	188	79											0.3308 ± 0.0001	5
	188	79											0.3404 ± 0.0001	24
	188	79											0.4055 ± 0.0001	9.1%
	188	79											0.5334 ± 0.0003	5.9
	188	79											0.6061 ± 0.0001	16
	188	79											0.6708 ± 0.0001	8
	188	79											0.6791 ± 0.0001	2
	188	79											0.9491 ± 0.0001	2.4
	188	79											1.0470 ± 0.0001	1.9
	188	79											1.0843 ± 0.0001	6.6
	188	79											1.1153 ± 0.0001	5
	188	79											1.1705 ± 0.0001	2.6
	188	79											1.3126 ± 0.0001	3.0
	188	79											1.3601 ± 0.0001	4.1
	188	79											1.5104 ± 0.0001	2.7
	188	79											1.5450 ± 0.0001	2.3
	188	79											1.8825 ± 0.0002	1.5
	188	79											2.0300 ± 0.0001	2.5
	188	79											2.2319 ± 0.0001	2.6
	188	79											2.4469 ± 0.0002	1.4
	188	79											2.6268 ± 0.0003	1.8
	188	79											2.7810 ± 0.0002	2.2
$_{79}Au^{189m}$	189	79			4.6 m.	$\beta+$,E.C.					11/2-		0.1667 ± 0.0002	100 +
	189	79											0.3211 ± 0.0005	19
$_{79}Au^{189}$	189	79		188.963720	28.7 m.	E.C.(96%)	2.700				1/2+		ann.rad.	
	189	79				$\beta+$(4%)							Pt k x-ray	53%
	189	79											0.2157 ± 0.0001	3.2%
	189	79											0.2220 ± 0.0001	5.0%
	189	79											0.2257 ± 0.0001	2.1%
	189	79											0.2537 ± 0.0002	2.0%
	189	79											0.2975 ± 0.0003	3.0%
	189	79											0.3482 ± 0.0001	8.4%
	189	79											0.44121 ± 0.0001	704%
	189	79											0.4478 ± 0.0001	11%
	189	79											0.5295 ± 0.0001	6.5%
	189	79											0.5295 ± 0.0001	6.5%
	189	79											0.6311 ± 0.0005	2.4%
	189	79											0.7133 ± 0.0001	21%
	189	79											0.8128 ± 0.0003	13%
	189	79											1.0715 ± 0.0003	5.5%
	189	79											1.1605 ± 0.0003	7.0%
	189	79											1.1769 ± 0.0007	3.3%
$_{79}Au^{190}$	190	79		189.964685	43 m.	$\beta+$(2%)	4.442				1-	0.066	ann.rad.	4%
	190	79				E.C.(98%)							Pt k x-ray	41%
	190	79											0.2958 ± 0.0001	72%
	190	79											0.3018 ± 0.0001	24%
	190	79											0.3189 ± 0.0001	4.7%
	190	79											0.4412 ± 0.0001	3.8%
	190	79											0.5977 ± 0.0001	9.0%
	190	79											0.6208 ± 0.0001	2.4%
	190	79											0.6250 ± 0.0002	3.0%
	190	79											1.0575 ± 0.0004	3.2%
	190	79											1.3953 ± 0.0002	2.1%
	190	79											1.44131 ± 0.0004	2.9%
	190	79											1.7849 ± 0.0004	2.1%
	190	79											2.3824 ± 0.0002	3.8%
	190	79											2.4163 ± 0.0002	3.3%
	190	79											2.7527 ± 0.0002	2.7%

Isotope	A	Z	% Natural abundance	Atomic mass	Half-life	Decay mode	Decay energy (MeV)	Particle energy (MeV)	Particle intensity	Thermal neutron cross section	Spin (h/2π)	μ Nucl. mag. moment	Gamma-ray energy (MeV)	Gamma-ray intensity
	190	79											2.9598 ± 0.0009	1.0%
$_{79}Au^{191m}$	191	79			0.9 s.	I.T.	0.2663				(11/2-		Au k x-ray	9%
	191	79											0.2414 ± 0.0005	13%
	191	79											0.2526 ± 0.0004	61%
$_{79}Au^{191}$	191	79		190.963630	3.2 h.	E.C.	1.830				3/2 +	0.138	Pt k x-ray	52%
	191	79											0.1665 ± 0.0001	3.1%
	191	79											0.1941 ± 0.0001	2.6%
	191	79											0.2064 ± 0.0001	2.1%
	191	79											0.2540 ± 0.0001	2.4%
	191	79											0.2716 ± 0.0001	2.4%
	191	79											0.2779 ± 0.0001	6.8%
	191	79											0.2804 ± 0.0001	2.8%
	191	79											0.2389 ± 0.0001	6.3%
	191	79											0.2935 ± 0.0001	2.7%
	191	79											0.3539 ± 0.0001	2.9%
	191	79											0.3869 ± 0.0001	3.4%
	191	79											0.3903 ± 0.0001	2.6%
	191	79											0.3998 ± 0.0001	4.5%
	191	79											0.4137 ± 0.0001	3.5%
	191	79											0.4214 ± 0.0001	3.2%
	191	79											0.4780 ± 0.0001	3.7%
	191	79											0.4876 ± 0.0001	2.6%
	191	79											0.5864 ± 0.0001	16%
	191	79											0.6742 ± 0.0001	6.4%
	191	79											(0.08 - 1.3)weak	
$_{79}Au^{192}$	192	79		191.964793	5.0 h.	β+(5%)	3.515	2.19			1/2-	0.0079	ann.rad.	10%
	192	79				E.C.(95%)		2.49					Pt k x-ray	41%
	192	79											0.2959 ± 0.0001	30%
	192	79											0.3084 ± 0.0001	4.6%
	192	79											0.3165 ± 0.0001	78%
	192	79											0.4681 ± 0.0001	2.3%
	192	79											0.4772 ± 0.0002	1.5%
	192	79											0.5826 ± 0.0001	3.6%
	192	79											0.6043 ± 0.0001	1.4%
	192	79											0.6124 ± 0.0001	5.9%
	192	79											0.7591 ± 0.0002	2.2%
	192	79											0.8787 ± 0.0002	1.1%
	192	79											1.0615 ± 0.0003	1.2%
	192	79											1.1209 ± 0.00023	2.0%
	192	79											1.1402 ± 0.0002	3.5%
	192	79											1.4229 ± 0.0002	4.0%
	192	79											1.5766 ± 0.0003	3.0%
	192	79											1.6244 ± 0.0003	2.3%
	192	79											1.7066 ± 0.0003	2.6%
	192	79											1.7231 ± 0.0002	4.2%
$_{79}Au^{193m}$	193	79			3.9 s.	I.T.	0.2901				11/2-		Au k x-ray	11%
	193	79											0.2197 ± 0.0001	3.8%
	193	79											0.2580 ± 0.0001	66%
$_{79}Au^{193}$	193	79		192.964050	17.6 h.	E.C.	1.000				3/2 +	0.140	Pt k x-ray	55%
	193	79											0.1125 ± 0.0001	2.1%
	193	79											0.1735 ± 0.0001	2.9%
	193	79											0.1862 ± 0.0001	10%
	193	79											0.2556 ± 0.0001	6.7%
	193	79											0.2682 ± 0.0001	3.9%
	193	79											0.4390 ± 0.0001	1.9%
	193	79											0.4913 ± 0.0001	0.7%
$_{79}Au^{194}$	194	79		193.965348	39.5 h.	β+(3%)	2.609	1.49			1-	0.074	ann.rad.	6%
	194	79				E.C.(97%)							Pt k x-ray	39%
	194	79											0.2935 ± 0.0001	11%
	194	79											0.3284 ± 0.0001	63%
	194	79											0.3649 ± 0.0001	1.5%
	194	79											0.4828 ± 0.0001	1.2%
	194	79											0.5288 ± 0.0001	1.7%
	194	79											0.6220 ± 0.0001	1.8%
	194	79											0.6451 ± 0.0001	2.3%
	194	79											0.9387 ± 0.0001	1.2%
	194	79											0.9483 ± 0.0001	2.3%
	194	79											1.1041 ± 0.0001	2.1%
	194	79											1.1508 ± 0.0001	1.4%
	194	79											1.1753 ± 0.0001	2.1%
	194	79											1.2188 ± 0.0001	1.2%
	194	79											1.3422 ± 0.0001	1.2%
	194	79											1.4689 ± 0.0001	6.7%
	194	79											1.5924 ± 0.0002	1.1%
	194	79											1.5958 ± 0.0001	1.8%
	194	79											1.8559 ± 0.0002	1.8%
	194	79											1.8570 ± 0.0002	1.6%
	194	79											1.9242 ± 0.0001	2.1%
	194	79											2.0437 ± 0.0001	3.8%
$_{79}Au^{195m}$	195	79			30.5 s.	I.T.	0.3186				11/2-		Au k x-ray	11%
	195	79											0.2004 ± 0.0001	1.7%
	195	79											0.2617 ± 0.0001	68%
$_{79}Au^{195}$	195	79		194.965013	186.1 d.	E.C.	0.230				3/2 +	0.148	Pt k x-ray	48%
$_{79}Au^{196m2}$	196	79			9.7 h.	I.T.	0.5954				12-	5.35	Au k x-ray	42%
	196	79											0.1478 ± 0.0001	42%

TABLE OF THE ISOTOPES (Continued)

Isotope	A	Z	% Natural abundance	Atomic mass	Half-life	Decay mode	Decay energy (MeV)	Particle energy (MeV)	Particle intensity	Thermal neutron cross section	Spin (h/2π)	μ Nucl. mag. moment	Gamma-ray energy (MeV)	Gamma-ray intensity
	196	79											0.1684 ± 0.0001	7.6%
	196	79											0.1883 ± 0.0001	37%
	196	79											0.2855 ± 0.0001	4.3%
	196	79											0.3162 ± 0.0001	2.9%
$_{79}Au^{196m1}$	196	79			8.1 s.	I.T.	0.0846				8+		0.0847 ± 0.0001	0.3%
$_{79}Au^{196}$	196	79		195.966544	6.18 d.	E.C.(92%)	0.507				2-	+0.514	Pt k x-ray	37%
$_{79}Au^{197m}$					7.8 s.	I.T.	0.4094				11/2-		Au k x-ray	12%
	197	79				β-(8%)	0.686						0.1302 ± 0.0001	3.1%
	197	79											0.2018 ± 0.0001	1.1%
	197	79											0.2790 ± 0.0001	71%
	197	79											0.4091 ± 0.0001	0.1%
$_{79}Au^{197}$	197	79	100%	196.966543						(0 + 98.7) b.	3/2+	+0.1457		
$_{79}Au^{198m}$	198	79			2.30 d.	I.T.	0.812				(12-)		Au k x-ray	43%
	198	79											0.0972 ± 0.0001	69%
	198	79											0.1803 ± 0.0001	51%
	198	79											0.2041 ± 0.0001	41%
	198	79											0.2419 ± 0.0001	77%
	198	79											0.3338 ± 0.0002	15%
$_{79}Au^{198}$	198	79		197.968217	2.693 d.	β-	1.372	0.290	1%	26 x 10^3 b.	2-	+0.5934	Hg k x-ray	1.4%
	198	79						0.961	99%				0.411794 ± 0.00001	95.5%
	198	79											0.67587 ± 0.00002	1.1%
	198	79											1.08766 ± 0.00002	0.2%
$_{79}Au^{199}$	199	79		198.968740	3.14 d.	β-	0.453	0.25	22%	30 b.	3/2+	+0.2715	Hg k x-ray	7.6%
	199	79					0.296	72%					0.15837 ± 0.00001	37%
	199	79						0.462	6%				0.20820 ± 0.00001	8.4%
$_{79}Au^{200m}$	200	79			18.7 h.	β-(84%)	1.0	0.56			12-	6.10	Au k x-ray	
	200	79				I.T.(16%)							0.1111 ± 0.0001	1.8%
	200	79											0.1203 ± 0.0001	1.0%
	200	79											0.1332 ± 0.0002	2.8%
	200	79											0.1373 ± 0.0003	1.2%
	200	79											0.1446 ± 0.0003	1.0%
	200	79											0.1461 ± 0.0002	3.5%
	200	79											0.1812 ± 0.0001	55%
	200	79											0.2185 ± 0.0001	1.6%
	200	79											0.2559 ± 0.0001	71%
	200	79											0.3328 ± 0.0004	12%
	200	79											0.36797 ± 0.0001	77%
	200	79											0.4978 ± 0.0001	73%
	200	79											0.5793 ± 0.0001	72%
	200	79											0.7595 ± 0.0001	66%
	200	79											0.9042 ± 0.0001	7.7%
$_{79}Au^{200}$	200	79		199.970670	48.4 m.	β-	2.210	0.7	15%		1-		0.3679 ± 0.0001	19%
	200	79						2.2	77%				1.2254 ± 0.0001	11%
	200	79											1.2629 ± 0.0001	3.1%
	200	79											(0.3 - 1.6)weak	
$_{79}Au^{201}$	201	79		200.971645	26 m.	β-	1.275	1.27	82%		3/2+		0.1674 ± 0.0001	1.0%
	201	79											0.3851 ± 0.0002	0.6%
	201	79											0.5170 ± 0.0003	1.3%
	201	79											0.5426 ± 0.0002	1.9%
	201	79											0.6132 ± 0.0003	1.2%
	201	79											0.6450 ± 0.0004	0.7%
$_{79}Au^{202}$	202	79		201.973840	28 s.	β-	3.00				(1-)		0.4396 ± 0.0001	10%
	202	79											0.9086 ± 0.0004	2%
	202	79											1.1254 ± 0.0004	2.5%
	202	79											1.2037 ± 0.0004	2.1%
	202	79											1.3065 ± 0.0005	2.3%
$_{79}Au^{203}$	203	79		202.975145	53 s.	β-	2.14	≈ 1.9			3/2+		0.690	10%
$_{79}Au^{204}$	204	79		203.978300	40 s.	β-	4.5						0.4366 ± 0.0002	91%
	204	79											0.6919 ± 0.0002	23%
	204	79											0.7230 ± 0.0003	22%
	204	79											1.3921 ± 0.0004	24%
	204	79											1.4048 ± 0.001	4%
	204	79											1.4148 ± 0.0005	8%
	204	79											1.5113 ± 0.0004	28%
	204	79											1.5531 ± 0.0004	8%
	204	79											1.7042 ± 0.0006	5%
	204	79											1.8283 ± 0.001	3%
	204	79											1.8416 ± 0.001	3%
Hg				200.59						374 b.				
$_{80}Hg^{178}$	178	80		177.982476	0.26 s.	E.C.(50%)	6.25				0+			
	178	80				α(50%)		6.43						
$_{80}Hg^{179}$	179	80		178.981630	1.09 s.	E.C.	7.88							
	179	80				α		6.29						
$_{80}Hg^{180}$	180	80		179.978250	2.9 s.	E.C.	5.54				0+		0.1250 ± 0.0004	10 +
	180	80				α		6.12					0.3005 ± 0.0003	100
	180	80											0.3812 ± 0.0004	69
	180	80											0.4050 ± 0.0005	17
	180	80											0.4505 ± 0.0005	16

Isotope	A	Z	% Natural abundance	Atomic mass	Half-life	Decay mode	Decay energy (MeV)	Particle energy (MeV)	Particle intensity	Thermal neutron cross section	Spin (h/2π)	μ Nucl. mag. moment	Gamma-ray energy (MeV)	Gamma-ray intensity
	180	80											0.4799 ± 0.0004	23
$_{80}Hg^{181}$	181	80		180.977720	3.6 s.	β+, EC (74%)	7.07				(1/2-)	+0.5071	0.0663 ± 0.001	
	181	80				α(26%)							0.0811 ± 0.001	
	181	80											0.0924 ± 0.001	
	181	80											0.1474 ± 0.001	
	181	80											0.1587 ± 0.001	
	181	80											0.2142 ± 0.001	
	181	80											0.2390 ± 0.001	
$_{80}Hg^{182}$	182	80		181.974750	11 s.	β+, EC (85%)	4.81				0+		0.1289 ± 0.0001	100 +
	182	80				α(15%)		5.87					0.2168 ± 0.001	75
	182	80											0.4126 ± 0.001	53
$_{80}Hg^{183}$	183	80		182.974350	8.8 s.	β+, EC (77%)	6.24				1/2	+0.524	0.0714 ± 0.001	
	183	80				α		5.83					0.0874 ± 0.001	
	183	80						5.91					0.1538 ± 0.001	
$_{80}Hg^{184}$	184	80		183.971810	30.9 s.	β+, EC (99%)	3.98				0+		0.0915 ± 0.0003	4.7 +
	184	80				α(1%)		5.54					0.1265 ± 0.0003	5
	184	80											0.1460 ± 0.0003	5
	184	80											0.1560 ± 0.0003	91
	184	80											0.1591 ± 0.0003	7
	184	80											0.1701 ± 0.0003	2
	184	80											0.2362 ± 0.0002	100
	184	80											0.2590 ± 0.0001	8
	184	80											0.2623 ± 0.0001	7
	184	80											0.2951 ± 0.0001	16
	184	80											0.2951 ± 0.0001	16
	184	80											0.3924 ± 0.0002	11
	184	80											0.4219 ± 0.0002	6
$_{80}Hg^{185m}$	185	80			21 s.	β+, EC, IT, α		5.37			13/2+		0.211	
	185	80											0.292	
$_{80}Hg^{185}$	185	80		184.971900	50 s.	β+, EC (95%)	5.68				1/2-		0.0236	
	185	80											0.0358	
	185	80											0.0958	
	105	00											0.1074	
	185	80											0.1078	
	185	80											0.1291	
	185	80											0.1810	
	185	80											0.1937	
	185	80											0.2052	
	185	80											0.2112	
	185	80											0.2125	
	185	80											0.2229	
	185	80											0.2442	
	185	80											0.2587	
	185	80											0.2701	
	185	80											0.2701	
	185	80											0.2887	
	185	80											0.2924	
	185	80											0.2924	
	185	80											(0.02 - 0.55)weak	
$_{80}Hg^{186}$	186	80		185.969350	1.4 m.	β+,E.C.	3.03				0+		0.1119 ± 0.0004	87 +
	186	80											0.2278 ± 0.0004	5
	186	80											0.2518 ± 0.0004	100
	186	80											0.3496 ± 0.0005	2.3
$_{80}Hg^{187m}$	187	80			1.7 m.	β+,E.C.					13/2+		See Hg^{187}	
$_{80}Hg^{187}$	187	80		186.969760	2.4 m.	β+,E.C.	4.94				3/2-	-0.593	0.1034 ± 0.0002	32 +
	187	80											0.2034 ± 0.0002	19
	187	80											0.2055 ± 0.0002	10
	187	80											0.2208 ± 0.0002	24
	187	80											0.2334 ± 0.0002	100
	187	80											0.2403 ± 0.0002	33
	187	80											0.27151 ± 0.0002	31
	187	80											0.2985 ± 0.0002	11
	187	80											0.3229 ± 0.0002	12
	187	80											0.3347 ± 0.0002	16
	187	80											0.3763 ± 0.0002	38
	187	80											0.4387 ± 0.0002	11
	187	80											0.4495 ± 0.0002	29
	187	80											0.4620 ± 0.0002	10
	187	80											0.4703 ± 0.0002	29
	187	80											0.4727 ± 0.0002	11
	187	80											0.4758 ± 0.0002	22
	187	80											0.4766 ± 0.0002	11
	187	80											0.4996 ± 0.0002	19
	187	80											0.5254 ± 0.0002	30
	187	80											0.6250 ± 0.0002	14
	187	80											1.9981 ± 0.0008	11
	187	80											2.0126 ± 0.0008	10
	187	80											2.0754 ± 0.0009	13

TABLE OF THE ISOTOPES (Continued)

Isotope	A	Z	% Natural abundance	Atomic mass	Half-life	Decay mode	Decay energy (MeV)	Particle energy (MeV)	Particle intensity	Thermal neutron cross section	Spin (h/2π)	μ Nucl. mag. moment	Gamma-ray energy (MeV)	Gamma-ray intensity
	187	80											2.1765 ± 0.001	20
$_{80}Hg^{188}$	188	80		187.967580	3.2 m.	β+,E.C.	2.340				0+		0.0988 ± 0.00074	12 +
	188	80											0.1148 ± 0.0007	37
	188	80											0.1346 ± 0.0007	11
	188	80											0.1424 ± 0.0007	20
	188	80											0.1824 ± 0.0007	18
	188	80											0.1858 ± 0.0007	13
	188	80											0.1900 ± 0.0007	100
	188	80											0.2540 ± 0.0007	10
	188	80											0.3360 ± 0.0007	10
	188	80											0.3453 ± 0.0007	10
	188	80											0.5238 ± 0.0007	19
$_{80}Hg^{189m}$	189	80			8.6 m.	E.C.					13/2+		0.0780 ± 0.0001	63 +
	189	80											0.1665 ± 0.0001	36
	189	80											0.2366 ± 0.0002	29
	189	80											0.2976 ± 0.0002	34
	189	80											0.3210 ± 0.0002	100
	189	80											0.3876 ± 0.0002	36
	189	80											0.3987 ± 0.0002	17
	189	80											0.4345 ± 0.0002	47
	189	80											0.4591 ± 0.0002	10
	189	80											0.4839 ± 0.0004	11
	189	80											0.4996 ± 0.0002	10
	189	80											0.5026 ± 0.0002	20
	189	80											0.5124 ± 0.0004	38
	189	80											0.5399 ± 0.0003	16
	189	80											0.5655 ± 0.0002	48
	189	80											0.6000 ± 0.0002	22
	189	80											0.7370 ± 0.0002	17
	189	80											1.2793 ± 0.0008	14
	189	80											2.0214 ± 0.0004	11
	189	80											2.0252 ± 0.0005	16
	189	80											2.0339 ± 0.0003	19
	189	80											(0.08 - 2.10)many	
$_{80}Hg^{189}$	189	80		188.968230	7.6 m.	E.C>	4.20				3/2-	-0.6086	0.2005 ± 0.0002	7 +
	189	80											0.2038 ± 0.0002	71
	189	80											0.2291 ± 0.0002	13
	189	80											0.2386 ± 0.0002	100
	189	80											0.2485 ± 0.0002	95
	189	80											0.2790 ± 0.0002	15
$_{80}Hg^{190}$	190	80		189.966400	20 m.	E.C.	1.60				0+		0.1296	1%
	190	80											0.1426	52%
	190	80											0.1547	2%
	190	80											0.1715	3.6%
$_{80}Hg^{191m}$	191	80			51 m.	β+(6%)					13/2+		ann.rad.	12%
	191	80				E.C.(94%)							Au k x-ray	50%
	191	80											0.2741 ± 0.001	13%
	191	80											0.3316 ± 0.0009	4%
	191	80											0.3570 ± 0.0004	5%
	191	80											0.3710 ± 0.0004	6%
	191	80											0.4097 ± 0.0009	3%
	191	80											0.4203 ± 0.0006	18%
	191	80											0.5215 ± 0.0007	4%
	191	80											0.5361 ± 0.0006	8%
	191	80											0.5787 ± 0.0004	17%
	191	80											0.6106 ± 0.0006	5%
	191	80											0.6710 ± 0.0007	3%
	191	80											0.7180 ± 0.0007	3%
	191	80											0.8870 ± 0.0008	2%
	191	80											0.9964 ± 0.0008	2%
	191	80											1.1093 ± 0.001	2%
	191	80											1.2844 ± 0.001	2%
	191	80											1.3298 ± 0.001	3%
	191	80											1.4478 ± 0.001	1.5%
	191	80											1.5036 ± 0.001	2%
	191	80											1.5488 ± 0.001	2%
	191	80											1.7394 ± 0.001	1.4%
	191	80											1.9081 ± 0.001	1.5%
	191	80											(0.07 - 1.9)many	
$_{80}Hg^{191}$	191	80			49 m.	β+,E.C.					(3/2-)		0.1963 ± 0.0002	67 +
	191	80											0.2247 ± 0.0002	60
	191	80											0.2408 ± 0.0002	44
	191	80											0.2524 ± 0.0002	100
	191	80											0.3314 ± 0.0005	39
	191	80											0.5214 ± 0.0006	9
	191	80											0.778 ± 0.001	6
$_{80}Hg^{192}$	192	80		191.965650	4.9 h.	E.C.	0.800				0+		Au k x-ray	53%
	192	80											0.1019	1%
	192	80											0.1572	6%
	192	80											0.1864	2.9%
	192	80											0.2454	1.5%
	192	80											0.2748	45%
	192	80											0.3065	4.8%

Isotope	A	Z	% Natural abundance	Atomic mass	Half-life	Decay mode	Decay energy (MeV)	Particle energy (MeV)	Particle intensity	Thermal neutron cross section	Spin (h/2π)	μ Nucl. mag. moment	Gamma-ray energy (MeV)	Gamma-ray intensity
$_{80}Hg^{193m}$	193	80			11.8 h.	β+, EC (91%)					13/2+	-1.0584	Hg k x-ray	53%
	193	80				I.T.(9%)	0.2901						0.1866 ± 0.0001	2.2%
	193	80											0.2181 ± 0.0001	6%
	193	80											0.2198 ± 0.0001	3.3%
	193	80											0.2580 ± 0.0001	58%
	193	80											0.2908 ± 0.0001	1.9%
	193	80											0.3419 ± 0.0001	3.0%
	193	80											0.3453 ± 0.0002	2.0%
	193	80											0.4076 ± 0.0001	37%
	193	80											0.4996 ± 0.0001	5.5%
	193	80											0.5351 ± 0.0001	4.5%
	193	80											0.5371 ± 0.0001	4.7%
	193	80											0.5733 ± 0.0001	31%
	193	80											0.6006 ± 0.0001	4.7%
	193	80											0.8701 ± 0.0002	2.9%
	193	80											0.8778 ± 0.0002	4.8%
	193	80											0.9324 ± 0.0002	15%
	193	80											0.9946 ± 0.0002	3.5%
	193	80											1.2413 ± 0.0002	5.6%
	193	80											1.3255 ± 0.0002	4.9%
	193	80											1.3395 ± 0.0002	4.7%
	193	80											1.3651 ± 0.0002	3.1%
	193	80											1.4861 ± 0.0003	3.4%
	193	80											1.6394 ± 0.0003	3.4%
	193	80											1.6485 ± 0.0003	2.6%
	193	80											(0.1 - 1.96)many	
$_{80}Hg^{193}$	193	80		192.966560	3.8 h.	2.339					3/2-	-0.6276	0.1866 ± 0.0001	16%
	193	80											0.2181 ± 0.0001	3.7%
	193	80											0.2580 ± 0.0001	14%
	193	80											0.3816 ± 0.0001	11%
	193	80											0.4295 ± 0.0001	11%
	193	80											0.5810 ± 0.0001	4%
	193	80											0.7461 ± 0.0002	2.3%
	193	80											0.7892 ± 0.0002	4.7%
	193	80											0.8278 ± 0.0002	4.0%
	193	80											0.8611 ± 0.0002	13%
	193	80											1.0405 ± 0.0003	2.3%
	193	80											1.0807 ± 0.0003	3.8%
	193	80											1.1188 ± 0.0002	8.3%
	193	80											1.2764 ± 0.0003	2.2%
	193	80											1.6034 ± 0.0003	2.1%
	193	80											1.8156 ± 0.0004	2.6%
	193	80											1.8622 ± 0.0004	1.5%
	193	80											1.9766 ± 0.0004	1.8%
$_{80}Hg^{194}$	194	80		193.965391	520 y.	E.C.	0.040				0+		Au L x-rays	
$_{80}Hg^{195m}$	195	80			40.0 h.	I.T.(54%) E.C.(46%)	0.3186				13/2+	-1.044	Hg k x-ray	2.3%
	195	80											Au k x-ray	51%
	195	80											0.2617 ± 0.0001	33%
	195	80											0.3879 ± 0.0001	2.3%
	195	80											0.5603 ± 0.0001	7.5%
	195	80											0.7798 ± 0.0001	5.0%(D)
$_{80}Hg^{195}$	195	80		194.966640	9.5 h.	E.C.	1.520				1/2-	+0.542	Au k x-ray	40%
	195	80											0.0614 ± 0.0001	6.4%
	195	80											0.1801 ± 0.0001	2%
	195	80											0.2071 ± 0.0001	1.6%
	195	80											0.2617 ± 0.0001	1.6%
	195	80											0.5851 ± 0.0001	2.0%
	195	80											0.5997 ± 0.0001	1.8%
	195	80											0.7798 ± 0.0001	7.0%
	195	80											1.1110 ± 0.0001	1.5%
	195	80											1.1724 ± 0.0001	1.3%
$_{80}Hg^{196}$	196	80	0.15%	195.965807						(120 + 3100) b.	0+			
$_{80}Hg^{197m}$	197	80			23.8 h.	I.T.(93%)	0.2989				13/2+	-1.0277	Hg k x-ray	17%
	197	80											Au k x-ray	13%
	197	80											0.13398 ± 0.00005	34%
	197	80											0.1650 ± 0.0001	0.3%
$_{80}Hg^{197}$	197	80		196.967187	64.1 h.	E.C.	0.600				1/2-	+0.5274	Au k x-ray	36%
	197	80											0.07735 ± 0.00002	18%
	197	80											0.19136 ± 0.00004	0.5%
	197	80											0.26871 ± 0.00003	0.04%
$_{80}Hg^{198}$	198	80	10.1%	197.966743						(0.01 + 1.9) b.	0+			
$_{80}Hg^{199m}$	199	80			42.6 m.	I.T.	0.532				13/2+	-1.015	Hg k x-ray	32%
	199	80											0.15841 ± 0.00002	52%
	199	80											0.37386 ± 0.00003	14%
	199	80											0.4134 ± 0.0002	0.03%
$_{80}Hg^{199}$	199	80	17%	198.968254						2.2 x 10^3 b.	1/2-	+0.5059		
$_{80}Hg^{200}$	200	80	23.1%	199.968300						< 60 b.	0+			
$_{80}Hg^{201}$	201	80	13.2%	200.970277						8 b.	3/2-	-0.5602		
$_{80}Hg^{202}$	202	80	29.65%	201.970617						4.9 b.	0+			

TABLE OF THE ISOTOPES (Continued)

Isotope	A	Z	% Natural abundance	Atomic mass	Half-life	Decay mode	Decay energy (MeV)	Particle energy (MeV)	Particle intensity	Thermal neutron cross section	Spin (h/2π)	μ Nucl. mag. moment	Gamma-ray energy (MeV)	Gamma-ray intensity
$_{80}Hg^{203}$	203	80		202.972848	46.6 ± 0.02 d.	β-	0.492	0.213	100%		5/2-	+0.8489	Tl k x-ray	6.3%
	203	80											0.279188 ± 0.00001	81.5%
$_{80}Hg^{204}$	204	80	6.8%	203.973467						0.4 b.	0+			
$_{80}Hg^{205}$	205	80		204.976047	5.2 ± 0.1 m.	β-	1.534	1.33	4%		1/2-	+0.601	0.20378 ± 0.00003	2.2%
	205	80											(0.2 - 1.4)weak	
$_{80}Hg^{206}$	206	80		205.977489	8.5 ± 0.1 m.	β-	1.309	0.935	34%		0+		Tl k x-ray	4%
	206	80						1.3	63%				0.3052 ± 0.0002	27%
	206	80											0.6502 ± 0.0002	2.5%
Tl		81		204.383						3.4 b.				
$_{81}Tl^{184}$	184	81		183.981670	11 s.	β+, EC (98%)	9.19						0.2868 ± 0.0003	39 +
	184	81				α(2%)		6.16					0.3399 ± 0.0003	25
	184	81						6.16					0.3667 ± 0.0003	100
	184	81											0.4188 ± 0.0003	9
	184	81											0.5342 ± 0.0003	17
	184	81											0.5541 ± 0.0003	5
	184	81											0.6083 ± 0.0003	11
	184	81											0.6168 ± 0.0003	8
	184	81											0.7222 ± 0.0003	3.5
$_{81}Tl^{185m}$	185	81			1.8 s.	I.T.	0.453				(9/2-)		0.1688 ± 0.0005	13 +
	185	81				α	5.97						0.2840 ± 0.0005	100
	185	81						6.01						
$_{81}Tl^{186m}$	186	81			4 s.	I.T.	0.374						0.3738 ± 0.0003	79%
$_{81}Tl^{186}$	186	81		185.978510	28 s.	β+,E.C.	8.53						0.3567 ± 0.0003	29%
	186	81											0.4026 ± 0.0003	45%
	186	81											0.4053 ± 0.0002	91%
	186	81											0.4241 ± 0.0002	13%
	186	81											0.4592 ± 0.0003	2.5%
	186	81											0.5975 ± 0.0002	4.2%
	186	81											0.6075 ± 0.0003	6.1%
	186	81											0.6755 ± 0.0003	14%
	186	81											0.7702 ± 0.0003	4.6%
	186	81											0.7884 ± 0.0004	2.6%
	186	81											0.8111 ± 0.0004	2.1%
	186	81											0.8264 ± 0.0004	2.2%
	186	81											1.2097 ± 0.0005	1.0%
	186	81											1.2476 ± 0.0005	1.3%
	186	81											1.2726 ± 0.0005	1.3%
$_{81}Tl^{187m}$	187	81			16 s.	I.T.	≈ 0.33				(9/2+)		0.2995 ± 0.0003	
$_{81}Tl^{187}$	187	81		186.976240	45 s.	β+,E.C.	6.04							
$_{81}Tl^{188m}$	188	81			71 s.	β+,E.C.					(7+)		Hg k x-ray	37%
	188	81											0.2917 ± 0.0001	3.5%
	188	81											0.3012 ± 0.0001	4.8%
	188	81											0.3269 ± 0.0001	9.4%
	188	81											0.3858 ± 0.0001	3.2%
	188	81											0.4129 ± 0.0001	88%
	188	81											0.4241 ± 0.0001	3.4%
	188	81											0.4527 ± 0.0001	2.5%
	188	81											0.4607 ± 0.0001	7.2%
	188	81											0.4682 ± 0.0001	5.0%
	188	81											0.5043 ± 0.0001	23%
	188	81											0.5693 ± 0.0001	3.4%
	188	81											0.5740 ± 0.0001	3.9%
	188	81											0.5921 ± 0.0001	61%
	188	81											0.7724 ± 0.0001	12%
	188	81											0.7952 ± 0.0001	10%
	188	81											0.8811 ± 0.0001	7.5%
	188	81											0.9048 ± 0.0001	11%
	188	81											1.0420 ± 0.0001	3.0%
	188	81											1.1705 ± 0.0004	2.1%
	188	81											1.4456 ± 0.00001	1.0%
$_{81}Tl^{188}$	188	81		187.975880	≈ 70 s.	β+,E.C.	7.73				(2-)		See Tal[188m]	
	188	81											0.4129 ± 0.0001	
$_{81}Tl^{189m}$	189	81			1.4 m.	β+,E.C.					(9/2-)		0.2156	90 +
	189	81											0.2284	50
	189	81											0.3175	100
	189	81											0.4452	14
$_{81}Tl^{189}$	189	81		188.980780	2.3 m.	β+,E.C.	5.20				(1/2+)		0.3337	100 +
	189	81											0.4510	49
	189	81											0.5223	27
	189	81											0.9422	69
$_{81}Tl^{190m}$	190	81			3.7 m.	β+,E.C.		4.2			(7+)		0.1968	4.5 +
	190	81											0.2401	3.4
	190	81											0.3053	15
	190	81											0.4164	91
	190	81											0.5439	5.6
	190	81											0.5570	6.3
	190	81											0.6154	4.2
	190	81											0.6254	82
	190	81											0.6838	7.0

TABLE OF THE ISOTOPES (Continued)

Isotope	A	Z	% Natural abundance	Atomic mass	Half-life	Decay mode	Decay energy (MeV)	Particle energy (MeV)	Particle intensity	Thermal neutron cross section	Spin (h/2π)	μ Nucl. mag. moment	Gamma-ray energy (MeV)	Gamma-ray intensity
	190	81											0.6921	4.7
	190	81											0.7311	37
	190	81											0.8397	24
	190	81											1.0999	4.3
$_{81}Tl^{190}$	190	81		189.973490	2.6 m.	β+,E.C.	6.60	5.7			(2-)		0.4164	87 +
	190	81											0.6254	12
	190	81											0.6838	9.3
	190	81											1.0999	8.1
$_{81}Tl^{191m}$	191	81			5.2 m.	β+, EC (98%)					(9/2+)		0.2157 ± 0.0001	100
	191	81											0.2647 ± 0.0001	51
	191	81											0.3256 ± 0.0003	67
	191	81											0.3359 ± 0.0003	45
	191	81											0.3743 ± 0.0005	16
	191	81											0.3781 ± 0.0003	27
	191	81											0.4775 ± 0.0003	11
	191	81											0.5351 ± 0.0004	10
	191	81											0.5631 ± 0.0003	20
	191	81											0.5797 ± 0.0004	35
	191	81											0.6153 ± 0.0004	12
	191	81											0.6390 ± 0.0004	16
$_{81}Tl^{192m}$	192	81			10.8 m.	β+,E.C.					(7+)		0.1740 ± 0.0001	12%
	192	81											0.3839 ± 0.0002	3.0%
	192	81											0.4228 ± 0.0001	96%
	192	81											0.4517 ± 0.0003	2.8%
	192	81											0.5841 ± 0.0001	2.1%
	192	81											0.6348 ± 0.0001	88%
	192	81											0.7455 ± 0.0001	32%
	192	81											0.7863 ± 0.0001	38%
	192	81											1.1130 ± 0.0002	4.2%
	192	81											1.2505 ± 0.0003	1.3%
	192	81											1.3451 ± 0.0003	1.0%
	192	81											1.3655 ± 0.0003	1.3%
	192	81											1.4218 ± 0.0002	1.9%
$_{81}Tl^{192}$	192	81		191.972120	9.4 m.	β+,E.C.	6.02				(2-)		0.3975 ± 0.0003	6.5%
	192	81											0.4228 ± 0.0001	81%
	192	81											0.6908 ± 0.0001	12%
	192	81											0.7967 ± 0.0003	2.4%
	192	81											1.1711 ± 0.0004	2.0%
	192	81											1.4218 ± 0.0002	2.6%
	192	81											1.6335 ± 0.0002	2.0%
	192	81											1.6589 ± 0.0002	2.2%
	192	81											1.9084 ± 0.0003	1.6%
	192	81											2.0545 ± 0.001	1.2%
	192	81											2.1163 ± 0.0003	2.6%
	192	81											2.1674 ± 0.0004	1.2%
	192	81											2.2627 ± 0.0004	1.4%
	192	81											2.3000 ± 0.0004	1.9%
$_{81}Tl^{193m}$	193	81			2.1 m.	I.T.(75%)					(9/2-)		0.3650 ± 0.0001	67%
$_{81}Tl^{193}$	193	81		192.970520	22 m.	β+,E.C.	3.68				($^1/_2$+)		0.2077 ± 0.0002	20 +
	193	81											0.2744 ± 0.0002	13
	193	81											0.2849 ± 0.0002	22
	193	81											0.3244 ± 0.0001	100
	193	81											0.3351 ± 0.0001	26
	193	81											0.3440 ± 0.0001	42
	193	81											0.4935 ± 0.0002	12
	193	81											0.6364 ± 0.0003	18
	193	81											0.6529 ± 0.0003	10
	193	81											0.6761 ± 0.0002	48
	193	81											0.6923 ± 0.0003	10
	193	81											0.7525 ± 0.0004	12
	193	81											0.7704 ± 0.0004	13
	193	81											0.9947 ± 0.0003	11
	193	81											1.0447 ± 0.0003	59
	193	81											1.1303 ± 0.0003	12
	193	81											1.2054 ± 0.0003	10
	193	81											1.2560 ± 0.0003	10
	193	81											1.5793 ± 0.001	45
$_{81}Tl^{194m}$	194	81			32.8 m.	β+(20%) E.C.(80%)	≈ 0.30				(7+)		ann.rad.	40%
	194	81											Hg k x-ray	49%
	194	81											0.1110 ± 0.0001	6.4%
	194	81											0.2554 ± 0.0001	9.2%
	194	81											0.3198 ± 0.0001	3.9%
	194	81											0.4282 ± 0.0003	96%
	194	81											0.6363 ± 0.0003	98%
	194	81											0.6503 ± 0.0003	7%
	194	81											0.7350 ± 0.0003	22%
	194	81											0.7490 ± 0.0003	77%
$_{81}Tl^{194}$	194	81		193.970920	33 m.	β+,E.C.	5.15.				2-		0.3955 ± 0.0005	1.9%
	194	81											0.4039 ± 0.0007	2.3%
	194	81											0.4282 ± 0.0003	92%
	194	81											0.6363 ± 0.0003	21%
	194	81											0.6452 ± 0.0002	12%
	194	81											1.0403 ± 0.0005	5%
	194	81											1.0733 ± 0.0005	4%
$_{81}Tl^{195m}$	195	81			3.6 s.	I.T.	0.483				9/2-		Tl k x-ray	3.2%

TABLE OF THE ISOTOPES (Continued)

Isotope	A	Z	% Natural abundance	Atomic mass	Half-life	Decay mode	Decay energy (MeV)	Particle energy (MeV)	Particle intensity	Thermal neutron cross section	Spin (h/2π)	μ Nucl. mag. moment	Gamma-ray energy (MeV)	Gamma-ray intensity
	195	81											0.0990 ± 0.0001	0.6%
	195	81											0.3836 ± 0.0001	91%
$_{81}Tl^{195}$	195	81		194.969630	1.13 h.	E.C.(97%)	7.78	≈ 1.8			1/2+		ann.rad.	6%
	195	81				β+(3%)							Hg k x-ray	41%
	195	81											0.2422 ± 0.0001	4.5%
	195	81											0.2792 ± 0.0001	3.9%
	195	81											0.3006 ± 0.0001	2.5%
	195	81											0.5584 ± 0.0001	2.7%
	195	81											0.5635 ± 0.0001	11%
	195	81											0.8147 ± 0.0001	2%
	195	81											0.8845 ± 0.0001	10.5%
	195	81											0.9216 ± 0.0001	2.3%
	195	81											0.9675 ± 0.0001	2.2%
	195	81											1.1003 ± 0.0001	2.4%
	195	81											1.1217 ± 0.0001	2.6%
	195	81											1.2695 ± 0.0001	2.5%
	195	81											1.3639 ± 0.0001	9.1%
	195	81											1.5116 ± 0.0001	1.5%
	195	81											1.7059 ± 0.0002	1.7%
	195	81											1.7782 ± 0.0003	1.1%
	195	81											1.9778 ± 0.0002	1.8%
	195	81											2.0148 ± 0.0002	1.0%
	195	81											2.285 ± 0.001	0.64%
	195	81											2.5133 ± 0.0002	0.5%
	195	81											(0.13 - 2.5)many	
$_{81}Tl^{196m}$	196	81			1.4 h.	β+, EC (95%)	4.9				(7+)		0.0840 ± 0.0001	7%
	196	81											0.3015 ± 0.0002	4%
	196	81											0.4261 ± 0.0001	91%
	196	81											0.6353 ± 0.0001	51%
	196	81											0.6954 ± 0.0005	41%
	196	81											1.0364 ± 0.001	2%
	196	81											(0.08 - 1.0)	
$_{81}Tl^{196}$	196	81		195.970460	1.8 h.	β+(15%)	4.34				2-		ann.rad.	30%
	196	81				E.C.(85%)							Hg k x-ray	34%
	196	81											0.4257 ± 0.0002	84%
	196	81											0.6105 ± 0.0005	12%
	196	81											0.6352 ± 0.0005	9.8%
	196	81											0.9646 ± 0.001	3.6%
	196	81											1.0362 ± 0.001	2.6%
	196	81											1.3890 ± 0.0005	2.5%
	196	81											1.4958 ± 0.0005	8.2%
	196	81											1.5530 ± 0.0007	4.8%
	196	81											1.6214 ± 0.002	4.9%
	196	81											1.6967 ± 0.002	3.0%
	196	81											1.7755 ± 0.001	2.8%
	196	81											2.0113 ± 0.0015	2.8%
	196	81											2.1278 ± 0.0025	2.8%
	196	81											2.2120 ± 0.002	3.4%
	196	81											(0.03 - 2.4)weak	
$_{81}Tl^{197m}$	197	81			0.54 s.	I.T.(53%)	0.608						Tl k x-ray	33%
	197	81				β+, EC (47%)							0.2262 ± 0.0003	5.2%
	197	81											0.2596 ± 0.0003	2.9%
	197	81											0.2828 ± 0.0002	28%
	197	81											0.4118 ± 0.0001	55%
	197	81											0.4418 ± 0.0003	2.1%
	197	81											0.4896 ± 0.0003	4.4%
	197	81											0.5192 ± 0.0003	3.5%
	197	81											0.5872 ± 0.0002	51%
	197	81											0.6367 ± 0.0002	55%
	197	81											0.7673 ± 0.0003	1.1%
$_{81}Tl^{197}$	197	81		196.969498	2.83 h.	β+(1%)	2.15				1/2+	+1.58	Hg k x-ray	43%
	197	81				E.C.(99%)							0.1522 ± 0.0001	7.2%
	197	81											0.3086 ± 0.0002	2.2%
	197	81											0.4258 ± 0.0001	13%
	197	81											0.4331 ± 0.0001	2.5%
	197	81											0.5780 ± 0.0001	4.4%
	197	81											1.6743 ± 0.0002	1.4%
	197	81											0.7015 ± 0.0001	1.0%
	197	81											0.7921 ± 0.0001	1.7%
	197	81											0.8572 ± 0.0001	2.0%
	197	81											0.9827 ± 0.0001	1.2%
	197	81											1.3853 ± 0.0001	1.2%
	197	81											1.4113 ± 0.0001	4.5%
$_{81}Tl^{198m}$	198	81			1.87 h.	β+, EC (53%)					7+	0.64	Hg k x-ray	33%
	198	81				I.T.(47%)	0.5347						Tl k x-ray	14%
	198	81											0.2262 ± 0.0003	5.2%
	198	81											0.2596 ± 0.0003	2.9%
	198	81											0.2828 ± 0.0002	28%
	198	81											0.4118 ± 0.0001	55%
	198	81											0.4896 ± 0.0003	4.4%
	198	81											0.5192 ± 0.0003	3.5%
	198	81											0.5872 ± 0.0002	57%
	198	81											0.6367 ± 0.0002	55%

TABLE OF THE ISOTOPES (Continued)

Isotope	A	Z	% Natural abundance	Atomic mass	Half-life	Decay mode	Decay energy (MeV)	Particle energy (MeV)	Particle intensity	Thermal neutron cross section	Spin (h/2π)	μ Nucl. mag. moment	Gamma-ray energy (MeV)	Gamma-ray intensity
	198	81											0.7673 ± 0.0003	1.1%
$_{81}Tl^{198}$	198	81		197.940460	5.3 h.	β+(1%)	3.46	1.4			2-	0.00	Hg k x-ray	40%
	198	81						2.1					0.4118 ± 0.0001	82%
	198	81						2.4					0.6367 ± 0.0002	10%
	198	81											0.6759 ± 0.0001	11%
	198	81											1.0076 ± 0.0003	2.7%
	198	81											1.0876 ± 0.0003	2.4%
	198	81											1.2006 ± 0.0002	9.7%
	198	81											1.3122 ± 0.0002	4.7%
	198	81											1.4206 ± 0.0003	8.0%
	198	81											1.4354 ± 0.0003	3.5%
	198	81											1.4470 ± 0.0003	4.3%
	198	81											1.4896 ± 0.0003	2.6%
	198	81											1.5936 ± 0.0002	2.1%
	198	81											1.7208 ± 0.0003	2.8%
	198	81											1.8326 ± 0.0003	4.2%
	198	81											1.8993 ± 0.0003	2.2%
	198	81											2.0402 ± 0.0002	8.4%
	198	81											2.1905 ± 0.0003	2.7%
	198	81											2.4682 ± 0.0003	1.1%
	198	81											(0.23 - 2.8)weak	
$_{81}Tl^{199}$	199	81		198.969870	7.4 h.	E.C.	1.500				1/2-	+1.60	Hg k x-ray	46%
	199	81											0.1584 ± 0.0001	4.9%
	199	81											0.2082 ± 0.0001	12%
	199	81											0.2473 ± 0.0001	9.2%
	199	81											0.2841 ± 0.0001	2.2%
	199	81											0.3339 ± 0.0001	1.7%
	199	81											0.4034 ± 0.0001	1.5%
	199	81											0.4555 ± 0.0001	12%
	199	81											0.4923 ± 0.0001	1.5%
	199	81											0.7504 ± 0.0001	1.0%
	199	81											1.0129 ± 0.0001	1.7%
$_{81}Tl^{200}$	200	81		199.970934	1.09 ± 0.01 d.	E.C.	2.454	1.07			2-	0.04	Hg k x-ray	39%
	200	81						1.44					0.36799 ± 0.00001	87%
	200	81											0.57932 ± 0.00005	14%
	200	81											0.66145 ± 0.00004	2.3%
	200	81											0.8284 ± 0.0001	11%
	200	81											0.88618 ± 0.00005	2.0%
	200	81											1.2057 ± 0.0001	30%
	200	81											1.2255 ± 0.0001	3.4%
	200	81											1.3631 ± 0.0001	3.4%
	200	81											1.5150 ± 0.0001	4.0%
	200	81											1.6045 ± 0.0001	1.2%
	200	81											(0.11 - 2.3)many	
$_{81}Tl^{201}$	201	81		200.970794	3.05 ± 0.01 d.	E.C.	0.482				1/2+	+1.61	Hg k x-ray	38%
	201	81											0.13528 ± 0.00003	2.7%
	201	81											0.16582 ± 0.00004	0.2%
	201	81											0.16740 ± 0.00004	9.4%
$_{81}Tl^{202}$	202	81		201.972085	12.23 ± 0.02 d.	E.C.	1.367				2-	0.06	Hg k x-ray	38%
	202	81											0.43957 ± 0.00001	91%
	202	81											0.52014 ± 0.00007	0.9%
	202	81											0.95971 ± 0.00007	0.1%
$_{81}Tl^{203}$	203	81	29.52%	202.972320						11.4 b.	1/2+	+1.6222		
$_{81}Tl^{204}$	204	81			3.78 ± 0.02 y.	β-(97%)	0.763	0.763	97%		2-	0.0908	Hg k x-ray	0.7%
	204	81				E.C.(3%)	0.345							
$_{81}Tl^{205}$	205	81	70.476%	204.974401						0.10 b.	1/2+	+1.6382		
$_{81}Tl^{206m}$	206	81			3.76 ± 0.02 m.	I.T.	2.644				12-		Tl k x-ray	20%
	206	81											0.2166 ± 0.001	89%
	206	81											0.2477 ± 0.001	9.7%
	206	81											0.2661 ± 0.0002	86%
	206	81											0.4534 ± 0.0008	94%
	206	81											0.4576 ± 0.0008	22%
	206	81											0.5644 ± 0.0008	13%
	206	81											0.6866 ± 0.0006	100%
	206	81											1.0219 ± 0.0008	76%
	206	81											1.1400 ± 0.0008	7.5%
$_{81}Tl^{206}$	206	81		205.976084	4.20 ± 0.02 m.	β-	1.531	1.53	99.9%		0-		Pb k x-ray	0.03%
	206	81											0.80313 ± 0.00005	0.005%

TABLE OF THE ISOTOPES (Continued)

Isotope	A	Z	% Natural abundance	Atomic mass	Half-life	Decay mode	Decay energy (MeV)	Particle energy (MeV)	Particle intensity	Thermal neutron cross section	Spin (h/2π)	μ Nucl. mag. moment	Gamma-ray energy (MeV)	Gamma-ray intensity
$_{81}Tl^{207m}$	207	81			1.33 ± 0.1 s.	I.T.	1.350				11/2-		Tl k x-ray	12.5%
	207	81											0.3501 ± 0.002	79%
	207	81											1.0000 ± 0.002	87%
$_{81}Tl^{207}$	207	81		206.977404	4.77 ± 0.02 m.	β-	1.427	1.43	99.8%		1/2+		0.89723 ± 0.00007	0.24%
$_{81}Tl^{208}$	208	81		207.981988	3.052 ± 0.003 m.	β-	4.994	1.28	23%		(5+)		Pb k x-ray	3.6%
	208	81						1.52	22%				0.27728 ± 0.00006	6.8%
	208	81						1.796	51%				0.51061 ± 0.00002	22%
	208	81											0.58302 ± 0.00002	86%
	208	81											0.86030 ± 0.00006	12%
	208	81											2.61448 ± 0.00005	99.8%
$_{81}Tl^{209}$	209	81		208.985334	2.20 ± 0.07 m.	β-	3.976	1.8	100%		(1/2+)		Pb k x-ray	10%
	209	81											0.1172 ± 0.0001	81%
	209	81											0.4651 ± 0.0002	81%
	209	81											1.5669 ± 0.0001	98%
$_{81}Tl^{210}$	210	81		209.990056	1.30 ± 0.03 m.	β-	5.490	1.3	25%		(5+)		Pb k x-ray	4.6%
	210	81						1.9	56%				0.081 ± 0.003	2.0%
	210	81											0.2981 ± 0.001	79%
	210	81											0.79788 ± 0.0001	99%
	210	81											0.860 ± 0.002	7%
	210	81											1.068 ± 0.001	12%
	210	81											1.110 ± 0.001	7%
	210	81											1.208 ± 0.001	17%
	210	81											1.315 ± 0.001	21%
	210	81											1.408 ± 0.001	5%
	210	81											2.008 ± 0.001	7%
	210	81											2.268 ± 0.001	3%
	210	81											2.358 ± 0.002	8%
	210	81											2.428 ± 0.002	9%
Pb		82		207.2						0.171 b.				
$_{82}Pb^{184}$	184	82		183.988120	0.6 s.	α		6.63			0+			
$_{82}Pb^{185}$	185	82		184.987490	4.1 s.	α		6.34						
	185	82						6.40						
	185	82						6.40						
	185	82						6.48						
$_{82}Pb^{186}$	186	82		185.984300	8 s.	β+, EC (95%)	5.39				0+			
	186	82				α(5%)		6.32						
$_{82}Pb^{187m}$	187	82			18.3 s.	E.C.					13/2+		0.1930 ± 0.0003	15 +
	187	82				α		6.08					0.3314 ± 0.0003	60
	187	82											0.3435 ± 0.0003	75
	187	82											0.3934 ± 0.0003	100
$_{82}Pb^{187}$	187	82		186.983830	15.2 s.	β+,E.C.	7.07	5.99			(1/2-)		0.0674 ± 0.0003	
	187	82						6.19					0.2080 ± 0.0003	
	187	82											0.2755 ± 0.0003	
	187	82											0.2995 ± 0.0003	
	187	82											0.4487 ± 0.0003	
	187	82											0.7477 ± 0.0003	
$_{82}Pb^{188}$	188	82		187.980970	24 s.	E.C.(78%)	4.740				0+		0.1850	49%
	188	82				α(22%)		5.98					0.7582	29%
$_{82}Pb^{189}$	189	82		188.980780	51 s.	E.C.	6.50							
	189	82				α		5.58						
$_{82}Pb^{190}$	190	82		189.978070	β+(13%)	4.27					0+		ann.rad.	26%
	190	82				E.C.(86%)							Tl k x-ray	47%
	190	82				α(0.9%)		5.58					0.1415 ± 0.0005	11%
	190	82											0.1512 ± 0.0001	9.0%
	190	82											0.1582 ± 0.0002	1.7%
	190	82											0.1932 ± 0.0002	1.3%
	190	82											0.2105 ± 0.0002	3.6%
	190	82											0.2742 ± 0.0001	3.1%
	190	82											0.3627 ± 0.0002	1.9%
	190	82											0.3764 ± 0.0001	7.1%
	190	82											0.3817 ± 0.0002	1.8%
	190	82											0.5660 ± 0.0002	4.7%
	190	82											0.5983 ± 0.0002	8.1%
	190	82											0.7394 ± 0.0002	4.1%
	190	82											0.7909 ± 0.0002	3.0%
	190	82											0.9422 ± 0.0001	34%
	190	82											1.2355 ± 0.0002	4.6%
	190	82											1.8545 ± 0.0003	0.7%
$_{82}Pb^{191m}$	191	82			2.2 m.	β+,E.C.					13/2+		ann.rad.	
	191	82											0.3250 ± 0.0002	27 +
	191	82											0.34112 ± 0.0002	20

Isotope	A	Z	% Natural abundance	Atomic mass	Half-life	Decay mode	Decay energy (MeV)	Particle energy (MeV)	Particle intensity	Thermal neutron cross section	Spin (h/2π)	μ Nucl. mag. moment	Gamma-ray energy (MeV)	Gamma-ray intensity
	191	82											0.3871 ± 0.0002	100
	191	82											0.4040 ± 0.0002	11
	191	82											0.5606 ± 0.0002	27
	191	82											0.6135 ± 0.0002	40
	191	82											0.7057 ± 0.0002	16
	191	82											0.7122 ± 0.0002	46
	191	82											0.8739 ± 0.0002	23
	191	82											1.0933 ± 0.0002	15
$_{82}Pb^{191}$	191	82		190.978160	1.3 m.	β+,E.C.	5.90						ann rad	
	191	82											0.9368 ± 0.0002	
$_{82}Pb^{192}$	192	82		191.975790	≈ 3.5 m.	β+,E.C.	3.42				0+		ann.rad.	
	192	82											0.1675 ± 0.0001	14 +
	192	82											0.2131 ± 0.0003	4
	192	82											0.2149 ± 0.0003	5
	192	82											0.2507 ± 0.0002	5
	192	82											0.3710 ± 0.0002	8
	192	82											0.4141 ± 0.0003	6
	192	82											0.6082 ± 0.0001	18
	192	82											0.7816 ± 0.0003	9
	192	82											1.1954 ± 0.0002	48
$_{82}Pb^{193}$	193	82		192.976120	5.8 m.	β+,E.C.	5.22				13/2+		ann.rad.	
	193	82											0.3650	100 +
	193	82											0.3922	21
	193	82											0.7165	7
	193	82											0.7361	5
	193	82											0.7558	3
$_{82}Pb^{194}$	194	82		193.973980	10 m.	β+,E.C.	2.85				0+		ann.rad.	
	194	82											0.2036 ± 0.0005	
$_{82}Pb^{195m}$	195	82			15 m.	β+(8%)					13/2+		ann.rad.	16%
	195	82				E.C.(92%)							Tl k x-ray	44%
	195	82											0.3132 ± 0.0001	7%
	195	82											0.3836 ± 0.0001	92%(D)
	195	82											0.3942 ± 0.0001	44%
	195	82											0.4284 ± 0.0001	4%
	195	82											0.6076 ± 0.0002	8%
	195	82											0.7077 ± 0.0002	14%
	195	82											0.7422 ± 0.0002	4.2%
	195	82											0.8784 ± 0.0002	24%
	195	82											1.0679 ± 0.0002	6.3%
$_{82}Pb^{195}$	195	82		194.974480	37 m.	β+,E.C.	4.52						ann.rad.	
	195	82											0.0.3836 ± 0.0001	100 +
	195	82											0.3937 ± 0.0003	7
	195	82											0.7776 ± 0.0002	6
	195	82											0.8354 ± 0.0002	3
	195	82											0.8712 ± 0.0002	2
	195	82											0.8834 ± 0.0003	4
$_{82}Pb^{196}$	196	82		195.972680	37 m.	β+,E.C.	2.060				0+		Tl k x-ray	51%
	196	82											0.1977 ± 0.0005	11%
	196	82											0.2400 ± 0.0005	8%
	196	82											0.2531 ± 0.0005	27%
	196	82											0.3022 ± 0.0005	4%
	196	82											0.3665 ± 0.0005	11%
	196	82											0.4939 ± 0.0005	6%
	196	82											0.5021 ± 0.0005	26%
	196	82											0.9541 ± 0.0008	4%
$_{82}Pb^{197m}$	197	82			43 m.	E.C.(79%)					13/2+		Tl k x-ray	43%
	197	82				β+(2%)							0.3079 ± 0.0002	3%
	197	82				I.T.(19%)	0.3193						0.3877 ± 0.0001	25%
	197	82											0.4162 ± 0.0001	2%
	197	82											0.5578 ± 0.0001	3.5%
	197	82											0.6956 ± 0.0001	9.5%
	197	82											0.7241 ± 0.0001	3.7%
	197	82											0.7743 ± 0.0001	14%
	197	82											0.8933 ± 0.0001	2.1%
	197	82											0.9577 ± 0.0001	5.8%
	197	82											1.1177 ± 0.0001	3.2%
	197	82											1.4972 ± 0.0001	1.0%
	197	82											(0.2 - 2.2)weak	
$_{82}Pb^{197}$	197	82		196.973360	≈ 8 m.	E.C.(97%)	3.60				(3/2-)		Tl k x-ray	42%
	197	82				β+(3%)							0.3755 ± 0.0001	14%
	197	82											0.3858 ± 0.0001	56%
	197	82											0.7611 ± 0.0001	15%
	197	82											0.8716 ± 0.0001	6.8%
	197	82											0.8961 ± 0.0001	5.6%
	197	82											0.9017 ± 0.0001	3.5%
	197	82											1.0928 ± 0.0001	3.7%
	197	82											1.1561 ± 0.0001	4.2%
	197	82											1.2612 ± 0.0001	9.2%
	197	82											1.2889 ± 0.0001	5.4%
	197	82											1.6746 ± 0.0001	3.3%
	197	82											1.8540 ± 0.0001	6.8%
	197	82											2.3455 ± 0.0001	4.9%
$_{82}Pb^{198}$	198	82		197.971960	2.4 h.	E.C.	1.44				0+		Tl k x-ray	49%
	198	82											0.1734 ± 0.0001	18%
	198	82											0.2595 ± 0.0001	5.8%

TABLE OF THE ISOTOPES (Continued)

Isotope	A	Z	% Natural abundance	Atomic mass	Half-life	Decay mode	Decay energy (MeV)	Particle energy (MeV)	Particle intensity	Thermal neutron cross section	Spin ($h/2\pi$)	μ Nucl. mag. moment	Gamma-ray energy (MeV)	Gamma-ray intensity
	198	82											0.2903 ± 0.0001	36%
	198	82											0.3654 ± 0.0001	19%
	198	82											0.3820 ± 0.0001	5.6%
	198	82											0.3977 ± 0.0001	2.9%
	198	82											0.5750 ± 0.0001	3.1%
	198	82											0.6490 ± 0.0001	1.8%
	198	82											0.7430 ± 0.0003	1.5%
	198	82											0.8653 ± 0.0001	5.9%
$_{82}Pb^{199m}$	199	82			12.2 m.	I.T.(93%)	0.4248				13/2 +		Pb k x-ray	21%
	199	82				β +, EC (7%)							0.4255 ± 0.0005	18.5%
$_{82}Pb^{199}$	199	82		198.972870	1.5 h.	E.C.(99%) β + (1%)	2.800				5/2-		Tl k x-ray	42%
	199	82											0.3534 ± 0.0001	14%
	199	82											0.4005 ± 0.0001	1.9%
	199	82											0.7202 ± 0.0001	9.5%
	199	82											0.7539 ± 0.0001	2.3%
	199	82											0.7620 ± 0.0001	3.3%
	199	82											0.7815 ± 0.0001	2.7%
	199	82											0.8748 ± 0.0001	2.4%
	199	82											0.9379 ± 0.0001	3.1%
	199	82											1.0292 ± 0.0001	2.4%
	199	82											1.1210 ± 0.0001	2.2%
	199	82											1.1350 ± 0.0001	11.5%
	199	82											1.2391 ± 0.0001	3.1%
	199	82											1.3827 ± 0.0001	4.2%
	199	82											1.5020 ± 0.0001	3.1%
	199	82											1.6584 ± 0.0001	8.2%
	199	82											1.7497 ± 0.0001	3.4%
	199	82											(0.22 - 2.4)weak	
$_{82}Pb^{200}$	200	82		199.971790	21.5 ± 0.4 h.	E.C.	0.800				0+		Tl k x-ray	51%
	200	82											0.14763 ± 0.00002	38%
	200	82											0.23562 ± 0.00002	4.3%
	200	82											0.25717 ± 0.00002	4.5%
	200	82											0.26837 ± 0.00002	4.0%
	200	82											0.28916 ± 0.00005	1.1%
	200	82											0.28992 ± 0.00002	1.7%
	200	82											0.45052 ± 0.00003	3.3%
$_{82}Pb^{201m}$	201	82			1.02 ± 0.03 m.	I.T.	0.6291				13/2 +		Pb k x-ray	15%
	201	82											0.6288 ± 0.0005	54%
$_{82}Pb^{201}$	201	82		200.972830	9.33 ± 0.03 h.	E.C.	1.90				5/2-		Tl k x-ray	44%
	201	82											0.33120 ± 0.00003	79%
	201	82											0.36131 ± 0.00003	9.9%
	201	82											0.40607 ± 0.00004	2.0%
	201	82											0.58462 ± 0.00004	3.6%
	201	82											0.69252 ± 0.00003	4.3%
	201	82											0.76738 ± 0.00004	3.2%
	201	82											0.90764 ± 0.00005	5.7%
	201	82											0.94594 ± 0.00004	7.4%
	201	82											1.07009 ± 0.00005	1.1%
	201	82											1.09858 ± 0.00004	1.8%
	201	82											1.23884 ± 0.00005	1.2%
	201	82											1.27714 ± 0.00004	1.6%
	201	82											(0.11 - 1.8)weak	
$_{82}Pb^{202m}$	202	82			3.62 ± 0.03 h.	I.T.(90%)	2.170				9-		Pb k x-ray	2.3%
	202	82				β + (10%)							Tl k x-ray	3%
	202	82											0.42219 ± 0.00003	86%
	202	82											0.45979 ± 0.00007	8.6%
	202	82											0.49055 ± 0.00007	9.1%

TABLE OF THE ISOTOPES (Continued)

Isotope	A	Z	% Natural abundance	Atomic mass	Half-life	Decay mode	Decay energy (MeV)	Particle energy (MeV)	Particle intensity	Thermal neutron cross section	Spin (h/2π)	μ Nucl. mag. moment	Gamma-ray energy (MeV)	Gamma-ray intensity
	202	82											0.65753 ± 0.00003	32%
	202	82											0.78700 ± 0.00006	50%
	202	82											0.96271 ± 0.00005	92%
$_{82}Pb^{202}$	202	82		201.972134	5.3 x 10^4 y.	E.C.	0.046				0+		Tl L x-ray	16%
$_{82}Pb^{203m}$	203	82			6.3 ± 0.2 s.	I.T.	0.8252				13/2+		Pb k x-ray	7.6%
	203	82											0.0203 ± 0.0002	0.4%
	203	82											0.8252 ± 0.0001	71%
$_{82}Pb^{203}$	203	82		202.973365	2.169 ± 0.004d	E.C.	0.974				5/2-		Tl k x-ray	42%
	203	82											0.279188 ± 0.00003	80.1%
	203	82											0.40131 ± 0.00001	3.4%
	203	82											0.68050 ± 0.00001	0.7%
$_{82}Pb^{204m}$	204	82			1.12 ± 0.01 h.	I.T.	2.185				9-		Pb k x-ray	4.5%
	204	82											0.37481 ± 0.00006	89%
	204	82											0.89922 ± 0.00007	99%
	204	82											0.91175 ± 0.00007	94%
$_{82}Pb^{204}$	204	82	1.4%	203.973020						0.66 b.	0+			
$_{82}Pb^{205}$	205	82		204.974458	1.51 x 10^7y.	E.C.	0.053						Tl L x-ray	16%
$_{82}Pb^{206}$	206	82	24.1%	205.974440						(0.006 + 0.025) b.	0+			
$_{82}Pb^{207m}$	207	82			0.796 s.	I.T.	1.632				13/2+		Pb k x-ray	4.8%
	207	82											0.56915 ± 0.00002	98%
	207	82											1.06310 ± 0.00002	89%
$_{82}Pb^{207}$	207	82	22.1%	206.975872						0.70 b.	1/2-	+0.5926		
$_{82}Pb^{208}$	208	82	52.4%	207.976627						0.5 mb.	0+			
$_{82}Pb^{209}$	209	82		208.981065	3.25 ± 0.02 h.	β-	0.644	0.645	100%		9/2+			
$_{82}Pb^{210}$	210	82		209.984163	22.3 ± 0.2 y.	β-	0.063	0.017	81%		0+			
	210	82						0.061	19%					
$_{82}Pb^{211}$	211	82		210.988735	36.1 ± 0.2 m.	β-	1.379	0.57	5%		(9/2+)		0.40486 ± 0.00003	3.8%
	211	82						1.36	92%				0.42700 ± 0.00003	1.7%
	211	82											0.83186 ± 0.00003	3.8%
	211	82											(0.09 - 1.27)weak	
$_{82}Pb^{212}$	212	82		211.991871	10.64±0.01 h.	β-	0.574	0.28	83%		0+		Bi k x-ray	18%
	212	82						0.57	12%				0.23858 ± 0.00001	43.6%
	212	82											0.30003 ± 0.00001	3.3%
$_{82}Pb^{213}$	213	82		212.996510	10.2 ± 0.3 m.	β-	2.00							
$_{82}Pb^{214}$	214	82		213.999798	26.8 ± 0.9 m.	β-	1.032	0.67	48%		0+		Bi k x-ray	11%
	214	82						0.73	42%				0.24192 ± 0.00003	7.5%
	214	82											0.29509 ± 0.00002	19.2%
	214	82											0.35187 ± 0.00004	37%
	214	82											0.78583 ± 0.00002	1.1%
Bi		83		208.9804						0.034 b.				
$_{83}Bi^{190}$	190	83		189.988480	5.4 s.	β+, EC (10%)	9.70							
	190	83				α(90%)		6.45						
$_{83}Bi^{191}$	191	83		190.986110	13 s.	β+, EC (60%)	0.370							
	191	83				α(40%)		6.32						
$_{83}Bi^{192}$	192	83		191.985400	42 s.	β+, EC (80%)	8.95							
	192	83				α(20%)		6.06						
$_{83}Bi^{193m}$	193	83			3.5 s.	β+,E.C.								
	193	83				α		6.48						
$_{83}Bi^{193}$	193	83		192.983180	64 s.	β+, EC (40%)	6.58							
	193	83				α(60%)		5.91						
$_{83}Bi^{194}$	194	83		193.982540	1.8 m.	β+, EC (99.9%)	7.98				(10-)		0.1661	46 +
	194	83				α(0.1%)							0.1740	28

TABLE OF THE ISOTOPES (Continued)

Isotope	A	Z	% Natural abundance	Atomic mass	Half-life	Decay mode	Decay energy (MeV)	Particle energy (MeV)	Particle intensity	Thermal neutron cross section	Spin (h/2π)	μ Nucl. mag. moment	Gamma-ray energy (MeV)	Gamma-ray intensity
	194	83											0.2802	70
	194	83											0.421	55
	194	83											0.5754	87
	194	83											0.9650	100
$_{83}Bi^{195m}$	195	83			1.5 m.	β+, EC (94%)								
	195	83				α(6%)		6.11						
$_{83}Bi^{195}$	195	83		194.980700	2.8 m.	β+, EC (99.8%)	5.80							
	195	83				α(0.2%)		5.45						
$_{83}Bi^{196}$	196	83		195.980690	4.5 m.	E.C.	7.46						0.1376 ± 0.0003	10 +
	196	83											0.3368 ± 0.0003	16
	196	83											0.3720 ± 0.0006	46
	196	83											0.6880 ± 0.0005	62
	196	83											1.0486 ± 0.0005	100
$_{83}Bi^{197}$	197	83		196.978880	≈ 10 m.	β+,E.C.					1/2+			
$_{83}Bi^{198m}$	198	83			7.7 s.	I.T.	0.2485				(10-)		0.2485 ± 0.0005	38%
$_{83}Bi^{198}$	198	83		197.979000	β+,E.C.	6.55					(7+)		0.0900	8 +
	198	83											0.1381	2
	198	83											0.1976	80
	198	83											0.3179	37
	198	83											0.4343	7
	198	83											0.5465	3
	198	83											0.5624	79
	198	83											0.9173	5
	198	83											1.0635	100
$_{83}Bi^{199m}$	199	83			24.7 m.	β+,E.C.							ann.rad.	
$_{83}Bi^{199}$	199	83		198.977520	27 m.	β+,E.C.	4.32				9/2-		0.7203 ± 0.0005	1.9%
	199	83											0.7794 ± 0.0005	1.9%
	199	83											0.8374 ± 0.0005	13%
	199	83											0.8417 ± 0.0005	16%
	199	83											0.9141 ± 0.0005	2.1%
	199	83											0.9264 ± 0.0005	7.6%
	199	83											0.9460 ± 0.0005	15%
	199	83											0.9661 ± 0.0005	2.8%
	199	83											0.9775 ± 0.0005	2.3%
	199	83											1.0228 ± 0.0005	6.3%
	199	83											1.0340 ± 0.0005	8.4%
	199	83											1.0528 ± 0.0005	10%
	199	83											1.1370 ± 0.0005	7.9%
	199	83											1.1464 ± 0.0005	6.4%
	199	83											1.2122 ± 0.0005	6.2%
	199	83											1.3056 ± 0.0005	10%
	199	83											1.4486 ± 0.0005	2.5%
	199	83											1.5059 ± 0.0005	7.8%
	199	83											1.5173 ± 0.0005	2.1%
	199	83											1.5405 ± 0.0005	1.3%
	199	83											1.7808 ± 0.0005	1.4%
	199	83											1.9216 ± 0.0005	1.3%
	199	83											2.0215 ± 0.0005	1.5%
	199	83											2.0587 ± 0.0005	1.9%
	199	83											2.6669 ± 0.0005.	1.0%
	199	83											(0.12 - 3.2)many	
$_{83}Bi^{200m}$	200	83			31 m.	β+,E.C.					(2+)		0.2453 ± 0.0001	4.3%
	200	83											0.4198 ± 0.0001	20%
	200	83											0.4624 ± 0.0001	36%
	200	83											0.7127 ± 0.0001	1.5%
	200	83											1.0265 ± 0.0002	85%
	200	83											1.7395 ± 0.0002	3.7%
$_{83}Bi^{200}$	200	83		199.978090	36 m.	E.C.(90%) β+(10%)	5.86				7+		ann.rad.	20%
	200	83											Pb k x-ray	49%
	200	83											0.2452 ± 0.0001	46%
	200	83											0.4198 ± 0.0001	91%
	200	83											0.4623 ± 0.0001	98%
	200	83											0.5455 ± 0.0002	4.5%
	200	83											0.6478 ± 0.0004	2.6%
	200	83											0.7810 ± 0.0005	2.0%
	200	83											0.9316 ± 0.0005	2.6%
	200	83											1.0265 ± 0.0002	100%
$_{83}Bi^{201m}$	201	83			59.1 ± 0.6 m.	I.T.	0.846				(1/2+)		Bi k x-ray	9 +
	201	83				β+,E.C.							0.8464	100
$_{83}Bi^{201}$	201	83		200.976930	1.80 ± 0.05 h.	E.C.	3.810				9/2-		Pb k x-ray	51%
	201	83											0.6288 ± 0.0005	24%
	201	83											0.7859 ± 0.0004	9.7%
	201	83											0.9015 ± 0.0005	8.5%
	201	83											0.9357 ± 0.0004	11%
	201	83											1.0138 ± 0.0007	11%
	201	83											1.3255 ± 0.001	6.1%
	201	83											1.5030 ± 0.0007	1.0%
	201	83											1.6509 ± 0.0005	5.9%
	201	83											(0.13 - 2.4)weak	
$_{83}Bi^{202}$	202	83		201.977660	1.72 h.	β+(3%) E.C.(97%)	5.15				5+		ann.rad.	6%
	202	83											Pb k x-ray	43%

TABLE OF THE ISOTOPES (Continued)

Isotope	A	Z	% Natural abundance	Atomic mass	Half-life	Decay mode	Decay energy (MeV)	Particle energy (MeV)	Particle intensity	Thermal neutron cross section	Spin (h/2π)	μ Nucl. mag. moment	Gamma-ray energy (MeV)	Gamma-ray intensity
	202	83											0.16815 ± 0.00004	4.8%
	202	83											0.24896 ± 0.00004	3.1%
	202	83											0.32018 ± 0.00005	3.1%
	201	83											0.34651 ± 0.00003	4.6%
	202	83											0.30531 ± 0.00003	4.8%
	202	83											0.57860 ± 0.00004	7.3%
	202	83											0.67622 ± 0.00003	1.9%
	202	83											0.85261 ± 0.00007	2.3%
	202	83											0.85848 ± 0.00004	1.6%
	202	83											0.92734 ± 0.00004	7.1%
	202	83											1.24553 ± 0.00003	2.8%
	202	83											1.55667 ± 0.00005	1.9%
	202	83											1.78056 ± 0.00006	0.7%
	202	83											2.3226 ± 0.000106	0.2%
	202	83											2.3408 ± 0.0007	0.2%
	202	83											(0.08 - 3.5)weak	
$_{83}Bi^{203}$	203	83		202.976830	11.76 ± 0.05h.	E.C. (99.8%)	3.22				9/2-	+4.62	Pb k x-ray	42%
	203	83				β+(0.2%)		1.35					0.1865 ± 0.0002	3.1%
	203	83											0.2642 ± 0.0003	5.2%
	203	83											0.3818 ± 0.0004	1.3%
	203	83											0.5694 ± 0.0004	1.2%
	203	83											0.6337 ± 0.0004	1.3%
	203	83											0.7224 ± 0.0003	4.8%
	203	83											0.8162 ± 0.0005	4.0%
	203	83											0.8203 ± 0.0002	30%
	203	83											0.8473 ± 0.0003	8.5%
	203	83											0.8666 ± 0.0005	1.5%
	203	83											0.8969 ± 0.0004	13%
	203	83											0.9334 ± 0.0004	1.4%
	203	83											1.0339 ± 0.0003	8.8%
	203	83											1.1986 ± 0.0004	2.0%
	203	83											1..2031 ± 0.0004	1.5%
	203	83											1.2537 ± 0.0005	1.3%
	203	83											1.5067 ± 0.0003	3.7%
	203	83											1.5365 ± 0.0004	7.5%
	203	83											1.5523 ± 0.0004	1.5%
	203	83											1.5929 ± 0.0005	1.09%
	203	83											1.6796 ± 0.0003	8.85%
	203	83											1.7198 ± 0.0003	3.4%
	203	83											1.7485 ± 0.0004	1.9%
	203	83											1.8475 ± 0.0003	11.4%
	203	83											1.8882 ± 0.0003	1.9%
	203	83											1.8931 ± 0.0003	8.2%
	203	83											1.9282 ± 0.001	1.1%
	203	83											2.0114 ± 0.0005	1.8%
	203	83											(0.1 - 2.9)many	
$_{83}Bi^{204}$	204	83		203.977740	11.2 ± 0.1 h.	E.C.	4.39				6+	+4.28	Pb k x-ray	44%
	204	83											0.17617 ± 0.00006	1.1%
	204	83											0.21608 ± 0.00007	1.4%
	204	83											0.21951 ± 0.00008	2.3%c
	204	83											0.24909 ± 0.00006	2.1%
	204	83											0.28926 ± 0.00007	2.8%
	204	83											0.37481 ± 0.00006	81%
	204	83											0.44056 ± 0.00005	2.5%
	204	83											0.53284 ± 0.00007	1.3%
	204	83											0.66161 ± 0.00007	2.6%
	204	83											0.67085 ± 0.00006	10.6%
	204	83											0.79130 ± 0.00007	3.2%

Isotope	A	Z	% Natural abundance	Atomic mass	Half-life	Decay mode	Decay energy (MeV)	Particle energy (MeV)	Particle intensity	Thermal neutron cross section	Spin (h/2π)	μ Nucl. mag. moment	Gamma-ray energy (MeV)	Gamma-ray intensity
	204	83											0.89922 ± 0.00007	98%
	204	83											0.91175 ± 0.00007	14%
	204	83											0.91231 ± 0.00008	11.1%
	204	83											0.91834 ± 0.00008	10.8%
	204	83											0.98409 ± 0.00005	58%
	204	83											1.11140 ± 0.00006	1.4%
	204	83											1.20392 ± 0.00008	2.1%
	204	83											1.21181 ± 0.00006	3.1%
	204	83											1.27481 ± 0.00007	2.2%
	204	83											1.70355 ± 0.00009	2.0%
	204	83											1.75534 ± 0.00006	1.2%
	204	83											1.89640 ± 0.00009	1.3%
	204	83											2.6808 ± 0.0002	0.4%
	204	83											2.8374 ± 0.0001	0.2%
$_{83}Bi^{205}$	205	83		204.977365	15.31 ± 0.04d.	E.C.	2.708				9/2-	4.16	Pb k x-ray	35%
	205	83											0.54986 ± 0.00003	0.3%
	205	83											0.57060 ± 0.00003	0.43%
	205	83											0.57974 ± 0.00003	0.54%
	205	83											0.70347 ± 0.00003	31%
	205	83											0.98764 ± 0.00003	1.6%
	205	83											1.04375 ± 0.00003	0.75%
	205	83											1.19004 ± 0.00004	0.23%
	205	83											1.61435 ± 0.00004	0.23%
	205	83											1.76435 ± 0.00004	32.5%
	205	83											1.86171 ± 0.00003	0.61%
	205	83											1.90345 ± 0.00004	0.25%
$_{83}Bi^{206}$	206	83		205.978478	6.243 ± 0.003d	E.C.	3.761				6+		Pb k x-ray	54%
	206	83											0.18403 ± 0.00002	15.8%
	206	83											0.34353 ± 0.00002	23.4%
	206	83											0.39803 ± 0.00002	10.7%
	206	83											0.49700 ± 0.00003	15.3%
	206	83											0.51619 ± 0.00003	40.7%
	206	83											0.53748 ± 0.00003	30.4%
	206	83											0.62053 ± 0.00003	5.8%
	206	83											0.63228 ± 0.00003	4.5%
	206	83											0.65721 ± 0.00003	1.9%
	206	83											0.80313 ± 0.00005	98.9%
	206	83											0.88100 ± 0.00003	66.2%
	206	83											0.89503 ± 0.00003	15.7%
	206	83											1.01856 ± 0.00003	7.6%
	206	83											1.09825 ± 0.00003	13.5%
	206	83											1.40510 ± 0.00004	1.4%
	206	83											1.59525 ± 0.00003	5.0%
	206	83											1.71878 ± 0.00003	31.8%

Isotope	A	Z	% Natural abundance	Atomic mass	Half-life	Decay mode	Decay energy (MeV)	Particle energy (MeV)	Particle intensity	Thermal neutron cross section	Spin (h/2π)	μ Nucl. mag. moment	Gamma-ray energy (MeV)	Gamma-ray intensity
	206	83											1.87898 ± 0.00005	2.0%
$_{83}Bi^{207}$	207	83		206.978446	32.2 ± 0.1 y.	E.C.	2.398				9/2-	4.10	Pb k x-ray	36%
	207	83											0.56915 ± 0.00002	97.8%
	207	83											1.06310 ± 0.00002	74.9%
	207	83											1.76971 ± 0.00004	6.9%
$_{83}Bi^{208}$	208	83		207.979717	3.68 x 10^5y.	E.C.	2.878				5+		Pb k x-ray	32%
	208	83											2.61435 ± 0.0001	99.8%
$_{83}Bi^{209}$	209	83	100%	208.980374						(10 mb + 24 mb)	9/2-	+4.110	6	
$_{83}Bi^{210m}$	210	83			3.0 x 10^6 y.	α		4.420(3)	0.29%		9-		Tl k x-ray	6.6%
	210	83						4.569(3)	3.9%				0.2661 ± 0.0002	50%
	210	83						4.584(3)	1.4%				0.3052 ± 0.0002	28%
	210	83						4.908(4)	39%				0.6502 ± 0.0002	3.6%
	210	83						4.946(3)	55%					
$_{83}Bi^{210}$	210	83		209.984095	5.01 ± 0.01 d.	β-	1.16	1.16	99%		1-	-0.044	6 0.2661 ± 0.0002	4x10-
	210	83											0.3.52 ± 0.0002	6x10-
$_{83}Bi^{211}$	211	83		210.987255	2.14 ± 0.02 m.	α(99.7%)		6.279	16%		9/2-		Tl k x-ray	1.3%
	211	83				β-(0.3%)	0.584	6.623	84%				0.3501 ± 0.0002	12.8%
$_{83}Bi^{212m2}$	212	83			9 ± 1 m.	β-					(15-)			
$_{83}Bi^{212m1}$	212	83			25 ± 1 m.	α(93%)		6.300	40%		(9-)		0.120 ± 0.001	
	212	83				β-(7%)		6.340	53%				0.233 ± 0.001	
	212	83											0.275 ± 0.001	
	212	83											0.404 ± 0.001	
	212	83											0.727 ± 0.001	
$_{83}Bi^{212}$	212	83		211.991255	1.009 ± 0.001 h.	β-(64%)	2.248				(1-)		Tl k x-ray	0.13%
	212	83				α(36%)		6.051	25%				Po k x-ray	0.1%
	212	83						6.090	9.6%				0.2881 ± 0.0001	0.34%
	212	83											0.4528 ± 0.0001	0.36%
	212	83											0.72725 ± 0.00005	6.6%
	212	83											0.78551 ± 0.00005	1.1%
	212	83											0.87342 ± 0.00006	0.37%
	212	83											0.89342 ± 0.00006	0.37%
	212	83											0.9522 ± 0.0001	0.18%
	212	83											1.0787 ± 0.0001	0.53%
	212	83											1.51275 ± 0.00006	0.31%
	212	83											1.62066 ± 0.00006	1.5%
	212	83											1.8059 ± 0.0001	0.1%
$_{83}Bi^{211}$	211	83		212.994359	45.6 ± 0.1 m.	β-(98%)	1.422	1.02	31%		9/2-		Po k x-ray	1.3%
	213	83				α(2%)		1.42	66%				0.3288 ± 0.0001	0.2%
	213	83						5.549	0.16%				0.44034 ± 0.00002	16%
	213	833						5.869	2.0%				0.80727 ± 0.00004	0.26%
	213	83											1.10006 ± 0.00005	0.28%
$_{83}Bi^{214}$	214	83		213.998691	19.9 ± 0.4 m.	β-	3.27						0.60931 ± 0.00001	46.1%
	214	83											0.66544 ± 0.00002	1.6%
	214	83											0.76835 ± 0.00001	4.9%
	214	83											0.80615 ± 0.00002	1.2%
	214	83											0.93404 ± 0.00002	3.2%
	214	83											1.12027 ± 0.00002	15%
	214	83											1.15518 ± 0.00002	1.7%
	214	83											1.23810 ± 0.00002	5.96%
	214	83											1.28095 ± 0.00002	1.5%
	214	83											1.37766 ± 0.00002	4.0%
	214	83											1.40148 ± 0.00004	1.4%
	214	83											1.40797 ± 0.00004	2.5%
	214	83											1.50922 ± 0.00003	2.2%

Isotope	A	Z	% Natural abundance	Atomic mass	Half-life	Decay mode	Decay energy (MeV)	Particle energy (MeV)	Particle intensity	Thermal neutron cross section	Spin (h/2π)	μ Nucl. mag. moment	Gamma-ray energy (MeV)	Gamma-ray intensity
	214	83											1.66126 ± 0.00002	1.1%
	214	83											1.72958 ± 0.00002	3.0%
	214	83											1.76449 ± 0.00002	15.9%
	214	83											1.84741 ± 0.00003	2.1%
	214	83											2.20409 ± 0.00011	4.99%
	214	83											2.44768 ± 0.00004	1.55%
	214	83											(0.19 - 3.27)many	
$_{83}Bi^{215}$	215	83		215.001930	7.4 ± 0.6 m.	β-	2.250							
Po		84												
$_{84}Po^{194}$	194	84		193.988180	0.7 s.	α		6.85			0+			
$_{84}Po^{195m}$	195	84			2.0 s.	α		6.70						
$_{84}Po^{195}$	195	84		194.988010	≈ 4.5 s.	α		6.61						
$_{84}Po^{196}$	196	84		195.985540	≈ 5.5 s.	α(95%)		6.520			0+			
	196	84				β+, EC (5%)								
$_{84}Po^{197m}$	197	84			26 s.	α(84%)		6.385			13/2+			
	197	84				β+, EC (16%)								
$_{84}Po^{197}$	197	84		196.985600	56 s.	α(44%)		6.282			(3/2-)			
	197	84				β+, EC (56%)								
$_{84}Po^{198}$	198	84		197.983360	1.76 m.	α(70%)		6.182			0+			
	198	84				β+, EC (30%)								
$_{84}Po^{199m}$	199	84			4.2 m.	β+, EC (51%)					13/2+		ann.rad.	
	199	84				α(39%)		6.059					0.2745	7.4%
	199	84											0.4998 ± 0.0005	25%
	199	84											1.0020 ± 0.0005	60%
$_{84}Po^{199}$	199	84		198.983610	5.2 m.	β+, EC (88%)	5.67				(3/2-)		Bi k x-ray	38%
	199	84				α(12%)		5.952					0.1877 ± 0.0005	7.5%
	199	84											0.2291 ± 0.0005	4.8%
	199	84											0.2335 ± 0.0005	5.5%
	199	84											0.2460 ± 0.0006	4.2%
	199	84											0.2607 ± 0.0005	4.0%
	199	84											0.3616 ± 0.0005	22%
	199	84											0.3978 ± 0.0004	4%
	199	84											0.4749 ± 0.0005	7%
	199	84											0.5068 ± 0.0005	3.7%
	199	84											0.9984 ± 0.0005	15%
	199	84											1.0214 ± 0.0006	24%
	199	84											1.0344 ± 0.0005	47%
$_{84}Po^{200}$	200	84		199.981700	11.5 ± 0.1 m.	β+, EC (85%)	3.370				0+		0.14748 ± 0.00001	4.4%
	200	84				α(15%)		5.863					0.32792 ± 0.00009	2.6%
	200	84											0.43007 ± 0.00012	4.8%
	200	84											0.4343 ± 0.00001	9.3%
	200	84											0.6176 ± 0.0001	19.7%
	200	84											0.6709 ± 0.0001	34%
	200	84											0.6956 ± 0.0002	5.5%
	200	84											0.7966 ± 0.0001	7.9%
	200	84											0.8499 ± 0.0001	4.9%
	200	84											0.8758 ± 0.0001	1.8%
	200	84											0.8958 ± 0.0002	1.5%
	200	84											0.9146 ± 0.0002	1.2%
	200	84											0.9455 ± 0.0001	1..1%
	200	84											1.0845 ± 0.0002	3.8%
	200	84											1.1729 ± 0.0002	1.1%
	200	84											1.2856 ± 0.0001	1.2%
	200	84											1.3876 ± 0.0002	1.0%
	200	84											1.8019 ± 0.0002	1.3%
$_{84}Po^{201m}$	201	84			8.9 ± 0.2 m.	β+, EC (57%)					13/2+		Bi k x-ray	19%
	201	84				I.T.(40%)	0.418						Po k x-ray	44%
	201	84				α(3%)		5.786					0.2726 ± 0.0004	2.8%
	201	84											0.4123 ± 0.0005	15.1%
	201	84											0.4179 ± 0.0003	33%
	201	84											0.9670 ± 0.0005	34%
$_{84}Po^{201}$	201	84		200.982190	15.3 ± 0.2 m.	β+, EC (98%)	4.90				3/2-		Bi k x-ray	38%
	201	84				α(2%)		5.683(3)					0.2056 ± 0.0003	8.0%
	201	84											0.2229 ± 0.0004	5.5%
	201	84											0.2250 ± 0.0004	12%
	201	84											0.2390 ± 0.0005	8.1%
	201	84											0.4285 ± 0.0004	8.9%

Isotope	A	Z	% Natural abundance	Atomic mass	Half-life	Decay mode	Decay energy (MeV)	Particle energy (MeV)	Particle intensity	Thermal neutron cross section	Spin (h/2π)	μ Nucl. mag. moment	Gamma-ray energy (MeV)	Gamma-ray intensity
	201	84											0.5375 ± 0.0004	5.5%
	201	84											0.5520 ± 0.0004	6.3%
	201	84											0.6390 ± 0.0004	5.4%
	201	84											0.8483 ± 0.0005	13%
	201	84											0.8904 ± 0.0004	5.4%
	201	84											0.9048 ± 0.0005	29%
	201	84											1.1639 ± 0.0005	3.7%
	201	84											1.2060 ± 0.0005	3.3%
$_{84}Po^{202}$	202	84		201.980680	44.7 ± 0.5 m.	β+, EC (90%)	2.820				0+		0.0410 ± 0.0001	3.0%
	201	84				α(2%)		5.588					0.1656 ± 0.0001	8.7%
	202	84											0.2135 ± 0.0002	3.4%
	202	84											0.3158 ± 0.0002	14%
	202	84											0.3365 ± 0.0002	2%
	202	84											0.4275 ± 0.0002	1.6%
	202	84											0.4581 ± 0.0002	3.8%
	202	84											0.5061 ± 0.0002	4.4%
	202	84											0.5977 ± 0.0002	2.6%
	202	84											0.6433 ± 0.0002	3.6%
	202	84											0.6884 ± 0.0003	51%
	202	84											0.7126 ± 0.0002	4.6%
	202	84											0.7168 ± 0.0002	6.1%
	202	84											0.7903 ± 0.0005	7.2%
	202	84											0.9736 ± 0.0002	4.9%
	202	84											1.1684 ± 0.0005	2.0%
	202	84											1.2148 ± 0.0005	1.7%
$_{84}Po^{203m}$	203	84			1.2 ± 0.2 m.	I.T.(96%)	0.6414				13/2+		Bi k x-ray	2.0%
	203	84				β,EC(4%)							Po k x-ray	14%
	203	84											0.5770 ± 0.0005	2.4%
	203	84											0.6414 ± 0.0001	50%
	203	84											0.9049 ± 0.0005	4.4%
$_{84}Po^{203}$	203	84		202.981370	34.8 ± 0.1 m.	β+,E.C.	4.24				5/2-		0.17516 ± 0.00006	3.0%
	203	84											0.18951 ± 0.00008	3.9%
	203	84											0.21477 ± 0.00006	14.5%
	203	84											0.41942 ± 0.00006	2.5%
	203	84											0.4861 ± 0.0001	2.1%
	203	84											0.64776 ± 0.00007	2.1%
	203	84											0.82291 ± 0.00007	2.4%
	203	84											0.8835 ± 0.001	2.0%
	203	84											0.89350 ± 0.00008	19.0%
	203	84											0.90863 ± 0.00007	56%
	203	84											1.09095 ± 0.00007	19.6%
	203	84											1.20233 ± 0.00007	4.7%
	203	84											1.33758 ± 0.00009	3.0%
	203	84											1.35282 ± 0.00008	1.4%
	203	84											1.8175 ± 0.0001	1.1%
	203	84											1.9308 ± 0.0005	0.9%
	203	84											2.0295 ± 0.0003	0.6%
	203	84											2.2369 ± 0.0001	0.6%
$_{84}Po^{204}$	204	84		203.980280	3.53 ± 0.02 h.	E.C.	2.37				0+		Bi k x-ray	65%
	204	84											0.1370 ± 0.0003	12.0%
	204	84											0.20368 ± 0.00007	3.4%
	204	84											1.2300 ± 0.0001	3.0%
	204	84											0.2702 ± 0.0001	31%
	204	84											0.3049 ± 0.0003	3.4%
	204	84											0.3167 ± 0.0001	4.9%
	204	84											0.4270 ± 0.0008	2.2%
	204	84											0.45194 ± 0.00007	2.7%
	204	84											0.4601 ± 0.0001	1.5%
	204	84											0.5349 ± 0.0003	12.9%
	204	84											0.5401 ± 0.0004	1.8%
	204	84											0.6807 ± 0.0001	9.2%
	204	84											0.6951 ± 0.0004	2.5%
	204	84											0.76265 ± 0.00007	11.4%
	204	84											0.8844 ± 0.0001	34%
	204	84											1.0162 ± 0.0001	24.6%
	204	84											1.0400 ± 0.0003	10.8%
	204	84											(0.11 - 1.9)weak	

Isotope	A	Z	% Natural abundance	Atomic mass	Half-life	Decay mode	Decay energy (MeV)	Particle energy (MeV)	Particle intensity	Thermal neutron cross section	Spin (h/2π)	μ Nucl. mag. moment	Gamma-ray energy (MeV)	Gamma-ray intensity
$_{84}Po^{205}$	205	84		204.981150	1.80 ± 0.04 h.	β+,E.C.	3.53				5/2-	+0.26	Bi k x-ray	45%
	205	84											0.21202 ± 0.00007	3.6%
	205	84											0.26108 ± 0.00007	4.0%
	205	84											0.59983 ± 0.00009	2.6%
	205	84											0.61426 ± 0.00007	1.6%
	205	84											0.62478 ± 0.0006	1.0%
	205	84											0.83681 ± 0.00006	19%
	205	84											0.84983 ± 0.00007	25%
	205	84											0.87241 ± 0.00007	37%
	205	84											1.00124 ± 0.00007	29%
	205	84											1.2391 ± 0.0001	4.6%
	205	84											1.5137 ± 0.0002	2.1%
	205	84											1.5519 ± 0.0001	2.9%
	205	84											1.7297 ± 0.0001	1.6%
	205	84											1.8112 ± 0.0001	1.2%
	205	84											2.1689 ± 0.0002	0.4%
	205	84											(0.12 - 2.77)weak	
$_{84}Po^{206}$	206	84		205.980456	8.8 ± 0.1 d.	E.C.(95%)	1.843				0+		Bi k x-ray	46%
	206	84				α(5%)		5.223					0.28644 ± 0.00002	24%
	206	84											0.31156 ± 0.00002	4.2%
	206	84											0.33844 ± 0.00002	19.2%
	206	84											0.46334 ± 0.00002	1.8%
	206	84											0.51134 ± 0.00002	24%
	206	84											0.52252 ± 0.00003	15.7%
	206	84											0.80737 ± 0.00002	22.7%
	206	84											0.86096 ± 0.00002	3.5%
	206	84											0.98027 ± 0.00003	7.1%
	206	84											1.00716 ± 0.00002	3.1%
	206	84											1.03228 ± 0.00002	32.9%
	206	84											1.19102 ± 0.00003	0.5%
	206	84											0.31871 ± 0.00002	0.65%
	206	84											(0.11 - 1.5)weak	
$_{84}Po^{207m}$	207	84			2.8 ± 0.2 s.	I.T.	1.383				19/2		Po k x-ray	28%
	207	84											0.2682 ± 0.001	45%
	207	84											0.30074 ± 0.00006	33%
	207	84											0.81448 ± 0.00006	99%
$_{84}Po^{207}$	207	84		206.981570	5.83 ± 0.07 h.	2.910					5/2-	+0.27	Bi k x-ray	41%
	207	84											0.24962 ± 0.00007	1.6%
	207	84											0.34521 ± 0.00002	2.0%
	207	84											0.36953 ± 0.00008	1.9%
	207	84											0.40570 ± 0.00006	10.1%
	207	84											0.68764 ± 0.00007	2.0%
	207	84											0.74263 ± 0.00006	29.2%
	207	84											0.91176 ± 0.00007	18.0%
	207	84											0.99225 ± 0.00007	60%
	207	84											1.14833 ± 0.00006	6.1%
	207	84											1.37244 ± 0.00008	1.4%
	207	84											2.06008 ± 0.00008	1.44%
$_{84}Po^{208}$	208	84		207.981222	2.898 y.	α	5.213	4.233	0.00	02	0+			

Isotope	A	Z	% Natural abundance	Atomic mass	Half-life	Decay mode	Decay energy (MeV)	Particle energy (MeV)	Particle intensity	Thermal neutron cross section	Spin (h/2π)	μ Nucl. mag. moment	Gamma-ray energy (MeV)	Gamma-ray intensity
	208	84						5.1158	100%					
$_{84}Po^{209}$	209	84		208.982404	105 ± 5 y.	α	4.976	4.624	0.56	%	½-	+0.77	0.26049 ± 0.00003	0.17%
	208	84						4.879	99.2	%			0.8964 ± 0.0002	0.25%
$_{84}Po^{210}$	210	84		209.982848	138.4 d.	α	5.407	4.516	0.00	1%	0+		0.80313 ± 0.00005	0.001%
	210	84						5.304	100%					
$_{84}Po^{211m}$	211	84			25.5 ± 0.3 s.	α		7.273	91%		25/2	+	Pb k x-ray	4.5%
	211	84						7.994	1.7%				0.32808 ± 0.00007	0.01%
	211	84						8.316	0.25	%			0.56915 ± 0.00002	92%
	211	84						8.875	7.0%				0.89723 ± 0.00007	1.6%
	211	84											1.06310 ± 0.00002	83.2%
$_{84}Po^{211}$	211	84		210.986627	0.52 s.	α	7.594	6.570	0.54	%	9/2+		0.56915 ± 0.00002	0.53%
	211	84						6.892	0.55	%			0.89723 ± 0.00007	0.52%
	211	84						7.450	98.9	%				
$_{84}Po^{212m}$	212	84			45.1 ± 0.6 s.	α		8.514	2.0%		16+			
	212	84						9.086	1.0%					
	212	84						11.650	97%					
$_{84}Po^{212}$	212	84		211.988842	0.3 μs.	α	8.953	8.784			0+			
$_{84}Po^{213}$	213	84		212.992833	≈ 4.2 μs.	α	8.537	7.614	0.00	3%	9/2+			
	213	84						8.375	100%					
$_{84}Po^{214}$	214	84		213.995176	163 μs.	α	7.833	6.904	0.01	%	0+			
	214	84						7.686	99.9	9%				
$_{84}Po^{215}$	215	84		214.999419	1.78 ms.	α	7.526	6.950	0.02	%	(9/2	+)		
	215	84						6.957	0.03	%				
	215	84						7.386	100%					
$_{84}Po^{216}$	216	84		216.001889	0.15 s.	α	6.906	5.895	0.00	2%	0+			
	216	84						6.778	99.9	9%				
$_{84}Po^{217}$	217	84		217.006260	<10 s.	α	6.662	6.539						
$_{84}Po^{218}$	218	84			3.11 ± 0.02 m.	α	6.114	5.181	1.00	%	0+			
	218	84						6.002	100%					
At		85												
$_{85}At^{196}$	196	85		195.995730	0.3 s.	α		7.06						
$_{85}At^{197}$	197	85		196.993410	0.4 s.	β+,E.C.	7.28				(9/2-)			
	197	85				α		6.96						
$_{85}At^{198m}$	198	85			1.5 s.	β+, EC (75%)								
	198	85				α(25%)		6.85						
$_{85}At^{198}$	198	85		197.992550	≈ 4.9 s.	α		6.75						
$_{85}At^{199}$	199	85		198.990580	7.0 s.	β+, EC (8%)	6.500				9/2-			
	199	85				α(92%)		6.64						
$_{85}At^{200m}$	200	85			4.3 ± 0.3 s.	β+, EC (80%)					10-			
	200	85				α(20%)		6.536						
$_{85}At^{200}$	200	85		199.990370	43 ± 5 s.	β+, EC (65%)	8.08				5+			
	200	85				α(35%)		6.412	21%					
	200	85						6.465	14%					
$_{85}At^{201}$	201	85		200.988440	1.48 ± 0.05 s.	β+, EC (29%)	5.82				9/2-			
	201	85				α(71%)	6.474	6.344						
$_{85}At^{202m}$	202	85			1.1 s.	I.T.	0.391							
$_{85}At^{202}$	202	85		201.988420	3.02 ± 0.05 m.	β+, EC (88%)	7.21				5+		ann.rad.	
	202	85				α(12%)		6.135	7.7%				0.4413 ± 0.0003	41%
	202	85						6.225	4.3%				0.5697 ± 0.0004	81%
	202	85											0.6753 ± 0.0005	87%
$_{85}At^{203}$	203	85		202.986790	7.4 ± 0.2 m.	β+, EC (69%)	5.040				9/2-		0.1458 ± 0.0001	14 +
	203	85				α(31%)	6.210	6.088					0.2459 ± 0.0002	48
	203	85											0.3616 ± 0.0003	23
	203	85											0.4169 ± 0.0001	14
	203	85											0.5319 ± 0.0001	18
	203	85											0.6088 ± 0.0001	20
	203	85											0.6414 ± 0.0001	53
	203	85											0.6562 ± 0.0001	30
	203	85											0.7379 ± 0.0001	42
	203	85											0.8458 ± 0.0001	30
	203	85											0.8804 ± 0.0001	41
	203	85											1.0020 ± 0.0001	86
	203	85											1.0340 ± 0.0001	100
$_{85}At^{204}$	204	85		203.987210	9.2 ± 0.2 m.	β+, EC (95%)	6.45				(5+)		Po k x-ray	27%
	204	85				α(5%)		5.951					0.3271 ± 0.0007	4.7%
	204	85											0.3367 ± 0.0007	5.6%

Isotope	A	Z	% Natural abundance	Atomic mass	Half-life	Decay mode	Decay energy (MeV)	Particle energy (MeV)	Particle intensity	Thermal neutron cross section	Spin (h/2π)	μ Nucl. mag. moment	Gamma-ray energy (MeV)	Gamma-ray intensity
	204	85											0.4254 ± 0.0003	66%
	204	85											0.4904 ± 0.001	4.7%
	204	85											0.5156 ± 0.0003	90%
	204	85											0.5888 ± 0.001	8.5%
	204	85											0.6084 ± 0.0006	19.8%
	204	85											0.6837 ± 0.0005	94%
	204	85											0.7621 ± 0.0007	4.7%
	204	85											0.8427 ± 0.0007	8.5%
$_{85}At^{205}$	205	85		204.986000	26.2 ± 0.5 m.	β+, EC (90%)	4.51				(9/2	-)	Po k x-ray	47%
	205	85				α(10%)	6.020	5.902					0.1543 ± 0.0002	2.4%
	205	85											0.1610 ± 0.0001	1.2%
	205	85											0.3114 ± 0.0001	3.4%
	205	85											0.4488 ± 0.0001	1.4%
	205	85											0.5207 ± 0.0001	3.5%
	205	85											0.6179 ± 0.0001	1.9%
	205	85											0.6209 ± 0.0001	4.6%
	205	85											0.6596 ± 0.0001	2.0%
	205	85											0.6696 ± 0.0001	8.1%
	205	85											0.6729 ± 0.0001	3.0%
	205	85											0.7194 ± 0.0001	27%
	205	85											0.7832 ± 0.0001	1.6%
	205	85											0.8724 ± 0.0005	2.0%
	205	85											1.3254 ± 0.0003	1.1%
	205	85											1.4758 ± 0.0003	0.6%
	205	85											1.4792 ± 0.0003	0.6%
$_{85}At^{206}$	206	85		205.986580	29.4 ± 0.3 m.	β+, EC (99%)	5.70				5+		Po k x-ray	35%
	206	85				α(1%)	5.881	5.703					0.20186 ± 0.0006	5.4%
	206	85											0.23354 ± 0.00006	3.1%
	206	85											0.25658 ± 0.00004	4.4%
	206	85											0.2756 ± 0.0001	2.0%
	206	85											0.27899 ± 0.00004	2.6%
	206	85											0.38678 ± 0.00006	2.6%
	206	85											0.39561 ± 0.00004	48%
	206	85											0.47716 ± 0.00003	86%
	206	85											0.52742 ± 0.00006	2.9%
	206	85											0.56562 ± 0.00006	3.2%
	206	85											0.70071 ± 0.00003	98%
	206	85											0.70464 ± 0.00006	6.0%
	206	85											0.73375 ± 0.00004	10.1%
	206	85											0.86831 ± 0.00004	7.6%
	206	85											0.92303 ± 0.00006	5.6%
	206	85											1.01394 ± 0.00007	2.9%
	206	85											1.04811 ± 0.00007	2.2%
	206	85											1.05938 ± 0.00004	3.4%
	206	85											1.12489 ± 0.00004	1.8%
	206	85											0.1969 ± 0.0001	1.5%
	206	85											0.2576 ± 0.0001	1.2%
	206	85											1.44616 ± 0.00007	1.3%
	206	85											1.63746 ± 0.00009	1.2%
	206	85											1.93813 ± 0.00007	1.3%
$_{85}At^{207}$	207	85		206.985730	1.81 h.	β+, EC (90%)	3.88				9/2-		Po k x-ray	46%
	207	85				α(10%)	5.873	5.758					0.16801 ± 0.00007	1.1%
	207	85											0.30074 ± 0.00006	9.7%
	207	85											0.35733 ± 0.00007	1.8%
	207	85											0.4220 ± 0.0001	1.4%
	207	85											0.45681 ± 0.00007	1.3%
	207	85											0.45966 ± 0.00008	1.1%

TABLE OF THE ISOTOPES (Continued)

Isotope	A	Z	% Natural abundance	Atomic mass	Half-life	Decay mode	Decay energy (MeV)	Particle energy (MeV)	Particle intensity	Thermal neutron cross section	Spin (h/2π)	μ Nucl. mag. moment	Gamma-ray energy (MeV)	Gamma-ray intensity
	207	85											0.46720 ± 0.00005	5.3%
	207	85											0.52991 ± 0.00006	2.5%
	207	85											0.58842 ± 0.00006	14.7%
	207	85											0.61720 ± 0.00001	1.1%
	207	85											0.62600 ± 0.00007	1.7%
	207	85											0.63740 ± 0.00007	1.7%
	207	85											0.64809 ± 0.00006	3.3%
	207	85											0.65848 ± 0.00006	5.0%
	207	85											0.67065 ± 0.00007	2.8%
	207	85											0.67521 ± 0.00007	4.9%
	207	85											0.6936 ± 0.0001	1.6%
	207	85											0.72119 ± 0.00005	4.9%
	207	85											0.81448 ± 0.00006	33%
	207	85											0.90721 ± 0.00007	4.0%
	207	85											0.96059 ± 0.00009	1.7%
	207	85											0.99401 ± 0.00006	1.7%
	207	85											1.07780 ± 0.00007	1.4%
	207	85											1.11529 ± 0.00006	3.3%
	207	85											1.22582 ± 0.00007	1.1%
	207	85											1.39640 ± 0.00006	1.0%
	207	85											1.67682 ± 0.00007	2.0%
	207	85											1.71270 ± 0.00009	1.0%
	207	85											1.7310 ± 0.0001	2.8%
	207	85											2.7122 ± 0.0002	0.9%
$_{85}At^{208}$	208	85		207.986510	1.63 ± 0.03 h.	β+, EC (99%)	4.93				(6+)		Po k x-ray	38%
	208	85				α(1%)	5.752	5.626	0.01 %				0.1770 ± 0.0006	46%
	208	85						5.641	0.53 %				0.2060 ± 0.0008	5.3%
	208	85											0.5170 ± 0.0007	7.0%
	208	85											0.6311 ± 0.0008	4.3%
	208	85											0.6601 ± 0.0001	90%
	208	85											0.6852 ± 0.001	98%
	208	85											0.8081 ± 0.0008	8.4%
	208	85											0.8450 ± 0.0007	21%
	208	85											0.8961 ± 0.001	6.0%
	208	85											0.9861 ± 0.0002	9%
	208	85											0.9931 ± 0.0007	14%
	208	85											1.0281 ± 0.001	27%
	208	85											1.2311 ± 0.002	3.3%
	208	85											1.2801 ± 0.0008	3.8%
	208	85											1.5391 ± 0.0007	1.6%
	208	85											2.0281 ± 0.0002	1.5%
	208	85											2.6361 ± 0.002	2.3%
$_{85}At^{209}$	209	85		208.986149	5.41 ± 0.05 h.	β+, EC (96%)	4.93				(6+)		Po k x-ray	38%
	209	85				α(4%)	5.757	5.647	4.1%				0.10422 ± 0.00007	2.4%
	209	85											0.19505 ± 0.00006	22.6%
	209	85											0.23916 ± 0.00006	12.4%
	209	85											0.54503 ± 0.00007	91%
	209	85											0.55103 ± 0.00005	4.9%
	209	85											0.78189 ± 0.00006	83%
	209	85											0.79020 ± 0.00006	63%
	209	85											0.86399 ± 0.00006	2.1%
	209	85											0.90315 ± 0.00006	3.6%

Isotope	A	Z	% Natural abundance	Atomic mass	Half-life	Decay mode	Decay energy (MeV)	Particle energy (MeV)	Particle intensity	Thermal neutron cross section	Spin (h/2π)	μ Nucl. mag. moment	Gamma-ray energy (MeV)	Gamma-ray intensity
	209	85											1.10351 ± 0.00007	5.4%
	209	85											1.17076 ± 0.00006	3.1%
	209	85											1.26261 ± 0.0006	1.9%
	209	85											1.58168 ± 0.00006	1.8%
	209	85											1.76713 ± 0.00005	0.5%
	209	85											(0.1 - 2.6)many	
$_{85}At^{210}$	210	85		209.987126	8.1 ± 0.4 h.	E.C. (99.8%)	3.98				5+		Po k x-ray	40%
	210	85				α(0.2%)	5.632	5.361	0.05%				0.24535 ± 0.00008	79%
	210	85						5.442	0.05%				0.52758 ± 0.00007	1.1%
	210	85											0.81723 ± 0.00009	1.7%
	210	85											0.8527 ± 0.0001	1.4%
	210	85											0.95577 ± 0.00007	1.8%
	210	85											1.18143 ± 0.00009	99%
	210	85											1.43678 ± 0.00006	29%
	210	85											1.48335 ± 0.00005	46%
	210	85											1.59956 ± 0.00006	13%
	210	85											2.25401 ± 0.00009	1.5%
	210	85											(0.04 - 2.4)weak	
$_{85}At^{211}$	211	85		210.987469	7.21 ± 0.01 h.	E.C.(58%)	0.784				9/2-		Po k x-ray	19%
	211	85				α(42%)	5.980	5.211	0.00	4%			0.66956 ± 0.00007	0.003%
	211	85						5.868	42%				0.6870 ± 0.0001	0.25%
	211	85											0.74263 ± 0.00006	0.0009
$_{85}At^{212m}$	212	85			0.12 s.	α		7.837	65%		(9-)			
	212	85						7.897	33%					
$_{85}At^{212}$	212	85		211.990725	0.3 s.	α	7.828	7.058	0.4%		(1-)			
	212	85						7.088	0.6%					
	212	85						7.618	15%					
	212	85						7.681	84%					
$_{85}At^{213}$	213	85		212.992911	0.11 μs.	α	9.254	9.080			9/2-			
$_{85}At^{214m}$	214	85			0.7 μs.	α		8.762			(9-)			
$_{85}At^{214}$	214	85		213.976347	0.56 μs.	α	8.987	8.819	100%		(1-)			
$_{85}At^{215}$	215	85		214.998638	100 μs.	α	8.178	7.626	0.04	5%	(9/2-)		0.40486 ± 0.00003	0.045%
	215	85						8.023	99.9%					
$_{85}At^{216}$	216	85		216.002390	300 μs.	α	7.947	7.595	0.2%		(1-)			
	216	85						7.697	2.1%					
	216	85						7.800	97%					
$_{85}At^{217}$	217	85		217.004694	32.3 ± 0.4 μs.	α	7.202	6.812	0.06%	(9/2-)		0.2595 ± 0.0008		
	217	85						7.067	99.9%				0.3345 ± 0.0008	
	217	85											0.5940 ± 0.0008	
$_{85}At^{218}$	218	85		218.008684	1.6 ± 0.4 s.	α	6.883	6.654	6%					
	218	85						6.695	90%					
	218	85						6.748	4%					
$_{85}At^{219}$	219	85		219.011300	54 s.	α	6.390	6.275						
Rn		86												
$_{86}Rn^{200}$	200	86		199.995700	1.0 ± 0.2 s.	α(98%)		6.909			0+			
	200	86				E.C.(2%)	8.080							
$_{86}Rn^{201m}$	201	86			3.8 ± 0.4 s.	E.C.(10%)					13/2+			
	201	86				α(90%)		6.770						
$_{86}Rn^{201}$	201	86		200.995570	7.0 ± 0.4 s.	α(80%)	6.860	6.721			(3/2-)			
	201	86				E.C.(20%)	6.65							
$_{86}Rn^{202}$	202	86		201.993230	9.9 ± 0.2 s.	α(12%)	6.771	6.636(3)			0+			
	202	86				E.C.(88%)	4.48							
$_{86}Rn^{203m}$	203	86			28 s.	α		6.548(3)			13/2+			
$_{86}Rn^{203}$	203	86		202.993330	45 ± 3 s.	α(66%)	6.629	6.498			0			
	203	86				E.C.(34%)	6.09							
$_{86}Rn^{204}$	204	86		203.991330	1.24 ± 0.03 m.	α(68%)	6.546	6.417(3)			0+			
	204	86				E.C.(32%)	3.84							
$_{86}Rn^{205}$	205	86		204.991650	2.83 ± 0.1 m.	α(23%)	6.390	6.123(3)	0.02%		(5/2-)		0.2652 ± 0.0007	100 +
	205	86				E.C.(77%)	5.27	6.262(3)	23%				0.3553 ± 0.0008	4
	205	86											0.4648 ± 0.0008	25
	205	86											0.6205 ± 0.0008	25
	205	86											0.6753 ± 0.001	20
	205	86											0.7300 ± 0.0008	20

Isotope	A	Z	% Natural abundance	Atomic mass	Half-life	Decay mode	Decay energy (MeV)	Particle energy (MeV)	Particle intensity	Thermal neutron cross section	Spin (h/2π)	μ Nucl. mag. moment	Gamma-ray energy (MeV)	Gamma-ray intensity
$_{86}Rn^{206}$	206	86		205.990140	5.67 ± 0.2 m.	α(68%)	6.384	6.258(3)			0+		0.06170 ± 0.0009	14 +
	206	86				E.C.(32%)	3.32						0.0968 ± 0.0001	5
	206	86											0.1009 ± 0.0002	4
	206	86											0.1337 ± 0.0001	5
	206	86											0.1862 ± 0.0003	7
	206	86											0.1954 ± 0.0001	12
	206	86											0.2080 ± 0.0001	26
	206	86											0.2131 ± 0.0004	11
	206	86											0.2906 ± 0.0003	7
	206	86											0.3019 ± 0.0002	53
	206	86											0.3245 ± 0.0001	100
	206	86											0.3504 ± 0.0003	13
	206	86											0.3711 ± 0.0002	52
	206	86											0.3862 ± 0.0001	63
	206	86											0.4356 ± 0.0001	5
	206	86											0.4439 ± 0.0003	28
	206	86											0.4582 ± 0.0006	5
	206	86											0.4654 ± 0.0002	4
	206	86											0.4822 ± 0.0002	59
	206	86											0.4853 ± 0.0003	31
	206	86											0.4973 ± 0.0001	104
	206	86											0.5271 ± 0.0002	25
	206	86											0.5363 ± 0.0003	17
	206	86											0.6318 ± 0.0003	15
	206	86											0.6429 ± 0.0007	6.7
	206	86											0.7166 ± 0.0007	6.8
	206	86											0.7382 ± 0.0006	15
	206	86											0.7568 ± 0.0006	11
	206	86											0.7728 ± 0.0003	60
	206	86											0.7948 ± 0.0004	10
$_{86}Rn^{207}$	207	86		206.990690	9.3 ± 0.2 m.	β+, EC (77%)	4.62				5/2-		At k x-ray	21%
	207	86				α(23%)	6.252	5.995(4)	0.02%				0.32947 ± 0.00004	3.0%
	207	86						6.068(3)	0.15%				0.34455 ± 0.00004	45%
	207	86						6.126(3)	22.8%				0.36767 ± 0.00008	2.5%
	207	86											0.40267 ± 0.00004	11.8%
	207	86											0.55323 ± 0.0001	1.2%
	207	86											0.62873 ± 0.00007	1.1%
	207	86											0.63159 ± 0.00009	2.9%
	207	86											0.6433 ± 0.0002	1.2%
	207	86											0.6472 ± 0.0001	1.8%
	207	86											0.6858 ± 0.0001	1.2%
	207	86											0.6971 ± 0.0001	2.4%
	207	86											0.74723 ± 0.00005	14.1%
	207	86											0.7754 ± 0.0001	2.0%
	207	86											0.85343 ± 0.00009	2.3%
	207	86											0.8927 ± 0.0007	1.0%
	207	86											0.9086 ± 0.0001	1%
	207	86											0.97328 ± 0.00008	2.5%
	207	86											0.9992 ± 0.0002	1.2%
	207	86											(0.18 - 1.47)weak	
$_{86}Rn^{208}$	208	86		207.989610	24.4 ± 0.1 m.	α(60%)	6.260	5.469(2)	0.00	3%	0+			
	208	86				E.C.(40%)	2.88	6.140(2)	60%					
$_{86}Rn^{209}$	209	86		208.990370	28.5 ± 0.1 m.	β+(83%)	3.930	2.16	2.3%		5/2-		At k x-ray	36%
	209	86				α(17%)		5.887(3)	0.04%				0.27933 ± 0.00007	1.1%
	209	86						5.898(3)	0.02%				0.33753 ± 0.00003	14.7%
	209	86						6.039(2)	16.9%				0.40841 ± 0.00003	51%
	209	86											0.46154 ± 0.00006	1.5%
	209	86											0.67293 ± 0.00004	3.3%
	209	86											0.68942 ± 0.00004	9.8%
	209	86											0.74594 ± 0.00003	23.1%
	209	86											0.79481 ± 0.00005	3.4%
	209	86											0.85590 ± 0.00004	4.9%
	209	86											1.03811 ± 0.00005	4.2%

TABLE OF THE ISOTOPES (Continued)

Isotope	A	Z	% Natural abundance	Atomic mass	Half-life	Decay mode	Decay energy (MeV)	Particle energy (MeV)	Particle intensity	Thermal neutron cross section	Spin (h/2π)	μ Nucl. mag. moment	Gamma-ray energy (MeV)	Gamma-ray intensity
	209	86											1.05460 ± 0.00005	1.7%
	209	86											1.06567 ± 0.00007	1.7%
	209	86											1.15893 ± 0.00006	0.85%
	209	86											1.5432 ± 0.0001	0.8%
	209	86											1.9258 ± 0.0003	0.3%
	209	86											2.1142 ± 0.0002	0.33%
	209	86											2.6428 ± 0.0002	0.32%
	209	86											(0.18 - 3.2)many	
$_{86}Rn^{210}$	210	86		209.989669	2.4 ± 0.1 h.	α(96%)	6.157	5.351(2)	0.00	5%	0+		At k x-ray	2.3%
	210	86				E.C.(4%)	2.368	6.039(2)	96%				0.19625 ± 0.00007	0.32%
	210	86											0.23324 ± 0.00006	0.5%
	210	86											0.45824 ± 0.00006	1.6%
	20	86											0.57104 ± 0.00006	0.8%
	210	86											0.64868 ± 0.00006	0.82%
	210	86											0.76148 ± 0.00006	0.52%
	210	86											0.95773 ± 0.00007	0.3%
	210	86											(0.14 - 1.7)weak	
$_{86}Rn^{211}$	211	86		210.990576	14.6 ± 0.2 h.	β+, EC (74%)	2.894				1/2-		At k x-ray	30%
	211	86				α(26%)	5.964	5.619(1)	0.7%				0.16877 ± 0.00006	6.8%
	211	86						5.784(1)	16.4%				0.25022 ± 0.00006	6.1%
	211	86						5.851(1)	8.8%				0.37049 ± 0.00008	1.4%
	211	86											0.41632 ± 0.00006	3.5%
	211	86											0.44209 ± 0.00006	23%
	211	86											0.67412 ± 0.00007	46%
	211	86											0.67839 ± 0.00006	29%
	211	86											0.85377 ± 0.00006	4.7%
	211	86											0.86600 ± 0.00007	8.0%
	211	86											0.93481 ± 0.00006	3.7%
	211	86											0.94666 ± 0.00006	3.7%
	211	86											0.94744 ± 0.00007	5.1%
	211	86											1.12668 ± 0.00006	22.5%
	211	86											1.36298 ± 0.00005	33.1%
	211	86											1.53985 ± 0.00008	4.8%
	211	86											1.9926 ± 0.0001	0.5%
	211	86											(0.11 - 2.7)weak	
$_{86}Rn^{212}$	212	86		211.990697	24 ± 2 m.	α	6.385	5.587(4)	0.05%		0+			
	212	86						6.260(4)	99.95%					
$_{86}Rn^{213}$	213	86		212.993856	25.0 ± 0.2 ms.	α	8.243	7.552(8)	1.0%		9/2+			
	213	86						8.087(8)	99%					
$_{86}Rn^{214m}$	214	86			7.3 ns.	α(4%)		10.63(3)			(8+)			
	214	86				I.T.	1.626							
$_{86}Rn^{214}$	214	86		213.996347	0.27 μs.	α	9.209	9.037(9)			0+			
$_{86}Rn^{215}$	215	86		214.998720	2.3 μs.	α	8.840	8.674(8)			(9/2+)			
$_{86}Rn^{217}$	217	86		217.003902		α	7.885	7.500	0.1%		9/2+			
	215	86						7.742(4)	100%					
$_{86}Rn^{218}$	218	86	218.005580		35 ± 6 ms.	α	7.267	6.534(1)	0.16%		0+			
	218	86						7.133(1)	99.8%					
$_{86}Rn^{219}$	219	86		219.009479	3.96 s.	α	6.946(1)	6.3130(5)	0.05%		(5/2+)		Po k x-ray	0.9%
	219	86						6.425(3)	7.5%				0.13057 ± 0.00006	0.13%
	219	86						6.5309(4)	0.12%				0.27113 ± 0.00005	9.9%
	219	86						6.5531(3	) 12.2%				0.40170 ± 0.00006	6.6%
	219	86						6.8193(3)	81%				(0.1 - 1.05)weak	
$_{86}Rn^{220}$	220	86		220.011368	55.6 ± 0.1 s.	α	6.404	5.7486(5)	0.07%		0+			
	220	86						6.2883(1)	99.9%					
$_{86}Rn^{221}$	221	86		221.015470	25 m.	α(22%)	6.148	5.778(3)	1.8%				Fr L x-ray	1.6%

Isotope	A	Z	% Natural abundance	Atomic mass	Half-life	Decay mode	Decay energy (MeV)	Particle energy (MeV)	Particle intensity	Thermal neutron cross section	Spin (h/2π)	μ Nucl. mag. moment	Gamma-ray energy (MeV)	Gamma-ray intensity
	221	86				β-(78%)	1.150	5.788(3)	2.2%				0.07384 ± 0.00004	0.53%
	221	86						6.037(3)	18%				0.08323	8.2%
	221	86											0.0610	13.6%
	221	86											0.09727	4.9%
	221	86											0.09982	2.8%
	221	86											0.10060	1.6%
	221	86											0.10836 ± 0.00003	2.2%
	211	86											0.11156 ± 0.00003	2.1%
	211	86											0.15008 ± 0.00003	4.4%
	211	86											0.18639 ± 0.00004	20.4%
	211	86											0.21686 ± 0.00004	2.3%
	211	86											0.254 ± 0.0003	2%
	211	86											0.26468 ± 0.00004	1.1%
	211	86											0.27927 ± 0.00003	1.8%
$_{86}Rn^{222}$	222	86		222.017570	3.82 d.	α	5.590	4.987(1)	0.08%	0.7 b.	0+		0.510 ± 0.002	0.07%
	222	86						5.4897(3)	99.9%					
$_{86}Rn^{223}$	223	86			43 m.	β-								
$_{86}Rn^{224}$	224	86			1.78 ± 0.005h.	β-					0+		0.1085 ± 0.0005	3.3 +
	224	86											0.1132 ± 0.0003	3.3
	224	86											0.2026 ± 0.0003	4.4
	224	86											0.2562 ± 0.0003	3.0
	224	86											0.2601 ± 0.0001	23
	224	86											0.2655 ± 0.0001	21
	224	86											0.3719 ± 0.0003	3.5
	224	86											0.398 ± 0.001	5.6
	224	86											0.402 ± 0.001	5.7
$_{86}Rn^{225}$	225	86			4.5 m.	β-								
$_{86}Rn^{226}$	226	86			6.0 m.	β-								
Fr		87												
$_{87}Fr^{201}$	201	87		201.004110	48 ms.	α	7.54	7.388(15)						
$_{87}Fr^{202}$	202	87		202.003300	0.34 s.	α	7.590	7.250(20)						
$_{87}Fr^{203}$	203	87		203.001000	0.55 s.	α	7.280	7.132(5)						
$_{87}Fr^{204}$	204	87		204.000670	2.1 s.	α	7.170	6.967(5)	30 +					
	204	87						7.027(5)	70					
$_{87}Fr^{205}$	205	87		204.998610	3.96 s.	α	7.050	6.914(5)						
$_{87}Fr^{206}$	206	87		205.99846s	16.0 ± 0.1 s.	α	7.416	6.789(5)						
$_{87}Fr^{207}$	207	87		206.996800	14.8 ± 0.1 s.	α	6.900	6.766(5)			9/2-			
$_{87}Fr^{208}$	208	87		207.997080	59 s.	α(77%)	6.770	6.636(5)			7			
	208	87				E.C.(23%)	6.960							
$_{87}Fr^{209}$	209	87		208.995878	50.0 ± 0.3 s.	α(89%)	5.130	6.646(3)			9/2-			
	209	87				E.C.(11%)	6.778							
$_{87}Fr^{210}$	210	87		209.996340	3.2 m.	α	6.670	6.543(5)			6+		0.2030 ± 0.0008	35 +
	210	87											0.2562 ± 0.0001	11
	210	87											0.4252 ± 0.0001	10
	210	87											0.461 ± 0.001	11
	210	87											0.6438 ± 0.0008	100
	210	87											0.733 ± 0.001	10
	210	87											0.8175 ± 0.0008	60
	210	87											0.9008 ± 0.0007	30
$_{87}Fr^{211}$	211	87		210.995490	3.1 m.	α	6.660	6.534(5)			9/2-		0.220 ± 0.0008	9 +
	211	87				E.C.	4.570						0.2799 ± 0.001	34
	211	87											0.4389 ± 0.001	20
	211	87											0.5389 ± 0.001	100
	211	87											0.9169 ± 0.001	55
	211	87											0.9819 ± 0.0008	20
$_{87}Fr^{212}$	212	87		211.996130	20.0 ± 0.6 m.	E.C.(57%)	5.070	6.076(3)	0.17%		(5+)		Rn x-ray	14%
	212	87				α(43%)	6.529	6.127(3)	0.43%				0.0789	2.4%
	212	87						6.173(4)	0.5%				0.08107	14%
	212	87						6.183(3)	0.6%				0.08152	4%
	212	87						6.261(1)	16%				0.08378	24%
	212	87						6.335(1)	4%				0.09468	8%
	212	87						6.343(1)	1.3%				0.1383 ± 0.0001	7.7%
	212	87						6.383(1)	10%				0.3091 ± 0.0002	1.0%
	212	87						6.406(1)	9.5%				0.3115 ± 0.0002	1.3%
	212	87											0.5320 ± 0.0005	2.8%
	212	87											0.8019 ± 0.0015	3.5%
	212	87											1.0473 ± 0.0014	7.2%
	212	87											1.1784 ± 0.0020	1.3%
	212	87											1.1856 ± 0.0014	14%
	212	87											1.2748 ± 0.0020	46%
$_{87}Fr^{213}$	213	87		212.996165	34.6 ± 0.3 s.	α	6.905	6.775(2)			9/2-			

TABLE OF THE ISOTOPES (Continued)

Isotope	A	Z	% Natural abundance	Atomic mass	Half-life	Decay mode	Decay energy (MeV)	Particle energy (MeV)	Particle intensity	Thermal neutron cross section	Spin (h/2π)	μ Nucl. mag. moment	Gamma-ray energy (MeV)	Gamma-ray intensity
$_{87}Fr^{214m}$	214	87			3.4 ms.	α		7.594(5)	0.5%		9-			
	214	87						7.708(5)	1.1%					
	214	87						7.963(5)	0.7%					
	214	87						8.046(5)	0.9%					
	214	87						8.476(4)	51%					
	214	87						8.547(4)	46%					
$_{87}Fr^{214}$	214	87		213.998948	5.1 ms.	α	8.587	7.409(3)	0.3%		(1-)			
	214	87						7.605(8)	1.0%		214	87		
	214	87						7.940(3)	1.0%					
	214	87						8.355(3)	4.7%					
	214	87						8.427(3)	93%					
$_{87}Fr^{215}$	215	87		215.000310	0.12 μs.	α	9.537	9.360(8)			(9/2-)			
$_{87}Fr^{217}$	217	87		216.003178	0.7 μs.	α	9.175	9.005(10)						
$_{87}Fr^{217}$	217	87		217.004609	22 μs.	α	8.471	8.315(8)			(9/2-)			
$_{87}Fr^{218}$	218	87		218.007553	0.7 ms.	α	8.014	7.384(10)	0.5%					
	218	87						7.542(15)	1.0%					
	218	87						7.572(10)	5%					
	218	87						7.732(10)	0.5%					
	218	87						7.867(2)	93%					
$_{87}Fr^{219}$	219	87		219.009242	21 ± 1 s.	α	8.132	6.802(2)	0.25%		(9/2-)			
	219	87						6.967(2)	0.6%					
	219	87						7.146(2)	0.25%					
	219	87						7.313(2)	99%					
$_{87}Fr^{220}$	220	87		220.012293	27.4 ± 0.3 s.	α	6.800	6.389(1)	0.3%		1		0.0450 ± 0.0003	2.3 +
	220	87						6.413(1)	1.2%				0.061 ± 0.004	0.4%
	220	87						6.438(2)	0.24%				0.1060 ± 0.0004	1.7%
	220	87						6.483(1)	1.3%				0.1539 ± 0.0004	1.0%
	220	87						6.490(2)	0.6%				0.1617 ± 0.0004	1.5%
	220	87						6.519(1)	0.6%					
	220	87						6.527(1)	3%					
	220	87						6.535(1)	2.5%					
	220	87						6.582(1)	10%					
	220	87						6.630(2)	6%					
	220	87						6.641(1)	12%					
	220	87						6.686(1)	61%					
$_{87}Fr^{221}$	221	87		221.014230	4.9 ± 0.2 m.	α	6.457	5.9393(7)	0.17%		(5/2-)		At k x-ray	1.4%
	221	87						5.9797(7)	0.49%				0.0995 ± 0.0001	0.10%
	221	87						6.0751(7)	0.15%				0.21798 ± 0.00004	10.9%
	221	87						6.1270(7)					0.4091 ± 0.0002	0.13%
	221	87						6.2433(3)	1.3%					
	221	87						6.3410(7)	83.4%					
$_{87}Fr^{222}$	222	87		14.4 m.	β-	2.060	1.78				2			
	222	87				α	5.850							
$_{87}Fr^{223}$	223	87		223.019733	21.8 ± 0.4 m.	β-	1.147	1.17	65%		(3/2+)		0.05014 ± 0.00004	33%
	223	87											0.07972 ± 0.00003	8.9%
	223	87											0.08543	2.4%
	223	87											0.08543	2.4%
	223	87											0.08847	4.0%
	223	87											0.09991	1.5%
	223	87											0.20495 ± 0.00005	1.1%
	223	87											0.23482 ± 0.00005	3.7%
	223	87											0.31918 ± 0.00007	0.54%
	223	87											0.3693 ± 0.0001	0.11%
	223	87											0.7758 ± 0.0001	0.40%
	223	87											(0.13 - 0.93)weak	
$_{87}Fr^{224}$	224	87		224.023220	2.7 ± 0.2 m.	β-	2.830				1		0.13150 ± 0.00006	83 +
	224	87											0.2057 ± 0.0002	14
	224	87											0.21575 ± 0.00006	180
	224	87											0.7625 ± 0.0002	12
	224	87											0.8018 ± 0.0003	6
	224	87											0.8367 ± 0.0002	58
	224	87											0.8810 ± 0.0002	5
	224	87											0.9683 ± 0.0002	5
	224	87											1.1619 ± 0.0002	5
	224	87											1.2983 ± 0.0002	5
	224	87											1.3402 ± 0.0003	25
	224	87											1.3777 ± 0.0002	17
	224	87											1.4356 ± 0.0002	10
	224	87											1.6521 ± 0.0005	6
	224	87											(0.1 - 2.21)weak	
$_{87}Fr^{225}$	225	87		225.025590	3.9 m.	β-	1.850							
$_{87}Fr^{226}$	226	87		226.029200	48 s.	β-	3.540						0.18606 ± 0.00004	66 +
	226	87											0.25373 ± 0.00004	83
	226	87											1.0069 ± 0.0002	15

Isotope	A	Z	% Natural abundance	Atomic mass	Half-life	Decay mode	Decay energy (MeV)	Particle energy (MeV)	Particle intensity	Thermal neutron cross section	Spin (h/2π)	μ Nucl. mag. moment	Gamma-ray energy (MeV)	Gamma-ray intensity
	226	87											1.0489 ± 0.0002	15
	226	87											1.3219 ± 0.0002	8.5
	226	87											1.3889 ± 0.002	4.4
$_{87}Fr^{227}$	227	87		227.031770	2.4 m.	β-	2.420							
$_{87}Fr^{228}$	228	87		228.035570	39 s.	β-	4.200							
$_{87}Fr^{229}$	229	87			50 s.	β-								
Ra		88												
$_{88}Ra^{206}$	206	88		206.003800	0.4 s.	α	7.416	7.272(5)			0+			
$_{88}Ra^{207}$	207	88		207.003740	1.3 ± 0.2 s.	α	7.270	7.133(3)						
$_{88}Ra^{208}$	208	88		208.001750	1.4 ± 0.4 s.	α	7.273	7.133(5)			0+			
$_{88}Ra^{209}$	209	88		209.001930	4.6 ± 0.2 s.	α	7.150	7.008(5)						
$_{88}Ra^{210}$	210	88		210.000430	3.7 ± 0.2 s.	α	7.610	7.020(5)			0+			
$_{88}Ra^{211}$	211	88		211.000860	13 ± 2 s.	α	7.046	6.912(5)			(5/2-)			
	211	88				E.C.								
$_{88}Ra^{212}$	212	88		211.999760	13.0 ± 0.5 s.	α	7.033	6.901(2)			0+			
$_{88}Ra^{213}$	213	88		213.000330	2.7 m.	E.C.(20%)	3.880				(1/2-)		0.1024 ± 0.0001	0.3%
	213	88				α(80%)	6.860	6.521(3)	4.8%				0.11010 ± 0.00009	6.4%
	213	88						6.622(3)	39%				0.2125 ± 0.0001	1.1%
	213	88						6.730(3)	36%					
$_{88}Ra^{214}$	214	88		214.000079	2.46 ± 0.03 s.	α	7.272	7.136(4)			0+			
$_{88}Ra^{215}$	215	88		215.002695	1.59 ± 0.09 ms.	α	8.864	7.883(6)	2.8%		(9/2+)			
	215	88						8.171(3)	1.4%					
	215	88						8.700(3)	95.9%					
$_{88}Ra^{216}$	216	88		216.003509	0.18 μs.	α	9.526	9.349(8)			0+			
$_{88}Ra^{217}$	217	88		217.006294	1.6 ± 0.2 μs.	α	9.161	8.992(8)			9/2-			
$_{88}Ra^{218}$	218	88		218.007117	14 ± s μs.	α	8.547	8.390(8)			0+			
$_{88}Ra^{219}$	219	88		219.010053		α	8.132	7.680(10)	65%					
	219	88						7.982(9)	35%					
$_{88}Ra^{220}$	220	88		220.011004	23 ± 5 ms.	α	7.593	6.998(7)	1.0%		0+		0.465 ± 0.004	1.0%
	220	88						7.455(7)	99%					
$_{88}Ra^{221}$	221	88		221.013889	28 ± 2 s.	α	6.879	6.254(10)	0.7%					
	221	88						6.578(5)	3%					
	221	88						6.585(3)	8%					
	221	88						6.608(3)	35%					
	221	88						6.669(3)	21%					
	221	88						6.758(3)	31%					
$_{88}Ra^{222}$	22	88		222.015353	38.0 ± 0.5 s.	α	5.590	6.237(2)	3.0%		0+			
	222	88						6.556(2)	97%					
$_{88}Ra^{223}$	223	88		223.018501	11.43 ± 0.02d.	α	5.979	5.287(1)	0.15%		(1/2+)		Rn k x-ray	25%
	223	88						5.338(1)	0.13%				0.12231 ± 0.00006	1.2%
	223	88						5.365(1)	0.13%				0.14418 ± 0.00003	3.3%
	223	88						5.433(5)	2.3%				0.15418 ± 0.00003	5.6%
	223	88						5.502(1)	1.0%				0.15859 ± 0.00003	0.67%
	223	88						5.540(1)	9.2%				0.26939 ± 0.00003	14%
	233	88						5.607(3)	24%				0.32388 ± 0.00003	3.9%
	233	88						5.716(3)	52%				0.33328 ± 0.00004	2.8%
	233	88						5.747(1)	9%				0.44494 ± 0.00005	1.3%
	223	88						5.857(1)	0.32%				(0.10 - 0.71)weak	
	223	88						5.872(1)	0.85%					
$_{88}Ra^{224}$	224	88		224.020186	3.66 ± 0.04 d.	α	5.789	5.034(10)	0.00	3%	0+		Rn k x-ray	0.2%
	224	88						5.047(1)	0.007%				0.2407 ± 0.0001	3.9%
	224	88						5.164(5)	0.007%				0.4093 ± 0.0007	0.004%
	224	88						5.449(2)	4.9%				0.6501 ± 0.0007	0.007%
	224	88						5.685(2)	95%					
$_{88}Ra^{225}$	225	88		225.023604	14.8 ± 0.2 d.	β-	0.371	0.32	100%		(3/2+)		Ac k x-ray	6%
	225	88											0.0434 ± 0.0017	29%
$_{88}Ra^{266}$	226	88		226.025402	1600 ± 7 y.	α	4.870	4.194(1)	0.001%		0+		Rn k x-ray	0.3%
	226	88						4.343(1)	0.006%				0.1861 ± 0.0001	3.3%
	226	88						4.601(1)	5.5%				0.2624 ± 0.0002	0.005%
	226	88						4.784(1)	94%					
$_{88}Ra^{227}$	227	88		227.029170	42.2 ± 0.5 m.	β-	1.324	1.03			(3/2+)		Ac L x-ray	25%
	227	88						1.30					Ac k x-ray	43%
	227	88											0.02739 ± 0.00001	17%
	227	88											0.08767	2.6
	227	88											0.0988	4.3%

Isotope	A	Z	% Natural abundance	Atomic mass	Half-life	Decay mode	Decay energy (MeV)	Particle energy (MeV)	Particle intensity	Thermal neutron cross section	Spin (h/2π)	μ Nucl. mag. moment	Gamma-ray energy (MeV)	Gamma-ray intensity
	227	88											0.10261	1.5%
	227	88											0.23062 ± 0.00007	1.4%
	227	88											0.25843 ± 0.00006	2.0%
	227	88											0.27743 ± 0.00006	2.9%
	227	88											0.28367 ± 0.00002	3.5%
	227	88											0.30007 ± 0.00002	5.3%
	227	88											0.30267 ± 0.00002	3.8%
	277	88											0.33007 ± 0.00002	2.9%
	227	88											0.40789 ± 0.00004	2.5%
	227	88											0.48703 ± 0.00009	2.5%
	227	88											0.5013 ± 0.0001	1.0%
	227	88											0.5164 ± 0.0001	1.5%
	227	88											0.6117 ± 0.0002	1.3%
$_{88}Ra^{228}$	228	88		228.031064	5.75 ± 0.03 y.	β-	0.045				0+		0.0135 ± 0.001	100 +
	228	88											0.016	
$_{88}Ra^{229}$	229	88		229.034870	4.0 ± 0.2 m.	β-	1.760	1.76			(3/2+)			
$_{88}Ra^{230}$	230	88		230.036990	1.55 ± 0.05 h.	β-	0.700	0.7			0+		0.0631 ± 0.0001	40 +
	230	88											0.0720 ± 0.0001	113
	230	88											0.0921 ± 0.0001	21
	230	88											0.1011 ± 0.0001	16
	230	88											0.1107 ± 0.0001	3
	230	88											0.1343 ± 0.0001	4.5
	230	88											0.1479 ± 0.0001	5.6
	230	88											0.1841 ± 0.0001	11
	230	88											0.1892 ± 0.0001	11
	230	88											0.2028 ± 0.0001	31
	230	88											0.2118 ± 0.0001	11
	230	88											0.2516 ± 0.0001	10
	230	88											0.2852 ± 0.0001	18
	230	88											0.44898 ± 0.00007	15
	230	88											0.4580 ± 0.0001	18
	230	88											0.4698 ± 0.0001	29
	230	88											0.4787 ± 0.0001	24
	230	88											0.5092 ± 0.0001	6
Ac		89												
$_{89}Ac^{210}$	210	89		210.009230	0.35 s.	α	7.610	7.462(8)						
$_{89}Ac^{211}$	221	89		211.007590	≈ 0.25 s.	α	7.620	7.480(8)						
$_{89}Ac^{212}$	212	89		212.007760	0.93 s.	α	7.520	7.379(8)						
$_{89}Ac^{213}$	213	89		213.006530	0.8 s.	α	7.500	7.364(8)			(9/2-)			
$_{89}Ac^{214}$	214	89		214.006840	8.2 ± 0.2 s.	α(86%)	7.350	7.007(8)	3 +		(5+)			
	214	89				E.C.(14%)		7.082(5)	38					
	214	89						7.214(5)	45					
$_{89}Ac^{215}$	215	89		215.006410	0.17 ± 0.01 s.	α	7.750	7.604(5)			(9/2-)			
$_{89}Ac^{216m}$	216	89			0.33 ms.	α		8.198(8)	1.7%		(9-)			
	216	89						8.283(8)	2.5%					
	216	89						9.028(5)	49%					
	216	89						9.106(5)	46%					
$_{89}Ac^{216}$	216	89		216.008650	≈ 0.33 ms.	α	9.241	8.990(2)	10%		(1)			
	216	89						9.070(8)	90%					
$_{89}Ac^{217m}$	217	89			0.4 ± 0.1 μs.	α		10.540	100%					
$_{89}Ac^{217}$	217	89		217.009322	0.11 μs.	α	9.832	9.650(10)	100%		9/2-			
$_{89}Ac^{218}$	218	89		218.001620	0.27 ± 0.04 μs	α	9.380	9.205(15)						
$_{89}Ac^{219}$	219	89		219.012390	7 ± 2 μs.	α	8.830	8.664(10)			(9/2-)			
$_{89}Ac^{220}$	220	89		220.014740	26.1 ± 0.5 ms.	α	8.350	7.610(20)	23%					
	220	89						4.680(20)	21%					
	220	89						7.790(10)	13%					
	220	89						7.850(10)	24%					
	220	89						7.985(10)	4%					
	220	89						8.005(10)	5%					
	220	89						8.060(10)	6%					
	220	89						8.195(10)	3%					
$_{89}Ac^{221}$	221	89		221.015570	52 ± 2 ms.	α	7.790	7.170(10)	2%					
	221	89						7.375(10)	10%					
	221	89						7.440(15)	20%					
	220	89						7.645(10)	70%					
$_{89}Ac^{222m}$	222	89			1.10 ± 0.05 m.	α(>89%)		6.710(20)	7%					
	222	89				E.C.(1%)		6.750(20)	13%					

Isotope	A	Z	% Natural abundance	Atomic mass	Half-life	Decay mode	Decay energy (MeV)	Particle energy (MeV)	Particle intensity	Thermal neutron cross section	Spin (h/2π)	μ Nucl. mag. moment	Gamma-ray energy (MeV)	Gamma-ray intensity
	222	89				I.T.(<10%)		6.810(20)	24%					
	222	89						6.840(20)	9%					
	222	89						6.890(20)	13%					
	222	89						6.970(20)	7%					
	222	89						7.000(20)	13%					
$_{89}Ac^{222}$	222	89		222.017824	4.2 s.	α	7.141	6.967(10)	6%					
	222	89						7.013(2)	94%					
$_{89}Ac^{223}$	223	89		223.019128	2.2 ± 0.1 m.	α(99%)	6.783	6.131(2)	0.12%		(5/2-)		0.0725 ± 0.0009	0.2%
	223	89				E.C.(1%)	0.584	6.177(2)	0.94%				0.0830 ± 0.0007	0.2%
	223	89						6.293(1)	0.47%				0.0927 ± 0.0007	0.2%
	223	89						6.326(1)	0.3%				0.0990 ± 0.0006	0.2%
	223	89						6.332(2)	0.14%				0.1917 ± 0.0007	0.25%
	223	89						6.360(1)	0.22%				0.2158 ± 0.0009	0.15%
	223	89						6.397(1)	0.13%				0.3588 ± 0.0011	0.10%
	233	89						6.448(1)	0.2%				0.4768 ± 0.0015	0.14%
	223	89						6.473(1)	3.1%					
	223	89						6.523(2)	0.6%					
	223	89						6.528(1)	3.1%					
	223	89						6.563(1)	13.6%					
	223	89						6.582(3)	0.3%					
	223	89						6.646(1)	44%					
	223	89						6.661(1)	31%					
$_{89}Ac^{224}$	224	89		224.021685	2.9 ± 0.2 h.	E.C.(90%)	1.397	5.841(1)	0.5	+			Ra L kx-ray	18%
	224	89				α(10%)	6.323	5.860(1)	0.75				Ra k x-ray	35%
	224	89						5.875(1)	1.7				0.08426 ± 0.00005	1.1%
	224	89						5.941(1)	4.4				0.13150 ± 0.00006	20%
	224	89						6.000(1)	6.7				0.1571 ± 0.0003	0.5%
	224	89						6.013(1)	1.4				0.21575 ± 0.00006	44%
	224	89						6.056(1)	22				0.2619 ± 0.0003	0.2%
	224	89						6.138(1)	26				(0.03 - 0.37)weak	
	224	89						6.154(1)	1.0					
	224	89						6.204(1)	12					
	224	89						6.210(1)	20					
$_{89}Ac^{225}$	225	89		225.023205	10.0 ± 0.1 d.	α	5.935	5.286(1)	0.2%		3/2		Fr k x-ray	2.1%
	225	89						5.444(3)	0.1%				0.9958 ± 0.0004	0.6%
	225	89						5.554(1)	0.1%				0.9982 ± 0.0006	1.7%
	225	89						5.608(1)	1.1%				0.1084 ± 0.0001	0.3%
	225	89						5.636(1)	4.5%				0.1116 ± 0.0001	0.33%
	225	89						5.681(1)	1.4%				0.1451 ± 0.0001	0.13%
	225	89						5.722(1)	2.9%				0.1539 ± 0.0001	0.15%
	225	89						5.731(1)	10%				0.15724 ± 0.00003	0.31%
	225	89						5.791(1)	9%				0.18799 ± 0.00005	0.46%
	225	89						5.793(1)	18%				0.19575 ± 0.00003	0.14%
	225	89											0.2162 ± 0.001	0.34%
	225	89											0.21686 ± 0.00004	0.42%
	225	89											0.25351 ± 0.00004	0.10%
	225	89											0.4524 ± 0.0001	0.11%
	225	89											(0.025 - 0.52)weak	
$_{89}Ac^{226}$	226	89		226.026084	1.2 d.	E.C.(17%)	0.635				(1-)		Ra k x-ray	6.0%
	226	89				β-(83%)	1.117						Th k x-ray	1.8%
	226	89				α(0.006%)	5.510	5.399(5)	0.00	6%			0.07218 ± 0.0000	3 0.6%
	226	89											0.15816 ± 0.00003	17.3%
	226	89											0.18606 ± 0.0001	4.6%
	226	89											0.23034 ± 0.00003	29.6%
	226	89											0.25373 ± 0.00001	5.8%
$_{89}Ac^{227}$	227	89		227.027750	21.77 ± 0.03y.	β-(98.6%)	0.041	β0.0455	54%		(3/2-)	+1.1	0.01520 ± 0.00009	0.035%
	227	89				α(1.4%)	5.043	α4.869(1)	0.09%				0.0698 ± 0.0001	0.02%
	227	89						4.938(1)	0.52%				0.0997 ± 0.0005	0.03%
	227	89						4.951(1)	0.65%				0.1600 ± 0.0004	0.02%
	227	89											(0.009 - 0.17)weak	
$_{89}Ac^{228}$	228	89		228.031015	6.13 h.	β-	2.142	1.11	32%		(3+)		Th L x-ray	20%
	228	89						1.85	12%				Th k x-ray	5.6%
	228	89						2.18	11%				0.12903 ± 0.00007	2.9%
	228	89											0.20939 ± 0.00007	4.1%
	228	89											0.27026 ± 0.00008	3.8%
	228	89											0.32807 ± 0.00009	3.5%
	228	89											0.33842 ± 0.00006	12.4%

TABLE OF THE ISOTOPES (Continued)

Isotope	A	Z	% Natural abundance	Atomic mass	Half-life	Decay mode	Decay energy (MeV)	Particle energy (MeV)	Particle intensity	Thermal neutron cross section	Spin (h/2π)	μ Nucl. mag. moment	Gamma-ray energy (MeV)	Gamma-ray intensity
	228	89											0.40962 ± 0.00008	2.2%
	228	89											0.46310 ± 0.00007	4.6%
	228	89											0.5815 ± 0.0002	3%
	228	89											0.75528 ± 0.00008	1.3%
	228	89											0.77228 ± 0.00007	1.1%
	228	89											0.7948 ± 0.0001	4.6%
	228	89											0.83560 ± 0.00009	1.7%
	228	89											0.91116 ± 0.00003	29%
	228	89											0.96897 ± 0.00005	17%
	228	89											1.4592 ± 0.0001	1.1%
	228	89											1.4960 ± 0.0003	1.0%
	228	89											1.5882 ± 0.0001	3.6%
	228	89											1.6304 ± 0.0002	1.9%
	228	89											(0.2 - 1.96)many	
$_{89}$Ac^{229}	229	89		229.032980	1.05 ± 0.01 h.	β-	1.140	1.1			(3/2+)		0.07450 ± 0.00002	8 +
	229	89											0.11715 ± 0.00002	15
	229	89											0.13533 ± 0.00001	34
	229	89											0.14635 ± 0.00002	35
	229	89											0.16451 ± 0.00001	100
	229	89											0.2392 ± 0.0002	4
	229	89											0.24529 ± 0.00002	9
	229	89											0.24866 ± 0.00001	9.2
	229	89											0.25201 ± 0.00008	24
	229	89											0.26188 ± 0.00008	39
	229	89											0.2747 ± 0.0002	1.2
	229	89											0.27806 ± 0.00001	2.5
	229	89											0.2849 ± 0.0002	4.3
	229	89											0.2878 ± 0.0001	6
	229	89											0.28795 ± 0.00001	2.7
	229	89											0.29132 ± 0.00001	11
	229	89											0.31713 ± 0.00001	22
	229	89											0.32051 ± 0.00001	6.4
	229	89											0.3228 ± 0.0002	3
	229	89											0.3320 ± 0.0001	2.0
	229	89											0.3656 ± 0.0002	2.7
	229	89											0.4046 ± 0.0001	9
	229	89											0.4065 ± 0.0001	6
	229	89											0.4228 ± 0.0001	7
	229	89											0.4359 ± 0.0001	6.3
	229	89											0.4492 ± 0.0001	16
	229	89											0.4784 ± 0.0001	17
	229	89											0.5085 ± 0.0002	37
	229	89											0.5267 ± 0.0001	6
	229	89											0.5399 ± 0.0001	20
	229	89											0.5635 ± 0.0001	6
	229	89											0.56916 ± 0.00008	91%
	229	89											0.5758 ± 0.0001	5
	229	89											0.60497 ± 0.0012	23
$_{89}$Ac^{230}	230	89		230.936240	2.03 ± 0.05 m.	β-	2.900	1.4			1+		Th k x-ray	0.8%
	230	89											0.12091 ± 0.00002	0.3%
	230	89											0.39769 ± 0.00005	0.4%
	230	89											0.45497 ± 0.00003	8.9%
	230	89											0.50820 ± 0.00003	5.1%
	230	89											0.58178 ± 0.00008	0.5%
	230	89											0.62885 ± 0.00009	0.24%
	230	89											0.72820 ± 0.00004	0.5%

Isotope	A	Z	% Natural abundance	Atomic mass	Half-life	Decay mode	Decay energy (MeV)	Particle energy (MeV)	Particle intensity	Thermal neutron cross section	Spin (h/2π)	μ Nucl. mag. moment	Gamma-ray energy (MeV)	Gamma-ray intensity
	230	89											0.78143 ± 0.00004	0.4%
	230	89											0.78899 ± 0.00007	0.53%
	230	89											0.8167 ± 0.0001	0.32%
	230	89											0.8671 ± 0.0001	0.47%
	230	89											0.89275 ± 0.00005	0.7%
	230	89											0.95100 ± 0.00004	0.83%
	230	89											1.22681 ± 0.00005	0.96%
	230	89											1.24396 ± 0.00007	3.5%
	230	89											1.30258 ± 0.00006	0.54%
	230	89											1.34772 ± 0.00005	1.6%
	230	89											1.37535 ± 0.00006	1.2%
	230	89											1.69170 ± 0.00008	0.6%
	230	89											1.77525 ± 0.00007	1.1%
	230	89											1.89666 ± 0.00007	0.52%
	230	89											1.90273 ± 0.00009	0.75%
	230	89											1.9138 ± 0.0001	0.56%
	230	89											1.94989 ± 0.00007	1.25%
	230	89											2.00094 ± 0.00008	0.4%
	230	89											2.0986 ± 0.0001	0.52%
	230	89											2.12280 ± 0.00009	0.58%
	230	89											(0.12 - 2.5)many	
$_{89}Ac^{231}$	231	89		231.038550	7.5 ± 0.1 m.	β-	2.100	2.1	100%		(1/2+)		0.14379 ± 0.00001	9 +
	231	89											0.18574 ± 0.00001	45
	231	89											0.19893 ± 0.00002	6
	231	89											0.22140 ± 0.00002	52
	231	89											0.22088 ± 0.00002	11
	231	89											0.2721 ± 0.0001	7
	231	89											0.28250 ± 0.00001	100
	231	89											0.3070 ± 0.0001	80
	231	89											0.3688 ± 0.0001	38
	231	89											0.3722 ± 0.0001	4.4
	231	89											0.3759 ± 0.0002	4.1
	231	89											0.4003 ± 0.0002	2.6
	231	89											0.4079 ± 0.0001	8.5
	231	89											0.5282 ± 0.0002	2.3
	231	89											0.5546 ± 0.0001	3.9
$_{89}Ac^{232}$	232	89		232.042130	35 ± 5 s.	β-	3.800				(2-)			
Th		90		232.0381						7.4 b.				
$_{90}Th^{212}$	212	90		212.012890	≈ 30 ms.	α		7.80			0+			
$_{90}Th^{213}$	213	90		213.012940	0.14 ± 0.04 s.	α	7.840	7.692(10)						
$_{90}Th^{214}$	214	90		214.011430	0.86 ± 0.1 s.	α	7.825	7.677(10)			0+			
$_{90}Th^{215}$	215	90		215.011690	1.2 ± 0.2 s.	α	7.660	7.33(10)	8%		(1/2-)			
	215	90						7.395(8)	52%					
	215	90						7.524(8)	40%					
$_{90}Th^{216}$	216	90		216.011030	28 ± 2 ms.	α	8.071	7.921(8)			0+			
$_{90}Th^{217}$	217	90		217.013050	252 ± 7 μs.	α	9.424	9.250(10)						
$_{90}Th^{218}$	218	90		218.013252	0.11 μs.	α	9.847	9.665(10)			0+			
$_{90}Th^{219}$	219	90		219.015510	1.05 ± 0.03 μs	α	9.510	9.340(20)						
$_{90}Th^{220}$	220	90		220.015724	9.7 ± 0.6 μs.	α	8.953	8.790(20)			0+			
$_{90}Th^{221}$	221	90		221.018160	1.68 ± 0.06 ms	α	8.628	7.743(8)	6%					
	221	90						8.146(5)	56%					
	221	90						8.4272(5)	39%					
$_{90}Th^{222}$	222	90		222.018447	2.8 ± 0.3 ms.	α	8.129	7.982(8)			0+			
$_{90}Th^{223}$	223	90		223.020659	0.66 ± 0.02 s.	α	7.454	7.287(10)	60%					
	223	90						7.317(10)	40%					

TABLE OF THE ISOTOPES (Continued)

Isotope	A	Z	% Natural abundance	Atomic mass	Half-life	Decay mode	Decay energy (MeV)	Particle energy (MeV)	Particle intensity	Thermal neutron cross section	Spin (h/2π)	μ Nucl. mag. moment	Gamma-ray energy (MeV)	Gamma-ray intensity
$_{90}Th^{224}$	224	90		224.021449	1.04 ± 0.5 s.	α	7.305	6.768(5)	1.2%					
	224	90						6.997(5)	19%					
	224	90						7.170(5)	79%					
$_{90}Th^{225}$	225	90		225.023922	8.0 m.	E.C.(10%)	0.668				(3/2+)			
	225	90				α(90%)	6.920	6.441(2)	15 +					
	225	90						6.479(2)	43					
	225	90						6.501(3)	14					
	225	90						6.627(3)	3					
	225	90						6.650(5)	3					
	225	90						6.700(5)	2					
	225	90						6.743(3)	7					
	225	90						6.796(2)	9					
$_{90}Th^{226}$	226	90		226.024885	31 m.	α	6.454	6.026(1)	0.2%		0+		Ra k x-ray	0.5%
	226	90						6.041(1)	0.19	%			0.11110 ± 0.00003	3.3%%
	226	90						6.098(1)	1.3%				0.13100 ± 0.00004	0.28%
	226	90						6.2283(4)	23%				0.19028 ± 0.00005	0.11%
	216	90						6.3375(4)	75%				0.20621 ± 0.00005	0.19%
	216	90											0.24210 ± 0.00004	0.87%
	216	90											(0.1 - 0.8)weak	
$_{90}Th^{227}$	227	90		227.027703	18.72 d.	α	6.146				(3/2+)		Ra L x-ray	21%
	227	90											Ra k x-ray	3.1%
	227	90											0.02987 ± 0.00002	0.1%
	227	90											0.04373 ± 0.00004	0.23%
	227	90											0.04985 ± 0.00003	0.2%
	227	90											0.05014 ± 0.00004	8.5%
	227	90											0.06236 ± 0.00004	0.24%
	227	90											0.7972 ± 0.00003	2.1%
	227	90											0.09393 ± 0.00004	1.4%
	227	90											0.11312 ± 0.00005	0.15%
	227	90											0.00319 ± 0.00006	0.56%
	227	90											0.00717 ± 0.00006	0.17%
	227	90											0.14144 ± 0.00007	0.13%
	227	90											0.20420 ± 0.00005	0.23%
	227	90											0.20604 ± 0.00005	0.23%
	227	90											0.21058 ± 0.00005	1.1%
	227	90											0.23597 ± 0.00004	11.2%
	227	90											0.25012 ± 0.00005	0.37%
	225	90											0.25246 ± 0.00006	0.11%
	227	90											0.25466 ± 0.00005	0.8%
	227	90											0.25624 ± 0.00003	6.7%
	227	90											0.26272 ± 0.00005	0.10%
	227	90											0.27295 ± 0.00004	0.49%
	227	90											0.28131 ± 0.00005	0.16%
	227	90											0.28611 ± 0.00003	1.59%
	227	90											0.29654 ± 0.00005	0.43%
	227	90											0.29997 ± 0.00004	2.1%
	227	90											0.30034 ± 0.00009	0.20%
	227	90											0.30451 ± 0.00004	1.09%
	227	90											0.31257 ± 0.00005	0.47%
	227	90											0.31482 ± 0.00005	0.46%
	277	90											0.32984 ± 0.00004	2.73%

Isotope	A	Z	% Natural abundance	Atomic mass	Half-life	Decay mode	Decay energy (MeV)	Particle energy (MeV)	Particle intensity	Thermal neutron cross section	Spin (h/2π)	μ Nucl. mag. moment	Gamma-ray energy (MeV)	Gamma-ray intensity
	227	90											0.34244 ± 0.00005	0.38%
	227	90											0.35048 ± 0.00008	0.11%
	227	90											(0.02 - 1.02)weak	
$_{90}Th^{228}$	228	90		228.028715	1.913 y.	α	5.520	5.1770(2)	0.18%		0+			
	228	90						5.2114(1)	0.4%					
	228	90						5.3405(1)	26.7%					
	228	90						5.4233(1)	73%					
$_{90}Th^{229}$	229	90		229.031755	7.3 x 10^3 y	α	5.168	4.680(1)	0.15%		5/2+	+0.46		
	229	90						4.7618(5)	0.63%					
	229	90						4.7979(5)	1.27%					
	229	90						4.809(20)	0.22%					
	229	90						4.8140(5)	9.3%					
	229	90						4.833(25)	0.29%					
	229	90						4.838(5)	4.8%					
	229	90						4.845(5)	56%					
	229	90						4.861(25)	0.2%					
	229	90						4.9008(5)	10.2%					
	229	90						4.9301(5)	0.11%					
	229	90						4.9678(5)	5.97					
	229	90						4.9786(5)	3.2%					
	229	90						5.0354(5)	0.24%					
	229	90						5.050(25)	5.2%					
	229	90						5.0525(5)	1.6%					
	229	90						5.0774(5)	0.01					
$_{90}Th^{230}$	230	90		230.033127	7.54 x 10^4y.	α	4.771	4.4383(6)	0.03%		0+			
	230	90						4.4798(6)	0.12%					
	230	90						4.6211(6)	23.4%					
	230	90						4.6876(6)	76.3%					
$_{90}Th^{231}$	231	90		231.036298	25.2 h.	β-	0.389	0.138	22%		5/2+		Pa L x-ray	37%
	231	90						0.218	20%				Pa k x-ray	0.6%
	231	90						0.305	52%				0.02564 ± 0.00001	15%
	231	90											0.084203 ± 0.000	096.6%
	231	90											0.08995 ± 0.00001	0.9%
	231	90											0.10225 ± 0.00001	0.4%
	231	90											0.10816 ± 0.	0.23%
	231	90											0.16311 ± 0.00001	0.15%
	231	90											(0.02 - 0.35)weak	
$_{90}Th^{232}$	232	90	100%	232.038054	1.4 x 10^{10}y.	α	4.081	3.830(10)	0.2%	7	4 b.	0+	0.0590 ± 0.0001	0.19%
	232	90						3.952(5)	23%				0.124 ± 0.001	0.04%
	232	90						4.010(5)	77%					
$_{90}Th^{233}$	233	90		233.041577	22.3 m.	β-	1.243	1.245		1500 b.	1/2+		Pa L x-ray	4.6%
	233	90								σf 15 b.			Pa k x-ray	0.8%
	233	90											0.02938 ± 0.00001	2.6%
	233	90											0.08653 ± 0.00001	2.6%
	233	90											0.08805 ± 0.00004	0.21%
	233	90											0.09288 ± 0.00004	0.51%
	233	90											0.09472 ± 0.00002	0.9%
	233	90											0.09586 ± 0.00004	0.8%
	233	90											0.10816 ± 0.00004	0.30%
	233	90											0.11189 ± 0.00004	0.10%
	233	90											0.16251 ± 0.00004	0.17%
	233	90											0.16258 ± 0.00008	0.15%
	233	90											0.16918 ± 0.00004	0.15%
	233	90											0.17078 ± 0.00008	0.13%
	233	90											0.19054 ± 0.00007	0.13%
	233	90											0.3598 ± 0.0001	0.12%
	233	90											0.44117 ± 0.00009	0.23%
	233	90											0.44784 ± 0.00009	0.15%
	233	90											0.45930 ± 0.00001	1.4%
	233	90											0.4907 ± 0.0001	0.17%
	233	90											0.4989 ± 0.0001	0.21%
	233	90											0.5953 ± 0.0009	0.16%
	233	90											0.66978 ± 0.00008	0.12%

Isotope	A	Z	% Natural abundance	Atomic mass	Half-life	Decay mode	Decay energy (MeV)	Particle energy (MeV)	Particle intensity	Thermal neutron cross section	Spin (h/2π)	μ Nucl. mag. moment	Gamma-ray energy (MeV)	Gamma-ray intensity
	233	90											0.7645 ± 0.0008	0.12%
	233	90											0.8901 ± 0.0001	0.14%
	233	90											(0.02 - 1.2)many	
$_{90}Th^{234}$	234	90		234.043593	24.10 d.	β-	0.270	0.102	20%		0+		Pa L x-ray	4%
	234	90						0.198	72%				0.06329 ± 0.00002	3.8%
	234	90											0.09235 ± 0.00003	2.7%
	234	90											0.09278 ± 0.00003	2.7%
	234	90											0.11280 ± 0.00003	0.24%
$_{90}Th^{235}$	235	90		235.047510	6.9 ± 0.2 m.	β-	1.940						0.4162 ± 0.0010	
	235	90											0.6594 ± 0.001	
	235	90											0.7272 ± 0.001	
	235	90											0.747 ± 0.001	
	235	90											0.9318 ± 0.001	
$_{90}Th^{236}$	236	90			37.1 ± 0.2 m.	β-	≈ 1						Pa k x-ray	3.3%
	236	90											0.1107 ± 0.0005	2.4%
	236	90											0.11189	0.40%
	236	90											0.1127 ± 0.0005	0.6%
	236	90											0.1316 ± 0.001	0.5%
	236	90											0.2296 ± 0.001	0.4%
Pa		91												
$_{91}Pa^{216}$	216	91		216.018960	0.20 s.	α	8.010	7.720						
	216	91						7.820						
	216	91						7.920						
$_{91}Pa^{217m}$	217	91			1.6 ms.	α		10.160(20)						
$_{91}Pa^{217}$	217	91		217.018250	4.9 ms.	α	8.490	8.340(10)						
$_{91}Pa^{218}$	218	91		218.019960	0.12 ms.	α		9.54						
	218	91						9.61						
$_{91}Pa^{222}$	222	91		222.023560	≈ 4.3 ms.	α	8.700	8.180	50%					
	222	91							8.33	0 20%				
	222	91						8.540	30%					
$_{91}Pa^{223}$	223	91		223.023950	6 ms.	α	8.340	8.006(10)	55%					
	223	91						8.196(10)	45%					
$_{91}Pa^{224}$	224	91		224.025530	0.95 s.	α	7.630	7.490(10)	100%					
$_{91}Pa^{225}$	225	91		225.026090	1.8 ± 0.3 s.	α	7.380	7.195(10)	30%					
	225	91						7.245(10)	70%					
$_{91}Pa^{226}$	226	91		226.027928	1.8 ± 0.2 s.	α(74%)	6.987	6.728(10)	0.7%					
	226	91				E.C.(26%)	2.834	6.823(10)	35%					
	226	91						6.863(10)	39%					
$_{91}Pa^{227}$	227	91		227.028797	38.3 m.	α(85%)	6.582	6.357(4)	7%		(5/2-)		0.0649 ± 0.001	5.3%
	227	91				E.C.(15%)	1.020	6.376(10)	2.2%				0.0669 ± 0.001	1.0%
	227	91						6.401(4)	8%				0.1100 ± 0.001	1.7%
	227	91						6.416(4)	13%					
	227	91						6.423(10)	10%					
	227	91						6.465(4)	43%					
$_{91}Pa^{228}$	228	91		228.030773	22 ± 1 h.	E.C.(98%)					(3+)		Th k x-ray	35%
	228	91				α(2%)		5.779	0.23%				0.20939 ± 0.00007	1.7%
	228	91						5.805	0.15%				0.27026 ± 0.00008	2.1%
	228	91						6.078	0.4%				0.28202 ± 0.00008	1.2%
	228	91						6.105	0.25%				0.32767 ± 0.00009	2%
	288	91						6.118	0.22%				0.32807 ± 0.00009	1.9%
	228	91											0.33248 ± 0.00009	1.6%
	228	91											0.33842 ± 0.00006	5.1%
	228	91											0.40962 ± 0.00008	6.4%
	228	91											0.46310 ± 0.00007	13.2%
	228	91											0.5815 ± 0.0002	1.0%
	228	91											0.7553 ± 0.0001	1.2%
	228	91											0.77228 ± 0.00007	1.2%
	228	91											0.7948 ± 0.0001	2.0%
	228	91											0.8306 ± 0.0001	1.9%
	228	91											0.8356 ± 0.0001	2.8%
	228	91											0.8404 ± 0.0001	1.0%
	228	91											0.8944 ± 0.0001	2.6%
	228	91											0.9043 ± 0.0001	2.8%
	228	91											0.91116 ± 0.00003	16.0%
	228	91											0.9457 ± 0.0008	1.8%
	228	91											0.96464 ± 0.00008	9.4%

Isotope	A	Z	% Natural abundance	Atomic mass	Half-life	Decay mode	Decay energy (MeV)	Particle energy (MeV)	Particle intensity	Thermal neutron cross section	Spin (h/2π)	μ Nucl. mag. moment	Gamma-ray energy (MeV)	Gamma-ray intensity
	228	91											0.96897 ± 0.00005	9.7%
	228	91											0.9757 ± 0.0001	1.6%
	228	91											1.2466 ± 0.0002	0.9%
	228	91											1.4592 ± 0.0001	0.72%
	228	91											1.5882 ± 0.0001	2.4%
	228	91											1.7385 ± 0.0002	0.64%
	228	91											1.7579 ± 0.0001	0.5%
	228	91											1.8351 ± 0.0001	0.64%
	228	91											1.8869 ± 0.0001	1.5%
	228	91											(0.1 - 1.96)many	
$_{91}Pa^{229}$	229	91		229.032073	1.4 ± 0.4 d.	E.C. (99.8%)	0.296				(5/2)		0.04244 ± 0.00001	
	229	91				α(0.2%)	5.836	5.536(2)	0.02%				(0.024 - 0.18)weak	
	229	91						5.579(2)	0.09%					
	229	91						5.668(2)	0.05%					
$_{91}Pa^{230}$	230	91		230.034527	17.4 ± 0.5 d.	E.C.(90%)		0.51		σf 1500 b.	(2-)		Th L x-ray	25%
	230	91				β-(10%)							Th k x-ray	30%
	230	91											0.39769 ± 0.00005	1.8%
	230	91											0.39996 ± 0.00006	0.61%
	230	91											0.44379 ± 0.00003	5.4%
	230	91											0.45477 ± 0.00003	6.1%
	230	91											0.46359 ± 0.00005	0.8%
	230	91											0.50820 ± 0.00003	3.5%
	230	91											0.51860 ± 0.00005	1.9%
	230	91											0.57116 ± 0.00006	1.0%
	230	91											0.72820 ± 0.00004	1.8%
	230	91											0.78143 ± 0.00004	1.4%
	230	91											0.89876 ± 0.00003	6.0%
	230	91											0.91856 ± 0.00005	8.0%
	230	91											0.95199 ± 0.00004	28%
	230	91											0.9591 ± 0.0002	0.5%
	230	91											0.95651 ± 0.00005	1.7%
	230	91											1.00974 ± 0.00005	1.7%
	230	91											1.02613 ± 0.00006	1.4%
	230	91											1.07467 ± 0.00006	0.73%
$_{91}Pa^{231}$	231	91		231.035880	3.27 x 10^4y.	α	5.148	4.6781(5)	1.5%		3/2-	2.01	Ac L x-ray	22%
	231	91						4.7102(5)	1.0%				Ac k x-ray	0.8%
	231	91						4.7343(5)	8.4%				0.01899 ± 0.00002	0.33%
	231	91						4.8513(5)	1.4%				0.027396 ± 0.00009	9.3%
	231	91						4.9339(5)	3%				0.03823 ± 0.00001	0.15%
	231	91						4.9505(5)	22.8%				0.04639 ± 0.00001	0.21%
	231	91						4.9858(5)	1.4%				0.25586 ± 0.00002	0.1%
	231	91						5.0131(5)	25.4%				0.26029 ± 0.00003	0.18%
	231	91						5.0292(5)	20%				0.28367 ± 0.00002	1.6%
	231	91						5.0318(5)	2.5%				0.30007 ± 0.00002	2.4%
	231	91						5.0587(5)	11%				0.30264 ± 0.00002	0.6%
	231	91											0.31306 ± 0.00002	0.13%
	231	91											0.33007 ± 0.00002	1.3%
	231	91											0.34087 ± 0.00002	0.17%
	231	91											0.35727 ± 0.00002	0.15%
	231	91											0.4271 ± 0.0001	0.10%
	231	91											(0.02 - 0.61)many	
$_{91}Pa^{232}$	232	91		232.038565	1.31 d.	β-	1.34			500 b.	(2-)		U k x-ray	1.8%

Isotope	A	Z	% Natural abundance	Atomic mass	Half-life	Decay mode	Decay energy (MeV)	Particle energy (MeV)	Particle intensity	Thermal neutron cross section	Spin (h/2π)	μ Nucl. mag. moment	Gamma-ray energy (MeV)	Gamma-ray intensity
	232	91								σf 700 b.			0.10900 ± 0.00001	2.8%
	232	91											0.15009 ± 0.00001	11%
	232	91											0.18414 ± 0.00001	1.3%
	232	91											0.38792 ± 0.00001	7.0%
	232	91											0.42196 ± 0.00001	2.5%
	232	91											0.45369 ± 0.00001	8.6%
	232	91											0.47243 ± 0.00001	4.2%
	232	91											0.51565 ± 0.00001	5.5%
	232	91											0.56323 ± 0.00001	3.7%
	232	91											0.58142 ± 0.00001	6.0%
	232	91											0.81925 ± 0.00001	7.5%
	232	91											0.86386 ± 0.00003	2.2%
	232	91											0.86683 ± 0.00001	5.8%
	232	91											0.89439 ± 0.00001	20%
	232	91											0.96934 ± 0.00001	42%
	232	91											(0.10 - 1.17)weak	
$_{91}Pa^{233}$	233	91		233.040242	27.0 ± 0.1 d.	β-	0.572	0.15	40%		3/2-	+3.5	U L x-ray	19%
	233	91						0.256	60%				U k x-rau	16%
	233	91											0.07534 ± 0.00001	1.2%
	233	91											0.08665 ± 0.00002	1.8%
	233	91											0.30017 ± 0.00002	6.2%
	233	91											0.31201 ± 0.00002	36%
	233	91											0.34059 ± 0.00002	4.2%
	233	91											0.39866 ± 0.00002	1.2%
	233	91											0.41593 ± 0.00002	1.5%
$_{91}Pa^{234m}$	234	91			1.17 m.	β-(99.9%)	2.29				(0-)		U k x-ray	0.2%
	234	91				I.T.(0.13%)							0.25818 ± 0.00003	0.06%
	234	91											0.74282 ± 0.00002	0.056%
	234	91											0.76641 ± 0.00001	0.21%
	234	91											0.78629 ± 0.00002	0.03%
	234	91											1.00100 ± 0.00003	0.65%
	234	91											1.7378 ± 0.0003	0.014%
	234	91											1.8317 ± 0.0004	0.011%
	234	91											(0.06 - 1.96)many	
$_{91}Pa^{234}$	234	91		234.043303	6.70 ± 0.05 h.	β-	2.199	0.51			(4+)		U L x-ray	52%
	234	91											U k x-ray	25%
	234	91											0.06278 ± 0.00004	3.2%
	234	91											0.09985 ± 0.00001	4.8%+
	234	91											0.1255 ± 0.0002	1.0%
	234	91											0.1312 ± 0.0002	20%
	234	91											0.15269 ± 0.00003	6.7%
	234	91											0.1859 ± 0.0002	2.0%
	234	91											0.20096 ± 0.00005	1.1%
	234	91											0.20320 ± 0.00003	1.1%
	234	91											0.2266 ± 0.0002	5.9%
	234	91											0.2272 ± 0.0002	5.5%
	234	91											0.2489 ± 0.0002	2.8%
	234	91											0.2721 ± 0.0002	1.0%
	234	91											0.2938 ± 0.0002	3.9%
	234	91											0.3700 ± 0.0002	2.9%
	234	91											0.4586 ± 0.0002	1.5%
	234	91											0.5067 ± 0.0002	1.6%

Isotope	A	Z	% Natural abundance	Atomic mass	Half-life	Decay mode	Decay energy (MeV)	Particle energy (MeV)	Particle intensity	Thermal neutron cross section	Spin (h/2π)	μ Nucl. mag. moment	Gamma-ray energy (MeV)	Gamma-ray intensity
	234	91											0.5136 ± 0.0003	1.3%
	234	91											0.5652 ± 0.0004	1.4%
	234	91											0.5683 ± 0.0002	3.0%
	234	91											0.5695 ± 0.0002	11%
	234	91											0.6648 ± 0.001	1.3%
	234	91											0.6666 ± 0.0002	1.6%
	234	91											0.6697 ± 0.0003	1.4%
	234	91											0.6927 ± 0.0004	1.5%
	234	91											0.6988 ± 0.0002	1.6%
	234	91											0.70602 ± 0.00009	3.0%
	234	91											0.7332 ± 0.0002	8.6%
	234	91											0.7380 ± 0.0004	1.0%
	234	91											0.7428 ± 0.0002	2.4%
	234	91											0.7804 ± 0.0004	1.1%
	234	91											0.78629 ± 0.00002	1.4%
	234	91											0.7936 ± 0.001	1.5%
	234	91											0.7953 ± 0.0003	3.8%
	234	91											0.80587 ± 0.00009	3.4%
	234	91											0.8258 ± 0.0002	4.0%
	234	91											0.8314 ± 0.0002	5.5%
	234	91											0.8805 ± 0.0002	4.0%
	234	91											0.88053 ± 0.00004	9.0%
	234	91											0.88324 ± 0.00004	12.0%
	234	91											0.8986 ± 0.0002	4.0%
	234	91											0.9256 ± 0.0002	11%
	234	91											0.9258 ± 0.0002	2.9%
	234	91											0.92671 ± 0.00004	9.0%
	234	91											0.94602 ± 0.00003	8.1%
	234	91											0.9841 ± 0.0002	1.9%
	234	91											1.3528 ± 0.0002	1.7%
	234	91											1.3941 ± 0.0002	3.0%
	234	91											1.4527 ± 0.0002	1.0%
	234	91											1.6682 ± 0.0007	1.2%
	234	91											1.6942 ± 0.0004	1.2%
	234	91											1.9260 ± 0.0006	1.5%
	234	91											(0.02 - 1.99)many	
$_{91}Pa^{235}$	235	91		235.045430	24.1 m.	β-	1.4	1.4	97%		(3/2-)		0.0308 ± 0.0003	
	235	91											0.0367 ± 0.0002	
	235	91											0.05162 ± 0.00002	
	235	91											0.12928 ± 0.00002	
	235	91											0.34494 ± 0.00003	
	235	91											0.37502 ± 0.00002	
	235	91											0.38017 ± 0.00003	
	235	91											0.39312 ± 0.00004	
	235	91											0.41369 ± 0.00002	
	235	91											0.63796 ± 0.00008	
	235	91											0.64598 ± 0.00007	
	235	91											0.65218 ± 0.00007	
	235	91											0.65893 ± 0.00007	
$_{91}Pa^{236}$	236	91		236.048890	9.1 m.	β-	3.100	1.1	40%		(1-)		U k x-ray	2.6%
	236	91						2.0	50%				0.64235 ± 0.00005	29%
	236	91						3.1	10%				0.68759 ± 0.00005	7.8%
	236	91											1.5601 ± 0.001	1.9%
	236	91											1.7630 ± 0.0001	5.4%
	236	91											1.8082 ± 0.0007	2.0%
	236	91											2.0416 ± 0.0007	1.6%
	236	91											(0.04 - 2.18)weak	
$_{91}Pa^{237}$	237	91		237.051140	8.7 ± 0.2 m.	β-	2.25	1.1	60%		(1/2+)		0.4986 ± 0.0001	2.4%
	237	91						1.6	30%				0.5293 ± 0.0001	14.8%
	237	91						2.3	10%				0.5407 ± 0.0001	9.3%
	237	91											0.5549 ± 0.0001	1.5%
	237	91											0.8536 ± 0.0001	34%
	237	91											0.8650 ± 0.0001	15%
	237	91											(0.04 - 1.4)weak	

Isotope	A	Z	% Natural abundance	Atomic mass	Half-life	Decay mode	Decay energy (MeV)	Particle energy (MeV)	Particle intensity	Thermal neutron cross section	Spin (h/2π)	μ Nucl. mag. moment	Gamma-ray energy (MeV)	Gamma-ray intensity
$_{91}Pa^{238}$	238	91		238.055040	2.3 m.	β-	3.960	1.2			(3-)		0.10350 ± 0.00004	12 +
	238	91						1.7					0.1785 ± 0.0005	11
	238	91											0.2179 ± 0.0005	14
	238	91											0.3698 ± 0.0005	12
	238	91											0.2930 ± 0.001	12
	238	91											0.3964 ± 0.0004	18
	238	91											0.4369 ± 0.0004	16
	238	91											0.4484 ± 0.0004	76
	238	91											0.4961 ± 0.0005	19
	238	91											0.4889 ± 0.0004	20
	238	91											0.5019 ± 0.0005	26
	238	91											0.5471 ± 0.0004	40
	238	91											0.6057 ± 0.0005	10
	238	91											0.6236 ± 0.001	19
	238	91											0.6350 ± 0.0004	88
	238	91											0.6800 ± 0.0004	73
	238	91											0.6870 ± 0.0004	54
	238	91											0.8058 ± 0.0004	44
	238	91											0.8491 ± 0.0005	14
	238	91											0.8637 ± 0.0005	54
	238	91											0.8857 ± 0.0004	45
	238	91											0.9049 ± 0.0005	23
	238	91											0.9111 ± 0.0004	19
	238	91											0.9526 ± 0.0005	21
	238	91											0.9571 ± 0.0005	18
	238	91											1.01446 ± 0.0004	100
	238	91											1.0602 ± 0.0005	45
	238	91											1.0834 ± 0.0003	50
	238	91											1.8892 ± 0.0004	17
	238	91											(0.04 - 2.5)many	
U		92		238.0289						7.57 b. σf 4.20 b.				
$_{92}U^{226}$	226	92		226.029170	0.50 ± 0.2 s.	α	7.560	7.430			0+			
$_{92}U^{227}$	227	92		227.030990	1.1 ± 0.3 m.	α	7.200	6.870						
$_{92}U^{228}$	228	92		228.031356	9.1 ± 0.2 m.	α	6.803	6.404(6)	0.6+		0+		0.095	1.6%
	228	92						6.440(5)	0.7				0.152	0.2%
	228	92						6.589(5)	29				0.187	0.3%
	228	92						6.681(6)	70				0.246	0.4%
$_{92}U^{229}$	229	92		229.033474	58 ± 3 m.	E.C.(80%)	1.305	6.223	3 +		(3/2+)			
	229	92				α(20%)	6.473	6.297(3)	11					
	229	92						6.332(3)	20					
	229	92						6.360(3)	64					
$_{92}U^{230}$	230	92		230.033921	20.8 d.	α	5.992	5.5866(3)	0.01%		0+		Th L x-ray	6.0%
	230	92						5.6624(3)	0.26%				0.07218 ± 0.00003	0.60%
	230	92						5.6663(3)	0.38%				0.15421 ± 0.00003	0.12%
	230	92						5.8178(3)	32%				0.23034 ± 0.00004	0.12%
	230	92						5.8887(3)	67%					
$_{92}U^{231}$	231	92		231.036270	4.2 ± 0.1 d.	E.C>					(5/2-)		Pa L x-ray	41%
	231	92											Pa k x-ray	27%
	231	92											0.02564 ± 0.00001	13%
	231	92											0.08420 ± 0.00001	6.0%
	231	92											0.21793 ± 0.00002	0.8%
	231	92											0.23598 ± 0.00002	0.2%
	231	92											0.31100 ± 0.00003	0.06%
$_{92}U^{232}$	231	92		232.037130	68.9 ± 0.1 y.	α	5.414	4.9979(1)	0.003%	73 b.	0+			
	232	92						5.1367(1)	0.3%					
	232	92						5.2635(1)	31%					
	232	92						5.3203(1)	69%					
$_{92}U^{233}$	233	92		233.039628	1.59 x 10^5y.	α	4.909	4.5097(8)	0.01%	466 b.	5/2+	+0.55	Th L x-ray	3.3%
	233	92						4.5130(8)	0.02%	σf 529 b.			0.04244 ± 0.00001	0.06%
	233	92						4.6323(8)	0.01%				0.09714 ± 0.00002	0.02%
	233	92						4.6642(8)	0.04%					
	233	92						4.6809(8)	0.01%					
	233	92						4.7014(8)	0.06%					
	233	92						4.7292(8)	1.6%					
	233	92						4.7541(8)	0.16%					
	233	92						4.758(10)	0.02%					
	233	92						4.7830(8)	13.2%					
	233	92						4.7960(8)	0.3%					
	233	92						4.804(8)	0.05%					

Isotope	A	Z	% Natural abundance	Atomic mass	Half-life	Decay mode	Decay energy (MeV)	Particle energy (MeV)	Particle intensity	Thermal neutron cross section	Spin (h/2π)	μ Nucl. mag. moment	Gamma-ray energy (MeV)	Gamma-ray intensity
	233	92						4.8247(8)	84.4%					
$_{92}U^{234}$	234	92		234.040946	2.45 x 10^5y.	α	4.856	4.604(1)	0.24%	100 b.	0+		0.05323 ± 0.00003	0.12%
	234	92						4.7231(1)	27.5%				0.12091 ± 0.00002	0.04%
	234	92						4.776(1)	72.5%					
$_{92}U^{235m}$	235	92			26 ± 2 m.	I.T.	0.0007				1/2+			
$_{92}U^{235}$	235	92	0.720%	235.043924	7.04 x 10^8y.	α	4.6793	4.1525(9)	0.9%	98 b.	7/2-	-0.35	Th L x-ray	15%
	235	92						4.2157(9)	5.7%	σf 583 b.			Th k x-ray	5.5%
	235	92						4.3237(9)	4.6%				0.10917 ± 0.00001	1.5%
	235	92						4.3641(9)	11%				0.14378 ± 0.00001	10.5%
	235	92						4.370(4)	6%				0.16338 ± 0.00001	4.7%
	235	92						4.3952(9)	55%				0.18574 ± 0.00001	53%
	235	92						4.4144(9)	2.1%				0.20213 ± 0.00001	1.0%
	235	92						4.5025(9)	1.7%				0.20533 ± 0.00001	4.7%
	235	92						4.5558(9)	4.2%				0.22140 ± 0.00002	0.1%
	235	92						4.5970(9)	5.0%				(0.03 - 0.79)weak	
$_{92}U^{236}$	236	92		236.045562	2.34 x 10^7y.	α	4.569	4.332(8)	0.26%	5.1 b.	0+		Th L x-ray	3.4%
	236	92						4.445(5)	26%				0.04937 ± 0.00001	0.08%
	236	92						4.494(3)	74%				0.11275 ± 0.00001	0.02%
$_{92}U^{237}$	237	92		237.048724	6.75 ± 0.01 d.	β-	0.519	0.24		400 b.	1/2+		Np L x-ray	30%
	237	92						0.25					Np k x-ray	26%
	237	92											0.02634 ± 0.00001	2.3%
	237	92											0.05953 ± 0.00001	33%
	237	92											0.06482 ± 0.00001	1.2%
	237	92											0.16459 ± 0.00001	1.8%
	237	92											0.20801 ± 0.00001	22%
	237	92											0.33236 ± 0.00001	1.2%
	237	92											0.37092 ± 0.00002	0.11%
$_{92}U^{238}$	238	92		238.050784	4.46 x 10^9y.	α		4.039(5)	0.23	% 2.68 b.	0+		Th L x-ray	4.1%
	238	92						4.147(5)	23%				0.04955 ± 0.00006	0.07%
	238	92						4.196(5)	77%					
$_{92}U^{239}$	239	92		239.054289	23.54 ± 0.05m.	β-	1.264	1.2		22 b.	5/2+		0.04354 ± 0.00001	4.4%
	239	92						1.3		σf 15 b.			0.07467 ± 0.00001	52%
	239	92											0.11770 ± 0.00003	0.12%
	239	92											0.66225 ± 0.00002	0.20%
	239	92											0.74805 ± 0.00003	0.10%
	239	92											0.81927 ± 0.00003	0.15%
	239	92											0.84410 ± 0.00003	0.17%
$_{92}U^{240}$	240	92		240.056587	14.1 ± 0.2 h.	β-	0.50	0.36			0+		Np L x-ray	18%
	240	92											0.04410 ± 0.00007	1.7%
$_{92}U^{242}$	242	92			16.8 m.	β-							0.05558 ± 0.0005	3.7%
	242	92											0.06760 ± 0.00005	9.2%
	242	92											0.1604 ± 0.0001	0.75%
	242	92											0.1820 ± 0.0001	0.70%
	242	92											0.32972 ± 0.00009	0.75%
	242	92											0.5729 ± 0.0001	1.8%
	242	92											0.5849 ± 0.0001	1.8%
Np		93												
$_{93}Np^{228}$	228	93			1.0 m.	S.F.								
$_{93}Np^{229}$	229	93		229.036230	4.0 ± 0.2 m.	α	7.010	6.890(20)						
$_{93}Np^{230}$	230	93		230.937810	4.6 m.	E.C.(97%)	3.620							
	230	93				α(3%)		6.660(20)						
$_{93}Np^{231}$	231	93		231.038240	48.8 ± 0.2 m.	E.C.(98%)	1.84				5/2		0.2629 ± 0.0003	2.8 +
	231	93				α(2%)	6.368	6.280	2%				0.3475 ± 0.0003	3.6

Isotope	A	Z	% Natural abundance	Atomic mass	Half-life	Decay mode	Decay energy (MeV)	Particle energy (MeV)	Particle intensity	Thermal neutron cross section	Spin (h/2π)	μ Nucl. mag. moment	Gamma-ray energy (MeV)	Gamma-ray intensity
	231	93											0.3703 ± 0.0002	9.8
	231	93											0.4201 ± 0.0003	1.0
	231	93											0.4838 ± 0.0005	1.6
	231	93											0.7369 ± 0.0002	1.2
	231	93											0.8364 ± 0.0004	0.4
	231	93											0.8511 ± 0.0003	0.7
	231	93											1.1072 ± 0.0002	0.5
$_{93}Np^{232}$	232	93		232.040020	14.7 m.	E.C.(99%)	2.70				(4-)		U L x-ray	35%
	232	93											U k x-ray	38%
	232	93											0.2229 ± 0.0002	2.2%
	232	93											0.2822 ± 0.0002	19.8%
	232	93											0.3268 ± 0.0002	52%
	232	93											0.75486 ± 0.00003	4.5%
	232	93											0.81415 ± 0.00007	4.1%
	232	93											0.81925 ± 0.00001	32.6%
	232	93											0.86386 ± 0.00003	19.7%
	232	93											0.86683 ± 0.00001	25.1%
	232	93											1.0371 ± 0.0002	3.3%
	232	93											1.1255 ± 0.0002	1.5%
$_{93}Np^{233}$	233	93		233.040800	36.2 ± 0.1 m.	E.C.	1.090				(5/2+)		U L x-ray	18%
	233	93											U k x-ray	35%
	233	93											0.2344 ± 0.0002	0.15%
	233	93											0.25846 ± 0.00003	0.10%
	233	93											0.2804 ± 0.0001	0.13%
	233	93											0.29887 ± 0.00003	0.48%
	233	93											0.31201 ± 0.00002	0.70%
	233	93											0.5061 ± 0.0002	0.15%
	233	93											0.5465 ± 0.0002	0.28%
$_{93}Np^{234}$	234	93		234.042888	4.4 ± 0.1 d.	β+,E.C.	1.808	0.79		σf 900 b.	(0+)		U L x-ray	17%
	234	93											U k x-ray	29%
	234	93											0.45092 ± 0.00004	1.3%
	234	93											0.6255 ± 0.0003	1.1%
	234	93											0.74282 ± 0.00002	5.0%
	234	93											0.78629 ± 0.00002	3.0%
	234	93											1.00100 ± 0.00003	1.4%
	234	93											1.19374 ± 0.00003	5.5%
	234	93											1.23721 ± 0.00003	2.2%
	234	93											1.3920 ± 0.0003	2.0%
	234	93											1.4354 ± 0.0003	6.3%
	234	93											1.5272 ± 0.0002	11.5%
	234	93											1.5587 ± 0.0003	18.4%
	234	93											1.5707 ± 0.0002	5.6%
	234	93											1.6022 ± 0.0003	9.7%
$_{93}Np^{235}$	235	93		235.044056	1.08 y.	E.C. (99.9%)	0.123			150 b.	5/2+		U k x-ray	16%
	235	93				α(0.001%)	5.191							
$_{93}Np^{236m}$	236	93			22.5 h.	E.C.(52%)					(1-)		U L x-ray	10%
	236	93				β-(48%)							Pu L x-ray	1.5%
	236	93											U k x-ray	18%
	236	93											0.64235 ± 0.00005	0.9%
	236	93											0.68759 ± 0.00005	0.25%
$_{93}Np^{236}$	236	93		236.046550	1.2 x 10^5 y.	E.C.(91%)	0.99			σf 2500 b.	(6-)		U L x-ray	57%
	236	93				β-(9%)	0.54						U k x-ray	33%
	236	93											0.10423 ± 0.00001	7.5%
	236	93											0.16031 ± 0.00001	28%
$_{93}Np^{237}$	237	93		237.048167	2.14 x 10^6y.	α	4.957	4.5779(5)	0.4%	180 b.	5/2+	+3.14	Pa L x-ray	25%
	237	93						4.6395(5)	6.2%	σf 0.02 b.			Pa k x-ray	2.6%
	237	93						4.659(2)	0.6%				0.029378 ± 0.00009	13%
	237	93						4.6645(5)	3.3%				0.08653 ± 0.00001	13%
	237	93						4.6971(7)	0.5%				0.09472 ± 0.00002	0.8%
	237	93						4.7071(5)	1.0%				0.11758 ± 0.00002	0.16%
	237	93						4.766(5)	8%				0.14323 ± 0.00002	0.40%

Isotope	A	Z	% Natural abundance	Atomic mass	Half-life	Decay mode	Decay energy (MeV)	Particle energy (MeV)	Particle intensity	Thermal neutron cross section	Spin (h/2π)	μ Nucl. mag. moment	Gamma-ray energy (MeV)	Gamma-ray intensity
	237	93						4.7715(5)	25%				0.15142 ± 0.00002	0.24%
	237	93						4.7884(5)	47%				0.19504 ± 0.00003	0.20%
	237	93						4.8040(5)	1.6%				0.21241 ± 0.00002	0.15%
	237	93						4.8173(5)	2.5%				(0.03 - 0.28)weak	
	237	93						4.8734(5)	2.6%					
$_{93}Np^{238}$	238	93		238.050941	2.117 d.	β-	1.291	1.2		σf 2100 b.	2+		Pu L x-ray	16%
	238	93											Pu k x-ray	0.3%
	238	93											0.1019 ± 0.0001	0.27%
	238	93											0.11990 ± 0.00007	0.11%
	238	93											0.56103 ± 0.00003	0.11%
	238	93											0.88258 ± 0.00002	0.9%
	238	93											0.91870 ± 0.00002	0.6%
	238	93											0.92339 ± 0.00002	2.9%
	238	93											0.98447 ± 0.00002	28%
	238	93											1.02588 ± 0.00002	9.6%
	238	93											1.02855 ± 0.00002	20%
$_{93}Np^{239}$	239	93		239.052933	2.35 d.	β-	0.721	0.341	30%	(30 + 30)b.	5/2+		Pu L x-ray	25%
	239	93						0.438	48%				Pu k x-ray	24%
	239	93											0.10613 ± 0.00001	23%
	239	93											0.20975 ± 0.00001	3.3%
	239	93											0.228186 ± 0.00002	10.7%
	239	93											0.27760 ± 0.00002	14.2%
	239	93											0.31588 ± 0.00001	1.6%
	239	93											0.33431 ± 0.00001	2.0%
	239	93											(0.04 - 0.50)weak	
$_{93}Np^{240m}$	240	93			7.22 m.	β-(99.9%)		2.18			(1-)		0.25143 ± 0.00005	0.9%
	240	93				I.T.(0.1%)							0.26333 ± 0.00006	1.1%
	240	93											0.30296 ± 0.00004	1.2%
	240	93											0.55454 ± 0.00003	22%
	240	93											0.59735 ± 0.00003	12.6%
	240	93											0.75864 ± 0.00004	1.2%
	240	93											0.81787 ± 0.00006	1.3%
	240	93											0.85750 ± 0.00004	0.5%
	240	93											0.91604 ± 0.00005	1.0%
	240	93											0.93805 ± 0.00007	1.3%
	240	93											0.96158 ± 0.00007	0.14%
	240	93											1.44533 ± 0.00007	0.36%
	240	93											1.49685 ± 0.00006	1.3%
	240	93											1.53967 ± 0.00006	0.79%
	240	93											1.63326 ± 0.00009	0.14%
$_{93}Np^{240}$	240	93		240.056050	1.03 h.	β-	2.090	0.89			5+		0.1471 ± 0.0003	1.5%
	240	93											0.15262 ± 0.00002	9.0%
	240	93											0.175 ± 0.001	6.5%
	240	93											0.1930 ± 0.0002	7.3%
	240	93											0.2708 ± 0.0003	9.0%
	240	93											0.4482 ± 0.0002	18%
	240	93											0.4669 ± 0.0002	2.2%
	240	93											0.5664 ± 0.0002	29%
	240	93											0.6008 ± 0.0002	22%
	240	93											0.847 ± 0.001	5%
	240	93											0.8674 ± 0.0002	9.0%
	240	93											0.8449 ± 0.001	4.0%

Isotope	A	Z	% Natural abundance	Atomic mass	Half-life	Decay mode	Decay energy (MeV)	Particle energy (MeV)	Particle intensity	Thermal neutron cross section	Spin (h/2π)	μ Nucl. mag. moment	Gamma-ray energy (MeV)	Gamma-ray intensity
	240	93											0.8963 ± 0.0003	14%
	240	93											0.9591 ± 0.0002	2.5%
	240	93											0.9741 ± 0.0002	23%
	240	93											0.98764 ± 0.00004	4.0%
	240	93											1.1672 ± 0.0002	5%
	240	93											1.1802 ± 0.0002	0.7%
$_{93}Np^{241}$	241	93		241.058250	13.9 ± 0.2 m.	β-	1.5	1.3			5/2+		0.1330 ± 0.0009	98 +
	241	93											0.1740 ± 0.0009	100
	241	93											0.280	
$_{93}Np^{242m}$	242	93			5.5 ± 0.1 m.	β-					6+		0.15910 ± 0.00008	32 +
	242	93											0.2651 ± 0.0001	24
	242	93											0.78570 ± 0.00008	100
	242	93											0.9448 ± 0.0001	63
	242	93											1.104 ± 0.0001	0.6
$_{93}Np^{242}$	242	93		242.061640	2.2 ± 0.2 m.	β-	2.7	2.7			(1+)		0.6209 ± 0.0001	0.9%
	242	93											0.6477 ± 0.0002	0.27%
	242	93											0.6853 ± 0.0001	0.35%
	242	93											0.73620 ± 0.00004	5.0%
	242	93											0.78074 ± 0.00004	2.6%
	242	93											0.81395 ± 0.00008	1.2%
	242	93											1.0076 ± 0.0002	0.15%
	242	93											1.0346 ± 0.0002	0.3%
	242	93											1.0938 ± 0.0001	1.1%
	242	93											1.1103 ± 0.0002	0.35%
	242	93											1.1374 ± 0.0001	1.2%
	242	93											1.47340 ± 0.00006	2.2%
	242	93											1.51794 ± 0.00006	1.2%
	242	93											1.55114 ± 0.00008	0.35%
	242	93											1.8596 ± 0.0002	0.55%
	242	93											1.9500 ± 0.0001	1.75%
	242	93											1.9702 ± 0.0001	0.52%
	242	93											1.9924 ± 0.0003	0.2%
	242	93											(0.04 - 2.37)weak	
Pu		94												
$_{94}Pu^{232}$	232	94		232.041169	34 m.	E.C.>80%	1.070				0+			
	232	94				α <20%	6.716	6.542(10)	38 +					
	232	94						6.600(10)	62					
$_{94}Pu^{233}$	233	94		233.042970	E.C.(99.9%)	2.020							0.1503 ± 0.0003	15 +
	233	94				α(0.1%)	6.416	6.300(20)	0.1%				0.1804 ± 0.0003	12
	233	94											0.1911 ± 0.0002	13
	233	94											0.2076 ± 0.0002	24
	233	94											0.2218 ± 0.0003	12
	233	94											0.2353 ± 0.0002	100
	233	94											0.4571 ± 0.0002	10
	233	94											0.4781 ± 0.0002	14
	233	94											0.5002 ± 0.0002	39
	233	94											0.5038 ± 0.0002	21
	233	94											0.5125 ± 0.0002	13
	233	94											0.5243 ± 0.0002	13
	233	94											0.5346 ± 0.0002	90
	233	94											0.5587 ± 0.0002	27
	233	94											0.6880 ± 0.0003	33
	233	94											0.8308 ± 0.0003	11
	233	94											0.9779 ± 0.0002	13
	233	94											0.9917 ± 0.0002	23
	233	94											1.0008 ± 0.0002	18
	233	94											1.0039 ± 0.0003	31
	233	94											1.0123 ± 0.0002	28
	233	94											1.0281 ± 0.0003	6.6
	233	94											1.0352 ± 0.0003	5.7
$_{94}Pu^{234}$	234	94		234.043299	8.8 h.	E.C.(94%)	0.383				0+			
	234	94				α(6%)	6.310	6.035(3)	0.024%					
	234	94						6.149(3)	1.9%					
	234	94						6.200(3)	4.0%					
$_{94}Pu^{235}$	235	94		235.045260	25.6 m.	E.C. (99+%)	1.13				(5/2+)			
	235	94				α(0.003%)	5.957	5.850(20)	0.003					
$_{94}Pu^{236}$	236	94		236.046032	2.85 ± 0.01 y.	α	5.867	5.4519(8)	0.002%	σf 160 b.	0+			
	236	94						5.6138(7)	0.18%					
	236	94						5.7210(7)	32%					
	236	94						5.7677(7)	68%					
$_{94}Pu^{237}$	237	94		237.048401	45.1 d.	E.C. (99.9%)	0.218				7/2-		Np L x-ray	21%

Isotope	A	Z	% Natural abundance	Atomic mass	Half-life	Decay mode	Decay energy (MeV)	Particle energy (MeV)	Particle intensity	Thermal neutron cross section	Spin (h/2π)	μ Nucl. mag. moment	Gamma-ray energy (MeV)	Gamma-ray intensity
	237	94				α(0.003%)	5.747	5.334(4)	0.0015				Np k x-ray	20%
	237	94						5.356(4)	0.0006				0.026344 ± 0.00001	0.23%
	237	94						5.650(4)	0.0007				0.03319 ± 0.00001	0.08%
	237	94											0.05954 ± 0.00001	3.3%
	237	94											(0.03 - 0.5)weak	
$_{94}Pu^{238}$	238	94		238.049554	87.74 y.	α	5.593	5.3583(1)	0.10%	540 b.	0+		U k x-ray	5.2%
	238	94						5.463(1)	28.3%	σf 18 b.			0.04347 ± 0.00001	0.04%
	238	94						5.4992(1)	71.6%				(0.04 - 1.1)weak	
$_{94}Pu^{239}$	239	94		239.052157	2.411 x 10^4y	α	5.244	5.0542(4)	0.023%	269 b.	$^1/_2$+	+0.203	U k x-ray	0.006%
	239	94						5.0752(4)	0.056%	σf 742 b.			0.05162 ± 0.00002	0.03%
	239	94						5.1047(4)	10.6%				0.05682 ± 0.00003	0.001%
	239	94						5.1428(4)	15.1%				0.12928 ± 0.00002	0.006%
	239	94						5.1555(4)	73.2%				0.37502 ± 0.00002	0.002%
	239	94											0.41369 ± 0.00002	0.0015%
$_{94}Pu^{240}$	240	94		240.053808	6537 ± 10 y.	α	5.255	5.0212(1)	0.07%	290 b.	0+		U L x-ray	5%
	240	39						5.1237(1)	26.4%				0.04524 ± 0.00001	0.045%
	240	94						5.1681(1)	73.5%				0.10423 ± 0.00001	0.007%
	240	94											(0.04 - 0.97)weak	
$_{94}Pu^{241}$	241	94		241.056845	14.4 ± 0.2 y.	β-(99+%)	0.021	4.8532(7)	3x10^{-4}%	360 b.	5/2+	-0.683	0.14854 ± 0.00001	0.00018%
	241	94				α(0.002%)	5.139	4.8966(7)	0.002%	σf 1010 b.				
$_{94}Pu^{242}$	242	94		242.058737	3.76 x 10^5y.	α	4.983	4.7546(7)	0.098%	19 b.	0+		U L x-ray	4.1%
	242	94						4.8564(7)	22.4%				0.04491 ± 0.00001	0.04%
	242	94						4.9006(7)	78%				0.10350 ± 0.00004	0.008%
	242	94											0.15880 ± 0.00008	0.0004%
$_{94}Pu^{243}$	243	94		243.061998	4.95 h.	β-	0.580	0.49	21%	90 b.	7/2+		Am L x-ray	6.2%
	243	94						0.58	60%	σf 200 b.			0.0417 ± 0.0002	0.8%
	243	94											0.06710 ± 0.0002	0.2%
	243	94											0.0839 ± 0.0002	23%
	243	94											0.10547 ± 0.00005	0.18%
	243	94											0.3564 ± 0.0002	0.13%
	243	94											0.3817 ± 0.0002	0.55%
	243	94											0.4232 ± 0.0002	0.01%
$_{94}Pu^{244}$	244	94		244.064199	8.2 x 10^7y.	α(99.9%)	4.665	4.546(1)	19.4%	1.7 b.	0+		U L x-ray	0.35%
	244	94				S.F.(0.1%)		4.589(1)	80.5%				0.0439 ± 0.0008	0.003%
$_{94}Pu^{245}$	245	94		245.067820	10.5 ± 0.1 h.	β-	1.28	0.93	57%		(9/2-)		Am L x-ray	6.6%
	245	94						1.21	11%				Am k x-ray	12%
	245	94											0.2804 ± 0.0001	1.3%
	245	94											0.30832 ± 0.00008	4.9%
	245	94											0.32752 ± 0.00001	25%
	245	94											0.37677 ± 0.00001	3.2%
	245	94											0.49169 ± 0.00001	2.7%
	245	94											0.56014 ± 0.00001	5.4%
	245	94											0.6302 ± 0.0001	2.7%
	245	94											0.8000 ± 0.0001	1.6%
	245	94											0.91063 ± 0.00002	1.4%
	245	94											0.93852 ± 0.00002	1.0%
	245	94											0.98770 ± 0.00004	1.3%
	245	94											1.0183 ± 0.0001	1.0%
	245	94											(0.03 - 1.2)weak	
$_{94}Pu^{246}$	246	94		246.070171	10.85 ± 0.02d.	β-	0.374	0.150	85%		0+		Am L x-ray	23%
	246	94						0.35	10%				Am k x-ray	22%
	246	94											0.02756 ± 0.00002	3.5%
	246	94											0.04379 ± 0.00002	25%
	246	94											0.17992 ± 0.00002	9.7%
	246	94											0.22371 ± 0.00002	23%

TABLE OF THE ISOTOPES (Continued)

Isotope	A	Z	% Natural abundance	Atomic mass	Half-life	Decay mode	Decay energy (MeV)	Particle energy (MeV)	Particle intensity	Thermal neutron cross section	Spin (h/2π)	μ Nucl. mag. moment	Gamma-ray energy (MeV)	Gamma-ray intensity
	246	94											0.25553 ± 0.00002	0.2%
Am		95												
$_{95}Am^{237}$	237	95		237.050050	1.22 h.	E.C.(99.98)	1.540				(5/2-)		Pu k x-ray	40%
	237	95				α(0.02%)	6.20	6.042(5)	0.02%				0.14559 ± 0.00001	0.5%
	237	95											0.28026 ± 0.00001	47%
	237	95											0.32101 ± 0.00002	1.4%
	237	95											0.42585 ± 0.00007	1.9%
	237	95											0.43845 ± 0.00007	8.3%
	237	95											0.47355 ± 0.00007	4.3%
	237	95											0.6553 ± 0.0002	1.3%
	237	95											0.9089 ± 0.0001	2.6%
$_{95}Am^{238}$	238	95		238.051980	1.63 h.	E.C.	2.26				1+		Pu L x-ray	27%
	238	95				α(0.0001%)	6.04	5.940	0.0001%				Pu k x-ray	36%
	238	95											0.35767 ± 0.00003	2.1%
	238	95											0.56103 ± 0.00003	10.9%
	238	95											0.60511 ± 0.00003	7.6%
	238	95											0.91870 ± 0.00002	23%
	238	95											0.94137 ± 0.00004	2.2%
	238	95											0.96278 ± 0.00002	28%
	238	95											1.2662 ± 0.0003	1.7%
	238	95											1.5772 ± 0.0001	2.9%
	238	95											1.6364 ± 0.0001	1.3%
$_{95}Am^{239}$	239	95		239.053016	11.901 h.	E.C.(99.99)	0.800				5/2-		Pu L x-ray	45%
	239	95				α(0.01%)	5.924	5.734(2)	0.001%				Pu k x-ray	61%
	239	95						5.776(2)	0.008%				0.18172 ± 0.00001	1.1%
	239	95											0.20975 ± 0.00001	3.5%
	239	95											0.22638 ± 0.00001	3.3%
	239	95											0.22818 ± 0.00001	11.4%
	239	95											0.27760 ± 0.00001	15%
$_{95}Am^{240}$	240	95		240.055278	50.9 h.	E.C.	1.369				(3-)		Pu L x-ray	38%
	240	95				α	5.592	5.378(1)	$16x10^{-4}$				Pu k x-ray	28%
	240	95											0.09558 ± 0.00001	1.5%
	240	95											0.88878 ± 0.00004	25%
	240	95											0.98764 ± 0.00004	73%
	240	95											(0.1 - 1.3)weak	
$_{95}Am^{241}$	241	95		241.056823	432.2 y.	α	5.637	5.2443(1)	0.002%	(50 + 550) b.	5/2-	+1.61	Np L x-ray	20%
	241	95						5.3221(1)	0.015%	σf 3.2 b.			0.02634 ± 0.00001	2.4%
	241	95						5.3884(1)	1.4%				0.033192 ± 0.00001	0.12%
	241	95						5.4431(1)	12.8%				0.059536 ± 0.00001	35.7%
	241	95						5.4857(1)	85.2%				(0.03 - 0.95)weak	
	241	95						5.5116(1)	0.20%					
	241	95						5.5442(1)	0.34%					
$_{95}Am^{242m}$	242	95			141 ± 2 y.	I.T.(99.5%)	0.048				5-		Am L x-ray	12%
	242	95				α(0.5%)	5.62	5.1413(4)	0.026%				0.04863 ± 0.00005	0.00013%
	242	95						5.2070(2)	0.4%				0.08648 ± 0.00003	0.04%
	242	95											0.10944 ± 0.00009	0.024%
	242	95											0.13497 ± 0.00006	0.01%
	242	95											0.16304 ± 0.00004	0.023%
$_{95}Am^{242}$	242	95		242.059541	16.01 ± 0.02h.	β-(83%)	0.661	0.63	46%		1-	+0.3878	Pu L x-ray	4.9%
	242	95				E.C.(17%)		0.67	37%				Cm L x-ray	8.6%
	242	95											Pu k x-ray	5.8%
	242	95											0.0422 ± 0.0001	0.04%
	242	95											0.04453 ± 0.00001	0.014%
$_{95}Am^{243}$	243	95		243.061375	$7.37 x 10^3$y.	α	5.438	5.1798(5)	1.1%	(74 + 4) b.	5/2-	+1.61	0.04354 ± 0.00001	5.1%

Isotope	A	Z	% Natural abundance	Atomic mass	Half-life	Decay mode	Decay energy (MeV)	Particle energy (MeV)	Particle intensity	Thermal neutron cross section	Spin (h/2π)	μ Nucl. mag. moment	Gamma-ray energy (MeV)	Gamma-ray intensity
	243	95						5.2343(5)	11%	σf 0.2 b.			0.07467 ± 0.00001	60%
	243	95						5.2766(5)	88%				0.08657 ± 0.00003	0.30%
	243	95						5.394(5)	0.12%				0.11770 ± 0.00003	0.6%
	243	95						5.3500(5)	0.16%				0.14197 ± 0.00004	0.11%
$_{95}Am^{244m}$	244	95			26 m.	β-	1.498			σf 1600 b.	(1-)		0.0429 ± 0.0001	0.02%
$_{95}Am^{244}$	244	95		244.064279	10.1 h.	β-	1.427			σf 2300 b.			Am L x-ray	50%
	244	95											Cm k x-ray	3.6%
	244	95											0.0994 ± 0.0001	5%
	244	95											0.1540 ± 0.0008	18%
	244	95											0.7460 ± 0.0008	67%
	244	95											0.9000 ± 0.0008	28%
$_{95}Am^{245}$	245	95		245.066444	2.05 ± 0.01 h.	β-	0.894	0.65	19%		(5/2+)		Cm L x-ray	3.4%
	245	95						0.90	77%				Cm k x-ray	5.7%
	245	95											0.04287 ± 0.00002	0.06%
	245	95											0.24106 ± 0.00003	0.33%
	245	95											0.25299 ± 0.00002	6.1%
	245	95											0.29587 ± 0.00002	0.23%
$_{95}Am^{246m}$	246	95			25.0 m.	β-	1.31	79%			2-		Cm L x-ray	13%
	246	95						1.60	14%				Cm k x-ray	1.9%
	246	95						2.1	7%				0.27002 ± 0.00003	1.0%
	246	95											0.73442 ± 0.00002	1.2%
	246	95											0.79881 ± 0.00002	25%
	246	95											0.83358 ± 0.00002	1.8%
	246	95											1.03600 ± 0.00002	13%
	246	95											1.06201 ± 0.00002	17%
	246	95											1.07885 ± 0.00002	28%
	246	95											1.08517 ± 0.00002	1.5%
	246	95											1.27472 ± 0.00004	0.27%
	246	95											1.59068 ± 0.00003	0.5%
	246	95											1.66165 ± 0.00003	0.22%
	246	95											(0.04 - 2.29)weak	
$_{95}Am^{246}$	246	95		246.069770	39 m.	β-	2.38	1.2			(7-)		Cm L x-ray	63.%
	246	95											Cm k x-ray	4.6%
	246	95											0.1529 ± 0.0001	25%
	246	95											0.2046 ± 0.0008	36%
	246	95											0.6289 ± 0.001	2.7%
	246	95											0.6786 ± 0.0008	53%
	246	95											0.7558 ± 0.0004	13%
	246	95											0.78131 ± 0.00003	4.0%
	246	95											0.8389 ± 0.0014	2%
$_{95}Am^{247}$	247	95		247.072170	22 ± 3 m.	β-	1.700						Cm L x-ray	13%
	247	95											Cm k x-ray	21%
	247	95											0.2267 ± 0.0007	5.8%
	247	95											0.2853 ± 0.0002	23%
Cm		96												
$_{96}Cm^{238}$	238	96		238.053020	2.4 ± 0.1 h.	E.C. (>90%)	0.970				0+			
	238	96				α(<10%)	6.632	6.520(50)	<10%					
$_{96}Cm^{239}$	239	96		239.054840	≈ 3 h.	E.C.	1.700						0.0407 ± 0.0005	
	239	96											0.1466 ± 0.0005	
	239	96											0.1874 ± 0.0004	
$_{96}Cm^{240}$	240	96		240.055503	27 ± 1 d.	α	6.397	5.989	0.014%		0+			
	240	96						6.147	0.05%					
	240	96						6.2478(6)	28.8%					
	240	96						6.2906(6)	70.6%					
$_{96}Cm^{241}$	241	96		241.057645	32.8 ± 0.2 d.	E.C.(99%)	0.765				1/2+		Am k x-ray	35%
	241	96				α(1%)	6.184	5.8842(4)	0.12%				0.13241 ± 0.00001	3.9%
	241	96						5.9291(4)	0.18%				0.16505 ± 0.00001	3.0%
	241	96						5.9389(4)	0.69%				0.18028 ± 0.00001	0.5%

Isotope	A	Z	% Natural abundance	Atomic mass	Half-life	Decay mode	Decay energy (MeV)	Particle energy (MeV)	Particle intensity	Thermal neutron cross section	Spin (h/2π)	μ Nucl. mag. moment	Gamma-ray energy (MeV)	Gamma-ray intensity
	241	96											0.20588 ± 0.00001	2.8%
	241	96											0.43063 ± 0.00001	4.1%
	241	96											0.46327 ± 0.00001	1.2%
	241	96											0.47181 ± 0.00001	71%
	241	96											0.63686 ± 0.00001	1.5%
$_{96}Cm^{242}$	242	96		242.058830	462.8 d.	α.	6.216	5.9694(1)	0.035%	20 b.	0+		Pu L x-ray	4.8%
	242	96						6.069(1)	25%				0.04408 ± 0.00002	0.03%
	242	96						6.1129(1)	74%				0.10189 ± 0.00002	0.002%
	242	96											0.15742 ± 0.00006	0.001%
	242	96											(0.04 - 1.2)weak	
$_{96}Cm^{243}$	243	96		243.061381	28.5 ± 0.2 y.	α	6.167	5.6815(5)	0.2%		5/2+	0.41	Pu L x-ray	19%
	243	96						5.6856(5)	1.6%				Pu k x-ray	23%
	243	96						5.7420(5)	10.6%				0.10612 ± 0.00001	0.26%
	243	96						5.7859(5)	73.3%				0.20975 ± 0.00001	3.3%
	243	96						5.9922(5)	6.5%				0.22819 ± 0.00001	10.6%
	243	96						6.0103(5)	1.0%				0.27760 ± 0.00001	14%
	243	96						6.0589(5)	5%				0.28546 ± 0.00001	0.73%
	243	96						6.0666(5)	1.5%				0.33431 ± 0.00001	0.023%
	243	96											(0.04 - 0.7)weak	
$_{96}Cm^{244}$	244	96		244.062747	18.11 y.	α	5.902	5.6656(1)	0.02%	15 b.	0+		Pu L x-ray	4%
	244	96						5.7528(1)	24%	σf 1.0 b.			0.04282 ± 0.00001	0.02%
	244	96						5.8050(1)	76%				0.09885 ± 0.00001	0.001%
	244	96											0.15262 ± 0.00002	0.001%
$_{96}Cm^{245}$	245	96		245.065483	8.5 x 10^3 y.	α	5.623	5.235(10)	0.3%	360 b.	7/2+	0.5	Pu L x-ray	55%
	245	96						5.3038(10)	5.0%	σf 2100 b.			Pu k x-ray	34%
	245	96						5.3620(7)	93%				0.04195 ± 0.00003	0.35%
	245	96						5.4927(11)	0.8%				0.13299 ± 0.00003	2.8%
	245	96						5.5331(11)	0.6%				0.13606 ± 0.00006	0.11%
	245	96											0.17494 ± 0.00004	9.5%
	245	96											0.18982 ± 0.00006	0.2%
$_{96}Cm^{246}$	246	96		246.067218	4.78 x 10^3y.	α	5.476	5.343(3)	21%	1.2 b.	0+		Pu L x-ray	3.9%
	246	96						5.386(3)	79%	σf 0.2 b.			0.04453 ± 0.00001	0.027%
$_{96}Cm^{247}$	247	96		247.070347	1.56 x 10^7y.	α	5.352	4.818(4)	4.7%	60 b.	9/2-	0.37	Pu k x-ray	2.1%
	247	96						4.8690(20)	71%	σf 80 b.	9/2-		0.2792 ± 0.0008	3.4%
	247	96						4.941(4)	1.6%				0.2886 ± 0.0007	2.0%
	247	96						4.9820(20)	2.0%				0.3471 ± 0.0008	1%
	247	96						5.1436(20)	1.2%				0.4035 ± 0.0005	72%
	247	96						5.2104(20)	5.7%					
	247	96						5.2659(20)	13.8%					
$_{96}Cm^{248}$	248	96		248.072343	3.4 x 10^5 y.	α(92%)	5.162	4.931(5)	0.07%	2.6 b.	0+			
	248	96				S.F.(8%)		5.0349(2)	16.5%	σf 0.4 b.				
	248	96						5.0784(2)	75	1%				
$_{96}Cm^{249}$	249	96		249.075948	64.15 m.	β-	0.902	0.9		2 b.	1/2+		Bk k x-ray	0.2%
	249	96											0.36897 ± 0.00004	0.35%
	249	96											0.51846 ± 0.00004	0.09%
	249	96											0.56039 ± 0.00005	0.84%
	249	96											0.62191 ± 0.00005	0.18%
	249	96											0.63431 ± 0.00005	1.5%
	249	96											0.65277 ± 0.00005	0.14%
$_{96}Cm^{250}$	250	96		250.078352	≈ 7.4x10^3 y.	S.F.				80 b.	0+			
	250	96				α	5.27							
$_{96}Cm^{251}$	251	96		251.082290	16.8 m.	β-	1.420	0.90	16%		(1/2+)		0.3896 ± 0.0002	1.3%
	251	96											0.4381 ± 0.0003	1.2%
	251	96											0.5299 ± 0.0002	1.6%
	251	96											0.5425 ± 0.0002	10.9%

Isotope	A	Z	% Natural abundance	Atomic mass	Half-life	Decay mode	Decay energy (MeV)	Particle energy (MeV)	Particle intensity	Thermal neutron cross section	Spin (h/2π)	μ Nucl. mag. moment	Gamma-ray energy (MeV)	Gamma-ray intensity
	251	96											0.5624 ± 0.0002	1.0%
	251	96											0.9782 ± 0.0002	1.0%
Bk		97												
$_{97}Bk^{242}$	242	97		242.061940	7.0 ± 0.1 m.	E.C.	2.900							
$_{97}Bk^{243}$	243	97		243.062997	4.5 h.	E.C. (99.8%)	1.505	6.542(4)	0.03%		(3/2-)		0.1466 ± 0.0005	0.01%
	243	97				α(0.15%)	6.871	6.5738(2)	0.04%				0.1874 ± 0.0004	0.06%
	243	97						6.7180(22)	0.02%				0.755 ± 0.002	10%
	243	97						6.7581(20)	0.02%				0.840 ± 0.004	3.0%
	243	97											0.946 ± 0.002	8%
$_{97}Bk^{244}$	244	97		244.065160	4.4 h.	E.C. (99.99%)	2.250				(4-)		0.1445 ± 0.001	7 +
	244	97				α(0.01%)	6.778	6.625(4)	0.003%				0.1876 ± 0.0003	16
	244	97						6.667(4)	0.003%				0.2176 ± 0.0003	100
	244	97											0.3335 ± 0.0005	10
	244	97											0.4905 ± 0.0005	18
	244	97											0.7461 ± 0.0008	8
	244	97											0.9815 ± 0.001	114
	244	97											0.9215 ± 0.001	22
	244	97											0.988 ± 0.001	5
	244	97											1.178 ± 0.001	5
	244	97											1.233 ± 0.001	4
	244	97											1.252 ± 0.001	3
	244	97											1.505 ± 0.005	3
$_{97}Bk^{245}$	245	97		245.066357	4.94 d.	E.C. (99.9%)	0.814				3/2-		Cm L x-ray	32%
	245	97				α(0.1%)	6.453	5.8851(5)	0.03%				Cm k x-ray	56%
	245	97						6.1176(9)	0.01%				0.25299 ± 0.00002	29%
	245	97						6.1467(5)	0.02%				0.3809 ± 0.0001	2.4%
	245	97						6.3087(5)	0.014%				0.3851 ± 0.0001	0.6%
	245	97						6.3492(5)	0.018%					
$_{97}Bk^{246}$	246	97		246.068720	1.80 d.	E.C>	1.400				(2-)		Cm L x-ray	38%
	246	97											Cm k x-ray	30%
	246	97											0.73442 ± 0.00002	3.2%
	246	97											0.79881 ± 0.00002	61%
	246	97											0.83358 ± 0.00002	4.9%
	246	97											1.06201 ± 0.00002	2.9%
	246	97											1.07885 ± 0.00002	3.8%
	246	97											1.08142 ± 0.00002	5.8%
	246	97											1.12427 ± 0.00002	4.3%
$_{97}Bk^{247}$	247	97		247.070300	1.4 x 10^3 y.	α	5.889	5.465(5)	1.5%		(3/2-)		0.04175 ± 0.0002	1%
	247	97						5.501(5)	7%				0.0839 ± 0.0002	40%
	247	97						5.532(5)	45%				0.268 ± 0.005	30%
	247	97						5.6535(20)	5.5%					
	247	97						5.678(2)	13%					
	247	97						5.712(2)	17%					
	247	97						5.753(2)	4.3%					
	247	97						5.794(2)	5.5%					
$_{97}Bk^{248}$	248	97		248.073106	23.7 ± 0.2 h.	β-(70%)	0.710	0.86			(1-)		Cm L x-ray	5.7%
	248	97				E.C.(30%)	0.860						Cf L x-ray	5.0%
	248	97											Cm k x-ray	9.8%
	248	97											Cf k x-ray	0.02%
	248	97											0.5507 ± 0.0001	5.0%
$_{97}Bk^{249}$	249	97		249.074980	320 ± 6 d.	β-	0.125	100%		710 b.	7/2+	2.0		
	249	97				α(0.001%)	5.525	5.3899(6)	0.0002%					
	249	97						5.4174(6)	0.001%					
$_{97}Bk^{250}$	250	97		250.078312	3.22 h.	β-	1.781	0.74		σf 1000 b.	2-		Cf L x-ray	10%
	250	97											Cf k x-ray	0.5%
	250	97											0.88996 ± 0.00001	1.5%
	250	97											0.92947 ± 0.00002	1.2%
	250	97											0.98912 ± 0.0001	45%
	250	97											1.02863 ± 0.00002	4.9%
	250	97											1.03184 ± 0.00001	36%
	250	97											(0.04 - 1.6)weak	
$_{97}Bk^{251}$	251	97		251.080760	57 m.	β-	1.100				(3/2-)		0.02481 ± 0.00001	+
	251	97											0.1528 ± 0.0001	39
	251	97											0.1776 ± 0.0001	100
Cf		98												

Isotope	A	Z	% Natural abundance	Atomic mass	Half-life	Decay mode	Decay energy (MeV)	Particle energy (MeV)	Particle intensity	Thermal neutron cross section	Spin (h/2π)	μ Nucl. mag. moment	Gamma-ray energy (MeV)	Gamma-ray intensity
$_{98}Cf^{240}$	240	98		240.062280	1.06 ± 0.1 m.	α	7.719	7.590(10)			0+			
$_{98}Cf^{241}$	241	98		241.063520	≈ 3.8 ± 0.7 m.	E.C.	3.080							
	241	98				α	7.60	7.335(5)						
$_{98}Cf^{242}$	242	98		242.063690	3.5 ± 0.2 m.	α	7.509	7.351(6)	20%		0+			
	242	98						7.385(4)	80%					
$_{98}Cf^{243}$	243	98		243.065390	10.7 ± 0.5 m.	E.C.(86%)	2.230	7.060(6)	20%		0+			
	243	97				α(14%)	7.40	7.170	4%					
$_{98}Cf^{244}$	244	98		244.065979	19.4 ± 0.6 m.	α	7.328	7.168(5)	25%		0+			
	244	98						7.210(5)	75%					
$_{98}Cf^{245}$	245	98		245.068037	43.6 ± 0.8 m.	α(30%)	7.255	6.886						
	245	98				E.C.(70%)	1.565	6.983						
	245	98						7.036						
	245	98						7.084						
	245	98						7.137(2)						
$_{98}Cf^{246}$	246	98		246.068800	36 h.	α	6.869	6.6156(10)	0.18%		0+		Cm L x-ray	4.2%
	246	98						6.7086(7)	21.8%				0.04221 ± 0.00001	0.02%
	246	98						6.7501(7)	78.0%				0.0945 ± 0.001	0.008%
	246	98											0.147 ± 0.004	0.004%
$_{98}Cf^{247}$	247	98		247.071020	3.11 ± 0.03 h.	E.C. (99.96%)	0.670				7/2+		Bk k x-ray	32%
	247	98				α(0.04%)	6.55	6.301(5)					0.2941 ± 0.0001	1.0%
	247	98											0.4070 ± 0.0001	0.2%
	247	98											0.4179 ± 0.0001	0.34%
	247	98											0.4778 ± 0.0001	0.55%
$_{98}Cf^{248}$	248	98		248.072183	334 ± 3 d.	α	6.369	6.220(5)	17%		0+			
	248	98						6.262(5)	83%					
$_{98}Cf^{249}$	249	98		249.074844	351 y.	α	6.295	5.7582(2)	3.7%	500 b.	9/2-		Cm L x-ray	10%
	249	98						5.8119(2)	84%	σf 1600 b.			Cm k x-ray	3.3%
	249	98						5.8488(2)	1.0%				0.25299 ± 0.00002	2.5%
	249	98						5.9029(2)	2.8%				0.33351 ± 0.00003	14.4%
	249	98						5.9451(2)	4.0%				0.38832 ± 0.00003	66%
	249	98						6.1401(2)	1.1%					
	249	98						6.1940(2)	2.2%					
$_{98}Cf^{250}$	250	98		250.076400	13.1 ± 0.1 y.	α	6.129	5.8913(4)	0.3%	2 x 10^3 b.	0+		Cm L x-ray	2.9%
	250	98						5.9889(4)	15%				0.04285 ± 0.00001	0.014%
	250	98						6.0310(4)	84.5%					
$_{98}Cf^{251}$	251	98		251.079580	8.9 x 10^2 y.	α	6.172	5.56448(7)	1.5%	2.9 x 10^3 b.	1/2+			
	251	98						5.632(1)	4.5%	σf 4.8 x 10^3 b.				
	251	98						5.648(1)	3.5%					
	251	98						5.6773(6)	35%					
	251	98						5.762(3)	3.8%					
	251	98						5.7937(7)	2.0%					
	251	98						5.8124(8)	4.2%					
	251	98						5.8514(6)	27%					
	251	98						6.0140(7)	11.6%					
	251	98						6.0744(7)	2.7%					
$_{98}Cf^{252}$	252	98		252.081621	2.64 y.	α(96.9%)	6.217	5.7977(1)	0.23%	20 b.	0+		Cm L x-ray	2.9%
	251	98				S.F.(3.1%)		6.0756(4)	15.2%	σf 32 b.			0.04339 ± 0.00002	0.015%
	252	98						6.1184(4)	81.6%				0.1002 ± 0.0001	0.01%
$_{98}Cf^{253}$	253	98		253.085127	17.8 d.	β-(99.7%)	0.29	0.27	100%	18 b.	(7/2+)			
	253	98				α(0.3%)	6.126	5.921(5)	0.02%	f 1300 b.				
	253	98						5.979(5)	0.29%					
$_{98}Cf^{254}$	254	98		254.087318	60.5 d.	S.F. (99.7%)				5 b.	0+			
	254	98				α(0.3%)	5.930	5.792(5)	0.05%					
	254	98						5.834(5)	0.26%					
$_{98}Cf^{255}$	255	98			1.4 h.	β-								
Es		99												
$_{99}Es^{243}$	243	99		243.069470	21 s.	α(>30%)	8.10	7.890(20)	>30%					
	243	99				E.C. (<70%)	3.810							
$_{99}Es^{244}$	244	99		244.070810	37 ± 4 s.	E.C.(76%)	4.500							
	244	99				α(4%)	7.84	7.570(20)	4%					
$_{99}Es^{245}$	245	99		245.071260	1.33 ± 0.1 m.	α(40%)	7.858	7.730(20)						
	245	99				E.C.(60%)	3.000							
$_{99}Es^{246}$	246	99		246.072920	≈ 7.7 ± 0.5 m.	E.C.(90%)	3.840							
	246	99				α(10%)	7.70	7.350(20)						
$_{99}Es^{247}$	247	99		247.073590	4.7 ± 0.3 m.	E.C.(93%)	2.400							
	247	99				α(7%)	7.441	7.320(20)						

TABLE OF THE ISOTOPES (Continued)

Isotope	A	Z	% Natural abundance	Atomic mass	Half-life	Decay mode	Decay energy (MeV)	Particle energy (MeV)	Particle intensity	Thermal neutron cross section	Spin (h/2π)	μ Nucl. mag. moment	Gamma-ray energy (MeV)	Gamma-ray intensity
$_{99}Es^{248}$	248	99		248.075440	27 ± 3 m.	E.C. (99.7%)	3.030							
	248	99				α(0.3%)	7.15	6.870(10)						
$_{99}Es^{249}$	249	99		249.076340	1.70 ± 0.1 h.	E.C. (99.4%)	1.395				(7/2+)			
	249	99				α(0.6%)	6.881	6.770(5)						
$_{99}Es^{250m}$	250	99			2.2 h.	E.C.							Cf L x-ray	26%
	250	99											Cf k x-ray	34%
	250	99											0.30395 ± 0.00003	0.1%
	250	99											0.62614 ± 0.00002	1.0%
	250	99											0.82883 ± 0.00002	5.5%
	250	99											0.98912 ± 0.00001	13.5%
	250	99											1.03184 ± 0.00001	10.6%
	250	99											1.16725 ± 0.00003	3.0%
	250	99											1.17550 ± 0.00002	1.5%
	250	99											1.20177 ± 0.00003	1.2%
	250	99											1.61525 ± 0.00002	1.8%
	250	99											1.65797 ± 0.00002	1.1%
$_{99}Es^{250}$	250	99		250.078660	8.6 ± 0.1 h.	E.C.	2.100						Cf L x-ray	120%
	250	99											Cf k x-ray	71%
	250	99											0.082283 ± 0.00006	2.6%
	250	99											0.08509 ± 0.00006	1.1%
	250	99											0.14069 ± 0.00006	4.7%
	250	99											0.22297 ± 0.00001	1.8%
	250	99											0.24686 ± 0.00005	3.8%
	250	99											0.30339 ± 0.00005	22.3%
	250	99											0.34948 ± 0.00005	20.4%
	250	99											0.38381 ± 0.00005	14%
	250	99											0.71228 ± 0.00005	1.3%
	250	99											0.76399 ± 0.00002	4.0%
	250	99											0.81009 ± 0.00002	9.1%
	250	99											0.82883 ± 0.00002	74%
	250	99											0.86315 ± 0.00002	5.1%
	250	99											0.88662 ± 0.00002	1.3%
$_{99}Es^{251}$	251	99		251.079986	1.38 d.	E.C. (99.5%)	0.379				(3/2-)			
	251	99				α(0.5%)	6.597	6.4625(13)	0.05%					
	251	99						6.492(1)	0.4%					
$_{99}Es^{252}$	252	99		252.082944	1.29 y.	α(76%)	6.739	6.051(1)	0.8%		(5-)			
	252	99				E.C.(24%)	1.12	6.238(1)	0.43%					
	252	99						6.266(3)	0.57%					
	252	99						6.4224(12)	0.34%					
	252	99						6.4827(12)	1.66%					
	252	99						6.5621(12)	10.3%					
	252	99						6.6316(12)	61.0%					
$_{99}Es^{253}$	253	99		253.084818	20.47 d.	α	6.739	6.2497(1)	0.04%	160 b.	7/2+	4.10	0.04180 ± 0.00004	0.05%
	253	99						6.4314(1)	0.06%				0.3871 ± 0.0001	0.018%
	253	99						6.4793(1)	0.08%				0.3892 ± 0.0001	0.026%
	253	99						6.4972(1)	0.26%				(0.03 - 1.1)weak	
	253	99						6.5405(1)	0.85%					
	253	99						6.5514(1)	0.7%					
	253	99						6.5916(1)	6.6%					
	253	99						6.6241(1)	0.8%					
	253	99						6.6327(5)	89.8%					
$_{99}Es^{254m}$	254	99			1.64 d.	β-(99.6%)	1.16	1.127		1 b.	2+		Fm L x-ray	13%
	254	99				α(0.3%)	6.67	6.3821(9)	0.25%	σf 1800 b.	2+		Fm k x-ray	0.8%
	254	99						6.5558(9)	0.02%				0.58432 ± 0.00002	2.9%
	254	99						6.5907(9)	0.01%				0.64879 ± 0.00002	28.9%

TABLE OF THE ISOTOPES (Continued)

Isotope	A	Z	% Natural abundance	Atomic mass	Half-life	Decay mode	Decay energy (MeV)	Particle energy (MeV)	Particle intensity	Thermal neutron cross section	Spin (h/2π)	μ Nucl. mag. moment	Gamma-ray energy (MeV)	Gamma-ray intensity
	254	99											0.68867 ± 0.00002	12.4%
	254	99											0.69377 ± 0.00002	24.7%
$_{99}Es^{254}$	254	99		254.088019	275 d.	α	6.617	6.3475(10)	0.75%	< 40 b.	(7+)		0.0426 ± 0.0001	0.15%
	254	99						6.3573(10)	2.6%	σf 2900 b.			0.0650 ± 0.0012	2.0%
	254	99						6.4161(10)	1.8%				0.3167 ± 0.0013	0.15%
	254	99						6.4266(12)	93.1%					
	254	99						6.4801(13)	0.23%					
$_{99}Es^{255}$	255	99		255.090270	39.8 ± 0.1 d.	β-(92%)	0.300			60 b.	(7/2+)			
	255	99				α(8%)	6.436	6.213(10)	0.20%					
	255	99				S.F. (0.004%)		6.260(10)	0.8%					
	255	99						6.2995(15)	7.0%					
$_{99}Es^{256m}$	256	99			7.6 h.	β-					(8-)		0.1114	5.7%
	256	99											0.1726	30%
	256	99											(0.05 - 1.1)weak	
$_{99}Es^{256}$	256	99		256.093560	25 m.	β-	1.676				(1+)			
Fm		100												
$_{100}Fm^{243}$	243	100		243.074460	0.18 s.	α		8.546(25)						
$_{100}Fm^{244}$	244	100		244.074120	3.7 ms.	S.F.					0+			
$_{100}Fm^{245}$	245	100		245.076250	4 s.	α	8.40	8.150(20)						
$_{100}Fm^{246}$	246	100		246.075290	1.1 ± 0.2 s.	α(92%)	8.373	8.240(20)			0+			
	246	100				S.F.(8%)								
$_{100}Fm^{247m}$	247	100			9.2 ± 1.0 s.	α		8.180(30)						
$_{100}Fm^{247}$	247	100		247.076800	35 ± 4 s.	α	8.20	7.870(50)	70 +					
	247	100						7.930(50)	30					
$_{100}Fm^{248}$	248	100		248.077171	36 ± 3 s.	α(99.9%)	8.001	7.830(20)	20%		0+			
	248	100				S.F.(0.1%		7.870(20)	80%					
$_{100}Fm^{249}$	249	100		249.078910	≈ 2.6 m.	E.C.					(7/2+)			
	249	100				α	7.70	7.530(20)						
$_{100}Fm^{250m}$	250	100			1.8 ± 0.1 s.	I.T.								
$_{100}Fm^{250}$	250	100		250.079509	30 m.	α	7.548	7.430(30)			0+			
$_{100}Fm^{251}$	251	100		251.081590	5.3 h.	E.C.(98%)	1.49				(9/2-)			
	251	100				α(2%)	7.424	6.7817(8)	0.09%					
	251	100						6.8324(8)	1.57%					
	251	100						7.3051(8)	0.02%					
$_{100}Fm^{252}$	252	100		252.082466	25.4 h.	α	7.154	6.999(8)	15%		0+			
	252	100						7.040(9)	85%					
$_{100}Fm^{253}$	253	100		253.085173	E.C.(88%)	0.334					$^1/_2$+		Es k x-ray	20%
	253	100				α(12%)	7.200	6.544(2)	0.18%				0.14497 ± 0.00005	0.2%
	253	100						6.633(4)	0.31%				0.2719 ± 0.0004	2.6%
	253	100						6.653(4)	0.29%					
	253	100						6.6757(14)	2.78%					
	253	100						6.8467(13)	1.0%					
	253	100						6.9010(13)	1.18%					
	253	100						6.9433(13)	5.12%					
	253	100						7.0245(13)	0.8%					
$_{100}Fm^{254}$	254	100		254.086846	3.24 h.	α	7.303	7.050(4)	0.9%	80 b.	0+			
	254	100				S.F. (0.06%)		7.147(4)	14%					
	254	100						7.189(4)	84.9%					
$_{100}Fm^{255}$	255	100		255.089948	20.1 h.	α	7.240	6.8916(5)	0.62%	26 b.	7/2+			
	255	100						6.9635(5)	5.0%	σf 3400 b.				
	255	100						7.0225(5)	93.4%					
	255	100						7.0800(5)	0.40%					
$_{100}Fm^{256}$	256	100		256.091767	2.63 h.	S.F.(92%)				50 b.	0+			
	256	100				α(18%)	7.025	6.92						
$_{100}Fm^{257}$	257	100		257.075099	100.5 d.	α(99.8%)				5800 b.	(9/2+)		0.0616 ± 0.0001	1.4%
	257	100				S.F.(0.2%)				σf 3000 b.			0.1794 ± 0.0001	8.7%
	257	100											0.2410 ± 0.0001	11%
$_{100}Fm^{258}$	258	100			0.38 ms.	S.F.								
Md		101												
$_{101}Md^{248}$	248	101		248.082750	7 ± 3 s.	E.C.(80%)	5.210							
	248	101				α(20%)	8.60	8.320(20)	15%					
	248	101						8.360(30)	5%					
$_{101}Md^{249}$	249	101		249.082950	24 ± 4 s.	E.C > (<80%)	3.760							
	249	101				α(>20%)	8.46	8.030(20)						
$_{101}Md^{250}$	250	101		250.084380	≈ 52 ± 6 s.	E.C.(94%)	4.54							
	250	101				α(6%)	8.25	7.750(20)	4 +					
	250	101						7.820(30)	2					
$_{101}Md^{251}$	251	101		251.084830	4.0 m.	E.C. (>94%)	3.020							
	251	101				α(<6%)	8.05	7.550(20)						
$_{101}Md^{252}$	252	101		252.086470	≈ 2.3 m.	E.C. (>50%)	3.73							
	252	101				α(<50%)	7.85	7.73						
$_{101}Md^{254m}$	254	101			≈ 28 m.	E.C.								
$_{101}Md^{254}$	254	101		254.089630	10 ± 3 m.	E.C.	2.600							
$_{101}Md^{255}$	255	101		255.091081	27 ± 2 m.	E.C.(92%)	1.055				(7/2-)			
	255	101				α(8%)	7.911	7.326(5)						

Isotope	A	Z	% Natural abundance	Atomic mass	Half-life	Decay mode	Decay energy (MeV)	Particle energy (MeV)	Particle intensity	Thermal neutron cross section	Spin (h/2π)	μ Nucl. mag. moment	Gamma-ray energy (MeV)	Gamma-ray intensity
$_{101}Md^{256}$	256	101		256.093960	76 m.	E.C.(90%)	2.041							
	256	101				α(10%)	7.483	7.14	16 +					
	256	101						7.22	63					
$_{101}Md^{257}$	257	101		257.095580	≈ 5.2 h.	E.C.(90%)	0.450				(7/2-)			
	257	101				α(10%)	7.60	7.068(5)						
$_{101}Md^{258m}$	258	101			43 m.	E.C.					(1-)			
$_{101}Md^{258}$	258	101		258.098570	56 d.	α	7.40	6.716(5)	72%		(8-)			
	258	101						6.79(1)	28%					
$_{101}Md^{259}$	259	101			1.6 h.	S.F.								
No		102												
$_{102}No^{250}$	250	102			250 μs.	S.F.					0+			
$_{102}No^{251}$	251	102		251.088870	0.8 s.	α		8.600(20)	80%					
	251	102						8.680(20)	20%					
$_{102}No^{252}$	252	102		252.088949	2.3 ± 0.2 s.	α(73%)	8.551	8.372(8)	18%		0+			
	252	102				S.F.(27%)		8.415(6%)	55%					
$_{102}No^{253}$	253	102		253.090530	1.7 ± 0.3 m.	α	8	40	8.010(20)		(9/2-)			
$_{102}No^{254m}$	254	102			0.28 s.	I.T.								
$_{102}No^{254}$	254	102		254.090953	≈ 55 s.	α	8.235	8.100(20)			0+			
$_{102}No^{255}$	255	102		255.093260	3.1 m.	α(62%)	8.445	7.620(10)	1.7%		½+			
	255	102				E.C.(38%)		7.717(11)	1.5%					
	255	102						7.771(7)	5.5%					
	255	102						7.879(11)	2.6%					
	255	102						7.927(7)	7.3%					
	255	102							8.007(11)	3.9%				
	255	102						8.077(9)	7.3%					
	255	102						8.121(6)	27.9%					
	255	102						8.266(8)	3.1%					
	255	102						8.312(9)	1.2%					
$_{102}No^{256}$	256	102		256.094252	3.2 s.	α	8.554	8.430			0+			
$_{102}No^{257}$	257	102		257.096850	25 s.	α	8.452	8322(2)	55%		(7/2+)			
	257	102						8.27(2)	26%					
	257	102						8.32(2)	19%					
$_{102}No^{258}$	258	102		258.098150	≈ 1.2 ms.	S.F.					0+			
$_{102}No^{259}$	259	102		259.100931	≈ 58 m.	α(78%)	7.794	7.443(10)	13 +		(9/2+)			
	259	102				E.C.(22%)		7.488(10)	39					
	259	102						7.521(10)	23					
	259	102						7.593(10)	14					
	259	102						7.673(10)	11					
Lr		103												
$_{103}Lr^{253}$	253	103		253.095150	≈ 1.4 s.	α		8.721(20)						
	253	103						8.805(20)						
$_{103}Lr^{254}$	254	103		254.096320	≈ 20 ± 10 s.	α		8.455(20)						
$_{103}Lr^{255}$	255	103		255.096670	22 ± 4 s.	α	8.80	8.370(13)	60%					
	255	103						8.429(20)	40%					
$_{103}Lr^{256}$	256	103		256.098490	28 s.	α(99.7%)	8.554	8.43						
	256	103				S.F.(0.3%)								
$_{103}Lr^{257}$	257	103		257.099480	0.65 s.	α	9.30	8.796(13)	15%		7/2+			
	257	103						8.861(12)	85%					
$_{103}Lr^{258}$	258	103		258.101710	4.3 s.	α	9.00	8.54(2)	10%					
	258	103						8.589(10)	45%					
	258	103						8.614(10)	35%					
	258	103						8.648(10)	10%					
$_{103}Lr^{259}$	259	103		259.102900	≈ 5.4 s.	α	8.70	8.46(2)	100%					
$_{103}Lr^{260}$	260	103		260.105320	3 m.	α	8.30	8.04(2)	100%					
Rf		104												
$_{104}Rf^{257}$	257	104		257.102950	4.8 s.	α	9.20	8.663						
	257	104						8.720						
	257	104						8.778						
	257	104						8.824						
	257	104						8.870						
	257	104	8.951											
	257	104						9.016						
$_{104}Rf^{258}$	258	104		258.103430	11 ms.	S.F.								
$_{104}Rf^{259}$	259	104		259.105530	≈ 3.1 s.	α	9.20	8.77(2)						
	259	104						8.86						
$_{104}Rf^{260}$	260	104		260.106300	≈ 20 ms.	S.F.								
$_{104}Rf^{261}$	261	104		261.108690	≈ 65 s.	α	8.60	8.29						
$_{104}Rf^{262}$	262	104			≈ 63 ms.									
Ha		105												
$_{105}Ha^{257}$	257	105		257.107770	≈ 1 s.	α		8.96						
	257	105						9.08						
	257	105						9.16						
$_{105}Ha^{258}$	258	105		258.109020	4 s.	α		9.02						
	258	105						9.09						
	258	105						9.18						
$_{105}Ha^{259}$	259	105		259.109580	≈ 1.2 s.	S.F.								
$_{105}Ha^{260}$	260	105		260.111040	1.5 s.	α		9.05						
	260	105				S.F.		9.08						
	260	105						9.13						
$_{105}Ha^{261}$	261	105		261.111820	≈ 1.8 s.	α		8.93						

TABLE OF THE ISOTOPES (Continued)

Isotope	A	Z	% Natural abundance	Atomic mass	Half-life	Decay mode	Decay energy (MeV)	Particle energy (MeV)	Particle intensity	Thermal neutron cross section	Spin (h/2π)	μ Nucl. mag. moment	Gamma-ray energy (MeV)	Gamma-ray intensity
	261	105				S.F.								
$_{105}Ha^{262}$	262	105		262.113760	34 s.	S.F.								
	262	105				α		8.45						
	262	105						8.53						
	262	105						8.67						
106		106												
	259	106			≈ 7 ms.	S.F.								
	263	106			0.8 s.	S.F.								
	261	106			≈ 1 ms.	α								
	262	106			≈ 115 ms.	α		9.70						
109		109												
	266	109			5 ms.			11.10						

PERMISSIBLE QUARTERLY INTAKES OF RADIONUCLIDES

Karl Z. Morgan

The following table lists the permissible quarterly intakes (oral and inhalation) or radionuclides and critical body organs recommended by NCPR for occupational exposure.

PERMISSIBLE QUARTERLY INTAKES (ORAL AND INHALATION) OF RADIONUCLIDES AND CRITICAL BODY ORGANS RECOMMENDED BY NCRP FOR OCCUPATIONAL EXPOSURE (1)

Radionuclide	Radioactive half-life days	Solubility state	Oral intake		Intake by inhalation	
			Critical organ	Permissible quarterly intake microcuries	Critical organ	Permissible quarterly intake microcuries
3-H	4.50E 03	Sol.	Body tissue	6.4E 03(2)	Body tissue	3.1E 03(2)
7-BE	5.36E 01	Sol.	GI LLI	3.6E 03	Tot. body	3.5E 03
		Insol.	GI LLI	3.6E 03	Lungs	7.5E 02
14-C	2.00E 06	Sol.	Fat	1.6E 03	Fat	2.2E 03
18-F	7.80E-02	Sol.	GI SI	1.7E 03	GI SI	3.3E 03
		Insol.	GI ULI	1.0E 03	GI ULI	1.6E 03
22-NA	9.50E 02	Sol.	Tot. body	8.0E 01	Tot. body	1.1E 02
		Insol.	GI LLI	6.0E 01	Lungs	5.3E 00
24-NA	6.30E-01	Sol.	GI SI	3.8E 02	GI SI	7.6E 02
		Insol.	GI LLI	5.6E 01	GI LLI	8.9E 01
31-SI	1.10E-01	Sol.	GI S	1.8E 03	GI S	3.5E 03
		Insol.	GI ULI	3.8E 02	GI ULI	6.2E 02
32-P	1.43E 01	Sol.	Bone	3.8E 01	Bone	4.4E 01
		Insol.	GI LLI	4.6E 01	Lungs	4.9E 01
35-S	8.71E 01	Sol.	Testes	1.3E 02	Testes	1.7E 02
		Insol	GI LLI	5.4E 02	Lungs	1.6E 02
36-CL	1.20E 08	Sol.	Tot. body	1.6E 02	Tot. body	2.2E 02
		Insol.	GI LLI	1.2E 02	Lungs	1.4E 01
38-CL	2.60E-02	Sol.	GI S	8.0E 02	GI S	1.6E 03
		Insol.	GI S	8.0E 02	GI S	1.3E 03
42-K	5.20E-01	Sol.	GI S	6.2E 02	GI-S	1.2E 03
		Insol.	GI LLI	4.0E 01	GI LLI	6.7E 01
45-CA	1.64E 02	Sol.	Bone	1.8E 01	Bone	2.0E 01
		Insol.	GI LLI	3.6E 02	Lungs	7.5E 01
47-CA	4.90E 00	Sol.	Bone	1.0E 02	Bone	1.1E 02
		Insol.	GI LLI	6.6E 01	GI LLI	1.1E 02
					Lungs	1.2E 02
46-SC	8.50E 01	Sol.	GI LLI	7.6E 01	Liver	1.5E 02
					GI LLI	1.5E 02
		Insol.	GI LLI	7.6E 01	Lungs	1.4E 01
47-SC	3.43E 00	Sol.	GI LLI	1.8E 02	GI LLI	3.6E 02
		Insol.	GI LLI	1.8E 02	GI LLI	2.9E 02
48-SC	1.83E 00	Sol.	GI LLI	5.4E 01	GI LLI	1.1E 02
		Insol.	GI LLI	5.4E 01	GI LLI	8.7E 01
48-V	1.61E 01	Sol.	GI LLI	5.8E 01	GI LLI	1.2E 02
		Insol.	GI LLI	5.8E 01	Lungs	3.5E 01
51-CR	2.78E 01	Sol.	GI LLI	3.2E 03	GI LLI	6.4E 03
					Tot. body	6.7E 03
		Insol.	GI LLI	3.0E 03	Lungs	1.4E 03
52-MN	5.55E 00	Sol.	GI LLI	6.6E 01	GI LLI	1.3E 02
		Insol.	GI LLI	6.0E 01	Lungs	8.7E 01
					GI LLI	9.5E 01
54-MN	3.00E 02	Sol.	GI LLI	2.6E 02	Liver	2.4E 02
		Insol.	GI LLI	2.4E 02	Lungs	2.2E 01
56-MN	1.10E-01	Sol.	GI LLI	2.4E 02	GI LLI	4.7E 02
		Insol.	GI LLI	2.0E 02	GI LLI	3.3E 02
55-FE	1.10E 03	Sol.	Spleen	1.6E 03	Spleen	5.3E 02
		Insol.	GI LLI	4.6E 03	Lungs	6.4E 02
59-FE	4.51E 01	Sol.	GI LLI	1.2E 02	Spleen	9.3E 01
		Insol.	GI LLI	1.1E 02	Lungs	3.3E 01
57-CO	2.70E 02	Sol.	GI LLI	1.1E 03	GI LLI	2.2E 03
		Insol.	GI LLI	7.6E 02	Lungs	1.0E 02
58M-CO	3.80E-01	Sol.	GI LLI	5.6E 03	GI LLI	1.1E 04
		Insol.	GI LLI	4.0E 03	Lungs	5.5E 03
58-CO	7.20E 01	Sol	GI LLI	2.6E 02	GI LLI	5.3E 02

PERMISSIBLE QUARTERLY INTAKES (ORAL AND INHALATION) OF RADIONUCLIDES AND CRITICAL BODY ORGANS RECOMMENDED BY NCRP FOR OCCUPATIONAL EXPOSURE (1) (Continued)

Radionuclide	Radioactive half-life days	Solubility state	Oral intake: Critical organ	Oral intake: Permissible quarterly intake microcuries	Intake by inhalation: Critical organ	Intake by inhalation: Permissible quarterly intake microcuries
					Tot. body	6.0E 02
		Insol	GI LLI	1.8E 02	Lungs	3.5E 01
60-CO	1.90E 03	Sol	GI LLI	9.8E 01	GI LLI	2.0E 02
					Tot. body	2.2E 02
		Insol.	GI LLI	7.0E 01	Lungs	5.5E 00
59-NI	2.90E 07	Sol.	Bone	4.0E 02	Bone	2.9E 02
		Insol	GI LLI	1.0E 03	Lungs	4.7E 02
63-NI	2.90E 04	Sol	Bone	5.4E 01	Bone	4.0E 01
		Insol	GI LLI	1.4E 03	Lungs	1.7E 02
65-NI	1.10E-01	Sol.	GI ULI	2.8E 02	GI ULI	5.6E 02
		Insol.	GI ULI	2.0E 02	GI ULI	3.3E 02
64-CU	5.30E-01	Sol.	GI LLI	6.6E 02	GI LLI	1.3E 03
		Insol.	GI LLI	4.2E 02	GI LLI	6.6E 02
65-ZN	2.45E 02	Sol.	Tot. body	2.0E 02	Tot. body	6.6E 01
			Prostate	2.4E 02	Prostate	8.0E 01
			Liver	2.6E 02		
		Insol.	GI LLI	3.6E 02	Lungs	3.6E 01
69M-ZN	5.80E-01	Sol.	GI LLI	1.4E 02	Prostate	2.4E 02
		Insol.	GI LLI	1.2E 02	GI LLI	2.0E 02
69-ZN	3.60E-02	Sol.	GI S	3.6E 03	Prostate	4.4E 03
		Insol.	GI S	3.6E 03	GI S	5.8E 03
72-GA	5.90E-01	Sol.	GI LLI	7.4E 01	GI LLI	1.5E 02
		Insol.	GI LLI	7.4E 01	GI LLI	1.2E 02
71-GE	1.20E 01	Sol.	GI LLI	3.2E 03	GI LLI	6.4E 03
		Insol.	GI LLI	3.2E 03	Lungs	4.0E 03
73-AS	7.60E 01	Sol.	GI LLI	9.6E 02	Tot. body	1.3E 03
		Insol.	GI LLI	9.2E 02	Lungs	2.4E 02
74-AS	1.75E 01	Sol.	GI LLI	1.1E 02	GI LLI	2.2E 02
		Insol.	GI LLI	1.0E 02	Lungs	7.8E 01
76-AS	1.10E 00	Sol.	GI LLI	4.0E 01	GI LLI	8.0E 01
		Insol.	GI LLI	3.8E 01	GI LLI	6.2E 01
77-AS	1.60E 00	Sol.	GI LLI	1.6E 02	GI LLI	3.3E 02
		Insol.	GI LLI	1.6E 02	GI LLI	2.5E 02
75-SE	1.27E 02	Sol.	Kidneys	6.0E 02	Kidneys	7.8E 02
			Tot. body	6.8E 02		
		Insol.	GI LLI	5.4E 02	Lungs	7.6E 01
82-BR	1.50E 00	Sol.	Tot. body	5.2E 02	Tot. body	7.1E 02
			GI SI	5.6E 02		
		Insol.	GI LLI	7.4E 01	GI LLI	1.2E 02
86-RB	1.86E 01	Sol.	Tot. body	1.3E 02	Tot. body	1.8E 02
			Pancreas	1.3E 02	Pancreas	1.8E 02
					Liver	2.5E 02
		Insol.	GI LLI	4.8E 01	Lungs	4.2E 01
87-RB	1.80E 13	Sol.	Pancreas	2.2E 02	Pancreas	3.1E 02
					Tot. body	4.0E 02
					Liver	4.2E 02
		Insol.	GI LLI	3.4E 02	Lungs	4.0E 01
85M-SR	4.90E-02	Sol.	GI SI	1.3E 04	GI SI	2.5E 04
		Insol.	GI SI	1.3E 04	GI SI	2.2E 04
85-SR	6.50E 01	Sol.	Tot. body	1.9E 02	Tot. body	1.4E 02
		Insol.	GI LLI	3.4E 02	Lungs	6.6E 01
89-SR	5.05E 01	Sol.	Bone	2.4E 01	Bone	1.7E 01
		Insol.	GI LLI	5.6E 01	Lungs	2.2E 01
90-SR	1.00E 04	Sol.	Bone	8.0E-01	Bone	7.3E-01
		Insol.	GI LLI	7.0E 01	Lungs	3.5E 00
91-SR	4.00E-01	Sol.	GI LLI	1.4E 02	GI LLI	2.7E 02
		Insol.	GI LLI	9.8E 01	GI LLI	1.6E 02
92-SR	1.10E-01	Sol.	GI ULI	1.4E 02	GI ULI	2.7E 02
		Insol.	GI ULI	1.2E 02	GI ULI	1.8E 02
90-Y	2.68E 00	Sol.	GI LLI	4.0E 01	GI LLI	8.0E 01
		Insol.	GI LLI	4.0E 01	GI LLI	6.4E 01
91M-Y	3.50E-02	Sol.	GI SI	6.8E 03	GI SI	1.4E 04
		Insol.	GI SI	6.8E 03	GI SI	1.1E 04
91-Y	5.80E 01	Sol.	GI LLI	5.2E 01	Bone	2.2E 01
		Insol.	GI LLI	5.2E 01	Lungs	2.0E 01
92-Y	1.50E-01	Sol.	GI ULI	1.2E 02	GI ULI	2.4E 02
		Insol.	GI ULI	1.2E 02	GI ULI	1.8E 02
93-Y	4.20E-01	Sol.	GI LLI	5.4E 01	GI LLI	1.1E 02
		Insol.	GI LLI	5.4E 01	GI LLI	8.6E 01
93-ZR	4.00E 08	Sol.	GI LLI	1.6E 03	Bone	8.0E 01
		Insol.	GI LLI	1.6E 03	Lungs	2.0E 02
95-ZR	6.33E 01	Sol.	GI LLI	1.3E 02	Tot. body	8.0E 01
		Insol.	GI LLI	1.3E 02	Lungs	2.0E 01
97-ZR	7.10E-01	Sol.	GI LLI	3.6E 01	GI LLI	7.3E 01
		Insol.	GI LLI	3.6E 01	GI LLI	5.6E 01
93M-NB	3.70E 03	Sol.	GI LLI	8.0E 02	Bone	7.6E 01
		Insol.	GI LLI	8.0E 02	Lungs	1.0E 02
95-NB	3.50E 01	Sol.	GI LLI	1.9E 02	Tot. body	2.9E 02
					GI LLI	3.8E 02
		Insol.	GI LLI	1.9E 02	Lungs	6.2E 01
97-NB	5.10E-02	Sol.	GI ULI	1.9E 03	GI ULI	3.8E 03
		Insol.	GI ULI	1.9E 03	GI ULI	2.9E 03
99-MO	2.79E 00	Sol.	Kidneys	3.6E 02	Kidneys	4.5E 02
			GI LLI	4.8E 02		
		Insol.	GI LLI	7.8E 01	GI LLI	1.3E 02
96M-TC	3.60E-02	Sol.	GI LLI	2.4E 04	GI LLI	4.7E 04
		Insol.	GI LLI	2.0E 04	Lungs	1.8E 04
96-TC	4.30E 00	Sol.	GI LLI	2.0E 02	GI LLI	4.0E 02
		Insol.	GI LLI	9.4E 01	GI LLI	1.5E 02

PERMISSIBLE QUARTERLY INTAKES (ORAL AND INHALATION) OF RADIONUCLIDES AND CRITICAL BODY ORGANS RECOMMENDED BY NCRP FOR OCCUPATIONAL EXPOSURE (1) (Continued)

Radionuclide	Radioactive half-life days	Solubility state	Oral intake: Critical organ	Oral intake: Permissible quarterly intake microcuries	Intake by inhalation: Critical organ	Intake by inhalation: Permissible quarterly intake microcuries
97M-TC	9.20E 01	Sol.	GI LLI	7.2E 02	GI LLI	1.4E 03
		Insol.	GI LLI	3.4E 02	Lungs	9.5E 01
97-TC	3.70E 06	Sol.	GI LLI	3.4E 03	GI LLI	6.7E 03
					Kidneys	8.0E 03
		Insol.	GI LLI	1.6E 03	Lungs	1.8E 02
99M-TC	2.50E-01	Sol.	GI ULI	1.1E 04	GI ULI	2.4E 04
		Insol.	GI ULI	5.6E 03	GI ULI	8.7E 03
99-TC	7.70E 07	Sol.	GI LLI	6.6E 02	GI LLI	1.3E 03
		Insol.	GI LLI	3.2E 02	Lungs	3.8E 01
97-RU	2.80E 00	Sol.	GI LLI	7.2E 02	GI LLI	1.4E 03
		Insol.	GI LLI	7.0E 02	GI LLI	1.1E 03
103-RU	4.10E 01	Sol.	GI LLI	1.6E 02	GI LLI	3.3E 02
		Insol.	GI LLI	1.6E 02	Lungs	5.3E 01
105-RU	1.90E-01	Sol.	GI ULI	2.2E 02	GI ULI	4.4E 02
		Insol.	GI ULI	2.0E 02	GI ULI	3.3E 02
106-RU	3.65E 02	Sol.	GI LLI	2.4E 01	GI LLI	4.7E 01
		Insol.	GI LLI	2.4E 01	Lungs	3.5E 00
103M-RH	3.80E-02	Sol.	GI S	2.4E 04	GI S	4.7E 04
		Insol.	GI S	2.4E 04	GI S	3.8E 04
105-RH	1.52E 00	Sol.	GI LLI	2.6E 02	GI LLI	5.3E 02
		Insol.	GI LLI	2.0E 02	GI LLI	3.3E 02
103-PD	1.70E 01	Sol.	GI LLI	6.8E 02	Kidneys	8.4E 02
		Insol.	GI LLI	5.4E 02	Lungs	4.7E 02
109-PD	5.70E-01	Sol.	GI LLI	1.8E 02	GI LLI	3.5E 02
		Insol.	GI LLI	1.4E 02	GI LLI	2.2E 02
105-AG	4.00E 01	Sol.	GI LLI	1.9E 02	GI LLI	3.8E 02
		Insol.	GI LLI	1.9E 02	Lungs	5.1E 01
110M-AG	2.70E 02	Sol.	GI LLI	6.0E 01	GI LLI	1.2E 02
		Insol.	GI LLI	6.0E 01	Lungs	6.4E 00
111-AG	7.50E 00	Sol.	GI LLI	8.8E 01	GI LLI	1.8E 02
		Insol.	GI LLI	8.6E 01	GI LLI	1.4E 02
109-CD	4.75E 02	Sol.	GI LLI	3.4E 02	Liver	3.3E 01
					Kidneys	3.5E 01
		Insol.	GI LLI	3.4E 02	Lungs	4.5E 01
115M-CD	4.30E 01	Sol.	GI LLI	5.0E 01	Liver	2.2E 01
		Insol.	GI LLI	5.0E 01	Lungs	2.2E 01
115-CD	2.20E 00	Sol.	GI LLI	6.8E 01	GI LLI	1.4E 02
		Insol.	GI LLI	7.2E 01	GI LLI	1.1E 02
134-CS	8.40E 02	Sol.	Tot. body	1.7E 01	Tot. body	2.4E 01
		Insol.	GI LLI	8.0E 01	Lungs	8.0E 00
135-CS	1.10E 09	Sol.	Liver	2.2E 02	Liver	2.9E 02
			Spleen	2.4E 02	Spleen	3.3E 02
			Tot. body	2.6E 02	Tot. body	3.6E 02
		Insol.	GI LLI	4.6E 02	Lungs	5.6E 01
136-CS	1.30E 01	Sol.	Tot. body	1.7E 02	Tot. body	2.4E 02
		Insol.	GI LLI	1.3E 02	Lungs	1.1E 02
137-CS	1.10E 04	Sol.	Tot. body	3.0E 01	Tot. body	4.0E 01
			Liver	3.6E 01		
			Spleen	4.4E 01		
			Muscle	4.8E 01		
		Insol.	GI LLI	8.8E 01	Lungs	9.1E 00
131-BA	1.16E 01	Sol.	GI LLI	3.6E 02	GI LLI	7.3E 02
		Insol.	GI LLI	3.4E 02	Lungs	2.2E 02
140-BA	1.28E 01	Sol.	GI LLI	5.2E 01	Bone	8.0E 01
		Insol.	GI LLI	5.0E 01	Lungs	2.7E 01
140-LA	1.68E 00	Sol.	GI LLI	4.8E 01	GI LLI	9.6E 01
		Insol.	GI LLI	4.8E 01	GI LLI	7.6E 01
141-CE	3.20E 01	Sol.	GI LLI	1.8E 02	Liver	2.7E 02
					GI LLI	3.6E 02
					Bone	3.6E 02
		Insol.	GI LLI	1.8E 02	Lungs	9.8E 01
143-CE	1.33E 00	Sol.	GI LLI	8.0E 01	GI LLI	1.6E 02
		Insol.	GI LLI	8.0E 01	GI LLI	1.3E 02
144-CE	2.90E 02	Sol.	GI LLI	2.4E 01	Bone	6.0E 00
		Insol.	GI LLI	2.4E 01	Lungs	4.0E 00
142-PR	8.00E-01	Sol.	GI LLI	6.0E 01	GI LLI	1.2E 02
		Insol.	GI LLI	6.0E 01	GI LLI	9.8E 01
143-PR	1.37E 01	Sol.	GI LLI	9.8E 01	GI LLI	2.0E 02
		Insol.	GI LLI	9.8E 01	Lungs	1.1E 02
144-ND	7.30E 17	Sol.	Bone	1.3E 02	Bone	5.1E-02
		Insol.	GI LLI	1.6E 02	Lungs	1.8E-01
147-ND	1.13E 01	Sol.	GI LLI	1.2E 02	Liver	2.2E 02
					GI LLI	2.4E 02
		Insol.	GI LLI	1.2E 02	Lungs	1.4E 02
149-ND	8.30E-02	Sol.	GI LLI	5.6E 02	GI LLI	1.1E 03
		Insol.	GI ULI	5.6E 02	GI ULI	9.1E 02
147-PM	9.20E 02	Sol.	GI LLI	4.4E 02	Bone	4.0E 01
		Insol.	GI LLI	4.4E 02	Lungs	6.0E 01
149-PM	2.20E 00	Sol.	GI LLI	8.8E 01	GI LLI	1.8E 02
		Insol.	GI LLI	8.8E 01	GI LLI	1.4E 02
147-SM	4.80E 13	Sol.	Bone	1.1E 02	Bone	4.4E-02
		Insol.	GI LLI	1.4E 02	Lungs	1.6E-01
151-SM	3.70E 04	Sol.	GI LLI	7.4E 02	Bone	4.0E 01
		Insol.	GI LLI	7.4E 02	Lungs	8.7E 01
153-SM	1.96E 00	Sol.	GI LLI	1.5E 02	GI LLI	3.1E 02
		Insol.	GI LLI	1.5E 02	GI LLI	2.5E 02
152M-EU	3.80E-01	Sol.	GI LLI	1.3E 02	GI LLI	2.5E 02
		Insol.	GI LLI	1.3E 02	GI LLI	2.0E 02
152-EU	4.70E 03	Sol.	GI LLI	1.5E 02	Kidneys	7.6E 00
		Insol.	GI LLI	1.5E 02	Lungs	1.1E 01

PERMISSIBLE QUARTERLY INTAKES (ORAL AND INHALATION) OF RADIONUCLIDES AND CRITICAL BODY ORGANS RECOMMENDED BY NCRP FOR OCCUPATIONAL EXPOSURE (1) (Continued)

Radionuclide	Radioactive half-life days	Solubility state	Oral intake: Critical organ	Oral intake: Permissible quarterly intake microcuries	Intake by inhalation: Critical organ	Intake by inhalation: Permissible quarterly intake microcuries
154-EU	5.80E 03	Sol.	GI LLI	4.4E 01	Kidneys	2.4E 00
					Bone	2.4E 00
		Insol.	GI LLI	4.4E 01	Lungs	4.4E 00
155-EU	6.21E 02	Sol.	GI LLI	4.0E 02	Kidneys	5.8E 01
					Bone	6.2E 01
		Insol.	GI LLI	4.0E 02	Lungs	4.5E 01
153-GD	2.36E 02	Sol.	GI LLI	4.2E 02	Bone	1.4E 02
		Insol.	GI LLI	4.2E 02	Lungs	5.6E 01
159-GD	7.50E-01	Sol.	GI LLI	1.6E 02	GI LLI	3.1E 02
		Insol.	GI LLI	1.6E 02	GI LLI	2.5E 02
113M-IN	7.30E-02	Sol.	GI ULI	2.6E 03	GI ULI	5.3E 03
		Insol.	GI ULI	2.6E 03	GI ULI	4.2E 03
114M-IN	4.90E 01	Sol.	GI LLI	3.4E 01	Kidneys	6.4E 01
					GI LLI	6.7E 01
					Spleen	7.1E 01
		Insol.	GI LLI	3.4E 01	Lungs	1.3E 01
115M-IN	1.90E-01	Sol.	GI ULI	7.4E 02	GI ULI	1.5E 03
		Insol.	GI ULI	7.4E 02	GI ULI	1.2E 03
115-IN	2.20E 17	Sol.	GI LLI	1.8E 02	Kidneys	1.5E 02
		Insol.	GI LLI	1.8E 02	Lungs	2.2E 01
113-SN	1.12E 02	Sol.	GI LLI	1.7E 02	Bone	2.2E 02
		Insol.	GI LLI	1.6E 02	Lungs	3.3E 01
125-SN	9.50E 00	Sol.	GI LLI	3.6E 01	GI LLI	7.3E 01
		Insol.	GI LLI	3.4E 01	Lungs	5.3E 01
					GI LLI	5.5E 01
122-SB	2.80E 00	Sol.	GI LLI	5.8E 01	GI LLI	1.2E 02
		Insol.	GI LLI	5.8E 01	GI LLI	9.1E 01
124-SB	6.00E 01	Sol.	GI LLI	4.6E 01	GI LLI	9.3E 01
		Insol.	GI LLI	4.6E 01	Lungs	1.2E 01
125-SB	8.77E 02	Sol.	GI LLI	2.0E 02	Lungs	3.3E 02
					Tot. body	3.6E 02
					GI LLI	4.0E 02
					Bone	4.5E 02
		Insol.	GI LLI	2.0E 02	Lungs	1.7E 01
125M-TE	5.80E 01	Sol.	Kidneys	3.2E 02	Kidneys	2.2E 02
			GI LLI	3.6E 02		
			Testes	4.4E 02		
		Insol.	GI LLI	2.4E 02	Lungs	8.0E 01
127M-TE	1.05E 02	Sol.	Kidneys	1.2E 02	Kidneys	8.2E 01
					Testes	8.7E 01
		Insol.	GI LLI	1.1E 02	Lungs	2.5E 01
127-TE	3.90E-01	Sol.	GI LLI	5.2E 02	GI LLI	1.0E 03
		Insol.	GI LLI	3.4E 02	GI LLI	5.3E 02
129M-TE	3.30E 01	Sol.	GI LLI	6.6E 01	Kidneys	4.9E 01
					Testes	5.6E 01
		Insol.	GI LLI	4.0E 01	Lungs	2.0E 01
129-TE	5.10E-02	Sol.	GI S	1.6E 03	GI S	3.3E 03
		Insol.	GI ULI	1.6E 03	GI ULI	2.5E 03
131M-TE	1.25E 00	Sol.	GI LLI	1.2E 02	GI LLI	2.4E 02
		Insol.	GI LLI	7.4E 01	GI LLI	1.2E 02
132-TE	3.20E 00	Sol.	GI LLI	6.4E 01	GI LLI	1.3E 02
		Insol.	GI LLI	4.2E 01	GI LLI	6.6E 01
126-I	1.33E 01	Sol.	Thyroid	3.4E 00	Thyroid	4.5E 00
		Insol.	GI LLI	1.8E 02	Lungs	2.0E 02
129-I	6.30E 09	Sol.	Thyroid	7.6E-01	Thyroid	1.0E 00
		Insol.	GI LLI	4.2E 02	Lungs	4.5E 01
131-I	8.00E 00	Sol.	Thyroid	4.0E 00	Thyroid	5.3E 00
		Insol.	GI LLI	1.3E 02	GI LLI	2.0E 02
					Lungs	2.0E 02
132-I	9.70E-02	Sol.	Thyroid	1.1E 02	Thyroid	1.5E 02
		Insol.	GI ULI	3.6E 02	GI ULI	5.8E 02
133-I	8.70E-01	Sol.	Thyroid	1.5E 01	Thyroid	2.0E 01
		Insol.	GI LLI	8.2E 01	GI LLI	1.3E 02
134-I	3.60E-02	Sol.	Thyroid	2.4E 02	Thyroid	3.1E 02
		Insol.	GI S	1.2E 03	GI S	2.0E 03
135-I	2.80E-01	Sol.	Thyroid	4.8E 01	Thyroid	6.4E 01
		Insol.	GI LLI	1.4E 02	GI LLI	2.2E 02
131-CS	1.00E 01	Sol.	Tot. body	4.8E 03	Tot. body	6.6E 03
					Liver	8.0E 03
		Insol.	GI LLI	1.9E 03	Lungs	2.0E 03
134M-CS	1.30E-01	Sol.	GI S	1.1E 04	GI S	2.2E 04
		Insol.	GI ULI	2.2E 03	GI ULI	3.6E 03
160-TB	7.30E 01	Sol.	GI LLI	8.8E 01	Bone	6.2E 01
		Insol.	GI LLI	9.0E 01	Lungs	2.0E 01
165-DY	9.70E-02	Sol.	GI ULI	8.0E 02	GI ULI	1.6E 03
		Insol.	GI ULI	8.0E 02	GI ULI	1.3E 03
166-DY	3.40E 00	Sol.	GI LLI	7.6E 01	GI LLI	1.5E 02
		Insol.	GI LLI	7.6E 01	GI LLI	1.2E 02
166-HO	1.10E 00	Sol.	GI LLI	6.2E 01	GI LLI	1.2E 02
		Insol.	GI LLI	6.2E 01	GI LLI	1.0E 02
169-ER	9.40E 00	Sol.	GI LLI	1.9E 02	GI LLI	3.8E 02
		Insol.	GI LLI	1.9E 02	Lungs	2.4E 02
171-ER	3.10E-01	Sol.	GI ULI	2.2E 02	GI ULI	4.4E 02
		Insol.	GI ULI	2.2E 02	GI ULI	3.6E 02
170-TM	1.27E 02	Sol.	GI LLI	9.2E 01	Bone	2.2E 01
		Insol.	GI LLI	9.2E 01	Lungs	2.2E 01
171-TM	6.94E 02	Sol.	GI LLI	1.0E 03	Bone	6.9E 01
		Insol.	GI LLI	1.0E 03	Lungs	1.4E 02
175-YB	4.10E 00	Sol.	GI LLI	2.2E 02	GI LLI	4.4E 02
		Insol.	GI LLI	2.2E 02	GI LLI	3.6E 02

PERMISSIBLE QUARTERLY INTAKES (ORAL AND INHALATION) OF RADIONUCLIDES AND CRITICAL BODY ORGANS RECOMMENDED BY NCRP FOR OCCUPATIONAL EXPOSURE (1) (Continued)

Radionuclide	Radioactive half-life days	Solubility state	Oral intake		Intake by inhalation	
			Critical organ	Permissible quarterly intake microcuries	Critical organ	Permissible quarterly intake microcuries
177-LU	6.80E 00	Sol.	GI LLI	2.0E 02	GI LLI	4.0E 02
		Insol.	GI LLI	2.0E 02	GI LLI	3.3E 02
					Lungs	4.4E 02
181-HF	4.60E 01	Sol.	GI LLI	1.4E 02	Spleen	2.4E 01
		Insol.	GI LLI	1.4E 02	Lungs	4.5E 01
182-TA	1.12E 02	Sol.	GI LLI	8.0E 01	Liver	2.4E 01
		Insol.	GI LLI	8.0E 01	Lungs	1.4E 01
181-W	1.40E 02	Sol.	GI LLI	7.2E 02	GI LLI	1.4E 03
		Insol.	GI LLI	6.4E 02	Lungs	7.8E 01
185-W	7.40E 01	Sol.	GI LLI	2.4E 02	GI LLI	4.7E 02
		Insol.	GI LLI	2.2E 02	Lungs	6.9E 01
187-W	1.00E 00	Sol.	GI LLI	1.4E 02	GI LLI	2.7E 02
		Insol.	GI LLI	1.2E 02	GI LLI	2.0E 02
183-RE	7.30E 01	Sol.	GI LLI	1.1E 03	Tot. body	1.6E 03
		Insol.	GI LLI	5.6E 02	Lungs	9.8E 01
186-RE	3.79E 00	Sol.	GI LLI	1.9E 02	GI LLI	3.8E 02
		Insol.	GI LLI	9.4E 01	GI LLI	1.5E 02
187-RE	1.80E 13	Sol.	GI LLI	5.0E 03	Skin	5.6E 03
			Skin	5.6E 03		
		Insol.	GI LLI	5.0E 03	Lungs	3.1E 02
188-RE	7.10E-01	Sol.	GI LLI	1.3E 02	GI LLI	2.5E 02
		Insol.	GI LLI	6.2E 01	GI LLI	1.0E 02
185-OS	9.50E 01	Sol.	GI LLI	1.5E 02	GI LLI	2.9E 02
		Insol.	GI LLI	1.3E 02	Lungs	2.9E 01
191M-OS	5.80E-01	Sol.	GI LLI	5.0E 03	GI LLI	1.0E 04
		Insol.	GI LLI	4.8E 03	Lungs	5.8E 03
191-OS	1.60E 01	Sol.	GI LLI	3.4E 02	GI LLI	6.7E 02
		Insol.	GI LLI	3.2E 02	Lungs	2.5E 02
193-OS	1.30E 00	Sol.	GI LLI	1.2E 02	GI LLI	2.4E 02
		Insol.	GI LLI	1.1E 02	GI LLI	1.7E 02
190-IR	1.20E 01	Sol.	GI LLI	4.0E 02	GI LLI	8.0E 02
		Insol.	GI LLI	3.6E 02	Lungs	2.5E 02
192-IR	7.45E 01	Sol.	GI LLI	8.0E 01	Kidneys	7.8E 01
		Insol.	GI LLI	7.2E 01	Lungs	1.6E 01
194-IR	7.90E-01	Sol.	GI LLI	6.8E 01	GI LLI	1.4E 02
		Insol.	GI LLI	6.0E 01	GI LLI	9.6E 01
191-PT	3.00E 00	Sol.	GI LLI	2.4E 02	GI LLI	4.7E 02
		Insol.	GI LLI	2.2E 02	GI LLI	3.5E 02
193M-PT	3.40E 00	Sol.	GI LLI	2.2E 03	GI LLI	4.4E 03
		Insol.	GI LLI	2.0E 03	GI LLI	3.3E 03
					Lungs	4.0E 03
193-PT	1.80E 05	Sol.	Kidneys	1.9E 03	Kidneys	6.4E 02
		Insol.	GI LLI	3.0E 03	Lungs	2.0E 02
197M-PT	5.60E-02	Sol.	GI ULI	2.0E 03	GI ULI	4.0E 03
		Insol.	GI ULI	1.9E 03	GI ULI	2.9E 03
197-PT	7.50E-01	Sol.	GI LLI	2.4E 02	GI LLI	4.7E 02
		Insol.	GI LLI	2.2E 02	GI LLI	3.5E 02
196-AU	5.60E 00	Sol.	GI LLI	3.2E 02	GI LLI	6.4E 02
		Insol.	GI LLI	3.0E 02	Lungs	3.8E 02
198-AU	2.70E 00	Sol.	GI LLI	1.0E 02	GI LLI	2.0E 02
		Insol.	GI LLI	9.2E 01	GI LLI	1.5E 02
199-AU	3.15E 00	Sol.	GI LLI	3.4E 02	GI LLI	6.7E 02
		Insol.	GI LLI	3.2E 02	GI LLI	5.1E 02
197M-HG	1.00E 00	Sol.	Kidneys	3.8E 02	Kidneys	4.5E 02
		Insol.	GI LLI	3.4E 02	GI LLI	5.3E 02
197-HG	2.70E 00	Sol.	Kidneys	6.0E 02	Kidneys	7.3E 02
		Insol.	GI LLI	9.8E 02	GI LLI	1.5E 03
203-HG	4.58E 01	Sol.	Kidneys	3.6E 01	Kidneys	4.4E 01
		Insol.	GI LLI	2.2E 02	Lungs	7.8E 01
200-TL	1.13E 00	Sol.	GI LLI	8.8E 02	GI LLI	1.6E 03
		Insol.	GI LLI	4.4E 02	GI LLI	7.1E 02
201-TL	3.00E 00	Sol.	GI LLI	6.2E 02	GI LLI	1.2E 03
		Insol.	GI LLI	3.4E 02	GI LLI	5.5E 02
202-TL	1.20E 01	Sol.	GI LLI	2.4E 02	GI LLI	4.7E 02
		Insol.	GI LLI	1.4E 02	Lungs	1.5E 02
204-TL	1.10E 03	Sol.	GI LLI	2.2E 02	Kidneys	3.8E 02
					GI LLI	4.4E 02
		Insol.	GI LLI	1.2E 02	Lungs	1.6E 01
203-PB	2.17E 00	Sol.	GI LLI	7.8E 02	GI LLI	1.6E 03
		Insol.	GI LLI	7.0E 02	GI LLI	1.1E 03
210-PB	7.10E 03	Sol.	Tot. body	2.4E-01	Kidneys	7.8E-02
			Kidneys	2.8E-01		
		Insol.	GI LLI	3.6E 02	Lungs	1.5E-01
212-PB	4.40E-01	Sol.	GI LLI	3.8E 01	Kidneys	1.1E 01
			Kidneys	4.0E 01		
		Insol.	GI LLI	3.4E 01	Lungs	1.2E 01
206-BI	6.40E 00	Sol.	GI LLI	7.6E 01	Kidneys	1.2E 02
		Insol.	GI LLI	7.6E 01	Lungs	9.1E 01
207-BI	2.90E 03	Sol.	GI LLI	1.3E 02	Kidneys	1.1E 02
		Insol.	GI LLI	1.3E 02	Lungs	8.6E 00
210-BI	5.00E 00	Sol.	GI LLI	8.2E 01	Kidneys	4.0E 00
		Insol.	GI LLI	8.2E 01	Lungs	3.6E 00
212-BI	4.20E-02	Sol.	GI S	7.0E 02	Kidneys	6.0E 01
		Insol.	GI S	7.0E 02	Lungs	1.3E 02
210-PO	1.38E 02	Sol.	Spleen	1.4E 00	Spleen	3.1E-01
					Kidneys	3.1E-01
		Insol.	GI LLI	5.8E 01	Lungs	1.3E-01
211-AT	3.00E-01	Sol.	Thyroid	3.4E 00	Thyroid	4.4E 00
			Ovaries	3.4E 00		
		Insol.	GI ULI	1.5E 02	Lungs	2.2E 01

PERMISSIBLE QUARTERLY INTAKES (ORAL AND INHALATION) OF RADIONUCLIDES AND CRITICAL BODY ORGANS RECOMMENDED BY NCRP FOR OCCUPATIONAL EXPOSURE (1)
(Continued)

			Oral intake		Intake by inhalation	
Radionuclide	Radioactive half-life days	Solubility state	Critical organ	Permissible quarterly intake microcuries	Critical organ	Permissible quarterly intake microcuries
220-RN[3]	6.40E-04				Lungs	1.8E 02
222-RN[3]	3.82E 00				Lungs	1.8E 01
223-RA	1.17E 01	Sol.	Bone	1.4E 00	Bone	1.1E 00
		Insol.	GI LLI	8.4E 00	Lungs	1.5E-01
224-RA	3.64E 00	Sol.	Bone	4.6E 00	Bone	3.5E 00
		Insol.	GI LLI	1.1E 01	Lungs	4.5E-01
226-RA	5.90E 05	Sol.	Bone	2.4E-02	Bone	1.8E-02
		Insol.	GI LLI	6.4E 01	Lungs	3.3E-02
228-RA	2.40E 03	Sol.	Bone	5.6E-02	Bone	4.2E-02
		Insol.	GI LLI	5.0E 01	Lungs	2.4E-02
227-AC	8.00E 03	Sol.	Bone	3.8E 00	Bone	1.5E-03
		Insol.	GI LLI	6.0E 02	Lungs	1.6E-02
228-AC	2.60E-01	Sol.	GI ULI	1.7E 02	Liver	4.7E 01
					Bone	5.3E 01
		Insol.	GI ULI	1.7E 02	Lungs	1.1E 01
227-TH	1.84E 01	Sol.	GI LLI	3.6E 01	Bone	2.2E-01
		Insol.	GI LLI	3.6E 01	Lungs	1.1E-01
228-TH	7.00E 02	Sol.	Bone	1.5E 01	Bone	5.6E-03
		Insol.	GI LLI	2.6E 01	Lungs	3.8E-03
230-TH	2.90E 07	Sol.	Bone	3.4E 00	Bone	1.4E-03
		Insol.	GI LLI	6.4E 01	Lungs	6.4E-03
231-TH	1.07E 00	Sol.	GI LLI	4.6E 02	GI LLI	9.3E 02
		Insol.	GI LLI	4.6E 02	GI LLI	7.5E 02
232-TH	5.10E 12	Sol.	Bone	3.0E 00	Bone	1.2E-03[4]
		Insol.	GI LLI	7.6E 01	Lungs	7.3E-03
234-TH	2.41E 01	Sol.	GI LLI	3.4E 01	Bone	3.6E 01
		Insol.	GI LLI	3.4E 01	Lungs	2.0E 01
NAT-TH[5]		Sol.	Bone	2.6E 00	Bone	1.0E-03[4]
		Insol.	GI LLI	1.9E 01	Lungs	2.5E-03
230-PA	1.77E 01	Sol.	GI LLI	4.8E 02	Bone	1.1E 00
		Insol.	GI LLI	5.0E 02	Lungs	4.9E-01
231-PA	1.30E 07	Sol.	Bone	1.7E 00	Bone	7.1E-04
		Insol.	GI LLI	5.4E 01	Lungs	6.7E-02
233-PA	2.74E 01	Sol.	GI LLI	2.4E 02	Kidneys	3.8E 02
		Insol.	GI LLI	2.4E 02	Lungs	1.1E 02
230-U	2.08E 01	Sol.	Kidneys	4.8E 00	Kidneys	1.8E-01
		Insol.	GI LLI	9.2E 00	Lungs	7.1E-02
232-U	2.70E 04	Sol.	Bone	1.7E 00	Bone	6.4E-02
		Insol.	GI LLI	5.8E 01	Lungs	1.7E-02
233-U	5.90E 07	Sol.	Bone	8.4E 00	Bone	3.3E-01
		Insol.	GI LLI	6.4E 01	Lungs	7.5E-02
234-U	9.10E 07	Sol.	Bone	8.6E 00	Bone	3.5E-01
		Insol.	GI LLI	6.4E 01	Lungs	7.5E-02
235-U[5]	2.60E 11	Sol.	Kidneys	7.4E 00	Kidneys	2.9E-01
					Bone	3.6E-01
		Insol.	GI LLI	5.6E 01	Lungs	8.0E-02
236-U[5]	8.70E 09	Sol.	Bone	9.0E 00	Bone	3.6E-01
		Insol.	GI LLI	6.8E 01	Lungs	7.8E-02
238-U[5]	1.60E 12	Sol.	Kidneys	1.2E 00	Kidneys	4.5E-02
		Insol.	GI LLI	7.0E 01	Lungs	8.6E-02
240-U	5.88E-01	Sol.	GI LLI	6.8E 01	GI LLI	1.4E 02
		Insol.	GI LLI	6.8E 01	GI LLI	1.1E 02
NAT-U[5]		Sol.	Kidneys	1.2E 00	Kidneys	4.5E-02
		Insol.	GI LLI	3.2E 01	Lungs	4.0E-02
237-NP	8.00E 08	Sol.	Bone	6.2E 00	Bone	2.5E-03
		Insol.	GI LLI	7.0E 01	Lungs	7.5E-02
239-NP	2.33E 00	Sol.	GI LLI	2.6E 02	GI LLI	5.3E 02
		Insol.	GI LLI	2.6E 02	GI LLI	4.2E 02
238-PU	3.30E 04	Sol.	Bone	1.0E 01	Bone	1.2E-03
		Insol.	GI LLI	5.6E 01	Lungs	2.2E-02
239-PU	8.90E 06	Sol.	Bone	9.0E 00	Bone	1.1E-03
		Insol.	GI LLI	5.8E 01	Lungs	2.4E-02
240-PU	2.40E 06	Sol.	Bone	9.0E 00	Bone	1.1E-03
		Insol.	GI LLI	5.8E 01	Lungs	2.4E-02
241-PU	4.80E 03	Sol.	Bone	4.6E 02	Bone	5.6E-02
		Insol.	GI LLI	2.6E 03	Lungs	2.4E 01
242-PU	1.40E 08	Sol.	Bone	9.4E 00	Bone	1.1E-03
		Insol.	GI LLI	6.2E 01	Lungs	2.4E-02
243-PU	2.08E-01	Sol.	GI ULI	6.8E 02	GI ULI	1.1E 03
		Insol.	GI ULI	6.8E 02	GI ULI	1.4E 03
244-PU	2.80E 10	Sol.	Bone	8.6E 00	Bone	1.0E-03
		Insol.	GI LLI	2.2E 01	Lungs	2.0E-02
241-AM	1.70E 05	Sol.	Kidneys	7.6E 00	Bone	3.6E-03
					Kidneys	3.8E-03
		Insol.	GI LLI	5.4E 01	Lungs	6.4E-02
242M-AM	5.60E 04	Sol.	Bone	8.8E 00	Bone	3.5E-03
		Insol.	GI LLI	1.8E 02	Lungs	1.6E-01
242-AM	6.77E-01	Sol.	GI LLI	2.6E 02	Liver	2.4E 01
		Insol.	GI LLI	2.6E 02	Lungs	2.9E 01
243-AM	2.90E 06	Sol.	Bone	8.8E 00	Bone	3.5E-03
					Kidneys	3.8E-03
		Insol.	GI LLI	5.6E 01	Lungs	6.7E-02
244-AM	1.81E-02	Sol.	GI SI	9.6E 03	Bone	2.5E 03
					Kidneys	2.7E 03
		Insol.	GI SI	9.6E 03	Lungs	1.5E 04
					GI(SI)	1.5E 04
242-CM	1.62E 02	Sol.	GI LLI	4.8E 01	Liver	7.5E-02
		Insol.	GI LLI	5.0E 01	Lungs	1.0E-01
243-CM	1.30E 04	Sol.	Bone	1.0E 01	Bone	4.0E-03
		Insol.	GI LLI	5.0E 01	Lungs	6.2E-02

PERMISSIBLE QUARTERLY INTAKES (ORAL AND INHALATION) OF RADIONUCLIDES AND CRITICAL BODY ORGANS RECOMMENDED BY NCRP FOR OCCUPATIONAL EXPOSURE (1) (Continued)

Radionuclide	Radioactive half-life days	Solubility state	Oral intake: Critical organ	Oral intake: Permissible quarterly intake microcuries	Intake by inhalation: Critical organ	Intake by inhalation: Permissible quarterly intake microcuries
244-CM	6.70E 03	Sol.	Bone	1.4E 01	Bone	5.6E-03
		Insol.	GI LLI	5.2E 01	Lungs	6.2E-02
245-CM	7.30E 06	Sol.	Bone	7.0E 00	Bone	2.9E-03
		Insol.	GI LLI	5.6E 01	Lungs	6.7E-02
246-CM	2.40E 06	Sol.	Bone	7.2E 00	Bone	2.9E-03
		Insol.	GI LLI	5.6E 01	Lungs	6.6E-02
247-CM	3.30E 10	Sol.	Bone	7.2E 00	Bone	2.9E-03
		Insol.	GI LLI	4.4E 01	Lungs	6.7E-02
248-CM	1.70E 08	Sol.	Bone	8.8E-01	Bone	3.6E-04
		Insol.	GI LLI	2.6E 00	Lungs	8.2E-03
249-CM	4.40E-02	Sol.	GI S	4.4E 03	Bone	7.8E 03
		Insol.	GI S	4.4E 03	GI S	7.1E 03
249-BK	2.90E 02	Sol.	GI LLI	1.2E 03	Bone	5.8E-01
		Insol.	GI LLI	1.2E 03	Lungs	7.5E 01
250-BK	1.34E-01	Sol.	GI ULI	4.4E 02	Bone	9.1E 01
		Insol.	GI ULI	4.4E 02	GI ULI	7.1E 02
249-CF	1.70E 05	Sol.	Bone	8.2E 00	Bone	9.8E-04
		Insol.	GI LLI	4.8E 01	Lungs	6.2E-02
250-CF	3.70E 03	Sol.	Bone	2.6E 01	Bone	3.1E-03
		Insol.	GI LLI	5.0E 01	Lungs	6.2E-02
251-CF	2.90E 05	Sol.	Bone	8.6E 00	Bone	1.0E-03
		Insol.	GI LLI	5.2E 01	Lungs	6.2E-02
252-CF	8.03E 02	Sol.	GI LLI	1.5E 01	Bone	4.0E-03
		Insol.	GI LLI	1.5E 01	Lungs	2.0E-02
253-CF	1.80E 01	Sol.	GI LLI	2.8E 02	Bone	5.3E-01
		Insol.	GI LLI	2.8E 02	Lungs	4.7E-01
254-CF	5.60E 01	Sol.	GI LLI	2.4E-01	Bone	3.3E-03
		Insol.	GI LLI	2.4E-01	Lungs	3.1E-03
253-ES	2.00E 01	Sol.	GI LLI	4.6E 01	Bone	4.7E-01
		Insol.	GI LLI	4.6E 01	Lungs	3.8E-01
254M-ES	1.60E 00	Sol.	GI LLI	3.8E 01	Bone	3.3E 00
		Insol.	GI LLI	3.8E 01	Lungs	3.6E 00
254-ES	4.80E 02	Sol.	GI LLI	2.8E 01	Bone	1.2E-02
		Insol.	GI LLI	2.8E 01	Lungs	6.7E-02
255-ES	3.00E 01	Sol.	GI LLI	5.6E 01	Bone	3.1E-01
		Insol.	GI LLI	5.6E 01	Lungs	2.5E-01
254-FM	1.35E-01	Sol.	GI ULI	2.4E 02	Bone	4.0E 01
		Insol.	GI ULI	2.4E 02	Lungs	4.4E 01
255-FM	8.96E-01	Sol.	GI LLI	6.6E 01	Bone	1.0E 01
		Insol.	GI LLI	6.6E 01	Lungs	6.7E 00
256-FM	1.11E-01	Sol.	GI ULI	1.8E 00	Bone	1.7E 00
		Insol.	GI ULI	1.8E 00	Lungs	1.1E 00

[1] These quarterly intakes are calculated from values of $(MPC)_a$ and $(MPC)_w$ recommended by the NCRP for occupational exposure (see NBS Handbook 69 and NCRP Report No. 32)* except for uranium, ^{90}Sr, and certain transuranic isotopes given by the ICRP (see ICRP Publications 2 and 6).** Specifically, except where indicated in the footnotes, the quarterly intakes are calculated to two-digit accuracy as $91 \times 2 \times 10^7 \times (MPC)_a$ and $91 \times 2200 \times (MPC)_w$ where 91 ~ days/quarter, 2×10^7 ~ cc of air breathed/day, 2200 ~ cc of water intake/day, and $(MPC)_a$, $(MPC)_w$ are μCi/cc of air and water recommended as limits for occupational exposure for the 168-hour week. These quarterly intakes are calculated for standard adult man and would be expected to deliver a total dose (= dose commitment) to the critical organ of 1/4 the maximum permissible annual dose (MPD) recommended as limiting by the NCRP for occupational exposure. These limiting MPD are the following: Bone, skin, thyroid—MPD = 30 rem; blood-forming organs, lenses of the eyes, gonads—MPD = 5 rem/yr; other organs—MPD = 15 rem/yr. These limits are not appropriate for exposure of the population or, specifically, for persons below the age of 18 years. For more detailed information on appropriate exposure limits recommended by the NCRP, see NBS Handbook 69 and NCRP Report No. 32.*

The abbreviations GI, S, SI, ULI, and LLI refer to gastrointestinal tract, stomach, small intestine, upper large intestine, and lower large intestine, respectively.

With few exceptions, the above intakes do not take account of chemical toxicity which should be considered separately.

[2] This intake includes absorption through the skin (see ICRP Publication 2, p. 22).**

[3] The daughter isotopes of ^{220}Rn and ^{222}Rn are assumed present to the extent they occur in unfiltered air (see NBS Handbook 69, p. 12).* For all other isotopes, the daughter elements are not considered as part of the intake and if present must be considered on the basis of the rules for mixtures.

[4] These are provisional values for ^{232}Th and nat-Th (see footnote, p. 83, NBS Handbook 69).*

[5] A curie of natural uranium is here considered to correspond to 2.94 g and a curie of natural thorium to 9.0 g of these materials (see NBS Handbook 69, p. 14).* For soluble nat-U, ^{238}U, ^{236}U, and ^{235}U these limits are based on considerations of chemical toxicity (see ICRP Publication 6, p. 13).**

* NBS Handbook 69, *Maximum Permissible Body Burdens and Maximum Permissible Concentrations of Radionuclides in Air and Water for Occupational Exposure*, 1959 (available from the Superintendent of Documents, U.S. Government Printing Office, Washington, D.C. 20402). NCRP Report No. 32, *Radiation Protection in Educational Institutions*, 1966 (NCRP Publications, P.O. Box 4867, Washington, D.C. 20008).

** *Recommendations of the International Commission on Radiological Protection*, ICRP Publication 2 (Pergamon Press, London, 1959) or with the bibliography, *Healthy Phys.* **3**, 1–380 (June 1960); ICRP Publication 6 (A Pergamon Press Book, 1964, distributed by The Macmillan Co., New York).

CRYOGENIC PROPERTIES OF GASES

Property	Conditions	He	Ne	Ar	Kr	Xe	H_2	CH_4	NH_3	N_2	O_2	F_2
Density	32°F, 1 atm, lb/ft³	0.01114	0.0562	0.1113	0.234	0.368	0.00561	0.0448	0.0481	0.0781	0.0892	0.106
	0°C, 1 atm, kg/m³	0.1784	0.9002	1.783	3.748	5.895	0.0899	0.718	0.770	1.251	1.429	1.698
Boiling point	°F, 1 atm	-452.08	-410.89	-302.3	-242.1	-160.8	-423.2	-263.2	-28.03	-320.4	-297.35	-306.7
	°C, 1 atm	-268.934	-246.048	-185.7	-152.90	-107.1	-252.87	-164.0	-33.35	-195.8	-182.97	-188.14
	°K, 1 atm	4.216	27.10	87.45	120.25	166.05	20.28	109.15	239.80	77.35	90.18	85.01
Melting point	°F, 1 atm	-458.0[a]	-415.6	-308.6	-249.9	-169.4	-434.5	-296.46	-107.9	-345.87	-361.1	-363.3
	°C, 1 atm	-272.2[a]	-248.67	-189.2	-156.6	-111.9	-259.14	-182.48	-77.7	-209.86	-218.4	-219.62
	°K, 1 atm	0.95[a]	24.48	83.95	116.55	161.25	14.01	90.67	195.45	63.29	54.75	53.53
Vapor density at boiling point	lb/ft³	0.999	0.593	0.368	0.518	0.606	0.0830	0.1124	0.0556	0.288	0.279	
	kg/m³	16.002	9.499	5.895	8.298	9.707	1.329	1.8004	0.8906	4.613	4.4692	
Liquid density at boiling point	lb/ft³	7.803	74.91	86.77	149.8	193.5	4.37	26.47	42.58	50.19	71.23	94.4
	kg/m³	125.	1200.	1390.	2400.	3100.	70.0	424.	682.1	804.	1142	1512
Vapor pressure of solid at melting point	lb/in.²		6.25	9.98	10.6	11.8	1.04	1.35	0.87	1.86	0.038	0.002
	kg/m²		323.	516	549.	612.	54.	70.	45.2	96.4	2.0	0.12
	(N/m²) × 10⁴		4.34	6.93	7.36	8.20	0.723	0.938	0.604	1.29	0.0026	0.00014
Heat of vaporization at boiling point	Btu/lb	10.3	37.4	70.0	46.4	41.4	194.4	248.4	588.6	85.7	91.588	73.7
	kcal/kg	5.72	20.8	38.9	25.8	23.0	108	138	327	47.6	50.88	40.9
	(J/kg) × 10³	23.932	87.027	162.76	107.95	96.23	451.9	577.4	1368.2	199.2	212.9	171.1
Heat of fusion at melting point	Btu/lb	1.8	7.2	12.1	7.0	5.9	25.2	26.1	152.1	11.0	5.9	5.8
	kcal/kg	1.0	4.0	6.7	3.9	3.3	14.0	14.5	84.0	6.1	3.27	3.2
	(J/kg) × 10³	4.184	16.74	28.03	16.3	13.8	58.6	60.7	351.5	25.5	13.7	13.4
C_p	59°F, 1 atm, Btu/lb-°F or 15°C, 1 atm, kcal/kg-°C	1.25[b]	0.25[c]	0.125	0.06[c]	0.04[c]	3.39	0.528	0.523	0.248	0.220	0.180
	288.15°K, 1 atm, (J/kg) × 10³	5.23[b]	1.05[c]	0.523	0.251[c]	0.167[c]	14.2	2.21	2.188	1.038	0.9205	0.753
C_p/C_v	15–20°C, 1 atm 288–293°K, 1 atm	1.66[b]	1.64[c]	1.67	1.68[c]	1.66[c]	1.41	1.31	1.31	1.40	1.40	
Critical temperature	°F	-450.2	-397.7	-188.5	-82.7	61.9	-399.8	-116.5	270.3	-232.8	-181.3	-200.2
	°C	-267.9	-228.7	-122.5	-63.7	16.6	-239.9	-82.5	132.4	-147.1	-118.57	-129.0
	°K	5.25	44.45	150.65	209.45	289.75	33.25	190.65	405.55	126.05	154.58	144.15
Critical pressure	lb/in.² (absolute)	33.2	394.6	705.2	798	855	188.1	672	1639	492.3	731.4	808.3
	kg/cm²	2.33	27.7	49.6	46.1	60.1	13.2	47.2	115.5	34.6	51.4	56.8
	(kg/m²) × 10³	23.3	277	496	561	601	132	472	1155	346	514	568
	(N/m²) × 10⁴	23.1	274.1	489.9	554.4	594	130.7	466.9	1139	342.	508.1	561.6

Note: For conversion factors see p. B-439.

[a] At 26 atmospheres.
[b] At -292°F or -180°C.
[c] Approximate.

CONVERSION FACTORS FOR TABLE OF CRYOGENIC PROPERTIES OF GASES

To convert from	To	Multiply by
lbs ft^{-3}	kg m^{-3}	16.018
lbs $in.^{-2}$	N m^{-2}	6894.8
lbs ft^{-2}	N m^{-2}	47.880
BTU lb^{-1}	J kg^{-1}	2324.4
cal g^{-1}	J kg^{-1}	4184
cal g^{-1} °F	J kg^{-1} °C	4184

VISCOSITY AND THERMAL CONDUCTIVITY OF NITROGEN AT CRYOGENIC TEMPERATURES

The viscosity and thermal conductivity of nitrogen gas for the temperature range 5 K–135 K have been computed from the second Chapman-Enskog approximation. Quantum effects, which become appreciable at the lower temperatures, are included by utilizing collision integrals based on quantum theory. A Lennard-Jones (12-6) potential was assumed. The computations yield viscosities about 20% lower than those predicted for the high end of this temperature range by the method of corresponding states, but the agreement is excellent when the computed values are compared with existing experimental data.

T, °K	η, micropoise	λ, $\frac{\mu cal}{cm\ sec\ °K}$
4.575	4.16639	1.10990
5.49	4.85525	1.29307
6.405	5.54773	1.47715
7.320	6.24722	1.63156
8.235	6.95025	1.85020
9.15	7.65067	2.03666
13.725	10.95231	2.91729
18.3	13.85738	3.69313
22.875	16.55204	4.41068
27.45	19.21670	5.11847
32.025	21.9423	5.84208
36.60	24.7606	6.59077
41.175	27.67445	7.36545
45.75	30.67422	8.16348
54.9	36.8747	9.81383
64.05	43.2469	11.51005
82.35	56.14634	14.94298
91.5	62.54767	16.64610
137.25	92.96598	24.74580

From Pearson, W. E., NASA Technical Note D-7565, National Aeronautics and Space Administration, 1974 (available from Superintendent of Documents, U.S. Government Printing Office, Washington, D.C.).

DEFINITIVE RULES FOR NOMENCLATURE OF ORGANIC CHEMISTRY

IUPAC Rules

These rules were taken from the IUPAC's *Nomenclature of Organic Chemistry*, 1979 Edition, published by Pergamon Press, Inc., Maxwell House, Fairview Park, Elmford, New York 10523, U.S.A. Permission to reproduce this information was granted by IUPAC. Such permission is gratefully acknowledged.

The IUPAC rules cover a wide variety of organic compounds. Those types of compounds for which rules of nomenclature are presented in the IUPAC publication are:

Acyclic hydrocarbons
Monocyclic hydrocarbons
Fused polycyclic hydrocarbons
Bridged hydrocarbons
Spiro hydrocarbons
Hydrocarbon ring assemblies
Cyclic hydrocarbons with side chains
Terpene hydrocarbons
Heterocyclic spiro compounds
Heterocyclic ring assemblies
Bridged heterocyclic systems
Halogen derivatives
Alcohols, phenols and their derivatives
Aldehydes, ketones and their derivatives
Carboxylic acids and their derivatives
Compounds containing bivalent sulfur
Sulfur halides, sulfoxides, sulfones, and sulfur acids and their derivatives
Compounds containing selenium or tellurium linked to an organic radical
Groups containing one nitrogen atom
Groups containing more than one nitrogen atom
Radical names
Coordination compounds
Organometallic compounds
Chains and rings with regular patterns of hetero atoms
Organic compounds containing phosphorus, arsenic, antimony, or bismuth
Organosilicon compounds
Organoboron compounds
Sterochemistry:
- Types of isomerism
- *cis-trans* isomerism
- Fused rings
- Chirality
- Conformations
- Stereoformulae

General rules for naming natural products and related compounds
Isotopically modified compounds, symbols, definitions and formulae for
Names for isotopically modified compounds
Numbering of isotopically modified compounds
Locants for nuclides in isotopically modified compounds

The information which follows and which was taken from the IUPAC publication deals nearly exclusively with the nomenclature of hydrocarbons.

A. HYDROCARBONS

ACYCLIC HYDROCARBONS

Rule A-1. Saturated Unbranched-chain Compounds and Univalent Radicals

1.1—The first four saturated unbranched acyclic hydrocarbons are called methane, ethane, propane, and butane. Names of the higher members of these series consist of a numerical term, followed by "-ane" with elision of terminal "a" from the numerical term. Examples of these names are shown in the table below. The generic name of saturated acyclic hydrocarbons (branched or unbranched) is "alkane".

Examples of names:

(n = total number of carbon atoms)

n	*n*	*n*
1 Methane	15 Pentadecane	29 Nonacosane
2 Ethane	16 Hexadecane	30 Triacontane
3 Propane	17 Heptadecane	31 Hentriacontane
4 Butane	18 Octadecane	32 Dotriacontane
5 Pentane	19 Nonadecane	33 Tritriacontane
6 Hexane	20 Icosane*	40 Tetracontane
7 Heptane	21 Henicosane	50 Pentacontane
8 Octane	22 Docosane	60 Hexacontane
9 Nonane	23 Tricosane	70 Heptacontane
10 Decane	24 Tetracosane	80 Octacontane
11 Undecane	25 Pentacosane	90 Nonacontane
12 Dodecane	26 Hexacosane	100 Hectane
13 Tridecane	27 Heptacosane	132 Dotriacontahectane
14 Tetradecane	28 Octacosane	

1.2—Univalent radicals derived from saturated unbranched acyclic hydrocarbons by removal of hydrogen from a terminal carbon atom are named by replacing the ending "-ane" of the name of the hydrocarbon by "-yl". The carbon atom with the free valence is numbered as 1. As a class, these radicals are called normal, or unbranched chain, alkyls.

Examples:

Pentyl $\overset{5}{CH_3}-\overset{4}{CH_2}-\overset{3}{CH_2}-\overset{2}{CH_2}-\overset{1}{CH_2}-$

Undecyl $\overset{11}{CH_3}-\overset{10-2}{[CH_2]_9}-\overset{1}{CH_2}-$

Rule A-2. Saturated Branched-chain Compounds and Univalent Radicals

2.1—A saturated branched acyclic hydrocarbon is named by prefixing the designations of the side chains to the name of the longest chain present in the formula.

Example:

$$\overset{5}{CH_3}-\overset{4}{CH_2}-\overset{3}{CH}(CH_3)-\overset{2}{CH_2}-\overset{1}{CH_3}$$

3-Methylpentane

The following names are retained for unsubstituted hydrocarbons only:

Isobutane $(CH_3)_2CH-CH_3$
Isopentane $(CH_3)_2CH-CH_2-CH_3$
Neopentane $(CH_3)_4C$
Isohexane $(CH_3)_2CH-CH_2-CH_2-CH_3$

2.2—The longest chain is numbered from one end to the other by Arabic numerals, the direction being so chosen as to give the lowest numbers possible to the side chains. When series of locants containing the same number of terms are compared term by term, that series is "lowest" which contains the lowest number on the occasion of the first difference. This principle is applied irrespective of the nature of the substituents.

Examples:

$$\overset{5}{C}H_3-\overset{4}{C}H_2-\overset{3}{C}H-\overset{2}{C}H_2-\overset{1}{C}H_3$$
$$|$$
$$CH_3$$

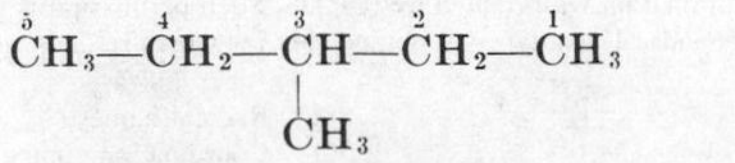

3-Methylpentane

$$\overset{6}{C}H_3-\overset{5}{C}H-\overset{4}{C}H_2-\overset{3}{C}H-\overset{2}{C}H-\overset{1}{C}H_3$$
$$|\qquad\qquad|\quad\ |$$
$$CH_3\qquad CH_3\ CH_3$$

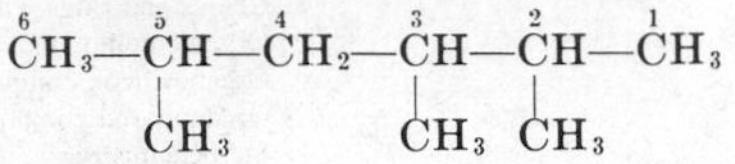

2,3,5-Trimethylhexane (not 2,4,5-Trimethylhexane)

$$\overset{10}{C}H_3-\overset{9}{C}H_2-\overset{8}{C}H-\overset{7}{C}H-\overset{6}{C}H_2-\overset{5}{C}H_2-\overset{4}{C}H_2-\overset{3}{C}H_2-\overset{2}{C}H-\overset{1}{C}H_3$$
$$|\quad\ |\qquad\qquad\qquad\qquad\qquad\ \ |$$
$$CH_3\ CH_3\qquad\qquad\qquad\qquad\quad CH_3$$

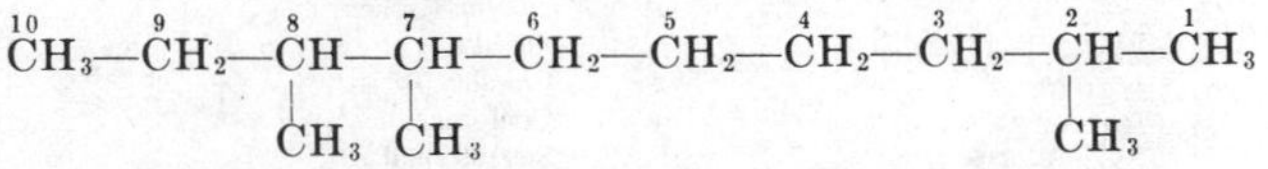

2,7,8-Trimethyldecane (not 3,4,9-Trimethyldecane)

$$\overset{9}{C}H_3-\overset{8}{C}H_2-\overset{7}{C}H_2-\overset{6}{C}H_2-\overset{5}{C}H-\overset{4}{C}H-\overset{3}{C}H_2-\overset{2}{C}H_2-\overset{1}{C}H_3$$
$$|\qquad\ |$$
$$CH_3\quad CH_2-CH_2-CH_3$$

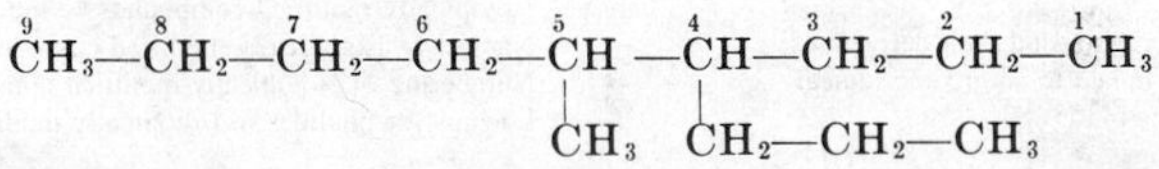

5-Methyl-4-propylnonane (not 5-Methyl-6-propylnonane since 4,5 is lower than 5,6)

2.25—Univalent branched radicals derived from alkanes are named by prefixing the designation of the side chains to the name of the unbranched alkyl radical possessing the longest possible chain starting from the carbon atom with the free valence, the said atom being numbered as 1.

Examples:

1-Methylpentyl	$\overset{5}{C}H_3\overset{4}{C}H_2\overset{3}{C}H_2\overset{2}{C}H_2\overset{1}{C}H(CH_3)-$
2-Methylpentyl	$CH_3CH_2CH_2CH(CH_3)CH_2-$
5-Methylhexyl	$(CH_3)_2CHCH_2CH_2CH_2CH_2-$

The following names may be used for the unsubstituted radicals only:

Isopropyl	$(CH_3)_2CH-$
Isobutyl	$(CH_3)_2CHCH_2-$
sec-Butyl	CH_3CH_2CH-
	$\qquad\qquad\ \ |$
	$\qquad\qquad\ CH_3$
tert-Butyl	$(CH_3)_3C-$
Isopentyl	$(CH_3)_2CHCH_2CH_2-$
Neopentyl	$(CH_3)_3CCH_2-$
	$\qquad\qquad\ \ |$
	$\qquad\qquad\ CH_3$
tert-Pentyl	CH_3CH_2C-
	$\qquad\qquad\ \ |$
	$\qquad\qquad\ CH_3$
Isohexyl	$(CH_3)_2CHCH_2CH_2CH_2-$

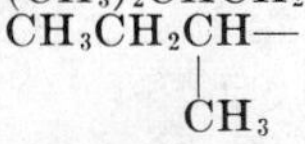

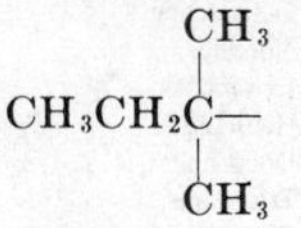

2.3—If two or more side chains of different nature are present, they are cited in alphabetical order.* The alphabetical order is decided as follows:
(i) The names of simple radicals are first alphabetized and the multiplying prefixes are then inserted.

Example:

$$\qquad\qquad\qquad\qquad CH_3-CH_2\ \ CH_3$$
$$\qquad\qquad\qquad\qquad\qquad\quad |\qquad |$$
$$\overset{7}{C}H_3-\overset{6}{C}H_2-\overset{5}{C}H_2-\overset{4}{C}H-\overset{3}{C}-\overset{2}{C}H_2-\overset{1}{C}H_3$$
$$\qquad\qquad\qquad\qquad\qquad |$$
$$\qquad\qquad\qquad\qquad\quad CH_3$$

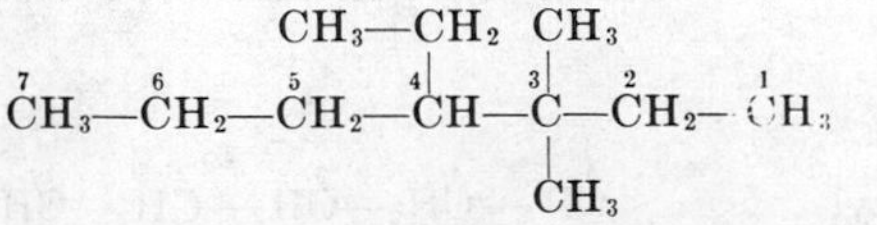

ethyl is cited before methyl, thus 4-Ethyl-3.3-dimethylheptane

(ii) The name of a complex radical is considered to begin with the first letter of its complete name.

Example:

$$\qquad\qquad\qquad\qquad\qquad CH_3$$
$$\qquad\qquad\qquad\qquad\qquad |$$
$$\qquad\qquad CH_3-\overset{1}{C}H-\overset{2}{C}H-\overset{3}{C}H_2-\overset{4}{C}H_2-\overset{5}{C}H_3$$
$$\overset{13}{C}H_3-[\overset{12-8}{C}H_2]_5-\overset{7}{C}H-\overset{6}{C}H_2-\overset{5}{C}H-\overset{4}{C}H_2-\overset{3}{C}H_2-\overset{2}{C}H_2-\overset{1}{C}H_3$$
$$\qquad\qquad\qquad\qquad\qquad\quad |$$
$$\qquad\qquad\qquad\qquad\quad CH_2-CH_3$$

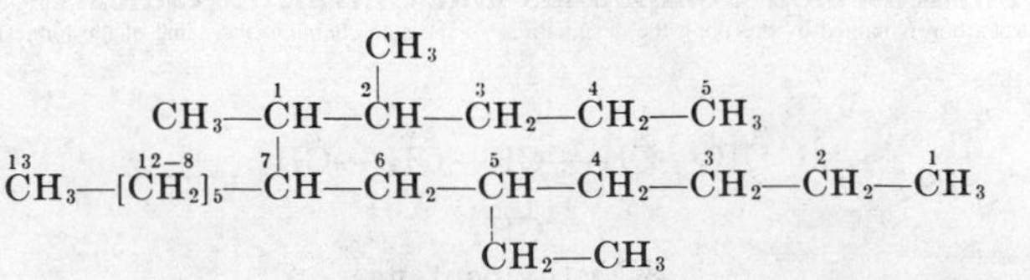

dimethylpentyl (as a complete single substituent) is alphabetized under "d", thus 7-(1.2-Dimethylpentyl)-5-ethyltridecane

(iii) In cases where names of complex radicals are composed of identical words, priority for citation is given to that radical which contains the lowest locant at the first cited point of difference in the radical.

* Use of an order of complexity given as alternative in the First and Second Editions is abandoned.

Example:

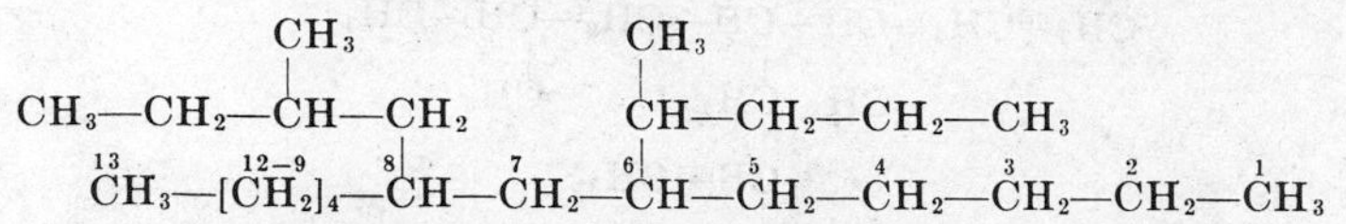

6-(1-Methylbutyl)-8-(2-Methylbutyl)tridecane

2.3—If two or more side chains are in equivalent positions, the one to be assigned the lower number is that cited first in the name.

Examples:

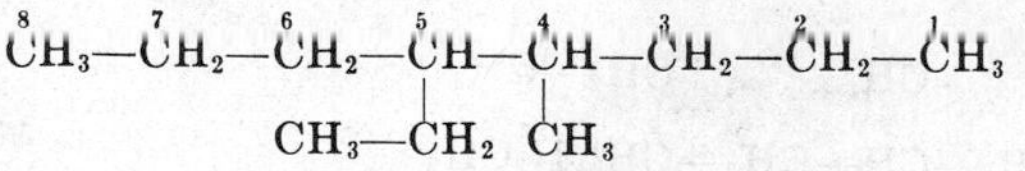

4-Ethyl-5-methyloctane

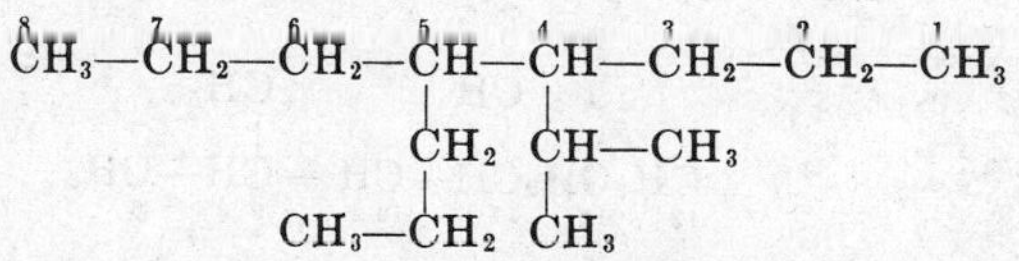

4-Isopropyl-5-propyloctane

2.5—The presence of identical unsubstituted radicals is indicated by the appropriate multiplying prefix di-, tri-, tetra-, penta-, hexa-, hepta-, octa-, nona-, deca-, undeca-, etc.

Example:

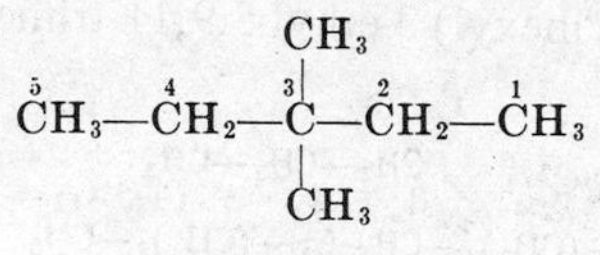

3,3-Dimethylpentane

The presence of identical radicals each substituted in the same way may be indicated by the appropriate multiplying prefix bis-, tris-, tetrakis-, pentakis-, etc. The complete expression denoting such a side chain may be enclosed in parentheses or the carbon atoms in side chains may be indicated by primed numbers.

Examples:

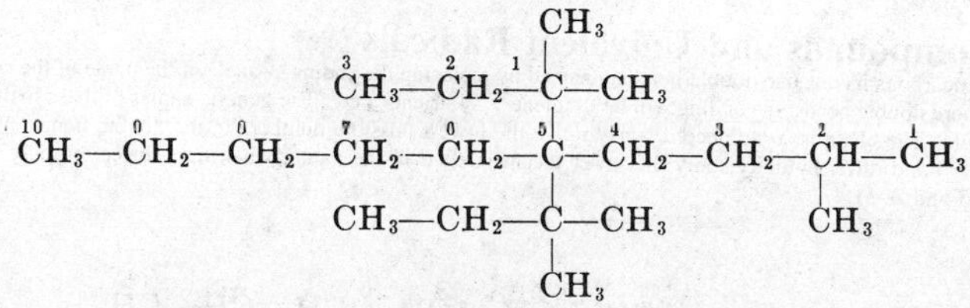

(a) Use of parentheses and unprimed numbers: 5,5-Bis(1,1-dimethylproply)-2-methyldecane
(b) Use of primes: 5,5-Bis-1′,1′-dimethylpropyl-2-methyldecane

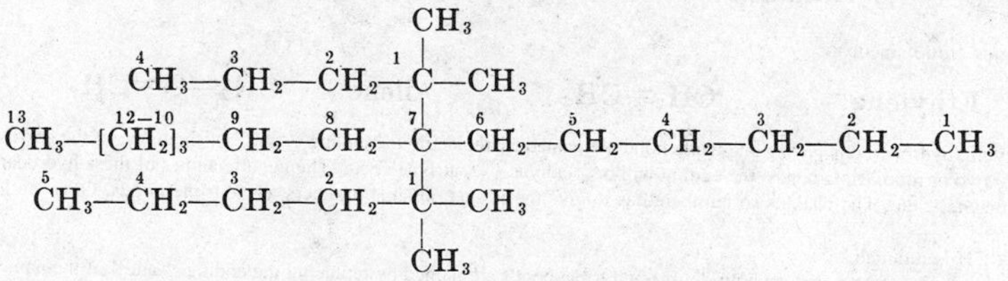

(a) Use of parentheses and unprimed numbers: 7-(1,1-Dimethylbutyl)-7-(1,1-dimethylpentyl)tridecane
(b) Use of primes: 7-1′,1′-Dimethylbutyl-7-1″,1″-dimethylpentyltridecane

2.6—If chains of equal length are competing for selection as main chain in a saturated branched acyclic hydrocarbon, then the choice goes in series to:

(a) The chain which has the greatest number of side chains.

Example:

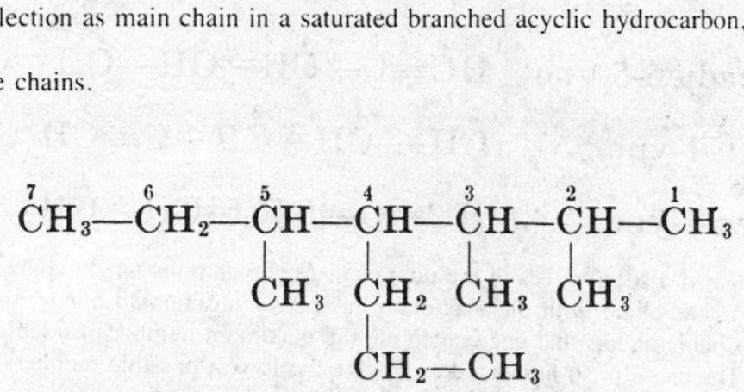

2,3,5-Trimethyl-4-propylheptane

(b) The chain whose side chains have the lowest-numbered locants.

Example:

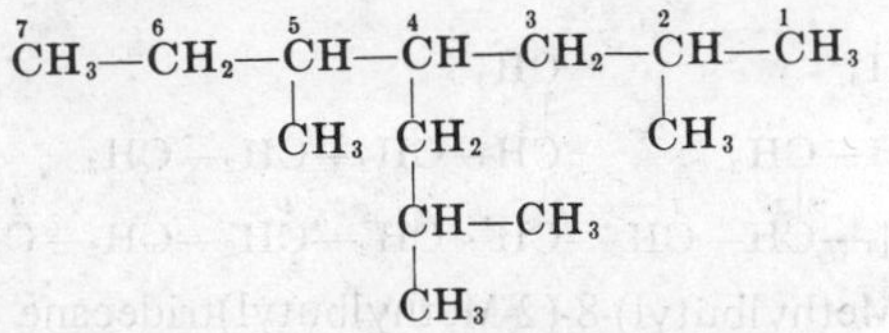

4-Isobutyl-2,5-dimethylheptane

(c) The chain having the greatest number of carbon atoms in the smaller side chains.

Example*:

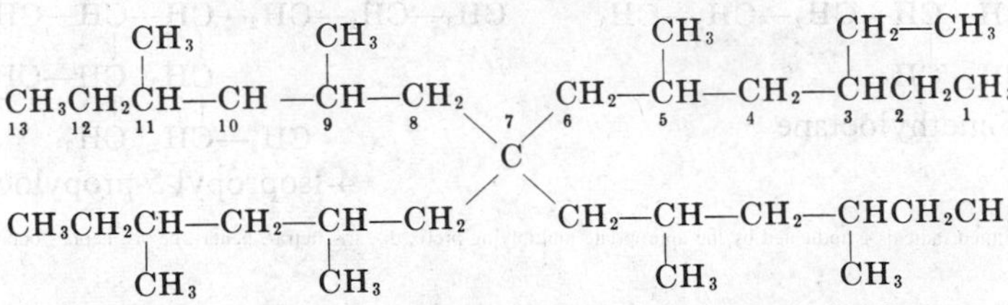

7,7-Bis(2,4-dimethylhexyl)-3-ethyl-5,9,11-trimethyltridecane

(d) The chain having the least branched side chains.

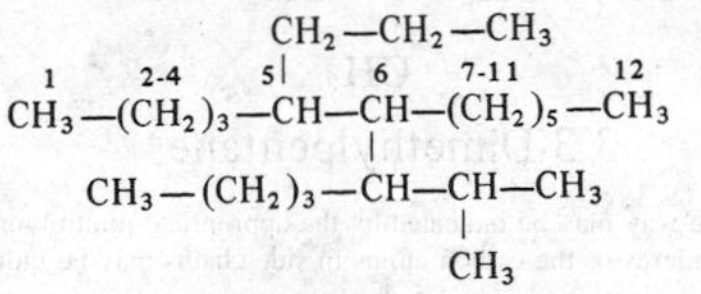

6-(1-Isopropylpentyl)-5-propyldodecane

Rule A-3. Unsaturated Compounds and Univalent Radicals

3.1—Unsaturated unbranched acyclic hydrocarbons having one double bond are named by replacing the ending "-ane" of the name of the corresponding saturated hydrocarbon with the ending "-ene". If there are two or more double bonds, the ending will be "-adiene", "-atriene", etc. The generic names of these hydrocarbons (branched or unbranched) are "alkene", "alkadiene", "alkatriene", etc. The chain is so numbered** as to give the lowest possible numbers to the double bonds. When, in cyclic compounds or their substitution products, the locants of a double bond differ by unity, only the lower locant is cited in the name; when they differ by more than unity, one locant is placed in parentheses after the other (see Rules **A-31.3** and **A-31.4**).

Examples:

2-Hexene	$\overset{6}{C}H_3-\overset{5}{C}H_2-\overset{4}{C}H_2-\overset{3}{C}H=\overset{2}{C}H-\overset{1}{C}H_3$
1,4-Hexadiene	$\overset{6}{C}H_3-\overset{5}{C}H=\overset{4}{C}H-\overset{3}{C}H_2-\overset{2}{C}H=\overset{1}{C}H_2$

The following nonsystematic names are retained:

Ethylene $CH_2=CH_2$ Allene $CH_2=C=CH_2$

3.2—Unsaturated unbranched acyclic hydrocarbons having one triple bond are named by replacing the ending "-ane" of the name of the corresponding saturated hydrocarbon with the ending "-yne". If there are two or more triple bonds, the ending will be "-adiyne", "-atriyne", etc. The generic names of these hydrocarbons (branched or unbranched) are "alkyne", "alkadiyne", "alkatriyne", etc. The chain is so numbered as to give the lowest possible numbers to the triple bonds. Only the lower locant for a triple bond is cited in the name of a compound.

The name "acetylene" for $HC\equiv CH$ is retained.

3.3—Unsaturated unbranched acyclic hydrocarbons having both double and triple bonds are named by replacing the ending "-ane" of the name of the corresponding saturated hydrocarbon with the ending "-enyne", "-adienyne", "-atrienyne", "-enediyne", etc. Numbers as low as possible are given to double and triple bonds even though this may at times give "-yne" to lower number than "-ene". When there is a choice in numbering, the double bonds are given the lowest numbers.

Examples:

1,3-Hexadien-5-yne	$H\overset{6}{C}\equiv\overset{5}{C}-\overset{4}{C}H=\overset{3}{C}H-\overset{2}{C}H=\overset{1}{C}H_2$
3-Penten-1-yne	$\overset{5}{C}H_3-\overset{4}{C}H=\overset{3}{C}H-\overset{2}{C}\equiv\overset{1}{C}H$
1-Penten-4-yne	$H\overset{5}{C}\equiv\overset{4}{C}-\overset{3}{C}H_2-\overset{2}{C}H=\overset{1}{C}H_2$

3.4—Unsaturated branched acyclic hydrocarbons are named as derivatives of the unbranched hydrocarbons which contain the maximum number of double and triple bonds. If there are two or more chains competing for selection as the chain with the maximum number of unsaturated bonds, then the choice goes to (1) that one with the greatest number of carbon atoms; (2) the number of carbon atoms being equal, that one containing the maximum number of double bonds. In other respects, the same principles apply as for naming saturated branched acyclic hydrocarbons. The chain is so numbered as to give the lowest possible numbers to double and triple bonds in accordance with Rule **A-3.3.**

* Here the choice lies between two possible main chains of equal length, each containing six side chains in the same positions. Listing in increasing order, the number of carbon atoms in the several side chains of the first choice as shown and of the alternate second choice results as follows:

first choice	1, 1, 1, 2, 8, 8
second choice	1, 1, 1, 1, 8, 9

** The expression, "the greatest number of carbon atoms in the smaller side chains", is taken to mean the largest side chain at the first point of difference when the size of the side chains is examined step by step. Thus, the selection in this case is made at the fourth step where 2 is greater than 1.
Only the lower locant for a double bond is cited in the name of an acyclic compound.

Examples:

3,4-Dipropyl-1,3-hexadien-5-yne

$$CH_2-CH_2-CH_3$$
$$\underset{6}{CH}\equiv \underset{5}{C}-\underset{4}{C}=\underset{3}{C}-\underset{2}{CH}=\underset{1}{CH_2}$$
$$CH_2-CH_2-CH_3$$

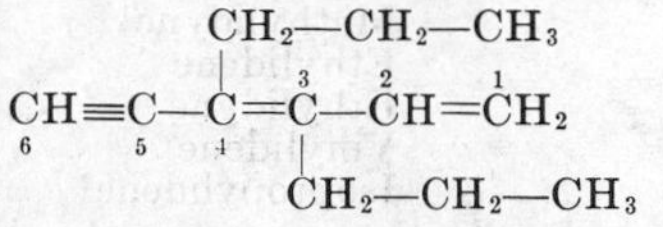

5-Ethynyl-1,3,6-heptatriene

$$\underset{7}{CH_2}=\underset{6}{CH}-\underset{5}{CH}-\underset{4}{CH}=\underset{3}{CH}-\underset{2}{CH}=\underset{1}{CH_2}$$
$$C\equiv CH$$

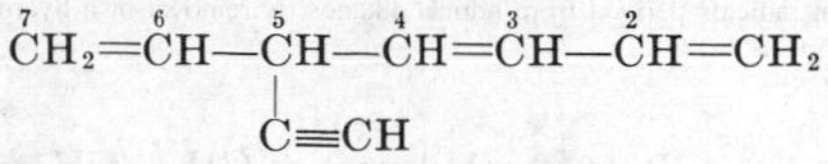

5,5-Dimethyl-1-hexene

$$CH_3$$
$$\underset{6}{CH_3}-\underset{5}{C}-\underset{4}{CH_2}-\underset{3}{CH_2}-\underset{2}{CH}=\underset{1}{CH_2}$$
$$CH_3$$

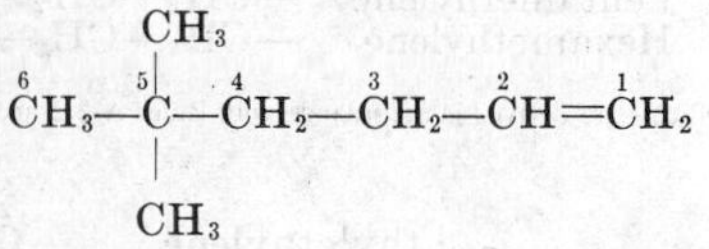

4-Vinyl-1-hepten-5-yne

$$CH_3$$
$$\underset{7}{CH_3}-\underset{6}{C}\equiv \underset{5}{C}-\underset{4}{CH}-\underset{3}{CH_2}-\underset{2}{CH}=\underset{1}{CH_2}$$
$$CH=CH_2$$

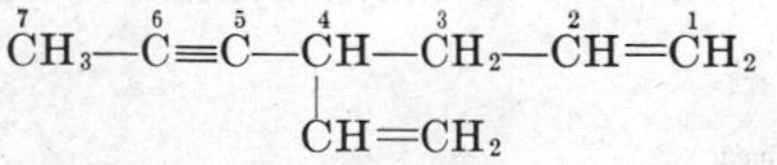

The name "isoprene" is retained for the unsubstituted compound only:

$$CH_3$$
$$CH_2=CH-C=CH_2$$

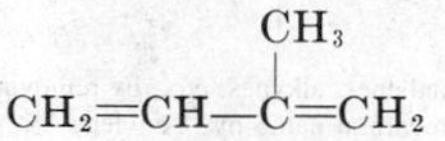

3.5—The names of univalent radicals derived from unsaturatd acyclic hydrocarbons have the endings "-enyl", "-ynyl", "-dienyl", etc., the positions of the double and triple bonds being indicated where necessary. The carbon atom with the free valence is numbered as 1.

Examples:

Ethynyl	$CH\equiv C-$
2-Propynyl	$CH\equiv C-CH_2-$
1-Propenyl	$CH_3-CH=CH-$
2-Butenyl	$CH_2=CH-CH_2-$
1,3-Butadienyl	$CH_2=CH-CH=CH-$
2-Pentenyl	$CH_3-CH_2-CH=CH-CH_2-$
2-Penten-4-ynyl	$CH\equiv C-CH=CH-CH_2-$

Exceptions: The following names are retained:

Vinyl (for ethenyl)	$CH_2=CH-$
Allyl (for 2-propenyl)	$CH_2=CH-CH_2-$
Isopropenyl	$CH_2=C-$ (for unsubstituted radical only)
(for 1-methylvinyl)	CH_3

3.6—When there is a choice for the fundamental chain of a radical, that chain is selected which contains (1) the maximum number of double and triple bonds; (2) the largest number of carbon atoms; and (3) the largest number of double bonds.

Examples:

$$\underset{10}{CH_3}-\underset{9}{CH}=\underset{8}{CH}-\underset{7}{CH}=\underset{6}{CH}-\underset{5}{CH}-\underset{4}{CH}=\underset{3}{CH}-\underset{2}{C}\equiv \underset{1}{C}-$$
$$CH_2-CH_2-CH=CH-CH_3$$

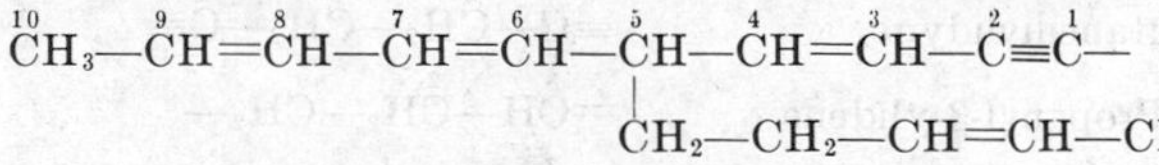

5-(3-Pentenyl)-3,6,8-decatrien-1-ynyl

$$\underset{12}{CH_3}-\underset{11}{CH_2}-\underset{10}{C}\equiv \underset{9}{C}-\underset{8}{CH}=\underset{7}{CH}-\underset{6}{CH}-\underset{5}{CH}=\underset{4}{CH}-\underset{3}{CH}=\underset{2}{CH}-\underset{1}{CH_2}-$$
$$CH=CH-CH=CH-CH_3$$

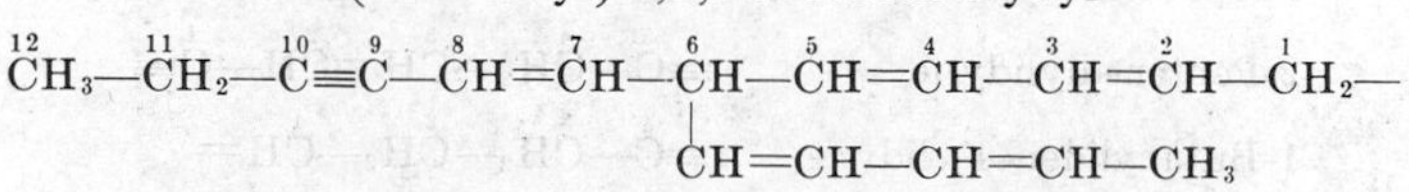

6-(1,3-Pentadienyl)-2,4,7-dodecatrien-9-ynyl

$$\underset{11}{CH_3}-\underset{10}{CH}=\underset{9}{CH}-\underset{8}{CH}=\underset{7}{CH}-\underset{6}{CH}-\underset{5}{CH}=\underset{4}{CH}-\underset{3}{CH}=\underset{2}{CH}-\underset{1}{CH_2}$$
$$CH=CH-C\equiv C-CH_3$$

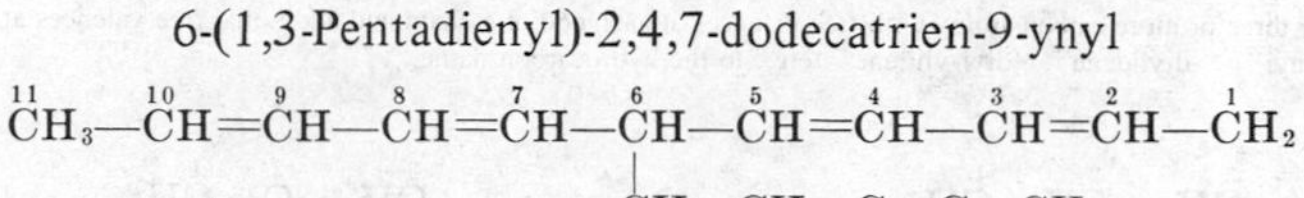

6-(1-Penten-3-ynyl)-2,4,7,9-undecatetraenyl

$$\underset{4}{CH_3}-\underset{3}{CH}=\underset{2}{C}-\underset{1}{CH_2}-$$
$$CH_2-CH_2-CH_2-CH_2-CH_2-CH_2-CH_2-CH_2-CH_3$$

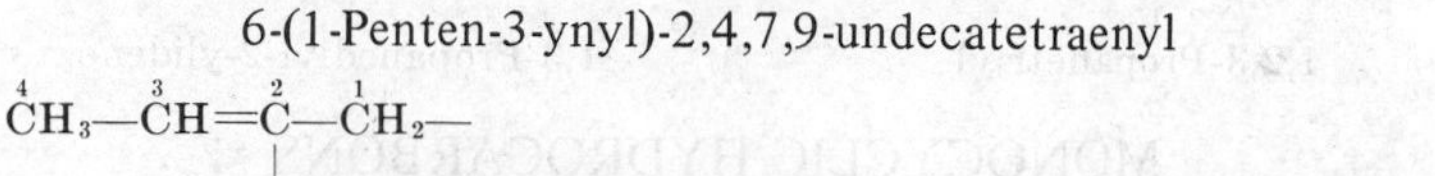

2-Nonyl-2-butenyl

Rule A-4. Bivalent and Multivalent Radicals*

4.1—Bivalent and trivalent radicals derived from univalent acyclic hydrocarbon radicals whose authorized names end in "-yl" by removal of one or two hydrogen atoms from the carbon atom with the free valences are named by adding "-idene" or "-idyne", respectively, to the name of the corresponding univalent radical. The carbon atom with the free valence is numbered as 1.

The name "methylene" is retained for the radical $CH_2=$.

* Rule **D-4.14** introduces an alternate method of naming radicals derived from any position of unbranched chains or ring systems by adding "-yl", "-diyl", "-triyl", etc. to the name of the chain or ring system with elision of "e" before "-yl". Examples: 2-pentanyl $CH_3-CH_2-CH_2-CH-CH_3$; 1,6-hexanediyl $-CH_2-(CH_2)_4-CH_2-$.

Examples:

Methylidyne[1]	$CH\equiv$
Ethylidene	$CH_3-CH=$
Ethylidyne	$CH_3-C\equiv$
Vinylidene	$CH_2=C=$
Isopropylidene[2]	$(CH_3)_2C=$

4.2—The names of bivalent radicals derived from normal alkanes by removal of a hydrogen atom from each of the two terminal carbon atoms of the chain are ethylene, trimethylene, tetramethylene, etc.

Examples:

Pentamethylene	$-CH_2-CH_2-CH_2-CH_2-CH_2-$
Hexamethylene	$-CH_2-CH_2-CH_2-CH_2-CH_2-CH_2-$

Names of the substituted bivalent radicals are derived in accordance with Rules **A-2.2** and **A-2.25**.

Example:

Ethylethylene 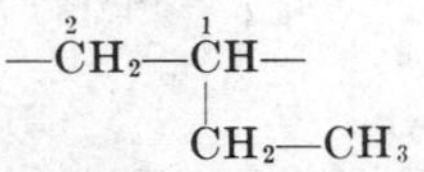

The name "propylene" is retained:

4.3—Bivalent radicals similarly derived from unbranched alkenes, alkadienes, alkynes, etc., by removing a hydrogen atom from each of the terminal carbon atoms are named by replacing the endings "-ene", "-diene", "-yne", etc., of the hydrocarbon name by "-enylene", "-dienylene", "-ynylene", etc., the positions of the double and triple bonds being indicated where necessary.

Example:

Propylene $-\overset{3}{C}H_2-\overset{2}{C}H=\overset{1}{C}H-$

The name "vinylene" is retained (for ethenylene):

$$-CH=CH-$$

Names of the substituted bivalent radicals are derived in accordance with Rule **A-3.4**.

Example:

4 Propyl-2-pentenylene 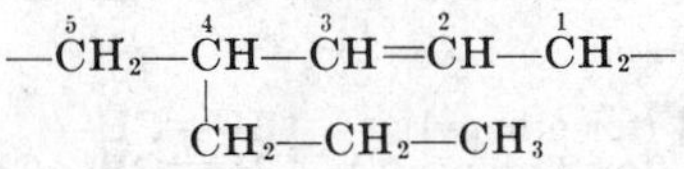

4.4—Trivalent, quadrivalent, and higher-valent acyclic hydrocarbon radicals of two or more carbon atoms with the free valences at each end of a chain are named by adding to the hydrocarbon name the terminations "-yl" for single free valence, "-ylidene" for a double, and "-ylidyne" for a triple free valence on the same atom (the final "e" in the name of the hydrocarbon is elided when followed by a suffix beginning with "-yl"). If different types are present in the same radical, they are cited and numbered in the order "-yl", "-ylidene", "-ylidyne".

Examples:

Butanediylidene	$=\overset{4}{C}H-\overset{3}{C}H_2-\overset{2}{C}H_2-\overset{1}{C}H=$
Butanediylidyne	$\equiv\overset{4}{C}-\overset{3}{C}H_2-\overset{2}{C}H_2-\overset{1}{C}\equiv$
1-Propanyl-3-ylidene	$=\overset{3}{C}H-\overset{2}{C}H_2-\overset{1}{C}H_2-$
Propadienediylidene	$=\overset{3}{C}=\overset{2}{C}=\overset{1}{C}=$
2-Pentenediylidyne	$\equiv\overset{5}{C}-\overset{4}{C}H_2-\overset{3}{C}H=\overset{2}{C}H-\overset{1}{C}\equiv$
1-Butanyliden-4-ylidyne	$\equiv\overset{4}{C}-\overset{3}{C}H_2-\overset{2}{C}H_2-\overset{1}{C}H=$

4.5—Multivalent radicals containing three or more carbon atoms with free valences at each end of a chain and additional free valences at intermediate carbon atoms are named by adding the endings "-triyl", "-tetrayl", "-diylidene", "diyl-ylidene", etc., to the hydrocarbon name.

Examples:

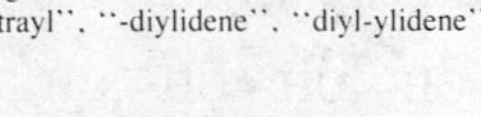

1,2,3-Propanetriyl

1,3-Propanediyl-2-ylidene

MONOCYCLIC HYDROCARBONS

Rule A-11. Unsubstituted Compounds and Radicals*

11.1—The names of saturated monocyclic hydrocarbons (with no side chains) are formed by attaching the prefix "cyclo" to the name of the acyclic saturated unbranched hydrocarbon with the same number of carbon atoms. The generic name of saturated monocyclic hydrocarbons (with or without side chains) is "cycloalkane".

1 The group =CH– may be referred to as the "methine" group.
2 For unsubstituted radical only.
* See footnote to Rule A-4.

Examples:

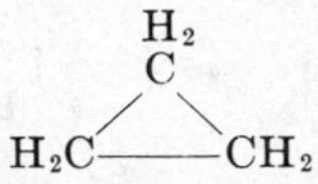

Cyclopropane

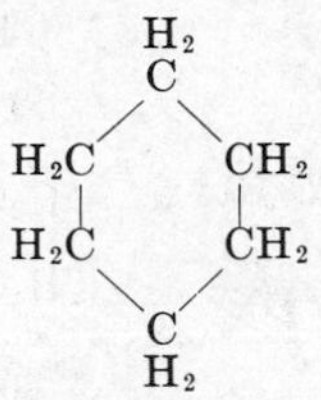

Cyclohexane

11.2—Univalent radicals derived from cycloalkanes (with no side chains) are named by replacing the ending "-ane" of the hydrocarbon name by "-yl", the carbon atom with the free valence being numbered as 1. The generic name of these radicals is "cycloalkyl".

Examples:

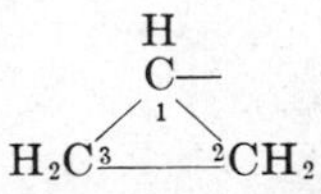

Cyclopropyl

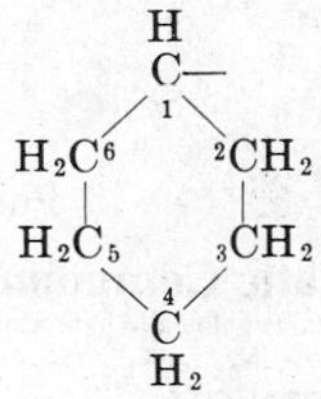

Cyclohexyl

11.3—The names of unsaturated monocyclic hydrocarbons (with no side chains) are formed by substituting "-ene", "-adiene", "-atriene", "-yne", "-adiyne", etc., for "-ane" in the name of the corresponding cycloalkane. The double and triple bonds are given numbers as low as possible as in Rule **A-3.3**.

Examples:

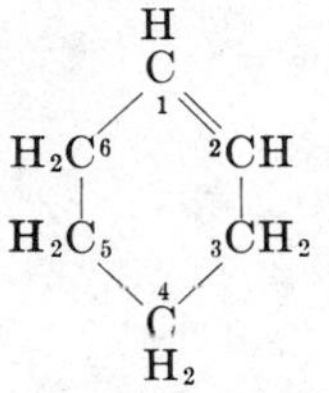

Cyclohexene

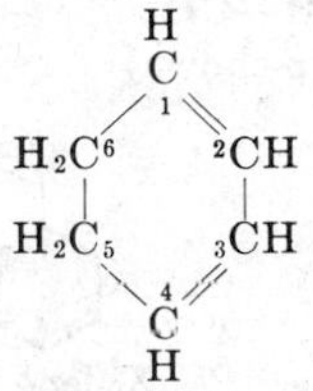

1,3-Cyclohexadiene

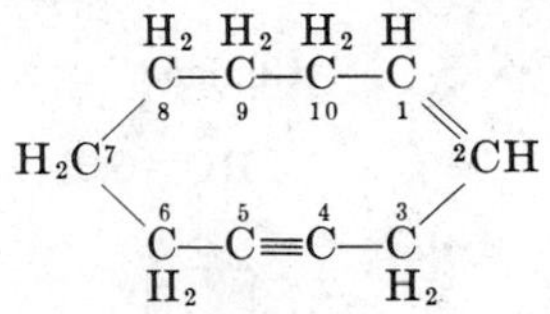

1-Cyclodecen-4-yne

The name "benzene" is retained.

11.4—The names of univalent radicals derived from unsaturated monocyclic hydrocarbons have the endings "-enyl", "-ynyl", "-dienyl", etc., the positions of the double and triple bonds being indicated according to the principles of Rule **A-3.3**. The carbon atom with the free valence is numbered as 1, except as stated in the rules for terpenes (see Rules **A-72** to **A-75**).

Examples:

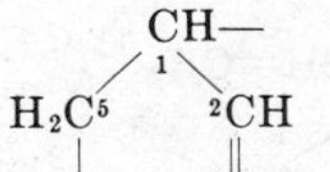

2-Cyclopenten-1-yl

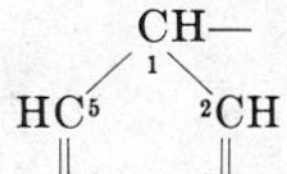

2,4-Cyclopentadien-1-yl

The radical name "phenyl" is retained.

11.5—Names of bivalent radicals derived from saturated or unsaturated monocyclic hydrocarbons by removal of two atoms of hydrogen from the same carbon atom of the ring are obtained by replacing the endings "-ane", "-ene", "-yne", by "-ylidene", "-enylidene" and "-ynylidene", respectively. The carbon atom with the free valences is numbered as 1, except as stated in the rules for terpenes.

Examples:

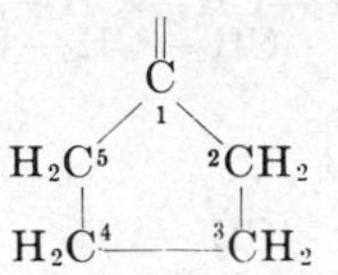

Cyclopentylidene

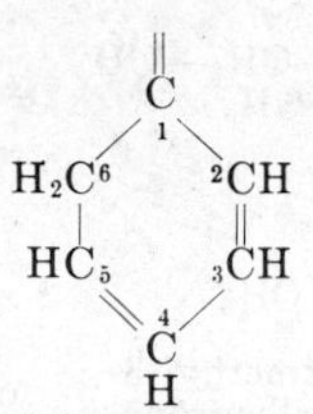

2,4-Cyclohexadien-1-ylidene

11.6—Bivalent radicals derived from saturated or unsaturated monocyclic hydrocarbons by removing a hydrogen atom from each of two different carbon atoms of the ring are named by replacing the endings "-ane", "-ene", "-diene", "-yne", etc., of the hydrocarbon name by "-ylene", "-enylene", "-dienylene", "-ynylene", etc., the positions of the double and triple bonds and of the points of attachment being indicated. Preference in lowest numbers is given to the carbon atoms having the free valences.

Examples:

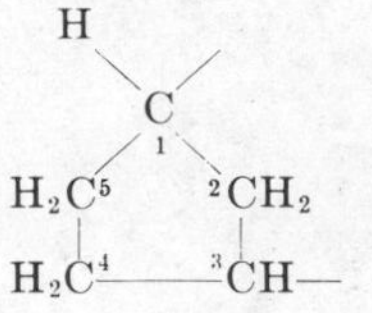

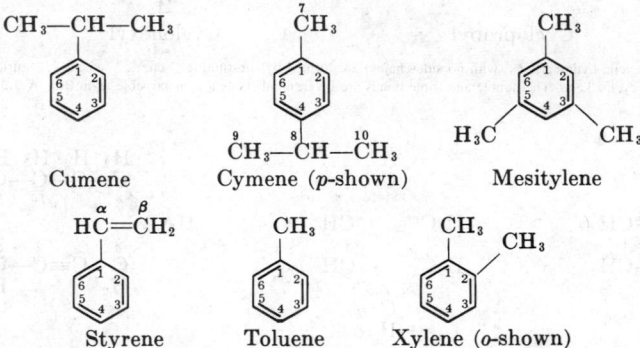

1,3-Cyclopentylene

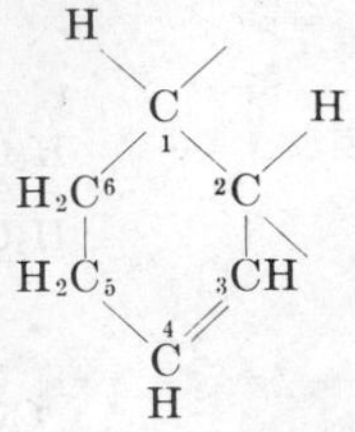

3-Cyclohexen-1,2-ylene

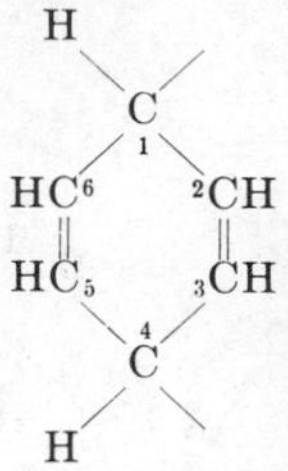

2,5-Cyclohexadien-1,4-ylene

The name "phenylene" is retained:

Phenylene (*p*-shown)

Rule A-12. Substituted Aromatic Compounds

12.1—The following names for monocyclic substituted aromatic hydrocarbons are retained:

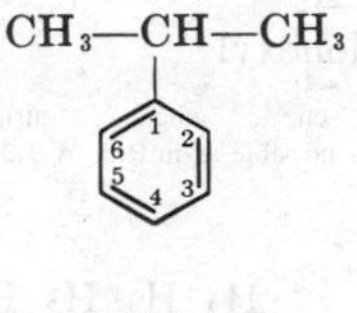

Cumene

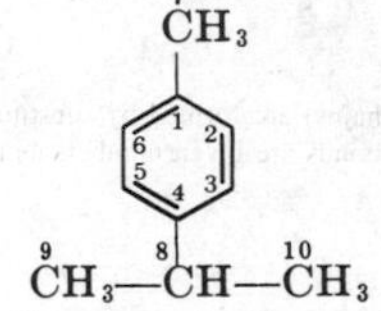

Cymene (*p*-shown)

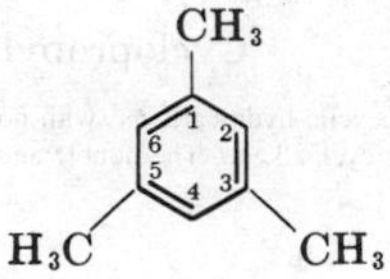

Mesitylene

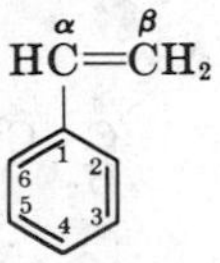

Styrene

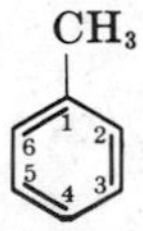

Toluene

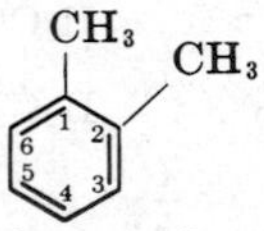

Xylene (*o*-shown)

12.2—Other monocyclic substituted aromatic hydrocarbons are named as derivatives of benzene or of one of the compounds listed in Part **.1** of this rule. However, if the substituent introduced into such a compound is identical with one already present in that compound, then the substituted compound is named as a derivative of benzene (see Rule **61.4**).

12.3—The position of substituents is indicated by numbers except that *o-(ortho),m-(meta)* and *p-(para)* may be used in place of 1,2-, 1,3-, and 1,4-, respectively, when only two substituents are present. The lowest numbers possible are given to substituents, choice between alternatives being governed by Rule **A-2** so far as applicable, except that when names are based on those of compounds listed in Part **.1** of this rule the first priority for lowest numbers is given to the substituent(s) already present in those compounds.

Examples:

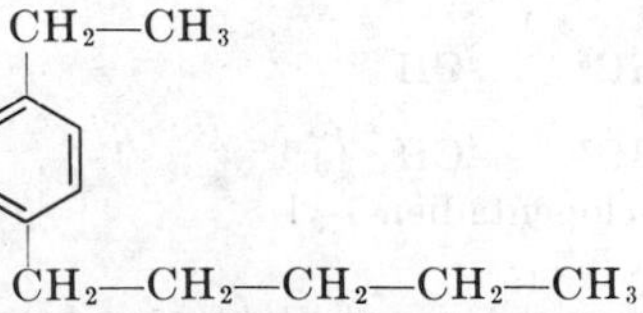

1-Ethyl-4-pentylbenzene or *p*-Ethylpentylbenzene

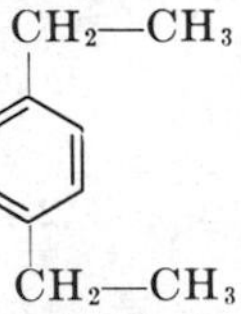

1,4-Diethylbenzene or *p*-Diethylbenzene

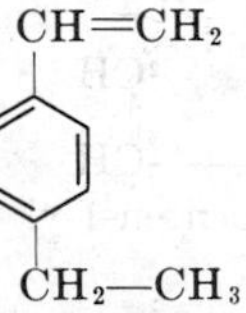

4-Ethylstyrene or *p*-Ethylstyrene

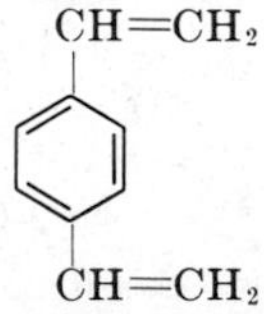

1,4-Divinylbenzene or *p*-Divinylbenzene, not *p*-Vinylstyrene

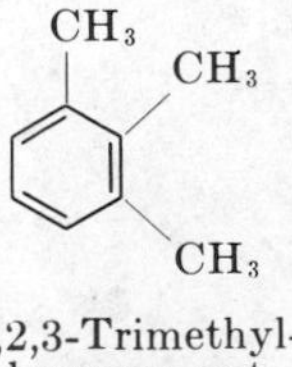

1,2,3-Trimethylbenzene, not Methylxylene nor Dimethyltoluene

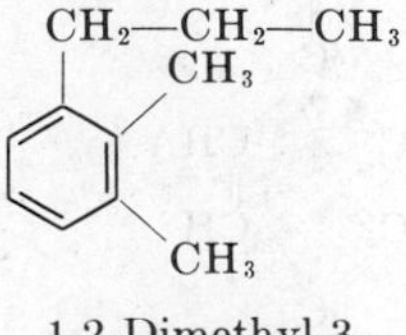

1,2-Dimethyl-3-propylbenzene or 3-Propyl-*o*-xylene

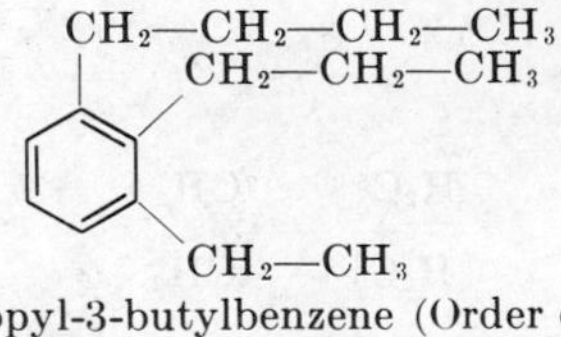

1-Ethyl-2-propyl-3-butylbenzene (Order of complexity) or 1-Butyl-3-ethyl-2-propylbenzene (Alphabetical order)

12.4—The generic name of monocyclic and polycyclic aromatic hydrocarbons is "arene".

Rule A-13. Substituted Aromatic Radicals

13.1—Univalent radicals derived from monocyclic substituted aromatic hydrocarbons and having the free valence at a ring atom are given the names listed below. Such radicals not listed below are named as substituted phenyl radicals. The carbon atom having the free valence is numbered as 1.

Phenyl C_6H_5—

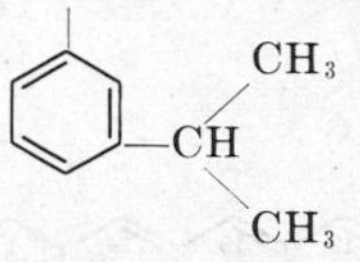

Cumenyl (*m*-shown)

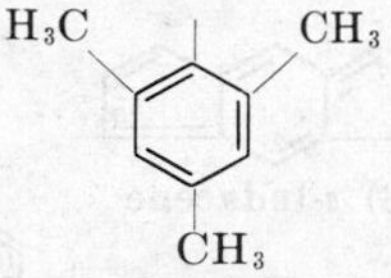

Mesityl

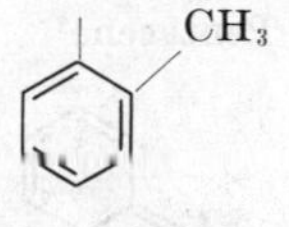

Tolyl (*o*-shown)

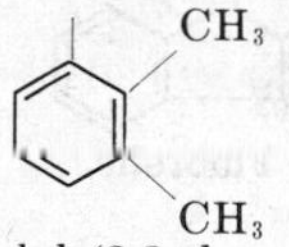

Xylyl (2,3-shown)

13.2—Since the name phenylene (*o*-, *m*- or *p*-) is retained for the radical $-C_6H_4-$ (exception to Rule **A-11.6**), bivalent radicals formed from substituted benzene derivatives and having the free valences at ring atoms are named as substituted phenylene radicals. The carbon atoms having the free valences are numbered 1,2-, 1,3-, or 1,4- as appropriate.

13.3—The following trivial names for radicals having a single free valence in the side chain are retained:

Benzyl	$C_6H_5-\overset{\alpha}{C}H_2-$
Benzhydryl (alternative to Diphenylmethyl)	$(C_6H_5)_2\overset{\alpha}{C}H-$
Cinnamyl	$C_6H_5-\overset{\gamma}{C}H=\overset{\beta}{C}H-\overset{\alpha}{C}H_2-$
Phenethyl	$C_6H_5-\overset{\beta}{C}H_2-\overset{\alpha}{C}H_2-$
Styryl	$C_6H_5-\overset{\beta}{C}H=\overset{\alpha}{C}H-$
Trityl	$(C_6H_5)_3C-$

13.4—Multivalent radicals of aromatic hydrocarbons with the free valences in the side chain are named in accordance with Rule **A-4**.

Examples:

Benzylidyne	$C_6H_5-C\equiv$
Cinnamylidene	$C_6H_5-\overset{\gamma}{C}H=\overset{\beta}{C}H-\overset{\alpha}{C}H=$

13.5—The generic names of univalent and bivalent aromatic hydrocarbon radicals are "aryl" and "arylene", respectively.

FUSED POLYCYCLIC HYDROCARBONS

Rule A-21. Trivial and Semi-trivial names

21.1—The names of polycyclic hydrocarbons with maximum number of noncumulative* double bonds end in "-ene". The names listed on pp. C-9 and C-10 are retained.

21.2—The names of hydrocarbons containing five or more fused benzene rings in a straight linear arrangement are formed from a numerical prefix as specified in Rule **A-1.1** followed by "-acene". (Examples on pp. C-9 and C-10).

Examples:

$CH_2=C=C=C=CH_2$

Cumulative

$CH_3-CH=CH-CH=CH-CH=CH_2$

or

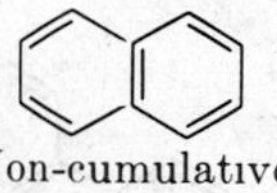

Non-cumulative

Examples (to Rule **A-21.2**):

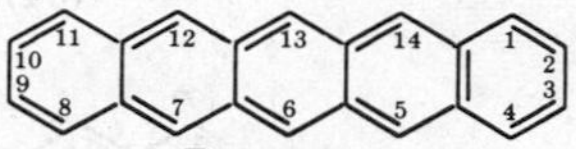

Pentacene

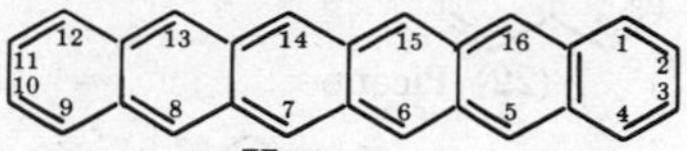

Hexacene

The following list contains the names of polycyclic hydrocarbons which are retained (see Rule **A-21.1**). This list is not limiting.

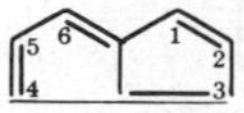

(1) Pentalene

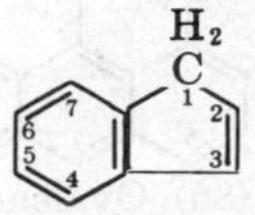

(2) Indene

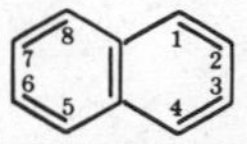

(3) Naphthalene

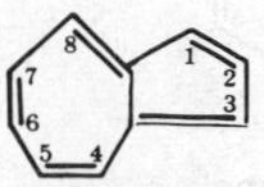

(4) Azulene

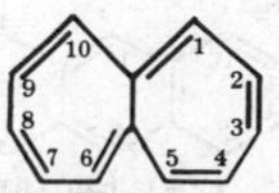

(5) Heptalene

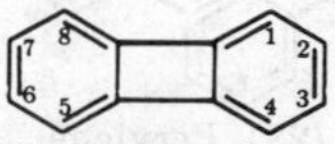

(6) Biphenylene

* Cumulative double bonds are those present in a chain in which at least three contiguous carbon atoms are joined by double bonds; non-cumulative double bonds comprise every other arrangement of two or more double bonds in a single structure. The generic name "cumulene" is given to compounds containing three or more cumulative double bonds.

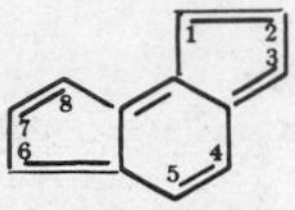

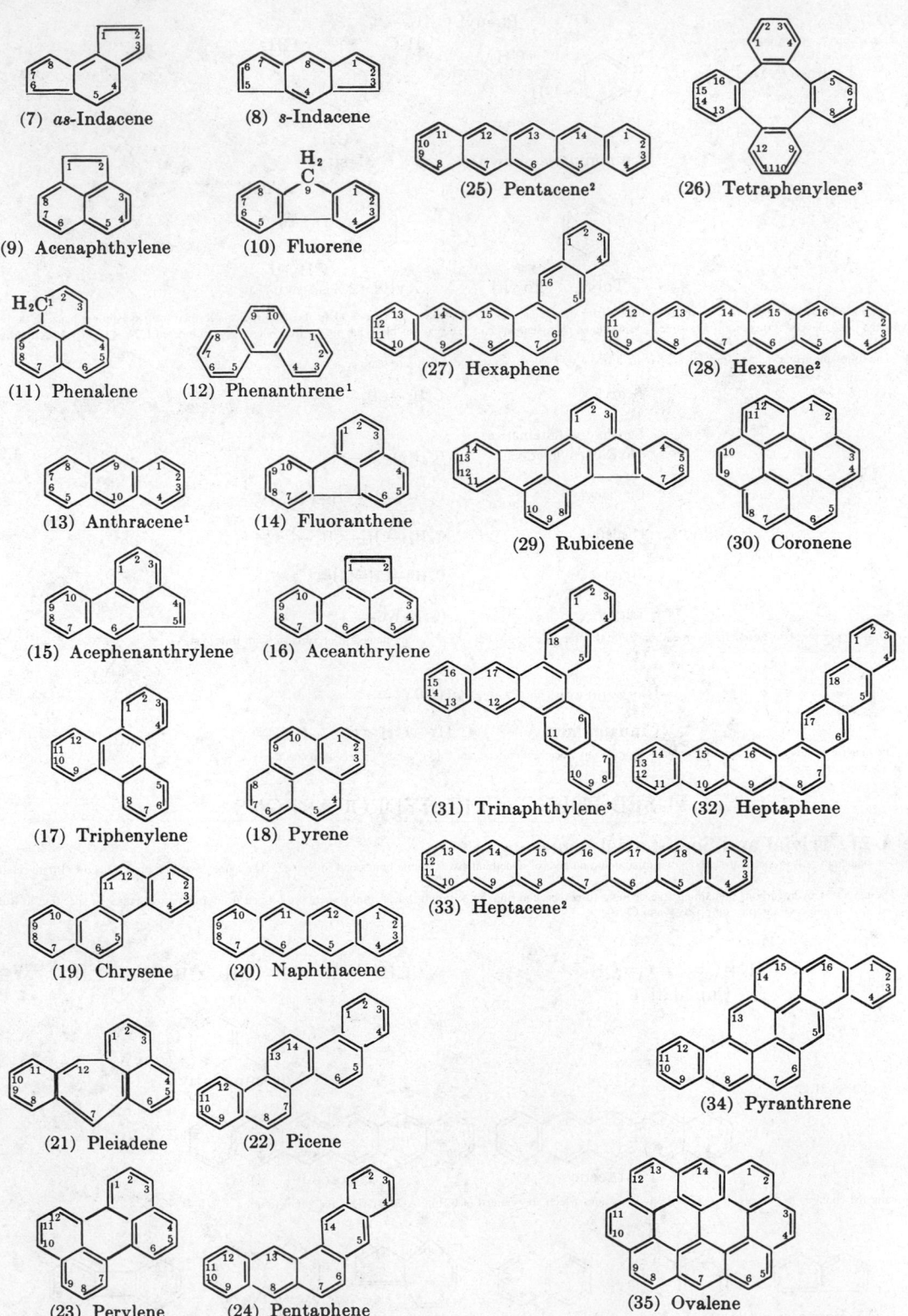

(7) *as*-Indacene

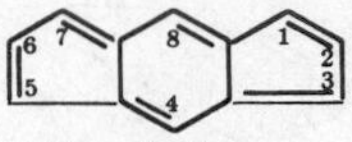

(8) *s*-Indacene

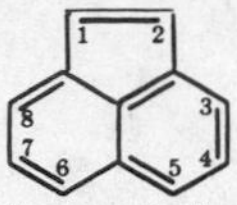

(9) Acenaphthylene

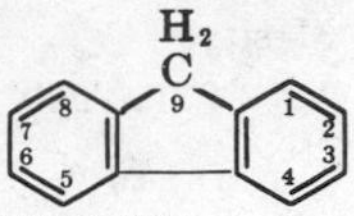

(10) Fluorene

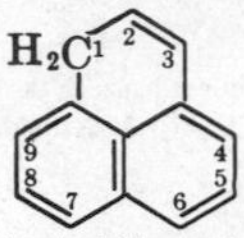

(11) Phenalene

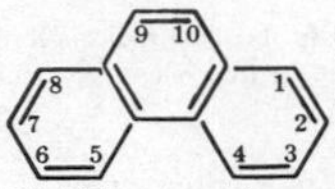

(12) Phenanthrene[1]

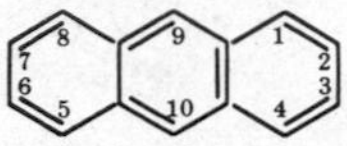

(13) Anthracene[1]

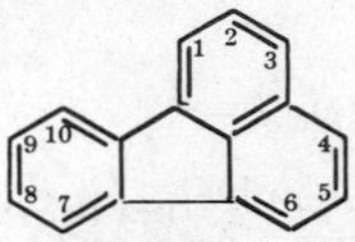

(14) Fluoranthene

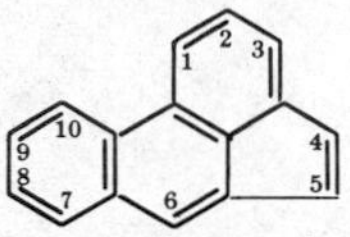

(15) Acephenanthrylene

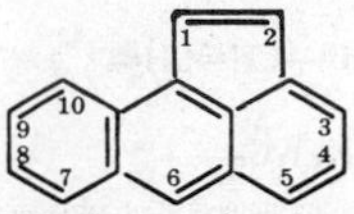

(16) Aceanthrylene

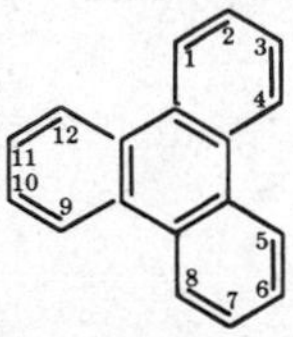

(17) Triphenylene

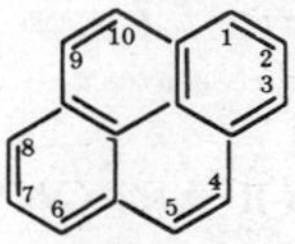

(18) Pyrene

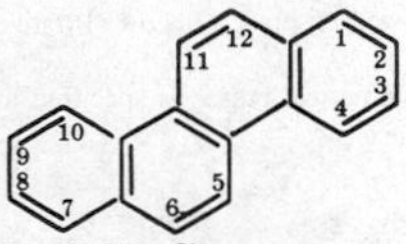

(19) Chrysene

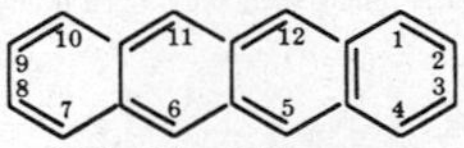

(20) Naphthacene

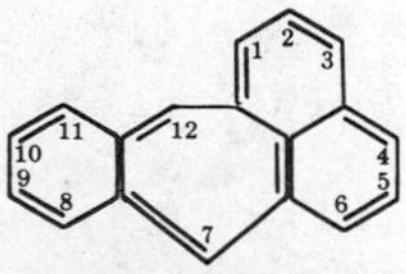

(21) Pleiadene

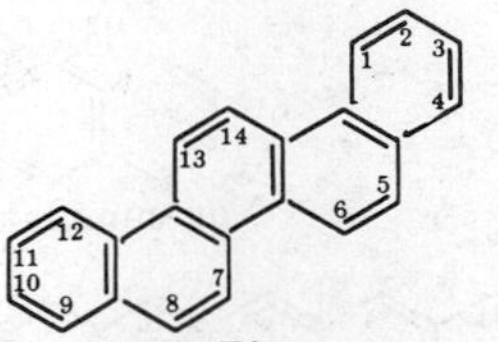

(22) Picene

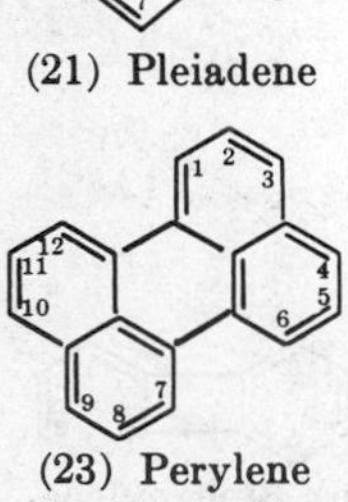

(23) Perylene

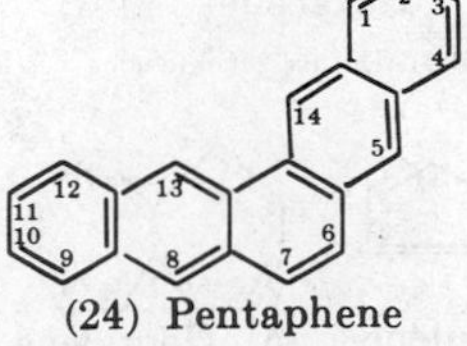

(24) Pentaphene

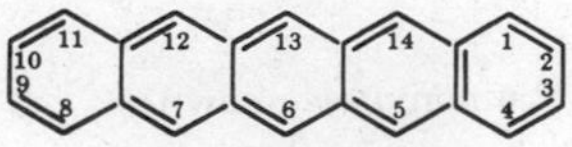

(25) Pentacene[2]

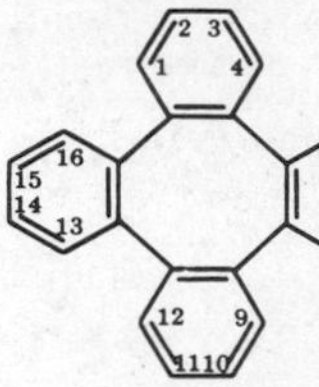

(26) Tetraphenylene[3]

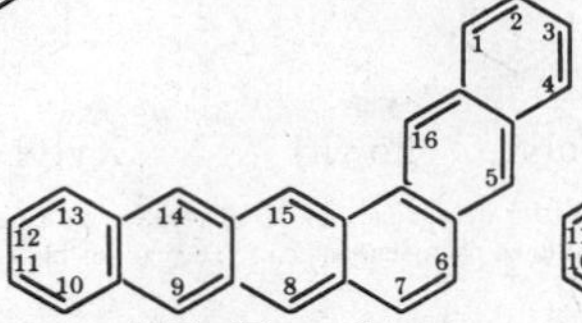

(27) Hexaphene

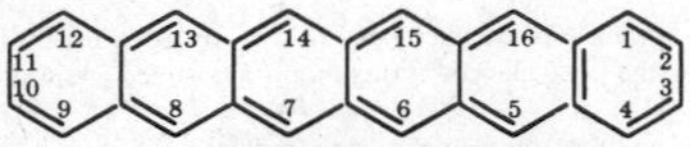

(28) Hexacene[2]

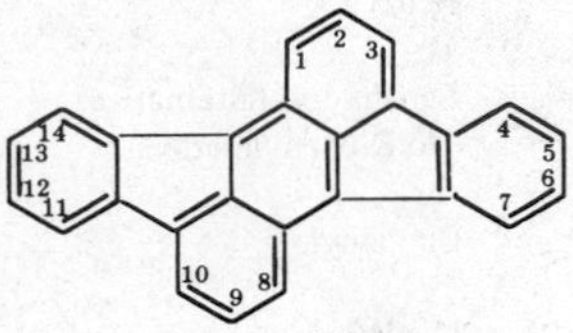

(29) Rubicene

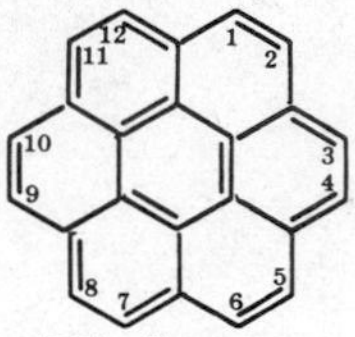

(30) Coronene

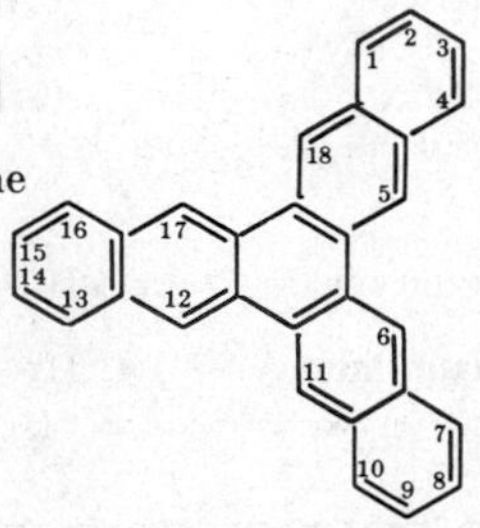

(31) Trinaphthylene[3]

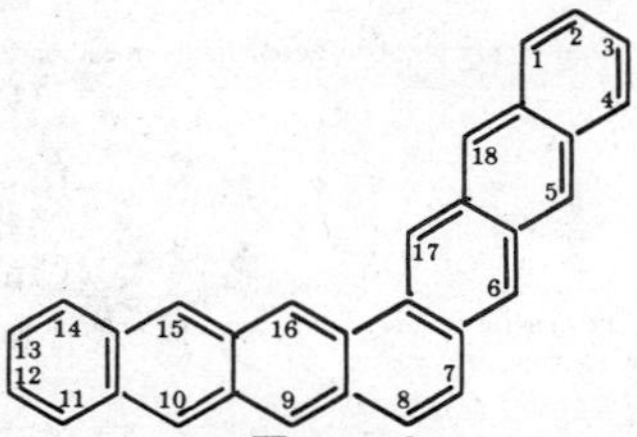

(32) Heptaphene

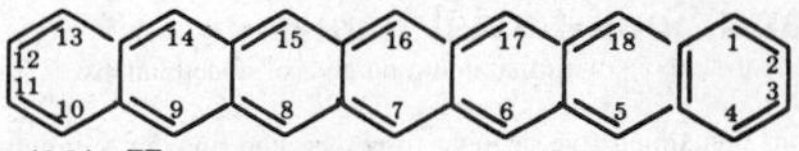

(33) Heptacene[2]

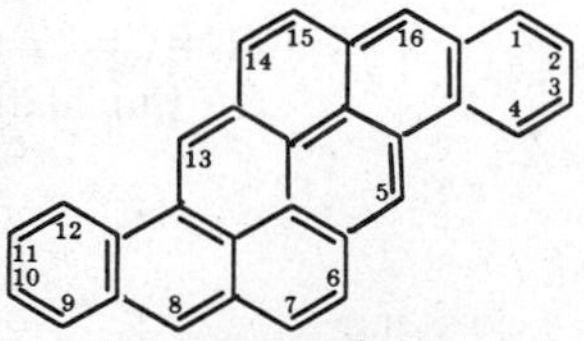

(34) Pyranthrene

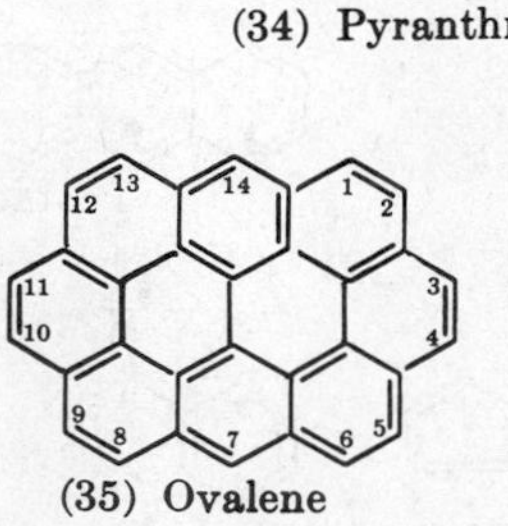

(35) Ovalene

1 Denotes exception to systematic numbering.
2 See Rule **A-21.2**
3 For isomer shown only.

21.3—"*Ortho*-fused"* or "*ortho-* and *peri*-fused"** polycyclic hydrocarbons with maximum number of noncumulative double bonds which contain at least two rings of five or more members and which have no accepted trivial name such as those of Part **.1** of this rule, are named by prefixing to the name of a component ring or ring system (the base component) designations of the other components. The base component should contain as many rings as possible (provided it has a trivial name), and should occur as far as possible from the beginning of the list of Rule **A-21.1.** The attached components should be as simple as possible.

Example:

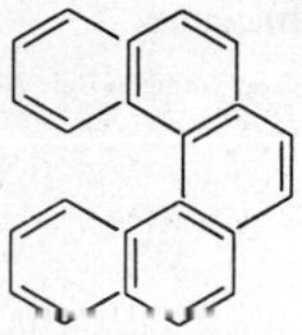

(not Naphthophenanthrene; benzo is "simpler" than naphtho, even though there are two benzo rings and only one naphtho)

Dibenzophenanthrene

21.4—The prefixes designating attached components are formed by changing the ending "-ene" of the name of the component hydrocarbon into "-eno"; e.g., "pyreno" (from pyrene). When more than one prefix is presented, they are arranged in alphabetical order. The following common abbreviated prefixes are recognized (see list in Part **.1** of this rule):

Acenaphtho	from	Acenaphthylene	Naphtho	from	Naphthalene
Anthra	from	Anthracene	Perylo	from	Perylene
Benzo	from	Benzene	Phenanthro	from	Phenanthrene

For monocyclic prefixes other than "benzo", the following names are recognized, each to represent the form with the maximum number of noncumulative double bonds: cyclopenta, cyclohepta, cycloocta, cyclonona, etc. When the base component is a monocyclic system, the ending "-ene" signifies the maximum number of noncumulative double bonds, and thus does not denote one double bond only.***

Examples:

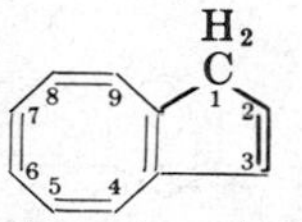

1*H*-Cyclopentacyclooöctene

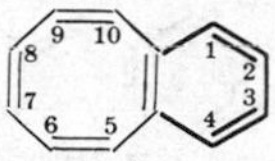

Benzocyclooöctene

21.5—Isomers are distinguished by lettering the peripheral sides of the base component *a, b, c*, etc., beginning with "*a*" for the side "1,2", "*b*" for "2,3" (or in certain cases "2,2*a*") and lettering every side around the periphery. To the letter as early in the alphabet as possible, denoting the side where fusion occurs, are prefixed, if necessary, the numbers of the positions of attachment of the other component. These numbers are chosen to be as low as is consistent with the numbering of the component, and their order conforms to the direction of lettering of the base component (see Examples II and IV). When two or more prefixes refer to equivalent positions so that there is a choice of letters, the prefixes are cited in alphabetical order according to Rule A-**21.4** and the location of the first cited prefix is indicated by a letter as early as possible in the alphabet (see Example V). The numbers and letters are enclosed in square brackets and placed immediately after the designation of the attached component. This expression merely defines the manner of fusion of the components.

Examples:

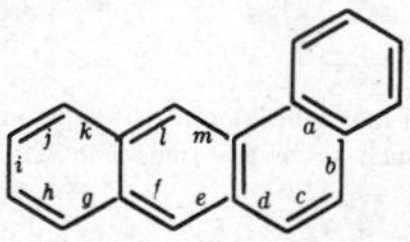

I

Benz[*a*]anthracene

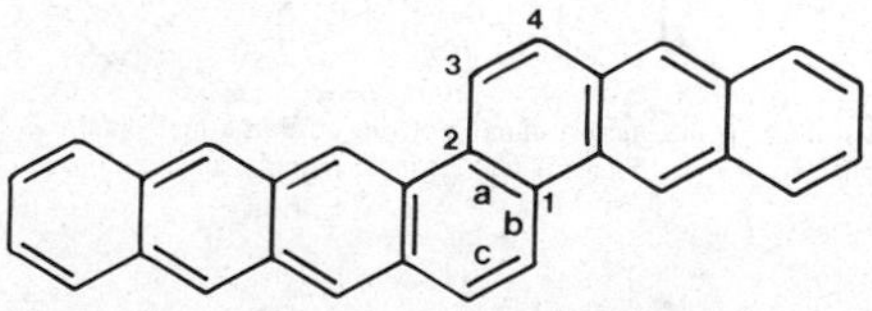

II

Anthra[2,1-*a*]naphthacene

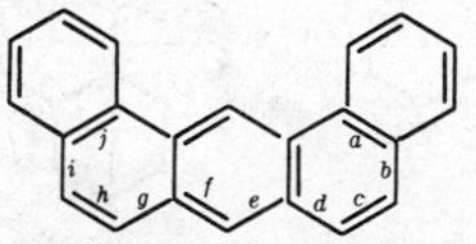

III

Dibenz[*a,j*]anthracene
(not Naphtho[2,1-*b*]phenanthrene)

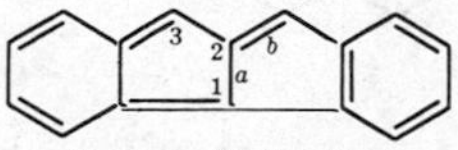

IV

Indeno[1,2-*a*]indene

* Polycyclic compounds in which two rings have two, and only two, atoms in common are said to be "*ortho*-fused". Such compounds have *n* common faces and 2*n* common atoms (Example I).

** Polycyclic compounds in which one ring contains two, and only two, atoms in common with each of two or more rings of a contiguous series of rings are said to be "*ortho*- and *peri*-fused". Such compounds have *n* common faces and less than 2*n* common atoms (Examples II and III).

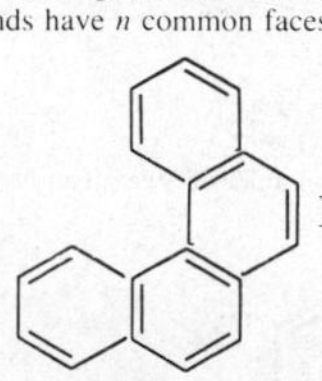

I

3 common faces
6 common atoms
"Ortho-fused" system

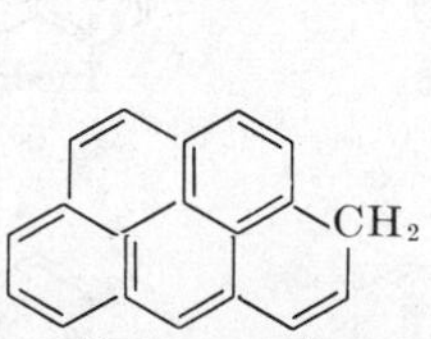

II

7 common faces
8 common atoms

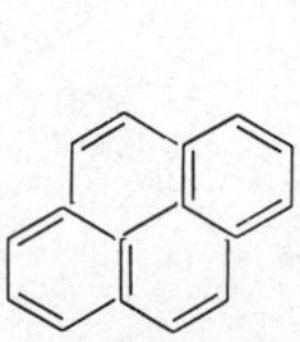

III

5 common faces
6 common atoms

"Ortho- and peri-fused" systems

*** The final "o" of acenaphtho, benzo, naphtho and perylo and the "a" of the monocyclic prefixes cyclopropa, cyclopenta, cyclohepta, etc. are elided before another vowel, as benz(o)[*a*]anthracene. In all other cases the final "o" or "a" is retained.

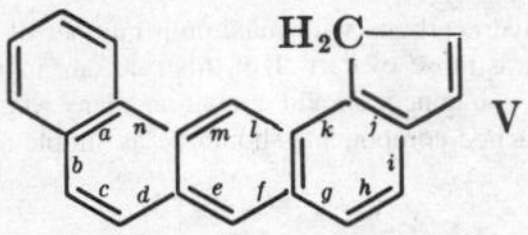

1*H*-Benzo[*a*]cyclopent[*j*]anthracene

The completed system consisting of the base component and the other components is then renumbered according to Rule **A-22**, the enumeration of the component parts being ignored.

Example:

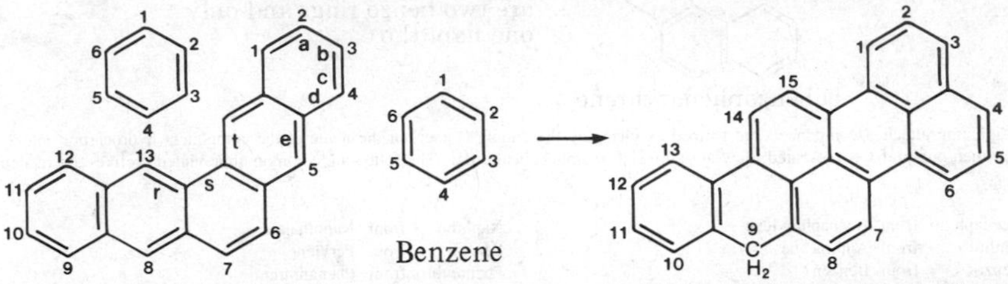

Benzene Pentaphene Benzene 9-*H*-Dibenzo[*de*,*rst*] pentaphene

21.6—When a name applies equally to two or more isomeric condensed parent ring systems with the maximum number of noncumulative double bonds and when the name can be made specific by indicating the position of one or more hydrogen atoms in the structure, this is accomplished by modifying the name with a locant, followed by italic capital *H* for each of these hydrogen atoms. Such symbols ordinarily precede the name. The said atom or atoms are called "indicated hydrogen". The same principle is applied to radicals and compounds derived from these systems.*

Examples:

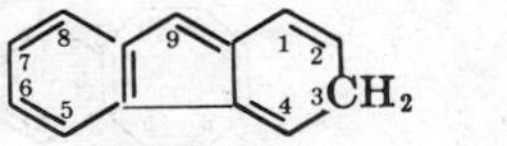

3*H*-Fluorene

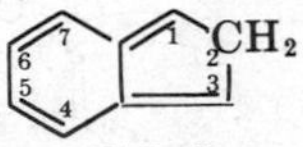

2*H*-Indene

Rule A-22. Numbering

22.1—For the purposes of numbering, the individual rings of a polycyclic "*ortho*-fused" or "*ortho*- and *peri*-fused" hydrocarbon system are normally drawn as follows:

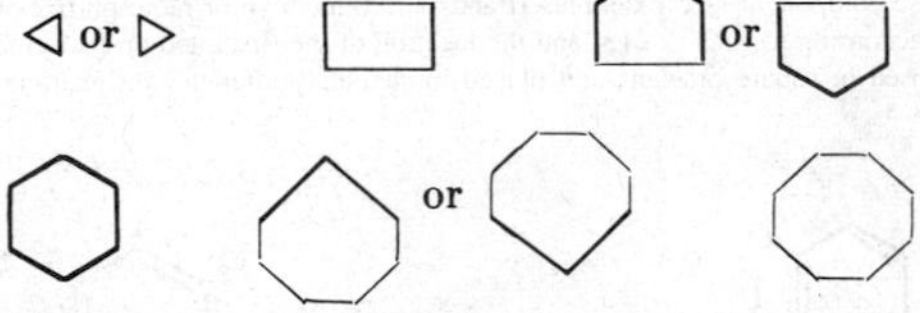

and the polycyclic system is oriented so that (a) the greatest number of rings are in a horizontal row and (b) a maximum number of rings are above and to the right of the horizontal row (upper right quadrant). If two or more orientations meet these requirements, the one is chosen which has as few rings as possible in the lower left quadrant.

Example:

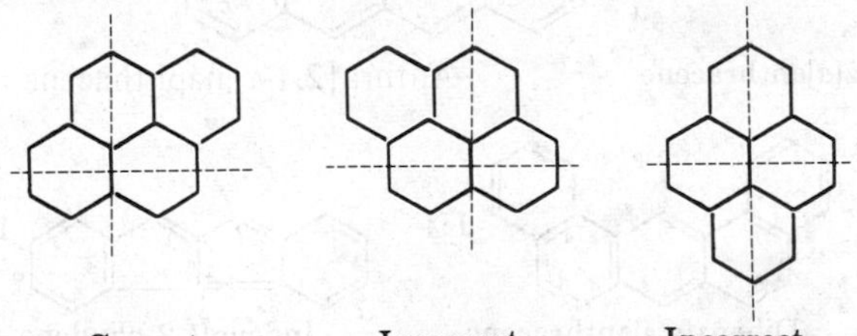

Correct orientation | Incorrect orientation | Incorrect orientation

The system thus oriented is numbered in a clockwise direction commencing with the carbon atom not engaged in ring-fusion in the most counter-clockwise position of the uppermost ring, or if there is a choice, of the uppermost ring farthest to the right, and omitting atoms common to two or more rings.

Example:

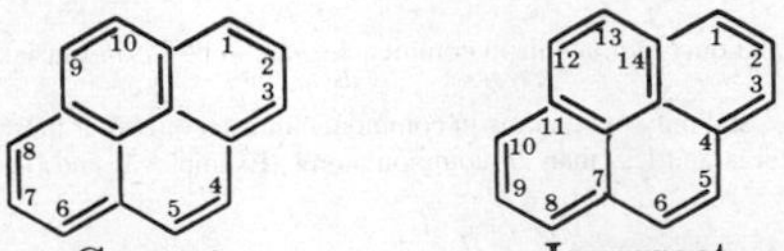

Correct | Incorrect

22.2—Atoms common to two or more rings are designated by adding roman letters "a", "b", "c", etc., to the number of the position immediately preceding. Interior atoms follow the highest number, taking a clockwise sequence wherever there is a choice.

Example:

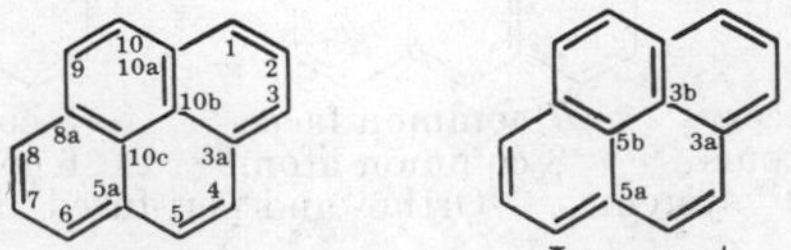

Correct | Incorrect

* See Rule B-5.12 for examples of "indicated hydrogen" in monocyclic rings.

22.3—When there is a choice,* carbon atoms common to two or more rings follow the lowest possible numbers.

Examples

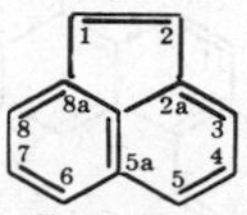

Correct

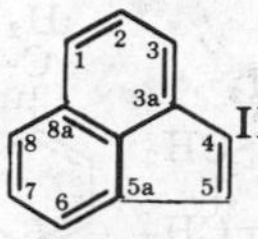

II

Incorrect

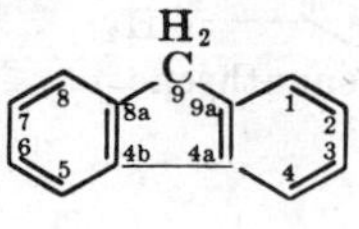

Correct

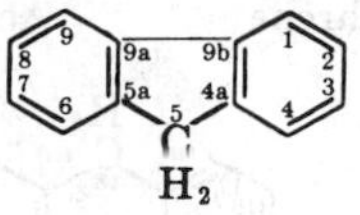

I

Incorrect

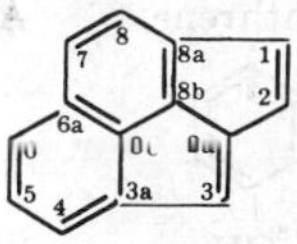

Correct

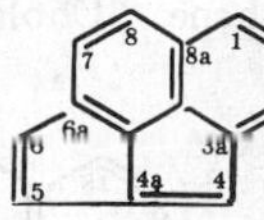

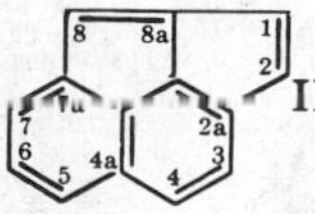

III

Incorrect

Note: I. 4, 4, 8, 9 is lower than 4, 5, 9, 9.
II. 2, 5, 8 is lower than 3, 5, 8.
III. 2, 3, 6, 8 is lower than 3, 4, 6, 8 or 2, 4, 7, 8.

22.4—When there is a choice, the carbon atoms which carry an indicated hydrogen atom are numbered as low as possible.

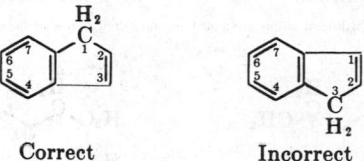

Example:

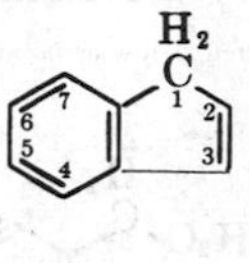

Correct

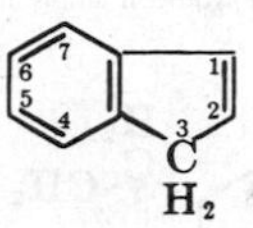

Incorrect

22.5—The following are recommended exceptions to the above rules on numbering:

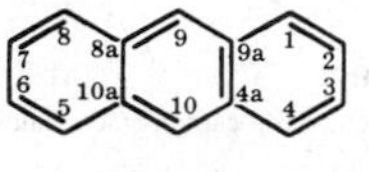

Anthracene

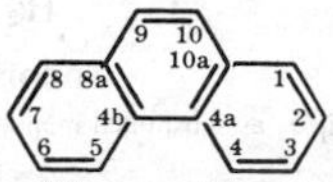

Phenanthrene

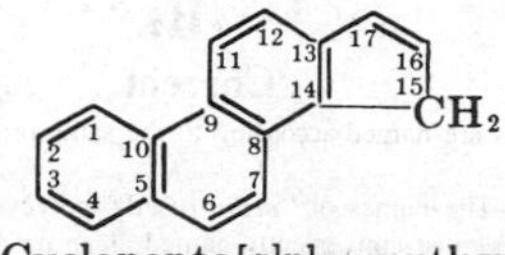

Cyclopenta[a]phenanthrene
(15*H*- shown)
See also rules on steroids**

Rule A-23. Hydrogenated Compounds

23.1—The names of "*ortho*-fused" or "*ortho*- and *peri*-fused" polycyclic hydrocarbons with less than maximum number of noncumulative double bonds are formed from a prefix "dihydro-", "tetrahydro-", etc., followed by the name of the corresponding unreduced hydrocarbon. The prefix "perhydro-" signifies full hydrogenation. When there is a choice for *H* used for indicated hydrogen it is assigned the lowest available number.

Examples:

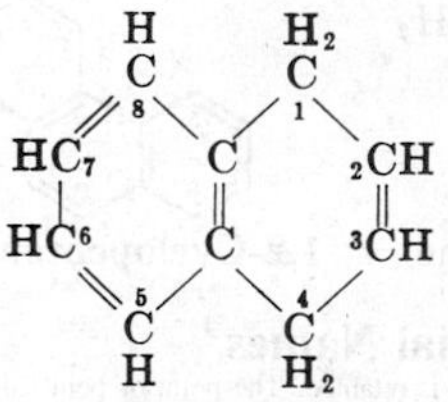

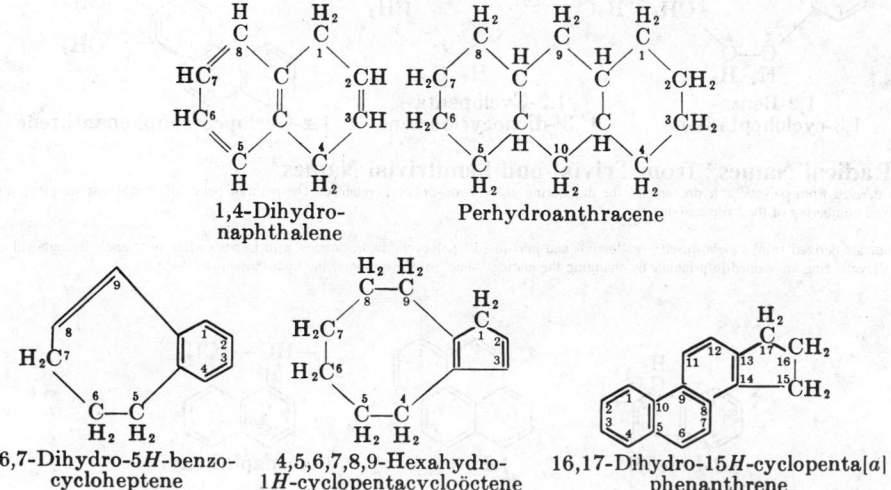

1,4-Dihydro-naphthalene

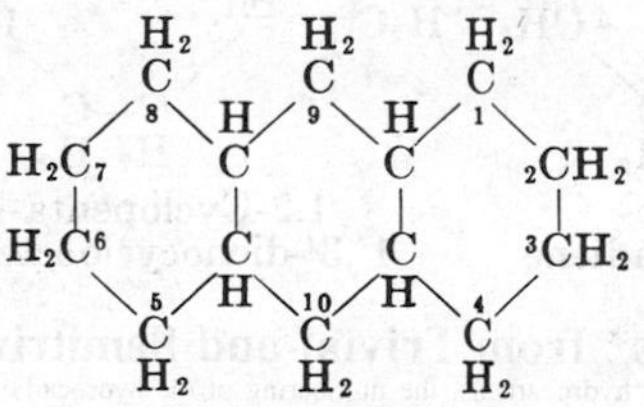

Perhydroanthracene

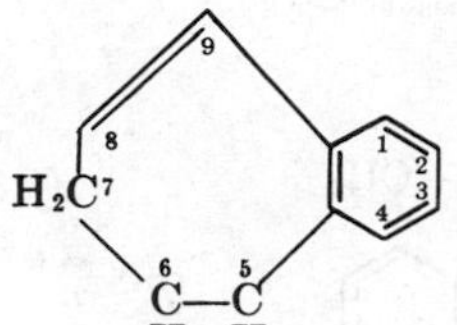

6,7-Dihydro-5*H*-benzo-cycloheptene

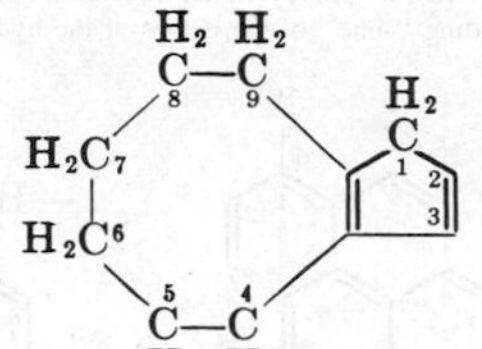

4,5,6,7,8,9-Hexahydro-1*H*-cyclopentacycloöctene

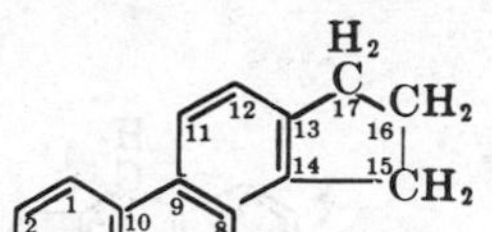

16,17-Dihydro-15*H*-cyclopenta[*a*]phenanthrene

* If after the requirements of the rules on orientation (cf Rule **A-22.1**) are met a choice remains, Rule **A-22.3** is applied.

** Definitive Rules for Nomenclature of Steroids, *Pure and Applied Chemistry*, Vol. 31, Nos. 1—2, 1972, pp. 285—322.

Exceptions: The following names are retained:

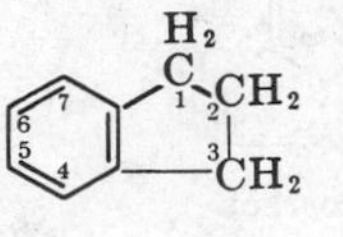
Indan

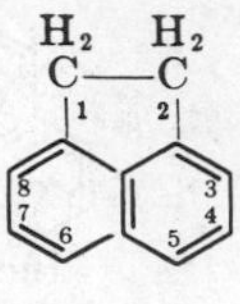
Acenaphthene

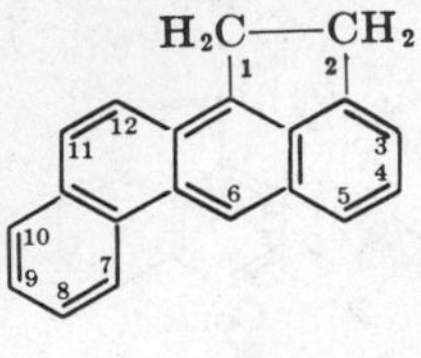
Cholanthrene

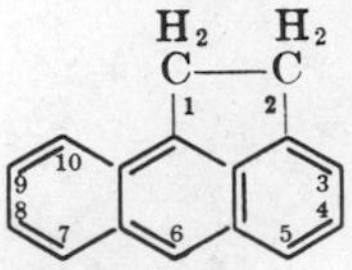
Aceanthrene

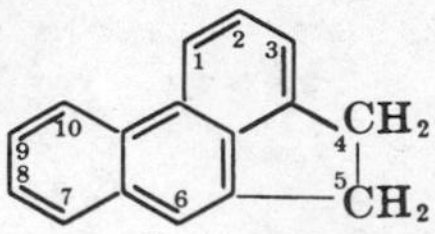
Acephenanthrene

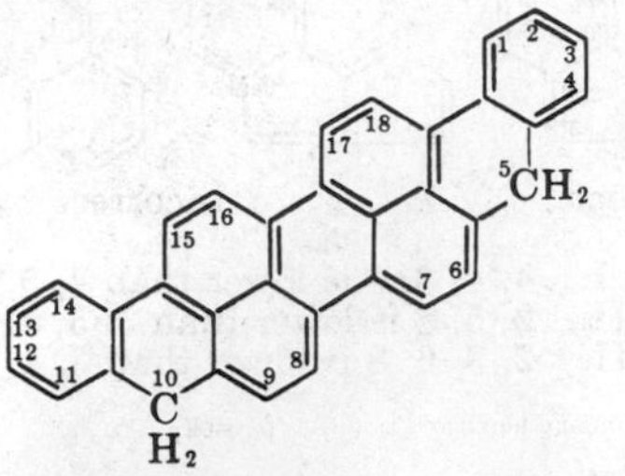
Violanthrene

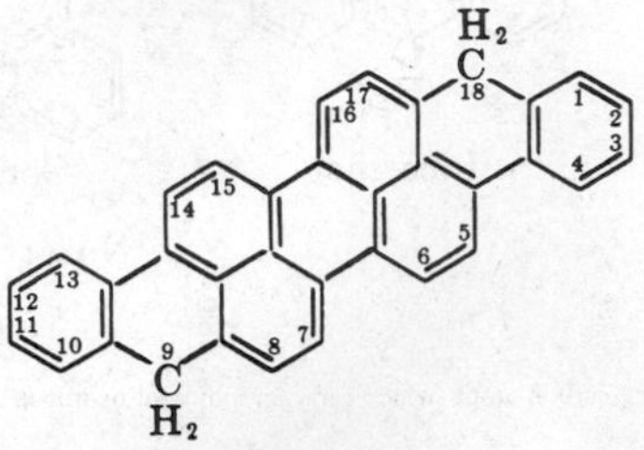
Isoviolanthrene

23.2—When there is a choice, the carbon atoms to which hydrogen atoms are added are numbered as low as possible.

Example:

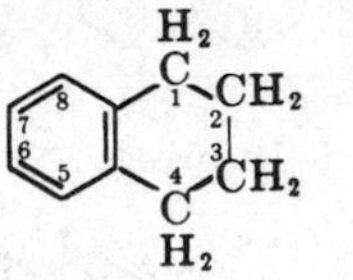
Correct

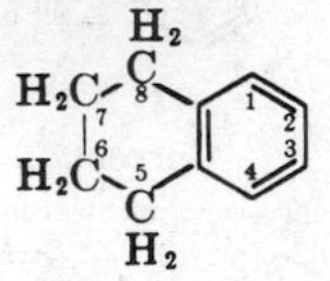
Incorrect

23.3—Substituted polycyclic hydrocarbons are named according to the same principles as substituted monocyclic hydrocarbons (see Rules **A-12** and **A-61**).

23.5 (Alternate to part of Rule A-23.1)—The names of "*ortho*-fused" polycyclic hydrocarbons which have (a) less than the maximum number of noncumulative double bonds, (b) at least one terminal unit which is most conveniently named as an unsaturated cycloalkane derivative, and (c) a double bond at the positions where rings are fused together, may be derived by joining the name of the terminal unit to that of the other component by means of a letter "o" with elisions of a terminal "e". The abbreviations for fused aromatic systems laid down in Rule **A-21.4** are used, and the exceptions of Rule **A-23.1** apply.

Examples:

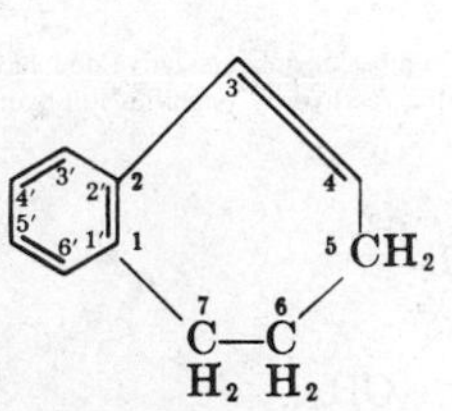
1,2-Benzo-
1,3-cycloheptadiene

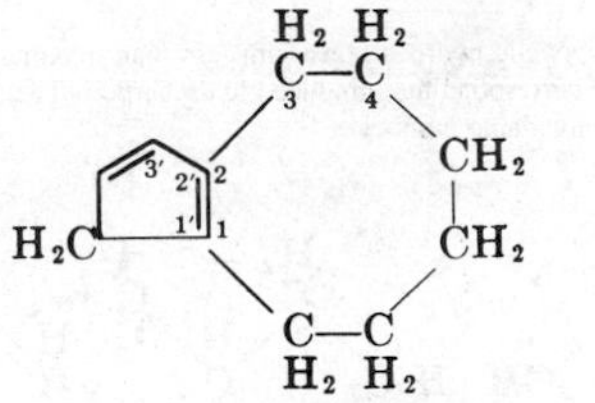
1,2-Cyclopenta-
1′,3′-dienocycloöctene

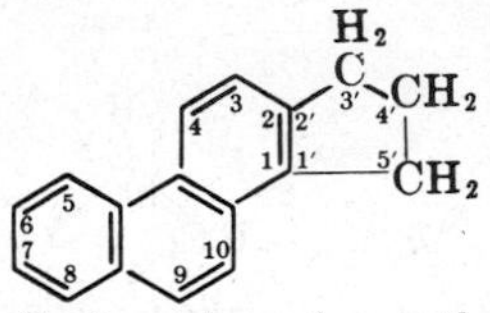
1,2-Cyclopentenophenanthrene

Rule A-24. Radical Names* from Trivial and Semitrivial Names

24.1—For radicals derived from polycyclic hydrocarbons, the numbering of the hydrocarbon is retained. The point or points of attachment are given numbers as low as is consistent with the fixed numbering of the hydrocarbon.

24.2—Univalent radicals derived from "*ortho*-fused" or "*ortho*- and *peri*-fused" polycyclic hydrocarbons with names ending in "-ene" by removal of a hydrogen atom from an aromatic or alicyclic ring are named in principle by changing the ending "-ene" of the names of the hydrocarbons to "-enyl".

Examples:

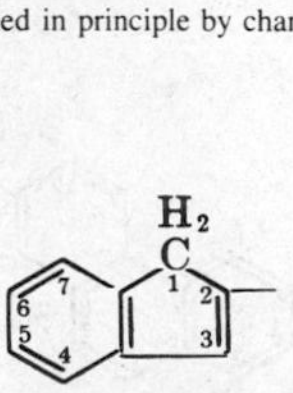
2-Indenyl

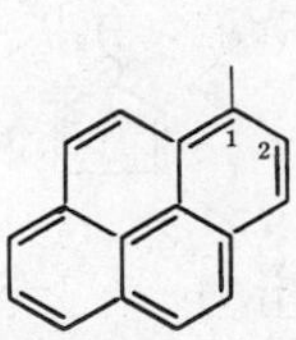
1-Pyrenyl

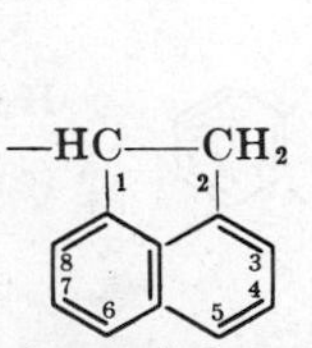
1-Acenaphthenyl

Exceptions:

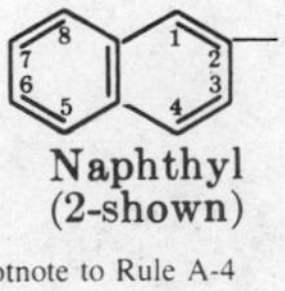
Naphthyl
(2-shown)

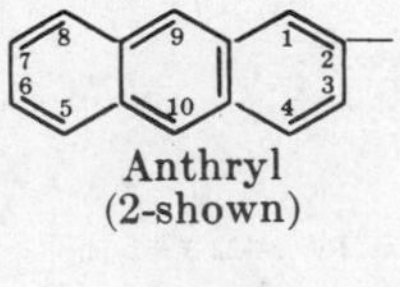
Anthryl
(2-shown)

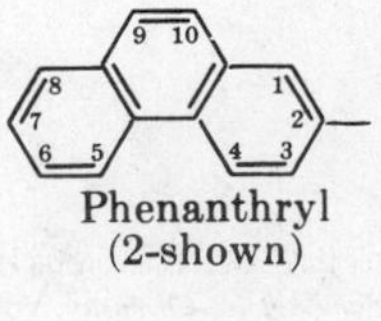
Phenanthryl
(2-shown)

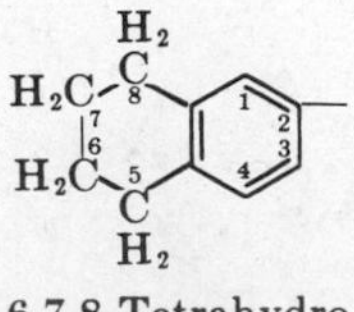
5,6,7,8-Tetrahydro-
2-naphthyl

* See footnote to Rule A-4

24.3—Bivalent radicals derived from univalent polycyclic hydrocarbon radicals whose names end in "-yl" by removal of one hydrogen atom from the carbon atom with the free valence are named by adding "-idene" to the name of the corresponding univalent radical.

Examples:

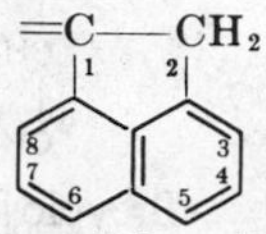

1-Acenaphthenylidene

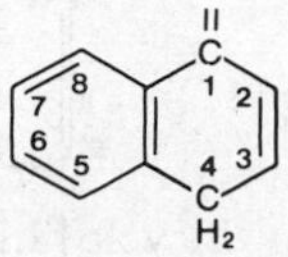

1(4*H*)-Naphthylidene
(for 4*H* see Rule **A-21.6**)
or 1,4-Dihydro-1-naphthylidene

24.4—Bivalent radicals derived from "*ortho*-fused" or "*ortho*- and *peri*-fused" polycyclic hydrocarbons by removal of a hydrogen atom from each of two different carbon atoms of the ring are named by changing the ending "yl" of the univalent radical name to "-ylene" or by adding "-diyl" to the name of the ring system Multivalent radicals, similarly derived, are named by adding "-triyl", "-tetrayl", etc., to the name of the ring system.

Examples:

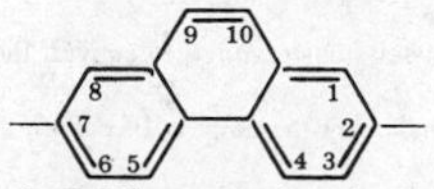

2,7-Phenanthrylene
or 2,7-Phenanthrenediyl

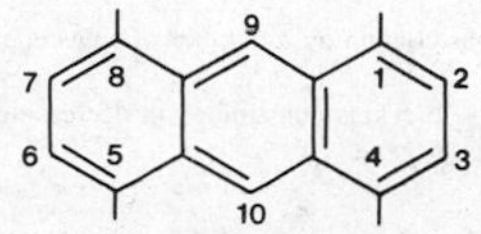

1,4,5,8-Anthracenetetrayl

Rule A-28. Radical Names for Fused Cyclic Systems with Side Chains

28.1—Radicals formed from hydrocarbons consisting of polycyclic systems and side chains are named according to the principles of the preceding rules.

BRIDGED HYDROCARBONS

EXTENSION OF THE VON BAEYER SYSTEM

Rule A-31. Bicyclic Systems

31.1—Saturated alicyclic hydrocarbon systems consisting of two rings only, having two or more atoms in common, take the name of an open chain hydrocarbon containing the same total number of carbon atoms preceded by the prefix "bicyclo-". The number of carbon atoms in each of the three bridges* connecting the two tertiary carbon atoms is indicated in brackets in descending order.

Examples:

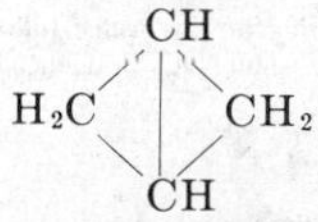

Bicyclo [1.1.0]-butane

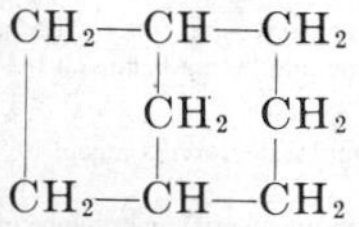

Bicyclo [3.2.1]-octane

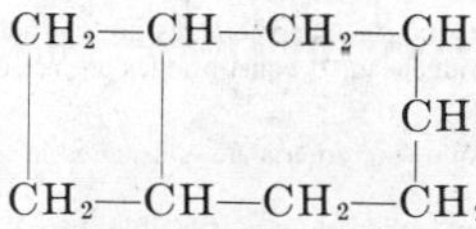

Bicyclo [5.2.0]nonane

31.2—The system is numbered commencing with one of the bridgeheads, numbering proceeding by the longest possible path to the second bridgehead; numbering is then continued from this atom by the longer unnumbered path back to the first bridgehead and is completed by the shortest path from the atom next to the first bridgehead.

Examples:

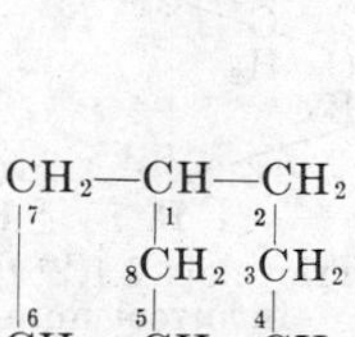

Bicyclo [3.2.1]octane

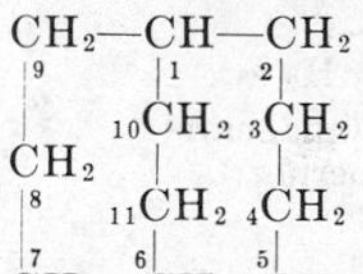

Bicyclo [4.3.2]undecane

Note: Longest path 1, 2, 3, 4, 5
Next longest path 5, 6, 7, 1
Shortest path 1, 8, 5

31.3—Unsaturated hydrocarbons are named in accordance with the principles set forth in Rule **A-11.3**. When after applying Rule **A-31.2** a choice in numbering remains unsaturation is given the lowest numbers.

Examples:

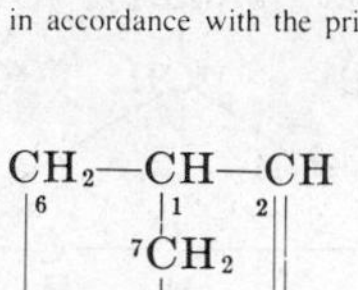

Bicyclo [2.2.1]hept-2-ene

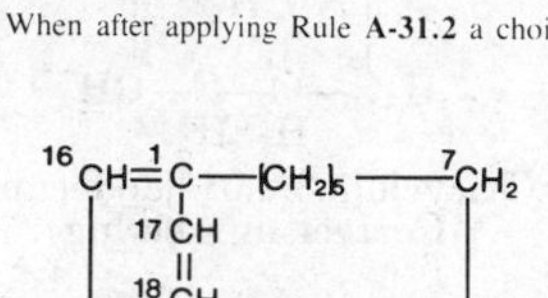

Bicyclo[12.2.2] octadeca-1(16),14,17-triene
or Bicyclo[12.2.2] octadeca-14,16(1),17-triene
(See Rule **A-3.1** for double locants)

31.4—Radicals derived from bridged hydrocarbons are named in accordance with the principles set forth in Rule **A-11**. The numbering of the hydrocarbon is retained and the point or points of attachment are given numbers as low as is consistent with the fixed numbering of the saturated hydrocarbon.

* A bridge is a valence bond or an atom or an unbranched chain of atoms connecting two different parts of a molecule. The two tertiary carbon atoms connected through the bridge are termed "bridgeheads".

Examples:

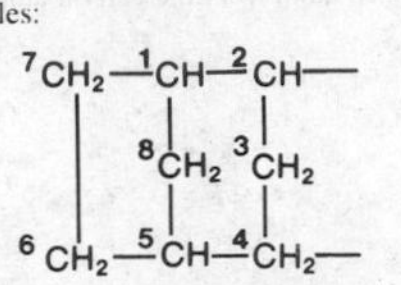

Bicyclo [3.2.1] oct-2-yl

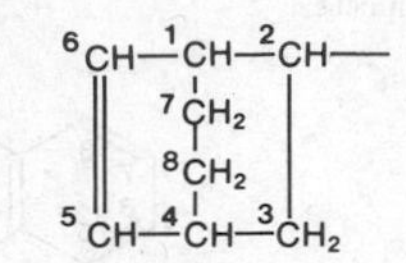

Bicyclo [2.2.2] oct-5-en-2-yl

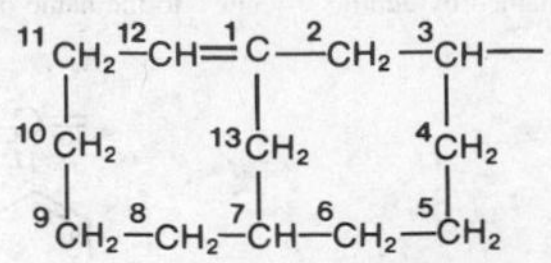

Bicyclo [5.5.1] tridec-1(12)-en-3-yl
or Bicyclo [5.5.1] tridec-12(1)-en-3-yl
(See Rule **A-3.1** for double locants)

Rule A-32. Polycyclic Systems

32.11—Cyclic hydrocarbon systems consisting of three or more rings may be named in accordance with the principles stated in Rule **A-31**. The appropriate prefix "tricyclo-", "tetracyclo-", etc., is substituted for "bicyclo-" before the name of the open-chain hydrocarbon containing the same total number of carbon atoms. Radicals derived from these hydrocarbons are named according to the principles set forth in Rule **A-31.4**.

32.12—A polycyclic system is regarded as containing a number of rings equal to the number of scissions required to convert the system into an open-chain compound.

32.13—The word "cyclo" is followed by brackets containing, in decreasing order, numbers indicating the number of carbon atoms in: the two branches of the main ring, the main bridge, and the secondary bridges.[1]

Examples:

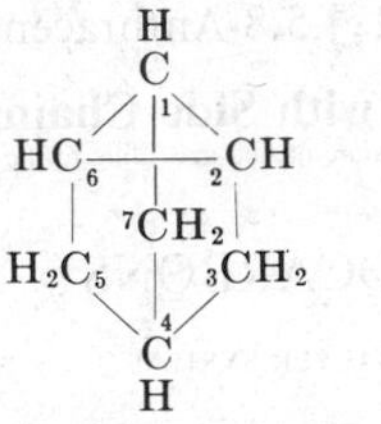

Tricyclo[2.2.1.0^{1}]heptane

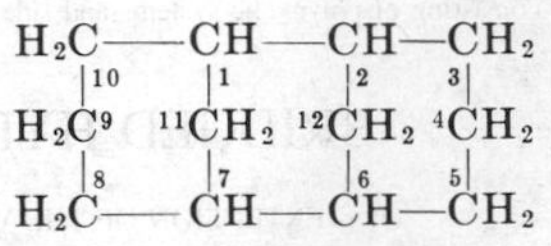

Tricyclo[5.3.1.1^{1}]dodecane

32.21—The main ring and the main bridge form a bicyclic system whose numbering is made in compliance with Rule **A-31**.

32.22—The location of the other or so-called secondary bridges is shown by superscripts following the number indicating the number of carbon atoms in the said bridges.

32.23—For the purpose of numbering, the secondary bridges are considered in decreasing order. The numbering of any bridge follows from the part already numbered, proceeding from the highest-numbered bridgehead. If equal bridges are present, the numbering begins at the highest-numbered bridgehead.

32.31—When there is a choice, the following criteria are considered in turn until a decision is made:

(a) the main ring shall contain as many carbon atoms as possible, two of which must serve as bridgeheads for the main bridge.

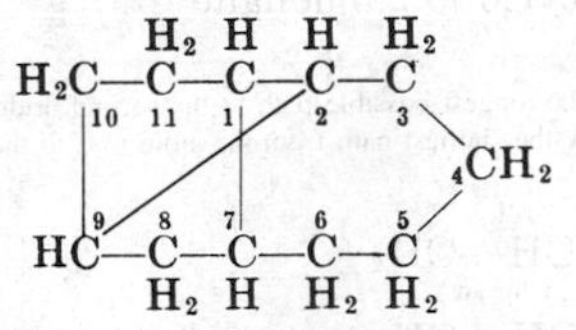

Tricyclo[5.4.0.$0^{2,9}$]undecane
Correct numbering

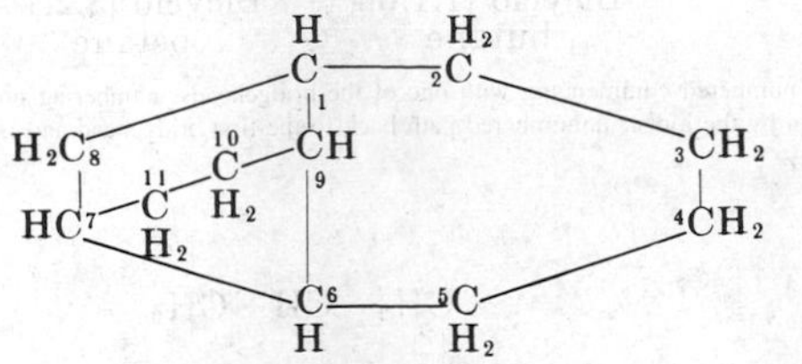

Tricyclo[4.2.1.$2^{7,9}$]undecane
Incorrect numbering

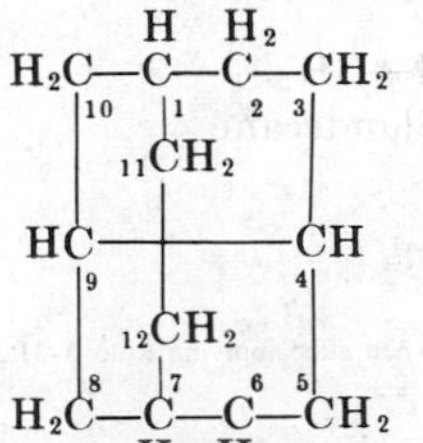

Tricyclo[5.3.2.$0^{4,9}$]dodecane
Correct numbering

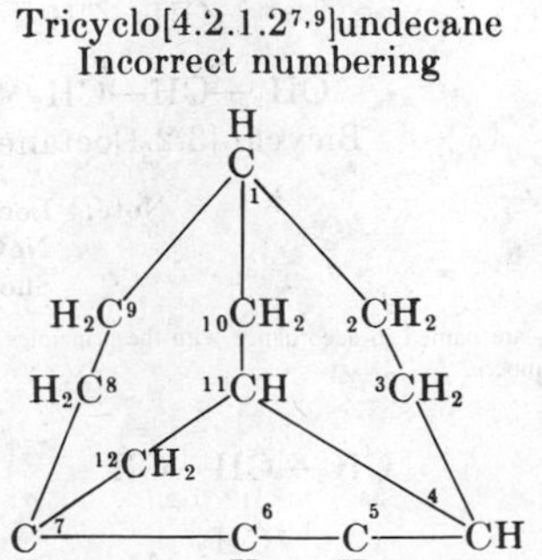

Tricyclo[5.2.3.$0^{4,11}$]dodecane
Incorrect numbering

[1] For location and numbering of the secondary bridge see Rules **A-32.22**, **A-32.23**, **A-32.31**.

(b) The main bridge shall be as large as possible.

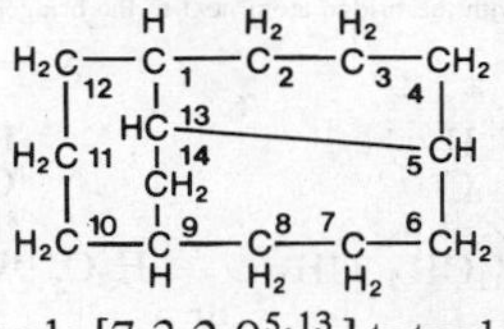

Tricyclo[7.3.2.$0^{5,13}$] tetradecane
Correct numbering

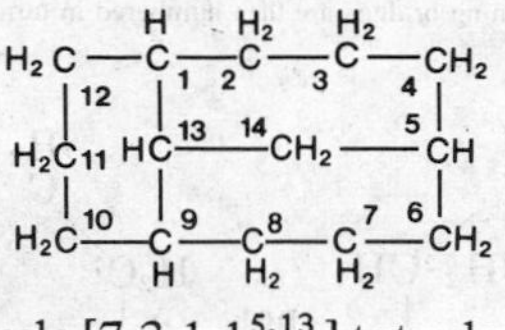

Tricyclo[7.3.1.$1^{5,13}$] tetradecane
Incorrect numbering

(c) The main ring shall be divided as symmetrically as possible by the main bridge.

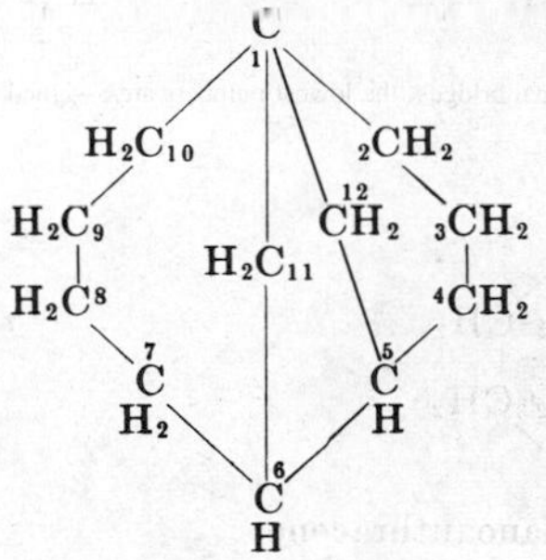

Tricyclo[4.4.1.$1^{1,5}$]dodecane:
Correct numbering

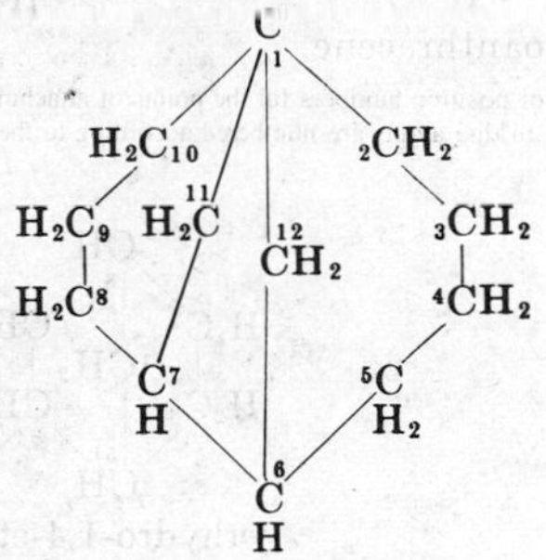

Tricyclo[5.3.1.$1^{1,6}$]dodecane:
Incorrect numbering

(d) The superscripts locating the other bridges shall be as small as possible (in the sense indicated in Rule **A-2.2**).

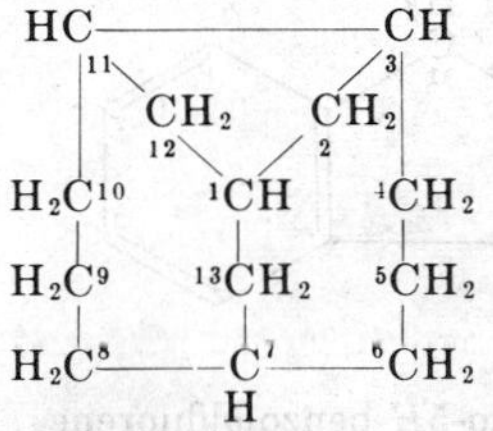

Tricyclo[5.5.1.$0^{3,11}$]tridecane
Correct numbering

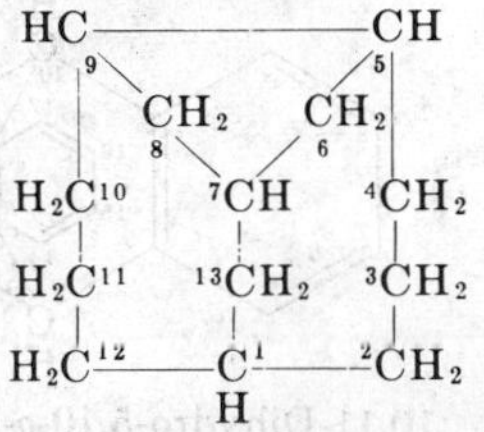

Tricyclo[5.5.1.$0^{5,9}$]tridecane
Incorrect numbering

Rule A-34. Hydrocarbon Bridges

34.1—Polycyclic hydrocarbon systems which can be regarded as "*ortho*-fused" or "*ortho*- and *peri*-fused" systems according to Rule **A-21** and which, at the same time, have other bridges*, are first named as "*ortho*-fused" or "*ortho*- and *peri*-fused" systems. The other bridges are then indicated by prefixes derived from the name of the corresponding hydrocarbon by replacing the final "-ane", "-ene", etc., by "-ano", "-eno", etc., and their positions are indicated by the points of attachment in the parent compound. If bridges of different types are present, they are cited in alphabetical order.

Examples of bridge names:

Butano	$—CH_2—CH_2—CH_2—CH_2—$	Etheno	$—CH{=}CH—$
Benzeno (*o*-, *m*-, *p*-)	$—C_6H_4—$	Methano	$—CH_2—$
Ethano	$—CH_2—CH_2—$	Propano	$—CH_2—CH_2—CH_2—$

Examples

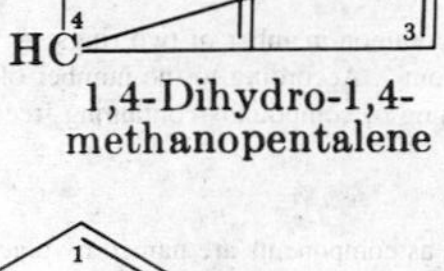

1,4-Dihydro-1,4-methanopentalene

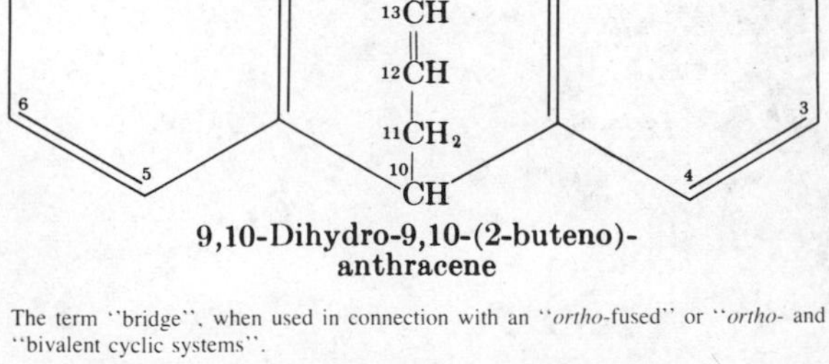

9,10-Dihydro-9,10-(2-buteno)-anthracene

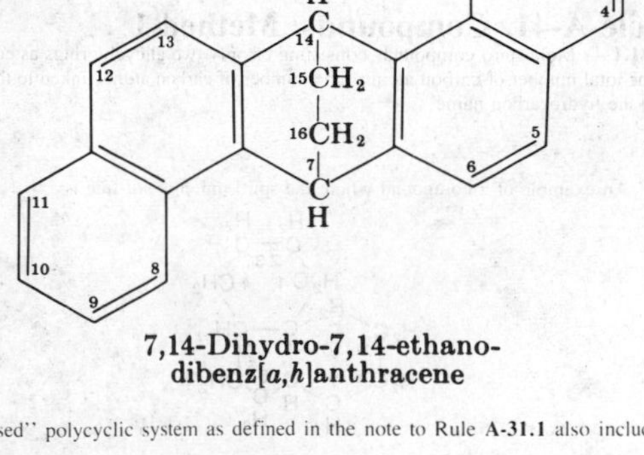

7,14-Dihydro-7,14-ethano-dibenz[*a*,*h*]anthracene

* The term "bridge", when used in connection with an "*ortho*-fused" or "*ortho*- and *peri*-fused" polycyclic system as defined in the note to Rule **A-31.1** also includes "bivalent cyclic systems".

34.2—The parent "*ortho*-fused" or "*ortho*- and *peri*-fused" system is numbered as prescribed in Rule **A-22**. Where there is a choice, the position numbers of the bridgeheads should be as low as possible. The remaining bridges are then numbered in turn starting each time with the bridge atom next to the bridgehead possessing the highest number.

Example:

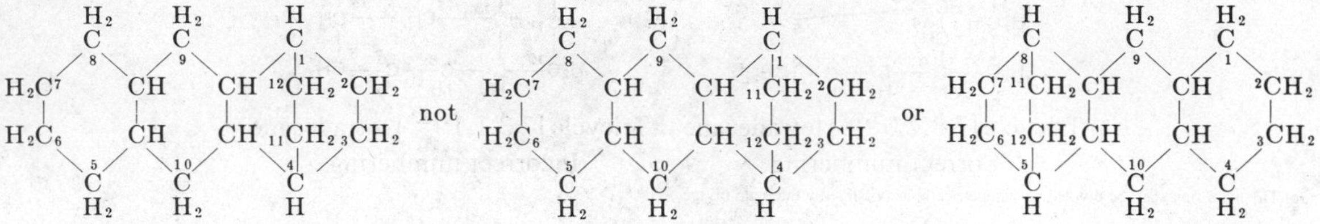

Perhydro-1,4-ethanoanthracene

34.3—When there is a choice of position numbers for the points of attachment for several individual bridges, the lowest numbers are assigned to the bridgeheads in the order of citation of the bridges and the bridge atoms are numbered according to the preceding rule.

Example:

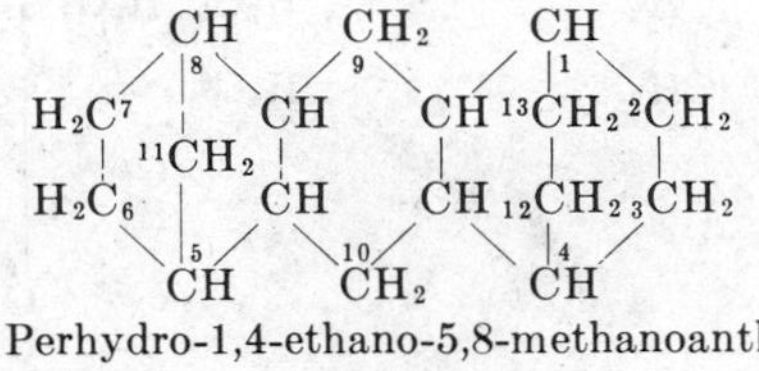

Perhydro-1,4-ethano-5,8-methanoanthracene

34.4—When the bridge is formed from a bivalent cyclic hydrocarbon radical, low numbers are given to the carbon atoms constituting the shorter bridge and numbering proceeds around the ring.

Example:

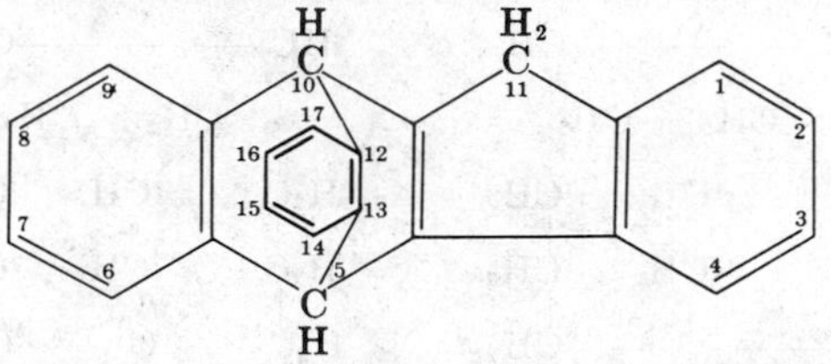

10,11-Dihydro-5,10-*o*-benzeno-5*H*-benzo[*b*]fluorene

34.5—Names for radicals derived from the bridged hydrocarbons considered in Rule **A-34.1** are constructed in accordance with the principles set forth in Rule **A-24**. The abbreviated radical names naphthyl, anthryl, phenanthryl, naphthylene, etc., permitted as exceptions to Rules **A-24.2** and **A-24.4** are replaced in such cases by the regularly formed names naphthalenyl, anthracenyl, phenanthrenyl, naphthalenediyl, etc.

Examples:

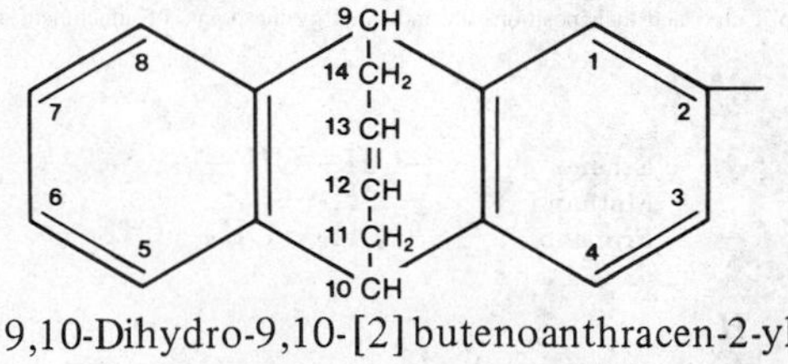

9,10-Dihydro-9,10-[2]butenoanthracen-2-yl

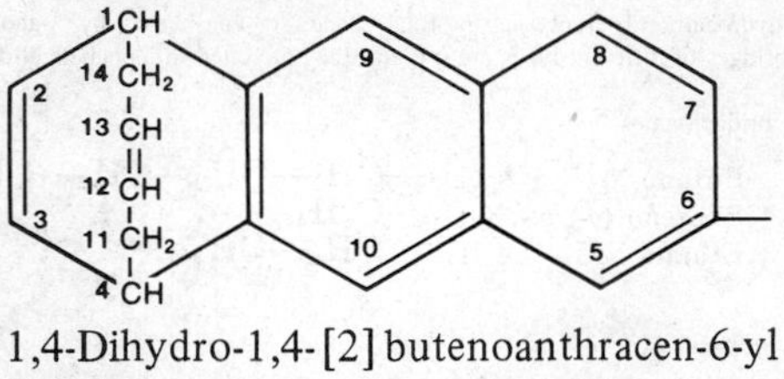

1,4-Dihydro-1,4-[2]butenoanthracen-6-yl

SPIRO HYDROCARBONS

A "spiro union" is one formed by a single atom which is the only common member of two rings. A "free spiro union" is one constituting the only union direct or indirect between two rings.* The common atom is designated as the "spiro atom". According to the number of spiro atoms present, the compounds are distinguished as monospiro-, dispiro-, trispirocompounds, etc. The following rules apply to the naming of compounds containing free spiro unions.

Rule A-41. Compounds: Method 1

41.1 — Monospiro compounds consisting of only two alicyclic rings as components are named by placing "spiro" before the name of the normal acyclic hydrocarbon of the same total number of carbon atoms. The number of carbon atoms linked to the spiro atom in each ring is indicated in ascending order in brackets placed between the spiro prefix and the hydrocarbon name.

* An example of a compound where the spiro union is *not* free is:

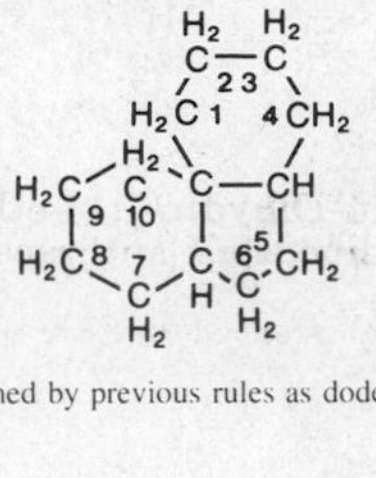

This compound is named by previous rules as dodecahydrobenz[*c*]indene.

Examples:

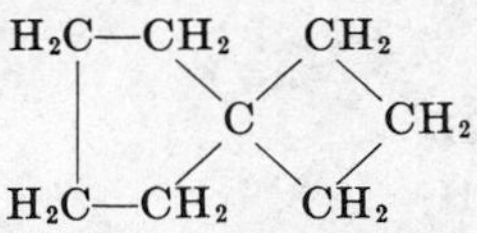

Spiro[3.4]octane

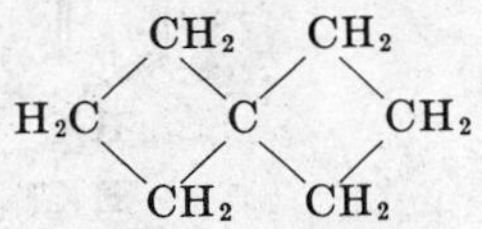

Spiro[3.3]heptane

41.2 — The carbon atoms in monospiro hydrocarbons are numbered consecutively starting with a ring atom next to the spiro atom, first through the smaller ring (if such be present) and then through the spiro atom and around the second ring.

Example:

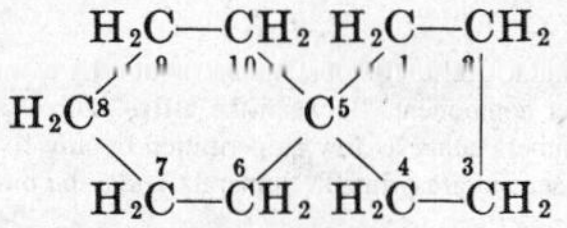

Spiro[4.5]decane

41.3 — When unsaturation is present, the same enumeration pattern is maintained, but in such a direction around the rings that the double and triple bonds receive numbers as low as possible in accordance with Rule **A-11.**

Example:

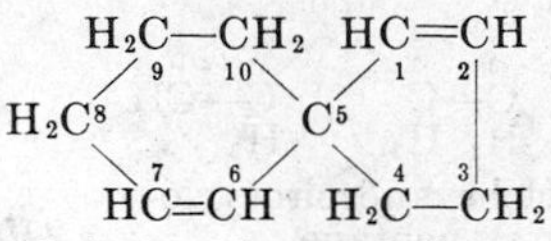

Spiro[4.5]deca-1,6-diene

41.4 — If one or both components of the monospiro compound are fused polycyclic systems, "spiro" is placed before the names of the components arranged in alphabetical order and enclosed in brackets. Established numbering of the individual components is retained. The lowest possible number is given to the spiro atom, and the numbers of the second component are marked with primes. The position of the spiro atom is indicated by placing the appropriate numbers between the names of the two components.

Example:

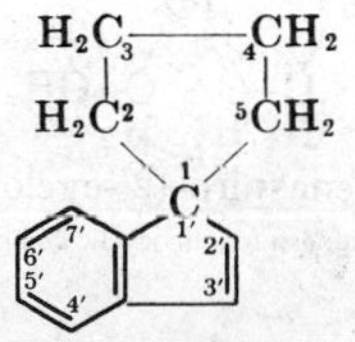

Spiro[cyclopentane-1,1′-indene]

41.5 — Monospiro compounds containing two similar polycyclic components are named by placing the prefix "spirobi" before the name of the component ring system. Established enumeration of the polycyclic system is maintained and the numbers of one component are distinguished by primes. The position of the spiro atom is indicated in the name of the spiro compound by placing the appropriate locants before the name.

Example:

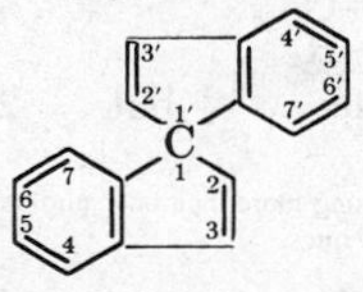

1,1′-Spirobiindene

41.6 — Polyspiro compounds consisting of a linear assembly of three or more alicyclic systems are named by placing "dispiro-", "trispiro-", "tetraspiro-", etc., before the name of the unbranched-chain acyclic hydrocarbon of the same total number of carbon atoms. The numbers of carbon atoms linked to the spiro atoms in each ring are indicated in brackets in the same order as the numbering proceeds about the ring. Numbering starts with a ring atom next to a terminal spiro atom and proceeds in such a way as to give the spiro atoms as low numbers as possible after numbering all the carbon atoms of the first ring linked to the terminal spiro atom.

Example:

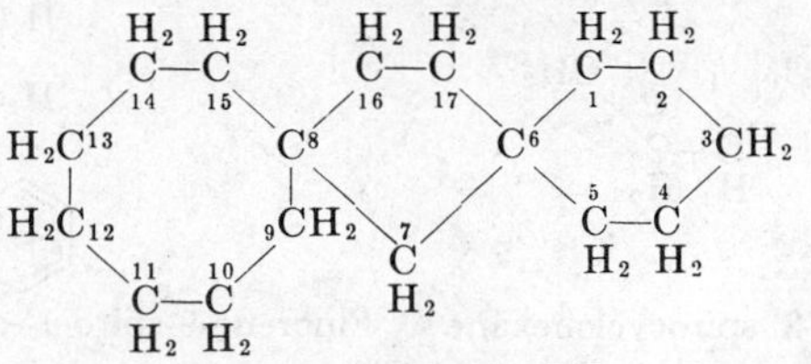

Dispiro[5.1.7.2]heptadecane

41.7 — Polycyclic compounds containing more than one spiro atom and at least one fused polycyclic component are named in accordance with Part **.4** of this rule by replacing "spiro" with "dispiro", "trispiro", etc., and choosing the end components by alphabetical order.

Example:

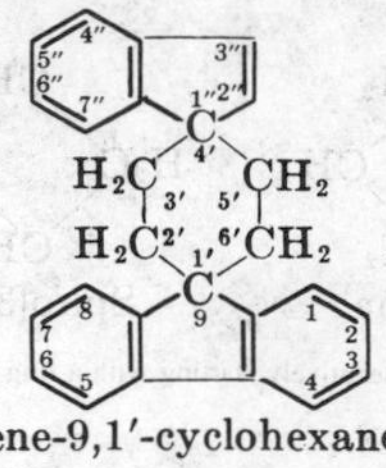

Dispiro[fluorene-9,1′-cyclohexane-4′,1″-indene]

Rule A-42. Compounds: Method 2

42.1 (Alternate to Rules **A-41.1** and **A-41.2**) — When two dissimilar cyclic components are united by a spiro union, the name of the larger component is followed by the affix "spiro" which, in turn, is followed by the name of the smaller component. Between the affix "spiro" and the name of each component system is inserted the number denoting the spiro position in the appropriate ring system, these numbers being as low as permitted by any fixed enumeration of the component. The components retain their respective enumerations but numerals for the component mentioned second are primed. Numerals 1 may be omitted when a free choice is available for a component.

Examples:

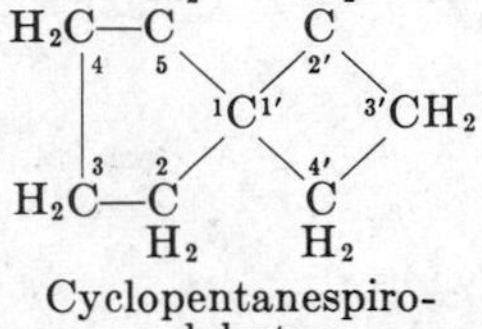

Cyclopentanespiro-cyclobutane

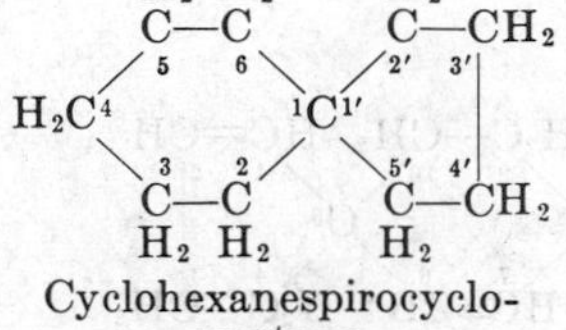

Cyclohexanespirocyclo-pentane

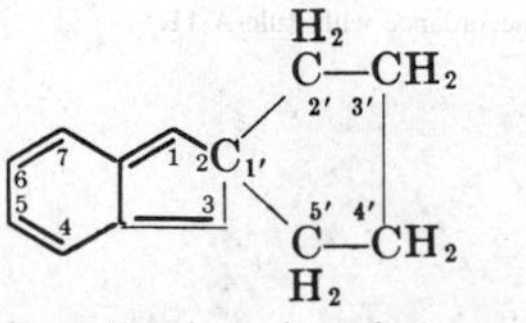

2*H*-Indene-2-spiro-1′-cyclopentane

42.2 (Alternate to **A-41.3**) — Rule **A-41.3** applies also with appropriate different enumeration, where nomenclature is according to Rule **A-42.1** but the spiro junction has priority for lowest numbers over unsaturation.

Example:

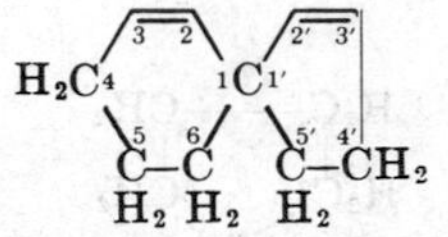

2-Cyclohexenespiro-(2′-cyclopentene)

42.3 (Alternate to **A-41.5**) — The nomenclature of Rule **A-41.5** is applied also to monocyclic components with identical saturation, the spiro union being numbered 1.

Example:

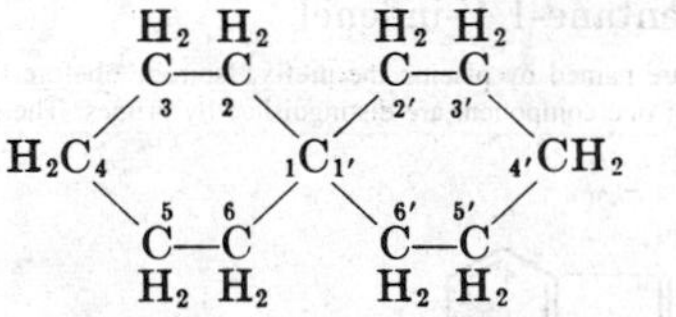

Spirobicyclohexane

but

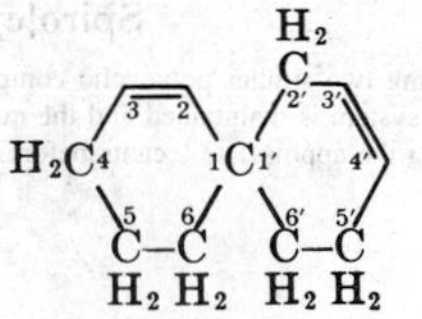

2-Cyclohexenespiro-(3′-cyclohexene)

42.4 (Alternate to **A-41.6** and **A-41.7**) — Polycyclic compounds containing more than one spiro atom are named in accordance with Rule **A-42.1** starting from the senior* end-component irrespective of whether the components are simple or fused rings.

Examples:

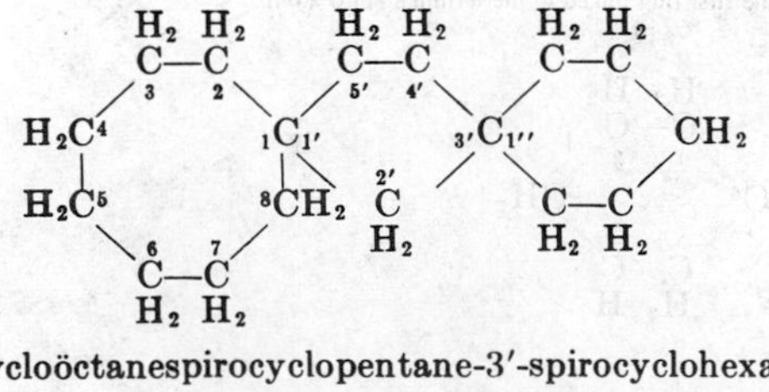

Cycloöctanespirocyclopentane-3′-spirocyclohexane

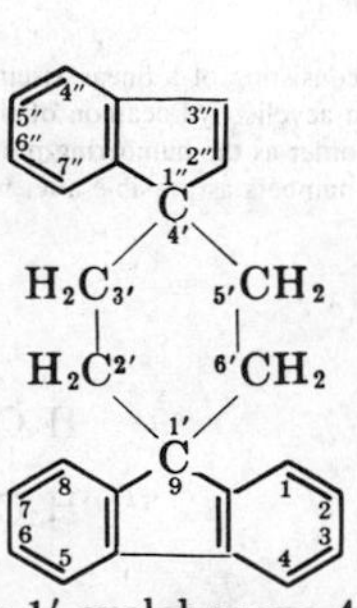

Fluorene-9-spiro-1′-cyclohexane-4′-spiro-1″-indene

Rule A-43. Radicals

43.1 — Radicals derived from spiro hydrocarbons are named according to the principles set forth in Rules **A-11** and **A-24**.

Examples:

* "Seniority" in respect to spiro compounds is based on the principles: (i) an aggregate is senior to a monocycle; (ii) of aggregates, the senior is that containing the largest number of individual rings; (iii) of aggregates containing the same number of individual rings, the senior is that containing the largest ring; and (iv) if aggregates consist of equal numbers of equal rings the senior is the first occurring in the alphabetical list of names.

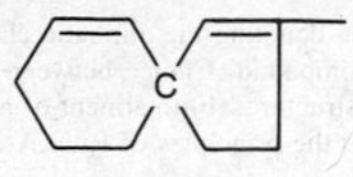

Spiro[4.5]deca-1,6-dien-2-yl
(cf. Rules **A-41.3** and **A-11**)
or 2-Cyclohexenespiro-2′-cyclopenten-3′-yl (cf. Rule **A-42.2**)

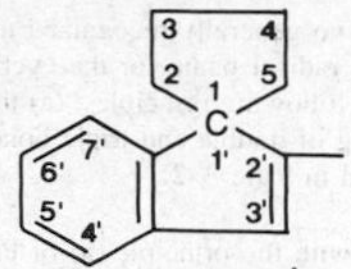

Spiro[cyclopentane-1,1′-inden]-2′-yl
(cf. Rules **A-41.4** and **A-24**)

HYDROCARBON RING ASSEMBLIES

Rule A-51. Definition

51.1—Two or more cyclic systems (single rings or fused systems) which are directly joined to each other by double or single bonds are named "ring assemblies" when the number of such direct ring junctions is one less than the number of cyclic systems involved.

Examples:

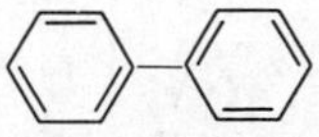

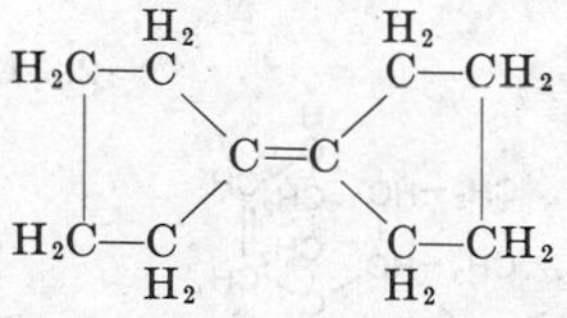

Ring assemblies

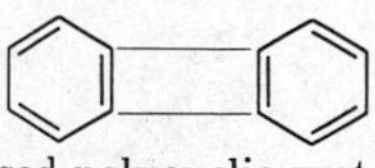

Fused polycyclic system

Rule A-52. Two Identical Ring Systems

52.1—Assemblies of two identical cyclic hydrocarbon systems are named in either of two ways: (a) by placing the prefix "bi-" before the name of the corresponding radical, or (b) for systems joined by a single bond by placing the prefix "bi-" before the name of the corresponding hydrocarbon. In each case, the numbering of the assembly is that of the corresponding radical or hydrocarbon, one system being assigned unprimed numbers and the other primed numbers. The points of attachment are indicated by placing the appropriate locants before the name.

Examples:

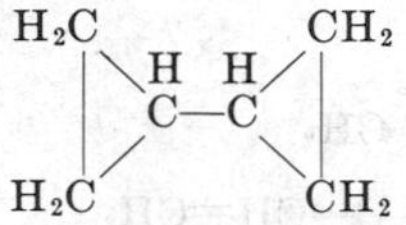

1,1′-Bicyclopropyl
or 1,1′-Bicyclopropane

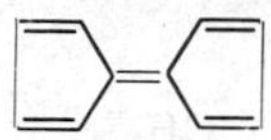

1,1′-Bicyclopentadienylidene
or $\Delta^{1,1'}$-Bicyclopentadienylidene*
(cf. footnote to Rule **B-1.2**)

52.2—If there is a choice in numbering, unprimed numbers are assigned to the system which has the lower-numbered point of attachment.

Example:

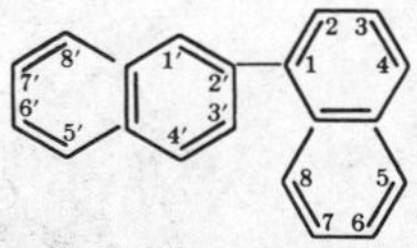

1,2′-Binaphthyl
or 1,2′-Binaphthalene

52.3—If two identical hydrocarbon systems have the same point of attachment and contain substituents at different positions, the locants of these substituents are assigned according to Rule **A-2.2**; for this purpose an unprimed number is considered lower than the same number when primed. Assemblies of primed and unprimed numbers are arranged in ascending numerical order.

Examples:

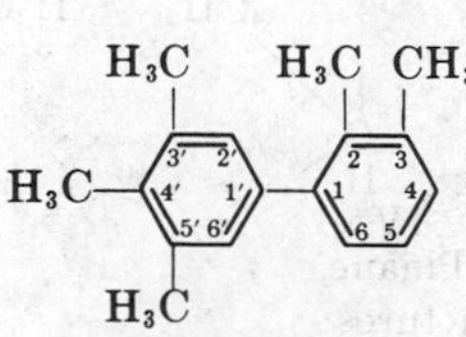

2,3,3′,4′,5′-Pentamethylbiphenyl
(not 2′,3,3′,4,5-Pentamethylbiphenyl)

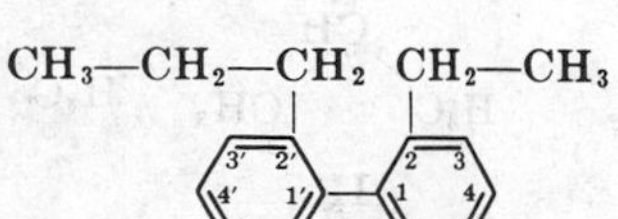

2-Ethyl-2′-propylbiphenyl

52.4—The name "biphenyl" is used for the assembly consisting of two benzene rings.

Biphenyl

CYCLIC HYDROCARBONS WITH SIDE CHAINS**

Rule A-61. General Principles

61.1—Hydrocarbons more complex than those envisioned in Rule **A-12**, composed of cyclic nuclei and aliphatic chains, are named according to one of the methods given below. Choice is made so as to provide the name which is the simplest permissible or the most appropriate for the chemical intent.

* A Greek capital delta (Δ) followed by superscript locants is used to denote the double bond.
** Note: cf. Rules **A-12** and **A-13**.

61.2—When there is no generally recognized trivial name for the hydrocarbon, then (1) the radical name denoting the aliphatic chain is prefixed to the name of the cyclic hydrocarbon, or (2) the radical name for the cyclic hydrocarbon is prefixed to the name of the aliphatic compound. Choice between these methods is made according to the more appropriate of the following principles: (a) the maximum number of substitutions into a single unit of structure; (b) treatment of a smaller unit of structure as a substituent into a larger. Numbering of double and triple bonds in chains or nonaromatic rings is assigned according to the principles of Rule **A-3**; numbering and citation of substituents are effected as described in Rule **A-2**.

61.3—In accordance with the principle (a) of Part .2 of this rule, hydrocarbons containing several chains attached to one cyclic nucleus are generally named as derivatives of the cyclic compound; and compounds containing several side chains and/or cyclic radicals attached to one chain are named as derivatives of the acyclic compound.

Examples:

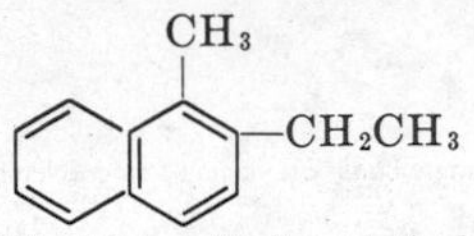

2-Ethyl-1-methylnaphthalene

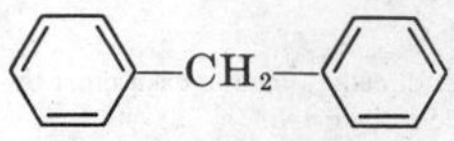

Diphenylmethane

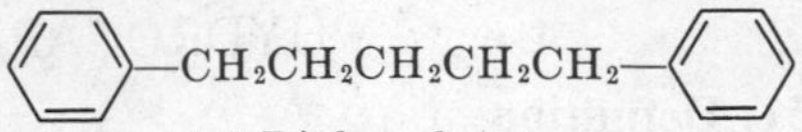

1,5-Diphenylpentane

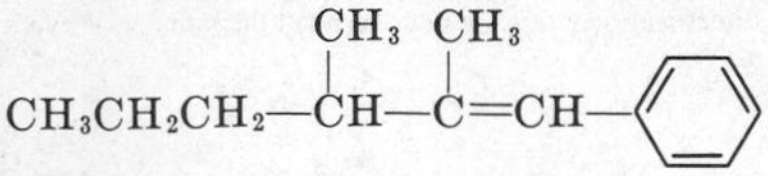

2,3-Dimethyl-1-phenyl-1-hexene

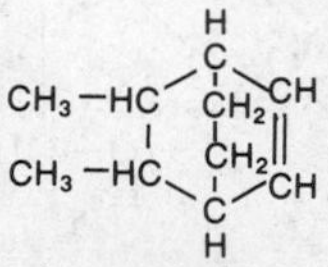

5,6-Dimethylbicyclo[2.2.2]oct-2-ene

TERPENE HYDROCARBONS

Owing to long-established custom, terpenes are given exceptional treatment in these rules.

Rule A-71. Acyclic Terpenes

71.1—The acyclic terpene hydrocarbons are named in a manner similar to that used for other unsaturated acyclic hydrocarbons when compounds with known structures are involved.

Example:

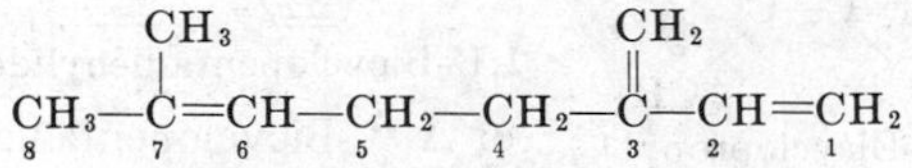

7-Methyl-3-methylene-1,6-octadiene

Rule A-72. Cyclic Terpenes

72.1—The following structural types with their special names and special systems of numbering are used as the basis for the specialized nomenclature of monocyclic and bicyclic terpene hydrocarbons. The name "bornane" replaces camphane and bornylane; "norbornane" replaces norcamphane and norborylane.*

Fundamental terpene types:

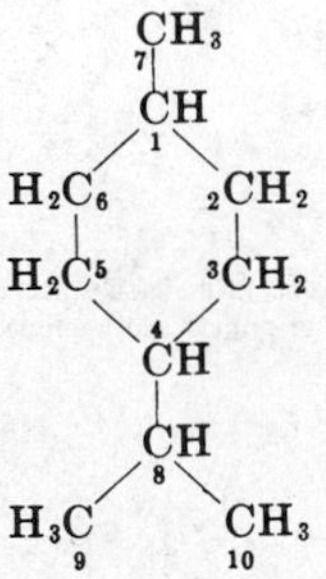

I
Menthane (*p*-form)

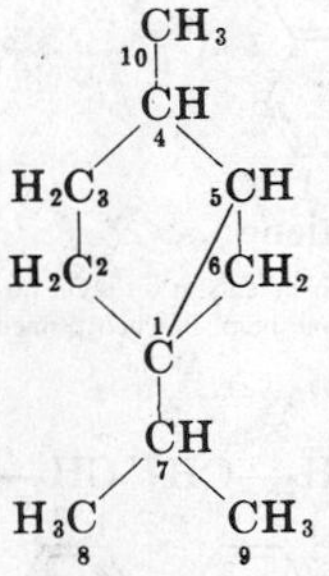

II
Thujane

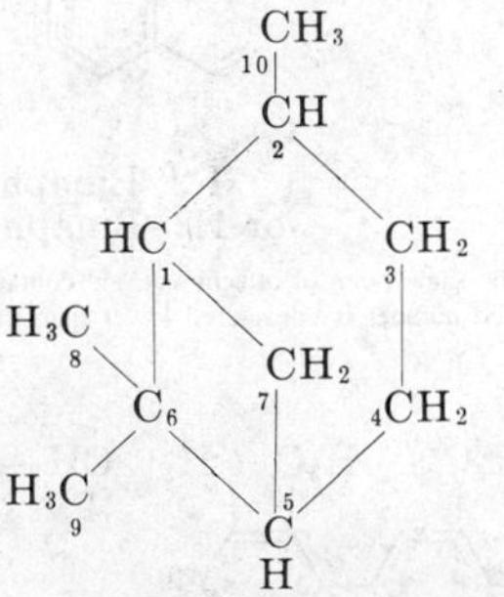

IV
Pinane

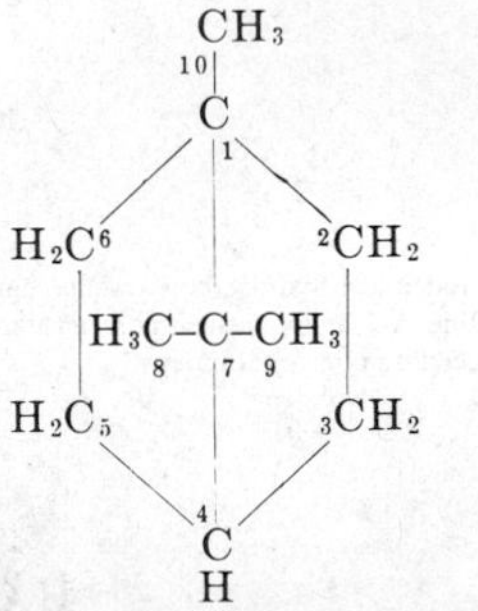

V
Bornane

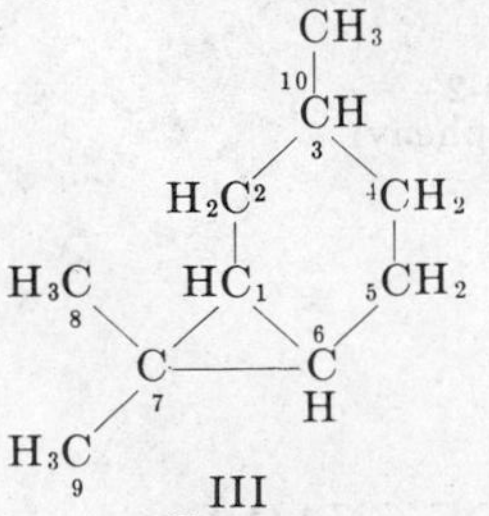

III
Carane

Nor-structures:

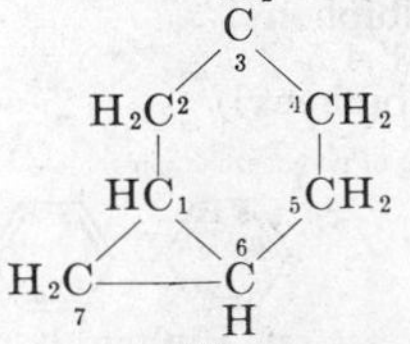

VI
Norcarane

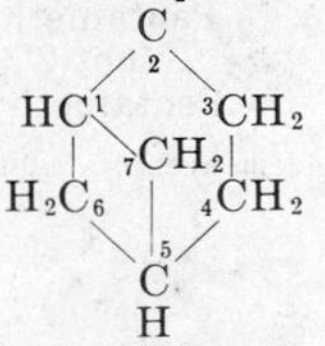

VII
Norpinane

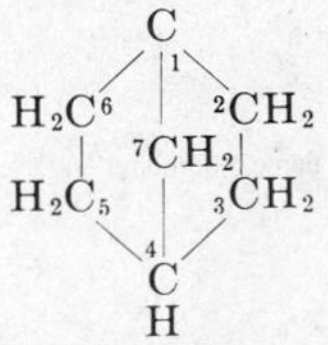

VIII
Norbornane

Rule A-73. Monocyclic Terpenes

73.1—Menthane Type: Monocyclic terpene hydrocarbons of this type (*ortho*-, *meta*-, and *para*-isomers) are named menthane, menthene, menthadiene, etc., and are given the fixed numbering of menthane (Formula I). Such compounds substituted by additional alkyl groups are named in accordance with Rules **A-11** and **A-61**.

* These names have been superseded (cf. Rule **F-4.2**).

Examples:

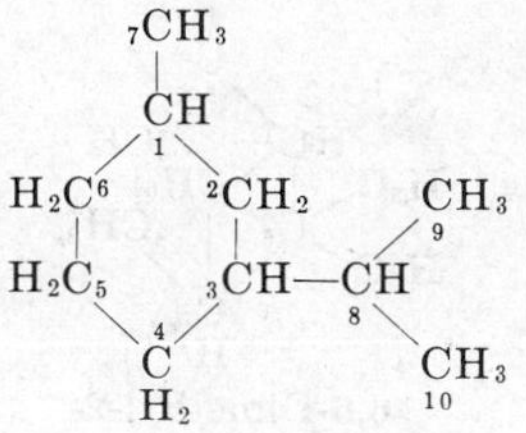

m-Menthane

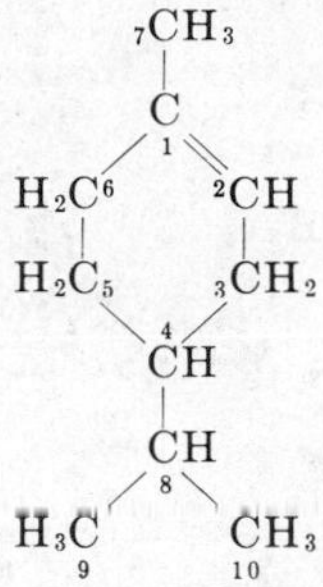

1-*p*-Menthene

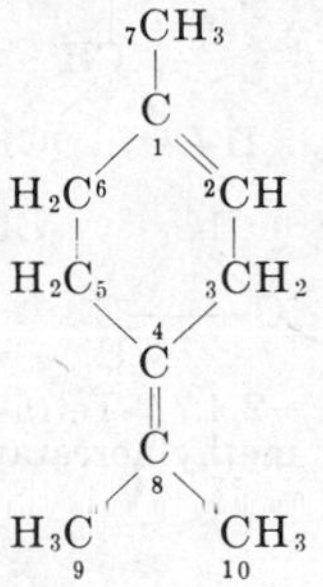

1,4(8)-*p*-Menthadiene

73.2—Tetramethylcyclohexane Type: Monocyclic terpene hydrocarbons of this type are named systematically as derivatives of cyclohexane, cyclohexene, and cyclohexadiene (see Rule **A-11**).

Examples:

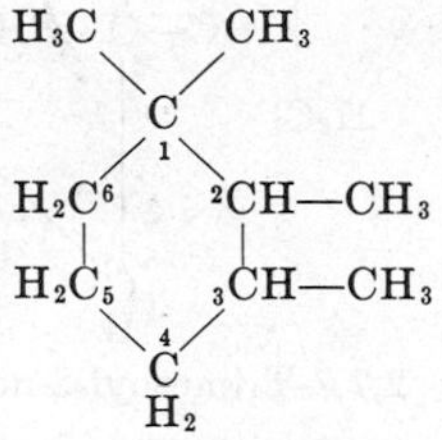

1,1,2,3-Tetramethylcyclohexane

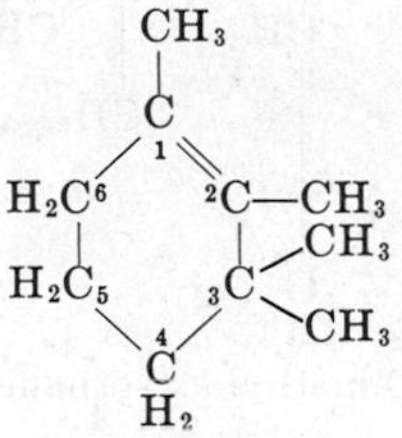

1,2,3,3-Tetramethylcyclohexene

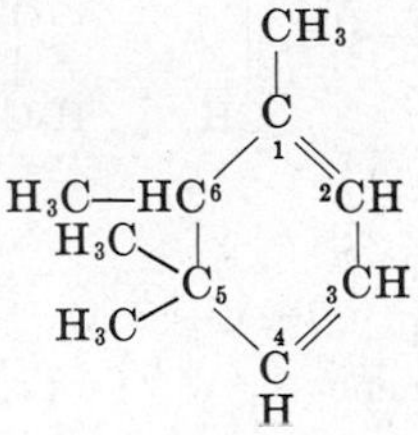

1,5,5,6-Tetramethyl-1,3-cyclohexadiene

Rule A-74. Bicyclic Terpenes

74.1—Bicyclic terpene hydrocarbons having the skeleton of Formula II or this skeleton and additional side chains except methyl or isopropyl (or methylene if one methylene group is already present) are named as thujane, thujene, thujadiene, etc., and are given the fixed numbering shown for thujane (Formula II). Other hydrocarbons containing the thujane ring-skeleton are named from bicyclo[3.1.0]hexane and are given systematic bicyclo numbering (cf. Rule **A-31**).

Examples:

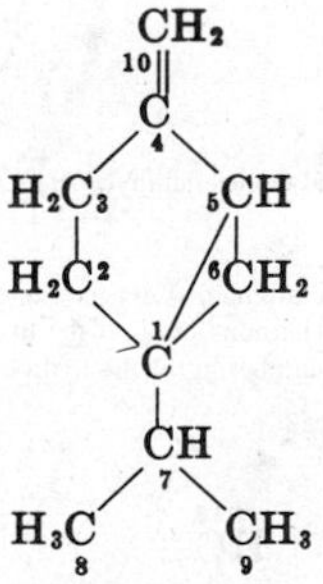

4(10)-Thujene

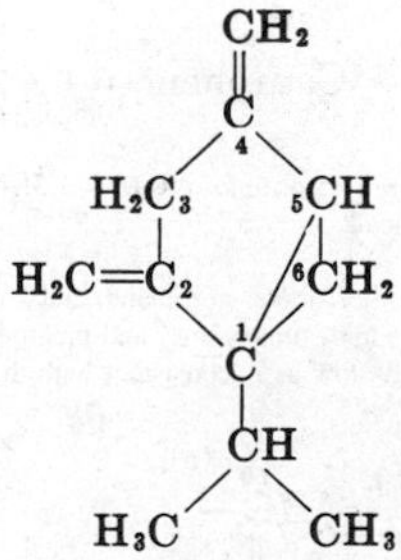

1-Isopropyl-2,4-dimethylenebicyclo-[3.1.0]hexane

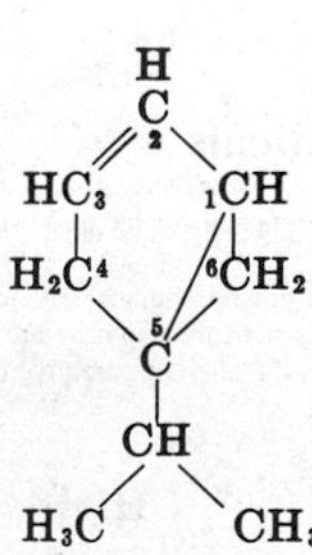

5-Isopropylbicyclo[3.1.0]hex-2-ene

74.2—Bicyclic terpene hydrocarbons having the skeleton of Formula III, IV, or V and additional side chains except methyl (or methylene if one methylene group is already present) are named, respectively, as carane, carene, caradiene, etc.; pinane, pinene, pinadiene, etc.; bornane, bornene, bornadiene, etc. They are given, respectively, the fixed numbering shown for carane (Formula III), pinane (Formula IV), and bornane (Formula V). Other hydrocarbons containing the ring-skeleton of carane, pinane, orbornane are named, respectively, from norcarane (Formula VI), norpinane (Formula VII), or norbornane (Formula VIII). These names are preferred to those from bicyclo[4.1.0]heptane, bicyclo[3.1.1]heptane, or bicyclo[2.2.1]heptane. The nor-names* are given systematic bicyclo numbering (cf. Rule **A-31**).

Examples:

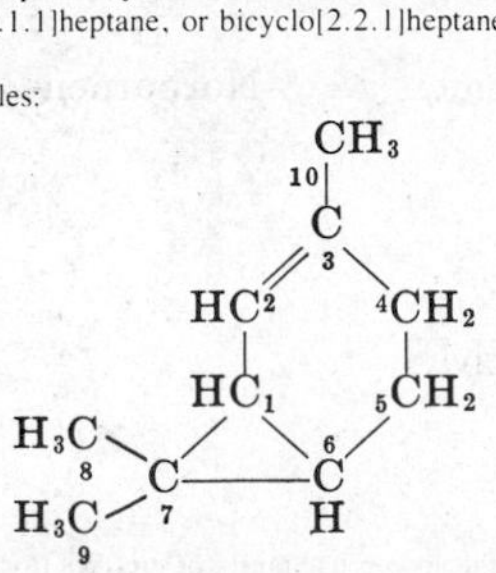

2-Carene

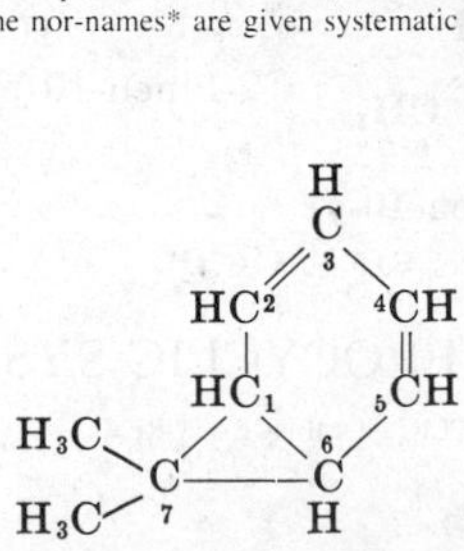

7,7-Dimethyl-2,4-norcaradiene

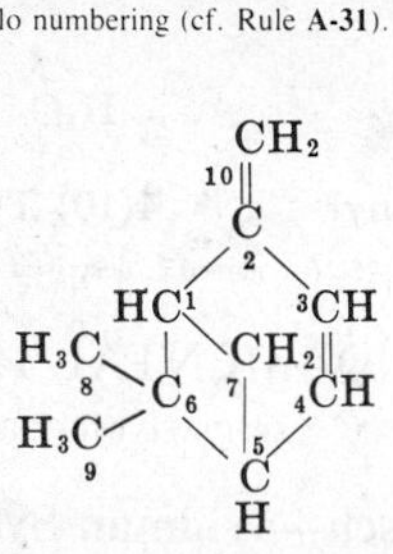

2(10),3-Pinadiene

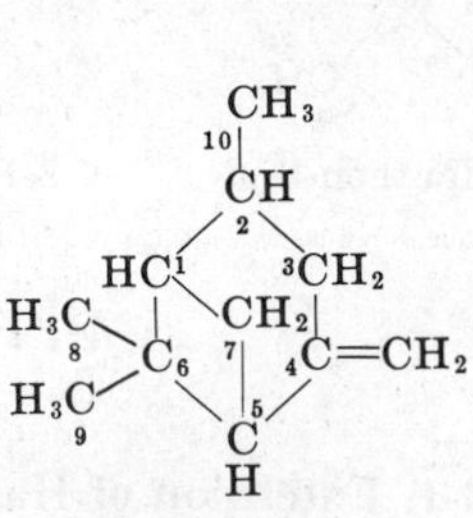

4-Methylenepinane

* These names have been superseded (cf. Rule **F-4.1**).

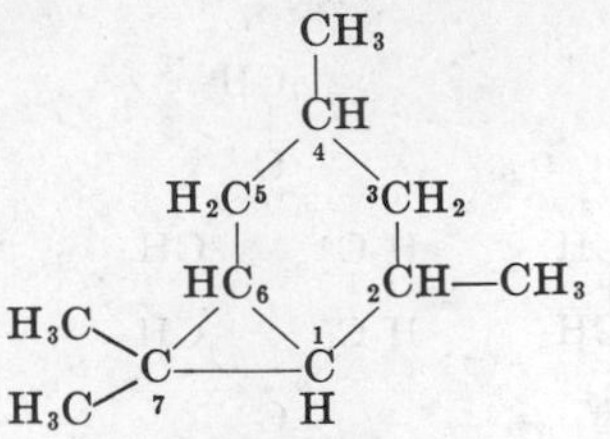

2,4,7,7-Tetra-
methylnorcarane

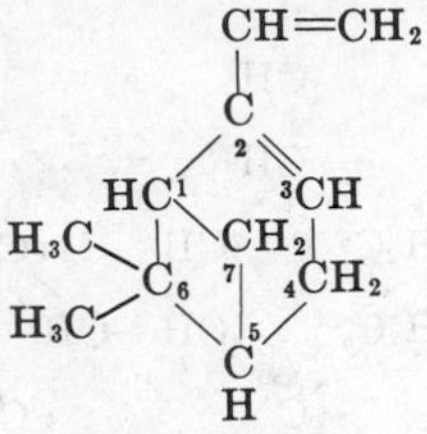

6,6-Dimethyl-2-
vinyl-2-norpinene

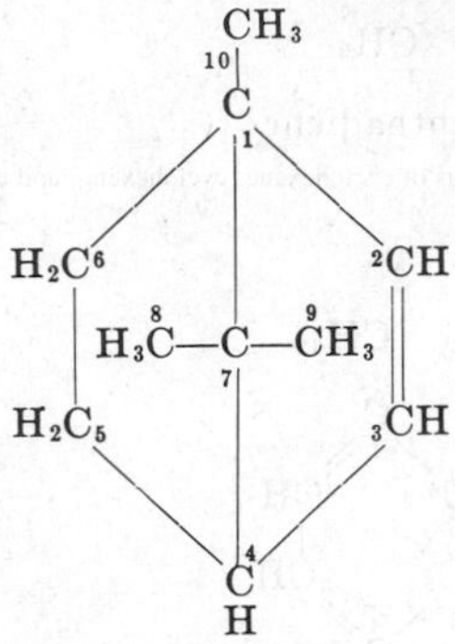

2-Bornene

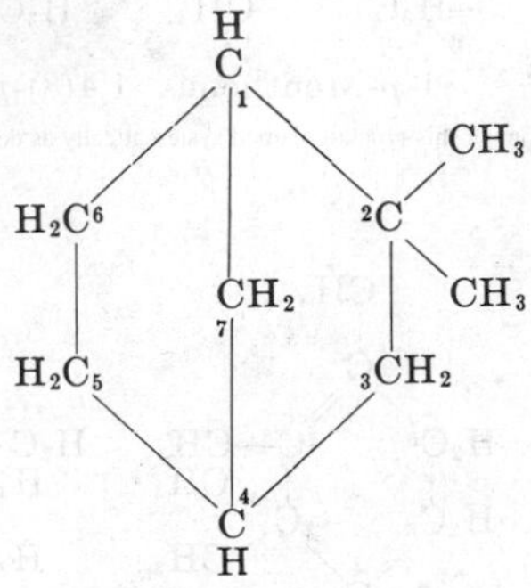

2,2-Dimethylnorbornane

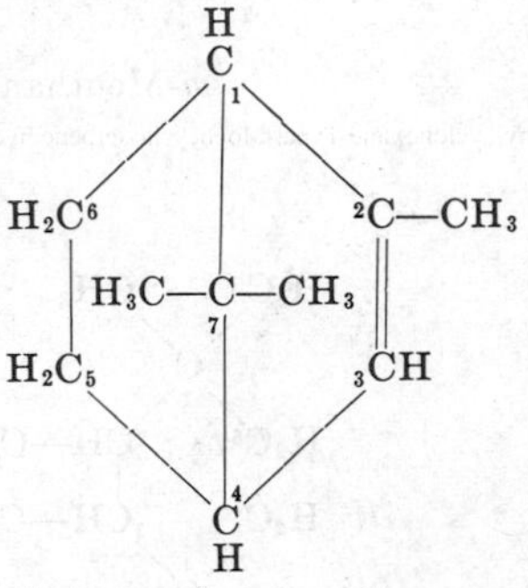

2,7,7-Trimethyl-2-norbornene

74.3—The name "camphene" is retained for the unsubstituted compound 2,2-dimethyl-3-methylenenorbornane.

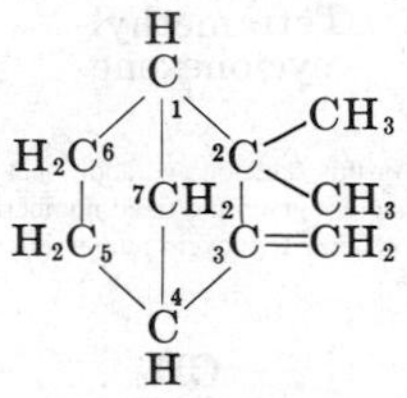

Camphene

Rule A-75. Terpene Radicals

75.1—Simple acyclic hydrocarbon terpene radicals are named and numbered according to Rule **A-3.5**. The trivial names geranyl, neryl, linalyl and phytyl [for (E)-(7R, 11R)-3,7,11,15-tetramethyl-2-hexadecenyl] are retained for the unsubstituted radicals.

75.2—Radicals derived from menthane, pinane, thujane, carane, bornane, norcarane, norpinane, and norbornane are named in accordance with the principles set forth in Rule **A-1.2** and **A-11.4** except that the saturated radicals of pinane are named pinanyl, pinanylene, and pinanylidene. The numbering of the hydrocarbon is retained and the point or points of attachment, whether in the ring or side chain, are given numbers as low as is consistent with the fixed numbering of the hydrocarbon.

Examples:

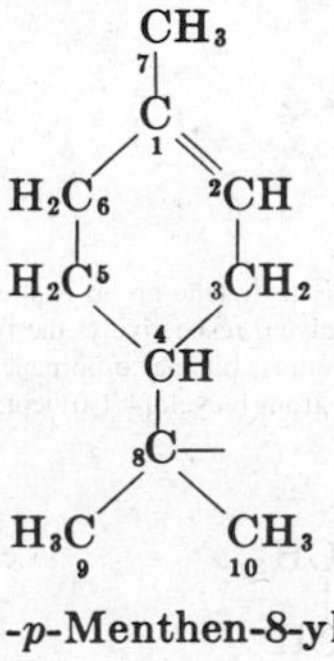

-p-Menthen-8-yl

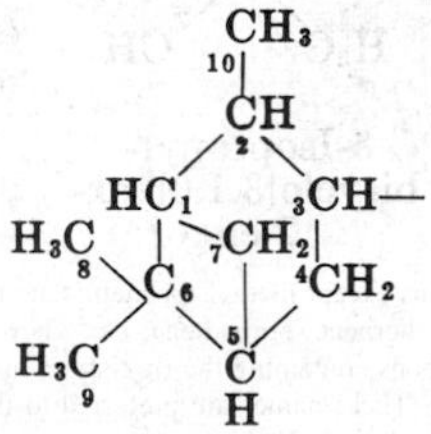

3-Pinanyl

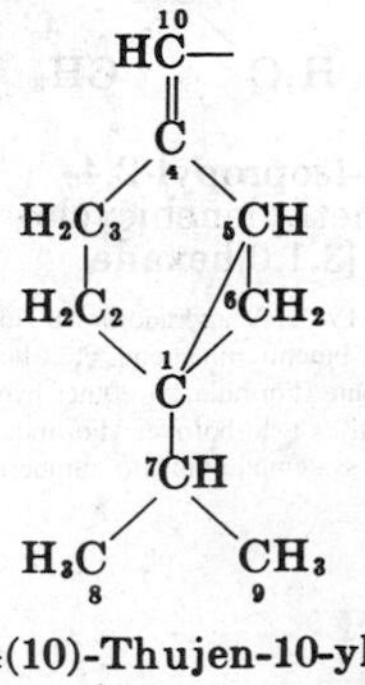

4(10)-Thujen-10-yl

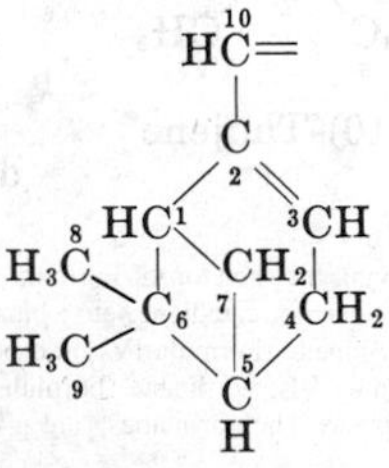

2-Pinen-10-ylidene

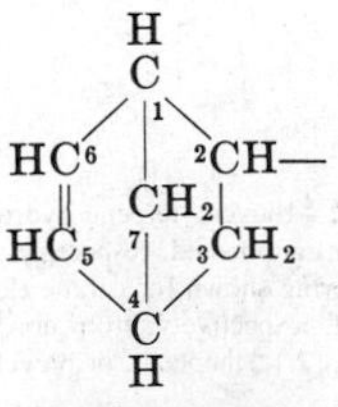

5-Norbornen-2-yl

75.3—Radicals not named in Rules **A-75.1** and **A-75.2** are named as described in Rules **A-11** and **A-31.4**.

B. FUNDAMENTAL HETEROCYCLIC SYSTEMS

SPECIALIST HETEROCYCLIC NOMENCLATURE

Rule B-1. Extension of Hantzsch—Widman System

1.1—Monocyclic compounds containing one or more hetero atoms in a three- to ten-membered ring are named by combining the appropriate prefix or prefixes from Table I (eliding "a" where necessary) with a stem from Table II.* The state of hydrogenation is indicated either in the stem, as shown in Table II, or by the prefixes "dihydro-", "tetrahydro-", etc., according to Rule **B-1.2**.

* It is necessary to elide the final "a" of the prefix when followed immediately by a vowel, e.g. ox(a)azole.

Table 1
IN DECREASING ORDER OF PRIORITY

Element	Valence	Prefix	Element	Valence	Prefix
Oxygen	II	Oxa	Antimony	III	Stiba[a]
Sulfur	II	Thia	Bismuth	III	Bisma
Selenium	II	Selena	Silicon	IV	Sila
Tellurium	II	Tellura	Germanium	IV	Germa
Nitrogen	III	Aza	Tin	IV	Stanna
Phosphorus	III	Phospha[a]	Lead	IV	Plumba
Arsenic	III	Arsa[a]	Boron	III	Bora
			Mercury	II	Mercura

[a] When immediately followed by "-in" or "-ine", "phospha-" should be replaced by "phosphor-", "arsa-" should be replaced by "arsen-" and "stiba-" should be replaced by "antimon-". In addition, the saturated six-membered rings corresponding to phosphorin and arsenin are named phosphorinane and arsenane. Further exceptions: borin is replaced by borinane.

Table II

No. of members in the ring	Rings containing nitrogen		Rings containing no nitrogen	
	Unsaturation[a]	Saturation	Unsaturation[a]	Saturation
3	-irine	-iridine	-irene	-irane[c]
4	-ete	-etidine	-ete	-etane
5	-ole	-olidine	-ole	-olane
6	-ine[b]	[c]	-in[b]	-ane[d]
7	-epine	[c]	-epine	-epane
8	-ocine	[c]	-ocin	-ocane
9	-onine	[c]	-onin	-onane
10[f]	-ecine	[c]	-ecin	-ecane

[a] Corresponding to the maximum number of noncumulative double bonds, the hetero elements having the normal valences shown in Table I.
[b] For phosphorus, arsenic, antimony and boron, see the special provisions of Table 1.
[c] Expressed by prefixing "perhydro" to the name of the corresponding unsaturated compound.
[d] Not applicable to silicon, germanium, tin and lead. In this case, "perhydro-" is prefixed to the name of the corresponding unsaturated compound.
[e] The syllables denoting the size of rings containing 3, 4 or 7-10 members are derived as follows: "ir" from *tri*, "et" from *tetra*, "ep" from *hepta*, "oc" from *octa*, "on" from *nona*, and "ec" from *deca*.
[f] Rings with more than ten members are named by replacement nomenclature (cf. Rule **B-4**).

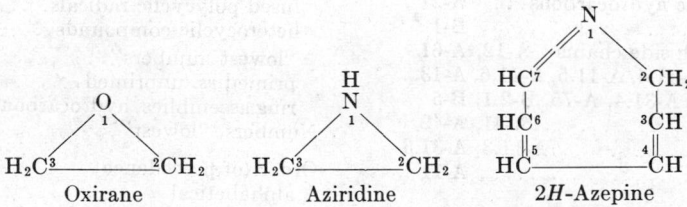

Examples:

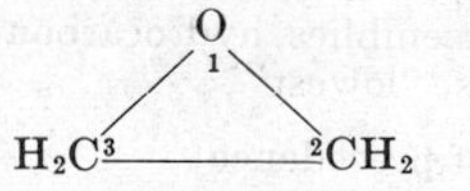
Oxirane

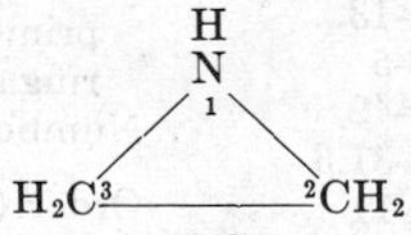
Aziridine

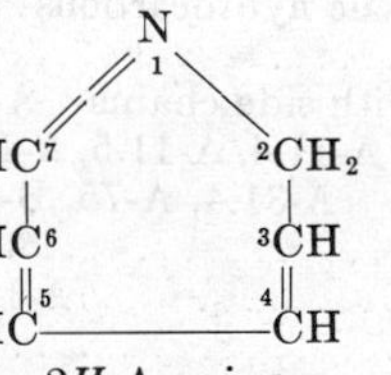
2H-Azepine

1.2—Heterocyclic systems whose unsaturation is less than the one corresponding to the maximum number of noncumulative double bonds are named by using the prefixes "dihydro-", "tetrahydro-", etc.

In the case of 4- and 5-membered rings, a special termination is used for the structures containing one double bond, when there can be more than one non-cumulative double bond.

No. of members of the partly saturated rings	Rings containing nitrogen	Rings containing no nitrogen
4	-etine	-etene
5	-oline	-olene

Examples:

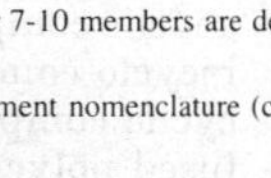
Δ^3-1,2-Azarsetine*

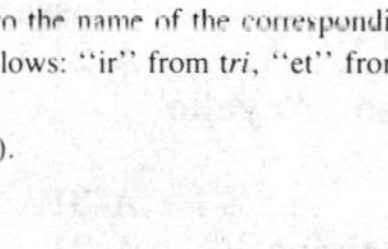
3-Silolene

* As exceptions, Greek capital delta (Δ), followed by superscript locant(s), is used to denote a double bond in a compound named according to Rule **B-1.2** if its name is preceded by locants for hetero atoms; and also to denote a double bond uniting components in an assembly of rings (cf. Examples to Rules **A-52.1** and **C-71.1**) or in conjunctive names (cf. Rule **C-55.1**).

INDEX

ILLUSTRATIVE PREFIXES

acetamido (acetylamino)	CH_3CONH-
acetimido (acetylimino)	$CH_3C(=NH)-$
acetoacetamido	CH_3COCH_2CONH-
acetoacetyl	CH_3COCH_2CO-
acetonyl	CH_3COCH_2
acetonylidene	$CH_3COCH=$
acetyl	CH_3CO-
acrylyl	$CH_2=CHCO-$
adipyl (from adipic acid)	$-OC(CH_2)_4CO-$
alanyl (from alanine)	$CH_3CH(NH_2)CO-$
β-alanyl	$H\,N(CH_2)_2CO-$
allophanoyl	$H_2NCONHCO-$
allyl (2-propenyl)	$CH_2=CHCH_2-$
allylidene (2-propenylidene)	$CH_2=CHCH=$
amidino (aminoiminomethyl)	$H_2NC(=NH)-$
amino	H_2N-
amyl (pentyl)	$CH_3(CH_2)_4-$
anilino (phenylamino)	C_6H_5NH-
anisidino	$CH_3OC_6H_4NH-$
anisyl (from anisic acid)	$CH_3OC_6H_4CO-$
anthranoyl (2-aminobenzoyl)	$2-H_2NC_6H_4CO-$
arsino	AsH_2-
azelaoyl (from azelaic acid)	$-OC(CH_2)_7CO-$
azido	N_3-
azino	$=NN=$
azo	$-N=N-$
azoxy	$-N(O)N-$
benzal	$C_6H_5CH=$
benzamido (benzylamino)	C_6H_5CONH-
benzhydryl (diphenylmethyl)	$(C_6H_5)_2CH-$
benzimido (benzylimino)	$C_6H_5C(=NH)-$
benzoxy (benzoyloxy)	C_6H_5COO-
benzoyl	C_6H_5CO-
benzyl	$C_6H_5CH_2-$
benzylidine	$C_6H_5CH=$
benzyldyne	$C_6H_5C\equiv$
biphenylyl	$C_6H_5C_6H_5-$
biphenylene	$-C_6H_4C_6H_4-$
butoxy	C_4H_9O-
sec-butoxy	$C_2H_5CH(CH_3)O-$
tert-butoxy	$(CH_3)_3CO-$
butyl	$CH_3(CH_2)_3-$
iso-butyl (3-methylpropyl)	$(CH_3)_2(CH_2)_2-$
sec-butyl (1-methylpropyl)	$C_2H_5CH(CH_3)-$
tert-butyl (1,1, dimethylethyl)	$(CH_3)_3C-$
butyryl	C_3H_7CO-
caproyl (from caproic acid)	$CH_3(CH_2)_4CO-$
capryl (from capric acid)	$CH_3(CH_2)_8CO-$
caprylyl (from caprylic acid)	$CH_3(CH_2)_6CO-$
carbamido	$H_2NCONH-$
carbamoyl (aminocarbonyl)	H_2NCO-
carbamyl (aminocarbonyl)	H_2NCO-
carbazoyl (hydrazinocarbonyl)	$H_2NNHCO-$
carbethoxy	$C_2H_5O_2C-$
carbobenzoxy	$C_6H_5CH_2O_2C-$
carbonyl	$-C=O-$
carboxy	$HOOC-$
cetyl	$CH_3(CH_2)_{15}-$
chloroformyl (chlorocarbonyl)	$ClCO-$
cinnamyl (3-phenyl-2-propenyl)	$C_6H_5CH=CHCH_2-$
cinnamoyl	$C_6H_5CH=CHCO-$
cinnamylidene	$C_6H_5CH=CHCH=$
cresyl (hydroxymethylphenyl)	$HO(CH_3)C_6H_4-$
crotoxyl	$CH_3CH=CHCO-$
crotyl (2-butenyl)	$CH_3CH=CHCH_2$
cyanamido (cyanoamino)	$NCNH-$
cyanato	$NCO-$
cyano	$NC-$
decanedioyl	$-OC(CH_2)_8CO-$
decanoly	$CH_3(CH_2)_8CO-$
diazo	$N_2=$
diazoamino	$-NHN-N-$
disilanyl	H_3SiSiH_2-
disiloxanoxy	$H_3SiOSiH_2O-$
disulfinyl	$-S(O)S(O)-$
dithio	$-SS-$
enanthyl	$CH_3(CH_2)_5CO-$
epoxy	$-O-$
ethenyl (vinyl)	$CH_2=CH-$
ethinyl	$HC\equiv C-$
ethoxy	C_2H_5O-
ethyl	CH_3CH_2-
ethylthio	C_2H_5S-
formamido (formylamino)	$HCONH-$
formyl	$HCO-$
fumaroyl (from fumaric acid)	$-OCCH=CHCO-$
furfuryl (2-furanylmethyl)	$OC_4H_3CH_2-$
furfurylidene (2-furanylmethylene)	$OC_4H_3CH=$
furyl (furanyl)	OC_4H_3-
glutamyl (from glutamic acid)	$-OC(CH_2)_2CH(NH_2)CO-$
glutaryl (from glutaric acid)	$-OC(CH_2)CO-$
glycidyl (oxiranylmethyl)	CH_2-CHCH_2-
glycinamido	H_2NCH_2CONH-
glycolyl (hydroxyacetyl)	$HOCH_2CO-$
glycyl (aminoacetyl)	H_2NCH_2CO-
glyoxylyl (oxoacetyl)	$HCOCO-$
guanidino	$H_2NC(=NH)NH-$
guanyl	$H_2NC(=NH)-$
heptadecanoyl	$CH_3(CH_2)_{15}CO-$
heptanamido	$CH_3(CH_2)_{15}CONH-$
heptanedioyl	$-OC(CH_2)_5CO-$
heptanoyl	$CH_3(CH_2)_5CO-$
hexadecanoyl	$CH_3(CH_2)_4CO-$
hexamethylene	$-(CH_2)_6-$
hexanedioyl	$-OC(CH_2)_4CO-$
hippuryl (N-benzoylglycyl)	$C_6H_5CONHCH_2CO-$
hydantoyl	$H_2NCONHCH_2CO-$
hydrazino	N_2NNH-
hydrazo	$-HNNH-$
hydrocinnamoyl	$C_6H_5(CH_2)_2CO-$

hydroperoxy	$HOO-$
hydroxamino	$HONH-$
hydroxy	$HO-$
imino	$HN=$
iodoso	$OI-$
isoamyl (isopentyl)	$(CH_3)_2CH(CH_2)_2-$
isobutenyl (2-methyl-1-propenyl)	$(CH_3)_2C=CH-$
isobutoxy	$(CH_3)_2CHCH\ O-$
isobutyl	$(CH_3)_2CHCH_2-$
isobutylidene	$(CH_3)_2CHCH=$
isobutyryl	$(CH_3)_2CHCO-$
isocyanato	$OCN-$
isocyano	$CN-$
isohexyl	$(CH_3)_2CH(CH_2)_3-$
isoleucyl (from isoleucine)	$C_2H_5CH(CH_3)CH(NH_2)CO-$
isonitroso	$HON=$
isopentyl	$(CH_3)_2CH(CH_2)_2-$
isopentylidene	$(CH_3)_2CHCH_2CH=$
isopropenyl	$H_2C=C(CH_3)-$
isopropoxy	$(CH_3)_2CHO-$
isopropyl	$(CH_3)_2CH-$
isopropylidene	$(CH_3)_2C=$
isothiocyanato (isothiocyano)	$SCN-$
isovaleryl (from isovaleric acid)	$(CH_3)_2CHCH_2CO-$
keto (oxo)	$O=$
lactyl (from lactic acid)	$CH_3CH(OH)CO-$
lauroyl (from lauric acid)	$CH_3(CH_2)_{10}CO-$
leucyl (from leucine)	$(CH_3)_2CHCH_2CH(NH_2)CO-$
levulinyl (From levulinic acid)	$CH_3CO(CH_2)_2CO-$
malonyl (from malonic acid)	$-OCCH_2CO-$
mandelyl (from mandelic acid)	$C_6H_5CH(OH)CO-$
mercapto	$HS-$
methacrylyl (from methacrylic acid)	$CH_2=C(CH_3)CO-$
methallyl	$CH_2=C(CH_3)CH_2-$
methionyl (from methionine)	$CH_3SCH_2CH_2CH(NH_2)CO-$
methoxy	CH_3O-
methyl	H_3C-
methylene	$H_2C=$
methylenedioxy	$-OCH_2O-$
methylenedisulfonyl	$-O_2SCH_2SO_2-$
methylol	$HOCH_2-$
methylthio	CH_3S-
myristyl (from myristic acid)	$CH_3(CH_2)_{12}CO-$
naphthal	$(C_{10}H_7)CH=$
naphthobenzyl	$(C_{10}H_7)CH_2-$
naphthoxy	$(C_{10}H_7)O-$
naphthyl	$(C_{10}H_7)-$
naphthylidene	$(C_{10}H_6)=$
neopentyl	$(CH_3)_3CCH_2-$
nitramino	O_2NNH-
nitro	O_2N-
nitrosamino	$ONNH-$
nitrosimino	$ONN=$
nitroso	$ON-$
nonanoyl (from nonanoic acid)	$CH_3(CH_2)_7CO-$
oleyl (from oleic acid)	$CH_3(CH_2)_7CH=CH(CH_2)_7CO-$
oxalyl (from oxalic acid)	$-OCCO-$
oxamido	$H_2NCOCONH-$
oxo (keto)	$O=$
palmityl (from palmitic acid)	$CH_3CH_2)_{14}CO-$
pelargonyl (from pelargonic acid)	$CH_3(CH_2)_7CO-$
pentamethylene	$-(CH_2)_5-$
pentyl	$CH_3(CH_2)_4-$
phenacyl	$C_6H_5COCH_2-$
phenacylidene	$C_6H_5COCH=$
phenanthryl	$(C_{14}H_9)-$
phenethyl	$C_6H_5CH_2CH_2-$
phenoxy	C_6H_5O-
phenyl	C_6H_5-
phenylene	$-C_6H_4-$
phenylenedioxy	$-OC_6H_4O-$
phosphino	H_2P-

phosphinyl	$H_2P(O)-$
phospho	O_2P-
phosphono	$(HO)_2P(O)-$
phthalyl (from phthalic acid)	$1,2-C_6H_4(CO-)_2$
picryl (2,4,6-trinitrophenyl)	$2,4,6-(NO_2)_3C_6H_2-$
pimelyl (from pimelic acid)	$-OC(CH_2)_5CO-$
piperidino	$C_5H_{10}N-$
piperidyl (piperidinyl)	$(C_5H_{10}N)-$
piperonyl	$3,4-(CH_2O_2)C_6H_3CH_2-$
pivalyl (from pivalic acid)	$(CH_3)_3CCO-$
prenyl (3-methyl-2-butenyl)	$(CH_3)_2C=CHCH_2-$
propargyl (2-propynyl)	$HC\equiv CCH_2-$
propenyl	$CH_2=CHCH_2-$
iso-propenyl	$(CH_3)_2C=$
propionyl	CH_3CH_2CO-
propoxy	$CH_3CH_2CH_2O-$
propyl	$CH_3CH_2CH_2-$
iso-propyl	$(CH_3)_2CH-$
propylidene	$CH_3CH_2CH=$
pyridino	C_5H_5N-
pyridyl (pyridinyl)	$(C_5H_4N)-$
pyrryl (pyrrolyl)	$(C_4H_4N)-$
salicyl (2-hydroxybenzoyl)	$2-HOC_6H_4CO-$
selenyl	$HSe-$
seryl (from serine)	$HOCH_2CH(NH_2)CO-$
siloxy	H_3SiO-
silyl	H_3Si-
silylene	$H_2Si=$
sorbyl (from sorbic acid)	$CH_3CH=CHCH=CHCO-$
stearyl (from stearic acid)	$CH_3(CH_2)_{16}CO-$
styryl	$C_6H_5CH=CH-$
suberyl (from suberic acid)	$-OC(CH_2)_6CO-$
succinamyl	$H_2NCOCH_2CH_2CO-$
succinyl (from succinic acid)	$-OCCH_2CH_2CO-$
sulfamino	$HOSO_2NH-$
sulfamyl	H_2NSO-
sulfanilyl	$4-H_2NC_6H_4SO_2-$
sulfeno	$HOS-$
sulfhydryl (mercapto)	$HS-$
sulfinyl	$OS=$
sulfo	HO_3S-
sulfonyl	$-SO_2-$
terephthalyl	$1,4-C_6H_4(CO-)_2$
tetramethylene	$-(CH_2)_4-$
thenyl	$(C_4H_3S)CH-$
thienyl	$(C_4H_3S)-$
thiobenzoyl	C_6H_5CS-
thiocarbamyl	H_2NCS-
thiocarbonyl	$-CS-$
thiocarboxy	$HOSC-$
thiocyanato	$NCS-$
thionyl (sulfinyl)	$-SO-$
thiophenacyl	$C_6H_5CSCH_2-$
thiuram (aminothioxomethyl)	H_2NCS-
threonyl (from threonine)	$CH_3CH(OH)CH(NH_2)CO-$
toluidino	$CH_3C_6H_4NH-$
toluyl	$CH_3C_6H_4CO-$
tolyl (methylphenyl)	$CH_3C_6H_4-$
α-tolyl	$C_6H_5CH_2-$
tolylene (methylphenylene)	$(CH_3C_6H_3)=$
α-tolylene	$C_6H_5CH=$
tosyl [(4-methylphenyl) sulfonyl)]	$4-CH_3C_6H_4SO_2-$
triazano	$H_2NNHNH-$
trimethylene	$-(CH_2)_3-$
triphenylmethyl (trityl)	$(C_6H_5)_3C-$
tyrosyl (from tyrosine)	$4-HOC_6H_4CH_2CH(NH_2)CO-$
ureido	$H_2NCONH-$
valeryl (from valeric acid)	C_4H_9CO
valyl (from valine)	$(CH_3)_2CHCH(NH_2)CO-$
vinyl	$CH_2=CH-$
vinylidene	$CH_2=C=$
xenyl (biphenylyl)	$C_6H_5C_6H_4-$
xylidino	$(CH_3)_2C_6H_3NH-$
xylyl (dimethylphenyl)	$(CH_3)_2C_6H_3-$
xylylene	$-CH_2C_6H_4CH_2-$

ORGANIC RING COMPOUNDS

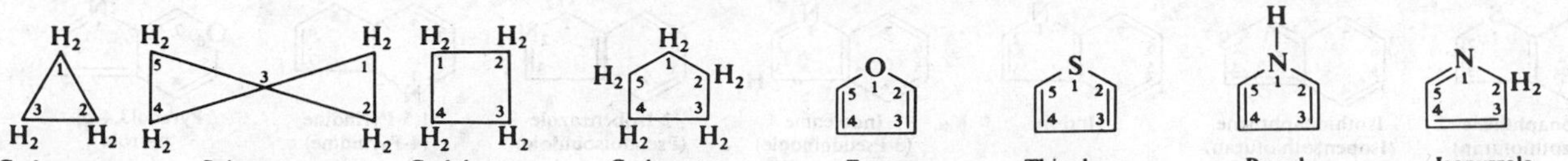

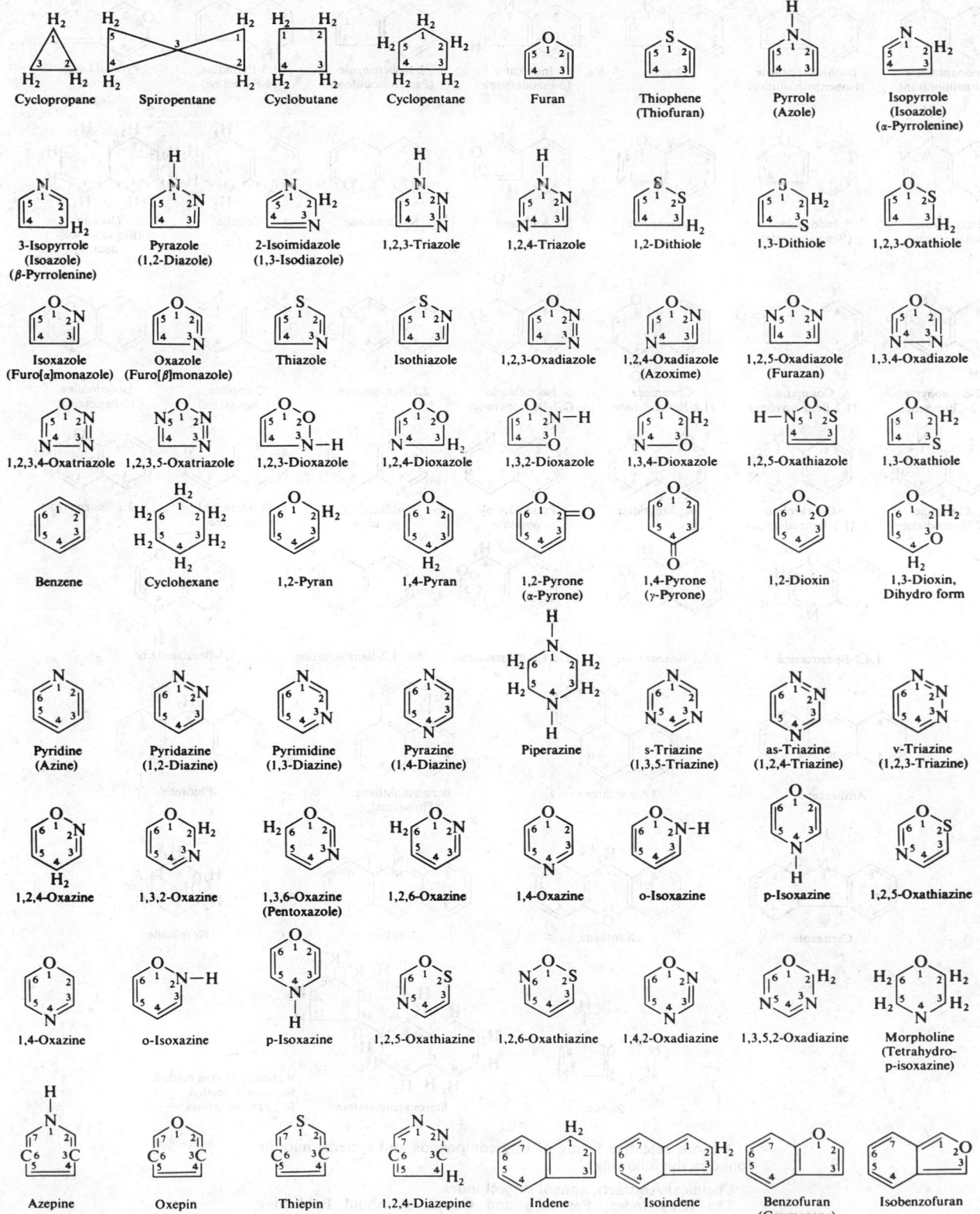

Cyclopropane | Spiropentane | Cyclobutane | Cyclopentane | Furan | Thiophene (Thiofuran) | Pyrrole (Azole) | Isopyrrole (Isoazole) (α-Pyrrolenine)

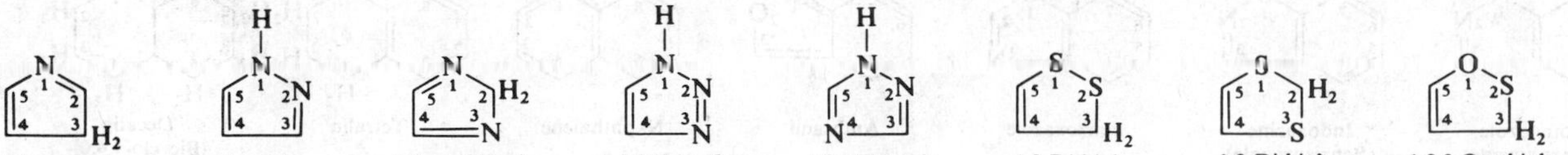

3-Isopyrrole (Isoazole) (β-Pyrrolenine) | Pyrazole (1,2-Diazole) | 2-Isoimidazole (1,3-Isodiazole) | 1,2,3-Triazole | 1,2,4-Triazole | 1,2-Dithiole | 1,3-Dithiole | 1,2,3-Oxathiole

Isoxazole (Furo[α]monazole) | Oxazole (Furo[β]monazole) | Thiazole | Isothiazole | 1,2,3-Oxadiazole | 1,2,4-Oxadiazole (Azoxime) | 1,2,5-Oxadiazole (Furazan) | 1,3,4-Oxadiazole

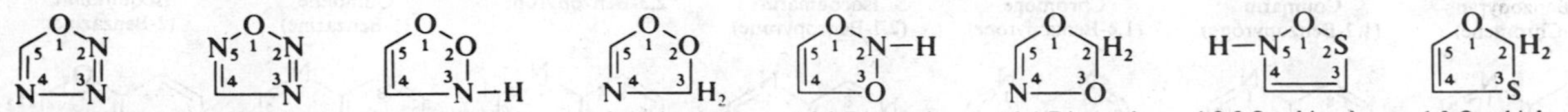

1,2,3,4-Oxatriazole | 1,2,3,5-Oxatriazole | 1,2,3-Dioxazole | 1,2,4-Dioxazole | 1,3,2-Dioxazole | 1,3,4-Dioxazole | 1,2,5-Oxathiazole | 1,3-Oxathiole

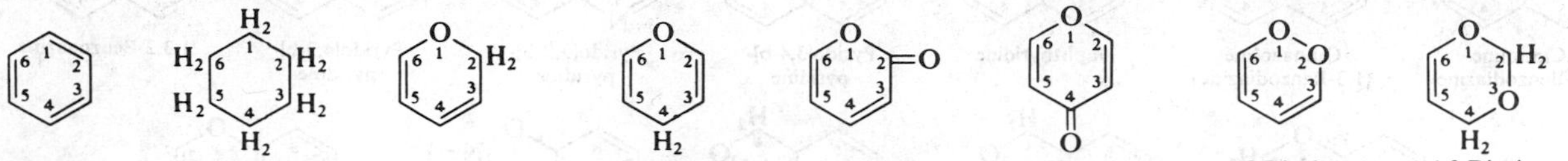

Benzene | Cyclohexane | 1,2-Pyran | 1,4-Pyran | 1,2-Pyrone (α-Pyrone) | 1,4-Pyrone (γ-Pyrone) | 1,2-Dioxin | 1,3-Dioxin, Dihydro form

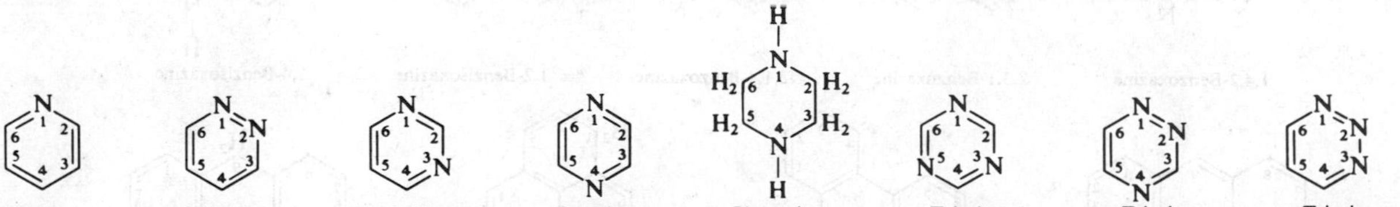

Pyridine (Azine) | Pyridazine (1,2-Diazine) | Pyrimidine (1,3-Diazine) | Pyrazine (1,4-Diazine) | Piperazine | s-Triazine (1,3,5-Triazine) | as-Triazine (1,2,4-Triazine) | v-Triazine (1,2,3-Triazine)

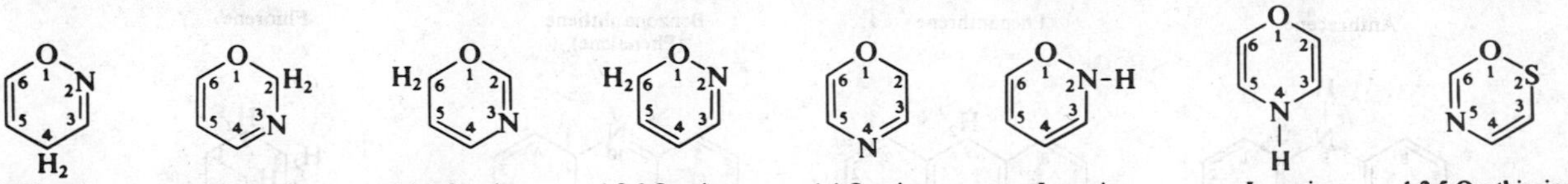

1,2,4-Oxazine | 1,3,2-Oxazine | 1,3,6-Oxazine (Pentoxazole) | 1,2,6-Oxazine | 1,4-Oxazine | o-Isoxazine | p-Isoxazine | 1,2,5-Oxathiazine

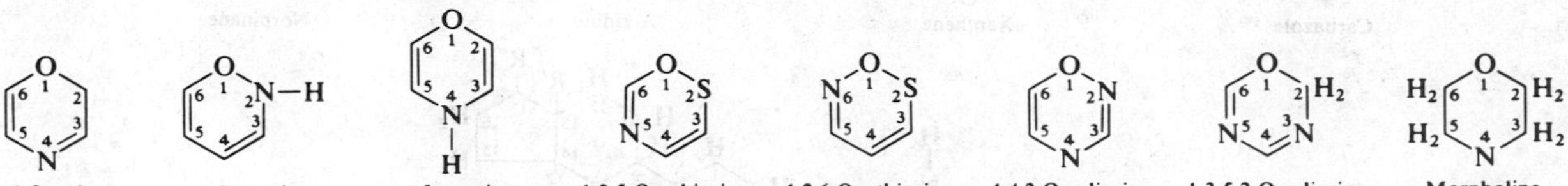

1,4-Oxazine | o-Isoxazine | p-Isoxazine | 1,2,5-Oxathiazine | 1,2,6-Oxathiazine | 1,4,2-Oxadiazine | 1,3,5,2-Oxadiazine | Morpholine (Tetrahydro-p-isoxazine)

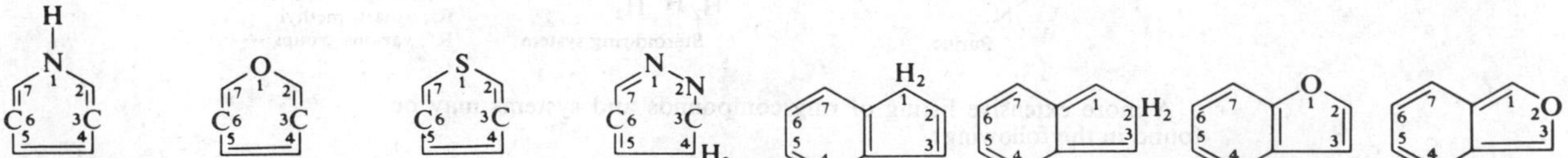

Azepine | Oxepin | Thiepin | 1,2,4-Diazepine | Indene | Isoindene | Benzofuran (Coumarone) | Isobenzofuran

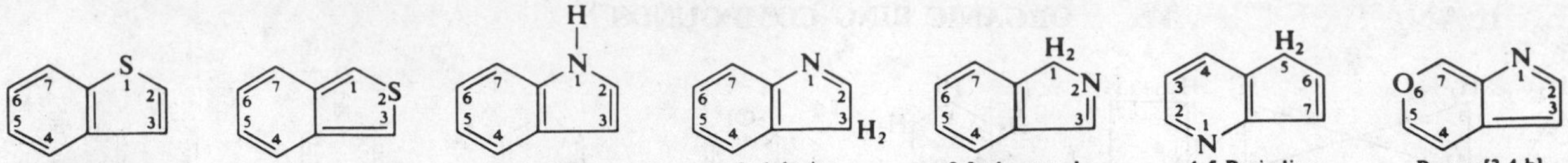

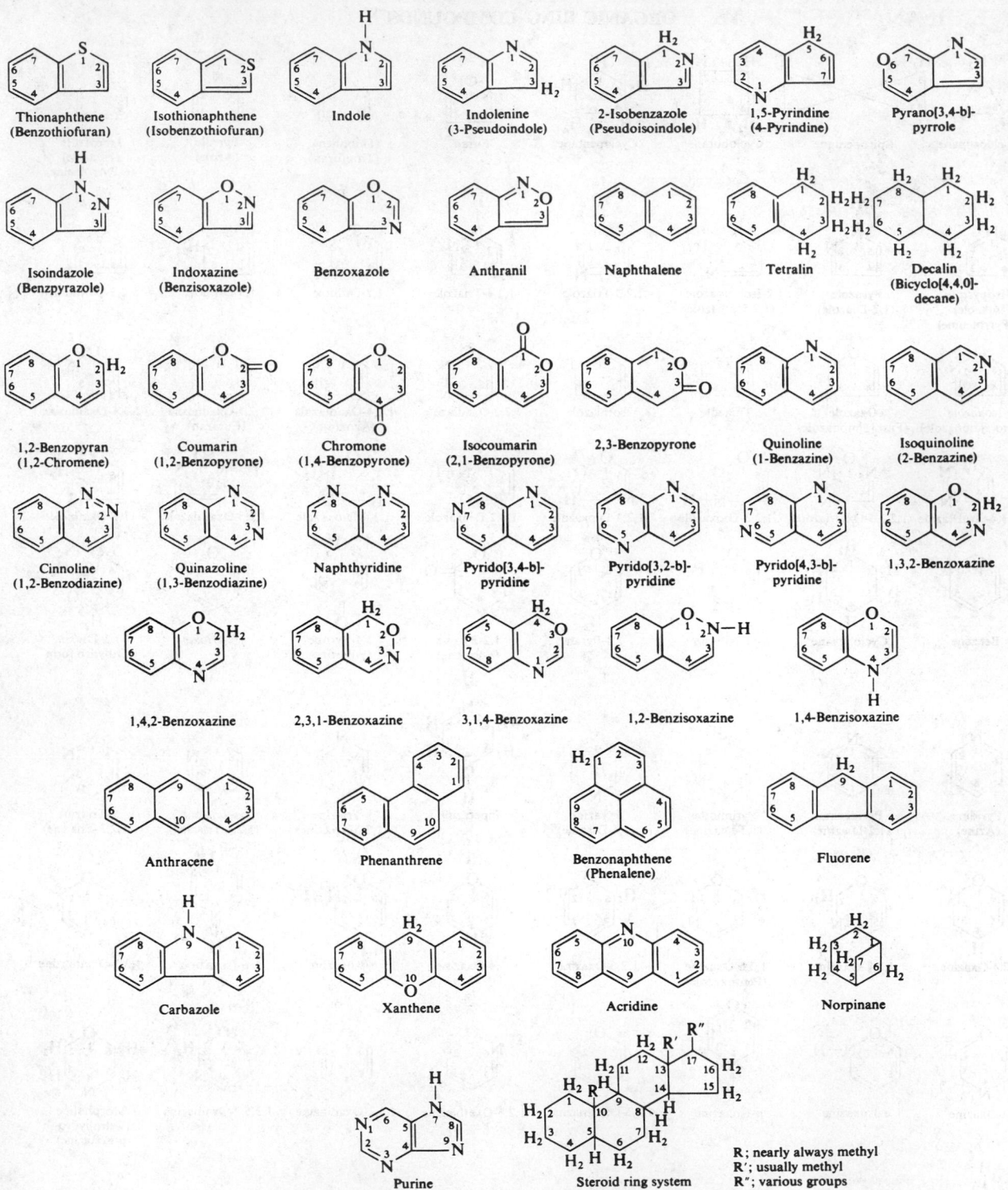

Thionaphthene (Benzothiofuran) · Isothionaphthene (Isobenzothiofuran) · Indole · Indolenine (3-Pseudoindole) · 2-Isobenzazole (Pseudoisoindole) · 1,5-Pyrindine (4-Pyrindine) · Pyrano[3,4-b]-pyrrole

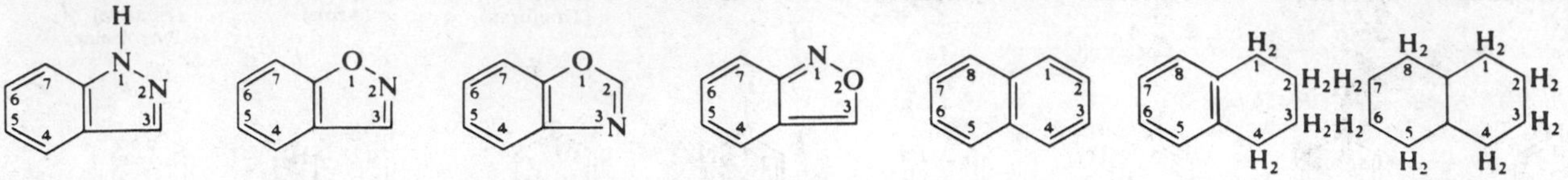

Isoindazole (Benzpyrazole) · Indoxazine (Benzisoxazole) · Benzoxazole · Anthranil · Naphthalene · Tetralin · Decalin (Bicyclo[4,4,0]-decane)

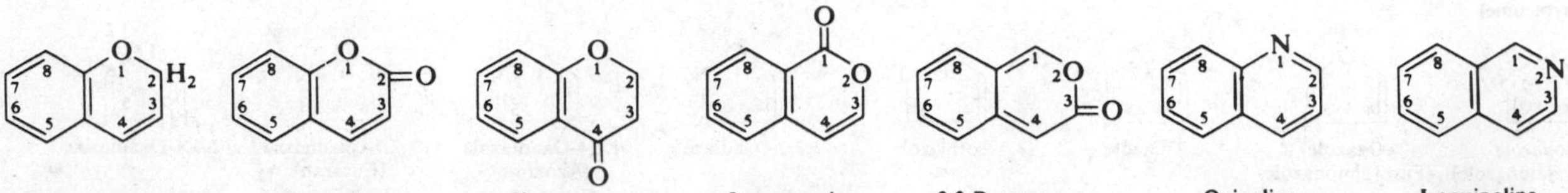

1,2-Benzopyran (1,2-Chromene) · Coumarin (1,2-Benzopyrone) · Chromone (1,4-Benzopyrone) · Isocoumarin (2,1-Benzopyrone) · 2,3-Benzopyrone · Quinoline (1-Benzazine) · Isoquinoline (2-Benzazine)

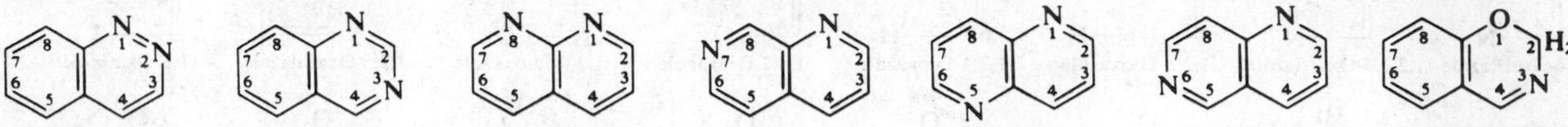

Cinnoline (1,2-Benzodiazine) · Quinazoline (1,3-Benzodiazine) · Naphthyridine · Pyrido[3,4-b]-pyridine · Pyrido[3,2-b]-pyridine · Pyrido[4,3-b]-pyridine · 1,3,2-Benzoxazine

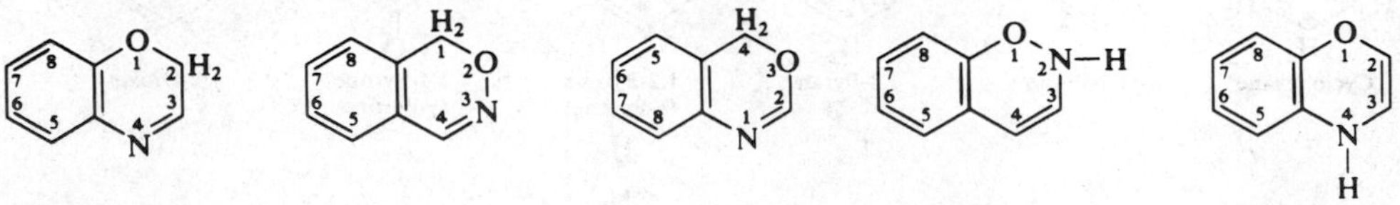

1,4,2-Benzoxazine · 2,3,1-Benzoxazine · 3,1,4-Benzoxazine · 1,2-Benzisoxazine · 1,4-Benzisoxazine

Anthracene · Phenanthrene · Benzonaphthene (Phenalene) · Fluorene

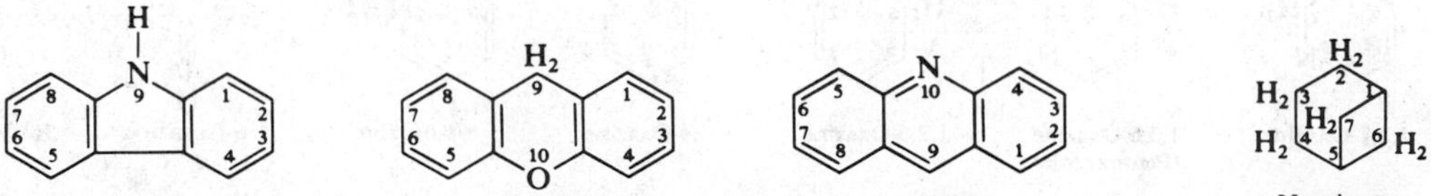

Carbazole · Xanthene · Acridine · Norpinane

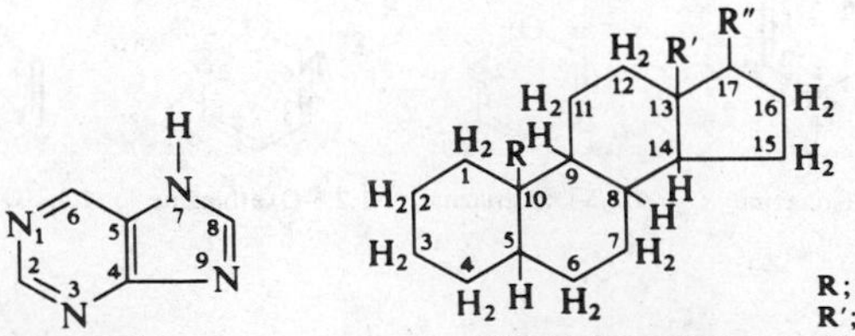

Purine · Steroid ring system

R; nearly always methyl
R′; usually methyl
R″; various groups

A more extensive listing of ring compounds and systems may be found in the following:

Chemical Abstracts, annual subject index.
The Ring Index, Patterson and Capell, Reinhold Publishing Company, 1940.
Lexikon der Kohlenstoffverbindungen, Richter, Leopold Voss, 1910.

The numbering system for the compounds listed above is that used in The Ring Index.

EXPLANATION OF TABLE PHYSICAL CONSTANTS OF ORGANIC COMPOUNDS

This table is a compilation of data on some 15,000 compounds of interest to chemists working in industrial or academic research as well as those in other areas having an occasional need for physical constant data.

An effort has been made to use names which are in most common usage with cross-reference to I.U.C. derived names. As far as possible, compounds are listed alphabetically, but derivatives are listed under the parent compound heading. Thus esters are listed under the name of the parent acid.

Frequent use is made of synonyms. As an example $C_3H_7CO_2H$ is listed under both butyric acid and butanoic acid but the derivatives, for the most part, are listed under butyric acid. As a general rule, if a derivative of a compound is not listed under the name in the first column, look for it to be listed as a derivative of the synonym.

Each compound has been given a number which is used in the formula, melting point, and boiling point indexes found at the end of the table.

References to original literature or other Compendia are given whenever possible (see Table of Abbreviations). The most important of these are, B (Berichte der Deutsche Chemische Gesselschaft), and A (Atlas for Spectral Data and Physical Constants).

Structural Formulas are given whenever possible. More complex formulas are given at the end of the Table.

Molecular Weights are obtained from recently published values of Atomic Weights.

The color of crystalline compounds is listed by abbreviations (see Table of Abbreviations) associated with the crystalline form. Solid compounds are considered as crystalline if not otherwise stated. Data on solvents of crystallization are given in brackets. (see Table of Abbreviations)

The following are specific examples:

pl	**Plates**
pl (al)	Plates obtained from alcohol as solvent
pl (al + 1½)	Plates obtained from alcohol with 1½ mol of alcohol of crystallization
pl (aq ace + 2w)	Solvent of crystallization is one of components of solvent, e.g. 2 mol of water
(al)	Crystalized from alcohol; crystalline form not reported.

Specific Rotation follows common usage e.g.

$$[\alpha]^{25}_{D} = -25.8 \text{ (w,c = 4)} \qquad \text{w = water} \quad \text{c = 4\%}$$

Melting Points are rounded to the nearest 0.1°C. A second, but probably less reliable, melting point is listed in brackets for a few compounds.

Boiling Points are at atmospheric pressure (760 mm) unless otherwise indicated by a superscript.

Density (specific Gravity). In most cases specific gravities are given. Thus $0.86^{20}{}_4$ indicates the density of the liquid at 20°C relative to the density of water at 4°C. Where only one temperature is given as a superscript, the value is in grams per milliliter at the indicated temperature.

The **Refractive Index** is reported for the D line of the sodium spectrum (η_D) at the temperature indicated by the superscript. In a few cases other spectral lines are indicated.

Common Solvents for each compound are given as a guide to solubility. No distinction is made between soluble and very soluble primarily because definitions for these terms differ widely among different investigators. Solvents in which the compounds are soluble in all proportions are underlined.

Three indexes are located at the end of the Table. These are:

1. **Empirical Formula Index** which is arranged according to increasing C, H, and remaining elements in alphabetical order. Hydrates are entered under the formula of the anhydrous compound. Salts, complexes, etc. are found under their total formula.

2. **Melting Point Index** which is arranged according to increasing melting point or freezing point. Only the lower temperature of the melting point range is given. Melting points are rounded off to the nearest unit.

3. **Boiling Point Index** which is arranged according to increasing boiling point of compounds of which the boiling point is given at a pressure near to 1 atmosphere, (i.e., 700 to 780 mm Hg). Only the lower temperature of the boiling range is given. No special entries are made for the remarks on the boiling point contained in the main table. Boiling points listed in the Index are rounded off to the nearest unit.

RULES FOR THE NAMING OF COMPOUNDS

1. Compounds are given common names when these are widely used. These compounds are usually cross-indexed to I.U.C. derived names. e.g., $HO_2CCH_2CH_2CO_2H$ is listed as succinic acid and also as butandioic acid. Physical constants are listed for both entries.

2. Derivatives of a parent compound having one principal function are listed under this compound; e.g. Ethyl butyrate is listed under butyric acid. Chloroacetic acid has a separate entry with many derivatives listed under it. However, 2-chlorobutyric acid is listed as a derivative of butyric acid because there are a limited number of chlorobutyric acid derivatives.

3. Whenever a derivative is not found under the name of the parent compound, look for it under the name of the synonym, or as a principal entry. Derivatives of derivatives are not in the table.

Aliphatic Hydrocarbons

4. I.U.C. nomenclature is used.

a) Saturated hydrocarbons are named as alkanes. Branched chain compounds are named as derivatives of the longest chain.

b) Hydrocarbons having carbon-carbon double bonds are named as alkenes, alkadienes, alkatrienes, etc., with the position of the double bonds indicated by suitable numbers. Branched chain compounds are named as derivatives of the longest chain that contains the maximum number of double bonds, e.g.

$CH_3CH{=}CHCH_3$	2-Butene
$CH_2{=}CHCH{=}CH_2$	1.3-Butadiene
$CH_2{=}CHCH(C_3H_7)CH{=}CH_2$	3-Propyl+1.4-pentadiene

c) Hydrocarbons having carbon-carbon triple bonds are named as alkynes. Branched chain alkynes are named as derivatives of the longest chain containing the triple bonds. The position of the triple bond is indicated by suitable numbers, e.g.

$CH_3CH_2C{\equiv}CH$ 1-Butyne

$CH_3CH_2CHCH_2CH_3$ (with $C{\equiv}CH$ on the central carbon) 3-Ethyl-1-pentyne

d) When both double and triple bonds are found in the same molecule, the double bond takes precedence in naming.

$CH_3CH_2CH = CHC \equiv CH$ 3 - Pentene - 1 - yne

5. **Cyclic Hydrocarbons** — Naphthenes — are given I.U.C. derived names.

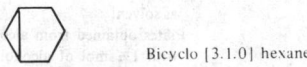

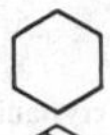

Cyclohexane

Cyclohexene

Cyclopentadiene

Cyclobutane

6. **Bridged Hydrocarbon** ring systems appear under the headings bicyclo-, tricyclo-. Most such compounds however, are listed under common names.

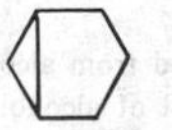

Bicyclo [3.1.0] hexane

7. **Spiro-, dispiro-**, etc. hydrocarbons and derivatives appear under the spiro-headings.

Spiro-[4,5-decane]

8. **Aromatic Hydrocarbons** are listed under their common names, e.g. Benzene (C_6H_6), Toluene ($C_6H_5CH_3$), Xylene [$C_6H_4(CH_3)_2$,] Naphthalene ($C_{10}H_8$) etc. Aromatic hydrocarbons having aliphatic side chains are listed as derivatives, e.g.

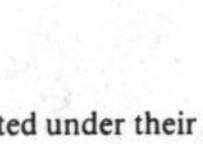

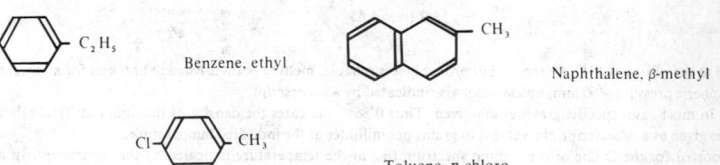

Benzene, ethyl

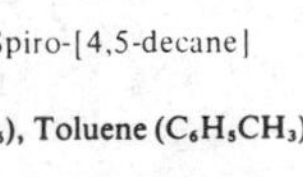

Naphthalene, β-methyl

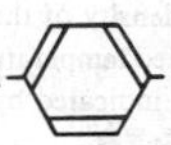

Toluene, *p*-chloro

When two or more benzene rings, phenyl groups, are attached to a single carbon atom or to more than one carbon atom of a hydrocarbon molecule, they are named as phenyl derivatives.

$(C_6H_5)_2CH_2$ Diphenyl methane

$(C_6H_5)_3CH$ Triphenyl methane

$C_6H_5CH_2CH_2C_6H_5$ 1 2,-Diphenyl ethane

9. **Heterocyclic Systems.** Common names are used or when these are lacking, systematic names using *oxo, azo, thio,* etc. to indicate an oxygen, nitrogen or sulfur atom in the cyclic structure.

The following common names are used; furan, pyran, pyrrole, pyrazole, imidazole, piperidine, pyridine, pyrazine, pyrimidine, pyrazidine, thiophene, etc.

10. **Radicals.** Radical names such as methyl, ethyl, isopropyl, phenyl, tolyl, benzyl, xylyl, allyl, styryl, etc. are commonly used. (See List of Substituent Prefixes)

Compounds Containing Functional Groups

Order of Precedence of Functions. When a compound contains one or more functional groups, the parent compound is designated according to the following order of precedence: acid, aldehyde, ketone, alcohol, amine, ether, sulfide, sulfone, and sulfoxide.

ACIDS AND DERIVATIVES

1. **Common names** have long been used in naming acids and well established names are used in this table, e.g. Acetic acid, Caprioc acid, Benzoic acid, Palmitic acid, Oleic acid, Stearic acid, Succinic acid, Tartaric acid, etc.

2. **Acids having no well-established common names** are names as alkanoic acids, alkenoic acids, alkane dioic acids, alkene dioic acids, etc., e.g.

$CH_3(CH_2)_7CO_2H$	Nonanoic acid	$CH_2=CHCH=CHCO_2H$	2.4-Pentadienoic acid
$CH_3CH=CHCH_2CO_2H$	3-Pentenoic acid	$HO_2CCH_2C(C_4H_9)=CHCO_2H$	2-Pentenedioic acid, -3-butyl
$(CH_3)_2CH=CHCO_2H$	2-Butenoic acid,-3-methyl		

Use the synonym name to find derivatives not listed under the name of the acid in the first column.

The term *-iso* is used to indicate branching at the end of the hydrocarbon chain. Thus $(CH_3)_2CHCO_2H$ is named Isobutyric acid. Listing of this acid and its derivative is found following Butyric acid — alphabetically under "B".

Polycarboxylic acids use the carboxylic acid suffix added to the longest carbon chain that does not contain these groups, e.g.

$HO_2CCH_2CH(CO_2H)CH_2CO_2H$ 1.2.3-Propane tricarboxylic acid

3. **Cyclic Acids.** After the name of the ring, the suffix carboxylic acid is attached.

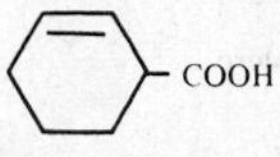

2-Cyclohexene carboxylic acid

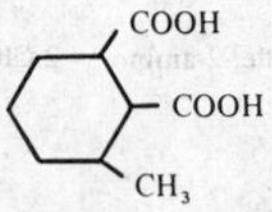

1.2-Cyclohexane dicarboxylic acid, -3-methyl

4. **Thioacids** are named as thio-, dithio-, thiolo-, or thiono-acids. The latter two prefixes are used only when it is certain that the hydroxyl or ketonic oxygen is replaced by the sulfur atom, e.g.,

CH_3CSOH	Thioacetic acid	CH_3CH_2COSH	Thiolopropionic acid
$(CH_3)_2CHCS_2H$	Dithioisobutyric acid	$CH_3CH_2C(S)OH$	Thionopropionic acid

5. **Imidic, Hydroxamic Acids.** Only simple compounds of these types appear in the Table. The following are examples

$CH_3CH_2CH(=NH)OH$	Butanimidic acid	$CH_3CH_2C(=NOH)OH$	Propanehydroxamic acid
$C_6H_5C(=NH)OH$	Benzimidic acid	$C_6H_5C(=NOH)OH$	Benzohydroxamic acid

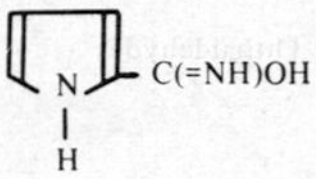

2-Pyrrolecarboxyimidic acid

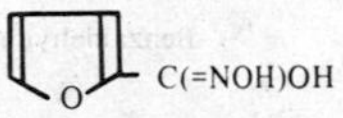

2-Furancarbohydroxamic acid

6. **Sulfonic and Sulfinic Acids.** Compounds containing the $-SO_3H$ group are named as sulfonic acids. When a carboxyl group is also present this group is designated by the prefix sulfo-.

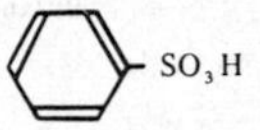

Benzene sulfonic acid

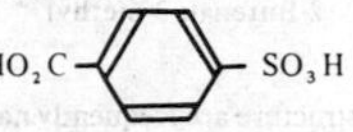

Benzoic acid, 4-sulfo

Compounds containing the $-SO_2H$ group are named as sulfinic acids.

7. **Keto Acids.** When the carbonyl function is present in the principal chain, the compounds are designated as oxo- acids.

$CH_3COCH_2CH_2CO_2H$ Valeric acid, 4-oxo or. Pentanoic acid, 4-oxo

Common names are listed when practical, e.g., $CH_3COCH_2CO_2H$ — Acetoacetic acid.

8. **Amino acids.** In most cases common names are used. Otherwise they are named as aminoalkanoic acids, aminoalkenoic acids, etc.

9. **Ortho acids (or esters)** are listed under ortho-, e.g.,

$HC(OCH_3)_3$ Orthoformic acid, trimethyl ester

10. **Acid halides and amides** are listed independently from the acids.

CH_3COCl	Acetyl chloride	$C_6H_5CON(C_2H_5)_2$	Benzamide, -*N,N*-diethyl
C_3H_7COBr	Butyryl bromide or Butanoyl bromide	$C_6H_5SO_2NH_2$	Benzenesulfonamide
C_6H_5COCl	Benzoyl chloride	$C_6H_5SO_2NHCH_3$	Benzenesulfonamide, -*N*-Methyl
$C_2H_5CONH_2$	Propionamide	$CH_3CONHCOCH_3$	Acetamide, -*N*-acetyl
$C_6H_5CONHCH_3$	Benzamide, -*N*-Methyl		

N-substituted amides in which the substituent is phenyl are named as amides. e.g.,

$CH_3CONHC_6H_5$	Acetanilide	$C_3H_7CONHC_6H_5$	Butyranalide

11. **Esters.** Esters are listed under the name of the acid, usually the common name.

$CH_3CO_2C_2H_5$	Acetic acid. ethyl ester	$C_2H_5O_2CCH_2CH_2CO_2C_2H_5$	Succinic acid. diethyl ester
$C_6H_5CO_2C_4H_9$	Benzoic acid. butyl ester	$(CH_3)_2C{=}CHCO_2CH$	2-Butenoic acid, 4-Methyl-methyl ester

Esters of complicated or polyhydric alcohols may appear under the name of the alcohol.

$CH_2O_2CC_{17}H_{35}$
|
$CHO_2CC_{17}H_{35}$ Glyceryl tristearate or Stearin
|
$CH_2O_2CC_{17}H_{35}$

12. **Lactams and Lactones** are listed under the names of the corresponding amino- or hydroxy acids.
13. **Nitriles** are listed independently from the corresponding acids and are designated as nitriles, e.g.,

C_3H_7CN	Butyronitrile	$NCCH_2CH_2CN$	succinonitrile
$CH_3CH(NH_2)CN$	Propionitrile,-2-amino	$2\text{-}ClC_6H_4CN$	Benzonitrile,-*o*-chloro

14. **Imides** are listed independently from the corresponding acids, e.g.,

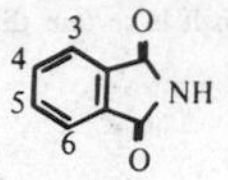
Phthalimide

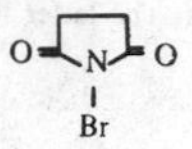
Succinimide, N-bromo

ALDEHYDES

1. Common names are listed for aldehydes. These are derived by dropping the -ic ending of the related acid and adding aldehyde, e.g.,

CH_3CHO	Acetaldehyde	$2\text{-}CH_3C_6H_4CHO$	*o*-Tolualdehyde
$(CH_3)_2CHCHO$	Isobutyraldehyde	$CH_3CH{=}CHCHO$	Crotonaldehyde
C_6H_5CHO	Benzaldehyde	$C_6H_4(CHO)_2$	Phthaldehyde

2. Where common names are not widely used, aldehydes are named as alkanals, alkenals, etc. e.g.,

$CH_3CH_2CH_2CHO$	Butanal	$CH_2{=}CHCH{=}CHCHO$	2,4-Butadienal
$(CH_3)_2C{=}CHCHO$	2-Butenal,-3-methyl	$OHCCH(CH_3)CHCHO$	Butanedial, 2-methyl

3. Compounds having a -CHO group attached to a cyclic structure are frequently named as carboxaldehydes, e.g.,

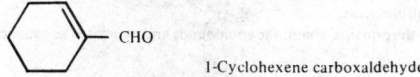

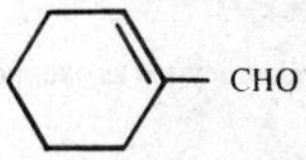
1-Cyclohexene carboxaldehyde

4. The -CHO group is frequently referred to as formyl when a carboxyl group is also present. The $-C{=}O$ group may also be designated as oxo, e.g.,

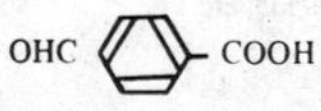
Benzoic acid,-4-formyl

$OHCCH_2CH_2COOH$ — Butanoic acid,-4-oxo

5. Acetals and hemi-acetals are listed under the corresponding aldehyde or ketone, e.g.,

$CH_3CH(OC_2H_5)_2$	Acetaldehyde, diethyl acetal	$CH_3CH_2CH(OH)OCH_3$	Propionaldehyde, methyl-hemi-acetal
$CH_3CH_2CH(CH_3)CH(OCH_3)_2$	Butanal,-2-methyl, dimethyl acetal		

KETONES

1. Ketones having only aliphatic groups are listed as dialkyl ketones, or as alkanones, alkenones, etc., e.g.,

CH_3COCH_3	Dimethyl ketone or acetone	$CH_2{=}CHCOCH_3$	Methyl vinyl ketone or 3-butene-2-one
$(CH_3)_2CHCOCH_3$	Methyl isopropyl ketone or 2-butanone,-3-methyl		

When the derivative being sought is not found under the name appearing in the first column, look for it under the synonym name or as a principal entry.

2. When one of the groups attached to the carbonyl group ($-C{=}O$) is phenyl the compound is named as phenyl ketone or as a phenone, e.g.,

$C_6H_5COCH_3$	Acetophenone or methyl phenyl ketone	$C_6H_5COC_3H_7$	Butyrophenone or propyl phenyl ketone
$C_6H_5COC_6H_5$	Benzophenone or diphenyl ketone		

3. When cyclic structures are attached to the carbonyl group, the ketones may be named as acetyl-, propionyl-, butyryl-, etc. derivatives of the cyclic compound.

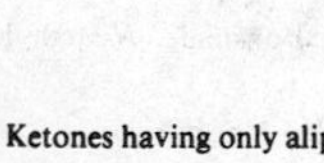
Furan, -2-acetyl

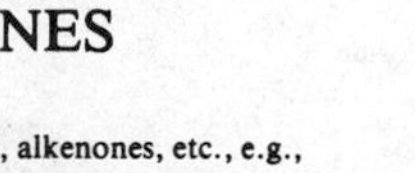
Pyrrole, -2-propionyl

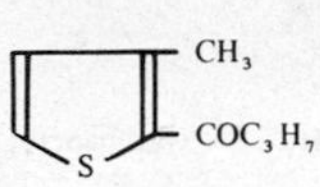

Thiophene, -3-methyl-2-butyryl

4. **Thioketones are named as thiones.**

$C_6H_5CSCH_3$	Acetothiophenone	$C_6H_5CSC_6H_5$	Dibenzothiophenone

5. **Cyclic ketones and thiones are given the name of the parent compound followed by the ending -one or -thione.**

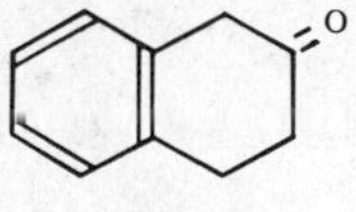

2-Tetralone

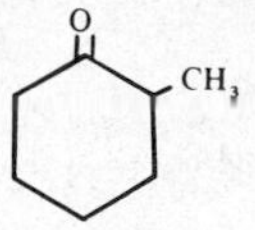

Cyclohexanone, 2-methyl

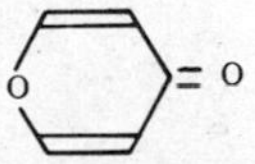

4-Pyrone

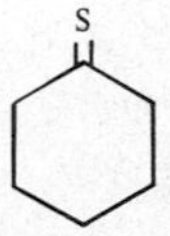

Cyclohexanethione

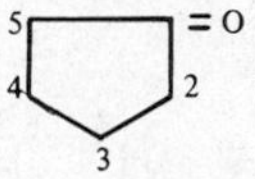

Cyclopentanone

6. **Quinones, the name is reserved for structures of the type**

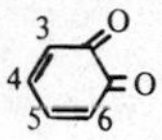

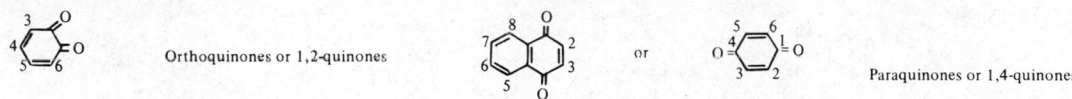

Orthoquinones or 1,2-quinones

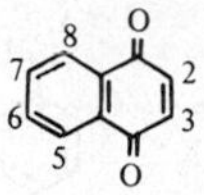

or

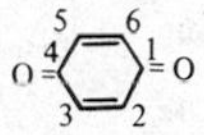

Paraquinones or 1,4-quinones

Examples are:

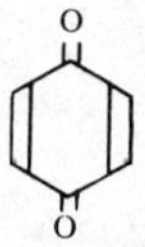

1,4-Benzoquinone

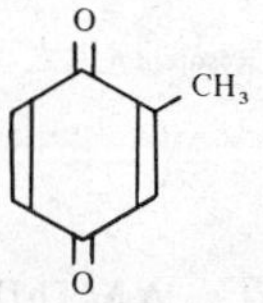

1,4-Benzoquinone, -2-methyl

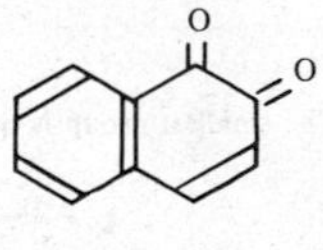

1,2-Naphthoquinone

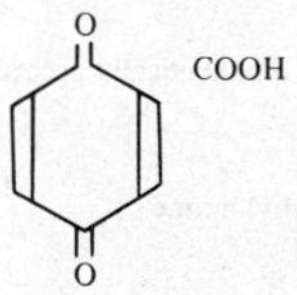

1,4-Benzoquinone, -2-carboxylic acid

ALCOHOLS

1. **Alcohols are named as alkyl alcohols, or as alkanols, alkenols, alkanediols, etc.**

CH_3CH_2OH	Ethyl alcohol ethanol	$(CH_3)_3COH$	Tert-butyl alcohol 2-Propanol, 2-methyl
$(CH_3)_2CHCH_2OH$	Isobutyl alcohol 1-propanol,-2-methyl		
$CH_3CH_2CH(OH)CH_3$	Sec-butyl alcohol 2-butanol		

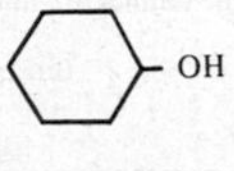

Cyclohexanol

$CH_3CH(OH)CH(OH)CH_3$ 2,3-Butanediol

Phenyl substituted alcohols are named as phenyl-, diphenyl-, alkanols, etc.

$(C_6H_5)_2CHOH$ Methanol,–diphenyl

2. **Iso-, sec-, and tert-alcohols and derivatives are listed sequentially after the name of the normal alcohol, e.g. isobutyl, sec-butyl and tert-butyl alcohols are listed as butyl alcohols.**
3. **Common names are frequently used in naming alcohols, e.g.**

$CH_2{=}CHCH_2OH$	Allyl alcohol	$C_{11}H_{23}CH_2OH$	Lauryl alcohol
$HOCH_2CH_2OH$	Ethylene glycol	$C_{17}H_{35}CH_2OH$	Stearyl alcohol
$HOCH_2CH(OH)CH_2OH$	Glycerol		

4. Thioalcohols are names as thiols or mercaptans

C_2H_5SH	Ethanethiol ethyl mercaptan	$(CH_3)_2CHCH_2SH$	1-Propanethiol,-3-methyl isobutyl mercaptan
$HOCH_2CH_2SH$	Ethanedithiol		

PHENOLS

1. Hydroxyaromatic compounds are named as phenols. They are much more acidic than alcohols and deserve separate listing, e.g.,

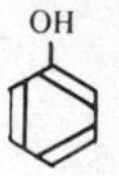

Phenol

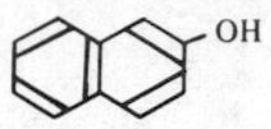

β-Naphthol

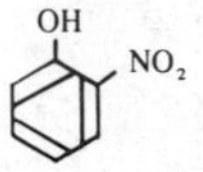

Phenol, *o*-nitro

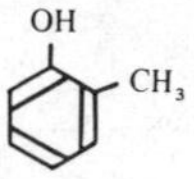

o-Cresol

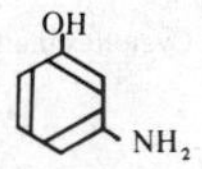

Phenol, *m*-amino

2. Common names are given to aromatic diphenols or triphenols and some mono phenols, e.g.,

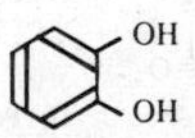

Catechol

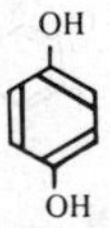

Hydroquinone

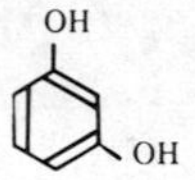

Resorcinol

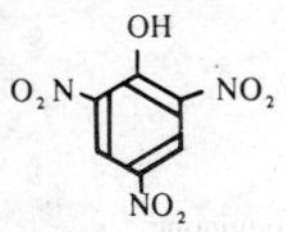

Picric acid

AMINES

1. Alkyl amines, dialkyl amines and trialkyl amines are listed alphabetically according to the alkyl groups attached. The smallest group is named first and thus determines the listing. Refer to the synonym if necessary, e.g.,

CH_3NH_2	Methyl amine	$(CH_3)_2NH$	Dimethyl amine
$C_4H_9NH_2$	Butyl amine	$C_3H_7NHC_2H_5$	Ethyl propyl amine
		$(C_2H_5)_3N$	Triethyl amine
		$(C_2H_5)_2NCH_3$	Methyl diethyl amine
		$H_2NCH_2CH_2CH_2NH_2$	1.3-Propane diamine

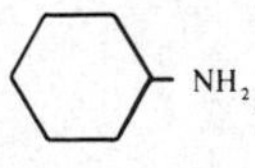

Cyclohexyl amine

2. When more complex radical names are found the amines are names as amino alkanes, amino alkenes, etc.

$CH_3CH_2CH(NH_2)CH_3$	Butane,-2-amino	$(CH_3)_2CH_2CH(NH_2)CH_3$	Butane,-4-methyl-2-amino

3. Amino substituted heterocyclic compounds are names as amino- derivatives, e.g.,

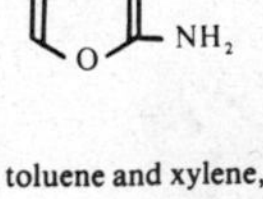

Furan,-2-amino

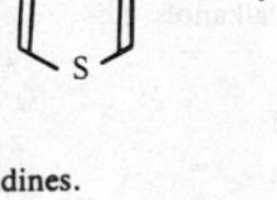

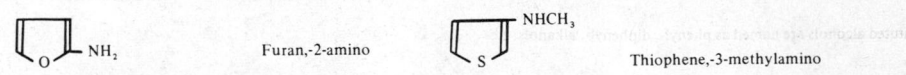

Thiophene,-3-methylamino

4. Amino substituted toluene and xylene, respectively are named as toluidines and xylidines.

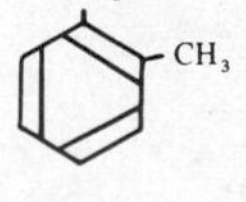

o-Toluidine

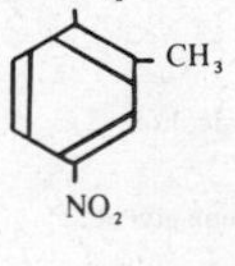

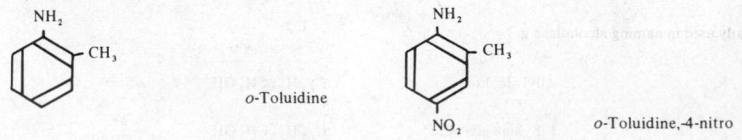

o-Toluidine,-4-nitro

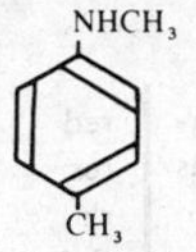

p-Toluidine,-N-Methyl

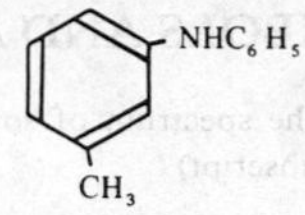

m-Toluidine, -N-phenyl

ETHERS

1. Dialkyl ethers are listed alphabetically with the smaller group being named first, e.g.,

$C_2H_5-O-C_2H_5$	Diethyl ether	$CH_3-O-C_6H_5$	Methyl phenyl ether
$C_2H_5-O-i-C_3H_7$	Ethyl isopropyl ether	$CH_3-O-CH=CH_2$	Methyl vinyl ether
$(CH_3)_3C-O-C_5H_{11}$	Tert-butyl pentyl ether	$C_6H_5CH_2-O-CH_2C_6H_5$	Dibenzyl ether

2. Alkyl ethers of aromatic or heterocyclic compounds are named as alkoxides, e.g.,

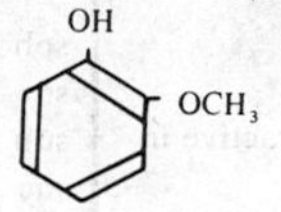

Phenol, *o*-methoxy

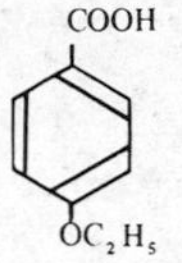

Benzoic acid, *p*-ethoxy

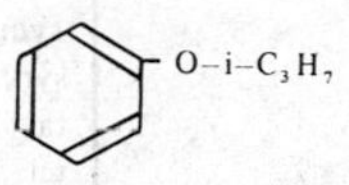

Benzene, isopropoxy

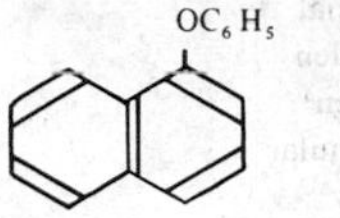

Naphthalene, α-paenoxy

SULFIDES, DISULFIDES, SULFOXIDES AND SULFONES

1. These compounds are named in the same way as ethers. The alkyl, or aryl groups are named alphabetically, beginning with the smaller group, e.g.,

$C_2H_5-S-C_2H_5$	Diethyl sulfide	$CH_3-SS-C_2H_5$	Methyl ethyl disulfide
$CH_3-S-C_4H_9$	Methyl butyl sulfide	$C_2H_5-SO_2-C_2H_5$	Diethyl sulfone
		$C_2H_5-SO_2-C_6H_5$	Ethyl phenyl sulfone

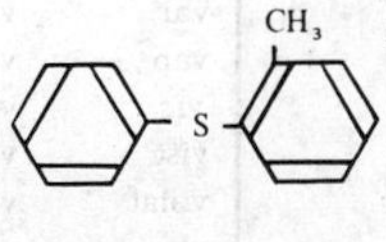

Phenyl-*o*- tolyl sulfide

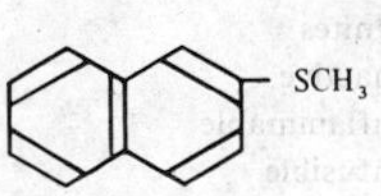

Naphthalene, -β-methylthio

$CH_3C-SO-C_2H_5$ Methyl ethyl sulfoxide

There may be times when the compound can be located more readily by the name of the synonym.

SYMBOLS AND ABBREVIATIONS

$[\alpha]$	specific rotation
>	above, more than
<	below, less than
?	unkown
aa	acetic acid
abs	absolute
ac	acid
Ac	acetyl
ace	acetone
AFCL	Aliphatic Fluorine Compounds
al	alcohol[1]
ALD	Aldrich Handbook of Organic Chemicals and Biochemicals
alk	alkali
Am	J. Am. Chem. Soc.
Am	amyl (pentyl)
amor	amorphous
anh	anhydrous
aqu	aqueous
as	asymmetric
Atlas	Atlas of Spectral Data and Physical Constants for Organic Compounds
atm	atmospheres
b	boiling
B	Beilstein
Ber	Chem. Ber.
bipym	bipyramidal
bk	black[2]
bl	blue[2]
BOSC	Bayant, et al., Organosilicon Compounds
br	brown[2]
bt	bright
Bu	butyl
bz	Benzene
CAS	Chemical Abstracts
c	percentage concentration
ca	about (circa)
chl	chloroform
co	columns
col	colorless
con	concentrated
cor	corrected
cr	crystals
cy	cyclohexane
d	decomposes
D	line in the spectrum of sodium (subscript)
D, d	dextro[3]
dd	slight decomposition
dil	diluted
diox	dioxane
distb	distillable
dk	dark
Dl, dl	racemic[3]
dlq	deliquescent
DMF	dimethyl formamide
E	Elsevier's
eff	efforescent
Et	ethyl
eth	ether[4]
exp	explodes
extrap	extrapolated
fl	flakes
flam	flammable
flr	fluorescent
fr	freezes
fr. p.	freezing point
fum	fuming
gel	gelatinous
gl	glacial
gold	golden
gr	green[3]
gran	granular
gy	gray[3]
h	hot
H	Helv. Chim. Acta
hex	hexagonal
HDOC	Helibron Dictionary of Organic Compounds
hp	heptane
htng	heating
hx	hexane
hyd	hydrate
hyg	hygroscopic
i	insoluble
i-	iso-
ign	ignites
in	inactive
inflam	inflammable
infus	infusible
irid	iridescent
iso	isooctaine
J	J. Chem. Soc.
JOC	J. Org. Chem.
KHOC	Kaufman Handbook of Organometallic Compounds
L, l	levo[3]
la	large
lf	leaf
liq	ligroin
liq	liquid
lo	long
lt	light
m	melting
m-	meta-
M	molar (concentration)
M	Merck Index, 7th Edition
mcl	monoclinic
Me	methyl
met	metallic
micr	microsopic
min	mineral
mod	modification
mut	mutarotatory
n	normal chain, refractive index
N	normal (concentration)
N	nitrogen[5]
nd	needles
o-	ortho-
oct	octahedral
og	orange[2]
ord	ordinary
org	organic
orh	orthorhombic
p-	para-
pa	pale
par	partial
PCHE	Egloff Physical Constants of Hydrocarbons
peth	petroleum ether
pk	pink[2]
Ph	phenyl
pl	plates
pr	prisms
Pr	propyl
Prak	J. Prak. Chem.
purp	purple[2]
pw	powder
Py	pyrimidine
pym	pyramids
rac	racemic
rect	rectangular
red	red
res	resinous
rh	rhombic
rhd	rhombodohedral
s	soluble
sc	scales
sec	secondary
sf	softens
sh	shoulder
silv	silvery
sl	slightly
so	solid
sol	solution
solv	solvent
Sol-vents	Entries in this column medium type means sol ble; entries in boldfa means very soluble
sph	sphenoidal
st	stable
sub	sublimes
suc	supercooled
sulf	sulfuric acid
sym	symmetrical
syr	syrup
ta	tablets
tcl	triclinic
tert	tertiary
Tet	Tetrahedron
tetr	tetragonal
THF	tetrahydrofuran
to	toluene
tr	transparent
trg	trigonal
undil	undiluted
uns	unsymmetrical
unst	unstable
v	very
vac	vacuum
var	variable
vap	vapor
vic	vicinal
visc	viscous
volat	volatile or volatilises
vt	violet[2]
w	water
wh	white[1]
wr	warm
wx	waxy
ye	yellow[2]
xyl	xylene

[1] Generally means ethyl alcohol.

[2] The abbreviation of a color ending in "sh" is to be read as ending with the suffix "-ish," e.g., grsh means greenish.

[3] *D, L* generally means configuration and *d, l* generally mean optical rotation, but there are many examples in the chemical literature for which the meaning of these symbols is ambiguous and/or interchangeable.

[4] Generally means diethyl ether.

[5] *N* indicates a position in the molecule.

BEILSTEIN REFERENCES

"Beilstein is the short name for "Beilsteins Handbuch der Organischen Chemie". This Handbuch contains the most complete set of data on organic compounds in the world. The 1st edition was published as two volumes during the years 1881 to 1882. That edition contained data for about 15,000 compounds described on 2200 pages. The 2nd edition of Beilstein consisted of 3 volumes and 4080 pages. The 2nd edition appeared in the years 1885 to 1889. The 3rd edition was published over the years 1892 to 1906. It consisted of 8 volumes published on approximately 11,000 pages. The 4th edition of Beilstein began to be published in 1909. This 4th edition may be described as follows:

The Series of the Beilstein Handbook (4th edition)

Series	Abbreviation	Period of literature completely covered	Colour of label on spine
Basic Series	H	up to 1909	green
Supplementary Series I	E I	1910—1919	red
Supplementary Series II	E II	1920—1929	white
Supplementary Series III	E III	1930—1949	blue
Supplementary Series III/IV	E III/IV*	1930—1959	blue/black
Supplementary Series IV	E IV	1950—1959	black

* Volumes 17 to 27 of Supplementary Series III and IV, covering the heterocyclic compounds are combined in a joint issue.

Preparations are currently in progress for the Fifth Supplementary Series (E V), which will cover the literature from 1960 to 1979.

Each of these series of the 4th edition comprises 27 volumes (or groups of volumes) in which the individual compounds are arranged according to the "Beilstein System". Which compounds are dealt with in which Beilstein volumes is shown in the following two tables:

The Main Divisions of the Beilstein Handbook

Main Division	Volume No.	System No.
A. *Acyclic* Compounds	1—4	1—449
B. *Isocyclic* Compounds	5—16	450—2358
C. *Heterocyclic* Compounds	17—27	2359—4720

Contents of the 27 Volumes of the Beilstein Handbook

	Type of registry compound	Feature of the functional group		Beilstein Volume No. for Main Division A (Acyclics)	B (Isocyclics)	C (Heterocyclics) Type and number of ring heteroatoms: 1O*	2O*, 3O*,...	1N	2N	3N, 4N,...	1N, 1O*, 1N, 2O*, 2N, 1O*, 2N, 2O*, further heteroatoms**
1	Compounds without functional groups		–		5			20			
2	Hydroxy-compounds		–OH	1	6	17			23		
3	Oxo-compounds	=O	=O		7			21	24		
			=O + –OH		8						
4	Carboxylic acids	$-C(=O)OH$	$-C(=O)OH$	2	9						
			$-C(=O)OH$ + –OH; $-C(=O)OH$ + =O; $-C(=O)OH$ + =O + –OH	3	10						
5	Sulfinic acids		$-SO_2H$								
6	Sulfonic acids		$-SO_3H$								
7	Seleninic acids, Selenonic acids, Tellurinic acids		$-SeO_2H$ and $-SeO_3H$ $-TeO_2H$	4	11	18	19	22	25	26	27
8	Amines	$-NH_2$	$-NH_2$		12						
			$[-NH_2]_n$; $-NH_2$ + –OH		13						
			$-NH_2$ + =O; $-NH_2$ + $-C(=O)OH$; $-NH_2$ + ...		14						
9	Hydroxylamines and Dihydroxyamines		–NH–OH $-N(OH)_2$								
10	Hydrazines		$-NH-NH_2$		15						
11	Azo-compounds		–N=NH								
12	Diazonium compounds		$-N{\equiv}N]^{\oplus}$								
13	Compounds with groups of 3 or more N-atoms		$-NH-NH-NH_2$, $-N(NH_2)_2$, $-N=N-NH_2$, etc.								
14	Compounds containing carbon directly bonded to P, As, Sb, and Bi	e.g.	$-PH_2$, PH–OH, $-P(OH)_2$, $-PH_4$, ..., $-PO(OH)_2$		16						
15	Compounds containing carbon directly bonded to Si, Ge, and Sn	e.g.	$-SiH_3$, $-SiH_2(OH)$, ...								
16	Compounds containing carbon directly bonded to elements of the 3rd–1st A-groups of the periodic table	e.g.	$-BH_2$, –BH(OH), ..., $-Mg^{\oplus}$								
17	Compounds containing carbon directly bonded to elements of the 1st–8th B-groups of the periodic table	e.g.	–HgH, $-Hg^{\oplus}$, ...								

* Instead of O also S, Se, Te.
** e.g. B, Si, P, but not S, Se, Te.

Additional informational material on Beilstein is available upon request free of charge from

Springer-Verlag KG
Heidelberger Platz 3
D-1000 Berlin 33

Springer Verlag New York Inc.
175 Fifth Avenue
New York, New York 10010

Included in such information are:

- Beilstein Reference Chart ("Contents of Beilstein Handbook of Organic Chemistry")
- Brochure, "What is Beilstein"
- How to Use Beilstein
- Beilstein Wordbook (German/English)
- Beilstein Outline (List of all available volumes; revised annually)

The Beilstein references listed in this edition of the *CRC Handbook of Chemistry and Physics* were revised and updated early in 1980 by personnel of Beilstein.

PHYSICAL CONSTANTS OF ORGANIC COMPOUNDS

No.	Name, Synonyms, and Formula	Mol. wt.	Color, crystalline form, specific rotation and λ_{max} (log ε)	b.p. °C	m.p. °C	Density	n_D	Solubility	Ref.
1	Abietic acid or sylvic acid $C_{20}H_{30}O_2$	302.46	mcl pl (al-w)	250	173-4			al, et, ace, bz	B9^{3}, 2904
2	Abietic acid, methyl ester $C_{21}H_{32}O_2$	316.48		225-6^{16}		1.0491$^{20/4}$	1.5344	al	B9^{3}, 2907
3	Acenaphthanthracene or Naphtho-2′,3′,4,5-acenaphthene $C_{20}H_{14}$	254.33	pa ye lf (lig)	192.5-3.5				bz, lig	B5^{3}, 2468
4	Acenaphthene $C_{12}H_{10}$	154.21		279	96.2	1.0242$^{90/4}$	1.6048^{95}	al, bz	B5^{4}, 1834
5	Acenaphthene, 1-amino $C_{12}H_{11}N$	169.23	cr (peth)	135	sub			al, bz, CS_2	B12^{2}, 764
6	Acenaphthene, 3-amino $C_{12}H_{11}N$	169.23	pl (al), nd (peth)		81.5			al, chl	B12^{3}, 3210
7	Acenaphthene, 4-amino $C_{12}H_{11}N$	169.23	nd (al,w)		87			bz, lig, w	B12^{3}, 3212
8	Acenaphthene, 5-amino $C_{12}H_{11}N$	169.23	nd (lig) red in air		108			al	B12^{3}, 3212
9	Acenaphthene, 5-bromo $C_{12}H_9Br$	233.11	pl (al)	335	52	1.4392$^{54/4}$	1.6565^{54}	al	B5^{4}, 1839
10	Acenaphthene, 5-chloro $C_{12}H_9Cl$	188.66	pl or nd (al)	319.2^{770},163^{13}	70.5	1.1954$^{20/4}$	1.6288^{7}	al	B5^{4}, 1837
11	Acenaphthene, 5-iodo $C_{12}H_9I$	280.11	nd (al)		65	1.6738$^{62/4}$	1.6909^{65}	al, bz	B5^{1}, 276
12	Acenaphthene, 5,6-dinitro $C_{12}H_8N_2O_4$	244.21			220-4				B5^{3}, 1784
13	Acenaphthene, 3-nitro $C_{12}H_9NO_2$	199.21	gr-ye nd (aa)		151.5			aa	B5^{3}, 1783
14	Acenaphthene, 5-nitro $C_{12}H_9NO_2$	199.21			103-4			w, al, eth, lig	B5^{4}, 1840
15	Acenaphthene, 2a,3,4,5-tetrahydro $C_{12}H_{14}$	158.24		246	12			chl	B5^{3}, 1385
16	Acenaphthene, 5-carboxylic acid or 5-Acenaphthoic acid $C_{13}H_{10}O_2$	198.22	nd (bz or lig)		220-1			bz, lig	B9^{3}, 3289
17	Acenaphthenequinone $C_{12}H_{16}O_2$	182.18	ye nd (aa)		261			al, bz	B7^{3}, 3796
18	3-Acenaphthene sulfonic acid $C_{12}H_9SO_3H$	234.27	hyg nd (bz)		87-9				B11^{3}, 435
19	1-Acenaphthenone or 1-oxoacenaphthene $C_{12}H_8O$	168.19	nd (al)		121			w, al, bz, chl	B7^{3}, 2046
20	Acenaphthylene $C_{12}H_8$	152.20	pr (eth), pl (al)	265-75	92-3	0.8988$^{16/2}$		al, eth, bz	B5^{4}, 2138
21	Acetaldehyde CH_3CHO	44.05		20.8	-121	0.7834$^{18/4}$	1.3316^{20}	**w, al, eth, ace, bz**	B1^{4}, 3094
22	Acetaldehyde, 2,4-dinitrophenylhydrazone $CH_3CH{=}NNHC_6H_3(NO_2)_2$-2,4	224.18	ye sc (al)		168.5			eth, ace, bz, ch	B15^{3}, 426
23	Acetaldehyde, amino, diethyl acetal $H_2NCH_2CH(OC_2H_5)_2$	133.19		163		0.9159^{25}	1.4170^{20}	w, al, eth, chl	B4^{4}, 1918
24	Acetaldehyde, ammonia or 1-aminoethanol $CH_3CH(OH)NH_2$	61.08	rh (eth-al)	110d	97			w	C5^{2}
25	Acetaldehyde, bromo, diethyl acetal $BrCH_2CH(OC_2H_5)_2$	197.07		180, 66^{18}		1.280$^{20/4}$	1.4376^{20}	al, eth	B1^{4}, 3151
26	Acetaldehyde, bromo, dimethyl acetal $BrCH_2CH(OCH_3)_2$	169.02		148.50-50^{15}		1.5049^{184}	1.4450^{20}	eth, ace, chl	B1^{3}, 2672
27	Acetaldehyde, chloro $ClCH_2CHO$	78.50		85-5.5^{748}				eth	B1^{4}, 3134
28	Acetaldehyde, chloro, diethyl acetal $ClCH_2CH(OC_2H_5)_2$	152.62		157.4, 71-2^{35}		1.068$^{20/4}$	1.4150^{20}	**al, eth,** bz	B1^{4}, 3134
29	Acetaldehyde, chloro, dimethyl acetal $ClCH_2CH(OCH_3)_2$	124.57		127-8		1.068$^{20/4}$	1.4150^{20}	al, bz	B1^{4}, 3134
30	Acetaldehyde, *bis*(2-chloroethyl) acetal $CH_3CH(OCH_2CH_2Cl)_2$	187.07		194-6, 106-7^{14}		1.1737$^{20/4}$	1.45266^{20}	al, eth	B1^{4}, 3104
31	Acetaldehyde, diacetate or Ethylidene diacetate $CH_3CH(O_2CCH_3)_2$	146.14		169, 65-7^{10}	18.9	1.3985^{25}	1.070^{25}	al, eth	B2^{4}, 282
32	Acetaldehyde, dichloro Cl_2CHCHO	112.94		90-1				al	B1^{4}, 3140
33	Acetaldehyde, dichloro, diethyl acetal $Cl_2CHCH(OC_2H_5)_2$	187.07		183-4, 67-71^{12}		1.1381^{14}		al, eth	B1^{4}, 3140

No.	Name, Synonyms, and Formula	Mol. wt.	Color, crystalline form, specific rotation and λ_{max} (log ε)	b.p. °C	m.p. °C	Density	n_D	Solubility	Ref.
34	Acetaldehyde, dichloro, hydrate $Cl_2CHCH(OH)_2$	130.96	cr (bz), ta	97	56-7			w, al, eth	B1, 614
35	Acetaldehyde, diethyl acetal $CH_3CH(OC_2H_5)_2$	118.18		$103.2, 21^{22}$		$0.8314^{20/4}$	1.3834^{20}	w, al, eth, ace	$B1^4$, 3103
36	Acetaldehyde, diethyl mercaptal $CH_3CH(SC_2H_5)_2$	150.30		$183\text{-}5, 73^{19}$		0.9706^{20}	1.5025^{20}	eth, ace, bz	$B1^4$, 3162
37	Acetaldehyde, dimethyl acetal $CH_3CH(OCH_3)_2$	90.12		64.5	−113.2	$0.85015^{20/4}$	1.3668^{20}	w, al, eth, ace, bz	$B1^4$, 3103
38	Acetaldehyde, dimethylamino, diethyl acetal $(CH_3)_2NCH_2CH(OC_2H_5)_2$	161.24	cr (bz), ye	170-1		0.885^{7}	1.4129^{20}	w, al, eth, ace	$B1^3$, 870
39	Acetaldehyde, 2,4-dinitrophenylhydrazone $CH_3CH{=}NNHC_6H_3(NO_2)_2$-2,4	224.18	ye sc (al)		168.5			eth, ace, bz	$B15^3$, 426
40	Acetaldehyde, diphenyl $(C_6H_5)_2CHCHO$	196.25		157.5^{5}		$1.1061^{21/4}$	1.5920^{21}	al, eth, ace	$B7^3$, 2117
41	Acetaldehyde, ethoxy $C_2H_5OCH_2CHO$	88.11		71-3		$0.942^{20/4}$	1.3956^{20}	w, al, ace	$B1^3$, 3181
42	Acetaldehyde, hydroxy or Glycolaldehyde $HOCH_2CHO$	60.05	pl		97	1.366^{100}	1.4772^{19}	w, al	$B1^4$, 3955
43	Acetaldehyde, hydroxy, diethyl acetal $HOCH_2CH(OC_2H_5)_2$	134.18		$167, 57\text{-}8^{8}$		$0.888^{24/4}$	1.4073^{20}	al, eth	$B1^4$, 3958
44	Acetaldehyde, methoxy CH_3OCH_2CHO	74.08		92.3^{770}		$1.005^{25/4}$	1.3950^{20}	w, al, eth, ace	$B1^4$, 3955
45	Acetaldehyde, methoxy, diethyl acetal $CH_3OCH_2CH(OC_2H_5)_2$	148.20		144-8			$1.3990^{20/D}$	al, eth, ace	$B1^4$, 3958
46	Acetaldehyde, oxime or Acetaldoxime $CH_3CH{=}NOH$	59.07	nd	115	47	$0.9656^{20/4}$	1.42567^{20}	al, eth	$B1^4$, 3121
47	Acetaldehyde, phenyl or α-Tolualdehyde $C_6H_5CH_2CHO$	120.15		$195, 88^{18}$	33-4	$1.0272^{20/4}$	1.5255^{20}	al, eth, bz	$B7^3$, 1003
48	Acetaldehyde, semicarbazone $CH_3CH{=}NNHCONH_2$	101.11			163	$1.300^{0/4}$		al	$B3^4$, 178
49	Acetaldehyde, phenyl, dimethyl acetal $C_6H_5CH_2CH(OCH_3)_2$	166.22		193-4					$B7^3$, 1006
50	Acetaldehyde, phenylhydrazone $CH_3CH{=}NNHC_6H_5$	134.18		$133\text{-}6^{21}$	98-101			al	$B15^3$, 79
51	Acetaldehyde, tribromo or bromal Br_3CCHO	280.74		$174, 61^{9}$		$2.6650^{25/4}$	1.5939^{20}	al, eth, ace	$B1^4$, 3155
52	Acetaldehyde, tribromo, hydrate or Bromal hydrate $Br_3CCH(OH)_2$	298.76	mcl pr (w + l)	d	53.5	$2.5662^{40/4}$		al, eth	$B1^4$, 3155
53	Acetaldehyde, trichloro or chloral Cl_3CCHO	147.39		97.8	-57.5	$1.5121^{20/4}$	1.45572^{20}	w, al, eth	$B1^4$, 3142
54	Acetaldehyde, trichloro, diethyl acetal $Cl_3CCH(OC_2H_5)_2$	221.51		$205, 84\text{-}5^{10}$		$1.266^{25/4}$	1.4586^{25}	al, eth	$B1^4$, 3144
55	Acetaldehyde, trichloro, ethylhemiacetal $Cl_3CCH(OH)OC_2H_5$	193.46		$115\text{-}6^{760}$	56-7	1.143^{40}		w, al, eth	$B1^4$, 3144
56	Acetaldehyde, trichloro, hydrate or chloral hydrate $Cl_3CCH(OH)_2$	165.40	mcl pl (w)	$96.3^{764/d}$	57	$1.9081^{20/4}$		w, al, eth, ace, bz	$B1^4$, 3143
57	Acetaldehyde, trimethyl or Pivaldehyde $(CH_3)_3CCHO$	86.13		77-8	6	0.7923^{17}	1.3791^{20}	al, eth	$B1^4$, 3295
58	Acetamide CH_3CONH_2	59.07	trg mcl (al-eth)	$221.2, 120^{20}$	82.3	$0.9986^{85/4}$	1.4278^{78}	w, al	$B2^4$, 399
59	Acetamide, *N*-acetyl or diacetamide $CH_3CONHCOCH_3$	101.11	nd (eth)	$223.5, 113^{12}$	79			w, al, eth	$B2^4$, 416
60	Acetamide, *N*-acetyl-*N*-ethyl or Diacetylethylamine $CH_3CON(C_2H_5)COCH_3$	129.16		195-9		1.0092^{20}	1.4513^{20}	al	$B4^4$, 352
61	Acetamide, *N*-acetyl-*N*-methyl or Diacetylmethylamine $CH_3CON(CH_3)COCH_3$	115.31		$194.5, 114.5^{61}$	-25	$1.0663^{25/4}$	1.4502^{25}	w	$B4^4$, 183
62	Acetamide, N-acetyl-N-phenyl or Diacetanilide $(CH_3CO)_2NC_6H_5$	177.20	ta (lig)	$200^{100}\ 142^{41}$	37-8			al, bz	$B12^3$, 472
63	Acetamide, *N*-(2-aminoethyl) $CH_3CONHCH_2CH_2NH_2$	102.14		128^{3}	51			w, al, bz	$B4^4$, 1193
64	Acetamide, *N*-benzyl $CH_3CONHCH_2C_6H_5$	149.19		157^{2}	61			al, eth	$B12^5$, 2255
65	Acetamide, *N*-bromo, hydrate $CH_3CONHBr \cdot H_2O$	155.98	bt ye pl (w + l)		108-9				$B2^1$, 82
66	Acetamide, *N*-butyl $CH_3CONHC_4H_9$	115.18		229			1.4388^{25}		$B4^4$, 565

PHYSICAL CONSTANTS OF ORGANIC COMPOUNDS (Continued)

No.	Name, Synonyms, and Formula	Mol. wt.	Color, crystalline form, specific rotation and λ_{max} (log ε)	b.p. °C	m.p. °C	Density	n_D	Solubility	Ref.
67	Acetamide, *N*-butyl-*N*-phenyl $CH_3CON(C_4H_9)C_6H_5$	191.27		$281, 141^{10}$	24.5	$0.9912^{20/4}$	1.5146^{20}	chl	$B12^{3}$, 467
68	Acetamide, chloro or chloroacetamide $ClCH_2CONH_2$	93.51		$224\text{-}5^{747}$	121			w, al	$B2^{4}$, 490
69	Acetamide, *N*-cyanomethyl $CH_3CONHCH_2CN$	98.10			79-81			w	$B4^{2}$, 790
70	Acetamide, *N,N*-diacetyl or Triacetamide $CH_3CON(COCH_3)_2$	143.14	nd (eth)		79			eth	$B2^{4}$, 416
71	Acetamide, dibenzyl $(C_6H_5CH_2)_2CHCONH_2$	239.32	nd (al, w)	259^{18}	129			al, eth	$B9^{3}$, 3356
72	Acetamide, dichloro $Cl_2CHCONH_2$	127.96		$233\text{-}4^{745}$	99.4			w, al, eth	$B2^{4}$, 505
73	Acetamide, *N,N*-diethyl $CH_3CON(C_2H_5)_2$	115.18		185-6		$0.9130^{17.4/D}$	$1.4374^{17.4/D}$	w, al, eth, ace, bz	$B4^{4}$, 349
74	Acetamide, *N,N*-dimethyl $CH_3CON(CH_3)_2$	87.12		$165^{758}, 84^{22}$	-20	$0.9366^{25/4}$	1.4380^{20}	w, al, ace, bz, eth	$B4^{4}$, 180
75	Acetamide, *N,N*-dipropyl $CH_3CON(C_3H_7)_2$	143.23		$209\text{-}10, 101^{16}$		$0.8992^{17.4/4}$	$1.4419^{17.4}$	al	$B4^{4}$, 476
76	Acetamide, *N*-ethyl $CH_3CONHC_2H_5$	87.12		$205, 104\text{-}5^{18}$		$0.942^{4.5/4}$		w, al, chl	$B4^{4}$, 347
77	Acetamide, hydroxy $HOCH_2CONH_2$	75.07		120				w	$B3^{4}$, 597
78	Acetamide, *N*-(2-hydroxyethyl) or *N*-acetylethanolamine $CH_3CONHCH_2CH_2OH$	103.12		$166\text{-}7^{8}$	63.5	$1.1079^{25/4}$	1.4674^{20}	w, ace	$B4^{4}$, 1535
79	Acetamide, *N*-methyl $CH_3CONHCH_3$	73.09	nd	$204\text{-}6, 95^{14}$	28	$0.9571^{25/4}$	1.4301^{20}	w, al, eth, ace, bz	$B4^{4}$, 176
80	Acetamide, *N*-methyl-*N*-α-naphthyl $CH_3CON(CH_3)\alpha C_{10}H_7$	199.25	nd (lig)		95-7			al, eth	$B12^{3}$, 2867
81	Acetamide, *N*-methyl-*N*-(4-nitrophenyl) $CH_3CON(CH_3)(C_6H_4NO_2\text{-}4)$	194.19	pl (w)		95-7			al, eth	$B12^{3}$, 1595
82	Acetamide, phenoxy $C_6H_5OCH_2CONH_2$	151.16			101.5			al	$B6^{4}$, 638
83	Acetamide, phenyl $C_6H_5CH_2CONH_2$	135.17			157			al, eth	$B9^{3}$, 2193
84	Acetamide, *N*-phenyl or Acetanilide $CH_3CONHC_6H_5$	135.17		304	114.3	1.2190^{15}		al, eth, ace, bz, chl	$B12^{3}$, 459
85	Acetamide, thiono CH_3CSNH_2	75.13			115-6			w, al	$B2^{4}$, 565
86	Acetamide, trichloro Cl_3CCONH_2	162.40		$238\text{-}9^{740}$	142			al, eth	$B2^{4}$, 520
87	Acetamidine $CH_3C(NH_2)=NH$	58.08		80^{10}	63-65			al	$B2^{4}$, 428
88	Acetamidine, *N,N'*-diphenyl $CH_3C(NC_6H_5)NHC_6H_5$	210.28	nd (al)		131-2			eth	$B12^{3}$, 471
89	Acetamidine, hydrochloride $CH_3C(NH_2)NH \cdot HCl$	94.54	nd or pr (al)		177-8			w	$B2^{4}$, 429
90	Acetanilide or N-phenylacetamide $CH_3CONHC_6H_5$	135.17	rh or pl (w)	304	114.3	1.2190^{15}		al, eth, ace, bz	$B12^{3}$, 459
91	Acetoacetanilide $CH_3COCH_2CONHC_6H_5$	177.20			86			al, eth	$B12^{3}$, 993
92	Acetanilide, *o*-amino $2\text{-}H_2NC_6H_4NHCOCH_3$	150.18			132			w, al, eth	$B13^{3}$, 41
93	Acetanilide, *m*-amino $3\text{-}H_2NC_6H_4NHCOCH_3$	150.18	nd or pl (bz)		87-9			w, al, bz, eth, ace	$B13^{3}$, 75
94	Acetanilide, *p*-amino $4\text{-}H_2NC_6H_4NHCOCH_3$	150.18	nd (w)	267	165-8			al, eth, w	$B13^{3}$, 166
95	Acetanilide, *N*-bromo $C_6H_5N(Br)COCH_3$	214.06	ye pl (peth)		88			chl, lig	$B12^{3}$, 1078
96	Acetanilide, *o*-bromo $2\text{-}Br\text{-}C_6H_4NHCOCH_3$	214.06	nd (al)		99			al, eth	$B12^{3}$, 1417
97	Acetanilide, *m*-bromo $3\text{-}BrC_6H_4\text{-}NHCOCH_3$	214.06	nd (aq,al)		87.5			al, eth	$B12^{3}$, 1424
98	Acetanilide, *p*-bromo $4\text{-}BrC_6H_4NHCOCH_3$	214.06	nd (60% al)		168	1.717		al	$B12^{3}$, 1437
99	Acetanilide, *N*-butyl $C_6H_5N(C_4H_9)COCH_3$	191.27		$281, 140^{10}$	24.5	$0.9912^{20/4}$	1.5146^{20}		$B12^{3}$, 467
100	Acetanilide, *p*-butyl $4\text{-}C_4H_9C_6H_4NHCOCH_3$	191.27	wh, pl (al)		105			al	$B12^{3}$, 2715

No.	Name, Synonyms, and Formula	Mol. wt.	Color, crystalline form, specific rotation and λ_{max} (log ε)	b.p. °C	m.p. °C	Density	n_D	Solubility	Ref.
101	**Acetanilide, N-chloro** $C_6H_5N(Cl)COCH_3$	169.61	nd (dil aa) pl (peth-chl)		91				B12[3], 1077
102	Acetanilide, *o*-chloro 2-$ClC_6H_4NHCOCH_3$	169.61	nd (dil aa)		87-8 (sub)			al, eth, bz	B12[3], 1288
103	Acetanilide, 2-chloro-3-methyl or 3-acetamino-2-chloro-toluene 2-Cl-3-CH_3-$C_6H_3NHCOCH_3$	183.64	nd (al)		133-4			al, bz	B12[3], 1991
104	Acetanilide, 2-chloro-4-nitro 2-Cl-4-$NO_2C_6H_3NHCOCH_3$	214.61	pr (al)		139-40				B12[3], 1663
105	Acetanilide, 2-chloro-5-nitro 2-Cl-5-$NO_2C_6H_3NHCOCH_3$	214.61	nd (al)		156				B12[2], 398
106	**Acetanilide, *m*-chloro** 3-$ClC_6H_4NHCOCH_3$	169.61	nd		79			al, eth, bz	B12[3], 1309
107	Acetanilide, 3-chloro-2-methyl or 2-acetamino-6-chloro-toluene 3-Cl-6-$CH_3C_6H_3NHCOCH_3$	183.64	nd (dil al)		157-9			al, bz	B12[2], 1919
108	Acetanilide, 3-chloro-4-methyl or 4-acetamino-2-chloro-toluene 3Cl-4-CH_3-C_6H_3-$NHCOCH_3$	183.64	tcl cr		105			al	B12[2], 530
109	Acetanilide, 3-chloro-5-methyl 3-Cl-5-$CH_3C_6H_3NHCOCH_3$	183.64	nd (al)		151			al	B12, 871
110	Acetanilide, 3-chloro-6-methyl or 2-acetamino-4-chloro-toluene 3-Cl-6-$CH_3C_6H_3NHCOCH_3$	183.64	nd (w)		139-40			al, eth, w	B12[3], 1912
111	Acetanilide, 3-chloro-4-nitro 3-Cl-4-$NO_2C_6H_3NHCOCH_3$	214.61	pa ye nd (al)		145				B12[3], 1666
112	**Acetanilide, *p*-chloro** 4-$ClC_6H_4NHCOCH_3$	169.61	nd (aq aa) ta (al ace) cr (w)		179	1.385^{224}		al, eth	B12[3], 1336
113	Acetanilide, 4-chloro-2-methyl or 2-acetamino-5-chloro-toluene 4-Cl-2-$CH_3C_6H_3NHCOCH_3$	183.64	lf (al)		140			al, bz	B12[3], 1914
114	Acetanilide, 4-chloro-3-methyl or 5-acetamino-2-chloro-toluene 4-Cl-3-$CH_3C_6H_3NHCOCH_3$	183.64	lf (al)		91.2-.7			al, bz	B12[2], 473
115	Acetanilide, 4-chloro-2-nitro 4-Cl-2-$NO_2C_6H_3NHCOCH_3$	214.61	ye nd (al)		104				B12[3], 1652
116	Acetanilide, 4-chloro-3-nitro 4-Cl-3-$NO_2C_6H_3NHCOCH_3$	214.61	ye nd (al)		150				B12[2], 357
117	Acetanilide, *α*-cyano $C_6H_5NHCOCH_2CN$	160.18	nd (al)		199-200				B12[3], 558
118	Acetanilide, 4-cyclohexyl 4-C_6H_{11}-$C_6H_4NHCOCH_3$	217.31	nd (peth)		129			al, eth, chl	B12[3], 2819
119	Acetanilide, 2,4-dimethyl or 2,4-Acetoxylide 2,4-$(CH_3)_2C_6H_3NHCOCH_3$	163.22	nd (al)	170^{10}	128-9			al	B12[3], 2474
120	Acetanilide, 4,6-dimethyl-2-nitro 2-NO_2-4,6-$(CH_3)_2C_6H_2NHCOCH_3$	208.22	yesh nd (w)		172-3			al, bz, w	B12[3], 2491
121	Acetanilide, 2,3-dinitro 2,3-$(NO_2)_2$-C_6H_3-$NHCOCH_3$	225.16	nd (al)		187			al, eth, bz	B12[3], 1680
122	Acetanilide, 2,4-dinitro 2,4-$(NO_2)_2C_6H_3NHCOCH_3$	225.16	ye nd (al or bz)		125-6			al, eth, bz	B12[3], 1686
123	Acetanilide, 2,5-dinitro 2,5-$(NO_2)_2C_6H_3NHCOCH_3$	225.16	nd (al)		121			al	B12[3], 1704
124	Acetanilide, 2,6-dinitro 2,6-$(NO_2)_2C_6H_3NHCOCH_3$	225.16	nd (aa)		197			al, aa	B12[3], 1705
125	Acetanilide, 3,4-dinitro 3,4-$(NO_2)_2C_6H_3NHCOCH_3$	225.16	ye cr (al) nd (w)		144-5			al, w	B12[3], 1705
126	Acetanilide, 3,5-dinitro 3,5-$(NO_2)_2C_6H_3NHCOCH_3$	225.16	ye nd (dil aa, w)		191			al, w, aa	B12[3], 1706
127	Acetamide, *N,N*-dipropyl $CH_3CON(C_3H_7)_2$	143.23		209-10, 101^{18}		$0.8992^{17.4/4}$	$1.4419^{17.4}$	al	B4[4], 476
128	Acetanilide, 2-ethoxy or *o*-Acetophenetidide 2-(C_2H_5O)-$C_6H_4NHCOCH_3$	179.22	lf (dil al)	>250	79			al, eth	B13[3], 779
129	Acetanilide, 2-ethoxy-4-nitro 2-(C_2H_5O)4-$(NO_2)C_6H_3NHCOCH_3$	224.22	nd (al)		202			al	B13[2], 194
130	Acetanilide, 2-ethoxy-5-nitro 2(C_2H_5O)-5$(NO_2)C_6H_3NHCOCH_3$	224.22	ye nd (al)		199			bz, al	B13[3], 881

PHYSICAL CONSTANTS OF ORGANIC COMPOUNDS (Continued)

No.	Name, Synonyms, and Formula	Mol. wt.	Color, crystalline form, specific rotation and λ_{max} (log ε)	b.p. °C	m.p. °C	Density	n_D	Solubility	Ref.
131	Acetanilide, 3-ethoxy or *m*-accetophenetidide $3\text{-}(C_2H_5O)\text{-}C_6H_4NHCOCH_3$	179.22	gy pl (w)	...	97-9	...	...	al, eth, w	B13³, 950
132	Acetanilide, 4-ethoxy or Phenacitin $4\text{-}(C_2H_5O)C_6H_4NHCOCH_3$	179.22	mcl pr	dic	137-8	...	1.571	al, ace	B13³, 1057
133	Acetanilide, 4-ethoxy-2-nitro $4\text{-}C_2H_5O\text{-}2NO_2C_6H_3NHCOCH_3$	224.22	ye nd (w)	...	104	...	...	al, eth, ace, chl	B13³, 1208
134	Acetanilide, 4-ethoxy-3-nitro $4\text{-}(C_2H_5O)\text{-}3\text{-}NO_2C_6H_3NHCOCH_3$	224.22	nd (dil al)	...	123	...	...	al, ace, bz	B13³, 1202
135	Acetanilide, 5-ethoxy-2-nitro $5\text{-}(C_2H_5O)\text{-}2\text{-}NO_2C_6H_3NHCOCH_3$	224.22	nd (al)	...	95	...	...	al, eth, ace, bz, chl	B13³, 976
136	Acetanilide, *N*-ethyl $C_6H_5N(C_2H_5)COCH_3$	163.22	(w) rh (eth)	258^{731}	55	...	...	w, eth	B12³, 466
137	Acetanilide, *N*-ethyl-3-nitro $3\text{-}NO_2\text{-}C_6H_4N(C_2H_5)COCH_3$	208.22	pa ye nd (dil al)	...	88-9	...	...	al, bz	B12, 704
138	Acetanilide, *N*-ethyl-4-nitro $4\text{-}NO_2\text{-}C_6H_4\text{-}N(C_2H_5)COCH_3$	208.22	lf or pr (dil al)	...	118-9	...	...	al, bz, eth	B12³, 1596
139	Acetanilide, *o*-hydroxy or *o*-Acetamidophenol $2\text{-}HOC_6H_4NHCOCH_3$	151.16	pl (dil al)	...	208	...	...	al, eth, bz	B13³, 778
140	Acetanilide, *m*-hydroxy or *m*-Acetamidophenal $3\text{-}HO\text{-}C_6H_4\text{-}NHCOCH_3$	151.16	nd (w)	...	148-9	...	...	w, al, eth, bz	B13³, 950
141	Acetanilide, *p*-hydroxy or p-acetamidophenol $4\text{-}HOC_6H_4NHCOCH_3$	151.17	nd (bz-peth)	...	151	...	...	al, eth, bz	B13³, 1056
142	Acetanilide, *p*-iodo $4\text{-}I\text{-}C_6H_4NHCOCH_3$	261.06	ta (w) pr (w, al)	...	184.5	...	...	al, eth, bz	B12³, 1496
143	Acetanilide, *o*-methyl or *o*-acetotoluidide $2\text{-}CH_3\text{-}C_6H_4NHCOCH_3$	149.19	nd (al)	296	110	1.168^{15}	...	al, eth, ace, bz	B12³, 1853
144	Acetanilide, *m*-methyl or *m*-acetotoluidide $3\text{-}CH_3\text{-}C_6H_4NHCOCH_3$	149.19	nd (w)	303, $182\text{-}3^{14}$	65.5	1.141^{15}	...	al, eth	B12³, 1962
145	Acetanilide, *p*-methyl or *p*-acetotoluidide $4\text{-}CH_3\text{-}C_6H_4NHCOCH_3$	149.19	mcl cr or nd (dil al)	307 sub	148.5	1.212^{15}	...	al, eth, bz	B12³, 2051
146	Acetanilide, *N*-methyl-*o*-hydroxy $2\text{-}HOC_6H_4N(CH_3)COCH_3$	165.19	nd (bz-peth)	...	151	...	...	al, eth, bz	B13³, 783
147	Acetanilide, *N*-methyl-*p*-hydroxy $4\text{-}HOC_6H_4N(CH_3)COCH_3$	165.19	sc (w)	...	245	...	...	al, eth	B13¹, 162
148	Acetanilide, *o*-methoxy or *o*-Acetanisidine $2\text{-}CH_3OC_6H_4\text{-}NHCOCH_3$	163.19	nd (w)	303-5	87-8	...	...	al, eth, ace, w	B13³, 778
149	Acetanilide, 2-methoxy-3-nitro $2\text{-}CH_3O\text{-}3\text{-}NO_2C_6H_3NHCOCH_3$	210.19	pa ye pr (dil aa MeOH)	...	103.4	...	...	MeOH, al	B13², 195
150	Acetanilide, 2-methoxy-4-nitro $2\text{-}CH_3O\text{-}4\text{-}NO_2C_6H_3NHCOCH_3$	210.19	pa ye cr (AcOEt)	...	153-4	...	...	al, bz	B13³, 890
151	Acetanilide, 2-methoxy-5-nitro $2\text{-}CH_3O\text{-}5\text{-}NO_2C_6H_3NHCOCH_3$	210.19	nd (w)	...	178	...	...	...	B13³, 881
152	Acetanilide, 2-methoxy-6-nitro $2\text{-}CH_3O\text{-}6\text{-}NO_2C_6H_3NHCOCH_3$	210.19	pa ye nd (dil al)	...	158-9	...	...	bz, al	B13³, 876
153	Acetanilide, *m*-methoxy or *m*-acetanisidine $3\text{-}CH_3OC_6H_4NHCOCH_3$	165.19	nd or pl (w)	...	81	...	...	w, al, eth, ace	B13³, 950
154	Acetanilide, 3-methoxy-2-nitro $3\text{-}CH_3O\text{-}2NO_2C_6H_3NHCOCH_3$	210.19	br (al) amor	...	265 sub	...	...	al, bz	B13¹, 136
155	Acetanilide, 3-methoxy-4-nitro $3\text{-}CH_3O\text{-}4NO_2C_6H_3NHCOCH_3$	210.19	ye nd (w)	...	165	...	...	al	B13³, 978
156	Acetanilide, 3-methoxy-5-nitro $3\text{-}CH_3O\text{-}5\text{-}NO_2C_6H_3NHCOCH_3$	210.19	nd (aa)	...	201	...	...	al, eth, bz, aa	B13², 216
157	Acetanilide, *p*-methoxy or *p*-acetanisidine $4\text{-}CH_3OC_6H_4NHCOCH_3$	165.19	pl (w)	...	130-2	...	...	al, ace, eth, w	B13³, 1056
158	Acetanilide, 4-methoxy-2-nitro $4\text{-}CH_3O\text{-}2\text{-}NO_2C_6H_3NHCOCH_3$	210.19	ye nd (al)	...	116.5-7	...	...	al, eth, bz	B13³, 1208
159	Acetanilide, 4-methoxy-3-nitro $4\text{-}CH_3O\text{-}3\text{-}NO_2C_6H_3NHCOCH_3$	210.19	og-ye nd (w or dil al)	...	153	...	...	al, lig	B13¹, 186
160	Acetanilide, 5-methoxy-2-nitro $5\text{-}CH_3O\text{-}2\text{-}NO_2C_6H_3NHCOCH_3$	210.19	wh nd (al)	...	125	...	...	al, lig, w	B13³, 976
161	Acetanilide, *N*-methyl or Exalgin $C_6H_5N(CH_3)COCH_3$	149.19	nd (eth) pr (al) lf (lig)	253^{712}	102-4	$1.0036^{105/4}$	1.576	w, al, eth, lig	B12³, 465
162	Acetanilide, *N*-methyl-4-hydroxy-3-nitro $3\text{-}NO_2\text{-}4\text{-}HOC_6H_3N(CH_3)NOCH_3$	210.19	(al)	...	161-2	...	...	al	B13¹, 186
163	Acetanilide, *N*-methyl-4-nitro $4\text{-}NO_2\text{-}C_6H_4N(CH_3)COCH_3$	194.19	pl (w)	...	153	...	...	al, eth	B12¹, 352

No.	Name, Synonyms, and Formula	Mol. wt.	Color, crystalline form, specific rotation and λ_{max} (log ε)	b.p. °C	m.p. °C	Density	n_D	Solubility	Ref.
164	Acetanilide, *N*-methyl-2-methyl or *N*-methyl-*o*-acetotoluidide $2-CH_3C_6H_4N(CH_3)COCH_3$	163.22	....	260	55-6	....	....	al	$B12^3$, 1854
165	Acetanilide, *N*-methyl-3-methyl or *N*-methyl-*m*-acetotoluidide $3-CH_3-C_6H_4N(CH_3)COCH_3$	163.22	cr	....	64	....	....	al	$B12^3$, 1963
166	Acetanilide, *N*-methyl-4-methyl or *N*-methyl-*p*-acetotoluidide $4-CH_3-C_6H_4N(CH_3)COCH_3$	163.22	lf (eth-al)	263	83	....	....	al, eth, lig	$B12^3$, 2052
167	Acetanilide, 4-methyl-2-nitro $4-CH_3-2-NO_2C_6H_3NHCOCH_3$	194.19	dk ye nd (peth)	....	96	....	....	....	$B12^3$, 2178
168	Acetanilide, *o*-nitro $2-NO_2C_6H_4NHCOCH_3$	180.16	ye pr (lig) lf (dil al)	$100^{0.1}$	94	1.419^{15}	....	w, al, eth, lig	$B12^3$, 1523
169	Acetanilide, *m*-nitro $3-NO_2C_6H_4NHCOCH_3$	180.16	wh lf (al)	$100^{0.008}$	154-6	....	....	w, al, chl	$B12^3$, 1551
170	Acetanilide, *p*-nitro $4-NO_2C_6H_4NHCOCH_3$	180.16	ye pr (w)	$100^{0.008}$	216 (217)	....	....	al, eth, aa, lig	$B12^3$, 1594
171	Acetanilide, *p*-octyl $4-(C_8H_{17})C_6H_4NHCOCH_3$	247.38	lf or pr (al)	....	94	....	....	al, eth	B12, 1185
172	Acetanilide, *N*-phenyl or *N*-acetyldiphenylamine $(C_6H_5)_2NCOCH_3$	211.26	rh or nd (w or lig)	sub	103	....	....	al	$B12^3$, 468
173	Acetanilide, *N*-propyl $C_6H_5N(C_3H_7)COCH_3$	177.25	mcl lf (eth, lig)	266^{712}	49 (56)	....	....	al, eth	$B12^3$, 466
174	Acetanilide, *N*-isopropyl $C_6H_5N(i-C_3H_7)COCH_3$	177.22	lf (lig)	$262-3^{712}$	39	....	....	lig	$B12^3$, 466
175	Acetic acid or Ethanoic acid CH_3CO_2H	60.05	rh (hyg)	117.9, 17^{10}	16.6	$1.0492^{20/4}$	1.3716^{20}	**w, al, ace, bz**	$B2^4$, 94
176	Acetic acid, allyl ester or Allyl acetate $CH_3CO_2CH_2CH=CH_2$	100.12	....	103.5	......	$0.9276^{20/4}$	1.4049^{20}	**al, eth,** ace, w	$B2^4$, 180
177	Acetic acid, anhvdride or Acetic anhydride $(CH_3CO)_2O$	102.09	....	139.55, 44^{15}	-73.1	$1.0820^{20/4}$	1.39006^{20}	al, **eth**, bz, w	$B2^4$, 386
178	Acetic acid, anhydride, trifluoro or Trifluoro acetic anhydride $(F_3CO)_2O$	210.03	....	39.5-40.1	-65	$1.490^{25/4}$	1.269^{25}	eth, aa	$B2^4$, 469
179	Acetic acid, benzyl ester or Benzyl acetate $CH_3CO_2CH_2C_6H_5$	150.18	....	215.5, $93-4^{10}$	-51.3	$1.0550^{20/4}$	1.5232^{20}	**al**, eth, ace	$B6^4$, 2262
180	Acetic acid, dibenzyl or Dibenzylacetic acid $(C_6H_5CH_2)_2CHCO_2H$	240.3	pl (peth), dil (aa), nd (w)	235^{18}	89	....	....	al, eth, bz	$B9^3$, 3356
181	Acetic acid, dibenzyl, methyl ester or Methyl dibenzylacetate $(C_6H_5CH_2)_2CHCO_2CH_3$	254.33	nd (al)	....	42-3	....	....	al, peth	$B9^2$, 475
182	Acetic acid, 2-bromo ethyl ester or (2-Bromoethyl) acetate $CH_3CO_2CH_2CH_2Br$	167.00	....	162-3	-13.8	$1.514^{20/4}$	1.457^{25}	w, al, **eth**, chl	$B2^4$, 136
183	Acetic acid, bromomethyl ester $CH_3CO_2CH_2Br$	152.98	....	$130-3^{750}$	......	$1.6350^{20/4}$	1.4520^{20}	al, eth, chl	$B2^4$, 280
184	Acetic acid, 2-butenyl ester or Crotyl acetate $CH_3CO_2CH_2CH=CHCH_3$	114.14	....	132	......	$0.9192^{20/4}$	1.4181^{20}	al, eth, bz	$B2^4$, 183
185	Acetic acid, butyl ester or Butyl acetate $CH_3CO_2C_4H_9$	116.16	....	126.5	-77.9	$0.8825^{20/4}$	1.3941^{20}	**al, eth**, bz	$B2^4$, 143
186	Acetic acid, *iso*-butyl ester or *iso*-Butyl acetate $CH_3CO_2-i-C_4H_9$	116.16	....	117.2	-98.58	$0.8712^{20/4}$	1.3902^{20}	**al, eth**, ace	$B2^4$, 149
187	Acetic acid, *sec*-butyl ester *d* or *sec*-butyl acetate $CH_3CO_2CH(CH_3)CH_2CH_3$	116.16	$[\alpha]^{20/D}$ + 25.43	112	......	$0.8758^{16/4}$	1.3877^{20}	al, eth, ace	$B2^3$, 241
188	Acetic acid, *sec*-butyl ester *dl* or *sec*-butyl acetate *dl* $CH_3CO_2CH(CH_3)CH_2CH_3$	116.16	....	112.2	......	$0.8716^{20/4}$	1.3888^{20}	al, eth, ace	$B2^4$, 148
189	Acetic acid, *sec*-butyl ester *l* or *sec*-butyl acetate *l* $CH_3CO_2CH(CH_3)CH_2CH_3$	116.16	$[\alpha]^{19/546}$ − 20.2	116-7	......	$0.87301^{19/4}$	1.3899^{18}	al, eth, ace	$B2^3$, 241
190	Acetic acid, *tert*-butyl ester or *tert*-butyl acetate $CH_3CO_2C(CH_3)_3$	116.16	....	97-8	......	$0.8665^{20/4}$	1.3855^{20}	al, eth, aa	$B2^4$, 151
191	Acetic acid, (1-chloroethyl) ester $CH_3CO_2CHCl-CH_3$	122.55	....	121.3^{746}	......	$1.110^{20/4}$	1.409^{20}	eth	$B2^4$, 282
192	Acetic acid, (2 chloroethyl) ester $CH_3CO_2CH_2CH_2Cl$	122.55	....	145, 50^{18}	......	$1.178^{20/4}$	1.4234^{20}	**al, eth**	$B2^4$, 136
193	Acetic acid, (2-chloro-*iso*-propyl) ester $CH_3CO_2CCl(CH_3)_2$	136.58	....	149-50	......	1.0788^{20}	1.4223^{20}	al, eth	$B2^4$, 286
194	Acetic acid, chloromethyl ester $CH_3CO_2CH_2Cl$	108.52	....	$115-6^{751}$	......	$1.194^{20/4}$	1.409^{20}	al, eth	$B2^4$, 280

PHYSICAL CONSTANTS OF ORGANIC COMPOUNDS (Continued)

No.	Name, Synonyms, and Formula	Mol. wt.	Color, crystalline form, specific rotation and λ_{max} (log ε)	b.p. °C	m.p. °C	Density	n_D	Solubility	Ref.
195	Acetic acid, (3-chloropropyl) ester $CH_3CO_2CH_2CH_2CH_2Cl$	136.58		163-5^{747}, 62-3^{10}		1.250^{19}	1.431^{20}	al, eth, ace	B2^4, 140
196	Acetic acid, cyclohexyl ester or Cyclohexyl acetate $CH_3CO_2C_6H_{11}$	142.20		173, 62-3^{12}		0.9698$^{20/4}$	1.4401^{20}	al, eth	B6^4, 36
197	Acetic acid, cyclopentyl ester or Cyclopentyl acetate $CH_3CO_2C_5H_9$	128.17		51.5-2.5^{12}		0.9522^{16}			B6^4, 7
198	Acetic acid, decyl ester or Decyl acetate $CH_3CO_2C_{10}H_{21}$	220.32		244, 125.8^{155}	-15.05	0.8671$^{20/4}$	1.4273^{20}	al, eth, bz, aa	B2^4, 168
199	Acetic acid, 1,3-dichloro-*iso*-propyl ester $CH_3CO_2CH(CH_2Cl)_2$	171.02		202-8, 81^{15}		1.281^{20}	1.4542^{20}	al, eth, chl	B2^3, 233
200	Acetic acid, 1,1-dimethyl butyl ester $CH_3CO_2C(CH_3)_2C_3H_7$	144.21		152-3^{755}, 34.5^{10}		0.8798$^{18/4}$	1.4068^{10}	al, eth	B2^3, 257
201	Acetic acid, (2,4-dinitrophenyl) ester $CH_3CO_2[C_6H_3(NO_2)_2$-2.4]	226.15	cr (MeOH)		72-3				B6^4, 1380
202	Acetic acid, 3,5,-dinitro phenyl ester $CH_3CO_2[C_6H_3(NO_2)_2$-3,5]	226.15	cr (bz-peth)		126-7			bz, aa	B6, 258
203	Acetic acid, dodecyl ester $CH_3CO_2C_{12}H_{25}$	228.38		260-70			1.4439^{20}		B2^4, 170
204	Acetic acid, ethyl ester or Ethyl acetate $CH_3CO_2C_2H_5$	88.11		77.06	-83.6	0.9003$^{20/4}$	1.3723^{20}	w, al, eth, ace, bz	B2^4, 127
205	Acetic acid, 2-ethylbutyl ester $CH_3CO_2CH_2CH(C_2H_5)_2$	144.21		162-3,63^{20}		0.8790$^{20/4}$	1.4109^{20}	al, eth	B2^4, 161
206	Acetic acid, 2-ethyl hexyl ester $CH_3CO_2CH_2CH(C_2H_5)C_4H_9$	172.27		199,95^{25}	-93	0.8734$^{20/20}$	1.4204^{20}	al, eth	B2^4, 166
207	Acetic acid, 2-ethoxyethyl ester or Cellosolve acetate $CH_3CO_2CH_2CH_2OH$	132.16		156.4,49^{12}	-61.7	0.9740$^{20/4}$	1.4054^{20}	w, al, eth, ace	B2^4, 214
208	Acetic acid, 3-ethyl-3-pentyl ester $CH_3CO_2C(C_2H_5)_3$	158.24		160-3				al	B2, 134
209	Acetic acid, 9-fluorenyl $C_{13}H_9CH_2CO_2H$	224.26	mcl nd (al)	218-20^{ll}	138-9				B9^3, 3448
210	Acetic acid, furfuryl ester or Furfuryl acetate $CH_3CO_2CH_2(C_4H_3O)$	140.14		177		1.1175$^{20/4}$	1.4327^{20}	al, eth	B17^4, 1246
211	Acetic acid, 2-furylester $CH_3CO_2(C_4H_3$-O)	126.11		102-4$^{0.4}$	68-9			w, bz, peth, MeOH	B17^4, 1220
212	Acetic acid, 1-heptylester $CH_3CO_2C_7H_{15}$	158.24		192.4,96^{28}	-50.2	0.8750$^{15/4}$	1.4150^{20}	al, eth	B2^4, 162
213	Acetic acid, 4-heptyl ester $CH_3CO_2CH(C_3H_7)_2$	158.24		170-2,60-70^{17}		0.8742$^{0/0}$	1.4105^{19}	eth	B2^4, 163
214	Acetic acid, hexadecylester or Cetyl acetate $CH_3CO_2C_{16}H_{33}$	284.48		220-5^{205}	α,-18.5, β, 24.2	0.8574$^{25/4}$	1.4438^{20}		B2^4, 171
215	Acetic acid, hexylester or Hexyl acetate $CH_3CO_2C_6H_{13}$	144.21		171.5,61.5^{12}	-80.9	0.8779$^{15/4}$	1.4092^{20}	al, eth	B2^4, 159
216	Acetic acid, 2-hexyl ester *d* $CH_3CO_2CH(CH_3)C_4H_9$	144.21		158,57^{20}		0.8658$^{18/4}$	1.4014^{25}	al, eth	B2^2, 145
217	Acetic acid, 2-hexyl ester *dl* $CH_3CO_2CH(CH_3)C_4H_9$	144.21		157-8		0.8651$^{15/15}$	1.4014^{25}	al, eth	B2^3, 256
218	Acetic acid, 3-hexyl ester *d* $CH_3CO_2CH(C_2H_5)C_3H_7$	144.21	$[\alpha]^{20/}_{D}$ + 0.55	149-51		0.8672$^{20/4}$	1.4037^{20}	al, eth	B2^4, 160
219	Acetic acid, hydrazide $CH_3CONHNH_2$	74.08		129^{18}	67			w, al, eth	B2^4, 435
220	Acetic acid, (2-hydroxyethyl) ester or Glycol mono acetate $CH_3CO_2CH_2CH_2OH$	104.11		187-9		1.108^{15}		w, al, eth	B2^4, 214
221	Acetic acid, menthyl ester *l* or Menthyl acetate *l* $CH_3CO_2C_{10}H_{19}$	198.31	$[\alpha]^{20/D}$- 79.42	109^{10}		0.9185$^{20/4}$	1.4469^{20}		B6^4, 153
222	Acetic acid, (4-methoxybenzyl) ester $CH_3CO_2CH_2C_6H_4OCH_3$-4	180.20		110-7		1.104-7$^{25/25}$			B6^3, 4549
223	Acetic acid, (2-methoxyethyl) ester $CH_3CO_2CH_2CH_2OCH_3$	118.13		144-5,40-1^{12}		1.0090$^{19/19}$	1.4002^{20}	w, al, eth	B2^4, 214
224	Acetic acid, (2-methoxy phenyl) ester $CH_3CO_2C_6H_4OCH_3$-2	166.18		134-3^{18}	31-2			al, eth	B6^3, 4227
225	Acetic acid, methyl ester or Methyl acetate $CH_3CO_2CH_3$	74.08		57	-98.1	0.9330$^{20/4}$	1.3595^{20}	al, eth, w, ace, bz, chl	B2^4, 122
226	Acetic acid, 2-methyl-2-butyl ester or *tert*-amyl acetate $CH_3CO_2C(CH_3)_2CH_2CH_3$	130.19		124-4.5		0.8740^{20}	1.4010^{20}	al, eth, ace	B2^4, 156
227	Acetic acid, 3-methyl butyl ester or *iso* Pentyl acetate $CH_3CO_2CH_2CH_2CH(CH_3)_2$	130.19		142	-78.5	0.8670$^{20/4}$	1.4003^{20}	al, eth, ace	B2^4, 157

PHYSICAL CONSTANTS OF ORGANIC COMPOUNDS (Continued)

No.	Name, Synonyms, and Formula	Mol. wt.	Color, crystalline form, specific rotation and λ_{max} (log ε)	b.p. °C	m.p. °C	Density	n_D	Solubility	Ref.
228	Acetic acid, methylene diester or Methanediol diacetate $CH_2(O_2CCH_3)_2$	132.13		164-5	-23	$1.136^{20/4}$	1.4025^{24}	al, eth	$B2^4$, 280
229	Acetic acid, (2-methyl-3-heptyl) ester $CH_3CO_2\text{-}CH(C_4H_9)CH(CH_3)_2$	172.27		172		6.875^{20}	1.4166^{20}	al	B2, 134
230	Acetic acid, 3-methyl-2-heptyl ester $CH_3CO_2CH(CH_3)CH(CH_3)C_4H_9$	172.27		185		$0.8545^{21/4}$	1.418^{21}	al	$B2^2$, 146
231	Acetic acid, 6-methyl-2-heptyl ester $CH_3CO_2\text{-}CH(CH_3)(CH_2)_3CH(CH_3)_2$	172.27		$187\text{-}8^{768}$		0.8474^{20}	1.4137^{20}	al	B2, 135
232	Acetic acid, 6-methyl-3-heptyl ester $CH_3CO_2CH(C_2H_5)CH_2CH_2CH(CH_3)_2$	172.27		184-5		0.8554^{20}	1.41602^{20}	al	B2, 135
233	Acetic acid, (2-methyl-1-naphthyl) ester $1\text{-}CH_3CO_2\text{-}C_{10}H_6\text{-}CH_3\text{-}2$	200.24	nd (eth-peth)		81-2			eth	$B6^3$, 3027
234	Acetic acid, (4-methyl-l-naphthyl) ester $1\text{-}CH_3CO_2C_{10}H_6CH_3\text{-}4$	200.24	nd (eth-peth)		86.8				C64, 19519
235	Acetic acid, (6-methyl-1-naphthyl) ester or 5-acetoxy-2-methyl naphthalene. $5\text{-}CH_3CO_2C_{10}H_6CH_3\text{-}(2)$	200.24		124^2					$B6^3$, 3028
236	Acetic acid, (7-methyl-l-naphthyl) ester or l-Acetoxy-7-methyl naphthalene $1\text{-}CH_3CO_2C_{10}H_6\text{-}CH_3\text{-}7$	200.24		188^{16}	39.41				$B6^3$, 3030
237	Acetic acid, (2-methyl-3-pentyl) ester $CH_3CO_2CH(C_2H_5)CH(CH_3)_2$	144.21		148.5^{747}		0.8688^{20}		al, eth	$B2^4$, 160
238	Acetic acid, (3-methyl-3-pentyl) ester $CH_3CO_2C(C_2H_5)_2CH_3$	144.21		148		$0.8834^{20/20}$	1.4109^{18}	al, eth	$B2^4$, 161
239	Acetic acid, (4-methyl-2-pentyl) ester $CH_3CO_2CH(CH_3)CH_2CH(CH_3)_2$	144.21		147-8		0.8805^0	1.3980^{20}	al, eth	$B2^4$, 161
240	Acetic acid, 2-methylpentyl ester $CH_3CO_2CH_2CH(CH_3)C_3H_7$	144.21		162.2^{740}		$0.8717^{25/25}$		al, eth	$B2^4$, 160
241	Acetic acid, α-naphthyl ester or α-naphthyl acetate $CH_3CO_2\text{-}\alpha\text{-}C_{10}H_7$	186.21	nd or pl (al)		49			al, eth	$B6^3$, 2928
242	Acetic acid, β-naphthyl ester or β-naphthyl acetate $CH_3CO_2\text{-}\beta\text{-}C_{10}H_7$	186.21	nd (al)		70			al, eth	$B6^3$, 2982
243	Acetic acid, (2-nitrophenyl) ester $CH_3CO_2C_6H_4NO_2\text{-}2$	181.15	nd or pr (lig)	253d	40-1			al, eth, ace, bz, w	$B6^4$, 1256
244	Acetic acid, 3-nitrophenyl ester $CH_3CO_2\text{-}C_6H_4NO_2\text{-}3$	181.15	nd (peth)		55-6			al, w, lig	$B6^4$, 1273
245	Acetic acid, (4-nitro phenyl) ester $CH_3CO_2C_6H_4NO_2\text{-}4$	181.15	lf (dil al)		81-2			al, bz, w, lig	$B6^4$, 1298
246	Acetic acid, octadecyl ester or Stearyl acetate $CH_3CO_2C_{18}H_{36}$	313.51		$222\text{-}3^{15}$	34.5			al	$B2^4$, 171
247	Acetic acid, octyl ester or Octyl acetate $CH_3CO_2C_8H_{17}$	172.27		$210, 112\text{-}3^{30}$	-38.5	$0.8705^{20/4}$	1.4150^{20}	al, eth	$B2^4$, 165
248	Acetic acid, 2-octylester *d* $CH_3CO_2CH(CH_3)C_6H_{13}$	172.27	$[\alpha]^{20/}{}_D$ + 7.00	196		$0.8606^{15/4}$	1.4141^{20}	al, eth, bz	$B2^3$, 260
249	Acetic acid, 2-octylester *dl* $CH_3CO_2CH(CH_3)C_6H_{13}$	172.27		194.5^{749}		$0.8626^{14/4}$	1.4146^{20}	al, eth	$B2^4$, 165
250	Acetic acid, 2-octyl ester *l* $CH_3CO_2CH(CH_3)C_6H_{13}$	172.27	$[\alpha]^{14.5/D}$ - 6.0	$196, 89\text{-}90^{17}$		$0.8570^{20/4}$	1.4140^{20}	al, eth	$B2^4$, 165
251	Acetic acid, 3-octyl ester *l* $CH_3CO_2CH(C_2H_5)C_5H_{11}$	172.27	$[\alpha]^{20/D}$ – 4.3		191-1.5, 56.5^2	$0.8641^{20/4}$	1.4152^{20}	al, eth, CS_2	$B2^4$, 166
252	Acetic acid, pentyl ester or Pentyl acetate $CH_3CO_2C_5H_{11}$	130.19		149.25	-70.8	$0.8756^{20/4}$	1.4023^{20}	al, eth	$B2^4$, 152
253	Acetic acid, 2-pentyl ester *d* or See-pentyl acetate *d*. $CH_3CO_2CH(CH_3)C_3H_7$	130.19	$[\alpha]^{20/D}$ + 17.16 (undil)	130-1		$0.8692^{18/4}$	1.3960^{20}	al, eth, ace	$B2^1$, 60
254	Acetic acid, 2-pentyl ester *dl* $CH_3CO_2CH(CH_3)C_3H_7$	130.19		134		$0.8692^{18/4}$	1.3960^{20}	al, eth	$B2^4$, 155
255	Acetic acid, 2-pentyl ester *l* $CH_3CO_2CH(CH_3)C_3H_7$	130.19	$[\alpha]^{20/D}$ + 3.30	142		0.8803^{15}	1.4012^{20}	al, eth	B2, 132
256	Acetic acid, 3-pentyl ester $CH_3CO_2CH(C_2H_5)_2$	130.19		132^{741}		$0.8712^{20/4}$	1.4005^{20}	al, eth	$B2^4$, 155
257	Acetic acid, phenacyl ester or ω-aceto aceto phenone $CH_3CO_2CH_2COC_6H_5$	178.19	rh pl	270	49-9.5	$1.1169^{65/4}$	1.5036^{65}	al, eth	$B8^3$, 301
258	Acetic acid, (phenoxyethyl) ester $CH_3CO_2CH_2CH_2OC_6H_5$	180.20		$109\text{-}12^3$		1.1082^{20}	1.5080^{20}		$B6^4$, 575
259	Acetic acid, phenyl ester $CH_3CO_2C_6H_5$	136.15		$195.7, 75\text{-}6^8$		$1.0780^{20/4}$	1.5035^{20}	al, eth	$B6^4$, 611

No.	Name, Synonyms, and Formula	Mol. wt.	Color, crystalline form, specific rotation and λ_{max} (log ε)	b.p. °C	m.p. °C	Density	n_D	Solubility	Ref.
260	Acetic acid, 2-phenyl ethyl ester or β-phenyl ethyl acetate $CH_3CO_2CH_2CH_2C_6H_5$	164.20		232.6, 153^{76}	-31.1	$1.0883^{20/4}$	1.5171^{20}	al, eth	$B6^3$, 1709
261	Acetic acid, *iso*-propenyl ester $CH_3CO_2\text{-}C(CH_3)=CH_2$	100.12		$92\text{-}4^{732}$	-92.9	0.9090^{20}	1.4033^{20}	al, eth, ace	$B2^4$, 179
262	Acetic acid, propyl ester or Propyl acetate $CH_3CO_2C_3H_5$	102.12		101.6	-95	$0.8878^{20/4}$	1.3842^{20}	al, eth	$B2^4$, 138
263	Acetic acid, *iso*-propyl-ester or *iso*-propyl acetate $CH_3CO_2CH(CH_3)_2$	102.13		90	-73.4	$0.8718^{20/4}$	1.3773^{20}	al, eth, ace, w	$B2^4$, 141
264	Acetic acid, tetrahydro furfuryl ester $CH_3CO_2CH_2(C_4H_7O)$	143.17		204-7		$1.0624^{20/4}$	1.4350^{25}	w, al, eth, chl	$B17^4$, 1103
265	Acetic acid, 2-thionyl or 2-Thiophene acetic acid $(2\text{-}C_4H_3S)CH_2CO_2H$	142.18	cr (w)		76			w, al, eth	$B18^4$, 4062
266	Acetic acid, 2-toly estor or *o*-Cresyl acetate $CH_3CO_2\text{-}C_6H_4CH_3\text{-}2$	150.18		208, 89^{10}		1.0533^{15}	1.5002^{20}	al, eth	$B6^4$, 1960
267	Acetic acid, 3-tolyl ester or *m*-Cresyl acetate $CH_3CO_2C_6H_4CH_3\text{-}3$	150.18		212		$1.043^{20/4}$	1.4978^{20}	al, eth, bz	$B6^4$, 2047
268	Acetic acid, 4-tolyl ester or *p*-Cresyl acetate $CH_3CO_2C_6H_4CH_3\text{-}4$	150.18		212.5, $107\text{-}9^{5}$		$1.0512^{17/4}$	1.5163^{22}	al, eth, chl	$B6^4$, 2112
269	Acetic acid, 1,2,2-trimethyl propyl ester *d* $CH_3CO_2CH(CH_3)C(CH_3)_3$	144.21	$[\alpha]^{25/D}$ + 16.2	141^{756}		$0.856^{20/4}$	1.4001^{25}	eth	$B2^3$, 258
270	Acetic acid, vinyl ester or Vinyl acetate $CH_3CO_2CH=CH_2$	86.09		72.3	-93.2	$0.9317^{20/4}$	1.3959^{20}	al, eth, ace, bz, chl	$B2^4$, 176
271	Aceto acetamide, *n*-phenyl or Acetoacetanilide $CH_3COCH_2CONHC_6H_5$	177.20	pl or nd (bz or lig)		86			al, eth, bz, chl, liq	$B12^3$, 993
272	Aceto acetamide, *N*-2-tolyl or Acetoacet-2-toluide $CH_3COCH_2CONH(C_6H_4CH_3\text{-}2)$	191.23	pr (AcOEt)		107-8			al, bz	$B12^3$, 1895
273	Acetoacetamide, *N*-3-tolyl or Acetoacet-3-toluide $CH_3COCH_2CONH(C_6H_4CH_3\text{-}3)$	191.23	pl (bz-peth)		57-8			al, bz	$B12^3$, 1983
274	Aceto acetamide, *N*-4-tolyl $CH_3COCH_2CONH(C_6H_4CH_3\text{-}4)$	191.23	pr (AcOEt)		95			al, eth	$B12^3$, 2129
275	Aceto acetanilide, α-bromo $CH_3COCHBrCONHC_6H_5$	256.10	lf (al)		138d			al, eth, chl	$B12^4$, 276
276	Aceto acetanilide, α-chloro $CH_3COCHClCONHC_6H_5$	211.65	nd (al)		137.5			al, bz, chl, MeOH	$B12^3$, 995
277	Acetoacetic acid or 3-Oxobutanoic acid CH_3COCH_2COOH	102.09	syr	<100d				w, al, eth	$B3^4$, 1527
278	Acetoacetic acid-α-acetyl, ethyl ester or Ethyl α-acetyl ace-toacetate $(CH_3CO)_2CHCO_2C_2H_5$	172.18		209-11, 104^{16}		$1.1045^{20/4}$	1.4690^{20}	al, eth, bz	$B3^4$, 1781
279	Acetoacetic acid, allyl ester or Allyl acetoacetate $CH_3COCH_2CO_2CH=CH_2$	128.13		$194\text{-}5^{737}$, 66.5^{14}	-85	$1.0385^{20/20}$	1.4398^{20}	al, bz, w, lig	$B3^4$, 1538
280	Aceto acetic acid, α-benzylidene, ethyl ester or Ethyl α-benzylidine acetoacetate $CH{:}COC(=CHC_6H_5)CO_2C_2H_5$	216.24	rh pl or pyr (dil al)	295-7	60-1			chl, al, eth, bz	$B10^3$, 3158
281	Aceto-acetic acid, α-bromo, ethyl ester or Ethyl α-bro-moaceto acetate $CH_3COCHBrCO_2C_2H_5$	209.04		210-5d, 104-10^{15}		$1.4294^{16/4}$	1.463^{14}	al, eth	$B3^4$, 1551
282	Aceto acetic acid, γ-bromo, ethyl ester $BrCH_2COCH_2CO_2H_5$	209.04		$114\text{-}7^{14}$, $56^{0.005}$		$1.4840^{20/4}$	1.5281^{20}	al, eth	$B3^4$, 1551
283	Acetoacetic acid, butyl ester or Butyl aceto acetate $CH_3COCH_2CO_2C_4H_9$	158.20		127^{50}, 85^{8}	-35.6	$0.9671^{25/4}$	1.4137^{20}	al, bz, lig	$B3^4$, 1536
284	Acetoacetic acid, *iso*-butyl ester or *iso*-Butyl acetoacetate $CH_3COCH_2CO_2CH_2CH(CH_3)_2$	158.20		198-202		0.932^{23}			$B3^4$, 1536
285	Acetoacetic acid, α-chloro, ethyl ester $CH_3COCHClCO_2C_2H_5$	164.59		197^{748}, $86\text{-}9^{12}$		$1.191^{14/17}$	1.4414^{20}	al, eth	$B3^4$, 1549
286	Acetoacetic acid, 2-chloro,ethyl estor $ClCH_2COCH_2CO_2C_2H_5$	164.59		220d, 115^{14}	-8	$1.2157^{20/4}$	1.4546^{17}	al, eth, ace, bz, chl	$B3^4$, 1550
287	Acetoacetic acid, α,α-dibromo, ethylester or Ethyl α, α-dibromoaceto acetate $CH_3COCBr_2CO_2C_2H_5$	287.94		$120\text{-}4^{13}$				al, eth	$B3^2$, 427
288	Acetoacetic acid, α,α-dichloro, ethyl ester $CH_3COCCl_2CO_2C_2H_5$	199.03		$205\text{-}7^{756}$, 91^{11}		$1.293^{16/17}$	1.4492^{17}	al	$B3^2$, 427
289	Aceto acetic acid, α,α-dimethyl, ethyl ester $CH_3COC(CH_3)_2CO_2C_2H_5$	158.20		184^{760}		$0.9777^{20/20}$	1.4180^{20}	al, eth	$B3^4$, 1594
290	Acetoacetic acid, ethyl ester or Ethylacetoacetate $CH_3COCH_2COOC_2H_5$	130.14		180.4, 74^{14}	<-80	$1.0282^{20/4}$	1.4194^{20}	w, al, eth. bz, chl	$B3^4$, 1528

No.	Name, Synonyms, and Formula	Mol. wt.	Color, crystalline form, specific rotation and λ_{max} (log ε)	b.p. °C	m.p. °C	Density	n_D	Solubility	Ref.
291	Aceto acetic acid, ethylester (enol form) or Ethyl aceteo-acetate $CH_3C(OH)=CHCOOC_2H_5$	130.14				1.0119^{10}	1.4432^{20}		$B3^4$, 1528
292	Acetoacetic acid, ethyl ester (keto form) or Ethylaceto acetate $CH_3COCH_2COOC_2H_5$	130.14			-39	$1.0368^{10/4}$	1.4171^{20}		$B3^4$, 1528
293	Acetoacetic acid, α-ethyl, ethyl ester or Ethyl ethyl aceto acetate $CH_3COCH(C_2H_5)CO_2C_2H_5$	158.20		$190, 58^3$		$0.9847^{16/4}$	1.4214^{25}	al, eth	$B3^4$, 1592
294	Acetoacetic acid, α-ethyl, methyl ester $CH_3COCH(C_2H_5)CO_2CH_3$	144.17		$182, 79\text{-}80^{14}$		0.995^{14}		al, eth, ace	$B3^4$, 1592
295	Aceto acetic acid, methyl ester or Methyl aceto acetate $CH_3COCH_2CO_2CH_3$	116.12		$171.7, 60^8$	27-8	$1.0762^{20/4}$	1.4184^{20}	w, al, eth	$B3^4$, 1527
296	Aceto acetic acid, α-methyl, ethyl ester or Methyl ethyl aceto acetate $CH_3COCH(CH_3)CO_2C_2H_5$	144.17		$187, 44^2$		$0.9941^{20/4}$	1.4185^{20}	al, eth, ace	$B3^4$, 1573
297	Aceto acetic acid, α-methyl, methyl ester $CH_3COCH(CH_3)CO_2CH_3$	130.14		$177.4, 80^{20}$		$1.0247^{25/25}$	1.416^{24}	al, eth	$B3^3$, 1225
298	Aceto acetic acid, α-phenyl, ethyl ester $CH_3COCH(C_6H_5)CO_2C_2H_5$	206.24		156^{22}		$1.0855^{20/4}$	1.5176^{20}	al, eth	$B10^3$, 3047
299	Acetoacetic acid, *iso*-propyl ester or *iso*-Propyl aceto acetate $CH_3COCH_2CO_2CH(CH_3)_2$	144.17		$185\text{-}7, 75\text{-}6^{15}$	-27.3	$0.9835^{20/4}$	1.4173^{20}	**al, eth**, lig, w	$B3^4$, 1535
300	Acetoacetic acid, α-*iso*-propyl, ethyl ester $CH_3COCH(iC_3H_7)CO_2C_2H_5$	172.22		$201^{758}, 97\text{-}8^{20}$		$0.9648^{18/4}$	$1.4256^{18/4}$	**al, eth**	$B3^4$, 1608
301	Acetoacetic acid, 4,4,4-trifluoro, ethyl ester or ethyl-tri-fluoro aceto acetate $F_3CCOCH_2CO_2C_2H_5$	184.11		131.5^{757}		$1.2586^{15.5}$	1.37830	al, eth	$B3^4$, 1548
302	Acetoacetonitrile CH_3COCH_2CN	83.09		120-5				al, ace	$B3^4$, 1545
303	Acetoacetonitrile, α-phenyl $CH_3COCH(C_6H_5)CN$	159.19	pr (bz), cr (dil al or ace-peth)		90-1			al, eth, bz, chl	$B10^3$, 3048
304	Aceto acetyl chloride CH_3COCH_2COCl	120.54			-50			eth	$B3^4$, 1545
305	Acetoin or 3-hydroxy-2-butanone $CH_3CHOHCOCH_3$	88.11		143,37.1	-72	$1.0062^{20/20}$	1.4171^{20}	**w**, ace	$B1^4$, 3991
306	Acetone or 2-Propanone CH_3COCH_3	58.08		56.2	-95.35	$0.7899^{20/4}$	1.3588^{20}	**w, al, eth, bz, chl**	$B1^4$, 3180
307	Acetone azine $(CH_3)_2C=NN=C(CH_3)_2$	112.17		133	-12.5	0.83899^{20}	1.4535^{20}	**w, al, eth,** ace	$B1^4$, 3207
308	Acetone, amino, hydrochloride or Acetonylamine hydrochloride $CH_3COCH_2NH_2{:}HCl$	109.56			75			w, al, eth	$B4^3$, 877
309	Acetone, bromo or Bromoacetone CH_3COCH_2Br	136.98		$136.5^{725}, 31.5^8$	-36.5	1.634^{23}	1.4697^{15}	al, eth, ace	$B1^4$, 3223
310	Acetone, chloro or Chloroacetone CH_3COCH_2Cl	92.53		119^{763}	-44.5	1.15^{20}		w, al, eth, chl	$B1^4$, 3215
311	Acetone, 1-chloro-3-phenyl $C_6H_5CH_2COCH_2Cl$	168.62	nd (chl)	$159\text{-}61^{17}$	72-3				$B7^3$, 1041
312	Acetone, 1,3-diamino,dihydrochloride $(H_2NCH_2)_2CO\cdot 2HCl$	161.03	(dil al, dil aa), pr (w + 1)		180d			w	$B4^2$, 763
313	Acetone, 1,1-dichloro $CH_3COCHCl_2$	126.97		$120, 47^{76}$		$1.3051^{18/15}$		al, **eth**	$B1^4$, 3218
314	Acetone, 1,3-dichloro $(ClCH_2)_2CO$	126.97	pr or nd	$173.4, 86\text{-}8^{12}$	45	$1.3826^{46/4}$	1.4716^{46}	w, al, eth	$B1^4$, 3219
315	Acetone, diethyl acetal or 2,2-Diothoxypropane $(CH_3)_2C(OC_2H_5)_2$	132.20		$114, 48^{60}$		$0.8200^{21/4}$	1.3891^{20}	al, eth, ace, bz	$B1^4$, 3200
316	Acetone, diethyl amino $CH_3COCH_2N(C_2H_5)_2$	129.20		$155\text{-}6d, 64^{16}$		$0.8620^{20/4}$	1.4249^{20}	**w, al, eth**	$B4^3$, 877
317	Acetone, 1,3-dihydroxy $(HOCH_2)_2CO$	90.08			89-91			**w**, al, ace	$B1^4$, 4119
318	Acetone, 2,4-dinitrophenyl hydrazone $(CH_3)_2C=NNH[C_6H_3(NO_2)_2\text{-}2,4]$	238.20	ye nd or pl (al)		128			bz, al, eth	$B15^3$, 427
319	Acetone, 1,1-diphenyl $CH_3COCH(C_6H_5)_2$	210.28		$306\text{-}7^{750}, 174\text{-}6^{10}$	46		1.5361^{16}		$B7^3$, 2171

No.	Name, Synonyms, and Formula	Mol. wt.	Color, crystalline form, specific rotation and λ_{max} (log ε)	b.p. °C	m.p. °C	Density	n_D	Solubility	Ref.
320	Acetone, 1,3-diphenyl or Dibenzyl ketone $(C_6H_5CH_2)_2CO$	210.28	cr (al, peth)	331, 112-25^{01}	35	1.195$^{0/4}$		al, eth, peth	$B7^3$, 2160
321	Acetone, dipropylamino $CH_3COCH_2N(C_3H_7)_2$	157.26		70-2^{6}			1.4267$^{25/D}$		$B4^3$, 877
322	Acetone, fluoro CH_3COCH_2F	76.07		75			1.3700$^{20/D}$		$B1^4$, 3213
323	Acetone, hexachloro or Perchloroacetone $(Cl_3C)_2CO$	264.75		202-4, 110^{40}	-2	1.444$^{12/12}$	1.5112^{20}	bz	$B1^4$, 3223
324	Acetone, hexafluoro $(CF_3)_2CO$	166.02		-28	-129				$B1^4$, 3215
325	Acetone, hydroxy or Acetol CH_3COCH_2OH	74.08		145-6d, 5 4^{18}	-7	1.0824$^{20/20}$	1.4295^{20}	w, al, eth	$B1^4$, 3977
326	Acetone, iodo or Iodoacetone CH_3COCH_2I	183.98	yesh (lig)	62^{12}		2.17^{15}		al	$B1^4$, 3226
327	Acetone, iodo, oxime or Iodoacetoxime $CH_3(C{=}NOH)CH_2I$	198.99	pr (peth)		64.5				B1, 660
328	Acetone, 3-methoxyphenyl $(3\text{-}CH_3OC_6H_4)CH_2COCH_3$	164.20		258-60, 95-7$^{0.7}$		1.0812^{0}	1.5230^{25}		$B8^3$, 397
329	Acetone, 4-methoxyphenyl $(4\text{-}CH_3OC_6H_4)CH_2COCH_3$	164.20		267-9, 142^{14}	<-15	1.0670$^{18/4}$	1.5233^{20}	al, eth	$B8^3$, 398
330	Acetone, 4-nitrophenylhydrazone $(CH_3)_2C{=}NNHC_6H_4NO_2\text{-}4$	193.21			152 (149)			al, eth	$B15^3$, 427
331	Acetone, oxime or Acetoxime $(CH_3)_2C{=}NOH$	73.09	pr (al)	134.8^{728}, 61^{20}	61	0.9113$^{62/4}$	1.4156^{20}	al, ace, w, liq	$B1^4$, 3202
332	Acetone, pentabromo $CHBr_2COCBr_3$	372.40			79-80				$B1^4$, 3226
333	Acetone, pentachloro $Cl_3CCOCHCl_2$	230.31	cr (w + 4)	192^{253}, 98^{40}	2.1 (an-kyd)	1.69$^{15/15}$			$B1^4$, 3222
334	Acetone, phenacyl or Phenacylacetone $C_6H_5COCH_2CH_2COCH_3$	176.22	ye oil	162^{12}			1.5250^{10}	ace	$B7^3$, 3509
335	Acetone, phenoxy $C_6H_5OCH_2COCH_3$	150.18		229-30, 117-9^{20}		1.0903$^{20/4}$	1.5228^{2}	eth, ace	$B6^4$, 604
336	Acetone, phenyl or methyl benzyl ketone $CH_3COCH_2C_6H_5$	134.18		216.5, 101^{14}	-15	1.0157$^{20/4}$	1.5168^{20}	al, eth, bz	$B7^3$, 1036
337	Acetone, phenyl hydrazone $(CH_3)_2C{=}NNHC_6H_5$	148.21	rh	163^{50}	42			al, eth	$B15^3$, 88
338	Acetone, semi carbazone $(CH_3)_2C{=}NNHCONH_2$	115.13	nd (w, ace)		190-1d			al, eth	$B3^4$, 179
339	Acetone, 1,1,1,3-tetrachloro $CH_2ClCOCCl_3$	195.86	liq (an hyd) pr (w + 4)	183, 71-2^{13}	46(+ 4 w) 65	1.624$^{15/4}$	1.497^{18}	eth, ace	$B1^4$, 3222
340	Acetone, 1,1,3,3-tetrachloro $(Cl_2CH)_2CO$	195.86		180-2^{718}				al, eth, ace, bz	$B1^4$, 3222
341	Acetone, 1,1,1-trichloro CH_3COCCl_3	161.42		149^{764}, 28^{10}		1.435$^{20/4}$	1.4635^{17}	al, eth	$B1^4$, 3222
342	Acetonitrile CH_3CN	41.05		81.6	-45.7	0.7857^{20}	1.34423^{20}	**w, al, eth, ace, bz**	$B2^4$, 419
343	Acetonitrile, cyclohexylidone $C_6H_{10}{=}CHCN$	121.18		107-8^{22}		0.9483$^{15/4}$	1.4832^{25}	al, eth	$B9^3$, 163
344	Acetonitrile, dibenzyl $(C_6H_5CH_2)_2CHCN$	221.30	lf or pr (al)		89-91			al, eth	$B9^3$, 3359
345	Acetonitrile, dichloro Cl_2CHCN	109.94		112-3		1.369^{20}	1.4391^{25}	al	$B2^4$, 506
346	Acetonitrile, diphenyl $(C_6H_5)_2CHCN$	193.25	pr (eth), lf (di al)	181-4^{12}	72-3			al, eth	$B9^3$, 3304
347	Acetonitrile, methoxy CH_3OCH_2CN	71.08		118.1		0.9492$^{20/4}$	1.3831^{20}	al, eth, ace	$B3^3$, 399
348	Acetonitrolic acid $CH_3C({=}NOH)NO_2$	104.07	ye rh (w, al, eth)		87-8d			w, al, eth, ace	$B2^4$, 434
349	Acetonyl acetate or 2-oxopropyl acetate $CH_3CO_2CH_2COCH_3$	116.12		170-1^{755}, 63^{11}		1.0757$^{20/4}$	1.4141^{20}	w, al eth	$B2^4$, 297
350	Acetonyl acetone or 2,5-Hexanedione $CH_3COCH_2CH_2COCH_3$	114.14		194^{754}, 89^{25}	-5.5	0.9737$^{20/4}$	1.4421^{20}	w, **al, eth**, ace, bz	$B1^4$, 3688
351	Acetophenone or Methyl phenyl ketone $C_6H_5COCH_3$	120.15	mcl pr or pl	202.6, 79^{10}	20.5	1.0281$^{20/4}$	1.53718^{20}	al, eth, ace, bz, chl	$B7^3$, 936
352	Acetophenone, 2-amino or o-acetyl aniline $2\text{-}H_2NC_6H_4COCH_3$	135.17	ye cr	250-2d, 135^{17}	20		1.6160^{20}	eth	$B14^3$, 80

PHYSICAL CONSTANTS OF ORGANIC COMPOUNDS (Continued)

No.	Name, Synonyms, and Formula	Mol. wt.	Color, crystalline form, specific rotation and λ_{max} (log ε)	b.p. °C	m.p. °C	Density	n_D	Solubility	Ref.
353	Acetophenone, 3-amino or *m*-acetyl aniline $3\text{-}H_2NC_6H_4COCH_3$	135.17	pa ye pl (al), lf (eth)	289-90	98-9				B14^3, 88
354	Acetophenone, 4-amino or *pp*-acetyl aniline $4\text{-}H_2NC_6H_4COCH_3$	135.17	ye mcl pr (al)	293-5, 195-200^{15}	106			al, eth	B14^3, 93
355	Acetophenone, *p*-amino, α-chloro or *p*-Amino phenacyl chloride $4\text{-}H_2NC_6H_4COCH_2Cl$	169.61	ye pl		148				B14^3, 99
356	Acetophenone, 4-amino-3-chloro- $4\text{-}H_2N\text{-}3\text{-}Cl\text{-}C_6H_3COCH_3$	169.61	pr (chl-peth)		92				B14, 49
357	Acetophenone, 3-amino-4-methoxy $3(H_2N)\text{-}4(CH_3O)C_6H_3COCH_3$	165.19	pr (al)		102			al, eth, bz	B14^3, 548
358	Acetophenone, α-bromo or Phenacyl bromide $C_6H_5COCH_2Br$	199.05	nd (al), rh pr (dil al), pl (peth)	135^{18}	50-1	$1.647^{20/4}$		al, eth, bz, chl	B7^3, 979
359	Acetophenone, 2-bromo or 2-Bromophenyl methyl ketone $3\text{-}BrC_6H_4COCH_3$	199.05	ye	131-5^{20}			1.5678^{20}		B7^3, 976
360	Acetophenone, 3-bromo or -3-bromophenyl methyl ketone $3\text{-}BrC_6H_4OCH_3$	199.05		131^{16}	7-8		1.5755^{20}	ace, bz	B7^3, 977
361	Acetophenone, 4-bromo or 4-Bromo phenyl methyl ketone $4\text{-}BrC_6H_4COCH_3$	199.05	lf (al)	255.5^{736}, 130^{11}	50-1	1.647		al, eth, bz, aa	B7^3, 977
362	Acetophenone, α-bromo-3-chloro or 3-Chlorophenacyl bromide $3\text{-}ClC_6H_4COCH_2Br$	233.49	nd		395-40			al	B7^3, 982
363	Acetophenone, α-bromo-4-chloro or 4-Chlorophenacyl bromide $4\text{-}ClC_6H_4COCH_2Br$	233.49	nd		96-7				B7^3, 983
364	Acetophenone, 4-bromo-α-chloro or 4-Bromophenacyl chloride $4\text{-}BrC_6H_4COCH_2Cl$	233.49	nd (al)		116-7			al	B7^3, 982
365	Acetophenone, α bromo-4-methyl or 4-Methyl phenacyl bromide $4\text{-}CH_3C_6H_4COCH_2Br$	213.07	nd or lf (al)	155-9^{14}	51			al, eth	B7^3, 1060
366	Acetophenone, 4-*tert*-butyl $4\text{-}(CH_3)_3CC_6H_4COCH_3$	176.26		136-8^{20}		0.9705^{0}	$1.518^{15/D}$		B7^3, 1171
367	Acetophenone, 4-*tert* butyl-2,6 dimethyl, 3.5-dinitro or Musk ketone $C_{14}H_{18}N_2O_5$	294.31	ye		134.5-6.5				B7^3, 1228
368	Acetophenone, α-chloro or Phenacyl chloride $C_6H_5COCH_2Cl$	154.60	pl (dil al), rh, lf (peth)	247, 139-41^{14}	56.5	$1.324^{15/4}$		al, eth, ace, bz	B7^3, 967
369	Acetophenone, 2-chloro $2\text{-}ClC_6H_4COCH_3$	154.60		227-8^{738}, 113^{18}		$1.2016^{17/4}$	1.685^{25}	eth	B7^3, 1008
370	Acetophenone, 3-chloro $3\text{-}ClC_6H_4COCH_3$	154.60		241-5^{744}, 127-31^{30}		1.2130^{04}	1.5494^{20}	al, eth, ace	B7^2, 218
371	Acetophenone, 4-chloro $4\text{-}ClC_6H_4COCH_3$	154.60		273, 106^{10}	20	$1.1922^{20/4}$	1.5550^{20}	al, eth	B7^3, 1008
372	Acetophenone, α-chloro-4-methyl or 4-Methyl phenacyl chloride $4\text{-}CH_3C_6H_4COCH_2Cl$	168.62	nd (al)	260-3, 113^{4}	57-8			al, eth	B7^3, 1065
373	Acetophenone, α-chloro-2,4-dimethyl $2,4(CH_3)_2C_6H_3COCH_2Cl$	182.65	nd		62			al, eth, ace	B7^3, 1106
374	Acetophenone, αα-dibromo or Phenacylidene bromide $C_6H_5COCHBr_2$	277.94		159-60^{13}	36-7			al, eth, chl	B7^3, 985
375	Acetophenone, α, 4-dibromo or 4-Bromophenacyl bromide $4\text{-}BrC_6H_4COCH_2Br$	277.94	nd (al)		110-2			eth, al	B7^3, 985
376	Acetophenone, αα-dichloro or Phenacylidine chloride $C_6H_5COCHCl_2$	189.04	amor	249, 143^{25}	20-1	1.340^{16}	1.5686^{20}	al, bz	B7^3, 972
377	Acetophenone, α,4-dichloro or *p*-Chorophenacyl chloride $4\text{-}ClC_6H_4COCH_2Cl$	189.04	nd (al)	270	101-2			al, bz	B7^3, 972
378	Acetophenone, 2,4-dichloro $2,4\text{-}Cl_2\text{-}C_6H_3COCH_3$	189.04		245-7, 140-50^{15}	33-4		1.5640^{20}		B7^3, 969
379	Acetophenone, 3,4-dichloro $3,4\text{-}Cl_2\text{-}C_6H_3COCH_3$	189.04	nd (peth)	135^{12}	76				B7^3, 971

No.	Name, Synonyms, and Formula	Mol. wt.	Color, crystalline form, specific rotation and λ_{max} (log ε)	b.p. °C	m.p. °C	Density	n_D	Solubility	Ref.
380	Acetophenone, 2,3-dihydroxy or 3-Acetylcatechol $2,3(HO)_2C_6H_3COCH_3$	152.15	ye pr (bz-lig)		97-8			al	$B8^3$, 2080
381	Acetophenone, 2,4-dihydroxy or Resacetophenone $2,4\text{-}(HO)_2C_6H_3COCH_3$	152.15	nd or lf		147		1.1800^{141}	al, bz	$B8^3$, 2082
382	Acetophenone, 2,5-dihydroxy or 2- Acetyl hydroquinone $2,5(HO)_2C_6H_3COCH_3$	152.15	ye-gr nd (dil al or w)		204-5			al	$B8^3$, 2100
383	Acetophenone, 3,4-dihydroxy or 4-Acetvl catechol $3,4(HO)_2C_6H_3COCH_3$	152.15	nd (w or chl)		115-6				$B8^3$, 2108
384	Acetophenone, 3,5-dihydroxy or 5-Acetyl resorcinol $3,5(HO)_2C_6H_3COCH_3$	152.15	cr (w)		147-8			al, eth, ace, w	$B8^2$, 301
385	Acetophenone, 3,4-dimethoxy or Aceto veratrone $3,4(CH_3O)_2C_6H_3COCH_3$	180.20	pr (dil al)	286-8, $160\text{-}2^{15}$	51			w, al, chl, bz	$B8^3$, 2110
386	Acetophenone, 3,5-dimethoxy-4-hydroxy or Aceto syringone $3,5(CH_3O)_2\text{-}4\text{-}(HO)C_6H_2COCH_3$	196.20	nd (w) pr (peth)		122-3			al, eth, ace	$B8^3$, 2118
387	Acetophenone, 2,4-dimethyl $2,4(CH_3)_2C_6H_3COCH_3$	148.20		228, 110^{13}		1.0121^{15}	1.5340^{20}	al	$B7^3$, 1105
388	Acetophenone, 2,5-dimethyl $2,5,(CH_3)_2C_6H_3COCH_3$	148.20		232-3, 107^{13}		$0.9963^{19/4}$	1.5291^{20}	al, eth, bz	$B7^3$, 1104
389	Acetophenone, 3,4-dimethyl $3,4\text{-}(CH_3)_2C_6H_3COCH_3$	148.20		246-7, 213^{310}		$1.0090^{14/4}$	1.5413^{15}	al, eth, bz	$B7^3$, 1104
390	Acetophenone, 3-dimethylamino $3\text{-}(CH_3)_2NC_6H_4COCH_3$	163.22		148^{13}	43			ace	B14, 45
391	Acetophenone, 4-dimethylamino $4(CH_3)_2NC_6H_4COCH_3$	163.22	nd (w, Peth)	$172\text{-}5^{11}$	105.5			eth, w, lig	$B14^3$, 94
392	Acetophenone, 2-ethoxy or 2-Acetylphenetole $2\text{-}C_2H_5OC_6H_4COCH_3$	164.20	pr (dil al), pl (lig)	243-4	43	1.0036^{78}		al, eth, lig	B885
393	Acetophenone, 4-ethoxy or 4-Acetylphenetole $4\text{-}C_2H_5OC_6H_4COCH_3$	164.20	pl (eth)		39			al, eth	$B8^3$, 280
394	Acetophenone, 4 fluoro $4\text{-}F\text{-}C_6H_4COCH_3$	138.14		196^{760}, 79^{10}	-45	$1.1382^{25/4}$	1.5081^{25}		$B7^3$, 961
395	Acetophenone, furfurylidene $C_6H_5COCH{=}CH(C_4H_3O)$	198.22		317, 181^{29}		1.1140^{20}		al, eth	$B17^2$, 377
396	Acetophenone, α-hydroxy or Phenacylalcohol $C_6H_5COCH_2OH$	136.15	hex pl (al or eth), pl (w or dil al)	$124\text{-}6^{12}$ (sub), 56^1	90	$1.0963^{99/4}$		al, eth, lig	$B8^3$, 298
397	Acetophenone, α-hydroxy, acetate or Phenacyl acetate $C_6H_5COCH_2O_2CCH_3$	178.19	rh pl (eth, lig or peth)	270, $150\text{-}2^{19}$	49.6	$1.1169^{65/4}$		al, eth, bz, chl	$B8^3$, 301
398	Acetophenone, 2-hydroxy or 2-Acetylphenol $2\text{-}HOC_6H_4COCH_3$	136.15		218, 106^{17}	4-6	$1.1307^{20/4}$	1.5584^{20}	al, eth, aa	$B8^3$, 261
399	Acetophenone, 3-hydroxy or 3-Acetylphenol $3\text{-}HOC_6H_4COCH_3$	136.15	nd or lf	296^{756}, 153^5	96	1.0992^{109}	1.5348^{109}	al, eth, bz	$B8^3$, 272
400	Acetophenone, 4-hydroxy or 4 Acetylphenol $4\text{-}HOC_6H_4COCH_3$	136.15	nd (eth, dil al)	$147\text{-}8^3$	109-10	1.1090^{109}	1.5577^{109}	al, eth	$B8^3$, 276
401	Acetophenone, α hydroxy-4-methoxy or 4-methoxy phenacyl alcohol $4\text{-}CH_3OC_6H_4COCH_2OH$	166.18	pl (dil al)		104			al, bz	$B8^3$, 2120
402	Acetophenone, 2-hydroxy-3-methoxy or *o* Aceto vanillon $2(HO)\text{-}3(CH_3O)C_6H_3COCH_3$	166.18	pa ye nd (peth or eth-peth)		53-4			eth, bz	$B8^3$2, 081
403	**Acetophenone, 2-hydroxy-4-methoxy or Peonol** $2\text{-}HO\text{-}4\text{-}CH_3OC_6H_3COCH_3$	166.18	nd (al)	158^{20}	52-3	1.3102^{81}	1.5452^{81}	al, eth, bz, chl	$B8^3$, 2084
404	Acetophenone, 2-hydroxy- 5-methoxy $2\text{-}HO\text{-}5\text{-}CH_3OC_6H_3COCH_3$	166.18	pa ye pr (dil al)		52			bz, al	B14, 477
405	Acetophenone, 3-hydroxy-4-methoxy or Iso aceto-vanillone $3\text{-}HO\text{-}4\text{-}CH_3OC_6H_3COCH_3$	166.18	cr (eth lig) or cr (w + 1)		67-8 (+1w) 91 anhyd			eth, w	$B8^3$, 2109
406	Acetophenone, 4-hydroxy-α-methoxy $4\text{-}HO\text{-}C_6H_4COCH_2OCH_3$	166.18	nd (bz), pr (w + 1)		130-1			al, ace, eth, bz	$B8^2$, 302
407	Acetophenone, 4-hydroxy-3-methoxy or Resacetophenone-2-methylether iso peonol $4\text{-}HO\text{-}3\text{-}CH_3O\text{-}C_6H_3COCH_3$	166.18	nd (w)		138			bz, w	$B8^1$, 617
408	Acetophenone, 4-hydroxy-3-methoxy or Acetovanillon $4\text{-}HO\text{-}3\text{-}CH_3OC_6H_3COCH_3$	166.18	pr (w)	295-300, 233-$5^{15\text{-}20}$	115			al, eth, ace, bz, chl	$B8^3$, 2108
409	Acetophenone, α-iodo or Phenacyl iodide $C_6H_5COCH_2I$	246.05		$158\text{-}60^{15}$	34.4			ace	$B7^3$, 989

No.	Name, Synonyms, and Formula	Mol. wt.	Color, crystalline form, specific rotation and λ_{max} (log ε)	b.p. °C	m.p. °C	Density	n_D	Solubility	Ref.
410	Acetophenone, 2-iodo $2I-C_6H_4COCH_3$	246.05	...	139-140^{12}	...	1.746$^{20/4}$	1.6180^{20}	bz	B7^3, 988
411	Acetophenone, 3-iodo $3-I-C_6H_4COCH_3$	246.05	...	129^{B}	...	...	1.622^{20}	bz	B7^1, 988
412	Acetophenone, 4-iodo $4-I-C_6H_4COCH_3$	246.05	...	153^{18}	85	...	...	al, bz, eth, lig	B7^3, 989
413	Acetophenone, 2-Methoxy or 2,Acetylanisole $2-CH_3OC_6H_4COCH_3$	150.18	ye	245	...	1.0897$^{20/4}$	1.5393^{20}	al, ace	B8^3, 263
414	Acetophenone, 3-methoxy $3-CH_3OC_6H_4COCH_3$	150.18	...	240, 125-6^{12}	95-6	1.0343^{19}	1.5410^{20}	al, ace, w, eth	B8^3, 273
415	Acetophenone, 4-methoxy $4-CH_3O_6H_4COCH_3$	150.18	...	258, 138-9^{15}	38-9	1.0818$^{41/4}$	1.547^{41}	al, eth, ace	B8^3, 277
416	Acetophenone, 2-methyl or 2-acetyl toluene $2CH_3C_6H_4COCH_3$	134.18	...	214, 89-92^{10}	...	1.026$^{20/4}$	1.5276^{20}	...	B7^3, 1052
417	Acetophenone, 3-methyl or 3 Acetyltoluene $3-CH_3C_6H_4COCH_3$	134.18	...	220^{766}, 109^{12}	...	1.0070$^{20/4}$	1.5270^{20}	al, eth, ace	B7^3, 1057
418	Acetophenone, 4-methyl or 4-acetyl toluene $4-CH_3C_6H_4COCH_3$	134.18	nd	226, 113^{11}	28	1.0051$^{20/4}$	1.5335^{20}	al, cth, bz	B7^3, 1060
419	Acetophenone, 2-nitro $2-NO_2C_6H_4COCH_3$	165.15	...	158^{16}	28-9	...	1.5468^{20}	al, eth	B7^3, 990
420	Acetophenone, 3-nitro $3-NO_2C_6H_4COCH_3$	165.15	nd (al)	202, 167^{18}	81	...	...	eth, w	B7^3, 991
421	Acetophenone, 4-nitro $4-NO_2C_6H_4COCH_3$	165.15	yesh pr (al)	...	80-2	...	...	al, eth	B7^3, 993
422	Acetophenone oxime $C_6H_5C(:NOH)CH_3$	135.17	nd (w)	245, 119^{20}	60	...	...	al, eth, ace, bz, lig	B7^3, 954
423	Acetophenone, 5-*iso*-propyl-2-methyl or Carvacryl methyl ketone $5-i-C_3H_7-2-CH_3C_6H_3COCH_3$	176.26	...	249.50	<-20	0.956$^{20/4}$	1.5181^{20}	...	B7^3, 1176
424	Acetophenone, ααα-trichloro $C_6H_5COCCl_3$	223.49	...	256-7, 145^{25}	...	1.425^{16}	...	al, eth	B7^3, 974
425	Acetophenone, ααα trifluoro $C_6H_5COCF_3$	174.12	...	152^{730}	-40	1.279^{20}	1.4583^{20}	...	B7^3, 962
426	Acetophenone, 2,4,5-trimethyl $2,4,5-(CH_3)_3C_6H_2COCH_3$	162.23	...	246-7, 137-8^{20}	10-1	1.0039$^{15/4}$	1.541^{15}	al, eth, bz, aa	B7^3, 1145
427	Acetophenone, 2,4,6-trimethyl $2,4,6-(CH_3)_3C_6H_2COCH_3$	162.23	...	240.5^{735}, 120^{12}	...	0.9754$^{20/4}$	1.5175^{20}	al, eth, ace, bz	B7^3, 1137
428	Acetoxyacetic acid $CH_3CO_2CH_2CO_2H$	118.09	nd (bz)	144-5^{12}	67-8	...	...	w, al, eth, ace, chl	B3^4, 576
429	Acetoxy acetic acid, ethylester or Ethyl acetoxyacetate $CH_3CO_2CH_2CO_2H_5$	146.14	...	179	...	1.0880$^{20/4}$	1.4112^{20}	al, eth, aa	B3^4, 580
430	(2-Acetoxypropyl) trimethyl ammonium chloride *D* or *o*-acetyl-B-methyl choline chloride $CH_3CH(O_2CCH_3)CH_2N+(CH_3)_3Cl-$	195.69	hyg. $[\alpha]_D+41.9$	d	172-3	...	...	w, al	B4^4, 1671
431	(2-Acetoxypropyl) trimethyl ammonium chloride *L* or *o*-Acetyl-β-methyl choline chloride $CH_3CH(O_2CCH_3)CH_2N+(CH_3)_3Cl-$	195.69	$[\alpha]_D$-41.9	d	172-3	...	...	...	B4^4, 1671
432	Acetyl acetone or 2-4-Pentanedione $CH_3COCH_2COCH_3$	100.12	...	139^{746}	-23	0.9721$^{25/4}$	1.4494^{20}	w, **al, eth, ace, chl**	B1^4, 3662
433	Acetyl benzoyl peroxide $CH_3CO-OO-COC_6H_5$	180.16	wb nd (lig)	85-100 (exp)	40-1	...	...	eth	B9^3, 1051
434	Acetyl bromide CH_3COBr	122.95	ye in air	76	-98	1.6625$^{16/4}$	1.45376^{16}	**eth**, ace, **bz, chl**	B2^4, 398
435	Acetyl chloride CH_3COCl	78.50	...	50.9	-112	1.1051$^{20/4}$	1.38976^{20}	**eth, ace, bz, chl**	B2^4, 395
436	Acetyl chloride, dichloro $Cl_2CHCOCl$	147.39	...	108-10	...	1.5315$^{16/4}$	1.4591^{20}	eth	B2^4, 504
437	Acetyl chloride, ethoxy $C_2H_5OCH_2COCl$	122.55	...	123-4, 49-50^{37}	...	1.1170	1.4204^{20}	eth, ace	B3^3, 396
438	Acetyl chloride, methoxy CH_3OCH_2COCl	108.52	...	99, 46-9^{62}	...	1.1871$^{20/4}$	1.4196^{20}	eth, ace, chl	B3^3, 396
439	Acetyl fluoride CH_3COF	62.04	...	20.8	...	1.002$^{15/4}$	...	**al, eth**, ace, bz	B2^4, 393
440	Acetyl iodide CH_3COI	169.95	...	108, 36^{50}	...	2.0674$^{20/4}$	1.5491^{20}	eth	B2^4, 399
441	Acetyl isothiocyanate CH_3CONCS	101.12	...	132-3, 30^{9}	...	1.1523$^{13/4}$	1.5231^{18}	eth	B3^3, 278

PHYSICAL CONSTANTS OF ORGANIC COMPOUNDS (Continued)

No.	Name, Synonyms, and Formula	Mol. wt.	Color, crystalline form, specific rotation and λ_{max} (log ε)	b.p. °C	m.p. °C	Density	n_D	Solubility	Ref.
442	Acetyl peroxide $CH_3CO\text{-}OO\text{-}COCH_3$	118.09	nd (eth), lf	63^{21}	30 (26)			al, eth	$B2^4$, 392
443	Acetylene $HC{\equiv}CH$	26.04		-84.0	-80.8	$0.6208^{-82/4}$	1.00051^0	ace, bz, chl	$B1^4$, 939
444	Acetylene-bromo $BrC{\equiv}CH$	104.93		4.7				eth	$B1^3$, 919
445	Acetylene-chloro $ClC{\equiv}CH$	60.48		-30	-126				$B1^4$, 957
446	Acetylene-dibromo $BrC{\equiv}CBr$	183.83	nd	76 (exp)	-25			al, eth, ace, bz	$B1^3$, 919
447	Acetylene-dichloro $ClC{\equiv}CCl$	94.93		exp	-66			al, eth, ace	$B1^4$, 957
448	Acetylene-diido $IC{\equiv}CI$	277.83	rh nd (lig)	32-3 (exp)	81-2			al, eth, ace, bz	$B1^4$, 958
449	Acetylene-diphenyl $C_6H_5C{\equiv}CC_6H_5$	178.23	mcl pr or pl (al)	300, 170^{19}	62.5	$0.9657^{100/4}$		eth	$B5^3$, 2119
450	Acetylene, *bis*(l-hydroxycyclohexyl) $(C_6H_{10}OHC)_2$	222.33	nd (ccl_4)	182^{13}	112-3			al, eth, ace	$B6^3$, 4741
451	Acetylene, l-phenyl-2-methyl $C_6H_5C{\equiv}CCH_3$	116.16		181		0.942^{15}	$1.563^{15/}{}_D$		$B5^2$, 408
452	Aconic acid $C_5H_4O_4$	127.08	lf (eth), rh (w, al)	d	164			w, al	$B18^4$, 5333
453	Aconine $C_{25}H_{41}NO_9$	499.60	amor, $[\alpha]^{20/D}$ + 23		132			w, al	$B21^4$, 2899
454	Aconitomide $H_2NCOCH_2C(CONH_2){=}CHCONH_2$	171.16	ye nd (w)		sinters 260			w	B2, 853
455	Aconitic acid, *cis* or 1, 2, 3-Propene tricarboxylic acid $HO_2CCH_2C(CO_2H){=}CHCO_2H$	174.11	nd (w)		130			w	$B2^4$, 2405
456	Aconitic acid, *trans* or 1, 2, 3-Propene tricarboxylic acid $HO_2CCH_2C(CO_2H){=}CHCO_2H$	174.11	lf (w), nd (w, eth)		198-9			w, al	$B2^4$, 2405
457	Aconitic acid, triethyl ester *trans* $C_2H_5O_2CH_2C(CO_2C_2H_5){=}CHCO_2C_2H_5$	258.27		275d, 159^9		$1.1064^{20/4}$	1.4556^{20}	al, eth	$B2^4$, 2405
458	Aconitic acid, trimethylester $CH_3O_2CH_2C(CO_2CH_3){=}CHCO_2CH_3$	216.19		270-1, 161^{14}				al, eth	$B2^4$, 2405
459	Aconitic acid, tripropylester $C_3H_7O_2CH_2C(CO_2C_3H_7){=}CHCO_2C_3H_7$	300.35		195^{13}		$1.050^{25/4}$	1.4521^{20}	al, eth	$B2^4$, 2405
460	Aconitine $C_{34}H_{47}NO_{11}$	645.75	rh lf $[\alpha]^{20/D}$ + 19 (chl)		204			al, bz, chl	$B21^4$, 2901
461	Aconitine, hydrobromide $C_{34}H_{47}NO_{11}\cdot HBr\cdot 1\frac{1}{2}H_2O$	753.68	yesh pr (w), $[\alpha]_D$-31 (w,c=2.3)		209.10			w, al, eth	$B21^4$, 2903
462	Aconitine, hydrochloride $C_{34}H_{47}NO_{11}\cdot HCl\cdot 3\frac{1}{2}H_2O$	745.26	(w + 3 1/2), $[\alpha]_D$-30.5 (w)		170-2			w, al, eth	$B21^4$, 2902
463	Aconitine, nitrate $C_{34}H_{47}NO_{11}HNO_3$	708.76	$[\alpha]^{20/D}$-35 (2% aq soln)		ca. 200d			w	$B21^4$, 2903
464	Acridine or 2,3,5,6-Dibenzopyridine $C_{13}H_9N$	179.22	rh nd or pr (al)	345-6	111	$1.005^{20/4}$		al, eth, bz	$B20^4$, 3987
465	Acridine, 2-amino $C_{13}H_{10}N_2$,2-$H_2NC_{13}H_8N$	194.24	ye nd (w or al)		213-4			al, eth	$B22^4$, 4984
466	Acridine, 2-amino-5(4-aminophenyl) or Chrysaniline $C_{19}H_{15}N_3.2H_2O$	321.38	ye nd (95% al)		260-7				$B22^4$, 5513
467	Acridine, 3-amino $C_{13}H_{10}N_2$,3-$NH_2C_{13}H_8N$	194.24	ye nd (w or al + 1)		213-4			al, eth	$B22^4$, 4987
468	Acridine, 4-amino $C_{13}H_{10}N_2$,4-$NH_2C_{13}H_8N$	194.24	red-br pr (peth), og nd (MeOH)	183-4	108			al, eth, ace, bz	$B22^4$, 4989
469	Acridine, 4 amino,-hydrochloride $C_{11}H_{10}N_2.HCl$	230.70	ye nd (dil al)		234d			w, al	$B22^2$, 376
470	Acridine, 9-amino $C_{11}H_{10}N_2$	194.24	ye nd (ace or al)		241 (cor)				$B21^4$, 4174
471	Acridine, 9-chloro $C_{11}H_8ClN$	213.67	nd (al)	sub	122			al, w	$B20^4$, 3995
472	Acridine, 3,6-diamino $C_{11}H_{11}N_3$	209.25	ye nd (al or w)		284-6			al, eth, w	$B22^4$, 5487

No.	Name, Synonyms, and Formula	Mol. wt.	Color, crystalline form, specific rotation and λ_{max} (log ε)	b.p. °C	m.p. °C	Density	n_D	Solubility	Ref.
473	Acridine, 6,9-diamino-2-ethoxy or Rivanol $C_{15}H_{15}N_3O$	255.30	ye nd		123-44d				B22^{4}, 6679
474	Acridine, 6,9-dichloro-2-methoxy $C_{14}H_9NOCl_2$	278.14			163-5				B21^{4}, 1553
475	Acridine, 9,10-dihydro or Acridan, carbazine $C_{13}H_{11}N$	181.24	pl or pr (al)	sub 300d	169-71			al, eth, ace	B20^{4}, 3885
476	Acridine, 9,10-dihydro-9-oxo or Acridone $C_{13}H_9NO$	195.22	ye lf (al)	>354	154			ace	B21^{4}, 4171
477	Acridine, 4-hydroxy $C_{13}H_9NO$	195.22			117			al, eth	B21^{4}, 1562
478	Acridine, 2-methyl $C_{13}H_{11}N$	193.25	ye nd (dil al)		134			al, eth, bz	B20^{4}, 4037
479	Acridine, 9 phenyl $C_{19}H_{13}N$	255.32	lf ye nd (al)	401-4 sub	184				B20^{4}, 4355
480	Acrolein or Propenal $CH_2{=}CHCHO$	56.06		52.5-3.5	-86.9	$0.8410^{20/4}$	1.4017^{20}	w, al, eth, ace	B1^{4}, 3435
481	Acrolein, 2-chloro $CH_2{=}C(Cl)CHO$	90.51		40^{10}		1.199^{20}	1.463^{20}	al, eth	B1^{4}, 3440
482	Acrolein, diethylacetal or 3.3-diethoxy-1-propene $CH_2{=}CHCH(OC_2H_5)_2$	130.19		123.5		0.8543^{15}	1.4000^{20}	eth, CCl_4	B1^{4}, 3437
483	Acrolein 3(2-furyl) or Furacrolein $(2\text{-}C_4H_3O)\text{-}CH{=}CHCHO$	122.12	ye or wh nd	> 200d	54			al, eth	B17^{4}, 4695
484	Acrolein, 2-methyl or methacrolein $CH_2{=}C(CH_3)CHO$	70.09		68.4		$0.837^{20/4}$	1.4144^{20}	w, al, eth	B1^{4}, 3455
485	Acrylamide $CH_2{=}CHCONH_2$	71.08	lf (bz)		84-5			w, al, eth, chl	B2^{4}, 1471
486	Acrylic acid or Propenoic acid $CH_2{=}CHCO_2H$	72.06		141.6, 48.5^{15}	13	$1.0511^{20/4}$	1.4224^{20}	w, al, eth, ace, bz	B2^{4}, 1455
487	Acrylic acid, allyl ester or Allyl acrylate $CH_2{=}CHCO_2CH_2CH{=}CH_2$	112.13		172-4		$0.9441^{20/4}$	1.4320^{20}	al, eth	B2^{4}, 1468
488	Acrylic acid, β-benzoyl $C_6H_5COCH{=}CHCO_2H$	176.17	nd or pr (to)		99			al, eth, to	B10^{1}, 3144
489	Acrylic acid, benzyl ester or Benzyl acrylate $CH_2{=}CHCO_2CH_2C_6H_5$	162.19		228		$1.0573^{20/4}$	1.5143^{20}	al, eth, ace	B6^{1}, 1481
490	Acrylic acid, 2-bromo, ethyl ester or Ethyl 2-bromo acrylate $CH_2{=}C(Br)CO_2C_2H_5$	179.01		155-8		1.4581^{25}	1.4660^{25}		B2^{4}, 1487
491	Acrylic acid, butyl ester or Butylacrylate $CH_2{=}CHCO_2C_4H_9$	128.17		146-8, 39^{10}	-64.6	$0.8898^{20/4}$	1.4185^{20}	al, eth, ace	B2^{4}, 1463
492	Acrylic acid, *iso* butylester or *iso*-Butylacrylate $CH_2{=}CHCO_2CH_2CH(CH_3)_2$	128.17		132		$0.8896^{20/4}$	1.4150^{20}	al, eth	B2^{4}, 1465
493	Acrylic acid, 2 chloro, ethyl ester or Ethyl-2-chloroacrylate $CH_2{=}C(Cl)CO_2C_2H_5$	134.56		$51\text{-}3^{18}$		$1.1404^{20/4}$	1.4384^{20}	al, eth	B2^{4}, 1482
494	Acrylic acid, 2-chloro, methyl ester or Methyl-2-chloroacrylate $CH_2{=}C(Cl)CO_2CH_3$	120.54		52^{51}		$1.189^{20/4}$	1.4420^{20}	eth	B2^{4}, 1482
495	Acrylic acid, 3-chloro-*cis* or *cis*-3-chloropropenoic acid $ClCH{=}CHCO_2H$	106.51	lf or nd (HCl)	$107^{17.5}$	63-4			al, eth	B2^{4}, 1481
496	Acrylic acid, 3-chloro-*trans* or *trans*-3-chloropropenoic acid $ClCH{=}CHCO_2H$	106.51	lf	94^{18}	86			al, eth	B2^{4}, 1481
497	Acrylic acid, cyclohexyl ester or Cyclohexyl acrylate $CH_2{=}CHCOOC_6H_{11}$	154.21		$182\text{-}4^{750}$, 88^{20}		$1.0275^{20/4}$	1.4673^{20}	al, eth, chl	B6^{4}, 38
498	Acrylic acid, 2,3-dichloro $ClCH{=}C(Cl)CO_2H$	140.95	mcl pr (chl)		87-8			w, al, eth, ace, chl	B2^{4}, 1484
499	Acrylic acid, 3,3-dichloro $Cl_2C{=}CHCO_2H$	140.95	nd (peth) pr (chl)	sub	76-7			eth, chl	B2^{4}, 1484
500	Acrylic acid, 2,3-diphenyl or α-phenyl cinnamic acid $C_6H_5CH{=}C(C_6H_5)CO_2H$	224.26	nd (lig, dil al)	sub	172.5-3			al, eth	B9^{3}, 3414
501	Acrylic acid, ethyl ester or Ethyl acrylate $CH_2{=}CH_1CO_2C_2H_5$	100.12		99.8	-71.2	$0.9234^{20/4}$	1.4068^{20}	al, eth, chl	B2^{4}, 1460
502	Acrylic acid, 3-(2-furyl)-*cis* or 2- Furanacrylic acid *cis* $(2\text{-}C_4H_3O)CH{=}CHCO_2H$	138.12	wh pr or pl		103-4			eth	B18^{4}, 4143
503	Acrylic acid, 3-(2-furyl) *trans* or 2-Furanacrylic acid *trans* $(2\text{-}C_4H_3O)CH{=}CHCO_2H$	138.12	nd (w)	286 (sub)	141		1.5286^{20}	al, eth, bz	B18^{4}, 4173

No.	Name, Synonyms, and Formula	Mol. wt.	Color, crystalline form, specific rotation and λ_{max} (log ε)	b.p. °C	m.p. °C	Density	n_D	Solubility	Ref.
504	Acrylic acid, 3-(2-furyl) *cis*, benzylester or Benzyl-2-furanacrylate.......... $(2\text{-}C_4H_3O)CH=CHCH_2C_6H_5$	138.12	pa ye	201-3^{12}	42-5	....	1.5872^{25}	eth, ace, bz, al	B18^4, 4147
505	Acrylic acid, 3-(2-furul) *cis*, butyl ester or Butyl-2-furanacrylate.......... $(2\text{-}C_4H_3O)CH=CHCO_2C_4H_9$	194.23	....	147-50^{15}, 117-8^{3}	......	1.045^{20}	1.5129^{20}	al, ace	B18^4, 4146
506	Acrylic acid, 3-(2-furyl) *cis*, pentylester or Pentyl-2-furanacrylate.......... $(2C_4H_3O)CH=CHCO_2C_5H_{11}$	208.26	....	116-8^{2}	......	1.0322$^{20/4}$	1.5289^{24}	al, eth	B18^4, 4146
507	Acrylic acid, 3-(2-furyl) *cis*, propyl ester or Propyl-2-furan-acrylate .... $(2\text{-}C_4H_3O)CH=CHCO_2C_3H_7$	180.20	....	91-4^{3}	......	1.07 44$^{20/4}$	1.5392^{24}	al, eth, bz	B18^4, 4146
508	Acrylic acid, 3-(2-furyl) *cis*, methyl ester or Methyl-2-furanacrylate.......... $(2\text{-}C_4H_3O)CH=CHCO_2CH_3$	152.15	....	227.5^{774}, 112^{15}	35-7	....	1.4447^{20}	al, eth, bz	B18^4, 4144
509	Acrylic acid, methyl ester or Methyl acrylate .... $CH_2=CHCO_2CH_3$	86.09	....	80.5	< - 75	0.9535$^{20/4}$	1.4040^{20}	al, eth, ace, bz	B2^4, 1457
510	Acrylic acid, 2 methylbutyl ester or (2-methylbutyl) acrylate .... $CH_2=CHCO_2CH_2CH(CH_3)CH_2CH_3$	142.20	....	160, 52^{11}	......	0.8936$^{20/4}$	1.4240^{20}	al, eth	B2^4, 1525
511	Acrylic acid, 3-(α-naphthyl) *cis* .... $\alpha\text{-}C_{10}H_7CH=CHCO_2H$	198.22	cal pl (al)	....	156	....	....	al	B9^3, 3284
512	Acrylic acid, 3-(α-naphthyl) *trans* .... $\alpha\text{-}C_{10}H_7CH=CHCO_2H$	198.22	nd (al,w,aa)	sub	211-2	....	....	eth, chl	B9^3, 3284
513	Acrylic acid, 3 (β-naphthyl) *trans* .... $\beta\text{-}C_{10}H_7CH=CHCO_2H$	198.22	nd (al, w)	....	210	....	....	al	B9^3, 3288
514	Acrylic acid, 2-phenyl or 2-Phenylpropenoic acid, Atropic acid .... $CH_2=C(C_6H_5)CO_2H$	148.17	lf (al), nd (w)	267d	106-7	....	....	al, eth, bz, chl	B9^3, 2751
515	Acrylic acid, trichloro .... $CCl_2=CClCO_2H$	175.40	....	158	......	....	1.5271$^{18.5}$	bz	B2^4, 1486
516	Acrylonitrile .... $CH_2=CHCN$	53.06	....	77.5-9	......	0.8060$^{20/4}$	1.3911^{20}	**al, eth**, ace, bz	B2^4, 1473
517	Acrylonitrile, 3,3 diphenyl .... $(C_6H_5)_2C=CHCN$	205.26	....	125-8$^{0.15}$	......	....	1.6352^{20}	....	B9^3, 3439
518	Acrylonitrile, triphenyl .... $(C_6H_5)_2C=C(C_6H_5)CN$	281.36	nd (al), pr	....	166-7	....	....	eth, al	B9^3, 3630
519	Acrylylchloride .... $CH_2=CHCOCl$	90.51	....	75-6	......	1.1136$^{20/4}$	1.4343^{20}	chl	B2^4, 1471
520	Acrylyl chloride, trichloro .... $CCl_2=CClCOCl$	193.84	....	158	......	....	1.5271$^{18.5}$	bz	B2^4, 1486
521	Actidione *l* or Cycloheximide .... $C_{15}H_{23}NO_4$	281.33	pl (al) $[\alpha]^{25/D}$-33 (chl C=l)	....	144-5	....	....	al	B20^4, 1406
522	Adamantane or Tricyclo *3,3,1,1^{3,7}*decane .... $C_{10}H_{16}$	136.24	nd (sub)	sub	268 (sealed)	1.07	1.568	bz	B5^4, 469
523	Adenosine or Adenine, 9-β-D-ribofuranosyladenine .... $C_{10}H_{13}N_5O_4$	267.24	nd (w + 1 1/2), $[\alpha]^{20/D}$ -60.0 (w,c=1)	....	235-6 (anh)	....	....	w	B31, 27
524	5-Adenylic acid or Adenosine-5-phosphate.......... $C_{10}H_{14}N_5O_7P$	347.22	pw, nd, (w, dil al) cr $[\alpha]^{50/D}$ -41.78	....	195-208d (sealed tube)	....	....	w, al	B31, 27
525	Adipaldehyde or 1,6-Hexanedial .... $OHC(CH_2)_4CHO$	114.14	....	92-4^{9}	-8	1.003$^{19/4}$	1.4350^{20}	al, eth, bz, aa	B1^4, 3686
526	Adipamic acid or Adipic acid monoamide .... $H_2NCO(CH_2)_4CO_2H$	145.16	nd (w)	....	161-2	....	....	....	B2^4, 1972
527	Adipamide .... $H_2NCO(CH_2)_4CONH_2$	144.17	pl	....	220	....	....	al	B2^4, 1972
528	Adipic acid or Hexanedioc acid .... $HO_2C(CH_2)_4CO_2H$	146.14	mcl pr (w, ace, lig)	265^{100}	153	1.360$^{25/4}$	....	al, eth	B2^4, 1956
529	Adipic acid, 2-amino -*dl* .... $HO_2C(CH_2)_3CH(NH_2)CO_2H$	161.16	pl (w)	....	206 (anh)	....	....	....	B4^4, 1555
530	Adipic acid, dibutylester or di-butyl adipate .... $C_4H_9O_2C(CH_2)_4CO_2C_4H_9$	258.36	....	165^{10}	-32.4	0.9615$^{20/4}$	1.4369^{20}	**al, eth**	B2^4, 1961
531	Adipic acid, diisobutylester or Di-*iso*butyl adipate.......... $i\text{-}C_4H_9O_2C(CH_2)_4CO_2\text{-}i\text{-}C_4H_9$	258.36	....	186-8$^{15.7}$	......	0.9543^{19}	1.4301^{20}	....	B2^4, 1962

PHYSICAL CONSTANTS OF ORGANIC COMPOUNDS (Continued)

No.	Name, Synonyms, and Formula	Mol. wt.	Color, crystalline form, specific rotation and λ_{max} (log ε)	b.p. °C	m.p. °C	Density	n_D	Solubility	Ref.
532	Adipic acid, diethyl ester or diethyl adipate .... $C_2H_5O_2C(CH_2)_4CO_2C_2H_5$	202.25	....	245	-19.8	$1.0076^{20/4}$	1.4272^{20}	al, eth	$B2^4$, 1960
533	Adipic acid, di-2 ethyl butyl ester or di-2-ethyl butyl adipate .... $(C_2H_5)_2CHCH_2O_2C(CH_2)_4CO_2CH_2CH(C_2H_5)_2$	314.47	....	200^{10}	-15	$0934^{25/4}$	1.4434^{20}	al, ace, aa	$B2^4$, 1964
534	Adipic acid, di-(2 ethylhexyl) ester or di-(2-ethylhexyl) adipate .... $C_8H_{17}O_2C(CH_2)_4CO_2C_8H_{17}$	370.57	....	214^5	-67.8	$0.922^{25/4}$	1.4474^{20}	al, eth, ac, aa	$B2^4$, 1964
535	Adipic acid, dimethyl ester or dimethyl adipate .... $CH_3O_2C(CH_2)_4CO_2CH_3$	174.20	cr	115^{13}	10.3	$1.0600^{20/4}$	1.4283^{20}	al, eth, aa	$B2^4$, 1959
536	Adipic acid, dipropyl ester or dipropyl adipate .... $C_3H_7O_2C(CH_2)_4CO_2C_3H_7$	230.30	....	151^{11}	-15.7	$0.9790^{20/4}$	1.4314^{20}	al, eth, chl	$B2^4$, 1961
537	Adipic acid, di *iso*-propyl ester or di *iso*-propyladipate.... i-$C_3H_7O_2C(CH_2)_4CO_2$-i-C_3H_7	230.30	....	$120^{6.5}$	-1.1	0.9569^{20}	1.4247^{20}	al, eth, ace, aa	$B2^4$, 1961
538	Adipic acid, monoethyl ester or Ethyladipate .... $HO_2C(CH_2)_4CO_2C_2H_5$	174.20	hyg (eth-peth)	285, 170^{17}	29	$0.9796^{20/4}$	1.4311^{20}	al, eth	$B2^4$, 1960
539	Adipic acid, mono methyl ester or methyl adipate........ $HO_2C(CH_2)_4CO_2CH_3$	160.17	lf(Me_3N-MeOH)	158^{10}	9	1.0623^{20}	1.4283^{20}	al	$B2^3$, 1717
540	Adipic acid, 2-methyl .... $HO_2C(CH_2)_3CH(CH_3)CO_2H$	160.17	cr (peth-bz)	209^{13}	93	....	....	w, al, eth, chl	$B2^4$, 2010
541	Adipic acid, 3-methyl *d* .... $HO_2C(CH_2CH(CH_3)CH_2CO_2H$	160.17	cr (chl-bz)	230^{30}	94-4.5	....	....	w, al, eth, ace, bz	$B2^4$, 2014
542	Adipic acid, 3-methyl *dl* .... $HO_2CCH_2CH_2CH(CH_3)CH_2CO_2H$	160.17	nd (bz), cr (ace-bz)	190-200^{12}	97	....	....	w, al, eth, ace, bz	$B2^4$, 2014
543	Adipic acid, 2-oxo .... $HO_2C(CH_2)_3COCO_2H$	160.13	cr (al or eth)	....	127	....	....	w, al, ace	$B3^4$, 1821
544	Adipic acid, 3-oxo .... $HO_2CCH_2CH_2COCH_2CO_2H$	160.13	pl (ace - chl)	....	123	....	....	ace	$B3^4$, 1822
545	Adiponitrile .... $NC(CH_2)_4CN$	108.14	nd (eth)	295, 180^{20}	1	0.9676^{20}	1.4380^{20}	al, chl	$B2^4$, 1975
546	Adipoyl chloride or Adipilacid, dichloride............ $ClCO(CH_2)_4COCl$	183.03	....	126^{12}	....	....	....	....	$B2^4$, 1972
547	Adonitol or Adonite, Ribitol .... $HOCH_2(CHOH)_3CH_2OH$	152.15	pr (w), nd (al)	....	104	....	....	w, al	$B1^4$, 2832
548	Adrenaline, *d* or *d*-epinephrine .... $C_9H_{13}NO_3$	183.21	$[\alpha]^{20/D}$ + 50.5 (HCl)	....	215d	....	....	aa	$B13^3$, 2384
549	Adrenaline, *l* or *l*-epinephrine .... $C_9H_{13}NO_3$	183.21	br (in air) pw $[\alpha]^{20/D}$-53 (aq. Hcl)	....	211-2	....	....	aa	$B13^3$, 2384
550	Adrenalone............ $C_9H_{11}NO_3$ 3,4$(HO)_2C_6H_3COCH_2NHCH_3$	181.19	nd	....	235-6d	....	....	....	$B14^3$, 614
551	Adrenochrome *dl* .... $C_9H_9NO_3$	179.18	red-br rods (MeOH-HCO_2H)	....	125 (anh)	....	....	w, al	$B21^4$, 6436
552	Ajmalicene or Py-tetrahydro serpentine .... $C_{21}H_{24}N_2O_3$	352.43	pr (MeOH) $[\alpha]^{24/D}$ -60 (chl, C=0.5)	....	258-d 261-3(vac)	....	....	MeOH	J76, 1332
553	Ajmaline .... $C_{20}H_{26}N_2O_2$	326.44	pl (+ 3.5 w, aq AcOEt) $[\alpha]^{20/D}$ + 144(Chl)	....	158-60 (hyd) 250-7 (anh)	....	....	al, chl, eth	J1954, 1242
554	α-Alanine-*D* or l-α-aminopropionic acid .... $CH_3CH(NH_2)CO_2H$	89.09	nd (w, al) $[\alpha]^{25/D}$-13.6 (6N HCl,C=1)	sub	314d	....	....	w, al	$B4^3$, 1219
555	α Alanine-*DL* or dl-α-aminopropionic acid .... $CH_3CH(NH_2)CO_2H$	89.09	orh pr or nd (w)	sub 258	295-6d	1.424	....	w, al	$B4^3$, 1222
556	α-Alanine-*L* or d-α amino propioniuc acid............ $CH_3CH(NH_2)CO_2H$	89.09	rh (w), $[\alpha]^{25/D}$ + 2.8 (w, C=6)	sub 160-5	314d	1.432^{22}	....	w, al	$B4^3$, 1208
557	α-Alanine, *N*-alanyl-*l* .... $H_2NCH(CH_3)CONHCH(CH_3)CO_2H$	160.17	lf $[\alpha]^{20/D}$-21.6 (w)	....	298	....	....	w	$B4^3$, 1218

PHYSICAL CONSTANTS OF ORGANIC COMPOUNDS (Continued)

No.	Name, Synonyms, and Formula	Mol. wt.	Color, crystalline form, specific rotation and λ_{max} (log ε)	b.p. °C	m.p. °C	Density	n_D	Solubility	Ref.
558	α-Alanine, N-benzoyl-*d* $CH_3CH(NHCOC_6H_5)CO_2H$	193.20	pl (w), $[\alpha]^{20/D}$ -2.4 (w,c=l)	152-4					$B9^3$, 1141
559	α-Alanine, N-benzoyl-*dl* $CH_3CH(NHCOC_6H_5)CO_2H$	193.20	pl or pr, lf (eth)	d	165-6			w, al	$B9^3$, 1142
560	α-Alanine, N-benzoyl-*l* $CH_3CH(NHCOC_6H_5)CO_2H$	193.20	$[\alpha]^{20/D}$ + 24 (w, c=1)		151				$B9^3$, 1141
561	α-Alanine, N-(Carboxymethyl)-*dl* $HO_2CCH_2NHCH(CH_3)CO_2H$	147.13	cr (dil al or w)		222-3			w	$B4^3$, 1250
562	α-Alanine, N-(4-Chlorophenyl), nitrile $CH_3CH[NH(C_6H_4Cl\text{-}4)]CN$	180.64	lf (eth-peth)		114.5			al, eth, bz, chl	B12, 617
563	**α-Alanine, N,N-diethyl, nitrile** $CH_3CH[N(C_2H_5)_2]CN$	126.20		81^{27}		$0.857^{10/4}$		w, al, eth	$B4^3$, 1238
564	α-Alanine, 3(3,4 dihydroxyphenyl)-*L* or *l*-Dopa $3,4\text{-}(HO)_2C_6H_3CH_2CH(NH_2)COOH$	197.19	pl (dil al), pr or nd $(w + SO_2)$ $[\alpha]^{15/D}$ −39.5 (w, p = 1.3)		285.5d			w	$B14^3$, 1629
565	α-Alanine, ethyl ester, hydrochloride $CH_3CH(NH_2)CO_2C_2H_5.HCl$	153.61	pr or hyg nd (al)	d	87-8			w, al, eth	$B4^3$, 1231
566	α-Alanine, N-fumaryl-*DL* or Fumaro alanide $HO_2CCH{=}CHCONHCH(CH_3)CO_2H$	187.15	nd		229d			al	$B4^3$, 1248
567	α-Alanine, N- methyl-*dl* $CH_3CH(NHCH_3)CO_2H$	103.12	rh pr (aks al)	sub 292	280 (d)				$B4^3$, 1235
568	β Alanine $H_2NCH_2CH_2CO_2H$	89.09	nd, rh pr (al)		207d	1.437^{19}		w	$B4^3$, 1259
569	β Alanine, ethyl ester, hydrochloride $N_2NCH_2CH_2CO_2C_2H_5HCl$	153.61			65.3			w	$B4^1$, 499
570	β Alanine, N-methyl $CH_3NHCH_2CH_2CO_2H$	103.12	pl (al, + 1w)		146 (anh)			w, al	$B4^3$, 1264
571	β Alanine, N-methyl, ethyl ester $CH_3NHCH_2CH_2CO_2C_2H_5$	131.17		80^{21}		$1.0082^{20/20}$	1.4443^{20}	al, eth	$B4^3$, 1264
572	β Alanine, N-methyl, nitrile or β Methylamino propionitrile $CH_3NHCH_2CH_2CN$	84.13		$101\text{-}4^{49}$		$0.8992^{20/4}$	1.4320^{20}	w, ace, bz, chl	$B4^3$, 1264
573	β Alanine, N-phenyl $C_6H_5NHCH_2CH_2CO_2H$	165.19	nd (bz or dil al), lf (dil, al, bz, eth)		92			al, eth, bz	$B12^3$, 933
574	**Aldrin of Octalene** $C_{12}H_8Cl_6$	364.91			104			al, eth, ace, bz	$B5^3$, 1385
575	Alizarin or 1,2-dihydroxy anthraquinone $1,2\text{-}(HO)_2C_{14}H_6O_2$	240.22	og or red tcl nd or pr (al, sub)	430 sub	289-90 (cor)			al, eth, ace, bz	$B8^3$, 3767
576	Alkanin-*l* $C_{16}H_{16}O_5$	288.30	red,br,pr, (bz,sub), nd(eth-al) $[\alpha]^{20/cd}$ -157 (bz)	sub $140^{0.001}$	149			ace, al, eth	$B8^3$, 4089
577	Allantoic acid or Dicarbamoacetic acid $(H_2NCONH)_2CHCO_2H$	176.13	nd, lf (MeOH)		173d			al	$B3^4$, 1492
578	Allantoin or Glyoxyldiureide $C_4H_6N_4O_3$	158.12	mcl pl or pr (w)		238.40			al, w	$B25^2$, 379
579	Allanturic acid or Glyoxalurea $C_3H_5N_2O_2$	101.08	amor, hyg pw		180 turns br				$B25^2$, 388
580	Allene or Propadiene $CH_2{=}C{=}CH_2$	40.06		-34.5	-136		1.4168	bz, peth	$B1^4$, 966
581	Allene, tetra fluoro $F_2C{=}C{=}CF_2$	112.03		-38					$B1^4$, 968
582	Allicin or S-oxodiallyl disulfide $CH_2{=}CHCH_2SS(O)CH_2CH{=}CH_2$	162.26		d		1.112^{20}	1.561^{20}	w	$B4^4$, 7
583	D-Allitol or D-Allodulcitol $HOCH_2(CHOH)_4CH_2OH$	182.17			150-1			w	$B1^4$, 2839

No.	Name, Synonyms, and Formula	Mol. wt.	Color, crystalline form, specific rotation and λ_{max} (log ε)	b.p. °C	m.p. °C	Density	n_D	Solubility	Ref.
584	Alloluecine-*l* $CH_3CH_2CH(CH_3)CH(NH_2)COOH$	131.18	lf (w) $[\alpha]^{20/D}$ -14.2 (w,c=2)		280-1d			w	$B4^{3}$, 1462
585	Allomucic acid or Tetrahydroxy adipic acid $HOOC(CHOH)_4COOH$	210.14	nd or pr (w)		198-200(d)			w	$B3^{3}$, 1116
586	α-Allonic acid-γ-lactone $C_6O_6H_{10}$	178.14	pr (al) $[\alpha]^{20/D}$ -6.8 (w,c=11)					w, al	$B18^{4}$, 3024
587	Alloocimene-A or 2,6-Dimethyl-2,3,6-Octateiene (4-trans, 6-trans) $CH_3CH=C(CH_3)CH=CHCH=C(CH_3)_2$	136.24		188^{750}, 91^{20}	-35.4	$0.8118^{20/4}$	1.5446^{20}		$B1^{4}$, 1106
588	Alloocimene-B or 2,6-Dimethyl-2,4,6-Octatriene (4-*trans*-6-*cis*) $CH_3CH=C(CH_3)CH=CHCH=C(CH_3)_2$	136.24		89^{20}	-20.6	0.8060^{20}	1.5446^{20}		$B1^{4}$, 1106
589	Allophanic acid, ethyl ester $H_2NCONHCO_2C_2H_5$	132.12	nd (w), (bz)	d	195				$B3^{4}$, 127
590	Allophanonitrile, 3-phenyl or 1-Cyano-3-phenyl urea $C_6H_5NHCONHCN$	161.16	nd		125			al	$B12^{3}$, 781
591	D-Allose (β-anomer) $C_6H_{12}O_6$,$HOCH_2(CHOH)_4CHO$	180.16	cr (w) $[\alpha]_D$ 0.58→14.4 (w,c=5) (mut)		128			w	$B1^{4}$, 4299
592	L-Allose $C_6H_{12}O_6$	180.16	pr (dil al) $[\alpha]^{20/D}$ -0.58 (al)		128-9			w	$B1^{4}$, 4300
593	Alloxan or Mesoxalurea $C_4H_2N_2O_4$	142.07		sub (vac)	256 d (anh)			w, al, ace, bz, aa	$B24^{2}$, 301
594	Alloxanic acid $C_4H_4N_2O_5$	160.09	tcl pr (eth)		162-3d			w, al	$B25^{2}$, 266
595	Alloxantin $C_8H_6N_4O_8$	286.16	rh pr (w + 2)		253-5d				$B26^{2}$, 335
596	Allyl alcohol or 1-Propene-3-ol $CH_2=CHCH_2OH$	58.08		97.1	-129	$0.8540^{20/4}$	$1.4135^{20/}{}_D$	**w, al, eth,** chl	$B1^{4}$, 2079
597	Allyl alcohol, 2-bromo $CH_2=CBr-CH_2OH$	136.98		$153\text{-}4^{755}$, 62^{11}		1.621^{18}	1.500^{18}	eth, chl	$B1^{4}$, 2094
598	Allyl alcohol, 2-chloro $CH_2=C(Cl)CH_2OH$	92.53		$136\text{-}40$, 47^{10}		$1.1618^{20/4}$	1.4588^{20}		$B1^{4}$, 2091
599	Allyl alcohol, 3-chloro $ClCH=CHCH_2OH$	92.53		(i) 146^{746}, (ii) 153^{756}		(i) $1.1769^{20/4}$, (ii) $1729^{20/4}$	(i) 1.4738^{20}, (ii) 1.4664^{20}		$B1^{4}$, 2090
600	Allyl alcohol, 3-(3,5-dimethoxy-4-hydroxyphenyl) or Sinaphyl alcohol. Syringenin. [*3,5(CH₃O)₂-4-HOC₆H₂*] $CH=CHCH_2OH$	210.23	nd (eth-peth)		66-7			eth	$B6^{3}$, 6690
601	Allyl alcohol, 3-(4-hydroxy phenyl) or *P*-coumaryl alcohol $4\text{-}HOC_6H_4CCH=CHCH_2OH$	150.18	pr (dil al)		124			al, eth, ace, bz	C45, 3359
602	Allyl alcohol, 1-phenyl $CH_2=CH-CH(C_6H_5)OH$	134.18		215-6, 111^{18}		$1.0251^{21/0}$	1.5406^{20}	al, eth, bz, chl	$B6^{3}$, 2417
603	Allyl alcohol, 3-phenyl *trans* or Cinnamyl alcohol $C_6H_5CH=CHCH_2OH$	134.18	wh nd (eth-peth)	257.5, $127\text{-}8^{10}$	34	$1.0440^{20/4}$	1.5819^{20}	al, eth	$B6^{3}$, 2401
604	Allylamine or 3-Aminopropene $H_2NCH_2CH=CH_2$	57.10		58		$0.7621^{20/4}$	1.4205^{20}	**w, al, eth,** chl	$B4^{4}$, 1057
605	Allylamine, *N*-methyl or Allylmethyl amine $CH_2CH_2NHCH_3$	58.10		65			1.4065^{20}	w, al, eth, ace	$B4^{4}$, 1058
606	Allylamine, *N*-phenyl or *N*-allylaniline $CH_2CH_2NHC_6H_5$	120.17		$217\text{-}9^{736}$, $105\text{-}8^{12}$		$0.982^{25/4}$	1.563^{20}	al, eth	$B12^{3}$, 277
607	Allylamine, *N*-iso-Propyl or Allyl-iso-propyl amine $CH_2CH_2NHCH(CH_3)_2$	86.16		96-7			$1.4140^{25/}$		$B4^{4}$, 1059
608	Allyl benzene, α-chloro $C_6H_5CHClCH=CH_2$	152.62		212-4, 97^{18}		$1.073^{14/4}$	1.545^{14}	eth, ace, bz	$B5^{4}$, 1363
609	Allylbromide or 3-Bromopropene $BrCH_2CH=CH_2$	120.98		70^{752}	119.4	$1.398^{20/4}$	1.4697^{20}	**al, eth**	$B1^{4}$, 754
610	Allyl chloride or 3-chloropropene $ClCH_2CH=CH_2$	76.53		45	134.5	$0.9376^{20/4}$	1.4157^{20}	**al, eth, ace, bz, lig**	$B1^{4}$, 738
611	Allyl ether or Diallyl ether $(CH_2=CHCH_2)_2O$	98.14		94		$0.8260^{20/4}$	1.4163^{20}	**al, eth,** ace	$B1^{4}$, 2086

PHYSICAL CONSTANTS OF ORGANIC COMPOUNDS (Continued)

No.	Name, Synonyms, and Formula	Mol. wt.	Color, crystalline form, specific rotation and λ_{max} (log ε)	b.p. °C	m.p. °C	Density	n_D	Solubility	Ref.
612	Allyl 4-chlorophenyl ether $4\text{-}ClC_6H_4OCH_2CH{=}CH_2$	168.62		$106\text{-}7^{12}$		1.131^{15}	1.5348^{25}	al, eth, bz	$B6^4$, 825
613	Allyl ethyl ether $CH_2{=}CHCH_2OC_2H_5$	86.13		66		$0.7651^{20/4}$	1.3881^{20}	al, eth, ace	$B1^4$, 2083
614	Allyl isocyanide $CH_2{=}CHCH_2NC$	67.09		98		$0.794^{17/4}$		al, eth	B4, 208
615	Allyl mercaptan $CH_2{=}CHCH_2SH$	74.14		67-8		$0.925^{23/4}$	1.4832^{20}	al, eth, Chl	$B1^4$, 2095
616	Allyl methyl ether $CH_2{=}CHCH_2OCH_3$	72.11		46		$0.77^{11/11}$	1.3778^{20}	al, eth, ace	$B1^4$, 2083
617	Allyl *iso*pentyl ether or Allyl-isoamyl-ether $CH_2{=}CHCH_2OCH_2CH_2CH(CH_3)_2$	128.21		120-2				al, eth	$B1^2$, 477
618	Allyl phenyl ether $CH_2{=}CHCH_2OC_6H_5$	134.17		$191.7, 74^5$		$0.9811^{20/4}$	1.5223^{20}	al, eth	$B6^4$, 562
619	Allyl propyl ether $CH_2{=}CHCH_2OC_3H_7$	100.16		90-2		0.7764^{20}	1.3919^{20}	al, eth, ace	$B1^4$, 2083
620	Allyl *iso*propyl ether $CH_2{=}CHCH_2OCH(CH_3)_2$	100.16		83-4		0.7764^{20}	1.3946^{20}	al, eth, ace	$B1^4$, 2083
621	Allyl 2-tolyl ether $(2\text{-}CH_3C_6H_4)OCH_2CH{=}CH_2$	148.20		$205\text{-}8, 85^{12}$		$0.9698^{15/4}$	1.5188^{15}		$B6^4$, 1946
622	Allyl 3-tolyl ether $(3\text{-}CH_3C_6H_4)OCH_2CH{=}CH_2$	148.20		$211\text{-}4, 93.5^{13}$		0.9564^{20}	1.5179^{20}	bz	$B6^4$, 2041
623	Allyl 4-tolyl ether $(4\text{-}CH_3C_6H_4)OCH_2CH{=}CH_2$	148.20		$214.5, 97\text{-}8^{16}$		$0.9728^{15/15}$	1.5157^{24}	bz	$B6^4$, 2101
624	Allyl vinyl ether $CH_2{=}CHCH_2\text{-}O\text{-}CH{=}CH_2$	84.12		65^{740}		$0.8050^{20/4}$	1.4062^{20}	eth, ace	$B1^4$, 2085
625	Allyl isothiocyanate or Allyl mustard oil $CH_2{=}CHCH_2NCS$	99.15		$152, 44^{12}$	−80	$1.0126^{20/4}$	1.5306^{20}	al, eth, bz	$B4^4$, 1081
626	Allyl sulfide $(CH_2{=}CHCH_2)_2S$	114.21		$139^{768}, 35^6$	−85	$0.8877^{27/4}$	1.4870^{25}	al, eth	$B1^4$, 2097
627	Allyl sulfoxide $(CH_2{=}CHCH_2)_2SO$	130.20		$112\text{-}5^{12}$	23.5	$1.0261^{20/4}$	1.5115^{20}	al, eth, ace	$B1^4$, 2097
628	Allyl thio cyanate $CH_2{=}CHCH_2SCN$	99.15		161		1.056^{15}		al, eth	$B3^4$, 332
629	Allyl (2,4,6-tribromophenyl) ether $(2,4,6\text{-}Br_3C_6H_2)OCH_2CH{=}CH_2$	370.87			33-4				$B6^2$, 194
630	Allyl trisulfide $(CH_2{=}CHCH_2)_2S_3$	178.33		$112\text{-}22^{16}$		1.0845^{15}		eth	B1, 441
631	Aloetic Acid $C_{15}H_4N_4O_{11}$	416.22	og-ye nd (aa), ye cr (w + l)		285d				B8, 525
632	Aloin or Barbaloin $C_{21}H_{22}O_9$	418.40	ye nd (al), $[\alpha]_D$ −8.3 (dil al)		148.5			w, al, ace	$B18^4$, 3630
633	Alstonine $C_{21}H_{20}N_2O_3$	348.40	ye nd (ace)		205-10d				$B27^2$, 824
634	D-Altrose (β-anomer) $C_6H_{12}O_6$	180.16	pr (MeOH-al), $[\alpha]^{20}{}_D$ + 11.7 → 33.1 (w, c = 7) (mut)		103.5			w	$B1^4$, 4300
635	L-Altrose (β-anomer) $C_6H_{12}O_6$	180.16	pr (al, aa) −28.55→ −32.30 (w) (mut)		107-9.5				$B1^4$, 4301
636	Amalic acid or Tetramethyl alloxantin $C_{12}H_{14}N_4O_8$	324.27	cr (w)		245d				$B26^2$, 336
637	Amarine $C_{21}H_{18}N_2$	298.39	pr (eth, bz-lig)	198d	136			al, eth, bz	$B23^2$, 274
638	Amarine hydrate $C_{21}H_{18}N_2 \cdot \frac{1}{2}H_2O$	307.39	pr (aq al + ½ w)		106			w	$B23^2$, 274
639	Amaron or Benzoin-imide $C_{28}H_{20}N_2$	384.48	tcl nd or pr (ace, al), nd (aa)	sub	2-46			ace, bz, chl	$B23^2$, 304
640	Aminoacetamide or Glycinamide $H_2NCH_2CONH_2$	74.08	hyg nd (chl)		67-8			w, al, ace	$B4^3$, 1118

PHYSICAL CONSTANTS OF ORGANIC COMPOUNDS (Continued)

No.	Name, Synonyms, and Formula	Mol. wt.	Color, crystalline form, specific rotation and λ_{max} (log ε)	b.p. °C	m.p. °C	Density	n_D	Solubility	Ref.
641	Aminoacetamide, *N'*(4-ethoxyphenyl) or Glycine-*p*-phenetidide $H_2NCH_2CONH\text{-}C_6H_4\text{-}OC_2H_5\text{-}4)$	194.23	nd (+lw)		100.5 (anh)			al, eth	B13[1], 179
642	Aminoacetamide Hydrochloride $H_2NCH_2CONH_2\text{-}HCl$	110.54	nd (al)		203-4			w, al	B4[3], 1119
643	Aminoacetamide, *N*-phenyl $C_6H_5NHCH_2CONH_2$	151.19			127-8			w, al	B12[3], 915
644	Aminoacetic acid or Glycine $H_2NCH_2CO_2H$	75.07	mcl or trg pr (dil al)		262d	1.607		w	B4[3], 1097
645	Aminoacetic acid, *N*-acetyl or Aceturic acid $CH_3CONHCH_2CO_2H$	117.10	lo nd (w, MeOH)		206			al, ace, w	B4[3], 1150
646	Aminoacetic acid, *N*-acetyl-*N*-phenyl or *N*-phenyl aceturic acid $CH_3CON(C_6H_5)CH_2CO_2H$	193.25	lf (w)		194			w	B12[3], 919
647	Aminoacetic acid, *N*-(4 aminophenyl) hydrate $(4\text{-}H_2N\text{-}C_6H_4)NHCH_2CO_2H\text{-}H_20$	184.20	pl (dil aa)		222-3d			w	B13[1], 34
648	Aminoacetic acid, *N*-benzoyl or Hippuric acid $C_6H_5CONHCH_2COOH$	179.18	pr (w or al)		190-3	$1.371^{20/4}$		w, al	B9[3], 1123
649	Aminoacetic acid, *N*-benzyl or *N*-benzyl glycine $C_6H_5CH_2NHCH_2CO_2H$	165.19	nd (w)		198-9				B12[3], 2283
650	Aminoacetic acid, *N*-benzyl, ethyl ester or Ethyl-*N*-benzyl glycinate $C_6H_5CH_2NHCH_2CO_2C_2H_5$	193.25		$175\text{-}9^{50}$			1.5041^{20}		B12[3], 2283
651	Aminoacetic acid, *N*-bromoacetyl-*N*-phenyl $BrCH_2CON(C_6H_5)CH_2COOH$	272.10	pl (w)		153d			al, bz	B12, 477
652	Aminoacetic acid, *N*-(2-carboxyphenyl) or Anthranilidoacetic acid $(2\text{-}HO_2CC_6H_4)NHCH_2CO_2H$	179.18	nd (MeOH)		218-20			al, eth	B14[3], 938
653	Aminoacetic acid, *N*-chloroacetyl-*N*-phenyl $ClCH_2CON(C_6H_5)CH_2CO_2H$	227.65	pl or pr (bz)		132-3			al, bz	B12, 476
654	Aminoacetic acid, *N*-chloroacetyl-*N*-phenyl,methyl ester $ClCH_2CON(C_6H_5)CH_2CO_2CH_3$	241.67	pr (lig)		59-60			al, eth, bz	B12, 477
655	**Aminoacetic acid, *N*-ethyl of Ethyl aminoacetic acid** $C_2H_5NHCH_2CO_2H$	103.12	pl (al)		181-2d			w, al	B4[3], 1134
656	Aminoacetic acid, ethyl ester or Ethyl glycinate $H_2NCH_2CO_2C_2H_5$	103.12		$148\text{-}9^{750}$, $57\text{-}8^{18}$		$1.0275^{10/4}$	1.4242^{10}	w, al, eth, ace bz, lig	B4[3], 1116
657	Aminoacetic acid, ethyl ester,hydrochloride or Ethyl glycinate hydrochloride $H_2NCH_2CO_2C_2H_5\text{-}HCl$	139.58	nd (al)	sub	144			w, al	B4[3], 1117
658	Aminoacetic acid, *N,N*-(dicarbethoxy),ethyl ester $(C_2H_5O_2C)_2NCH_2CO_2C_2H_5$	247.25	pr (peth)	$152\text{-}3^{10}$	36.5			al, eth, bz	B4, 365
659	Aminoacetic acid, *N,N*-dimethyl $(CH_3)_2NCH_2CO_2H$	103.12	hyg nd (PrOH)		185-6			w, al, eth	B4[3], 1124
660	Aminoacetic acid hydrazide $H_2NCH_2CONHNH_2$	89.10		dec 150	80.5			chl	B4[3], 1121
661	Aminoacetic acetic acid, hydrochloride or Glycine hydrochloride $H_2NCH_2CO_2H\text{-}HCl$	111.53	hyg rh nd (w)		200-1			w	B4[3], 1111
662	Aminoacetic acid, *N,N*-bis(2-hydroxyethyl) $(HOCH_2CH_2)_2NCH_2CO_2H$	163.17	nd (al)		193-5d			w	B4[3], 1145
663	Aminoacetic acid, *N*-(4-hydroxyphenyl) $(4\text{-}HOC_6H_4)NHCH_2CO_2H$	167.16	pl (w)		245-7d				B13[3], 1132
665	Aminoacetic acid, *N*-methyl,hydrochloride $CH_3NHCH_2CO_2H \cdot HCl$	125.56			166-8d			w	B4, 345
666	Aminoacetic acid, methyl ester or Methyl glycinate, Methylamine acetate $NH_2CH_2CO_2CH_3$	89.09		130d, 45^{20}				al	B4[3], 1115
667	Aminoacetic acid, *N,N*-methylene di or *N,N*-methylene diamino acetic acid $CH_2(NHCH_2CO_2H)_2$	162.15	pl		199d			al	B4[3], 1148
668	Aminoacetic acid, *N*-(2-naphthyl) $2\text{-}C_{10}H_7\text{-}NHCH_2COOH$	201.22	(w)		134-5			al, ace, w, aa eth	B12, 1298
669	Aminoacetic acid, *N*-(2-nitro phenyl) $(2\text{-}O_2N\text{-}C_6H_4)\text{-}NH\text{-}CH_2CO_2H$	196.16	dk red pr (al)		192-3d			al	B12, 695
670	Aminoacetic acid, *N*-phenyl or Anilino-acetic acid $C_6H_5NHCH_2CO_2H$	151.16			127-8			w, al	B12[3], 914

PHYSICAL CONSTANTS OF ORGANIC COMPOUNDS (Continued)

No.	Name, Synonyms, and Formula	Mol. wt.	Color, crystalline form, specific rotation and λ_{max} (log ε)	b.p. °C	m.p. °C	Density	n_D	Solubility	Ref.
671	Aminoacetic acid, *N*-phenyl,ethyl ester or Ethyl-*N*-phenyl amino acetate $C_6H_5NHCH_2CO_2C_2H_5$	179.22	lf (dil al)	273-4, 163^{18}	58			al, eth	$B12^3$, 915
672	Aminoacetic acid, *N*-phenyl,methyl ester or Methyl-*N*-phenyl amino acetate $C_6H_5NHCH_2CO_2CH_3$	165.19	nd (al)		48			al, eth	$B12^1$, 263
673	Amino acetic acid, *N*-phthaloyl or Phthalimide-*N*-acetic acid $C_{10}H_7NO_4$	205.17	nd or pr (w or al)		193			al, eth	$B21^4$, 5176
674	Aminoacetic acid, *N*-phthaloyl,ethyl ester or Ethyl-*N*-phthaloyl amino acetate o-$C_6H_4(CO)_2N$-$CH_2CO_2C_2H_5$	233.23	nd (w, al or eth)	300	112-3			al, eth, bz, chl	$B21^4$, 5177
675	Aminoacetic acid, *N*-succinyl,ethyl ester or Ethyl succinimide-*N*-acetate $[C_2H_4(CO)_2N]$-$CH_2CO_2C_2H_5$	185.18	nd (eth)	198^{32}	67			w, al	$B21^2$, 305
676	Amino acetonitrile H_2NCH_2CN	56.07		58^{15}				al	$B4^3$, 1120
677	Amino acetonitrite-*N,N*-diethyl $(C_2H_5)_2NCH_2CN$	112.17		170, 70^{24}			1.4260^{20}	w, al	$B4^3$, 1136
678	Amino acetonitrile-*N,N*-dimethyl $(CH_3)_2NCH_2CN$	84.12		137-8, 42^{22}		$0.8650^{20/0}$	1.4095^{20}	w, al	$B4^3$, 1126
679	Amino acetonitrile-*N*-ethyl $C_2H_5NHCH_2CN$	84.12		166-7, $81\text{-}3^{29}$				al, eth	$B4^2$, 787
680	Aminoacetonitrile-*N*-phenyl $C_6H_5NHCH_2CN$	132.16	pl (lig-eth)		48			al, bz	$B12^3$, 916
681	Aminoacetyl chloride-*N*-phthaloyl or *O-phthalinido-N*-acetyl chloride $[o$-$C_6H_4(CO)_2N]CH_2COCl$	223.62	nd (lig)		84-5			al, bz	$B21^4$, 5180
682	2-Aminobenzoic acid or Anthanilic acid 2-$H_2NC_6H_4CO_2H$	137.14	lf (al)	sub	146-7	1.412^{20}		w, al, eth	$B14^3$, 879
683	3-Aminobenzamide 3-$H_2NC_6H_4CONH_2$	136.15	ye mcl nd (+lw), nd (bz)		79—80 (hyd), 113—4 (anh)			w, al, eth	$B14^3$, 998
684	3-Aminobenzoic acid-*N*-acetyl 3-$CH_3CONHC_6H_4CO_2H$	179.18	nd (al)		248-50			al	$B14^3$, 1006
685	3-Amino benzoic acid-*N*-acetyl-6-ethoxy 3-(CH_3CONH)-6-$(C_2H_5O)C_6H_3CO_2H$	223.23	nd (w)		190			al	B14, 583
686	3-Aminobenzoic acid-6-chloro 3-H_2N-6-Cl-$C_6H_3CO_2H$	171.58			188	1.519^{15}		al	$B14^3$, 1017
687	3-Aminobenzoic acid-*N,N*-dimethyl 3-$(CH_3)_2NC_6H_4CO_2H$	165.19	nd (w)		151			w, al, eth	B14, 392
688	3-Aminobenzoic acid-*N*-ethyl 3-$(C_2H_5NH)C_6H_4CO_2H$	165.19	nd or pr (dil al)		154			al, eth, ace	B14, 393
689	3-Amino benzoic acid, ethyl ester or Ethyl 3-Amino benzoate 3-$H_2NC_6H_4CO_2C_2H_5$	165.19		294, $160\text{-}1^5$		$1.1248^{22/4}$	1.5600^{22}	al, eth	$B14^3$, 993
690	3-Amino benzoic acid, 2-hydroxy or 3-Amino salicylic acid 3-H_2N-2-$HOC_6H_3CO_2H$	153.14			235d				$B14^3$, 1434
691	3-Aminobenzoic acid, 4-hydroxy 3-H_2N-4-$HOC_6H_3CO_2H$	153.14	pr (w + l)		210 (anh)			w	$B14^2$, 360
692	3-Aminobenzoic acid, 4-hydroxy,methyl ester 3-H_2N-4-$HOC_6H_3CO_2CH_3$	167.16	(i) nd (bz or aa) (st) (ii) nd (chl) (unst)		(i)143, (ii)111			al, eth	$B14^3$, 1477
693	3-Aminobenzoic acid, methyl ester or Methyl 3-amino benzoate 3-$H_2NC_6H_4CO_2CH_3$	151.16		$152\text{-}3^{11}$	39	1.232^{20}		eth, al, bz, lig	$B14^3$, 993
694	3-Amino benzoic acid, *N*-methyl 3-$CH_3NHC_6H_4CO_2H$	151.16	pl (peth)		127			al, ace, bz, chl	$B14^1$, 559
695	3-Amino benzoic acid, *N*-methyl,methyl ester 3-$CH_3NHC_6H_4CO_2CH_3$	165.19	cr (al)		72			al, eth	B14, 392
696	3-Amino benzoic acid, 4-methyl 3-H_2N-4-CH_3-$C_6H_3CO_2H$	151.16	nd (al)		164-6			w, al	$B14^2$, 291
697	3-Aminobenzoic acid, 6-methyl 3-H_2N-6-$CH_3C_6H_3CO_2H$	151.16	pr (w)		196			w, al	$B14^2$, 290

No.	Name, Synonyms, and Formula	Mol. wt.	Color, crystalline form, specific rotation and λ_{max} (log ε)	b.p. °C	m.p. °C	Density	n_D	Solubility	Ref.
698	3-Aminobenzoic acid, 2-nitro $3\text{-}H_2N\text{-}2\text{-}NO_2C_6H_3CO_2H$	182.14	ye nd (w, dil al)		156-7			al, eth, ace	B14, 414
699	3-Aminobenzoic acid, 4-nitro $3\text{-}H_2N\text{-}4\text{-}NO_2C_6H_3CO_2H$	182.14	red pl or nd (al)		298d			al, eth, ace	B14, 415
700	3-Amino benzoic acid, 5-nitro $3\text{-}H_2N\text{-}5\text{-}NO_2C_6H_3CO_2H$	182.14	ye pr (w)		209-10			al	$B14^3$, 565
701	3-Aminobenzoic acid, 6-nitro $3\text{-}H_2N\text{-}6\text{-}NO_2C_6H_3CO_2H$	182.14	ye nd or pr (w)		235d			al, **ace**	$B14^3$, 1021
702	3-Aminobenzoic acid, 2,4,6-tribromo $3\text{-}H_2N\text{-}2,4,6\text{-}Br_3C_6HCO_2H$	373.83	nd (w)		171.5-3			**al**	$B14^3$, 1019
703	3-Aminobenzomitrile or m-Cyanoaniline $3\text{-}H_2NC_6H_4CN$	118.14	nd (dil al or CCl_4)	286-96	53-4			**al, eth, ace**	$B14^3$, 1001
704	3-Aminobenzonitrile, 2-methyl $3\text{-}NH_2\text{-}2\text{-}CH_3C_6H_3CN$	132.16	rh cr (al)		90			w	$B1^4$, 477
705	3-Amino benzonitrile, 4-methyl $3NH_2\text{-}4\text{-}CH_3C_6H_3N$	132.16	pr (al)		81-2			al, eth, **ace, bz**	B14, 487
706	3-Aminobenzo nitrile, 5-methyl $3\text{-}H_2N\text{-}5\text{-}CH_3C_6H_3CN$	132.16	nd (lig)		75			al	$B14^1$, 600
707	3-Aminobenzonitrile, 6-methyl $3\text{-}H_2N\text{-}6\text{-}CH_3C_6H_3CN$	132.16	nd (peth)	$100\text{-}10^{22}$	88			al	$B14^2$, 290
708	4-Aminobenzamide $4\text{-}H_2NC_6H_4CONH_2$	136.15	ye cr (+ ¼ w)		183			al, eth	$B14^3$, 1061
709	**4-Aminobenzamide, N(2-diethyl aminoehtyl) hydrochloride or Procainamide, hydrochloride** **$4\text{-}H_2NC_6H_4CONH[CH_2CH_2N(C_2H_5)]_2 \cdot HCl$**	313.85	cr		165-9			**w**, al	C54, 428
710	4-Amino benzamide, N phenyl $4\text{-}H_2NC_6H_4CONHC_6H_5$	212.25			135-6			al	$B14^3$, 1062
711	4-Aminobenzoic acid $4\text{-}H_2NC_6H_4CO_2H$	137.14	mcl pr (w)		188-9	$1.374^{20/4}$		al, eth, w	$B14^3$, 1023
712	4-Aminobenzoic acid, N-acetyl $4\text{-}(CH_3CONH)C_6H_4CO_2H$	179.18	nd (aa)		256.5			al	$B14^3$, 1112
713	4-Aminobenzoic acid, butyl ester or Butesin $4\text{-}H_2NC_6H_4CO_2C_4H_9$	193.25	(al or bz)	$173\text{-}4^8$	58			al, eth, bz, chl	$B14^3$, 1027
714	4-Aminobenzoic acid, *iso*-butyl ester $4\text{-}H_2NC_6H_4CO_2\text{-}i\text{-}C_4H_9$	193.25			64.5				$B14^3$, 1027
715	4-Aminobenzoic acid, 2-chloro or 4-Amino-2-chlorobenzoic acid $4\text{-}H_2N\text{-}2\text{-}ClC_6H_3CO_2H$	171.58			dec 213			al	$B14^3$, 1153
716	4-Aminobenzoic acid, 3,5-dichloro $4\text{-}H_2N\text{-}3,5\text{-}Cl_2C_6H_2CO_2H$	206.03	(al)		291			al, eth, bz	$B14^3$, 1156
717	4-Aminobenzoic acid, (2-diethylaminoethyl) ester or Novocaine $4\text{-}H_2NC_6H_4CO_2[CH_2CH_2N(C_2H_5)_2]$	236.31	nd (w + 2), pl (lig or eth)		51, 61 (anh)			al, eth, bz, chl	$B14^3$, 1037
718	**4-Aminobenzoic acid, 2-(diethylaminoethyl) ester, hydrochloride or Novocaine hydrochloride** **$4\text{-}H_2N\text{-}C_6H_4CO_2[CH_2CH_2N(C_2H_5)_2] \cdot HCl$**	272.77	nd (al), mcl or tcl pl (w)		156	0.707^{17}		**w**, al	$B14^3$, 1037
719	4-Aminobenzoic acid, 3,5-diiodo $4\text{-}H_2N\text{-}3,5\text{-}I_2C_6H_2CO_2H$	388.93	nd (aa-NH_3)		> 350				$B14^3$, 1161
720	4-Aminobenzoic acid, 3,5-diido, ethyl ester $4\text{-}H_2N\text{-}3,5\text{-}I_2C_6H_2CO_2C_2H_5$	416.98	nd (al)		148				B14, 439
721	4-Aminobenzoic acid, N,N-dimethyl or 4-(Dimethyl amino) benzoic acid $4\text{-}(CH_3)_2NC_6H_4CO_2H$	165.19	nd (al)		242.5—3.5			al	$B14^3$, 1082
722	4-Aminobenzoic acid, ethyl ester or Benzocaine $4\text{-}H_2NC_6H_4CO_2C_2H_5$	165.19	nd (w), rh (eth)	310	92			**al, eth**, chl	$B14^3$, 1025
723	4-Aminobenzoic acid, N-ethyl $4\text{-}(C_2H_5NH)C_6H_4CO_2H$	165.19	cr (bz)		177-8			al, eth, ace, bz	$B14^3$, 1086
724	4-Aminobenzoic acid, 2-hydroxy or 4-Aminosalicylic acid $4\text{-}H_2N\text{-}2\text{-}HOC_6H_3CO_2H$	153.14	nd pl (al-eth)		dec 150-1			w, al, eth, ace	$B14^3$, 1436
725	4-Aminobenzoic acid, 2-hydroxy, methyl ester $2\text{-}HO\text{-}4\text{-}H_2NC_6H_3CO_2CH_3$	167.16			120-1				$B14^3$, 1437
726	4-Aminobenzoic acid, methyl ester or Methyl-4-aminobenzoate $4\text{-}H_2NC_6H_4CO_2CH_3$	151.16	lf or nd (aq MeOH)		114			Chl	$B14^3$, 1025
727	4-Aminobenzoic acid, N-methyl $4\text{-}CH_3NHC_6H_4CO_2H$	151.16	nd (bz w or dil al)		168 (159)			**al, eth**, w, bz	$B14^3$, 1081
728	4-Aminobenzoic acid, N-methyl,methyl ester $4\text{-}CH_3NHC_6H_4CO_2CH_3$	165.19	pl (dil al or lig)		95.5			al, eth	$B14^3$, 1081

No.	Name, Synonyms, and Formula	Mol. wt.	Color, crystalline form, specific rotation and λ_{max} (log ε)	b.p. °C	m.p. °C	Density	n_D	Solubility	Ref.
729	4-Aminobenzoic acid, 2-methyl $4\text{-}H_2N\text{-}2\text{-}CH_3C_6H_3CO_2H$	151.16	nd (al)		dec 165			al	B14³,1200
730	**4-Aminobenzoic acid, 3-methyl** $4\text{-}H_2N\text{-}3\text{-}CH_3C_6H_3CO_2H$	151.16	nd (w)		170			w	B14², 290
731	4-Aminobenzoic acid, 2-nitro $4\text{-}H_2N\text{-}2\text{-}NO_2C_6H_3CO_2H$	182.14	rd nd (w), pr (dil aa)		239d			al, aa	B14³, 1161
732	4-Aminobenzoic acid, 3-nitro $4\text{-}H_2N\text{-}3\text{-}NO_2C_6H_3CO_2H$	182.14	rd-ye nd (al)		284d			ace, aa	B14³, 1161
733	4-Aminobenzoic acid, phenyl ester $4\text{-}H_2NC_6H_4CO_2C_6H_5$	213.24	nd (al)		173			al, eth	B14¹, 568
734	4-Aminobenzoic acid, propyl ester or Propaesin $4\text{-}H_2NC_6H_4CO_2C_3H_7$	179.22	pr		75			al, eth, bz, chl	B14³, 1026
735	4-Aminobenzonitrile or *p*-Cyanoaniline $4\text{-}H_2NC_6H_4CN$	118.14	pr or pl (w)		86			al, eth, ace, bz, aa	B14³, 1079
736	4-Aminobenzonitrile, 2-methyl $2\text{-}CH_3\text{-}4\text{-}H_2NC_6H_3CN$	132.16	rh cr (al)		90			al	B14¹, 598
737	4-Aminobenzonitrile, 3-methyl $4\text{-}H_2N\text{-}3\text{-}CH_3C_6H_3CN$	132.16	nd (w)		95				B14¹, 598
738	Ammelide or Cyanuromonoamide $C_3H_4N_4O_2$	128.09	mcl pr (w)		d				B26², 132
739	Ammeline or Cyanurodiamide $C_3H_5N_5O$	127.11	nd (aq Na_2CO_3)		d				B26², 132
740	Amphetamine (d) or Dexedrine, β-phenyl-*iso*-propyl amine $C_6H_5CH_2CH(NH_2)CH_3$	135.21	$[\alpha]^{12}{}_D$ + 37.6	203-4, 80^{10}		$0.949^{15/4}$	1.4704^{20}	al, eth	B12³, 2664
741	Amphetamine (dl) or Benzedrine $C_6H_5CH_2CH(NH_2)CH_3$	135.21		203, 97^{20}		$0.9306^{25/4}$	1.518^{26}	al, chl	B12³, 2664
742	**Amphetamine, sulfate (d) or Dexedrine sulfate** $[C_6H_5CH_2CH(NH_2)CH_3]_2\cdot H_2SO_4$	368.49	$[\alpha]^{20}{}_D$ + 22 (w, c=8)		> 300	$1.15^{25/4}$		w	B12³, 2665
743	**Amphetamine, sulfate (dl) or Benzedrine sulfate** $[C_6H_5CH_2CH(NH_2)CH_3]_2\cdot H_2SO_4$	368.49			280—1	$1.15^{25/4}$		w	B12³, 2665
744	Amygdalin or Mandelonitrile-β-gentiobioside $C_{20}H_{27}NO_{11}$	457.43			223—6			w	B17⁴, 3614′
745	**Amygdalin trihydrate** $C_{20}H_{27}NO_{11}\cdot 3H_2O$	511.48	rh (w + 3) $[\alpha]^{20}{}_D$ −40 (w, c=1)		200 (re-melts 125—30)			w, al	B31, 400
746	α-Amyrin or α-Amyrenol $C_{30}H_{50}O$	426.73	nd (al), $[\alpha]^{17}{}_D$ + 91.6 (bz, c = 1.3) + 83.5 (chl)	$243^{0.5}$	186			al, eth, bz, aa	B6³, 2889
747	β-Amyrin or β-Amyrenol $C_{30}H_{50}O$	426.73	nd (lig or al), $[\alpha]^{19}{}_D$ + 99.3 (bz, c = 1.3) + 88.4 (chl)	$260^{0.5}$	197				B6³, 2894
748	Anabasine (l) or l-2-(3-Pyridyl)piperidine $C_{10}H_{14}N_2$	162.23	$[\alpha]^{20}{}_D$ −83.1	276, 105^2	9	$1.0455^{20/20}$	1.5430^{20}	w, al, eth, bz	B23², 113
749	Anagyrine or Monolupine $C_{15}H_{20}N_2O$	244.34	pa ye glass, $[\alpha]^{25}{}_D$ −168 (al, c = 4.8)	260—70^{12}				al, w, eth, bz	B24², 84
750	**Analgen or 5-benxamide-8-ethoxyquinoline** $C_{18}O_2N_2H_{16}$	292.34	ye nd (al)		206				B22, 503
751	Anatabine *l* or 2-(3-Pyridyl)-1,2,3,6-tetrahydropyridine $C_{10}H_{12}N_2$	160.22	$[\alpha]^{17}{}_D$ − 177.8	145-6^{10}		$1.091^{19/4}$	1.5676^{20}	w, al, eth, bz	C31, 3055
752	Androstane or Etiocholane $C_{19}H_{32}$	260.46	lf (ace-MeOH), or (hl-ace)	75-$80^{0.01}$(sub)	50-2			al, eth, ace, chl, lig	B5⁴, 1211
753	Androsterone or 3-α-hydroxy-17-keto androstane $C_{19}H_{30}O_2$	290.45	lf or nd (al, ace), $[\alpha]^{20}{}_D$ + 94.6 (abs al, c=0.71)		185 (cor)			al,eth, ace, bz	B7³, 586

PHYSICAL CONSTANTS OF ORGANIC COMPOUNDS (Continued)

No.	Name, Synonyms, and Formula	Mol. wt.	Color, crystalline form, specific rotation and λ_{max} (log ε)	b.p. °C	m.p. °C	Density	n_D	Solubility	Ref.
754	Anemonin or Anemone camphor $C_{10}H_8O_4$	192.17	rh pl (chl), nd (al or bz)		158			chl	$B19^4$, 1986
755	Anethole or 43Propenyl anisole $4\text{-}CH_3OC_6H_4CH{=}CHCH_3$	148.20	lf (al)	234.5^{763}, 115^{12}	21.35	$0.9882^{20/4}$	1.5615^{20}	al, eth, ace, bz	$B6^3$, 2395
756	Angelic acid or *cis*-2-Methyl-2-butenoic acid $CH_3CH{=}C(CH_3)CO_2H$	100.12	mcl pr or nd	185, $88\text{-}9^{10}$	45-6	$0.9834^{49/4}$	1.4434^{47}	eth, al	$B2^4$, 1551
757	Anhalamine $C_{11}H_{15}NO_3$	209.25	nd (al)		187-8			al, ace	$B21^4$, 2521
758	Anhalonidine $C_{12}H_{17}NO_3$	223.27	oct (bz, eth)		160-1			w, al	$B21^4$, 2524
759	Anhalonine *d* $C_{12}H_{15}NO_3$	221.26	$[\alpha]^{25}{}_D$ + 56.7 (chl)	$140^{0.02}$	84.5-5			w, al, eth, peth	$B27^2$, 542
760	Anhalonine *dl* $C_{12}H_{15}NO_3$	221.26	rh nd (peth)		85-5.5			al, w, eth, peth	$B27^2$, 542
761	Anhalonine *l* $C_{12}H_{15}NO_3$	221.26	nd (peth), $[\alpha]^{25}{}_D$− 56.3 (chl, c=4)	$140^{0.02}$				w, al, eth, peth	$B27^2$, 542
762	Anhalonine-hydrochloride *l* $C_{12}H_{15}NO_3HCl$	257.72	rh pr, $[\alpha]^{25}{}_D$− 40.5 (al)		254-5d			al, w	$B27^2$, 542
763	Anhydroecgonine *dl* or Ecgonidine $C_9H_{13}NO_2$	167.21	cr (MeOH, MeOH-eth)		226-30d			w	$B22^4$, 284
764	Anhydroecgonine hydrochloride or Ecgonidine hydrochloride $C_9H_{13}NO_2\cdot HCl$	203.67	rh nd (al)		240 1			al, eth, w	$B22^4$, 284
765	Aniline or Phenylamine $C_6H_5NH_2$	93.13		184, 68.3^{10}	−6.3	$1.02173^{20/4}$	1.5863^{20}	al, eth, ace, bz lig	$B12^3$, 217
766	Aniline, benzene sulfonate $C_6H_5NH_2\text{-}C_6H_5SO_3H$	251.30	nd		240d			w, al	$B12^2$, 75
767	Aniline, *m*-benzyl or (3-aminophenyl)phenyl methane $3\text{-}H_2NC_6H_4CH_2C_6H_5$	183.25	cr (lig)		46			lig	$B12^3$, 3215
768	Aniline, *p*-benzyl or (4-aminophenyl)phenyl methane $4\text{-}(C_6H_5CH_2)C_6H_4NH_2$	183.25	mcl (lig)	300	34-5	1.038^{25}		al, eth, lig	$B12^3$, 3215
769	Aniline, *N*-benzylidene or Benzalaniline $C_6H_5N{=}CHC_6H_5$	181.24	pa ye nd (CS_2), pl (dil al)	310	54	$1.038^{55/4}$	1.600^{100}	al, eth	$B12^3$, 319
770	Aniline, 2-bromo $2\text{-}BrC_6H_4NH_2$	172.02		229, 110.5^{19}	32	$1.578^{20/4}$	1.6113^{20}	al, eth	$B12^3$, 1415
771	Aniline, 2-bromo-4,6-dichloro $2\text{-}Br\text{-}4,6\text{-}Cl_2C_6H_2NH_2$	240.91	nd (al)	273	83.5			al, bz, chl	$B12^2$, 335
772	Aniline, 2-bromo-3,5-dinitro $2\text{-}Br\text{-}3,5\text{-}(NO_2)_2C_6H_2NH_2$	262.02	ye lf (al)		181			al	$B12^3$, 1728
773	Aniline, 2-bromo-4,5-dinitro $2\text{-}Br\text{-}4,5\text{-}(NO_2)_2C_6H_2NH_2$	262.02	pa ye (al)		186			al	B12, 762
774	Amiline, 2-bromo-4,6-dinitro $2\text{-}Br\text{-}4,6\text{-}(NO_2)_2C_6H_2NH_2$	262.02	ye nd (aa or al)	sub	153-4			w, ace	$B12^3$, 1724
775	Aniline, 2-bromo-4-nitro $2\text{-}Br\text{-}4\text{-}(NO_2)C_6H_3NH_2$	217.02	ye nd (al)		104.5			al, aa	$B12^2$, 403
776	Aniline, 2-bromo-5-nitro $2\text{-}Br\text{-}5\text{-}(NO_2)C_6H_3NH_2$	217.02	pa ye nd (al)		141			w, al, eth	$B12^3$, 1674
777	Aniline, 3-bromo $3\text{-}BrC_6H_4NH_2$	172.02		251, 130^{12}	18.5	$1.5793^{20.4/4}$	$1.6260^{20.4}$	al, eth	$B12^3$, 1421
778	Aniline, 3-bromo-4-nitro $3\text{-}Br\text{-}4\text{-}(NO_2)_2C_6H_3NH_2$	217.02	ye nd (al)		175-6			al	$B12^3$, 1676
779	Aniline, 4-bromo $4\text{-}BrC_6H_4NH_2$	172.02	rh bip yrm nd (60 % al)	d	66.4	$1.4970^{100/4}$		al, eth	$B12^3$, 1429
780	Aniline, 4-bromo-2,3-dichloro $4\text{-}Br\text{-}2,3\text{-}Cl_2C_6H_2NH_2$	240.91	nd		77.5			al, eth, bz	B12, 653
781	Aniline, 4-bromo-2,5-dichloro $4\text{-}Br\text{-}2,5\text{-}Cl_2C_6H_2NH_2$	240.91	nd		91			al, eth	$B12^3$, 1471
782	Aniline, 4-bromo-3,5-dichloro $4\text{-}Br\text{-}3,5\text{-}Cl_2C_6H_2NH_2$	240.91	nd (w + al)		129			al, bz, chl	B12, 654

PHYSICAL CONSTANTS OF ORGANIC COMPOUNDS (Continued)

No.	Name, Synonyms, and Formula	Mol. wt.	Color, crystalline form, specific rotation and λ_{max} (log ε)	b.p. °C	m.p. °C	Density	n_D	Solubility	Ref.
783	Aniline, 4-bromo-2,5-dinitro $4\text{-Br-}2,5\text{-}(NO_2)_2C_6H_2NH_2$	262.02	ye cr (al)		186			al	B12, 761
784	Aniline, 4-bromo-2,6-dinitro $4\text{-Br-}2,6\text{-}(NO_2)_2C_6H_2NH_2$	262.02	og-red pl (abs.al)		163 (cor)			al	B12³, 1726
785	Aniline, 4-bromo-2-nitro $4\text{-Br-}2\text{-}(NO_2)C_6H_3NH_2$	217.02	og ye nd (w)	sub	111.5			al	B12³, 1670
786	Aniline, 4-bromo-3-nitro $4\text{-Br-}3\text{-}(NO_2)C_6H_3NH_2$	217.02	nd (al)		132			al, eth, chl, aa	B12³, 1674
787	Aniline, 5-bromo-2,4-dinitro $5\text{-Br-}2,4\text{-}(NO_2)_2C_6H_2NH_2$	262.02	pa ye nd (dil al)		178.4			al, eth, bz	B12³, 1723
788	Aniline, 5-bromo-2-nitro $5\text{-Br-}2\text{-}(NO_2)C_6H_3NH_2$	217.02	rd-ye nd (dil al)		151-2			al	B12³, 1671
789	Aniline, 6-bromo-2,3-dinitro $6\text{-Br-}2,3\text{-}(NO_2)_2C_6H_2NH_2$	262.02	dk rd cr (al)		158				B12, 760
790	Aniline, 6-bromo-2-nitro $6\text{-Br-}2\text{-}(NO_2)C_6H_3NH_2$	217.02	og or ye nd (dil al)		74.5	1.988		al	B12², 402
791	Aniline, *N*-butyl or Butyl phenyl amine $C_6H_5NHC_4H_9$	149.24		241.6, $118\text{-}20^{15}$	-14.4	$0.93226^{20/4}$	1.53412^{20}	al, eth	B12³, 267
792	Aniline, *N*-isobutyl or Isobutyl phenyl amine $C_6H_5NH\text{-i-}C_4H_9$	149.24		231-3, 109^{13}		$0.940^{15/4}$	1.5328^{20}	bz, eth	B12³, 269
793	Aniline, *N*-*sec*-butyl or *Sec*-butyl phenyl amine $C_6H_5NHCH(CH_3)C_2H_5$	149.24		225, $112\text{-}4^{22}$			1.5333^{20}		B12³, 269
794	Aniline, *N*-*tert*-butyl or *tert*-Butyl phenyl amine $C_6H_5NHC(CH_3)_3$	149.24		$214\text{-}6^{753}$, $93\text{-}8^{14}$		1.5270^{20}		al, ace, bz	B12³, 270
795	Aniline, 2-butyl or *o*-Butyl aniline $2\text{-}C_4H_9C_6H_4NH_2$	149.24	ye oil	$122\text{-}5^{12}$		$0.953^{20/4}$		al, ace, bz	B12³, 2714
796	Aniline, 2-Sec-butyl $2\text{-}CH_3CH_2CH(CH_3)C_6H_4NH_2$	149.24		$120\text{-}2^{16}$		0.9574^{20}		al, ace, bz	B12³, 2721
797	Aniline, 2-*tert*-butyl $2\text{-}(CH_3)_3CC_6H_4NH_2$	149.24		233-5		0.977^{15}	1.5453^{20}	al, eth, bz	B12³, 2726
798	Aniline, 4-butyl or *p*-Butyl aniline $4\text{-}C_4H_9C_6H_4NH_2$	149.24	pa ye	261, 133^{14}		$0.945^{20/4}$		al, aa, bz	B12³, 2715
799	Aniline, 4-*iso*-butyl $4\text{-}tert\text{-}C_4H_9C_6H_4NH_2$	149.24	pa ye	238^{762}, 112^{11}		$0.949^{15/4}$		al, ace bz	B12³, 2715
800	Aniline, 4-*sec*-butyl $4\text{-}Sec\text{-}C_4H_9C_6H_4NH_2$	149.24		238^{762}, 118^{15}		$0.949^{15/4}$	1.5360^{20}	eth, bz	B12², 635
801	Aniline, 4-*tert*-butyl $4\text{-}(CH_3)_3CC_6H_4NH_2$	149.24	ye rd (peth)	240^{740}	17	$0.9525^{15/4}$	1.5380^{20}	al, eth, bz	B12³, 2728
802	Aniline, 3-*tert*-butyl $3\text{-}(CH_3)CC_6H_4NH_2$	149.24		229^{708}				al, eth, bz	B12², 637
803	Aniline, 2-chloro-(α) $2\text{-}ClC_6H_4NH_2$	127.57		208.8	-14	$1.21253^{20/4}$	1.58951^{2}ls0	al, eth, ace	B12³, 1281
804	Aniline, 2-chloro (β) $2\text{-}ClC_6H_4NH_2$	127.57		208.8, 84.6^{10}	-1.9	$1.21266^{20/4}$	1.5889^{20}	al eth, ace, bz	B12³, 1281
805	Aniline, 2-chloro,hydrochloride $2\text{-}ClC_6H_4NH_2\text{-}HCl$	164.03	pl(w,aq.al)		235	1.505^{18}		w	B12², 315
806	Aniline, 2-chloro-4-nitro $2\text{-Cl-}4\text{-}(NO_2)C_6H_3NH_2$	172.57	ye nd (lig-CS_2)		108			al, eth	B12³, 1662
807	Aniline, 3-chloro $3\text{-}ClC_6H_4NH_2$	127.57		229.9, 101.2^{10}	-10.3	$1.21606^{20/4}$	1.59414^{20}	al, eth, ace, bz	B12³, 1303
808	Aniline, 3-chloro-2,6-dinitro $3\text{-Cl-}2,6\text{-}(NO_2)_2C_6H_2NH_2$	217.57	og-ye nd (al)		112.			al	B12³, 1719
809	Aniline, 3-chloro, hydrochloride $3\text{-}ClC_6H_4NH_2\cdot HCl$	164.03	pl		222			w, al	B12², 320
810	Aniline, 3-chloro-4-nitro $3\text{-Cl-}4\text{-}(NO_2)C_6H_3NH_2$	172.57	ye lf (bz)		156-7			al, eth	B12³, 1665
811	Aniline, 4-chloro $4\text{-}ClC_6H_4NH_2$	127.57	rh pr	232	72.5	$1.429^{19/4}$	1.5546^{87}	al, eth, w	B12³, 1325
812	Aniline, 4-chloro-2,6-dinitro $4\text{-Cl-}2,6\text{-}(NO_2)_2C_6H_2NH_2$	217.57	og-ye nd (al)		147 (cor)			al	B12³, 1720
813	Aniline, 4-chloro-2-methoxy. or 4-Chloro-*o*-anisidine $4\text{-Cl-}2\text{-}(CH_3O)C_6H_3NH_2$	157.60	nd or pr (dil.al)	260	52			al, eth, bz	B13², 184
814	Aniline, 4-chloro-2-nitro $4\text{-Cl-}2\text{-}(NO_2)C_6H_3NH_2$	172.57	dk og-ye pr (dil al)		116-7			al, eth, aa	B12³, 1649
815	Aniline, 4-chloro-3-nitro $4\text{-Cl-}3\text{-}NO_2C_6H_3NH_2$	172.57	ye nd or pr (w) nd (peth)		103			al, eth, ace	B12³, 1660

PHYSICAL CONSTANTS OF ORGANIC COMPOUNDS (Continued)

No.	Name, Synonyms, and Formula	Mol. wt.	Color, crystalline form, specific rotation and λ_{max} (log ε)	b.p. °C	m.p. °C	Density	n_D	Solubility	Ref.
816	Aniline, 5-chloro-2-ethoxy or 5-Chloro-*o*-phenetidine 5-Cl-2-$(C_2H_5O)C_6H_3NH_2$	171.63	nd (dil.al)		42			al, bz, aa	B13, 383
817	Aniline, 5-chloro-2-methoxy or 5-Chloro-*o*-anisidine 5-Cl-2-$(CH_3O)C_6H_3NH_2$	157.60	nd (dil al)		84			al	B13^{2}, 183
818	Aniline, 5-chloro-2-nitro 5-Cl-2-$(NO_2)C_6H_3NH_2$	172.57	gold ye nd (CS_2) ye lf (al, bz)	sub	126.5			al, eth	B12^{3}, 1655
819	Aniline, 5-chloro-3-nitro 5-Cl-3-$(NO_2)C_6H_3NH_2$	172.57	og.ye nd (al)		133-4			al, eth	B12, 732
820	Aniline, 6-chloro-2-nitro 6-Cl-2-$(NO_2)C_6H_3NH_2$	172.57	ye nd (dil al)		76			al	B12^{3}, 1658
821	Aniline, 6-chloro-3-nitro 6-Cl-3-$(NO_2)C_6H_3NH_2$	172.57	ye nd (lig)		121			al, eth, ace	B12^{3}, 1661
822	Aniline, *N*-cyclohexyl or Phenylcyclohexylamine *N*-cyclohexyl aniline $C_6H_5NHC_6H_{11}$	175.27	mcl pr	279^{764}	16	$1.0155^{20/4}$	1.5610^{20}	al, eth, bz	B12^{2}, 98
823	Aniline, 4-cyclohexyl 4-$C_6H_{11}C_6H_4NH_2$	175.27	pl (lig)		55			al,bz	B12^{3}, 2819
824	Aniline, 2,3-dibromo 2,3-$Br_2C_6H_3NH_2$	250.92	pl (dil al)		43			al, eth, aa	B12, 655
825	Aniline, 2,4-dibromo 2,4-$Br_2C_6H_3NH_2$	250.92	rh bi pym (chl)nd or lf (dil al)	156^{24}	79.5-80.5	2.260^{20}		al, eth	B12^{3}, 1471
826	**Aniline, 2,4-dibromo-6-nitro** **$Br_2NO_2C_6H_2NH_2$**	295.92	ye cr		128				B12^{2}, 403
827	Aniline, 2,5-dibromo 2,5-$Br_2C_6H_3NH_2$	250.92	pr (al)		53-5			al, eth	B12^{3}, 1474
828	Aniline, 2,6-dibromo 2,6-$Br_2C_6H_3NH_2$	250.92	nd (al)	262-4	87-8			al, eth, bz, chl	B12^{3}, 1474
829	Aniline, 2,6-dibromo-4-nitro 2,6-Br_2-4-$(NO_2)C_6H_2NH_2$	295.92	ye nd (al or aa)	207				aa	B12^{3}, 1678
830	Aniline, 3,4-dibromo 3,4-$Br_2C_6H_3NH_2$	250.92	lf (dil al)	100 sub	81			al	B12^{1}, 329
831	Aniline, 3,5-dibromo 3,5-$Br_2C_6H_3NH_2$	250.92	nd (dil al)		57			al, eth, bz	B12^{3}, 1475
832	Aniline, 3,5-dibromo-4-methoxy 3,5-Br_2-4-$CH_3OC_6H_2NH_2$	280.95	pl (lig)		66			al, eth, ace, bz	B13^{3}, 1195
833	Aniline, 4,6-dibromo-2-ethoxy 4,6-Br_2-2$(C_2H_5O)C_6H_2NH_2$	294.97	pr (dil al)		52			al, lig	B13^{3}, 867
834	Aniline, *N,N*-dibutyl or Dibutylphenylamine $C_6H_5N(C_4H_9)_2$	205.34		274.8, 138.8^{10}	−32.2	$0.9037^{20/4}$	1.5186^{20}	al, eth, ace, bz	B12^{3}, 269
835	Aniline, 2,3-dichloro 2,3-$Cl_2C_6H_3NH_2$	162.02	nd (lig)	252	24			al, eth, ace	B12, 621
836	Aniline, 2,4-dichloro 2,4-$Cl_2C_6H_3NH_2$	162.02	pr (ace) nd (dil al) (lig)	245	63-4			al, eth	B12^{3}, 1389
837	Aniline, 2,5-dichloro 2,5-$Cl_2C_6H_3NH_2$	162.02	nd (lig)	251	50			al, eth, bz	B12^{3}, 1397
838	Aniline, 2,6-dichloro 2,6-$Cl_2C_6H_3NH_2$	162.02			39			al, eth	B12^{2}, 337
839	Aniline, 2,6-dichloro-4-ethoxy or 4-Amino-3,5-dichloro-phenetole 2,6-Cl_2-4-$(C_2H_5O)C_6H_2NH_2$	206.07	nd (dil al)	275	46			al, eth, bz	B13^{2}, 276
840	Aniline, 2,6-dichloro-4-nitro or Dichloran 2,6-Cl_2-4-$(NO_2)C_6H_2NH_2$	207.02	ye nd (al, aa)		191			al	B12^{3}, 1669
841	Aniline, 3,4-dichloro 3,4-$Cl_2C_6H_3NH_2$	162.02	nd (lig)	272, 145^{15}	72			al, eth	B12^{3}, 1402
842	Aniline, 3,5-dichloro 3,5-$Cl_2C_6H_3NH_2$	162.02	nd (lig, dil al)	260^{741}	51-3			al, eth, bz, chl	B12^{2}, 337
843	Aniline, 3,5-dichloro-4-ethoxy or 4-Amino-2,6-dichloro-phenetole 3,5-Cl_2-4$(C_2H_5O)C_6H_2NH_2$	206.07	nd (peth)		105-7			al, eth, bz	B13^{2}, 275
844	Aniline, *N,N*-diethyl or Diethylphenylamine $C_6H_5N(C_2H_5)_2$	149.24	ye oil	216.27, 92^{10}	−38.8	$0.93507^{20/4}$	1.5409^{20}	al, eth, ace	B12^{3}, 260
845	Aniline, 2-diethylamino 2-$[(C_2H_5)_2N]C_6H_4NH_2$	164.25	oil	312^{244}, 127^{25}				al, ace, bz	B13^{3}, 34

No.	Name, Synonyms, and Formula	Mol. wt.	Color, crystalline form, specific rotation and λ_{max} (log ε)	b.p. °C	m.p. °C	Density	n_D	Solubility	Ref.
846	Aniline, 3-diethylamino $3\text{-}[(C_2H_5)_2N]C_6H_4NH_2$	164.25	...	276-8, 117^{4}	...	...	...	al	$B13^{2}$, 26
847	Aniline, 4-diethylamino $4\text{-}[(C_2H_5)_2N]C_6H_4NH_2$	164.25	ye oil	260-2, $139\text{-}40^{10}$	...	...	...	bz	$B13^{3}$, 113
848	Aniline, *N,N*-diethyl-3-bromo $3\text{-}BrC_6H_4N(C_2H_5)_2$	228.13	...	142^{10}	...	...	...	al, ace	$B12^{1}$, 315
849	Aniline, *N,N*-diethyl-4-bromo $4\text{-}BrC_6H_4N(C_2H_5)_2$	228.13	nd or pr	270	33	...	...	al, eth	$B12^{2}$, 347
850	**Aniline, *N,N*-diethyl-4-chloro** $4\text{-}ClC_6H_4N(C_2H_5)_2$	183.68	nd (al)	251-3, $95\text{-}6^{1.5}$	45.5-6.5	...	...	al	$B12^{3}$, 1329
851	Aniline, *N,N*-diethyl-2-ethoxy or *N,N*-diethyl-*o*-phentidine $2\text{-}(C_2H_5O)C_6H_4N(C_2H_5)_2$	193.29	...	231-3	...	...	...	**al, eth, bz**	B13, 365
852	Aniline, *N,N*-diethyl-3-ethoxy or *N,N*-diethyl-*m*-phenetidine $3\text{-}(C_2H_5O)C_6H_4N(C_2H_5)_2$	193.29	...	268-70, 145^{14}	...	...	1.5325^{25}	al, bz, aa	$B13^{3}$, 942
853	Aniline, *N,N*-diethyl-3-nitro $3\text{-}NO_2C_6H_4N(C_2H_5)_2$	194.23	ye	288-90	...	...	...	...	$B12^{3}$, 1545
854	Aniline, *N,N*-diethyl-4-nitro $4\text{-}NO_2C_6H_4N(C_2H_5)_2$	194.23	ye nd (lig) pl (al)	...	77-8	1.225	...	al	$B12^{3}$, 1585
855	Aniline, *N,N*-diethyl-4-nitroso $4\text{-}ONC_6H_4N(C_2H_5)_2$	178.23	gr mcl pr (eth) gr lf (ace)	...	87-8	$1.24^{15/4}$	...	al, eth, ace	$B12^{3}$, 1585
856	Aniline, 3,4-dihydroxy $3,4\text{-}(HO)_2C_6H_3NH_2$	125.13	br-vt nd (al-bz)	...	124-5d	...	...	w, al, eth	$B13^{2}$, 464
857	Aniline, 3,5-dihydroxy or Phloramin $3,5\text{-}(HO)_2C_6H_3NH_2$	125.13	nd	...	146-52	...	...	al	B13, 787
858	Aniline, 2,4-diiodo $2,4\text{-}I_2C_6H_3NH_2$	344.92	br nd or rh cr (al)	...	95-6	2.748	...	al, eth, ace, bz, chl	$B12^{3}$, 1508
859	Aniline, 2,4-diiodo-5-nitro $2,4\text{-}I_2\text{-}5\text{-}(NO_2)C_6H_2NH_2$	389.92	pa ye mcl pl	...	125	...	...	...	B12, 747
860	Aniline, 2,5-diiodo $2,5\text{-}I_2C_6H_3NH_2$	344.92	nd (al)	...	88-9	...	...	al, eth, ace	B12, 675
861	Aniline, 2,6-diiodo $2,6\text{-}I_2C_6H_3NH_2$	344.92	nd (al)	...	122	...	...	al, eth, ace, bz	B12, 675
862	Aniline, 2,6-diiodo-3-nitro $2,6\text{-}I_2\text{-}3\text{-}(NO_2)C_6H_2NH_2$	389.92	ye nd (dil al)	...	145.5	...	...	al	B12, 747
863	Aniline, 2,6-diiodo-4-nitro $2,6\text{-}I_2\text{-}4\text{-}(NO_2)C_6H_2NH_2$	389.92	pa ye lf or nd (bz)	...	245	...	...	bz	$B12^{3}$, 1680
864	Aniline, 3,4-diiodo $3,4\text{-}I_2C_6H_3NH_2$	344.92	pa ye lf or pr (bz-peth)	...	74.5	...	...	al, eth, bz	B12, 675
865	Aniline, 3,5-diiodo $3,5\text{-}I_2C_6H_3NH_2$	344.92	nd (al)	...	110	...	...	al, eth, chl	$B12^{1}$, 337
866	Aniline, 4,6-diiodo-2-nitro $4,6\text{-}I_2\text{-}2\text{-}(NO_2)C_6H_2NH_2$	389.92	ye nd (ace)	...	154	...	...	eth, ace, bz, chl	$B12^{1}$, 361
867	Aniline, 4,6-diiodo-3-nitro $4,6\text{-}I_2\text{-}3\text{-}(NO_2)C_6H_2NH_2$	389.92	pa ye nd (eth-al)	...	149	...	...	eth	B12, 747
868	Aniline, 2,3-dimethoxy or 3-Amino veratrole $2,3\text{-}(CH_3O)_2C_6H_3NH_2$	153.18	...	137^{15}	...	...	...	w	$B13^{3}$, 2104
869	Aniline, 2,4-dimethoxy $2,4\text{-}(CH_3O)_2C_6H_3NH_2$	153.18	pl (lig)	...	33.5	...	...	al, eth, lig	$B13^{3}$, 2131
870	Aniline, 2,6-dimethoxy $2,6\text{-}(CH_3O)_2C_6H_3NH_2$	153.18	pl (al) lf (peth)	146^{23}	75	...	...	al, eth, bz, lig	$B13^{3}$, 2128
871	Aniline, 3,4-dimethoxy or 4-Amino veratroile $3,4\text{-}(CH_3O)_2C_6H_3NH_2$	153.18	lf (eth)	...	87-8	...	...	eth	$B13^{3}$, 2105
872	Aniline, *N,N*-dimethyl $C_6H_5N(CH_3)_2$	121.18	pa ye	194, 77^{13}	2.45	$0.9557^{20/4}$	1.5582^{20}	al, eth, ace, bz, chl	$B12^{3}$, 245
873	Aniline, 2-dimethylamino $2\text{-}[(CH_3)_2N]C_6H_4NH_2$	136.20	oil	218^{751}, 117^{25}	...	0.995^{22}	...	al, eth, ace, bz	$B13^{3}$, 33
874	Aniline, 3-dimethylamino $3\text{-}[(CH_3)_2N]C_6H_4NH_2$	136.20	...	$268\text{-}70^{740}$, 138^{10}	<-20	0.995^{25}	...	al, eth	$B13^{3}$, 68
875	Aniline, 4-dimethylamino $4\text{-}[(CH_3)_2N]C_6H_4NH_2$	136.20	nd (bz)	263, 158^{11}	53	$1.036^{20/4}$	...	al, eth, bz, w, chl	$B13^{3}$, 109
876	Aniline, *N,N*-dimethyl-2-bromo $2\text{-}BrC_6H_4N(CH_3)_2$	200.08	...	$107\text{—}8^{14}$	...	$1.3880^{25/25}$	1.5768^{25}	al	$B12^{3}$, 1416

PHYSICAL CONSTANTS OF ORGANIC COMPOUNDS (Continued)

No.	Name, Synonyms, and Formula	Mol. wt.	Color, crystalline form, specific rotation and λ_{max} (log ε)	b.p. °C	m.p. °C	Density	n_D	Solubility	Ref.
877	Aniline, *N,N*-dimethyl-3-bromo $3BrC_6H_4N(CH_3)_2$	200.08		$239, 126^{14}$	11			al, aa	$B12^3$, 1422
878	Aniline, *N,N*-dimethyl-4-bromo $4\text{-}BrC_6H_4N(CH_3)_2$	200.08	lf (al)	264	55			al, eth	$B12^3$, 1432
879	Aniline, *N,N*-dimethyl-2-chloro $2\text{-}ClC_6H_4N(CH_3)_2$	155.63		$203, 98\text{-}9^{18}$		$1.1067^{20/4}$	1.5578^{20}	al, bz	$B12^3$, 1283
880	Aniline, *N,N*-dimethyl-4-chloro $4\text{-}ClC_6H_4N(CH_3)_2$	155.63	nd (al)	231	35.5			al	$B12^3$, 1329
881	Aniline, *N,N*-dimethjyl-4-chloro-3-nitro $4\text{-}Cl\text{-}3(NO_2)C_6H_3N(CH_3)_2$	200.62	ye nd (dil al)		81.5-2.5			al, lig	$B12^3$, 1660
882	Aniline, *N,N*-dimethyl, hydrochloride $C_6H_5N(CH_3)_2\text{-}HCl$	157.66	hyg pl (w) (bz)		85-95	$1.1156^{19/4}$		w, al, chl	$B12^3$, 251
883	Aniline, *N,N*-dimethyl-2-nitro $2\text{-}(NO_2)C_6H_4N(CH_3)_2$	166.18	ye-og	146^{20}	fp-20	$1.1794^{20/4}$	1.6102^{20}	w, al, eth, chl	$B12^3$, 1516
884	Aniline, *N,N*-dimethyl-3-nitro $3\text{-}NO_2C_6H_4N(CH_3)_2$	166.18	og-ye or red nd pr (eth or eth-al)	280-5	60-1			al, eth	$B12^3$, 1544
885	Aniline, *N,N*-dimethyl-4-nitro $4\text{-}NO_2C_6H_4N(CH_3)_2$	166.18	ye nd (al)		164.5			al, eth, aa	$B12^3$, 1583
886	Aniline, *N,N*-dimethyl-4-nitroso $4\text{-}ON\text{-}C_6H_4N(CH_3)_2$	150.18	gr-pl (eth)		92.5-3.5	1.145^{20}		al, eth	$B12^3$, 1509
887	Aniline, 2,3-dimethyl or 2,3-xylidine $2,3\text{-}(CH_3)_2C_6H_3NH_2$	121.18		$221\text{-}2, 106^{15}$	<−15	0.9931^{20}	1.5684^{20}	al, eth	$B12^3$, 2438
888	Aniline, 2,3-dimethyl-4-nitro or 4-Nitro-2,3-xylidine $2,3\text{-}(CH_3)_2\text{-}4\text{-}(NO_2)C_6H_2NH_2$	166.18	ye pr (al)		114			al	$B12^3$, 2442
889	Aniline, 2,3-dimethyl-5-nitro or 5-Nitro-2,3-xylidine $2,3\text{-}(CH_3)_2\text{-}5NO_2C_6H_2NH_2$	166.18	pa ye nd (al)		111-2			al	$B12^1$, 479
890	Aniline, 2,3-dimethyl-6-nitro or 6-Nitro-2,3-xylidine $2,3\text{-}(CH_3)_2\text{-}6\text{-}NO_2C_6H_2NH_2$	168.18	red pl (al)		118-9			al	$B12^3$, 2442
891	Aniline, 2,4-dimethyl or 2,4-xylidine $2,4\text{-}(CH_3)_2C_6H_3NH_2$	121.18		$214, 91^{10}$	−14.3	$0.9723^{20/4}$	1.5569^{20}	al, eth, bz	$B12^3$, 2469
892	Aniline, 2,4-dimethyl-3-nitro or 3-Nitro-2,4-xylidine $2,4\text{-}(CH_3)_2\text{-}3\text{-}NO_2C_6H_2NH_2$	166.18	ye nd		81-2			al, lig	$B12^2$, 612
893	Aniline, 2,4-dimethyl-5-nitro or 5-Nitro-2,4-xylidine $2,4\text{-}(CH_3)_2\text{-}5\text{-}NO_2C_6H_2NH_2$	166.18	og ye nd (al)		123			al	$B12^2$, 612
894	Aniline, 2,4-dimethyl-6-nitro or 6-Nitro-2,4-xylidine $2,4\text{-}(CH_3)_2\text{-}6\text{-}NO_2C_6H_2NH_2$	166.18	og red nd or pl (lig)		76			al	$B12^2$, 612
895	Aniline, 2,5-dimethyl or 2,5-Xylidine $2,5(CH_3)_2C_6H_3NH_2$	121.18	ye lf (lig)	$214, 97\text{-}100^{10}$	15.5	$0.9790^{21/4}$	1.5591^{21}	eth	$B12^3$, 2503
896	Aniline, 2,6-dimethyl or 2,6-Xylidine $2,6(CH_3)_2C_6H_3NH_2$	121.18		214^{739}	11.2	0.9842^{20}	1.5610^{20}	al, eth	$B12^3$, 2462
897	Aniline, 2,6-dimethyl-3-nitro or 3-Nitro-2,6-xylidine $2,6\text{-}(CH_3)_2\text{-}3\text{-}NO_2C_6H_2NH_2$	166.18	ye nd (dil al)		81-2			al	$B12^2$, 605
898	Aniline, 2,6-dimethyl-4-nitro or 4-Nitro-2,6-xylidine $2,6\text{-}(CH_3)_2\text{-}4\text{-}NO_2C_6H_2NH_2$	166.18	og-red nd or pl (lig)		76			al	$B12^3$, 2469
899	Aniline, 3,4-dimethyl or 3,4-Xylidine $3,4\text{-}(CH_3)_2C_6H_3NH_2$	121.18	pl or pr (lig)	228	51	1.076^{18}		eth, lig	$B12^3$, 2443
900	Aniline, 3,4-dimethyl-2-nitro or 2-Nitro-3,4-xylidine $3,5\text{-}(CH_3)_2\text{-}2\text{-}NO_2C_6H_2NH_2$	166.18	red pr (al)		65-6			al	$B12^3$, 2454
901	Aniline, 3,4-dimethyl-5-nitro or 5-Nitro-3,4-xylidine $3,4\text{-}(CH_3)_2\text{-}5\text{-}NO_2C_6H_2NH_2$	166.18	og lf (al)		74-5			al	B12, 1106
902	Aniline, 3,5-dimethyl or 3,5-Xylidine $3,5\text{-}(CH_3)_2C_6H_3NH_2$	121.18		$220\text{-}1, 99\text{-}100^{20}$	9.8	$0.9706^{20/4}$	1.5581^{20}	eth	$B12^3$, 2495
903	Aniline, 3,5-dimethyl-2-nitro or 2-Nitro-3,5-xylidine $3,5\text{-}(CH_3)_2\text{-}2\text{-}NO_2C_6H_2NH_2$	166.18	ye nd (lig)		56				$B12^2$, 613
904	Aniline, 3,5-dimethyl-4-nitro or 4-Nitro-3,5-xylidine $3,5\text{-}(CH_3)_2\text{-}4\text{-}NO_2C_6H_2NH_2$	166.18	og pr (bz), lf (lig)		133			al, bz	$B12^3$, 2500
905	Aniline, 4,5-dimethyl-2-nitro or 2-Nitro-4,5-xylidine $4,5\text{-}(CH_3)_2\text{-}2\text{-}NO_2C_6H_2NH_2$	166.18	red-br pr (al)		140			ace, bz	$B12^3$, 2455
906	Aniline, 2,3-dinitro $2,3\text{-}(NO_2)_2C_6H_3NH_2$	183.12			178	1.646^{50}		al, eth	$B12^3$, 1680
907	Aniline, 2,4-dinitro $2,4\text{-}(NO_2)_2C_6H_3NH_2$	183.12	ye nd (dil ace), gr-ye ta (al)		180 (188)	1.615^{14}			$B12^3$, 1681
908	Aniline, 2,6-dinitro $2,6\text{-}(NO_2)_2C_6H_3NH_2$	183.12	gold lf (50% aa), ye nd (al)		141-2			eth, bz	$B12^3$, 1704

No.	Name, Synonyms, and Formula	Mol. wt.	Color, crystalline form, specific rotation and λ_{max} (log ε)	b.p. °C	m.p. °C	Density	n_D	Solubility	Ref.
909	Aniline, 3,5-dinitro $3,5\text{-}(NO_2)_2C_6H_3NH_2$	183.12			16.3	1.601^{50}		al, eth	$B12^3$, 1705
910	Aniline, *N,N*-dipropyl $C_6H_5N(C_3H_7)_2$	177.29	lf ye	$243, 127^{10}$		0.9104^{20}	1.5271^{20}	al, eth, ace, bz	$B12^3$, 266
911	Aniline, 2-ethoxy or *o*-Phenetidine $2\text{-}(C_2H_5O)C_6H_4NH_2$	137.18		$232.5, 127\text{-}8^{14}$	<-21		1.5560^{20}	al, eth	$B13^3$, 756
912	Aniline, 3-ethoxy or *m*-Phenetidine $3\text{-}(C_2H_5O)C_6H_4NH_2$	137.18		$248, 127\text{-}8^{11}$				al, eth	$B13^3$, 932
913	Aniline, 4-ethoxy or *p*-Phenetidine $4\text{-}(C_2H_5O)C_6H_4NH_2$	137.18		$254, 125^{11}$	2.4	$1.06521^{16/4}$	1.5528^{20}	al, eth	$B13^3$, 996
914	Aniline, *N*-Ethyl or Ethylphenylamine $C_6H_5NHC_2H_5$	121.18		$204.7, 97.5^{18}$	-63.5	$0.9625^{20/4}$	1.5559^{20}	al, eth, ace, bz	$B12^3$, 255
915	Aniline, *N*-ethyl-2-chloro $2\text{-}ClC_6H_4NHC_2H_5$	155.63		219^{726}		$1.104^{20/4}$			$B12^3$, 1284
916	Aniline, *N*-ethyl, hydrochloride $C_6H_5NHC_2H_5\text{-}HCl$	157.64	nd		178.5	1.0085^{182}		w, al, chl	$B12^3$, 257
917	Aniline, *N*-ethyl-*N*-methyl $C_6H_5N(CH_3)(C_2H_5)$	135.21		$203\text{-}5, 93\text{-}5^{13}$		$0.9193^{55/4}$		al, eth	$B12^3$, 258
918	Aniline, *N*-ethyl-*N-nitroso* or *N*-ethylphenylnitrosamine $C_6H_5N(NO)C_2H_5$	150.18	yesh	$119\text{-}20^{15}$		1.0874^{20}		aa	$B12^3$, 1119
919	Aniline, *o*-ethyl $2\text{-}C_2H_5C_6H_4NH_2$	121.18		209-10	-43	$0.983^{22/4}$	1.5584^{22}	al, eth	$B12^3$, 2374
920	Aniline, *m*-ethyl $3\text{-}C_2H_5C_6H_4NH_2$	121.18		$214\text{-}5^{764}, 93\text{-}5^{6}$	-64	0.9896		al, eth	$B12^3$, 2379
921	Aniline, *p*-ethyl $4\text{-}C_2H_5C_6H_4NH_2$	121.18		$217\text{-}8, 92.3^{10}$	-4.87	$0.9679^{20/4}$	1.5554^{20}	al, eth	$B12^3$, 2380
922	Aniline, *o*-fluoro $2\text{-}FC_6H_4NH_2$	111.12	pa ye	$174\text{-}6^{757}$	-28.5	1.1513^{21}	1.5421^{20}	al, eth	$B12^3$, 1273
923	Aniline, *m*-fluoro $3\text{-}FC_6H_4NH_2$	111.12		$187\text{-}9, 82\text{-}3^{18}$		1.1561^{19}	1.5436^{20}	al, eth	$B12^3$, 1274
924	Aniline, *p*-fluoro $4\text{-}FC_6H_4NH_2$	111.12	pa ye	$180.5\text{-}2.5^{757}$, 85^{19}	-0.8	$1.1725^{20/4}$	$1.5195^{20/4}$	al, eth	$B12^3$, 1276
925	Anilinehydrochloride $C_6H_5NH_2\text{-}HCl$	129.59	lf or nd	245	198	1.2215^{4}		w, al	$B12^3$, 232
926	Aniline, *o*-iodo $2\text{-}IC_6H_4NH_2$	219.02	nd (dil al)		60-1			al, eth, ace	$B12^3$, 1487
927	Aniline, *m*-iodo $3\text{-}IC_6H_4NH_2$	219.02	lf or nd	280	33		1.6811^{20}	al, chl	$B12^3$, 1489
928	Aniline, *p*-iodo $4\text{-}IC_6H_4NH_2$	219.02	nd (w)		67-8			al, eth	$B12^3$, 1493
929	Aniline, *o*-methoxy or *o*-Anisidine $2\text{-}(CH_3O)C_6H_4NH_2$	123.15		$224, 90^{4}$	6.2	$1.0923^{20/4}$	1.5715^{10}	al, eth, ace, bz	$B13^3$, 754
930	Aniline, 2-methoxy-3-nitro $2\text{-}(CH_3O)\text{-}3\text{-}NO_2\text{-}C_6H_3NH_2$	168.15	pa ye nd (lig)		67			al, lig	$B13^2$, 195
931	Aniline, *m*-methoxy or *m*-Anisidine $3\text{-}(CH_3O)C_6H_4NH_2$	123.15		251 (cor)	-1	$1.096^{20/4}$	1.5794^{20}	al, eth, ace, bz	$B13^3$, 932
932	Aniline, *p*-methoxy or *p*-Anisidine $4\text{-}(CH_3O)C_6H_4NH_2$	123.15	ta (w), rh pl	$243, 115^{13}$	57.2	$1.071^{57/4}$	1.5559^{60}	w, al, eth, ace, bz	$B13^3$, 994
933	Aniline, *N*-methyl or Methylphenyl amine $C_6H_5NHCH_3$	107.16		$196.25, 86^{15}$	-57	$0.9891^{20/4}$	1.5684^{20}	al, eth	$B12^3$, 240
934	Aniline, *N*-methyl, hydrochloride or Methylphenyl amine hydrochloride $C_6H_5NHCH_3\text{-}HCl$	143.62	nd (chl-eth)	122-3		$1.0660^{131/4}$		w, al	$B12^3$, 244
935	Aniline, *N*-(3-methylbutyl) or *N*-isopentylaniline $C_6H_5NH\text{-}i\text{-}C_5H_{11}$	163.26		$254\text{-}5, 126\text{-}7^{14}$		$0.8912^{55/4}$	1.5305^{20}	al, eth	$B12^3$, 272
936	Aniline, *N*-methyl-2-chloro $2\text{-}ClC_6H_4NHCH_3$	141.60		$218, 106^{20}$		1.1735^{12}	1.5780^{25}	al, ace, bz	$B12^3$, 1283
937	Aniline, *N*-methyl-4-chloro $4\text{-}ClC_6H_4NHCH_3$	141.60		$239^{764}, 120^{20}$		1.614^{20}	1.5835^{20}	al, bz	$B12^3$, 1329
938	Aniline, *N*-methyl,-2-nitro or Methyl-2-nitrophenylamine $2\text{-}O_2NC_6H_4NHCH_3$	152.15	red or og nd (peth)	d	38			al, eth, ace, bz, lig	$B12^3$, 1516
939	Aniline, *N*-methyl,3-nitro $3\text{-}NO_2C_6H_4NHCH_3$	152.15	red-ye nd or pr (al), cr (lig)		68			al, eth, bz	$B12^3$, 1544
940	Aniline, *N*-methyl-4-nitro $4\text{-}NO_2C_6H_4NHCH_3$	152.15	br-ye pr (al), cr (eth)	d	152	$1.201^{155/4}$		al, bz	$B12^3$, 1584

No.	Name, Synonyms, and Formula	Mol. wt.	Color, crystalline form, specific rotation and λ_{max} (log ε)	b.p. °C	m.p. °C	Density	n_D	Solubility	Ref.
941	Aniline, *N*-methyl-*N*-nitroso or *N*-Nitroso methylphenylamine $C_6H_5N(NO)CH_3$	136.15	ye	225d, 121^{13}	14.7	$1.1240^{20/4}$	1.57688^{20}	al, eth	$B12^3$, 1119
942	Aniline, *N*-methyl-4-nitroso 4-$ONC_6H_4NHCH_3$	136.15	bl pl (bz)		118			al, eth, chl	$B7^3$, 3370
943	Aniline, *N*-methyl-4-*iso*-Propyl or Cuminyl amine 4-$(CH_3)_2CHC_6H_4NHCH_3$	149.24	oil	$225\text{-}7^{734}$, $111\text{-}2^{11}$				al, eth	$B12^3$, 2741
944	Aniline, *N*-methyl-2,4,5-trimethyl 2,4,5-$(CH_3)_3C_6H_2NHCH_3$	149.24	nd (dil al)		85			al, chl	B12, 1152
945	Aniline, *N*-methyl-3,4,5-trimethyl 3,4,5-$(CH_3)_3C_6H_2NHCH_3$	149.24	lf (w)		123				B12, 1176
946	Aniline, *N*-methyl, *N*,2,4,6-tetranitro or Tetryl 2,4,6$(NO_2)_3C_6H_2N(NO_2)CH_3$	287.15	ye pr (al)	exp 187	131-2	1.57^{10}		ace, bz	$B12^3$, 1738
947	Aniline, 2-methyl-5-*iso*-propyl or Carvacryl amine 2-CH_3-5-*i*-$C_3H_7C_6H_3NH_2$	149.24		241, 118^{12}	-16	$0.9942^{20/4}$	1.5387^{20}	al, eth	$B12^3$, 2733
948	Aniline, 4-methylamino 4-$(CH_3NH)C_6H_4NH_2$	122.17	lf (eth-peth)	257-9, 162^{20}	36			w, al, eth, ace, bz	$B13^3$, 108
949	Aniline, 4-methylthio 4-$(CH_3S)C_6H_4NH_2$	139.22		272-3, 140^{15}		$1.1379^{20/4}$	1.6395^{20}	al, eth, ace, bz	$B13^3$, 1221
950	Aniline nitrate $C_6H_5NH_2\text{-}HNO_3$	156.15	rh		d190	1.356^4		w, al, eth	$B12^3$, 232
951	Aniline, *N*-nitro $C_6H_5NHNO_2$	138.13	lf (peth)	exp	43			w, al, eth, ace, bz	$B16^2$, 343
952	Aniline, 2-nitro 2-$(NO_2)C_6H_4NH_2$	138.13	gold-ye pl or nd	284, $165\text{—}6^{28}$	71.5	1.442^{15}		al, eth, ace, bz	$B12^3$, 1513
953	Aniline, 2-nitro-5-methoxy 2-NO_2-5-$(CH_3O)C_6H_3NH_2$	168.16	br nd	sub	131			al	$B13^3$, 975
954	Aniline, 3-nitro 3-$(NO_2)C_6H_4NH_2$	138.13	ye nd, rh bipym (w)	305-7d, $100^{0.16}$	114	$1.1747^{160/4}$		al, eth, ace	$B12^3$, 1541
955	Aniline, 4-nitro 4-$(NO_2)C_6H_4NH_2$	138.13	pa ye mcl nd (w)	331.7, $106^{0.03}$	148-9	$1.424^{20/4}$		al, eth, ace, chl	$B12^3$, 1580
956	Aniline, 5-nitro-2-propoxy 5-(NO_2)-2-$(C_3H_7O)C_6H_3NH_2$	196.21	og (PrOH-peth)		49			al	$B13^3$, 879
957	Aniline, 4-nitroso 4-$(ON)C_6H_4NH_2$	122.12	bl nd (bz)	dec	173-4			w, al	$B7^3$, 3370
958	Aniline, 4-octyl 4-$C_8H_{17}C_6H_4NH_2$	205.35		310-1, $170\text{-}2^{17}$	20			eth	$B12^3$, 2770
959	Aniline oxalate(mono) $C_6H_5NH_2\text{-}H_2C_2O_4$	183.16	ta (aq al)		150-1			w, al, bz	$B12^3$, 236
960	Aniline oxalate(di) $(C_6H_5NH_2)_2\text{-}H_2C_2O_4$	276.29	tcl pr (w)		174-5d			w	$B12^3$, 236
961	Aniline, pentabromo $C_6Br_5NH_2$	487.61	nd (al-to)		265-6			al	$B12^3$, 1487
962	Aniline, pentachloro $C_6Cl_5NH_2$	265.35	nd (al)		232			al, eth, lig	$B12^3$, 1415
963	Aniline, pentafluoro $C_6F_5NH_2$	183.08		153-4	34				CAS53, 3112
964	Aniline, pentamethyl $(CH_3)_5C_6NH_2$	163.26	mcl pr (al)	277-8	152-3			al, eth	$B12^3$, 2758
965	Aniline, *N*-pentyl or Pentylphenyl amine $C_6H_5NHC_5H_{11}$	163.26		260-2					$B12^3$, 271
966	Aniline, 2-phenoxy 2-$(C_6H_5O)C_6H_4NH_2$	185.23	cr (lig)	$307\text{-}8^{728}$, $172\text{-}3^{14}$	44-5			al, eth, ace, bz	$B13^3$, 758
967	Aniline, 3-phenoxy 3-$(C_6H_5O)C_6H_4NH_2$	185.23	pr (lig)	315, $190\text{-}1^{14}$	37			al, eth, ace, bz	$B13^3$, 933
968	Aniline, 4-phenoxy or 4-Aminodiphenylether 4-$(C_6H_5O)C_6H_4NH_2$	185.23	nd (w), cr (dil al)	$187\text{-}9^{14}$	85-6			al, eth, w	$B13^3$, 999
969	Aniline phosphate $C_6H_5NH_2\text{-}H_3PO_4$	191.12	nd (al), lf (w-al)		180			w, al	B12, 117
970	Aniline-picrate $C_{12}H_{10}N_4O_7$	322.23	ye or red mcl pr (w), dk gr		181	1.558		al	$B12^3$, 236
971	Aniline, *N*-propyl $C_6H_5NHC_3H_7$	135.21		222 (cor), 100^{11}		$0.9443^{20/4}$	1.5428^{20}	al, eth	$B12^3$, 264
972	Aniline, *o*-propyl 2-$C_3H_7C_6H_4NH_2$	135.21		226^{758}, 116^{15}		$0.9602^{20/4}$	1.5427^{20}	al, eth	$B12^3$, 2657
973	Aniline, *p*-propyl 4-$C_3H_7C_6H_4NH_2$	135.21		224-6					$B12^3$, 2658

PHYSICAL CONSTANTS OF ORGANIC COMPOUNDS (Continued)

No.	Name, Synonyms, and Formula	Mol. wt.	Color, crystalline form, specific rotation and λ_{max} (log ε)	b.p. °C	m.p. °C	Density	n_D	Solubility	Ref.
974	Aniline, *N*-isopropyl or Phenylisopropyl amine $C_6H_5NH\text{-}i\text{-}C_3H_7$	135.21		203, 72-5^{14}			1.5380^{20}	al, eth	B12^{3}, 266
975	Aniline, 2-isopropyl $2\text{-}i\text{-}C_3H_7C_6H_4NH_2$	135.21		270-1^{745}, 95^{13}		0.9760$^{12/4}$		eth, bz	B12^{3}, 2683
976	Aniline, 2-isopropyl-5-methyl or Thymylamine $2\text{-}i\text{-}C_3H_7\text{-}5\text{-}CH_3C_6H_3NH_2$	149.24	oil	238-42				al, eth	B12^{3}, 2741
977	Aniline, 4-isopropyl or Cumidine $4\text{-}(CH_3)_2CHC_6H_4NH_2$	135.21		225	-63	0.953$^{20/4}$	1.3415^{20}	al, eth, bz	B12^{3}, 2684
978	Aniline sulfate(mono) $C_6H_5NH_2\text{-}H_2SO_4$	191.20	lf (w + 1/2)		162				B12^{3}, 232
979	Aniline sulfate *(di)* $(C_6H_5NH_2)_2\text{-}H_2SO_4$	284.33	lf (al)		d	1.377^{4}		w	B12^{3}, 232
980	Aniline, 2,3,4,5-tetrachloro $2,3,4,5\text{-}Cl_4\text{-}C_6HNH_2$	230.91	nd (al)		118-20			al, eth, bz	B12^{2}, 340
981	Aniline, 2,3,5,6-tetrachloro $2,3,5,6\text{-}Cl_4C_6HNH_2$	230.91	nd (lig, al)		108			al, eth	B12^{3}, 1414
982	Aniline, 2,3,4,5-tetramethyl or Prehnidine $2,3,4,5\text{-}(CH_3)_4C_6HNH_2$	149.24	lf (w)	259-60	70			al, eth, lig	B12^{3}, 2743
983	Aniline, 2,3,4,6-tetramethyl or isoduridine $2,3,4,6\text{-}(CH_3)_4C_6HNH_2$	149.24		255	23-4	0.978^{24}		al	B12^{3}, 2744
984	Aniline, 2,4,5-tribromo $2,4,5\text{-}Br_3C_6H_2NH_2$	329.83	nd (al)		80-1 (85)			al, eth, bz	B12, 662
985	Aniline, 2,4,6-tribromo $2,4,6\text{-}Br_3\text{-}C_6H_2NH_2$	329.83	nd (al, bz)	300	122	2.35$^{20/20}$		ace, chl	B12^{3}, 1477
986	Aniline, 2,4,6-tribromo,hydrobromide $2,4,6\text{-}Br_3C_6H_2NH_2\text{-}HBr$	410.75	nd	sub	195-6				B12^{1}, 330
987	Aniline, 3,3,5-tribromo $3,4,5\text{-}Br_3C_6H_2NH_2$	329.83	nd (al)		123			al, eth	B12^{3}, 1487
988	Aniline, 2,3,4-trichloro $2,3,4\text{-}Cl_3C_6H_2NH_2$	196.46	nd (lig)	292^{774}	73			al	B12, 626
989	Aniline, 2,4,5-trichloro $2,4,5\text{-}Cl_3C_6H_2NH_2$	196.46	nd(lig or 50% al)	ca 270	96.5			al, eth	B12^{3}, 1409
990	Aniline, 2,4,6-trichloro or *sym*-Trichloroaniline $2,4,6\text{-}Cl_3C_6H_2NH_2$	196.46	cr (al), nd (lig or peth)	262^{746}	78.5			al, eth	B12^{3}, 1409
991	Aniline, 2,3,5-triido $2,3,5\text{-}I_3C_6H_2NH_2$	470.82	nd		116			al, bz	B12, 676
992	Aniline, 2,3,6-triido $2,3,6\text{-}I_3C_6H_2NH_2$	470.82	nd (al or al-eth)		116.8			al	B12, 676
993	Aniline, 2,4,6-triido $2,4,6\text{-}I_3C_6H_2NH_2$	470.82	ye nd or pl (al), pr (aa)		185.5			aa	B12^{3}, 1508
994	Aniline, 3,4,5-triido $3,4,5\text{-}I_3C_6H_2NH_2$	470.82	nd (al-ace)		174.5d			al, eth, ace,bz	B12, 676
995	Aniline, 2,4,5-trimethyl or Pseudocumidine $2,4,5,\text{-}(CH_3)_3C_6H_2NH_2$	135.21	nd (w)	234-5	68	0.957		al, w	B12^{3}, 2700
996	Aniline, 2,4,6-trimethyl or Mesidine $2,4,6\text{-}(CH_3)_3C_6H_2NH_2$	135.21		232-3	-5	1.5495^{20}			B12^{3}, 2708
997	Aniline, 2,4,6-trinitro or Pieramide $2,4,6\text{-}(NO_2)_3C_6H_2NH_2$	228.12	dk ye pr (aa)	exp	192-5	1.762^{12}		ace, bz	B12^{3}, 1737
998	*o*-Anisaldehyde or 2-Methoxybenzaldehyde $2\text{-}(CH_3O)C_6H_4CHO$	136.16	pr	243-4, 124-5^{18}	37-8	1.1326$^{20/4}$	1.5600^{20}	al, eth, ace, bz, chl	B8^{3}, 143
999	*m*-Anisaldehyde or 3-Methoxybenzaldehyde $3\text{-}(CH_3O)C_6H_4CHO$	136.16		230, 62^{1}		1.1187$^{20/4}$	1.5530^{20}	al, eth, ace, bz, chl	B8^{3}, 199
1000	*p*-Anisaldehyde or 4-Methoxtbenzaldehyde $4\text{-}(CH_3O)C_6H_4CHO$	136.16		249.5, 83^{2}	0	1.1191$^{15/4}$	1.5730^{20}	w, eth, ace, bz chl	B8^{3}, 218
1001	*o*-Anisamide $2\text{-}(CH_3O)C_6H_4CONH_2$	151.16	nd (bz), pr (eth), pl (w)		129			eth, bz	B10^{3}, 155
1002	*p*-Anisamide $4\text{-}(CH_3O)C_6H_4CONH_2$	151.16	nd or ta (w)		295, 183^{40}	166-7	16.5	w, al	B10^{3}, 341
1003	*o*-Anisic acid or 2-Methoxybenzoic acid $2\text{-}(CH_3O)C_6H_4CO_2H$	152.15	pl (w)	200	101			al, eth, bz, chl	B10^{3} 97
1004	*o*-Anisic acid, ethyl ester $2\text{-}(CH_3O)C_6H_4CO_2C_2H_5$	180.20		261, 134-6^{12}		1.1124$^{20/4}$	1.5224^{20}	al, eth	B10^{3}, 116

No.	Name, Synonyms, and Formula	Mol. wt.	Color, crystalline form, specific rotation and λ_{max} (log ε)	b.p. °C	m.p. °C	Density	n_D	Solubility	Ref.
1005	*o*-Anisic acid, methyl ester $2-(CH_3O)C_6H_4CO_2CH_3$	166.18		$245, 127^{11}$		$1.1511^{19/4}$	$1.534^{19.5}$	al	B10³, 109
1006	*m*-Anisic acid or 3-Methoxybenzoic acid $3-(CH_3O)C_6H_4CO_2H$	152.15							B10³, 244
1007	*m*-Anisic acid, ethyl ester $3-(CH_3O)C_6H_4CO_2C_2H_5$	180.20		$260-1, 110^5$		$1.0993^{20/4}$	1.5161^{20}	al, eth	B10³, 250
1008	*m*-Anisic acid, methyl ester $3-(CH_3O)C_6H_4CO_2CH_3$	166.18		$252, 121-4^{10}$		$1.13101^{20/4}$	1.5224^{20}	al	B10³, 250
1009	*p*-Anisic acid or 4-Methoxybenzoic acid $4-(CH_3O)C_6H_4CO_2H$	152.15	nd (w)	$170-2^{10}$	184			al, eth, bz	B10³, 280
1010	*p*-Anisic acid, butyl ester $4-(CH_3O)C_6H_4CO_2C_4H_9$	208.26		183^{40}		$1.054^{16.5}$	$1.5141^{16.5}$		B10³, 307
1011	*p*-Anisic acid, ethyl ester $4-(CH_3O)C_6H_4CO_2C_2H_5$	180.20		$269-70, 136-7^{13}$	7-8	$1.1038^{20/4}$	1.5254^{20}	al, eth	B10³, 300
1012	*p*-Anisic acid, methyl ester $4-(CH_3O)C_6H_4CO_2CH_3$	166.18	fl (al or eth)	$256, 160^{20}$	49			al, eth	B10³, 297
1013	*o*-Anisidine or 2-Methoxy aniline $2-(CH_3O)C_6H_4NH_2$	123.15		$224, 90^4$	6.2	$1.0923^{20/4}$	1.5713^{20}	al, eth, ace, bz	B13³, 754
1014	*o*-Anisidine, 4-nitro or 2-Methoxy-4-nitro aniline $2-(CH_3O)-4(NO_2)C_6H_3NH_2$	168.15	pa ye nd (dil al)		139-40	1.2112^{150}		al, ace	B13³, 887
1015	**o-Anisidine, 5-nitro or 2-Methoxy-5-nitro aniline** **$2-(CH_3O)-5-NO_2-C_6H_3NH_2$**	168.15	og-red nd (al, eth, w)		118	1.2068^{156}		al, eth, ace, bz, aa	B13³, 878
1016	**o-Anisidine, 6-nitro or 2-Methoxy-6-nitro aniline** **$2-(CH_3O)-6-NO_2-C_6H_3NH_2$**	168.15	ye pa red nd (al)		76			al, bz	B13³, 875
1017	*m*-Anisidine or Methoxy aniline $3-(CH_3O)C_6H_4NH_2$	123.15		251		$1.096^{20/4}$	1.5794^{20}	al, ace, eth, bz	B13³, 932
1018	*m*-Anisidine, 2-nitro or 3-Methoxy-2-nitro aniline $3-(CH_3O)-2-(NO_2)C_6H_3NH_2$	168.15	ye nd (bz)		124 (143)			al, ace	B13³, 974
1019	*m*=Anisidine, 4-nitro or 3-Methoxy-4-nitro aniline $3(CH_3O)-4(NO_2)C_6H_3NH_2$	168.15	ye nd (al)	sub	169			al, ace	B13³, 975
1020	*m*-Anisidine, 5-nitro or 3-Methoxy-5-nitro aniline $3-(CH_3O)-5(NO_2)C_6H_3NH_2$	168.15	og cr (w)		120	1.2034^{150}		al, ace, bz	B13³, 977
1021	*p*-Anisidine or 4-Methoxy aniline $4-(CH_3O)C_6H_4NH_2$	123.15	ta (w), rh pl	$243 (cor), 115^{13}$	57.2	$1.071^{57/4}$	1.5559^{67}	al, eth, w, ace, bz	B13³, 994
1022	*p*-Anisidine hydrochloride or 4-Methoxy aniline,hydrochloride $4-(CH_3O)C_6H_4NH_2-HCl$	159.62	lf, nd		236			w, al	B13², 223
1023	*p*-Anisidine, 2-nitro or 4-Methoxy-2-nitro aniline $4(CH_3O)-2(NO_2)C_6H_3NH_2$	168.15	dk red pr (w or al)		129			w, al, eth, ace	B13³, 1203
1024	*p*-Anisidine, 3-nitro or 4-Methoxy-3-nitro aniline $4(CH_3O)-3(NO_2)C_6H_3NH_2$	168.15	red (eth), og pr or pl (eth-lig)		57			al, eth, ace, bz	B13², 284
1025	Anisole or Methoxybenzene, Methyl phenyl ether $C_6H_5OCH_3$	108.14		155	-37.5	$0.9961^{20/4}$	1.5179^{20}	al, eth, ace, bz	B6⁴, 548
1026	Anisole, 2acetyl or 2-Methoxy acetophenone $2-(CH_3CO)C_6H_4OCH_3$	150.18	ye	245		$1.0897^{20/4}$	1.5393^{20}	al, ace	B8³, 263
1027	Anisole, 3-acetyl or 3-Methoxy acetophenone $3-(CH_3CO)C_6H_4OCH_3$	150.18		$240, 125-6^{12}$	95-6	1.0343^{19}	1.5410^{20}	w, al, ace	B8³, 273
1028	Anisole, 4-acetyl or 4-Methoxy acetophenone $4-(CH_3CO)C_6H_4OCH_3$	150.18	pl (eth)	$258,138-9^{15}$	38-9	$1.0818^{41/4}$	1.547^{41}	al, eth, ace	B8³ 277
1029	Anisole, 4-acetyl-2-nitro or 4-Methoxy-3-nitro acetophen-one $4-(CH_3CO)-2(NO_2)C_6H_3OCH_3$	195.17	nd (al)		99.5			al, eth, ace, bz	B8³, 295
1030	Anisole, 2-bromo $2-BrC_6H_4OCH_3$	187.04		$216 (223), 94^{10}$	2.5	$1.5018^{20/4}$	1.5727^{20}	al, eth	B6³, 736
1031	Anisole, 3-bromo $3-Br-C_6H_4OCH_3$	187.04		$210-1^{752}, 105^{16}$		1.5635^{20}		al, eth, bz	B6⁴, 1043
1032	Anisole, 4-bromo $4-BrC_6H_4OCH_3$	187.04		$215, 100^{16}$	13-4	$1.4564^{20/4}$	1.5642^{20}	al, eth, chl	B6⁴, 1044
1033	Anisole, 2-chloro $2-ClC_6H_4OCH_3$	142.58		$198.5, 90-1^{16}$	-26.8	$1.1911^{20/4}$	1.5480^{20}		B6⁴, 785
1034	Anisole, 3-chloro-2-nitro $3-Cl-2-(NO_2)C_6H_3OCH_3$	187.58			56			eth, bz, chl	B6⁴, 1347
1035	Anisole, 3-chloro $3-ClC_6H_4OCH_3$	142.58		$193-4, 70^9$		$1.1759^{12/4}$	1.5365^{20}	al, eth	B6⁴, 811
1036	Anisole, 4-chloro $4-ClC_6H_4OCH_3$	142.58		$197.5, 75^{10}$	−18	$1.201^{20/4}$	1.5390^{20}	al, eth, chl	B6⁴, 822

PHYSICAL CONSTANTS OF ORGANIC COMPOUNDS (Continued)

No.	Name, Synonyms, and Formula	Mol. wt.	Color, crystalline form, specific rotation and λ_{max} (log ε)	b.p. °C	m.p. °C	Density	n_D	Solubility	Ref.
1037	Anisole, 4-chloro-2-nitro $4\text{-Cl-2-}(NO_2)C_6H_3OCH_3$	187.58	ye nd or pr (al)		98			al, MeOH	B6⁴, 1348
1038	Anisole, 2-cyclohexyl $2\text{-}C_6H_{11}C_6H_4OCH_3$	190.29		267-8.5		1.007^{18}	1.5365^{18}		B6³, 2493
1039	Anisole, 4-cyclohexyl $4\text{-}C_6H_{11}C_6H_4OCH_3$	190.29		$275\text{-}6^{748}$	57-8				B6³, 2502
1040	Anisole, 2,4-dichloro $2,4\text{-}Cl_2C_6H_3OCH_3$	177.03	pr	$235, 125^{10}$	28-9			al, eth	B6⁴, 855
1041	Anisole, 2,3-dimethyl or 3-Methoxy-*o*-xylene $2,3\text{-}(CH_3)_2C_6H_3OCH_3$	136.19		$199, 85^{18}$	29	0.9596^{40}	1.5120^{40}	al, eth, ace, bz	B6³, 1723
1042	Anisole, 2,4-dimethyl $2,4\text{-}(CH_3)_2C_6H_3OCH_3$	136.19		$192, 83\text{-}4^{15}$		$0.9740^{16/4}$	1.5190^{16}	al, eth, bz	B6³, 1744
1043	Anisole, 2,5-dimethyl $2,5\text{-}(CH_3)_2C_6H_3OCH_3$	136.19		194^{773}		$0.9693^{13/4}$	1.5182^{15}	al, eth, bz, peth	B6³, 1772
1044	Anisole, 2,6-dimethyl $2,6\text{-}(CH_3)_2C_6H_3OCH_3$	136.19		182—3		$0.9619^{14/4}$	1.5053^{14}	al, eth, bz	B6³, 1737
1045	Anisole, 3,4-dimethyl $3,4\text{-}(CH_3)_2C_6H_3OCH_3$	136.19		$204\text{-}5, 96\text{-}7^{17}$		$0.9744^{14/4}$	1.5198^{14}	al, eth, bz	B6³, 1727
1046	Anisole, 3,5-dimethyl $3,5\text{-}(CH_3)_2C_6H_3OCH_3$	136.19		$194.5, 89^{15}$		$0.9627^{15/4}$	1.5110^{20}	al, eth, bz, aa	B6³, 1756
1047	Anisole, 2,3-dinitro $2,3\text{-}(NO_2)_2C_6H_3OCH_3$	198.14	nd (al), pl (to)		119	1.2290^{137}		al, lig	B6⁴, 1369
1048	Anisole, 3,4-dinitro $3,4\text{-}(NO_2)_2C_6H_3OCH_3$	198.14	ye nd (dil al)		71	1.3332^{110}		al, MeOH	B6⁴, 1384
1049	Anisole, 2,6-dinitro $2,6\text{-}(NO_2)_2C_6H_3OCH_3$	198.14	nd (al)		118	1.3000^{128}		al	B6³, 868
1050	Anisole, 2,4-dinitro $2,4\text{-}(NO_2)_2C_6H_3OCH_3$	198.14	nd (al or w)	$206\text{-}7^{12}$, sub	94.5-5.5	1.3364^{131}		al, eth, ace, bz, to	B6⁴, 1372
1051	Anisole, 2,5-dinitro $2,5\text{-}(NO_2)_2C_6H_3OCH_3$	198.14	nd (bz-lig)	$136\text{-}8^{2}$	97	1.476^{18}		ace, bz, al	B6⁴, 1383
1052	Anisole, 3,5-dinitro $3,5\text{-}(NO_2)_2C_6H_3OCH_3$	198.14	nd (al)		105.5	1.558^{12}		ace, bz, MeOH	B6⁴, 1385
1053	Anisole, 2-ethyl $2\text{-}C_2H_5C_6H_4OCH_3$	136.19		$186\text{-}8^{755}, 80^{14}$		$0.9636^{19/4}$	1.5142^{20}	eth, bz	B6³ 1656
1054	Anisole, 3-ethyl $3\text{-}C_2H_5C_6H_4OCH_3$	136.19		$196\text{-}7^{758}, 74^{10}$		$0.9575^{18/4}$	1.5102	eth, bz	B6³, 1662
1055	Anisole, 4-ethyl $4\text{-}C_2H_5C_6H_4OCH_3$	136.19		$195\text{-}6, 83\text{-}4^{16}$		$0.9624^{15/4}$	1.5120^{20}	eth, bz	B6³, 1665
1056	Anisole, 2-fluoro $2\text{-}FC_6H_4OCH_3$	126.13		$154\text{-}5, 59^{11}$	-39	1.5489^{17}	$1.4969^{17.5}$	eth	B6⁴, 771
1057	Anisole, 4-fluoro $4\text{-}FC_6H_4OCH_3$	126.13		154	-45	$1.1781^{18/4}$	1.4886^{18}	eth	B6⁴, 773
1058	Anisole, *o*-(1-hydroxyethyl) $2\text{-}[CH_3CH(OH)]C_6H_4OCH_3$	152.19		125^{768}	-85.1	$0.9647^{20/4}$	1.4024^{20}	al, eth, ace, bz	B6³, 4563
1059	Anisole, *m*-(1-hydroxyethyl) $3\text{-}[CH_3CH(OH)]C_6H_4OCH_3$	152.19		133^{15}		$1.0781^{19/4}$	1.5325^{20}	al, eth	B6³, 4564
1060	Anisole, *p*-(1-hydroxyethyl) $4\text{-}[CH_3CH(OH)]C_6H_4OCH_3$	152.19		$310d, 140\text{-}1^{17}$		$1.0794^{20/4}$	1.5310^{25}	al, eth	B6³, 4565
1061	Anisole, 2-iodo $2\text{-}IC_6H_4OCH_3$	234.04		$239\text{-}40^{730}, 91\text{-}2^{2}$		1.8^{20}		al, eth, ace, bz, lig	B6⁴, 1070
1062	Anisole, 3-iodo $3\text{-}IC_6H_4OCH_3$	234.04		$244\text{-}5, 123^{14}$				al, eth	B6⁴, 1073
1063	Anisole, 4-iodo $4\text{-}IC_6H_4OCH_3$	234.04	lf (al), nd (MeOH)	$237^{726}, 139^{25}$	53			eth, al	B6⁴, 1075
1064	Anisole, 4-(3-methylbutyl) or *p*-Isopentyl anisole $4\text{-}i\text{-}C_5H_{11}C_6H_4OCH_3$	178.28		121^{14}				bz	B6³, 1960
1065	Anisole, 2-nitro $2\text{-}NO_2C_6H_4OCH_3$	153.14		$276.8^{752}, 144^{4}$	10.5	$1.2540^{20/4}$	1.5161^{20}	al, eth	B6⁴, 1249
1066	Anisole, 3-nitro $3\text{-}NO_2C_6H_4OCH_3$	153.14	nd (al), pl (bz-lig)	258	38-9	1.373^{18}		al, eth	B6⁴, 1270
1067	Anisole, 4-nitro $4\text{-}NO_2C_6H_4OCH_3$	153.14	pr (al), nd (dil al)	274	54	$1.2192^{60/4}$	1.5070^{60}	al, eth	B6⁴, 1282
1068	Anisole, 4-propenyl or Anethole $4\text{-}(CH_3CH{=}CH)_2\text{-}C_6H_4OCH_3$	148.20	lf (al)	$234.5^{763}, 115^{12}$	21.3	$0.9882^{20/4}$	1.5615^{20}	al, eth, ace, bz	B6³, 2395
1069	Anisole, 4-propyl $4\text{-}C_3H_7C_6H_4OCH_3$	150.22		$210, 98^{17}$		$0.94718^{20/4}$	1.5045^{20}	al, eth, ace, bz, chl	B6³, 1789

PHYSICAL CONSTANTS OF ORGANIC COMPOUNDS (Continued)

No.	Name, Synonyms, and Formula	Mol. wt.	Color, crystalline form, specific rotation and λ_{max} (log ε)	b.p. °C	m.p. °C	Density	n_D	Solubility	Ref.
1070	Anisole, 2,3,4,6-tetrabromo $2,3,4,6\text{-}Br_4C_6HOCH_3$	423.72	nd (dil al or aa)	340	113-4			al	$B6^1$, 766
1071	Anisole, 2,3,4,5-tetrachloro $2,3,4,5\text{-}Cl_4C_6HOCH_3$	245.92	cr (MeOH)		83			al, bz	$B6^2$, 182
1072	Anisole, 2,3,5,6-tetrachloro $2,3,5,6\text{-}Cl_4C_6HOCH_3$	245.92	nd (al)		89-90			al, eth, bz	$B6^2$, 182
1073	Anisole, 2,3,4,6-tetrachloro $2,3,4,6\text{-}Cl_4C_6HOCH_3$	245.92	nd (MeOH), pr (al)		64-5			al, eth, bz	$B6^2$, 182
1074	Anisole, 2,3,4-tribromo $2,3,4\text{-}Br_3C_6H_2OCH_3$	344.83	nd (al)		106			al, ace, bz	$B6^4$, 1066
1075	Anisole, 2,3,5-tribromo $2,3,5\text{-}Br_3C_6H_2OCH_3$	344.83	pr (dil al)	305-12	82			al, ace, bz	$B6^2$, 192
1076	Anisole, 1,3,5-tribromo $1,3,5\text{-}Br_3C_6H_2OCH_3$	344.83	nd (al)	297-9	88	2.491		ace, bz	$B6^4$, 1067
1077	Anisole, 2,4,5-tribromo $2,4,5\text{-}Br_3C_6H_2OCH_3$	344.83	nd (al)	$306\text{-}9^{775}$	105			al, ace, bz	$B6^4$, 1067
1078	Anisole, 3,4,5-tribromo $3,4,5\text{-}Br_3C_6H_2OCH_3$	344.83	cr (al)	300-10	91-4			al, ace, bz	$B6^2$ 195
1079	Anisole, 2,3,5-trichloro $2,3,5\text{-}Cl_3C_6H_2OCH_3$	211.48	nd (al)		84			ace, al	$B6^2$, 180
1080	Anisole, 2,4,5-trichloro $2,4,5\text{-}Cl_3C_6H_2OCH_3$	211.48	nd (dil al)	$252\text{-}5^{742}$	77.5			al, ace	$B6^2$, 180
1081	Anisole, 2,3,6-trichloro $2,3,6\text{-}Cl_3C_6H_2OCH_3$	211.48	pr (al)	$227\text{-}9^{754}$	45			al, ace	$B6^2$, 180
1082	Anisole, 2,4,6-trichloro $2,4,6\text{-}Cl_3C_6H_2OCH_3$	211.48	mcl nd (al)	240^{718}	61-2 (65)	1.640		al, ace	$B6^1$, 723
1083	Anisole, 2,4,6-triido $2,4,6\text{-}I_3C_6H_2OCH_3$	485.83	lf (bz), nd (eth or al)		98-9			al, ace	$B6^2$, 204
1084	Anisole, 2,3,4-trinitro $2,3,4\text{-}(NO_2)_3C_6H_2OCH_3$	243.13	pa ye lf (al)	exp	155			ace, aa, al	$B6^1$, 129
1085	Anisole, 2,3,5-trinitro $2,3,5\text{-}(NO_2)_3C_6H_2OCH_3$	243.13	ye nd (al)		106.8	1.618^{15}		al, bz, ace	$B6^3$, 873
1086	Anisole, 2,4,5-trinitro $2,4,5\text{-}(NO_2)_3C_6H_2OCH_3$	243.13	ye (al)		106-7			eth, bz, w	$B6^2$, 253
1087	Anisole, 2,4,6-trinitro $2,4,6\text{-}(NO_2)_3C_6H_2OCH_3$	243.13	nd (dil MeOH)		(i) 69, (ii) 56-7	1.4947^{80}		al, eth, bz	$B6^4$, 1456
1088	Anisole, 2-vinyl or 2-Methoxystyrene $2\text{-}(CH_2{=}CH)C_6H_4OCH_3$	134.18	nd	195-200, $83\text{-}4^{12}$	29	$1.0049^{17/4}$	1.5388^{20}	al, eth, ace, bz	$B6^3$, 2383
1089	Anisole, 3-vinyl or 3-Methoxystyrene $3\text{-}(H_2C{=}CH)C_6H_4OCH_3$	134.18		$114\text{-}6^{16\text{-}7}$, $90\text{-}3^{15}$		$0.999^{16/4}$	1.5586^{23}	al, eth, bz	$B6^3$, 2385
1090	Anisole, 4-vinyl or 4-Methoxystyrene $4\text{-}(CH_2{=}CH)C_6H_4OCH_3$	134.18		$204\text{-}5^{756}$, 91^{13}		$1.0001^{13/4}$	1.5642^{13}	al, eth, bz	$B6^3$, 2386
1091	*o*-Anisonitrile or *o*-Anisic acid nitrile $2\text{-}NCC_6H_4OCH_3$	133.15		255-6, 146^{20}	24.5	$1.1063^{20/4}$		al, eth	$B10^3$, 159
1092	*p*-Anisonitrile or *p*-Anisic acid nitrile $4\text{-}CH_3OC_6H_4CN$	133.15	nd (w), lf (al)	256-7	61-2			al, eth, bz	$B10^3$, 344
1093	*o*-Anisyl chloride $2\text{-}CH_3OC_6H_4COCl$	170.60		254, 128^{11}					$B10^3$, 151
1094	*m*-Anisyl chloride $3\text{-}CH_3OC_6H_4COCl$	170.60		243-4, $123\text{-}5^{15}$				eth, ace, bz	$B10^3$, 253
1095	*p*-Anisyl chloride $4\text{-}CH_3OC_6H_4COCl$	170.60	nd	262-3, 91^1	24-5	$1.261^{20/4}$	1.580^{20}	eth, ace, bz	$B10^3$, 337
1095a	*o*-Anisyl ether or bis(2-methoxy phenyl) ether $(2\text{-}CH_3OC_6H_4)_2O$	230.26	pl (lig)	330-1	79-80			al, eth	
1096	Anthracene $C_{14}H_{10}$	178.23	ta or mcl pr (al)	340 (cor), 226.5^{53} sub	216-4	$1.283^{25/4}$		ace, bz	$B5^3$, 2123
1097	Anthracene, 1-acetyl $1\text{-}CH_3COC_{14}H_9$	220.27	pa ye (al)		107.5-9			al	$B7^3$, 2538
1098	Anthracene, 2-acetyl $2\text{-}CH_3COC_{14}H_9$	220.27	ye (al), cr (AcOEt-peth)		190-2			al	$B7^3$, 2538
1099	Anthracene, 9-acetyl $9\text{-}CH_3COC_{14}H_9$	220.27	pa ye (al)		76			al	$B7^3$, 2539
1100	Anthracene, 1-amino or 1-Anthrylamine $1\text{-}H_2NC_{14}H_9$	193.25	gold-ye nd (al)		130			al	$B12^3$, 3335

PHYSICAL CONSTANTS OF ORGANIC COMPOUNDS (Continued)

No.	Name, Synonyms, and Formula	Mol. wt.	Color, crystalline form, specific rotation and λ_{max} (log ε)	b.p. °C	m.p. °C	Density	n_D	Solubility	Ref.
1101	Anthracene, 2-amino or 2-Anthrylamine $2\text{-}H_2NC_{14}H_9$	193.25	ye lf (al)	sub	238-41			al	$B12^{3}$, 3355
1102	Anthracene, 9-amino or 9-Anthrylamine $9\text{-}H_2NC_{14}H_9$	193.25	ye lf (dil al), br (bz)		145-50			al, eth, bz, chl	$B7^{3}$, 2361
1103	Anthracene, 9-benzoyl or 9-Anthraphenone $9\text{-}C_6H_5COC_{14}H_9$	282.35	ye nd (bz, aa)		148			ace, bz, aa	$B7^{2}$, 502
1104	Anthracene, 1-chloro $1\text{-}ClC_{14}H_9$	212.68	lf (aa)		83.5	$1.1707^{100/4}$	1.6959^{100}	al, eth, bz	$B5^{3}$, 2132
1105	Anthracene, 2-chloro $2\text{-}ClC_{14}H_9$	212.68	nd or lf		215 (223)			CCl_4	$B5^{3}$, 2133
1106	Anthracene, 9-chloro $9\text{-}ClC_{14}H_9$	212.68	gold-ye nd (al)		106			al, eth, bz	$B5^{3}$, 2133
1107	Anthracene, 9,10-diamino $9,10\text{-}(NH_2)_2C_{14}H_8$	208.26	red cr		196				$B14^{3}$, 291
1108	Anthracene, 9,10-dibenzyl $9,10(C_6H_5CH_2)_2C_{14}H_8$	358.48			245			lig, bz	$B5^{3}$, 2640
1109	Anthracene, 9,10-dibromo $9,10\text{-}Br_2C_{14}H_8$	336.03	ye nd (to or xyl)	sub	226			chl, bz	$B5^{3}$, 2135
1110	Anthracene, 9,10-dichloro $9,10\text{-}Cl_2C_{14}H_8$	247.12	ye nd (MeCOEt or CCl_4)		212			bz	$B5^{3}$, 2134
1111	Anthracene, 9,10-dihydro or 9,10-Dihydro anthracene $C_{14}H_{12}$	180.25	ta or pr	305, 165-70^{13}, sub	111	$0.8976^{11/4}$		al, eth, bz	$B5^{3}$, 1987
1112	Anthracene, 9,10-dihydro-10-chloro-9-NO_2 $10\text{-}Cl\text{-}9\text{-}NO_2\text{-}C_{14}H_{10}$	259.69	nd (bz)		163			bz	$B5^{3}$, 1989
1113	Anthracene, 9,10-dihydro-9-ethyl $9\text{-}C_2H_5C_{14}H_{11}$	208.30		320-3 (cor)		$1.049^{18/18}$		al, eth, bz, aa	$B5^{2}$, 560
1114	Anthracene, 9,10-dihydro-1-hydroxy $1\text{-}HOC_{14}H_{11}$	196.25	sl grsh flr lf or nd (bz, peth)		94			al, eth, aa	$B6^{2}$, 660
1115	Anthracene, 9,10-dihydro-2-hydroxy $2\text{-}HOC_{14}H_{11}$	196.25	(bz-peth), lf (dil al)		129				$B6^{2}$, 660
1116	Anthracene, 9,10-dihydro-9-hydroxy or 9-Hydroanthrol $9\text{-}HO\text{-}C_{14}H_{11}$	196.25	nd (peth)		76			al, eth, bz	$B6^{3}$, 3504
1117	Anthracene, 9,10-dihydro-9-oxo or 9-Anthrone $C_{14}H_{10}O$	194.23	nd (bz-lig, aa)		155			ace, bz	$B7^{3}$, 2359
1118	Anthracene, 9,10-dihydro-9-oxo-10-nitro $C_{14}H_9NO_3$	239.23	cr (bz-lig), nd (CS_2)		140 (148d)			al, bz	$B7^{3}$, 2371
1119	Anthracene, 1,2-dihydroxy or 1,2-Anthradiol $1,2\text{-}(HO)_2C_{14}H_8$	210.23	pa gr lf		160-2			al, eth, aa	$B6^{2}$, 998
1120	Anthracene, 1,5-dihydroxy or 1,5 -Anthradiol, Ruful $1,5\text{-}(HO)_2C_{14}H_8$	210.23	ye nd		265d			al, eth, bz	$B6^{3}$, 5683
1121	Anthracene, 1,8-dihydroxy or 1,8-Anthradiol, Chrysazol $1,8\text{-}(HO)_2C_{14}H_8$	210.23	ye nd (dil al), lf (al-aa)		225d			al, eth, bz, AcOEt	B6, 1033
1122	Anthracene, 2,6-dihydroxy or 2,6-Anthradiol, Flavol $2,6\text{-}(HO)_2C_{14}H_8$	210.23	pa ye lf (al)		295-300d			al, eth, aa	$B6^{3}$, 5684
1123	Anthracene, 9,10-dihydroxy or 9,10-Anthradiol, Anthraquinol $9,10\text{-}(HO)_2C_{14}H_8$	210.23	br or ye nd		180			al, eth	$B6^{3}$, 5685
1124	Anthracene, 1,3-dimethyl $1,3\text{-}(CH_3)_2C_{14}H_8$	206.29	pa bl fir lf (eth), cr (al)	140-5^{2}	83			al, eth	$B5^{3}$, 2165
1125	Anthracene, 2,3-dimethyl $2,3\text{-}(CH_3)_2C_{14}H_8$	206.29	bl gr fir lf (bz)		252			al, bz	$B5^{3}$, 2166
1126	Anthracene, 9-ethyl $9\text{-}C_2H_5C_{14}H_9$	206.29	bl flr lf (al or MeOH)		59	$1.0413^{99/4}$	1.6762^{99}	al, eth	$B5^{3}$, 2165
1127	Anthracene, 1,2,3,4,5,6-hexahydro $C_{14}H_{16}$	184.28	lf (MeOH), cr (al)	160^{15}	67 (70)			bz	$B5^{2}$, 472
1128	Anthracene, 1-hydroxy or 1-Anthrol $1\text{-}(HO)C_{14}H_9$	194.23	cr (bz), br nd or lf (al or aa)	234^{15}	158			al, eth	$B6^{3}$, 3551
1129	Anthracene, 2-hydroxy or 2-Anthrol $2\text{-}(HO)C_{14}H_9$	194.23	ye (bz), br lf or nd (dil al)		253			al, eth, ace, bz	$B6^{3}$, 3552

No.	Name, Synonyms, and Formula	Mol. wt.	Color, crystalline form, specific rotation and λ_{max} (log ε)	b.p. °C	m.p. °C	Density	n_D	Solubility	Ref.
1130	Anthracene, 9-hydroxy or 9-Anthrol 9-$HOC_{14}H_9$	194.23	ye red lf (dil al), pa ye nd (aa)		160-4			al, bz	B6^{3}, 3554
1131	Anthracene, 1-methyl 1-$CH_3C_{14}H_9$	192.26	bl nd (MeOH), lf (al)	199—200	85-6	$1.0471^{99/4}$	$1.6802^{99/4}$	al, eth, bz, aa	B5^{3}, 2149
1132	Anthracene, 2-methyl 2-$CH_3C_{14}H_9$	192.26	gr-bl flr lf (sub)	sub	209	$1.81^{0/4}$		bz, chl	B5^{3}, 2149
1133	Anthracene, 9-methyl 9-$CH_3C_{14}H_9$	192.26	yesh nd (dil al), pr (bz, al)	$196\text{-}7^{12}$	81.5	$1.065^{99/4}$	1.6959^{99}	al, eth, ace, bz	B5^{3}, 2150
1134	Anthracene, 9-nitro 9-$O_2NC_{14}H_9$	223.23	ye nd (al), pr (aa or xyl)	ca 275^{17}	146			bz, al	B5^{3}, 2136
1135	Anthracene, octahydro or Octhracene $C_{14}H_{18}$	186.30	pl (al)	293-5, 167^{12}	78	$0.9703^{80/4}$	1.5372^{80}	al, bz, aa	B5^{4}, 1584
1136	Anthracene, 2-phenyl 2-$C_6H_5C_{14}H_9$	254.33			207				B5^{3}, 2462
1137	Anthracene, 9-phenyl 9-$C_6H_5C_{14}H_9$	254.33	bl flr in sol, lf (al) (aa)		207			al, ace, bz, chl	B5^{3}, 2462
1138	Anthracene, tetradecahydro $C_{14}H_{24}$	192.34		128^{11}	93				B5^{4}, 492
1139	Anthracene, 1,2,3,4-tetrahydro $C_{14}H_{14}$	182.27			106-7				B5^{4}, 1909
1140	Anthracene, 1,2,9-trihydroxy 1,2,9-$(HO)_3C_{14}H_7$	226.23	og ye lf		149-51				B6^{3}, 2728
1141	Anthracene, 1,2,10-trihydroxy 1,2,10-$(HO)_3C_{14}H_7$	226.23	ye lf, nd (al-w)		208			al, eth, ace, bz, aa	B8^{2}, 372
1142	Anthracene, 1,4,9-trihydroxy 1,4,9-$(HO)_3C_{14}H_7$	226.23	og-red nd (al)		156			al	B6^{3}, 2800
1143	Anthracene, 1,5,9-trihydroxy 1,5,9-$(HO)_3C_{14}H_7$	226.23	gold lf (al)		200d without melt			al	B6^{3}, 2801
1144	Anthracene, 1,8,9-trihydroxy or Anthralin 1,8,9-$(HO)_3C_{14}H_7$	226.23	ye pl or nd (lig)		178-80			al, eth, ace, bz	B6^{3}, 2802
1145	Anthracene, 1,9,10-trihydroxy (enol form) 1,9,10-$(HO)_3C_{14}H_7$	226.23	gr nd (eth)		204-6			al, eth	B8^{2}, 372
1146	Anthracene, 1,9,10-trihydroxy (keto form) 1,9,10-$(HO)_3C_{14}H_7$	226.23	ye nd (lig)		135-7			al	B8^{2}, 372
1147	Anthracene, 2,3,9-trihydroxy 2,3,9-$(HO)_3C_{14}H_7$	226.23	ye br nd (al)		288-9			al, eth, ace, aa	B6^{3}, 2803
1148	Anthracene-2-vinyl 2-$CH_2{=}CHC_{14}H_9$	204.27			125				
1149	Anthracene-9-vinyl 9-$CH_2{=}CHC_{14}H_9$	204.27			64-5				
1150	9-Anthraldehyde or 9-Anthracene carboxaldehyde 9-$C_{14}H_9CHO$	206.24	og nd (dil al)		104-5			bz, aa	B7^{3}, 2527
1151	Anthranil or 3,4-Benzo isoxazol C_7H_5NO	119.12		215d, 99^{13}	−18	$1.8127^{20/4}$	1.5845^{20}	al, ace	B27^{2}, 17
1152	Anthranil amide 2-$H_2NC_6H_4CONH_2$	136.15	lf (chl, w)	300	110-1			al, AcOEt	B14^{3}, 889
1153	Anthranilic acid or 2-Aminobenzoic acid 2-$H_2NC_6H_4CO_2H$	137.14	lf (al)	sub	146-7	1.412^{20}		w, al, eth, chl	B14^{3}, 879
1154	Anthranilic acid, *N*-acetyl 2-$CH_3CONHC_6H_4CO_2H$	179.18	nd (aa)		185			al, eth, ace, bz, aa	B14^{3}, 922
1155	Anthranilic acid, *N*-acetyl-4-ethoxy 2-CH_3CONH-4-C_2H_5O-$C_6H_3CO_2H$	223.23	nd (al or MeOH)		199d			al	B14^{1}, 657
1156	Anthanilic acid, butyl ester or Butyl anthranilate 2-$H_2NC_6H_4CO_2C_4H_9$	193.25		182					B14^{2}, 209
1157	Anthranilic acid, iso--Butyl ester or iso-Butyl anthranilate 2-$H_2NC_6H_4CO_2$-i-C_4H_9	193.25		$156\text{-}7^{13.5}$					B14^{2}, 209
1158	Anthranilic acid, 3,4-dichloro 3,5-Cl_2-2-$H_2NC_6H_2CO_2H$	206.03	nd (aa)		237-8			al, eth, chl	B14^{1}, 549
1159	Anthranilic acid, 3,5-dichloro 3,5-Cl_2-2-$H_2NC_6H_2CO_2H$	206.03	nd or lf (al)		231-2d			al, ace, eth, bz	B14^{3}, 966
1160	Anthranilic acid, 3,6-dichloro 3,6-Cl_2-2-$H_2NC_6H_2CO_2H$	206.03	nd (w or aa)	sub	155			al, eth, ace, w, bz, aa	B14, 367

PHYSICAL CONSTANTS OF ORGANIC COMPOUNDS (Continued)

No.	Name, Synonyms, and Formula	Mol. wt.	Color, crystalline form, specific rotation and λ_{max} (log ε)	b.p. °C	m.p. °C	Density	n_D	Solubility	Ref.
1161	Anthranilic acid, 4,5-dichloro $4,5\text{-}Cl_2\text{-}2\text{-}H_2NC_6H_4CO_2H$	206.03	nd (aa)		213-4			al, eth, aa	B14^1, 549
1162	Anthranilic acid, 5,6-dichloro $5,6\text{-}Cl_2\text{-}2\text{-}H_2NC_6H_2CO_2H$	206.03	nd (MeOH)		176-7d			al, eth, aa	B14, 368
1163	Anthranilic acid, 3,5-diido $3,5\text{-}I_2\text{-}2\text{-}H_2NC_6H_2CO_2H$	388.93	pr (al)		232-3			eth	B14^3, 973
1164	Anthranilic acid, 3,5-diido,ethyl ester $3,5\text{-}I_2\text{-}2\text{-}H_2NC_6H_2CO_2C_2H_5$	416.98	pr (al)		101				B14^1, 555
1165	Anthranilic acid, 4,5-diido $4,5\text{-}I_2\text{-}2\text{-}H_2NC_6H_2CO_2H$	388.93	cr (dil NH_3)		210-2d			al, eth, bz	B14^1, 555
1166	Anthranilic acid, 4,5-diido,ethyl ester $4,5\text{-}I_2\text{-}2\text{-}H_2NC_6H_2CO_2C_2H_5$	416.99	pr (al)		137			al, eth, bz	B14^1, 555
1167	Anthranilic acid, *N,N*-dimethyl $2\text{-}(CH_3)_2NC_6H_4CO_2H$	165.19	pr nd (eth)	sub d	72			w, al, eth	B14^3, 896
1168	Anthranilic acid, *N*-ethyl $2\text{-}(C_2H_5NH)C_6H_4CO_2H$	165.19	pr or nd (dil al)		154			al, eth, ace	B14^3, 897
1169	Anthranilic acid, ethyl ester or Ethyl anthranilate $2\text{-}H_2NC_6H_4CO_2C_2H_5$	165.19		268, 145-7^{15}	13	1.1374$^{20/4}$	1.5646^{20}	al, eth	B14^3, 885
1170	Anthranilic acid hydrochloride or 2-Aminobenzoic acid hydrochloride $2\text{-}H_2NC_6H_4CO_2H\text{-}HCl$	173.60						w	B14^2, 207
1171	Anthranilic acid, 3-hydroxy or 2-Amino-3-hydroxy benzoic acid $3\text{-}HO\text{-}2\text{-}H_2NC_6H_3CO_2H$	153.14	lf(w)		164			al, eth, chl	B14^3, 1463
1172	Anthranilic acid, 3 hydroxy or 2-Amino-3-hydroxy benzoic acid $3\text{-}HO\text{-}2\text{-}H_2NC_6H_3CO_2H$	153.14	lf (w)		164			al, eth, chl	B14^2, 1463
1173	Anthranilic acid, 4-hydroxy or 2-Amino-4-hydroxy benzoic acid $4\text{-}HO\text{-}2\text{-}H_2NC_6H_3CO_2H$	153.14			148^d			w, al, eth, ace	B14^3, 1475
1174	Anthranilic acid, 5-hydroxy $5\text{-}HO\text{-}2\text{-}H_2NC_6H_3CO_2H$	153.14	vt pr (w)		252d			al, eth, ace, bz	B14^3, 1468
1175	Anthranilic acid, *N*-methyl $2\text{-}CH_3NHC_6H_4CO_2H$	151.16	pl (al or lig)	80$^{0.01}$	179			al, eth, bz, chl, lig	B14^3, 895
1176	Anthranilic acid, *N*-methyl,ethyl ester or Ethyl-*N*-methyl anthranilate $2\text{-}CH_3NHC_6H_4CO_2C_2H_5$	179.22		266	39			eth	B14^2, 213
1177	Anthranilic acid, *N*-methyl,methyl ester $2\text{-}CH_3NHC_6H_4CO_2CH_3$	165.01	cr (peth)	255, 130-1^{13}	19	1.120^{15}	1.5839^{15}	al, eth	B14^3, 895
1178	Anthranilic acid, methyl ester or Methyl anthranilate $2H_2NC_6H_4CO_2CH_3$	151.16		256, 135.5^{15}	24—5	1.1682$^{10/4}$	1.5810	al, eth	B14^3, 894
1179	Anthranilic acid, 3-methyl or 2-Amino-3-methyl benzoic acid $3\text{-}CH_3\text{-}2\text{-}H_2NC_6H_3CO_2H$	151.16	nd (al), pr (w)		172			al, eth	B14^2, 290
1180	Anthranilic acid, 5-methyl $5\text{-}CH_3\text{-}2\text{-}H_2NC_6H_3CO_2H$	151.16	lf (al), nd (w)		175			al, eth	B14^2, 291
1181	Anthranilic acid, 6-methyl $6\text{-}CH_3\text{-}2\text{-}H_2NC_6H_3CO_2H$	151.16	nd (MeOH)		125-6d			MeOH	B14^3, 1502
1182	Anthranilic acid, 3-nitro $3\text{-}NO_2\text{-}2\text{-}H_2NC_6H_3CO_2H$	182.14	ye nd (w)		208-9	1.558^{15}		al, eth	B14^3, 973
1183	Anthranilic acid, 4-nitro $4\text{-}NO_2\text{-}2\text{-}H_2NC_6H_3CO_2H$	182.14	og pr (dil al)		269			al, eth, ace, xyl	B14^3, 975
1184	Anthranilic acid, 5-nitro $5\text{-}NO_2\text{-}2\text{-}H_2NC_6H_3CO_2H$	182.14	lf (al), ye nd (w, dil al)			268-70, (280)		al, eth	B14^3, 980
1185	Anthranilic acid, 6-nitro $6\text{-}NO_2\text{-}2\text{-}H_2NC_6H_3CO_2H$	182.14	ye nd or lf (w)		184			w, al, eth, ace, aa	B14^3, 988
1186	Anthranilic acid, *N*-phenyl $2\text{-}C_6H_5NHC_6H_4CO_2H$	213.24	lf, nd or pr (al)		184d			al	B14^3, 898
1187	Anthranilic acid, phenyl ester or Phenyl anthranilate $2\text{-}H_2NC_6H_4CO_2C_6H_5$	213.24	nd (al)		70			al, eth	B14^3, 885
1188	Anthranilic acid, propyl ester or Propyl anthranilate $2\text{-}H_2NC_6H_4CO_2C_3H_7$	179.22		270				al, eth	B14^2, 209
1189	Anthranilonitrile or *o*-Cyanoaniline $2\text{-}H_2NC_6H_4CN$	118.14	ye pr (CS_2), nd (peth)	263^{751}	51			al, eth, ace, bz	B14^2, 210
1190	Anthranilonitrile, 4-methyl $4\text{-}CH_3\text{-}2\text{-}H_2NC_6H_3CN$	132.16	lf (dil al)		94			al, ace, bz, chl	B14^3, 1208

PHYSICAL CONSTANTS OF ORGANIC COMPOUNDS (Continued)

No.	Name, Synonyms, and Formula	Mol. wt.	Color, crystalline form, specific rotation and λ_{max} (log ε)	b.p. °C	m.p. °C	Density	n_D	Solubility	Ref.
1191	Anthanilonitrile, 5-methyl 5-CH_3-2-$H_2NC_6H_3CN$	132.16	cr (dil al)		63			al, eth, ace, bz	B14, 482
1192	Anthanilonitrile, 6-methyl 6-CH_3-2-$H_2NC_6H_3CN$	132.16	ye pr (bz) (w)		128				$B14^2$, 290
1193	Anthranylamide 2-$H_2NC_6H_4CONH_2$	136.15	lf (chl or w)	300	110-1			al, AcOEt, w	$B14^3$, 889
1194	β-Anthraquinoline or Naphtho(2,3:5,6)quinoline $C_{17}H_{11}N$	229.28	lf or ta (al)	446	170			al, eth, bz	$B20^4$, 4290
1195	Anthraquinone or 9,10-Dioxoanthracene $C_{14}H_8O_2$	208.22	ye rh nd (al, bz)	379.8	286 (sub)	1.438^4			$B7^3$, 4059
1196	Anthraquinone, 1-amino 1-$H_2NC_{14}H_7O_2$	223.23	red nd (al), gl (aa)	sub	253-4			eth, ace, bz, aa, chl	$B14^3$, 407
1197	Anthraquinone, 1-amino-2-benzoyl 1-NH_2-2-$(C_6H_5CO)C_{14}H_6O_2$	327.34	red nd (aa)		190			aa	B14, 482
1198	Anthraquinone, 1-amino-2-bromo 1-NH_2-2-$BrC_{14}H_6O_2$	302.13	ye, red nd (aa), nd (xyl)		182			to, aa	$B14^3$, 423
1199	Anthraquinone, 1-amino-3-bromo 3-Br-1-$H_2NC_{14}H_6O_2$	302.13	red nd (to)		243				$B14^3$, 423
1200	Anthraquinone, 1-amino-4-bromo 1-H_2N-4-Br-$C_{14}H_6O_2$	302.13			170				$B14^3$, 423
1201	Anthraquinone, 1-amino-4-bromo-2-methyl $C_{15}H_{10}NO_2Br$	316.15			232			al, eth	$B14^3$, 494
1202	Anthraquinone, 1-amino-4-hydroxy $C_{14}H_9NO_3$	239.23			215			al, ace	$B14^3$, 652
1203	Anthraquinone, 1-amino-2-methyl $C_{15}H_{11}NO_2$	237.26			205-6			al, eth, bz, chl, aa	$B14^3$, 492
1204	Anthraquinone, 1-amino-4-methyl $C_{15}H_{11}NO_2$	237.26			181				$B14^2$, 123
1205	Anthraquinone, 1-amino-4-nitro $C_{14}H_8N_2O_4$	268.23			298-300			al	$B14^3$, 427
1206	Anthraquinone, 2-amino 2-$NH_2C_{14}H_8O_2$	223.23	red nd (al aa)	sub	303-6			ace, bz, chl	$B14^3$, 429
1207	Anthraquinone, 2-amino-3-benzoyl 2-NH_2-3$(C_6H_5CO)C_{14}H_6O_2$	327.34	ye pl (Py)		331				$B14^1$, 482
1208	Anthraquinone, 2-amino-3-chloro $C_{14}H_8NO_2Cl$	257.68			280-3				$B14^3$, 435
1209	Anthraquinone, 3-amino-1,2-dihydroxy or 3-Aminoalizarin 3-NH_2-1,2-$(HO_2)_2C_{14}H_5O_2$	255.23	dk red pr (aa)	sub d	>300			aq NH_3	$B14^2$, 185
1210	Anthraquinone, 2-amino-1-hydroxy 2-NH_2-1-$HOC_{14}H_6O_2$	239.23	red br nd (al)	sub	226-7			al, eth, bz	$B14^3$, 651
1211	Anthraquinone, 4-amino-1,2-dihydroxy or 4-Aminoalizarin 4-NH_2-1,2-$(HO)_2C_{14}H_5O_2$	255.23	gr bl nd (al)		d			al	$B14^2$, 185
1212	Anthraquinone, 1-bromo 1-$BrC_{14}H_7O_2$	287.11	ye nd (bz)	sub	188			al, bz	$B7^3$, 4076
1213	Anthraquinone, 1-bromo-4-methylamino 1-Br-4$(CH_3NH)C_{14}H_6O_2$	316.15	br-red nd (Py)		194				$B14^3$, 424
1214	Anthraquinone, 2-bromo 2-$BrC_{14}H_7O_2$	287.11		sub	204-5			bz	$B7^3$, 4076
1215	Anthraquinone, 2-bromo-1-methylamino 2-Br-1-$(CH_3NH)C_{14}H_6O_2$	316.15	br nd (aa)		170-2				$B14^1$, 446
1216	Anthraquinone, 3-bromo-1,2-dihydroxy or 3-Bromoalizarin 3-Br-1,2-$(HO)_2C_{14}H_5O_2$	319.11	br-red nd (to)	sub	260-1			w	$B8^3$, 3772
1217	Anthraquinone, 1-chloro 1-$ClC_{14}H_7O_2$	242.66	ye nd (to or al)	sub	162			eth, bz, aa	$B7^3$, 4064
1218	Anthraquinone, 1-chloro-2-methyl 1-Cl-2-$CH_3C_{14}H_6O_2$	256.69			170-1				$B7^3$, 4105
1219	Anthraquinone, 2-chloro 2-$ClC_{14}H_7O_2$	242.66	pa ye nd (aa or al)	sub	211			bz, al	$B7^3$, 4066
1220	Anthraquinone, 1,2-diamino 1,3-$(NH_2)_2C_{14}H_6O_2$	238.25	vt nd		303-4				$B14^3$, 438
1221	Anthraquinone, 1,3-diamino 1,3-$(NH_2)_2C_{14}H_6O_2$	238.25	red		290			Py	$B14^3$, 439

No.	Name, Synonyms, and Formula	Mol. wt.	Color, crystalline form, specific rotation and λ_{max} (log ε)	b.p. °C	m.p. °C	Density	n_D	Solubility	Ref.
1222	Anthraquinone, 1,4-diamino 1,4-$(NH_2)_2C_{14}H_6O_2$	238.25	dk vt nd (Py), vt cr		268			al, bz, Py	B14^3, 439
1223	Anthraquinone, 1,5-diamino 1,5-$(NH_2)_2C_{14}H_6O_2$	238.25	dk red nd (al, aa)	sub	319 (cor)				B14^3, 466
1224	Anthraquinone, 1,6-diamino 1,6-$(NH_2)_2C_{14}H_6O_2$	238.25	red nd (aa)		297				B14^2, 119
1225	Anthraquinone, 1,7-diamino 1,7-$(NH_2)_2C_{14}H_6O_2$	238.25	red, nd		290				B14^1, 470
1226	Anthraquinone, 1,8-diamino 1,8-$(NH_2)_2C_{14}H_6O_2$	238.25	red (al or aa)	265				al, aa	B14^3, 478
1227	Anthraquinone, 2,3-diamino 2,3-$(NH_2)_2C_{14}H_6O_2$	238.25	red		353			Py	B14^1, 480
1228	Anthraquinone, 2,6-diamino 2,6-$(NH_2)_2C_{14}H_6O_2$	238.25	red-br pr (aq Py)		320d			Py, xy, al	B14^3, 480
1229	Anthraquinone, 2,7-diamino 2,7-$(NH_2)_2C_{14}H_6O_2$	238.25	og-ye nd (al), dk red nd (sub)	sub	>320				B14^1, 483
1230	Anthraquinone, 2,3-dibromo 2,3-$Br_2C_{14}H_6O_2$	366.01	ye nd (to)	sub	283			bz, chl	B7^3, 4077
1231	Anthraquinone, 2,7-dibromo 2,7-$Br_2C_{14}H_6O_2$	366.01	lt ye lf	sub	248			bz, aa	B7^2, 718
1232	Anthraquinone, 1,3-dichloro 1,3-$Cl_2C_{14}H_6O_2$	277.11	ye nd (aa)		209-10			aa	B7^3, 4068
1233	Anthraquinone, 1,4-dichloro 1,4-$Cl_2C_{14}H_6O_2$	277.11	og-ye nd (aa)		187-8			Py, aa	B7^3, 4068
1234	Anthraquinone, 1,5-dichloro 1,5-$Cl_2C_{14}H_6O_2$	277.11	yesh (to), ye nd		252			aa	B7^3, 4068
1235	Anthraquinone, 1,6-dichloro 1,6-$Cl_2C_{14}H_6O_2$	277.11	pa ye nd (aa)		203-4			ace, to	B7^3, 4070
1236	Anthraquinone, 1,7-dichloro 1,7-$Cl_2C_{14}H_6O_2$	277.11	ye nd (aa)		213-4			bz, to	B7^3, 4070
1237	Anthraquinone, 1,8-dichloro 1,8-$Cl_2C_{14}H_6O_2$	277.11	ye nd (aa)		202-3			bz	B7^3, 4070
1238	Anthraquinone, 2,3-dichloro 2,3-$Cl_2C_{14}H_6O_2$	277.11	ye nd (aa)		271			bz, aa	B7^3, 4071
1239	Anthraquinone, 2,6-dichloro 2,6-$Cl_2C_{14}H_6O_2$	277.11	ye nd (aa or al)		291			bz, aa, al	B7^3, 4071
1240	Anthraquinone, 2,7-dichloro 2,7-$Cl_2C_{14}H_6O_2$	277.11	yesh nd		212 (231)			eth	B7^3, 4072
1241	Anthraquinone, 1,2-dihydroxy or Alizarin 1,2-$(HO)_2C_{14}H_6O_2$	240.22	og or red tcl nd or pr (al, sub)	430 (sub)	289-90 (cor)			al, eth, ace, bz, Py	B8^3, 3767
1242	Anthraquinone, 1,2-dihydroxy-3-iodo or β-iodo alizarin 1,2-$(HO)_2$-3-$IC_{14}H_5O_2$	366.11	og-red nd (xyl)		229			w	B8^3, 3773
1243	Anthraquinone, 1,2-dihydroxy-3-methyl or β-methylalizarin 1,2$(HO)_2$-3-CH_3-$C_{14}H_5O_2$	254.24	og nd	sub	245			al, eth, ace	B8^3, 3808
1244	Anthraquinone, 1,2-dihydroxy-3-nitro or β-nitroalizarin, Alizarin orange 1,2-$(HO)_2$-3$(NO_2)C_{14}H_5O_2$	285.21	og-ye nd (bz), ye, pl (gl aa, al)	sub, d	244d			al, bz, aa	B8^3, 3774
1245	Anthraquinone, 1,2-dihydroxy-4-nitro or 4-Nitroalizarin 1,2-$(HO)_2$-3$(NO_2)C_{14}H_5O_2$	285.21	gold-ye nd (aa or al)	sub, d	289d			al, bz, aa, chl	B8^2, 491
1246	Anthraquinone, 1,3-dihydroxy or Purpuro xanthin 1,3-$(HO)_2C_{14}H_6O_2$	240.22	ye-red nd (sub)		268-70			al, ace, bz, aa	B8^3, 3774
1247	Anthraquinone, 1,4-dihydroxy or Quinizarin 1,4-$(HO)_2C_{14}H_6O_2$	240.22	ye red lf (eth), dk red nd		200			w, al, bz, eth	B8^3, 3775
1248	Anthraquinone, 1,4-dihydroxy-5,6,7,8-tetrachloro or 5,6,7,8-Tetrachloroquinizarin 1,4-$(HO)_2$-5,6,7,8-$Cl_4C_{14}H_2O_2$	378.00	red pl (aa)		270			bz, aa	B8^3, 3783
1249	Anthraquinone, 1,5-dihydroxy or Anthrarufin 1,5-$(HO)_2C_{14}H_6O_2$	240.22	pa ye pl (gl aa)	sub	280			bz	B8^3, 3787
1250	Anthraquinone, 1,6-dihydroxy 1,6-$(HO)_2C_{14}H_6O_2$	240.22	og-ye nd (gl aa)		276				B8^3, 3791
1251	Anthraquinone, 1,7-dihydroxy 1,7-$(HO)_2C_{14}H_6O_2$	240.22	ye nd (sub)	sub	292-3			al, eth, bz, chl	B8^3, 3791

No.	Name, Synonyms, and Formula	Mol. wt.	Color, crystalline form, specific rotation and λ_{max} (log ε)	b.p. °C	m.p. °C	Density	n_D	Solubility	Ref.
1252	Anthraquinone, 1,8-dihydroxy or Chrysazin $1,8\text{-}(HO)_2C_{14}H_6O_2$	240.22	red or redsh-ye nd or lf (al)	sub	193			al, eth, ace, chl	B8[3], 3792
1253	Anthraquinone, 1,8-dihydroxy-3-hydroxymethyl or Aloe emedin $1,8\text{-}(HO)_2\text{-}3(HOCH_2)C_{14}H_5O_2$	270.24	og-ye nd (to, al)	sub	223-4			al, eth, bz	B8[3], 4160
1254	Anthraquinone, 1,8-dihydroxy-3-methyl or Chrysophenol, Chrysophanic acid $1,8\text{-}(HO)_2\text{-}3\text{-}CH_3C_{14}H_5O_2$	254.24	ye hex or mcl nd (sub)	sub	196	0.92		ace, bz, aa	B8[2], 510
1255	Anthraquinone, 1,8-dihydroxy-2,4,5,7-tetrabromo or 2,4,5,7-Tetrabromoquinizarin $1,8\text{-}(HO)_2\text{-}2,4,5,7\text{-}Br_4C_{14}H_2O_2$	555.80	og-ye nd (bz)		312				B8[1], 722
1256	Anthraquinone, 1,8-dihydroxy-2,4,5,7-tetranitro or Chrysammic acid, Chrysamminic acid $1,8\text{-}(HO)_2\text{-}2,4,5,7\text{-}(NO_2)_4C_{14}H_2O_2$	420.21	ye pl or lf	d	exp			al, eth	B8[1], 723
1257	Anthraquinone, 2,3-dihydroxy or Hystazin $2,3\text{-}(HO)_2C_{14}H_6O_2$	240.23	ye-br nd (aa), ye nd (sub)	sub	>330				B8[3], 3794
1258	Anthraquinone, 2,6 dihydroxy or Anthraflavin $2,6\text{-}(HO)_2C_{14}H_6O_2$	240.22	ye nd (al)		360d				B8[3], 3796
1259	Anthraquinone, 2,7-dihydroxy or Isoanthraflavin $2,7\text{-}(HO)_2C_{14}H_6O_2$	240.22	ye nd (+1 w, dil al), nd (sub)	sub	350-5			al, aa	B8[3], 3798
1260	Anthraquinone, 1,8-dimethoxy $1,8(CH_3O)_2C_{14}H_6O_2$	268.27			221			al, bz, lig	B8[3], 3793
1261	Anthraquinone, 2,6-dimethoxy $2,6\text{-}(CH_3O)_2C_{14}H_6O_2$	268.27			256				B8[1], 3797
1262	Anthraquinone, 1,2-dimethyl $1,2\text{-}(CH_3)_2C_{14}H_6O_2$	236.27	nd (ace or aa)		156			al, eth, ace, bz, aa	B7[3], 4130
1263	Anthraquinone, 1,3-dimethyl $1,3\text{-}(CH_3)_2C_{14}H_6O_2$	236.27	nd (aa)		162			aa	B7[3], 4130
1264	Anthraquinone, 1,4-dimethyl $1,4\text{-}(CH_3)_2C_{14}H_6O_2$	236.27	ye nd (al, sub)	sub	140-1			bz, xyl, aa	B7[3], 4131
1265	Anthraquinone, 2,3-dimethyl $2,3\text{-}(CH_3)_2C_{14}H_6O_2$	236.27	ye nd (al, to or xyl), cr (aa)	sub	210			al, bz, xyl	B7[3], 4131
1266	Anthraquinone, 2,6-dimethyl $2,6\text{-}(CH_3)_2C_{14}H_6O_2$	236.27	ye nd (aa or al)	sub	242			to	B7[3], 4132
1267	Anthraquinone, 2,7-dimethyl $2,7\text{-}(CH_3)_2C_{14}H_6O_2$	236.27	yesh nd (al)		170			al	B7[3], 4133
1268	Anthraquinone, 1,3-dinitro $1,3\text{-}(NO_2)_2C_{14}H_6O_2$	298.21	ye nd (HNO_3)		246-50				E13, 436
1269	Anthraquinone, 1,5-dinitro $1,5\text{-}(NO_2)_2C_{14}H_6O_2$	298.21	pa ye nd (xyl), ye-cr (sub)	sub	422 (385)				B7[3], 4081
1270	Anthraquinone, 1,6-dinitro $1,6\text{-}(NO_2)_2C_{14}H_6O_2$	298.21			257-9				B7[3], 4082
1271	Anthraquinone, 1,7-dinitro $1,7(NO_2)_2C_{14}H_6O_2$	298.21			295				B7[3], 4082
1272	Anthraquinone, 1,8-dinitro $1,8\text{-}(NO_2)_2C_{14}H_6O_2$	298.21			312				B7[3], 4082
1273	Anthraquinone, 1,5-disulfonic acid $1,5\text{-}(HO_3S)_2C_{14}H_6O_2$	368.33	ye nd (HCl + 4w), pl (dil aa + 4w)		310-1d			w, al, aa	B11[3], 634
1274	Anthraquinone, 1,6-disulfonic acid $1,6\text{-}(HO_3S)_2C_{14}H_6O_2$	368.33	ye nd (HCl + 5w), gold pr (dil aa + 5w)		215-7d			w, al, aa	B11[3], 634
1275	Anthraquinone, 1,7-disulfonic acid $1,7\text{-}(HO_3S)_2C_{14}H_6O_2$	368.33	ye hyg pw (dil aa + 4w)		d at 120			w, al, aa	B11[3], 634
1276	Anthraquinone, 1,8-disulfonic acid $1,8\text{-}(HO_3S)_2C_{14}H_6O_2$	368.33	ye nd (+5w)		293-4d			w, al	B11[3], 635
1277	Anthraquinone, 2-ethyl-1-nitro $1\text{-}(NO_2)\text{-}2(C_2H_5)C_{14}H_6O_2$	281.27	yesh br (aa)		226				B7[2], 743

PHYSICAL CONSTANTS OF ORGANIC COMPOUNDS (Continued)

No.	Name, Synonyms, and Formula	Mol. wt.	Color, crystalline form, specific rotation and λ_{max} (log ε)	b.p. °C	m.p. °C	Density	n_D	Solubility	Ref.
1278	Anthraquinone, 1,2,3,5,6,7-hexahydroxy or Rufigallol $1,2,3,5,6,7-(HO)_6C_{14}H_2O_2$	304.21	red rh, red, ye nd (sub)		sub, d			ace	$B8^3$, 4401
1279	Anthraquinone, 1-hydroxy or Erythroxyanthraquinone $1-HOC_{14}H_7O_2$	224.22	red-og nd (al)	sub	194-5			al, eth, bz	$B8^3$, 2906
1280	Anthraquinone, 2-hydroxy $2-HOC_{14}H_7O_2$	224.22	ye pl or nd (al or aa)	sub	306			al, eth	$B8^3$, 2921
1281	Anthraquinone, 2-methoxy $2-CH_3OC_{14}H_7O_2$	238.24			196			bz, al	$B8^3$, 2922
1282	Anthraquinone, 2-methyl $2-CH_3C_{14}H_7O_2$	222.24	yesh nd (al aa)	sub	182-3			al, bz, aa	$B7^3$, 4104
1283	Anthraquinone, 1-methylamino $1-(CH_3NH)C_{14}H_7O_2$	237.26	ye-red nd		170			al, to, aa	$B14^3$, 408
1284	Anthraquinone, 2-methylamino $2-(CH_3NH)C_{14}H_7O_2$	237.26	red nd (aa)		226-7			al, eth, aa, to	$B14^3$, 430
1285	Anthraquinone, 2-methyl-1-nitro $2-CH_3-1-NO_2C_{14}H_6O_2$	267.24	pa ye nd (aa)		270-1				$B7^3$, 4110
1286	Anthraquinone, 6-methyl-1,2,5-trihydroxy or Morindone $6-CH_3-1,2,5-(HO)_3C_{14}H_4O_2$	270.24	og-red nd (to)		282			al, eth, bz, Py	$B8^3$, 4151
1287	Anthraquinone, 6-methyl-1,3,8-trihydroxy or Emodin $6-CH_3-1,3,8-(HO)_3C_{14}H_4O_2$	270.24	og-red mcl nd (aa), cr (dil aa + 1w)	sub	256-7			al, eth	$B8^3$, 4154
1288	Anthraquinone, 1-nitro $1-NO_2C_{14}H_7O_2$	253.21	yesh pr (ace), nd (aa)	$270-1^{7}$	232-3			ace, bz, aa	$B7^3$, 4080
1289	Anthraquinone, 2-nitro $2-NO_2C_{14}H_7O_2$	253.21	ye nd (aa or al)	sub	184-5			bz, chl	$B7^3$, 4080
1290	Anthraquinone, 1,2,4,5,8-pentahydroxy or Alizarin cyanine R $1,2,4,5,8-(HO)_5C_{14}H_3O_2$	288.21							$B8^3$, 4379
1291	Anthraquinone, 2-sulfonamide $2-(H_2NO_2S)C_{14}H_7O_2$	287.29	ye nd (aa)		261				B11, 339
1292	Anthraquinone, 1-sulfonic acid $1-HO_3SC_{14}H_7O_2$	288.27	lf (aa), ye lf (con HCl + 3w)		214 (cor) anh			w, al	$B11^3$, 626
1293	Anthraquinone, 1-sulfonic acid-5-chloro $1-HO_3S-5-Cl-C_{14}H_6O_2$	322.72	ye rh pr (HCl or aa + 5w)		236-7			w	$B11^3$, 627
1294	Anthraquinone, 2-sulfonic acid $2-HO_3SC_{14}H_7O_2$	288.27	ye lf (+ 3w)					w, al	$B11^3$, 628
1295	Anthraquinone-2-sulfonic acid-5-nitro $2-HO_3S-5-NO_2C_{14}H_6O_2$	333.27	yesh pl (dil HNO_3)		255d			w	$B11^3$, 629
1296	Anthraquinone, 1,2-4,6-tetrahydroxy or Hydroxyflavopurpurin $1,2,4,6-(HO)_4C_{14}H_4O_2$	272.21	dk red nd (sub)					al, Py	$B8^3$, 4288
1297	Anthraquinone, 1,2,4,7-tetrahydroxy or 4-Hydroxyanthrapurpurin $1,2,4,7-(HO)_4C_{14}H_4O_2$	272.21	red-ye (al, Py or aa)					al	$B8^3$, 4288
1298	Anthraquinone, 1,2,5,6-tetrahydroxy or Rufiopin $1,2,5,6-(HO)_4C_{14}H_4O_2$	272.21	og-red nd (Py)	sub	340			w, al, aa	$B8^3$, 4288
1299	Anthraquinone, 1,2,5,8-tetrahydroxy or Quinalizarin $1,2,5,8-(HO)_4C_{14}H_4O_2$	272.21	og nd		>275				$B8^3$, 4289
1300	Anthraquinone, 1,2,6,7-tetrahydroxy $1,2,6,7-(HO)_4C_{14}H_4O_2$	272.21			>330				$B8^3$, 584
1301	Anthraquinone, 1,2,7,8-tetrahydroxy $1,2,7,8-(HO)_4C_{14}H_4O_2$	272.21	red nd or pr (aa)		318d				$B8^2$, 585
1302	Anthraquinone, 1,3,5,7-tetrahydroxy or Anthrachrysone $1,3,5,7-(HO)_4C_{14}H_4O_2$	272.21	yesh nd (al + 2w)	sub	150—60d (+ 2w), <360 (anh)			al, aa, bz, ace	$B8^3$, 4289
1303	Anthraquinone, 1,4,5,8-tetrahydroxy $1,4,5,8-(HO)_4C_{14}H_4O_2$	272.21	gr nd (aa), br nd (bz-lig)	sub	>300			al	$B8^3$, 4290
1304	Anthraquinone, 1,2,3-trihydroxy or Anthragallol $1,2,3-(HO)_3C_{14}H_5O_2$	256.21	ye nd (dil al), br (aa), og nd	sub (290)	313			al, eth, aa	$B8^3$, 4140

No.	Name, Synonyms, and Formula	Mol. wt.	Color, crystalline form, specific rotation and λ_{max} (log ε)	b.p. °C	m.p. °C	Density	n_D	Solubility	Ref.
1305	Anthraquinone, 1,2,4-trihydroxy or Purpurin $1,2,4-(HO)_3C_{14}H_5O_2$	256.21	og red, dk red or og-ye nd (al)	sub	259			al, eth, bz	$B8^3$, 4141
1306	Anthraquinone, 1,2,5-trihydroxy or 2-Hydroxyanthrarufin $1,2,5-(HO)_3C_{14}H_5O_2$	256.21	red nd (gl aa)		278			eth	$B8^2$, 554
1307	Anthraquinone, 1,2,6-trihydroxy or Flavopurpurin $1,2,6-(HO)_3C_{14}H_5O_2$.	256.21	ye nd (al)	459d	330 (sub)			al, bz, w	$B8^3$, 4144
1308	Anthraquinone, 1,2,7-trihydroxy or Anthrapurpurin $1,2,7-(HO)_3C_{14}H_5O_2$	256.21	og nd (al)	462	374			al, bz, aa	$B8^3$, 4145
1309	Anthraquinone, 1,2,8-trihydroxy or 2-Hydroxychrysazin $1,2,8-(HO)_3C_{14}H_5O_2$	256.21	red nd (aa-sub)	sub	239-40				$B8^3$, 4145
1310	Anthraquinone, 1,3,8-trihydroxy $1,3,8-(HO)_3C_{14}H_5O_3$	256.21	bt red-br nd (bz), gold-ye pl (AcOEt)		287-8				$B8^3$, 4145
1311	Anthraquinone, 1,4,5-trihydroxy or 5-Hydroxyquinizarin $1,4,5-(HO)_3C_{14}H_5O_2$	256.21	red-br nd or lf dk red nd (Py)		271				$B8^3$, 4146
1312	Anthraquinone, 1,4,6-trihydroxy or 6-Hydroxyquinizarin $1,4,6-(HO)_3C_{14}H_5O_2$	256.21	bt red nd (al), br pw		> 300			al, Py	$B8^2$, 558
1313	Anthraquinone-2-carboxylic acid $2-HO_2CC_{14}H_7O_2$	252.23	ye nd (aa)	sub	290-2			ace	$B10^3$, 3640
1314	Anthraquinone-2-carboxylic acid, 1,3-dihydroxy or Munjiston $1,3-(HO)_2C_{14}H_5O_2(CO_2H)-2$	284.24	ye nd (al-w + 1w), lf anh					eth	$B10^3$, 4787
1315	Anthraquinone-2-carboxylic acid, 1,4-dihydroxy $1,4-(HO)_2C_{14}H_5O_2(CO_2H)-2$	284.24	ye br or red nd		249-50			al, ace	$B10^3$, 4787
1316	Anthraquinone-2-carboxylic acid, 4,5-dihydroxy or Cassic acid $4,5-(HO)_2C_{14}H_5O_2(CO_2H)-2$	284.24	ye or og nd (MeOH, Py)	sub	321			Py	$B10^3$, 4789
1317	Anthraquinone-2-carboxylic acid, 4,5-dihydroxy-7-methoxy or Parietinic acid $4,5-(HO)_2-7-(CH_3O)C_{14}H_4O_2(CO_2H)-2$	314.25	red-br or ye nd (sub)		300				$B10^3$, 4829
1318	Anthraquinone-2-carboxylic acid, 1-nitro $1-(NO_2)C_{14}H_6O_2(CO_2H)-2$	297.22	nd (al or aa)		288d			al, ace	$B10^3$, 3650
1319	Anthraquinone-2-carboxylic acid, 5-nitro $5-(NO_2)-C_{14}H_6O_2(CO_2H)-2$	297.22	yesh nd (aa)					aa	$B10^3$, 3653
1321	1-Anthroic acid or 1-Anthracene carboxylic acid $1-HO_2CC_{14}H_9$	222.24	ye nd (aa), ye pr (al, AcOEt)	sub	251-2			eth	$B9^3$, 3492
1322	2-Anthroic acid or 2-Anthracene carboxylic acid $2-HO_2CC_{14}H_9$	222.24	ye lf (al), nd, lf (sub)	sub	281			aa	$B9^3$, 3493
1323	9-Anthroic acid or 9-Anthracene carboxylic acid $9-HO_2CC_{14}H_9$	222.24	pa ye nd (bz, al)	sub	217d			al	$B9^3$, 3494
1324	Anthrone or 9-Oxodihydroanthracene $C_{14}H_{10}O$	194.23	nd (bz-lig, aa)		155			ace, bz	$B7^3$, 2359
1325	Anthrone-10-nitro $10-NO_2-C_{14}H_9O$	239.23	cr (bz-lig), nd (CS_2)		140(148 d)			al, bz	$B7^3$, 2371
1326	Antimalarine or Plasmocid $C_{17}H_{25}N_3O$	287.41		$182^{1.0}$		$1.0569^{24/4}$	1.5855^{24}	dil HCl	$B22^4$, 5787
1327	Antipyrine (α - form) or Analgesine, Phenazone $C_{11}H_{12}N_2O$	188.23	mcl lf or sc (w, bz, eth)	319^{741}, $211-2^{10}$	114	$1.0747^{20/4}$	1.5697	w, al, bz, chl	$B24^2$, 11
1328	Antipyrene (β) $C_{11}H_{12}N_2O$	188.23	cr (unst)		109(β→ α at 18)			bz	$B24^2$, 11
1329	Antipyrine, 4-acetamido $4-(CH_3CONH)C_{11}H_{11}N_2O$	245.28			201			w, al	$B24^2$, 152
1330	Antipyrene, o-amino $C_{11}H_{13}N_3O$	203.24	nd (AcOEt-eth)		165			w, al	$B24^1$, 210
1331	Antipyrene, m-amino $C_{11}H_{13}N_3O$	203.24	redsh in air (bz)		148			w, al, chl	$B24^1$, 210
1332	Antipyrine, p-amino $C_{11}H_{13}N_3O$	203.24			109			w, al, eth, bz	$B24^2$, 151

No.	Name, Synonyms, and Formula	Mol. wt.	Color, crystalline form, specific rotation and λ_{max} (log ε)	b.p. °C	m.p. °C	Density	n_D	Solubility	Ref.
1333	Antipyrene, *p*-bromo $C_{11}H_{11}BrN_2O$	267.13	nd (w)	300°	122			w, al, eth, to	B24, 33
1334	Antipyrene, *p*-dimethylamino $C_{13}H_{17}N_3O$	231.30	pr or pl (lig or AcOEt)		134—5			w, al, bz	B24, 46
1335	Antipyrene, 2-hydroxybenzoate or Salazolon $C_{11}H_{12}N_2O\text{-}C_7H_6O_2$	310.35	pw		92			al, chl	$B24^2$, 16
1336	Aphanin $C_{40}H_{54}O$	550.87	bl blk lf (bz-MeOH)		178			chl	$B7^3$, 2858
1337	Aphyiline $C_{15}H_{24}N_2O$	248.37	$[\alpha]^{20}_D$ + 10.3 (MeOH, c = 2)	200d	52-7			al, eth, ace, bz	C26, 2742
1338	Apiol or 2,5-Dimethoxy saffrole parsley camphor $C_{12}H_{14}O_4$	222.24	nd	294, 179^{35}	29.55	$1.015^{20/4}$	1.5360^{20}	al, eth, ace, bz, lig	$B19^4$, 1033
1339	Apoatropine or Atropamine $C_{17}H_{21}NO_2$	271.36	pr (chl)		62			al, eth, ace, bz, chl	$B21^4$, 175
1340	Apoatropine hydrochloride $C_{17}H_{21}NO_2\text{-}HCl$	307.82	lf (w)		239			w	$B21^4$, 175
1341	Apocinchonidine $C_{19}H_{22}N_2O$	294.40	lf (al), $[\alpha]^{20}_D$ − 139.3 (chl-al, c = 2)		252			al, chl	$B23^1$, 131
1342	Apocinchonine or Allocinchonine $C_{19}H_{22}N_2O$	294.40	pr (al), $[\alpha]^{20}_D$ + 167.4 (abs al, c=3)		219			al, bz	$B23^2$, 367
1343	Apocinchonine hydrochloride $C_{19}H_{22}N_2O\text{-}HCl\text{-}2H_2O$	366.89	nd (+ 2w), $[\alpha]^{16}_D$ + 139 (w, c=0.006)					w	$B23^2$, 367
1344	Apocodeine or Apomorphine-3-methyl ether $C_{18}H_{19}NO_2$	281.35	pr (MeOH), $[\alpha]^{23}_D$ − 90 (abs al, c=0.449)		123.4(anh)			al, eth, aa, lig	$B21^4$, 2420
1345	Apocodeine, ethanol solvate $C_{18}H_{19}NO_2\text{-}C_2H_5OH$	327.43	lf (al + l)		104.5-6.5			al	$B21^4$, 2420
1346	Apocodeine, methanol solvate $C_{18}H_{19}NO_2\text{-}CH_3OH$	313.40	nd (MeOH + l)		85			MeOH	$B21^4$, 2420
1347	Apocyclene C_9H_{14}	122.21	cr (al)	$138\text{-}9^{764}$	42.5-4.3	$0.8710^{40/4}$	1.4514^{40}	al, eth, bz	$B5^4$, 430
1348	Apocynin or 4-Hydroxy-3-methoxy acetophenone $4\text{-}HO\text{-}3\text{-}(CH_3O)C_6H_3COCH_3$	166.18	pr (w)	295-300, $233\text{-}5^{15\text{-}20}$	115			al, eth, ace, bz, chl	$B8^3$, 2108
1349	Apofenchocamphoric acid or 4,4-Dimethyl-1,3-cyclopentane dicarboxylic acid $4,4\text{-}(CH_3)_2\text{-}1,3\text{-}(HO_2C)_2C_5H_6$	186.21	mcl		144-5			al, eth	$B9^3$, 3849
1350	Apomorphine $C_{17}H_{17}NO_2$	267.33	hex pl (chl-peth), rods (eth + l)		195d			al, eth, ace, bz	$B21^4$, 2419
1351	Apomorphine-hydrochloride $C_{17}H_{17}NO_2\text{-}HCl\text{-}1/2H_2O$	312.80	gr in air, mcl pr, $[\alpha]^{25}_D$ − 48 (w, c = 1.2)		200-10				$B2^4$, 2419
1352	Aponal or *tert*-Pentylcarbamate $H_2NCOOCH_2C(CH_3)_3$	131.17	nd (dil al)		86—7			ace, bz	$B3^1$, 14
1353	Apoquinine (α) or Apocupreine $C_{19}H_{22}N_2O_2$	310.40	pr (eth), $[\alpha]^{20}_D$ − 214.8 (al)		190d			al	$B23^2$, 412
1354	Apoquinine(β) $C_{19}H_{22}N_2O_2\text{-}2H_2O$	346.43	$[\alpha]^{20}_D$ − 194 (al)		190d			al	$B23^2$, 412
1355	Aposafranone or Benzeneindone,10-phenyl-2-phenazinone $C_{18}H_{12}N_2O$	272.31	br, gr (red in sl), nd (al)						$B23^2$, 364

No.	Name, Synonyms, and Formula	Mol. wt.	Color, crystalline form, specific rotation and λ_{max} (log ε)	b.p. °C	m.p. °C	Density	n_D	Solubility	Ref.
1356	D-Arabinose (α-anomer) $C_5H_{10}O_5$	150.13	(cr, MeOH)		155.5—6.5 (cor)	1.585		w	B1[4], 4215
1357	D-Arabinose (β-anomer) $C_5H_{10}O_5$	150.13	cr (MeOH), $[\alpha]^{20}_D$ − 175→-108 (mut)		155.5—6.5 (cor)	1.625		w	B1[4], 4215
1358	L-Arabinose or pectinose $C_5H_{10}O_5$	150.13	pr nd (al)		164—5 (cor)	$1.585^{20/4}$		w	B1[4], 4223
1359	L-Arabinose (α-anomer) $C_5H_{10}O_5$	150.13	orth cr (pr), cr (dil al)		159-60	$1.585^{20/4}$			B1[4], 4218
1360	L-Arabinose (β-anomer) $C_5H_{10}O_5$	150.13	orth, $[\alpha]^{20}_D$ + 190.5→ + 104.5 (c = 3) (mut)		159-60	$1.625^{20/4}$		w	B1[4], 4218
1361	D-Arabinose-diphenylhydrazone $HOCH_2(CHOH)_3CH = NN(C_6H_5)_2$	316.36	orth pr (dil al)		207				B31, 33
1362	DL-Arabinose-diphenylhydrazone $HOCH_2(CHOH)_3CH = NN(C_6H_5)_2$	316.36	nd (aq Py)		206			Py	B31, 47
1363	L-Arabinose-diphenylhydrazone $HOCH_2(CHOH)_3CH = NN(C_6H_5)_2$	316.36	orth pr, nd (dil al), $[\alpha]^{20}_D$ + 18.5 Py		204.5			al	B31, 44
1364	D-Arabitol or Arabite, D-Lyxitol 1,2,3,4,5-pentanepentol $HOCH_2(CHOH)_3CH_2OH$	152.15	pr (dil al), $[\alpha]^{20}_D$ + 11.8 (borax sol, c = 9.5)					w	B1[4], 2832
1365	DL-Arabitol $HOCH_2(CHOH)_3CH_2OH$	152.15	pr (90% al), cr (al-ace)		106			w	B1[4], 2832
1366	L-Arabitol $HOCH_2(CHOH)_3CH_2OH$	152.15	$[\alpha]_D$ − 5.4 (borax sol)		102-3			w	B1[4], 2832
1367	D-Arabonic acid $HOCH_2(CHOH)_3CO_2H$	166.13	(dil aa), $[\alpha]^{25}_D$ + 10.5 (c=6)		114-16			w, al	B3[4], 1205
1368	DL-Arabonic acid $HOCH_2(CHOH)_3CO_2H$	166.13						w	B3[2], 474
1369	L-Arabonic acid $HOCH_2(CHOH)_3CO_2H$	166.13	cr (al), $[\alpha]^{20}_D$ − 9.6 (w, c=2.5)						B3[1], 979
1370	Arachidic acid or Eicosanoic acid $CH_3(CH_2)_{18}CO_2H$	312.54	pl (al)	328d, 203—5[1]	77	$0.8240^{100/4}$	1.425^{100}	eth, bz, chl	B2[4], 1275
1371	Arachidic acid-ethyl ester or Ethyl arachidate $CH_3(CH_2)_{18}CO_2C_2H_5$	340.69		295-7[100], 186-7[2]	50			al, eth, bz, chl	B2[3], 1067
1372	Arachidic acid, methyl ester or methyl arachidate $CH_3(CH_2)_{18}CO_2CH_3$	326.56	lf (MeOH)	215-6[10], 188[2]	54.5		1.4317^{60}	al, eth, bz, chl	B2[4], 1276
1373	Arachidic alcohol or l-Eicosanol $CH_3(CH_2)_{18}CH_2OH$	298.55	wx (al), cr (chl)	309, 220-5[3]	72-3	$0.8405^{20/4}$	1.4550^{20}	ace, bz	B1[4], 1900
1374	Arachidonic acid $CH_3(CH_2)_3(CH_2CH{=}CH)_4(CH_2)_2CO_2H$	304.44		d	− 49.5		1.4824^{20}	al, eth, ace	B2[4], 1802
1375	Aramite or Niagaramite $C_{15}H_{23}ClO_4S$	334.86		195[2]	− 37.3	$1.145^{20/20}$	1.5100^{20}	al, eth, ace, bz	M, 97
1376	Arbutin or Hydroquinone-β-d-glucopyranoside $C_{12}H_{16}O_7$	272.24	nd (w + l), $[\alpha]^{25}_D$ − 64 (w, c=3)		199.5-200(st, anh)			w, al	B17[4], 2983
1377	Arbutin hydrate $C_{12}H_{16}O_7$-H_2O	290.27	$[\alpha]^{17}_D$ − 60.3 (w, p=5)		142			al	B31, 210
1378	Arecaidine or l-Methyl-l,2,5,6-tetrahydroricotinic acid $C_7H_{11}NO_2$	141.17	pl (dil al), ta (dil al + lw)		232 (anh)			w	B22[3], 184
1379	Arecoline or Arecaidine methyl ester $C_8H_{13}NO_2$	155.20		209 (220), 94[1]		$1.0504^{20/20}$	1.4860^{20}	w, al, eth chl	B22[3], 185
1380	Arecoline hydrobromide $C_8H_{13}NO_2HBr$	236.11	mcl pr (al)		172 (177)			w, al	B22[3], 185

PHYSICAL CONSTANTS OF ORGANIC COMPOUNDS (Continued)

No.	Name, Synonyms, and Formula	Mol. wt.	Color, crystalline form, specific rotation and λ_{max} (log ε)	b.p. °C	m.p. °C	Density	n_D	Solubility	Ref.
1381	Arecoline hydrochloride $C_8H_{13}NO_2$-HCl	191.66	nd (al)		157-8			w, al	$B22^3$, 185
1382	Arginine (*DL*) or *DL*-2-Amino-5-guanido pentanoic acid $H_2NC(=NH)NH(CH_2)_3CH(NH_2)CO_2H$	174.20			217-8			w	$B4^3$, 1359
1383	Arginine (*L*) $C_6H_{14}N_4O_2$	174.20			244d			w	$B4^3$, 1348
1384	Arginine, benzylidene (*L*) $C_{13}H_{18}N_4O_2$	262.31	lf (w)		204-5			MeOH	$B7^3$, 837
1385	Arginine diflavinate (*L*) $C_6H_{14}N_4O_2(C_{10}H_6N_2O_8S)_2$	802.66	ye nd		202d				$B11^3$, 547
1386	Arginine dipicrate (*DL*) $C_6H_{14}N_4O_2(C_6H_3N_3O_7)_2$	632.41			196d				$B6^4$, 1450
1387	Arginine dipicrate (*L*) $C_6H_{14}N_4O_2(C_6H_3N_3O_7)_2$	632.41			200d				$B6^4$, 1450
1388	Arginine picrate (*D*) $C_6H_{14}N_4O_2{\cdot}C_6H_3N_3O_7{\cdot}2H_2O$	439.34	ye-og lf		258-60d				$B11^3$, 547
1389	Arginine picrate (*D*) $C_6H_{14}N_4O_2$-$C_6H_3N_3O_7{\cdot}2H_2O$	439.34	nd (w)		217-8d				$B6^4$, 1450
1390	Arginine picrate (*DL*) $C_6H_{14}N_4O_2$-$C_6H_3N_3O_7{\cdot}2H_2O$	439.34	pr (w)		200 (223d)				$B6^4$, 1450
1391	Arginine picrate (*L*) $C_6H_{14}N_4O_2$-$C_6H_3N_3O_7$-$2H_2O$	439.34	ye nd (w)		217-8d				$B6^4$, 1450
1392	Arsenic acid, diphenyl $(C_6H_5)_2AsO_2H$	262.14	nd pr		178			al, chl	$B16^2$, 443
1393	Arsenic acid, triethyl ester or Triethyl arsenate $(C_2H_5O)_3AsO$	226.10		235-8, 118-20^{15}		$1.3023^{20/0}$	1.4343^{20}		$B1^4$, 1358
1394	Arsenobenzene $C_{36}H_{30}As_6$	912.16			212			bz, al	$B16^3$, 1157
1395	Arsenous acid, triethyl ester or Treiethyl arsenite $(C_2H_5O)_3As$	210.10		165-6, 66-7^{12}		$1.2239^{20/4}$	1.4369^{13}		$B1^4$, 1358
1396	Arsine, *bis*(trifluoromethyl)iodo $(CF_3)_2AsI$	339.84	ye oil	92			1.425^{23}	eth	$B3^4$, 267
1397	Arsine, chloro,diphenyl $(C_6H_5)_2AsCl$	264.59	rh pl (peth)	337, 193^{20}	44	$1.48204^{16/4}$	1.6332^{56}	al, eth, ace, bz	$B16^3$, 944
1398	Arsine, dibromo,trifluoromethyl CF_3AsBr_2	303.75		118^{745}		1.528^{20}			$B3^4$, 267
1399	Arsine, dichloro,trifluoromethyl CF_3AsCl_2	214.83		133, 37^{25}	− 42.5	1.8358^{20}	1.5677^{15}	al, eth	$B3^4$, 267
1400	Arsine, dichloro,phenyl $C_6H_5AsCl_2$	222.93		254-7, 131^{14}		$1.6516^{20/4}$	1.6386^{15}	al, eth, ace, bz	$B16^3$, 958
1401	Arsine, dichloro (trifluoromethyl) CF_3AsCl_2	214.83		71		1.431^{20}		al, ace	$B3^4$, 268
1402	Arsine, diethyl $(C_2H_5)_2AsH$	134.05		105 (97)	$1.1338^{24/4}$	1.4709		al, eth, ace, bz	$B4^3$, 1797
1403	Arsine, difluoro,ethyl $C_2H_5AsF_2$	141.98	fume in air	94.3, 74^{100}	-38.7	1.708^{17}			$B4^3$, 1799
1404	Arsine, difluoro,methyl CH_3AsF_2	127.95	fume in air	76.5	-29.7	1.924^{18}			$B4^3$, 1796
1405	Arsine, difluoro,phenyl $C_6H_5AsF_2$	190.03	wax	110^{48}	42				$B16^3$, 959
1406	Arsine, diido,trifluoromethyl F_3CAsI_2	397.74		183d, 100^{48}					$B3^4$, 268
1407	Arsine, dimethyl or Cacodylhydride $(CH_3)_2AsH$	106.00	ign in air	36		$1.213^{29/29}$		al, eth, ace, bz	$B4^2$, 978
1408	Arsine, diphenyl $(C_6H_5)_2AsH$	230.14	oil	174^{25}		$1.30^{25/25}$		al, eth	$B16^3$, 918
1409	Arsine, ethyl $C_2H_5AsH_2$	106.00		36		$1.217^{22/22}$		al, eth	$B4^3$, 1797
1410	Arsine, 4-methoxyphenyl,oxide $(4\text{-}CH_3OC_6H_4)AsO$	198.05	cr (chl-eth)- or bz-peth		114-6 (anh)			bz, chl	$B16^2$, 448
1411	Arsine, methyl CH_3AsH_2	91.97		2	-143			al, eth, ace	$B4^3$, 1795
1412	Arsine, methyl, oxide CH_3AsO	105.96	pr (CS_2) (al)	275d	95			al, bz, chl	$B4^3$, 1817
1413	Arsine, pheny,oxide C_6H_5AsO	168.03	cr (bz-eth or chl-eth)		144-6			bz, chl	$B16^3$, 956

No.	Name, Synonyms, and Formula	Mol. wt.	Color, crystalline form, specific rotation and λ_{max} (log ε)	b.p. °C	m.p. °C	Density	n_D	Solubility	Ref.
1414	Arsine, triethyl $(C_2H_5)_3As$	162.11		138-9		$1.150^{20/4}$	1.467^{20}	al, eth, ace	B4[3], 1798
1415	Arsine, trifluoromethyl CF_3AsH_2	145.94		-11.6^{781}					B3[4], 266
1416	Arsine, trimethyl $(CH_3)_3As$	120.03		52	-87.3	1.144^{15}		al, eth, bz	B4[3], 1795
1417	Arsine, triphenyl $(C_6H_5)_3As$	306.24			61	$1.2634^{18/4}$	1.6888^{21}	eth, bz, chl	B16[3], 921
1418	Arsine, triphenyl, oxide $(C_6H_5)_3AsO$	322.24			192				B16[3], 1021
1419	Arsine, triphenyl, sulfide $(C_6H_5)_3AsS$	338.20			162				B16[3], 1026
1420	Arsine, *tris*-(pentafluoroethyl) $(C_2F_5)_3As$	431.96		96					C49, 14635
1421	Arsine, *tris*-(trifluoromethyl) $(CF_3)_3As$	281.94		33.3				eth	B3[4], 266
1422	Artemsic acid $C_{15}H_{16}O_3$	244 29	nd (dil aa)		135 6			al, eth, aa, lig	B10[3], 220
1423	Ascaridole $C_{10}H_{16}O_2$	168.24	unst, $[\alpha]_D$ − 4.14	exp^{760}, 115^{15}, $39—40^{0.2}$	3.3	$1.0103^{20/4}$	1.4769^{20}	al, ace, bz	B19[4], 164
1424	D-Ascorbic acid $C_6H_8O_6$	176.13	pl or mcl nd, $[\alpha]^{18}{}_D$ − 48 (MeOH, c=1), $[\alpha]^{20}{}_D$ − 23.8 (w, c=3)		192d			w, al	B18[4], 3046
1425	DL-Ascorbic acid $C_6H_8O_6$	176.13			168-9			w, al	B18[4], 3047
1426	L-Ascorbic acid or Vitamin C $C_6H_8O_6$	176.13	pl or mcl nd, $[\alpha]^{20/}{}_D$ + 24 (w, c = 1)		192d	1.65			B18[4], 3038
1427	L-Ascorbic acid, 6-desoxy $C_6H_8O_5$	160.13	pr (AcOEt), $[\alpha]^{22/}{}_D$ + 36.7 (O.1N HCl, c = 1)	sub, $160^{0.001}$	168			w, al, ace	B18[4], 2301
1428	β-Asparagine (*D*) $H_2NCOCH_2CH(NH_2)CO_2H{\cdot}H_2O$	150.13	$[\alpha]^{20/}{}_D$ + 5.4 (w, c = 1.3)		234.5 (anh)	$1.543^{15/4}$		w	B4[3], 1522
1429	β-Asparagine (*DL*) $H_2NCOCH_2CH(NH_2)CO_2H{\cdot}H_2O$	150.13	tcl cr (w + 1), pr (w al)	213-5d	182-3	$1.4540^{15/4}$		w	B4[3], 1524
1430	β-Asparagine (*L*) $H_2NCOCH_2CH(NH_2)CO_2H$	132.12	rh (w + 1), $[\alpha]^{20/}{}_D$ −5.42 (w, c = 1.3)		236 (anh)	$1.543^{15/4}$		w	B4[3], 1513
1431	Aspartic acid (*D*) or Aminosuccinic acid (*D*) $HO_2CCH_2CH(NH_2)CO_2H$	133.10	$[\alpha]^{20/}{}_D$ −25.5 (HCl)		269-71	$1.6613^{13/13}$		w, dil, HCl	B4[3], 1522
1432	Aspartic acid (*DL*) or Aminosuccinic acid $HO_2CCH_2CH(NH_2)CO_2H$	133.10	mcl pr (w)		338-9 (275 sealed tube)	$1.6632^{13/13}$		w	B4[3], 1523
1433	Aspartic acid (*L*) or Aminosuccinic acid $C_4H_7NO_4$	133.10	rh lf (w)		324d, (270 sealed tube)	$1.6613^{13/13}$		w, dil, HCl	B4[3], 1506
1434	Aspartic acid, *N*-benzoyl $HO_2CCH_2CH(NHCOC_6H_5)CO_2H$	237.21	nd or lf, $[\alpha]^{20/}{}_D$ + 37 (0.76N NaOH)		171-3				B9[3], 1189
1435	Aspidospermine or Vallesine $C_{22}H_{30}N_2O_2$	354.49	nd or pr (al), nd (peth)					al, bz, chl	J1940, 1051

PHYSICAL CONSTANTS OF ORGANIC COMPOUNDS (Continued)

No.	Name, Synonyms, and Formula	Mol. wt.	Color, crystalline form, specific rotation and λ_{max} (log ε)	b.p. °C	m.p. °C	Density	n_D	Solubility	Ref.
1436	Atabrin or Quinacrine dihydrochloride (*DL*) $C_{23}H_{30}ClN_3O\text{-}2HCl\text{-}2H_2O$	508.92	yesh nd (w), ye cr pw		248-50d			w, al, MeOH	$B21^4$, 6248
1437	Atisine or Anthorine $C_{22}H_{33}NO_2$	343.51	rh bipym		57-60			al, eth, chl	Am78, 4139
1438	Atisine-hydrochloride $C_{22}H_{33}NO_2HCl$	377.97	nd (dil al), $[\alpha]_D$ +28, (w, c=1.1)		340			w, al	C36, 4826
1439	Atrolactic acid or 2-hydroxy-2-phenyl propanoic acid $CH_3C(OH)(C_6H_5)CO_2H$	166.18	pr (w), $[\alpha]^{16.5/}{}_D$ +37.7 (al, c=3.5)		116.5-7			w, ace, bz, al	$B10^3$, 560
1440	Atropic acid or 2-phenylpropenoic acid $CH_2{=}C(C_6H_5)CO_2H$	148.16	lf (al), nd (w)	267d	106-7			al, eth, bz, chl	$B9^3$, 2751
1441	Atropine or (*DL*)-hyoscyamine $C_{17}H_{23}NO_3$	289.37	rh nd (dil al), orth pr (ace)	sub(vac) 93-100	118-9			al, eth, bz, chl	$B21^4$, 183
1442	Atropine hydrochloride $C_{17}H_{23}NO_3\text{-}HCl$	325.84	nd (al)		165			w, al	$B21^4$, 184
1443	Atropine pentanoate $C_{17}H_{23}NO_3\text{-}C_5H_{10}O_2\text{-}\frac{1}{2}H_2O$	400.52	cr		42			w, al, eth	M, 111
1444	Atropine sulfate, hydrate $(C_{17}H_{23}NO_3)_2\cdot H_2SO_4\cdot H_2O$	694.82	nd (al eth or al ace)	sub	194 (anh)			w, al	$B21^4$, 182
1445	Auramine or *bis*(*p*-dimethylaminophenyl)methylene imine $C_{17}H_{21}N_3$	267.37	ye or col pl (al)		136			al	$B14^3$, 227
1446	Auramine hydrochloride or Auramine O $C_{17}H_{21}N_3\text{-}HCl$	303.83	ye nd (w)		267			al, chl	$B14^2$, 58
1447	Auramine-*N* $C_{18}H_{23}N_3$	281.40	ye cr (al)		133			al, ace, aa	B14, 93
1448	Aurin or Rosolic acid $C_{19}H_{14}O_3$	290.32	dk red lf or rh		308-10d			al, eth	$B8^3$, 3024
1449	Auroemycin or Chlorotetracycline $C_{22}H_{23}N_2O_8Cl$	478.89	gold ye, ye flr, $[\alpha]^{25/}{}_D$ −275 (MeOH)		168-9				$B14^3$, 1710
1450	Auroemycin hydrochloride $C_{22}H_{23}N_2O_8Cl\cdot HCl$	515.35	ye or th, $[\alpha]^{20/}{}_D$ −106.5 (dil al)		216d				$B14^3$, 1710
1451	Auxin A or Auxenetriolic acid $C_{18}H_{32}O_5$	328.45	hex (al-lig), $[\alpha]^{20/}{}_D$ −3.19 (al)		196	$1.292^{19/4}$		al	$B10^3$, 2045
1452	Auxin B or Auxenolonic acid $C_{18}H_{30}O_4$	310.43	cr (al-lig), $[\alpha]^{20/}{}_D$ −2.8 (al)		183	$1.269^{20/4}$		al, eth	$B10^3$, 4194
1453	Azelaic acid or Nonanedioic acid $HO_2C(CH_2)_7CO_2H$	188.22	lf or nd	287^{100}, 225^{10}	106.5	$1.225^{25/4}$	1.4303^{111}	al, eth	$B2^4$, 2055
1454	Azelaic acid, diethylester or Diethyl azelate $C_2H_5O_2C(CH_2)_7CO_2H_5$	244.33		291-2, $174\text{-}5^{20}$	-18.5	$0.97294^{20/4}$	1.43509^{20}	al, eth	$B2^3$, 1787
1455	Azelaic acid, di(2-ethylbutyl) ester $H_2C[(CH_2)_3{}'CO_2CH_2CH(C_2H_5)]_2$	356.54		230^5	-45	$0.928^{25/4}$	1.443^{25}	al, ace, bz	$B2^3$, 1787
1456	Azelaic acid, di-(2-ethylhexyl) ester $CH_2[(CH_2)_3CO_2CH_2CH(C_2H_5)C_4H]_2$	412.65		237^5	-78	1.446^{25}		al, ace, bz	$B2^3$, 1787
1457	Azelaic acid, dimethylester or Dimethyl azelate $CH_3O_2C(CH_2)_7CO_2CH_3$	216.28		156^{20}		$1.0082^{20/4}$	1.4367^{20}	al, ace, bz	$B2^3$, 1786
1458	Azelaic acid, diphenylester or Diphenyl azelate $C_6H_5O_2C(CH_2)_7CO_2C_6H_5$	340.42	nd (al)		59-60			eth, bz	$B6^3$, 606
1459	Azelanitrile or Azelaic acid, dinitrile $NC(CH_2)_7CN$	150.22		$198\text{-}9^{25}$		0.9200^{19}	1.4518^{19}	al, eth, bz	$B2^4$, 2059
1460	Azelayl chloride $ClOC(CH_2)_7COCl$	225.11		166^{18}		1.4680^{20}		eth, bz	$B2^3$, 1789
1461	l-Azacyclooctane, 2-methyl or α-Methyl heptamethylenimine $C_8H_{17}N$	139.24	nd (ace)	$162\text{-}3^{746}$	156-7	0.853^{20}	1.4620^{21}		$B20^1$, 30
1462	Azetidine or Trimethyleneimine C_3H_7N	57.10		63^{748}		0.8436^{20}	1.4287^{25}	w, al, eth, ace, bz	$B20^4$, 53

No.	Name, Synonyms, and Formula	Mol. wt.	Color, crystalline form, specific rotation and λ_{max} (log ε)	b.p. °C	m.p. °C	Density	n_D	Solubility	Ref.
1463	Aziridine or Ethyleneimine C_2H_5N	43.07		56^{756}		$0.8321^{20/4}$		w, al, eth, ace, bz	$B20^4$, 3
1464	Azobenzene (*cis*) $C_6H_5N{=}NC_6H_5$	182.22	og-red pl (peth)		71			al, eth, bz, aa	$B16^3$, 4
1465	Azobenzene (*trans*) $C_6H_5N{=}NC_6H_5$	182.22	og red mcl lf (al)	29.3	68.5	$1.203^{20/4}$	1.6266^{78}	al, eth, bz, aa	$B16^3$, 5
1466	Azobenzene, 2-acetamide-4′,5′-dimethyl $2(CH_3CONH)\text{-}5\text{-}CH_3\text{-}C_6H_3N{=}N(C_6H_4CH_3\text{-}4)$	267.33	ye nd (al-aa)		157			al, eth, chl	$B16^2$, 182
1467	Azobenzene, acetamido $4\text{-}(CH_3CONH)C_6H_4N{=}NC_6H_5$	239.28	gold-ye nd (al)		144-6				$B16^3$, 346
1468	Azobenzene, acetamido-2′,3-dimethyl $4(CH_3CONH)\text{-}3\text{-}CH_3\text{-}C_6H_3N{=}N(C_6H_4CH_3\text{-}2)$	267.33	red nd (al)		186-7			eth, chl	$B16^3$, 387
1469	Azobenzene, 4-acetoxy-2′-methyl $4(CH_3CO_2)C_6H_4N{=}N(C_6H_4\text{-}CH_3\text{-}2)$	254.29	red-ye lf (al)		68			al	B16, 105
1470	Azobenzene, 4-acetoxy-3-methyl $4\text{-}(CH_3CO_2)\text{-}3\text{-}(CH_3)C_6H_3N{=}NC_6H_5$	254.29	ye pl (dil al)		81-2			al, eth, bz, chl	B16, 130
1471	Azobenzene, 4-acetoxy-4′-methyl $4\text{-}(CH_3CO_2)C_6H_4N{=}N(C_6H_4CH_3\text{-}4)$	254.29	og nd (al, bz)		98			al	$B16^2$, 42
1472	Azobenzene, 2-amino or 2-Benzeneazoaniline $2\text{-}H_2NC_6H_4N{=}N\text{-}C_6H_5$	197.24	red nd or pr (al)		59			eth, ace, bz	$B16^3$, 334
1473	Azobenzene, 3-amino $3\text{-}H_2NC_6H_4N{=}NC_6H_5$	197.24	(i) og-ye nd (peth), (ii) br-red cr		(i) 69-70, (ii) 90-d1			al, eth, ace, bz, chl	$B16^3$, 335
1474	Azobenzene, amino $4\text{-}H_2NC_6H_4N{=}NC_6H_5$	197.24	og mcl nd (al)	>360	127			al, eth, bz, chl	$B16^2$, 149
1475	Azobenzene, 4-amino-2,2′-dimethyl $4(H_2N)\text{-}2\text{-}(CH_3)C_6H_3N{=}N(C_6H_4CH_3\text{-}2)$	225.29	ye nd (lig)		116-7			al, lig	$B16^3$, 180
1476	Azobenzene, 4-amino-2,3′-dimethyl $4\text{-}(H_2N)\text{-}2\text{-}(CH_3)\text{-}C_6H_3N{=}N(C_6H_4CH_3\text{-}3)$	225.29	ye-gold nd (al), ye-br nd (lig)		80			al	$B16^3$, 392
1477	Azobenzene, 4-amino-2′,3-dimethyl $4\text{-}(H_2N)\text{-}3\text{-}CH_3C_6H_3N{=}N(C_6H_4CH_3\text{-}2)$	225.29	ye lf (al)		101.5-3			al, eth	$B16^3$, 386
1478	Azobenzene, 4-amino-2,4′-dimethyl $4(H_2N)\text{-}2CH_3C_6H_3N{=}N(C_6H_4CH_3\text{-}4)$	225.29	ye pl (al)-gold-ye (lig)		127			al, lig	B16, 348
1479	Azobenzene, 4-amino-3,3′-dimethyl $4(H_2N)\text{-}3\text{-}(CH_3)C_6H_3N{=}N(C_6H_4CH_3\text{-}3)$	225.29	ye br lf or nd (lig)		124			al, lig	B16, 345
1480	Azobenzene, 4-amino-3,4′-dimethyl $4(H_2N)\text{-}3\text{-}(CH_3)C_6H_3N{=}N(C_6H_4CH_3\text{-}4)$	225.29	og-ye nd (lig), ye pl (al)		128				$B16^3$, 389
1481	Azobenzene, 4-amino, hydrobromide $4\text{-}H_2NC_6H_4N{=}NC_6H_5\text{-}HBr$	278.15	bk-vt nd (dil al)		206-7				B16, 307
1482	Azobenzene, 4-amino, hydrochloride $4\text{-}H_2NC_6H_4N{=}NC_6H_5\text{-}HCl$	233.70	bl-vt or pa red, nd or pw		240			w, al	$B16^2$, 149
1483	Azobenzene, 4-benzoxy-4′-methyl $4\text{-}(C_6H_5CH_2O)C_6H_4N{=}N(C_6H_5CH_3\text{-}4)$	302.38	pa ye lf (lig)		128			al, eth, bz	B16, 107
1484	Azobenzene, 3,3′-*bis*(dimethylamino) $3\text{-}(CH_3)_2NC_6H_4N{=}N(C_6H_4N(CH_3)_2\text{-}3)$	268.36	red nd (al)		118			al, bz	B16, 305
1485	Azobenzene, 4,4′-*bis*-(dimethylamino) $4\text{-}(CH_3)_2NC_6H_4N{=}N[C_6H_4N(CH_3)_2\text{-}4]$	268.36	og nd (bz)	sub	273			eth, bz, chl	$B16^3$, 375
1486	Azobenzene, 4-bromo $4\text{-}BrC_6H_4N{=}NC_6H_5$	261.12	og nd (lig)		90-1			al, eth, ace, lig	$B16^3$, 29
1487	Azobenzene, 2,2′-diamino $2\text{-}H_2NC_6H_4N{=}N(C_6H_4NH_2\text{-}2)$	212.25	red pl (al or bz)		134			eth	$B16^2$, 148
1488	Azobenzene, 2,4-diamino or Chrysoidine $2,4\text{-}(H_2N)_2C_6H_3N{=}NC_6H_5$	212.25	pa ye nd (w)		117.5			al, eth, bz	$B16^3$, 436
1489	Azobenzene, 4,4′-diamino or *o*-Azodianiline $4\text{-}H_2NC_6H_4N{=}N(C_6H_4NH_2\text{-}4)$	212.25	gold-ye nd (al), og-ye pr (al)		250-1			al, bz, chl	$B16^3$, 375
1490	Azobenzene, 2,2′-diethoxy or *o*-Azophenetole $2\text{-}C_2H_5OC_6H_4N{=}N(C_6H_4OC_2H_5\text{-}2)$	270.33	red pr (al)	240d	131			al, eth	$B16^3$, 83
1491	Azobenzene, 4,4′-diethoxy or *p*-Azophenetole $4\text{-}C_2H_5OC_6H_4N{=}N(C_6H_4OC_2H_5\text{-}4)$	270.33	ye-lf (al)	d	162			eth, bz, chl, aa	$B16^3$, 93

No.	Name, Synonyms, and Formula	Mol. wt.	Color, crystalline form, specific rotation and λ_{max} (log ε)	b.p. °C	m.p. °C	Density	n_D	Solubility	Ref.
1492	Azobenzene, 2,2′-dihydroxy or *o*-Azophenol $2-HOC_6H_4N=N(C_6H_4OH-2)$	214.22	gold-ye lf (bz), nd (al)	sub	172			eth, al, bz	B16^{2}, 33
1493	Azobenzene, 2,4-dihydroxy $2,4(HO)_2C_6H_3N=NC_6H_5$	214.22	dk red nd (dil al)		170 (anh)			al, eth, bz, aa	B16^{3}, 166
1494	Azobenzene, 2,4-dihydroxy-4′-nitro $2,4-(HO)_2C_6H_3N=N(C_6H_4NO_2-4)$	259.22	red pw (al or MeOH)		200				B16^{3}, 162
1495	Azobenzene, 2,5-dihydroxy $2,5(HO)_2C_6H_3N=NC_6H_5$	214.22			149			eth, ace, bz	B16^{3}, 176
1496	Azobenzene, 3,3′-dihydroxy or *m*-Azophenol $3-HOC_6H_4N=N(C_6H_4OH-3)$	214.22	ye lf (dil al)		207			eth, ace	B16^{3}, 85
1497	Azobenzene, 4-dimethylamino or Butteryellow $4-(CH_3)_2NC_6H_4=NC_6H_5$	225.30	ye lf (al)	d	117			al, eth, aa, Py	B16^{3}, 340
1498	Azobenzene, 4,4′-dihydroxy or *p*-Azophenol $4-HOC_6H_4N=N(C_6H_4OH-4)$	214.22	og-ye pl (dl al, + 1w)		216-6.5 anh			al, eth, ace	B16^{3}, 92
1499	Azobenzene, 2,2′-dimethyl or *o*-Azotoluene $2-CH_3C_6H_4N=N(C_6H_4CH_3-2)$	210.28	dk red mcl pr (eth)		55-6	$1.0215^{65/4}$	1.6180^{65}	al, eth, bz	B16^{2}, 19
1500	Azobenzene, 2,2′-dimethyl-4-ethoxy $2-CH_3-4-C_2H_5OC_6H_3N=N(C_6H_4CH_3-2)$	254.34	dk red nd (al)		64			al, eth, ace, bz	B16, 134
1501	Azobenzene, 2,2′-dimethyl-4-hydroxy or 4-*o*-Tolueneazo-*m*-cresol $2-CH_3-4-HOC_6H_3N=N(C_6H_4CH_3-2)$	226.28	ag-red pl (bz), red cr (w + l)		113 (anh)			al, eth, bz	B16^{2}, 61
1502	Azobenzene, 2,3′-dimethyl-4-ethoxy $2-CH_3-4-C_2H_5OC_6H_3N=N(C_6H_4CH_3-3)$	254.33	red pr (al)	73				al, eth, ace, bz	B16, 135
1503	Azobenzene, 2′,3-dimethyl-4-ethoxy $3-CH_3-4-C_2H_5OC_6H_3N=N(C_6H_4CH_3-2)$	254.33	red cr (lig)		35-7			al, bz, lig	B16, 131
1504	Azobenzene, 2,3-dimethyl-4-hydroxy or 4-Benzeneazo-*o*-xylenol $2,3(CH_3)_2-4-HO-C_6H_2N=NC_6H_5$	226.28	red pr (al, lig)		132			al, eth, ace, bz	B16^{2}, 63
1505	Azobenzene, 2,3′-dimethyl-4-hydroxy or 4-*m*-Tolueneazo-*m*-cresol $2-CH_3-4-HOC_6H_3N=N(C_6H_4CH_3-3)$	226.28	og-ye pl (bz)		106-7			eth, bz, lig	B16^{3}, 107
1506	Azobenzene, 2,4′-dimethyl-4-ethoxy $2-CH_3-4-C_2H_5OC_6H_3N=N(C_6H_4CH_3-4)$	254.33	og-red pl (al)		64			al, eth, bz, lig	B16, 135
1507	Azobenzene, 2,4-dimehyl-5-hydroxy or 5-benzeneazo-2,4-xylenol $2,4-(CH_3)_2-5-HOC_6H_2N=NC_6H_5$	226.28	og-ye nd (lig-peth)		114			al, eth, ace, bz, aa	B16, 146
1508	Azobenzene, 2,4′dimethyl-4-hydroxy or 4-*p*-Tolueneazo-*m*-cresol $2-CH_3-4-HOC_6H_3N=N(C_6H_4CH_3-4)$	226.28	og-ye pr (bz)		135			al, eth, bz	B16^{2}, 62
1509	Azobenzene, 3,3′-dimethyl *cis* or *m*-Azotoluene $3-CH_3C_6H_4N=N(C_6H_4CH_3-3)$	210.28	red (peth)		46				B16^{3}, 47
1510	Azobenzene, 3,3′-dimethyl *trans* or *m*-Azotoluene $3-CH_3C_6H_4N=N(C_6H_4CH_3-3)$	210.28	og-red orh		84-4.5	1.0123^{66}	1.6152^{66}	al, eth, bz	B16^{3}, 48
1511	Azobenzene, 3,3′-bis(dimethylamino) $3(CH_3)_2N-C_6H_4N=NC_6H_4[N(CH_3)_2]-3$	268.36	og nd (bz)	sub	273			eth, bz, chl	B16^{3}, 339
1512	Azobenzene, 3,3′-dimethyl-4-ethoxy $3-CH_3-4-C_2H_5OC_6H_3N=N(C_6H_4CH_3-3)$	254.33	red-ye pl (al)		46-7			al, eth, ace, bz, lig	B16, 131
1513	Azobenzene, 3,3′-dimethyl-4-hydroxy or 4-*m*-Tolueneazo-*o*-cresol $3-CH_3-4-HOC_6H_3N-N(C_6H_4CH_3-3)$	226.28	gold-ye nd (bz)		115			al, eth, bz	B16^{2}, 60
1514	Azobenzene, 3,4′-dimethyl-4-ethoxy $3-CH_3-4-C_2H_5OC_6H_3N=N(C_6H_4CH_3-4)$	254.33	og-ye nd (al)	251^{42}	73-4			al, eth, ace, bz	B16, 131
1515	Azobenzene, 3,4′-dimethyl-4-hydroxy or 4-*p*-Tolueneazo-*o*-cresol $3-CH_3-4-HOC_6H_3N=N(C_6H_4CH_3-4)$	226.28	og nd (bz)		163			al, eth, bz	B16^{3}, 106
1516	Azobenzene, 3,5-dimethyl-2-hydroxy or 6-Benzeneazo-*m*-4-xylenol $3,5-(CH_3)_2-2-HOC_6H_2N=NC_6H_5$	226.28	dk red nd (al, lig)		90			al, eth, bz, lig	B16, 145
1517	Azobenzene, 3,5-dimethyl-4-hydroxy or 4-Benzeneazo-2,6-xylenol $3,5(CH_3)_2-4-HOC_6H_2N=NC_6H_5$	226.28	og-red cr (lig)		150-1			al, eth, bz	B16, 145
1518	Azobenzene, 4,4′-dimethyl *cis* or *p*-Azotoluene $4-CH_3C_6H_4N=N(C_6H_4CH_3-4)$	210.28	dk red		104			bz, lig	B16^{3}, 49
1519	Azobenzene, 4,4′-dimethyl *trans* or *p*-Azotoluene $4-CH_3C_6H_4N=N(C_6H_4CH_3-4)$	210.28	og ye nd (al)		144			eth, bz, lig	B16^{3}, 49
1520	Azobenzene, 4,4′-bis(dimethyl amino) $4-(CH_3)_2N-C_6H_4-N=NC_6H_4[N(CH_3)_2-4]$	268.36			273			eth, bz	B16^{3}, 375

No.	Name, Synonyms, and Formula	Mol. wt.	Color, crystalline form, specific rotation and λ_{max} (log ε)	b.p. °C	m.p. °C	Density	n_D	Solubility	Ref.
1521	Azobenzene, 4,4′-dimethyl-2-hydroxy or 6-*p*-Toluenеazo-*m*-cresol $4\text{-}CH_3\text{-}2\text{-}HOC_6H_3N{=}N(C_6H_4CH_3\text{-}4)$	226.28	og red cr (lig)	...	150-1	...	...	al, eth, bz, lig	B16^3, 106
1522	Azobenzene, 4′,5-dimethyl-2-ethoxy $5\text{-}CH_3\text{-}2\text{-}C_2H_5OC_6H_3N{=}N(C_6H_4CH_3\text{-}4)$	254.33	pa red nd (abs al)	253-5^{61}	43	...	...	al	B16, 141
1523	Azobenzene, 4′,5-dimethyl-2-hydroxy or 2-*p*-Tolueneazo-P-cresol $5\text{-}CH_3\text{-}2\text{-}HOC_6H_3N{=}N(C_6H_4CH_3\text{-}4)$	226.28	red cr, ye pl (to)	...	112-3	...	...	al, bz, to, chl	B16^3, 109
1524	Azobenzene, 2-ethoxy or *o*-Benzeneazo phenetole $2\text{-}C_2H_5OC_6H_4N{=}NC_6H_5$	226.28	red pl or mcl pr (peth)	...	44	...	...	al, eth, ace, bz, peth	B16^3, 81
1525	Azobenzene, 3-ethoxy or *m*-Benzeneazo phenetole $3\text{-}C_2H_5OC_6H_4N{=}NC_6H_5$	226.28	pl (peth)	208^{22}	63.5-4	...	...	al, eth, ace, bz	B16, 95
1526	Azobenzene, 4-ethoxy or *p*-Benzenazo phenetole $4\text{-}C_2H_5OC_6H_4N{=}NC_6H_5$	226.28	og nd (60-70% al)	339-40	85	1.0400$^{100/}{}_4$	1.6419$^{100/}{}_{5875}$	al, eth, ace, bz	B16^2, 40
1527	Azobenzene, 4-ethoxy-2-methyl $2\text{-}CH_3\text{-}4\text{-}C_2H_5OC_6H_3N{=}NC_6H_5$	240.30	or red nd (dil al)	...	51.5-2.0	...	...	al, eth, lig	B16, 134
1528	Azobenzene, 4-ethoxy-2′-methyl $4\text{-}C_2H_5OC_6H_4N{=}N(C_6H_4CH_3\text{-}2)$	240.30	og pl (al)	...	53	...	...	al, eth, bz, chl	B16, 105
1529	Azobenzene, 4-ethoxy-3-methyl $3\text{-}CH_3\text{-}4\text{-}C_2H_5OC_6H_3N{=}NC_6H_5$	240.30	og nd or mcl pr (al)	...	60	...	...	al, eth, bz	B16, 130
1530	Azobenzene, 4-ethoxy-3′-methyl $4\text{-}C_2H_5OC_6H_4N{=}N(C_6H_4CH_3\text{-}3)$	240.30	og-red pr (al)	...	65	...	...	al, eth, bz	B16, 106
1531	Azobenzene, 4-ethoxy-4′-methyl $4\text{-}C_2H_5OC_6H_4N{=}(C_6H_4CH_3\text{-}4)$	240.30	red lf (al)	...	121-2	...	...	al, bz, chl	B16, 107
1532	Azobenzene, 2-hydroxy or *o*-Benzeneazo phenol $2\text{-}HOC_6H_4N{=}N\text{-}C_6H_5$	198.22	og-red nd (eth)	...	82.5-3	...	...	al, eth, ace, bz	B16^1, 80
1533	Azobenzene, 2-hydroxy, benzoate $2.C_6H_5CO_2C_6H_4N{=}NC_6H_5$	302.33	og-red nd (lig)	...	93	...	...	al, bz	B16^3, 81
1534	Azobenzene, 2-hydroxy-4-methyl $4\text{-}CH_3\text{-}2\text{-}HOC_6H_3N{=}NC_6H_5$	212.25	red pl (lig)	...	122	...	...	al, ace, bz, lig	B16^1, 241
1535	Azobenzene, 2-hydroxy-4-methyl,benzoate $2\text{-}C_6H_5CO_2\text{-}4\text{-}CH_3C_6H_3N{=}NC_6H_5$	316.36	...	...	98	...	...	bz	B16^1, 241
1536	Azobenzene, 2-hydroxy-5-methyl or 2-benzeneazo-*p*-cresol $2\text{-}HO\text{-}5\text{-}CH_3C_6H_3N{=}NC_6H_5$	212.25	og-ye lf (bz, bz-lig, gold lf) (w-al)		108-9	...	...	al, eth, bz, chl	B16, 108
1537	Azobenzene, 3-hydroxy or *m*-Benzeneazo phenol $3\text{-}HOC_6H_4N{=}NC_6H_5$	198.22	ye nd (w, lig), pr (bz)	...	116.5-7	...	...	al, eth, ace, bz	B16^3, 85
1538	Azobenzene, 3-hydroxy, acetate $3\text{-}CH_3CO_2C_6H_4N{=}NC_6H_5$	240.26	og pl (peth)	...	67.5	...	...	peth	B16, 95
1539	Azobenzene, 3-hydroxy, benzoate $3\text{-}C_6H_5CO_2\text{-}C_6H_4N{=}N\text{-}C_6H_5$	302.33	og-red pl (peth)	...	92	...	...	peth	B16, 95
1540	Azobenzene, 4-hydroxy or *p*-Benzeneazo phenol $4\text{-}HOC_6H_4N{=}NC_6H_5$	198.22	ye lf (bz), og pr (al)	220-30$^{20/d}$	155-7	...	...	al, eth, bz	B16^2, 38
1541	Azobenzene, 4-hydroxy acetate $4\text{-}CH_3CO_2C_6H_4N{=}N\text{-}C_6H_5$	240.26	og lf (al), nd (al, lig)	> 360 d	89	...	...	lig, al	B16^1, 236
1542	Azobenzene, 4-hydroxy,benzoate $4\text{-}C_6H_5CO_2C_6H_4N{=}NC_6H_5$	302.33	og-ye nd (al, lig), ye-red pr (eth, al)	...	138	...	...	to, lig, al, eth	B16^3, 89
1543	Azobenzene, 4-hydroxy-2′-methoxy-2-methyl $2\text{-}CH_3\text{-}4\text{-}HOC_6H_3N{=}N(C_6H_4OCH_3\text{-}2)$	242.28	og	...	161	...	...	al, eth	B16, 135
1544	Azobenzene, 4-hydroxy-2′-methoxy-3-methyl $3\text{-}CH_3\text{-}4\text{-}HO\text{-}C_6H_3N{=}N(C_6H_4OCH_3\text{-}2)$	242.28	og red pl (dil al)	...	68	...	...	al, eth, bz, lig	B16, 131
1545	Azobenzene, 4-hydroxy-2-methyl or 4-Benzeneazo-*m*-cresol $4\text{-}HO\text{-}2\text{-}CH_3C_6H_3N{=}NC_6H_5$	212.25	ye nd (lig)	...	109	...	...	al, eth, bz, chl	B16^2, 61
1546	Azobenzene, 4-hydroxy-2′-methyl or *o*-Toluenwazo-*p*-phenol $4\text{-}HOC_6H_4N{=}N(C_6H_4CH_3\text{-}2)$	212.25	red pl or lf (bz-lig), og-ye nd (bz)	...	107-8	...	...	al, eth, bz, chl	B16^3, 91
1547	Azobenzene, 4-hydroxy-3-methyl or 4-Benzeneazo-*o*-cresol $4\text{-}HO\text{-}3\text{-}CH_3C_6H_3N{=}NC_6H_5$	212.25	gold-ye lf or nd (al)	...	128-30	...	...	al, eth, bz, lig	B16^3, 104
1548	Azobenzene, 4-hydroxy-3-methyl, benzoate $4\text{-}C_6H_5CO_2\text{-}3\text{-}CH_3C_6H_3N{=}NC_6H_5$	316.36	ye nd (al)	...	110-1	...	...	eth, ace, chl	B16, 130

No.	Name, Synonyms, and Formula	Mol. wt.	Color, crystalline form, specific rotation and λ_{max} (log ε)	b.p. °C	m.p. °C	Density	n_D	Solubility	Ref.
1549	Azobenzene, 4-hydroxy-3′-methyl or *m*-Tolueneazo-*p*-phenol $4-HO-C_6H_4N=N(C_6H_4CH_3-3)$	212.25	ye-pl (al)		144-5			al, bz	$B16^2$, 41
1550	Azobenzene, 4-hydroxy-4′-methyl or *p*-Tolueneazo-*p*-phenol $4-HOC_6H_4N=N(C_6H_4CH_3-4)$	212.25	og red mcl (bz-lig)		152			al, eth, bz	$B16^2$, 42
1551	Azobenzene, 4-hydroxy-4′-methyl, benzoate $4-C_6H_5CO_2C_6H_4N=N(C_6H_4CH_3-4)$	316.36	og-red pr (bz), redsh-ye nd (lig)		178			al, bz, eth	$B16^1$, 237
1552	Azobenzene, 2-methoxy or *o*-Benzeneazo anisole $2-CH_3OC_6H_4N=NC_6H_5$	212.25	og-red nd (dil al)	$195-7^{14}$	41			al, bz	$B16^3$, 81
1553	Azobenzene, 3-methoxy or *m*-Benzeneaz anisole $3-CH_3OC_6H_4N=NC_6H_5$	212.25	og-red pl (MeOH)	193^{15}	32-3			al, ace	$B16^3$, 85
1554	Azobenzene, 4-methoxy or *p*-Benzeneazo anisole $4-CH_3OC_6H_4N=NC_6H_5$	212.25	og-red pl, lf (al) (peth)	340	56 (64)	1.12^{75}		al, eth, ace	$B16^2$, 40
1555	Azobenzene, 4-methoxy-4′-methyl $4-CH_3OC_6H_4N-N(C_6H_4CH_3-4)$	226.28	og-ye pr (al)		110-1			al, eth, ace, bz	$B16^2$, 236
1556	Azobenzene, 4-nitro $4-NO_2C_6H_4N=NC_6H_5$	227.22	lf, nd, (al, lig)		135 (155)			al, ace, bz, lig	$B16^2$, 17
1557	Azobenzene, 4-phenylamino or 4-Anilineazobenzene $4-C_6H_5NHC_6H_4N=NC_6H_5$	273.34	ye pl or pr		82			al, eth, lig	$B16^3$, 343
1558	Azobenzene, 2,4,3′-triamino or Bismark's brown $2,4(H_2N)_2C_6H_3N=N(C_6H_4NH_2-3)$	226.23	og lf (w), red (bz)		143-5			al,eth	B16, 386
1559	2-Azobenzene carboxylic acid or *o*-Benzeneazobenzoic acid $2-HO_2C-C_6H_4N=NC_6H_5$	226.24	og-red nd or pl (al)		97-8			al, eth, ace, bz, aa	$B16^3$, 213
1560	2-Azobenzene carboxylic acid, 4′-dimethylamino or Methyl red $2-HO_2CC_6H_4N=N-[C_6H_4N(CH_3)_2-4]$	269.30	vt or red pr (to or bz), nd (ag aa)		183			al, ace, bz, aa	$B16^2$, 164
1561	3-Azobenzene carboxylic acid or *m*-Benzeneazobenzoic acid $3-HO_2CC_6H_4N=NC_6H_5$	226.23	og-red pl (w), red lf (al)		170-1			al, eth, bz, chl	B16, 229
1562	3-Azobenzene carboxylic acid, 4-hydroxy or 5-Benzeneazosalicylic acid $3-HO_2C-4-HOC_6H_3NC_6H_5$	242.23	nd (bz)		230d			al, eth, ace	$B16^3$, 237
1563	3-Azobenzene carboxylic acid, 4-hydroxy-2′-nitro or 5-*o*-Nitrobenzeneazosalicylic acid $3-(HO_2C)-4-HOC_6H_3N=N(C_6H_4NO_2-2)$	287.23	br-red cr (al)		215-7			al, aa	$B16^2$, 101
1564	3-Azobenzene carboxylic acid, 4-hydroxy-3′-nitro or 5-*m*-Nitrobenzeneazosalicylic acid $3-HO_2C-4-HOC_6H_3N=N[C_6H_4NO_2-3]$	287.23	red-br nd (al)					al, eth, bz, aa	$B16^3$, 238
1565	3-Azobenzene carboxylic acid, 4-hydroxy, 4′-nitro or 5-*p*-nitro benzeneazosalicylic acid $3-(HO_2C)-4-HOC_6H_3N=N(C_6H_4NO_2-4)$	287.23	og br nd (dil aa)		257			al, aa	$B16^2$, 101
1566	3-Azobenzene carboxylic acid, 3′-sulfo $3(HO_2C)C_6H_4N=(C_6H_4SO_3H-3)$								B16268
1567	4-Azobenzene carboxylic acid or *p*-Benzeneazobenzoic acid $4-HO_2CC_6H_4N=NC_6H_5$	226.23	red pl or lf (al)		249			al, eth, ace, bz	$B16^3$, 218
1568	2,2′-Azobenzenedicarboxylic acid or *o*-Azobenzoic acid $2-HO_2CC_6H_4N=N(C_6H_4CO_2H-2)$	270.24	dk red nd (al)		245 (cor)			eth, al	$B16^3$, 216
1569	3,3′-Azobenzenedicarboxylic acid or *m*-Azobenzoic acid $3-HO_2CC_6H_4N=NC_6H_4CO_2H-3)$	270.24	ye nd (aa)		340d			al	$B16^3$, 217
1570	4,4′-Azobenzeenedicarboxylic acid or *p*-Azobenzoic acid $4-HO_2CC_6H_4N=N(C_6H_4CO_2H-4)$	270.24	dk ye nd (al), red nd (aa)		330d				$B16^3$, 228
1571	3,3′-Azobenzenedisculfonamide $3-(H_2NO_2S)C_6H_4N=N(C_6H_4SO_2NH_2-3)$	340.37	ye nd (al)		305			al	B16, 268
1572	3,4′-Azobenzene disulfonamide $3-H_2NO_2SC_6H_4N=N(C_6H_4SO_2NH_2-4)$	340.37	ye nd (al)		288			al	B16, 279
1573	4,4′-Azobenzene disulfonamide $4-H_2NO_2SC_6H_4N=N(C_6H_4SO_2NH_2-4)$	340.37	og nd (al)		> 250d			al, eth	$B16^3$, 299
1574	3,3′- Azobenzene disulfonic acid or *p*-Azobenzoic acid $3-HO_3SC_6H_4N=N(C_6H_4SO_3H-3)$	342.34	ye lf (w + 5)					w, al, eth	$B16^3$, 281
1575	3,4′-Azobenzne disulfonic acid $3-HO_3SC_6H_4N=N(C_6H_4SO_3H-4)$	342.34	syr						$B16^3$, 299

No.	Name, Synonyms, and Formula	Mol. wt.	Color, crystalline form, specific rotation and λ_{max} (log ε)	b.p. °C	m.p. °C	Density	n_D	Solubility	Ref.
1576	4,4′-Azobenzene disulfonic acid $4\text{-}HO_3SC_6H_4N{=}N(C_6H_4SO_3H\text{-}4)$	342.34	red nd (w + 5)		169d (anh)			w	$B16^3$, 299
1577	3,3′-Azobenzene disulfonyl chloride $3\text{-}(ClO_2S)C_6H_4N{=}N(C_6H_4SO_2Cl\text{-}3)$	379.23	red nd (eth)		166-7			eth	B16, 268
1578	3,4′-Azobenzene disulfonyl chloride $3\text{-}ClO_2SC_6H_4N{=}N(C_6H_4SO_2Cl\text{-}4)$	379.23	red nd (eth)		123-5			eth	B16, 279
1579	4,4′-Azobenzene disulfonyl chloride $4\text{-}ClO_2SC_6H_4N{=}N(C_6H_4SO_2Cl\text{-}4)$	379.23	br-red nd (eth-bz)					bz, chl	B16, 280
1580	3-Azobenzene sulfonic acid-4′-hydroxy-3′-nitro $3\text{-}NO_2\text{-}4\text{-}HOC_6H_3N{=}N(C_6H_4SO_3H\text{-}3)$	323.28	gold-ye pr (dil HCl + w)		235 (anh), 116 (+ w)			w, al	B16, 267
1581	4-Azobenzene sulfonic acid or 4-phenyl azobenzene sulfonic acid $4\text{-}HO_3SC_6H_4N{=}N(C_6H_5)$	262.28	og-red lf (w + 3)		127 (hyd)			w	$B16^2$, 281
1582	4-Azobenzene sulfonylchloride $4\text{-}ClO_2SC_6H_4N{=}NC_6H_5$	280.73	og-red nd or pl (bz)		82			al, bz	$B16^3$, 282
1583	*p,p*′-Azobiphenyl or Di-*p*-xenyldiimide $4{-}C_6H_5C_6H_4N{=}NC_6H_4C_6H_5{-}4$	334.42	og-red pl (bz)		256			eth	$B16^3$, 65
1584	Azomethane or Dimethyl diimide $CH_3N = NCH_3$	58.08	col or pa ye gas	1.5	-78	$0.744^{0/15}$	1.4199^{19}	al, eth, ace	$B4^2$, 1747
1585	Azomethane, hexafluoro $F_3CN{=}NCF_3$	166.03	pa grsh gas	-31.6	-133				$B3^4$, 246
1586	α,α′-Azonaphthalene or Di-α-naphthyldiimide $\alpha\text{-}C_{10}H_7N{=}NC_{10}H_7\text{-}\alpha$	282.34	red nd (gl aa)	sub	190			ace, bz	$B16^2$, 26
1587	α,α′-Azonaphthalene, 4-amino $4\text{-}H_2NC_{10}H_6N{=}NC_{10}H_7$	297.36	red-br nd		183 (cor)				
1588	1,2′-Azonaphthalene or α,β-Naphthyldiimide $\alpha\text{-}C_{10}H_7N = N\text{-}\beta\text{-}C_{10}H_7\text{-}(2)$	282.34	red nd, br lf (aa)		136			al, bz, aa	B16, 80
1589	2,2′-Azonaphthalene or Di-β-Naphthyldiimide $\beta\text{-}C_{10}H_7N{=}N\text{-}\beta\text{-}C_{10}H_7$	282.34	red lf (bz), og-ye nd (chl, al)	sub	208			bz	$B16^2$, 26
1590	Azoxybenzene *cis* $C_6H_5N(O){=}NC_6H_5$	198.22			87	$1.166^{20/4}$	1.633^{20}		$B16^3$, 580
1591	Azoxybenzene *trans* $C_6H_5N(O){=}C_6H_5$	198.22	bt ye nd	d	36	$1.1590^{26/4}$	1.652^{20}	al, eth, lig	$B16^3$, 579
1592	Azoxybenzene, 4-bromo $4\text{-}BrC_6H_4N(O){=}NC_6H_5$	277.12	dk ye nd		93.5-4.5	1.4138^{100}		al, bz	$B16^3$, 584
1593	Azoxybenzene, 2,2′-dimethoxy or *o*-Azoxyanisole $2\text{-}CH_3OC_6H_4N(O){=}N(C_6H_4OCH_3\text{-}2)$	258.28	ye-og cr (al)		81			al, eth, ace, bz, chl	B16, 635
1594	Azoxybenzene, 3,3′-dimethoxy or *m*-Azoxyanisole $3\text{-}CH_3OC_6H_4N(O){=}N(C_6H_4OCH_3\text{-}3)$	258.28	ye cr (al)		51-2			al, eth	$B16^2$, 325
1595	Azoxybenzene, 4,4′-dimethoxy or *p*-Azoxyanisole $4\text{-}CH_3OC_6H_4N(O){=}N(C_6H_4OCH_3\text{-}4)$	258.28	ye mcl nd (al)		119-20	$1.1711^{115/4}$		ace, bz, al	$B16^3$, 597
1596	Azoxybenzene, 2,2′-dimethyl *trans* $2\text{-}CH_3C_6H_4N(O){=}N(C_6H_4CH_3\text{-}2)$	226.28			60	$1.0215^{65/4}$	1.61804^{65}	al, eth, bz	$B16^3$, 587
1597	Azoxybenzene, 3,3′-dimethyl *trans* $3\text{-}CH_3C_6H_4N(O){=}N(C_6H_4CH_3\text{-}3)$	226.28			39	1.0123^{66}	1.6152^{66}	al, eth, bz, lig	$B16^3$, 588
1598	2,2-Azoxybenzene dicarboxylic acid or *o*-Azoxybensoic acid $2\text{-}HO_2CC_6H_4N(O){=}N(C_6H_4CO_2H\text{-}2)$	286.24	ye pl or pr (al)		254.5			al, ace, aa	$B16^3$, 609
1599	3,3′-Azoxybenzene dicarboxylic acid or *m*-Azoxybenzoic acid $3\text{-}HO_2CC_6H_4N(O){=}N(C_6H_4CO_2H\text{-}3)$	286.24	pa ye nd or lf (gl aa)		320d				$B16^3$, 610
1600	4,4′-Azoxybenzene dicarboxylic acid or *p*-Azoxybenzoic acid $4\text{-}HO_2CC_6H_4N(O)\text{-}N(C_6H_4CO_2H\text{-}4)$	286.24	ye am or pw		360d			Py	$B16^3$, 615
1601	1,1′-Azoxynaphthalene $\alpha\text{-}C_{10}H_7N(O){=}N\text{-}\alpha\text{-}C_{10}H_7$	298.34	ye or red orh pl (al)		127			al	$B16^3$, 591
1602	2,2′-Azoxynaphthalene $\beta\text{-}C_{10}H_7N(O){=}N\text{-}\beta\text{-}C_{10}H_7$	298.34	ye or red nd (al, eth, gl aa)		167-8			bz, aa, chl	$B16^3$, 592
1603	Azulene $C_{10}H_8$	128.17	bl or grsh-bk lf (al)	270d, $115\text{-}35^{10}$	99-100			al, eth, ace	B541636
1604	Azulene, 1,4-dimethyl-7-isopropyl or Guaiazulene $C_{15}H_{18}$	198.31	bl-vt pl (al)	$167\text{-}8^{12}$	31.5			al, eth	$B5^4$, 1751
1605	Azulene, 1,5-dimethyl-8-isopropyl or Chamazulene $C_{15}H_{18}$	198.31	bl lig	161^{12}		$0.9883^{20/4}$		al, eth	$B5^2$, 474

PHYSICAL CONSTANTS OF ORGANIC COMPOUNDS (Continued)

No.	Name, Synonyms, and Formula	Mol. wt.	Color, crystalline form, specific rotation and λ_{max} (log ε)	b.p. °C	m.p. °C	Density	n_D	Solubility	Ref.
1606	Azulene, 4,8-dimethyl-2-isopropyl or Elamazulene, Vetivazulene $C_{15}H_{18}$	198.31	red-vt nd (al)	140-60²	31-2			al, eth	B5⁴, 1754
1607	Azulene, 4,6,8-trimethyl $C_{13}H_{14}$	170.25			79-81			chl	B5⁴, 1721

No.	Name, Synonyms, and Formula	Mol. wt.	Color, crystalline form, specific rotation and λ_{max} (log ε)	b.p. °C	m.p. °C	Density	n_D	Solubility	Ref.
1608	Barbituric acid or Pyrimidinetrione $C_4H_4N_2O_3$	128.09	rh pr (w + 2)	260d	248			eth, w	B24², 292
1609	Barbituric acid, 5,allyl-5-butyl $C_{11}H_{16}N_2O_3$	224.26	(w, dil al)		128			w, al	B24², 292
1610	Barbituric acid, 5-allyl-5(2-cyclopenten-l-yl) or Cycclopal $C_{12}H_{14}N_2O_3$	234.25	(w, dil al)		139-40			w, al	C24, 5308
1611	Barbituric acid, 5-allyl-5-*iso*-butyl or Sandoptal $C_{11}H_{16}N_2O_3$	224.26	(w, dil al)		138			w, al, eth, ace, chl	B24², 292
1612	Barbituric acid, 5- allyl-5-*iso*-propyl $C_{10}H_{14}N_2O_3$	210.23			140-1				B24², 290
1613	Barbituric acid, 5-allyl-5-*iso*-propyl-l-methyl or Narconumal $C_{11}H_{16}N_2O_3$	224.26	(w, dil al)	176-8¹²	56-7			al, eth, chl, ace, bz	C32, 1052
1614	Barbituric acid, 5- allyl-5-phenyl or Alphenal $C_{13}H_{12}N_2O_3$	244.25			156-7			al, eth, chl	C26, 4828
1615	Barbituric acid, 5- amino or Uramil $C_4H_5N_3O_3$	143.10	nd or pl (w)		>400			w, chl	B25¹, 704
1616	Barbituric acid, 5-benzylidene $C_{11}H_8N_2O_3$	216.20	pr (aa)		256			ace	B24², 299
1617	Barbituric acid, 5-(2-bromoallyl)-5-*iso*-propyl or Propallylonal $C_{10}H_{13}BrN_2O_3$	289.13	(dil aa, dil al)		181			al, eth, ace, aa	B24², 291
1618	Barbituric acid, 5-butyl-5-ethyl $C_{10}H_{16}N_2O_3$	212.25			128-9				B24², 285
1619	Barbituric acid, 5-(l-cyclohexenyl)-5-ethyl or Cyclobarbitone $C_{12}H_{16}N_2O_3$	236.27	lf (w)		172-4			al, eth	B24², 294
1620	Barbituric acid, 5,5-diallyl or Allobarbitone $C_{10}H_{12}N_2O_3$	208.22	pl (w or 50% al)		174			al, eth, bz	B24², 293
1621	Barbituric acid, 5,5-dibromo or Dibromin $C_4H_2Br_2N_2O_3$	285.88	pl (MeOH-bz), lf (dil HNO_3)	d	235			al, eth	B24², 272
1622	Barbituric acid, 5,5-diethyl or barbital, veronal $C_8H_{12}N_2O_3$	184.19	(i) trig (w) (ii) mcl pr (iii) mcl nd (iv) tcl		(i) 190 (ii) 183 (iii) 181 (iv) 176	1.220		al, eth, ace, aa	B24², 279
1623	Barbituric acid, 5,5-diethyl-l-methyl or Metharbital $C_9H_{14}N_2O_3$	198.22	nd		150-1			w	B24², 281
1624	Barbituric acid, 5,5-dipropyl or Proponal $C_{10}H_{16}N_2O_3$	212.25	pl (w)		146 (166)			al, eth, bz, chl	B24², 286
1625	Barbituric acid, 5-ethyl-5-hexyl $C_{12}H_{20}N_2O_3$	240.30	nd (w)		112-3			al, eth, bz	B24², 288
1626	Barbituric acid, 5-ethyl-5(2-methallyl)-2-thio $C_{10}H_{14}N_2O_2S$	226.29			160-1				C35, 5599
1627	Barbituric acid, 5-ethyl-5-(1-methylbutyl) $C_{11}H_{18}N_2O_3$	226.28			130			al, eth	B24², 287
1628	Barbituric acid, 5-ethyl-5-(1-methylbutyl)-2-thio $C_{11}H_{18}N_2O_2S$	242.34			158-9				C29, 8237
1629	Barbituric acid, 5-ethyl-5-(3-methylbutyl) $C_{11}H_{18}N_2O_3$	226.28			157-9.5				
1630	Barbituric acid, 5-ethyl-(3-methylbutyl) or Amytal $C_{11}H_{18}N_2O_3$	226.28	lf (w, dil al)		156-8			al, eth, bz, chl	B24², 287
1631	Barbituric acid, 5-ethyl-l-methyl-5-phenyl or Methobarbital, Prominal $C_{13}H_{14}N_2O_3$	246.27	wh cr (w)		176			al, chl	C28, 281
1632	Barbituric acid, 5-ethyl-5-pental $C_{11}H_{18}N_2O_3$	226.28	cr (dil al)		135-6			al, eth, chl	B24², 286
1633	Barbituric acid, 5-ethyl-5-(2-pentyl) or Nembutal, Pentobarbital $C_{11}H_{18}N_2O_3$	226.28	nd (w)		130			al, eth	B24², 287
1634	Barbituric acid, 5-ethyl-5-phenyl or Phenobarbital, Luminal $C_{12}H_{12}N_2O_3$	232.24	pl (w)		174			al, eth	B24², 297
1635	Barbituric acid, 5-ethyl-5-(1-piperidyl) or Eldoral $C_{11}H_{17}N_3O_3$	239.28	wh cr (dil al)		215			al, eth, ace	C36, 3160
1636	Barbituric acid, 5-ethyl-5-*iso*-propyl or Ipral, Probarbital $C_9H_{14}N_2O_3$	198.22	nd (w)		203			eth	B24², 284

PHYSICAL CONSTANTS OF ORGANIC COMPOUNDS (Continued)

No.	Name, Synonyms, and Formula	Mol. wt.	Color, crystalline form, specific rotation and λ_{max} (log ε)	b.p. °C	m.p. °C	Density	n_D	Solubility	Ref.
1637	Barbituric acid, 5-(2-furfurylidine) $C_9H_6N_2O_4$	206.16			315				B27^{1}, 607
1638	Barbituric acid, 5-(2-furfurylidine)-2-thio $C_9H_6N_2O_3S$	222.22	ye pl		> 280d				B27^{1}, 607
1639	Barbituric acid, 5-hydroxy or Dialuric acid $C_4H_4N_2O_4$	144.09	pr or pl nd (+ lw, w)		224 (anh)			al, ace, w, aa	B25^{2}, 61
1640	Barbituric acid, l-methyl $C_5H_6N_2O_3$	142.11			132-3			al, ace	B24^{2}, 270
1641	Barbituric acid, 5-methyl-5-phenyl or Rufinal $C_{11}H_{10}N_2O_3$	218.21	cr		220			al, eth	B24^{2}, 296
1642	Barbituric acid, 5-nitro $C_4H_3N_3O_5$	173.08	pr, lf (w + 3)		180-1			al, w	B24^{2}, 273
1643	Barbituric acid, 5-(2-phenylethyl) $C_{12}H_{12}N_2O_3$	232.24	cr		212-3			al	B24^{2}, 296
1644	Barbituric acid, 2-thio $C_4H_4N_2O_2S$	144.15	pl (w)		235d			al, w	B24^{2}, 275
1645	Bebeerine *d* or Chondrodendrin $C_{36}H_{38}N_2O_6$	594.71	cr (bz, eth, chl-MeOH), $[\alpha]^{25}_D$ + 297 (al)		221			al, eth, ace, chl	B27^{2}, 896
1646	Berbeerine *l* or Curine $C_{36}H_{38}N_2O_6$	594.71	pr, nd (chl-MeOH), $[\alpha]^{20}_D$ −328 p_4, $[\alpha]^{28}_D$ −298 (al)		221			ace, bz, Py	B27^{2}, 894
1647	Behenic acid or Docosanoic acid $CH_3(CH_2)_{20}CO_2H$	340.59	nd	306^{60}	80	0.8223^{90}	1.4270^{100}		B2^{4}, 1290
1648	Behenic acid, ethyl ester or Ethyl behenate $CH_3(CH_2)_{20}CO_2C_2H_5$	368.64	nd (al), or (ace)	$240\text{-}2^{10}$	50			al, eth	B2^{4}, 1292
1649	Behenic acid, methyl ester or Methylbehenate $CH_3(CH_2)_{20}CO_2CH_3$	354.62	nd (ace)	$224\text{-}5^{15}$	54		1.4339^{60}	al, eth	B2^{4}, 1291
1650	Benzal acetate, 2-acetoxy $2\text{-}CH_3CO_2C_6H_4CH(O_2CCH_3)_2$	266.25	nd or pl (al), pr (al)		107			eth, bz, chl	B8^{3}, 148
1651	Benzal acetate, 3- acetoxy $3\text{-}CH_3CO_2C_6H_4CH(O_2CCH_3)_2$	266.25	lf (w-al)		76			al, eth	B8, 60
1652	Benzal acetate, 4-acetoxy $4\text{-}CH_3CO_2C_6H_4CH(O_2CCH_3)_2$	266.25	pr (eth or lig)		94			eth, lig, al	B8^{1}, 530
1653	Benzalacetophenone or *trans* Chalcone $C_6H_5COCH{=}CHC_6H_5$	208.26	pa ye lf, pr, nd (peth)	345-8d, 208^{25}	(i) 59 (ii) 57 (iii) 49	$1.0712^{62/4}$		eth, bz, chl	B7^{3}, 2380
1654	Benzaldehyde C_6H_5CHO	106.12		178, 62^{10}	-26(fr-56)	$1.0415^{10/4}$	1.5463^{20}	al, eth, ace, bz, lig	B7^{3}, 805
1655	Benzaldehyde, 2-acetamido $2\text{-}CH_3CONHC_6H_4CHO$	163.18			70-1			w, eth, ace	B14^{3}, 51
1656	Benzaldehyde, 3-acetamido $3\text{-}CH_3CONHC_6H_4CHO$	163.18	pl (bz)		84			al, eth, bz	B14, 29
1657	Benzaldehyde, 4-acetamido $4\text{-}CH_3CONHC_6H_4CHO$	163.18	pr (w)		156			w, bz	B14^{2}, 25
1658	Benzaldehyde, 3-allyl-2-hydroxy $3\text{-}(CH_2{=}CHCH_2)\text{-}2\text{-}HOC_6H_3CHO$	162.19		$245\text{-}6^{755}$	1.098^{15}				B8^{1}, 559
1659	Benzaldehyde, 2-amino or Anthranil aldehyde $2\text{-}H_2NC_6H_4CHO$	121.14	silv lf		39-42			al, eth, bz, chl	B14^{3}, 47
1660	Benzaldehyde, 2-amino, oxime $2\text{-}H_2NC_6H_4CH{=}NOH$	136.17	nd (bz)		135-6			al, eth, bz, aa	B14^{3}, 50
1661	Benzaldehyde, 3-amino $3\text{-}H_2NC_6H_4CHO$	121.14	nd (AcOEt)		28-30			eth	B14^{3}, 53
1662	Benzaldehyde, 3-amino, oxime $3\text{-}H_2NC_6H_4CH{=}NOH$	136.15	nd (bz)		88			al, eth	B14, 28
1663	Benzaldehyde, 4-amino $4\text{-}H_2NC_6H_4CHO$	121.14	pl (w)		71-2			w, al, eth	B14^{3}, 57
1664	Benzaldehyde, 4-amino, oxime $4\text{-}H_2NC_6H_4CH{=}NOH$	136.15	ye cr (w)		124			al, eth	B14^{3}, 58
1665	Benzaldehyde azine or Dibenzal hydrazine $C_6H_5CH{=}NN{=}CHC_6H_5$	208.26	ye pr (al)		93			al, eth, ace, bz	B7^{3}, 844
1666	Benzaldehyde, 2-bromo $2\text{-}BrC_6H_4CHO$	185.02		230, 118^{12}	21-2		1.5925^{20}	al, bz	B7^{3}, 882

No.	Name, Synonyms, and Formula	Mol. wt.	Color, crystalline form, specific rotation and λ_{max} (log ε)	b.p. °C	m.p. °C	Density	n_D	Solubility	Ref.
1667	Benzaldehyde, 2-bromo, diacetate $2\text{-}BrC_6H_4CH(O_2CCH_3)_2$	287.11			84-6			al, eth	$B7^2$, 181
1668	Benzaldehyde, 2-bromo-4-hydroxy $2\text{-}Br\text{-}4\text{-}HOC_6H_3CHO$	201.02	pa ye nd (w)		159.5				$B8^2$, 74
1669	Benzaldehyde, 2-bromo-5-hydroxy $2\text{-}Br\text{-}5\text{-}HOC_6H_3CHO$	201.02	nd (w)		135			al, eth, ace, bz	$B8^2$, 57
1670	Benzaldehyde, 3-bromo $3\text{-}BrC_6H_4CHO$	185.02		233-6			1.5915^{20}	al, eth	$B7^3$, 882
1671	Benzaldehyde, 3-bromo-2-hydroxy or 3-Bromosalicylaldehyde $3\text{-}Br\text{-}2\text{-}HOC_6H_3CHO$	201.02	nd (dil al)		49			al, eth, ace, bz	B8, 54
1672	Benzaldehyde, 3-bromo-4-hydroxy $3\text{-}Br\text{-}4\text{-}HOC_6H_3CHO$	201.02	lf (w)		124			al, eth, ace, bz	$B8^3$, 290
1673	Benzaldehyde, 4-bromo $4\text{-}BrC_6H_4CHO$	185.02	lf (dil al)	$66\text{-}8^2$	67			al, bz	$B7^3$, 883
1674	Benzaldehyde, 4-bromo, diacetate $4\text{-}BrC_6H_4CH(O_2CCH_3)_2$	287.11	ye		94-5			eth, al	$B7^2$, 182
1675	Benzaldehyde, 4-bromo-2-hydroxy or 4-Bromosalicylaldehyde $4\text{-}Br\text{-}2\text{-}HOC_6H_3CHO$	201.02	nd (dil al)		52			al, eth, ace, bz	B8, 54
1676	Benzaldehyde, 4-bromo-3-hydroxy $4\text{-}Br\text{-}3\text{-}HOC_6H_3CHO$	201.02			131.5			ace, bz, chl	$B8^2$, 56
1677	Benzaldehyde, 5-bromo-2-hydroxy or 5-Bromosalicylaldehyde $5\text{-}Br\text{-}2\text{-}HOC_6H_3CHO$	201.02	nd (al), lf (eth)		105-6			al, eth	$B8^3$, 269
1678	Benzaldehyde, 5-bromo-4-hydroxy-3-methoxy or 5-Bromovanillan $5\text{-}Br\text{-}4\text{-}HO\text{-}3\text{-}CH_3OC_6H_2CHO$	231.05	pl (aa), nd, pl (al)		164-6			al	$B8^3$, 2050
1679	Benzaldehyde, 2-chloro $2\text{-}ClC_6H_4CHO$	140.57	nd	211.9, 84.3^{10}	12.4	$1.2483^{20/4}$	1.5662^{20}	al, eth, ace, bz	$B7^3$, 864
1680	Benzaldehyde, 2-chloro-3-hydroxy $2\text{-}Cl\text{-}3\text{-}HOC_6H_3CHO$	156.57	cr (ag aa)		139.5			al, aa	$B8^2$, 54
1681	Benzaldehyde, 2-chloro-4-hydroxy $2\text{-}Cl\text{-}4\text{-}HOC_6H_3CHO$	156.57	nd (w or aa)		147-8			al, eth, aa	B8, 81
1682	Benzaldehyde, 2-chloro-5-hydroxy $2\text{-}Cl\text{-}5\text{-}HOC_6H_3CHO$	156.57	nd (w or aa)		110-1			al, aa	$B8^1$, 526
1683	Benzaldehyde, 2-chloro, phenyl hydrazone $2\text{-}ClC_6H_4CH{=}NNHC_6H_5$	230.70			84				$B15^3$, 86
1684	Benzaldehyde, 3-chloro $3\text{-}ClC_6H_4CHO$	140.57	pr	213-4, 55^1	17-8	$1.2410^{20/4}$	1.5650^{20}	al, eth, ace, bz	$B7^3$, 869
1685	Benzaldehyde, 3-chloro-2-hydroxy or 3-Chlorosalicylaldehyde $3\text{-}Cl\text{-}2\text{-}HOC_6H_3CHO$	156.57	nd (MeOH)		55			al, eth, ace, bz	$B8^3$, 179
1686	Benzaldehyde, 3-chloro-4-hydroxy $3\text{-}Cl\text{-}4\text{-}HOC_6H_3CHO$	156.57	nd (w)	$149\text{-}50^{14}$	139			al, eth	$B8^3$, 286
1687	Benzaldehyde, 4-chloro $4\text{-}ClC_6H_4CHO$	140.57	pl	213-4, $72\text{-}5^3$	47.5	$1.196^{61/4}$	1.5552^{61}	w, al, eth, ace, bz	$B7^3$, 872
1688	Benzaldehyde, 4-chloro-2-hydroxy or 4-chlorosalicylaldehyde $4\text{-}Cl\text{-}2\text{-}HOC_6H_3CHO$	156.57	nd (al or ag aa)		52.5			w, eth, ace, bz	$B8^2$, 44
1689	Benzaldehyde, 4-chloro-3-hydroxy $4\text{-}Cl\text{-}3\text{-}HOC_6H_3CHO$	156.57	nd (al)		121			w, al, eth, ace, bz	$B8^2$, 55
1690	Benzaldehyde, 4-chloro-3-nitro $4\text{-}Cl\text{-}3\text{-}NO_2C_6H_3CHO$	185.57			64.5			chl	$B7^3$, 919
1691	Benzaldehyde, 5-chloro-2-hydroxy or 5-Chlorosalicylaldehyde $5\text{-}Cl\text{-}2\text{-}HOC_6H_3CHO$	156.57	pl (al)	105^{12}	99-100			al, eth	$B8^2$, 45
1692	Benzaldehyde, 5-chloro-2-hydroxy, oxime $5\text{-}Cl\text{-}2\text{-}HOC_6H_3CH{=}NOH$	171.58	nd (w)		128			al	B8, 53
1693	Benzaldehyde, 5-chloro-4-hydroxy-3-methoxy or 5-Chlorovanillan $5\text{-}Cl\text{-}4\text{-}HO\text{-}3\text{-}CH_3OC_6H_2CHO$	186.60	tetr		165			al, aa	$B8^3$, 2063
1694	Benzaldehyde diacetate or Benzylidene diacetate $C_6H_5CH(O_2CCH_3)_2$	208.22	pl (eth)	220, 154^{20}	46	1.11^{20}		al, eth, bz	$B7^3$, 827
1695	Benzaldehyde, 3,5-dibromo-2-hydroxy or 3,5-Dibromosalicylaldehyde $3,5\text{-}Br_2\text{-}2\text{-}HOC_6H_2CHO$	279.92			82.0-3.5				$B8^3$, 269
1696	Benzaldehyde, 2,3-dichloro $2,3\text{-}Cl_2\text{-}C_6H_3CHO$	175.01	cr (dil al)		65-7			al, eth	$B7^3$, 878

No.	Name, Synonyms, and Formula	Mol. wt.	Color, crystalline form, specific rotation and λ_{max} (log ε)	b.p. °C	m.p. °C	Density	n_D	Solubility	Ref.
1697	Benzaldehyde, 2,4-dichloro 2,4-$Cl_2C_6H_3CHO$	175.01	pr		72			al, eth, bz	B7^{3}, 878
1698	Benzaldehyde, 2,4-dichloro-3-hydroxy 2,4-Cl_2-3-HOC_6H_2CHO	191.01	cr (aa)		141			lig	B8^{2}, 55
1699	Benzaldehyde, 2,5-dichloro 2,5-$Cl_2C_6H_3CHO$	175.01	nd (al)	231-3	58			al, eth, bz, chl	B7^{3}, 878
1700	Benzaldehyde, 2,6-dichloro 2,6-Cl_2-C_6H_3CHO	175.01	nd (lig)		71			al, eth, lig	B7^{3}, 878
1701	Benzaldehyde, 2,6-dichloro-3-hydroxy 2,6-Cl_2-3-HOC_6H_2CHO	191.01	cr (w)		142			aa	B8^{2}, 204
1702	Benzaldehyde, 3,4-dichloro 3,4-$Cl_2C_6H_3CHO$	175.01		247-8	44			al, eth	B7^{3}, 880
1703	Benzaldehyde, 3,5-dichloro 3,5-$Cl_2C_6H_3CHO$	175.01	nd or lf (dil al)	235-40^{748}	65			al, eth, ace, bz	B7^{3}, 880
1704	Benzaldehyde, 3,5-dichloro-2-hydroxy or 3,5-dichlorsalicylaldehyde 3,5-Cl_2-2-HOC_6H_2CHO	191.01	ye rh (aa)		95				B8^{3}, 183
1705	Benzaldehyde, 3,5-dichloro-2-hydroxy, oxime 3,5-Cl_2-2-$HOC_6H_2CH{=}NOH$	206.03	nd (dil al)		195-6			al, eth, bz	B8, 54
1706	Benzaldehyde, 3,5-dichloro-4-hydroxy 3,5-Cl_2-4-HOC_6H_2CHO	191.01	nd (chl, dil al)		158-9			al, eth	B8, 81
1707	Benzaldehyde, 4,6-dichloro-3-hydroxy 4,6-Cl_2-3-HOC_6H_2CHO	191.01	nd		130			w, al, eth, bz	B8^{1}, 526
1708	Benzaldehyde, 2,3-diethoxy 2,3-$(C_2H_5O)_2C_6H_3CHO$	194.23		169^{37}				al, eth	B8^{3}, 1982
1709	Benzaldehyde, 3,4-diethoxy 3,4-$(C_2H_5O)_2C_6H_3CHO$	194.23		278-30				al	B8^{2}, 2025
1710	Benzaldehyde, 4-diethylamino 4-$(C_2H_5)_2NC_6H_4CHO$	177.23	ye nd (w)	174^{7}	41			w, al, eth, bz	B14^{3}, 71
1711	Benzaldehyde, 2,4-dihydroxy or β-Resorcyl aldehyde 2,4-$(HO)_2C_6H_3CHO$	138.12	nd (eth-lig)	220-8^{22}	201-2			w, al, eth, chl	B8^{3}, 1989
1712	Benzaldehyde, 2,4-dihydroxy, oxime 2,4-$(HO)_2C_6H_3CH{=}NOH$	153.14			192			w, al, eth	B8^{2}, 274
1713	Benzaldehyde, 2,5-dihydroxy or Gentisic aldehyde 2,5-$(HO)_2C_6H_3CHO$	138.12	ye nd (bz)		99			w, al, eth, chl	B8^{3}, 2003
1714	Benzaldehyde, 2,6-dihydroxy-4-methyl or Atranol 2,6-$(HO)_2$-4-$CH_3C_6H_2CHO$	152.16	ye nd (w)		124			w, al, eth, chl	B8^{3}, 2006
1715	Benzaldehyde, 3,4-dihydroxy or Protocatechualdehyde 3,4-$(HO)_2C_6H_3CHO$	138.12	lf (w, to)		153-4			w, al, eth	B8^{3}, 2009
1716	Benzaldehyde, 3,4-dihydroxy, oxime 3,4-$(HO)_2C_6H_3CH{=}NOH$	153.14	nd (w)		157d			w, al	B8^{2}, 285
1717	Benzaldehyde, 3,5-diido-4-hydroxy 3,5-I_2-4-HOC_6H_2CHO	373.91	nd (al)		206.5			al, ace, aa	B8^{3}, 251
1718	Benzaldehyde, 2,4-dimethoxy 2,4-$(CH_3O)_2C_6H_3CHO$	166.18	nd (al or lig)	165^{10}	72			al, eth, bz, lig	B8^{3}, 1992
1719	Benzaldehyde, 2,5-dimethoxy 2,5-$(CH_3O)_2C_6H_3CHO$	166.18		146^{10}	52			al, eth	B8^{3}, 2004
1720	Benzaldehyde, 2,4-dimethoxy-6-hydroxy 2,4-$(CH_3O)_2$-6-HOC_6H_2CHO	182.18	nd, pl (dil al)	190-5^{25}, 165^{10}	71			al, eth, bz, aa	B8^{3}, 3364
1721	Benzaldehyde, 2,6-dimethoxy-4-hydroxy 2,6-$(CH_3O)_2$-4-HOC_6H_2CHO	182.18	nd or pl (MeOH), pr (bz)		70-1				B8^{3}, 3365
1722	Benzaldehyde, 3,4-dimethoxy or Veratraldehyde 3,4-$(CH_3O)_2C_6H_3CHO$	166.18	nd (eth, lig, to)	258, 172-5^{18}	44 (58)			al, eth	B8^{3}, 2020
1723	Benzaldehyde, 3,4-dimethoxy-5-hydroxy 3,4-$(CH_3O)_2$-5-HOC_6H_2CHO	182.18		177-80^{12}	62-3			ual, bz, aa, lig	B8^{2}, 437
1724	Benzaldehyde, 3,5-dimethoxy 3,5-$(CH_3O)_2C_6H_3CHO$	166.18		151^{16}	45.5			al, bz	B8^{3}, 2073
1725	Benzaldehyde, 3,5-dimethoxy-4-hydroxy 3,5-$(CH_3O)_2$-4-HOC_6H_2CHO	182.18	br nd (lig)	192-3^{14}	113			al, eth, bz, aa	B8^{3}, 3368
1726	Benzaldehyde, 2,4-dimethyl 2,4-$(CH_3)_2C_6H_3CHO$	134.18		218, 99^{10}	-9			al, eth, ace, bz	B7^{3}, 1072
1727	Benzaldehyde, 2,5-dimethyl 2,5-$(CH_3)_2C_6H_3CHO$	134.18		220, 100^{10}				w, al, eth, ace, bz	B7^{3}, 1072
1728	Benzaldehyde, 3,4-dimethyl 3,4-$(CH_3)_2C_6H_3CHO$	134.18		223-5				al, eth, ace	B7^{3}, 1073
1729	Benzaldehyde, 3,5-dimethyl 3,5-$(CH_3)_2C_6H_3CHO$	134.18		220-2	9			al, eth, ace, bz	B7^{3}, 1073

No.	Name, Synonyms, and Formula	Mol. wt.	Color, crystalline form, specific rotation and λ_{max} (log ε)	b.p. °C	m.p. °C	Density	n_D	Solubility	Ref.
1730	Benzaldehyde, 4-dimethylamino 4-(CH$_3$)$_2$NC$_6$H$_4$CHO	149.19	lf (w)	176-7^{17}	74			al, eth, ace, bz	B14^3, 58
1731	Benzaldehyde, 2,4-dinitro 2,4-(NO$_2$)$_2$C$_6$H$_3$CHO	196.12	pa ye pr (al), pl (bz)	190-210$^{10-20}$	72			al, eth, bz	B7^3, 923
1732	Benzaldehyde, 2,6-dinitro 2,6-(NO$_2$)$_2$C$_6$H$_3$CHO	196.12	lf (dil al)		123			w, al, eth, bz, aa	B7^2, 206
1733	Benzaldehyde, 2-ethoxy 2-C$_2$H$_5$OC$_6$H$_4$CHO	150.18		247-9, 143-7^{25}	20-2			al, eth	B8^3, 144
1734	Benzaldehyde, 3-ethoxy 3-C$_2$H$_5$OC$_6$H$_4$CHO	150.18		245.5		1.0768$^{20/4}$	1.5408^{20}	al, eth, bz	B8, 60
1735	Benzaldehyde, 3-ethoxy-2-hydroxy or 3-Ethoxysalicylaldehyde 3-C$_2$H$_5$O-2-HOC$_6$H$_3$CHO	166.18		263-4^{740}	64-5				B8^3, 1981
1736	Benzaldehyde, 3-ethoxy-4-hydroxy or Ethylvanillan 3-C$_2$H$_5$O-4-HOC$_6$H$_3$CHO	166.18			77-8			al, eth, bz, chl	B8^3, 2022
1737	Benzaldehyde, 4-ethoxy 4-C$_2$H$_5$OC$_6$H$_4$CHO	150.18		249, 140^{20}	13-4	1.08$^{21/21}$		al, eth, bz	B8^3, 223
1738	Benzaldehyde, 4-ethoxy-2-hydroxy or 4-Ethoxysalicylaldehyde 4-C$_2$H$_5$O-2-HOC$_6$H$_3$CHO	166.18	nd (dil al)		35			al	B8^3, 1992
1739	Benzaldehyde, 4-ethoxy-3-methoxy 4-C$_2$H$_5$O-3-CH$_3$OC$_6$H$_3$CHO	180.21	mcl pr	sub	64-5 (73-4)			al, eth, bz, aa	B8^3, 2024
1740	Benzaldehyde, 5-ethoxy-2-hydroxy or 5-Hydroxysalicylaldehyde 5-C$_2$H$_5$O-2-HOC$_6$H$_3$CHO	166.18	ye pr	230	51.5			al, eth, chl	B8, 245
1741	Benzaldehyde hydrazone C$_6$H$_5$CH=NNH$_2$	120.15	lf	140^{14}	16			al	B7^3, 863
1742	Benzaldehyde, 2-hydroxy or Salicylaldehyde 2-HOC$_6$H$_4$CHO	122.12		197, 93^{25}	-7	1.1674$^{20/4}$	1.5740^{20}	al, eth, ace, bz	B8^3, 135
1743	Benzaldehyde-2-hydroxy-3-methoxy or o-Vanillan 2-HO-3-CH$_3$OC$_6$H$_3$CHO	152.15	lt ye-lt gr nd (w, lig)	256-6, 128^{10}	44-5			al, eth, lig	B8^3, 1979
1744	Benzaldehyde, 2-hydroxy-4-methoxy or 4-Methoxysalicylaldehyde 2-HO-4-CH$_3$OC$_6$H$_3$CHO	152.15	nd (w), cr (al)		40-2			al, eth, bz, lig	B8^3, 1991
1745	Benzaldehyde, 2-hydroxy-5-methoxy or 5-Methoxysalicylaldehyde 2-HO-5-CH$_3$OC$_6$H$_3$CHO	152.15	ye (w)	247-8	4			al, eth	B8^3, 2004
1746	Benzaldehyde, 2-hydroxy-3-nitro or 3-Nitrosalicylaldehyde 2-HO-3-NO$_2$ C$_6$H$_3$CHO	167.12	nd (aa)		109-10			al, bz	B8^3, 190
1747	Benzaldehyde, 2-hydroxy-5-nitro or 5-Nitrosalicylaldehyde 5-NO$_2$-2--HO-C$_6$H$_3$CHO	167.12	(dil aa)		128			al, eth	B8^3, 192
1748	Benzaldehyde, 2-hydroxy-5-nitro, oxime 2-HO-5-NO$_2$C$_6$H$_3$CH=NOH	182.14			225				B8^3, 193
1749	Benzaldehyde, 3-hydroxy 3-HOC$_6$H$_4$CHO	122.12	nd (w)	240, 161^{20}	108			al, eth, ace, bz	B8^3, 197
1750	Benzaldehyde, 3,hydroxy, azine 3-HOC$_6$H$_4$CH=NN=CHC$_6$H$_4$OH-(3)	240.26	ye pl (al)		162				B8^3, 203
1751	Benzaldehyde, 3-hydroxy-4-methoxy or Isovanillan 3-HO-4-CH$_3$OC$_6$H$_3$CHO	152.15	pl (w)	179^{15}	116-7	1.198		al, eth, bz, chl, aa	B8^3, 2019
1752	Benzaldehyde, 3-hydroxy-5-methoxy 3-HO-5-CH$_3$OC$_6$H$_3$CHO	152.15		110-20, sub	130-1			al, bz, aa	B8^3, 2073
1753	Benzaldehyde, 3-hydroxy-2-nitro 3-HO-2-NO$_2$C$_6$H$_3$CHO	167.12	nd or pl (bz, lig), d		152			al, bz	B8^3, 210
1754	Benzaldehyde, 3-hydroxy-2-nitro, oxime 3-HO-2-NO$_2$C$_6$H$_3$CH=NOH	182.14	pa ye nd		172.5				B8^2, 58
1755	Benzaldehyde, 3-hydroxy-4-nitro 4-NO$_2$-3-HOC$_6$H$_3$CHO	167.12	ye lf		128			al, eth, bz	B8^3, 211
1756	Benzaldehyde, 3-hydroxy-4-nitro, oxime 4-NO$_2$-3-HOC$_6$H$_3$CH=NOH	182.14	ye nd (chl)		164			al, chl	B8^2, 58
1757	Benzaldehyde, 3-hydroxy, oxime 3-HOC$_6$H$_4$CH=NOH	137.14	(bz)		90			w, al, eth	B8, 61
1758	Benzaldehyde, 4-hydroxy 4-HOC$_6$H$_4$CHO	122.12	nd (w)		117	1.129$^{130/4}$	1.5705^{130}	al, eth, bz	B8^3, 215

PHYSICAL CONSTANTS OF ORGANIC COMPOUNDS (Continued)

No.	Name, Synonyms, and Formula	Mol. wt.	Color, crystalline form, specific rotation and λ_{max} (log ε)	b.p. °C	m.p. °C	Density	n_D	Solubility	Ref.
1759	Benzaldehyde, 4-hydroxy, azine $4\text{-}HOC_6H_4CH{=}NN{=}CHC_6H_4OH(\text{-}4)$	240.26	ye (al)	......	239-40, 268 (d)	......	......	al, bz	$B8^1$, 531
1760	Benzaldehyde, 4-hydroxy-2-iodo-3-methoxy or 2-Iodo vanillan $4\text{-}HO\text{-}3\text{-}CH_3O\text{-}2IC_6H_2CHO$	278.05	......	......	155-6	......	......	al	$B8^3$, 2058
1761	Benzaldehyde, 4-hydroxy-5-iodo-3-methoxy or 5-Iodo vanillan $5\text{-}I\text{-}4\text{-}HO\text{-}3\text{-}CH_3OC_6H_2CHO$	278.05	pa ye	......	180	......	......	......	$B8^3$, 2059
1762	Benzaldehyde, 4-hydroxy-2-methoxy $4\text{-}HO\text{-}2\text{-}CH_3OC_6H_3CHO$	152.15	lf (bz), nd (w)	......	153	......	......	al, eth, chl	$B8^3$, 1990
1763	Benzaldehyde, 4-hydroxy-3-methoxy or Vanillan $4\text{-}HO\text{-}3\text{-}CH_3O\text{-}C_6H_3CHO$	152.15	(i) nd (w,lig) (ii) tetr (w, lig)	285, 170^{15}	(i) 77—9 (ii) 81—2	1.056	......	al, eth, ace, bz, chl	$B8^3$, 2011
1764	Benzaldehyde, 4-hydroxy-2-nitro $4\text{-}HO\text{-}2\text{-}NO_2C_6H_3CHO$	167.12	ye nd	......	67	......	......	al, eth, bz	$B8^3$, 253
1765	Benzaldehyde, 4-hydroxy-3-nitro $4\text{-}HO\text{-}3\text{-}NO_2C_6H_3CHO$	167.12	dk ye nd (al, w)	......	144.3	......	......	al, w	$B8^3$, 253
1766	Benzaldehyde, 4-hydroxy-3-nitro, oxime $4\text{-}HO\text{-}3\text{-}NO_2C_6H_3CH{=}NOH$	182.14	pr or nd (al-chl)	......	189	......	......	al, eth, aa	$B8^3$, 254
1767	Benzaldehyde imine, *N*-ethyl or *N*-Benzalethylamine $C_6H_5CH{=}NC_2H_5$	133.19	......	195^{745}	......	$0.937^{20/4}$	1.5378^{15}	al, eth	$B7^3$, 831
1768	Benzaldehyde imine,2-hydroxy-*N*-phenyl $2\text{-}HOC_6H_4CH{=}NC_6H_5$	197.24	......	......	47-9	1.087	......	al	$B12^3$, 369
1769	Benzaldehyde imine, *N*-methyl $C_6H_5CH{=}NCH_3$	119.17	......	185^{760}, $90\text{-}1^{30}$	......	$0.9672^{14/2}$	1.5526^{20}	al, eth, ace, chl	$B7^3$, 831
1770	Benzaldehyde imine, *N*-phenyl $C_6H_5CH{=}NC_6H_5$	181.24	......	310	54	$1.038^{55/4}$	1.600^{99}	al, eth, chl	$B12^3$, 319
1771	Benzaldehyde imine, *N*-(2-tolyl) $C_6H_5CH{=}NC_6H_4CH_3\text{-}(2)$	195.26	......	314, 176^{15}	......	$1.041^{20/4}$	1.6310^{25}	ace	$B12^3$, 1848
1772	Benzaldehyde imine, *N*-(3-tolyl) $C_6H_5CH{=}N(C_6H_4CH_3\text{-}3)$	195.26	......	315^{775}	30-2	......	1.6353^{2}	ace	$B12^3$, 1958
1773	Benzaldehyde imine, *N*-(4-tolyl) or *N*-Benzal-p-toluidine. $C_6H_5CH{=}N(C_6H_4CH_3\text{-}4)$	195.26	ye cr	318^{755}, 178^{11}	35	......	......	ace	$B12^3$, 2096
1774	Benzaldehyde, 2-methoxy or *o*-Anisaldehyde $2\text{-}CH_3OC_6H_4CHO$	136.15	pr	243-4, $124\text{-}5^{18}$	37-8	$1.1326^{20/4}$	1.5600^{20}	al, eth, ace, bz, chl	$B8^3$, 143
1775	Benzaldehyde, 2-methoxy-3-nitro or 3-Nitro-*o*-anisaldehyde $2\text{-}CH_3O\text{-}3\text{-}NO_2C_6H_3CHO$	181.15	ye pr (dil al), nd (bz)	......	89-90	......	......	al, eth	$B8^3$, 190
1776	Benzaldehyde, 3-methoxy or *m*-Anisaldehyde $3\text{-}CH_3OC_6H_4CHO$	136.15	......	230, 62^{1}	......	$1.1187^{20/4}$	1.5330^{20}	al, eth, ace, bz	$B8^3$, 199
1777	Benzaldehyde, 4-methoxy or *p*-Anisaldehyde $p\text{-}CH_3OC_6H_4CHO$	136.15	......	249.5, 83^{2}	0	$1.1191^{15/4}$	1.5730^{20}	al, eth, ace, bz	$B8^3$, 218
1778	Benzaldehyde, 4-methoxy-3-methyl or 4-methoxy-*m*-tolualdehyde $4\text{-}CH_3O\text{-}3\text{-}CH_3C_6H_3CHO$	150.18	......	$80\text{-}5^{1}$	......	......	1.5670^{20}	......	$B8^3$, 367
1779	Benzaldehyde, 4-methoxy, oxime $4\text{-}CH_3OC_6H_4CH{=}NOH$	151.16	nd (bz)	......	64—5 (anti), 133 (syn)	......	......	......	$B8^3$, 235
1780	Benzaldehyde, 2-methyl or *o*-Tolualdehyde $2\text{-}CH_3C_6H_4CHO$	120.15	......	200, 94^{10}	......	$1.0386^{19/4}$	1.5481^{20}	al eth, ace, bz, chl	$B7^3$, 1011
1781	Benzaldehyde, 3-methyl or *m*-Tolualdehyde $3\text{-}CH_3C_6H_4CHO$	120.15	......	199, $93\text{-}4^{17}$	......	$1.0189^{21/4}$	1.5413^{21}	**al, eth,** ace, bz, chl	$B7^3$, 1013
1782	Benzaldehyde, 4-methyl or *p*-Tolualdehyde $4\text{-}CH_3C_6H_4CHO$	120.15	......	204-5, 106^{10}	......	$1.0194^{17/4}$	1.5454^{20}	**al, eth, ace,** chl	$B7^3$, 1016
1783	Benzaldehyde, 2-nitro $2\text{-}NO_2C_6H_4CHO$	151.12	ye nd (w)	153^{23}	43.5-4	$1.2844^{20/4}$	......	al, eth, ace, bz	$B7^3$, 889
1784	Benzaldehyde, 2-nitro, diacetate $2\text{-}NO_2C_6H_4CH(O_2CCH_3)_2$	253.21	pr (lig)	......	90	......	......	eth, ace, bz	$B7^3$, 892
1785	Benzaldehyde, 2-nitro, dimethyl acetal $2\text{-}NO_2C_6H_4CH(OCH_3)_2$	197.19	gr-ye	274-6, $138\text{-}9^{11}$	......	......	......	bz	$B7^3$, 891
1786	Benzaldehyde, 3-nitro $3\text{-}NO_2C_6H_4CHO$	151.12	lt ye nd (w)	164^{23}	58	$1.2792^{20/4}$	......	al, eth, ace, bz	$B7^3$, 897
1787	Benzaldehyde, 3-nitro, diacetate $3\text{-}NO_2C_6H_4CH(O_2CCH_3)_2$	253.21	pr or nd (al)	......	72	1.393	......	al, eth, ace, bz	$B7^3$, 900

No.	Name, Synonyms, and Formula	Mol. wt.	Color, crystalline form, specific rotation and λ_{max} (log ε)	b.p. °C	m.p. °C	Density	n_D	Solubility	Ref.
1788	Benzaldehyde, 3-nitro, dimethyl acetal $3\text{-}NO_2C_6H_4CH(OCH_3)_2$	197.19		162-4[19]		1.209[15]		bz	B7[3], 899
1789	Benzaldehyde, 4-nitro $4\text{-}NO_2C_6H_4CHO$	151.12	lf, pr (w)	sub	106	1.496		al, bz, aa	B7[3], 907
1790	Benzaldehyde, 4-nitro, diacetate $4\text{-}NO_2C_6H_4CH(O_2CCH_3)_2$	253.21	pr (al)		127			al, eth, ace, bz	B7[3], 909
1791	Benzaldehyde, 4-nitro, dimethylacetal $4\text{-}NO_2C_6H_4CH(OCH_3)_2$	197.19		294-6[774]	23-5			bz	B7[3], 909
1792	Benzaldehydeoxime *anti* or *anti*-Benzaldoxime $C_6H_5CH{=}NOH$	121.14	nd (eth)		130	1.145[20/4]		al, eth, w	B7[3], 840
1793	Benzaldehydeoxime *syn* or *syn*-Benzaldoxime $C_6H_5CH{=}NOH$	121.14	pr	200,118-9[10]	36-7	1.1111[20/4]	1.5908[20]	al, eth, bz	B7[3], 840
1794	Benzaldehyde, pentachloro C_6Cl_5CHO	278.35	nd (bz, al)		202.5			eth, ace, bz	B7[3], 881
1795	Benzaldehyde, phenylhydrazone $C_6H_5CH{=}NNHC_6H_5$	196.26	nd (lig), pr		156			ace, bz	B15[2], 57
1796	Benzaldehyde, 4-isopropyl or Cumaldehyde $4\text{-}(CH_3)_2CHC_6H_4CHO$	148.21		235-6, 103-4[10]		0.9755[20/4]	1.5301[20]	al, eth	B7[3], 1095
1797	Benzaldehyde, semicarbazone $C_6H_5CH{=}NNHCONH_2$	163.18			224				B7[3], 854
1798	Benzaldehyde, 2,3,4,5-tetrachloro $2,3,4,5Cl_4C_6HCHO$	243.90			106				B7[2], 181
1799	Benzaldehyde, 2,3,4-trichloro $2,3,4\text{-}Cl_3C_6H_2CHO$	209.46	nd (al)		91			al	B7, 238
1800	Benzaldehyde, 2,3,5- trichloro $2,3,5\text{-}Cl_3C_6H_2CHO$	209.46	nd (al)		56			al, eth, ace, bz	B7[2], 180
1801	Benzaldehyde, 2,3,6-trichloro $2,3,6\text{-}Cl_3\text{-}C_6H_2CHO$	209.46	nd (lig)		86-7			eth, ace, bz	B7[3], 881
1802	Benzaldehyde, 2,4,5-trichloro $2,4,5\text{-}Cl_3C_6H_2CHO$	209.46	nd (al)		112-3			al, eth, ace, bz	B7[2], 180
1803	Benzaldehyde, 2,4,6, trichloro $2,4,6\text{-}Cl_3\text{-}C_6H_2CHO$	209.46			58-9			eth, lig	B7[3], 881
1804	Benzaldehyde, 3,4,5-trichloro $3,4,5\text{-}Cl_3\text{-}C_6H_2CHO$	209.46	nd (al)		90-1			al, eth, ace, bz, peth	B7[2], 180
1805	Benzaldehyde, 2,4,5-trimethoxy $2,4,5\text{-}(CH_3O)_3C_6H_2CHO$	196.20		140[4]-sub	114			w, eth, chl, lig	B8[3], 3361
1806	Benzaldehyde, 3,4,5-trimethoxy $3,4,5\text{-}(CH_3O)_3C_6H_2CHO$	196.20			71-4			chl	B8[3], 3368
1807	Benzaldehyde, 2,4,6-trimethyl or Mesityl aldehyde $2,4,6\text{-}(CH_3)_3C_6H_2CHO$	148.20		237-40, 192[50]	14			al, eth, ace, bz	B7[3], 1111
1808	Benzaldehyde, 2,4,6-trinitro $2,4,6\text{-}(NO_2)_3C_6H_2CHO$	241.12	dk gr pl (bz)		119			al, eth, ace, bz,aa	B7[3], 926
1809	Benzamide or Benzoic acid amide $C_6H_5CONH_2$	121.14	mcl pr or pl (w)	290	132-3	1.0792[130/4], 1.341[4]		al, bz, w	B9[3], 1064
1810	Benzamide, 2-amino or Anthranilamide $2\text{-}H_2NC_6H_4CONH_2$	136.15	lf (chl or w)	300	110-1.5			w, al	B14[3], 889
1811	Benzamide, 3-amino $3\text{-}H_2NC_6H_4CONH_2$	136.15	ye mcl nd (+lw) nd		79—80 (hyd), 113—4 (anh)			w, al, eth	B14[3], 998
1812	Benzamide, 4-amino $4\text{-}H_2NC_6H_4CONH_2$	136.15	ye cr (+¼w)		183			al, eth	B14[3], 1061
1813	Benzamide, 2-benzoyl $2\text{-}C_6H_5COC_6H_4COCNH_2$	225.25	nd (to)		165 (cor)			al, bz, to	B10[3], 3295
1814	Benzamide, 2-bromo $2\text{-}BrC_6H_4COCNH_2$	200.03	nd (w)	sub	160-1			al	B9[3], 1387
1815	Benzamide, 3-bromo $3\text{-}BrC_6H_4CONH_2$	200.03	lf (w or al)	sub	155.3			al	B9[3], 1399
1816	Benzamide, 4-bromo $4\text{-}BrC_6H_4CONH_2$	200.03	nd or pl (w)		192			al, w, aa	B9[3], 1418
1817	Benzamide, 2-chloro $2\text{-}ClC_6H_4CONH_2$	155.58	rh nd (w)		142.4			al, eth	B9[3], 1338
1818	Benzamide, 3-chloro $3\text{-}ClC_6H_4CONH_2$	155.58	nd		135-7			al, eth	B9[3], 1349

PHYSICAL CONSTANTS OF ORGANIC COMPOUNDS (Continued)

No.	Name, Synonyms, and Formula	Mol. wt.	Color, crystalline form, specific rotation and λ_{max} (log ε)	b.p. °C	m.p. °C	Density	n_D	Solubility	Ref.
1819	Benzamide, 4-chloro $4\text{-ClC}_6H_4CONH_2$	155.58			179-80			al, eth, w	B9^3, 1363
1820	Benzamide, 2,5-dimethoxy $2,5(CH_3O)_2C_6H_3CONH_2$	181.19	lf (hz-peth), nd (w)		141-2			al, eth, ace, bz, chl	B10^2, 258
1821	Benzamide, 3,4-dimethoxy or Veratramide $3,4\text{-}(CH_3O)_2C_6H_3CONH_2$	181.19	cr (w)		164			w, eth, bz	B10^2, 261
1822	Benzamide, 3,5-dimethoxy or 3,5-Dimethoxy benzamide $3,5\text{-}(CH_3O)_2C_6H_3CONH_2$	181.19	nd (bz)		148—9			w, al, eth, bz, lig	B10^3, 1449
1823	Benzamide, *N,N*-diphenyl or *N*-Benzoyl diphenylamine $C_6H_5CON(C_6H_5)_2$	273.33	rh pr (al), nd		180				B12^3, 511
1824	Benzamide, 2-ethoxy $2\text{-}C_2H_5OC_6H_4CONH_2$	165.19	nd (w, al)		132-4			al, eth	B10^2, 58
1825	Benzamide, 3-ethoxy $3\text{-}C_2H_5OC_6H_4CONH_2$	165.19	nd (w)		139			al, eth, ace, chl	B10^1, 64
1826	Benzamide, 4-ethoxy $4\text{-}C_2H_5OC_6H_4NH_2$	165.19	pr (dil al)		206			al	B10^3, 342
1827	Benzamide, 3-hydroxy $3\text{-HOC}_6H_4CONH_2$	137.14	pl (w)		170.5			al, eth	B10^3, 254
1828	Benzamide, 3-hydroxy $3\text{-HOC}_6H_4CONH_2$	137.14	pl (w)		170.5			al, eth	B10^2, 82
1829	Benzamide, 3-hydroxy-*N*-phenyl $3\text{-HOC}_6H_4CONHC_6H_5$	213.24	nd or pl (w, dil al)		156			al	B12^1, 269
1830	Benzamide, 4-hydroxy $4\text{-HOC}_6H_4CONH_2$	137.14	nd (w + l)		162 (hyd)			al, eth, chl	B10^3, 340
1831	Benzamide, 4-hydroxy-*N*-phenyl $4\text{-HOC}_6H_4CONHC_6H_5$	213.24	pl or nd (w)		201-2			al	B12^3, 947
1832	Benzamide, 2-methoxy or *o*-Anisamide $2\text{-CH}_3OC_6H_4CONH_2$	151.16	nd (bz), pr (eth), pl (w)		129			eth, bz	B10^2, 58
1833	Benzamide, 4-methoxy $4\text{-CH}_3OC_6H_4CONH_2$	151.16	nd or ta (w)	295	166-7			al	B10^2, 100
1834	Benzamide, *N*-methyl $C_6H_5CONHCH_3$	135.17		291^{765}, 167^{12}	82			al, ace	B9^3, 1068
1835	Benzamide, *N*-α-naphthyl $C_6H_5CONHC_{10}H_7\text{-}\alpha$	247.30	nd (al or aa)		161-2			bz, aa	B12^3, 2870
1836	Benzamide, 2-nitro $2\text{-NO}_2C_6H_4CONH_2$	166.14		317	176.6			al, eth	B9^3, 1477
1837	Benzamide, 2-nitro-*N*-phenyl $2\text{-NO}_2C_6H_4CONHC_6H_5$	242.23	nd (al or bz)		155			al, eth, bz, chl	B12^3, 506
1838	Benzamide, 3-nitro $3\text{-NO}_2C_6H_4CONH_2$	166.14		310—5	142.7			w, al, eth	B9^3, 1515
1839	Benzamide, 4-nitro $4\text{-NO}_2C_6H_4CONH_2$	166.14			201.4			al, eth	B9^3, 1710
1840	Benzamide, *N*-(2-nitrophenyl) $C_6H_5CONH(C_6H_4NO_2\text{-}2)$	242.23	gold-ye nd (al)		98			al, eth	B12^3, 1525
1841	Benzamide, *N*-(3-nitrophenyl) $C_6H_5CONH(C_6H_4NO_2\text{-}3)$	242.23	pl		157			eth, chl	B12^3, 1552
1842	Benzamide, *N*-(4-nitrophenyl) $C_6H_5CONH(C_6H_4NO_2\text{-}4)$	242.23	ye nd (AcOEt)		199				B12^3, 1597
1843	Benzamide, *N*-phenyl or Benzanilide $C_6H_5CONHC_6H_5$	197.24	lf (al)	$117\text{-}9^{10}$, sub	163	1.315			B17^3, 502
1844	Benzamide, thiono, *N*-phenyl $C_6H_5CSNHC_6H_5$	213.30	ye pl or pr (al)	d	10^2			al, eth, bz	B12^3, 507
1845	Benzamide, *N*-(2-tolyl) $C_6H_5CONH(C_6H_4CH_3\text{-}2)$	211.26	rh nd (AcOEt-ace)		145-6	1.205^{15}		al, eth	B12^3, 1858
1846	Benzamide, *N*-(3-tolyl) $C_6H_5CONH(C_6H_4CH_3\text{-}3)$	211.26	mcl pr (dil al)		125	1.170^{15}		al	B12^3, 1964
1847	Benzamide, *N*-(4-tolyl) $C_6H_5CONH(C_6H_4CH_3\text{-}4)$	211.26	rh nd (al)	232	158	1.202^{15}		al, eth	B12^3, 2062
1848	Benzamide, 3,4,5-trihydroxy or Gallamide $3,4,5\text{-(HO)}_3C_6H_2CONH_2$	169.14	lf (w + 1)		244-5d			al, w, chl	B10^2, 346
1849	Benzomide, 3,4,5-trihydroxy-*N*-phenyl or Gallanilide $3,4,5\text{-(HO)}_3C_6H_2CONHC_6H_5$	245.24	lf (+ 2w, dil al)		207			al, aa, w	B12^2, 263
1850	Benzamidine $C_6H_5C(=NH)NH_2$	120.15	lf (al)		80			w, al	B9^3, 1264

No.	Name, Synonyms, and Formula	Mol. wt.	Color, crystalline form, specific rotation and λ_{max} (log ε)	b.p. °C	m.p. °C	Density	n_D	Solubility	Ref.
1851	Benzamidine hydrochloride $C_6H_5C(:NH)NH_2 \cdot HCl$	156.61	rh pr (w + 2)		169			w, al	$B9^2$199
1852	Benzamidine, *N,N*-diphenyl $C_6H_5C(=NH)N(C_6H_5)_2$	272.35	rh pl (eth)		112			al, eth, bz	$B12^3$, 511
1853	Benzamidine, *N,N*-diphenyl, hydrochloride $C_6H_5C(=NH)N(C_6H_5)_2.HCl$	308.81	mcl pr or nd (al)		223d			w, al	B12, 270
1854	Benzamidine, *N,N'*-diphenyl $C_6H_5C(=NC_6H_5)NHC_6H_5$	272.35	nd (bz or al), pr (al)	146-7				al, eth	$B12^3$, 514
1856	Benzamidine, *N*-(1-naphthyl) $C_6H_5C(=NH)NHC_{10}H_7$-(α)	246.31	pl (al)		141			al, eth	B12, 1233
1857	Benzamidoxime $C_6H_5C(NH_2)=NOH$	136.15	mcl pr (w)		80			al, eth, bz, chl	$B9^3$, 1308
1858	1,2-Benzanthracene or Naphthanthracene $C_{18}H_{12}$	228.29	ye-br flr pl or lf (al-aa)	435 (sub)	162			al, eth, ace, bz	$B5^3$, 2374
1859	1,2-Benzanthracene, 9,10-dihydro $C_{18}H_{14}$	230.31	lf (bz, al, aa)		112				$B5^3$, 2305
1860	1,2-Benzanthracene, 9,10-dimethyl $9,10(CH_3)_2C_{18}H_{10}$	256.35	pa ye pl (al, aa)		122-3			ace, bz	$B5^3$, 2413
1861	1,2-Benzanthracene, 4'-hydroxy $4\text{-}HOC_{18}H_{11}$	244.29	pa ye or og nd (bz)		225			bz, to, aa	$B6^3$, 3725
1862	1,2-Benzanthracene, 3-methyl $3\text{-}CH_3C_{18}H_{11}$	242.32	pl (bz, al), nd (bz-lig)	$160^{0.01}$, (sub)	156-7			al, ace, bz, aa	$B5^3$, 2388
1863	1,2-Benzanthracene, 4-methyl $4\text{-}CH_3C_{18}H_{11}$	242.32	nd (al)	1	126-7			al, eth	$B5^3$, 2388
1864	1,2-Benzanthracene, 5-methyl $5\text{-}CH_3C_{18}H_{11}$	242.32	ye mcl pl (bz-al)		160			al, eth, bz, xyl	$B5^3$, 2388
1865	1,2-Benzanthracene, 6-methyl $6\text{-}CH_3C_{18}H_{11}$	242.32	pl (al)		150-1			al, eth, chl	$B5^3$, 2388
1866	1,2-Benzanthracene, 7-methyl $7\text{-}CH_3C_{18}H_{11}$	242.32	ye pl (al)		183-3.6			al, aa	2389
1867	1,2-Benzanthracene, 8-methyl $8\text{-}CH_3C_{18}H_{11}$	242.32	pl or nd (al, aa)		118			al, aa	$B5^3$, 2391
1868	1,2-Benzanthracene, 9-methyl $9\text{-}CH_3C_{18}H_{11}$	242.32	pa ye nd or pl (aa, MeOH)		138			al, aa	$B5^3$, 2391
1869	1,2-Benzanthracene, 10-methyl $10\text{-}CH_3C_{18}H_{11}$	242.32			141			al, eth, ace	$B5^3$, 2392
1870	1,2-Benzanthracene, 10-carboxaldehyde $10\text{-}OHCC_{18}H_{11}$	256.30	ye pr or nd		148			al, eth, ace, bz	$B7^3$, 2807
1871	1,2-Benz-3,4-anthraquinone $C_{18}H_{10}O_2$	258.28	red nd (to)		262-3d				$B7^2$, 759
1872	1,2-Benz-9,10-anthraquinone $C_{18}H_{10}O_2$	258.28	ye pr (aa)	sub	169			ace, bz, chl	$B7^3$, 4278
1873	Benzanthrene $C_{17}H_{12}$	216.28	lf (al)		84			al	$B5^3$, 2289
1874	1,9-Benzanthr-10-one $C_{17}H_{10}O$	230.27	ye nd (xyl or al)		170				$B7^3$, 2694
1875	1,9-Benzanthr-10-one, *bz*-1-bromo $C_{17}H_9BrO$	309.17	ye nd (aa)		178			al, MeOH	$B7^3$, 2706
1876	1,9-Benzanthr-10-one, *bz*-1-chloro $C_{17}H_9ClO$	264.71	ye nd (aa)		182-3				$B7^3$, 2700
1877	1,9-Benzanthr-10-one, *bz*-2-hydroxy $C_{17}H_{10}O_2$	246.27	ye nd		294				$B8^3$, 1629
1878	1,9-Benzanthr-10-one, 4-hydroxy $C_{17}H_{10}O_2$	246.27	ye nd (aa)		178-9			eth, bz	$B8^3$, 1638
1879	1,2-Benzazulene $C_{14}H_{10}$	178.23			176				$B5^3$, 2122
1880	Benzedrine or Amphetamine *dl* $C_6H_5CH_2CH(NH_2)CH_3$	135.21		203, 97^{20}		$0.9306^{25/4}$	1.518^{26}	al, chl	$B12^3$, 2665
1881	Benzestrol or 2,4-Bis(4-hydroxyphenyl)-3-ethyl hexane $C_{20}H_{26}O_2$	298.43	cr (al)		162-6			al, eth, ace, aa	$B6^3$, 5538
1882	Benzene C_6H_6	78.11	rh pr	80.1	5.5	$0.8765^{20/4}$	1.5011^{20}	al, eth, ace, aa	$B5^4$, 583

No.	Name, Synonyms, and Formula	Mol. wt.	Color, crystalline form, specific rotation and λ_{max} (log ε)	b.p. °C	m.p. °C	Density	n_D	Solubility	Ref.
1883	Benzene, allyl or Allylbenzene $CH_2=CHCH_2C_6H_5$	118.18		156, 47^{13}	-40	$0.8920^{20/4}$	1.5131^{20}	al, eth, bz	$B5^4$, 1362
1884	Benzene, 1-allyl-4-bromo $4\text{-}BrC_6H_4CH_2CH=CH_2$	197.07		222-2, 96^{12}		$1.324^{15/4}$	1.559^{20}	eth, bz, chl	$B5^4$, 1363
1885	Benzene, 1-allyl-3,4-dimethoxy $3,4\text{-}(CH_3O)_2C_6H_3CH_2CH=CH_2$	178.23	cr (hx)	254.7, 104^2	-4	$1.0396^{20/4}$	1.5340^{20}	al, eth	$B6^3$, 5024
1886	Benzene, 1-allyl-2,3,4,5-tetramethoxy $2,3,4,5\text{-}(CH_3)C_6HCH_2CH=CH_2$	238.28			25	1.087^{25}	1.51462^{25}		$B6^3$, 6691
1887	Benzene, (3-aminobutyl) $C_6H_5(CH_2CH_2CH(NH_2)CH_3)$	149.24		221-3^{750}, 101-2^{14}	148	$0.9289^{15/4}$	1.51520^{20}	al	$B12^3$, 2717
1888	Benzene, (1-aminoethyl) *d* $C_6H_5CH(NH_2)CH_3$	121.18	$[\alpha]^{25/}{}_D$ + 39.3 (MeOH)	187		09651^{15}		w, al, eth	$B12^3$, 2386
1889	Benzene, (1-aminoethyl) *dl* $C_6H_5CH(NH_2)CH_3$	121.18		187, 87^{24}		0.9395^{15}	1.5238^{25}	w, al, eth	$B12^3$, 2386
1890	Benzene, (1-aminoethyl) *l* $C_6H_5CH(NH_2)CH_3$	121.18	$[\alpha]^{18/}{}_D$ −38 (MeOH)	184-6, 77^{16}		$0.9520^{20/4}$		w, al, eth	$B12^3$, 2386
1891	Benzene, (2-aminoethyl) or β-Phenethylamine $C_6H_5CH_2CH_2NH_2$	121.18		197-8		$0.9580^{24/4}$	1.5290^{25}	w, al, eth	$B12^3$, 2408
1892	Benzene, 1-(2-aminoethyl)-3,4-dimethoxy $3,4\text{-}(CH_3O)_2C_6H_3CH_2NH_2$	181.23		163-5^{14}			1.5464^{20}		$B13^3$, 2207
1893	Benzene, 1-benzyl-4-ethyl or *p*-Ethyl diphenylmethane $4\text{-}C_2H_5C_6H_4CH_2C_6H_5$	196.29		297, $85^{0.2}$	−24	0.9777^{20}	1.5616^{20}	al, eth, chl	$B5^4$, 1926
1894	Benzene, 1,2-bis (bromomethyl) or *o*-Xylyene bromide $1,2\text{-}(BrCH_2)_2C_6H_4$	263.96	rh (chl)	128-$30^{4.5}$	95	1.988^0		al, eth, peth	$B5^4$, 929
1895	Benzene, 1,3-bis (bromomethyl) or *m*-Xylyene bromide $1,3\text{-}(BrCH_2)_2C_6H_4$	263.96	nd (chl), pr (ace)	135-40^{20}	77	1.959^0		al, eth, lig	$B5^4$, 946
1896	Benzene, 1,4-bis (bromomethyl) or *p*-Xylyene bromide $1,4\text{-}(BrCH_2)_2C_6H_4$	263.96	mcl pr (al), cr (chl, bz)	245, 155-8^{14}	145-7			al, bz, chl	$B5^4$, 970
1897	Benzene, 1,3,-bis (bromomethyl)-5-methyl $1,3\text{-}(BrCH_2)_2\text{-}5\text{-}CH_3C_6H_3$	278.99	pr		66			al, eth, bz, lig	$B5^4$, 1027
1898	Benzene, 1,3-bis(carboethoxy) $1,3\text{-}C_6H_4(OCH_2CO_2H_5)_2$	226.19	nd (aa, w)		195			w, aa	$B6^3$, 4324
1899	Benzene, 1,3-bis (carbethoxymethoxy) $1,3\text{-}C_6H_4(OCH_2CO_2H_5)_2$	282.30	nd (eth)	278^{32}	42				B6, 818
1900	Benzene, 1,2-bis (chloromethyl) or *o*Xylyene chloride $1,2\text{-}(ClCH_2)_2C_6H_4$	175.06	mcl (lig)	239-41, 130-5^{10}	55	1.393^0		al, eth, lig	$B5^4$, 927
1901	Benzene, 1,3-bis (chloromethyl or *m*-Xylyene chloride $1,3\text{-}(ClCH_2)_2C_6H_4$	175.06		250-5	34.2			al, eth	$B5^4$, 944
1902	Benzene, 1,4-bis (chlormethyl) or *p*-Xylyene chloride $1,4\text{-}(ClCH_2)_2C_6H_4$	175.06	pl (al)	240-50d, 135^{16}	100	1.417^0		al, eth, ace, chl	$B5^4$, 967
1903	Benzene, 1,2-bis (cyanomethyl) or *o*-Xylyene dicyanide $1,2\text{-}(NCCH_2)_2C_6H_4$	156.19	(i) nd or pr (MeOH), (ii) pr (al)		(i) 18 (unst), (ii) 60 (st)			al, eth	$B9^3$, 4293
1904	Benzene, 1,3-bis (cyanomethyl) or *m* - Xylyene dicyanide $1,3\text{-}(NCCH_2)_2C_6H_4$	156.19	nd (w), pr (eth)	305-10^{100}, $170^{20\text{-}30}$	30			al, eth, bz, chl	$B9^3$, 4294
1905	Benzene, 1,4-bis (cyanomethyl) or *p*-Xylyene dicyanide $1,4\text{-}(NCCH_2)_2C_6H_4$	156.19	pr (al), nd (w)		98			al, eth, chl	$B9^3$, 4295
1906	Benzene, 1,2-bi(dibromomethyl) $1,2\text{-}(Br_2CH)_2C_6H_4$	421.75	mcl		116-7			chl	$B5^4$, 929
1907	Benzene, 1,3-bis (dibromomethyl) $1,3\text{-}(Br_2CH)_2C_6H_4$	421.75	nd (al), pr (chl)		107			al, bz, chl, lig	$B5^3$, 839
1908	Benzene, 1,4-bis (dibromomethyl) $1,4\text{-}(Br_2CH)_2C_6H_4$	421.75	mcl pr (chl)		172			bz	$B5^3$, 860
1909	Benzene, 1,2-bis (diethylamino) $1,2\text{-}[(C_2H_5)_2N]_2C_6H_4$	220.36		119^{10}		$0.9267^{13/4}$	1.5213^{13}	al, eth	$B13^2$, 12
1910	Benzene, 1,3-bis (diethylamino) $1,3[(C_2H_5)_2N]_2C_6H_4$	220.36		148^9		$0.9522^{12/4}$	1.5537^{12}	al, eth	$B13^2$, 26
1911	Benzene, 1,4-bis (diethylamino) $1,4\text{-}[(C_2H_5)_2N]_2C_6H_4$	220.36	mcl pr, pl (al-w)	280	52			al, eth, bz, lig	$B13^2$, 40
1912	Benzene, 1,4-bis (dimethylamino) $1,4\text{-}[(CH_3)_2N]_2C_6H_4$	164.25	lf (dil al or lig)	260	51			al, eth, bz, lig	$B13^3$, 111
1913	Benzene, 1,2-bis (hydroxymethyl) or *o* -Xylylene glycol $1,2\text{-}C_6H_4(CH_2OH)_2$	138.17	pl (eth, peth)		65-6			w, al, eth	$B6^3$, 4587
1914	Benzene, 1,3-bis (hydroxymethyl) or Isophthalyl glycol $1,3\text{-}(HOCH_2)_2C_6H_4$	138.17	nd (bz)	154-9^{13}	57			w, al, eth	$B6^3$, 4600

PHYSICAL CONSTANTS OF ORGANIC COMPOUNDS (Continued)

No.	Name, Synonyms, and Formula	Mol. wt.	Color, crystalline form, specific rotation and λ_{max} (log ε)	b.p. °C	m.p. °C	Density	n_D	Solubility	Ref.
1915	Benzene, 1,4-bis (hydroxymethyl) or Terephthalyl glycol $1,4\text{-}(HOCH_2)_2C_6H_4$	138.17	nd (w)	....	115-6	....	....	w, al, eth, ace	$B6^3$, 4608
1916	Benzene, 1,2-bis (methylamino)-4-chloro $4\text{-}Cl\text{-}1,2(CH_3NH)_2C_6H_3$	170.64	pr (lig)	....	61	....	....	w	B13, 25
1917	Benzene, 1,3-bis (3-methylbutoxy) $[(CH_3)_2CHCH_2CH_2O]_2C_6H_4$	250.39	....	....	47	....	....	....	B6, 815
1918	Benzene, 1,3-bis((4-tolylamino) $1,3\text{-}(4\text{-}CH_3C_6H_4NH)_2C_6H_4$	288.40	nd (al)	....	139-9	....	....	....	B13, 42
1919	Benzene, 1,2-bis (trimethylsiloxy) $1,2\text{-}[(CH_3)_3SiO]_2C_6H_4$	254.48	....	235^{760}	....	....	$1.4685^{20/}{}_D$	....	C^2, 249
1920	Benzene, 1,3-bis (trimethylsiloxy) $1,3\text{-}[(CH_3)_3SiO)]_2C_6H_4$	254.48	....	$237\text{-}40^{740}$	....	$0.950^{20/4}$	$1.4748^{20/}{}_D$	....	C^2, 249
1921	Benzene, bromo or phenylbromide C_6H_5Br	157.01	....	156, 43^{18}	-30.8	$1.4950^{20/4}$	1.5597^{20}	al, eth, bz	$B5^4$, 670
1922	Benzene, (4-bromobutoxy) $Br(CH_2)_4OC_6H_5$	229.12	cr (al)	$153\text{-}6^{18}$	41	....	....	al	$B6^4$, 558
1923	Benzene, 1-bromo-4-*tert* butyl $4\text{-}[(CH_3)_3C]C_6H_4Br$	213.12	....	231-2, 103^{10}	19	$1.2286^{20/4}$	1.5436^{20}	eth, bz, chl	$B5^4$, 1050
1924	Benzene, 1-bromo-2-chloro $2\text{-}ClC_6H_4Br$	191.45	....	204^{765}	-12.3	$1.6387^{25/4}$	1.5809^{20}	bz	$B5^4$, 680
1925	Benzene, 1-bromo-3-chloro $3\text{-}ClC_6H_4Br$	191.45	....	196	-21.5	$1.6302^{20/4}$	1.5771^{20}	al, eth	$B5^4$, 680
1926	Benzene, 1-bromo-4-chloro $4ClC_6H_4Br$	191.45	nd or pl (al, eth)	196^{756}	68	$1.576^{71/4}$	1.5531^{70}	eth, bz, chl	$B5^4$, 681
1927	Benzene, 1-bromo 4 cyclohexyl $4\text{-}C_6H_{11}\text{-}C_6H_4Br$	239.16	....	160^{25}	....	$1.283^{25/4}$	1.5584^{20}	bz, chl	$B5^3$, 1258
1928	Benzene, 1-bromo-2,3-dichloro $2,3\text{-}Cl_2C_6H_3Br$	225.90	pl or lf (al)	243^{705}	60	....	....	eth, bz, chl	B5, 209
1929	Benzene, 1-bromo-3,5-dichloro $3,5\text{-}Cl_2C_6H_3Br$	225.90	pr (al)	232^{757}	82-4	....	....	eth, bz, chl, al	$B5^2$, 162
1930	Benzene, 1-bromo-2,6-dichloro $2,6\text{-}Cl_2C_6H_3Br$	225.90	pr (al)	242^{765}	65	....	....	eth, bz, chl	B5, 210
1931	Benzene, 1-bromo-2,5-dichloro $2,5\text{-}Cl_2C_6H_3Br$	225.90	pr or nd (al)	235^{731}, 119^{20}	35	....	....	al, eth, bz, lig	$B5^3$, 564
1932	Benzene, 1-bromo-3,4-dichloro $3,4\text{-}Cl_2C_6H_3Br$	225.90	pr	237, 124^{33}	....	....	....	al, eth, bz, chl	$B5^3$, 564
1933	Benzene, 1-bromo-2,3-dimethyl $2,3\text{-}(CH_3)_2C_6H_3Br$	185.06	....	214, 83^{11}	....	$1.365^{20/4}$	....	eth, ace, bz	$B5^4$, 928
1934	Benzene, 1-bromo-2,4-dimethyl $2,4\text{-}(CH_3)_2C_6H_3Br$	185.06	....	205, 84^{13}	0	....	1.5501^{20}	al, eth, bz	$B5^4$, 945
1935	Benzene, 1-bromo-2,5-dimethyl $2,5\text{-}(CH_3)_2C_6H_3Br$	185.06	lf or pl	199-200, $88\text{-}9^{13}$	9	1.3582^{18}	1.5514^{18}	al, bz	$B5^4$, 969
1936	Benzene, 1-bromo-2,6-dimethyl $2,6(CH_3)_2C_6H_3Br$	185.06	....	203-4, $98\text{-}9^{20}$	....	....	1.5552^{20}	eth, ace, bz	$B5^4$, 945
1937	Benzene, 1-bromo-3,4-dimethyl $3,4\text{-}(CH_3)_2C_6H_3Br$	185.06	....	214.5	-0.2	$1.3708^{20/4}$	1.5530^{20}	al, eth	$B5^4$, 928
1938	Benzene, 1-bromo-3,5-dimethyl $3,5\text{-}(CH_3)_2C_6H_3Br$	185.06	....	204, $83\text{-}9^{12}$	....	1.342^{20}	1.5462^{22}	eth, ace, bz	$B5^4$, 945
1939	Benzene, 1-bromo-2,3-dinitro $2,3\text{-}(NO_2)_2C_6H_3Br$	247.00	pa ye pl (al)	320	101-2	....	....	al	$B5^4$, 749
1940	Benzene, 1-bromo-2,4-dinitro $2,4\text{-}(NO_2)_2C_6H_3Br$	247.00	ye nd (al)	....	75	....	....	al	$B5^4$, 749
1941	Benzene, 1-bromo-2,5-dinitro $2,5\text{-}(NO_2)_2C_6H_3Br$	247.00	nd (al), pr (al-eth)	....	70	....	....	eth, al	$B5^4$, 750
1942	Benzene, 1-bromo-2,6-dinitro $2,6\text{-}(NO_2)_2C_6H_3Br$	247.00	ye pr (al)	....	107	....	....	al	$B5^4$, 749
1943	Benzene, 1-bromo-3,4-dinitro $3,4\text{-}(NO_2)_2C_6H_3Br$	247.00	nd (al), pl (al-eth)	....	34-5 (unst)	....	....	eth, al	$B5^3$, 640
1944	Benzene, 1-bromo-2-ethoxy $2\text{-}C_2H_5OC_6H_4Br$	201.06	....	222-6	....	....	....	al, eth	$B6^4$, 1038
1945	Benzene, 1-bromo-3-ethoxy $3\text{-}C_2H_5OC_6H_4Br$	201.06	....	222	....	....	....	al, eth	$B6^3$, 739
1946	Benzene, 1-bromo-4-ethoxy $4\text{-}C_2H_5OC_6H_4Br$	201.06	....	230-2, 109^{17}	4	$1.4071^{25/4}$	1.5517^{20}	al, eth	$B6^4$, 1045
1947	Benzene, (1-bromoethyl) *d* $C_6H_5(CHBrCH_3)$	185.06	$[\alpha]^{14/}{}_D$ +1.5	203, $86\text{-}8^{15}$	....	1.3108^{23}	1.5612^{20}	al, eth	$B5^4$, 906

No.	Name, Synonyms, and Formula	Mol. wt.	Color, crystalline form, specific rotation and λ_{max} (log ε)	b.p. °C	m.p. °C	Density	n_D	Solubility	Ref.
1948	Benzene, (1-bromoethyl) *dl* $C_6H_5(CHBrCH_3)$	185.06		202-3, 85^{13}		$1.3605^{20/4}$	1.5612^{20}	al, eth, bz	$B5^4$, 906
1949	Benzene, (2-bromoethyl) $C_6H_5CH_2CH_2Br$	185.06		$217\text{-}8^{734}$, 92^{11}		$1.3587^{20/4}$	1.5572^{20}	eth, bz	$B5^4$, 907
1950	Benzene, 1-bromo-2-ethyl 2-$C_2H_5C_6H_4Br$	185.06		199			$1.5473^{20/}{}_D$		$B5^4$, 906
1951	Benzene, 1-bromo-4-ethyl 4-$C_2H_5C_6H_4Br$	185.06		204			1.5445^{20}		$B5^4$, 906
1952	Benzene, 1-bromo-4-fluoro 4-F-C_6H_4-Br	175.00		152^{764}	-8 (fr-17)	$1.4946^{20/4}$	1.5604^{20}	al, eth	$B5^4$, 678
1953	Benzene, (2-bromo-1-hydroxyethyl) $C_6H_5CH(OH)CH_2Br$	201.06		$109\text{-}10^{2}$		$1.4994^{20/4}$	$1.5800^{17/}{}_D$		$B6^3$, 1690
1954	Benzene, 1-bromo-2-iodo 2-I-C_6H_4-Br	282.91		257^{754}, $120\text{-}1^{15}$	9-10	$2.2571^{25/4}$	1.6618^{25}	ace	$B5^4$, 698
1955	Benzene, 1-bromo-3-iodo 3-I-C_6H_4Br	282.91		252^{754}, 120^{18}	-9.3				$B5^4$, 698
1956	Benzene, 1-bromo-4-iodo 4-I-C_6H_4Br	282.91	pr or pl (eth-al)	252^{754}	92			eth	$B5^4$, 698
1957	Benzene, 1-(bromomethyl)-2-methyl or *o*-Xylylbromide 2-$CH_3C_6H_4CH_2Br$	185.06	pr	$216\text{-}7^{742}$, 108^{16}	21	1.3811^{23}	1.5730^{20}	al, eth, ace, bz	$B5^4$, 928
1958	Benzene, 1-(bromomethyl)-3-methyl or *m*-Xylylbromide 3-$CH_3C_6H_4CH_2Br$	185.06		212-3, 105^{11}		1.3711^{23}	1.5660^{20}	al, eth, chl	$B5^2$, 293
1959	Benzene, 1-(bromomethyl)-4-methyl or *p*-Xylylbromide 4-$CH_3C_6H_4CH_2Br$	185.06	nd (al)	$218\text{-}20^{740}$	35	1.324		al, eth, chl	$B5^3$, 858
1960	Benzene, 1-(bromomethyl)-3,5-dimethyl or Mesityl bromide 3,5-$(CH_3)_2C_6H_3CH_2Br$	199.09	nd (eth)	$229\text{-}31^{740}$(d), 118^{22}	40			al, eth, bz, chl	$B5^4$, 1027
1961	Benzene, 1-bromo-2-nitro 2-$NO_2C_6H_4Br$	202.01	pa, ye (al)	258^{756}	43	$1.6245^{80/4}$		al, eth, ace, bz	$B5^4$, 728
1962	Benzene, 1-bromo-3-nitro 3-$NO_2C_6H_4Br$	202.01	rh	265	17 (unst), 56 (st)	$1.7036^{20/4}$	1.5979^{20}	al, eth, bz	$B5^4$, 729
1963	Benzene, 1-bromo-4-nitro 4-$NO_2C_6H_4Br$	202.01	rh or mcl pr (al)	256	127	1.948		al, eth, bz	$B5^4$, 729
1964	Benzene, 1-bromo-2-nitroso 2-ONC_6H_4Br	186.01	nd		98			bz, chl	$B5^4$, 706
1965	Benzene, 1-bromo-4-nitroso 4-ONC_6H_4Br	186.01	nd (al)		95			al, bz, chl	$B5^4$, 706
1966	Benzene, 3-bromo-1-propenyl) or Cinnamyl bromide $BrCH_2CH{=}CHC_6H_5$	197.07	nd (al, eth)	130^{10}	34	1.3428^{30}	1.610—1.613^{20}	al	$B5^3$, 1188
1967	Benzene, (3-bromopropoxy) or 2-Phenoxypropyl bromide $C_6H_5\text{-}OCH_2CH_2CH_2Br$	215.09		127^{18}	7-8	$1.365^{16/16}$		eth	$B6^3$, 549
1968	Benzene, (1-bromopropyl)-(*DL*) $C_6H_5CHBrCH_2CH_3$	199.09	*d*$[\alpha]_D$ + 5.9 (al, c = 1), *l*$[\alpha]_D$ −5.7 (eth, c = 10)	105^{17}		$1.3098^{19/4}$	1.5517^{19}	eth, bz	$B5^4$, 981
1969	Benzene, (2-bromopropyl) $C_6H_5CH_2CHBrCH_3$	199.09		$107\text{-}9^{16}$		1.2908^{16}	1.5450^{20}	chl	$B5^4$, 981
1970	Benzene, (3-bromopropyl) $C_6H_5CH_2CH_2CH_2Br$	199.09		110^{12}		$1.3106^{25/4}$	1.5440^{25}	eth	$B5^4$, 982
1971	Benzene, (3-bromoisopropyl) $BrCH_2CH(CH_3)C_6H_5$	199.09	$[\alpha]^{25/}{}_D$*(d)* + 15.6	188-9, $106\text{-}8^{18}$		1.3155^{20}	1.5548^{28} (d)	bz, chl	$B5^4$, 995
1972	Benzene, 1-bromo-4-isopropyl 4-i-$C_3H_7C_6H_4Br$	199.09		218.7, $97\text{-}8^{5}$	-22.5	$1.3145^{20/4}$	1.5569^{20}	eth, bz, chl	$B5^4$, 994
1973	Benzene, 2-bromo-4-*iso*-propyl-1-methyl 4-i-C_3H_7-2-$BrC_6H_3CH_3$	213.12		234.3, 99^{9}		$1.2689^{18/4}$	1.5360^{20}	al, eth, chl	$B5^4$, 1063
1974	Benzene, 1-bromo-4-triazo 4-$N_3C_6H_4Br$	198.02	pl	105^{10}	20			eth, bz	$B5^4$, 761
1975	Benzene, 1-bromo-2,3,5-trimethyl 2,3,5-$(CH_3)_3C_6H_2Br$	199.09		238, 117^{17}	<-15		1.5516^{20}	bz	$B5^4$, 1014
1976	Benzene, 1-bromo-2,4,5-trimethyl 2,4,5-$(CH_3)_3C_6H_2Br$	199.09	nd (al)	233-5	73			al	$B5^4$, 1014
1977	Benzene, 1-bromo-2,4,6-trimethyl 2,4,6-$(CH_3)_3C_6H_2Br$	199.09		225, 117^{25}	-1	1.3191^{10}	1.5510^{20}	eth, bz	$B5^4$, 1027
1978	Benzene, butoxy or Butyl phenyl ether $C_6H_5OC_4H_9$	150.22		210, 95^{17}	-19.4	$0.9351^{20/4}$	1.4969^{20}	ace, bz	$B6^4$, 558

PHYSICAL CONSTANTS OF ORGANIC COMPOUNDS (Continued)

No.	Name, Synonyms, and Formula	Mol. wt.	Color, crystalline form, specific rotation and λ_{max} (log ε)	b.p. °C	m.p. °C	Density	n_D	Solubility	Ref.
1979	Benzene, isobutoxy or Isobutyl phenyl ether $i\text{-}C_4H_9OC_6H_5$	150.22		196		$0.9240^{24/15}$	1.4932^{14}	ace, bz, peth	$B6^4$, 559
1980	Benzene, 1-*iso*-butoxy-2-nitro $2\text{-}NO_2C_6H_4O\text{-}i\text{-}C_4H_9$	195.22	ye nd	275-80		1.1361^{20}		eth	B6, 218
1981	Benzene, *sec*-butoxy $C_6H_5OCH(CH_3)CH_2CH_3$	150.22		194.5, $70\text{-}2^{5}$		$0.9415^{20/4}$	1.4926^{25}	eth	$B6^4$, 558
1982	Benzene, butyl $C_6H_5C_4H_9$	134.22		183, 62.2^{10}	-88	$0.8601^{20/4}$	1.4898^{20}	al, eth, ace, bz	$B5^4$, 1033
1983	Benzene, 1-butyl-2-methyl $2\text{-}CH_3C_6H_4C_4H_9$	148.25		208, 81^{10}		$0.8710^{20/4}$	1.4960^{20}	eth, ace, bz	$B5^3$, 999
1984	Benzene, 1-butyl-3-methyl $3\text{-}CH_3C_6H_4C_4H_9$	148.25		205, 79^{10}		$0.8590^{20/4}$	1.4910^{20}	eth, ace, bz	$B5^3$, 999
1985	Benzene, 1-butyl-4-methyl $4\text{-}CH_3C_6H_4C_4H_9$	148.25		207, 80.6^{10}	-85	$0.8586^{20/4}$	1.4916^{20}	eth, ace, bz	$B5^4$, 1094
1986	Benzene, *iso*-butyl $C_6H_5CH_2CH_2(CH_3)_2$	134.22		172.8, 53.1^{10}	-51.5	$0.8532^{20/4}$	1.4866^{20}	al, eth, ace, bz, peth	$B5^4$, 1042
1987	Benzene, *sec*-butyl(*d*) $C_6H_5CH(CH_3)CH_2CH_3$	134.22	$[\alpha]^{20}/_D$ +27.3 (undil)	173, 63^{15}	-75	$0.8621^{20/4}$	1.4895^{20}	al, eth, ace, bz	$B5^4$, 1038
1988	Benzene, *sec*-butyl(*dl*) $C_6H_5CH(CH_3)CH_2CH_3$	134.22		173, 53.6^{10}	-75.5	$0.8621^{20/4}$	1.4902^{20}	al, eth, ace, bz	$B5^4$, 1038
1989	Benzene, *sec*-butyl *l*) $C_6H_5CH(CH_3)CH_2CH_3$	134.22	$[\alpha]^{25}/_D$ −17.9 (al, c = 5)	61^{18}		$0.868^{24/4}$	1.4891^{20}	al, eth, ace, bz	$B5^4$, 1038
1990	Benzene, 1-*sec*-butyl-4-methyl $4\text{-}CH_3C_6H_4CH(CH_3)C_2H_5$	148.25		197, 73^{10}		0.8640^{19}	1.493^{20}	eth, bz, chl	$B5^4$, 1095
1991	Benzene, *tert*-butyl $C_6H_5C(CH_3)_3$	134.22		169, 30.7^{18}	-57.8	$0.8665^{20/4}$	1.4927^{20}	al, eth, ace, bz	$B5^4$, 1045
1992	Benzene, 1-*tert*-butyl-3,5-dimethyl,2,4,6-trinitro or Musk xylene $3,5\text{-}(CH_3)_2\text{-}2,4,6\text{-}(NO_2)_3C_6\text{-}tert\text{-}C_4H_9$	297.27	pl, nd (al)		110			eth, al	$B5^4$, 1132
1993	Benzene, 2-*tert*-butyl-3,5-dinitro-1-*iso*-propyl-4-methyl or Moskene $2\text{-}tert\text{-}C_4H_9\text{-}3,5(NO_2)_2\text{-}4\text{-}CH_3C_6H(i\text{-}C_3H_7)$	280.32	Pa ye cr		132-3				C49, 3044
1994	Benzene, 2-*tert*-butyl-1,3-dinitro-4,5,6-trinitro or Musk tebetine $2\text{-}(CH_3)_3C\text{-}1,3\text{-}(NO_2)_2\text{-}4,5,6\text{-}(CH_3)_3C_6$	266.30	pa ye pr (al)		135-6			eth	$B5^3$, 1055
1995	Benzene, 5-*tert*-butyl-1,3-dinitro-4-methoxy-2-methyl or Musk ambrette $5\text{-}(CH_3)_3C\text{-}1,3\text{-}(NO_2)_2\text{-}4\text{-}CH_3O\text{-}2\text{-}CH_3\text{-}C_6H$	268.25	pa ye lf (al)	135^{16}	84-6			eth	$B6^3$, 1984
1996	Benzene, 1-*tert*-butyl-2-methyl $2\text{-}CH_3\text{-}C_6H_4C(CH_3)_3$	148.25		170^{743}		$1.49423^{17/D}$			$B5^4$, 1096
1997	Benzene, 1-*tert*-butyl-3-methyl-2,4,6-trinitro or Artificial musk $3\text{-}CH_3\text{-}2,4,6\text{-}(NO_2)_3C_6H\text{-}tertC_4H_9$	283.24	ye nd (al)		96-7			al, eth, bz, chl	$B5^3$, 1003
1998	Benzene, 1-*tert*-butyl-4-methyl $4\text{-}CH_3C_6H_4C(CH_3)_3$	148.25		193, 70^{10}	-52	$0.8612^{20/4}$	1.4918^{20}	eth, ace, bz	$B5^4$, 1097
1999	Benzene, 2-*tert*-butyl-4-methyl-1,3,5-trinitro or Musk baur $2\text{-}(CH_3)_3C\text{-}4\text{-}CH_3\text{-}1,3,5\text{-}(NO_2)_3C_6H$	283.24	ye nd (al)		97			al, eth, bz, peth	$B5^3$, 1003
2000	Benzene, chloro or Phenyl chloride C_6H_5Cl	112.56		132, 22^{10}	-45.6	$1.1058^{20/4}$	1.5241^{20}	al, eth, bz	$B5^4$, 640
2001	Benzene, (chloro-*tert*-butyl) $C_6H_5C(CH_3)_2CH_2Cl$	168.67		222^{741}, $104\text{-}5^{18}$		$1.047^{20/4}$	$1.5247^{20/D}$	al, eth, ace, bz	$B5^4$, 1048
2002	Benzene, 1-chloro-2,5-diacetamido $2,5\text{-}(CH_3CONH)_2\text{-}C_6H_3Cl$	226.66	nd		196-7			al	B13, 118
2003	Benzene, 1-chloro-2,4-diamino $2,4\text{-}(NH_2)_2C_6H_3Cl$	142.59	pl or nd		91			al	$B13^3$, 97
2004	Benzene, 1-chloro-2,5-diamino $2,5\text{-}(NH_2)_2C_6H_3Cl$	142.59	nd (bz-lig)		64			w, bz	$B13^2$, 58
2005	Benzene, 1-chloro-2,6-diamino $2,6\text{-}(NH_2)_2C_6H_3Cl$	142.29			85-6			eth	$B13^1$, 15
2006	Benzene, 1-chloro-3,4-diamino $3,4\text{-}(NH_2)_2C_6H_3Cl$	142.59	pl (bz-lig), lf (w)		76			al, eth, bz, lig	$B13^3$, 20
2007	Benzene, 1-chloro-3,5-diamino $3,5\text{-}(NH_2)_2C_6H_3Cl$	142.59	rh pr (al), nd (to-al)		105-6			al, eth, ace	$B13^2$, 29
2008	Benzene, 1-chloro-2,4-dimethyl or 4-Chloro-*m*-xylene $2,4\text{-}(CH_3)_2\text{-}C_6H_3Cl$	140.61		187-8, 89^{24}		$1.0598^{20/20}$	1.5230^{25}	ace, bz	$B5^4$, 943

No.	Name, Synonyms, and Formula	Mol. wt.	Color, crystalline form, specific rotation and λ_{max} (log ε)	b.p. °C	m.p. °C	Density	n_D	Solubility	Ref.
2009	Benzene, 1-chloro-2,5-dimethyl or 2-Chloro-*p*-xylene $2,5\text{-}(CH_3)_2C_6H_3Cl$	140.61		187 (192)	1.6	$1.0589^{15/4}$		ace, bz	$B5^4$, 965
2010	Benzene, 1-chloro-2,6-dimethyl $2,6\text{-}(CH_3)_2C_6H_3Cl$	140.61		$185\text{-}7, 62\text{-}3^{12}$		1.053^{20}	1.526^{20}	ace, bz, chl	$B5^4$, 943
2011	Benzene, 1-chloro-3,4-dimethyl $3,4\text{-}(CH_3)_2C_6H_3Cl$	140.61		194^{755}	-6	$1.0692^{15/15}$		ace, bz	$B5^3$, 815
2012	Benzene, 1-chloro-3,5-dimethyl or 5-Chloro-*m*-xylene $3,5\text{-}(CH_3)_2C_6H_3Cl$	140.61		$187\text{-}8, 66^{12}$				ace, bz	$B5^3$, 834
2013	Benzene, 1-chloro-2,3-dinitro $2,3\text{-}(NO_2)C_6H_3Cl$	202.55	pr (al), cr (MeOH)		78			eth	$B5^4$, 744
2014	Benzene, 1-chloro-2,4-dinitro $2,4\text{-}(NO_2)_2C_6H_3Cl$	202.55	α: ye rh (eth), β: ye rh (eth), nd (al)	$315, 158\text{-}60^2$	α: 53 (st), β: 43 (unst)	$1.4982^{75/4}$	1.5857^{60}	eth, bz	$B5^4$, 744
2015	Benzene, 1-chloro-2,6-dinitro $2,6\text{-}(NO_2)_2C_6H_3Cl$	202.55	ye nd (al, aa)	315	88	1.6867^{16}		al, eth, to	$B5^4$, 744
2016	Benzene, 1-chloro-3,4-dinitro $3,4\text{-}(NO_2)C_6H_3Cl$	202.55	α-cr (al), β-pr (lig), γ-mcl or rh nd (eth or lig)	$315d, 160^4$	α 36, β 37, γ 40-1	$1.6867^{20/4}$		eth, bz	$B5^4$, 744
2017	Benzene, 1-chloro,3,5-dinitro $3,5\text{-}(NO_2)_2C_6H_3Cl$	202.55	nd (al, peth)		59			al, eth	$B5^3$, 637
2018	Benzene, 1-chloro-2-ethoxy $2\text{-}C_2H_5OC_6H_4Cl$	156.61		$210, 97\text{-}8^{15}$		$1.1288^{15/4}$	1.5284^{25}	al, eth, bz	$B6^4$, 785
2019	Benzene, 1-chloro-3-ethoxy $3\text{-}C_2H_5OC_6H_4Cl$	156.61		$204\text{-}5^{747}$		$1.1712^{20/4}$		al, eth, bz	$B6^4$, 811
2020	Benzene, 1-chloro-4-ethoxy $4\text{-}C_2H_5OC_6H_4Cl$	156.61		$212\text{-}4, 98^{17}$	21	$1.1254^{20/4}$	1.5252^{20}	al, eth, bz, aa	$B6^4$, 823
2021	Benzene, (2-chlorethoxy) $C_6H_5OCH_2CH_2Cl$	156.61		$217\text{-}20, 100\text{-}2^{12}$	28			al, eth, ace, bz	$B6^4$, 556
2022	Benzene, (1-chloroethyl)(*d*) or *d*-α-Phenethyl chloride $C_6H_5CHClCH_3$	140.61	$[\alpha]^{20}{}_D$ + 50.6 (undil)	85^{20}		$1.0631^{20/4}$	1.5250^{25}	al, eth, bz	$B5^4$, 898
2023	Benzene, (1-chloroethyl)(*dl*) or α-Phenethyl chloride $C_6H_5CHClCH_3$	140.61		$81\text{-}2^{17}$		$1.0620^{20/4}$	1.5276^{20}	al, eth, bz	$B5^4$, 898
2024	Benzene, (1-chloroethyl)(*l*) or *l*-α-Phenethyl chloride $C_6H_5CHClCH_3$	140.61	$[\alpha]^{20}{}_D$ −30.1 (undil)	85^{20}		$1.0632^{20/4}$		al, eth, bz	$B5^4$, 898
2025	Benzene, (2-chloroethyl) or β-Phenethyl chloride $C_6H_5CH_2CH_2Cl$	140.61		$197\text{-}8, 92^{20}$		$1.069^{25/4}$	1.5276^{20}	eth, al, ace, bz, lig	$B5^4$, 899
2026	Benzene, (2-chloroethylthio) $2,C_6H_5SCH_2CHCl$	172.67		$117\text{-}8^{11}$		$1.1769^{25/4}$	1.5828^{20}	eth, chl	$B6^4$, 1469
2027	Benzene, 1-chloro-2-ethyl $2\text{-}C_2H_5C_6H_4Cl$	140.61		178.4	-82.7	$1.0569^{20/4}$	1.5218^{20}	ace, bz, chl	$B5^4$, 897
2028	Benzene, 1-chloro-4-ethyl $4\text{-}C_2H_5C_6H_4Cl$	140.61		$184.4, 63.5^{10}$	-62.6	$1.0455^{20/4}$	1.5175^{20}	al, eth, ace, peth, bz	$B5^4$, 897
2029	Benzene, 1-chloro-2-fluoro $2\text{-}FC_6H_4Cl$	130.55		137.6	-43	$1.2233^{30/4}$	1.4968^{30}	ace, bz	$B5^4$, 652
2030	Benzene, 1-chloro-3-fluoro $3\text{-}FC_6H_4Cl$	130.55		127.6		1.221^{25}	1.4911		$B5^4$, 652
2031	Benzene, 1-chloro-4-fluoro $4\text{-}FC_6H_4Cl$	130.55		130^{757}	-26.8	1.4990^{15}		al, eth, bz	$B5^4$, 652
2032	Benzene, 1-chloro-3-hydroxylamino $3\text{-}(HONH)C_6H_4Cl$	143.57	pl (bz)		49			w, bz, peth	$B15^2$, 8
2033	Benzene, 1-chloro-4-hydroxylamino $4\text{-}(HONH)C_6H_4Cl$	143.57	lf (dil al)		87-8			al, bz, chl	$B15^2$, 9
2034	Benzene, 1-chloro-2-iodo $2\text{-}IC_6H_4Cl$	238.46		$234\text{-}5, 110^{18}$		1.9515^{25}	1.6331^{25}	ace	$B5^4$, 695
2035	Benzene, 1-chloro-3-iodo $3\text{-}IC_6H_4Cl$	238.46		230				bz	$B5^4$, 695
2036	Benzene, 1-chloro-4-iodo $4\text{-}IC_6H_4Cl$	238.46	lf (ace) (al)	$227, 108^{13}$	57	$1.886^{27/4}$		al	$B5^4$, 696
2037	Benzene, 1-(chloromethyl)-2,4-dimethyl $2,4\text{-}(CH_3)_2C_6H_3CH_2Cl$	154.64		$215\text{-}6, 86\text{-}7^{12}$				al, eth, bz	$B5^4$, 1012

No.	Name, Synonyms, and Formula	Mol. wt.	Color, crystalline form, specific rotation and λ_{max} (log ε)	b.p. °C	m.p. °C	Density	n_D	Solubility	Ref.
2038	Benzene, 1-(chloromethyl)-4-ethyl or 4-Ethylbenzyl chloride $4\text{-}C_2H_5C_6H_4CH_2Cl$	154.64	....	95-6^{15}	......	....	1.5290^{25}	al, bz, chl	B5^4, 1004
2039	Benzene, 1-(chloromethyl)-2-methyl or o-xylylchloride $2\text{-}CH_3C_6H_4CH_2Cl$	140.61	....	197-9, 80^{12}	......	....	1.5410^{25}	**al, eth**	B5^4, 926
2040	Benzene, 1-chloromethyl-3-methyl or *m*-xylylchloride $3\text{-}CH_3C_6H_4CH_2Cl$	140.61	....	195-6, 101-2^{30}	......	1.064^{20}	1.5345^{20}	al, eth	B5^4, 943
2041	Benzene, 1-(chloromethyl)-4-methyl or *p*-xylylchloride $4\text{-}CH_3C_6H_4CH_2Cl$	140.61	....	200-2, 81^{15}	......	1.0512$^{20/4}$	1.5380	al, **eth**	B5^4, 966
2042	Benzene, 1-chloroisopropyl $C_6H_5CCl(CH_3)_2$	154.64	....	98^{1}	......	1.192^{25}	1.5290^{25}	....	B5^4, 992
2043	Benzene, 1-chloro-2-nitro $2\text{-}NO_2C_6H_4Cl$	157.56	mcl nd	246, 119^{8}	34-5	....	....	al, eth, ace, bz	B5^4, 721
2044	Benzene, 1-chloro-3-nitro $3\text{-}NO_2C_6H_4Cl$	157.56	pa ye rh pr (al)	235-6	24 (unst) 46 (st)	1.343$^{50/4}$	1.5374$^{80/\alpha}$	al, eth, bz	B5^4, 722
2045	Benzene, 1-chloro-4-nitro $4\text{-}NO_2C_6H_4Cl$	157.56	mcl pr	242, 113^{8}	83.6	1.2979$^{90.5}$	1.5376$^{100/\alpha}$	al, ace	B5^4, 723
2046	Benzene, 1-chloro-2-nitroso $2\text{-}ONC_6H_4Cl$	141.56	nd (al)	....	65-6	....	....	al, eth, bz, peth	B5^4, 705
2047	Benzene, 1-Chloro-3-nitroso $3\text{-}ONC_6H_4Cl$	141.56	nd (bz)	....	72 (77)	....	....	al, eth, ace, bz, chl	B5^4, 705
2048	Benzene, 1-chloro-4-nitroso $4\text{-}ONC_6H_4Cl$	141.56	(aa or al)	....	92-3	....	....	al	B5^4, 705
2049	Benzene, chloro-pentafluoro C_6ClF_5	202.51	....	122-3^{750}	......	1.568^{25}	1.4256^{20}	....	B5^4, 654
2050	Benzene, (3-chloropropoxy) $C_6H_5OCH_2CH_2CH_2Cl$	170.64	....	245-55, 139^{25}	12	1.1167^{20}	1.5235^{25}	eth, ace	B6^4, 557
2051	Benzene, (3-chloropropylthio) $C_6H_5SCH_2CH_2CH_2Cl$	186.70	....	116-7^{4}	......	1.1536$^{20/4}$	1.5752^{20}	ace, Py	B6^3, 981
2052	Benzene, 1-chloro-4-*iso*-propyl or 4-Chlorocumene $4\text{-}(CH_3)_2CHC_6H_4Cl$	154.64	....	198.3, 74^{10}	-12.3	1.0208$^{20/4}$	1.5117^{20}	**al, eth, ace,** bz	B5^4, 992
2053	Benzene, 1-chloro-4-triazo $4\text{-}N_3C_6H_4Cl$	153.57	....	96^{20}	20	....	....	eth	B5^4, 760
2054	Benzene, 1-chloro 2,4,6-trimethyl $2,4,6(CH_3)_3C_6H_2Cl$	154.64	....	204-6, 104^{25}	<-20	1.0337^{30}	1.5212^{30}	al, eth	B5^4, 1026
2055	Benzene, 1-chloro-2,4,5-trinitro $2,4,5\text{-}(NO_2)_3C_6H_2Cl$	247.55	ye lf (al)	....	116	....	....	al, bz, aa	B5^4, 757
2056	Benzene, 1-chloro-2,4,6-trinitro or Picryl chloride $2,4,6\text{-}(NO_2)_3C_6H_2Cl$	247.55	wh nd or pl (chl, al-lig)	....	83	1.797^{20}	....	al, ace, bz, to	B5^4, 757
2057	Benzene, cyclohexyl or Phenylcyclohexane $C_6H_5C_6H_{11}$	160.26	pl	235-6, 127-8^{30}	7-8	0.9502$^{20/4}$	1.5329^{20}	al, eth	B5^4, 1424
2058	Benzene, cyclopentyl or Phenylcyclopentane $C_6H_5C_5H_9$	146.23	....	219, 102^{18}	......	0.9462$^{20/4}$	1.5280^{20}	eth	B5^4, 1409
2059	Benzene, 1,2-diacetamido $1,2\text{-}(CH_3CONH)_2C_6H_4$	192.22	nd (w)	....	185-6	....	....	al, ace, aa, w	B13^3, 42
2060	Benzene, 1,4-diacetyl $1,4\text{-}(CH_3CO)_2C_6H_4$	162.19	pr (al, eth)	sub	114	....	....	al	B7^3, 3504
2061	Benzene, 1,2-diamino or *o*-Phenylenediamine $1,2\text{-}(H_2N)_2C_6H_4$	108.14	brsh ye lf (w), pl (chl)	256-8	102-3	....	....	al, eth, bz	B13^3, 28
2062	Benzene, 1,3-diamino or *m*-Phenylenediamine $1,3,\text{-}(H_2N)_2C_6H_4$	108.14	rh (al)	282-4	63-4	1.0696$^{58/4}$	1.6339^{58}	w, al, eth, bz	B13^3, 65
2063	Benzene, 1,4-diamino or *p*-Phenylenediamine $1,4\text{-}(H_2N)_2C_6H_4$	108.14	wh pl (bz, eth)	267	140	....	....	eth, chl, w, al	B13^3, 104
2064	Benzene, 1,2-dibenzoxy or Catechol dibenzyl ether $1,2\text{-}(C_6H_5CH_2O)_2C_6H_4$	290.36	yesh nd or pr (al)	....	63-4	....	....	al, eth, peth	B6^3, 4219
2065	Benzene, 1,4-dibenzoxy or Hydroquinone dibenzyl ether $1,4\text{-}(C_6H_5CH_2O)_2C_6H_4$	290.36	pl (al)	130	......	....	....	al, eth, aa	B6^3, 4402
2066	Benzene, 1,4-dibenzoyl $1,4\text{-}(C_6H_5CO)_2C_6H_4$	286.33	....	....	161	....	....	....	B7^3, 4299
2067	Benzene, 1,4-dibenzyl $1,4\text{-}(C_6H_5CH_2)_2C_6H_4$	258.36	....	225^{18}	87-8	....	....	al, bz	B5^3, 2334
2068	Benzene, 1,2-dibromo or *o*-Dibromobenzene $1,2\text{-}Br_2C_6H_4$	235.91	....	225, 92^{10}	7.1	1.9843$^{20/4}$	1.6155^{20}	al, **eth, ace, bz**	B5^4, 682

No.	Name, Synonyms, and Formula	Mol. wt.	Color, crystalline form, specific rotation and λ_{max} (log ε)	b.p. °C	m.p. °C	Density	n_D	Solubility	Ref.
2069	Benzene, (1,2-dibromoethyl) $C_6H_5CHBrCH_2Br$	263.96		139-41^{15}	74			al, eth, bz, lig	B5^{4}, 909
2070	Benzene, 1,2-dibromo-3-nitro $1,2\text{-}Br_2\text{-}3\text{-}NO_2C_6H_3$	280.90	mcl pr		85			eth, ace, chl	B5^{4}, 731
2071	Benzene, 1,2-dibromo-4-nitro $1,2\text{-}Br_2\text{-}4\text{-}NO_2C_6H_3$	280.90	nd (al, aa) mcl pr	296, 180^{20}	58-9	2.354^{8}	1.9835^{111}	al, bz, aa	B5^{4}, 732
2072	Benzene, 1,3-dibromo or *m*-Dibromobenzene $1,3\text{-}Br_2C_6H_4$	235.91		218, 66^{5}	-7	1.9523$^{20/4}$	1.6083^{17}	al, eth	B5^{4}, 682
2073	Benzene, 1,3-dibromo-2-nitro $1,3\text{-}Br_2\text{-}2\text{-}NO_2C_6H_3$	280.91	nd (al), mcl pr (al)	sub	84	1.9211^{111}, 2.211^{8}		al, ace, bz	B5^{3}, 621
2074	Benzene, 1,3-dibromo-5-nitro $1,3\text{-}Br_2\text{-}5\text{-}NO_2C_6H_3$	280.91	pl or pr (eth), lf or nd (al)		106	1.9341^{111}, 2.363^{8}		al, eth, bz	B5^{3}, 621
2075	Benzene, 1,4-dibromo or *p*-Dibromobenzene $1,4\text{-}Br_2C_6H_4$	235.91	pl	218-9	87.3	1.5742		al, eth, ace, bz	B5^{4}, 683
2076	Benzene, 1,5-dibromo-2,4-dimethyl or 4,6-Diamino-*m*-xylene $1,5\text{-}Br_2\text{-}2,4\text{-}(CH_3)_2C_6H_2$	263.98	pl (al)	255-6	68			al	B5^{3}, 839
2077	Benzene, 1,4-dibromo-2-nitro $1,4\text{-}Br_2\text{-}2\text{-}NO_2C_6H_3$	280.90	yesh pl (ace)		85-6	1.9416^{111}, 2.368^{8}		bz, ace	B5^{4}, 739
2078	Benzene, 2,3-dibromo 1,4,5-trimethyl $2,3\text{-}Br_2\text{-}1,4,5\text{-}(CH_3)_3C_6H$	277.99	nd (al)		63-4			al, eth, bz	B5^{3}, 910
2079	Benzene, 2,4-dibromo-1-nitro $2,4\text{-}Br_2C_6H_3NO_2$	280.90	ye pl or pr (al)		62	1.9581^{111}, 2.356^{8}		ace, bz	B5^{3}, 620
2080	Benzene, 2,4-dibromo-1,3,5-trimethyl $2,4\text{-}Br_2\text{-}1,3,5\text{-}(CH_3)_3C_6H$	277.99	nd (al)	285	65.5			al, bz	B5^{3}, 921
2081	Benzene, 1,4-di-*tert*-butyl or *p*-Di-*tert*-butyl benzene $1,4\text{-}[(CH_3)_3C]_2C_6H_4$	190.33	nd (MeOH)	237^{743}, 109^{15}	80-1			al, eth	B5^{4}, 1163
2082	Benzene, 1,2-dichloro or *o*-Dichlorobenzene $1,2\text{-}Cl_2\text{-}C_6H_4$	147.00		180.5, 86^{18}	-17	1.3048$^{20/4}$	1.5515^{20}	al, eth, ace, bz	B5^{4}, 654
2083	Benzene, (1,2-dichloroethyl) $C_6H_5CHClCH_2Cl$	175.06		233-4, 115^{15}		1.240$^{15/4}$	1.5544^{15}	ace, bz	B5^{4}, 901
2084	Benzene, 1,2-dichloro-3-nitro $1,2\text{-}Cl_2\text{-}3NO_2C_6H_3$	192.00	mcl nd (peth, aa)	257-8	61-2	1.721^{14}		al, eth, ace, bz	B5^{4}, 725
2085	Benzene, 1,2-dichloro-4-nitro $1,2\text{-}Cl_2\text{-}4\text{-}NO_2C_6H_3$	192.00	nd (al)	255-6, 189^{100}	43	1.4558$^{75/4}$		al, eth	B5^{4}, 726
2086	Benzene, 1,3-dichloro or *m*-Dichlorobenzene $1,3\text{-}Cl_2C_6H_4$	147.00		173, 53^{10}	-24.7	1.2884$^{20/4}$	1.5459^{20}	al, eth, ace, bz	B5^{4}, 657
2087	Benzene, 1,3-dichloro-2-nitro $1,3\text{-}Cl_2\text{-}2\text{-}NO_2C_6H_3$	192.00	nd or pr (al, CS_2)	130^{8}	72.5	1.603^{17}, 1.4094^{80}		eth, al	B5^{4}, 726
2088	Benzene, 1,3-dichloro-5-nitro $1,3\text{-}Cl_2\text{-}5\text{-}NO_2C_6H_3$	192.00	mcl pr or lf (aa, al)		65.4	1.692^{14}, 1.4000^{100}		eth, al	B5^{4}, 727
2089	Benzene, 1,4-dichloro or *p*-Dichlorobenzene $1,4\text{-}Cl_2\text{-}C_6H_4$	147.00	mcl pr, lf (ace)	174, 55^{10}	53.1	1.2475$^{20/4}$	1.5285^{20}	al, eth, ace, bz	B5^{4}, 685
2090	Benzene, 1,4-dichloro-2-iodo $1,4\text{-}Cl_2\text{-}2\text{-}IC_6H_3$	272.90	pl (al)	255-6, 134^{12}	23			al, eth, bz, chl	B5^{4}, 580
2091	Benzene, 1,4-dichloro-2-nitro $1,4\text{-}Cl_2\text{-}2\text{-}NO_2C_6H_3$	192.00	pl or pr (al), pl (AcOEt)	267	56	1.669^{22}, 1.439^{75}		al, eth, bz, chl	B5^{4}, 726
2092	Benzene, 2,4-dichloro-1-ethoxy $2,4\text{-}Cl_2C_6H_3OC_2H_5$	191.06		237				al, eth, bz	B6^{4}, 885
2093	Benzene, 2,4-dichloro-1-nitro $2,4\text{-}Cl_2\text{-}C_6H_3NO_2$	192.00	nd (al)	258.5, 100-1^{4}	34	1.4790^{80}	1.5512$^{78/o}$	al, eth	B5^{4}, 726
2094	Benzene, 2,4-dichloro-1-triazo $2,4\text{-}Cl_2\text{-}C_6H_3N_3$	188.02	ye nd (al), pr (ace or bz)		54			al, eth, bz, peth	B5^{3}, 208
2095	Benzene, 2,4-dichloro-1,3,5-trimethyl $2,4\text{-}Cl_2\text{-}1,3,5\text{-}(CH_3)_3C_6H$	189.08		243.4	59			al	B5^{3}, 919
2096	Benzene, 2,5-dichloro-1-ethyl $2,5\text{-}Cl_2\text{-}C_6H_3C_2H_5$	175.06		213.5		1.239^{0}		bz	B5^{4}, 901
2097	Benzene, 1-dichlorophosphino-4-isopropyl $C_6H_4PCl_2\text{-}4\text{-}(CH_3)_2CH$	221.07		268-70, 129-30^{16}		1.1917^{25}	1.5677^{25}	ace	B16, 773
2098	Benzene, 1,2-diethyl or *o*-Diethyl benzene $1,2\text{-}(C_2H_5)_2C_6H_4$	134.22		183.4, 63^{10}	-31.2	0.8800^{20}	1.5035^{20}	al, eth, ace, bz, lig	B5^{4}, 1065
2099	Benzene, 1,3-diethyl or *m*-Diethyl benzene $1,3\text{-}(C_2H_5)_2C_6H_4$	134.22		181, 60^{10}	-83.9	0.8602$^{20/4}$	1.4955^{20}	al, eth, ace, bz	B5^{4}, 1066
2100	Benzene, 1,3-diethyl-5-methyl $1,3\text{-}(C_2H_5)_2\text{-}5\text{-}CH_3C_6H_3$	148.25		205, 79^{10}	-74.1	0.8748$^{20/4}$	1.5027^{20}	al, eth, ace, bz	B5^{4}, 1106

No.	Name, Synonyms, and Formula	Mol. wt.	Color, crystalline form, specific rotation and λ_{max} (log ε)	b.p. °C	m.p. °C	Density	n_D	Solubility	Ref.
2101	Benzene, 1,4-diethyl or *p*-Diethylbenzene $1,4\text{-}(C_2H_5)_2C_6H_4$	134.22	....	$183.8, 63^{10}$	−42.8	$0.8620^{20/4}$	1.4967^{20}	al, eth, ace, bz	$B5^4$, 1067
2102	Benzene, 1,2-difluoro or *o*-Difluorobenzene $1,2\text{-}F_2C_6H_4$	114.09	....	$91\text{-}2^{751}$	−34	$1.1599^{18/4}$	1.44506^{18}	ace, bz, chl	$B5^4$, 637
2103	Benzene, 1,2-dihydroxy or Catechol $2\text{-}HOC_6H_4OH$	110.11	cr	245	105	1.1493^{22}	1.604	w, al, eth, ace, bz	$B6^2$, 4187
2104	Benzene, 1,3-dihydroxy or Resorcinal $1,3\text{-}(HO)_2C_6H_4$	110.11	nd (bz), pl (w)	178^{16}	111	1.2717	....	w, al, eth, bz	$B6^3$, 4292
2105	Benzene, 1,4-dihydroxy or Hydroquinone $1,4\text{-}(HO)_2C_6H_4$	110.11	mcl pr (sub), nd (w), Pr (MeOH)	285^{730}	173-4	1.328^{15}	....	w, al, eth, ace	$B6^3$, 4374
2106	Benzene, 1,4-dicyclohexyl or *p*-Dicyclohexyl benzene $1,4\text{-}(C_6H_{11})_2C_6H_4$	242.40	....	120-3	......	....	....	....	$B5^4$, 1607
2107	Benzene, 1,4-difluoro or *p*-Difluorobenzene $1,4\text{-}F_2C_6H_4$	114.09	....	89^{760}	-13	$1.1688^{20/4}$	1.4422^{20}	ace, bz	$B5^4$, 637
2108	Benzene, 1,3-difluoro-4-nitro $1,3\text{-}F_2\text{-}4\text{-}NO_2C_6H_3$	159.09	....	207^{760}	9.8	1.4571^{14}	1.5149^{14}	chl	$B5^4$, 720
2109	Benzene, 1,2-diiodo or *o*-Diidobenzene $1,2\text{-}I_2C_6H_4$	329.91	pl or pr (lig)	$286^{750}, 109^{3}$	27	2.54^{20}	1.7179^{20}	eth	$B5^4$, 700
2110	Benzene, 1,3-diiodo or *m*-Diidobenzene $1,3\text{-}I_2C_6H_4$	329.91	rh pl or pr (eth-al)	285	40.4	2.47^{25}	....	al, eth, chl	$B5^4$, 700
2111	Benzene, 1,4-diiodo or *p*-Diidobenzene $1,4\text{-}I_2C_6H_4$	329.91	rh lf (al)	285 sub	131-2	....	....	al, eth	$B5^4$, 700
2112	Benzene, 1,2-dimethoxy-4,5-dinitro $4,5\text{-}(NO_2)_2\text{-}1,2\text{-}(CH_3O)_2C_6H_2$	228.16	ye nd (al)	....	130-2	$1.3164^{140/4}$	....	....	$B6^3$, 4274
2113	Benzene, 1,2-dimethoxy-4-iodo or 4-Iodoveratrole $1,2\text{-}(CH_3O)_2\text{-}4\text{-}IC_6H_3$	264.06	nd (dil MeOH)	$163\text{-}4^{26}, 90^{0.2}$	35	....	....	al, bz	$B6^4$, 4262
2114	Benzene, 1,2-dimethoxy-3-nitro $1,2\text{-}(CH_3O)_2\text{-}3\text{-}NO_2C_6H_3$	183.16	nd (al)	....	64-5	$1.1404^{132/4}$	....	al, eth, bz, aa	$B6^3$, 4263
2115	Benzene, 1,2-dimethoxy-4-nitro $1,2\text{-}(CH_3O)_2\text{-}4\text{-}NO_2C_6H_3$	183.16	ye nd (al-w)	$230^{15\text{-}20}$	98	$1.1888^{133/4}$	....	al, eth, chl	$B6^3$, 4264
2116	Benzene, 1,2-dimethoxy-4-propenyl(*cis*) or Isoengenol methyletha $1,2\text{-}(CH_3O)_2\text{-}4\text{-}CH_3CH{=}CH\text{-}C_6H_3$	178.23	....	$270.5, 138\text{-}40^{12}$	......	$1.0521^{20/4}$	1.5616^{20}	ace, bz	$B6^3$, 4995
2117	Benzene, 1,3-dimethoxy-2-nitro $1,3\text{-}(CH_3O)_2\text{-}2\text{-}NO_2C_6H_3$	183.16	ye nd (al or aa)	....	131	....	....	bz, chl, al	$B6^3$, 4344
2118	Benzene, 1,3-dimethoxy 5-nitro $1,3\text{-}(CH_3O)_2\text{-}5\text{-}NO_2C_6H_3$	183.16	pa ye nd (AcOEt)	....	89	$1.1693^{133/4}$	....	al, bz	$B6^3$, 4347
2119	Benzene, 1,4-dimethoxy-2-nitro $1,4\text{-}(CH_3O)_2\text{-}2\text{-}NO_2C_6H_3$	183.16	gold-ye nd (dil al)	169^{13} sub	72-3	$1.1666^{132/4}$	....	al, bz, chl	$B6^3$, 4442
2120	Benzene, 1,4-dimethoxy-2-*iso*-propyl $1,4\text{-}(CH_3O)_2\text{-}2\text{-}(CH_3)_2CHC_6H_3$	180.25	....	$114\text{-}6^{15}$	......	$1.0129^{17/4}$	1.5103^{17}	eth, bz	B6, 929
2121	Benzene, 2,4-dimethoxy-l-nitro $2,4\text{-}(CH_3O)_2\text{-}1\text{-}NO_2C_6H_3$	183.16	nd (al)	....	76-7	$1.1876^{132/4}$	....	al	$B6^2$, 822
2122	Benzene, 1,2-dimethyl-4-*iso*-propyl $1,2(CH_3)_2\text{-}4\text{-}(CH_3)_2CHC_6H_3$	148.25	....	$199, 86\text{-}7^{16}$	......	$0.8710^{20/4}$	$1.4951^{20/D}$	....	$B5^4$, 1103
2123	Benzene, 1,3-dimethyl-5-*iso*-propyl $1,3\text{-}(CH_3)_2\text{-}5\text{-}(CH_3)_2CHC_6H_3$	148.25	....	$83\text{-}5^{17}$	......	....	$1.4935^{25/D}$	....	$B5^4$, 1105
2124	Benzene, 1,2-dinitro or *o*-Dinitrobenzene $1,2\text{-}(NO_2)_2C_6H_4$	168.11	nd (bz) pl (al)	$319^{775}, 194^{30}$	118.5	$1.3119^{120/4}$, 1.565^{17}	....	al, bz, chl	$B5^4$, 738
2125	Benzene, 1,3-dinitro or *m*-Dinitrobenzene $1,3\text{-}(NO_2)_2C_6H_4$	168.11	rh pl (al)	$291^{756}, 167^{14}$	90	$1.5751^{18/4}$	....	al, eth, ace, bz, Py	$B5^4$, 379
2126	Benzene, 1,4-dinitro or *p*-Dinitro benzene $1,4\text{-}(NO_2)_2C_6H_4$	168.11	nd (al)	$298^{777}, 183^{34}$, sub	174	$1.625^{18/4}$	....	ace, bz, to, aa	$B5^4$, 741
2127	Benzene, 1,4-dinitro-2,3,5,6-tetramethyl or Dinitrodurene $1,4\text{-}(NO_2)_2\text{-}2,3,5,6\text{-}(CH_3)_4C_6$	224.22	pr (al)	sub	211-2	....	....	eth, bz	$B5^4$, 1080
2128	Benzene, 2,4-dinitro-1-fluoro $2,4\text{-}(NO_2)_2C_6H_3F$	186.10	....	$296, 178^{25}$	25.8	1.4718^{84}	....	al	$B5^4$, 742
2129	Benzene, 2,4-dinitro-1,3,5-trimethyl $2,4\text{-}(NO_2)_2\text{-}1,3,5\text{-}(CH_3)_3C_6H$	210.19	rh (al)	418 exp	86	....	....	al	$B5^4$, 1029
2130	Benzene, 1,4-dipropyl or *p*-Dipropyl benzene $1,4\text{-}(C_3H_7)_2C_6H_4$	162.27	....	109^{23}	......	$0.8563^{10.4/4}$	$1.4917^{19.4}$	....	$B5^4$, 1124
2131	Benzene, 1,2-diisopropyl or *o*-Diisopropyl benzene $1,2\text{-}(i\text{-}C_3H_7)_2C_6H_4$	162.27	....	$204, 115.4^{50}$	-57	$0.8701^{20/4}$	1.4960^{20}	al, eth, ace, bz	$B5^4$, 1125

PHYSICAL CONSTANTS OF ORGANIC COMPOUNDS (Continued)

No.	Name, Synonyms, and Formula	Mol. wt.	Color, crystalline form, specific rotation and λ_{max} (log ε)	b.p. °C	m.p. °C	Density	n_D	Solubility	Ref.
2132	Benzene, 1,3-di*iso*propyl or *m*-Diisopropyl benzene $1,3\text{-}(i\text{-}C_3H_7)_2C_6H_4$	162.27		$203.2, 75^{9}$	-61	$0.8559^{20/4}$	1.4883^{20}	al, eth, ace, bz	B5[4], 1125
2133	Benzene, 1,4-di*iso*-propyl or *p*-Diisopropyl benzene $1,4\text{-}(i\text{-}C_3H_7)_2C_6H_4$	162.27		$210.3, 120^{50}$	-17	$0.8568^{20/4}$	1.4898^{20}	al, eth, ace, bz	B5[4], 1126
2134	Benzene, 1,2-divinyl or *o*-Divinyl benzene $1,2(CH_2{=}CH)_2C_6H_4$	130.19		76^{14}		$0.9325^{22/4}$	1.5767^{20}	ace, bz	B5[4], 1540
2135	Benzene, 1,3-divinyl or *m*-Divinyl benzene $1,3\text{-}(CH_2{=}CH)_2C_6H_4$	130.19		$121^{76}, 52^{3}$	-52.3	$0.9294^{20/4}$	1.5760^{20}	ace, bz	B5[2], 1367
2136	Benzene, 1,4-divinyl or *p*-Divinyl benzene $1,4\text{-}(CH_2{=}CH)_2C_6H_4$	130.19		$95\text{-}6^{18}, 34^{0.2}$	31	0.913^{40}	1.5835^{25}	ace, bz	B5[4], 1541
2137	Benzene, dodecyl $C_6H_5C_{12}H_{25}$	246.44		$331, 185\text{-}8^{15}$	-7	$0.8551^{20/4}$	1.4824^{20}		B5[4], 1200
2138	Benzene, (epoxy isopropyl) $C_6H_5OCH(CH_3)_2$	136.19		$83\text{-}5^{15}$		1.0280^{20}	1.5232^{20}		B17[4], 410
2139	Benzene, ethoxy or Phenetole $C_6H_5OC_2H_5$	122.17		$170, 60^{9}$	-29.5	$0.9666^{20/4}$	1.5076^{20}	al, eth	B6[4], 554
2140	Benzene, ethyl or ethylbenzene $C_6H_5C_2H_5$	106.17		$136.2, 25.8^{10}$	-95	$0.8670^{20/4}$	1.4959^{20}	al, eth	B5[4], 885
2141	Benzene, 1-ethyl-4-*iso*-butyl $4\text{-}i\text{-}C_4H_9C_6H_4C_2H_5$	162.27		211				eth, ace	B5[4], 1123
2142	Benzene, 1-ethyl-2-Iodo $2\text{-}IC_6H_4C_2H_5$	232.06		226		$1.6189^{10/4}$	1.5941^{22}	ace, bz	B5[4], 970
2143	Benzene, 1-ethyl-4-iodo $4\text{-}IC_6H_4C_2H_5$	232.06		209	-17	$1.6095^{10/4}$	1.5909^{22}	ace, bz	B5[4], 801
2144	Benzene, 1-ethyl-2-methyl or 2-Ethyl toluene $2\text{-}CH_3C_6H_4C_2H_5$	120.19		$165.2, 62.3^{20}$	-80.8	0.8807^{20}	1.5046^{20}	al, eth, ace, bz	B5[4], 999
2145	Benzene, 1-ethyl-3-methyl or 3-Ethyl toluene $3\text{-}CH_3C_6H_4C_2H_5$	120.19		$161.3, 45.6^{10}$	-95.5	$0.8645^{20/4}$	1.4966^{20}	al, eth, ace, bz peth	B5[4], 1001
2146	Benzene, 1-ethyl-4-methyl or 4-Ethyl toluene $4\text{-}CH_3C_6H_4C_2H_5$	120.19		$162, 45.6^{10}$	-62.3	$0.8614^{20/4}$	1.4959^{20}	al, eth, ace, bz peth	B5[4], 1003
2147	Benzene, 1-ethyl-2-nitro $2\text{-}NO_2C_6H_4C_2H_5$	151.16		$228, 116^{22}$	-23	1.1345^{0}		al, eth, ace	B5[4], 911
2148	Benzene, 1-ethyl-3-nitro $3\text{-}NO_2C_6H_4C_2H_5$	151.16		242-3		1.1345^{0}		al, eth, ace	B5[4], 911
2149	Benzene, 1-ethyl-4-nitro $4\text{-}NO_2C_6H_4C_2H_5$	151.16		$245\text{-}6, 134\text{-}6^{23}$	-12.3	$1.1192^{20/4}$	1.5455^{20}	al, eth, ace	B5[4], 912
2150	Benzene, 1-ethyl-4-propyl $4\text{-}C_3H_7C_6H_4C_2H_5$	148.25		205		$0.8594^{20/4}$	1.4921^{20}	al, eth, ace, bz	B5[4], 1100
2151	Benzene, 1-ethyl-3-*iso*-propyl $3\text{-}i\text{-}C_3H_7C_6H_4C_2H_5$	148.25		$192, 69.2^{10}$	<-20	$0.859^{22/4}$	1.4921^{20}	al, eth, bz	B5[3], 1105
2152	Benzene, 1-ethyl-4-*iso*-propyl $4\text{-}i\text{-}C_3H_7C_6H_4C_2H_5$	148.25		$136.6, 73^{10}$	<-20	$0.8585^{20/4}$	1.4923^{20}	eth, bz	B5[4], 1100
2153	Benzene, fluoro or Phenyl fluoride C_6H_5F	96.10		85.1	-41.2	$1.0225^{20/4}$	1.4684^{40}	al, eth, ace, bz, lig	B5[4], 632
2154	Benzene, 1-fluoro-2-iodo $2\text{-}IC_6H_4F$	222.00		$188.6, 78^{20}$			1.5910^{20}	ace, bz, chl	B5[4], 693
2155	Benzene, 1-fluoro-4-iodo $4\text{-}IC_6H_4F$	222.00		$182\text{-}4, 67\text{-}9^{11}$	(i)-27 (ii)-18	1.9523^{15}	1.5270^{22}	al, eth, ace	B5[4], 694
2156	Benzene, 1-fluoro-2-nitro $2\text{-}NO_2C_6H_4F$	141.10	ye	$214.6d, 86\text{-}7^{11}$	-6	$1.3285^{18/4}$	1.5489^{17}	al, eth	B5[4], 718
2157	Benzene, 1-fluoro-3-nitro $3\text{-}NO_2C_6H_4F$	141.10	ye	198-200	41	$1.3254^{19/4}$	1.5262^{15}	al, eth	B5[4], 719
2158	Benzene, 1-fluoro-4-nitro $4\text{-}NO_2C_6H_4F$	141.10	ye nd	$206\text{-}7, 87^{14}$	(i) 27 (st), (ii) 3 (unst)	$1.3300^{20/4}$	1.5316^{20}	al, eth	B5[4], 719
2159	Benzene, 1-fluoro-2,4,6-trimethyl $2,4,6\text{-}(CH_3)_3C_6H_2F$	138.18		168.7	-36.7	$0.9745^{25/4}$	1.4809^{25}	ace, bz, chl	B5[4], 1225
2160	Benzene, hexabromo C_6Br_6	551.49	mcl nd (bz)		327			bz, chl, peth	B5[4], 687
2161	Benzene, hexachloro C_6Cl_6	284.78	nd (bz-al)	322 sub	230	$1.5691^{23.6}$		eth, bz, chl	B5[4], 670
2162	Benzene, hexaethyl $C_6(C_2H_5)_6$	246.44	mcl pr (al or bz)	298	129	0.8305^{130}	1.4736^{130}	al, eth, bz	B5[4], 1208
2163	Benzene, hexafluoro C_6F_6	186.06		80.5	5.3	1.6184^{20}	$1.37774^{20/D}$		B5[4], 640
2164	Benzene, heptyl $C_6H_5C_7H_{15}$	176.30		$245.5, 116^{12}$	−48	$0.8567^{20/4}$	1.4865^{20}	bz, chl	B5[4], 1143

No.	Name, Synonyms, and Formula	Mol. wt.	Color, crystalline form, specific rotation and λ_{max} (log ε)	b.p. °C	m.p. °C	Density	n_D	Solubility	Ref.
2165	Benzene, hexahydroxy or Hexaphenol $C_6(OH)_6$	174.11	nd (w)	...	>300	...	...	...	B6³, 6922
2166	Benzene, hexaiodo C_6I_6	833.49	red-br nd or mcl pr (bz)	...	350d	...	...	...	B5³, 585
2167	Benzene, hexamethyl or Mellitene $C_6(CH_3)_6$	162.27	rh pr or nd (al)	265	166-7	1.0630^{25}	...	al, eth, ace, bz, aa	B5⁴, 1137
2168	Benzene, 1-hydroxymethyl-4-*iso*-propyl or Cumic alcohol $4\text{-}i\text{-}C_3H_7C_6H_4CH_2OH$	150.22	...	249	28	$0.9818^{20/4}$	1.5210^{20}	al, eth	B6³, 1911
2169	Benzene, iodo or Phenyl iodide C_6H_5I	204.01	...	$188.3, 75^{10}$	−31.3	$1.8308^{20/4}$	1.6200^{20}	al, **eth, ace, bz**	B5⁴, 688
2170	Benzene, (2-iodoethoxy) $C_6H_5OCH_2CH_2I$	248.06	cr (dil al)	...	31-2	...	...	...	B6³, 548
2171	Benzene, 1-iodomethyl-2-methyl $2\text{-}CH_3C_6H_4CH_2I$	232.06	nd (eth, (peth)		33-4			eth	B5², 286
2172	Benzene, 1-iodomethyl-4-methyl $4\text{-}CH_3C_6H_4CH_2I$	232.06	nd (eth) (peth)	...	46-7	...	...	bz, eth, peth	B5⁴, 970
2173	Benzene, 1-iodo-2-nitro $2\text{-}NO_2C_6H_4I$	249.01	ye rh nd (al)	$288\text{-}9^{729}$, 162^{18}	54	1.9186^{75}		al, eth	B5², 190
2174	Benzene, 1-iodo-3-nitro $3\text{-}NO_2C_6H_4I$	249.01	mcl pr	$280, 153^{14}$	(i) 38.5 (st), (ii)10 (unst)	$1.9477^{50/4}$	...	al, eth	B5⁴, 733
2175	Benzene, 1-iodo-4-nitro $4\text{-}NO_2C_6H_4I$	249.01	ye nd (al)	289^{772}	174			al, aa	B5², 191
2176	Benzene, 1-iodo-2-triazo $2\text{-}N_3C_6H_4I$	245.02	ye	$90\text{-}100^{0.9}$		$1.8893^{25/4}$	1.6631^{25}	al, ace	B5, 278
2177	Benzene, iodoso C_6H_5OI	220.01	ye pw	...	210 exp	...	...	al, w	B5⁴, 692
2178	Benzene, iodoxy $C_6H_5IO_2$	236.01	nd (w)	...	236-7 exp	...	...	w, aa	B5⁴, 693
2179	Benzene, *isocyano or Phenylisocyanate* C_6H_5NCO	119.12	...	$162\text{-}3^{751}, 55^{13}$	...	$1.0956^{19.6/4}$	$1.5368^{19.6}$	eth, ace	B12³, 903
2180	Benzene, methoxy or Anisole $C_6H_5OCH_3$	108.14	...	155	−37.5	$0.9961^{20/4}$	1.5179^{20}	al, eth, ace, bz	B6⁴, 548
2181	Benzene, (3-methylbutoxy) or Phenyl isopentyl ether $C_6H_5O\text{-}i\text{-}C_5H_{11}$	164.25	...	225	...	$0.9198^{22/4}$	1.4872^{20}	w	B6³, 553
2182	Benzene, (1-methylbutyl) $C_6H_5CH(CH_3)C_3H_7$	148.25	...	$198\text{-}9^{757}$	...	$0.8594^{21/4}$	1.4875^{25}	al, eth	B5⁴, 1087
2183	Benzene, (2-methylbutyl) $C_6H_5CH_2CH(CH_3)CH_2CH_3$	148.25	...	$193.8, 102^{15}$	...	$0.8584^{20/4}$	1.4873^{20}	...	B5⁴, 1089
2184	Benzene, (2-methyl-l-propenyl) $C_6H_5CH{=}C(CH_3)_2$	132.21	...	$99^{43.5}$	...	$0.9029^{20/4}$	$1.5388^{20/D}$	...	B5⁴, 1380
2185	Benzene, (2-methyl-2-butyl) $C_6H_5C(CH_3)_2CH_2CH_3$	148.25	...	$192\text{-}4, 71\text{-}2^{12}$	...	$0.8587^{20/4}$	1.4934^{20}	**al, eth**	B5⁴, 1090
2186	Benzene, (3-methylbutyl) or isopentyl benzene $i\text{-}C_5H_{11}C_6H_5$	148.25	...	199	...	$0.8558^{20/4}$	1.4853^{20}	al, eth, bz	B5⁴, 1089
2187	Benzene, (4-methylpentyl) or Isohexyl benzene $i\text{-}C_6H_{13}C_6H_5$	162.27	...	$214\text{-}5^{740}$	...	0.8568^{16}	...	al, eth, bz	B5⁴, 1116
2188	Benzene, 1-methyl-4-(nitromethyl) or 4-Nitromethyl toluene $4\text{-}(O_2NCH_2)C_6H_4CH_3$	151.16	...	$150\text{-}1^{35}$	11-2	$1.1234^{20/4}$	1.5278^{20}	bz	B5⁴, 972
2189	Benzene, 1-methyl-2-*iso*-propenyl or 2-*iso*-Propenyl toluene $2\text{-}[CH_2{=}C(CH_3)]C_6H_4CH_3$	132.21	...	$175, 59\text{-}62^{11}$	...	$0.9181^{15/0}$	$1.5112^{30/0}$	...	B5³, 1214
2190	Benzene, 1-methyl-2-*iso*-propyl or l-cymene $2\text{-}i\text{-}C_3H_7C_6H_4CH_3$	134.22	...	$178.1, 57.3^{10}$	−71.5	0.8766^{2014}	1.5006^{20}	al, eth, ace, bz	B5⁴, 1383
2191	Benzene, 1-methyl-3-*iso*-propyl or *m*-Cymene $3\text{-}i\text{-}C_3H_7C_6H_4CH_3$	134.22	...	$175.1, 55^{10}$	-63.7	$0.8610^{20/4}$	1.4930^{20}	**al, eth, ace, bz**	B5⁴, 1383
2192	Benzene, 1-methyl-4-*iso*-propyl or *p*-Cymene $4\text{-}i\text{-}C_3H_7C_6H_4CH_3$	134.22	...	$177.1, 56.3^{10}$	-68	$0.8573^{20/4}$	1.4909^{20}	**al, eth, ace, bz**	B6⁴, 1466
2193	Benzene, methylthio or Thioanisole $CH_3SC_6H_5$	124.20	...	$193, 74^{10}$	...	$1.0579^{20/4}$	1.5868^{20}	al, bz	B5⁴, 708
2194	Benzene, nitro $C_6H_5NO_2$	123.11	...	210.8	5.7	$1.2037^{20/4}$	1.5562^{20}	al, eth, ace, bz	B5⁴, 998
2195	Benzene, (α-nitro-*iso*-propyl) $C_6H_5C(NO_2)(CH_3)_2$	165.19	...	$224d, 125\text{-}7^{15}$	...	1.1025^{20}	1.5209^{20}	ace, bz	B5⁴, 728

No.	Name, Synonyms, and Formula	Mol. wt.	Color, crystalline form, specific rotation and λ_{max} (log ε)	b.p. °C	m.p. °C	Density	n_D	Solubility	Ref.
2196	Benzene, nitro,pentachloro or Brassicol $C_6Cl_5NO_2$	295.34	cr (al)		144	$1.718^{25/4}$		bz, chl	$B5^4$, 7286
2197	Benzene, 1-nitro-2,3,5,6-tetrachloro 2,3,5,6-$Cl_4C_6HNO_2$	260.89			99-100	$1.744^{25/4}$		al, bz, chl	$B5^4$, 728
2198	Benzene, 1-nitro-2-triazo 2-$N_3C_6H_4NO_2$	164.12	ye nd (bz-al), pr (al)		53-5			al, bz, aa, chl	$B5^4$, 761
2199	Benzene, 1-nitro-3-triazo 3-$N_3C_6H_4NO_2$	164.12	wh nd (dil al or peth)		56			al, eth, bz	$B5^4$, 761
2200	Benzene, 1-nitro-4-triazo 4-$N_3C_6H_4NO_2$	164.12	wh pl (dil al)		75			al, eth, bz, aa	$B5^4$, 762
2201	Benzene, 1-nitro-2,3,4-tribromo 2,3,4-$Br_3C_6H_2NO_2$	359.80	cr (al)		85.4			al, eth, bz	B5, 251
2202	Benzene, 1-nitro-2,3,5-tribromo 2,3,5-$Br_3C_6H_2NO_2$	359.80	nd		119			eth, bz	B5, 251
2203	Benzene, 1-nitro-2,3,6-tribromo 2,3,6-$Br_3C_6H_2NO_2$	359.80	pl or pr (eth-al)		185 (sub)			al, eth, bz	B5, 251
2204	Benzene, 1-nitro-2,4,5-tribromo 2,4,5-$Br_3C_6H_2NO_2$	359.80	nd (al)		95 (sub)			al, eth	$B5^3$, 621
2205	Benzene, 1-nitro-2,4,6-tribromo 2,4,6-$Br_3C_6H_2NO_2$	359.80	mcl pr (chl)	177^{11}	125			eth, chl, aa	$B5^4$, 732
2206	Benzene, 1-nitro-3,4,5-tribromo 3,4,5-$Br_3C_6H_2NO_2$	359.80	tcl (eth-al)	sub	112	2.645		eth, bz	$B5^4$, 732
2207	Benzene, 1-nitro-2,3,4-trichloro 2,3,4-$Cl_3C_6H_2NO_2$	226.45	nd (al)		55.5			CS_2	$B5^2$, 186
2208	Benzene, 1-nitro-2,4,5-trichloro 2,4,5-$Cl_3C_6H_2NO_2$	226.45	pr (al), nd (al)	288	57-8	1.790^{23}		al, eth, bz	$B5^4$, 728
2209	Benzene, 1-nitro-2,3,6-trichloro 2,3,6-$Cl_3C_6H_2NO_2$	226.45	nd (al)		89			al	$B5^2$, 187
2210	Benzene, 1-nitro-2,4,6-trichloro 2,4,6-$Cl_3C_6H_2NO_2$	226.45	nd (al)		71			al, lig	$B5^4$, 728
2211	Benzene, 1-nitro-3,4,5-trichloro 3,4,5-$Cl_3C_6H_2NO_2$	226.45	pa ye tcl (al)		72.5	1.807		al	$B5^2$, 187
2212	Benzene, 1-nitro-2,3,5-triiodo 2,3,5-$I_3C_6H_2NO_2$	500.80	ye pr		124			bz	B5, 256
2213	Benzene, 1-nitro-2,3,6-triiodo 2,3,6-$I_3C_6H_2NO_2$	500.80	nd (aa)		137			aa	B5, 256
2214	Benzene, 1-nitro-2,4,5-triiodo 2,4,5-$I_3C_6H_2NO_2$	500.80	pa ye nd (CS_2)		178				B5, 256
2215	Benzene, 1-nitro-3,4,5-triiodo 3,4,5-$I_3C_6H_2NO_2$	500.80	ye pr (chl), nd (al)		167	3.256		eth, bz, aa, chl	$B5^3$, 626
2216	Benzene, 1-nitro-2,3,5-trimethyl 2,3,5-$(CH_3)_3$-$C_6H_2NO_2$	165.19	pr (al)	$139\text{-}40^7$	20			al	$B5^4$, 1015
2217	Benzene, 1-nitro-2,4,5-trimethyl 2,4,5-$(CH_3)_3C_6H_2NO_2$	165.19	ye nd (al)	265	71			al, peth	$B5^4$, 1015
2218	Benzene, 1-nitro-2,4,6-trimethyl 2,4,6-$(CH_3)_3C_6H_2NO_2$	165.19	rh pr (al)	255	44	1.51		al	$B5^4$, 1028
2219	Benzene, nitroso C_6H_5NO	107.11	rh or mcl (al-eth)	$57\text{-}9^{18}$	68-9			al, eth, bz, lig	$B5^4$, 702
2220	Benzene, nonyl $C_6H_5C_9H_{19}$	204.36		280.1		$0.8584^{20/4}$	$1.4816^{20/}$		$B5^3$, 1075
2221	Benzene, pentamino $C_6H(NH_2)_5$	153.19	dk red nd		228d			w	$B13^2$, 155
2222	Benzene, pentabromo C_6HBr_5	472.59	wh nd (aa or al)	sub	160-1			bz, chl	$B5^4$, 687
2223	Benzene, pentachloro C_6HCl	250.34	nd (al)	277	86	$1.8342^{16.5}$			$B5^4$, 669
2224	Benzene, pentaethyl $C_6H(C_2H_5)_5$	218.38		277	<−20	$0.895^{19/19}$	1.5127^{20}		$B5^4$, 1197
2225	Benzene, pentaiodo C_6HI_5	707.60	nd (al)	sub	172			bz, chl, aa	$B5^4$, 702
2226	Benzene, pentamethyl $C_6H(CH_3)_5$	148.25	pr (al)	232, 100^{10}	54.5	$0.917^{20/4}$	1.527^{20}	al, bz	$B5^4$, 1109
2227	Benzene, pentoxy or Phenyl pentyl ether $C_6H_5OC_5H_{11}$	164.25		200^{756}, 111^{17}		$0.9270^{20/4}$	1.4947^{20}	al, ace	$B6^3$, 552
2228	Benzene, pentyl $C_6H_5C_5H_{11}$	148.25		205.4, 80.6^{10}	-75	$0.8585^{20/4}$	1.4878^{20}	al, eth, ace, bz	$B5^4$, 1085

PHYSICAL CONSTANTS OF ORGANIC COMPOUNDS (Continued)

No.	Name, Synonyms, and Formula	Mol. wt.	Color, crystalline form, specific rotation and λ_{max} (log ε)	b.p. °C	m.p. °C	Density	n_D	Solubility	Ref.
2229	Benzene, (2-pentyl) $C_6H_5CH(CH_3)C_3H_7$	148.25		$198\text{-}9^{757}$		$0.8594^{21/4}$	1.4875^{21}	al, eth	$B5^4$, 1087
2230	Benzene, propenyl (*cis*) $C_6H_5CH{=}CHCH_3$	118.18		69^{28}	-60.5	$0.9088^{20/4}$	1.5420^{20}	al, eth, ace, bz	$B5^4$, 1359
2231	Benzene, propenyl (*trans*) $C_6H_5CH{=}CHCH_3$	118.18		175.6	-27^{1}	0.9019^{25}	1.5508^{20}	al, eth, ace, bz	$B5^4$, 1360
2232	Benzene, 1-propenyl-2,4,5-trimethoxy (α form) or α Asaron $2,4,5\text{-}(CH_3O)_3\text{-}C_6H_2CH{=}CHCH_3$	208.34	mcl nd (w)	296, $167\text{-}8^{12}$	67	$1.165^{20/4}$	1.5683^{20}	al, eth, peth	$B6^3$, 6441
2233	Benzene, *iso*-propyl $C_6H_5C(CH_3){=}CH_2$	118.18		165.4, 48.5^{10}	-23.2	$0.9106^{20/4}$	1.5386^{20}	al, eth, ace, bz	$B5^4$, 1364
2234	Benzene, propoxy or Phenyl propyl ether $C_6H_5OC_3H_7$	136.19		189.9	-27	$0.9474^{20/4}$	1.5014^{20}	al, eth	$B6^4$, 556
2235	Benzene, *iso*-propoxy or Phenyl-*iso*-propyl ether $C_6H_5OCH(CH_3)_2$	136.19		176.8	-33	0.9408	1.4975^{20}	w, al, ace, bz	$B6^3$, 549
2236	Benzene, propyl $C_6H_5C_3H_7$	120.19		159.2, 43.3^{10}	-99.5	$0.8620^{20/4}$	1.4920^{20}	al, eth, ace, bz	$B5^4$, 977
2237	Benzene, *iso*-propyl or Cumene $(CH_3)_2CHC_6H_5$	120.19		152.4, 38.2^{10}	-96	$0.8618^{20/4}$	1.4915^{20}	al, eth, ace, bz	$B5^4$, 985
2238	Benzene, 1-*iso*-propyl-2,4,5-trimethyl $2,4,5\text{-}(CH_3)_3C_6H_2\text{-}i\text{-}C_3H_7$	162.27		218-20		$0.8795^{21/4}$	$1.50648^{21/D}$		$B5^4$, 1038
2239	Benzene, 1,2,3,5-tetrabromo $1,2,3,5\text{-}Br_4C_6H_2$	393.70	nd (al)	329	99-100			eth, bz, al	$B5^4$, 686
2240	Benzene, 1,2,4,5-tetrabromo $1,2,4,5\text{-}Br_4C_6H_2$	393.70	mcl pr (CS_2)		182	3.072^{20}		eth	$B5^4$, 687
2241	Benzene, 1.2.3.4-tetrachloro $1,2,3,4\text{-}Cl_4C_6H_2$	215.89	nd (al)	254	47.5			eth, lig, aa	$B5^3$, 667
2242	Benzene, 1,2,3,5-tetrachloro $1,2,3,5\text{-}Cl_4C_6H_2$	215.89	nd (al)	246	54.5			eth, bz, lig, al	$B5^4$, 686
2243	Benzene, 1,2,4,5-tetrachloro $1,2,4,5\text{-}Cl_4C_6H_2$	215.89	nd mcl pr (eth, al or bz)	243-6	139-40			eth, bz, chl	$B5^4$, 686
2244	Benzene, 1,2,3,4-tetraethyl $1,2,3,4\text{-}(C_2H_5)_4C_6H_2$	190.33		251^{734}, 121.7^{14}	11.8	$0.8875^{20/4}$	1.5125^{20}	al, eth	$B5^4$, 1168
2245	Benzene, 1,2,4,5-tetraethyl $1,2,4,5\text{-}(C_2H_5)_4C_6H_2$	190.33		250	10	$0.8788^{20/4}$	1.5054^{20}	al, eth	$B5^4$, 1168
2246	Benzene, 1,2,3,4-tetrafluoro $1,2,3,4\text{-}F_4C_6H_2$	150.08		93-4			$1.4054^{20/D}$		$B5^4$, 639
2247	Benzene, 1,2,4,5-tetrafluoro $1,2,4,5\text{-}F_4C_6H_2$	150.08		89-90	4-5	1.4255^{20}	1.4075^{20}		$B5^4$, 639
2248	Benzene, 1,2,3,5-tetrahydroxy $1,2,3,5\text{-}(HO)_4C_6H_2$	142.11	nd (w), lf (eth-AcOEt)		165-7			w, al, ace	$B6^1$, 6652
2249	Benzene, 1,2,4,5-tetrahydroxy $1,2,4,5\text{-}(HO)_4C_6H_2$	142.11	lf (w or aa)		232.5			w, al, ace	$B6^1$, 6655
2250	Benzene, 1,2,3,4-tetraiodo $1,2,3,4\text{-}I_4C_6H_2$	581.70	pr (eth-aa)	sub	136			al, eth, aa	B5, 229
2251	Benzene, 1,2,3,5-tetraiodo $1,2,3,5\text{-}I_4C_6H_2$	581.70	pr (eth, aa)	sub	148			aa	B5, 229
2252	Benzene, 1,2,4,5-tetraiodo $1,2,4,5\text{-}I_4C_6H_2$	581.70	pr (bz), nd (eth)	sub (vac)	254 (165)			aa	B5, 229
2253	Benzene, 1,2,3,4-tetramethyl or Prebnitene $1,2,3,4\text{-}(CH_3)_4C_6H_2$	134.22	cr (peth)	205, 79.4^{10}	-6.2	$0.9052^{20/4}$	1.5203^{20}	al, eth, ace, bz	$B5^4$, 1072
2254	Benzene, 1,2,3,5-tetramethyl or Isodurene $1,2,3,5\text{-}(CH_3)_4C_6H_2$	134.22		198, 74.4^{10}	-23.7	$0.8903^{20/4}$	1.5130^{20}	al, eth, ace, bz	$B5^4$, 1073
2255	Benzene, 1,2,4,5-tetramethyl or Durene $1,2,4,5\text{-}(CH_3)_4C_6H_2$	134.22		196.8, 73.5^{10}	79.2	$0.8380^{81/4}$	1.4790^{81}	al, eth, ace, bz	$B5^4$, 1076
2256	Benzene, 1,2,4,5-tetra-*iso*propyl $1,2,4,5\text{-}(i\text{-}C_3H_7)_4C_6H_2$	246.44		260^{775}, 133^{17}	118.4	$0.758^{150/4}$			$B5^4$, 1207
2257	Benzene, 1,3,5-triacetyl $1,3,5\text{-}(CH_3CO)_3C_6H_3$	204.23	nd (w,aa or al)		163			aa	$B7^3$, 4579
2258	Benzene, 1,2,3-triamino $1,2,3\text{-}(H_2N)_3C_6H_3$	123.16	cr (dil HCl)	336	103			w, al, eth	$B13^3$, 551
2259	Benzene, 1,2,4-triamino $1,2,4\text{-}(H_2N)_3C_6H_3$	123.16	pl or lf (chl)	340	95-8			w, al, chl	$B13^3$, 552
2260	Benzene, triazo or Phynylazide $C_6H_5N_3$	119.13	pa ye oil	70^{11}	-27.5	$1.0880^{20/20}$	1.5589^{25}		$B5^3$, 759

No.	Name, Synonyms, and Formula	Mol. wt.	Color, crystalline form, specific rotation and λ_{max} (log ε)	b.p. °C	m.p. °C	Density	n_D	Solubility	Ref.
2261	Benzene, 1,2,3-tribromo 1,2,3-$Br_3C_6H_3$	314.80	pl (al)		87.8	2.658		eth	$B5^3$, 569
2262	Benzene, 1,2,4-tribromo 1,2,4-$Br_3C_6H_3$	314.80	nd (al or eth)	275	44-5			al, eth, ace	$B5^4$, 685
2263	Benzene, 1,3,5-tribromo 1,3,5-$Br_3C_6H_3$	314.80	nd or pr (al)	271^{765}	121-2			eth, bz, chl	$B5^4$, 685
2264	Benzene, 2,4,6-tribromo-1,3,5-trimethyl 1,3,5-$(CH_3)_3C_6(Br_3)$	356.90	tcl nd (al or bz)		227-8			bz	$B5^3$, 921
2265	Benzene, 1,2,3-trichloro 1,2,3-$Cl_3C_6H_3$	181.45	pl (al)	218-9	53-4			eth, bz	$B5^4$, 664
2266	Benzene, 1,2,4-trichloro 1,2,4-$Cl_3C_6H_3$	181.45	rh	213.5, 84.8^{10}	17	$1.4542^{20/4}$	1.5717^{20}	eth	$B5^4$, 664
2267	Benzene, 1,3,5-trichloro 1,3,5-$Cl_3C_6H_3$	181.45	nd	208^{763}	63-4			eth, bz, lig	$B5^4$, 666
2268	Benzene, 1,2,3-trichloro-4,5,6-trihydroxy 1,2,3-$Cl_3C_6(OH)_3$	229.45	nd (al or bz), nd (+3w, w)		185			w, al, eth	B6, 1084
2269	Benzene, 1,2,4-trichloro-3,5,6-trihydroxy 1,2,4-$Cl_3C_6(OH)_3$	229.45	nd (bz or aa)		160			al, eth	$B6^2$, 1072
2270	Benzene, 1,3,5-trichloro-2,4,6-trihydroxy 1,3,5-$Cl_3C_6(OH)_3$	229.45	cr (al)	sub	136			al	B6, 1104
2271	Benzene, 1,3,5-triethoxy or Phloroglucinol triethylether 1,3,5-$(C_2H_5O)_3C_6H_3$	210.27	cr (al, dil al)	175^{24}	43.5			al, eth	$B6^3$, 6306
2272	Benzene, 1,2,4-triethyl 1,2,4-$(C_2H_5)_3C_6H_3$	162.27		217.5^{755}, 99^{15}		$0.8738^{20/4}$	1.5024^{20}	al, eth	$B5^4$, 1133
2273	Benzene, 1,3,5-triethyl 1,3,5-$(C_2H_5)_3C_6H_3$	162.27		216	-66.5	1.4969^{20}		al, eth	$B5^4$, 1133
2274	Benzene, 1,2,3-trihydroxy or Pyrogallol 1,2,3-$(HO)_3C_6H_3$	126.11	lf or nd (bz)	309, 171^{12}	133-4	$1.453^{4/4}$	1.561^{134}	w, al, eth	$B6^3$, 2620
2275	Benzene, 1,2,5-trihydroxy, triacetate 1,2,5-$(CH_3CO_2)_3C_6H_3$	252.22	pr (al)		165			al	$B6^3$, 6282
2276	Benzene, 1,2,4-trihydroxy or Hydroxyquinol 1,2,4-$(HO)_3C_6H_3$	126.11	pl (eth), lf or pl (w)		140-1			w, al, eth	$B6^3$, 6272
2277	Benzene, 1,2,4-trihydroxy,triacetate 1,2,4 $(CH_3CO_2)C_6H_3$	252.22	nd (MeOH)	>300	97-8			al	$B6^3$, 6282
2278	Benzene, 1,3,5-trihydroxy or Phloroglucinol 1,3,5-$(HO)_3C_6H_3$	126.11	lf or pl (w + 2)	sub	117 (hyd), 218-9 (anh)	1.46		al, eth, bz	$B6^3$, 6301
2279	Benzene, 1,3,5-trihydroxy, triacetate or Phloroglucinol triacetate 1,3,5-$(CH_3CO_2)C_6H_3$	252.22	pr (w), hd (dil al)		105-6			al	$B6^3$, 6306
2280	Benzene, 1,2,3-triiodo 1,2,3-$I_3C_6H_3$	455.80	nd (al), pr (bz)	sub	116			al, eth, chl	$B5^1$, 122
2281	Benzene, 1,2,4-triiodo 1,2,4-$I_3C_6H_3$	455.80	nd (al)	sub	91.5			eth, chl	$B5^3$, 585
2282	Benzene, 1,3,5-triiodo 1,3,5-$I_3C_6H_3$	455.80	nd (aa)	sub	184.2			aa	$B5^4$, 702
2283	Benzene, 1,2,3-trimethoxy or Pyrogallol trimethyl ether 1,2,3-$(CH_3O)_3C_6H_3$	168.19	rh nd (al)	235, 140^{12}	48-9	$1.1118^{45/45}$		al, eth, bz	$B6^3$, 6265
2284	Benzene, 1,3,5-trimethoxy or Phloroglucinol trimethyl ether 1,3,5-$(CH_3O)_3C_6H_3$	168.19	pr (al), lf (peth)	255.5	54-5			al, eth, bz	$B6^3$, 6305
2285	Benzene, 1,2,3-trimethyl or Hemimellitene 1,2,3-$(CH_3)_3C_6H_3$	120.19		176.1, 56.7^{10}	-25.4	$0.8944^{20/4}$	1.5139^{20}	**al, eth, ace, bz**	$B5^4$, 1007
2286	Benzene, 1,2,4-trimethyl or Pseudocumene 1,2,4-$(CH_3)_3C_6H_3$	120.19		169.3, 51.6^{10}	-43.8	$0.8758^{20/4}$	1.5048^{20}	**al, eth, ace, bz**	$B5^4$, 1010
2287	Benzene, 1,3,5-trimethyl or Mesitylene 1,3,5-$(CH_3)_3C_6H_3$	120.19		164.7, 48.7^{10}	-44.7	$0.8652^{20/4}$	1.4994^{20}	**al, eth, ace, bz**	$B5^4$, 1016
2288	Benzene, 1,2,3-trimethyl-4,5,6-trinitro 4,5,6-$(NO_2)_3C_6(CH_3)_3$	255.19	pr (al)		209			al, ace, bz	B5, 400
2289	Benzene, 1,2,4-trimethyl-3,5,6-trinitro 3,5,6-$(NO_2)_3C_6(CH_3)_3$	255.19	rh pr (al)		185			bz, to	$B5^4$, 1015
2290	Benzene, 1,3,5-trimethyl-2,4,6-trinitro or Trinitromesitylene 2,4,6-$(NO_2)_3C_6(CH_3)_3$	255.19	tcl nd (al), pr (ace)	415 exp	238.2			ace, bz	$B5^4$, 1029
2291	Benzene, 1,2,3-trinitro 1,2,3-$(NO_2)_3C_6H_3$	213.11	ye nd or pr (MeOH)		127.5			al	$B5^4$, 754

No.	Name, Synonyms, and Formula	Mol. wt.	Color, crystalline form, specific rotation and λ_{max} (log ε)	b.p. °C	m.p. °C	Density	n_D	Solubility	Ref.
2292	Benzene, 1,2,4-trinitro $1,2,4-(NO_2)_3C_6H_3$	213.11	lf (eth) pa ye pr (al)		61.2			al, eth, ace, bz, chl	$B5^{4}$, 754
2293	Benzene, 1,3,5-trinitro $1,3,5-(NO_2)_3C_6H_3$	213.11	rh pl (bz), lf (w)	315, 175^{2}	(i) 121-2, (ii) 61			ace, bz	$B5^{4}$, 755
2294	Benzene, 1,3,5-triphenyl $1,3,5-(C_6H_5)_3C_6H_3$	306.41	rh nd (al or aa)	459^{17}	176 (cor)	$1.199^{10/4}$		al, eth, bz	$B5^{3}$, 2563
2295	Benzene, 1,3,5-tri*iso*-propyl $1,3,5(i-C_3H_7)_3C_6H_3$	204.36		238	-7.4	0.8545^{20}	1.4882^{20}	ace, bz, chl	$B5^{4}$, 1178
2296	Benzene, 1,3,5-tris(phenyl amino) or *Sym*-trianilino benzene $1,3,5-(C_6H_5NH)_3C_6H_3$	351.45	nd (al)		193			eth	$B13^{2}$, 147
2297	Benzene, 1,3,5-tris-(4-tolylamino) $1,3,5-(4-CH_3C_6H_4NH)_3C_6H_3$	393.53	nd (al)		186-7				B13, 299
2298	Benzene arsonic acid or Phenylarsonic acid $C_6H_5AsO(OH)_2$	202.04	cr (w)		158-62d			w, al	$B16^{3}$, 1057
2299	Benzene arsonic acid-2-amino $2-H_2NC_6H_4AsO(OH)_2$	217.06	nd (al-eth)		153-4			w, al, aa	$B16^{3}$, 1099
2300	Benzene arsonic acid-3-amino $3-H_2NC_6H_4AsO(OH)_2$	217.06	pr (w)		213			w	$B16^{3}$, 1104
2301	Benzene arsonic acid, 4-amino or Arsanilic acid $4H_2NC_6H_4AsO(OH)_2$	217.06	mcl, nd, (w or al)		232	1.9571^{10}		w, al, eth	$B16^{3}$
2302	Benzenearsonic acid, 2-chloro $2-ClC_6H_4AsO(OH)_2$	236.49	nd (w or dil al)		186-7			al, w	$B16^{2}$, 457
2303	Benzenearsonic acid, 3-chloro $3-ClC_6H_4AsO(OH)_2$	236.49	cr (w)		175			w, al	$B16^{2}$, 457
2304	Benzenearsonic acid, 4-chloro $4-ClC_6H_4AsO(OH)_2$	236.49	wh nd (al)		283-5d			al	$B16^{3}$, 1058
2305	Benzenearsonic acid, 2-hydroxy $2-HOC_6H_4AsO(OH)_2$	218.04	nd (w)		190-1			w, al, bz, ace	$B16^{2}$, 464
2306	Benzenearsonic acid, 3-hydroxy $3-HOC_6H_4AsO(OH)_2$	218.04	cr (w)		162-73d			w, al	$B16^{1}$, 454
2307	Benzenearsonic acid, 4-hydroxy $4-HOC_6H_4AsO(OH)_2$	218.04	nd (aa)		170-4			w, al	$B16^{3}$, 1070
2308	Benzenearsonic acid, 2-nitro $2-NO_2C_6H_4AsO(OH)_2$	246.03			233d			aa	$B16^{3}$, 1059
2309	Benzenearsonic acid, 3-nitro $3-NO_2C_6H_4AsO(OH)_2$	246.03	lf ye lf (w)		200				$B16^{3}$, 1059
2310	Benzenearsonic acid, 4-nitro $4-NO_2C_6H_4AsO(OH)_2$	246.03	lf or nd (w)		>310				$B16^{3}$, 1059
2311	Benzenearsonic acid, 4-ureido or Carbarsone $C_7H_9AsN_2O_4$	260.08	nd (w)		174			w	$B16^{2}$, 497
2312	Benzeneazoethane or Ethylphenyl diimide $C_6H_5N{=}NC_2H_5$	134.18	bt ye oil	175-85, $82-3^{20}$		$0.9628^{22/4}$	α-1.5313 β-1.5579	al, eth, bz	$B16^{3}$, 4
2313	Benzeneazomethane or Methylphenyl diimide $C_6H_5N{=}NCH_3$	120.15	ye oil	150d, 60^{15}				al, eth	$B16^{3}$, 3
2314	Benzeneazo-α-naphthylene $C_6H_5N{=}N-\alpha-C_{10}H_7$	232.28	dk red lf (al)		70			al, eth, bz, lig	$B16^{3}$, 57
2315	Benzene-azo-α-naphthalene, 2′-amino $C_6H_5N{=}N-(\alpha-C_{10}H_6NH_2-2)$	247.30	red pl (al)		102-4			al, aa	$B16^{3}$, 417
2316	Benzeneazo, α-naphthalene, 4-amino or Naphthyl red $C_6H_5N{=}N-(\alpha-C_{10}H_6NH_2-4)$	247.30	red lf (dil al), nd (dil al)		125-6			al, eth, bz	$B16^{3}$, 406
2317	Benzenazo-α-naphthalene, 4-amino,hydrochloride $C_6H_5N{=}N-(\alpha-C_{10}H_6NH_2-4).HCl$	283.76	gr nd, pr (al, aa)		205-6			al, aa	$B16^{1}$, 324
2318	Benzeneazo-α-naphthalene, 2,3-dimethyl-2′-hydroxy $2,3-(CH_3)_2C_6H_3N{=}N-(\alpha-C_{10}H_6OH-2)$	276.34	pa ye amor (al-bz)		125-30			al, to	$B16^{2}$, 71
2319	Benzeneazo-α-naphthalene, 2,4-dimethyl-2′-hydroxy $2,4-(CH_3)_2C_6H_3N{=}-(\alpha C_{10}H_6OH-2)$	276.34	red nd (al)		166			al, eth	$B16^{2}$, 72
2320	Benzeneazo-α-naphthalene, 2,5-dimethyl-2′-hydroxy $2,5-(CH_3)_2C_6H_3N{=}N-(\alpha-C_{10}H_6OH-2)$	276.34	nd (al)		153			al	$B16^{3}$, 137
2321	Benzeneazo-α-naphthalene, 3,4-dimethyl,2′-hydroxy $3,4-(CH_3)_2C_6H_3N{=}N(\alpha-C_{10}H_6OH-2)$	276.34	red nd (al)		146			bz, chl	$B16^{1}$, 260
2322	Benzeneazo-α-naphthalene, 2′-hydroxy or Sudan yellow $C_6H_5N{=}N-(\alpha-C_{10}H_6OH-2)$	248.28	red-gold lf or nd (al)	133-4				al, eth, bz	$B16^{3}$, 129
2323	Benzeneazo-α-naphthylene, 4′-hydroxy $C_6H_5N{=}N-(\alpha-C_{10}H_6OH-4)$	248.28	vt-br lf (bz)		205-6d			al, bz	$B16^{3}$, 125

PHYSICAL CONSTANTS OF ORGANIC COMPOUNDS (Continued)

No.	Name, Synonyms, and Formula	Mol. wt.	Color, crystalline form, specific rotation and λ_{max} (log ε)	b.p. °C	m.p. °C	Density	n_D	Solubility	Ref.
2324	Benzeneazo-N = N-α-naphthalene, 2-nitro-2′-hydroxy $2\text{-}NO_2C_6H_4N{=}N(\alpha\text{-}C_{10}H_6OH\text{-}2)$	293.28	og-red nd (gl aa)		294			aa	B16^{2}, 70
2325	Benzeneazo-α-naphthalene, 4-nitro-2′-hydroxy or Para red Paranitroaniline red $4\text{-}NO_2C_6H_4N{=}N(\alpha\text{-}C_{10}H_6OH\text{-}2)$	293.28	br-og pl (to or bz)		257			al, bz	B16^{2}, 70
2326	Benzene azo-β-naphthalene $C_6H_5\text{-}N{=}N\beta\text{-}C_{10}H_7$	232.28	ye (al)		131			al, eth, bz, lig	B16^{3}, 60
2327	Benzeneazo-β-naphthalene, 2,4-dimethyl, 1′-hydroxy $2,4\text{-}(CH_3)_2C_6H_3N{=}N(\beta\text{-}C_{10}H_6OH\text{-}1)$	276.34	gold-red lf or nd (al-chl)		186			bz, chl	B16^{1}, 249
2328	Benzeneazo-β-naphthalene, 4-dimethylamino, 1′-hydroxy $4\text{-}(CH_3)_2NC_6H_4N{=}N(\beta\text{-}C_{10}H_6OH\text{-}1)$	275.33	ye-br (bz-lig)		174			bz	B16, 323
2329	Benzeneazo-β-naphthalene, 1′-hydroxy $C_6H_5N{=}N(\beta C_{10}H_6OH\text{-}1)$	248.28	red nd (al)	sub	138			al, aa	B16^{3}, 123
2330	Benzene diazonium chloride $C_6H_5N_2Cl$	140.57	nd (al)		exp			w, al, ace, aa	B16^{3}, 506
2331	Benzene diazonium cyanide $C_6H_5N_2CN$	131.14	ye pr (w)		69				B16, 432
2332	Benzene diazonium nitrate $C_6H_5N_2ONO_2$	167.12	nd (al-eth)		90 exp			w, al	B16^{2}, 268
2333	Benzene hexacarboxylic acid or Mellitic acid $C_6(CO_2H)_6$	342.17	nd (al)		286-8d			w, al	B9^{3}, 4914
2334	Benzene pentacarboxylic acid $C_6H(CO_2H)_5$	298.16	nd (w + 5)		228-30 (hyd), 238 (anh)			w, al	B9^{3}, 4908
2335	Benzene phosphinic acid $C_6H_5P(OH)_2$	142.09			82-3			w, al	B16^{3}, 874
2336	Benzene phosphinic acid, diethylester $C_6H_5P(OC_2H_5)_2$	198.09		235		1.032^{16}			B16^{3}, 847
2337	Benzene phosphonic acid $C_6H_5P(O)(OH)_2$	158.09	lf (w)		160			w, al, eth	B16^{3}, 883
2338	Benzenephosphonyl chloride $C_6H_5POCl_2$	194.98		258, $137\text{-}8^{15}$		1.197^{25}	1.5581^{25}		B16^{3}, 885
2339	Benzenephosphonic acid, tetrachloride $C_6H_5PCl_4$	249.89			73				B16^{3}, 885
2340	Benzenephosphothionic acid, ethyl-4-nitro phenylester or EPN $C_6H_5PS(OC_2H_5)OC_6H_4NO_2\text{-}4$	323.31			36	$1.27^{25/4}$	1.5978^{30}	al, eth, bz	C48, 2742
2341	Benzeneseleninic acid $C_6H_5SeO_2H$	189.07	pl (w)		124-5	$1.652^{125/4}$			B11^{3}, 716
2342	Benzenesiliconic acid $C_6H_5SiO_2H$	138.20	glassy (eth)		92			al, eth	B16, 911
2343	Benzene stibonic acid $C_6H_5SbO(OH)_2$	248.87	nd (aa)		139				B16^{3}, 1178
2344	Benzene sulfenyl chloride C_6H_5SCl	144.62	red liq	$73\text{-}5^{9}$				bz	B6^{4}, 1564
2345	Benzene sulfenyl chloride, 2,4-dinitro $2,4\text{-}(NO_2)_2C_6H_3SCl$	234.61	ye pr (bz-peth)		99			dz, chl, aa	B6^{4}, 1772
2346	Benzene sulfenyl chloride, 2-nitro $2\text{-}NO_2C_6H_4SCl$	189.62	ye nd (bz)		75			eth, bz, chl	B6^{4}, 1676
2347	Benzene sulfenyl chloride, 4-nitro $4\text{-}NO_2C_6H_4SCl$	189.62	ye lf (peth)	$125^{0.1}$	52			bz	B6^{4}, 1717
2348	Benzenesulfinic acid $C_6H_5SO_2H$	142.17	pr (w)	d at 100	84			al, eth, bz	B11^{3}, 3
2349	Benzenesulfinic acid, 3-acetamido $3\text{-}(CH_3CONH)C_6H_4SO_2H$	199.22	cr (w)		145			w	B14^{2}, 427
2350	Benzenesulfinic acid, 4-acetamido $4\text{-}(CH_3CONH)C_6H_4SO_2H$	199.22	cr (w)		160			w	B14^{3}, 1893
2351	Benzenesulfinic acid, 4-bromo $4\text{-}BrC_6H_4SO_2H$	221.07	nd (w)		114			al, eth	B11^{3}, 5
2352	Benzenesulfinic acid, 4-chloro $4\text{-}ClC_6H_4SO_2H$	176.62	lf or nd (w)		99			al, eth, w	B11^{3}, 5
2353	Benzenesulfinic acid, ethyl ester $C_6H_5SO_2C_2H_5$	170.23	liq	d				al, eth, bz	B11^{3}, 4
2354	Benzenesulfinic acid, 3-nitro $3\text{-}NO_2C_6H_4SO_2H$	187.17	nd		98			w, al, eth, ace	B11^{3}, 5

PHYSICAL CONSTANTS OF ORGANIC COMPOUNDS (Continued)

No.	Name, Synonyms, and Formula	Mol. wt.	Color, crystalline form, specific rotation and λ_{max} (log ε)	b.p. °C	m.p. °C	Density	n_D	Solubility	Ref.
2355	Benzenesulfinic acid, 4-nitro 4-$NO_2C_6H_4SO_2H$	187.17	pr or nd (w)		159			al, eth, aa	$B11^{3}$, 6
2356	Benzenesulfinyl chloride C_6H_5SOCl	160.62	pl (peth)	$71\text{-}2^{1.5}$	38	1.3469^{25}	1.3470^{25}	eth, chl	$B11^{2}$, 4
2357	Benzenesulfonamide $C_6H_5SO_2NH_2$	157.19	lf, nd (w)		156			al, eth, w	$B11^{3}$, 52
2358	Benzenesulfonamide, *N,N*-dichloro or Dichloramine B $C_6H_5SO_2NCl_2$	226.08	ye mcl or pl		76			al	$B11^{3}$, 79
2359	Benzenesulfonamide, *N*-hydroxy $C_6H_5SO_2NHOH$	173.19	pl (w), rh		126d			w, al, eth, ace, aa	$B11^{3}$, 83
2360	Benzenesulfonamide, *N*-phenyl $C_6H_5SO_2NHC_6H_5$	233.28	tetr pr (al)		110			al, eth	$B12^{3}$, 1079
2361	Benzene sulfonamido, 3-acetamido 3-$(CH_3CONH)C_6H_4SO_2NH_2$	214.26	(aa)		216-9			ace, aa	$B14^{1}$, 718
2362	Benzenesulfonamido, 4-acetamido 4-$(CH_3CONH)C_6H_4SO_2NH_2$	214.24	nd (aa)		219-20			w, al, ace	B14, 702
2363	Benzenesulfonamide, 2-amino 2-$H_2NC_6H_4SO_2NH_2$	172.20	mcl pl, nd or pr (w)		152-3			al, ace, aa	$B14^{3}$, 1898
2364	Benzenesulfonamide, 3-amino 3-$H_2NC_6H_4SO_2NH_2$	172.20	lf or nd (w)		142			al	$B14^{3}$, 1908
2365	Benzenesulfonamide, 4-amino or Sulfanilamide 4-$H_2NC_6H_4SO_2NH_2$	172.20	lf (aq al)		165-6	1.08		w, al, eth, ace, MeOH	$B14^{3}$, 1919
2366	Benzenesulfonamide, 2-bromo or 2-bromo benzenesulfonamide 2-$BrC_6H_4SO_2NH_2$	236.08	nd (w), pr (al)		186				B11, 56
2367	Benzenesulfonamide, 3-bromo or 3-bromo sulfonamide 3-$Br\text{-}C_6H_4SO_2NH_2$	236.08	nd or lf (w), pr (al)		154			w, al	B11, 57
2368	Benzenesulfonamide, 4-bromo or 4-bromo sulfonamide 4-$Br\text{-}C_6H_4SO_2NH_2$	236.08	nd (w or dil al)		166			al	$B11^{3}$, 104
2369	Benzene sulfonamide, 2-carboxy or *o*-sulfamyl benzoic acid 2-$HO_2CC_6H_4SO_2NH_2$	201.20	pl or nd (w)		165-7			w, al, eth, bz	$B11^{2}$, 216
2370	Benzenesulfonamide, 3-carboxy or *m*-sulfamyl benzoic acid 3-$HO_2CC_6H_4SO_2NH_2$	201.20	pl (w)		246			w, al	$B11^{3}$, 663
2371	Benzenesulfonamide, 4-carboxy or *p*-sulfamyl benzoic acid 4-$HO_2CC_6H_4SO_2NH_2$	201.20	pr or lf (w)		290-2d			al	$B11^{3}$, 672
2372	Benzenesulfonamide, 2-chloro 2-$Cl\text{-}C_6H_4SO_2NH_2$	191.63	lf (al)		188			al	$B11^{3}$, 88
2373	Benzenesulfonamide, 3-chloro 3-$ClC_6H_4SO_2NH_2$	191.63	lf (dil al)		148			w, al, eth	$B11^{3}$, 88
2374	Benzenesulfonamide, 4-chloro 4-$ClC_6H_4SO_2NH_2$	191.63	pr or pl (eth)	141^{15}	55			eth, bz	$B11^{3}$, 90
2375	Benzenesulfonamide, 4-fluoro 4-$FC_6H_4SO_2NH_2$	175.18	pl or nd (w, al)		126			al, eth, ace	$B11^{3}$,88
2376	Benzenesulfonamide, 4-hydroxy 4-$HOC_6H_4SO_2NH_2$	173.19	cr (al or w)		176-7			w, al	$B11^{3}$, 506
2377	Benzenesulfonamide, 4-ureido,amide 4-$H_2NCONHC_6H_4SO_2NH_2$	215.23	nd (al)		206-7			w, al	$B14^{3}$, 2142
2378	Benzenesulfonic acid $C_6H_5SO_3H$	158.17	nd (bz)		65-6 (anh)			w, al, aa	$B11^{3}$, 32
2379	Benzenesulfonic acid, hydrate $C_6H_5SO_3H.1.5H_2O$	185.19	pl (w)		45-6			w, al	$B11^{2}$, 18
2380	Benzenesulfonic acid, 2-amino or orthanilic acid 2-$H_2NC_6H_4SO_3H$	173.19	pr (+ ½ w)		>320d				$B14^{3}$, 1896
2381	Benzenesulfonic acid, 2-amino-4-chloro 2-H_2N-4-$Cl\text{-}C_6H_3SO_3H$	207.63	nd or pl (w)		310-30d				$B14^{3}$, 1900
2382	Benzenesulfonic acid, 2-amino-5-chloro 2-H_2N-5-$ClC_6H_3SO_3H$	207.63	nd (w)		280d			w	$B14^{3}$, 1901
2383	Benzenesulfonic acid, 2-amino-3,4-dimethyl 2-H_2N--3,4-$(CH_3)_2C_6H_2SO_3H$	201.24	pr (w + 1)		300d				$B14^{1}$, 731
2384	Benzenesulfonic acid, 2-amino-3,5,-dimethyl 2-H_2N-3,5-$(CH_3)_2C_6H_2SO_3H$	201.24		d					$B14^{2}$, 452
2385	Benzenesulfonic acid, 2-amino-3,6-dimethyl 2-H_2N-3,6-$(CH_3)_2C_6H_2SO_3H$	201.24	amor		260				$B14^{1}$, 732
2386	Benzenesulfonic acid, 2-amino-4,5-dimethyl 2-H_2N-4,5-$(CH_3)_2C_6H_2SO_3H$	201.24	pl		>300				$B14^{1}$, 731

PHYSICAL CONSTANTS OF ORGANIC COMPOUNDS (Continued)

No.	Name, Synonyms, and Formula	Mol. wt.	Color, crystalline form, specific rotation and λ_{max} (log ε)	b.p. °C	m.p. °C	Density	n_D	Solubility	Ref.
2387	Benzenesulfonic acid, 3-amino or Metanilic acid $3\text{-}H_2NC_6H_4SO_3H$	173.19	nd, pr (w + l)		d				$B14^3$, 1907
2388	Benzenesulfonic acid, 3-amino-2,5-dimethyl $3\text{-}H_2N\text{-}2,5\text{-}(CH_3)_2C_6H_2SO_3H$	201.24	nd (w + l)						$B14^3$, 2237
2389	Benzenesulfonic acid, 3-amino-4,5-dimethyl $3\text{-}H_2N\text{-}4,5\text{-}(CH_3)_2C_6H_2SO_3H$	201.24	red nd (w + l)		315				$B14^1$, 731
2390	Benzenesulfonic acid, 3-amino-4-hydroxy $3\text{-}H_2N\text{-}4\text{-}HOC_6H_3SO_3H$	189.19	rh (w + l)		>300				$B14^3$, 2275
2391	Benzenesulfonic acid, 4-amino or Sulfanilic acid $4\text{-}H_2NC_6H_4SO_3H$	173.19	rh pl or mcl (w + 2)		288	$1.485^{25/4}$		w	$B14^3$, 1916
2392	Benzenesulfonic acid, 4-amino-2,3-dimethyl $4\text{-}H_2N\text{-}2,3\text{-}(CH_3)_2C_6H_2SO_3H$	201.24	nd		305				$B14^2$, 452
2393	Benzenesulfonic acid, 4-amino-2,5-dimethyl $4\text{-}H_2N\text{-}2,5\text{-}(CH_3)_2C_6H_2SO_3H$	201.24	pl or nd		>300			w	$B14^3$, 2236
2394	Benzenesulfonic acid, 5-amino-2,3-dimethyl $5\text{-}H_2N\text{-}2,3\text{-}(CH_3)_2C_6H_2SO_3H$	201.24	pl (w + 2)		294d			w	$B14^1$, 731
2395	Benzenesulfonic acid, 5-amino-2,4-dimethyl $5\text{-}H_2N\text{-}2,4\text{-}(CH_3)_2C_6H_2SO_3H$	201.24	pr or nd		290d				$B14^3$, 2235
2396	Benzenesulfonic acid, 2-bromo $2\text{-}BrC_6H_4SO_3H$	237.07	nd					w, al	B11, 56
2397	Benzenesulfonic acid, 4-bromo $4\text{-}BrC_6H_4SO_3H$	237.07	nd (al)	155^{25}	102-3			w, al	$B11^3$, 97
2398	Benzenesulfonic acid, 2-carboxy or 2-sulfobenzoic acid $2\text{-}HO_2CC_6H_4SO_3H$	202.18	nd (w + 3)		141 (anh), 70 (+ 3w)			w, al	$B11^3$, 658
2399	Benzenesulfonic acid, 2-carboxamide $2\text{-}H_2NCOC_6H_4SO_3H$	201.20	pr (w + 1)		193-4 (anh)			w, al, eth	$B11^2$, 215
2400	Benzenesulfonic acid, 2-carboxy, anhydride (endo) $C_7H_4SO_4$	184.17	nd or pr (bz)	$184\text{-}6^{18}$	129.5			eth, bz, chl	$B19^3$, 1641
2401	Benzenesulfonic acid, 2-carboxy, imide or Saccharin, 2-Sulfobenzoic acid imide $C_7H_5NSO_3$	183.18	mcl (ace), pr (al), lf (w)	sub(vac)	228-9d	0.828		al, ace	$B27^2$, 217
2402	Benzenesulfonic acid, 3-carboxy or 3-Sulfobenzoic acid $3\text{-}HO_2CC_6H_4SO_3H$	202.18	cr (w + 2)		191 (anh), 98 (hyd)			w, al, eth	$B11^3$, 662
2403	Benzenesulfonic acid, 4-carboxy or 4-Sulfobenzoic acid $4\text{-}HO_2CC_6H_4SO_3H$	202.18	nd (w + 3)		259-60 (anh)			w, al, eth	$B11^3$, 671
2404	Benzenesulfonic acid, 4-chloro $4\text{-}ClC_6H_4SO_3H$	192.62	nd (w + 1)	$147\text{-}8^{25}$	67			w, al	$B11^3$, 89
2405	Benzenesulfonic acid, 4-chloro,phenylester or Ovotran $4\text{-}ClC_6H_4SO_3C_6H_5$	268.71			62			ace	C51, 1264
2406	Benzenesulfonic acid, 3,4-diamino $3,4\text{-}(H_2N)_2C_6H_3SO_3H$	188.20	nd		d			w	B14, 717
2407	Benzenesulfonic acid, 3,5-diamino $3,5\text{-}(NH_2)_2C_6H_3SO_3H$	188.20			d				$B14^2$, 446
2408	Benzenesulfonic acid, 2,5-dichloro, dihydrate $2,5\text{-}Cl_2C_6H_3SO_3H\text{-}2H_2O$	263.09	nd (w + 2)		<100			w, al	$B11^2$, 30
2409	Benzenesulfonic acid, 3,4-dichloro, dihydrate $3,4\text{-}Cl_2C_6H_3SO_3H\text{-}2H_2O$	263.09	nd		71-2			w, al, eth	B11, 55
2410	Benzenesulfonic acid, 3,5-diido-4-hydroxy, trihydrate or Sozoiodolic acid $3,5\text{-}I_2\text{-}4\text{-}HO\text{-}C_6H_2SO_3H\text{-}3H_2O$	480.02	nd (eth), pr (w + 3)		120 (hyd), 190d			w, al, eth	$B11^3$, 514
2411	Benzenesulfonic acid, 3,4-dimethyl $3,4\text{-}(CH_3)_2C_6H_3SO_3H$	186.23	pl or pr (chl + 2w)		63-4			w	$B11^3$, 337
2412	Benzenesulfonic acid, 2,4-dinitro $2,4\text{-}(NO_2)_2C_6H_3SO_3H$	248.17	nd (w + 3)		130 (anh), 108			w, al	$B11^3$, 159
2413	Benzenesulfonic acid, 3,5-dinitro $3,5\text{-}(NO_2)_2C_6H_3SO_3H$	248.17	ye red (dil al)		235			al, w	$B11^2$, 36
2414	Benzenesulfonic acid, 3-ethylamino $3\text{-}(C_2H_5NH)C_6H_4SO_3H$	201.24	nd (w)		294d			w	$B14^2$, 435
2415	Benzenesulfonic acid, 4-ethylamino $4\text{-}(C_2H_5NH)C_6H_4SO_3H$	201.24	pl (w)		258d			w	$B14^3$, 2025
2416	Benzenesulfonic acid, ethyl ester $C_6H_5SO_3C_2H_5$	186.23		156^{15}		$1.2167^{20/2}$	1.5081^{20}	al, eth, chl	$B11^3$, 36

No.	Name, Synonyms, and Formula	Mol. wt.	Color, crystalline form, specific rotation and λ_{max} (log ε)	b.p. °C	m.p. °C	Density	n_D	Solubility	Ref.
2417	Benzenesulfonic acid, 2-hydroxy or *o*-phenolsulfonic acid $2\text{-}HOC_6H_4SO_3H$	174.18	cr (w + 1)		145d			w, al	$B11^3$, 488
2418	Benzenesulfonic acid, 4-hydroxy or *p*-phenolsulfonic acid $4\text{-}HOC_6H_4SO_3H$	174.18	nd					w, al	$B11^3$, 498
2419	Benzenesulfonic acid, 4-hydroxy-3-nitro $4\text{-}HO\text{-}3\text{-}NO_2C_6H_3SO_3H$	219.17	ye pl (AcOEt-bz), nd (w + 3)		141-2 (anh), 51.5 (hyd)			w, al	$B11^3$, 515
2420	Benzenesulfonic acid, 4-hydrazino $4\text{-}H_2NNHC_6H_4SO_3H$	188.22	nd or lf (w)		286				$B15^3$, 865
2421	Benzenesulfonic acid, methyl ester $C_6H_5SO_3CH_3$	172.20							$B11^3$, 36
2422	Benzenesulfonic acid, 2-methyl-4-isopropyl $4\text{-}i\text{-}C_3H_7\text{-}2\text{-}CH_3C_6H_3SO_3H$	214.28	pl or pr		88-90			w	B11, 139
2423	Benzenesulfonic acid, 2-methyl-5-isopropyl $5\text{-}i\text{-}C_3H_7\text{-}2\text{-}CH_3C_6H_3SO_3H$	214.28	pl (+ 2w) mcl pr		220, 78-9 (+ 2w)			w	$B11^3$, 349
2424	Benzenesulfonic acid, 5 methyl 2 isopropyl $2\text{-}i\text{-}C_3H_7\text{-}5\text{-}CH_3C_6H_3SO_3H$	214.28			130-1			w, al	$B11^3$, 350
2425	Benzenesulfonic acid, 2-nitro $2\text{-}NO_2C_6H_4SO_3H$	203.17			85			w, al	$B11^2$, 31
2426	Benzenesulfonic acid, 3-nitro $3\text{-}NO_2C_6H_4SO_3H$	203.17	pl		48			w, al	$B11^3$, 118
2427	Benzenesulfonic acid, 4-nitro $4\text{-}NO_2C_6H_4SO_3H$	203.17	(+ 2w)		95, (109-11)			w	$B11^3$, 134
2428	Benzenesulfonic acid, 4-phenylamino or 4 Sulfodiphenylamine $4\text{-}C_6H_5NHC_6H_4SO_3H$	249.29	pl (al eth)		206			w, al	$B14^1$, 721
2429	Benzenesulfonic acid, propylester $C_6H_5SO_3C_3H_7$	200.26		$162\text{-}3^{15}$		$1.1804^{17/4}$	1.5035^{25}	al, eth, chl	$B11^2$, 20
2430	Benzenesulfonic acid, *iso*-propyl ester $C_6H_5SO_3CH(CH_3)_2$	200.26					1.5003^{20}	al, eth	$B11^2$, 20
2431	Benzenesulfonyl chloride $C_6H_5SO_2Cl$	176.62		251-2d, 120^{10}	14.5	$1.3482^{15/15}$		al, eth	$B11^3$, 51
2432	Benzenesulfonyl chloride, 3-acetamido $3\text{-}(CH_3CONH)C_6H_4SO_2Cl$	233.67	nd (bz-peth)		88			al, eth, aa	$B14^2$, 435
2433	Benzenesulfonyl chloride, 4-acetamide $4\text{-}(CH_3CONH)C_6H_4SO_2Cl$	233.67	nd (bz), pr (bz-chl)		149			al, eth	$B14^3$, 2043
2434	Benezenesulfonyl chloride, 2-bromo $2\text{-}BrC_6H_4SO_2Cl$	225.51	pr (eth)		51				B11, 56
2435	Benzenesulfonyl chloride, 4-bromo $4\text{-}BrC_6H_4SO_2Cl$	255.51	tcl or mcl pr (eth)	153^{15}	76			eth	$B11^3$, 104
2436	Benzenesulfonyl chloride, 3-carboxy $3\text{-}HO_2CC_6H_4SO_2Cl$	220.63	pr (bz)		133-4			eth, bz	$B11^3$, 663
2437	Benzenesulfonyl chloride, 4-carboxy $4\text{-}HO_2CC_6H_4SO_2Cl$	220.63	nd (ace)		237-8d			ace, to	$B11^3$, 672
2438	Benzenesulfonyl chloride, 4-chloro $4\text{-}ClC_6H_4SO_2Cl$	211.06	pr or pl (eth)	141^{15}	55			eth, bz	$B11^3$, 90
2439	Benzenesulfonyl chloride, 2,5-dichloro or 2,5-dichlorobenzene sulfonyl chloride $2,5\text{-}Cl_2C_6H_3SO_2Cl$	245.51	mcl pr (bz)		38				$B11^3$, 93
2440	Benzenesulfonyl chloride, 3,4-dichloro or 3,4-Dichlorobenzenesulfonyl chloride $3,4\text{-}Cl_2C_6H_3SO_2Cl$	245.51	mcl pr		fr. p. 22.4				$B11^3$, 94
2441	Benzenesulfonyl chloride, 2,3-dimethyl or 2,3-Dimethylsulfonyl chloride $2,3\text{-}(CH_3)_2C_6H_3SO_2Cl$	204.67	pr (peth)		47			peth	B11, 120
2442	Benzenesulfonyl chloride, 3,4-dimethyl or 3,4-Dimethylbenzenesulfonyl chloride $3,4\text{-}(CH_3)_2C_6H_3SO_2Cl$	204.67	pr (peth)		51-2			eth	$B11^3$, 338
2443	Benzenesulfonyl chloride, 3,5-dimethyl or 3,5-Dimethylbenzenesulfonyl chloride $3,5\text{-}(CH_3)_2C_6H_3SO_2Cl$	204.67	nd (peth or bz)		94			eth	$B11^1$, 34
2444	Benzenesulfonyl chloride, 4-fluoro or 4-Fluorobenzenesulfonyl chloride $4\text{-}FC_6H_4SO_2Cl$	194.61	pl or nd	$95\text{-}6^{2.2}$	36			eth, bz, chl	$B11^3$, 88

No.	Name, Synonyms, and Formula	Mol. wt.	Color, crystalline form, specific rotation and λ_{max} (log ε)	b.p. °C	m.p. °C	Density	n_D	Solubility	Ref.
2445	Benzenesulfonyl chloride, 2-methoxy or 2-Methoxybenzenesulfonyl chloride $2\text{-}CH_3OC_6H_4SO_2Cl$	206.64	nd (peth)	126-9$^{0.1}$	56			eth, peth	B11, 235
2446	Benzenesulfonyl chloride, 4-methoxy or 4-Methoxybenzenesulfonyl chloride $4\text{-}CH_3OC_6H_4SO_2Cl$	206.64	nd or pr (bz)		42-3			al, eth, bz	B11[3], 504
2447	Benzenesulfonyl chloride, 2-nitro or 2-Nitrobenzenesulfonyl chloride $2\text{-}NO_2C_6H_4SO_2Cl$	221.62	pr (lig, eth-peth)		68-9			eth	B11[3], 114
2448	Benzenesulfonyl chloride, 3-nitro or 3-Nitrobenzenesulfonyl chloride $3\text{-}NO_2C_6H_4SO_2Cl$	221.62	mcl pr (eth), nd (lig)		64			al	B11[3], 126
2449	Benzenesulfonyl chloride, 4-nitro or 4-Nitrobenzenesulfonyl chloride $4\text{-}NO_2C_6H_4SO_2Cl$	221.62	mcl pr (peth, lig)		79-80			peth	B11[3], 136
2450	Benzenesulfonyl fluoride $C_6H_5SO_2F$	160.16		203-4, 90-1^{4}		1.3286$^{20/4}$	1.4932^{18}	al, eth	B11[3], 51
2451	1,3-Benzenedisulfonic acid, 4-amino $4\text{-}H_2NC_6H_3(SO_3H)_2\text{-}(1,3)$	253.24	nd (w + 2)		120d			w, al	B14[2], 470
2452	1,3-Benzenedisulfonic acid, 4-hydroxy $4\text{-}HOC_6H_3(SO_3H)_2\text{-}(1,3)$	254.23	nd (w)		>100d			w, al	B11[3], 522
2453	1,2,3,4-Benzenetetracarboxylic acid or Prehnitic acid $1,2,3,4\text{-}C_6H_2(COOH)_4$	254.15	pr (+ 6w)		241d			w, ace	B9[3], 4872
2454	1,2,3,4-Benzenetetracarboxylic acid, tetramethyl ester $1,2,3,4\text{-}C_6H_2(CO_2CH_3)_4$	310.26	nd (MeOH or w)		133.5			al, bz	B9[3], 4872
2455	1,2,3,5-Benzenetetracarboxylic acid, tetramethyl ester or Tetramethyl mellophanate $1,2,3,5\text{-}C_6H_2(CO_2CH_3)_4$	310.26	nd (MeOH, al) MeOH		111			al	B9[3], 4872
2456	1,2,4,5-Benzenetetracarboxylic acid or Puromellitic acid $1,2,4,5\text{-}C_6H_2(CO_2H)_4$	254.15	tcl pr (w + 2)		276 (anh), 242 (+ 2w)			al	B9[3], 4873
2457	1,2,4,5-Benzenetetracarboxylic acid, tetraethyl ester $1,2,4,5\text{-}C_6H_2(CO_2C_2H_5)_4$	366.37	nd (al)	sub	54			al	B9[2], 731
2458	1,2,4,5-Benzenetetracarboxylic acid, tetramethyl ester $1,2,4,5\text{-}C_6H_2(CO_2CH_3)_4$	310.26	lf (al)		143-4				B9[3], 4873
2459	1,2,3-Benzenetricarboxylic acid or Hemimellitic acid $1,2,3\text{-}C_6H_3(CO_2H)_3$	210.14	tcl pl (+ 2w), nd (w)		197d (anh), 223-4 (hyd)	1.546^{20}		w	B9[3], 4791
2460	1,2,4-Benzenetricarboxylic acid or Trimellitic acid $1,2,4\text{-}C_6H_3(CO_2H)_3$	210.24	nd (w), cr (aa or al)		238d			w, al, eth	B9[3], 4792
2461	1,2,4-Benzenetricarboxylic acid, 6-hydroxy $6\text{-}HOC_6H_2(CO_2H)_3\text{-}(1,2,4)$	226.14	pr (w + 2)		278-81d			w, al	B10, 580
2462	1,3,5-Benzenetricarboxylic acid or Trimesic acid $1,3,5\text{-}C_6H_3(CO_2H)_3$	210.14	pr or nd (w + 1)		380 (anh)			al, eth	B9[3], 4793
2463	1,3,5-Benzenetricarboxylic acid, 2-chloro $2\text{-}Cl\text{-}C_6H_2(CO_2H)_3\text{-}(1,3,5)$	244.59	nd or pl (w + 1)	sub	285 (anh)			al, eth, w	B9[2], 713
2464	1,3,5-Benzenetricarboxylic acid, 2-hydroxy $2\text{-}HOC_6H_2(CO_2H)_3\text{-}(1,3,5)$	226.14	pr (w + 1), nd (w + 2)		306 (anh)			al, w	B10, 580
2465	1,3,5-Benzenetricarboxylic acid, triethyl ester $1,3,5\text{-}C_6H_3(CO_2C_2H_5)_3$	294.30	pr or nd (al)		133-4			al, eth, bz, aa	B9[3], 4794
2466	1,3,5-Benzenetricarboxylic acid, trimethyl ester $1,3,5\text{-}C_6H_3(CO_2CH_3)_3$	252.22	nd (dil al)		144			al	B9[3], 4794
2467	1,3,5-Benzenetrisulfonic acid $1,3,5\text{-}C_6H_3(SO_3H)_3$	318.29	cr (w + 3)		d>100			w, al	B11[3], 483
2468	Benzhydrol or Diphenylmethanol $(C_6H_5)_2CHOH$	184.24	nd (lig)	297-8^{748}, 180^{20}	69			al, eth	B6[3], 3364
2469	Benzhydrol, *p*-amino $4\text{-}H_2NC_6H_4CH(OH)C_6H_5$	199.25	nd (w, bz)		121			al, ace	B13, 696
2470	Benzhydrol, 4,4′-dimethyl $(4\text{-}CH_3C_6H_4)_2CHOH$	212.39	nd (al)		69			al, eth, ace, chl	B6[3], 3422
2471	Benzhydrol, bis(2-dimethylamino)-α-phenyl $[2\text{-}(CH_3)_2NC_6H_4]_2C(OH)C_6H_5$	346.47	pr (lig)		105				B13, 741
2472	Benzhydrol, bis(3-dimethylamino)-α-phenyl $[3\text{-}(CH_3)_2NC_6H_4]_2C(OH)C_6H_5$	346.47	cr (eth)		128-9			eth	B13, 696

No.	Name, Synonyms, and Formula	Mol. wt.	Color, crystalline form, specific rotation and λ_{max} (log ε)	b.p. °C	m.p. °C	Density	n_D	Solubility	Ref.
2473	Benzhydrol, bis(4-dimethylamino)-α-phenyl $[4-(CH_3)_2NC_6H_4]_2C(OH)C_6H_5$	346.48	cr (eth, bz, lig, MeOH)	....	121-3	....	....	eth, bz	B13³, 1957
2474	Benzhydrol, 4-hydroxy or Benzaurin $4-HOC_6H_4CH(OH)C_6H_5$	200.24	ye red pw	....	110-20	....	....	....	B8³, 1644
2475	Benzhydrol, 4-methoxy $4-CH_3OC_6H_4CH(OH)C_6H_5$	214.27	nd (w, lig, dil, al)	....	66-8	....	....	al, bz, chl	B6³, 5417
2476	Benzhydrol, α-naphthyl $(C_6H_5)_2C(OH)$-α-$C_{10}H_7$	310.40	cr (lig, bz)	d	136.5	....	....	al, eth, bz	B6³, 3817
2477	Benzhydrol, β-naphthyl $(C_6H_5)_2C(OH)$-β-$C_{10}H_7$	310.40	pr (eth-lig)	....	118	....	....	al, eth, ace, bz	B6³, 3818
2478	Benzhydryl amine $(C_6H_5)_2CHNH_2$	183.25	hex pl	304^{763}, 176^{23}	34	$1.0635^{20/20}$	1.5963	bz	B12³, 3221
2479	o-Benzidine or 2,2′Diamino biphenyl $2-H_2NC_6H_4-C_6H_4NH_2-2$	184.24	mcl pr or nd (al)	162^{4}	81	....	....	w, bz	B13³, 410
2480	m-Benzidine or 3,3′-Diamino biphenyl $3-H_2NC_6H_4C_6NH_2-3$	184.24	nd (w), pr (bz)	205	93—4	....	....	eth, bz	B13³, 422
2481	p-Benzidine or 4,4′-Diamino biphenyl $(4-H_2NC_6H_4-)_2$	189.24	nd (w)	400^{740}	128	....	....	al	B12³, 425
2482	Benzil or Diphenylglyoxal $C_6H_5COCOC_6H_5$	210.23	ye pr (al)	346—8d, 188^{12}	95-6	$1.084^{102/4}$	....	al, eth, ace, bz	B7³, 3804
2483	Benzil dioxime (anti) or α Benzil dioxime $C_6H_5C{=}(NOH)C{=}(NOH)C_6H_5$	240.26	lf (al), or (ace)	....	238d	....	....	....	B7², 680
2484	Benzil dioxime (syn) or β-Benzil dioxime $C_6H_5C{=}(NOH)C{=}(NOH)C_6H_5$	240.26	nd (al)	....	207d	....	....	al, eth	B7², 681
2485	Benzil dioxime (amphi) or α-Benzil dioxime $C_6H_5-C{=}(NOH)C{=}(NOH)C_6H_5$	240.26	nd (al + 1), (aa)	....	164-6	....	....	al, ace, bz	B7¹, 681
2486	Benzil monoxime (α) $C_6H_5C{=}(NOH)COC_6H_5$	225.25	lf (dil al, bz)	200d	137-8	....	....	al, eth, aa, chl	B7², 678
2487	Benzil monoxime (β) $C_6H_5(C{=}NOH)COC_6H_5$	225.25	....	....	70 (+ ½ bz), 113-4	....	....	al, eth, ace, bz, chl	B7², 679
2488	Benzil osazone (anti) $C_{26}H_{22}N_4$	390.49	ye nd (al)	....	230-2	....	....	bz, chl	B15³, 114
2489	Benzil osazone (syn) $C_{26}H_{22}N_4$	390.49	ye nd (bz-al)	....	210	....	....	al, bz	B15³, 114
2490	Benzilic acid, α-hydroxydiphenyl acetic acid $(C_6H_5)_2C(OH)CO_2H$	228.25	mcl nd (w)	d180	151	....	....	al, eth	B10³, 1168
2491	Benzilic acid, ethyl ester or Ethyl benzilate $(C_6H_5)_2C(OH)COOC_2H_5$	256.30	pr or nd	201^{21}	34	....	1.5620^{20}	al, eth	B10³, 1171
2492	Benzilic acid, methyl ester or Methyl benzilate $(C_6H_5)_2C(OH)COOCH_3$	242.27	mcl or tcl cr (al)	187^{13}	75	....	....	....	B10³, 1170
2493	Benzimidazole or 1,3-Benzodiazole $C_7H_6N_2$	118.14	rh bipym pl (w)	>360	170.5	....	....	al	B23², 151
2494	Benzimidazole, 2-amino $2-H_2N-(C_7H_5N_2)$	133.15	pl (w)	....	224	....	....	w, al, ace	B24¹, 116
2495	Benzimidazole, 5,6-dimethyl $5,6-(CH_3)_2(C_7H_4N_2)$	146.19	(eth)	140^{2} sub	205-6	....	....	w, eth, al, chl	C50, 1087
2496	Benzimidazole, 2-hydroxy or o-Phenylene urea $2-HO(C_7H_5N_2)$	134.14	lf (w or al)	....	318d	....	....	al, ace	B24², 62
2497	Benzimidazole, 2-mercapto $2-HS(C_7H_5N_2)$	150.20	pl (dil al, aq NH_3)	....	298	....	....	al	B24², 65
2498	Benzimidazole, 1-methyl $1-CH_3(C_7H_5N_2)$	132.16	nd (peth), pl (al)	286^{756}	66	$1.1254^{20/4}$	1.6013^{7}	al, peth	B23², 152
2499	Benzimidazole, 2-methyl $2-CH_3(C_7H_5N_2)$	132.16	pr or nd (w)	....	176-7	....	....	w	B23², 160
2500	Benzimidazole, 2-methyl-5-nitro-1-phenyl $1-C_6H_5-2-CH_3-5-NO_2(C_7H_3N_2)$	253.26	nd (al)	....	170-1	....	....	al	B23², 161
2501	Benzimidazole, 4-methyl $4-CH_3(C_7H_5N_2)$	132.16	pl (w), nd (bz)	....	145	....	....	al, w	B23¹, 38
2502	Benzimidazole, 5-methyl $5-CH_3(C_7H_5N_2)$	132.16	(w)	....	114	....	....	w	B23², 157
2503	Benzimidazole, 6-nitro $6-NO_2(C_7H_5N_2)$	163.14	nd (w)	....	209—10	....	....	al	B23², 154
2504	Benzimidazole, 1-phenyl $1-C_6H_5(C_7H_5N_2)$ $C_{13}H_{10}N_2$	194.24	....	$210\text{-}2^{14}$	98	....	....	al, w	B23², 153

No.	Name, Synonyms, and Formula	Mol. wt.	Color, crystalline form, specific rotation and λ_{max} (log ε)	b.p. °C	m.p. °C	Density	n_D	Solubility	Ref.
2505	Benzimidazole, 2-phenyl $2\text{-}C_6H_5(C_7H_5N_2)$	194.24	pl (aa), (al-w), nd (bz w)		293			al, aa	B23^{2}, 238
2506	4,5-Benzindane or 1,2-Cyclopentanonaphthalene $C_{13}H_{12}$	168.24	oil	294-5, 170^{5}		1.066$^{20/4}$	1.6290^{20}	aa	B5^{4}, 1865
2507	1,2-Benzisothiazole C_7H_5NS	135.18		116-8^{18}					B27^{2}, 16
2508	1,2-Benzisothiazole, 5-hydroxy-3-phenyl $3\text{-}C_6H_5\text{-}5\text{-}HO(C_7H_3NS)$	227.28	nd (dil al, aa)		159-60			al, ace, bz, aa	B27^{2}, 87
2509	Benzisoxazole, 3-methyl C_8H_7NO	133.15		108-10^{16}				eth	B27^{2}, 19
2510	1,2-Benzocarbazole, 3,4-dihydro $C_{16}H_{13}N$	219.29	(lig or MeOH)		163-4			MeOH, lig	B20^{4}, 4187
2511	1,2-Benzo-1-cycloactene-3-one $C_{12}H_{14}O$	174.24		87-8$^{0.001}$			1.5577^{25}		AM76, 5462
2512	Benzodichlorofluoride $C_6H_5CCl_2F$	179.02		178-80		1.3138^{11}	1.5180^{11}	al	B5^{4}, 818
2513	5,6-Benzoflavone or β-Naphthoflavone $C_{19}H_{12}O_2$	272.30	nd (al)		167-8			eth, bz	B17^{4}, 5555
2514	7,8-Benzoflavone or α-Naphthoflavone $C_{19}H_{12}O_2$	272.30	ye pl (al), lf or nd (dil al)		157-9 (167)				B17^{4}, 5551
2515	3,4-Benzofluoranthene $C_{20}H_{12}$	252.32	nd (bz)		168				B5^{3}, 2516
2516	10,11-Benzofluoranthene $C_{20}H_{12}$	252.32	ye pl (al), nd (aa)		166				B5^{3}, 2517
2517	11,12-Benzofluoranthene $C_{20}H_{12}$	252.32	pa ye nd (bz)	480	217			al, bz, aa	B5^{3}, 2516
2518	1,2-Benzofluorene or Chrysofluorene $C_{17}H_{12}$	216.28	pl (ace or aa)	413	189-90			eth, chl	B5^{3}, 2288
2519	1,2-Benzofluorene, 9-phenyl $9\text{-}C_6H_5(C_{17}H_{11})$	292.38	nd (aa)		195.5			eth, bz	B5^{3}, 2558
2520	Benzofuran or Coumaron C_8H_6O	118.14		174, 62-3^{15}	<−18	1.0913^{25}	1.5615^{17}	al, eth	B17^{4}, 478
2521	Benzofuran, 2-acetyl $2\text{-}CH_3CO(C_8H_5O)$	160.17		136^{11}	76			w	B17^{4}, 5080
2522	Benzofuran, 2-benzoyl $2\text{-}C_6H_5CO(C_8H_5O)$	222.24		360	91				B17^{4}, 5426
2523	Benzofuran, 2-(chloromethyl)-2,3-dihydro $2\text{-}ClCH_2\text{-}2,3\text{-}H_2\text{-}(C_8H_5O)$	168.62		118-9^{11}	41-2	1.2196$^{7.5/16}$	1.5620$^{7.5}$	eth, ace	B17^{4}, 421
2524	Benzofuran, 2-methyl $2\text{-}CH_3(C_8H_5O)$	132.16		197-8, 93-4^{20}		1.0540$^{20/4}$	1.5495^{22}	al, eth	B17^{4}, 497
2525	Benzofuran, 3-methyl $3\text{-}CH_3(C_8H_5O)$	132.16		196-7^{742}, 86^{20}		1.0540$^{25/4}$	1.5536^{16}	al, eth	B17^{4}, 500
2526	Benzofuran, 5-methyl $5\text{-}CH_3(C_8H_5O)$	132.16		197-9, 83.5^{17}		1.0603^{19}	1.5570^{19}	al, eth	B12^{4}, 502
2527	Benzofuran, 7-methyl $7\text{-}CH_3(C_8H_5O)$	132.16		190-1		1.0400^{19}	1.5525^{19}	al, eth	B17^{4}, 503
2528	Benzoguananine or 2,4-Diamino-6-phenyl-1,3,5-triazine $2,4\text{-}(NH_2)_2\text{-}6\text{-}C_6H_5(C_3N_3)$	187.20	nd or pl (al)		226.4			al, eth	B26^{1}, 69
2529	Benzohydroxamic acid $C_6H_5CO(NHOH)$	137.14	rh ta, lf (eth)	exp	121—2			al, w	B9^{3}, 1304
2530	Benzohydroxamic acid, 2-hydroxy or Salicylhydroxamic acid $2\text{-}HO(C_6H_4CO(NHOH)$	153.14	nd (aa)		168			al, eth	B10^{3}, 160
2531	Benzoic acid $C_6H_5CO_2H$	122.12	mcl lf or nd	249, 133^{10}	122.13	1.0749^{130}, 1.2659$^{15/4}$	1.504^{12}	al, eth, ace, bz, chl	B9^{3}, 360
2532	Benzoic acid, 2-acetamido $2\text{-}(CH_3CONH)C_6H_4CO_2H$	179.18	nd (aa)		185			eth, ace, bz	B14^{3}, 922
2533	Benzoic acid, 3-acetimido $3\text{-}CH_3CONHC_6H_4CO_2H$	179.18	nd (al)		248-50			al	B14^{2}, 241
2534	Benzoic acid, 4-acetamido $4\text{-}(CH_3CONH)C_6H_4CO_2H$	179.18	nd (aa)		256.5			al	B14^{3}, 1112
2535	Benzoic acid, (4-acetimidophenyl) ester $C_6H_5CO_2(C_6H_4NHCOCH_3\text{-}4)$	255.28	nd (al)		171			al, bz, aa	B13^{3}, 1065
2536	Benzoic acid, 2-acetyl $2\text{-}CH_3CO\text{-}C_6H_4CO_2H$	164.16	nd (w), pr (bz)	110-2^{2}	114-5			al, w	B10^{3}, 102

PHYSICAL CONSTANTS OF ORGANIC COMPOUNDS (Continued)

No.	Name, Synonyms, and Formula	Mol. wt.	Color, crystalline form, specific rotation and λ_{max} (log ε)	b.p. °C	m.p. °C	Density	n_D	Solubility	Ref.
2537	Benzoic acid (-2-acetylphenyl) ester $C_6H_5CO_2(C_6H_4COCH_3\text{-}2)$	240.26	nd (al)	....	88	....	....	al, eth, aa	$B9^3$, 729
2538	Benzoic acid, 3-acetyl $3\text{-}CH_3COC_6H_4CO_2H$	164.16	....	....	210	....	....	w	$B10^3$, 248
2539	Benzoic acid, 4-acetyl $4\text{-}CH_3COC_6H_4CO_2H$	164.16	nd (w)	sub	210	....	....	w	$B10^3$, 294
2540	Benzoic acid, 4-acetyl, methyl ester $4\text{-}CH_3COC_6H_4CO_2CH_3$	178.19	nd (w)	$140\text{-}5^{4}$ sub	95	....	....	w	B10, 695
2541	Benzoic acid, (4-acetylphenyl) ester $C_6H_5CO_2(C_6H_4COCH_3\text{-}4)$	240.26	nd (al, aq MeOH)	....	135-6	....	....	al, eth, bz, aa	$B9^3$, 730
2542	Benzoic acid, allyl ester or Allylbenzoate $C_6H_5CO_2CH_2CH{=}CH_2$	162.19	....	242	......	$1.0578^{15/15}$	1.5178^{20}	al, eth, ace	$B9^3$, 402
2543	Benzoic acid, 5-allyl-2-hydroxy-3-methoxy or Eugenic acid .... $5\text{-}(CH_2{=}CHCH_2)\text{-}2\text{-}HO\text{-}3\text{-}CH_3OC_6H_2CO_2H$	208.21	pl (w + l)	....	85—8 (+lw), 127 (anh)	....	....	al, eth	$B10^3$, 1855
2544	Benzoic acid, 2-amino or Anthranilic acid $2\text{-}H_2NC_6H_4CO_2H$	137.14	lf (al)	sub	146-7	1.412^{20}	....	w, al, eth, chl	$B14^3$, 879
2545	Benzoic acid, anhydride or Benzoic anhydride $(C_6H_5CO)_2O$	226.23	pr (eth)	360	42-3	$1.989^{15/4}$	1.5767^{15}	al, eth	$B9^3$, 852
2546	Benzoic acid, 2-benzamido $2\text{-}(C_6H_5CONH)C_6H_4CO_2H$	241.25	nd (al or dz)	....	181	....	....	al, eth	$B14^3$, 925
2547	Benzoic acid, 3-Benzamido $3\text{-}(C_6H_5CONH)C_6H_4CO_2H$	241.25	red pr (al)	....	252-3	$1.510^{4/4}$	....	al	$B14^1$, 562
2548	Benzoic acid, 4-benzamido $4\text{-}(C_6H_5CONH)C_6H_4CO_2H$	241.25	nd (al)	....	278	....	....	al, eth	$B14^1$, 577
2549	Benzoic acid, 2-Benzoyl or 2-Benzophenone carboxylic acid .... $2\text{-}(C_6H_5CO)C_6H_4CO_2H$	226.23	tcl nd (w + l)	....	127-9 (anh)	....	....	al, eth, bz	$B10^3$, 3289
2550	Benzoic acid, 2-benzoyl-4-chloro $2\text{-}C_6H_5CO\text{-}4\text{-}ClC_6H_3CO_2H$	260.68	nd (MeOH)	....	92	....	....	al	$B10^3$, 3296
2551	Benzoic acid, 2-benzoyl, ethyl ester $2\text{-}(C_6H_5CO)C_6H_4CO_2C_2H_5$	254.28	rh pl (dil al)	....	58	$1.221^{64/4}$	1.560^{64}	al, eth	$B10^3$, 3297
2552	Benzoic acid, 2-benzoyl, methyl ester $2\text{-}(C_6H_5CO)C_6H_4CO_2CH_3$	240.26	pl or mcl pr (dil al)	350-2	52	$1.1903^{19/4}$	1.591^{20}	al, eth	$B10^3$, 3291
2553	Benzoic acid, benzoylmethyl ester $C_6H_5CO_2(CH_2COC_6H_5)$	240.26	pl (dil al)	....	118.5	....	....	eth, bz, chl, al	$B9^3$, 730
2554	Benzoic acid, 3-benzoyl or 3-Benzophenonecarboxylic acid .... $3\text{-}(C_6H_5CO)C_6H_4CO_2H$	226.23	nd (w), fl (dil al)	sub	161-2	....	....	al, eth	$B10^3$, 3304
2555	Benzoic acid, 4-Benzoyl or 4-Benzophenonecarboxylic acid .... $4\text{-}(C_6H_5CO)C_6H_4CO_2H$	226.23	nd (dil aa), pl (al), mcl lf (w)	sub	198—200 (226—7)	....	....	al, eth, aa	$B10^3$, 3305
2556	Benzoic acid, benzyl ester or Benzyl benzoate $C_6H_5CO_2CH_2C_6H_5$	212.25	nd or lf	323—4 (cor), $170\text{—}1^{11}$	21	$1.1121^{25/4}$	1.5680^{20}	al, eth, ace, bz	$B9^3$, 428
2557	Benzoic acid, 2-benzyl $2\text{-}(C_6H_5CH_2)C_6H_4CO_2H$	212.25	nd (dil al)	sub	118	....	....	al, eth, bz, chl	$B9^3$, 3317
2558	Benzoic acid, 3-benzyl $3\text{-}(C_6H_5CH_2)C_6H_4CO_2H$	212.25	nd (w), lf (dil al)	sub	157-8	....	....	al, eth, bz, chl	$B9^3$, 676
2559	Benzoic acid, 4-benzyl $4\text{-}(C_6H_5CH_2)C_6H_4CO_2H$	212.25	nd (w), lf (dil al)	sub	157-8	....	....	al, eth, bz, chl	$B9^3$, 3319
2560	Benzoic acid, 4-(benzylsulfonamido) or Caronamide $4\text{-}(C_6H_5CH_2SO_2NH)C_6H_4CO_2H$	291.33	....	....	229-30	....	....	al	C45, 3418
2561	Benzoic acid, 2-bromo or 2-Bromobenzoic acid $2\text{-}BrC_6H_4CO_2H$	201.02	mcl pr (w), nd	sub	150	$1.929^{25/4}$	....	al, eth, bz, chl	$B9^3$, 1383
2562	Benzoic acid, 2-bromo, anhydride $(2\text{-}BrC_6H_4CO)_2O$	384.02	nd (al)	....	79.6	....	....	al, bz, chl	$B9^1$, 142
2563	Benzoic acid, 2-bromo-3-chloro $2\text{-}Br\text{-}3\text{-}ClC_6H_3CO_2H$	235.46	cr (bz)	....	224-6	....	....	bz	B9, 355
2564	Benzoic acid, 2-bromo-5-chloro $2\text{-}Br\text{-}5\text{-}ClC_6H_3CO_2H$	235.46	cr (bz)	....	153	....	....	al, bz	$B9^3$, 1426
2565	Benzoic acid, 2-bromo-3,5-dinitro $3,5\text{-}(NO_2)_2\text{-}2\text{-}BrC_6H_2CO_2H$	291.01	ye nd (w)	....	213	....	....	al, bz, aa	$B9^3$, 1954
2566	Benzoic acid, 2-bromo, ethyl ester $2\text{-}BrC_6H_4CO_2C_2H_5$	229.07	....	254-5, 135^{15}	......	$1.4438^{15/4}$	1.5455^{15}	al, eth, ace, bz	$B9^3$, 1385
2567	Benzoic acid, 2-bromo, methyl ester $2\text{-}BrC_6H_4CO_2CH_3$	215.05	....	244, 122^{17}	......	....	....	al	$B9^3$, 1385

No.	Name, Synonyms, and Formula	Mol. wt.	Color, crystalline form, specific rotation and λ_{max} (log ε)	b.p. °C	m.p. °C	Density	n_D	Solubility	Ref.
2568	Benzoic acid, 2-bromo-3-nitro $3-NO_2-2-BrC_6H_3CO_2H$	246.02	(dil al)	...	191	...	...	al	$B9^2$, 277
2569	Benzoic acid, 2-bromo-4-nitro $4-NO_2-2-BrC_6H_3CO_2H$	246.02	nd (w or dil al)	sub>155	166-7	...	...	al, eth	$B9^3$, 1771
2570	Benzoic acid, 2-bromo-5-nitro $5-NO_2-2-BrC_6H_3CO_2H$	246.02	nd (w)	sub	180-1	...	...	al, eth	$B9^3$, 1771
2571	Benzoic acid, 3-bromo $3-BrC_6H_4CO_2H$	201.02	mcl nd (dil al)	>280	155	1.845^{20}	k ...	al, eth	$B9^3$, 1392
2572	Benzoic acid, 3-bromo-2-chloro $3-Br-ClC_6H_3CO_2H$	235.46	cr (al)	...	165	...	...	al	$B9^3$, 1426
2573	Benzoic acid, 3-bromo-4-chloro $3-Br-4-ClC_6H_3CO_2H$	235.46	pl (dil aa), cr (al)	...	215-6	...	...	al, aa	$B9^3$, 1427
2574	Benzoic acid, 3-bromo-6-chloro $3-Br-6-ClC_6H_3CO_2H$	235.46	nd (w), cr (aa)	...	155-6	...	...	al, aa	$B9^3$, 1427
2575	Benzoic acid, 3-bromo, ethyl ester or Ethyl-3-bromobenzoate $3-BrC_6H_4CO_2C_2H_5$	229.07	...	261, 133^{15}	...	$1.4308^{19/4}$	1.5430^{19}	al, eth, ace, bz	$B9^3$, 1393
2576	Benzoic acid, 3-bromo, methyl ester $3-BrC_6H_4CO_2CH_3$	215.05	pl	122.5^{15}	32	...	...	al, eth	$B9^3$, 1393
2577	Benzoic acid, 3-bromo-2-nitro $2-NO_2-3-BrC_6H_3CO_2H$	246.02	(eth)	...	250	...	...	eth, bz	$B9^3$, 1770
2578	Benzoic acid, 3-bromo-4-nitro $4-NO_2-3-BrC_6H_3CO_2H$	246.02	nd (dil al)	...	197	...	...	al, eth, chl	B9, 408
2579	Benzoic acid, 3-bromo-5-nitro $5-NO_2-3-BrC_6H_3CO_2H$	246.02	nd (w, bz or eth), pl (al)	...	159-60	...	...	al, eth, bz, chl	$B9^3$, 1771
2580	Benzoic acid, 4-bromo $4-BrC_6H_4CO_2H$	201.02	nd (eth), lf (w), mcl pr	...	254.5	1.894^{20}	...	al, eth	$B9^3$, 1403
2581	Benzoic acid, 4-bromo-3-chloro $4-Br-3-ClC_6H_3CO_2H$	235.46	pl (dil aa), cr (al)	...	218	...	...	al, aa	$B9^2$, 236
2582	Benzoic acid, 4-bromo, ethyl ester or Ethyl-4-bromo benzoate $4-BrC_6H_4CO_2C_2H_5$	229.07	...	262^{737}, 125^{15}	...	$1.4332^{17/4}$	1.5438^{17}	al, eth, ace, bz	$B9^3$, 1405
2583	Benzoic acid, 4-bromo, methyl ester or Methyl-4-bromobenzoate $4-BrC_6H_4CO_2CH_3$	215.05	lf (dil al), nd (eth)	...	81	1.689	...	al, eth, ace, bz, chl	$B9^3$, 1405
2584	Benzoic acid, 4-bromo-2-nitro $2-NO_2-4-BrC_6H_3CO_2H$	246.02	nd (w)	...	163	...	...	al, eth, bz, chl	$B9^3$, 1770
2585	Benzoic acid, 4-bromo-3-nitro $3-NO_2-4-BrC_6H_3CO_2H$	246.02	nd (dil aa)	sub	203-4	...	...	al	$B9^3$, 1770
2586	Benzoic acid, 5-bromo-2-nitro $2-NO_2-5-BrC_6H_3CO_2H$	246.02	cr (w, al, bz, to)	...	140	1.920^{18}	...	al, bz	$B9^1$, 165
2587	Benzoic acid, butyl ester or Butylbenzoate $C_6H_5CO_2C_4H_9$	178.23	...	250.3	-22.4	1.000^{20}	1.4940^{25}	al, eth, ace	$B9^3$, 392
2588	Benzoic acid, *iso*-butyl ester or Isobutyl benzoate $C_6H_5CO_2\text{-}i\text{-}C_4H_9$	178.23	...	242	...	$0.9990^{20/4}$	...	al, eth, ace	$B9^3$, 394
2589	Benzoic acid, 2-*tert*-butyl $2-[(CH_3)_3C]C_6H_4CO_2H$	178.23	pl (dil al)	...	80-1	...	...	al	$B9^2$, 365
2590	Benzoic acid, 3-*tert*-butyl $3-[(CH_3)_3C]C_6H_4CO_2H$	178.23	nd (peth)	...	127-8	...	...	al, peth	$B9^3$, 2525
2591	Benzoic acid, 4-*tert*-butyl $4-[(CH_3)_3C]C_6H_4CO_2H$	178.23	nd (dil al)	...	164-5	...	...	al, bz	$B9^3$, 2525
2592	Benzoic acid, (1-chloroethyl) ester or (1-Chloroethyl) benzoate $C_6H_5CO_2CHClCH_3$	184.62	...	134^{30}	...	1.172^{20}	...	al, eth	$B9^2$, 127
2593	Benzoic acid, (2-chloroethyl) ester or (2-Chloroethyl) benzoate $C_6H_5CO_2CH_2CH_2Cl$	184.62	...	254^{729}	$118-20^2$	...	...	al, eth	$B9^3$, 388
2594	Benzoic acid, 2-chloro $2-ClC_6H_4CO_2H$	156.57	mcl pr (w)	sub	142	1.544^{20}	...	al, eth, ace, bz	$B9^3$, 1330
2595	Benzoic acid, 2-chloro, ethyl ester $2-ClC_6H_4CO_2C_2H_5$	184.62	...	243, $122-5^{15}$	...	$1.1942^{15/4}$	1.5247^{15}	al, eth	$B9^3$, 1333
2596	Benzoic acid, 2-chloro, methyl ester $2-ClC_6H_4CO_2CH_3$	170.60	...	$234-5^{762}$	...	...	...	al	$B9^3$, 1333
2597	Benzoic acid, 2-chloro-4-methyl $2-Cl-4-CH_3C_6H_3CO_2H$	170.60	nd (al)	...	155-6	...	...	al, eth, bz, chl	B9, 497

No.	Name, Synonyms, and Formula	Mol. wt.	Color, crystalline form, specific rotation and λ_{max} (log ε)	b.p. °C	m.p. °C	Density	n_D	Solubility	Ref.
2598	Benzoic acid, 2-chloro-5-methyl $2\text{-Cl-5-}CH_3C_6H_3CO_2H$	170.60	nd (w or al)		167			w, al	B9[3], 2327
2599	Benzoic acid, 2-chloro-6-methyl $2\text{-Cl-6-}CH_3C_6H_3CO_2H$	170.60	nd (w)		102				B9[3], 2310
2600	Benzoic acid, 2-chloro-4-nitro $4\text{-}NO_2\text{-2-}ClC_6H_3CO_2H$	201.57	nd (w)		140-2			al, eth, w	B9[3], 1766
2601	Benzoic acid, 2-chloro-5-nitro $5\text{-}NO_2\text{-2-}ClC_6H_3CO_2H$	201.57	nd or pr (w)		165	1.608^{18}		al, eth, bz	B9[3], 1765
2602	Benzoic acid, 3-chloro $3\text{-}ClC_6H_4CO_2H$	156.57	pr (w)	sub	158	$1.496^{25/4}$		al, eth	B9[3], 1345
2603	Benzoic acid, 3-chloro, anhydride $(3\text{-}ClC_6H_4CO)_2O$	295.14	nd (al or peth)		95.5			al, bz, chl	B9[2], 224
2604	Benzoic acid, 3-chloro, ethyl ester $3\text{-}ClC_6H_4CO_2C_2H_5$	184.62		243, 121^{20}		$1.1859^{15/4}$	1.5223^{20}	al, eth	B9[3], 1347
2605	Benzoic acid, 3-chloro, methyl ester $3\text{-}ClC_6H_4CO_2CH_3$	170.60		231, 114^{18}	21			al	B9[3], 1346
2606	Benzoic acid, 3-chloro-2-methyl $3\text{-Cl-2-}CH_3C_6H_3CO_2H$	170.60	nd (al)		159			al, eth	B9[3], 2309
2607	Benzoic acid, 3-chloro-4-methyl $3\text{-Cl-4-}CH_3C_6H_3CO_2H$	170.60	nd or lf (dil al)		200—2			al	B9[3], 2355
2608	Benzoic acid, 3-chloro-5-methyl $3\text{-Cl-5-}CH_3C_6H_3CO_2H$	170.60	nd (dil al)		178			al	B9, 479
2609	Benzoic acid, 3-chloro-2-nitro $2\text{-}NO_2\text{-3-}ClC_6H_3CO_2H$	201.57	nd or pl (w)		237-9	1.566^{18}		al, eth	B9, 400
2610	Benzoic acid, 3-chloro-5-nitro $5\text{-}NO_2\text{-3-}ClC_6H_3CO_2H$	201.57	nd (w)		147			al, eth, aa	B9[1], 165
2611	Benzoic acid, 4-chloro $4\text{-}ClC_6H_4CO_2H$	156.57	tcl pr (al-eth)		243			al	B9[3], 1354
2612	Benzoic acid, 4-chloro, anhydride $(4\text{-}ClC_6H_4CO)_2O$	295.14	nd or lf (bz)		193-4				B9[3], 1361
2613	Benzoic acid, 2-(4-chlorobenzoyl) $2\text{-}(4\text{-}ClC_6H_4CO)\text{-}C_6H_4CO_2H$	260.68	cr (bz, aa)		150			al, eth, bz	B10[2], 518
2614	Benzoic acid, 4-chloro, ethyl ester $4\text{-}ClC_6H_4CO_2C_2H_5$	184.62		237-8, 122^{15}				al	B9[3], 1356
2615	Benzoic acid, 4-chloro, methyl ester $4\text{-}ClC_6H_4CO_2CH_3$	170.60	nd or mcl pr		44	1.382^{20}		al	B9[3], 1356
2616	Benzoic acid, 4-chloro-2-methyl $4\text{-Cl-2-}CH_3C_6H_3CO_2H$	170.60	nd (w, al, dil aa, bz)		173			al, aa	B9[2], 321
2617	Benzoic acid, 4-chloro-3-methyl $4\text{-Cl-3-}CH_3C_6H_3CO_2H$	170.60	nd (w)		209-10				B9[3], 2327
2618	Benzoic acid, 4-chloro-2-nitro $2\text{-}NO_2\text{-4-}ClC_6H_3CO_2H$	201.57	pl (bz-lig), pr (bz), nd (w)		142.3			al, eth, bz, w	B9[3], 1763
2619	Benzoic acid, 4-chloro-3-nitro $3\text{-}NO_2\text{-4-}ClC_6H_3CO_2H$	201.57	nd or pl (w)		181-2	1.645^{18}			B9[3], 1764
2620	Benzoic acid, 4-chloro-3-nitro, ethyl ester $3\text{-}NO_2\text{-4-}ClC_6H_3CO_2C_2H_5$	229.60	ye nd (al)		59			al, bz, aa	B9, 402
2621	Benzoic acid, 4-chloro-3-nitro, methyl ester $3\text{-}NO_2\text{-4-}ClC_6H_3CO_2CH_3$	215.57	nd (MeOH)		83	1.522^{18}		al	B9, 402
2622	Benzoic acid, 5-chloro-2-methyl $5\text{-Cl-2-}CH_3C_6H_3CO_2H$	170.60	nd (al)		168-9			al	B9[2], 320
2623	Benzoic acid, 5-chloro-2-nitro, methyl ester $2\text{-}NO_2\text{-5-}ClC_6H_3CO_2CH_3$	215.57	pl (MeOH)		48.5	1.453^{18}		MeOH	B9, 401
2624	Benzoic acid, cyclohexyl ester or Cyclohexylbenzoate $C_6H_5CO_2C_6H_{11}$	204.27		285	<-10	1.0429^{20}	1.5200^{20}	al, eth	B9[3], 404
2625	Benzoic acid, 3-cyano or Isophthalic acid mononitrile $3\text{-}NCC_6H_4CO_2H$	147.13	nd (w)	sub d	217			al, eth	B9[3], 4243
2626	Benzoic acid, 2,3-diamino $2,3\text{-}(H_2N)_2C_6H_3CO_2H$	152.15	nd (dil al)	d	190-1d			al, aa	B14[3], 1172
2627	Benzoic acid, 2,4-diamino $2,4\text{-}(H_2N)_2C_6H_3CO_2H$	152.15		>200d	140			al, aa	B14, 448
2628	Benzoic acid, 2,5-diamino $2,5\text{-}(H_2N)_2C_6H_3CO_2H$	152.15	br pr (w)		darkens				B14, 448
2629	Benzoic acid, 3,4-diamino $3,4\text{-}(H_2N)_2C_6H_3CO_2H$	152.15	lf (w)		215-8d				B14[3], 1177

No.	Name, Synonyms, and Formula	Mol. wt.	Color, crystalline form, specific rotation and λ_{max} (log ε)	b.p. °C	m.p. °C	Density	n_D	Solubility	Ref.
2630	Benzoic acid, 3,5-diamino $3,5-(H_2N)_2C_6H_3CO_2H$	152.15	nd (+ lw)		240 (rapid)			al, eth	$B14^3$, 1179
2631	Benzoic acid, 2,3-dibromo $2,3-Br_2C_6H_3CO_2H$	279.92	nd (w)		149-50				$B9^1$, 146
2632	Benzoic acid, 2,4-dibromo $2,4-Br_2C_6H_3CO_2H$	279.92	lf (w)	sub	174			al, eth	$B9^2$, 237
2633	Benzoic acid, 2,5-dibromo $2,5-Br_2C_6H_3CO_2H$	279.92	nd (al or w)	sub	157			al, eth, chl, aa	$B9^3$, 1428
2634	Benzoic acid, 2,6-dibromo $2,6-Br_2C_6H_3CO_2H$	279.92	nd (w), (lig)	$209—10^{16}$	150-1			al, eth, ace, chl	$B9^3$, 1428
2635	Benzoic acid, 2,6-dibromo-3,4,5-trihydroxy or Gallobromol $2,6-Br_2C(OH)_3CO_2H$	327.91	nd, pr or lf (w + 1)		139 (hyd)			w, al, eth	$B10^2$, 347
2636	Benzoic acid, 3,4-dibromo $3,4-Br_2C_6H_3CO_2H$	279.92	nd (w), pl (al)		234-5			al, eth, MeOH	$B9^3$, 1428
2637	Benzoic acid, 2-(*N,N*-dichlorosulfamyl) $2-(Cl_2NO_2S)C_6H_4CO_2H$	270.09	yesh gr pl (chl)		146-8 exp				B11, 377
2638	Benzoic acid, 4-(*N,N*-dichlorosulfamyl) or Halazone $4-(Cl_2NO_2S)C_6H_4CO_2H$	270.09	pr (aa)		213			aa	$B11^2$, 230
2639	Benzoic acid, 2,3-dichloro $2,3-Cl_2C_6H_3CO_2H$	191.01	nd (w)		168.3			al, eth, w	$B9^2$, 228
2640	Benzoic acid, 2,4-dichloro $2,4-Cl_2C_6H_3CO_2H$	191.01	nd (w or bz)	sub	164.2			al, eth, bz, chl	$B9^3$, 1374
2641	Benzoic acid, 2,5-dichloro $2,5-Cl_2C_6H_3CO_2H$	191.01	nd (w)	301	154.4			al, eth, w	$B9^3$, 1376
2642	Benzoic acid, 2,6-dichloro $2,6-Cl_2C_6H_3CO_2H$	191.01	nd (al), pr (w)	sub	144			al, eth, bz, w	$B9^3$, 1377
2643	Benzoic acid, 3,4-dichloro $3,4-Cl_2C_6H_3CO_2H$	191.01	nd (w, al, bz)		208-9			al, eth, w	$B9^3$, 1378
2644	Benzoic acid, 3,5-dichloro $3,5-Cl_2C_6H_3CO_2H$	191.01	nd (al or w)	sub	188			al, eth	$B9^3$, 1379
2645	Benzoic acid, 2,3-dihydroxy or *o*-Pyrocatechuic acid $2,3(HO)_2C_6H_3CO_2H$	154.12	pr or nd (w + 1)		204 (anh)	$1.542^{20/4}$		w, al, eth	$B10^3$, 1363
2646	Benzoic acid, 2,4-dihydroxy or β-Resorcylic acid $2-4-(HO)_2C_6H_3CO_2H$	154.12	cr + w (w)					al, eth, bz	$B10^3$, 1370
2647	Benzoic acid, 2,4-dihydroxy-6-methyl or 4,6-dihydroxy-*o*-toluic acid $2,4(HO)_2-6-CH_3C_6H_2CO_2H$	168.15	nd (dil aa + 1w)		176d			al, eth	$B10^2$, 272
2648	Benzoic acid, 2,4-dihydroxy-6-methyl, ethyl ester $6-CH_3-2,4-(HO)_2C_6H_2CO_2C_2H_5$	196.20	lf (aa), pr (al)	sub	132			al, eth	$B10^3$, 1482
2649	Benzoic acid, 2,4-dihydroxy-6-pentyl $6-C_5H_{11}-2,4-(HO)_2C_6H_2CO_2H$	224.26	wh nd		147			al, eth	$B10^3$, 1577
2650	Benzoic acid, 2,5-dihydroxy or Gentisic acid $2,5-(HO)_2C_6H_3CO_2H$	154.12	nd or pr (w)		205			w, al, eth	$B10^3$, 1384
2651	Benzoic acid, 2,6-dihydroxy or γ-Resorcylic acid $2,6-(HO)_2C_6H_3CO_2H$	154.12	nd (+ w)		167d			al, eth, w	$B10^3$, 1401
2652	Benzoic acid, 3,4-dihydroxy or Protocatechuic acid $3,4-(HO)_2C_6H_3CO_2H$	154.12	mcl nd (w + 1)		200-2d	1.524^4		w, al, eth	$B10^3$, 1403
2653	Benzoic acid, 3,5-dihydroxy or α-Resorcylic acid $3,5-(HO)_2C_6H_3CO_2H$	154.12	pr or nd		238-40			al, eth, w	$B10^3$, 1446
2654	Benzoic acid, 3,4-dimethoxy or Veratric acid $3,4-(CH_3O)_2C_6H_3CO_2H$	182.18	nd (w or aa), rh (sub)	sub	181-2 (sub)			al, eth	$B10^3$, 1404
2655	Benzoic acid, 2,3-dimethyl or Hemimellitic acid $2,3-(CH_3)_2C_6H_3CO_2H$	150.18	pr (al)		144			al, eth	$B9^3$, 2434
2656	Benzoic acid, 2(dimethylamino) or *N,N*-Dimethylanthranilic acid $2-(CH_3)_2NC_6H_4CO_2H$	165.19	pr nd (eth)	sub d	72			w, al, eth	$B14^3$, 896
2657	Benzoic acid, 3-(dimethylamino) $3-(CH_3)_2NC_6H_4CO_2H$	165.19	nd (w)		151			w, al, eth	$B14^3$, 1002
2658	Benzoic acid, 4-(dimethylamino) $4-(CH_3)_2NC_6H_4CO_2H$	165.19	nd (al)		242-3			al	$B14^3$, 1082
2659	Benzoic acid, 2,4-dimethyl or 2,4-Xylylic acid $2,4-(CH_3)_2C_6H_3CO_2H$	150.18	mcl or tcl nd (w)	267^{727}, sub	127 (anh), 90 (hyd)			al, ace, bz	$B9^2$, 350
2660	Benzoic acid, 2,5-dimethyl or Isoxylylic acid $2,5-(CH_3)_2C_6H_3CO_2H$	150.18	nd (al)	268 sub	132	$1.069^{21/4}$		al, eth, ace, bz	$B9^1$, 210

PHYSICAL CONSTANTS OF ORGANIC COMPOUNDS (Continued)

No.	Name, Synonyms, and Formula	Mol. wt.	Color, crystalline form, specific rotation and λ_{max} (log ε)	b.p. °C	m.p. °C	Density	n_D	Solubility	Ref.
2661	Benzoic acid, 2,6-dimethyl or 2,6-Xylylic acid $2,6\text{-}(CH_3)_2C_6H_3CO_2H$	150.18	nd (lig)	274-5	116			al, eth	$B9^{3}$, 2435
2662	Benzoic acid, 3,4-dimethyl or Paraxylylic acid $3,4\text{-}(CH_3)_2C_6H_3CO_2H$	150.18	pr (al)		166			al, eth, bz	$B9^{3}$, 2441
2663	Benzoic acid, 3,5-dimethyl or Mesitylenic acid $3,5\text{-}(CH_3)_2C_6H_3CO_2H$	150.18	nd (w or al)	sub	170-1			al, eth	$B9^{3}$, 2444
2664	Benzoic acid, 2,4-dinitro $2,4\text{-}(NO_2)_2C_6H_3CO_2H$	212.12	nd (w)						$B9^{3}$, 1776
2665	Benzoic acid, 2,4-dinitro-3-hydroxy $2,4\text{-}(NO_2)_2\text{-}3\text{-}HOC_6H_2CO_2H$	228.12	(w)		204			al, eth	$B10^{3}$, 264
2666	Benzoic acid, 2,5-dinitro $2,5\text{-}(NO_2)_2C_6H_3CO_2H$	212.12	pr (w)		177			al, eth	$B9^{3}$, 1778
2667	Benzoic acid, 2,6-dinitro $2,6\text{-}(NO_2)_2C_6H_3CO_2H$	212.12	nd (w)		202.3			al, eth	$B9^{3}$, 1778
2668	Benzoic acid, 3,4-dinitro $3,4\text{-}(NO_2)_2C_6H_3CO_2H$	212.12	nd (w)		165			al, eth	$B9^{3}$, 177
2669	Benzoic acid, 3,5-dinitro $3,5\text{-}(NO_2)_2C_6H_3CO_2H$	212.12	mcl pr (al)		205			al, aa	$B9^{3}$, 1779
2670	Benzoic acid, 3,5-dinitro, benzyl ester $3,5\text{-}(NO_2)_2C_6H_3CO_2CH_2C_6H_5$	302.25	nd (lig)		112				$B9^{3}$, 1848
2671	Benzoic acid, 3,5-dinitro, butyl ester $3,5\text{-}(NO_2)_2C_6H_3CO_2C_4H_9$	268.23	mcl nd (al)		62.5			al	$B9^{3}$, 1782
2672	Benzoic acid, 3,5-dinitro, isobutyl ester $3,5\text{-}(NO_2)_2C_6H_3CO_2\text{-}i\text{-}C_4H_9$	268.23			87-8			w, al	$B9^{2}$, 281
2673	Benzoic acid, 3,5-dinitro, ethyl ester $3,5\text{-}(NO_2)_2C_6H_3CO_2C_2H_5$	240.18	nd (al)		72.9	1.295^{111}	1.560	al	$B9^{3}$, 1781
2674	Benzoic acid, 3,5-dinitro-2-hydroxy $3,5\text{-}(NO_2)_2\text{-}2\text{-}HOC_6H_2CO_2H$	228.12	ye nd or pl (+1w)		182 (anh)			w, al, eth, bz	$B10^{3}$, 207
2675	Benzoic acid, 3,5-dinitro-4-hydroxy $3,5\text{-}(NO_2)_2\text{-}4\text{-}HOC_6H_2CO_2H$	228.12	ye lf (al)		248-9			al, eth	$B10^{3}$, 383
2676	Benzoic acid, 3,5-dinitro, methyl ester $3,5\text{-}(NO_2)_2C_6H_3CO_2CH_3$	226.15	nd (w)		112			w, al	$B9^{3}$, 1781
2677	Benzoic acid, 3,5-dinitro, pentyl ester $3,5\text{-}(NO_2)_2C_6H_3CO_2C_5H_{11}$	282.25			46.4			al	$B9^{2}$, 281
2678	Benzoic acid, 3,5-dinitro, phenyl ester $3,5\text{-}(NO_2)_2C_6H_3CO_2C_6H_5$	288.22	rods (al)		145-6			al, bz	$B9^{3}$, 1846
2679	Benzoic acid, 3,5-dinitro, propyl ester $3,5\text{-}(NO_2)_2C_6H_3CO_2C_3H_7$	254.20	mcl pl (al)		73			al	$B9^{3}$, 1781
2680	Benzoic acid, 3,5-dinitro, isopropyl ester $3,5\text{-}(NO_2)_2C_6H_3CO_2\text{-}i\text{-}C_3H_7$	254.20	nd (al)		122			al	$B9^{3}$, 1782
2681	Benzoic acid, 3,5-dinitro, tetrahydrofurfuryl ester $3,5\text{-}(NO_2)_2C_6H_3CO_2CH_2(C_4H_7O)$	296.24	nd (al)		83-4			al	$B17^{4}$, 1106
2682	Benzoic acid, 2-ethoxy $2\text{-}C_2H_5OC_6H_4CO_2H$	166.18		$211\text{-}2^{39}$	20.7				$B10^{3}$, 98
2683	Benzoic acid, (2-ethoxyethyl)ester or Ethylcellosolve benzoate $C_6H_5CO_2CH_2CH_2OCH_2CH_3$	194.23		$260\text{-}1^{739}$		$1.0585^{25/25}$	1.4969^{25}	al, eth, ace, bz	$B9^{3}$, 533
2684	Benzoic acid, 2-ethoxy, ethyl ester $2\text{-}C_2H_5OC_6H_4CO_2C_2H_5$	194.23		251, $180\text{-}5^{111}$		1.005^{20}		al, eth	$B10^{3}$, 117
2685	Benzoic acid, 3-ethoxy $3\text{-}C_2H_5OC_6H_4CO_2H$	166.18	nd (w)	sub	137			w, al, eth, bz	$B10^{3}$, 245
2686	Benzoic acid, 3-ethoxy, ethyl ester $3\text{-}C_2H_5OC_6H_4CO_2C_2H_5$	194.23		264, $172\text{-}3^{50}$		$1.0725^{20/20}$		al, eth	$B10^{3}$, 251
2687	Benzoic acid, 4-ethoxy $4\text{-}C_2H_5OC_6H_4CO_2H$	166.18	nd (w)		198.5			al, eth, bz	$B10^{3}$, 282
2688	Benzoic acid, 4-ethoxy, ethyl ester $4\text{-}C_2H_5OC_6H_4CO_2C_2H_5$	194.23		275, $148\text{-}9^{14}$		1.076^{12}		al, eth	$B10^{3}$, 301
2689	Benzoic acid, 4-ethoxy-2-hydroxy or 4-Ethoxysalicylic acid $4\text{-}C_2H_5O\text{-}2\text{-}HOC_6H_3CO_2H$	182.18	nd (w or bz)		154			al, eth, bz	B10, 379
2690	Benzoic acid, ethyl ester or Ethyl benzoate $C_6H_5CO_2C_2H_5$	150.18		213, 87^{10}	-34.6	$1.0468^{20/4}$	1.5007^{20}	al, eth, ace, bz, peth	$B9^{3}$, 384
2691	Benzoic acid, 2-ethyl $2\text{-}C_2H_5C_6H_4CO_2H$	150.18	nd (w)	259	68	$1.0431^{100/4}$	1.5099^{100}	al, eth	$B9^{3}$, 2425
2692	Benzoic acid, 3-ethyl $3\text{-}C_2H_5C_6H_4CO_2H$	150.18	nd (w or dil al)		47	1.042^{100}	1.5345^{100}	al, eth	$B9^{3}$, 2429

PHYSICAL CONSTANTS OF ORGANIC COMPOUNDS (Continued)

No.	Name, Synonyms, and Formula	Mol. wt.	Color, crystalline form, specific rotation and λ_{max} (log ε)	b.p. °C	m.p. °C	Density	n_D	Solubility	Ref.
2693	Benzoic acid, 4-ethyl $4-C_2H_5C_6H_4CO_2H$	150.18	pr (al), pr or lf (w)	....	113.5	....	....	al, bz, chl	B9[3], 2430
2694	Benzoic acid, 2-ethylamino $2-(C_2H_5NH)C_6H_4CO_2H$	165.19	pr or nd (dil al)	....	154	....	....	al, eth, ace	B14[3], 897
2695	Benzoic acid, 3-ethylamino $3-(C_2H_5NH)C_6H_4CO_2H$	165.19	nd or pl (dil al)	sub	112	....	....	al, eth, ace	B14, 393
2696	Benzoic acid, 4-ethylamino $4-(C_2H_5NH)C_6H_4CO_2H$	165.19	cr (bz)	....	177-8	....	....	al, eth, ace, bz	B14[3], 1086
2697	Benzoic acid, 2-fluoro $2-FC_6H_4CO_2H$	140.11	nd (w)	....	1.265	$1.460^{25/4}$	....	al, eth, chl	B9[3], 1324
2698	Benzoic acid, 3-fluoro $3-FC_6H_4CO_2H$	140.11	lf (w)	....	124	$1.474^{25/4}$	....	eth	B9[3], 1327
2699	Benzoic acid, 4-fluoro $4-FC_6H_4CO_2H$	140.11	pr (w)	....	185	$1.479^{25/4}$	....	al, eth	B9[3], 1327
2700	Benzoic acid, 2-formamido $2-(HCONH)C_6H_4CO_2H$	165.15	nd (w + l)	....	169	....	....	al, eth	B14[3], 921
2701	Benzoic acid, 2-formyl or Phthalaldehydic acid $2-(HCO)C_6H_4CO_2H$	150.13	lf (+ w)	....	98-9	1.404	....	w, al, eth	B10[1], 2986
2702	Benzoic acid, 3-formyl or Isophthalaldehydic acid $3-(HCO)C_6H_4CO_2H$	150.13	nd (w)	....	175	....	....	al, eth, w	B10[3], 2988
2703	Benzoic acid, 4-formyl or Terephthalaldehydic acid $4-(HCO)C_6H_4CO_2H$	150.13	nd (w)	sub	256	....	....	al, eth, chl	B10[3], 2989
2704	Benzoic acid, hexyl ester or Hexylbenzoate $C_6H_5CO_2C_6H_{13}$	206.28	....	272^{770}, $139\text{-}140^{8}$	113-7	....	....	al, ace	B9[3], 398
2705	Benzoic acid, hydrazide or Benzoyl hydrazine $C_6H_5CONHNH_2$	136.15	pl (w)	....	113-7	....	....	w, al	B9[3], 1312
2706	Benzoic acid, 2-hydrazino $2-H_2NNHC_6H_4CO_2H$	152.15	nd (w)	....	250-1	....	....	w, al	B15[1], 831
2707	Benzoic acid, 2-hydrazino, hydrochloride $2-H_2NNHC_6H_4CO_2H.HCl$	188.61	nd (w)	....	194-5d	....	....	w	B15[2], 295
2708	Benzoic acid, 3-hydrazino $3-H_2NNHC_6H_4CO_2H$	152.15	pa ye lf (w)	....	186d	....	....		B15[1], 836
2709	Benzoic acid, 4-hydrazino $4-H_2NNHC_6H_4CO_2H$	152.15	ye nd or pl (w)	....	220-5d	....	....		B15[1], 837
2710	Benzoic acid, 2-hydroxy or Salicylic acid $2-HOC_6H_4CO_2H$	138.12	nd (w), nd pr (al)	211^{20}, sub	158	$.443^{20/4}$	1.565	al, eth, ace	B10[3], 87
2711	Benzoic acid, 3-hydroxy $3-HOC_6H_4CO_2H$	138.12	nd (w), pl, pr (al)	....	202-3	....	....	eth, ace	B10[1], 2421
2712	Benzoic acid, 3-hydroxy, ethyl ester $3-HOC_6H_4CO_2C_2H_5$	166.18	pl (bz)	295, 211^{65}	73.5	....	....	al, eth	B10[3], 250
2713	Benzoic acid, 3-hydroxy, methyl ester $3-HOC_6H_4CO_2CH_3$	152.15	nd (bz-peth)	280^{709}, 178^{17}	71.5	....	....	al, bz	B10[3], 249
2714	Benzoic acid, 4-hydroxy $4-HOC_6H_4CO_2H$	138.12	pr or pl (w, al), cr (dil al, ace)	....	214-5	....	....	al, eth, ace	B10[3], 277
2715	Benzoic acid, 4-hydroxy, butyl ester $4-HOC_6H_4CO_2C_4H_9$	194.23	....	....	68-9	....	....	al	B10[3], 307
2716	Benzoic acid, 4-hydroxy, ethyl ester $4-HOC_6H_4CO_2C_2H_5$	166.18	cr (dil al)	297-8	116-8	....	....	al, eth	B10[3]300
2717	Benzoic acid, 4-hydroxy, methyl ester $4-HOC_6H_4CO_2H_3$	152.15	nd (dil al)	270-80	131	....	....	al, eth, ace	B10[3], 297
2718	Benzoic acid, 4-hydroxy, propyl ester $4-HOC_6H_4CO_2C_3H_7$	180.20	pr (eth)	....	96-8	$1.0630^{102/4}$	1.5050^{102}	al, eth	B10[3], 306
2719	Benzoic acid, 4-(α-hydroxybenzyl) $4-[C_6H_5CH(OH)]C_6H_4CO_2H$	228.25	nd (w)	....	164-5	....	....	al, eth, w	B10[3], 1184
2720	Benzoic acid, 2-hydroxyethyl $2-(HOCH_2CH_2)C_6H_4CO_2H$	166.18	....	$260\text{-}1^{739}$	$1.0585^{25/25}$	1.4969^{25}	....	al, eth, ace, bz	B10[3], 572
2721	Benzoic acid, 2-hydroxyethyl ester or Ethylene glycol monobenzoate $C_6H_5CO_2CH_2CH_2OH$	166.18	....	$150\text{-}1^{10}$	45	....	....	al	B9[3], 532
2722	Benzoic acid, 2-(hydroxymethyl) $2-(HOCH_2)C_6H_4CO_2H$	152.15	nd (w)	....	128	....	....	al, eth, w	B10[3], 500
2723	Benzoic acid, 2-iodo $2-IC_6H_4CO_2H$	248.02	nd (w)	233 exp	163	....	....	al, eth	B9[3], 1432
2724	Benzoic acid, 2-iodo, methyl ester $2-IC_6H_4CO_2CH_3$	262.05	....	$277\text{-}8^{720}$, 146^{16}	....	....	1.6052^{20}	al	B9[3], 1434
2725	Benzoic acid, 2-iodo-4-nitro $4-NO_2-2-IC_6H_3CO_2H$	293.02	pa ye pr (w)	....	146-7	....	....	al, eth	B9[3], 1774

No.	Name, Synonyms, and Formula	Mol. wt.	Color, crystalline form, specific rotation and λ_{max} (log ε)	b.p. °C	m.p. °C	Density	n_D	Solubility	Ref.
2726	Benzoic acid, 3-iodo $3\text{-}IC_6H_4C_2H$	248.02	mcl pr (ace)	sub	187-8			al	$B9^3$, 1437
2727	Benzoic acid, 3-iodo, methyl ester $3\text{-}IC_6H_4CO_2CH_3$	262.05	nd (dil al)	$276\text{-}7^{739}$, 50^{18}	54-5			al, eth, ace	$B9^3$, 1438
2728	Benzoic acid, 4-iodo $4\text{-}IC_6H_4CO_2H$	248.02	mcl pr (dil al), lf (sub)	sub	270	2.184^{20}			$B9^3$, 1442
2729	Benzoic acid, 4-iodo, methyl ester $4\text{-}IC_6H_4CO_2CH_3$	262.05	nd (eth-al)	sub	114			al, eth	$B9^3$, 1443
2730	Benzoic acid, 4-iodo-2-nitro $2\text{-}NO_2\text{-}4\text{-}IC_6H_3CO_2H$	293.02	ye lf or pr (dil al)		192-3			al, eth, bz	$B9^2$, 278
2731	Benzoic acid, 4-iodo-3-nitro $3\text{-}NO_2\text{-}4\text{-}IC_6H_3CO_2H$	293.02	ye pr (al)		213			al	$B9^2$, 278
2732	Benzoic acid, 5-iodo-3-nitro $3\text{-}NO_2\text{-}5\text{-}IC_6H_3CO_2H$	293.02	nd (al), pr (peth)		167			al, w	$B9^2$, 278
2733	Benzoic acid, 2-iodoso $2\text{-}IOC_6H_4CO_2H$	264.02	lf (w)		223-5d				$B9^1$, 1433
2734	Benzoic acid, 3-iodoso $3\text{-}IOC_6H_4CO_2H$	264.02	ye amor		175-80				B9, 365
2735	Benzoic acid, 4-iodoso $4\text{-}IOC_6H_4CO_2H$	264.02	amor		212d				B9, 366
2736	Benzoic acid, 2-mercapto or Thiosalicylic acid $2\text{-}HSC_6H_4CO_2H$	154.18	lf or nd (al, w, aa)	sub	168-9			al, eth, w, aa	$B10^3$, 212
2737	Benzoic acid, (2-methoxyethyl) ester $C_6H_5CO_2CH_2CH_2OCH_3$	180.20		254-6		$1.0891^{25/25}$	1.5040^{25}	al, eth, ace, bz	$B9^3$, 533
2738	Benzoic acid, 3-methoxy-2-methylamino or Damascenine $3\text{-}CH_3O\text{-}2\text{-}CH_3NHC_6H_3CO_2H$	181.19	pr (al)	270^{750}d, $147\text{-}8^{10}$	27-9			al, eth, bz, lig	$B14^1$, 654
2739	Benzoic acid, (2-methoxyphenyl) ester or Guaiacyl benzoate $C_6H_5CO_2C_6H_4OCH_3\text{-}2$	228.25			58			al, eth, ace, chl	$B9^3$, 551
2740	Benzoic acid, methyl ester or Methyl benzoate $C_6H_5CO_2CH_3$	136.15		199.6, $96\text{-}8^{24}$	−12.3	$1.0888^{20/4}$	1.5164^{20}	al, eth, MeOH	$B9^3$, 381
2741	Benzoic acid, 3-methylbutyl ester or *iso* Pentyl benzoate $C_6H_5CO_2(CH_2)_2CH(CH_3)_2$	192.26		262.3, 133^{14}		$1.0040^{20/4}$	1.4950^{20}	al, eth	$B9^3$397
2742	Benzoic acid, methylene diester or Methanediol dibenzoate $(C_6H_5CO_2)_2CH$	256.26	nd or pr (eth)	225d	99	1.275^{22}		eth, ace, bz	$B9^3$, 715
2743	Benzoic acid, 2-methyl or *o*-Toluic acid $2\text{-}CH_3C_6H_4CO_2H$	136.15	pr or nd (w)	$258\text{-}9^{751}$	107-8	1.062^{115}	1.512^{115}	al, eth, chl	$B9^3$, 2298
2744	Benzoic acid, 2-methylamino $2\text{-}CH_3NHC_6H_4CO_2H$	151.16	pl (al or lig)	$80^{0.01}$	179			al, eth, bz	$B14^3$, 895
2745	Benzoic acid, 2-methylamino, ethyl ester $2\text{-}CH_3NHC_6H_4CO_2C_2H_5$	179.22		266, $141\text{-}3^{15}$	39			eth	$B14^2$, 213
2746	Benzoic acid, 2-methylamino, methyl ester $2\text{-}(CH_3NH)C_6H_4CO_2CH_3$	165.19	cr (peth)	255, $130\text{-}1^{13}$	18-19	1.120^{15}	1.5839^{12}	al, eth	$B14^3$, 895
2747	Benzoic acid, 3-methylamino $3\text{-}(CH_3NH)C_6H_4CO_2H$	151.16	pl (peth)		127			al, ace, bz, chl	$B14^1$, 559
2748	Benzoic acid, 3-methylamino, methyl ester $3\text{-}(CH_3NH)C_6H_4CO_2CH_3$	165.19	cr (al)		72			al, eth	B14, 392
2749	Benzoic acid, 4-methylamino $4\text{-}(CH_3NH)C_6H_4CO_2H$	151.16	nd (bz, w or dil al)		168			al, eth, w, bz	$B14^3$, 1081
2750	Benzoic acid, 4-methylamino, methyl ester $4\text{-}(CH_3NH)C_6H_4CO_2CH_3$	165.19	pl (dil al or lig)		93.5			al, eth	$B14^3$, 1081
2751	Benzoic acid, 3-(3-methylbutoxy) $3\text{-}[(CH_3)_2CH(CH_2)_2O]C_6H_4CO_2H$	208.26	cr (al)		74-5				$B10^1$, 64
2752	Benzoic acid, 4-(3-methylbutoxy) $4\text{-}[(CH_3)_2CH(CH_2)_2O]C_6H_4CO_2H$	208.26	nd		141-2				$B10^1$, 70
2753	Benzoic acid, 2-(α naphthoyl) $2\text{-}(\alpha\text{-}C_{10}H_7CO)C_6H_4CO_2H$	276.29	pl (al-w)		176.4			al, ace, bz, chl	$B10^3$, 3426
2754	Benzoic acid, 2-(β-naphthoyl) $2\text{-}(\beta\text{-}C_{10}H_7CO)C_6H_4CO_2H$	276.29	nd (to)		168			al, eth, ace, bz	$B10^3$, 3429
2755	Benzoic acid, α napthyl ester $C_6H_5CO_2\text{-}\alpha\text{-}C_{10}H_7$	248.28	pl or pr (al-eth)		56			al, eth	$B9^3$, 491
2756	Benzoic acid, β-naphthyl ester $C_6H_5CO_2\text{-}\beta\text{-}C_{10}H_7$	248.28	nd or pr (al)		107			al	$B9^3$, 492
2757	Benzoic acid, 2-nitro or 2-Nitrobenzoic acid $2\text{-}NO_2C_6H_4CO_2H$	167.12	tcl nd (w)		147-8	$1.575^{20/4}$		al, eth, ace	$B9^3$, 1466

No.	Name, Synonyms, and Formula	Mol. wt.	Color, crystalline form, specific rotation and λ_{max} (log ε)	b.p. °C	m.p. °C	Density	n_D	Solubility	Ref.
2758	Benzoic acid, (2-nitrobenzyl) ester $C_6H_5CO_2(CH_2C_6H_4NO_2\text{-}2)$	257.25	nd (dil al)	....	101-2	....	....	al, eth, bz, aa	B9, 121
2759	Benzoic acid, 2-nitro, ethyl ester $2\text{-}NO_2C_6H_4CO_2C_2H_5$	195.17	tcl (dil al)	275, 173^{18}	30	....	....	al, eth	$B9^3$, 1469
2760	Benzoic acid, 3-nitro or 3-Nitrobenzoic acid $3\text{-}NO_2C_6H_4CO_2H$	167.12	mcl pr (w)	....	140-2	1.494^{20}	....	al, eth, ace, chl	$B9^3$, 1489
2761	Benzoic acid, (3-nitrobenzyl) ester $C_6H_5CO_2(CH_2C_6H_4NO_2\text{-}3)$	257.25	....	....	71-2	....	....	al, eth	$B9^3$, 431
2762	Benzoic acid, 4-nitro $4\text{-}NO_2C_6H_4CO_2H$	167.12	mcl lf (w)	sub	242	1.610^{20}	....	al, eth, chl	$B9^3$, 1537
2763	Benzoic acid, (4-nitrobenzyl) ester $C_6H_5CO_2(CH_2C_6H_4NO_2\text{-}4)$	257.25	....	....	94-5	....	....	al, eth	$B9^3$, 431
2764	Benzoic acid, 2-nitroso $2\text{-}ONC_6H_4CO_2H$	151.12	cr (al or aa)	....	210d	....	....	....	$B9^3$, 1462
2765	Benzoic acid, 3-nitroso $3\text{-}ONC_6H_4CO_2H$	151.12	cr	....	23d	....	....	al	B9, 369
2766	Benzoic acid, 4-nitroso $4\text{-}ON_6H_4CO_2H$	151.12	ye pw	....	<350	....	....	al	$B9^3$, 1465
2767	Benzoic acid, 4-octyl $4\text{-}C_8H_{17}C_6H_4CO_2H$	234.34	lf (al)	....	139	....	....	al	$B9^3$, 2611
2768	Benzoic acid, pentachloro $C_6Cl_5CO_2H$	294.35	nd or pl (bz or dil aa)	sub (vac)	208	....	....	al, to	$B9^3$, 1383
2769	Benzoic acid, pentamethyl $C_6(CH_3)_5CO_2H$	192.26	nd (w), lf or nd (dil al)	sub	210.5	....	....	al	$B9^3$, 2564
2770	Benzoic acid, per or Perbenzoic acid C_6H_5COOOH	138.12	mcl pl (peth)	$97\text{-}110^{13.5}$ sub	41-3	....	....	al, eth, ace, bz	$B9^3$, 1049
2771	Benzoic acid, 2-phenoxy $2\text{-}C_6H_5OC_6H_4CO_2H$	214.22	lf (dil al)	355d	113-4	....	....	al, eth, chl	$B10^3$, 99
2772	Benzoic acid, 3-phenoxy $3\text{-}C_6H_5OC_6H_4CO_2H$	214.22	....	....	145	....	....	al, eth	$B10^3$, 247
2773	Benzoic acid, 4-phenoxy $4\text{-}C_6H_5OC_6H_4CO_2H$	214.22	....	....	161	....	....	al, eth	$B10^3$, 289
2774	Benzoic acid, phenyl ester or Phenylbenzoate $C_6H_5CO_2C_6H_5$	198.22	mcl pr (eth-al)	314	71	$1.235^{20/4}$	....	al, eth	$B9^3$, 415
2775	Benzoic acid, (1-phenylethyl) ester $C_6H_5CO_2CH(C_6H_5)CH_3$	226.27	....	189^{21}	....	1.1108^{18}	1.5588^{21}	al	$B9^3$, 431
2776	Benzoic acid, 2-phoshono $2\text{-}(HO_3P)C_6H_4CO_2H$	202.10	nd (w or al)	....	>300	....	....	w	B16, 820
2777	Benzoic acid, propyl ester or Propyl benzoate $C_6H_5CO_2C_3H_7$	164.20	....	211	-51.6	$1.0230^{20/4}$	....	al, eth	$B9^3$, 389
2778	Benzoic acid, *iso*-propyl ester or Isopropyl benzoate $C_6H_5CO_2CH(CH_3)_2$	164.20	....	218	....	$1.0172^{15/15}$	1.4890^{20}	al, eth, ace	$B9^3$, 391
2779	Benzoic acid, 2-propyl $2\text{-}C_3H_7C_6H_4CO_2H$	164.20	lf (dil al)	272^{739}, $164\text{-}5^{20}$	58	....	....	al, eth, w	$B9^1$, 213
2780	Benzoic acid, 2-*iso*-propyl or *o*-Cuminic acid $2\text{-}[(CH_3)_2CH]C_6H_4CO_2H$	164.20	pr (w or peth)	$160\text{-}1^{25}$	64	....	....	al, eth, bz, peth	$B9^3$, 2481
2781	Benzoic acid, 4-propyl $4\text{-}C_3H_7C_6H_4CO_2H$	164.20	pr or lf (w)	....	141	....	....	al, eth, w, bz, lig	$B9^3$, 2479
2782	Benzoic acid, 4-*iso*-propyl or Cumic acid $4\text{-}[(CH_3)_2CH]C_6H_4CO_2H$	164.20	tcl pl (al)	sub	117-8	1.162^4	....	al, eth, peth	$B9^3$, 2482
2783	Benzoic acid, 2,3,4,5-tetrachloro $2,3,4,5\text{-}Cl_4C_6HCO_2H$	259.90	nd (al), cr (ace-w)	....	194-5	....	....	al, eth	$B9^3$, 1381
2784	Benzoic acid, tetrahydrofurfuryl ester $C_6H_5CO_2(C_4H_7O)$	192.21	....	$300\text{-}2^{750}$, $138\text{-}40^2$	....	$1.137^{20/4}$	....	al, eth, chl	$B17^4$, 1105
2785	Benzoic acid, 2,3,4,5-tetrahydroxy $2,3,4,5\text{-}(HO)_4C_6HCO_2H$	186.12	pr	....	84-5	....	....	w	C50, 14644
2786	Benzoic acid, thiolo $C_6H_5C(=O)SH$	138.18	ye pl (aa)	$85\text{-}7^{10}$	24	....	1.6040^{20}	al, eth, ace, bz	$B9^3$, 1961
2787	Benzoic acid, 2-tolylester $C_6H_5CO_2C_6H_4CH_3\text{-}2$	212.25	....	$307\text{-}8^{728}$, $154\text{-}6^{815}$	....	1.114^{19}	....	al, eth	$B9^3$, 423
2788	Benzoic acid, 3-tolylester $C_6H_5CO_2C_6H_4CH_3\text{-}3$	212.25	....	314, $168\text{-}70^8$	55-6	....	....	al, eth	$B9^3$, 424
2789	Benzoic acid, 4-tolylester $C_6H_5CO_2C_6H_4CH_3\text{-}4$	212.25	pl (eth-al)	316	71.5	....	....	al, eth	$B9^3$, 427
2790	Benzoic acid, 2-(2-tolyl) $2\text{-}(2\text{-}CH_3C_6H_4)C_6H_4CO_2H$	212.25	nd (w + l), cr (bz)	....	130-2 (anh)	....	....	al, eth, w	$B9^3$, 3323
2791	Benzoic acid, 2-(3-tolyl) $2\text{-}(\text{-}3\text{-}CH_3C_6H_4)C_6H_4CO_2H$	212.25	nd (w + l)	....	162	....	....	eth, ace, w	$B9^3$, 3326

No.	Name, Synonyms, and Formula	Mol. wt.	Color, crystalline form, specific rotation and λ_{max} (log ε)	b.p. °C	m.p. °C	Density	n_D	Solubility	Ref.
2792	Benzoic acid, 2-(4-tolyl) 2-(4-$CH_3C_6H_4)C_6H_4CO_2H$	212.25	pr (+lw, al-to), nd (al)		146			al, eth, ace, bz	B9, 677
2793	Benzoic acid, 2-(4-tolyl), methyl ester 2-(4-$CH_3C_6H_4)C_6H_4CO_2CH_3$	226.27	pl (MeOH)		66			al, bz	B10, 759
2794	Benzoic acid, 4-(2-tolyl) 4-(2-$CH_3C_6H_4)C_6H_4CO_2H$	212.25			177				B9^{3}, 3328
2795	Benzoic acid, 4-(4-tolyl) 4-(4-$CH_3C_6H_4)C_6H_4CO_2H$	212.25	nd (MeOH or ace)		228			al, ace	B9, 677
2796	Benzoic acid, 2,3,5-triamino 2,3,5-$(H_2N)_3C_6H_2CO_2H$	167.17	cr (w)		d			w	B14, 455
2797	Benzoic acid, 3,4,5-triamino 3,4,5-$(H_2N)_3C_6H_2CO_2H$	167.17	nd (w + ½)		d			w	B14, 455
2798	Benzoic acid, 2,3,4-tribromo 2,3,4-$Br_3C_6H_2CO_2H$	358.81	nd (bz)		197-8			al, eth, bz	B9^{1}, 147
2799	Benzoic acid, 2,3,5-tribromo 2,3,5-$Br_3C_6H_2CO_2H$	358.81	nd (al)		193-4			al, eth, ace, bz	B9^{1}, 147
2800	Benzoic acid, 2,4,5-tribromo 2,4,5-$Br_3C_6H_2CO_2H$	358.81	nd (al or bz)		195-6			al, eth	B9^{3}, 1430
2801	Benzoic acid, 2,4,6-tribromo 2,4,6-$Br_3C_6H_2CO_2H$	358.81	pr (w)		198			al, eth, bz	B9^{3}, 1430
2802	Benzoic acid, 3,4,5-tribromo 3,4,5-$Br_3C_6H_2CO_2H$	358.81	nd (bz or al)		240			al, eth	B9^{3}, 1431
2803	Benzoic acid, 2,3,4-trichloro 2,3,4-$Cl_3C_6H_2CO_2H$	225.46	nd (w)		187-8				B9, 345
2804	Benzoic acid, 2,3,5-trichloro 2,3,5-$Cl_3HC_6H_2CO_2H$	225.46	nd (w)		163			al, eth, ace, bz	B9^{3}, 1380
2805	Benzoic acid, 2,3,6-trichloro 2,3,6-$Cl_3C_6H_2CO_2H$	225.46			124-5			eth	B9^{3}, 1380
2806	Benzoic acid, 2,4,5-trichloro 2,4,5-$Cl_3C_6H_2CO_2H$	225.46	nd (w or sub)	sub	168			al, eth, w	B9^{3}, 1380
2807	Benzoic acid, 2,4,6-trichloro 2,4,6,$Cl_3C_6H_2CO_2H$	225.46	nd (w)		164			al, eth, chl	B9^{3}, 1381
2808	Benzoic acid, 3,4,5-trichloro 3,4,5-$Cl_3C_6H_2CO_2H$	225.46	nd (dil al)		210			al, eth, ace, bz	B9^{2}, 230
2809	Benzoic acid, 2,3,4-trihydroxy 2,3,4-$(HO)_3C_6H_2CO_2H$	170.12	nd (+w)	sub	207-8d			al, eth, ace	B10^{3}, 2057
2810	Benzoic acid, 2,3,4-trihydroxy, ethyl ester 2,3,4-$(HO)_3C_6H_2CO_2C_2H_5$	198.18	cr (w + l)		102, (anh), 86 (+w)			al, eth, w	B10, 467
2811	Benzoic acid, 2,3,4-trihydroxy, methyl ester 2,3,4-$(HO)_3C_6H_2CO_2CH_3$	184.15	nd (w + 2)		151-2			al, w	B10^{3}, 2058
2812	Benzoic acid, 2,4,5-trihydroxy 2,4,5-$(HO)_3C_6H_2CO_2H$	170.12	nd (w + ½), nd		217-8d			al, w	B10^{3}, 2065
2813	Benzoic acid, 2,4,6-trihydroxy 2,4,6-$(HO)_3C_6H_2CO_2H$	170.12	(w + l)		100d			al, eth, w	B10^{2}, 334
2814	Benzoic acid, 2,4,6-trihydroxy, ethyl ester 2,4,6-$(HO)_3C_6H_2CO_2H_5$	198.18	pr or nd (w + l), pr (lig)		129			al, eth, w, bz	B10^{1}, 236
2815	Benzoic acid, 2,4,6-trihydroxy, methyl ester 2,4,6-$(HO)_3C_6H_2CO_2CH_3$	184.15	cr (dil al)		174-6			al, eth	B10^{3}, 2069
2816	Benzoic acid, 3,4,5-trihydroxy or Gallic acid 3,4,5-$(HO)_3C_6H_2CO_2H$	170.12	pr (w + l)		253d	$1.694^{6/4}$		al, ace	B10^{3}, 2070
2817	Benzoic acid, 3,4,5-trihydroxy, ethyl ester 3,4,5-$(HO)_3C_6H_2CO_2C_2H_5$	198.18	mcl pr (w + 2½), nd (chl)		160-2			al, eth, w	B10^{2}, 343
2818	Benzoic acid, 3,4,5-trihydroxy, methyl ester or Gallicin 3,4,5-$(HO)_3C_6H_2CO_2CH_3$	184.15	mcl pr (MeOH)		202			al	B10^{3}, 2076
2819	Benzoic acid, 3,4,5-trihydroxy, propylester 3,4,5-$(HO)_3C_6H_2CO_2C_3H_7$	212.20	nd (w)		130				B10^{3}, 2078
2820	Benzoic acid, 3,4,5-trihydroxy, *iso*-propylester 3,4,5-$(HO)_3C_6H_2CO_2CH(CH_3)_2$	212.20			123-4			w, al, eth	B10^{2}, 343
2821	Benzoic acid, 2,3,5-triido 2,3,5-$I_3C_6H_2CO_2H$	499.81	pr (al)		224-6			al, eth	B9^{3}, 1456
2822	Benzoic acid, 2,4,5-triido 2,4,5-$I_3C_6H_2CO_2H$	499.81	nd (al)		248			eth, al	B9^{1}, 150

No.	Name, Synonyms, and Formula	Mol. wt.	Color, crystalline form, specific rotation and λ_{max} (log ε)	b.p. °C	m.p. °C	Density	n_D	Solubility	Ref.
2823	Benzoic acid, 3,4,5-triido $3,4,5\text{-}I_3C_6H_2CO_2H$	499.81	pr (al)		292-3			al	B9³, 1457
2824	Benzoic acid, 2,3,4-trimethoxy $2,3,4\text{-}(CH_3O)_3C_6H_2CO_2H$	212.20	cr (w or peth)		100			w, al, eth	B10³, 2058
2825	Benzoic acid, 2,4,5-trimethoxy or Asaronic acid $2,4,5\text{-}(CH_3O)_3C_6H_2CO_2H$	212.20	nd (al or bz-peth)	300	144			w, al, bz, peth	B10³, 2065
2826	Benzoic acid, 3,4,5-trimethoxy $3,4,5\text{-}(CH_3O)_3C_6H_2CO_2H$	212.20	mcl nd (w)	$225\text{-}7^{10}$	171-2			al, eth, chl	B10³, 2073
2827	Benzoic acid, 2,3,4-trimethyl or Prehnitylic acid $2,3,4\text{-}(CH_3)_3C_6H_2CO_2H$	164.20	pr (al)		167.5			w, al, eth	B9³, 2489
2828	Benzoic acid, 2,3,5-trimethyl or α-Isodurylic acid $2,3,5\text{-}(CH_3)_3C_6H_2CO_2H$	164.20	pl (lig)		127			al	B9³, 2489
2829	Benzoic acid, 2,3,6-trimethyl $2,3,6\text{-}(CH_3)_3C_6H_2CO_2H$	164.20	nd (w or peth)		110-1			w, al, eth	B9³, 2489
2830	Benzoic acid, 2,4,5-trimethyl or Durylic acid $2,4,5\text{-}(CH_3)_3C_6H_2CO_2H$	164.20	nd (bz)		152-3			al, eth	B9³, 2501
2831	Benzoic acid, 2,4,6-trimethyl or Mesitoic acid $2,4,6\text{-}(CH_3)_3C_6H_2CO_2H$	164.20	pr (lig)		155			al, eth, ace, chl	B9³, 2489
2832	Benzoic acid, 3,4,5-trimethyl or α-Isodurylic acid $3,4,5\text{-}(CH_3)_3C_6H_2CO_2H$	164.20	nd (w)		215-6			al, eth	B9, 554
2833	Benzoic acid, 2,3,6-trinitro $2,3,6\text{-}(NO_2)_3C_6H_2CO_2H$	257.12	wh nd (w + 2)		160d, 55 (+ 2w)			al	B9¹, 168
2834	Benzoic acid, 2,4,5-trinitro $2,4,5\text{-}(NO_2)_3C_6H_2CO_2H$	257.12	ye lf or pl (w)		194.5d			al, eth, bz, w	B9³, 1956
2835	Benzoic acid, 2,4,6-trinitro $2,4,6\text{-}(NO_2)_3C_6H_2CO_2H$	257.12	rh (w)		228d			al, eth, ace	B9³, 1956
2836	Benzoic acid, 3,4,5-trinitro $3,4,5\text{-}(NO_2)_3C_6H_2CO_2H$	257.12	ye nd (eth + l)		168d			eth	B9¹, 168
2837	Benzoin *(d)* or α-Hydroxybensyl phenyl ketone $C_6H_5CH(OH)COC_6H_5$	212.25	$[\alpha]^{15}{}_D$ + 92.8 (Py. c=1)		133-4			al, ace, Py	B8³, 1272
2838	Benzoin *(dl)* $C_6H_5CH(OH)COC_6H_5$	212.25		344^{768}, 194^{12}	137	$1.310^{20/4}$		al, chl, aa	B8³, 1273
2839	Benzoin *(l)* $C_6H_5CH(OH)COC_6H_5$	212.25	nd (MeOH), $[\alpha]^{12}{}_D$ −117.5 (ace, c=1.25)		133-4			al, bz	B8³, 1272
2840	Benzoin, acetate *(dl)* $C_6H_5CH(O_2CCH_3)COC_6H_5$	254.29	pr or pl (eth)		83			al, eth	B8², 196
2841	Benzoin, 4,4′-dimethoxy or Anisoin $4\text{-}CH_3OC_6H_4CH(OH)COC_6H_4OCH_3\text{-}4$	272.30	pr (dil al)		113			al, ace	B8³, 3655
2842	Benzoin, ethyl ether $C_6H_5CH(COC_2H_5)COC_6H_5$	252.31	nd (lig)	$194\text{-}5^{20}$	62	$1.1016^{17/4}$	1.5727^{17}	al, eth, bz, lig	B8², 195
2843	Benzoin hydrazone *(dl)* $C_6H_5CH(OH)C(=NNH_2)C_6H_5$	226.28	pr (al)		75			al	B8, 176
2844	Benzoin methyl ether *(dl)* $C_6H_5CH(OCH_3)COC_6H_5$	226.28	nd (lig)	$188\text{-}9^{15}$	49-50	$1.1278^{14/4}$		al, eth, bz	B8², 195
2845	Benzoin oxime *(l)* $C_6H_5CH(OH)C(=NOH)C_6H_5$	227.26	amor or pr, (bz) $[\alpha]^{24}{}_D$ -3.2 (chl, c=0.85)		163-4			al, eth, ace	B8, 167
2846	Benzoin oxime *(dl, anti)* $C_6H_5CH(OH)C(=NOH)C_6H_5$	227.26	pr (bz)		151-2			al, eth, ace	B8², 196
2847	Benzoin oxime *(dl, syn)* $C_6H_5CH(OH)C(=NOH)C_6H_5$	227.26	pr (eth)		99			al, eth, ace	B8², 196
2848	Benzonitrile or Benzoic acid nitrile C_6H_5CN	103.12		190.7	−13	$1.0102^{15/15}$	1.5289^{20}	al, eth, ace, bz	B9³, 1255
2849	Benzonitrile, 2-amino $2\text{-}H_2NC_6H_4CN$	118.14		263^{751}	51			al, eth, ace, bz, chl	B14², 210
2850	Bensonitrile, 3-amino $3\text{-}H_2NC_6H_4CN$	118.14		288-90	53-4			al, eth, ace, chl	B14³, 1001
2851	Benzonitrile, 4-amino $4\text{-}H_2NC_6H_4CN$	118.14			86			al, eth, ace, bz, aa	B14³, 1079
2852	Benzonitrile, 2-bromo $2\text{-}BrC_6H_4CN$	182.02	nd (w)	$251\text{-}3^{754}$	55.5			al	B9³, 1387
2853	Benzonitrile, 3-bromo $3\text{-}BrC_6H_4CN$	182.03	(al)	225	39-40			al, eth	B9³, 1399

No.	Name, Synonyms, and Formula	Mol. wt.	Color, crystalline form, specific rotation and λ_{max} (log ε)	b.p. °C	m.p. °C	Density	n_D	Solubility	Ref.
2854	Benzonitrile, 4-bromo 4-BrC_6H_4CN	182.03	nd (w or al)	235-7	114	...	...	al, eth	B9³, 1420
2855	Benzonitrile, 2-chloro 2-ClC_6H_4CN	137.57	nd	232	43-6	...	...	al, eth	B9³, 1339
2856	Benzonitrile, 3-chloro 3-ClC_6H_4CN	137.57	...	99-100^{15}	40-2	...	...	al, eth	B9³, 1349
2857	Benzonitrile, 4-chloro 4-ClC_6H_4CN	137.57	nd (al)	223^{750}, 95^{5}	94-6	...	...	al, eth, bz, chl	B9³, 1366
2858	Benzonitrile, 3 chloromethyl or 3-Chloromethyl benzonitrile 3-$ClCH_2C_6H_4CN$	151.60	pr (al)	258-60	67	...	...	...	B9³, 2328
2859	Benzonitrile, 4-chlormethyl or 4-Chloromethyl benzonitrile 4-$ClCH_2C_6H_4CN$	151.60	pr (al)	263^{756}	79.5	...	...	...	B9³, 2357
2860	Benzonitrile, 2,4-dimethoxy or 2,4-Dimethoxy benzonitrile 2,4-$(CH_3O)_2C_6H_3CN$	163.18	cr (al, lig)	...	96	...	...	aa	B10³, 1366
2861	Benzonitrile, 2,5-dimethoxy 2,5-$(CH_3O)_2C_6H_3CN$	163.18	nd (al)	...	82	...	...	al, bz, chl	B10³, 1388
2862	Benzonitrile, 2,6-dimethoxy or 2,6-Dimethoxy benzonitrile 2,6-$(CH_3O)_2C_6H_3CN$	163.18	nd or pl	310	118-20	...	...	al, ace, bz, chl	B10², 260
2863	Benzonitrile, 3,4-dimethoxy or Veratronitrile 3,4-$(CH_3O)_2C_6H_3CN$	163.18	nd (w)	...	67-8	...	...	bz, al, w	B10², 264
2864	Benzonitrile, 2-ethoxy 2-$C_2H_5OC_6H_4CN$	147.18	...	260.7, 153^{15}	5	...	...	al, eth, lig	B10, 97
2865	Benzonitrile, 4-ethoxy 4-$C_2H_5OC_6H_4CN$	147.18	nd (lig)	238	61-2	...	...	al, eth, lig	B10³, 344
2866	Benzonitrile, 4-fluoro or 4-Fluoro benzonitrile 4-FC_6H_4CN	121.11	nd (peth)	188.8	34.8	1.1070^{55}	1.4925^{55}	peth	B9², 221
2867	Benzonitrile, 3-formyl or 3-Cyanobenzaldehyde 3-$HCOC_6H_4CN$	131.13	...	210	79-81	...	...	al, eth, w, chl	B10, 671
2868	Benzonitrile, 4-formyl or 4-Cyanobenzaldehyde 4-$HCOC_6H_4CN$	131.13	nd (w), pr (eth, or dil al)	133^{12}	101-2	...	...	al, eth, w, chl	B10³, 2990
2869	Benzonitrile, 2-hydroxy 2-HOC_6H_4CN	119.12	...	149^{14}	98	1.1052$^{100/4}$	1.5372^{100}	al, eth, bz, chl	B10², 60
2870	Benzonitrile, 3-hydroxy 3-HOC_6H_4CN	119.12	pr (al, eth), lf (w)	...	83-4	...	...	al, eth, bz, chl	B10³, 255
2871	Benzonitrile, 4-hydroxy 4-HOC_6H_4CN	119.12	lf (w)	...	113	...	...	al, eth, chl	B10³, 344
2872	Benzonitrile, 2-methoxy 2-$CH_3OC_6H_4CN$	133.15	...	255-6, 146^{20}	24.5	1.1063$^{20.5/4}$	...	al, eth	B10²,,60
2873	Benzonitrile, 4-methoxy 4-$CH_3C_6H_4CN$	133.15	nd (w), lf (al)	256-7, 106-8^{6}	61-2	...	...	al, eth, bz	B10², 101
2874	Benzonitrile, 2-methyl 2-$CH_3C_6H_4CN$	117.15	...	205, 90^{15}	-13.5	0.9955$^{20/4}$	1.5279^{20}	al, eth	B9³, 2307
2875	Benzonitrile, 2-methyl-5-nitro 5-NO_2-2$CH_3C_6H_3CN$	162.15	nd (95% al)	174-5^{18}	106	...	...	w, al, eth, ace, bz	B9³, 2314
2876	Benzonitrile, 3-methyl 3-$CH_3C_6H_4CN$	117.15	...	213, 84.5^{10}	-23	1.0316$^{20/4}$	1.5252^{20}	al, eth	B9³, 2324
2877	Benzonitrile, 4-methyl 4-$CH_3C_6H_4CN$	117.15	...	217.6, 91^{11}	29.5	0.9805$^{30/30}$	...	al, eth	B9³, 2348
2878	Benzonitrile, 3-(trifluoromethyl) 3-$CF_3C_6H_4CN$	171.12	...	189	14.5	1.28126^{20}	1.4508^{20}	...	B9³, 2327
2879	3,4-Benzophenanthrene or Benzo (c) phenanthrene $C_{18}H_{12}$	228.29	...	...	68	...	...	...	B9², 2379
2880	9,10-Benzophenanthrene or Triphenylene $C_{18}H_{12}$	228.29	nd (al, chl, bz)	425	199	...	...	al, bz, chl	B9³, 2380
2881	1,2-Benzophenazine $C_{16}H_{10}N_2$	230.27	pr (al), ye nd (bz)	> 360	142.5	...	...	al, eth, ace, bz, aa	B23², 259
2882	Benzophenone or Diphenyl ketone $C_6H_5COC_6H_5$	182.22	(α) rh pr (al, eth); (β) mcl pr	305.9	(α) 48.1, (β) 26	(α) 1.146^{20}, (β) 1.1076	α 1.6077^{19} β 1.6059^{23}	al, eth, ace, bz	B7³, 2048
2883	Benzophenone, 2-amino 2-$H_2NC_6H_4COC_6H_4$	197.24	pa ye lf or pr al	...	110-1	...	...	al, eth	B14³, 213
2884	Benzophenone, 2-amino-5-chloro 5-Cl-2-$H_2NC_6H_3COC_6H_5$	231.68	...	...	100-1	...	...	w, al, chl, peth	B14³, 214

PHYSICAL CONSTANTS OF ORGANIC COMPOUNDS (Continued)

No.	Name, Synonyms, and Formula	Mol. wt.	Color, crystalline form, specific rotation and λ_{max} (log ε)	b.p. °C	m.p. °C	Density	n_D	Solubility	Ref.
2885	Benzophenone, 2-amino-4′-methyl $2\text{-}H_2NC_6H_4CO(C_6H_4CH_3\text{-}4)$	211.26	ye pr or pl (al)		96			al, eth, bz	$B14^3$, 248
2886	Benzophenone, 2-amino-5-methyl $2\text{-}H_2N\text{-}5\text{-}CH_3C_6H_3COC_6H_5$	211.26	ye nd or pl (al)		66			al, eth, ace, aa, lig	$B14^3$, 246
2887	Benzophenone, 2-amino-5-nitro $5\text{-}NO_2\text{-}2H_2NC_6H_3COC_6H_5$	242.23			161.5			al, aa	B14,79
2888	Benzophenone, 3-amino $3\text{-}H_2NC_6H_4COC_6H_5$	197.24	ye nd (w)					al, eth	$B14^1$, 388
2889	Benzophenone, 3-amino-4-methyl $3\text{-}H_2N\text{-}4\text{-}CH_3C_6H_3COC_6H_5$	211.26	pa ye nd (MeOH)		109			al, eth, ace, bz	$B14^3$, 247
2890	Benzophenone, 3-amino-4′-methyl $3\text{-}H_2NC_6H_4CO(C_6H_4CH_3\text{-}4)$	211.26	pr (al)		111			al, eth	B14, 107
2891	Benzophenone, 4-amino $4\text{-}H_2NC_6H_4COC_6H_5$	197.24	lf (dil al)		124			al, eth, aa	$B14^3$, 217
2892	Benzophenone, 4-amino-3-methyl $4\text{-}H_2N\text{-}3\text{-}CH_3C_6H_3COC_6H_5$	211.26	ye pr (w)		112			al, eth	$B14^3$, 245
2893	Benzophenone, 4-amino-4′-methyl $4\text{-}H_2NC_6H_4CO(C_6H_4CH_3\text{-}4)$	211.26	nd (bz)		186-7			al, eth, chl	$B14^3$, 248
2894	Benzophenone, 4,4′-bis(diethylamino) $[4\text{-}(C_2H_5)_2NC_6H_4]_2CO$	324.47	lf (al)		95-6				$B14^2$, 59
2895	Benzophenone, 4,4′-bis(dimethylamino) or Michler's ketone $[4\text{-}(CH_3)_2NC_6H_4]_2CO$	268.36	lf (al), nd (bz)	>360d	179			bz, Py	$B14^3$, 226
2896	Benzophenone, 2-bromo $2\text{-}BrC_6H_4COC_6H_5$	261.12	pl (al), nd (lig)	345	42			ace, al, lig	$B7^3$, 2079
2897	Benzophenone, 3-bromo $3\text{-}BrC_6H_4COC_6H_5$	261.12	nd (al)	$185\text{-}7^{5}$	81			al	$B7^3$, 2079
2898	Benzophenone, 4-bromo $4\text{-}BrC_6H_4COC_6H_5$	261.12	lf (al)	350^{757}	82.5				$B7^3$, 2079
2899	Benzophenone, 5-*tert* butyl-2-methoxy $5\text{-}t\text{-}C_4H_9\text{-}2CH_3OC_6H_3COC_6H_5$	268.35			62-3			chl	B8,187
2900	2-Benzophenone Carboxylic acid $2\text{-}HO_2CC_6H_4COC_6H_5$	226.23			127-9			al, eth, bz	$B10^3$, 3289
2901	2-Benzophenone carboxylic acid, ethyl ester $2\text{-}(C_2H_5O_2C)C_6H_4COC_6H_5$	254.28			58	$1.221^{64/4}$	1.560^{64}	al, eth	$B10^2$, 517
2902	2-Benzophenone carboxylic acid, methyl ester $2\text{-}(CH_3O_2C)C_6H_4COC_6H_5$	240.26		350-2	52	$1.1903^{19/4}$	1.591^{20}	al, eth	$B10^2$, 517
2903	Benzophenone, 2-chloro $2\text{-}ClC_6H_4COC_6H_5$	216.67	pl (chl-lig)	330, $185\text{-}8^{13}$	52-6				$B7^3$, 2071
2903a	Benzophenone, 2-chloro-3,5-dinitro $2\text{-}Cl\text{-}3,5\text{-}(NO_2)_2C_6H_2COC_6H_5$	306.66	ye nd (aa)		148			chl, aa	B7, 428
2904	Benzophenone, 3-chloro $3\text{-}ClC_6H_4COC_6H_5$	216.67	nd		82-3			bz	$B7^3$, 2071
2905	Benzophenone, 4-chloro $4\text{-}ClC_6H_4COC_6H_5$	216.67	nd (al)	332^{771}	77-8			w, al, eth	$B7^3$, 2072
2906	Benzophenone, 2,2′-diamino $(2H_2NC_6H_4)_2CO$	212.25	lf (dil al), pr (bz)		134-5			al, eth	B14,87
2907	Benzophenone, 3,3′-diamino $(3\text{-}H_2NC_6H_4)_2CO$	212.25	nd (al)	285^{11}	173-4			al, eth	$B14^3$, 225
2908	Benzophenone, 4,4′-diamino $(4\text{-}H_2NC_6H_4)_2CO$	212.25	nd (al)		244-5			al, eth	$B14^3$, 226
2909	Benzophenone, 3,3′-dibromo $(3\text{-}BrC_6H_4)_2CO$	340.01			141			al, eth	$B7^2$, 361
2910	Benzophenone, 4,4′-dibromo $(4\text{-}BrC_6H_4)_2CO$	340.01	pl (al)	395	177			al, ace, bz, chl	$B7^3$, 2081
2911	Benzophenone, 4,4′-dicarboxylic acid $(4\text{-}HO_2CC_6H_4)_2CO$	270.24	nd (al)	sub	>360			aa	$B10^3$, 4010
2912	Benzophenone, 2,4′-dichloro $2\text{-}ClC_6H_4CO(C_6H_4Cl\text{-}4)$	251.11		$214\text{-}5^{22}$	67	1.393^{14}		al, chl	$B7^3$, 2075
2913	Benzophenone, 3,3′-dichloro $(3\text{-}ClC_6H_4)_2CO$	251.11		$160\text{-}6^{2}$	124			al, eth	$B7^3$, 2076
2914	Benzophenone, 3,4′-dichloro $3\text{-}ClC_6H_4CO(C_6H_4Cl\text{-}4)$	251.11			112-3				$B7^3$, 2076
2915	Benzophenone, 4,4′-dichloro $(4\text{-}ClC_6H_4)_2CO$	251.11	pl (al)	353^{757}	147-8			al, eth, ace, chl, aa	$B7^3$, 2076
2916	Benzophenone, 3,5-dichloro-2-hydroxy $3,5\text{-}Cl_2\text{-}2HOC_6H_2COC_6H_5$	267.11			116			al, bz	$B8^3$, 1233

No.	Name, Synonyms, and Formula	Mol. wt.	Color, crystalline form, specific rotation and λ_{max} (log ε)	b.p. °C	m.p. °C	Density	n_D	Solubility	Ref.
2917	Benzophenone, 2,2′-dihydroxy $(2\text{-}HOC_6H_4)_2CO$	214.22	lf or pr (lig)	330-40	59.5			al, eth, chl	$B8^3$, 2644
2918	Benzophenone, 2,3′-dihydroxy $2\text{-}HOC_6H_4CO(C_6H_4OH\text{-}3)$	214.22	nd (w)		126			al, eth	B8, 315
2919	Benzophenone, 2,4-dihydroxy $2,4\text{-}(HO)_2C_6H_3COC_6H_5$	214.22	nd (w)		144			al, eth, aa	$B8^3$, 2640
2920	Benzophenone, 2,4-dihydroxy--6-methoxy or Isocotoin $2,4\text{-}(HO)_2\text{-}6\text{-}CH_3OC_6H_2COC_6H_5$	244.25	ye nd (lig)		162			w, ace	$B8^2$, 467
2921	Benzophenone, 2,4′-dihydroxy $2\text{-}HOC_6H_4CO(C_6H_4OH\text{-}4)$	214.22	pl (w)		150-1			al, eth, bz	$B8^3$, 2646
2922	Benzophenone, 2,5-dihydroxy $2,5\text{-}(HO)_2C_6H_3COC_6H_5$	214.22	ye nd (dil al)		125-6			al, eth, bz, w	$B8^3$, 2643
2923	Benzophenone, 2,6-dihydroxy-4-methoxy or Cotoin $2,6\text{-}(HO)_2\text{-}4\text{-}CH_3OC_6H_2COC_6H_5$	244.25	yesh pr (chl), lf or nd (w)		130-1			al, eth, ace, bz	$B8^3$, 3638
2924	Benzophenone, 3,3′-dihydroxy $(3\text{-}HOC_6H_4)_2CO$	214.22	nd (w)		170			al, w	$B8^3$, 2647
2925	Benzophenone, 3,4′-dihydroxy $3\text{-}HOC_6H_4CO(C_6H_4OH\text{-}4)$	214.22	nd (w)		208			al, eth, w	$B8^3$, 2648
2926	Benzophenone, 2,4′-dihydroxy $(4\text{-}HOC_6H_4)_2CO$	214.22	nd (lig), cr (w)		210	1.133^{111}		al, eth, ace	$B8^3$, 2646
2927	Benzophenone, 4,4′-diido $(4\text{-}IC_6H_4)_2CO$	434.01	pl (to), nd (bz)	281^{12}	238.5				B7,425
2928	Benzophenone, 2,4-dimethoxy $2,4\text{-}(CH_3O)_2C_6H_3COC_6H_5$	242.27	pr (dil al)	218^{10}	87-8			al, chl	$B8^3$, 2641
2929	Benzophenone, 2,4′-dimethoxy $2\text{-}CH_3OC_6H_4CO(C_6H_4OCH_3\text{-}4)$	242.27	nd (al)		100			al, eth, bz, aa	$B8^1$, 640
2930	Benzophenone, 2,5-dimethoxy $2,5(CH_3O)_2C_6H_3COC_6H_5$	242.27	(lig)	225^{38}	51				$B8^3$, 2644
2931	Benzophenone, 3,4-dimethoxy $3,4\text{-}(CH_3O)_2C_6H_3COC_6H_5$	242.27	nd or pl (al)		103-4			al	$B8^2$, 354
2932	Benzophenone, 3,4′-dimethoxy $3\text{-}CH_3OC_6H_4CO(C_6H_4OCH_3\text{-}4)$	242.27	pr (al)		58-9			al	$B8^2$, 354
2933	Benzophenone, 4,4′-dimethoxy $(4\text{-}CH_3OC_6H_4)_2CO$	242.27	nd (al)		148			al, eth, ace, bz	$B8^2$, 355
2934	Benzophenone, 2,2′-dimethyl $(2\text{-}CH_3C_6H_4)_2CO$	210.28		310^{740}, $175\text{-}80^{17}$	72			al, ace, bz	$B7^3$, 2178
2935	Benzophenone, 4,4′-dimethyl $(4\text{-}CH_3C_6H_4)_2CO$	210.28	rh (al)	333^{725}	95			al, eth, ace, bz	$B7^3$, 2181
2936	Benzophenone, 3-(dimethylamino) $3\text{-}(CH_3)_2NC_6H_4COC_6H_5$	225.29	pa ye pl (al)	216^{15}	47			al	$B14^1$, 388
2937	Benzophenone, 4-(dimethylamino) $4\text{-}(CH_3)_2NC_6H_4COC_6H_5$	225.29	ye lf (al), nd (peth)		92-3			al, eth, peth, chl	$B14^3$, 218
2938	Benzophenone, 2,2′-dinitro $(2\text{-}NO_2C_6H_4)_2CO$	272.21	nd (to or aa)		188-9			to, aa	$B7^3$, 2085
2939	Benzophenone, 3,3′-dinitro $(3\text{-}NO_2C_6H_4)_2CO$	272.22			157			ace	$B7^3$, 2085
2940	Benzophenone, 2-hydroxy $2\text{-}HOC_6H_4COC_6H_5$	198.22	pl (dil al)	250^{560}	39			al, eth, bz, aa	$B8^3$, 1227
2941	Benzophenone, 3-hydroxy $3\text{-}HOC_6H_4COC_6H_5$	198.22	lf or pl (al)		116			al, eth	$B8^3$, 1235
2942	Benzophenone, 4-hydroxy $4\text{-}HOC_6H_4COC_6H_5$	198.22	nd (al), pr (dil al)		135 (st), 122 (unst)			al, eth, aa, w	$B8^3$, 1237
2943	Benzophenone, 2-hydroxy-4-methoxy $4\text{-}CH_3O\text{-}2\text{-}HOC_6H_3COC_6H_5$	228.25			65.6				$B8^3$, 2640
2944	Benzophenone, 2-hydroxy-4′-methoxy $2\text{-}HOC_6H_4CO(C_6H_4OCH_3\text{-}4)$	228.25			98				
2945	Benzophenone, 2-hydroxy-5-methoxy $5\text{-}CH_3O\text{-}2\text{-}HOC_6H_3COC_6H_5$	228.25			84				$B8^3$, 2644
2946	Benzophenone, 2-hydroxy-4′-methyl $2\text{-}HOC_6H_4CO(C_6H_4CH_3\text{-}4)$	212.25			40-1				$B8^3$, 1301
2947	Benzophenone, 2-hydroxy-4-methyl $4\text{-}CH_3\text{-}2\text{-}HOC_6H_3COC_6H_5$	212.25			63				$B8^3$, 1300
2948	Benzophenone, 2-hydroxy-5-methyl $5\text{-}CH_3\text{-}2\text{-}HOC_6H_3COC_6H_5$	212.25			84				$B8^3$, 1295

PHYSICAL CONSTANTS OF ORGANIC COMPOUNDS (Continued)

No.	Name, Synonyms, and Formula	Mol. wt.	Color, crystalline form, specific rotation and λ_{max} (log ε)	b.p. °C	m.p. °C	Density	n_D	Solubility	Ref.
2949	Benzophenone, 2-hydroxy-5-nitro $5\text{-}NO_2\text{-}2\text{-}HOC_6H_3COC_6H_5$	243.22			124-5				B8, 157
2950	Benzophenone, 4-hydroxy-3-methyl $3\text{-}CH_3\text{-}4\text{-}HOC_6H_3COC_6H_5$	212.25			173-4				$B8^3$, 1294
2951	Benzophenone, 4-hydroxy-4′-nitro $4\text{-}HOC_6H_4CO(C_6H_4NO_2\text{-}4)$	243.22			190-2			al, eth, aa	B8,163
2952	Benzophenone imine $(C_6H_5)_2C{=}NH$	181.24		$282, 158^{12}$		$1.0847^{19/4}$	1.6191^{19}	eth	$B7^3$, 2061
2953	Benzophenone, 2-methoxy $2\text{-}CH_3OC_6H_4COC_6H_5$	212.25		$194\text{-}6^{18}$	41			al, bz, aa	$B8^3$, 1227
2954	Benzophenone, 3-methoxy $3\text{-}CH_3OC_6H_4COC_6H_5$	212.25		$342\text{-}3^{730}, 201^{17}$	44			al, bz, aa	$B8^3$, 1236
2955	Benzophenone, 4-methoxy $4\text{-}CH_3OC_6H_4COC_6H_5$	212.25	pr (eth)	$354\text{-}5^{729}, 168^{12}$	61-2			al, eth, ace, bz, aa	$B8^3$, 1238
2956	Benzophenone, 2-methyl $2\text{-}CH_3C_6H_4COC_6H_5$	196.25		$309.5^{762}, 128^{12}$	< −18			al	$B7^3$, 2123
2957	Benzophenone, 3-methyl $3\text{-}CH_3C_6H_4COC_6H_5$	196.25	oil	$314\text{-}5^{725}, 170^{9}$		$1.088^{17.5}$		al, eth, bz, chl, aa	$B7^3$, 2126
2958	Benzophenone, 4-methyl $4\text{-}CH_3C_6H_4COC_6H_5$	196.25	mcl pr	327-8	59-60			eth, bz, chl	$B7^3$, 2127
2959	Benzophenone, 4-methyl, diphenyl acetal $4\text{-}CH_3C_6H_4C(OC_6H_5)_2C_6H_5$	366.47	(dil al, eth-peth)		134				$B7^2$, 372
2960	Benzophenone, 4-methyl, imine $4\text{-}CH_3C_6H_4C({=}NH)C_6H_5$	195.26		147^{5}	37	$1.0617^{20/4}$	1.6097^{20}		$B7^3$, 2129
2961	Benzophenone, 4-methyl-2-nitro $2\text{-}NO_2\text{-}4\text{-}CH_3C_6H_3COC_6H_5$	241.25	nd or pl (al)	sub	126-7			al, bz, chl	B7, 442
2962	Benzophenone, 4-methyl-2′-nitro $4\text{-}CH_3C_6H_4CO(C_6H_4NO_2\text{-}2)$	241.25	pr (aa or al)		155			bz, chl, ace	$B7^3$, 2132
2963	Benzophenone, 4-methyl-3-nitro $3\text{-}NO_2\text{-}4\text{-}CH_3C_6H_3COC_6H_5$	241.25	pa ye pl (al or aa)		130-2			al, eth, ace, bz, aa	$B7^3$, 2131
2964	Benzophenone, 4-methyl-3′-nitro $4\text{-}CH_3C_6H_4CO(C_6H_4NO_2\text{-}3)$	241.25	lf (al)		111			eth, bz, chl	$B7^2$, 375
2965	Benzophenone, 4-methyl-4′-nitro $4\text{-}CH_3C_6H_4CO(C_6H_4NO_2\text{-}4)$	241.25	nd (al)	sub	122-4			al, eth, bz, chl, aa	$B7^2$, 375
2966	Benzophenone, 4-methyl-4′-nitro, oxime $4\text{-}CH_3C_6H_4C({=}NOH)(C_6H_4NO_2\text{-}4)$	256.26	nd (eth-lig)		145			al, eth, bz	B7, 443
2967	Benzophenone, 2-nitro $2\text{-}NO_2C_6H_4COC_6H_5$	227.22	mcl (al)		105			al	$B7^3$, 2082
2968	Benzophenone, 3-nitro $3\text{-}NO_2C_6H_4COC_6H_5$	227.22	ye nd (al)	234^{18}	95			al	$B7^3$, 2082
2969	Benzophenone, 4-nitro $4\text{-}NO_2C_6H_4COC_6H_5$	227.22	nd or lf (al)		138	1.406		bz	$B7^3$, 2082
2970	Benzophenone oxime $(C_6H_5)_2C{=}NOH$	197.24	nd (al)		144			al, eth, ace, chl	$B7^3$, 2063
2971	Benzophenone phenyl hydrazone $(C_6H_5)_2C{=}NNHC_6H_5$	272.35	pr or nd (al)		137			eth, bz, aa	$B15^2$, 63
2972	Benzophenone, 2,2′,3,4-tetrahydroxy $2,3,4\text{-}(HO)_3C_6H_2CO(C_6H_4OH\text{-}2)$	246.22	ye lf or pl (w + l)		149, 102 (+ w)			al, eth, w, aa	$B8^2$, 539
2973	Benzophenone, 2,2′,4,4′-tetrahydroxy $[2,4\text{-}(HO)_2C_6H_3]_2CO$	246.22	ye nd (w + l)		196-8			al, eth, ace, w, bz, aa	$B8^3$, 4064
2974	Benzophenone, 2,2′,4,6′-tetrahydroxy or Isoeuxanthonic acid $2,4\text{-}(HO)_2C_6H_3CO[C_6H_3(OH)_2\text{-}2,6]$	246.22	(w + l)		200d			w, al, eth	B8, 496
2975	Benzophenone, 2,2′,5,6′-tetrahydroxy or Euxanthoic acid $2,5\text{-}(HO)_2C_6H_3CO[C_6H_3(OH)_2\text{-}2,6]$	246.22	ye nd (w)		200-2d			w, al	$B8^2$, 541
2976	Benzophenone, 2,3′,4,4′-tetrahydroxy $2,4\text{-}(HO)_2C_6H_3CO[C_6H_3(OH)_2\text{-}3,4]$	246.22	nd (w + 2)		202 (anh)			w, al, eth, ace, aa	$B8^2$, 541
2977	Benzophenone, 2,3′,4,6-tetrahydroxy $2,4,6(HO)_3C_6H_2CO(C_6H_4OH\text{-}3)$	246.22	pa ye lf (w)		246d			w, al	$B8^2$, 540
2978	Benzophenone, 2,4,4′,6-tetrahydroxy $2,4,6\text{-}(HO)_3C_6H_2CO(C_6H_4OH\text{-}4)$	246.22	pr or nd (w + 2)		210			w, al, eth	$B8^2$, 540
2979	Benzophenone, 3,3′,4,4′-tetrahydroxy $[3,4\text{-}(HO)_2C_6H_3]_2CO$	246.22	(w)		227-8			w, al, bz	$B8^2$, 541
2980	Benzophenone, 2,2′,4,4′-tetramethoxy $(2,4\text{-}(CH_3O)_2C_6H_3)_2CO$	302.33			130			al, bz	$B8^3$, 4064
2981	Benzophenone, 2,2′,5,5′-tetramethoxy $[2,5\text{-}(CH_3O)_2C_6H_3]_2CO$	302.33	ye (aa or al)		109			al, eth, bz, chl, aa	$B8^2$, 541

PHYSICAL CONSTANTS OF ORGANIC COMPOUNDS (Continued)

No.	Name, Synonyms, and Formula	Mol. wt.	Color, crystalline form, specific rotation and λ_{max} (log ε)	b.p. °C	m.p. °C	Density	n_D	Solubility	Ref.
2982	Benzophenone, 2,2′,6,6′-tetramethoxy ... $[2,6\text{-}(CH_3O)_2C_6H_3]_2CO$	302.33	pl (bz)	...	204	...	...	chl	B8³, 4065
2983	Benzophenone, 2,3′,4,4′-tetramethoxy ... $2,4\text{-}(CH_3O)_2C_6H_3CO[C_6H_3(OCH_3)_2\text{-}3,4]$	302.33	nd, lf or pr (al)	...	126	...	...	al	B8³, 4065
2984	Benzophenone, 2,3′,4,5′-tetramethoxy ... $2,4\text{-}(CH_3O)_2C_6H_3CO[C_6H_3(OCH_3)_2\text{-}3,5]$	302.33	nd (bz-peth)	...	73-4	...	...	al, eth	B8¹, 735
2985	Benzophenone, 2,3′4′,5-tetramethoxy ... $2,5\text{-}(CH_3O)_2C_6H_3CO[C_6H_3\text{-}(OCH_3)_2\text{-}3,4]$	302.33	pr (dil al)	...	101-2	...	...	al, eth	B8, 497
2986	Benzophenone, 2,3,4,6-tetramethoxy ... $2,3,4,6\text{-}(CH_3O)_4C_6HCOC_6H_5$	302.33	nd (lig)	...	125-6	...	...	al, ace, bz	B8¹, 734
2987	Benzophenone, 2,4,4′,5-tetramethoxy ... $2,4,5\text{-}(CH_3O)_3C_6H_2CO(C_6H_4OCH_3\text{-}4)$	302.33	ye pw (al)	...	122-4	...	...	al, ace, bz, chl	B8¹, 734
2988	Benzophenone, 2,4,4′,6-tetramethoxy ... $2,4,6(CH_3O)_3C_6H_2CO(C_6H_4OCH_3\text{-}4)$	302.33	pr (al)	...	146	...	...	al, eth	B8, 496
2989	Benzophenone, 3,3′,4,4′-tetramethoxy or Veratrophenone $[3,4\text{-}(CH_3O)_2C_6H_3]_2CO$	302.33	pr (al)	...	145	...	...	w, bz	B8³, 4066
2990	Benzophenone, 3,3′,4,5′-tetramethoxy ... $3,4\text{-}(CH_3O)_2C_6H_3CO[C_6H_3(OCH_3)_2\text{-}3,5]$	302.33	nd (bz)	...	114-5	...	...	al, eth, bz, chl	B8¹, 735
2991	Benzophenone, 2,2′,4,4′-tetramethyl ... $[2,4\text{-}(CH_3)_2C_6H_3]_2CO$	238.33	...	190^{10}	1.043^{15}	1.5790^{25}	...	...	B7³, 2247
2992	Benzophenone, thio or Thiobenzophenone ... $C_6H_5CSC_6H_5$	198.28	...	174^{14}	53-4	...	...	bz	B7³, 2087
2993	Benzophenone, 2,2′,6-trihydroxy ... $2,6\text{-}(HO)_2C_6H_3CO(C_6H_4OH\text{-}2)$	230.22	ye nd (dil al)	...	133-4	...	...	al, eth, bz	B8², 468
2994	Benzophenone, 2,3,4-trihydroxy or Callobenzophenone .. $2,3,4\text{-}(HO)_3C_6H_2COC_6H_5$	230.22	ye nd (dil al)	...	140-1	...	...	w, al, eth, ace, aa	B8³, 3635
2995	Benzophenone, 2,4,4′-trihydroxy ... $2,4\text{-}(HO)_2C_6H_3CO(C_6H_4OH\text{-}4)$	230.22	ye nd (w + 2)	...	200-1	...	...	w, al	B8³, 3639
2996	Benzophenone, 2,4,6-trihydroxy ... $2,4,6\text{-}(HO)_3C_6H_2COC_6H_5$	230.22	ye nd (w + l, dil al)	...	165	...	...	w, al, eth	B8³, 3637
2997	Benzophenone, 3,4,5-trihydroxy ... $3,4,5\text{-}(HO)_3C_6H_2COC_6H_5$	230.22	ye pl (+ lw), col (chl)	...	177-8 (anh)	...	...	w, al, eth, ace	B8, 422
2998	Benzophenone, 2,3,4-trimethoxy ... $2,3,4\text{-}(CH_3O)_3C_6H_2COC_6H_5$	272.30	pr (dil al)	...	55	...	...	al, eth	B8,418
2999	Benzophenone, 2,4,4′-trimethoxy ... $2,4\text{-}(CH_3O)_2C_6H_3CO(C_6H_4OCH_3\text{-}4)$	272.30	nd (al)	...	73-4	...	...	al, eth, bz, aa	B8¹, 702
3000	Benzophenone, 2,4,5-trimethoxy ... $2,4,5\text{-}(CH_3O)_3C_6H_2COC_6H_5$	272.30	ye nd (w)	...	97	...	...	w, al, ace, bz	B8¹, 701
3001	Benzophenone, 2,4,6-trimethoxy ... $2,4,6\text{-}(CH_3O)_3C_6H_2COC_6H_5$	272.30	mcl pr or rh pl (al)	...	115	...	...	al, eth, chl	B8³, 3638
3002	Benzophenone, 3,3′,4-trimethoxy ... $3,4\text{-}(CH_3O)_2C_6H_3CO(C_6H_4OCH_3\text{-}3)$	272.30	nd (MeOH)	...	83-4	...	...	al, ace, bz	B8², 468
3003	Benzophenone, 3,4,4′-trimethoxy ... $3,4\text{-}(CH_3O)_2C_6H_3CO(C_6H_4OCH_3\text{-}4)$	272.30	nd (al)	...	98-9	...	...	al, ace, bz	B8², 469
3004	Benzophenone, 3,4′,5-trimethoxy ... $3,5\text{-}(CH_3O)_2C_6H_3CO(C_6H_4OCH_3\text{-}4)$	272.30	nd (bz)	...	97-8	...	...	al, eth, bz, chl	B8¹, 702
3005	Benzopinacolone (α-form) or α,α,α-triphenylacetophenone ... $C_6H_5COC(C_6H_5)_3$	348.44	nd	...	206-7	...	...	bz, chl	C55, 22234
3006	Benzopinacolone (β form) or α,α,α-triphenylacetophenone ... $C_6H_5COC(C_6H_5)_3$	348.44	nd (al)	...	182	...	...	eth, bz, chl	B7³, 2941
3007	1,2-Benzopyrene or Benzo[a]pyrene ... $C_{20}H_{12}$	252.32	ye pl (bz-lig)	$310\text{-}12^{10}$	179-179.3	...	...	...	B5³, 2520
3009	3,4-Benzopyrene ... $C_{20}H_{12}$	252.31	...	$310\text{-}12^{10}$	176.5-7.5	...	...	...	B5³, 2517
3010	3,4-Benzopyrene, 5-amino ... $C_{20}H_{13}N$	267.33	ye pl (bz-lig)	...	239-41	...	...	al, ace, bz, chl	B12³, 3404
3011	3,4-Benzopyrene, 5-hydroxy ... $C_{20}H_{12}O$	268.31	nd (eth-lig)	...	207-9	...	...	al, bz, aa	B6³, 3810
3012	3,4-Benzopyrene, 8-hydroxy ... $C_{20}H_{12}O$	268.31	ye nd (bz-peth)	...	226-7d	...	...	al, eth, ace, bz	E14s, 704
3013	5,6-Benzoquinoline or β-Naphthoquinoline ... $C_{11}H_9N$	179.22	lf (peth or w)	350^{721}, $202—5^{8}$	94	...	...	al, eth, ace, bz	B20⁴, 4009

No.	Name, Synonyms, and Formula	Mol. wt.	Color, crystalline form, specific rotation and λ_{max} (log ε)	b.p. °C	m.p. °C	Density	n_D	Solubility	Ref.
3014	7,8-Benzoquinoline or α-Naphthoquinoline $C_{13}H_9N$	179.22	lf (eth), pl (peth)	338^{719}, 233^{47}	52			al, eth, ace, bz	B20⁴, 4003
3015	7,8-Benzoquinoline, 2-methyl $C_{14}H_{11}N$	193.25		324-6		1.1464^{20}	1.6738^{20}	al	B20⁴, 4043
3016	1,2-Benzoquinone or o-Quinone 1,2-O=C_6H_4=O	108.10	red pl or pr		60-70d			eth, ace, bz	B7³, 3352
3017	1,2-Benzoquinone, 3-chloro 3-Cl-(1,2-O=C_6H_3=O)	142.54	pa ye-red pr (hx)		68d			al, eth	B7¹, 338
3018	1,2-Benzoquinone, 4-chloro 4-Cl-(1,2-O=C_6H_3=O)	142.54	pa ye-red nd (hx)		78			eth	B7³, 3354
3019	1,2-Benzoquinone, 4,5-dichloro 4,5-Cl_2-(1,2-O=C_6H_2=O)	176.99	yesh red pr or pl		94d			eth, bz	B7¹, 338
3020	1,2-Benzoquinone, 4-methoxy,1-oxime 4-(H_3O-(1-HON=C_6H_3=O-2)	153.14	ye pr		158-9			al, bz, aa	B8³, 1966
3021	1,2-Benzoquinone, 3-methoxy 3-CH_3O(1,2-O=C_6H_3=O)	138.12	br-red pl, pr, nd		115-20			w, al, eth, bz, chl	B8³, 1964
3022	1,2-Benzoquinone, 5-isopropyl-4-methyl-1-oxime $C_{10}H_{13}NO_2$	179.22	nd (bz-chl)		165-7d			bz, chl	B7², 595
3023	1,4-Benzoquinone or p-Quinone 1,4-O=C_6H_4=O	108.10	ye mcl pr (w)	sub	115-7	1.318^{20}		al, eth	B7³, 3356
3024	1,4-Benzoquinone, 2-bromo-6-methyl 2-Br-6-CH_3-(1,4-O=C_6H_2=O)	201.02	ye nd (al), pr (eth or lig)	sub	95			al, eth, chl	B7³, 3393
3025	1,4-Benzoquinone, 5-bromo-2-methyl 5-Br-2-CH_3(1,4-O=C_6H_2=O)	201.02	ye lf (lig)		106			al	B7², 591
3026	1,4-Benzoquinone, 2-chloro 2-Cl-(1,4-O=C_6H_3=O)	142.54	ye-red rh (hx)		57			w, al, eth, chl	B7³, 3373
3027	1,4-Benzoquinone, 2-chloro,oxime 2-Cl-1,4(1-HON=C_6H_3=O)	157.56	gr-ye nd (bz-aa)		184d			al, eth	B7³, 3374
3028	1,4-Benzoquinone, 2-chloro-3-methyl 2-Cl-3-CH_3(1,4-O=C_6H_2=O)	156.57	cr (lig)		55			al, chl, lig	B7³, 3391
3029	1,4-Benzoquinone, 2-chloro-5-methyl 2-Cl-5-CH_3-(1,4-O=C_6H_2=O)	156.57	ye nd (w or al)		105			al, eth, chl	B7³, 3391
3030	1,4-Benzoquinone, 2-chloro-6-methyl 2-Cl-6-CH_3(1,4-O=C_6H_2=O)	156.57	ye nd (w)		90			al, eth, chl	B7³, 3392
3031	1,4-Benzoquinone, 2,5-diaminoanil or Bandrowski's base $C_{18}H_{18}N_6$	318.38	dk red or br lf		238			dil HCl	B14³, 363
3032	1,4-Benzoquinone, 2,6-dibromo-1-imine, N-chloro $C_6H_2Br_2ClNO$	299.36	ye pr (aa, al)		85-6				B7², 584
3033	1,4-Benzoquinone, 2,6-dibromo-4-imine-N-chloro 2,6-Br_2-(1-O=C_6H_2=NCl-4)	299.36	ye pr (al or aa)		83			al	B7², 584
3034	1,4-Benzoquinone, 3,5-dibromo,2,6-dimethyl 2,6-$(CH_3)_2$(1,4-O=C_6Br_2=O)	293.94	ye lf (al)	sub	176			al	B7³, 3401
3035	1,4-Benzoquinone, 3,5-dibromo-2-methyl 3,5-Br_2-2-CH_3(1,4-O=C_6H=O)	279.92	ye		117			al, eth, chl	B7³, 3393
3036	1,4-Benzoquinone, 2,5-di-*tert*-butyl 2,5-*t*-C_4H_9(1,4-O=C_6H_2=O)	220.31	ye (al)		152.5			eth, bz, aa	B7³, 3428
3037	1,4-Benzoquinone, 2,6-di-*tert*-butyl 2,6-di-*t*-C_4H_9(1,4-O=C_6H_2=O)	220.31			102-4				
3038	1,4-Benzoquinone, 2,3-dichloro 2,3-Cl_2-(1,4-O=C_6H_2=O)	176.99	ye lf		100-1			eth, bz	B7³, 3375
3039	1,4-Benzoquinone, 2,5-dichloro 2,5-Cl_2(1,4-O=C_6H_2=O)	176.99	pa ye mcl pr (al)		161-2			eth, chl	B7³, 3376
3040	1,4-Benzoquinone, 2,5-dichloro-3,6-dihydroxy or Chloranilic acid 3,6-$(HO)_2$(1,4-O=C_6Cl_2=O)	208.99	red lf (w + 2)		283-4			w	B8³, 3350
3041	1,4-Benzoquinone, 2,6-dichloro 2,6-Cl_2(1,4-O=C_6H_2=O)	176.99	ye rh (lig, bz)		120-1			chl	B7³, 3376
3042	1,4-Benzoquinone, 2,6-dichloro-1-imine, N-chloro 2,6-Cl_2-(1-ClN=C_6H_2=O-4)	210.45	ye nd (al)		67-8			eth, chl	B7², 581
3043	1,4-Benzoquinone, 2,5-dihydroxy 2,5-$(HO)_2$-(1,4-O=C_6H_2=O)	140.10	dk ye nd	215d sub	211			aa	B8³, 3348
3044	1,4-Benzoquinone, 2,5-dihydroxy-3-methoxy-6-methyl or Spinulosin $C_8H_8O_5$	184.16	red-bl	120^1 sub	202-3				B8³, 3983
3045	1,4-Benzoquinone, 2,5-dihydroxy-3-undecyl or Embelin 3-$C_{11}H_{23}$-2,5-$(HO)_2$-(1,4-O=C_6H=O)	294.39	og red pl (al-bz)	sub	143				B8², 452

No.	Name, Synonyms, and Formula	Mol. wt.	Color, crystalline form, specific rotation and λ_{max} (log ε)	b.p. °C	m.p. °C	Density	n_D	Solubility	Ref.
3046	1,4-Benzoquinone, diimine $1,4\text{-}HN{=}C_6H_4{=}NH$	106.13	ye nd		124			bz, chl	$B7^2$, 574
3047	1,4-Benzoquinone, diimine, *N,N*-dichloro $1,4\text{-}ClN{=}C_6H_4{=}NCl$	175.02	nd (w)		126d			al, eth, bz	$B7^2$, 574
3049	1,4-Benzoquinone, 2,6-dimethoxy $2,6\text{-}(CH_3O)_2\text{-}(1,4\text{-}O{=}C_6H_2{=}O)$	168.15	ye mcl pr (aa)	sub	256			aa	$B8^3$, 3354
3050	1,4-Benzoquinone, 2,3-dimethyl or *o*-Xyloquinone $2,3\text{-}(CH_3)_2\text{-}(1,4\text{-}O{=}C_6H_2{=}O)$	136.15	ye nd	sub	55			al, eth	$B7^3$, 3397
3051	1,4 Benzoquinone, 2,5-dimethyl $2,5\text{-}(CH_3)_2\text{-}(1,4\text{-}O{=}C_6H_2{=}O)$	136.15			125			eth, bz	$B7^3$, 3397
3052	1,4-Benzoquinone, 2,6-dimethyl $2,6\text{-}(CH_3)_2\text{-}(1,4\text{-}O{=}C_6H_2{=}O)$	136.15	ye nd	sub	72-3	$1.0479^{78/4}$			$B7^3$, 3399
3053	1,4-Benzoquinone dioxime $1,4\text{-}C_6H_4({=}NOH)_2$	138.13	pa ye nd (w)		240d				B7,627
3054	1,4-Benzoquinone, 2,5-diphenyl $2,5\text{-}(C_6H_5)_2\text{-}(1,4\text{-}O{=}C_6H_2{=}O)$	260.30	og-ye pl (bz or aa)		214			bz, aa	$B7^3$, 4251
3055	1,4-Benzoquinone, 2-hydroxy-3,5,6-trimethyl $C_9H_{10}O_3$	166.18			90-2			chl	$B8^3$, 2183
3056	1,4-Benzoquinone, 3-hydroxy-2-methoxy-5-methyl or Fumigatin $C_8H_8O_4$	168.16	br nd or pl (peth)		116			al, eth, ace, bz, chl	$B8^3$, 3374
3057	1,4-Benzoquinone, 2-methoxy $2\text{-}CH_3O\text{-}(1,4\text{-}O{=}C_6H_3{=}O)$	138.12	ye nd (w)	sub	145			al	$B8^3$, 1969
3058	1,4-Benzoquinone, 2-methyl or Toluquinone $2\text{-}CH_3\text{-}(1,4\text{-}O{=}C_6H_3{=}O)$	122.12	ye pl or nd	sub	69	$1.08^{75.5/4}$		al, eth	$B7^3$, 3387
3059	1,4-Benzoquinone, 2-methyl-6-propyl-4-oxime $2\text{-}CH_3\text{-}6\text{-}C_3H_7\text{-}(1,4\text{-}O{=}C_6H_2{=}NOH)$	179.22	br nd (lig)		93-4			al, bz, aa	$B7^1$, 595
3060	1,4-Benzoquinone, 2-methyl-3,5,6-tribromo $2\text{-}CH_3\text{-}(1,4\text{-}O{=}C_6Br_3{=}O)$	358.81	ye pl (al)		235-6			eth, bz	$B7^2$, 592
3061	1,4-Benzoquinone monoimine, *N*-chloro $1\text{-}ClN{=}C_6H_4{=}O\text{-}4$	141.56	ye nd (peth)		85			w, al, eth	B7,619
3062	1,4-Benzoquinone, 2-phenyl $2\text{-}C_6H_5\text{-}(1,4\text{-}O{=}C_6H_3{=}O)$	184.21	ye lf (peth, al)		114			al, bz, chl	$B7^3$, 3764
3063	1,4-Benzoquinone, 2-isopropyl-5-methyl $2\text{-}i\text{-}C_3H_7\text{-}5\text{-}CH_3[1,4\text{-}O{=}C_6H_2{=}O]$	164.20			45-7			chl	$B7^3$, 3411
3064	1,4-Benzoquinone, 2-isopropyl-5-methyl-dioxime or Thymoquinone dioxime $2\text{-}i\text{-}C_3H_7\text{-}5\text{-}CH_3[1,4\text{-}HON{=}C_6H_2{=}NOH]$	179.22	nd (bz-chl)		165-7d			bz, chl	$B7^3$, 3414
3065	1,4-Benzoquinone, tetrachloro or Chloranil $1,4\text{-}O{=}C_6Cl_4{=}O$	245.88	ye mcl pr (bz), ye lf (aa)	sub	290 (sealed tube)			eth	$B7^3$, 3378
3066	1,4-Benzoquinone, tetrahydroxy $1,4\text{-}O{=}C_6(OH)_4{=}O$	172.09	bl-bk cr					al, w	$B8^3$, 4204
3067	1,4-Benzoquinone, tetramethyl or Duroquinone $1,4\text{-}O{=}C_6(CH_3)_4{=}O$	164.20	ye nd (al or lig)		111-2			al, eth, ace, bz, aa	$B7^3$, 3417
3068	1,4-Benzoquinone, trichloro $1,4\text{-}O{=}C_6Cl_3H{=}O$	211.43	ye pl (al)		169-70			al, eth	$B7^3$, 3377
3069	*α*-Benzosuberone $C_{11}H_{12}O$	160.22		$124\text{-}5^7$, 108^1		$1.0780^{20/4}$	1.5698^{20}	al	$B7^3$, 1442
3070	Benzotrichloride $C_6H_5CCl_3$	195.48		220.6, 150^{100}	− 4.75	1.3723^{20}	1.5580^{20}	al, eth, bz	$B5^4$, 820
3071	Benzotrichloride, 2-chloro $2\text{-}ClC_6H_4CCl_3$	229.92		264.3, 129.5^{13}	30	$1.5187^{20/4}$	1.5836^{20}	eth, ace	$B5^4$, 823
3072	Benzotrichloride, 3-chloro $3\text{-}ClC_6H_4CCl_3$	229.92		255		1.495^{14}	1.4461^{20}	eth, ace	$B5^4$, 823
3073	Benzotrichloride, 4-chloro $4\text{-}ClC_6H_4CCl_3$	229.92		245, $108\text{-}12^8$			1.4463^{20}	eth, ace	$B5^4$, 823
3074	Benzotrifluoride $C_6H_5CF_3$	146.11		102, 10^{10}	− 29.1	1.1884^{20}	1.4146^{20}	**al, eth, ace, bz**	$B5^4$, 802
3075	Benzotrifluoride, 3-amino $3\text{-}H_2NC_6H_4CF_3$	161.13		187.5^{764}, $74\text{-}5^{10}$			1.4787^{20}	al, eth	$B12^3$, 1988
3076	Benzotrifluoride, 2-nitro $2\text{-}NO_2C_6H_4CF_3$	191.11	cr (al)	216.3^{765}	32.5			al, bz, aa	$B5^3$, 744
3077	Benzotrifluoride, 3-nitro $3\text{-}NO_2C_6H_4CF_3$	191.11		202.8, 81.6^{10}	− 2.4	$1.4357^{15/4}$	1.4719^{20}	al, eth	$B5^3$, 744
3078	Benzothiazole C_7H_5NS	135.18		231, 131^{34}	2	$1.2460^{20/4}$	1.6379^{20}	al, eth, ace, bz	$B27^2$, 17

PHYSICAL CONSTANTS OF ORGANIC COMPOUNDS (Continued)

No.	Name, Synonyms, and Formula	Mol. wt.	Color, crystalline form, specific rotation and λ_{max} (log ε)	b.p. °C	m.p. °C	Density	n_D	Solubility	Ref.
3079	Benzothiazole, 2-amino $C_7H_6N_2S$	150.20	pl (w)		132			al, eth, chl	$B27^2$, 225
3080	Benzothiazole, 2-amino-6-chloro $C_7H_5N_2ClS$	184.64			199-201				$B27^2$, 230
3081	Benzothiazole, 2-amino-6-ethoxy $C_9H_{10}N_2OS$	194.25	nd (al)		163-4			al	$B27^2$, 335
3082	Benzothiazole, 2-amino-4-methyl $C_8H_8N_2S$	164.22	nd (w), pl (al)		145				$B27^2$, 237
3083	Benzothiazole, 2-amino-5-methyl $C_8H_8N_2S$	164.22	pl (dil al)		171-2			al	$B27^2$, 240
3084	Benzothiazole, 2-amino-6-methyl $C_8H_8N_2S$	164.22	nd (w), pr (dil al)		142			al, w	$B27^2$, 241
3085	Benzothiazole, 5-amino-2-mercapto $C_7H_6N_2S_2$	182.26	nd		216			aniline	$B27^2$, 475
3086	Benzothiazole, 6-amino $C_7H_6N_2S$	150.20	pr (w)		87			al	B27,366
3087	Benzothiazole, 6-amino-2-mercapto $C_7H_6N_2S_2$	182.26			263				$B27^2$, 475
3088	Benzothiazole, 2-chloro C_7H_4ClNS	169.63		248, $136\text{-}6^{28}$	24	$1.3715^{10/}{}_{4}$	1.6338^{10}	al, eth, ace	$B27^2$, 18
3089	Benzothiazole, 5-chloro-2-methyl C_8H_6ClNS	183.66	pl (eth)	$70\text{-}90^{0.5}$ sub	69			al, peth	$B27^2$, 22
3090	Benzothiazole, 6-dimethylamino-2-mercapto $C_9H_{10}N_2S_2$	210.32	ye nd (bz)		230d			ace	$B27^2$, 475
3091	Benzothiazole, 2-(2,4-dinitrophenylthio) $C_{13}H_7N_3O_4S_2$	333.35	ye		162	$1.24^{20/4}$			C48,1443
3092	Benzothiazole, 2-hydroxy C_7H_5NOS	151.18	pr (dil al) nd	360	138			al, eth	$B27^2$, 225
3093	Benzothiazole, 2-(2-hydroxyphenyl) $C_{13}H_9NOS$	227.28	nd or lf (al)	$175\text{-}93^{3}$	132-3			al	$B27^2$, 91
3094	1,2-Benzisothiazole, 5-hydroxy-3-phenyl $C_{13}H_9NOS$	227.28	nd (dil al or aa)		159-60			al, ace, bz, aa	$B27^2$, 87
3095	Benzothiazole, 6-hydroxy-2-phenyl $C_{13}H_9NOS$	227.28	wh nd (dil al)		227 (cor)			al, ace, bz, aa	$B27^2$, 88
3096	Benzothiazole, 2-mercapto $C_7H_5NS_2$	167.24	nd (al or dil MeOH)		180-2	$1.42^{20/4}$		al	B27, 185
3097	Benzothiazole, 2-mercapto,benzoate $C_{14}H_9NOS_2$	271.35	ye		132				C51,3491
3098	Benzothiazole, 2-mercapto-4-methyl $C_8H_7NS_2$	181.27	nd (dil aa)		186			aa	$B27^2$, 240
3099	Benzothiazole, 2-mercapto-5-methyl $C_8H_7NS_2$	181.27	nd (to)		171-3			al, ace, bz	$B27^2$, 241
3100	Benzothiazole, 2-mercapto-6-methyl $C_8H_7NS_2$	181.27			181			al, ace, bz	$B27^2$, 242
3101	Benzothiazole, 2-mercapto-7-methyl $C_8H_7NS_2$	181.27			184			ace, bz, to	$B27^2$, 242
3102	Benzothiazole, 2-mercapto-6-nitro $C_7H_4N_2O_2S_2$	212.24	ye nd (aa)		255-7			ace, aa	$B27^2$, 234
3103	Benxothiazole, 2-methyl C_8H_7NS	149.21		238, $150\text{-}1^{15}$	14	$1.1763^{19/4}$	1.6092^{19}	al	$B27^2$, 21
3104	Benzothiazole, 2-methyl-6-nitro $C_8H_6N_2O_2S$	194.21			175				B27, 47
3105	Benzothiazole, 2-(methylthio) $C_8H_7NS_2$	181.28	pr (dil al)		52			al	$B27^2$, 71
3106	Benzothiazole, 2-phenyl $C_{13}H_9NS$	211.28	nd (dil al)	>360	114			al, eth	$B27^2$, 37
3107	Benzothiazole, 2-phenylamino $C_{13}H_{10}N_2S$	226.30	nd (al)		161			AcOEt	$B27^2$, 226
3108	Benzothiazoline, 3-methyl-2-imino $C_8H_8N_2S$	164.22	pl (w), nd (al)		128			al, eth, chl	$B27^2$, 228
3109	2-Benzothiazolinethione, 3-methyl $C_8H_7NS_2$	181.27	nd (al), pr (aa)	335^{757}	90			al, bz, chl	$B27^2$, 233
3110	Benzothiophene or Thionaphthene C_8H_6S	134.20	lf	221, $103\text{-}5^{20}$	32	$1.1484^{32/4}$	1.6374^{37}	al, eth, ace, bz	$B17^3$, 482
3111	Benzothiophene, 3-hydroxy C_8H_6OS	150.20	nd (w)		71			al, eth, ace, bz	$B17^4$, 1458

PHYSICAL CONSTANTS OF ORGANIC COMPOUNDS (Continued)

No.	Name, Synonyms, and Formula	Mol. wt.	Color, crystalline form, specific rotation and λ_{max} (log ε)	b.p. °C	m.p. °C	Density	n_D	Solubility	Ref.
3112	Benzothiophene, 4-hydroxy C_8H_6OS	150.20	nd (sub), cr (peth)	sub	78-9			al	B17^4, 1467
3113	2,3-Benzothiophene, 5-methyl C_9H_8S	148.22		111-5^{12}	19-22	1.111$^{22/D}$	1.615$^{22/D}$		B17^4, 502
3114	2,3-Benzothiophene quinone or Thioisatin $C_8H_4O_2S$	164.18	gold-ye pr (al)	247	121			al, bz, aa	B17^4, 6129
3115	1,2,3-Benzotriazole or Azimidobenzene $C_6H_5N_3$	119.13	nd (chl or bz)	204^{15}	100			al, bz, chl	B26^2, 17
3116	1,2,3-Benzoxadiazole, 5,7-dinitro $C_6H_2N_4O_5$	210.11	ye pl (al)		158			al	B16^3, 549
3117	2,3-Benzoxazin-1-one $C_8H_5NO_2$	147.13	cr (bz)		120d				B27^2, 249
3118	Benzoxazole C_7H_5NO	119.12	pr (dil al)	182.5, 45^4	31		1.5594^{20}	al	B27^2, 17
3119	Benzoxazole, 2-chloro C_7H_4ClNO	153.57		201-2	7	1.3453$^{18/4}$	1.5678^{20}		B27^2, 17
3120	Benzoxazole, 2,5-dimethyl C_9H_9NO	147.18		218-9			1.5412^{20}		B27^2, 25
3121	Benzoxazole, 2-hydroxy or 2(3)-Benzoxazolone $C_7H_5NO_2$	135.12	ng (bz or w + 1)	230^{30}	141—2 (anh), 97—8 (+w)			al, eth	B27^2, 223
3122	Benzoxazole, 2-(2-hydroxyphenyl) $C_{13}H_9NO_2$	211.22	pink nd (al or aa)	338	123-4			al, eth, ace, bz	B27^2, 91
3123	Benzoxazole, 4-hydroxy-2-phenyl $C_{13}H_9NO_2$	211.22	nd (bz)		138-9			al, eth, aa	B27^2, 88
3124	Benzoxazole, 5-hydroxy-2-phenyl $C_{13}H_9NO_2$	211.22	nd (lig or dil al)		175			al, bz, chl, aa	B27^2, 88
3125	Benzoxazole, 6-hydroxy-2-phenyl $C_{13}H_9NO_2$	211.22	nd		216-7			ace, bz, aa	B27^2, 88
3126	Benzoxazole, 7-hydroxy-2-phenyl $C_{13}H_9NO_2$	211.22	nd (bz or dil al)		191-2			al, eth, bz, chl, lig	B27^2, 91
3127	Benzoxazole, 2-mercapto C_7H_5NOS	151.18	nd (w)		196			eth, aa	B27^2, 224
3128	Benzoxazole, 2-methyl C_8H_7NO	133.15		200—1, 59-60^{12}	8.5-10	1.1211$^{20/4}$	1.5497^{20}	al, eth	B27^2, 20
3129	Benzoyl acetamide or α-Benzoyl acetanilide $C_6H_5COCH_2CONHC_6H_5$	239.27	lf (bz)		108			al	B12^3, 1005
3130	Benzoyl acetic acid $C_6H_5COCH_2CO_2H$	164.16	nd (bz-peth)		103-4d			al, eth, bz	B10^3, 2990
3131	Benzoyl acetic acid, ethyl-ester or Ethylbenzoyl acetate $C_6H_5COCH_2CO_2C_2H_5$	192.21		265-70, 165^{14}	<0	1.1220$^{20/4}$	1.5312^{16}	al, eth	B10^3, 2991
3132	Benzoyl acetic acid, methyl-ester $C_6H_5COCH_2CO_2CH_3$	178.19	pa ye	265d, 151.5^{12}		1.158$^{29/4}$	1.537^{20}	al, eth, ace	B10^3, 2991
3133	Benzoyl acetone or 1-Phenyl-1,3-butanedione $C_6H_5COCH_2COCH_3$	162.19	pr	261-2	56	1.0599$^{74/4}$	1.5678^{78}	eth	B7^3, 34821
3134	Benzoyl acetonitrite $C_6H_5COCH_2CN$	145.16	pr or lf (w)	160^{10}	80-1			al, eth, bz	B10^3, 2994
3135	Benzoyl azide $C_6H_5CON_3$	147.14	pl (ace)	exp	32			al, eth	B9^3, 1324
3136	Benzoyl bromide C_6H_5COBr	185.02		218-9, 48-50$^{0.05}$	−24	1.570^{15}	1.5868^{25}	eth	B9^3, 1064
3137	Benzoyl chloride C_6H_5COCl	140.57		197.2, 71^9		1.2120$^{20/4}$	1.5537^{20}	eth	B9^3, 1058
3138	Benzoyl chloride, 2-bromo $2\text{-}BrC_6H_4COCl$	219.47	nd	245, 118^{10}	11		1.5963^{20}		B9^3, 1387
3139	Benzoyl chloride, 4-bromo $4\text{-}Br\text{-}C_6H_4COCl$	219.47	nd (peth)	245-7, 123-6^{15}	42			al, eth, bz, lig	B9^3, 1418
3140	Benzoyl chloride, 2-chloro $2\text{-}ClC_6H_4COCl$	175.01		238, 110^{15}	−4		1.5726^{20}		B9^3, 1338
3141	Benzoyl chloride, 3-chloro $3\text{-}ClC_6H_4COCl$	175.01		225, 103-4^{14}			1.5677^{20}		B9^3, 1348
3142	Benzoyl chloride, 4-chloro $4\text{-}ClC_6H_4COCl$	175.01		222, 111^{18}	16	1.3770$^{20/4}$	1.5756^{20}		B9^3, 1362
3143	Benzoyl chloride, 2,4-dichloro $2,4\text{-}Cl_2C_6H_3COCl$	209.46	lig	150^{34}, 111$^{7.5}$	15-8		1.5895^{20}		B9^3, 1375
3144	Benzoyl chloride, 3,4-dichloro $3,4\text{-}Cl_2C_6H_3COCl$	209.46	lig	242, 160^{42}	24-6				B9^3, 1379

No.	Name, Synonyms, and Formula	Mol. wt.	Color, crystalline form, specific rotation and λ_{max} (log ε)	b.p. °C	m.p. °C	Density	n_D	Solubility	Ref.
3145	Benzoyl chloride, 3,5-dinitro $3,5\text{-}(NO_2)_2C_6H_3COCl$	230.56	ye nd (bz)		74			eth	B9^3, 1779
3146	Benzoyl chloride, 3-hydroxy $3\text{-}HOC_6H_4COCl$	156.57		110-2$^{0.5}$	<−15			chl	B10^2, 82
3147	Benzoyl chloride, 2-methoxy or o-Anisoyl chloride $2\text{-}CH_3OC_6H_4COCl$	170.60		254, 128^{11}					B10^3, 151
3148	Benzoyl chloride, 3-methoxy $3\text{-}CH_3OC_6H_4COCl$	170.60		243-4, 123-5^{15}				eth, ace, bz	B10^3, 252
3149	Benzoyl chloride, 4-methoxy $4\text{-}CH_3OC_6H_4COCl$	170.60	nd	262-3, 91^1	24-5	1.261$^{20/4}$	1.580^{20}	eth, ace, bz	B10^3, 337
3150	Benzoyl chloride, 2-nitro $2\text{-}NO_2C_6H_4COCl$	185.57		275-8, 154^{18}				eth	B9^3, 1477
3151	Benzoyl chloride, pentachloro C_6Cl_5COCl	312.79	pl (al)	5	87			al	B9^2, 230
3152	Benzoyl chloride, 2,4,6-trimethyl or Mesitoyl chloride $2,4,6\text{-}(CH_3)_3C_6H_2COCl$	182.65		143-6^{60}				ace	B9^3, 2469
3153	Benzoyl fluoride C_6H_5COF	124.11		154-5		>1		al, eth	B9^3, 1058
3154	Benzoyl glycolic acid $C_6H_5COCH(OH)CO_2H$	180.17	lo pr (lig)		112			al, eth, chl	B10^2, 677
3155	Benzoyl iodide C_6H_5COI	232.02	nd	128^{20}	3	1.748$^{18/18}$	1.137^{20}	al, eth	B9^3, 1064
3156	Benzoyl peroxide or Dibenzoyl peroxide $C_6H_5CO\text{-}OO\text{-}COC_6H_5$	242.23	rh (eth), pr	exp	106-8		1.543	al, eth, ace, bz	B9^3, 1052
3157	Benzoyl peroxide, 3,3′-dinitro $3\text{-}O_2NC_6H_4CO\text{-}OO(COC_6H_4NO_2\text{-}3)$	332.23	nd (al)		139-40d			eth, ace, bz	B9^2, 252
3158	Benzoyl peroxide, 4,4′-dinitro $4\text{-}O_2NC_6H_4CO\text{-}OO(COC_6H_4NO_2\text{-}4)$	332.23	ye cr (ace), nd (to)		156d			to	B9^2, 270
3159	β-Benzoyl propionic acid $C_6H_5COCH_2CH_2CO_2H$	178.19	lf (dil al)		116			al, eth, bz, chl	B10^2, 482
3160	Benzyl alcohol or α-Hydroxy toluene $C_6H_5CH_2OH$	108.14		205.3, 93^{10}	−15.3	1.0419$^{24/4}$	1.5396^{20}	w, al, eth, ace, bz	B6^4, 2222
3161	Benzyl alcohol, 2-amino $2\text{-}H_2NC_6H_4CH_2OH$	123.15		270-80, 160$^{5-10}$	83-4			w, al, eth, bz, chl	B13^3, 1615
3162	Benzyl alcohol, 2-amino-3-methyl $3\text{-}CH_3\text{-}2\text{-}H_2NC_6H_3CH_2OH$	137.18	nd (bz)	135-45^{12}	71				B13^2, 367
3163	Benzyl alcohol, 2-amino-4-methyl $4\text{-}CH_3\text{-}2\text{-}H_2NC_6H_3CH_2OH$	137.18	nd (bz)	140-50^{13}	141				B13^2, 369
3164	Benzyl alcohol, 2-amino-5-methyl $2\text{-}NH_2\text{-}5\text{-}CH_3C_6H_3CH_2OH$	137.18	nd (bz)	145-50^{12}	123				B13^2, 367
3165	Benzyl alcohol, 3-amino $3\text{-}H_2NC_6H_4CH_2OH$	123.15			97			w, al, eth, bz, chl	B13^3, 1617
3166	Benzyl alcohol, 4-amino $4\text{-}H_2NC_6H_4CH_2OH$	123.15			65			al, eth, bz	B13^3, 1619
3167	Benzyl alcohol, 2-bromo $2\text{-}BrC_6H_4CH_2OH$	187.04	nd (lig)		80			w, al, eth	B6^4, 2600
3168	Benzyl alcohol, 4-bromo $4\text{-}BrC_6H_4CH_2OH$	187.04	nd (lig)		77			al, eth, bz	B6^4, 2602
3169	Benzyl alcohol, 5-bromo-2-hydroxy or 5-Bromosaligenin $5\text{-}Br\text{-}2\text{-}HOC_6H_3CH_2OH$	203.04	lf (bz)		113			al, eth, bz, chl	B6^3, 4541
3170	Benzyl alcohol, 2-chloro $2\text{-}ClC_6H_4CH_2OH$	142.58	lf or nd (dil al)	230, 100-5^{28}	74			al, eth, lig	B6^4, 2589
3171	Benzyl alcohol, 4-chloro $4\text{-}ClC_6H_4CH_2OH$	142.58	nd (w), pr (bz or bz-lig)	235	75			al, eth, bz	B6^4, 2593
3172	Benzyl alcohol, 2,5-dihydroxy or Gentisyl alcohol $2,5\text{-}(HO)_2C_6H_3CH_2OH$	140.14	nd (chl)	sub vac	100			w, al, eth, chl	B6^3, 6322
3173	Benzyl alcohol, 2,3-dimethoxy $2,3\text{-}(CH_3O)_2C_6H_3CH_2OH$	168.19		257-8, 155-60^{17}	50				B6^2, 1082
3174	Benzyl alcohol, 3,4-dimethoxy $3,4\text{-}(CH_3O)_2C_6H_3CH_2OH$	168.19		135-8$^{0.3}$		1.179$^{17/17}$	1.555$^{17'}_{0}$	al, w	B6^3, 6324
3175	Benzyl alcohol, α,α-dimethyl $C_6H_5C(OH)(CH_3)_2$	136.19	pr	202, 93^{11}	35-7	0.9735$^{20/4}$	1.5325^{20}	al, eth, bz, aa	B6^3, 1813
3176	Benzyl alcohol, α-ethyl $C_6H_5CH(OH)C_2H_5$	136.19		213-5, 98^{10}		0.9938$^{22/4}$	1.5210^{22}	al, eth	B6^3, 1792
3177	Benzyl alcohol, hexahydro $C_6H_{11}CH_2OH$	114.19		183, 83^{14}	−43	0.9297$^{20/4}$	1.4644^{20}	al, eth	B6^4, 106

PHYSICAL CONSTANTS OF ORGANIC COMPOUNDS (Continued)

No.	Name, Synonyms, and Formula	Mol. wt.	Color, crystalline form, specific rotation and λ_{max} (log ε)	b.p. °C	m.p. °C	Density	n_D	Solubility	Ref.
3178	Benzyl alcohol, 2-hydroxy or Saligenin Salicyl alcohol $2\text{-}HOC_6H_4CH_2OH$	124.14	lf (bz), nd or pl (w, eth)	sub	87	1.1613^{25}	....	w, al, eth, bz, chl	$B6^3$, 4537
3179	Benzyl alcohol, 2-hydroxy, glucoside or Saligenin-β-D-glucoside $C_{13}H_{18}O_7$	286.28	rh nd or lf (w), $[\alpha]^{20/}{}_D$ +62.6 (w, c=3)	240d	205—9	1.434^{26}	....	w, al, aa	$B6^3$, 4537
3180	Benzyl alcohol, 3-hydroxy $3\text{-}HO\text{-}C_6H_4CH_2OH$	124.14	nd (bz), cr (CCl_4)	300d	73	1.161^{25}	....	al, eth, w	$B6^3$, 4545
3181	Benzyl alcohol, 4-hydroxy $4\text{-}HOC_6H_4CH_2OH$	124.14	pr or nd (w)	252	124-5	....	....	w, al, eth	$B6^3$, 4546
3182	Benzyl alcohol, 4-hydroxy-3-methoxy or Vanillyl alcohol $3\text{-}CH_3O\text{-}4\text{-}HOC_6H_3CH_2OH$	154.17	pr (w), nd (bz)	d	115	....	....	....	$B6^3$, 6323
3183	Benzyl alcohol, 2-methoxy or Saligenin, 2-methylether $2\text{-}CH_3OC_6H_4CH_2OH$	138.17	....	249, 119^8	......	$1.0395^{25/15}$	1.5455^{20}	al, eth	$B6^3$, 4538
3184	Benzyl alcohol, 4-methoxy or Anisyl alcohol $4\text{-}CH_3OC_6H_4CH_2OH$	138.17	nd	259.1, $134\text{-}5^{12}$	25	$1.109^{26/4}$	1.5420^{25}	al, eth, w	$B6^3$, 4547
3185	Benzyl alcohol, 2-methyl $2\text{-}CH_3C_6H_4CH_2OH$	122.17	nd	223^{750}, $117\text{-}9^{20}$	37-9	1.023^{40}	....	al, eth, chl	$B6^3$, 1733
3186	Benzyl alcohol, 3-methyl $3\text{-}CH_3C_6H_4CH_2OH$	122.17	....	215-6	<−20	0.9157^{17}	....	al, eth	$B6^3$, 1768
3187	Benzyl alcohol, 4-methyl $4\text{-}CH_3C_6H_4CH_2OH$	122.17	nd (hp)	217, $116\text{-}8^{20}$	61-2	$0.978^{22/4}$	....	al, eth	$B6^3$, 1779
3188	Benzyl alcohol, α-1-naphthyl $C_6H_5CH(OH)\text{-}1\text{-}C_{10}H_7$	234.30	cr (al, lig)	ca 360	86.5	....	....	al, eth, bz	$B6^3$, 3608
3189	Benzyl alcohol, α-2-naphthyl $C_6H_5CH(OH)\text{-}2\text{-}C_{10}H_7$	234.30	nd (al, lig)	....	87-8	....	....	al, eth, bz	$B6^3$, 3609
3191	Benzyl alcohol, 2-nitro $2\text{-}NO_2C_6H_4CH_2OH$	153.14	nd (w)	270, 168^{20}	74	....	....	al, eth	$B6^3$, 1563
3192	Benzyl alcohol, 3-nitro $3\text{-}NO_2C_6H_4CH_2OH$	153.14	rh nd (w)	$175\text{-}80^3$	30.5	$1.296^{19/15}$	....	al, eth, w	$B6^3$, 1565
3193	Benzyl alcohol, 4-nitro $4\text{-}NO_2C_6H_4CH_2OH$	153.14	nd (w)	250-60d, 185^{22}	96-7	....	....	al, eth	$B6^3$, 1567
3194	Benzyl alcohol, α-isopropyl $C_6H_5CH(OH)CH(CH_3)_2$	150.22	*d*: $[\alpha]^{20/}{}_D$ +47.7, *l*: $[\alpha]^{20/}{}_D$ -25.2	222.4, $112\text{-}3^{15}$	......	0.9869^{14}	1.5193^{14}	al, ace	$B6^3$, 1859
3195	Benzyl alcohol, 4-isopropyl $4\text{-}i\text{-}C_3H_7C_6H_4CH_2OH$	150.22	....	246, 122.5^{13}	28	$0.9401^{20/4}$	1.519^{20}	al, eth, bz	$B6^3$, 1911
3196	Benzyl amine or α-Aminotoluene $C_6H_5CH_2NH_2$	107.16	....	185^{770}, 90^{12}	......	$0.9813^{20/4}$	1.5401^{20}	w, al, eth, ace, bz	$B12^3$, 2194
3197	Benzyl amine, 3-bromo $3\text{-}BrC_6H_4CH_2NH_2$	186.05	....	244-5, 84^{15}	....	....	....	eth	$B12^3$, 2349
3198	Benzyl amine, 4-bromo $4\text{-}BrC_6H_4CH_2NH_2$	186.05	....	127^{15}	20	....	....	eth	$B12^3$, 2351
3199	Benzyl amine, N-*tert*-butyl or Benzyl *tert*-butyl amine $C_6H_5NHC(CH_3)_3$	163.26	....	73.7^{44}	......	....	1.4951^{25}	....	$B12^3$, 2208
3200	Benzyl amine, N-(3-butynyl)-N-methyl $C_6H_5CH_2CH_2N(CH_3)CH_2CH_2C{=}CH$	173.25	....	127^{16}	......	0.9372^{20}	1.5202^{20}	....	C51,7295
3201	Benzyl amine, 2-chloro $2\text{-}ClC_6H_4CH_2NH_2$	141.60	....	$72\text{-}3^2$	......	....	$1.5594^{25/}{}_D$	....	$B12^3$, 2340
3202	Benzyl amine, 3-chloro $3\text{-}ClC_6H_4CH_2NH_2$	141.60	....	89^2	......	....	$1.5570^{25/}{}_D$	....	$B12^3$, 2342
3203	Benzyl amine, 4-chloro $4\text{-}ClC_6H_4CH_2NH_2$	141.60	....	$109\text{-}10^{13}$	......	....	1.5566^{25}	....	$B12^3$, 2343
3204	Benzyl amine, N-cyclopropyl $C_6H_5CH_2NHC_3H_5$	147.22	....	$80\text{-}1^5$	......	1.5222^{25}	....	....	C59,509
3205	Benzyl amine, 2,4-dichloro $2,4\text{-}Cl_2C_6H_3CH_2NH_2$	176.05	....	$124\text{-}6^{13}$	......	....	$1.5762^{25/}{}_D$	chl	....
3206	Benzyl amine, 3,4-dimethoxy $3,4\text{-}(CH_3O)_2C_6H_3CH_2NH_2$	167.21	....	$154\text{-}8^{12}$	1.43^0	....	....	chl	$B13^3$, 2183
3207	Benzyl amine, N,N-dimethyl or Benzyl dimethyl amine $C_6H_5CH_2N(CH_3)_2$	135.21	....	180-2, $73\text{-}4^{15}$	......	$0.915^{0/4}$	1.5011^{20}	al, eth	$B12^3$, 2203
3208	Benzyl amine, 2,4-dimethyl $2,4\text{-}(CH_3)_2C_6H_3CH_2NH_2$	135.21	....	$86\text{-}90^{10}$	......	....	....	....	$B12^3$, 2709
3209	Benzyl amine, 2,5-dimethyl $2,5\text{-}(CH_3)_2C_6H_3CH_2NH_2$	135.21	....	225-6	......	....	$1.5377^{20/}{}_D$	....	ALD 12695-C

No.	Name, Synonyms, and Formula	Mol. wt.	Color, crystalline form, specific rotation and λ_{max} (log ε)	b.p. °C	m.p. °C	Density	n_D	Solubility	Ref.
3210	Benzylamine, *N,N*-diphenyl or Benzyl diphenyl amine $C_6H_5CH_2N(C_6H_5)_2$	259.36	nd (al)		95			eth, ace, bz	$B12^3$, 2220
3211	Benzylamine, *N*-ethyl or Benzyl ethyl amine $C_6H_5CH_2NHC_2H_5$	135.21		$194, 82^{15}$		$0.9350^{17/15}$	1.5117^{20}	al, eth, bz	$B12^3$, 2204
3212	Benzyl amine, *N*-ethyl-*N*-phenyl or Benzyl ethyl phenyl amine $C_6H_5CH_2N(C_2H_5)C_6H_5$	211.31	pa ye	285-6d	34-6	$1.034^{19/4}$	1.5930^{20}	al, eth	$B12^3$, 2218
3213	Benzyl amine hydrochloride $C_6H_5CH_2NH_2$-HCl	143.62			255-8			w, al	$B12^3$, 2197
3214	Benzyl amine, 2-hydroxy--5-nitro 5-NO_2-2-$HOC_6H_3CH_2NH_2$	168.16	ye nd or lf (w)		253d			w, al	B13,587
3215	Benzyl amine, 4-hydroxy-3-nitro 3-NO_2-4-$HOC_6H_3CH_2NH_2$	168.16	og red nd (w + l)		225d			w	B13, 610
3216	Benzyl amine, *N*-(3-hydroxy propyl)-*N*-methyl $C_6H_5CH_2N(CH_3)CH_2CH_2CH_2OH$	179.26		$132\text{-}5^4$			1.425^{20}	CS_2	$B12^3$, 2234
3217	Benzyl amine, 3-methoxy 3-$CH_3OC_6H_4CH_2NH_2$	137.18		141^{30}				al, eth	$B13^3$, 1569
3218	Benzyl amine, 4-methoxy 4-$CH_3OC_6H_4CH_2NH_2$	137.18		$236\text{-}7, 133\text{-}4^{33}$		1.050^{15}	1.5462^{20}	w, al, eth	$B13^3$, 1594
3219	Benzyl amine, 2-methyl 2-$CH_3C_6H_4CH_2NH_2$	121.18		$66\text{-}8^2$			1.5408^{25}		$B12^3$, 2460
3220	Benzyl amine, *N*-nitroso-*N*-phenyl $C_6H_5CH_2N(NO)C_6H_5$	212.25	ye nd (al)		58			al, eth, chl, lig	$B12^3$, 2335
3221	Benzyl amine, *N*-phenyl or *N*-Benzyl aniline $C_6H_5CH_2NHC_6H_5$	183.25		$306\text{-}7, 171.5^{10}$	37-8	$1.0298^{65/4}$	1.6118^{25}	al, eth	$B12^3$, 2215
3222	Benzyl amine, *N*--2-tolyl $C_6H_5CH_2NH(C_6H_4CH_3\text{-}2)$	197.28		$300\text{-}5, 176^{10}$	60	$1.0142^{65/4}$	1.5861^{65}	al, ace, chl	$B12^3$, 2220
3223	Benzyl arsonic acid $C_6H_5CH_2AsO(OH)_2$	216.07	nd (al)		167-8			chl	$B16^3$, 1057
3224	Benzyl azide $C_6H_5CH_2N_3$	133.15		$108^{23}, 74^{11}$		$1.0655^{25/4}$	1.53414^{25}	**al, eth**	$B5^4$, 759
3225	Benzyl boric acid $C_6H_5CH_2B(OH)_2$	135.96	cr (w, bz)		140, 104 (hyd)			al, bz	$B16^3$, 1278
3226	Benzyl bromide $C_6H_5CH_2Br$	171.04	pr	$201, 114^{15}$	-3	1.4380^{25}	1.5752^{20}	**al, eth**	$B5^4$, 829
3227	Benzyl bromide, 2-bromo 2-$BrC_6H_4CH_2Br$	249.93	cr (al, lig)	129^{19}	31			**al, eth, aa**	$B5^4$, 836
3228	Benzyl bromide, 3-bromo 3-$BrC_6H_4CH_2Br$	249.93	nd or lf		41			al, eth, aa	$B5^4$, 836
3229	Benzyl bromide, 4-bromo 4-$BrC_6H_4CH_2Br$	249.93	nd (al)		63			al, eth, bz, aa	$B5^4$, 836
3230	Benzyl bromide, 2-chloro 2-$ClC_6H_4CH_2Br$	205.48		120^{10}				bz, chl	$B5^4$, 833
3231	Benzyl bromide, 2,4-dibromo 2,4-$Br_2C_6H_3CH_2Br$	328.83			40-1			al	$B5^2$, 240
3232	Benzyl bromide, 3,5-dibromo 3,5-$Br_2C_6H_3CH_2Br$	328.83	pl or nd (al)	173^{19}	96			al	$B5^3$, 719
3233	Benzyl bromide, 2-nitro 2-$O_2NC_6H_4CH_2Br$	216.03	pl (dil al)		46-7			al, eth, bz, lig	$B5^3$, 752
3234	Benzyl bromide, 3-nitro 3-$O_2NC_6H_4CH_2Br$	216.03	nd or pl (al)	$153\text{-}4^8$	58-9			al	$B5^4$, 860
3235	Benzyl bromide, 4-nitro 4-$O_2NC_6H_4CH_2Br$	216.03	nd (al)		99-100			al, eth, aa	$B5^4$, 861
3236	Benzyl butyl ether $C_6H_5CH_2C_4H_9$	164.25		$223, 92^{10}$		$0.9227^{20/4}$	1.4833^{20}	**al, eth, ace**	$B6^4$, 2231
3237	Benzyl-*iso*-butyl ether $C_6H_5CH_2O$-*i*-C_4H_9	164.25		$211\text{-}2^{743}$		$0.9233^{20/4}$	1.4826^{20}	eth, chl	$B6^3$, 1457
3238	Benzyl chloride or α-chloro toluene $C_6H_5CH_2Cl$	126.59		$179.3, 66^{11}$	-39	$1.1002^{20/20}$	1.5391^{20}	**al, eth, chl**	$B5^4$, 809
3239	Benzyl chloride, 2-bromo 2-$BrC_6H_4CH_2Cl$	205.48		$124\text{-}6^{20}$				al, eth	$B5^4$, 832
3240	Benzyl chloride, 3-bromo 3-$BrC_6H_4CH_2Cl$	205.48	nd (al or peth)	119^{18}	22-3			al	$B5^4$, 832
3241	Benzyl chloride, 4-bromo 4-$BrC_6H_4CH_3$	205.48	nd (al or peth)	$236, 110\text{-}1^9$	50			al, eth, peth	$B5^4$, 832
3242	Benzyl chloride, 2-chloro 2-$ClC_6H_4CH_2Cl$	161.03		$217, 94\text{-}5^{10}$	-17	$1.2699^{0/4}$	1.5530^{20}		$B5^4$, 816

No.	Name, Synonyms, and Formula	Mol. wt.	Color, crystalline form, specific rotation and λ_{max} (log ε)	b.p. °C	m.p. °C	Density	n_D	Solubility	Ref.
3243	Benzyl chloride, 3-chloro $3\text{-ClC}_6\text{H}_4\text{CH}_2\text{Cl}$	161.03		$215\text{-}6^{753}$, $110\text{-}1^{25}$		$1.2695^{15/4}$	1.5554^{20}	al	B5[4], 816
3244	Benzyl chloride, 4-chloro $4\text{-ClC}_6\text{H}_4\text{CH}_2\text{Cl}$	161.03	nd (dil al)	222, 117^{20}	31			eth, bz, aa	B5[4], 816
3245	Benzyl chloride, 2,6-dichloro $2,6\text{-Cl}_2\text{C}_6\text{H}_3\text{CH}_2\text{Cl}$	195.48	cr (lig, eth, al-eth)	$117\text{-}9^{14}$	39-40			al, eth, lig	B5[4], 820
3246	Benzyl chloride, 3,4-dichloro $3,4\text{-Cl}_2\text{C}_6\text{H}_3\text{CH}_2\text{Cl}$	195.48		241	37.5			al	B5[4], 820
3247	Benzyl chloride, 3,5-dichloro $3,5\text{-Cl}_2\text{C}_6\text{H}_3\text{CH}_2\text{Cl}$	195.48	cr(MeOH)	$60^{0.35}$	36			al	B5[3], 699
3248	Benzyl chloride, 4-hydroxy-3-nitro $3\text{-NO}_2\text{-4-HOC}_6\text{H}_3\text{CH}_2\text{Cl}$	187.58	ye nd (lig or al), lf (peth)		75			al, bz	B6, 413
3249	Benzyl chloride, 4-methoxy $4\text{-CH}_3\text{OC}_6\text{H}_4\text{CH}_2\text{Cl}$	156.61		$116\text{-}20^{15}$, $83\text{-}4^{2}$		1.159^{20}	1.553		B6[4], 2137
3250	Benzyl chloride, 2-nitro $2\text{-NO}_2\text{C}_6\text{H}_4\text{CH}_2\text{Cl}$	171.58	cr (lig)		50-2		1.5557^{62}	al, eth, ace, bz, aa	B5[4], 854
3251	Benzyl chloride, 3-nitro $3\text{-NO}_2\text{C}_6\text{H}_4\text{CH}_2\text{Cl}$	171.58	pa ye nd (lig)	$175\text{-}83^{30-5}$	45-7		1.5577^{62}	al, eth, ace, bz, aa	B5[4], 855
3252	Benzyl chloride, 4-nitro $4\text{-NO}_2\text{C}_6\text{H}_4\text{CH}_2\text{Cl}$	171.58	pl or nd (al)		71		1.5647^{62}	al, eth, ace, bz	B5[4], 856
3253	Benzyl chloride, 2,4,5-trichloro $2,4,5\text{-Cl}_3\text{C}_6\text{H}_2\text{CH}_2\text{Cl}$	229.92		273		1.547^{20}		al, eth, ace	B5[4], 822
3254	Benzyl chloromethyl ether $\text{C}_6\text{H}_5\text{CH}_2\text{OCH}_2\text{Cl}$	156.61		103^{13}		$1.1350^{20/4}$	1.5192^{20}		B6[4], 2253
3255	Benzyl ether or Dibenzyl ether $(\text{C}_6\text{H}_5\text{CH}_2)_2\text{O}$	198.26		298, 160^{11}	3.6	$1.0428^{20/4}$	1.5168^{20}	al, eth	B6[4], 2240
3256	Benzyl diphenyl methanol $(\text{C}_6\text{H}_5)_2\text{C(OH)CH}_2\text{C}_6\text{H}_5$	274.36	nd (bz-lig), pr (peth)	222^{11}	89-90			al	B6[3], 3680
3257	Benzyl ethyl ether $\text{C}_6\text{H}_5\text{CH}_2\text{OC}_2\text{H}_5$	136.29		185, 70^{15}		$0.9490^{20/4}$	1.4955^{20}	al, eth	B6[4], 2229
3258	Benzyl ethyl sulfide $\text{C}_6\text{H}_5\text{CH}_2\text{SC}_2\text{H}_5$	152.25		218-20					B6[4], 2635
3259	Benzyl fluoride $\text{C}_6\text{H}_5\text{CH}_2\text{F}$	110.13	nd (fr)	139.8^{753}, 40^{14}	-35	$1.0228^{25/4}$	1.4892^{25}		B5[4], 800
3260	*N*-Benzyl hydroxy, 1-amine $\text{C}_6\text{H}_5\text{CH}_2\text{NHOH}$	123.15	nd (peth or lig)		57			al, lig	B15[3], 20
3261	Benzyl iodidie or α-Iodo toluene $\text{C}_6\text{H}_5\text{CH}_2\text{I}$	218.04	col or ye nd (MeOH)	93^{10}	24.5	1.7335^{25}	1.6334^{25}	al, eth, bz	B5[4], 842
3262	Benzyl isocyanide $\text{C}_6\text{H}_5\text{CH}_2\text{NC}$	117.15		198-200d, $93\text{-}4^{55}$		0.972^{15}			B12[3], 2241
3263	Benzyl isothiocyanate or Benzyl mustard oil $\text{C}_6\text{H}_5\text{CH}_2\text{NCS}$	149.21	ye oil	243, $124\text{-}5^{12}$		$1.1246^{16/4}$	1.6049^{15}	al, eth	B12[2], 567
3264	Benzylmercapton or α- Mercapto toluene $\text{C}_6\text{H}_5\text{CH}_2\text{SH}$	124.20		194-5		1.058^{20}	1.5751^{20}	al, eth	B6[4], 2632
3265	Benzyl methyl ether $\text{C}_6\text{H}_5\text{CH}_2\text{OCH}_3$	122.17		170, $59\text{-}60^{12}$	-52.6	$0.9634^{20/4}$	1.5008^{20}	al, eth, bz	B6[4], 2229
3266	Benzyl (2-methylbutyl) ether $\text{C}_6\text{H}_5\text{CH}_2\text{OCH}_2\text{CH(CH}_3)\text{C}_2\text{H}_5$	178.28		231^{722}		$0.911^{22/4}$	1.4854^{22}	al, eth	B6, 341
3267	Benzyl methyl sulfide $\text{C}_6\text{H}_5\text{CH}_2\text{SCH}_3$	138.23		195-8			$1.5620^{20/}{}_D$		B6[4], 2633
3268	Benzyl α-naphthyl ketone $\alpha\text{-C}_{10}\text{H}_7\text{COCH}_2\text{C}_6\text{H}_5$	246.31	ta, lf (al)	$194\text{-}6^{0.05}$	66-7			eth, chl	B7[2], 461
3269	Benzyl β-naphthyl ketone $\beta\text{-C}_{10}\text{H}_7\text{COCH}_2\text{C}_6\text{H}_5$	246.31	nd (al)		99.5			al, eth, bz, chl	B7[2], 461
3270	Benzyl phenyl ketone $\text{C}_6\text{H}_5\text{COCH}_2\text{C}_6\text{H}_5$	196.25	pl (al)	320, 177^{20}	60	$1.201^{0/4}$		al, eth, chl	B7[2], 368
3271	Benzyl isopentyl ether $\text{C}_6\text{H}_5\text{CH}_2\text{O-i-C}_5\text{H}_{11}$	178.27		$236\text{-}7^{748}$, $117\text{-}9^{19}$		$0.9098^{20/4}$	1.4792^{20}	al, eth	B6[2], 410
3272	Benzyl phenyl methanol *d* $\text{C}_6\text{H}_5\text{CH}_2\text{CH(OH)C}_6\text{H}_5$	198.26	nd (eth-peth or dil al), $[\alpha]^{25/}{}_D$ + 53 (al)	$167\text{-}70^{10}$	67-8	$1.0358^{70/4}$		al, eth	B6[3], 3390
3273	Benzyl phenyl methanol *dl* $\text{C}_6\text{H}_5\text{CH}_2\text{CH(OH)C}_6\text{H}_5$	198.26	nd (bz-peth)	177^{15}	69			al, eth	B6[3], 3390
3274	Benzyl phenyl methanol *l* $\text{C}_6\text{H}_5\text{CH}_2\text{CH(OH)C}_6\text{H}_5$	198.26	$[\alpha]^{20/}{}_D$ -9.4 (w, c=10)		67	$1.0358^{70/4}$		al, eth	B6[3], 3390

No.	Name, Synonyms, and Formula	Mol. wt.	Color, crystalline form, specific rotation and λ_{max} (log ε)	b.p. °C	m.p. °C	Density	n_D	Solubility	Ref.
3275	Benzyl phenyl ketone or α-Phenylacetophenone $C_6H_5CH_2COC_6H_5$	196.25	pl (al)	320, 177^{20}	60	$1.201^{0/4}$		al, eth, chl	B7^{2}, 368
3276	Benzyl phenyl ketone, α-chloro or Desyl chloride $C_6H_5COCHClC_6H_5$	230.69	nd (al)	d	68.5				B7^{2}, 369
3277	Benzyl phenyl sulfide $C_6H_5CH_2SC_6H_5$	200.30	lf (al)	197^{27}	42-3.5			al, eth	B6^{4}, 2644
3278	Benzyl phenyl sulfone $C_6H_5CH_2SO_2C_6H_5$	232.30	nd (al)		146	$1.1261^{153/4}$			B6^{4}, 2647
3279	Benzyl-4-tolyl sulfone $(4\text{-}CH_3C_6H_4)SO_2CH_2C_6H_5$	246.32	nd (al)		144-5			al, bz, aa	B6^{4}, 2649
3280	Benzyl sulfide or Dibenzyl sulfide $(C_6H_5CH_2)_2S$	214.33	pl (eth or chl)	d	49-50	$1.071^{50/50}$		al, eth	B6^{4}, 2649
3281	Benzyl sulfonamide $C_6H_5CH_2SO_2NH_2$	171.21	pr or nd (w), nd (al)		105			w, al	B11^{3}, 331
3282	Benzyl sulfonamide, *N*-methyl $C_6H_5CH_2SO_2NHCH_3$	185.25	nd or lf (aa-lig)		108-9			al, eth	B11^{2}, 73
3283	Benzyl sulfonamide, *N*-2-tolyl $C_6H_5CH_2SO_2NH(C_6H_4CH_3\text{-}2)$	261.34	cr (dil al)		83			al	B12^{2}, 452
3284	Benzyl sulfonamide, *N*-3-tolyl $C_6H_5CH_2SO_2NH(C_6H_4CH_3\text{-}3)$	261.34	cr (al)		75			al	B12^{2}, 473
3285	Benzyl sulfonamide, *N*-4-tolyl $C_6H_5CH_2SO_2NH(C_6H_4CH_3\text{-}4)$	261.35	pr (al)		113			al, eth	B12^{2}, 528
3286	Benzyl sulfonyl chloride $C_6H_5CH_2SO_2Cl$	190.64	pr (eth), nd (bz)		93			eth, bz	B11^{3}, 331
3287	Benzyl sulfoxide or Dibenzyl sulfoxide $(C_6H_5CH_2)_2SO$	230.32	lf (al, w)	210d	134-5			al, eth	B6^{4}, 2651
3288	Benzyl thiocyanate $C_6H_5CH_2SCN$	149.21	pr (al)	256	43			al, eth	B6^{4}, 2680
3289	Benzyl 4-tolyl sulfide $C_6H_5CH_2S(C_6H_4CH_3\text{-}4)$	214.33	nd (al)		144-5			al, bz, aa	B6^{4}, 2649
3290	Benzyltrimethylammonium bromide $C_6H_5CH_2N^+(CH_3)_3Br^-$	230.15	pl (al-lig), (w)		235			w, al	B12^{3}, 2204
3291	Benzyltrimethylammonium chloride $C_6H_5CH_2N^+(CH_3)_3Cl^-$	185.70	(ace)		243			w	B12^{3}, 2203
3292	Benzyltrimethylammonium iodide $C_6H_5CH_2N^+(CH_3)_3I^-$	277.15	(al)		180			al	B12^{3}, 2204
3293	Benzyltrimethylammonium nitrate $C_6H_5CH_2N^+(CH_3)_3NO_3$	212.25			151-60			w, al	B12^{2}, 546
3294	Benzyldimethylphenylammonium chloride $C_6H_5CH_2N^+(CH_3)_2C_6H_5Cl^-$	247.77			134-8			w	B12^{3}, 2218
3295	Benzyltriethylammonium bromide $C_6H_5CH_2N^+(C_2H_5)_3Br^-$	272.23			194d			w	B12^{2}, 547
3296	Benzylidene bromide or α,α-Dibromotoluene, Benzal bromide $C_6H_5CHBr_2$	249.93		156^{23}		1.51^{15}	1.6147^{20}	al, eth	B5^{4}, 836
3297	Benzylidene bromide, 4-nitro $4\text{-}NO_2C_6H_4CHBr_2$	294.93	nd (al)		84			al, eth	B5^{4}, 862
3298	Benzylidene chloride or α,α-Dichlorotoluene $C_6H_5CHCl_2$	161.03		205.2	-16.4	1.2557^{14}	1.5502^{20}	al, eth	B5^{4}, 817
3299	Benzylidene chloride, 2-chloro $2\text{-}ClC_6H_4CHCl_2$	195.48		228.5		1.399^{15}			B5^{3}, 699
3300	Benzylidene chloride, 2,5-dichloro $2,5\text{-}Cl_2C_6H_3CHCl_2$	229.92	cubic cr (chl)		42			al, eth, bz	B5^{3}, 702
3301	Benzylidene chloride, 3,4-dichloro $3,4\text{-}Cl_2C_6H_3CHCl_2$	229.92		257		$1.518^{22/22}$		al, eth, bz, aa	B5^{3}, 702
3302	Benzylidene chloride, 3,5-dichloro $3,5\text{-}Cl_2C_6H_3CHCl_2$	229.92	cr (MeOH or dil aa)		36.5			eth, ace	B5^{3}, 703
3303	Benzylidene chloride, 3-nitro $3\text{-}NO_2C_6H_4CHCl_2$	206.03	mcl (al)		65			al, eth	B5^{3}, 750
3304	Benzylidene chloride, 4-nitro $4\text{-}NO_2C_6H_4CHCl_2$	206.03	pr (al)		46			al, eth	B5^{4}, 859
3305	Benzylidene chloride, pentachloro $C_6Cl_5CHCl_2$	333.26	lf (al)	334, 199^{13}	119.5				B5^{4}, 824
3306	Benzylidene chloride, 2,3,4-trichloro $2,3,4\text{-}Cl_3C_6H_2CHCl_2$	264.37	cr (lig)	275-85	84			bz	B5^{1}, 153

No.	Name, Synonyms, and Formula	Mol. wt.	Color, crystalline form, specific rotation and λ_{max} (log ε)	b.p. °C	m.p. °C	Density	n_D	Solubility	Ref.
3307	Benzylidene chloride, 2,3,6-trichloro 2,3,6-$Cl_3C_6H_2CHCl_2$	264.37	nd (MeOH)	$145-50^{12}$	83			bz	$B5^4$, 823
3308	Benzylidene chloride, 2,4,5-trichloro 2,4,5-$Cl_3C_6H_2CHCl_2$	264.37		280-1, $153-5^{15}$	<0	$1.5956^{20/4}$	1.5992^{20}	bz	$B5^4$, 824
3309	Benzylidene chloride, 2,4,6-trichloro 2,4,6-$Cl_3C_6H_2CHCl_2$	264.37	cr (MeOH)	158^{15}	27				$B5^3$, 704
3310	Benzylidene diacetate or Benzaldehyde diacetate $C_6H_5CH(O_2CCH_3)_2$	208.22	pl (eth)	220, 154^{20}	46	1.11^{20}		al, eth, bz	$B7^2$, 161
3311	Benzylidene ethyl amine $C_6H_5CH{=}NC_2H_5$	133.20		195^{740}, 117^{12}		$0.9370^{20/4}$	1.5365^{20}	al, eth, ace	$B7^2$, 163
3312	Benzylidene methylamine $C_6H_5CH{=}N\text{-}CH_3$	119.17		185, $90-1^{30}$		$0.9672^{14/2}$	1.5526^{20}	al, eth, ace	$B7^2$, 162
3313	Benzylidene fluoride or α,α-Difluoro toluene $C_6H_5CHF_2$	128.12		139.9		1.1357^{20}	1.4577^{20}	al	$B5^4$, 801
3314	Berbamine $C_{37}H_{40}N_2O_6$	608.73	lf (+2w, al), cr (peth), $[\alpha]^{20}_D$ +109 (chl)		197-200 (anh), 156 (hyd)			al, eth, peth, chl	$B27^2$, 891
3315	Berberine $C_{20}H_{19}NO_5$	353.37	red-ye nd (w+6), cr (chl+1)		145 (anh), 110 (+6w)			al, eth	$B27^2$, 567
3316	Berberine hydrochloride $C_{20}H_{20}ClNO_5$	389.84	ye cr (w+2), nd (w+4)					w	$B27^1$, 514
3317	Berberine nitrate $C_{20}H_{18}N_2O_7$	398.37	red-ye nd (al)		155d			w	B27, 500
3318	Berberine sulfate,trihydrate $C_{40}H_{42}N_2O_{15}S$	822.84	red-ye nd					w, al, chl	B27, 500
3319	Berberine, tetrahydro *(dl)* $C_{20}H_{21}NO_4$	339.39	mcl nd (al)		173-4			al, chl	$B27^2$, 557
3320	Berbine *(dl)* $C_{17}H_{17}N$	235.33	nd (eth or MeOH)		89			al, eth, ace	$B20^4$, 4108
3321	Betaine $(CH_3)_3N^+CH_2COO^-$	117.15	(w+1), pr or lf (al)		293d			w, al	$B4^3$, 1127
3322	Betonicine $C_7H_{13}NO_3$	159.19	pr (dil al +1w), $[\alpha]^{21}_D$ -37 (w, c=4.8)		252d			al	$B22^4$, 2054
3323	Betulin or Lupenediol $C_{30}H_{50}O_2$	442.73	nd (al+1), $[\alpha]^{15}_D$ +20 (Py, c=2)	$170-80^{0.08/d}$	251-2			eth, aa	$B6^3$, 5234
3324	Betulinic acid or Betulic acid $C_{30}H_{48}O_3$	456.71	pr or nd (al+1), $[\alpha]^{22/546}$ +7.9 (Py)					Py	$B10^3$, 1059
3325	Biacene or Biacenaphthene $C_{24}H_{16}$	304.39	red-ye pl or nd (bz)		277			bz	$B5^3$, 2595
3326	Biacetyl or 2,3-Butanedione $CH_3COCOCH_3$	86.09		88	− 2.4	$0.9808^{18.5/4}$	1.3951^{20}	w, al, eth, ace, bz	$B1^4$, 3644
3327	Biacetyl dioxime or Dimethyl glyoxime $CH_3C(:NOH)C(:NOH)CH_3$	116.12	nd (to or dil al)	sub 234-5	245-6			al, eth	$B1^4$, 3647
3328	Biacetyl monoxime $CH_3COC(:NOH)CH_3$	101.11	pr (chl), lf (w)	185-6	77-8			al, eth, chl	$B1^4$, 3646
3329	10,10′-Bianthronyl or Bianthrone $C_{28}H_{18}O_2$	386.45	pl (ace)		256-8d			chl	$B7^3$, 4486
3330	Biarsine, tetraethyl or Ethylcacodyl $(C_2H_5)_2As\text{-}As(C_2H_5)_2$	266.09		185-7		$1.1388^{24/4}$	1.4709	al, eth	$B4^3$, 1832
3331	Biarsine, tetrakis(trifluoromethyl) $(CF_3)_2AsAs(CF_3)_2$	425.87		106-7			1.372^{19}		$B3^4$, 269
3332	Bibenzyl, 4,4′-diamino 4-$H_2NC_6H_4CH_2CH_2(C_6H_4NH_2\text{-}4)$	212.29	pl (w)	sub	135-6 sub			al	$B13^3$, 470
3333	Bicyclo[2,2,1]-hepta-2,5-diene C_7H_8	92.14		89.5	− 19.1	$0.9064^{20/4}$	1.4702^{20}	al, eth, ace, bz lig	$B5^4$, 879

No.	Name, Synonyms, and Formula	Mol. wt.	Color, crystalline form, specific rotation and λ_{max} (log ε)	b.p. °C	m.p. °C	Density	n_D	Solubility	Ref.
3334	Bicyclo[2,2,1]heptane or Norbornane C_7H_{12}	96.17		sub	87.5			al, eth, ace, bz	B5[4], 258
3335	Bicyclo[4,1,0]-heptane, 7-azo $C_6H_{11}N$	97.16		48-51[22]	20-2			al, eth, bz	B20[4], 1937
3336	Bicyclo[2,2,1]-heptane-2,3-dicarboxylic acid or 2,3-Norcamphane dicarboxylic acid $C_9H_{12}O_4$	184.19			192-3				B9[3], 3970
3337	Bicyclo[2,2,1]-heptane-2-carboxaldehyde $C_8H_{10}O$	122.17	cr (w)	70-2[22]		1.0227[19/4]	1.4760[25]	eth	B7[3], 267
3338	Bicycloheptyl $C_{14}H_{26}$	194.36		290-1[728]		0.9069[20/0]			B5[4], 344
3339	Bicyclo[3,1,0]hex-2-ene-4-one, 5-*iso*-propyl-2-methyl $C_{10}H_{14}O$	150.22	$[\alpha]_D$ − 36.5	219.20[749]		0.9581[15/15]	1.48325		B7[3], 582
3340	Bicyclohexyl or Dodecahydrobiphenyl *(cis,cis)* C_6H_{11}-C_6H_{11}	166.31		238	4	0.8914[20/4]	1.4766[20]	al, eth	B5[3], 273
3341	Bicyclohexyl *(trans,trans)* $C_{12}H_{22}$	166.31		217-8, 95-6[9]	4.2	0.8592[20/4]	1.4663[20]	al, eth	B5[4], 334
3342	Bicyclo[3,1,1]-heptane, 2,4,6-trimethyl $C_{10}H_{18}$	138.25		169-70[768]		0.8467[21/4]	1.4605[21]		PCHE2, 238
3343	Bicyclo[3,3,1]-nonane C_9H_{16}	124.23	cr (MeOH)	169-70 sub	145-6			al, aa	B5[4], 293
3344	Bicyclo[2,2,2]-octane C_8H_{14}	110.20			169-71				B5[4], 279
3345	Bicyclo[3,2,1]octane C_8H_{14}	110.20			139-41				B5[4], 278
3346	Bicyclo[3,3,0]octane *(cis)* C_8H_{14}	110.20		137[765]	< − 80	0.8638[25/4]	1.4595[25]	al	B5[4], 277
3347	Bicyclo[3,3,0]octane *(trans)* C_8H_{14}	110.20		132[755]	− 30	0.8624[18/4]	1.4625[18]	al	B5[4], 277
3348	Bicyclo[3,3,0]octane, 2,6-dione $C_8H_{10}O_2$	138.17		86-8[0.2]	45	1.1290[60/4]	1.4877[54]		B7[3], 3279
3349	Bicyclo[2,2,2]octane, 2-methyl $C_8H_{13}CH_3$	124.23		158[740]	33-4	0.8664[40.5/4]	1.4608[40.5]		B5[4], 295
3350	Bicyclo[3,3,0]octane, 2-one *(cis)* $C_8H_{12}O$	124.18		72[13]		1.0097[20/4]	1.4790[20]	al, ace	B7[3], 264
3351	9.9′-Bifluorenyl $C_{26}H_{18}$	330.43	nd (bz-al)		247			aa, Py	B5[3], 2626
3352	9,9′-Bifluorenyl, 9,9′-diphenyl $C_{38}H_{26}$	482.62	pl (bz)		256 (under CO_2)	1.266[0/4]			B5[3], 2782
3353	Bifluorenylidene $C_{26}H_{16}$	328.41	red nd (bz)		194-5			eth, bz, chl	B5[3], 2652
3354	Biguanide $H_2NC(:NH)NHC(:NH)NH_2$	101.11	pr or nd (al)	d 142	136			w, al	B3[2], 76
3355	Biguanide, 1-phenyl $C_6H_5NHC(=NH)NHC(=NH)NH_2$-HCl	177.21			143				B12[3], 807
3356	Biguanide, 1-(2-tolyl) $C_9H_{13}N_5$	191.24	nd or pl (w + 1)		144			al, ace	B12[1], 1873
3357	2-2′-Biindane, 1,1′,3,3′-tetraoxo or Bisdiketohydrindene $C_{18}H_{10}O_4$	290.28	red nd (bz)		297			bz	B7[2], 863
3358	Bikhaconitine $C_{36}H_{51}NO_{11}$	673.80	$[\alpha]_D$ + 12 (al)		118-23			al, eth, chl	B21[4], 2869
3359	Bilifucsin $C_{16}H_{20}N_2O_4$	304.35	dk br pw		183			al, ace	C30,1936
3360	Bilirubin or Haematoidine $C_{33}H_{36}N_4O_6$	584.67	red mcl pr or pl (chl)					bz, chl	C38,1230
3361	Biliverdin or Dehydrobilirubin $C_{33}H_{34}N_4O_6$	582.66	dk gr pl or pr (MeOH)		>300			al, bz	J 1961,2284
3362	α,α′-Binaphthyl or α,α′-Dinaphthyl α-$C_{10}H_7$-α-$C_{10}H_7$, $C_{10}H_{14}$	254.33	(i) pl (aa), (ii) rh (peth)	>360, 240-2[12]	(i) 144, (ii) 160			eth, ace, bz	B5[3], 2465
3363	α,α′-Binaphthyl, 4,4′-diamino-3,3′-dimethyl (3-CH_3-4-H_2N-α-$C_{10}H_5$)$_2$	312.41			213			al, bz	B13[3], 542
3364	α,α′-Binaphthyl, 2,2′-dihydroxy or β-Dinaphthol [2-HO-α-$C_{10}H_6$]$_2$	286.33	nd (al), cr (w)		220			al, eth	B6[3], 5877
3365	α,α′-Binaphthyl, 4,4′-dihydroxy or α-Dinaphthol [4-HO-α-$C_{10}H_6$]$_2$	286.33	pl	sub	300			al, eth	B6[1], 5878

No.	Name, Synonyms, and Formula	Mol. wt.	Color, crystalline form, specific rotation and λ_{max} (log ε)	b.p. °C	m.p. °C	Density	n_D	Solubility	Ref.
3366	β,β′-Binaphthyl or β,β′-Dinaphthyl $\beta-C_{10}H_7-\beta-C_{10}H_7$	254.33	bl fluor pl (al)	452^{753} sub	187-8	....	....	eth, bz	B5³, 2467
3367	Biotin or Vitamin H-Coenzyme R $C_{10}H_{16}N_2O_3S$	244.31	nd (w)	....	232d	....	....	w	Am67, 2096
3368	Biotin, methyl ester $C_{11}H_{18}N_2O_3S$	258.34	pl (MeOH-eth), $[\alpha]^{22}_D$ + 57 chl	sub	166.7	....	....	al, ace, chl	C43,1810
3369	Biphenyl or Phenylbenzene $C_6H_5-C_6H_5$	154.21	lf (dil al)	255.9, 145^{22}	71	$0.8660^{20/4}$	1.475^{20}, 1.588^{75}	al, eth, bz	B5⁴, 1807
3370	Biphenyl, 2-acetamide $2-(CH_3CONH)C_6H_4-C_6H_5$	211.26	pr or nd (dil al or peth)	355	121	....	....	al, eth	B12³, 3125
3371	Biphenyl, 3-acetamido $3-(CH_3CONH)C_6H_4-C_6H_5$	211.26	nd (al)	....	149	....	....	al, bz	B12², 751
3372	Biphenyl, 4-acetamido-3-NO_2 $3-NO_2-4-(CH_3CONH)C_6H_3-C_6H_5$	211.27	cr (dil MeOH)	....	172	....	....	al, ace	B12³, 3156
3373	Biphenyl, 4-acetamido,3-nitro $4-(CH_3CONH)-3-NO_2C_6H_3C_6H_5$	256.26	ye nd (al)	....	132	....	....	al, eth, aa	B12³, 3199
3374	Biphenyl, 4-acetyl or 4-Phenylacetophenone $4-(CH_3CO)C_6H_4-C_6H_5$	196.25	pr (ace), cr (al)	325-7	121	....	....	al, bz	B7², 377
3375	Biphenyl, 2-amino $2-H_2NC_6H_4-C_6H_5$	169.23	lf (dil al)	299, 170^{15}	51-3	....	....	al, eth, bz	B12³, 3124
3376	Biphenyl, 2-amino-4′-nitro $2-H_2NC_6H_4-C_6H_4NO_2-4$	214.22	og-red nd (al)	....	159	....	....	al	B12³, 3141
3377	Biphenyl, 2-amino-5-nitro $5-NO_2-2-H_2NC_6H_3-C_6H_5$	214.22	ye nd (al)	....	125	....	....	al	B12³, 3140
3378	Biphenyl, 3-amino $3-H_2NC_6H_4-C_6H_5$	169.23	nd	254^{135}	30	....	....	al, eth, ace, bz	B12³, 3146
3379	Biphenyl, 3-amino-4-hydroxy $3-H_2N-4-HO-C_6H_3-C_6H_5$	185.23	nd (chl)	....	208	....	....	al, eth, bz	B13³, 1946
3380	Biphenyl, 3-amino-4-nitro $4-NO_2-3-H_2NC_6H_3-C_6H_5$	214.22	og nd (dil al)	....	116	....	....	al	B12³, 3149
3381	Biphenyl, 3-amino-4′-nitro $3-H_2NC_6H_4-C_6H_4NO_2-4$	214.22	og nd (al)	....	137	....	....	al, aa	B12², 753
3382	Biphenyl, 4-amino or Xenylamine $4-H_2NC_6H_4-C_6H_5$	169.23	lf (dil al)	302, 191^{15}	53-4	....	....	al, eth, chl	B12³, 3152
3383	Biphenyl, 4-amino-2′-hydroxy $4-H_2NC_6H_4-C_6H_4OH-2$	185.23	nd (to)	....	181-2	....	....	al, aa	B13³, 1943
3384	Biphenyl, 4-amino-4′-hydroxy $4-H_2NC_6H_4-C_6H_4OH-4$	185.23	pl (dil al)	....	275	....	....	....	B13², 420
3385	Biphenyl, 4-amino-2′-nitro $4-H_2NC_6H_4-C_6H_4NO_2-2$	214.22	red mcl pr (al)	....	99	....	....	al	B12³, 3199
3386	Biphenyl, 4-amino-3-nitro $3-NO_2-4-H_2NC_6H_3-C_6H_5$	214.22	red nd (al)	....	170-1	....	....	al, eth, chl, aa	B12³, 3198
3387	Biphenyl, 4-amino-4′-nitro $4-H_2NC_6H_4-C_6H_4NO_2-4$	214.22	red nd (al)	....	203-4	....	....	al, aa	B12³, 3200
3388	Biphenyl, 5-amino-2-hydroxy $5-H_2N-2-HOC_6H_3-C_6H_5$	185.23	nd (al or bz)	....	201	....	....	al, bz	B13³, 1940
3389	Biphenyl, 2-benzyl or o-Biphenylphenyl methane $2-(C_6H_5CH_2)C_6H_4-C_6H_5$	244.34	mcl nd (al)	$283-7^{110}$	54-6	....	....	al, eth, bz	B5³, 2323
3390	Biphenyl, 4-benzyl or p-Biphenylphenyl methane $4-(C_6H_5CH_2)C_6H_4-C_6H_5$	244.34	lf	$285-6^{110}$	85	$1.171^{0/4}$	....	al, eth, bz	B5³, 2324
3391	Biphenyl, 4,4′-bis(diethylamino) $[4-(C_2H_5)_2NC_6H_4-]_2$	296.46	nd (al)	....	85	....	....	al, eth	B13³, 430
3392	Biphenyl, 2,4′-bis(dimethylamino) $2-(CH_3)_2N-C_6H_4C_6H_4[N(CH_3)_2-4]$	240.35	pl (al)	$206-7^{11}$	51-2	....	....	al, eth	B13², 88
3393	Biphenyl, 4,4′-bis(dimethylamino) $4-(CH_3)_2N-C_6H_4-C_6H_4[N(CH_3)_2-4]$	240.35	nd (al or bz-lig)	>360	198	....	....	bz, chl	B13³, 429
3394	Biphenyl, 4,4′-bis(ethylamino) $4-C_2H_5NHC_6H_4-C_6H_4(NHC_2H_5-4)$	240.35	nd or pl (al)	....	120.5	....	....	al, eth, bz	B13, 222
3395	Biphenyl, 4,4′-bis(methylamino) $4-CH_3NHC_6H_4-C_6H_4(NHCH_3-4)$	212.29	lf (al, w, or lig)	....	91	....	....	al, lig	B13², 97
3396	Biphenyl, 4,4′-bis(phenylamino) $4-C_6H_5NHC_6H_4-C_6H_4(NHC_6H_5-4)$	336.24	lf (to)	....	244-5	....	....	aa, to	B13³, 431
3397	Biphenyl, 2-bromo $2-BrC_6H_4-C_6H_5$	233.11	....	296-8, 160^{11}	1-2	1.2175^{26}	1.6248^{25}	al, eth	B5⁴, 1818
3398	Biphenyl, 3-bromo $3-BrC_6H_4-C_6H_5$	233.11	....	$299-30^{1}$, $169-73^{17}$	....	....	1.6411^{20}	....	B5⁴, 1818

No.	Name, Synonyms, and Formula	Mol. wt.	Color, crystalline form, specific rotation and λ_{max} (log ε)	b.p. °C	m.p. °C	Density	n_D	Solubility	Ref.
3399	Biphenyl, 3-bromo-4-hydroxy $3\text{-Br-4-HOC}_6\text{H}_3\text{-C}_6\text{H}_5$	249.11	nd (chl-peth)		96			al, chl, aa	$B6^3$, 3333
3400	Biphenul, 4-bromo $4\text{-BrC}_6\text{H}_4\text{-C}_6\text{H}_5$	233.11	pl (al)	310	91.2	$0.9327^{25/4}$		al, eth, bz, aa	$B5^4$, 1819
3401	Biphenyl, 4-(bromoacetyl) $4\text{-(BrCH}_2\text{CO)C}_6\text{H}_4\text{-C}_6\text{H}_5$	275.15	nd (95% al)		127				$B7^3$, 2137
3402	Biphenyl, 4-bromo-4′-hydroxy $4\text{-BrC}_6\text{H}_4\text{-C}_6\text{H}_4\text{OH-4)}$	249.11	pl (al)		164-6			al, eth, ace, bz	$B6^3$, 3334
3403	Biphenyl, 2-chloro $2\text{-ClC}_6\text{H}_4\text{-C}_6\text{H}_5$	188.66	nd (dil al)	274, 154^{12}	34	$1.1499^{32.5}$		al, eth, lig	$B5^4$, 1816
3404	Biphenyl, 3-chloro $3\text{-ClC}_6\text{H}_4\text{-C}_6\text{H}_5$	188.66		284-5, 150-60^6	16	$1.1579^{25/4}$	1.6181^{25}	al, eth, ace	$B5^4$, 1816
3405	Biphenyl, 3-chloro-2-hydroxy $3\text{-Cl-2-HOC}_6\text{H}_3\text{-C}_6\text{H}_5$	204.66		317-8d	6	$1.24^{25/4}$	1.6237^{30}	al, eth, ace, bz	$B6^3$, 3297
3406	Biphenyl, 4-chloro $4\text{-ClC}_6\text{H}_4\text{-C}_6\text{H}_5$	188.66	lf (lig or al)	291, 180-95^{20-30}	77.7			al, eth, lig	$B5^4$, 1816
3407	Biphenyl, 4-(chloroacetyl) $4\text{-(ClCH}_2\text{CO)C}_6\text{H}_4\text{-C}_6\text{H}_5$	230.69	pl (al)		125			al	B7, 443
3408	Biphenyl, 4-chloro-4′-hydroxy $4\text{-Cl-C}_6\text{H}_4\text{-(C}_6\text{H}_4\text{OH-4)}$	204.66	cr (dil al)		146-7			al, eth, ace, bz	$B6^3$, 3332
3409	Biphenyl, 5-chloro-2-hydroxy $5\text{-Cl-2-HOC}_6\text{H}_3\text{C}_6\text{H}_5$	204.66		319^{745}, 128-30^2	11				$B6^3$, 3299
3410	Biphenyl, 2,2′-diacetamido $[2\text{-(CH}_3\text{CONH)C}_6\text{H}_4\text{-]}_2$	268.32	pr (al)		164-5			bz, aa	$B13^3$, 412
3411	Biphenyl, 2,4-diacetamido $2,4\text{-(CH}_3\text{CONH)}_2\text{C}_6\text{H}_3\text{-C}_6\text{H}_5$	268.32	nd (al)		202			al	$B13^3$, 417
3412	Biphenyl, 4,4′-diacetamido $[4\text{-(CH}_3\text{CONH)C}_6\text{H}_4\text{-]}_2$	268.32	nd (aa)		328.3				$B13^3$, 437
3413	Biphenyl, 2,2′-diamino or *o*-Benzidine $2\text{-H}_2\text{NC}_6\text{H}_4\text{-C}_6\text{H}_4\text{NH}_2\text{-2}$	184.24	mcl pr or nd (al)	162^4	81			w, bz	$B13^3$, 410
3414	Biphenyl, 2,4′-diamino or Diphenyline $2\text{-H}_2\text{NC}_6\text{H}_4\text{-C}_6\text{H}_4\text{NH}_2\text{-4}$	184.24	nd (dil al)	363	54.5			al, eth	$B13^3$, 416
3415	Biphenyl, 3,3′-diamino or *m*-Benzidine $3\text{-H}_2\text{NC}_6\text{H}_4\text{-C}_6\text{H}_4\text{NH}_2\text{-3}$	184.24	nd (w), pr (bz)	205	93-4			eth, bz	$B13^3$, 422
3416	Biphenyl, 3,4-diamino $3,4\text{-(H}_2\text{N)}_2\text{C}_6\text{H}_3\text{-C}_6\text{H}_5$	184.24	lf (eth or al)		103			al, eth	$B13^2$, 89
3417	Biphenyl, 4,4′-diamino or Benzidine $4\text{-H}_2\text{NC}_6\text{H}_4\text{-C}_6\text{H}_4\text{NH}_2\text{-4}$	184.24	nd (w)	400^{740}	125			al	$B13^3$, 425
3418	Biphenyl, 4,4′-diamino-3,3′-dimethoxy $[3\text{-CH}_3\text{O-4-H}_2\text{NC}_6\text{H}_3\text{-]}_2$	244.29	lf or nd (w)		137			al, eth, ace, bz, chl	$B13^3$, 2310
3419	Biphenyl, 4,4 -diamino-2,2 -dimethyl $[2\text{-CH}_3\text{-4-H}_2\text{NC}_6\text{H}_3\text{-]}_2$	212.29	pr (w)		108-9			al, eth	B13, 255
3420	Biphenyl, 4,4′-diamino-3,3′-dimethyl $[3\text{-CH}_3\text{-4-H}_2\text{NC}_6\text{H}_3\text{-]}_2$	212.29	lf (dil al)		131-2			al, eth	$B13^3$, 484
3421	Biphenyl, 4,4′-diamino-3-ethoxy $3\text{-C}_2\text{H}_5\text{O-4-H}_2\text{NC}_6\text{H}_3\text{-C}_6\text{H}_4\text{-NH}_2\text{-4}$	228.29	nd (w)		134			al	$B13^2$, 419
3422	Biphenyl, 4,4′-dibromo $4\text{-BrC}_6\text{H}_4\text{-C}_6\text{H}_4\text{Br-4}$	312.00	mcl pr (MeOH)	355-60	164			bz	$B5^4$, 1820
3423	Biphenyl, 3,3′-dichloro $3\text{-ClC}_6\text{H}_4\text{-C}_6\text{H}_4\text{Cl-3}$	223.10		322-4	29			al, eth, bz	$B5^3$, 1739
3424	Biphenyl, 4,4′-dichloro $4\text{-ClC}_6\text{H}_4\text{-C}_6\text{H}_4\text{Cl-4}$	223.11	pr or nd (al or to-peth)	315-9	148-9			bz	$B5^4$, 1817
3425	Biphenyl, 4,4′-dichloro-2,2′-dinitro $[4\text{-Cl-2-NO}_2\text{C}_6\text{H}_3\text{-]}_2$	313.10	ye cr (al)		140			bz, aa	$B5^4$, 1828
3426	Biphenyl, 2,2′-diethoxy-3,3′-dimethyl $[3\text{-CH}_3\text{-2-C}_2\text{H}_5\text{O-C}_6\text{H}_3\text{-]}_2$	270.37	lf (al)		85				$B6^2$, 974
3427	Biphenyl, 2,4′-diethoxy-3,3′-dimethyl $3\text{-CH}_3\text{-2-C}_2\text{H}_5\text{O-C}_6\text{H}_3\text{-C}_6\text{H}_3\text{-OC}_2\text{H}_5\text{(4)-CH}_3\text{-3}$	270.37	nd (al)		53				$B6^2$, 974
3428	Biphenyl, 4,4′-diethoxy-3,3′-diethyl $[3\text{-C}_2\text{H}_5\text{-4-C}_2\text{H}_5\text{OC}_6\text{H}_3\text{-]}_2$	298.43	lf (al)		120				B6, 1015
3429	Biphenyl, 4,4′-diethoxy-3,3′-dimethyl $[3\text{-CH}_3\text{-4-C}_2\text{H}_5\text{OC}_6\text{H}_3\text{-]}_2$	270.37	pl (al)		156				B6, 1010
3430	Biphenyl, 3,3′-diethyl-6,6′-dihydroxy $[3\text{-C}_2\text{H}_5\text{-6-HOC}_6\text{H}_3\text{-]}_2$	242.32	nd (dil al)		131				$B6^2$, 981
3431	Biphenyl, 2,2′-difluoro $2\text{-F-C}_6\text{H}_4\text{-C}_6\text{H}_4\text{-F-2}$	190.19			118.5-9.5	$1.393^{20/4}$		al	$B5^3$, 1735

No.	Name, Synonyms, and Formula	Mol. wt.	Color, crystalline form, specific rotation and λ_{max} (log ε)	b.p. °C	m.p. °C	Density	n_D	Solubility	Ref.
3432	Biphenyl, 3,3′-difluoro $3\text{-}FC_6H_4C_6H_4F\text{-}3$	190.19	....	130^{14}	8	$1.192^{25/4}$	1.5678^{20}	....	B5[3], 1736
3433	Biphenyl, 4,4′-difluoro $4\text{-}FC_6H_4\text{-}C_6H_4F\text{-}4$	190.19	mcl pr (al), lf (w)	254—5, 119^{14}	94-5	....	....	al, eth, ace, bz, chl	B5[3], 1736
3434	Biphenyl, 2,2′-dihydroxy or *o,o′*-Biphenol $2\text{-}HOC_6H_4\text{-}C_6H_4OH\text{-}2$	186.21	lf (w + 1), pr (to)	325-6	110-2 (anh)	....	....	al, eth, ace, bz, aa	B6[3], 5374
3435	Biphenyl, 2,2′-dihydroxy-3,3′-dimethyl $[3\text{-}CH_3\text{-}2\text{-}HOC_6H_3\text{-}]_2$	214.26	nd (peth)	sub	113	....	....	al, eth, bz	B6[3], 5445
3436	Biphenyl, 2,2′-dihydroxy-5,5′-dimethyl $[5\text{-}CH_3\text{-}2\text{-}HOC_6H_3\text{-}]_2$	214.26	nd (bz or w)	sub	153.5	....	....	al, eth, ace, bz	B6[3], 5447
3437	Biphenyl, 2,2′-dihydroxy-6,6′-dimethyl $[6\text{-}CH_3\text{-}2\text{-}HOC_6H_3\text{-}]_2$	214.26	pl (dil al)	....	164	....	....	al	B6[3], 5445
3438	Biphenyl, 2,2′-dihydroxy-3,3′,5,5′-tetramethyl $[3,5\text{-}(CH_3)_2\text{-}2\text{-}HOC_6H_2\text{-}]_2$	242.32	nd or pl (eth or lig)	$140\text{-}60^{0.05}$	137-8	....	....	al, eth, lig	B6[3], 5483
3439	Biphenyl, 2,4′-dihydroxy or *o,p′*-Biphenol $2\text{-}HOC_6H_4\text{-}C_6H_4OH\text{-}4$	186.21	mcl pr or nd (dil al)	342, $206\text{-}10^{11}$	162-3	....	....	eth	B6[3], 5387
3440	Biphenyl, 2,5-dihydroxy $2,5\text{-}(HO)_2C_6H_3C_6H_5$	186.21	nd (dil al)	....	97-8	....	....	al	B6[3], 5371
3441	Biphenyl, 2,5′-dihydroxy-2′,5-dimethyl $5\text{-}CH_3\text{-}2\text{-}HOC_6H_3C_6H_3OH(5)\text{-}CH_3\text{-}2$	214.26	....	....	158	....	....	al, eth	B6[3], 5445
3442	Biphenyl, 3,3′-dihydroxy or *m,m′*-Biphenol $3\text{-}HOC_6H_4\text{-}C_6H_4OH\text{-}3$	186.21	nd (w)	247^{18}	123-4	....	....	al, eth, bz, chl	B6[3], 5388
3443	Biphenyl, 3,4-dihydroxy $3\text{-}HOC_6H_4\text{-}C_6H_4OH\text{-}4$	186.21	....	>360	145	....	....	al, eth, ace, bz, chl	B6[3], 5387
3444	Biphenyl, 4,4′-dihydroxy or *p,p′*-Biphenol $4\text{-}HOC_6H_4\text{-}C_6H_4OH\text{-}4$	186.21	nd or pl (al)	sub	274-5	....	....	al, eth	B6[3], 5389
3445	Biphenyl, 4,4′-dihydroxy-3,3′-diethyl $[3\text{-}C_2H_5\text{-}4\text{-}HOC_6H_3\text{-}]_2$	242.32	nd (aa)	....	148	....	....	al, ace, bz, aa	B6[3], 5482
3446	Biphenyl, 4,4′-dihydroxy-3,3′-dimethyl $[3\text{-}CH_3\text{-}4\text{-}HOC_6H_3\text{-}]_2$	214.26	lf (w), nd	....	161	....	....	al, eth	B6[3], 5445
3447	Biphenyl, 4,4′-diacetoxy-3,3′-dimethyl $[3\text{-}CH_3\text{-}4\text{-}CH_3CO_2C_6H_3\text{-}]_2$	298.34	wh nd (al, aa)	....	135.3	....	....	....	B6[3], 5446
3448	Biphenyl, 4,4′-dihydroxy-3,3′,5,5′-tetramethoxy or Hydrocerulignone $[3,5\text{-}(CH_3)_2\text{-}4\text{-}HOC_6H_2\text{-}]_2$	306.32	mcl pr (al)	....	190	....	....	al	B6[1], 593
3449	Biphenyl, 4,4′-dihydroxy-3,3′,5,5′-tetramethyl $[3,5\text{-}(CH_3)_2\text{-}4\text{-}HOC_6H_2\text{-}]_2$	242.32	pa ye nd or pr (aa)	sub	222-3	....	....	al, aa	B6, 1015
3450	Biphenyl, 5,5′-dihydroxy-2,2′-dimethyl $[2\text{-}CH_3\text{-}5\text{-}HOC_6H_3\text{-}]_2$	214.26	pr (al)	....	229	....	....	al, eth	B6[3], 5444
3451	Biphenyl, 4,4′-dihydroxy-3,3′,5,5′-tetranitro $[3,5\text{-}(NO_2)_2\text{-}4\text{-}HOC_6H_2\text{-}]_2$	366.20	ye nd	....	223	....	....	....	B6[3], 5399
3452	Biphenyl, 2,2′-dimethoxy or *o,o′*-Bianisole $(2\text{-}CH_3OC_6H_4\text{-})_2$	214.26	rh bipyr pr (al)	$307\text{-}8^{766}$	155	1.268	....	al, bz, chl	B6[3], 5375
3453	Biphenyl, 3,3′-dimethoxy or *m, m′* Bianisole $(3\text{-}CH_3OC_6H_4\text{-})_2$	214.26	nd (dil al)	328, $211\text{-}20^{15}$	36	....	....	al, eth, ace, bz, chl	B6[3], 5388
3454	Biphenyl, 4,4′-dimethoxy or *p, p′*-Bianisole $(4\text{-}CH_3OC_6H_4\text{-})_2$	214.26	lf (bz)	sub	173	....	....	al, bz, chl	B6[3], 5391
3455	Biphenyl, 2,2′-dimethoxy-5,5′-dimethyl $[5\text{-}CH_3\text{-}2\text{-}CH_3OC_6H_3\text{-}]_2$	242.32	nd (dil al)	188^{12}	71	....	....	al, eth, ace, bz	B6[3], 5447
3456	Biphenyl, 2,5′-dimethoxy-2′,5-dimethyl $5\text{-}CH_3\text{-}2\text{-}CH_3OC_6H_3\text{-}C_6H_3OCH_3(5)\text{-}CH_3\text{-}2$	242.32	pr (al)	168^{4}	86	....	....	bz, peth	B6[2], 973
3457	Biphenyl, 4,4′-dimethoxy-3,3′-dimethyl $[3\text{-}CH_3\text{-}4\text{-}CH_3OC_6H_3\text{-}]_2$	242.32	pr (al)	....	145.5	....	....	....	B6[3], 5445
3458	Biphenyl, 2,2′-dimethyl or *o, o′*-Bitolyl $(2\text{-}CH_3C_6H_4\text{-})_2$	182.27	cr (al)	256	19—20	0.9906^{20}	1.5752^{20}	al, eth, ace, bz	B5[4], 1897
3459	Biphenyl-2,3-dimethyl $2,3\text{-}(CH_3)_2C_6H_3C_6H_5$	182.26	....	141^{14}	42	....	1.5845^{23}	eth	B5[4], 1897
3460	Biphenyl, 2,3′-dimethyl or *o,m′*-Bitolyl $2\text{-}CH_3C_6H_4\text{-}C_6H_4CH_3\text{-}3$	182.27	....	270	....	0.9924^{20}	1.5810^{20}	al, eth, ace, bz	B5[4], 1902
3461	Biphenyl, 2,4-dimethyl $2,4\text{-}(CH_3)_2C_6H_3\text{-}C_6H_5$	182.27	....	$270\text{-}6^{767}$	0.9947 20/4	1.5844^{20}	....	....	B5[4], 1897
3462	Biphenyl, 2,4′-dimethyl or *o,p′*-Bitolyl $2\text{-}CH_3C_6H_4\text{-}C_6H_4CH_3\text{-}4$	182.27	....	273-6, $137^{12.5}$	....	0.9924^{20}	1.5826^{20}	al, eth, ace, bz	B5[4], 1903
3463	Biphenyl, 2,5-dimethyl $2,5\text{-}(CH_3)_2C_6H_3\text{-}C_6H_5$	182.27	....	$140^{14.5}$	....	$0.9931^{20/4}$	1.5819^{20}	....	B5[4], 1897
3464	Biphenyl, 2,6-dimethyl $2,6\text{-}(CH_3)_2C_6H_3\text{-}C_6H_5$	182.27	....	260-5, $132^{16.5}$	− 5	$0.9907^{20/4}$	1.5745^{20}	....	B5[4], 1897

PHYSICAL CONSTANTS OF ORGANIC COMPOUNDS (Continued)

No.	Name, Synonyms, and Formula	Mol. wt.	Color, crystalline form, specific rotation and λ_{max} (log ε)	b.p. °C	m.p. °C	Density	n_D	Solubility	Ref.
3465	Biphenyl, 3,3′-dimethyl or *m, m*′-Bitolyl $(3-CH_3C_6H_4-)_2$	182.27		$280, 150^{18}$	9	$0.9995^{20/4}$	1.5946^{20}	al, eth, ace, bz	$B5^4$, 1903
3466	Biphenyl, 3,3′-dimethyl-4,4′-dipropoxy $(3-CH_3-4-C_3H_7OC_6H_3-)_2$	298.43	lf		115				B6, 1010
3467	Biphenyl, 3,4-dimethyl $3,4-(CH_3)_2C_6H_3C_6H_5$	182.27		$281-3, 139-40^{8}$	29.2-9.7	$1.0087^{20/4}$	1.6036^{20}	bz	$B5^4$, 1903
3468	Biphenyl, 3,4′-dimethyl $3-CH_3C_6H_4-C_6H_4CH_3-4$	182.27		$288-9^{752}, 153^{15}$	14-15	$0.9978^{20/4}$	1.5968^{20}	bz	$B5^4$, 1905
3469	Biphenyl, 3,5-dimethyl $3,5-(CH_3)_2C_6H_3-C_6H_5$	182.27		273-6	22-3	$0.9990^{20/4}$	1.5952^{20}		$B5^4$, 1903
3470	Biphenyl, 4,4′-dimethyl $(4-CH_3C_6H_4-)_2$	182.27	mcl pr (eth)	295	125	$0.917^{121/4}$		eth, ace, bz	$B5^4$, 1906
3471	Biphenyl, 2,2′-dinitro $(2-NO_2C_6H_4-)_2$	244.21	ye mcl pr or nd (al)		127-8	1.45 (sol)		al, eth, bz, aa	$B5^4$, 1826
3472	Biphenyl, 2,3′-dinitro $2-NO_2C_6H_4-C_6H_4NO_2-3$	244.21			118-9.5			al	$B5^4$, 1826
3473	Biphenyl, 2,4′-dinitro $2-NO_2C_6H_4-C_6H_4NO_2-4$	244.21	mcl pr (al)		93-4	1.474		al, eth, bz, aa	$B5^4$, 1827
3474	Biphenyl, 3,3′-dinitro $(3-NO_2C_6H_4-)_2$	244.21	ye og nd (al or aa)		200			bz, aa	$B5^4$, 1827
3475	Biphenyl, 4,4′-dinitro $(4-NO_2C_6H_4-)_2$	244.21	nd (al)		240-3			bz, aa	$B5^4$, 1827
3476	Biphenyl, 2-ethoxy $2-C_2H_5OC_6H_4-C_6H_5$	198.26	pr (peth)	$276, 132^{6}$	34			al, eth, ace, bz, chl	$B6^3$, 3284
3477	Biphenyl, 3-ethoxy $3-C_2H_5OC_6H_4-C_6H_5$	198.26	(peth)	$305, 158^{8}$	35			al, eth, bz, ace	$B6^3$, 3313
3478	Biphenyl, 3-ethyl $3-C_2H_5C_6H_4-C_6H_5$	182.27		$283-4^{763}$		1.043^{0}			$B5^4$, 1896
3479	Biphenyl, 2-fluoro $2-F-C_6H_4-C_6H_5$	172.20		248	73.5	$1.2452^{25/4}$		al, eth	$B5^4$, 1815
3480	Biphenyl, 4-fluoro $4-F-C_6H_4-C_6H_5$	172.20		253	74.5			eth	$B5^4$, 1815
3481	Biphenyl, 2,2′,4,4′,6,6′-hexamethyl $[2,4,6-(CH_3)_3C_6H_2-]_2$	238.37		296^{735}	103—4	1.023^{50}		eth, bz	$B5^4$, 1989
3482	Biphenyl, 2-hydroxy-2′-methoxy-5,5′-dimethyl $C_{15}H_{16}O_2$	228.29		205^{12}				al, eth, bz, chl	$B6^2$, 974
3483	Biphenyl, 2-iodo $2-IC_6H_4-C_6H_5$	280.11		$189-92^{36}$		$1.6038^{25/25}$	1.6620^{20}	al, eth, bz, aa	$B5^4$, 1820
3484	Biphenyl, 4-iodo $4-IC_6H_4-C_6H_5$	280.11	nd (al or aa)	$320d, 183^{11}$	113-4			al, eth, bz, aa	$B5^4$, 1821
3485	Biphenyl, 2-methoxy or 2-Phenylanisole $2-CH_3OC_6H_4-C_6H_5$	184.24	pr (peth)	$274, 150^{13}$	29	$1.0233^{99/4}$	1.5641^{99}	al, peth	$B6^3$, 3284
3486	Biphenyl, 4-methoxy or 4-Phenylanisole $4-CH_3OC_6H_4-C_6H_5$	184.24	pl (al)	157^{10}	90	$1.0278^{100/4}$	1.5744^{100}	al, eth	$B6^3$, 3321
3487	Biphenyl, 2-methyl or 2-Phenyl toluene $2-CH_3C_6H_4-C_6H_5$	168.24		$255.5, 130-6^{27}$	-0.2	$1.010^{22/4}$	1.5914^{20}	al, eth	$B5^4$, 1855
3488	Biphenyl, 3-methyl $3-CH_3C_6H_4C_6H_5$	168.24		$272.7, 148.50^{20}$	4.5	$1.0182^{17/4}$	1.5972^{20}	al, eth	$B5^4$, 1858
3489	Biphenyl, 4-methyl $4-CH_3C_6H_4C_6H_5$	168.24	pl (lig, MeOH)	$267-8, 134-6^{15}$	49-50	1.015^{27}		al, eth	$B5^4$, 1860
3490	Biphenyl, 2-nitro $2-NO_2C_6H_4-C_6H_5$	199.21	pl (al or MeOH)	$320, 201^{30}$	37.2	1.44		al, eth	$B5^4$, 1823
3491	Biphenyl, 3-nitro $3-NO_2C_6H_4-C_6H_5$	199.21	ye pl or nd (dil al)	$225-30^{35}$	62			al, eth, aa, lig	$B5^4$, 1823
3492	Biphenyl, 4-nitro $4-NO_2C_6H_4-C_6H_5$	199.21	ye nd (al)	$340, 224^{30}$	114			eth, bz, chl, aa	$B5^4$, 1823
3493	Biphenyl, 2,2′,4,4′-tetrahydroxy $[2,4-(HO)_2C_6H_3-]_2$	218.21			226-7			w, al, eth, ace	$B6^3$, 6705
3494	Biphenyl, 3,3′,5,5′-tetrahydroxy or Diresorcinol $[3,5-(HO)_2C_6H_3-]_2$	218.21	pl or nd (w + 2)		310 (anh)			al, eth, w	$B6^2$, 1129
3495	Biphenyl, 2,2′,4,4′-tetramethyl $[2,4-(CH_3)_2C_6H_3-]_2$	210.32			41			al	$B5^3$, 1891
3496	Biphenyl, 2,2′,5,5′-tetramethyl $[2,5-(CH_3)_2C_6H_3-]_2$	210.32		284^{732}	50			eth, bz	$B5^3$, 1892
3497	Biphenyl, 2,2′,4,4′-tetranitro $[2,4-(NO_2)_2C_6H_3]_2$	334.20	ye pr (bz)		165-6			bz, aa	$B5^3$, 1772

No.	Name, Synonyms, and Formula	Mol. wt.	Color, crystalline form, specific rotation and λ_{max} (log ε)	b.p. °C	m.p. °C	Density	n_D	Solubility	Ref.
3498	Biphenyl, 2,4,4′-triamino $2,4-(H_2N)_2C_6H_3-C_6H_4NH_2-4$	199.26	nd		134				B13^{3}, 560
3499	Biphenyl, 4-vinyl $4-(CH_2=CH)C_6H_4-C_6H_5$	180.25		136-8^{6}	119				B5^{3}, 1987
3500	2-Biphenyl carboxylic acid $2-(HO_2C)C_6H_4-C_6H_5$	198.22	lf (dil al)	343-4, 199^{10}	113-4			al, bz, aa	B9^{3}, 3268
3501	2-Biphenylcarboxylonitrile $2-NCC_6H_4-C_6H_5$	179.22	nd	170-2^{15}	41			al, eth	B9^{3}, 3269
3502	3-Biphenylcarboxylic acid $3-(HO_2C)C_6H_4C_6H_5$	198.22	lf (al)		165—6			al, eth, bz, aa	B9^{3}, 3274
3503	4-Biphenylcarboxylic acid $4-(HO_2C)C_6H_4C_6H_5$	198.22	nd (bz or al)	sub	228			al, eth, bz	B9^{3}, 3276
3504	3-Biphenylcarboxylic acid, 2-hydroxy $2-HO-3-(HO_2C)C_6H_3-C_6H_5$	214.22	cr (bz)		186-7			al	B10^{3}, 1159
3505	2,2′-Biphenyldicarboxylic acid or Diphenic acid $[2-(HO_2C)C_6H_4-]_2$	242.23	mcl pr or lf (w), cr (aa)	sub	233.5			al, eth	B9^{3}, 4496
3506	2,2′-Biphenyldicarboxylic acid anhydride $C_{14}H_8O_3$	224.22	nd (aa or bz)	sub	217				B17^{4}, 6425
3507	2,2′-Biphenyldicarboxylyl chloride $[2-(ClCO)C_6H_4-]_2$	279.14		sub	94 (97)				B9^{2}, 657
3508	2,2′-Biphenyldicarboxylic acid, diethyl ester $[2-(C_2H_5O_2C)C_6H_4-]_2$	298.34			42			eth	B9^{2}, 656
3509	2,2′-Biphenyldicarboxylic acid, dimethyl ester $[2-(CH_3O_2C)C_6H_4-]_2$	270.28	mcl pr (MeOH)	204-6^{14}	74			al, eth, bz	B9^{3}, 4497
3510	2,2′-Biphenyldicarboxylic acid, imide $C_{14}H_9NO_2$	223.23	nd (al)		219-20			chl	B21^{4}, 5601
3511	2,3′-Biphenyldicarboxylic acid or Isodiphenic acid $2-(HO_2C)-C_6H_4-C_6H_4(CO_2H)-3$	242.22	nd (w or dil aa)		216			al	B9^{2}, 663
3512	2,4′-Biphenyldicarboxylic acid $2-(HO_2C)-C_6H_4-C_6H_4(CO_2H)-4$	242.23	lf (al)		272-3			al, bz, aa	B9^{3}, 4514
3513	3,3′-Biphenyldicarboxylic acid $[3-(HO_2C)-C_6H_4-]_2$	242.23	lf (al)		356-7			chl	B9^{3}, 4517
3514	3,3′-Biphenyl dicarboxylic acid, dimethyl ester $[3-CH_3O_2C-C_6H_4-]_2$	270.29	lf (MeOH)		104			al, eth, bz	B9^{3}, 4517
3515	3,4′-Biphenyl dicarboxylic acid $[3-(HO_2C)-C_6H_4-C_6H_4-(CO_2H)-4]$	242.23	nd		334-5				B9^{3}, 4518
3516	3,4′-Biphenyldicarboxylic acid, dimethyl ester $3-(CH_3O_2C)-C_6H_4\cdot C_6H_4CO_2CH_3-4$	270.28	nd (lig or MeOH)		98-9			lig, MeOH	B9, 927
3517	3,5-Biphenyldicarboxylic acid $3,5-(HO_2C)_2C_6H_3C_6H_5$	242.23	lf (aa)		>310			al, eth, ace, bz	B9, 926
3518	3,5-Biphenyldicarboxylic acid, dimethyl ester $3-5-(CH_3O_2C)_2C_6H_3C_6H_5$	270.28	lf (MeOH)		214				B9^{2}, 665
3519	2,2′-Biphenyldicarboxylic acid, 3,3′-dimethyl-5-nitro $C_{16}H_{13}NO_6$	315.28	lf (w or dil al)		267			al, eth, bz	B9^{2}, 659
3520	2,2′-Biphenyldicarboxylic acid, 3,3′-dimethyl-6-nitro *dl* $C_{16}H_{13}NO_6$	315.28	lf (w)		248-50d			al, eth, ace, aa	B9^{2}, 659
3521	2,2′-Biphenyldicarboxylic acid, 3,3′-dimethyl-4-nitro $C_{16}H_{13}NO_6$	315.28	lf, pr or wh nd (w)		217 (250)			al, eth	B9^{2}, 659
3522	2,2′-Biphenyldicarboxylic acid, 3,3′-dimethyl $[3-CH_3-2-(HO_2C)C_6H_3-]_2$	270.28			230			al, eth, bz	B9^{1}, 407
3523	2,2′-Biphenyl disulfonyl chloride $[2-ClO_2S-C_6H_4-]_2$	351.24	pr (chl), cr (aa)		142-4			eth, bz, chl	B11^{3}, 469
3524	3,3′-Biphenyl disulfonamide $[3-(H_2NO_2S)-C_6H_4-]_2$	312.37	nd (ace)		285			al	B11, 219
3525	3,3′-Biphenyl disulfonyl chloride $[3-(ClO_2S)-C_6H_4-]_2$	351.24	nd (chl)		128			eth, bz	B11^{3}, 471
3526	4,4′-Biphenyl disulfonic acid $[4-(HO_3S)-C_6H_4-]_2$	314.34	pr	>200	72.5			w	B11^{3}, 472
3527	4,4′-Biphenyl disulfonamide $[4-(H_2NO_2S)-C_6H_4-]_2$	312.37	nd (w)		300			w, eth	B11, 220
3528	4,4′-Biphenyl disulfonyl chloride $[4-(ClO_2S)C_6H_4-]_2$	351.24	pr (aa)		205—7			eth, bz, aa	B11^{3}, 472
3529	2,2′-Biphenyl disulfonic acid, 4,4′-diamino $[2-(HO_3S)-4-H_2N-C_6H_3-]_2$	344.37	lf		175d				B14^{3}, 2264
3530	2,2′-Bipyridyl $2-NC_5H_4-C_5H_4N-2$	156.19	pr (peth)	272-5	71-3			al, eth, bz, chl, lig	B23^{2}, 211

No.	Name, Synonyms, and Formula	Mol. wt.	Color, crystalline form, specific rotation and λ_{max} (log ε)	b.p. °C	m.p. °C	Density	n_D	Solubility	Ref.
3531	2,3′-Bipyridyl or Isonicoteine $2\text{-}NC_5H_4\text{-}C_5H_4N\text{-}(3)$	156.19		295-6		$1.140^{20/4}$	1.6223^{20}	al, eth, bz, chl	$B23^2$, 212
3532	2,4′-Bipyridyl $2\text{-}NC_5H_4\text{-}C_5H_4N\text{-}(4)$	156.19		280-2, $148\text{-}50^1$	61.5			al, eth, chl	B23, 200
3533	3,3′-Bipyridyl $3\text{-}NC_5H_4\text{-}C_5H_4N\text{-}(3)$	156.19		291-2, $190\text{-}2^{25}$	68	$1.1635^{20/20}$		w, al	$B23^2$, 212
3534	3,4′-Bipyridyl $3\text{-}NC_5H_4\text{-}C_5H_4N\text{-}(4)$	156.19	lf (peth)	297	62			w, al, peth	B23, 212
3534a	4,4′-Bipyridyl $4\text{-}NC_5H_4\text{-}C_5H_4N\text{-}(4)$	156.19	nd (w + 2)	305 sub	114, 171-2			al, eth, bz, chl, lig	$B23^2$, 212
3535	2,2′-Biquinolyl $(C_9H_6N)_2$	256.31	pl or lf (al)		196			al, eth, ace, bz	$B23^2$, 267
3536	2,3′-Biquinolyl $(C_9H_6N)_2$	256.31	lf (al), ye pl or nd (bz)	>400	176-7			al, eth, bz, chl	$B23^2$, 267
3537	2,6′-Biquinolyl $(C_9H_6N)_2$	256.31	pl (al)		144			al, ace, bz	B23, 294
3538	2,7′-Biquinolyl(higher melting) $(C_9H_6N)_2$	256.31	mcl pl (al)		160			al	B23, 294
3539	2,7′-Biquinolyl(lower melting) $C_{18}H_{12}N_2$	256.31	tcl		115			al, eth, bz	B23, 294
3540	3,4′-Biquinolyl $(C_9H_6N)_2$	256.31	pw (peth)		83-4			al, bz	$B23^2$, 267
3541	3,7′-Biquinolyl $(C_9H_6N)_2$	256.31	lf or nd (al or bz)		190			al, bz, chl	$B23^2$, 268
3542	4,4′-Biquinolyl $(C_9H_6N)_2$	256.31	pr (peth)		171			al, ace, bz	$B23^2$,268
3543	4,6′-Biquinolyl $(C_9H_6N)_2$	256.31	cr (bz)		122			al, bz, chl	B23,294
3544	6,6′-Biquinolyl $(C_9H_6N)_2$	256.31	lf (al)		181			al, eth, bz	B23, 295
3545	6,8′-Biquinolyl $(C_9H_6N)_2$	256.31	lf (al)		148			al, bz	B23, 296
3546	8,8′-Biquinolyl $(C_9H_6N)_2$	256.31	lf or pl (al or aa)		205-7			al, ace, bz, chl	$B23^2$, 268
3547	Bismuthine, triphenyl $(C_6H_5)_3Bi$	440.30		242^{14}	77.6	$1.715^{75/4}$	1.7040^{75}	eth, ace, bz, lig	$B16^3$, 1188
3548	Biquinone, 3,3′-dihydroxy-5,5′-dimethyl or Phenicin $(C_7H_5O_3)_2$	274.23	yesh-br (al)		230-1			al, chl, aa	$B8^3$, 4251
3549	2,2-Bithiophene or 2,2-Bithienyl $(C_4H_3S)_2$	166.26	lf (al)	260, 103^3	33			al, eth, aa	$B19^4$, 265
3550	2,2-Bithiophene, hexabromo $C_8Br_6S_2$	639.63	nd (bz)		257-8			bz	$B19^4$, 266
3551	3,3′-Bithiophene $(3\text{-}C_4H_3S\text{-}3)_2$	166.26			132			al, eth, bz, chl, lig	$B19^4$, 267
3552	Biuret or Carbamoylurea $H_2NCONHCONH_2$	103.08	pl (al), nd (w + 1)		190d			al, w	$B3^4$, 141
3553	Biuret, acetyl $CH_3CONHCONHCONH_2$	145.12	nd (w or al)		193-4			w, al	$B3^4$, 142
3554	Biuret, 1,5-diamino $H_2NNHCONHCONHNH_2$	133.11	pr (dil al), nd (aa)		199-200d			w, aa	$B3^4$, 178
3555	Bixin $C_{25}H_{30}O_4$	394.51	vt pr (ace)		198			al, ace, chl	$B2^3$, 2020
3556	Boric acid, tributyl ester or Tributoxyborine $B(OC_4H_9)_3$	230.15	oil	230—1, $114\text{-}5^{25}$		$0.8567^{20/4}$	1.4106^{18}	al, eth, bz	$B1^4$, 1544
3557	Boric acid, isobutyl $i\text{-}C_4H_9B(OH)_2$	101.94	lo pl (w)		112			al, eth	$B4^3$, 1965
3558	Boric acid, isopentyl $i\text{-}C_5H_{11}B(OH)_2$	115.97	pl (w)		169			al, eth, ace	$B4^2$, 1023
3559	Boric acid, triethyl ester or Triethoxyborine $B(OC_2H_5)_3$	146.00		120		$0.8546^{20/4}$	1.3749^{20}	al, eth	$B1^4$, 1365
3560	Boric acid, trimethyl ester or Trimethoxyborine $B(OCH_3)_3$	103.92		67-9	−29.3	0.915^{20}	1.3568^{20}	al, eth, bz	$B1^4$, 1269
3561	Boric acid, tri isopentyl ester or Triisopentyl borate $B\text{-}(O\text{-}i\text{-}C_5H_{11})_3$	272.24		254-5, $132\text{-}3^{12}$		$0.8518^{20/4}$	1.4156^{20}	al, eth	$B1^4$, 1688
3562	Boric acid, tripropyl ester or Tripropoxyborine $B(OC_3H_7)_3$	188.07		179-80, 64^9		$0.8576^{20/4}$	1.3948^{20}	al, eth	$B1^4$, 1436
3563	Boric acid, tri-*iso*-propyl ester or Tri-*iso*-propoxyborine $B\text{-}(O\text{-}i\text{-}C_3H_7)_3$	188.07		164, $52\text{-}3^{32}$		$0.8251^{20/4}$	1.3772^{20}	al, eth, bz	$B1^4$, 1488

PHYSICAL CONSTANTS OF ORGANIC COMPOUNDS (Continued)

No.	Name, Synonyms, and Formula	Mol. wt.	Color, crystalline form, specific rotation and λ_{max} (log ε)	b.p. °C	m.p. °C	Density	n_D	Solubility	Ref.
3564	Boric acid, tri-isopentyl ester $B\text{-}(O\text{-}i\text{-}C_5H_{11})_3$	272.24		$254\text{-}5, 132\text{-}3^{12}$		$0.8518^{20/4}$	1.4156^{20}	al, eth	$B1^4$, 1688
3565	Borine, bis(dimethylamino)-fluoro $[(CH_3)_2N]_2BF$	117.96		106	−44.3			eth, ace	$B4^4$, 303
3566	Borine, bis(methylthio) methyl $(CH_3S)_2BCH_3$	120.05		100^{147}	−59			eth, ace	Am78,1523
3567	Borine, tri-(methylthio) $B(SCH_3)_3$	152.11		218.2	5	1.126^{20}	1.5788^{20}	eth, ace	C55,9346
3568	Borine, difluoro-(dimethyl amino) $(CH_3)_2NBF_3$	92.89	rh	sub 132	165-8d				$B4^4$303
3569	Borine, phenyl-difluoro $C_6H_5BF_2$	125.91		$97\text{-}8^{747}$	−36.2	1.087^{25}	1.4441^{25}	eth, bz	$B16^2$, 638
3570	Borine, 4-tolyl, difluoro $(4\text{-}CH_3C_6H_4)BF_2$	139.94		$127\text{-}8^{747}$		1.055^{25}	1.4535^{25}	eth, bz	C18,992
3571	Borine, dimethyl-(dimethyl amino) $(CH_3)_2BN(CH_3)_2$	84.96		65	−92			eth, ace	$B4^3$, 1960
3572	Borine, dimethyl,methoxy $(CH_3)_2BOCH_3$	71.91		21				eth, ace	Am75,3872
3573	Borine, (methylthio)-dimethyl $CH_3SB(CH_3)_2$	87.97		71	−84			eth, ace	Am76,3307
3574	Borine, triethyl $B(C_2H_5)_3$	98.00		95-6	−92.9	0.6961^{23}		al, eth	$B4^3$, 1957
3575	Borine, tri-*iso*-butyl $B(i\text{-}C_4H_9)_3$	182.16		$188^{766}, 86^{20}$		$0.7380^{25/4}$	1.4188^{23}	al, eth, bz	$B4^3$, 1958
3576	Borine, trimethyl $B(CH_3)_3$	55.91		20	−161.5			al, eth	$B4^3$, 1955
3577	Borine, triisopentyl $(i\text{-}C_5H_{11})_3B$	224.24		119.14		$0.7600^{25/4}$	1.4321	al, eth, ace	$B4^2$, 1023
3578	Borine, triphenyl $B(C_6H_5)_3$	242.13	wh cr	$245—50^{15}$	142			bz, lig	$B16^3$, 1271
3579	Borine, tri-isopentyl $(i\text{-}C_5H_{11})_3B$	224.24		119.14		$0.7600^{23/4}$	1.4321	al, eth, ace, w	$B4^2$, 1023
3580	Borine, tripropyl $B(C_3H_7)_3$	140.08		$159, 43\text{-}4^{17}$	−56	$0.7204^{25/4}$	$1.4135^{22.5}$		$B4^3$, 1957
3581	3-Bornanone-*(d)* or Epicamphor $C_{10}H_{16}O$	152.24	$[\alpha]^{17}{}_D$ +45.4 (bz)		182			al, eth, peth	$B7^3$, 420
3582	3-Bornanone *(dl)* $C_{10}H_{16}O$	152.24	cr (peth)		175-7			al, eth, peth	$B7^3$, 421
3583	3-Bornanone oxime-*(d)* $C_{10}H_{17}NO$	167.25	nd (MeOH), $[\alpha]_D$ −98.9		103			eth, ace	$B7^1$, 86
3584	3-Bornanone oxime-*(dl)* $C_{10}H_{17}NO$	167.25	nd (dil al)		98-100			eth, ace	$B7^1$, 87
3585	3-Bornanone oxime-*(l)* $C_{10}H_{17}NO$	167.25	nd (dil MeOH), $[\alpha]_D$ +100.5		103-4			eth, ace	$B7^1$, 86
3586	3-Bornanone semicarbazide *(d)* $C_{11}H_{19}N_3O$	209.29	nd (al)		237-8				$B7^3$, 421
3587	3-Bornanone semicarbazide *(l)* $C_{11}H_{19}N_3O$	209.29	nd (al), $[\alpha]^{20}{}_D$ +145 (MeOH, c=0.73)		237-8d				$B7^3$, 421
3588	Borneol *(d)* $C_{10}H_{18}O$	154.25	lf or hex pl (peth), $[\alpha]^{20}{}_D$ +37.7 (al)		208	$1.011^{20/4}$		al, eth, bz, lig	$B6^4$, 281
3589	Borneol *(dl)* $C_{10}H_{18}O$	154.25	lf (lig)	sub	210.5	$1.011^{20/4}$		al, eth, bz	$B6^4$, 281
3590	Borneol *(l)* $C_{10}H_{18}O$	154.25	hex pl, $[\alpha]^{20}{}_D$ −37.74 (al)	210^{779}	208.6	$1.1011^{20/4}$		al, eth, ace, bz	$B6^4$, 281
3591	Borneol acetate *(d)* $C_{12}H_{20}O_2$	196.29	rh, $[\alpha]^{20}{}_D$ +44.4 (al)	$223\text{-}4, 107^{15}$	29	$0.9920^{20/4}$			$B6^3$, 302
3592	Borneal acetate *(dl)* $C_{12}H_{20}O_2$	196.29		223-4	<−17	1.4630^{20}			$B6^3$,303

No.	Name, Synonyms, and Formula	Mol. wt.	Color, crystalline form, specific rotation and λ_{max} (log ε)	b.p. °C	m.p. °C	Density	n_D	Solubility	Ref.
3593	Borneol acetate *(l)* $C_{12}H_{20}O_2$	196.29	$[\alpha]^{20/D}$ −44.45 (undil)	223-4, 107^{15}	29	$0.9920^{20/4}$	1.4634^{20}	al, eth	$B6^{3}303$
3594	Borneol formate $C_{11}H_{18}O_2$	182.26	$[\alpha]_D$ +48.75 (undil)	90^{10}		1.009^{22}	1.4700^{15}		$B6^{3}$, 301
3595	Bornylamine *(d)* $C_{10}H_{17}NH_2$	153.27	$[\alpha]^{20/}_D$ +47.2 (al)	200 sub	163			al, eth, ace, bz	$B12^{3}$, 193
3596	Bornyl chloride *(d)* $C_{10}H_{17}Cl$	172.70	nd	207-8 sub	132			al, eth, bz, peth	$B5^{4}$, 319
3597	Bornylene *(d)* or 2-Bornene $C_{10}H_{16}$	136.24	cr (al), $[\alpha]_D$ +30.5 (to)	146^{750} sub	109-10				$B5^{4}$, 460
3598	Bornylene *(l)* $C_{10}H_{16}$	136.24	cr (al), $[\alpha]^{19/}_D$ −23.9 (bz)	146^{746}	113			al, eth, bz	$B5^{4}$, 460
3599	Brazilein $C_{16}H_{12}O_5$	284.27	red-br nd or lf (w + 1)		250			al, w, bz, chl, aa	$B18^{4}$, 2770
3600	Bromcresolgreen or 3,3′,5,5′-Tetrabromo-*m*-Cresol sulfophthalein $C_{21}H_{14}O_5BrS$	698.02	wh or red (+7w), ye (aa)		218-9			al, eth, bz, aa	$B19^{4}$, 1133
3601	Bromo acetamide or Bromo acetamide $BrCH_2CONH_2$	137.96	nd (al or bz)		91			w, bz	$B2^{4}$, 530
3602	Bromoacetic acid $BrCH_2CO_2H$	138.95	hex or rh	208, 127.5^{30}	50	$1.9335^{50/4}$	1.4804^{50}	w, **al**, **eth**, ace, bz	$B2^{4}$, 526
3603	Bromoacetic acid, *iso*-butyl ester or *iso*-Butyl bromoacetate $BrCH_2CO_2$-*i*-C_4H_9	195.06		188^{752}, 74.5^{10}		$1.3269^{20/4}$		al, eth, ace	$B2^{3}$, 482
3604	Bromoacetic acid, *tert*-butyl ester or *tert*-Butyl bromoacetate $BrCH_2CO_2C(CH_3)_3$	195.06		$73\text{-}4^{25}$		1.4430^{20}		al, eth	$B2^{3}$, 482
3605	Bromoacetic acid, ethyl ester or Ethyl bromoacetate $BrCH_2CO_2C_2H_5$	167.00		168-9, 58.9^{15}		$1.5059^{20/20}$	1.4489^{20}	**al**, **eth**, ace	$B2^{4}$, 527
3606	Bromoacetic acid, methyl ester or Methyl bromoacetate $BrCH_2CO_2CH_3$	152.98		144d, 64^{33}				al, eth, ace	$B2^{4}$, 527
3607	Bromoacetic acid, phenyl ester $BrCH_2CO_2C_6H_5$	215.05	pl (al)	140^{20}	32			al, eth	$B6^{1}$, 87
3608	Bromoacetic acid, propyl ester or Propyl bromoacetate $BrCH_2CO_2C_3H_7$	181.03		176^{762}		$1.4099^{20/4}$	1.4518^{20}	al, eth, ace	$B2^{4}$, 528
3609	Bromoacetonitrile $BrCH_2CN$	119.95	pa ye	$150\text{-}1^{752}$, 46^{13}				eth	$B2^{4}$, 531
3610	Bromo acetyl bromide $BrCH_2COBr$	201.85		130		$2.317^{22/22}$	1.5449^{20}	ace	$B2^{4}$, 530
3611	Bromochloro acetic acid $BrClCHCO_2H$	173.39		215d, $103\text{-}4^{11}$	38	$1.9848^{31/4}$	1.5014^{31}	w, al, eth, ace	$B2^{4}$, 532
3612	Bromochloro acetic acid, ethyl ester or Ethyl bromochloro acetate $BrClCHCO_2C_2H_5$	201.45		174d		$1.5890^{22/4}$	1.4639^{24}	al, eth	$B2^{4}$, 532
3613	Bromo difluoro acetic acid BrF_2CCO_2H	174.93	lf (chl)	145-6, 87^{82}	40			w, al, chl	$B2^{4}$, 532
3614	Bromo diphenyl acetic acid $(C_6H_5)_2CBrCO_2H$	291.14	(chl-peth)		133-4			chl, to	$B9^{2}$, 471
3615	Bromodiphenyl acetyl bromide $(C_6H_5)_2CBrCOBr$	354.04	nd (lig)		65-6			al, eth, bz, chl	$B9^{1}$, 283
3616	bis-(2-Bromoethyl) ether $(BrCH_2CH_2)_2O$	231.91		115^{32}		$1.8222^{27/4}$	$1.5131^{27/D}$		$B1^{4}$, 1386
3617	Bromofluoro acetamide $BrCHFCONH_2$	155.96	nd (CCl_4)		44			w, al, eth	B2,217
3618	Bromofluoro acetic acid $BrCHFCO_2H$	156.94		183, 102^{20}	48			w, al, chl	$B2^{4}531$
3619	Bromofluoro acetic acid, ethyl ester $BrFCHCO_2C_2H_5$	184.99		154		1.5587^{17}			$B2^{4}$, 531
3620	Bromophenol blue or 3,3′,5,5′-Tetrabromophenol sulfonphthalein $C_{19}H_{10}O_5Br_4S$	669.96	hex pr (aa-ace)	279d				al, bz, aa	$B19^{4}$, 1129
3621	bis-(4-Bromophenyl) ether $(4\text{-}BrC_6H_4)_2O$	328.00	lf (al)	338-40, 210^{11}	60.5	1.8 (sol)		al, eth, bz	$B6^{4}$, 1048

No.	Name, Synonyms, and Formula	Mol. wt.	Color, crystalline form, specific rotation and λ_{max} (log ε)	b.p. °C	m.p. °C	Density	n_D	Solubility	Ref.
3622	4-Bromophenyl isocyanate $(4\text{-}BrC_6H_4)NCO$	198.02	nd	226, 158^{14}				eth	B12[1], 321
3623	4-Bromphenyl isothiocyanate $4\text{-}BrC_6H_4NCS$	214.08	nd		60-1			al	B12[3], 1463
3624	Brucine $C_{23}H_{26}N_2O_4$	394.47	mcl pr (w + 4), $[\alpha]^{20/}_{546.1}$ −149.5 (chl, c=1		178 (anh) 105 (hyd)			al, chl	B27[2], 797
3625	Brucine hydrochloride $C_{23}H_{26}N_2O_4 \cdot HCl$	430.93	pr					w, al	B27[2], 801
3626	Brucine nitrate, dihydrate $C_{23}H_{26}N_2O_4.HNO_3.2H_2O$	493.51	pr	sub	230d				B27[2], 797
3627	Brucine sulfate, heptahydrate $C_{46}H_{54}N_4O_{16}S \cdot 7H_2O$	1013.12	nd $[\alpha]_D$ −24.4 (w)					w, MeOH	B27[2], 797
3628	Bufotalin $C_{26}H_{36}O_6$	444.57	cr (+ 1 al), $[\alpha]^{20/}_D$ + 5.4 (chl, c=0.5)		223d			al, chl	B18[4], 2557
3629	Bulbocapnine *(d)* $C_{19}H_{19}NO_4$	325.36	pr (al), $[\alpha]_D$ + 237.1 (chl, c=4)					al, chl	B27[2], 554
3630	Bulbocapnine *(l)* $C_{19}H_{19}NO_4$	325.36		209-10					B27[2], 554
3631	1,2-Butadiene or Methylallene $CH_2{=}C{=}CHCH_3$	54.09		10.8	−136.2	$0.676^{0\ 4}$	$1.4205^{1\ 3}$	**al, eth**, bz	B1[4], 975
3632	1,2-Butadiene, 4-bromo $CH_2{=}C{=}CH\text{-}CH_2Br$	132.99		109-11		$1.4255^{20/4}$	1.5248^{20}	ace	B1[3], 929
3633	1,2-Butadiene, 4-chloro $CH_2{=}C{=}CH\text{-}CH_2Cl$	88.54		88		$0.9891^{20/4}$	1.4775^{20}	eth, ace, bz	B1[4], 975
3634	1,2-Butadiene, 4-iodo $CH_2{=}CCCH\text{-}CH_2I$	179.99		130		$1.7129^{20/4}$	1.5709^{20}		B1[3], 929
3635	1,2-Butadiene, 4-methoxy $CH_2{=}C{=}CH\text{-}CH_2OCH_3$	84.12		87-9		$0.8286^{20/4}$	1.435^{20}	al	B1[4], 2221
3636	1,2-Butadiene, 3-methyl $CH_2{=}C{=}C(CH_3)_2$	68.12		40	−120	$0.6804^{20/4}$	1.4166^{20}	**al, eth, ace, bz, peth**	B1[4], 1006
3637	1,2-Butadiene, 4-ol $CH{=}C{=}CH\text{-}CH_2OH$	70.09		126-8, $68\text{-}9^{45}$		$0.9164^{20/4}$	1.4759^{20}	w, al, ace, eth, bz	B1[4], 2221
3638	1,3-Butadiene or Bivinyl $CH_2{=}CH\text{-}CH{=}CH_2$	54.09		−4.4	−108.9	$0.6211^{20/4}$	1.4292^{-25}	al, eth, ace, bz	B1[4], 976
3639	1,3-Butadiene, 2-bromo or Bromoprene $CH_2{=}CBr\text{-}CH{=}CH_2$	132.99		$42\text{-}3^{165}$		$1.397^{20/4}$	1.4988^{20}	al, eth	B1[4], 989
3640	1,3-Butadiene, 1-chloro $CH_2{=}CH\text{-}CH{=}CHCl$	88.54		68		$0.9606^{20/4}$	1.4712^{20}	al, eth, chl	B1[3], 949
3641	1,3-Butadiene, 1-chloro-2-methyl $ClCH{=}C(CH_3)CH{=}CH_2$	102.56		107, 50.4^{100}		$0.9710^{20/4}$	1.4792^{20}	al, ace	B1[3], 974
3642	1,3-Butadiene, 1-chloro-3-methyl $ClCH{=}CH\text{-}C(CH_3){=}CH_2$	102.56		99-100		$0.9543^{20/4}$	1.4719^{20}	al, eth, ace, chl	B1[3], 975
3643	1,3-Butadiene, 2-chloro or Chloroprene $CH_2{=}C(Cl)CH{=}CH_2$	88.54		59.4, 6.4^{100}		$0.9583^{20/4}$	1.4583^{20}	**eth, ace, bz**	B1[4], 984
3644	1,3-Butadiene, 2-chloro-3-methyl $CH_2{=}CCl\text{-}C(CH_3){=}CH_2$	102.56		93		$0.9593^{20/4}$	1.4686^{20}	al, eth, ace, chl	B1[4], 1004
3645	1,3-Butadiene, 4-cyano or 2,4-Pentadieno nitrile $CH_2{=}CH\text{-}CH{=}CHCN$	79.10		135-8	0.8444^{20}	1.4880			B2[4], 1692
3646	1,3-Butadiene, 1,1-dichloro $Cl_2C{=}CH\text{-}CH{=}CH_2$	122.98		$42\text{-}3^{90}$		$1.1831^{20/4}$	1.5022^{20}	eth, bz	B1[4], 985
3647	1,3-Butadiene, 1,2-dichloro $ClCH{=}CCl\text{-}CH{=}CH_2$	122.98		$60\text{-}5^{105}$, 35^{40}		$1.1991^{20/4}$	1.4960^{20}	CCl_4	B1[4], 985
3648	1,3-Butadiene, 2,3-dichloro $CH_2{=}CCl\text{-}CCl{=}CH_2$	122.98		98		$1.1829^{20/4}$	1.4890^{20}	chl	B1[4], 986
3649	1,3-Butadiene, 2,3-dimethyl *(cis,cis)* or Biisopropenyl $CH_2{=}C(CH_3{-}C)CH_3{=}CH_2$	82.15		68-78	−76	$1.4394^{20/4}$			B1[4], 1023
3650	Butadiene dioxide $CH_2CHCHCH_2$ (with two O bridges)	86.09		144	4	1.113^{20}	1.435^{20}	w, al	B19[4], 111
3651	1,3-Butadiene, 1,4-diphenyl *(cis,cis)* or *cis,cis*-Bistyryl $C_6H_5CH{=}CH\text{-}CH{=}CHC_6H_5$	206.29	lf or nd (al or MeOH)		70.5	0.9697^{100}	1.6183^{100}	al, eth, bz, chl, peth	B5[3], 2159

No.	Name, Synonyms, and Formula	Mol. wt.	Color, crystalline form, specific rotation and λ_{max} (log ε)	b.p. °C	m.p. °C	Density	n_D	Solubility	Ref.
3652	1,3, Butadiene, 1,4-diphenyl *(trans,trans)* $C_6H_5CH=CH\text{-}CH=CHC_6H_5$	206.29	lf (al or aa)	350^{720}	152.5	...	...	al, eth, bz, chl, peth	$B5^3$, 2159
3653	1,3-Butadiene, 1-ethoxy $C_2H_5OCH=CH\text{-}CH=CH_2$	98.15	...	109-12	...	$0.8154^{20/4}$	1.4529^{20}	al, eth, ace, bz, chl	$B1^4$, 2222
3654	1,3-Butadiene, 2-ethoxy $CH_2=C(OC_2H_5)\text{-}CH=CH_2$	98.14	...	94-5	...	$0.8177^{20/4}$	1.4400^{20}	al, eth, ace, bz	$B1^4$, 2223
3655	1,3-Butadiene, 2-fluoro or Fluoroprene $CH_2=CF\text{-}CH=CH_2$	72.08	...	12	...	$0.843^{4/4}$	1.4004	...	$B1^4$, 982
3656	1,3-Butadiene, hexachloro $CCl_2=CCl\text{-}CCl=CCl_2$	260.76	...	215, 101^{20}	−21	1.5542^{20}	...	al, eth	$B1^1$, 955
3657	1,3-Butadiene, hexafluoro $CF_2=CF\text{-}CF=CF_2$	162.03	...	6	−132	$1.553^{20/4}$	1.378^{-20}	...	$B1^4$, 983
3658	1,3-Butadiene, 2-iodo or Iodoprene $CH_2=CI\text{-}CH=CH_2$	179.99	...	111-3	...	$1.7278^{20/4}$	1.5616	...	$B1^3$, 956
3659	1,3-Butadiene, 1-methoxy $CH_3OCH=CH\text{-}CH=CH_2$	84.12	...	91-2	...	$0.8296^{20/4}$	1.4594^{20}	w, al	$B1^4$, 2221
3660	1,3-Butadiene, 2-methoxy $CH_2=C(OCH_3)\text{-}CH=CH_2$	84.12	...	75	...	$0.8272^{20/4}$	1.4442^{20}	al, eth, ace, bz	$B1^4$, 2223
3661	1,3-Butadiene, 2-methyl or Isoprene $CH_2=C(CH_3)\text{-}CH=CH_2$	68.12	...	34	−146	$0.6810^{20/4}$	1.4219^{20}	al, eth, ace, bz	$B1^4$, 1001
3662	1,3-Butadiene, pentafluoro-2-trifluoro methyl $CF_2=C(CF_3)\text{-}CF=CF_2$	212.04	...	39	...	$1.527^{0/4}$	1.3000^{0}	eth, ace, bz	C49,2479
3663	1,3-Butadiene, 1-phenyl *(trans)* $C_6H_5CH=CH\text{-}CH=CH_2$	130.19	...	76^{11}	4.5	$0.9286^{20/4}$	1.6089^{25}	al, eth, ace, bz	$B5^4$, 1536
3664	1,3-Butadiene, 2-phenyl $CH_2=C(C_6H_5)\text{-}CH=CH_2$	130.19	...	$60\text{-}1^{17}$	...	$0.9266^{20/4}$	1.5489^{20}	eth, bz	$B5^4$, 1539
3665	1,3-Butadiene, 1,2,3,4-tetrachloro (liquid) $ClCH=CCl\text{-}CCl=CHCl_2$	191.87	...	188, 67^{10}	−4	$1.516^{15/15}$	1.5455^{20}	al, eth, ace, bz	$B1^4$, 987
3666	1,3-Butadiene, 1,2,3,4-tetrachloro (solid) $ClCH=CCl\text{-}CCl=CHCl$	191.87	...	...	52	1.4961^{20}	1.5438^{20}	al, eth, ace, bz, chl	$B1^4$, 987
3667	1,3-Butadiene, 1,2,3-trichloro $ClCH=CCl\text{-}CCl=CH_2$	157.43	...	$33\text{-}4^{7}$	...	$1.4060^{20/4}$	1.5262^{20}	eth, chl	$B1^3$, 954
3668	Butadiyne or Biacetylene $CH\equiv C\text{-}C\equiv CH$	50.06	...	10.3	−36.4	$0.7364^{0/4}$	1.4189^{5}	al, eth, ace, chl	$B1^4$, 1116
3669	Butadiyne, 1,4-dichloro $ClC\equiv C\text{-}C\equiv CCl$	118.95	nd	...	1-3	...	...	chl	$B1^3$, 1057
3670	Butadiyne, 1,4-bis(1-hydroxycyclohexyl) $(C_6H_{11}O)_2$	246.35	cr (bz)	174	...	...	...	MeOH	$B6^3$, 5178
3671	Butanal or Butyraldehyde $CH_3CH_2CH_2CHO$	72.11	...	75.7	−99	$0.8170^{20/4}$	1.3843^{20}	w, **al, eth, ace,** bz	$B1^4$, 3229
3672	Butane C_4H_{10}	58.12		−0.5	−138.4	$0.6012^{0/4}$, $0.5788^{20/4}$	1.3543^{-13}, 1.3326^{20}	al, eth, chl, w	$B1^4$, 236
3673	Butane, 1-amino or *n*-Butylamine $CH_3CH_2CH_2CH_2NH_2$	73.13	...	77.8	−49.1	$0.7414^{20/4}$	1.4031^{20}	**w**, al, eth	$B4^4$, 540
3674	Butane, 1-amino-3-methyl or *iso*-Pentylamine $(CH_3)_2CH_2CH_2CH_2NH_2$	87.16	...	$95\text{-}7^{761}$	...	$0.7505^{20/4}$	1.4083^{20}	**w, al, eth,** ace, chl	$B4^4$, 696
3675	Butane, 2-amino or *sec*-Butylamine *(d)* $CH_3CH_2CH(NH_2)CH_3$	73.13	$[\alpha]^{20/}_{D}$ + 7.4 (w)	63	−104.5	$0.724^{20/4}$	1.344^{20}	w, **al, eth,** ace, chl	$B4^4$, 617
3676	Butane, 2-amino *(dl)* or *sec*-Butylamine $CH_3CH_2CH(NH_2)CH_3$	73.13	...	63.5^{764}	<-72	$0.7246^{20/4}$	1.3932^{20}	**w, al, eth,** ace, chl	$B4^4$, 617
3677	Butane, 2-amino *(l)* or *sec*-Butylamine $CH_3CH_2CH(NH_2)CH_3$	73.13	$[\alpha]^{20/}_{D}$ -7.4 (w, c=4.7)	63	...	$0.7205^{20/4}$	...	w, al, eth, ace, chl	$B4^4$, 617
3678	Butane, 2-amino-2,3-dimethyl $(CH_3)_2CH\text{-}C(NH_2)(CH_3)_2$	101.19	...	104-5	...	$0.7683^{0/4}$	1.4096^{17}	...	$B4^4$, 733
3679	Butane, 2-amino-2-methyl $CH_3CH_2C(CH_3)_2NH_2$	87.16	...	77	-105	$0.731^{25/4}$	1.3954^{25}	**w, al, eth,** ace	$B4^4$, 694
3680	Butane, 2-amino-3-methyl $(CH_3)_2CH\text{-}CH(CH_3)NH_2$	87.16	...	84-7	...	0.7574^{19}	1.4096^{18}	w, al	$B4^4$, 695
3681	Butane, 3-amino-2,2 dimethyl $CH_3CH(NH_2)\text{-}C(CH_3)_2CH_3$	101.19	...	102	-20	...	...	w	$B4^4$, 730
3682	Butane, 3-amino 2,2-dimethyl, hydrochloride $(CH_3)_3CCH(NH_2)CH_3.HCl$	137.65	nd	sub 245	300-1 (cor)	...	...	w	$B4^4$, 730
3683	Butane, 1-bromo $CH_3CH_2CH_2CH_2Br$	137.02	...	101.6, 18.8^{30}	-112.4	$1.2758^{20/4}$	1.4401^{20}	**al, eth, ace,** chl	$B1^4$, 258
3684	Butane, 1-bromo-4-chloro $BrCH_2CH_2CH_2CH_2Cl$	171.46	...	$174\text{-}5^{756}$, $63\text{-}4^{10}$	...	$1.488^{20/4}$	1.4885^{20}	al, eth, chl	$B1^4$, 264

No.	Name, Synonyms, and Formula	Mol. wt.	Color, crystalline form, specific rotation and λ_{max} (log ε)	b.p. °C	m.p. °C	Density	n_D	Solubility	Ref.
3685	Butane, 1-bromo-3,3-dimethyl $BrCH_2CH_2C(CH_3)_3$	165.07	...	$138, 54^{40}$	...	$1.556^{20/4}$	1.4440^{20}	al, eth, chl	$B1^3$, 409
3686	Butane, 1-bromo-2,4-diphenyl *(d)* $C_6H_5CH_2CH_2CH(C_6H_5)CH_2Br$	289.21	$[\alpha]^{20}$ 16.8 (chl, c=10)	$122\text{-}30^{0.02}$	...	...	1.5812^{25}	...	...
3687	Butane, 1-bromo-2,4-diphenyl *(dl)* $C_6H_5CH_2CH_2CH(C_6H_5)CH_2Br$	289.21	...	$126\text{-}8^{0.01}$	...	...	1.5812^{25}	...	...
3688	Butane, 1-bromo-4-fluoro $Br(CH_2)_4F$	155.01	...	$134\text{-}5^{740}$	...	...	1.4370^{25}	al, eth	$B1^4$, 263
3689	Butane, 1-bromo-2-methyl *(d)* or *act*-Amylbromide $BrCH_2CH(CH_3)CH_2CH_3$	151.05	$[\alpha]^{20}$ + 3.68	121.6	...	$1.2234^{20/4}$	1.4451^{20}	al, eth, chl	$B1^4$, 327
3690	Butane, 1-bromo-3-methyl *(d)* $BrCH_2CH(CH_3)CH_2CH_3$	151.05	...	$120.5, 12.3^{10}$	...	$1.2205^{20/4}$	1.4452^{20}	al, eth, chl	$B1^4$, 327
3691	Butane, 1-bromo-3-methyl or Isoamylbromide $BrCH_2CH_2CH(CH_3)_2$	151.05	...	$120.4, 12.3^{10}$	-112	$1.2071^{20/4}$	1.4420^{20}	al, eth, chl	$B1^4$, 328
3692	Butane, 2-bromo *(dl)* or *sec*-butylbromide $CH_3CH_2CHBrCH_3$	137.02	...	91.2	-111.9	$1.2585^{20/4}$	1.4366^{20}	**eth, ace,** chl	$B1^4$, 261
3693	Butane, 2-bromo *(l)* or *sec*-butylbromide $CH_3CH_2CHBrCH_3$	137.02	$[\alpha]^{22/}{}_D$ -23.13 (undil)	90-1	...	$1.2536^{25/4}$	1.4359^{19}	al, **eth, ace**, chl	$B1^4$, 261
3694	Butane, 2-bromo-1-chloro $ClCH_2CHBrCH_2CH_3$	171.46	...	$146\text{-}7^{758}$	...	$1.468^{20/4}$	1.4880^{20}	al, eth, bz, chl	$B1^4$, 264
3695	Butane, 2-bromo-2,3-dimethyl $CH_3CHBr(CH_3)CH(CH_3)_2$	165.07	...	$132\text{-}3^{742}, 87^{180}$	24-5	1.1772^{10}	1.4517	eth, chl	$B1^4$, 374
3696	Butane, 2-bromo-2-methyl $(CH_3)_2CBrCH_2CH_3$	151.05	...	108^{765}	...	$1.198^{18}f^{5/15}$	1.44207	...	$B1^4$, 327
3697	Butane, 1,4-bis(dicarbethoxy amino) $C_2H_5O_2CNH\text{-}(CH_2)_4NHCO_2C_2H_5$	232.28	nd (lig)	...	85-6	...	...	al, eth, bz, chl	$B4^4$, 1292
3698	Butane, 1-chloro or *n*-Butylchloride $CH_3CH_2CH_2CH_2Cl$	92.57	...	78.4	-123.1	$0.8862^{20/4}$	1.4021^{20}	**al, eth**	$B1^4$, 246
3699	Butane, 1-chloro-2,3-dimethyl $ClCH_2CH(CH_3CH(CH_3)_2$	120.62	...	116^{735}	...	1.4200^{20}	...	al, eth, chl	$B1^4$, 373
3700	Butane, 1-chloro-3,3-dimethyl $ClCH_2CH_2C(CH_3)_3$	120.62	...	$115,41^{50}$	...	$0.8670^{20/4}$	1.4161^{20}	al, eth, chl	$B1^4$, 369
3701	Butane, 1-chloro-4-fluoro $Cl(CH_2)_4F$	110.56	...	114.7	...	$1.0627^{25/4}$	1.4020^{25}	al, eth	$B1^4$, 249
3702	Butane, 1-chloro-2-methyl *(d)* or *act*-Amyl chloride $ClCH_2CH(CH_3)CH_2CH_3$	106.60	$[\alpha]^{20/}{}_{5892}$ + 1.64	$100.5, 43^{100}$	...	$0.8857^{20/4}$	1.4126^{20}	al, eth	$B1^4$, 324
3703	Butane, 1-chloro-2-methyl *(dl)* $ClCH_2CH(CH_3)CH_2CH_3$	106.60	...	$99.9, 52.2^{50}$	...	$0.8818^{15/15}$	1.4102^{25}	al, eth	$B1^4$, 324
3704	Butane, 1-chloro-3-methyl or Isopentyl chloride $(CH_3)_2CHCH_2CH_2Cl$	106.60	...	98.5	-104.4	$0.8704^{20/4}$	1.4084^{20}	**al, eth**, chl	$B1^4$, 325
3705	Butane, 1-chloro-2,2,3,3-tetramethyl $(CH_3)_3CC(CH_3)_2CH_2Cl$	148.68	...	$80\text{-}1^{40}$	52-3	...	...	eth	$B1^3$, 502
3706	Butane, 2-chloro *(d)* or *sec*-Butyl chloride $CH_3CH_2CHClCH_3$	92.57	...	68.2	-131.3	$0.8732^{20/4}$	1.3971^{20}	**al, eth**, bz, chl	$B1^4$, 248
3707	Butane, 2-chloro *(l)* $CH_3CH_2CHClCH_3$	92.57	$[\alpha]^{20/}{}_D$ -8.48	68	-140.5	$0.8950^{0/4}$	...	**al, eth**, bz, chl	$B1^4$, 248
3708	Butane, 2-chloro-2,3-dimethyl $CH_3CCl(CH_3)CH(CH_3)_2$	120.62	...	112	-10.4	$0.8780^{20/4}$	1.4191^{20}	al, ace	$B1^4$, 373
3709	Butane, 2-chloro-2-methyl or *tert*-Amyl chloride $CH_3CH_2CCl(CH_3)_2$	106.60	...	85.6	-73.5	$0.8653^{20/4}$	1.4055^{20}	al, eth	$B1^4$, 324
3710	Butane, 2-chloro-3-methyl *(dl)* $CH_3CHClCH(CH_3)_2$	106.60	...	$92.8, 25.7^{60}$	...	$0.8620^{20/4}$	1.4020^{20}	al, eth	$B1^3$, 358
3711	Butane, 2-(chloromethyl)-1,3-dichloro *(dl)* $CH_3CHClCH(CH_2Cl)_2$	175.49	...	$79\text{-}81^{15}$	...	$1.2793^{15/4}$	...	al, eth, chl	C31, 1003
3712	Butane, 2-(chloromethyl)-1,2,3-trichloro $CH_3CHClCCl(CH_2Cl)_2$	209.93	...	$102\text{-}3^{13}$	...	$1.3977^{18/4}$	1.5012^{18}	eth, chl	$B1^3$, 362
3713	Butane, 2-chloro-2,3,3-trimethyl $(CH_3)_2CClC(CH_3)_3$	134.65	...	sub	136	...	...	eth	$B1^4$, 411
3714	Butane, 3-chloro-2,3-dimethyl or Pinacolyl chloride $(CH_3)_3CCHClCH_3$	120.62	...	$111, 7^{10}$	0.9	$0.8767^{20/4}$	1.4182^{20}	eth	$B1^4$, 373
3715	Butane, decafluoro or Perfluoro butane C_4F_{10}	238.03	...	3.96	-128	1.6484	...	bz, chl	$B1^4$, 245
3716	Butane, 1,4-diamino or Putrescine $H_2N(CH_2)_4NH_2$	88.15	lf	158-9	27-8	0.877^{25}	1.4969^{20}	w	$B4^4$, 1283
3717	Butane, 1,4-diamino, dihydrochloride $H_2N(CH_2)_4NH_2 \cdot 2HCl$	161.07	nd or lf (al or w)	sub	315d	...	...	w, al	$B4^4$, 1284

PHYSICAL CONSTANTS OF ORGANIC COMPOUNDS (Continued)

No.	Name, Synonyms, and Formula	Mol. wt.	Color, crystalline form, specific rotation and λ_{max} (log ε)	b.p. °C	m.p. °C	Density	n_D	Solubility	Ref.
3718	Butane, 1,2-dibromo $CH_3CH_2CHBrCH_2Br$	215.92		166.3	-65.4	$1.7915^{20/4}$	1.4025^{20}	eth, chl	B1⁴, 266
3719	Butane, 1,3-dibromo $CH_3CHBrCH_2CH_2Br$	215.92		174, 72^{20}		1.800^{20}	1.507^{20}	eth, chl	B1⁴, 266
3720	Butane, 1,4-dibromide or Tetramethylene dibromide $Br(CH_2)_4Br$	215.92		197, 79^{10}	-16.5	$1.7890^{20/4}$	1.5190^{20}	chl	B1⁴, 267
3721	Butane, 2,3-dibromo $CH_3CH_2BrCH_2BrCH_3$	215.92		161	<-80	$1.7893^{22/4}$	1.5133^{22}	eth, chl	B1⁴, 268
3722	Butane, 2,3-dibromo *(meso)* $CH_3CHBrCHBrCH_3$	215.92		157.3		$1.7913^{15/4}$	$1.5132^{15.4}$		B1⁴, 268
3723	Butane, 1,1-dichloro or Butylidene dichloride $CH_3CH_2CH_2CHCl_2$	127.01		113.8		$1.0863^{20/4}$	1.4355^{20}	chl	B1⁴, 250
3724	Butane, 1,1-dichloro-3-methyl *(dl)* $(CH_3)_2CHCH_2CHCl_2$	141.04		130, 48-9^{40}		1.0473^{20}	1.4344^{20}	al, eth	B1⁴, 326
3725	Butane, 1,2-dichloro $CH_3CH_2CHClCH_2Cl$	127.01		124		$1.1116^{25/4}$	1.4450^{20}	eth, chl	B1⁴, 250
3726	Butane, 1,2-dichloro-2-methyl $CH_3CH_2CClCl(CH_3)CH_2Cl$	141.04		133-5, 71.5^{100}		$1.0785^{20/4}$	$1.4432^{21.5}$	al, eth, chl	B1⁴, 325
3727	Butane, 1,3-dichloro $CH_3CHClCH_2CH_2Cl$	127.01		134		$1.1158^{20/4}$	1.4445^{20}	eth, chl	B1⁴, 250
3728	Butane, 1,3-dichloro-3-methyl *(d)* $(CH_3)_2CClCH_2CH_2Cl$	141.04		145-6, 39^{10}		$1.0654^{20/4}$	1.4455^{20}	eth, chl	B1⁴, 326
3729	Butane, 1,4-dichloro or Tetramethylene dichloride $Cl(CH_2)_4Cl$	127.01		153.9, 39.7^{10}	-37.3	$1.1408^{20/4}$	1.4542^{20}	chl	B1⁴, 250
3730	Butane, 1,4-dichloro-2-methyl $ClCH_2CH_2CH(CH_3)CH_2Cl$	141.04		168-9, 50^{12}		$1.1003^{25/4}$	1.4562^{21}	chl	B1³, 360
3731	Butane, 2,2-dichloro $CH_3CH_2CCl_2CH_3$	127.01		104	-74		1.4295	chl	B1⁴, 251
3732	Butane, 2,2-dichloro-3,4-dimethyl $(CH_3)_3CCCl_2CH_3$	155.07			151-2			al, eth	B1⁴, 369
3733	Butane, 2,3-dichloro $CH_3CHClCHClCH_3$	127.01		116, 49.5^{80}	-80	$1.1134^{20/4}$	1.4420^{20}	chl	B1⁴, 251
3734	Butane, 2,3-dichloro-2,3-dimethyl $(CH_3)_2CClCCl(CH_3)_2$	155.07	pr (dil al)		164			al, eth	B1⁴, 374
3735	Butane, 2,3-dichloro-2-methyl $CH_3CHClCHClCH(CH_3)_2$	141.04		129, 37.5^{20}		$1.0696^{15/4}$	1.4450^{18}	al, eth	B1⁴, 325
3736	Butane, 1,1-dicyclohexyl $CH_3CH_2CH_2CH(C_6H_{11})_2$	222.41		280.2		$0.8842^{16/0}$	1.485^{16}		B5⁴, 359
3737	Butane, 1,2-dicyclohexyl $CH_3CH_2CH(C_6H_{11})CH_2C_6H_{11}$	222.41		276.8		$0.9084^{18/0}$	1.475^{21}	chl, eth	B5⁴, 358
3738	Butane, 1,2,3,4-diepoxy *(dl)* or Butadiene dioxide $CH_2CHCHCH_2$ (O bridging C1–C2 and C3–C4)	86.09		144	4	1.113^{20}	1.435^{20}	w, al	B19⁴, 111
3739	Butane, 1,2,3,4-diepoxy *(meso)*	86.09		138^{767}	-16	$1.1157^{20/4}$	1.4330^{20}	w, al	B19⁴, 110
3740	Butane, 1,1-diethoxy $CH_3CH_2CH_2CH(OC_2H_5)_2$	146.23		143		$0.841^{25/4}$			B1⁴, 3232
3741	Butane, 1,4-diethoxy $C_2H_5O(CH_2)_4OC_2H_5$	146.23		155-7^{730}					B1⁴, 2517
3742	Butane, 1,4-difluoro-octachloro $FCCl_2CCl_2CCl_2CCl_2F$	369.66		152.5^{20}	4-5	$1.9272^{20/4}$	1.5256^{20}		B1⁴, 258
3743	Butane, 2,2-di(2-furyl) $CH_3CH_2C(C_4H_3O)_2CH_3$	190.24		64-6^{1}		$1.0330^{20/4}$	1.4970^{20}	al, eth	B19⁴, 287
3744	Butane, 1,4-diido or Tetramethylene diiodide $I(CH_2)_4I$	309.92		125-$6^{15}/_d$	5.8	$2.349^{26/4}$	1.619^{25}		B1⁴, 276
3745	Butane, 2,2-dimethyl or *neo*-Hexane $CH_3CH_2C(CH_3)_3$	86.18		49.7	-99.9	$0.6485^{20/4}$	1.3688^{20}	al, eth, ace, bz	B1⁴, 367
3746	Butane, 2,3-dimethyl $(CH_3)_2CH\text{-}CH(CH_3)_2$	86.18		58	-128.5	$0.6616^{20/4}$	1.3750^{20}	al, eth, ace, bz	B1⁴, 371
3747	Butane, 2,3-dimethyl-2,3-epoxy $(CH_3)_2C\text{-}C(CH_3)_2$ (O bridging C2–C3)	100.16		90-4^{745}		$0.8156^{16/4}$	$1.3984^{16.4}{}_D$	w	B17⁴, 89
3748	Butane, 1,4-bis (dimethylamino) or Tetramethyl putrescine $(CH_3)_2N(CH_2)_4N(CH_3)_2$	144.26		168, 78-80^{28}		0.7942^{15}	1.4621^{25}	w, al, eth	B4⁴, 1284
3749	Butane, 1,4-dinitro $O_2N(CH_2)_4NO_2$	148.12	pl (al)	176-8^{13}	33-4			eth, bz	B1⁴, 280
3750	Butane, 2,3-dinitro *(dl)* $CH_3CH(NO_2)CH(NO_2)CH_3$	148.12			48-9			eth	B1⁴, 280

No.	Name, Synonyms, and Formula	Mol. wt.	Color, crystalline form, specific rotation and λ_{max} (log ε)	b.p. °C	m.p. °C	Density	n_D	Solubility	Ref.
3751	Butane, 2,3-dinitro *(meso)* $CH_3CH(NO_2)CH(NO_2)CH_3$	148.12	pr (eth)	$76\text{-}7^{1}$				eth	$B1^4$, 280
3752	Butane, 1,1-diphenyl $(C_6H_5)_2CH\text{-}CH_2CH_2CH_3$	210.32		$286\text{-}8, 161\text{-}3^{20}$	27	$0.9928^{20/4}$	1.5664^{20}	al, eth, bz, chl	$B5^4$, 1944
3753	Butane, 1,2-diphenyl $C_6H_5CH_2CH(C_6H_5)CH_2CH_3$	210.32		$289^{750}, 152^{11}$		0.9777^{20}	1.5554^{20}	al, eth, bz, chl	$B5^4$, 1939
3754	Butane, 1,3-diphenyl *(l)* $C_6H_5CH_2CH_2CH(C_6H_5)CH_3$	210.32	$[\alpha]^{20/}{}_D$ -15.6 (chl, c=10)	$68\text{-}70^{0.01}$			1.5503^{25}	chl	$B5^4$, 1938
3755	Butane, 1,4-diphenyl $C_6H_5(CH_2)_4C_6H_5$	210.32		$317, 108\text{-}9^{0.1}$	52.5			al, eth, chl	$B5^4$, 1937
3756	1,4-Butane dithiol $HS(CH_2)_4SH$	122.24		195-6	-53.9	$1.0621^{0/4}$	1.5290^{20}	al	$B1^4$, 2523
3757	Butane, 1,2-epoxy or 1,2-Butylene oxide $CH_3CH_2CHCH_2$	72.11		63.3		$0.837^{17/4}$	1.3851^{20}	al, **eth**, ace	$B17^4$, 45
3758	Butane, 2,3-epoxy *(cis)* $CH_3CHCHCH_3$	72.11		59.7^{742}	-80	$0.8226^{25/4}$	1.3802^{20}	eth, ace, bz	$B17^4$, 48
3759	Butane, 2,3-epoxy *(trans)* $CH_3CHCHCH_3$	72.11		56-7	-85	$0.8010^{25/4}$	1.3736^{20}	eth, ace, bz	$B17^4$, 49
3760	Butane, 2,2-bis(ethylsulfonyl) $CH_3CH_2C(SO_2C_2H_5)_2CH_3$	274.35	pl (w)	d	76	$1.199^{85/4}$		al, eth, bz, peth, lig	$B1^3$, 2790
3761	Butane, 1-fluoro or *n*-Butyl fluoride $CH_3CH_2CH_2CH_2F$	76.11		32.5	-134	$6.7789^{20/4}$	1.3396^{20}	al	$B1^4$, 244
3762	Butane, 1,1,2,2,3,3,4,4,-heptachloro $Cl_2CHCHClCCl_2CHCl_2$	299.24		$137.5^{13.5}$		$1.742^{20/20}$	1.5407^{20}	ace, CCl_4	$B1^4$, 257
3763	Butane, 1,1,2,3,4,4-hexachloro (liquid) $Cl_2CHCHClCHClCHCl_2$	264.79		111^{10}		$1.6460^{20/4}$	1.5258^{20}	bz, CCl_4	$B1^3$, 288
3764	Butane, 1,1,2,3,4,4-hexachloro (solid) $Cl_2CHCHClCHClCHCl_2$	264.79	nd (al), pr (bz, aa)		109-10			bz, CCl_4	$B1^3$, 288
3765	Butane, 1,1-bis(4-hydroxyphenyl) $(4\text{-}HOC_6H_4)_2CHC_3H_7$	242.32	nd (to)	270^{12}	137			al, bz, to	$B6^3$, 5476
3766	Butane, 2,2-bis(4-hydroxyphenyl) $(4\text{-}HOC_6H_4)_2CH(CH_3)C_2H_5$	242.32	nd or pr (w)	250-3	133-4			al, eth, ace, bz	$B6^3$, 5477
3767	Butane, 1-iodo or *m*-Butyl iodide $CH_3CH_2CH_2CH_2I$	184.02		$130.5, 19.2^{10}$	-103	$1.6154^{20/4}$	1.5001^{20}	al, eth, chl	$B1^4$, 271
3768	Butane, 1-iodo-2-methyl *(d)* or *act*-Amyliodide $CH_3CH_2CH(CH_3)CH_2I$	198.05	$[\alpha]^{15}{}_D$ + 5.78 (undil)	$148^{20}, 47.1^{20}$		$1.5253^{20/4}$	1.4977^{20}	al, eth	$B1^3$, 366
3769	Butane, 1-iodo-2-methyl *(dl)* $CH_3CH_2CH(CH_3)CH_2I$	198.05		144-7			1.497^{20}	al, eth	$B1^3$, 366
3770	Butane, 1-iodo-3-methyl or Isopentyl iodide $(CH_3)_2CHCH_2CH_2I$	198.05		147		$1.5118^{20/4}$	1.4939^{20}	**al, eth**	$B1^4$, 331
3771	Butane, 2-iodo *(dl)* or *sec*-Butyl iodide $CH_3CH_2CHICH_3$	184.02		$120, 33^{45}$	-104.2	$1.5920^{20/4}$	1.4991^{20}	**al, eth**, chl	$B1^4$, 272
3772	Butane, 2-iodo *(l)* $CH_3CH_2CHICH_3$	184.02	$[\alpha]_D$ -12.15 (al, c=20)	117-8		$1.585^{20/4}$	1.4945^{19}	**al, eth**, chl	$B1^4$, 272
3773	Butane, 2-iodo-2-methyl or *tert*-Amyliodide $(CH_3)_2CICH_2CH_3$	198.05		124.5		$1.4937^{20/4}$	1.4981^{20}	**al, eth**	$B1^4$, 331
3774	Butane, 2-iodo-3-methyl $(CH_3)_2CHCHICH_3$	198.05		138-9				al, eth, ace	$B1^3$, 367
3775	Butane, 1-isocyano or Butyl carbylamine $CH_3CH_2CH_2CH_2NC$	83.13		118		1.4061^{20}		**al, eth**	$B4^4$, 562
3776	Butane, 1-isocyano-3-methyl or Isoamylcarbylamine $(CH_3)_2CHCH_2CH_2NC$	97.16		140		$0.806^{20/4}$	1.406^{20}	**al, eth**	B4, 184
3777	Butane, 2-methyl or Isopentane $CH_3CH_2CH(CH_3)_2$	72.15		27.8	-159.9	$0.6201^{20/4}$	1.3537^{20}	al, eth	$B1^4$, 320
3778	Butane, 3-methyl-2-phenyl $CH_3CH(C_6H_5)CH(CH_3)_2$	148.25		186-8		$0.8672^{16/4}$	1.4972^{16}		$B5^4$, 1118
3779	Butane, 2-methyl-1,2,3-trichloro $CH_3CHClCCl(CH_3)CH_2Cl$	175.49		$183\text{-}5^{762}, 65.5^{11}$		$1.2527^{20/4}$		chl	$B1^3$, 361
3780	Butane, 2-methyl-2,3,3-trichloro $(CH_3)_2CClCCl_2CH_3$	175.49		182-3		$1.215^{15/4}$	1.472^{21}	chl, aa	$B1^3$, 361
3781	Butane, 1-nitro $CH_3CH_2CH_2CH_2NO_2$	103.12		153		$0.9710^{20/4}$	1.4303^{20}	**al, eth**	$B1^4$, 277
3782	Butane, 2-nitro *(dl)* $CH_3CH_2CH(NO_2)CH_3$	103.12		140	-132	$0.9854^{17/4}$	1.4044^{20}		$B1^4$,,278
3783	Butane, 1,1,2,2,3,3,4,4-octachloro $Cl_2CHCCl_2CCl_2CHCl_2$	333.68	cr (al)		81			al, eth, ace, bz	$B1^4$, 257

No.	Name, Synonyms, and Formula	Mol. wt.	Color, crystalline form, specific rotation and λ_{max} (log ε)	b.p. °C	m.p. °C	Density	n_D	Solubility	Ref.
3784	Butane, 1,1,2,3,4-pentachloro (liquid) $ClCH_2(CHCl)_2CHCl_2$	230.35		95.5^{11}		1.561^{18}	1.5140^{18}	al, CCl_4	B1⁴, 256
3785	Butane, 1,1,2,3,4-pentachloro (solid) $ClCH_2CHClCHClCHCl_2$	230.35	lf (al)	$230, 102^{11}$	49	1.539^{53}	1.5065^{53}	al, CCl_4	B1⁴, 256
3786	Butane, 1,2,2,3,4-pentachloro $ClCH_2CHClCCl_2CH_2Cl$	230.35		85^{10}		$1.5543^{20/4}$	1.5157^{20}	ace, chl	B1⁴, 256
3787	Butane, 1,1,4,4-tetrabromo $Br_2CHCH_2CH_2CHBr_2$	373.71		$138\text{-}45^{10}$		$2.529^{20/4}$	1.6077^{20}	eth, bz, chl	B1³, 298
3788	Butane, 1,2,2,3-tetrabromo $CH_3CHBrCBr_2CH_2Br$	373.71		$128\text{-}30^{14}$	-2	$2.5100^{20/4}$	1.6070^{20}	ace	B1³, 298
3789	Butane, 1,2,2,4-tetrabromo $BrCH_2CH_2CBr_2CH_2Br$	373.71	nd (lig)		72-3			lig	B1³, 298
3790	Butane, 1,2,3,4-tetrabromo *(dl)* $BrCH_2CHBrCHBrCH_2Br$	373.71	lf (peth)		40-1			al, eth, ace, lig	B1⁴, 271
3791	Butane, 1,2,3,4-tetrabromo *(meso)* $BrCH_2CHBrCHBrCH_2Br$	373.71	nd (al or lig)	$180\text{-}1^{60}$	118-9			al, ace, chl	B1⁴, 271
3792	Butane, 2,2,3,3-tetrabromo $CH_3CBr_2CBr_2CH_3$	373.71	lf (lig), pr (eth-lig)		243			eth, ace, bz, chl lig	B1³, 298
3793	Butane, 1,1,1,2-tetrachloro $CH_3CH_2CHClCCl_3$	195.90		$134\text{-}5^{742}$		$1.3932^{20/20}$	1.4920^{25}	ace, chl	B1⁴, 254
3794	Butane, 1,2,2,3-tetrachloro $CH_3CHClCCl_2CH_2Cl$	195.90		$182, 85^{10}$	-48	$1.4276^{18/4}$	1.491^{20}	ace, chl	B1³, 286
3795	Butane, 1,2,3,3-tetrachloro $CH_3CCl_2CHClCH_2Cl$	195.90		$90^{32}, 55\text{-}7^{10}$		$1.4204^{20/4}$	1.4958^{20}	eth, ace, chl	B1⁴, 254
3796	Butane, 2,2,3,3-tetramethyl $CH_3C(CH_3)_2C(CH_3)_2CH_3$	114.23	lf (eth)	$106.5, 13.1^{10}$	100.7	0.8242^{20}, $0.6485^{110}{}_4$	1.4695^{20}	eth	B1⁴, 447
3797	Butane, 1,1,2-tribromo $CH_3CH_2CHBrCHBr_2$	294.81		$216.2, 98^{14}$		$2.1836^{20/4}$	1.5626^{17}	al, eth, chl	B1², 84
3798	Butane, 1,2,2-tribromo $CH_3CH_2CBr_2CH_2Br$	294.81		$213.8, 90.1^{14}$		$2.1692^{20/4}$	1.5624^{10}	al, eth, chl	B1², 85
3799	Butane, 1,2,3-tribromo $CH_3CHBrCHBrCH_2Br$	294.81		$220, 97^{10}$	-19	$2.1908^{20/4}$	1.5680^{20}	al, eth, chl	B1⁴, 271
3800	Butane, 1,2,4-tribromo $BrCH_2CH_2CHBrCH_2Br$	294.81		$215, 93^{10}$	-18	$2.170^{20/4}$	1.5608^{20}	al, eth, chl	B1⁴, 271
3801	Butane, 1,3,3-tribromo $CH_3CBr_2CH_2CH_2Br$	294.81		$200\text{-}5, 70^{8}$		$2.1446^{20/4}$	1.5564^{20}	al, eth, chl	B1³, 298
3802	Butane, 2,2,3-tribromo $CH_3CHBrCBr_2CH_3$	294.81		$200, 86^{10}$	1.8	$2.1724^{20/4}$	1.5602^{20}	al, eth, chl	B1³, 298
3803	Butane, 1,1,3-trichloro $CH_3CHClCH_2CHCl_2$	161.46		152^{753}-		1.317^{15}-	1.4600^{15}	al, eth, chl	B1⁴, 252
3804	Butane, 1,2,3-trichloro $CH_3CHClCHClCH_2Cl$	161.46		$165\text{-}8^{725}, 63^{28}$		$1.3164^{20/4}$	1.4790^{20}	al, eth, chl	B1³, 285
3805	Butane, 1,2,4-trichloro $ClCH_2CHClCH_2CH_2Cl$	161.46		$61\text{-}2.5^{10}$		1.3175^{20}	1.4820^{20}	bz, chl	B1⁴, 253
3806	Butane, 2,2,3-trichloro $CH_3CHClCCl_2CH_3$	161.46		143-5		$1.2699^{20/4}$	1.4645^{20}	chl	B1³, 285
3807	Butane, 2,2,3-trimethyl or Triptane $(CH_3)_2CHC(CH_3)_3$	100.20		80.9	-24.2	$0.6901^{20/4}$	1.3864^{20}	al, eth, ace, bz	B1⁴, 410
3808	Isobutane or 2-Methylpropane $(CH_3)_2CHCH_3$	58.12		-11.7	-159.4	0.549^{10}		al, eth, chl	B1⁴, 282
3809	Isobutane, 1,2-dibromo or 1,2-dibromo-2-methylpropane $(CH_3)_2CBrCH_2Br$	215.92		$149\text{-}51, 61^{40}$	9-12	$1.759^{20/4}$	1.509	al, eth, bz, chl	B1⁴, 298
3810	Isobutane, 1,1-dichloro or 1,1-dichloro-2-methyl propane $(CH_3)_2CHCHCl_2$	127.01		105-6		$1.0111^{12/12}$	1.4330^{25}	al, eth, bz, chl	B1⁴, 292
3811	Isobutane, 1,2-dichloro or 1,2-dichloro-2-methyl propane $(CH_3)_2CClCH_2Cl$	127.01		$108, 38\text{-}9^{70}$		$1.093^{20/4}$	1.4370^{20}	al, eth, ace, bz	B1⁴, 292
3812	Isobutane, 1,3-dichloro or 1,3-dichloro-2-methyl propane $CH_3CH(CH_2Cl)_2$	127.01		$134.6, 60^{49}$		$1.1325^{25/4}$	1.4488^{25}	al, eth, bz	B1⁴, 293
3813	Isobutane, 1,2-epoxy or Isobutylene oxide $(CH_3)_2C\text{-}CH_2$ (O bridge)	72.11		52		0.8650^{0}	1.3712^{22}	al, eth	B17⁴, 46
3814	Isobutane, 1,2-epoxy-3-chloro $ClCH_2C(CH_3)CH_2$ (O bridge)	106.55		$12^{2}, 51^{55}$		$1.1011^{20/4}$	1.4340^{20}	eth, w	B17⁴, 47
3815	Isobutane, 1-nitro $(CH_3)_2CHCH_2NO_2$	103.12		$140.5, 61\text{-}2^{45}$		$0.9625^{25/25}$	1.4066^{20}	al, eth	B1⁴, 301
3816	Isobutane, 2-nitro $(CH_3)_3CNO_2$	103.12		127.2	26.23	0.9501^{28}	1.4015^{20}	al, eth, ace, bz	B1⁴, 301

No.	Name, Synonyms, and Formula	Mol. wt.	Color, crystalline form, specific rotation and λ_{max} (log ε)	b.p. °C	m.p. °C	Density	n_D	Solubility	Ref.
3817	Isobutane, 1,1,1,2,3-pentachloro $ClCH_2CCl(CH_3)CCl_3$	230.35	(al)	215^{757}, $90\text{-}3^{10}$	73.5	$1.5686^{25/4}$	1.5165^{25}	chl	$B1^4$, 294
3818	Isobutane, 1,1,1,2-tetrabromo $(CH_3)_2CBrCBr_3$	373.71	lf	...	217	...	...	al, eth	B1, 128
3819	Isobutane, 1,1,2,3-tetrabromo $BrCH_2CBr(CH_3)CHBr_2$	373.71	...	134^{11}	...	2.4545^{20}	1.5990^{20}	chl	$B1^3$, 325
3820	Isobutane, 1,1,1,2-tetrachloro $(CH_3)_2CClCCl_3$	195.90	cr (al)	$192^{117.5}$ sub	178-9	...	...	al, eth, chl	$B1^4$, 293
3821	Isobutane, 1,1,2,3-tetrachloro $ClCH_2CCl(CH_3)CHCl_2$	195.90	...	190-1, 69^{12}	-46	$1.4393^{25/4}$	1.4963^{20}	eth, bz, chl	$B1^4$, 294
3822	Isobutane, 1,1,2-tribromo $(CH_3)_2CBrCHBr_2$	294.81	...	208-15d, 96^{14}	...	$2.0169^{20/4}$	...	eth, chl	$B1^3$, 325
3823	Isobutane, 1,2,3-tribromo $(BrCH_2)_2CBrCH_3$	294.81	...	88.5^{9}	...	$2.1750^{20/4}$	1.5652^{20}	eth, chl	$B1^4$, 299
3824	Isobutane, 1,1,2-trichloro $(CH_3)_2CClCHCl_2$	161.46	...	145-6, $46\text{-}7^{18}$	6	$1.2588^{20/4}$	1.4666^{20}	eth, chl	$B1^4$, 293
3825	Isobutane, 1,2,3-trichloro $(ClCH_2)_2CClCH_3$	161.46	...	162-3, 81^{50}	...	$1.3012^{25/4}$	1.4765^{20}	chl	$B1^4$, 293
3826	1,2-Butanediol-*(d)* $CH_3CH_2CHOHCH_2OH$	90.12	$[\alpha]^{20/}{}_D$ + 14.5 (al, c=6)	192.4, $68^{0.4}$	...	$1.0059^{17.5/0}$	1.4375^{20}	w, al, ace	$B1^4$, 2507
3827	1,2-Butanediol *(dl)* or α-Butylene glycol $CH_3CH_2CHOHCH_2OH$	90.12	...	190.5, 96.5^{10}	...	$1.0024^{20/4}$	1.4378^{20}	w, al, ace	$B1^4$, 2507
3828	1,2-Butanediol *(l)* $CH_3CH_2CHOHCH_2OH$	90.12	$[\alpha]^{22/}{}_D$ -7.4 (al, c=4)	$94\text{-}6^{12}$	...	...	...	w, al, ace	$B1^4$, 2507
3829	1,3-Butanediol *(d)* or β-Butyleneglycol $CH_3CH(OH)CH_2CH_2OH$	90.12	$[\alpha]^{22/}{}_D$ + 18.5 (al, c=4)	204, $60\text{-}5^{0.8}$	...	$1.0053^{20/4}$	1.4418^{20}	w, al	$B1^4$, 2508
3830	1,3-Butanediol *(dl)* $CH_3CH(OH)CH_2CH_2OH$	90.12	...	207.5, $103\text{-}4^{8}$	...	$1.0053^{20/4}$	1.4410^{20}	w, al	$B1^4$, 2508
3831	1,3-Butanediol *(l)* $CH_3CH(OH)CH_2CH_2OH$	90.12	$[\alpha]^{25/}{}_D$ -18.8 (al, c=4)	$107\text{-}10^{23}$	...	$1.005^{20/4}$	...	w, al	$B1^4$, 2508
3832	1,3-Butanediol sulfite $C_4H_8O_3S$	136.17	...	185, $76\text{-}7^{17}$	5	$1.2352^{20/4}$	1.4661^{20}	al, eth, ace, bz	$B1^4$, 2510
3833	1,4-Butanediol or Tetramethylene glycol $HOCH_2CH_2CH_2CH_2OH$	90.12	...	235, 120^{10}	20.1	$1.0171^{20/4}$	1.4460^{20}	w, al	$B1^4$, 2515
3834	1,4-Butanediol diacetate $CH_3CO_2(CH_2)_4CO_2CH_3$	174.20	...	229^{765}	12	1.0479^{15}	1.4251^{15}	...	$B2^4$, 224
3835	1,4-Butanediol dibenzoate $C_6H_5CO_2(CH_2)_4CO_2C_6H_5$	298.34		...	81-2	...	...	eth	$B9^3$, 540
3836	1,4-Butanediol, 2-hexyl $HOCH_2CH_2CH(C_6H_{13})CH_2OH$	174.28	...	$118^{0.05}$	...	...	...	...	...
3837	1,4-Butanediol, 2-methyl $HOCH_2CH_2CH(CH_3)CH_2OH$	104.15	$[\alpha]^{22/}{}_D$ + 11.65	131-3	...	$0.9929^{22/4}$	...	...	$B1^4$, 2546
3838	2,3-Butanediol *(d)* $CH_3CHOHCHOHCH_3$	90.12	$[\alpha]^{25/}{}_D$ + 12.5 (undil)	180-2	34 (anh), 16.86 (+ sw)	0.9872^{25}	1.4306^{25}	**w, al**, eth, ace	$B1^4$, 2524
3839	2,3-Butyldiol *(dl)* $CH_3CHOHCHOHCH_3$	90.12	cr (i-$Pr_2$0)	182.5, 86^{10}	7.6	$1.0033^{20/4}$	1.4310^{25}	**w, al, eth, ace**	$B1^4$, 2524
3840	2,3-Butanediol *(l)* $CH_3CHOHCHOHCH_3$	90.12	$[\alpha]^{25/}{}_D$ -13.0 (undil)	178-81, 77.5^{10}	19.7	$0.9869^{25/4}$	1.4340^{18}	**w, al**, eth, ace	$B1^4$, 2524
3841	2,3-Butanediol *(meso)* $CH_3CHOHCHOHCH_3$	90.12	cr (i-$Pr_2$0)	181.7, 83.5^{10}	34.4	$1.0003^{20/4}$	1.4367^{20}	w, al, eth	$B1^4$, 2524
3842	2,3-Butanediol, 2,3-dimethyl or Pinacol $(CH_3)_2C(OH)C(OH)(CH_3)_2$	118.18	nd (al or eth)	174.4	43	...	...	al, eth	$B1^4$, 2575
3843	2,3-Butanediol, 2,3-dimethyl, hexahydrate or Pinacol hydrate $(CH_3)_2C(OH)C(OH)(CH_3)_2 \cdot 6H_2O$	226.27	pl (w + 6)	...	47	0.967^{15}	...	al, eth, w	$B1^4$, 2575
3844	2,3-Butanediol, 2,3-diphenyl or Acetophenone Pinacol $C_6H_5COH(CH_3)COH(CH_3)C_6H_5$	242.32	pr (dil al)	...	121-2	...	...	al, eth	$B6^3$, 5474
3845	1,2-Butanediol, 3-methyl or α-Isopentylene glycol $(CH_3)_2CHCHOHCH_2OH$	104.15	...	206, $81\text{-}3^{5}$	...	$0.9987^{0/4}$	...	al, eth	$B1^4$, 2549
3846	1,3-Butanediol, 3-methyl or α-Isopentylene glycol $(CH_3)_2C(OH)CH_2CH_2OH$	104.15	...	202-3, 108^{16}	...	$0.9448^{20/4}$	1.4452^{20}	w, al	$B1^4$, 2549
3847	2,3-Butanediol, 2-methyl or β-Isopentyl glycol $CH_3CHOHC(CH_3)OHCH_3$	104.15	...	175, 68^{5}	...	$0.9920^{25/4}$	1.4375^{20}	w, al, eth	$B1^4$, 2547

PHYSICAL CONSTANTS OF ORGANIC COMPOUNDS (Continued)

No.	Name, Synonyms, and Formula	Mol. wt.	Color, crystalline form, specific rotation and λ_{max} (log ε)	b.p. °C	m.p. °C	Density	n_D	Solubility	Ref.
3848	2,3-Butanedione or Biacetyl, Dimethyl glycol $CH_3COCOCH_3$	86.09		88	-2.4	$0.9808^{18.5/4}$	1.3951^{20}	w, al, eth, ace, bz	B1[4], 3644
3849	1,4-Butanedithiol $HS(CH_2)_4SH$	122.24		195-6, $110\text{-}12^{50}$	-53.9	$1.0621^{0/4}$	1.5290^{20}	al	B1[4], 2523
3850	1,2,3,4-Butane tetracarboxylic acid *(dl)* $HO_2CCH_2CH(CO_2H)CH(CO_2H)CH_2CO_2H$	234.16	lf (w), cr (ace)		236-7			w, al	B2[4], 2419
3851	1-Butanethiol or *n*-Butyl mercaptan $CH_3CH_2CH_2CH_2SH$	90.18		98.4	-115.7	$0.8337^{20/4}$	1.4440^{20}	al, eth	B1[4], 1555
3852	1-Butanethiol, 2-methyl-*(d)* or *act*-Amyl mercaptan $CH_3CH_2CH(CH_3)CH_2SH$	104.21	$[\alpha]^{23/}_D$ + 3.21	118.2		$0.8420^{20/4}$	1.4440^{20}		B1[4], 1668
3853	1-Butanethiol, 3-methyl or Isopentyl mercaptan $(CH_3)_2CHCH_2CH_2SH$	104.21		118		$0.8350^{20/4}$	1.4418^{20}	al, eth	B1[4], 1688
3854	2-Butanethiol, *(d)* or *sec*-Butyl mercaptan $CH_3CH_2CHSHCH_3$	90.18	$[\alpha]^{20/}_D$ + 15.7	85-95, 37.4^{134}		$0.8299^{20/4}$	1.43385^{25}	al, eth, bz, peth	B1[3], 1549
3855	2-Butanethiol, *(dl)* or *sec*-Butyl mercaptan $CH_3CH_2CHSHCH_3$	90.18	$[\alpha]^{17/}_D$ - 17.35	85		$0.8295^{20/4}$	1.4366^{20}	al, eth, bz, peth	B1[4], 1584
3856	2-Butanethiol, *(l)* $CH_3CH_2CHSHCH_3$	90.18	$[\alpha]^{17/}_D$ - 17.35	83-4		$0.8300^{17/4}$		al, eth, bz, peth	B1[3], 1549
3857	2-Butanethiol, 2-methyl $CH_3CH_2C(SH)(CH_3)_2$	104.21		99-100			1.4385^{20}		B1[4], 1674
3858	1,2,3-Butanetriol or 1-Methyl glycerol $CH_3CHOHCHOHCH_2OH$	106.12		170^{20}			1.4462^{20}	w, al	B1[4], 2774
3859	1,2,4-Butanetriol $HOCH_2CH_2CH(OH)CH_2OH$	106.12		$172\text{-}4^{12}$		1.018^{20}	1.4688^{20}	w, al	B1[4], 2775
3860	1-Butanol or *n*-Butyl alcohol $CH_3CH_2CH_2CH_2OH$	74.12	$[\alpha]^{20/}_D$ + 9.8 (w)	117.2	-89.5	$0.8098^{20/4}$	1.3993^{20}	w, al, eth, ace, bz	B1[4], 1506
3861	1-Butanol, 2-amino-*(d)* $CH_3CH_2CHNH_2CH_2OH$	89.14	$[\alpha]^{20/}_D$ + 9.8 (w)	80^{11}		0.947^{20}	1.4518^{20}	w, al, eth	B4[4], 1705
3862	1-Butanol, 2-amino-*(dl)* $CH_3CH_2CHNH_2CH_2OH$	89.14		178	-2	0.9162^{20}	1.4489^{25}	w, al, eth	B4[4], 1705
3863	1-Butanol, 2-amino-1-phenyl $CH_3CH_2CH(NH_2)CH(OH)C_6H_5$	165.24	pl (bz-eth)		79-80			al, bz, chl	B13[3], 1791
3864	1-Butanol, 3-amino $CH_3CHNH_2CH_2CH_2OH$	89.14		$82\text{-}5^{19}$			1.4534^{25}	w, al	B4[4], 1710
3865	1-Butanol, 4-amino $H_2NCH_2CH_2CH_2CH_2OH$	89.14		206^{776}		0.967^{12}	1.4625^{20}	w, al	B4[4], 1711
3866	1-Butanol, 2-chloro $CH_3CH_2CHClCH_2OH$	108.57		$74\text{-}6^{25}$		$1.062^{25/4}$	1.4438^{20}	al, eth	B1[4], 1548
3867	1-Butanol, 3-chloro $CH_3CHClCH_2CH_2OH$	108.57		170-80, 73^{20}		$1.0883^{20/4}$	1.4518^{20}	al, eth	B1[4], 1549
3868	1-Butanol, 4-chloro $ClCH_2CH_2CH_2CH_2OH$	108.57		$84\text{-}5^{16}$		$1.0883^{20/4}$	1.4518^{20}	al, eth	B1[4], 1550
3869	1-Butanol, 2,2-dimethyl $CH_3CH_2C(CH_3)_2CH_2OH$	102.18		136.7	<-15	$0.8283^{20/4}$	1.4208^{20}	al, eth	B1[4], 1726
3870	1-Butanol, 2,3-dimethyl-*(d)* $CH_3CH(CH_3)CH(CH_3)CH_2OH$	102.18	$[\alpha]^{25}_D$ + 1.9	142		$0.823^{25/4}$		al, eth, ace	B1[4], 1729
3871	1-Butanol, 2,3-dimethyl-*(dl)* $CH_3CH(CH_3)CH(CH_3)CH_2OH$	102.18		144-5		$0.8297^{20.5/4}$	$1.4195^{20.5}$	al, eth, ace	B1[4], 1729
3872	1-Butanol, 3,3-dimethyl $(CH_3)_3CCH_2CH_2OH$	102.18		143	-60	1.4323^{15}		al, eth, ace	B1[4], 1729
3873	1-Butanol, 2,4-diphenyl $C_6H_5CH_2CH_2CH(C_6H_5)CH_2OH$	226.32	$[\alpha]^{20}_D$ −17.5 (chl, c=10)	$145\text{-}6^{0.1}$	51-2		1.5686^{25}	chl	B6[3], 3428
3874	1-Butanol, 2-diethyl $(C_2H_5)_3ChCH_2OH$	130.23		146.3	<-15	$0.8326^{20/4}$	1.4220^{20}	al, eth	B1[4], 1725
3875	1-Butanol, 4-fluoro $FCH_2CH_2CH_2CH_2OH$	92.11		58^{15}			1.3942^{15}	al, eth, ace	B1[4], 1546
3876	1-Butanol, 2,2,3,3,4,4,4-heptafluoro $CF_3CF_2CF_2CH_2OH$	200.06		95		$1.600^{20/4}$	1.294^{20}	al, ace	B1[4], 1547
3877	1-Butanol, 3-methoxy $CH_3CH(OCH_3)CH_2CH_2OH$	104.15		160		0.923^{23}	1.4148^{25}	al, eth, ace	B1[4], 2509
3878	1-Butanol, 2-methyl-*(d)* or *act*-Amyl alcohol-*(d)* $CH_3CH_2CH(CH_3)CH_2OH$	88.15		128, 65.7^{50}		$1.8191^{20/4}$	1.4102^{20}	al, eth, ace	B1[4], 1666
3879	1-Butanol, 2-methyl-*(dl)* $CH_3CH_2CH(CH_3)CH_2OH$	88.15		127-8, 70^{60}		$0.8152^{25/4}$	1.4092^{20}	al, eth, ace	B1[4], 1666
3880	1-Butanol, 2-methyl-*(l)* $CH_3CH_2CH(CH_3)CH_2OH$	88.15	$[\alpha]^{18/}_D$ + 3.75	129		$0.816^{18/4}$	1.4098^{20}	al, eth, ace	B1[4], 1666

PHYSICAL CONSTANTS OF ORGANIC COMPOUNDS (Continued)

No.	Name, Synonyms, and Formula	Mol. wt.	Color, crystalline form, specific rotation and λ_{max} (log ε)	b.p. °C	m.p. °C	Density	n_D	Solubility	Ref.
3881	1-Butanol, 2-methyl-4-phenyl *(l)* $C_6H_5CH_2CH_2CH(CH_3)CH_2OH$	164.25	...	135^{11}	...	$0.9719^{20/4}$	1.5173^{16}	al eth, ace, bz	$B6^3$, 1962
3882	1-Butanol, 3-methyl or *iso*-Butyl alcohol $(CH_3)_2CH_2CH_2CH_2OH$	88.15	...	128.5^{750}	...	$0.8092^{20/4}$	1.4053^{20}	al, eth, ace	$B1^4$, 1677
3883	1-Butanol, 3-methyl-1-phenyl or α-Hydroxy isopentyl benzene $(CH_3)_2CHCH_2CH(OH)C_6H_5$	164.25	...	$235\text{-}6^{746}$, 112^9	...	$0.9537^{19/4}$	1.5080^{18}	al, eth, ace, bz	$B6^2$, 505
3884	1-Butanol, 3-methyl -2-phenyl $(CH_3)_2CHCH(C_6H_5)CH_2OH$	164.25	...	130^{15}	...	$0.9694^{25/4}$	1.5137^{20}	al, eth, ace, bz	$B6^2$, 506
3885	1-Butanol, 2-nitro $CH_3CH_2CH(NO_2)CH_2OH$	119.12	...	105^{10}	-47	$1.1332^{25/4}$	1.4390^{20}	w, al, eth, ace, aa	$B1^4$, 1555
3886	1-Butanol, 1-phenyl $CH_3CH_2CH_2CH(C_6H_5)OH$	150.22	...	232, $113\text{-}5^{17}$	16	$0.9740^{20/4}$	1.5139^{20}	al, eth	$B6^3$, 1845
3887	1-Butanol, 2,2,3-trichloro $CH_3CHClCCl_2CH_2OH$	177.46	pr (dil al)	199-200, $97\text{-}8^{18}$	62	...	...	al, eth, ace	$B1^3$, 1518
3888	2-Butanol-*(d)* or *sec*-Butyl alcohol $CH_3CH_2CH(OH)CH_3$	74.12	$[\alpha]^{20/}_D$ + 13.9	99.5	...	$0.8080^{20/4}$	1.3954^{20}	al, eth, ace, bz	$B1^4$, 1566
3889	2-Butanol-*(dl)* or *sec*-Butyl alcohol $CH_3CH_2CHOHCH_3$	74.12	...	99.5, 45.5^{60}	...	$0.8063^{20/4}$	1.3978^{20}	al, eth, ace, bz	$B1^4$, 1567
3890	2-Butanol-*(l)* $CH_3CH_2CHOHCH_3$	74.12	$[\alpha]^{20/}_D$ + 13.9	99.5	...	$0.8070^{20/4}$	1.3975^{20}	al, eth, ace, bz	$B1^4$, 1566
3891	2-Butanol, 3-amino $CH_3CHNH_2CHOHCH_3$	89.14	...	$159\text{-}60^{745}$, 70^{20}	18-20 (44)	$0.9299^{25/4}$	1.4502^{20}	w, al, ace	$B1^4$, 1726
3892	2-Butanol, 3-bromo-*(dl)* $CH_3CHBrCHOHCH_3$	153.02	...	154, $46\text{-}50^8$	...	$1.4550^{20/4}$	1.4786^{20}	al, eth	$B1^4$, 1580
3893	2-Butanol, 1-chloro $CH_3CH_2CHOHCH_2Cl$	108.57	...	141, 52^{15}	...	$1.068^{25/4}$	1.4400^{20}	al, eth	$B1^4$, 1578
3894	2-Butanol, 1-chloro-2-methyl $CH_3CH_2C(CH_3)OHCH_2Cl$	122.59	...	150-2	...	1.0161^{20}	1.4469^{20}	al	$B1^4$, 1673
3895	2-Butanol, 3-chloro $CH_3CHClCHOHCH_3$	108.57	...	138-40, $52\text{-}4^{30}$	...	$1.0669^{20/4}$	1.4432^{20}	al, eth	$B1^3$, 1538
3896	2-Butanol, 3-chloro-*(erythro, dl)* $CH_3CHClCHOHCH_3$	108.57	...	135.4^{748}, 56.1^{30}	...	$1.0610^{25/4}$	1.4397^{25}	al, eth, chl	$B1^3$, 1538
3897	2-Butanol, 3-chloro-*(threo)* $CH_3CH(OH)CHClCH_3$	108.57	...	130.8^{748}	...	$1.0586^{25/4}$	1.4386^{25}	...	$B1^3$, 1538
3898	2-Butanol, 3-chloro-2-methyl $CH_3CHClC(CH_3)OHCH_3$	122.59	...	141-2, $55\text{-}6^{30}$	...	$1.0295^{20/4}$	1.4436^{20}	al, eth	$B1^3$, 1627
3899	2-Butanol, 4-chloro $ClCH_2CH_2CHOHCH_3$	108.57	...	67^{20}	...	...	1.4408^{20}	al, eth	$B1^4$, 1578
3900	2-Butanol, 4-cyclohexyl $C_6H_{11}CH_2CH_2CH(OH)CH_3$	156.27	...	112^{14}	...	0.903^{21}	1.464^{21}	...	$B6^3$, 123
3901	2-Butanol, 1,4-dibromo $BrCH_2CH_2CH(OH)CH_2Br$	231.91	...	$114\text{-}5^{15}$	...	2.023^0	1.544^{20}	...	$B1^4$, 1581
3902	2-Butanol, 1,3-dichloro $CH_3CHClCHOHCH_2Cl$	143.01	...	$63\text{-}4^{10}$	...	$1.2860^{15/4}$	1.4766^{20}	al, eth	$B1^3$, 1541
3903	2-Butanol, 2,3-dimethyl $(CH_3)_2C(CH_3)(OH)CH_3$	102.18	...	118.4	-14	$0.8236^{20/4}$	1.4176^{20}	al, eth	$B1^4$, 1729
3904	2-Butanol, 3,3-dimethyl or Pinacolyl alcohol $(CH_3)_3CCHOHCH_3$	102.18	...	120.4	5.6	0.8122^{25}	1.4148^{20}	al, eth	$B1^4$, 1727
3905	2-Butanol, 2-methyl $CH_2CH_2C(CH_3)(OH)CH_3$	88.15	...	102, 50^{60}	-8.4	$0.8059^{25/4}$	1.4052^{20}	al, eth, ace, bz	$B1^4$, 1668
3906	2-Butanol, 2-methyl-1-phenyl-*(dl)* $CH_3CH_2C(CH_3)(OH)CH_2C_6H_5$	164.25	...	$215\text{-}25^{747}$, $103\text{-}5^{11}$	...	$0.9754^{20/0}$	1.5182^{20}	al, eth, lig	$B6^3$, 1958
3907	2-Butanol, 2-methyl-3-phenyl $CH_3CH(C_6H_5)C(OH)(CH_3)_2$	164.25	...	196-8, 118^{24}	...	$0.9794^{20/4}$	1.5193^{20}	al, eth, ace, bz	$B6^3$, 1971
3908	2-Butanol, 2-methyl-4-phenyl $C_6H_5CH_2CH_2CH(OH)(CH_3)_2$	164.25	nd	121^{13}	24.5	$0.9626^{21/4}$	1.5077^{21}	al, eth, ace, bz	$B6^3$, 1962
3909	2-Butanol, 3-methyl-*(d)* $(CH_3)_2CHCHOHCH_3$	88.15	$[\alpha]^{20/}_D$ + 5.34 (al)	112^{734}	...	$0.8225^{16/4}$	1.4089^{20}	al, eth, ace, bz, chl	$B1^1$, 196
3910	2-Butanol, 3-methyl-*(dl)* $(CH_3)_2CHCHOHCH_3$	88.15	...	112.9	...	$0.8180^{20/4}$	1.4089^{20}	al, eth, ace, bz	$B1^4$, 1675
3911	2-Butanol, 3-methyl-3-phenyl $(CH_3)_2C(C_6H_5)CHOHCH_3$	164.25	...	196-8	...	$0.9653^{13/4}$	1.5161^{13}	al, eth, ace, bz	$B6^1$, 269
3912	2-Butanol, 1-nitro $CH_3CH_2CH(OH)CH_2NO_2$	119.12	...	204^{767}, 75^2	...	$1.1353^{20/4}$	1.4435^{20}	al, eth, ace, bz	$B1^4$, 1582
3913	2-Butanol, 3-nitro $CH_3CH(NO_2)CH(OH)CH_3$	119.12	...	$55^{0.5}$	...	...	1.4414^{20}	...	$B1^4$, 1583

PHYSICAL CONSTANTS OF ORGANIC COMPOUNDS (Continued)

No.	Name, Synonyms, and Formula	Mol. wt.	Color, crystalline form, specific rotation and λ_{max} (log ε)	b.p. °C	m.p. °C	Density	n_D	Solubility	Ref.
3914	2-Butanol, 2-phenyl-*(d)* $CH_3CH_2C(OH)(C_6H_5)CH_3$	150.22	$[\alpha]^{22/}{}_D$ + 17.5	$112-4^{21}$	-13	$0.984^{25/4}$	1.5185^{20}	al, eth	$B6^3$, 185
3915	2-Butanol, 2-phenyl-*(dl)* $CH_3CH_2C(OH)(C_6H_5)CH_3$	150.22		211-2, $90-1^4$		$0.984^{25/4}$	1.5150^{20}	al, eth	$B6^3$, 1854
3916	2-Butanol, 1,1,1-trichloro $CH_3CH_2CHOHCCl_3$	177.46		$169-71^{738}$, $82-4^{22}$		$1.3670^{25/25}$	1.4800^{20}	al, eth, ace, bz, chl	$B1^4$, 1579
3917	2-Butanol, 2,3,3-trimethyl $CH_3C(CH_3)_2C(CH_3)(OH)CH_3$	116.20	cr (dil al + ½ w)	131-2, $40-1^{15}$	83-4	$0.8380^{25/4}$	1.4233^{22}	al, eth, ace	$B1^4$, 1755
3918	2-Butanone or Methyl ethyl ketone $CH_3CH_2COCH_3$	72.11		79.6, 30^{119}	-86.3	$0.8054^{20/4}$	1.3788^{20}	w, al, eth, ace, bz	$B1^4$, 3243
3919	2-Butanone, 1-chloro $CH_3CH_2COCH_2Cl$	106.55		137-8, $34-5^{10}$		$1.0850^{20/4}$	1.4372^{20}	MeOH	$B1^4$, 3255
3920	2-Butanone, 3-chloro $CH_3CHClCOCH_3$	106.55		115, 40^{30}		1.0554^0	1.4219^{20}	al, eth	$B1^4$, 3256
3921	2-Butanone, 3-chloro-3-methyl $(CH_3)_2CHClCOCH_3$	120.58		117.2^{758}		$1.0083^{20/4}$	1.4204^{20}	al, eth	$B1^3$, 2818
3922	2-Butanone, 4-chloro $ClCH_2CH_2COCH_3$	106.55		120-1d, 48^{15}		1.0680^{23}	1.4284^{23}	al, eth	$B1^4$, 3256
3923	2-Butanone, 3,4-dibromo-4-phenyl or Benzalacetone dibromide $C_6H_5CHBrCHBrCOCH_3$	307.00	nd (al)		124-5			al, chl	$B7^3$, 1085
3924	2-Butanone, 1,3-dichloro $CH_3CHClCOCH_2Cl$	141.00		166-7, 55.5^{10}		$1.3116^{20/4}$	1.4686^{20}	al, eth, ace, bz	$B1^3$, 2786
3925	2-Butanone, 4-(diethylamino) $(C_2H_5)_2NCH_2CH_2COCH_3$	143.23		84^{30}		$0.8630^{20/4}$	1.4333^{24}	al, eth, ace, bz	$B4^4$, 1930
3926	2-Butanone, 3,3-dimethyl or Pinacolone $(CH_3)_3CCOCH_3$	100.16		106	-49.8	$0.8012^{25/4}$	1.3952^{20}	al, eth, ace	$B1^4$, 3310
3927	2-Butanone, 3,3-diphenyl $CH_3C(C_6H_5)_2CH_2COCH_3$	224.30	pr (al)	310-1, 176^{14}	41	$1.069^{20/4}$	1.5748^{20}	al, eth, chl, aa	$B7^3$, 2209
3928	2-Butanone, 1-hydroxy $CH_3CH_2COCH_2OH$	88.11		160, 48^9		$1.0272^{20/4}$	1.4189^{20}	w, al, eth	$B1^4$, 3989
3929	2-Butanone, 3-hydroxy or Acetoin $CH_3CHOHCOCH_3$	88.11		143, 37^{11}	-72	$1.0062^{20/20}$	1.4171^{20}	w, ace	$B1^4$, 3991
3930	2-Butanone, 3-methyl or Methyl isopropyl ketone $(CH_3)_2CHCOCH_3$	86.13		94-5	-92	$0.8051^{20/4}$	1.3880^{20}	al, eth, ace	$B1^4$, 3287
3931	2-Butanone, 3-methyl, oxime or Methyl isopropyl ketoxime $(CH_3)_2CHC(:NOH)CH_3$	101.15			157-8			al, eth	$B1^4$, 3289
3932	2-Butanone oxime $CH_3CH_2C(:NOH)CH_3$	87.12		152-3, $59-60^{15}$	-29.5	$0.9232^{20/4}$	1.4410^{20}	w, al, eth	$B1^4$, 3250
3933	2-Butanone, 1-phenyl or Ethyl benzyl ketone $CH_3CH_2COCH_2C_6H_5$	148.20		230^{755}		$1.002^{20/4}$		al, eth, ace	$B7^3$, 1080
3934	2-Butanone, 4-phenyl or Benzylacetone $C_6H_5CH_2CH_2COCH_3$	148.20		233-4, 115^{13}		$0.9849^{22/4}$	1.511^{22}	al, eth, ace	$B7^3$, 1081
3935	2-Butenal or Crotonaldehyde $CH_3CH=CHCHO$	70.09		104-5	-74	$0.8495^{25/4}$	1.4366^{20}	al, eth, ace, bz	$B1^4$, 3447
3936	2-Butenal diethylacetal or Crotonaldehyde diethylacetal $CH_3CH=CHCH(OC_2H_5)_2$	144.21		147-8		$0.8473^{18/4}$	1.4097^{20}	al, eth, ace, bz	$B1^4$, 3450
3937	2-Butenal, 2-bromo, diethyl acetal $CH_3CH=CBrCH(OC_2H_5)_2$	223.11		86^{15}		1.2255^{21}	1.4565^{21}		$B1^4$, 3452
3938	2-Butenal, 2-chloro $CH_3CH=CClCHO$	104.54		147-8, $53-4^{20}$		$1.1404^{23/4}$	1.4780^{25}	al, eth, chl	$B1^3$, 2981
3939	2-Butenal, 3-ethoxy, diethylacetal or 1,1,3-triethoxy-2-butene $CH_3C(OC_2H_5)=CH-CH(OC_2H_5)_2$	188.27		190-5, $79-82^{10}$		$0.908^{21/0}$	1.430^{21}	al	$B1^4$, 4082
3940	2-Butenal, 2-methyl or Tiglaldehyde $CH_3CH=C(CH_3)CHO$	84.12		116.7^{738}, $63-5^{110}$		$0.8710^{20/4}$	1.4475^{20}	w, al, eth	$B1^4$, 3464
3941	2-Butenal, 3-methyl or β,β-Dimethylacrolein $(CH_3)_2C=CHCHO$	84.12		133^{730}		$0.8722^{20/4}$	1.4528^{20}	w, al, eth	$B1^4$, 3464
3942	1-Butene $CH_3CH_2CH=CH_2$	56.11		-6.3	-185.3	$0.5951^{20/4}$	1.3962^{20}	al, eth, bz	$B1^4$, 765
3943	1-Butene, 1-bromo-*(cis)* $CH_3CH_2CH=CHBr$	135.00		86.15		$1.3265^{15/4}$	1.4536^{20}	eth, ace, bz	$B1^2$, 174
3944	1-Butene, 1-bromo-*(trans)* $CH_3CH_2=CHBr$	135.00		94.7	-100.3	$1.3209^{15/4}$	1.4527^{20}	eth, ace, bz	$B1^3$, 726
3945	1-Butene, 2-bromo $CH_3CH_2CBr=CH_2$	135.00		88, 25^{10}	-133.4	$1.3209^{15/4}$	1.4527^{20}	eth, ace, bz	$B1^4$, 775

PHYSICAL CONSTANTS OF ORGANIC COMPOUNDS (Continued)

No.	Name, Synonyms, and Formula	Mol. wt.	Color, crystalline form, specific rotation and λ_{max} (log ε)	b.p. °C	m.p. °C	Density	n_D	Solubility	Ref.
3946	1-Butene, 2-bromo-3-methyl $(CH_3)_2CHCBr=CH_2$	149.03		105^{757}		$1.2328^{20/4}$	1.4504^{20}	eth, bz, chl	B1³, 800
3947	1-Butene, 2-bromo-4-phenyl $C_6H_5CH_2CH_2CBr=CH_2$	211.10		$117\text{-}8^{21}$, $90\text{-}1^{5}$		$1.2907^{20/4}$	1.5450^{20}	ace	B5³, 1209
3948	1-Butene, 4-bromo $BrCH_2CH_2CH=CH_2$	135.00		98.5		$1.3230^{20/4}$	1.4622^{20}	al, eth, bz	B1⁴, 775
3949	1-Butene, 1-chloro-*(cis)* $CH_3CH_2CH=CHCl$	90.55		63.5		$0.9153^{15/4}$	1.4194^{15}	al, eth, ace, chl	B1³, 723
3950	1-Butene, 1-chloro-*(trans)* $CH_3CH_2\text{-}CH=CHCl$	90.55		68		$0.9205^{15/4}$	1.4223^{15}	al, eth, ace, chl	B1³, 723
3951	1-Butene, 1-chloro-2-methyl $CH_3CH_2C(CH_3)=CHCl$	104.58		96-7		$0.9170^{20/4}$	1.4141^{20}	eth, ace	B1³, 787
3952	1-Butene, 1-chloro-3-methyl $(CH_3)_2CH\text{-}CH=CHCl$	104.58		$86\text{-}8^{756}$			1.4229^{20}	eth, ace, chl	B1³, 799
3953	1-Butene, 2-chloro $CH_3CH_2CCl=CH_2$	90.55		58.5		$0.9107^{15/4}$	1.4165^{21}	al, eth, ace, bz	B1⁴, 769
3954	1-Butene, 2(chloromethyl)-1,3-dichloro $CH_3CHClC(CH_2Cl)=CHCl$	174.48		$68\text{-}70^{8}$		$1.2775^{19/4}$		CCl_4	B1³, 788
3955	1-Butene, 3-chloro $CH_3CHClCH=CH_2$	90.55		64-5		$0.8978^{20/4}$	1.4149^{20}	eth, ace, chl	B1⁴, 769
3956	1-Butene, 3-chloro-2-chloromethyl $CH_3CHClC(CH_2Cl)=CH_2$	139.02		155, $31\text{-}3^{7}$		$1.1233^{20/4}$	1.4724^{20}	ace, chl	B1³, 787
3957	1-Butene, 3-chloro-2-methyl $CH_3CHClC(CH_3)=CH_2$	104.58		94		$0.9088^{20/4}$	1.4304^{20}	eth, ace, chl	B1⁴, 819
3958	1-Butone, 4-chloro $ClCH_2CH_2CH=CH_2$	90.55		75^{773}		$0.9211^{20/4}$	1.4233^{20}	eth, ace, chl	B1⁴, 771
3959	1-Butene, 1,3-dichloro $CH_3CHClCH=CHCl$	125.00		125, $58\text{-}60^{25}$		$1.1341^{24/4}$	1.4647^{20}	al, eth, ace, chl	B1⁴, 772
3960	1-Butene, 2,3-dichloro $CH_3CHClCCl=CH_2$	125.00		112		$1.1340^{20/4}$	1.4580^{20}	eth, ace, chl	B1⁴, 772
3961	1-Butene, 3,3-dichloro-2-methyl $CH_3CCl_2C(CH_3)=CH_2$	139.02		151-3		$1.1276^{18/4}$	1.4737^{18}	eth, ace, chl	B1³, 787
3962	1-Butene, 3,4-dichloro $ClCH_2CHClCH=CH_2$	125.00		115-7, 45.5^{40}		1.1170^{20}	1.4475^{20}, 1.4641^{20}	al, eth, bz	B1⁴, 772
3963	1-Butene, 2,3-dimethyl $CH_3CH(CH_3)C(CH_3)=CH_2$	84.16		55.67	-157.3	$0.6803^{20/4}$	1.3995^{20}	al, eth, ace	B1⁴, 852
3964	1-Butene, 3,3-dimethyl $(CH_3)_3CCH=CH_2$	84.16		41.2	-115.2	$0.6529^{20/4}$	1.3763^{20}	al, eth	B1⁴, 850
3965	1-Butene-3,4-diol $HOCH_2CHOHCH=CH_2$	88.11		196.5, 98^{16}		$1.0470^{20/4}$	1.4628^{21}	w, al	B1⁴, 2658
3966	1-Butene, 1,3-diphenyl $CH_3CH(C_6H_5)CH=CHC_6H_5$	208.30		$175\text{-}6^{14}$		1.016^{15}	1.590		B5⁴, 2205
3967	1-Butene, 3,4-epoxy or Butadiene monoxide $CH_2CHCH=CH_2$	70.09		70		0.9006^{0}	1.4168^{20}	al, eth, bz	B17⁴, 145
3968	1-Butene, 2-ethyl $CH_3CH_2C(C_2H_5)=CH_2$	84.16		64.7	-131.5	$0.6894^{20/4}$	1.3969^{20}	eth, ace, bz, chl	B1⁴, 850
3969	1-Butene, 2-ethyl-3-methyl $(CH_3)_2CH\text{-}C(C_2H_5)=CH_2$	98.19		89		$0.7150^{20/4}$	1.410^{20}	eth, ace, bz, chl	B1⁴, 871
3970	1-Butene, 2-methyl $CH_3CH_2CCH_3=CH_2$	70.13		31.2	-137.5	$0.6504^{20/4}$	1.3778^{20}	al, eth, bz	B1⁴, 818
3971	1-Butene, 3-methyl $(CH_3)_2CHCH=CH_2$	70.13		20	-168.5	$0.6272^{20/4}$	1.3643^{20}	al, eth, bz	B1⁴, 825
3972	1-Butene, perfluoro $F_3CCF_2CF=CF_2$	200.03		4.8		$1.615^{-20/4}$, 1.5443^{0}			B1⁴, 769
3973	1-Butene, 1-phenyl *(cis)* $CH_3CH_2CH=CHC_6H_5$	132.21		196.2^{755}, $84\text{-}5^{23}$		$0.9106^{20/4}$	1.5381^{16}	al, eth, bz	B5³, 1205
3974	1-Butene, 1-phenyl-*(trans)* $CH_3CH_2CH=CHC_6H_5$	132.21		198.7, $91\text{-}2^{23}$	-43.1	$0.9019^{20/4}$	1.5420^{20}	al, eth, bz	B5⁴, 1374
3975	1-Butene, 4-phenyl $C_6H_5CH_2CH_2CH=CH_2$	132.21		177, 64^{10}	-70	$0.8831^{20/4}$	1.5059^{20}	eth, bz	B5⁴, 1378
3976	1-Butene, 1,3,4,4-tetrachloro $Cl_2CHCHClCH=CHCl$	193.89		88^{20}		$1.0711^{20/4}$	1.4773^{20}	chl	B1⁴, 774
3977	1-Butene, 2,3,3,4-tetrachloro $ClCH_2CCl_2CCl=CH_2$	193.89		$41\text{-}2^{7}$		$1.4602^{20/4}$	1.5135^{20}	ace, chl	B1³, 726
3978	1-Butene, 2,3,4-trichloro $ClCH_2CHClCCl=CH_2$	159.44		60^{20}		$1.3430^{20/4}$	1.4944^{20}	ace, chl	B1³, 725

No.	Name, Synonyms, and Formula	Mol. wt.	Color, crystalline form, specific rotation and λ_{max} (log ε)	b.p. °C	m.p. °C	Density	n_D	Solubility	Ref.
3979	1-Butene, 2,3,3-trimethyl or Triptene $(CH_3)_3CC(CH_3)=CH_2$	98.19		77.9	-109.9	$0.7050^{20/4}$	1.4025^{20}	eth, bz, MeOH	B1⁴, 873
3980	2-Butene *(cis)* $CH_3CH=CHCH_3$	56.11		3.7	-138.9	$0.6213^{20/4}$	1.3931^{-25}	al, eth, bz	B1⁴, 778
3981	2-Butene *(trans)* $CH_3CH=CHCH_3$	56.11		0.9	-105.5	$0.6042^{20/4}$	1.3848^{-25}	al, eth, bz	B1⁴, 781
3982	2-Butene, 1-bromo $CH_3CH=CHCH_2Br$	135.00		$103\text{-}6, 13^{10}$		$1.3371^{25/4}$	1.4822^{20}	al, eth, bz	B1⁴, 789
3983	2-Butene, 1-bromo-3-methyl $(CH_3)_2C=CH\text{-}CH_2Br$	149.03		$129\text{-}33d, 50\text{-}1^{40}$		$1.2819^{20/0}$	1.4930^{15}	al, eth, ace, bz	B1⁴, 824
3984	2-Butene, 1-bromo-4-phenyl $C_6H_5CH_2CH=CHCH_2Br$	211.10		$126\text{-}30^{10.5}$		$1.2660^{20/4}$	1.5678^{20}	eth	B5⁴, 1377
3985	2-Butene, 2-bromo *(cis)* $CH_3CH=CBrCH_3$	135.00		93.9	-111.5	$1.3416^{15/4}$	1.4631^{19}	al, eth, bz	B1⁴, 790
3986	2-Butene, 2-bromo-3-methyl $(CH_3)_2C=CBrCH_3$	149.03		119-20		$1.2773^{20/4}$	1.4738^{20}	eth, ace, chl	B1⁴, 824
3987	2-Butene, 2-bromo-3-phenyl $CH_3C(C_6H_5)=CBr\text{-}CH_3$	211.10		$120\text{-}30^{11}$		$1.3348^{20/4}$	1.5811^{20}	eth, bz, chl	B5⁴, 1379
3988	2-Butene, 1-chloro *(cis)* $CH_3CH=CHCH_2Cl$	90.55		84.1		$0.9426^{20/4}$	1.4390^{20}	al, ace, chl	B1⁴, 783
3989	2-Butene, 1-chloro *(trans)* $CH_3CH=CHCH_2Cl$	90.55		84.8^{752}		$0.9295^{20/4}$	1.4350^{20}	ace, chl	B1⁴, 784
3990	2-Butene, 1-chloro-2,3-dimethyl $(CH_3)_2C=C(CH_3)CH_2Cl$	118.61		$111\text{-}2^{756}$		$0.9355^{20/4}$	1.4605^{20}	eth, ace, chl	B1⁴, 856
3991	2-Butene, 1-chloro-2-methyl $CH_3CH=C(CH_3)CH_2Cl$	104.58		$110, 26.4^{25}$		$0.9327^{20/4}$	1.4481^{20}	al, eth, ace, chl	B1⁴, 822
3992	2-Butene, 1-chlor-3-methyl $CH_3C(CH_3)=CHCH_2Cl$	104.58		$109, 54.3^{95}$		$0.9273^{20/4}$	1.4485^{20}	al, eth, ace, chl	B1⁴, 823
3993	2-Butene, 2-chloro *(cis)* $CH_3CH=CClCH_3$	90.55		70.6	-117.3	$0.9239^{20/4}$	1.4240^{20}	al, ace, chl	B1⁴, 785
3994	2-Butene, 2-chloro *(trans)* $CH_3CH=CClCH_3$	90.55		62.8	-105.8	$0.9138^{20/4}$	1.4190^{20}	al, ace, chl	B1⁴, 785
3995	2-Butene, 2-chloro-3-methyl $(CH_3)_2C=CClCH_3$	104.58		94		$0.9324^{20/4}$	1.4320^{20}	al, eth, ace, chl	B1³, 794
3996	2-Butene, 1,4-dibromo *(trans)* $BrCH_2CH=CHCH_2Br$	213.90	pl (peth)	$203, 85^{10}$	53.4			al, ace, peth	B1⁴, 791
3997	2-Butene, 1,1-dichloro $CH_3CH=CHCHCl_2$	125.00		125-7		1.1310^{20}	1.466^{18}	eth, ace, chl	B1⁴, 787
3998	2-Butene, 1,2-dichloro (high b.p.) $CH_3CH=CClCH_2Cl$	125.00		130-1		$1.1601^{20/4}$	1.4734^{20}	eth, ace, chl	B1³, 741
3999	2-Butene, 1,2-dichlor (low b.p.) $CH_3CH=CClCH_2Cl$	125.00		$116\text{-}8^{765}$		$1.1544^{20/4}$	1.4642^{20}	bz, CCl_4	B1³, 741
4000	2-Butene, 1,3-dichloro *(cis)* $CH_3CCl=CHCH_2Cl$	125.00		$129.9^{745}, 34^{20}$		$1.1605^{20/4}$	1.4735^{20}	al, eth, ace, bz	B1⁴, 786
4001	2-Butene, 1,3-dichloro *(trans)* $CH_3CCl=CHCH_2Cl$	125.00		$130^{745}, 53^{50}$		$1.1585^{20/4}$	1.4719^{20}	al, eth, ace, bz	B1⁴, 786
4002	2-Butene, 1,3-dichloro-2-methyl $CH_3CCl=C(CH_3)CH_2Cl$	139.02		151-3		$1.1293^{20/4}$		ace, chl	B1³, 795
4003	2-Butene, 1,4-dichloro *(cis)* $ClCH_2.CH=CHCH_2Cl$	125.00		$152.5, 22.5^{3}$	-48	$1.188^{25/4}$	1.4887^{25}	al, eth, ace, bz	B1⁴, 787
4004	2-Butene, 1,4-dichloro *(trans)* $ClCH_2CH=CHCH_2Cl$	125.00		$155.5^{758}, 55.5^{20}$	1-3	$1.183^{25/4}$	1.4871^{25}	al, eth, ace, bz	B1⁴, 787
4005	2-Butene, 1,4-dichloro-2-methyl $ClCH_2CH=C(CH_3)CH_2Cl$	139.02		93^{50}	1	$1.1526^{20/4}$	1.4932^{20}	ace, chl	B1³, 795
4006	2-Butene, 2,3-dichloro *(cis)* $CH_3CCl=CClCH_3$	125.00		$125\text{-}6^{758}$		$1.1618^{20/4}$	1.4590^{20}	al, eth, ace, bz	B1⁴, 787
4007	2-Butene, 2,3-dichloro *(trans)* $CH_3CCl=CClCH_3$	125.00		$101\text{-}3^{758}$		$1.1416^{20/4}$	1.4582^{20}	al, eth, ace, bz	B1⁴, 787
4008	2-Butene, 2,3-dimethyl $(CH_3)_2=C(CH_3)_2$	84.16		73.2	-74.3	$0.7080^{20/4}$	1.4122^{20}	al, eth, ace, chl	B1⁴, 853
4009	2-Butene-1,4-diol *(cis)* $HOCH_2CH=CHCH_2OH$	88.11		$235, 132^{16}$	4	$1.0698^{20/4}$	1.4782^{20}	w, al	B1⁴, 2660
4010	2-Butene-1,4-diol *(trans)* $HOCH_2CH=CHCH_2OH$	88.11		131^{13}	25	$1.0700^{20/4}$	1.4755^{20}	w, al	B1⁴, 2660
4011	2-Butene, 1,1,2,3,4,4-hexachloro (liquid) $Cl_2CHCCl=CClCHCl_2$	262.78		$97\text{-}8^{10}$	-19	$1.651^{15/15}$	1.5331	chl	B1 , 789

No.	Name, Synonyms, and Formula	Mol. wt.	Color, crystalline form, specific rotation and λ_{max} (log ε)	b.p. °C	m.p. °C	Density	n_D	Solubility	Ref.
4012	2-Butene, 1,1,2,3,4,4-hexachloro (solid) $Cl_2CHCCl{=}CClCHCl_2$	262.78	lf (al)		80			al, eth, bz, chl	$B1^4$, 789
4013	2-Butene, 2-methyl $CH_3CH{=}C(CH_3)_2$	70.13		38.6	-133.8	$0.6623^{20/4}$	1.3874^{20}	al, eth, bz, lig	$B1^4$, 820
4014	2-Butene, perfluoro $CF_3CF{=}CFCF_3$	200.03		0—3	-129	1.5297^{0}			$B1^4$, 783
4015	2-Butene, 1,1,1,4,4-pentachloro $Cl_2CHCH{=}CHCCl_3$	228.33		$78\text{-}80^{11}$		$1.612^{21/21}$	1.5538^{21}	al, chl	$B1^3$, 744
4016	2-Butene, 1,2,4-trichloro $ClCH_2CH{=}CClCH_2Cl$	159.44		$67\text{-}9^{10}$		$1.3843^{20/4}$	1.5175^{20}	al, eth, bz, chl	$B1^3$, 744
4017	2-Butenoic acid *(cis)* or Isocrotonic acid $CH_3CH{=}CHCO_2H$	86.09	nd or pr (peth)	169.3, 74^{15}	15.5	$1.0267^{20/4}$	1.4483^{14}	w, al	$B2^4$, 1497
4017a	2-Butenoic acid *(trans)* or Crotonic acid $CH_3CH{=}CHCO_2H$	86.09	mcl pr or nd (w or lig)	185	71.5	$1.1018^{15/4}$	1.4249^{77}	w, al, eth, ace	$B2^4$, 1509
4018	2-Butenoic acid, 2-chloro *(cis)* $CH_3CH{=}CClCO_2H$	120.54	nd (w)		67			w, al, lig	$B2^4$, 1510
4019	2-Butenoic acid, 2-chloro, ethyl ester *(cis)* $CH_3CH{=}CClCO_2C_2H_5$	148.59		75^{30}		$1.1021^{18/4}$		al, eth	$B2^4$, 1510
4020	2-Butenoic acid, 3-chloro *(cis)* $CH_3CCl{=}CHCO_2H$	120.54		195 sub	61	$1.1995^{66/4}$	1.4704^{66}	al, peth	$B2^4$, 1511
4021	2-Butenoic acid, 3-chloro, ethyl ester *(cis)* $CH_3CCl{=}CHCO_2C_2H_5$	148.59		161.4, 50^{10}		$1.0860^{20/4}$	$1.4542^{19/4}$	al, eth	$B2^4$, 1510
4022	2-Butenoic acid, 3-chloro, methyl ester *(cis)* $CH_3CCl{=}CHCO_2CH_3$	134.56		142.4, $42\text{-}3^{13}$		$1.138^{20/4}$	1.4573^{19}	eth, MeOH	$B2^4$, 1500
4023	2-Butenoic acid, ethyl ester *(cis)* or Ethyl isocrotonate $CH_3CH{=}CHCO_2C_2H_5$	114.14		136		$0.9182^{20/4}$	1.4242^{20}	al, eth, ace	$B2^4$, 1551
4024	2-Butenoic acid, 2-methyl *(cis)* or Angelic acid $CH_3CH{=}C(CH_3)CO_2H$	100.12	mcl pr or nd	185, $88\text{-}9^{10}$	45-6	$0.983^{49/4}$, $0.9539^{76/4}$	1.4434^{47}	al, eth	$B2^4$, 1497
4026	2-Butenoic acid, allyl ester *(trans)* $CH_3CH{=}CHCO_2CH_2{=}CH_2$	126.16		$88\text{-}9^{70}$		$0.9440^{20/4}$	1.4465^{20}		$B2^4$, 1503
4027	2-Butenoic acid, 3-amino, ethyl ester *(trans)* $CH_3C(NH_2){=}CHCO_2C_2H_5$	129.16	mcl pr	210-5d, 105^{15}	34 (st), 20-1 (unst)	$1.0219^{19/4}$	1.4988^{22}	al, eth, bz, chl, lig	$B3^3$, 1199
4028	2-Butenoic acid, 3-bromo *(trans)* or 3-Bromo-crotonic acid $CH_3CBr{=}CHCO_2H$	164.99	nd (lig), lf (w)		97			al, eth, bz, aa	$B2^3$, 1275
4029	2-Butenoic acid, 4-bromo, ethyl ester *(trans)* or Ethyl-4-bromocrotonate $BrCH_2CH{=}CHCO_2C_2H_5$	193.04		$97\text{-}8^{15}$		$1.402^{16/4}$	1.4925^{20}	al	$B2^4$, 1517
4030	2-Butenoic acid, 2-chloro *(trans)* or 2-Chlorocrotonic acid $CH_3CH{=}CClCO_2H$	120.54	nd (w or peth)	212, $111\text{-}2^{14}$	100.5			al, eth	$B2^4$, 1509
4031	2-Butenoic acid, 2-chloro, ethyl ester *(trans)* or Ethyl 2-chlorocrotonate $CH_3CH{=}CHCClCO_2C_2H_5$	148.59		176-8, 61^{10}		$1.1135^{20/4}$	1.4538^{20}	al, eth	$B2^4$, 1510
4032	2-Butenoic acid, 2-chloro, methyl ester *(trams)* or Methyl 2-chlorocrotonate $CH_3CH{=}CClCO_2CH_3$	134.56		161.5^{762}, 59.5^{16}		$1.160^{20/4}$	1.4569^{21}	eth	$B2^4$, 1510
4033	2-Butenoic acid, 3-chloro *(trans)* or 3-Chlorocrotonic acid $CH_3CCl{=}CHCO_2H$	120.54		206-11d	94-5			al, CS_2	$B2^4$, 1510
4034	2-Butenoic acid, 3-chloro, ethyl ester *(trans)* $CH_3CCl{=}CHCO_2C_2H_5$	148.59		184, 66^{10}		$1.1062^{20/4}$	1.4592^{20}	al, eth	$B2^4$, 1511
4035	2-Butenoic acid, 3-chloro, methyl ester *(trans)* $CH_3CCl{=}CHCO_2CH_3$	134.56		$64\text{-}7^{14}$		$1.157^{20/4}$	1.4630^{20}	al, eth	$B2^4$, 1510
4036	2-Butenoic acid, 4-chloro-*(trans)* $ClCH_2CH{=}CHCO_2H$	120.54	cr (peth-eth)	$117\text{-}8^{13}$	83			eth	$B2^4$, 1511
4037	2-Butenoic acid, ethyl ester *(trans)* or Ethyl crotonate $CH_3CH{=}CHCO_2C_2H_5$	114.14		136.5, $58\text{-}9^{48}$		$0.9175^{20/4}$	1.4243^{20}	al, eth	$B2^4$, 1500
4038	2-Butenoic acid, 2-ethyl *(trans)* $CH_3CH{=}C(C_2H_5)CO_2H$	114.14	mcl pr (peth)	209,109^{13}	45-6	$0.9578^{50/4}$	1.4475^{50}	al, eth	$B2^3$, 1329
4039	2-Butenoic acid, methyl ester *(trans)* or Methyl crotonate $CH_3CH{=}CHCO_2CH_3$	100.12		121	-42	$0.9444^{20/4}$	1.4242^{20}	al, eth	$B2^4$, 1500
4040	2-Butenoic acid, 2-methyl *(trans)* or Tiglic acid $CH_3CH{=}C(CH_3)CO_2H$	100.12	ta (w)	198.5	64.5-5	$0.9641^{76/4}$	1.4330^{76}	al, eth	$B2^4$, 1552
4041	2-Butenoic acid, 2-methyl, ethyl ester *(trans)* or Ethyl tiglate $CH_3CH{=}C(CH_3)CO_2C_2H_5$	128.17		156, 55.5^{11}		$0.9200^{20/4}$	1.4340^{20}	al, bz	$B2^4$, 1553

No.	Name, Synonyms, and Formula	Mol. wt.	Color, crystalline form, specific rotation and λ_{max} (log ε)	b.p. °C	m.p. °C	Density	n_D	Solubility	Ref.
4042	3-Butenoic acid or Vinyl acetic acid $CH_2=CHCH_2CO_2H$	86.09	...	$169, 69\text{-}70^{12}$	-35	$1.0091^{20/4}$	1.4239^{20}	w, al, eth	B2^4, 1491
4043	3-Butenoic acid-ethyl ester or Ethyl, 3-butenoate $CH_2=CHCH_2CO_2C_2H_5$	114.14	...	119	...	0.9122^{20}	1.4105^{20}	al	B2^4, 1491
4044	3-Butenoic acid, 2-hydroxy, ethyl ester $CH_2=CHCHOHCO_2C_2H_5$	130.14	...	$173d, 68^{15}$	...	$1.0470^{15/4}$	1.436^{13}	w, al, eth	B2^3, 685
4045	3-Butenoic acid, 2-hydroxy-4-phenyl or Benzollactic acid $C_6H_5CH=CHCHOHCO_2H$	178.19	nd (w)	...	137	...	...	...	B10^3, 862
4046	3-Butenoic acid, 4-phenyl or Styrylacetic acid $C_6H_5CH=CHCH_2CO_2H$	162.19	nd (w), pr (CS_2)	302	87	...	...	al, eth	B9^3, 2756
4047	3-Butenonitrile or Allyl cyanide $CH_2=CHCH_2CN$	67.09	...	119	-84	$0.8329^{20/4}$	1.4060^{20}	al, eth	B2^4, 1491
4048	2-Butene-1-ol or Crotyl alcohol $CH_3CH=CHCH_2OH$	72.11	...	121.2	<-30	$0.8521^{20/4}$	1.4288^{20}	al, eth	B1^4, 2107
4049	2-Butene-1-ol, 2-chloro $CH_3CH=CClCH_2OH$	106.55	...	159	...	$1.1180^{20/4}$	1.4682^{20}	w, al	B1^3, 1900
4050	2-Butene-1-ol, 4-chloro $ClCH_2CH=CHCH_2OH$	106.55	...	$64\text{-}5^{2}$	...	...	1.4845^{20}	al, eth	B1^4, 2110
4051	3-Butene-1-ol $CH_2=CHCH_2CH_2OH$	72.11	...	113.5	...	$0.8424^{20/4}$	1.4224^{20}	w, al, eth, ace	B1^4, 2105
4052	3-Buten-1-ol, 2-chloro $CH_2=CHCHClCH_2OH$	106.55	...	$66\text{-}7^{30}$	...	$1.1044^{20/4}$	1.4665^{20}	al, eth	B1^4, 2106
4053	3-Buten-2-ol *(d)* $CH_2=CH\text{-}CHOHCH_3$	72.11	$[\alpha]^{20/}_{D}$ + 33.9 (undil)	96.5^{745}	...	$0.8362^{15/4}$	1.4120^{20}	...	B1^4, 2102
4054	3-Buten-2-ol *(dl)* $CH_2=CH\text{-}CHOHCH_3$	72.11	...	97.3	<-100	$0.8318^{20/4}$	1.4137^{20}	...	B1^4, 2102
4055	3-Buten-2-ol, l-chloro $CH_2=CH\text{-}CHOHCH_2Cl$	106.55	...	$144\text{-}7, 63.3^{30}$	...	$1.111^{20/4}$	1.4643^{20}	chl	B1^3, 1893
4056	3-Buten-2-ol, 3-chloro $CH_2=CClCHOHCH_3$	106.55	...	$53\text{-}7^{19}$	...	$1.1138^{23/4}$	...	al, eth	B1^3, 1893
4057	3-Buten-2-one or Methyl vinyl ketone $CH_2=CHCOCH_3$	70.09	...	$81.4, 33\text{-}4^{130}$	...	$0.8636^{20/4}$	1.4081^{20}	w, al, eth, ace, bz	B1^4, 3444
4058	3-Butene-2-one, 4-bromo-4-phenyl $C_6H_5CBr=CHCOCH_3$	225.08	...	$150\text{-}1^{10}$	...	...	...	bz	B7^3, 1407
4059	3-Butene-2-one, 4-(2-hydroxyphenyl) $2\text{-}HOC_6H_4CH=CHCOCH_3$	162.19	nd (al or lig), pr (bz)	...	140	...	...	eth, al, bz	B8^3, 810
4060	3-Butene-2-one, 4-(3-hydroxyphenyl) $3\text{-}HOC_6H_4CH=CHCOCH_3$	162.19	ye pr (bz)	...	97-8	...	...	bz	B8^3, 812
4061	3-Butene-2-one, 4-(4-hydroxyphenyl) $4\text{-}HOC_6H_4CH=CHCOCH_3$	162.19	nd (w)	...	114-5	...	...	al, aa	B8^3, 812
4062	3-Butene-2-one, 4-(4-methoxy phenyl) or Anisylidene acetone $4\text{-}CH_3OC_6H_4CH=CHCOCH_3$	176.22	lf (al, eth, aa)	...	73	...	...	al, eth, bz, aa	B8^3, 812
4063	3-Butene-2-one, 3-methyl or Isopropenyl methyl ketone $CH_2=C(CH_3)COCH_3$	84.12	...	98	-54	$0.8527^{20/4}$	1.4220^{20}	al	B1^4, 3462
4064	3-Butene-2-one, 4-phenyl *(trans)* or Benzal acetone $C_6H_5CH=CHCOCH_3$	146.19	pl	$26^{2}, 140^{16}$	42	$1\ \ 097^{45}$	1.5836^{45}	al, eth, ace, bz, chl	B7^3, 1399
4065	1-Butene-3-yne or Vinyl acetylene $CH_2=CH\text{-}C\equiv CH$	52.08	...	5.1	...	$0.7095^{0/0}$	1.4161^{1}	bz	B1^4, 1083
4066	1-Butene-3-yne, 4-chloro $CH_2=CH\text{-}C\equiv CCl$	86.52	...	55-7	...	$1.0022^{20/4}$	1.4656^{20}	chl	B1^4, 1085
4067	1-Butene-3-yne, 4-methoxy $CH_3OCH=CH\text{-}C\equiv CH$	82.10	...	$122\text{-}5d, 30\text{-}2^{15}$	...	$0.906^{20/4}$	1.4818^{20}	...	B1^4, 2300
4068	1-Butene-3-yne, 2-methyl $CH_2=C(CH_3)C\equiv CH$	66.10	...	34	...	$0.6801^{11/4}$	1.4105^{20}	...	B1^4, 1089
4069	*n*-Butyl alcohol or 1-Butanol $CH_3CH_2CH_2CH_2OH$	74.12	...	117.2	-89.5	$0.8098^{20/4}$	1.3993^{20}	w, al, eth, ace, bz	B1^4, 1506
4070	Isobutyl alcohol or 2-methyl-1-propanol $(CH_3)CHCH_2OH$	74.12	...	108.1	...	$0.8018^{20/4}$	1.3955^{20}	al, eth, ace	B1^4, 1588
4071	Isobutyl alcohol, 2-amino $(CH_3)_2C(NH_2)CH_2OH$	89.14	...	$165.5, 69\text{-}70^{10}$	25-6	$0.934^{20/4}$	1.449^{20}	w	B4^4, 1740
4072	Isobutyl alcohol, 2-chloro or β Isobutylene chlorohydrin $(CH_3)_2CClCH_2OH$	108.57	...	$132\text{-}3d, 59\text{-}61^{50}$	...	$1.0472^{20/4}$	1.4388^{20}	...	B1^4, 1603
4073	Isobutyl alcohol, 2-nitro $(CH_3)_2C(NO_2)CH_2OH$	119.12	nd or pl (MeOH)	$94\text{-}5^{10}$	89-90	...	...	al, eth	B1^4, 1604

No.	Name, Synonyms, and Formula	Mol. wt.	Color, crystalline form, specific rotation and λ_{max} (log ε)	b.p. °C	m.p. °C	Density	n_D	Solubility	Ref.
4074	Isobutyl alcohol, 3-chloro $ClCH_2CH(CH_3)CH_2OH$	108.57	....	76-8^{11}	......	1.083$^{25/4}$	1.4460^{25}	al, eth	B1^3, 1564
4075	*sec*-Butyl alcohol or 2-Butanol $CH_3CH_2CH(OH)CH_3$	74.12	....	99.5, 45.5^{60}	......	0.8063$^{20/4}$	1.3978^{20}	**al, eth,** ace, bz	B1^4, 1566
4076	*tert*-Butyl alcohol or 2-Methyl-2-propanol $(CH_3)_3COH$	74.12	....	82.3, 20^{31}	25.5	0.7887$^{20/4}$	1.3878^{20}	**w, al, eth**	B1^4, 1609
4077	*tert*-Butyl alcohol, methoxy $(CH_3)_2C(OH)CH_2OCH_3$	104.15	....	116.6^{747}	......	0.9021$^{15/15}$	....	....	B1^4, 1615
4078	*tert*-Butyl alcohol, 1,1,1-tribromo or Brometone $(CH_3)_2C(OH)CBr_3$	310.81	nd (lig), cr (dil al)	sub	168-70	....	....	al, eth	B1^3, 1588
4079	*tert*-Butyl alcohol, 1,1,1-trichloro or Chloreton $(CH_3)_2C(OH)CCl_3$	177.46	hyg nd (w + 1)	167	98-9, 77 (hyd)	....	....	al, eth, ace, bz	B1^4, 1629
4080	*n*-Butyl amine $CH_3CH_2CH_2CH_2NH_2$	73.14	....	77.8	-49.1	0.7414$^{20/4}$	1.4031^{20}	w, al, eth	B4^4, 540
4081	Butyl dimethyl amino $C_4H_9N(CH_3)_2$	101.19	....	95	......	0.7206$^{20/4}$	1.3970^{20}	**w, al, eth, ace, bz**	B4^4, 546
4082	Butyl ethyl amine $C_4H_9NHC_2H_5$	101.19	....	108-9	......	0.7398$^{20/4}$	1.4040^{20}	**al, eth, ace, bz**	B4^4, 547
4083	Butyl bis(2-hydroxyethyl) amine $C_4H_9N(CH_2CH_2OH)_2$	161.25	....	273-5^{741}, 80^{35}	......	0.9692$^{20/4}$	1.4625^{20}	**w, al, eth, ace**	B4^4, 1520
4084	Butyl phenyl amine or *N*-butylanilene $C_4H_9NHC_6H_5$	149.24	....	241.6, 118-20^{15}	-14.4	0.9322$^{20/4}$	1.5341^{20}	al, eth	B12^3, 267
4085	*sec*-Butyl amine *(d)* or 2-Aminobutane *(d)* $CH_3CH_2CH(NH_2)CH_3$	73.14	$[\alpha]^{20/}{}_D$ + 7.4 (w)	63	-104.5	0.724$^{20/4}$	1.344^{20}	w, **al, eth,** ace	B4^4, 617
4086	*sec*-Butyl amine *(dl)* or 2-Amino butane *(dl)* $CH_3CH_2CH(NH_2)CH_3$	73.14	....	63.5^{764}	<-72	0.7246$^{20/4}$	1.3932^{20}	**al, eth,** ace	B4^4, 618
4087	*sec*-Butyl amine *(l)* or 2-Amino butane *(l)* $CH_3CH_2CH(NH_2)CH_3$	73.14	$[\alpha]^{20/}{}_D$ -7.4 (w, c=4.7)	63	......	0.7205$^{20/4}$	....	w, **al, eth,** ace	B4^4, 617
4088	*sec*-Butyl ethyl amine *(d)* *sec*-$C_4H_9NC_2H_5$	101.19	$[\alpha]^{15/}{}_D$ + 18	98	......	0.7396$^{15/4}$	1.4043^{15}	**al, eth, ace, bz**	B4^3, 307
4089	*sec*-Butyl ethyl amine *(dl)* or Ethyl-*sec*-butyl lamine *(dl)* *sec*-$C_4H_9NHC_2H_5$	101.19	....	97-8^{741}	-104.3	0.7358$^{20/4}$	....	**al, eth, ace, bz**	B4^2, 636
4090	*sec*-Butyl phenyl amine or *N*- *sec*-Butylaniline $CH_3CH_2CH(CH_3)NHC_6H_5$	149.24	....	223, 112-4^{22}	......	....	1.5333^{20}	....	B12^3, 269
4091	*tert*-Butyl amine $(CH_3)_3CNH_2$	73.14	....	44.4	-67.5	0.6958$^{20/4}$	1.3784^{20}	**w, al, eth**	B4^4, 657
4092	*tert*-Butyl phenyl amine or *N*- *tert*-Butylaniline $(CH_3)_3CNHC_6H_5$	149.24	....	214-6^{753}, 93-8^{19}	......	....	1.5270^{20}	al, ace, bz, chl	B12^3, 270
4093	Butyl 4-aminophenyl ketone $(4\text{-}H_2NC_6H_4)COC_4H_9$	177.25	cr (bz-peth)	160-3^3	74-5	....	....	al, eth	B14^2, 43
4094	Butylarsonic acid $C_4H_9AsO(OH)_2$	182.05	....	....	160	....	....	w, al	B4^3, 1824
4095	*n*-Butyl bromide or 1-Bromobutane $CH_3CH_2CH_2CH_2Br$	137.02	....	101.6, 18.8^{30}	-112.4	1.2758$^{20/4}$	1.4401^{20}	**al, eth, ace,** chl	B1^4, 258
4096	Isobutylbromide or 1-Bromo-2-methyl propane $(CH_3)_2CHCH_2Br$	137.02	....	91.7, 41-3^{135}	-117.4	1.2532$^{20/4}$	1.4348^{20}	al, eth, ace, bz	B1^4, 294
4097	*sec*-Butyl bromide *(dl)* or 2-Bromobutane $CH_3CH_2CHBrCH_3$	137.02	....	91.2	-111.9	1.2585$^{20/4}$	1.4366^{20}	eth, **ace,** chl	B1^4, 261
4098	*tert*-Butyl bromide or 2-Bromo-2-methylpropane $(CH_3)_3CBr$	137.02	....	73.25	-16.2	1.2209$^{20/4}$	1.4278^{20}	....	B1^4, 295
4099	*n*-Butyl chloride or 1-Chlorobutane $CH_3CH_2CH_2CH_2Cl$	92.57	....	78.44	-123.1	0.8862$^{20/4}$	1.4021^{20}	**al, eth**	B1^4, 246
4100	Isobutyl chloride or 1-Chloro-2-methylpropane $(CH_3)_2CHCH_2Cl$	92.57	....	68-70	-130.3	0.8810$^{20/4}$	1.39841^{20}	eth, ace, chl	B1^4, 287
4101	*sec*-Butyl chloride or 2-Chlorobutane $CH_3CH_2CHClCH_3$	92.57	....	68.2	-131.3	0.8732$^{20/4}$	1.3971^{20}	**al, eth,** bz, chl	B1^4, 248
4102	*tert*-Butyl chloride or 2-Chloro-2-methyl propane $(CH_3)_3CCl$	92.57	....	....	-25.4	0.8420$^{20/4}$	1.3857^{20}	**al, eth,** bz, chl	B1^4, 288
4103	Butyl (β-chloroethyl) ether $C_4H_9OCH_2CH_2Cl$	136.62	....	154.5d, 49-50^{11}	......	0.9335$^{20/4}$	1.4155^{20}	eth	B1^4, 1519
4103a	1,2-Butylene oxide $CH_3CH_2CH\text{-}CH_2$ (—O—)	72.11	....	63.3	......	0.837$^{17/4}$	1.3851^{20}	al, **eth, ace**	B1^4, 796
4104	Isobutylene or 2-methylpropene $(CH_3)_2C{=}CH_2$	56.11	....	-6.9	-140.3	0.5942$^{20/4}$	1.3926^{-25}	al, eth, bz	B1^3, 762
4105	Isobutylene (trimer) or Trisobutylene $(C_4H_8)_3$	168.32	....	174-81, 56^{10}	-76	0.7590$^{20/4}$	1.4314^{20}	....	B1^3, 763

No.	Name, Synonyms, and Formula	Mol. wt.	Color, crystalline form, specific rotation and λ_{max} (log ε)	b.p. °C	m.p. °C	Density	n_D	Solubility	Ref.
4106	Isobutylene (tetramer) $(C_4H_8)_4$	224.44		$242\text{-}6, 109.5^{15}$	-98	$0.7944^{20/4}$	1.4482^{20}		$B1^4$, 803
4107	Isobutylene, 1-chloro or 1-chloro-2-methylpropene $(CH_3)_2C{=}CHCl$	90.55		68^{754}		$0.9186^{20/4}$	1.4221^{20}	al, eth, ace, chl	$B1^4$, 803
4108	Isobutylene, 3-chloro or 3-chloro-2-methyl propene $ClCH_2C(CH_3){=}CH_2$	90.55		71-2		$0.9165^{20/4}$	1.4291^{20}	al, eth, ace, chl	$B1^4$, 804
4109	Isobutylene, 1,1-dichloro or 1,1-dichloro-2-methylpropene $(CH_3)_2C{=}CCl_2$	125.00		$108\text{-}9, 42\text{-}3^{75}$		$1.1449^{20/0}$	1.4580^{20}	eth, bz, chl	$B1^4$, 2533
4110	Isobutylene glycol or 2-methyl-1,2-propane diol $(CH_3)_2C(OH)CH_2OH$	90.12		$176, 79\text{-}80^{12}$		$1.0024^{20/4}$	1.4350^{20}	w, al, eth	$B1^4$, 2536
4111	Isobutylene glycol, 3-chloro or 3-chloro-2-methyl propylene glycol $ClCH_2C(OH)(CH_3)CH_2OH$	124.57		$114\text{-}7^{20}$		$1.2362^{20/4}$	1.4788^{20}	w, al, eth	$B1^4$, 805
4112	Isobutylene, 3,3-dichloro or 3,3-dichloro-2-methyl propene $Cl_2CHC(CH_3){=}CH_2$	125.00		$108\text{-}12, 49\text{-}50^{120}$		$1.3631^{24/4}$	1.4523^{24}	eth, bz, chl	$B1^3$, 1904
4113	Isobutylene, 3-methoxy or 3-methoxy-2-methyl propene $CH_3OCH_2C(CH_3){=}CH_2$	86.13		68^{773}		$0.7698^{20/4}$	1.3964^{20}		$B17^4$, 45
4115	Isobutylene, 1,1,3-trichloro or 1,1,3-trichloro-2-methyl propene $ClCH_2C(CH_3){=}CCl_2$	159.44		$156, 45\text{-}6^{12}$		1.346^{20}	1.4990^{20}	ace, bz, chl	$B1^4$, 805
4116	Isobutylene, 3,3,3-trichloro or 3,3,3-trichloro-2-methyl propene $Cl_3CC(CH_3){=}CH_2$	159.44			132-4	1.293^{20}	1.4770^{20}	ace, bz, chl, aa	$B1^4$, 805
4117	Butyl ether or Dibutylether $C_4H_9OC_4H_9$	130.23		142	-95.3	$0.7689^{20/4}$	1.3992^{20}	al, eth, ace	$B1^4$, 1520
4118	*sec*-Butyl ether *(dl)* or Di-*sec*-Butyl ether $(sec\text{-}C_4H_9)_2O$	130.23		120-1		0.756^{25}	1.393^{25}	al, eth, ace	$B1^3$, 1533
4119	Isobutyl ether α, β-dichloro or α, β-Diclorodiisobutyl ether $(CH_3)_2CCCHCl\text{-}O\text{-}i\text{-}C_4H_9$	199.12		$192.5, 83^{15}$		$1.031^{5/4}$		eth, ace	B1, 675
4120	Butyl isobutyl ether $(CH_3)_2CHCH_2OC_4H_9$	130.23		$148\text{-}52^{730}$		$0.7980^{22/4}$	1.4077^{21}	al, eth, ace	$B1^4$, 1594
4121	Butyl ethyl ether $C_4H_9OC_2H_5$	102.18		96		$0.7490^{20/4}$	1.3818^{20}	al, eth, ace	$B1^4$, 1518
4122	*sec*-Butyl ethyl ether *sec*-$C_4H_9OC_2H_5$	102.18		81		$0.7503^{20/4}$	1.3802^{20}	al, eth	$B1^4$, 1572
4123	*tert*-Butyl ethyl ether $(CH_3)_3COC_2H_5$	102.18		73.1	-94	0.7519^{25}	1.3794^{20}	al, eth	$B1^4$, 1615
4124	Butyl ethynyl ether or Butoxy acetylene $C_4H_9OC{\equiv}CH$	98.14		102-4,exp ca 100		$0.8200^{20/4}$	1.4020^{20}	al, eth	$B1^4$, 2213
4125	Butyl ethyl sulfide $C_4H_9SC_2H_5$	118.24		$144.2, 33.3^{10}$	-95.1	$0.8376^{20/4}$	1.4491^{20}	al, chl	$B1^4$, 1558
4126	Butyl furfuryl ether $C_4H_9OCH_2(C_4H_3O)$	154.21		189.9^{765}		$0.9516^{20/4}$	1.4522^{20}	al, eth	$B17^2$, 115
4127	*tert*-Butyl hydro peroxide $(CH_3)_3COOH$	90.12		$d\ at89, 35\text{-}7^{17}$	6	0.8960^{20}	1.4015^{20}	w, al, eth, chl	$B1^4$, 1616
4128	*tert*-Butyl hypochlorite $(CH_3)_3COCl$	108.57	ye lig	77-8		$0.9583^{18/4}$	1.403^{20}	eth, ace, bz	$B1^4$, 1621
4129	*tert*-Butyl iodide or 2-Iodo-2-methyl propane $(CH_3)_3CI$	184.02		$100, 20.8^{30}$	-38.2	$1.5445^{20/4}$	1.4918^{20}	al, eth	$B1^4$, 300
4130	Isobutyl isocyanate *i*-C_4H_9NCO	99.13		106					$B4^4$, 653
4131	*tert*-Butyl isocyanate $(CH_3)_3CNCO$	99.13		85.5		0.8670^{9}	1.4061^{20}		$B4^4$, 669
4132	*tert*-Butyl isocyanide or 2-Isocyano-2-methyl propane $(CH_3)_3CNC$	83.13		$167\text{-}70, 91^{38}$				eth, al	$B4^4$, 562
4133	Butyl isothiocyanate C_4H_9NCS	115.19		$168, 64\text{-}6^{12}$		$0.9546^{20/4}$	1.501^{20}	al, eth	$B4^4$, 596
4134	*sec*-Butyl isothiocyanate *(d)* or *sec*-Butyl mustard oil *sec*-C_4H_9NCS	115.19	$[\alpha]^{20/}_{D}$ +61.88	159		$0.943^{20/4}$		al, eth	$B4^1$, 372
4135	*sec*-Butyl isothiocyanate *(dl)* *sec*-C_4H_9NCS	115.19		159.5		0.944^{12}		al, eth	$B4^4$, 624
4136	*sec*-Butyl isothiocyanate *(l)* *sec*-C_4H_9NCS	115.19	$[\alpha]^{20/}_{D}$ -61.8	159		$0.942^{20/4}$		al, eth	B4, 161
4137	Isobutyl isothiocyanate or Isobutyl mustard oil *i*-C_4H_9NCS	115.19		160 (cor)		$0.9638^{14/4}$	1.5005^{14}	al, eth	$B4^4$, 653

PHYSICAL CONSTANTS OF ORGANIC COMPOUNDS (Continued)

No.	Name, Synonyms, and Formula	Mol. wt.	Color, crystalline form, specific rotation and λ_{max} (log ε)	b.p. °C	m.p. °C	Density	n_D	Solubility	Ref.
4138	*tert*-Butyl isothiocyanate $(CH_3)_3CNCS$	115.19		140^{770}	10-11	$0.9187^{10/4}$		eth	$B4^4$, 669
4139	*n*-Butyl mercaptan or l-Butanethiol $CH_3CH_2CH_2CH_2SH$	90.18		98.4	-115.7	$0.8337^{20/4}$	1.4440^{20}	al, eth	$B1^4$, 1555
4140	Isobutyl mercaptan or 2-Methyl-l-propanethiol $(CH_3)_2CHCH_2SH$	90.18		88.7	<-70	$0.8339^{20/4}$	1.4387^{20}	al, eth, ace	$B1^4$, 1605
4141	*sec*-Butyl mercaptan *(d)* or 2-Butanethiol $CH_3CH_2CHSHCH_3$	90.18	$[\alpha]^{20/}{}_D$ + 15.7	85-95, 37.4^{134}		$0.8299^{20/4}$	1.4338^{25}	al, eth, bz, peth	$B1^3$, 1549
4142	*sec*-Butyl mercaptan *(dl)* $CH_3CH_2CH(SH)CH_3$	90.18		85		$0.8295^{20/4}$	1.4366^{20}	al, eth, bz, peth	$B1^4$, 1584
4143	*sec*-Butyl mercaptan *(l)* $CH_3CH_2CH(SH)CH_3$	90.18	$[\alpha]^{17/}{}_D$ - 17.35	83-4		$0.8300^{17/4}$		al, eth, bz, peth	$B1^3$, 1549
4144	*tert*-Butyl mercaptan or 2-Methyl-2-propanethiol $(CH_3)_3CSH$	90.18		64.2	1.11	$0.8002^{20/4}$	1.4232^{20}		$B1^4$, 1634
4145	Butyl methyl ether $C_4H_9OCH_3$	88.15		71	-115.5	$0.7443^{20/4}$	1.3736^{20}	al, eth, ace	$B1^4$, 1518
4146	Isobutyl methyl ether *i*-$C_4H_9OCH_3$	88.15		58		$0.7311^{20/4}$		al, eth	$B1^4$, 1593
4147	*sec*-Butyl methyl ether *sec*-$C_4H_9OCH_3$	88.15		60		$0.7415^{20/4}$	1.3680^{25}	al, eth, ace	$B1^4$, 1572
4148	*tert*-Butyl methyl ether $(CH_3)_3COCH_3$	88.15		55.2	-109	$0.7405^{20/4}$	1.3690^{20}	al, eth	$B1^4$, 1615
4149	Butyl methyl sulfide $C_4H_9SCH_3$	104.21		123.2	-97.8	$0.8426^{20/4}$	1.4477^{20}	al	$B1^4$, 1557
4150	Isobutyl methyl sulfide *i*-$C_4H_9SCH_3$	104.21		112.5		$0.8335^{20/4}$	1.4433^{20}	al, eth, ace	$B1^4$, 1606
4151	*n*-Butyl nitrate $C_4H_9ONO_2$	119.12		135.5, $70\text{-}1^{88}$		1.0228^{30}	1.4013^{23}	al, eth	$B1^4$, 1524
4152	*sec*-Butyl nitrate *sec*-$C_4H_9ONO_2$	119.12		124, 59^{80}		$1.0264^{20/4}$	1.4015^{20}	al, eth	$B1^4$, 1573
4153	iso-butyl nitrite *i*-C_4H_9ONO	103.12		67		$0.8699^{22/4}$	1.3715^{22}	al, eth	$B1^4$, 1595
4154	*tert*-Butyl nitrite $(CH_3)_3CONO$	103.12	pa ye	63, 34^{250}		$0.8670^{20/4}$	1.368^{20}	al, eth, chl	$B1^4$, 1622
4155	*tert*-Butyl peroxide or Di-*tert*-butyl peroxide $(CH_3)_3COOC(CH_3)_3$	146.23		111, 70^{197}	-40	0.704^{20}	1.3890^{20}	ace, lig	$B1^4$, 1619
4156	Butyl phenyl ether $C_4H_9OC_6H_5$	150.22		210, 95^{17}	-19.4	$0.9351^{20/4}$	1.4969^{20}	eth, ace	$B6^4$, 558
4157	*sec*-Butyl phenyl ether *sec*-$C_4H_9OC_6H_5$	150.22		194-5, $70\text{-}2^{5}$		$0.9415^{20/4}$	1.4926^{25}	eth	$B6^4$, 558
4158	*tert*-Butyl phenyl ether $(CH_3)_3COC_6H_5$	150.22		185-6		0.9214^{20}			$B6^4$, 559
4159	Butyl phenyl ketone or Valerophenone $C_4H_9COC_6H_5$	162.23		248.5, $131\text{-}3^{13}$		$0.988^{20/20}$	1.5158^{20}	al, eth	$B7^3$, 1114
4160	Butylphosphonic acid $C_4H_9PO(OH)_2$	138.10	pl (bz)	d	106			w, al, eth	$B4^3$, 1782
4161	Isobutyl phosphonic acid *i*-$C_4H_9PO(OH)_2$	138.10	pl (xyl)	d	119			w, al, eth, xyl	$B4^1$, 573
4162	*sec*-Butyl phosphonic acid *sec*-$C_4H_9PO(OH)_2$	138.10	lf (eth-lig)		48			w, al, eth, bz	Am75, 3379
4163	*tert*-Butyl phosphonic acid *tert*-$C_4H_9PO(OH)_2$	138.10	wh nd (xyl, aa-lig)	d	192			w, al	Am75, 3379
4164	Butyl propyl ether $C_4H_9OC_3H_7$	116.20		117.1		$0.7773^{0/0}$		al, eth	$B1^4$, 1519
4165	Isobutyl propyl ether *i*-$C_4H_9OC_3H_7$	116.20		$105\text{-}6^{720}$		0.7549^{20}	1.3852^{25}	al, eth	$B1^4$, 1594
4166	Butyl isopropyl ether *i*-$C_3H_7OC_4H_9$	116.20		108^{738}		$0.7594^{15/4}$	1.3870^{15}	al, eth, ace	$B1^4$, 1519
4167	Butyl sulfate $(C_4H_9O)_2SO_2$	210.29		109.5^{4}	1.4192^{20}	1.0616^{20}			$B1^4$, 1523
4168	Butyl sulfide (α-form) or α-Dibutyl ssulfide $(C_4H_9)_2S$	146.29		185	-79.7	$0.8386^{20/4}$	1.4530^{20}	al, eth, chl	$B1^4$, 1559
4169	Butyl sulfide (β-form) or β-Dibutyl sulfide $(C_4H_9)_2S$	146.29		190-230d				al, eth, ace	Ber62, 2168
4170	Butyl sulfide-2,2′-dimethyl *(d)* $[C_2H_5CH(CH_3)CH_2]_2S$	146.29		165		$0.8348^{20/4}$	1.4506^{20}	al, eth	B1, 387

No.	Name, Synonyms, and Formula	Mol. wt.	Color, crystalline form, specific rotation and λ_{max} (log ε)	b.p. °C	m.p. °C	Density	n_D	Solubility	Ref.
4171	Isobutyl sulfide or Di isobutyl sulfide $(i\text{-}C_4H_9)_2S$	146.29		170.5^{752}	-105.5	0.8363^{10}			$B1^4$, 1607
4172	sec-Butyl sulfide or Di-*sec*-Butyl sulfide $(sec\text{-}C_4H_9)_2S$	146.29		165		$0.8348^{20/4}$	1.4506^{20}	al, eth	$B1^4$, 1586
4173	Butyl sulfite $(C_4H_9O)_2SO$	194.30		230, 116^{19}		$0.9957^{20/4}$	0.4310^{20}	al, eth	$B1^4$, 1522
4174	Isobutyl sulfite $(i\text{-}C_4H_9O)_2SO$	194.30		209^{741}, $92\text{-}4^{13}$		$0.9862^{20/4}$	1.4268^{20}		$B1^4$, 1607
4175	*n*-Butyl sulfonamide $C_4H_9SO_2NH_2$	137.20	lf (eth-lig)		48			w, al, eth, bz	$B4^4$, 45
4176	*n*-Butyl sulfonyl chloride $C_4H_9SO_2Cl$	156.63		75^{10}			1.4559^{20}		$B4^4$, 45
4177	Isobutyl sulfonyl chloride $i\text{-}C_4H_9SO_2Cl$	156.63		189-91, 87^{15}			1.4520^{25}		$B4^4$, 49
4178	Butyl sulfoxide or Di butyl sulfoxide $(C_4H_9)_2SO$	162.29	nd (dil al)	d	32.6	$0.8317^{23/4}$	1.4669^{20}	al, eth	$B1^4$, 1561
4179	Butyl thiocyanate C_4H_9SCN	115.19		185^{743}		0.9563^{15}	1.4360^{20}	al, eth	$B3^4$, 329
4180	Isobutyl thiocyanate $(CH_3)_2CHCH_2SCN$	115.19		175.4, 66^{15}	− 59			al, eth	$B3^4$, 330
4181	*tert*-butyl thiocyanate $(CH_3)_3CSCN$	115.20		140^{770}d, $39\text{-}40^{10}$	10.5	0.9187^{10}			$B3^4$, 330
4182	Butyl 2-tolyl ether $(2\text{-}CH_3C_6H_4)OC_4H_9$	164.25		223		$0.943^{0/0}$			$B6^3$, 1247
4183	Butyl 3-tolyl ether or 3-Butoxy toluene $(3\text{-}CH_3C_6H_4)OC_4H_9$	164.25		229.2		$0.9407^{70/0}$	1.4970^{20}	eth	$B6^4$, 2040
4184	Butyl 4-tolyl ether or 4-Butoxy toluene $(4\text{-}CH_3C_6H_4)OC_4H_9$	164.25		229.5, 88^3		$0.9232^{25/25}$	1.4970^{20}	eth	$B6^3$, 1354
4185	Butyl vinyl ether $C_4H_9OCH{=}CH_2$	100.16		93.8	− 92	$6.7888^{20/4}$	1.4026^{20}	al, **eth**, ace, bz	$B1^4$, 2052
4186	Isobutyl vinyl ether $i\text{-}C_4H_9OCH{=}CH_2$	100.16		83	− 112	$0.7645^{20/4}$	1.3966^{20}	al, **eth**, ace, bz	$B1^4$, 2054
4187	Butyl nitrite C_4H_9ONO	103.12		− 77.8, 27^{88}		$0.8823^{20/4}$	1.3762^{20}	**al, eth**	$B1^4$, 1523
4188	*sec*-Butyl nitrite $sec\text{-}C_4H_9ONO$	103.12		68-9, 28^{180}		$0.8726^{20/4}$	1.3710^{20}	al, eth, chl	$B1^4$, 1573
4189	Isobutylene or 2-Methylpropene $(CH_3)_2C{=}CH_2$	56.11	gas	− 6.9	− 140.3	$0.5942^{20/4}$	1.3926^{25}	al, eth, bz	$B1^4$, 796
4190	Isobutylene trimer or Triisobutylene $(C_4H_8)_3$	168.32		179.8, 56^{16}	− 76	0.7590^{204}	1.4314^{20}		$B1^3$, 762
4191	Isobutylene tetramer or Tetraiso butylene $(C_4H_8)_4$	224.43		243-6, 109.5^{22}	− 98	0.7444^{20}	1.4482^{20}		$B1^3$, 763
4192	Isobutylene, l-chloro $(CH_3)_2C{=}CHCl$	90.55		68^{754}		$0.9186^{20/4}$	1.4221^{20}	**al, eth**, ace, chl	$B1^4$, 803
4193	Isobutylene, 3-chloro or 3-Chloro-2-methylpropene $ClCH_2C(CH_3){=}CH_2$	90.55		71-2		$0.9165^{20/4}$	1.4291^{20}	**al, eth**, ace, chl	$B1^4$, 803
4194	Isobutylene oxide or l,2-Epoxy-2-methylpropane $(CH_3)_2\underset{O}{C}CH_2$	72.11		52		0.8650^0	1.3712^{22}	al, eth	$B17^4$, 46
4195	1-Butyne $CH_3CHC{\equiv}CH$	54.09		8.1	− 125.7	$0.6784^{0/0}$	1.3962^{20}	al, eth	$B1^4$, 969
4196	1-Butyne, 3-chloro $CH_3CHClC{\equiv}CH$	88.54		68.5		0.9466^{25}	1.4218^{25}		$B1^4$, 970
4197	1-Butyne, 3-chloro-3-methyl $(CH_3)_2CClC{\equiv}CH$	102.56		77-9		$0.9061^{20/4}$			$B1^4$, 1000
4198	1-Butyne, 3,3-dimethyl or *tert*-Butyl acetylene $(CH_3)_3CC{\equiv}CH$	82.15		39-40	− 81.2	$0.6695^{20/4}$	1.3738^{20}		$B1^4$, 1022
4199	1-Butyne, 3-methyl or Isopropyl acetylene $(CH_3)_2CHC{\equiv}CH$	68.12		29.5	− 89.7	$0.6660^{20/4}$	1.3723^{20}	**al, eth**	$B1^4$, 999
4200	1-Butyne, 4-phenyl $C_6H_5CH_2CH_2C{\equiv}CH$	130.19		190		$0.9258^{20/4}$	1.5208^{20}		$B5^4$, 1535
4201	2-Butynal $CH_3C{\equiv}CCHO$	68.08		106-7, $27\text{-}8^{34}$	− 26	$0.9265^{17/0}$	1.446^{19}	eth, ace	$B1^4$, 3540
4202	2-Butyne or Dimethyl acetylene $CH_3C{\equiv}CCH_3$	54.09		27	− 32.2	$0.6910^{20/4}$	1.3921^{20}	al, eth	$B1^4$, 971
4203	2-Butyne, 1-chloro $CH_3C{\equiv}CCH_2Cl$	88.54		104-6		1.0152^{20}	1.4581^{20}	al, eth, ace	$B1^4$, 973

No.	Name, Synonyms, and Formula	Mol. wt.	Color, crystalline form, specific rotation and λ_{max} (log ε)	b.p. °C	m.p. °C	Density	n_D	Solubility	Ref.
4204	2-Butyne, 1,4-dibromo $BrCH_2C{\equiv}CCH_2Br$	211.88	....	92^{15}	......	2.014^{18}	1.588^{18}	eth, ace, chl	$B1^{4}$, 974
4205	2-Butyne, 1,4-dichloro $ClCH_2C{\equiv}CCH_2Cl$	122.98	....	165-6, 73^{24}	......	$1.258^{20/4}$	1.5058^{20}	eth, ace, chl	$B1^{4}$, 973
4206	2-Butyne, 1,4-diiodo $ICH_2C{\equiv}CCH_2I$	305.88	nd (al)	$70\text{-}2^{0.1}$	53	....	....	al, eth, ace, chl	$B1^{4}$, 975
4207	2-Butyne dinitrile $NCC{\equiv}CCN$	76.06	....	$76\text{-}6.5^{753}$	20.5-1	$0.9703^{25/4}$	1.46471^{25}	....	$B2^{4}$, 2295
4208	2-Butyne, 1,4-diol $HOCH_2C{\equiv}CCH_2OH$	86.09	pl (bz, AcOEt)	238, 145^{15}	58	....	1.4804^{20}	w, al, ace	$B1^{4}$, 2687
4209	2-Butyne, 1,4-diol,diacetate $CH_3CO_2CH_2\text{-}C{\equiv}C\text{-}CH_2O_2CCH_3$	170.17	....	$122\text{-}3^{10}$	......	....	1.4611^{20}	....	$B2^{4}$, 244
4210	3-Butyne, 1,2-diol $HC{=}CCHOHCH_2OH$	86.09	....	$64\text{-}6^{0.2}$	40	....	....	w, al	$B1^{4}$, 2689
4211	2-Butyne, perfluoro $CF_3C{\equiv}CCF_3$	162.03	....	− 24.6	− 117.4	....	....	al, eth, ace, aa	$B1^{4}$, 972
4212	2-Butynedioic acid or Acetylene dicarboxylic acid $HO_2CC{\equiv}CCO_2H$	114.06	pl (eth)	....	179	....	....	w, al, eth	$B2^{4}$, 2290
4213	2-Butynedioic acid, diethyl ester $H_5C_2O_2CC{\equiv}CCO_2C_2H_5$	170.17	....	184^{200}	1-2	$1.0675^{20/4}$	1.4425^{20}	al, eth	$B2^{4}$, 2294
4214	2-Butynedioic acid, dimethyl ester $CH_3O_2CC{\equiv}CCO_2CH_3$	142.11	....	195-8d, 98^{20}	......	$1.1564^{20/4}$	1.4434^{20}	al, eth	$B2^{4}$, 2291
4215	2-Butynoic acid or Tetrolic acid $CH_3C{\equiv}CCO_2H$	84.07	pl (eth, peth)	203, $99\text{-}100^{18}$	78	$0.9641^{20/4}$	....	w, al, eth, chl	$B2^{4}$, 1690
4216	2-Butynoic acid, ethyl ester $CH_3C{\equiv}CCO_2C_2H_5$	112.13	....	163, 105^{190}	......	$0.9641^{20/4}$	1.4372^{20}	....	$B2^{4}$, 1691
4217	2-Butyn-1-ol $CH_3C{\equiv}CCH_2OH$	70.09		143, $52\text{-}3^{14}$	− 2.2	$0.9370^{20/4}$	1.4530^{20}	al, eth	$B1^{4}$, 2220
4218	3-Butyn-1-ol $HC{\equiv}CCH_2CH_2OH$	70.09	....	129	− 63.6	$0.9257^{20/4}$	1.4409^{20}	w, al	$B1^{4}$, 2219
4219	3-Butyn-2-ol $HC{\equiv}C\text{-}CHOHCH_3$	70.09	....	107	......	0.8858^{20}	1.4265^{20}	w, al, eth	$B1^{4}$, 2218
4220	3-Butyn-2-ol-2, methyl $(CH_3)_2COHC{\equiv}CH$	84.12	....	104, 56^{97}	+ 3	$0.8618^{20/4}$	1.4207^{20}	w, al	$B1^{4}$, 2229
4221	2-Butynyl methyl ether $CH_3C{\equiv}CCH_2OCH_3$	84.12	....	99-100, 33^{27}	......	$0.8496^{20/4}$	1.4262^{20}	al, eth, bz	$B1^{3}$, 1973
4222	Butyraldehyde or Butanal $CH_3CH_2CH_2CHO$	72.11	....	75.7	− 99	$0.8170^{20/4}$	1.3843^{20}	w, al, eth, ace, bz	$B1^{4}$, 3229
4223	Butyraldehyde phenyl hydrazone $C_3H_7CH{=}NNHC_6H_5$	162.23	....	$190\text{-}5^{80}$, 152^{14}	93-5	....	....	....	$B15^{3}$, 80
4224	Butyraldehyde oxime or Butyraldoxime $CH_3CH_2CH_2CH{=}NOH$	87.12	....	152^{715}	− 29.5	$0.923^{20/4}$	....	al, eth, ace, bz	$B1^{4}$ 3234
4225	Butyraldehyde, 2-bromo $CH_3CH_2CHBrCHO$	151.00	....	33^{17}	......	1.469^{20}	1.4683^{20}	eth, ace, bz	$B1^{4}$, 3241
4226	Butyraldehyde, 3-chloro, diethyl acetal $CH_3CHClCH_2CH(OC_2H_5)_2$	180.67	....	$70\text{-}1^{12}$	......	$0.9709^{20/4}$	1.4210^{20}	al, eth, ace, bz	$B1^{4}$, 3239
4227	Butyraldehyde, 4-chloro $ClCH_2CH_2CH_2CHO$	106.55	....	$50\text{-}1^{13}$	......	$1.107^{8.15/15}$	$1.4466^{8.5}$	al, eth, ace	$B1^{4}$, 3240
4228	Butyraldehyde, 2,3-dichloro $CH_3CHClCHClCHO$	141.00	....	$58\text{-}60^{20}$	......	$1.2666^{21/4}$	1.4618^{21}	al, eth, ace, chl	$B1^{4}$, 3240
4229	Butyraldehyde, 2-ethyl $(C_2H_5)_2CHCHO$	100.16	....	$117\text{-}9^{160}$	......	$0.8110^{20/4}$	1.4025^{20}	al, eth	$B1^{4}$, 3310
4230	Butyraldehyde, 3-hydroxy or Aldol $CH_3CH(OH)CH_2CHO$	88.11	....	83^{20}, d85	......	$1.103^{20/4}$	1.4238^{20}	w, al, eth, ace	$B1^{4}$, 3984
4231	Butyraldehyde, 2-methyl *(dl)* $CH_3CH_2CH(CH_3)CHO$	86.14	....	92-3, 54^{200}	......	$0.8029^{20/4}$	1.3869^{20}	al, eth, ace	$B1^{4}$, 3286
4232	Butyraldehyde, 3-methyl or Isovaleraldehyde $(CH_3)_2CH_2CH_2CHO$	86.14	....	92.5	− 51	$0.7977^{20/4}$	1.3902^{20}	al, eth	$B1^{4}$, 3291
4233	Butyraldehyde, 3-methyl,oxime or Isovaleraldoxime $(CH_3)_2CH_2CH_2CH{=}NOH$	101.15	....	161.3	48.5	$0.8934^{20/4}$	1.4367^{20}	al, eth, ace	$B1^{4}$, 3293
4234	Butyraldehyde, 2,2,3-trichloro or *n*-Butyl chloral $CH_3CHClCCl_2CHO$	175.44	....	163-5, 49^{8}	......	$1.3956^{20/4}$	1.4755^{20}	w, al eth	$B1^{4}$, 3241
4235	Butyraldehyde, 2,2,3-trichloro,hydrate $CH_3CHClCCl_2CH(OH)_2$	192.45	rh pl or lf (w)	d	78	$1.694^{20/4}$	....	w, al, eth	$B1^{3}$, 2768
4236	Isobutyraldehyde $(CH_3)_2CHCHO$	72.11	....	64.2-4.6	......	$0.7938^{20/4}$	1.3730^{20}	w, eth, ace, chl	$B1^{4}$, 3262

No.	Name, Synonyms, and Formula	Mol. wt.	Color, crystalline form, specific rotation and λ_{max} (log ε)	b.p. °C	m.p. °C	Density	n_D	Solubility	Ref.
4237	Isobutyraldehyde, 2-chloro $(CH_3)_2CClCHO$	106.55		90		$1.053^{15/4}$	1.4160^{16}	al, eth	B1⁴, 3267
4238	Isobutyraldehyde, 3-(4-isopropylphenyl) or Cyclamenaldehyde $4\text{-}(CH_3)_2CHC_6H_4CH_2CH(CH_3)CHO$	190.29		$133\text{-}7^{99}$, 115^{9}		0.951^{15}	1.5068^{20}	al, eth, bz	B7³, 1200
4239	Isobutyraldehyde, 2-chloro $(CH_3)_2CClCHO$	106.55							B1⁴, 3267
4240	Butyramide or Butanoic acid,amide $CH_3CH_2CH_2CONH_2$	87.12	lf (bz)	216	114.8	0.8850^{120}	1.4087^{130}	al	B2⁴, 804
4241	Butyramide, α-bromo $CH_3CH_2CHBrCONH_2$	166.02	lf (bz), nd (ace)		112—3			w, al, eth, ace, bz	B2⁴, 834
4242	Butyramide, 2-bromo-2-ethyl or Neuronal $(C_2H_5)_2CBrCONH_2$	194.07			67			al, eth, bz	B2³, 755
4243	Butyramide, 2-bromo-3-methyl *(dl)* $(CH_3)_2CHCHBrCONH_2$	180.04	lf (bz)		133			w, lig	B2⁴, 906
4244	Butyramide, 2-bromo-2-*iso*-propyl or Neodorme $(CH_3)_2CHCBr(C_2H_5)CONH_2$	208.10	nd (sub)	sub	50-1			al, eth, ace, bz	B2⁴, 979
4244a	Butyramide, 2-bromo-*N*-methyl-*N*-phenyl $CH_3CH_2CHBrCON(CH_3)C_6H_5$	256.14	cr (lig)		44			to, lig	B12, 254
4245	Butyramide, *N,N*-diethyl $CH_3CH_2CH_2CON(C_2H_5)_2$	143.23		206, 97^{16}			1.4403^{25}	w, al	B4⁴, 354
4246	Butyramide, *N,N*-dimethyl $CH_3CH_2CH_2CON(CH_3)_2$	115.18		185-8, $124\text{-}5^{100}$	− 40	$0.9064^{25/4}$	1.4391^{25}	w, al, eth, ace, bz	B4⁴, 185
4247	Butyramide, 2,3-dimethyl $(CH_3)_2CHCH(CH_3)CONH_2$	115.18	pl (ace-peth)		130.9			al, eth	B2³, 761
4248	Butyramide, 3,3-dimethyl or *tert*-Butyl acetamide $(CH_3)_3CCH_2CONH_2$	115.18	lf (w or ace-peth)		134			al	B2⁴, 956
4249	Butyramide, 2-ethyl-*N,N*-diethyl $(C_2H_5)_2CHCON(C_2H_5)_2$	171.28		220-1, 108^{12}				al, eth, bz	B4, 111
4250	Butyramide, 3-methyl $(CH_3)_2CHCH_2CONH_2$	101.15	mcl lf (al)	224-8	137			w, al, eth, peth	B2⁴, 902
4251	Butyramide, 3-methyl-2-phenyl $(CH_3)_2CHCH(C_6H_5)CONH_2$	177.25	nd (dil al)	$180\text{—}2^{14}$	111-2			al, eth	B9³, 2518
4252	Butyramide, 2-phenoxy $CH_3CH_2CH(OC_6H_5)CONH_2$	179.22	nd (w or al)		123			al, eth, ace, chl	B6⁴, 645
4253	Butyramide, 4-phenoxy $C_6H_5OCH_2CH_2CH_2CONH_2$	179.22	lf (dil al), nd (bz)		80			al	B6³, 617
4254	Butyramide, *N*-phenyl or Butyranilide $C_3H_7CONHC_6H_5$	163.22	mcl pr (al, bz, eth)	189^{15}	97	1.134		al, eth	B12³, 474
4255	Butyramide, 2-phenyl $CH_3CH_2CH(C_6H_5)CONH_2$	163.22	cr	185^{16}	86			al	B9³, 2465
4256	Butyramide, 3-phenyl $CH_3CH(C_6H_5)CH_2CONH_2$	163.22	nd (dil al)		106-7			al	B9³, 2459
4257	Butyramide, 4-phenyl $C_6H_5CH_2CH_2CH_2N$	163.22	pl (w)		84.5			al, eth	B9³, 2453
4258	Butyramide, 2,2,3,3-tetramethyl $CH_3C(CH_3)_2C(OH_3)_2CONH_2$	143.23	nd (peth-al)		201-2			al	B2⁴, 1018
4259	Isobutyramide $(CH_3)_2CHCONH_2$	87.12			127-9			chl	B2⁴, 852
4260	Isobutyramide, 2-bromo or 2-Bromo-2-methylpropionamide $(CH_3)_2CBrCONH_2$	166.02	pr (chl)	145^{17}	148			al, chl	B2⁴, 863
4262	Isobutyramide, *N*-phenyl $(CH_3)_2CHCONHC_6H_5$	163.22	mcl pr (al, eth), nd (lig)		106-7			al, eth	B12³, 475
4263	Butyric acid or Butanoic acid $CH_3CH_2CH_2CO_2H$	88.11		165.5	− 4.5	$0.9577^{20/4}$	1.3980^{20}	al, eth	B2⁴, 779
4264	Butyric acid, 2-acetyl,ethyl ester $CH_3CH_2CH(OCCH_3)CO_2C_2H_5$	158.20		$85\text{-}7^{13}$		$0.9924^{15/15}$	1.4237^{15}	al, eth	B2⁴, 1592
4265	Butyric acid, 2-acetyl-3-oxo-ethyl ester or Ethyl-α-acetyl aceto acetate $(CH_3CO)_2CHCO_2C_2H_5$	172.18		209^{11}, 104^{16}		$1.1045^{20/4}$	1.4690^{20}	al, eth, bz	B2⁴, 1781
4266	Butyric acid, allyl ester or Allylbutyrate, allylbutanoate $CH_3CH_2CH_2CO_2CH_2CH{=}CH_2$	128.17		$142\text{-}3^{772}$, 44.5^{15}		$0.9017^{20/4}$	1.4158^{20}	al, eth	B2⁴, 793
4267	Butyric acid, 2-amino- *(d)* or 2-Aminobutanoic acid $C_2H_5CHNH_2CO_2H$	103.12	lf (dil al), $[\alpha]^{16}{}_D$ + 8.4		292d			w	B4³, 1294

No.	Name, Synonyms, and Formula	Mol. wt.	Color, crystalline form, specific rotation and λ_{max} (log ε)	b.p. °C	m.p. °C	Density	n_D	Solubility	Ref.
4268	Butyric acid, 2-amino- *(dl)* $C_2H_5CHNH_2CO_2H$	103.12	lf (w)	sub	304d	....	....	w	$B4^3$, 1296
4269	Butyric acid, 2-amino-*(l)* $C_2H_5CHNH_2CO_2H$	103.12	lf (w-al), cr (al), $[\alpha]^{20}_D$ − 14.9	....	292d	....	....	w	$B4^3$, 1296
4270	Butyric acid, 3-amino- *(d)* $CH_3CH(NH_2)CH_2CO_2H$	103.12	pr (MeOH), $[\alpha]^{20}_D$ + 35.3	....	d at 220	....	....	w	$B4^3$, 1312
4271	Butyric acid, 3-amino-*(dl)* $CH_3CH(NH_2)CH_2CO_2H$	103.12	nd (al)	....	193-4	....	....	w	$B4^3$, 1312
4272	Butyric acid, 3-amino-*(l)* $CH_3CHNH_2CH_2CO_2H$	103.12	pr (MeOH), $[\alpha]^{20}_D$ − 35.2 (w, c=10)	....	d at 220	....	....	w	$B4^3$, 1312
4273	Butyric acid, 3-amino-2-hydroxy or 3-Methyl *iso*-serine $CH_3CHNH_2CHOHCO_2H$	119.12	pr (dil al)	....	200d	....	....	w	B4, 513
4274	Butyric acid, 3-amino-3-methyl $(CH_3)_2CNH_2CO_2H$	117.15	pr (w + 1), cr (dil al), nd (eth-al)	....	217	....	....	w	$B4^3$, 1364
4275	Butyric acid, 4-amino or Piperidinic acid $H_2NCH_2CH_2CH_2CO_2H$	103.12	pr or nd (dil al), lf (MeOH-eth)	....	203d	....	....	w	$B4^3$, 1316
4276	Butyric acid, 4-amino-2-hydroxy $H_2NCH_2CH_2CHOHCO_2H$	119.12	pr (w or dil al)	....	214	....	....	w	$B4^1$, 548
4277	Butyric acid, 4-amino-3-hydroxy *(dl)* $H_2NCH_2CHOHCH_2CO_2H$	119.12	pr (w), cr (dil al)	....	218	....	....	....	$B4^3$, 1635
4278	Butyric acid anhydride or Butyric anhydride $(C_3H_7CO)_2O$	158.20	....	199-201	− 75	$0.9668^{20/4}$	1.4070^{20}	eth	$B2^4$, 802
4279	Butyric acid, 2-benzoyl,ethyl ester $CH_3CH_2CH(COC_6H_5)CO_2C_2H_5$	220.27	....	152^7	......	$1.0706^{15/4}$	1.509^{15}	eth	$B10^3$, 3065
4280	Butyric acid, 4-benzoyl $C_6H_5CO(CH_2)_3CO_2H$	192.21	pl (w)	....	128-9	....	....	w	$B10^3$, 3060
4281	Butyric acid, benzyl ester or Benzyl butyrate $C_3H_7CO_2CH_2C_6H_5$	178.23	....	238-40, 105^7	......	$1.0111^{20/4}$	1.4920^{20}	al, eth	$B6^4$, 2266
4282	Butyric acid, 2-bromo *(d)* or α-Bromobutyric acid $CH_3CH_2CHBrCO_2H$	167.00	$[\alpha]^{20}_D$ + 35.2 (eth, c=20)	105-7^{15}	......	$1.568^{20/4}$	1.4483^{20}	al	$B2^3$, 630
4283	Butyric acid, 2-bromo *(dl)* $CH_3CH_2CHBrCO_2H$	167.00	....	217d, 108^{13}	− 4	$1.5669^{20/20}$	....	al, eth	$B2^4$, 833
4284	Butyric acid, 2-bromo,ethyl ester $CH_3CH_2CHBrCO_2C_2H_5$	195.06	....	177.5^{765}, 43-$4^{5.5}$	......	$1.3297^{20/20}$	1.4475^{20}	al, eth	$B2^4$, 834
4285	Butyric acid, 2-bromo,methyl ester $CH_3CH_2CHBrCO_2CH_3$	181.03	....	170-2, 75-8^{18}	......	1.4528^{20}	1.4029^{25}	al	$B2^4$, 834
4286	Butyric acid, 2-bromo-3-methyl-*(d)* $(CH_3)_2CHCHBrCO_2H$	181.03	pr (peth), $[\alpha]^{20}_D$ + 22.8 (bz, p=4)	230, 95-100^2	44-5	....	....	al, eth, ace, bz	$B2^3$, 705
4287	Butyric acid, 2-bromo-3-methyl-*(dl)* $(CH_3)_2CHCHBrCO_2H$	181.03	pr (eth or chl)	230d, 136-40^{25}	44	1.459^{20}	....	al, eth, ace, bz	$B2^4$, 905
4288	Butyric acid, 2-bromo-3-methyl-*(l)* $(CH_3)_2CHCHBrCO_2H$	181.03	cr (peth), $[\alpha]^{20}_D$ − 21.6 (bz, p=4)	150^{40}, 119-20^{14}	43-4	....	....	al, eth, ace, bz	$B2^4$, 905
4289	Butyric acid, 2-bromo-3-methyl, ethyl ester $(CH_3)_2CHCHBrCO_2C_2H_5$	209.08	....	186, 73-4^{12}	......	$1.2760^{20/4}$	1.4496^{20}	al, eth	$B2^4$, 906
4290	Butyric acid, 2-bromo-3-methyl, methyl ester $(CH_3)_2CHCHBrCO_2CH_3$	195.06	....	176-8, 64-5^{11}	......	$1.353^{13/13}$	1.4530^{20}	al, eth	$B2^3$, 706
4291	Butyric acid, 3-bromo-3-methyl $(CH_3)_2CBrCH_2CO_2H$	181.03	nd (lig)	....	73.5	....	....	al, eth, bz	$B2^4$, 905
4292	Butyric acid, 4-bromo $BrCH_2CH_2CH_2CO_2H$	167.00	....	124-7^7	33	....	....	....	$B2^4$, 835
4293	Butyric acid, 4-bromo, methyl ester $BrCH_2CH_2CH_2CO_2CH_3$	181.03	....	186-7, 86^{15}	......	1.371^{25}	1.4567^{25}	al	$B2^4$, 835
4294	Butyric acid, butyl ester or Butyl butyrate $C_3H_7CO_2C_4H_9$	144.21	....	166.6, 55^{11}	− 91.5	$0.8700^{20/4}$	1.4075^{20}	al, eth	$B2^4$, 789

No.	Name, Synonyms, and Formula	Mol. wt.	Color, crystalline form, specific rotation and λ_{max} (log ε)	b.p. °C	m.p. °C	Density	n_D	Solubility	Ref.
4295	Butyric acid, *sec*-butyl ester-*(d)* or *sec*-Butyl butyrate $C_3H_7CO_2$-*sec*-C_4H_9	144.21	$[\alpha]^{20}_D + 22$	151.5^{747}, 54^{18}	− 91.5	$0.8737^{13/4}$	1.4011^{20}	al, bz, Py	$B2^3$, 600
4296	Butyric acid, *iso*-butyl ester or *iso*-Butyl butyrate $C_3H_7CO_2$-*i*-C_4H_9	144.21		157		$0.8364^{18/4}$	1.4032^{20}	al, eth	$B2^4$, 790
4297	Butyric acid, *sec*-butyl ester-*(dl)* $C_3H_7CO_2$-*sec*-C_4H_9	144.21		152.5, 52^{16}		$0.8609^{20/4}$	1.4019^{20}		$B2^4$, 790
4298	Butyric acid, *tert*-butyl ester or *tert*-Butyl butyrate $C_3H_7CO_2C(CH_3)_3$	144.21		145-7			$1.4007^{17.5}$	al, eth, ace	$B2^4$, 790
4299	Butyric acid, 2-chloro $CH_3CH_2CHClCO_2H$	122.55		189^{627}, 101^{15}		$1.1796^{20/4}$	1.4411^{20}	al, eth	$B2^4$, 821
4300	Butyric acid, 2-chloro, ethyl ester $CH_3CH_2CHClCO_2C_2H_5$	150.61		163-4, 63^{70}		1.0560^{20}	1.4248^{20}	al, eth	$B2^4$, 822
4301	Butyric acid, 2-chloro, methyl ester or Methyl 2-chlorobutyrate $CH_3CH_2CHClCO_2CH_3$	136.58		$145\text{-}6^{756}$		1.0979^{14}	1.4247^{20}	al, eth	$B2^4$, 821
4302	Butyric acid, 2-chloro-2-methyl $CH_3CH_2CCl(CH_3)CO_2H$	136.58		$200\text{-}5^{754/d}$		$1.1204^{20/4}$	1.4445^{20}	al, eth	$B2^3$, 688
4303	Butyric acid, 2-chloro-2-methyl, ethyl ester $CH_3CH_2C(CH_2)ClCO_2C_2H_5$	164.63		175^{747}		1.069^{14}	1.4388^{11}	al, eth	$B2^3$, 688
4304	Butyric acid, 2-chloro-2-methyl-3-oxo, ethyl ester or Ethyl α-chloro-α-acetopropionate $CH_3COC(CH_3)ClCO_2C_2H_5$	164.63		$178\text{-}9^{756}$		1.021^{13}		al, eth	$B3^2$, 433
4305	Butyric acid, 2-chloro-3-methyl or 2-chloroisovaleric acid $(CH_3)_2CHCHClCO_2H$	136.58		$210\text{-}2^{756}$, 126^{32}	20—2	1.135^{13}	1.4450^{11}	al, eth	$B2^4$, 904
4306	Butyric acid, 3-chloro *(dl)* $CH_3CHClCH_2CO_2H$	122.55	cr (eth)	116^{22}	16	$1.1898^{20/4}$	1.4221^{20}	al, eth	$B2^4$, 823
4307	Butyric acid, 3-chloro, ethyl ester $CH_3CHClCH_2CO_2C_2H_5$	150.61		109, 65^{15}		$1.0517^{20/4}$	1.4246^{20}	al	$B2^4$, 824
4308	Butyric acid, 3-chloro, methyl ester $CH_3CHClCH_2CO_2CH_3$	136.58		155-6		$1.0996^{20/4}$	1.4258^{20}	eth	$B2^4$, 824
4309	Butyric acid, 4-chloro $ClCH_2CH_2CH_2CO_2H$	122.55		196^{22}, $68^{0.2}$	16	$1.2236^{20/4}$	1.4642^{20}	eth	$B2^4$, 825
4310	Butyric acid, 4-chloro, ethyl ester $ClCH_2CH_2CH_2CO_2C_2H_5$	150.61		186, 77^{10}		$1.0756^{20/4}$	1.4311^{20}	al, eth, ace	$B2^4$, 825
4311	Butyric acid, 4-chloro, methyl ester $ClCH_2CH_2CH_2CO_2CH_3$	136.58		$175\text{-}6^{764}$, 55^4		$1.1201^{20/4}$	1.4321^{20}	al, eth, ace	$B2^4$, 825
4312	Butyric acid, cyclohexyl ester or Cyclohexyl butyrate $C_3H_7CO_2C_6H_{11}$	170.25		212^{750}		$0.9572^{0/4}$		al	$B6^4$, 37
4313	Butyric acid, 2,3-dibromo (high m.p.) $CH_3CHBrCHBrCO_2H$	245.90	nd (eth)	$100\text{-}10^{20}$	87			al, eth, bz	$B2^4$, 837
4314	Butyric acid, 2,3-dibromo (low m.p.) $CH_3CHBrCHBrCO_2H$	245.90	nd (lig)		59-60			al, eth	$B2^4$, 837
4315	Butyric acid, 2,3-dibromo, ethyl ester $CH_3CHBrCHBrCO_2C_2H_5$	273.95	nd	113^{30}	58-9			al, eth	$B2^4$, 837
4316	Butyric acid, 2,4-dibromo, ethyl ester $BrCH_2CH_2CHBrCO_2C_2H_5$	273.95		$149\text{-}50^{52}$		$1.6990^{20/0}$	1.4960^{20}	al, eth	$B2^4$, 838
4317	Butyric acid, 2,2-dichloro $CH_3CH_2CCl_2CO_2H$	157.00		$107\text{-}10^{14}$		$1.389^{20/4}$		al, eth	$B2^2$, 254
4318	Butyric acid, 2,3-dichloro (high m.p.) $CH_3CHClCHClCO_2H$	157.00	pr (dil al)	131.5^{20}	78			al, eth	$B2^4$, 827
4319	Butyric acid, 2,3-dichloro (low m.p.) $CH_3CHClCHClCO_2H$	157.00	pr (dil al)	$124\text{-}5^{20}$	63			al, eth, bz, chl	$B2^4$, 827
4320	Butyric acid, 2,2-diethyl-3-oxo, ethyl ester $CH_3COC(C_2H_5)_2CO_2C_2H_5$	186.25		$215\text{-}6^{744}$, 64^3		$0.9717^{18/4}$	1.4326^{17}	al, eth	$B3^4$, 1625
4321	Butyric acid, 2,2-dimethyl $CH_3CH_2C(CH_3)_2CO_2H$	116.16		186, 80^{11}	− 14	$0.9276^{20/4}$	1.4145^{20}		$B2^4$, 954
4322	Butyric acid, (2,2-dimethylpropyl) ester or *neo*-pentyl butyrate $C_3H_7CO_2CH_2C(CH_3)_3$	158.24		165-6		0.8719^0			B2, 272
4323	Butyric acid, 2,2-dimethyl-3-oxo, ethyl ester $CH_3COC(CH_3)_2CO_2C_2H_3$	158.20		184, $40\text{-}1^3$		$0.9773^{20/20}$	1.4180^{20}	al, eth	$B3^4$, 1594
4324	Butyric acid, 2,3-dimethyl $(CH_3)_2CH\text{-}CH(CH_3)CO_2H$	116.16		191.7	− 1.5	$0.9275^{20/4}$	1.4146^{20}	al, eth	$B2^4$, 958
4325	Butyric acid, 3,3-dimethyl or *tert*-Butylacetic acid $(CH_3)_3CCH_2CO_2H$	116.16		190, 96^{26}	6-7	$0.9124^{20/4}$	1.4096^{20}	al, eth	$B2^4$, 955
4326	Butyric acid, 3,3-dimethyl-2-oxo $(CH_3)_3CCOCO_2H$	130.14		189^{747}, 80^{15}	90-1			eth, bz, chl	$B3^4$, 1595

No.	Name, Synonyms, and Formula	Mol. wt.	Color, crystalline form, specific rotation and λ_{max} (log ε)	b.p. °C	m.p. °C	Density	n_D	Solubility	Ref.
4327	Butyric acid, 2,3-epoxy-3-phenyl, ethyl ester $CH_3C(C_6H_5)CH\text{-}CO_2C_2H_5$	206.24	...	$272—5, 147\text{-}9^{22}$	...	1.0442^{20}	1.5182^{20}	...	$B18^2$, 275
4328	Butyric acid, ethyl ester or Ethylbutyrate $C_3H_7CO_2C_2H_5$	116.16	...	$121\text{-}6, 48.8^{50}$	− 100.8	$0.8785^{20/4}$	1.4000^{20}	al, eth	$B2^4$, 787
4329	Butyric acid, 2-ethyl or Diethyl acetic acid $(C_2H_5)_2CHCO_2H$	116.16	...	$194, 90^{13}$	− 31.8	$0.9239^{20/4}$	1.4132^{20}	al, eth	$B2^4$, 951
4330	Butyric acid, 2-ethyl-2-methyl $(C_2H_5)_2C(CH_3)CO_2H$	130.19	...	$208.5, 104^{13}$	<−20	...	1.4250^{20}	al	$B2^4$, 981
4331	Butyric acid, 4-fluoro $FCH_2CH_2CH_2CO_2H$	106.10	...	$76\text{-}8^{5}$	...	...	1.3993^{25}	al, eth	$B2^4$, 809
4332	Butyric acid, furfuryl ester $C_3H_7CO_2CH_2(C_4H_3O)$	168.19	...	$212\text{-}3^{764}$	...	$1.0530^{20/4}$	...	al, eth	$B17^4$, 1247
4333	Butyric acid, heptafluoro $F_3CCF_2CF_2CO_2H$	214.04	...	120^{735}	− 17.5	$1.651^{20/4}$	1.295^{15}	w, eth, to	$B2^4$, 810
4334	Butyric acid, heptafluoro, ethyl ester $F_3CCF_2CF_2CO_2C_2H_5$	242.09	...	95	...	...	1.3011^{20}	eth, ace	$B2^4$, 813
4335	Butyric acid, heptafluoro, methyl ester $CF_3CF_2CF_2CO_2CH_3$	228.07	...	80	...	$1.483^{20/4}$	1.295^{20}	eth, ace	$B2^4$, 812
4336	Butyric acid, heptyl ester or Heptyl butyrate $C_3H_7CO_2C_7H_{15}$	186.29	...	$225.8, 105^{10}$	− 57.5	0.8637^{20}	1.4231^{20}	al	$B2^4$, 791
4337	Butyric acid, hexyl ester or Hexyl butyrate $C_3H_7CO_2C_6H_{13}$	172.27	...	208	− 78	0.8652^{20}	1.4160^{15}	al	$B2^4$, 791
4338	Butyric acid, 2-hydroxy *(dl)* $CH_3CH_2CHOHCO_2H$	104.11	nd (CCl_4)	$266d, 140^{14}$	44-4.5	1.125^{20}	...	w, al, eth	$B3^4$, 754
4339	Butyric acid, 2-hydroxy, ethyl ester *(d)* $CH_3CH_2CHOHCO_2C_2H_5$	132.16	$[\alpha]^{22}{}_D + 8.4$	165-70	...	0.978^{15}	1.4101	al	$B3^4$, 756
4340	Butyric acid, 2-hydroxy, ethyl ester *(dl)* $CH_3CH_2CHOHCO_2C_2H_5$	132.16	...	$167, 74.5^{24}$	...	$1.0069^{20/4}$	1.4179^{20}	al	$B3^4$, 756
4341	Butyric acid, 2-hydroxy-3-methyl *(d)* $(CH_3)_2CHCHOHCO_2H$	118.13	cr (eth-peth), $[\alpha]^{20}{}_D$ − 1.81 (w, c=12)	$124\text{-}5^{13}$	69.5	...	...	w, al, eth, ace	$B3^4$, 830
4342	Butyric acid, 2-hydroxy-3-methyl *(dl)* $(CH_3)_2CHOHCO_2H$	118.13	rh bipyr	...	86	...	...	w, al, eth, ace	$B3^4$, 830
4343	Butyric acid, 3-hydroxy-*(dl)* $CH_3CHOHCH_2CO_2H$	104.11	...	$130^{12\text{-}14}, 94\text{-}6^{0.1}$	48-50	...	1.4424^{20}	w, al, eth	$B3^4$, 760
4344	Butyric acid, 3-hydroxy-*(l)* $CH_3CHOHCH_2CO_2H$	104.11	$[\alpha]^{25}{}_D$ − 24.5 (w, c=5)	...	49-50	...	...	w, al, eth	$B3^4$, 760
4345	Butyric acid, 3-hydroxy, ethyl ester *(dl)* $CH_3CHOHCH_2CO_2C_2H_5$	132.16	...	$184\text{-}5^{755}, 76\text{-}7^{15}$	...	$1.017^{20/4}$	1.4182^{20}	al, w	$B3^4$, 762
4346	Butryic acid, 3-hydroxy, methyl ester *(l)* $CH_3CHOHCH_2CO_2CH_3$	118.13	$[\alpha]^{20}{}_D$ − 21.09	$76\text{-}7^{20}$	...	$1.058^{20/20}$	...	w, al, eth, bz	$B3^3$, 569
4347	Butyric acid, 3-hydroxy-3-methyl $(CH_3)_2COHCH_2CO_2H$	118.13	...	162^{12}	< − 32	$0.9384^{20/4}$	1.5081^{20}	w, al, eth	$B3^4$, 827
4348	Butryic acid, 4-hydroxy $4HOCH_2CH_2CH_2CO_2H$	104.11	...	d at 178-80	< − 17	...	...	...	$B3^4$, 774
4349	Butyric acid, 4-hydroxy, lactone or γ-Butyrolactone $CH_2CH_2CH_2CO$ (ring, O)	86.09	...	$206, 89^{12}$	− 42	$1.1286^{16/0}$	1.4341^{20}	w, al, eth, ace, bz	$B17^4$, 4159
4350	Butyric acid, 4-hydroxy-2-methylene, lactone or γ-Butyrolactone $CH_2CH_2C(=CH_2)CO$ (ring, O)	98.10	...	$85\text{-}6^{10}$	...	1.1206^{20}	1.4650^{20}	w, al, eth, ace, bz	$B17^4$, 4304
4351	Butyric acid, methyl ester or Methyl butyrate $C_3H_7CO_2CH_3$	102.13	...	102.3	− 84.8	$0.8984^{20/4}$	1.3878^{20}	al, eth	$B2^4$, 786
4352	Butyric acid, 2-methyl *(d)* $CH_3CH_2CH(CH_3)CO_2H$	102.13	$[\alpha]^{15}{}_D$ + 19.2 (w)	$176, 77^{12}$	...	$0.9419^{20/4}$	1.4058^{20}	al, eth	$B2^4$, 888
4353	Butyric acid, 2-methyl *(dl)* $CH_3CH_2CH(CH_3)CO_2H$	102.13	...	177	<− 80	$0.9410^{20/4}$	1.4051^{20}	al, eth	$B2^4$, 889
4354	Butyric acid, 2-methyl *(l)* $CH_3CH_2CH(CH_3)CO_2H$	102.13	$[\alpha]^{20}{}_D$ − 24 (w, c=0.9)	$176\text{-}7, 71\text{-}2^{12}$	...	$0.9340^{20/4}$	1.4042^{25}	al, eth	$B2^4$, 888
4355	Butytic acid, 2-methyl butyl ester *(d)* $C_3H_7CO_2CH_2CH(CH_3)C_2H_5$	158.24	$[\alpha]^{20}{}_D + 3.5$	179^{765}	...	$0.8620^{20/4}$	1.4135^{20}	...	$B2^3$, 304
4356	Butyric acid, 2-methyl, ethyl ester *(d)* $CH_3CH_2CH(CH_3)CO_2C_2H_5$	130.19	$[\alpha]^{26}{}_{5892}$ + 5.16	$131\text{-}3^{730}, 35^{16}$	...	$0.8689^{25/4}$	1.3964^{20}	al, bz	$B2^4$, 890
4357	Butyric acid, 2-methyl butyl ester *(dl)* $C_3H_7CO_2CH_2CH(CH_3)C_2H_5$	158.24	...	166-7	...	$0.862^{20/4}$	1.4100^{25}	...	B2, 304

No.	Name, Synonyms, and Formula	Mol. wt.	Color, crystalline form, specific rotation and λ_{max} (log ε)	b.p. °C	m.p. °C	Density	n_D	Solubility	Ref.
4358	Butyric acid, 2-(2-methyl butyl) ester or *tert*-Pentyl butyrate $C_3H_7CO_2C(CH_3)_2C_2H_5$	158.24		164		$0.8646^{15/0}$			B2[3], 602
4359	Butyric acid, 3-methyl or Isovaleric acid $(CH_3)_2CHCH_2CO_2H$	102.13		176.7	− 29.3	$0.9286^{20/4}$	1.4033^{20}	al, eth, chl	B2[4], 895
4360	Butyric acid, 3-methyl,anhydride $[(CH_3)_2CHCH_2CO]_2O$	186.25		215^{762}, $102\text{-}3^{15}$		$0.9327^{20/4}$	1.4043^{20}	eth	B2[4], 901
4361	Butyric acid, 3-methyl, isopropyl ester $(CH_3)_2CHCH_2CO_2CH(CH_3)_2$	144.21		142^{756}, $69\text{-}70^{55}$		0.8538^{17}	$1.3960^{20.1}$	al, eth, ace	B2[3], 698
4362	Butyric acid, 3-methylbutyl ester or Isopentyl butyrate $C_3H_7CO_2CH_2CH_2CH(CH_3)_2$	158.24		178.5, $65\text{-}8^{12}$		$0.8651^{20/4}$	1.4110^{20}	al, eth	B2[4], 791
4363	Butyric acid, 3-methyl, isobutyl ester $(CH_3)_2CHCH_2CO_2\text{-}i\text{-}C_4H_9$	158.24		171.4, $60\text{-}2^{12}$		$0.8736^{20/4}$	1.4057^{20}	al, eth, ace	B2[4], 790
4364	Butyric acid, 3-methyl, ethyl ester or Ethyl isovalerate $(CH_3)_2CHCH_2CO_2C_2H_5$	130.19		134.7	− 99.3	$0.8656^{20/4}$	1.3962^{20}	al, eth	B2[4], 898
4365	Butyric acid, 3-methyl, methyl ester or Methyl isovalerate $(CH_3)_2CHCH_2CO_2CH_3$	116.16		116.7		$0.8808^{20/4}$	1.3927^{20}	al, eth, ace	B2[4], 897
4366	Butyric acid, 3-methyl-2-oxo $(CH_3)_2CHCOCO_2H$	116.12		170.5, 73^{11}	31.0 — 1.5	$0.9968^{20/4}$	1.3850^{16}	w, al, eth	B3[4], 1577
4367	Butyric acid, 3-methyl, propyl ester or Propyl isovalerate $(CH_3)_2CHCH_2CO_2C_3H_7$	144.21		155.7, 40.5^{13}		$0.8617^{20/4}$	1.4031^{20}	al, eth	B2[4], 898
4368	Butyric acid, 3-methyl-2-phenyl $(CH_3)_2CHCH(C_6H_5)CO_2H$	178.23	pr (lig)	$159\text{-}60^{14}$	63			al, eth	B9[3], 2518
4369	Butyric acid, octyl ester or Octyl butyrate $C_3H_7CO_2C_8H_{17}$	200.32		244.1	− 55.6	0.8629^{20}	$1.4267^{15/}{}_{H9}$	al	B2[4], 791
4370	Butyric acid, 2-oxo or α-Ketobutyric acid $CH_3CH_2COCO_2H$	102.09	pl	$80\text{-}2^{16}$	31-2	$1.200^{17/4}$	1.3972^{20}	w, al	B3[4], 1524
4371	Butyric acid, 2-oxo,oxime $CH_3CH_2C(NOH)CO_2H$	117.10	nd (w), tcl (diox)		164			al	B3[4], 1525
4372	Butyric acid, pentyl ester or Pentyl butyrate, amyl butyrate $C_3H_7CO_2C_5H_{11}$	158.24		186.4	− 73.2	$0.8713^{15/4}$	1.4123^{20}	al, eth	B2[4], 790
4373	Butyric acid, 1-methylpentyl ester *(d)* or *sec*-Hexyl butyrate $C_3H_7CO_2CH(CH_3)C_4H_9$	172.27	$[\alpha]_D$ + 10.16	85^{20}		$0.8744^{21/4}$			B2[1], 120
4374	Butyric acid, 2-phenoxy $CH_3CH_2CH(OC_6H_5)CO_2H$	180.20	nd (w), pl (lig)	258	98			al, eth, ace, bz	B6[4], 644
4375	Butyric acid, 2-phenoxy, ethyl ester $CH_3CH_2CH(OC_6H_5)CO_2C_2H_5$	208.26		$250\text{-}1^{749}$, $87\text{-}90^{2}$		1.0388^{21}		al, eth, ace, chl	B6[4], 645
4376	Butyric acid, 4-phenoxy $C_6H_5OCH_2CH_2CH_2CO_2H$	180.20	pl (lig), cr (w)	$192\text{-}7^{15}$	64-5			al, eth, ace, bz	B6[4], 645
4377	Butyric acid, 4-phenoxy, ethyl ester $C_6H_5OCH_2CH_2CH_2CO_2C_2H_5$	208.26		$170\text{-}3^{25}$		$1.045^{35/25}$	1.491^{33}	al, eth	B6[2],159
4378	Butyric acid, phenyl ester or Phenyl butyrate $C_3H_7CO_2C_6H_5$	164.20		227-8, 85^{8}		$1.0382^{15/4}$	$1.0267^{15/}{}_{15}$	al, eth	B6[4], 615
4379	Butyric acid, 2-phenyl $CH_3CH_2CH(C_6H_5)CO_2H$	164.20	pl (eth)	270-2, $145\text{-}50^{14}$	47.5			eth, bz	B9[3], 2461
4380	Butyric acid, 2-phenyl, methyl ester $CH_3CH_2CH(C_6H_5)CO_2CH_3$	178.23	nd (dil al)	228	77-8			al, eth	B9[3], 2461
4381	Butyric acid, 3-phenyl *(dl)* $CH_3CH(C_6H_5)CH_2CO_2H$	164.20		$140\text{-}5^{3}$	46-7	1.0701^{20}	1.5155^{20}		B9[3], 2457
4382	Butyric acid, 4-phenyl $C_6H_5CH_2CH_2CH_2CO_2H$	164.20	lf (w)	290, 171^{15}	52			al, eth	B9[3], 2451
4383	Butyric acid, propyl ester or Propyl butyrate $C_3H_7CO_2C_3H_7$	130.19		143, 39.2^{14}	− 97.2	$0.8730^{20/4}$	1.4001^{20}	al, eth	B2[4], 788
4384	Butyric acid, *iso*-propyl ester or *iso*-Propyl butyrate $C_3H_7CO_2CH(CH_3)_2$	130.19		130-1		$0.8588^{20/4}$	1.3936^{20}	al	B2[4], 789
4385	Butyric acid, 3-isopropyl-3-oxo,ethyl ester $CH_3COCH(i\text{-}C_3H_7)CO_2C_2H_5$	172.22		201^{758}, $97\text{-}8^{20}$		$0.9648^{18/4}$	$1.4256^{18.5}$	al, eth	B3[4], 1608
4386	Butyric acid, 3-thioxo, ethyl ester $CH_3CSCH_2CO_2C_2H_5$	146.20	dk red	75^{15}		$1.0554^{31/4}$	1.4712^{26}	al, eth	B3[4], 1552
4387	Butyric acid, 2,2,3-trichloro $CH_3CHClCCl_2CO_2H$	191.44	lf or nd (peth)	236-8	60			eth	B2[3], 629
4388	Butyric acid, 2,2,3-trichloro, ethyl ester $CH_3CHClCCl_2CO_2C_2H_5$	219.50		212, 101.9^{17}		$1.3138^{20/20}$		al, eth	B2, 281
4389	Butyric acid, 2,2,4-trichloro $ClCH_2CH_2CCl_2CO_2H$	191.44	cr (peth)		73-5			al, eth	B2, 281

PHYSICAL CONSTANTS OF ORGANIC COMPOUNDS (Continued)

No.	Name, Synonyms, and Formula	Mol. wt.	Color, crystalline form, specific rotation and λ_{max} (log ε)	b.p. °C	m.p. °C	Density	n_D	Solubility	Ref.
4390	Butyric acid, 2,3,3-trichloro $CH_3CCl_2CHClCO_2H$	191.44	pl (lig)	....	52	....	....	al, eth, ace, bz	B2, 281
4391	Butyric acid, 4,4,4-trichloro $Cl_3CCH_2CH_2CO_2H$	191.44	nd (w)	....	35	....	....	al, eth, chl	$B2^4$, 830
4392	Isobutyric acid or 2-Methyl propionic acid $(CH_3)_2CHCO_2H$	88.11	....	153.2, 53.7^{10}	− 46.1	$0.9681^{20/4}$	1.3930^{20}	al, eth, w	$B2^4$, 843
4393	Isobutyric acid, allyl ester or Allyl isobutyrate $(CH_3)_2CHCO_2CH_2CH{=}CH_2$	128.17	....	$133\text{-}5^{755}$	....	....	....	al, eth, ace	$B2^3$, 650
4394	Isobutyric acid anhydride or Isobutyric anhydride $[(CH_3)_2CHCO]_2O$	158.20	....	181.5^{734}, $89\text{-}90^{32}$	− 53.5	$0.9535^{20/4}$	1.4061^{19}	eth, chl	$B2^4$, 851
4395	Isobutyric acid, 2-amino $(CH_3)_2C(NH_2)CO_2H$	103.12	ta or pr (w)	sub 280	337 (cor)	....	....	w	$B4^3$, 1322
4396	Isobutyric acid, benzyl ester $(CH_3)_2CHCO_2CH_2C_6H_5$	178.23	....	$114\text{-}5^{20}$	....	1.0075^{15}	1.4883^{20}	....	$B6^4$, 2267
4397	Isobutyric acid, 2-bromo or 2-Bromo-2-methyl propionic acid $(CH_3)_2CBrCO_2H$	167.00	cr (peth)	198-200, 115^{24}	48-9	$1.5225^{60/60}$	....	al, eth	$B2^4$, 862
4398	Isobutyric acid, 2-bromo, ethyl ester $(CH_3)_2C(Br)CO_2C_2H_5$	195.06	....	164^{762}, 70^{20}	....	$1.3182^{20/4}$	1.4446^{20}	al, eth	$B2^4$, 862
4399	Isobutyric acid, isobutyl ester or Isobutyl isobutyrate $(CH_3)_2CHCO_2CH_2CH(CH_3)_2$	144.21	....	148.6, $36\text{-}40^{11}$	− 80.6	$0.8750^{0/4}$	1.3999^{20}	al, eth, ace	$B2^4$, 847
4400	Isobutyric acid, *tert*-butyl ester or *tert*-Butyl isobutyrate $(CH_3)_2CHCO_2C(CH_3)_3$	144.21	....	126.7	....	....	1.3921^{20}	al, eth, ace	$B2^3$, 648
4401	Isobutyric acid, 2-chloro $(CH_3)_2CClCO_2H$	122.55	....	118^{50}	31	....	1.450^{20}	al	$B2^4$, 858
4402	Isobutyric acid, 2-chloro,ethyl ester $(CH_3)_2CClCO_2C_2H_5$	150.61	....	148-9	....	1.062^{0}	1.4109^{16}	al, eth	$B2^4$, 859
4403	Isobutyric acid, 2-chloro, methyl ester $(CH_3)_2CClCO_2CH_3$	136.58	....	135, $42\text{-}4^{17}$	....	$1.0893^{15/15}$	1.4122^{21}	eth	$B2^4$, 859
4404	Isobutyric acid, 3-chloro $ClCH_2CH(CH_3)CO_2H$	122.55	....	$128\text{-}33^{50}$	....	1.0153^{20}	1.4310^{20}	al, eth	$B2^4$, 860
4405	Isobutyric acid, cyclohexyl ester or Cyclohexyl isobutyrate $(CH_3)_2CHCO_2C_6H_{11}$	170.25	....	204^{750}	....	$0.9489^{0/4}$	....	al, eth	$B6^3$, 24
4406	Isobutyric acid, 3,3-dichloro-2-hydroxy $Cl_2CHC(CH_3)(OH)CO_2H$	173.00	pr (al-eth)	d	82-3	....	....	w, al, eth	$B3^2$, 224
4407	Isobutyric acid, ethyl ester or Ethyl isobutyrate $(CH_3)_2CHCO_2C_2H_5$	116.16	....	111.0	− 88.2	$0.8693^{20/4}$	1.3869^{18}	al, eth, ace	$B2^4$, 846
4408	Isobutyric acid, furfuryl ester or Furfuryl isobutyrate $(CH_3)_2CHCO_2CH_2(C_4H_3O)$	168.19	....	$85\text{-}6^{15}$	....	$1.0313^{20/4}$	....	al, eth	$B17^2$, 115
4409	Isobutyric acid, 2-hydroxy or Acetonic acid $(CH_3)_2C(OH)CO_2H$	104.11	hyg pr (eth), nd (bz)	212, $108\text{-}11^{8}$	82-3	....	....	al, eth, w	$B3^4$, 782
4410	Isobutyric acid, 2-hydroxy, ethyl ester $(CH_3)_2C(OH)CO_2C_2H_5$	132.16	....	150 (cor), 46^{14}	....	0.987^{20}	1.4080^{20}	w, al	$B3^4$, 783
4411	Isobutyric acid, 2-hydroxy, methyl ester $(CH_3)_2C(OH)CO_2CH_3$	118.13	....	137, $62\text{-}4^{12}$	....	....	1.4056^{20}	w, al	$B3^4$, 783
4412	Isobutyric acid, methyl ester or Methyl isobutyrate $(CH_3)_2CHCO_2CH_3$	102.13	....	92.3	− 84.7	$0.8906^{20/4}$	1.3840^{20}	al, eth, ace	$B2^4$, 846
4413	Isobutyric acid, 2-nitropentyl ester or 2-Nitropentyl isobutyrate $(CH_3)_2CHCO_2CH_2CH(NO_2)C_3H_7$	202.23	....	248-51, 122^{10}	....	$1.0329^{20/20}$	1.4315^{20}	eth, ace	$B2^3$, 648
4414	Isobutyric acid, isopentyl ester or Isopentyl isobutyrate $(CH_3)_2CHCO_2\text{-i-}C_5H_{11}$	158.24	....	168.9		0.8627^{20}	....	al, eth, ace	$B2^3$, 649
4415	Isobutyric acid, 3-phenyl $C_6H_5CH_2CH(CH_3)CH_2CO_2H$	164.20	pl (dil al)	272, $155\text{-}6^{11}$	36.5	....	....	al, eth	$B9^3$, 2472
4416	Isobutyric acid, propyl ester or Propyl isobutylrate $(CH_3)_2CHCO_2C_3H_7$	130.19		135-6	....	$0.8843^{0/4}$	1.3955^{20}	al, eth, ace	$B2^4$, 847
4417	Isobutyric acid, isopropyl ester or Isopropyl isobutylrate $(CH_3)_2CHCO_2CH(CH_3)_2$	130.19		120.7	....	$0.8471^{21/4}$		al, eth, ace	$B2^4$, 847
4418	Isobutyric acid, vinyl ester or Vinyl isobutyrate $(CH_3)_2CHCO_2CH{=}CH_2$	114.14		104-5	....	0.8932^{20}	1.4061^{20}	....	$B2^4$, 848
4419	γ-Butyrolactone $CH_2CH_2CH_2CO$ 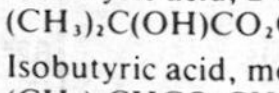	86.09		206, 89^{12}	− 42	$1.1286^{16/0}$	1.4341^{20}	w, al, eth, ace, bz	$B17^4$, 4159

No.	Name, Synonyms, and Formula	Mol. wt.	Color, crystalline form, specific rotation and λ_{max} (log ε)	b.p. °C	m.p. °C	Density	n_D	Solubility	Ref.
4420	γ-Butyrolactone, 2-methylene or Butanoic acid-4-hydroxy-2-methylene, lactone $CH_2CH_2C(=CH_2)CO$ (ring through O)	98.10	...	85-6^{10}	...	1.1206^{20}	1.4650^{20}	w, al, eth, ace, bz	B17^{4}, 4304
4421	Butyronitrile $CH_3CH_2CH_2CN$	69.11	...	118	− 112	0.7936$^{20/4}$	1.3842^{10}	**al, eth**, bz	B2^{4}, 806
4422	Butyronitrile, 4-bromo $BrCH_2CH_2CH_2CN$	148.00	...	205-7, 91^{12}	...	1.4967$^{20/4}$	1.4818^{20}	al, eth	B2^{4}, 836
4423	Butyronitrile, 4-chloro $ClCH_2CH_2CH_2CN$	103.55	...	189-91, 75^{11}	...	1.0934^{15}	1.4413^{20}	al, eth	B2^{4}, 827
4424	Butyronitrile, 3,4-epoxy or Epicyanohydrin CH_2CHCH_2CN (epoxide ring through O)	83.09	pr	...	162	...	...	al	B18^{4}, 3822
4425	Butyronitrile, 2-ethyl $(C_2H_5)_2CHCN$	97.16	...	145-6	...	...	1.3891^{24}	**al, eth**	B2^{4}, 953
4426	Butyronitrile, 2-methyl $CH_3CH_2CH(CH_3)CN$	83.13	...	125	...	0.7913$^{15/4}$	1.3933^{20}	al, eth	B2^{4}, 892
4427	Butyronitrile, 3-methyl or Isovaleronitrile $(CH_3)_2CHCH_2CN$	83.13	...	130.5, 53^{50}	− 100.8 f.p.	0.7914$^{20/4}$	1.3927^{20}	**al, eth**, ace	B2^{4}, 902
4428	Butyronitrile, 3-methyl-2-phenyl $(CH_3)_2CHCH(C_6H_5)CN$	159.23	...	245-9^{765}	...	0.967$^{15.5}$	1.5038^{25}	al, bz	B9^{2}, 364
4429	Butyronitrile, 2-phenoxy $CH_3CH_2CH(OC_6H_5)CN$	161.20	...	228-30^{748}/$_{d}$	...	...	...	al, eth	B6, 164
4430	Butyronitrile, 4-phenoxy $C_6H_5OCH_2CH_2CH_2CN$	161.20	nd	287-9^{765}, 170.5^{22}	45-6	...	...	eth	B6^{4}, 646
4431	Butyronitrile, 2-phenyl $CH_3CH_2CH(C_6H_5)CN$	145.20	...	238-40^{763}, 141-3^{8}	...	...	...	al, eth, bz	B9^{3}, 2468
4432	Butyronitrile, 4-phenyl $C_6H_5CH_2CH_2CH_2CN$	145.20	...	142-5^{16}	...	...	...	...	B9^{3}, 2454
4433	Isobutyronitrile $(CH_3)_2CHCN$	69.11	...	103.8	− 71.5	0.7608$^{30/4}$	1.3720^{20}	al, eth, ace, chl	B2^{4}, 853
4434	Isobutyronitrile, 2-bromo $(CH_3)_2CBrCN$	148.00	...	139-40, 61-2^{5}	...	1.4796$^{15/4}$	1.4739^{15}	al, eth, bz	B2^{4}, 863
4435	Isobutyronitrile, 2-hydroxy or Acetone cyanohydrin $(CH_3)C(OH)CN$	85.11	...	82^{23}	− 19	0.932$^{20/4}$	1.3996^{20}	w, al, eth, ace, bz	B3^{4}, 785
4436	Isobutyronitrile, 2-hydroxy-3-chloro $ClCH_2C(OH)(CH_3)CN$	119.55	...	110^{27}	...	1.2027^{15}	1.4356^{11}	w, al, ace	B3^{3}, 599
4437	Butyrophenone or *n*-Propyl phenyl ketone $CH_3CH_2CH_2COC_6H_5$	148.20	...	228-9	11-3	0.988$^{20/4}$	1.5203^{20}	**al, eth**, ace	B7^{3}, 1075
4438	Butyrophenone, *p*-methyl or *p*-Totyl-propyl ketone $(4\text{-}CH_3C_6H_4)COC_3H_7$	162.23	...	251.5^{758}	12	0.9745$^{20/4}$	1.5232^{20}	al, eth	B7^{3}, 1127
4439	Isobutyrophenone $C_6H_5COCH(CH_3)_2$	148.20	...	221, 86^{4}	...	0.9863$^{11/4}$	1.5172^{20}	al, eth	B7^{3}, 1088
4440	Butyryl bromide C_3H_7COBr	151.00	...	128	...	1.4162$^{17/4}$	1.1596^{17}	...	B2^{4}, 804
4441	Butyrylbromide, 2-bromo $CH_3CH_2CHBrCOBr$	229.90	...	172-4, 57-60^{10}	...	...	...	eth	B2^{4}, 834
4442	Butyrylbromide, 3-methyl or Isovaleryl bromide $(CH_3)_2CHCH_2COBr$	165.03	mcl pl (al)	143	...	...	...	eth	B2^{4}, 902
4443	Isobutyryl bromide $(CH_3)_2CHCOBr$	151.00	...	116-8	...	1.4067$^{15/4}$	1.4552^{15}	...	B2^{4}, 852
4444	Isobutyryl bromide, 2-bromo $(CH_3)_2CBrCOBr$	229.90	...	162-4, 91-8^{100}	...	1.4067$^{14/4}$	1.4552^{14}	ace, CS_2	B2^{3}, 661
4445	Butyryl chloride $CH_3CH_2CH_2COCl$	106.55	...	102	− 89	1.0277$^{20/4}$	1.4121^{20}	**eth**	B2^{4}, 803
4446	Butyryl chloride, 2-bromo $CH_3CH_2CHBrCOCl$	185.45	...	150-2, 41^{12}	...	1.5320^{20}	...	eth	B2^{4}, 834
4447	Butyryl chloride, 2-chloro $CH_3CH_2CHClCOCl$	141.00	...	130-1, 51-2^{41}	...	1.2360^{17}	1.4475^{20}	eth	B2^{4}, 822
4448	Butyryl chloride, 2-chloro-2-methyl $CH_3CH_2C(CH_3)ClCOCl$	155.02	...	144^{750}	...	1.187^{14}	...	eth	B2^{4}, 893
4449	Butyryl chloride, 2-chloro-3-methyl $(CH_3)_2CHCHClCOCl$	155.02	...	148-9	...	1.135^{11}	...	eth	B2^{4}, 904
4450	Butyryl chloride, 3-chloro $CH_3CHClCH_2COCl$	141.00	...	40-1^{12}	...	1.2163$^{20/4}$	1.4509^{20}	CS_2	B2^{4}, 824
4451	Butyryl chloride, 4-chloro $ClCH_2CH_2CH_2COCl$	141.00	...	173-4, 60-1^{12}	...	1.2581$^{20/4}$	1.4616^{20}	eth	B2^{4}, 826
4452	Butyryl chloride, 4-chloro-3-oxo $ClCH_2COCH_2COCl$	154.98	...	117-9^{17}	...	1.4397$^{20/4}$	1.4860^{20}	bz	B3^{3}, 1207

No.	Name, Synonyms, and Formula	Mol. wt.	Color, crystalline form, specific rotation and λ_{max} (log ε)	b.p. °C	m.p. °C	Density	n_D	Solubility	Ref.
4453	Butyryl chloride, 2,2-dimethyl $CH_3CH_2C(CH_3)_2COCl$	134.61	....	132, 27^{11}	......	$0.9801^{20/4}$	1.4245^{20}	eth	$B2^4$, 955
4454	Butyryl chloride, 2,3-dimethyl $(CH_3)_2CHCH(CH_3)COCl$	134.61	....	$135\text{-}6^{751}$, $38\text{-}9^{18}$	......	$0.9795^{20/4}$	....	eth	$B2^3$, 761
4455	Butyryl chloride, 3,3-dimethyl $(CH_3)_3CCH_2COCl$	134.61	....	$128\text{-}30^{745}$, 68^{100}	......	$0.9696^{20/4}$	1.4210^{20}	eth	$B2^4$, 956
4456	Butyryl chloride, 2-ethyl or Diethylacetyl chloride $(C_2H_5)_2CHCOCl$	134.61	....	140, 40^{20}	......	$0.9825^{20/4}$	1.4234^{20}	eth	$B2^4$, 952
4457	Butyryl chloride, 3-methyl or Isovaleryl chloride $(CH_3)_2CHCH_2COCl$	120.58	....	$114\text{-}5^{771}$	......	$0.9844^{20/4}$	1.4149^{20}	eth	$B2^4$, 901
4458	Butyryl chloride, 2-methyl-*(dl)* $CH_3CH_2CH(CH_3)COCl$	120.58	....	116	......	$0.9917^{20/4}$	1.4170^{20}	....	$B2^4$, 891
4459	Butyryl chloride, 2,2,3-trimethyl $(CH_3)_2CHC(CH_3)_2COCl$	148.63	....	148-50	......	....	....	eth	$B2^4$, 982
4460	Isobutyryl chloride $(CH_3)_2CHCOCl$	106.55	....	92	− 90	$1.0174^{20/4}$	1.4079^{20}	eth	$B2^4$, 852
4461	Isobutyryl chloride, 2-chloro $(CH_3)_2CClCOCl$	141.00	....	126-7	......	....	1.4369^{20}	eth	$B2^4$, 859
4462	Isobutyryl chloride. 3-chloro $CH_3CH_2ClCHCOCl$	141.00	....	92	− 90.0	$1.0174^{20/4}$	1.4079^{20}	eth	$B2^4$, 860

PHYSICAL CONSTANTS OF ORGANIC COMPOUNDS (Continued)

No.	Name, Synonyms, and Formula	Mol. wt.	Color, crystalline form, specific rotation and λ_{max} (log ε)	b.p. °C	m.p. °C	Density	n_D	Solubility	Ref.
4463	Cacodyl or Tetramethyl biarsine $(CH_3)_2AsAs(CH_3)_2$	209.98	pl	165	− 6	1.447^{15}		al, eth	$B4^3$, 1831
4464	Cacodyl chloride or Dimethyl chloraisine $(CH_3)_2AsCl$	140.45		109	< − 45	$1.5046^{12/4}$	1.5203^{12}	al	$B4^3$, 1797
4465	Cacodyl oxide or bis-Dimethyl arsenous oxide $(CH_3)_2AsOAs(CH_3)_2$	225.98		150	− 25	1.4816^{15}	1.5225^{9}	al, eth	$B4^3$, 1814
4466	β-Cadinene-*(l)* or 3,9-Cadinadiene $C_{15}H_{24}$	204.36	$[\alpha]^{27}_D$ − 15.9 (chl, c=1)	274, 149^{20}		$0.9230^{20/4}$	1.5059^{20}	eth, lig	$B5^3$, 1086
4467	Caffeine or 1,3,7-Trimethylxanthine $C_8H_{10}N_4O_2$	194.19	wh nd (w + 1), hex pr (sub)	sub 178, sub 89^{15}	238 (anh)	1.23^{19}		Py, chl	$B26^2$, 266
4468	Caffeine benzoate $C_8H_{10}N_4O_2$-$C_6H_5CO_2H$	316.32	wh so pw					al, w	$B26^2$, 268
4469	Caffeine citrate $C_8H_{10}N_4O_2$-$C_6H_8O_7$	386.32	mcl cr						$B26^2$, 269
4470	Caffeine hydrobromide $C_8H_{10}N_4O_2$-HBr-$2H_2O$	311.14	ye		d 80-100			w	$B26^1$, 137
4471	Caffeine hydrochloride $C_8H_{10}N_4O_2 \cdot HCl \cdot 2H_2O$	266.68	mcl pr		d 80-100				$B26^2$, 268
4472	Caffeine, 2-hydroxybenzoate or Caffeine salicylate $C_8H_{10}N_4O_2$-$C_7H_6O_3$	322.32	wh nd (w)		137			al	$B26^2$, 269
4473	Caffeine, 3-methylbutanoate or Caffeine isovalerate $C_{10}H_8N_4O_2$-$C_5H_{10}O_2$	296.33	unst nd					w	B26, 467
4474	Caffeine sulfate $C_8H_{10}N_4O_2$-H_2SO_4	292.27	wh nd						B26, 466
4475	Caffeine, 8-ethoxy or 1,3,7-Trimethyl-2,6-dioxo-8-ethoxy purine $C_{10}H_{14}N_4O_3$	238.25	wh or yesh nd (w)		143			al	$B26^2$, 322
4476	Caffeine, 8-methoxy $C_9H_{12}N_4O_3$	224.22	wh nd (al or w)		176	$1.399^{25/4}$		al, bz	$B26^2$, 322
4477	Calciferol or Vitamin D2 $C_{28}H_{44}O$	396.66	pr (ace), $[\alpha]^{20}_D$ + 102.5 (al)	sub	115—8			al, eth, ace	$B6^3$, 3089
4478	Camphane or Bornane $C_{10}H_{18}$	138.25	hex pl (al), pr (MeOH)	sub 161	158-9			al, eth	$B5^4$, 319
4479	Camphane-3-carboxylic acid $C_{10}H_{17}CO_2H$	182.26	pl (dil aa), $[\alpha]^{20}_D$ + 56 (al, p=8)	153^{13}	90-1			al, aa	$B9^3$, 244
4480	Camphene-*(d)* $C_{10}H_{16}$	136.24	nd $[\alpha]^{17}_D$ + 103.9 (eth, c=4)	160-2, 52^{17}	52	$0.8450^{50/4}$	1.4570^{25}	eth	$B5^4$, 461
4481	Camphene *(dl)* $C_{10}H_{16}$	136.24	nd (sub)	158-9	51-2	$0.879^{20/4}$	1.4551^{54}	al, eth	$B5^4$, 462
4482	Camphene-*(l)* $C_{10}H_{16}$	136.24	$[\alpha]^{19}_D$ − 106.1 (eth, c=4)	158	52	$0.8446^{50/4}$	1.4564^{54}	eth	$B5^4$, 461
4483	Camphenilone $C_9H_{14}O$	138.21	$[\alpha]^{20}_D$ + 70.4 (al)	193^{751}, 76^{12}	41			eth	$B7^3$, 306
4484	3-Camphanol or Epiborneol $C_{10}H_{15}OH$	154.24	nd $[\alpha]^{17}$ + 11.1 (al)	213^{742}	181-2			peth	$B6^4$, 288
4485	3-Camphanol acetate $C_{12}H_{20}O_2$	196.29	$[\alpha]^{10}_D$ + 15.63	101^{11}	< − 15	$0.9872^{14/4}$	1.4651^{14}		$B6^2$, 92
4486	Campholic acid-*(d)* $C_{10}H_{18}O_2$	170.25	pr $[\alpha]^{20}_D$ + 59.3 (bz)	255^{768}, 146^{12}	106			al, eth	$B9^3$, 98
4487	Campholic acid-*(dl)* $C_{10}H_{18}O_2$	170.25	tcl pr		109			al, eth	$B9^3$, 99
4488	Campholic acid-*(l)* $C_{10}H_{18}O_2$	170.25	pr (dil al), $[\alpha]^{15}_D$ − 49.1 (al)	250	106-7			al, eth	$B9^3$, 99
4489	Campholytic acid-*(α,l)* $C_9H_{14}O_2$	154.21	$[\alpha]^{13}_D$ − 60.4	240-3, 140^{15}		1.0145^{18}	1.4712^{17}	lig	$B9^3$, 187
4490	Camphor-*(d)* or 2-Camphanone *(d)* $C_{10}H_{16}O$	152.24	pl $[\alpha]^{20}_D$ + 44.26 (al)	sub 204	179.8	$0.990^{25/4}$	1.5462	al, eth, ace, bz	$B7^3$, 400

No.	Name, Synonyms, and Formula	Mol. wt.	Color, crystalline form, specific rotation and λ_{max} (log ε)	b.p. °C	m.p. °C	Density	n_D	Solubility	Ref.
4491	Camphor-*(dl)* $C_{10}H_{16}O$	152.24	wh	sub	178.8			al, eth, ace, bz	B7³, 406
4492	Camphor-*(l)* $C_{10}H_{16}O$	152.24	$[\alpha]^{16}{}_D$ − 44.2 (al, c=16.5)	204 sub	178.6	0.9853[18]		al, eth, ace	B7³, 405
4493	Camphor oxime-*(d)* $C_{10}H_{16}NOH$	167.25	pr (lig-eth), $[\alpha]^{22}{}_D$ + 42.5 (al)		115			al, eth	B7³, 408
4494	Camphor oxime-*(dl)* $C_{10}H_{16}NOH$	167.25	cr (peth)		118			al, eth	B7³, 409
4495	Camphor oxime-*(l)* $C_{10}H_{16}NOH$	167.25	mcl nd or pr (dil al), $[\alpha]^{20}{}_D$ − 42.4 (al)	249-54d	118	1.01[116/4]		al, eth	B7³, 408
4496	Camphor, 3-amino-*(d)* or 3-Camphoryl amine $C_{10}H_{17}NO$	167.25	wx	244	110-15d			al, eth	B14³, 15
4497	Camphor, 3-bromo-*(d,α)* $C_{10}H_{15}BrO$	231.13	pr (al), $[\alpha]^{20}{}_D$ + 129.3 (MeOH, c=4.6)	274(d) sub	76			al, eth, bz, chl	B7³, 414
4498	Camphor, 3-bromo-*(d,α′)* $C_{10}H_{15}BrO$	231.13	$[\alpha]^{20}{}_D$ 29.4	d265	78	1.484[14]		al, chl	B7³, 415
4499	Camphor, 3-bromo-*(dl,α)* $C_{10}H_{15}BrO$	231.13			51			al, eth, bz, chl	B7³, 415
4500	Camphor, 3-bromo-*(l,α)* $C_{10}H_{15}BrO$	231.13	mcl nd (al), $[\alpha]^{18}{}_D$ − 138.8 (ace, c=6)		76			al, eth, bz, chl	B7³, 415
4501	Camphor, 3-bromo-*(d,α′)* $C_{10}H_{15}BrO$	231.13	nd (dil al)	265d	78	1.484[14]		al, chl, aa	B7², 101
4502	Camphor, 5-bromo-*(exo)* $C_{10}H_{15}BrO$	231.13	$[\alpha]^{16}{}_D$ bz 2%	100[15]	114			al, eth, ace, bz	B7³, 415
4503	Camphor, 8-bromo-*(d)* $C_{10}H_{15}BrO$	231.13	tetr pr (lig), $[\alpha]^{19/}{}_D$ + 122.2 (chl)	sub	93			al, eth, ace, bz	B7³, 416
4504	Camphor, 8-bromo-(dl) $C_{10}H_{15}BrO$	231.13	pr (eth-peth), pym (eth)	sub	92.7			al, eth, ace, bz	B7³, 417
4505	Camphor, 10-bromo-*(d)* $C_{10}H_{15}BrO$	231.13	pr (peth), $[\alpha]^{20}{}_D$ + 19.2 (abs al), + 15.7 (bz)	265d	78			al, eth, ace, bz	B7³, 416
4506	Camphor, 10-bromo-*(dl)* $C_{10}H_{15}BrO$	231.13	cr		77				B7³, 416
4507	Camphor, 3-chloro-*(d,α)* $C_{10}H_{15}ClO$	186.68	lf $[\alpha]^{20}{}_D$ + 71.1	244-7	94			al, eth, ace, bz	B7³, 411
4508	Camphor, 3-chloro-*(d,α′)* $C_{10}H_{15}ClO$	186.68	$[\alpha]^{20}{}_D$ + 35 (al, c=5)	231d	118			eth, chl	B7³, 411
4509	Camphor, 8-chloro-*(d)* $C_{10}H_{15}ClO$	186.68	pr (al), $[\alpha]_D$ + 99.9 (chl)		139				B7³, 414
4510	Camphor, 8-chloro-*(dl)* $C_{10}H_{15}ClO$	186.68		sub	138				B7, 136
4511	Camphor, 10-chloro-*(d)* $C_{10}H_{15}ClO$	186.68	pr (al), $[\alpha]^{14}{}_D$ + 40.7 (al)		132-5			al, eth, bz, chl	B7², 100
4512	Camphor, 3,3-dibromo-*(d)* $C_{10}H_{14}Br_2O$	310.03	wh ye rh pr (al, peth), $[\alpha]^{20}{}_D$ + 40 (chl) + 39.2 (al)	sub	64	1.8954[21.6/4]		al, eth, bz, chl	B7², 101
4513	Camphor, 3-nitro-*(l)* $C_{10}H_{15}NO_3$	197.23	mcl pr (bz), $[\alpha]^{13}{}_D$ - 26→-9 (al), (mut)					al, eth, bz, chl	B7³, 419

PHYSICAL CONSTANTS OF ORGANIC COMPOUNDS (Continued)

No.	Name, Synonyms, and Formula	Mol. wt.	Color, crystalline form, specific rotation and λ_{max} (log ε)	b.p. °C	m.p. °C	Density	n_D	Solubility	Ref.
4514	Camphor-3-carboxylic acid-*(d)* $C_{11}H_{16}O_3$	196.25	pr (eth, 50% al), $[\alpha]^{20}_D$ +34.9 (bz)		128d			w, al, eth, bz	B10[3], 2925
4515	Camphor-3-carboxylic acid-*(dl)* $C_{11}H_{16}O_3$	196.25	cr (bz)		136-7			al, eth, bz	B10[3], 2927
4516	Camphor-3-carboxylic acid $C_{11}H_{16}O_3$	196.25	pr (eth), cr (bz), $[\alpha]^{20}_D$ -64 (al)		127-8d			al, eth, bz	B10[3], 2927
4517	β-Camphor-*(l)* or *l*-EpiCamphor, 1,3-Bornanone $C_{10}H_{16}O$	152.24	$[\alpha]^{19}_D$ -58.21 (bz, c=13)	213	184			al, eth, peth	B7[3], 421
4518	Camphoric acid-*(d)* or 1,2,2-Trimethylcyclopentane-1,3-dicarboxylic acid $C_{10}H_{16}O_4$	200.23	Pr, lf (w), $[\alpha]^{20}_D$ +47.7 (al)		188.2	$1.186^{20/4}$		al, eth, ace	B9[3], 3876
4519	Camphoric acid-*(dl)* $C_{10}H_{16}O_4$	200.23	pr (al, aa), mcl nd		208	$1.228^{20/4}$		al, ace	B9[3], 3878
4519a	Camphoric acid-*(l)* $C_{10}H_{16}O_4$	200.23	Cr (w), $[\alpha]^{18/}_D$ -48.1 (abs. al, c=8)	223	270d	1.190		al, eth, ace	B9[3], 3878
4520	Camphoric anhydride-*(d)* $C_{10}H_{14}O_3$	182.22	rh (al), pr (bz)	7270d	223.5	1.194^{20}		bz	B17[4], 5957
4521	Camphoric anhydride-*(dl)* $C_{10}H_{14}O_3$	182.22	rh (al)	270	221	$1.194^{20/4}$		bz	B17[1], 238
4522	Camphoric anhydride-*(l)* $C_{10}H_{14}O_3$	182.22	$[\alpha]_D$ -77 (bz)	>270	221	$1.194^{20/4}$		bz	B17[4], 5958
4523	Camphoric acid, diethyl ester-*(d)* $C_{14}H_{24}O_4$	256.34	$[\alpha]^{15/}_D$ +7.5 (al), +9 (bz)	286^{752}, 164^{20}		$1.0298^{20/4}$	1.4535^{20}	al, eth, bz	B9[2], 536
4524	Camphoric acid, dimethyl ester-*(d)* $C_{12}H_{20}O_4$	228.29	$[\alpha]_D$ +49.07 (al)	264^{758}, 155^{15}	<-16	$1.0747^{20/4}$	1.4627^{19}	al, eth	B9[3], 3880
4525	Camphoric acid, 1-monoamide or β-Camphoramic acid $C_{10}H_{17}NO_3$	199.25	pl (nd), $[\alpha]^{20}_D$ +74 (al)		183			al, ace	B9[3], 3886
4526	Camphoric acid, 3-monoamide-*(d)* or α-Camphoramic acid $C_{10}H_{17}NO_3$	199.25	nd or lf (w), $[\alpha]^{20/}_D$ +25 (al)		176-7				B9[3], 3887
4527	Camphoric acid, 3-monoamide-*(dl)* or α-Camphoramic acid $C_{10}H_{17}NO_3$	199.25	nd (w)		198				B9, 761
4528	Camphoronic acid-*(d)* or 2,4-Dimethylpentane-1,2,3-tricarboxylic acid $(CH_3)_2C(CO_2H)C(CH_3)(CO_2H)CH_2CO_2H$	218.21	nd (w), $[\alpha]^{19}$ +27.05 (w)					w, al, eth, ace	B2, 837
4529	Camphoronic acid-*(dl)* $(CH_3)_2C(CO_2H)C(CH_3)(CO_2H)CH_2CO_2H$	218.21	nd or pr (w)		172d			al, eth, ace	B2[3], 2045
4530	Camphoronic acid-*(l)* $(CH_3)_2C(CO_2H)C(CH_3)(CO_2H)CH_2CO_2H$	218.21	nd (w), $[\alpha]^{19/}_D$ -26.9		164-5d			w, al, eth, ace	B2[3], 2045
4531	Camphorpinacol-*(l)* or 2,2′-Bicamphane-2,2′-diol $(C_{10}H_{17}O)_2$	306.49	rh $[\alpha]_D$ −27.2 (bz)		158			al, eth	B6[3], 4767
4532	Camphor quinone or 2,3-Camphor dione $C_{10}H_{14}O_2$	166.22	ye nd (dil al, w), pr (eth) $[\alpha]^{20/}_D$ −113.2 (bz)	sub	199			al, eth, bz, chl	B7[3], 3297
4533	Camphor-3-sulfonic acid, methyl ester-*(d)* $C_{11}H_{18}SO_4$	246.32	cr (MeOH), nd (peth), $[\alpha]^{20/}_D$ +98.6 (chl, c=5)					al	B11[2], 179
4534	Camphor-10-sulfonic acid-*(d)* $C_{10}H_{16}O_4S$	232.29	pr (aa), $[\alpha]^{20/}_D$ +32.8 (AcOEt, + c=3), +24 (w)		195d			w	B11[3], 585

No.	Name, Synonyms, and Formula	Mol. wt.	Color, crystalline form, specific rotation and λ_{max} (log ε)	b.p. °C	m.p. °C	Density	n_D	Solubility	Ref.
4535	Camphor-10-sulfonic acid-*(dl)* $C_{10}H_{16}O_4S$	232.29	cr (aa)		202d			w	B11[3], 587
4536	Camphor-10-sulfonic acid-*(l)* $C_{10}H_{16}O_4S$	232.29	cr (aa), nd (AcoEt), $[\alpha]^{20}_D$ −20.75 (w)		194-5d			w	B11[3], 585
4537	α-Camphyl amine or 4-(2-Aminoethyl)-1,5,5-trimethyl cyclopentene $C_{10}H_{19}N$	153.27	$[\alpha]_D + 6$	194-6, 95^{12}		0.8688^{20}	1.4728^{18}		B12[3], 176
4538	β-Camphyl amine or 2-(2-Aminoethyl)-1,5,5-trimethyl cyclopentene $C_{10}H_{19}N$	153.27		206		$0.8697^{20/20}$			B12, 40
4539	Canadine-*(d)* or *d*-Tetrahydrobeberine $C_{20}H_{21}NO_4$	339.39	ye nd (dil al), $[\alpha]^{29/}_D$ + 299 (chl, c=1)		132 (140)			al, eth, bz, chl	B27[2], 557
4540	Canadine-*(dl)* $C_{20}H_{21}NO_4$	339.39	mcl nd (al)		174			chl	B27[2], 557
4541	Canadine-*(l)* $C_{20}H_{21}NO_4$	339.39	ye nd (al) $[\alpha]^{20/}_D$- 299.2 (chl)		134			al, eth, bz, chl	B27[2], 557
4542	Canaline $H_2NOCH_2CH_2CH(NH_2)CO_2H$	134.14	nd (al), $[\alpha]^{23/}_D$ -8.31 (w)		214d			w, al	B4[3], 1636
4543	Canavanine-*(L)*. $HN{:}C(NH_2)NHOCH_2CH_2CH(NH_2)CO_2H$	176.18	cr (al), $[\alpha]^{20/D}$ + 7.9 (w, C=2)					w	B4[3], 1636
4544	Cannabidiol $CH_3C_6H_7CCH_3CH_2C_6H_2(OH)_2(CH_2)_4CH_3$	314.47	rods (peth)	$187\text{-}90^{2}$	67			al, eth, bz, chl	B6[3], 5362
4545	Cannabinol $C_{21}H_{26}O_2$	310.44	pl, lf (peth) $[\alpha]^{20/}_D$ -148 (al)	$185^{0.05}$	77			al, eth, ace, bz	B17[4], 1652
4546	Cantharidin $C_{10}H_{12}O_4$	196.20	rh pl or sc	sub 84	218				B19[4], 1958
4547	Capraldehyde or Decanal $CH_3(CH_2)_8CHO$	156.27		208-9, 81^{9}	−5	$0.830^{15/4}$	1.4287^{20}	al, eth, ace	B1[4], 3366
4548	Capradehyde oxime or Capraldoxime $CH_3(CH_2)_8CHOH$	171.28	lf (dil MeOH)		69			al, eth	B1, 711
4549	Capramide $CH_3(CH_2)_8CONH_2$	171.28	lf (eth)		108	$0.999^{20/4}$	1.4261^{110}	al, eth, ace	B2[4], 1050
4550	Capric acid or Decanoil acid. $CH_3(CH_2)_8CO_2H$	172.27	nd	270, $148\text{-}50^{11}$	fr 31.5	$0.8858^{40/4}$	1.4288^{40}	al, eth, ace, bz, peth	B2[4], 1041
4551	Capric acid, 2-acetyl,ethyl ester $C_8H_{15}CH(COCH_3)CO_2C_2H_5$	242.36		280-2		$0.9354^{18.5/17.5}$			B3[3], 1278
4552	Capric anhydride $(C_9H_{19}CO)_2O$	326.52	lf		24.7	$0.8865^{25/4}$	1.400^{25}	al, eth	B2[4], 1049
4553	Capric acid, 2-bromo $C_8H_{17}CHBrCO_2H$	251.16		$140\text{-}1^{2}$	4	1.1912^{24}	1.4595^{24}	eth	B2[4], 1054
4554	Capric acid, decyl ester $C_9H_{19}CO_2C_{10}H_{21}$	312.54		219^{15}	9.7	0.8586^{20}	1.4423^{20}	eth	B2[4], 1045
4555	Capric acid, ethyl ester or Ethyl caprate $C_9H_{19}CO_2C_2H_5$	200.32		241.5, $122\text{-}4^{11}$	−20	$0.8650^{20/4}$	1.4256^{20}	al, eth, chl	B2[4], 1044
4556	Capric acid, 10-fluoro $F(CH_2)_9CO_2H$	190.26		$135\text{-}8^{10}$	49			al, eth, lig	B2[4], 1051
4557	Capric acid, methyl ester or Methyl caprate $C_9H_{19}CO_2CH_3$	186.29		224, 114^{15}	−18	$0.8730^{20/4}$	1.4259^{20}	al, eth, chl	B2[4], 1044
4558	Capric acid, 2-octyl or 9-Heptadecane carboxylic acid $(C_8H_{17})_2CHCO_2H$	284.48	nd or lf (al)	$212\text{-}8^{13}$	38.5			al, eth	B2[4], 1254
4559	Capric acid, 4-oxo or γ-Ketocapric acid $C_6H_{13}COCH_2CH_2CO_2H$	186.25	(dil al)		70-1			al	B3[4], 1642
4560	Capric acid, propyl ester or Propyl caprate $C_9H_{19}CO_2C_3H_7$	214.35		128.5^{10}	0.8623^{20}	1.4280^{20}			B2[4], 1045
4561	Capric acid, isopropyl ester or Isoporpyl caprate $C_9H_{19}CO_2CH(CH_3)_2$	214.35		121^{10}		0.8543^{20}	1.4221^{25}		B2[4], 1045
4562	Caprinitrile $C_9H_{19}CN$	153.27		243, 106^{10}	fr−17.9	$0.8199^{20/4}$	1.4296^{20}	al, eth, ace, chl	B2[4], 1051

No.	Name, Synonyms, and Formula	Mol. wt.	Color, crystalline form, specific rotation and λ_{max} (log ε)	b.p. °C	m.p. °C	Density	n_D	Solubility	Ref.
4563	Capryl chloride $C_9H_{19}COCl$	190.71		232, 114^{15}	−34.5	0.973$^{8/4}$		eth	B2^4, 1050
4564	Caproaldehyde or Hexanal $CH_3(CH_2)_4CHO$	100.16		128, 28^{12}	−56	0.8139$^{20/4}$	1.4039^{20}	al, eth, ace, bz	B1^4, 3296
4565	Caproaldehyde oxime or Capraldoxime, Hexanaldoxime $CH_3(CH_2)_4CH{=}NOH$	115.18	cr (MeOH)		51				B1^2, 745
4566	Caproaldehyde, 2-ethyl $CH_3(CH_2)_3CH(C_2H_5)CHO$	128.21		163, 65^{15}	<−100	0.8540^{20}	1.4142^{20}	al, eth	B1^4, 3345
4567	Caproaldehyde, 3-methyl or 3-Methylhexanal $CH_3(CH_2)_2CH(CH_3)CH_2CHO$	114.19		142-3^{755}		0.8203$^{20/4}$	1.4122^{20}	al, eth	B1^2, 756
4568	Caproamide or Hexananoamide $CH_3(CH_2)_4CONH_2$	115.18	cr (ace)	255	101	0.999$^{20/4}$	1.4200^{110}	al, eth, bz, chl	B2^4, 929
4569	Caproamide, 2-ethyl $C_4H_9CH(C_2H_5)CONH_2$	143.23	nd (w)		102-3			w	B2^4, 1007
4570	Caproamide, *N*-phenyl or Capranilide $CH_3(CH_2)_4CONHC_6H_5$	191.27	nd (peth), pr (al)		95	1.112		al, eth	B12^3, 478
4571	Caproic acid or Hexanoic acid $CH_3(CH_2)_4CO_2H$	116.16		205	−2	0.9274$^{20/4}$	1.4163^{20}	al, eth	B2^4, 917
4572	Caproic acid, 2-acetyl, ethyl ester or Ethyl 2-acetyl caproate $C_4H_9CH(COCH_3)CO_2C_2H_5$	186.25		219-24, 104^{12}		0.9523$^{20/4}$	1.4301^{20}	eth, ace	B3^4, 1616
4573	Caproic acid, 2-amino $C_4H_9CH(NH_2)CO_2H$	131.17			297-300			w	B4^3, 1386
4574	Caproic acid, 6-amino $H_2NCH_2(CH_2)_4CO_2H$	131.17	lf (eth)	202-3				w	B4^3, 1393
4575	Caproic acid, 6-amino-ε-lactam or ε-Caprolactam $C_6H_{11}NO$	113.16	lf (lig)	139^{12}	69-71			w, al, bz, chl	B21^4, 3196
4576	Caproic acid, 6-amino-3-methyl, lactam-(*l*) $CH_2(CH_2)_2CH(CH_3)CH_2CO$ (ring closed through NH)	127.19	cr (bz-peth), $[\alpha]^{20/}{}_D$ −36.1		105-6			w, bz	B21, 243
4577	Caproic acid, 6-amino-5-methyl, lactam-(*l*) $CH_2CH(CH_3)(CH_2)_3CO$ (ring closed through NH)	127.19	cr (peth, bz-lig), $[\alpha]^{20/}{}_D$ −22.2		68-9			w, eth	B21, 242
4578	Caproic acid anhydride $(C_5H_{11}CO)_2O$	214.30		254-7d, 143^{15}	−41	0.9240$^{15/4}$	1.4297^{20}	al, eth	B2^4, 928
4579	Caproic acid, 6-benzoylamino $C_6H_5CONH(CH_2)_5CO_2H$	235.28	nd (al-eth)		79-80			AcOEt	B9^3, 1157
4580	Caproic acid, 6-benzoylamino-2-bromo-(*dl*) $C_6H_5CONH(CH_2)_4CHBrCO_2H$	314.18	cr (dil al)		166			al	B9^3, 1158
4581	Caproic acid, 6-benzoylamino-2-bromo-(*l*) $C_6H_5CONH(CH_2)_4CHBrCO_2H$	314.18	nd (dil al), $[\alpha]^{18/D}$ −29.2 (al)		129			al	B9^3, 1157
4582	Caproic acid, 2-bromo-(*dl*) $CH_3(CH_2)_3CHBrCO_2H$	195.06		240, 140-2^{23}	4			al, eth	B2^4, 938
4583	Caproic acid, 2-bromo-(*l*) $CH_3(CH_2)_3CHBrCO_2H$	195.06	$[\alpha]^{20/D}$ −27 (eth, C=5)	129^{14}				al, eth	B2^3, 736
4584	Caproic acid, 2-bromo, ethyl ester-(*dl*) $CH_3(CH_2)_3CHBrCO_2C_2H_5$	223.12		205-10, 95-6^9				al	B2^4, 938
4585	Caproic acid, 3-bromo $C_3H_7CHBrCH_2CO_2H$	195.06	nd (dil al)		35			al, bz, chl, lig	B2^4, 939
4586	Caproic acid, 6-bromo $Br(CH_2)_5CO_2H$	195.06	cr (peth)	165-70^{20}	35			peth	B2^4, 940
4587	Caproic acid, butyl ester or Butyl caproate $C_5H_{11}CO_2C_4H_9$	172.27		208	−64.3	0.8653$^{20/4}$	1.4152^{20}	al, eth	B2^4, 922
4588	Caproic acid, 6-cyclohexyl $C_6H_{11}(CH_2)_5CO_2H$	198.31		180^{11}	33-5	0.9626$^{20/4}$	1.4750^{20}	eth	B9^3, 112
4589	Caproic acid, ethyl ester or Ethyl caproate $C_5H_{11}CO_2C_2H_5$	144.21		168	−67	0.8710$^{20/4}$	1.4073^{20}	al, eth	B2^4, 921
4590	Caproic acid, 2-ethyl $C_4H_9CH(C_2H_5)CO_2H$	144.21		228^{755}, 120^{13}		0.9031$^{25/4}$	1.4241^{20}	eth	B2^4, 1003
4591	Caproic acid, 6-fluoro $F(CH_2)_5CO_2H$	134.15		138^{28}, 67-8$^{0.6}$			1.4166^{25}	al, eth	B2^4, 932
4592	Caproic acid, heptyl ester $C_5H_{11}CO_2C_7H_{15}$	214.35		261	−34.4	0.8611^{20}	1.4293^{15}	al, eth, ace, bz	B2^4, 923

No.	Name, Synonyms, and Formula	Mol. wt.	Color, crystalline form, specific rotation and λ_{max} (log ε)	b.p. °C	m.p. °C	Density	n_D	Solubility	Ref.
4593	Caproic acid, hexyl ester or Hexyl caproate $C_5H_{11}CO_2C_6H_{13}$	200.32		246	−55	0.865^{18}	1.4264^{15}	al, eth, ace, bz	$B2^4$, 922
4594	Caproic acid, 2-hydroxy-*(d)* $C_4H_9CH(OH)CO_2H$	132.16	$[\alpha]^{20/D}$ +0.7 (w, C=14)		60			w, al, eth	$B3^4$, 838
4595	Caproic acid, 2-hydroxy-*(dl)* $CH_3(CH_2)_3CH(OH)CO_2H$	132.16	pr (eth-al, peth)		60-1			w, al, eth, chl	$B3^4$, 838
4596	Caproic acid, 2-hydroxy-*(l)* $C_4H_9CH(OH)CO_2H$	132.16	pr (eth), $[\alpha]^{20/D}$ −3.8 (w, C=4.5)		60-1			w, al, eth, chl	$B3^4$, 838
4597	Caproic acid, 4-hydroxy, lactone or γ-Caprolactone $CH_3CH_2CHCH_2CH_2CO$ (—O— ring)	114.14		215-6, 103^{14}	−18		1.4495^{20}	w, al	$B17^4$, 4194
4598	Caproic acid, 6-hydroxy-ε-lactone $C_6H_{10}O_2$	114.14		108^{10}	−1.3	$1.0693^{20/4}$	1.4611^{20}	al, eth, ace	$B17^4$, 4186
4599	Caproic acid, methyl ester or Methyl caproate $C_5H_{11}CO_2CH_3$	130.19		151, 52^{15}	−71	$0.8846^{20/4}$	1.4049^{20}	al, eth ace, bz	$B2^4$, 921
4600	Caproic acid, 2-methyl-(d) or 2-methyl hexanoic acid $C_4H_9CH(CH_3)CO_2H$	130.19	$[\alpha]^{22/D}$ + 19.6 (eth)	105^5		0.909^{25}	1.4189^{20}	al, eth, ace, bz	$B2^3$, 773
4601	Caproic acid, 2-methyl-*dl* or 2-Methylhexanoic acid $C_4H_9CH(CH_3)CO_2H$	130.19		215-6, 100^{11}		$0.9612^{20/4}$	1.4195^{20}	al, eth, ace, bz	$B2^4$, 969
4602	Caproic acid, 2-methyl-*(l)* or 2-Methylhexanoic acid $C_4H_9CH(CH_3)CO_2H$	130.19	$[\alpha]^{25/}{}_D$ -4.3 (w, c=26)	121^{20}		$0.909^{25/4}$	1.4189^{25}	al, eth, ace, bz	$B2^3$, 773
4603	Caproic acid, 4-methyl-*(d)* or 4-Methylhexanoic acid $C_2H_5CH(CH_3)CH_2CH_2CO_2H$	130.19	$[\alpha]^{20/}{}_D$ + 7.6 (MeOH)	221, 115^{16}		$0.9228^{20/4}$	1.4198^{20}	al, ace, bz	$B2^4$, 973
4604	Caproic acid, 4-methyl-*(dl)* or 4-Methylhexanoic acid $C_2H_5CH(CH_3)(CH_2)_2CO_2(CH_2)_2CO_2H$	130.19		217-8, 85^2	-80	$0.9215^{20/4}$	1.4211^{20}	al, eth, ace, bz	$B2^4$, 974
4605	Caproic acid, 5-methyl or 5-Methylhexanoic acid $(CH_3)_2CH(CH_2)_3CO_2H$	130.19		216, 109^{16}	<-25	$0.9138^{21/4}$	1.4220^{20}	al, eth, ace, bz	$B2^4$, 970
4606	Caproic acid, octyl ester or Octyl caproate $C_5H_{11}CO_2C_8H_{17}$	228.38		275	-28	0.8603^{20}	1.4326^{15}	al, eth, ace, bz	$B2^3$, 727
4607	Caproic acid, 4-oxo or Homolevulinic acid $C_2H_5COCH_2CH_2CO_2H$	130.14	hyg ta or lf (eth-peth)	183^{20}, $89^{0.4}$	41-2			w, al, eth	$B3^4$, 1581
4608	Caproic acid, pentyl ester or Pentyl caproate $C_5H_{11}CO_2C_5H_{11}$	186.29		226, 116.6^{20}	-47	$0.8612^{25/4}$	1.4202^{25}	al, eth, ace	$B2^4$, 922
4609	Caproic acid, isopentyl ester or Isopentyl caproate $C_5H_{11}CO_2CH_2CH_2CH(CH_3)_2$	186.29		224-7		$0.861^{20/4}$		al, eth	$B2^3$, 727
4610	Caproic acid, propyl ester or Propyl caproate $C_5H_{11}CO_2C_3H_7$	158.24		187	-68.7	$0.8672^{20/4}$	1.4170^{20}	al, eth	$B2^4$, 922
4611	Caprolactam or Hexanoic acid-6-amino-ε-lactam $CH_2(CH_2)_4CO$ (—NH— ring)	113.16	lf (lig)	139^{12}	69-71			w, al, bz, chl	$B21^4$, 3196
4612	Caprolactam, 6-amino-3-methyl $C_7H_{13}NO$	127.19	cr (bz-peth), $[\alpha]^{20/}{}_D$ - 36.1		105-6			w, bz	B21, 243
4613	Caprolactam, 6-amino-5-methyl $C_7H_{13}NO$	127.19	cr (peth, bz lig), $[\alpha]^{20/}{}_D$ -22.2		68-9			w, eth	B21, 242
4614	2-Caprolactone $C_6H_{10}O_2$	114.14		215-6, 103^{14}	-18		1.4495^{20}	w, al	$B17^4$, 4194
4615	Capronitrile $CH_3(CH_2)_4CN$	97.16		163.6, 47.3^{10}	-80.3	$0.8051^{20/4}$	1.4068^{20}	al, eth, chl	$B2^4$, 930
4616	Caprophenone or Pentyl phenyl ketone $C_5H_{11}COC_6H_5$	176.26	fl	265, 122-4^{15}	27	$0.9576^{20/4}$	1.5027^{25}	al, eth, ace	$B7^3$, 1151
4617	Caproyl chloride $C_5H_{11}COCl$	134.61		153	-87	$0.9754^{20/4}$	1.4264^{20}	eth, ace	$B2^4$, 928
4618	Caproyl chloride, 3-methyl $C_4H_9CH(CH_3)CH_2COCl$	148.63		163^{751}, 82^{50}		$0.967^{20/4}$	1.4293^{25}	bz	$B2^3$, 777
4619	Caproyl chloride, 4-methyl $C_2H_5CH(CH_3)(CH_2)_2COCl$	148.63		167^{767}		$0.9677^{20/4}$		eth	$B2^2$, 299
4620	Caproyl chloride, 5-methyl $(CH_3)_2CH(CH_2)_3COCl$	148.63		168^{739}, 76-82^{34}				eth	$B2^4$, 971
4621	Caprylaldehyde or Octanal $CH_3(CH_2)_6CHO$	128.21		171, 72^{20}		$0.8211^{20/4}$	1.4217^{20}	al, eth, ace, bz	$B1^4$, 3337
4622	Caprylaldehyde oxime or Caprylaldoxime $CH_3(CH_2)_6CH{=}NOH$	143.23	nd (peth, dil al)	112^9	60			al, ace	$B1^2$, 758

PHYSICAL CONSTANTS OF ORGANIC COMPOUNDS (Continued)

No.	Name, Synonyms, and Formula	Mol. wt.	Color, crystalline form, specific rotation and λ_{max} (log ε)	b.p. °C	m.p. °C	Density	n_D	Solubility	Ref.
4623	Caprylamide $C_7H_{15}CONH_2$	143.23	lf, pl	239	fr 106-10	0.8450^{110}	....	al, eth, ace	$B2^4$, 992
4624	Caprylic acid or Octanoic acid $CH_3(CH_2)_6CO_2H$	144.21	....	239.3, 140^{23}	16.5	0.9088^{20}	1.4285^{20}	al, chl	$B2^4$, 982
4625	Caprylic acid-2-amino-*(d)* $C_6H_{13}CH(NH_2)CO_2H$	159.23	$[\alpha]^{26}{}_D$ + 23.5 (6N HCl, c=1)	....	......	....	....	aa	$B4^2$, 886
4626	Caprylic acid, 2-amino-*(dl)* $C_6H_{13}CH(NH_2)CO_2H$	159.23	lf (w)	(sub, d)	270	....	....	aa	$B4^3$, 1472
4627	Caprylic acid, 2-amino-*(l)* $C_6H_{13}CH(NH_2)CO_2H$	159.23	$[\alpha]_D$ -23 (5N HCl)	....	276	....	....	aa	$B4^2$, 886
4628	Caprylic acid, 8-amino $H_2N(CH_2)_7CO_2H$	159.23	....	....	172	....	....	al	$B4^1$, 527
4629	Caprylic anhydride $(C_7H_{15}CO)_2O$	270.41	....	280-5, 186^{15}	-1	$0.9065^{18/4}$	1.4358^{18}	al, eth, ace	$B2^3$, 796
4630	Caprylic acid, 2-bromo $C_6H_{13}CHBrCO_2H$	223.11	....	140^5	......	1.2785^{24}	1.4613^{24}	....	$B2^4$, 1000
4631	Caprylic acid, butyl ester or Butyl caprylate $C_7H_{15}CO_2C_4H_9$	200.32	....	240.5, $121\text{-}2^{20}$	-42.9	0.8628^{20}	1.4232^{25}	al, eth, ace	$B2^4$, 987
4632	Caprylic acid, ethyl ester or Ethyl caprylate $C_7H_{15}CO_2C_2H_5$	172.27	....	208.5, 104^{80}	-43.1	$0.8693^{20/4}$	1.4178^{20}	al, eth	$B2^4$, 987
4633	Caprylic acid, 8-fluoro $F(CH_2)_7CO_2H$	162.21	....	$132\text{-}3^4$	35	....	....	al, eth	$B2^4$, 994
4634	Caprylic acid, heptyl ester or Heptyl caprylate $C_7H_{15}CO_2C_7H_{15}$	242.40	....	290.5, 160^{14}	-10.6	0.8596^{20}	1.4340^{20}	al, eth, ace	$B2^3$, 794
4635	Caprylic acid, hexyl ester or Hexyl caprylate $C_9H_{15}CO_2C_6H_{13}$	228.38	....	277.4	-30.6	0.8603^{20}	1.4323^{25}	al, eth, ace	$B2^3$, 794
4636	Caprylic acid, 2-hydroxy $C_6H_{13}CH(OH)CO_2H$	160.21	pl	$160\text{-}5^{10}$	70	....	....	al, eth	$B3^4$, 874
4637	Caprylic acid, 4-hydroxy, lactone or 2-Caprylolactone $C_4H_9\underbrace{CH-CH_2CH_2CO}_{O}$	142.20	....	$132\text{-}3^{20}$	......	$0.9796^{19/4}$	1.4451^{19}	al	$B17^4$, 4228
4638	Caprylic acid, methyl ester or Methyl caprylate $C_7H_{15}CO_2CH_3$	158.24	....	192.9, 83^{15}	-40	$0.8775^{20/4}$	1.4170^{20}	al, eth	$B2^4$, 986
4639	Caprylic acid, 2-methyl-3-oxo,ethyl ester $C_5H_{11}COCH(CH_3)CO_2C_2H_5$	200.28	....	$128\text{-}9^{12}$	......	$0.963^{0/4}$	....	al	B3, 713
4640	Caprylic acid, 2-methallyl ester $C_7H_{15}CO_2CH_2(CH_3C{=}CH_2$	198.31	....	147.8^{50}	......	0.8703	1.4308	al	$B2^3$, 795
4641	Caprylic acid, octyl ester or Octyl caprylate $C_7H_{15}CO_2C_8H_{17}$	256.43	....	306.8, 192.5^{30}	-18.1	$0.8554^{20/4}$	1.4352^{20}	al, eth, ace	$B2^4$, 988
4642	Caprylic acid, pentyl ester or Pentyl caprylate $C_7H_{15}CO_2C_5H_{11}$	214.35	....	260.2, $124\text{-}6^{20}$	-34.8	0.8613^{20}	1.4262^{25}	al, eth, ace	$B2^3$, 794
4643	Caprylic acid, perfluoro $CF_3(CF_2)_6CO_2H$	414.07	....	187-9	53	....	....	....	$B2^4$, 994
4644	Caprylic acid, perfluoro, methyl ester $C_7H_{15}CO_2CH_3$	428.10	....	158	......	$1.684^{20/4}$	1.304^{27}	....	$B2^4$, 995
4645	Caprylic acid, propyl ester or Propyl caprylate $C_7H_{15}CO_2C_3H_7$	186.29	(peth)	226.4, 112^{20}	-46.2	0.8659^{20}	1.4191^{25}	al, eth, ace	$B2^4$, 987
4646	Caprylic acid, ispropyl ester or Isopropyl caprylate $C_7H_{15}CO_2CH(CH_3)_2$	186.29	....	93.8^{10}	......	0.8555^{20}	1.4147^{25}	....	$B2^4$, 987
4647	Caprylonitrile $C_7H_{15}CN$	125.21	....	205.2, $77\text{-}8^{10}$	-45.6	$0.8136^{20/4}$	1.4203^{20}	eth	$B2^4$, 993
4648	Caprylyl chloride $C_7H_{15}COCl$	162.66	....	195.6, 89^{20}	-63	$0.9535^{15/4}$	1.4335^{20}	eth	$B2^4$, 992
4649	Capsaicin $C_{18}H_{27}NO_3$	305.42	mcl pr or sc (peth)	$210\text{-}20^{0.01}$	65	....	....	al, eth, bz, peth	$B13^3$, 2192
4650	Carbamic acid, benzyl ester $H_2NCO_2CH_2C_6H_5$	151.16	pl (to), lf (w)	220d	91	....	....	al	$B6^3$, 1485
4651	Carbamic acid, *N*-benzyl, ethyl ester $C_6H_5CH_2NHCO_2C_2H_5$	179.22	lf (lig)	230d	49	....	....	al, eth, bz, chl	$B12^3$, 2271
4652	Carbamic acid, N-nitro, benzyl ethylester $C_6H_5CH_2N(NO_2)CO_2C_2H_5$	224.22	ye	d	......	$1.213^{20/20}$	1.5203^{20}	**al, eth**	C55, 24616
4653	Carbamic acid, butyl ester $H_2NCO_2C_4H_9$	117.15	pr	204d	54	....	....	al	$B3^4$, 54
4654	Carbamic acid, *N*-butyl, butyl ester $C_4H_9NHCO_2C_4H_9$	173.26	....	88^3	......	$0.9238^{20/20}$	1.4359^{20}	**al, eth**	$B4^4$, 577

No.	Name, Synonyms, and Formula	Mol. wt.	Color, crystalline form, specific rotation and λ_{max} (log ε)	b.p. °C	m.p. °C	Density	n_D	Solubility	Ref.
4655	Carbamic acid, *N*-butyl-*N*-nitro, butyl ester $C_4H_9N(NO_2)CO_2C_4H_9$	218.26		98[1]		1.048[20/20]	1.4359[20]	**al, eth**	Am73, 5449
4656	Carbamic acid, isobutyl ester or Isobutyl carbamate $H_2NCO_2CH_2CH(CH_3)_2$	117.15	lf	207	67		1.4098[76]	al, eth	B3[4], 56
4657	Carbamic acid, *N*-isobutyl, ethyl ester or Isobutyl urethane $(CH_3)_2CHCH_2NHCO_2C_2H_5$	145.20		110[30]	>-65	0.9432[20/4]	1.4288[20]	al, eth	B4[4], 647
4658	Carbamic acid, *N*-*tert*-butyl-*N*-nitro, ethyl ester $(CH_3)_3CN(NO_2)CO_2C_2H_5$	190.20		56[2]d		1.051[20/20]	1.4331[20]	al, eth	Am83, 1191
4659	Carbamic acid, *N,N*-diethyl $(C_2H_5)_2NCO_2H$	117.15	nd (eth)	171	-15d	0.9276[20/4]	1.4206[20]	w, al, eth	B4[3], 222
4660	Carbamic acid, *N,N*-diphenyl, ethyl ester or Diphenyl urethane $(C_6H_5)_2NCO_2C_2H_5$	241.29	pr (lig)	360	72			w, eth, bz	B12[3], 888
4661	Carbamic acid, ethyl ester or Urethane $H_2NCO_2C_2H_5$	89.09	pr (bz, to)	185	48-50	0.9862[21/4]	1.4144[51]	al, w, eth, bz	B3[4], 40
4662	Carbamic acid, *N*-ethyl,butyl ester $C_2H_5NHCO_2C_4H_9$	145.20		66[3]		0.9413[20/20]	1.4301[20]	w, al, eth	C55, 2334
4663	Carbamic acid, *N*-ethyl, ethyl ester or Ethyl urethane $C_2H_5NHCO_2C_2H_5$	117.15		176, 75[14]		0.9813[20/4]	1.4215[20]	w, al, eth	B4[4], 365
4664	Carbamic acid, *N*-ethyl-*N*-nitro,butyl ester $C_2H_5N(NO_2)CO_2C_4H_9$	190.20		79[3]		1.091[20/20]	1.4455[20]	w, **al, eth**	Am73, 5043
4665	Carbamic acid, *N*-ethyl-*N*-nitro,ethyl ester $C_2H_5N(NO_2)CO_2C_2H_5$	162.15		107[31]		1.163[20/20]	1.4432[20]	**al, eth**	Am73, 5449
4666	Carbamic acid, *N*-ethyl-*N*-nitro,methyl ester $C_2H_5N(NO_2)CO_2CH_3$	148.12		72[11]		1.233[20/20]	1.4483[20]	**al, eth**	Am73, 5449
4667	Carbamic acid, *N*-ethylidine, diethyl ester or Ethylidine diurethane $CH_3CH(NHCO_2C_2H_5)_2$	204.23	nd (eth)	170-8[20]	126			w, al, ace, chl	B3[1], 11
4668	Carbamic acid, methyl ester or Urethylan $H_2NCO_2CH_3$	75.07	nd	177, 82[14]	54	1.1361[56/4]	1.4125[56]	w, al, eth	B3[4], 37
4669	Carbamic acid, 2-methyl-2-butyl ester or *tert*-Pentylcarbamate, Aponal $H_2NCO_2C(CH_3)_2CH_2CH_3$	131.17	nd (dil al)		85-7			ace, bz	B3[1], 14
4670	Carbamic acid, 3-methylbutyl ester or Isopentyl carbamate $H_2NCO_2CH_2CH_2CH(CH_3)_2$	131.17	nd (w)	220, 114-5[16]	64	0.9438[71/4]	1.4175[71]	al, eth	B3[4], 58
4671	Carbamic acid, *N*-methyl, ethyl ester or Methyl urethane $CH_3NHCO_2C_2H_5$	103.12		170, 80[15]		1.0115[20/4]	1.4183[20]	w, al	B4[4], 200
4672	Carbamic acid, *N*-nitro,ethyl ester $O_2NNHCO_2C_2H_5$	134.09	pl (eth, lig)	140d	64	1.0074[20/4]		w, al, ace, lig	B3[4], 247
4673	Carbamic acid, *N*-phenyl,ethyl ester or Ethylcarbanilate $C_6H_5NHCO_2C_2H_5$	165.19	wh nd (w), pl (dil al)	237d	53	1.1064[30/4]	1.5376[30]	al, eth, bz	B12[3], 612
4674	Carbamic acid, *N*-phenyl,isobutyl ester $C_6H_5NHCO_2CH_2CH(CH_3)_2$	193.25	nd (dil al)	216	86			al, eth, bz	B12[3], 614
4675	Carbamic acid, *N*-phenyl,propyl ester $C_6H_5NHCO_2C_3H_7$	179.22	wh nd (dil al)		57-9			al, eth, bz	B12[3], 613
4676	Carbamic acid, *N*-phenyl, *iso*propyl ester $C_6H_5NHCO_2CH(CH_3)_2$	179.22	wh nd (al)		90	1.09[20]	1.4989[91]	al, bz	B12[3], 613
4677	Carbamic acid, propyl ester $H_2NCO_2C_3H_7$	103.12	pr	196, 92[12]	60			w, al, eth, ace	B3[4], 52
4678	Carbamic acid, isopropyl ester $H_2NCO_2CH(CH_3)_2$	103.12	nd	181[711]	92-4	0.9951[66]			B3[4], 53
4679	Carbamic acid, *N*-propyl,ethyl ester $C_3H_7NHCO_2C_2H_5$	131.17		192[751], 92[22]		0.9921[15]		al	B4[4], 480
4680	Carbamic acid, *N*-propyl-*N*-nitro, ethyl ester $C_3H_7N(NO_2)CO_2C_2H_5$	176.17		66[3]		1.123[20/20]	1.4431[20]	**al, eth**	Am73, 5449
4681	Carbamic acid, *N*-propyl-*N*-nitro,methyl ester $C_3H_7N(NO_2)CO_2CH_3$	162.15				1.2585[15/15]		al, **w, eth**	B4, 146
4682	Carbamic acid, *N*-*iso*propyl, ethyl ester or Isopropylurethane $(CH_3)_2CHNHCO_2C_2H_5$	131.17		79[15]		0.9548[20/20]	1.4229[20]	**al, eth**	B4[4], 520
4683	Carbamic acid, *N*-isopropyl-*N*-nitro, ethyl ester $(CH_3)_2CHN(NO_2)CO_2C_2H_5$	176.17		72[7]		1.112[20/20]	1.4381[20]	**al, eth**	Am73, 5449
4684	Carbamic acid, thiolo, ethyl ester $H_2NCOSC_2H_5$	105.15	pl (w)	sub d	109			al, eth	B3[4], 294
4685	Carbamic acid, thiono, ethyl ester or Thiourethane $H_2NCSOC_2H_5$	105.15	nd lf or pyr	d	41	1.069[20/4]	1.520[20]	**al, eth**, chl	B3[4], 294

No.	Name, Synonyms, and Formula	Mol. wt.	Color, crystalline form, specific rotation and λ_{max} (log ε)	b.p. °C	m.p. °C	Density	n_D	Solubility	Ref.
4686	Carbamonitrile, *N*-ethyl-*N*-phenyl $C_6H_5N(CH_2CH_3)CN$	146.19		271,153^{19}					B12^{3}423
4687	Carbamyl chloride H_2NCOCl	79.49		62d					B3^{3}, 65
4688	Carbamyl chloride, *N,N*-diethyl $(C_2H_5)_2NCOCl$	135.59		186					B4^{4}, 379
4689	Carbamyl chloride, *N,N*-diphenyl $(C_6H_5)_2NCOCl$	231.68	lf (al)		85				B12^{3}, 893
4690	Carbamyl chloride, *N*-methyl-*N*-phenyl $C_6H_5N(CH_3)COCl$	169.61	pl (al)	280	88-9			al, eth	B12^{3}, 874
4691	Carbazic acid, methyl ester $H_2NNHCO_2CH_3$	90.08		108^{12}	73			w, al, bz	B3^{2}, 78
4692	Carbazide $H_2NNHCONHNH_2$	90.08	nd (dil al)		154	1.616^{20}		w, al	B3^{4}, 240
4693	Carbazide, 1,5-diphenyl $(C_6H_5NHNH)_2CO$	242.28	cr (al + 1), cr (aa)	d	170			bz, aa	B15^{3}, 187
4694	Carbazide, 1-phenyl $C_6H_5NHNHCONHNH_2$	166.18	nd (al)		151			w	B15^{3}, 187
4695	Carbazide, 1,1,5,5-tetraphenyl $[(C_6H_5)_2NNH]_2CO$	394.48	(al), nd (aa)		242			al, aa	B15^{2}, 115
4696	Carbazide, 3-thio $(H_2NNH)_2CS$	106.15	nd, pl (w)		170d			w	B3^{3}, 319
4697	Carbazole or Dibenzopyrrole $C_{12}H_9N$	167.21	pl or lf	355, 200^{147}	247-8			ace	B20^{4}, 3824
4698	Carbazole, 9-acetyl $C_{14}H_{11}NO$	209.25	(eth), nd (w)	190^{6}	69	1.161$^{100/24}$	1.640^{100}		B20^{4}, 3836
4699	Carbazole, 9-benzoyl $C_{19}H_{13}NO$	271.32	nd or pr (al)		98.5			al, bz	B20^{4}, 3838
4700	Carbazole, 9-benzyl $C_{19}H_{15}N$	257.33	nd (al)	267-8^{24}	118-20			bz	B20^{4}, 3831
4701	Carbazole, 9-butyl $C_{16}H_{17}N$	223.32	nd (al)	218-9^{19}	58			eth	B20^{4}, 3829
4702	Carbazole, 9-ethenyl $C_{14}H_{11}N$	193.25	cr (al)		66			eth	B20^{4}, 3830
4703	Carbazole, 9-ethyl $C_{14}H_{13}N$	195.27	nd (al)	190^{10}	68	1.059$^{80/4}$	1.6394^{80}	al, eth	B20^{4}, 3829
4704	Carbazole, 9-methyl $C_{13}H_{11}N$	181.24	nd, lf (al)	195^{12}	88			eth	B20^{4}, 3828
4705	Carbazole, 1-nitro $C_{12}H_8N_2O_2$	212.21	ye nd (aa)		187			aa	B20^{4}, 3863
4706	Carbazole, 3-nitro $C_{12}H_8N_2O_2$	212.21	ye		214				B20^{4}, 3865
4707	Carbazole, 3-nitro-9-nitroso $C_{12}H_7N_3O_3$	241.21	ye nd (al)		169			chl	B20^{4}, 3868
4708	Carbazole, 1-oxo-1,2,3,4-tetrahydro $C_{12}H_{11}NO$	185.23	nd (dil al)		170			al, bz, aa	B21^{4}, 4075
4709	Carbazole, 9-phenyl $C_{18}H_{13}N$	243.31	nd or pl (al)		95			al, eth, bz, aa	B20^{4}, 3830
4710	Carbazole, 9-propionyl $C_{15}H_{13}NO$	223.27			90			al, eth	B20^{4}, 3837
4711	Carbazole, 9-propyl $C_{15}H_{15}N$	209.29	nd (al)		50			eth	B20^{4}, 3829
4712	Carbazole, 1,2,3,4-tetrahydro $C_{12}H_{13}N$	171.24	lf (dil al)	325-30, 190^{10}	120			al, eth, bz	B20^{4}, 3566
4713	Carbazone, 1,5-diphenyl $C_6H_5N{=}NCONHNHC_6H_5$	240.26	og nd (bz), pr (al)		157d			al, bz, chl	B16^{3}, 18
4714	Carbodiimide, diphenyl $C_6H_5N{=}C{=}NC_6H_5$	194.24		331, 218^{21}	168-70			bz	B12^{3}, 906
4715	Carbon dioxide CO_2	44.01		−78.6 sub	−56.6 (5.2 atm)	1.0310^{-20}			B3, 4
4716	Carbon diselenide CSe_2	169.93	ye	126, 46^{50}	−45.5	2.6824$^{20/4}$	1.8454^{20}		B3^{4}, 436
4717	Carbon disulfide CS_2	76.13		46.2	−111.5	1.2632$^{20/4}$	1.6319^{20}	al, eth, chl	B3^{4}, 395
4718	Carbonic acid, bis(2-chloroethyl) ester or bis-(2-Chloroethyl) carbonate $(ClCH_2CH_2O)_2CO$	187.02		241	8	1.3506$^{20/4}$	1.461^{20}		B3^{4}, 6

PHYSICAL CONSTANTS OF ORGANIC COMPOUNDS (Continued)

No.	Name, Synonyms, and Formula	Mol. wt.	Color, crystalline form, specific rotation and λ_{max} (log ε)	b.p. °C	m.p. °C	Density	n_D	Solubility	Ref.
4719	Carbonic acid, bis (3-chloropropyl) ester $(ClCH_2CH_2CH_2O)_2CO$	215.08	...	$265\text{-}70^{740}$	...	...	...	...	$B3^3$, 8
4720	Carbonic acid, bis(2-ethoxyethyl) ester $(C_2H_5OCH_2CH_2O)_2CO$	206.24	...	$245\text{-}6^{758}$, $112\text{-}3^{5}$	...	$1.0439^{20/4}$	1.4227^{20}	al, eth, ace	$B3^3$, 17
4721	Carbonic acid, bis(2 methoxyethyl) ester $(CH_3OCH_2CH_2O)_2CO$	178.19	...	230-2, $99\text{-}100^{5}$	...	$1.0988^{20/4}$	1.4204^{20}	al, eth, ace	$B3^3$, 17
4722	Carbonic acid, bis(2-methoxyphenyl) ester or Duotal $(2\text{-}CH_3OC_6H_4O)_2CO$	274.28	cr (al)	...	89	...	...	eth, chl	$B6^3$, 4233
4723	Carbonic acid, bis(trichloromethyl) ester $[Cl_3CO]_2CO$	296.75	cr (eth, peth)	203d	79	...	...	...	$B3^4$, 33
4724	Carbonic acid, dibutyl ester or Dibutyl carbonate $(C_4H_9O)_2CO$	174.24	...	207, $96\text{-}7^{16}$	...	$0.9251^{20/4}$	1.4117^{20}	al, eth	$B3^4$, 8
4725	Carbonic acid, diisabutyl ester $(i\text{-}C_4H_9O)_2CO$	174.24	...	190, 85^{16}	...	$0.9138^{20/4}$	1.4072^{20}	al, eth	$B3^4$, 9
4726	Carbonic acid, di-*tert*-butyl ester or di-*tert*-Butyl carbonate $[(CH_3)_3CO]_2CO$	174.24	cr (al)	158^{767}	40 (sub)	...	...	al	$B3^4$, 9
4727	Carbonic acid, diethyl ester or Diethyl carbonate $(C_2H_5O)_2CO$	118.13	...	126	-43	$0.9752^{20/4}$	1.3845^{20}	al, eth	$B3^4$, 5
4728	Carbonic acid, dimethyl ester or Dimethyl carbonate $(CH_3O)_2CO$	90.08	...	90-1	2-4	$1.0694^{20/4}$	1.3687^{20}	...	$B3^4$, 3
4729	Carbonic acid, diphenyl ester or Diphenyl carbonate $(C_6H_5O)_2CO$	214.22	nd (al, bz)	306, 168^{15}	83 (88)	$1.1215^{87/4}$	...	eth	$B6^4$, 629
4730	Carbonic acid, di-isopentyl ester or Diisopentyl carbonate $(i\text{-}C_5H_{11}O)_2CO$	202.29	...	$232\text{-}5^{751}$, 122^{16}	...	$0.9067^{20/4}$	1.4174^{20}	...	$B3^3$, 11
4731	Carbonic acid, di-propyl ester or Dipropyl carbonate $(C_3H_7O)_2CO$	146.19	...	168, 59.5^{15}	...	$0.9435^{20/4}$	1.4008^{20}	al, eth	$B3^4$, 6
4732	Carbonic acid, Diisopropyl ester $(i\text{-}C_3H_7O)_2CO$	146.19	...	147, 43^{12}	...	$0.9162^{20/4}$	1.3932^{20}	al	$B3^3$, 9
4733	Carbonic acid, dithiolo, diethyl ester $(C_2H_5S)_2CO$	150.25	ye	197, $85\text{-}7^{19}$	...	1.085^{20}	1.5237^{18}	al, eth	$B3^3$, 339
4734	Carbonic acid, di-2-tolyl ester $(2\text{-}CH_3C_6H_4O)_2CO$	242.27	nd (al)	$144\text{-}5^{0.5}$	60	...	...	aa	$B6^3$, 1256
4735	Carbonic acid, di-3-tolyl ester $(3\text{-}CH_3C_6H_4O)_2CO$	242.27	cr (al)	...	50-1	...	...	bz, chl	$B6^3$, 1307
4736	Carbonic acid, di-4-tolyl ester $(4\text{-}CH_3C_6H_4O)_2CO$	242.27	...	...	115	...	...	chl	$B6^3$, 1366
4737	Carbonic acid, ethyl-2-butoxyethyl ester $C_2H_5O\text{-}CO\text{-}OCH_2CH_2OC_4H_9$	190.24	...	224^{759}	...	$0.9756^{25/4}$	1.4143^{25}	al, eth, ace	$B3^3$, 14
4738	Carbonic acid, ethyl-methyl ester $CH_3O\text{-}CO\text{-}OC_2H_5$	104.11	...	107-8	−14	$1.012^{20/4}$	1.3778^{20}	al, eth	$B3^4$, 4
4739	Carbonic acid, trithio $(HS)_2CS$	110.21	red	57d	−30	$1.47^{17/4}$	...	al, chl, to	$B3^4$, 428
4740	Carbon suboxide or 1,3-Dioxoallene $OC{=}C{=}CO$	68.03	gas	6.8	−107	$1.114^{0/4}$	1.4538^{0}	eth, bz	$B1^4$, 3764
4741	Carbonyl fluoride COF_2	66.01	...	−83	−114	1.139^{-114}	...	...	$B3^4$, 21
4742	Carbonyl sulfide COS	60.08	...	−50	−138	$1.028^{17/4}$	...	al	$B3^4$, 271
4743	Carbothialdine $C_5H_{10}N_2S_2$	162.27	cr (al)	...	120d	...	...	al	$B27^2$, 687
4744	Δ^3-Carene-*(dl)* $C_{10}H_{16}$	136.24	...	167^{732}, $44\text{-}5^{8}$	...	$0.8602^{20/4}$	1.4759^{20}	eth, ace, bz, aa	$B5^2$, 362
4745	Δ^3-Carene-*(l)* $C_{10}H_{16}$	136.24	$[\alpha]^{20}_{D}$ −5.72	$168\text{-}9^{705}$, $123\text{-}4^{200}$	...	$0.8586^{30/30}$	1.4684^{30}	eth, ace, bz, aa	$B5^4$, 449
4746	Δ^4-Carene-*(l)* $C_{10}H_{16}$	136.24	$[\alpha]^{20}_{D}$ + 62.2	167^{707}, 64^{20}	...	$0.8441^{30/4}$	1.4740^{30}	eth, ace, bz, aa	$B5^4$, 451
4747	Carminic acid $C_{22}H_{20}O_{13}$	492.39	red mcl pr (ag MeOH)	...	d 136	...	...	al	$B10^3$, 4874
4748	Carnaubyl alcohol $CH_3(CH_2)_{20}OH$	438.82	lf (al)	...	69	...	...	al, ace	$B1^2$, 472
4749	Carnosine-*(D,-)* or Ignotine. β-Alanyl-*(D,-)*-histidine $C_9H_{14}N_4O_3$	226.24	$[\alpha]^{18}_{D}$ −20.4 (w, c=1.5)	...	260	...	...	w	$B25^2$, 408

No.	Name, Synonyms, and Formula	Mol. wt.	Color, crystalline form, specific rotation and λ_{max} (log ε)	b.p. °C	m.p. °C	Density	n_D	Solubility	Ref.
4750	Carnosine-*(L+)* $C_9H_{14}N_4O_3$	226.24	nd (w-al), $[\alpha]^{20}_D$ +24.1 (w, c=1.5)		246-50d			w	B25[2], 408
4751	α-Carotene or α-Carotin $C_{40}H_{56}$	536.88	red pl or pr (peth, bz-MeOH)		187.5	$1.00^{20/20}$		eth, bz, chl	B5[3], 2457
4752	β-Carotene or Provitamin A $C_{40}H_{56}$	536.88	red br hex pr (bz-MeOH)		184	$1.00^{20/20}$		eth, ace, bz, peth	B5[3], 2453
4753	γ-Carotene or γ-Carotin $C_{40}H_{56}$	536.88	red br (bz-MeOH) vt pr (bz-eth)		178			bz, chl	B5[3], 2451
4754	Carpaine-*(d)* $C_{28}H_{50}N_2O_4$	478.72	mcl pr (al or ace) $[\alpha]^{21}_D$ +24.7 (al, c=1.07)	sub $120^{0.01}$	121			al, eth, ace, bz, chl	B27[2], 209
4755	Carpaine hydrochloride $C_{28}H_{50}N_2O_4$-HCl	515.18	wh mcl nd or pl		225d			w, al, eth	B27[2], 210
4756	Carpiline or Carpidine, Pilosine $C_{16}H_{18}N_2O_3$	286.33	pl (al), pr (dil al or w), $[\alpha]^{20}_D$ +35.9 (al)		187			al	B27[1], 612
4757	Carvacrol or 2-Methyl-5-*iso*-propyl phenol 2-CH_3-5-$(CH_3)_2CHC_6H_3OH$	150.22	nd	237.7, $101\text{-}2^{10}$	1	$0.9772^{20/4}$	1.5230^{20}	al, eth, ace	B6[3], 1885
4758	Carvacrol-acetate or 2-Methyl-5-*iso*-propyl phenyl acetate 2-CH_3-5-$(CH_3)_2CHC_6H_3O_2CCH_3$	192.24		245-8		0.9896^{25}	1.4913^{28}	al, eth	B6[2], 494
4759	Carvacrol, 4-amino or 2-Methyl-5-*iso*-propyl-4-aminophenol 2-CH_3-4-H_2N-5-$CH(CH_3)_2$-C_6H_2OH	165.24	cr (MeOH)		134			MeOH	B13[3], 1801
4760	Carvenone-*(dl)* $C_{10}H_{16}O$	152.24		$235\text{-}6^{762}$, 104^{10}		$0.9263^{20/4}$	1.4826^{20}	ace	B7[3], 332
4761	Carvenone-*(l)* $C_{10}H_{16}O$	152.24	$[\alpha]_D$ −2.08	232-4		$0.9290^{20/4}$	1.4805	ace	B7[1], 66
4762	Carveol, dihydro-*(d)* $C_{10}H_{18}O$	154.25	$[\alpha]^{18}_D$ + 34.2	225, 107^{15}		$0.9274^{20/4}$	1.4780^{20}	eth	B6[3], 256
4763	Carveol, dihydro-*(l)* $C_{10}H_{18}O$	154.25	$[\alpha]^{20}_D$ −33.3	107^{14}		0.9368^{15}	1.4836^{20}	ace	B6[3], 256
4764	β-Carveol, dihydro $C_{10}H_{18}O$	154.25	$[\alpha]_D$ + 7.64	130^{20}		$0.9266^{20/4}$	1.4809^{20}		B6, 64
4765	Carvomenthane-*(d)* $C_{10}H_{18}$	138.25	$[\alpha]_{5780}$ + 118	175-7, 77^{24}		$0.8246^{18/4}$	1.4563^{18}	al, bz, peth	B5[4], 301
4766	Carvomenthol-*(d)* or Hexahydrocarvacrol $C_{10}H_{20}O$	156.27	$[\alpha]^{21}_D$31.4	222, 102^{14}		$0.8995^{20/4}$	1.4617^{20}	al, eth	B6[4], 148
4767	Carvomenthol-*(l-neo)* $C_{10}H_{20}O$	156.27	$[\alpha]^{21}_D$ −41.7	217-8, 102^{18}		$0.9012^{20/4}$	1.4632^{20}	al, eth	B6[4], 148
4768	Carvomenthone $C_{10}H_{18}O$	154.25	$[\alpha]^{21}_D$ + 17.15	$218\text{-}20^{745}$, $95\text{-}9^{15}$		0.9075^{15}	1.4544^{20}	al, ace, chl	B7[3], 146
4769	Carvone-*(d)* or Carvol $C_{10}H_{14}O$	150.22	$[\alpha]^{20}_D$ + 69.1	231, 104^{11}		$0.9608^{20/4}$	1.4999^{18}	al, eth, chl	B7[3], 561
4770	Carvone-*(dl)* $C_{10}H_{14}O$	150.22		231, 85^5		$0.9645^{15/15}$	1.5003^{20}	al, eth, chl	B7[3], 564
4771	Carvone-*(l)* $C_{10}H_{14}O$	150.22	$[\alpha]^{20}_D$ −62.46	231, 98^9		$0.9593^{20/4}$	1.4988^{20}	al, eth, chl	B7[3], 561
4772	Carvone oxime-*(α,d)* or *D*-Carotime-*(α)* $C_{10}H_{15}NO$	165.24	lf (al) $[\alpha]^{17}_D$ + 39.71 (al, p=8.45)		72			al, eth, ace	B7[3], 564
4773	Carvone oxime-*(l,α)* or *L*-Carvoximine-(α) $C_{10}H_{15}NO$	165.24	mcl (dil al) $[\alpha]^{18}_D$ −39.43 (al, p=4.33)		73.5	1.0140^{73}		al, eth, ace	B7[3], 564

PHYSICAL CONSTANTS OF ORGANIC COMPOUNDS (Continued)

No.	Name, Synonyms, and Formula	Mol. wt.	Color, crystalline form, specific rotation and λ_{max} (log ε)	b.p. °C	m.p. °C	Density	n_D	Solubility	Ref.
4776	Carvone oxime-*(dl)* or *dl*-Carvoxime $C_{10}H_{15}NO$	165.24			93-4			al, eth, ace	$B7^3$, 564
4777	Carvone, dihydro-*(d)* $C_{10}H_{16}O$	152.24	$[\alpha]_D$ +17.5	221-2		0.928^{19}	1.4724	eth, ace	$B7^3$, 337
4778	Carvone, dihydro-*(l)* $C_{10}H_{16}O$	152.24	$[\alpha]^{20}/_D$−19	221-2, 104^{18}		$0.9253^{20/4}$	1.4717^{20}	eth, ace	$B7^3$, 337
4779	α-Caryophyllene $C_{15}H_{24}$	204.36	$[\alpha]^{20/4}$ +1±0.3 (chl, c=9.26)	123^{10}		$0.8905^{20/4}$	1.5038^{20}		$B5^4$, 1171
4780	β-Caryophyllene $C_{15}H_{24}$	204.36	$[\alpha]^{20}/_D$ −9.08	$122^{13.5}$		$0.9075^{20/4}$	1.4988^{20}	bz	$B5^4$, 1182
4781	γ-Caryophyllene or Isocaryophyllene $C_{15}H_{24}$	204.36	$[\alpha]^{19}/_D$ −26.2	130-1^{24}		$0.8953^{20/4}$	1.4967^{19}	bz	$B5^3$, 1083
4782	Caryophyllenic acid-*(l,cis)* $C_9H_{14}O_4$	186.21	pr (w) $[\alpha]_{546}$−7.4		77-8			w, al, eth, ace, bz	$B9^3$, 3850
4783	Caryophyllenic acid-*(d,trans)* $C_9H_{14}O_4$	186.21	nd $[\alpha]^{20}/_D$+35.3 (bz)		81-2			w, al, eth, ace, bz	$B9^3$, 3850
4784	Caryophyllin or Oleanolic acid $C_{30}H_{48}O_3$	456.71	nd or pr (al) $[\alpha]^{20}/_D$ +83.3 (chl, c=0.9)	280-308, sub (vac)	310d			aa, Py	$B10^3$, 1049
4785	Catechin-*(cis'd)* or 3,5,7,3′,4′-Flavanpentol $C_{15}H_{14}O_6$	290.27	nd (w+4) $[\alpha]^{18}/_D$ +18.4 (w, c=0.9)	240-5	96 (hyd) 177 (anh)	1.344^{44}		al, ace, aa	$B17^4$, 3841
4786	Catechin-*(cis,dl)* $C_{15}H_{14}O_6$	290.27	nd (w+3)		212-4d			al, ace	$B17^4$, 3842
4787	Catechin-*(cis,l)* $C_{15}H_{14}O_6$	290.27	nd (+4, w), $[\alpha]_{5780}$ (w ace, p=3)		96 (hyd) 177 (anh)			al, ace, aa	$B17^2$, 255
4788	Catechol or 1,2-Dihydroxybenzene 2-HOC_6H_4OH	110.11	cr	245^{750}	105	1.1493^{21}	1.604	w, al, eth, ace	$B6^3$, 4187
4789	Catechol, 3-acetyl or 2,3-Dihydroxyacetophenone 2, 3-$(HO)_2C_6H_3COCH_3$	152.15	ye pr (bz-lig)		97-8			al	$B8^3$, 2080
4790	Catechol, 4-acetyl or 3,4-Dihydroxyacetophenone 3,4$(HO)_2C_6H_3COCH_3$	152.15	nd (w or chl)		115-6				$B8^3$, 2108
4791	Catechol, 4-bromo or 4-Bromo-1,2-dihydroxybenzene 1,2-$(HO)_2$-4-Br-C_6H_3	189.01	pr or nd (chl)		87			w, al, eth, bz	$B6^3$, 4253
4792	Catechol, 3-chloro or 3-Chloro-1,2-dihydroxybenzene 3-Cl-1,2-$(HO)_2C_6H_3$	144.56	cr (lig)	110-1^{11}	46-8			lig	$B6^3$, 4249
4793	Catechol, 4-chloro 4-Chloro-1,2-$(HO)_2C_6H_3$	144.56	lf (bz-peth)	$139^{10.5}$	90-1			w, al, eth, ace, aa	$B6^3$, 4249
4794	Catechol, diacetate 2-CH_3CO_2-C_6H_4-O_2CCH_3	194.19	nd (al)	142-3^9	64-5			al, eth, chl	$B6^3$, 4228
4795	Catechol, dibenzoate 1,2-$(C_6H_5CO_2)C_6H_4$	318.33	lf (eth-al)		86			al, eth, bz	$B9^3$, 552
4796	Catechol, dibenzyl ether 1,2-$(C_6H_5CH_2O)_2C_6H_4$	290.37	yesh nd or pr (al)		63-4			eth, peth	$B6^3$, 4219
4797	Catechol, dibutyl ether or 1,2-Dibutoxybenzene 1,2-$(C_4H_9O)_2C_6H_4$	222.23	ye	241^{765}, 135-8^{12}					$B6^3$, 4210
4798	Catechol, 3,5-dichloro 3,5-$Cl_2C_6H_2(OH)_2$-1,2	179.00	pr		83-4			al, ace	B6, 783
4799	Catechol, 4,5-dichloro 4,5-$Cl_2C_6H_2(OH)_2$-1,2	179.00	pr (chl-CS_2), nd (bz-peth)		116-7			w, al, bz	$B6^3$, 4252
4800	Catechol, diethyl ether or 1,2-Diethoxybenzene 1,2-$(C_2H_5O)_2C_6H_4$	166.22	pr (peth, dil al)	219	43-5	$1.0075^{20/4}$	1.5083^{25}	al, eth	$B6^3$, 4208
4801	Catechol, dimethylether or Veratrole-1-2 dimethoxy benzene 1,2-$(CH_3O)_2C_6H_4$	138.17	cr (lig)	206^{750}, 90^{10}	22.5	$1.0842^{25/25}$	$1.5827^{21}/_4$	al, eth	$B6^3$, 4205
4802	Catechol, 3,5-dimethyl 3,5-$(CH_3)_2C_6H_2(OH)_2$-1,2	138.17	pr (w), nd (peth-bz)		73-4			w, al, eth	$B6^3$, 4591
4803	Catechol, 4,5-dimethyl 4,5-$(CH_3)_2C_6H_2(OH)_2$-1,2	138.17	mcl pr or nd (peth)	sub	87-8			w, al, eth	$B6^3$, 4584

No.	Name, Synonyms, and Formula	Mol. wt.	Color, crystalline form, specific rotation and λ_{max} (log ε)	b.p. °C	m.p. °C	Density	n_D	Solubility	Ref.
4804	Catechol, dipropylether or 1,2-Dipropoxy benzene $1,2\text{-}(C_3H_7O)_2C_6H_4$	194.27		234-7, 117-20^{12}		0.9554$^{33/4}$	1.4950^{27}		B6^3, 4209
4805	Catechol, dithio or 1,2-Dimercaptobenzene $1,2\text{-}(HS)_2C_6H_4$	142.25		238-9, 120^{17}	28-9			al, eth, bz	B6^3, 4286
4806	Catechol, 4-iodo $4\text{-}IC_6H_3(OH)_2\text{-}1,2$	236.01	lf (CCl_4)	sub	92			al, eth, ace, bz	B6^3, 4262
4807	Catechol, 3-methoxy $3\text{-}CH_3OC_6H_4(OH)_2\text{-}1,2$	140.14	nd	129^{10}	43-4				B6^3, 6264
4808	Catechol, 3-methyl $3\text{-}CH_3C_6H_3(OH)_2\text{-}1,2$	124.14	lf (bz)	241, 127^{12}	68			w, al, bz, chl	B6^3, 4492
4809	Catechol, monoacetate $2\text{-}(CH_3CO_2)C_6H_4OH$	152.15	pl	189-91^{102}, 148^{25}	57-8			w, al, ace, peth	B6^3, 4227
4810	Catechol, monobenzoate $2\text{-}(C_6H_5CO_2)C_6H_4OH$	214.22	nd (w)		130-1			al	B9^3, 551
4811	Catechol, monobenzylether $2\text{-}(C_6H_5CH_2O)C_6H_4OH$	200.24		173-4^{13}		1.154^{22}	1.5906^{18}	al, eth	B6^3, 4218
4812	Catechol, monobutyl ether $2\text{-}(C_4H_9O)C_6H_4OH$	166.22		231-4, 159^{69}		1.026^{25}	1.5113^{25}		B6^3, 4209
4813	Catechol, monoethyl ether or 2-Ethoxy phenol $2\text{-}(C_2H_5O)C_6H_4OH$	138.17		217, 68^{4}	29			al, eth	B6^3, 4207
4814	Catechol, monomethyl ether or 2-Methoxy phenol, Guaiacol $2\text{-}(CH_3O)C_6H_4OH$	124.14	hex pr	205	32	1.1287$^{21/4}$	1.5429^{20}	al, eth, chl	B6^3, 4200
4815	Catechol, monomethyl ether, acetate $2\text{-}CH_3OC_6H_4OCOCH_3$	166.18		123-4^{13}		1.1285$^{25/4}$	1.5101^{25}	al, eth	B6^3, 4227
4816	Catechol, monopropryl ether or 2-Propoxy phenol $2\text{-}(C_3H_7O)C_6H_4OH$	152.19		228-9, 80-3^{4}		1.0523^{25}	1.5176^{25}	al	B6^3, 4209
4817	Catechol, 4-nitro $4\text{-}NO_2C_6H_3(OH)_2\text{-}1,2$	155.11		174				w, al, eth, chl	B6^3, 4263
4818	Catechol, 3-*iso*-propyl-6-methyl $3\text{-}i\text{-}C_3H_7\text{-}6\text{-}CH_3\text{-}C_6H_2(OH)_2\text{-}1,2$	166.22		270	48			eth, ace	B6^3, 4673
4819	Catechol, 4-propyl $4\text{-}C_3H_7C_6H_3(OH)_2\text{-}1,2$	152.19	pr (w, bz)	152^{13}	60	1.100$^{18/4}$	1.4440^{18}	al, eth, ace	B6^3, 4613
4820	Catechol, 4-*iso*-propyl $4\text{-}i\text{-}C_3H_7C_6H_3(OH)_2\text{-}1,2$	152.29	lf (lig)	270-2, 168^{26}	78				B6^3, 4632
4821	Catechol, tetrabromo $Br_4C_6(OH)_2\text{-}1,2$	425.70	nd (bz, al) lf (bz-peth)		192-3			al, bz	B6^3, 4261
4822	Catechol, tetrachloro $Cl_4C_6(OH)_2\text{-}1,2$	247.89	cr (dil al, bz) cr (+3w) (aq aa)		110 (anh), 94 (+3w)				B6^3, 4253
4823	Catechol, 3,4,5-trichloro $3,4,5\text{-}Cl_3C_6H(OH)_2\text{-}1,2$	213.45	(i) pr (+1w, aa), (ii) pr (+½w, bz)	115, 134-5				al, eth, aa	B6^1, 389
4824	Cedrene $C_{15}H_{24}$	204.36	$[\alpha]^{20/}{}_D$ −91.3	262-3, 124-6^{12}		0.9342$^{20/4}$	1.5034^{20}	bz, lig	B5^3, 1095
4825	Cedrol $C_{15}H_{26}O$	222.37	$[\alpha]$ +10.08 (chl, c=10)		86	0.9496$^{90/20}$	1.4824$^{90/0}$		B6^3, 424
4826	*β*-Cellobiose or 4-*α*-*β*-D-Glucopyranosyl-*β*-D-glucose $C_{12}H_{22}O_{11}$	342.30	cr (dil al), $[\alpha]^{20/}{}_D$ +14.2→ +34.6 (mut) (w, c=8, 15 hr)		d 225			w	B17^4, 3061
4827	Cellobiose, octa-acetate (*α*-anomer) $C_{28}H_{38}O_{19}$	678.60	nd (al), $[\alpha]^{20/D}$ +43.6 (chl, c=6)		229.5			chl, aa	B17^4, 3589
4828	Cellobiose, octa-acetate (*β*-anomer) $C_{28}H_{38}O_{19}$	678.60	nd (al), $[\alpha]^{20/}{}_D$ −14.7 (chl, c=5)		202			chl	B17^4, 3590
4829	Cellulose or Polycellobioso $(C_6H_{10}O_5)_x$	(162.14)$_x$	wh amor		260-70d	1.27-1.60			

No.	Name, Synonyms, and Formula	Mol. wt.	Color, crystalline form, specific rotation and λ_{max} (log ε)	b.p. °C	m.p. °C	Density	n_D	Solubility	Ref.
4830	Cellulose, hexanitrate or Gun cotton $(C_{12}H_{14}N_6O_{22})_x$	(594.27)$_x$	wh amor		160-70 (ign)	1.66		ph NO_2	
4831	Cellulose, pentanitrate $(C_{12}H_{15}N_5O_{20})_x$	(549.28)$_x$	wh amor		1.66			eth-al	
4832	Cellulose, tetranitrate or in Collodion $(C_{12}H_{16}N_4O_{18})_x$	(504.28)$_x$	wh amor			1.66		eth-al	
4833	Cellulose, triacetate $(C_{12}H_{16}O_8)_x$	(288.25)$_x$	yesh fl $[\alpha]_D$ −22.5 (chl)					aa	
4834	Cellulose, triethylether or Ethylcellulose $(C_{12}H_{22}O_5)_x$	(246.30)$_x$	wh nd (bz) $[\alpha]^{20/}{}_D$ +26.1 (bz)		240-55			eth	
4835	Cellulose, trinitrate or in Collodion $(C_{21}H_{17}N_3O_{16})_x$	(459.28)$_x$	wh			1.66		ace, aa	
4836	Cepharanthine $C_{37}H_{38}N_2O_6$	606.72	ye amor pw $[\alpha]^{20/}{}_D$ +277 (chl)		145-55			al, eth, ace, bz	C49, 1745
4837	Cerane or Isohexacosane $CH_3(CH_2)_{24}CH_3$	366.71	pl (eth), sc (w)	$207^{0.7}$	61			al, eth	$B1^2$, 143
4838	Cerulignone $[(C_6H_2O(CH_3O)_2]_2$	304.30	bl gr						$B8^2$, 573
4839	Cetane or Hexadecane $CH_3(CH_2)_{14}CH_3$	226.45	lf (ace)	287, 149^{10}	18.2	$0.7733^{20/4}$	1.4345	**eth**	$B1^4$, 537
4840	Cetene or 1-Hexadecene $CH_3(CH_2)_{13}CH=CH_2$	224.43	lf	284.4, 155^{15}	4.1	$0.7811^{20/4}$	1.4412^{20}	al, eth, peth	$B1^4$, 927
4841	Cetyl alcohol or 1-Hexadecanol $CH_3(CH_2)_{14}CH_2OH$	242.45	fl (AcOEt)	344, 190^{15}	50	$0.8176^{50/4}$	1.4283^{79}	eth, ace, bz, chl	$B1^4$, 1876
4842	Cetylamine or 1-Amino hexadecane $CH_3(CH_2)_{14}CH_2NH_2$	241.46	lf	322.5, 144^2	46.8	$0.8129^{20/4}$	1.4496^{20}	al, eth, ace, bz, chl	$B4^4$, 818
4843	Cetyl Phenyl Ether or Hexadecyl phenyl ether $C_{16}H_{33}OC_6H_5$	318.54	lf (al)	200^1	41.8	0.8434^{82}	1.4556^{82}		$B6^3$, 555
4844	Cetyl sulfate $(C_{16}H_{33}O)_2SO_2$	546.93			66.2			w	$B1^4$, 1879
4845	Cevagenine $C_{27}H_{43}NO_8$	509.64	nd (MeOH-eth), $[\alpha]_{20/D}$ −47.5 (al)		246-8				$B21^4$, 6815
4846	Chalcone dibromide-*(threo)* $C_6H_5CHBrCHBrCOC_6H_5$	368.07	nd (al)		122-3			al	$B7^3$, 2155
4847	Chalcone dibromide-*(erythro)* $C_6H_5CHBrCHBrCOC_6H_5$	368.07	pr or nd (al)		159-60				$B7^3$, 2154
4848	Chalcone-*(trans)* or Benzalacetophenone $C_6H_5COCH=CHC_6H_5$	208.26	pa ye lf, pr, nd (peth)	345-8d, 208^{25}	(i)59 (ii)57 (iii)49	$1.0712^{62/4}$		eth, bz, chl	$B7^3$, 2380
4849	Chalcone, 4,4-dimethyl $(4-(CH_3C_6H_4)COCH=CH(C_6H_4CH_3-4)$	236.32	cr (MeOH)		127-9			al	$B7^2$, 441
4850	Chalcone, 3,3′-dinitrotro $(3-O_2NC_6H_4)COCH=CH(C_6H_4NO_2-3)$	298.25	pa ye nd (aa)		210-1			bz	$B7^3$, 2407
4851	Chalcone, 2-methoxy or 2-Anisylidene acetophenone $(2-CH_3OC_6H_4)CH=CHCOC_6H_5$	238.29	yesh nd (peth or eth-lig)		64-5			al, eth, bz, chl	$B8^3$, 1456
4852	Chalcone, 3-methoxy $(3-CH_3OC_6H_4)CH=CHCOC_6H_5$	238.29	yesh pl or pr (MeOH)	247^{12}	65			al, eth, ace, bz	$B8^3$, 1463
4853	Chalcone, 4-methoxy $(4-CH_3OC_6H_4)CH=CHCOC_6H_5$	238.29	ye nd (al)	$187\text{-}8^{18}$	79			al, eth, chl, aa	$B8^3$, 1464
4854	Chalcone, 3,4-methylene dioxy or Piperonylidine acetophenone $[3,4-(CH_2O_2)C_6H_3]CH=CHCOC_6H_5$	252.27	ye nd (al)		128			al, aa	$B19^4$, 1866
4855	Chalcone, 2-nitro $(2-O_2NC_6H_4)CH=CHCOC_6H_5$	253.26	pa br nd (al)		125			al, eth, aa	$B7^3$, 2399
4856	Chalcone, 2′-nitro $C_6H_5CH=CHCO(C_6H_4NO_2-2)$	253.26	nd (al)		128-9			al, eth	$B7^3$, 2402
4857	Chalcone, 3-nitro $(3-O_2NC_6H_4)CH=CHCOC_6H_5$	253.26	ye nd (al or bz)		145-6			al, bz, chl, aa	$B7^3$, 2400
4858	Chalcone, 4-nitro $(4-O_2NC_6H_4)CH=CHCOC_6H_5$	253.26	pa ye nd (al), pl (bz)		164			al, chl	$B7^3$, 2401

PHYSICAL CONSTANTS OF ORGANIC COMPOUNDS (Continued)

No.	Name, Synonyms, and Formula	Mol. wt.	Color, crystalline form, specific rotation and λ_{max} (log ε)	b.p. °C	m.p. °C	Density	n_D	Solubility	Ref.
4859	Chalcone, α-nitro $C_6H_5C(NO_2)=CHCOC_6H_5$	253.26	ye pl (eth or bz-lig) cr aa		90			eth, ace, bz	$B7^3$, 2403
4860	Chaulmoogric acid-(d) or d-13(2-cyclopentenyl)tridecanoic acid $C_{18}H_{32}O_2$	280.46	pl or lf (al, aa), $[\alpha]_D$ + 62 (chl)	$247\text{-}8^{20}$	68.5			eth, chl	$B9^3$, 284
4861	Chaulmoogric acid-(dl) $C_{18}H_{32}O_2$	280.45	cr (peth)	$247\text{-}8^{20}$				eth, chl	$B9^3$, 285
4862	Chloroacetaldehyde, trimer or 2,4,6-tris (chloromethyl) 1,3,5-trioxane $C_6H_9Cl_3O_3$	235.49	nd (eth)	$142\text{-}4^{10}$	87			eth	$B19^4$, 4718
4863	Chelerythrine $C_{21}H_{19}NO_5$	365.39	cr (chl-MeOH), cr (al + 1)	207				chl	$B27^2$, 563
4864	Chelidonic acid or 4-Pyrone-2,6-dicarboxylic acid $C_7H_4O_6$	184.11	rose mcl nd (al-w + 1w)		262				$B18^4$, 6136
4865	Chelidonic acid, diethyl ester $C_{11}H_{12}O_6$	240.21	pr, nd		69			eth	$B18^4$, 6137
4866	Chelidonine-(d) $C_{20}H_{19}NO_5$	353.37	mcl pl (al + 1w), $[\alpha]_D$ + 151 (al, 1%)		136-40d			al, eth, chl	$B27^2$, 615
4867	Chelidonine, hydrochloride-(d) $C_{20}H_{19}NO_5\text{-}HCl$	389.84	wh (w)						B27, 557
4868	Chloral or Trichloroacetaldehyde Cl_3CCHO	147.39		97.8	−57.5	$1.51214^{20/4}$	1.4557^{20}	w, al, eth	$B1^4$, 3142
4869	Chloral ammonia $Cl_3CCH(OH)NH_2$	164.42	nd (al)	100d	72-4			al, eth, bz	$B1^4$, 3147
4870	Chloral hydrate $Cl_3CCH(OH)_2$	165.40	mcl pl (w)	$96.3^{764}{}_d$	57	$1.9081^{20/4}$		w, al, eth, ace, bz, chl, Py	$B1^4$, 3143
4871	Chloralide or 2,5-bis(trichloromethyl)1,3-dioxolan-4-one $C_5H_2Cl_6O_3$	322.79	pr (al or eth)	272-3, $147\text{-}8^{12}$	116			eth, aa	$B19^4$, 1571
4872	Chloroacetamide $ClCH_2CONH_2$	93.51	mcl pr	224.5^{743}	121			w, al	$B2^4$, 490
4873	Chloroacetamide, *N*-allyl-*N*-phenyl $ClCH_2CON(C_6H_5)CH_2CH=CH_2$	217.74		$119^{0.15}$			1.5079^{25}		Am78, 2556
4874	Chloroacetamide, *N,N*-bis(2-chloroallyl) $ClCH_2CON(CH_2C(Cl)=CH_2)_2$	242.53		$161\text{—}3^{12}$			1.5220^{25}		$B4^4$, 1089
4875	Chloroacetamide-*N,N-bis*(3-chloroallyl) $ClCH_2CON[CH_2C(Cl)=CH_2]_2$	242.53		140.5^1			1.5220^{25}		Am78, 2556
4876	Chloroacetamide, *bis*(2-chloropropyl) $ClCH_2CON(CH_2CHClCH_3)_2$	246.57		$134^{0.7}$			1.5018^{25}		$B4^4$, 499
4877	Chloroacetamide, *N,N-bis*(2-ethylhexyl) $ClCH_2CON[CH_2CH(C_2H_5)C_4H_9]_2$	217.94		$154^{0.8}$			1.4622^{25}		Am78, 2556
4878	Chloroacetamide, *N,N-bis*(2-methylallyl) $ClCH_2CON[CH_2C(CH_3)=CH_2)]_2$	201.70		$133\text{-}5^{20}$			1.4882^2		$B4^4$, 1105
4879	Chloroacetamide, *N,N-bis*(3-Methylbutyl) $ClCH_2CON[CH_2CH_2CH(CH_3)_2]_2$	233.78		$109^{0.6}$			1.4625^{25}		$B4^4$, 702
4880	Chloroacetamide, *N*-butyl $ClCH_2CONHC_4H_9$	149.62		110^7			1.4665^{25}		$B4^4$, 567
4881	Chloroacetamide, *N*-butyl-*iso*-propyl $ClCH_2CON(C_4H_9)CH(CH_3)_2$	191.70		$101^{1.4}$			1.5078^{25}		$B4^4$, 567
4882	Chloroacetamide, *N*-*sec*-butyl $ClCH_2CONHCH(CH_3)C_2H_5$	149.62		$68^{0.7}$	45-6			peth	$B4^4$, 621
4883	Chloroacetamide, *N*-*tert*-butyl $ClCH_2CONHC(CH_3)_3$	149.62	cr (peth)		84			peth	$B4^4$, 662
4884	Chloroacetamide, *N*-butyl-*N*-ethyl $ClCH_2CON(C_2H_5)C_4H_9$	177.67		$90^{1.5}$			1.4665^{25}		$B4^4$, 567
4885	Chloroacetamide, *N*-2-chloroallyl $ClCH_2CONHCH_2CCl=CH_2$	168.02		$101^{1.4}$			1.5078^{25}		$B4^4$, 1089
4886	Chloroacetamide, *N*-2-chloroallyl-*N*-phenyl $ClCH_2CON(C_6H_5)CH_2CCl=CH_2$	243.11		$138^{0.7}$			1.5602^{25}		Am78, 2556
4887	Chloroacetamide, *N*-3-chloroallyl $ClCH_2CONHCH_2CH=CHCl$	168.02	cr (peth)	$112^{0.5}$	52-3			peth	$B4^4$, 1087
4888	Chloroacetamide, *N*-(2-chloro-4-nitrophenyl) $(4\text{-}NO_2\text{-}2\text{-}Cl\text{-}C_6H_3)NCOCH_2Cl$	248.05	cr (peth)		118-9				$B12^3$, 1664

No.	Name, Synonyms, and Formula	Mol. wt.	Color, crystalline form, specific rotation and λ_{max} (log ε)	b.p. °C	m.p. °C	Density	n_D	Solubility	Ref.
4889	Chloroacetamide, *N*-(4-chlorophenyl)-*N*-ethyl $(4\text{-}ClC_6H_4)N(C_2H_5)COCH_2Cl$	232.11	cr (peth)		70-1			peth	Am78, 2557
4890	Chloroacetamide, *N*-2-chloropropyl $ClCH_2CONHCH_2CHClCH_3$	170.04		$88^{1.5}$			1.4942^{25}		B4^4, 499
4891	Chloroacetamide, *N*-3-chloropropyl $ClCH_2CONHCH_2CH_2CH_2Cl$	170.04			36-7				B4^4, 501
4892	Chloroacetamide, *N*-*N*-diallyl $ClCH_2CON(CH_2CH{=}CH_2)_2$	173.64		$92^{0.7}$			1.4932^{25}		B4^4, 1064
4893	Chloroacetamide, *N,N*-dibenzyl $ClCH_2CON(CH_2C_6H_5)_2$	273.76		$190^{1.8}$			1.5837^{25}		Am78, 2556
4894	Chloroacetamide, *N,N-di-iso*-butyl $ClCH_2CON(i\text{-}C_4H_9)_2$	205.73		99^{2}			1.4642^{25}		B4^4, 633
4895	Chloroacetamide, *N,N-di-sec*-butyl $ClCH_2CON(sec\text{-}C_4H_9)_2$	205.73		$92^{0.7}$			1.4681^{25}		Am78, 2556
4896	Chloroacetamide, *N*-(2,3-dichloroallyl) $ClCH_2CONHCH_2CCl{=}CHCl$	202.47		$126\text{-}31^{1.8}$			1.5311^{25}		Am78, 2556
4897	Chloroacetamide, N-(2,4-dichlorobenzyl) $ClCH_2CONH\text{-}(CH_2C_6H_3Cl_2\text{-}2,4)$	252.53	cr (bz)		96-7			bz	Am78,2556
4898	Chloroacetamide, *N*-(2,4-dichlorobenzyl) $ClCH_2CONH\text{-}(CH_2C_6H_3Cl_2\text{-}2,4)$	252.53	cr (dil al)		105-6			al	Am78, 2556
4899	Chloroacetamide, *N*-(2,4-dichlorophenyl) $ClCH_2CONH(C_6H_3Cl_2\text{-}2,4)$	238.50	cr (dil al)		101-2			al, peth	Am78, 2556
4900	Chloroacetamide, *N*-(2,5-dichlorophenyl) $ClCH_2CONH(C_6H_3Cl_2\text{-}2,5)$	238.50	cr (dil al)		116-7			al	Am78, 2556
4901	Chloroacetamide, *N*(2,3-dichloropropyl) $ClCH_2CONHCH_2CHClCH_2Cl$	204.48	cr (peth)		65-6			peth	B4^4, 502
4902	Chloroacetamide, *N,N*-diethyl $ClCH_2CON(C_2H_5)_2$	149.62		$190\text{-}5^{25}$					B4^4, 350
4903	Chloroacetamide, *N,N*-dihexyl $ClCH_2CON(C_6H_{13})_2$	261.84	cr (peth)		114-5				B4^4, 713
4904	Chloroacetamide, *N*-(2,4-dinitrophenyl) $[2,4\text{-}(NO_2)_2C_6H_3]NHCOCH_2Cl$	259.61	nd (al)		114-5				B9^2, 315
4905	Chloroacetamide, *N,N*-dipentyl $ClCH_2CON(C_5H_{11})_2$	233.78		126^{1}			1.4651^{25}		B4^4, 679
4906	Chloroacetamide, *N,N*-dipropyl $ClCH_2CON(C_3H_7)_2$	177.67		120^{8}, $90\text{-}2^{0.8}$			1.4670^{20}		B4^4, 476
4907	Chloroacetamide, *N,N*-di-*iso*-propyl $ClCH_2CON(i\text{-}C_3H_7)_2$	177.67	cr (peth)	$86^{2.7}$	48-9		1.4619^{25}		B4^4, 516
4908	Chloroacetamide, *N*-ethyl-*N*-hexyl $ClCH_2CON(C_2H_5)C_6H_{13}$	205.73		120^{11}			1.4978^{25}		Am78, 2556
4909	Chloroacetamide, *N*-furfuryl $ClCH_2CONH(CH_2C_4H_3O)$	173.60	cr(peth)		58			peth	B18^4, 7080
4910	Chloroacetamide, *N*-hexyl $ClCH_2CONHC_6H_{13}$	177.67	cr (peth)	$95\text{-}105^{0.2}$	108-9			peth	B4^4, 713
4911	Chloroacetamide, *N*-hexyl-*N*-methyl $ClCH_2CON(CH_3)C_6H_{13}$	191.70		$134^{3.8}$			1.5005^{25}		Am78, 2556
4912	Chloroacetamide, 3-methoxypropyl $ClCH_2CONH(CH_2)_3COCH_3$	165.62		$88^{0.5}$	30		1.4712^{25}		B4^4, 1645
4913	Chloroacetamide, *N*-(2-methylallyl) $ClCH_2CONHCH_2C(CH_3){=}CH_2$	147.61		96^{1}			1.4860^{25}		B4^4, 1105
4914	Chloroacetamide, *N*-(3-methylbutyl) $ClCH_2CONH\text{-}CH_2CH_2CH(CH_3)_2$	163.65		$134\text{-}5^{13}$	−15				B4^2, 647
4915	Chloroacetamide, *N*-pentyl $ClCH_2CONHC_5H_{11}$	163.65		$82^{0.5}$			1.4665^{25}		B4^4, 678
4916	Chloroacetamide, *N*-propyl $ClCH_2CONHC_3H_7$	135.59		$105\text{-}6^{10.5}$	62			al, eth, ace, chl	B4^4, 476
4917	Chloroacetamide, *N-iso*-propyl $ClCH_2CONHCH(CH_3)_2$	135.59	cr (peth)		62			peth	B4^4, 515
4918	Chloroacetamide, *N*-tetradecyl $ClCH_2CONHC_{14}H_{29}$	289.89	cr(peth)		64-5				B4^4, 814
4919	Chloroacetamide, *N*-tetrahydrofurfuryl $ClCH_2CONH(C_4H_7O)$	177.63	cr (peth)		62-3			peth	B18^4, 7039
4920	Chloroacetic acid (α) $ClCH_2CO_2H$	94.50	mcl pr	187.8, 104^{20}	63	$1.4043^{40/4}$	1.4351^{55}	w, al, eth, bz, chl	B2^4, 474
4921	Chloroacetic acid (β) $ClCH_2CO_2H$	94.50	mcl pr	187.9, 104^{20}	56.2	$1.4043^{40/4}$	1.4351^{50}	w, al, eth, bz, chl	B2^4, 474

No.	Name, Synonyms, and Formula	Mol. wt.	Color, crystalline form, specific rotation and λ_{max} (log ε)	b.p. °C	m.p. °C	Density	n_D	Solubility	Ref.
4922	Chloroacetic acid (γ) $ClCH_2CO_2H$	94.50		$187.8, 104^{20}$	52.5	$1.4043^{40/4}$	1.4351^{55}	w, al, eth, bz, chl	$B2^4$, 474
4923	Chloroacetic acid, anhydride $(ClCH_2CO)_2O$	170.98	pr (bz)	203	46	1.5497^{20}			$B2^4$, 487
4924	Chloroacetic acid, benzyl ester or Benzyl chloracetate $ClCH_2CO_2CH_2C_6H_5$	184.62		147.5^9, $84\text{-}6^{0.4}$		$1.2223^{4/4}$	1.5426^{18}	al, eth	$B6^3$, 1479
4925	Chloroacetic acid, butyl ester or Butyl chloroacetate $ClCH_2CO_2C_4H_9$	150.61		$183, 94^{38}$		$1.0704^{20/4}$	1.4297^{20}	al, eth	$B2^4$, 482
4926	Chloroacetic acid, *iso*-butyl ester or *iso*-Butyl chloroacetate $ClCH_2CO_2CH_2CH(CH_3)_2$	150.61		170		$1.0612^{20/4}$	1.4255^{20}	eth, ace	$B2^4$, 483
4927	Chloroacetic acid, *sec*-butyl ester or *sec*-Butylchloroacetate $ClCH_2CO_2\text{-}sec\text{-}C_4H_9$	150.61		163-4		$1.062^{20/20}$	1.4251^{19}	al, eth	$B2^3$, 443
4928	Chloroacetic acid, 2-chloroethyl ester or 2-chloroethyl chloroacetate $ClH_2CO_2CH_2CH_2Cl$	157.00		$202, 89^{10}$		$1.3600^{25/4}$	1.4619^{25}	eth	$B2^4$, 481
4929	Chloroacetic acid, ethyl ester or Ethyl chloroacetate $ClCH_2CO_2C_2H_5$	122.55		$144^{740}, 52^{20}$	−26	$1.1585^{20/4}$	1.4215^{20}	**al, eth, ace,** bz	$B2^4$, 481
4930	Chloroacetic acid, 2-hydroxyethyl ester or 2-Hydroxyethyl chloroacetate $ClCH_2CO_2CH_2CH_2OH$	138.55		240d, $86^{1.5}$		$1.3300^{20/4}$	1.4609^{20}	**w, al**	$B2^3$, 447
4931	Chloroacetic acid, hydrazide $ClCH_2CONHNH_2$	108.53			93				
4932	Chloroacetic acid, 2-methoxyethyl ester or 2-Methoxyethyl chloroacetate $ClCH_2CO_2CH_2CH_2OCH_3$	152.57		$85\text{-}6^9$		$1.2015^{20/4}$	1.4382^{20}	eth	$B2^3$, 447
4933	Chloroacetic acid, methyl ester or Methyl chloroacetate $ClCH_2CO_2CH_3$	108.52		129.8, 29^{10}	−32.1	$1.2337^{20/4}$	1.4218^{20}	**al, eth, ace, bz**	$B2^4$, 480
4934	Chloroacetic acid, phenyl ester or Phenyl chloroacetate $ClCH_2CO_2C_6H_5$	170.60	nd or pl (al)	230-5, 114^8	44-5	$1.2202^{44/4}$	1.5146^{44}	al, eth	$B6^3$, 598
4935	Chloroacetic acid, propylester or Propyl chloroacetate $ClCH_2CO_2C_3H_7$	136.58		161^{764}		$1.1033^{20/4}$	1.4261^{20}	eth	$B2^4$, 482
4936	Chloroacetic acid, *iso*-propyl ester or *iso*-Propyl chloroacetate $ClCH_2CO_2\text{-}i\text{-}C_3H_7$	136.58		150-1		$1.0888^{20/4}$	1.4382^{20}	eth	$B2^4$, 482
4937	Chloroacetic acid, 4-tolyl ester $ClCH_2CO_2(C_6H_4CH_3\text{-}4)$	184.62	pl	162^{41}	32			al, eth	$B6^2$, 378
4938	Chloroacetone cyanohydrin $ClCH_2C(CH_3)(OH)CN$	119.55		110^{27}		1.2027^{15}	1.4536^{11}	w, al, ace	$B3^3$, 599
4939	Chloroacetomitrile $ClCH_2CN$	75.50		126-7, $30\text{-}2^{15}$		1.1930^{20}	1.4202^{25}	al, eth	$B2^4$, 492
4940	Chloroacetyl chloride $ClCH_2COCl$	112.94		107		$1.4202^{20/4}$	1.4541^{20}	**eth,** ace	$B2^4$, 488
4941	2-Chloroally isothiocyanate $CH{=}CClCH_2NCS$	133.60		182		1.27^{12}			B4, 219
4942	Chlorodifluoro acetic acid ClF_2CCOOH	130.48			22.9			chl	$B2^4$, 497
4943	Chlorodiphenyl acetamide $(C_6H_5)_2CClCONH_2$	245.71	cr (to)		115			al, eth, bz, chl	$B9^3$, 3308
4944	Chlorodiphenyl acetic acid $(C_6H_5)_2CClCO_2H$	246.79	pl (bz-lig)	118-9d				al, eth, ace, bz	$B9^3$, 3307
4945	Chlorodiphenyl acetic acid, ethyl ester $(C_6H_5)_2CClCO_2C_2H_5$	274.75	pl (chl), cr (al)	185^{14}	43-4			al, eth	$B9^3$, 3307
4946	Chlorodiphenyl acetyl chloride $(C_6H_5)_2CClCOCl$	265.14	cr (lig)	180^{14}	50-1				$B9^3$, 3308
4947	bis-(2-Chloroethyl) methylamine, hydrochloride $(ClCH_2CH_2)_2NCH_3 \cdot HCl$	192.52	hyg nd		111-2			w, al	$B4^4$, 446
4948	tris-(2-Chloroethyl) amine or Nitrogen mustard gas $(ClCH_2CH_2)_3N$	204.53	Pa ye	$143\text{-}4^{15}$	−4			**al, eth, bz**	$B4^4$, 447
4949	(α-Chloroethyl)ether or bis-(α-chloroethyl) ether $(CH_3CHCl)_2O$	143.02		116-7		$1.1060^{25/4}5$	1.4186^{25}	al, **eth,** chl	$B1^4$, 3120
4950	(α-Chloroethyl)(β-Chloroethyl) ether $CH_3CHClOCH_2CH_2Cl$	143.01		d^{760}, $55\text{-}7^{17}$		$1.1867^{20/4}$	1.4473^{20}	al, eth, chl	$B1^3$, 2655
4951	(β-Chloroethyl)diethyl amminium chloride $ClCH_2CH_2NH(C_2H_5)_2Cl^-$	172.10	nd (al-eth)		210-1			w, al	$B4^4$, 447

No.	Name, Synonyms, and Formula	Mol. wt.	Color, crystalline form, specific rotation and λ_{max} (log ε)	b.p. °C	m.p. °C	Density	n_D	Solubility	Ref.
4952	α-Chloroethyl methyl ether $CH_3OCHClCH_3$	94.54		$72\text{-}3^{751}$		$0.9902^{20/4}$	1.4004^{20}	eth	$B1^4$, 3119
4953	α-Chloroethyl pentyl ether $CH_3CH_2Cl\text{-}O\text{-}C_5H_{11}$	150.65		$63\text{-}6^{8}$		$0.9200^{20/4}$	1.4218^{20}	eth	$B1^3$, 2656
4954	α Chloroethyl propyl ether $CH_3CHClOC_3H_7$	122.59		$112\text{-}5^{731}$		$0.9322^{20/4}$	1.4013^{20}	eth	$B1^3$, 2655
4955	(β-Chloroethyl)ether or bis-(β-Chloroethyl)ether $(ClCH_2CH_2)_2O$	143.01		178, 75^{20}	−24.5	$1.2199^{20/4}$	1.4575^{20}	al, eth, ace, bz	$B1^4$, 1375
4956	α-Chloroethyl methyl ether $ClCH_2CH_2OCH_3$	94.54		92-3		$1.0345^{20/4}$	1.4111^{20}	w, eth	$B1^4$, 1375
4957	β-Chloroethyl vinyl ether $ClCH_2CH_2OCH{=}CH_2$	106.55		108		$1.0475^{20/4}$	1.4378^{20}	al, eth	$B1^4$, 2051
4958	Chloroformic acid, benzyl ester or Carbobenzoxy chloride $ClCO_2CH_2C_6H_5$	170.60		103^{20}		1.20	1.5150^{20}	eth, ace, bz	$B6^3$, 1485
4959	Chloroformic acid, butyl ester $ClCO_2C_4H_9$	136.58		138^{750}, 35.5^{11}		$1.0513^{20/4}$	1.4121^{20}	eth, ace	$B3^4$, 25
4960	Chloroformic acid, isobutyl ester or Isobutyl chloroformate $ClCO_2CH_2CH(CH_3)_2$	136.58		128.8		$1.0426^{18/4}$	$1.4071^{18/He}$	eth, bz, chl	$B3^4$, 26
4961	Chloroformic acid, β-chloroethyl ester $ClCO_2CH_2CH_2Cl$	142.97		156		$1.3847^{20/4}$	1.4483^{20}	al, eth, ace, bz	$B3^4$, 24
4962	Chloroformic acid, chloromethyl ester $ClCO_2CH_2Cl$	128.54		107		1.465^{15}	1.4286^{22}	eth, ace	$B3^4$, 30
4963	Chloroformic acid, (3-chloropropyl) ester $ClCO_2CH_2CH_2CH_2Cl$	157.00		177		$1.2949^{25/20}$	1.4456^{20}		$B3^3$, 25
4964	Chloroformic acid, cyclohexyl ester $ClCO_2C_6H_{11}$	162.62		87.5^{27}				eth	$B6^4$, 43
4965	Chloroformic acid, dichloromethyl ester $ClCO_2CHCl_2$	163.39		110-1				ace	$B3^4$, 31
4966	Chloroformic acid, β-ethoxyethyl ester or Cellosolve chloroformate $ClCO_2CH_2CH_2OC_2H_5$	152.58		67.2^{14}		1.1341^{25}	1.4169^{25}	al	$B3^2$, 29
4967	Chloroformic acid, ethyl ester or Ethyl chloroformate $ClCO_2C_2H_5$	108.52		95	−80.6	$1.1352^{20/4}$	1.3974^{20}	eth, bz, chl	$B3^4$, 23
4968	Chloroformic acid, 2-methoxyethyl ester $ClCO_2CH_2CH_2OCH_3$	138.55		58.7^{13}		1.1905^{25}	1.4163^{20}	eth	$B3^2$, 29
4969	Chloroformic acid, methyl ester or Methyl chloroformate $ClCO_2CH_3$	94.50		70-1		$1.2231^{20/4}$	1.3868^{20}	al, eth, bz, chl	$B3^4$, 23
4970	Chloroformic acid, isopentyl ester or Isopentyl chloroformate $ClCO_2\text{-}i\text{-}C_5H_{11}$	150.61		154.3, 60^{15}		$1.0288^{17/4}$	1.4176^{20}	al, eth	$B3^4$, 26
4971	Chloroformic acid, pentyl ester $ClCO_2C_5H_{11}$	150.61		$60\text{-}2^{15}$			1.4181^{18}	eth	$B3^4$, 26
4972	Chloroformic acid, isopropenyl ester $ClCO_2C(CH_3){=}CH_2$	120.54		100		$1.103^{20/20}$		eth	$B3^3$, 28
4973	Chloroformic acid, propyl ester or Propyl chloroformate $ClCO_2C_3H_7$	122.55		115.2		$1.0901^{20/4}$	1.4035^{20}	al, eth	$B3^4$, 24
4974	Chloroformic acid, isopropyl ester or Isopropyl chloroformate $ClCO_2CH(CH_3)_2$	122.55		105, 66.3^{200}			1.4013^{20}		$B3^4$, 24
4975	Chloroformic acid, trichloro methyl ester or Diphosgene $ClCO_2CCl_3$	197.83		128, 49^{50}	−57	1.6525^{14}	1.4566^{22}	al, eth	$B3^4$, 33
4976	Chloromethyl ether or bis-(Chloromethyl) ether $(ClCH_2)_2O$	114.96		104	-41.5	$1.328^{15/4}$	1.435^{21}	al. eth	$B1^4$, 3051
4977	Chloromethyl propyl ether $ClCH_2OC_3H_7$	108.57		109		$0.9884^{20/4}$	1.4125^{20}	al, eth	$B1^3$, 2589
4978	Chloromethyl thiocyanate $ClCH_2SCN$	107.56		185		1.37^{15}			$B3^2$, 124
4979	Chloromycetin or Chloroamphenicol $C_{11}H_{12}Cl_2N_2O_5$	323.14	pa ye pl or nd (w), $[\alpha]^{25}_{D}$ + 19 (al, c=5) -25.5 (AcOEt)	sub vac	150-1			al, ace, chl	$B13^3$, 2268
4980	(4-Chlorophenyl) ether or bis-(4-Chlorophenyl) ether $(4\text{-}ClC_6H_4)_2O$	239.10	nd (al)	312-4, 168.7^{7}	30	1.1231^{20}	1.611^{20}		$B6^4$, 826

PHYSICAL CONSTANTS OF ORGANIC COMPOUNDS (Continued)

No.	Name, Synonyms, and Formula	Mol. wt.	Color, crystalline form, specific rotation and λ_{max} (log ε)	b.p. °C	m.p. °C	Density	n_D	Solubility	Ref.
4981	4-Chlorophenyl phenyl ether $(4\text{-}ClC_6H_4)OC_6H_5$	204.66		284-5		1.2026^{15}	1.599_D		$B6^4$, 826
4982	2-Chlorophenyl isocyanate $2\text{-}ClC_6H_4NCO$	153.57		$115\text{-}7^{43}$	30-1				$B12^3$, 1296
4983	3-Chlorophenyl isocyanate $3\text{-}ClC_6H_4NCO$	153.57		$113\text{-}4^{43}$					$B12^3$, 1316
4984	4-Chlorophenyl isocyanate $4\text{-}ClC_6H_4NCO$	153.57		$115\text{-}7^{45}$					$B12^3$, 1376
4985	4-Chlorophenyl isothiocyanate $4\text{-}ClC_6H_4NCS$	169.63	nd (al)	249-50	45				$B12^3$, 1376
4986	Chlorophyll-a $C_{55}H_{72}MgN_4O_5$	893.51	bl bk hex pl		150-3			al, eth, lig	M, 245
4987	Chlorophyll b $C_{55}H_{70}MgN_4O_6$	907.49	bl bk gr pw		120-30			al, eth, lig	M, 245
4988	Chloroprene or 2-Chloro-1,3-Butadiene $CH_2{=}CCl\text{-}CH{=}CH_2$	88.54		59.4, 6.4^{100}		$0.9583^{20/4}$	1.4583^{20}	**eth, ace, bz**	$B1^4$, 984
4989	Chlorpromazine or Thorazine $C_{17}H_{19}ClN_2S$	318.86		$200\text{-}5^{0.8}$				al, eth, bz, chl	C50, 1951
4990	Chlorpromazin, hydrochloride $C_{17}H_{19}ClN_2S\text{-}HCl$	355.33		194-7d				w, al, chl	C50, 1931
4991	(β-Chloropropyl) ether or bis-(β-chloropropyl) ether $CH_3CHClCH_2OCH_2CHClCH_3$	171.07		188		$1.109^{20/4}$	1.4467^{20}	al, eth	$B1^4$, 1442
4992	(γ-Chloropropyl) ether or bis-(α-Chloropropyl)ether $(ClCH_2CH_2CH_2)_2O$	171.07		215^{745}, 90.5^{11}		$1.140^{20/20}$	1.4158^{20}	al, eth	$B1^2$, 370
4993	(β-Chloroisopropyl) ether or bis-(β-Chloroisopropyl) ether $[ClCH_2CH(CH_3)]_2O$	171.07		187		$1.103^{20/4}$	1.4505^{20}	**al, eth,** ace, bz	$B1^3$, 1470
4994	Bis-(1,2,2,2-Tetrachloroethyl) ether $(Cl_3CCHCl)_2O$	349.68	cr (al, MeOH)	$130\text{-}1^{11}$	40.2			bz, peth	$B1^3$, 2672
4995	Chlorosulfinic acid, ethyl ester or Ethyl chlorosulfinate $(C_2H_5O)SOCl$	128.57		52.5^{44}, 32^{16}		$1.2766^{25/4}$	1.4550^{25}	eth	$B1^3$, 1316
4996	Chlorosulfonic acid, ethyl ester or Ethyl chlorosulfonate $C_2H_5OSO_2Cl$	144.57		151-4, 52^{14}		$1.3502^{25/4}$	1.416^{20}	eth, chl, lig	$B1^4$, 1326
4997	Chlorosulfonic acid, methyl ester or Methyl chlorosulfonate CH_3OSO_2Cl	130.55		133-5, 48^{29}		$1.4805^{25/4}$	1.4138^{18}	eth, ace, bz	$B1^4$, 1252
4998	Cholanic acid or Ursocholamic acid $C_{24}H_{40}O_2$	360.58	nd (al), cr (aa), $[\alpha]^{20/}{}_D$ + 21.7 (chl)		163-4			al, chl, aa	$B9^3$, 2656
4999	Cholanic acid, 3α, 6α-dihydroxy or Hyodeoxycholic acid $C_{24}H_{40}O_4$	392.58	cr (AcOEt), $[\alpha]^{20/}{}_D$ + 37.2 (MeOH)		198-9			al, aa	$B10^3$, 1631
5000	Cholanic acid, 3α,7α-dihydroxy or Chenodeoxycholic acid $C_{24}H_{40}O_4$	392.58	nd (AcOEt-hp), $[\alpha]^{20/}{}_D$ + 11.1 (al, c=2.1)		143			al, eth, ace, aa	$B10^3$, 1635
5001	Cholanic acid, ethyl ester $C_{26}H_{44}O_2$	388.63	lf, nd (dil al), $[\alpha]^{19/}{}_D$ + 21 (chl)	273^{12}	93-4			chl	$B9^3$, 2658
5002	Cholanic acid, methyl ester $C_{25}H_{42}O_2$	374.61	nd $[\alpha]_D$ + 23±2 (diox)		87-8			diox	$B9^3$, 2658
5003	Clolanthrene or Benz *(j)* aceanthrylene $C_{20}H_{14}$	254.33	pa ye lf (bz-al)	sub, $210^{0.2}$	174-5d			al, bz, aa, lig	$B5^3$, 2469
5004	Cholanthrene, 6,7-dihydro-20-methyl $C_{21}H_{18}$	270.37	lf (MeOH)		155			MeOH	$B5^3$, 2429
5005	Cholanthrene, 11,14-dihydro-20-methyl $C_{21}H_{18}$	270.37	nd (PrOH)		138 9				$B5^3$, 2429
5006	Cholanthrene, 20-methyl $C_{21}H_{16}$	268.36	yesh nd (bz)		180				$B5^3$, 2484
5007	$\Delta^{2,4}$-Cholestadiene $C_{27}H_{44}$	368.65	cr (eth-ace), $[\alpha]^{23/}{}_D$ + 168.5 (eth, c=1.5)		68.5			al, eth, chl	$B5^3$, 1428

No.	Name, Synonyms, and Formula	Mol. wt.	Color, crystalline form, specific rotation and λ_{max} (log ε)	b.p. °C	m.p. °C	Density	n_D	Solubility	Ref.
5008	$\Delta^{3,5}$-Cholestadiene or Cholesterilene $C_{27}H_{44}$	368.65	wh nd (al), $[\alpha]^{20/}{}_D$ -129.6 (chl, c=3)	260^{11}	80	$0.925^{100/4}$		al, eth, bz, chl	$B5^4$, 1620
5009	$\Delta^{5,7}$-Cholestadiene-3β-ol or 7-Dehydrocholesterol $C_{27}H_{44}O$	384.65	pl (+1w, eth-MeOH), $[\alpha]^{25/}{}_D$ -115 (chl, C=2.5)		150-1			eth, ace	$B6^3$, 2819
5010	$\Delta^{4,6}$-Cholestadiene-3-one $C_{27}H_{42}O$	382.63	ye pr $[\alpha]^{20/}{}_D$ +31 (chl)		78				$B7^3$, 1760
5011	Cholestane $C_{27}H_{48}$	372.68	sc or pl (eth-al, ace), $[\alpha]^{20/}{}_D$ +30.2 (chl, c=2)	250^1	80	$0.9090^{88/4}$	1.4887^{88}	eth, bz, chl	$B5^4$, 1227
5012	3β-Cholestane carboxylic acid $C_{28}H_{48}O_2$	416.69	nd $[\alpha]^{25/}{}_D$ +28.8 (chl, c=1.7)		210-1				$B9^3$, 2668
5013	3,6-Cholestane dione $C_{27}H_{44}O_2$	400.65	nd $[\alpha]^{20/}{}_D$ +8.9		171-2				$B7^3$, 3282
5014	3α-Cholestanol or Epidihydro cholesterol $C_{27}H_{48}O$	388.68	nd (al), $[\alpha]^{20/}{}_D$ +34 (chl, c=1.1)		188				$B6^3$, 2135
5015	3β-Cholestanol or Dihydrocholesterol $C_{27}H_{48}O$	388.68	lf (al+1w), pr, pl (MeOH), $[\alpha]^{22/}{}_D$ +24.2 (chl, c=1.3		141-3			al, eth, chl	$B6^3$, 2131
5016	3-Cholestanone or Zymostanone $C_{27}H_{46}O$	386.66	nd or lf (al), $[\alpha]_D$ +42 (chl, c=2.12)		128-30			al	$B7^3$, 1330
5017	6-Cholestanone, 3β-hydroxy or 6-Keto Cholestanol $C_{27}H_{46}O_2$	402.66	nd (al), $[\alpha]'_D$ -3.0 (chl)		150-1			al	$B8^3$, 658
5018	7-Cholestanone, 3β-hydroxy or 7-Oxocholestanol $C_{27}H_{46}O_2$	402.66	pl $[\alpha]^{22/}{}_D$ -34 (chl)		165-8			al	$B8^3$, 665
5019	2-Cholestene or Neocholestene $C_{27}H_{46}$	370.66	nd (eth-ace or al), $[\alpha]_D$ +66 (c=1.65)		75-6				$B5^4$, 1507
5020	3-Cholestene $C_{27}H_{46}$	370.66	wh pr or nd (al), $[\alpha]^{18/}{}_D$ -56.3 (chl)		93-4				$B5^4$, 1508
5021	5-Cholestene, 3β-bromo or Cholestyryl bromide $C_{27}H_{45}Br$	449.56	mcl lf (al), $[\alpha]^{20/}{}_D$ -19 (bz, c=0.4)		100-2			bz, chl	$B5^4$, 1512
5022	5-Cholestene, 3β-chloro or Cholesteryl chloride $C_{27}H_{45}Cl$	405.11	nd (al or ace), $[\alpha]^{20/}{}_D$ -26.4 (bz) -33.3 (chl)		96			bz, chl	$B5^4$, 1510
5023	5-Cholestene, 3β,7β-diol or 7β-Hydroxycholesterol $C_{27}H_{46}O_2$	402.66	wh nd (eth), $[\alpha]^{20/}{}_D$ +7.2 (chl, c=2)	sub, $145^{0.005}$	177-8				$B6^3$, 5130
5024	5-Cholestene, 3β-iodo or Cholestyryl iodide $C_{27}H_{45}I$	496.56	nd (ace or AcOEt)		106-7			bz, chl, lig	$B5^4$, 1512
5025	1-Cholesten-3-one $C_{27}H_{44}O$	384.65	nd (dil ace or al), $[\alpha]^{25/}{}_D$ +88.2 (chl)		99-101			bz	$B7^3$, 1592

No.	Name, Synonyms, and Formula	Mol. wt.	Color, crystalline form, specific rotation and λ_{max} (log ε)	b.p. °C	m.p. °C	Density	n_D	Solubility	Ref.
5026	4-Cholestene-3-one $C_{27}H_{44}O$	384.65	nd or pl (al), $[\alpha]^{25/}{}_D$ +92 (chl, c=2.01)		81-2			eth, bz, peth	$B7^3$, 1594
5027	5-Cholesten-3-one $C_{27}H_{44}O$	384.65	lf (al), $[\alpha]^{20/}{}_D$ -4.3 (chl)		127			eth, al	$B7^3$, 1608
5028	Cholesterol $C_{27}H_{45}OH$	386.66	rh or tcl lf (al + 1w) nd (eth), $[\alpha]^{20/}{}_D$ -31.5 (eth), $[\alpha]^{20/}{}_D$ -31.5 (eth, c=2) -39.5 (chl, c=2)	360d, $233^{0.5}$	148.5 (anh)	$1.067^{20/4}$		eth, bz, chl, aa	$B6^2$, 2607
5029	Cholesterol, acetate or Cholestyryl acetate $C_{29}H_{48}O_2$	428.70	wh nd (ace or al), $[\alpha]^{20/}{}_D$ -47.7 (chl, c=2)		115-6			eth, ace, bz, chl	$B6^3$, 2630
5030	Cholesterol, benzoate $C_{34}H_{50}O_2$	490.77	wh nd, $[\alpha]$ -13.7 (chl, c=0.9)		152-3			eth, chl	$B9^3$, 460
5031	Cholesterol, hexadecanoate or Cholesteryl palmitate $C_{43}H_{76}O_2$	625.08	wh nd (eth or al), $[\alpha]^{20}$ -25.4 (chl, c=2)		80			bz, chl	$B6^3$, 2640
5032	Cholestrophane or Dimethylparabanic acid Oxalyldimethylurea $C_5H_6N_2O_3$	142.11	lf or pl (w or al)	275-7, $148\text{-}50^{13}$	155.5			w, eth	$B24^2$, 265
5033	Cholic acid $C_{24}H_{40}O_5$	408.58	rh (eth), tetr rh (w or dil al), cr (al + 1), $[\alpha]^{20/}{}_D$ +37 (al, c=0.6)		198 (anh)			al, eth, ace, chl, aa	$B10^3$, 2162
5034	Choline or Trimethyl(2-hydroxymethyl) ammonium hydroxide $(CH_3)_3N^+CH_2CH_2OH \cdot OH^-$	121.18	syr					w, al	$B4^4$, 1443
5035	Choline, *O*-acetyl, bromide or Acetylcholine bromide $(CH_3)_3N^+CH_2CH_2O_2CCH_3\text{-}Br^-$	226.11	hex pr (dil al)	d	143			w, al	$B4^4$, 1446
5036	Choline, *O*-acetyl, chloride or Acetyl choline chloride $(CH_3)_3N^+CH_2CH_2O_2CCH_3\text{-}Cl^-$	181.66	yesh nd		153			w, al	$B4^4$, 1446
5037	Choline, *O*-benzyl, chloride $(CH_3)_3N^+CH_2CHOO_2CC_6H_5Cl^-$	243.73	pr (ace-al)		200			al, ace	$B9^3$, 877
5038	Choline, carbonate, chloride $(CH_3)_3N^+CH_2CH_2O_2CNH_2\text{-}Cl^-$	182.65	hyg pr or pw		210-2			w	$B4^4$, 1455
5039	Chroman or Dihydrobenzopyran $C_9H_{10}O$	134.18		$214\text{-}5^{742}$, $98\text{-}9^{18}$		1.0610^{20}	1.5444^{20}		$B17^4$, 413
5040	Chroman, 2,2-dimethyl $C_{11}H_{14}O$	162.23		225^{769}, 98^{11}		$1.009^{15/15}$			$B17^4$, 445
5041	Chromanone $C_9H_8O_2$	148.16		160^{50}, 128^{13}	38-9	$1.1291^{100/4}$	1.5460^{100}	al, eth, ace, bz	$B17^4$, 4957
5042	3-Chromene C_9H_8O	132.16		91^{13}, 50^1			1.5879^{20}		$B17^4$, 496
5043	Chromone or α-Benzopyrone $C_9H_6O_2$	146.15	nd (peth or w)	sub	59			al, eth, bz, chl	$B17^4$, 5052
5044	*cis*-Chrysanthemucic acid-*(d)* or Cyclopropane carboxylic acid-2,2-dimethyl-3-(2-methyl propenyl) $C_{10}H_{16}O_2$	168.24	pr $[\alpha]^{22/}{}_D$ +83.3 (chl, c=1.6)	$95^{0.1}$	40-2			al, eth	$B9^3$, 210
5045	*trans*-Chrysanthemucic acid-*(d)* or Cyclopropane carboxylic acid-2,2-dimethyl-3-(2 methyl propenyl*(trans)* $C_{10}H_{16}O_2$	168.24	pr $[\alpha]^{19/}{}_D$ +25.8 (chl, c=2.5)	245d, 135^{12}	18-21			al, eth, chl	$B9^3$, 211
5046	Chrysazin or 1,8-Dihydroxyanthraquinone $1,8\text{-}(HO)_2C_{14}H_6O_2$	240.22	red or redsh-ye nd or lf (al)	sub	193			al, eth, ace, chl	$B8^3$, 3792

PHYSICAL CONSTANTS OF ORGANIC COMPOUNDS (Continued)

No.	Name, Synonyms, and Formula	Mol. wt.	Color, crystalline form, specific rotation and λ_{max} (log ε)	b.p. °C	m.p. °C	Density	n_D	Solubility	Ref.
5047	Chrysene or 1,2-Benzophenanthrene $C_{18}H_{12}$	228.29	red bl flr rh pl (bz-aa)	448	255-6	1.274^{20}			$B5^3$, 2380
5048	Chrysene, 5,6-dimethyl $C_{20}H_{16}$	256.35	pl or nd (bz-al)	$200^{0.5}$ (sub 140 vac)	128-9			al, aa	$B5^3$, 2418
5049	Chrysene, 1-methyl $C_{19}H_{14}$	242.32	lf (hx, bz, to)	sub 130- 140 (vac)	256-7			al	$B5^3$, 2395
5050	Chrysene, 2-methyl $C_{19}H_{14}$	242.32	lf (bz-al)		229-30			al, aa	$B5^3$, 2395
5051	Chrysene, 3-methyl $C_{19}H_{14}$	242.32	lf (bz-peth)		172-3			al	$B5^3$, 2395
5052	5,6-Chrysoquinone $C_{18}H_{10}O_2$	258.28	red nd (bz or to), lf or pl (aa)	sub	239.5			al, bz	$B7^3$, 4285
5053	6,12-Chrysoquinone $C_{18}H_{10}O_2$	258.25	red ye nd (aa)		288-90 d			al, aa	$B7^3$, 4285
5054	Cinchonamine $C_{19}H_{24}N_2O$	296.41	rh nd (al), orh pr (MeOH), $[\alpha]^{20}/_D$ + 123 (al, c=0.66		186 (194)			al, eth, bz, chl	$B23^2$, 358
5055	Cinchonicine or Cinchotoxine $C_{19}H_{22}N_2O$	294.40	nd or pr (eth), $[\alpha]^{15}/_D$ + 48 (al, c=1)		58-60			al, eth, ace, bz, chl	$B24^2$, 100
5056	Cinchonidine or Cinchovatine $C_{19}H_{22}N_2O$	294.40	orh pl or pr (al), $[\alpha]^{20}/_D$ - 109.2 (al, p=1)	sub	210.5			al	$B23^2$, 373
5057	Cinchonidine, hydrochloride $C_{19}H_{22}N_2O\text{-}HCl\text{-}H_2O$	348.87	wh pr (w), $[\alpha]^{20}/_D$ - 117.6 (w, c=1.2)		242d (anh)			w, al, chl	$B23^2$, 373
5058	Cinchonidine, sulfate $(C_{19}H_{22}N_2O)_2\text{-}H_2SO_4\cdot 6H_2O$	794.97	mcl nd (al + 2w), $[\alpha]^{18.5}/_D$ - 97.9 (w, c=1.2)		205 (anh)				$B23^2$, 373
5059	β-Cinchonidine $C_{19}H_{22}N_2O$	294.40	pr or lf (al or dil al)		241			al, chl	$B23^1$, 131
5060	Cinchonine $C_{19}H_{22}N_2O$	294.40	nd or mcl cr (al)	sub	255 (265)			chl	$B23^2$, 369
5061	Cinchonine, dihydrochloride $C_{19}H_{22}N_2O\cdot 2HCl$	367.32	pl $[\alpha]^{24}/_D$ + 205.5 (w, c=3.6)					w, al	$B23^1$, 133
5062	Cinchonine, hydrochloride $C_{19}H_{22}N_2O\text{-}HCl\text{-}2H_2O$	366.89	mcl $[\alpha]^{25}/_D$ + 133.6 (chl, c=1.4)		ca 215 (anh)	1.234		w, al, chl	$B23^2$, 370
5063	Cinchonine, sulfate $(C_{19}H_{22}N_2O)_2\text{-}H_2SO_4$	686.87	rh pr (w), $[\alpha]^{15}/_D$ + 169 (w, c=1.4)		206-7			al, w	$B23^2$, 371
5064	Cinchotine or Hydrocinchonine $C_{19}H_{24}N_2O$	296.41	pr $[\alpha]^{21}/_D$ + 203.4		268-9			w	$B23^2$, 356
5065	1,4-Cineole or 1,4-epoxy-p-menthane $C_{10}H_{16}O$	154.24		173-4	1	0.8997^{20}	1.4562^{20}	al, eth, bz, lig	$B17^4$, 213
5066	1,8-Cineole or Eucalyptol $C_{10}H_{16}O$	154.24		176.4, 61^{14}	1.5	0.9267^{20}	1.4586^{20}	al, eth, chl	$B17^4$, 273
5067	Cineolic acid-*(d)* $C_{10}H_{16}O_5$	216.23	cr (w + 1), $[\alpha]^{20}/_D$ + 18.6 (w, p=8.21)		79 (+w), 138-9 (anh)			al	B18, 322
5068	Cineolic acid-*(dl)* $C_{10}H_{16}O_5$	216.23			(i)197.5 (ii)208			al, eth	$B18^4$, 4446
5069	Cineolic acid-*(l)* $C_{10}H_{16}O_5$	234.23	rh (w + 1), $[\alpha]^{20}/_D$ - 19.1(w)		79 (+w), 138.9 (anh)			al	B18, 322

PHYSICAL CONSTANTS OF ORGANIC COMPOUNDS (Continued)

No.	Name, Synonyms, and Formula	Mol. wt.	Color, crystalline form, specific rotation and λ_{max} (log ε)	b.p. °C	m.p. °C	Density	n_D	Solubility	Ref.
5070	Cinnamaldehyde-*(trans)* or β-Phenyl acrolein $C_6H_5CH=CHCHO$	132.16	yesh	253d, 127^{16}	-7.5	$1.0497^{20/4}$	1.6195^{20}	al, eth, chl	$B7^3$, 1364
5071	Cinnamaldehyde, β-bromo-*(cis)* $C_6H_5CBr=CHCHO$	211.06	...	$144\text{-}6^{12}$	...	$1.492^{20/4}$	1.6368^{20}	eth	$B7^3$, 1383
5072	Cinnamaldehyde, α-ethyl $C_6H_5CH=C(C_2H_5)CHO$	160.22	...	$157\text{-}8^{5}$	...	$1.0201^{22/4}$	1.578^{20}	...	$B7^3$, 1435
5073	Cinnamaldehyde, 4-hydroxy-3-methoxy or Coniferaldehyde $[4\text{-}(HO)\text{-}3\text{-}(CH_3O)C_6H_3]CH=CHCHO$	178.19	cr (bz)	$157^{2.5}$	84	...	...	al, eth, bz	$B8^3$, 2331
5074	Cinnamaldehyde, α-methyl $C_6H_5CH=C(CH_3)CHO$	146.19	ye	150^{100}	...	$1.0407^{17/4}$	1.6057^{17}	...	$B7^3$, 1412
5075	Cinnamaldehyde, 4-methyl $4\text{-}CH_3C_6H_4CH=CHCHO$	146.19	ye lf (dil al)	154^{25}	41.5	...	...	al	$B7^3$, 1414
5076	Cinnamaldehyde, 2-nitro $2\text{-}O_2NC_6H_4CH=CHCO$	177.16	nd (eth or al)	...	127.5	...	...	al, eth, chl	$B7^3$, 1387
5077	Cinnamaldehyde, 3-nitro $3\text{-}O_2NC_6H_4CH=CHCHO$	177.16	ye nd (w), pr (al), cr (aa)	...	116	...	...	bz, aa	$B7^2$, 282
5078	Cinnamaldehyde, 4-nitro $4\text{-}O_2NC_6H_4CH=CHCHO$	177.16	nd (w or al)	...	141-2	...	...	al, eth, ace, bz	$B7^3$, 1388
5079	Cinnamaldehyde, oxime-*(trans)* $C_6H_5CH=CHCH=NOH$	147.18	nd	...	138.5	...	...	bz	$B7^3$, 1376
5080	Cinnamamide-*(trans)* $C_6H_5CH=CHCONH_2$	147.18	nd (bz)	...	148	...	...	al, eth	$B9^3$, 2711
5081	Cinnamic acid-*(cis)*-(1st form) or Phenyl acrylic acid $C_6H_5CH=CHCO_2H$	148.16	mcl pr (w)	...	42	...	...	al, aa, lig	$B9^3$, 2670
5082	Cinnamic acid-*(cis)*-(2nd form) $C_6H_5CH=CHCO_2H$	148.16	mcl pr (lig)	265	58	...	...	al, eth, ace, chl	$B9^3$, 2670
5083	Cinnamic acid-*(cis)*-(3rd form) or Allocinnamic acid $C_6H_5CH=CHCO_2H$	148.16	mcl pr	...	68	...	...	al, eth, lig	$B9^3$, 2670
5084	Cinnamic acid -*(trans)* $C_6H_5CH=CHCO_2H$	148.16	mcl pr (dil al)	300 (cor)	135-6	$1.2475^{4/4}$	...	al, eth, ace, bz, chl	$B9^3$, 2671
5085	Cinnamic acid, α-acetamido $C_6H_5CH=C(NHOCH_3)CO_2H$	205.21	cr (+ 2w)	...	193-4	...	...	w	$B10^3$, 3001
5086	Cinnamic acid, allyl ester or Allylcinnamate $C_6H_5CH=CHCO_2CH_2CH=CH_2$	188.23	...	268d, 163^{17}	...	1.048^{23}	1.530^{20}	al, eth	$B9^2$, 387
5087	Cinnamic acid, 2-amino-*(trans)* $2\text{-}H_2NC_6H_4CH=CHCO_2H$	163.18	ye nd (w)	...	158-9d	...	...	al, eth	$B14^3$, 1304
5088	Cinnamic acid, 3-amino-*(trans)* $3\text{-}H_2NC_6H_4CH=CHCO_2H$	163.18	nd (al)	...	191-3	...	...	aa	$B14^3$, 1305
5089	Cinnamic acid, 4-amino $4\text{-}H_2NC_6H_4CH=CHCO_2H$	163.18	ye nd (w or al)	...	175-6d	...	...	al, eth	$B14^3$, 1305
5090	Cinnamic acid-*(trans)*, anhydride $(C_6H_5CH=CHCO)_2O$	278.31	nd (bz or al) pr, (al)	...	138	...	...	bz	$B9^3$, 2703
5091	Cinnamic acid, 2-benzamido-*(trans)* $2\text{-}(C_6H_5CONH)C_6H_4CH=CHCO_2H$	267.28	nd (al)	...	191-3	...	...	aa	$B14^3$, 1304
5092	Cinnamic acid, 3-benzamido-*(trans)* $3\text{-}(C_6H_5CONH)C_6H_4CH=CHCO_2H$	267.28	nd (AcOEt)	...	229	...	...	ace, aa	$B14^3$, 1305
5093	Cinnamic acid, 4-benzamido-*(trans)* $4\text{-}(C_6H_5CONH)C_6H_4CH=CHCO_2H$	267.28	lf (aa), nd (ace)	...	274d	...	...	ace, bz, aa	$B14^3$, 1312
5094	Cinnamic acid, benzyl ester-*(trans)* or Benzyl cinnamate $C_6H_5CH=CHCO_2CH_2C_6H_5$	238.29	pr	350d, 244^{5}	39	1.109^{15}	...	al, eth	$B9^3$, 2691
5095	Cinnamic acid, α-bromo-*(cis)* $C_6H_5CH=CBrCO_2H$	227.06	lf (w), pr (chl)	...	120-1	...	...	al, bz	$B9^3$, 2734
5096	Cinnamic acid, α-bromo-*(trans)* $C_6H_5CH=CBrCO_2H$	227.06	nd (w)	...	131-2	...	...	al, eth, bz	$B9^3$, 2734
5097	Cinnamic acid, β-bromo-*(cis)* $C_6H_5CBr=CHCO_2H$	227.06	nd (bz), pl (al)	...	159-60	...	...	eth, chl	$B9^3$, 2732
5098	Cinnamic acid, β-bromo-*(trans)* $C_6H_5CBr=CHCO_2H$	227.06	pa ye nd or pl (w), pr (chl)	...	135	...	...	al, bz	$B9^3$, 2733
5099	Cinnamic acid, 2-carboxy-*(trans)* $(2\text{-}HO_2CC_6H_4)CH=CHCO_2H$	192.17	pr or nd (w)	...	208-9	...	...	al	$B9^3$, 4384
5100	Cinnamic acid, 3-carboxy-*(trans)* $(3\text{-}HO_2CC_6H_4)CH=CHCO_2H$	192.17	nd (ace-lig)	...	275	...	...	aa	$B9^2$, 642
5101	Cinnamic acid, 4-carboxy-*(trans)* $4\text{-}HO_2CC_6H_4CH=CHCO_2H$	192.17	pw	sub> 350	358d	...	...	...	$B9^2$, 642

PHYSICAL CONSTANTS OF ORGANIC COMPOUNDS (Continued)

No.	Name, Synonyms, and Formula	Mol. wt.	Color, crystalline form, specific rotation and λ_{max} (log ε)	b.p. °C	m.p. °C	Density	n_D	Solubility	Ref.
5102	Cinnamic acid, α-chloro-*(cis)* $C_6H_5CH{=}CClCO_2H$	182.61			111			al	B9^3, 2729
5103	Cinnamic acid, α-chloro,methyl ester-*(trans)* $C_6H_5CH{=}CHClCO_2CH_3$	196.63		108-9$^{0.5}$	33			ace	B9^2, 396
5104	Cinnamic acid, β-chloro-*(cis)* $C_6H_5CCl{=}CHCO_2H$	182.61			133			al, eth	B9^3, 2728
5105	Cinnamic acid, β-chloro-*(trans)* $C_6H_5CCl{=}CHCO_2H$	182.61			143			al, eth	B9^3, 2728
5106	Cinnamic acid, β-chloro,methyl ester $C_6H_5CCl{=}CHCO_2CH_3$	196.63		113-4$^{0.5}$	29	1.2248$^{21/4}$	1.5781^{21}	ace	B9^2, 396
5107	Cinnamic acid, α,β-dibromo-*(cis)* $C_6H_5CBr{=}CBrCO_2H$	305.95	ye pr or pl (chl or lig)	124$^{0.5}$	100			eth, chl, aa	B9^3, 2736
5108	Cinnamic acid, 2,4-dihydroxy-*(trans)* or Umbellic acid $2,4\text{-}(HO)_2C_6H_3CH{=}CHCO_2H$	180.16	ye nd or pl		260d 240 (darkens)			al	B10^3, 1830
5109	Cinnamic acid, 2,5-dihydroxy-*(trans)* $2,5\text{-}(HO)_2C_6H_3CH{=}CHCO_2H$	180.16	ye cr (w), cr (dil al + 1w)		207d			al	B10^3, 1833
5110	Cinnamic acid, 3,4-dihydroxy-*(trans)* or Caffeic acid $3,4\text{-}(HO)_2C_6H_3CH{=}CHCO_2H$	180.17	ye pr or pl (w)		225d			al	B10^3, 1834
5111	Cinnamic acid, 3,5-dihydroxy-*(trans)* $3,5\text{-}(HO)_2C_6H_3CH{=}CHCO_2H$	180.16	nd (w + ½)		245-6			al, eth	B10^2, 297
5112	Cinnamic acid, 2,3-dimethoxy-*(trans)* $2,3\text{-}(CH_3O)_2C_6H_3CH{=}CHCO_2H$	208.21			180-1				B10^3, 1829
5113	Cinnamic acid, 2,4-dimethoxy-*(cis)* $2,4\text{-}(CH_3O)_2C_6H_3CH{=}CHCO_2H$	208.21	nd (al)		138			al, eth, bz	B10^3, 1831
5114	Cinnamic acid, 2,4-dimethoxy-*(trans)* $2,4\text{-}(CH_3O)_2C_6H_3CH{=}CHCO_2H$	208.21	nd (w or dil al)		187-9			al, eth, bz, chl	B10^3, 1831
5115	Cinnamic acid, 2,5-dimethoxy-*(trans)* $2,5\text{-}(CH_3O)_2C_6H_3CH{=}CHCO_2H$	208.21	pa ye or ye-gr nd (w)		148-9			al, eth	B10^3, 1833
5116	Cinnamic acid, 3,4-dimethoxy-*(trans)* $3\text{-}4\text{-}(CH_3O)_2C_6H_3CH{=}CHCO_2H$	208.21	nd (w, dil al), pa ye pw (dil aa)		183			al, eth	B10^3, 1835
5117	Cinnamic acid, 3,4-dimethoxy, methyl ester $3,4\text{-}(CH_3O)_2C_6H_3CH{=}CHCO_2CH_3$	222.24			68-70			chl	B10^3, 1838
5118	Cinnamic acid, 3,5-dimethoxy-*(trans)* $3,5\text{-}(CH_3O)_2C_6H_3CH{=}CHCO_2H$	208.22	nd (w)		175-6			al, bz	B10^2, 297
5119	Cinnamic acid, 3,5-dimethoxy-4-hydroxy-*(trans)* or 5-Methoxyferulic acid $4\text{-}HO\text{-}3,5\text{-}(CH_3O)_2C_6H_2CH{=}CHCO_2H$	224.21	pa ye nd (al)		192			al	B10, 508
5120	Cinnamic acid, α-ethyl-*(cis)* or α-Benzal butyric acid $C_6H_5CH{=}C(C_2H_5)CO_2H$	176.22	nd (w)		82			al, eth, bz	B9^3, 2783
5121	Cinnamic acid, α-ethyl-*(trans)* or α-Benzal butyric acid $C_6H_5CH{=}C(C_2H_5)CO_2H$	176.22	nd (w)		106 (114)			al, eth	B9^3, 2783
5122	Cinnamic acid, ethyl ester-*(trans)* or Ethyl cinnamate $C_6H_5CH{=}CHCO_2C_2H_5$	176.22		271.5, 144^{15}	12	1.0491$^{20/4}$	1.5598^{20}	al, eth, ace, bz	B9^3, 2682
5123	Cinnamic acid, α-fluoro $C_6H_5CH{=}CFCO_2H$	166.15		290	157.6			al, eth	B9^1, 237
5124	Cinnamic acid, 2-hydroxy-*(trans)* or *o*-Coumaric acid $2\text{-}HOC_6H_4CH{=}CHCO_2H$	164.16	nd (w)		217d			al	B10^3, 833
5125	Cinnamic acid, 3-hydroxy-*(trans)* or *m*-Coumaric acid $3\text{-}HOC_6H_4CH{=}CHCO_2H$	164.16	pr (w)		193			al, eth, bz	B10^3, 840
5126	Cinnamic acid-4-hydroxy-*(trans)* or *p*-Coumaric acid $4\text{-}HOC_6H_4CH{=}CHCO_2H$	164.16	nd (w + 1), cr (w)		215d			eth	B10^3, 844
5127	Cinnamic acid-4-hydroxy-3-methoxy or Ferulic acid $4\text{-}(HO)\text{-}3\text{-}CH_3OC_6H_3CH{=}CHCO_2H$	194.19	pr or nd (w)		171			al, chl	B10^3, 1834
5128	Cinnamic acid-2-methoxyphenyl ester-*(trans)* or *o*-Anisylcinnamate $C_6H_5CH{=}CHCO_2(C_6H_4OCH_3\text{-}2)$	254.29	wh nd (al)		130			ace, bz, chl	B9^3, 2698
5129	Cinnamic acid-4-methoxy $4\text{-}CH_3OC_6H_4CH{=}CHCO_2H$	178.19	wh nd (al)		172-5			aa	B10^3, 845
5130	Cinnamic acid-methyl ester-*(trans)* $C_6H_5CH{=}CHCO_2CH_3$	162.19	cr (peth or dil al)	261.9, 127^{10}	36.5	1.0911$^{20/4}$	1.5766^{22}	al, eth, ace, bz	B9^3, 2680
5131	Cinnamic acid-α-methyl-*(cis)* or α-Benzal propionic acid $C_6H_5CH{=}C(CH_3)CO_2H$	162.19	nd (bz)	288, 190^{21}	74			al, eth, bz, peth	B9^3, 2764

No.	Name, Synonyms, and Formula	Mol. wt.	Color, crystalline form, specific rotation and λ_{max} (log ε)	b.p. °C	m.p. °C	Density	n_D	Solubility	Ref.
5132	Cinnamic acid-α-methyl-*(trans)* or α-Benzal propionic acid $C_6H_5CH=C(CH_3)CO_2H$	162.19	pr (aa, eth or dil al)	...	81-2	...	...	al, eth, bz, peth	B9^3, 2764
5133	Cinnamic acid-α-methyl-2-nitro $2\text{-}O_2NC_6H_4CH=C(CH_3)CO_2H$	207.19	mcl pr (al)	...	164-5	...	...	al, eth	B9^3, 2766
5134	Cinnamic acid-α-methyl-3-nitro $3\text{-}O_2NC_6H_4CH=C(CH_3)CO_2H$	207.19	nd or pw	...	203.5	...	...	eth, bz, aa	B9^3, 2767
5135	Cinnamic acid-α-methyl-4-nitro $4\text{-}O_2NC_6H_4CH=C(CH_3)CO_2H$	207.19	ye rh (aa), tcl pym (aa, al-eth)	...	208	...	...	al, eth, bz	B9^3, 2767
5136	Cinnamic acid-2-methyl-4-nitro $[2\text{-}(CH_3)\text{-}4\text{-}(O_2N)C_6H_3]CH=CHCO_2H$	207.19	nd (al)	...	256	...	...	...	B9^1, 256
5137	Cinnamic acid-4-methyl-3-nitro-*(trans)* $[4\text{-}(CH_3)\text{-}3\text{-}(O_2N)C_6H_3]CH=CHCO_2H$	207.19	ye pl or nd (al)	...	173.5	...	...	al, eth	B9^3, 2770
5138	Cinnamic acid-2-nitro-*(cis)* $2\text{-}O_2NC_6H_4CH=CHCO_2H$	193.16	yesh (bz or chl)	...	146-7	...	...	al, bz, chl	B9^1, 246
5139	Cinnamic acid-2-nitro-*(trans)* $2\text{-}O_2NC_6H_4CH=CHCO_2H$	193.16	nd (al)	sub	242-3	...	...	...	B9^3, 2739
5140	Cinnamic acid-2-nitro, ethyl ester-*(trans)* $2\text{-}O_2NC_6H_4CH=CHCO_2C_2H_5$	221.21	ye rh bipym (al)	...	44	...	...	al, eth, bz	B9^3, 2739
5141	Cinnamic acid-2-nitro,methyl ester-*(trans)* $2\text{-}O_2NC_6H_4CH=CHCO_2CH_3$	207.19	wh nd (w)	187-9^{15}	73	...	...	al	B9^3, 2739
5142	Cinnamic acid-3-nitro-*(cis)* $3\text{-}O_2NC_6H_4CH=CHCO_2H$	193.16	nd	...	158	...	...	...	B9^1, 247
5143	Cinnamic acid-3-nitro-*(trans)* $3\text{-}NO_2C_6H_4CH=CHCO_2H$	193.16	nd (al)	...	204-5	...	...	al	B9^3, 2741
5144	Cinnamic acid-3-nitro,ethyl ester-*(trans)* $3\text{-}NO_2C_6H_4CH=CHCO_2C_2H_5$	221.21	nd (al), pr (aa)	...	78-9	...	...	...	B9^3, 2742
5145	Cinnamic acid-3-nitro,methyl ester-*(trans)* $3\text{-}O_2NC_6H_4CH=CHCO_2CH_3$	207.19	pa ye pr (MeOH)	d	123-4	...	...	eth, bz, chl	B9^3, 2742
5146	Cinnamic acid-4-nitro-*(trans)* $4\text{-}O_2NC_6H_4CH=CHCO_2H$	193.16	yesh-wh pr (al)	...	286	...	...	...	B9^3, 2744
5147	Cinnamic acid-4-nitro,ethyl ester-*(trans)* $4\text{-}O_2NC_6H_4CH=CHCO_2C_2H_5$	221.21	pl (aa)	...	141-2	...	...	...	B9^3, 2744
5148	Cinnamic acid-4-nitro,methyl ester-*(trans)* $4\text{-}O_2NC_6H_4CH=CHCO_2CH_3$	207.19	wh nd (al)	281-6	162	...	...	al	B9^3, 2744
5149	Cinnamic acid-2-octyl ester-*(d)* $C_6H_5CH=CHCO_2CH(CH_3)C_6H_{13}$	260.38	$[\alpha]^{17}{}_D$ +40.2	218^{28}	...	0.9645$^{20/4}$	1.5145^{20}	bz, chl	B9^3, 8687
5150	Cinnamic acid-2-octyl ester-*(trans,dl)* $C_6H_5CH=CHCO_2CH(CH_3)C_6H_{13}$	260.38	...	240^{60}	...	0.9715$^{17/4}$	...	bz, chl	B9^1, 230
5151	Cinnamic acid, 2-octyl ester*(trans,l)* $C_6H_5CH=CHCO_2CH(CH_3)C_6H_{13}$	260.38	$[\alpha]^{17}{}_D$ -39.78	211^{28}	...	0.9692$^{17/4}$	...	bz, chl	B9^3, 2687
5152	Cinnamic acid, α-phenyl or α-Phenyl cinnamic acid $C_6H_5CH=C(C_6H_5)CO_2H$	224.26	nd (lig, dil al)	sub	134-5	...	...	al, eth	B9^3, 3414
5153	Cinnamic acid, phenylester-*(trans)* $C_6H_5CH=CHCO_2H_5$	224.26	...	205-7^{15}	72.5	...	...	eth, bz	B9^3, 2689
5154	Cinnamic acid, propyl ester-*(trans)* $C_6H_5CH=CHCO_2C_3H_7$	190.24	...	285	...	1.0435$^{0/0}$	...	...	B9^3, 2684
5155	Cinnamic acid, *iso*-propyl ester-*(trans)* or Isopropyl cinnamate $C_6H_5CH=CHCO_2CH(CH_3)_2$	190.24	...	268-70, 153-5^{20}	...	1.0320^{20}	1.5455^{20}	al, eth, ace	B9^3, 2685
5156	Cinnamic acid, 4-isopropyl or Cumiliden acetic acid $4\text{-}i\text{-}C_3H_7C_6H_4CH=CHCO_2H$	190.24	pr (bz)	...	165	...	...	al	B9^3, 2814
5157	Cinnamic acid, 3,4,5-trimethoxy $3,4,5\text{-}(CH_3O)_3C_6H_2CH=CHCO_2H$	238.24	nd (w)	...	126-7	...	...	chl	B10^3, 2200
5158	Cinnamonitrile, *(cis)* or Allocinnamonitrile $C_6H_5CH=CHCN$	129.16	...	249, 139^{30}	-4.4	...	1.5843^{20}	al, bz	B9^3, 2720
5159	Cinnamonitrile, *(trans)* $C_6H_5CH=CHCN$	129.16	...	263.8, 134-6^{28}	22	1.0304$^{20/4}$	1.6013^{20}	al, ace	B9^3, 2721
5160	Cinnamylalcohol, (cis) or 3-phenyl-2-propen-2-ol $C_6H_5CH=CHCH_2OH$	134.18	wh nd (eth-peth)	257.5, 127-8^{10}	34	1.0440$^{20/4}$	1.5819^{20}	al, eth	B6^3, 2401
5161	Cinnamyl alcohol, acetate-*(trans)* or Cinnamyl acetate $C_6H_5CH=CHCH_2O_2CCH_3$	176.22	...	145-6^{15}, 114^{1}	...	1.0567^{20}	1.5425^{20}	al, eth, ace, bz	B6^3, 2406
5162	Cinnamyl chloride, *(trans)* $C_6H_5CH=CHCOCl$	166.61	...	257.5, 131^{11}	37-8	1.1617$^{45/4}$	1.614$^{42.5}$	lig	B9^3, 2710

No.	Name, Synonyms, and Formula	Mol. wt.	Color, crystalline form, specific rotation and λ_{max} (log ε)	b.p. °C	m.p. °C	Density	n_D	Solubility	Ref.
5163	Citraconic acid or Methyl maleic acid $HO_2CC(CH_3)=CHCO_2H$	130.10	tcl pr or pl (eth-bz) nd (eth-lig)		93-4	1.617		w	$B2^4$, 2230
5164	Citraconic anhydride or Methylmaleic anhydride $C_5H_4O_3$	112.08		213-4, $99\text{-}100^{15}$	7-8	$1.2469^{16/4}$	1.4716^{21}	al, eth, ace	$B17^4$, 5912
5165	Citraconic acid, diethyl ester or Diethyl citraconate $C_2H_5O_2CC(CH_3)=CHCO_2C_2H_5$	186.21		228^{766}, 120^{20}		$1.0491^{20/4}$	1.4467^{20}	al, eth, aa	$B2^4$, 2232
5166	Citraconic acid, dimethyl ester or Dimethyl citraconate $CH_3O_2CC(CH_3)=CHCO_2CH_3$	158.15		210.5^{758}, 92.8^{10}		$1.1153^{20/4}$	1.4473^{20}	al, eth, ace, aa	$B2^4$, 2232
5167	Citral a or Geranial $C_{10}H_{16}O$	152.24		229, $118\text{-}9^{20}$		0.8888^{20}	1.4898^{20}	al, eth	$B1^4$, 3569
5168	Citral b or Neral $C_{10}H_{16}O$	152.24		120^{20}		0.8869^{20}	1.4869^{20}	al, eth	$B1^4$, 3569
5169	β-Citraurin $C_{30}H_{40}O_2$	432.65	pl (bz-peth), cr (al)		147			al, eth, ace, bz	$B8^3$, 1599
5170	Citric acid or 2-Hydroxy-1,2,3-propane tricarboxylic acid $HOC(CH_2CO_2H)_2CO_2H$	192.13	rh (w + 1)	d	153 anh	$1.665^{20/4}$		w, al, eth	$B3^4$, 1272
5171	Citric acid, anhydro-methylene or 1,3-Dioxolan-4-one-5,5-diacetic acid $C_9H_{12}O_7$	220.18	(w)		298			bz, chl	$B19^2$, 324
5172	Citric acid, tribenzyl ester or Benzyl citrate $HOC(CH_2CO_2CH_2C_6H_5)_2CO_2CH_2C_6H_5$	462.50	nd (al)		51			al	$B6^3$, 1537
5173	Citric acid, triethyl ester or Ethyl citrate $HOC(CH_2CO_2C_2H_5)_2CO_2C_2H_5$	276.29		294, 185^{17}		$1.1369^{20/4}$	1.4455^{20}	al, eth	$B3^4$, 1276
5174	Citric acid, trimethyl ester or Methyl citrate $HOC(CH_2CO_2CH_3)_2CO_2CH_3$	234.21	tol	287d, 176^{16}	78.9			al, eth	$B3^4$, 1276
5175	Citric acid, triphenyl ester or Phenyl citrate $HOC(CH_2CO_2C_6H_5)_2CO_2C_6H_5$	420.42	nd (al)		124.5			eth	B6, 170
5176	Citric acid, tripropyl ester or Propyl citrate $HOC(CH_2CO_2C_3H_7)_2CO_2C_3H_7$	318.37		198^{18}				al, eth	B3, 568
5177	Citric acid, triethyl ester, acetate $CH_3CO_2C(CH_2CO_2C_2H_5)CO_2C_2H_5$	318.32		131-2		1.135	1.4380		
5178	Citric triamide $HOC(CH_2CONH_2)_2CONH_2$	189.17	(w)		210-5d			w	$B3^3$, 197
5179	Citrinin $C_{13}H_{14}O_5$	250.25	ye nd (MeOH), $[\alpha]^{11'}_D$ -37		178-9d			ace, bz, chl	$B18^4$, 6329
5180	Citronellal-*(d)* or *d*-Rhodinal $(CH_3)_2C=CH(CH_2)_2CH(CH_3)CH_2CHO$	154.25	$[\alpha]^{18'}_D$ + 13.09	207.8, 92^{14}		$0.8573^{20/4}$	1.4456^{20}	al, eth	$B1^4$, 3515
5181	Citronellal-*(dl)* or *dl*-Rhodinal $C_{10}H_{18}O$	154.25		207-8, $79\text{-}81^{10}$		$0.8535^{17/4}$	1.4473^{20}		$B1^4$, 3515
5182	Citronellal-*(l)* or *l*-Rhodinal $C_{10}H_{18}O$	154.25	$[\alpha]^{20'}_D$ -2.5	205-6, 87^{10}		$0.8567^{17/4}$	1.4479^{20}	al, eth	$B1^4$, 3515
5183	Citronellol-*(d)* or *d*-Rhodinol $(CH_3)_2C=CH(CH_2)_2CH(CH_3)CH_2CH_2OH$	156.27	$[\alpha]^{17'}_D$ + 6.8	244.4, 118^{17}		0.8590^{20}	1.4565^{20}	al, eth	$B1^4$, 2188
5184	Citronellol-*(dl)* or Dihydrogeranoil $C_{10}H_{20}O$	156.27		99^{10}		$0.8560^{20/4}$	1.4543^{20}	al, eth	$B1^4$, 2188
5185	Citronellol-*(l)* or *l*-Rhodinol $C_{10}H_{20}O$	156.27	$[\alpha]^{18'}_D$ -5.3	$108\text{-}9^{10}$		$0.859^{18/4}$	1.4576^{18}	al, eth	$B1^4$, 2188
5186	Citrulline-*(L)* or α-Amino-δ-ureido valeric acid $H_2NCONH(CH_2)_3CH(NH_2)CO_2H$	175.19	pr (aq MeOH), $[\alpha]^{20'}_D$ + 3.7 (w, c=2)		234-7			w	$B4^3$, 1347
5187	Clovene $C_{15}H_{24}$	204.36	$[\alpha]_D$ + 2.84	259-60		0.9241^{18}	1.4999^{18}		$B5^2$, 356
5188	Cocaine-*(d)* or Benzoyl methyl ecgonine $C_{17}H_{21}NO_4$	303.36	mcl pr (eth), $[\alpha]^{20'}_D$ + 15.8 (chl, p=10)		98			al, eth, ace, bz	$B22^2$, 150
5189	Cocaine-*dl* $C_{17}H_{21}NO_4$	303.36	rh bipym pr (peth)		79-80			al, eth, ace, bz, lig	$B22^4$, 2103

PHYSICAL CONSTANTS OF ORGANIC COMPOUNDS (Continued)

No.	Name, Synonyms, and Formula	Mol. wt.	Color, crystalline form, specific rotation and λ_{max} (log ε)	b.p. °C	m.p. °C	Density	n_D	Solubility	Ref.
5190	Cocaine-*(l)* $C_{17}H_{21}NO_4$	303.36	mcl pr (al), $[\alpha]^{20}_{D}$ -16.3 (chl, c=4)	$187\text{-}8^{0.1}$, (sub, vac)	98		1.5022^{98}	al, eth, ace, bz, chl	$B22^4$, 2101
5191	Cocaine, hydrochloride-*(dl)* $C_{17}H_{21}NO_4$-HCl	339.82	pl (al)		187 (cor)			w, al, ace	$B22^2$, 156
5192	Cocaine, hydrochloride-*(l)* $C_{17}H_{21}NO_4$-HCl	339.82	mcl pr (al), cr (w + 2), $[\alpha]^{20}_{D}$ -71.95		197			w, al, ace	$B22^4$, 2102
5193	Coclaurine-*(l)* $C_{17}H_{19}NO_3$	285.34	pl (al), $[\alpha]^{20}_{D}$ -17.01		220-1				$B21^4$, 2605
5194	Codamine $C_{20}H_{25}NO_4$	343.42	pr (bz or eth)		127			al, eth, chl	$B21^4$, 2704
5195	Codeine or Morphine-3-methylether $C_{18}H_{21}NO_3$	299.37	rh oct (+ 1w, w or dil al), cr (eth)	250^{22}, sub $140^{1.5}$	157-8	1.32		al, eth, bz, chl	$B27^2$, 137
5196	Codeine, hydrate $C_{18}H_{21}NO_3$-H_2O	317.38	rh oct (w, aq al), $[\alpha]^{25}_{D}$ -136 (al, c=2.8)			1.31		al, eth, ace, bz, chl	$B27^2$, 137
5197	Codeine, hydrochloride $C_{18}H_{21}NO_3$-HCl-$2H_2O$	371.86	nd, $[\alpha]^{15}_{D}$ -108.2		287d			w, al	$B27^2$, 143
5198	Codeine, phosphate $C_{18}H_{21}NO_3 \cdot H_3PO_4 \cdot 1\frac{1}{2}H_2O$	424.39	lf or pr (dil al)		220-35d			eth, chl	$B27^2$, 144
5199	Codeine, sulfate $(C_{18}H_{21}NO_3)_2$-$H_2SO_4 5H_2O$	786.89	pr, $[\alpha]^{15}_{D}$ -100.9 (w, p=3)		278 (anh)			w	$B27^2$, 144
5200	β-Codiene or Neopine $C_{18}H_{21}NO_3$	299.37	nd (peth), $[\alpha]^{23}_{D}$ -28 (chl, c=7.5)		127.5			al, eth, bz, chl	$B27^2$, 176
5201	Codeine, dihydro-*(d)* or Dihydroneopine $C_{18}H_{23}NO_3$	301.39	cr (+ 1w, dil MeOH)	248^{15}	112-3				$B27^2$, 103
5202	Colchiceine or *N*-Acetyltrimethyl colchicinic acid $C_{21}H_{23}NO_6$	385.42	pa ye nd (diox)		178-9	1.24		al, chl	$B14^3$, 692
5203	Colchicine $C_{22}H_{25}NO_6$	399.44	ye pl (w + 1½), pa ye nd (AcOEt), ye cr (bz), $[\alpha]^{17}_{D}$ -121 (chl, c=0.9), -429 (w, c=1.72)		155-7			w, al	$B14^3$, 693
5204	Colchicinic acid, trimethyl $C_{19}H_{21}NO_5$	342.38	pa ye nd (al), $[\alpha]^{25}_{D}$ -184.5 (chl, c=1)		155-7			w, al	Am75, 5292
5205	Conessine or Neriine $C_{24}H_{40}N_2$	356.60	lf or pl (ace), $[\alpha]^{20}_{D}$ + 25.3 (al, c=0.7)	$165\text{-}7^{0.1}$	125-6			aa	$B22^4$, 4382
5206	Congo red $C_{32}H_{22}N_6Na_2O_6S_2$	696.66	pw					al	$B16^3$, 474
5207	Conhydrine-*(d)* or 2(1-Hydroxypropyl)piperidine $C_8H_{17}NO$	143.23	lf (eth), $[\alpha]_D$ + 10 (w)	226	121			w, al, eth, chl	$B21^4$, 122
5208	Conhydrine-*(dl)*(lower m.p.) $C_8H_{17}NO$	143.23	nd (peth)	sub	69-70			w, al, eth, bz	$B21^4$, 122
5209	Conhydrine-*(dl)*(higher m.p.) or 2-(1-Hydroxypropyl) piperidine $C_8H_{17}NO$	143.23	nd (eth)	sub	98-9			w, al, eth, bz	$B21^4$, 123

PHYSICAL CONSTANTS OF ORGANIC COMPOUNDS (Continued)

No.	Name, Synonyms, and Formula	Mol. wt.	Color, crystalline form, specific rotation and λ_{max} (log ε)	b.p. °C	m.p. °C	Density	n_D	Solubility	Ref.
5210	α-Coniceine-*(d)* or 2-Methylconidine $C_8H_{15}N$	125.21	$[\alpha]^{15}_{D}$ +18.4 (al, c=2)	158	-16	$0.891^{15.5/4}$	...	al	$B20$, 152
5211	Corticosterone, 11-dehydro $C_{21}H_{28}O_4$	344.45	pr (ace-w, al or ace-eth), $[\alpha]^{25}_{D}$ +258 (al)	...	183-4	...	...	al, ace, bz	$B8^3$, 3624
5212	α-Coniceine-*(dl)* or 2-Methyl conidine $C_8H_{15}N$	125.21	...	156-9	...	$0.890^{15.5/4}$	...	al	B20, 153
5213	β-Coniceine-*(dl)* or 2-Propenyl piperidine $C_8H_{15}N$	125.21	nd $[\alpha]^{45}_{D}$ +49.9	168-9	38-9	...	...	al, eth	B20, 146
5214	β-Coniceine-*(dl)* or 2-Propenylpiperidine $C_8H_{15}N$	125.21	nd	$168\text{-}70^{753}$	8	$0.8716^{15/4}$	...	al, eth	B20, 146
5215	β-Coniceine-*(l)* or 2-Propenylpiperidine $C_8H_{15}N$	125.21	nd $[\alpha]^{45}_{D}$ -50.5	168-9	41	$0.8520^{50/4}$	...	al, eth	B20, 146
5216	γ-Coniceine $C_8H_{15}N$	125.21	...	173-4, $64\text{-}5^{14}$	...	$0.8720^{20/4}$	1.4607^{18}	al	$B20^4$, 1970
5217	ε-Coniceine-*(d)* or *d*-2-Methyl coniceine $C_8H_{15}N$	125.21	$[\alpha]^{15}_{D}$ +67.4	152-4	...	$0.8856^{15/4}$	...	...	B20, 151
5218	ε-Coniceine-*(dl)* $C_8H_{15}N$	125.21	...	150-1	...	$0.8836^{15/4}$	...	...	B20, 152
5219	ε-Coniceine-*l* $C_8H_{15}N$	125.21	$[\alpha]^{15}_{D}$ -87.34	151-3.5	...	$0.8642^{15/4}$	...	...	B20, 151
5220	α-Conidendrin or Tsugalactone $C_{20}H_{20}O_6$	356.38	cr (al), $[\alpha]^{20}_{D}$ -51.5 (ace, c=2.1)	...	255-6	...	...	al, eth, bz	$B18^4$, 3346
5221	α-Conidine, 3-methyl-*(l)* $C_8H_{15}N$	125.21	$[\alpha]^{8}_{D}$ +16	158	...	$0.8856^{15/4}$	...	al, eth	B20, 153
5222	α-Conidine, 3-methyl-*(dl)* $C_8H_{15}N$	125.21	...	158	...	$0.8946^{15/4}$	...	al, eth	B20, 153
5223	Conidine, 3-methyl-*(l)* $C_8H_{15}N$	125.21	$[\alpha]^{17}_{D}$ -17.1	158	...	$0.8856^{15/4}$	...	al, eth	B20, 153
5224	Coniferin or 4-(3-hydroxypropenyl)-3-Methoxyphenyl-D-glucoside $C_{16}H_{22}O_8$	342.35	nd (w + 2), $[\alpha]^{20}_{D}$ -68 (w, c=0.5)	...	186 (anh)	...	...	...	$B17^4$, 2999
5225	Coniferyl alcohol $C_{10}H_{12}O_3$	180.20	pr (eth-lig)	$163\text{-}5^{3}$	74	...	...	al, eth	$B6^3$, 6442
5226	Coniine-*(d)* or 2-Propyl piperidine $C_8H_{17}N$	127.23	$[\alpha]^{20}_{D}$ +15.6	166-7, 64^{18}	-2	0.8440^{20}	1.4512^{22}	al, eth, bz	$B20^4$, 1611
5227	Coniine-*(dl)* or 2-Propyl piperidine $C_8H_{17}N$	127.23	...	$166\text{-}7^{745}$, $59\text{-}63^{17}$	...	0.8447^{20}	1.4513^{23}	al, eth, bz	$B20^4$, 1611
5228	Coniine-*(l)* $C_8H_{17}N$	127.23	$[\alpha]^{15}_{D}$ -15.6	166, 64^{18}	...	$0.845^{15/4}$	1.4512^{22}	al, eth, bz	$B20^2$, 62
5229	Coniine, hydrobromide-*(d)* $C_8H_{17}N\text{-}HBr$	208.14	pr	...	211	...	...	w, al, eth, chl	B20, 112
5230	Coniine, hydrochloride-*(d)* $C_8H_{17}N\text{-}HCl$	163.69	orth (w), $[\alpha]^{20}_{D}$ +10.1 (lig, NH_3)	...	221	...	...	w, al	$B20^4$, 1611
5231	Coniine, hydrochloride-*(dl)* $C_8H_{17}N\text{-}HCl$	163.69	nd (al-eth)	...	216-7	...	...	w, al	$B20^1$, 31
5232	Coniine, hydrochloride-*(l)* $C_8H_{17}N\text{-}HCl$	163.69	nd	...	220-1	...	...	w, al	B20, 118
5233	Coniine, *N*-methyl-*(d)* $C_9H_{19}N$	141.26	$[\alpha]^{24}_{D}$ +82.4	$173\text{-}4^{757}$	...	$0.8326^{23/4}$	1.4538^{13}	al, ace	$B20^4$, 1612
5234	Coniine, picrate-*(d)* $C_8H_{17}N\text{-}C_6H_3N_3O_7$	356.34	ye pr (w)	...	75	...	...	al, eth	B20, 112
5235	Coniine, picrate-*(l)* $C_8H_{17}N\text{-}C_6H_3N_3O_7$	356.34	ye pr (w)	...	74	...	...	al, eth	$B20^2$, 62
5236	Conquinamine $C_{19}H_{24}N_2O_2$	312.41	ye tetr, $[\alpha]^{15}_{D}$ +200 (al, c=0.5)	...	123	...	...	al, eth, chl	$B27^2$, 667
5237	Copaene $C_{15}H_{24}$	204.36	$[\alpha]_D$ -25.8	246-51, $119\text{-}20^{10}$	...	$0.8996^{20/4}$	1.4894^{20}	eth, ace, aa, lig	$B5^4$, 1189

No.	Name, Synonyms, and Formula	Mol. wt.	Color, crystalline form, specific rotation and λ_{max} (log ε)	b.p. °C	m.p. °C	Density	n_D	Solubility	Ref.
5238	Coproergostane or Pseudoergostane $C_{28}H_{50}$	386.71	nd (ace), $[\alpha]^{19/}{}_D$ + 25.3 (chl, c=2)		64			eth, chl	B5³, 1143
5239	Coprostane or Pseudocholestane $C_{27}H_{48}$	372.68	orth nd (al, ace or eth-al), $[\alpha]^{20/}{}_D$ + 25.1 (chl, c=2)		72	$0.9119^{87.7/4}$	$1.4884^{88/4}$	eth, chl	B5⁴, 1226
5240	3β-Coprostanol or Coprosterol $C_{27}H_{48}O$	388.68	nd (MeOH), $[\alpha]^{18/}{}_D$ + 28 (chl, c=1.8)		102 (105)			al, eth, bz, chl	B6³, 2128
5241	Coprostenol or Allocholesterol $C_{27}H_{46}O$	386.66	nd (eth-MeOH), $[\alpha]_D$ + 43.7 (bz, c=1)					al, eth, ace, bz, chl	B6³, 2604
5242	Coprostenone or Δ^4-Cholesten-3-one $C_{27}H_{44}O$	384.65	pl (MeOH), $[\alpha]^{/}{}_D$ + 88.6 (chl)		81-2			eth, bz, lig	B7³, 1592
5243	Coronene or Hexabenzobenzene $C_{24}H_{12}$	300.36	ye nd (bz)	525	438-40 (cor)	1.371			B5³, 2651
5244	Corticosterone or 11,21-Dihydroprogesterone $C_{21}H_{30}O_4$	346.47	nd (al), pl (ace), $[\alpha]^{15/}{}_D$ + 223 (al, c=1.1)	sub $190^{0.01}$	180-2			al, eth, ace	B8³, 3574
5245	Corticosterone, 17-hydroxy or Cortisol Hydrocortisone $C_{21}H_{30}O_5$	362.47	pr(al or i-prOH), $[\alpha]^{22/}{}_D$ + 167 (al)		220			aa, al	B8³, 4036
5246	Cortisone-*(d)* $C_{21}H_{28}O_5$	360.45	$[\alpha]^{25/}{}_D$ + 209 E + OH, c=1.2)		220-4			al, ace	B8³, 4057
5247	Cortisone, 21-acetate $C_{23}H_{30}O_6$	402.49	nd (ace), rods (chl), $[\alpha]^{24/}{}_D$ + 164 (ace, c=0.5)		239-40			ace, chl	B8³, 4058
5248	Corybulbine-*(d)* or Corydalis-6 $C_{21}H_{25}NO_4$	355.43	nd (al), $[\alpha]^{20/}{}_D$ 303 (chl, c=1.4)		237-8			ace, chl	B21⁴, 2779
5249	Corybulbine-*(dl)* $C_{21}H_{25}NO_4$	355.43	cr (chl-al)		220-2			al, ace, chl, bz	B21, 217
5250	Corycavamine $C_{21}H_{21}NO_5$	367.40	pr (eth or al), $[\alpha]^{20/}{}_D$ + 166.6 (chl, c=2.2)		149			al, chl	B27², 621
5251	Corycavine $C_{21}H_{21}NO_5$	367.40	orh pl (al)		221-2			chl	B27², 621
5252	Corydaldine $C_{11}H_{13}NO_3$	207.23	mcl pr (w or al)		175			w, al, eth, bz, chl	B21⁴, 6443
5253	Corydaline-*(d)* or Corydolis-A $C_{22}H_{27}NO_4$	369.46	pr (al), $[\alpha]^{20/}{}_D$ + 311 (al, c=0.8)		136			al, eth	B21⁴, 2779
5254	Corydaline-*(dl)* $C_{22}H_{27}NO_4$	369.46	cr (al)		135-6			eth, al, w	B21⁴, 2780
5255	Corydaline-*(meso,d)* $C_{22}H_{27}NO_4$	369.46	pr (eth), $[\alpha]_D$ + 180 (chl, c=3)		155-6			eth, al	B21⁴, 2779
5256	Corydaline-*(meso,dl)* $C_{22}H_{27}NO_4$	369.46	cr (al)		163-4			eth, al, w	B21⁴, 2779

No.	Name, Synonyms, and Formula	Mol. wt.	Color, crystalline form, specific rotation and λ_{max} (log ε)	b.p. °C	m.p. °C	Density	n_D	Solubility	Ref.
5257	Corydaline-*(meso,l)* $C_{22}H_{27}NO_4$	369.46	pr (eth), $[\alpha]^{20/}_D$ -181 (chl, c=3)		155-6			eth, al, w	B21^1, 257
5258	Corynantheine $C_{22}H_{26}N_2O_3$	366.46	$[\alpha]^{18/}_D$ +28.8 (MeOH, c=28.8)		165-6			al	B25^2, 212
5259	Cotarnine $C_{12}H_{15}NO_4$	237.26	nd (bz), cr (eth)		132-3d			al, eth, bz, chl	B27^2, 543
5260	Cotarnine, chloride or Stypticin $C_{12}H_{14}ClNO_3 \cdot 2H_2O$	291.73	ye pw or nd (al-AcOEt)		197			w, al	B27^1, 456
5261	Cotarnine, *O*-phthalate or Styptol $(C_{12}H_{14}NO_3)_2C_6H_4(CO_2)_2$	604.61	og cr or pw		103-5			w	B27, 476
5262	Coumalic acid or α-Pyrone-5-carboxylic acid $C_6H_4O_4$	140.10	pr (MeOH)	218^{120}	205-10d			al, aa	B18^4, 5382
5263	Coumaran or 2,3-Dihydrobenzofuran C_8H_8O	120.15		188-9, 76^{14}	-21.5	1.0576$^{24/4}$	1.5426^{20}	al, eth, chl	B17^4, 404
5264	3-Coumaranone $C_8H_6O_2$	134.13	red nd (al)	152-4^{10}	102-3			bz	B17^4, 1456
5265	Coumarin or 1,2-Benzopyrone $C_9H_6O_2$	146.15	rh pym (eth)	301.7	71	0.935$^{20/4}$		al, eth, chl	B17^4, 5055
5266	Coumarin, 6-amino $C_9H_7NO_2$	161.16			168-70			w, al	B18^4, 7920
5267	Coumarin, 4-chloro $C_9H_5ClO_2$	180.59	nd (al)		165			al, eth, bz	B17^4, 5058
5268	Coumarin, 7-diethylamino-4-methyl $C_{14}H_{17}NO_2$	231.29	cr (al, bz-lig)		89 (al), 135 (bz-lig)			al, eth, ace	B18, 612
5269	Coumarin, 3,4-dihydro or Hydrocoumarin $C_9H_8O_2$	148.16	lf	272, 145^{18}	25	1.169^{18}	1.5563^{20}	chl	B17^4, 4956
5270	Coumarin, 5,7-dihydroxy-4-methyl $C_{10}H_8O_4$	192.17	nd (al), lf (aa)		282-4			al	B18^4, 1367
5271	Coumarin, 6,7-dihydroxy or Aesculetin $C_9H_6O_4$	178.14	nd (w + 1), pr (aa), lf (sub)	sub	276			al, ace, chl	B18^4, 1322
5272	Coumarin, 7,8-dihydroxy or Daphnetin $C_9H_6O_4$	178.14	yesh (dil al)	sub	261-3			al	B18^4, 1330
5273	Coumarin, 6,7-dihydroxy-4-methyl or 4-Methyl aesculetin $C_{10}H_8O_4$	192.17	ye nd (dil al)		274-6			al	B18^4, 1371
5274	Coumarin, 7,8-dihydroxy-6-methoxy or Fraxetin $C_{10}H_8O_5$	208.17	pl (dil al)		230-2			al	B18^4, 2371
5275	Coiumarin, 5,7-dimethoxy or Citropten $C_{11}H_{10}O_4$	206.20	pr or nd (al)	200d	148-50			al, ace, chl, aa	B18^4, 1322
5276	Coumarin, 7-ethoxy-4-methyl or Maraniol $C_{12}H_{12}O_3$	204.23	wh		114			al	B18^4, 334
5277	Coumarin, 3-hydroxy $C_9H_6O_3$	162.14			154			w, al, eth, ace	B17^4, 6152
5278	Coumarin, 4-hydroxy or Benzotetronic acid $C_9H_6O_3$	162.14	nd (w)		213-4 (232)			al, eth	B17^4, 6153
5279	Coumarin, 4-hydroxy-3-(1-phenyl-3-oxobutyl) or Coumadin, Warfarin $C_{19}H_{16}O_4$	308.33	cr (al)		161			al, bz, diox	B17^4, 6794
5280	Coumarin, 5-hydroxy $C_9H_6O_3$	162.14			224-7			al, eth, bz	B18^3, 291
5281	Coumarin, 6-hydroxy $C_9H_6O_3$	162.14	nd (dil HCl)		250	1.25		al	B18^4, 293
5282	Coumarin, 7-hydroxy or Umbelliferone $C_9H_6O_3$	162.14	nd (w)	sub	230-1			al, chl, aa	B18^4, 294
5283	Coumarin, 7-hydroxy-6-methoxy or Chrysatropic acid $C_{10}H_8O_4$	192.17	nd or pr (al)		204			chl, aa	B18^4, 1323
5284	Coumarin, 7-hydroxy-4-methyl or 4-Methyl umbelliferone $C_{10}H_8O_3$	176.17	nd (al)		185-7			al, aa	B18^4, 332
5285	Coumarin, 8-hydroxy $C_9H_6O_3$	162.14	nd (dil al)		160			al, aa	B18^4, 304

PHYSICAL CONSTANTS OF ORGANIC COMPOUNDS (Continued)

No.	Name, Synonyms, and Formula	Mol. wt.	Color, crystalline form, specific rotation and λ_{max} (log ε)	b.p. °C	m.p. °C	Density	n_D	Solubility	Ref.
5286	Coumarin, 7-methoxy or Herniarin $C_{10}H_8O_3$	176.17	lf (w or MeOH)		117-8			al, eth	B18^{4}, 295
5287	Coumarin, 8-methoxy $C_{10}H_8O_3$	176.17			89			al, eth, bz	B18^{4}, 304
5288	Coumarin, 3-methyl $C_{10}H_8O_2$	160.17	rh bipym (al)	292.5	91			al	B17^{4}, 5073
5289	Coumarin, 4-methyl $C_{10}H_8O_2$	160.17	nd (w), pr (bz)		83-4			al, bz	B17^{4}, 5074
5290	Coumarin, 5-methyl $C_{10}H_8O_2$	160.17		$173\text{-}4^{12}$	65.8			al, eth, bz	B17^{4}, 5075
5291	Coumarin, 6-methyl $C_{10}H_8O_2$	160.17		303^{725}, 174^{14}	75-6			al, eth, bz	B17^{4}, 5076
5292	Coumarin, 7-methyl $C_{10}H_8O_2$	160.17		171.5^{11}	128			al, eth	B17^{4}, 5077
5293	Coumarin, 8-methyl $C_{10}H_8O_2$	160.17		178^{20}	109-10			al, eth, bz	B17^{4}, 5079
5294	Coumarin, 6,7,8-trimethoxy or Fraxetin dimethyl ether $C_{12}H_{12}O_5$	236.22	rh bipym pl (dil al)	$90\text{-}100^{0.2}$	103-4			al, eth	B18^{4}, 2372
5295	3-Coumarin carboxylic acid $C_{10}H_6O_4$	190.16	nd (w or bz)		190d			al	B18^{4}, 5569
5296	2-Coumarone carboxylic acid or Coumarilic acid $C_9H_6O_3$	162.14	nd (w)	310-5d	192-3			al	B18^{4}, 4247
5297	2-Coumarone carboxylic acid, ethyl ester $C_{11}H_{10}O_3$	190.20		274^{720}, 161^{15}	30-1	$1.1656^{28.5/4}$	$1.564^{27.6}$		B18^{1}, 442
5298	2-Coumarone carboxylic acid, 3-methyl $C_{10}H_8O_3$	176.18	nd (dil al)		188-9			al	B17^{4}, 5074
5299	*P*-Coumaryl alcohol or -3(4 Hydroxyphenyl)-2-propene-1-ol $4\text{-}HOC_6H_4CH{=}CH\text{-}CH_2OH$	150.18	pr (dil al)		124			al, eth, ace, bz	C45, 3359
5300	Coumestrol $C_{15}H_8O_5$	268.23	gy micr rods		385d				B19^{4}, 2870
5301	Coumestrol, diacetate $C_{19}H_{12}O_7$	352.30	pl		234				B19^{4}, 2872
5302	Creatine or (α-Methylguanido) acetic acid $H_2NC({=}NH)N(CH_3)CH_2CO_2H$	131.13	mcl pr (w + 1)		303	1.33		w	B4^{3}, 1170
5303	Creatinine or l-Methylglylocy amidine $C_4H_7N_3O$	113.12	rh pr (w + 2), lf (w)		ca 300d				B24^{2}, 128
5304	*o*-Cresol or 2-Hydroxytoluene $2\text{-}CH_3C_6H_4OH$	108.14		191, 74.9^{10}	30.9	$1.0273^{20/4}$	1.5361^{20}	al, eth, ace, bz	B6^{4}, 1940
5305	*o*-Cresol, 3-amino $3\text{-}H_2N\text{-}2\text{-}CH_3C_6H_3OH$	123.15	nd (w)		129			al	B13, 579
5306	*o*-Cresol, 4-amino $4\text{-}H_2N\text{-}2\text{-}CH_3C_6H_3OH$	123.15	nd or lf (bz)	sub	175			al, eth	B13^{3}, 1531
5307	*o*-Cresol, 4-amino-6-nitro $6\text{-}NO_2\text{-}4\text{-}NH_2\text{-}2\text{-}CH_3C_6H_2OH$	168.15	br-red nd (al)		118			al	B13, 578
5308	*o*-Cresol, 6-amino $6\text{-}H_2N\text{-}2\text{-}CH_3C_6H_3OH$	123.15	pl (w)		89			al, eth, ace, bz	B13^{3}, 1527
5309	*o*-Cresol, 6-amino-4-nitro $4\text{-}NO_2\text{-}6\text{-}H_2N\text{-}2\text{-}CH_3C_6H_2OH$	168.15	red-br nd (bz)		176			al, bz	B13^{3}, 1528
5310	*o*-Cresol, 3-bromo $3\text{-}Br\text{-}2\text{-}CH_3C_6H_3OH$	187.04	nd (peth)	$55\text{-}7^{4}$	95			al, eth, ace, bz	B6, 360
5311	*o*-Cresol, 4-bromo $4\text{-}Br\text{-}2\text{-}CH_3C_6H_3OH$	187.04	nd (al or peth)	235 sub, $137\text{-}43^{18}$	64			al, eth, ace	B6^{4}, 2006
5312	*o*-Cresol, 5-bromo $5\text{-}Br\text{-}2\text{-}CH_3C_6H_3OH$	187.04	nd (lig or peth)		80			al, eth, ace	B6^{2}, 333
5313	*o*-Cresol, 3-chloro $3\text{-}Cl\text{-}2\text{-}CH_3C_6H_3OH$	142.58	lo nd (w)	225	86			al, eth, bz	B6^{4}, 2000
5314	*o*-Cresol, 4-chloro $4\text{-}Cl\text{-}2\text{-}CH_3C_6H_3OH$	142.58	nd (peth)	223	51				B6^{4}, 1987
5315	*o*-Cresol, 5-chloro $5\text{-}Cl\text{-}2\text{-}CH_3C_6H_3OH$	142.58	nd (peth)		73-4			al, bz	B6^{4}, 1986
5316	*o*-Cresol, 6-chloro $6\text{-}Cl\text{-}2\text{-}CH_3H_6H_3OH$	142.58		$188\text{-}9^{740}$, $80\text{-}1^{20}$			1.5449^{20}	eth	B6^{4}, 1984
5317	*o*-Cresol, 3,5-dibromo $3,5\text{-}Br_2\text{-}2\text{-}CH_3C_6H_2OH$	265.93	nd (peth)	$283\text{-}7^{758}$	98-101			peth	B6^{2}, 334
5318	*o*-Cresol, 3,6-dibromo $3,6\text{-}Br_2\text{-}2\text{-}CH_3C_6H_2OH$	265.93	cr	255-60	38				B6^{1}, 176

No.	Name, Synonyms, and Formula	Mol. wt.	Color, crystalline form, specific rotation and λ_{max} (log ε)	b.p. °C	m.p. °C	Density	n_D	Solubility	Ref.
5319	*o*-Cresol, 4,6-dibromo $4,6\text{-}Br_2\text{-}2\text{-}CH_3C_6H_2OH$	265.93	nd (peth)	$263\text{-}6^{745}/_d$	58	...	...	al, eth, bz	$B6^3$, 1271
5320	*o*-Cresol, 4,5-dichloro $4,5\text{-}Cl_2\text{-}2\text{-}CH_3C_6H_2OH$	177.03	nd (peth)	...	101	...	...	al, bz, aa	$B6^2$, 333
5321	*o*-Cresol, 4,5dichloro $4,6\text{-}Cl_2\text{-}2\text{-}CH_3C_6H_2OH$	177.03	nd (w or peth)	$266.5, 73\text{-}8^4$	55	...	...	al, eth, chl	$B6^4$, 2001
5322	*o*-Cresol, 3,5-dinitro $3,5\text{-}(NO_2)_2\text{-}2\text{-}CH_3C_6H_2OH$	198.14	ye pr (al)	...	85.8	...	...	al, eth, ace	C51, 10414
5323	*o*-Cresol, 4,6-dinitro $4,6\text{-}(NO_2)_2\text{-}2\text{-}CH_3C_6H_2OH$	198.14	ye pr or nd (al)	...	86.5	...	...	al, eth, ace	$B6^4$, 2014
5324	*o*-Cresol, 3-nitro $3\text{-}NO_2\text{-}2\text{-}CH_3C_6H_3OH$	153.14	pa ye nd (w)	...	147	...	...	al, eth	$B6^1$, 178
5325	*o*-Cresol, 4-nitro $4\text{-}NO_2\text{-}2\text{-}CH_3C_6H_3OH$	153.14	ye or col nd (w or aq al)	$186\text{-}90^9$	96 (anh)	...	...	al, eth, bz, aa	$B6^4$, 2011
5326	*o*-Cresol, 5-nitro $5\text{-}NO_2\text{-}2\text{-}CH_3C_6H_3OH$	153.14	ye nd (lig)	...	118	...	...	al, eth, bz	$B6^4$, 2010
5327	*o*-Cresol, 6-nitro $6\text{-}NO_2\text{-}2\text{-}CH_3C_6H_3OH$	153.14	ye pr (dil al or peth)	$250\text{-}60d, 185^{12}$	70	...	...	al, eth	$B6^4$, 2009
5328	*o*-Cresol, 4-nitroso $4\text{-}ON\text{-}2\text{-}CH_3C_6H_3OH$	137.14	nd (w)	...	134-5d	...	...	al, eth, bz, chl	$B7^3$, 3388
5329	*o*-Cresol-4-isopropyl $4\text{-}i\text{-}C_3H_7\text{-}2\text{-}CH_3C_6H_3OH$	150.22	...	$230^{766}, 83^3$	8.6	0.9793^{25}	1.5253^{20}	al, bz, chl	$B6^3$, 1884
5330	*o*-Cresol, 5-isopropyl $5\text{-}i\text{-}C_3H_7\text{-}2\text{-}CH_3\text{-}C_6H_3OH$	150.22	...	117-20	0-3	$0.976^{20/4}$	1.523^{20}	...	$B6^3$, 1885
5331	*o*-Cresol, 6-isopropyl $6\text{-}i\text{-}C_3H_7\text{-}2\text{-}CH_3C_6H_3OH$	150.22	...	$225, 104^{14}$	-14.5	0.9789^{25}	1.5239^{20}	al, bz, chl	$B6^3$, 1882
5332	*o*-Cresol-3,4,5,6-tetrabromo $2\text{-}CH_3C_6Br_4OH$	423.72	ye nd (chl or aa)	d	208	...	...	al, eth, bz, chl, aa	$B6^3$, 1272
5333	*o*-Cresol-3,4,5,6-tetrachloro $2\text{-}CH_3C_6Cl_4OH$	245.92	nd (lig)	...	190	...	...	al, eth, bz, aa	$B6^3$, 333
5334	*o*-Cresol-4,5,6-trinitro $4,5,6\text{-}(NO_2)_3\text{-}2\text{-}CH_3C_6HOH$	243.13	og-ye pr (ace)	...	102	...	...	al, eth, ace, chl	B6, 369
5335	*o*-Cresylphosphate or Tri-*o*-cresylphosphate $(2\text{-}CH_3C_6H_4O)_3PO$	368.37	col or pa ye	$410, 283\text{-}5^{20}$	11	$1.1955^{20/4}$	1.5575^{20}	al, eth, aa	$B6^4$, 1979
5336	*m*-Cresol or 3-Hydroxy toluene $3\text{-}CH_3C_6H_4OH$	108.14	...	202.2, 86^{10}	11.5	$1.0336^{20/4}$	1.5438^{20}	al, eth, ace, bz	$B6^4$, 2035
5337	*m*-Cresol-2-amino $2\text{-}NH_2\text{-}3\text{-}CH_3C_6H_3OH$	123.15	pl (w)	sub	150	...	...	eth	$B13^2$, 324
5338	*m*-Cresol-4-amino $4\text{-}H_2N\text{-}3\text{-}CH_3C_6H_3OH$	123.15	pr (dil al), cr (bz)	...	179	...	...	al, eth	$B13^3$, 1559
5339	*m*-Cresol-5-amino $5\text{-}H_2N\text{-}3\text{-}CH_3C_6H_3OH$	123.15	cr (dil MeOH)	245	139	...	...	al	C47, 9300
5340	*m*-Cresol-5-amino-2-nitro $2\text{-}NO_2\text{-}H_2\text{-}5\text{-}H_2N\text{-}3CH_3C_6H_2OH$	168.15	red-br nd (al)	...	201	...	...	al	B13, 595
5341	*m*-Cresol-6-amino $6\text{-}H_2N\text{-}3\text{-}CH_3C_6H_3OH$	123.15	nd (bz, dil al)	...	162d	...	...	al, eth, ace	$B13^3$, 1552
5342	*m*-Cresol-4-bromo $4\text{-}Br\text{-}3\text{-}CH_3C_6H_3OH$	187.04	nd (peth or w)	$137\text{-}43^{16}$	63.5	...	...	eth, Py	$B6^4$, 2072
5343	*m*-Cresol-5-bromo $5\text{-}Br\text{-}3\text{-}CH_3C_6H_3OH$	187.04	nd (w)	$161\text{-}2^{28}$	56-7	...	...	al, eth	$B6^2$, 357
5344	*m*-Cresol-6-bromo $6\text{-}Br\text{-}3\text{-}CH_3C_6H_3OH$	187.04	cr (peth)	$206\text{-}8^{731}$, $81\text{-}2^4$	38	...	...	al, eth, ace	$B6^3$, 1320
5345	*m*-Cresol-2-chloro $2\text{-}Cl\text{-}3\text{-}CH_3C_6H_3OH$	142.58	lf or nd (dil al)	$230, 100\text{-}5^{28}$	74	...	...	al, eth	$B6^4$, 2064
5346	*m*-Cresol-4-chloro $4\text{-}Cl\text{-}3\text{-}CH_3C_6H_3OH$	142.58	nd (peth)	235	66-8	...	...	al, eth	$B6^4$, 2064
5347	*m*-Cresol-6-chloro $6\text{-}Cl\text{-}3\text{-}CH_3C_6H_3OH$	142.58	pr (peth)	196	45-6	1.215^{15}	...	al, w	$B6^3$, 1315
5348	*m*-Cresol-2,4-dichloro $2,4\text{-}Cl_2\text{-}3\text{-}CH_3C_6H_3OH$	177.03	pr (peth)	241-2	27	...	...	eth, chl	$B6^3$, 1319
5349	*m*-Cresol-2,6-dichloro $2,6\text{-}Cl_2\text{-}3\text{-}CH_3C_6H_3OH$	177.03	...	$235\text{-}6^{745}, 75\text{-}80^4$	58-9	...	...	eth, chl	$B6^3$, 1319
5350	*m*-Cresol-4,6-dichloro $4,6\text{-}Cl_2\text{-}3\text{-}CH_3C_6H_2OH$	177.03	pr (peth)	$235\text{-}6, 110^{18}$	72-4	...	1.572^{20}	chl, peth	$B6^4$, 2069
5351	*m*-Cresol, 4-nitro $4\text{-}NO_2\text{-}3\text{-}CH_3C_6H_3OH$	153.14	nd or pr (w)	...	129	...	...	al, eth, bz, chl	$B6^4$, 2075

No.	Name, Synonyms, and Formula	Mol. wt.	Color, crystalline form, specific rotation and λ_{max} (log ε)	b.p. °C	m.p. °C	Density	n_D	Solubility	Ref.
5352	*m*-Cresol, 5-nirto 5-NO_2-3-$CH_3C_6H_3OH$	153.14	pa ye cr (bz)		90-1			eth, bz	B6^{2}, 361
5353	*m*-Cresol, 6-nitro 6-NO_2-3-$CH_3C_6H_3OH$	153.14	ye mcl nd (eth or bz)		56			al, eth, bz	B6^{3}, 1326
5354	*m*-Cresol, 4-nitroso 4-ON-3-$CH_3C_6H_3OH$	137.14	nd (w or bz), pr (aa)		165d			al, eth, bz, aa	B7^{3}, 3389
5355	*m*-Cresol, 2,4,5,6-tetrabromo 3-$CH_3C_6Br_4OH$	423.72	nd (chl aa)		194			eth	B6^{3}, 1324
5356	*m*-Cresol, 2,4,5,6-tetrachloro 3-$CH_3C_6Cl_4OH$	245.92	nd (peth)		189-90			al, eth, ace, bz	B6^{4}, 2071
5357	*m*-Cresol, 2,4,6-tribromo 2,4,6-Br_3-3-CH_3C_6HOH	344.83			84				B6^{3}, 1324
5358	*m*-Cresol, 2,4,6-trinitro or Methyl picric acid 2,4,6-$(NO_2)_3$-3-CH_3C_6HOH	243.13	pa ye nd (w or al)	150 exp	109-10			al, eth, ace, bz, chl	B6^{4}, 2079
5359	*m*-Cresyl phosphate or Tri-*m*-Cresyl phosphate $(3\text{-}CH_3C_6H_4O)_3PO$	368.37	wax	260^{15}	25-6	1.150^{25}	1.5575^{20}	eth, aa	B6^{4}, 2057
5360	*p*-Cresol or 4-Hydroxytoluene 4-$CH_3C_6H_4OH$	108.14	pr	201.9, 85.7^{10}	34.8	$1.0178^{20/4}$	1.5312^{20}	**al, eth, ace, bz**	B6^{4}, 2093
5361	*p*-Cresol, 2-amino 2-NH_2-4-$CH_3C_6H_3OH$	123.15	cr (w), rh (bz), lf or nd (sub)	sub	137			al, eth, chl	B13^{3}, 1576
5362	*p*-Cresol, 2-amino-5-nitro 5-NO_2-2-NH_2-4-$CH_3C_6H_2OH$	168.15	ye-og cr (al)		199-200d			al, dil HCl	B13^{2}, 346
5363	*p*-Cresol, 2-amino-6-nitro 6-NO_2-2-NH_2-4-$CH_3C_6H_2OH$	168.15	red-br (al)		119			al	B13^{2}, 345
5364	*p*-Cresol, 3-amino 3-H_2N-4-$CH_3C_6H_3OH$	123.15	cr (w or eth), lf (sub)	sub	156-7			eth	B13^{2}, 337
5365	*p*-Cresol, 2-bromo 2-Br-4-$CH_3C_6H_3OH$	187.04	nd (peth)	213-4	56-7	$1.5468^{25/25}$	1.5772^{20}	al, bz	B6^{4}, 2143
5366	*p*-Cresol, 3-bromo 3-Br-4-CH_3-C_6H_3OH	187.04	nd (peth)	245-7	56			al, eth, ace, bz	B6^{4}, 2143
5367	*p*-Cresol, 2-chloro 2-Cl-4-$CH_3C_6H_3OH$	142.58		195-6		$1.1785^{25/4}$	1.5200^{27}	al, eth, bz, aa	B6^{4}, 2135
5368	*p*-Cresol, 3-chloro 3-Cl-4-$CH_3C_6H_3OH$	142.58	nd (al)	228	55-6			al, eth, bz, aa	B6^{3}, 1374
5369	*p*-Cresol, 2,6-dichloro 2,6-Cl_2-4-$CH_3C_6H_2OH$	177.03	nd (lig)	$138\text{-}9^{28}$	39 (42)			al, eth, aa	B6^{4}, 2141
5370	*p*-Cresol, 2,6-dinitro 2,6-$(NO_2)_2$-4-$CH_3C_6H_2OH$	198.14	ye nd (eth or peth)		85			al, eth, bz	B6^{4}, 2152
5371	*p*-Cresol, 2-methoxy or Cresolol 2-CH_3O-4-$CH_3C_6H_3OH$	138.17	pr	221, 113.5^{22}	5.5	$1.098^{20/4}$	1.5353^{25}	al, eth	B6^{2}, 865
5372	*p*-Cresol, 3-methylamino 3(CH_3NH)-4-$CH_3C_6H_3OH$	137.18	cr (bz-lig)		108			al, eth, bz	B13, 599
5373	*p*-Cresol, 2-nitro 2-NO_2-4-$CH_3C_6H_3OH$	153.14	ye nd (al or w)	125^{22}	36.5	$1.2399^{20/4}$	1.574^{40}	al, eth, ace, bz	B6^{4}, 2149
5374	*p*-Cresol, 3-nitro 3-NO_2-4-$CH_3C_6H_3OH$	153.14	ye pr (eth)		79			al, eth	B6^{3}, 1384
5375	*p*-Cresol, 2-isopropyl 2-*p*-C_3H_7-4-$CH_3C_6H_3OH$	150.22		228-9, 82^{3}	36-7	$0.9910^{20/4}$	1.5275^{20}	al, bz, chl	B6^{3}, 1882
5376	*p*-Cresol, 2,3,5,6-tetrabromo 4-$CH_3C_6Br_4OH$	423.72	nd (al or chl)		198-9			al, eth, chl	B6^{3}, 1383
5377	*p*-Cresol, 2,3,5,6-tetrachloro 4-$CH_3C_6Cl_4OH$	245.92	nd (dil al, aa bz-lig)		190			al, bz, chl, aa	B6^{4}, 2142
5378	*p*-Cresyl phosphate or Tri-*p*-Cresyl phosphate $(4\text{-}CH_3C_6H_4O)_3PO$	368.37	nd (al), ta (eth)	$224^{3.5}$	77-8	1.247^{25}		al, eth, bz, chl	B6^{4}, 2130
5379	Crocetin-*(trans)* or Gardenin $C_{20}H_{24}O_4$	328.41	brick red rh		285-7 (cor)			Py	B2^{3}, 2018
5380	Croconic acid or Crocic acid $C_5H_2O_5$	142.07	pa ye nd (+3w al-diox)		sub d> 150			w, al	B8^{3}, 3977
5381	Crotonaldehyde or 2-Butenal $CH_3CH{=}CHCHO$	70.09		104-5	-74	$0.8495^{25/4}$	1.4355^{20}	al, eth, ace, **bz**	B1^{4}, 3447
5382	Crotonaldehyde, diethylacetal or 2-Butenaldiethylacetal $CH_3CH{=}CHCH(OC_2H_5)_2$	144.21		147-8, 49^{17}		$0.8473^{18/4}$	1.4097^{20}	**al, eth, ace, bz**	B1^{4}, 3450
5383	Crotonamide $CH_3CH{=}CHCONH_2$	85.11	nd (ace)	sub at 140^{13}	161.5 (cor)		1.4420^{165}	al, bz	B2^{4}, 1506

No.	Name, Synonyms, and Formula	Mol. wt.	Color, crystalline form, specific rotation and λ_{max} (log ε)	b.p. °C	m.p. °C	Density	n_D	Solubility	Ref.
5384	Crotonic acid or *(trans)*-2-Butenoic acid $CH_3CH{=}CHCO_2H$	86.09	mcl pr or nd (w or lig)	185	71.5	$1.018^{15/4}$	1.4249^{77}	w, al, eth, ace	$B2^4$, 1498
5385	Crotonic acid, allyl ester $CH_3CH{=}CHCO_2CH_2CH{=}CH_2$	126.15	....	$88\text{-}9^{70}$	......	$0.9440^{20/4}$	1.4465^{20}	....	$B2^4$, 1503
5386	Crotonic acid, anhydride or Crotonic anhydride $(CH_3CH{=}CHCO)_2O$	154.17	....	246-8, 129^{19}	......	1.0397^{20}	1.4745^{20}	eth	$B2^4$, 1505
5387	Crotonic acid, ethyl ester $CH_3CH{=}CHCO_2C_2H_5$	114.14	....	136.5, $58\text{-}9^{48}$	......	$0.9175^{20/4}$	1.4243^{20}	al, eth	$B2^4$, 1500
5388	Crotonic acid, methyl ester or Methyl crotonate $CH_3CH{=}CHCO_2CH_3$	100.13	....	121	-42	$0.9444^{20/4}$	1.4242^{20}	al, eth	$B2^4$, 1500
5389	Crotononitrile or 2-Butenonitrile $CH_3CH{=}CHCN$	67.09	....	$12\text{-}1^{762}$	-51.5	$0.8239^{20/4}$	1.4225^{20}	eth, ace	$B2^4$, 1507
5390	Crotonyl acetate $CH_3CO_2CH_2CH{=}CHCH_3$	114.15	....	132	......	$0.9192^{20/4}$	1.4181^{20}	al, eth, ace	$B2^4$, 183
5391	Crotonyl chloride $CH_3CH{=}CHCOCl$	104.54	....	124-5, 35^{18}	......	1.0905^{20}	1.460^{18}	ace	$B2^4$, 1506
5392	Cryptopine or Cryptocavine $C_{21}H_{23}NO_5$	369.42	pr or pl (bz), nd (chl-MeOH)	....	223 (cor)	$1.315^{20/4}$	....	chl, aa	$B27^2$, 578
5393	Cryptoxanthin or β-Caroten-3-ol $C_{40}H_{56}O$	552.88	garnet red pr (bz-MeOH)	....	169	....	....	bz, chl	$B6^3$, 3772
5394	Cumene or *iso*-Propylbenzene $(CH_3)_2CHC_6H_5$	120.19	....	152.4, 38.2^{10}	-96	$0.8618^{20/4}$	1.4915^{20}	al, eth, ace, bz	$B5^4$, 985
5395	Cumene, 2-nitro or 1-*iso*-Propyl-2-nitro benzene $2\text{-}NO_2C_6H_4CH(CH_3)_2$	165.19	pa ye	....	103^9	1.101^{12}	1.5259^{20}	ace, bz, aa	$B5^4$, 997
5396	Cumene, 4-nitro or 4-Nitro-1-*iso*-propyl benzene $4\text{-}NO_2C_6H_4CH(CH_3)_2$	165.19	pa ye oil	122^9	......	$1.0830^{20/4}$	1.5367^{20}	ace, bz, lig	$B5^4$, 997
5397	Cubebin $C_{20}H_{20}O_6$	356.38	nd (al or bz), $[\alpha]^{25}{}_D$ -45.6 (chl, c=5)	....	131-2	....	....	al, eth, chl	$B19^4$, 5967
5398	Cumic alcohol or 1-Hydroxymethyl-4-isopropylbenzene $4\text{-}i\text{-}C_3H_7C_6H_4CH_2OH$	150.22	....	289	28	$0.9818^{18/4}$	1.5210^{20}	al, eth	**$B6^3$, 1911**
5399	Cupreine or Hydroxycinchonine $C_{19}H_{22}N_2O_2$	310.40	pr (eth), $[\alpha]^{17}{}_D$ -175.5 (al)	....	198 (anh)	....	....	al	$B23^4$, 416
5400	C-Curarine-III-hydroxide or C-flurorcuraninehydroxide $C_{20}H_{28}N_2O_2$	328.46	cr (MeOH-eth)	....	212	....	....	w, al	H36, 102
5401	ar-Curcumene $C_{15}H_{22}$	202.34	$[\alpha]^{18}{}_D$ +35.8	140^{19}	......	$0.8821^{20/20}$	1.4989^{20}	bz	$B5^4$, 1465
5402	Curcumin $C_{21}H_{20}O_6$	368.39	or ye pr, rh pr (MeOH)	....	183	....	....	al, aa	$B8^3$, 4312
5403	Cuscohygrine or α,α' Bis(*N*-methyl-α-pyrrolidyl) acetone $C_{13}H_{24}N_2O$	224.35	....	185^{22}	......	$0.9782^{16/4}$	....	w, al, eth, bz	$B24^2$, 36
5404	Cuscohygrine, hydrate $C_{13}H_{24}N_2O.3\frac{1}{2}H_2O$	287.40	nd (a)	....	40-1	....	....	eth, bz	B24, 78
5405	Cusparine $C_{19}H_{17}NO_3$	307.35	(i) wh, nd (peth), (ii) ye, nd, (iii) pr	....	(i) 92, (ii) 92, (iii) 110-22	....	....	al, eth, ace, bz, chl	$B27^2$, 545
5406	Cyamelide or *sym*-Trioxane triimine $C_3H_3N_3O_3$	129.08	am or pw	d	......	$1.127^{15/4}$	....	....	$B3^3$, 30
5407	Cyanamide or Carbamonitrile H_2NCN	42.04	nd	140^{19}	42 (46)	$1.282^{20/4}$	1.4418^{48}	w, al, eth, ace, bz, chl	$B3^4$, 145
5408	Cyanamide, benzyl $C_6H_5CH_2NHCN$	132.16	pl (eth)	....	43	....	....	al, eth	B12, 1051
5409	Cyanamide, diallyl $(CH_2{=}CHCH_2)_2NCN$	122.17	....	$140\text{-}5^{90}$, 95^9	......	....	....	al, eth, ace, bz	$B4^4$, 1078
5410	Cyanamide, dibutyl $(C_4H_9)_2NCN$	154.26	....	$187\text{-}91^{190}$, $146\text{-}51^{15}$	......	....	....	al, eth, ace, bz	$B4^4$, 592
5411	Cyanamide, diethyl $(C_2H_5)_2NCN$	98.15	....	188, 62^{10}	......	$0.854^{20/4}$	1.4126^{25}	al, eth	$B4^4$381
5412	Cyanamide, dimethyl $(CH_3)_2NCN$	70.09	....	163.5, 56^{15}	......	....	1.4089^{19}	al, eth, ace	$B4^4$, 226

No.	Name, Synonyms, and Formula	Mol. wt.	Color, crystalline form, specific rotation and λ_{max} (log ε)	b.p. °C	m.p. °C	Density	n_D	Solubility	Ref.
5413	Cyanamide, diphenyl $(C_6H_5)_2NCN$	194.24	pr (al)	$235\text{-}40^{60}$	73-4	...	...	al, lig	$B12^3$, 895
5414	Cyanamide, methyl-α-naphthyl $\alpha\text{-}C_{10}H_7N(CH_3)CN$	182.22	yesh	$185\text{-}7^2$	...	...	...	al, eth	$B12^2$, 697
5415	Cyanamide, phenyl or Carbanilonitrile C_6H_5NHCN	118.14	cr (w, eth), lf (aa)	...	47 (hyd)	...	...	al, eth	$B12^3$, 805
5416	Cyanic acid HOCN	43.03	gas	23.5	-81	$1.140^{20/4}$	...	w, eth, bz, chl, aa	$B3^4$, 80
5417	Cyanic acid, ethyl ester or Ethyl cyanate C_2H_5OCN	71.08	...	$162d, 30^{12}$	...	$0.89^{20/4}$	1.3788^{25}	al, eth	Tet, 964, 2829
5418	Cyano acetamide $NCCH_2CONH_2$	84.08	pl (w)	...	121-2	...	...	w	$B2^4$, 1891
5419	Cyano acetamide, *N*-phenyl or α-Cyano acetanilide $NCCH_2CONHC_6H_5$	160.18	nd (al)	...	199-200	...	...	...	$B12^2$, 167
5420	Cyanoacetic acid $NCCH_2CO_2H$	85.06	...	$108^{0.15}/_d$	70-1	...	...	w, al, eth	$B2^4$, 1888
5421	Cyano acetic acid, benzal or α-Cyanocinnamic acid $C_6H_5CH = C(CN)CO_2H$	173.17	cr (al)	...	183	...	...	...	$B9^3$, 4379
5422	Cyanoacetic acid, benzal, ethyl ester $C_6H_5CH=C(CN)CO_2C_2H_5$	201.22	(i) nd, (al), (ii) oil	(ii) 188^{15}	(i) 51	(ii) 1.1076	(ii) 1.5033	eth, bz, chl	B9, 4380
5423	Cyano acetic acid, 1-cyclo hexenyl $C_6H_9CH(CN)CO_2H$	165.19	nd (bz)	...	109-10	...	...	al, ace, bz	$B9^2$, 560
5424	Cyanoacetic acid, diethyl, ethyl ester $(C_2H_5)_2C(CN)CO_2C_2H_5$	169.22	...	$214\text{-}5, 100\text{-}1^{15}$	...	...	1.4200^{27}	al, eth	$B2^3$, 1761
5425	Cyanoacetic acid, ethyl ester or Ethyl cyanoacetate $NCCH_2CO_2C_2H_5$	113.12	...	$205, 99^{15}$	-22.5	$1.0654^{20/4}$	1.4175^{20}	al, eth	$B2^4$, 1889
5426	Cyanoacetic acid, methyl ester or Methyl cyanoacetate $NCCH_2CO_2CH_3$	99.09	...	$200\text{-}1, 115^{36}$	-22.5	1.1128^{20}	1.4176^{20}	al, eth	$B2^4$, 1889
5427	Cyanoacetic acid, phenyl, ethyl ester or Ethyl phenyl cyen-ate $C_6H_5CH(CN)CO_2C_2H_5$	189.21	oil	$275d, 165^{20}$	...	$1.091^{20/4}$	1.5012^{25}	al, eth, ace, bz	$B9^3$, 4262
5428	α-Cyanocaproic acid, ethyl ester $C_4H_9CH(CN)CO_2C_2H_5$	169.22	...	$245\text{-}50^{762}$, 105^9	...	0.988^{15}	1.4248^{20}	al, eth	$B2^3$, 1746
5429	Cyanogen or Oxalodinitrile NCCN	52.04	gas	-21.2	-27.9	0.9537^{-21}	...	w, al, eth	$B2^4$, 1863
5430	Cyanogen bromide BrCN	105.92	nd	61.4	52	$2.015^{20/4}$	...	w, al, eth	$B3^4$, 92
5431	Cyanogen chloride ClCN	61.47	gas	12.7	-6	$1.186^{20/4}$	...	w, al, eth	$B3^4$, 90
5432	Cyanogen iodide ICN	152.92	nd (al or eth)	sub>45	146-7	2.84^{18}	...	al, eth	$B3^4$, 93
5433	Cyanogen sulfide $S(CN)_2$	84.10	rh pl	sub 30-40	65	...	...	w, al, eth	$B3^4$, 339
5434	Cyanuric acid, dihydrate or 2,4,6,-Triazinetriol $C_3H_3N_3O_3 \cdot 2H_2O$	165.11	mcl (w + 2)	d	>360d	$2.500^{20/4}$	...	w	$B26^2$, 131
5435	Cyanuric acid, tribenzyl ester or Benzyl cyanurate $C_{24}H_{21}N_3O_3$	399.45	nd (al)	>320	159	...	...	al	$B26^1$, 76
5436	Cyanuric chloride $C_3N_3Cl_3$	184.41	cr (eth or bz)	190^{720}	154	...	...	al	$B26^2$, 16
5437	Cyclamen aldehyde $4\text{-}i\text{-}C_3H_7C_6H_4CH_2CH(CH_3)CHO$	190.29	...	$133\text{-}7^{99}, 115^5$	...	0.951^{15}	1.5068^{20}	al, eth, bz	$B7^3$, 1200
5438	Cyclobutane or Tetramethylene C_4H_8	56.11	...	12	-50	$0.720^{5/4}$	1.4260^{20}	al, eth, ace, bz	$B5^4$, 6
5439	Cyclobutane, benzoyl or Cyclobutyl phenyl ketone $C_6H_5COC_4H_7$	160.22	...	$260, 122^{10}$	...	$1.0457^{25/25}$	1.5472^{20}	...	B7, 374
5440	Cyclobutane, ethyl $C_2H_5C_4H_7$	84.16	...	70.7	-142.9	$0.7284^{20/4}$	1.4020^{20}	al, eth, ace, bz	$B5^4$, 87
5441	Cyclobutane, methyl $CH_3C_4H_7$	70.13	...	36.3	...	$0.6884^{20/4}$	1.3866^{20}	al, eth, ace, bz, peth	$B5^4$, 21
5442	Cyclobutane, octafluoro or Perfluro cyclobutane C_4F_8	200.03	...	-4^{764}	-38.7	...	...	eth	$B5^4$, 8
5443	Cyclobutane carboxylic acid $C_4H_7CO_2H$	100.12	...	$190^{754}, 74\text{-}5^2$	-2	$1.0599^{20/4}$	1.4400^{20}	al, eth	$B9^3$, 6
5444	1,1-Cyclobutane dicarboxylic acid $1,1\text{-}C_4H_6(CO_2H)_2$	144.13	pr (eth or w)	...	156.6 (cor)	...	...	w, al, eth, bz	$B9^3$, 3797
5445	1,1-Cyclobutane dicarboxylic acid, diethyl ester $1,1\text{-}C_6H_4(CO_2C_2H_5)_2$	200.23	...	$229^{735}, 104^{12}$	...	$1.0456^{20/4}$	1.4344^{20}	al	$B9^3$, 3798

No.	Name, Synonyms, and Formula	Mol. wt.	Color, crystalline form, specific rotation and λ_{max} (log ε)	b.p. °C	m.p. °C	Density	n_D	Solubility	Ref.
5446	1,2-Cyclobutane dicarboxylic acid *(cis, dl)* $1,2\text{-}C_4H_6(CO_2H)_2$	144.13	pl (w), pr (bz)		138			w, al, eth	B9[3], 3798
5447	1,2-Cyclobutane dicarboxylic acid *(trans, d)* $1,2\text{-}C_4H_6(CO_2H)_2$	144.13	$[\alpha]^{30}{}_D$ +123.3 (w, c=2)		105			w, al	B9[3], 3799
5448	1,2-Cyclobutane dicarboxylic acid *(trans-dl)* $1,2\text{-}C_4H_6(CO_2H)_2$	144.13	rh nd (bz)		131			w, al	B9[3], 3799
5449	1,2-Cyclobutane dicarboxylic acid *(trans-l)* $1,2\text{-}C_4H_6(CO_2H_2$	144.13	nd (HCl), $[\alpha]^{30}{}_D$ +123.3 (w, c=0.8)		105			w, al	B9[3], 3798
5450	1,3-Cyclobutane dicarboxylic acid *(cis)* $1,3\text{-}C_4H_6(CO_2H)_2$	114.13	pr (w)	252	143-4			w, al	B9[3], 3801
5451	1,3-Cyclobutane dicarboxylic acid *(trans)* $1,3\text{-}C_4H_6(CO_2H)_2$	144.13	pr (w), nd (sub)	sub	171			w, al	B9[3], 3802
5452	Cyclobutane, acetyl $CH_3COC_4H_7$	98.14		137-9		0.9020^{20}	1.4322^{19}		B7[3], 45
5453	Cyclobutane, 1,2,bis (amino methyl) $1,2\text{-}(CH_3NH)_2C_4H_6$	114.19			193-4		$1.4778^{27.5}$		
5454	Cyclobutanone C_4H_6O	70.09		96-7		$0.9548^{0/0}$	1.4215^{20}	al, eth, bz, chl	B7[3], 4
5455	Cyclobutene C_4H_6	54.09			2	$0.733^{0/4}$		ace, bz, peth	B5[4], 207
5456	Cyclobutene, perfluoro C_4F_6	162.03		3	-60	$1.602^{-20/4}$	1.298^{-20}		B5[4], 208
5457	Cyclobutyl phenyl ketone $C_4H_7COC_6H_5$	160.22		260,121-2^{10}		$1.0457^{25/25}$	1.5472^{20}		B7, 374
5458	Cyclocamphene or Epicyclene $C_{10}H_{16}$	136.24		150-1	117-8	0.7948^{121}		al, aa	B5[3], 393
5459	1,6-Cyclodecanediol *(trans)* $C_{10}H_{20}O_2$	172.27	cr (chl, AcOEt)		151-3			eth	B6[3], 4106
5460	1,6-Cyclodecadione $C_{10}H_{16}O_2$	168.24	cr (eth)		100			ace, aa	B7[3]3246
5461	Cyclodecanol $C_{10}H_{19}OH$	156.27		125^{12}	40-1	$0.9606^{20/4}$	1.4926^{20}	al	B6[4], 138
5462	Cyclodecanone $C_{10}H_{18}O$	154.25	amor pw	106-7^{13}	28	$0.9654^{20/4}$	1.4806^{20}	eth, bz, chl	B7[3], 134
5463	Cyclofenchene $C_{10}H_{16}$	136.24		144-6		$0.859^{20/4}$	1.4503^{22}		B5[4], 468
5464	9-Cycloheptadecen-1-one or Civetone $C_{17}H_{30}O$	250.42		342^{742}, 159^{2}	32.5	$0.9170^{33/4}$	1.4830^{33}	al, bz	B7[3], 524
5465	Cycloheptane or Suberane C_7H_{14}	98.19		118.5	-12	$0.8098^{20/4}$	1.4436^{20}	al, eth, bz, lig, chl	B5[4], 92
5466	Cycloheptane, 1-aza $C_6H_{13}N$	99.18		138^{749}		$0.8643^{22/4}$	1.4631^{20}	al, eth	B20[4], 1406
5467	Cycloheptane, bromo or Suberyl bromide $C_7H_{13}Br$	177.08		101.5^{40}, 75^{12}		$1.2887^{22/4}$	1.4996^{20}	eth, chl	B5[4], 93
5468	Cycloheptane, methyl $C_7H_{13}CH_3$	112.22		134		$0.8001^{20/4}$	1.4401^{20}	al, eth, bz, peth	B5[4], 114
5469	1,3-Cycloheptanedione $C_7H_{10}O_2$	126.16	ye	107-9^{17}	-40	$1.0607^{22/22}$	1.4689^{22}	al	C50, 6327
5470	Cycloheptanol or Suberol $C_7H_{13}OH$	114.19		185, 95^{24}	2	0.9554^{20}	1.4705^{20}	al, eth	B6[4], 94
5471	Cycloheptanone or Suberone $C_7H_{12}O$	112.17		178-9, 71^{9}		$0.9508^{20/4}$	1.4608^{20}	al, eth	B7[3], 46
5472	Cycloheptane carboxylic acid $C_7H_{13}CO_2H$	142.20		254-8^{711}, 130-1^{8}		$1.0423^{20/4}$	1.4753^{20}	al	B9[3], 47
5473	Cycloheptasiloxane, tetradicamethyl $C_{14}H_{42}O_7Si_7$	519.08		154^{20}	-26	$0.9703^{20/4}$	1.4040^{20}		B4[3], 1886
5474	1,,3,5-Cycloheptatriene or Tropilidene C_7H_8	92.14	cubic (at-80)	117, 60.5^{122}	-79.5	$0.8875^{19/4}$	1.5343^{20}	al, eth, bz, chl	B5[4], 765
5475	2,4,6-Cycloheptatriene-1-one or Tropone C_7H_6O	106.12		113^{15}	-7	$1.095^{22/4}$	1.6172^{22}		Am73, 876
5476	2,4,6-Cycloheptatriene-1-one, 2-amino $2\text{-}H_2N(C_7H_5O)$	121.14	ye pl (bz)		106-7			al, bz, chl	C46, 7559
5477	2,4,6-Cycloheptatriene-1-one, 3-bromo-2-hydroxy $3\text{-}Br\text{-}2\text{-}HO(C_7H_4O)$	201.02	ye pl or nd		107-8			al, eth	C49, 2405

PHYSICAL CONSTANTS OF ORGANIC COMPOUNDS (Continued)

No.	Name, Synonyms, and Formula	Mol. wt.	Color, crystalline form, specific rotation and λ_{max} (log ε)	b.p. °C	m.p. °C	Density	n_D	Solubility	Ref.
5478	2,4,6-Cycloheptatrien-1-one, 2-hydroxy or Tropolone 2-HO(C_7H_6O)	122.12	nd	sub 40^{4}	51-2			w, eth, ace	Am74, 4456
5479	2,4,6-Cycloheptatriene-1-one, 2-hydroxy-4-isopropyl $C_{10}H_{12}O_2$	164.20	pa ye (peth)		50-1				$B8^{3}$, 440
5480	2,4,6-Cycloheptatriene-1-one, 2-hydroxy-4-methyl 2-HO-4-CH_3(C_7H_4O)	136.15	nd (peth)		75-6			eth, chl	$B8^{3}$, 260
5481	2,4,6-Cycloheptatriene-1-one, 2-methoxy 2-CH_3O-(C_7H_5O)	136.15	pa, ye nd (+ ½ w)	128^{5}	41(+ ½ w)			al, bz	C46, 4521
5482	Cycloheptene or Suberene C_7H_{12}	96.17		115	-56	$0.8228^{20/4}$	1.4552^{20}	al, eth, bz, peth	$B5^{4}$, 244
5483	1,3-Cyclohexadiene or 1,2-Dihydrobenzene C_6H_8	80.13		80.5	-89	$0.8405^{20/4}$	1.4755^{20}	al, eth, bz, chl	$B5^{4}$382
5484	1,3-Cyclohexadiene, 5-methyl *(dl)* 5-$CH_3C_6H_7$	94.16		101.5^{762}		$0.8354^{20/4}$	1.4763^{20}	al, eth, bz, lig	$B5^{3}$, 318
5485	1,3-Cyclohexadiene, perfluoro C_6F_8	224.05			62-4	1.601^{20}	1.3149^{20}		$B5^{4}$384
5486	1,4-Cyclohexadiene or 1,4-Dihydrobenzene C_6H_8	80.13		85.6	-49.2	$0.8471^{20/4}$	1.4725^{20}	**al, eth**, bz, peth, chl	$B5^{4}$, 385
5487	1,4-Cyclohexadiene, octafluoro C_6F_8	224.05			57-8		1.318^{18}		$B5^{4}$, 386
5488	1,4-Cyclohexadiene-1,2-dicarboxylic acid or 3,6-Dihydrophthalic acid 1,4-$(HO_2C)_2C_6H_6$	168.15	mcl pr (w)		153			al	$B9^{3}$, 4047
5489	2,4-Cyclohexadiene-1,2-dicarboxylic acid or 2,3-Dihydrophthalic acid 1,2-$(HO_2C)_2C_6H_6$	168.15	pr (w or al)		179.80			al	$B9^{3}$, 4047
5490	2,6-Cyclohexadiene-1,2-dicarboxylic acid or 4,5-Dihydrophthalic acid 1,2-$(HO_2C)_2C_6H_6$	168.15	tcl (w)		215			al, ace	$B9^{2}$, 575
5491	Cyclohexane C_6H_{12}	84.16		80.7	6.5	$0.7785^{20/4}$	1.4266^{20}	**al, eth, ace, bz, lig**	$B5^{4}$, 27
5492	Cyclohexane, acetyl or Cyclohexyl methyl ketone $CH_3COC_6H_{11}$	126.20		180-1, 69^{12}		$0.9176^{20/4}$	1.4565^{16}	eth	$B7^{3}$, 84
5493	Cyclohexane, allyl (CH_2=$CHCH_2$)C_6H_{11}	124.23		131.5^{757}		0.8135^{20}	1.4500^{20}	al, eth, ace, bz, chl	$B5^{4}$, 283
5494	Cyclohexane, amino or Cyclohexyl amine $C_6H_{11}NH_2$	99.18		134.5, 30.5^{15}	-17.7	$0.8191^{20/4}$	1.4372^{20}	w, al, **eth, ace, bz**	$B12^{3}$, 10
5495	Cyclohexane, bromo or Cyclohexyl bromide $C_6H_{11}Br$	163.06		166.2, 45.5^{10}	-56.5	$1.3359^{20/4}$	1.4957^{20}	**al, eth, ace, bz**	$B5^{4}$, 67
5496	Cyclohexane, 1-bromo-1-methyl 1-Br-1-CH_3(C_6H_{10})	177.08		156-60, $65\text{-}9^{10}$		1.2510^{20}	1.4866^{20}	al, chl	$B5^{4}$, 100
5497	Cyclohexane, 1-bromo-2-methyl 1-Br-2-CH_3(C_6H_{10})	177.08		$90\text{-}2^{10}$					$B5^{2}$, 12
5498	Cyclohexane, 1-bromo-3-methyl *(dl)* 1-Br-3-CH_3(C_6H_{10})	177.08		181, 60^{11}		$1.275^{25/4}$	1.4979^{20}	eth, bz	$B5^{3}$, 76
5499	Cyclohexane, 1-bromo-4-methyl 1-Br-4-CH_3(C_6H_{10})	177.08		130^{200}, 55^{15}				eth, bz	$B5^{4}$, 100
5500	Cyclohexane, (bromomethyl) $BrCH_2C_6H_{11}$	177.08		$76\text{-}7^{26}$		$1.2763^{25/4}$	1.4907^{20}	eth, bz, chl	$B5^{4}$, 100
5501	Cyclohexane, butyl or Cyclohexyl butane $C_4H_9C_6H_{11}$	140.27		181, 59^{10}	−74.7	$0.7992^{20/4}$	1.4408^{20}		$B5^{4}$, 146
5502	Cyclohexane, isobutyl *i*-$C_4H_9C_6H_{11}$	140.27		171.3	−95	$0.7952^{20/4}$	1.4386^{20}	al, eth, ace, bz, chl	$B5^{4}$, 147
5503	Cyclohexane, *sec*-butyl $CH_3CH_2CH(CH_3)C_6H_{11}$	140.27		179.3		$0.8131^{20/4}$	1.4467^{20}	ace	$B5^{4}$, 146
5504	Cyclohexane, *tert*-butyl $(CH_3)_3CC_6H_{11}$	140.27		171.5	−41.2	$0.8127^{20/4}$	1.4469^{20}		$B5^{4}$, 147
5505	Cyclohexane, butylamino $C_4H_9NHC_6H_{11}$	155.28		207				al, eth	$B12^{3}$, 15
5506	Cyclohexane, chloro or Cyclohexyl chloride $C_6H_{11}Cl$	118.61		143	−43.9	$1.000^{20/4}$	1.4626^{20}	**al, eth, ace, bz**, chl	$B5^{4}$, 48
5507	Cyclohexane, cyclopentyl $C_5H_9C_6H_{11}$	152.28		215.1		$0.8758^{20/4}$	1.4725^{20}		$B5^{4}$, 328
5508	Cyclohexane, 1,2-dibromo *(cis)* 1,2-$Br_2C_6H_{10}$	241.95		115^{14}	9.7	$1.803^{25/25}$	1.5514^{25}	eth, ace, bz, chl, lig	$B5^{4}$, 70
5509	Cyclohexane, 1,2-dibromo *(trans, dl)* 1,2-$Br_2C_6H_{10}$	241.95		$145\text{—}6^{100}$, 105^{20}	−4	$1.7759^{20/4}$	1.5445^{19}	al, eth, ace, bz	$B5^{4}$, 71

PHYSICAL CONSTANTS OF ORGANIC COMPOUNDS (Continued)

No.	Name, Synonyms, and Formula	Mol. wt.	Color, crystalline form, specific rotation and λ_{max} (log ε)	b.p. °C	m.p. °C	Density	n_D	Solubility	Ref.
5510	Cyclohexane, 1,3-dibromo *(cis)* $1,3\text{-}Br_2C_6H_{10}$	241.95	rods (al)	...	112	...	...	bz, al	$B5^4$, 71
5511	Cyclohexane, 1,3-dibromo *(trans)* $1,3\text{-}Br_2C_6H_{10}$	241.95	...	116^{16}	1	...	1.5480^{20}	al, bz	$B5^4$, 72
5512	Cyclohexane, 1,4-dibromo *(cis)* $1,4\text{-}Br_2C_6H_{10}$	241.95	...	$137\text{-}8^{25}$	...	$1.7834^{20/4}$	1.5531^{20}	eth	$B5^4$, 72
5513	Cyclohexane, 1,4-dichloro *(cis)* $1,4\text{-}Cl_2C_6H_{10}$	153.05	...	80.3^{25}	18	$1.1900^{20/4}$	1.4942^{20}	...	$B5^4$, 51
5514	Cyclohexane, 1,4-dibromo *(trans)* $1,4\text{-}Br_2C_6H_{10}$	241.95	cr (eth)	...	...	...	...	eth	$B5^4$, 72
5515	Cyclohexane, 1,.2-dichloro *(cis)* $1,2\text{-}Cl_2C_6H_{10}$	153.05	...	$206\text{-}9^{762}$, 91^{20}	−1.5	$1.2021^{20/4}$	1.4967^{20}	bz	$B5^4$, 50
5516	Cyclohexane, 1,2-dichloro-*(trans,dl)* $1,2\text{-}Cl_2C_6H_{10}$	153.05	...	189, 78^{20}	−6.3	$1.1839^{20/4}$	1.4902^{20}	...	$B5^4$, 51
5517	Cyclohexane, (diethyl amino) $(C_2H_5)_2NC_6H_{11}$	155.28	...	$192\text{-}3^{740}$, $85\text{-}6^{20}$	...	$0.872^{0/0}$	...	al	$B12^3$, 14
5518	Cyclohexane, (difluoramino) decafluro $C_6F_{11}NF_2$	333.05	...	75-6	...	$1.787^{25/4}$	1.286^{25}	...	J1950, 1966
5519	Cyclohexane, 1,1-dimethyl $1,1\text{-}(CH_3)_2C_6H_{10}$	112.22	...	119.5, 10^{10}	−33.5	$0.7809^{20/4}$	1.4290^{20}	al, eth, ace, bz, lig	$B5^4$, 117
5520	Cyclohexane, 1,2-dimethyl *(cis)* $1,2\text{-}(CH_3)_2C_6H_{10}$	112.22	...	129.7, 18.3^{10}	−50.1	$0.7963^{20/4}$	1.4360^{20}	al, eth, ace, bz, lig	$B5^4$, 118
5521	Cyclohexane, 1,2-dimethyl *(trans)* $1,2\text{-}(CH_3)_2C_6H_{10}$	112.22	...	123.4, 12.9^{10}	−89.2	$0.7760^{20/4}$	1.4270^{20}	al, eth, ace, bz, lig	$B5^4$, 118
5522	Cyclohexane, 1,2-dimethyl,perfluoro $1,2\text{-}(CF_3)_2C_6F_{10}$	400.06	...	101.5	−56	$1.829^{25/4}$	1.283^{25}	...	$B5^3$, 98
5523	Cyclohexane, 1,3-dimethyl *(cis)* $1,3\text{-}(CH_3)_2C_6H_{10}$	112.22	...	120.1, 11.1^{10}	−75.6	$0.7660^{20/4}$	1.4229^{20}	al, eth, ace, bz	$B5^4$, 121
5524	Cyclohexane, 1,3-dimethyl *(trans, d)* $1,3\text{-}(CH_3)_2C_6H_{10}$	112.22	$[\alpha]_{549}$ + 1.33	124.4, 15^{10}	−90	$0.7847^{20/4}$	1.4309^{20}	al, eth, ace, bz, lig	$B5^4$, 121
5525	Cyclohexane, 1,4-dimethyl *(cis)* $1,4\text{-}(CH_3)_2C_6H_{10}$	112.22	...	124.3, 14.4^{10}	−87.4	$0.7829^{20/4}$	1.4230^{20}	al, eth, ace, bz, lig	$B5^4$, 122
5526	Cyclohexane, 1,4-dimethyl *(trans, dl)* $1,4\text{-}(CH_3)_2C_6H_{10}$	112.22	...	119.3, 10^{10}	−37.0	$0.7626^{20/4}$	1.4209^{20}	al, eth, ace, bz, lig	$B5^4$, 123
5527	Cyclohexane, 1,2-dimethylene $(1,2\text{-}CH_2)_2C_6H_8$	108.18	...	124^{740}, $60\text{—}1^{90}$	...	$0.8229^{25/4}$	1.4718^{25}	al, eth, ace, bz, chl	$B5^4$, 409
5528	Cyclohexane, 1,2-epoxy or Cyclohexene oxide $C_6H_{10}O$	98.14	...	131.5, $54\text{-}5^{10}$	<−10	0.9663^{20}	1.4519^{20}	al, eth, ace, bz	$B17^4$, 164
5529	Cyclohexane, 1,2-epoxy-4 (epoxy ethyl) or 4-Vinylcyclohexene dioxide $C_8H_{12}O_2$	140.18	...	227, 92^5	<−55	$1.0986^{20/20}$	1.4787^{20}	w	$B19^4$, 161
5530	Cyclohexane, 1,2-epoxy-4-vinyl $C_8H_{12}O$	124.18	...	169, 20^2	<−100	$0.9598^{20/20}$	1.4700^{20}	...	$B17^4$, 314
5531	Cyclohexane, ethyl $C_2H_5C_6H_{11}$	112.22	...	131.8, 20.5^{10}	−111.3	$0.7880^{20/4}$	1.4330^{20}	al, eth, ace, bz, lig	$B5^4$, 115
5532	Cyclohexane, ethylamino $C_2H_5NHC_6H_{11}$	127.23	...	164, $62\text{-}5^{15}$	...	$0.868^{0/0}$	...	al, eth	$B12^3$, 14
5533	Cyclohexane, fluoro or Cyclohexyl fluoride $C_6H_{11}F$	102.15	...	100.2, 48^{100}	13	$0.9279^{20/4}$	1.4146^{20}	Py	$B5^4$, 44
5534	Cyclohexane, 1,1,2,3,4,5,6-heptachloro $C_6H_5Cl_7$	325.28	rods	...	55-6	...	...	...	$B5^4$, 63
5535	Cyclohexane, 1,2,3,4,5,6-hexabromo or Benzene-β-hexabromide *(β or cis)* $C_6H_6Br_6$	557.54	pr	...	253d	...	...	...	$B5^4$, 78
5536	Cyclohexane, 1,2,3,4,5,6-hexabromo *(α or trans)* $C_6H_6Br_6$	557.54	mcl pr (xyl)	...	212	...	...	...	$B5^4$, 78
5537	Cyclohexane, 1,2,3,4,5,6-hexachloro *(α dl)* or Benzene-*trans*-hexachloride $C_6H_6Cl_6$	290.83	mcl pr (al or aa)	288	159-60	...	...	al, bz, chl	$B5^4$, 60
5538	Cyclohexane, 1,2,3,4,5,6-hexachloro-*(β)* or Benzene-*cis*-hexachloride $C_6H_6Cl_6$	290.83	cr (bz, al or xyl)	$60^{0.50}$	314-5 sub	1.89^{19}	...	...	$B5^4$, 61
5539	Cyclohexane, 1,2,3,4,5,6-hexachloro -γ- or Benzene-γ-hexachloride, Lindane $C_6H_6Cl_6$	290.83	nd (al)	323.4, 176.2^{10}	112-3	...	...	ace, bz	$B5^4$, 58
5540	Cyclohexane, 1,2,3,4,5,6-hexachloro-(δ) $C_6H_6Cl_6$	290.83	pl	$60^{0.36}$	141-2	...	...	...	$B5^4$, 57

No.	Name, Synonyms, and Formula	Mol. wt.	Color, crystalline form, specific rotation and λ_{max} (log ε)	b.p. °C	m.p. °C	Density	n_D	Solubility	Ref.
5541	Cyclohexane, α-hydroxyethyl $C_6H_{11}[CH(OH)CH_3]$	128.21	....	189, 81-2^{15}	......	0.9250$^{20/4}$	1.4677^{20}	al, eth	B6^4, 117
5542	Cyclohexane, β-hydroxy ethyl $C_6H_{11}CH_2CH_2OH$	128.21	....	207-9^{757}, 97-9^{15}	......	0.9229$^{20/4}$	1.4641^{20}	al, eth, bz	B6^4, 119
5543	Cyclohexane, hydroxymethyl $C_6H_{11}CH_2OH$	114.19	....	183, 83^{14}	−43	0.9297$^{20/4}$	1.4644^{20}	al, eth	B6^4, 106
5544	Cyclohexane, iodo or Cyclohexy iodide $C_6H_{11}I$	210.06	....	180d, 81.5^{20}	......	1.6244$^{20/4}$	1.5477^{20}	al, eth, ace, bz, lig, chl	B5^4, 78
5545	Cyclohexane, methyl or Hexahydrotoluene $CH_3C_6H_{11}$	98.19	....	100.9, 16.3^{10}	−126.6	0.7694$^{20/4}$	1.4231^{20}	al, eth, ace, **bz**	B5^4, 94
5546	Cyclohexane, methylamino $CH_3NHC_6H_{11}$	113.20	....	145-7, 76-7^{18}	......	0.8660^{23}	1.4530^{23}	al, eth	B12^3, 13
5547	Cyclohexane, 1-methyl-4-ethyl $4\text{-}C_2H_5(C_6H_{10})CH_3$	126.24	....	150-1	......	0.791$^{20/20}$	1.435^{20}	....	B5^4, 136
5548	Cyclohexane, 1-methyl-4-methylene $4\text{-}CH_2{=}(C_6H_9)CH_3$	110.20	....	122	......	0.7923$^{19/19}$	1.4465^{18}	....	B5^4, 271
5549	Cyclohexane, 1-methyl-2-pentyl $2\text{-}C_5H_{11}(C_6H_{10})CH_3$	168.32	....	216-9	......	0.816$^{20/20}$	1.4487^{20}	....	B5^4, 171
5550	Cyclohexane, methyl, perfluoro $CF_3C_6F_{11}$ $CF_3C_6F_{11}$	350.05	....	76.1	−44.7	1.7878$^{25/4}$	1.285^{17}	ace, bz	B5^4, 97
5551	Cyclohexane, methylene $CH_2{:}(C_6H_{10})$	96.17	....	102—3^{764}	−106.7	0.8074$^{20/4}$	1.4523^{20}	eth, bz, lig	B5^4, 250
5552	Cyclohexane, nitro $O_2NC_6H_{11}$	129.16	....	205.5^{768}, 95^{22}	fr-34	1.0610$^{20/4}$	1.4612^{19}	al, lig	B5^4, 81
5553	Cyclohexane, 1,2,3,4,5-pentahydroxy *(d)* or *d*-Quercitol $1,2,3,4,5\text{-}(HO)_5C_6H_7$	164.16	pr (w or dil al), $[\alpha]^{15/}{}_D$ +25 (w,c=10)	....	235-7	1.5845^{13}	....	w	B6^3, 6873
5554	Cyclohexane, 1,2,3,4,5-pentahydroxy *(l)* or Viboquercetol $1,2,3,4,5\text{-}(HO)_5C_6H_7$	164.16	pr (w), nd (al), nd (w + 1), $[\alpha]^{20/}{}_D$ −50 (w,c=4)	....	180-1 (anh)	....	....	w	B6^3, 6873
5555	Cyclohexane, pentyl $C_5H_{11}C_6H_{11}$	154.30	....	202.8, 75.3^{10}	−57.5	0.8037$^{20/4}$	1.4437^{20}	al, eth, ace, bz	B5^4, 164
5556	Cyclohexane, *iso*-pentyl $i\text{-}C_5H_{11}C_6H_{11}$	154.30	....	196.5	......	0.8023$^{20/4}$	1.4420^{20}	bz, lig	B5^3, 143
5557	Cyclohexane, phenyl or Cyclohexyl benzene $C_6H_5C_6H_{11}$	160.26	....	235-6, 127-8^{30}	7-8	0.9502$^{20/4}$	1.5329^{20}	al, eth	B5^4, 1424
5558	Cyclohexane, propyl $C_3H_7C_6H_{11}$	126.24	....	156.7, 40.1^{10}	−94.9	0.7936$^{20/4}$	1.4370^{20}	**al**, eth, **ace**, bz, peth	B5^4, 134
5559	Cyclohexane, *iso*-propyl or Hexahydrocumene $(CH_3)_2CHC_6H_{11}$	126.24	....	154.5, 38.3^{10}	90	0.8023$^{20/4}$	1.4410^{20}	al, eth, **ace, bz**	B5^4, 134
5560	Cyclohexane, 1,1,3-trimethyl $1,1,3\text{-}(CH_3)_3C_6H_9$	126.24	....	138-94	......	0.7664$^{20/0}$	1.4237^{15}	....	B5^4, 137
5561	Cyclohexane, 1,3,5-trimethyl *(cis)* or Hexahydromesitylene $1,3,5\text{-}(CH_3)_3C_6H_9$	126.24	....	138.5	−49.7	0.7708$^{20/4}$	1.4269^{20}	eth, bz, lig	B5^4, 138
5562	Cyclohexane, 1,3,5-trimethyl *(trans)* or Hexahydromesitylene $1,3,5\text{-}(CH_3)_3C_6H_9$	126.24	....	140.5	−107.4	0.7794$^{20/4}$	1.4307^{20}	eth, bz, lig	B5^4, 138
5563	Cyclohexanecarboxaldehyde or Hexahydrobenzaldehyde $C_6H_{11}CHO$	112.17	....	159.3, 36^{10}	......	0.9035$^{20/4}$	1.4496^{20}	eth	B7^3, 66
5564	Cyclohexane carboxylic acid or Hexahydrobenzoic acid $C_6H_{11}CO_2H$	128.17	mcl pr	232-3, 120-1^{13}	31-2	1.0334$^{22/4}$	1.4599^{22}	al, bz, chl	B9^3, 15
5565	Cyclohexane carboxylic acid, ethyl ester $C_6H_{11}CO_2C_2H_5$	156.22	....	196, 63^{12}	......	0.9362$^{20/4}$	1.4501^{15}	al, eth, ace, chl	B9^3, 17
5566	Cyclohexane carboxylic acid, 2-hydroxy or Hexahydrosalicylic acid $2\text{-}HOC_6H_{10}CO_2H$	144.17	nd (AeOEt)	....	111	....	....	w, al, eth	B10^3, 14
5567	Cyclohexane carboxylic acid, methyl ester $C_6H_{11}CO_2CH_3$	142.20	....	183, 73^{15}	......	0.9954$^{15/4}$	1.4433^{20}	al, eth, ace, chl	B9^3, 16
5568	Cyclohexane carboxylic acid, propyl ester $C_6H_{11}CO_2C_3H_7$	170.25	....	215.5	......	0.9530$^{15/4}$	1.4486^{15}	al, eth, ace, chl	B9^3, 17

No.	Name, Synonyms, and Formula	Mol. wt.	Color, crystalline form, specific rotation and λ_{max} (log ε)	b.p. °C	m.p. °C	Density	n_D	Solubility	Ref.
5569	Cyclohexane carboxylic acid, 1,3,4,5-tetrahydroxy *(d)* or *d*-Quinic acid .... 1,3,4,5-$(HO)_4C_6H_7CO_2H$	192.17	mcl pr (w), $[\alpha]^{20/D}$ + 44 (w,c=10)	d	164	1.637	....	w	B10, 538
5570	Cyclohexane carboxylic acid, 1,3,4,5-tetra hydroxy *(dl)* or *dl*-Quinic acod .... 1,3,4,5-$(HO)_4C_6H_7CO_2H$	192.17	pr (w)	....	142	....	....	w	B10³, 2408
5571	Cyclohexane carboxylic acid, 1,3,4,5-tetrahydroxy *(l)* or *l*-Quinic acid .... 1,3,4,5-$(HO)_4(C_6H_7)CO_2H$	192.17	pr (w), $[\alpha]^{18/D}$ −44.1 (w, c=12)	d	172	1.64	....	w	B10³, 2407
5572	Cyclohexane carboxylonitrile, epoxy .... C_7H_9NO	123.15	....	244.5, 110^{10}	−33	$1.0929^{20/20}$	1.4763^{20}	w, eth	B18⁴, 3891
5573	Cyclohexanecarboxylyl chloride .... $C_6H_{11}COCl$	146.62	....	180, $75\text{-}7^{15}$	......	$1.0962^{15/4}$	1.4711^{20}	....	B9³, 27
5574	1,2-Cyclohexanedicarboxylic acid *(cis)* or Hexahydrophthalic acid .... 1,2-$C_6H_{10}(CO_2H)_2$	172.18	tcl nd (al)	d	192	....	....	al, eth, ace, bz	B9³, 3812
5575	1,2-Cyclohexane dicarboxylic acid *(trans, d)* .... 1,2-$C_6H_{10}(CO_2H)_2$	172.18	pw (w), $[\alpha]_D$ + 18.2	....	179-83	....	....	w	B9³, 3812
5576	Cyclohexanehexone, octahydrate .... $C_6O_6 \cdot 8H_2O$	312.18	mic nd (dil HNO_3)	....	100-1	....	....	....	B7³, 4857
5577	Cyclohexanol .... $C_6H_{11}OH$	100.16	hyg nd	161.1	25.1	$0.9624^{20/4}$	1.4641^{20}	w, al, ace, eth, bz	B6⁴, 20
5578	Cyclohexanol, 1-acetyl .... 1-$CH_3COC_6H_{10}OH$	142.20	....	125-6, 91^{11}	......	$1.0248^{25/4}$	1.4670^{25}	al, eth	B8³, 15
5579	Cyclohexanol, 2-allyl *(trans)* .... 2-$(CH_2{=}CHCH_2)C_6H_{10}OH$	140.23	....	$94\text{-}6^{15}$	......	$0.947^{20/4}$	1.4778^{20}	aa	B6⁴, 235
5580	Cyclohexanol, 2-amino *(trans, dl)* .... 2-$H_2NC_6H_{10}OH$	115.18	hyg	105^{10}	68	....	....	bz, chl, aa	B13³, 704
5581	Cyclohexanol, 2-butyl *(trans)* .... 2-$C_4H_9C_6H_{10}OH$	156.27	....	$111\text{-}2^{16}$	......	$0.9020^{20/4}$	1.4641^{20}	eth, ace, bz	B6³, 121
5582	Cyclohexanol, 2-chloro *(cis, dl)* .... 2-$ClC_6H_{10}OH$	134.61	hyg (peth)	$93\text{-}4^{26}$	36-7	1.1261^{25}	1.4894^{25}	al, bz, chl	B6⁴, 64
5583	Cyclohexanol, 2-chloro *(cis, l)* .... 2-$ClC_6H_{10}OH$	134.61	hyg $[\alpha]_{549}$ 19.5	87^{15}	......	1.137^{15}	1.4894^{25}	w, al, bz, chl	B6³, 39
5584	Cyclohexanol, 2-chloro *(trans)* .... 2-$ClC_6H_{10}OH$	134.61	pr (bz-lig)	93^{26}	29	$1.146^{16/4}$	1.4899^{20}	al, eth, bz, chl	B6⁴, 64
5585	Cyclohexanol, 4-chloro *(trans)* .... 4-$ClC_6H_{10}OH$	134.16	pl (cy)	106^{14}	82-3	$1.1435^{17/4}$	1.4930^{17}	al, eth, bz, chl	B6⁴, 68
5586	Cyclohexanol, 3-dimethylamino .... 3-$(CH_3)_2NC_6H_{10}OH$	143.23	....	231, $126\text{-}7^{22}$	73	$0.9766^{25/25}$	1.4852^{20}	al	B13³, 719
5587	Cyclohexanol, 1-ethyl .... 1-$C_2H_5C_6H_{10}OH$	128.21	pr	166, 67^{10}	34-5	0.9227^{25}	1.4633^{20}	bz, peth	B6⁴, 115
5588	Cyclohexanol, 2-ethyl *(cis, dl)* .... 2-$C_2H_5C_6H_{10}OH$	128.21	....	180-2, 74^{12}	......	$0.9274^{20/4}$	1.4655^{21}	eth, ace, bz, peth	B6³, 85
5589	Cyclohexanol, 2-ethyl *(trans, dl)* .... 2-$C_2H_5C_6H_{10}OH$	128.21	....	79^{12}	......	$0.9193^{21/4}$	1.4640^{21}	eth, ace, bz, peth	B6⁴, 117
5590	Cyclohexanol, 1-ethynyl .... 1-$(HC{\equiv}C)C_6H_{10}OH$	124.18	cr (peth)	174, 73^{12}	31-2	$0.9873^{20/4}$	1.4822^{20}	al, bz, peth	B6⁴, 348
5591	Cyclohexanol, 2-(l-hydroxyethyl) .... 2$[CH_3CH(OH)]C_6H_{10}OH$	144.21	....	140^{12}	......	$0.976^{20/7}$	1.4900^{20}	....	C50, 3299
5592	Cyclohexanol, 1-methyl .... 1-$CH_3C_6H_{10}OH$	114.19	....	155, 70^{25}	25	$0.9194^{20/4}$	1.4595^{20}	al, bz, chl	B6⁴, 95
5593	Cyclohexanol, 2-methyl *(cis, dl)* .... 2-$CH_3C_6H_{10}OH$	114.19	....	165, 60^{12}	7	$0.9360^{20/4}$	1.4640^{20}	al, eth	B6⁴, 100
5594	Cyclohexanol, 2-methyl *(trans, d)* .... 2-$CH_3C_6H_{10}OH$	114.19	$[\alpha]^{20/}{}_D$ + 17.19 (undil)	166, 78^{20}	......	0.9454^{20}	1.4610^{20}	al, eth	B6³, 62
5595	Cyclohexanol, 2-methyl *(trans, dl)* .... 2-$CH_3C_6H_{10}OH$	114.19	....	$167\text{-}8, 78^{20}$	−4	$0.9247^{20/4}$	1.4616^{20}	al, eth	B6⁴, 100
5596	Cyclohexanol, 2-methyl-*(trans, l)* .... 2-$CH_3C_6H_{10}OH$	114.19	$[\alpha]^{20}{}_D$ − 35.5 (undil)	166, 78^{20}	......	0.9454^{20}	1.4610^{20}	al, eth	B6³, 62
5597	Cyclohexanol, 3-methyl *(cis, l)* .... 3-$CH_3C_6H_{10}OH$	114.19	$[\alpha]^{22/}{}_D$ − 4.75 (undil)	174-5, 94^{12}	−4.7	$0.9155^{20/4}$	1.4574^{20}	al, eth	B6³, 67

PHYSICAL CONSTANTS OF ORGANIC COMPOUNDS (Continued)

No.	Name, Synonyms, and Formula	Mol. wt.	Color, crystalline form, specific rotation and λ_{max} (log ε)	b.p. °C	m.p. °C	Density	n_D	Solubility	Ref.
5598	Cyclohexanol, 3-methyl *(trans, l)* $3\text{-}CH_3C_6H_{10}OH$	114.19	$[\alpha]^{20}/_D - 7.3$ (undil)	174-5, 84^{13}	−1	$0.9214^{20/4}$	1.4590^{20}	al, eth	$B6^3$, 68
5599	Cyclohexanol, 4-methyl *(cis)* $4\text{-}CH_3C_6H_{10}OH$	114.19		173-4, $78\text{-}9^{20}$	−9.2	$0.9170^{20/4}$	1.4614^{20}	al, eth	$B6^4$, 105
5600	Cyclohexanol, 4-methyl *(trans)* $C_7H_{14}O$	114.19		173-4, 54^3		$0.9118^{20/4}$	1.4561^{20}	al, eth	$B6^4$, 105
5601	Cyclohexanol, 1-phenyl $1\text{-}C_6H_5C_6H_{10}OH$	176.26		157.5^{28}, $112\text{-}3^5$	63-3.5	1.035^{16}	1.5415^{16}		$B6^3$, 2510
5602	Cyclohexanol, 2-phenyl *(cis, dl)* $2\text{-}C_6H_5C_6H_{10}OH$	176.26		$140\text{-}1^{16}$	41-2 (56)	1.035^{16}	1.5415^{16}		$B6^3$, 2510
5603	Cyclohexanol, 2-phenyl *(trans, dl)* $2\text{-}C_6H_5C_6H_{10}OH$	176.26	cr (peth)	$152\text{-}5^{16}$	56-7			al, chl	$B6^3$, 2511
5604	Cyclohexanol, 2-isopropyl *(cis)* $2\text{-}(CH_3)_2CHC_6H_{10}OH$	142.24		77^{13}	52-3	0.9223^{25}	1.4665^{25}		$B6^4$, 131
5605	Cyclohexanol, 2,2,6,6-tetrakis (hydroxymethyl) $2,2,6,6,(HOCH_2)_4C_6H_7OH$	220.27	pl (al)		131			w, al	$B6^3$, 6877
5606	Cyclohexanol, 1,2,2-trimethyl *(dl)* $1,2,2\text{-}(CH_3)_3C_6H_8OH$	142.24	cr (+ ½ w)	$81\text{-}2^{20}$	41 (hyd)	$0.9230^{20/4}$	1.4682^{20}	al, eth, bz	$B6^3$, 114
5607	Cyclohexanol, 1,2,6-trimethyl $1,2,6\text{-}(CH_3)_3C_6H_8OH$	142.24		78^{22}		$0.9126^{15/4}$	1.4598^{15}	al, eth, ace, bz	$B6^1$, 17
5608	Cyclohexanol, 1,3,3-trimethyl $1,3,3\text{-}(CH_3)_3C_6H_8OH$	142.24	pr (dil al)		74			al, eth, ace, bz	$B6^1$, 16
5609	Cyclohexanol, 1,3,5-trimethyl $1,3,5\text{-}(CH_3)_3C_6H_8OH$	142.24		181, $82\text{-}3^{19}$		$0.8876^{17/4}$	$1.454^{16.3}$	al, eth, chl	$B6^3$, 117
5610	Cyclohexanol, 1,4,4-trimethyl $1,4,4\text{-}(CH_3)_3C_6H_8OH$	142.24	hyg nd (dil al)	$79\text{-}80^{15}$	58			al, eth, chl	$B6^1$, 16
5611	Cyclohexanol, 2,2,3-trimethyl $2,2,3\text{-}(CH_3)_3C_6H_8OH$	142.24		$85\text{-}7^{15}$				al, eth	$B6^1$, 16
5612	Cyclohexanol, 2,2,5-trimethyl or Pulenol $2,2,5\text{-}(CH_3)_3C_6H_8OH$	142.24		187-9, $90\text{-}2^{23}$		$0.8955^{22/4}$	1.4569^{20}	al	B6, 22
5613	Cyclohexanol, 2,2,6-trimethyl (liquid) $2,2,6\text{-}(CH_3)_3C_6H_8OH$	142.24		$186\text{-}7^{752}$		$0.9128^{20/4}$	1.4600^{20}	al, eth, chl	$B6^4$, 135
5614	Cyclohexanol, 2,2,6-trimethyl (solid) $2,2,6\text{-}(CH_3)_3C_6H_8OH$	142.24	cr (peth or al)	87^{28}	51			al, eth, chl	$B6^4$, 135
5615	Cyclohexanol, 2,3,3-trimethyl $2,3,3\text{-}(CH_3)_3C_6H_8OH$	142.24	nd	197, 97^{19}	28			al, ace	$B6^1$, 16
5616	Cyclohexanol, 2,3,6-trimethyl $2,3,6\text{-}(CH_3)_3C_6H_8OH$	142.24		$193\text{-}5^{747}$		$0.9117^{17/4}$		al, chl	B6, 22
5617	Cyclohexanol, 2,4,5-trimethyl *(cis)* $2,4,5\text{-}(CH_3)_3C_6H_8OH$	142.24	hyg	191-3, 84^{17}		$0.9120^{20/4}$	1.463^{20}	al, eth, chl	$B6^2$, 36
5618	Cyclohexanol, 2,4,5-trimethyl *(trans)* $2,4,5\text{-}(CH_3)C_6H_8OH$	142.24	hyg	196, 112^{35}		$0.906^{20/4}$	1.461^{20}	al, eth, chl	$B6^2$, 36
5619	Cyclohexanol, 3,3,5-trimethyl *(cis)* $3,3,5\text{-}(CH_3)_3C_6H_8OH$	142.24		$201\text{-}3^{750}$, 92^{12}	37.3	$0.9006^{16/4}$	1.4550^{16}	al, eth, chl	$B6^4$, 135
5620	Cyclohexanol, 3,3,5-trimethyl *(trans)* $3,3,5\text{-}(CH_3)_3C_6H_8OH$	142.24	cr (eth)	189.2	55.8	$0.8647^{60/20}$		al, eth, chl	$B6^4$, 135
5621	Cyclohexanone $C_6H_{10}O$	98.14		155.6, 47^{15}	−16.4	$0.9478^{20/4}$	1.4507^{20}	al, eth, ace, bz, chl	$B7^3$, 14
5622	Cyclohexanone, 2-acetyl $2\text{-}CH_3CO(C_6H_9O)$	140.18		$111\text{-}2^{18}$		1.0782^0	1.5138^{20}		$B7^3$, 3223
5623	Cyclohexanone, 2,6-dibromo *(cis)* $2,6\text{-}Br_2(C_6H_8O)$	255.94	cr (eth, aa)		106-7				$B7^3$, 38
5624	Cyclohexanone, 2-butyl $2\text{-}C_4H_9(C_6H_9O)$	154.25		70^2		$0.905^{20/4}$	1.4545^{20}		$B7^3$, 140
5625	Cyclohexanone, 2-butylidene $C_{10}H_{16}O$	152.24		$98\text{-}100^{10}$		$0.935^{20/4}$	1.4800^{20}	al, eth, ace, bz	C49, 1598
5626	Cyclohexanone, 2-chloro $2\text{-}Cl(Cl_6H_9O)$	132.59		82^{15}	23	$1.161^{20/15}$	1.4825^{20}	eth, bz	$B7^3$, 36
5627	Cyclohexanone, 3-chloro $3\text{-}Cl(C_6H_9O)$	132.59		$91\text{-}2^{14}$				eth	B7, 10
5628	Cyclohexanone, 4-chloro $4\text{-}Cl(C_6H_9O)$	132.59		95^{17}			1.4867^{20}	eth	$B7^2$, 11
5629	Cyclohexanone, 2-β-cyanoethyl $2\text{-}NCCH_2CH_2\text{-}(C_6H_9O)$	151.21		$138\text{-}42^{10}$		$1.0181^{20/4}$	1.4755^{20}		$B10^3$, 2835
5630	Cyclohexanone, cyanohydrin $C_6H_{10}(OH)CN$	125.17		$109\text{-}13^9$	34-6		1.4643^{20}	eth, w	$B10^3$, 11

PHYSICAL CONSTANTS OF ORGANIC COMPOUNDS (Continued)

No.	Name, Synonyms, and Formula	Mol. wt.	Color, crystalline form, specific rotation and λ_{max} (log ε)	b.p. °C	m.p. °C	Density	n_D	Solubility	Ref.
5631	Cyclohexanone, 2-cyclohexyl 2-$C_6H_{11}(C_6H_9O)$	180.29		264	−32	0.9752$^{25/25}$	1.4877^{25}		B7^3, 474
5632	Cyclohexanone, 2,6-dibenzyl 2,6-$(C_6H_5CH_2)_2(C_6H_8O)$	278.39	ye nd (al)	185-95^{20}	117-8			bz, aa	B7^3, 2661
5633	Cyclohexanone, 2,4-dimethyl *(trans, d)* 2,4-$(CH_3)_2(C_6H_8O)$	126.20	$[\alpha]^{24}{}_D$ 64.8 al 6%	178.7^{766}, 69^{17}		0.9004$^{16/4}$	1.4488^{22}	eth, ace, bz	B7^3, 93
5634	Cyclohexanone, 2,5-dimethyl *(d)* 2,5-$(CH_3)_2(C_6H_8O)$	126.20	$[\alpha]^{20}$ + 11.5 (undil)	172-4^{750}, 51^{10}		0.8985$^{20/4}$	1.4445^{20}	al, eth	B7^3, 97
5635	Cyclohexanone, 2,5-dimethyl *(trans, dl)* 2,5-$(CH_3)_2(C_6H_8O)$	126.20		171-3, 76-7^{27}		0.9025^{20}	1.4446^{20}	al, eth	B7^3, 97
5635a	Cyclohexanone, 2-(dimethylaminomethyl) 2-$(CH_3)_2NCH_2(C_6H_9O)$	155.24		92$^{10.5}$		0.9504$^{20/4}$	1.4672^{20}	al, eth	B14^3, 5
5636	Cyclohexanone, 2-ethylidene (2-CH_3CH=)(C_6H_8O)	124.18		92^{20}		0.962$^{20/4}$	1.4882^{20}		B7^2, 58
5637	Cyclohexanone, 2-hydroxy or Adipoin 2-$HO(C_6H_9O)$	114.14	nd (al or MeOH)		113		1.4785^{21}	w, al	B8^3, 4
5638	Cyclohexanone, 2-methyl *(d)* 2-$CH_3(C_6H_9O)$	112.17	$[\alpha]^{25}{}_D$ + 14.21 (chl)	167-8^{735}		0.9262$^{18/4}$	1.4440^{25}		B7^3, 49
5639	Cyclohexanone, 2-methyl *(dl)* 2-$CH_3(C_6H_9O)$	112.17		165^{757}, 90^{20}	−13.9	0.9250$^{20/4}$	1.4483^{25}	al, eth	B7^3, 49
5640	Cyclohexanone, 2-methyl *(l)* 2-$CH_3(C_6H_9O)$	112.17	$[\alpha]^{25}{}_D$ − 15.22 (undil)	59-60^{20}		0.9230$^{25/4}$	1.4440^{25}	al, eth	B7^3, 49
5641	Cyclohexanone, 3-methyl *(d)* 3-$CH_3(C_6H_9O)$	112.17	$[\alpha]^{20}{}_D$ + 12.7 (undil)	169		0.9155$^{20/4}$	1.4493^{20}	al, eth	B7^3, 55
5642	Cyclohexanone, 3-methyl *(dl)* 3-$CH_3(C_6H_9O)$	112.17		168-9^{738}, 65^{15}	−73.5	0.9136$^{20/4}$	1.4456^{20}	al, eth	B7^3, 57
5643	Cyclohexanone, 4-methyl 4-$CH_3(C_6H_9O)$	112.17		170	−40.6	0.9138$^{20/4}$	1.4451^{20}	al, eth	B7^3, 63
5644	Cyclohexanone, oxime C_6H_{10}=NOH	113.16	hex pr (lig)	206-10	90			w, al, eth	B7^3, 32
5645	Cyclohexanone, 2-propyl 2-$C_3H_7(C_6H_9O)$	140.23		195, 70^6		0.927$^{20/4}$	1.4538^{20}	al, eth, ace, bz	B7^3, 115
5646	Cyclohexanone, 2-isopropyl 2-*i*-$C_3H_7(C_6H_9O)$	140.23			72-3	0.922$^{16/4}$	1.4564^{15}	al, eth, ace, bz	B7^3, 117
5647	Cyclohexanethiol or Cyclohexyl mercaptan $C_6H_{11}SH$	116.22		158, 41^{12}		0.9782$^{20/4}$	1.4921^{20}	al, eth, ace, bz, chl	B6^4, 72
5648	Cyclohexanethione $C_6H_{10}S$	114.21		74^{11}			1.5375^{20}	al, eth, ace	B7^3, 39
5649	1,2-Cyclohexane dicarboxylic acid *(trans, dl)* 1,2-$C_6H_{10}(CO_2H)_2$	172.19	lf or pr (w)		222			w	B9^3, 3813
5650	1,2-Cyclohexane dicarboxylic acid *(trans, l)* 1,2-$C_6H_{10}(CO_2H)_2$	172.19	pw (w)		179-82				B9, 732
5651	1,2-Cyclohexane dicarboxylic acid, diethyl ester *(cis)* 1,2-$C_6H_{10}(CO_2HC_{25})_2$	228.29		133^{10}		1.0540$^{22/4}$	1.45512^{14}	eth	B9^3, 3813
5652	1,2-Cyclohexane dicarboxylic acid,diethyl ester *(trans, dl)* 1,2-$C_6H_{10}(CO_2C_2H_5)_2$	228.29		135^{11}		1.040$^{20/4}$	1.4522^{13}	eth	B9^3, 3813
5653	1,3-Cyclohexane dicarboxylic acid *(cis)* or *cis*-Hexadhydro isophthalic acid 1,3-$C_6H_{10}(CO_2H)_2$	172.18	nd (con HCl), cr (w)		167-8			w, al, eth, bz	B9^3, 3817
5654	1,3-Cyclohexane dicarboxylic acid *(trans, d)* 1,3-$C_6H_{10}(CO_2H)_2$	172.18	cr (w), $[\alpha]^{22}{}_D$ + 23.8 (w, c=4)		134			w, al, eth	B9^2, 523
5655	1,3-Cyclohexane dicarboxylic acid *(trans, dl)* 1,3-$C_6H_{10}(CO_2H)_2$	172.18	nd (w)		150.5			w, al, eth	B9^3, 3817
5656	1,3-Cyclohexane dicarboxylic acid *(trans, l)* 1,3-$C_6H_{10}(CO_2H)_2$	172.18	w, $[\alpha]^{22}{}_D$ − 23.2 (w, c=2)		134			w, al, eth	B9^3, 523
5657	1,3-Cyclohexane dicarboxylic acid, diethyl ester *(cis)* 1,3-$C_6H_{10}(CO_2C_2H_5)_2$	228.29		288, 142^{11}		1.0450$^{20/4}$	1.4521^{20}		B9^3, 3817
5658	1,3-Cyclohexane dicarboxylic acid, diethyl ester *(trans, dl)* 1,3-$C_6H_{10}(CO_2C_2H_5)_2$	228.29		286^{756}, 142^{12}		1.0485$^{21/4}$	1.4530^{20}		B9^3, 3817
5659	1,4-Cyclohexane dicarboxylic acid *(cis)* or *cis*-Hexahydro terephthalic acid 1,4-$C_6H_{10}(CO_2H)_2$	172.18	lf (w)		170-1			al, eth, chl	B9^3, 3818

PHYSICAL CONSTANTS OF ORGANIC COMPOUNDS (Continued)

No.	Name, Synonyms, and Formula	Mol. wt.	Color, crystalline form, specific rotation and λ_{max} (log ε)	b.p. °C	m.p. °C	Density	n_D	Solubility	Ref.
5660	1,4-Cyclohexane dicarboxylic acid *(trans)* $1,4-C_6H_{10}(CO_2H)_2$	172.19	pr (w), pl (ace)	300 sub	312-3			al, ace	$B9^3$, 3818
5661	1,4-Cyclohexane dicarboxylic acid, diethyl ester *(cis)* $1,4-C_6H_{10}(CO_2C_2H_5)_2$	228.29		151^{13}		$1.0516^{21/4}$	1.4522^{21}	eth	$B9^2$, 524
5662	1,4-Cyclohexane dicarboxylic acid, diethyl ester *(trans)* $1,4-C_6H_{10}(CO_2C_2H_5)_2$	228.29	nd		43-4	$1.0110^{20/4}$	1.4337^{64}	eth	$B9^2$, 524
5663	1,2-Cyclohexanediol *(cis)* $1,2-(HO)_2C_6H_{10}$	116.16	cr (eth), pl (bz)	120^{15}	99-101	$1.0297^{101/4}$		al, ace, bz	$B6^3$, 4058
5664	1,2-Cyclohexanediol *(trans)* $1,2-(HO)_2C_6H_{10}$	116.16	cr (ace)	117^{13}	105	$1.147^{24/4}$		w, al	$B6^3$, 4060
5665	1,4-Cyclohexanediol *(cis)* or *cis*-Quinitol $1,4-(HO)_2C_6H_{10}$	116.16	pr (ace)		113-4			w, al	$B6^3$, 4080
5666	1,4-Cyclohexanediol *(trans)* $1,4-(HO)_2C_6H_{10}$	116.16	mcl pr (ace)		143	$1.18^{20/4}$		w, al	$B6^3$, 4081
5667	1,2-Cyclohexanedione $C_6H_8O_2$	112.13	cr (peth)	193-5, $96\text{-}7^{25}$	38-40		1.4995^{20}	w, al, eth, bz	$B7^3$, 3209
5668	1,2-Cyclohexanedione, 3,5-dimethyl $C_8H_{12}O_2$	140.18	cr (dil MeOH)		71-2			w	$B7^1$, 314
5669	1,2-Cyclohexanedione, 5,5-dimethyl or Dimedone $C_8H_{12}O_2$	140.18	yesh nd (w, aq ace), mcl pr (al-eth)		150			ace, chl, aa	$B7^3$, 3225
5670	1,2-Cyclohexanedione, dioxime or Nioxime $C_6H_{10}N_2O_2$	142.16	nd (w or ace)		191-3			ace	$B7^3$, 3210
5671	1,3-Cyclohexanedione $C_6H_8O_2$	112.13	pr (bz)		105-6	1.0861^{91}	1.4576^{102}	w, al, ace, chl	$B7^3$, 3210
5672	1,3-Cyclohexanedione, 2-bromo $C_6H_7BrO_2$	191.02	micr nd		169-70			al	B7, 556
5673	1,3-Cyclohexanedione, dioxime $C_6H_{10}N_2O_2$	142.16	cr (w)		156-7			w, al, aa	$B7^3$, 3211
5674	1,4-Cyclohexanedione $C_6H_8O_2$	112.13	mcl pl (w), nd (peth)	sub 100	78			w, al, eth, ace, bz	$B7^3$, 3211
5675	1,4-Cyclohexanedione, dioxime $C_6H_{10}N_2O_2$	142.16	cr (w)		188			w	$B7^3$,3212
5676	Cyclohexasiloxane, dodecamethyl $C_{12}H_{36}O_6Si_6$	444.93		245, 128^{20}	−3	0.9672	1.4015^{20}		$B4^3$, 1886
5677	Cyclohexasiloxane, 2,4,6,8,10,12-hexamethyl $C_6H_{24}O_6Si_6$	360.77			−79	$1.006^{20/4}$	1.3944^{20}		$B4^3$, 1874
5678	Cyclohexene C_6H_{10}	82.15		83	−103.5	$0.8102^{20/4}$	1.4465^{20}	al, eth, ace, bz	$B5^4$, 218
5679	Cyclohexene, 1-acetyl $1-(CH_3CO)C_6H_9$	124.18		201-2, $63\text{-}4^6$		$0.9655^{20/4}$	1.4881^{20}	al, eth	$B7^4$, 244
5680	1-Cyclohexene, 1-bromo $1-BrC_6H_9$	161.04		164-6, 69^{35}		$1.3901^{20/4}$	1.5134^{20}	eth, ace, bz	$B5^4$, 236
5681	Cyclohexene, 3-bromo $3-BrC_6H_9$	161.04		$80\text{-}2^{40}$		$1.3890^{20/4}$	1.5230^{20}	eth, bz, chl	$B5^4$, 237
5682	Cyclohexene, 1-chloro $1-ClC_6H_9$	116.59		142-3, 35^{13}		$1.0361^{19/4}$	1.4797^{20}	eth, ace, chl	$B5^4$, 230
5683	Cyclohexene, 1,2-dimethyl $1,2-(CH_3)_2C_6H_8$	110.20		136		$0.823^{20/4}$	1.4580^{21}		$B5^4$, 268
5684	Cyclohexene, 1-ethyl $1-C_2H_5C_6H_9$	110.20		$135\text{-}6^{755}$		0.8238^{19}	1.4567^{19}		$B5^4$, 266
5685	Cyclohexene, 1-methyl $1-CH_3C_6H_9$	96.17		110, 24.6^{30}	−121	$0.8102^{20/4}$	1.4503^{20}	eth, bz	$B5^4$, 245
5686	Cyclohexene, 3-methyl *(d)* $3-CH_3C_6H_9$	96.17	$[\alpha]^{20/}{}_D$ + 110	104		$0.8010^{20/4}$	1.4414^{20}	eth, bz, peth, chl	$B5^3$, 200
5687	Cyclohexene, 3-methyl *(dl)* $3-CH_3C_6H_9$	96.17		104	−115.5	$0.7990^{20/4}$	1.4414^{20}	eth, bz, peth, chl	$B5^4$, 247
5688	Cyclohexene, 4-methyl $4-CH_3C_6H_9$	96.17		102.7, 19^{30}	−115.5	$0.7991^{20/4}$	1.4414^{20}	al, eth	$B5^4$, 248
5689	Cyclohexene, 1,3,4,5,6-pentachloro *(γ)* $C_6H_5Cl_5$	254.37		$115\text{-}6^4$			1.5630^{20}		$B5^4$, 234
5690	Cyclohexene, 1,3,4,5,6-pentachloro *(d)* $C_6H_5Cl_5$	254.37			68-9	1.80		al	$B5^4$, 234
5691	Cyclohexene-per fluoro C_6F_{10}	262.05		$52\text{-}3^{750}$			1.293^{20}		$B5^4$, 229

No.	Name, Synonyms, and Formula	Mol. wt.	Color, crystalline form, specific rotation and λ_{max} (log ε)	b.p. °C	m.p. °C	Density	n_D	Solubility	Ref.
5692	Cyclohexene, 1-phenyl $1\text{-}C_6H_5C_6H_9$	158.24		251-3, 125-6^{14}	−11	0.9939$^{20/4}$	1.5718^{20}	MeOH	B5^4, 1557
5693	Cyclohexene, 1-isopropyl-4-methylene $i\text{-}C_3H_7\text{-}4(CH_2{=})C_6H_7$	136.24		173-4		0.838^{22}	1.4754^{22}		B5^4, 437
5694	Cyclohexene, 4-isopropyl-1-methyl $4\text{-}i\text{-}C_3H_7C_6H_8CH_3$	136.24		174.5		0.8465$^{15.5/15.5}$	1.4735^{20}		B5^4, 300
5695	Cyclohexene, 1-vinyl $1\text{-}(CH_2{=}CH)C_6H_9$	108.18		145, 50-2^{22}		0.8623$^{15/4}$	1.4915^{20}	eth, bz, MeOH	B5^4, 405
5696	Cyclohexene, 4-vinyl $4\text{-}(CH_2{=}CH)C_6H_9$	108.18		128.9, 66-7^{100}		0.8299$^{20/4}$	1.4639^{20}	eth, bz, peth	B5^4, 406
5697	Cyclohexene, 1-carboxaldehyde C_6H_9CHO	110.16		72^{15}		0.9694$^{20/4}$	1.5005^{20}	al, eth	B7^3, 234
5698	3-Cyclohexene, 1-carboxaldehyde C_6H_9CHO	110.16		164, 52^{13}	fr−96.1	0.9709$^{20/4}$	1.4725^{19}	ace, MeOH	B7^3, 237
5699	Cyclohexene, 1-carboxylic acid $2\text{-}C_6H_9CO_2H$	126.16		240-2, 138^{14}	38	1.109$^{20/4}$	1.4902^{20}	al, ace	B9^3, 144
5700	3-Cyclohexene, 1-carboxylic acid $4\text{-}C_6H_9CO_2H$	126.16		237^{748}, 132-3^{20}	17	1.0815$^{20/4}$	1.4812^{20}	w, al, ace	B9^3, 148
5701	1-Cyclohexene-1,2-dicarboxylic acid or Δ′-Tetrahydra-phthalic acid $C_8H_{10}O_4$	170.17	nd pr (w)		126			w	B9^3, 3939
5702	1-Cyclohexene-1,2-dicarboxylic anhydride $C_8H_8O_3$	152.15	pl (eth)		74			al, eth, ace, chl	B17^4, 5995
5703	2-Cyclohexene-1,2-dicarboxylic anhydride $C_8H_8O_3$	152.15	pr (eth)		78-9			al, eth, chl	B17^4, 5994
5704	4-Cyclohexene-1,2-dicarboxylic anhydride *(cis)* $C_8H_8O_3$	152.15	pl (eth or lig)		103-4			al, ace, chl	B17^4, 5996
5705	4-Cyclohexene-1,2-dicarboxylic anhydride *(trans, d)* $C_8H_8O_3$	152.15	lf $[\alpha]^{25}{}_D$ + 6.6 (al)		128			al, bz	B17, 462
5706	4-Cyclohexene-1,2-dicarboxylic anhydride *(trans, dl)* $C_8H_8O_3$	152.15	cr (bz-lig)		141			al, bz, chl	B17^4, 5996
5707	2-Cyclohexene-1-ol C_6H_9OH	98.14		164-6, 63-5^{12}		0.9923$^{15/4}$	1.4790^{22}	al, ace	B6^4, 196
5708	2-Cyclohexene-1-ol, 5-methyl *(cis, d)* $CH_3C_6H_8OH$	112.17	$[\alpha]^{30}{}_D$ + 6.95	83^{25}		0.9391$^{25/4}$	1.4727^{25}	eth, lig	B6^4, 206
5709	2-Cyclohexene-1-ol, 5-methyl-*(cis, l)* $CH_3C_6H_8OH$	112.17	$[\alpha]^{25}{}_D$ − 7	82^{25}		0.9391$^{25/4}$	1.4727^{25}	eth, lig	B6^4, 206
5710	2-Cyclohexene-1-ol, 5-methyl *(trans, d)* $CH_3C_6H_8OH$	112.17	$[\alpha]^{27}{}_D$ + 127 (ace, c=19.4)	68-9^{24}		0.9430$^{20/4}$	1.4737^{25}	eth, lig	B6^4, 207
5711	2-Cyclohexene-1-ol, 5-methyl-*(trans, l)* $CH_3C_6H_8OH$	112.17	$[\alpha]^{27}{}_D$ − 163.9	82-3^{24}		0.9430$^{20/4}$	1.4737^{25}	eth, lig	B6^4, 207
5712	3-Cyclohexene-1-ol C_6H_9OH	98.14		164, 68-9^{16}		0.9845$^{20/4}$	1.4851^{20}	eth, ace	B6^4, 200
5713	2-Cyclohexene-1-one C_6H_8O	96.13		169-71, 61-2^{10}		0.9620^{25}	1.4883^{20}	al, ace	B7^3, 224
5714	2-Cyclohexene-1-one, 2,3-dimethyl $(CH_3)_2(C_6H_6O)$	124.18		93-6^{20}		0.9695^{20}	1.4995^{20}	al, eth	B7^3, 250
5715	2-Cyclohexene-1-one, 2,5-dimethyl $(CH_3)_2(C_6H_6O)$	124.18		189-90		0.938^{22}	1.4753^{22}	al, eth	B7^1, 51
5716	2-Cyclohexene-1-one, 3,5-dimethyl $(CH_3)_2(C_6H_6O)$	124.18		208-9, 94^{17}		0.9400$^{20/4}$	1.4812^{20}	al, eth	B7^3, 255
5717	2-Cyclohexene-1-one, 3,5-dimethyl-4-carboxyethyl $C_{11}H_{16}O_3$	196.25		157-8^{18}		1.0493$^{20/4}$	1.4773^{20}	ace	B10^3, 2902
5718	2-Cyclohexene-1-one, 3,6-dimethyl $(CH_3)_2(C_6H_6O)$	124.18		75^{19}		1.008$^{18/18}$	1.4805^{18}	al, eth	B7^3, 257
5719	2-Cyclohexene-1-one, 2-methyl $CH_3(C_6H_7O)$	110.16		178-9, 56^9		0.9667$^{20/4}$	1.4833^{20}	bz	B7^3, 233
5720	2-Cyclohexene-1-one, 3-methyl $CH_3(C_6H_7O)$	110.16		200-2, 78-9^{12}	−21	0.9693$^{20/4}$	1.4947^{20}	bz	B7^3, 230
5721	3-Cyclohexene-1-one, 4-methyl $CH_3(C_6H_7O)$	110.16		169-72^{755}, 74^{17}		0.9551$^{20/4}$	1.4652^{20}	al, ace, bz	B7^3, 232
5722	2-Cyclohexene-1-one, 5-methyl *(dl)* $CH_3(C_6H_7O)$	110.16		179-83, 60^8		0.947$^{20/4}$	1.4739^{25}	al, ace, bz	B7^3, 236
5723	2-Cyclohexene-1-one, 5-isopropyl-3-methyl or Hexeton $C_{10}H_{16}O$	152.24	pa ye	244, 124^{15}		0.9340^{21}	1.4865^{21}	al, ace	B7^3, 323

No.	Name, Synonyms, and Formula	Mol. wt.	Color, crystalline form, specific rotation and λ_{max} (log ε)	b.p. °C	m.p. °C	Density	n_D	Solubility	Ref.
5724	3-Cyclohexene-1-one, 4,6-dimethyl $(CH_3)_2(C_6H_6O)$	124.18		194		0.9539^{0}		eth	$B7^{1}$, 256
5725	Cycloheximide $C_{15}H_{23}NO_4$	281.35	wh pl (w), $[\alpha]^{29}{}_D$ − 3.4 (al)		119-21			al, eth, ace	$B21^{4}$, 6632
5726	Cyclohexyl acetic acid $C_6H_{11}CH_2CO_2H$	142.20	nd (HCO₂H)	224-6, 135^{13}	33	$1.0423^{18/4}$	1.4775^{20}	al, eth	$B9^{1}$, 47
5727	Cyclohexylamine $C_6H_{11}NH_2$	99.18		134.5, 30.5^{15}	−17.7	$0.8191^{20/4}$	1.4372^{20}	al, eth, ace, bz	B12 , 10
5728	Cyclohexylamine, hydrochloride $C_6H_{11}NH_2.HCl$	135.64	nd (w or al-eth)		206-7			w, al	$B12^{1}$, 12
5729	Cyclohexyl bromide $C_6H_{11}Br$	163.06		166.2, 45.5^{10}	−56.5	$1.3359^{20/4}$	1.4957^{20}	al, eth, ace, bz, lig	$B5^{4}$, 67
5730	Cyclohexyl chloride $C_6H_{11}Cl$	118.61		143	$1.000^{20/}$	1.4626^{20}		al, eth, ace, bz, chl	$B5^{4}$, 48
5731	Cyclohexyl ether or Dicyclohexyl ether $(C_6H_{11})_2O$	182.31		242-3		$0.9227^{20/4}$	1.4741^{20}		$B6^{3}$, 19
5732	Cyclohexyl 2-furyl ether $C_6H_{11}O(C_4H_3O)$	166.22		118.9^{28}		$1.0200^{28/4}$	1.4861^{28}	al, eth, ace	$B17^{4}$, 1219
5733	Cyclohexyl hydro peroxide $C_6H_{11}OOH$	116.16		$42^{0.1}$	−20	$1.019^{20/4}$	1.4645^{25}	al, eth, aa	$B6^{4}$, 53
5734	Cyclohexyl methyl ether or Hexahydro anisole $C_6H_{11}OCH_3$	114.19		133	−74.4	$0.8756^{20/4}$	1.4355^{20}	al, eth	$B6^{4}$, 26
5735	Cyclohexyl methyl ether, 2-bromo *(dl, trans)* 2-$BrC_6H_{10}OCH_3$	193.08		$78\text{-}9^{12}$		$1.3314^{20/4}$	1.4871^{20}	al, eth	$B6^{3}$, 43
5736	Cyclohexyl phenyl ether $C_6H_{11}OC_6H_5$	176.26		128^{15}		$1.0077^{20/4}$	1.520^{22}	ace, bz	$B6^{4}$, 565
5737	Cyclohexyl isothiocyanate C_5H_9NCS	141.23		219^{746}			1.5375^{20}	al, eth	$B12^{3}$, 53
5738	Cyclononanone $C_9H_{16}O$	140.23		148.5^{24}	34	$0.9560^{20/4}$	1.4729^{20}	al	$B7^{3}$, 111
5739	Cyclononasiloxane, octadecamethyl $C_{18}H_{54}O_9Si_9$	667.39		188^{20}			1.4070^{20}	bz, lig	$B4^{3}$, 1886
5740	Cyclononene *(cis)* C_9H_{16}	124.23		167-9, $73\text{-}4^{30}$		$0.8671^{20/4}$	1.4805^{20}	bz	$B5^{4}$, 280
5741	Cyclononene-*(trans)* C_9H_{16}	124.23		$94\text{-}6^{30}$		$0.8615^{20/4}$	1.4799^{20}	bz	$B5^{4}$, 281
5742	1,5-Cyclooctadiene *(cis, cis)* C_8H_{12}	108.18		150.8^{757}, $51\text{-}2^{25}$	−70	$0.8818^{25/4}$	1.4905^{25}	bz	$B5^{4}$, 403
5743	1,3-Cyclooctadiene *(cis, cis)* C_8H_{12}	108.18		$54\text{-}5^{35}$	−57	$0.8699^{25/4}$	1.4940^{25}	eth, bz, chl	$B5^{4}$, 401
5744	Cyclooctane C_8H_{16}	112.22		$148\text{-}9^{749}$, 63^{45}	14.3	$0.8349^{20/4}$	1.4586^{20}	bz, lig	$B5^{4}$, 111
5745	Cyclooctanol $C_8H_{15}OH$	128.21		99^{16}	25.1	$0.9740^{20/4}$	1.4871^{20}	al	$B6^{4}$, 113
5746	Cyclooctanone or Azelaone $C_8H_{14}O$	126.20		194-8, 74^{12}	28-30	$0.9581^{20/4}$	1.4694^{20}	al, ace, bz	$B7^{3}$, 77
5747	Cyclooctanone, semicarbazone $C_8H_{14}NHCONHNH_2$	183.25	lf (dil MeOH)		170-1				$B7^{3}$, 78
5748	Cyclooctasiloxane, hexadecamethyl $C_{16}H_{48}O_8Si_8$	593.24		290, 175^{20}	31.5	1.177	1.4060^{20}	bz, lig	$B4^{3}$, 1886
5749	Cyclooctatetraene C_8H_8	104.16	ye or wh	140.5, 29.1^{10}	−4.7	$0.9206^{20/4}$	1.5381^{20}	al, eth, ace, bz	$B5^{4}$, 1331
5750	Cyclooctatetraene, chloro C_8H_7Cl	138.60		$50\text{-}1^{5.5}$		$1.1199^{25/4}$	1.5542^{25}	ace, bz	$B5^{4}$, 1334
5751	Cyclooctatetraene, methyl $CH_3C_8H_7$	118.18		84.5^{67}		$0.8978^{25/4}$	1.5249^{25}	al, eth, bz, chl	$B5^{4}$, 1358
5752	Cyclooctene *(cis)* C_8H_{14}	110.20		138, 42^{18}	−12	$0.8472^{20/4}$	1.4698^{20}	al, eth	$B5^{4}$, 262
5753	Cyclooctene *(trans)* C_8H_{14}	110.20		143, 75^{78}	−59	$0.8483^{20/4}$	1.4741^{25}	al, chl	$B5^{4}$, 263
5754	Cyclopentadecanone or Exhaltone $C_{15}H_{28}O$	224.39		$120^{0.3}$	63	0.8895	1.4637^{66}	al, ace	$B7^{3}$, 203
5755	Cyclopentadecanone, 3-methyl or Muscone, muskone 3-$CH_3C_{15}H_{27}O$	238.41	ye $[\alpha]^{17}{}_D$ − 13 (undil)	$327\text{-}30^{752}$, $130^{0.5}$		$0.9221^{17/4}$	1.4802^{17}	al, eth, ace	$B7^{3}$, 208

No.	Name, Synonyms, and Formula	Mol. wt.	Color, crystalline form, specific rotation and λ_{max} (log ε)	b.p. °C	m.p. °C	Density	n_D	Solubility	Ref.
5756	Cyclopentadiene C_5H_6	66.10		40.0	−97.2	$0.8021^{20/4}$	1.4440^{20}	**al, eth**, ace, bz	$B5^4$, 377
5757	Cyclopentadiene, perchloro C_5Cl_6	272.77	ye gr lig	239^{753}, $48\text{-}9^{0.3}$	−9	$1.7019^{25/4}$	1.5658^{20}		$B5^4$, 381
5758	1,3-Cyclopentadiene, 5-isopropylidene $[CH_2{=}C(CH_3)]C_5H_5$	106.17		$49\text{-}50^{11}$		$0.881^{20/4}$	1.5474^{20}		$B5^4$, 974
5759	Cyclopentadiene benzoquinone $C_{11}H_{10}O_2$	174.20	gr-ye lf (MeOH)		77-8			al, eth, ace, bz	$B6^3$, 5311
5760	Cyclopentadienone, tetraphenyl or Tetracyclone $(C_6H_5)_4C_5O$	384.48	bk-vt lf, cr (aa or xyl)		220-1			al, bz	$B7^3$, 2997
5761	Cyclopentasiloxane, 2,4,6,8,10-pentamethyl $C_5H_{20}O_5Si_5$	300.64		169	−108	$0.9985^{20/4}$	1.3912^{20}		$B4^3$, 1874
5762	Cyclopentane or Pentamethylene C_5H_{10}	70.13		49.2	−93.9	$0.7457^{20/4}$	1.4065^{20}	**al, eth, ace, bz**	$B5^4$, 14
5763	Cyclopentane, acetyl or Cyclopentyl methyl ketone $CH_3COC_5H_9$	112.17		158-9		$0.918^{20/20}$	1.4409^{20}	eth	$B7^3$, 71
5764	Cyclopentane, allyl $(CH_2{=}CHCH_2)C_5H_9$	110.20		124-6		$0.793^{25/4}$	1.4412^{20}		$B5^4$, 272
5765	Cyclopentane, amino or Cyclopentyl amine $H_2NC_5H_9$	85.15		108	−85.7	$0.8689^{20/4}$	1.4778^{25}	ace, bz	$B12^3$, 5
5766	Cyclopentane, bromo or Cyclopentyl bromide C_5H_9Br	149.03		137-8, 56^{49}		$1.3873^{20/4}$	1.4886^{20}		$B5^4$, 19
5767	Cyclopentane, butyl $C_4H_9C_5H_9$	126.24		156.7, 41.6^{10}	−108	$0.7846^{20/4}$	1.4316^{20}	**al, eth, ace, bz, peth**	$B5^4$, 139
5768	Cyclopentane, chloro or Cyclopentyl chloride C_5H_9Cl	104.58		$113\text{-}4^{752}$		$1.0051^{20/4}$	1.4510^{20}	eth, ace, bz	$B5^4$, 18
5769	Cyclopentane, 1,2-diethyl *(trans)* $1,2\text{-}(C_2H_5)_2C_5H_8$	126.24		153.6	−95.6	$0.7832^{20/4}$	1.4295^{20}	eth, bz, peth	$B5^4$, 141
5770	Cyclopentane, 1,1-dimethyl $1,1\text{-}(CH_3)_2C_5H_8$	98.19		87.5		$0.7552^{20/0}$	1.4139^{20}		$B5^4$, 105
5771	Cyclopentane, 1,2-dimethyl *(cis)* $1,2\text{-}(CH_3)_2C_5H_8$	98.19		99.25	−62	$0.7718^{20/4}$			$B5^4$, 106
5772	Cyclopentane, 1,2-dimethyl *(trans)* $1,2\text{-}(CH_3)_2C_5H_8$	98.19		91.8	−120	$0.7495^{20/4}$			$B5^4$, 107
5773	Cyclopentance, ethyl $C_2H_5C_5H_9$	98.19		103.5, 19.4^{30}	−138.4	$0.7665^{20/4}$	1.4198^{20}	**al, eth, ace, bz, peth**	$B5^4$, 104
5774	Cyclopentane, β-hydroxyethyl $C_5H_9CH_2CH_2OH$	114.19		$183\text{-}4^{770}$, $96\text{-}7^{24}$		$0.9180^{20/4}$	1.4577^{20}	eth	$B6^4$, 110
5775	Cyclopentane, iodo or Cyclopentyl iodide C_5H_9I	196.03		166-7, 52^{12}		$1.7096^{20/4}$	1.5447^{20}	eth, bz	$B5^4$, 20
5776	Cyclopentane, methyl $CH_3C_5H_9$	84.16		71.8	−142.4	$0.7486^{20/4}$	1.4097^{20}	**al, eth, ace, bz**	$B5^4$, 84
5777	Cyclopentane, 1-methyl-2-propyl (trans) $1\text{-}CH_3\text{-}2\text{-}C_3H_7(C_5H_8)$	126.24		152.6	−104.9	$0.7921^{20/4}$	1.4321^{20}		$B5^4$, 140
5778	Cyclopentane, nitro $C_5H_9NO_2$	115.13		$90\text{-}1^{40}$		$1.0776^{23/4}$	1.4538^{20}	bz	$B5^4$, 20
5779	Cyclopentane, phenyl $C_6H_5C_5H_9$	146.23		219, 102^{18}		$0.9462^{20/4}$	1.5309^{20}		$B5^4$, 1409
5780	Cyclopentane, propyl $C_3H_7C_5H_9$	112.22		131, 21.2^{10}	−117.3	$0.7763^{20/4}$	1.4266^{20}	**al, eth, ace, bz**	$B5^4$, 125
5781	Cyclopentane, isopropyl $(CH_3)_2CHC_5H_9$	112.22		126.4, 16.3^{10}	−111.4	$0.7765^{20/4}$	1.4258^{20}	**al, eth, ace, bz**	$B5^4$, 125
5782	Cyclopentane, 1,1,2-trimethyl $1,1,2\text{-}(CH_3)_3C_5H_7$	112.22		$113\text{-}4^{749}$		$0.7661^{20/D}$	1.4199^{20}		$B5^4$, 127
5783	Cyclopentane, 1,1,3-trimethyl $1,1,3\text{-}(CH_3)_3C_5H_7$	112.22		115-6		$0.7703^{20/4}$	1.4223^{20}		$B5^4$, 128
5784	Cyclopentane carboxaldehyde C_5H_9CHO	98.14		133-4		$0.9371^{20/4}$	1.1432^{20}	w, al, eth	$B7^3$, 43
5785	Cyclopentane carboxylic acid $C_5H_9CO_2H$	114.14		$212\text{-}3^{752}$, 104^{11}	−7	$1.0527^{20/4}$	1.4532^{20}		$B9^3$, 11
5786	Cyclopentanecarboxylic acid, 3-formyl-2,2,3-trimethyl,methyl ester *(d)* $C_{11}H_{18}O_3$	198.26	$[\alpha]_D$ + 51.4 (al)	$130\text{-}2^{8}$		$1.048^{20/4}$	1.4160^{22}	al	$B10^3$, 2868
5787	Cyclopentanecarboxylic acid, 2,oxo,ethyl ester $C_8H_{12}O_3$	156.18		218^{704}, 110^{16}		$1.0781^{21/4}$	1.4519^{20}	eth, bz	$B10^3$, 2808
5788	Cyclopentanecarboxylic acid, 3-oxo $C_6H_8O_3$	128.13	197^{30}	64-5					$B10^3$, 2812

PHYSICAL CONSTANTS OF ORGANIC COMPOUNDS (Continued)

No.	Name, Synonyms, and Formula	Mol. wt.	Color, crystalline form, specific rotation and λ_{max} (log ε)	b.p. °C	m.p. °C	Density	n_D	Solubility	Ref.
5789	1,2-Cyclopentanedicarboxylic acid *(cis)* $1,2\text{-}C_5H_8(CO_2H)_2$	158.15	nd (w)	...	140	...	...	w	$B9^{3}$, 3807
5790	1,2-Cyclopentanedicarboxylic acid *(trans d)* $C_7H_{10}O_4$	158.15	cr (w), $[\alpha]_D$ + 87.6 (w, c=0.9)	...	181	...	...	w, al	$B9^{3}$, 3807
5791	1,2-Cyclopentanedicarboxylic acid *(trans, dl)* $C_7H_{10}O_4$	158.15	cr (w)	...	162-3	...	...	w, al	$B9^{3}$, 3807
5792	1,2-Cyclopentanedicarboxylic acid *(trans,l)* $C_7H_{10}O_4$	158.15	cr (w)$[\alpha]_D$ −85.9 (w, c=1.2)	...	180-1	...	...	w, al	$B9^{3}$, 3807
5793	1,3-Cyclopentanedicarboxylic acid *(cis)* or Norcamphoric acid $1,3\text{-}C_5H_8(CO_2H)_2$	158.15	pr (w)	>300d	121	...	...	al, eth, ace, chl	$B9^{3}$, 3808
5794	1,3-Cyclopentanedicarboxylic acid *(trans, d)* $1,3\text{-}C_5H_8(CO_2H)_2$	158.15	cr (CCl_4), $[\alpha]_D$ + 5.9 (w, c=5)	...	93.5	...	...	w	$B9^{2}$, 519
5795	1,3-Cyclopentanedicarboxylic acid *(trans, dl)* $C_7H_{10}O_4$	158.15	pr (CCl_4)	...	88	...	...	w	$B9^{3}$, 3808
5796	1,3-Cyclopentanedicarboxylic acid *(trans, l)* $C_7H_{10}O_4$	158.15	cr (CCl_4), $[\alpha]_D$ − 5.3 (w, c=5)	...	93	...	...	w	$B9^{2}$, 519
5797	1,3-Cyclopentanedicarboxylic acid, 4,4-dimethyl or Apofenchocamphoric acid $4,4\text{-}(CH_3)_2C_5H_6(CO_2H)_2\text{-}1,3$	186.21	mcl	...	144-5	...	...	w, al, eth	$B9^{3}$, 3849
5798	Cyclopentanol C_5H_9OH	86.13	...	140.8, 53^{10}	−19	$0.9478^{20/4}$	1.4530^{20}	al, eth, ace	$B6^{4}$, 5
5799	Cyclopentanol, 2,acetyl-1,3,3,4,4-pentamethyl *(a)* or Desoxymesityl oxide $C_{12}H_{22}O_2$	198.31	cr (peth-eth)	...	45	...	...	bz	$B8^{3}$, 32
5800	Cyclopentanol, 1-methyl $1\text{-}CH_3C_5H_8OH$	100.16	...	$53\text{-}4^{30}$	35-7	$0.9044^{23.5/4}$	$1.4429^{23.5}$	...	$B6^{4}$, 86
5801	Cyclopentanone C_5H_8O	84.12	...	130.6	−51.3	$0.9487^{20/4}$	1.4366^{20}	al, **eth**, ace	$B7^{3}$, 5
5802	Cyclopentanone, 2-methyl $2\text{-}CH_3(C_5H_7O)$	98.14	...	139.5, 44^{18}	−75	0.9139^{20}	1.4364^{20}	al, eth, ace	$B7^{3}$, 40
5803	Cyclopentanone, 3-methyl *(d)* $3\text{-}CH_3(C_5H_7O)$	98.14	$[\alpha]^{25/}{}_D$ + 143.7 (undil)	143.5^{742}, $43\text{-}4^{12}$	−58.4	$0.9140^{19/4}$	1.4340^{19}	w, al, eth, ace, aa	$B7^{3}$, 42
5804	Cyclopentanone, 3-methyl *(dl)* $3\text{-}CH_3(C_5H_7O)$	98.14	...	144, 38^{11}	...	0.913^{22}	1.4329^{20}	w, al, eth, ace	$B7^{3}$, 43
5805	Cyclopentasiloxane, decamethyl $C_{10}H_{30}O_5Si_5$	370.77	...	210, 101^{20}	−38	$0.9593^{20/4}$	1.3982^{20}	...	$B4^{3}$, 1885
5806	Cyclopentene C_5H_8	68.12	...	44.2	−135	$0.7720^{20/4}$	1.4225^{20}	al, eth, bz, peth	$B5^{4}$, 209
5807	Cyclopentene, 3-chloro $3\text{-}ClC_5H_7$	102.56	...	$25\text{-}31^{30}$	...	1.0577^{15}	1.4708^{26}	al, eth, chl	$B5^{4}$, 212
5808	Cyclopentene, perchloro C_5Cl_8	343.68	nd (al)	283, 140^{10}	41	$1.8200^{50/4}$	1.5660^{50}	al	$B5^{4}$, 213
5809	Cyclopentene, 1,2-deimthyl $1,2(CH_3)_2C_5H_6$	96.17	...	103^{757}	...	$0.7992^{13.5}$	$1.4447^{13.5}$	...	$B5^{4}$, 254
5810	Cyclopentene, 2,3-dimethyl-4-isopropyl $2,3\text{-}(CH_3)_2\text{-}4\text{-}(CH_3)_2CHC_5H_5$	138.25	...	164-6	...	$0.8085^{22/4}$	1.4503^{22}	...	PCHE 2, 319
5811	Cyclopentene, 2,4-dimethyl $2,4\text{-}(CH_3)_2C_5H_6$	96.17	...	93.2	...	...	1.4283^{20}	...	$B5^{4}$, 255
5812	Cyclopentene, 1-ethyl $1\text{-}C_2H_5C_5H_7$	96.17	...	108	−123.3	$0.8000^{20/4}$	1.4429^{21}	...	$B5^{4}$, 252
5813	Cyclopentene, 3-ethyl $3\text{-}C_2H_5C_5H_7$	96.17	...	$99\text{-}103^{758}$	...	$0.7874^{20/4}$	1.4303^{20}	...	$B5^{4}$, 252
5814	Cyclopentene, 1,2-dichloro,perfluoro $C_5Cl_2F_6$	244.95	...	90.7	−105.8	$1.6546^{20/4}$	1.3676^{20}	...	$B5^{4}$, 213
5815	Cyclopentene, 1-methyl $1\text{-}CH_3C_5H_7$	82.15	...	75.5	−127.2	$0.7851^{15/4}$	1.4347^{15}	...	$B5^{4}$, 239
5816	Cyclopentene, 3-methyl $3\text{-}CH_3C_5H_7$	82.15	...	69-71	...	$0.9705^{20/4}$	1.42476^{20}	...	$B5^{4}$, 240
5817	Cyclopentene, 1,2,3-trimethyl $1,2,3(CH_3)_3C_5H_5$	110.20	...	121.6	...	$0.8039^{15/4}$	$1.4464^{16.5}$	...	$B5^{3}$, 219
5818	2-Cyclopentene-1-one, 3-phenyl $C_{11}H_{10}O$	158.20	...	234.2	−23	0.9711^{20}	1.5440^{20}	al, ace, chl	$B5^{3}$, 1654

PHYSICAL CONSTANTS OF ORGANIC COMPOUNDS (Continued)

No.	Name, Synonyms, and Formula	Mol. wt.	Color, crystalline form, specific rotation and λ_{max} (log ε)	b.p. °C	m.p. °C	Density	n_D	Solubility	Ref.
5819	3-Cyclopentene-1-one, 3,4-bis(4-methoxyphenyl) $(CH_3OC_6H_4)_2C_5H_4O$	294.35	ye br		129			al, eth, ace	B8, 355
5820	1,2-Cyclopentenophenanthrene $C_{17}H_{14}$	218.30	nd (al), cr (peth)		135-6			al	$B5^{3}$, 2241
5821	1,2-Cyclopentenophenanthrene, 3-methyl $C_{18}H_{16}$	232.33	cr (aa)		126-7				$B5^{3}$, 2252
5822	2,3-Cyclopentenophenanthrene $C_{17}H_{14}$	218.30	pl or pr (al), nd (MeOH)		84-5			al	$B5^{3}$, 2240
5823	9,10-Cyclopentenophenanthrene $C_{17}H_{14}$	218.30	pl (xyl), nd (i-ProH)		155-6			al, bz	$B5^{3}$, 2242
5824	Cyclopropane C_3H_6	42.08		−32.7	−127.6	$0.720^{-79/4}$	$1.3799^{-42.5}$	al, eth, bz, peth	$B5^{4}$, 3
5825	Cyclopropane, acetyl or Cyclopropyl methyl ketone $CH_3COC_3H_5$	84.12		114^{772}	fp − 68.4	$0.8984^{20/4}$	1.4251^{20}	w, al, eth	$B7^{3}$, 13
5826	Cyclopropane, amino or Cyclopropyl amine $H_2NC_3H_5$	57.10		50-1		$0.8240^{20/4}$	1.4210^{20}	w, al, eth	$B12^{3}$, 3
5827	Cyclopropane, 1,1-dimethyl 1,1-$(CH_3)_2C_3H_4$	70.13		20.6	−109	$0.6589^{20/4}$	1.3668^{20}	al, eth	$B5^{4}$, 24
5828	Cyclopropane, 1,2-dimethyl *(cis)* 1,2-$(CH_3)_2C_3H_4$	70.13		37		0.6928^{20}	1.3822^{20}	al, eth	$B5^{4}$, 25
5829	Cyclopropane, 1,2-dimethyl *(trans, dl)* 1,2-$(CH_3)_2C_3H_4$	70.13		29		0.6769^{20}	1.3713^{20}	al, eth	$B5^{4}$, 26
5830	Cyclopropane, 1,2-dimethyl *(trans, l)* 1,2-$(CH_3)_2C_3H_4$	70.13	$[\alpha]^{22/}{}_D$ − 2.39 dig-lyme 13.9%	28-9		1.3699^{16}		al, eth	TETRA 20, 1965
5831	Cyclopropane, ethyl $C_2H_5C_3H_5$	70.13		34.5		$0.677^{20/4}$	1.379^{20}		$B5^{4}$, 23
5832	Cyclopropane, methoxy $CH_3OC_3H_5$	72.11		44.7	−119	$0.8100^{20/4}$	1.3802^{20}	w, al, eth, bz	$B6^{3}$, 3
5833	Cyclopropane, methyl $CH_3C_3H_5$	56.11		4-5	−117.2	$0.6912^{-20/4}$		al, eth	$B5^{4}$, 13
5834	Cyclopropane, 1-methyl-1-phenyl 1-CH_3-1-$C_6H_5C_3H_4$	132.21		91^{50}			1.5160^{20}	ace, bz, chl	$B5^{4}$, 1388
5835	Cyclopropane, phenyl $C_6H_5C_3H_5$	118.18		173.6^{758}, 80^{37}	−31	$0.9317^{20/4}$	1.5285^{20}	eth, ace, chl	$B5^{3}$, 1200
5836	Cyclopropane, isopropenyl $[CH{=}CH(CH_3)]C_3H_5$	82.15		$69.5\text{-}70^{751}$		$0.7500^{20/4}$	1.4252^{20}		$B5^{4}$, 243
5837	Cyclopropane, 1,1,2-trimethyl *(dl)* 1,1,2-$(CH_3)_3C_3H_3$	84.16		52.6	− 138.3	$0.6974^{20/4}$	1.3864^{20}	eth, bz	$B5^{4}$, 91
5838	Cyclopropane carbonitrile C_3H_5CN	67.09		135, $69\text{-}70^{88}$		$0.8946^{20/4}$	1.4229^{20}	eth	$B9^{3}$, 6
5839	Cyclopropane carboxylic acid $C_3H_5CO_2H$	86.09		182-4	18-9	$1.0885^{20/4}$	1.4390^{20}	al, eth	$B9^{3}$, 3
5840	Cyclopropane carboxylic acid, 2,2-dimethyl-3-(2-methyl-propenyl) or *d-cis*-Chrysanthemamic acid $C_{10}H_{16}O_2$	168.24	pr $[\alpha]^{22/}{}_D$ + 83.3 (chl, c=1.6)	$95^{0.1}$	40-2			al, eth, chl	$B9^{3}$, 210
5841	Cyclopropane carboxylic acid, 2,2-dimethyl-3-(2-methyl-propenyl) *cis,dl* $C_{10}H_{16}O_2$	168.24	pr (AcOEt), cr (peth)		115-6			al, eth	$B9^{3}$, 211
5842	Cyclopropane carboxylic acid, 2,2-dimethyl-3-(2-methyl-propenyl) *(cis,l)* $C_{10}H_{16}O_2$	168.24	pr $[\alpha]^{19/}{}_D$ − 83.3 (chl, c=1.6)	$95^{0.1}$	41-3			al, eth	$B9^{3}$, 211
5843	Cyclopropane carboxylic acid, 2,2-dimethyl-3-(2-methyl-propenyl) *trans,d* or *trans*-Chrysanthemucic acid *(d)* $C_{10}H_{16}O_2$	168.24	pr $[\alpha]^{20/}{}_D$ + 25.8 (chl, c=2.5)	245d, 135^{12}	18-21			al, eth, chl	$B9^{3}$, 211
5844	Cyclopropane carboxylic acid, 2,2-dimethyl-3-(2-methyl-propenyl) *trans,dl* $C_{10}H_{16}O_2$	168.24	pr (AcOEt)	$145\text{-}6^{13}$	54			al, eth, chl	$B9^{3}$, 212
5845	Cyclopropane carboxylic acid, 2,2-dimethyl-3-(2-methyl-propenyl) *trans,l* $C_{10}H_{16}O_2$	168.24	$[\alpha]^{20/}{}_D$ − 25.8 (chl 2.9%)	$99\text{-}100^{0.2}$	17-21			al	$B9^{3}$, 211
5846	Cyclopropane carboxylic acid, methyl ester $C_3H_5CO_2CH_3$	100.12		119		$0.9848^{20/4}$	1.4144^{19}		$B9^{3}$, 3
5847	1,1-Cyclopropane dicarboxylic acid 1,1-$C_3H_4(CO_2H)_2$	130.10	pr or nd (chl), pr (w + 1)		140-1			w, eth	$B9^{3}$, 3795

No.	Name, Synonyms, and Formula	Mol. wt.	Color, crystalline form, specific rotation and λ_{max} (log ε)	b.p. °C	m.p. °C	Density	n_D	Solubility	Ref.
5848	1,1-Cyclopropane dicarboxylic acid, diethyl ester $1,1\text{-}C_3H_4(CO_2C_2H_5)_2$	186.21		$214\text{-}6^{748}$, $99\text{-}100^{12}$		$1.0566^{25/25}$	1.4345^{18}	al, eth	$B9^3$, 3795
5849	1,2-Cyclopropane dicarboxylic acid *(cis)* $1,2\text{-}C_3H_4(CO_2H)_2$	130.10	pr (eth or w)		139			w, al, eth	$B9^3$, 3796
5850	1,2-Cyclopropane dicarboxylic acid *(trans, d)* $1,2\text{-}C_3H_4(CO_2H)_2$	130.10	$[\alpha]^{27/}{}_D$ + 84.87 (w)		175			w, al, eth	$B9^3$, 3796
5851	1,2-Cyclopropane dicarboxylic acid *(trans, dl)* $1,2\text{-}C_3H_4(CO_2H)_2$	130.10	nd (eth), pl (ace-bz)	210^{30}	175			w, al, eth	$B9^3$, 3797
5852	1,2-Cyclopropane dicarboxylic acid *(trans, l)* $1,2\text{-}C_3H_4(CO_2H)_2$	130.10	$[\alpha]^{27/}{}_D$ − 84.40 (w)		175			w, al, eth	$B9^3$, 3797
5853	1,2-Cyclopropane dicarboxylic acid, 1-bromo $1\text{-}BrC_3H_3(CO_2H)_2\text{-}1,2$	209.00	pr (eth-chl), cr (ace-bz)		175			eth, ace	$B9^2$, 514
5854	1,2-Cyclopropane dicarboxylic acid, diethyl ester-*(cis)* $1,2\text{-}C_3H_4(CO_2C_2H_5)_2$	186.21		106-7		$1.062^{12/4}$	1.4450^{20}	al, eth	$B9^2$, 513
5855	1,2-Cyclopropane dicarboxylic acid, dimethyl ester $C_7H_{20}O_4$	158.15		219-20, 110^3		$1.1584^{16/4}$	1.4472^{14}	al, eth	$B9^2$, 513
5856	1,2,3-Cyclopropane tricarboxylic acid $C_3H_3(CO_2H)_3$	174.11	nd (w), cr (HCl)		220			w, al	$B9^3$, 4746
5857	Cyclopropyl methyl ketone $CH_3COC_3H_5$	84.12		114^{772}	fp − 68.4	$0.8984^{20/4}$	1.4251^{20}	w, al, eth	$B7^3$, 13
5858	Cyclotetrasiloxane, octamethyl $C_8H_{24}O_4Si_4$	296.62		175.8, 74^{20}	17.5	0.9561^{20}	1.3968^{20}		$B4^3$, 1885
5859	Cyclotetrasiloxane, octaphenyl $(C_6H_5)_8O_4Si_4$	793.19	nd (bz-al or aa)	$330\text{-}4^1$	200-1			bz, aa	$B16^3$, 1213
5860	Cyclotetrasiloxane, 1,3,5,7-tetramethyl-1,3,5,7-tetraphenyl $C_{28}H_{32}O_4Si_4$	544.90	cr (aa)	$237^{1.5}$	99	$1.1183^{20/4}$	1.5461^{20}	ace	$B16^3$, 1211
5861	Cyclotetrasiloxane, 2,4-6,8-tetramethyl $C_4H_{16}O_4Si_4$	240.51		134-5	−65	$0.9912^{20/4}$	1.3870^{20}		$B4^3$, 1874
5862	Cyclotrisiloxane, hexaphenyl $(C_6H_5)_6O_3Si_3$	594.89	pl (bz-al or aa)	$290\text{-}300^1$	190	$1.23^{25/4}$		bz, aa	$B16^3$, 1213
5863	Cyclotrisiloxane, 1,3,5-triethyl,1,3,5-triphenyl $(C_2H_5)_3(C_6H_5)_3O_3Si_3$	450.76		$166^{0.025}$	177.5	$1.0952^{25/4}$	1.5402^{25}	bz	$B16^3$, 1211
5864	Cymarose or 4,5-Dihydroxy-3-methoxyhexanal or 2,6-didesoxy-3-*O*-methyl-D-allose $CH_3(CHOH)_2CH(OCH_3)CH_2CHO$	162.19	pr (eth-peth), nd (ace), $[\alpha]^{21/}{}_D$ + 53.4 (w, c=2.2)		100-2			w, al, ace	$B1^4$, 4193
5865	*o*-Cymene or 1-Methyl-2-*iso*-propyl benzene $2\text{-}(CH_3)_2CHC_6H_4CH_3$	134.22		178.1, 57.3^{10}	−71.5	$0.8766^{20/4}$	1.5006^{20}	**al, eth, ace, bz**	$B5^4$, 1057
5866	*m*-Cymene or 1-Methyl-3-*iso*-propyl benzene $3\text{-}(CH_3)_2CHC_6H_4CH_3$	134.22		175.1, 55^{10}	−63.7	$0.8610^{20/4}$	1.4930^{20}	**al, eth, ace, bz**	$B5^4$, 1058
5867	*p*-Cymene or 1-Methyl-4-*iso*-propyl benzene $4\text{-}(CH_3)_2C\ HC_6H_4CH_3$	134.22		177.1, 56.3^{10}	−67.9	$0.8573^{20/4}$	1.4909^{20}	**al, eth, ace, bz**	$B5^4$, 1060
5868	*p*-Cymene, 2-chloro or 2-Chloro-4-isopropyltoluene $2\text{-}Cl\text{-}4\text{-}i\text{-}C_3H_7C_6H_3CH_3$	168.67		216-7, 104^{20}		$1.0104^{25/4}$	1.50782^{20}	ace, bz	$B5^4$, 1062
5869	*p*-Cymene, 2-nitro $4\text{-}(CH_3)_2CH\text{-}2\text{-}NO_2\text{-}C_6H_3CH_3$	179.22		126^{10}		$1.0744^{20/4}$	1.5301^{20}	al, eth	$B5^4$, 1064
5870	Cysteic acid *(d)* or 1,2-amino-3-sulfopropanoic acid $HO_2CCH(NH_2)CH_2SO_3H$	169.15	oct cr or nd (dil al), pr or nd (w + 1), $[\alpha]^{20/}{}_D$ + 8.66 D (w)		260d			w	$B5^3$, 1713
5871	Cysteic acid *(dl)* $HO_2CCH(NH_2)CH_2SO_3H$	169.15	pr (w)		272-4d			w	$B4^3$, 1714
5872	Cysteine *(L)* or *L*-β-Mercaptoalanine $HSCH_2CH(NH_2)CO_2H$	121.15	cr (w), $[\alpha]^{30;/}{}_D$ + 9.8 (w, c=1.3)		240d			w, al, aa	$B4^3$, 1580
5873	Cystine *(D)* or *D*-Dicystine $[HO_2CCH(NH_2)CH_2S]_2$	240.29	cr (dil NH3), $[\alpha]^{20/}{}_D$ + 224 (1N HCl, c=1)		247-9				$B4^3$, 1618
5874	Cystine *(DL)* $[HO_2CCH(NH_2)CH_2S\text{-}]_2$	240.29			260				$B4^3$, 1621

No.	Name, Synonyms, and Formula	Mol. wt.	Color, crystalline form, specific rotation and λ_{max} (log ε)	b.p. °C	m.p. °C	Density	n_D	Solubility	Ref.
5875	Cystine *(L)* $[HO_2CCH(NH_2)CH_2S-]_2$	240.29	hex pl or pr (w), $[\alpha]^{20/}{}_D$ − 223.4 (1N HCl, c=1)		260-1d	1.677			B4[1], 1593
5876	Cystine *(meso)* $[HO_2CCH(NH_2)CH_2S-]_2$	240.29			200-21d				B4[1], 1593
5877	Cytidine or 1-β-D-ribofuranosylcytosine $C_9H_{13}N_3O_5$	243.22	nd (dil al), $[\alpha]^{20.5/}{}_D$ + 35.3 (w, c=1)		230-1d			w	B31, 24
5878	Cytidylic acid or 3-Cytosylic acid $C_9H_{14}N_3O_8P$	323.20	orh nd $[\alpha]^{20/}{}_D$ + 49.4 (w, c=1)		233-4d			w, al	B31, 25
5879	Cytisine *(l)* or Sophorine, Baptitoxine $C_{11}H_{14}N_2O$	190.24	orh pr (aa or ace)	218^2	154.5			w, al, ace, bz, chl	B24[2], 70
5880	Cytisine, *N*-methyl or Caulophylline $C_{12}H_{16}N_2O$	204.27	cr (w + 2), nd (al, bz or lig), pr (al)		137			w, al, ace, bz	B24[2], 70
5881	Cytosine or 4-amino-1,2-dihydro-1,3-diazin-2-one $C_4H_5N_3O$	111.10	mcl or tcl pl (w + 1)		320-5d			w	B24, 314
5882	Cytosine, 5-methyl $C_5H_7N_3O$	125.13	pr (w + ½)		270d			w	B24, 355

No.	Name, Synonyms, and Formula	Mol. wt.	Color, crystalline form, specific rotation and λ_{max} (log ε)	b.p. °C	m.p. °C	Density	n_D	Solubility	Ref.
5883	1,3-Decadiene $CH_3(CH_2)_5CH=CHCH=CH_2$	138.25	...	168-70	...	0.752^{20}	...	bz	B1^4, 1056
5884	3,7-Decadiene-5-yne, 4,7-dipropyl $C_2H_5CH=C(C_3H_7)C{\equiv}CCH=C(C_3H_7)C_2H_5$	218.38	...	125-7^{18}	...	0.8131$^{19/4}$	1.4890^{20}	bz	B1^4, 1134
5885	2,4-Decadienoic acid, ethyl ester $C_5H_{11}CH=CHCH=CHCO_2C_2H_5$	196.29	...	130$^{0.6}$	...	...	1.5020^{26}	...	B2^4, 17314
5886	4,6-Decadiyne or Dipentyne $C_3H_7C{\equiv}C\text{-}C{\equiv}CC_3H_7$	134.22	...	88^{12}	...	0.8695$^{19/4}$	...	...	B1^3, 1064
5887	Decalin *(cis)* or Decahydronaphthalene $C_{10}H_{18}$	138.25	...	195.6, 69.4^{10}	−43	0.8965$^{20/4}$	1.4810^{20}	al, eth, ace, bz, chl	B5^4, 310
5888	Decalin *(trans)* or Decahydronaphthalene $C_{10}H_{18}$	138.25	...	187.2, 63^{10}	−30.4	0.8699$^{20/4}$	1.4695^{20}	al, eth, ace, bz, chl	B5^4, 311
5889	Decalin, 1-amino *(cis)* $1\text{-}H_2NC_{10}H_{17}$	153.27	...	100^{12}	(i) 8, (ii) − 2	...	...	...	B12^3, 178
5890	Decalin, 1-amino *(trans)* $1\text{-}H_2NC_{10}H_{17}$	153.27	...	106^{16}	(i) − 18, (ii) − 1	...	...	...	B12^2, 35
5891	Decalin, 1-chloro $C_{10}H_{17}Cl$	172.70	...	d^{760}, 114-6^{20}	...	...	...	...	B5^4, 313
5892	Decalin, 2-methylene *(trans)* $2\text{-}CH_2{:}C_{10}H_{16}$	150.26	...	200-1^{756}, 82^{10}	...	0.8897$^{20/4}$	1.4841^{22}	...	B5^3, 397
5893	Decalin-1,3-dione *(cis)* $C_{10}H_{14}O_2$	166.22	nd (bz, ace, dil al)	...	124-5	...	...	al, ace, bz	B7^3, 3288
5894	Decalin-1,3-dione *(trans)* $C_{10}H_{14}O_2$	166.22	nd (bz, dil al)	...	152-3	...	...	al	B7^3, 3289
5895	Decalin-2,3-dione *(cis)* $C_{10}H_{14}O_2$	166.22	rh (al, ace, lig)	...	88-9	...	...	al	B7^2, 552
5896	Decalin-2,3-dione *(trans)* $C_{10}H_{14}O_2$	166.22	nd (w), lf (dil al)	...	100-1	...	...	al, lig	B7^3, 3291
5897	2-Decalincarboxylic acid *(cis)* or Decahydro-2-naphthoic acid $2\text{-}C_{10}H_{17}CO_2H$	182.26	cr (hx)	150^{15}	81	...	...	al, eth, bz, chl	B9^3, 233
5898	Decanal or Capraldehyde $CH_3(CH_2)_8CHO$	156.27	...	208-9, 81^7	*ca* −5	0.830$^{15/4}$	1.4287^{20}	al, eth, ace	B1^4, 3366
5899	Decane $C_{10}H_{22}$	142.28	...	174.1, 57.6^{10}	−29.7	0.7300$^{20/4}$	1.4102^{20}	al, eth	B1^4, 464
5900	Decane, 1-amino or Decylamine $CH_3(CH_2)_8CH_2NH_2$	157.30	...	220.5, 95.8^{10}	17	0.7936$^{20/4}$	1.4369^{20}	al, eth, ace, bz, chl	B4^4, 783
5901	Decane, 1-bromo or Decyl bromide $CH_3(CH_2)_8CH_2Br$	221.18	...	240.6, 110^{10}	−29.2	1.0702$^{20/4}$	1.4557^{20}	eth, chl	B1^4, 470
5902	Decane, 1-bromo-10-fluoro $F(CH_2)_{10}Br$	239.17	...	131-2^{11}	...	1.152$^{20/4}$	1.4512^{25}	...	**B1^4, 471**
5903	Decane, 2-bromo *(dl)* $CH_3(CH_2)_7CHBrCH_3$	221.18	...	111^{11}	...	1.0512^{20}	1.4526^{25}	eth, chl	B1^3, 523
5904	Decane, 1-chloro or Decyl chloride $CH_3(CH_2)_8CH_2Cl$	176.73	...	223.4, 97^{10}	−31.3	0.8705$^{20/4}$	1.4379^{20}	eth, chl	B1^4, 469
5905	Decane, 1-chloro-10-fluoro $F(CH_2)_{10}Cl$	194.72	...	115^9	...	0.957$^{20/4}$	1.4333^{25}	al, eth	B1^4, 469
5906	Decane, 1,10-dibromide or Decamethylene dibromide $Br(CH_2)_{10}Br$	300.08	pl (al)	160^{15}d, 127-30^4	28	1.335^{30}	1.4905^{20}	eth	B4^4, 1368
5907	Decane, 1,10-dibromide or Decamethylene dibromide $Br(CH_2)_{10}Br$	300.09	pr (al)	160^{15}d, 127-30^4	28	1.335^{30}	1.4905^{20}	eth	B1^4, 471
5908	Decane, 2,5-dimethyl $(CH_3)_2CH(CH_2)_2CH(CH_3)C_5H_{11}$	170.34	$[\alpha]^{25}{}_D$ −0.05	122^{100}	...	0.739$^{25/4}$	...	...	**B1^4, 506**
5909	Decane, 1-fluoro or Decyl fluoride $CH_3(CH_2)_8CH_2F$	160.28	...	186.2, 69^{10}	−35	0.8194$^{20/4}$	1.4085	eth	B1^4, 468
5910	Decane, 1-iodo or Decyl iodide $CH_3(CH_2)_8CH_2I$	268.18	...	132^{15}	−16.3	1.2546$^{20/4}$	1.4858^{20}	al, eth	B1^4, 472
5911	Decane, 2-methyl $(CH_3)_2CH(CH_2)_7CH_3$	156.31	...	189.2	−48.86	0.7368^{20}	1.4154^{20}	...	**B1^4, 491**
5912	Decane, 3-methyl $C_7H_{15}CH(CH_3)C_2H_5$	156.31	...	188.1, 71-2^{12}	−92.9	0.7422$^{20/4}$	1.4177^{20}	...	**B1^4, 491**
5913	Decane, 1-nitro $CH_3(CH_2)_8CH_2NO_2$	187.28	...	86^1	...	...	1.4387^{20}	...	**B1^4, 473** 8536
5914	Decandioic acid or Sebacic acid $HO_2C(CH_2)_8CO_2H$	202.25	lf	295^{100}, 232^{10}	134.5	1.2705$^{20/4}$	1.422^{133}	al, eth	B2^4, 2078
5915	1-10-Decanediol or Decamethylene glycol $HO(CH_2)_{10}OH$	174.28	nd (w)	175-6^{14}	72-5	...	...	al	B1^4, 2613

No.	Name, Synonyms, and Formula	Mol. wt.	Color, crystalline form, specific rotation and λ_{max} (log ε)	b.p. °C	m.p. °C	Density	n_D	Solubility	Ref.
5916	1-Decanethiol or Decylmercaptan $CH_3(CH_2)_8CH_2SH$	174.34		240.6, 125-7^{19}	−26	0.8443$^{20/4}$	1.4569^{20}	al, eth	B1^4, 1821
5917	Decanoic acid or Capric acid $CH_3(CH_2)_8CO_2H$	172.27	nd	270, 148-50^{11}	fr 31.5	0.8858$^{40/4}$	1.4288^{40}	al, eth, ace, bz, chl	B2^4, 1041
5918	Decanoic acid, 10-fluoro or 10-Fluoro capric acid $F(CH_2)_9CO_2H$	190.26		135-8^{10}	49			al, eth, lig	B2^4, 1052
5919	Decanoic acid, 2-octyl or 9-Heptadecane carboxylic acid $(C_8H_{17})_2CHCO_2H$	284.49	nd or lf (al)	212-8^{13}	38.5			al, eth	B2^4, 1254
5920	Decanoic acid, 4-oxo or γ-Ketocapric acid $C_6H_{13}COCH_2CH_2CO_2H$	186.25	(dil al)		70-1			al	B3^4, 1642
5921	1-Decanol or Decyl alcohol $CH_3(CH_2)_8CH_2OH$	158.28		229, 107-8^7	fr 7	0.8297$^{20/4}$	1.4372^{20}	al, eth, ace, bz, chl	B1^4,1815
5922	1-Decanol, 10-chloro $Cl\text{-}(CH_2)_{10}OH$	192.73		185-9^{15}	12-3	0.9630^{25}	1.4578^{20}	al, eth	B1^4, 1821
5923	1-Decanol, 10-fluoro $F(CH_2)_{10}OH$	176.27		136-7^{15}	*ca* 22	0.919$^{20/4}$	1.4322^{25}	al, eth	B1^4, 1821
5924	2-Decanol *(dl)* $CH_3(CH_2)_7CHOHCH_3$	158.28		211, 110-11^{10}	−2.4	0.8250$^{20/4}$	1.4326^{25}	al, eth, ace, bz	B1^4, 1823
5925	4-Decanol $C_3H_7CHOHC_6H_{13}$	158.28		210-1, 96^{13}	−11	0.8262$^{20/0}$	1.4320^{20}	al	B1^4, 1824
5926	2-Decanone or Methyl-*m*-octyl ketone $CH_3(CH_2)_7COCH_3$	156.27	nd	210-1^{767}, 95-7^{12}	14	0.8248^{20}	1.4255^{20}	al, eth	B1^4, 3367
5927	3-Decanone or Ethyl-*M*-heptyl ketone $CH_3(CH_2)_6COCH_2CH_3$	156.27		203^{754}	1-4	0.8251$^{20/4}$	1.4252^{20}	al, eth	B1^4, 3368
5928	4-Decanone or Propyl hexyl ketone $C_6H_{13}COC_3H_7$	156.27	nd	206-7, 87-9^{11}	-9	0.824$^{20/D}$	1.4240^{21}	al, eth	B1^4, 3368
5929	Decasiloxane, dicosamethyl $CH_3[Si(CH_3)_2O]_9Si(CH_3)_3$	755.62		183^4		0.925$^{20/4}$	1.3988^{20}	bz, lig	C47, 4679
5930	1,4,9-Decatriene *(trans)* $CH_2{=}CHCH_2CH{=}CH(CH_2)_3CH{=}CH_2$	136.24		164-6			1.4496^{20}		
5931	1-Decene $CH_3(CH_2)_7CH{=}CH_2$	140.27		170.5, 54.3^{10}	fr-66.3	0.7408$^{20/4}$	1.4215^{20}	al, eth	B1^3, 858
5932	1-Decene, 2-bromo $CH_3(CH_2)_7CBr{=}CH_2$	219.16		115-6^{22}		1.0844$^{20/4}$	1.4629^{20}		B1^3, 859
5933	1-Decene-3-yne $CH_3(CH_2)_5C{\equiv}C\text{-}CH{=}CH_2$	136.24		76^{20}		0.7873^{20}	1.4620^{20}		B1^4, 1105
5934	1-Decene-4-yne $CH_3(CH_2)_4C{\equiv}CCH_2CH{=}CH_2$	136.24		73-4^{22}		0.7880^{20}	1.445^{20}		B1^3, 1049
5935	2-Decene, 1-bromo $CH_3(CH_2)_6CH{=}CHCH_2Br$	219.16		121^{17}		1.074$^{18/4}$	1.4716^{18}	lig	B1^4, 902
5936	2-Decen-4-yne $CH_3(CH_2)_4C{\equiv}C\text{-}CH{=}CHCH_3$	136.24		55^5		0.7850$^{25/4}$	1.4609^{25}		B1^3, 1049
5937	5-Decene *(cis)* $CH_3(CH_2)_3CH{=}CH(CH_3)_3CH_3$	140.27		170^{739}, 73^{20}	−112	0.7445$^{20/4}$	1.4258^{20}	al, eth	B1^4, 902
5938	5-Decene *(trans)* $CH_3(CH_2)_3CH{=}CH(CH_2)_3CH_3$	140.27		170.2^{739}	−73	0.7401$^{20/4}$	1.4243^{20}	al, eth	B1^4, 902
5939	*n*-Decyl nitrate $C_{10}H_{21}ONO_2$	203.28		127-8^{11}, 88-9^1		0.951$^{0/4}$		al, eth	B1^4, 1819
5940	Decyl nitrite $C_{10}H_{21}ONO$	187.28	yesh	105-8^{12}			1.4247^{20}	al, eth	B1^4, 1819
5941	Decyl sulfate $(C_{10}H_{21}O)_2SO_2$	378.61			37-8				B1^4, 1819
5942	1-Decyne or *n*-Octyl acetylene $CH_3(CH_2)_7C{\equiv}CH$	138.25		174, 57^{10}	−36	0.7655$^{20/4}$	1.4265^{20}	al, eth	B1^4, 1054
5943	3-Decyne $CH_3CH_2C{\equiv}C(CH_2)_5CH_3$	138.25		175-6		0.765$^{21/4}$	1.433^{21}		B1^4, 1055
5944	4-Decyne $C_4H_9C{\equiv}CC_3H_7$	138.25		74.5^{19}		0.772$^{17/4}$	1.436^{17}		B1^3, 1017
5945	4-Decyne, 3,3-dimethyl $CH_3CH_2C(CH_3)_2C{\equiv}C(CH_2)_4CH_3$	166.31		86^{20}		0.7731$^{20/4}$	1.4399^{20}		B1^3, 1026
5946	5-Decyne or Dibutyl acetylene $C_4H_9C{\equiv}CC_4H_9$	138.25		177^{751}, 788^{25}	−73	0.7690$^{20/4}$	1.4331^{20}	al, eth	B1^4, 1055
5947	Dehydroacetic acid $C_8H_8O_4$	168.15	nd (w), rh nd or pr (al)	270, 132-3^5	109			eth, w	B17^2, 524

No.	Name, Synonyms, and Formula	Mol. wt.	Color, crystalline form, specific rotation and λ_{max} (log ε)	b.p. °C	m.p. °C	Density	n_D	Solubility	Ref.
5948	Dehydrochloric acid or 3,7,12-Trioxocholanic acid $C_{24}H_{34}O_5$	402.53	(ace), $[\alpha]^{20}/_D$ +26 (al, c=1.4)		237			ace, chl	B10^3, 3986
5949	Dehydro ergosterol $C_{28}H_{42}O$	394.64	lf (al + 1w), pl (al), nd (eth), $[\alpha]^{15}_D$ +149.2 (chl,c=1.9	$230^{0.5}$	146			al, eth, ace, bz, chl	B6^3, 3479
5950	Delphinidine chloride or 3,3′,4′,5′,7′-hexahydroxy flavinium chloride $C_{15}H_{11}ClO_7$	338.70	br pr, nd or pl (HCl)		>350			w, al	B18^2, 247
5951	Delphinine $C_{33}H_{45}NO_9$	599.72	orh (al), $[\alpha]^{25}/_D$ +25 (al)		198-200d			al, eth, ace, chl	B2^4, 2867
5952	Demissine, solamine-d $C_{50}H_{30}NO_{20}$.	1018.21	nd (al), $[\alpha]$ −20 Py		276-9			al	B21^4, 844
5953	Derritol $C_{21}H_{22}O_6$	370.40	ye nd (MeOH)	$220\text{-}5^{0.06}$	164				B18^4, 3352
5954	Deserpidine or Canescine $C_{32}H_{38}N_2O_8$	578.66	nd or pr $[\alpha]^{24.5}/_D$ −137 (chl)		229-32			al, chl	Am77, 4335
5955	Desoxycholic acid or 3,12-Dihydroxy cholamic acid $C_{24}H_{40}O_4$	392.58	(al) $[\alpha]^{20}/_D$ +57 (al)		176			al	B10^3, 1641
5956	Desoxycorticosterone or Δ^4-Pregnene-3,20-dion-21-ol $C_{19}H_{30}O_3$	306.45	pl (eth), $[\alpha]^{20}/_D$ +178 (al,c=1)		141-2			al, eth, ace	B8^3, 2506
5957	Desthiobiotin $C_{10}H_{18}N_2O_3$	214.26	lo nd (w), $[\alpha]^{21}/_D$ +10.7 (w,c=2)		156-8			w	J1948, 1552
5958	Desthiobiotin, methyl ester $C_{11}H_{20}N_2O_3$	228.29	cr (MeOH), $[\alpha]^{28}/_D$ +2.6 (chl,c=2)	$194\text{-}7^{0.03}$	69-70			al	C44, 4934
5959	Dextrin (starch) or Amylin $(C_6H_{10}O_5)_x$	$(162.14)_x$	amor $[\alpha]'_D$ > +200		chars	$1.0384^{20/4}$		w	J1925, 636
5960	Dextropimaric acid, methyl ester $C_{21}H_{32}O_2$	316.48	$[\alpha]_D$ +60.5 (MeOH)	$140^{0.03}$	69	$1.030^{19/4}$	1.5208^{19}	al, eth	B9^3, 2912
5961	Diacetone alcohol or 4-Methyl-2-pentanon-4-ol $(CH_3)_2C(OH)CH_2COCH_3$	116.16		164, $67\text{-}9^{19}$	−44	$0.9387^{20/4}$	1.4213^{20}	w, al, eth	B1^4, 4023
5962	Diacetyl disulfide $CH_3COSSCOCH_3$	150.21		$105\text{-}6^{18}$	20			al, eth	B2^4, 564
5963	Diallyl amine $(CH_2{=}CHCH_2)_2NH$	97.16		111			1.4387^{20}	al, eth	B4^4, 1060
5964	Diallyl trisulfide $(CH_2{=}CHCH_2)_2S_3$	178.33		$112\text{-}22^{16}$		1.0845^{15}		eth	B1, 441
5965	Di(4-aminophenyl)amine $(4\text{-}H_2NC_6H_4)_2NH$	199.26	lf (w)	d	158			al, eth	B13^3, 256
5966	Diaziridine, 3-ethyl-3-methyl $C_4H_{10}N_2$	86.14		32^{17}			1.4390^{20}		
5967	Diazoacetic acid, ethyl ester or Ethyl diazo acetate $N_2CH_2CO_2C_2H_5$	114.10	ye rh	$140\text{-}1^{750}/_d$	−22	$1.0852^{18/4}$	1.4605^{20}	al, eth, bz, lig	B3^4, 1495
5968	Diazoamino benzene or 1,3-Diphenyltriazine $C_6H_5N{=}N\text{-}NHC_6H_5$	197.24	ye lf or pr (al)		98			al, eth, bz, Py	B16^3, 643
5969	Diazoamino benzene, 2,2′-dimethyl $(2\text{-}CH_3C_6H_4)N{=}N\text{-}NH(C_6H_4CH_2\text{-}2)$	225.29	og (al)		51			al, eth, lig	B16^3, 653
5970	Diazoamino benzene, 2′,3-dimethyl $(3\text{-}C_6H_3C_6H_4)N{=}N\text{-}NH(C_6H_4CH_3\text{-}2)$	225.29	ye cr (lig)		74			ace, bz, lig	B16^3, 654
5971	Diazoamino benzene, 2,4′-dimethyl $(2\text{-}CH_3C_6H_4)N{=}N\text{-}NH(C_6H_4\text{-}CH_3\text{-}4)$	225.29	ye nd (lig)		120			lig	B16^3, 655
5972	Diazoamino benzene, 3,3′-dimethyl $(3\text{-}CH_3C_6H_4)N{=}N\text{-}NH(C_6H_4CH_3\text{-}3)$	225.29	ye nd (peth)		52			eth, ace, bz	B16^3, 655

No.	Name, Synonyms, and Formula	Mol. wt.	Color, crystalline form, specific rotation and λ_{max} (log ε)	b.p. °C	m.p. °C	Density	n_D	Solubility	Ref.
5973	Diazoamino benzene, 3,4′-dimethyl $(3\text{-}CH_3C_6H_4)N{=}NNH(C_6H_4CH_3\text{-}4)$	225.29	ye nd (lig)		97				B16³, 656
5974	Diazoamino benzene, 4,4′-dimethyl $(4\text{-}CH_3C_6H_4)N{=}N\text{-}NH(C_6H_4CH_3\text{-}4)$	225.29	red-ye nd (lig) pr (al)		118			al, lig	B16³, 656
5975	Diazoamino benzene, 4,4′-dinitro $(4\text{-}O_2NC_6H_4)N{=}NNH(C_6H_4NO_2\text{-}4)$	287.23	ye nd (al), lf (bz)		240d			eth	B16³, 650
5976	Diazoamino benzene, 3-methyl $(3\text{-}CH_3C_6H_4)N{=}N\text{-}NHC_6H_5$	211.27	ye nd (lig)		86				B16³, 654
5977	Diazoamino benzene, 4-methyl $(4\text{-}CH_3C_6H_4)N{=}NNHC_6H_5$	211.27	ye pl (lig)		86-7				B16³, 655
5978	α,α′-Diazoamino naphthalene $\alpha\text{-}C_{10}H_7N{=}N\text{-}NH\text{-}\alpha\text{-}C_{10}H_7$	297.36	ye lf (al)		exp > 100				B16, 716
5979	β,β′-Diazoamino naphthalene $\beta\text{-}C_{10}H_7N{=}N\text{-}NH\text{-}\beta\text{-}C_{10}H_7$	297.36	red nd (xyl)		156				Ber19, 1282
5980	Diazoethane, 1,1,1-trifluoro CF_3CHN_2	110.04	ye	13^{752}				w, eth	B1³, 2659
5981	1,4-Diazopine, 1-methyl-perhydro $C_6H_{14}N_2$	114.19		$152\text{-}5^{742}$		0.9111^{20}	1.4769^{20}		CAS 59, 7370
5982	1,2:3,4-Dibenzanthracene $C_{22}H_{14}$	278.35	nd (aa or al)		205			bz	B5³, 2555
5983	1,2:5,6-Dibenzanthracene $C_{22}H_{14}$	278.35	pl (dil ace)		269-70			ace, bz, aa	B5³, 2553
5984	1,2:5,6-Dibenzanthracene, 4′, 4″ dihydroxy $C_{22}H_{14}O_2$	310.35	og (bz)	sub	415-8				B6³, 5914
5985	1,2:6,7-Dibenzanthracene $C_{22}H_{14}$	278.35	ye lf or nd (xyl)	sub 275^{2-4}	263-4				D5³, 2552
5986	1,2:7,8-Dibenzanthracene $C_{22}H_{14}$	278.35	og lf or nd (bz)		197-8			peth	B5³, 2553
5987	Dibenzanthrone or Violanthrone $C_{34}H_{16}O_2$	456.50	vt-bl or bk nd ($PhNO_2$)		490-5d				B7³, 4539
5988	2,3:6,7-Dibenzocycloheptadiene-5-one $C_{23}H_{16}O$	308.38		$203\text{-}4^{7}$	30	1.1635^{20}	1.6324^{20}		
5989	1,2:5,6-Dibenzofluorene $C_{21}H_{14}$	266.34	pl (bz-al)	$195\text{-}200^{0.3}$	174-5			bz	B5³, 2527
5990	1,2:7,8-Dibenzo fluorene $C_{21}H_{14}$	266.34	lf (bz), pl (aa)		234			bz	B5³, 2527
5991	1,2:6,7-Dibenzo-9-fluorenone $C_{21}H_{12}O$	280.33	og-ye pl or pr (aa or xyl)	sub $190\text{-}200^{0.04}$	214				B7³, 2897
5992	Dibenzofuran or Diphenylene oxide $C_{12}H_8O$	168.19	lf or nd (al)	287	86-7	$1.0886^{99/4}$	1.6079^{99}	al, eth, ace, aa	B17⁴, 585
5993	Dibenzofuran, 1-amino $1\text{-}H_2N(C_{12}H_7O)$	183.21	br nd (dil MeOH)		85				B18⁴7183
5994	Dibenzofuran, 2-amino $2\text{-}H_2N(C_{12}H_7O)$	183.21	pl (dil al)		128			al, eth	B18⁴7184
5995	Dibenzofuran, 3-amino $3\text{-}H_2N(C_{12}H_7O)$	183.21	(dil al)		94 (99)				B18⁴, 7191
5996	Dibenzofuran, 4-amino $4\text{-}H_2N(C_{12}H_7O)$	183.21	(al)		85				B18⁴, 7211
5997	Debenzofuran, 2-bromo $2\text{-}Br(C_{12}H_7O)$	247.09	nd (al), lf (aa)	220^{40}	110			al	B17⁴, 588
5998	Dibenzofuran, 3-bromo $3\text{-}Br(C_{12}H_7O)$	247.09	lf (al)	220^{40}	120			al, eth	B17⁴, 588
5999	Dibenzofuran, 4-bromo $4\text{-}Br(C_{12}H_7O)$	247.09			67				B17⁴, 589
6000	Dibenzofuran, 2,8-dibromo $2,8\text{-}Br_2(C_{12}H_6O)$	325.98	lf (al)		199-200			al, eth, bz, aa	B17⁴, 589
6001	Dibenzofuran, 3,6-dinitro $3,6\text{-}(NO_2)_2(C_{12}H_6O)$	258.19			245				B17⁴, 595
6002	Dibenzofuran, 3,8-dinitro $3,8\text{-}(NO_2)_2(C_{12}H_6O)$	258.19			225-6			ace, bz	B17⁴, 594
6003	Dibenzofuran, 1-nitro $1\text{-}O_2N(C_{12}H_7O)$	213.19	lt ye nd (al)		120-1				B17⁴, 591
6004	Dibenzofuran, 3-nitro $3\text{-}O_2N(C_{12}H_7O)$	213.19	ye nd (aa)	$180\text{-}5^{3}$	181-2			aa	B17⁴, 591

PHYSICAL CONSTANTS OF ORGANIC COMPOUNDS (Continued)

No.	Name, Synonyms, and Formula	Mol. wt.	Color, crystalline form, specific rotation and λ_{max} (log ε)	b.p. °C	m.p. °C	Density	n_D	Solubility	Ref.
6005	Dibenzofuran, 4-nitro $4\text{-}O_2N(C_{12}H_7O)$	213.19	ye nd	$190\text{-}205^{15}$	138-9				$B17^4$, 591
6006	1-Dibenzofuran carboxylic acid $1\text{-}(C_{12}H_7O)CO_2H$	212.20	nd (50% al)		232-3				$B18^4$, 4340
6007	2-Dibenzo furan carboxylic acid $2\text{-}(C_{12}H_7O)CO_2H$	212.20	nd (dil al aa)		246-7 (252)			al, eth	$B18^4$, 4341
6008	4-Dibenzofuran carboxylic acid $4\text{-}(C_{12}H_7O)CO_2H$	212.20	nd (al)		209-10				$B18^4$, 4346
6009	1,2:6,7-Dibenzophenanthrene $C_{22}H_{14}$	278.35	pa gr-ye lf (xyl)		294			bz, diox	$B5^3$, 2552
6010	2,3:6,7-Dibenzophenanthrene $C_{22}H_{14}$	278.35	ye-gr nd or lf (xyl)		257			al, bz	$B5^3$, 2552
6011	1,2:4,5-Dibenzopyrene $C_{24}H_{14}$	302.38	pa ye nd (xyl)		233-4				$B5^3$, 2621
6012	Dibenzothiophene or Diphenylene sulfide $C_{12}H_8S$	184.26	nd (dil al or lig)	332-3, $152\text{-}4^3$	99-100			al, bz	$B17^4$, 601
6013	Dibenzothiophene, 2-amino $2\text{-}H_2N(C_{12}H_7S)$	199.27	lf (dil al)		122-3				$B18^4$, 7186
6014	Dibenzothiophene, 3-amino $3\text{-}H_2N(C_{12}H_7S)$	199.27	(dil al)		129-31				$B18^4$, 7204
6015	Dibenzothiophene, 2-bromo $2\text{-}Br(C_{12}H_7S)$	263.15	nd (al)		125-6				$B17^4$, 605
6016	Dibenzothiophene, 2-bromo-monoxide $2\text{-}Br(C_{12}H_7SO)$	279.15			171-2				$B17^2$, 71
6017	Dibenzothiophene, 3-bromo, dioxide $3\text{-}Br\ (C_{12}H_7SO_2)$	295.15			224-5				$B17^4$, 605
6018	Dibenzothiophene, 4-bromo $4\text{-}Br(C_{10}H_7S)$	263.15	(al)		84				$B17^4$, 605
6019	Dibenzothiophene, 2,8-diamino $2,8\text{-}(NH_2)_2(C_{12}H_6S)$	214.28	nd (al)		194-6			al	$B18^4$, 7286
6020	Dibenzothiophene, 3,7-diamino $3,7\text{-}(NH_2)_2(C_{12}H_6S)$	214.28	pa ye cr		169-70				$B18^4$, 7289
6021	Dibenzothiophene, 3,7-diamino, dioxide $3,7\text{-}(NH_2)_2(C_{12}H_6SO_2)$	246.28	ye nd (al)		327-8				$B18^4$, 7289
6022	Dibenzothiophene, 2,8-dibromo $2,8\text{-}Br_2(C_{12}H_6S)$	342.05	(aa)		229				$B17^4$, 605
6023	Dibenzothiophene, 2,8-dibromo, dioxide $2,8\text{-}Br_2(C_{12}H_6SO_2)$	374.05	(aa)		361-2				$B17^4$, 606
6024	Dibenzothiophene, 3,7-dinitro, dioxide $3,7\text{-}(NO_2)_2(C_{12}H_6SO_2)$	306.25	(ace)		273-5				$B17^4$, 608
6025	Dibenzothiophene, 2-nitro $2\text{-}O_2N(C_{12}H_7S)$	229.25	pa ye nd		186				$B17^4$, 606
6026	Dibenzothiophene, 2-nitro, dioxide $2\text{-}O_2N(C_{12}H_7SO_2)$	261.25	(ace)		257-8				$B17^4$, 607
6027	Dibenzothiophene, 3-nitro $3\text{-}O_2N(C_{12}H_7S)$	229.25	pa ye (dil al)		153-4				$B17^4$,607
6028	Dibenzothiophene, 3-nitro, monoxide $3\text{-}O_2N(C_{12}H_7SO)$	245.25	(al)		210				$B17^4$, 607
6029	2-Dibenzothiophene carboxylic acid $2\text{-}(C_{12}H_7S)CO_2H$	228.27	(al)		255				$B18^4$, 4344
6030	4-Dibenzothiophene carboxylic acid $4\text{-}(C_{12}H_7S)CO_2H$	228.27	(dil MeOH)		261-2				$B18^4$, 4349
6031	Dibenzoyl disulfide $C_6H_5COSSCOC_6H_5$	274.35	pr (al) sc (chl-peth)	d	136				$B9^3$, 1977
6032	Dibenzoylmethane $(C_6H_5CO)_2CH_2$	224.26			70-1			eth, chl	$B7^3$, 3838
6033	Dibenzoyl methane, (enol form) $C_6H_5COCH{=}C(OH)C_6H_5$	224.26	rh bipym (eth)	$219\text{-}21^{18}$	78-9			al, eth, chl	$B7^3$, 3838
6034	Dibenzoylmethane, keto form $(C_6H_5CO)_2CH_2$	224.26	nd or pl (eth)		81			al, eth, chl	$B7^3$, 3838
6035	Dibenzoylmethane, α,α-dibromo $(C_6H_5CO)_2CBr$	382.05	pr (eth)		95				$B7^3$, 3846
6036	Dibenzoylmethane, oxo or Diphenyl triketone $(C_6H_5CO)_2CO$	238.24	ye nd (lig)	289^{175}, 248^{60}	68-70			eth	$B7^3$, 4620
6037	Dibenzyl acetamide $(C_6H_5CH_2)_2CHCONH_2$	239.32	nd (al, w)	259^{18}	129			al, eth	$B9^2$, 476

PHYSICAL CONSTANTS OF ORGANIC COMPOUNDS (Continued)

No.	Name, Synonyms, and Formula	Mol. wt.	Color, crystalline form, specific rotation and λ_{max} (log ε)	b.p. °C	m.p. °C	Density	n_D	Solubility	Ref.
6038	Dibenzylacetic acid $(C_6H_5CH_2)_2CHCO_2H$	240.30	pl (peth, dil aa) nd (w)	235^{18}	89	...	...	al, eth, bz, chl, aa	$B9^3$, 3356
6039	Dibenzyl acetic acid, methyl ester or Methyldibenzyl acetate $(C_6H_5CH_2)_2CHCO_2CH_3$	254.33	nd (al)	...	42-3	...	...	...	$B9^2$, 475
6040	Dibenzyl acetonitrile $(C_6H_5CH_2)_2CHCN$	221.30	lf or pl (al)	...	89-91	...	...	...	$B9^3$, 3359
6041	Dibenzyl amine $(C_6H_5CH_2)_2NH$	197.28	...	300d, 270^{250}	-26	$1.0256^{22/4}$	1.5731^{20}	al, eth	$B12^2$, 2221
6042	Dibenzyl disulfide $C_6H_5CH_2SSCH_2C_6H_5$	246.39	lf (MeOH or al) nd (aa)	...	71-2	...	...	al, eth, bz, MeOH	$B6^4$, 2760
6043	Dibenzylphenyl amine $(C_6H_5CH_2)_2NC_6H_5$	273.38	nd or pr (al)	> 300d, 226^{10}	71-2	$1.0444^{80/4}$	1.6065^{80}	eth, bz	$B12^3$, 2225
6044	Dibenzyl sulfone or Benzylsulfene $(C_6H_5CH_2)_2SO_2$	246.32	nd (al-bz)	290d	155	...	...	ace, bz, aa	$B6^4$, 2651
6045	Dibenzyl sulfoxide $(C_6H_5CH)_2SO$	230.32	lf (al or w)	210d	134-5	...	...	al, eth	$B6^4$, 2651
6046	Diborane, methylthio $CH_3SB_2H_5$	73.75	...	53	-101.5	...	...	ace, bz	$B1^4$, 1287
6047	Dibromo acetamide, N,N-dimethyl $Br_2CHCON(CH_3)_2$	244.91	pr (w or eth)	128^{16}	79-80	...	...	...	B4, 59
6048	Dibromoacetic acid Br_2CHCO_2H	217.84	dlq cr	195^{250}	48	...	...	w, al, eth	$B2^4$, 533
6049	Dibromoacetic acid, ethyl ester or Ethyl dibromo acetate $Br_2CHCO_2C_2H_5$	245.90	...	194, 121^{74}	...	$1.9025^{20/20}$	1.5017^{13}	al, eth	$B2^4$, 533
6050	Dibromoacetic acid, methyl ester or Methyl dibromo acetate $Br_2CHCO_2CH_3$	231.87	...	182-3				al, eth	B2, 219
6051	Dibutyl amine $(C_4H_9)_2NH$	129.25	...	159, 48^{13}	−60	$0.7670^{20/4}$	1.4177^{20}	w, al, eth, ace, bz	$B4^4$, 550
6052	Diisobutyl amine $(i\text{-}C_4H_9)_2NH$	129.25	...	139-40	−73.5	...	1.4090^{20}	al, eth, ace, bz	$B4^4$, 630
6053	Di-*sec*-butyl amine $(Sec\text{-}C_4H_9)_2NH$	129.25	...	135^{765}	...	$0.7534^{20/4}$	1.4162^{20}	w, al	$B4^4$, 620
6054	Dibutyl disulfide $C_4H_9SSC_4H_9$	178.35	...	226, 85^3	...	$0.9383^{20/4}$	1.4926^{20}	al, eth	$B1^4$, 1562
6055	Dibutyl sulfone or Butyl sulfone $(C_4H_9)_2SO_2$	178.29	pl (w or al)	...	46	...	...	al, eth	$B1^4$, 1561
6056	Dibutyl sulfoxide $(C_4H_9)_2SO$	162.29	nd (dil al)	d	32.6	$0.8317^{23/4}$	1.4669^{20}	al, eth	$B1^4$, 1561
6057	Dichloroacetaldehyde Cl_2CHCHO	112.94	...	90-1	...	...	...	al	$B1^4$, 3140
6058	Dichloroacetaldehyde, diethyl acetal $Cl_2CHCH(OC_2H_5)_2$	187.07	...	183-4, 67-71^{12}	...	1.1383^{14}	...	al, eth	$B1^4$, 3140
6059	Dichloroacetamide $Cl_2CHCONH_2$	127.96	mcl pr (w)	233-4^{745} sub	99.4	...	...	al, eth	$B2^4$, 505
6060	Dichloroacetic acid Cl_2CHCO_2H	128.94	...	194	13.5	$1.5634^{20/4}$	1.4658^{20}	w, al, eth, ace	$B2^4$, 498
6061	Dichloroacetic acid, anhydride or Dichloroacetic anhydride $(Cl_2CHCO)_2O$	239.87	...	214-6d	...	1.574^{24}	...	...	$B2^4$, 503
6062	Dichloroacetic acid, butyl ester or Butyl dichloroacetate $Cl_2CHCO_2C_4H_9$	185.05	...	193-4, 102^{37}	...	$1.1820^{20/4}$	1.4420^{20}	al, eth	$B2^4$, 502
6063	Dichloroacetic acid, ethyl ester or Ethyl dichloroacetate $Cl_2CHCO_2C_2H_5$	157.00	...	155.5^{764}, 56^{10}	...	$1.2827^{20/4}$	1.4386^{20}	al, eth, ace	$B2^4$, 501
6064	Dichloroacetic acid, 2-hydroxy ethyl ester or (2-Hydroxyethyl) dichloroacetate $Cl_2CHCO_2CH_2CH_2OH$	173.00	...	81-$2^{0.5}$	...	$1.438^{20/4}$	1.4735^{20}	al	$B2^3$, 460
6065	Dichloroacetic acid, methyl ester or Methyl dichloroacetate $Cl_2CHCO_2CH_3$	142.97	...	142.8, 38^{10}	−51.9	$1.3774^{20/4}$	1.4429^{20}	al	$B2^4$, 501
6066	Dichloroacetic acid, propyl ester or Propyl dichloro acetate $Cl_2CHCO_2C_3H_7$	171.02	...	176	...	1.2240^{20}	1.4398^{20}	al, eth	$B2^4$, 502
6067	Dichloroacetic acid, *iso*-propyl ester or *iso*-Propyl dichloro acetate $Cl_2CHCO_2CH(CH_3)_2$	171.02	...	163-4	...	$1.2053^{20/4}$	1.4328^{20}	al, eth	$B2^4$, 502

PHYSICAL CONSTANTS OF ORGANIC COMPOUNDS (Continued)

No.	Name, Synonyms, and Formula	Mol. wt.	Color, crystalline form, specific rotation and λ_{max} (log ε)	b.p. °C	m.p. °C	Density	n_D	Solubility	Ref.
6068	Dichloro acetonitrile Cl_2CHCN	109.94		112-3		1.369^{20}	1.4391^{25}	al	$B2^4$, 506
6069	Dichloroacetyl chloride $Cl_2CHCOCl$	147.39		108-10		1.5315$^{16/4}$	1.4591^{20}	eth	$B2^4$, 504
6070	Di-(β-chloroethyl) methyl amine, hydrochloride $(ClCH_2CH_2)_2NCH_3HCl$	192.52	hyg nd		111-2			w, al	$B4^4$, 446
6071	Dichloro fluoro acetic acid Cl_2FCCO_2H	146.93			162.5				$B2^4$, 507
6072	Dicoumarin-(cis) $C_{18}H_{10}O_4$	290.28	lf (aa)		262				$B19^4$, 2106
6073	Dicoumarin-(trans) $C_{18}H_{10}O_4$	290.28	nd or pl (aa)		>275				$B19^4$, 2106
6074	Dictamnine $C_{12}H_9NO_2$	199.21	pr (al)		133-4			al, eth, chl	$B27^2$, 79
6075	Dicumarol or 4,4′-Dihydroxy-3,3′-methylene bis coumarin $C_{19}H_{12}O_6$	336.30	nd		288-92				$B19^4$, 2261
6076	Dicyclohexadiene $C_{12}H_{16}$	160.26		229-30, 104^{16}		0.9950$^{20/4}$	1.5267$^{20.5}$	eth, ace, bz, aa	$B5^3$, 1267
6077	Dicyclohexylamine $(C_6H_{11})_2NH$	181.32		255.8d, 113-5^9	−0.1	0.9123$^{20/4}$	1.4842^{20}	al, eth, bz	$B12^1$, 19
6078	Dicyclohexyl ketone $(C_6H_{11})_2CO$	194.32		159^{20}		0.986%	1.4860^{20}	eth, ace	$B7^1$, 494
6079	Dicyclohexyl methane, 4,4′-diamino *(cis,cis)* $(4\text{-}H_2NC_6H_{10})_2CH_2$	210.36		141^2	60-2		1.5014^{27}		Am73, 641
6080	Dicyclohexyl methane, 4,4′-diamino (cis,trans) $(4\text{-}H_2NC_6H_{10})_2CH_2$	210.36	cr	127-8$^{1.2}$	36-7	0.9608$^{25/4}$	1.5046^{25}		Am73, 741
6081	Dicyclohexyl methane, 4,4′-diamino (trans,trans) $C_{13}H_{26}N_2$	210.36	cr (peth)	130-1$^{0.3}$	64-5		1.5032^{25}		Am73, 741
6082	α-Dicylopentadiene *(endo form)* $C_{10}H_{12}$	132.21		170d, 64-5^{14}	32	0.9302$^{35/4}$	1.5050^{35}	al, eth, aa	$B5^4$, 1399
6083	α-Dicyclopentadiene, 3,4,5,6,7,8,8a-heptachloro or Heptachlor $C_{10}H_5Cl_7$	373.32	wh		95-6	1.57^9		al, eth, bz, lig	$B5^1$, 1236
6084	α-Dicyclopentadiene, tetrahydro or Tricyclodecane $C_{10}H_{16}$	136.24	(al or aa)	193^{769}, 86-7^{12}	77	0.9128^{79}	1.4726^{79}	al, aa	$B5^4$, 467
6085	Didodecylamine or Dilaurylamine $(C_{12}H_{25})_2NH$	353.68			55-6			al, eth, bz, chl	$B4^4$, 801
6086	Dieldrin or Octalox $C_{12}H_8OCl_6$	380.91			175-6	1.75		ace, bz	$B17^4$, 526
6087	Diethanol amine $(HOCH_2CH_2)_2NH$	105.14		271, 154-5^{10}	28	1.0966$^{20/4}$	1.4776^{20}	w, al	$B4^4$, 1514
6088	Diethanol amine, *N*-phenyl $(HOCH_2CH_2)_2NC_6H_5$	181.23	pl (al)	228^{15}	58			al, eth, ace, bz	$B12^1$, 299
6089	Diethoxyacetic acid, ethyl ester or Ethyldiethoxy acetate $(C_2H_5O)_2CHCO_2C_2H_5$	176.21		199, 83-5^{13}		0.994^{18}	1.4089^{25}	al, eth	$B3^4$, 1494
6090	Diethoxy disulfide $C_2H_5OSSOC_2H_5$	154.24		67-8^{16}		1.0913$^{20/4}$	1.4766^{20}		$B1^4$, 1324
6091	Diethoxy sulfide $(C_2H_5O)_2S$	122.18		117^{733}		0.9940$^{20/4}$	1.4234^{20}	w, al, eth, ace, bz	$B1^3$, 1314
6092	Diethyl amine $(C_2H_5)_2NH$	73.14		56.3	−48	0.7056$^{20/4}$	1.3864^{20}	w, al, eth	$B4^4$, 313
6093	Diethyl amine, hydrochloride $(C_2H_5)_2NH.HCl$	109.60	lf (al-eth)	320-30	227-30	1.0477$^{22/4}$		w, al	$B4^4$, 318
6094	Diethyl amine, *N*-nitro $(C_2H_5)_2N.NO_2$	118.14		206.5^{757}, 93^{16}		1.057^{15}		al, eth	$B4^3$, 233
6095	Diethyl amine, *N*-nitroso $(C_2H_5)_2N.NO$	102.14	ye	1 76.9		0.9422$^{20/4}$	1.4386^{20}	w, al, eth	$B4^3$, 233
6096	Diethyl (B-bromo ethyl) amine, hydrobromide $(C_2H_5)_2NC\ H_2CH_2Br.HBr$	261.00	nd (al-eth)		209			w	$B4^3$, 249
6097	Diethyl (2,2-diethoxy ethyl) amine or Diethylamino acetal $(C_2H_5)_2NCH_2CH(OC_2H_5)_2$	189.30		194-5		0.863	1.4189^{20}	w, al, eth	$B4^4$, 1919
6098	Diethyl disulfide $C_2H_5SSC_2H_5$	122.24		154	−101.5	0.9931$^{20/4}$	1.5073^{20}	al, eth	$B1^4$, 1397
6099	Diethyl (methoxymethyl) amine $(C_2H_5)_2NCH_2OCH_3$	117.19		117^{763}				w, al, eth	$B4^3$, 203
6100	Diethyl methyl amine $(C_2H_5)_2NCH_3$	87.16		66		0.703$^{25/4}$	1.3879^{25}	w, al, eth	$B4^4$, 321

No.	Name, Synonyms, and Formula	Mol. wt.	Color, crystalline form, specific rotation and λ_{max} (log ε)	b.p. °C	m.p. °C	Density	n_D	Solubility	Ref.
6101	Diethyl selenide $(C_2H_5)_2Se$	137.08	pa ye	108		$1.2300^{20/4}$	1.4768^{20}	al, eth, bz	$B1^4$, 1411
6102	Diethyl sulfide $(C_2H_5)_2S$	90.18		92.1	-103.8	$0.8362^{20/4}$	1.4430^{20}	al, eth	$B1^4$, 1394
6103	Diethyl sulfide, 2,2-diamino or bis(2-Aminoethyl) sulfide $(H_2NCH_2CH_2)_2S$	120.21	ye	$231\text{-}3^{755}$, $118\text{-}20^{17}$					$B4^4$, 1577
6104	Diethyl sulfide, 2,2′-dichloro or bis-(2-Chloroethyl) sulfide mustard gas $(ClCH_2CH_2)_2S$	159.07	ye pr	217, 95^{10}	13-4	$1.2741^{20/4}$	1.5312^{20}	al, eth, ace, bz	$B1^4$, 1407
6105	Diethyl sulfide, 2,2′-dihydroxy or bis-(2-Hydroxyethyl) sulfide $(HOCH_2CH_2)_2S$	122.18		$164\text{-}6^{20}$	−10	$1.1819^{20/4}$	1.5203^{20}	w, al, eth, chl	$B1^4$, 2437
6106	Diethyl sulfide, 2,2′-diphenoxy or bis-(2-Phenoxyethyl) sulfide $(C_6H_5OCH_2CH_2)_2S$	274.38	nd (al)		42			al, eth	$B6^4$, 581
6107	Diethyl sulfone or Ethyl sulfone $(C_2H_5)_2SO_2$	122.18	rh pl	248	73-4	$1.357^{20/4}$		w, bz	$B1^4$, 1396
6108	Diethyl sulfoxide $(C_2H_5)_2SO$	106.18	syr	104^{25}	14			w, al, eth	$B1^4$, 1395
6109	bis (Diethylthiocarbamyl) disulfide or Antabuse $[(C_2H_5)_2NCS]_2S_2$	296.52	(al)	117^{17}	71-2			al, chl	$B4^4$, 398
6110	Diethyl trisulfide $C_2H_5S_3C_2H_5$	154.30	ye	$96\text{-}7^{26}$		1.114^{20}	1.5689^{13}		$B1^4$, 1398
6111	Diethylene glycol $HOCH_2CH_2OCH_2CH_2OH$	106.12		245, 133^{14}	-10.5	$1.1197^{15/4}$	1.4472^{20}	w, al, eth	$B1^4$, 2390
6112	Diethylene glycol, monobutyl ether or Butylcarbitol $C_4H_9OCH_2CH_2OCH_2CH_2OH$	162.23		231, 118^{12}	-68.1	$0.9553^{20/4}$	1.4321^{20}	w, al, eth, ace, bz	$B1^4$, 2394
6113	Diethylene glycol, monobutyl ether, acetate $C_4H_9OCH_2CH_2OCH_2CH_2O_2CCH_3$	204.27		245	−32	$0.985^{20/4}$	1.4262^{20}	w, al, eth, ace	$B2^3$, 308
6114	Diethylene glycol, diacetate $CH_3CO_2CH_2CH_2OCH_2CH_2O_2CCH_3$	190.20		245-51, $110\text{-}35^{16}$		$1.1078^{15/15}$	1.4348^{20}	w, al, eth	$B2^4$, 216
6115	Diethylene glycol, dibenzyl ether $(C_6H_5CH_2OCH_2CH_2)_2O$	286.37		$279\text{-}81^{24}$, 250^1	33.5	$1.1701^{15/15}$		w, al	$B9^3$, 535
6116	Diethylene glycol, diisobutyrate $(i\text{-}C_3H_7CO_2CH_2CH_2)_2O$	246.30		$178\text{-}80^{40}$			1.4282^{20}		$B2^3$, 652
6117	Diethylene glycol, diethyl ether or Diethyl carbitol $(C_2H_5OCH_2CH_2)_2O$	162.23		189		$0.9063^{20/4}$	1.4115^{20}	w, al, eth	$B1^4$, 2394
6118	Diethylene glycol, dimethyl ether $(CH_3OCH_2CH_2)_2O$	134.18		162	-68	$0.9451^{20/20}$		w, al, eth	$B1^4$, 2393
6119	Diethylene glycol, di-octadecanoate or Diethylene glycol disfearate $(C_{17}H_{35}CO_2CH_2CH_2)_2O$	639.07	wax		54-5	$0.9333^{20/4}$			$B2^4$, 1223
6120	Diethylene glycol, dioleate $(C_{17}H_{33}CO_2CH_2CH_2)_2O$	635.04	pa ye oil			$0.9310^{20/4}$		al, eth	C47, 4618
6121	Diethylene glycol, monododecanoate or Diethylene glycol monolaurate $C_{11}H_{23}CO_2CH_2CH_2OCH_2CH_2OH$	288.43	lt ye	> 270	17-8	$0.96^{25/25}$		al, eth, ace, bz	$B2^3$, 887
6122	Diethylene glycol monoethyl ether or Ethylcellosolve $C_2H_5OCH_2CH_2OCH_2CH_2OH$	134.19	hyg liq	195		$0.9881^{20/4}$	1.4300^{20}	w, al, eth, ace, bz	$B1^4$, 2377
6123	Diethylene glycol, monoethyl ether, acetate or Carbitol acetate $C_2H_5OCH_2CH_2OCH_2CH_2O_2CCH_3$	176.21		218	−25	$1.0096^{20/4}$	1.4230^{25}	w, al, eth, ace	$B2^3$, 308
6124	Diethylene glycol, (2-hydroxypropyl) ether $[(CH_3CH(OH)CH_2]OCH_2CH_2OCH_2CH_2OH$	164.20		277-9		$1.0789^{20/4}$	1.4498^{20}	w, al, bz	C52, 17693
6125	Diethylene glycol, methyl ether or Methyl carbitol $CH_3OCH_2CH_2OCH_2CH_2OH$	120.15		193		$1.0270^{20/4}$	1.4264^{20}	w, al, eth, ace	$B1^4$, 2392
6126	Diethylene triamine or bis-(2-Aminoethyl) amine $H_2NCH_2CH_2NHCH_2CH_2NH_2$	103.17	ye hyg liq	207	−39	$0.9586^{20/20}$	1.4810^{25}	w, al, lig	$B4^3$, 1238
6127	Difluoroacetic acid F_2CHCO_2H	96.03		134.2, $67\text{-}70^{20}$	−0.3	1.5255^{20}	1.3420^{20}	w, al, eth, ace, bz	$B2^4$, 455
6128	Difluoroacetic acid, ethyl ester or Ethyl difluoro acetate $F_2CHCO_2C_2H_5$	124.09		99.2^{750}		$1.1893^{9.8}$			$B2^4$, 455
6129	Difurfuryl amine $(OC_4H_3CH_2)_2NH$	177.20		$135\text{-}42^{15}$		$1.1045^{20/4}$	1.5168^{20}	eth	$B18^4$, 7092
6130	Di(2-furfuryl) disulfide $[(2\text{-}OC_4H_3)CH_2]_2S_2$	226.31		$112\text{-}3^{0.5}$	10			al	$B17^4$, 1258
6131	*m*-Digallic acid or Gallicacid-3-monogallate $C_{14}H_{10}O_9$	332.23	nd (dil al + 1w)		268-70d			al, ace	$B10^3$, 2086

PHYSICAL CONSTANTS OF ORGANIC COMPOUNDS (Continued)

No.	Name, Synonyms, and Formula	Mol. wt.	Color, crystalline form, specific rotation and λ_{max} (log ε)	b.p. °C	m.p. °C	Density	n_D	Solubility	Ref.
6132	Digitalose or 3-*o*-Methyl-D-fucose $C_7H_{14}O_5$	178.19	nd (AcOEt) $[\alpha]^{22}_D$ +109→ +126 (mut)		106→ 119			w	B1[4], 4270
6133	Digitogenin or 5α,22α-Spirostan-2,3,15-triol $C_{27}H_{44}O_5$	448.64	nd (al) $[\alpha]^{20}_D$ −18 (chl, c=1.4)		280-3			chl	B19[4], 1242
6134	Digitoxigenin $C_{23}H_{34}O_4$	374.52	(dil MeOH) $[\alpha]^{20}_D$ +119.1 (MeOH, c=1.36)		253 (256)			al	B18[4], 1468
6135	Digitoxin $C_{41}H_{64}O_{13}$	764.95	wh (chl-eth) pr (dil al) $[\alpha]^{20}_D$ +4.8 (diox, c=1.2)		255-6			al, eth, chl	B18[4], 1478
6136	Digitoxose or 2-Deoxy-D-*altro*-Methylose $C_6H_{12}O_4$	148.16	cr (MeOH-eth) $[\alpha]^{15}_D$ +27.9→ 43.3 (Pyr=1, mut)		112			w, ace, py	B1[4], 4191
6137	Diglycolic acid or Oxydiethanoic acid $O(CH_2CO_2H)_2$	134.09	mcl pr (w + 1)	d	148			w, al, eth	B3[4], 577
6138	Diglycolyl chloride $O(CH_2COCl)_2$	146.96		116^{15}				chl	B3, 240
6139	Diglycolic acid, hydrate $O(CH_2CO_2H)_2 \cdot H_2O$	152.10	mcl pr (w + 1)	d	148			w, al, eth	B3, 234
6140	Digoxigenin $C_{23}H_{34}O_5$	390.52	pr (AcoEt)		222			al, MeOH	B18[4], 2450
6141	Diheptyl amine $(C_7H_{15})_2NH$	213.41	nd	271^{750}, $134\text{-}6^{9}$	30			al, eth	B4[4], 736
6142	Diheptyl sulfide $(C_7H_{15})_2S$	230.45		298, 164^{20}		$0.8416^{20/4}$	1.4606^{20}	eth	B1[4], 1739
6143	Dihexyl amine $(C_6H_{13})_2NH$	185.35		192-5, $112\text{-}4^{12}$			1.4339^{20}	al, eth	B4[4], 711
6144	Dihexyl sulfide $(C_6H_{13})_2S$	202.40		230, 113.5^{4}		$0.8411^{20/4}$	1.4586^{20}		B1[4], 1706
6145	Dihydrosamidin $C_{21}H_{24}O_7$	388.42	$[\alpha]_D$ + 19(al)		117-9			al, eth	B19[4], 2788
6146	3,4-Dihydroxybenzoic acid, 5-bromo 5-Br-3,4-$(HO)_2C_6H_2CO_2H$	233.02	nd (w, aa, dil al)		230			aa	B10[1], 192
6147	2,3-Dihydroxybenzoic acid, 5-bromo 5-Br-2,3-$(HO)_2C_6H_2CO_2H$	233.02	pr (w + 1) nd (w)		187 (pr) 215 (nd)			al, eth, aa	B10[3], 1367
6148	2,4-Dihydroxybenzoic acid, 3-bromo 3-Br-2,4-$(HO)_2C_6H_2CO_2H$	233.02	br or ye nd (w)		202			aa	B10[2], 254
6149	2,4-Dihydroxybenzoic acid, 5-bromo 5-Br-2,4-$(HO)_2C_6H_2CO_2H$	233.02	micr pr (w + 1)		212			al, eth	B10[3], 1378
6150	Di-(β-Hydroxyethyl) methyl amine $(HOCH_2CH_2)_2NCH_3$	119.16		$246\text{-}8^{747}$, $123\text{-}5^{4}$		1.0377^{20}	1.4642^{20}	w, al	B4[4], 1571
6151	Diiodoacetic acid I_2CHCO_2H	311.85	lf ye cr or wh nd (bz)		110			w, al, eth, bz	B2[4], 537
6152	Diisoeugenol $C_{20}H_{24}O_4$	328.41	nd(bz liq or al)		180-1			al, eth, chl	B6[3], 6765
6153	Dilactic acid $[HO_2CCH(CH_3)]_2O$	162.14	rh		112-3			w, eth	B3[3], 468
6154	Dilauryl amine $(C_{12}H_{25})_2NH$	353.68			55-6			al, eth, bz, chl	B4[4], 801
6155	Dimethisoquin hydrochloride or Quotane $C_{17}H_{24}N_2OHCl$	308.85			146			w, al	B21[4], 1335
6156	Dimethyl amine $(CH_3)_2NH$	45.08		7.4	−93	$0.6804^{0/4}$	1.350^{17}	w, al, eth	B4[4], 128

No.	Name, Synonyms, and Formula	Mol. wt.	Color, crystalline form, specific rotation and λ_{max} (log ε)	b.p. °C	m.p. °C	Density	n_D	Solubility	Ref.
6157	Dimethyl amine hydrochloride or Dimethylammonium chloride $(CH_3)_2NH.HCl$	81.55	rh, nd (al)		171			w, al, chl	$B4^4$, 132
6158	Dimethyl amine, hexafluoro $(CF_3)_2NH$	153.03		−6.7	−130				$B3^4$, 78
6159	Dimethylamine, perfluoro $(CF_3)_2NF$	171.02		−37					$B3^4$, 170
6160	Dimethylamine, *N*-nitro $(CH_3)_2N\text{-}NO_2$	90.08	nd(eth)	187	58	$1.1090^{72/4}$	1.4462^{72}	w, al, eth, ace, bz	$B4^3$, 167
6161	Dimethyl amine, *N*-nitroso or Dimethylnitrosamine $(CH_3)_2\text{-}N\text{-}NO$	74.08	ye	154		$1.0059^{20/4}$	1.4358^{20}	w, al, eth	$B4^3$, 166
6162	3,4-Dimethylbenzoic acid, ethyl ester $3,4\text{-}(CH_3)_2C_6H_3CO_2C_2H_5$	178.23		$127\text{-}8^{10}$			1.5144^{20}		$B9^2$, 353
6163	Dimethyl isobutyl amine $i\text{-}C_4H_9N(CH_3)_2$	101.19		80-1		$0.7097^{20/4}$	1.3907^{20}	w	$B4^4$, 627
6164	Dimethyl disulfide CH_3SSCH_3	94.19		109.7, 6.4^{10}	−84.7	$1.0625^{20/4}$	1.5289^{20}	al, eth	$B1^4$, 1281
6165	Dimethyl disulfide, hexafluoro CF_3SSCF_3	202.13		34.6	>1			al, peth	$B3^4$, 278
6166	Dimethyl ethyl amine $(CH_3)_2NC_2H_5$	73.14		36-7	−36	0.675	1.3705		$B4^4$, 312
6167	Dimethyl glyoxime $CH_3C(:NOH)C(:NOH)CH_3$	116.12	nd (to or dil al)	sub 234	245-6			al eth	$B1^4$, 3647
6168	Dimethyl hexesterol or 3,4-bis(4-hydroxy-3-methylphenyl) hexane $(C_{10}H_{13}O)_2$	298.43	cr (dil al)		145				$B6^3$, 5541
6169	Dimethyloxonium bromide $[(CH_3)_2OH]^+Br^-$	126.98			−13			eth	$B1^4$, 1247
6170	Dimethyloxonium chloride $[(CH_3)_2OH]^+Cl^-$	82.53	gas	−2	−97			eth, lig, HCl	$B1^4$, 1247
6171	Dimethyl pentyl amine $(CH_3)_2NC_5H_{11}$	115.22		123		$0.743^{20/4}$	1.4083^{20}	eth	$B4^4$, 675
6172	Dimethyl phosphinic acid, ethyl ester $(CH_3)_2PO_2C_2H_5$	122.10		89^{15}		$1.0278^{25/4}$	1.4281^{25}	al, eth	Am73, 5466
6173	bis-(Dimethyl phosphino) amine or Amino-bis(dimethylphosphine) $[(CH_3)_2P]_2NH$	137.10		$33.5^{5.4}$ sub	39.5				Am75, 3869
6174	Dimethyl sulfone $(CH_3)_2SO_2$	94.13	pr	238	110	$1.1702^{110/0}$	1.4226	w, al, bz	$B1^4$, 1279
6175	Dimethyl sulfoxide $(CH_3)_2SO$	78.13		189, $85\text{-}7^{20}$	18.4	$1.1014^{20/4}$	1.4770^{20}	w, al, eth, ace	$B1^4$, 1277
6176	bis-(Dimethylthiocarbamyl) disulfide or Arasan $[(CH_3)_2NCS]_2S_2$	240.41	wh or ye mcl (chl-al)	129^{20}	155.6			chl	$B4^4$, 242
6177	Di-α-Naphthyl amine $(\alpha\text{-}C_{10}H_7)_2NH$	269.35		$310\text{-}5^{15}$	115			al, eth, ace, bz, chl	$B12^3$, 2859
6178	Diβ-Naphthyl amine $(\beta\text{-}C_{10}H_7)_2NH$	269.35	lf (bz)	471	172.2			eth	$B12^3$, 3003
6179	α,α′-Dinaphthyl disulfide $(\alpha\text{-}C_{10}H_7)_2S_2$	318.45	pl (al) nd (lig)		91	1.144^{20}		eth	$B6^3$, 2947
6180	β,β′-Dinaphthyl disulfide $(\beta\text{-}C_{10}H_7)_2S_2$	318.45	nd		139-40	$0.8409^{20/4}$	1.4555^{20}	al, eth	$B6^3$, 3013
6181	α,α′-Dinaphthyl ketone $\alpha\text{-}C_{10}H_7CO\text{-}\alpha\text{-}C_{10}H_7$	282.34	wh nd yesh pr (aa)		104			al, eth, bz	$B7^2$, 503
6182	α,β′-Dinaphthyl ketone $\alpha\text{-}C_{10}H_7CO\text{-}\beta\text{-}C_{10}H_7$	282.34	nd (al bz-lig)	$235^{0.06}$	136-7			chl	$B6^3$, 2869
6183	β,β′-Dinaphthyl ketone $\beta\text{-}C_{10}H_7CO\text{-}\beta\text{-}C_{10}H_7$	282.34	(i) nd (eth), (ii) lf (chl-eth)		(i) 125.5, (ii) 164.5			chl	$B7^2$, 504
6184	Di(α-naphthyl)methane $(\alpha\text{-}C_{10}H_7)_2CH_2$	268.36	pr or nd (al)	>360, 270^{14}	109			eth, bz, chl	$B5^3$, 2480
6185	Di(β-Naphthyl)methane $(\beta\text{-}C_{10}H_7)_2CH_2$	268.36	nd (al eth)		93			bz	$B5^1$, 360
6186	2,4-Dinitrobenzoic acid $2,4\text{-}(NO_2)_2C_6H_3CO_2H$	212.12	nd(w)		183			bz	$B9^3$, 1776
6187	2,5-Dinitrobenzoic acid $2,5\text{-}(NO_2)_2C_6H_3CO_2H$	212.12	pr (w)		177			al, eth	$B9^3$, 1778

No.	Name, Synonyms, and Formula	Mol. wt.	Color, crystalline form, specific rotation and λ_{max} (log ε)	b.p. °C	m.p. °C	Density	n_D	Solubility	Ref.
6188	2,6-Dinitrobenzoic acid $2,6-(NO_2)_2C_6H_3CO_2H$	212.12	nd(w)		202-3			al, eth	$B9^{3}$, 1778
6189	3,4-Dinitrobenzoic acid $3,4-(NO_2)_2C_6H_3CO_2H$	212.12	nd (w)		165			al, eth	$B9^{3}$, 1778
6190	3,5-Dinitrobenzoic acid $3,5-(NO_2)_2C_6H_3CO_2H$	212.12	mcl pr (al)		205			al, aa	$B9^{3}$, 1779
6191	3,5-Dinitrobenzoic acid, benzyl ester $3,5-(NO_2)_2C_6H_3CO_2CH_2C_6H_5$	302.24	nd (lig)		112				$B9^{3}$, 1848
6192	3,5-Dinitrobenzoic acid, 2-bromo $3,5-(NO_2)_2-2-BrC_6H_2CO_2H$	291.01	ye nd (w)		213			al, bz, aa, lig	$B9^{3}$, 1954
6193	3,5-Dinitrobenzoic acid, butyl ester $3,5-(NO_2)_2C_6H_3CO_2C_4H_9$	268.23	mcl nd (al)		62.5		1.488		$B9^{3}$, 1782
6194	3,5-Dinitrobenzoic acid, isobutyl ester $3,5-(NO_2)_2C_6H_3CO_2-i-C_4H_9$	268.23	mcl pl, nd (al)		87-8			al	$B9^{3}$, 1783
6195	3,5-Dinitrobenzoic acid, ethyl ester $3,5-(NO_2)_2C_6H_3CO_2C_2H_5$	240.17	nd (al)		92.9	1.295^{111}	1.560	al	$B9^{3}$, 1781
6196	3,5-Dintrobenzoic acid, furfuryl ester $3,5-(NO_2)_2C_6H_3CO_2CH_2(C_4H_3O)$	292.20	(bz-py)		78-81				$B17^{4}$, 1249
6197	3,5-Dinitrobenzoic acid, methyl ester $3,5-(NO_2)_2C_6H_3CO_2CH_3$	226.15	nd(w)		112			al, w	$B9^{3}$, 1781
6198	3,5-Dinitrobenzoic acid, tetrahydrofurfuryl ester $3,5-(NO_2)_2C_6H_3CO_2CH_2(C_4H_7O)$	296.24	nd (al)		83-4			al	$B17^{4}$, 1106
6199	3,5-Dinitrobenzoic acid, pentyl ester $3,5-(NO_2)_2C_6H_3CO_2C_5H_{11}$	282.25			46.4			al	$B9^{2}$, 281
6200	3,5-Dinitrobenzoic acid, phenyl ester $3,5-(NO_2)_2C_6H_3CO_2C_6H_5$	288.22	rods (al)		145-6			al, bz	$B9^{3}$, 1846
6201	3,5-Dinitrobenzoic acid, propyl ester $3,5-(NO_2)_2C_6H_3CO_2C_3H_7$	254.20	mcl pl (al)		73			al	$B9^{3}$, 1781
6202	3,5-Dintrobenzoic acid, isopropyl ester $3,5-(NO_2)_2C_6H_3CO_2-i-C_3H_7$	254.20	nd (al)		122			al	$B9^{3}$, 1782
6203	3,5-Dinitrobenzoyl chloride $3,5-(NO_2)_2C_6H_3COCl$	230.56	ye nd bz	196^{12}	74			eth	$B9^{3}$, 1936
6204	Dioctadecyl amine or Distearyl amine $(C_{18}H_{37})_2NH$	522.00			73-4			chl	$B4^{4}$, 829
6205	Dioctyl amine $(C_8H_{17})_2NH$	241.46	nd	297-8, 175^{14}	35.6	$0.7968^{26/4}$	1.4415^{26}	al, eth	$B4^{4}$, 753
6206	Di-2-Octyl amine $[C_6H_{13}CH(CH_3)]_2NH$	241.46		281.5^{739}		$0.7948^{20/4}$		al, eth	$B4^{4}$, 765
6207	1,3-Dioxane or *m*-Dioxane $C_4H_8O_2$	88.11		105^{755}	−42	$1.0342^{20/4}$	1.4165^{20}	w, al, eth, ace, bz	$B19^{4}$, 8
6208	1,3-Dioxane, 2,4-dimethyl $C_6H_{12}O_2$	116.16		115-8		$0.9392^{20/4}$	1.4136^{20}		$B19^{4}$, 61
6209	1,3-Dioxane, 5-ethyl-4-propyl $C_9H_{18}O_2$	158.24		196		$0.9305^{20/4}$	1.4370^{20}		C52, 2795
6210	1,3-Dioxane, 5-hydroxy-2-methyl $C_5H_{10}O_3$	118.13		176		$1.0705^{17/4}$	1.4375^{17}	w	$B19^{4}$, 624
6211	1,3-Dioxane, 4-methyl $C_5H_{10}O_2$	102.13		114		0.9758^{20}	1.4159^{20}		$B19^{4}$, 49
6212	1,3-Dioxane, 4-methyl-4-phenyl $C_{11}H_{14}O_2$	178.23		256, 102^{4}	35-40	1.0864^{20}	1.5240^{20}		$B19^{4}$, 233
6213	1,3-Dioxane, 2-phenyl $C_{10}H_{12}O_2$	164.20	nd (peth)	252-4, $98-9^{6}$	41			al, eth	$B19^{4}$ 215
6214	1,3-Dioxane, 4-phenyl $C_{10}H_{12}O_2$	164.20		245, $128-30^{10}$		$1.1038^{20/4}$	1.5306^{18}		$B19^{4}$, 218
6215	1,4-Dioxane or Diethylene dioxide $C_4H_8O_2$	88.11		101^{750}	11.8	$1.0337^{20/4}$	1.4224^{20}	w, al, eth, ace, bz	$B19^{4}$, 9
6216	1,4-Dioxane, -2,3-dichloro $C_4H_6Cl_2O_2$	157.00		$80-2^{10}$	30	$1.468^{20/4}$	1.4928^{20}	eth, ace, bz	$B19^{4}$, 30
6217	1,4-Dioxane, heptachloro $C_4HCl_7O_2$	329.22		$123-8^{8}$	54-6			eth, ace, bz, lig	$B19^{4}$, 32
6218	1,4-Dioxene $C_4H_6O_2$	86.09		94.1		$1.0836^{20/4}$	1.4372^{20}	eth, ace, bz	$B19^{4}$, 108
6219	1,4-Dioxine $C_4H_4O_2$	84.07		74.6^{748}		$1.115^{20/4}$	1.4350^{20}	eth, ace, bz	$B19^{4}$, 154
6220	1,3-Dioxolane or Glycol methylene ether $C_3H_6O_2$	74.08		78^{765}	−95	$1.0600^{20/4}$	1.3974^{20}	w, al, eth, ace	$B19^{4}$, 5

No.	Name, Synonyms, and Formula	Mol. wt.	Color, crystalline form, specific rotation and λ_{max} (log ε)	b.p. °C	m.p. °C	Density	n_D	Solubility	Ref.
6221	1,3-Dioxolane, 4-(hydroxymethyl)-2-methyl or Glycerol-ethylidine ether $C_5H_{10}O_3$	118.13		$187, 68\text{-}70^{1}$		$1.1243^{17/4}$	1.4413^{17}	al	$B19^{4}$, 631
6222	1,3-Dioxolane, 2-methyl or Glycol ethylidine ether $C_4H_8O_2$	88.11	81-2			$0.9811^{20/4}$	1.4035^{17}	w, al, eth	$B19^{4}$, 42
6223	1,3-Dioxlane-4-carboxaldehyde, 2,2-dimethyl $C_6H_{10}O_3$	130.14		74^{50}			1.4189^{25}	w	$B19^{4}$, 1579
6224	1,3-Dioxolan-2-one or 1,2-Ethanediol carbonate $C_3H_4O_3$	88.06	mcl pl (al)	248^{760}	39-40	$1.3214^{39/4}$	1.4158^{50}	w, al, eth, bz, chl	$B19^{4}$, 1556
6225	1,3-Dioxolane-2-one, 4-methyl or Propylene carbonate $C_4H_6O_3$	102.09		24.2	−48.8	$1.2069^{20/20}$	1.4189^{20}	w, al, eth, ace, bz	$B19^{4}$, 1564
6226	Dipentyl amine $(C_5H_{11})_2NH$	157.30		$202\text{-}3, 91\text{-}3^{14}$		$0.7771^{20/4}$	1.4272^{20}	al, %eth,] ace	$B4^{4}$, 676
6227	Dipentylamine, 2,2′-dimethyl-2,2′-dihydroxy $[C_3H_7C(OH)(CH_3)CH_2]_2NH$	217.35		$165\text{-}70^{15}$		$0.9264^{20/4}$	1.4585^{20}	al, %eth,] ace	C53, 16829
6228	Diisopentyl amine $(i\text{-}C_5H_{11})_2NH$	157.30		188	−44	$0.7672^{21/4}$	1.4235^{20}	al, %eth]	$B4^{4}$, 699
6229	Dipentyl disulfide or [m] -Amyldi sulfide $C_5H_{11}SSC_5H_{11}$	206.40		119^{7}		$0.9221^{20/4}$	1.4889^{20}		$B1^{4}$, 1654
6230	Di-[iso] -Pentyl disulfide or Isoamylsulfide $([i]\text{-}C_5H_{11})_2S_2$	206.40		250		$0.9192^{20/4}$	1.4864^{20}		$B1^{4}$, 1689
6231	Diphenadione or 2-Diphenyl acetyl-1,3-indanedione $C_{23}H_{16}O_3$	340.38	pa ye mcl (al)		146-7		1.670	ace, aa	C49, 3264
6232	Diphenyl acetamide $(C_6H_5)_2CHCONH_2$	211.26	pl (al)		167-8			al	$B9^{3}$, 3301
6232a	Diphenyl acetamide, α-chloro $(C_6H_5)_2CClCONH_2$	245.71	cr (to)		115			al, eth, bz, chl	
6233	Diphenyl acetic acid $(C_6H_5)_2CHCO_2H$	212.25	nd (w) lf (al)	194^{25} sub	148	$1.258^{15/15}$		al, eth, chl	$B9^{3}$, 3290
6234	Diphenyl acetic acid-anhydride $[(C_6H_5)_2CHCO]_2O$	406.48	nd (eth,)	$220\text{-}5^{15}$	98			bz, chl	$B9^{1}$, 281
6235	Diphenyl acetic acid, α-bromo $(C_6H_5)_2C(Br)CO_2H$	291.14	(chl-peth)		133-4			chl, to	$B9^{2}$, 471
6236	Diphenyl acetic acid, α-chloro $(C_6H_5)_2C(Cl)CO_2H$	246.69	pl (bz-liq)		118-9d			al, eth, ace, bz	$B9^{3}$, 3307
6237	Diphenyl acetic acid, α-chloro, ethyl ester or Ethyl-diphenyl chloroacetate $(C_6H_5)_2C(Cl)CO_2C_2H_5$	274.75	pl (chl) cr (al)	185^{14}	43-4			al, eth	$B9^{3}$, 3307
6238	Diphenl acetic acid, ethyl ester or Ethyldiphenyl acetate $(C_6H_5)_2CHCO_2C_2H_5$	240.30	nd (al) rh (AcOEt)	195^{25}	59			al, eth	$B9^{3}$, 3291
6239	Diphenyl acetic acid, α-hydroxy or Benzilic acid $(C_6H_5)_2C(OH)CO_2H$	228.25	mcl nd (w)	d180	151			al, eth	$B10^{3}$, 1168
6240	Diphenyl acetic acid, α-hydroxy, ethyl ester $(C_6H_5)_2C(OH)CO_2C_2H_5$	256.30	pr or nd	201^{21}	34		1.5620^{20}	al, eth	$B10^{3}$, 1171
6241	Diphenyl acetic acid, α-hydroxy, methyl ester or Methyl benzilate $(C_6H_5)_2C(OH)CO_2CH_3$	242.27	mcl or tcl cr (al)	187^{13}	75			al, eth, aa	$B10^{3}$, 1170
6242	Diphenyl acetic acid, methyl ester or Methyl diphenyl acetate $(C_6H_5)_2CHCO_2CH_3$	226.27	mcl pl (AcOEt) lf (dil al)		60			al, eth	$B9^{3}$, 3291
6243	Diphenyl acetic acid, 2,2′,4,4′-tetranitro,ethyl ester $[2,4\text{-}(NO_2)C_6H_3]_2CHCO_2C_2H_5$	420.29	lf (al) nd (bz, aa)		154				$B9^{3}$, 3315
6244	Diphenyl acetic acid, 2,2′,4,4′-tetranitro,methyl ester $[2,4\text{-}(NO_2)C_6H_3]_2CHCO_2CH_3$	406.27	lf (chl-MeOH)		159			chl	B9, 675
6245	Diphenyl acetonitrile $(C_6H_5)_2CHCN$	193.25	pr (eth) lf (dil al)	$181\text{-}4^{12}$	72-3			al, eth	$B9^{3}$, 3304
6246	Diphenyl acetyl bromide, α-bromo $(C_6H_5)_2CBrCOBr$	354.04	nd (lig)		65-6			al, eth, bz, chl	$B9^{1}$, 283
6247	Diphenyl acetyl chloride $(C_6H_5)_2CHCOCl$	230.69	pl (lig)	178^{15}	56-7				$B9^{3}$, 3300
6248	Diphenyl acetyl chloride, α-chloro $(C_6H_5)_2CClCOCl$	265.14	cr (lig)	180^{14}	50-1				$B9^{3}$, 3308
6249	Diphenyl amine $(C_6H_5)_2NH$	169.23	mcl lf (dil al)	$302, 179^{22}$	54-5	$1.160^{22/20}$		al, eth, ace, bz	$B12^{3}$, 284
6250	Diphenyl amine, 4,4′-dimethylamino $[4\text{-}(CH_3)_2NC_6H_4]_2NH$	255.36	tetr pl (CS_2)		119			eth	$B13^{2}$, 56

No.	Name, Synonyms, and Formula	Mol. wt.	Color, crystalline form, specific rotation and λ_{max} (log ε)	b.p. °C	m.p. °C	Density	n_D	Solubility	Ref.
6251	Diphenyl amine, 2,2′-dinitro $(2\text{-}NO_2C_6H_4)_2NH$	259.23	lf (al ace) ye cr (al aa)		169			al, ace, aa	B12^3, 1518
6252	Diphenyl amine, 2,4-dinitro $2,4\text{-}(NO_2)C_6H_3NHC_6H_4$	259.23	ye red nd (al)		157			al, ace, chl, py	B12^3, 1683
6253	Diphenyl amine, 2,4′-dinitro $2\text{-}NO_2C_6H_4NH(C_6H_4NO_2\text{-}4)$	259.23	red nd (aa)		222-3			chl, to	B12^3, 1587
6254	Diphenyl amine, 2,6-dinitro $2,6\text{-}(NO_2)_2C_6H_3NHC_6H_5$	259.23	og lf (al, aa)		107-8			al, aa	B12^3, 1705
6255	Diphenyl amine, 3,4′-dinitro $3\text{-}NO_2C_6H_4NH(C_6H_4NO_2\text{-}4)$	249.23	pa ye (chl aq py)		217			ace, chl	B12^3, 1587
6256	Diphenyl amine, 4,4′-dinitro $(4\text{-}NO_2C_6H_4)_2NH$	259.23	ye nd (al)		216			ace, aa	B12^3, 1587
6257	Diphenyl amine, 2,4-dinitro-4′-hydroxy $2,4\text{-}(NO_2)_2C_6H_3NH(C_6H_4OH\text{-}4)$	275.22	red lf		195-6				B13^3,1019
6258	Diphenyl amine, 2,4-dinitro-5-hydroxy $2,4\text{-}(NO_2)\text{-}5\text{-}HOC_6H_2NHC_6H_5$	275.22	dk ye nd (al)		139			al	B13^1, 138
6259	Diphenyl amine, 2,6-dinitro-2′-hydroxy $2,6\text{-}(NO_2)C_6H_3NH(C_6H_4OH\text{-}2)$	275.22	red vt nd (al)		191			al, eth, bz	B13, 365
6260	Diphenyl amine, 2,6-dinitro-3-hydroxy $2,6\text{-}(NO_2)_2\text{-}3\text{-}HOC_6H_2NHC_6H_5$	275.22	ye br nd (MeOH)		124-5			al	B13^2, 216
6261	Diphenyl amine, 2,2′,4,4′,6,6′-hexanitro or Dipicryi amine $[2,4,6\text{-}(NO_2)_3C_6H_2]_2NH$	439.22	pa ye pr (aa)		244d			py	B12^3, 1734
6262	Diphenyl amine, hydrobromide $(C_6H_5)_2NH.HBr$	250.14	pl (dil al)		230d			w	B12, 180
6263	Diphenyl amine, *N*-nitroso $(C_6H_5)_2N\text{-}NO$	198.23	ye pl (lig)		66.5			al, bz	B12^3, 1120
6264	Diphenyl amine, 2,2′,4,4′-tetrabromo $(2,4\text{-}Br_2C_6H_3)_2NH$	484.81	nd (chl or bz)		187.5			bz, chl	B12^3, 1472
6265	Diphenyl amine, 2,2′,4,4′-tetranitro $[2,4\text{-}(NO_2)_2C_6H_3]_2NH$	349.22	ye nd or cr (aa)		201			py	B12^3, 1684
6266	Diphenyl diselenide $(C_6H_5)_2Se_2$	312.13	ye nd		63-4	$1.557^{80/4}$		al, eth, xyl	B6^3, 1110
6267	Diphenyl disulfide $(C_6H_5)_2S_2$	218.33	nd (al) or rh	310, 192^{15}	61-2	$1.353^{20/4}$		al, eth, bz	B6^3, 1027
6268	1,2-Diphenylethane or Bibenzyl $C_6H_5CH_2CH_2C_6H_5$	182.27	nd (al)	285, $95\text{-}6^1$	52.2	$0.9583^{60/4}$	1.5478^{60}	al, eth	B5^4, 1868
6269	1,2-Diphenylethane, 4,4′-diamino $(4\text{-}H_2NC_6H_4)CH_2CH_2(C_6H_4NH_2\text{-}4)$	212.29	pl (w)	sub	135-6 sub			al	B13^3, 470
6270	1,2-Diphenylethane, 2,2′-dibromo $(2\text{-}BrC_6H_4)CH_2CH_2(C_6H_4Br\text{-}2)$	340.06	pl (al)	$138\text{-}40^{0.012}$	84.5			al	B5^4, 1875
6271	1,2-Diphenylethane, 4,4′-dibromo $(4\text{-}BrC_6H_4)CH_2CH_2(C_6H_4Br\text{-}4)$	340.06	pr (al)	ca 198^{10}	115				B5^4, 1875
6272	1,2-Diphenyl ethane, α,α′-dinitro*(dl)* $C_6H_5CH(NO_2)CH(NO_2)C_6H_5$	272.26	(al)pr(aa)		154-5			al, eth, ace, bz, chl	B5^4, 1878
6273	1,2-Diphenyl ethane, α,α′-dinitro-*(meso)* $C_6H_5CH(NO_2)CH(NO_2)C_6H_5$	272.26	nd (aa)		235-6			ace	B5^4, 1878
6274	1,2-Diphenyl ethane, 2,2′-dinitro $(2\text{-}O_2NC_6H_4)CH_2CH_2(C_6H_4NO_2\text{-}2)$	272.26	pr (aa)		127			eth, bz, aa	B5^4, 1878
6275	1,2-Diphenyl ethane, 4,4′-dinitro $(4\text{-}O_2NC_6H_4)CH_2CH_2(C_6H_4NO_2\text{-}4)$	272.26	yesh nd (al or bz)		180.5				B5^4,1878
6276	1,2-Diphenyl ethane, α-hydroxy-*(l)* $C_6H_5CH(OH)CH_2C_6H_5$	198.26	nd (eth-peth), $[\alpha]^{20/}{}_D$ -9.4 (w, c=10)		67	$1.0358^{20/4}$		al, eth	B6^2, 637
6277	1,2-Diphenyl ethane, α-hydroxy-*(dl)* $C_6H_5CH(OH)CH_2C_6H_5$	198.26	nd (bz-peth)	177^{15}	69			al, eth	B6^3, 3390
6278	1,2-Diphenyl ethane, α-hydroxy-*(d)* $C_6H_5CH(OH)CH_2C_6H_5$	198.26	nd (eth-peth or dil al), $[\alpha]^{25/}{}_D$ + 53 (al)	$167\text{-}70^{10}$	67-8	$1.0358^{20/4}$		al, eth	B6^3, 3390
6279	Diphenyl ether, 2,3′-dimethoxy $(2\text{-}CH_3OC_6H_4)O(C_6H_4OCH_3\text{-}3)$	230.26	pr(bz-peth)	326-9, 152^2	54			al, eth, bz	B6^3, 4318
6280	Diphenyl glyoxal or Benzil $C_6H_5COCOC_6H_5$	210.23	ye pr (al)	346-8d, 188^{12}	95-6	$1.084^{102/4}$		al, eth, ace, bz	B7^3, 3804
6281	Diphenyl glyoxime (anti) or Benzildioxime (anti) $C_6H_5C(=NOH)C(=NOH)C_6H_5$	240.26	lf (al or ace)		238d				B7^3, 3816

PHYSICAL CONSTANTS OF ORGANIC COMPOUNDS (Continued)

No.	Name, Synonyms, and Formula	Mol. wt.	Color, crystalline form, specific rotation and λ_{max} (log ε)	b.p. °C	m.p. °C	Density	n_D	Solubility	Ref.
6282	Diphenyl methane or Ditan $(C_6H_5)_2CH_2$	168.24	pr nd	264.3, 125.5^{10}	25.3	$1.0060^{20/4}$	1.5753^{20}	al, eth, chl	$B5^4$, 1841
6283	Diphenyl methane, α-amino $(C_6H_5)_2CHNH_2$	183.25	hex pl	304^{763}, 176^{23}	34	$1.0675^{20/0}$	1.5963	bz	$B12^3$, 3221
6284	Diphenyl methane, 3-amino $(3\text{-}H_2NC_6H_4)CH_2C_6H_5$	183.25	cr(lig)		46			lig	$B12^3$, 3215
6285	Diphenyl methane, 4-amino $(4\text{-}H_2NC_6H_4)CH_2C_6H_5$	183.25	mcl (lig)	300	34-5	1.038^{55}		al, eth, lig	$B12^3$, 3215
6286	Diphenyl methane, α-bromo $(C_6H_5)_2CHBr$	247.13	tcl (peth)	193^{26}, $111^{0.3}$	45			al, bz	$B5^4$, 1850
6287	Diphenyl methane, α-chloro or Benzydryl chloride $(C_6H_5)_2CHCl$	202.68	nd	173^{19}	20.5	$1.1398^{20/4}$	1.5959^{20}		$B5^4$, 1847
6288	Diphenyl methane, 4-chloro $(4\text{-}ClC_6H_4)CH_2C_6H_5$	202.68		298^{742}, $147\text{-}8^{8}$	7.5	$1.1247^{20/4}$		ace	$B5^4$, 1847
6289	Diphenyl methane, 4,4′-diamino or 4,4′-Diamino ditan $(4\text{-}H_2NC_6H_4)_2CH_2$	198.27	pl or nd (w) pl (bz)	$398\text{-}9^{768}$, 257^{18}	92-3			al, eth, bz	$B13^3$, 454
6290	Diphenyl methane, diazo $(C_6H_5)_2CN_2$	194.24	bl-red nd (peth)	exp	30-2			al, eth, ace	$B7^3$, 2068
6291	Diphenyl methane, α,α-dichloro or Benzophenone dichloride $(C_6H_5)_2CCl_2$	237.13		305d, 190^{21}		1.235^{18}		eth, bz	$B5^4$, 1848
6292	Diphenyl methane, 4,4′-dichloro $(4\text{-}ClC_6H_4)_2CH_2$	237.13		$186\text{-}90^{18}$	55-6	1.365^{17}		al	$B5^4$, 1848
6293	Diphenyl methane, 5,5′-dichloro-2,2′-dihydroxy or Dichlorophene $(5\text{-}Cl\text{-}2\text{-}HOC_6H_3)_2CH_2$	269.13	cr (bz peth)		177-8			al, ace	$B6^3$, 5406
6294	Diphenyl methane, 2,4′-dihydroxy $(2\text{-}HOC_6H_4)CH_2(C_6H_4OH\text{-}4)$	200.24	nd (dil al, bz or w)		119-20			al, eth	$B6^3$, 5409
6295	Diphenyl methane, 3,3′-dihydroxy or *m,m′*-Methylene diphenol $[3\text{-}HOC_6H_4]_2CH_2$	200.24	nd (dil aa)	$230\text{-}40^{3}$	102-3			al, eth, aa	$B6^3$, 5411
6296	Diphenyl methane, 4,4′-dihydroxy or *p,p′*-Methylene diphenol $(4\text{-}HOC_6H_4)_2CH_2$	200.24	lf or nd (w)	sub	162-3			al, eth, chl	$B6^3$, 5412
6297	Diphenyl methane, 4,4′-dimethylamino $[4\text{-}(CH_3)_2NC_6H_4]_2CH_2$	254.38	pl or ta (al lig)	390d, $182\text{-}5^{3}$	91-2			eth, bz	$B13^3$, 454
6298	Diphenyl methane, α,α-dinitro $(C_6H_5)_2C(NO_2)_2$	258.23	pl (dil al)		79-80			al, eth, bz, chl	$B5^3$, 1797
6299	Diphenyl methane, 2,2′-dinitro $(2\text{-}O_2NC_6H_4)_2CH_2$	258.23	cr		83.5			al, eth	$B5^3$, 1796
6300	Diphenyl methane, 2,2′-dinitro,4,4′-diamino $[2\text{-}NO_2\text{-}4\text{-}H_2NC_6H_3]_2CH_2$	288.26	og pl (al)		205			al, aa	$B13^2$, 113
6301	Diphenyl methane, 2,4′-dinitro $(2\text{-}O_2NC_6H_4)CH_2(C_6H_4NO_2\text{-}4)$	258.23	ye mcl pr(bz)		118				$B5^3$, 1797
6302	Diphenyl methane, 3,3′-dinitro $(3\text{-}O_2NC_6H_4)_2CH_2$	258.23	lf(aa)		175.5			al, bz, aa	$B5^3$, 1797
6303	Diphenyl methane, 3,3′-dinitro-4,4′-diamino $[3\text{-}NO_2\text{-}4\text{-}NH_2C_6H_3]_2CH_2$	288.26	red nd		232-3				$B13^3$, 113
6304	Diphenyl methane, 3,4′-dinitro $(3\text{-}O_2NC_6H_4)CH_2(C_6H_4NO_2\text{-}4)$	258.23	nd(al)		103-4				$B5^3$, 1797
6305	Diphenyl methane, 4,4′-dinitro $(4\text{-}O_2NC_6H_4)_2CH_2$	258.23	nd (bz, peth, aa)		188			bz, aa	$B5^3$, 1797
6306	Diphenyl methane, 4-ethyl $(4\text{-}C_2H_5\text{-}C_6H_4)CH_2C_6H_5$	196.29		297 $85^{0.2}$	−24	0.9777^{20}	1.5618^{20}	al, eth, chl	$B5^3$, 1870
6307	Diphenyl methane, 4-hydroxy or 4-Benzylphenol $(4\text{-}HOC_6H_4)CH_2C_6H_5$	184.24	nd or pl(al)	325-30, $198\text{-}200^{10}$	84			al, eth, bz, chl, aa	$B6^3$, 3357
6308	Diphenyl methane, 2-methyl or Phenyl-2-tolyl methane $(2\text{-}CH_3H_5)CH_2C_6H_5$	182.27		280.5	6.6		$1.5763^{20/}{}_D$		$B5^1$, 1855
6309	Diphenyl methane, 3-methyl or Phenyl-3-tolyl methane $(3\text{-}CH_3C_6H_4)CH_2C_6H_5$	182.27		279.2, $120^{0.2}$	−28	$0.9913^{20/4}$	1.5712^{20}	al, eth, bz, aa, chl	$B5^4$, 1858
6310	Diphenyl methane, 4-methyl $(4\text{-}CH_3C_6H_4)CH_2C_6H_5$	182.27		286, $114\text{-}5^{3}$	−30	$0.9976^{20/4}$	1.5712^{20}	al, eth, bz, chl, aa	$B5^4$, 1860
6311	Diphenyl methane, 4-propyl $(4\text{-}C_3H_7C_6H_4)CH_2C_6H_5$	210.32		$152\text{-}5^{10}$		$0.9739^{18/4}$	1.5552^{20}		$B5^4$, 1950
6312	Diphenyl methane, 2,2′,4,4′-tetramethyl $[2,4\text{-}(CH_3)_2C_6H_3]_2CH_2$	224.35		$140\text{-}2^{3}$			1.5635^{15}		$B5^4$, 1971
6313	Diphenyl methane, 2,2′,4,4′-tetranitro $[2,4\text{-}(NO_2)_2C_6H_3]_2CH_2$	348.23	ye pr (aa)		181				$B5^3$, 1798

No.	Name, Synonyms, and Formula	Mol. wt.	Color, crystalline form, specific rotation and λ_{max} (log ε)	b.p. °C	m.p. °C	Density	n_D	Solubility	Ref.
6314	Diphenyl methanol or Benzhydrol $(C_6H_5)_2CHOH$	184.24	nd(lig)	297-8^{748}, 180^{20}	69			al, eth, chl	B6^3, 3364
6315	Diphenyl methanol, 4,4′-dimethyl or di-(4-tolyl)methanol $(4\text{-}CH_3C_6H_4)_2CHOH$	212.29	nd(al)		69			al, eth, ace, chl, aa	B6^3, 3422
6316	Di-(α-Phenylethyl) amine $[CH_3CH(C_6H_5)]_2NH$	225.23	ye	295-8, 190^{10}		1.018^{13}	1.573		B12^2, 589
6317	Di-(β-Phenylethyl)amine $(C_6H_5CH_2CH_2)_2NH$	225.33		335-7^{603}, 190^{15}	28-30		1.5550^{25}	al, eth	B12^2, 593
6318	Diphenyl ethyl amine $(C_6H_5)_2NC_2H_5$	197.28		295-6, 148^{11}		1.0396$^{20/20}$	1.6095^{20}	al, eth	B12^3, 291
6319	Diphenyl methyl amine $(C_6H_5)_2NCH_3$	183.25		293-4, 145^{10}	−7.5	1.0476$^{20/4}$	1.6193^{20}		B12^3, 290
6320	Diphenyl propanetrione or Diphenyl triketone $C_6H_5COCOCOC_6H_5$	238.24	ye nd(lig)	289^{175}, 248^{60}	68-70			eth	B7^3, 4620
6321	Diphenyl selenide $(C_6H_5)_2Se$	233.17	ye nd(bz)	301-2, 126-7^{5}	2.5	1.351$^{20/4}$	1.5500^{20}	al, eth, bz, xyl	B6^4, 1779
6322	Diphenyl selenonium dichloride $(C_6H_5)_2SeCl_2$	304.08	pa ye pr(xyl, ace)nd(al)		183			w, al, ace	B6^3, 1107
6323	Diphenyl sulfide $(C_6H_5)_2SO_2$	218.27	mcl pr (bz) pl (al) nd (w)	379, 232^{18}	128-9	1.252$^{20/4}$		eth, bz	B6^4, 1490
6324	Diphenyl sulfoxide $(C_6H_5)_2SO$	202.27	pr(lig)	340d, 210^{15}	70.5			al, eth, bz, aa	B6^4, 1489
6325	Diphosphine, tetrakis(trifluoromethyl) $(F_3C)_2PP(CF_3)_2$	337.97		84		>1.0			B3^4, 265
6326	Dipicryl amine $[2,4,6\text{-}(NO_2)_3C_6H_2]_2NH$	439.21	pa ye pr (aa)		244d			py	B12^2, 422
6327	Diploicin $C_{16}H_{10}Cl_4O_5$	424.06	(bz)		232				B19^4, 2347
6328	Dipropanol amine, *N*-methyl $(HOCH_2CH_2CH_2)_2NCH_3$	147.22	hyg	164-5^{12}				w, al, bz	B4^4, 1644
6329	Di-*iso*-Propanol amine $[CH_3CH(OH)CH_2]_2NH$	133.19	cr	249-50^{745}, 151^{23}	44-5			w, al	B4^3, 761
6330	Di-*iso*-Propanol amine, *N*-2-hydroxy ethyl $[CH_3CH(OH)CH_2]_2NCH_2CH_2OH$	177.24		155-6^{1}		1.0458$^{20/4}$	1.4708^{20}	w, al, ace	B4^3, 764
6331	Dipropyl amine $(C_3H_7)_2NH$	101.19		109-10	−39.6	0.7400$^{20/4}$	1.4050^{20}	w, al, eth, ace, bz	B4^4, 469
6332	Dipropyl amine, *N*-nitroso $(C_3H_7)_2N\text{-}NO$	130.19	gold	206, 89^{13}		0.9163$^{20/4}$	1.4437^{20}	al, eth	B4^3, 264
6333	Di-*m*-Propyl-*iso*-butyl amine, perfluoro $i\text{-}C_4F_9N(C_3F_7)_2$	571.08		146-8		1.84$^{25/4}$	1.283^{25}	al, eth	B2^4, 858
6334	Di-*iso*-Propyl amine $(i\text{-}C_3H_7)_2NH$	101.19		84	−61	0.7169$^{20/4}$	1.3924^{20}	al, eth, ace, bz	B4^4, 510
6335	Di-*iso*-Propyl amine, *N*-nitroso $[(CH_3)_2CH]_2N\text{-}NO$	130.19	cr(eth w)	194.5, 76-81^{14}	48	0.9422$^{20/4}$		al, eth, bz	B4^3, 281
6336	Dipropyl disulfide $(C_3H_7)_2S_2$	150.30		193.5		0.9599$^{20/4}$	1.4981^{20}		B1^4, 1454
6337	Diisopropyl disulfide $(i\text{-}C_3H_7)_2S_2$	150.30		177.2, 56.8^{10}		0.9435$^{20/4}$	1.4916^{20}		B1^4, 1503
6338	Dipropylene glycol $(CH_3CH(OH)CH_2)_2O$	134.18		229-32		1.0224$^{20/20}$		w, al	B1^4, 2473
6339	Dipropyl sulfide $(C_3H_7)_2SO_2$	150.24	sc		29-30	1.0278$^{50/4}$	1.4456^{30}	al, eth	B1^4, 1453
6340	Di-isopropyl sulfone $(i\text{-}C_3H_7)_2SO_2$	150.24	eth		36			w, eth	B1^4, 1502
6341	Diisopropyl ketone, cyanohydrin $(i\text{-}C_3H_7)_2C(OH)CN$	141.21	rh(eth or peth)	111^{18}	59			al, eth, ace, bz	B3^2, 239
6342	Di-(α-Pyrryl)methane $[2\text{-}C_4H_3NH]_2CH_2$	146.19	lf or nd(al)	163-7^{12}	73			al, eth, bz	B23, 167
6343	Disilane, 1,2-dichloro-1,1,2,2-tetramethyl $(CH_3)_2SiClSiCl(CH_3)_2$	187.22		49-50^{18}		1.010^{20}	1.4548^{20}		KHOC, 366
6344	Disilane, 1,2-difluoro-1,1,2,2-tetramethyl $(CH_3)_2SiFSiF(CH_3)_2$	154.31		92-9		0.9120$^{20/4}$	1.3837^{20}		BOSC2^2, 183
6345	Disilane, 1,2-diphenyl-1,1,2,2-tetramethyl $C_6H_5Si(CH_3)_2Si(CH_3)C_6H_5$	270.52		111^{1}, 73$^{0.1}$	34.5	0.9892$^{20/4}$	1.5161^{20}		BOSC2^2, 285
6346	Disilane, hexamethyl $(CH_3)_3SiSi(CH_3)_3$	116.31		112.5-4.3	12.8-14	0.7247$^{22.5/4}$	1.4229^{20}		KHOC, 597

No.	Name, Synonyms, and Formula	Mol. wt.	Color, crystalline form, specific rotation and λ_{max} (log ε)	b.p. °C	m.p. °C	Density	n_D	Solubility	Ref.
6347	Disiloxane, 1,3-dichlo-1,1,3,3,-tetramethyl $(CH_3)_2SiClOSiCl(CH_3)_2$	203.22		138	−37.5	$1.038^{20/4}$			KHOC, 593
6348	Disiloxane, dichloromethyl-pentamethyl $Cl_2CHSi(CH_3)_2OSi(CH_3)_3$	231.27		200-5		1.046^{20}	1.4382^{20}		KHOC, 596
6349	Disiloxane, 1,3-diethenyl-1,1,3,3-tetramethyl $CH_2{=}CHSi(CH_3)_2\text{-}O\text{-}Si(CH_3)_2CH{=}CH_2$	186.40		39	−99.7	0.811^{20}	1.4123^{20}		KHOC, 602
6350	Disiloxane, 1,3-dimethoxy-1,1,3,3-tetramethyl $CH_3OSi(CH_3)_2O\text{-}Si(CH_3)_2OCH_3$	194.38		139		$0.9048^{20/4}$	1.3835^{20}		$BOSC2^2$, 197
6351	Disiloxane, 1,3-diphenyl-1,1,3,3-tetramethyl $C_6H_5Si(CH_3)_2\text{-}O\text{-}Si(CH_3)_2C_6H_5$	286.53		110^2		$0.9763^{20/4}$	$1.5176^{2}3^{0}$		Am79, 1437
6352	Disiloxane, hexaethyl $[(C_2H_5)_3Si]_2O$	246.54		233^{756}, 129^{30}		$0.8590^{0/0}$	1.4340^{20}		B4, 627
6353	Disiloxane, hexakis (2-ethylbutoxy) $([C_2H_5)_2CHCH_2O]_3Si)_2O$	679.18		220^1	>-54	$0.9219^{20/20}$	1.4330^{20}	eth, bz	C48, 3761
6354	Disiloxane, hexakis(2-ethyhexoxy) $([CH_3(CH_2)_3CH(C_2H_5)CH_2O]_3Si)_2O$	847.50		$253^{0.9}$		$0.9044^{20/20}$	1.4402^{20}	eth, bz	C48, 3761
6355	Disiloxane, hexamethyl $(CH_3)_3SiOSi(CH_3)_3$	162.38		99.5-100	-66	0.7638^{20}	$1.3774^{20/}{}_D$		KHOC, 598
6356	Disiloxane, 1,1,3,3-tetramethyl $(CH_3)_2SiH\text{-}O\text{-}SiH(CH_3)_2$	134.33		$70.5\text{-}71^{731}$		$0.7572^{20/4}$	$1.3700^{20/}{}_D$		$BOSC2^2$, 184
6357	1,2-Dithiane $C_4H_8S_2$	120.23		89^{14}	32-3		1.5981^{25}	eth, bz, chl	$B19^4$, 8
6358	1,4-Dithiane or Diethylene disulfide $C_4H_8S_2$	120.23	mcl pr	199-200	111-2			al, eth, aa	$B19^4$, 35
6359	1,3,5-Dithiazine, 4,5-dihydro-5-methyl or Methylthioformaldine $C_4H_9NS_2$	135.24	nd(eth)	185d	65			al, eth, aa	$B27^2$, 524
6360	1,4-Dithiine, 2,5-diphenyl $C_{16}H_{12}S_2$	268.39	ye pr (al)		118-9			al, bz	$B19^4$, 404
6361	1,3-Dithiolane or Trimethylene-1,3-disulfide $C_3H_6S_2$	106.20		175	−50	1.259^{17}	1.5975^{15}	al, eth, xyl	$B19^4$, 6
6362	di(2-Thienyl)ketone or Thienone $C_4H_3SC0C_4H_3S$	194.27	nd(al)	326	90			eth, ace	$B19^4$, 1745
6363	Dithizone or Diphenyl thiocarbazone $C_6H_5N{=}NCSNHNHC_6H_5$	256.33	bl-bk(chl-al)		165-9d			chl	$B16^3$, 19
6364	Dithioacetic acid or Thiolo-thionoacetic acid CH_3CS_2H	92.17	ye-red oil	66^{85}, 37^{15}		1.24^{20}		w, al, eth, ace, bz	$B2^4$, 572
6365	Di-2-tolylamine $(2\text{-}CH_3C_6H_4)_2NH$	197.28	bl flr, wh cr	312^{727}, 192^{23}	52-3				$B12^2$, 437
6366	Di-3-tolylamine $(3\text{-}CH_3C_6H_4)_2NH$	197.28	pa ye (peth)	319-20	53			al, eth, peth	$B12^2$, 467
6367	Di-4-tolylamine $(4\text{-}CH_3C_6H_4)_2NH$	197.28	nd(peth)	330.5	79			eth, peth	$B12^3$, 2033
6368	3,4'-Ditolylamine hydrochloride $(3\text{-}CH_3C_6H_4)NH(C_6H_4CH_3\text{-}4).HCl$	233.74	cr		202-3			al, ace, bz, chl	$B12^1$, 414
6369	Di-2-tolyl disulfide $(2\text{-}CH_3C_6H_4)_2S_2$	246.39	lf(al)		38-9			al, eth, ace	$B6^4$, 2027
6370	Di-4-tolyl disulfide $(4\text{-}CH_3C_6H_4)_2S_2$	246.39	nd or lf(al)	$210\text{-}15^{20}$	47-8	1.114^{51}		al, eth, ace	$B6^4$, 2206
6371	2,2'-Ditolylsulfide or *o*-Totylsulfide $(2\text{-}CH_3C_6H_4)_2S$	214.33	pl(al)	285, 174^{15}	64			al, eth, chl	$B6^4$, 2019
6372	3,4'-Ditolylsulfide $(3\text{-}CH_3C_6H_4)S(C_6CH_3\text{-}4)$	214.33	nd(al)	179^{11}	28			al, eth	$B6^4$, 2173
6373	4,4'-Ditolylsulfide or *p*-Totylsulfide $(4\text{-}CH_3C_6H_4)_2S$	214.33	nd(al)	>300, 179^{11}	57.3			al, eth, ace, bz	$B6^4$, 2173
6374	2,4'-Ditolylsulfide $(2\text{-}CH_3C_6H_4)S(C_6H_4CH_3\text{-}4)$	214.33		173^{11}		$1.0774^{15/4}$		al, eth	$B6^4$, 2172
6375	2,2'-Ditolylsulfone or *o*-Tolylsulfone $(2\text{-}CH_3C_6H_4)_2SO_2$	246.32	nd(al)		134-5			al, eth, bz, chl	$B6^4$, 2020
6376	3,4'-Ditolylsulfide $(3\text{-}CH_3C_6H_4)SO_2(C_6H_4CH_3\text{-}4)$	246.32			116				$B6^3$, 1404
6377	4,4'-Ditolylsulfide $(4\text{-}CH_3C_6H_4)_2SO_2$	246.32	pr(bz)nd(w, al)pl(al)	405^{714}	159			bz, chl	$B6^4$, 2174
6378	4,4'-Ditolylsulfoxide or bis-(4-tolyl)sulfoxide $(4\text{-}CH_3C_6H_4)_2SO$	230.32	cr(lig)		94			al, eth, bz, chl, aa	$B6^4$, 2173
6379	Diurea or Dicarbamide $H_2NCONHNHCONH_2$	116.08	pr(w)		270				$B3^4$, 236

PHYSICAL CONSTANTS OF ORGANIC COMPOUNDS (Continued)

No.	Name, Synonyms, and Formula	Mol. wt.	Color, crystalline form, specific rotation and λ_{max} (log ε)	b.p. °C	m.p. °C	Density	n_D	Solubility	Ref.
6380	Djenkoic acid or β,β′-Methylene dithio dialanine $[HO_2CCH(NH_2)CHS]_2CH_2$	254.32	nd(w)	....	300-50d	....	....	w	B4^3, 1591
6381	Docosane $CH_3(CH_2)_{20}CH_3$	310.61	pl(to) cr(eth)	368.6, 213^{10}	44.4	0.7944$^{20/4}$	1.4455^{20}	al, eth, chl	B1^4, 572
6382	Docosanoic acid or Behenic acid $CH_3(CH_2)_{20}CO_2H$	340.59	nd	306^{60}	80	0.8223^{90}	1.4270^{100}	....	B2^4, 1290
6383	Docosanoic acid, ethyl ester or Ethyl behenate $C_{21}H_{43}CO_2C_2H_5$	368.64	nd (al) cr (ace)	240-2^{10}	50	0.8820^{51}	....	al, eth	B2^4, 1292
6384	Docosanoic acid, methyl ester or Methyl behenate $C_{21}H_{43}CO_2CH_3$	354.62	nd(ace)	224-5^{12}	54	....	1.4339^{60}	al, eth	B2^4, 1291
6385	1-Docosanol or Docosyl alcohol $CH_3(CH_2)_{20}CH_2OH$	326.61	(ace chl)	180$^{0.22}$	71(87)	....	....	al, chl	B1^4, 1906
6386	4,7,11-Docosatriene-18-ynoic acid or Clupanodonic acid $C_{22}H_{34}O_2$	330.51	pa ye	236^5	<−78	0.9290^{20}	1.4868^{20}	eth	B1^3, 1528
6387	13-Docosenoic acid-*(cis)* or Erucic acid $CH_3(CH_2)_7CH{=}CH(CH_2)_{11}CO_2H$	338.57	nd(al)	265^{15}	33-4	0.860$^{55/4}$	1.4758^{20}	al, eth	B2^4, 1676
6388	13-Docosenoic acid-*(trans)* or Brassidic acid $CH_3(CH_2)_7CH{=}CH(CH_2)_{11}CO_2H$	338.57	pl(al)	282^{30}	61.5	0.8585$^{57/4}$	1.4472^{64}	....	B2^4, 1677
6389	13-Docosenoic acid, anhydride-*(trans)* or Brassidic anhydride $(C_{21}H_{41}CO)_2O$	659.12	nd (al) pl (eth)	....	64	0.835$^{70/4}$	1.4366^{100}	eth, ace, peth	B2^2, 448
6390	13-Docosynoic acid or Behenolic acid $CH_3(CH_2)_7C{\equiv}C(CH_2)_{11}CO_2H$	336.56	mcl pr or nd(al)	....	59.5	....	....	al, eth, chl	B2^4, 1764
6391	1,11-Dodecadiene $CH_2{=}CH(CH_2)_8CH{=}CH_2$	166.31	....	208-9^{758}	....	0.7702$^{20/4}$	1.4400^{20}	....	B1^4, 1067
6392	Dodecanal or Lauraldehyde $CH_3(CH_2)_{10}CHO$	184.32	lf	185^{100}, 100$^{3.5}$	44.5	0.8352$^{15/4}$	1.435^{22}	al, eth	B1^4, 3380
6393	Dodecane $C_{12}H_{26}$	170.34	....	216.3, 91.5^{10}	−9.6	0.7487$^{20/4}$	1.4216^{20}	al, eth, ace, chl	B1^4, 498
6394	Dodecane, 1-amino or Lauryl amine $CH_3(CH_2)_{10}CH_2NH_2$	185.35	....	259, 126.5^{10}	28.3	0.8015$^{20/4}$	1.4421^{20}	al, eth, bz, chl	B4^4, 1794
6395	Dodecane, 1-bromo or Lauryl bromide $CH_3(CH_2)_{10}CH_2Br$	249.23	....	276, 139^{10}	−9.5	1.0399$^{20/4}$	1.4583^{20}	al, eth, ace	B1^4, 502
6396	Dodecane, 1-bromo-12-fluoro $F(CH_2)_{12}Br$	267.22	....	85.8$^{0.15}$	....	....	1.4524^{25}	al, eth, ace	B1^4, 502
6397	Dodecane, 1-chloro or Lauryl chloride $CH(CH_2)_{10}CH_2Br$	204.78	....	260, 126.4^{10}	fr−9.3	0.8687$^{20/4}$	1.4433^{20}	al, ace, bz, lig	B1^4, 501
6398	Dodecane, 1,12-dibromo $Br(CH_2)_{12}Br$	328.13	nd(aa al)	215^{15}	41	....	....	al, eth, chl, aa	B1^4, 503
6399	Dodecane, 1-iodo or Lauryl iodide $CH_3(CH_2)_{10}CH_2I$	296.24	....	298.2, 153^{10}	0.3	1.1999$^{20/4}$	1.4840^{20}	al, eth, ace, chl	B1^4, 503
6400	6-Dodecanol $C_5H_{11}CH(OH)C_6H_{13}$	186.34	peth	119^9	30	....	....	al, eth	B1^3, 1794
6401	2-Dodecanone or *n*-Decyl methyl Ketone $CH_3COC_{10}H_{21}$	184.32	....	246-7, 144^{11}	21	0.8198$^{20/4}$	1.4330^{20}	al, eth, ace	B1^4, 3382
6402	6-Dodecanone $C_6H_{13}COC_5H_{11}$	184.32	....	112^9	9	....	1.4302$^{20/}{}_D$	....	B1^4, 3383
6403	1-Dodecene $CH_3(CH_2)_9CH{=}CH_2$	168.32	....	213.4, 88.7^{10}	−35.2	0.7584$^{20/4}$	1.4300^{20}	al, eth, ace, bz	B1^4, 914
6404	Dodecanedioic acid $HO_2C(CH_2)_{10}CO_2H$ $HO_2C(CH_2)_{10}CO_2H$	230.30	....	....	128	....	....	....	B2^4, 2126
6405	Dodecanedioic acid, dimethyl ester $CH_3O_2C(CH_2)_{10}CO_2CH_3$	258.36	pr	167-9^9	31.3	....	....	....	B2^4, 2126
6406	1-Dodecanethiol or Lauryl mercapton $CH_3(CH_2)_{10}CH_2SH$	202.40	....	142-5^{15}	....	0.8450$^{20/20}$	1.4589^{20}	al, eth	B1^4, 1851
6407	Dodecanoic acid or Lauric acid $CH_3(CH_2)_{10}CO_2H$	200.32	nd(al)	131^1	44	0.8679$^{50/4}$	1.4304^{50}	al, eth, ace, bz, peth	B2^4, 1082
6408	Dodecanoic acid, 2-bromo $CH_3(CH_2)_9CHBrCO_2H$	279.22	pl	157-9^2	32	1.1474^{74}	1.4585^{24}	al, eth, bz, chl, lig	B2^4, 1106
6409	1-Dodecanol or Lauryl alcohol $CH_3(CH_2)_{10}CH_2OH$	186.34	lf(dil al)	255-9, 150^{20}	26	0.8309$^{24/4}$	....	al, eth	B1^4, 1844
6410	2-Dodecanol $CH_3(CH_2)_9CH(OH)CH_3$	186.34	....	252	19	0.8286^{20}	1.4400$^{20/}{}_D$	....	B1^3, 1793
6411	2-Dodecenedioic acid-*(cis)* or Traumatic acid $HO_2CCH{=}CH(CH_2)_8CO_2H$	228.29	(al, ace)	....	67-8	....	....	al, eth, bz, chl	B2^3, 1979
6412	1-Dodecene-3-yne $CH_3(CH_2)_7C{\equiv}C\text{-}CH{=}CH_2$	164.29	....	78^4	....	0.7858$^{25/4}$	1.4510^{25}	....	B1^4, 1112

No.	Name, Synonyms, and Formula	Mol. wt.	Color, crystalline form, specific rotation and λ_{max} (log ε)	b.p. °C	m.p. °C	Density	n_D	Solubility	Ref.
6413	2-Dodenedioic acid-*(trans)* $HO_2CCH{=}CH(CH_2)_8CO_2H$	228.29	(al, ace)		165-6			al, eth, chl	B2[4], 2279
6414	Dodecyl sulfate $(C_{12}H_{25}O)_2SO_2$	434.72			48.5				B1[4], 1849
6415	Dodecyltrimethyl ammonium chloride $C_{12}H_{25}N^+(CH_3)_3Cl^-$	263.89			246d			w, al, ace, chl	B4[4], 798
6416	1-Dodecyne $CH_3(CH_2)_9C{\equiv}CH$	166.31		215, 89^{10}	−19	$0.7788^{20/4}$	1.4340^{20}		B1[4], 1066
6417	2-Dodecyne $CH_3(CH_2))_8C{\equiv}CCH_3$	166.31		105^{15}	−9	$0.7917^{15/4}$	1.4828^{20}		B1, 261
6418	3-Dodecyne $CH_3(CH_2)_7C{\equiv}CCH_2CH_3$	166.31		95^{12}		$0.7871^{20/4}$	1.4442^{20}	eth, ace	B1[3], 1025
6419	6-Dodecyne $CH_3(CH_2)_4C{\equiv}C(CH_2)_4CH_3$	166.31		209^{745}, 100^{14}		$0.7871^{20/4}$	1.4442^{20}	al, eth, ace	B1[4], 1067
6420	Dotriacontane or Dicetyl $CH_3(CH_2)_{30}CH_3$	450.88	pl (bz, chl, aa, eth)	467, 292.7^{10}	69.7	$0.8124^{20/4}$	1.4550^{20}	eth, bz	B1[4], 595
6421	1-Dotriacontanol $CH_3(CH_2)_{30}CH_2OH$	466.88	pl(bz)	sub 200-50^{1}	89.4				B1[4], 1919
6422	Durene or 1,2,4,5-tetramethyl benzene $1,2,4,5\text{-}(CH_3)_4C_6H_2$	134.22		196.8, 73.5^{10}	79.2	$0.8380^{81/4}$, $0.8875^{20/4}$	1.4790^{81}, 1.5116^{20}	**al, eth, ace, bz**	B5[4], 1076

No.	Name, Synonyms, and Formula	Mol. wt.	Color, crystalline form, specific rotation and λ_{max} (log ε)	b.p. °C	m.p. °C	Density	n_D	Solubility	Ref.
6423	Ecgonidine-*(l)* or Anhydroecgonine $C_9H_{13}NO_2$	167.21	cr(MeOH-eth), $[\alpha]^{14/}_D$ -84.6 (w, p=1.7)		225d			al	B22^{4}, 284
6424	Ecgonine-*(dl)* $C_9H_{15}NO_3$	185.22	pl(w + 3)		203			w, al	B22^{2}, 156
6425	Ecgonine, Ecqonine $C_9H_{15}NO_3$	185.23	mcl pr, $[\alpha]^{15/}_D$ - 45.5 (w, c=5)		205			w, al	B22^{4}, 2097
6426	Ecgonine hydrate-*(l)* $C_9H_{15}NO_3 \cdot H_2O$	203.24	mcl pr (al) eff 120-30		198				B22, 196
6427	Ecgonine benzoate-*(l)* or O-Benzoyl ecqonine $C_{16}H_{19}NO_4$	289.34	nd(w), $[\alpha]^{14/}_D$ -63.5 (w, p=1.7)		195			al, bz	B22^{4}, 2098
6428	Ecgonine benzoate, ethyl ester (l) or Homococaine $C_{18}H_{23}NO_4$	317.38	pr(eth)		109			al, eth	B22, 202
6429	Ecognine benzoate, tetrahydrate (l) $C_{16}H_{19}NO_4 \cdot 4H_2O$	361.39	pr(w)		92			al, bz	B22, 197
6430	Ecgonine hydrochloride (l) $C_9H_{15}NO_3.HCl$	221.68	rh(al), $[\alpha]^{25/}_D$ -59 (w, c=10)		246			w, al	B22^{4}, 2098
6431	Echinochrome A $C_{12}H_{10}O_7$	266.21	dk red nd (to)	sub 120$^{.0001}$	220d			al, eth, ace	B8^{3}, 4365
6432	Echinochrome A, 3,6,7-trimethyl ether $C_{15}H_{16}O_7$	308.29	dk red lf (aq diox)		133 vac			al	B8^{3}, 4361
6433	Echinopsine or N-Methyl-α-quinoline $C_{10}H_9NO$	159.19	α-nd(bz) β-cr(al)		α152 β135			w, al, bz, chl	B21^{4}, 3722
6434	Echitamidine $C_{20}H_{26}N_2O_3$	342.44	pl(eth), $[\alpha]^{16/}_D$ -515 (al)		244d			w, al	J1932, 2628
6435	Echitamine or Ditaine $C_{22}H_{28}N_2O_4 \cdot 4H_2O$	456.54	pr (al + 4w) eff −105 (-3w), $[\alpha]^{20/}_D$ -29 (al)		206 (+ 1w)			w, al, eth	J1925, 1640
6436	Echitamine, hydrochloride $C_{22}H_{28}N_2O_4.HCl$	420.94	nd(w), $[\alpha]^{15/}_D$ -58 (w, c=1)		295-300d			al	J1925, 1640
6437	Echitin $C_{32}H_{52}O_2$	468.76	lf, $[\alpha]_D$ + 73 (eth)		170			al, eth, ace, bz	M, 397
6438	Egonol $C_{19}H_{18}O_5$	326.35	pl(BuOH)	228-30$^{0.15}$	118			chl	B19^{4}, 4882
6439	Eicosane or Didecyl $C_{20}H_{42}$	282.55	lf(al)	343, 195.7^{10}	36.8	$0.7886^{20/4}$	1.4425^{20}	eth, ace, bz, peth	B1^{4}, 563
6440	Eicosane, 1-cyclohexyl $C_{26}H_{52}$	364.70		422	48.5	0.8318^{20}	$1.4622^{20/}_D$		B5^{4}, 198
6441	Eicosane, 9-octyl $C_{11}H_{23}CH(C_8H_{15})_2$	394.77		257^{10}, 199$^{0.5}$	0.5	$0.8075^{20/4}$	$1.4515^{20/}_D$		B1^{4}, 590
6442	Eicosane, 1-phenyl $CH_3(CH_2)_{19}C_6H_5$	358.65		212^{1}	42.3	$0.8235^{60/4}$	$1.4725^{40/}_D$		B5^{4}, 1222
6443	Eicosane, 2-phenyl $CH_3(CH_2)_{17}CH(CH_3)C_6H_5$	358.65		204.5^{1}	29.0	$0.8547^{20/4}$	1.4795^{20}		B5^{4}, 1222
6444	Eicosone, 3-phenyl $CH_3(CH_2)_{16}CH(C_2H_5)C_6H_5$	358.65		202.0^{1}	29.3	$0.8546^{20/4}$	$1.4796^{20/}_D$		B5^{4}, 1222
6445	Eicosane, 4-phenyl $CH_3(CH_2)_{15}CH(C_3H_7)C_6H_5$	358.65		199.0^{1}	31.4	$0.8546^{20/4}$	$1.4794^{20/}_D$		B5^{4}, 1223
6446	Eicosane, 5-phenyl $CH_3(CH_2)_{14}CH(C_4H_9)C_6H_5$	358.65		197^{1}	30.2	$0.8549^{20/4}$	$1.4796^{20/}_D$		B5^{4}, 1223
6447	Eicosane, 9-phenyl $CH_3(CH_2)_7CH(C_6H_5)(CH_2)_{10}CH_3$	358.65		196^{10}	17.9	$0.8534^{20/4}$	1.4790^{20}		B5^{4}, 1223
6448	Eicosanedioic acid or Octadecane dicarboxylic acid $HO_2C(CH_2)_{18}CO_2H$	342.52	cr(bz or al)	233-4^{2}	125-6			eth	B2^{4}, 2185
6449	Eicosanedioic acid, diethyl ester or Diethyl eicosanedioate $C_2H_5O_2C(CH_2)_{18}CO_2C_2H_5$	398.63		240^{12}	54-5			al, eth	B2^{3}, 1881
6450	Eicosanoic acid or Arachidic acid $CH_3(CH_2)_{18}CO_2H$	312.54	pl(al)	328d, 203-5^{1}	77	$0.8240^{100/4}$	1.425^{100}	eth, chl	B2^{4}, 1275

No.	Name, Synonyms, and Formula	Mol. wt.	Color, crystalline form, specific rotation and λ_{max} (log ε)	b.p. °C	m.p. °C	Density	n_D	Solubility	Ref.
6451	Eicosanoic acid, ethyl ester or Ethyleicosonoate $C_{19}H_{39}CO_2C_2H_5$	340.59	....	$295\text{-}7^{100}$, $186\text{-}7^{2}$	50	....	....	al, eth, bz, chl	B2[3], 1067
6452	Eicosanoic acid, methyl ester or Methyl eicosanoate $C_{19}H_{39}CO_2CH_3$	326.56	lf(MeOH)	$215\text{-}6^{10}$	54.5	....	1.4317^{60}	al, eth, bz, chl	B2[4], 1276
6453	1-Eicosanol or Arachidic alcohol $CH_3(CH_2)_{18}CH_2OH$	298.55	wx(al) cr(chl)	309, $220\text{-}5^{3}$	72-3	$0.8405^{20/4}$	1.4550^{20}	ace, bz	B1[4], 1900
6454	2-Eicosonal $CH_3(CH_2)_{17}CH(OH)CH_3$	298.56	cr(MeOH)	357	63-4	$0.8378^{20/4}$	1.4312^{80}	ace, bz	B1[4], 1901
6455	2-Eicosanone or Methyl octadecyl ketone $C_{18}H_{35}COCH_3$	296.54	lf(MeOH)	....	58	....	....	al, ace, bz	B1[4], 3402
6456	3-Eicosanone or Ethyl heptadecyl ketone $C_{17}H_{35}C(NOH)C_2H_5$	296.54	lf(al)	....	60-1	....	....	eth, ace, bz, chl, aa	B1[4], 3403
6457	3-Eicosanone oxime $C_{17}H_{35}C(H)C_2H_5$	311.55	nd(al)	....	α55-6, β64-5	....	....	al, eth	B1[3], 2932
6458	7-Eicosanone or Hexyl tridecyl ketone $C_{13}H_{27}COC_6H_{13}$	296.54	cr	$210\text{-}1^{11}$	52-3	....	1.4258^{20}	eth, ace	B1[4], 3403
6459	5,8,11,14-Eicosatetraenoic acid or Arachidonic acid $CH_3(CH_2)_3(CH_2CH{=}CH)_4(CH_2)_3CO_2H$	304.47	....	d	−49.5	....	1.4824^{20}	al, eth, ace, chl	B2[4], 1802
6460	1-Eicosene $CH_3(CH_2)_{17}CH{=}CH_2$	280.54	....	341, $151^{1.5}$	28.5	$0.7882^{30/4}$	1.440^{30}	bz, peth	B1[4], 934
6461	1-Eicosyne $CH_3(CH_2)_{17}C{\equiv}CH$	278.52	....	340, 191.8^{10}	36	$0.8073^{20/4}$	1.4501^{20}	bz, peth	B1[4], 1077
6462	Elaidamide $CH_3(CH_2)_7CH{=}CH(CH_2)_7CONH_2$	281.48	....	....	93-4	....	....	al	B2[4], 1668
6463	Elaidic acid or 9-Octadecenoic acid (trans) $CH_3(CH_2)_7CH{=}CH(CH_2)_7CO_2H$	282.47	pl(al)	288^{100}, 234^{15}	45	0.8734^{45}	1.4499^{45}	al, eth, bz, chl	B2[4], 1647
6464	Elaidic acid, dibromide or 9,10-Dibromo octadecanedioic acid $CH_3(CH_2)_7CHBrCHBr(CH_2)_7CO_2H$	442.27	col or ye	....	29-30	$1.2458^{30/4}$	1.4893^{42}	eth	B2[3], 1048
6465	Elaidic acid, ethyl ester or Ethylelaidate $C_{17}H_{33}CO_2C_2H_5$	310.52	....	$217\text{-}9^{15}$	5.8	0.8664^{25}	1.4480^{25}	al, eth	B2[4], 1652
6466	Elaidic acid, methyl ester $C_{17}H_{33}CO_2CH_3$	296.49	....	$213\text{-}5^{15}$	.....	0.8730^{20}	1.4513^{20}	al, eth	B2[4], 1651
6467	Elaidyl alcohol or 9-Octadecen-1-ol *(trans)* $C_8H_{17}CH{=}CH(CH_2)_7CH_2OH$	268.48	....	333, 198^{10}	36-7	$0.8338^{40/4}$	1.4552^{40}	al, eth, ace	B1[4], 2204
6468	α-Elaterin $C_{32}H_{44}O_8$	556.70	cr(chl-MeOH or al), $[\alpha]^{20}{}_D$ -64.3 (chl, c=1.6)	....	234	....	....	eth, bz, chl	B8[3], 4377
6469	β-Elaterin $C_{20}H_{28}O_5$	348.44	nd(al), $[\alpha]^{25}{}_D$ + 13.9	....	195.5	....	....	chl	B8[3], 4377
6470	Elemane or Dihydroelemene $C_{15}H_{30}$	210.40	....	$115\text{-}9^{10}$	.....	$0.8509^{20/4}$	1.4640^{20}	eth, bz, peth	B5[2], 117
6471	α-Elemene (d) $C_{15}H_{24}$	204.36	$[\alpha]_D$ 116 (chl, 14.85%)	$120\text{-}30^{7}$	.....	$0.8782^{20/4}$	1.5130^{26}	ace, bz	B5[3], 1083
6472	Elemenonic acid or Dihydro-β-elemonic acid $C_{30}H_{48}O_3$	456.71	nd(al or AcOEt)	....	249-50	....	....	....	B10[3], 3227
6473	α-Elemol $C_{15}H_{26}O$	222.37	cr, $[\alpha]_D$ + 43.7 (chl, c=1.9)	$142\text{-}3^{12}$	52-3	$0.9345^{18/4}$	1.4980^{18}	....	B6[3], 410
6474	α-Eleostearic acid or 9,11,13-Octadecatrienoic acid (cis) $C_4H_9[CH{=}CH]_3(CH_2)_7CO_2H$	278.44	$[\alpha]_D$ -5.8 (chl, c=3.4) nd (al)	235^{12}d, 170^{1}	49	$0.9028^{50/4}$	1.5112^{50}	al, eth	B2[4], 1787
6475	β-Eleostearic acid or 9,11,13-Octadecatrienoic acid (trans) $C_4H_9[CH{=}CH]_3(CH_2)_7CO_2H$	278.44	lf (al, MeOH)	188^{1}	71-2	$0.8839^{80/4}$	1.5000^{80}	MeOH, al	B2[4], 1787
6476	Eluetherin $C_{16}H_{16}O_4$	272.30	$[\alpha]^{15}{}_D$ 346 (chl)	....	175	....	....	....	B18[4], 1642
6477	Ellagene or Indino-2′,3′:2,3-fluorene $C_{20}H_{14}$	254.33	pl(bz)	....	216	....	....	....	E14s, 498
6478	Ellagic acid, dihydrate $C_{14}H_6O_8.2H_2O$	338.23	pa ye nd(py)	....	450-80d	....	....	....	B19[4], 3164
6479	Elliptic acid $C_{20}H_{18}O_8$	386.36	nd(aq al)	....	190	....	....	....	B18[4], 3342

No.	Name, Synonyms, and Formula	Mol. wt.	Color, crystalline form, specific rotation and λ_{max} (log ε)	b.p. °C	m.p. °C	Density	n_D	Solubility	Ref.
6480	Emeraldine $C_{48}H_{40}N_8$	728.90	indigo-bl pw					aa(80)	B13^{1}, 257
6481	Emetine (l) or Cephaline-O-methyl ether $C_{29}H_{40}N_2O_4$	480.65	amor pw $[\alpha]^{20/}{}_D$ -50 (chl, c=2)		74 (cor)			al, eth, ace	B23^{2}, 449
6482	Emetine, hydrochloride (l) $C_{29}H_{40}N_2O_4 \cdot 2HCl \cdot 7H_2O$	679.68	nd(w), $[\alpha]_D$ + 11 (w, c=1)		269-70d			w, al	B23^{2}, 451
6483	Emicymarin $C_{30}H_{46}O_9$	550.69	nd or pr (+ MeOH), $[\alpha]^{20/}{}_D$ + 12.5 (al, c=2.5)		Ca 207				B18^{4}, 2440
6484	Enneaphyllin $C_{90}H_{154}$	1236.21	rods (bz)		295-6			al	C32, 2686
6485	Eosin or 2,4,5,7-tetrabromfluorescein $C_{20}H_8O_5Br_4$	647.90	ye-red		295-6			al	B19^{4}, 2917
6486	Ephedrine (d) $C_6H_5CH(OH)CH(CH_3)NHCH_3$	165.24	pl(w)	225	40			w, al, eth, bz, chl	B13^{3}, 1723
6487	Ephedrine, hydrochloride-(d) $C_{10}H_{15}NO \cdot HCl$	201.70	pl(abs al), $[\alpha]^{20/}{}_D$ + 35.8 (w, c=11.5)		218			w, al	B13^{3}, 1723
6488	Ephedrine (dl) or Racephedrine $C_6H_5CH(OH)CH(CH_3)NHCH_3$	165.24	nd(eth or peth)	135-7^{12}	76-7			w, al, eth, bz, chl	B13^{3}, 1723
6489	Ephedrine, p-amino dihydrochloride $CH_3CH(NH_2)CH(OH)C_6H_4NH_2$-4.2HCl	239.15	lf(al-eth)		192-3d			w, al	C27, 2762
6490	Ephedrine, hydrochloride (dl) or Ephotonin $C_{10}H_{15}NO.HCl$	201.70	pl(al)		189-90			w, al	B13^{3}, 1724
6491	Ephedrine (l) or Natural ephedrine $C_6H_5CH(OH)CH(CH_3)NHCH_3$	183.25	pl(w + 1)	225	40			w, al, eth, bz, chl	B13^{3}, 1720
6492	Ephedrine, hydrochloride (l) $C_{10}H_{15}NO.HCl$	201.70	orb nd, $[\alpha]^{20/}{}_D$ - 36.6		218-20			w, al	B13^{3}, 1721
6493	Ephedrine, N-methyl $C_6H_5CH(OH)CH(CH_3)N(CH_3)_2$	179.27	nd or pl (al or eth), $[\alpha]_D$ -29.5 (MeOH, c=4.5)		87-8			al, eth, MeOH	B13^{3}, 1726
6494	Ephedrine, N-(4-nitrobenzoyl) (dl) $C_6H_5CH(OH)CH(CH_3)NH(CH_3)COC_6H_4NO_2$-(4)	314.34	pa ye pl(al)		162				B13^{2}, 384
6495	Ephedrine sulfate (l) $(C_{10}H_{15}NO)_2.H_2SO_4$	428.54	hex pl or orh nd(w)		245-8d			w	C21, 2169
6496	Epi-β-amyrin acetate $C_{30}H_{50}O$	426.73	cr(MeOH)		225 (cor)				B6^{3}, 2896
6497	Epi-α-amyrin acetate $C_{32}H_{52}O_2$	468.76	nd(chl-MeOH), $[\alpha]_D$ + 39 (chl)		135 (cor)				B6^{3}, 2890
6498	Epiandrosterone $C_{10}H_{17}OH$	290.45	cr(bz-peth ace), $[\alpha]^{20/}{}_D$ + 108 (MeOH)		177-9				B8^{3}, 584
6499	Epiborneol (l) or 3-Camphanol $C_{10}H_{15}OH$	154.25	nd(peth)	213^{742}	181-2				B6^{4}, 288
6500	Epiborneol, acetate or 3-Camphanol acetate $C_{12}H_{20}O_2$	196.29	$[\alpha]^{19/}{}_D$ + 15.63	101^{11}	<-15	0.9872$^{14/0}$	1.4651^{14}		B6^{3}, 318
6501	Epibreinonol or Breinonol A $C_{30}H_{48}O_2$	440.71	pl (MeOH) nd (chl), $[\alpha]^{17/}{}_D$ + 37 (chl)		204			chl	B8^{3}, 1093
6502	Epicamphor (d) or 3-Bornanone $C_{10}H_{16}O$	152.24	$[\alpha]^{17/}{}_D$ + 45.4 (bz)		182			al, eth, peth	B7^{3}, 420
6503	Epicamphor, oxime (d) or d-3-Bornanone oxime $C_{10}H_{16}NOH$	167.25	nd(MeOH)		103			eth, ace	B7^{1}, 86
6504	Epicamphor, semicarbazone (d) $C_{10}H_{16}NNHCONH_2$	209.29	nd(al)		237-8			al	B7^{3}, 421

No.	Name, Synonyms, and Formula	Mol. wt.	Color, crystalline form, specific rotation and λ_{max} (log ε)	b.p. °C	m.p. °C	Density	n_D	Solubility	Ref.
6505	Epicamphor (dl) or 3-Bornanone $C_{10}H_{16}O$	152.24	cr(peth)		175-7			al, eth, peth	$B7^{3}$, 421
6506	Epicamphor, oxime (dl) $C_{10}H_{16}NOH$	167.25	nd(dil al)		98-100			eth, ace	$B7^{1}$, 87
6507	Epicamphor-(l) or β-Camphor $C_{10}H_{16}O$	152.24		213	184 (187)			al, eth, bz, peth	$B7^{3}$, 421
6508	Epicamphor, bromo (l) $BrC_{10}H_{15}O$	231.13	nd or pw (peth), $[\alpha]_D$ -86.6 (AcOGt ,c=3.6)		133-4			bz, chl, lig	$B7^{3}$, 422
6509	Epicamphor, oxime (l) or β-camphor oxime $C_{10}H_{16}NOH$	167.25	nd(dil MeOH), $[\alpha]_D$ + 100.5 (bz, c=6.3)		103-4			al, eth, ace	$B7^{1}$, 86
6510	Epicamphor, semicarbazone (l) $C_{10}H_{16}NNHCONH_2$	209.29	nd(al)		237-8			al	$B7^{3}$, 421
6511	α-Epicamphyl amine $C_{10}H_{17}NH_2$	153.27	$[\alpha]_D$ + 17.6 (bz, c=6.5)	$127\text{-}8^{100}$				w	$B12^{3}$, 176
6512	Epicatechin (dl) or Epicatechol $C_{15}H_{14}O_6$	290.27	nd(w + 1)pr (w + 4)		224-6d			al, ace	$B17^{2}$, 258
6513	Epicatechin (l) or Epicatechol $C_{15}H_{14}O_6$	290.27	cr(w + 4), $[\alpha]^{25}{}_D$ -69 (al)		245d			al, ace	$B17^{4}$, 3841
6514	Epicholestan-3-ol or 3-α-hydroxycholestane $C_{27}H_{48}O$	388.68	nd (al, MeOH), $[\alpha]^{20}{}_D$ + 34 (chl)		185-6			eth, chl	$B6^{3}$, 2135
6515	Epicholestan-4-ol or 4α-Hydroxycholestane $C_{27}H_{48}O$	388.68	lf(al MeOH-ace), $[\alpha]^{21}{}_D$ + 29.0 (chl)		187-8			eth, chl	$B6^{3}$, 2155
6516	Epicholesterol or 5-Cholesten-3α-ol $C_{27}H_{46}O$	386.66	cr(al chl MeOH), $[\alpha]^{30}{}_D$ - 37.5 (al)		141.5				$B6^{3}$, 2622
6517	Epicoprostanol or 3-α-Coprostanol $C_{27}H_{48}O$	388.68	cr(al ace), $[\alpha]^{20}{}_D$ + 31.6 (chl)		117-8			al, eth, bz, chl	$B6^{3}$, 2130
6518	Epicoprostenol or 4-Cholesten-3α-ol $C_{27}H_{46}O$	386.66	nd(ace), $[\alpha]^{24}{}_D$ + 120.8 (bz)		84			al, eth, bz, chl	$B6^{3}$, 2605
6519	Epidicentrin (dl) or dl-Domesticine methyl ether $C_{20}H_{21}NO_4$	339.39	pr(MeOH)		142			al, eth, chl	$B27^{2}$, 553
6520	Epidicentrin (l) or l-Domesticin methyl ether $C_{20}H_{21}NO_4$	339.39	pr(MeOH), $[\alpha]^{18}{}_D$ - 101.3 (chl, c=0.5)		138-9			al, eth, chl	$B27^{2}$, 533
6521	Epiergosterol *(d)* or Δ-7,9,(11),22-Ergostatrien-3α-ol $C_{28}H_{44}O$	396.66	nd(eth-MeOH), $[\alpha]^{19}{}_D$ + 36.2 (chl)		203-4				$B6^{3}$, 3096
6522	D-*epi*-Fucitol or 6-deoxy-D-glucitol $C_6H_{14}O_5$	166.17	cr(eth w), $[\alpha]^{21}{}_D$ + 2.2 (w, c=1)		105-7			w, eth	Am74, 4373
6523	L-*epi*-Fucitol $C_6H_{14}O_5$	166.17	cr(w), $[\alpha]^{20}{}_D$ -2.3 (w, c=1)		105-7			w, eth	C24, 2431
6524	L-*epi*-Fucose or L-Quinovose $CH_3(CHOH)_4CHO$	164.16	cr(AcOEt), $[\alpha]_D$ -36.9 (w, c=6)		135-45			w	$B1^{4}$, 4267

PHYSICAL CONSTANTS OF ORGANIC COMPOUNDS (Continued)

No.	Name, Synonyms, and Formula	Mol. wt.	Color, crystalline form, specific rotation and λ_{max} (log ε)	b.p. °C	m.p. °C	Density	n_D	Solubility	Ref.
6525	Epiisofenchol or 4,6,6-Trimethyl-2-norbornanol $C_{10}H_{17}OH$	154.25	nd(sub), $[\alpha]^{20/}{}_D$ -7.35		71-2				$B6^3$, 292
6526	Epiisofenchone or 4,6,6-Trimethyl norbornanone (d) $C_{10}H_{16}O$	152.24	$[\alpha]_D$ + 19.6	195		$0.934^{20/4}$	1.459^{20}	eth, ace	$B7^3$, 396
6527	Epiisofenchone (dl) or 4,6,6,-Trimethyl norbornanone $C_{10}H_{16}O$	152.24		195-8, $90\text{-}3^{21}$			1.4625^{25}	eth, ace	$B7^3$, 397
6528	Epilupinine or d-isolupinine $C_{10}H_{19}NO$	169.27	nd(peth), $[\alpha]^{17/}{}_D$ + 32 (al, c=1.5)		76-8			eth, ace, bz	$B21^4$, 290
6529	Epinine $C_9H_{13}NO_2$	167.21	nd(al)		188-9				$B13^3$, 2209
6530	Epiquinidine $C_{20}H_{24}N_2O_2$	324.42	cr (AcOEt) lf (eth), $[\alpha]^{20/}{}_D$ + 103.7 (al, c=1.86)		113			al, eth	B23, 505
6531	Epirhodanhydrin or 2,3-Epoxy propyl rhodanine CH_2CHCH_2SCN (O bridging CH_2–CH)	115.15	dk red liq- (garlic odor)	d				al, chl	B17, 106
6532	Episarsapogenin $C_{27}H_{44}O_3$	416.64	nd(ace), $[\alpha]_D$ -71		2 04-6				$B19^4$, 825
6533	Epitruxillic acid or 2,4-*cis*-Diphenyl cyclobutane-1,3-*trans*-dicarboxylic acid $C_{18}H_{16}O_4$	296.32	cr(dil al or bz-aa)		285-7				$B9^3$, 4625
6534	2,3-Epoxypropyl ethyl ether $CH_2CHCH_2OC_2H_5$ (O bridging CH_2–CH)	102.13		128		0.9700^{20}	1.4320^{20}	w, al, eth	$B17^4$, 987
6535	2,3-Epoxypropyl phenyl ether $CH_2CHCH_2OC_6H_5$ (O bridging CH_2–CH)	150.18		242.5^{755}		$1.1109^{21\cdot 2/4}$	$1.5307^{21/}{}_D$		$B17^4$, 990
6536	Equilenin (d) $C_{18}H_{18}O_2$	266.34	$[\alpha]^{16/}{}_D$ + 87 (diox)	sub $170\text{-}80^{0\cdot 1}$	258-9				$B8^3$, 1523
6537	Equilenin (dl) $C_{18}H_{18}O_2$	266.34	cr(bz)		276-8				$B8^3$, 1525
6538	Equilenin (l) $C_{18}H_{18}O_2$	266.34	$[\alpha]^{20/}{}_D$ -85 (diox)	sub $170\text{-}80^1$	258-9			al	$B8^3$, 1522
6539	Equilin $C_{18}H_{20}O_2$	268.36	orh sph pl(AcOEt)	170-200 (sub vac)	238-40			al, ace	$B8^3$, 1415
6540	Equilin, α-dihydro $C_{18}H_{22}O_2$	270.37	cr(ace), $[\alpha]_D$ + 220 (diox, c=1)		174-6			al, ace, diox	$B6^3$, 5530
6541	Equisetrin $C_{27}H_{30}O_{16}$	610.53	ye nd (+ 2w)		195-6			al	B18, 3296
6542	Equol or 4,7-Isoflavandiol $C_{15}H_{14}O_3$	242.27	cr(aq al)		189-90				$B17^4$, 2186
6543	Eremophilol $C_{15}H_{24}O$	220.35	visc oil	164.5^{13}			1.5202^{20}		$B6^3$, 2080
6544	Eremophilone $C_{15}H_{22}O$	218.34	nd(MeOH)	171^{15}	42-3	$0.9994^{25/25}$	1.5182^{25}		$B7^3$, 1266
6545	Eremophilone, 8,9-epoxy $C_{15}H_{22}O_2$	234.34	nd(peth)		63-4			al, eth, ace, bz	$B17^4$, 4767
6546	Ergine or Ergonovine $C_{16}H_{17}N_3O$	267.33	cr (MeOH) Pr (aq ace)		135-40				H32, 506
6547	Ergocornine $C_{37}H_{39}N_5O_5$	561.68	cr(MeOH), $[\alpha]^{20/}{}_D$ -188 (chl, c=1)		182-4d			al, ace, bz, chl	H52, 1549
6548	Ergocorninine $C_{31}H_{39}N_5O_5$	561.68	lo pr (al), $[\alpha]^{20/}{}_D$ + 409 (chl, c=1)		228d			al, ace, bz, chl	H52, 1549
6549	Ergocristine $C_{35}H_{39}N_5O_5$	609.73	rh (bz + 2), $[\alpha]^{20/}{}_D$ -183 (chl, c=1) -93 Py		175d			al, ace, chl	H34, 1944

No.	Name, Synonyms, and Formula	Mol. wt.	Color, crystalline form, specific rotation and λ_{max} (log ε)	b.p. °C	m.p. °C	Density	n_D	Solubility	Ref.
6550	Ergocristinine $C_{35}H_{39}N_5O_5$	609.73	pr(al), $[\alpha]^{20}/_D$ +366 (chl, c=0.68)		237-8d				$B27^{2}$, 860
6551	Ergocryptine $C_{32}H_{41}N_5O_5$	575.71	pr(al), $[\alpha]^{20}/_D$ -187 (chl, c=1)		212-4d			al, chl	$B27^{2}$, 860
6552	Ergocryptinine $C_{32}H_{41}N_5O_5$	575.71	lo pr (al) $[\alpha]^{20}/_D$ +408 (chl, c=1) +479 Py		245d,			ace, chl	$B27^{2}$, 860
6553	Ergometrine (l) or Ergobasine $C_{19}H_{23}N_3O_2$	325.41	nd(bz), $[\alpha]^{20}/_D$ -89 (w)		159-62d			al, ace	Am60, 1701
6554	Ergometrinine (d) or Ergobasinine $C_{19}H_{23}N_3O_2$	325.41	pr(ace), $[\alpha]^{20}/_D$ +416 (chl, c=0.26)		195-7d			chl, py	Am60, 1701
6555	Ergometrinine (l) $C_{19}H_{23}N_3O_2$	325.41	pr(ace), $[\alpha]^{20}/_D$ -41.5 (chl, c=0.26)		196			chl, py	Am60, 1701
6556	Ergopinacol II or Bisergostadienol $C_{56}H_{36}O_2$	791.30	nd(bz-al), $[\alpha]_D$ -155 (Py, c=0.8)		205				$B6^{1}$, 5897
6557	Ergosine $C_{30}H_{37}N_5O_5$	547.65	pr(MeOH, AcOEt)		228d			ace, chl	H34, 1544
6558	Ergosinine $C_{30}H_{37}N_5O_5$	547.65	Pr (al), (aq ace, bz) nd MeOH, $[\alpha]^{20}/_D$ +420 (chl, c=1)		220d			ace, chl	J1937, 396
6559	Δ^{5}:^{6,7}:^{8}-Ergostadien-3β-ol or Provitamin D_4 $C_{28}H_{46}O$	398.67	nd(MeOH-AcOEt), $[\alpha]^{19}/_D$ -109 (chl)		152-3				$B6^{1}$, 2836
6560	$\Delta^{14,22}$-Ergostadien-3β-ol $C_{28}H_{46}O$	398.67	cr(MeOH), $[\alpha]^{20}/_D$ -9 (chl)		116				E14s, 1762
6561	α-Ergostadienone $C_{28}H_{44}O$	396.66	lf(al), $[\alpha]^{19}/_D$ +2 (chl)		182-3				$B7^{3}$, 1767
6562	Ergostane or 24-Methyl-5α-cholestane $C_{28}H_{50}$	386.71	lf or pl(eth-MeOH or ace), $[\alpha]^{25}/_D$ +21 (chl, c=2)		85			eth, ace, chl	$B5^{4}$, 1234
6563	Ergostanol or Ergostan-3β-ol $C_{28}H_{50}O$	402.70	nd(MeOH-eth), $[\alpha]_D$ +15.4 (chl, c=1.8)		144-5			eth, chl	$B6^{3}$, 2161
6564	$\Delta^{3,5,7,22}$-Ergostatetraene $C_{28}H_{42}$	378.64	pl(al)$[\alpha]^{20}/_D$ -40.5 (chl)		104				$B5^{3}$, 1927
6565	$\Delta^{4,6,22}$-Ergostatrienone or Isoergosterone $C_{28}H_{42}O$	394.64	nd(ace-eth)		110				$B7^{3}$, 2031
6566	α-Ergostenol or α-Tetrahydroergosterol $C_{28}H_{48}O$	400.69	lf or nd (MeOH) nd (gl aa), $[\alpha]^{16}/_D$ +11 (MeOH, c=0.9)		131 (135)			eth, bz, chl	$B6^{3}$, 2685

No.	Name, Synonyms, and Formula	Mol. wt.	Color, crystalline form, specific rotation and λ_{max} (log ε)	b.p. °C	m.p. °C	Density	n_D	Solubility	Ref.
6567	β-Ergostenol or β-Tetrahydroergosterol $C_{28}H_{48}O$	400.69	pl or ta(al), $[\alpha]^{20/}{}_D$ +21.2 (chl, c=0.9)	...	141-2	...	...	...	$B6^{3}$, 2689
6568	α-Ergostenol or α-Tetrahydroergosterol $C_{28}H_{48}O$	400.69	nd (MeOH) cr (ProH) lf (w)	...	148 (152)	...	...	...	$B6^{3}$, 2683
6569	δ-Δ-^{8}:^{9}-Ergostenol $C_{28}H_{46}O$	400.69	$[\alpha]_D$ +39	...	155	...	...	chl	$B6^{3}$, 2685
6570	Ergosterol or Δ ^{5},^{7},^{22}-Ergostatrien-2β-ol $C_{28}H_{44}O$	396.66	pl (+w, al) nd (eth), $[\alpha]^{20/}{}_D$ -135 (chl, c=1.2)	$250^{0.01}$	168 (+w)	...	...	bz, chl	$B6^{3}$, 3099
6571	Ergosterol D $C_{28}H_{44}O$	396.76	nd(al), $[\alpha]^{17/}{}_D$ +24.6	...	167	...	...	bz, chl	$B6^{3}$, 3120
6572	Ergosterol-5,6-dihydro or α-Dihydroergosterol $C_{28}H_{46}O$	398.67	lf (ace) pl (chl-MeOH), $[\alpha]^{20/}{}_D$ -19 (chl)	...	176-7	...	...	...	$B6^{3}$, 2841
6573	Ergosterone $C_{28}H_{42}O$	394.64	nd(ace-MeOH), $[\alpha]^{20/}{}_D$ -4.52 (chl)	...	132	...	...	...	$B7^{3}$, 2032
6574	Ergotamine $C_{33}H_{35}N_5O_5$	581.67	nd (al) pr (bz) pl (aq ace), $[\alpha]^{20/}{}_D$ -160 (chl, c=1)	...	213-4d	...	...	eth, bz, chl	$B27^{2}$, 860
6575	Ergotaminine $C_{33}H_{35}N_5O_5$	581.67	rh pl (MeOH) pl (al), $[\alpha]^{20/}{}_D$ +369 (chl, c=0.5)	...	252d	...	...	chl, py	$B27^{2}$, 860
6576	Ergothioneine or Thiasine $C_9H_{15}N_3O_2S$	229.30	pl(w+2)nd or lf (dil al), $[\alpha]_D$ +115 (w, c=1)	...	290d	...	...	...	$B25^{2}$, 413
6577	Erucic acid or *cis*-13-Docosenoic acid $CH_3(CH_2)_7CH{=}CH(CH_2)_{11}CO_2H$	338.57	nd(al)	265^{15}	33-4	$0.860^{55/4}$	1.4758^{20}	al, eth	$B2^{4}$, 1676
6578	Erysocine $C_{18}H_{21}NO_3$	299.37	nd(eth), $[\alpha]_D$ +238.1	...	162	...	...	al, eth, chl	$B21^{4}$, 2624
6579	Erysodine $C_{18}H_{21}NO_3$	299.37	nd(al), $[\alpha]^{27/}{}_D$ +248 (al)	...	204-5	...	...	al, eth	$B21^{4}$, 2623
6580	Erysonine $C_{17}H_{19}NO_3$	285.34	cr(al)$[\alpha]^{25/}{}_D$ +285 (aq.HCl)	...	236-7d	...	...	...	$B21^{4}$, 2623
6581	Erysopine $C_{17}H_{19}NO_3$	285.34	cr(al), $[\alpha]^{25/}{}_D$ +265.2 (al-glycerol)	...	241-2	...	...	...	$B21^{4}$, 2622
6582	Erysothiopine $C_{10}H_{21}NO_7S$	299.37	cr(al-w), $[\alpha]^{25/}{}_D$ +194 (al)	...	168-9	...	...	...	$B21^{4}$, 2624
6583	Erysovine $C_{18}H_{21}NO_3$	299.37	pr(eth), $[\alpha]_D$ +252 (al)	...	178-9	...	...	al, eth, chl	$B21^{4}$, 2623
6584	Erythraline $C_{18}H_{19}NO_3$	297.35	cr(al), $[\alpha]^{27/}{}_D$ +211.8 (al)	...	106-7	...	...	al, chl	Am73, 589

PHYSICAL CONSTANTS OF ORGANIC COMPOUNDS (Continued)

No.	Name, Synonyms, and Formula	Mol. wt.	Color, crystalline form, specific rotation and λ_{max} (log ε)	b.p. °C	m.p. °C	Density	n_D	Solubility	Ref.
6585	Erythramine or Dihydroerythraline $C_{18}H_{21}NO_3$	299.37	cr(eth-peth), $[\alpha]^{20/}{}_{D}$ + 228 (al, c=0.19)	$125^{4\times10^{-4}}$	103-4			al, eth, ace, bz	Am73, 589
6586	Erythraline $C_{18}H_{19}NO_4$	315.33	cr(eth-peth), $[\alpha]^{28/}{}_{D}$ + 145.5 (al)		170				Am73, 589
6587	*meso*-Erythritol $HOCH_2(CHOH)_2CH_2OH$	122.12	bipym tetr pr,	329-31	121.5	$1.451^{20/4}$		w, py	B1⁴, 2807
6588	Erythritol anhydride $C_4H_6O_2$	86.09		138		$1.113^{18/4}$			
6589	Erythritol, tetranitrate or Cardilate $C_4H_6N_4O_{12}$	302.11	leaflets		61			al, w	
6590	β-Erythroidine $C_{16}H_{19}NO_3$	273.33	cr(abs al), $[\alpha]^{25/}{}_{D}$ + 88.8 (w)		99-100			w, al, eth, bz, chl	Am80, 3905
6591	D-Erythronic acid-γ-lactone $C_4H_6O_4$	118.09	pr, $[\alpha]^{20/}{}_{D}$ -73.2 (w, c=4)		104-5			w, al	B18⁴, 1099
6592	L-Erythronic acid-γ-lactone $C_4H_6O_4$	118.09	nd (AcOEt), $[\alpha]^{30/}{}_{D}$ + 73 (w, c=4)		105			al	B18⁴, 1099
6593	D-Erythrose $C_4H_8O_4$	120.11	syr $[\alpha]^{20/}{}_{D}$ + 1→-14.3 (w, c=11)					w, al	B1⁴, 4172
6594	L-Erythrose $C_4H_8O_4$	120.11	syr $[\alpha]^{24/}{}_{D}$ + 11.5→ + 30.5 (w, c=3)						B1⁴, 4172
6595	Erythrosin or 2,4,5,7-Tetraiodo fluorescein $C_{20}H_8O_5I_4$	835.90	og-ye (eth)					al, eth	B19⁴, 2923
6596	L-Erythrulose $HOCH_2CH(OH)COCH_2OH$	120.11	syr [α]D + 11.2	d				w, al	B1⁴, 4176
6597	Escholerine $C_{41}H_{61}NO_{13}$	775.93	pl (ace-w), $[\alpha]^{25/}{}_{D}$ -30 (Py, c=1)		235d			ace	B21⁴, 6844
6598	Esculin or Aesculin (αβ-glucoside) $C_{15}H_{16}O_9$	340.29	pr (w + 2) $[\alpha]^{18/}$D-78.4 (50% aq diox)	230d	205d			aa, py	B18⁴, 1326
6599	α-Estradiol $C_{18}H_{24}O_2$	272.39	nd(+ ½ w) (80% al), $[\alpha]^{20/}{}_{D}$ + 56 (diox, c=0.9)		220-3			al, ace, peth	B6³, 5332
6600	β-Estradiol $C_{18}H_{24}O_2$	272.39	pr (80% al), $[\alpha]^{25/}{}_{D}$ + 76 (diox)		178-9			al, ace, diox	B6³, 5337
6601	Estriol $C_{18}H_{24}O_3$	288.39	lf(al) mcl (dil al), $[\alpha]_D$ + 61 (al) + 30 (Py)		288d	1.27			B6³, 6520
6602	Estrone $C_{18}H_{22}O_2$	270.37	α mcl(al) βγ orth(al)		260.2	β-1.236, γ-1.228		diox, py	B8³, 1171
6603	Ethane CH_3CH_3	30.07	gas, hex cr	−88.6	−183.3	$0.572^{-100/4}$	$1.0377^{0/}{}_{546}$ mm	bz	B1⁴, 108
6604	Ethane, 1,2-bis (ethylthio) $C_2H_5SCH_2CH_2SC_2H_5$	150.30		217, 95.5^{12}		$0.9815^{20/4}$	1.5118^{20}	al, eth	B1⁴, 2452

PHYSICAL CONSTANTS OF ORGANIC COMPOUNDS (Continued)

No.	Name, Synonyms, and Formula	Mol. wt.	Color, crystalline form, specific rotation and λ_{max} (log ε)	b.p. °C	m.p. °C	Density	n_D	Solubility	Ref.
6605	Ethane, 2,2-bis (4-methoxyphenyl)-1,1,1-trichloro or Methoxychlor $(4-CH_3OC_6H_4)_2CHCCl_3$	345.65	cr (dil al)	...	94	...	...	al, eth, bz	$B6^3$, 5436
6606	Ethane, 1,2-bis(methyl thio) $CH_3SCH_2CH_2SCH_3$	122.24	...	182.5^{750}, 78-80^{11}	...	1.0371$^{20/4}$	1.5292^{20}	w, al, eth, ace, chl	$B1^4$, 2457
6607	Ethane, 1,2-bis (phenyl sulfonyl) $C_6H_5SO_2CH_2CH_2O_2SC_6H_5$	310.38	nd or lf (al)	...	180	...	...	al, bz, aa	$B6^4$, 1495
6608	Ethane, 1,2-bis (phenylthio) $C_6H_5SCH_2CH_2SC_6H_5$	246.39	ta (al)	...	70	...	...	ace	$B6^4$, 1493
6609	Ethane, 2,2-bis (4 chlorophenyl)-1,11-trichloro or DDT, Dichloro diphenyl trichloro ethane $C_{14}H_9Cl_5$	354.49	nd (al)	260	108-9	...	...	eth, ace, bz, chl, peth	$B5^3$, 1833
6610	Ethane, bromo or Ethyl bromide C_2H_5Br	108.97	...	38.4	−118.6	1.4604$^{20/4}$	1.4239^{20}	al, eth, chl	$B1^4$, 150
6611	Ethane, 1-bromo-2-chloro or Ethylene chloro bromide $ClCH_2CH_2Br$	143.41	...	107	−16.7	1.7392$^{20/4}$	1.4908^{20}	al, eth	$B1^4$, 155
6612	Ethane, 1-bromo-2-fluoro FCH_2CH_2Br	126.96	...	71-2	...	1.7044$^{25/4}$	1.4236^{20}	al, eth	$B1^4$, 154
6613	Ethane, 1-bromo-2-methoxy $BrCH_2CH_2OCH_3$	140.00	...	110.3	...	1.4623$^{20/4}$	1.44753^{20}	...	$B1^4$, 1386
6614	Ethane, chloro or Ethyl chloride C_2H_5Cl	64.51	...	12.3	−136.4	0.8978$^{20/4}$	1.3676^{20}	al, eth	$B1^4$, 124
6615	Ethane, 1-chloro-2-fluoro FCH_2CH_2Cl	82.51	...	59^{750}	...	1.1747$^{20/4}$	1.3775^{20}	al, eth	$B1^4$,127
6616	Ethane, 1-chloro-2-iodo ICH_2CH_2Cl	190.41	...	140	−15.6	2.16439^{0}	...	...	$B1^4$, 167
6617	Ethane, 1-chloro-1-nitro $CH_3CHCl(NO_2)$	109.51	...	124-5	...	1.2860$^{20/20}$	...	...	$B1^4$, 172
6618	Ethane, chloro pentafluoro $ClCF_2CF_3$	154.47	gas	−38	−106	...	...	al, eth	$B1^4$, 129
6619	Ethane, 2-chloro-1,1,1-trifluoro $ClCH_2CF_3$	118.49	...	6.93	−105.5	1.389$^{0/4}$	1.3090^{0}/$_{D}$	...	$B1^4$, 128
6620	Ethane, 1-(2 chloroethoxy)-2-phenoxy $ClCH_2CH_2OCH_2CH_2OC_6H_5$	200.67	...	149^{10}	...	1.149$^{15/15}$	...	...	$B6^3$, 568
6621	Ethane, 1,1-dibromo or Ethylidene bromide CH_3CHBr_2	187.86	...	108, 9.0^{10}	−63	2.0555$^{20/4}$	1.5128^{20}	al, eth, ace, bz	$B1^4$, 157
6622	Ethane, 1,2-dibromo or Ethylene dibromide CH_2BrCH_2Br	187.86	...	131.3, 29.1^{10}	9.8	2.1792$^{20/4}$	1.5387^{20}	al, eth, ace, bz	$B1^4$, 158
6623	Ethane, 1,2-dibromo-1,1-dichloro $BrCH_2CBrCl_2$	256.75	...	178.3, 58.8^{10}	−66.8	2.2623$^{20/4}$	1.5567^{20}	al, eth, ace, bz	$B1^4$, 161
6624	Ethane, 1,2-dibromo-1,2-dichloro BrCHCl-CHBrCl	256.75	...	195, 84^{45}	−26	2.135$^{20/4}$	1.5662^{20}	al, eth, ace, bz	$B1^4$, 161
6625	Ethane, 1,2-dibromo-tetrafluoro $BrCF_2CF_2Br$	259.82	...	46.4	−112	2.149$^{25/4}$	...	...	$B1^4$, 160
6626	Ethane, 1,1-dichloro or Ethylidene chloride CH_3CHCl_2	98.96	...	57.3	−97	1.1757$^{20/4}$	1.4164^{20}	al, eth, ace, bz	$B1^4$, 130
6627	Ethane, 1,1-dichloro-1-fluoro CH_3CCl_2F	116.95	...	32	−103.5	1.250$^{10/4}$	1.3600^{10}/$_{D}$	...	$B1^4$, 134
6628	Ethane, 1,1-dichloro-1,2,2,2 tetrafluoro F_3CCCl_2F	170.92	...	3.6	−94	1.455$^{25/4}$	1.3092^{0}	al, eth, bz, chl	$B1^4$, 136
6629	Ethane, 1,2-dichloro or Ethylene dichloride $ClCH_2CH_2Cl$	98.96	...	83.5	−35.3	1.2351^{20}	1.4448^{20}	al, eth, ace, bz	$B1^4$, 131
6630	Ethane, 1,2-dichloro-1,1-difluoro $CH_2Cl_2ClCClF_2$	134.94	...	46.8	−1012	1.4163$^{20/4}$	1.36193^{20}/$_{D}$	...	$B1^4$, 135
6631	Ethane, 1,2-dichloro-1-fluoro $ClCH_2CHClF$	116.95	...	73.7	...	1.3814$^{20/4}$	1.41132^{20}/$_{D}$	...	$B1^4$, 134
6632	Ethane, 1,1-difluoro or Ethylidene fluoride CH_3CHF_2	66.05	gas	−24.7	−117	0.95^{20} (sat pr)	1.3011^{-72}	...	$B1^4$, 120
6633	Ethane, 1,1-difluoro-1,2,2,2-tetrachloro Cl_3CCClF_2	203.83	...	91.5	40.6	...	...	al, eth, chl	$B1^4$, 146
6634	Ethane, 1,2-difluoro or Ethylene difluoride CH_2FCH_2F	66.05	...	30.7	...	...	...	eth, bz, chl	$B1^4$, 121
6635	Ethane, 1,2-difluoro-1,1,2,2-tetrachloro $FCCl_2-CFCl_2$	203.83	...	93	25	1.6447$^{25/4}$	1.4130^{25}	al, eth, chl	$B1^4$, 146
6636	Ethane, 1,1-diiodo or Ethylidine iodide CH_3CHI_2	281.86	...	179-80, 60-1^{12}	...	2.84^{0}	1.673^{20}	al, eth, chl, ace	$B1^4$, 169

No.	Name, Synonyms, and Formula	Mol. wt.	Color, crystalline form, specific rotation and λ_{max} (log ε)	b.p. °C	m.p. °C	Density	n_D	Solubility	Ref.
6637	Ethane, 1,2-diido or Ethylene diiodo CH_2ICH_2I	281.86	ye mcl pr or rh (eth) d in lt	200, 74^{10}	83	$3.325^{20/4}$	1.871^{20}	al, eth, ace, chl	B1[4], 169
6638	Ethane, 1,2-di-N-morpholyl $C_{10}H_{20}N_2O_2$	200.28	wh-yesh (eth or lig)	$160\text{-}3^{25}$	75	....	....	w, al, ace, bz	B27, 7
6639	Ethane, 1,1-dinitro $CH_3CH(NO_2)_2$	120.06	ye mcl (bz or MeOH)	185-6, 72^{12}	......	$1.3503^{24/24}$	....	al, eth	B1[4], 174
6640	Ethane, 1,2-dinitro-1,1,2,2-tetrafluoro $O_2NCF_2CF_2(NO_2)$	192.03	....	58-9	−41.5	$1.6024^{25/4}$	1.3265^{25}	ace	B1[4], 175
6641	Ethane, 1,1-diphenyl or α-Methylditan $CH_3CH(C_6H_5)_2$	182.27	....	286, 148^{15}	−21.5	$0.9997^{20/4}$	1.5756^{20}	al, eth, bz	B5[4], 1880
6642	Ethane, 1,2-diphenyl or Bibenzyl $C_6H_5CH_2CH_2C_6H_5$	182.27	nd (al)	285, $95\text{-}6^{1}$	52.2	$0.9583^{60/4}$	1.5478^{60}	al, eth	B5[4], 1868
6643	Ethane, 1,2-di-(4-tolyl) $(4\text{-}CH_3C_6H_4)CH_2CH_2(C_6H_4CH_3\text{-}4)$	210.32	lf(MeOH or dil al) pl (lig)	296-8, 178^{18}	82-3	....	....	bz, peth	B5[4], 1943
6644	Ethane, 2,2-di-4-tolyl-1,1,1-trichloro $(4\text{-}CH_3C_6H_4)_2CHCCl_3$	313.65	mcl pr (al eth-al)	....	92	....	....	al, eth, ace	B5[4], 1949
6645	Ethane, 1-ethoxy-2-methylamino or Ethyl-β-methyl amino ethyl ester $C_2H_5OCH_2CH_2NHCH_3$	103.17	....	$114\text{-}5^{744}$	......	$0.8363^{20/4}$	1.4147^{20}	w, al, eth, ace, bz	B4[3], 647
6646	Ethane, fluoro or Ethyl fluoride CH_3CH_2F	48.06	gas	−37.7	−143.2	$0.7182^{20/4}$ (liq)	1.2656^{20}	al, eth	B1[4], 120
6647	Ethane, 1-fluoro-1,2,2-trichloro $Cl_2CHCHFCl$	151.40	....	101-3	......	1.54968^{17}	$1.4390^{20/D}$	....	B1[4], 141
6648	Ethane, fluoro penta chloro FCl_2CCCl_3	220.29	....	134-6	101.3	....	....	al, eth	B1[4], 148
6649	Ethane, hexabromo or Perbromo ethane C_2Br_6	503.45	rh pr (bz)	d200-10	d	$2.823^{20/4}$	1.863	....	B1[3], 193
6650	Ethane, hexachloro or Perchloro ethane C_2Cl_6	236.74	rh (al-eth)	186^{777}	186-7 (sealed tube)	$2.091^{20/4}$	....	al, eth, bz	B1[4], 148
6651	Ethane, hexafluoro or Perfluoro ethane C_2F_6	138.01	gas	−79	−94	1.590^{-78}	....	....	B1[4], 123
6652	Ethane, hexaphenyl $(C_6H_5)_3CC(C_6H_5)_3$	486.66	cr (ace)	d	145-7d	....	....	eth, ace, chl, MeOH	B5[3], 2746
6653	Ethane, iodo or Ethyl iodide C_2H_5I	155.97	....	72.3	−108	$1.9358^{20/4}$	1.5133^{20}	al, eth	B1[4], 163
6654	Ethane, isocyano or Ethyl carbylamine CH_3CH_2NC	55.08	....	79^{775}	<-66	$0.7402^{20/4}$	1.3622^{20}	al, eth, ace	B4[4], 342
6655	Ethane-1-(4-methoxyphenyl)-1-phenyl or 1-p-Anisyl-1-phenyl ethane $CH_3CH(C_6H_5)(C_6H_4OCH_3\text{-}4)$	212.29	....	$180\text{-}2^{19}$	......	$1.0473^{20/4}$	1.5725^{20}	eth, ace, bz	B6[2], 639
6656	Ethane, nitro $C_2H_5NO_2$	75.07	....	115	−50	$1.0448^{25/4}$	1.3917^{20}	al, eth, ace	B1[4], 170
6657	Ethane, nitro-pentafluoro $CF_3CF_2NO_2$	165.02	....	0	......	....	....	eth	B1[4], 172
6658	Ethane, 1,nitro-2,2,2-trifluoro $F_3CCH_2NO_2$	129.04	....	96	......	$1.3914^{20/4}$	1.3394^{20}	eth	B1[4], 172
6659	Ethane, nitroso-pentafluoro CF_3CF_2NO	149.02	....	−42	......	....	....	....	B1[4], 169
6660	Ethane, pentabromo Br_2CHBr_3	424.55	mcl pr (dil al)	210^{300}	56-7	$3.312^{20/4}$	....	al, eth	B1[3], 193
6661	Ethane, pentachloro $CHCl_2CCl_3$	202.29	....	162	−29	$1.6796^{20/4}$	1.5025^{20}	al, eth	B1[4], 147
6662	Ethane, pentaiodo CHI_2CI_3	659.55	mcl pr (aa)	....	182-4	....	....	al, eth, bz, aa	B1[1], 31
6663	Ethane, perfluoro CF_3CF_3	138.01	....	−79	−100.6	1.590^{-98}	....	....	B1[4], 123
6664	Ethane, 1,1,1,2-tetrabromo CH_2BrCBr_3	345.65	....	$112^{18}{}_d$	0.0	$2.8748^{20/4}$	1.6277^{20}	al, eth, ace, bz, chl	B1[4], 162
6665	Ethane, 1,1,2,2,-tetrabromo $CHBr_2CHBr_2$	345.65	yesh	243.5, 114.8^{10}	0	$2.9656^{20/4}$	1.6353^{20}	al, eth, ace, bz, aa	B1[4], 162
6666	Ethane, 1,1,1,2--tetrachloro CH_2ClCCl_3	167.85	yesh red	130.5, 22.1^{10}	-70.2	$1.5406^{20/4}$	1.4821^{20}	al, eth, ace, bz, chl	B1[4], 143
6667	Ethane, 1,1,2,2-tetrachloro $CHCl_2CHCl_2$	167.85	....	146.2, 33.9^{10}	−36	$1.5953^{20/4}$	1.4940^{20}	al, eth, ace, bz	B1[4], 144

No.	Name, Synonyms, and Formula	Mol. wt.	Color, crystalline form, specific rotation and λ_{max} (log ε)	b.p. °C	m.p. °C	Density	n_D	Solubility	Ref.
6668	Ethane, 1,1,1,2-tetrafluoro CH_2FCF_3	102.03		-26.5^{736}				eth	B1⁴, 123
6669	Ethane, 1,1,1,2-tetraphenyl $C_6H_5CH_2C(C_6H_5)_3$	334.46	mcl (eth-peth)	$277\text{-}80^{21}$	143-4			bz	B5³, 2575
6670	Ethane, 1,1,2,2-tetraphenyl $(C_6H_5)_2CHCH(C_6H_5)_2$	334.46	cr (bz + l) rh nd (chl)	358-62, 260^{16}	214-5			aa	B5³, 2574
6671	Ethane, 1,1,2-tribromo $CH_2BrCHBr_2$	266.76		188.9, 73.1^{10}	−29.3	$2.6211^{20/4}$	1.5933^{20}	al, eth, bz, chl	B1⁴, 161
6672	Ethane, 1,1,1-trichloro or Methyl chloroform CH_3CCl_3	133.40		74.1	−30.4	$1.3390^{20/4}$	1.4379^{20}	**al, eth,** chl	B1⁴, 138
6673	Ethane, 1,1,1-trichloro-2,2,2-trifluoro CF_3CCl_3	187.38		45.8	14.2	$1.5790^{20/4}$	1.3610^{25}	al, eth, chl	B1⁴, 142
6674	Ethane, 1,1,2-trichloro $CH_2ClCHCl_2$	133.40		113.8, 9.5^{10}	−36.5	$1.4397^{20/4}$	1.4714^{20}	al, **eth,** chl	B1⁴, 139
6675	Ethane, 1,1,2-trichloro-1,2,2-trifluoro CF_2ClCCl_2F	187.38		47.7	−36.4	$1.5635^{25/4}$	1.3557^{25}	al, **eth, bz**	B1⁴, 142
6676	Ethane, 1,1,1-trifluoro CH_3CF_3	84.04	gas	−47.3	−111.3			eth, chl	B1⁴, 122
6677	Ethane, 1,1,1-triido CH_3CI_3	407.76	ye oct (al)		95			eth, bz	B1³, 199
6678	Ethane, 1,1,1-triphenyl or α-Methyl tritan $CH_3C(C_6H_5)_3$	258.36	nd (al, eth)	$205\text{-}10^{18}$	95			eth, bz	B5³, 2331
6679	Ethane, 1,1,2-triphenyl $C_6H_5CH_2CH(C_6H_5)_2$	258.36	mcl lf (dil al) nd (al)	$348\text{-}9^{751}$	57			al, eth, bz	B5³, 2329
6680	Ethanediol or Ethylene glycol $HOCH_2CH_2OH$	62.07		198, 93^{13}	−11.5	$1.1088^{20/4}$	1.4318^{20}	**w, al,** eth, **ace**	B1⁴, 2369
6681	1,2-Ethanediol, 1,2-dicylohexyl (dl) or Cyclohoxanone pinacol $C_6H_{11}CH(OH)\text{-}CH(OH)C_6H_{11}$	226.36	nd		129-30			bz, peth	B6³, 4156
6682	1,2-Ethanediol, 1,2-diphenyl (d) $C_6H_5CH(OH)CH(OH)C_6H_5$	214.26	nd (w) lf or pr (abs al) [α] + 9_2 (abs al c=1.2) + 128 Cbz, c=0.3		148-9d			al	B6³, 5431
6683	1,2-Ethanediol, 1,2-diphenyl (dl) $C_6H_5CH(OH)CH(OH)C_6H_5$	214.26	nd (w or al) ta (eth)	>300, $133^{0.023}$	122-3			al, eth, chl	B6³, 5431
6684	1,2-Ethanediol, 1,2-diphenyl (l) $C_6H_5CH(OH)CH(OH)C_6H_5$	214.26	lf (eth abs al or bz) pr(bz or abs al)		148-9			al, eth, ace, bz	B6³, 5431
6685	1,2-Ethanediol, 1,2-diphenyl-*(meso)* $C_6H_5CH(OH)CH(OH)C_6H_5$	214.26	nd or lf (w or bz-peth) mcl lf (al, w)	>300, $139^{0.023}$	139-40			al, chl	B6³, 5429
6686	1,2-Ethanediol, phenyl-(dl) or Phenyl ethylene glycol $C_6H_5CH(OH)CH_2OH$	138.17	nd (lig)	$272\text{-}4^{755}$	69-70			w, al, eth, bz	B6³, 4572
6687	1,2-Ethanol, tetraphenyl or Bensopinacol $(C_6H_5)_2C(OH)C(OH)(C_6H_5)_2$	366.46	pr (bz + l) cr (ace)		182			eth, ace, bz, chl	B6³, 5923
6688	Ethanedione, di-(2-furyl) $(OC_4H_3)COCO(C_4H_3O)$	190.16	ye nd (al) cr (bz)		165-6			al, eth, bz, chl	B19⁴, 2007
6689	1,2-Ethane disulfonic acid $HO_3SCH_2CH_2SO_3H$	190.19	hyg nd (gl aa)		111-2 (+ 2w) 174 (anh)			w, al, diox	B4⁴, 78
6690	1,2-Ethanedithiol or Dithioglycol $HSCH_2CH_2SH$	94.19		146, $46\text{-}7^{6}$	−41.2	$1.1243^{20/4}$	1.5590^{20}	al, eth, ace, bz	B1⁴, 2450
6691	Ethane dithiolic acid, diethyl ester or Diethyl dithio oxalate $C_2H_5SCO\text{-}COSC_2H_5$	178.26	ye nd (eth)	235, $80\text{-}2^{32}$	27	$1.0565^{21/4}$		eth	B1¹, 244
6692	Ethane phosphonic acid $C_2H_5PO(OH)_2$	110.05	pl or nd (w)		61-2			w, al, eth	B4³, 1779
6693	Ethane phosphonic acid, diethyl ester $C_2H_5PO(OC_2H_5)_2$	166.16		198, 83^{13}		$1.0259^{20/4}$	1.4163^{20}	al, eth	B4³, 1779
6694	Ethane phosphonic acid, dimethyl ester $C_2H_5PO(OCH_3)_2$	138.10		82^{18}		$1.1029^{30/4}$	1.4128^{30}	w, al, bz	J1954, 3222
6695	Ethane sulfonic acid or Ethyl sulfonic acid $CH_3CH_2SO_3H$	110.13	hyg	123^{1}	−17	1.3341^{25}	1.4335^{20}	w, al	B4⁴, 33

No.	Name, Synonyms, and Formula	Mol. wt.	Color, crystalline form, specific rotation and λ_{max} (log ε)	b.p. °C	m.p. °C	Density	n_D	Solubility	Ref.
6696	Ethane sulfonyl chloride $C_2H_5SO_2Cl$	128.57 etc.	pa ye	171, 65^{13}		1.357$^{22.5}$	1.4531^{20}	eth	B4^4, 34
6697	Ethane sulfonyl chloride, 2-bromo $BrCH_2CH_2SO_2Cl$	207.47	pa ye	102^{13}		1.921^{20}	1.5242^{20}		B4^4, 37
6698	Ethane sulfonyl chloride, 1-chloro $CH_3CHClSO_2Cl$	163.02		80-1^{22}			1.4782^{20}		B1^4, 3121
6699	Ethane sulfonyl chloride, 2-chloro $ClCH_2CH_2SO_2Cl$	163.02		200-3, 93-7^{17}		1.555$^{20/4}$	1.4920^{20}		B4^4, 36
6700	Ethane sulfonic acid, 2-hydroxy, dihydrate or Isethionic acid $HOCH_2CH_2SO_3H_2H_2O$	162.16	hyg cr (aa-AC_2O)		111-2			w, al	B4^4, 84
6701	1,1,2,2-Ethane tetracarboxylic acid, 1,2-diethyl ester $C_2H_5O_2CCH(CO_2H)CH(CO_2H)CO_2C_2H_5$	262.22	hyg lf (+ 1/2 w)		132-3d			w, al, eth	B2, 858
6702	1,1,2,2-Ethane tetracarboxylic acid, tetraethyl ester $(C_2H_5O_2C)_2CHCH(CO_2C_2H_5)_2$	318.32		d 305	77	1.064^{80}	1.4105^{80}	al	B2^4, 2415
6703	1,1,2,2-Ethane tetracarboxylic acid, tetramethyl ester $(CH_3O_2C)_2CHCH(CO_2CH_3)_2$	262.22	cr (eth al bz)		138			al	B2^3, 2076
6704	Ethanethiol or Ethyl mercaptan C_2H_5SH	62.13		35	−144.4	0.8391$^{20/4}$	1.4310^{20}	al, eth, ace	B1^4, 1390
6705	Ethanethiol, 2-amino or Cysteamine $H_2NCH_2CH_2SH$	77.14	cr (sub)	d^{760} Sub (vac)	99-100			w, al	B4^4, 1570
6706	Ethanethiol, 2-chloro $ClCH_2CH_2SH$	96.57		113		1.1826$^{20/4}$	1.4929^{20}	al, eth, diox	B1^4, 1406
6707	Ethanethiol, 1-phenyl (l) $CH_3CH(C_6H_5)SH$	138.23	$[\alpha]^{20}_D$ −89 al, c=6	199-200, 83^{10}		1.022$^{20/4}$	1.5593^{20}	al, eth, bz	B6^3, 1697
6708	Ethanol or Ethyl alcohol CH_3CH_2OH	46.07		78.5	-117.3	0.7893$^{20/4}$	1.3611^{20}	**w, eth, ace, bz**	B1^4, 1289
6709	Ethanol, β-(4-amino phenyl) $(4\text{-}NH_2C_6H_4)CH_2CH_2OH$	137.18	nd (al)		108				B13^3, 1679
6710	Ethanol, 2-bromo or Ethylene bromohydrin $BrCH_2CH_2OH$	124.97		149-50^{750}, 51^4		1.7629$^{20/4}$	1.4915^{20}	**w, al, eth**	B1^4, 1385
6711	Ethanol-2-chloro or Ethylene chlorohydrin $ClCH_2CH_2OH$	80.51		128, 44^{20}	-67.5	1.2003$^{20/4}$	1.4419^{20}	**w, al**	B1^4, 1372
6712	Ethanol, 2-chloro-1-phenyl or Styrene Chlorohydrin $ClCH_2CH(OH)C_6H_5$	156.61		128^{17}		1.1926$^{20/4}$	1.5523^{20}	al, eth	B6^3, 1683
6713	Ethanol, 2,2-dichloro Cl_2CHCH_2OH	114.96		146, 37-8^6		1.4040$^{25/4}$	1.4626^{25}	al, eth	B1^4, 1383
6714	Ethanol, 2,2-diphenyl $(C_6H_5)_2CHCH_2OH$	198.26		195^{20}, 144-5^1	64.5			eth, al, ace	B6^3, 3397
6715	Ethanol, 2-ethylthio $C_2H_5SCH_2CH_2OH$	106.18		184	Ca-100	1.0166$^{20/4}$	1.4867^{20}	al, ace	B1^4, 2430
6716	Ethanol, 2-fluoro or Ethylene fluorohydrin FCH_2CH_2OH	64.06		103.5	-26.4	1.1040$^{20/4}$	1.3647^{18}	**w, al, eth**, ace	B1^4, 1366
6717	Ethanol, 2-iodo or Ethylene iodohydrin ICH_2CH_2OH	171.97		176-7d, 85-8^{25}		2.1968$^{20/4}$	1.5713^{20}	w, al, eth	B1^4, 1387
6718	Ethanol, 2-mercapto or Monothio ethylene glycol $HSCH_2CH_2OH$	78.13		157-8^{742}, 55^{13}		1.1143$^{20/4}$	1.4996^{20}	w, al, eth, bz	B1^4, 2428
6719	Ethanol, 2-methylthio $CH_3SCH_2CH_2OH$	92.16		68-70^{20}		1.6640$^{20/20}$	1.4867^{30}	w, al, eth	B1^4, 2429
6720	Ethanol, 2-nitro or β-Nitro ethanol $O_2NCH_2CH_2OH$	91.07		194^{765}, 102^{10}	-80	1.270$^{15/4}$	1.4438^{19}	**w, al, eth**	B1^4, 1388
6721	Ethanol, 2-triazo $N_3CH_2CH_2OH$	87.08		75^{40}		1.149$^{24/24}$		w	B1^4, 1389
6722	Ethanol, 2,2,2-tribromo or Avertin, bromethol CBr_3CH_2OH	282.76	nd or pr (peth)	92-3^{10}	81			al, eth, bz	B1^3, 1362
6723	Ethanol, 2,2,2-trichloro CCl_3CH_2OH	149.40	hyg rh ta or pl	151^{737}, 52^{11}	19		1.4861^{20}	**al, eth**	B1^4, 1383
6724	Ethanol, 2,2,2-trifluoro CF_3CH_2OH	100.04		74	-43.5	1.4680^{20}, 1.3739$^{22/4}$	1.2907^{22}	al, eth, ace, bz	B1^4, 1370
6725	Ethanol, 1,1,2-triphenyl $C_6H_5CH_2C(OH)(C_6H_5)_2$	274.36	nd (bz-lig) pr (peth)	222^{11}	89-90			al	B6^3, 3680
6726	Ethanol, 2,2,2-triphenyl $(C_6H_5)_3CCH_2OH$	274.36	cr (al eth lig)		110.5d			al, eth, bz, lig	B6^3, 3683
6727	Ethanolamine or 2-Amino ethanol $H_2NCH_2CH_2OH$	61.08		170, 58^5	10.3	1.0180$^{20/4}$	1.4541^{20}	**w, al, chl**	B4^4, 1406

No.	Name, Synonyms, and Formula	Mol. wt.	Color, crystalline form, specific rotation and λ_{max} (log ε)	b.p. °C	m.p. °C	Density	n_D	Solubility	Ref.
6728	Ethanolamine, N-acetyl $CH_3CONHCH_2CH_2OH$	103.12	nd (ace)	$166\text{-}7^{8}$	63-5	$1.1079^{25/4}$	1.4674^{20}	w	$B4^{4}$, 1535
6729	Ethanolamine, N-butyl $C_4H_9NHCH_2CH_2OH$	117.19	......	199-200, $91\text{-}2^{11}$	......	$0.8907^{20/4}$	1.4437^{20}	w, al, eth	$B4^{3}$, 682
6730	Ethanolamine, N-isobutyl $i\text{-}C_4H_9NHCH_2CH_2OH$	117.19	......	199-200, 90^{16}	......	$0.8818^{20/4}$	1.4402^{20}	w, al, eth	$B4^{3}$, 683
6731	Ethanolamine, N,N-diethyl $(C_2H_5)_2NCH_2CH_2OH$	117.19	hyg	163, $56\text{-}7^{15}$	......	$0.8921^{20/4}$	1.4412^{20}	w, al, eth, ace, bz	$B4^{4}$, 1471
6732	Ethanolamine, N,N-dimethyl $(CH_3)_2NCH_2CH_2OH$	89.14	......	134	......	$0.8866^{20/4}$	1.4300^{20}	**w, al, eth**	$B4^{4}$, 1424
6733	Ethanolamine, N-ethyl $C_2H_5NHCH_2CH_2OH$	89.14	......	169-70, $78\text{-}80^{27}$	......	$0.914^{20/4}$	1.444^{20}	w, al, eth	$B4^{4}$, 1465
6734	Ethanolamine, N-methyl $CH_3NHCH_2CH_2OH$	75.11	......	158, 52^{6}	......	0.937^{20}	1.4385^{20}	**w, al, eth**	$B4^{4}$, 1422
6735	Ethanolamine, N-methyl-N-phenyl $C_6H_5N(CH)_3CH_2CH_2OH$	151.21	yesh	150^{14}	......	$0.9995^{15/O}$	......	al, eth, ace, bz	$B12^{3}$, 296
6736	Ethanolamine, N-β-naphthyl $\beta\text{-}C_{10}H_7NHCH_2CH_2OH$	187.24	lf (eth or al)	$197\text{-}8^{3}$	52	......	......	eth	$B12^{2}$, 717
6737	Ethanolamine, N-phenyl $C_6H_5NHCH_2CH_2OH$	137.18	......	286, 167^{17}	......	1.0945^{20}	1.5760^{20}	al, eth, chl	$B12^{3}$, 293
6738	Ethanolamine, 1-phenyl $H_2NCH_2CH(OH)C_6H_5$	137.18	nd (al-eth-peth)	160^{17}	56-7	......	......	w, al	$B13^{4}$, 1657
6739	Ethanolamine, N-isopropyl $i\text{-}C_3H_7NHCH_2CH_2OH$	103.16	......	172-4, $76\text{-}7^{15}$	......	$0.8970^{20/4}$	1.4395^{20}	**w, al, eth**	$B4^{3}$, 681
6740	Ethoxyacetic acid $C_2H_5OCH_2CO_2H$	104.11	......	206-7, 111^{25}	......	$1.1021^{20/4}$	1.4194^{20}	w, al, eth	$B2^{4}$, 574
6741	Ethoxyacetic acid, ethyl ester or Ethyl ethoxy acetate $C_2H_5OCH_2CO_2C_2H_5$	132.16	......	158, 52^{12}	......	$0.9702^{20/4}$	1.4029^{20}	al, eth, ace	$B3^{4}$, 581
6742	Ethoxyacetic acid, l-menthyl ester or l-Menthyl ethoxy acetate $C_2H_5OCH_2CO_2C_{10}H_{19}$	242.36	$[\alpha]^{20}_{D}$ −66.35	155^{20}	......	$0.9545^{20/4}$	......	al, eth, chl	$B6^{2}$, 47
6743	Ethoxyacetic acid, methyl ester or Methylethoxy acetate $C_2H_5OCH_2CO_2CH_3$	118.13	......	$147\text{-}8^{734}$	......	1.0112^{15}	......	al, eth, ace	$B3^{4}$, 578
6744	2-Ethoxybenzoic acid $2\text{-}C_2H_5OC_6H_4CO_2H$	166.18	......	$211\text{-}2^{35}$	20.7	......	......	......	$B10^{3}$, 98
6745	2-Ethoxybenzoic acid, ethyl ester $2\text{-}C_2H_5OC_6H_4CO_2C_2H_5$	194.23	......	251, $180\text{-}5^{113}$	......	1.005^{20}	......	al, eth	$B10^{3}$, 117
6746	3-Ethoxybenzoic acid $3\text{-}C_2H_5OC_6H_4CO_2H$	166.18	nd (w or sub)	sub	137	......	......	al, eth, bz	$B10^{3}$, 245
6747	3-Ethoxybenzoic acid, ethyl ester or Ethyl-3-ethoxy benzoate $3\text{-}C_2H_5OC_6H_4CO_2C_2H_5$	194.23	......	264, $172\text{-}3^{50}$	......	$1.0725^{20/20}$	......	al, eth	$B10^{3}$, 251
6748	4-Ethoxybenzoic acid $4\text{-}C_2H_5OC_6H_4CO_2H$	166.18	nd (w)	......	198.5	......	......	al, eth, bz	$B10^{3}$, 282
6749	4-Ethoxybenzoic acid, ethyl ester or Ethyl-4-ethoxy benzoate $4\text{-}C_2H_5OC_6H_4CO_2C_2H_5$	194.23	......	275 $148\text{-}9^{14}$	: lf	1.076^{14}	......	al, eth	$B10^{3}$, 301
6750	Ethoxyl amine or α-Ethyl hydroxyl amine $C_2H_5ONH_2$	61.08	......	......	68	$0.8872^{8/8}$	......	**w, al, eth**	$B1^{4}$, 1326
6751	Ethylaceto acetate $CH_3COCH_2CO_2C_2H_5$	130.14	......	180.4, 74^{14}	<-80	$1.0282^{20/4}$	1.4194^{25}	**al, eth,** bz, chl	$B3^{4}$, 1528
6752	Ethylacetoacetate (enol form) $CH_3C(OH){=}CHCO_2C_2H_5$	130.14	......	......	......	1.0119^{10}	1.4432^{20}	......	$B3^{4}$, 1528
6753	Ethylacetoacetate (keto form) $CH_3COCH_2CO_2C_2H_5$	130.14	......	......	-39	$1.0368^{10/4}$	1.4171^{20}	......	$B3^{4}$, 1528
6754	Ethyl amine $C_2H_5NH_2$	45.08	......	16.6	-81	$0.6829^{20/4}$	1.3663^{20}	**w, al, eth**	$B4^{4}$, 307
6755	Ethyl amine, hydrobromide $C_2H_5NH_2\text{-}HBr$	126.00	mcl nd or pl (al)	......	159.3	......	......	w, al	$B4^{4}$, 310
6756	Ethyl amine, hydrochloride $C_2H_5NH_2.HCl$	81.55	mcl pl (al)	d 315	109-10	......	......	w, al	$B4^{4}$, 310
6757	Ethyl amine, hydroiodide $C_2H_5NH_2HI$	173.00	mcl nd (w)	......	188.5	2.100	......	w	$B4^{4}$, 310
6758	Ethyl amine, 2-bromo, hydrochloride $BrCH_2CH_2NH_2.HCl$	160.44	lf (al-AcOEt)	......	174-5d	......	......	w, al, bz	$B4^{3}$, 248
6759	Ethyl amine, 2-chloro, hydrochloride $ClCH_2CH_2NH_2.HCl$	115.99	hyg cr (al-eth)	......	144 (148)	......	......	w, al, ace	$B4^{3}$, 236

No.	Name, Synonyms, and Formula	Mol. wt.	Color, crystalline form, specific rotation and λ_{max} (log ε)	b.p. °C	m.p. °C	Density	n_D	Solubility	Ref.
6760	Ethyl amine, 2-ethoxy $C_2H_5OCH_2CH_2NH_2$	89.14		108^{758}		$0.8512^{20/4}$	1.4101^{20}	w, al, eth, ace, bz	$B4^{4}$, 1411
6761	Ethyl amine, 2-methoxy $CH_3OCH_2CH_2NH_2$	75.11		95^{756}				w, al	$B4^{4}$, 1411
6762	Ethyl amime, perfluoro $CF_3CF_2NF_2$	171.02		−35					$B2^{4}$, 473
6763	Ethyl β-amino, ethyl ether $C_2H_5OCH_2CH_2NH_2$	89.14		108^{758}		$0.8512^{20/4}$	1.4101^{20}	w, al, eth, ace, bz	$B4^{4}$, 1411
6764	Ethyl benzene $C_6H_5CH_2CH_3$	106.17		136.2, 25.8^{10}	-95	$0.8670^{20/4}$	1.4959^{20}	al, eth	$B5^{3}$, 776
6765	2-Ethyl benzoic acid $2\text{-}C_2H_5C_6H_4CO_2H$	150.18	nd (w)	259	68	$1.0413^{100/4}$	1.5099^{100}	al, eth	$B9^{3}$, 2425
6766	3-Ethyl benzoic acid $3\text{-}C_2H_5C_6H_4CO_2H$	150.18	nd (w or dil al)		47	1.042^{100}	1.5345^{100}	al, eth	$B9^{1}$, 208
6767	4-Ethyl benzoic acid $4\text{-}C_2H_5C_6H_4CO_2H$	150.18	pr (al) pl or lf (w)		113.5			al, bz, chl	$B9^{2}$, 349
6768	Ethyl benzyl ketone or 1-Phenyl-2-butanone $CH_3CH_2COCH_2C_6H_5$	148.29		230^{755}		$1.002^{20/4}$		al, eth, ace	$B7^{3}$, 1080
6769	Ethylboric acid $C_2H_5B(OH)_2$	73.89	pl (eth)	d	40 (sub)			w, al, eth	$B4^{3}$, 1964
6770	Ethyl bromide CH_3CH_2Br	108.97		38.4	-118.6	$1.4604^{20/4}$	1.4239^{20}	al, eth, chl	$B1^{4}$, 150
6771	Ethyl 2-bromoethyl ether $C_2H_5OCH_2CH_2Br$	153.02	ye nd (al)	$127\text{-}8^{755}$, 40^{24}		$1.3572^{20/4}$	1.4447^{20}	al, eth	$B1^{4}$, 1386
6772	Ethyl bromomethyl ether $C_2H_5OCH_2Br$	138.99		109^{746}		$1.4402^{20/4}$	1.4515^{20}	eth	$B1^{3}$, 2594
6773	Ethyl isobutyl ether $C_2H_5\text{-O-}i\text{-}C_4H_9$	102.18		81		$0.751^{20/4}$	1.3739^{25}	al, eth, ace, chl	$B1^{4}$, 1593
6774	Ethyl isobutyl sulfide $i\text{-}C_4H_9SC_2H_5$	118.24		134.2, 24.8^{10}		$0.8306^{20/4}$	1.4450^{20}	al, eth	$B1^{4}$, 1606
6775	Ethyl chloride CH_3CH_2Cl	64.51		12.3	-136.4	$0.8978^{20/4}$	1.3676^{20}	al, eth	$B1^{4}$, 124
6776	Ethyl (l-chloroethyl) ether $C_2H_5OCHClCH_3$	108.57		92-5		$0.950^{20/4}$	1.4053^{20}		$B1^{4}$, 3119
6777	Ethyl β-chloroethyl ether $C_2H_5OCH_2CH_2Cl$	108.57		107-8		$0.9894^{20/4}$	1.4113^{20}	eth, chl	$B1^{4}$, 1375
6778	Ethyl β-chloroethyl sulfide $C_2H_5SCH_2CH_2Cl$	124.63		156, $63\text{-}5^{47}$		$1.0663^{20/4}$	1.4878^{20}	chl	$B1^{4}$, 1707
6779	Ethyl chloromethyl ether $C_2H_5OCH_2Cl$	94.54		83^{763}		$1.0372^{0/4}$	1.4040^{20}	al, eth	$B1^{4}$, 3047
6780	Ethyl chloro sulfinate C_2H_5OSOCl	128.57		52.5^{44}		$1.2766^{25/4}$	1.4550^{25}	eth	$B1^{3}$, 1316
6781	Ethyl chloro sulfonate $C_2H_5OSO_2Cl$	144.57		151-4, 52^{14}		$1.3502^{25/4}$	1.416^{20}	eth, chl, lig	$B1^{4}$, 1326
6782	Ethyl cyanoacetate $NCCH_2CO_2C_2H_5$	113.12		205, 99^{15}	-22.5	$1.0654^{20/4}$	1.4175^{20}	al, eth	$B2^{4}$, 1889
6783	Ethyl cyanoformate $NCCO_2C_2H_5$	99.09		$116\text{-}8^{765}$		$1.0034^{20/4}$	1.3821^{20}	al, eth	$B2^{4}$, 1862
6784	Ethyl di-benzylamine $C_2H_5N(CH_2C_6H_5)_2$	225.33		306, 131^{11}				al, eth	$B12^{3}$, 2222
6785	Ethyl (1,2-dibromoethyl) ether $C_2H_5OCH_2BrCH_2Br$	231.91		80^{20}		$1.7320^{20/4}$	1.5044^{20}	al, chl	$B1^{4}$, 3152
6786	Ethyl (1,2-dichloroethyl) ether $C_2H_5OCHClCH_2Cl$	143.01		145, $66\text{-}8^{45}$		$1.1370^{20/4}$	1.4435^{20}	al, eth	$B1^{4}$, 3136
6787	Ethyl (1,2-dichlorovinyl) ether $C_2H_5OCCl{=}CHCl$	141.00		128.2		$1.1972^{25/4}$	1.4558^{17}		$B1^{4}$, 3425
6788	Ethyl (diethylaminomethyl) ether $(C_2H_5)_2NCH_2OC_2H_5$	131.22		136, 76^{11}				al, eth, ace	$B4^{4}$, 336
6789	Ethyl-di-isopropylamine $C_2H_5N(i\text{-}C_3H_7)_2$	129.25		126.5			1.4138^{20}		$B4^{4}$, 511
6790	Ethyl 2,3-epoxypropyl ether $C_2H_5OCH_2CHCH_2$ (epoxide O bridging CH and CH_2)	102.13		128		0.9700^{20}	1.4320^{20}	w, al, eth	B17, 105
6791	Ethyl ether or Diethyl ether $C_2H_5OC_2H_5$	74.12		34.5	fr−116.2	$0.7138^{20/4}$	1.3526^{20}	al, ace, bz, chl	$B1^{4}$, 1314
6792	Ethyl ether borofluoride $(C_2H_5)_2O.BF_3$	141.93		125-6 60^{20}	-60.4	$1.3572^{20/4}$	1.4447^{20}	al, eth	$B1^{4}$, 1321

PHYSICAL CONSTANTS OF ORGANIC COMPOUNDS (Continued)

No.	Name, Synonyms, and Formula	Mol. wt.	Color, crystalline form, specific rotation and λ_{max} (log ε)	b.p. °C	m.p. °C	Density	n_D	Solubility	Ref.
6793	Ethyl ethynyl ether or Ethoxy acetylene $C_2H_5OC{\equiv}CH$	70.09		50 exp 100		$0.8000^{20/4}$	1.3796^{20}		$B1^{4}$, 2211
6794	Ethyl furfuryl ether $C_2H_5OCH_2(C_4H_3O)$	126.16		$149\text{-}50^{770}$		$0.9844^{20/4}$	1.4523^{20}	al, eth	$B17^{2}$, 114
6795	Ethyl-2-furyl ether $(2\text{-}C_4H_3O)OC_2H_5$	112.13		125-6		$0.9849^{23/4}$	1.4500^{23}		$B17^{4}$, 1219
6796	Ethyl heptyl ether $C_2H_5OC_7H_{15}$	144.26		166.6		$0.790^{16/4}$	1.4111^{20}	al, eth	$B1^{4}$, 1733
6797	Ethyl hexyl ether $C_2H_5OC_6H_{13}$	130.23		$142\text{-}3^{773}$, 42^{14}		$0.7722^{20/4}$	1.4008^{20}	al, eth	$B1^{4}$, 1697
6798	Ethyl hydrogensulfate $C_2H_5OSO_3H$	126.13		280d		$1.3657^{20/4}$	1.4105^{20}	w	$B1^{4}$, 1324
6799	Ethyl hydroperoxide C_2H_5OOH	62.07		93-7 ext >100	-100	$0.9332^{20/4}$	1.3800^{20}	w, **al, eth**, bz	$B1^{4}$, 1323
6800	Ethyl-bis (2-hydroxyethyl) amine $C_2H_5N(CH_2CH_2OH)_2$	133.19	ye	246-8, 118^{1}		$1.0135^{20/4}$	1.4663^{20}	al	$B4^{3}$, 693
6801	Ethyl (2-hydroxy--5-methylphenyl) ketone $(5\text{-}CH_3\text{-}2\text{-}HOC_6H_3)COC_2H_5$	164.20		$129\text{-}30^{16.5}$	2	$1.0841^{14/4}$	$1.549^{13.8}$	chl	$B8^{2}$, 120
6802	Ethyl (2-hydroxy--3-methylphenyl) ketone $(3\text{-}CH_3\text{-}2\text{-}HOC_6H_3)COC_2H_5$	164.20		$127\text{-}9^{15}$	22-3			chl	$B8^{2}$, 119
6803	Ethyl (2-hydroxy--4-methylphenyl) ketone $(4\text{-}CH_3\text{-}2\text{-}HOC_6H_3)COC_2H_5$	164.20		$115\text{-}20^{10}$	41.5-2.5			chl	$B8^{3}$, 471
6804	Ethyl hypochlorite C_2H_5OCl	80.51	ye lig	36^{732}		$1.013^{-6/4}$		al, **eth, bz, chl**	$B1^{4}$, 1324
6805	Ethyl iodide CH_3CH_2I	155.97		72.3	-108	$1.9358^{20/4}$	1.5133^{20}	**al**, eth	$B1^{4}$, 163
6806	Ethyl isocyanate C_2H_5NCO	71.08		60		$0.9031^{20/4}$	1.3808^{20}	**al, eth**	$B4^{4}$, 402
6807	Ethyl methyl amine or Methylethylamine $CH_3NHC_2H_5$	59.11		36.7				w, al, eth, ace	$B4^{4}$, 312
6808	Ethyl methyl amine, hydrochloride $C_2H_5NHCH_3.HCl$	95.57	pl (al-eth)		126-30	$1.0874^{20/4}$		w, al, eth, ace	$B4^{2}$, 589
6809	Ethyl methyl ether $C_2H_5OCH_3$	60.10		10.8		$0.7252^{0/0}$	1.3420^{4}	w, **al, eth**, ace, chl	$B1^{4}$, 1314
6810	Ethyl methyl sulfide $C_2H_5SCH_3$	76.16		66.6	-105.9	$0.8422^{20/4}$	1.4404^{20}	**al**, eth	$B1^{4}$, 1392
6811	Ethyl α-naphthyl ether or 1-Ethoxynaphthalene α-$C_{10}H_7OC_2H_5$	172.23	nd	280.5, $136\text{-}8^{14}$	5.5	$1.060^{20/4}$	1.5953^{25}	al, eth	$B6^{3}$, 2924
6812	Ethyl β-naphthyl ether β-$C_{10}H_7OC_2H_5$	172.23	pl (al)	282, 148^{10}	37-8	$1.0640^{20/20}$	1.5975^{36}	al, eth, lig, to	$B6^{3}$, 2972
6813	Ethyl nitrate $C_2H_5ONO_2$	91.07	flam	87.2	-94.6	$1.1084^{20/4}$	1.3852^{20}	w, **al, eth**	$B1^{4}$, 1327
6814	Ethyl nitrite C_2H_5ONO	75.07	yesh	$16\text{-}7^{725}$		$0.90^{15/15}$	1.3418^{10}	**al, eth**	$B1^{4}$, 1327
6815	Ethyl octyl ether $C_2H_5OC_8H_{17}$	158.28		186.3, 74^{9}	12.5	$0.7847^{20/4}$	1.4127^{20}	al	$B1^{4}$, 1759
6816	Ethyl pentyl ether $C_2H_5OC_5H_{11}$	116.20		119-20		$0.7622^{20/4}$	1.3927^{20}	**al, eth**	$B1^{3}$, 1602
6817	Ethyl *iso*pentyl ether $C_2H_5O\text{-}i\text{-}C_5H_{11}$	116.20		112-3		$0.7695^{21/15}$		**al, eth**	$B1^{4}$, 1681
6818	Ethyl *tert*-pentyl ether $C_2H_5O\text{-}tert\text{-}C_5H_{11}$	116.20		101		$0.7657^{20/4}$	1.3912^{20}	al, **eth**	$B1^{3}$, 1626
6819	Ethylarsonic acid $C_2H_5AsO(OH)_2$	154.00	nd (al) rh nd (w)	$209\text{-}11^{12}$	99.5			w, al	$B4^{3}$, 1823
6820	Ethyl perchlorate $C_2H_5OClO_3$	128.51	oil		89			**al, eth**	$B1^{3}$, 1314
6821	Ethyl Diethylperoxide $C_2H_5OOC_2H_5$	90.12		65	-70	0.8240^{19}	1.3715^{17}	**al, eth**	$B1^{4}$, 1323
6822	Ethyl phenyl sulfide or (Ethylthio)benzene $C_6H_5SC_2H_5$	138.23		205, 84^{10}		$1.0211^{20/4}$	1.5670^{20}	al	$B6^{4}$, 1468
6823	Ethyl propyl amine $C_2H_5NHC_3H_7$	87.16		61-2		0.7204^{17}	1.3858^{25}	w, al, ace	$B4^{4}$, 468
6824	Ethyl isopropyl amine i-$C_3H_7NHC_2H_5$	87.16		70-1			1.3872^{25}		$B4^{4}$, 508
6825	Ethyl propyl ether $C_2H_5OC_3H_7$	88.15		63.6	<-79	$0.7386^{20/4}$	1.3695^{20}	**al, eth, aa**	$B1^{4}$, 1421

No.	Name, Synonyms, and Formula	Mol. wt.	Color, crystalline form, specific rotation and λ_{max} (log ε)	b.p. °C	m.p. °C	Density	n_D	Solubility	Ref.
6826	Ethyl iso propyl ether $C_2H_5OCH(CH_3)_2$	88.15		63-4		$0.720^{25/4}$	1.3698^{25}	al, eth, ace, chl	$B1^4$, 1471
6827	Ethyl propyl sulfide $C_2H_5SC_3H_7$	104.21		118.5, 13.5^{10}	-117	$0.8370^{20/4}$	1.4462^{20}	al	$B1^4$, 1451
6828	Ethyl 1-propynyl ether $C_2H_5OC{=}CCH_3$	84.12		84		$0.8276^{20/4}$	1.4039^{20}	al, eth	$B1^3$, 1969
6829	Ethyl 2-propynyl ether or Ethyl propargyl ether $C_2H_5OCH_2C{\equiv}CH$	84.12		82		$0.8326^{20/4}$	1.4039^{20}	al, eth	$B1^4$, 2215
6830	N-Ethyl quinolinium iodide $C_9H_7N^+(C_2H_5)I^-$	285.13	ye pr (al)		158			w, al, chl	$B20^4$, 3358
6831	Ethyl pyridinium bromide $(C_5H_5N^+)C_2H_5Br^-$	188.07	cr (al)		111-2			w, al	$B20^4$, 2309
6832	Ethyl sulfate $(C_2H_5O)_2SO_2$	154.18		208d, 96^{15}	-24.5	$1.1774^{20/4}$	1.4004^{20}	al, eth	$B1^4$, 1326
6833	Ethyl sulfite $(C_2H_5O)_2SO$	138.18		230, 116^{10}		$0.9957^{20/4}$	1.4310^{20}	al, eth	$B1^4$, 1324
6834	Ethyl sulfoxide or Diethyl sulfoxide $(C_2H_5)_2SO$	106.18	syr	104^{25}	14			w, al, eth	$B1^4$, 1395
6835	Ethyl telluride or Diethyl telluride $(C_2H_5)_2Te$	185.72	red-ye	137-8		$1.599^{15/4}$	1.5182^{15}	al	$B1^4$, 1412
6836	Ethyl thioacetic acid $C_2H_5SCH_2CO_2H$	120.17		164^{83}, 109^5	-8.5	$1.1497^{20/4}$		w, al, eth	$B3^4$, 603
6837	Ethyl thiocyanate C_2H_5SCN	87.14		145^{758}	-85.5	$1.0071^{22/4}$	1.4684^{15}	al, eth	$B3^4$, 328
6838	Ethyl isothiocyanate or Ethyl mustard oil C_2H_5NCS	87.14		131-2	-5.9	$0.9990^{20/4}$	1.5130^{20}	al, eth	$B4^4$, 403
6839	Ethyl vinyl ether $C_2H_5OCH{=}CH_2$	72.11		35-6	-115.8	$0.7589^{20/4}$	1.3767^{20}	al, eth	$B1^4$, 2049
6840	Ethylene $CH_2{=}CH_2$	28.05	gas, mcl pr	-103.7	-169		1.363^{100}	eth	$B1^4$, 677
6841	Ethylene, amino or Vinylamine $CH_2{=}CHNH_2$	43.07		$55\text{-}6^{750}$		0.8321^{24}		w, al, eth	B4, 203
6842	Ethylene, 1,2-bis(trimethylsilyl)-*(trans)* $(CH_3)_3SiCH{=}CHSi(CH_3)_3$	172.42		145.5		0.7589^{20}	$1.4310^{20}/_D$		KHOC, 603
6843	Ethylene, l-bromo-2-chloro $BrCH{=}CHCl$	141.39		84.6	-86.7	1.7972^{15}	1.4982		$B1^3$, 671
6844	Ethylene, 1-bromo-1,2,2-triphenyl $C_6H_5CBr{=}C(C_6H_5)_2$	335.24	nd (aa)		116-7			aa	$B5^3$, 2400
6845	Ethylene, 1-chloro-2-dichloroarsino (trans) or Lewisite $ClCH{=}CHAsCl_2$	207.32		196d, 93^{20}	0.1	$1.888^{20/4}$		al, eth	$B4^3$, 1810
6846	Ethylene chlorohydrin $ClCH_2CH_2OH$	80.51		128, 44^{20}	-67.5	$1.2003^{20/4}$	1.4419^{20}	w, al	$B1^4$, 1372
6847	Ethylene, 1-chloro-1,2,2-trifluoro $FCCl{=}CF_2$	116.47		-26.2	-157.5	$1.54^{-60/4}$	1.38^0	bz	$B1^4$, 704
6848	Ethylene, 1-chloro-1,2,2-triphenyl $(C_6H_5)_2C{=}CClC_6H_5$	290.79			117			al, eth, ace, bz, chl	$B5^3$, 2400
6849	Ethylenediamine $H_2NCH_2CH_2NH_2$	60.10		116.5	8.5	$0.8995^{20/20}$	1.4568^{20}	w, al	$B4^4$, 1166
6850	Ethylene diamine, N,N′-dibenzoyl $C_6H_5CONHCH_2CH_2NHCOC_6H_5$	268.32	pr or nd (al)		247			aa	$B9^3$, 1210
6851	Ethylenediamine, N,N′-diethyl $C_2H_5NHCH_2CH_2NHC_2H_5$	116.21		144, $38\text{-}40^{15}$		$0.8280^{20/4}$	1.4340^{20}	w, al, eth, to	$B4^4$, 1174
6852	Ethylenediamine, N,N′-dimethyl $CH_3NHCH_2CH_2NHCH_3$	88.15		120		$0.828^{15/4}$		al, eth, dil HCl	$B4^4$, 1171
6853	Ethylenediamine, N,N′-diphenyl $C_6H_5NHCH_2CH_2NHC_6H_5$	212.29	lf (dil al)	$178\text{-}82^2$	74			al, eth	$B12^3$, 1042
6854	Ethylenediamine, hydrate $H_2NCH_2CH_2NH_2.H_2O$	78.11		118	10	$0.964^{20.5/4}$	$1.4500^{20.5}$	w	$B4^4$, 1168
6855	Ethylenediamine, hydrochloride $H_2NCH_2CH_2NH_2.2HCl$	133.02	mcl pr (w)	sub	300-30 sub		1.633	w	$B4^4$, 1168
6856	Ethylenediamine, N-β-hydroxyethyl $HOCH_2CH_2NHCH_2CH_2NH_2$	104.15		238-40, 123^{10}		$1.0254^{25/D}$	1.4861^{20}	w, al, ace	$B4^4$, 1558
6857	Ethylene, dibenzoyl *(cis)* $C_6H_5COCH{=}CHCOC_6H_5$	236.27	nd (al)		134			eth, ace, bz, chl	$B3^7$, 4115
6858	Ethylene, dibenzoyl *(trans)* $C_6H_5COCH{=}CHCOC_6H_5$	236.27	ye nd (al or bz)		111			bz, chl, aa	$B3^7$, 4116
6859	Ethylene, 1,1-dibromo or Vinylidene bromide $CH_2{=}CBr_2$	185.85		92		$2.1780^{21/4}$		al, eth, ace, bz	$B1^4$, 720

No.	Name, Synonyms, and Formula	Mol. wt.	Color, crystalline form, specific rotation and λ_{max} (log ε)	b.p. °C	m.p. °C	Density	n_D	Solubility	Ref.
6860	Ethylene, 1,1-dibromo--2-ethoxy $C_2H_5OCH=CBr_2$	229.90		170-2^{747}, 73-5^{15}		1.7697$^{18/4}$		eth	B1^2, 473
6861	Ethylene, 1,2-dibromo *(cis)* $CHBr=CHBr$	185.85		112.5^{760}	-53	2.2464^{20}	1.5428^{20}	al, eth, ace, bz, chl	B1^4, 720
6862	Ethylene, 2,2-dibromo *(trans)* $CHBr=CHBr$	185.85		108	-6.5	2.2308^{20}	1.5505^{18}	al, eth, ace, bz, chl	B1^4, 721
6863	Ethylene, 1,1-dichloro or Vinylidene chloride $CH_2=CCl_2$	96.94		37	-122.1	1.218^{20}	1.4249^{20}	al, eth, ace, bz, chl	B1^4, 706
6864	Ethylene, 1,1-dichloro-2-fluoro $CCl_2=CHF$	114.93		37.5		1.37324$^{16.4}$	1.4031$^{16.4/}{}_D$		B1^4, 711
6865	Ethylene, 1,2-dichloro *(cis)* $CHCl=CHCl$	96.94		60.3	-80.5	1.2837$^{20/4}$		**al, eth, ace,** bz, chl	B1^4, 707
6866	Ethylene, 1,2-dichloro *(trans)* $CHCl=CHCl$	96.94		47.5	-50	1.2565$^{20/4}$	1.4454^{20}	**al, eth, ace,** bz, chl	B1^4, 709
6867	Ethylene, 1,2-dichloro-1,2-difluoro $CFCl=CFCl$	132.92		21.1	-130.5	1.4950$^{0/4}$	1.3777$^{0/}{}_D$		B1^4, 712
6868	Ethylene, 1,1-difluoro or Vinylidene fluoride $CH_2=CF_2$	64.03	gas		<-84			al, eth	B1^4, 696
6869	Ethylene, 1,2-diiodo $CHI=CHI$	279.85		72.5^{16}	-14	3.0625^{20}		eth, chl	B1^4, 724
6870	Ethylene, 1,1-diphenyl $CH_2=C(C_6H_5)_2$	180.25		277, 94-5^{11}	8.2	1.0281$^{16/4}$	1.6100^{20}	eth, chl	B5^3, 1975
6871	Ethylene, fluoro-trichloro $CFCl=CCl_2$	149.38		71	-108.9	1.5460$^{20/4}$	1.4379$^{20/}{}_D$	chl	B1^4, 715
6872	Ethylene, nitro $CH_2=CHNO_2$	73.05		98.5, 38-9^{80}	-55.5	1.2212$^{14/4}$	1.4282^{20}	al, eth, ace, bz, chl	B1^4, 725
6873	Ethylene, tetrabromo $Br_2C=CBr_2$	343.64	pl (dil al) nd (al)	225-7, 100^{15}	56.5			al, eth, ace, bz	B1^4, 722
6874	Ethylene, tetrachloro $Cl_2C=CCl_2$	165.83		121, 14^{10}	-19	1.6227$^{20/4}$	1.5053^{20}	**al, eth, bz**	B1^4, 715
6875	Ethylene, tetracyano $(NC)_2C=C(CN)_2$	128.09		223	198-200	1.348^{25}	1.560^{25}	ace	B2^4, 2450
6876	Ethylene, tetrafluoro $F_2C=CF_2$	100.02	gas	-76.3	-142.5	1.519$^{-76.3}$			B1^4, 698
6877	Ethylene, tetraiodo $I_2C=CI_2$	531.64	ye lf pr (eth)	sub	192	2.983^{20}		bz, chl	B1^4, 724
6878	Ethylene, tetraphenyl $(C_6H_5)_2C=C(C_6H_5)_2$	332.44	mcl or rh (bz-eth or chl-al)	415-25	225	1.155$^{0/4}$		bz	B5^3, 2598
6879	Ethylene tetracarboxylic acid, tetraethyl ester $(C_2H_5O_2C)_2C=C(CO_2C_2H_5)_2$	316.31	tcl pr (eth)	325-8d, 210^{22}	58			al, eth	B2^4, 2450
6880	Ethylene, tribromo $BrCH=CBr_2$	264.74		163-4, 75^{15}		2.708$^{20.5/4}$	1.6045^{16}	al, eth, ace, chl	B1^4, 722
6881	Ethylene, trichloro $ClCH=CCl_2$	131.39		87	−73	1.4642$^{20/4}$	1.4773^{20}	al, eth, ace, chl	B1^4, 712
6882	Ethylene, triphenyl $C_6H_5CH=C(C_6H_5)_2$	256.35	lf (al or MeOH)	220-1^{14}	72-3	1.0373$^{78/4}$	1.6292^{78}	al, eth	B5^3, 2398
6883	Ethylene glycol or Ethanediol $HOCH_2CH_2OH$	62.07		198, 93^{13}	-11.5	1.1088$^{20/4}$	1.4318^{20}	**w, al, eth, ace**	B1^4, 2369
6884	Ethylene glycol, bis(chloroacetate) $ClCH_2CO_2CH_2CH_2O_2CCH_2Cl$	215.03	pr (eth-peth)	142-4^{2}	45-6			eth	B2^3, 448
6885	Ethylene glycol, bis(2-chloroethyl)ether $ClCH_2CH_2OCH_2CH_2OCH_2CH_2Cl$	187.07		230, 118^{10}		1.197$^{20/20}$	1.4592^{25}		B1^4, 2379
6886	Ethylene glycol, diacetate or Ethylene glycol diacetate $CH_3CO_2CH_2CH_2O_2CCH_3$	146.14		190	-31	1.1063$^{20/20}$	1.4159^{20}	**w, al, eth, ace, bz**	B2^4, 217
6887	Ethylene glycol, dibenzoate $C_6H_5CO_2CH_2CH_2O_2CC_6H_5$	270.28	rh pr (eth)	>360d	73-4			eth	B9^3, 536
6888	Ethylene glycol, dibutyrate $C_3H_7CO_2CH_2CH_2O_2CC_3H_7$	202.25		240, 118-21^{11}		1.0005$^{20/4}$	1.4262^{20}	al, eth	B2^4, 796
6889	Ethane-1,1-di-4-tolyl $(4CH_3C_6H_4)_2CHCH_3$	210.32		298-9, 153-6^{11}	<-20	0.974$^{20/4}$		bz	B5^4, 1948
6890	Ethylene glycol, diethylether or 1,2-diethoxyethane $C_2H_5OCH_2CH_2OC_2H_5$	118.18		123.5		0.8484^{20}	1.3860^{20}	al, eth, ace, bz	B1^4, 2379
6891	Ethylene glycol, diformate $HCO_2CH_2CH_2O_2CH$	118.09		174		1.193$^{0/4}$	1.3580	al, eth	B2^4, 37
6892	Ethylene glycol, dilaurate or Ethyleneglycoldidoclecanoate $C_{11}H_{23}CO_2CH_2CH_2O_2CC_{11}H_{23}$	426.68	pl (al)	188^{20}	56.6			al, eth	B2^4, 1094

PHYSICAL CONSTANTS OF ORGANIC COMPOUNDS (Continued)

No.	Name, Synonyms, and Formula	Mol. wt.	Color, crystalline form, specific rotation and λ_{max} (log ε)	b.p. °C	m.p. °C	Density	n_D	Solubility	Ref.
6893	Ethylene glycol, dimethylether or 1,2-dimethoxyethane $CH_3OCH_2CH_2OCH_3$	90.12		83-4	-58	$0.8628^{20/4}$	1.3796^{20}	w, al, eth, ace, bz	$B1^4$, 2376
6894	Ethylene glycol, dimyristate or Ethylene glycol ditetradecylate $C_{13}H_{27}CO_2CH_2CH_2O_2CC_{13}H_{27}$	482.79	cr (eth or ace)		65			eth, ace, bz	$B2^4$, 1133
6895	Ethylene glycol, dinitrate $O_2NOCH_2CH_2ONO_2$	152.06	ye	197-200	-22.3	$1.4918^{20/4}$		al, eth	$B1^4$, 2413
6896	Ethylene glycol, dinitrite $ONOCH_2CH_2ONO$	120.06		98	<-15	$1.2156^{0/4}$		al, eth	$B1^4$, 2411
6897	Ethylene glycol, dipalmitate or Ethanediol dihexadecylate $C_{15}H_{31}CO_2CH_2CH_2O_2CC_{15}H_{31}$	538.90	lf or nd (al-chl)	226 (vac)	72	0.8594^{78}		eth, ace	$B2^4$, 1169
6898	Ethylene glycol, di-phenyl ether $C_6H_5OCH_2CH_2OC_6H_5$	214.26	lf (al)	$180\text{-}5^{12}$	98			eth, chl	$B6^4$, 573
6899	Ethylene glycol, dipropionate $C_2H_5CO_2CH_2CH_2O_2CC_2H_5$	174.20			211	1.020^{15}		al, eth, ace	$B2^4$, 715
6900	Ethylene glycol, distearate or Ethylene glycol di-octadecylate $C_{17}H_{35}CO_2CH_2CH_2O_2CC_{17}H_{35}$	595.00	lf	241^{20}	79			eth, ace	$B2^4$, 1223
6901	Ethylene glycol, dithiocyanate $NCSCH_2CH_2SCN$	144.21	rh pl or nd (w) ta (al or eth)	d	90			al, eth, ace	$B3^4$, 333
6902	Ethylene glycol, methyl, ethyl ether $CH_3OCH_2C_2OC_2H_5$	104.15		102		$0.8529^{20/4}$	1.3868^{20}	w, al, eth ace, bz	$B1^3$, 2078
6903	Ethylene glycol, monoacetate or 2-Hydroxyethyl acetate $CH_3CO_2CH_2CH_2OH$	104.11		187-9		1.108^{15}		w, al, eth	$B2^4$, 154
6904	Ethylene glycol, monoallyl ether $HOCH_2CH_2OCH_2CH{=}CH_2$	102.13		$159^{755}, 64^{15}$		$0.9580^{20/4}$	1.4358^{20}	w, al, bz	$B1^4$, 2388
6905	Ethylene glycol, monobenzoate $C_6H_5CO_2CH_2CH_2OH$	166.18		$150\text{-}1^{10}$	45			al	$B9^3$, 532
6906	Ethylene glycol, monobenzyl ether or Benzyl cellosolve $HOCH_2CH_2OCH_2C_6H_5$	152.19		$256, 138^{15}$	<-75	$1.0640^{20/4}$	1.5233^{20}	w, al, eth	$B6^4$, 2241
6907	Ethylene glycol, monobutyl ether or Butyl cellosolve $HOCH_2CH_2OC_4H_9$	118.18		$171, 50^4$		$0.9015^{20/4}$	1.4198^{20}	w, al, eth	$B1^4$, 2380
6908	Ethylene glycol, mono-*iso*-butyl ether or Isobutyl cellosolve $HOCH_2CH_2OCH_2CH(CH_3)_2$	118.18		159^{745}		$0.8900^{20/4}$	1.4143^{20}		$B1^4$, 2382
6909	Ethylene glycol, mono-β-chloroethyl ether or β-Chloroethyl cellosolve $HOCH_2CH_2OCH_2CH_2Cl$	124.57		180-85, $91\text{-}2^{13}$			1.4805^{19}	w, al, eth	$B1^3$, 2078
6910	Ethylene glycol, monoethyl ether or Ethyl cellosolve $HOCH_2CH_2OC_2H_5$	90.12		135, 35^{10}		$0.9297^{20/4}$	1.4080^{20}	w, al, eth, ace	$B1^4$, 2377
6911	Ethylene glycol, monohexyl ether or Hexyl cellosolve $HOCH_2CH_2OC_6H_{13}$	146.23		$208, 96^{13}$	-45.1	$0.8894^{20/20}$	1.4291^{20}	al, eth	$B1^4$, 2383
6912	Ethylene glycol, monomethyl ether or Methyl cellosolve $HOCH_2CH_2OCH_3$	76.10		125^{768}	-85.1	$0.9647^{20/4}$	1.4024^{20}	w, al, eth, ace, bz	$B1^4$, 2375
6913	Ethylene glycol, monophenyl ether or Phenyl cellosolve $HOCH_2CH_2OC_6H_5$	138.17		237, $134\text{-}5^{18}$		1.1020^{20}	1.5340^{20}	al, eth	$B6^4$, 571
6914	Ethylene glycol, monopropyl ether or Propyl cellosolve $HOCH_2CH_2OC_3H_7$	104.15		150^{743}		$0.9112^{20/4}$	1.4133^{20}	w, al, eth	$B1^4$, 2379
6915	Ethylene glycol, mono isopropyl ether or Isopropyl cellosolve $i\text{-}C_3H_7OCH_2CH_2OH$	104.15		144^{743}		$0.9030^{20/4}$	1.4095^{20}	w, al, eth, ace	$B1^4$, 2380
6916	Ethylene glycol, monostearate $C_{17}H_{35}CO_2CH_2CH_2OH$	328.54	(peth)	$189\text{-}91^3$	60-1	$0.8780^{60/4}$	1.4310^{60}	eth	$B4^2$, 1222
6917	Ethylene glycol sulfate $C_2H_4O_2SO_2$	124.1	nd or pr (bz -lig)	sub	99			al, eth, ace, bz	$B1^3$, 2110
6918	Ethylene glycol sulfite $C_2H_4O_2SO$	108.11		$173, 70\text{-}1^{20}$	-11	$1.4402^{20/4}$	1.4463^{20}	w, al, eth, ace, bz	$B1^4$, 2409
6919	Ethylene imine or Aziridine C_2H_5N	43.07		50^{756}		$0.8321^{20/4}$		w, al, eth, ace, bz	$B20^4$, 3
6920	Ethylene oxide or Epoxyethane C_2H_4O	44.06		13.2^{746}	-111	$0.8824^{10/10}$	1.3597^7	w, al, eth, ace, bz	$B17^4$, 3
6921	Ethylidene diacetate $CH_3CH(O_2CCH_3)_2$	146.14		$169, 65\text{-}7^{10}$	18.9	1.070^{25}	1.3985^{25}	al, eth	$B2^4$, 282
6922	Ethynyl methyl ether $CH_3OC{\equiv}CH$	56.06		50		$0.8001^{20/4}$	1.3812^{20}	al, eth	$B1^4$, 2211
6923	Ethynyl phenyl ether C_6H_5OCCH	118.14		$61\text{-}2^{25}$	-36	$1.0614^{20/4}$	1.5125^{20}	al, eth	$B6^4$, 565

No.	Name, Synonyms, and Formula	Mol. wt.	Color, crystalline form, specific rotation and λ_{max} (log ε)	b.p. °C	m.p. °C	Density	n_D	Solubility	Ref.
6924	Ethynyl propyl ether $C_3H_7OC{\equiv}CH$	84.12	....	75	......	$0.8080^{20/4}$	1.3935^{20}	al, 6924	$B1^4$, 2213
6925	α-Eucaine $C_{19}H_{27}NO_4$	333.43	pr (eth or al)	....	104-5	....	....	w, al, eth, bz, chl	B22, 194
6926	α-Eucaine, hydrochloride $C_{19}H_{27}NO_4 \cdot HCl$	369.89	pl (w + 1) pr	....	*ca* 200d	....	....	w, al	B22, 194
6927	β-Eucaine-*(d)* $C_{15}H_{21}NO_2$	247.34	pr (peth)	....	57-8	....	....	al, eth, bz, chl, peth	$B21^2$, 14
6928	β-Eucaine *(dl)* or Betacaine $C_{15}H_{21}NO_2$	247.34	pr (peth)	....	57-8	....	....	al, eth, bz, chl, peth	$B21^2$, 13
6929	β-Eucaine *(l)* $C_{15}H_{21}NO_2$	247.34	pr (peth)	....	57-8	....	....	al, eth, bz, chl, peth	$B21^2$, 14
6930	β-Eucaine, hydrochloride-*(dl)* $C_{15}H_{21}NO_2.HCl$	283.80	pl (w)	....	277-9	....	....	w, al, eth, chl	$B21^2$, 13
6931	Eugenol or 5-Allylguaiacol 4-$(CH_2{=}CHCH_2)$-2-$CH_3OC_6H_3OH$	164.20	cr (hx)	253.2, 130.5^{10}	−7.5	$1.0652^{20/4}$	1.5405^{20}	al 6931	$B6^3$, 5021
6932	Eugenol-acetate 4-$(CH_2{=}CHCH_2)$-2-$CH_3OC_6H_3O_2CCH_3$	206.24	pr (al)	$127\text{-}8^6$	30-1	$1.0806^{20/4}$	1.5205^{20}	al	$B6^3$, 5029
6933	Eupitone or Eupittonic acid $C_{25}H_{26}O_9$	470.48	nd (al-eth)	....	200	....	....	a: 6933	$B8^3$, 4427
6934	Euxanthic acid $C_{19}H_{16}O_{16}$	404.33	ye nd (w + 1) α−108 (+ lw)	....	130d (+ w), 162d (anh)	....	....	al	B31, 277
6935	Eucarvone $C_{10}H_{14}O$	150.22	....	$99\text{-}100^{22}$, 88^{16}	......	$0.9490^{20/4}$	1.50872^{20}	eth, ace	JOCEA 26, 1609
6936	Evernic acid $C_{17}H_{16}O_7$	332.31	nd (w or ace) pr (al)	....	170	....	....	....	$B10^3$, 1488
6937	Evodiamine *(d)* or Rhetsine $C_{19}H_{17}N_3O$	303.36	yesh lf (al) $[\alpha]^{15}{}_D$ + 352 (ace,c=0.5)	....	278	....	....	....	$B26^2$, 103
6938	Evodiamine, hydrate *(d)* $C_{19}H_{17}N_3O.H_2O$	321.38	pl (al)	....	146-7	....	....	....	$B24^2$, 72

No.	Name, Synonyms, and Formula	Mol. wt.	Color, crystalline form, specific rotation and λ_{max} (log ε)	b.p. °C	m.p. °C	Density	n_D	Solubility	Ref.
6939	Fagaramide $C_{14}H_{17}NO_3$	247.30	nd (bz, dil al or peth)		119.5			al, bz	B19^2, 299
6940	β-Fagarine or Skimmianine $C_{14}H_{13}NO_4$	259.27	pym oct (al)		177			al, chl	B27, 134
6941	α-Fagarine or Haplophine $C_{13}H_{11}NO_3$	229.24	pr (al)		142			al, eth, bz, chl	C51, 4402
6942	α-Farnesine or 3,7,11-trimethyl-1,3,6,10-dodecatetracene $C_{15}H_{24}$	204.36		$129\text{-}32^{12}$		$0.8410^{20/4}$	1.4836^{20}	eth, ace, peth, lig	B1^3, 1067
6943	β-Farnesine $C_{15}H_{24}$	204.36		$121\text{-}2^{9}$		$0.8363^{20/4}$	1.4899^{20}	eth, ace, chl, aa	B1^4, 1133
6944	Farnesol *(trans, trans)* $(CH_3)_2C{=}CH(CH_2)_2C(CH_3){=}CH(CH_2)_2C(CH_3){=}CHCH_2OH$	222.37		160^{10}		$0.8846^{20/4}$	1.4877^{20}	al, eth, ace	B1^4, 2335
6945	Farnesol *(cis, trans)* $C_{15}H_{26}O$	222.37		$120^{0.3}$		$0.8846^{20/4}$	1.4877^{20}	al, eth, ace	B1^4, 2335
6946	Fenchane *(d)* $C_{10}H_{18}$	138.25		$151\text{-}2^{765}$		$0.8345^{20/4}$	1.44714^{20}	al, eth	B5^3, 256
6947	Fenchane *(dl)* $C_{10}H_{18}$	138.25		$151\text{-}2^{265}$		$0.8345^{20/4}$	1.4471^{20}	al, eth	B5^3, 256
6948	α-Fenchene *(d)* $C_{10}H_{16}$	136.24	$[\alpha]^{14/}{}_D+29$	155-6		$0.8660^{20/4}$	1.4713^{20}	al, eth, ace	B5^1, 86
6949	α-Fenchene *(dl)* or Isopinene $C_{10}H_{16}$	136.24		154-6		$0.8660^{20/4}$	1.4705^{20}	al, eth, ace	B5^3, 389
6950	α-Fenchene *(l)* $C_{10}H_{16}$	136.24	$[\alpha]^{20/}{}_D$-43.8	158-9		$0.8670^{20/4}$	1.4713^{20}	al, eth, ace	B5^4, 466
6951	β-Fenchene $C_{10}H_{16}$	136.24	$[\alpha]'_D$ 62.91 (al)	151-3		$0.8591^{20/4}$	$1.4645^{25/}{}_D$		B5^4, 465
6952	Fenchone *(d)* or Trimethyl norcamphor $C_{10}H_{16}O$	152.24	$[\alpha]^{20/}{}_D$ + 66.9 (al)	193.5, 80^{20}	6	$0.9465^{20/4}$	1.4623^{20}	al, eth, ace	B7^3, 392
6953	Fenchone-*(dl)* $C_{10}H_{16}O$	152.24		193-4, $72\text{-}3^{12}$	-18	$0.9501^{15/15}$	1.4702^{20}	al, eth, ace	B7^3, 393
6954	Fenchone *(l)* $C_{10}H_{16}O$	152.24	$[\alpha]^{23}{}_D$ −66.94 (al)	192-4	5(8.5)	0.948^{20}	1.4636^{20}	al, eth, ace	B7^3, 392
6955	Fenchyl alcohol *(dl)* $C_{10}H_{18}O$	154.25		α202-301, β201	α38-9, β6			al, eth	B6;3, 288
6956	Filixic acid BBB or Filicin $C_{36}H_{44}O_{12}$	668.74	cr (AcOEt-ace)		172-4			bz, chl	B8^3, 4436
6957	Ferrocene or Dicyclopentadienyl iron $C_{10}H_{10}Fe$	186.04		249	172.5-3.0				KHOC, 1525
6958	Flavaniline or 2-(*p*-Aminophenyl) lepidine $C_{16}H_{14}N_2$	234.30	pr (bz)	$133\text{-}41^{15}$	97			al, bz	B22, 469
6959	Flavanone or 2,3-dihydro-2-phenyl-1,4-benzopyrone $C_{15}H_{12}O_2$	224.26	nd (lig)		76			ace, bz	B17^4, 5338
6960	Flavanone, 4′-methoxy-3′,5,7-trihydroxy or Hesperetin $C_{16}H_{14}O_6$	302.28	pl (dil al + ½ w)	sub $205^{0.004}$	227-8			al, eth	B18^4, 3215
6961	Flavanone, 3′,4′,5,7-tetra hydroxy or Eriodictyol $C_{15}H_{12}O_6$	288.26	Pa br nd (dil al + 1.5 w)		267 d			al, aa	B18^4, 3214
6962	Flavone or 2-Phenyl-α-benzopyrone $C_{15}H_{10}O_2$	222.24	nd (lig)cr (30% al)		100			al, eth, ace, bz, chl	B17^4, 5413
6963	Flavone, 6-bromo $C_{15}H_9BrO_2$	301.14	nd (al)		191-2			al	B17^4, 5417
6964	Flavone, 5,7-dihydroxy or Chrysin $C_{15}H_{10}O_4$	254.24	Pa ye pl or pr (MeOH) nd (sub)	sub	275			al, ace	B18^4, 1766
6965	Flavone, 5,7-dihydroxy 4′-methoxy or Acacetin $C_{16}H_{12}O_5$	284.27	Pa ye nd (al)		261			ace	B18^4, 2683
6966	Flavone, 5,7-dihydroxy-6-methoxy $C_{16}H_{12}O_5$	284.27	ye nd (al)		231-2			al, eth, ace, aa	B18^4, 2671
6967	Flavone, 3-hydroxy or Flavanol $C_{15}H_{10}O_3$	238.24	Pa ye nd (al)		169-70			al	B17^4, 6428
6968	Flavone, 2′,3,3′,5,7-pentahydroxy $C_{15}H_{10}O_7$	302.24	ye nd (aa + 1.5 w)		300			al	B18^4, 3468
6969	Flavone, 2′,3,4′,5,7-pentahydroxy or Morin $C_{15}H_{10}O_4$	302.24	Pa ye nd (+1w, dil al)		303-4			al, bz	B18^4, 3468

No.	Name, Synonyms, and Formula	Mol. wt.	Color, crystalline form, specific rotation and λ_{max} (log ε)	b.p. °C	m.p. °C	Density	n_D	Solubility	Ref.
6970	Flavone, 2′,3,5,5′,7-pentahydroxy $C_{15}H_{10}O_7$	302.24	red ye cr (dil al + 1 w)		306-8				B18^{4}, 3470
6971	Flavone, 3,3′,4′,5,7-pentahydroxy or Quercitin $C_{15}H_{10}O_7$	302.24	ye rd (dil al + 2w)	sub	316-7			al, ace, aa	B18^{4}, 3470
6972	Flavone, 3,3′,4′,7,8-pentahydroxy $C_{15}H_{10}O_7$	302.24	ye nd (dil al + 1w)		308d			al, ace	B18^{4}, 3506
6973	Flavone, 3,3′,5,5′,7-pentahydroxy $C_{15}H_{10}O_7$	302.24	ye nd		>300				B18^{2}, 239
6974	Flavone, 3′,4′,5,5′,7-pentahydroxy or Tricetin $C_{15}H_{10}O_7$	302.24	ye nd(dil al + w)		>330d				B18^{4}, 3454
6975	Flavone, 2′,3,5,7-tetrahydroxy or Datiscetin $C_{15}H_{10}O_6$	286.24	Pa ye nd (al, aq aa)		277-8			al, eth, ace	B18^{4}, 3281
6976	Flavone, 3,3′,4′,7-tetrahydroxy or Fisetin $C_{15}H_{10}O_6$	286.24	lf ye nd (dil al + 1w)	sub ^{760}	330			al, ace	B18^{4}, 3304
6977	Flavone, 3′,4′,5,7-tetrahydroxy or Luteolin $C_{15}H_{10}O_6$	286.24	ye nd (dil al + 1w)	sub ^{760}	329-30 d			al, eth	B18^{4}, 3261
6978	Flavone, 3,4′,5,7-tetrahydroxy or Kaempferol $C_{15}H_{10}O_6$	286.24	ye nd (al + 1w) (aa)		276-8			al, ace	B18^{4}, 3283
6979	Flavone, 4′,5,7-trihydroxy or Apigenin $C_{15}H_{10}O_5$	270.24	ye nd (py-w) lf (al))	sub ^{760}	347-8			al, py	B18^{4}, 2682
6980	Flavone, 5,6,7-trihydroxy or Baicalein $C_{15}H_{10}O_5$	270.24	ye pr (al)		264-5d			al, eth, ace	B18^{4}, 2671
6981	Floridoside or 2-o-α-D-Galactosyl glycerol $C_9H_{18}O_8$	254.24	pr (al) $[\alpha]'_D$ + 15.1 (w)		86-7			w	B17^{4}, 2995
6982	Fluoran or 9-Hydroxy-9-xanthene-o-benzoic acid acetone $C_{20}H_{12}O_3$	300.31	nd (al + 2 al)		182-3			al	B19^{4}, 2903
6983	Fluoran, 1,6-dihydroxy $C_{20}H_{12}O_5$	332.31	ye nd		>260				B19^{4}, 2902
6984	Fluoran, 2,6-dihydroxy $C_{20}H_{12}O_5$	332.31	ye nd (al)		177				B19^{4}, 2902
6985	Fluoran, 3,5-dihydroxy $C_{20}H_{12}O_5$	332.31	ye gr nd or pl (aa)		179				B19^{4}, 2903
6986	Fluoranthene or 1,2-Benzacenaphthene $C_{16}H_{10}$	202.26	pa ye nd or pl (al)	375, 217^{50}		1.252$^{0/4}$		al, eth, bz, aa	B5^{3}, 2276
6987	Fluorene or 2,3-Benzindene $C_{13}H_{10}$	166.22	lf (al)	293-5	116-7	1.203$^{0/4}$		eth, ace, bz	B5^{3}, 1936
6988	Fluorene, 2-acetamido or N(2-Fluorenyl) acetamide $C_{15}H_{13}NO$	223.27	nd (50% al or 50% aa)		194			al, eth, aa	B12^{3}, 3287
6989	Fluorene, 9-acetamido $C_{15}H_{13}NO$	223.27	nd (aa)		262				B12^{3}, 3299
6990	Fluorene, 1-amino-9-hydroxy $C_{13}H_{11}NO$	197.24	dk red rd (w)		142			al, eth, bz, aa	B13^{2}, 435
6991	Fluorene, 2-amino $C_{13}H_{11}N$	181.24	lo pl or nd (dil al)		131-2			al, eth	B12^{3}, 3285
6992	Fluorene, 2-amino-9-hydroxy $C_{13}H_{11}NO$	197.24	irid nd (al)		200-1			al	B13^{3}, 2023
6993	Fluorene, 4-amino-9-hydroxy $C_{13}H_{11}NO$	197.24	ye (60% al)		183-4			al	B13^{3}, 2023
6994	Fluorene, 9-amino $C_{13}H_{11}N$	181.24	nd (lig)		64-5			al, eth, ace, bz, chl	B12^{3}, 3297
6995	Fluorene, 9-benzhydrylidene or ω, ω -Diphenyl dibenzo fulvene $C_{26}H_{18}$	330.43	ye (bz)		229.5			chl	B5^{3}, 2625
6996	Fluorene, 9-benzylidene or ω -Phenyl dibenzo fulvene $C_{20}H_{14}$	254.33	lf (al)		76			al, bz	B5^{3}, 2464
6997	Fluorene, 2-bromo $C_{13}H_9Br$	245.12	nd or pl (al)	185^{135}	113-4			al, chl, aa	B5^{3}, 1943
6998	Fluorene, 9-bromo $C_{13}H_9Br$	245.12	(lig or al)		104-5			al, ace	B5^{3}, 1944
6999	Fluorene, 9 (3-bromobenzylidene) $C_{20}H_{13}Br$	333.23	ye nd (aa)		92-3			al, MeOH	B5^{1}, 358
7000	Fluorene, 9 (4-bromobenzylidene) $C_{20}H_{13}Br$	333.23	ye nd (aa, AcOEt-chl)		147-8			aa	B5^{2}, 640

No.	Name, Synonyms, and Formula	Mol. wt.	Color, crystalline form, specific rotation and λ_{max} (log ε)	b.p. °C	m.p. °C	Density	n_D	Solubility	Ref.
7001	Fluorene, 9 (2-chlorobenzylidene) $C_{20}H_{13}Cl$	288.78	ye nd (aa, MeOH)	$180^{0.7}$	69-70			al, aa	B5³, 2464
7002	Fluorene, 9-(3-chlorobenzylidene) $C_{20}H_{13}Cl$	288.78	Pa ye pr or pym (MeOH)		90.5			al, aa	B5¹, 358
7003	Fluorene, 9-(4 chlorobenzylidene) $C_{20}H_{13}Cl$	288.78	ye nd (aa,al)		151			al, aa	B5¹, 358
7004	Fluorene, 9-Cinnamylidene *(trans)* $C_{22}H_{16}$	280.37	pa ye nd (aa)		155			al, chl	B5³, 2533
7005	Fluorene, 2,7-diamino $C_{13}H_{12}N_2$	196.25	nd (w) pr (bz) pl (eth)		165-7			al, chl	B13³, 507
7006	Fluorene, 2,7-dichloro $C_{13}H_8Cl_2$	235.11	pl or nd (bz)	sub	128			bz, chl	B5³, 1942
7007	Fluorene, 9,9-dichloro $C_{13}H_8Cl_2$	235.11	rh pr (bz-eth) nd (peth)		103			al, eth, ace, bz	B5³, 1942
7008	Fluorene, 1,8-dimethyl-9-(2-tolyl) $C_{22}H_{20}$	284.40			168-9			al, eth, ace	C50, 11293
7009	Fluorene, 2-hydroxy $C_{13}H_{10}O$	182.22	lf (w) nd (chl)		171-4			al, eth, ace, aa	B6³, 3487
7010	Fluorene, 9-hydroxy or 9-Fluorenol $C_{13}H_{10}O$	182.22	hex nd (w or peth)		154			eth, ace, bz	B6³, 3489
7011	Fluorene, 9-hydroxy-9-phenyl $C_{19}H_{14}O$	258.32	ye or col pr (lig)		108-9			bz, aa	B6³, 3732
7012	Fluorene, 9-methyl $C_{14}H_{12}$	180.25	pr	$154\text{-}6^{15}$	46-7	$1.0263^{66/4}$	1.610^{66}	al, eth, ace, bz, chl	B5³, 1992
7013	Fluorene, 9-(2-methylbenzylidene) $C_{21}H_{16}$	268.36	nd or pr (aa)		109.5			al, bz	B5¹, 359
7014	Fluorene, 9-(4-methylbenzylidene) $C_{21}H_{16}$	268.36	nd (aa) pr (al)		97.5			al, bz	B5¹, 359
7015	Fluorene, 9-methylene or Dibenzofulvene $C_{14}H_{10}$	178.23			53			al, eth, ace, bz	B5³, 2147
7016	Fluorene, 2-nitro $C_{13}H_9NO_2$	211.22	nd (50% aa or ace)		158			ace, bz	B5³, 1948
7017	Fluorene, 3-nitro $C_{13}H_9NO_2$	211.22	pa ye nd (al, chl-peth)		106			al, ace	B5³, 1949
7018	Fluorene, 9-nitro $C_{13}H_9NO_2$	211.22	gr ye nd (al) lf (bz)		181-2d			eth, ace,bz, chl, aa	B5³, 1949
7019	Fluorene, 9-phenyl $C_{19}H_{14}$	242.32	nd or lf (al or bz)		148			al, bz, aa, chl	B5³, 2385
7020	2-Fluorene sulfonic acid $C_{13}H_{10}SO_3$	246.28	nd (aa)		155 (+1w)			w, al, ace, chl	B11³, 440
7021	9-Fluorenone or 9-Oxofluorene $C_{13}H_8O$	180.21	ye rh bipym (al, bz-peth)	341.5	84	$1.1300^{99/4}$	1.6369^{99}	al, eth, ace, bz	B7³, 2330
7022	9-Fluorenone, 1-amino $C_{13}H_9NO$	195.22	ye nd (dil al)		118-20			al, eth, ace	B14³, 285
7023	9-Fluorenone, 2-amino $C_{13}H_9NO$	195.22	red vt pr (al)		163			al, eth, bz, aa	B14³, 286
7024	9-Fluorenone, 3-amino $C_{13}H_9NO$	195.22	ye nd (w or dil al)		158-9			al	B14³, 289
7025	9-Fluorenone, 4-amino $C_{13}H_9NO$	195.22	red nd (al)		145			al, eth, ace, chl, aa	B14³, 289
7026	9-Fluorenone, 2-bromo $C_{13}H_7BrO$	259.10	ye nd (al or aa)		149			ace, bz, chl, aa	B7³, 2340
7027	9-Fluorenone, 2-chloro $C_{13}H_7ClO$	214.65	og ye nd (dil al)	sub	125-6			al	B7³, 2338
7028	9-Fluorenone, 1,8-dimethyl $C_{15}H_{12}O$	208.26	ye		197-8			chl, aa	C62, 7611
7029	9-Fluorenone, 2-nitro $C_{13}H_7NO_3$	225.20	ye nd or lf (aa)	sub	222-3			ace	B7³, 2344
7030	9-Fluorenone, oxime $C_{13}H_9NO$	195.22	nd (chl-peth or bz)		195-6			al, chl	B7³, 2335
7031	Fluorene, 2,3,7-trinitro $C_{13}H_7N_3O_7$	315.20	Pa ye nd (aa)		180-1			bz, chl	B7³, 2348

PHYSICAL CONSTANTS OF ORGANIC COMPOUNDS (Continued)

No.	Name, Synonyms, and Formula	Mol. wt.	Color, crystalline form, specific rotation and λ_{max} (log ε)	b.p. °C	m.p. °C	Density	n_D	Solubility	Ref.
7032	9-Fluorenone, 2,4,7-trinitro $C_{13}H_5N_3O_7$	315.20	pa ye nd (aa or bz)	...	176	...	...	ace, bz, chl	B7³, 2348
7033	9-Fluorenone-1-carboxamide $(C_{13}H_7O)CONH_2$	223.23	ye nd (al)	...	229-30	...	...	ace	B10³, 3369
7034	9-Fluorenone-1-carboxylic acid $(C_{13}H_7O)CO_2H$	224.22	og, red nd (dil al)	...	192-4	...	...	al, eth	B10³, 3368
7035	9-Fluorenone-1-carboxylic acid, ethyl ester $(C_{13}H_7O)CO_2C_2H_5$	252.27	ye nd (dil al)	...	84-5	...	...	al, eth	B10³, 3369
7036	9-Fluorenone-1-carboxylyl chloride $(C_{13}H_7O)COCl$	242.66	pa ye nd (bz)	...	140	...	...	al, eth, bz	B10, 774
7037	9-Fluorenone-2-carboxylic acid $(C_3H_7O)_2CO_2H$	224.22	ye nd (al or aa)	sub 340	338	...	...	al, aa	B10³, 3370
7038	9-Fluorenone-2-carboxylic acid, methyl ester $C_{15}H_{10}O_3$ 2-$(C_3H_7O)CO_2CH_3$	238.24	ye nd (MeOH)	...	181	...	...	al, eth, ace	B10³, 3370
7039	9-Fluorenone-3-carboxylic acid $C_{14}H_8O_3$	224.22	ye (aa, MeOH)	...	299	...	...	al	B10³, 3372
7040	9-Fluorenone 4-carboxylic acid $C_{14}H_7O_3$	224.22	ye nd (al)	...	227	...	...	al, eth	B10³, 3372
7041	Fluorescein or 3′,4′-Dehydroxy fluoran $C_{20}H_{12}O_5$	332.31	red rh pr	...	314-6d (sealed tube)	...	...	ace, py, MeOH	B19⁴, 2904
7042	Fluorescin or 2 (3,6-Dihydroxyxanthyl) benzoic acid $C_{20}H_{14}O_5$	334.33	col or ye nd (aa or eth) pl (bz)	...	125-7	...	...	al, eth, ace	B19⁴, 2904
7043	Fluoroacetic acid FCH_2CO_2H	78.04	nd	165	35.2	1.3693^{36}	...	al, w	B2⁴, 446
7044	Fluorophosphoric acid, diisopropyl ester $(i\text{-}C_3H_7O)_2POF$	184.15	...	62^{9}	...	1.055	1.3830^{25}	eth	B1⁴, 1480
7045	Folic acid or Pteroylglutamic acid, Vitamin Bc $C_{19}H_{19}N_7O_6$	441.40	ye og nd (w) $\alpha^{25}/_D$ + 23 (0.1 N NaOH, c=0.5)	...	250d (darkens)	...	...	al, aa, py	Am69, 1476
7046	Folinic acid or 5-Formyl-5,6,7,8-tetrahydropteroyl-*L*-glutamic acid $C_{20}H_{23}N_7O_7$	473.45	cr (w + 3) $[\alpha]^{25}/_D$ + 16.76 (5% Na_2CO_3, c=3.5)	...	248-50d	...	...	...	Am73, 1979
7047	Formaldehyde or Methanal HCHO	30.03	gas	−21	−92	$0.815^{20/4}$	...	w, al, **eth, ace,** bz	B1⁴, 3017
7048	Formaldehyde, bis-4-chlorophenyl acetal $H_2C(OC_6H_4Cl\text{-}4)_2$	269.13	...	$189\text{-}94^{6}$	70	...	...	eth, ace, bz	B6⁴, 833
7049	Formaldehyde, dibutyl acetal $H_2C(OC_4H_9)_2$	160.26	...	179.2	−58.1	$0.834^{20/D}$	$1.4072^{17.2}/_D$	...	B1⁴, 3029
7050	Formaldehyde, β,β′-(dichloro isopropyl) ethyl acetal $[(ClCH_2)_2CHO]CH_2[OC_2H_5]$	187.07	...	$96\text{-}8^{16}$	...	$1.182^{17/17}$	1.4491^{17}	eth	B1³, 2574
7051	Formaldehyde, diethyl acetal or Ethylal $CH_2(OC_2H_5)_2$	104.15	...	89	−66.5	$0.8319^{20/4}$	1.3748^{18}	w, **al, eth,** ace bz	B1⁴, 3027
7052	Formaldehyde, dimethyl acetal or Methylal $CH_2(OCH_3)_2$	76.10	...	45.5	−104.8	$0.8593^{20/4}$	1.3513^{20}	w,**al, eth, ace,** bz	B1⁴, 3026
7053	Formaldehyde, 2,4-dinitrophenyl hydrazone $[2,4\text{-}(NO_2)_2C_6H_3]NHN{=}CH_2$	210.15	ye cr (al) pr (lig)	...	167	...	...	...	B15³, 426
7054	Formaldehyde, dipropyl acetal $CH_2(OC_3H_7)_2$	132.20	...	140.5	−97.3	0.8345^{20}	1.3939^{19}	w, al, eth, ace, bz	B1⁴, 3029
7055	Formaldehyde, fluoro or Formyl fluoride FCHO	48.02	gas	−24	...	...	...	...	B2⁴, 42
7056	Formaldehyde, oxime or Formaldoxime $H_2C{=}NOH$	45.04	...	109^{15}	2.5	1.133	...	w, al, eth	B1⁴, 3055
7057	Formaldomedone $C_{17}H_{22}O_4$	292.36	nd (al or bz)	...	189-90	...	...	...	B7³, 4736
7058	Formamide $HCONH_2$	45.04	...	111^{20}	2.5	$1.1334^{20/4}$	1.4472^{20}	w, **al,** ace	B2⁴, 45
7059	Formamide, *N,N*-diethyl $HCON(C_2H_5)_2$	101.15	...	177-8, 68^{15}	...	0.9080^{19}	1.4321^{25}	w, al, eth, **ace,** bz	B4⁴, 346
7060	Formamide, *N,N*-dimethyl $HCON(CH_3)_2$	73.09	...	149-56, 39.9^{10}	−60.5	$0.9487^{20/4}$	1.4305^{20}	**w, al, eth, ace,** bz, chl	B4⁴, 171

No.	Name, Synonyms, and Formula	Mol. wt.	Color, crystalline form, specific rotation and λ_{max} (log ε)	b.p. °C	m.p. °C	Density	n_D	Solubility	Ref.
7061	Formamide, *N,N*-diphenyl $HCON(C_6H_5)_2$	197.24	rh (dil al)	337.5, 190^{13}	73-4	...	...	al, eth, bz	B12^{3}, 455
7062	Formamide, *N*-ethyl $HCONHC_2H_5$	73.09	...	197-9, 109.6^{30}	...	$0.9552^{20/4}$	1.4320^{20}	w, al, eth	B4^{4}, 346
7063	Formamide, *N*-(1-hydroxy-2,2,2-trichloro ethyl) or Chloral formamide $HCONH[CH(OH)CCl_3]$	192.43	cr	118	...	...	...	w, al, eth, ace	B2^{2}, 37
7064	Formamide, *N*-methyl $HCONHCH_3$	59.07	...	180-5, $102\text{-}3^{20}$	...	1.011^{19}	1.4319^{20}	w, al, ace	B4^{4}, 170
7065	Formamide, *N*-phenyl or Formanilide $HCONHC_6H_5$	121.14	mcl pr (lig-xyl)	271, 166^{14}	50	$1.1322^{50/50}$	...	al, eth, bz	B12^{3}, 453
7066	Formamide, *N*-2-tolyl $HCONH(C_6H_4CH_3\text{-}2)$	135.17	lf (al)	288	62	$1.086^{55/4}$	...	al	B12^{3}, 1852
7067	Formamide, *N*-3-tolyl $HCONH(C_6H_4CH_3\text{-}3)$	135.17	...	278^{724}d, $176\text{-}8^{17}$	<-18	...	...	w	B12^{3}, 1962
7068	Formamide, *N*-4-tolyl $HCONH(C_6H_4CH_3\text{-}4)$	135.17	nd	...	53	...	...	al, ace	B12^{3}, 2050
7069	Formamidine $HN{=}CHNH_2$	44.06	pr	d	81	...	...	w, al	B2^{4}, 82
7070	Formamidine, *N,N′*-diphenyl $C_6H_5N{=}CHNHC_6H_5$	196.25	nd (al)	> 250	142	...	...	al, eth, ace, bz, chl	B12^{3}, 456
7071	Formamidoxime or Isoretin $HON{=}CHNH_2$	60.06	rh nd (al)	d	114-5	...	...	al	B2^{4}, 84
7072	Formic acid or Methanoic acid HCO_2H	46.03	...	100.7, 50^{120}	8.4	$1.220^{20/4}$	1.3714^{20}	w, al, eth, ace bz	B2^{4}, 3
7073	Formic acid, allyl ester or Allyl formate $HCO_2CH_2CH{=}CH_2$	86.09	...	83.6	...	$0.9460^{20/4}$	...	al, eth,	B2^{3}, 46
7074	Formic acid, benzyl ester or Benzyl formate $HCO_2CH_2C_6H_5$	136.15	...	$202\text{-}3^{747}$, $84\text{-}5^{10}$	...	$1.081^{20/4}$	1.5154^{20}	al, eth, ace	B6^{4}, 2262
7075	Formic acid, butyl ester or Butyl formate $HCO_2C_4H_9$	102.13	...	106.8	−91.3	$0.8885^{20/4}$	1.3912^{20}	al, eth, ace	B2^{4}, 28
7076	Formic acid, *iso*butyl ester or Isobutyl formate $HCO_2\text{-}i\text{-}C_4H_9$	102.13	...	98.4	−95.8	$0.8854^{20/4}$	1.3857^{20}	al, eth, ace	B2^{4}, 29
7077	Formic acid, *sec*-butyl ester-*(dl)* or *sec*-Butyl formate $HCO_2\text{-}sec\text{-}C_4H_9$	102.13	...	97	...	$0.8846^{20/4}$	1.3865^{20}	al, eth, ace	B2^{4}, 29
7078	Formic acid, cyclohexyl ester or Cyclohexyl formate $HCO_2C_6H_{11}$	128.17	...	162.5^{750}	...	$1.0057^{0/4}$	1.4430^{20}	al, eth, aa	B6^{4}, 35
7079	Formic acid, ethyl ester or Ethyl formate $HCO_2C_2H_5$	74.08	...	54.5	−80.5	0.9168^{20}	1.3598^{10}	w, al, eth, ace	B2^{4}, 23
7080	Formic acid, heptyl ester or Heptyl formate $HCO_2C_7H_{15}$	144.22	...	178.1, 83^{30}	...	0.8784^{20}	1.4140^{20}	al, eth	B2^{4}, 31
7081	Formic acid, hexyl ester or Hexyl formate $HCO_2C_6H_{13}$	130.19	...	155.5	−62.6	$0.8813^{20/4}$	1.4071^{20}	al, eth	B2^{4}, 31
7082	Formic acid, hydrazide $HCONHNH_2$	60.06	ye lf or nd (al)	...	54	...	...	al, eth, bz, chl	B2^{4}, 85
7083	Formic acid, methyl ester or Methyl formate HCO_2CH_3	60.05	...	31.5	−99	$0.9742^{20/4}$	1.3433^{20}	w, al, eth	B2^{4}, 20
7084	Formic acid, octyl ester or Octyl formate $HCO_2C_8H_{17}$	158.24	...	198.8	−39.1	0.8744^{20}	1.4208^{15}	al, eth	B2^{4}, 31
7085	Formic acid, pentyl ester $HCO_2C_5H_{11}$	116.16	...	132.1	−73.5	$0.8853^{20/4}$	1.3992^{20}	al, eth	B2^{4}, 30
7086	Formic acid, isopentyl ester $HCO_2\text{-}i\text{-}C_5H_{11}$	116.16	...	124.2	−93.5	$0.8857^{20/4}$	1.3976^{20}	al, eth	B2^{4}, 30
7087	Formic acid, propyl ester or Propyl formate $HCO_2C_3H_7$	88.11	...	81.3	−92.9	$0.9058^{20/4}$	1.3779^{20}	al, eth	B2^{4}, 26
7088	Formic acid, isopropyl ester or Isopropyl formate $HCO_2CH(CH_3)_2$	88.11	...	68.2	...	$0.8728^{20/4}$	1.3678^{20}	al, eth, ace	B2^{4}, 27
7089	Formimidic acid, *N*-phenyl, ethyl ether $C_6H_5N{=}CHOC_2H_5$	149.19	...	213-5	...	$1.0051^{20/4}$	1.5279^{20}	eth, bz	B12^{3}, 455
7090	Formonitrolic acid or Methyl nitrolic acid $HON{=}CHNO_2$	90.04	nd (eth or eth-peth)	...	68d	...	...	w, al, eth	B2^{4}, 85
7091	Frangulin A (αβ-L-Rhamnoside) $C_{21}H_{20}O_9$	416.38	ye or red (al or AcOEt)	...	228	...	...	al, bz, aa	B17^{4}, 2535
7092	Fraxin or αβ-Glucoside of fraxetin $C_{16}H_{18}O_{10}$	370.31	ye nd (al) (w + 3)	...	205	...	...	...	B18^{4}, 2373

PHYSICAL CONSTANTS OF ORGANIC COMPOUNDS (Continued)

No.	Name, Synonyms, and Formula	Mol. wt.	Color, crystalline form, specific rotation and λ_{max} (log ε)	b.p. °C	m.p. °C	Density	n_D	Solubility	Ref.
7093	β-D-Fructose or Levulose $C_6H_{12}O_6$	180.16	pr or nd (w) orh, pr (al) $[\alpha]^{20/}{}_D$(mut) -133→-92 (w, c=2)		103-5d	$1.60^{20/4}$		w, al, ace	B1[4], 4401
7094	α-L-Fucose or 6-Deoxy-L-galactose $C_6H_{12}O_5$	164.16	nd (al) $[\alpha]^{20/}{}_D$(mut) -124.1→-75.6 (w, c=9)						B1[4], 4265
7095	Fucoxanthin $C_{42}H_{58}O_6$	658.92	red br pl (eth-peth) hex pl c + 2w, dil al) $[\alpha]^{18/}{}_D$ + 72.5 (chl)		168			al, eth	B18[4], 2820
7096	Fulvene C_6H_6	78.11		$7\text{-}8^{56}$			1.4920^{20}	bz	B5[4], 764
7097	Fulvene-6-vinyl *(trans)* 6-$(CH_2{=}CH)C_6H_5$	104.15		45^{12}	−35	0898^{20}		bz, chl	HCACA 47, 102
7098	Fumagacin $C_{32}H_{41}O_7$	537.68	nd (dil aa) $[\alpha]^{18/}{}_D$ -125 (chl, c=1)		212 d			eth, ace, bz, aa	Am78, 5275
7099	Fumaric acid or *trans*-Butenedioic acid $HO_2CCH{=}CHCO_2H$	116.08	nd mcl pr or lf (w)	$165^{1.7}$ sub	300-2 (sealed tube)	$1.635^{20/4}$		al	B2[4], 2202
7100	Fumaric acid, bromo or Bromofumaric acid $HO_2CCBr{=}CHCO_2H$	194.97	pr (AcOEt)	d 200	185-6			w, al	B2[4], 2224
7101	Fumaric acid, dibutyl ester or Dibutyl fumarate $C_4H_9O_2CCH{=}CHCO_2C_4H_9$	228.29		150^4		$0.9869^{20/4}$	1.4469^{20}	ace, chl	B2[4], 2210
7102	Fumaric acid, diisobutyl ester *i*-$C_4H_9O_2CCH{=}CHCO_2$-*c*-C_4H_9	228.29		170^{160}, 122^5		$0.9760^{20/4}$	1.4432^{20}	al, eth, ace	B2[4], 2211
7103	Fumaric acid, chloro $HO_2CCCl{=}CHCO_2H$	150.52	pl (aa)	sub	192-3			al, eth	B2[4], 2221
7104	Fumaric acid, chloro, diethyl ester $C_2H_5O_2CCCl{=}CHCO_2C_2H_5$	206.63		250d, 127^{10}		$1.1880^{20/4}$	1.4571^{20}	al, eth	B2[3], 1909
7105	Fumaric acid, chloro, dimethyl ester $CH_3O_2CCCl{=}CHCO_2CH_3$	178.57		224, 108^{15}		$1.2899^{25/4}$	1.4720^{18}	al, eth	B2[3], 1909
7106	Fumaric acid, diethyl ester $C_2H_5O_2CCH{=}CHCO_2C_2H_5$	172.18		214, 98^{10}	1-2	$1.0452^{20/4}$	1.4412^{20}	ace, chl	B2[4], 2207
7107	Fumaric acid, dimethyl $HO_2CC(CH_3){=}C(CH_3)CO_2H$	144.13	nd (w)		241				B2[4], 2243
7108	Fumaric acid, dimethyl ester $CH_3O_2CCH{=}CHCO_2CH_3$	144.13		193	103-4	$1.37^{20/4}$	1.40625^{111}	ace, chl	B2[4], 2205
7109	Fumaric acid, diphenyl ester $C_6H_5O_2CCH{=}CHCO_2C_6H_5$	268.27	nd (al)	219^{14}	161-2				B6[4], 628
7110	Fumaric acid, dipropyl ester $C_3H_7O_2CCH{=}CHCO_2C_3H_7$	200.24		110^5		$1.0129^{20/4}$	1.4435^{20}	al, eth	B2[4], 2209
7111	Fumaric acid, diisopropyl ester $(CH_3)_2CHO_2CCH{=}CHCO_2CH(CH_3)_2$	200.24		225-6				al, eth, ace	B2[4], 2209
7112	Fumaric acid, methyl or Mesaconic acid $HO_2CC(CH_3){=}CHCO_2H$	130.10	rh nd or mcl pr (eth, AcOEt)	sub	204-5	$1.466^{20/4}$		al, eth	B2[4], 2231
7113	Fumaronitrile $NCCH{=}CHCN$	78.07	nd (bz-peth)	186	96.8	0.9416^{111}	1.4349^{111}	w, al, eth, ace, bz	B2[4], 2219
7114	Fumaryl chloride $ClOCCH{=}CHCOCl$	152.97	pa ye lig	158-60, 63^{13}		1.408^{20}	1.5004^{18}		B2[4], 2217
7115	Fumaryl chloride, chloro $ClOCCCl{=}CHCOCl$	187.41	pa gr	184-7d, $73\text{-}5^{20}$		$1.564^{20/4}$	1.5206^{20}	eth, aa	B2[3], 1909
7116	Furan C_4H_4O	68.08		31.4	−85.6	$0.9514^{20/4}$	1.4214^{20}	al, eth, ace, bz	B17[4], 225
7117	Furan, 2-acetyl or 2-Furyl methyl ketone 2-$CH_3CO(C_4H_3O)$	110.11	cr (lig)	175, 67^{10}	33	1.098^{20}	1.5017^{20}	al, eth	B17[4], 4500
7118	Furan, 2-benzoyl or 2-Furyl phenyl ketone $C_6H_5CO(2\text{-}C_4H_3O)$	172.18		285, 164^{19}	<-15	1.1732^{20}	1.6055^{20}	al, eth	B17[4], 5184

No.	Name, Synonyms, and Formula	Mol. wt.	Color, crystalline form, specific rotation and λ_{max} (log ε)	b.p. °C	m.p. °C	Density	n_D	Solubility	Ref.
7119	Furan, 2-bromo or α-Furyl bromide $2\text{-}Br(C_4H_3O)$	146.97		102^{744}		1.6500^{20}	1.4980^{20}	al, eth, ace, bz	$B17^4$, 231
7120	Furan, 3-bromo or β Furyl bromide $3\text{-}Br(C_4H_3O)$	146.97		103		$1.6606^{20/4}$	1.4958^{20}	al, eth, ace, bz	$B17^4$, 232
7121	Furan, 2-*tert* butyl $2\text{-}(CH_3)_3C(C_4H_3O)$	124.19		119-20		$0.869^{20/4}$	1.4373^{20}	al, eth, ace	$B17^4$, 307
7122	Furan, 2-chloro or α-Furyl chloride $2\text{-}Cl(C_4H_3O)$	102.52		77.5^{744}		$1.1923^{70/4}$	1.4569^{20}	al, eth, ace	$B17^4$, 230
7123	Furan, 3-chloro or β-Furyl chloride $3\text{-}Cl(C_4H_3O)$	102.52		79^{742}		$1.2094^{20/4}$	1.4601^{20}	eth, ace	$B17^4$, 230
7124	Furan, 2,5-dibromo $2,5\text{-}Br_2(C_4H_2O)$	225.87	pl	$164\text{-}5^{764}$, 62^{13}		$2.27^{20/20}$	1.5455^{20}		$B17^4$, 232
7125	Furan, 2,5-di-*tert*-butyl $2,5\text{-}[tert\text{-}(C_4H_9)_2](C_4H_2O)$	180.29		210, $61\text{-}2^{17}$		$0.837^{20/4}$	1.4369^{20}	al, eth,	$B17^4$, 336
7126	Furan, 2,5-dichloro $2,5\text{-}Cl_2(C_4H_2O)$	136.97		115		1.371^{25}			$B17^4$, 230
7127	Furan, 2,5-dimethoxy $2,5(CH_3O)_2(C_4H_2O)$	128.13		145-7			$1.4168^{20/D}$		$B17^4$, 1993
7128	Furan, 2,4-dimethyl $2,4(CH_3)_2(C_4H_2O)$	96.13		94		$0.8993^{20/4}$	1.4371^{20}		$B17^4$, 287
7129	Furan, 2,5-dimethyl $2,5\text{-}(CH_3)_2(C_4H_2O)$	96.13		93-4	−62.8	$0.8883^{20/4}$	1.4363^{20}	al, eth, ace, bz	$B17^4$, 289
7130	Furan, 2,5-dinitro $2,5\text{-}(NO_2)_2(C_4H_2O)$	158.07	nd (w) pr (al)		102			eth	$B17^4$, 234
7131	Furan, 2,5-diphenyl $2,5\text{-}(C_6H_5)_2(C_4H_2O)$	220.27	nd or lf (dil al)	343-5	91			al, eth, ace, bz	$B17^4$, 682
7132	Furan, 2-ethyl $2\text{-}C_2H_5(C_4H_3O)$	96.14		$92\text{-}3^{768}$		$0.912^{15/15}$	1.4466^{23}	al, eth, ace, bz	$B17^4$, 284
7133	Furan, 2-furfuryl $(C_4H_3O)CH_2(C_4H_3O)$	148.16		94^{22}		$1.102^{20/4}$	1.5049^{20}	al, eth, ace	$B19^4$, 269
7134	Furan, 2-iodo or α-Furyl iodide $2\text{-}I\text{-}(C_4H_3O)$	193.97		$43\text{-}5^{15}$		$2.024^{20/4}$	1.5661^{20}	eth	$B17^4$, 232
7135	Furan, 3-iodo or β Furyl iodide $3\text{-}I(C_4H_3O)$	193.97		132.2^{732}, $37\text{-}8^{22}$		$2.045^{20/4}$	1.5610^{20}	eth	$B17^4$, 232
7136	Furan, 2-methyl or Sylvan $2\text{-}CH_3(C_4H_3O)$	82.10		63^{737}		$0.9132^{20/4}$	1.4342^{20}	al, eth	$B17^4$, 265
7137	Furan, 3-methyl $3\text{-}CH_3(C_4H_3O)$	82.10		65.5^{749}		$0.923^{18/4}$	1.4330^{19}	al, eth	$B17^4$, 276
7138	Furan, 2-nitro $2\text{-}O_2N(C_4H_3O)$	113.07	yesh mcl cr (peth)	$133\text{-}5^{123}$	29			al, eth	$B17^4$, 233
7139	Furan, 2-phenyl $2\text{-}C_6H_4(C_4H_3O)$	144.17		$107\text{-}8^{18}$		$1.083^{20/4}$	1.5920^{20}	ace, bz	$B17^4$, 542
7140	Furan, 2-propionyl or Ethyl Furyl Ketone $2\text{-}CH_3CH_2CO(C_4H_3O)$	124.14	cr	88^{14}	28	1.0626^{28}	1.4922^{25}	eth	$B17^4$, 4537
7141	Furan, 2-propyl $2\text{-}C_3H_7(C_4H_3O)$	110.16		$114\text{-}5^{750}$		$0.8876^{20/4}$	1.4549^{20}	al, eth, ace	$B17^4$, 296
7142	Furan, tetrahydro C_4H_8O	72.11		67	fr-108	$0.8892^{20/4}$	1.4050^{20}	al, eth, ace, bz	$B17^4$, 24
7143	Furan, tetrahydro-2,2-diethyl $2,2\text{-}(C_2H_5)_2(C_4H_6O)$	128.21		146		$0.8703^{20/4}$	1.4317^{20}	al, eth, ace,,bz	$B17^4$, 107
7144	Furan, tetrahydro,2-ethyl $2\text{-}C_2H_5(C_4H_7O)$	100.16		109		$0.8570^{19/4}$	1.4147^{19}	al, eth, ace, bz	$B17^4$, 78
7145	Furan, tetrahydro-2-methyl $2\text{-}CH_3(C_4H_7O)$	86.13		80		$0.8552^{20/4}$	1.4059^{21}	al, eth, ace, bz, chl	$B17^4$, 60
7146	Furan, tetrahydro-3-methyl $3\text{-}CH_3(C_4H_7O)$	86.13		86-7		$0.8642^{20/4}$	1.4122^{20}	al, eth, ace, bz	$B17^2$, 21
7147	Furan, tetrahydro-2-propyl $2\text{-}C_3H_7(C_4H_7O)$	114.19		139.4		$0.8547^{20/4}$	1.4242^{20}		$B17^4$, 93
7148	Furan, tetraiodo C_4I_4O	571.66	nd		165			al	$B17^4$, 233
7149	Furan, tetraphenyl $(C_6H_5)_4(C_4O)$	372.47		220	175			al, eth, ace,bz, aa	$B17^4$, 810
7150	Furan, 2,3,5-trichloro $2,3,5\text{-}Cl_3(C_4HO)$	171.41		147		1.50^{25}			$B17^4$, 231
7151	2-Furancarboxylic acid or α-Furoic acid $2\text{-}(C_4H_3O)CO_2H$	112.09	mcl nd or lf (w)	230-2, $141\text{-}4^{20}$	133-4			w, al, eth	$B18^4$, 3914

No.	Name, Synonyms, and Formula	Mol. wt.	Color, crystalline form, specific rotation and λ_{max} (log ε)	b.p. °C	m.p. °C	Density	n_D	Solubility	Ref.
7152	2-Furancarboxylic acid, 5-ethoxy $5\text{-}C_2H_5O(C_4H_2O)CO_2H\text{-}2$	156.14			140-1			al	B18^4, 4827
7153	2-Furancarboxylic acid, 5-iodo $5\text{-}I(C_4H_2O)CO_2H\text{-}2$	237.99			197d				B18^4, 3992
7154	2-Furancarboxylic acid, 5-methoxy $5\text{-}CH_3O(C_4H_2O)CO_2H\text{-}2$	142.11			136-8d			al	B18^4, 4827
7155	3-Furancarboxylic acid or β-Furoic acid $3\text{-}(C_4H_3O)CO_2H$	112.09	nd (w)	105-10^{12} sub	122-3			al, eth	B18^4, 4052
7156	3-Furancarboxylic acid, 4-methyl $4\text{-}CH_3(C_4H_2O)CO_2H\text{-}3$	126.11	nd (bz-peth)		138-9			w	B18^4, 4089
7157	3-Furancarboxylic acid, 5-methyl $5\text{-}CH_3(C_4H_2O)CO_2H\text{-}3$	126.11	(w)	sub	119			w, eth	B18^4, 4076
7158	2,3-Furandicarboxylic acid $2,3\text{-}(C_4H_2O)(CO_2H)_2$	156.10	pr (aa or sub)	sub	226			w, al	B18^4, 4477
7159	2,3-Furandicarboxylic acid, dimethyl ester $2,3\text{-}(C_4H_2O)(CO_2CH_3)_2$	184.15	(MeOH)		39			al, eth	B18^4, 4477
7160	2,4-Furandicarboxylic acid $2,4\text{-}(C_4H_2O)(CO_2H)_2$	156.10	lf (w + 1)	sub	266			al, ace	B18^4, 4479
7161	2,4-Furandicarboxylic acid, dimethyl ester $2,4\text{-}(C_4H_2O)(CO_2CH_3)_2$	184.15	pr (MeOH)		109-10				B18^4, 4480
7162	2,5-Furandicarboxylic acid or Dehydromucic acid $2,5\text{-}(C_4H_2O)(CO_2H)_2$	156.10	nd (w) lf (al)	sub	> 320				B18^4, 4481
7163	2,5-Furandicarboxylic acid, dimethyl ester $2,5\text{-}(C_4H_2O)(CO_2CH_3)_2$	184.15	nd (w) cr (MeOH)	154-6^{15}	112			al, eth, chl	B18^4, 4482
7164	3,4-Furandicarboxylic acid $3,4\text{-}(C_4H_2O)(CO_2H)_2$	156.10			217-8				B18^4, 4497
7165	2-Furanone, tetrahydro $C_4H_6O_2$	86.09		206, 89^{12}	−42	1.1286$^{16/0}$	1.4341^{20}	w, al, eth, ace, bz	B17^4, 4159
7166	2-Furanone, 5-methyl (d), tetrahydro $C_5H_8O_2$	100.12	$[\alpha]^{20/}{}_D$ + 13.5 undil	86-90^{14}					B17^2, 288
7167	2-Furanone, 5-methyl (dl), tetrahydro $C_5H_8O_2$	100.12		206, 83-4^{13}	−31	1.0465^{23}	1.4328^{20}		B17^4, 4176
7168	2-Furanone, 5-methyl (l), tetrahydro $C_5H_8O_2$	100.12	$[\alpha]^{20/}{}_D$ −4.6 eth 10%	78-80^8			1.4322^{20}		B17^4, 4176
7169	Furazan, 3,4-dimethyl or 3,4-dimethyl-1,2,5-oxadiazole $C_4H_6N_2O$	98.10		156^{764}	−7	1.0528$^{14/4}$	1.4237^{20}	al, eth	B27^2, 628
7170	Furfural or α Furaldehyde $(OC_4H_3)CHO$	96.09		161.7, 90^{65}	−38.7	1.1594$^{20/4}$	1.5261^{20}	al, eth, ace, bz, chl	B17^4, 4403
7171	Furfural acetone $(C_4H_3O)CH{=}CHCOCH_3$	136.15	nd	229d, 112-3^{10}	39-40	1.0496$^{57/4}$	1.5788^{25}	al, eth, chl, peth	B17^4, 4714
7172	Furfural, 5-bromo $5\text{-}Br(OC_4H_3)CHO\text{-}2$	174.98	cr (50% al)	112^{16}	82			al, eth	B17^4, 4456
7173	Furfural, 5-chloro $5\text{-}Cl(C_4H_2O)CHO\text{-}2$	130.53		70^1	31-3				B17^4, 4454
7174	Furfural, diacetate or Furfurylidene diacetate $(C_4H_3O)CH(O_2CCH_3)_2$	198.18	nd or pl (eth-peth)	220, 143-4^{20}	52-3			al, eth, bz	B17^4, 4413
7175	Furfural, diethyl acetal $(C_4H_3O)CH(OC_2H_5)_2$	170.21		191-2, 62-4^5		0.9994$^{20/20}$	1.4451^{20}	al	B17^4, 4412
7176	Furfural, 5-hydroxymethyl $5\text{-}HOCH_2(C_4H_2O)CHO\text{-}2$	126.11	nd (eth-peth)	114-6$^{0.5}$	35	1.2062$^{25/4}$	1.5627^{18}	w, al, eth, bz, chl	B18^4, 100
7177	Furfural, 5-methyl $5\text{-}CH_3(C_4H_2O)CHO\text{-}2$	110.11		187, 79-81^{12}		1.1072$^{18/4}$	1.5262^{20}	w, al, eth	B17^4, 4523
7178	Furfural, 5-nitro $5\text{-}NO_2(C_4H_2O)CHO\text{-}2$	141.08	pa ye (peth)	128-32^{10}	35-6			peth	B17^4, 4459
7179	Furfural, 5-nitro, semicarbazone $C_6H_6N_4O_4$	198.14	pa ye nd (w) darkens in light		237d				B17^4, 4467
7180	Furfural, oxime, *(anti)* or α-Furfuraldoxime $2\text{-}(C_4H_3O)CH{=}NOH$	111.10	nd (lig)		75-6			al, eth, bz, aa	B17^4, 4430
7181	Furfural, oxime *(syn)* or β Furfuraldoxime $2\text{-}(C_4H_3O)CH{=}NOH$	111.10	nd (lig)	201-8d, 98^9	91-2			al, eth, bz, chl	B17^4, 4428
7182	Furfural, phenylhydrazone $2\text{-}(C_4H_3O)CH{=}NNHC_6H_5$	186.21	ye lf (al)		97-8			al, eth	B17^4, 4432
7183	Furfural, tetrahydro $2\text{-}(C_4H_7O)CHO$	100.12		142-3^{779}, 45-7^{29}		1.0727$^{20/4}$	1.4366^{20}	w, eth,	B17^4, 4179
7184	Furfuryl alcohol or 2-Hydroxymethyl furan $2\text{-}(C_4H_3O)CH_2OH$	98.10	col ye	171^{750}, 68-9^{20}		1.1296$^{20/4}$	1.4868^{20}	w, al, eth	B17^4, 1242

PHYSICAL CONSTANTS OF ORGANIC COMPOUNDS (Continued)

No.	Name, Synonyms, and Formula	Mol. wt.	Color, crystalline form, specific rotation and λ_{max} (log ε)	b.p. °C	m.p. °C	Density	n_D	Solubility	Ref.
7185	Furfuryl alcohol, 5- methyl $5\text{-}CH_3(C_4H_2O)CH_2OH\text{-}2$	112.13		194-6$^{744}_{d}$ 81^{23}		1.0769$^{20/4}$	1.4853^{20}	al, eth	B17^4, 1279
7186	Furfuryl alcohol, tetrahydro $2\text{-}(C_4H_7O)CH_2OH$	102.13		177-8^{750}, 80-2^{20}		1.0544$^{20/4}$	1.4517^{20}	eth, ace	B17^4, 1095
7187	Furfuryl amine $2\text{-}(C_4H_3O)CH_2NH_2$	97.12		145-6, 80^{84}		1.0995$^{20/4}$	1.4908^{20}	w, al, eth	B18^4, 7068
7188	Furfuryl amine, tetrahydro $2\text{-}(C_4H_7O)CH_2NH_2$	101.15		151-2^{735}		0.9770$^{20/20}$	1.4551^{20}	w, al, eth	B18^4, 7034
7189	Furfuryl bromide $2\text{-}(C_4H_3O)CH_2Br$	161.00	pa ye	33-4^{2}		1.560$^{20/20}$	1.5380^{20}	eth	B17^4, 268
7190	2-Furfuryl bromide, tetrahydro $2\text{-}(C_4H_7O)CH_2Br$	165.03		168-70^{744}, 69-70^{22}		1.3653$^{20/4}$	1.4850^{20}	al, eth	B17^4, 62
7191	Furfuryl chloride $2\text{-}(C_4H_3O)CH_2Cl$	116.55		49^{26}		1.1783$^{20/4}$	1.4941^{20}	al, eth, ace, bz	B17^4, 268
7192	Furfuryl mercaptan $2\text{-}(C_4H_3O)CH_2SH$	114.16		155, 47^{12}		1.1319$^{20/4}$	1.5329^{20}		B17^4, 1255
7193	Furfurin $C_{15}H_{12}N_2O_3$	268.27	lt br nd or rh pr (w or eth)		116-7			al, eth	B27^2, 918
7194	Furfuryl ether or Difurfuryl ether $[(C_4H_3O)CH_2]_2O$	178.19		101^{2}		1.1405$^{20/4}$	1.5088^{20}		B17^2, 116
7196	Furfuryl methyl ether $(C_4H_3O)CH_2OCH_3$	112.14		131-3		1.0163$^{20/4}$	1.4570^{20}	al, eth	B17^4, 1243
7197	α -Furoic acid or 2-Furancarboxylic acid $2\text{-}(C_4H_3O)CO_2H$	112.09	mcl nd or lf (w)	230-2, 141-4^{20}	133-4			w, al, eth	B18^4, 3914
7198	α -Furoic acid, allyl ester $2\text{-}(C_4H_3O)CO_2(CH_2CH{=}CH_2)$	152.15		206-9		1.118$^{23/23}$	1.4945^{20}	eth, ace	B18^4, 3919
7199	α -Furoic acid, benzyl ester $2\text{-}(C_4H_3O)CO_2CH_2C_6H_5$	202.21	ye	179-81^{18}		1.1623$^{22/4}$	1.5550^{20}	eth, ace	B18^4, 3921
7200	α -Furoic acid, butyl ester or Butyl furoate $2\text{-}(C_4H_3O)CO_2C_4H_9$	168.19		233, 83-4^{1}		1.0555$^{20/4}$	1.4740	al, eth, bz, peth	B18^4, 3918
7201	α-Furoic acid, *iso*butyl ester or Isobutyl α-furoate $2\text{-}(C_4H_3O)CO_2\text{-i-}C_4H_9$	168.19		221-3, 97$^{13.5}$		1.0388$^{20/4}$	1.4676^{20}	al, eth, ace, bz	B18^2, 266
7202	α-Furoic acid, *sec*butyl ester $2\text{-}(C_4H_3O)CO_2CH(CH_3)C_2H_5$	168.19		67-9^{1}		1.0465$^{20/4}$		al, eth	B18^2, 266
7203	α-Furoic acid, 5-ethoxy $5\text{-}C_2H_5O(C_4H_2O)CO_2H\text{-}2$	156.14			140-1			al	B18^4, 4827
7204	α-Furoic acid, ethyl ester or Ethyl-α-furoate $2\text{-}(C_4H_3O)CO_2C_2H_5$	140.14	lf or pr	196.8, 128^{95}	34-5	1.1174$^{21/4}$	1.4797^{21}	al, eth, ace, bz, peth	B18^4, 3917
7205	α-Furoic acid, furfuryl ester $2\text{-}(C_4H_3O)CO_2CH_2(C_4H_3O)$	192.17	dimorphic	122^{2}	27.5	1.2384$^{25/25}$	1.5280^{20}	al, eth, ace, bz, peth	B18^4, 3936
7206	α-Furoic acid, heptyl ester or Heptyl-α-furoate $2\text{-}(C_4H_3O)CO_2C_7H_{15}$	210.27		116-7^{1}		1.0005$^{20/4}$		al, eth	B18^2, 267
7207	α-Furoic acid, hexylester or Hexyl-α-Furoate $2\text{-}(C_4H_3O)CO_2C_6H_{13}$	196.25		105-7^{1}		1.0170$^{20/4}$		al, eth, bz	B18^2, 267
7208	α-Furoic acid, 5-iodo $5\text{-}I(C_4H_2O)CO_2H\text{-}2$	237.98			197d				B18^4, 3992
7209	α-Furoic acid, 5-methoxy $5\text{-}CH_3O(C_4H_2O)CO_2H\text{-}2$	142.11			136-8d			al	B18^4, 4827
7210	α-Furoic acid, methyl ester or Methyl-α-Furoate $2\text{-}(C_4H_3O)CO_2CH_3$	126.11		181.3		1.1786$^{21/4}$	1.4860^{20}	al, eth, bz	B18^4, 3916
7211	α-Furoic acid, 3-methyl $3\text{-}CH_3(C_4H_2O)CO_2H\text{-}2$	126.11	nd (w)	sub	134				B18^4, 4067
7212	α-Furoic acid, 3-methyl, ethyl ester $3\text{-}CH_3(C_4H_2O)(CO_2C_2H_5)\text{-}2$	154.17	pl	205	47-8				B18^4, 4067
7213	α-Furoic acid, 3-methyl, methyl ester $3\text{-}CH_3(C_4H_2O)(CO_2CH_3)\text{-}2$	140.14	pl (al)	72-6^{2}	36-8				B18^4, 4067
7214	α-Furoic acid, 4-methyl $4\text{-}CH_3(C_4H_2O)CO_2H\text{-}2$	126.11	nd (bz-peth)		131-2				B18^4, 4073
7215	α-Furoic acid, 5-methyl $5\text{-}CH_3(C_4H_2O)CO_2H\text{-}2$	126.11	pl or nd (w)	105^{1}	109-10			al, eth, chl	B18^4, 4076
7216	α-Furoic acid, 5-methyl, methyl ester $5\text{-}CH_3(C_4H_2O)CO_2CH_3\text{-}2$	140.14		205, 98^{15}					B18^4, 4076
7217	α-Furoic acid, 5-nitro $5\text{-}O_2N(C_4H_2O)CO_2H\text{-}2$	157.09	pa ye pl (w)	sub	184			al, eth	B18^4, 3992
7218	α-Furoic acid, octyl ester or Octyl-α-furoate $2\text{-}(C_4H_3O)CO_2C_8H_{17}$	224.30		126-7^{1}		0.9885$^{20/4}$		al	B18^2, 267

No.	Name, Synonyms, and Formula	Mol. wt.	Color, crystalline form, specific rotation and λ_{max} (log ε)	b.p. °C	m.p. °C	Density	n_D	Solubility	Ref.
7219	α-Furoic acid, pentyl ester or Pentyl-α-Furoate 2-$(C_4H_3O)CO_2C_5H_{11}$	182.22		95-7^{1}		1.0335$^{20/4}$		al	B18^{2}, 266
7220	α-Furoic acid, isopentyl ester or Isopentyl-α-Furoate 2-$(C_4H_3O)CO_2$-i-C_5H_{11}	182.22		282		1.030$^{20/4}$	1.4274^{20}	al, bz, peth	B18^{2}, 226
7221	α-Furoic acid, 5-phenoxy 5-$C_6H_5O(C_4H_2O)CO_2H$-2	204.19			122-3			al	B18^{4}, 4827
7222	α-Furoic acid, propylester or Propyl-α-Furoate 2-$(C_4H_3O)CO_2C_3H_7$	154.17		210.9		1.0745$^{20/4}$	1.4737^{20}	al, eth, ace, bz, peth	B18^{4}, 3918
7223	α-Furoic acid, isopropyl ester or isopropyl-α-furoate 2-$(C_4H_3O)CO_2CH(CH_3)_2$	154.17		198-9		1.0655$^{24/4}$	1.4682^{24}	al, eth, ace, bz	B18, 275
7224	α-Furoic acid, tetrahydro 2-$(C_4H_7O)CO_2H$	116.12		145^{25}	21	1.1933$^{20/20}$	1.4612^{20}	w	B18^{4}, 3824
7225	α-Furonitrile 2-$(C_4H_3O)CN$	93.09		146^{738}		1.0822$^{20/4}$	1.4798^{20}	al, eth	B18^{4}, 3964
7226	α-Furoyl chloride or 2-Furan carboxylyl chloride 2-$(C_4H_3O)COCl$	130.53		173, 66^{10}	-2			eth, chl	B18^{4}, 3938
7227	β-Furoic acid or 3-Furan carboxylic acid 3-$(C_4H_3O)CO_2H$	112.09	nd (w)	105-10^{12}, sub	122-3			al, eth	B18^{4}, 4052
7228	β-Furoic acid, 4-bromo 4-$Br(C_4H_2O)CO_2H$-3	190.98	nd (w)		129			al, eth, bz, chl	B18^{4}, 3980
7229	β-Furoic acid, 5-bromo 5-$Br(C_4H_2O)CO_2H$-3	190.98	lf (w)		190-1			al, eth, chl	B18^{4}, 3980
7230	β-Furoic acid, 5-bromo, ethyl ester 5-$Br(C_4H_2O)(CO_2C_2H_5)$-3	219.04	pr	235^{767}, 134-6^{34}	17	1.528^{20}		al, eth	B18^{4}, 3981
7231	β-Furoic acid, 4-chloro 4-$Cl(C_4H_2O)CO_2H$-3	146.53	pl or pr (w)		149			al, eth, bz	B18^{4}, 3977
7232	β-Furoic acid, 5-chloro 5-$Cl(C_4H_2O)CO_2H$-3	146.53	lf (w)		179-80			al, eth, bz	B18^{4}, 3977
7233	β-Furoic acid, 5-chloro, methyl ester 5-$Cl(C_4H_2O)CO_2CH_3$-3	160.56			40-1			eth	B18^{4}, 3978
7234	β-Furoic acid, 2,5-dimethyl or Pyrotritaric acid 2,5-$(CH_3)_2(C_4HO)CO_2H$-3	140.14	nd (w)	sub	135			al, eth	B18^{4}, 4099
7235	β-Furoic acid, methyl ester 3-$(C_4H_3O)CO_2CH_3$	126.11		160, 79^{42}		1.1744$^{15/15}$	1.4676^{20}	al, ace	B18^{4}, 4052
7236	β-Furoic acid, 2-methyl 2-$CH_3(C_4H_2O)CO_2H$-3	126.11	cr (w)		102-3			al, eth	B18^{4}, 4072
7237	β-Furoic acid, 2-methyl, ethyl ester 2-$CH_3(C_4H_2O)CO_2C_2H_5$-3	154.17		85-7^{20}		1.0102$^{25/4}$	1.4620^{25}	eth	B18^{4}, 4072
7238	β-Furoic acid, 4-methyl 4-$CH_3(C_4H_2O)CO_2H$-3	126.11	nd (bz-peth)		138-9				B18^{4}, 4089
7239	β-Furoic acid, 5-methyl 5-$CH_3(C_4H_2O)CO_2H$-3	126.11	(w)	sub	119			w, eth	B18^{4}, 4076
7240	Furoin $C_{10}H_8O_4$	192.17	nd (al)		138-9			eth, MeOH	B19^{4}, 2543
7241	2-Furylacetic acid or 2-Furanacetic acid (2-$C_4H_3O)CH_2CO_2H$	126.11	lf (bz peth)	102-4$^{0.4}$	68-9			w, MeOH	B18^{4}, 4061
7242	2-Furylacetonitrile or Furfuryl cyanide (2-$C_4H_3O)CH_2CN$	107.11		75-80^{20}		1.0854$^{25/4}$	1.4693^{20}	al, eth	B18^{4}, 4062
7243	2-Furyl octyl ether (2-$C_4H_3O)OC_8H_{17}$	196.29		129-30^{18}		0.9214$^{28/4}$	1.4520^{28}	eth	B17^{4}, 1219
7244	2-Furyl phenyl ether (2-$C_4H_3O)OC_6H_5$	160.17		105-6^{18}		1.1010$^{23/4}$	1.5418^{23}	eth	B17^{4}, 1219
7245	2-Furyl phenyl ketone or 2-Benzoyl furan (2-$C_4H_3O)COC_6H_5$	172.18		285, 164^{19}	<-15	1.1732^{20}	1.6055^{20}	al, eth, bz	B17^{4}, 5184
7246	2-Furyl phenyl sulfide (2-$C_4H_3O)SC_6H_5$	176.23		119-20^{8}		1.1341^{26}	1.5976^{20}	al, eth	B17^{4}, 1220
7247	2-Furyl isopropyl ether or 2-isopropoxy furan (2-$C_4H_3O)O$-i-C_3H_7	126.16		135-6		0.9689$^{20/4}$	1.4419^{20}		B17^{4}, 1219
7248	2-Furyl methyl ether (2-$C_4H_3O)OCH_3$	98.10		110-1		1.0646$^{25/4}$	1.4468^{25}		B17^{4}, 1219

No.	Name, Synonyms, and Formula	Mol. wt.	Color, crystalline form, specific rotation and λ_{max} (log ε)	b.p. °C	m.p. °C	Density	n_D	Solubility	Ref.
7249	Galacitol or D-dulcitol $CH_2OH(CHOH)_4CH_2OH$	182.18	mcl pr (w)	275-8^1 sub	189	1.466	...	w	$B1^4$, 2844
7250	D-Galactonic acid-γ-lactone $C_6H_{10}O_6$	178.14	nd (w + 1) nd (al or AcOEt) $[\alpha]^{20}{}_D$ −65.5 (w)	...	112 (anh), 66 (hyd)	...	...	w	$B18^4$, 3026
7251	D-Galactose $C_6H_{12}O_6$	180.16	pl or pr (al) pr or nd (w + 1) $[\alpha]$ + 83.3 (w)	...	...	...	...	w, py	$B1^4$, 4336
7252	D-Galactose, 3,6-anhydro- $C_6H_{10}O_5$	162.14	lf (PrOH) $[\alpha]_D$ + 24 (w)	...	123-5	...	...	w	$B18^4$, 2278
7253	α-D-Galactose, 2,3,4,6-tetra-O-methyl $C_{10}H_{20}O_6$	236.27	$[\alpha]^{20}{}_D$ + 150 → + 114 (w)	172^{12}	71-3	...	...	w, al, eth	$B1^4$, 4371
7254	D-Galacturonic acid $C_6H_{10}O_7$	194.14	nd (w + 1) β-$[\alpha]^{20}{}_D$ + 27 (w) α-$[\alpha]_D$ + 98 → + 53 (w, c=10)	...	α 156, β 150	...	...	w, al	$B3^4$, 2000
7255	Galegine or 4-Guanidino-2-methyl-2-butene $(CH_3)_2C{=}CHCH_2N{=}C(NH_2)_2$	127.19	hyg	d	60-5	...	...	w, al	$B4^3$, 465
7256	Galipine or Galipoline methyl ether $C_{20}H_{21}NO_3$	323.39	pr (al eth) nd (peth)	...	115.5	...	...	al, eth, ace, bz	$B21^4$, 2656
7257	Gallein or 4,5-Dihydro fluorescein $C_{20}H_{12}O_5$	332.31	br red pw (+ 1.5w) red (anh)	...	>300	...	...	al, ace	$B19^4$, 3147
7258	Gallin or 4,5-Dihydroxyfluorescin $C_{20}H_{14}O_7$	366.33	nd (eth) turns red in air	...	...	...	...	al, ace, aa	B18, 368
7259	Gelsemine (d) or Gelseminine $C_{20}H_{22}N_2O_2$	322.41	cr (ace)	...	178	...	...	al, eth, ace, bz,chl	$B27^2$, 720
7260	Gelsemine, hydrochloride $C_{20}H_{22}N_2O_2.HCl$	358.87	pr (w)	...	326	...	...	w	$B27^2$, 720
7261	Geranial or Citral a $C_{10}H_{16}O$	152.24	$[\alpha]^{20}{}_D$ + 2.5 (w)	229, 118-9^{20}	...	0.8888^{20}	1.4898^{20}	**al, eth**	$B1^4$, 3569
7262	Geraniol or 2,7-Dimethyl-2,6-octadiene-1-ol $(CH_3)_2C{=}CHCH_2CH_2C(CH_3)CH_2CH_2OH$	154.25	...	230, 121^{18}	>-15	0.8894^{20}	1.4766^{20}	al, eth, ace	$B1^4$, 2277
7263	Geraniol formate $C_{11}H_{18}O_2$	182.26	...	229d, 113-4^{25}	...	$0.9086^{25/4}$	1.4659^{20}	al, eth, ace	$B2^4$, 35
7264	Geraniol, tetrahydro (d) or 3,7-Dimethyl-1-octanol $(CH_3)_2CH(CH_2)_3CH(CH_3)CH_2CH_2OH$	158.28	$[\alpha]^{20}{}_D$ + 4.09	212-3, 105-6^{10}	...	$0.8285^{20/4}$	1.4355^{20}	eth	$B1^4$, 1830
7265	Geraniol, tetrahydro or 3,7-Dimethyl-1-octanol $(CH_3)_2CH(CH_2)_3CH(CH_3)CH_2CH_2OH$	158.28	...	202-3, 106^{12}	...	$0.8308^{10/4}$	1.4367^{20}	al, eth, ace, bz	$B1^4$, 1830
7266	Geraniol, tetrahydro (l) or 3,7-Dimethyl-1-octanol $(CH_3)_2CH(CH_2)_3CH(CH_3)CH_2CH_2OH$	158.28	$[\alpha]^{27}{}_{548}$ − 3.67	212-3, 109^{15}	...	0.830^{18}	1.4370^{15}	eth	$B1^3$, 1768
7267	Geranyl bromide $(CH_3)_2C{=}CHCH_2CH_2C(CH_3)CH_2CH_2Br$	217.15	...	101-2^{12}, 47-8$^{0.005}$	...	$1.0940^{22/4}$	1.5027^{20}	al, eth	$B1^4$, 1059
7268	Germanidine $C_{37}H_{57}NO_{10}$	675.87	nd $[\alpha]^{24}{}_D$ − 30 (al)		221-2 (vac)	...	...	...	$B21^4$, 6806
7269	Germanitrine $C_{39}H_{59}NO_{11}$	717.90	nd (aq aw) $[\alpha]^{24}{}_D$ − 61 (Py, c=1)	...	228-9 (vac)	...	...	...	$B21^4$,6807
7270	Germerine $C_{37}H_{59}NO_{11}$	693.88	cr (ace) lf (bz) $[\alpha]^{25}{}_D$ + 15.7 (chl, c=1.02)	...	193-5d	...	...	al, ace, bz, chl	$B21^4$, 6808
7271	Germidine $C_{34}H_{53}NO_{10}$	635.80	pl (aq MeOH) $[\alpha]^{25}{}_D$ + 13 (chl)	...	242-4, 203-3	...	...	al, bz	$B21^4$, 6802

No.	Name, Synonyms, and Formula	Mol. wt.	Color, crystalline form, specific rotation and λ_{max} (log ε)	b.p. °C	m.p. °C	Density	n_D	Solubility	Ref.
7272	Germine $C_{27}H_{43}NO_8$	509.64	pr or cr (MeOH) $[\alpha]^{25}_{D}$ + 5 (95% al)		220			bz	B21[4], 6796
7273	Germinitrine $C_{39}H_{57}NO_{11}$	715.89	pr (dil ace) $[\alpha]^{24}_{D}$ −36 (Py, c=1.12)		175 (vac)				Am75, 4925
7274	Germitrine $C_{39}H_{61}NO_{12}$	735.92	cr (dil al) $[\alpha]^{25}_{D}$ + 11 (chl)		197-9			bz, chl	B21[4], 6809
7275	Gitogenin or Digine $C_{27}H_{44}O_4$	432.65	lf (bz) nd (eth) $[\alpha]^{20}_{D}$ −75 (chl, c=1)		271-2			al, chl	B19[4], 1050
7276	Gitoxigenin or Hydroxy digitoxin $C_{23}H_{34}O_5$	390.52	pr (AcOEt) pr (+w, dil al) $[\alpha]^{20}_{546}$ + 38.5 (MeOH, c=0.7)		234 (anh)			chl	B18[4], 2456
7277	Gitoxin or Anhydrogitalin $C_{41}H_{64}O_{14}$	780.96	pr (chl MeOH) $[\alpha]^{24}_{D}$ + 5 (py, c=1)		285d				B18[4], 2462
7278	D-Glucoascorbic acid or 3-keto-D-*keto*-heptonofuranolactone $C_7H_{10}O_7$	206.16	reds (+w) (ace MeOH Peth) $[\alpha]^{20}_{D}$ −22 (MeOH, c=1)		191 (anh) 140 (hyd)			w, al	B18[4], 3400
7279	D-*gluco*-Heptose $C_7H_{14}O_7$	210.19	rh pl (w) $[\alpha]^{20}_{D}$ −19.7 (w) (mut)		193 (210)			w	B1[4], 4436
7280	D-gluco-Methylose or D-Isorhamnose $C_6H_{12}O_5$	164.16	cr (AcOEt) $[\alpha]^{20}_{D}$ + 73→ + 29.7 (w, mut)		139.40			w, al	B1[4], 4260
7281	D-Gluconic acid or Dextronic acid $C_6H_{12}O_7$	196.16	nd (al eth) $[\alpha]^{25}_{D}$ −3.49→ + 12.95 (w, mut)		131			w	B3[4], 1255
7282	D-Gluconic acid-γ-lactone $C_6H_{10}O_6$	178.14	nd (al) $[\alpha]_D$ + 67.5→ + 6.2 (w)		134-6			al	B18[4], 3024
7283	D-Gluconic acid-δ-lactone $C_6H_{10}O_6$	178.14	nd (al) $[\alpha]^{25}_{D}$ + 63.5→ + 6.2 (w)						B18[4], 3018
7284	D-Gluconitrile $C_6H_{11}NO_5$	177.16	(i) cr (al, aa) (ii) pl (al) (i) $[\alpha]^{24}_{D}$ + 10.0 (w, c=1.8) (ii) $[\alpha]^{21}_{D}$ + 8.8 (w)		(i)146.8, (ii)120.5			w, py	B18[4], 3024
7285	D-Gluconic acid, phenylhydrazide $C_{12}H_{18}N_2O_6$	286.29	pr (w) $[\alpha]$ + 12(w)		204-5			w	B15[2], 122
7286	D-Gluconic acid, 5-oxo or 5-keto-D-gluconic acid $C_6H_{10}O_7$	194.14	cr or syr $[\alpha]_D$ −14.5 (w)		125-6			w, al	B3[4], 1993
7287	Gluconol (d) $C_6H_{10}O_4$	146.15	hyg nd $[\alpha]^{22}_{D}$ −7 (w)		60				B17[4], 2332

No.	Name, Synonyms, and Formula	Mol. wt.	Color, crystalline form, specific rotation and λ_{max} (log ε)	b.p. °C	m.p. °C	Density	n_D	Solubility	Ref.
7288	D-Glucose (equilib mixt) or Dextrose $C_6H_{12}O_6$	180.16	$[\alpha]^{20/}{}_D$ + 52.7(w)		146 (150)			w	$B1^4$, 4302
7289	α-D-Glucose $C_6H_{12}O_6$	180.16	reds cubes orh nd (al) $[\alpha]^{20/}{}_D$ + 112.2→ + 52.7 (w, c=4)		146d	$1.5620^{18/4}$		w	$B1^4$, 4304
7290	α-D-Glucose, monohydrate $C_6H_{12}O_6.H_2O$	198.18	lf pl or orh (w) $[\alpha]^{20/}{}_D$ + 102→ + 47.9		86	$1.54^{25/4}$		w	$B1^4$, 4306
7291	β-D-Glucose $C_6H_{12}O_6$	180.16	nd (al) (w + 1) $[\alpha]^{20/}{}_D$ + 17.5→ + 52.7 (w, c=4)		150	$1.5620^{18/4}$		w	$B1^4$, 4306
7292	α-D-Glucose, pentaacetate $C_{16}H_{22}O_{11}$	390.35	pl or nd (al) $[\alpha]^{20/}{}_D$ + 100.9 (al, c=0.5)	sub	112-3			eth	$B17^4$, 3276
7293	β-D-Glucose, pentaacetate $C_{16}H_{22}O_{11}$	390.35	nd (al) $[\alpha]'_D$ + 3.9 (chl, c=6)	sub (vac)	134			bz, chl	$B17^4$, 3278
7294	D-Glucose, pentamethyl ether $C_{11}H_{22}O_6$	250.30	$[\alpha]^{20/}{}_D$ + 147.4 (w, p=10)	$180^{0.4}$		$1.0944^{20/4}$	1.4466^{20}	w, al, eth, ace	$B17^4$, 2928
7295	D-Glucose, phenylhydrazone (α type) $C_{12}H_{18}N_2O_5$	270.29	pl (al) $[\alpha]^{25/}{}_D$ − 8.7→− 52.2 (w, c=2) (mut)		160			w	B31, 173
7296	D-Glucose, phenylhydrazone (β type) $C_{12}H_{18}N_2O_5$	270.29	pr nd (al) $[\alpha]^{19/}{}_D$ -4.5→- 53.7 (w py) (mut)		140-1			w	B31, 173
7297	D-Glucose, phenylosazone or D-Glucosazone, D-Fructosazone, D-Mannosazone $C_{18}H_{22}N_4O_4$	358.40	ye nd (dil al) $[\alpha]'_D$-41 (MeOH)	d 213	210				B31, 350
7298	DL-Glucose, phenylosazone $C_{18}H_{22}N_4O_4$	358.40	ye nd (al)						B31, 355
7299	β-D-Glucose, 2-amino or D-Glucosamine $C_6H_{13}NO_5$	179.18	nd (al MeOH) $[\alpha]_D$ + 28→ + 47.5 (w, c=0.4)		110d			w	$B4^4$, 2019
7300	D-Glucose, 2-amino, hydrochloride $C_6H_{13}NO_5 \cdot HCl$	215.64	mcl (w or dil al) $[\alpha]_{20D}$ + 25→ + 72.6 (w)					w	$B4^4$, 2019
7301	D-Glucose, 2-(methylamino) or N-methyl-D-glucosamine $C_7H_{15}NO_5$	193.20	gummy, $[\alpha]^{25/}{}_4$ -65 (MeOH, c=1)		130-2d			w	$B18^4$, 7526
7302	D-Glucothiose $C_6H_{12}SO_5$	196.22	hyg pw(w + 1), $[\alpha]^{30/}{}_D$ + 48.7 (w, c=1.4)		70			w	$B1^4$, 4391
7303	D-Glucoside, α-methyl or Methyl-α-D-glucopyranoside $C_7H_{14}O_6$	194.19	rh nd, (al), $[\alpha]^{20/}{}_D$ + 158.9 (w)	$200^{0.2}$	168	$1.46^{30/4}$		w	$B17^4$, 2909

No.	Name, Synonyms, and Formula	Mol. wt.	Color, crystalline form, specific rotation and λ_{max} (log ε)	b.p. °C	m.p. °C	Density	n_D	Solubility	Ref.
7304	D-Glucoside, β-methyl or Methyl-β-D-glucopyranoside $C_7H_{14}O_6$	194.18	tetr pr (al), $[\alpha]^{20}/_D$ -34.2 (w, p=10)		115-6			w	B17[4], 2911
7305	β-D-Glucuronic acid $C_6H_{10}O_7$	194.14	nd(al AcOEt), $[\alpha]^{20}/_D$ +11.7 + 36.3 (w, c=6) (mut)		165			w, al	B3[4], 1996
7306	D-Glucuronic acid, γ-lactone or D-glucurone $C_6H_8O_6$	176.13	mcl pl (w) cr (al), $[\alpha]^{25}/_D$ + 19.8 (w, c=5.2)		177-8	$1.76^{20/4}$		w	B18[4], 3055
7307	Glutaconic acid, diethyl ester or Ethyl glutaconate $C_2H_5O_2CCH_2CH=CHCO_2C_2H_5$	186.21		236-8, 125^{12}		$1.0496^{20/4}$	1.4411^{20}	al, eth	B2[4], 2227
7308	Glutamic acid (D) or 2-Aminopentanedioic acid $HO_2CCH_2CH_2CH(NH_2)CO_2H$	147.13	lf (w), $[\alpha]^{25}/_D$ -31.7 (1.7 NHCl)		213d				B4[3], 1549
7309	Glutamic acid (DL) $HO_2CCH_2CH_2CH(NH_2)CO_2H$	147.13	rh(al w)		199d, (225-7)	$1.4601^{20/4}$			B4[3], 1550
7310	Glutamic acid (L +) $HO_2CCH_2CH_2CH(NH_2)CO_2H$	147.13	orh(dil al), $[\alpha]^{22}/_D$ + 31.4 (6N HCl, c=1)	sub 175^{10}	224-5d	$1.538^{20/4}$			B4[3], 1530
7311	Glutamic acid, hydrochloride (L) or Acidulin $HO_2CCH_2CH_2CH(NH_2)CO_2H.HCl$	183.59	rh pl (w), $[\alpha]^{19}/_D$ + 31.1 (dil HCl)		214d			w, al	B4[3], 1537
7312	Glutamic acid, N-acetyl (L) $HO_2CCH_2CH_2CH(NHCOCH_3)CO_2H$	189.17	pr (w), $[\alpha]_D$ -15.3 (w, c=2)		199				B4[3], 1544
7313	Glutamic acid, 3-hydroxy (D) $HO_2CCH_2CH(OH)CH(NH_2)CO_2H$	163.13	hyg pr(w)	d	135			w, aa	B4[3], 1676
7314	Glutamic acid, 3-hydroxy (DL) $HO_2CCH_2CH(OH)CH(NH_2)CO_2H$	163.13	rh pr or nd(w)	d	198d			w	B4[3], 1676
7315	Glutamine (L +) or α-Amino glutaramic acid $HO_2CCH(NH_2)CH_2CH_2CONH_2$	146.15	nd (w or dil al), $[\alpha]^{25}/_D$ + 6.5 (w, c=2)		185-6d			w	B4[3], 1540
7316	Glutyraldehyde or 1,5-Pentanedial $OHC(CH_2)_3CHO$	100.12		187-9d, 71-2^{10}				w, al, bz	B1[4], 3111
7317	Glutyraldehyde, dioxime or Glutyraldoxime $HON=CH(CH_2)_3CH=NOH$	130.15	nd(w or Py)	sub	178				B1[4], 3660
7318	Glutaric acid or Pentanedioic acid $HO_2C(CH_2)_3CO_2H$	132.12	nd(bz)	302-4d, 200^{20}	99	$1.424^{25/4}$	1.4188^{106}	w, al, eth, chl	B2[4], 1934
7319	Glutaric acid, 2-acetyl, diethyl ester $C_2H_5O_2CCH_2CH_2CH(COCH_3)CO_2C_2H_5$	230.26		271-2d, 119-$2^{0.1}$		$1.0712^{20/4}$	1.4420^{15}	al, eth	B3[4], 1835
7320	Glutaric anhydride, 2-phenyl $C_{11}H_{10}O_3$	190.20	nd(eth)	218-30^{13}	95			al	B17[4], 6185
7321	Glutaric anhydride, 3-phenyl $C_{11}H_{10}O_3$	190.20	cr(bz)	217-9^{15}	105			eth, bz, chl	B17[4], 6185
7322	Glutaric acid, diethyl ester or Diethylglutarate $C_2H_5O_2C(CH_2)_3CO_2C_2H_5$	188.22	syr	236-7, 103-4^7	−24.1	$1.0220^{20/4}$	1.4241^{20}	eth	B2[4], 1937
7323	Glutaric acid, dimethyl ester or Dimethyl glutarate $CH_3O_2C(CH_2)_3CO_2CH_3$	160.17		214^{751}, 109^{21}		$1.0876^{20/4}$	1.4242^{20}	al, eth	B2[4], 1937
7324	Glutaric acid, 2,3-dimethyl $HO_2CCH_2CH_2C(CH_3)_2CO_2H$	160.17	nd(bz-lig)		85			al, chl, aa	B2[4], 2018
7325	Glutaric acid, 3,3-dimethyl $(CH_3)_2C(CH_2CO_2H)_2$	160.17	mcl pl nd(bz)	126-$7^{4.5}$	103-4	$1.4278^{20/4}$		w, al, eth	B2[4], 2023
7326	Glutaric acid, diphenyl ester $C_6H_5O_2C(CH_2)_3CO_2C_6H_5$	284.31	nd(lig)	236.5^{15}	54			al, ace, lig	B6[4], 626
7327	Glutaric acid, 2-ethyl-3-methyl $HO_2CCH_2CH(CH_3)CH(C_2H_5)CO_2H$	174.20	(i) pr (w), (ii) pr (chl-lig)		(i)100 (ii)88			w, eth	B2[4], 2047

PHYSICAL CONSTANTS OF ORGANIC COMPOUNDS (Continued)

No.	Name, Synonyms, and Formula	Mol. wt.	Color, crystalline form, specific rotation and λ_{max} (log ε)	b.p. °C	m.p. °C	Density	n_D	Solubility	Ref.
7328	Glutaric acid, 2-ethyl-4-methyl (dl) or Paramethyl ethyl glutaric acid $HO_2CCH(CH_3)CH_2CH(C_2H_5)CO_2H$	174.20	nd(w)		107			eth, lig	$B2^4$, 2042
7329	Glutaric acid, 2-ethyl-4-methyl (meso) $HO_2CCH(CH_3)CH_2CH(C_2H_5)CO_2H$	174.20	nd(al)		83-4			eth	$B2^4$, 2042
7330	Glutaric acid, 3-ethyl-3-methyl $(HO_2CCH_2)_2C(CH_3)C_2H_5$	174.20	pl(bz-peth)	260^{740}	87			al, eth, bz	$B2^3$, 1781
7331	Glutaric acid, 2-hydroxy (d) $HO_2CCH_2CH_2CH(OH)CO_2H$	148.12	$[\alpha]^{19/}_D$ + 1.76 (w, c=1)		72				$B3^4$, 1146
7332	Glutaric acid, 2-hydroxy (dl) $HO_2CCH_2CH_2CH(OH)CO_2H$	148.12	pr(AcOEt)		72			w, al	$B3^4$, 1146
7333	Glutaric acid, 2-hydroxy (l) $HO_2CCH_2CH_2CH(OH)CO_2H$	148.12	$[\alpha]_D$ -1.98 (w)		72-3			w, al	$B3^4$,1146
7334	Glutaric acid, 3-methyl $CH_3CH(CH_2CO_2H)_2$	146.14		$165\text{-}7^{0.5}$	87			w, al, eth	$B2^4$, 1992
7335	Glutaric acid, 2-oxo $HO_2CCH_2CH_2COCO_2H$	146.10	cr(ace-bz)		115-6			w, al, eth, ace	$B3^4$, 1813
7336	Glutaric acid, 3-oxo or Acetone dicarboxylic acid $(HO_2CCH_2)_2C{=}O$	146.10	nd(al AcOEt) rh(w)	d	135d			w, al	$B3^4$, 1816
7337	Glutaric acid, 3-oxo, diethyl ester $(C_2H_5O_2CCH_2)_2C{=}O$	202.21		250, 140^{12}		$1.113^{20/4}$		al	$B3^4$, 1817
7338	Glutaric acid, 2-phenyl $HO_2CCH_2CH_2CH(C_6H_5)CO_2H$	208.21	cr(bz or eth-peth)		82-3				$B9^3$, 4298
7340	Glutaric acid, 2,3,4-trihydroxy (d) $HO_2C(CHOH)_3CO_2H$	180.11	cr(ace) pl(w), $[\alpha]^{20/}_D$ + 22.2		128			w, al	$B3^4$, 1265
7341	Glutaric acid, 2,3,4-trihydroxy (dl) $HO_2C(CHOH)_3CO_2H$	180.11	cr(ace)		154d			w, al, ace	B3, 553
7342	Glutaronitrile $NC(CH_2)_3CN$	94.12		286, 160.4^{22}	−29	$0.9911^{15/4}$	1.4295^{20}	al, chl	$B2^4$, 1941
7343	Glutaryl chloride $ClCO(CH_2)_3COCl$	169.01		216-8		$1.324^{30/4}$	1.4728^{20}	eth	$B2^4$, 1939
7344	Glutathione or α-Glutamyl cysteinyl glycine $C_{10}H_{17}N_3O_6S$	307.32	rh (w) cr (50% al), $[\alpha]2^{7/}_D$ -21.3 (w, c=2)		195			w, DMF	$B4^3$, 1612
7345	D-Glyceraldehyde or D-2,3-Dihydroxy proponal $HOCH_2CH(OH)CHO$	90.08	syr, $[\alpha]_D$ + 14 (w)					w	$B1^4$, 4114
7346	DL-Glyceraldehyde $HOCH_2CH(OH)CHO$	90.08	nd or pr(40% MeOH)	$140\text{-}50^{0.8}$	145	$1.455^{18/18}$		w	$B1^4$, 4114
7347	L-Glyceraldehyde $HOCH_2CH(OH)CHO$	90.08	$[\alpha]$ -13.8 (w, c=1)					w	$B1^4$, 4114
7348	D-Glyceraldehyde, diethyl acetal or 3,3-Diethoxy-1,2- $HOCH_2CH(OH)CH(OC_2H_5)_2$	164.20	$[\alpha]^{15/}_D$ +21.2 (w, c=18)	$127\text{-}9^{17}$				w, al, eth, ace	$B1^2$, 888
7349	DL-Glyceric acid or 2,3-Dihydroxy propionic acid $HOCH_2CH(OH)CO_2H$	106.08	syr	d				w, al, eth	$B3^4$, 1050
7350	D-Glyceric acid, ethyl ester or D-Ethyl glycerate $HOCH_2CH(OH)CO_2C_2H_5$	134.13	$[\alpha]^{11/}_D$ -22.73					w, al, eth	$B3^4$, 1052
7351	DL-Glyceric acid, ethyl ester $HOCH_2CH(OH)CO_2C_2H_5$	134.13		230-40, $120\text{-}1^{16}$		$1.1908^{15/15}$		w, al, eth	$B3^4$, 1052
7352	DL-Glyceric acid, methyl ester or DL-Methyl glycerate $HOCH_2CH(OH)CO_2CH_3$	120.11		239-44, $119\text{-}20^{14}$		$1.2814^{15/15}$	1.4502^{20}	w, al	$B3^4$, 1052
7353	L-Glyceric acid, methyl ester or L-Methyl glycerate $HOCH_2CH(OH)CO_2CH_3$	120.11	$[\alpha]^{15/}_D$ −6.44	$119\text{-}20^{14}$, 74-$5^{0.2}$		$1.2798^{15/15}$		w, al	$B3^4$, 1052
7354	Glycerol or Glycerin, 1,2,3-trihydroxy propane $HOCH_2CH(OH)CH_2OH$	92.09	syr rh pl	290d, 182^{20}	20	$1.2613^{20/4}$	1.4746^{20}	w, al	$B1^4$, 2751
7355	Glycerol, 1-monoacetate or α-Monoacetin $HOCH_2(CHOH)CH_2O_2CCH_3$	134.13		158^{165}, $129\text{-}31^3$		$1.2060^{20/4}$	1.4157^{20}	w, al	$B2^4$, 251
7356	Glycerol borate $(C_3H_5BO_3)n$	(99.88)n	glass		Ca 150				B1, 519
7357	Glycerol, 1-monobutyrate (dl) or α-Monobutyrin $HOCH_2CH(OH)CH_2O_2CC_3H_7$	162.19		289-71, 163^{16}		1.129^{18}	1.4531^{20}	w, al	$B2^4$, 798

PHYSICAL CONSTANTS OF ORGANIC COMPOUNDS (Continued)

No.	Name, Synonyms, and Formula	Mol. wt.	Color, crystalline form, specific rotation and λ_{max} (log ε)	b.p. °C	m.p. °C	Density	n_D	Solubility	Ref.
7358	Glycerol, 1(2-chlorophenyl) ether $(2\text{-}ClC_6H_4)OCH_2CH(OH)CH_2OH$	202.64	nd (bz)	250^{19}	71-2	...	...	eth	$B6^4$, 790
7359	Glycerol, 1-(4-chlorophenyl) ether $(4\text{-}ClC_6H_4)OCH_2CH(OH)CH_2OH$	202.64	nd (eth peth) lf (bz)	$214\text{-}5^{19}$	76	...	...	al, eth, ace	$B6^4$, 831
7360	Glycerol, 1,3-diacetate or Diacetin $HOCH(CH_2O_2CCH_3)_2$	176.17	...	$280, 155\text{-}6^{15}$	40	$1.1779^{15/4}$	1.4395^{20}	w, al	$B2^4$, 252
7361	Glycerol, 1,2-dibutyrate (d) or α,β-Dibutyrin $HOCH_2CH(O_2CC_3H_7)CH_2(O_2CC_3H_7)$	232.28	$[\alpha]_D$ +1.7 (py,c=7)	$273.5, 167^{20}$	...	...	1.4422^{20}	al	$B2^4$, 799
7362	Glycerol, 1,3-dilaurate or α,γ-Dilarurin $HOCH(CH_2O_2CC_{11}H_{23})_2$	456.71	pl (al) nd (eth-al)	...	[α] −49.5 (unst) β −56.5 (st)	...	...	al, eth, bz, chl, lig	$B2^4$, 1098
7363	Glycerol, 1,2-dimethyl ether (dl) or 2,3-Dimethoxy-1-propanol $HOCH_2CH(OCH_3)CH_2OCH_3$	120.15	...	$180, 100^{40}$	...	$1.016^{25/4}$	1.4200^{20}	w, al, eth	$B1^3$, 2317
7364	Glycerol, 1,3-dimethyl ether or 1,3-Dimethoxy-2-propanol $HOCH(CH_2OCH_3)_2$	120.15	...	$169, 88^{40}$	...	$1.0085^{20/4}$	1.4192^{20}	w, al, eth	$B1^3$, 2318
7365	Glycerol, 1,3-dinitrate $HOCH_2(CH_2ONO_2)_2$	182.09	pr (w) cr (eth + 1w)	$148^{15}, 116^{0.6}$	26 (hyd)	$1.523^{20/4}$	1.4715^{20}	w, al, eth	$B1^2$, 591
7366	Glycerol, 1,3-dipalmitate $HOCH(CH_2O_2CC_{15}H_{31})_2$	568.92	cr (al, chl)	...	72-4	...	...	eth	$B2^4$, 1174
7367	Glycerol, 1,3-diphenylether $HOCH(CH_2OC_6H_5)_2$	244.29	lf (al)	$224.5^{17.5}$, 175^2	81-2	$1.179^{24/4}$	...	al, eth, bz, chl	$B6^4$, 590
7368	Glycerol, 1,3-diphenylether-2-acetate $CH_3CO_2CH(CH_2OC_6H_5)_2$	286.33	(dil al)	190^{160}	70-1	...	...	al, eth, bz, chl	$B6^3$, 583
7369	Glycerol, 1,3-dipropionate $HOCH(CH_2O_2CC_2H_5)_2$	204.22	...	$170\text{-}3^{10}$	...	...	...	al	$B2^4$, 717
7370	Glycerol, 1,3-distearate or α,γ-Distearin $HOCH(CH_2O_2CC_{17}H_{35})_2$	625.03	nd or pl (eth chl lig)	...	79.1	...	...	eth	$B2^4$, 1231
7371	Glycerol, 1-mono (2-hydroxybenzoate) or α-Glyceryl salicylate $(2\text{-}HOC_6H_4)CO_2CH_2CH(OH)CH_2OH$	212.20	nd (eth)	...	76	...	...	al, bz	$B10^3$, 142
7372	Glycerol, 1-monolaurate (dl) or α-Monolaurin (dl) $HOCH_2CH(OH)CH_2O_2CC_{11}H_{23}$	274.40	lf (peth)	186^2	63	0.9248^{97}	1.4350^{86}	eth, ace, bz, chl	$B2^4$, 1096
7373	Glycerol, 1-monolaurate (l) or α-Monolaurin (l) $HOCH_2CH(OH)CH_2O_2CC_{11}H_{23}$	274.40	cr (eth or peth)	...	54-5	...	...	eth, ace, bz, chl	$B2^4$, 1096
7374	Glycerol, 1-linoleate (dl) or Glycerol-1-(9,12-actadecadienoate) $C_{17}H_{31}CO_2CH_2CH(OH)CH_2OH$	354.53	cr (bz)	...	14-5	...	1.4758^{20}	eth, bz, chl	$B2^4$, 1758
7375	Glycerol, 1-(2-methoxyphenyl) ether or Guaiacol-α-glyceryl ether $(2\text{-}CH_3OC_6H_4)OCH_2CH(OH)CH_2OH$	198.22	rh pr (eth, eth-peth)	$215^{19}, 126^{0.2}$	78-9	...	...	al, bz, chl, w	$B6^3$, 4224
7376	Glycerol, 1-methyl ether or 3-Methoxy-1,2-propanediol $HOCH_2CH(OH)CH_2OCH_3$	106.12	hyg liq	$220, 110\text{-}2^{13}$	...	$1.830^{20/4}$	1.442^{25}	w, al, eth, ace	$B1^4$, 2755
7377	Glycerol, 2-methyl ether or 2-Methoxy-1,3-propandiol $CH_3OCH(CH_2OH)_2$	106.12	hyg liq	$232, 119\text{-}20^9$	...	$1.124^{25/4}$	1.4505^{12}	w, al, eth, ace	$B1^3$, 2317
7378	Glycerol, 1-mononitrate $HOCH_2CH(OH)CH_2ONO_2$	137.09	pr (w, al, eth)	155-60, 102^1	61	$1.4164^{20/4}$	1.4698^{20}	w, al	$B1^4$, 2761
7379	Glycerol, 2-mononitrate $ONO_2CH(CH_2OH)_2$	137.09	lf (w)	155-60	54	$1.40^{22/4}$	...	w, al, eth	$B1^2$, 591
7380	Glycerol, 1-octadecyl ether (d) or Batyl alcohol $HOCH_2CH(OH)CH_2OC_{18}H_{37}$	344.58	pl (bz, aa)	$215\text{-}20^2$	70-1	...	...	eth	$B1^4$, 2758
7381	Glycerol, 1-oleate or α-Monoolein Glycerol-1-(9-octadeceneoate) $C_{17}H_{33}CO_2CH_2CH(OH)CH_2OH$	356.55	pl (al)	$238\text{-}40^3$	35	$0.9420^{20/4}$	1.4626^{20}	al, eth, chl	$B2^4$, 1657
7382	Glycerol, 1-oleate or α-Monoolein Glycerol-1 mono hexadicaneate $C_{15}H_{31}CO_2CH_2CH(OH)CH_2OH$	330.51	pl or lf(eth lig)	...	77	...	...	al	$B2^4$, 1170
7383	Glycerol, monopalmitate (l) or α-Monopalmitate (l) $HOCH_2CH(OH)CH_2O_2CC_{15}H_{31}$	330.51	$[\alpha]_D$ −4.37 (py)	...	71-2	...	...	...	$B2^4$, 1170
7384	Glycerol, 1-palmityl ether or Glycerol-1-hexadecylether $HOCH_2CH(OH)CH_2OC_{16}H_{33}$	316.52	lf (hex) $[\alpha]^{20}_D$ +3 (chl)	$120^{0.005}$	64	...	...	ace, chl, peth	$B1^3$, 2322
7385	Glycerol, 1-phenyl ether or Antodyne $HOCH_2CH(OH)CH_2OC_6H_5$	168.19	nd (eth peth)	200^{22}	67-8	$1.225^{20/4}$	...	w, al, eth, bz	$B6^4$, 589

PHYSICAL CONSTANTS OF ORGANIC COMPOUNDS (Continued)

No.	Name, Synonyms, and Formula	Mol. wt.	Color, crystalline form, specific rotation and λ_{max} (log ε)	b.p. °C	m.p. °C	Density	n_D	Solubility	Ref.
7386	Glycerol, 1-ricinoleate or 1-Mono (12-hydroxy-9-octadecanoate) $HOCH_2CH(OH)CH_2[O_2C(CH_2)_7CH{=}CHCH_2(OH)CHC_6H_{13}]$	372.55	ye			$1.028^{20/4}$		al, eth, ace, bz	B3[4], 1030
7387	Glycerol, 1-stearate or α-Monostearin $C_{17}H_{35}CO_2CH_2CH(OH)CH_2OH$	358.36	pl (MeOH)		81	$0.9841^{20/4}$	1.4400^{86}	lig	B2[4], 1225
7388	Glycerol, 1-stearate (l) $C_{17}H_{35}CO_2CH(OH)CH_2OH$	358.56	cr (eth or peth) $[\alpha]_D$ −3.58 (py)		76-7				B2[4], 1225
7389	Glycerol, 1-(2-tolyl) ether $(2\text{-}CH_3C_6H_4)OCH_2CH(OH)CH_2OH$	182.22	nd (bz-peth)		70-1d			al	B6[3], 1952
7390	Glycerol, triacetate or Triacetin $(CH_3CO_2CH_2)_2CHO_2CCH_3$	218.21	cr (al)	258-60, 130.5^7	4.1	$1.1596^{20/4}$	1.4301^{20}	**al, eth,** ace, **bz, chl**	B2[4], 253
7391	Glycerol, tribenzoate or Tribenzoin $C_6H_5CO_2CH(CH_2O_2CC_6H_5)_2$	404.43	nd (MeOH)		76	$1.228^{12/4}$		eth, ace, bz, chl	B9[3], 666
7392	Glycerol, tributyrate or Tributyrin $C_3H_7CO_2CH(CH_2O_2CC_3H_7)_2$	302.37		305-10, 190^{15}	−75	$1.0350^{20/4}$	1.4359^{20}	al, eth, ace, bz	B2[4], 799
7393	Glycerol, tricaproate or Tricaproin Glycerol trihexanoate $C_5H_{11}CO_2CH(CH_2O_2CC_5H_{11})_2$	386.54		>200	−60	$0.9867^{20/4}$	1.4427^{20}	**al, eth,** ace, **bz,** peth	B2[4], 926
7394	Glycerol, tricaprylate or Tricuprylin Glycerol trioctoate $C_7H_{15}CO_2CH(CH_2O_2CC_7H_{15})_2$	470.70		233.1	10 (st) −22 (unst)	$0.9540^{20/4}$	1.4482^{20}	**al,** eth, bz, chl, lig	B2[4], 991
7395	Glycerol, trielaidate or Trielaidin Glycerol tri (*trans*-9-octacteceneate) $C_{17}H_{33}CO_2CH(CH_2O_2CC_{17}H_{33})_2$	885.47			α-16.6β-42.8			eth, bz, chl	B2[4], 1664
7396	Glycerol, trilaurate or Trilaurin Glycerol tridodicanoate $C_{11}H_{23}CO_2CH(CH_2O_2CC_{11}H_{23})_2$	639.03	nd(al)		46	0.8986^{55}	1.4404^{60}	al, eth, ace, bz, chl	B2[4], 1098
7397	Glycerol, trimethylether or Trimethoxy propane $CH_3OCH(CH_2OCH_3)_2$	134.18			148	$0.9460^{15/4}$	1.4055^{15}	w, eth, ace, bz	B1[4], 2755
7398	Glycerol, trimyristate or Glycerol tritetradecanoate $C_{13}H_{27}CO_2CH(CH_2O_2CC_{13}H_{27})_2$	768.28	poly-morphic (al-eth)	311	56.5 (st) 32 (unst)	$.08848^{60/4}$	1.4428^{60}	eth, ace, bz, chl	B2[4], 1134
7399	Glycerol, trinitrate or Nitroglycerin $O_2NOCH(CH_2ONO_2)_2$	227.09	pa ye fcl or rh	256 exp, 125^2	13	$1.5931^{20/4}$	1.4786^{12}	al, eth, ace, bz, chl	B1[4], 2762
7400	Glycerol, trioleate or Trioleen, Glycerol-(*cis*-9-octodeceneate) $C_{17}H_{33}CO_2CH(CH_2O_2CC_{17}H_{33})_2$	885.47	poly-morphic	$235\text{-}40^{18}$		0.8988^{40}	1.4621^{40}	eth, chl, peth	B2[4], 1664
7401	Glycerol, tripalmatate or Tripalmitin Glycerol trihexadecanoate $C_{15}H_{31}CO_2CH(CH_2O_2CC_{15}H_{31}$	807.35	nd (eth)	310-20	66 (st) 44.7 (unst)	$0.8752^{70/4}$	1.4381^{80}	eth, bz, chl	B2[4], 1176
7402	Glycerol, tripropionate $C_2H_5CO_2CH(CH_2O_2CC_2H_5)_2$	260.29		$175\text{-}6^{20}$		$1.100^{20/18}$	1.4318^{19}	al, eth, chl	B2[4], 717
7403	Glycerol, tristearate or Tristearin Glycerol triactadecanoate $C_{17}H_{35}CO_2CH(CH_2O_2CC_{17}H_{35})_2$	891.51			α-55, β-73	$0.8559^{90/4}$	1.4395^{80}	ace	B2[4], 1233
7404	Glycerol, triisovalerate or Glycerol tri (3-methylbutyrate) $i\text{-}C_4H_9CO_2CH(CH_2O_2C\text{-}i\text{-}C_4H_9)$	344.45		330-5, 194^{15}		$0.9984^{20/4}$	1.4354^{20}	al, eth	B2[4], 900
7405	Glycidic acid or 2,3-Epoxypropionic acid $CH_2CHCOOH$ O	88.06						**w, al, eth**	B18[1], 435
7406	Glycidol (d) or 3-Hydroxypropylene oxide CH_2CHCH_2OH O	74.08	α + 15 (undil)	56.5^{11}		$1.117^{20/4}$	1.4293^{16}	w, al, eth, ace, bz	B17[4], 985
7407	Glycidol dl or 2,3-Epoxy-1-propanol (dl) CH_2CHCH_2OH O	74.08		$65\text{-}6^{2.5}$		1.1143^{25}	1.4287^{20}	**w, al,** eth, ace, bz	B17[4], 985
7408	Glycidol (l) or 3-Hydroxy propylene oxide $HOCH_2CHCH_2$ O	74.08	$[\alpha]^{18}_D$ −8.6 (indil)	56^{11}		1.1050^{18}	1.4293^{16}	w, al, eth, ace, bz	B17[4], 985
7409	Glycidol, phenyl $C_6H_5CHCHCH_2OH$ O	150.18		138^3	26.5	1.512^{27}	1.5432^{27}	al, eth	B6[3], 1800
7410	Glycine or Aminoacetic acid $H_2NCH_2CO_2H$	75.07	mcl or trg pr (dil al)		262d	1.607		w	B4[3], 1097
7411	Glycine, N-Leucyl (DL) or DL-Leucyl Glycine $(CH_3)_2CHCH_2\text{-}CH(NH_2)CONHCH_2CO_2H$	188.23			243d			w	B4[3], 1434
7412	Glycine, -Leucyl (L) or L-Leucyl glycine $(CH_3)_2CHCH_2CH(NH_2)CONHCH_2CO_2H$	188.23	lf or nd (w al)		248d			w	B4[3], 1414
7413	Glycyl glycine $H_2NCH_2CONHCH_2CO_2H$	132.12	$[\alpha]^{20}_D$ + 85.8 (w,c=2)		dec 215				B4[3], 1191

PHYSICAL CONSTANTS OF ORGANIC COMPOUNDS (Continued)

No.	Name, Synonyms, and Formula	Mol. wt.	Color, crystalline form, specific rotation and λ_{max} (log ε)	b.p. °C	m.p. °C	Density	n_D	Solubility	Ref.
7414	Glycocholic acid or Cholylglycine $C_{26}H_{43}NO_6$	465.63	nd (w) $[\alpha]^{23}_D$ + 32.3 (al,c=1)		165-8 (anh) 132-4 (+w)				B10^{3}, 2176
7415	Glycocyamine or N-Guanyl glycine NH=C(NH$_2$)NHCH$_2$CO$_2$H	117.11	pl or nd (w)		>300				B4^{3}, 1165
7416	Glycogen or Animal starch $(C_6H_{10}O_5)n$	(162.14) n	wh pw $[\alpha]^{25}_D$ + 196.5 (w)					w	C25, 4940
7417	Glycolic acid or Hydroxyacetic acid HOCH$_2$COOH	76.05	rh nd (w) lf (eth)	d	80			w, al, eth	B3^{4}, 571
7418	Glycolic acid, benzoyl C_6H_5COCH(OH)CO$_2$H	180.16	lo pr (lig)		112			al, eth, chl	B9, 954
7419	Glycolide or 2,5-*p*-dioxane dione $C_4H_4O_4$	116.07	lf (al-chl or al)		86-7			ace	B19^{4}, 1922
7420	Glycoluril or Glyoxaldiurene $C_4H_6N_4O_2$	142.12	nd or pr (w)		300d			eth	B26^{2}, 260
7421	18α-Glycyrrhetinic acid or Glycyrhetic acid $C_{30}H_{46}O_4$	470.69	α-pl (dil al) β-nd (al peth)	α283 β296				al	B10^{3}, 4392
7422	Glycyrrhizic acid or Glycyrhizin $C_{42}H_{62}O_{16}$	822.94	pl or pr (aa)		220d			w	B18^{4}, 5156
7423	Glyoxal or Ethanedial OHCCHO	58.04	ye pr	50.4	15	1.14^{20}	1.3826^{20}	w, al, eth	B1^{4}, 3625
7424	Glyoxal, dioxine HON=CHCH=NOH	88.07	rh pl (w)	sub	178d			w, al, eth	B1^{4}, 3629
7425	Glyoxal, methyl, phenyl or Methylphenyl glyoxal CH$_3$COCOC$_6$H$_5$	148.16	ye oil	228, 101^{12}		$1.0065^{20/4}$	1.537^{10}	w, al, eth	B7^{3}, 3463
7426	Glyoxal, methylphenyl, dioxime or Methylphenyl glyoxime CH$_3$C(=NOH)C(=NOH)C$_6$H$_5$	178.19	nd (dil al)		140-1			al	B7^{3}, 3465
7427	Glyoxal, phenyl or Benzoyl formaldehyde C_6H_5COCHO	134.13	nd (+w)	142^{125}	91	(hyd)		w, al, eth, ace, bz	B7^{3}, 3443
7428	Glyoxal, phenyl, hydrate C_6H_5COCH(OH)$_2$	152.15	nd (w, chl, al, lig)	93.4				al, eth, chl	B7^{3}, 3443
7429	Glyoxal, phenyl-1-oxime C_6H_5COCH=NOH	149.15	mcl pr or lf (chl, w)		129			chl	B7^{3}, 3447
7430	Glyoxylic acid or Oxoacetic acid HCOCO$_2$H	74.04	rh pr (w + 1/2)		98			w	B3^{4}, 1489
7431	Glyoxylic acid, phenyl or Phenylglyoxylic acid, Benzoyl formic acid C_6H_5COCO$_2$H	150.13	pr (CCl$_4$)	147-51^{12}	66			w, al, eth	B10^{3}, 2972
7432	Glyoxylic acid, phenyl, methyl ester or Methyl phenyl glyoxylate C_6H_5COCO$_2$CH$_3$	164.16	ye	246-8, 137^{14}			1.5268^{20}		B10^{3}, 2973
7433	Glyoxylonitrile, phenyl or Benzoyl cyanide C_6H_5COCN	131.13	ta	206-8, 99^{19}	32-3			al, eth	B10^{3}, 2976
7434	Gramine or 3-(Dimethylaminomethyl indole) $C_{11}H_{14}N_2$	174.25	lf (eth) nd (ace)		138-9			al, eth, chl	B22^{4}, 4302
7435	Griseofulvin or Fulvicin $C_{17}H_{17}ClO_6$	352.77	oct or rh (bz) $[\alpha]^{17}D$ + 376 (chl,Sat sul)		220				B18^{4}, 3160
7435a	Guaiol or champicol $C_{15}H_{26}O$	222.37	trq Pr (al), $[\alpha]^{-30}$ c = 4	165_{17}	91	$0.9074^{100/4}$	$1.4716^{100/6}$	al, eth	
7436	Guanamine or 2,4-Diamino-1,3,5-triazine 2,4(H$_2$N)$_2$(C$_3$HN$_3$)	111.11	nd (w)		329d			w	B26^{1}, 65
7437	Guanidine or Aminomethanamidine, Carbaniedine HN=C(NH$_2$)$_2$	59.07	cr		Ca 50			al, w	B3^{4}, 148
7438	Guanidine acetate HN=C(NH$_2$)$_2$·CH$_3$CO$_2$H	119.12	nd (al-eth)		229-30			w, al	B3^{4}, 152
7439	Guanidine, amine or Guanyl hydrazine HN=C(NH$_2$)NHNH$_2$	74.09	cr		d			w, al	B3^{4}, 236
7440	Guanidine, carbonate [HN=C(NH$_2$)$_2$]$_2$.H$_2$CO$_3$	180.17	oct tetr pr (w)		198	1.24^{4}		w	B3^{4}, 152
7441	Guanidine, 1-cyano HN=C(NH$_2$)NHCN	84.08	rh lf or pl (al)	d	211-2	1.404^{14}		w, al, ace	B3^{4}, 160

PHYSICAL CONSTANTS OF ORGANIC COMPOUNDS (Continued)

No.	Name, Synonyms, and Formula	Mol. wt.	Color, crystalline form, specific rotation and λ_{max} (log ε)	b.p. °C	m.p. °C	Density	n_D	Solubility	Ref.
7442	Guanidine, 1,3-diphenyl or Melaniline $HN=C(NHC_6H_5)_2$	211.27	mcl nd (al or to)	d 170	150	$1.13^{20/4}$		al, eth	B12[3], 805
7443	Guanidine, 1,3-di(2-totyl) $HN=C[NH(C_6H_4CH_3\text{-}2)]$	239.32	cr (dil al)		179	$1.10^{20/4}$		eth, chl	B12[3], 1871
7444	Guanidine, hydrochloride $HN=C(NH_2)_2).HCl$	95.53	rh bipym (al)		178-85	$1.354^{20/4}$		w, al	B3[4], 150
7445	Guanidine, nitrate $HN=C(NH_2)_2.HNO_3$	122.08	lf (w)	d	217			w, al	B3[4], 151
7446	Guanidine, 1-nitro $HN=C(NH_2)NHNO_2$	104.07	nd or pr (w)		239d				B3[4], 249
7447	Guanidine, picrate $HN=C(NH_2)_2.C_6H_3N_3O_7$	288.18	og ye pl or nd (w)		333d				B6[3], 960
7448	Guanidine, thiocyanate $HN=C(NH_2)_2HCNS$	118.16	lf		118			w	B3[2], 121
7449	Guanidine, tetraphenyl $HN=C[N(C_6H_5)_2]_2$	363.46	rh (lig)		130-1			al, eth, bz	B12, 430
7450	Guanidine, 1,1,3-triphenyl $HN=C(NHC_6H_5)N(C_6H_5)_2$	287.37	ta (dil al)		134			al, eth	B12[3], 895
7451	Guanidine, 1,2,3-triphenyl $C_6H_5N=C(NHC_6H_5)_2$	287.37	nd or pr(al)	d	146-7	$1.163^{20/4}$		al	B12[3], 907
7452	Guanidine, 1-ureido or Dicyandiamidine $HN=C(NH_2)NHCONH_2$	102.10	pr(al)	d160	105			py	B3[4], 155
7453	Guanine or 2-Aminohypoxanthine $C_5H_5N_5O$	151.13	nd or pl(aq NH₃)	sub	360d				B26[2], 262
7454	Guanosine or 9-D-Ribosidoguanine $C_{10}H_{13}N_5O_5$	283.24	nd (w), $[\alpha]^{20/}{}_D$ -60.5 (0.1 NNaOH, p=3)		239d			aa	B31, 28
7455	Guanylic acid or Guanosine phosphoric acid $C_{10}H_{14}N_5O_8P$	363.22	nd or pr (w + 2), $[\alpha]^{20/}{}_D$ -7.5 (w, p=1) $^{25/}{}_D$ -65 (5% NaOH, c=2)		208d			w	B31, 29
7456	Guiaicol or Catechol monomethyl ether $2\text{-}CH_3OC_6H_4OH$	124.14	hex pr	205, 106.5^{24}	32	$1.1287^{21/4}$	1.5429^{20}	al, eth, chl	B6[3], 4200
7457	Guiaicol, 3-nitro or 2-Methoxy-6-nitro phenol $2\text{-}CH_3O\text{-}6\text{-}NO_2C_6H_3OH$	169.14	og ye nd(sub)	sub	62			w, al	B6[3], 4263
7458	Guiaicol, 4-nitro or 2-Methoxy-5-nitro phenol $2\text{-}CH_3O\text{-}5\text{-}NO_2C_6H_3OH$	169.14	pa ye nd(w)		105			al, eth	B6[3], 4264
7459	Guiaicol, 5-nitro or 2-Methoxy-4-nitro phenol $2\text{-}CH_3O\text{-}4\text{-}NO_2C_6H_3OH$	169.14	ye nd(w)		103-4			al, eth	B6[3], 4264
7460	Guiaicol, 6-nitro or 2-Methoxy-3-nitro phenol $2\text{-}OCH_3\text{-}3\text{-}NO_2C_6H_3OH$	169.14	yesh rh pr(peth)		102-3			al	B6[3], 4263
7461	Guaiacol, 5-vinyl or 2-Methoxy-4-vinyl phenol $2\text{-}CH_3O\text{-}4\text{-}(CH_2=CH)C_6H_3OH$	150.18	cr		57			al, eth	B6[3], 4981
7462	D-Gulonic acid-γ-lactone $C_6H_{10}O_6$	178.14	pr ta (w), $[\alpha]^{20}{}_D$ −57.1 (w)		180-1			w	B18[4], 3025
7463	L-Gulonic acid-γ-lactone $C_6H_{10}O_6$	178.14	rh pr (w), $[\alpha]^{20}{}_D$ + 55.1 (w)		185			w	B18[4], 3026
7464	D-Gulonic acid, phenylhydrazide $C_{12}H_{18}N_2O_6$	286.28	(w), $[\alpha]^{20/}{}_D$ + 13.45	d 195	147-9			w, al	B15[1], 82
7465	D-Gulose $C_6H_{12}O_6$	180.16	syr $[\alpha]^{20}{}_D$ −20.4 (w)	d				w	B1[4], 4333
7466	L-Gulose $C_6H_{12}O_6$	180.16	sy $[\alpha]^{20}{}_D$ + 61.6 (w)	d				w	B1[4], 4334
7467	Guvacine or 1,2,5,6-Tetrahydronicotinic acid $C_6H_9NO_2$	127.14	pr(w) rods(+ 1w dil al)		295d			w	B22[1], 489
7468	β-Gurjunene $C_{15}H_{24}$	204.36	$[\alpha]_D$ + 74.5	$120\text{-}3^{13}$		0.9348	1.5028		B5[3], 1093

PHYSICAL CONSTANTS OF ORGANIC COMPOUNDS (Continued)

No.	Name, Synonyms, and Formula	Mol. wt.	Color, crystalline form, specific rotation and λ_{max} (log ε)	b.p. °C	m.p. °C	Density	n_D	Solubility	Ref.
7469	Halostachine (l) $C_6H_5CH(OH)CH_2NHCH_3$	151.21	[α] -47		43-5			w, al, eth	$B13^{3}$, 1658
7470	Harmaline or 3,4-Dihydroharmine $C_{13}H_{14}N_2O$	214.27	ta(MeOH) rh pr(al)		250d				$B23^{2}$, 345
7471	Harmine or Banisterine, Telepathine $C_{13}H_{12}N_2O$	212.25	rh(al) pr(MeOH)	sub	272-4			py	$B23^{2}$, 348
7472	Hecogenin $C_{27}H_{42}O_4$	430.63	pl (eth), $[\alpha]^{22}/_D$ + 7		265-8			al, eth, ace	$B19^{4}$, 2581
7473	Hecogenin, acetate $C_{29}H_{44}O_5$	472.67	cr(MeOH)		243 (252)			al	$B19^{4}$, 2584
7474	Hederagenin $C_{30}H_{48}O_4$	472.71	pr (al), $[\alpha]^{20}/_D$ + 70.1 (chl-MeOH)		332-4			al	$B10^{3}$, 1923
7475	Helenine or Alantolactone $C_{15}H_{20}O_2$	232.32	nd	275, 197^{10}	76			al, eth, bz	$B17^{4}$, 5030
7476	Helicin or Salicylaldehyde-β-D- glucoside $C_{13}H_{16}O_7$	284.27	nd (a), $[\alpha]^{20}/_D$ -60.4 (w)		175			w, al	$B17^{4}$, 3010
7476a	Helvolic acid or Fumagacin $C_{32}H_{44}O_8$	554.68	nd (dil al), $[\alpha]^{18}/_D$ -125 (chl, c=1)		212d			eth, ace, bz, aa	Am 78, 5275
7477	Hematein or Haematein $C_{16}H_{12}O_6$	300.27	red br cr		250d				$B18^{4}$, 3343
7478	Hematin or Ferriporphyrin hydroxide $C_{34}H_{32}N_4O_4$.FeOH	633.51	br pw(py)		>200				M, 508
7479	Hematommic acid $C_9H_{14}O_5$	202.21	nd(aa)		172-3			aa	H16, 282
7480	Hematoporphyrin or Photodyn $C_{34}H_{38}N_4O_6$	598.70	red		172-3			aa	M, 509
7481	Hematoxylin $C_{16}H_{14}O_6 \cdot 3H_2O$	362.38	yesh cr, [α] + 11 (w, c=3.7)		140			al	$B17^{2}$, 273
7482	Hemimellitic acid or 1,2,3-Benzene tricarboxylic acid 1,2,3-$C_6H_3(CO_2H)_3$	210.14	tcl pl(+ 2w) nd(w)		197d	1.546^{20}		w	$B9^{3}$, 4791
7483	Heneicoscane or Uneicosane $CH_3(CH_2)_{19}CH_3$	296.58	cr (w)	356.5, 203^{10}	40.5	$0.7919^{20/4}$	1.4441^{20}	Peth	
7484	Heptachlor or 3,4,5,6,7,8,8a Heptachloro-α-Dicyclopentadiene $C_{10}H_5Cl_7$	373.32	wh		95-6	1.57^{9}		al, eth, bz, lig	$B5^{3}$, 1236
7485	Heptacosane $CH_3(CH_2)_{25}CH_3$	380.74	cr(al bz) lf(AcOEt)	442, 270^{15}	59.5	$0.7796^{60/4}$	1.4345^{65}		$B1^{4}$, 586
7486	7,10-Heptadecadıyne $CH_3(CH_2)_5C{\equiv}CCH_2C{\equiv}C(CH_2)_5CH_3$	232.41		150^{6}		$0.84^{19/4}$	1.4700^{19}		$B1^{3}$, 1068
7487	Heptadecanal or Margaraldehyde $CH_3(CH_2)_{15}CHO$	254.46	nd (peth) cr (al + 1)	204^{26}	36			eth, bz, aa	$B1^{4}$, 3395
7488	Heptadecane $CH_3(CH_2)_{15}CH_3$	240.47	hex lf	301.8, 161.7^{10}	22	$0.7780^{20/4}$	1.4369^{20}	eth	$B1^{4}$, 548
7489	Heptadecane, 1-amino or Heptadecyl amine $CH_3(CH_2)_{15}CH_2NH_2$	255.49		336, 189^{10}	49	$0.8510^{20/4}$	1.4510^{20}	al, eth	$B4^{4}$, 824
7490	Heptadecane, 1-bromo or Heptadecyl bromide $CH_3(CH_2)_{15}CH_2Br$	319.37		349, 199^{10}	32	$0.9916^{20/4}$	1.4625^{20}	chl	$B1^{4}$, 549
7491	Heptadecane, 1,17-dibromo $Br(CH_2)_{17}Br$	398.26	lf(al)	$208\text{-}10^{3}$	38			chl	$B1^{3}$, 564
7492	Heptadecane, 9-hexyl $CH_3(CH_2)_5CH[(CH_2)_7CH_3]_2$	324.63		213.0^{10}, $151^{0.5}$	-19.4	$0.7976^{20/4}$	$1.4465^{20}/_D$		$B1^{4}$, 578
7493	Heptadecane-9-octyl $CH[(CH_2)_7CH_3]_3$	352.69		231.5^{10}	-13.8	$0.8020^{20/4}$	$1.4487^{20}/_D$		$B1^{4}$, 583
7494	Heptadecane-9-Phenethyl $C_6H_5(CH_2)_2CH(C_8H_{17})_2$	344.62		189^{10}	-26.7	$0.8560^{20/4}$	$1.4806^{20}/_D$		PCHG 3, 180
7495	Heptadecanoic acid or Margaric acid $CH_3(CH_2)_{15}CO_2H$	270.46	pl(peth)	227^{100}	62-3	0.8532^{60}	1.4342^{60}	eth, ace, bz, chl	$B2^{4}$, 1193
7496	Heptadecanoic acid, ethyl ester or Ethyl heptadecanoate $CH_3(CH_2)_{15}CO_2C_2H_5$	298.51	pl(dil al)	185^{5}	28			al, eth, ace, bz	$B2^{4}$, 1194

No.	Name, Synonyms, and Formula	Mol. wt.	Color, crystalline form, specific rotation and λ_{max} (log ε)	b.p. °C	m.p. °C	Density	n_D	Solubility	Ref.
7497	Heptadecanoic acid, methyl ester $CH_3(CH_2)_{15}CO_2CH_3$	284.48	pl(al)	$184\text{-}7^{9}$	30	....	....	al, eth, ace, bz	$B2^4$, 1194
7498	Heptadecanonitrile or Margaronitrile $CH_3(CH_2)_{15}CN$	251.46	cr(al)	$349, 183^{10}$	34	$0.8315^{20/4}$	1.4467^{20}	eth	$B2^4$, 1195
7499	1-Heptadecanol $CH_3(CH_2)_{15}CH_2OH$	256.47	lf(al) cr(ace)	308	54	$0.8475^{20/4}$	....	al, eth	$B1^4$, 1884
7500	2-Heptadecanol $CH_3(CH_2)_{14}CH(OH)CH_3$	256.47	pl(dil al)	$140^{0.5}$	54	....	1.4407^{37}	al, eth	$B1^4$, 1885
7501	9-Heptadecanol $(C_8H_{17})_2CHOH$	256.47	pl(dil al)	174^{9}	61	....	1.4262^{80}	al, eth, ace, bz	$B1^4$, 1884
7502	2-Heptadecanone or Methyl penta decylketone $C_{15}H_{31}COCH_3$	254.46	pl(dil al)	$320, 246^{110}$	48	$0.8140^{48/48}$	....	eth, ace, bz	$B1^4$, 3395
7503	9-Heptadecanone or Pelargone $(C_8H_{17})_2CO$	254.46	pl(MeOH)	$250\text{-}3, 142^{1}$	53	....	....	....	$B1^4$, 3396
7504	1-Heptadecene $CH_3(CH_2)_{14}CH{=}CH_2$	238.46	....	$300, 160^{10}$	11.2	$0.7852^{20/4}$	1.4432^{20}	eth, bz, lig	$B1^4$, 927
7505	8-Heptadecene, 9-octyl $C_7H_{15}CH{=}C(C_8H_{17})_2$	350.67	....	227^{10}	......	$0.8086^{20/4}$	$1.4554^{20/}{}_D$	....	$B1^4$, 936
7506	Heptadecyl amine $CH_3(CH_2)_{15}CH_2NH_2$	255.49	....	$336, 189^{10}$	49	$0.8510^{20/4}$	1.4510^{20}	al, eth	$B4^4$, 827
7507	1,4-Heptadiene $CH_3CH_2CH{=}CHCH_2CH{=}CH_2$	96.17	....	93^{772}	......	$0.7270^{20/4}$	1.4370^{20}	eth, bz, peth	$B1^3$, 999
7508	1,5-Heptadiene $CH_3CH{=}CHCH_2CH_2CH{=}CH_2$	96.17	....	94	......	$0.7186^{20/4}$	1.4200^{20}	al, eth, ace, bz	$B1^3$, 999
7509	1-5-Heptadiene-4-ol $CH_3CH{=}CHCH(OH)CH_2CH{=}CH_2$	112.17	....	$155\text{-}6^{742}$, 68^{24}	......	$0.8598^{20/4}$	1.4510^{25}	ace	$B1^4$, 2248
7510	1,6-Heptadiene-3-yne $CH_2{=}CHCH_2C{\equiv}CCH{=}CH_2$	92.14	....	110^{950}	......	$0.787^{25/4}$	1.4694^{25}	bz, peth	$B1^3$, 1061
7511	2,4-Heptadiene $CH_3CH_2CH{=}CHCH{=}CHCH_3$	96.17	....	108	......	$0.7384^{20/4}$	1.4578^{20}	al, eth, ace, bz	$B1^4$, 1029
7512	2,5-Heptadien-4-one,2,6-dimethyl or Phorone $CH_3C(CH_3){=}CHCOCH{=}C(CH_3)_2$	138.21	ye gr pr	197.8	28	$0.8850^{20/4}$	1.4998^{20}	al, eth, ace	$B1^4$, 3051
7513	3,5-Heptadiene-2-one or Crotonylidene acetone $CH_3CH{=}CHCH{=}CHCOCH_3$	110.16	....	88^{28}	......	$0.8946^{19/4}$	1.5177^{19}	eth	$B1^4$, 3549
7514	1,5-Heptadiyne $CH_3C{\equiv}CCH_2CH_2C{\equiv}CH$	92.14	....	26^{30}	......	$0.8100^{21/4}$	1.4521^{21}	bz, peth	$B1^2$, 247
7515	1,6-Heptadiyne $HC{\equiv}C(CH_2)_3C{\equiv}CH$	92.14	....	$112, 36^{20}$	−85	$0.8164^{17/4}$	1.451^{17}	bz, aa	$B1^4$, 1121
7516	Heptamethyleneimine, α-methyl or 1-Azacyclooctane-2-methyl $C_8H_{17}N$	127.23	nd(ace)	$162\text{-}3^{746}$	156-7	0.853^{30}	1.4620^{21}	w, al, eth, ace, bz	$B20^1$, 30
7517	Heptanal or Enanthaldehyde. Heptaldehyde $CH_3(CH_2)_5CHO$	114.19	....	152.8, 59.6^{30}	−43.3	$0.8495^{20/4}$	1.4113^{20}	al, eth	$B1^4$, 3314
7518	Heptanal, 2-benzylidene or α-Pentylcinnamaldehyde $C_5H_{11}C({=}CHC_6H_5)CHO$	202.30	ye oil	$174\text{-}5^{20}$	80	0.9711^{20}	1.5381^{20}	ace	$B7^3$, 1517
7519	Heptanal, oxime or Heptaldoxime $CH_3(CH_2)_5CH{=}NOH$	129.20	pl(al)	$195, 100.5^{14}$	57-8	0.8583^{55}	1.4210^{20}	al, eth	$B1^4$, 3316
7520	1,6-Heptadiene-4-ol, 4-methyl $CH_3C(OH)(CH_2CH{=}CH)_2$	126.20	....	158.4, $54\text{-}6^{11}$	......	$0.86258^{20/20}$	1.4500^{23}	al, ace	$B1^4$, 2260
7521	1,6-Heptadiene-3-yne $CH_2{=}CHC{\equiv}CCH_2CH{=}CH_2$	92.14	....	110^{750}	......	$0.787^{25/4}$	1.4694^{25}	bz, peth	$B1^3$, 1061
7522	Heptane C_7H_{16}	100.20	....	98.4	−90.6	$0.6837^{20/4}$	1.3878^{20}	al, eth, ace, chl, peth	$B1^4$, 376
7523	Heptane, 1-amino or Heptyl amine $CH_3(CH_2)_5CH_2NH_2$	115.22	....	156.9, 45.6^{20}	−18	$0.7754^{20/4}$	1.4251^{20}	al, eth	$B4^4$, 734
7524	Heptane, 2-amino $CH_3(CH_2)_4CHNH_2CH_3$	115.22	....	142	......	0.7665^{19}	1.4199^{19}	al, eth, peth	$B4^4$, 743
7525	Heptane, 1-bromo or Heptyl bromide $CH_3(CH_2)_5CH_2Br$	179.11	....	$178\text{-}9, 59.7^{10}$	−56.1	$1.1400^{20/4}$	1.4502^{20}	al, eth, chl	$B1^4$, 391
7526	Heptane, 1-bromo-7-fluoro $Br(CH_2)F$	197.10	....	85^{11}	......	....	1.4463^{20}	al, eth	$B1^4$, 392
7527	Heptane, 2-bromo $C_5H_{11}CHBrCH_3$	179.10	....	165-7, 63.8^{20}	......	$1.1277^{20/4}$	1.4503^{20}	bz	$B1^3$, 431
7528	Heptane, 3-bromo $CH_3CH_2CHBr(CH_2)_3CH_3$	179.10	....	62^{18}	......	$1.1362^{20/4}$	1.4503^{20}	bz, chl	$B1^4$, 392

No.	Name, Synonyms, and Formula	Mol. wt.	Color, crystalline form, specific rotation and λ_{max} (log ε)	b.p. °C	m.p. °C	Density	n_D	Solubility	Ref.
7529	Heptane, 4-bromo $CH_3(CH_2)_2CHBrCH_2CH_2CH_3$	179.10		84.6^{72}		$1.1351^{20/4}$	1.4495^{20}	bz, chl	$B1^{4}$, 392
7530	Heptane, 1-chloro or Heptyl chloride $CH_3(CH_2)_5CH_2Cl$	134.65		159, 45^{10}	−69.5	$0.8758^{20/4}$	1.4256^{20}	**al, eth**	$B1^{4}$, 389
7531	Heptane, 1-chloro-7-fluoro $F(CH_2)_7Cl$	152.64		70^{10}		0.993^{20}	1.4222^{25}	al, eth, bz, chl	$B1^{4}$, 390
7532	Heptane, 2-chloro $CH_3(CH_2)_4CHClCH_3$	134.65		46^{19}		0.8672^{20}	1.4221^{20}	eth, bz, chl, aa	$B1^{4}$, 390
7533	Heptane, 2-chloro-2-methyl $C_5H_{11}CCl(CH_3)_2$	148.68		50^{15}		$0.8568^{25/4}$	1.4240^{25}	al, eth, bz, chl	$B1^{4}$, 428
7534	Heptane, 2-chloro-6-methyl $(CH_3)_2CH(CH_2)_3CHClCH_3$	148.68		74^{35}			1.4260^{15}	al, eth, bz, chl	$B1^{3}$, 472
7535	Heptane, 3-chloro $CH_3(CH_2)_3CHClCH_2CH_3$	134.65		144^{751}, 48.3^{20}		$0.8960^{20/4}$	1.4228^{20}	eth, bz	$B1^{4}$, 390
7536	Heptane, 3-chloro-2,3-dimethyl $C_4H_9CCl(CH_3)CH(CH_3)_2$	162.70		54^{8}		0.8395^{20}	1.4391^{20}	eth, chl	$B1^{3}$, 511
7537	Heptane, 3-chloro-3-ethyl $C_4H_9CCl(C_2H_5)_2$	162.70		46^{3}		$0.8856^{20/4}$	1.4400^{20}	eth, chl	$B1^{3}$, 510
7538	Heptane, 3-chloromethyl $CH_3CH_2CH(CH_2Cl)(CH_2)_3CH_3$	148.68		174		$0.8769^{20/4}$	1.4319^{20}	al, eth, ace, bz	$B1^{4}$, 430
7539	Heptane, 3-chloro-3-methyl $C_4H_9CCl(CH_3)C_2H_5$	148.68		64^{27}		$0.8764^{20/4}$	1.4317^{20}	al, eth, bz, chl	$B1^{4}$, 430
7540	Heptane, 4-chloro $(C_3H_7)_2CHCl$	134.65		144^{758}, 48.9^{21}		$0.8710^{20/4}$	1.4237^{20}	eth,bz	$B1^{4}$, 390
7541	Heptane, 4-chloro-4-ethyl $(C_3H_7)_2C(Cl)C_2H_5$	162.70		67^{12}		$0.8821^{20/4}$	1.4438^{20}	eth, bz, chl	$B1^{3}$, 511
7542	Heptane, 4-chloro-4-methyl $(C_3H_7)_2C(Cl)CH_3$	148.68		50^{12}		$0.8690^{20/4}$	1.4310^{15}	al, eth, bz, chl	$B1^{3}$, 477
7543	Heptane, 5-chloro-2,5-dimethyl $CH_3CH_2CCl(CH_3)CH_2CH_2CH(CH_3)_2$	162.70		63^{15}		$0.8692^{18/4}$	1.4346^{15}	bz, chl	$B1^{1}$, 64
7544	Heptane, 1,7-diamino or Heptamethylenediamine $H_2N(CH_2)_7NH_2$	130.23		223-5, 104-5^{12}	28.9			al, eth, ace, bz	$B4^{4}$, 1354
7545	Heptane, 1,7-dibromo $Br(CH_2)_7Br$	258.00		263, 132^{11}	−41.7	$1.5306^{20/4}$	1.5034^{20}	eth, ace, bz	$B1^{4}$, 393
7546	Heptane, 1,1-dichloro $CH_3(CH_2)_5CHCl_2$	169.09		187, 82^{20}		$1.0008^{20/4}$	1.4440^{20}	eth, bz, chl	$B1^{4}$, 390
7547	Heptane, 1,2-dichloro $CH_3(CH_2)_4CHClCH_2Cl$	169.09		68-72^{7}		$1.064^{20/4}$	1.4490^{20}	eth, bz, chl	$B1^{3}$, 430
7548	Heptane, 2,2-dichloro $CH_3(CH_2)_4CCl_2CH_3$	169.09		77^{25}		$1.012^{20/4}$	1.4440^{20}	eth, bz, chl	$B1^{1}$, 430
7549	Heptane, 2,6-dichloro-2,6-dimethyl $(CH_3)_2CCl(CH_2)_3CCl(CH_3)_2$	197.15		93^{16}	43			bz, chl	$B1^{3}$, 513
7550	Heptane, 4,4-dichloro $(C_3H_7)_2CCl_2$	169.09		86^{27}		1.0008^{17}	1.448^{17}	eth, chl	$B1^{2}$, 117
7551	Heptane, 3,3-di(hydroxymethyl) $C_2H_5C(CH_2OH)_2C_4H_9$	160.26	wh	262, 123^{15}	43.8	$0.929^{50/20}$	1.4587^{25}	al	$B1^{3}$, 2228
7552	Heptane, 2,2-dimethyl $(CH_3)_3C(CH_2)_4CH_3$	128.26		132.7	−113	$0.7105^{20/4}$	1.4016^{20}	eth, ace, bz	$B1^{4}$, 457
7553	Heptane, 2,3-dimethyl $CH_3(CH_2)_3CH(CH_3)CH(CH_3)_2$	128.26		140.5, 29.4^{10}	−116	$0.7260^{20/4}$	1.4088^{20}	**al, eth, ace, bz**	$B1^{4}$, 457
7554	Heptane, 2,4-dimethyl $C_3H_7CH(CH_3)CH_2CH(CH_3)_2$	128.26		133.5, 23.8^{10}		$0.7143^{20/4}$	1.4031^{20}	**al, eth, ace, bz**	$B1^{4}$, 457
7555	Heptane, 2,5-dimethyl (d) $C_2H_5CH(CH_3)(CH_2)_2CH(CH_3)_2$	128.26		136		$0.7198^{20/4}$	1.4033^{20}	**al, eth, ace, bz**	$B1^{4}$, 457
7556	Heptane, 2,6-dimethyl $(CH_3)_2CH(CH_2)_3CH(CH_3)_2$	128.26		135.2, 25.5^{10}	−102.9	$0.7089^{20/4}$	1.4011^{20}		$B1^{4}$, 458
7557	Heptane, 3,3-dimethyl $C_4H_9C(CH_3)_2CH_2CH_3$	128.26		137.3, 26^{10}		$0.7254^{20/4}$	1.4087^{20}	**al**, eth, ace, bz	$B1^{4}$, 458
7558	Heptane, 3,4-dimethyl $CH_3CH_2CH(CH_3)CH(CH_3)C_3H_7$	128.26		140.1		$0.7314^{20/4}$	1.4108^{20}	eth, ace, bz	$B1^{3}$, 514
7559	Heptane, 3,5-dimethyl $CH_3CH_2CH(CH_3)CH_2CH(CH_3)CH_2CH_3$	128.26		136		$0.7225^{20/4}$	1.4083^{20}	eth, ace, bz	$B1^{4}$, 458
7560	Heptane, 4,4-dimethyl $(C_3H_7)_2C(CH_3)_2$	128.26		135.2		$0.7221^{20/4}$	1.4076^{20}	eth, ace, bz	$B1^{4}$, 458
7561	Heptane, 4-ethyl $(C_3H_7)_2CHC_2H_5$	128.26		141.2, 31^{10}		$0.7270^{20/4}$	1.4096^{20}	**al, eth, ace, bz, chl**	$B1^{4}$, 457

PHYSICAL CONSTANTS OF ORGANIC COMPOUNDS (Continued)

No.	Name, Synonyms, and Formula	Mol. wt.	Color, crystalline form, specific rotation and λ_{max} (log ε)	b.p. °C	m.p. °C	Density	n_D	Solubility	Ref.
7562	Heptane, 1-fluoro or Heptyl fluoride $CH_3(CH_2)_5CH_2F$	118.12	......	117.9	−73	$0.8062^{20/4}$	1.3854^{20}	eth, ace, bz, peth	$B1^{4}$, 387
7563	Heptane, perfluoro C_7F_{16}	388.05	......	82.4	−78	1.7333^{20}	1.2618^{20}	al, eth, ace, chl	$B1^{4}$, 388
7564	Heptane, 1-iodo or Heptyl iodide $CH_3(CH_2)_5CH_2I$	226.10	......	204, 76.1^{10}	−48.2	$1.3791^{20/4}$	1.4904^{20}	al, eth, ace, chl	$B1^{4}$, 393
7565	Heptane, 2-iodo $CH_3CHI(CH_2)_4CH_3$	226.10	......	98^{50}	......	1.304^{20}	1.4826	ace, bz	$B1^{4}$, 393
7566	Heptane, 2-methyl $C_5H_{11}CH(CH_3)_2$	114.23	......	117.6, 12.3^{10}	−109	$0.6980^{20/4}$	1.3949^{20}	al eth, ace, bz, chl	$B1^{4}$, 428
7567	Heptane, 2- methylamino $C_5H_{11}CH(CH_3)NHCH_3$	129.25	......	155	......	......	......	......	$B4^{4}$, 743
7568	Heptane, 3-methyl (d) $C_4H_9CH(CH_3)C_2H_5$	114.23	$[\alpha]^{26}/_D$ + 9.34	115-8	......	$0.7075^{16/4}$	1.4002^{18}	al, eth, ace, bz, chl	$B1^{4}$, 429
7569	Heptane, 3-methyl (dl) $C_4H_9CH(CH_3)C_2H_5$	114.23	......	119, 13.3^{10}	−120.5	$0.7058^{20/4}$	1.3985^{20}	al, eth, ace, bz, chl	$B1^{4}$, 429
7570	Heptane, 3-methyl (l) $C_4H_9CH(CH_3)C_2H_5$	114.23	......	$117\text{-}8^{745}$	......	......	1.3990^{20}	al, eth, ace, bz, chl	$B1^{3}$, 476
7571	Heptane, 4-methyl $(C_3H_7)_2CHCH_3$	114.23	......	117.7, 12.4^{10}	−121	$0.7046^{20/4}$	1.3979^{20}	al, eth ace, bz, chl	$B1^{4}$, 431
7572	Heptane, 2,2,4,4,6-pentamethyl $(CH_3)_2CHCH_2CH(CH_3)_2CH_2C(CH_3)_3$	170.34	......	177.8	−67	$0.7463^{20/4}$	$1.4440^{20}/_D$	......	$B1^{4}$, 510
7573	Heptane, 2,2,4-trimethyl $(CH_3)_3CCH_2CH(CH_3)(C_3H_7)$	142.28	......	147.7, 32.9^{10}	......	$0.7275^{20/4}$	1.4092^{20}	bz, chl	$B1^{4}$, 481
7574	Heptane, 3,3,5-trimethyl $CH_3CH_2C(CH_3)_2CH_2CH(CH_3)CH_2CH_3$	142.28	......	155.7, 38.9^{10}	......	$0.7248^{20/4}$	1.4170^{20}	bz, chl	$B1^{4}$, 483
7575	Heptanedioic acid or Pimelic acid $HO_2C(CH_2)_5CO_2H$	160.17	pr(w)	272^{100}sub, 212^{10}	106	1.329^{15}	......	w, al, eth	$B2^{4}$, 2003
7576	1,7-Heptanediol or Heptamethylene glycol $HO(CH_2)_7OH$	132.20	......	262, 151^{14}	22	$0.9569^{25/4}$	1.4520^{25}	w, al	$B1^{4}$, 2580
7577	2,4-Heptanediol, 3-methyl $C_3H_7CH(OH)CH(CH_3)CH(OH)CH_3$	146.23	......	115^{3}	......	$0.928^{20/4}$	1.4459^{20}	al	B1, 491
7578	2,4-Heptanedione $C_3H_7COCH_2COCH_3$	128.17	......	174, 70^{20}	......	$0.9411^{25/4}$	......	......	$B1^{4}$, 3698
7579	1-Heptanethiol $CH_3(CH_2)_5CH_2SH$	132.26	......	177	−43	$0.8427^{20/4}$	1.4521^{20}	al, eth	$B1^{4}$, 1738
7580	1,4,7-Heptanetriol $(HOCH_2CH_2CH_2)_2CHOH$	148.20	......	$230\text{-}2^{25}$, 146^{1}	−35	1.075^{18}	1.4725^{20}	w, al, ace	$B1^{4}$, 2787
7581	2,4,6-Heptanetrione or Diacetyl acetone $(CH_3COCH_2)_2CO$	142.15	lf	121^{10}	49	$1.0681^{40/40}$	1.4930^{20}	w, al, eth	$B1^{4}$, 3783
7582	Heptano amide $CH_3(CH_2)_5CONH_2$	129.20	nd(al)lf(w)	250-8	96	$0.852^{110/4}$	1.4217^{110}	w, al, eth	$B2^{4}$, 963
7583	Heptanoic acid or Enanthic acid $CH_3(CH_2)_5CO_2H$	130.19	......	223, 116^{11}	−7.5	$0.9200^{20/4}$	1.4170^{20}	al, eth, ace	$B2^{4}$, 958
7584	Heptanoic acid, 7-amino $H_2N(CH_2)_6CO_2H$	145.20	cr (w, MeOH-peth)	......	195	......	......	w, al	$B4^{3}$, 1467
7585	Heptanoic anhydride $(C_6H_{13}CO)_2O$	242.36	......	268-71, $164^{12.5}$	−12.4	$0.9321^{20/4}$	1.4335^{15}	al, eth	$B2^{4}$, 962
7586	Heptanoic acid, 2-bromo $CH_3(CH_2)_4CHBrCO_2H$	209.08	......	250d, 147^{12}	......	1.319^{15}	1.471^{18}	eth, ace	$B2^{4}$, 967
7587	Heptanoic acid, 7-bromo $Br(CH_2)_6CO_2H$	209.08	wh cr(dil al)	280	31	......	......	al, eth, ace, bz	$B2^{4}$, 968
7588	Heptanoic acid, butyl ester $CH_3(CH_2)_5CO_2C_4H_9$	186.29	......	226.2	−67.5	0.8638^{20}	1.4204^{20}	al, eth, ace, bz	$B2^{3}$, 768
7589	Heptanoic acid, *iso*-butyl ester $C_6H_{13}CO_2\text{-}i\text{-}C_4H_9$	186.29	......	208	......	0.8593^{20}	......	al, eth, ace, bz	$B2^{1}$, 145
7590	Heptanoic acid, ethyl ester or Ethyl heptanoate $CH_3(CH_2)_5CO_2C_2H_5$	158.24	......	187, 78^{14}	−66.1	$0.8817^{20/4}$	1.4100^{20}	al, eth	$B2^{4}$, 960
7591	Heptanoic acid, 7-fluoro $F(CH_2)_6CO_2H$	148.18	......	133^{10}	......	1.039^{20}	1.4207^{25}	......	$B2^{4}$, 964
7592	Heptanoic acid, heptyl ester $C_6H_{13}CO_2C_7H_{15}$	228.38	......	276-8	−33	$0.8649^{20/4}$	1.4320^{20}	al, eth	$B2^{4}$, 961
7593	Heptanoic acid, hexyl ester $C_6H_{13}CO_2C_6H_{13}$	214.35	......	261	−48	0.8611^{20}	1.429^{15}	al, eth, ace, bz	$B2^{3}$, 768
7594	Heptanoic acid, 7-iodo $I(CH_2)_6CO_2H$	256.08	lf(dil al)	......	49-51	......	......	al, eth, ace, bz	$B2^{4}$, 969

No.	Name, Synonyms, and Formula	Mol. wt.	Color, crystalline form, specific rotation and λ_{max} (log ε)	b.p. °C	m.p. °C	Density	n_D	Solubility	Ref.
7595	Heptanoic acid, methyl ester $C_6H_{13}CO_2CH_3$	144.21		172	−56	$0.8815^{20/4}$	1.4152^{20}	al, eth, ace	$B2^4$, 960
7596	Heptanoic acid, octyl ester or Octyl heptanoate $C_6H_{13}CO_2C_8H_{17}$	242.40		290	−22.5	0.8596^{20}	1.4349^{15}	al, eth, ace, bz	$B2^3$, 768
7597	Heptanoic acid, 6-oxo $CH_3CO(CH_2)_4CO_2H$	144.17		$250\text{-}3^{280}$, 135^1	40.2		1.4306^{25}	w, al, eth, ace	$B3^4$, 1598
7598	Heptanoic acid, pentyl ester $C_6H_{13}CO_2C_5H_{11}$	200.32		245.4	−50	0.8623^{20}	1.4263^{15}	al, eth, ace, bz	$B2^4$, 960
7599	Heptanoic acid, propyl ester or Propyl heptanoate $C_6H_{13}CO_2C_3H_7$	172.27	cr(peth)	207.9	−63.5	$0.8641^{15/4}$	1.4183^{15}	al, eth, ace, bz	$B2^3$, 767
7600	Heptanonitrile $CH_3(CH_2)_5CN$	111.19		183^{765}, $70\text{-}2^{10}$		$0.8107^{20/0}$	1.4104^{30}	eth, ace, bz, aa	$B2^4$, 963
7601	Heptanoyl chloride $CH_3(CH_2)_5COCl$	148.63		125.2	−83.8	0.9590^{20}	1.4345^{18}	eth, lig	$B2^4$, 963
7602	1-Heptanol $CH_3(CH_2)_5CH_2OH$	116.20		176	−34.1	$0.8219^{20/4}$	1.4249^{20}	al, eth	$B1^4$, 1731
7603	1-Heptanol, 7-chloro $Cl(CH_2)_7OH$	150.65	cr(peth or bz)	150^{20}	11	$0.9998^{15/4}$	1.4537^{25}	al, peth	$B1^4$, 1738
7604	1-Heptanol, 7-fluoro $F(CH_2)_7OH$	134.19		$98\text{-}9^{12}$		$0.956^{20/4}$	1.4197^{25}	al, eth	$B1^4$, 17369
7605	1-Heptanol, 4-methyl $C_3H_7CH(CH_3)CH_2CH_2CH_2OH$	130.23		183, 71.6^{20}		$0.8065^{25/4}$	1.4258^{20}	al	$B1^4$, 1789
7606	1-Heptanol, 6-methyl $(CH_3)_2CH(CH_2)_5OH$	130.23		188^{764}, 95.8^{20}	−106	$0.8176^{25/4}$	1.4251^{25}	al, eth	$B1^4$, 1782
7607	1-Heptanol, 1-phenyl $CH_3(CH_2)_5CH(C_6H_5)OH$	192.30		275, $153\text{-}5^{18}$		0.946	1.5024^{20}		$B6^2$, 513
7608	2-Heptanol (d) $C_5H_{11}CH(OH)CH_3$	116.20	$[\alpha]^{20/}{}_D$ + 11.4 (al)	160-2, 73^{20}		$0.8190^{20/4}$	1.4209^{20}	al, eth	$B1^4$, 1740
7609	2-Heptanol (dl) $C_5H_{11}CH(OH)CH_3$	116.20		160, 66^{20}		$0.8167^{20/4}$	1.4210^{20}	al, eth	$B1^4$, 1740
7610	2-Heptanol (l) $C_5H_{11}CH(OH)CH_3$	116.20	$[\alpha]^{12/}{}_D$ -10.5	74^{23}		$0.8184^{20/4}$	1.4201^{20}	al, eth	$B1^3$, 1687
7611	2-Heptanol, 6-amino-2-methyl, hydrochloride $CH_3CH(NH_2)(CH_2)_3C(OH)(CH_3)_2 \cdot HCl$	181.71	cr		154-5			w, al	$B4^4$, 1809
7612	2-Heptanol, 1-chloro $C_5H_{11}CH(OH)CH_2Cl$	150.65		93^{13}		$0.9885^{20/4}$	1.4499^{20}	al, eth, ace	$B1^4$, 1741
7613	2-Heptanol, 2-methyl $C_5H_{11}C(OH)(CH_3)_2$	130.23		156, $66\text{-}8^{15}$		$0.8142^{20/4}$	1.4250^{20}	al, eth	$B1^4$, 1780
7614	2-Heptanol, 3-methyl $C_4H_9CH(CH_3)CH(OH)CH_3$	130.23		166.1, 68.1^{20}		$0.8177^{25/4}$	1.4199^{25}	al, eth	$B1^3$, 1729
7615	3-Heptanol (d) $C_4H_9CH(OH)C_2H_5$	116.20		157^{750}, 66^{18}	−70	$0.8227^{20/4}$	1.4201^{20}	al, eth	$B1^4$, 1741
7616	3-Heptanol, 2,6-dimethyl $(CH_3)_2CHCH(OH)CH_2CH_2CH(CH_3)_2$	144.26		175		$0.8212^{20/4}$	1.4246^{20}		$B1^3$, 1753
7617	3-Heptanol, 2-methyl (d) $C_4H_9CH(OH)CH(CH_3)_2$	130.23	$[\alpha]^{20/}{}_D$ + 27.7 (al)	72^{12}		$0.8235^{20/4}$	1.4265^{20}	al, eth	$B1^1$, 209
7618	3-Heptanol, 2-methyl (dl) $C_4H_9CH(OH)CH(CH_3)_2$	130.23		167.2, 73^{19}		$0.8235^{20/4}$	1.4265^{20}	al, eth	$B1^4$, 1781
7619	3-Heptanol, 2-methyl (l) $C_4H_9CH(OH)CH(CH_3)_2$	130.23	$[\alpha]_D$ -21.08	87^{36}		$0.8235^{20/4}$	1.4265^{20}	al, eth	$B1^1$, 209
7620	3-Heptanol, 3-methyl $C_4H_9C(OH)(CH_3)CH_2CH_3$	130.23		163, $64\text{-}5^{10}$	−83	$0.8282^{20/4}$	1.4279^{20}	al, eth	$B1^4$, 1783
7621	4-Heptanol $(C_3H_7)_2CHOH$	116.20		161, 63.8^{16}	−41.2	$0.8183^{20/4}$	1.4205^{20}	al, eth	$B1^4$, 1743
7622	4-Heptanol, 2,6-dimethyl $[(CH_3)_2CHCH_2]_2CHOH$	144.26		176-7		$0.809^{21/4}$	1.4242^{20}	al, eth	$B1^4$, 1810
7623	4-Heptanol, 4-ethyl $(C_3H_7)_2C(OH)C_2H_5$	144.26		182		0.8350^{20}	1.4332^{20}	al, eth	$B1^4$, 1809
7624	4-Heptanol, 3-methyl (dl) $C_2H_5CH(CH_3)CH(OH)C_3H_7$	130.23		164.7, 67.3^{20}		$0.8335^{25/4}$	1.4211^{25}	al, eth	$B1^3$, 1730
7625	4-Heptanol, 4-methyl $(C_3H_7)_2C(CH_3)OH$	130.23		161, $61\text{-}3^{12}$	-82	$0.8248^{20/4}$	1.4258^{20}	al, eth	$B1^4$, 1789
7626	4-Heptanol, 4-propyl $(C_3H_7)_3COH$	158.28		190-2, $89\text{-}90^{15}$		$0.8338^{21/0}$	1.4355^{21}	al, eth, bz	$B1^4$, 1831
7627	2-Heptanone or Methyl hexyl ketone $CH_3(CH_2)_4COCH_3$	114.19		151.4, 111^{21}	−35.5	$0.8111^{20/4}$	1.4088^{20}	al, eth	$B1^4$, 3318

No.	Name, Synonyms, and Formula	Mol. wt.	Color, crystalline form, specific rotation and λ_{max} (log ε)	b.p. °C	m.p. °C	Density	n_D	Solubility	Ref.
7628	2-Heptanone, 1-chloro $C_5H_{11}COCH_2Cl$	148.64	...	83^{16}	...	0.802^{20}	1.4371^{20}	**al, eth**	$B1^4$, 3320
7629	2-Heptanone, 3-methyl $C_4H_9CH(CH_3)COCH_3$	128.21	...	167	...	$0.8218^{20/4}$	1.4172^{20}	al, eth, ace, bz	$B1^3$, 2878
7630	2-Heptanone, 6-methyl $(CH_3)_2CH(CH_2)_3COCH_3$	128.21	...	167, $51\text{-}3^{12}$	...	$0.8151^{20/4}$	1.4162^{20}	al, eth, ace, bz	$B1^4$, 3344
7631	2-Heptanone, 4-isopropyl i-$C_3H_7CH(C_3H_7)CH_2COCH_3$	156.27	...	$82\text{-}4^{14}$	...	...	...	al, eth, ace, bz	$B1^3$, 2902
7632	3-Heptanone or Butyl ethyl ketone $C_4H_9COC_2H_5$	114.19	...	147^{765}	−39	$0.8183^{20/4}$	1.4057^{20}	**al, eth**	$B1^4$, 3321
7633	3-Heptanone, 6-dimethylamino-4,4-diphenyl (l) or l-Methadone $(CH_3)_2NCH(CH_3)CH_2C(C_6H_5)_2COC_2H_5$	309.45	$[\alpha]^{20/}{}_D$ -32 (al)	...	99-100	...	...	al	$B14^3$, 278
7634	3-Heptanone, 6-dimethylamino-4,4-diphenyl,hydrochloride (dl) or Physopeptone $(CH_3)_2NCH(CH_3)CH_2C(C_6H_5)_2COC_2H_5.HCl$	345.91	pl(al-eth)	...	236	...	...	w, al, eth, chl	$B14^3$, 279
7635	3-Heptanone, 6-dimethylamino-4,4-diphenyl hydrochloride (l) $(CH_3)_2NCH(CH_3)CH_2C(C_6H_5)_2COC_2H_5.HCl$	345.91	$[\alpha]^{20/}{}_D$ -169 (al, c=2.1)	...	245-6	...	...	w, al, chl	$B14^3$, 278
7636	3-Heptanone, 2-methyl $C_4H_9COCH(CH_3)_2$	128.21	...	158, $63\text{-}5^{25}$	...	$0.8163^{20/4}$	1.4115^{20}	al, eth, ace	$B1^4$, 3343
7637	3-Heptanone, 6-methyl $(CH_3)_2CHCH_2CH_2COC_2H_5$	128.21	...	163^{734}	...	0.8304^{20}	1.4209^{20}	al, eth, bz	$B1^4$, 3344
7638	4-Heptanone or Depropyl ketone $(C_3H_7)_2CO$	114.19	...	144	−33	$0.8174^{20/4}$	1.4069^{20}	**al, eth**	$B1^4$, 3323
7639	4-Heptanone, 2,6-dimethyl or Isovalerone $(i\text{-}C_4H_9)_2CO$	142.24	...	168, $60\text{-}1^{18}$	...	$0.8053^{20/4}$	1.412^{20}	**al, eth**	$B1^4$, 3360
7640	4-Heptanone, 2-methyl $C_3H_7COCH_2CH(CH_3)_2$	128.21	...	155^{750}	...	$0.813^{22/0}$	...	al, eth	$B1^4$, 3343
7641	Heptano phenone $C_6H_{13}COC_6H_5$	190.29	lf	283.3, 155^{15}	16.4	$0.9516^{20/4}$	1.5060^{20}	al, eth, ace	$B7^3$, 1188
7642	Heptasiloxane, hexadecamethyl $CH_3[Si(CH_3)_2O\text{-}]_6Si(CH_3)_3$	533.15	...	270, 165^{20}	−78	$0.9012^{20/4}$	1.3965^{20}	bz, lig	$B4^3$, 1880
7643	1-Heptene $CH_3(CH_2)_4CH{=}CH_2$	98.19	...	93.6	−119	$0.6970^{20/4}$	1.3998^{20}	al, eth	$B1^4$, 857
7644	1-Heptene, 1-chloro $C_5H_{11}CH{=}CHCl$	132.63	...	155, $78\text{-}82^{75}$	...	0.8948^{20}	1.4380^{20}	eth, ace, bz, chl	$B1^3$, 823
7645	1-Heptene, 2-chloro $C_5H_{11}CCl{=}CH_2$	132.63	...	138^{748}	...	$0.8895^{20/4}$	1.4349^{20}	al, eth, chl	$B1^3$, 823
7646	1-Heptene, 2-methyl $C_5H_{11}C(CH_3){=}CH_2$	112.22	...	118.2	−90.1	$0.2206^{20/4}$	$1.4120^{20/}{}_D$	...	$B1^4$, 881
7647	2-Heptene *(cis)* $CH_3(CH_2)_3CH{=}CHCH_3$	98.19	...	98.5	...	$0.708^{20/4}$	1.406^{20}	al, eth, ace, bz, chl	$B1^3$, 824
7648	2-Heptene (trans) $CH_3(CH_2)_3CH{=}CHCH_3$	98.19	...	98	−109.5	$0.7012^{20/4}$	1.4045^{20}	al, eth, ace, bz, chl	$B1^4$, 860
7649	2-Heptene, 4-chloro $C_3H_7CHClCH{=}CHCH_3$	132.63	...	140-5, 49^{21}	...	$0.879^{18/4}$	1.4430^{23}	al, eth, chl	$B1^3$, 825
7650	2-Heptene, 6-chloro-2-methyl $CH_3CHCl(CH_2)_2CH{=}C(CH_3)_2$	146.66	...	$60\text{-}1^{15}$	...	$0.8931^{18/4}$	1.4458^{18}	al, eth, ace, bz	$B1^2$, 200
7651	2-Heptene, 2-methyl $C_4H_9CH{=}C(CH_3)_2$	112.22	...	122.6	...	$0.7241^{20/0}$	1.4170^{20}	eth, bz	$B1^4$, 882
7652	2-Heptene, 2-methyl-6-methylamino $CH_3NHCH(CH_3)CH_2CH_2CH{=}C(CH_3)_2$	141.26	...	176-8, $58\text{-}9^{17}$	...	...	...	al, eth	$B4^3$, 467
7653	3-Heptene *(cis)* $C_3H_7CH{=}CHC_2H_5$	98.19	...	95.8	...	$0.7030^{20/4}$	1.4059^{20}	al, eth, ace, bz, chl	$B1^4$, 861
7654	3-Heptene (trans) $C_3H_7CH{=}CHC_2H_5$	98.19	...	95.7	−136.6	$0.6981^{20/4}$	1.4043^{20}	al, eth, ace, bz, peth	$B1^4$, 861
7655	3-Heptene, 4-chloro $C_3H_7CCl{=}CHC_2H_5$	132.63	...	139	...	0.883^{14}	1.437^{14}	al, eth, chl	$B1^3$, 827
7656	3-Heptene, 4-propyl $C_2H_5CH{=}C(C_3H_7)_2$	140.27	...	160.5	...	$0.7518^{17.8}$	$1.4302^{17.8}$	...	$B1^4$, 906
7657	1-Heptene-2-carboxaldehyde, 1-phenyl or Jasminaldehyde $C_5H_{11}C(CHO){=}CHC_6H_5$	203.30	...	$174\text{-}5^{20}$, 140^5	...	0.9718^{20}	1.5381^{20}	al, eth	$B7^3$, 1517
7658	1-Heptene-4-ol, 4-methyl $C_3H_7COH(CH_3)CH_2CH{=}CH_2$	128.21	...	159-60	...	$0.8345^{20/0}$	1.4479^{18}	al, eth	$B1^3$, 1945
7659	2-Heptene-1-ol *(trans)* $C_4H_9CH{=}CHCH_2OH$	114.19	...	177-9, 75^{10}	...	0.8516^{20}	1.4460^{20}	al, ace	$B1^3$, 1936

No.	Name, Synonyms, and Formula	Mol. wt.	Color, crystalline form, specific rotation and λ_{max} (log ε)	b.p. °C	m.p. °C	Density	n_D	Solubility	Ref.
7660	2-Heptene-4-ol (dl) $C_3H_7CH(OH)CH=CHCH_3$	114.19		152-4, 64^{14}		$0.8445^{20/4}$	1.4373^{20}	al, eth	$B1^4$, 2155
7661	3-Heptene-2-one *(trans)* $C_3H_7CH=CHCOCH_3$	112.17		62^{15}		$0.8496^{20/4}$	1.4436^{20}	al, eth	$B1^4$, 3481
7662	5-Heptene-2-one, 6-methyl $(CH_3)_2C=CHCH_2CH_2COCH_3$	126.20		173, 58.6^{10}	−67	$0.8546^{16/4}$	1.4445^{20}	**al, eth**	$B1^4$, 3493
7663	1-Heptene-3-yne- $C_3H_7C \equiv C\text{-}CH=CH_2$	94.16		110, 44^{75}		$0.7603^{20/4}$	1.4520^{25}	al, ace, bz, peth	$B1^4$, 1097
7664	1-Heptene-4-one, 6,6-dimethyl $(CH_3)_3CC \equiv CCH_2CH=CH_2$	122.21		125, 68^{100}		0.758^{20}	1.4312^{20}	al, ace, bz, peth	C55, 23329
7665	6-Heptene-4-one-3-ol, 3-ethyl $CH_2=CCHC \equiv CC(OH)(C_2H_5)_2$	138.21		62^4		$0.8875^{20/4}$	1.4800^{20}	al, eth	$B1^3$, 2034
7666	Heptyl amine $CH_3(CH_2)_5CH_2NH_2$	115.22		156.9, 45.6^{10}	−18	$0.7754^{20/4}$	1.4251^{20}	**al, eth**	$B4^4$, 734
7667	Heptyl ether or Diheptyl ether $(C_7H_{15})_2O$	214.39		258.5		$0.8008^{20/4}$	1.4275^{20}	al, eth	$B1^4$, 1733
7668	Heptyl methyl ether $C_7H_{15}OCH_3$	130.23		151		$0.7869^{15/15}$	1.4073^{20}	**al, eth**, ace	$B1^3$, 1682
7669	Heptyl nitrite $C_7H_{15}ONO$	145.20		155-8, 44^{18}		$0.8939^{0/4}$	1.4032^{20}	eth	$B1^4$, 1735
7670	Heptyl phenyl ether $C_7H_{15}OC_6H_5$	192.31		267, $128\text{-}30^{12}$		$0.9178^{15/15}$	1.4912^{20}	**al, eth**, ace	$B6^4$, 560
7671	Heptyl sulfate $(C_7H_{15}O)_2SO_2$	294.45	cr(peth)	$146.6^{1.5}$	13	$0.9819^{25/25}$	1.4362^{25}		$B1^3$, 1683
7672	Heptyl thiocyanate $C_7H_{15}SCN$	157.27		234-6, 136^{28}		0.92^{20}		**al, eth**	$B3^4$, 331
7673	1-Heptyne $CH_3(CH_2)_4C \equiv CH$	96.17		99.7, 6^{10}	−81	$0.7328^{20/4}$	1.4087^{20}	**al, eth**, bz, chl, peth	$B1^4$, 1025
7674	1-Heptyne, 1-bromo $CH_3(CH_2)_4C \equiv CBr$	175.07		164^{755}, 69^{25}		$1.2120^{22/4}$	1.4678^{22}	al, eth, ace, chl	$B1^3$, 998
7675	1-Heptyne, 1-chloro $CH_3(CH_2)_4C \equiv CCl$	130.62		141		$0.9250^{24/4}$	1.4411^{24}	al, eth	$B1^3$, 997
7676	1-Heptyne, 1-iodo $CH_3(CH_2)_4C \equiv CI$	222.07		$90\text{-}2^{17}$		$1.4701^{19/4}$	$1.5123^{19/}{}_D$		$B1^3$, 998
7677	2-Heptyne $CH_3(CH_2)_3C \equiv CCH_3$	96.17		112		$0.7480^{20/4}$	1.4230^{20}	**al, eth**, bz, chl, peth	$B1^4$, 1026
7678	2-Heptyne, 1-bromo $CH_3(CH_2)_3C \equiv CCH_2Br$	175.07		104^{55}, 84^{20}			1.4878^{25}	al, eth, ace	$B1^4$, 1026
7679	2-Heptyne, 1-chloro $CH_3(CH_2)_3C \equiv CCH_2Cl$	130.62		167, 73^{24}			1.4570^{25}		$B1^4$, 1026
7680	2-Heptyne, 7-chloro $Cl(CH_2)_4C \equiv CCH_3$	130.62		166			1.4507^{25}	al, eth	$B1^3$, 998
7681	3-Heptyne $C_3H_7C \equiv CC_2H_5$	96.17		105-6		$0.7527^{20/4}$	1.4220^{20}	**al, eth**, bz, chl, peth	$B1^4$, 1027
7682	3-Heptyne, 1-chloro $C_3H_7C \equiv CCH_2CH_2Cl$	130.62		162, $90\text{-}93^{20}$			1.4520^{25}		$B1^3$, 999
7683	3-Heptyne, 7-chloro $Cl(CH_2)_3C \equiv CC_2H_5$	130.62		164, $74\text{-}5^{31}$			1.4517^{20}	al, eth	$B1^3$, 999
7684	3-Heptyne, 2,6-dimethyl $(CH_3)_2CHCH_2C \equiv CCH(CH_3)_2$	124.23		130-6		$0.785^{20/4}$		eth, ace	$B1^3$, 1015
7685	3-Heptyne, 5,5-dimethyl $CH_3CH_2C(CH_3)_2C \equiv CC_2H_5$	124.23		69^{100}		$0.7610^{20/4}$	1.4360^{20}	eth, ace	$B1^3$, 1015
7686	3-Heptyne, 5-ethyl-5-methyl $(C_2H_5)_2C(CH_3)C \equiv CC_2H_5$	138.25		88^{100}		$0.7714^{20/4}$	1.4386^{20}	eth, ace	$B1^3$, 1021
7687	2-Heptyne-1-ol or Butyl propargyl alcohol $C_4H_9C \equiv CCH_2OH$	112.17		94^{22}			1.4523^{25}		$B1^4$, 2247
7688	Heroin or *o,o*-Diacetyl morphine $C_{21}H_{23}NO_5$	369.42	rh, $[\alpha]^{15/}{}_D$ -166 (MeOH)	$272\text{-}4^{12}$	173	1.56-1.61		bz, chl	$B27^2$, 151
7689	Hesperidin or Hesperitin-7-(6-*o*-L-rhamnopyransoyl)-β-D-glucoside $C_{28}H_{34}O_{15}$	610.57	wh nd(dil MeOH or aa), $[\alpha]^{20/}{}_D$ -77.5 (py, c=4.2		261-3			al, py, aa	$B18^4$, 3219
7690	Hexacene or Anthraceno-2′:3′,2:3-anthracene $C_{26}H_{16}$	328.41	dk bl-gr cr(sub)	sub	Ca 380				$B5^3$, 2654

No.	Name, Synonyms, and Formula	Mol. wt.	Color, crystalline form, specific rotation and λ_{max} (log ε)	b.p. °C	m.p. °C	Density	n_D	Solubility	Ref.
7691	Hexachlorophene or Bis(2-hydroxy-3,5,6-trichlorophenyl) methane $(2\text{-}HO\text{-}3,5,6\text{-}Cl_3C_6H)_2CH_2$	406.91	nd(bz)	...	166-7	...	...	al, eth, ace, chl	$B6^1$, 5407
7692	Hexacosane or Cerane $CH_3(CH_2)_{24}CH_3$	366.71	mcl tcl or rh (bz) cr (eth)	412.2, 248.2^{22}	56.4	0.7783^{60}, $0.8032^{20/4}$	1.4357^{60}	bz, lig, chl	$B1^4$, 583
7693	Hexacosane, 13-dodecyl $C_{13}H_{27}CH(C_{12}H_{25})_2$	353.03	...	272^1	13.7	$0.8188^{20/4}$	1.4577^{20}	...	$B1^4$, 600
7694	1-Hexacosanol or Ceryl alcohol Cerotin $CH_3(CH_2)_{24}CH_2OH$	382.71	rh pl (dil al)	305^{20} d	80	...	...	al, eth	$B1^4$, 1912
7695	1,15-Hexadecadiyne $HC{\equiv}C(CH_2)_{12}C{\equiv}CH$	218.38	fl(al)	$152\text{-}5^{12}$	44-5	...	...	...	$B1^2$, 249
7696	6,9-Hexadecadiyne $CH_3(CH_2)_5C{\equiv}CCH_2C{\equiv}C(CH_2)_5CH_3$	218.38	...	169^{15}	...	$0.845^{18/4}$	1.4694^{18}	...	$B1^3$, 1067
7697	6,10-Hexadecadiyne $CH_3(CH_2)_4C{\equiv}CCH_2CH_2C{\equiv}C(CH_2)_4CH_3$	218.28	...	157^{10}	...	$0.7907^{20/4}$	1.4523^{20}	...	$B1^3$, 1067
7698	Hexadecanal or Palmitaldehyde $CH_3(CH_2)_{14}CHO$	240.43	pl (eth) nd (peth)	$200\text{-}2^{29}$	34	...	...	al, eth, ace, bz	$B1^4$, 3393
7699	Hexadecanal, dimethyl acetal $CH_3(CH_2)_{14}CH(OCH_3)_2$	286.50	...	144^2	10	0.8542^{20}	1.4382^{25}	al, eth, ace	$B1^4$, 3393
7700	Hexadecanal, oxime $CH_3(CH_2)_{14}CH{=}NOH$	255.45	nd(dil al)	...	88	...	...	al, chl	$B1^3$, 2923
7701	Hexadecane or Cetane $CH_3(CH_2)_{14}CH_3$	226.45	lf(ace)	287, 149^{10}	18.2	$0.7733^{20/4}$	1.4345	eth	$B1^4$, 537
7702	Hexadecane, 1-amino or Cetyl amine $CH_3(CH_2)_{14}CH_2NH_2$	241.46	lf	322.5, 144^2	46.8	$0.8129^{20/4}$	1.4496^{20}	al, eth, ace, bz, chl	$B4^4$, 818
7703	Hexadecane, 1-bromo or Cetyl bromide $CH_3(CH_2)_{14}CH_2Br$	305.34	...	336, 188^{10}	17-9	$0.9991^{20/4}$	1.4618^{25}	...	$B1^4$, 542
7704	Hexadecane, 1-chloro or Cetyl chloride $CH_3(CH_2)_{14}CH_2Cl$	260.89	...	322, 177^{10}	17.9	$0.8652^{20/4}$	1.4505^{20}	...	$B1^4$, 542
7705	Hexadecane, 1,16-dibromo $Br(CH_2)_{16}Br$	384.24	lf(al)	204^4	56	...	...	chl	$B1^2$, 138
7706	Hexadecane, 6,11-dipentyl $(C_5H_{11})_2CH(CH_2)_4CH(C_5H_{11})_2$	366.71	...	231^{10}	−16.2	$0.8072^{20/4}$	$1.4502^{20/}{}_D$	...	$B1^4$, 586
7707	Hexadecane, 1-fluoro or Cetyl fluoride $CH_3(CH_2)_{14}CH_2F$	244.44	...	289, 152.6^{10}	18	$0.8321^{20/4}$	1.4317^{20}	eth, lig	$B1^4$, 542
7708	Hexadecane, 1-iodo or Cetyl iodide $CH_3(CH_2)_{14}CH_2I$	352.34	lf(al)	357, 202^{10}	24.7	$1.1257^{20/4}$	1.4818^{20}	eth, ace, bz, chl	$B1^4$, 543
7709	Hexadecane, 1-phenyl or Cetylbenzene $CH_3(CH_2)_{14}CH_2C_6H_5$	302.54	...	237^{16}	27	$0.8560^{20/4}$	1.4814^{20}	eth, bz, lig	$B5^4$, 1216
7710	Hexadecanedioic acid or Thapsic acid $HO_2C(CH_2)_{14}CO_2H$	286.41	pl(al, AcOEt)	...	126	...	...	al, ace	$B2^4$, 2162
7711	Hexadecanoic acid or Palmitic acid $CH_3(CH_2)_{14}CO_2H$	256.43	nd(al)	350, 267^{100}	63	$0.8527^{62/4}$	1.4335^{60}	al, eth, ace, bz, chl	$B2^4$, 1157
7712	1-Hexadecanol or Cetyl alcohol $CH_3(CH_2)_{14}CH_2OH$	242.45	fl(AcOEt)	344, 190^{15}	50	$0.8176^{50/4}$	1.4283^{20}	al, eth, ace, bz	$B1^4$, 1876
7713	2-Hexadecanol $CH_3(CH_2)_{13}CHOHCH_3$	242.45	...	314	44	0.8338^{20}	$1.4479^{20/}{}_D$	...	$B1^4$, 1882
7714	1-Hexadecanethiol or Cetyl mercapton $CH_3(CH_2)_{14}CH_2SH$	258.51	(lig)	$123\text{-}8^{0.5}$	18-20	...	...	eth	$B1^4$, 1881
7715	1-Hexadecene or Cetene $CH_3(CH_2)_{13}CH{=}CH_2$	224.43	lf	284.4, 155^{15}	4.1	$0.7811^{20/4}$	1.4412^{20}	al, eth, peth	$B1^4$, 927
7716	2-Hexadecenoic acid (form I) or Gaidic acid $CH_3(CH_2)_{12}CH{=}CHCO_2H$	254.41	lf(al)	...	39	...	...	al	B2, 461
7717	2-Hexadecenoic acid (form II) or Δ-α,β-Hypogeic acid $CH_3(CH_2)_{12}CH{=}CHCO_2H$	254.41	fl(al)	...	49	...	...	al, eth, chl, peth	$B2^4$, 1629
7718	7-Hexadecenoic acid $CH_3(CH_2)_7CH{=}CH(CH_2)_5CO_2H$	254.41	cr	230^{10}	33	...	...	al, eth	B2, 460
7719	Hexadecyl ether or Dicetyl ether $(C_{16}H_{33})_2O$	466.88	lf(al)	270d	55	0.978^{19}	...	al, eth	$B1^4$, 1878
7720	Hexadecylpyridenium chloride $(C_5H_5N^+)C_{16}H_{33}Cl^-$	340.00	wh pw	...	77-83	...	...	w, chl	$B20^4$, 2316
7721	1-Hexadecyne $CH_3(CH_2)_{13}C{\equiv}CH$	222.41	...	284, 147.8^{10}	15	$0.7965^{20/4}$	1.4440^{20}	bz	$B1^4$, 1073
7722	2-Hexadecyne $CH_3(CH_2)_{12}C{\equiv}CCH_3$	222.41	fl	160^{15}	20	$0.8039^{20/4}$	...	...	$B1^3$, 1028
7723	7-Hexadecynoic acid $C_8H_{17}C{\equiv}C(CH_2)_5CO_2H$	252.40	nd(w) cr(al)	214^{15}	47	...	...	al, eth	$B2^3$, 1474

PHYSICAL CONSTANTS OF ORGANIC COMPOUNDS (Continued)

No.	Name, Synonyms, and Formula	Mol. wt.	Color, crystalline form, specific rotation and λ_{max} (log ε)	b.p. °C	m.p. °C	Density	n_D	Solubility	Ref.
7724	2,4-Hexadienal or Sorbaldehyde $CH_3CH=CHCH=CHCHO$	96.13		$173\text{-}4^{756}$, 76^{30}		0.898^{20}	1.5384^{20}		$B1^{4}$, 3545
7725	1,2-Hexadiene or Propylallene $CH_3CH_2CH_2CH=C=CH_2$	82.15		76		0.7149^{20}	1.4282^{20}	eth, chl	$B1^{4}$, 1011
7726	1,3-Hexadiene $CH_3CH_2CH=CH\text{-}CH=CH_2$	82.15		73		$0.7050^{20/4}$	1.4380^{20}	eth	$B1^{4}$, 1011
7727	1,3-Hexadiene, 3-chloro $CH_3CH_2CH=CClCH=CH_2$	116.59		68^{117}		$0.9390^{20/4}$	1.4770^{20}	eth, chl	$B1^{4}$, 1012
7728	1,4-Hexadiene $CH_3CH=CHCH_2CH=CH_2$	82.15		65		$0.7000^{20/4}$	1.4150^{20}	eth	$B1^{4}$, 1013
7729	1,4-Hexadiene, 5-chloro-2-isopropyl $CH_3CCl=CHCH_2C(i\text{-}C_3H_7)=CH_2$	158.67		95^{18}		$0.9310^{25/4}$	1.4370^{25}	ace, chl	$B1^{3}$, 1014
7730	1,4-Hexadiene, 3,3,6-trichloro $ClCH_2CH=CHCCl_2CH=CH_2$	185.48		$100\text{-}3^{4}$		$1.3036^{20/4}$	1.5585^{20}	eth, chl	$B1^{3}$, 982
7731	1,5-Hexadiene or Biallyl $H_2C=CHCH_2CH_2CH=CH_2$	82.15		59.5	−141	$0.6880^{20/4}$	1.4042^{20}	al, eth, bz, chl	$B1^{4}$, 1013
7732	1,5-Hexadiene, 2,5-dimethyl $CH_2=C(CH_3)CH_2CH_2C(CH_3)=CH_2$	110.20		134	−75.6	0.7512^{20}	1.4399^{21}	ace, chl	$B1^{4}$, 1042
7733	1,5-Hexadiene, 2-methyl $CH_2=C(CH_3)CH_2CH_2CH=CH_2$	96.17		90-3	−128.8	$0.7198^{20/4}$	$1.4183^{20}/_D$		$B1^{4}$, 1030
7734	1,5-Hexadiene, perchloro $Cl_2C=CClCCl_2CCl_2CCl=CCl_2$	426.60	cr(ace)	$121^{0.03}$	49	$1.905^{52/4}$	1.6012^{51}	eth, ace, chl	$B1^{4}$, 1016
7735	1,5-Hexadeiene, 3,4-diol (d) or Divinyl glycol $CH_2=CHCH(OH)CH(OH)CH=CH_2$	114.14	$[\alpha]^{17}/_D$ +94.8 (al)	198, 97^{13}	−60	$1.006^{20/4}$	1.4700^{20}	al, eth, chl	$B1^{4}$, 2693
7736	1,5-Hexadiene-3,4-diol (dl) $CH_2=CHCH(OH)CH(OH)CH=CH_2$	114.14	hyg	$90\text{-}1^{8}$	21.7	$1.017^{19/4}$	1.4790^{19}	al, eth, chl	$B1^{3}$, 2272
7737	1,5-Hexadiene-3,4-diol *(meso)* $CH_2=CHCH(OH)CH(OH)CH=CH_2$	114.14	hyg	100^{14}	18	$1.023^{19/4}$	1.4810^{19}	w, al, eth, chl	$B1^{4}$, 2693
7738	2,4-Hexadiene $CH_3CH=CHCH=CHCH_3$	82.15		80	−79	$0.7196^{20/4}$	1.4500^{20}	al, eth, chl	$B1^{4}$, 1016
7739	2,4-Hexadiene, 6-chloro-2-methyl $ClCH_2CH=CHCH=C(CH_3)_2$	130.62		57^{11}		$0.9416^{20/4}$	1.5120^{20}	ace, chl	$B1^{3}$, 1000
7740	2,4-Hexadiene, 1,3-dichloro $CH_3CH=CHCCl=CHCH_2Cl$	151.04		$80\text{-}2^{17}$		$1.1456^{20/4}$	1.5271^{20}	bz, chl	$B1^{4}$, 1017
7741	2,4-Hexadiene, 2,5-dimethyl $(CH_3)_2C=CHCH=C(CH_3)_2$	110.20		134, 75^{100}	14-5	$0.7625^{20/4}$	1.4785^{20}	al, eth, bz, chl	$B1^{4}$, 1043
7742	2,4-Hexadiene, 1,3,4,6-tetrachloro $ClCH_2CH=CClCCl=CHCH_2Cl$	219.93		$84\text{-}9^{2}$		$1.4013^{20/4}$	1.5465^{20}	chl, MeOH	$B1^{3}$, 987
7743	2,4-Hexadienedioic acid or *cis*-Muconic acid $HO_2CCH=CHCH=CHCO_2H$	142.11		194-5				aa	$B2^{4}$, 2297
7744	2,4-Hexadienedioic acid *(trans)* or *trans*-Muconic acid $HO_2CCH=CHCH=CHCO_2H$	142.11	nd(al)	320	305d			aa, AcOEt	$B2^{4}$, 2298
7745	2,4-Hexadienoic acid or Sorbic acid $CH_3CH=CHCH=CHCO_2H$	112.13	nd(dil al)	228d, 153^{50}	134.5	$1.204^{19/4}$		al, eth	$B2^{4}$, 1701
7746	2,4-Hexadien-1-ol or Sorbyl alcohol $CH_3CH=CHCH=CHCH_2OH$	98.14	nd	76^{12}	30-1	$0.8967^{23/4}$	1.4981^{20}	al, eth	$B1^{4}$, 2239
7747	3,5-Hexadien-2-ol $CH_2=CHCH=CHCH(OH)CH_3$	98.14		$77\text{-}8^{26}$		0.8678^{20}	1.4816^{20}	al	$B1^{4}$, 2237
7748	3,5-Hexadiene, 2-one-6-phenyl or Cinnamylidene acetone $C_6H_5CH=CHCH=CHCOCH_3$	172.23	wh lf (eth)	$170\text{-}2^{15}$	68			al, eth, ace, bz, chl	$B7^{3}$, 1656
7749	1,5-Hexadiene-3-one or Divinyl acetylene $CH_2=CHC\equiv CCH=CH_2$	78.11		85	−88	$0.7851^{20/4}$	1.5035^{20}	bz	$B1^{4}$, 1120
7750	1,5-Hexadiene, 3-one-2,5-dimethyl $CH_2=C(CH_3)C\equiv CC(CH_3)=CH_2$	106.17	ye	123		$0.7863^{25/4}$	1.4845^{20}	bz, chl	$B1^{4}$, 1124
7751	3,5-Hexadiene, 1-yne $CH_2=CHCH=CH_2C\equiv CH$	78.11		83-4, 32^{100}		$0.7806^{20/4}$	1.5095^{20}	bz	$B1^{4}$, 1120
7752	1,4-Hexadiyne $CH_3C\equiv CCH_2C\equiv CH$	78.11		78-83	<-80	$0.825^{0/4}$		bz, chl	$B1^{3}$, 1057
7753	1,5-Hexadiyne or Dipropargyl $HC\equiv CCH_2CH_2C\equiv CH$	78.11		86 ,20^{46}	−6	$0.8049^{20/4}$	1.4380^{23}	al, eth, ace, bz	$B1^{4}$, 1118
7754	1,5-Hexadiyne, 1,6-diamino $H_2NC\equiv CCH_2CH_2C\equiv CNH_2$	108.14			104-5			bz	$B4^{4}$, 1399
7755	2,4-Hexadiyne $CH_3C\equiv C\text{-}C\equiv C\text{-}CH_3$	78.11	pr(sub)	129-30	68.5			al, eth	$B1^{4}$, 1119
7756	Hexaethyl tetraphosphate $[(C_2H_5O)_2P(O)O]_3PO$	506.26	hyg	>150d	ca−40	$1.2917^{27/4}$	1.4273^{27}	al, ace, bz	$B1^{3}$, 1331

No.	Name, Synonyms, and Formula	Mol. wt.	Color, crystalline form, specific rotation and λ_{max} (log ε)	b.p. °C	m.p. °C	Density	n_D	Solubility	Ref.
7757	Hexamethylene diamine or 1,6-Diamino hexane $H_2N(CH_2)_6NH_2$	116.21	rh bi pym pl	204-5, 100^{20}	41-2			w, al, bz	$B4^4$, 1320
7758	Hexamethylene tetramine or Hexamin. Urotropine $C_6H_{12}N_4$	140.19	rh (al)	sub	285-95 sub	1.331^{-5}		w, al, ace, chl	$B26^2$, 200
7759	Hexanal or Caproaldehyde $CH_3(CH_2)_4CHO$	100.16		128, 28^{12}	−56	$0.8139^{20/4}$	1.4039^{20}	al, eth, ace, bz	$B1^4$, 3296
7760	Hexane C_6H_{14}	86.18		69	−95	$0.6603^{20/4}$	1.3751^{20}	al, eth, chl	$B1^4$, 338
7761	Hexane, 1-amino or *n*-Hexyl amine $CH_3(CH_2)_4CH_2NH_2$	101.19		130	−19	0.7660^{20}	1.4180^{20}	al, eth	$B4^4$, 709
7762	Hexane, 2-amino (d) $CH_3(CH_2)_3CHNH_2CH_3$	101.19		114-5, 64^{90}		$0.755^{27/4}$		al, eth	$B4^3$, 361
7763	Hexane, 2-amino (dl) $CH_3(CH_2)_3CHNH_2CH_3$	101.19		117-8	−19	$0.7534^{20/0}$	1.4080^{25}	al, eth	$B4^4$, 721
7764	Hexane, 2-amino-4-methyl $CH_3CH_2CH(CH_3)CH_2CH(NH_2)CH_3$	115.22		130-5		0.7655^{20}	1.4150^{25}	al, eth, chl	$B4^4$, 747
7765	Hexane, 1-bromo or *n*-Hexyl bromide $CH_3(CH_2)_4CH_2Br$	165.07		155.3, 41^{10}	−84.7	$1.1744^{20/4}$	1.4478^{20}	al, eth, ace, chl	$B1^4$, 352
7766	Hexane, 1-bromo-6-fluoro $F(CH_2)_6Br$	183.06		$67\text{-}8^{11}$		$1.293^{20/4}$	1.4435^{25}	al, eth, ace, chl	$B1^4$, 353
7767	Hexane, 2-bromo or *sec*-Hexyl bromide $CH_3(CH_2)_3CHBrCH_3$	165.07		144^{749}, 78^{90}		$1.1658^{20/4}$	1.4832^{25}	al, eth, ace, chl	$B1^4$, 353
7768	Hexane, 3-bromo $C_3H_7CHBrC_2H_5$	165.07		141-3		$1.1799^{20/4}$	1.4472^{20}	al, eth, ace, chl	$B1^4$, 353
7769	Hexane, 1-chloro or *n*-Hexyl chloride $CH_3(CH_2)_4CH_2Cl$	120.62		134.5	−94	$0.8785^{20/4}$	1.4199^{20}	al, eth, ace, bz, chl	$B1^4$, 349
7770	Hexane, 1-chloro-3-ethyl (d) $C_3H_7CH(C_2H_5)CH_2CH_2Cl$	148.68	$[\alpha]^{27}{}_D$ +1.15	85^{40}		$0.879^{21/4}$	1.4335^{25}	eth	$B1^4$, 432
7771	Hexane, 1-chloro-6-fluoro $F(CH_2)_6Cl$	138.61		167^{740}, 62^{15}		$1.015^{20/4}$	1.4168^{25}	al, eth, chl	$B1^4$, 350
7772	Hexane, 1-chloro-3-methyl $C_3H_7CH(CH_3)CH_2CH_2Cl$	134.65		$150\text{-}2^{758}$		$0.8766^{20/4}$	1.4274^{20}	al, eth, chl	$B1^2$, 119
7773	Hexane, 2-chloro or *sec*-Hexyl chloride $CH_3(CH_2)_3CHClCH_3$	120.62		122-3, 61^{100}		$0.8694^{21/4}$	1.4142^{22}	al, eth, ace, bz, chl	$B1^4$, 349
7774	Hexane, 2-chloro-2,5-dimethyl $(CH_3)_2CHCH_2CH_2CCl(CH_3)_2$	148.68		86^{100}		$0.8476^{18/4}$	1.4232^{20}	al, eth, ace, bz	$B1^4$, 434
7775	Hexane, 2-chloro-2-methyl $C_4H_9CCl(CH_3)_2$	134.65		135d, 59.5^{52}		$0.8635^{20/4}$	1.4200^{20}	al, eth, chl	$B1^4$, 398
7776	Hexane-2-chloro-5-methyl $(CH_3)_2CHCH_2CH_2CHClCH_3$	134.65		138^{735}d		$0.863^{20/4}$		al, eth, chl	$B1^3$, 436
7777	Hexane-3-chloro $C_3H_7CHClC_2H_5$	120.62		123, 60^{95}		$0.8700^{20/20}$	1.4163^{20}	al, eth, ace, bz, chl	$B1^4$, 349
7778	Hexane-3-chloro-2,3-dimethyl $C_3H_7CCl(CH_3)CH(CH_3)_2$	148.68		$41\text{-}3^{12}$		$0.8869^{20/4}$	1.4333^{25}	al	$B1^3$, 481
7779	Hexane-3-chloro-3-ethyl $C_3H_7CCl(C_2H_5)_2$	148.68		155d, $62\text{-}3^{24}$		0.9018	1.4358^{20}	eth, chl	$B1^3$, 479
7780	Hexane-3-chloro-3-methyl $C_3H_7CCl(CH_3)C_2H_5$	134.65		135		$0.8787^{20/4}$	1.4250^{20}	al, eth, chl	$B1^3$, 438
7781	Hexane-3-chloro-2,2,3-trimethyl $C_3H_7CCl(CH_3)C(CH_3)_3$	162.70		$64\text{-}5^{13}$		$0.9010^{20/4}$	1.4465^{20}	al, eth, chl	$B1^3$, 515
7782	Hexane-1,6-diamino or Hexamethylene diamine $H_2N(CH_2)_6NH_2$	116.21	rh bipym pl	204-5, 100^{20}	41-2			w, al, bz	$B4^4$, 1320
7783	Hexane-1,6-diamino, dihydrochloride $H_2N(CH_2)_6NH_2.2HCl$	189.13	nd(al-eth)		248-50			w	$B4^4$, 1320
7784	Hexane-2,5-diamino $CH_3CH(NH_2)CH_2CH_2CH(NH_2)CH_3$	116.21		175				w, al, eth	B4, 269
7785	Hexane-1,2-dibromo $C_4H_9CHBrCH_2Br$	243.97		$103\text{-}5^{36}$		$1.5774^{20/4}$	1.5024^{20}	al, eth, bz, chl	$B1^3$, 392
7786	Hexane-1,6-dibromo $Br(CH_2)_6Br$	243.97		245-6, 110^{12}	−2.3	1.5948^{15}	1.5037^{20}	eth, ace	$B1^4$, 354
7787	Hexane-2,5-dibromo (dl) $CH_3CHBrCH_2CH_2CHBrCH_3$	243.97		$108\text{-}9^{30}$	−44.64	1.5788^{20}	1.5007^{20}	eth, ace, chl	$B1^4$, 354
7788	Hexane-1,2-dichloro $C_4H_9CHClCH_2Cl$	155.07		172-4, $73\text{-}4^{0.03}$		1.085^{15}		eth, chl	$B1^4$, 350
7789	Hexane-1,6-dichloro $Cl(CH_2)_6Cl$	155.07		203-5, 94^{22}		$1.0677^{20/4}$	1.4572^{20}	eth, chl	$B1^4$, 350

No.	Name, Synonyms, and Formula	Mol. wt.	Color, crystalline form, specific rotation and λ_{max} (log ε)	b.p. °C	m.p. °C	Density	n_D	Solubility	Ref.
7790	Hexane-2,2-dichloro $C_4H_9CCl_2CH_3$	155.07	...	68^{49}	...	$1.0150^{25/4}$	1.4353^{25}	eth, chl	B1[3], 390
7791	Hexane-2,3-dichloro $C_3H_7CHClCHClCH_3$	155.07	...	162-5	...	1.0527^{11}	...	eth, chl	B1[4], 350
7792	Hexane-2,5-dichloro (dl) $CH_3CHClCH_2CH_2CHClCH_3$	155.07	...	177^{751}, 106^{91}	fp-38.4	1.0474^{20}	1.4491^{20}	chl	B1[4], 350
7793	Hexane-2,5-dichloro *(meso)* $CH_3CHClCH_2CH_2CHClCH_3$	155.07	...	178^{752}, 109^{99}	19.9	$1.0474^{20/4}$	1.4484^{20}	chl	B1[4], 350
7794	Hexane-2,5-dichloro-2,5-dimethyl $(CH_3)_2CClCH_2CH_2CCl(CH_3)_2$	183.12	lf, nd	...	67-8	0.9543^{70}	...	al, eth, bz, chl	B1[4], 435
7795	Hexane-3,4-dichloro (dl) $CH_3CH_2CHClCHClCH_2CH_3$	155.07	...	167.7, 62^{20}	...	1.0617^{20}	1.4541^{20}	ace, chl	B1[4], 351
7796	Hexane-3,4-dichloro-3,4-dimethyl $C_2H_5CCl(CH_3)CCl(CH_3)C_2H_5$	183.12	...	165, $114\text{-}5^{18}$	...	...	...	chl	B1[4], 436
7797	Hexane-1,6-diido $I(CH_2)_6I$	337.97	nd	$141\text{-}2^{10}$	10	$2.03^{22/4}$	1.585^{20}	al, eth	B1[4], 356
7798	Hexane-2,2-dimethyl $C_4H_9C(CH_3)_3$	114.23	...	106.8	−121	$0.6953^{20/4}$	1.3935^{20}	al, eth, ace, bz, chl	B1[4], 432
7799	3-Hexane-2,3-dimethyl-(dl) $C_3H_7CH(CH_3)CH(CH_3)_2$	114.23	...	115.6, 9.9^{0}	...	$0.7121^{20/4}$	1.4011^{20}	al, eth, ace, bz, chl, lig	B1[4], 432
7800	Hexane-2,3-dimethyl (l) $C_3H_7CH(CH_3)CH(CH_3)_2$	114.23	$[\alpha]^{25/}{}_D$-0.92	113	...	...	...	al, eth, ace, bz, lig, chl	B1[3], 482
7801	Hexane, 2,4-dimethyl (d) $C_2H_5CH(CH_3)CH_2CH(CH_3)_2$	114.23	$[\alpha]^{30/}{}_D$ + 2.99	111	...	$0.696^{20/4}$	1.3810^{20}	al, eth, ace, bz, chl, lig	B1[3], 483
7802	Hexane, 2,4-dimethyl (dl) $C_2H_5CH(CH_3)CH(CH_3)_2$	114.23	...	109.4, 5.2^{10}	...	$0.7004^{20/4}$	1.3953^{20}	al, eth, ace, bz, lig, chl	B1[4], 433
7803	Hexane, 2,4-dimethyl (l) $C_2H_5CH(CH_3)CH(CH_3)_2$	114.23	$[\alpha]^{21/}{}_D$ -10.85	110	...	$0.703^{21/4}$	...	al, eth, ace, bz, chl, lig	B1[3], 483
7804	Hexane, 2,5-dimethyl $(CH_3)_2CHCH_2CH_2CH(CH_3)_2$	114.23	...	109, 5.3^{10}	−91.2	$0.6935^{20/4}$	1.3925^{20}	al, eth, ace, bz, lig, chl	B1[4], 434
7805	Hexane, 3,3-dimethyl $C_3H_7C(CH_3)_2C_2H_5$	114.23	...	112, 6.1^{10}	−126.1	$0.7100^{20/4}$	1.4001^{20}	al, eth, ace, bz	B1[4], 435
7806	Hexane, 3,4-dimethyl $C_2H_5CH(CH_3)CH(CH_3)C_2H_5$	114.23	...	117.7, 11.3^{10}	...	$0.7200^{20/4}$	1.4046^{20}	al, eth, ace, bz, chl, lig	B1[4], 436
7807	Hexane, 1,6-dinitro $O_2N(CH_2)_6NO_2$	176.17	cr(MeOH)	$100\text{-}3^{0.3}$	37.5	...	...	aa	B1[4], 357
7808	Hexane, 3-ethyl $C_3H_7CH(C_2H_5)_2$	114.23	...	118.5, 12.8^{10}	...	$0.7136^{20/4}$	1.4018^{20}	al, eth, ace, bz, lig	B1[4], 431
7809	Hexane, 3-ethyl-2-methyl $(CH_3)_2CHCH(C_2H_5)CH_2CH_2CH_3$	128.26	...	138	...	...	$1.4106^{20/}{}_D$	...	B1[4], 459
7810	Hexane, 3-ethyl-3-methyl $(CH_3CH_2)_2C(CH_3)C_3H_7$	128.26	...	140.6	...	...	$1.4140^{20/}{}_D$	...	B1[4], 459
7811	Hexane, 3-ethyl-4-methyl $(CH_3CH_2)_2CHCH(CH_3)C_2H_5$	128.26	...	140.4	...	...	$1.4134^{20/}{}_D$	...	B1[4], 459
7812	Hexane, 4-ethyl-2-methyl $(C_2H_5)_2CHCH_2CH(CH_3)_2$	128.26	...	133.8	...	...	$1.4063^{20/}{}_D$	...	B1[4], 459
7813	Hexane, 1-fluoro or *n*-Hexyl fluoride $CH_3(CH_2)_4CH_2F$	104.17	...	91.5	−103	0.7995^{20}	1.3738^{20}	eth, bz	B1[4], 348
7814	Hexane, 3,4-bis(4-hydroxy-3-methylphenyl) or Dimethylexesterol $[4\text{-}HO\text{-}3\text{-}CH_3C_6H_3CH(C_2H_5)\text{-}]_2$	298.43	cr(dil aa)	...	145	...	...	...	B6[3], 5541
7815	Hexane, 2,4-bis(4-hydroxyphenyl)-3-ethyl or Benzestrol $4\text{-}HOC_6H_4CH(C_2H_5)CH(C_2H_5)CH(CH_3)(C_6H_4OH\text{-}4)$	298.43	cr(al)	...	162-6	...	...	al, eth, ace, aa	B6[3], 5538
7816	Hexane, 1-iodo or *n*-Hexyl iodide $CH_3(CH_2)_4CH_2I$	212.07	...	181.3, 58.2^{10}	−75	$1.4397^{20/4}$	1.4929^{20}	...	B1[4], 355
7817	Hexane, 2-iodo (l) $CH_3CHIC_4H_9$	212.07	$[\alpha]^{17/}{}_D$ -38.35	$90\text{-}1^{70}$. 45^{9}	...	$1.4354^{17/4}$	1.4878^{25}	ace, chl	B1[3], 395
7818	Hexane, 2-methyl or Isoheptane $CH_3(CH_2)_3CH(CH_3)_2$	100.20	...	90	−118.3	$0.6787^{20/4}$	1.3848^{20}	al, eth, ace, bz, lig, chl	B1[4], 397
7819	Hexane, 3-methyl (d) $C_3H_7CH(CH_3)C_2H_5$	100.20	$[\alpha]^{20/}{}_D$ + 9.5	92	−119	$0.6860^{20/4}$	1.3887^{20}	al, eth, ace, bz, chl, lig	B1[4], 400
7820	Hexane, 3-methyl (dl) $C_3H_7CH(CH_3)C_2H_5$	100.20	...	92	ca−173	0.6872^{20}	1.3885^{20}	al, eth, ace, bz, lig, chl	B1[4], 400
7821	Hexane, 3-methyl (l) $C_3H_7CH(CH_3)C_2H_5$	100.20	$[\alpha]^{21/}{}_D$ -7.75	92	...	$0.687^{21/4}$	1.3854^{25}	al, eth, ace, bz, chl, lig	B1[3], 440
7822	Hexane, 1-nitro $CH_3(CH_2)_4CH_2NO_2$	131.17	...	$193\text{-}4^{765}$, 84^{21}	...	$0.9396^{20/4}$	1.4270^{20}	al, eth, ace, bz	B1[4], 356
7823	Hexane, 1,1,1,2,2,pentachloro $CH_3(CH_2)_3CCl_2CCl_3$	258.40	...	$129\text{-}31^{10}$	...	1.370^{25}	1.4872^{25}	eth, chl	B1[3], 390

No.	Name, Synonyms, and Formula	Mol. wt.	Color, crystalline form, specific rotation and λ_{max} (log ε)	b.p. °C	m.p. °C	Density	n_D	Solubility	Ref.
7824	Hexane, perfluoro C_6F_{14}	338.04		57.11	−87.1	$1.6995^{20/4}$	1.2515^{20}	eth, bz, chl	$B1^4$, 348
7825	Hexane, 1-phenyl or Hexylbenzene $CH_3(CH_2)_4CH_2C_6H_5$	162.27		227	−62	0.8613^{20}	1.4900^{20}	eth, bz, peth	$B5^4$, 1115
7826	Hexane, 2-phenyl $CH_3(CH_2)_3CH(C_6H_5)CH_3$	162.27		208		$0.869^{15/4}$	1.492^{15}		$B5^4$, 1116
7827	Hexane, 3-phenyl $CH_3CH_2CH(C_6H_5)C_3H_7$	162.27		209-12		$0.8254^{25/20}$	$1.4859^{20/}{}_D$		$B5^4$, 1117
7828	Hexane, 1,1,2,2,-tetrachloro $CH_3(CH_2)_3CCl_2CHCl_2$	223.96		$99\text{-}101^{14}$		$1.3096^{25/4}$	1.488^{25}	ace, bz, chl	$B1^3$, 390
7829	Hexane, 2,2,3-trimethyl $C_3H_7CH(CH_3)C(CH_3)_3$	128.26		131.7			1.4100^{20}		$B1^4$, 459
7830	Hexane, 2,2,4-trimethyl $C_2H_5CH(CH_3)CH_2C(CH_3)_3$	128.26		126.5	120	$0.711^{20/4}$	$1.40328^{20/}{}_D$		$B1^4$, 459
7831	Hexane, 2,2-5-trimethyl $(CH_3)_2CHCH_2CH_2C(CH_3)_3$	128.26		$124, 16.2^{10}$	−105.8	$0.7072^{20/4}$	1.3997^{20}	al, eth, ace, bz, lig	$B1^4$, 460
7832	Hexane, 2,3,3-trimethyl $C_3H_7C(CH_3)_2CH(CH_3)_2$	128.26		137.7	−116.8		$1.4141^{20/}{}_D$		$B1^4$, 461
7833	Hexane, 2,3,4-trimethyl $C_2H_5CH(CH_3)CH(CH_3)CH(CH_3)_2$	128.26		139			$1.4144^{20/}{}_D$		$B1^4$, 461
7834	Hexane, 2,3,5-trimethyl $CH_3CH(CH_3)CH(CH_3)CH_2CH(CH_3)_2$	128.26		131.3	−s127.9	$0.7818^{20/4}$	$1.4051^{20/}{}_D$		$B1^4$, 461
7835	Hexane, 2,4,4-trimethyl $C_2H_5C(CH_3)_2CH_2CH(CH_3)_2$	128.26		126.5	−123.4	$0.711^{20/4}$	$1.40328^{20/}{}_D$		$B1^4$, 461
7836	Hexane, 3,3,-4-trimethyl $CH_3CH_2C(CH_3)_2CH(CH_3)_2CH_2CH_3$	128.26		140.5	−101.2		$1.4178^{20/}{}_D$		$B1^4$, 462
7837	1,6-Hexanedial or Adipaldehyde $OHC(CH_2)_4CHO$	114.14		$92\text{-}4^9$	ms;8	$1.003^{19/4}$	1.4350^{20}	al, eth, bz, aa	$B1^4$, 3686
7838	Hexanedioic acid or Adipic acid $HOOC(CH_2)_4COOH$	146.14	mcl pr(w, ace-lig)	$265^{100}, 205^{10}$	153	$1.360^{25/4}$		al, eth	$B2^4$, 1956
7839	1,3-Hexanediol, 2-ethyl $C_3H_7CH(OH)CH(C_2H_5)CH_2OH$	146.23		244	−40	$0.9325^{22/4}$	1.4497^{20}	al, eth	$B1^4$, 2597
7840	1,6-Hexanediol or Hexamethylene glycol $HO(CH_2)_6OH$	118.18	nd(w)	$250, 132^9$	43(59)			w, al, ace	$B1^4$, 2556
7841	2,3-Hexanediol $C_3H_7CH(OH)CH(OH)CH_3$	118.18	cr	$204\text{-}6, 102^{0.8}$		0.9900^{15}	1.4510^{15}	w, al, eth	$B1^4$, 2561
7842	2,5-Hexanediol $CH_3CH(OH)CH_2CH_2CH(OH)CH_3$	118.18	cr(eth)	$216\text{-}8^{750}$, $85\text{-}7^1$	43	$0\ 9610^{20/4}$	1.4475^{20}	w, al, eth	$B1^4$, 2562
7843	3,4-Hexanediol, 3,4-diethyl $(C_2H_5)_2C(OH)C(OH)(C_2H_5)_2$	174.28	cr(eth)	$230, 112^{10}$	28	$0.9630^{13/25}$	1.467^{13}	al, eth	$B1^4$, 2621
7844	2,5-Hexanediol, 2,5-dimethyl $(CH_3)_2C(OH)CH_2CH_2C(OH)(CH_3)_2$	146.23	pr (AcOEt) fl (peth)	$214, 118^{15}$	92	0.898^{20}		w, al, bz, chl	$B1^4$, 2600
7845	2,3-Hexandione, 3-oxime $C_3H_7C(=NOH)COCH_3$	129.16	cr(al)		60			al	$B1^3$, 3127
7846	2,5-Hexanedione or Acetonyl acetone $CH_3COCH_2CH_2COCH_3$	114.14		$194^{754}, 89^{25}$	−5.5	$0.9737^{20/4}$	1.4421^{20}	w, al, eth, ace, bz	$B1^4$, 3688
7847	2,5-Hexanedione, dioxime $CH_3C(=NOH)CH_2CH_2C(=NOH)CH_3$	144.17	pl(bz)		137			al, eth	$B1^3$, 3130
7848	2,5-Hexanedione, 3-hydroxy $CH_3COCH_2CH(OH)COCH_3$	130.14		$62\text{-}7^{0.5}$			1.4497^{25}	eth, bz	$B1^3$, 3317
7849	3,4-Hexanedione, 2,2,5,5-tetramethyl or Di-*tert*-butyl glyoxal $(CH_3)_3CCOCOC(CH_3)_3$	170.25		168^{745}	−2	0.8776^{20}	1.4157^{20}	eth	$B1^4$, 3726
7850	1,2,3,4,5,6-Hexanehexol or Dulcitol $HOCH_2(CHOH)_4CH_2OH$	182.17	mcl pr	$257\text{-}80^1$	189	1.466^{15}		w	$B1^4$, 2844
7851	1,3,4,5-Hexanetetrol or Digitoxit $CH_3(CHOH)_3CH_2CH_2OH$	150.17	pr, $[\alpha]^{15/}{}_D$ -86.2		88			al	$B1^2$, 603
7852	1-Hexanethiol or Hexyl mercapton $CH_3(CH_2)_4CH_2SH$	118.24		151	−81	$0.8424^{20/4}$	1.4496^{20}	al, eth	$B1^4$, 1705
7853	2-Hexanethiol or *sec*-Hexyl mercaptan $CH_3(CH_2)_3CH(SH)CH_3$	118.24		$142, 60\text{-}6^{50}$	−147	$0.8345^{20/4}$	1.4451^{20}	al, eth, bz	$B1^4$, 1711
7854	3-Hexanethiol $C_3H_7CH(SH)C_2H_5$	118.24		57^{25}		0.9206^{20}	1.4496^{20}	eth	$B1^4$, 1713
7855	1,2,3-Hexanetriol (threo) $C_3H_7CH(OH)CH(OH)CH_2OH$	134.18		$130^{0.05}$	64-5	1.089^{26}	1.472^{26}	w, al	$B1^4$, 2784
7856	1,2,4-Hexanetriol $C_2H_5CH(OH)CH_2CH(OH)CH_2OH$	134.18		$190\text{-}2^{30}$				w, al	B1, 521

PHYSICAL CONSTANTS OF ORGANIC COMPOUNDS (Continued)

No.	Name, Synonyms, and Formula	Mol. wt.	Color, crystalline form, specific rotation and λ_{max} (log ε)	b.p. °C	m.p. °C	Density	n_D	Solubility	Ref.
7857	1,2,5-Hexanetriol $CH_3CH(OH)CH_2CH_2CH(OH)CH_2OH$	134.18		181^{10}		$1.1012^{20/4}$		w, al	$B1^4$, 2784
7858	2,3,4-Hexanetriol $CH_3CH_2(CHOH)_3CH_3$	134.18		256-7, $155\text{-}6^{20}$				w, al	$B1^3$, 2349
7859	Hexanoic acid or Caproic acid $CH_3(CH_2)_4CO_2H$	116.16		205	−2	$0.9274^{20/4}$	1.4163^{20}	al, eth	$B2^4$, 917
7860	Hexanoic acid, 2-acetyl, ethyl ester $C_4H_9CH(COCH_3)CO_2C_2H_5$	186.25		219-24, 104^{12}		$0.9523^{20/4}$	1.4301^{20}	eth, ace	$B3^4$, 1616
7861	Hexanoic acid, 2,4-dioxo or Propionyl pyruvic acid $CH_3CH_2COCH_2COCO_2H$	144.13	cr(al w + 1)		83			w, al, eth	$B3^4$, 1334
7862	Hexanoic acid, 2,4-dioxo, ethyl ester $CH_3CH_2COCH_2COCO_2C_2H_5$	172.18		163-5, $108\text{-}11^{11}$				al, eth, ace	$B3^4$, 1779
7863	Hexanoic acid, 2-ethyl $C_4H_9CH(C_2H_5)CO_2H$	144.21		228^{755}, 120^{13}		$0.9031^{25/4}$	1.4241^{20}	eth	$B2^4$, 1003
7864	Hexanoic acid, 2-methyl (dl) or 2-Methyl caproic acid $C_4H_9CH(CH_3)CO_2H$	130.19		215-6, 100^{12}		$0.9612^{20/4}$	1.4193^{20}	al, eth, ace, bz, chl	B24, 969
7865	Hexanoic acid, 4-methyl (dl) or 4-Methyl caproic acid $C_2H_5CH(CH_3)(CH_2)_2CO_2H$	130.19		217-8, 85^2	−80	$0.9215^{20/4}$	1.4211^{20}	al, eth, ace, bz	$B2^4$, 974
7866	Hexanoic acid, 5-methyl or 5-Methyl caproic acid $(CH_3)_2CH(CH_2)_3CO_2H$	130.19		216	<−25	$0.9138^{21/4}$	1.4220^{20}	al, eth, ace, bz	$B2^4$, 970
7867	Hexanoic acid, 5-oxo $CH_3CO(CH_2)_3CO_2H$	130.14		155^{12}	13-14		1.4445^{20}	w, al, eth	$B3^4$, 1583
7868	1-Hexanol or n-Hexyl alcohol $CH_3(CH_2)_4CH_2OH$	102.18		158	−46.7	$0.8136^{20/4}$	1.4178^{20}	al, eth, ace, bz, chl	$B1^4$, 1694
7869	1-Hexanol, 6-chloro $Cl(CH_2)_6OH$	136.62		107^{12}		$1.0241^{20/4}$	1.4550^{20}	al, eth	$B1^4$, 1704
7870	1-Hexanol, 3,5-dimethoxy $CH_3CH(OCH_3)CH_2CH(OCH_3)CH_2CH_2OH$	162.23		114^{13}		0.9631^{25}	1.4329^{25}		$B1^4$, 2785
7871	3-Hexanol, 2-ethyl $CH_3(CH_2)_3CH(C_2H_5)CH_2OH$	130.23		185, $84\text{-}6^{15}$	<−76	$0.8328^{20/4}$	1.4328^{20}	al, eth, ace, bz	$B1^4$, 1783
7872	1-Hexanol, 6-fluoro $F(CH_2)_6OH$	120.17		$85\text{-}6^{14}$		$0.975^{20/4}$	$1.4141^{2'}$	al, eth	$B1^4$, 1703
7873	1-Hexanol, 2-methyl (d) $C_4H_9CH(CH_3)CH_2OH$	116.20	$[\alpha]^{25/}{}_D$ + 2.45(al)	164-5, $70\text{-}2^{15}$		$0.8313^{13/4}$	1.4245^{17}	al, eth	$B1^2$, 444
7874	1-Hexanol, 2-methyl (dl) $C_4H_9CH(CH_3)CH_2OH$	116.20		164^{750}		$0.8270^{20/4}$	1.4226^{20}	al, eth	$B1^4$, 1745
7875	1-Hexanol, 3-methyl (dl) $C_3H_7CH(CH_3)CH_2CH_2OH$	116.20		$168\text{-}9^{754}$, 91^{13}		0.8258^{20}	1.4245^{20}	al, eth, ace	$B1^4$, 1748
7876	1-Hexanol, 3-methyl (l) $C_3H_7CH(CH_3)CH_2CH_2OH$	116.20	$[\alpha]^{27/}{}_D$ -1.67 (chl)	$161\text{-}2^{740}$, 80^{25}		$0.8208^{25/4}$	1.4204^{30}	al, eth, ace	$B1^4$, 1748
7877	1-Hexanol, 4-methyl (d) $C_2H_5CH(CH_3)(CH_2)_3OH$	116.20	$[\alpha]^{28/}{}_D$ + 2.2	77^{20}		$0.809^{23/4}$	1.4233^{25}	al, eth, ace, bz	$B1^4$, 1749
7878	1-Hexanol, 4-methyl (dl) $C_2H_5CH(CH_3)(CH_2)_3OH$	116.20		173, 83^{24}		$0.8239^{20/4}$	1.4219^{20}	al, eth, ace, bz	$B1^4$, 1749
7879	1-Hexanol, 5-methyl $(CH_3)_2CH(CH_2)_3CH_2OH$	116.20		170^{755}, $53\text{-}5^{15}$		$0.8119^{20/4}$	1.4175^{20}	al, eth	$B1^4$, 1748
7880	1-Hexanol, 1-phenyl $C_5H_{11}CH(OH)C_6H_5$	178.27		170^{50}		$0.9477^{25/4}$	1.5105^{20}	al, eth	$B6^3$, 1994
7881	1-Hexanol, 2-isopropyl-5-methyl $(CH_3)_2CHCH_2CH(i\text{-}C_3H_7)CH_2OH$	158.28		211		$0.8345^{20/4}$	1.4369^{20}	al, eth	$B1^4$, 1833
7882	2-Hexanol (d) $CH_3(CH_2)_3CH(OH)CH_3$	102.18	$[\alpha]^{25/}{}_D$ + 14.1 (eth, c=1)	138		$0.8104^{25/4}$	1.4126^{25}	al, eth	$B1^3$, 1663
7883	2-Hexanol (dl) $CH_3(CH_2)_3CH(OH)CH_3$	102.18		140		$0.8159^{20/4}$	1.4144^{20}	al, eth	$B1^4$, 1708
7884	2-Hexanol (l) $C_4H_9CH(OH)CH_3$	102.18	$[\alpha]^{18/}{}_{5780}$ -12.04	$136\text{-}8^{754}$		$0.8178^{18/4}$		al, eth	$B1^3$, 1663
7885	2-Hexanol, 1-chloro $C_4H_9CH(OH)CH_2Cl$	136.62		$73\text{-}5^{12}$		$0.0139^{20/4}$	1.4478^{20}	al, eth, ace	$B1^4$, 1710
7886	2-Hexanol, 2-methyl $C_4H_9C(OH)(CH_3)_2$	116.20		143, $53\text{-}5^{15}$		$0.8119^{20/4}$	1.4175^{20}	al, eth	$B1^4$, 1745
7887	2-Hexanol, 3-methyl $C_3H_7CH(CH_3)CH(OH)CH_3$	116.20		$79\text{-}81^{52}$		$0.8820^{25/4}$	1.4198^{18}	al, eth, ace	$B1^3$, 1693
7888	2-Hexanol, 5-methyl $(CH_3)_2CHCH_2CH_2CH(OH)CH_3$	116.20		150^{744}, 78^{28}		$0.814^{20/4}$	1.4180^{20}	al, eth	$B1^4$, 1747
7889	2-Hexanol, 2,3,4-trimethyl $C_2H_5CH(CH_3)CH(CH_3)C(OH)(CH_3)_2$	144.26		57^5		$0.853^{15/4}$	1.4415^{15}	al, eth, ace, peth	$B1^3$, 1756

No.	Name, Synonyms, and Formula	Mol. wt.	Color, crystalline form, specific rotation and λ_{max} (log ε)	b.p. °C	m.p. °C	Density	n_D	Solubility	Ref.
7890	2-Hexanol, 2,3,5-trimethyl $(CH_3)_2CHCH_2CH(CH_3)C(OH)(CH_3)_2$	144.26		171^{755}, 72^{21}		$0.8271^{20/20}$	1.4321^{20}	al, eth, ace	$B1^{1}$, 212
7891	3-Hexanol (d) $C_3H_7CH(OH)C_2H_5$	102.18	$[\alpha]^{20/}{}_D$ + 6.8 (chl)	131-3		$0.8213^{20/4}$	1.4150^{20}	al, eth, ace	$B1^{3}$, 1665
7892	3-Hexanol (dl) $C_3H_7CH(OH)C_2H_5$	102.18		135		$0.8182^{20/4}$	1.4167^{20}	al, eth, ace	$B1^{4}$, 1711
7893	3-Hexanol (l) $C_3H_7CH(OH)C_2H_5$	102.18	$[\alpha]^{20/}{}_D$-7.17	135		$0.8213^{20/4}$	1.4140^{20}	al, eth, ace	$B1^{3}$, 1665
7894	3-Hexanol, 1-chloro $C_3H_7CH(OH)CH_2CH_2Cl$	136.62		120^{35}, 78^{6}		$1.003^{25/4}$	1.446^{25}	al, eth, ace, bz	$B1^{4}$, 1712
7895	3-Hexanol, 2-chloro $C_3H_7CH(OH)CHClCH_3$	136.62		171		1.0143^{11}		al, eth, ace, bz	$B1^{2}$, 438
7896	3-Hexanol, 5-chloro $CH_3CHClCH_2CH(OH)C_2H_5$	136.62		$78\text{-}9^{13}$		$1.0012^{15/4}$	1.4433^{19}	al, eth, ace, bz	$B1^{4}$, 1712
7897	3-Hexanol, 2,4-dimethyl $C_2H_5CH(CH_3)CH(OH)CH(CH_3)_2$	130.23		$61\text{-}2^{18}$		$0.8371^{20/4}$	$1.4309^{20/}{}_D$		$B1^{4}$, 1791
7898	3-Hexanol, 3-ethyl $C_3H_7C(OH)(C_2H_5)_2$	130.23		160		$0.8373^{20/4}$	1.4300^{20}	al, eth, ace, bz	$B1^{4}$, 1790
7899	3-Hexanol, 3-ethyl-5-methyl $(CH_3)_2CHCH_2C(OH)(C_2H_5)_2$	144.26		272		$0.8396^{22/4}$	1.4346^{13}	al, eth	$B1^{3}$, 1755
7900	3-Hexanol, 3-methyl $C_3H_7C(OH)(CH_3)C_2H_5$	116.20		143, 56^{18}		$0.8234^{20/0}$	1.4231^{20}	al, eth	$B1^{4}$, 1749
7901	3-Hexanol, 5-methyl (d) $(CH_3)_2CHCH_2CH(OH)C_2H_5$	116.20	$[\alpha]^{35/}{}_D$ + 21.2	81^{60}			1.4171^{25}		$B1^{3}$, 1692
7902	3-Hexanol, 5-methyl (dl) $(CH_3)_2CHCH_2CH(OH)C_2H_5$	116.20		$147\text{-}8^{756}$		0.827^{0}	1.4128^{20}	al, eth	$B1^{4}$, 1747
7903	3-Hexanol, 5-methyl (l) $(CH_3)_2CHCH_2CH(OH)C_2H_5$	116.20	$[\alpha]^{23/}{}_D$ -3.88	$93\text{-}6^{105}$, 63^{19}			1.4171^{25}		$B1^{3}$, 1692
7904	3-Hexanol, 2,2,5,5-tetramethyl $(CH_3)_3CCH_2CH(OH)C(CH_3)_3$	158.28	cr(peth)	166-70	52-3			al, eth, ace, peth	$B1^{3}$, 1772
7905	3-Hexanol, 2,2,3-trimethyl $C_3H_7C(OH)(CH_3)C(CH_3)_3$	144.26		170		$0.8474^{20/4}$	1.4402^{20}	al, eth, ace	$B1^{3}$, 1755
7906	3-Hexanol, 2,3,5-trimethyl $(CH_3)_2CHCH_2C(OH)(CH_3)CH(CH_3)_2$	144.26		72^{21}		$0.8271^{20/20}$	1.4321^{20}	al, eth, ace	$B1^{4}$, 1813
7907	3-Hexanol, 2,4,4-trimethyl $C_2H_5C(CH_3)_2CH(OH)CH(CH_3)_2$	144.26		170		0.8489^{20}	1.4395^{20}	al, eth, ace	$B1^{3}$, 1756
7908	3-Hexanol, 2,5,5-trimethyl $(CH_3)_3CCH_2CH(OH)CH(CH_3)_2$	144.26		77^{32}		0.8250^{20}	1.4286^{20}	al, eth, ace	$B1^{4}$, 1813
7909	3-Hexanol, 3,4,4-trimethyl $C_2H_5C(CH_3)_2C(OH)(CH_3)C_2H_5$	144.26		165-6		$0.8323^{21/0}$	1.4341^{21}	al, eth, ace	B1, 425
7910	3-Hexanol, 3,5,5-trimethyl $(CH_3)_3CCH_2C(OH)(CH_3)C_2H_5$	144.26		62^{14}		0.8350^{20}	1.4352^{20}	al, eth, ace	$B1^{3}$, 1755
7911	2-Hexanone or Methyl butyl ketone $C_4H_9COCH_3$	100.16		128	−57	$0.8113^{20/4}$	1.4007^{20}	al, eth, ace	$B1^{4}$, 3298
7912	2-Hexanone, 3,3-dimethyl $C_3H_7C(CH_3)_2COCH_3$	128.21		149^{765}		$0.838^{0/4}$	1.4098^{20}	al, eth, ace	$B1^{4}$, 3350
7913	2-Hexanone, 3,4-dimethyl $C_2H_5CH(CH_3)CH(CH_3)COCH_3$	128.21		158		$0.8295^{22/4}$	1.4193^{20}	al, eth, ace	$B1^{3}$, 2882
7914	2-Hexanone, 3-methyl $C_3H_7CH(CH_3)COCH_3$	114.19		142-5		0.828^{25}	1.4035^{20}	al, eth, ace, bz	$B1^{3}$, 2863
7915	2-Hexanone, 4-methyl $C_2H_5CH(CH_3)CH_2COCH_3$	114.19		142, $35\text{-}7^{11}$			1.4081^{24}	al, eth, ace, bz	$B1^{4}$, 3329
7916	2-Hexanone, 5-methyl $(CH_3)_2CH(CH_2)_2COCH_3$	114.19		144		$0.888^{20/4}$	1.4062^{20}	al, eth, ace, bz	$B1^{4}$, 3329
7917	2-Hexanone, 5-methyl, oxime $(CH_3)_2CH(CH_2)_2C(=NOH)CH_3$	129.20		195-6		$0.8881^{20/4}$	1.4448^{20}		B1, 701
7918	3-Hexanone or Ethyl propyl ketone $C_3H_7COC_2H_5$	100.16		125		$0.8118^{20/4}$	1.4004^{20}	al, eth, ace	$B1^{4}$, 3301
7919	3-Hexanone, 2,2-dimethyl or *tert*-Butyl propyl ketone $C_3H_7COC(CH_3)_3$	128.21		$145\text{-}8^{745}$		$0.8105^{25/4}$	1.4119^{20}	al, eth, ace	$B1^{4}$, 3347
7920	3-Hexanone, 2,2-dimethyl,oxime $C_3H_7C(=NOH)C(CH_3)_3$	143.23	nd(al)		78				$B1^{3}$, 2881
7921	3-Hexanone, 2,5-dimethyl $(CH_3)_2CHCH_2COCH(CH_3)_2$	128.21		147-8		$0.8270^{0/0}$	1.4049^{20}	al, eth, ace	$B1^{4}$, 3349
7922	3-Hexanone, 4,4-dimethyl $C_2H_5C(CH_3)_2COC_2H_5$	128.21		151		0.8285^{20}	1.4203^{25}	al, bz, chl	$B1^{4}$, 3350

PHYSICAL CONSTANTS OF ORGANIC COMPOUNDS (Continued)

No.	Name, Synonyms, and Formula	Mol. wt.	Color, crystalline form, specific rotation and λ_{max} (log ε)	b.p. °C	m.p. °C	Density	n_D	Solubility	Ref.
7923	3-Hexanone, 6-dimethylamino-4,4-diphenyl-5-methyl (l) or l-Isomethadone. $(CH_3)_2NCH_2CH(CH_3)C(C_6H_5)_2COC_2H_5$	309.46		162-5$^{0.5}$				al	B14^{3}, 287
7924	3-Hexanone, 4-hydroxy or Propioin $CH_3CH_2CH(OH)COC_2H_5$	116.16		132-5^{227}, 73^{30}		0.956$^{21/4}$	1.4340^{21}	al, ace	B1^{4}, 4021
7925	3-Hexanone, 4-hydroxy-2,2,5,5-tetramethyl $(CH_3)_3CCH(OH)COC(CH_3)_3$	172.27		80^{10}	81(sub)			eth	B1^{4}, 4060
7926	3-Hexanone, 2-methyl or Propyl isopropyl ketone $C_3H_7COCH(CH_3)_2$	114.19		134-6		0.8091^{20}	1.4042^{20}	al, eth, ace, chl	B1^{4}, 3328
7927	3-Hexanone, 4-methyl $C_2H_5CH(CH_3)COC_2H_5$	114.19		134-5		0.8162^{20}	1.4069^{20}	al, eth, ace, bz	B1^{4}, 3329
7928	3-Hexanone, 5-methyl $(CH_3)_2CHCH_2COC_2H_5$	114.19		134^{735}		0.8090^{20}	1.4047^{20}	al, eth	B1^{4}, 3328
7929	Hexaphenyl ethane $(C_6H_5)_3CC(C_6H_5)_3$	486.66	cr(ace)	d	145-7d			eth, ace, chl, MeOH	B5^{3}, 2746
7930	Hexasiloxane, tetradecamethyl $(CH_3)_3SiO[-Si(CH_3)_2O-]_4Si(CH_3)_3$	443.97		245.5, 142^{20}	−59	0.8910$^{20/4}$	1.3948^{20}	bz	B4^{3}, 1880
7931	1,2,3,5 Hexatetracene, 4-chloro $CH_2=CHCCl=C=C=CH_2$	112.56		127d, 55^{54}		0.9997$^{20/4}$	1.5280^{20}	eth, ace, chl	B1^{3}, 1061
7932	1,2,4,5-Hexatetracene, 3,4-dichloro $CH_2=C=CClCCl=C=CH_2$	147.00		38-40^{8}		1.1819$^{20/4}$	1.5456^{20}	eth, ace, chl	B1^{3}, 1061
7933	1,2,4-Hexatriene, 3,4,6-trichloro $ClCH_2CH=CClCCl=C=CH_2$	183.46		50^{1}		1.3132$^{20/4}$	1.5517^{20}	ace, bz, chl	B1^{3}, 1041
7934	1,3,4-Hexatriene, 3,6-dichloro $ClCH_2CH=C=CClCH=CH_2$	149.02		45-6^{3}		1.1807$^{20/4}$	1.5195^{20}	al	B1^{3}, 1041
7935	1,3,5-Hexatriene *(cis)* or Divinyl ethylene $CH_2=CHCH=CHCH=CH_2$	80.13		78	−12	0.7175$^{20/4}$	1.4577^{20}	al, ace, chl, peth	B1^{3}, 1041
7936	1,3,5-Hexatriene *(trans)* $CH_2=CHCH=CHCH=CH_2$	80.13		78.5	−12	0.7369$^{15/4}$	1.5135^{20}	al, ace, chl, peth	B1^{4}, 1093
7937	1,3,5-Hexatriene, 2,5-dimethyl $CH_2=C(CH_3)CH=CHC(CH_3)=CH_2$	108.18		145^{747}	−9	0.7822$^{20/4}$	1.5122^{20}	ace, lig, MeOH	B1^{4}, 1102
7938	1,3,5-Hexatriene, 1,6-diphenyl $C_6H_5CH=CHCH=CHCH=CHC_6H_5$	232.33	lf(ace)		200-3				B5^{3}, 2243
7939	2-Hexenal *(trans)* $C_3H_7CH=CHCHO$	98.14		146-7, 43^{12}		0.8491$^{20/4}$	1.4480^{20}		B1^{4}, 3468
7940	3-Hexenal $CH_3CH_2CH=CHCH_2CHO$	98.14		42-3^{28}		0.8455$^{22/4}$	1.4275$^{21.5}$	eth, ace	B1^{4}, 3469
7941	1-Hexene $C_4H_9CH=CH_2$	84.16		63.3	−139.8	0.6731$^{20/4}$	1.3837^{20}	al, eth, bz, chl, peth	B1^{4}, 828
7942	1-Hexene, 5-amino-4-methyl $CH_3CH(NH_2)CH(CH_3)CH_2CH=CH_2$	113.20		133-6		0.793$^{15/0}$		w	B4, 226
7943	1-Hexene, 1-chloro $C_4H_9CH=CHCl$	118.61		121		0.8872^{22}	1.4300^{22}	eth, ace, bz, chl	B1^{4}, 831
7944	1-Hexene, 2-chloro $C_4H_9CCl=CH_2$	118.61		113, 63^{118}		0.8886$^{25/4}$	1.4278^{25}	ace, bz, chl	B1^{3}, 803
7945	1-Hexene, 5-chloro $CH_3CHCl(CH_2)_2CH=CH_2$	118.61		120.7, 28-30^{13}		0.8891$^{25/4}$	1.4305^{20}	eth, ace, bz, chl	B1^{4}, 832
7946	1-Hexene, 1,2-dichloro *(cis)* $C_4H_9CCl=CHCl$	153.05		88^{30}		1.0812$^{25/4}$	1.4631^{25}	bz, chl	B1^{3}, 803
7947	1-Hexene, 1,2-dichloro *(trans)* $C_4H_9CCl=CHCl$	153.05		63-5^{22}		1.1167$^{25/4}$	1.4576^{25}	bz, chl	B1^{3}, 803
7948	1-Hexene, 2-ethyl $C_4H_9C(C_2H_5)=CH_2$	112.22		120		0.7270$^{20/4}$	1.4157^{20}	eth, bz, peth	B1^{4}, 884
7949	1-Hexene, 2-methyl $C_4H_9C(CH_3)=CH_2$	98.19		91.1		0.7000$^{20/4}$	1.4040$^{20/}{}_{D}$		B1^{4}, 863
7950	1-Hexene, 3-methyl $C_3H_7CH(CH_3)CH=CH_2$	98.19		84		0.6945$^{20/4}$	1.3970^{20}		B1^{4}, 865
7951	1-Hexene, 4-methyl $C_2H_5CH(CH_3)CH_2CH=CH_2$	98.19		87.5		1.6969$^{20/4}$	1.3985$^{20/}{}_{D}$		B1^{4}, 867
7952	1-Hexene, perfluoro $C_4F_9CF=CF_2$	300.05		57				chl	B1^{4}, 831
7953	1-Hexene, 1,1,2-trichloro $C_4H_9CCl=CCl_2$	187.50		90-93^{10}		1.125^{25}	1.4760^{25}	eth	B1^{3}, 803
7954	2-Hexene *(cis)* $C_3H_7CH=CHCH_3$	84.16		68.8	−141.3	0.6869$^{20/4}$	1.3977^{20}	al, eth, bz, lig, chl	B1^{4}, 833
7955	2-Hexene *(trans)* $C_3H_7CH=CHCH_3$	84.16		68^{750}	−133	0.6784$^{20/4}$	1.3935^{20}	al, eth, bz, lig, chl	B1^{4}, 834

No.	Name, Synonyms, and Formula	Mol. wt.	Color, crystalline form, specific rotation and λ_{max} (log ε)	b.p. °C	m.p. °C	Density	n_D	Solubility	Ref.
7956	2-Hexene, 4-chloro $C_2H_5CHClCH{=}CHCH_3$	118.61	....	$123, 30^{10}$	......	$0.8934^{20/4}$	1.4400^{20}	eth, ace, bz, chl	B1⁴, 835
7957	2-Hexene, 2,3-dimethyl $C_3H_7C(CH_3){=}C(CH_3)CH_3$	112.22	....	122.1	......	$0.7405^{20/4}$	$1.4269^{20/}{}_D$	....	B1⁴, 887
7958	2-Hexene, 2,5-dimethyl $(CH_3)_2CHCH_2CH{=}C(CH_3)_2$	112.22	....	112.6	......	$0.7182^{20/4}$	$1.4135^{20/}{}_D$	....	B1⁴, 888
7959	2-Hexene, 2-methyl $C_3H_7CH{=}C(CH_3)_2$	98.19	....	95-8	......	....	$1.4040^{20.5/}{}_D$	....	B1⁴, 863
7960	3-Hexene *(cis)* $C_2H_5CH{=}CHC_2H_5$	84.16	....	66.4	−137.8	$0.6796^{20/4}$	1.3947^{20}	al, eth, bz, lig, chl	B1⁴, 837
7961	3-Hexene *(trans)* $C_2H_5CH{=}CHC_2H_5$	84.16	....	67.1	−113.4	$0.6772^{20/4}$	1.3943^{20}	al, eth, bz, lig, chl	B1⁴, 237
7962	3-Hexene, 1-chloro $C_2H_5CH{=}CH(CH_2)_2Cl$	118.61	....	61^{60}	......	$0.900^{24/4}$	1.435^{24}	eth, ace, bz, chl	B1⁴, 838
7963	3-Hexene, 1-chloro-4-ethyl $(CH_3CH_2)_2C{=}CHCH_2CH_2Cl$	146.66	....	173	......	$0.9102^{20/4}$	1.4524^{20}	bz, chl	B1², 201
7964	3-Hexene, 2-chloro,2,5-dimethyl $(CH_3)_2CHCH{=}CHCCl(CH_3)_2$	146.66	....	$45\text{-}60^{15}$	......	....	1.450^{20}	bz, chl	Am63, 3474
7965	3-Hexene, 3-chloro *(cis)* $C_2H_5CH{=}CClC_2H_5$	118.61	....	119.6	......	$0.9009^{20/4}$	1.4360^{20}	eth, ace, bz, chl	B1⁴, 838
7966	3-Hexene, 1,2,3,4,5,6-hexachloro $ClCH_2CHClCCl{=}CClCHClCH_2Cl$	290.83	cr (peth)	$110\text{-}12^{2}$	58-9	....	....	chl, MeOH	B1⁴, 839
7967	3-Hexene, 3-(4-hydroxyphenyl)-4-(4-methoxyphenyl) $CH_3CH_2C(4\text{-}CH_3OC_6H_4){=}C(4\text{-}HOC_6H_4)CH_2CH_3$	282.38	nd (bz-lig) lf (70% al)	$185\text{-}95^{0.3}$	117-8	....	....	al, eth, ace	B6³, 5623
7968	3-Hexene, perfluoro $C_2F_5CF{=}CFC_2F_5$	300.05	....	49	......	....	....	chl	B1⁴, 838
7969	2-Hexenoic acid, 2-methyl *(trans)* $C_3H_7CH{=}C(CH_3)CO_2H$	128.17	....	$204\text{-}6, 118^{11}$	......	$0.9627^{20/4}$	1.4601^{20}	eth, ace	B2⁴, 1581
7970	2-Hexenoic acid, 3-phenyl $C_3H_7C(C_6H_5){=}CHCO_2H$	190.24	cr (peth)	$183\text{-}4^{14}$	94	....	....	bz	B9³, 2811
7971	3-Hexenoic acid or Hydrosorbic acid $C_2H_5CH{=}CHCH_2CO_2H$	114.14	....	$208, 81\text{-}2^{2}$	12	$0.9640^{23/4}$	1.4935^{20}	....	B2⁴, 1566
7972	5-Hexenoic acid $CH_2{=}CH(CH_2)_3CO_2H$	114.14	....	$203, 107^{17}$	−37	$0.9610^{20/4}$	1.4343^{20}	al, eth	B2⁴, 1562
7973	1-Hexene-3-ol $C_3H_7CH(OH)CH{=}CH_2$	100.16	....	$134, 50^{20}$	......	$0.834^{22/4}$	1.4297^{18}	al, eth, ace	B1⁴, 2136
7974	2-Hexene-1-ol *(cis)* $C_3H_7CH{=}CHCH_2OH$	100.16	....	$156\text{-}8, 58\text{-}60^{15}$	......	$0.8472^{20/4}$	1.4397^{20}	al, eth, ace	B1⁴, 2138
7975	3-Hexene-1-ol *(cis)* $C_2H_5CH{=}CHCH_2CH_2OH$	100.16	....	$156\text{-}7, 58^{12}$	......	$0.8478^{22/4}$	1.4380^{20}	al, eth	B1⁴, 2139
7976	3-Hexene-2-one $C_2H_5CH{=}CHCOCH_3$	98.14	....	$140, 36\text{-}8^{18}$	......	$0.86554^{20/4}$	1.4418^{20}	al, eth, ace	B1⁴, 3468
7977	5-Hexene-2-one, 5-methyl $CH_2{=}C(CH_3)CH_2CH_2COCH_3$	112.17	....	150	......	$0.8475^{20/20}$	1.4348^{20}	al, eth, ace	B1⁴, 3482
7978	1-Hexene-3-one, 1-phenyl-5-methyl or Benzal pinacolone $(CH_3)_2CHCH_2COCH{=}CHC_6H_5$	188.26	cr	154^{25}	43	0.9509^{46}	1.5523^{25}	al, bz, chl	B7³, 1487
7979	4-Hexene-3-one, 2-methyl $CH_3CH{=}CHCO(CH_3)_2$	112.17	....	$147\text{-}8.5^{739}$	......	$0.843^{20/4}$	1.4345^{20}	al, eth, ace	B1⁴, 3483
7980	2-Hexene-4-one $CH_3CH_2COCH{=}CHCH_3$	98.14	....	138-9	......	$0.8559^{20/4}$	1.4388^{20}	al, eth, ace	B1⁴, 3468
7981	1-Hexene-3-yne $C_2H_5C{\equiv}CCH{=}CH_2$	80.13	....	85^{758}	......	$0.7492^{20/4}$	1.4522^{20}	eth, bz, peth, chl	B1⁴, 1091
7982	1-Hexene-3-yne, 5-chloro-5-methyl $(CH_3)_2CClC{\equiv}C\text{-}CH{=}CH_2$	128.60	....	48^{28}	......	0.9375^{15}	1.4778^{20}	al, eth, ace, bz, peth	B1⁴, 1098
7983	1-Hexene-5-yne $HC{\equiv}CCH_2CH_2CH{=}CH_2$	80.13	....	70	......	$0.7650^{20/4}$	1.4318^{20}	eth, bz, peth, chl	B1⁴, 1092
7984	3-Hexene-1-yne, 3-propyl $C_2H_5CH{=}C(C_3H_7)C{\equiv}CH$	122.21	....	136	......	$0.7799^{25/4}$	1.4432^{25}	bz, peth, chl	B1³, 1049
7985	4-Hexene-1-yn-3-ol (d) $CH_3CH{=}CHCH(OH)C{\equiv}CH$	96.13	$[\alpha]^{20/D}$ + 16.06	157-9	......	$0.9090^{20/4}$	1.4645^{17}	al, ace, bz	B1⁴, 2307
7986	4-Hexene-1-yn-3-ol (dl) $CH_3CH{=}CHCH(OH)C{\equiv}CH$	96.13	....	$154\text{-}6, 60^{18}$	......	$0.9148^{25/4}$	1.4651^{23}	al, ace, bz	B1⁴, 2307
7987	Hexylamine or 1-aminohexane $CH_3(CH_2)_4CH_2NH_2$	101.19	....	130	-19	0.7660^{20}	1.4180^{20}	al, eth	B4⁴, 709
7988	Hexyl nitrite $C_6H_{13}ONO$	131.17	ye	$129\text{-}30^{774}, 52^{44}$	......	$0.8778^{20/4}$	1.3987^{20}	al, eth	B1⁴, 1699

No.	Name, Synonyms, and Formula	Mol. wt.	Color, crystalline form, specific rotation and λ_{max} (log ε)	b.p. °C	m.p. °C	Density	n_D	Solubility	Ref.
7989	Hexyl phenyl ether $C_6H_{13}OC_6H_5$	178.27		240, 130^{22}	-19	$0.9174^{20/4}$	1.4921^{20}	eth	$B6^4$, 560
7990	Hexyl sulfate $(C_6H_{13}O)_2SO_2$	266.40		125.3^2		$1.0036^{21/0}$	1.433^{21}		$B1^4$, 1699
7991	1-Hexyne or *n*-Butyl acetylene $C_4H_9C{\equiv}CH$	82.15		71.3	-131.9	$0.7155^{20/4}$	1.3989^{20}	al, eth, bz, peth, chl	$B1^4$, 1006
7992	1-Hexyne, 5-methyl or isopentyl acetylene $(CH_3)_2CHCH_2CH_2C{\equiv}CH$	96.17		92	-125	$0.7274^{20/4}$	1.4059^{20}	al, eth, bz, peth, chl	$B1^3$, 1000
7993	2-Hexyne $C_3H_7C{\equiv}CCH_3$	82.15		84	-89.6	$0.7315^{20/4}$	1.4138^{20}	al, eth, bz, peth, chl	$B1^4$, 1009
7994	2-Hexyne, 5-methyl $(CH_3)_2CHCH_2C{\equiv}CH_3$	96.17		102.5	-92.9	$0.7378^{20/4}$	1.4176^{20}	eth, ace, bz, peth, chl	$B1^4$, 1030
7995	3-Hexyne or Diethyl acetylene $C_2H_5C{\equiv}CC_2H_5$	82.15		81.5^{764}	-103	$0.7231^{20/4}$	1.4115^{20}	al, eth, bz, peth, chl	$B1^4$, 1009
7996	3-Hexyne, 2,5-dimethyl-2,5-dichloro $(CH_3)_2CClC{\equiv}C\text{-}CCl(CH_3)_2$	181.11		$175\text{-}8^{745}$	29			chl	$B1^4$, 1042
7997	3-Hexyne, 1:2,5:6-diepoxy $CH_2CHC{\equiv}C\text{-}CH\text{-}CH_2$ (epoxy rings O-bridged)	110.11		98^{20}	-16	1.1189^{23}	1.4871^{23}	chl	$B19^2$, 19
7998	3-Hexyne, 2,5-dimethyl-2,5-diol or Acetylene pinacol $(CH_3)_2C(OH)C{\equiv}CC(OH)(CH_3)_2$	142.20	nd(w)	205	95	$0.949^{20/20}$		al, eth, ace, bz, chl	$B1^4$, 2699
7999	3-Hexyne, 2-methyl $CH_3CH_2C{\equiv}CCH(CH_3)_2$	96.17		95.2	-116.7	$0.7263^{20/4}$	1.4114^{20}	eth, bz, peth, chl	$B1^4$, 1029
8000	1-Hexyn-3-ol, 3-methyl $C_3H_7C(OH)(CH_3)C{\equiv}CH$	112.17		137		$0.8620^{20/4}$	1.4338^{20}	al, eth	$B1^4$, 2252
8001	3-Hexyne-2,5-diol $CH_3CH(OH)C{\equiv}CCH(OH)CH_3$	114.14		120^{11}		1.0180^{20}	1.4691^{20}		$B1^3$, 2271
8002	3-Hexyn-2-ol, 2-methyl $C_2H_5C{\equiv}CC(OH)(CH_3)_2$	112.17		145-7, $46\text{-}7^7$		0.962^0	1.4392^{25}	al, eth	$B1^4$, 2250
8003	Hippuric acid or N-benzoylaminoacetic acid $C_6H_5CONHCH_2CO_2H$	179.18	pr (w or al)		190-3	$1.371^{20/4}$		al, w	$B9^3$, 1123
8004	Hippuric acid-*p*-amino $(4\text{-}H_2NC_6H_4)CONHCH_2CO_2H$	194.19	pr or nd (w)		198-9			al	$B14^3$, 1069
8005	Hippuric acid, *o*-bromo $(2\text{-}BrC_6H_4)CONHCH_2CO_2H$	258.07	nd (w)		192-3			AcOEt	$B9^3$, 1387
8006	Hippuric acid, *m*-bromo $(3\text{-}BrC_6H_4)CONHCH_2CO_2H$	258.07	nd (w)		146-7			al, MeOH	$B9^2$, 233
8007	Hippuric acid-*p*-bromo $(4\text{-}BrC_6H_4)CONHCH_2CO_2H$	258.07	nd (w)		162			al	$B9^2$, 236
8008	Histamine or 4-Imidoazol ethylamine $C_5H_9N_3$	111.15	wh nd (chl)	209^{18}	86			w, al	$B25^2$, 302
8009	Histamine, dihydrochloride $C_5H_9N_3.2HCl$	184.07	pl (eth-ace) pr (w)		249-52			w, MeOH	$B25^2$, 303
8010	Histidine (d) $C_6H_9N_3O_2$	155.16	ta (w) $[\alpha]^{20/}{}_D$ +40.2 (w)		287d			w	$B25^2$, 404
8011	Histidine (dl) $C_6H_9N_3O_2$	155.16	ta, tetr pr (w)		285d			w	$B25^2$, 409
8012	Histidine (l) $C_6H_9N_3O_2$	155.16	nd or pl (dil al) $[\alpha]^{20/}{}_D$ −39.7 (w, c=1.13)		287d			w	$B25^2$, 404
8013	Histidine, bis(3,4-dichlorobenzene sulfonate) (d) $C_6H_6N_3O_2\text{-}2Cl_2C_6H_4O_3S$	609.28	rh nd(w)		280d				C51, 5184
8014	Histidine, diflavianate $C_6H_9N_3O_2.2C_{10}H_6N_2O_8S$	783.61	nd(w)		251-4d				$B25^2$, 407
8015	Histidine, dihydrochloride (dl) $C_6H_9N_3O_2.2HCl$	228.08			237d			w, al	$B25^1$, 718
8016	Histidine, dihydrochloride (l) $C_6H_9N_3O_2.2HCl$	228.08	rh pl		252d			w, al	B25, 513
8017	Histidine, monohydrochloride, hydrate $C_6H_9N_3O_2.HCl.H_2O$	209.63	pl (w) $[\alpha]^{26/}{}_D$ + 8.0 (3NHCl, c=2)		259d				$B25^2$, 407
8018	Holocaine or N,N-bis(p-ethoxyphenyl) acetamidine $C_{18}H_{22}N_2O_2$	298.38	nd (al)		117-8			al, eth, ace, bz	$B13^3$, 1069

PHYSICAL CONSTANTS OF ORGANIC COMPOUNDS (Continued)

No.	Name, Synonyms, and Formula	Mol. wt.	Color, crystalline form, specific rotation and λ_{max} (log ε)	b.p. °C	m.p. °C	Density	n_D	Solubility	Ref.
8019	Holocaine, hydrochloride or Phenacaine $C_{18}H_{22}N_2O_2.HCl$	334.85	cr(w + 1)		190-2			w, al, chl	B13^3, 1069
8020	Homatropine or Mandelyl tropine $C_{16}H_{21}NO_3$	275.35	pr (al or eth)		99-100			al, eth, ace, chl	B21^4, 179
8021	Homatropine, hydrobromide $C_{16}H_{21}NO_3 \cdot HB$-	356.26	rh pym or pl(w)		217-8d			w, al	B21^4, 179
8022	Homatropine, hydrochloride $C_{16}H_{21}NO_3 \cdot HCl$	311.81	wh pr(w)		220-7d			w, al	B21^4, 179
8023	Homocysteine (dl) $HSCH_2CH_2CH(NH_2)CO_2H$	135.18			270-5 d			w	B4^3, 1647
8024	Homocystine (d) $HO_2CCH(NH_2)CH_2CH_2SSCH_2CH_2CH(NH_2)CO_2H$	268.35	$_D$,$^{26/}{}_D$ −79 in HCl		281-4 d			w	B4^3, 1646
8025	Homogentisic acid or 2,5-dihydroxyphenyl acetic acid $(2,5\text{-}(HO)_2C_6H_3)CH_2CO_2H$	168.15	pr (w + 1) lf (al-chl)		152-4			w, al, eth	B10^3, 1456
8026	Homoveratric acid or 3,4-dimethoxyphenyl acetic acid $[3,4\text{-}(CH_3O)_2C_6H_3]CH_2CO_2H$	196.20	nd (w + 1) cr (bz-peth)		80-2 (hyd) 98-9 (anh)			w, al, eth	B10^3, 1459
8027	Hordenine or 1-(dimethylamino)-2-(4-hydroxyphenyl)ethane $4\text{-}HOC_6H_4CH_2CH_2N(CH_3)_2$	165.24	rh pr (al or bz-peth) nd (w)	173-4^{11} sub	117-8			al, eth, bz, lig, chl	B13^3, 1640
8028	Hordenine, sulfate $2(C_{10}H_{15}NO).H_2SO_4$	428.54	fl		210-11			w	B13^3, 1641
8029	Hordenine, sulfate, dihydrate $(C_{10}H_{15}NO)_2.H_2SO_4.2H_2O$	464.57	pr or pl		197			w	B13^3, 1641
8030	Humulon or α-Lupulic acid $C_{21}H_{30}O_5$	362.47	yesh cr (eth) $[\alpha]^{20/}{}_D$ −232 (bz)		66.5			al, eth, ace, bz	B8^3, 4034
8031	Hydantoic acid or N-Carbomoyl glycine $H_2NCONHCH_2CO_2H$	118.09	mcl pr		180d			w, al	B4^3, 1163
8032	Hydantoic acid, ethyl ester or Ethyl hydantoate $H_2NCONHCH_2CO_2C_2H_5$	146.15	nd (w)		135			w, al	B4^3, 1167
8033	Hydantoic acid, phenylthio or N-phenylpseudothio hydantoic acid $C_6H_5N{=}C(NH_2)SCH_2CO_2H$	210.26	wh (al)	175	159-61				B12^3, 868
8034	Hydantoin or Glycol urea $C_3H_4N_2O_2$	100.08	nd (MeOH) lf (w)		220			w, al	B24^2, 127
8035	Hydantoin, 1-acetyl-2-thio $C_5H_6N_2O_2S$	158.17	pl (al)		175-6			al	B24^1, 293
8036	Hydantoin, 1-benzoyl-2-thio $C_{10}H_8N_2O_2S$	220.25	pr (al)		165d			w	B24^1, 294
8037	Hydantoin, 5-benzylidene-2-thio $C_{10}H_8N_2OS$	204.25	ye nd (al)		258d			al	B24^1, 355
8038	Hydantoin, 5,5-dimethyl $C_5H_8N_2O_2$	128.13	pr (w-al)	sub	178			al, w, eth, ace, bz	B24^2, 157
8039	Hydantoin, 5,5-diphenyl $C_{15}H_{12}N_2O_2$	252.27	nd (al)		286			al, ace, aa	B24^2, 227
8040	Hydantoin, 3,5-diphenyl-2-thio $C_{15}H_{12}N_2OS$	268.33			233			al, eth, bz	B24, 385
8041	Hydantoin, 5-ethyl-5-phenyl (d) or d-Nirvanol $C_{11}H_{12}N_2O_2$	204.23	pl (10 al) $[\alpha]_D$ + 123 (al)		237			al	B24^2, 206
8042	Hydantoin, 5-ethyl-5-phenyl (dl) or dl-Nirvanol $C_{11}H_{12}N_2O_2$	204.23	pr (dil al)		199-200			al, aa	B24^2, 206
8043	Hydantoin, 5(2-hydroxybenzylidene)-2-thio $C_{10}H_8N_2O_2S$	220.25	nd (aa)		248			aa	B25^1, 502
8044	Hydantoin, 1-methyl $C_4H_6N_2O_2$	114.10	cr (w) pl (al)		157-9			w, al, chl	B24^2, 128
8045	Hydantoin, 5-methyl (dl) or α-Lactylurea $C_4H_6N_2O_2$	114.10	pr (w)		145-6 (anh)			w, al	B24^2, 155
8046	Hydantoin, 5-methyl (l) $C_4H_6N_2O_2$	114.10	(w) $[\alpha]^{20/}{}_D$ −50.6		175			w, al	B24^1, 304
8047	Hydantoin, 5-methyl,hydrate (dl) $C_4H_6N_2O_2.H_2O$	132.12	rh (w)		155-6			w, al, ace	B24^2, 155
8048	Hydantoin, 5-phenyl $C_9H_8N_2O_2$	176.17			184-5				B24^2, 201

PHYSICAL CONSTANTS OF ORGANIC COMPOUNDS (Continued)

No.	Name, Synonyms, and Formula	Mol. wt.	Color, crystalline form, specific rotation and λ_{max} (log ε)	b.p. °C	m.p. °C	Density	n_D	Solubility	Ref.
8049	Hydantoin, 2-thio or Glycol thiourea $C_3H_4N_2OS$	116.14	wh nd (w)		229-31d			w, al, eth	$B24^2$, 138
8050	β(-)Hydrastine $C_{21}H_{21}NO_6$	383.40	yesh pr (al) $[\alpha]^{17}/_D$ −67.8 (chl,c=2.5)		132 (135)			eth, ba, chl	$B27^2$, 603
8051	β(-)Hydrastine, hydrochloride $C_{21}H_{21}NO_6.HCl$	419.86	micr pw $[\alpha]/_D$ +158.0 (w, c=2)		116			w	$B27^2$, 604
8052	Hydrastinine $C_{11}H_{13}NO_3$	207.23	nd (lig) cr (eth)		116-7			al, eth, chl	$B27^2$, 530
8053	Hydrastinine, bisulfate $C_{11}H_{13}NO_3.H_2SO_4$	305.30	gr-flr ye cr (al)		216d			w, al	$B27^2$, 530
8054	Hydrastinine, hydrochloride $C_{11}H_{13}NO_3.HCl$	243.69	pa ye nd		212d			w, al	$B27^2$, 530
8055	Hydratropamide or 2-Phenyl propionamide $C_6H_5CH(CH_3)CONH_2$	149.19	lo nd (w dil al) $[\alpha]^{28}/_D$ +57.9 (chl, c=1.6)		100.5			al, chl	$B9^3$, 2420
8056	Hydratropic acid (d) or 2-Phenyl propionic acid $CH_3CH(C_6H_5)CO_2H$	150.18	$[\alpha]^{20}/_D$ +81.1 (al, c=3)	152^{16}					$B9^3$, 2417
8057	Hydratropic acid (dl) or 2-Phenyl propionic acid $C_6H_5CH(CH_3)CO_2H$	150.18		260-2, 160^{25}	<-20	$1.1^{0/4}$	1.5237^{20}		$B9^3$, 2418
8058	Hydratropic acid (l) or 2-Phenyl propionic acid. $C_6H_5CH(CH_3)CO_2H$	150.18	$[\alpha]^{20}/_D$ −58 (al)	152^{10}					$B9^3$, 2417
8059	Hydratroponitrile or 2-Phenyl propionitrile. $C_6H_5CH(CH_3)CN$	131.18		230-2, $116\text{-}7^{20}$		$0.9854^{20/4}$	1.5095^{25}	al, eth	$B9^3$, 2421
8060	Hydrazine, 1-acetyl-2-phenyl $CH_3CONHNHC_6H_5$	150.18	hex pr (eth)		130-2			w, al, bz, chl	$B15^2$, 92
8061	Hydrazine-allyl $CH_2{=}CHCH_2NHNH_2$	72.11		$122\text{-}4^{757}$				w, eth, chl	$B4^1$, 562
8062	Hydrazine, 1,2-bis-(3-aminophenyl) or m,m′-hydrazino dianiline $3\text{-}H_2NC_6H_4NHNHC_6H_4NH_2\text{-}(3)$	214.27	pym (al)		151				$B15^3$, 876
8063	Hydrazine, 1,2-bis-(4-aminophenyl) or p,p′-Hydrazino dianiline $4\text{-}H_2NC_6H_4NHNHC_6H_4NH_2\text{-}(4)$	214.27	ye cr		145			al, eth	$B15^3$, 879
8064	Hydrazine, benzyl $4\text{-}CH_3C_6H_4NHNHCH_2C_6H_5$	212.29	fl or pr (al)	103^{41}	26			w, al, eth	$B15^3$, 699
8065	Hydrazine, 1-benzyl-2-(4-tolyl) $4\text{-}H_2NC_6H_4NHNHCH_2C_6H_5$	212.29		212^{17}					B15, 533
8066	Hydrazine, 2-bromophenyl $2\text{-}BrC_6H_4NHNH_2$	187.04	nd		48				$B15^1$, 117
8067	Hydrazine, 4-bromophenyl $4\text{-}BrC_6H_4NHNH_2$	187.04	nd (w) lf (lig) cr (al)		108			al, eth, lig	$B15^3$, 289
8068	Hydrazine, 1-butyl-1-phenyl $C_6H_5N(C_4H_9)NH_2$	164.25		250^{763}					$B15^1$, 28
8069	Hydrazine, 1,2-diallyl $(CH_2{=}CHCH_2)NHNH(CH_2CH{=}CH_2)$	112.17		145^{752}					$B4^3$, 1737
8070	Hydrazine, 1,2-dibenzoyl $C_6H_5CONHNHCOC_6H_5$	240.26	nd (al)		241				$B9^3$, 1318
8071	Hydrazine, 1,2-dibenzoyl-1,2-dimethyl $C_6H_5CON(CH_3)N(CH_3)COC_6H_5$	268.32	pr (al)		85-6			al	$B9^2$, 217
8072	Hydrazine, 1,1-dibenzyl $(C_6H_5CH_2)_2NNH_2$	212.29	cr (peth)		65			al, eth	$B15^2$, 245
8073	Hydrazine, 1,2-dibenzyl $C_6H_5CH_2NHNHCH_2C_6H_5$	212.29	lf (dil al)		47				$B15^3$, 701
8074	Hydrazine, 1,2-diisobutyl $i\text{-}C_4H_9NHNH\text{-}i\text{-}C_4H_9$	144.26		170^{735}, 63^{10}		$0.8002^{20/4}$	1.4276	al, eth, ace, bz	$B4^2$, 962
8075	Hydrazine, (2,4-dichlorophenyl) $2,4\text{-}Cl_2C_6H_3NHNH_2$	177.03	nd (eth or peth)		94			al, eth, aa	$B15^2$, 152
8076	Hydrazine, 1,1-diethyl $(C_2H_5)_2NNH_2$	88.15		$98\text{-}9^{750}$		$0.8804^{20/4}$	1.4214^{20}	w, al, eth, bz, chl	$B4^2$, 959

No.	Name, Synonyms, and Formula	Mol. wt.	Color, crystalline form, specific rotation and λ_{max} (log ε)	b.p. °C	m.p. °C	Density	n_D	Solubility	Ref.
8077	Hydrazine, 1,2-diethyl $C_2H_5NHNHC_2H_5$	88.15		85-6		0.797^{26}	1.4204^{20}	al, eth, bz	$B4^3$, 1730
8078	Hydrazine, 1,1-dimethyl $(CH_3)_2NNH_2$	60.10		63^{752}		0.7914^{22}	1.4075^{22}	w, al, eth	$B4^3$, 1726
8079	Hydrazine, 1,2-dimethyl $CH_3NHNHCH_3$	60.10		81^{753}		$0.8274^{20/4}$	1.4209^{20}	w, al, eth	$B4^3$, 1727
8080	Hydrazine, 1,2-dimethyl, dihydrochloride $CH_3NHNHCH_3.2HCl$	133.02	pr (w)		170d			w, al	$B4^3$, 1727
8081	Hydrazine, (2,3-dimethylphenyl) $[2,3\text{-}(CH_3)_2C_6H_3]NHNH_2$	136.20	nd (al)		111-2			al, eth	$B15^3$, 718
8082	Hydrazine, (2,4-dimethylphenyl) $[2,4\text{-}(CH_3)_2C_6H_3]NHNH_2$	136.20	nd (eth)		85			al, eth	$B15^2$, 249
8083	Hydrazine, (2,5-dimethylphenyl) $[2,5\text{-}(CH_3)_2C_6H_3]NHNH_2$	136.20	nd (dil aa)		78			al, eth, ace, bz	$B15^3$, 719
8084	Hydrazine, (2,6-dimethylphenyl) $[2,6\text{-}(CH_3)_2C_6H_3]NHNH_2$	136.20	nd (lig)		46			lig	$B15^2$, 552
8085	Hydrazine, (3,4-dimethylphenyl) $[3,4\text{-}(CH_3)_2C_6H_3]NHNH_2$	136.20	yesh nd (eth)		57				$B15^2$, 249
8086	Hydrazine, 1,2-di-α-naphthyl or 1,1′-Hydrazonaphthalene $\beta\text{-}C_{10}H_7NHNH\text{-}\alpha\text{-}C_{10}H_7$	284.36	lf (bz) pl (peth)		153			eth, bz	$B15^3$, 729
8087	Hydrazine, 1,2-di-β-naphthyl or 2,2′-Hydrazonaphthalene $\beta\text{-}C_{10}H_7NHNH\text{-}\beta\text{-}C_{10}H_7$	284.36	red pl (bz)		140-1				$B15^3$, 734
8088	Hydrazine, (2,4-dinitrophenyl) $[2,4\text{-}(O_2N)_2C_6H_3]NHNH_2$	198.14	blsh-red (al)		194 (198d)				$B15^3$, 425
8089	Hydrazine, (2,6-dinitrophenyl) $[2,6\text{-}(O_2N)_2C_6H_3]NHNH_2$	198.14	red nd (dil al)		145				$B15^2$, 219
8090	Hydrazine, 1,1-diphenyl $(C_6H_5)_2NNH_2$	184.24	ta (lig)	220^{40-50}	49-52	$1.190^{16/4}$		al, eth, bz, chl	$B15^3$, 74
8091	Hydrazine, 1,2-diphenyl or Hydrazobenzene $C_6H_5NHNHC_6H_5$	184.24	ta (al-eth)		131	$1.158^{16/4}$		al	$B15^3$, 76
8092	Hydrazine, 1,2-diisopropyl $(CH_3)_2CHNHNHCH(CH_3)_2$	116.21		125, 63^{84}		$0.7894^{20/4}$	1.4173^{20}	al, eth, ace, bz	$B4^3$, 1732
8093	Hydrazine, 1,1-di-4-tolyl $(4\text{-}CH_3C_6H_4)_2NNH_2$	212.29	lf (al)		93			al	$B15^1$, 154
8094	Hydrazine, 1,2-di-(2-tolyl) or *o*-Hydrazotoluene $(2\text{-}CH_3C_6H_4)NHNH(C_6H_4CH_3\text{-}2)$	212.29	lf (al)		165			al, eth, bz	$B15^3$, 655
8095	Hydrazine, 1,2-di-(3-tolyl) or *m*-Hydrazotoluene $(3\text{-}CH_3C_6H_4)NHNH(C_6H_4CH_3\text{-}3)$	212.29	cr (peth)	224	38			al, eth, bz	$B15^3$, 670
8096	Hydrazine, 1,2-di-(4-tolyl) or *p*=Hydrazotoluene $(4\text{-}CH_3C_6H_4)NHNH(C_6H_4CH_3\text{-}4)$	212.29	lf (lig) cr (bz-al) pl (al-eth)		135	$0.957^{20/4}$		al, eth, bz	$B15^3$, 677
8097	Hydrazine, 1-isobutyl-1-phenyl $i\text{-}C_4H_9N(C_6H_5)NH_2$	164.25			240-5	$0.9633^{15/4}$			$B15^3$, 75
8098	Hydrazine, ethyl $C_2H_5NHNH_2$	60.11		101				w, al, eth, ace, bz	$B4^3$, 1730
8099	Hydrazine, 1-ethyl-1-phenyl $C_2H_5N(C_6H_5)NH_2$	136.20		237, 115-9^{19}		$1.0181^{21/4}$	1.5711^{21}	al, eth, ace, bz	$B15^3$, 74
8100	Hydrazine, 1-ethyl-2-phenyl $C_2H_5NHNHC_6H_5$	136.20		240^{750}, 110^{14}		$1.0150^{20/4}$	1.5676^{20}	al, eth, bz, chl	$B15^3$, 74
8101	Hydrazine, methyl CH_3NHNH_2	46.07		87.5^{760}	−52.4	0.874^{25}	$1.4325^{20}{}_{D}$	w, al, eth	$B4^3$, 1726
8102	Hydrazine-1-methyl, 1-phenyl $C_6H_5N(CH_3)NH_2$	122.17		227^{745}, 131^{35}		$1.0404^{20/4}$	1.5691^{20}	al, eth, bz, chl	$B15^3$, 73
8103	Hydrazine-1-methyl, 2-phenyl $CH_3NHNHC_6H_5$	122.17		230^{728}, 112^{14}		$1.0320^{20/4}$	1.5733^{20}	al, eth, bz, chl	$B15^3$, 73
8104	Hydrazine-1-methyl, 2-(3-tolyl) $(3\text{-}CH_3C_6H_4)NHNHCH_3$	136.20	ye		59-61	$1.0265^{100/4}$		al, bz	$B15^2$, 229
8105	Hydrazine-1-methyl, 2-(4-tolyl) $(4\text{-}CH_3C_6H_4)NHNHCH_3$	136.20	ye nd (eth) pl (lig)		91			al, eth, bz	$B15^2$, 154
8106	Hydrazine, α-naphthyl $\alpha\text{-}C_{10}H_7NHNH_2$	158.20		203^{20}	117			al, eth, bz, chl	$B15^3$, 728
8107	Hydrazine, β-naphthyl $\beta\text{-}C_{10}H_7NHNH_2$	158.20	lf (w)		124-5			al, bz	$B15^3$, 734
8108	Hydrazine, (2-nitrophenyl) $(2\text{-}O_2NC_6H_4)NHNH_2$	153.14	red nd (bz)		90-2			w	$B15^3$, 316

No.	Name, Synonyms, and Formula	Mol. wt.	Color, crystalline form, specific rotation and λ_{max} (log ε)	b.p. °C	m.p. °C	Density	n_D	Solubility	Ref.
8109	Hydrazine, (3-nitrophenyl) $(3\text{-}O_2NC_6H_4)NHNH_2$	153.14	red nd or pr (ace) ye nd (al)		93			chl, aa	$B15^3$, 326
8110	Hydrazine, (4-nitrophenyl) $(4\text{-}O_2NC_6H_4)NHNH_2$	153.14	og-red lf or nd (al)		158d			eth, chl	$B15^3$, 331
8111	Hydrazine, 1-pentyl-2-phenyl *(d)* $C_5H_{11}NHNHC_6H_5$	178.28	$[\alpha]^t_D$ + 4.45	$173\text{-}5^{50}$		0.986^{20}	1.5523^{20}	eth	B15, 121
8112	Hydrazine, 1-isopentyl-1-phenyl $i\text{-}C_5H_{11}N(C_6H_5)NH_2$	178.28			236	0.9588^{15}			B15, 121
8113	Hydrazine, phenyl $C_6H_5NHNH_2$	108.14	mcl pr or pl	243, 115^{10}	19.8	$1.0986^{20/4}$	1.6084^{10}	al, eth, ace, bz, chl	$B15^3$, 67
8114	Hydrazine, phenyl, hemihydrate $C_6H_5NHNH_2.½H_2O$	117.15	fl (w)	120^{12}	24	$1.0970^{25/25}$	1.6081^{20}		B15, 68
8115	Hydrazine, phenyl, hydrochloride $C_6H_5NHNH_2.HCl$	144.60	lf (al)	sub	243-6d			w, al	$B15^3$, 71
8116	Hydrazine, 1-phenyl-2-(2-tolyl) $(2\text{-}CH_3C_6H_4)NHNHC_6H_5$	198.27	pl (al)		101-2			eth, bz	$B15^3$, 655
8117	Hydrazine, 1-phenyl-2-(3-tolyl) $(3\text{-}CH_3C_6H_4)NHNHC_6H_5$	198.27	ye cr (peth)		61	$1.0265^{100/4}$		al, bz, lig	$B15^2$, 229
8118	Hydrazine, 1-phenyl-2-(4-tolyl) $(4\text{-}CH_3C_6H_4)NHNHC_6H_5$	198.27	pl (lig), cr (al)		91			al, bz	$B15^3$, 677
8119	Hydrazine, propyl $C_3H_7NHNH_2$	74.13			119				$B4^3$, 1731
8120	Hydrazine, *iso*-propyl $(CH_3)_2CHNHNH_2$	74.13		107^{750}				w, al, bz	$B4^3$, 1731
8121	Hydrazine, 1-isopropyl-2-methyl $(CH_3)_2CHNHNHCH_3$	88.15	cr (eth)	100					$B4^2$, 960
8122	Hydrazine, tetraphenyl $(C_6H_5)_2NN(C_6H_5)_2$	336.44	pr (chl-al)		149d			eth, ace, bz, chl	$B15^3$, 77
8123	Hydrazine, (2-tolyl) $(2\text{-}CH_3C_6H_4)NHNH_2$	122.17	nd (dil al)		59			al, eth, chl	$B15^3$, 654
8124	Hydrazine, (3-tolyl) $(3\text{-}CH_3C_6H_4)NHNH_2$	122.17		244d		$1.057^{20/4}$		al, eth, bz	$B15^3$, 669
8125	Hydrazine, (4-tolyl) $(4\text{-}CH_3C'6H_4)NHNH_2$	122.17	lf (w or eth)	244d	66			al, eth, bz	$B15^3$, 676
8126	Hydrazine, (2,4,6-tribromophenyl) $(2,4,6\text{-}Br_3C_6H_2)NHNH_2$	344.83	nd (peth) cr (lig)		146			bz, chl, lig	$B15^1$, 126
8127	Hydrazine, (2,4,6-trichlorophenyl) $(2,4,6\text{-}Cl_3C_6H_2)NHNH_2$	211.48	cr (bz)		143				$B15^3$, 281
8128	Hydrazine, (2,4,6-trinitrophenyl) $[2,4,6(O_2N)_3C_6H_2]NHNH_2$	243.14	red pl (al)		186			al, aa	$B15^3$, 652
8129	Hydrazine, triphenyl $(C_6H_5)_2NNHC_6H_5$	260.34	nd (bz-peth) cr (al)		142d	$0.869^{70/4}$		al, bz	$B15^2$, 54
8130	Hydrazine carboxylic acid, ethyl ester or *N*-Amino urethane $H_2NNHCO_2C_2H_5$	104.11	cr	198d, 93^9	46			al, eth	$B3^4$, 174
8131	1,1-Hydrazine dicarboxylic acid, diethyl ester $H_2NN(CO_2C_2H_5)_2$	176.17	pr (w)	138^{12}	29			al	$B3^3$, 79
8132	1,2-Hydrazine dicarbon amide $H_2NCONHNHCONH_2$	118.10	pl (w)		257-9				$B3^4$, 236
8133	1,2-Hydrazine dicarboxylic acid, diethyl ester $C_2H_5O_2CNHNH\ CO_2C_2H_5$	176.17	nd (chl) pr (w)	250d	135	1.324^8		al, eth	$B3^4$, 175
8134	Hydrazobenzene or 1,2-Diphenyl hydrazine $C_6H_5NHNHC_6H_5$	184.24	ta (al-eth)		131	$1.158^{16/4}$		al	$B15^3$, 76
8135	Hydrazo diformic acid OCHNHNHCHO	88.07	pr (al)		160			w	$B2^4$, 86
8136	α,α'-Hydrazo naphthalene $\alpha\text{-}C_{10}H_7NHNH\text{-}\alpha\text{-}C_{10}H_7$	284.36	lf (bz) pl (peth)		153			eth, bz	$B15^3$, 729
8137	β, β' -Hydrazo naphthalene $\beta\text{-}C_{10}H_7NHNH\text{-}\beta\text{-}C_{10}H_7$	284.36	red pl (bz)		140-1				$B15^3$, 734
8138	Hydrindane *(trans, l)* or Hexahydro indane C_9H_{16}	124.23		161, 71.7^{40}		$0.8627^{20/4}$	1.4636^{20}	eth, bz, peth	$B5^4$, 292
8139	2-Hydrindanone *(cis)* $C_9H_{14}O$	138.21		225^{754}, 108^{23}	10		1.4830^{20}	al, bz, lig	$B7^3$, 294
8140	2-Hydrindanone *(trans)* $C_9H_{14}O$	138.21		218^{754}	-12	$0.9807^{17/4}$	1.4769^{17}	al, bz, lig	$B7^3$, 294

No.	Name, Synonyms, and Formula	Mol. wt.	Color, crystalline form, specific rotation and λ_{max} (log ε)	b.p. °C	m.p. °C	Density	n_D	Solubility	Ref.
8141	Hydrobenzamide or Tribenzal diamine $C_6H_5CH(N{=}CHC_6H_5)_2$	298.39	nd (bz) cr (al w)	130	110			al, eth	$B7^3$, 838
8142	Hydroberberine *(d)* or *d*-Canadine $C_{20}H_{21}NO_4$	339.39	$[\alpha]^{20}_D$ +297.4 (chl, c=1)		132			al, eth, bz, chl	$B27^2$, 556
8143	Hydroberberine *(l)* $C_{20}H_{21}NO_4$	339.39	nd (al) $[\alpha]^{20/}_D$ −298.2 (chl,c=1)		154			al, eth, bz, chl	$B27^2$, 557
8144	Hydrocinchonidine or Cinchamidine $C_{19}H_{24}N_2O$	296.41	lf (al) $[\alpha]^{20/}_D$ −98.4 (al)	(al)	229			al	$B23^2$, 357
8145	Hydrocinnamaldehyde or 3-Phenylpropionaldehyde $C_6H_5CH_2CH_2CHO$	134.18	mcl	223^{745}, $104\text{-}5^{13}$	47			al, eth	$B7^3$, 1046
8146	Hydrocinnamide or 3-Phenylpropionamide $C_6H_5CH_2CH_2CONH_2$	149.19	nd (w)		106-8			al, eth	$B9^3$, 2393
8147	Hydrocinnamic acid or 3-Phenylpropionic acid $C_6H_5CH_2CH_2CO_2H$	150.18	pr (peth)	279.8, $169\text{-}70^{28}$	48.6	$1.0712^{49/4}$		al, eth, bz	$B9^3$, 2382
8148	Hydrocinnamic acid, 3-amino *(d)* or 3-Amino-3-phenylpropionic acid $C_6H_5CH(NH_2)CH_2CO_2H$	165.19	pl (w) $[\alpha]^{20/}_D$ + 7.0 (w,p=1)		234-5d				$B14^3$, 1218
8149	Hydrocinnamic acid, 3-amino *(dl)* $C_6H_5CH(NH_2)CH_2CO_2H$	165.19	cr (w)		231d				$B14^3$, 1218
8150	Hydrocinnamic acid, 3-amino *(l)* $C_6H_5CH(NH_2)CH_2CO_2H$	165.19	cr (w) $[\alpha]^{25/}_D$ −7.5 (w,c=1)		234-5d				$B14^3$, 1218
8151	Hydrocinnamic acid, benzyl ester or Benzyl hydrocinnamate $C_6H_5CH_2CH_2CO_2CH_2C_6H_5$	240.30		310-40, $198\text{-}9^{20}$		1.090^{15}		eth	$B9^2$, 339
8152	Hydrocinnamic acid, ethyl ester or Ethyl hydrocinnamate $C_6H_5CH_2CH_2CO_2C_2H_5$	178.23		247.2, 123^{16}		1.0147^{20}	1.4954^{20}	al, eth	$B9^3$, 2385
8153	Hydrocinnamic acid, methyl ester or Methyl hydrocinnamate $C_6H_5CH_2CH_2CO_2CH_3$	164.20		$238\text{-}9^{757}$		1.0455^{0}		al, eth, bz	$B9^3$, 2384
8154	Hydrocinnamic acid, propyl ester or Propyl hydrocinnamate $C_6H_5CH_2CH_2CO_2C_3H_7$	192.26		262.1, 135^{16}		1.008^{12}			$B9^3$, 2386
8155	Hydrocinnamic acid, isopropyl ester or Isopropyl hydrocinnamate $C_6H_5CH_2CH_2CO_2CH(CH_3)_2$	192.26		126^{11}		$0.9860^{25/4}$		al, eth	$B9^2$, 339
8156	Hydrocinnamonitrile or 3-Phenyl propionitrile $C_6H_5CH_2CH_2CN$	131.18		261, $125\text{-}6^{15}$		1.0016^{20}	1.5266^{28}	al, eth	$B9^3$, 2395
8157	Hydrocinnamyl chloride or 3-Phenyl propionyl chloride $C_6H_5CH_2CH_2COCl$	168.62		225d, 105^{10}		1.135^{21}		eth	$B9^3$, 2393
8158	Hydroconiferyl alcohol or 3-(4-hydroxy-3-methoxyphenyl)-l-propanol $(3\text{-}CH_3O\text{-}4\text{-}HOC_6H_3)CH_2CH_2CH_2OH$	182.22		197^{15}	65		1.5545^{25}	al, eth	$B6^3$, 6347
8159	Hydrocotamine, hemihydrate $C_{12}H_{15}NO_3.½H_2O$	230.26	pr (eth)		56			al, eth, ace, chl, aa	$B27^2$, 541
8160	Hydrocupreine $C_{19}H_{24}N_2O_2$	312.41	pl (dil al) $[\alpha]^{23/}_D$ −159.2 (abs, al)		230			eth, chl	$B23^2$, 399
8161	Hydrocyanic acid or Hydrogen cyanide HCN	27.03		25.7	−13.2	$0.6876^{20/4}$	1.2614^{20}	w, al, eth	$B2^4$, 50
8162	Hydrofuramide or Furfur amide $C_{15}H_{12}N_2O_3$	268.27	nd (al)		117			al, eth	$B17^4$, 4428
8163	Hydrohydrastinine $C_{11}H_{13}NO_2$	191.23	nd (lig) cr (peth)	303^{752}	66			al, eth, ace, bz, aa	$B27^2$, 528
8164	Hydrolaphacol $C_{15}H_{16}O_3$	244.29	ye nd (al) cr (peth)		94			al	$B8^3$, 2595
8165	Hydroquinidine $C_{20}H_{26}N_2O_2$	326.44	nd (al) $[\alpha]^{18/}_D$ + 229.26		168-9			al, eth, ace, chl	B23,411
8166	Hydroquinine *(d)* or Quinotine $C_{20}H_{26}N_2O_2$	326.44	$[\alpha]^{18/}_D$ + 143.5		171			al, eth, ace, chl	$B23^2$, 400

No.	Name, Synonyms, and Formula	Mol. wt.	Color, crystalline form, specific rotation and λ_{max} (log ε)	b.p. °C	m.p. °C	Density	n_D	Solubility	Ref.
8167	Hydroquinine *(dl)* $C_{20}H_{26}N_2O_2$	326.44	nd (al, chl)		175-7			al, eth, ace, chl	B23[2], 400
8168	Hydroquinine *(l)* $C_{20}H_{26}N_2O_2$	326.44	nd (eth, chl) $[\alpha]^{20}{}_D$ −142.2		172.3			al, eth, ace, chl	B23[2], 400
8169	Hydroquinone or 1,4-Dehydroxy benzene 1,4-$(HO)_2C_6H_4$	110.11	mcl pr (sub) nd (w)	285^{750}	173-4	1.328^{15}		w, al, eth, ace	B6[3], 4374
8170	Hydroquinone, 2-acetyl or 2,5-Dihydroxy acetophenone 2,5$(HO)_2C_6H_3COCH_3$	152.15	ye-gr nd (dil al or w)		204-5			al	B8[3], 2100
8171	Hydroquinone, monobenzoate 4-$(C_6H_5CO_2)C_6H_4OH$	214.22	nd (al)		163-4			al, eth, bz, chl	B9[3], 558
8172	Hydroquinone, 2-bromo or Adurol-1,4-dehydroxy-2-bromobenzene 1,4-$(HO)_2$-2-Br-C_6H_3	189.01	lf (lig) cr (chl)	sub	110-1			w, al, eth, bz	B6[3], 4436
8173	Hydroquinone, 2-chloro 2-Cl-1,4-$(HO)_2C_6H_3$	144.56	red lf (chl) nd (bz)	263	108			w, al, eth, bz	B6[3], 4432
8174	Hydroquinone, diacetate 1,4-$(CH_3CO_2)_2C_6H_4$	194.19	pl (w, al)		123-4	$0.8731^{25/4}$		al, eth, lig, chl	B6[3], 4414
8175	Hydroquinone, dibenzoate 1,4-$(C_6H_5CO_2)_2C_6H_9$	318.33	mcl nd (al or to)		204			w	B9[3], 559
8176	Hydroquinone, monobenzyl ether 4-$C_6H_5CH_2OC_6H_4OH$	200.24	pl (w)		122			al, eth, bz	B6[3], 4402
8177	Hydroquinone, dibenzyl ether 1,4-$(C_6H_5CH_2O)_2C_6H_4$	290.36	pl (al)		130			eth, aa	B6[3], 4402
8178	Hydroquinone, dibenzoyl 1,4-$(C_6H_5CO)_2C_6H_4$	286.33			161				B7[3], 4299
8179	Hydroquinone, 2,5-di-*tert*-butyl 2,5-$(t\text{-}C_4H_9)_2C_6H_2(OH)_2$-1,4	222.33	cr (aq aa)		213.4				B6[3], 4741
8180	Hydroquinone, 2,3-dichloro 2,3-$Cl_2C_6H_2(OH)_2$-1,4	179.00	cr (sub) nd (w + 2)		146-8 (anh)			al	B6[3], 4434
8181	Hydroquinone, 2,5-dichloro 2,5-Cl_2-C_6H_2-$(OH)_2$-1,4	179.00	nd or pr (w, ace, bz)		172.5			al, eth, ace	B6[3], 4434
8182	Hydroquinone, 2,6-dichloro 2,6-$Cl_2C_6H_2(OH)_2$-1,4	179.00	nd or lf (w, bz)		164			al, ace	B6[3], 4435
8183	Hydroquinone, diethyl ether or 1,4-diethoxybenzene 1,4-$(C_2H_5O)_2C_6H_4$	166.22	pl (dil al)	246	72			al, eth, bz, chl	B6[3], 4387
8184	Hydroquinone, 2,3-dimethyl 2,3-$(CH_3)_2C_6H_2(OH)_2$-1,4	138.17	cr (w)		224-5d			w, al, eth	B6[3], 4582
8185	Hydroquinone, 2,5-dimethyl 2,5-$(CH_3)_2C_6H_2(OH)_2$-1,4	138.17	lf (w, al, bz)		217 sub			al, eth, chl	B6[3], 4601
8186	Hydroquinone, 2,6-dimethyl 2,6-$(CH_3)_2$-$C_6H_2(OH)_2$-1,4	138.17	nd (xyl) cr (w)		153-4			w, al, eth	B6[3], 4588
8187	Hydroquinone, dimethyl ether or 1,4-dimethoxybenzene 1,4-$(CH_3O)_2C_6H_4$	138.17	lf (w)	212.6, 109^{20}	58-60	$1.0526^{55/55}$		al, eth, bz	B6[3], 4385
8188	Hydroquinone, dithio or 1,4-dimercapto benzene 1,4-$(HS)_2C_6H_4$	142.23	lf (al)		98			al, bz, aa	B6[3], 4472
8189	Hydroquinone, monoethyl ether or 4-ethoxy phenol 4-$C_2H_5OC_6H_4OH$	138.17	pr or lf (w)	246-7	66-7			al, eth	B6[3], 4387
8190	Hydroquinone, monoheptyl ether 4-$C_7H_{15}OC_6H_4OH$	208.30	cr (lig)		60				B6[3], 4391
8191	Hydroquinone, monohexyl ether 4-$C_6H_{13}OC_6H_4OH$	194.27	cr (lig)		48				B6[3], 4390
8192	Hydroquinone, 2-iodo 2,5-$(HO)_2C_6H_3I$	236.01			115-6				B6[3], 4440
8193	Hydroquinone, 2-methyl 2,5-$(HO)_2C_6H_3CH_3$	124.14		sub 163[11]	128			w, al, eth, ace	B6[3], 4498
8194	Hydroquinone, monomethyl ether or 4-methoxyphenol 4-$CH_3OC_6H_4OH$	124.14	pl	243	57			w, al, eth, bz	B6[3], 4383
8195	Hydroquinone, 2-nitro 2,5-$(HO)_2C_6H_3NO_2$	155.11	og-red rh (w)		133-4			al, eth, bz	B6[3], 4442
8196	Hydroquinon, monooctyl ether 4-$C_8H_7OC_6H_4OH$	222.33	cr (lig)		60-1			al	B6[3], 4391
8197	Hydroquinone, phenyl or 2,5-dihydroxybiphenyl 2,5-$(HO)_2C_6H_3.C_6H_5$	186.21	nd (dil al)		97-8			al	B6[3], 5371

No.	Name, Synonyms, and Formula	Mol. wt.	Color, crystalline form, specific rotation and λ_{max} (log ε)	b.p. °C	m.p. °C	Density	n_D	Solubility	Ref.
8198	Hydroquinone, monopropyl ether or 4-propoxyphenol $4\text{-}C_3H_7OC_6H_4OH$	152.19	cr (w, lig, al)		56-7				B6[3], 4388
8199	Hydroquinone, 2-*iso*-propyl $2,5\text{-}(HO)_2C_6H_3\text{-}i\text{-}C_3H_7$	152.19	nd (w)		130-1				B6[3], 4632
8200	Hydroquinone, 2-*iso*-propyl-5-methyl or Thynohydro quinone $[2,5\text{-}(HO)_2\text{-}4\text{-}CH_3]C_6H_2\text{-}i\text{-}C_3H_7$	166.22	pr (dil al)	290 sub	148			al, eth	B6[3], 4673
8201	Hydroquinone, tetrabromo $1,4\text{-}(HO)_2C_6Br_4$	425.70	mcl pr (al-eth)		244	3.023^{21}		al, eth, aa	B6[3], 4440
8202	Hydroquinone, tetrachloro $1,4\text{-}(HO)_2C_6Cl_4$	247.89	nd (aa)	sub	232			al, eth	B6[3], 4436
8203	Hydroquinone, tetraiodo $1,4\text{-}(HO)_2C_6I_4$	613.70	(aa)		258			eth, chl	B6[1], 417
8204	Hydroquinone, tetra methyl or Durohydroquinone $1,4\text{-}(HO)_2C_6(CH_3)_4$	166.22	nd (al)		233				B6[3], 4682
8205	Hydroquinone, 2,3,5-tribromo $2,3,5\text{-}Br_3C_6H(OH)_2\text{-}1,4$	346.80	nd (chl)		136-7			al, eth, bz, aa, chl	B6[3], 4439
8206	Hydroquinone, 2,3,5-trimethyl $2,3,5\text{-}(CH_3)_3C_6H(OH)_2\text{-}1,4$	152.19	nd (w)		168-70d			al, eth, bz	B6[2], 897
8207	Hydroquinonephthalein or 2,7-Dihydroxy fluoran $C_{20}H_{12}O_5$	332.31	nd (eth)		228-9			al, eth, ace, aa	B19[2], 247
8208	Hydroxyacetamide $HOCH_2CONH_2$	75.07	lf (al) rh (AcOEt)		120	1.415^{13}		w	B3[4], 597
8209	Hydroxyacetic acid or Glycolic acid $HOCH_2COOH$	76.05	rh, nd (w) lf (eth)	d	80			w, al, eth	B3[4], 571
8210	Hydroxyacetic acid, acetate or Acetoxyacetic acid $CH_3CO_2CH_2COOH$	118.19	nd (bz)	$144\text{-}5^{12}$	67-8			w, al, eth, ace, chl	B3[4], 576
8211	Hydroxyacetic acid, anhydride or Glycolic anhydride $(HOCH_2CO)_2O$	134.09	pw	d	128-30				B3[1], 92
8212	Hydroxyacetic acid, 4-bromophenyl ester *(dl)* $(4\text{-}BrC_6H_4)O_2CCH_2OH$	231.05	nd (bz)		118-20			al, eth, bz	B16[2], 125
8213	Hydroxyacetic acid, ethyl ester or Ethyl glycolate $HOCH_2CO_2C_2H_5$	104.11		$160, 69^{25}$		$1.0826^{23/4}$	1.4180^{20}	al, eth	B3[4], 580
8214	Hydroxyacetic acid, methyl ester or Methyl glycolate $HOCH_2CO_2CH_3$	90.08		151.1		$1.1677^{18/4}$		w, al, eth	B3[4], 578
8215	Hydroxyacetic acid, propyl ester or Propyl glycolate $HOCH_2CO_2C_3H_7$	118.14		170-1		$1.0631^{18/4}$	1.4231^{18}		B3[4], 588
8216	Hydroxyacetonitrile or Glycolonitrile $HOCH_2CN$	57.05		183d, 119^{24}	<-72		1.4117^{19}	w, al, eth	B3[4], 598
8217	Hydroxy amphetamine or Paredrine $C_9H_{13}NO$	151.21			125-6			al, chl	B13[3], 1709
8218	Hydroxy amphetamine, hydrobromide $C_9H_{13}NO.HBr$	232.12			189			w, al, ace	B13[3], 1709
8219	Hydroxycitronellal $(CH_3)_2COH(CH_2)_3CH(CH_3)CH_2CHO$	172.27		103^3		0.9220^{20}	1.4494^{20}	al, ace	B1[4], 4058
8220	bis [3-Hydroxy-*sec*-butyl] amine $[CH_3CH(OH)CH(CH_3)]_2NH$	161.24	yesh	$112\text{-}5^3$		$0.9775^{20/4}$	1.4162^{20}	w, al, ace	B4[4], 1726
8221	β-Hydroxyethyl 2-tolyl amine $(2\text{-}CH_3C_6H_4)NHCH_2CH_2OH$	151.21		$285\text{-}6, 149^4$		$1.0794^{20/4}$	1.5675^{20}	al, eth	B12[3], 1846
8222	bis (2-Hydroxyethyl) methyl amine $(HOCH_2CH_2)_2NCH_3$	119.17		$246\text{-}8^{747}$, $123\text{-}5^4$		1.0377^{20}	1.4642^{20}	w, al	B4[4], 1517
8223	bis-(2-Hydroxyethyl) ethyl amine $(HOCH_2CH_2)_2NC_2H_5$	133.19	ye	246-8, 118^3		$1.0135^{20/4}$	1.4663^{20}	w, al	B4[3], 693
8224	β-Hydroxyethyl-4-tolyl amine $(4\text{-}CH_3C_6H_4)NHCH_2CH_2OH$	151.21	pl (eth-lig)	286-8, $153\text{-}5^4$	42-3			al, eth, bz, chl	B12[3], 2034
8225	2-(Hydroxyethyl)-2-(hydroxybutyl) amine $(HOCH_2CH_2)NH(CH_2CH(OH)CH_2CH_3)$	133.19	yesh	137^9		1.0310^{204}	1.4690^{30}	w, al, ace	B4[4], 1725
8226	(2-Hydroxyethyl)-3-(hydroxybutyl) amine $(HOCH_2CH_2)NH(CH_2CH_2CH(OH)CH_3)$	133.19	yesh	$107\text{-}9^1$		$1.0331^{20/4}$	1.4718^{20}	w, ace	C59, 6397
8227	(2-Hydroxypropyl)trimethylammonium chloride or β-Methyl choline chloride $CH_3CH(OH)CH_2N^+(CH_3)_3Cl^-$	153.65	pr (BuOH)	d	165			w, al	B4[3], 754
8228	tris-(2-Hydroxyethyl) amine or Triethanolamine $(HOCH_2CH_2)_3N$	149.19	hyg cr	277^{150}	21.2	$1.1242^{20/4}$	1.4852^{20}	w, al, chl	B4[4], 1524
8229	2-Hydroxybenzoic acid or Salicylic acid $2\text{-}HOC_6H_4CO_2H$	138.12	nd (w), mcl pr (al)	211^{20} sub	159	$1.443^{20/4}$	1.565	al, eth, ace, bz	B10[3], 87

No.	Name, Synonyms, and Formula	Mol. wt.	Color, crystalline form, specific rotation and λ_{max} (log ε)	b.p. °C	m.p. °C	Density	n_D	Solubility	Ref.
8230	3-Hydroxybenzomide 3-HOC$_6$H$_4$CONH$_2$	137.14	pl (w)		170.5			al, eth	B10^3, 255
8231	3-Hydroxybenzamide-*N*-phenyl 3-HOC$_6$H$_4$CONHC$_6$H$_5$	213.24	nd or pl (w, dil al)		156			al	B12^1, 269
8232	3-Hydroxybenzoic acid 3-HOC$_6$H$_4$CO$_2$H	138.12	nd (w) pl or pr (al)		201-3			eth, ace, MeOH	B10^3, 242
8233	3-Hydroxybenzoic acid, 2-bromo 2-Br-3-HOC$_6$H$_3$CO$_2$H	217.02	nd (w)		160-1			eth	B10^2, 83
8234	3-Hydroxybenzoic acid, 4-bromo 4-Br-3-HOC$_6$H$_3$CO$_2$H	217.02	pl, nd (w)		214			al	B10^3, 258
8235	3-Hydroxybenzoic acid, 6-bromo 6-Br-3-HOC$_6$H$_3$CO$_2$H	217.02	(w)		185d			eth	B10^3, 258
8236	3-Hydroxybenzoic acid, 2-chloro 2-Cl-3-HO-C$_6$H$_3$CO$_2$H	172.57	lf (w or bz)		157-8			bz	B10^3, 257
8237	3-Hydroxybenzoic acid, 4-chloro 4-Cl-3-HOC$_6$H$_3$CO$_2$H	172.57	nd (w)		219-20				B10^2, 83
8238	3-Hydroxybenzoic acid, 6-chloro 6-Cl-3-HOC$_6$H$_3$CO$_2$H	172.57	(w)		178-9			al, ace	B10^3, 258
8239	3-Hydroxybenzoic acid, 4,5-dimethoxy 4,5-(CH$_3$O)$_2$-3-HOC$_6$H$_2$CO$_2$H	198.18	nd (aa or w)		197-8			al, aa	B10^3, 2073
8240	3-Hydroxybenzoic acid, 5,6-dimethoxy 5,6-(CH$_3$O)$_2$-3-HOC$_6$H$_2$CO$_2$H	198.18	pl or lf (w)		186-8			al	B10^3, 2060
8241	3-Hydroxybenzoic acid, 2,4-dinitro 2,4-(NO$_2$)$_2$-3-HOC$_6$H$_2$CO$_2$H	228.12	(w)		204			al, eth	B10^3, 264
8242	3-Hydroxybenzoic acid, ethyl ester or Ethyl 3-hydroxy-benzoate 3-HOC$_6$H$_4$CO$_2$C$_2$H$_5$	166.18	pl (bz)	295, 211^{65}	73.8			al, eth	B10^3, 250
8243	3-Hydroxybenzoic acid, 4-formyl 4-HCO-3-HOC$_6$H$_3$CO$_2$H	166.13	nd (w)	sub	234			al, eth	B10, 954
8244	3-Hydroxybenzoic acid, 2-iodo 2-I-3-HOC$_6$H$_3$CO$_2$H	264.02	nd (chl)	158-9				al, eth	B10^2, 84
8245	3-Hydroxybenzoic acid, 4-iodo 4-I-3-HOC$_6$H$_3$CO$_2$H	264.02	nd (w)		226-8			al, eth	B10^3, 261
8246	3-Hydroxybenzoic acid, 6-iodo 6-I-3-HOC$_6$H$_3$CO$_2$H	264.02	nd (w)	sub 160	198			al, eth	B10^3, 261
8247	3-Hydroxybenzoic acid, 4-methoxy or Isovanillic acid 4-CH$_3$O-3-HOC$_6$H$_3$CO$_2$H	168.15	nd, pr, pl (w)	sub	255-7			al, eth	B10^3, 1404
8248	3-Hydroxybenzoic acid, methyl ester 3-HOC$_6$H$_4$CO$_2$CH$_3$	152.15	nd (bz-peth)	280^{709}, 178^{17}	71.5			al	B10^3, 249
8249	3-Hydroxybenzoic acid, 2-methyl or 2,3-Cresotic acid 2-CH$_3$-3-HOC$_6$H$_3$CO$_2$H	152.15	nd (w, dil al)		145-6			al, eth	B10^3, 494
8250	3-Hydroxybenzoic acid, 4-methyl or 3,4-Cresotic acid 4-CH$_3$-3-HOC$_6$H$_3$CO$_2$H	152.15	nd or pr (w)	sub	208.5			al, eth	B10^3, 527
8251	3-Hydroxybenzoic acid, 5-methyl or 3,5-Cresotic acid 5-CH$_3$-3-HOC$_6$H$_3$CO$_2$H	152.15	nd (w)	sub	210			al, eth	B10, 227
8252	3-Hydroxybenzoic acid, 6-methyl or 3,6-Cresotic acid 6-CH$_3$-3-HOC$_6$H$_3$CO$_2$H	152.15	nd or pr (w)		185			al, eth	B10, 215
8253	3-Hydroxybenzoic acid, 2-nitro 2-NO$_2$-3-HOC$_6$H$_3$CO$_2$H	183.12	pl or pr (w + l)		180-1			al, eth	B10^2, 84
8254	3-Hydroxybenzoic acid, 4-nitro 4-NO$_2$-3-HOC$_6$H$_3$CO$_2$H	183.12	ye lf (w)		235			al, eth	B10^3, 263
8255	3-Hydroxybenzoic acid, 5-nitro 5-NO$_2$-3-HOC$_6$H$_3$CO$_2$H	183.12	ye lf or pl (25% HCl)		167			al, eth	B10^2, 85
8256	3-Hydroxybenzoic acid, 6-nitro 6-NO$_2$-3-HOC$_6$H$_3$CO$_2$H	183.12	ye nd or pr (w + 1)		172			al, eth	B10^3, 263
8257	3-Hydroxybenzoic acid, 4-sulfo 4-HO$_3$S-3-HOC$_6$H$_3$CO$_2$H	218.18	ye-gr nd (w + 2)		208 (213)			w, al	B11^3, 707
8258	3-Hydroxybenzoic acid, 5-sulfo 5-HO$_3$S-3-HOC$_6$H$_3$CO$_2$H	218.18	nd (w + 1)		120d			w, al, eth	B11^3, 708
8259	3-Hydroxybenzonitrile 3-HOC$_6$H$_4$CN	119.12	pr (al or eth) lf (w)		83-4			al, eth, bz, chl	B10^3, 255
8260	3-Hydroxybenzoyl chloride 3-HOC$_6$H$_4$COCl	156.57		110-3$^{0.5}$	<-15			chl	B10^2, 82
8261	4-Hydroxybenzamide 4-HOC$_6$H$_4$CONH$_2$	137.14	nd (w + 1)		162 (hyd)			al, eth	B10^3, 340

No.	Name, Synonyms, and Formula	Mol. wt.	Color, crystalline form, specific rotation and λ_{max} (log ε)	b.p. °C	m.p. °C	Density	n_D	Solubility	Ref.
8262	4-Hydroxybenzamide, *N*-phenyl $4\text{-}HOC_6H_4CONHC_6H_5$	213.24	pl or nd (w)		201-2			al	B12^3, 947
8263	4-Hydroxybenzoic acid $4\text{-}HOC_6H_4CO_2H$	138.12	pr or pl (w, al, xyl-al) cr (dil al or ace)		214-5			eth, ace	B10^3, 277
8264	4-Hydroxybenzoic acid, 2-bromo $2\text{-}Br\text{-}4\text{-}HOC_6H_3CO_2H$	217.02	nd (w)		151			eth	B10^2, 103
8265	4-Hydroxybenzoic acid, 3-bromo $3\text{-}Br\text{-}4\text{-}HOC_6H_3CO_2H$	217.02	nd or pr (+w) (w)		177			al, eth, aa	B10^3, 363
8266	4-Hydroxybenzoic acid, butyl ester or Butyl 4-hydroxy-benzoate $4\text{-}HOC_6H_4CO_2C_4H_9$	194.23			68-9			al	B10^3, 307
8267	4-Hydroxybenzoic acid, 2-chloro $2\text{-}Cl\text{-}4\text{-}HOC_6H_3CO_2H$	172.57	nd (w)		159			ace	B10^3, 360
8268	4-Hydroxybenzoic acid, 3-chloro $3\text{-}Cl\text{-}4\text{-}HOC_6H_3CO_2H$	172.57	nd (w)	sub	170-2			al, eth, ace	B10^2, 102
8269	4-Hydroxybenzoic acid, 3,5-dichloro $3,5\text{-}Cl_2\text{-}4\text{-}HOC_6H_2CO_2H$	207.01	nd (dil al or dil aa)	sub d	269			al, eth	B10^3, 362
8270	4-Hydroxybenzoic acid, 3,5-diido $3,5\text{-}I_2\text{-}4\text{-}HOC_6H_2CO_2H$	389.92	nd (dil al)	d 260	237			al, eth	B10^3, 370
8271	4-Hydroxybenzoic acid, 3,5-diido, ethyl ester $3,5\text{-}I_2\text{-}4\text{-}HOC_6H_2CO_2C_2H_5$	417.98	nd (dil al)		123			al	B10^3, 372
8272	4-Hydroxybenzoic acid, 2,3-dimethoxy $2,3\text{-}(CH_3O)_2\text{-}4\text{-}HOC_6H_2CO_2H$	198.18	pl or lf (w or al)		154-5			al, ace, chl	B10^2, 332
8273	4-Hydroxybenzoic acid, 2,6-dimethoxy $2,6\text{-}(CH_3O)_2\text{-}4\text{-}HOC_6H_2CO_2H$	198.18	pl (w)		175			bz, py	B10^1, 235
8274	4-Hydroxy benzoic acid, 3,5-dimethoxy $3,5\text{-}(CH_3O)_2\text{-}4\text{-}HOC_6H_2CO_2H$	198.18	nd (w)		204-5			al, eth, ace, chl	B10^3, 2073
8275	4-Hydroxybenzoic acid, 3,5-dinitro $3,5\text{-}(NO_2)_2\text{-}4\text{-}HOC_6H_2CO_2H$	228.12	ye lf (al)		248-9			al, eth	B10^3, 383
8276	4-Hydroxybenzoic acid, ethyl ester $4\text{-}HOC_6H_4CO_2C_2H_5$	166.18	cr (dil al)	297-8	116-8			al, eth	B10^3, 300
8277	4-Hydroxybenzoic acid, 3-formyl $3\text{-}HCO\text{-}4\text{-}HOC_6H_3CO_2H$	166.13	pr (w)	sub	244			al, eth	B10^2, 675
8278	4-Hydroxybenzoic acid, 2-iodo $2\text{-}I\text{-}4\text{-}HOC_6H_3CO_2H$	264.02	nd (w)		215d			al, eth	B10^2, 104
8279	4-Hydroxybenzoic acid, 3-iodo $3\text{-}I\text{-}4\text{-}HOC_6H_3CO_2H$	264.02	nd (w + ½)	sub	173-4			al, eth, aa	B10^3, 367
8280	4-Hydroxybenzoic acid, 3-methoxy or Vanillic acid $3\text{-}(CH_3O)\text{-}4\text{-}HOC_6H_3CO_2H$	168.15	nd (w)	sub	213-5			eth	B10^3, 1403
8281	4-Hydroxybenzoic acid, 3-methoxy, ethyl ester or Ethyl 3-methoxy-4-hydroxybenzoate $3\text{-}CH_3O\text{-}4\text{-}HOC_6H_3CO_2C_2H_5$	196.20	nd (dil al)	291-3	44			al, eth	B10, 397
8282	4-Hydroxybenzoic acid, 3-methoxy methyl ester $3\text{-}CH_3O\text{-}4\text{-}HOC_6H_3CO_2CH_3$	182.18	nd (dil al)	285-7, 118^2	64			al, chl	B10^3, 1410
8283	4-Hydroxybenzoic acid, methyl ester or Methyl 4-hy-droxy-benzoate $4\text{-}HOC_6H_4CO_2CH_3$	152.15	nd (dil al)	270-80d	131			al, eth, ace	B10^3, 296
8284	4-Hydroxybenzoic acid, 2-methyl or 4,2-Cresotic acid $2\text{-}CH_3\text{-}4\text{-}HOC_6H_3CO_2H$	152.15	nd (w + ½)	236-7 sub	177-8			al, eth	B10^3, 494
8285	4-Hydroxybenzoic acid, 3-methyl or 4,3-Cresotic acid $3\text{-}CH_3\text{-}4\text{-}HOC_6H_3CO_2H$	152.15	nd (w + ½)		174-5 sub			al, eth	B10^3, 512
8286	4-Hydroxybenzoic acid, 3-nitro $3\text{-}NO_2\text{-}4\text{-}HOC_6H_3CO_2H$	183.12	nd or lf (w)		186-7			al, eth	B10^3, 376
8287	4-Hydroxybenzoic acid, propyl ester $4\text{-}HOC_6H_4CO_2C_3H_7$	180.20	pr (eth)		96-8	$1.0630^{102/4}$	1.5050^{102}	eth	B10^3, 306
8288	4-Hydroxybenzoic acid, 5-*iso*-propyl-2-methyl or *p*-Thy-motinic acid $5\text{-}(CH_3)_2CH\text{-}2\text{-}CH_3\text{-}4\text{-}HOC_6H_2CO_2H$	194.23	pl (dil al)		157			al, eth, bz, chl	B10^3, 631
8289	4-Hydroxybenzoic acid, 3-sulfo $3\text{-}HO_3S\text{-}4\text{-}HOC_6H_3CO_2H$	218.18	nd or lf (w)		d			w, al	B11^3, 709
8290	4-Hydroxybenzonitrile $4\text{-}HOC_6H_4CN$	119.12	lf (w)		113			al, eth, chl	B10^3, 344
8291	Hydroxylamine, *N*-ethyl or *β*-Ethyl hydroxyl amine C_2H_5NHOH	61.08	nd (lig)		59-60d	$0.9079^{20/4}$	1.4152^{66}	w, al	B4^3, 1717

No.	Name, Synonyms, and Formula	Mol. wt.	Color, crystalline form, specific rotation and λ_{max} (log ε)	b.p. °C	m.p. °C	Density	n_D	Solubility	Ref.
8292	Hydroxylamine, *N*-methyl CH_3NHOH	47.06	hyg nd	62.5^{15}	87-8	$1.0003^{20/4}$	1.4164^{20}	w, al	B4^{3}, 1715
8293	Hydroxyquinol or 1,2,4-Trihydroxybenzene 1,2,4-$(HO)_3C_6H_3$	126.11	pl (eth), lf or pl (w)		140-1			w, al, eth	B6^{3}, 6276
8294	Hydroxyquinol, triacetate or 1,2,4-Triacetoxy benzene 1,2,4-$(CH_3CO_2)_3C_6H_3$	252.22	nd (MeOH)	>300	97-8			al	B6^{3}, 6282
8295	Hydroxyquinol, 3,5,6-trichloro or 3,5,6-trichloro-1,2,4-trihydroxybenzene $C_6H_3Cl_3O_3$	229.45	nd (bz, aa)		160			al, eth	B6^{2}, 1072
8296	Hyenic acid $C_{25}H_{50}O_2$	382.67	cr (eth) nd (bz)		77-8			eth	B2^{2}, 380
8297	Hygrine-*(l)* or 2-acetonyl-1-methylpyrrolidine $C_8H_{15}NO$	141.21	$[\alpha]_D$ −1.3	193-5, $92\text{-}4^{20}$		$0.935^{17/4}$		al, chl	B21^{4}, 3257
8298	Hyoscine *(dl)* or Scopalamine $C_{17}H_{21}NO_4$	303.36	syr					al, eth, ace, bz	B27^{1}, 248
8299	Hyoscine, hydrobromide *(d)* $C_{17}H_{21}NO_4.HBr$	384.27	$[\alpha]_D$ + 26.3 (w)		195			w, al	B27^{1}, 247
8300	Hyoscine, hydrobromide *(dl)* $C_{17}H_{21}NO_4.HBr$	384.27	eff (ace)		185			w, al	B27^{1}, 248
8301	Hyoscine, hydrobromide *(l)* $C_{17}H_{21}NO_4.HBr$	384.27	lf (al) $[\alpha]'_D$ −26		209			w	B27^{2}, 64
8302	Hyoscine, hydrobromide, trihydrate *(d)* $C_{17}H_{21}NO_4.HBr.\ 3\ H_2O$	438.32	ta (w) $[\alpha]'_D$ + 26.3 (w,c=3)		55				B27^{1}, 247
8303	Hyoscine, hydrobromide, trihydrate *(dl)* $C_7H_{21}NO_4.HBr\text{-}3\ H_2O$	438.32			55-8				B27^{1}, 248
8304	Hyoscine, hydrobromide, trihydrate-*(l)* $C_{17}H_{21}NO_4.HBr\text{-}3\ H_2O$	438.32	ta (w + 3) $[\alpha]_D$-22.8 (w,c=2)					w	B27^{2}, 64
8305	Hyoscine, hydrochloride, dihydrate *(l)* $C_{17}H_{21}NO_4.HCl.2H_2O$	375.85	pr (w)		80			w, al	B27, 101
8306	Hyoscine, monohydrate *(dl)* $C_{17}H_{21}NO_4.H_2O$	321.37	cr (w + 1)		56-7			al, eth, chl	B27, 102
8307	Hyoscine, monohydrate-*(l)* $C_{17}H_{21}NO_4H_2O$	321.37	cr (w + 1) $[\alpha]^{20/}_D$-28 (w,c=2.7)		59			al, eth, ace, bz, chl	B27^{2}, 63
8308	Hyoscine, hydrochloride *(l)* $C_{17}H_{21}NO_4.HCl$	339.82	cr (al)		200			w, al	B27^{2}, 64
8309	Hyoscyamine *(d)* $C_{17}H_{23}NO_3$	289.37	nd (dil al) $[\alpha]'_D$ + 31.3 (al,c=4)		106			al, eth, bz, chl	B21^{4}, 181
8310	Hyoscyamine *(l)* or Daturine $C_{17}H_{23}NO_3$	289.37	tetr nd (dil al)$[\alpha]^{20/}_D$-1 (al,c=1)		108.5			al, chl	B21^{4}, 181
8311	Hyoscyamine, hydrobromide-*(l)* $C_{17}H_{23}NO_3.HBr$	370.29	pr		152			w, al, chl	B21^{4}, 182
8312	Hyoscyamine, sulfate *(l)* $(C_{17}H_{23}NO_3)_2.H_2SO_4$	676.82	dlq nd (al or w)		206 anh			w, al	B21^{4}, 182
8313	Hyoscyamine, hydrochloride-*(l)* $C_{17}H_{23}NO_3.HCl$	325.84	$[\alpha]'_D$ −23.2 (w,c=0.5)		149-51			w, al	B21^{4}, 182
8314	Hyoscyamine, sulfate dihydrate *(l)* $(C_{17}H_{23}NO_3)_2.H_2SO_4.2H_2O$	712.85	dlq nd (al, w) $[\alpha]'_D$ −28.3 (w)		206			w, al, bz	B21^{4}, 182
8315	Hypaphorine *(d)* or *N,N*-Dimethyl-*L*-tryptophane betaine $C_{14}H_{18}N_2O_2$	246.31	cr (dil al) $[\alpha]^{25/}_D$ + 113.4 (w,c=1.6)		255d			w, al	B22^{2}, 469
8316	Hypnal or Antipyrine chloral hydrate $C_{11}H_{12}N_2O\cdot Cl_3CCH(OH)_2$	353.63	wh		68			w, al	B24^{1}, 196
8317	Hypoxanthene or 6-hydroxy purine $C_5H_4N_4O$	136.11	oct nd (w)		150 d				B26^{2}, 252

PHYSICAL CONSTANTS OF ORGANIC COMPOUNDS (Continued)

No.	Name, Synonyms, and Formula	Mol. wt.	Color, crystalline form, specific rotation and λ_{max} (log ε)	b.p. °C	m.p. °C	Density	n_D	Solubility	Ref.
8318	D-Iditol or 1,2,3,4,5,6-Hexane hexol $HOCH_2(CHOH)_4CH_2OH$	182.17	mcl pr (al) $[\alpha]^{20/}{}_D$ + 3.5 (w)		73.5			w	B1[4], 2843
8319	D-Idonic acid or 2,3,4,5,6-pentahydroxy hexanoic acid $HOCH_2(CHOH)_4CO_2H$	196.16	nd (w, al) $\alpha^{20/}{}_D$ + 5.2→- 13,7 (mut)		205d			w	B3[4], 1257
8320	D-Idonic acid-γ-lactone $C_6H_{10}O_6$	178.14	pl $\alpha^{20/}{}_D$ 52.6 (w)		174			w	B18[4], 3026
8321	D-Idonic acid, phenylhydrazide $HOCH_2(CHOH)_4CONHNHC_6H_5$	286.28	$[\alpha]^{20/}{}_D$ −15.´ (w,c=1)		115-7			w, al	B15[3], 208
8322	D-Idose $C_6H_{12}O_6$	180.16	syr $[\alpha]^{13/}{}_D$ + 16 (w)					w	B1[4], 4335
8323	L-Idose $C_6H_{12}O_6$	180.16	syr $[\alpha]^{20/}{}_D$ −17.4 (w,c = 6.2)					w	B1[4], 4336
8324	Imesatin or Isatin-3-imide $C_8H_6N_2O$	146.15	dk ye pr (dil al)		175-6			al	B21[4], 4984
8325	Imidazole or 1,3-Diazole. Glyoxaline $C_3H_4N_2$	68.08	ncl pr (bz)	257, 138.2^{12}	90-1	$1.0303^{101/4}$	1.4801^{101}	w, al, eth, ace, chl	B23[2], 34
8326	Imidazole, 2-mercapto-1-methyl or Methimazole $C_4H_6N_2S$	114.17	lf (al)	280d	142			w, al, chl	B24, 17
8327	Imidazole, l-ethyl $C_5H_8N_2$	96.13		209-10		0.999		w	B23, 46
8328	Imidazole, 1-methyl or Oxalmethylene $C_4H_6N_2$	82.11		195-6, .94-5	−6	1.0325^{20}	1.4970^{20}	w, al, eth, ace	B23[2], 35
8329	Imidazole, 4-methyl $C_4H_6N_2$	82.11		263, $120^{0.02}$	56	$1.0416^{14.3/4}$	$1.5037^{14.3}$	w, al	B23[2], 60
8330	Imidazole, 1-phenyl $C_9H_8N_2$	144.18		276,$153\text{-}4^{23}$	13		1.6025^{25}	eth, ace	B23[2], 36
8331	4,5-Imidazoledicarboxylic acid or 1,3-Diazole-4,5-dicarboxylic acid $C_5H_4N_2O_4$	156.10	pr	288d		1.749			B25[2], 159
8332	2-Imidazolidine thione or *N,N*-Ethylene-thiourea $C_3H_6N_2S$	102.15	nd (al), pr (al)		200-3			w, al	B24[2], 4
8333	2-Imidazolidone or Ethylene urea $C_3H_6N_2O$	86.09	nd (chl)		131-3			w, al	B24[2], 3
8334	2-Imidazoline, 2-methyl or 2-methyl-2-glyoxalidine $C_4H_8N_2$	84.12	hyg	195-8	107			w, al, chl	B23[2], 26
8335	Imino-di-acetic acid or Diglycolamidic acid $HN(CH_2CO_2H)_2$	133.10	rh pr		247.5				B4[3], 1176
8336	Imperatorin or Ammidin $C_{16}H_{14}O_4$	270.28	cr (al)		102			al, eth, ace, bz	B19[4], 2635
8337	Indaconitine or Acetylbenzoyl pseudoaconine $C_{34}H_{47}NO_{10}$	629.75	cr		202-3d			al, eth, chl	B21[4], 2889
8338	Indan or 2,3-dihydroindene C_9H_{10}	118.18		178, 73^{13}	−51.4	$0.9639^{20/4}$	1.5978^{20}	al, eth	B5[4], 1371
8339	Indan, 1-amino or *dl*-1-hydrindamine $C_9H_9NH_2$	133.19	oil	220.5^{747}, $96\text{-}7^{8}$		$1.038^{15/4}$	1.5613^{20}	eth, ace, bz	B12[3], 2798
8340	Indan-5-amino or 5-Hydrindamine $5\text{-}H_2NC_9H_9$	133.19	nd (peth)	$247\text{-}9^{745}$, 131^{15}	37-8			eth, ace, bz	B12[3], 2800
8341	Indan, 2,3-dibromo or Indene dibromide $2,3\text{-}Br_2C_9H_8$	275.97		144^{10}	31-2	$1.747^{25/4}$	1.6290^{25}	eth	B5[3], 1203
8342	Indan, 2,3-dichloro or Indene dichloride $2,3\text{-}Cl_2C_9H_8$	187.07		$87\text{-}90^{2}$		$1.254^{25/4}$	1.5715^{23}		B5[3], 1202
8343	Indan, 1,1-dimethyl $C_{11}H_{14}$	146.23		191		0.919^{20}	$1.5135^{25/}{}_D$		B5[4], 1415
8344	Indan, 1,2-dimethyl $1,2\text{-}(CH_3)_2C_9H_8$	146.23		$79\text{-}80^{10}$		$0.927^{20/4}$	$1.5186^{20/}{}_D$		B5[4], 1415
8345	Indan, 4,6-dimethyl $4,6\text{-}(CH_3)_2C_9H_8$	146.23		52.9^{1}			$1.5325^{20/}{}_D$		B5[2], 395
8346	Indan, 4,7-dimethyl $4,7\text{-}(CH_3)_2C_9H_8$	146.23		$94\text{-}7^{10}$		$0.949^{20/4}$	$1.5342^{20/}{}_D$		B5[4], 1416

No.	Name, Synonyms, and Formula	Mol. wt.	Color, crystalline form, specific rotation and λ_{max} (log ε)	b.p. °C	m.p. °C	Density	n_D	Solubility	Ref.
8347	Indan, 5,6-dimethyl $5\text{-}6\text{-}(CH_3)_2C_9H_8$	146.23		94^{10}		0.9449^{20}	$1.5360^{20/D}$		$B5^4$, 1416
8348	Indan, 1-ethyl $1\text{-}C_2H_5C_9H_9$	146.23		222		0.9348^{25}	$1.5121^{25/D}$		$B5^4$, 1414
8349	Indan, 5-hexyl $5\text{-}C_6H_{13}C_9H_9$	202.34		292.1		0.9114^{20}	1.5122^{2D}		$B5^4$, 1474
8350	Indan, 1-methyl $1\text{-}CH_3C_9H_9$	132.21		188-90, 60^{10}		$0.9402^{20/4}$	1.5260^{20}		$B5^4$, 1397
8351	Indan, 2-methyl $2\text{-}CH_3C_9H_9$	132.21		187, 70^{10}		$0.9034^{20/4}$	1.5070^{20}		$B5^4$, 1397
8352	Indan, 4-methyl $4\text{-}CH_3C_9H_9$	321.21		205.5		0.9577^{20}	$1.5356^{20/D}$		$B5^4$, 1397
8353	Indan, 5-methyl $5\text{-}CH_3C_9H_9$	132.21		$101\text{-}2^{32}$, 82^{17}		$0.9494^{20/4}$	1.5316^{20}		$B5^4$, 1398
8354	Indan, 4-nitro $4\text{-}O_2NC_9H_9$	163.18	wh cr (al)	139^{10}	44				$B5^3$, 1204
8355	Indan, per hydro C_9H_{16}	124.23		165-6		0.8334^{20}	$1.4629^{20/D}$		$B5^4$, 292
8356	Indan, 1-phenyl-1,3,3-trimethyl $1\text{-}C_6H_5\text{-}1,3,3\text{-}(CH_3)_3C_9H_6$	236.36	tcl pr (al)	307-10, $161\text{-}5^{12}$	52-3	$1.0009^{20/4}$	1.5681^{20}	bz, MeOH	$B5^4$, 2246
8357	Indan, 1,1,4,7-tetramethyl $1,1,4,7(CH_3)_4C_9H_6$	174.29		114.2^{15}		0.934^{25}	$1.5216^{25/D}$		$B5^3$, 1282
8358	Indan, 1,1,3-trimethyl $1,1,3\text{-}(CH_3)_3C_9H_7$	160.26		204^{748}			1.5082^{20}		$B5^4$, 1433
8359	Indan, 1,1,4-trimethyl $1,1,4\text{-}(CH_3)_3C_9H_7$	160.26		52.5^{1}			$1.5157^{20/D}$		$B5^4$, 1433
8360	Indan, 1,1,5-trimethyl $1,1,5\text{-}(CH_3)_3C_9H_7$	160.26		$86.^{10}$		0.9119^{20}	1.5126^{20}		$B5^4$, 1433
8361	Indan, 1,1,6-trimethyl $1,1,6\text{-}(CH_3)_3C_9H_7$	160.26		55^{1}			1.5134^{20}		API 23, 2(35.5202)
8362	Indan, 1,4,7-trimethyl $(CH_3)_3C_9H_7$	160.26		95^{10}		0.938^{20}	1.5252^{20}		$B5^4$, 1434
8363	Indan, 1,5,7-trimethyl $1,5,7\text{-}(CH_3)_3C_9H_7$	160.26		106.1^{15}			1.5231^{25}		$B5^3$, 1267
8364	Indan, 4,5,7-trimethyl $4,5,7\text{-}(CH_3)_3C_9H_7$	160.26		59.9^{1}			1.5322^{20}		API 23, 2(35.5202)
8365	1,2-Indandione or α, β-Dioxohydrindene $C_9H_6O_2$	146.15	gold ye pl or lf (bz, eth)		114-6			al, chl	$B7^3$, 3593
8366	1,2-Indandione, β-oxime $C_9H_6O(=NOH)$	161.16	nd (al), (bz)		215-20d			al, bz	$B7^3$, 3593
8367	1,3-Indandione or 1,3-Dioxohydrindene $C_9H_6O_2$	146.15	nd (eth, lig)		131-2d	1.37^{21}		al, eth, bz	$B7^3$, 3594
8368	1,3-Indandione, 2,2-dimethylpropoxy or Pivalyl indandione $C_{14}H_{14}O_3$	230.26	(dil al)		108-10			al, eth, ace	$B7^3$, 4599
8369	1,3-Indandione, dioxime $1,3\text{-}C_9H_6(=NOH)_2$	176.17	nd (w)		ca 225d				B7, 695
8370	1,3-Indandione, 2(3-methylbutoxy) or Valone $C_{14}H_{14}O_3$	230.26	ye (dil al)		67-8			al, eth, ace	$B7^3$, 4598
8371	1,3-Indandione, 2-phenyl or Danilone $C_{15}H_{10}O_2$	222.24	lf (al, bz)		149-51			al, eth, ace, bz, chl	$B7^3$, 4100
8372	1-Indanol or -1-hydroxyindan $1\text{-}HOC_9H_9$	134.18	pl (peth)	255, 128^{12}	54			al, eth, bz, chl	$B6^3$, 2423
8373	4-Indanol or 4-Hydroxy Indan $4\text{-}HOC_9H_9$	134.18	(i) tcl pr (peth) (ii) nd (peth)	120^{12}	(i) 50 (ii) 40				$B6^3$, 2427
8374	5-Indanol or 5-Hydroxy indan $5\text{-}HOC_9H_9$	134.18	nd (peth)	225, 110^{8}	56			al, eth	$B6^3$, 2428
8375	1-Indanone or α-Hydrindone C_9H_8O	132.16	ta, nd (w + 3)	$241\text{-}2^{739}$, 129^{12}	42	$1.1028^{40/40}$	1.561^{25}	al, eth, ace, chl, lig	$B7^3$, 1392
8376	1-Indanone, 3,3-dimethyl $C_{11}H_{12}O$	160.22		$130\text{-}1^{18}$		$1.0320^{14.5/4}$	$1.5453^{14.5}$		$B7^3$, 1450
8377	1-Indanone, 2-nitro $C_9H_7NO_3$	177.16	ye nd (bz-lig)		117d			w, al, eth, ace, bz	$B7^1$, 192
8378	1-Indanone, 6-nitro $C_9H_7NO_3$	177.16	ye lf or nd (peth, al)		74			al, eth, ace, bz, chl	$B7^2$, 285

No.	Name, Synonyms, and Formula	Mol. wt.	Color, crystalline form, specific rotation and λ_{max} (log ε)	b.p. °C	m.p. °C	Density	n_D	Solubility	Ref.
8379	2-Indanone or β-Hydrindone C_9H_8O	132.16	nd (al or eth)	218d	59(61)	$1.0712^{69/4}$	1.538^{67}	al, eth, ace, chl	$B7^3$, 1397
8380	2-Indanone, 5-nitro $C_9H_7NO_3$	177.16	br nd (al)		141			al, eth, aa	B7, 364
8381	Indanthrene or Dihydroanthraquinonazine-indanthrone $C_{28}H_{14}N_2O_4$	442.43	bl nd		470-500d				$B24^2$, 317
8382	Indazole or 1,2-Benzodiazole $C_7H_6N_2$	118.14	nd (al or w)	$267\text{-}70^{743}$	147-9			al, eth	$B23^2$, 117
8383	Indazole, 1-benzhydrylidene or ω,ω-Diphenyl benzofulvene $C_{22}H_{16}$	280.37	og-ye (al)		114.5			eth, ace, bz	$B5^3$, 2532
8384	Indazole, 3-chloro $C_7H_5ClN_2$	152.58	nd (w or lig)	sub	148			al, eth, bz	$B23^2$, 139
8385	Indazole, 4-chloro $C_7H_5ClN_2$	152.58	nd (to)		156			w, al, eth, ace	$B23^2$, 139
8386	Indazole, 2-methyl $C_8H_8N_2$	132.16		261, 135^{16}	56			al, eth, ace	$B23^2$, 118
8387	Indazole, 1-methyl $C_8H_8N_2$	132.16		231, 109^{17}	60-1			al, ace, eth	$B23^2$, 118
8388	Indazole, 3-methyl $C_8H_8N_2$	132.16		280-1, $169\text{-}74^9$	113			al, eth, ace	$B23^2$, 155
8389	Indazole, 5-methyl $C_8H_8N_2$	132.16		$293\text{-}4^{747}$	117				$B23^2$, 157
8390	Indazole, 4-nitro $C_7H_5N_3O_2$	163.14	nd (w)		205-7			al, eth, ace, bz, aa	$B23^2$, 144
8391	Indazole, 5-nitro $C_7H_5N_3O_2$	163.14	yesh nd or col nd (al)		208			al, eth, ace, bz, aa	$B23^2$, 145
8392	Indazole, 6-nitro $C_7H_5N_3O_2$	163.14	nd (w, al, aa)		181d			al, eth, ace, bz	$B23^2$, 146
8393	Indazole, 7-nitro $C_7H_5N_3O_2$	163.14	cr (al)		188-90 sub			eth, ace	$B23^2$, 150
8394	3-Indazolinone or Benzo pyrazolone $C_7H_6N_2O$	134.14	nd or lf (w or MeOH) pl or nd (al)		250-2				$B24^2$, 59
8395	Indene or Indonaphthene C_9H_8	116.16	(aa)	182.6	−1.8	$0.9960^{25/4}$	1.5768^{20}	al, eth, ace, bz, py	$B5^4$, 1532
8396	Indene, 1,2-Diphenyl 1,2-$(C_6H_5)_2C_9H_6$	268.36	lo nd (aa)		177-8			eth	$B5^3$, 2473
8397	Indene, 1,3-diphenyl 1,3-$(C_6H_5)_2C_9H_6$	268.36	(i) nd (aa), (ii) pym (aa)	230^{15}	(i) 68-9 (ii) 85			eth, ace	$B5^3$, 2474
8398	Indene, 2,3-diphenyl 2,3-$(C_6H_5)_2C_9H_6$	268.36	pr (aa)	$235\text{-}40^{12}$	108-9			eth, ace, bz	$B5^3$, 2473
8399	Indene, 2-methyl 2-$CH_3C_9H_7$	130.19	oil	187, $62\text{-}5^{20}$		$0.9034^{20/4}$	1.5070^{20}	eth, ace, bz	$B5^4$, 1545
8400	Indene, 3-methyl 3-$CH_3C_9H_7$	130.19		198.5d, 70^{10}		$0.9640^{20/4}$	1.5591^{27}	eth, ace, bz	$B5^4$, 1545
8401	1-Indenecarboxylic acid 1-$C_9H_7CO_2H$	160.17	pa ye nd(bz)	$193\text{-}5^{12}$	161			eth	$B9^3$, 3068
8402	2-Indenecarboxylic acid 2-$C_9H_7CO_2H$	160.17	nd or lf(bz)		234 sub			al, eth	$B9^3$, 3069
8403	1-Indenone, 2,3-diphenyl $C_{21}H_{14}O$	282.34	og-red (lig or al)		153-5			al, ace, bz	$B7^3$, 2861
8404	Indican or Indoxyl-β-glucoside $C_{14}H_{17}NO_6.3H_2O$	349.34	orh nd (w + 3), $[\alpha]^{19/}_{546}$ -65.6 (w, c=1)		57-8 (hyd) 178-80d (anh)			w, al, ace	$B21^4$, 748
8405	Indigo white or 2,2′-Diindoxyl leucoindigo $C_{16}H_{12}N_2O_2$	264.28	ye cr (dil al)					al, eth	$B23^2$, 429
8406	Indigotin $C_{16}H_{10}N_2O_2$	262.27			390-2d	1.35			$B24^2$, 233
8407	4,4′-Indigotin dicarboxylic acid $C_{18}H_{10}N_2O_6$	350.29	bl nd						B25, 273
8408	5,5′-Indigotin disulfonic acid, sodium Salt or Indigo carmine $C_{16}H_8N_2O_8Na_2S_2$	476.43	dk bl amor or br-red cr					w, al	$B25^2$, 298

No.	Name, Synonyms, and Formula	Mol. wt.	Color, crystalline form, specific rotation and λ_{max} (log ε)	b.p. °C	m.p. °C	Density	n_D	Solubility	Ref.
8409	Indigotin sulfonic acid $C_{16}H_{10}N_2S_2O_8$	422.38	amor		200d			w, al	B25[2], 246
8410	Indirubin or Indigo red $C_{16}H_{10}N_2O_2$	262.27	red or br rh nd (sub)		sub			eth	B24[2], 246
8411	Indole or 1-Benzo [b] pyrrole C_8H_7N	117.15	lf (w, peth) cr (eth)	254, $123\text{-}4^{5}$	52.5	1.22		al, eth, bz, lig	B20[4], 3176
8412	Indole, 1-acetyl 1-$CH_3CO(C_8H_6N)$ $C_{10}H_9NO$	159.19		$152\text{-}3^{14}$ $100^{0.001}$				eth, ace	B20[4], 3182
8413	Indole, 3-(2-aminoethyl) or Tryptamine $C_{10}H_{12}N_2$	160.22	nd (al-bz or liq)	$137^{0.15}$	120 (146)			al, ace	B22[4], 4319
8414	Indole, 1,3-dimethyl or N-Methyl skatole 1,3-$(CH_3)_2C_8H_5N$	145.20	nd	257-60, 119^{7}	141-3			eth	B20[4], 3208
8415	Indole, 2,3-dimethyl $C_{10}H_{11}N$ 2,3-$(CH_3)_2C_8H_5N$	145.20		285	105-7				B20[4], 3226
8416	Indole, 2-hydroxy-3-nitroso or β-Isatoxime $C_8H_6N_2O_2$	162.15	gold-ye nd		214			al	B21[4], 4988
8417	Indole, 3-hydroxy or Indoxyl 3-HOC_8H_6N	133.15	bt ye pr		85			w, al, eth, ace, bz	B21[3], 746
8418	Indole, 1-methyl 1-$CH_3C_8H_6N$	131.18		240-1, $70\text{-}5^{2}$		1.0707^{0}		al, eth, bz	B20[4], 3180
8419	Indole, 2-methyl 2-$CH_3C_8H_6N$	131.18	pl(dil al)nd or lf(w)	272	61	$1.07^{20/4}$		al, eth, ace	B20[4], 3202
8420	Indole, 3-methyl or Skatole 3-$CH_3C_8H_6N$	131.18	lf(liq)	$265\text{-}6^{755}$	97-8			w, al, eth, ace, bz	B20[4], 3206
8422	Indole, 3-methyl-2-phenyl 3-CH_3-2-$C_6H_5C_8H_5N$	207.27		$280\text{-}90^{120}$	91-2			al, bz	B20, 474
8423	Indole, 2-phenyl 2-$C_6H_5C_8H_6N$	193.25		250^{10}	189			eth, bz	B20[2], 302
8424	Indole, 1,2,3-trimethyl 1,2,3-$(CH_3)_3C_8H_4N$	159.23		$283\text{-}4^{750}$	18				B20[4], 3227
8425	3-Indolylacetic acid, 2-methyl $C_{11}H_{11}NO_2$	189.21	ace	195-200				al, eth, ace	B22[4], 1117
8426	2-Indolecarboxylic acid $C_9H_7NO_2$	161.16	ye pl(bz peth)		205-8			al, eth	B22[4], 1059
8427	2-Indolecarboxylic acid, 3-hydroxy or Indoxylic acid $C_9H_7NO_3$	177.16	cr(sub)	122-3 sub					B22[2], 168
8428	Indoline or 2,3-Dihydroindole C_8H_9N	119.17		228-30, $70\text{-}5^{2}$		$1.069^{20/4}$	1.5923^{20}	eth, ace, bz	B20[4], 2896
8429	3-Indolylacetic acid or Heteroauxin $(C_8H_6N)CH_2CO_2H$	175.19	lf(bz)pl(chl)		165-6			al, eth, ace	B20[4], 1088
8430	Indone-2,3-dibromo $C_9H_6Br_2O$	289.95	og-ye nd(al aa)		123			al, eth, chl	B7[3], 1647
8431	Indophenin $C_{24}H_{14}N_2O_2S_2$	426.51	bl nd or pw		d				B21[2], 330
8432	Indopenol $C_{12}H_9NO_2$	199.21	red br pl (ace-peth)		160			w, al, eth, bz, chl	B13[3], 1047
8433	Indoxazene or 4,5-Benzoisoxazol C_7H_5NO	119.12	oil	100^{26}		$1.1727^{21/4}$	1.5570^{20}	eth	B27[2], 15
8434	Inosine or Hyoxanthosine-β- -ribose $C_{10}H_{12}N_4O_5$	268.23	pl (w + 2) nd (80% al)		90 (+2w) 218d (anh)			al	B31, 25
8435	D-Inositol or 1,2,3,4,5,6-cyclohexanehexol $C_6H_{12}O_6$	180.16	pr (w + 2) (al), $[\alpha]_D$ + 65.0 (w, 12%)		249-50			w, aa	B6[3], 6925
8436	DL-Inositol or Phaseo mannitol $C_6H_{12}O_6$	180.16	mcl pr (w)cr(gl aa)	319(vac)	253	1.752^{15}		w, aa	B6[3], 6925
8437	L-Inositol $C_6H_{12}O_6$	180.16	nd(w + 2)	250 vac	247	1.598^{20}		w, aa	B6[3], 6925
8438	Inulin or Plant starch $(C_6H_{10}O_5)n$	~7000	wh amor or pw, $[\alpha]^{21}\Sigma_D$ - 38.3		178d	$1.35^{20/4}$			J1952, 2384

No.	Name, Synonyms, and Formula	Mol. wt.	Color, crystalline form, specific rotation and λ_{max} (log ε)	b.p. °C	m.p. °C	Density	n_D	Solubility	Ref.
8439	Iodoacetamide ICH_2CONH_2	184.96	cr(w)		95			w	$B2^{4}$, 536
8440	Iodoacetic acid ICH_2COOH	185.95	pl(w,peth)	d	83			w, al	$B2^{4}$, 534
8441	Iodoacetic acid, ethyl ester or Ethyliodoacetate $ICH_2COOC_2H_5$	214.00	oil	$178\text{-}80, 73^{16}$		$1.8173^{13/4}$	1.5079^{13}	al, eth	$B2^{4}$, 535
8442	Iodogorgoic acid (d) or 3,5-Diidotyrosine(d) $C_9H_9I_2NO_3$	432.98	yesh nd(w or 70% al), $[\alpha]^{20/}{}_D$ + 2.9 (HCl, c=5)		213				$B14^{3}$, 1563
8443	Iodogorgoic acid (dl) $C_9H_9I_2NO_3$	432.98	nd (50%) al, pl (w)		200d				$B14^{3}$, 1563
8444	Iodogorgoic acid (l) $C_9H_9I_2NO_3$	432.98	nd(w or70%), $[\alpha]^{20/}{}_D$ -2.98 (4% HCl, c=5)	213d				$B14^{2}$, 366	$B14^{3}$, 1563
8445	Ionene $C_{13}H_{18}$	174.29		$238\text{-}9^{730}$, $90\text{-}1^{4}$		$0.9356^{20/4}$	1.5257^{20}	al, eth, bz, chl	$B5^{4}$, 1445
8446	α-Ionol $C_{13}H_{22}O$	194.32	oil	127^{15}		$0.9474^{20/4}$	1.4735^{20}	al, eth, ace	$B6^{3}$, 402
8447	β-Ionol $C_{13}H_{22}O$	194.32		131^{15}, $89^{0.7}$		$0.9243^{20/4}$	1.4969^{20}	al, eth, ace	$B6^{3}$, 401
8448	α-Ionone (d,trans) or 4-(2,6,6-trimethyl-2-cyclohenyl)-3-buturic-2-one $C_{13}H_{20}O$	192.30	$[\alpha]^{25/}{}_D$ + 347				$1,5061^{20}$		$B7^{3}$, 640
8449	α-Ionone (dl) $C_{13}H_{20}O$	192.30		$146\text{-}7^{28}$		0.9298^{21}	1.5041^{20}	al, eth, ace	$B7^{3}$, 641
8450	α-Ionone (l) $C_{13}H_{20}O$	192.30	$[\alpha]^{27/}{}_D$ -406	$73\text{-}7^{0.15}$			1.5000^{25}	al, eth, ace	$B7^{3}$, 640
8451	α-Ionone, semicarbazone (dl) $C_{14}H_{23}N_3O$	249.36	(60% al)		(i) 107-8, (ii) 143				$B7^{3}$, 644
8452	β-Ionone $C_{13}H_{20}O$	192.30		140^{18}, $72\text{-}4^{0.1}$		$0.9462^{20/4}$	1.5198^{20}	al, eth	$B7^{3}$, 634
8453	β-Ionone, semicarbazone $C_{14}H_{23}N_3O$	249.36	nd(al)		149			al, eth, bz, chl	$B7^{3}$, 639
8454	Irene or anhydro irone (?) $C_{14}H_{20}$	188.31		$120\text{-}5^{10}$		$0.9332^{20/4}$	1.5217^{20}	ace, bz	$B5^{4}$, 1460
8455	β-Irone or 4-(2,5,6,6-tetramethyl-1-cyclohexenyl)-3-butene-2-one $C_{14}H_{22}O$	206.33		$85\text{-}90^{0.1}$		$0.9434^{21/4}$	1.5017^{20}	al, eth, bz, chl, lig	$B7^{3}$, 666
8456	Isatic acid or 2-Aminobenzoyl formic acid $H_2NC_6H_4COCO_2H$	165.15	pw		d			w	$B14^{3}$, 1650
8457	Isatin or 2,3-Indolinedione $C_8H_5NO_2$	147.13	yesh red pr(sub)	sub	203-5			al, ace, bz	$B21^{4}$, 4981
8458	Isatin, 1-acetyl $C_{10}H_7NO_3$	189.17	ye pr or nd		144-5			al, ace	$B21^{4}$, 4998
8459	Isatin, chloride or 2-chloro-3-indolone C_8H_4ClON	165.58	br nd		180d			al, eth, aa	$B21^{4}$, 3720
8460	Isatin, 1-methyl $C_9H_7NO_2$	161.16	red-ye orh nd (w)		134			al, eth, ace, bz	$B21^{4}$, 4991
8461	Isatin, 5-methyl $C_9H_7NO_2$	161.16	red pl (w)nd (w or al)		187			al	$B21^{4}$, 5451
8462	Isatin, 7-methyl $C_9H_7NO_2$	161.16			267				$B21^{4}$, 5454
8463	Isatin, 5-nitro $C_8H_4N_2O_4$	192.13	ye nd al		254-5d				$B21^{4}$, 5015
8464	Isatin, 2-oxime $C_8H_6N_2O_2$	162.15	ye-og nd al or w		198-200d			eth, ace, aa	$B21^{4}$, 4987
8465	Isatin, 3-oxime $C_8H_6N_2O_2$	162.15	gold ye nd		225d			al	$B21^{4}$, 4988
8466	Isatoic acid, anhydride or N-carboxyanthranilic anhydride $C_8H_5NO_3$	163.13	pr(al or gl aa)cr (al)		243d				$B27^{2}$, 299

No.	Name, Synonyms, and Formula	Mol. wt.	Color, crystalline form, specific rotation and λ_{max} (log ε)	b.p. °C	m.p. °C	Density	n_D	Solubility	Ref.
8467	d-α-Isatropic acid or d,α-1-phenyl-1,4-tetralindicarboxylic acid $C_{18}H_{16}O_4$	296.32	pr, $[\alpha]^{25}_D$ + 9.44 (al, c=12.6)		239d				$B9^1$, 417
8468	dl-α-Isatropic acid or dl-α-1-phenyl-1,4-tetralindicarboxylic acid $C_{18}H_{16}O_4$	296.32	cr(chl-peth)		238-9			aa	$B9^3$, 4635
8469	dl-β-Isatropic acid $C_{18}H_{16}O_4$	296.32	pl w		208-9			al, aa	$B9^3$, 4635
8470	l,β-Isatropic acid $C_{18}H_{16}O_4$	296.32	$[\alpha]_D$ -8.8 (al, c=5)		197			al, aa	$B9^3$, 4634
8471	Isoapiol $C_{12}H_{14}O_4$	222.24	mcl pr lf or nd (al)	303-4, 189^{33}	56			al, eth, ace, bz	$B19^4$, 1033
8472	Isobergaptene $C_{12}H_8O_4$	216.19	cr(al)		222-3			al, diox	$B19^4$, 2638
8473	Isoborneol (d) or 2-Hydroxybornane $C_{10}H_{18}O$	154.25	cr(peth)		212 (sealed tube)			al, eth, bz, chl	$B6^3$, 299
8474	Isoborneol (dl) or α,β-Camphol $C_{10}H_{18}O$	154.25	ta (peth)	sub	212(sealed tube)			al, eth, chl	$B6^4$, 282
8475	Isoborneol (l) or β-Camphol $C_{10}H_{18}$	154.25	(peth), $[\alpha]_D$ + 33.9 (al)		214 (218)			al, eth, chl	$B6^4$, 282
8476	Isoborneol, acetate (d) $CH_3CO_2C_{10}H_{17}$	196.29	$[\alpha]^{20}_D$ -50.2 (al)	112^{17}		$0.9905^{20/4}$	1.4633^{20}	al, ace	$B6^2$, 90
8477	Isoborneol, acetate (dl) $CH_3CO_2C_{10}H_{17}$	196.29		$115\text{-}7^{21}$		$0.9841^{20/4}$	1.4640^{20}	al, ace	$B6^4$, 283
8478	Isoborneol, acetate (l) $CH_3CO_2C_{10}H_{17}$	196.29		225, $123\text{-}7^{35}$	<-50	$1.002^{11/11}$		al, ace	B6, 89
8479	Isoborneol, formate (d) $HCO_2C_{10}H_{17}$	182.26	$[\alpha]^{20}_D$ + 129.5 (al, c=5)	94^{15}		$1.0136^{20/4}$	1.4678^{22}	al, ace	$B6^3$, 301
8480	Isobornylamine $C_{10}H_{17}NH_2$	153.27	pw, $[\alpha]_D$ -47.7 (4% al)		184			eth, ace	$B12^3$, 195
8481	Isocalycanthine $C_{22}H_{28}N_4.H_2O$	366.51	rh		235-6			al, eth, ace, chl	Am32, 1305
8482	Isocamphane (d) or Dihydrocamphene $C_{10}H_{18}$	138.25	cr (MeOH), $[\alpha]^{20}_D$ + 8.68 (bz, p=20)	166^{750}	62-3			al, ace	$B5^3$, 263
8483	Isocamphane (dl) $C_{10}H_{18}$	138.25	cr (MeOH)	$165\text{-}6^{730}$	65-7	$0.8276^{67/4}$	1.4419^{67}	al, ace, bz	$B5^3$, 263
8484	Isocamphane (l) $C_{10}H_{18}$	138.25	cr (al), $[\alpha]^{20}_D$ -8.5 (al)	$164\text{-}5^{757}$, $62\text{-}3^{17}$	64			al, ace	$B5^3$, 263
8485	Isocamphoric acid (d) or α-*trans*-1,2,2-Trimethyl-1,3-cyclopentane-dicarboxylic acid $C_{10}H_{16}O_4$	200.23	lf (w), $[\alpha]^{20}_D$ + 48.6		171-2			al, aa	B9, 762
8486	Isocamphoric acid (dl) $C_{10}H_{16}O_4$	200.23	pl(al or gl aa)cr(w)		197	1.249		eth	$B9^3$, 3879
8487	Isocamphoric acid (l) $C_{10}H_{16}O_4$	200.23	tetr, $[\alpha]^{17}_D$ -48.4 (MeOH, p=9.9)		173			al, aa	$B9^3$, 3878
8488	Isocarotene or Dehydro-β-carotene $C_{40}H_{54}$	534.87	vt pr(bz-MeOH) vt nd lf(bz)		192-3				$B5^3$, 2515
8489	Isocarvomenthol (d) $C_{10}H_{20}O$	156.27	$[\alpha]_D$ + 20.2	110^{20}		$0.904^{20/4}$	1.4669^{18}	al, eth, ace	$B6^3$, 131
8490	Isocarvomenthol (l) $C_{10}H_{20}O$	156.27	$[\alpha]^{16}_D$ -17.7	106^{17}		$0.9109^{20/4}$	1.4662^{20}	al, eth, ace	$B6^4$, 149
8491	Isocodeine $C_{18}H_{21}NO_3$	299.37	pl (bz) pr (AcoEt or aa), $[\alpha]^{15}_D$ -152 (chl, c=2)	d	171-2	1.87^4	1.675		$B27^2$, 175
8492	Isocorybulbine $C_{21}H_{25}NO_4$	355.43	lf (al), $[\alpha]^{15}_D$ + 301 (chl, c=1)		187-8	$1.045^{20/4}$		al, chl	$B21^4$, 4779

No.	Name, Synonyms, and Formula	Mol. wt.	Color, crystalline form, specific rotation and λ_{max} (log ε)	b.p. °C	m.p. °C	Density	n_D	Solubility	Ref.
8493	Isocorydine or Corytuberine methyl ether $C_{20}H_{23}NO_4$	341.41	pl, $[\alpha]^{20/}{}_D$ + 195.3 (chl)		185			chl	B21[4], 2755
8494	Isocoumarin or *o*-(β-Hydroxyvinyl) benzoic acid lactone $C_9H_6O_2$	146.15	pl(bz)	285-6[719]	47			al, eth, bz	B17[4], 5062
8495	Isocyanuric acid, trimethyl ester $C_6H_9N_3O_3$	171.16	mcl pr (w or al)	274	176-7			al	B26[2], 134
8496	Isoderritol $C_{21}H_{22}O_6$	370.40	ye lf (MeOH)		150				B18[4], 3351
8497	Isodurene or -1,2,3,5-tetramethylbenzene 1,2,3,5-$(CH_3)_4C_6H_2$	134.22		198, 24.4[10]	-23.7	0.8903[20/4]	1.5130[20]	al, eth, ace, bz	B5[4], 1073
8498	8-Isoestradiol $C_{18}H_{24}O_2$	272.39	cr (dil MeOH chl), $[\alpha]^{20/}{}_D$ + 18 (diox)		181			al, diox	B6[3], 5332
8499	8-Isoestrone or 8-Epiestrone $C_{18}H_{22}O_2$	270.37	cr (MeOH) $[\alpha]^{20/}{}_D$ + 94 (diox)		247			eth, diox	B8[3], 1170
8500	Iso-β-eucaine (dl) or 4-Benzoyloxy-2,2,6-trimethyl piperidine $C_{15}H_{21}NO_2$	247.34		188[19]	<-5	1.0467[22/22]		al	B21[4], 131
8501	Iso-β-eucaine, hydrochloride (l) $C_{15}H_{21}NO_2 \cdot HCl$	283.80	nd (w) $[\alpha]_{5461}$ + 17 (w,c=1)		271-3			w	B21[2], 16
8502	Iso-β-eucaine, hydrochloride (dl) $C_{15}H_{21}NO_2.HCl$	283.80	ta (w) pl (aq al or al-eth)		269-71			w	B21[2], 15
8503	Iso-β-eucaine, hydrochloride (l) $C_{15}H_{21}NO_2.HCl$	283.80	nd α_{5461} −16.3 (w,c=1)		271-3			w	B21[2], 15
8504	Isoeugenol *(cis)* or 2-Methoxy-4-propenylphenol 2-CH_3O-4-$(CH_2CH{=}CH)$-C_6H_3OH	164.20		134-5[13], 80-1[0.5]		1.0837[20/4]	1.5726[20]		B6[3], 4992
8505	Isoeugenol (trans) 2-CH_3O-4-$(CH_2CH{=}CH)C_6H_3OH$	164.20		141-2[13]	33-4	1.0852[20/4]	1.5784[20]	al, eth	B6[3], 4992
8506	Isoeugenol (cis), acetate 2-CH_3O-4-$(CH_2CH{=}CH)C_6H_3O_2CCH_3$	206.24		160-2[13]		1.0947[19/4]	1.5418[20]	eth	B6[3], 5007
8508	Isoeugenol, acetate (trans) 2-CH_3O-4-$(CH_2CH{=}CH)C_6H_3O_2CCH_3$	206.24	nd (al bz-lig)	282-3	80-1	1.0251[100/4]	1.5052[100]		B6[3], 5008
8509	Isoflavone, 4′,7-dihydroxy or Daidzein $C_{15}H_{10}O_4$	254.24	pa ye pr (50% al)	sub	323d			al, eth	B18[4], 1805
8510	Isoflavone, 4′,5,7-trihydroxy or Genistein $C_{15}H_{10}O_5$	270.24	nd (eth) pr (dil al)		301-2d				B18[4], 2724
8511	Isofurfurine $C_{15}H_{12}N_2O_3$	268.27	nd (w)		143				B27, 674
8512	Isogeraniolene or 2,6-Dimethyl-1,3-heptadiene $CH_2{=}C(CH_3)CH{=}CHCH_2CH(CH_3)_2$	124.23		143-4[755], 31[7]		0.7561[20/4]	1.4520[20]	eth, bz	B1, 1015
8513	Isoleucine (d) $C_2H_5CH(CH_3)CH(NH_2)CO_2H$	131.17	$[\alpha]^{20/}{}_D$ −12.2H₂O −(3,2%)		283-4 d			w	B4[3], 1458
8514	Isoleucine (l) $C_2H_5CH(CH_3)CH(NH_2)CO_2H$	131.17	$[\alpha]^{25/}{}_D$ + 12.2 (3.2% in H_2O) 36.7 (INHCl)		285-6 d			w	B4[3], 1454
8515	Isoleucine, allo (l) $C_2H_5CH(CH_3)CH(NH_2)CO_2H$	131.17	$[\alpha]^{20/}{}_D$ −14.2H_2O 2		dec 280-1			w	B4[3], 1462
8516	Isolysergic acid (d) $C_{16}H_{16}N_2O_2$	268.32	cr (w + 2) $[\alpha]^{20/}{}_D$ + 281 (py,c=1)		218d			py	B27[2], 860

PHYSICAL CONSTANTS OF ORGANIC COMPOUNDS (Continued)

No.	Name, Synonyms, and Formula	Mol. wt.	Color, crystalline form, specific rotation and λ_{max} (log ε)	b.p. °C	m.p. °C	Density	n_D	Solubility	Ref.
8517	D-Isomannide or 1,4,3,6-Dianhydro-D-mannitol $C_6H_{10}O_4$	146.14	mcl cr $[\alpha]^{26/}{}_D$ +62.2 ^{D} (chl) +91 (w)	274d	87-9			w	B19⁴, 990
8518	Isomenthol (d) or 2-isopropyl-5-methyl cyclohexanol $C_{10}H_{20}O$	156.27	nd (dil al) $[\alpha]^{20/}{}_D$ +26.5 (al,c=4)	218.6, 96.5^{10}	82 (85)			al, eth, aa	B6⁴, 151
8519	Isomenthol (dl) $C_{10}H_{20}O$	156.27	nd	218.5, $97.4^{16.5}$	53-4	0.9040^{30}	1.4510^{60}	al, eth, aa	B6⁴, 152
8520	Isomenthol (l) $C_{10}H_{20}O$	156.27	$[\alpha]^{25/}{}_D$ −24.1		82.5			al, eth, aa	B6⁴, 152
8521	Isomenthone $C_{10}H_{18}O$	154.25		$89\text{-}90^{15}$		$0.8995^{20/4}$	$1.4527^{20/}{}_D$		B7³, 151
8522	α-Isomorphine $C_{17}H_{19}NO_3$	285.34	nd (MeOH-AcOEt) $[\alpha]^{15/}{}_D$ −167 (MeOH,c=3)		248			al, MeOH	B27², 174
8523	Isonicotine or 4-(4-pyridyl) piperidine $C_{10}H_{14}N_2$	162.23	hyg wh nd	292	80			al, eth, bz, lig	B23, 119
8524	Isonicotinaldehyde or 4-Pyridine carboxaldehyde $4\text{-}C_5H_4NCHO$	107.11		$77\text{-}8^{12}$			1.5423^{20}	w, al, aa, chl	B21⁴, 3529
8525	Isonicotinic acid or 4-Pyridine carboxylic acid $4\text{-}C_5H_4NCO_2H$	123.11	nd (w)	sub 260^{15}	319				B22⁴, 518
8526	Isonicotinic acid-2,6-dihydro or Citrazinic acid $4\text{-}C_5H_6NCO_2H$	155.11	yesh-gr pw (w)		>330d				B22⁴, 2459
8527	Isonicotinic acid, ethyl betaine $C_8H_9NO_2$	151.16	nd		241d			w, al	B22, 47
8528	Isonicotinic acid, ethyl ester or Ethylisonicotinate $4\text{-}C_5H_4NCO_2H_5$	151.16	nd	220, 110^{15}	23	1.1052^{20}	1.5177^{20}	al, eth, bz, chl	B22⁴, 521
8530	Isonicotinic acid, hydrazide or Isoniazid $4\text{-}C_5H_4NCONHNH_2$	137.14	nd (al)		171				B22, 47
8531	Isonicotinic acid, methyl ester or Methyl isonicotinate $4\text{-}C_5H_4NCO_2CH_3$	137.14		209d, 104^{21}	8.5	$1.1599^{20/4}$	1.5315^{20}	al, eth, bz	B22⁴, 545
8532	Isonicotinonitrile $4\text{-}C_5H_4NCN$	104.11	nd (liq-eth)		83			w, al, eth, bz	B22⁴, 520
8533	Isopapaverine, N-benzyl $C_{27}H_{27}NO_4$	429.52	ye lf (al)		139-40			eth	B22⁴, 542
8534	Isopapaverine, N-ethyl $C_{22}H_{25}NO_4$	367.44	pr(al)		ca 101			eth, bz	B21, 229
8535	Isopapaverine, N-methyl $C_{21}H_{23}NO_4$	353.42	ye hg mcl pr (al)		129-31			w	B21⁴, 2799
8536	Isopelletriene or 2-Acetonyl piperidine $C_8H_{15}NO$	141.21	oil		$91\text{-}2^{14}$	$0.9624^{20/4}$	1.4683^{20}	al, chl	B21⁴, 3264
8537	Isopelletriene, N-methyl $C_9H_{17}NO$	155.24		$96\text{-}8^{13}$		$0.9478^{20/4}$	1.4674^{20}	w, lig	B21⁴, 3265
8538	Isophenolphthalein or 3-(*o*-hydroxy phenyl)-3-(*p*-hydroxy phenyl) phthalide $C_{20}H_{14}O_4$	318.33	(dil aa)		189-90			al, ace	B10³, 2013
8539	3-Isophenothiazin, 3-one or Azthione thiazone $C_{12}H_7NO$	181.19	red (dil al)		164			al, bz, chl	B27¹, 251
8540	3-Isophenothiazin-3-one, 7-hydroxy or Thionol $C_{12}H_7NO_2S$	229.25	red br pw or nd (aa)		>360			al	B27², 109
8541	Isophorone or 3,5,5-trimethyl-2-cyclohexene-1-one $C_9H_{14}O$	138.21		214^{754}, 99^{18}		0.9229^{20}	1.4759^{20}	al, eth, ace	B7³, 283
8542	Isophthaldehyde or 1,3-Benzene dicarboxaldehyde $1,3\text{-}(OHC)_2C_6H_4$	134.13	nd(dil al)	$245\text{-}8^{77}$	89-90			al, ace, bz	B7³, 3459
8543	Isophthalamide $1,3\text{-}(H_2NOC)_2C_6H_4$	164.16	pl (w)		280				B9¹, 372
8544	Isophthalamide, N,N,N′,N′-tetraethyl $1,3\text{-}C_6H_4[CON(C_2H_5)_2]_2$	276.38		242^{12}	85			al, eth, ace, bz	B9³, 4242
8545	Isophthalic acid or 1,3-Benzene dicarboxylic acid $1,3\text{-}C_6H_4(CO_2H)_2$	166.13	nd (w or al)	sub	348			al, aa	B9³, 4240
8546	Isophthalic acid, 2-amino $2\text{-}H_2NC_6H_3(CO_2H)_2\text{-}1,3$	181.15	pl (al), nd (aa)	sub 267	>260			al, eth	B14², 337

PHYSICAL CONSTANTS OF ORGANIC COMPOUNDS (Continued)

No.	Name, Synonyms, and Formula	Mol. wt.	Color, crystalline form, specific rotation and λ_{max} (log ε)	b.p. °C	m.p. °C	Density	n_D	Solubility	Ref.
8547	Isophthalic acid, 2-amino, dimethyl ester $2\text{-}H_2NC_6H_3(CO_2CH_3)_2\text{-}1,3$	209.20	nd (al)		103-4				B14[2], 337
8548	Isophthalic acid, 4-amino $4\text{-}H_2NC_6H_3(CO_2H)\text{-}1,3$	181.15	nd (w)		336-7			al, eth, ace, aa	B14[1], 633
8549	Isophthalic acid, 4-amino, dimethyl ester $4\text{-}H_2NC_6H_3(CO_2CH_3)_2\text{-}1,3$	209.20	nd(al)		131.5				B14[2], 337
8550	Isophthalic acid, 4-amino $5\text{-}H_2NC_6H_3(CO_2H)_2\text{-}1,3$	181.15	pr (al) pl (w)	sub	>360				B14[1], 636
8551	Isophthalic acid, 5-amino, dimethyl ester $5\text{-}H_2NC_6H_3(CO_2CH_3)_2\text{-}1,3$	209.20	lf or pl (MeOH)		176			eth	B14, 556
8552	Isophthalic acid, 4-bromo $4\text{-}BrC_6H_3(CO_2H)_2\text{-}1,3$	245.03	nd (al)		287			al	B9[3], 4247
8553	Isophthalic acid, 4-chloro $4\text{-}ClC_6H_3(CO_2H)_2\text{-}1,3$	200.58	nd (w)		295			al	B9[3], 4245
8554	Isophthalic acid, 5-chloro $5\text{-}ClC_6H_3(CO_2H)_2\text{-}1,3$	200.58	nd (w + ½)		278 (anh)			al	B9[3], 4246
8555	Isophthalic acid, 4,6-dichloro $4,6\text{-}Cl_2C_6H_2(CO_2H)_2\text{-}1,3$	235.02	nd (w, dil al)		280			al, eth, chl	B9[3], 4246
8556	Isophthalic acid, 4,5-dimethoxy or Isohemipinic acid $4,5\text{-}(CH_3O)_2C_6H_2(CO_2H)_2\text{-}1,3$	226.19	nd (w)		245-6			al, eth	B10[3], 2435
8557	Isophthalic acid, diethyl ester or Diethyl isophthalate $1,3\text{-}C_6H_4(CO_2C_2H_5)_2$	222.24		302, $170^{2.4}$	11.5	$1.1239^{17/4}$	1.508^{18}		B9[3], 4241
8558	Isophthalic acid, dimethyl ester or Dimethylisophthalate $1,3\text{-}C_6H_4(CO_2CH_3)_2$	194.19	nd (dil al)	282, 124^{12}	67-8	$1.194^{20/4}$	1.5168^{20}		B9[3], 4241
8559	Isophthalic acid, 4,6-dimethyl or α-Cumidic acid $4,6\text{-}(CH_3)_2C_6H_2(CO_2H)_2\text{-}1,3$	194.19	nd (w) pr (al-bz) lf (sub)	sub	266			al	B9[3], 4298
8560	Isophthalic acid, 2-hydroxy $2\text{-}HOC_6H_3(CO_2H)_2\text{-}1,3$	182.13	nd (w + 1)		244-5			al, eth, chl	B10[3], 2192
8561	Isophthalic acid, 4-hydroxy $4\text{-}HOC_6H_3(CO_2H)_2\text{-}1,3$	182.13	nd (w) lf (dil al)		310			al, eth	B10[3], 2193
8562	Isophthalic acid, 5-hydroxy $5\text{-}HOC_6H_3(CO_2H)_2\text{-}1,3$	182.13	nd (w + 2) cr (aq al)	sub	293			al, eth, bz	B10[3], 2195
8563	Isophthalic acid, 5-methyl or Uvitic acid $5\text{-}CH_3C_6H_3(CO_2H)_2\text{-}1,3$	180.16	nd (w)		298			al, eth, ace	B9[3], 4274
8564	Isophthalic acid, 2-nitro, dimethyl ester $2\text{-}NO_2\text{-}C_6H_3(CO_2CH_3)_2\text{-}1,3$	239.18	nd (w or al)		135			al, eth	B9[1], 373
8565	Isophthalic acid, 5-nitro $5\text{-}NO_2C_6H_3(CO_2H)_2\text{-}1,3$	211.13	gr lf (+ 3/2w)		260-1 (anh)			al, eth	B9[1], 373
8566	Isophthalic acid, 5-nitro, diethyl ester $5\text{-}NO_2C_6H_3(CO_2C_2H_5)_2\text{-}1,3$	267.24	nd (al)		83.5			al, eth	B9, 840
8567	Isophthalic acid, 5-nitro, dimethyl ester $5\text{-}NO_2C_6H_3(CO_2CH_3)_2\text{-}1,3$	239.18	nd (dil al)		123			al, eth	B9[2], 611
8568	Isophthalic acid, tetrabromo $1,3\text{-}(HO_2C)_2C_6Br_4$	481.71	nd (w)		288-92				B9, 839
8569	Isophthalonitrile $1,3\text{-}C_6H_4(CN)_2$	128.13	nd (al)		162			al, eth, bz, chl	B9[3], 4243
8570	Isophthalylalcohol or *m*-Xylylene glycol $1,3\text{-}(HOCH_2)_2C_6H_4$	138.17	nd (bz)	$154\text{-}9^{13}$	57	1.1359^{53}		w, al, eth	B6[3], 4600
8571	Isophthalyl chloride $1,3\text{-}C_6H_4(COCl)_2$	203.02	pr (eth)	276	43-4	$1.3880^{17/4}$	1.570^{47}	eth	B9[3], 4242
8572	Isopilocarpine or N-Methyl isopilocarpidine $C_{11}H_{16}N_2O_2$	208.26	pr	261^{10}				w, al, eth, bz, chl	B27[2], 697
8573	Isopimpinellin $C_{13}H_{10}O_5$	246.22	ye nd (MeOH)		151			MeOH	B19[4], 2811
8574	Isopomiferin $C_{25}H_{24}O_6$	420.46	nd (dil al)		265d			al	B19[4], 5231
8575	Isoprene or 2-Methyl-1,3-butadiene $CH_2{=}C(CH_3)\text{-}CH{=}CH_2$	68.12		34	−146	$0.6810^{20/4}$	1.4219^{20}	al, eth, ace, bz	B1[4], 1001
8576	Isopropenyl methyl ketone or 3-Methyl-3-buten-2- one $CH_2{=}C(CH_3)COCH_3$	84.12		98	-54	$0.8527^{20/4}$	1.4220^{20}	al	B1[4], 3462
8577	Isopulegol (d) or $\Delta^{8(9)}$-p-Menthenol-3 $C_{10}H_{18}O$	154.25	$[\alpha]_{5461}$ + 29.3	212, $93\text{-}4^{14}$		$0.9110^{20/4}$	1.4723^{20}	al, eth	B6[3], 257
8578	Isopulegol (l) $C_{10}H_{18}O$	154.25	$[\alpha]^{20}_{5461}$ −25.9	212, 94^{14}		$0.9110^{20/4}$	1.4723^{20}	al, eth	B6[3], 257

PHYSICAL CONSTANTS OF ORGANIC COMPOUNDS (Continued)

No.	Name, Synonyms, and Formula	Mol. wt.	Color, crystalline form, specific rotation and λ_{max} (log ε)	b.p. °C	m.p. °C	Density	n_D	Solubility	Ref.
8579	α-Isoquinine $C_{20}H_{24}N_2O_2$	324.42	(bz-peth) $[\alpha]^{18}_D$ −245 (al,c=1)		196.5			al, eth	B23^{2}, 414, 423
8580	β-Isoquinine $C_{20}H_{24}N_2O_2$	324.42	pr (dil al or amor) $[\alpha]^{17}_D$ −187 (97% al,c=1)		190-1			al, bz, chl	B23^{2}, 413
8581	Isoquinoline C_9H_7N	129.16	hyg pl	242.2^{743}, 142^{40}	26.5	1.0986^{20}	1.6148^{20}	al, eth, ace, bz	B20^{4}, 3410
8582	Isoquinoline, hydrochloride $C_9H_7N.HCl$	165.62	pr or pl (al)		209			w	B20^{4}, 3412
8583	Isoquinoline, hydrogen sulfate $C_9H_7N.H_2SO_4$	227.23	pr or pl (al)		209			w	B20^{4}, 3413
8584	Isoquinoline, 1-amino 1-$H_2N(C_9H_6N)$	144.18	pl (w)		123			al	B22^{4}, 4736
8585	Isoquinoline, 3-amino 3-$H_2N(C_9H_6N)$	144.18			178-9				B22^{4}, 4744
8586	Isoquinoline, 5-amino 5-$H_2N(C_9H_6N)$	144.18	pa ye nd (peth)	sub	128				B22^{4}, 4747
8587	Isoquinoline, 4-bromo 4-$Br(C_9H_6N)$	208.06	(peth)	280-5	40-3			eth	B20^{4}, 3448
8588	Isoquinoline, 1-chloro 1-$Cl(C_9H_6N)$	163.61		274-5, 135-40^{10}	37-8			bz	B20^{4}, 3444
8589	Isoquinoline, 6,7-dimethoxy-1,2-dimethyl-1,2,3,4-tetrahydro or Carnegine $C_{13}H_{19}NO_2$	221.30	pa br syr $[\alpha]^{25}_D$ +20 (w)	170^{1}				w, al, eth, chl	B21^{4}, 2125
8590	Isoquinoline, 1-hydroxy or Isocarbostyril 1-$HO(C_9H_6N)$	145.16	mcl (bz) nd (bz al w)	240 sub	209-10			al	B21^{4}, 1205
8591	Isoquinoline, 7-hydroxy 7-$HO(C_9H_6N)$	145.16			230			al	B21^{4}, 1214
8592	Isoquinoline, 7-methoxy 7-$CH_3O(C_9H_6N)$	159.19		182-6^{34}	49			al, lig	B21^{4}, 1214
8593	Isoquinoline, 1-methyl 1-$CH_3(C_9H_6N)$	143.19		248, 124-5^{10}	10	1.0777$^{20/4}$	1.6095^{20}	eth, ace, bz	B20^{4}, 3505
8594	Isoquinoline, 3-methyl 3-$CH_3(C_9H_6N)$	143.19	cr(eth)	246	68			eth, ace	B20^{4}, 3507
8595	Isoquinoline, 4-methyl 4-$CH_3(C_9H_6N)$	143.19		256				eth, ace, bz	B20^{4}, 3510
8596	Isoquinoline, 6-methyl 6-$CH_3(C_9H_6N)$	143.19	cr	265.5	85-6			al, eth, ace, bz	B20^{2}, 247
8597	Isoquinoline, 7-methyl 7-$CH_3(C_9H_6N)$	143.19		245	67-8			eth, ace	B20^{4}, 3511
8598	Isoquinoline, 8-methyl 8-$CH_3(C_9H_6N)$	143.19		258				eth, ace, bz	B20, 404
8599	Isoquinoline, 5-nitro 5-$O_2N(C_9H_6N)$	174.16	nd (w+1)	sub	110			al, eth, chl, aa	B20^{4}, 3450
8600	Isoquinoline, 1,2,3,4-tetrahydro or 2-Azatetralin $C_9H_{11}N$	133.19		232-3	<−15	1.0642$^{24/4}$	1.5668^{20}	al	B20^{4}, 2949
8601	1-Isoquinoline carboxylonitrile or 1-Cyanoisoquinoline 1-$NC(C_9H_6N)$	154.17	nd (peth MeOH)		78(93)			al, eth, bz	B22^{4}, 1205
8602	5-Isoquinoline carboxylonitrile or 5-Cyanoisoquinoline 5-$NC(C_9H_6N)$	154.17	nd (w or dil al)	sub 100-120	135			al, eth	B22^{4}, 1208
8603	Isoraunescine $C_{31}H_{36}N_2O_8$	564.64	wh nd		241-2			chl, aa	C50, 1267
8604	Isoreserpilline $C_{23}H_{28}N_2O_5$	412.49	wh pr $[\alpha]^{20}_D$ −84 (py)		210-2				C53, 14134
8605	Isorubijervine or Δ^5-β-18-Dihydroxy solanidene $C_{27}H_{43}NO_2$	413.64	dr (al) $[\alpha]_D$ +9.2 (al)		241-4			bz, chl	B21^{4}, 2312
8606	Isorubijervosine $C_{33}H_{53}NO_7$	575.79	wh nd $[\alpha]^{24}_D$ −20 (py,c=1.45)		279-80				B21^{4}, 2313
8607	Isosaccharic acid or 3,4-Dihydroxytetrahydro-2,5-furandicarboxylic acid $C_6H_8O_7$	192.13	rh $[\alpha]^{20}_D$ +46.7 (w,p=4.2)	d	185			w, al	B18^{2}, 309

No.	Name, Synonyms, and Formula	Mol. wt.	Color, crystalline form, specific rotation and λ_{max} (log ε)	b.p. °C	m.p. °C	Density	n_D	Solubility	Ref.
8608	Isosafrole *(trans)* $C_{10}H_{10}O_2$	162.19		253, 111-2[6]	fp 6.8	$1.1224^{20/4}$	1.5782^{20}	al, eth, ace, bz, chl	B19[4], 273
8609	Isoserine (l) $H_2NCH_2CH(OH)CO_2H$	105.09	cr (dil al w) $[\alpha]^{20}_D$ −32.6		199-201d			w	B4[3], 1566
8610	Isothebaine (d) $C_{19}H_{21}NO_3$	311.38	rh cr (al) $[\alpha]^{18}_D$ +285 (al,c=2)		203-4			al, chl	B21[4], 2646
8611	Isothebaine, sulfate $(C_{19}H_{21}NO_3)_2{,}H_2SO_4$	720.83	nd		120-1d			w	B21[1], 250
8612	Isovaleric acid or 3-Methylbutyric acid $(CH_3)_2CHCH_2CO_2H$	102.13		176.7	−29.3	$0.9286^{20/4}$	1.4033^{20}	al, eth, chl	B2[4], 895
8613	Isovalerophenone or Isobutyl phenyl ketone $(CH_3)_2CHCH_2COC_6H_5$	162.23		236.5, 137-8[38]		$0.9701^{16.4/4}$	$1.5139^{15.3}$	al, eth, ace	B7[3], 1121
8614	Isovaline (d) or 2-Amino-2-methyl butyric acid $CH_3CH_2C(CH_3)(NH_2)CO_2H$	117.15	nd (aq al) $[\alpha]^{t}_D$ +13 (w,c=2)	sub	ca 300			w	B7[3], 1361
8615	Isovaline (dl) or 2-Amino-2-methyl butyric acid $CH_3CH_2CH(CH_3)(NH_2)CO_2H$	117.15	rh nd (al-eth) mcl pr	sub 300	315 (sealed tube)			w, al	B4[3], 1361
8616	Isovaline (l) or 2-Amino-2-methyl butyric acid $CH_3CH_2CH(CH_3)(NH_2)CO_2H$	117.15	lo nd (w, ace) $[\alpha]^{20}_D$ −9.1 (w,c=2)					w, al	B4[2], 851
8617	Isoxanthen-3-one, 9-phenyl-2,6,7-trihydroxy or 9-Phenyl-2,6,7-trihydroxy-fluorone $C_{19}H_{12}O_5$	320.30	og red (al-HCl)		>300				B18[4], 2824
8618	Isoxazole C_3H_3NO	69.06		95-6		$1.078^{20/4}$	1.4298^{17}		B27[2], 9
8619	Isoxazole, 5-methyl 5-$CH_3(C_3H_2NO)$	83.09		122		$1.4386^{20/0}$			B27[2], 9
8620	Itaconic acid or Methylene succinic acid $CH_2{=}C(CO_2H)CH_2CO_2H$	130.10	rh (bz)	d	175	1.632		w, al, ace, chl	B2[4], 2228
8621	Itaconic anhydride or Methylene succinic anhydride $C_5H_4O_3$	112.08	rh bipym pr (eth chl)	139-40[30], 114-5[18]	68-70			chl	B17[4], 5913
8622	Itaconic acid, diethyl ester or Diethylitaconate $CH_2{=}C(CO_2C_2H_5)CH_2CO_2C_2H_5$	186.21		228, 111[13]	58-9	$1.0467^{20/4}$	1.4377^{20}	al, eth, ace, bz	B2[4], 2230
8623	Itaconic acid, dimethyl ester or Dimethylitaconate $CH_2{=}C(CO_2CH_3)CH_2CO_2CH_3$	158.15	hyg mcl (MeOH)	208, 108[11]	38	$1.1241^{18/4}$	1.4457^{20}	al, eth, ace	B2[4], 2229
8624	Iticonyl chloride or Methylene succienyl chloride $CH_2{=}C(COCl)CH_2COCl$	166.99		89[17]			1.4919^{20}	ace	B2[3], 1934

No.	Name, Synonyms, and Formula	Mol. wt.	Color, crystalline form, specific rotation and λ_{max} (log ε)	b.p. °C	m.p. °C	Density	n_D	Solubility	Ref.
8625	Jacareubin $C_{18}H_{14}O_6$	326.33	ye pr (MeOH)		256-7d			al, ace	B19[4], 3041
8626	Jaconecic acid $C_{10}H_{16}O_6$	232.23	nd (eth) $[\alpha]^{25}/_D$ +28.1 (95% al)		183-4				B18[4], 5079
8627	Japaconitine-A or Acetyl benzoyl aconitine $C_{34}H_{47}NO_{11}$	645.75	rh $[\alpha]^{25}/_D$ +20.7 (chl)		202-3d			ace, chl	B21[4], 2901
8628	Japaconitine-Al $C_{34}H_{47}NO_{11}$	645.75	rh (MeOH) $[\alpha]^{27}/_D$ +26.4 (chl)		208-9			al, eth, chl	Ber57, 1462
8629	Japaconitine-B $C_{34}H_{47}NO_{11}$	645.76	rh (MeOH) $[\alpha]^{17}/_D$ +26.9		208-9d			al, eth, chl	Ber57, 1462
8630	Jasmin aldehyde $C_5H_{11}C(CHO)=CHC_6H_5$	203.30		$174\text{-}5^{20}$, 140^{5}		0.9718^{20}	1.5381^{20}	al, eth	B7[3], 1517
8631	Jasmone or 3-Methyl-2-(2-pentenyl)-2-cyclopenten-1-one $C_{11}H_{16}O$	164.25	ye oil	$257\text{-}8^{755}$, $134\text{-}5^{12}$		$0.9437^{22/4}$	1.4979^{22}	al, eth, lig	B7[3], 601
8632	Javanicin $C_{15}H_{14}O_6$	290.27	red (al)		208d				B8[3], 4231
8633	Jervine $C_{27}H_{39}NO_3.2H_2O$	461.64	nd (w + 2) (MeOH w) $[\alpha]^{23}/_D$ −158.5 (al,c=0.99)		243-4d			al, ace, chl	Am73, 2970
8634	Julolidine $C_{12}H_{15}N$	173.26		280d, $155\text{-}6^{17}$	40	1.003^{20}	1.568^{25}		B20[4], 3281
8635	Julolidine-1,6-dioxo or 1,6-Diketo julolidine $C_{12}H_{11}NO_2$	201.22	ye (al)	$190\text{-}210^{0.3}$	145-6			al, MeOH	B21[4], 5525
8636	Junipal or 5-(α-propynyl)-2-formyl thiophene C_8H_6OS	150.20	nd (peth or dil al)		80				B17[4], 4946
8637	Juniperol or Macrocarpol $C_{15}H_{24}O$	220.35	tcl (al) $[\alpha]^{20}/_D$ +25.4 (al)	286-8d	112	$1.0460^{20/20}$	1.519		B6[3], 426
8638	Junipic acid or 5-(α-propynyl)-2-thiophencarboxylic acid $C_8H_6O_2S$	166.19	ye nd (peth) cr (aq al)		180			al	B18[4], 4198
8639	Junipic acid, methyl ester or Methyl janipate $C_9H_8O_2S$	180.22	(aq MeOH)	sub 50-5	62			ace	B18[4], 4199

No.	Name, Synonyms, and Formula	Mol. wt.	Color, crystalline form, specific rotation and λ_{max} (log ε)	b.p. °C	m.p. °C	Density	n_D	Solubility	Ref.
8640	Ketene $CH_2=CO$	42.04		−56	−151				$B1^4$, 3418
8641	Ketene, diethyl acetal or 1,1-diethoxyethylene $CH_2=C(OC_2H_5)_2$	116.16		68^{100}		$0.7932^{20/4}$	1.3643^{21}		$B1^4$, 3420
8642	Ketene, dimethyl $(CH_3)_2C=C=O$	70.09	ye	34	−97.5				$B1^4$, 3453
8643	Ketene, diphenyl $(C_6H_5)_2C=C=O$	194.23	ye-red	265-70d, 146^{12}		$1.1107^{14/4}$	1.615^{14}	eth, bz	$B7^3$, 2356
8644	Ketene, methyl $CH_3CH=C=O$	56.06			−80			eth	$B1^4$, 3433
8645	Khellin $C_{14}H_{12}O_5$	260.25	(MeOH or eth)	$180\text{-}200^{0.05}$	154-5 (vac)			ace, MeOH	$B19^4$, 2816
8646	Kynurenine or 3-Anthranyloyl alanine $C_{10}H_{12}N_2O_3$	208.22	lf (+ ½w) $[\alpha]^{20}{}_D$ −29 (w,c=4)		191d				$B14^4$, 1656

PHYSICAL CONSTANTS OF ORGANIC COMPOUNDS (Continued)

No.	Name, Synonyms, and Formula	Mol. wt.	Color, crystalline form, specific rotation and λ_{max} (log ε)	b.p. °C	m.p. °C	Density	n_D	Solubility	Ref.
8647	Lactamide (d) $CH_3CH(OH)CONH_2$	89.09	cr (AcOEt) $[\alpha]^{18/}_{578}$		49-51			w, al	$B3^{3}$, 450
8648	Lactamide (dl) $CH_3CH(OH)CONH_2$	89.09	pl (AcOEt)		75.5	$1.1381^{80/4}$		w, al	$B3^{4}$, 674
8649	Lactamide, N-(4-ethoxyphenyl) or N-Lactyl-β-phenetidide $CH_3CH(OH)CONH(C_6H_4OC_2H_5\text{-}4)$	209.25	nd (w)		118			al	$B13^{3}$, 1135
8650	D-Lactic acid or D-2-Hydroxypropionic acid $CH_3CH(OH)CO_2H$	90.08	pl (chl aa) $[\alpha]^{t}_{D}$ −2.26 (w,c=1.24)	103^{2}	53			w, al	$B3^{4}$, 633
8651	DL-Lactic acid $CH_3CH(OH)CO_2H$	90.08	ye	122^{15}	18	$1.2060^{21/4}$	1.4392^{20}	w, al, eth	$B3^{4}$, 633
8652	L-Lactic acid $CH_3CH(OH)CO_2H$	90.08	hyg pr (eth), $[\alpha]^{15/}_{D}$ +3.8 (w, c=10.5)					w, al	$B3^{4}$, 633
8653	DL-Lactic acid acetate $CH_3CO_2CH(CH_3)CO_2H$	132.13	dlq	$167\text{-}70^{78}$, 127^{11}	57-60	$1.1758^{20/4}$	1.4240^{20}	al, bz	$B3^{4}$, 638
8654	DL-Lactic acid, allyl ester or Allyl lactate $CH_3CH(OH)CO_2CH_2CH=CH_2$	130.14		$56\text{-}60^{8}$		$1.0452^{20/4}$	1.4369^{20}	py	$B3^{3}$, 486
8655	DL-Lactic acid anhydride $[CH_3CH(OH)CO]_2O$	162.14	pa ye amor or syr	250d				al, eth	$B3^{3}$, 494
8656	D-Lactic acid, butyl ester or D-Butyl lactate $CH_3CH(OH)CO_2C_4H_9$	146.19	$[\alpha]^{27/}_{D}$ +13.6	77^{10}		$0.9744^{27/4}$		al, eth	$B3^{2}$, 188
8657	DL-Lactic acid, butyl ester or DL-Butyl lactate $CH_3CH(OH)CO_2C_4H_9$	146.19		83^{13}	−49	$0.9807^{22/4}$	1.4217^{20}	al, eth	$B3^{4}$, 649
8658	D-Lactic acid, ethyl ester or D-Ethyl lactate $CH_3CH(OH)CO_2C_2H_5$	118.13	$[\alpha]^{19/}_{D}$ +14.5	58^{20}		$1.0324^{20.4/4}$	1.4125^{20}	w, al, eth	$B3^{3}$, 449
8659	DL-Lactic acid, ethyl ester $CH_3CH(OH)CO_2C_2H_5$	118.13		154.5, 58^{19}		$1.0302^{20/4}$	1.4124^{20}	w, al, eth	$B3^{4}$, 643
8660	L-Lactic acid, ethyl ester $CH_3CH(OH)CO_2C_2H_5$	118.13	$[\alpha]^{19/}_{D}$ -11.3	$69\text{-}70^{36}$		$1.0314^{20/4}$	1.4156^{20}	w, al, eth	$B3^{2}$, 446
8661	D-Lactic acid, methyl ester or D-Methyl lactate $CH_3CH(OH)CO_2CH_3$	104.11	$[\alpha]^{20/}_{D}$ +7.5	40^{11}		$1.0857^{25/4}$		w, al, eth	$B3^{3}$, 449
8662	DL-Lactic acid, Methyl ester $CH_3CH(OH)CO_2CH_3$	104.11		144.8		$1.0928^{20/4}$	1.4141^{20}	w, al, eth	$B3^{4}$, 640
8663	L-Lactic acid, Methyl ester $CH_3CH(OH)CO_2CH_3$	104.11	$[\alpha]^{20/}_{D}$ -8.3	58^{19}		$1.0895^{20/4}$	1.4139^{20}	w, al, eth	$B3^{3}$, 445
8664	Lactic acid, isopentyl ester or Isopentyl lactate $CH_3CH(OH)CO_2\text{-i-}C_5H_{11}$	160.21		202.4, 82^{7}		$0.9617^{25/25}$	1.4240^{25}	al, eth	$B3^{3}$, 483
8665	D-Lactic acid phenyl or Atrolatic acid $CH_3C(OH)(C_6H_5)CO_2H$	166.18	pr (w), $[\alpha]^{16.5/}_{D}$ +37.7 (al, c=3.5)		116-7			al, ace, bz	$B10^{3}$, 560
8666	DL-Lactic acid, 2-phenyl $CH_3C(OH)(C_6H_5)CO_2H$	166.18	nd pl(liq)		93-5			w, al, ace, bz	$B10^{3}$, 560
8667	L-Lactic acid, 2-phenyl $CH_3C(OH)(C_6H_5)CO_2H$	166.18	nd (bz, w), $[\alpha]^{13.8/}_{D}$ -37.7 (al, c=3.4)		116-7			w, al, ace, bz	$B10^{3}$, 560
8668	Lactic acid, 3-phenyl (d) or α-Hydroxy hydrocinnamic acid $C_6H_5CH_2CH(OH)CO_2H$	166.18	nd (w), $[\alpha]^{20/}_{D}$ +22.2 (w, c=2.2)		124-6			w, al, ace	$B10^{3}$, 554
8669	Lactic acid, 3-phenyl (dl) $C_6H_5CH_2CH(OH)CO_2H$	166.18	cr(chl bz) pr (w)	$148\text{-}50^{15}$	98			al, eth, ace	$B10^{3}$, 554
8670	Lactic acid, 3-phenyl (l) $C_6H_5CH_2CH(OH)CO_2H$	166.18	nd (w), $[\alpha]^{20/}_{D}$ -19.9 (w, c=3.2)		124-5			al, eth, ace	$B10^{3}$, 554
8671	Lactic acid, isopropyl ester or Isopropyl lactate $CH_3CH(OH)CO_2CH(CH_3)_2$	132.16		166-8, $75\text{-}80^{12}$		$0.9980^{20/4}$	1.4082^{25}	w, al, eth, bz	$B3^{3}$, 479
8672	Lactic acid, 3,3,3-trichloro $CCl_3CH(OH)CO_2H$	193.41	pr (eth)	$140\text{-}70^{45}$	125			w, al, eth, chl	$B3^{4}$, 680

No.	Name, Synonyms, and Formula	Mol. wt.	Color, crystalline form, specific rotation and λ_{max} (log ε)	b.p. °C	m.p. °C	Density	n_D	Solubility	Ref.
8673	Lactide (d) or 2,5-Dimethyl, 3,6-dioxo-1,4-dioxane $C_6H_8O_4$	144.13	hyg rh (eth), $[\alpha]^{18/}_D$ -298 (bz, c=1.17)	150^{25}	95				$B19^4$, 1927
8674	Lactide (dl) $C_6H_8O_4$	144.13	pa ye tcl pr or nd (al)	255^{757}, $138-42^8$	124.5	$0.862^{10/4}$		ace, bz	$B19^4$, 1927
8675	Lactide (l) $C_6H_8O_4$	144.13	rh (eth), $[\alpha]^{26/}_D$ + 281.6 (bz, c=0.82)	150^{25}	95				$B19^4$, 1927
8676	D-Lactobionic acid $C_{12}H_{22}O_{12}$	358.30	syr					w	$B17^4$, 3392
8677	D-Lactonitrile or Acetaldehyde cyanohydrin $CH_3CH(OH)CN$	71.08	ye lig	182-4d, 102^{30}	−40	$0.9877^{20/4}$	1.4058^{18}	w, al, eth	$B3^4$, 675
8678	D-Lactonitrile, acetate $CH_3CO_2CH(CH_3)CN$	113.12		172-3, $76-7^{25}$		$1.0278^{20/4}$	1.4027^{20}	w	$B3^4$, 676
8679	D-Lactonitrile-3,3,3-trichloro $Cl_3CCH(OH)CN$	174.41	pl(w)	215-20d	61			w, al, eth	$B3^4$, 680
8680	Lactose or Milk sugar (α-anomer) $C_{12}H_{22}O_{11}$	342.30	pw, $[\alpha]^{20/}_D$ + 92.6→ + 52.3 (w, c=4.5)		222.8			w	$B17^4$, 3066
8681	Lactose (β-anomer) $C_{12}H_{22}O_{11}$	342.30	$[\alpha]^{20/}_D$ + 34.2→ + 52.3 (mut, w)		253	1.59^{20}		w	$B17^4$, 3068
8682	Lactose, monohydrate (α-anomer) $C_{12}H_{22}O_{11}.H_2O$	360.32	mcl (w), $[\alpha]^{20/}_D$ + 83.5 (w, 10min)	d	201-2	1.525^{20}		w	$B17^4$, 3067
8683	Lactyl chloride, acetate (dl) $CH_3CO_2CH(CH_3)COCl$	150.56		150d, 56^{11}		$1.1920^{17/4}$	1.4241^{17}		$B3^4$, 674
8684	Lactyl chloride, acetate (d+) $CH_3CO_2CH(CH_3)COCl$	150.56	$[\alpha]^{18/}_{578}$ + 32.4	$51-3^{11}$		1.177^{20}			$B3^2$, 189
8685	Lanosterol or Isocholesterol $C_{30}H_{50}O$	426.73	nd (eth) cr (MeOH-ace), $[\alpha]^{20/}_D$ + 62 (chl, c=1)		140-1			al, eth, chl	$B6^3$, 2880
8686	Lanosterol, benzoate or Isocholesterol benzoate $C_{34}H_{50}O_2$	490.77	pw or nd (eth), $[\alpha]^{17/}_D$ + 72.2		191.5			al, eth	$B9^3$, 485
8687	Lanthionine (D) or d-β,β-Thio di-alanine $[HO_2CCH(NH_2)CH_2]_2S$	208.23	hex pl, $[\alpha]^{21/}_D$ -8.0 (2.4N NaOH, c=5)		293-5d darkens 245				$B4^3$, 1618
8688	Lanthionine (DL) $[HO_2CCH(NH_2)CH_2]_2S$	208.23	hex pl		282-95d chars 240				$B4^3$, 1620
8689	Lanthionine (L) $[HO_2CCH(NH_2)CH_2]_2S$	208.23	hex pl, $[\alpha]^{22/}_D$ + 8.6 (2.4N NaOH, c=5)		293-5d darkens 245				$B4^3$, 1593
8690	Lanthionine *(meso)* $[HO_2CCH(NH_2)]_2S$	208.23	hex pl (aq NH_3)		304d softens 270				$B4^3$, 1620
8691	Lanthopine $C_{23}H_{25}NO_4$	379.46	pw		200			chl	C20, 2715
8692	Lapachol or Taiguic acid $C_{15}H_{14}O_3$	242.27	ye pr (eth,bz) pl (aa,al)		139-40			al, eth, bz, aa	$B8^2$, 365
8693	Lapachol-δ-hydroxy $C_{15}H_{14}O_4$	258.27	ye nd (bz,w)		127			al, eth, bz	$B8^3$, 3661

No.	Name, Synonyms, and Formula	Mol. wt.	Color, crystalline form, specific rotation and λ_{max} (log ε)	b.p. °C	m.p. °C	Density	n_D	Solubility	Ref.
8694	Lappaconitine $C_{32}H_{44}N_2O_9$	600.71	hex pl (al), $[\alpha]^{18}{}_D$ + 27 (chl)	....	223	....	....	bz, chl	$B21^4$, 2850
8695	Laudanidine (d) $C_{20}H_{25}NO_4$	343.42	(MeOH), $[\alpha]^{17}{}_D$ + 93.5 (chl, c=1)	....	184-5	....	....	....	$B21^4$, 2703
8696	Laudanidine (l) or Tritopine $C_{20}H_{25}NO_4$	343.42	hex pr (al), $[\alpha]^{17}{}_D$ -94.8 (chl, c=2)	....	184-5	....	....	w, bz	$B21^4$, 2703
8697	Laudanine (dl) $C_{20}H_{25}NO_4$	343.42	ye wh pr (dil al or al-chl)	....	167	$1.26^{20/4}$	....	al, bz, chl	$B21^4$, 2703
8698	Laudanosine (d) or N-Methyl tetrahydro papaverine $C_{21}H_{27}NO_4$	357.45	nd (peth) pr (al), $[\alpha]^{16}{}_D$ + 106 (al, c=1.6)	....	89	....	....	al, eth, ace, chl	$B21^4$, 2704
8699	Laudanosine (dl) $C_{21}H_{27}NO_4$	357.45	nd(al)	....	115-6	....	....	al, eth, ace, bz, chl	$B21^4$, 2704
8700	Laudanosine (l) $C_{21}H_{27}NO_4$	357.45	cr (al), $[\alpha]^{15}{}_D$ -105.4 (al, c=31)	....	89	....	....	....	$B21^4$, 2704
8701	Laureline $C_{19}H_{19}NO_3$	309.36	ta (al) cubes (peth), $[\alpha]^{18}{}_D$ -99 (abs al,c=0.7)	....	114	....	....	al, eth	$B27^1$, 461
8702	Lauraldehyde or Dodecanal $CH_3(CH_2)_{10}CHO$	184.32	lf	185^{100}, $100^{3.5}$	44.5	$0.8352^{15/4}$	1.435^{22}	al, eth	$B1^4$, 3380
8703	Lauraldehyde, dimethylacetal $CH_3(CH_2)_{10}CH(OCH_3)_2$	230.39	....	$132\text{-}4^5$	....	....	1.4310^{25}	al, eth	$B1^4$, 3381
8704	Lauramide $CH_3(CH_2)_{10}CONH_2$	199.34	nd	199^{12}	110	....	1.4287^{110}	al, ace	$B2^4$, 1103
8705	Lauramide, N-Phenyl $CH_3(CH_2)_{10}CONHC_6H_5$	275.43	nd (dil al)	....	78	....	....	al, eth, ace, bz, chl	$B12^3$, 485
8706	Lauric acid or Dodecanoic acid $CH_3(CH_2)_{10}CO_2H$	200.32	nd (al)	131^1	44	$0.8679^{50/4}$	1.4304^{50}	al, eth, ace, bz, peth	$B2^4$, 1082
8707	Lauric anhydride $(C_{11}H_{23}CO)_2O$	382.63	lf (al,eth)	....	41.8	$0.8533^{70/4}$	1.4292^{70}	al	$B2^4$, 1100
8708	Lauric acid, benzyl ester or Benzyl laurate $C_{11}H_{23}CO_2CH_2C_6H_5$	290.45	....	$209\text{-}11^{12}$	8.5	$0.9457^{25/25}$	1.4812^{24}	al, eth, bz, chl, peth	$B6^4$, 2267
8709	Lauric acid, 2-bromo $CH_3(CH_2)_9CHBrCO_2H$	279.22	pl	$157\text{-}9^2$	32	1.1474^{24}	1.4585^{24}	al, eth, bz, chl, lig	$B2^4$, 1106
8710	Lauric acid, ethyl ester or Ethyl laurate $C_{11}H_{23}CO_2C_2H_5$	228.38	....	273^{764}, 154^{15}	fr−1.8	$0.8618^{20/4}$	1.4311^{20}	al, eth	$B2^4$, 1092
8711	Lauric acid, 12-fluoro $F(CH_2)_{11}CO_2H$	218.31	....	....	60-1	....	....	al, eth	$B2^4$, 1105
8712	Lauric acid, methyl ester or Methyl laurate $C_{11}H_{23}CO_2CH_3$	214.35	....	262^{766}, 141^{15}	fr 5.2	$0.8702^{20/4}$	1.4319^{20}	al, eth, ace, bz	$B2^4$, 1090
8713	Lauric acid, phenyl ester or Phenyl laurate $C_{11}H_{23}CO_2C_6H_5$	276.42	lf (al)	210^{15}	24.5	....	....	al, eth, ace	$B6^4$, 618
8714	Lauric acid, propyl ester or Propyl laurate $C_{11}H_{23}CO_2C_3H_7$	242.40	....	205^{60}, 124^2	....	0.8600^{20}	1.4335^{20}	....	$B2^4$, 1092
8715	Lauric acid, isopropyl ester or *iso*-Propyl laurate $C_{11}H_{23}CO_2CH(CH_3)_2$	242.40	....	196^{60}, 117^2	....	0.8536^{20}	1.4280^{25}	al, eth	$B2^4$, 1092
8716	Laurone or 12-Tricosanone Diundecyl ketone $(C_{11}H_{23})_2CO$	338.62	lf (al)	....	69.3	$0.8086^{69/4}$	1.4283^{80}	eth, bz, chl	$B1^4$, 3408
8717	Lauronitrile $C_{11}H_{23}CN$	181.32	....	277, 131^{10}	fr 4	$0.8240^{20/4}$	1.4361^{20}	al, eth, ace, bz, chl	$B2^4$, 1104
8718	Laurophenone or n-undecyl phenyl ketone $C_{11}H_{23}COC_6H_5$	260.42	og cr	$222\text{-}3^{21}$	46-7	$0.8969^{52/4}$	1.4850^{52}	ace	$B7^3$, 1291
8719	Lauroyl peroxide $C_{11}H_{23}CO\text{-}OO\text{-}COC_{11}H_{23}$	398.63	wh pl	....	49	....	....	....	$B2^4$, 1102
8720	Lauryl alcohol or 1-Dodecanol $CH_3(CH_2)_{10}CH_2OH$	186.34	lf (dil al)	255-9, 150^{20}	26	$0.8309^{24/4}$	....	al, eth	$B1^4$, 1844

No.	Name, Synonyms, and Formula	Mol. wt.	Color, crystalline form, specific rotation and λ_{max} (log ε)	b.p. °C	m.p. °C	Density	n_D	Solubility	Ref.
8721	Lauryl amine or 1-Amino dodecane $CH_3(CH_2)_{10}CH_2NH_2$	185.35		259, 126.5^{10}	28.3	$0.8015^{20/4}$	1.4421^{20}	al, eth, bz, chl	$B4^4$, 794
8722	Lauryl amine, acetate $C_{12}H_{25}NH_2.CH_3CO_2H$	245.41			fr 69.5			w, al	$B4^4$, 797
8723	Lauryl amine, hydrochloride $C_{12}H_{25}NH_2.HCl$	221.81			98			w, al	$B4^4$, 795
8724	Lauroyl chloride $C_{11}H_{23}COCl$	218.77		145^{18}	−17		1.4458^{20}	eth	$B2^4$, 1103
8725	Lauryl mercaptan $CH_3(CH_2)_{11}SH$	202.40		$142\text{-}5^{15}$		$0.8450^{20/20}$	1.4589^{20}	al, eth	$B1^4$, 1851
8726	Lauryl sulfate $(C_{12}H_{25}O)_2SO_2$	434.72			48.5				$B1^4$, 1849
8727	Lead, tetraphenyl $Pb(C_6H_5)_4$	515.61		126^{13}	227-8				$B16^3$, 1252
8728	Lecithin $RCO_2CH_2(CHOCOR)CH_2OP(O^-)(=O)OCH_2CH_2N^+(CH_3)_3$		$[\alpha]^{24/}_D$ + 7.0		236-7			eth, chl, peth	$B4^4$, 1462
8729	Ledol $C_{15}H_{26}O$	222.37	nd (al), $[\alpha]^{20/}_D$ + 28 (chl, c=10)	292 sub	105-6 sub	$0.9094^{100/20}$	1.4667^{110}	al, eth, ace	$B6^4$, 426
8730	*p*-Leucaniline, N,N,N′,N′-tetramethyl or bis(4-dimethylaminophenyl)4-aminophenylmethane $C_{23}H_{27}N_3$	345.49	(al)		151-2				$B13^3$, 566
8731	Leucic acid (D) or 2-Hydroxy-4-methyl valeric acid $(CH_3)_2CHCH_2CH(OH)CO_2H$	132.16	nd (bz) pr (eth-peth), $[\alpha]^{125/}_D$ + 10.7 (w, c=5)		80-1			w, al, eth	$B3^4$, 850
8732	Leucic acid-(DL) $(CH_3)_2CHCH_2CH(OH)CO_2H$	132.16	pl (eth-peth)		77			w, al, eth	$B3^4$, 851
8733	Leucic aicd (l) $(CH_3)_2CHCH_2CH(OH)CO_2H$	132.16	rh (eth), $[\alpha]^{20/}_D$ -11.3 (w, c=1)		81-2			w, al, eth	$B3^4$, 851
8734	Leucin amide (dl) or α-Amino isocaproamide $(CH_3)_2CHCH_2CH(NH_2)CONH_2$	130.19	pr (bz)		106-7			w, al, ace	$B4^3$, 1434
8735	Leucine (D) or D-α-Amino isocaproic acid $(CH_3)_2CHCH_2CH(NH_2)CO_2H$	131.17	pl (al), $[\alpha]^{20/}_D$ + 10.34	sub	293 (sealed tube)				$B4^3$, 1424
8736	Leucine (DL) $(CH_3)_2CHCH_2CH(NH_2)CO_2H$	131.17	lf (w)	sub	293-5 (sealed tube)	$1.293^{18/4}$		w	$B4^3$, 1430
8737	Leucine (L) $(CH_3)_2CHCH_2CH(NH_2)CO_2H$	131.17	hex pl (dil al), $[\alpha]^{25/}_D$ -10.4 (w, p=22)	sub	293-5 (sealed tube)	$1.293^{18/4}$		w	$B4^3$, 1408
8738	Leucine, N-acetyl-(dl) $(CH_3)_2CHCH_2CH(NHCOCH_3)CO_2H$	173.21	nd (dil al)		161			al	$B4^3$, 1439
8739	Leucine, N-benzoyl $(CH_3)_2CHCH_2CH(NHCOC_6H_5)CO_2H$	235.28	nd (dil al)		137-41			w, al, eth, chl	$B9^3$, 1159
8740	Leucine, N-glycyl (dl) $(CH_3)_2CHCH_2CH(NHOCH_2NH_2)CO_2H$	188.23	tetr (dil al)		242d	1.181		w	$B4^3$, 1443
8741	Leucine, N-glycyl (l) $(CH_3)_2CHCH_2CH(NHCONH_2)CO_2H$	188.23	pl (dil al), $[\alpha]^{20/}_D$ -35.2 (w)		256d			w	$B4^3$, 1420
8742	Leucomethylene blue $C_{16}H_{19}N_3S$	285.41	ye nd (eth,al)		185			al	$B27^2$, 448
8743	Levopimaric acid, methyl ester $C_{21}H_{32}O_2$	316.48	(MeOH or eth), $[\alpha]_D$ -190.4 (al) -268 (eth)	$166\text{-}9^{0.5}$	63-4	$1.0312^{22/4}$	1.5232^{22}	al	$B9^3$, 2904
8744	Levulin or Fructosin $(C_6H_{10}O_5)n$	(162.14)n	amor, $[\alpha]$ -52.1		140-5d			w, al	B1, 925
8745	Levulinaldehyde or 4-Oxovaleraldehyde $CH_3COCH_2CH_2CHO$	100.12		186-8d, 70^{12}	<−21	$1.0184^{21/4}$	1.4257^{22}	w, al, eth, ace, bz	$B1^4$, 3659
8746	Levulinic acid or 4-Oxovalericacid $CH_3COCH_2CH_2CO_2H$	116.12	lf or pl	245-6d, $139\text{-}40^8$	37.2	$1.1335^{20/4}$	1.4396^{20}	w, al, eth	$B3^4$, 1560

No.	Name, Synonyms, and Formula	Mol. wt.	Color, crystalline form, specific rotation and λ_{max} (log ε)	b.p. °C	m.p. °C	Density	n_D	Solubility	Ref.
8747	Levulinic acid, benzyl ester or Benzyl levulinate $CH_3COCH_2CH_2CO_2CH_2C_6H_5$	206.24	....	$132-4^{2}$	......	$1.0935^{20/4}$	1.5090^{20}	to	$B6^4$, 2481
8748	Levulinic acid, butyl ester or Butyl levulinate $CH_3COCH_2CH_2CO_2C_4H_9$	172.22	....	237.8	......	$0.9735^{20/4}$	1.4290^{20}	al, eth, ace, bz	$B3^4$, 1563
8749	Levulinic acid, iso butyl ester or Butyl levulinate $CH_3COCH_2CH_2CO_2$-i-C_4H_9	172.22	....	231	......	$0.9705^{20/4}$	1.4268^{20}	al, eth, ace, bz	$B3^4$, 1563
8750	Levulinic acid, *Sec*-butyl ester or *Sec*-Butyl levulinate $CH_3COCH_2CH_2CO_2CH(CH_3)C_2H_5$	172.22	....	225.8	......	$0.9669^{20/4}$	1.4249^{20}	al, eth, ace	$B3^3$, 1221
8751	Levulinic acid, methyl ester or Methyl levulinate $CH_3COCH_2CH_2CO_2CH_3$	130.14	....	196, $85-6^{14}$	......	$1.0511^{20/4}$	1.4233^{20}	al, eth, ace, bz	$B3^4$, 1562
8752	Levulinic acid-propyl ester or Propyl levulinate $CH_3COCH_2CH_2CO_2C_3H_7$	158.20	....	221.2	......	$0.9896^{20/4}$	1.4258^{20}	al, eth, ace, bz	$B3^4$, 1563
8753	Levulinic acid-isopropyl ester or Isopropyl levulinate $CH_3COCH_2CH_2CO_2CH(CH_3)_2$	158.20	....	209.3	......	$0.9842^{20/4}$	1.4420^{20}	al, eth, ace, bz	$B3^4$, 1563
8754	Limonene (d) or 4-isopropenyl-1-methyl cyclohexene $C_{10}H_{16}$	136.24	$[\alpha]^{20/}{}_D$ +125.6 (undil)	178, 61^{12}	-74.3	$0.8411^{20/4}$	1.4730^{20}	al, eth	$B5^4$, 438
8755	Limonene (dl) or Dipentene $C_{10}H_{16}$	136.24	....	178, 64.4^{15}	-95.5	$0.8402^{21/4}$	1.4727^{20}	....	$B5^4$, 440
8756	Limonene (l) $C_{10}H_{16}$	136.24	$[\alpha]^{20/}{}_D$ -122.1 (undil)	$177-8^{755}$, 64.4^{15}	......	$0.8422^{20/4}$	1.4746^{20}	al, eth	$B5^4$, 440
8757	Linalool (d) or 3,7-Dimethyl-1,6-octadien-3-ol $CH_2=CHC(OH)(CH_3)CH_2CH_2CH=C(CH_3)_2$	154.25	$[\alpha]^{20/}{}_D$ +19.18	198-200, $87-8^{12}$	......	$0.8700^{20/4}$	1.4636^{20}	al, eth	$B1^3$, 2012
8758	Linalool, acetate (l) or Linalyl acetate, Bergamol $C_{12}H_{20}O_2$	196.29	$[\alpha]^{20/}{}_D$ -9.45	220^{762}	......	$0.8951^{20/4}$	1.4544^{21}	al, eth	$B2^4$, 204
8759	Linalool, formate or Linalyl formate $CH_2=CHC(CH_3)(O_2CH)CH_2CH_2CH=C(CH_3)_2$	182.26	....	$100-3^{10}$	......	$0.915^{25/4}$	1.456^{20}	al	$B2^4$, 35
8760	Linalool, tetrahydro-(dl) or 3,7-dimethyl-3-octanol-(dl) $(CH_3)CH(CH_2)_3C(OH)(CH_3)CH_2CH_3$	158.28	....	196-7, $87-8^{10}$	31-2	0.8280^{20}	1.4335^{20}	al	$B1^4$, 1829
8761	Linaloolen or 2,6-Dimethyl-2,7-octadiene $(CH_3)_2C=CHCH_2CH_2CH(CH_3)CH=CH_2$	138.25	....	168, 58^{12}	......	0.7882^{20}	1.4561^{20}	al	B1, 261
8762	β-Linaloolene or Dihydromyrcene $CH_2=CHCH(CH_3)CH_2CH_2CH=C(CH_3)_2$	138.25	....	165-8	......	$0.7601^{20/4}$	1.4362^{20}	....	$B1^4$, 1060
8763	Linamarin or Acetone cyanohydrin-β-d-glycopyranoside $C_{10}H_{17}NO_6$	247.25	nd (w al), $[\alpha]^{18/}{}_D$ -29.1 (w, p=5)	....	145	....	....	ace	$B17^4$, 3340
8764	Linoleic acid or 9,12-Octadecadienoic acid $CH_3(CH_2)_4CH=CHCH_2CH=CH(CH_2)_7CO_2H$	280.45	....	$229-30^{16}$	-5	0.9022^{204}	1.4699^{20}	al, eth, ace, bz, chl	$B2^4$, 1754
8765	Linoleic acid, ethyl ester or Ethyl linoleate $CH_3(CH_2)_4CH=CH\text{-}CH_2\text{-}CH=CH\text{-}(CH_2)_7CO_2C_2H_5$	308.50	ye or col	$270-5^{180}$, 212^{12}	......	$0.8865^{20/4}$	....	al, eth	$B2^4$, 1757
8766	Linoleic acid, methyl ester or Methyl linoleate $CH_3(CH_2)_4CH=CHCH_2CH=CH(CH_2)_7CO_2CH_3$	294.48	....	215^{20}	-35	$0.8886^{18/4}$	1.4638^{20}	al, eth	$B2^4$, 1756
8767	10,12-Linoleic acid or 10,12-Octadecadienoic acid $CH_3(CH_2)_4CH=CHCH=CH(CH_2)_8CO_2H$	280.45	....	....	56-7	$0.8686^{70/4}$	1.4689^{60}	....	$B2^4$, 1752
8768	α-Linolenic acid or 9,12,15-Octadecatrienoic acid *(cis,cis,cis)* $CH_3[CH_2CH=CH]_3(CH_2)_7CO_2H$	278.44	....	$230-2^{17}$, $129^{0.05}$	-11.3	$0.9164^{20/4}$	1.4800^{20}	al, eth	$B2^4$, 1781
8769	α-Linolenic acid, ethyl ester or Ethyl linolenate $CH_3[CH_2CH=CH]_3(CH_2)CO_2C_2H_5$	306.49	....	218^{15}	......	$0.8919^{20/4}$	1.4694^{20}	....	$B2^4$, 1782
8770	α-Lipoic acid (d) or 6,8-Epidithiooctanoic acid $C_8H_{14}O_2S_2$	206.32	pa ye pl, $[\alpha]^{25/}{}_D$ +96.7	....	47.5	....	....	bz, MeOH	$B19^4$, 3459
8771	Lithocholic acid or 3α-Hydroxycholanic acid $C_{24}H_{40}O_3$	376.58	hex lf (al) pr (dil al or aa), $[\alpha]^{20/}{}_D$ +32.14 (al)	....	186	....	....	al, chl, aa	$B10^3$, 687
8772	Lobelanidine $C_{22}H_{29}NO_2$	339.48	sc(al eth)	distb vac	150	....	....	al, ace, bz, chl, py	$B21^4$, 2380
8773	Lobelanine $C_{22}H_{25}NO_2$	335.45	nd(peth or eth)	....	99	....	....	al, ace, bz, aa, chl	$B21^4$, 5630
8774	Lobeline (dl) $C_{22}H_{27}NO_2$	337.46	ye pl (eth) pr (al)	....	110	....	....	al, eth, bz, chl	$B21^4$, 6346

PHYSICAL CONSTANTS OF ORGANIC COMPOUNDS (Continued)

No.	Name, Synonyms, and Formula	Mol. wt.	Color, crystalline form, specific rotation and λ_{max} (log ε)	b.p. °C	m.p. °C	Density	n_D	Solubility	Ref.
8775	Lobeline (l) $C_{22}H_{27}NO_2$	337.46	nd (al eth bz), $[\alpha]^{15/}{}_D$ -42.8 (al, c=1)	...	130.1	...	...	al, eth, ace, bz, chl	$B21^4$, 6345
8776	Longifolene (d) $C_{15}H_{24}$	204.36	$[\alpha]^{18/}{}_D$ +42.7	$254\text{-}6^{706}$, $126\text{-}7^{15}$	...	$0.9319^{10/4}$	1.5040^{20}	bz	$B5^4$, 1192
8777	Lophine or 2,4,5-Triphenyl imidazole $C_{21}H_{16}N_2$	296.37	nd (al)	sub	275	...	...	al, eth	$B23^2$, 280
8778	Luciculine $C_{22}H_{38}NO_3$	361.52	cr (+1w ace), $[\alpha]^{11.5/}{}_D$ -11.4 (al)	$165^{0.02}$	148-50	...	...	al	$B21^4$, 2584
8779	Luciculine, hydrochloride $C_{22}H_{38}NO_3HCl$	397.99	cr (w + $^3/_2$), $[\alpha]_D$ -9.4 (w)	...	198-203	...	...	al	$B21^4$, 2584
8780	Luminol or 3-Aminophthalhydrazide $C_8H_7N_3O_2$	177.16	ye nd (al)	...	329-32	...	...	...	$B25^2$, 389
8781	Lumisterol $C_{28}H_{44}O$	396.66	nd (ace MeOH), $[\alpha]^{18/}{}_D$ +197 (chl)	...	118	...	...	al, eth, ace, bz, aa	$B6^3$, 3108
8782	Lupanine (d) $C_{15}H_{24}N_2O$	248.37	hyg nd, $[\alpha]^{20/}{}_D$ +82.4 (w, c=3)	$190\text{-}3^3$	40-4	...	1.544^{26}	w, al, eth, chl	$B24^2$, 53
8783	Lupanine (dl) $C_{15}H_{24}N_2O$	248.37	nd(peth) rh pr(ace)	$233\text{-}4^{18}$	98-9	...	...	w, al, eth, chl, peth	$B24^2$, 55
8784	Lupeol $C_{30}H_{50}O$	426.73	nd (al or ace), $[\alpha]^{20/}{}_D$ +27.2 (chl)	...	215-7	$0.9457^{218/4}$	1.4910^{218}	al, eth, ace, bz, chl	$B6^3$, 2901
8785	Lupinine (l) $C_{10}H_{19}NO$	169.27	rh (peth), $[\alpha]^{17/}{}_D$ -20.3 (al)	$269\text{-}70^{754}$	70	...	...	w, al, eth, bz, chl	$B21^4$, 291
8786	Lupinine, hydrochloride $C_{10}H_{19}NO.HCl$	205.73	pr (dil al), $[\alpha]_D$ -14 (w)	...	212-3	...	...	w, al	$B21^4$, 292
8787	Lupulone or β-Lupulinic acid $C_{26}H_{38}O_4$	414.59	pr (MeOH)	...	93	...	...	al, peth	$B7^3$, 4753
8788	2,3-Lutidine or 2,3-dimethyl pyridine 2,3-$(CH_3)_2(C_5H_3N)$	107.16	...	163-4	...	$0.9319^{25/4}$	1.5057^{20}	w, al, eth	$B20^4$, 2765
8789	2,4-Lutidine or 2,4-dimethyl pyridine 2,4-$(CH_3)_2(C_5H_3N)$	107.16	...	159	...	$0.9309^{20/4}$	1.5010^{20}	w, al, eth, ace	$B20^4$, 2768
8790	2,5-Lutidine or 2,5-dimethyl pyridine 2,5-$(CH_3)_2(C_5H_3N)$	107.16	...	157-9	−16	$0.9297^{20/4}$	1.5006^{20}	al, eth, ace	$B20^4$, 2774
8791	2,6-Lutidine or 2,6-dimethyl pyridine 2,6-$(CH_3)_2(C_5H_3N)$	107.16	...	145.7	−6.1	$0.9226^{20/4}$	1.4953^{20}	w, eth, ace	$B20^4$, 2776
8792	3,4-Lutidine or 3,4-dimethyl pyridine 3,4-$(CH_3)_2(C_5H_3N)$	107.16	...	163-4	...	$0.9281^{20/4}$	1.5096^{20}	al, eth, ace, chl	$B20^4$, 2787
8793	3,5-Lutidine or 3,5-dimethyl pyridine 3,5-$(CH_3)_2(C_5H_3N)$	107.16	...	171.6	...	$0.9419^{20/4}$	1.5061^{20}	w, al, eth, ace	$B20^4$, 2788
8794	Lycaconitine $C_{34}H_{34}N_2O_6$	566.65	amor, $[\alpha]_D$ +31.5	...	111-4	...	...	al, bz, chl, peth	$B21^4$, 4564
8795	Lycomarasmine $HO_2CCH_2CH(CO_2H)NHCH_2CH(CO_2H)NHCH_2CONH_2$	277.24	$[\alpha]^{20/}{}_D$ -48 (w)	...	227-9d	...	...	...	$B4^3$, 1521
8796	Lycopene $C_{40}H_{56}$	536.88	red pr or nd (peth)	...	175	...	...	eth, bz, chl	$B1^4$, 1166
8797	Lycorine (l) or Narcissine $C_{16}H_{17}NO_4$	287.32	pr (al py), $[\alpha]^{16/}{}_D$ -129 (al, c=0.16)	sub	280	...	...	...	$B27^2$, 547
8798	Lycoxanthin $C_{40}H_{56}O$	552.88	red pl (bz-MeOH)	...	168	...	...	bz	$B1^4$, 2368

No.	Name, Synonyms, and Formula	Mol. wt.	Color, crystalline form, specific rotation and λ_{max} (log ε)	b.p. °C	m.p. °C	Density	n_D	Solubility	Ref.
8799	Lysergic acid $C_{16}H_{16}N_2O_2$	268.32	lf or hex sc(w), $[\alpha]^{20/}{}_D$ + 40 (py, c=0.5)		240d			al, py	J1955, 1626
8800	Lysine(L) or L-α-ω-Diaminocaproic acid $H_2N(CH_2)_4CH(NH_2)CO_2H$	146.19	nd (w, dil al), $[\alpha]^{20/}{}_D$ + 14.6 (w, c=6)		224-5d darkens 210			w	B4[3], 1400
8801	Lysine, dihydro chloride (L) $H_2N(CH_2)_4CH(NH_2)CO_2H.2HCl$	219.11	(al-eth or aq HCl), $[\alpha]^{20/}{}_D$ + 15.3 (w)		201-2			w, al	B4[3], 1402
8802	Lysine, α-N-benzoyl (dl) $H_2N(CH_2)_4CH(NHCOC_6H_5)CO_2H$	250.30	nd (w)		235			w	B9[3], 1235
8803	Lysine, ω-N-benzoyl (dl) $C_6H_5CONH(CH_2)_4CH(NH_2)CO_2H$	250.30	cr (w)		268			w	B9[3], 1238
8804	Lysine, ω-N-benzoyl-(1) $C_6H_5CONH(CH_2)_4CH(NH_2)CO_2H$	250.30	lf (w), $[\alpha]^{19/}{}_D$ + 20.1 (aq HCl)		240d			w	B9[3], 1237

No.	Name, Synonyms, and Formula	Mol. wt.	Color, crystalline form, specific rotation and λ_{max} (log ε)	b.p. °C	m.p. °C	Density	n_D	Solubility	Ref.
8805	Malathion or S(1,2-Dicarboxymethyl)-0,0-dimethyldithiophosphate $C_{10}H_{19}PS_2O_6$	330.35	ye-br	$156\text{-}7^{0.7}/_d$	2.8	1.2076^{20}	1.4960^{20}	al, eth, bz	$B3^4$, 1136
8806	Maleicacid-*cis*-butenedioic acid $HO_2CCH{=}CHCO_2H$	116.07	mcl pr (w)		139-40	1.590^{20}		w, al, eth, ace, aa	$B2^4$, 2199
8807	Maleic acid, bromo or Bromo maleicacid $HO_2C\,CH{=}CHCO_2H$	194.97	nd or pr	d	136-40			w, al, eth	$B2^4$, 2224
8808	Maleic acid, chloro $HO_2CCl{=}CHCO_2H$	150.52	pr(eth-chl)		108 (114)			al, eth, ace, aa	$B2^4$, 2221
8809	Maleic acid, chloro,diethyl ester or Diethyl chloromaleate $C_2H_5O_2CCCl{=}CHCO_2C_2H_5$	206.63		235d, 125^{19}		$1.1741^{20/4}$		al, eth, ace, aa	$B2^3$, 1928
8810	Maleic acid, chloro, dimethyl ester $CH_3O_2CCCl{=}CHCO_2CH_3$	178.57		106.5^{18}		$1.2775^{25/4}$		al, eth	$B2^3$, 1928
8811	Maleic acid, dichloro $HO_2CCCl{=}CClCO_2H$	184.96	nd (liq eth)		119-20			w, al, eth	$B2^4$, 2223
8812	Maleic acid, diethyl ester or Diethyl maleate $C_2H_5O_2CCH{=}CHCO_2H_5$	172.18		223, $105\text{-}6^{14}$	−88	$1.0662^{20/4}$	1.4416^{20}	al, eth	$B2^4$, 2207
8813	Maleic acid, dihydroxy $HO_2CC(OH){=}C(OH)CO_2H$	148.07	pl (w + 2)		155 (anh)				$B3^4$, 1975
8814	Maleic acid, dimethyl ester or Dimethyl maleate $CH_3O_2CCH{=}CHCO_2CH_3$	144.13		202, 102^{17}	−19	$1.1606^{20/4}$	1.4416^{20}	eth	$B2^4$, 2204
8815	Maleic acid, diphenyl ester $C_6H_5O_2CCH{=}CHCO_2C_6H_5$	286.27	pl(lig)	226.15	73			al, eth, ace, bz, chl	$B6^4$, 628
8816	Maleic acid, dipropyl ester or Dipropylmaleate $C_3H_7O_2CCH{=}CHCO_2C_3H_7$	200.23		126^{12}		$1.0245^{20/4}$	1.4434^{20}	al, eth, ace, bz	$B2^4$, 2209
8817	Maleic acid, hydrazide $C_4H_4N_2O_2$	112.09	cr (w)		>300d				$B2^4$, 2220
8818	Maleic acid, methyl or Citraconic acid $HO_2CC(CH_3){=}CHCO_2H$	130.10	tcl pr or pl (eth-bz) nd (eth-lig)		93-4	1.617		w	$B2^4$, 2230
8819	Maleic acid, monoamide or Maleamic acid $HO_2CCH{=}CHCONH_2$	115.09	lf(w)		178			w, al	$B2^4$, 2218
8820	Maleic anhydride $C_4H_2O_3$	98.06	nd(chl eth)	197-9, 82^{14}	60	1.314^{60}		eth, ace, chl	$B17^4$, 5897
8821	Maleic anhydride, dimethyl or Pyrocinchonic anhydride $OC(CH_3){=}C(CH_3)CO$	126.11	pl or lf(dil al)	223, 105^{12}	96	$1.107^{100/4}$		al, eth, bz, chl	$B17^4$, 5919
8822	Maleic anhydride, methyl or Citraconic anhydride $C_5H_4O_3$	112.08		213-4, $99\text{-}100^{15}$	7-8	$1.2469^{16/4}$	1.4710^{21}	al, eth, ace	$B17^4$, 5912
8823	Maleic anhydride, phenyl $C_{10}H_7O_3$	174.16	ye nd (CS_2)		122			al, eth	$B17^4$, 6279
8824	Maleimide or Maleic acid, imide $OCHC{=}CH\text{-}(O\cdot NH)$ └─N──┘	97.07	pl (bz)	sub	93-5			w, al, eth	$B21^4$, 4627
8825	Maleimide, N-ethyl $C_6H_7NO_2$	125.13	cr (bz)		45.5			al, eth	$B21^4$, 4629
8826	Maleimide, N-phenyl or Maleanil $C_{10}H_7NO_2$	173.17	ye nd (bz-lig)	162^{12}	90-1			al, eth, bz	$B21^4$, 4631
8827	Malonamide $H_2C(CONH_2)_2$	102.09	mcl pr (w)		171-2			w	$B2^4$, 1887
8828	Malonamide, benzyl $C_6H_5CH_2CH(CONH_2)_2$	192.22	nd (al)		225				$B9^3$, 4285
8829	Malonamide, N,N′-diphenyl or Malonanilide $H_2C(CONHC_6H_5)_2$	254.29	nd (al)		226			al, bz, aa	$B12^3$, 558
8830	Malonic acid or Propanedioic acid $HO_2CCH_2CO_2H$	104.06	tcl(al)	d140	135.6	1.619^{16}		al, eth, py	$B2^4$, 1874
8831	Malonic acid, acetamide,diethyl ester $CH_3CONHCH(CO_2C_2H_5)_2$	217.22	cr(al bz-peth)	185^{20}	95-6			al	$B4^3$, 1503
8832	Malonic acid, acetyl, diethyl ester $(CH_3CO)CH(CO_2C_2H_5)_2$	202.21		232, 120^{17}		$1.0834^{26/4}$	1.4435^{25}	ace	$B3^4$, 1819
8833	Malonic acid, allyl or 3-Butene-1,1-dicarboxylic acid $(CH_2{=}CHCH_2)CH(CO_2H)_2$	144.13	tcl (eth)	>180d	105			w, al, eth	$B2^3$, 1945
8834	Malonic acid, allyl,diethyl ester $(CH_2{=}CHCH_2)CH(CO_2C_2H_5)_2$	200.23		222-3, 93^6		$1.0098^{20/4}$	1.4305^{20}	al, eth	$B2^4$, 2240
8835	Malonic acid, amino,monohydrate $H_2NCH(CO_2H)_2.H_2O$	137.09	pr (w + 1)		112d				$B4^3$, 1501
8836	Malonic acid, amino,diethyl ester $H_2NCH(CO_2C_2H_5)_2$	175.18		$122\text{-}3^{16}$		$1.100^{16/4}$	1.4353^{16}	w, al, eth, ace, bz	$B4^3$, 1501

PHYSICAL CONSTANTS OF ORGANIC COMPOUNDS (Continued)

No.	Name, Synonyms, and Formula	Mol. wt.	Color, crystalline form, specific rotation and λ_{max} (log ε)	b.p. °C	m.p. °C	Density	n_D	Solubility	Ref.
8837	Malonic acid, benzoyl amino,diethyl ester $(C_6H_5CONH)CH(CO_2C_2H_5)_2$	279.29	nd(peth)		61			al, eth	$B9^3$, 1186
8838	Malonic acid, benzyl $C_6H_5CH_2CH(CO_2H)_2$	194.19	pr(bz eth chl-peth)		121-2			w, al, eth, bz	$B9^3$, 4283
8839	Malonic acid, benzyl,diethyl ester $C_6H_5CH_2CH(CO_2C_2H_5)_2$	250.29		300, 169^{212}		$1.0750^{20/4}$	1.4872^{20}		$B9^3$, 4284
8840	Malonic acid, benzyl,hydroxy $C_6H_5CH_2C(OH)(CO_2H)_2$	210.19	pr		147d			w, al, eth	$B10^3$, 2216
8841	Malonic acid, benzylidene $C_6H_5CH=C(CO_2H)_2$	192.17	pr(w)		195-6d			al, ace	$B9^3$, 4378
8842	Malonic acid, benzylidene,diethyl ester $C_6H_5CH=C(CO_2C_2H_5)_2$	248.28		308-12d, 180^{10}	32	$1.1045^{20/4}$	1.5389^{20}	al, eth, ace, bz	$B9^3$, 4379
8843	Malonic acid, bromo $BrCH(CO_2H)_2$	182.96	nd(eth) pl(ace-bz)		113d			al, eth	$B2^4$, 1904
8844	Malonic acid, bromo,diethyl ester $BrCH(CO_2C_2H_5)_2$	239.07		253-5d, 123^{20}	−54	$1.4022^{25/4}$	1.4521^{20}	**al, eth**, ace	$B2^4$, 1904
8845	Malonic acid, butyl or 1,1-Pentane dicarboxylic acid $C_4H_9CH(CO_2H)_2$	160.17	pr(w)		104-5			w, al, eth	$B2^4$, 2011
8846	Malonic acid, butyl,diethyl ester $C_4H_9CH(CO_2C_2H_5)_2$	216.28		235-40, 122^{12}			1.4250^{20}	al, eth	$B2^4$, 2011
8847	Malonic acid, isobutyl $i-C_4H_9CH(CO_2H)_2$	160.17	cr(bz)	d	115d			w, al, eth	$B2^4$, 2023
8848	Malonic acid, isobutyl,diethyl ester $i-C_4H_9CH(CO_2C_2H_5)_2$	216.28		255, $119\text{-}20^{16}$		$0.9804^{20/4}$	1.4236^{20}	al, eth	$B2^3$, 1756
8849	Malonic acid, sec-Butyl,diethyl ester $C_2H_5CH(CH_3)CH(CO_2C_2H_5)_2$	216.28		245-50, 105^9		0.988^{15}	1.4248^{20}	al, eth	$B2^4$, 2019
8850	Malonic acid, cetyl or Hexadecyl malonic acid $C_{16}H_{33}CH(CO_2H)_2$	328.69	nd(lig) lf(aa)		121-2			eth, bz, aa	$B2^4$, 2183
8851	Malonic acid, cetyl, diethyl ester $C_{16}H_{33}CH(CO_2C_2H_5)_2$	384.60	amor	$238\text{-}40^{14}$	(i)25.1 (ii)12.7		1.4433^{20}	al	$B2^4$, 2183
8852	Malonic acid, chloro $ClCH(CO_2H)_2$	138.51	pr (w)	d	133			w, sl, eth	$B2^3$, 1637
8853	Malonic acid, chloro, diethyl ester or Ethyl chloromalonate $ClCH(CO_2C_2H_5)_2$	194.61		222, 118^{16}		$1.2040^{20/4}$	1.4327^{20}	**al, eth, chl**	$B2^4$, 1903
8854	Malonic acid, (3-chloropropyl), diethyl ester $Cl(CH_2)_3CH(CO_2C_2H_5)_2$	236.70		$147\text{-}9^{10}$			1.4429^{20}	al, eth, chl	$B2^4$, 1992
8855	Malonic acid, cinnamylidene-*(trans)* or Cinnamal malonic acid $C_6H_5CH=CHCH=C(CO_2H)_2$	218.21	dk ye nd		212d			al, chl	$B9^3$, 4432
8856	Malonic acid, cyclohexyl, diethyl ester $C_6H_{11}CH(CO_2C_2H_5)_2$	242.32		$163\text{-}5^{20}$		$1.0281^{19/4}$	1.4478^{25}	al, eth, ace, bz	$B9^3$, 3834
8857	Malonic acid, 2-cylopentenyl, diethyl ester $(2-C_5H_7)CH(CO_2C_2H_5)_2$	226.27		141^{10}		$1.0507^{20/4}$	1.4536^{20}	eth, ace	$B9^3$, 3951
8858	Malonic acid, cyclopentylidene, diethyl ester $(C_5H_8=)C(CO_2C_2H_5)_2$	226.27		140^{10}		$1.0616^{20/4}$	1.4724^{20}	eth, ace	$B9^3$, 3952
8859	Malonic acid, diallyl, diethyl ester $(CH_2=CHCH_2)_2C(CO_2C_2H_5)_2$	240.30		243-4, 128^{16}		$0.9943^{20/4}$	1.4445^{22}	al, eth, ace	$B2^3$, 2003
8860	Malonic acid, dibenzyl, diethyl ester $(C_6H_5CH_2)_2C(CO_2C_2H_5)_2$	340.42		$234\text{-}5^{23}$	14	$1.093^{20/4}$		al, eth	$B9^3$, 4553
8861	Malonic acid, dibromo, diethyl ester $Br_2C(CO_2C_2H_5)_2$	317.96		250-6d, 154^{28}				al, eth, ace	$B2^4$, 1905
8862	Malonic acid, dibutyl ester $CH_2(CO_2C_4H_9)_2$	216.28		251-2, 137^{14}	−83	$0.9824^{20/4}$	1.4262^{20}	al, eth, ace, bz, aa	$B2^4$, 1884
8863	Malonic acid, dibutyl,diethyl ester $(C_4H_9)_2C(CO_2C_2H_5)_2$	272.38		$153\text{-}4^{14}$		$0.9457^{20/4}$	1.4341^{20}	al, eth	$B2^4$, 2119
8864	Malonic acid, diethyl or 3,3-Pentanedicarboxylic acid $(C_2H_5)_2C(CO_2H)_2$	160.17	pr (w bz)	d	127			w, al, eth	$B2^4$, 2026
8865	Malonic acid, diethyl, diethyl ester $(C_2H_5)_2C(CO_2C_2H_5)_2$	216.28		230, 100^{12}		0.9643^{30}	1.4240^{20}	**al, eth**	$B2^4$, 2026
8866	Malonic acid, diethyl ester or Diethyl malonate $CH_2(CO_2C_2H_5)_2$	160.17		199.3, 96^{22}	−48.9	$1.0551^{20/4}$	1.4139^{20}	**al, eth**, ace, bz, chl	$B2^4$, 1881
8867	Malonic acid, dimethyl ester or Dimethyl malonate $CH_2(CO_2CH_3)_2$	132.12		181.4, 78.4^{15}	−61.9	1.156^{20}	1.4135^{20}	**al, eth**, ace, bz, chl	$B2^4$, 1880
8868	Malonic acid, dimethyl, diethyl ester $(CH_3)_2C(CO_2C_2H_5)_2$	188.22		197, $97\text{-}8^{22}$	−30.4	$0.9964^{20/4}$	1.4129^{20}	**al, eth**	$B2^4$, 1955

No.	Name, Synonyms, and Formula	Mol. wt.	Color, crystalline form, specific rotation and λ_{max} (log ε)	b.p. °C	m.p. °C	Density	n_D	Solubility	Ref.
8869	Malonic acid, dipropyl ester $CH_2(CO_2C_3H_7)_2$	188.22	glass	229, 113^{13}	−77.1	$1,0097^{20/4}$	1.4206^{20}	al, eth, ace, bz	$B2^{3}$, 1619
8870	Malonic acid, ethoxymethylene),diethyl ester $(C_2H_5OCH{=})C(CO_2C_2H_5)_2$	216.23		279-81d, 165^{19}			1.4600^{20}	al, eth	$B3^{4}$, 1192
8871	Malonic acid, ethyl or 1,1-Propanedicarboxylic acid $C_2H_5CH(CO_2H)_2$	132.12	pr (w + 1)	160d	114			w, al, eth, bz, chl	$B2^{4}$, 1952
8872	Malonic acid, ethyl, diethyl ester $C_2H_5CH(CO_2C_2H_5)_2$	188.22		$207\text{-}9^{755}$, $98\text{-}9^{12}$		$1.0047^{20/4}$	1.4166^{20}	al, eth, ace, chl	$B2^{4}$, 1953
8873	Malonic acid, ethyl (methyl) or 2,2-Butane dicarboxylic acid $C_2H_5C(CH_3)(CO_2H)_2$	146.14	pr or nd (eth)		122			w, al, eth, ace	$B2^{4}$, 1977
8874	Malonic acid, ethyl (phenyl), diethyl ester $C_6H_5C(C_2H_5)(CO_2C_2H_5)_2$	264.32		170^{19}		$1.071^{20/4}$	1.4896^{25}	al, eth	$B9^{3}$, 4304
8875	Malonic acid, ethyl (isopropyl), diethyl ester $(CH_3)_2CHC(C_2H_5)(CO_2C_2H_5)_2$	230.30		$232\text{-}4^{742}$, $108\text{-}10^{11}$			1.4280^{25}	al, eth, ace	$B2^{4}$, 2052
8876	Malonic acid, ethylidene,diethyl ester $(CH_3CH{=})C(CO_2C_2H_5)_2$	186.21		$115\text{-}8^{17}$		$1.0194^{17/4}$	1.4308^{17}	al, eth	$B2^{4}$, 2234
8877	Malonic acid, formamido,diethyl ester $HCONHCH(CO_2C_2H_5)_2$	203.20	pl (MeOH)	$173\text{-}4^{11}$	48-9			al	$B4^{4}$, 1503
8878	Malonic acid, heptyl or 1,1-Octanedicarboxylic acid $C_7H_{15}CH(CO_2H)_2$	202.25	pr (bz-peth)		96-8			al, eth, ace	$B2^{4}$, 2094
8879	Malonic acid, hexyl,diethyl ester $C_6H_{13}CH(CO_2C_2H_5)_2$	244.33		268-70, 143^{15}		$0.9577^{21/4}$	1.4278^{21}	al, eth, ace, bz	$B2^{3}$, 1791
8880	Malonic acid, hydroxy or Tartronic acid $HOCH(CO_2H)_2$	120.06	pr (w + 1)	sub	156-8			w, al	$B3^{4}$, 1120
8881	Malonic acid, hydroxy,diethyl ester $HOCH(CO_2C_2H_5)_2$	176.17		222-5, 121^{15}	−2.5			al, eth, ace, bz	$B3^{3}$, 905
8882	Malonic acid, hydroxy,dimethyl ester $HOCH(CO_2CH_3)_2$	148.12	cr (eth-peth)	122^{19}	45			w, al, eth, ace, bz	$B3^{3}$, 905
8883	Malonic acid, (2-hydroxycyclohexyl) lactone,ethyl ester $C_{11}H_{16}O_4$	212.25	pa ye	199^{30}		1.0735^{19}			$B18^{4}$, 5351
8884	Malonic acid, hydroxy-(methyl) or α-Isomalic acid $CH_3C(OH)(CO_2H)_2$	134.09		170d	142d			w, al, eth	$B3^{3}$, 927
8885	Malonic acid, methyl or 1,1-Ethane dicarbozylic acid $CH_3CH(CO_2H)_2$	118.09	nd (AcOEt-bz) pr(eth-bz)		135d	$1.455^{20/4}$		w, al, eth, aa	$B2^{4}$, 1932
8886	Malonic acid, methyl,diethyl ester or Diethyl methylmalonate $CH_3CH(CO_2C_2H_5)_2$	174.20		201, 94^{16}		$1.0225^{20/4}$	1.4126^{20}	al, eth, ace, chl	$B2^{4}$, 1932
8887	Malonic acid, methyl,dimethyl ester $CH_3CH(CO_2CH_3)_2$	146.14		176.5		$1.0977^{20/4}$	1.4128^{20}	al, eth, ace, chl	$B2^{3}$, 1681
8888	Malonic acid, monoamide-N-Phenyl or Malonanilic acid $HO_2CCH_2CONHC_6H_5$	179.18	cr(w eth al)		132			al, eth	$B12^{3}$, 557
8889	Malonic acid, monochloride,ethyl ester or Carbethoxy acetyl chloride $ClCOCH_2CO_2C_2H_5$	150.56		170-80d, $75\text{-}7^{15}$				eth	$B2^{4}$, 1887
8890	Malonic acid, monochloride,methyl ester $ClCOCH_2CO_2CH_3$	136.53		71^{15}				eth	$B2^{4}$, 1886
8891	Malonic acid, octyl or 1,1-Nonane dicarboxylic acid $C_8H_{17}CH(CO_2H)_2$	216.28	pr(bz-peth)		116	1.173^{17}		al, ace	$B2^{4}$, 2114
8892	Malonic acid, pentyl,diethyl ester $C_5H_{11}CH(CO_2C_2H_5)_2$	230.30		$134\text{-}6^{14}$		$0.9652^{20/4}$	1.4253^{20}	al, eth	$B2^{4}$, 2034
8893	Malonic acid, isopentyl,diethyl ester i-$C_5H_{11}CH(CO_2C_2H_5)_2$	230.30		240-2, 137.5^{19}		$0.9580^{25/4}$	1.4255^{25}	al, eth, ace	$B2^{3}$, 1777
8894	Malonic acid, (3-pentyl), diethyl ester $[(C_2H_5)_2CH]CH(CO_2C_2H_5)_2$	230.30		130^{16}			1.4291^{20}	al, eth	$B2^{3}$, 1781
8895	Malonic acid, phenyl,diethyl ester $C_6H_5CH(CO_2C_2H_5)_2$	236.27		205d, 168^{12}	16-7	$1.0950^{20/4}$	1.4977^{20}	al, ace	$B9^{3}$, 4260
8896	Malonic acid, phenyl,dimethyl ester $C_6H_5CH(CO_2CH_3)_2$	208.21	cr(lig)	$145\text{-}7^{13}$	51			al, eth	$B9^{3}$, 4260
8897	Malonic acid, phenyl amino,diethyl ester $C_6H_5NHCH(CO_2C_2H_5)_2$	251.28	cr(al or lig)		45			al, eth, bz, chl	$B12^{3}$, 971
8898	Malonic acid, phthalimido,diethyl ester $C_{15}H_{15}NO_6$	305.29	pr(al)		74			al, eth, ace, bz, chl	$B21^{4}$, 5264
8899	Malonic acid, propyl $C_3H_7CH(CO_2H)_2$	146.14	pl(bz)	d	96-7			w, al, eth, chl	$B2^{4}$, 1991
8900	Malonic acid, propyl,diethyl ester $C_3H_7CH(CO_2C_2H_5)_2$	202.25		221^{767}, 114^{22}		0.9873^{20}	1.4197^{20}	al, eth	$B2^{4}$, 1991

PHYSICAL CONSTANTS OF ORGANIC COMPOUNDS (Continued)

No.	Name, Synonyms, and Formula	Mol. wt.	Color, crystalline form, specific rotation and λ_{max} (log ε)	b.p. °C	m.p. °C	Density	n_D	Solubility	Ref.
8901	Malonic acid, isopropyl,diethyl ester $(CH_3)_2CHCH(CO_2C_2H_5)_2$	202.25		$235, 107\text{-}9^{18}$		$0.9970^{20/15}$	1.4188^{21}	al, eth, chl	$B2^4$, 2011
8902	Malonic acid, isopropylidene $(CH_3)_2C{=}C(CO_2H)_2$	144.13	cr (ace-chl)	d	170-1			al	$B2^3$, 1948
8903	Malonic acid, isopropylidene,diethyl ester $(CH_3)_2C{=}C(CO_2C_2H_5)_2$	200.23		$175\text{-}8, 140\text{-}1^{20}$		$1.0282^{18/4}$	1.4486^{17}	al, ace	$B2^4$, 2244
8904	Malononitrile $CH_2(CN)_2$	66.06		$218\text{-}9, 109^{20}$	32	$1.1910^{20/4}$	1.4146^{34}	w, al, eth, ace, bz	$B2^4$, 1892
8905	Malononitrile, benzyl $C_6H_5CH_2CH(CN)_2$	156.19	pl(al) nd(w lig)	174^{23}	91			al, eth, bz	B9, 870
8906	Malononitrile, phenyl $C_6H_5CH(CN)_2$	142.16	cr (dil al)	$152\text{-}3^{21}$	70-1			al	$B9^3$, 4263
8907	Malonyl chloride $CH_2(COCl)_2$	140.95		58^{26}		$1.4509^{20/4}$	1.4639^{20}	eth	$B2^4$, 1887
8908	Maltose (pyranose form, α-anomer, anhydrous) or 4-*O*-α-D-glucopyranosyl-D-glucose $C_{12}H_{22}O_{11}$	342.30	nd (abs al), $[\alpha]_D$ + 140.7 (w, c=10)		160-5			w	$B17^4$, 3057
8909	Maltose, monohydrate (β-anomer) $C_{12}H_{22}O_{11} \cdot H_2O$	360.32	nd (w), $[\alpha]^{20/}_D$ + 111.7→ + 130.4 (c=4, mut)		102-3	1.54		w	$B17^4$, 3057
8910	Malvidine chloride or Syringidine chloride $C_{17}H_{15}ClO_7$	366.75	rh ta or pr(al-aq HCl + 1w)		>300			al	$B18^4$, 3561
8911	Mandelic acid-(D) or α-Hydroxyphenyl acetic acid $C_6H_5CH(OH)CO_2H$	152.15	ta, $[\alpha]^{20/}_D$ -158 (w, c=2.5)		133-5	1.341		w, al, eth, aa, chl	$B10^3$, 445
8912	Mandelic Acid (DL) or α-Hyroxyphenyl acetic acid $C_6H_5CH(OH)COOH$	152.15	pl (wh) rh (bz)	d	121.3	$1.300^{20/4}$		w, al, eth	$B10^3$, 448
8913	Mandelic acid (L+) or L+-α-Hydroxyphenylacetic acid $C_6H_5CH(OH)CO_2H$	152.15	pl(w), $[\alpha]^{20/}_D$ + 156.6 (w, c=2.9)					w, al, eth, chl	$B10^3$, 447
8914	Mandelic acid, 2,4-dimethyl or (2,4-Dimethylphenyl)hydroxyacetic acid $[2,4\text{-}(CH_3)_2C_6H_3]CH(OH)CO_2H$	180.20	rh(w) nd(bz) lf(to peth al)		119			al, eth, chl	$B10^3$, 613
8915	Mandelic acid, 2,5-dimethyl or (2,5-Dimethylphenyl)-α-Hydroxyacetic acid $[2,5\text{-}(CH_3)_2C_6H_3]CH(OH)CO_2H$	180.20	nd or pr(bz)		116-7			al, eth, chl	$B10^3$, 612
8916	Mandelic acid, 3,4-dimethyl or (3,4-Dimethylphenyl)-α-hydroxyacetic acid $[3,4\text{-}(CH_3)_2C_6H_3]CH(OH)CO_2H$	180.20	lf(bz)		135			w, bz	$B10^3$, 611
8917	Mandelic acid, *p*-iodo-*(dl)* $4{=}IC_6H_4CH(OH)CO_2H$	278.05			135			w, al, eth	$B10^3$, 481
8918	Mandelic acid, *p*-isopropyl-*(dl)* $[4\text{-}i\text{-}C_3H_7C_6H_4]CH(OH)CO_2H$	194.23	nd(w)		159-60			al, eth	$B10^3$, 629
8919	Mandelic acid, *p*-isopropyl-*(dl)* $[4\text{-}i\text{-}C_3H_7C_6H_4]CH(OH)CO_2H$	194.23	lf (w), $[\alpha]^{17/}_D$ + 134.9 (abs, al, c=4)		153-4			al, eth	B10, 279
8920	Mandelic acid, *p*-isopropyl-*(l)* $[4\text{-}i\text{-}C_3H_7C_6H_4]CH(OH)CO_2H$	194.23	ta (20% al), $[\alpha]^{17/}_D$ -135 (abs, al, c=4.09)		153-4			al, eth	B10, 279
8921	D-Mannitane or 3,6-(1,4)Anhydro-D-mannitol $C_6H_{12}O_5$	164.16	amor pl		146-7			w, al	$B17^4$, 2641
8922	D-Mannitol or D-Mannite $C_6H_{14}O_6$	182.17	rh nd or pr (w), $[\alpha]^{25/}_D$ -0.49 (w)	$295^{3.5}$	168	$1.489^{20/4}$	1.3330	w	$B1^4$, 2841
8923	DL-Mannitol or α-Acritol $C_6H_{14}O_6$	182.17	cr(al)		168			w	$B1^3$, 2405
8924	L-Mannitol or L-Mannite $C_6H_{14}O_6$	182.17	nd(al)		163-4			w	$B1^4$, 2843
8925	Mannitol, hexanitrate or Nitro mannitol $O_2NOCH_2(CHONO_2)_4CH_2ONO_2$	452.16	nd (al), $[\alpha]^{25/}_{546}$ + 46.8	120 exp	112.3	$1.8^{20/4}$		al, eth, bz, aa	$B1^4$, 2849

No.	Name, Synonyms, and Formula	Mol. wt.	Color, crystalline form, specific rotation and λ_{max} (log ε)	b.p. °C	m.p. °C	Density	n_D	Solubility	Ref.
8926	α-D-Mannoheptose $C_7H_{14}O_7$	210.18	nd (al) $[\alpha]^{20/}{}_D$ +85→ +68.6 (w,c=1)		134-5			w	$B1^4$, 4438
8927	β-D-Mannoheptose, monohydrate $C_7H_{14}O_7 \cdot H_2O$	228.20	cr (w + 1) $[\alpha]^{20/}{}_D$ +45.7		83			w	$B1^4$, 4441
8928	D-Mannoic acid-γ-lactone $C_6H_{10}O_6$	178.14	pr (abs al) $[\alpha]^{20/}{}_D$ +54 (w,c=2)		151			w	$B18^4$, 3018
8929	D-Mannoheptose (β-anomer), hydrate $C_7H_{14}O_7$	210.18	rh nd (w) $[\alpha]_D$ −51.8		151			w	$B18^2$, 199
8930	D-Mannosaccharic acid, γ,γ′-dilactone $C_6H_6O_6$	174.11	nd (w + 2) $[\alpha]^{23/}{}_D$ +202 (w)		190d			al	$B19^4$, 2962
8931	L-Mannosaccharic acid, γ,γ′-dilactone $C_6H_6O_6$	174.11	nd (w al) $[\alpha]^{20/}{}_D$ −202.5 (w)		183-5d, 68 (hyd)			w	$B19^2$, 266
8932	D-Mannose (β-anomer) $C_6H_{12}O_6$	180.16	nd or orh pr (al or aa) $[\alpha]^{20/}{}_D$ −17→ +14.6 (w,p=3)		132d	$1.539^{20/4}$			$B1^4$, 4328
8933	DL-Mannose $C_6H_{12}O_6$	180.16	cr (al)		132-3			w	B31, 294
8934	L-Mannose (β-anomer) $C_6H_{12}O_6$	180.16	cr (al) $[\alpha]_D$ +14→ −14		132			w	$B1^4$, 4333
8935	D-Mannose, phenylhydrazone $C_{12}H_{18}N_2O_5$	270.29	ye pr (w) nd (al) $[\alpha]_D$ +26.3→ +33.8 (py)		199-200				B31, 290
8936	L-Mannose, phenylhydrazone $C_{12}H_{18}N_2O_5$	270.29	ye pr (w)		195				B31, 294
8937	D-Mannuronic acid (α-anomer), monohydrate $C_6H_{10}O_7$	194.14	hyg nd (al eth) $[\alpha]^{25/}{}_D$ +16→ −6 (w,mut)		120-30d			w	$B3^4$, 1998
8938	D-Mannuronic acid (β-anomer) $C_6H_{10}O_7$	194.14	cr (w ace eth) $[\alpha]^{25/}{}_D$ −47.9→ −23.9 (w)		165-7			w	$B3^4$, 1999
8939	Margaric acid or Heptadecanoic acid $CH_3(CH_2)_{15}CO_2H$	270.46	pl (peth)	227^{100}	62-3	0.8532^{60}	1.4342^{60}	eth, ace, bz, chl	$B2^4$, 1193
8940	Matrine or Sophocarpidine $C_{15}H_{24}N_2O$	248.37	α, nd or pl, β, orh prγ, lig;δ,pr or lf (peth), $[\alpha]^{15/}{}_D$ α40.9 (w), β −28.7 (w)	γ-223^6	α,76-7 β, 87 δ, 84	γ-$1.088^{20/4}$	γ-1.5286^{85}	al, eth, ace, bz	$B24^2$, 58
8941	Meconic acid or 3-hydroxy-γ-pyrone-2,6-dicarboxylic acid $C_7H_4O_7$	200.10	rh pl (w dilHCl) (+3w)	d at 120	-w at 100			al, bz	$B18^4$, 6203
8942	Meconidine $C_{21}H_{23}NO_4$	353.42	ye amor		58			al, eth, ace, bz, chl	M, 641
8943	Meconin or 6,7-Dimethoxyphthalide $C_{10}H_{10}O_4$	194.19	wh nd (w)	155sub	102-3			al, eth, ace, bz, chl	$B18^4$, 1226
8944	Medicagenic acid $C_{30}H_{46}O_6$	502.69	pr or nd		352-3				Am76, 2271
8945	Medicagenic acid, diacetate $C_{34}H_{50}O_8$	586.77	mcl $[\alpha]^{22/}{}_D$ +92 chl		210-2	$1.190^{25/25}$			Am79, 5292
8946	Melam $C_6H_9N_{11}$	235.21	pw						$B3^4$, 319
8947	Melamine or 2,4,6-triamino-1,3,5-triazine $C_3H_6N_6$	126.12	mcl pr (w)	sub	345d	1.573^{16}	1.872^{20}		$B26^2$, 132

PHYSICAL CONSTANTS OF ORGANIC COMPOUNDS (Continued)

No.	Name, Synonyms, and Formula	Mol. wt.	Color, crystalline form, specific rotation and λ_{max} (log ε)	b.p. °C	m.p. °C	Density	n_D	Solubility	Ref.
8948	Melene $C_{30}H_{60}$	420.81	nd (ace) cr (peth)	380	62-3	$0.9037^{25/25}$	1.4228^{90}	...	$B1^3$, 885
8949	Melissic acid $CH_3(CH_2)_{29}CO_2H$	466.83	sc or nd (al or ace)	...	93	...	...	bz	$B2^3$, 1097
8950	Melezitose $C_{18}H_{32}O_{16}$	504.44	cr (w + 2) $[\alpha]^{20/}_D$ + 88 (w,c=4)	...	153-4 (anh)	1.5565^0	...	w	$B17^4$, 3815
8951	Melibiose (α-anomer) or 6-*O*-α-D-galactopyranosyl-α-D-glucose $C_{12}H_{22}O_{11}$	342.30	amor (anh) $[\alpha]^{15}$ + 145.8 →141.6 (w,c=2,mut)	...	...	...	...	w	$B17^4$, 3075
8952	Melibiose (β-anomer), dihydrate $C_{12}H_{22}O_{11}.2H_2O$	378.33	mcl cr (w + 2) $[\alpha]^{20/}_D$ + 111.7 → + 129.5 (w,c=4)	...	84-5	...	...	w, MeOH	$B17^4$, 3075
8953	Mellitic acid or Benzene hexacarboxylic acid $C_6(CO_2H)_6$	342.17	nd (al)	...	286-8	...	...	w, al	$B9^3$, 4914
8954	P-Menthane *(cis)* or 1-isopropyl-4-methyl cyclohexane $C_{10}H_{20}$	140.27	...	170.9^{725}	−89.9	$0.8039^{20/4}$	1.4431^{20}	al, eth, bz, peth	$B5^4$, 151
8955	P-Menthane *(trans)* $C_{10}H_{20}$	140.27	...	170.6, 58.5^{15}	...	$0.7928^{20/4}$	1.4366^{20}	al, eth, bz, lig	$B5^4$, 151
8956	*p*-Menthane, 1,8-epoxy $C_{10}H_{18}O$	154.25	...	176-7	1.5	0.9267^{20}	$1.4584^{15/}_D$	al, eth, chl	$B17^4$, 213
8957	Δ^3-*p*-Menthene(d) (one form) or 1-isopropyl-4-methyl cyclohexene-(3) $C_{10}H_{18}$	138.25	$[\alpha]^{20/}_D$ + 115.6 (undil)	168	...	$0.8118^{20/4}$	1.4524^{20}	al, eth, bz, aa	$B5^4$, 301
8958	Δ^1-*p*-methene(d) (one form) $C_{10}H_{18}$	138.25	$[\alpha]_D$ + 29.6 → + 54.4	167-8	...	$0.8078^{20/4}$	...	al, eth, bz, peth	$B5^4$, 299
8959	Δ^3-*p*-methene(dl) $C_{10}H_{18}$	138.25	...	168^{754}, 60.5^{12}	...	$0.8069^{20/4}$	1.4503^{15}	al, eth, bz, peth	$B5^4$, 301
8960	Δ^3-*p*-Menthene $C_{10}H_{18}$	138.25	$[\alpha]_D$ −13.5 (al)	167-8	1	$0.8112^{19/19}$	1.4511^{20}	al, eth, bz, peth	$B5^4$, 301
8961	Menthol-(d) or 3-*p*-Menthol, Hexahydro thymol $C_{10}H_{20}O$	156.27	$[\alpha]_D$ + 49.2 (al,c=5)	$103-4^9$	42-3	...	...	al, eth, ace, bz	$B6^4$, 150
8962	Menthol-(dl) $C_{10}H_{20}O$	156.27	nd (peth)	216, $103-5^{16}$	(i)28 (ii)38	$0.904^{15/15}$	1.4615^{20}	al, eth, ace, bz	$B6^4$, 151
8963	Menthol-(l) $C_{10}H_{20}O$	156.27	nd (MeOH) $[\alpha]^{20/}_D$ −48 (al,c=2.5)	216.4, 111^{20}	44	$0.904^{15/15}$	1.460^{22}	al, eth, ace, chl	$B6^4$, 151
8964	Menthol, 3-isovalerate $C_{15}H_{28}O_2$	240.39	$[\alpha]^{20/}_D$ −64 (bz,c=10)	129^9	...	0.9089^{15}	1.4486^{20}	al, ace	$B6^3$, 145
8965	Menthol, acetate(l) $C_{12}H_{22}O_2$	198.31	$[\alpha]^{20/}_D$ −79.42	109^{10}	...	$0.9185^{20/4}$	1.4469^{20}	...	$B6^3$, 142
8966	Menthone(d) or 2-Isopropyl-5-methyl cyclohexanone $C_{10}H_{18}O$	154.25	$[\alpha]^{18/}_D$ + 24. 8	204^{750}, 85^{14}	...	$0.8963^{20/20}$	1.4503^{20}	al, eth, ace, bz, aa	$B7^3$, 154
8967	Menthone (dl) $C_{10}H_{18}O$	154.25	...	210.5	...	0.911^0	...	al, eth, ace, bz, aa	$B7^3$, 154
8968	Menthone (l) $C_{10}H_{18}O$	154.25	$[\alpha]^{20/}_D$-29.6	209.6, 96^{20}	...	$0.8954^{20/4}$	1.4505^{20}	al, eth, ace, bz	$B7^3$, 152
8969	Menthoxy acetic acid (l) $C_{12}H_{22}O_3$	214.30	cr (eth) $[\alpha]^{20/}_D$ −92.9 (MeOH)	171^{11}	53-5	...	...	al, eth, ace	$B6^3$, 155
8970	Menthoxyacetyl chloride (l) $C_{12}H_{21}ClO_2$	232.75	$[\alpha]^{13/}_D$ −84.8 (chl)	$128-31^{11}$	...	...	...	eth	$B6^3$, 156
8971	Mercaptoacetamide, N-Phenyl $HSCH_2CONHC_6H_5$	167.23	nd (al)	...	110-1	...	...	al, eth	$B12^3$, 925
8972	Mercaptoacetamide, N-β-naphthyl or Thionalide $HSCH_2CONH$-β-$C_{10}H_7$	217.29	nd	...	111-2	...	...	al	$B12^3$, 3052

PHYSICAL CONSTANTS OF ORGANIC COMPOUNDS (Continued)

No.	Name, Synonyms, and Formula	Mol. wt.	Color, crystalline form, specific rotation and λ_{max} (log ε)	b.p. °C	m.p. °C	Density	n_D	Solubility	Ref.
8973	Mercaptoacetic acid or Thioglycolic acid $HSCH_2COOH$	92.11	...	120^{20}	−16.5	1.3253^{20}	1.5030^{20}	**w, al, eth**	B3^4, 600
8974	Mercaptoacetic acid, acetate or Acetyl thioglycolic acid $(CH_3COS)CH_2CO_2H$	134.15	yo	$158\text{-}9^{17}$, $115\text{-}8^{2.5}$	...	...	...	w	B3^4, 610
8975	Mercaptoacetic acid, ethyl ester or Ethylmercapto acetate $HSCH_2CO_2C_2H_5$	120.17	...	156-8, 55^{17}	...	1.0964^{15}	1.4582^{20}	al, eth	B3^4, 617
8976	Mercaptoacetic acid, butyl ester or Butyl mercapto acetate $HSCH_2CO_2C_4H_9$	148.22	...	$85\text{-}8^{16}$	...	$1.03^{20/4}$	...	...	B3^4, 620
8977	Mercaptoacetic acid, dodecyl ester or Lauryl mercapto-acetate $HSCH_2CO_2C_{12}H_{25}$	260.44	...	...	3	$0.43^{20/4}$	...	...	B3^3, 432
8978	Mercaptoacetic acid, 2-ethylhexyl ester or 2-Ethylhexyl mercaptoacetate $HSCH_2CO_2[CH_2CH(C_2H_5)(CH_2)_3CH_3]$	204.33	...	133.5	...	$0.97^{20/4}$	...	...	...
8979	Mercaptoacetic acid, hexadecyl ester $HSCH_2CO_2C_{16}H_{33}$	316.54	...	...	20-5	...	...	chl	B3^4, 622
8980	Mercaptoacetic acid, isopropyl ester or isopropyl mercap-toacetate $HSCH_2CO_2CH(CH_3)_2$	134.19	...	$80\text{-}8^{45}$	...	$1.05^{20/4}$	...	...	B3^3, 430
8981	Mercaptoacetic acid, methyl ester or Methyl-mercaptoace-tate $HSCH_2CO_2CH_3$	106.14	...	$42\text{-}3^{10}$	...	...	1.4657^{20}	al, eth	B3^4, 614
8982	Mercaptoacetic acid, octyl ester or Octyl mercaptoacetate $HSCH_2CO_2C_8H_{17}$	204.33	...	125^{17}	...	...	1.4606^{21}	...	B3^3, 431
8983	Mesaconic acid or Methylfumaric acid $HO_2CC(CH_3)=CHCO_2H$	130.10	rh nd or mcl pr (eth, AcOEt)	sub	204-5	$1.466^{20/4}$	...	al, eth	B2^4, 2231
8984	Mesaconic acid, diethyl ester or Diethyl mesaconate $C_2H_5O_2CC(CH_3)=CCO_2C_2H_5$	186.21	...	229, $93\text{-}5^{10}$	...	$1.0453^{20/20}$	1.4488^{20}	al, eth, ace, bz	B2^4, 2232
8985	Mesaconic acid, dimethyl ester or Dimethyl mesaconate $CH_3O_2CC(CH_3)=CHCO_2CH_3$	158.15	...	203.5, 100^{10}	...	$1.0914^{20/4}$	1.4512^{20}	al, eth, ace	B2^4, 2232
8986	Mescaline or 3,4,5-trimethoxy-β-phenethylamine $C_{11}H_{17}NO_3$	211.26	cr	180^{12}	35-6	...	...	w, al, bz, chl	B13^3, 2375
8987	Mesitylene or 1,3,5-trimethylbenzene $1,3,5\text{-}(CH_3)_3C_6H_3$	120.19	...	164.7, 48.7^{10}	−44.7	$0.8652^{20/4}$	1.4994^{20}	**al, eth, ace, bz**	B5^4, 1016
8988	Mesitylene, acetyl or 2,4,6-trimethyl acetophenone $2,4,6\text{-}(CH_3)_3C_6H_2COCH_3$	162.23	...	240.5^{735}, 120^{12}	...	$0.9754^{20/4}$	1.5175^{20}	al, eth, ace, bz	B7^3, 1137
8989	Mesitylene, γ-bromo isobutylryl $2,4,6\text{-}(CH_3)_3C_6H_2COCBr(CH_3)_2$	269.18	gold-ye oil	$160\text{-}70^{24}$	27	...	...	eth	B7^3, 1209
8990	Mesitylene, vinyl $2,4,6(CH_3)_3C_6H_2CH=CH_2$	146.23	...	208-10, 83^{12}	...	$0.9057^{20/4}$	1.5296^{20}	...	B5^4, 1408
8991	Mesityl oxide or 4-Methyl-3-penten-2-one $(CH_3)_2C=CHCOCH_3$	98.14	...	129.7, 41^{25}	−52.9	$0.8653^{20/4}$	1.4440^{20}	w, **al, eth**, ace	B1^4, 3471
8992	Mesoxalic acid, hydrate or Dihydroxymalonic acid $(HO)_2C(CO_2H)_2$	136.06	dlq nd (w)	...	120-1	...	...	w, al, eth	B3^3, 1355
8993	Mesoxalic acid, diethyl ester $O=C(CO_2C_2H_5)_2$	174.15	pa ye gr oil	208^{220}, 105^{19}	ca −30	$1.1419^{16/4}$	1.4310^{22}	w, al, eth, chl	B3^3, 1356
8994	Mesoxalic acid, diethyl ester, hydrate or Ethyl dihydroxy malonate $(HO)_2C(CO_2C_2H_5)_2$	192.17	pl (bz)	ca 200	57	...	...	w, al, eth, ace, bz	B3^3, 1356
8995	Mesoxalic acid, diethyl ester, oxime $HON=C(CO_2C_2H_5)_2$	189.17	...	172^{12}	...	$1.1821^{18/4}$	1.4544^{18}	al, eth, ace, bz	B3^4, 1805
8996	Mesoxalonitrile or Oxomalonic acid, dinitrile $O{:}C(CN)_2$	80.05	...	65.5	−36	$1.124^{20/4}$	1.3919^{20}	eth, ace	B3^4, 1806
8997	Mestilbol or 3-(4-hydroxyphenyl)-4(4-methoxyphenyl)-3-Hexene $CH_3CH_2C(4\text{-}CH_3OC_6H_5)=C(4\text{-}HOC_6H_5)CH_2CH_3$	282.38	nd (bz-lig) lf (70% al)	$185\text{-}95^{0.3}$	117-8	...	...	al, eth, ace	B6^3, 5623
8998	Metacrolein or 2,4,6-triethynyl-1,3,5-trioxane $C_9H_{12}O_3$	168.19	pl (al)	170	50	...	...	al, eth	B1^4, 3435
8999	Metaldehyde or Metacetaldehyde $(C_2H_4O)_{4\text{-}6}$	(44.05)	tetr nd or pr (al)	sub 115	246.2 (sealed tube)	...	...	...	B19^4, 5643
9000	Metaldehyde II(tetramer) $(C_2H_4O)_4$	176.21	...	110, 65^{15}	47	...	...	al, eth, ace, bz	B19^4, 5643
9001	Metameconin or 5,6-Dimethoxy phthalide $C_{10}H_{10}O_4$	194.19	cr (dil al)	...	155-7	...	...	...	B18^4, 1223

PHYSICAL CONSTANTS OF ORGANIC COMPOUNDS (Continued)

No.	Name, Synonyms, and Formula	Mol. wt.	Color, crystalline form, specific rotation and λ_{max} (log ε)	b.p. °C	m.p. °C	Density	n_D	Solubility	Ref.
9002	Methacrolein $CH_2{=}C(CH_3)CHO$	70.09		68.4		$0.837^{20/4}$	1.4144^{20}	w, al, eth,	$B1^4$, 3455
9003	Methacryl amide $CH_2{=}C(CH_3)CONH_2$	85.11	cr (bz)		110-1			al	$B2^4$, 1538
9004	Methacrylic acid or 2-Methylpropenoic acid $CH_2{=}C(CH_3)COOH$	86.09	pr	$162\text{-}3^{757}$, 60^{12}	16	$0.0153^{20/4}$	1.4314^{20}	w, al, eth	$B2^4$, 1518
9005	Methacrylic acid, anhydride $[CH_2{=}C(CH_3)CO]_2O$	154.17		89^5			1.4540^{20}	al, eth	$B2^4$, 1537
9006	Methacrylic acid, butyl ester or Butyl methacrylate $CH_2{=}C(CH_3)CO_2C_4H_9$	142.20		160, 52^{11}		$0.9836^{20/4}$	1.4240^{20}	al, eth	$B2^4$, 1525
9007	Methacrylic acid, isobutyl ester or isobutyl methacrylate $CH_2{=}C(CH_3)CO_2\text{-}i\text{-}C_4H_9$	142.20		155, 45^{11}		$0.8858^{20/4}$	1.4199^{20}	al, eth	$B2^4$, 1526
9008	Methacrylic acid, ethyl ester or Ethyl methacrylate $CH_2{=}C(CH_3)CO_2C_2H_5$	114.14		117, 30^{18}		$0.9135^{20/4}$	1.4147^{20}	al, eth	$B2^4$, 1523
9009	Methacrylic acid, methyl ester or Methyl methacrylate $CH_2{=}C(CH_3)CO_2CH_3$	100.12		100-1, 24^{32}	−48	$0.9440^{20/4}$	1.4142^{20}	al, eth, ace	$B2^4$, 1519
9010	Methacrylic acid, propyl ester or Propyl methacrylate $CH_2{=}C(CH_3)CO_2C_3H_7$	128.17		141		$0.9022^{20/4}$	1.4190^{20}	al, eth	$B2^4$, 1524
9011	Methacrylic acid, isopropyl ester or Isopropyl methacrylate $CH_2{=}C(CH_3)CO_2\text{-}i\text{-}C_3H_7$	128.17		125		$0.8847^{20/4}$	1.4122^{20}	al, eth, ace, bz	$B2^4$, 1525
9012	Methacrylonitrile $CH_2{=}C(CH_3)CN$	67.09		90.3	−35.8	$0.7998^{20/4}$	1.4190^{20}	al, eth	$B2^4$, 1539
9013	Methacrylyl chloride $CH_2{=}C(CH_3)COCl$	104.54		96, 50^{135}		$1.0871^{20/4}$	1.4435^{20}	eth, ace, chl	$B2^4$, 1537
9014	Methallyl alcohol or 2-Methyl-2-propen-1-ol $CH_2{=}C(CH_3)CH_2OH$	72.11		114.5		0.8515^{20}	1.4255^{20}	w, al, eth	$B2^4$, 2114
9015	Methane CH_4	16.04	gas	−164	−182	0.466^{-164}, 0.5547^0		al, eth, bz	$B1^4$, 3
9016	Methane, bis (dimethyl amino) $[(CH_3)_2N]_2CH_2$	102.18		82.5		$0.7491^{18.7}$		w	$B4^4$, 153
9017	Methane, bis-(2-hydroxy-3,5,6-trichlorophenyl) or Hexachlorophene $(2\text{-}HO\text{-}3,5,6\text{-}Cl_3C_6H)_2CH_2$	406.91	nd (bz)		166-7			al, eth, ace, chl	$B6^3$, 5407
9018	Methane, bis (trichlorosilyl) $(Cl_3Si)_2CH_2$	282.92		182.7^{745}, 64^{10}		$1.5567^{20/4}$	1.4740^{20}	al, eth	$B1^4$, 3077
9019	Methane, bis (4-chlorophenoxy) or Oxythane $(4\text{-}ClC_6H_4O)_2CH_2$	269.13		$189\text{-}94^6$	69-70			eth, ace, bz	$B6^4$, 833
9020	Methane, bromo or Methyl bromide CH_3Br	94.94		3.6	−93.6	$1.6755^{20/4}$	1.4218^{20}	al, eth	$B1^4$, 68
9021	Methane, bromochloro CH_2BrCl	129.39		68.1	−86.5	$1.9344^{20/4}$	1.4838^{20}	al, eth, ace, bz	$B1^4$, 74
9022	Methane, bromochloro-dinitro $BrCCl(NO_2)_2$	219.38		$75\text{-}6^{15}$	9.3	$2.0394^{20/4}$	1.4793	al	$B1^3$, 115
9023	Methane, bromo-chloro-fluoro $BrCHClF$	147.37		36.1^{756}	−115	$1.9771^{0/4}$	1.4144^{25}	eth, ace, chl	$B1^4$, 75
9024	Methane, bromo-dichloro $BrCHCl_2$	163.83		90	−57.1	$1.980^{20/4}$	1.4964^{20}	al, eth, ace, bz, chl	$B1^4$, 76
9025	Methane, bromo-difluoro $BrCHF_2CHF_2$	130.92			−14.5	1.55^{16}		w, al	$B1^4$, 72
9026	Methane, bromo difluoro nitroso BrF_2CNO	159.92	bl gas	−12					$B1^4$, 100
9027	methane, bromo-diido $BrCHI_2$	346.73	ye (peth)	110^{25}	60				B1, 72
9028	Methane, bromo-diphenyl or Benzydryl bromide $(C_6H_5)_2CHBr$	247.13	tcl (peth)	193^{26}d, $111^{0.3}$	45			al, bz	$B5^4$, 1850
9029	Methane, bromo fluoro $BrCH_2F$	112.93		18-20				al, chl	$B1^4$, 72
9030	Methane, bromo-iodo $BrCH_2I$	220.84		138-41		2.926^{17}	1.6410^{20}	chl	$B1^4$, 95
9031	Methane, bromo-nitro $BrCH_2NO_2$	139.94		$148\text{-}9^{742}$			1.4880^{20}	al	$B1^4$, 106
9032	Methane, bromo-trichloro $BrCCl_3$	198.27		104.7, 0.6^{10}	−5.6	$2.0122^{20/4}$	1.5063^{20}	al, eth	$B1^4$, 77
9033	Methane, bromo-trifluoro $BrCF_3$	148.91	gas	-59^{740}				chl	$B1^4$, 73
9034	Methane, bromo-trinitro $BrC(NO_2)_3$	229.93		56^{10}	17-8	$2.0313^{20/4}$	1.4808^{20}	al, chl	$B1^3$, 116

No.	Name, Synonyms, and Formula	Mol. wt.	Color, crystalline form, specific rotation and λ_{max} (log ε)	b.p. °C	m.p. °C	Density	n_D	Solubility	Ref.
9035	Methane, chloro or Methyl chloride CH_3Cl	50.49	gas	−24.2	−97.1	$0.9159^{20/4}$	1.3389^{20}	al,eth, ace, bz, chl	$B1^4$, 28
9036	Methane, chloro-dibromo $ClCHBr_2$	208.28		$119\text{-}20^{748}$		$2.451^{20/4}$	1.5482^{20}	al, eth, ace, bz	$B1^4$, 81
9037	Methane, chloro-difluoro or Freon 22 $ClCHF_2$	86.47	gas	−40.8	−146			eth, ace, chl	$B1^4$, 32
9038	Methane, chloro difluoro nitro ClF_2CNO_2	131.47		25				chl	$B1^4$, 106
9039	Methane, chloro difluoro nitroso ClF_2CNO	115.47	bl gas	*ca* -35					$B1^4$, 99
9040	Methane, chloro diiodo $ClCHI_2$	302.28		200d, 88^{30}	−4			eth, ace, chl	$B1^4$, 97
9041	Methane, chloro dinitro $ClCH(NO_2)_2$	140.48		$34\text{-}6^{13}$		1.6125^{20}	1.4575^{20}		$B1^3$, 115
9042	Methane, chloro diphenyl or Benzhydryl chloride $(C_6H_5)_2CHCl$	202.68	nd	173^{19}	20.5	$1.1398^{20/4}$	1.5959^{20}		$B5^4$, 1847
9043	Methane, chloro-fluoro $ClCH_2F$	68.48	gas		−9.1			chl	$B1^4$, 32
9044	Methane, chloro-iodo $ClCH_2I$	176.38		109		$2.422^{20/4}$	1.5822^{20}	al, eth, ace, bz, chl	$B1^4$, 94
9045	Methane, chloro-nitro $ClCH_2NO_2$	95.49		122-3		1.466^{15}		w	$B1^4$, 106
9046	Methane, chloro tribromo $ClCBr_3$	287.18	lf (eth)	158-9	55	2.71^{15}		eth	$B1^4$, 85
9047	Methane, chloro trifluoro or Freon 13 $ClCF_3$	104.46	gas	−81.1	−181				$B1^4$, 34
9048	Methane, chloro trinitro $ClC(NO_2)_3$	185.48		133-5d, 56^{40}	4.5	$1.6769^{20/4}$	1.4500^{20}	al, eth, chl	$B1^3$, 116
9049	Methane, chloro triphenyl or Trityl chloride $(C_6H_5)_3CCl$	278.78	nd or pr (bz-peth)	310, $230\text{-}5^{20}$	113-4			eth, ace, bz, chl	$B5^3$, 2315
9050	Methane, deutero-trichloro $CDCl_3$	120.38		61-2	-64.1	$1.5004^{20/4}$	1.4450^{20}		$B1^4$, 54
9051	Methane, diazo or Acomethylene CH_2N_2	42.04	ye gas	*ca* -0	−145			eth	$B1^4$, 3056
9052	Methane, diazo diphenyl $(C_6H_5)_2CN_2$	194.24	bl-red nd (peth)	exp	30-2			al, eth, ace, bz	$B7^3$, 2068
9053	Methane, dibromo or Methylene bromide CH_2Br_2	173.83		97	−52.5	$2.4970^{20/4}$	1.5420^{20}	al, eth, ace	$B1^4$, 78
9054	Methane, dibromo chloro fluoro Br_2CClF	226.27		80.3		2.3173^{22}	$1.4570^{20/}{}_D$		$B1^4$, 82
9055	Methane, dibromo dichloro Br_2CCl_2	242.73		150.2	38			al, eth, ace, bz	$B1^4$, 82
9056	Methane, dibromo difluoro Br_2CF_2	209.82		24.5				al, eth, ace, bz	$B1^4$, 80
9057	Methane, dibromo dinitro $Br_2C(NO_2)_2$	263.83	nd	158d, 77^{21}	5.5	$2.4440^{20/4}$	1.5280^{25}	al	$B1^3$, 115
9058	Methane, dibromo fluoro $FCHBr_2$	191.83		64.9^{757}		$2.421^{20/4}$	1.4685^{20}	al, eth, ace, bz, chl	$B1^4$, 80
9059	Methane, dibromo iodo $ICHBr_2$	299.73	pl (peth)	91^{42}	22.5				$B1^3$, 99
9060	Methane, dichloro or Methylene chloride CH_2Cl_2	84.93		40	−95.1	$1.3266^{20/4}$	1.4242^{20}	al, eth	$B1^4$, 35
9061	Methane, dichloro difluoro or Freon 12 Cl_2CF_2	120.91		−29.8	−158	1.75^{-115}, 1.1834^{57}		al, eth, aa	$B1^4$, 40
9062	Methane, dichloro-dinitro $Cl_2C(NO_2)_2$	174.93		121-2, 46^{20}		$1.6124^{20/4}$	1.4575^{20}	al, eth, bz, chl	$B1^3$, 115
9063	Methane, dichloro diphenyl $(C_6H_5)_2CCl_2$	237.13		305d, 190^{21}		1.235^{18}			$B5^4$, 1848
9064	Methane, dichloro-fluoro or Freon 21 Cl_2CHF	102.92		9	−135	1.405^{9}	1.3724^{9}	al, eth, chl, aa	$B1^4$, 39
9065	Methane, dichloro-iodo Cl_2CHI	210.83		132, 40^{30}		$2.392^{20/4}$	1.5840^{20}	al, eth, ace, bz, chl	$B1^4$, 95
9066	Methane, dichloro-nitro Cl_2CHNO_2	129.93		107					$B1^3$, 113
9067	Methane, difluoro or Methylene fluoride CH_2F_2	52.02		−51.6		0.909^{20}	1.190^{20}	al	$B1^4$, 24

No.	Name, Synonyms, and Formula	Mol. wt.	Color, crystalline form, specific rotation and λ_{max} (log ε)	b.p. °C	m.p. °C	Density	n_D	Solubility	Ref.
9068	Methane, difluoro-iodo F_2CHI	177.92		21.6	−122	3.238^{-19}			B1^4, 92
9069	Methane, di-2-furyl $(2\text{-}C_4H_3O)_2CH_2$	148.16		$94^{22.5}$		$1.102^{20/4}$	1.5049^{20}	al, eth, ace	B19^4, 269
9070	Methane, diiodo or Methylene iodide CH_2I_2	267.84	ye nd or lf	182; 60^{10}	6.1	$3.3254^{20/4}$	1.7425^{20}	al, eth, bz, chl	B1^4, 96
9071	Methane, diiodo-fluoro $FCHI_2$	285.83	pa ye	100-1, 50^{50}	−34.5	3.1969^{22}		al, eth	B1^4, 97
9072	Methane, dinitro $CH_2(NO_2)_2$	106.04	ye nd	100 exp	<−15			al, eth	B1^4, 107
9073	Methane, dinitro diphenyl $(C_6H_5)_2C(NO_2)_2$	258.23	pl (dil al)		79-80			al, eth, bz, chl	B5^3, 1797
9074	Methane, diphenyl or Diphenyl nethane $(C_6H_5)_2CH_2$	168.24	pr nd	264.3, 125.5^{10}	25.3	$1.0060^{20/4}$	1.5753^{20}	al, eth, chl	B5^4, 1841
9075	Methane, diphenyl 3-tolyl $(3\text{-}CH_3C_6H_4)CH(C_6H_5)_2$	258.36	pr (al, MeOH)	354	62	1.07^{16}		eth, bz, chl, aa	B5^3, 2332
9076	Methane, di(α-pyrryl) or 2, 2′-Methylene dipyrrole $(2\text{-}C_4H_4N)_2CH_2$	146.19	lf or nd (al)	$163\text{-}7^{12}$	73			al, eth, bz	B23^2, 167
9077	Methane, disilano $(SiH_3)_2CH_2$	76.25		14.7^{754}		$0.6979^{4/4}$	1.4115^{4}		B1^4, 3072
9078	Methane, di-(4-tolyl) phenyl $(4\text{-}CH_3C_6H_4)_2CHC_6H_5$	272.39	nd (MeOH)	$218\text{-}20^{12}$	56			al, eth, bz, chl	B5^3, 2342
9079	Methane, fluoro or Methyl fluoride CH_3F	34.03		−78.4	−141.8	0.8428^{-60}	1.1727^{20}	al, eth, bz, chl	B1^4, 22
9080	Methane, fluoro-iodo FCH_2I	159.93		53.4		$2.366^{20/4}$	1.5256^{20}	eth, ace, bz, chl	B1^4, 92
9081	Methane, fluoro-tribromo $FCBr_3$	270.72	hyg nd	62.5^{15}	42	$1.0003^{20/4}$	1.4164^{20}	al	B1^4, 85
9082	Methane, iodo or Methyl iodide CH_3I	141.94		42.4	−66.4	$2.279^{20/4}$	1.5380^{20}	al, eth ace, bz	B1^4, 87
9083	Methane, iodo-trichloro $ICCl_3$	245.27		142		$2.355^{20/4}$	1.5854^{20}	eth, ace, bz, chl	B1^4, 95
9084	Methane, iodo-trifluoro CF_3I	195.91		−22.5		$2.3608^{-32/4}$	$1.3790^{-42}/_D$		B1^4, 92
9085	Methane, nitro CH_3NO_2	61.04		100.8	fr -17	$1.1371^{20/4}$	1.3817^{20}	al,eth,ace	B1^4, 100
9086	Methane, nitro-tribromo or Bromopicrin Br_3CNO_2	297.73		exp^{760}, $89\text{-}90^{20}$	10.2	$2.7930^{20/4}$	1.5790^{20}	al, eth, ace, bz, aa	B1^4, 106
9087	Methane, nitro-trichloro or Chloropicrin Cl_3CNO_2	164.38		111.8	−64.5	$1.6566^{20/4}$	1.4622^{20}	al, ace, bz, aa	B1^4, 106
9088	Methane, nitro-trifluoro or Fluoropicrin F_3CNO_2	115.01		−31.1					B1^4, 105
9089	Methane, nitroso-trifluoro F_3CNO	99.01		−84	−197				B1^4, 99
9090	Methane, (pentafluorothio)trifluoro $F_3C(SF_5)$	196.06		−20					B1^4, 35
9091	Methane, tetrabromo or Carbon tetrabromide CBr_4	331.63	mcl ta (dil al)	189-90, 102^{50}	90-4	$2.9609^{100/4}$	1.5942^{100}	al, eth	B1^4, 85
9092	Methane, tetrachloro or Carbon tetrachloride CCl_4	153.82		76.5	−23	$1.5940^{20/4}$	1.4601^{20}	al,eth, ace, bz, chl	B1^4, 56
9093	Methane, tetrafluoro or Carbon tetrafluoride CF_4	88.00		-129^{754}	−150	3.034^{0}		bz, chl	B1^4, 26
9094	Methane, tetraiodo or Carbon tetraiodide CI_4	519.63	red lf (bz,chl)	$130\text{-}40^{1\text{-}2}$	171d	4.23^{20}		chl, py	B1^4, 98
9095	Methane, tetranitro $C(NO_2)_4$	196.03		126, $21\text{-}3^{22}$	14.2	$1.6380^{20/4}$	1.4384^{20}	al, eth	B1^4, 107
9096	Methane, tetraphenyl $C(C_6H_5)_4$	320.43	rh nd (bz,sub)	431 sub					B5^3, 2568
9097	Methane, tribromo or Bromoform $CHBr_3$	252.73	hex sc	149.5, 46^{15}	8.3	2.8899	1.5976^{20}	al, eth, bz, chl, lig	B1^4, 82
9098	Methane, trichloro or Chloroform $CHCl_3$	119.38		61.7	−63.5	$1.4832^{20/4}$	1.4459^{20}	al, eth, ace, bz, lig	B1^4, 42
9099	Methane, tricyclohexyl $(C_6H_{11})_3CH$	262.48		322-9	48	$0.9274^{50/0}$	$1.4986^{40}/_D$	eth, bz	B5^4, 504
9100	Methane, trifluoro or Fluroform CHF_3	70.01		−82.2	−160	1.52^{-100}		al, ace, bz	B1^4, 24

No.	Name, Synonyms, and Formula	Mol. wt.	Color, crystalline form, specific rotation and λ_{max} (log ε)	b.p. °C	m.p. °C	Density	n_D	Solubility	Ref.
9101	Methane, triido or Iodoform CHI_3	393.73	ye hex pr or nd(ace)	*ca* 218	123	$4.008^{20/4}$		eth, ace, chl, aa	$B1^4$, 97
9102	Methane, trinitro or Nitroform $CH(NO_2)_3$	151.04		exp^{760}, $45\text{-}7^{22}$	19	$1.479^{20/4}$	1.4451^{24}	al, ace	$B1^4$, 107
9103	Methane, triphenyl or Triphenyl methane $(C_6H_5)_3CH$	244.34	rh (al)	$358\text{-}9^{754}$, $190\text{-}215^{10}$	94	$1.014^{99/4}$	1.5839^{99}	eth, ba, chl	$B5^3$, 2307
9104	Methane, tris (4-aminophenyl) or *p*-Leucaniline $(4\text{-}NH_2C_6H_4)_3CH$	289.38	lf (w,al,bz)		208			al, eth	$B13^3$, 566
9105	Methane, tris (4-dimethylaminophenyl) or Leucocrystal violet $[4\text{-}(CH_3)_2NC_6H_4]_3CH$	373.54	lf (al) nd (bz, lig)		175			eth, bz, chl, aa	$B13^3$, 566
9106	Methane, tris (2-tolyl) $(2\text{-}CH_3C_6H_4)_3CH$	286.42	nd (al)		130-1			al, eth	$B5^3$, 2347
9107	Methane arsonic acid $CH_3AsO(OH)_2$	139.97	lf (al)		160-1			w, al	$B4^3$, 1822
9108	Methanedisulfonic acid, dihydrate or Methionic acid $CH_2(SO_3H)_2.2H_2O$	212.18	hyg nd (w + 2)	$220\text{-}70^{15\text{-}20}$ d				w, al	$B1^4$, 3054
9109	Methane phosphonic acid or Methyl phosphonic acid $CH_3PO(OH)_2$	96.02	hyg pl	d	108-9			w, al, eth	$B4^3$, 1778
9110	Methanephosphonic acid, diethyl ester or Diethyl methylphosphonate $CH_3PO(OC_2H_5)_2$	152.13		$194, 85^{15}$		$1.0406^{30/4}$	1.4101^{30}	w, al, eth	$B4^4$, 1778
9111	Methanephosphonic acid, dimethyl ester or Dimethyl methylphosphonate $CH_3PO(OCH_3)_2$	124.08		181^{754}, 79.5^{20}				w	$B4^3$, 1778
9112	Methanesulfenyl chloride, trichloro or Perchloromethylmercaptan Cl_3CSCl	185.88	ye oil	147-8, 51^{25}		$1.6947^{20/4}$	1.5484^{20}	eth	$B3^4$, 290
9113	Methanesulfenylchloride, trifluoro or Perfluoromethylmercaptan F_3CSCl	136.52		-0.7					$B3^4$, 289
9114	Methane sulfinic acid, amino, imino $HN{=}C(NH_2)SO_2H$	108.11	nd (al)		144d			w	$B3^4$, 145
9115	Methane sulfonamide, trifluoro $F_3CSO_2NH_2$	149.09			119			w, chl	$B3^4$, 35
9116	Methanesulfonic acid CH_3SO_3H	96.10		167^{10}	20	$1.4812^{18/4}$	1.4317^{18}	w, al, eth	$B4^4$, 10
9117	Methanesulfonic acid, trifluoro F_3CSO_3H	150.07	hyg liq	$162, 81^{37.5}$				**eth**	$B3^4$, 34
9118	Methane sulfonamide, trifluoro-*N,N*-diethyl $F_3CSO_2N(C_2H_5)_2$	205.20		55^7				eth	$B4^4$, 415
9119	Methane sulfonyl chloride, trifluoro F_3CSO_2Cl	168.52		31.6					$B3^4$, 35
9120	Methane sulfonic acid, trifluoro,ethyl ester $F_3CSO_3C_2H_5$	178.13		$115, 42^{40}$				eth	$B3^4$, 34
9121	Methane sulfonyl chloride CH_3SO_2Cl	114.55		161^{730}, 55^{11}		$1.4805^{18/4}$	1.4573^{20}	al, eth	$B4^4$, 27
9122	Methane sulfonyl chloride, trichloro Cl_3CSO_2Cl	217.88	cr (al-w)	170 (sub)	140-1			al, eth	$B3^4$, 36
9123	Methane sulfonyl fluoride, trifluoro F_3CSO_2F	152.06		−21.7					$B3^4$, 34
9124	Methanethiol or Methyl mercaptan CH_3SH	48.10		6.2	−123	$0.8665^{20/4}$		al, eth	$B1^4$, 1273
9125	Methanetricarboxylic acid, trimethyl ester $HC(CP_2CH_3)_3$	190.15	pr (MeOH)	242.7, 128^{15}	46-7			al, eth, bz, chl	$B2^3$, 2023
9126	Methantheline bromide or Banthine bromide $C_{21}H_{26}NO_3Br$	420.35	cr (i-prOH)		172-7			w, al, eth	$B18^4$, 4352
9127	Methanol or Methyl alcohol CH_3OH	32.04		$65, 15^{73}$	−93.9	$0.7914^{20/4}$	1.3288^{20}	**w, al, eth, ace,** bz, chl	$B1^4$, 1227
9128	Methanol-*d* or *o*-Deutero methanol CH_3OD	33.05		65.5	−100	$0.8127^{20/4}$		**w, al, eth,** ace, bz	$B1^4$, 1244
9129	Methapyrilene (base) or Histadyl base $C_{14}H_{19}N_3S$	261.38		$173\text{-}5^3$			1.5915^{20}		$B22^4$, 3950
9130	Methionine *(DL)* or *dl*-2-amino-4-(methylthio) butyric acid $CH_3SCH_2CH_2CH(NH_2)CO_2H$	149.21	pl		281d	1.340		w	$B4^3$, 1647

No.	Name, Synonyms, and Formula	Mol. wt.	Color, crystalline form, specific rotation and λ_{max} (log ε)	b.p. °C	m.p. °C	Density	n_D	Solubility	Ref.
9131	Methionine *(L)* $CH_3SCH_2CH_2CH(NH_2)CO_2H$	149.21	hex pl (dil al $[\alpha]^{25}/_D$ −8.2 (w,c=1) +22.5 (INHCl)	sub 186	283d			w	B4^3, 1639
9132	Methoxyacetic acid or Methyl glycolic acid $CH_3OCH_2CO_2H$	90.08	hyg	203-4, 96^{13}		1.1768$^{20/4}$	1.4168^{20}	w, al, eth	B3^4, 574
9133	Methoxyacetic acid, ethyl ester or Ethyl methoxy acetate $CH_3OCH_2CO_2C_2H_5$	118.13		142, 44-5^9		1.0118^{15}	1.4050^{20}	al, eth	B3^3, 381
9134	Methoxyacetic acid, methyl ester or Methyl methoxy acetate $CH_3OCH_2CO_2CH_3$	104.11		131^{763}, 57^{50}		1.0511$^{20/4}$	1.3962^{20}	al, eth, ace	B3^4, 578
9135	2-Methoxybenzoic acid or *o*-Anisic acid 2-$CH_3OC_6H_4CO_2H$	152.15	pl (w), fl (al)	200	101			al, eth, bz, chl	B10^3, 97
9136	2-Methoxybenzoic acid, 5-chloro 5-Cl-2-$CH_3OC_6H_3CO_2H$	186.59	nd (w)		81-2			al, ace	B10^3, 165
9137	2-Methoxybenzoic acid, ethyl ester 2-$CH_3OC_6H_4CO_2C_2H_5$	180.20		261, 135-6^{12}		1.1124$^{20/4}$	1.5224^{20}	al, eth	B10^3, 116
9138	2-Methoxybenzoic acid, methyl ester 2-$CH_3OC_6H_4CO_2CH_3$	166.18		245, 127^{11}		1.1571$^{19/4}$	1.534$^{19.5}$	al	B10^3, 109
9139	3-Methoxybenzoic acid or *m*-Anisic acid 3-$CH_3OC_6H_4CO_2H$	152.15	nd (w)	170-2^{10}	110			al, eth, bz, chl	B10^3, 244
9140	3-Methoxybenzoic acid, ethyl ester 3-$CH_3OC_6H_4CO_2C_2H_5$	180.20		260-1, 110^5		1.0993$^{20/4}$	1.5161^{20}	al, eth	B10^3, 250
9141	3-Methoxybenzoic acid, methyl ester 3-$CH_3OC_6H_4CO_2CH_3$	166.18		252, 121-4^{10}		1.1310$^{20/4}$	1.5224^{20}	al	B10^3, 250
9142	4-Methoxybenzoic acid or *p*-Anisic acid 4-$CH_3OC_6H_4CO_2H$	152.15	pr or nd (w)	275-80	185			al, eth, chl	B10^3, 280
9143	4-Methoxybenzoic acid, butyl ester 4-$CH_3OC_6H_4CO_2C_4H_9$	208.26		183^{40}		1.054$^{16.5}$	1.5141$^{18.5}$		B10^3, 307
9144	4-Methoxybenzoic acid, ethyl ester 4-$CH_3OC_6H_4CO_2C_2H_5$	180.20		269-70	7-8	1.1038$^{20/4}$	1.5254^{20}	al, eth	B10^3, 300
9145	4-Methoxybenzoic acid, methyl ester 4-$CH_3OC_6H_4CO_2CH_3$	166.18	fl (al,eth)	256, 160^{20}	49			al, eth	B10^3, 297
9146	Methoxylamine, hydrochloride or *o*-Methyl hydroxylamine hydrochloride $CH_3ONH_2.HCl$	83.52	pr	149				w, al	B1^4, 1252
9147	2-Methoxyphenyl isothiocyanate 2-$CH_3OC_6H_4NCS$	165.21		131-2^{11}		1.1878^{20}	1.6458$^{20}/_D$		B13^3, 823
9148	Methylamine or Amino methane CH_3NH_2	31.06	gas	−6.3	−93.5	0.699^{-4}, 0.6628^{20}		w, al, eth, ace, bz	B4^4, 118
9149	Methylamine, hydrochloride $CH_3NH_2.HCl$	67.52	diq tetr ta (al)	sub 30^{15}	227-8			w, al	B4^4, 122
9150	Methyl isobutyl ketone or 4-methyl-2-pentanone *i*-$C_4H_9COCH_3$	100.16		116.8, 35-40^{16}	−84.7	0.7978^{20}	1.3962^{20}	al, eth, ace, bz, chl	B1^4, 3305
9151	β-Methyl chalcone $C_6H_5C(CH_3)=CHCOC_6H_5$	222.29		340-5d, 225^{22}		1.108$^{20/0}$	1.6312^{20}	eth	B7^3, 2422
9152	Methyl β-chloroethyl sulfide $CH_3SCH_2CH_2Cl$	110.60		140, 44^{20}		1.1097$^{25/4}$	1.4902^{20}	al, eth, ace	B1^4, 1406
9153	Methyl chlorosulfonate CH_3OSO_2Cl	130.55		133-5, 48^{29}		1.4805$^{25/4}$	1.4138^{18}	eth, ace, bz	B1^4, 1252
9154	β-Methyl choline chloride or (2-hydroxypropyl) trimethyl ammonium chloride $CH_3CH(OH)CH_2N^+(CH_3)_3\ Cl^-$	153.65	pr (BROH)	d	165			w, al	B4^3, 754
9155	Methyl ether or Dimethyl ether CH_3OCH_3	46.07	gas	−25	−138.5			w, al, eth ace, chl	B1^4, 1245
9156	Methyl ether, boro fluoride $(CH_3)_2O.BF_3$	113.88		127d	−14	1.2410$^{20/4}$	1.302^{20}		B1^4, 1248
9157	Methyl chloromethyl ether CH_3OCH_2Cl	80.51		59.1	−103.5	1.0605$^{20/4}$	1.3974^{20}	al., eth, ace, chl	B1^4, 3046
9158	Methyl (2,3-dibromopropyl) ether $CH_3OCH_2CHBrCH_2Br$	231.91		185, 84^{15}		1.8320$^{12/4}$	1.5123^{20}	eth	B1^4, 1447
9159	Methyl (2,3-epoxypropyl) ether or Epimethylin $CH_3OCH_2CHCH_2$ (O bridge)	89.11		115-8		0.9890^{20}	1.4320^{20}	w, al, eth, ace	B17^4, 986
9160	Methyl ethyl amine or Ethyl methyl amine $CH_3NHC_2H_5$	59.11		36.7				w, al, eth, ace	B4^4, 312

PHYSICAL CONSTANTS OF ORGANIC COMPOUNDS (Continued)

No.	Name, Synonyms, and Formula	Mol. wt.	Color, crystalline form, specific rotation and λ_{max} (log ε)	b.p. °C	m.p. °C	Density	n_D	Solubility	Ref.
9161	Methyl ethyl ketone or 2-Butanone $CH_3CH_2COCH_3$	72.11		$79.6, 30^{119}$	−86.3	$0.8054^{20/4}$	1.3788^{20}	w, **al**, **eth**, **ace**, bz	$B1^{4}$, 3243
9162	Methyl ethynyl ether or Methoxy acetylene $CH_3OC{\equiv}CH$	56.06		50		$0.8001^{20/4}$	1.3812^{20}	al, eth	$B1^{4}$, 2211
9163	Methyl glyoxal or 2-oxopropionaldehyde CH_3COCHO	72.06	ye hyg liq	72		1.0455^{24}	1.4002^{18}	al, eth, bz	$B1^{4}$, 3631
9164	Methylglyoxal, dioxime or Methyl glyoxime $CH_3C(NOH)CH{=}NOH$	102.09	nd (w, sub) Pr (al)	sub	157			al, eth	$B1^{4}$, 3633
9165	Methylglyoxal, 1-oxime $CH_3COCH{=}NOH$	87.08	nd (CCl_4) lf (eth-peth)	sub	69	1.0744^{67}		w, eth	$B1^{4}$, 3632
9166	Methyl green or Heptamethyl pararosaniline chloride $C_{26}H_{33}N_3Cl_2$	458.47	gr pw (al)					w	$B13^{3}$, 2074
9167	Methyl hydrogen sulfate CH_3OSO_3H	112.10		130-40d	<−30			w, al, **eth**	$B1^{4}$, 1250
9168	Methyl hydroperoxide CH_3OOH	48.04		$38\text{-}40^{65}$		$1.9967^{15/4}$	1.3641^{15}	w, **al**, **eth**, bz	$B1^{4}$, 1249
9169	Methyl isocyanate CH_3NCO	57.05		39.1-40.1	−45	$0.9230^{27/4}$	1.3419^{18}	w	$B4^{4}$, 247
9170	Methyl isothiocyanate or Methyl mustard oil CH_3NCS	73.11		119^{758}	36	$1.0691^{37/4}$	1.5258	**al**, eth	$B4^{4}$, 248
9171	Methyl *α*-naphthyl ether $CH_3O\text{-}\alpha\text{-}C_{10}H_7$	158.20		$269, 135^{10}$	<−10	$1.0964^{14/2}$	1.6227^{22}	al, eth, bz, chl	$B6^{3}$, 2922
9172	Methyl *β*-napthyl ether $CH_3O\text{-}\beta\text{-}C_{10}H_7$	158.20	lf (eth) pl (peth)	$274, 138^{10}$ sub	73-4			eth, bz, chl	$B6^{3}$, 2969
9173	Methyl nitrate CH_3ONO_2	77.04	exp.vapor	64.6 exp	−82.3	$1.2075^{20/4}$	1.3748^{20}	al, eth	$B1^{4}$, 1254
9174	Methyl nitrite CH_3ONO	61.04	gas	−12	−16	0.991^{15} (liq)		al, eth	$B1^{4}$, 1253
9175	Methyl nitro amine CH_3NHNO_2	76.05		$80\text{-}5^{10}$	38	1.2433^{49}	1.4616^{49}	w, al, eth, bz	$B4^{3}$, 1753
9176	Methyl (2-mitrophenyl) sulfide $(2\text{-}O_2NC_6H_4SCH_3$	169.20			64.5	$1.2628^{78/4}$	1.6246^{78}	al, bz, chl	$B6^{4}$, 1661
9177	Methyl (4-nitrophenyl) sulfide $(4\text{-}O_2NC_6H_4)SCH_3$	169.20		$135\text{-}40^{2}$	72	$1.2391^{80/4}$	1.64008^{20}		$B6^{4}$, 1687
9178	Methyl (2-octyn-1-yl) ether $CH_3O[CH_2C{\equiv}C(C_5H_{11})]$	140.23		77^{19}		$0.8370^{25/4}$	1.4380^{20}	eth	$B1^{3}$, 1996
9179	Methyl orange or Sodium 4′-dimethylamino azobenzene-4-sulfonate $C_{14}H_{14}N_3O_3NaS$	327.33	og,ye pl or sc (w)		d				$B16^{3}$, 371
9180	Methyl pentyl ether $CH_3OC_5H_{11}$	102.18		99-100		0.767^{19}	1.3855^{19}	al, eth, ace	$B1^{4}$, 1643
9181	Methyl *iso*pentyl ether $CH_3O\text{-}i\text{-}C_5H_{11}$	102.18		91^{765}		0.7517^{20}	1.3830^{20}	al, eth	$B1^{4}$, 1681
9182	Methyl-*tert*-pentyl ether $CH_3OC(CH_3)_2CH_2CH_3$	102.18		86.3		$0.7703^{20/4}$	1.3885^{20}	al, **eth**	$B1^{4}$, 1671
9183	Methyl pentyl sulfide $CH_3SC_5H_{11}$	118.24		145	−94	$0.8431^{20/4}$	1.4506^{20}	al, eth, ace, bz, chl	$B1^{3}$, 1608
9184	Methyl perchlorate CH_3OClO_3	114.49	oil	*ca* 52				**al**, **eth**	$B1^{4}$, 1249
9185	Methyl phenyl glyoxal $C_6H_5COCOCH_3$	148.16	ye oil	$222, 101^{12}$		$1.0065^{20/4}$	1.537^{10}	w, al, eth	$B7^{3}$, 3463
9186	Methylphenylglyoxal, dioxime or Methylphenylglyoxime $C_6H_5C({=}NOH)C({=}NOH)CH_3$	178.19	nd (dil al)		238-40			al	$B7^{3}$, 3465
9187	Methylphenylglyoxal, 1-oxime $C_6H_5C({=}NOH)COCH_3$	163.18	pa ye nd or lf (aa, al)		166-7				$B7^{3}$, 3464
9188	Methylphenylglyoxal, 2-oxime $C_6H_5COC({=}NOH)CH_3$	163.18	wh nd (w)		115				$B7^{3}$, 3464
9189	Methyl phenyl sulfide $CH_3SC_6H_5$	124.20		$193, 74^{10}$		$1.0579^{20/4}$	1.5868^{20}	al, bz	$B6^{4}$, 1466
9190	Methyl propyl ether $CH_3OC_3H_7$	74.12		38-9		$0.738^{20/4}$	1.3579^{25}	w, **al**, **eth**, ace	$B1^{4}$, 1421
9191	Methyl isopropyl ether $(CH_3)_2CHOCH_3$	74.12		32.5^{777}		$0.7237^{15/4}$	1.3576^{20}	al, eth	$B1^{4}$, 1471
9192	Methyl isopropyl ketone or 3-methyl-2-butanone $i\text{-}C_3H_7COCH_3$	86.13		94-5	−92	$0.8051^{20/4}$	1.3880^{20}	**al**, **eth**, ace	$B1^{4}$, 3287
9193	Methyl propyl sulfide $CH_3SC_3H_7$	90.18		95.5, -4^{10}	−113	0.8424^{20}	1.4442^{20}	w, al, eth, ace	$B1^{4}$, 1450

No.	Name, Synonyms, and Formula	Mol. wt.	Color, crystalline form, specific rotation and λ_{max} (log ε)	b.p. °C	m.p. °C	Density	n_D	Solubility	Ref.
9194	Methyl isopropyl sulfide $i\text{-}C_3H_7SCH_3$	90.18		84.7	−101.5	$0.8291^{20/4}$	1.4932^{20}	al, eth, ace	B1⁴, 1500
9195	*N*-Methyl quinolinium chloride $C_{10}H_9NCl$	178.64	cr (+w, al)		126			w, chl	B20⁴, 3357
9196	Methyl red or 4′ dimethylamino azobenzene-2-carboxylic acid $C_{15}H_{15}N_3O_2$	269.30	vt or red pr (to, bz) nd (aq, aa) lf (dil al)		183			al, ace, bz, chl, aa	B16³, 367
9197	Methyl selenide or Dimethyl selenide $(CH_3)_2Se$	109.03		$54\text{-}5^{753}$		$1.4077^{15/4}$		al, eth, chl	B1⁴, 1288
9198	Methyl sulfate or Dimethyl sulfate $(CH_3O)_2SO_2$	126.13		188.5d, 76^{15}	−31.7	1.3283^{20}	1.3874^{20}	w, al, eth, bz	B1⁴, 1251
9199	Methyl sulfide or Dimethyl sulfide $(CH_3)_2S$	62.13		37.3	−98.3	$0.8483^{20/4}$	1.4438^{20}	al, eth	B1⁴, 1275
9200	Methyl sulfite or Dimethyl sulfite $(CH_3O)_2SO$	110.13		126, 52^{45}		$1.2129^{20/4}$	1.4093^{20}	al, eth	B1⁴, 1250
9201	Methyl sulfone or Dimethyl sulfone $(CH_3)_2SO_2$	94.13	pr	238	110	$1.1707^{110/0}$	1.4226	w, al, bz	B1⁴, 1279
9202	Methyl sulfoxide or Dimethyl sulfoxide $(CH_3)_2SO$	78.13		189, $85\text{-}7^{20}$	18.4	$1.1014^{20/4}$	1.4770^{20}	w, al, eth, ace	B1⁴, 1277
9203	Methyl telluride or Dimethyl telluride $(CH_3)_2Te$	157.67	pa ye	93.5^{749}		>1		al	B1⁴, 1288
9204	Methyl thiocyanate CH_3SCN	73.11		132.9^{757}	−51	$1.0678^{25/4}$	1.4669^{25}	al, eth	B3⁴, 327
9205	Methyl vinyl ether $CH_3OCH{=}CH_2$	58.08		12	−122	$0.7725^{0/4}$	1.3730^{0}	al, eth, ace, bz	B1⁴, 2049
9206	Methyl vinyl ketone or 3-buten-2-one $CH_2{=}CH\text{-}COCH_3$	70.09		81.4, $33\text{-}4^{130}$		$0.8636^{20/4}$	1.4081^{20}	w, al, eth, ace, bz	B1⁴, 3444
9207	Methyl vinyl sulfide $CH_3SCH{=}CH_2$	74.14		69-70		$0.9026^{20/4}$	1.4837^{20}		B1⁴, 2065
9208	Methyl vinyl sulfone $CH_3SO_2CH{=}CH_2$	106.14		$122\text{-}4^{24}$		$1.2117^{20/4}$	1.4636^{20}	eth, ace	B1⁴, 2065
9209	Methylene blue or 3,9-bis Dimethylamino phenazothionium chloride $C_{16}H_{18}N_3ClS$	319.85	dk gr cr or pw (chl-eth)					w, al, chl	B27², 448
9210	5,5′-Methylene disalicylic acid or Bis-(3-carboxy-4 hydroxy phenyl)methane $[4\text{-}HO\text{-}3HO_2CC_6H_3]_2CH_2$	288.26	nd (bz)		243-4			al, eth, ace	B10³, 2507
9211	Methysticin or Kavatin $C_{15}H_{14}O_5$	274.27	nd (MeOH) pr (ace) $[\alpha]^{20}_D$ +94.34 (aa,p=5)		137			al, ace, bz, chl	B19⁴, 5161
9212	Metrazol or Cardiazole Leptazolo $C_6H_{10}N_4$	138.17	cr (bz-liq)	194^{12}	59-60			w, al, eth, ace, bz	B26², 213
9213	Metycaine or Piperocaine hydrochloride $C_{16}H_{19}NO_2Cl$	292.79			172-5			w, al, chl	B20⁴, 1460
9214	Michler's hydrol $(4\text{-}(CH_3)_2NC_6H_4)_2CHOH$	270.38	lt gr lf or pr (bz)		98			al, eth, bz, aa	B13³, 1958
9215	Mimosine *(l)* or *l*-Leucenol $C_8H_{10}N_2O_4$	198.18	ta (w) $[\alpha]^{22}_D$ −21 (w,c=0.5)	228-9d					B21⁴, 4648
9216	Morphine $C_{17}H_{19}NO_3$	285.34	pr		254-6			py, MeOH	B27², 118
9217	Morphine, acetate,trihydrate *(l)* $C_{17}H_{19}NO_3.CH_3CO_2H.3H_2O$	398.43	cr (dil al) $[\alpha]^{15}_D$ −77 (w)		200d			w	B27², 134
9218	Morphine, hydrate $C_{17}H_{19}NO_3.H_2O$	303.36	orh pr (dil al) $[\alpha]^{25}_D$ −132 (MeOH, c=1)	d	254-6 anh	$1.32^{20/4}$	1.56-1.64		B27², 122
9219	Morphine, hydrochloride,trihydrate $C_{17}H_{19}NO_3.HCl.3H_2O$	375.85	nd or fl (dil HCl) $[\alpha]^{25}_D$ −113.5 (w,c=2.2)		200d			w, al	B27², 132

PHYSICAL CONSTANTS OF ORGANIC COMPOUNDS (Continued)

No.	Name, Synonyms, and Formula	Mol. wt.	Color, crystalline form, specific rotation and λ_{max} (log ε)	b.p. °C	m.p. °C	Density	n_D	Solubility	Ref.
9220	Morphine, *N*-oxide or Genomorphine $C_{17}H_{19}NO_4$	301.34	pr (50% al)		274-5				B27[2], 159
9221	Morphine, sulfate,pentahydrate $2(C_{17}H_{19}NO_3).H_2SO_4.5H_2O$	758.84	pw or cubes $[\alpha]^{25/}_D$ −107.8 (w,c=4)		*ca* 250d			w	B27[2], 133
9222	Morphine, *o,o*-diacetyl or Heroin $C_{21}H_{23}NO_5$	369.42	rh$[\alpha]^{15/}_D$ −166 (MeOH)	272-4^{12}	173	1.56-1.61		bz, chl	B27[2], 151
9223	Morphine-*o,o*-diacetyl, hydrochloride, monohydrate $C_{21}H_{23}NO_5.HCl.H_2O$	423.89	$[\alpha]^{20/}_D$ −153 (w,c=1.17)		231-2			w, al, chl	B27[2], 153
9224	Morphine-3-ethyl ether hydrochloride, dihydrate or Dionin $C_{19}H_{23}NO_3.HCl.2H_2O$	385.89	cr		123-5d, 170 (anh)			w, al	B27[2], 148
9225	Morpholine or Tetrahydro-1,4-isoxazine C_4H_9NO	87.12	hyg	128.3, 24.8^{10}	−4.7	1.0005$^{20/4}$	1.4548^{20}	w, al, eth, ace, bz	B27[2], 3
9226	Morpholine, 4-acetyl 4-$CH_3CO(C_4H_8NO)$	129.16		152^{50}, 118^{12}	14.5	1.1165$^{20/20}$	1.4827^{20}	w, al, ace	C50, 7112
9227	Morpholine, 4-(2-aminoethyl) 4-$(H_2NCH_2CH_2)(C_4H_8NO)$	130.19		116^{50}	25.6	0.9915$^{20/20}$	1.4715^{20}	**w, al**, ace, **bz, lig**	C42, 6747
9228	Morpholine, 4-(3 aminopropyl) 4-$[H_2N(CH_2)_3](C_4H_8NO)$	144.22		219^{733}, 134^{50}	−15	0.9872$^{20/20}$	1.4762^{20}	**w, al**, ace, **bz, lig**	Am66, 725
9229	Morpholine, 4-benzyl 4-$C_6H_5CH_2(C_4H_8NO)$ $C_{11}H_{15}NO$	177.25		260-1, 128-9^{13}		1.0387$^{20/4}$	1.5302^{20}	ace, bz	B27[1], 203
9230	Morpholine, 4-butyl 4-$C_4H_9(C_4H_8NO)$ $C_8H_{17}NO$	143.23		213-4, 67-8^{10}	−57.1	0.9068$^{20/4}$	1.4451^{20}	w, al, ace, bz	Am61, 171
9231	Morpholine, 2,6,-dimethyl 2,6-$(CH_3)_2(C_4N_7NO)$	115.18		146.6, 58^{30}	fr -85	0.9346$^{20/20}$	1.4460^{20}	**w, al**, ace **bz, lig**	Am80, 1257
9232	Morpholine, 4-(2-ethoxy ethyl) 4-$(C_2H_5OCH_2CH_2)(C_4H_8NO)$ $C_8H_{17}NO_2$	159.23		206, 93-7^{14}	−100	0.963^{20}		**w**, eth, ace, bz	Am63, 298
9233	Morpholine, 4-ethyl 4-$C_2H_5(C_4H_8NO)$ $C_6H_{13}NO$	115.18		138-9^{763}		0.9886$^{20/4}$	1.4400^{20}	**w, al, eth,** ace, bz	B27[1], 203
9234	Morpholine, 4-(2-hydroxy ethyl) or 4-(*β*-morpholinol) ethanol $C_6H_{13}NO_2$	131.17		227^{757}		1.0710$^{20/4}$	1.4763^{20}	w, al	B27, 7
9235	Morpholine, 4-(2-hydroxy propyl) $C_7H_{15}NO_2$	145.20		92-4^{13}		1.0174$^{20/4}$	1.4638^{20}	w, al, eth, ace, bz	Am64, 970
9236	Morpholine, 4-methyl $C_5H_{11}NO$ 4-$CH_3(C_4H_8NO)$	101.15		115-6^{750}		0.9051$^{20/4}$	1.4332^{20}	w, al, eth	B27[1], 203
9237	Morpholine, 4-phenyl 4-$C_6H_5(C_4H_8NO)C_{10}H_{13}NO$	163.22	cr (al-eth)	259-60^{745}, 165-70^{45}	57-8			eth	B27[2], 3
9238	Morpholine, 4(4-tolyl) 4$(4-CH_3C_6H_4)(C_4H_8NO)$ $C_{11}H_{15}NO$	177.25	cr (dil al)	167^{30}	51			al, eth	B27[2], 4
9239	Mucic acid or 2,3,4,5-Tetrahydroxy hexanedioic acid or Galactaric acid $HO_2C(CHOH)_4CO_2H$	210.14	pr (w)		255 (rapid htng)				B3[4], 1292
9240	*cis*-Muconic acid or 2,4-Hexadiendioic acid $HO_2CCH{=}CHCH{=}CHCO_2H$	142.11			194-5			aa	B2[4], 2297
9241	*trans*-Muconic acid $HO_2CCH{=}CHCH{=}CHCO_2H$	142.11	nd (al)	320	305d				B2[4], 2298
9242	Murexide or Ammonium purpurate $C_8H_{10}N_6O_7$	302.20	red-gr pr (aq.NH_4Cl)						B25[1], 709
9243	Mycophenolic acid $C_{17}H_{20}O_6$	320.34	nd (w)		141			al, eth, chl	B18[4], 6513
9244	Myrcene, dihydro or *β*-Linaloolene $CH_2{=}CHCH(CH_3)CH_2CH_2CH{=}C(CH_3)_2$	138.25		165-8		0.7601$^{20/4}$	1.4362^{20}		B1[3], 1020
9245	Myrcene or -7-methyl-3-methylene-1,6-octadiene $(CH_3)_2C{=}CH(CH_2)_2\text{-}C({=}CH_2)CH{=}CH_2$	136.24		167, 65^{20}		0.8013$^{15/4}$	1.4722^{20}	al, eth, bz, chl, aa	B1[4], 1108
9246	Myricyl alcohol or 1-Triacontanol $CH_3(CH_2)_{28}CH_2OH$	438.82	nd (eth) pl (bz)		88	0.777^{95}		al, eth, bz	B1[3], 1850
9247	Myristaldehyde or Tetradecanal $C_{13}H_{27}CHO$	212.38	lf	166^{24}	30			al, eth, ace	B1[4], 3389
9248	Myristaldehyde, dimethyl acetal $C_{13}H_{27}CH(OCH_3)_2$	258.43		134-6^{4}			1.4342^{25}	al, eth	B1[4], 3389
9249	Myristaldehyde, oxime $C_{13}H_{27}CH{=}NOH$	227.39	lf or nd (al)		82-3			al	B1[2], 770
9250	Myristamide $C_{13}H_{27}CONH_2$	227.39	lf (ace)	217^{12}	105-7			al	B2[4], 1138

PHYSICAL CONSTANTS OF ORGANIC COMPOUNDS (Continued)

No.	Name, Synonyms, and Formula	Mol. wt.	Color, crystalline form, specific rotation and λ_{max} (log ε)	b.p. °C	m.p. °C	Density	n_D	Solubility	Ref.
9251	Myristic acid or Tetradecanoic acid $C_{13}H_{27}CO_2H$	228.38	lf (eth, 80% aa)	250.5^{100}, 149.3^{1}	58	$0.8439^{80/4}$	1.4305^{60}	al, ace, bz, chl	$B2^{4}$, 1126
9252	Myistic acid, anhydride $(C_{13}H_{27}CO)_2O$	438.73	lf (peth)	vac distb	53.4	$0.8502^{70/4}$	1.4335^{70}	al, eth	$B2^{4}$, 1138
9253	Myristic acid, benzyl ester or Benzyl myristate $C_{13}H_{27}CO_2CH_2C_6H_5$	318.50	....	229.3^{11}	20.5	$0.9321^{25/25}$	....	al, eth, bz, chl	$B6^{2}$, 417
9254	Myristic acid, ethyl ester or Ethyl myristate $C_{13}H_{27}CO_2H_5$	256.43	....	295, 162.5^{9}	12.3	$0.8573^{25/4}$	1.4362^{20}	al	$B2^{4}$, 1131
9255	Myristic acid, methyl ester or Methyl myristate $C_{13}H_{27}CO_2CH_3$	242.40	....	295^{751}, $155\text{-}7^{7}$	19	....	1.425^{45}	al, eth, ace, bz, chl	$B2^{4}$, 1131
9256	Myristic acid, propyl ester or Propyl myristate $C_{13}H_{27}CO_2C_3H_7$	270.46	....	147^{2}	......	0.8592^{20}	1.4356^{25}	al, eth, ace, bz	$B2^{4}$, 1132
9257	Myristic acid, isopropyl ester or Isopropyl myristate $C_{13}H_{27}CO_2CH(CH_3)_2$	270.46	....	192.6^{20}, 140.2^{2}	......	0.8532^{20}	1.4325^{25}	al, eth, ace, bz, chl	$B2^{4}$, 1132
9258	Myristicin or 1,3-Benzodioxide,4-methoxy-6-(2-propenyl) $C_{11}H_{12}O_3$	192.21	....	276-7, 157^{21}	<−20	$1.1437^{20/20}$	1.5403^{20}	eth, bz	$B19^{4}$, 801
9259	Myristonitrile $C_{13}H_{27}CN$	209.38	....	226.5^{100}, 119^{1}	19.2	$0.8281^{19/4}$	1.4392^{23}	al, eth, ace, bz, chl	$B2^{4}$, 1139
9260	Myristoyl chloride $C_{13}H_{27}COCl$	246.82	....	174^{16}	−1	....	....	eth	$B2^{4}$, 1138
9261	Myristyl alcohol or 1-Tetradecanol $C_{13}H_{27}CH_2OH$	214.39	lf	263.2, 167^{15}	39-40	$0.8236^{38/4}$	....	al, eth, ace, bz, chl	$B1^{4}$, 1864

PHYSICAL CONSTANTS OF ORGANIC COMPOUNDS (Continued)

No.	Name, Synonyms, and Formula	Mol. wt.	Color, crystalline form, specific rotation and λ_{max} (log ε)	b.p. °C	m.p. °C	Density	n_D	Solubility	Ref.
9262	Naphthacene or 2,3-Benzanthracene $C_{18}H_{12}$	228.29	og-ye lf (bz, xyl)	sub	357				$B5^2$, 2372
9263	Naphthacene, 9,10,-dihydro $C_{18}H_{14}$	230.31	nd (xyl) lf (bz)	*ca* 400	212			bz, aa	$B5^3$, 2304
9264	Naphthacene, 9,10,-diphenyl $C_{30}H_{20}$	380.49	og (eth)		207-8			eth	$B5^3$, 2703
9265	Naphthacene, 9,11-diphenyl $C_{30}H_{20}$	380.49	ye		301-2			eth, bz	$B5^3$, 2702
9266	Naphthacene, 9,10,11-triphenyl $C_{36}H_{24}$	456.59	og (eth) (bz)		236-7, 177 (+1 bz)			eth, bz	$B5^3$, 2766
9267	9,10-Naphthacene quinone or 2,3-Benzanthraquinone $C_{18}H_{10}O_2$	258.28	ye nd (aa)	sub	294			al	$B7^3$, 4273
9268	9,11-Naphthacene quinone $C_{18}H_{10}O_2$	258.28	dk red (aa,xyl)		322				$B7^3$, 4276
9269	1-Naphthaldehyde α-$C_{10}H_7CHO$	156.18	pa ye	292, 160^{15}	33-4	$1.1503^{20/4}$	1.6507^{20}	al, eth, ace, bz	$B7^3$, 1953
9270	1-Naphthaldehyde, 2-ethoxy 2-C_2H_5O-α-$C_{10}H_6CHO$	200.24	yesh nd (al,aa)	$185\text{-}7^{25}$	115			al, aa	$B8^3$, 1110
9271	1-Naphthaldehyde, 4-ethoxy 4-C_2H_5O-α-$C_{10}H_6CHO$	200.24	yesh cr (aa)		75			al, eth	$B8^2$, 174
9272	1-Naphthaldehyde, 2-hydroxy 2-HO-α-$C_{10}H_6CHO$	172.18	pr (al) nd (AcOEt)	192^{27}	82			al, eth, peth	$B8^3$, 1108
9273	2-Naphthaldehyde β-$C_{10}H_7CHO$	156.18	lf (w)	160^{19}	61-3	$1.0775^{99/4}$	1.6211^{99}	al, eth, ace	$B7^3$, 1957
9274	2-Naphthaldehyde, 1-hydroxy 1-HO-β-$C_{10}H_6CHO$	172.18	grsh-ye nd (dil al,dil aa,lig)		60			al, eth, aa	$B8^3$, 1118
9275	Naphthalene $C_{10}H_8$	128.17	mcl pl (al)	218, 87.5^{10}	80.5	$0.9625^{100/4}$, 1.0253^{20}	1.5898^{85}, 1.4003^{24}	al, eth, ace, bz	$B5^3$, 1549
9276	Naphthalene, 1-acetoxy or α-Naphthyl acetate α-$CH_3CO_2C_{10}H_7$	186.21	nd or pl (al)		49			al, eth	$B6^3$, 2928
9277	Naphthalene, 1-acetoxy-2-methyl 1-CH_3CO_2-2-$CH_3C_{10}H_6$	200.24	nd(eth-peth)		81-2			eth	$B6^3$, 3027
9278	Naphthalene, 1-acetoxy-4-methyl 1-CH_3CO_2-4-$CH_3C_{10}H_6$	200.24	nd(eth-peth)		86-8				C64, 19519
9279	Naphthalene, 1-acetoxy-7-methyl 1-CH_3CO_2-7-$CH_3C_{10}H_6$	200.24		188^{15}	38-41				$B6^3$, 3030
9280	Naphthalene, 2-acetoxy or β-Naphthyl acetate β-$CH_3CO_2C_{10}H_7$	186.21	nd (al)		70			al, eth, chl	$B6^3$, 2982
9281	Naphthalene, 5-acetoxy-2-methyl 5-CH_3CO_2-2-$CH_3C_{10}H_6$	200.24		124^3					$B6^3$, 3028
9282	Naphthalene, 1-acetyl or Methyl-α-naphthyl ketone α-$CH_3COC_{10}H_7$	170.21		296-8, 170^{20}	34	$1.1171^{21.5/4}$	1.6280^{22}	al, eth, ace	$B7^3$, 1960
9283	Naphthalene, 2-acetyl or Methyl-β-naphthyl ketone β-$CH_3COC_{10}H_7$	170.21	nd (liq, dil al)	301-3, $171\text{-}3^{11}$	56				$B7^3$, 1967
9284	Naphthalene, 1-allyl α-$(CH_2{=}CHCH_2)C_{10}H_7$	168.24		265-7, $129\text{-}30^{10}$		$1.0228^{20/4}$	1.6140^{20}	al, bz, chl	$B5^4$, 1740
9285	Naphthalene, 1(2-aminoethyl) α-$(H_2NCH_2CH_2)C_{10}H_7$	171.24		$182\text{-}3^{18}$				al, xyl	$B12^3$, 3112
9286	Naphthalene, β-(2-aminoethyl) β-$(H_2NCH_2CH_2)C_{10}H_7$	171.24		174^{25}				al, aa	$B12^3$, 3114
9287	Naphthalene, 1-aminomethyl α-$(H_2NCH_2)C_{10}H_7$	157.22	yesh turns red in air	294-5, $162\text{-}3^{12}$				al, eth	$B12^3$, 3097
9288	Naphthalene, 2-aminomethyl β-$(H_2NCH_2)C_{10}H_7$	157.22	pr (eth)	180^{24}	59-60			al, eth	$B12^3$, 3109
9289	Naphthalene, α-benzyl α-$(C_6H_5CH_2)C_{10}H_7$	218.30	mcl lf or ta (al)	350, $217\text{-}20^{19}$	59-60	1.166^{17}		al, eth, bz, chl	$B5^3$, 2236
9290	Naphthalene, β-benzyl or β-Benzyl naphthalene β-$(C_6H_5CH_2)C_{10}H_7$	218.30	mcl pr (al, MeOH)	350	58	1.176^0		eth, bz, chl,	$B5^3$, 2237
9291	Naphthalene, α-bromo α-$BrC_{10}H_7$	207.07	pr (β form)	281, 139^{16}	α-6.2, β2-3	$1.4826^{20/4}$	1.658^{20}	al, eth, ace, bz, chl	$B5^4$, 1665
9292	Naphthalene, 1-bromo-2-(bromomethyl) 2-$BrCH_2$-1-$BrC_{10}H_6$	299.99	nd (al), cr (peth)		107-8			al, bz, liq	$B5^3$, 1633
9293	Naphthalene, α-(bromo methyl) α-$(BrCH_2)C_{10}H_7$	221.10	cr (peth, al)	183^{18}, $145\text{-}50^2$	56			al, eth, ace, bz	$B5^4$, 1693

PHYSICAL CONSTANTS OF ORGANIC COMPOUNDS (Continued)

No.	Name, Synonyms, and Formula	Mol. wt.	Color, crystalline form, specific rotation and λ_{max} (log ε)	b.p. °C	m.p. °C	Density	n_D	Solubility	Ref.
9294	Naphthalene, β-bromo $\beta\text{-BrC}_{10}\text{H}_7$	207.07	pl or rh lf (al)	287-2, 147^{18}	59	1.605^{0}	1.6382^{60}	al, eth, bz, chl	$B5^4$, 1667
9295	Naphthalene, β-bromomethyl $\beta\text{-(BrCH}_2\text{)C}_{10}\text{H}_7$	221.10	lf(al)	213^{100}, $165\text{-}9^{14}$	56	...	...	al, eth, chl,	$B5^4$, 1698
9296	Naphthalene, α-butyl $\alpha\text{-C}_4\text{H}_9\text{C}_{10}\text{H}_7$	184.28	...	289.34, $151\text{-}2^{14}$	−19.76	$0.9738^{20/4}$	1.5819^{20}	al, eth, ace, bz,	$B5^4$, 1737
9297	Naphthalene, β-butyl $\beta\text{-C}_4\text{H}_9\text{C}_{10}\text{H}_7$	184.28	...	292, 146^{12}	−5	$0.9673^{20/4}$	1.57774^{20}	al, ace, bz	$B5^4$, 1738
9298	Naphthalene, β-*tert*-butyl $\beta\text{-[(CH}_3\text{)}_3\text{C]C}_{10}\text{H}_7$	184.28	...	$274\text{-}7^{756}$	−4	$0.9674^{20/4}$	$1.5685^{20/D}$	al, ace, bz	$B5^3$, 1805
9299	Naphthalene, α-chloro $\alpha\text{-ClC}_{10}\text{H}_7$	162.62	cr (al, ace)	258.8^{753}, 106.5^{5}	−2.3	$1.1938^{20/4}$	1.6326^{20}	al, eth, bz	$B5^4$, 1658
9300	Naphthalene, 1-chloro-2-nitro $\text{1-Cl-2-(O}_2\text{N)C}_{10}\text{H}_6$	207.62	pa ye nd (al, liq)	...	81	...	...	al	$B5^4$, 1677
9301	Naphthalene, 1-chloro-3-nitro $\text{1-Cl-3-(NO}_2\text{)C}_{10}\text{H}_6$	207.62	ye rd (al)	...	129.5	...	...	al	$B5^4$, 1677
9302	Naphthalene, 1-chloro-4-nitro $\text{1-Cl-4(NO}_2\text{)C}_{10}\text{H}_6$	207.62	br ye nd (peth,al)	...	87	...	...	al, eth	$B5^4$, 1676
9303	Naphthalene, 1-chloro-5-nitro $\text{1-Cl-5-(NO}_2\text{)C}_{10}\text{H}_6$	207.62	nd (dil al,aa)	>360, 181^{2}	111	...	...	al	$B5^3$, 1597
9304	Naphthalene, 1-chloro-6-nitro $\text{1-Cl-6-(NO}_2\text{)C}_{10}\text{H}_6$	207.62	ye nd (dil al)	...	131	...	...	al, eth, ace, bz, chl	$B5^3$, 1598
9305	Naphthalene, 1-chloro-8-nitro $\text{1-Cl-8-(NO}_2\text{)C}_{10}\text{H}_6$	207.62	lt ye nd (gl aa bz liq)	175^{2}	94-5	...	...	al, aa	$B5^4$, 1676
9306	Naphthalene, β-chloro $\beta\text{-ClC}_{10}\text{H}_7$	162.62	pl (dil al)lf	256, $121\text{-}2^{12}$	61	$1.1377^{71/4}$	1.6079^{13}	al, eth, bz	$B5^4$, 1660
9307	Naphthalene, 2-chloro-1-nitro $\text{1-NO}_2\text{-2-ClC}_{10}\text{H}_6$	207.62	nd (peth) ye nd (al)	>360	99-100	...	...	al, eth, ace, bz	$B5^4$, 1676
9308	Naphthalene, 2-chloro-3-nitro $\text{2-Cl-3(NO}_2\text{)C}_{10}\text{H}_6$	207.62	br cr or nd (al)	...	94.5	...	...	...	$B5^4$, 1677
9309	Naphthalene, 2-chloro-6-nitro $\text{2-Cl-6-(NO}_2\text{)C}_{10}\text{H}_6$	207.62	ye nd	$180\text{-}90^{15}$	170	...	...	...	$B5^3$, 1598
9310	Naphthalene, 2-chloro-7-nitro $\text{2-Cl-7(NO}_2\text{)C}_{10}\text{H}_6$	207.62	pa ye nd	...	136	...	...	...	$B5^4$, 1677
9311	Naphthalene, 2-chloro-8-nitro $\text{8-(NO}_2\text{)-2-ClC}_{10}\text{H}_6$	207.62	ye nd (al)	...	116	...	...	al, eth	$B5^4$, 1676
9312	Naphthalene, 3-chloro-1-nitro $\text{3-Cl-1-(NO}_2\text{)C}_{10}\text{H}_6$	207.62	grsh br nd	...	105	...	...	...	$B5^4$, 1676
9313	Naphthalene, 3-chloro-8-nitro $\text{8-(NO}_2\text{)-3-ClC}_{10}\text{H}_6$	207.62	nd (aq ace)	...	100.5	...	...	ace	$B5^4$, 1676
9314	Naphthalene, α-(chloromethyl) $\alpha\text{-(ClCH}_2\text{)C}_{10}\text{H}_7$	176.65	pr	291-2, $135\text{-}6^{6}$	32	...	...	al, bz	$B5^4$, 1692
9315	Naphthalene, β-(chloromethyl) $\beta\text{-(ClCH}_2\text{)C}_{10}\text{H}_7$	176.65	lf (al)	170^{20}	48-9	...	...	al, bz	$B5^4$, 1697
9316	Naphthalene, decahydro *(cis)* or Decalin $\text{C}_{10}\text{H}_{18}$	138.25	...	155.6, 69.4^{10}	−43	$0.8965^{22/4}$	1.4810^{20}	al, eth, ace, bz, chl	$B5^4$, 310
9317	Naphthalene, 1,2-diamino or 1,2-Naphthalenediamine $\text{1,2-(H}_2\text{N)}_2\text{C}_{10}\text{H}_6$	158.20	lf (w) (red in air)	214^{13}	98.5	...	...	al, eth, chl	$B13^3$, 377
9318	Naphthalene, 1,3-diamino-2-phenyl $\text{2-C}_6\text{H}_5\text{-1,3-(CH}_2\text{N)}_2\text{C}_{10}\text{H}_5$	234.30	pl, (MeOH, bz) (red in air)	...	116	...	...	al, bz	$B13^2$, 131
9319	Naphthalene, 1,4-diamino or 1,4-Nahthylene diamine $\text{1,4-(H}_2\text{N)}_2\text{C}_{10}\text{H}_6$	158.20	ye nd (w)	...	120	...	1.6441^{18}	...	$B13^3$, 383
9320	Naphthalene, 1,4-diamino-2-methyl $\text{2-CH}_3\text{-1,4-(H}_2\text{N)}_2\text{C}_{10}\text{H}_5$	172.23	ye cr (peth)	...	113-4	...	...	...	$B13^3$, 406
9321	Naphthalene, 1,4-diamino-2-methyl,dihydrochloride or Vitamin K_6 $\text{2-CH}_3\text{-1,4-(H}_2\text{N)}_2\text{C}_{10}\text{H}_5\text{.2HCl}$	245.15	cr (dil HCl)	...	300d	...	...	w	$B13^3$, 406
9322	Naphthalene, 1,5-diamino or 1,5-Naphthylenediamine $\text{1,5-(H}_2\text{N)}_2\text{C}_{10}\text{H}_6$	158.20	pr (eth,al,w)	sub	190	1.4	...	al, eth, chl	$B13^3$, 390
9323	Naphthalene, 1,5-diamino-2,methyl $\text{2-CH}_3\text{-1,5-(H}_2\text{N)}_2\text{C}_{10}\text{H}_5$	172.23	red-ye lf (dil al)	...	136	...	...	al, eth, bz	$B13^3$, 407
9324	Naphthalene, 1,6-diamino or 1,6-Naphthylene diamine $\text{1,6-(H}_2\text{N)}_2\text{C}_{10}\text{H}_6$	158.20	nd (w,eth)	...	85-6	$1.1477^{99/4}$	1.7083^{99}	al, bz	$B13^2$, 85
9325	Naphthalene, 1,7-diamino or 1,7-Naphthylenediamine $\text{1,7-(H}_2\text{N)}_2\text{C}_{10}\text{H}_6$	158.20	lf (bz)nd (w)	...	117.5	...	...	al, bz	$B13^3$, 398

No.	Name, Synonyms, and Formula	Mol. wt.	Color, crystalline form, specific rotation and λ_{max} (log ε)	b.p. °C	m.p. °C	Density	n_D	Solubility	Ref.
9326	Naphthalene, 1,8-diamino or 1,8-Naphthylenediamine $1,8-(CH_2N)_2C_{10}H_6$	158.20	nd(dil al)	205^{12},sub	66.5	$1.1265^{99/4}$	1.6828^{99}	al, eth	B13³, 398
9327	Naphthalene, 2,3-diamino or 2,3-Naphthylenediamine $2,3-(H_2N)_2C_{10}H_6$	158.20	lf (eth or w)		199	$1.0968^{26/4}$	1.6342^{26}	al, eth	B13³, 402
9328	Naphthalene, 2,6-diamino or 2,6-naphthylenediamine $2,6-(H_2N)_2C_{10}H_6$	158.20	nd or lf (w) lf (al)		222d				B13³, 402
9329	Naphthalene, 1,2-dichloro $1,2-Cl_2C_{10}H_6$	197.06	pl (al)	295-8, $151-3^{19}$	35-7	$1.3147^{49/4}$	1.5338^{49}	al, eth	B5⁴, 1661
9330	Naphthalene, 1,3-dichloro $1,3-Cl_2C_{10}H_6$	197.06	nd or pr (al)	291^{775}	61.5			al	B5⁴, 1661
9331	Naphthalene, 1,4-dichloro $1,4-Cl_2C_{10}H_6$	197.06	nd or pr (al,aa,ace)	$286-7^{740}$, 147^{12}	68	$1.2997^{76/4}$	1.6228^{76}	eth, ace, bz, aa	B5⁴, 1661
9332	Naphthalene, 1,5-dichloro $1,5-Cl_2C_{10}H_6$	197.06	nd or lf (al,aa) pr (sub)	sub	107				B5⁴, 1662
9333	Naphthalene, 1,6-dichloro $1,6-Cl_2C_{10}H_6$	197.06	nd or pr (al,peth,sub)	sub	49				B5⁴, 1662
9334	Naphthalene, 1,7-dichloro $1,7-Cl_2C_{10}H_6$	197.06	nd or pr (al, aa)	285-6	63-4	$1.2611^{100/4}$	1.6092^{100}	al, eth, bz, aa	B5⁴, 1662
9335	Naphthalene, 1,8-dichloro $1,8-Cl_2C_{10}H_6$	197.06	rh pl (hx) nd (al,sub)	d sub	89	$1.2924^{100/4}$	1.6236^{100}	al, peth	B5⁴, 1662
9336	Naphthalene, 2,3-dichloro $2,3-Cl_2C_{10}H_6$	197.06	rh lf (al)		120			eth	B5⁴, 1662
9337	Naphthalene, 2,6-dichloro $2,6-Cl_2C_{10}H_6$	197.06	pr (aa), nd or lf (al) pl (eth,bz)	285	140-1			eth, bz, chl, aa	B5⁴, 1662
9338	Naphthalene, 2,7-dichloro $2,7-Cl_2C_{10}H_6$	197.06	pl or lf (al)		114			al	B5⁴, 1662
9339	Naphthalene, 2,3-dihydrazino $2,3-(H_2NNH)_2C_{10}H_6$	188.23	red-br (al,w) col nd (bz)		167-8d			al	B15, 583
9340	Naphthalene, 1,2-dihydro $C_{10}H_{10}$	130.19	lf, pl	206-7, 78^{9}	−8	$0.9974^{20/4}$	1.5814^{20}		B5⁴, 1543
9341	Naphthalene, 1,2-dihydro-3-methyl $C_{11}H_{12}$	144.22		105^{13}		$0.9837^{20/20}$	1.5751^{20}	eth, bz	B5⁴, 1552
9342	Naphthalene, 1,2-dihydro-4-methyl $C_{11}H_{12}$	144.22		112^{18}		$0.9895^{20/4}$	1.5758^{20}	eth, bz	B5⁴, 1551
9343	Naphthalene, 1,4-dihydro $C_{10}H_{10}$	130.19	pl	211-2, 94^{17}	25(30)	$0.9928^{33/4}$	1.5577^{20}	aa	B5⁴, 1544
9344	Naphthalene, 2,3-dihydro-1,4-dimethyl $C_{12}H_{16}$	160.26		234		0.940^{20}	$1.528^{20/}{}_D$		35.5214) 35.5214)
9345	Naphthalene, 1,2-dihydroxy or 1,2-Naphthalenediol $1,2-(HO)_2C_{10}H_6$	160.17	lf or nd (CS_2) lf (w + 1) nd (liq)		103-4 (anh), 58-60 (+ 1w)			eth	B6³, 5240
9346	Naphthalene, 1,4-dihydroxy or 1,4-Naphthalenediol $1,4-(HO)_2C_{10}H_6$	160.17	mcl nd (bz,w)		192			al, eth, aa	B6³, 5260
9347	Naphthalene, 1,4-diacetoxy-2-methyl $2-CH_3-1,4-(CH_3CO_2)_2C_{10}H_5$	258.28	pr (al)		113			al	B6³, 5302
9348	Naphthalene, 1,5-dihydroxy or 1,5-Napthalene diol $1,5(HO)_2C_{10}H_6$	160.17	pr (w) nd (sub)	sub	265d			eth, ace, aa	B6³, 5265
9349	Naphthalene, 1,5-diacetoxy $1,5-(CH_3CO_2)_2C_{10}H_6$	244.25	nd (bz)		161			bz,aa	B6³, 5267
9350	Naphthalene, 1,6-dihydroxy or 1,6-Naphthylene diol $1,6-(HO)_2C_{10}H_6$	160.17	pr (bz)	sub	138			eth, ace, bz, MeOH	B6³, 5279
9351	Naphthalene, 1,7-dihydroxy $1,7(HO)_2C_{10}H_6$	160.17	nd (bz or sub)	sub	178-81			al, eth, bz, aa	B6³, 5281
9352	Naphthalene, 1,8-dihydroxy or 1,8-Napthalenediol $1,8-(HO)_2C_{10}H_6$	160.17	lf or nd (w)		144			al, eth, bz	B6³, 5283
9353	Naphthalene, 2,3-dihydroxy or 2,3 -Napthalenediol $2,3-(HO)_2C_{10}H_6$	160.17	lf(w)		163-4			al, eth, bz, aa, lig	B6³, 5287
9354	Naphthalene, 2,6-dihydroxy or 2,6-Naphthalenediol $2,6-(HO)_2C_{10}H_6$	160.17	rh pl (w)	sub	222			al, eth, ace, aa	B6³, 5287
9355	Naphthalene, 2,7-dihydroxy or 2,7-naphthalenediol $2,7-(HO)_2C_{10}H_6$	160.17	nd (w,dil al) pl (dil al)	sub	190-4			al, eth, bz, chl	B6³, 5291

No.	Name, Synonyms, and Formula	Mol. wt.	Color, crystalline form, specific rotation and λ_{max} (log ε)	b.p. °C	m.p. °C	Density	n_D	Solubility	Ref.
9356	Naphthalene, 1,5-dimercapto or 1,5-naphthylenedithiol $1,5\text{-}(HS)_2C_{10}H_6$	192.29	ye lf (al,eth,bz)		119			al, eth, bz	B6³, 5276
9357	Naphthalene, 1,2-dimethyl $1,2\text{-}(CH_3)_2C_{10}H_6$	156.23		266-7, 132^{12}	−1.6	$1.0179^{20/4}$	1.61656^{20}	eth, bz	B5⁴, 1708
9358	Naphthalene, 1,3-dimethyl $1,3\text{-}(CH_3)_2C_{10}H_6$	156.23		263, $138\text{-}40^{12}$	−6	$1.0144^{20/4}$	1.6140^{20}	eth, bz	B5⁴, 1708
9359	Naphthalene, 1,4-dimethyl $1,4\text{-}(CH_3)_2C_{10}H_6$	156.23		268, 129^{10}	7.6	$1.0166^{20/4}$	1.6127^{20}	al,eth, ace, bz, lig	B5⁴, 1709
9360	Naphthalene, 1,5-dimethyl $1,5\text{-}(CH_3)_2C_{10}H_6$	156.23		265	82			eth, bz	B5⁴, 1710
9361	Naphthalene, 1,6-dimethyl $1,6\text{-}(CH_3)_2C_{10}H_6$	156.23		264, 126^{13}	−16.9	1.0021^{20}	1.61656^{20}	eth, bz	B5⁴, 1711
9362	Naphthalene, 1,6-dimethyl-4-isopropyl or Cadalene $1,6\text{-}(CH_3)_2C_{10}H_5\text{-}i\text{-}C_3H_7\text{-}4$	198.31		$291\text{-}2^{720}$, 165^{20}		$0.9792^{19/4}$	1.5851^{19}		B5⁴, 1758
9363	Naphthalene, 1,7-dimethyl $1,7\text{-}(CH_3)_2C_{10}H_6$	156.23		263, 148^{15}	−13.9	$1.0115^{20/4}$	1.60831^{20}	eth, bz	B5⁴, 1711
9364	Naphthalene, 1,8-dimethyl $1,8\text{-}(CH_3)_2C_{10}H_6$	156.23		270, 140^{18}	65			eth, bz	B5⁴, 1712
9365	Naphthalene, 2,3-dimethyl or Guaiene $2,3\text{-}(CH_3)_2C_{10}H_6$	156.23	lf (al)	268, 128.7^{10}	105	$1.003^{20/4}$	1.5060^{20}	eth, bz	B5⁴, 1713
9366	Naphthalene, 1,2-dinitro $1,2\text{-}(NO_2)_2C_{10}H_6$	218.17			162-3				B5³, 1605
9367	Naphthalene, 1,3-dinitro $1,3\text{-}(NO_2)_2C_{10}H_6$	218.17	ye nd (bz,py-w)	sub	147-9			al, ace	B5⁴, 1680
9368	Naphthalene, 1,5-dinitro $1,5\text{-}(NO_2)_2C_{10}H_6$	218.17	hex nd (aa,ace)	sub	219			eth, bz	B5⁴, 1680
9369	Naphthalene, 1,7-dinitro $1,7\text{-}(NO_2)_2C_{10}H_6$	218.17			156			ace, bz	B5⁴, 1680
9370	Naphthalene, 1,8-dinitro $1,8\text{-}(NO_2)_2C_{10}H_6$	218.17	ye rh pl (chl)	445d	173			ace, py	B5⁴, 1681
9371	Naphthalene, 2,3-dinitro $2,3\text{-}(NO_2)_2C_{10}H_6$	218.17			172-4			al, bz	B5⁴, 1681
9372	Naphthalene, 2,4-dinitro-1-triazo $1\text{-}N_3\text{-}2,4\text{-}(NO_2)_2C_{10}H_5$	259.18	ye rh nd (al)		105d			eth, bz, chl	B5², 460
9373	Naphthalene, 2,6-dinitro $2,6\text{-}(NO_2)_2C_{10}H_6$	218.17			278			al, bz	B5³, 1609
9374	Naphthalene, 2,7-dinitro $2,7(NO_2)_2C_{10}H_6$	218.17			234			al, eth, bz	B5⁴, 1681
9375	Naphthalene, 1-ethoxy or Ethyl α-naphthyl ether $\alpha\text{-}C_{10}H_7OC_2H_5$	172.23	nd	280.5, $136\text{-}8^{14}$	5.5	$1.060^{20/4}$	1.5953^{25}	al, eth	B6³, 2924
9376	Naphthalene, 2-ethoxy or Ethyl β-naphthyl ether $\beta\text{-}C_{10}H_7OC_2H_5$	172.23	pl (al)	282, 148^{10}	37-8	$1.0640^{20/20}$	1.5975^{36}	al, eth, lig, to	B6³, 2972
9377	Naphthalene, α-ethyl $\alpha\text{-}C_2H_5C_{10}H_7$	156.23		258.6, 120^{10}	−13.9	$1.0082^{20/4}$	1.6062^{20}	al, eth	B5⁴, 1705
9378	Naphthalene, β-ethyl $\beta\text{-}C_2H_5C_{10}H_7$	156.23		258, 119^{10}	−7.4	$0,9922^{20/4}$	1.5999^{20}	al, eth	B5⁴, 1707
9379	Naphthalene, α-fluoro $\alpha\text{-}FC_{10}H_7$	146.16		215^{756}, 80^{11}	−9	1.1322^{20}	1.5939^{20}	al, eth, bz, chl, aa	B5⁴, 1657
9380	Naphthalene, β-fluoro $\beta\text{-}FC_{10}H_7$	146.16	nd (al)	211.5^{737}, 90^{16}	61			al, eth, bz, chl, aa	B5⁴, 1658
9381	Naphthalene, 1,2,3,4,9,10-hexahydro $C_{10}H_{14}$	134.22		200, 82^{2}		0.934^{23}	1.5260^{16}	eth, bz	B5⁴, 1082
9382	Naphthalene, 1-(α-hydroxyethyl)-*(dl)* $\alpha\text{-}C_{10}H_7CH(OH)CH_3$	172.23	nd (peth)	178^{15}	66	$1.1190^{14/4}$	1.6188^{25}	al, ace, bz, chl	B6³, 3034
9383	Naphthalene, 1-(α-hydroxy ethyl *(l)* $\alpha\text{-}C_{10}H_7\text{-}CH(OH)CH_3$	172.23	$[\alpha]^{20}{}_D$ -78.9 (al, c=5)	166^{11}	47	$1.1190^{14/4}$	1.6180^{25}	al, ace, bz, chl	B6³, 3034
9384	Naphthalene, α-hydroxymethyl $\alpha\text{-}(HOCH_2)C_{10}H_7$	158.20	nd (w,al) cr (bz-liq)	301^{715}, 163^{12}	64	$1.1039^{80/4}$		al, eth	B6³, 3024
9385	Naphthalene, α-iodo $\alpha\text{-}IC_{10}H_7$	254.07		302	4.2	$1.7399^{20/4}$	1.7026^{20}	al, eth, bz	B5⁴, 1670
9386	Naphthalene, β-iodo $\beta\text{-}IC_{10}H_7$	254.07	lf (dil al)	308, 172^{21}	54.5	$1.6319^{99/4}$	1.6662^{99}	al, eth, aa	B5⁴, 1671
9387	Naphthalene, α-mercapto or Naphthalene thiol $\alpha\text{-}C_{10}H_7SH$	160.23		285d, 161^{20}		$1.1607^{20/4}$	1.6802^{20}	al, eth	B6³, 2943
9388	Naphthalene, β-mercapto or β-Naphthalene thiol $\beta\text{-}C_{10}H_7SH$	160.23	pl(al)	288, 162.7^{20}	81	1.550		al, eth, lig	B6³, 3007

No.	Name, Synonyms, and Formula	Mol. wt.	Color, crystalline form, specific rotation and λ_{max} (log ε)	b.p. °C	m.p. °C	Density	n_D	Solubility	Ref.
9389	Naphthalene, α-methoxy or Methyl α-naphthyl ether α-$C_{10}H_7OCH_3$	158.20		269, 135^{10}	<−10	$1.0964^{14/2}$	1.6940^{25}	al, eth, bz, chl	$B6^{3}$, 2922
9390	Naphthalene, β-methoxy or Methyl β-naphthyl ether β-$C_{10}H_7OCH_3$	158.20	lf (eth) pl (peth)	274, 138^{10}	73-4			eth, bz, chl	$B6^{3}$, 2969
9391	Naphthalene, α-methyl α-$CH_3C_{10}H_7$	142.20		244.6, 107.4^{10}	−22	$1.0202^{20/4}$	1.6170^{20}	al, eth, bz	$B5^{4}$, 1687
9392	Naphthalene, 1-methyl-2-nitro 1-CH_3-2-$NO_2C_{10}H_6$	187.20	lt ye nd (al)		58-9			al	$B5^{4}$, 1693
9393	Naphthalene, 1-methyl-3-nitro 1-CH_3-3$NO_2C_{10}H_6$	187.20	ye nd (al)		81-2			al	$B5^{3}$, 1624
9394	Naphthalene, 1-methyl-4-nitro 1-CH_3-4-$NO_2C_{10}H_6$	187.20	pa ye nd (al)	$182\text{-}3^{18}$	71-2			al, eth, ace	$B5^{3}$, 1624
9395	Naphthalene, 1-methyl-5-nitro 1-CH_3-5-$NO_2C_{10}H_6$	187.20	brsh nd (al)		82-3			al, ace	$B5^{3}$, 1624
9396	Naphthalene, 1-methyl-6-nitro 1-CH_3-6$NO_2C_{10}H_6$	187.20	ye nd (dil al)		76-7			al	$B5^{3}$, 1625
9397	Naphthalene, 1-methyl-7-nitro 1-CH_3-7-$NO_2C_{10}H_6$	187.20	ye nd (al)		98-9			al	$B5^{3}$, 1625
9398	Naphthalene, 1-methyl-8-nitro 1-CH_3-8-$NO_2C_{10}H_6$	187.20	br lf (al)		65			al	$B5^{3}$, 1625
9399	Naphthalene, 1-methyl-7-isopropyl 1-CH_3-7-*i*-$C_3H_7C_{10}H_6$	184.28		152^{18}		$0.9740^{20/4}$	1.5833^{20}	eth, ace, bz	$B5^{4}$, 1741
9400	Naphthalene, β-methyl β-$CH_3C_{10}H_7$	142.20	mcl (al)	241, 104.7^{10}	34.6	$1.0058^{20/4}$	1.6015^{40}	al, eth, bz	$B5^{4}$, 1693
9401	Naphthalene, 2-methyl-1-nitro 2-CH_3-1-$NO_2C_{10}H_6$	187.20	yesh pr or nd (al)	188^{20}	81-2			al, ace	$B5^{4}$, 1698
9402	Naphthalene, 2-methyl-3-nitro 2-CH_3-3-$NO_2C_{10}H_6$	187.20	yesh pl (al)		117-8			al	$B5^{4}$, 1698
9403	Naphthalene, 2-methyl-4-nitro 2-CH_3-4-$NO_2C_{10}H_6$	187.20	pa ye nd (al)		49-50			al	$B5^{3}$, 1634
9404	Naphthalene, 2-methyl-5-nitro 2-CH_3-5-$NO_2C_{10}H_6$	187.20	yd nd (al)		61-2			al	$B5^{3}$, 1634
9405	Naphthalene, 2-methyl-6-nitro 2-CH_3-6-$NO_2C_{10}H_6$	187.20	ye nd (al)		119			al	$B5^{3}$, 1634
9406	Naphthalene, 2-methyl-7-nitro 2-CH_3-7-$NO_2C_{10}H_6$	187.20	yesh pl (al)		105			al	$B5^{3}$, 1634
9407	Naphthalene, 2-methyl-8-nitro 2-CH_3-8-$NO_2C_{10}H_6$	187.20	ye nd (al)		36-8			al	$B5^{3}$, 1635
9408	Naphthalene, α-nitramino α-$(O_2NNH)C_{10}H_7$	188.19	lt ye nd (w)		123-4			bz	$B16^{2}$, 346
9409	Naphthalene, β-nitramino β-$(O_2NNH)C_{10}H_7$	188.19	lf or nd		131-6				B16, 675
9410	Naphthalene, α-nitro α-$NO_2C_{10}H_7$	173.17	ye nd (al)	304 sub, $30\text{-}40^{0.01}$	61.5	$1.332^{20/4}$		al, eth, bz, py	$B5^{4}$, 1673
9411	Naphthalene, 1-nitro-5-triazo 1-NO_2-5-$N_3C_{10}H_6$	214.18	gold-ye nd (al)		121			al, ace	$B5^{2}$, 459
9412	Naphthalene, β-nitro β-$NO_2C_{10}H_7$	173.17	ye rh nd or pl (al)	312.5^{734}, 165^{15}	79			al, eth	$B5^{4}$, 1675
9413	Naphthalene, 2-nitro-1-triazo 2-NO_2-1-$N_3C_{10}H_6$	214.18	ye nd (dil ace)		103-4d			al, ace, bz, ace	B5, 565
9414	Naphthalene α-(nitrosohydroxyl amino) α-$C_{10}H_7(N(NO)OH)$	188.19	nd (peth)		54-5			chl	$B16^{3}$, 639
9415	Naphthalene, β-(nitrosohydroxylamino) β-$C_{10}H_7(N(NO)OH)$	188.19	nd (AcOEt-peth)		88-92			eth, aa	$B16^{1}$, 396
9416	Naphthalene, octachloro or Perchloro naphthalene $C_{10}Cl_8$	403.73	nd (bz-CCl_4)	$440\text{-}2^{7.4}$	197-8			bz, chl, lig	$B5^{4}$, 1665
9417	Naphthalene, α-isopentoxy or Isopentyl-α-naphthyl ether α-$C_{10}H_7O$-*i*-C_5H_{11}	214.31		$317\text{-}9^{742}$, $148\text{-}53^{3}$		$1.0069^{14/4}$	1.5705^{16}		$B6^{3}$, 2925
9418	Naphthalene, β-Isopentoxy or Isopentyl-β-naphthyl ether *i*-$C_5H_{11}O$-β-$C_{10}H_7$	214.31	lf	323-6d, $155\text{-}60^{6}$	26.5	$1.0155^{12/4}$	1.5768^{12}	al, eth	$B6^{3}$, 2974
9419	Naphthalene, α-phenyl α-$C_6H_5C_{10}H_7$	204.27	cr	334, 190^{12}	ca 45	$1.096^{20/4}$	1.6664^{20}	al, eth, bz, aa	$B5^{3}$, 2230
9420	Naphthalene, β-phenyl β-$C_6H_5C_{10}H_7$	204.27	lf (al)	345-6, $185\text{-}90^{5}$	103-4			al, eth, bz, aa, chl	$B5^{3}$, 2231

PHYSICAL CONSTANTS OF ORGANIC COMPOUNDS (Continued)

No.	Name, Synonyms, and Formula	Mol. wt.	Color, crystalline form, specific rotation and λ_{max} (log ε)	b.p. °C	m.p. °C	Density	n_D	Solubility	Ref.
9421	Naphthalene, picrate $C_{10}H_8.C_6H_3N_3O_7$	357.28	ye pr or pl (aa) cr (eth)		152	1.53		eth, bz	$B5^3$, 1568
9422	Naphthalene, 1-propoxy or Propyl α-naphthyl ether α-$C_{10}H_7OC_3H_7$	186.25		293.5, 167^{18}		$1.0447^{18/4}$	1.5928^{18}		$B6^3$, 2924
9423	Naphthalene, 2-propoxy or Propyl β-naphthyl ether β-$C_{10}H_7OC_3H_7$	186.25	nd (al)	305, 144^{10}	41			al	$B6^3$, 2973
9424	Naphthalene, α-propyl α-$C_3H_7C_{10}H_7$	170.25		274-5	−8.6				$B5^4$, 1721
9425	Naphthalene, β-isopropyl β-i-$C_3H_7C_{10}H_7$	170.25		268^2, $129\text{-}30^{14}$		$0.9753^{20/4}$	1.58482^{20}	al, eth, bz	$B5^4$, 1722
9426	Naphthalene, β-isopropyl β-i-$C_3H_7C_{10}H_7$	170.25			225			w, al, ace, bz	$B6^3$, 6699
9427	Naphthalene, 1,2,3,4-tetrahydroxy 1,2,3,4-$(HO)_4C_{10}H_4$	192.17			106-7			al, ace, bz	$B5^4$, 1745
9428	Naphthalene, 1,2,5,6-tetramethyl 1,2,5,6-$(CH_3)_4C_{10}H_4$	184.28		$150\text{-}5^{12}$	118			al, bz	$B5^4$, 1746
9429	Naphthalene, 1,3,6,8-tetramethyl 1,3,6,8-$(CH_3)_4C_{10}H_4$	184.28		$115\text{-}6^2$	84-5			al	$B5^4$, 1746
9430	Naphthalene, 1,3,5,8-tetranitro or γ-Tetranitro naphthalene. 1,3,5,8-$(NO_2)_4C_{10}H_4$	308.16	lt ye tetr (ace)		194-5			ace	$B5^4$, 1683
9431	Naphthalene, 1,3,6,8-tetranitro or β-tetranitro naphthalene. 1,3,6,8$(NO_2)_4C_{10}H_4$	308.16	ye nd (al,bz)	exp	207			bz, aa	$B5^4$, 1683
9432	Naphthalene, α-triazo or α-Naphthyl azide α-$C_{10}H_7N_3$	169.19	pa ye pr	d	12	1.1713^{25}	1.6550^{25}	al, eth, ace	$B5^4$, 1684
9433	Naphthalene, β-triazo or β-Napthyl azide β-$C_{10}H_7N_3$	169.19	pl (al nd (peth) ye in air		33			al, eth, ace. bz	$B5^3$, 1614
9434	Naphthalene, 1,2,3-trimethyl 1,2,3-$(CH_3)_3C_{10}H_5$	170.25		$125\text{-}30^{12}$	27-8				$B5^4$, 1725
9435	Naphthalene, 1,2,4-trimethyl 1,2,4-$(CH_3)_3C_{10}H_5$	170.25		146^{12}	55-6			al, eth, bz, chl	$B5^4$, 1725
9436	Naphthalene, 1,2,5-trimethyl 1,2,5-$(CH_3)_3C_{10}H_5$	170.25	nd (al)	140^{12}	33.5	$1.0103^{22/4}$	16093^{22}	eth, bz	$B5^4$, 1726
9437	Naphthalene, 1,2,6-trimethyl 1,2,6-$(CH_3)_3C_{10}H_5$	170.25	lf	146^{10}	14		1.6010^{20}	eth, bz	$B5^4$, 1726
9438	Naphthalene, 1,2,7-trimethyl 1,2,7-$(CH_3)_3C_{10}H_5$	170.25		$147\text{-}8^{16}$		$1.0087^{20/4}$	1.6097^{20}	eth, bz	$B5^4$, 1727
9439	Naphthalene, 2,3,6-trimethyl 2,3,6-$(CH_3)_3C_{10}H_5$	170.25		263-4, $146\text{-}8^{14}$	100-2			eth, bz	$B5^4$, 1730
9440	Naphthalene, 1,2,5-trinitro 1,2,5-$(NO_2)_3C_{10}H_5$	263.17	lt ye nd (al)		112-3			al	$B5^3$, 1614
9441	Naphthalene, 1,3,5-trinitro 1,3,5-$(NO_2)_3C_{10}H_5$	263.17	ye rh (chl)	364exp	122			al, ace, aa, chl	$B5^4$, 1682
9442	Naphthalene, 1,3,8-trinitro 1,3,8-$(NO_2)_3C_{10}H_5$	263.17	yesh mcl pr (al,ace,aa)		218			ace, py	$B5^4$, 1682
9443	Naphthalene, 1,4,5-trinitro 1,4,5-$(NO_2)_3C_{10}H_5$	263.17	ye lf or rh pl (al,aa,bz)		154			ace	$B5^2$, 458
9444	Naphthalene, β-vinyl β-$(CH_2{=}CH)C_{10}H_7$	154.21		$136\text{-}8^{17}$	66			al, ace, bz	$B5^4$, 1833
9445	1,2-Naphthalene dicarboxylic acid 1,2-$C_{10}H_6(CO_2H)_2$	216.19	nd (al) cr (w)		175d			al, eth, aa	$B9^3$, 4462
9446	1,2-Naphthalene dicarboxylic acid, 3,4-dihydro, anhydride 1,2-$C_{10}H_8(CO)_2O$	200.19	pa ye nd (lig,al)	$227\text{-}30^{23}$	126-7			bz, aa, MeOH	$B9^4$, 650
9447	1,4-Naphthalene dicarboxylic acid 1,4-$C_{10}H_6(CO_2H)_2$	216.19	rods (aa)		309 (320)			al, aa	$B9^3$, 4463
9448	1,5-Naphthalene dicarboxylic acid 1,5-$C_{10}H_6(CO_2H)_2$	216.19	nd		320-2d				$B9^3$, 4465
9449	1,6-Naphthalene dicarboxylic acid 1,6-$C_{10}H_6(CO_2H)_2$	216.19	nd (aa)		310 sinters			al,aa	$B9^2$, 651
9450	1,7-Naphthalene dicarboxylic acid 1,7-$C_{10}H_6(CO_2H)_2$	216.19	ye pw (dil al,aa)		308d			al, eth, ace, aa	$B9^3$, 4466
9451	2,3-Naphthalene dicarboxylic acid 2,3-$C_{10}H_6(CO_2H)_2$	216.19	pr (aa,w) pr (sub)		246				$B9^3$, 4470

PHYSICAL CONSTANTS OF ORGANIC COMPOUNDS (Continued)

No.	Name, Synonyms, and Formula	Mol. wt.	Color, crystalline form, specific rotation and λ_{max} (log ε)	b.p. °C	m.p. °C	Density	n_D	Solubility	Ref.
9452	2,6-Naphthalene dicarboxylic acid 2,6-$C_{10}H_6(CO_2H)_2$	216.19	nd (al or sub)		>300d			al	B9^{3}, 4471
9453	2,7-Naphthalene dicarboxylic acid 2,7-$C_{10}H_6(CO_2H)_2$	216.19	nd (w,al,dil HCl)		>300d			al	B9^{2}, 653
9454	1,3-Naphthalene disulfonic acid, 7-amino or Amino-G-acid 7-NH_2-1,3-$C_{10}H_5(SO_3H)_2$	303.30	mcl pr or nd (w + 2)		273-5			w, al	B14^{3}, 2263
9455	1,3-Naphthalene disulfonic acid, 7-hydroxy or G-acid 7-HO-1,3-$C_{10}H_5(SO_3H)_2$	304.29						w	B11^{3}, 560
9456	1,5-Naphthalene disulfonic acid 1,5-$C_{10}H_6(SO_3H)_2$	288.29	pl (+4w,dil aa)		240-5d	1.493		w, al	B11^{3}, 17
9457	1,6-Naphthalene disulfonic acid 1,6-$C_{10}H_6(SO_3H)_2$	288.29	og pr (+4w,aa or w)		125d (anh)			w, al	B11^{3}, 467
9458	2,7-Naphthalene disulfonic acid 2,7-$C_{10}H_6(SO_3H)_2$	288.29	hyg nd (con HCl)		199			w	B11^{3}, 468
9459	2,7-Naphthalene disulfonic acid, 4-amino-5-hydroxy or H acid 4-NH_2-5-HO-2,7-$C_{10}H_4(SO_3H)_2$	319.30							B14^{3}, 2292
9460	2,7-Naphthalene disulfonic acid, 4,5-dihydroxy or Chromotropic acid 4,5-$(HO)_2$-2,7-$C_{10}H_4(SO_3H)_2$	320.29	nd or lf (w + 2)					w	B11^{3}, 576
9461	2,7-Naphthalene disulfonic acid, 3-hydroxy or R-acid 3-HO-2,7-$C_{10}H_5(SO_3H)_2$	304.29	dlq nd		d			w, al	B11^{3}, 559
9462	α-Naphthalene phosphonic acid α-$C_{10}H_7PO(OH)_2$	208.15	cr (w)		189			al	B16^{3}, 896
9463	α-Naphthalene phosphonyl chloride α-$C_{10}H_7PO(Cl)_2$	245.04			*ca* 60				B16^{2}, 392
9464	α-Naphthalene sulfinic acid α-$C_{10}H_7SO_2H$	192.23	nd (w)		104			w, al	B11^{3}, 12
9465	β-Naphthalene sulfinic acid β-$C_{10}H_7SO_2H$	192.23	nd (w)		98(105)			w, al, eth	B11^{3}, 14
9466	α-Naphthalene sulfonic acid α-$C_{10}H_7SO_3H$	208.23	pr (+2w,dil HCl)		90 (hyd) 139-40 (anh)			w, al	B11^{3}, 382
9467	α-Naphthalene sulfonic acid, 4-amino or Naphthionic acid 4-H_2N-α-$C_{10}H_6SO_3H$	232.25	wh nd (w + ½) red-br cr		d	$1.6703^{25/4}$		py, MeOH	B14^{3}, 2241
9468	α-Naphthalene sulfonic acid, 4-amino-5-hydroxy or S-acid 4-H_2N-5-HO-α-$C_{10}H_5SO_3H$	239.25	nd						B14^{3}, 2290
9469	α-Naphthalene sulfonic acid, 7-amino or Bayer's acid, Cassela's acid 7-H_2N-α-$C_{10}H_6SO_3H$	223.25	nd (w + 1) pl (aq ace)					aa	B14^{3}, 2245
9470	α-Naphthalene sulfonic acid, 4-hydroxy or Nenile-winther acid 4-HO-α-$C_{10}H_6SO_3H$	224.23	ta or pl (w)		170d (rapid htg)				B11^{3}, 540
9471	α-Naphthalene sulfonic acid, 5-hydroxy or α-Naphtholsulfonic acid L 5-HO-α-$C_{10}H_6SO_3H$	224.23	dlq		120			w, aa	B11^{3}, 541
9472	α-Naphthalene sulfonic acid, 7-hydroxy or Croceic acid 7-HO-α-$C_{10}H_6SO_3H$	224.23						w	B11^{3}, 556
9473	α-Naphthalene sulfonic acid, 8-hydroxy or α-Naphthol sulfonic acid S 8-HO-α-$C_{10}H_6SO_3H$	224.23	cr (w + 1)		106-7			w	B11^{2}, 157
9474	α-Naphthalene sulfonic acid, 8-hydroxy,lactone or 1-Naphthol-8-sulfonic acid sultone $C_{10}H_6O_3S$	206.22	pr (bz)	>360	154			bz, chl	B19^{4}, 323
9475	α-Naphthalene sulfonyl chloride α-$C_{10}H_7SO_2Cl$	226.68	lf (eth)	195^{11}, $147.5^{0.9}$	68			al, eth, bz	B11^{3}, 383
9476	β-Naphthalene sulfonic acid β-$C_{10}H_7SO_3H$	208.23	dlq pl (+1w) cr (+3w,dil HCl)	d	124-5 (+1w) 83(+3 w)	$1.441^{25/4}$		w, al, eth	B11^{3}, 397
9477	β-Naphthalene sulfonic acid, 4-amino or Cleve's acid 4-H_2N-β-$C_{10}H_6SO_3H$	223.25	md (w + 1)						B14^{3}, 2248
9478	β-Naphthalene sulfonic acid, 5,7-dinitro-8-hydroxy or Flavianic acid 5,7-$(NO_2)_2$-8-HO-β-$C_{10}H_4SO_3H$	314.23	pa ye nd (con HCl, + 3w) cr (w)		100 (+3w) 151 (anh)			w, al	B11^{3}, 542

No.	Name, Synonyms, and Formula	Mol. wt.	Color, crystalline form, specific rotation and λ_{max} (log ε)	b.p. °C	m.p. °C	Density	n_D	Solubility	Ref.
9479	β-Naphthalene sulfonic acid, 1-hydroxy or α-Naphthol sulfonic acid 1-HO-β-$C_{10}H_6SO_3H$	224.23	pl (w)		>250			w, al	B11^{3}, 540
9480	β-Naphthalene sulfonic acid, 6-hydroxy or Schaeffer acid 6-HO-β-$C_{10}H_6SO_3H$	224.23	lf cr (w + 1)		167 (anh) 129 (+1w) 118 (+2w)			w, al, aa	B11^{3}, 553
9481	β-Naphthalene sulfonic acid, 7-hydroxy or F acid 7-HO-β-$C_{10}H_6SO_3H$	224.23	nd (HCl) cr (w + 1,2 or 4)		115-6 (anh) 108-9 (+1w)			w, al	B11^{3}, 555
9482	β-Naphthalene sulfonyl chloride β-$C_{10}H_7SO_2Cl$	226.68	pw or lf (bz-peth)	201^{13}, $148^{0.5}$	79			al, eth, bz, chl	B11^{3}, 399
9483	1,4,5,8-Naphthalene tetracarboxylic acid 1,4,5,8-$C_{10}H_4(CO_2H)_4$	304.21	lf or nd (w,dil HCl)		320			ace, aa	B9^{3}, 4889
9484	1,4,5,8-Naphthalene tetracarboxylic acid, 1,8: 4,5-dianhydride $C_{14}H_4O_6$	268.18	nd (al)	sub 320^{3}	>300				B19^{4}, 2258
9485	1,2,5-Naphthalene tricarboxylic acid 1,2,5-$C_{10}H_5(CO_2)_3$	260.20	nd (MeOH or sub)		270-2			w, MeOH	B9^{4}, 4835
9486	1,4,5-Naphthalene tricarboxylic acid 1,4,5-$C_{10}H_5(CO_2)_3$	260.20	cr (eth or con HCl)		266-8			al, eth	B9^{3}, 4835
9487	Naphthalic acid or 1,8-Naphthalene dicarboxylic acid 1,8-$C_{10}H_6(CO_2H)_2$	216.19	nd (al)		260				B9^{3}, 4466
9488	Naphthalic anhydride 1,8-$C_{10}H_6(CO)_2O$	198.18	nd (al) pr (aa) lf (sub)	sub	274			aa	B17^{4}, 6392
9489	Naphthalic acid, diethyl ester or Diethyl naphthalate 1,8-$C_{10}H_6(CO_2C_2H_5)_2$	272.30	yesh mcl or lf (dil al)	$238\text{-}9^{19}$	59-60	$1.1399^{70/4}$	1.5586^{70}	al, eth	B9^{3}, 4467
9490	Naphthalic acid, dimethyl ester or Dimethyl naphthalate 1,8-$C_{10}H_6(CO_2CH_3)_2$	244.25	nd (al) pr MeOH)		104			al, aa, MeOH	B9^{3}, 4467
9491	Naphthalic acid, 3,6-dinitro or 3,6-Dinitronaphthalic acid 3,6-$(NO_2)_2$-1,8-$C_{10}H_4(CO_2H)_2$	306.19	silvery lf (w)		212			al, aa	B9^{3}, 4470
9492	Naphthalic acid, imide 1,8-$C_{10}H_6(CO)_2NH$	197.19	nd (chl-al)		300				B21^{4}, 5557
9493	Naphthalic anhydride, 3-nitro 3-NO_2-1,8-$C_{10}H_5(CO)_2O$	243.18	yesh nd (aa)		252-3			aa	B17^{4}, 6396
9494	Naphthalic acid, 4-nitro 4-NO_2-1,8-$C_{10}H_5(CO_2H)_2$	261.19	ye nd		d140-50			aa	B9^{2}, 653
9495	Naphthaloyl chloride 1,8-$C_{10}H_6(COCl)_2$	253.08	pr (CS_2)	$195\text{-}200^{0.2}$	84-6			bz, chl	B9^{3}, 4468
9496	1-Naphthamidine $C_{11}H_{10}N_2$	170.21	lf (dil al)		154			al, ace, chl	B9^{3}, 3147
9497	2-Naphthamidine $C_{11}H_{10}N_2$	170.21	cr (bz)		133-6			al, bz, aa	B9^{3}, 3187
9498	α-Naphthamide α-$C_{10}H_7CONH_2$	171.20	nd or pl (al,gl aa)	sub	204-5			aa	B9^{3}, 3145
9499	Naphthocaine, hydrochloride $C_{17}H_{23}N_2O_2Cl$	322.83	pa ye		212-6				B14^{3}, 1333
9500	α-Naphthoic acid or α-Naphthalene carboxylic acid α-$C_{10}H_7CO_2H$	172.18	nd (aa-w,w,al)	>300, 231^{50}	161	1.398		al, eth, chl	B9^{3}, 3136
9501	α-Naphthoic acid-2-amino 2-H_2N-α-$C_{10}H_6CO_2H$	187.20	nd (dil al)		126d			al, eth, aa	B14^{3}, 1330
9502	α-Naphthoic acid-3-amino 3-H_2N-α-$C_{10}H_6CO_2H$	187.20	ye or pksh nd (eth)		181-2			al, MeOH	B14^{3}, 1330
9503	α-Naphthoic acid-4,amino 4-H_2N-α-$C_{10}H_6CO_2H$	187.20	brsh nd (w,al)		177			al, eth, ace, aa	B14^{3}, 1332
9504	α-Naphthoic acid-5-amino 5-H_2N-α-$C_{10}H_6CO_2H$	187.20	og nd (w, dil al) (sub) nd	211-2, 196 (sub)				al, aa	B14, 553
9505	α-Naphthoic acid-6-amino 6-H_2N-α-$C_{10}H_6CO_2H$	187.20	pa ye nd (al, w)		205-6			eth, ace, aa	B14^{3}, 1340
9506	α-Naphthoic aicd-7-amino 7-H_2N-α $C_{10}H_6CO_2H$	187.20	pa br pr (al)		223-4			al, eth, ace, aa	B14^{1}, 1340

PHYSICAL CONSTANTS OF ORGANIC COMPOUNDS (Continued)

No.	Name, Synonyms, and Formula	Mol. wt.	Color, crystalline form, specific rotation and λ_{max} (log ε)	b.p. °C	m.p. °C	Density	n_D	Solubility	Ref.
9507	α-Naphthoic anhydride $(\alpha\text{-}C_{10}H_7CO)_2O$	326.35	pr (bz)	....	145-6	....	....	eth, bz	B9^2, 450
9508	α-Naphthoic acid-4-bromo $4\text{-}Br\text{-}\alpha\text{-}C_{10}H_6CO_2H$	251.08	nd (aa, dil al, xyl)	....	220	....	....	al, eth, ace, bz	B9^3, 3153
9509	α-Naphthoic acid-5-bromo $5\text{-}Br\text{-}\alpha\text{-}C_{10}H_6CO_2H$	251.08	nd (aa, al)	sub	261	....	....	bz	B9^3, 3153
9510	α-Naphthoic acid-8-Bromo $8\text{-}Br\text{-}\alpha\text{-}C_{10}H_6CO_2H$	251.08	pr (w, bz)	....	178	....	....	al, eth, bz, aa	B9^3, 3154
9511	α-Naphthoic acid-2-chloro $2\text{-}Cl\text{-}\alpha\text{-}C_{10}H_6CO_2H$	206.63	cr (w, bz)	....	153	....	....	al, eth	B9^3, 3148
9512	α-Naphthoic acid-4-chloro $4\text{-}Cl\text{-}\alpha\text{-}C_{10}H_6CO_2H$	206.63	nd (al)	....	210 (223)	....	....	al, aa	B9^3, 3148
9513	α-Naphthoic acid-5-chloro $5\text{-}Cl\text{-}\alpha\text{-}C_{10}H_6CO_2H$	206.63	nd (dil al)	sub	245	....	....	al	B9^3, 3149
9514	α-Naphthoic acid-8-chloro $8\text{-}Cl\text{-}\alpha\text{-}C_{10}H_6CO_2H$	206.63	pl (al, w, bz)	sub	171-2	....	....	ace, aa	B9^3, 3151
9515	α-Naphthoic acid-1,2-dihydro $\alpha\text{-}C_{10}H_9CO_2H$	174.20	cr (50% al)	....	138	....	....	w	B9^3, 3076
9516	α-Naphthoic acid-1,4-dihydro $\alpha\text{-}C_{10}H_9CO_2H$	174.20	nd or pl (lig)	....	91	....	....	al, eth, ace	B9^3, 3075
9517	α-Naphthoic acid-3,4-dihydro $\alpha\text{-}C_{10}H_9CO_2H$	174.20	nd (w, AcOEt) cr (peth)	$305\text{-}6^{748}$	125	....	....	al, chl	B9^3, 3076
9518	α-Naphthoic acid-4,5-dinitro $4,5\text{-}(NO_2)_2\text{-}\alpha\text{-}C_{10}H_5CO_2H$	262.18	yesh nd or pl (al)	....	267 sub	....	....	al, eth	B9^3, 3168
9519	α-Naphthoic acid-ethyl ester or Ethyl α-naphthoate $\alpha\text{-}C_{10}H_7CO_2C_2H_5$	200.24	....	310, $183\text{-}6^{70}$	....	$1.1274^{15/15}$	1.5966^{15}	al	B9^3, 3139
9520	α-Naphthoic acid-5-hydroxy $5\text{-}HO\text{-}\alpha\text{-}C_{10}H_6CO_2H$	188.18	pr or nd (w) cr (bz)	....	236	....	....	al, eth, chl, aa	B10^3, 1070
9521	α-Naphthoic acid-6-hydroxy $6\text{-}HO\text{-}\alpha\text{-}C_{10}H_6CO_2H$	188.18	nd or pr (w)	....	212-3	....	....	al, eth, ace	B10^3, 1070
9522	α-Naphthoic acid-7-hydroxy $7\text{-}HO\text{-}\alpha\text{-}C_{10}H_6CO_2H$	188.18	nd (w, al aa)	....	256	....	....	al, aa	B10^3, 1072
9523	α-Naphthoic acid-8-iodo $8\text{-}I\text{-}\alpha\text{-}C_{10}H_6CO_2H$	298.08	br pr (w)	....	164-5	....	....	al, eth, bz, aa	B9^3, 3157
9524	α-Naphthoic acid-methyl ester or Methyl α-naphthoate $\alpha\text{-}C_{10}H_7CO_2CH_3$	186.21	....	$167\text{-}9^{20}$, $100\text{-}2^{0.04}$	59.5	$1.1290^{20/4}$	1.6086^{20}	al, bz	B9^3, 3138
9525	α-Naphthoic acid-3-nitro $3\text{-}NO_2\text{-}\alpha\text{-}C_{10}H_6CO_2H$	217.18	cr (al)	....	271-5	....	....	al	B9^3, 3158
9526	α-Naphthoic acid, 4-nitro $4\text{-}NO_2\text{-}\alpha\text{-}C_{10}H_6CO_2H$	217.18	yesh nd (al,aa)	....	225-6	....	....	al, chl, aa	B9^3, 3160
9527	α-Naphthoic acid, 5-nitro $5\text{-}NO_2\text{-}\alpha\text{-}C_{10}H_6CO_2H$	217.18	yesh nd (al)	sub	241-2	....	....	aa	B9^3, 3162
9528	α-Naphthoic acid, 8-nitro $8\text{-}NO_2\text{-}\alpha\text{-}C_{10}H_6CO_2H$	217.18	nd (w) pr (al)	....	217	....	....	al	B9^3, 3164
9529	α-Naphthoic acid, 1,2,3,4-tetrahydro $\alpha\text{-}C_{10}H_{11}CO_2H$	176.22	tcl pr (AcOEt)	....	85	....	....	al, eth, ace, bz	B9^3, 2801
9530	α-Naphthoic acid, 5,6,7,8-tetrahydro $\alpha\text{-}C_{10}H_{11}CO_2H$	176.22	pr (w) wh nd (dil aa)	....	150	....	....	al, bz, chl	B9^3, 2794
9531	α-Naphthonitrile $\alpha\text{-}C_{10}H_7CN$	153.18	nd (lig)	299, 148^{12}	37.5	$1.1113^{25/25}$	1.6298^{18}	al, eth, lig	B9^3, 3146
9532	α-Naphthoyl chloride $\alpha\text{-}C_{10}H_7COCl$	190.63	....	297.5, 172^{15}	20	....	....	....	B9^3, 3145
9533	β-Naphthamide $\beta\text{-}C_{10}H_7CONH_2$	171.20	lf (al)	....	195	....	....	al, eth, ace, chl, lig	B9^2, 454
9534	β-Naphthoic acid or 2-Naphthalene carboxylic acid $\beta\text{-}C_{10}H_7CO_2H$	172.20	nd (lig, chl, sub) pl (ace)	>300	185.5	$1.077^{100/4}$	....	al, eth, chl	B9^3, 3174
9535	β-Naphthoic acid, 4-acetyl-3-hydroxy $4\text{-}CH_3CO\text{-}3\text{-}HO\text{-}\beta\text{-}C_{10}H_5CO_2H$	230.22	ye pr (aa)	....	194	....	....	al	B10^3, 4408
9536	β-Naphthoic acid, 1-amino $1\text{-}H_2N\text{-}\beta\text{-}C_{10}H_6CO_2H$	187.20	nd (dil al,aa)	....	205 (rapid htq)	....	....	al, eth, bz	B14^3, 1340
9537	β-Naphthoic acid, 3-amino $3\text{-}H_2N\text{-}\beta\text{-}C_{10}H_6CO_2H$	187.20	ye lf (dil al)	....	216-7	....	....	al, eth	B14^3, 1341
9538	β-Naphthoic acid, 4-amino $4\text{-}H_2N\text{-}\beta\text{-}C_{10}H_6CO_2H$	187.20	nd (dil al)	....	215-6	....	....	al, eth, ace, bz	B14^3, 1345

No.	Name, Synonyms, and Formula	Mol. wt.	Color, crystalline form, specific rotation and λ_{max} (log ε)	b.p. °C	m.p. °C	Density	n_D	Solubility	Ref.
9539	β-Naphthoic acid, 5-amino $5\text{-}H_2N\text{-}\beta\text{-}C_{10}H_6CO_2H$	187.20	ye lf (al)		291-2			al, eth, ace, bz	$B14^3$, 1345
9540	β-Naphthoic acid, 6-amino $6\text{-}H_2N\text{-}\beta\text{-}C_{10}H_6CO_2H$	187.20	pa ye nd (w,dil al)		225			al, eth, ace, bz, aa	$B14^2$, 324
9541	β-Naphthoic acid, 7-amino $7\text{-}H_2N\text{-}\beta\text{-}C_{10}H_6CO_2H$	187.20	pa ye nd or lf (al)		243			al, eth, ace, aa	$B14^2$, 324
9542	β-Naphthoic acid, 8-amino $8\text{-}H_2N\text{-}\beta\text{-}C_{10}H_6CO_2H$	187.20	grsh-ye nd (aa)		220			al, eth, ace	$B14^2$, 324
9543	β-Naphthoic anhydride $(\beta\text{-}C_{10}H_7CO)_2O$	326.35	nd (eth)		135			bz, aa	$B9^3$, 3184
9544	β-Naphthoic acid, 1-bromo $1\text{-}Br\text{-}\beta\text{-}C_{10}H_6CO_2H$	251.08	nd (aa,bz)		191			al, bz, aa	$B9^3$, 3195
9545	β-Naphthoic acid, 5-bromo $5\text{-}Br\text{-}\beta\text{-}C_{10}H_6CO_2H$	251.08	nd (al,sub)	sub	270			al, eth, bz, aa	$B9^3$, 3196
9546	β-Naphthoic acid, 1-chloro $1\text{-}Cl\text{-}\beta\text{-}C_{10}H_6CO_2H$	206.63	nd (bz)		196			al, ace	$B9^3$, 3191
9547	β-Naphthoic acid, 3-chloro $3\text{-}Cl\text{-}\beta\text{-}C_{10}H_6CO_2H$	206.63	cr (dil MeOH)		216.5			al, eth, ace, bz, chl	$B9^3$, 3192
9548	β-Naphthoic acid, 5-chloro $5\text{-}Cl\text{-}\beta\text{-}C_{10}H_6CO_2H$	206.63	nd (al,aa)		270			bz, aa	$B9^3$, 3192
9549	β-Naphthoic acid, 1,2-dihydro $\beta\text{-}C_{10}H_9CO_2H$	174.20	nd or pr (dil al,w,peth)		105-6			al, chl, aa	$B9^3$, 3076
9550	β-Naphthoic acid, 1,4-dihydro $\beta\text{-}C_{10}H_9CO_2H$	174.20	pl (bz,w,dil al)		162-3			al, eth	$B9^3$, 3076
9551	β-Naphthoic acid, 3,4-dihydro $\beta\text{-}C_{10}H_9CO_2H$	174.20	nd (dil aa,dil al)		120			bz, chl, aa	$B9^3$, 3076
9552	β-Naphthoic acid, 1,3-dihydroxy,ethyl ester $1,3\text{-}(HO)_2\text{-}\beta\text{-}C_{10}H_5CO_2C_2H_5$	232.24	nd (dil al, dil aa)		83-4			al, eth, lig	$B10^3$, 1932
9553	β-Naphthoic acid, ethyl ester or Ethyl β-naphthoate $\beta\text{-}C_{10}H_7CO_2C_2H_5$	200.24		308-9, 224^{74}	32	$1.1143^{23/4}$	1.5951^{23}	al, eth, chl, aa	$B9^3$, 3177
9554	β-Naphthoic acid, 1-hydroxy $1\text{-}HO\text{-}\beta\text{-}C_{10}H_6CO_2H$	188.18	cr (dil al,w,aa) nd (al,eth,bz)		195			al, th, bz	$B10^3$, 1075
9555	β-Naphthoic acid, 1-hydroxy-4-chloro $1\text{-}HO\text{-}4\text{-}Cl\text{-}\beta\text{-}C_{10}H_5CO_2H$	222.63	nd (al,aa)		234			al, ace	$B10^3$, 1079
9556	β-Naphthoic acid, 1-hydroxy,phenyl ester $1\text{-}HO\text{-}\beta\text{-}C_{10}H_6CO_2C_6H_5$	264.28			96			al, bz	B10, 332
9557	β-Naphthoic acid, 3-hydroxy $3\text{-}HO\text{-}\beta\text{-}C_{10}H_6CO_2H$	188.18	ye lf (dil al) nd (dil al)		222.3			al, eth, bz, chl	$B10^3$, 1084
9558	β-Naphthoic acid, 3-hydroxy-4-chloro $4\text{-}Cl\text{-}3\text{-}HO\text{-}\beta\text{-}C_{10}H_5CO_2H$	222.63	ye nd		231d				$B10^3$, 1091
9559	β-Naphthoic acid, 3-hydroxy,ethyl ester $3\text{-}HO\text{-}\beta\text{-}C_{10}H_6CO_2C_2H_5$	216.24	nd or mcl pr (aa)	291	85			ace, chl	$B10^3$, 1086
9560	β-Naphthoic acid, 3-hydroxy,hexadecyl ester $3\text{-}HO\text{-}\beta\text{-}C_{10}H_6CO_2C_{16}H_{33}$	412.61	grsh-wh pr		72-3			bz, lig	$B10^3$, 1087
9561	β-Naphthoic acid, 3-hydroxy,methyl ester $3\text{-}HO\text{-}\beta\text{-}C_{10}H_6CO_2CH_3$	202.21	pa ye rh nd (dil MeOH)	205-7	75-6			al	$B10^3$, 1086
9562	β-Naphthoic acid, 5-hydroxy $5\text{-}HO\text{-}\beta\text{-}C_{10}H_6CO_2H$	188.18	wh nd (w,dil al)		213			al, eth, ace, aa	$B10^3$, 1098
9563	β-Naphthoic acid, 7-hydroxy $7\text{-}HO\text{-}\beta\text{-}C_{10}H_6CO_2H$	188.18	lf or pl (dil al) pa ye nd (al)		269-70			al, eth, ace, aa	$B10^3$, 1101
9564	β-Naphthoic acid, methyl ester or Methyl β-naphthoate $\beta\text{-}C_{10}H_7CO_2CH_3$	186.21	lf (MeOH)	290, $141\text{-}3^4$	77			al, eth, bz, chl	$B9^3$, 3176
9565	β-Naphthoic acid, 5-nitro $5\text{-}NO_2\text{-}\beta\text{-}C_{10}H_6CO_2H$	217.18	yesh nd (al)		295			ace	$B9^3$, 3203
9566	β-Naphthoic acid, 8-nitro $8\text{-}NO_2\text{-}\beta\text{-}C_{10}H_6CO_2H$	217.18	yesh nd (al)	sub	295				$B9^2$, 455
9567	β-Naphthoic acid, 1,2,3,4-tetra hydro $\beta\text{-}C_{10}H_{11}CO_2H$	176.22	nd (dil al)	$168\text{-}70^{15}$	97			al, eth, bz, chl	$B9^3$, 2805
9568	β-Naphthoic acid, 5,6,7,8-tetra hydro $\beta\text{-}C_{10}H_{11}CO_2H$	176.22	nd (al) cr (aa,bz)	216^{14}	154			al, bz	$B9^3$, 2802
9569	β-Naphthonitrile $\beta\text{-}C_{10}H_7CN$	153.18	lf (lig)	306.5, $156\text{-}8^{12}$	66	$1.0939^{60/60}$		al, eth, lig	$B9^3$, 3186
9570	β-Naphthoyl chloride $\beta\text{-}C_{10}H_7COCl$	190.63	cr (peth)	304-6, $142\text{-}3^5$	51			eth, bz, chl aa	$B9^3$, 3185

PHYSICAL CONSTANTS OF ORGANIC COMPOUNDS (Continued)

No.	Name, Synonyms, and Formula	Mol. wt.	Color, crystalline form, specific rotation and λ_{max} (log ε)	b.p. °C	m.p. °C	Density	n_D	Solubility	Ref.
9571	α-Naphthol or 1-Hydroxy naphthalene α-$C_{10}H_7OH$	144.17	ye mcl nd (w)	288 sub	96	$1.0989^{99/4}$	1.6224^{99}	al, eth, ace, bz, chl	$B6^3$, 2193
9572	α-Naphthol, 2-acetyl 2-CH_3CO-α-$C_{10}H_6OH$	186.21	(i)pr(bz,lig) (ii)gr-ye nd(al)	325d	(i)98 (ii)103			bz, aa	$B8^3$, 1130
9573	α-naphthol, 2-acetyl-4-bromo 2-CH_3CO-4-Br-α-$C_{10}H_5OH$	265.11	ye nd (al)		126-7			al, eth, bz, chl	$B8^3$, 1134
9574	α-Naphthol, 2-acetyl-4-nitro 2-CH_3CO-4-NO_2-α-$C_{10}H_5OH$	231.21	ye nd (al)		159			eth, bz	$B8^3$, 1135
9575	α-Naphthol, 3-acetyl 3-CH_3CO-α-$C_{10}H_6OH$	186.21	nd (bz)		173-4			al, aa	B8, 150
9576	α-naphthol, 4-acetyl 4-CH_3CO-$C_{10}H_6OH$	186.21	yesh pr (aa,al,to)		198			al, bz, aa	$B8^3$, 1124
9577	α-Naphthol, 2-benzyl 2-$C_6H_5CH_2$-α-$C_{10}H_6OH$	234.30	nd (lig) pr (bz)	$237\text{-}40^{12}$	73-4				$B6^3$, 3608
9578	α-Naphthol, 4-benzyl 4-$C_6H_5CH_2$-α-$C_{10}H_6OH$	234.30	pl or nd (aa-lig) pr (bz)	237^{10}	125-6				$B6^2$, 680
9579	α-Naphthol, 4-bromo 4-Br-α-$C_{10}H_6OH$	223.07	nd (dil al,hex)		128			chl, aa	$B6^3$, 2935
9580	α-Naphthol, 5-bromo 5-Br-α-$C_{10}H_6OH$	223.07	nd (w)		137			eth	$B6^3$, 2936
9581	α-Naphthol, 6-bromo 6-Br-α-$C_{10}H_6OH$	223.07	nd (w)		129-30			w	$B6^3$, 2936
9582	α-Naphthol, 7-bromo 7-Br-α-$C_{10}H_6OH$	223.07	cr (w)		105-6			w	$B6^2$, 583
9583	α-Naphthol, 8-bromo 8-Br-α-$C_{10}H_6OH$	223.07	pl (peth)		61			al	B6, 614
9584	α-Naphthol, 4-methyl 4-CH_3-α-$C_{10}H_6OH$	158.20		$165\text{-}7^{13}$	86-7			al, eth, ace, bz	$B6^3$, 3022
9585	α-Naphthol, 8-nitro or 8-nitro-1-naphthol 8-NO_2-α-$C_{10}H_6OH$	189.17	grsh ye nd (al,chl,bz-hx)		130-3			al, eth, ace, bz	$B6^3$, 2939
9586	α-Napthol, 2-propionyl 2-CH_3CH_2CO-α-$C_{10}H_6OH$	200.24	grsh ye lf or pl (al)		81			al, eth	$B8^3$, 1144
9587	α-Naphthol, 2,3,4-trichloro 2,3,4-Cl_3-α-$C_{10}H_4OH$	247.51	nd (aa or lig)		168			eth, al	$B6^2$, 582
9588	β-Naphthol or 2-Hydroxy naphthalene β-$C_{10}H_7OH$	144.17	mcl lf (w)	295	123-4	1.28^{20}		al, eth, bz, chl	$B6^3$, 2955
9589	β-Naphthol, 1-acetoamido 1-CH_3CONH-β-$C_{10}H_6OH$	201.22	lf (w,dil al)	sub	235d			al, eth, ace, bz	$B13^2$, 414
9590	β-Naphthol, acetate or 2-Acetoxy naphthalene β-$C_{10}H_7(O_2CCH_3)$	186.21	nd (al)	$132\text{-}4^2$	71-2			al, eth, chl	$B6^3$, 2982
9591	β-Naphthol, 1-acetyl 1-CH_3CO-β-$C_{10}H_6OH$	186.21	pa ye lf (peth) rh (lig)		64-5			al, eth, bz	$B8^3$, 1122
9592	β-Naphthol, 3-acetyl 3-CH_3CO-β-$C_{10}H_6OH$	186.21	ye lf or nd (al,peth)		112			ace, bz	$B8^3$, 1135
9593	β-Napthol, 1-amino 1-H_2N-β-$C_{10}H_6OH$	159.19	silvery lf (bz, eth)		150d			al	$B13^3$, 1891
9594	β-Naphthol, 3-amino 3-H_2N-β-$C_{10}H_6OH$	159.19	silvery lf (bz) nd (al,w)		235			al	$B13^3$, 1897
9595	β-Naphthol, 5-amino 5-H_2N-β-$C_{10}H_6OH$	159.19	nd or og pr (w)		190.6			al, eth, ace, w	$B13^3$, 1901
9596	β-Naphthol, 6-amino 6-H_2N-β-$C_{10}H_6OH$	159.19	pr (w)		192-4, 212d			al, w	$B13^3$, 1902
9597	β-Naphthol, 7-amino 7-H_2N-β-$C_{10}H_6OH$	159.19	nd or lf (al)		201 (208)			al, eth	$B13^2$, 416
9598	β-Naphthol, 8-amino 8-H_2N-β-$C_{10}H_6OH$	159.19	nd (w,al)	sub	205-7			eth	$B13^3$, 1907
9599	β-Naphthol, benzoate or 2-Benzoyloxy naphthalene β-$C_{10}H_7(O_2CC_6H_5)$	248.28	nd or pr (al) cr (lig)		108			al, eth	$B9^3$, 492
9600	β-Naphthol, 1-benzyl 1-$C_6H_5CH_2$-β-$C_{10}H_6OH$	234.30	nd (bz)	$247\text{-}50^{13}$	115			al, eth, ace, bz, chl	$B6^3$, 3607
9601	β-Naphthol, 1-bromo 1-Br-β-$C_{10}H_6OH$	223.07	rh pr(bz-lig) nd(aa lig)	d130	84			al, eth, bz, aa, lig	$B6^3$, 2994

PHYSICAL CONSTANTS OF ORGANIC COMPOUNDS (Continued)

No.	Name, Synonyms, and Formula	Mol. wt.	Color, crystalline form, specific rotation and λ_{max} (log ε)	b.p. °C	m.p. °C	Density	n_D	Solubility	Ref.
9602	β-Naphthol, 3-bromo 3-Br-β-$C_{10}H_6OH$	223.07	nd(lig)		84-5			al, bz	$B6^3$, 2995
9603	β-Naphthol, 5-Bromo 5-Br-β-$C_{10}H_6OH$	223.07	nd(w)		105			al	$B6^3$, 2996
9604	β-Naphthol, 6-bromo 6-Br-β-$C_{10}H_6OH$	223.07	nd(bz)		127			al, bz	$B6^3$, 3020
9605	β-Naphthol, 6-bromo-1-methyl 6-Br-1-CH_3-β-$C_{10}H_5OH$	262.31	nd(bz)		129			al, eth, ace, bz, chl	$B6^3$, 3020
9606	β-Naphthol, 7-bromo 7-Br-β-$C_{10}H_6OH$	223.07	cr(peth)		132-3			eth, ace	$B6^2$, 605
9607	β-Naphthol, 1-chloro 1-Cl-β-$C_{10}H_6OH$	178.62	nd(lig) pr(chl) pl(w)		71			al, bz, chl	$B6^3$, 2990
9608	β-Naphthol, 1,6-dibromo 1,6-Br_2-β-$C_{10}H_5OH$	301.97	nd(peth bz) lf(al)		106			al, eth, ace	$B6^3$, 2998
9609	β-Naphthol, 1,6-dinitro 1,6-$(NO_2)_2$-β-$C_{10}H_5OH$	234.17	pa ye nd(chl)		195d			al, eth, chl, py	$B6^3$, 3005
9610	β-Naphthol, 1-methyl 1-CH_3-β-$C_{10}H_6OH$	158.20	nd(w bz-lig dil aa)	180^{12} sub	112			al, eth, ace, bz, aa	$B6^3$, 3019
9611	β-Naphthol, 1-nitro or 1-nitro-2-naphthol 1-NO_2-β-$C_{10}H_6OH$	189.17	ye nd lf or pr(al)	$115^{0.05}$	104			al, eth	$B6^3$, 3002
9612	β-Naphthol, 5-nitro 5-NO_2-β-$C_{10}H_6OH$	189.17	lt ye nd(w)		147-9			eth, ace	$B6^3$, 3004
9613	β-Naphthol, 1,3,6-tribromo 1,3,6-Br_3-β-$C_{10}H_4OH$	380.86	nd(aa al)		133			al, bz	$B6^3$, 3000
9614	β-Naphthol, 1,4,6-tribromo 1,4,6-Br_3-β-$C_{10}H_4OH$	380.86	nd(bz)		157-8			al, bz, chl, aa	$B6^3$, 3000
9615	β-Naphthol, 3,4,6-tribromo 3,4,6-Br_3-β-$C_{10}H_4OH$	380.86	nd(bz)		127-8			al, bz	$B6^2$, 607
9616	β-Naphthol, 1,3,4-trichloro 1,3,4-Cl_3-β-$C_{10}H_4OH$	247.51	nd		162			al, aa	$B6^2$, 604
9618	1,2-Naphthoquinone $C_{10}H_6O_2$	158.16	ye-red nd (eth) og lf (bz)		146	1.450		w, a;, eth	$B7^3$, 3686
9619	1,2-Naphthoquinone, 3-bromo $C_{10}H_5BrO_2$	237.05	red nd or pl(aa al)	sub	178			al, bz	$B7^3$, 3692
9620	1,2-Naphthoquinone, 4-bromo $C_{10}H_5BrO_2$	237.05	red nd(bz-lig)		154			al, bz, aa	$B7^3$, 3691
9621	1,2-Naphthoquinone, 6-bromo $C_{10}H_5BrO_2$	237.05	og-red or ye pr or pl (bz) nd (w)		168d			al, ace	$B7^3$, 3691
9622	1,2-Naphthoquinone, 3-chloro $C_{10}H_5ClO_2$	192.60	red nd(al aa bz chl)		172d			bz	$B7^3$, 3690
9623	1,2-Naphthoquinone, 4-chloro $C_{10}H_5O_2Cl$	192.60			134-6			al	$B7^3$, 3690
9624	1,2-Naphthoquinone, 3,4-dibromo $C_{10}H_4Br_2O_2$	315.95	red lf or pl(aa bz)		172-4			bz	$B7^3$, 3693
9625	1,2-Naphthoquinone, 3,6-dibromo $C_{10}H_4Br_2O_2$	315.95	red pr(AcOEt)		176			al	$B7^3$, 3692
9626	1,2-Naphthoquinone, 4,6-dibromo $C_{10}H_4Br_2O_2$	315.95	og-red pr (bz peth) nd (AcOEt)		153			bz, aa	$B7^3$, 3692
9627	1,2-Naphthoquinone, 3,4-dichloro $C_{10}H_4Cl_2O_2$	227.05	red lf or pl(aa bz) nd(bz chl)	sub	184			bz, chl	$B7^3$, 3691
9628	1,2-Naphthoquinone, dioxime $C_{10}H_8N_2O_2$	188.19	ye nd(bz-lig dil al)		169			al, bz, diox	$B7^3$, 3690
9629	1,2-Naphthoquinone, 6-hydroxy $C_{10}H_6O_3$	174.16	red lf(ace)		165d			al, eth, ace, aa	$B8^3$, 2542
9631	1,2-Naphthoquinone, 7-hydroxy $C_{10}H_6O_3$	174.16	br nd		194			al	$B8^3$, 2542
9632	1,2-Naphthoquinone, 3-methyl $C_{11}H_8O_2$	172.18	red or og nd(abs al) lf(bz peth)		116 (122)			al	$B8^3$, 3709
9633	1,2-Naphthoquinone, 4-methyl $C_{11}H_8O_2$	172.18	og nd (MeOH) nd (aa)		248-50d			al	$B7^3$, 3708

No.	Name, Synonyms, and Formula	Mol. wt.	Color, crystalline form, specific rotation and λ_{max} (log ε)	b.p. °C	m.p. °C	Density	n_D	Solubility	Ref.
9634	1,2-Naphthoquinone, 3-nitro $C_{10}H_5NO_4$	203.15	red pl(aa)		158			bz, aa	B7^2, 651
9635	1,2-Naphthoquinone, 1-oxime $C_{10}H_7NO_2$	173.17	ye nd(bz) og pr or pl(al)		112			al, eth, ace, bz	B7^3, 3688
9636	1,4-Naphthoquinone $C_{10}H_6O_2$	158.16	bt ye nd(al peth) ye(sub)	sub	128.5			al, eth, bz, aa	B7^3, 3696
9637	1,4-Naphthoquinone, 2-amino $C_{10}H_7NO_2$	173.17			207			al, eth	B14^3, 388
9638	1,4-Naphthoquinone, 4-anil,2-anilino $C_{22}H_{16}N_2O$	324.38	ye red nd(bz al)		182-3			aa, bz	B14^1, 390
9639	1,4-Naphthoquinone, 2-bromo $C_{10}H_5BrO_2$	237.05	ye pl or nd (al, dil aa)		132			aa, bz, chl, aa	B7^3, 3705
9640	1,4-Naphthoquinone, 2-bromo-3-methyl $C_{11}H_7BrO_2$	251.08	ye-br nd(al)	sub 100	151			al, eth, ace, bz, chl	B7^3, 3714
9641	1,4-Naphthoquinone, 5-bromo-2,3-dichloro $C_{10}H_3BrCl_2O_2$	305.94	ye pr(al)		180			ace, bz	B7^2, 655
9642	1,4-Naphthoquinone, 6-bromo $C_{10}H_5BrO_2$	237.05	og-red or ye pr(bz AcOEt) nd(w)		168d			al, ace	B7, 722
9643	1,4-Naphthoquinone, 2-chloro $C_{10}H_5ClO_2$	192.60	ye nd(w al aa)		117-8			al, ace, bz	B7^3, 3702
9644	1,4-Naphthoquinone, 5-chloro $C_{10}H_5ClO_2$	192.60	ye nd(lig)	sub	163			al, aa, lig	B7^3, 3701
9645	1,4 Naphthoquinone, 6-chloro $C_{10}H_5ClO_2$	192.60	red-br(al eth) ye cr(dil MeOH)		109-10			bz	B7^3, 3702
9646	1,4-Naphthoquinone, 2,3-dibromo $C_{10}H_4Br_2O_2$	315.95	ye nd(aa)		218			aa	B7^3, 3705
9647	1,4-Naphthoquinone, 5,8-dibromo $C_{10}H_4Br_2O_2$	315.95	ye nd(al)		171-3				B7, 732
9648	1,4-Naphthoquinone, 2,3-dichloro $C_{10}H_4Cl_2O_2$	227.05	ye nd(al)		195			chl	B7^3, 3703
9649	1,4-Naphthoquinone, 2,6-dichloro $C_{10}H_4Cl_2O_2$	227.05	dk ye nd(al)		148-9			al	B7, 730
9650	1,4-Naphthoquinone, 5,6-dichloro $C_{10}H_4Cl_2O_2$	227.05	ye nd(al)	sub	181			eth	B7, 730
9651	1,4-Naphthoquinone, 5,8-dichloro $C_{10}H_4Cl_2O_2$	227.05	ye nd(al)	sub	173-4			eth	B7^3, 3702
9651a	1,4-Naphthoquinone, 2,3-dihydro $C_{10}H_8O_2$	160.17	lf (hx) nd (peth)		98-9			al	E12B, 2806
9652	1,4-Naphthoquinone, 2,3-dihydroxy or Isonaphthazarin $C_{10}H_6O_4$	190.16	red og nd or lf (sub)	sub	282			ace	B8^3, 3596
9653	1,4-Naphthoquinone, 3,5-dihydroxy-2-methyl or Proserone $C_{11}H_8O_4$	204.18	og-ye nd(al aa)	sub 100^3	181			al, eth, peth	B8^3, 3604
9654	1,4-Naphthoquinone, 5,8-dihydroxy or Naphthazarin $C_{10}H_6O_4$	190.16	dk red mcl pr (bz) red-br nd (al)	sub	276-80			aa	B8^3, 3600
9655	1,4-Naphthoquinone, 5,8-dihydroxy-2-methyl $C_{11}H_8O_4$	204.18	gr pl		173			al	B8^3, 3605
9656	1,4-Naphthoquinone, 2,3-dimethyl $C_{12}H_{10}O_2$	186.21	ye pr(al)		127			bz, aa	B7^3, 3717
9657	1,4-Naphthoquinone, 2,5-dimethyl $C_{12}H_{10}O_2$	186.21	ye nd(peth eth)		95			diox	B7^3, 3716
9658	1,4-Naphthoquinone, 2,6-dimethyl $C_{12}H_{10}O_2$	186.21	ye pr or nd (AcOEt)		136-7			al, eth, bz	B7^3, 3720
9659	1,4-Naphthoquinone, 2,8-dimethyl $C_{12}H_{10}O_2$	186.21	pr (peth) nd (MeOH)		135-6			al	B7^3, 3716
9660	1,4-Naphthoquinone, dioxime $C_{10}H_8N_2O_2$	188.19	nd(dil al)		207d			w, al	B7^2, 653
9661	1,4-Naphthoquinone, 2-ethyl $C_{12}H_{10}O_2$	186.21	pr(al) nd aa peth MeOH		88-9			al, aa	B7^3, 3715

No.	Name, Synonyms, and Formula	Mol. wt.	Color, crystalline form, specific rotation and λ_{max} (log ε)	b.p. °C	m.p. °C	Density	n_D	Solubility	Ref.
9662	1,4-Naphthoquinone, 2-ethyl-3-hydroxy $C_{12}H_{10}O_3$	202.21	ye nd(dil MeOH)		141			eth, ace	B8[3], 2586
9663	1,4-Naphthoquinone, 2-ethyl-3,5,6,7,8-pentahydroxy or Echinochrome A $C_{12}H_{10}O_7$	266.21	red nd(diox-w)	sub 120^{10}	220d			w, al, eth, ace, bz	B8[3], 4360
9664	1,4-Naphthoquinone, 2-hydroxy or Lawsone $C_{10}H_6O_3$	174.16	redsh-br(aa)		192d			al, aa	B8[3], 2543
9665	1,4-Naphthoquinone, 2-hydroxy,acetate $C_{12}H_8O_4$	216.19	ye lf(al)		131			al, eth, chl, aa	B8[3], 2547
9666	1,4-Naphthoquinone, 2-hydroxy-3-methyl or Phthiocol $C_{11}H_8O_3$	188.18	ye pr(eth-peth)	sub	173-4			eth, ace	B8[3], 2569
9667	1,4-Naphthoquinone, 2-hydroxy-3-phenyl $C_{16}H_{10}O_3$	250.25	gold-ye or og pr or nd(al bz MeOH)		147			al, eth, bz, chl, lig	B8[3], 2981
9668	1,4-Naphthoquinone, 3-hydroxy-2-bromo $C_{10}H_5BrO_3$	253.05	ye mcl pr (al) nd (al, w)	sub	202			ace	B8[3], 2552
9669	1,4-Naphthoquinone, 3-hydroxy-2-chloro $C_{10}H_5ClO_3$	208.60	ye nd(al aa)	sub	213			al, eth, bz	B8[2], 347
9670	1,4-Naphthoquinone, 5-hydroxy or Juglon, Nucin $C_{10}H_6O_3$	174.16	redsh-ye nd or pr(chl bz)	sub	154 (161)			al, eth, bz, chl, aa	B8[3], 2558
9671	1,4-Naphthoquinone, 5-hydroxy-2-methyl or Plumbagin $C_{11}H_8O_3$	188.18	gold pr or og-ye nd(dil al)	sub	78-9			al, eth, ace, bz, chl	B8[3], 2576
9672	1,4-Naphthoguinone, 6-hydroxy $C_{10}H_6O_3$	174.16	gold-ye or red ye nd(w bz al)		170d			al, eth, ace, MeOH	B8[2], 348
9673	1,4-Naphthoquinone, 2-methyl or Menadione, Vitamin K_3 $C_{11}H_8O_2$	172.18	ye nd(al peth)		107			ace, bz	B7[3], 3709
9674	1,4-Naphthoquinone, 2-phenyl. $C_{16}H_{10}O_2$	234.25	gold-ye nd(al)		111			al, eth, bz, chl	B7[3], 4209
9675	1,4-Naphthoquinone, 2-phenylamino or Lawsone anilide $C_{16}H_{11}NO_2$	249.27	red nd(dil al)	sub	193			eth, bz	B14[3], 389
9676	1,4-Naphthoquinone, 5,6,7,8-tetrahydro $C_{10}H_{10}O_2$	162.19	gold-ye nd(peth)		55-6			al, eth	B7[3], 3506
9677	1,4-Naphthoquinone, 2,5,8-trihydroxy $C_{10}H_6O_5$	206.15	red nd(bz MeOH)		195			al, aa	B8[3], 4043
9678	1,4-Naphthoquinone, 2-oxime $C_{10}H_7NO_2$	173.17	pa ye nd(bz) nd(dil al)		198			al, eth, ace	B7[3], 3700
9679	2,6-Naphthoquinone $C_{10}H_6O_2$	158.16	ye-red pr(bz bz-peth)		135d			al, MeOH	B7[3], 3706
9680	2,6-Naphthoquinone, 1,5-dichloro $C_{10}H_4Cl_2O_2$	227.05	og pr(chl) gold-ye nd(al)		206d			ace, bz, chl, aa	B7[3], 3706
9681	α-Naphthoxy acetic acid α-$C_{10}H_7OCH_2CO_2H$	202.21	pr		190			al, eth	B6[3], 2930
9682	β-Naphthoxy acetic acid β-$C_{10}H_7OCH_2CO_2H$	202.21	pr(w)		156			al, eth, aa	B6[3], 2985
9683	α-Naphthy acetamide α-$C_{10}H_7CH_2CONH_2$	185.23	nd(w al)		180-1 sub			eth, bz, aa	B9[3], 3208
9684	α-Naphthyl acetic acid α-$C_{10}H_7CH_2CO_2H$	186.21	wh nd(w)	d	133			eth, ace, bz, chl, aa	B9[3], 3206
9685	α-Naphthyl acetonitrile α-$C_{10}H_7CH_2CN$	167.21	wx	$162\text{-}4^{12}$	32-3		1.6192^{20}		B9[3], 3209
9686	β-Naphthyl acetamide β-$C_{10}H_7CH_2CONH_2$	185.23	lf(w)		202-4d			al, eth	B9[3], 3212
9687	β-Naphthyl acetonitrile β-$C_{10}H_7CH_2CN$	167.21	nd or lf(dil al)	145.50^2	85-6			eth, bz, chl	B9[3], 3212
9688	α-Naphthyl amine α-$H_2NC_{10}H_7$	143.19	nd(dil al eth)	300.8, 160^{12} sub	50	$1.1229^{25/25}$	1.6703^{51}	al, eth	B12[3], 2846
9689	α-Naphthyl amine, 2-acetyl,hydrochloride 2-CH_3CO-α-$C_{10}H_6NH_2.HCl$	221.69	cr(w)		220d			w, aa	B14[3], 208
9690	α-Naphthyl amine, N-acetyl α-$C_{10}H_7NHCOCH_3$	185.23	cr(al)		160				B12[3], 2866

No.	Name, Synonyms, and Formula	Mol. wt.	Color, crystalline form, specific rotation and λ_{max} (log ε)	b.p. °C	m.p. °C	Density	n_D	Solubility	Ref.
9691	α-Naphthyl amine, N-acetyl-2-nitro $2\text{-}O_2N\text{-}\alpha\text{-}C_{10}H_6NHCOCH_3$	230.22	lt ye nd(aa al)		200			al, aa	B12[3], 2969
9692	α-Naphthyl amine, N-acetyl-4-nitro $4\text{-}O_2N\text{-}\alpha\text{-}C_{10}H_6NHCOCH_3$	230.22	pa ye nd(ace)		192-3			al, ace	B12[3], 2972
9693	α-Naphthyl amine, N-acetyl-5-nitro $5\text{-}O_2N\text{-}\alpha\text{-}C_{10}H_6NHCOCH_3$	230.22	br pr(aa) ye cr(al)		220			al	B12[3], 2973
9694	α-Naphthyl amine, N-acetyl-8-nitro $8\text{-}NO_2\text{-}\alpha\text{-}C_{10}H_6NHCOCH_3$	230.22	nd(w)		191				B12[3], 2976
9695	α-Naphthyl amine, N-2-aminoethyl $\alpha\text{-}C_{10}H_7(NHCH_2CH_2NH_2)$	186.26	ye	320d, 204^{9}		$1.114^{25/4}$	1.6648^{25}	al, ace	B12[3], 2955
9696	α-Naphthyl amine, N-benzylidene $\alpha\text{-}C_{10}H_7N{=}CHC_6H_5$	231.30	ye lf(al)		73.5			al, eth, bz, MeOH	B12[3], 2861
9697	α-Naphthyl amine, 4-bromo $4Br\text{-}\alpha\text{-}C_{10}H_6NH_2$	222.08	nd(al bz peth)		102			al, ace, bz, lig	B12[3], 2964
9698	α-Naphthyl amine, 4-bromo-2-nitro $4\text{-}Br\text{-}2\text{-}NO_2\text{-}\alpha\text{-}C_{10}H_5NH_2$	267.08	og cr(al aa)		200			al, bz, chl	B12[3], 2979
9699	α-Naphthyl amine, 5-bromo $5\text{-}Br\text{-}\alpha\text{-}C_{10}H_6NH_2$	222.08	lf or pl(w lig)	sub	69			al, eth, ace, bz, lig	B12[3], 2966
9700	α-Naphthyl amine, 2-chloro $2\text{-}Cl\text{-}\alpha\text{-}C_{10}H_6NH_2$	177.63	nd(peth dil al)		60			al, ace	B12[3], 2960
9701	α-Naphthyl amine, 4-chloro $4\text{-}Cl\text{-}\alpha\text{-}C_{10}H_6NH_2$	177.63	nd(al bz lig)		99-100			al, eth, bz	B12[3], 2961
9702	α-Naphthyl amine, 2,4-dibromo $2,4\text{-}Br_2\text{-}\alpha\text{-}C_{10}H_5NH_2$	300.98	nd or pl(dil al)		118-9			al, eth, bz, chl, lig	B12[3], 2966
9703	α-Naphthyl amine, 2,4-dichloro $2,4\text{-}Cl_2\text{-}\alpha\text{-}C_{10}H_5NH_2$	212.08	nd(al)		83-4			al	B12[3], 2963
9704	α-Naphthyl amine, N,N-diethyl or 1-Diethylamino naphthalene $\alpha\text{-}(C_2H_5)_2NC_{10}H_7$	199.30		285, $155\text{-}65^{30}$		$1.015^{20/20}$	1.5961^{20}	al, eth, aa	B12[3], 2855
9705	α-Naphthyl amine, 5,8-dihydro $\alpha\text{-}C_{10}H_9NH_2$	145.20	pl or nd (to) (pink in air)	247^{408}	37.5			al, chl	B12[3], 2837
9706	α-Naphthyl amine, N,N-dimethyl $\alpha\text{-}C_{10}H_7N(CH_3)_2$	171.24	vt flr	274.5^{711}, $139\text{-}40^{13}$		$1.0423^{20/4}$	1.624^{15}	al, eth	B12[3], 2854
9707	α-Naphthyl amine, 2,4-dinitro $2,4\text{-}(NO_2)_2\text{-}\alpha\text{-}C_{10}H_5NH_2$	233.18	ye nd(al) pr(aa) cr(ace)		242			ace	B12[3], 2983
9708	α-Naphthyl amine, N-ethyl $\alpha\text{-}C_{10}H_7NHC_2H_5$	171.24		303^{723}, 191^{16}		1.060^{20}	1.6477^{15}		B12[3], 2854
9709	α-Naphthyl amine, 4-fluoro $4\text{-}F\text{-}\alpha\text{-}C_{10}H_6NH_2$	161.18	lt ye	162^{16}	48				B12[3], 2960
9710	α-Naphthyl amine, N-formyl $\alpha\text{-}C_{10}H_7NHCHO$	171.20	nd(w)		137.5			al, eth, ace	B12[3], 2866
9711	α-Naphthyl amine, hydrochloride $\alpha\text{-}C_{10}H_7NH_2.HCl$	179.65	nd	sub				al, w	B12[3], 2849
9712	α-Naphthyl amine, N-methyl $\alpha\text{-}C_{10}H_7NHCH_3$	157.22	oil	293, $165\text{-}7^{15}$			1.6722^{20}	al, eth	B12[3], 2854
9713	α-Naphthyl amine, 2-methyl $2\text{-}CH_3\text{-}\alpha\text{-}C_{10}H_6NH_2$	157.22	nd (peth) (turns red in air)		32			eth, lig	B12[3], 3702
9714	α-Naphthyl amine, 3-methyl $3\text{-}CH_3\text{-}\alpha\text{-}C_{10}H_6NH_2$	157.22	cr(peth)		51-2			al, eth, lig	B12[3], 3106
9715	α-Naphthyl amine, 4-methyl $4\text{-}CH_3\text{-}\alpha\text{-}C_{10}H_6NH_2$	157.22	nd(peth)	176^{12}	51-2			eth	B12[3], 3093
9716	α-Naphthyl amine, 2-nitro $2\text{-}O_2N\text{-}\alpha\text{-}C_{10}H_6NH_2$	188.19	ye-red mcl pr(al)		144			al	B12[3], 2968
9717	α-Naphthyl amine, 3-nitro $3\text{-}O_2N\text{-}\alpha\text{-}C_{10}H_6NH_2$	188.19	og-ye nd(50% al)		137			al, bz, chl	B12[3], 2969
9718	α-Naphthyl amine, 4-nitro $4\text{-}O_2N\text{-}\alpha\text{-}C_{10}H_6NH_2$	188.19	og-ye nd(al)		195			al, aa	B12[3], 2971
9719	α-Naphthyl amine, 5-nitro $5\text{-}O_2N\text{-}\alpha\text{-}C_{10}H_6NH_2$	188.19	red nd(w)		118-9			eth, aa	B12[3], 2973
9720	α-Naphthyl amine, 6-nitro $6\text{-}O_2N\text{-}\alpha\text{-}C_{10}H_6NH_2$	188.19	og-red nd(chl)		172-3			al	B12[3], 2974
9721	α-Naphthyl amine, 8-nitro $8\text{-}O_2N\text{-}\alpha\text{-}C_{10}H_6NH_2$	188.19	red lf(peth)		96-7			eth	B12[3], 2975

No.	Name, Synonyms, and Formula	Mol. wt.	Color, crystalline form, specific rotation and λ_{max} (log ε)	b.p. °C	m.p. °C	Density	n_D	Solubility	Ref.
9722	β-Naphthyl amine $\beta\text{-}C_{10}H_7NH_2$	143.19	lf(w)	306.1	113	$1.0614^{98/4}$	1.6493^{98}	al, eth	$B12^{3}$, 2989
9723	β-Naphthyl amine, N-acetyl $\beta\text{-}C_{10}H_7(NHCOCH_3)$	185.23	lf(al w)	....	134	....	....	....	$B12^{3}$, 3014
9724	β-Naphthyl amine, N-acetyl-1-nitro $1\text{-}NO_2\text{-}\beta\text{-}C_{10}H_6NHCOCH_3$	230.22	ye nd(al)	....	126	....	....	al, eth, bz, aa	$B12^{2}$, 731
9725	β-Naphthyl amine, N-acetyl-5-nitro $5\text{-}NO_2\text{-}\beta\text{-}C_{10}H_6NHCOCH_3$	230.22	ye-br rh(al) ye nd(bz)	....	186	....	....	al, aa	$B12^{3}$, 732
9726	β-Naphthyl amine, N-acetyl-6-nitro $6\text{-}NO_2\text{-}\beta\text{-}C_{10}H_6NHCOCH_3$	220.22	lf ye nd(dil al)	....	224	....	....	al, bz, aa	$B12^{2}$, 733
9727	β-Naphthyl amine, N-acetyl-8-nitro $8\text{-}NO_2\text{-}\beta\text{-}C_{10}H_6NHCOCH_3$	230.22	ye nd(al)	....	195.5	....	....	aa	$B12^{2}$, 733
9728	β-Naphthyl amine, N-benzylidene $\beta\text{-}C_{10}H_7N{=}CHC_6H_5$	231.30	yesh nd(al)	....	103	....	....	al, bz, chl, aa	$B12^{3}$, 3006
9729	β-Naphthyl amine, 1-bromo $1\text{-}Br\text{-}\beta\text{-}C_{10}H_6NH_2$	222.08	rh nd(dil al lig)	....	63-4	....	....	al, eth, bz, chl	$B12^{3}$, 3070
9730	β-Naphthyl amine, 3-bromo $3\text{-}Br\text{-}\beta\text{-}C_{10}H_6NH_2$	222.08	pl(al)	....	169	....	....	al, aa	$B12^{3}$, 3072
9731	β-Naphthyl amine, 4-bromo $4\text{-}Br\text{-}\beta\text{-}C_{10}H_6NH_2$	222.08	nd(bz-peth or 90% aa)	....	72	....	....	al, eth, bz	$B12^{3}$, 3073
9732	β-Naphthyl amine, 5-bromo $5\text{-}Br\text{-}\beta\text{-}C_{10}H_6NH_2$	222.08	cr	$207\text{-}10^{16}$	38	....	....	al, ace	$B12^{3}$, 3074
9733	β-Naphthyl amine, 6-bromo $6\text{-}Br\text{-}\beta\text{-}C_{10}H_6NH_2$	222.08	lf(al w peth)	....	128	....	....	al, ace, bz	$B12^{3}$, 3074
9734	β-Naphthyl amine, 1-chloro $1\text{-}Cl\text{-}\beta\text{-}C_{10}H_6NH_2$	177.63	nd(al peth)	....	60	....	....	al, ace	$B12^{3}$, 3065
9735	β-Naphthyl amine, 1,4-dibromo $1,4\text{-}Br_2\text{-}\beta\text{-}C_{10}H_5NH_2$	300.98	nd(al bz)	....	106-7	....	....	al, eth, bz	B12, 1311
9736	β-Naphthyl amine, 1,6-dibromo $1,6\text{-}Br_2\text{-}\beta\text{-}C_{10}H_5NH_2$	300.98	nd(al peth)	....	121	....	....	al, bz, aa	$B12^{1}$, 544
9737	β-Naphthyl amine, N,N-dimethyl $\beta\text{-}C_{10}H_7N(CH_3)_2$	171.24	dk red nd	$305, 160\text{-}2^{12}$	52-3	$1.0455^{60/60}$	1.6443^{53}	al, eth	$B12^{3}$, 2995
9738	β-Naphthyl amine, 1,4-dimethyl $1,4\text{-}(CH_3)_2\text{-}\beta\text{-}C_{10}H_5NH_2$	171.24	....	333	75	....	....	al, eth	B12, 1317
9739	β-Naphthyl amine, 1,6-dinitro $1,6\text{-}(NO_2)_2\text{-}\beta\text{-}C_{10}H_5NH_2$	233.18	gold-ye nd(al aa) ye pw(py)	....	248	....	....	....	$B12^{3}$, 3087
9740	β-Naphthyl amine, N-ethyl $\beta\text{-}C_{10}H_7NHC_2H_5$	171.24	....	$316\text{-}7, 191^{25}$	<15	1.0545^{21}	1.6544^{21}	....	$B12^{3}$, 2996
9741	β-Naphthyl amine, N-formyl $\beta\text{-}C_{10}H_7NHCHO$	171.20	lf(bz-peth)	....	129	....	....	al, bz	$B12^{3}$, 3013
9742	β-Naphthyl amine, hydrochloride $\beta\text{-}C_{10}H_7NH_2.HCl$	179.65	lf	....	254	....	....	w, al	$B12^{3}$, 2992
9743	β-Naphthyl amine, 3-iodo $3\text{-}I\text{-}\beta\text{-}C_{10}H_6NH_2$	269.08	cr(al)	....	137	....	....	al, bz, chl, aa	$B12^{3}$, 3078
9744	β-Naphthyl amine, N-methyl $\beta\text{-}C_{10}H_7NHCH_3$	157.22	dk in air	$317, 165\text{-}70^{12}$	....	....	1.6722^{20}	....	$B12^{3}$, 2995
9745	β-Naphthyl amine, 1-methyl $1\text{-}CH_3\text{-}\beta\text{-}C_{10}H_6NH_2$	157.22	nd(lig) pr(peth)	....	51	....	....	al, eth, ace, bz	$B12^{3}$, 3091
9746	β-Naphthyl amine, 6-methyl $6\text{-}CH_3\text{-}\beta\text{-}C_{10}H_6NH_2$	157.22	lf(peth w)(turns red in air)	....	129-30	....	....	....	$B12^{2}$, 743
9747	β-Naphthyl amine, 1-nitro $1\text{-}NO_2\text{-}\beta\text{-}C_{10}H_6NH_2$	188.19	og-ye nd(al)	....	126-7	....	....	al, ace, aa	$B12^{3}$, 3079
9748	β-Naphthyl amine, 5-nitro $5\text{-}NO_2\text{-}\beta\text{-}C_{10}H_6NH_2$	188.19	red nd(al)	....	143.5	....	....	al, bz, aa	$B12^{3}$, 3082
9749	β-Naphthyl amine, 6-nitro $6\text{-}NO_2\text{-}\beta\text{-}C_{10}H_6NH_2$	188.19	lt og pl(al) ye pl(aa)	....	207.5	....	....	bz, aa, diox	$B12^{3}$, 3082
9750	β-Naphthyl amine, 8-nitro $8\text{-}NO_2\text{-}\beta\text{-}C_{10}H_6NH_2$	188.19	red nd(aa)	....	104.5	....	....	al, eth, bz	$B12^{3}$, 3084
9751	β-Naphthyl amine, 1-nitroso $1\text{-}ON\text{-}\beta\text{-}C_{10}H_6NH_2$	172.19	gr nd(dil al bz)	....	150.2	....	....	al	$B7^{3}$, 3690
9752	β-Naphthyl amine, 1,3,6-tribromo $1,3,6\text{-}Br_3\text{-}\beta\text{-}C_{10}H_4NH_2$	379.88	pa red cr (chl al-eth)	....	143	....	....	eth, chl	$B12^{3}$, 3078
9753	α-Naphthyl ether or α-Naphthyl ether $(\alpha\text{-}C_{10}H_7)_2O$	270.33	lf(al al-eth)	$280\text{-}5^{22}$	110	....	....	eth, bz	$B6^{3}$, 2926

No.	Name, Synonyms, and Formula	Mol. wt.	Color, crystalline form, specific rotation and λ_{max} (log ε)	b.p. °C	m.p. °C	Density	n_D	Solubility	Ref.
9754	β-Naphthyl ether or bis-(2-Naphthyl) ether $(\beta\text{-}C_{10}H_7)_2O$	270.33	nd or lf(al)	250d	105			eth, bz	B6[3], 2976
9755	α-β-Naphthyl ether $\alpha\text{-}C_{10}H_7O\text{-}\beta\text{-}C_{10}H_7$	270.33	lf(al or al-eth)	264^{15}	81			eth, bz	B6[2], 600
9756	α-Naphthyl isocyanate $\alpha\text{-}C_{10}H_7NCO$	169.18		269-70		$1.1774^{20/4}$		eth, bz	B12[3], 2948
9757	β-Naphthyl isocyanate $\beta\text{-}C_{10}H_7NCO$	169.18	lf		55-6			eth, bz	B12[3], 3051
9758	α-Naphthyl isothiocyanate or α-Naphthyl mustard oil $\alpha\text{-}C_{10}H_7NCS$	185.24	wh nd(al)		58			al, eth, ace, bz, chl	B12[3], 2948
9759	β-Naphthyl isothiocyanate $\beta\text{-}C_{10}H_7NCS$	185.24	yesh nd(al)		62-3			al, eth, bz, chl	B12[3], 3052
9760	α-Naphthyl pentyl ether $\alpha\text{-}C_{10}H_7OC_5H_{11}$	214.31	nd(al)	322	30			al, eth, bz, chl	B6[3], 2925
9761	β-Naphthyl pentyl ether $\beta\text{-}C_{10}H_7OC_5H_{11}$	214.31	lf(al)	335	24.5		1.5587^{30}	al, eth, bz, chl	B6[3], 2973
9762	α-Naphthyl phenyl amine $\alpha\text{-}C_{10}H_7NHC_6H_5$	219.29	lf(lig) pr or nd(al)	335^{528}, 226^{8}	62			al, eth, bz, chl, aa	B12[3], 2856
9763	α-Naphthyl propyl amine $\alpha\text{-}C_{10}H_7NHC_3H_7$	185.27	ye	$316\text{-}8^{771}$					B12, 1224
9764	β-Naphthyl phenyl amine $\beta\text{-}C_{10}H_7NHC_6H_5$	219.29	nd(MeOH)	395-9, 237-13	108			al, eth, bz, aa	B12[3], 2999
9765	α-Naphthyl sulfide $(\alpha\text{-}C_{10}H_7)_2S$	286.39	nd or pr(al)	$289\text{-}90^{15}$	110			bz, aa	B6[3], 2945
9766	β-Naphthyl sulfide $(\beta\text{-}C_{10}H_7)_2S$	286.39	pl(al) lf(bz)	$295\text{-}6^{15}$	151			bz	B6[3], 3009
9767	α-Naphthyl thiocyanate $\alpha\text{-}C_{10}H_7SCN$	185.24	cr(peth)		55			peth	B6[2], 588
9768	β-Naphthyl thiocyanate $2\text{-}C_{10}H_7SCN$	185.24			35				B6[2], 611
9769	α-Naphthyl (2-tolyl)amine $(2\text{-}CH_3C_6H_4)NH\text{-}\alpha\text{-}C_{10}H_7$	233.31	nd(lig)	$198\text{-}202^{9}$	94-5			al, eth, bz	B12[3], 2857
9770	α-Naphthyl (4-tolyl) amine $(4\text{-}CH_3C_6H_4)NH\text{-}\alpha\text{-}C_{10}H_7$	233.31	pr(al)	360^{528}, 236^{15}	79			eth, bz	B12[3], 2857
9771	β-Naphthyl 2-tolyl amine or Yellow OB $(2\text{-}CH_3C_6H_4)NH\text{-}\beta\text{-}C_{10}H_7$	233.31	lf(lig)	400-5, $235\text{-}7^{14}$	95-6 (105)			al, eth, ace, bz, lig	B12[3], 3001
9772	β-Naphthyl 4-tolyl amine $(4\text{-}CH_3C_6H_4)NH\text{-}\beta\text{-}C_{10}H_7$	233.31	red lf(al)		103			eth, bz	B12[3], 3001
9773	α-Naphthyl 2-tolyl ketone $\alpha\text{-}C_{10}H_7CO(C_6H_4CH_3\text{-}2)$	246.31	(al)	365	64			al, eth	B7[3], 2639
9774	α-Naphthyl 3-tolyl ketone $\alpha\text{-}C_{10}H_7CO(C_6H_4CH_3\text{-}3)$	246.31			74-5			al	B7[3], 2639
9775	Narceine or Pseudonarcene $C_{23}H_{27}NO_8.3H_2O$	499.52	nd or pr (w + 3)		145.2 (+ 3w) 176-7 (anh)			al	B19[4], 4382
9776	Narceine, bisulfate,decahydrate $C_{23}H_{27}NO_8.H_2SO_4.10H_2O$	723.70	nd(sulf)		d			w, al, eth	B19, 372
9777	Narceine, hydrochloride,trihydrate $C_{23}H_{27}NO_8.HCl.3H_2O$	535.98	pr(HCl)		192 (anh)			al	B19[4], 4383
9778	α-Narcotine (dl) $C_{22}H_{23}NO_7$	413.43	nd(al chl MeOH)		232-3				B27[2], 607
9779	α-Narcotine (l) $C_{22}H_{23}NO_7$	413.43	pr or nd (al), [α] -200 (chl)		176			al, ace, bz, chl	B27[2], 605
9780	α-Narcotine, hydrochloride $C_{22}H_{23}NO_7 \cdot HCl$	449.89	(w + 3) [α] + 100		193 anh			w	B27[2], 606
9781	β-Narcotine-(dl) or β-Gnoscopine $C_{22}H_{23}NO_7$	413.43	nd pr(MeOH al)		180			al	B27[1], 559
9782	β-Narcotine, hydrochloride $C_{22}H_{23}NO_7.HCl$	449.89	pr		86-8, 224 (on standing)				B27[1], 559
9783	Naringin $C_{27}H_{32}O_{14} \cdot 2H_2O$	616.57	nd (w + 8), $[\alpha]^{11}{}_D$ -82.1 (al)		82 (+ 8w), 171 (+ 2w)				B18[4], 2637

No.	Name, Synonyms, and Formula	Mol. wt.	Color, crystalline form, specific rotation and λ_{max} (log ε)	b.p. °C	m.p. °C	Density	n_D	Solubility	Ref.
9784	Neoabietic acid, methyl ester or Methyl neoabietate $C_{21}H_{32}O_2$	316.48	cr(MeOH)		61-2			MeOH	B9^2, 433
9785	Neoamygdalin or L-Mandelonitrile-β-gentobioside $C_{20}H_{27}NO_{11}$	457.43	cr (al), $[\alpha]^{25/}_D$ -61.4 (w, c=8.5)		212			w, al	B31, 404
9786	Neobornyl amine or Isobornyl amine $C_{10}H_{19}N$	153.27	pw, $[\alpha]_D$ -47.7 (4% al)		184			eth, ace	B12^3, 195
9787	Neocarvomenthol (dl) $C_{10}H_{20}O$	156.27					1.4637^{20}	al	B6^3, 130
9788	Neocarvomenthol (l) $C_{10}H_{20}O$	156.27	$[\alpha]^{21/}_D$ -41.7	102^{18}		$0.9012^{20/4}$	1.4632^{20}	al	B6^4, 148
9789	Neoergesterol $C_{27}H_{40}O$	380.61	pr or nd (al), $[\alpha]^{17/}_D$ -12 (chl, c=2)		155-7			eth, ace	B6^3, 3474
9790	Neogermitrine $C_{36}H_{55}NO_{11}$	677.83	wh nd, $[\alpha]^{25/}_D$ -79.2 (py)		237-9			chl	B21^4, 6803
9791	Neoisocarvomenthol (l) $C_{10}H_{20}O$	156.27	$[\alpha]^{17/}_D$ -34.7	$87\text{-}8^{4}$	<−25	$0.9102^{20/4}$	1.4676^{20}	al, ace	B6^4, 148
9792	Neoisamenthol (d) or p-menthol-3. $C_{10}H_{20}O$	156.27	$[\alpha]^{15/}_D$ + 2.2 (al, c=2)	214.6, 91.5^{11}	−8	$0.9131^{18/4}$	1.4670^{20}	al, ace	B6^4, 149
9793	Neoisomenthol (dl) $C_{10}H_{20}O$	156.27		214.5, 81^{6}	14	$0.8854^{55/4}$	1.4649^{20}	al, ace	B6^4, 149
9794	Neomenthol (d) or *d*,β-Pulegomenthol $C_{10}H_{20}O$	156.27	$[\alpha]^{20/}_D$ + 19.6 (al)	211.7, 95^{12}	−15	$0.897^{22/4}$	1.4600^{20}	al, ace	B6^4, 149
9795	Neomenthol (dl) $C_{10}H_{20}O$	156.27	pl or pr(peth)	211.7, $103\text{-}5^{16}$	52	$0.903^{15/15}$	1.4600^{20}	al, ace	B6^4, 150
9796	Neomenthol-(l) $C_{10}H_{20}O$	156.27	$[\alpha]^{18/}_D$ -19.6 (al)	211.7, 97.6^{10}			1.4603^{20}	al, ace	B6^3, 140
9797	Neopentane or 2,2-Dimethylpropane $(CH_3)_4C$	72.15	gas	9.5	−16.5	0.6135^{20}	1.3476^{6}	al, eth	B1^4, 333
9798	Neopentyl alcohol or 2,2-Dimethyl-1-propanol $(CH_3)_3CCH_2OH$	88.15		113-4	52-3	0.812		al, eth	B1^4, 1690
9799	Neopentyl alcohol, 1-phenyl $(CH_3)_3CCH(OH)C_6H_5$	164.25	nd	$114\text{-}6^{16}$	45			al, eth	B6^3, 1972
9800	Neopentyl alcohol, 3-phenyl $C_6H_5CH_2C(CH_3)_2CH_2OH$	164.25	nd	$125\text{-}6^{14.5}$	34-5			al, eth	B6^2, 507
9801	Neopentyl amine or 1-amino-2,2-dimethyl propane $(CH_3)_3CCH_2NH_2$	87.16		$81\text{-}2^{741}$		$0.7455^{20/4}$	1.4023^{20}	eth	B4^4, 707
9802	Neopentyl bromide or 1-bromo-2,2-dimethyl propane $(CH_3)_3CCH_2Br$	151.05		106, 34.6^{100}		$1.1997^{20/4}$	1.4370^{20}	al, eth, ace, bz, chl	B1^4, 337
9803	Neopentyl chloride or 1-Chloro-2,2-dimethyl propane $(CH_3)_3CCH_2Cl$	106.60		84.3	−20	$0.86604^{20/4}$	1.4044^{20}	al, eth, bz, chl	B1^4, 336
9804	Nepentylene glycol or 2,2-Dimethyl-1,3-propanediol $(HOCH_2)_2C(CH_3)_2$	104.15	nd (bz)	206^{747}, $120\text{-}30^{15}$	130			w, al, eth	B1^4, 2551
9805	Neopentyl iodide or 1-Iodo-2,2-dimethyl propane $(CH_3)_3CCH_2I$	198.05		127-9d, $42\text{-}4^{20}$		1.4940^{20}	1.4890^{20}	al, eth	B1^4, 338
9806	Neral or Citral b $C_{10}H_{16}O$	152.24		120^{20}		0.8869^{20}	1.4869^{20}	al, eth	B1^4, 3569
9807	Nerol or 3,7-Dimethyl-2,6-octadiene-1-ol $(CH_3)_2C=CHCH_2CH_2C(CH_3)=CHCH_2OH$	154.25		$224\text{-}5^{745}$, 125^{25}	<-15	$0.8756^{20/4}$	1.4746^{20}	al	B1^4, 2276
9808	Nerolidol (d) or α-3,7,11-trimethyl-1,6,10-dodecatriene-3-ol $(CH_3)_2C=CHCH_2CH_2C(CH_3)=CH(CH_2)_2C(CH_3)(OH)CH=CH_2$	222.37	$[\alpha]^{20/}_D$ + 15.5 (undil)	276, $128\text{-}9^{6}$		$0.8778^{20/4}$	1.4898^{20}	al, eth, ace, aa	B1^4, 2336
9809	Nerolidol (dl) $C_{15}H_{26}O$	222.37		$145\text{-}6^{12}$, $75\text{-}6^{0.1}$		$0.8756^{19/4}$	1.4801^{16}	al, eth, ace	B1^4, 2336
9810	Nerolidol (l) $C_{15}H_{26}O$	222.37	$[\alpha]_D$-6.5 (undil)	$124\text{-}6^{3}$		$0.8881^{15/15}$	1.4799^{20}	al, eth, ace, aa	B1^3, 2042
9811	Neurine or Trimethyl ethenyl ammonium hydroxide $CH_2=CHN^+(CH_3)_3OH^-$	103.16	syr					w, al, eth	B4^4, 1053
9812	Nicotine (d) or α-N-Methyl-*d*-β-pyridylpyrrolidine $C_{10}H_{14}N_2$	162.23	hyg $[\alpha]^{20/}_D$ + 163.2	$245\text{-}6^{729}$		$1.0094^{20/4}$	1.5280^{20}	w, al, eth, chl, lig	B23, 117

No.	Name, Synonyms, and Formula	Mol. wt.	Color, crystalline form, specific rotation and λ_{max} (log ε)	b.p. °C	m.p. °C	Density	n_D	Solubility	Ref.
9813	Nicotine (dl) or Tetrahydronicotyrine $C_{10}H_{14}N_2$	162.23	....	242-3	......	$1.0082^{20/4}$	1.5289^{20}	w, al, eth, chl, lig	B23², 111
9814	Nicotine (l) $C_{10}H_{14}N_2$	162.23	hyg (br in air) $[\alpha]^{20/}{}_D$ −169	246.7^{745}, $124\text{-}5^{18}$	−79	$1.0097^{20/4}$	1.5282^{20}	w, al, eth, chl, lig	B23², 107
9815	Nicotine-hydrochloride (d) $C_{10}H_{14}N_2.HCl$	198.70	diq $[\alpha]^{20/}{}_D$ + 104 (w, p=10)	....	......	1.0337	....	w	B23, 114
9816	Nicotine amide or Niacin amide $3\text{-}C_5H_4NCONH_2$	122.13	wh pw nd (bz)	$150\text{-}60^{0.0005}$	129-31	1.400	1.466	w, al	B22⁴, 389
9817	Nicotinamide-N,N-diethyl or Coramine $3\text{-}C_5H_4NCON(C_2H_5)_2$	178.23	yesh	280d, 175^{25}	24-6	$1.060^{25/4}$	1.525^{20}	w, al, eth, ace, chl	B22⁴, 393
9818	Nicotinic acid or Niacin 3-Pyridine carboxylic acid $3\text{-}C_5H_4NCO_2H$	123.11	nd (w al)	sub	236-7	1.473	....	....	B22⁴, 348
9819	Nicotinic acid-6-amino $6\text{-}N_2N\text{-}(C_5H_3N)CO_2H\text{-}3$	138.13	cr (dil aa + 2w) aa	....	312	....	....	....	B22⁴, 6726
9820	Nicotinic acid-ethyl betaine $C_8H_9NO_2$	151.16	hyg pl	....	84-6	....	....	w	B22, 43
9821	Nicotinic acid-ethyl ester or Ethyl nicotinate $3\text{-}(C_5H_4N)CO_2C_2H_5$	151.16	....	224, $103\text{-}5^{5}$	8-9	1.1070^{20}	1.5024^{20}	w, al, eth, bz	B22⁴, 357
9823	Nictinic acid-hydrochloride $3\text{-}(C_5H_4N)CO_2H.HCl$	159.57	pr or pl rh bipym(w)	....	274	....	....	w, al	B22⁴, 354
9824	Nicotinic acid-2-hydroxy $2\text{-}HO(C_5H_3N)CO_2H\text{-}3$	139.11	nd (w)	....	α259-61d, β301-2d	....	....	....	B22⁴, 2139
9825	Nicotinic acid-4-hydroxy $4\text{-}HO(C_5H_3N)CO_2H\text{-}3$	139.11	nd (w + 2) cr (al)	....	254-5	....	....	....	B22⁴, 2145
9826	Nicotinic acid, 6-hydroxy $6\text{-}HO\text{-}3(C_5H_3N)CO_2H\text{-}3$	139.11	nd (w)	sub	304d	....	....	....	B22⁴, 2147
9827	Nicotinic acid, N-methyl or Trigoneiline $C_7H_7NO_2$	137.14	pr (aq al + 1w)	....	218d anh	....	....	w	B22⁴, 462
9828	Nicotinic acid, methyl ester or Methyl nicotinate $3\text{-}(C_5H_4N)CO_2CH_3$	137.14	cr	204, 118.5^{23}	42-3	....	....	w, al, bz	B22⁴, 356
9829	Nicotinonitrile $3\text{-}(C_5H_4N)CN$	104.11	nd (lig peth-eth)	240-5	50-2	....	....	w, al, eth, bz	B22⁴, 434
9830	2,2′-Nicotyrine (solid) or N-Methyl-2-(2-pyridyl)pyrrole $C_{10}H_{10}N_2$	162.19	cr, br in air	....	43-4	....	....	al, eth, bz, dil HCl	B23², 192
9831	2,2′-Nicotyrine (liquid) $C_{10}H_{10}N_2$	162.19	....	273^{764}, $149\text{-}50^{22}$	−28	....	....	ace, bz	B23², 191
9832	3,2′-Nicotyrine or N-Methyl-2-(3-pyridyl)pyrrole $C_{10}H_{10}N_2$	162.19	br in air	$280\text{-}1^{744}$, 150^{5}	......	$1.2111^{20/4}$	1.6057^{20}	al, eth, ace	B23², 192
9833	Ninhydrin or 1,2,3-triketo hydrindene monohydrate $C_9H_6O_4$	178.14	pr (w)	....	241-3d	....	....	w, al	B7³, 4592
9834	Nitranilic acid or 2,5-Dihydroxy-3,6-dinitro-p-benzoquinone $C_6H_2N_2O_8$	230.09	gold-ye pl (+ w dil HNO_3)	exp	100d	....	....	w, al	B8³, 3351
9835	Nitro acetic acid $O_2NCH_2CO_2H$	105.05	nd (chl)	....	92-3d	....	....	al, eth, bz, chl	B2⁴, 537
9836	Nitro acetic acid, ethyl ester or Ethyl nitro acetate $O_2NCH_2CO_2C_2H_5$	133.10	....	$105\text{-}7^{25}$	......	$1.1953^{20/4}$	1.4250^{20}	al, eth	B2⁴, 537
9837	2-Nitrobenzamide $2\text{-}NO_2C_6H_4CONH_2$	166.14	nd (dil al)	317	176-6	....	....	al, eth	B9³, 1477
9838	2-Nitrobenzamide, N-phenyl or N-Phenyl-2-nitrobenzamide $2\text{-}NO_2C_6H_4CONHC_6H_5$	242.23	nd (al bz)	....	155	....	....	al, bz, chl	B12³, 506
9839	2-Nitrobenzoic acid $2\text{-}NO_2C_6H_4CO_2H$	167.12	tcl nd (w)	....	147-8	$1.575^{20/4}$	....	al, eth, ace	B9³, 1466
9840	2-Nitrobenzoic acid, azide $2\text{-}NO_2C_6H_4CON_3$	192.13	ye pr (eth)	....	37.5	....	....	eth, bz, chl	B9³, 1489
9841	2-Nitrobenzoic acid, 3-bromo $3\text{-}Br\text{-}2\text{-}NO_2C_6H_3CO_2H$	246.02	(eth)	....	250	....	....	eth, bz	B9³, 1770
9842	2-Nitrobenzoic acid, 4-bromo $4\text{-}Br\text{-}2\text{-}NO_2C_6H_3CO_2H$	246.02	nd (w)	....	163	....	....	al, eth, bz, chl	B9³, 1770
9843	2-Nitrobenzoic acid, 5-bromo $5\text{-}Br\text{-}2\text{-}NO_2C_6H_3CO_2H$	246.02	cr (w al bz)	....	140	1.920^{18}	....	al, bz	B9¹, 165
9844	2-Nitrobenzoic acid, 3-chloro $3\text{-}Cl\text{-}2\text{-}NO_2C_6H_3CO_2H$	201.57	nd or pl (w)	....	237-9	1.566^{18}	....	al, eth	B9, 400

PHYSICAL CONSTANTS OF ORGANIC COMPOUNDS (Continued)

No.	Name, Synonyms, and Formula	Mol. wt.	Color, crystalline form, specific rotation and λ_{max} (log ε)	b.p. °C	m.p. °C	Density	n_D	Solubility	Ref.
9845	2-Nitrobenzoic acid, 4-chloro $4\text{-Cl-2-NO}_2C_6H_3CO_2H$	201.57	pl (bz-lig) pr (bz) nd (w)		142.3			al, eth	$B9^3$, 1763
9846	2-Nitrobenzoic acid, 5-chloro,methyl ester $5\text{-Cl-2-NO}_2C_6H_3CO_2CH_3$	215.59	pl (MeOH)		48.5	1.453^{18}		MeOH	B9, 401
9847	2-Nitrobenzoic acid, ethyl ester or Ethyl 2-nitrobenzoate $2\text{-NO}_2C_6H_4CO_2C_2H_5$	195.17	tcl (dil al)	275, 173^{18}	30			al, eth	$B9^3$, 1469
9848	2-Nitrobenzoic acid, hydrazide $2\text{-NO}_2C_6H_4CONHNH_2$	181.15	ye-br pr (w)		123			al	$B9^3$, 1481
9849	2-Nitrobenzoic acid, 4-iodo $4\text{-I-2-NO}_2C_6H_3CO_2H$	293.02	ye lf or pr (dil al)		192-3			al, eth	$B9^2$, 278
9850	2-Nitrobenzoic acid, methyl ester $2\text{-NO}_2C_6H_4CO_2CH_3$	181.15		275, 176^{21}	−13	1.2855^{20}		al, eth, bz, chl	$B9^3$, 1469
9851	2-Nitrobenzonitrile $2\text{-NO}_2C_6H_4CN$	148.12	nd (w, aa)	sub	111			al, eth, ace, bz, aa	$B9^3$, 1479
9852	2-Nitrobenzonitrile, 4-chloro $4\text{-Cl-2-NO}_2C_6H_3CN$	182.57	nd (w)		100-1			al, eth	$B9^3$, 1763
9853	3-Nitrobenzamide $3\text{-NO}_2C_6H_4CONH_2$	166.14	ye mcl nd (w)	310-5	142.7			al, eth	$B9^3$, 1515
9854	3-Nitrobenzamide, N-phenyl or N-Phenyl-3-nitrobenzamide $3\text{-NO}_2C_6H_4CONHC_6H_5$	242.23	lf (w, al)	sub	153-4			al, eth, bz	$B12^3$, 506
9855	3-Nitrobenzoic acid $3\text{-NO}_2C_6H_4CO_2H$	167.12	mcl pr (w)		140-2	$1.494^{20/4}$		al, eth, ace	$B9^3$, 1489
9856	3-Nitrobenzoic acid, azide $3\text{-NO}_2C_6H_4CON_3$	192.13	pl (dil al)		68			al, eth, bz, aa	$B9^3$, 1536
9857	3-Nitrobenzoic acid, 2-bromo $2\text{-Br-3-NO}_2C_6H_3CO_2H$	246.02	(dil al)		191			al	$B9^2$, 277
9858	3-Nitrobenzoic acid, 4-bromo $4\text{-Br-3-NO}_2C_6H_3CO_2H$	246.02	nd (dil aa)	sub	203-4			al	$B9^3$, 1770
9859	3-Nitrobenzoic acid, 5-bromo $5\text{-Br-3-NO}_2C_6H_3CO_2H$	246.02	nd (w, bz eth), pl (al)		159-60			al, eth, bz, aa, chl	$B9^3$, 1771
9860	3-Nitrobenzoic acid, 6-bromo $6\text{-Br-3-NO}_2C_6H_3CO_2H$	246.02	nd (w)	sub	180-1			al, eth, chl	$B9^3$, 1771
9861	3-Nitrobenzoic acid, 2-chloro $2\text{-Cl-3-NO}_2C_6H_3CO_2H$	201.57			185	1.662		al	$B9^2$, 275
9862	3-Nitrobenzoic acid, 4-chloro $4\text{-Cl-3-NO}_2C_6H_3CO_2H$	201.57	nd or pl (w)		181-2	1.645^{18}			$B9^3$, 1764
9863	3-Nitrobenzoic acid, 4-chloro, ethyl ester $4\text{-Cl-3-NO}_2C_6H_3CO_2C_2H_5$	229.62	ye nd (al)		59			al, bz, aa	B9, 402
9864	3-Nitrobenzoic acid, 4-chloro, methyl ester $4\text{-Cl-3-NO}_2C_6H_3CO_2CH_3$	215.59	nd (MeOH)		83	1.522^{18}		al	B9, 402
9865	3-Nitrobenzoic acid, 5-chloro $5\text{-Cl-3-NO}_2C_6H_3CO_2H$	201.57	nd (w)		147			al, eth, aa	$B9^1$, 165
9866	3-Nitrobenzoic acid, 6-chloro $6\text{-Cl-3-NO}_2C_6H_3CO_2H$	201.57	nd or pr (w)		165	1.608^{18}		al, eth, bz	$B9^3$, 1765
9867	3-Nitrobenzoic acid, ethyl ester or Ethyl-3-nitrobenzoate $3\text{-NO}_2C_6H_4CO_2C_2H_5$	195.17	mcl pr	296-8, 156^{10}	47			al, eth	$B9^3$, 1493
9868	3-Nitrobenzoic acid hydrazide $3\text{-NO}_2C_6H_4CONHNH_2$	181.15	nd (w)		153-4				$B9^3$, 1524
9869	3-Nitrobenzoicacid, 2-iodo $2\text{-I-3-NO}_2C_6H_3CO_2H$	293.02	pr (w)		206			al, eth	$B9^2$, 278
9870	3-Nitrobenzoic acid, 4-iodo $4\text{-I-3-NO}_2C_6H_3CO_2H$	293.02	ye pr (al)		213			al	$B9^2$, 278
9871	3-Nitrobenzoic acid, 5-iodo $5\text{-I-3-NO}_2C_6H_3CO_2H$	293.02	nd (al), pr (peth)		167			al	$B9^2$, 278
9872	3-Nitrobenzoic acid, methyl ester $3\text{-NO}_2C_6H_4CO_2CH_3$	181.15	nd		78				$B9^3$, 1493
9873	3-Nitrobenzonitrile $3\text{-NO}_2C_6H_4CN$	148.12	nd (w)	sub	118			al, eth, ace, aa	$B9^3$, 1521
9874	3-Nitrobenzoyl chloride $3\text{-NO}_2C_6H_4COCl$	185.57	ye cr	275-8, $154\text{-}5^{18}$	35			eth	$B9^3$, 1514
9875	4-Nitrobenzamide $4\text{-NO}_2C_6H_4CONH_2$	166.14	nd (w)		201-4			al, eth	$B9^3$, 1710

No.	Name, Synonyms, and Formula	Mol. wt.	Color, crystalline form, specific rotation and λ_{max} (log ε)	b.p. °C	m.p. °C	Density	n_D	Solubility	Ref.
9876	4-Nitrobenzamide, N-phenyl or N-Phenyl-4-nitrobenzamide $4-NO_2C_6H_4CONHC_6H_5$	242.23	lf (eth)	....	211	....	....	al, eth	$B12^{3}$, 506
9877	4-Nitrobenzoic acid $4-NO_2C_6H_4CO_2H$	167.12	mcl lf (w)	sub	242	1.610^{20}	....	al, eth, chl	$B9^{3}$, 1537
9878	4-Nitrobenzoic acid, 2-bromo $2-Br-4-NO_2C_6H_3CO_2H$	246.02	nd (w, dil al)	sub >155	166-7	....	....	al, eth	$B9^{3}$, 1771
9879	4-Nitrobenzoic acid, 3-bromo $3-Br-4-NO_2C_6H_3CO_2H$	246.02	nd (dil al)	....	197	....	....	al, eth, chl	B9, 408
9880	4-Nitrobenzoic acid, butyl ester or Butyl-4-nitrobenzoate $4-NO_2C_6H_4CO_2C_4H_9$	223.23	nd	160^{8}	35.3	....	....	eth, bz	$B9^{3}$, 1544
9881	4-Nitrobenzoic acid, 2-chloro $2-Cl-4-NO_2C_6H_3CO_2H$	201.57	nd (w)	....	140-2	....	....	al, eth	$B9^{3}$, 1766
9882	4-Nitrobenzoic acid, ethyl ester $4-NO_2C_6H_4CO_2C_2H_5$	195.17	tcl lf (al)	186.3	57	....	....	al, eth	$B9^{3}$, 1541
9883	4-Nitrobenzoic acid, hydrazide $4-NO_2C_6H_4CONHNH_2$	181.15	yesh nd (w)	....	214	....	....	....	$B9^{3}$, 1751
9884	4-Nitrobenzoic acid, 2-iodo $2-I-4-NO_2C_6H_3CO_2H$	293.02	pa ye pr (w)	....	146-7	....	....	al, eth	$B9^{2}$, 278
9885	4-Nitrobenzoic acid, methyl ester $4-NO_2C_6H_4CO_2CH_3$	181.15	ye mcl lf	....	96	....	....	al, eth, chl	$B9^{3}$, 1541
9886	4-Nitrobenzoic acid, 2,2,2-trichloroethyl ester $4-NO_2C_6H_4CO_2CH_2C(Cl)_3$	298.51	pr (al)	$106\text{-}7^{1}$	71	1.5343^{26}	....	....	$B9^{3}$, 1542
9887	4-Nitrobenzonitrile $4-NO_2C_6H_4CN$	148.12	lf (al), nd (bz)	sub	149	....	....	chl	$B9^{3}$, 1748
9888	4-Nitrobenzoyl chloride $4-NO_2C_6H_4COCl$	185.57	ye nd (lig)	$202\text{-}5^{105}$, $150\text{-}2^{15}$	75	....	....	eth	$B9^{3}$, 1709
9889	2-Nitrophenyl isocyanate $2-O_2NC_6H_4NCO$	164.12	wh nd (peth)	135^{7}	41	....	....	eth, bz, chl	$B12^{3}$, 1535
9890	3-Nitrophenyl isocyanate $3-O_2NC_6H_4NCO$	164.12	wh lf (lig)	$130\text{-}1^{11}$	51	....	....	eth, bz, chl	$B12^{3}$, 1573
9891	4-Nitrophenyl isocyanate $4-O_2NC_6H_4NCO$	164.12	pa ye nd	$137\text{-}8^{11}$	57	....	....	eth, bz, chl	$B12^{3}$, 1630
9892	bis-(2-nitrophenyl)trisulfide $(2-NO_2C_6H_4)_2S_3$	340.39	ye nd (al)	....	175-6	....	....	....	$B6^{3}$, 1062
9893	2-Nonanol (d) $CH_3CH(OH)C_7H_{15}$	144.26	$[\alpha]^{20/}{}_{D}$	105^{19}	....	$0.8230^{20/4}$	1.4299^{20}	al, eth	$B1^{1}$, 211
9894	2-Nonanol (dl) $CH_3CH(OH)C_7H_{15}$	144.26	....	193-4, 91^{12}	....	$0.84708^{20/4}$	1.43533^{20}	al, eth	$B1^{4}$, 1803
9895	3-Nonanol (s) $C_2H_5CH(OH)C_6H_{13}$	144.26	$[\alpha]^{25/}{}_{D}$ 7.08	97^{17}	....	$0.8281^{17/4}$	1.4308^{20}	al, eth	$B1^{3}$, 1746
9896	3-Nonanol (dl) $C_2H_5CH(OH)C_6H_{13}$	144.26	....	195^{750}, 93^{18}	−22	$0.8250^{20/4}$	1.4289^{20}	al, eth	$B1^{4}$, 1803
9897	4-Nonanol (s) $C_3H_7CH(OH)C_5H_{11}$	144.26	$[\alpha]^{25/}{}_{D}$ 0.57 eth 0.9%	192-3, $94\text{-}5^{18}$	....	$0.8282^{20/4}$	1.41971^{20}	al, eth	$B1^{3}$, 1747
9898	5-Nonanol $(C_4H_9)_2CHOH$	144.26	....	193^{750}, 97^{20}	....	$0.8356^{20/4}$	1.4289^{20}	al	$B1^{4}$, 1803
9899	Nitron or 4,5-Dihydro-1,4-diphenyl-3,5-phenylamino-1,2,4-triazole $C_{19}H_{16}N_4$	300.36	ye lf (al), nd (+ chl)	189d	....	....	....	al, ace, bz, chl	$B26^{2}$, 76,199
9900	Nonacosane $CH_3(CH_2)_{27}CH_3$	408.80	rh cr (peth)	440.8, 271.4^{10}	63.7	$0.8083^{20/4}$, $0.7630^{100/}{}_{4}$	1.4529^{20}	al, eth, ace, bz	$B1^{4}$, 591
9901	Nonacosane, 2-methyl $(CH_3)_2CH(CH_2)_{26}CH_3$	422.82	pl (lig)	$222^{0.3}$	73-4	....	....	eth, ace, bz	$B1^{1}$, 72
9902	1-Nonacosanol $CH_3(CH_2)_{27}CH_2OH$	424.79	cr (al)	sub $200\text{-}50^{1}$	84-5	....	....	....	$B1^{4}$, 1916
9904	Nonadecane $CH_3(CH_2)_{17}CH_3$	268.53	wax	329.7, 193^{15}	32.1	$0.7855^{20/}{}_{4}$	1.4409^{20}	eth, ace	$B1^{4}$, 560
9905	1,2,3-Nonadecane tricarboxylic acid, 2-hydroxy or Agaric acid, Laricic acid $CH_3(CH_2)_{15}CH(CO_2H)C(OH)(CO_2H)CH_2CO_2H$	416.56	lf (+ 3/2 w, dil al), $[\alpha]^{19/}{}_{D}$ -8.8 (NaOH)	....	142d	....	....	....	$B3^{4}$, 1284
9906	1,2,3-Nonadecane tricarboxylic acid, 2-hydroxy,triethyl ester $C_{28}H_{52}O_7$	500.72	nd	36-7	....	....	....	al, bz	$B3^{2}$, 373
9907	1,2,3-Nonadecane tricarboxylic acid, 2-hydroxy,trimethyl ester $C_{25}H_{46}O_7$	458.64	nd (al)	....	63-4	....	....	bz	$B3^{2}$, 373

No.	Name, Synonyms, and Formula	Mol. wt.	Color, crystalline form, specific rotation and λ_{max} (log ε)	b.p. °C	m.p. °C	Density	n_D	Solubility	Ref.
9908	Nonadecanoic acid or *n*-Nonadecylic acid $C_{18}H_{37}CO_2H$	298.51	lf (al)	$297-8^{100}$, $227-30^{10}$	69.4			al, eth, bz, chl, lig	$B2^4$, 1256
9909	1-Nonadecanol $C_{18}H_{37}CH_2OH$	284.53	cr (ace)	$166-7^{0.32}$	62-3		1.4328^{75}	eth, ace	$B1^4$, 1898
9910	2-Nonadecanone or Methyl-n-heptadecyl ketone $C_{17}H_{35}COCH_3$	282.51	pr (al)	266.5^{110}, 165^2	57	0.8108^{56}		eth, ace, chl	$B1^4$, 3400
9911	2-Nonadecanone oxime $C_{17}H_{35}C(=NOH)CH_3$	297.50	cr (al)		76-7			al	B1, 718
9912	4-Nonadecanone or Propyl pentadecyl ketone $C_{15}H_{31}COC_3H_7$	282.51	lf (al)	211^{11}d	50.5			eth, ace	$B1^4$, 3401
9913	10-Nonadecanone or Caprinone-dinonyl ketone $(C_9H_{19})_2CO$	282.51	lf (al)	>350, $155.6^{1.1}$	65.5			eth, ace, bz, chl, lig	$B1^4$, 3402
9914	1-Nonadecyne $C_{17}H_{35}C{\equiv}CH$	264.49		327, 181.6^{10}	37-8	$0.8054^{20/4}$	1.4488^{20}	eth, ace, bz	$B1^3$, 1030
9915	1,8-Nonadiyne $HC{\equiv}C(CH_2)_5C{\equiv}CH$	120.19		162, 55^{13}	−27.3	$0.8158^{20/4}$	1.4490^{20}	eth, ace	$B1^4$, 1125
9916	Nonanal $C_8H_{17}CHO$	142.24		190-2, 93.5^{23}		$0.8264^{22/4}$	1.4273^{20}	eth	$B1^4$, 3352
9917	Nonanal oxime $C_8H_{17}CH{=}NOH$	157.26	lf (dil al)		64			al, eth, ace	$B1^2$, 761
9918	Nonane $CH_3(CH_2)_7CH_3$	128.26		150.8, 39^{10}	−51	$0.7176^{20/4}$	1.4054^{20}	al, eth, ace, bz, chl	$B1^4$, 447
9919	Nonane, 1-amino or Nonyl amine $C_8H_{17}CH_2NH_2$	143.27		202.2, 80.8^{10}	−1	$0.7886^{20/4}$	1.4336^{20}	al, eth	$B4^4$, 777
9920	Nonane, 1-bromo $C_9H_{19}Br$	207.15		88^4		$1.0183^{20/20}$	1.4533^{20}		$B1^4$, 451
9921	Nonane, 5-butyl $(C_4H_9)_3CH$	184.37		217-8		$0.7635^{18.5/4}$	$1.4273^{18.5}$		$B1^4$, 517
9922	Nonane, 1-chloro or Nonyl chloride $CH_3(CH_2)_7CH_2Cl$	162.70		203.4, 80.5^{10}	−39.4	$0.8720^{20/4}$	1.4345^{20}	eth, chl	$B1^4$, 450
9923	Nonane, 1-chloro-9-fluoro $F(CH_2)_9Cl$	180.69		102^{11}		$0.966^{20/4}$	1.4301^{25}	al, eth	$B1^4$, 450
9924	Nonane, 2-chloro $CH_3(CH_2)_6CHClCH_3$	162.70		190^{764}		0.8790^{20}	1.4420^{20}	chl	B1, 166
9925	Nonane, 5-chloro $C_4H_9CHClC_4H_9$	162.70		$85-7^{14}$		$0.8639^{15/4}$	1.4314^{15}	eth	$B1^2$, 128
9926	Nonane, 2-methyl $CH_3(CH_2)_6CH(CH_3)_2$	142.28		166.8	−74.3	$0.7281^{20/4}$	1.4099^{20}	eth, bz, chl	$B1^4$, 473
9927	Nonane, 3-methyl (dl) $C_2H_5CH(CH_3)(CH_2)_5CH_3$	142.28		167.8	−84.6	$0.7354^{20/4}$	1.4125^{20}	eth, bz, chl	$B1^4$, 474
9928	Nonane, 4-methyl $C_3H_7CH(CH_3)(CH_2)_4CH_3$	142.28		165.7	−101.6	$0.7323^{20/4}$	1.4123^{20}	eth, bz, chl	$B1^4$, 475
9929	Nonane, 5-methyl $CH_3CH[(CH_2)_3CH_3]_2$	142.28		165.1	−86.5	$0.7326^{20/4}$	1.4116^{20}	eth, bz, chl	$B1^4$, 475
9930	Nonanedioic acid or Azelaic acid $HO_2C(CH_2)_7CO_2H$	188.22	lf or nd	>360d, 287^{100}, 225^{10}	106.5	$1.225^{25/4}$	1.4303^{111}	al	$B2^4$, 2055
9931	1,9-Nonanediol or Nonamethylene glycol $HOCH_2(CH_2)_7CH_2OH$	160.26	cr (bz)	$173-5^{20}$	45.8			al, eth, bz	$B1^4$, 2607
9932	Nonanoic acid or Pelargonic acid $CH_3(CH_2)_7CO_2H$	158.24		255, 150^{20}	15	$0.9057^{20/4}$	1.4343^{19}	al, eth, chl	$B2^4$, 1018
9933	1-Nonanol or Nonyl alcohol $CH_3(CH_2)_7CH_2OH$	144.26		213.5, 118^{15}	−5.5	$0.8273^{20/4}$	1.4333^{20}	al, eth	$B1^4$, 1798
9934	5-Nonanol, 5-butyl or Tri-*n*-butyl carbinol $(C_4H_9)_3COH$	200.36		230-5d, $118-20^{17}$	20	$0.8408^{20/4}$	1.4445^{20}	al	$B1^4$, 1863
9935	1-Nonanol, 9-chloro $Cl(CH_2)_8CH_2OH$	178.70		$146-8^{14}$	28		1.4575^{20}	al, eth	$B1^4$, 1802
9936	1-Nonanol, 9-fluoro $F(CH_2)_8CH_2OH$	162.25		$125-6^{15}$		$0.928^{20/4}$	1.4279^{25}	al, eth	$B1^4$, 1801
9937	2-Nonanone or Heptyl methyl ketone $C_7H_{15}COCH_3$	142.24		195.3, 73.8^{10}	−7.5	$0.8208^{20/4}$	1.4210^{20}	al, eth, ace, bz, chl	$B1^4$, 3353
9938	4-Nonanone or Propyl pentyl ketone $C_5H_{11}COC_3H_7$	142.24		187-8, $75-6^{20}$		$0.8190^{25/4}$	1.4189^{20}	al, eth, ace, chl	$B1^4$, 3354
9939	5-Nonanone or Dibutyl ketone $(C_4H_9)_2CO$	142.24		188.4, 88^{22}	−4.8	$0.8217^{20/4}$	1.4195^{20}	al, eth, chl	$B1^4$, 3355
9940	Nonasiloxane, eicosa methyl $CH_3[Si(CH_3)_2O]_8Si(CH_3)_3$	681.46		307.5, 198.8^{16}		0.9173^{20}	1.3980^{20}	bz	$B4^3$, 1881

PHYSICAL CONSTANTS OF ORGANIC COMPOUNDS (Continued)

No.	Name, Synonyms, and Formula	Mol. wt.	Color, crystalline form, specific rotation and λ_{max} (log ε)	b.p. °C	m.p. °C	Density	n_D	Solubility	Ref.
9941	1,3,6,8-Nonatetraen-5-one, 1,9-diphenyl or Dicinnamylidene acetone $(C_6H_5CH{=}CHCH{=}CH)_2CO$	286.33	ye nd (abs al)		144				B7^3, 2756
9942	1-Nonene $C_7H_{15}CH{=}CH_2$	126.24		146		0.730^{21}	1.414^{21}		B1^4, 894
9943	3-Nonene *(trans)* $C_2H_5CH{=}CHC_5H_{11}$	126.24		147-8		0.732$^{21/4}$	1.4181^{21}	eth, bz, chl	B1^4, 895
9944	3-Nonene, 2-methyl $CH_3CH(CH_3)CH{=}CHC_5H_{11}$	140.27		161.0		0.7340$^{20/4}$	1.4202^{20}		B1^4, 903
9945	4-Nonene *(trans)* $C_3H_7CH{=}CHC_4H_9$	126.24				0.7318$^{20/}{}_4$	1.4205^{20}	eth, bz, chl	B1^4, 896
9946	4-Nonene, 5-butyl $C_3H_7CH{=}C(C_4H_9)_2$	182.35		215-6		0.7745$^{20/4}$	1.4375^{20}		B1^4, 922
9947	1-Nonene-3-yne $C_5H_{11}C{\equiv}CCH{=}CH_2$	122.21		27-8^4		0.7602$^{25/4}$	1.4487^{25}	eth, ace	B1^4, 1103
9948	1-Nonene-4-yne $C_4H_9C{\equiv}C\text{-}CH_2CH{=}CH_2$	122.21		58^{22}		0.777$^{25/4}$	1.4413^{25}	eth, ace	B1^4, 1103
9949	2-Nonene-4-yne $C_4H_9C{\equiv}C\text{-}CH{=}CHCH_3$	122.21		70^{20}		0.7832$^{25/4}$	1.4590^{25}	eth, ace	B1^3, 1048
9950	1-Nonyne $C_7H_{15}C{\equiv}CH$	124.23		150.8, 33.3^{10}	−50	0.7568$^{20/4}$	1.4217^{20}	eth, bz	B1^4, 1047
9951	1-Nonyne, 1-chloro $C_7H_{15}C{\equiv}CCl$	158.67		75-7^{15}		0.906^{20}	1.450^{20}	eth	B1^3, 1012
9952	2-Nonyne $C_6H_{13}C{\equiv}C{-}CH_3$	124.23		158-9		0.7690$^{20/4}$	1.4337^{20}	eth, lig	B1^3, 1013
9953	3-Nonyne $C_5H_{11}C{\equiv}CC_2H_5$	124.23		153-5^{740}, 92^{97}		0.7616$^{20/4}$	1.4299^{20}	eth, lig	B1^3, 1013
9954	4-Nonyne $C_4H_9C{\equiv}CC_3H_7$	124.23		150-4^{750}		0.757$^{25/4}$	1.4296^{25}	eth, ace	B1^4, 1047
9955	4-Nonyne, 3,3-dimethyl $C_4H_9C{\equiv}C\text{-}C(CH_3)_2C_2H_5$	152.28		82^{40}		0.7667$^{20/4}$	1.4317^{20}	eth, bz	B1^3, 1024
9956	4-Nonyne, 8-methyl $(CH_3)_2CHCH_2CH_2C{\equiv}CC_3H_7$	138.25		104.5^{97}		0.7681$^{20/4}$	1.4311^{20}	eth, bz	B1^3, 1018
9957	*n*-Nonyl amine or 1-Amino nonane $C_8H_{17}CH_2NH_2$	143.27		202.2, 80.8^{10}	−1	0.7886$^{20/4}$	1.4336^{20}	al, eth	B4^4, 777
9958	Nopinone (d) $C_9H_{14}O$	138.21	$[\alpha]^{20/}{}_D$ + 34 (chl)	209, 87-8^{14}	0	0.9807$^{20/4}$	1.4787^{20}	w, al, eth	B7^3, 303
9959	Noradrenaline-(l) or 1-Norepinephrine $C_8H_{11}NO_3$	169.18	$[\alpha]^{25/}{}_D$ −37.5 (dil HCl)		216-8d			dil HCl	B13^3, 2382
9960	Norbornane or Norcamphane, bicyclo [2,2,1] heptane C_7H_{12}	96.17		sub	87-8			al, eth, ace, bz	B5^4, 258
9961	Norbornyl amine or Isobornyl amine $C_{10}H_{17}NH_2$	153.27	pw $[\alpha]'_D$ −47.7 (4% al)		184			eth, ace	B12^3, 195
9962	Norcamphane-2-carboxaldehyde $C_8H_{10}O$	122.17	cr (w)	70-2^{22}		1.0227$^{19/4}$	1.4760^{25}	eth	B7^3, 545
9963	2,3-Norcomphane dicarboxylic acid or Bicyclo [2,2,1]-heptane dicarboxylic acid $C_9H_{12}O_4$	184.19			192-5				B9^3, 3971
9964	2,3-Norcamphane dicarboxylic anhydride $C_9H_{10}O_3$	166.18			165-7			bz	B17^4, 6005
9965	Nordihydroguaiaretic acid or NDGA $C_{18}H_{22}O_4$	302.37	nd (w, al, aa)		185-6			al, eth, ace	B6^3, 6731
9966	Norephedrine hydrochloride-(dl) $C_6H_5CH(OH)CH(CH_3)NH_2.HCl$	187.67	pl (abs al), cr (dil HCl, al)		194			w, al	B13^2, 371
9967	Norephedrine, N,N-diethyl, hydrochloride $C_6H_5CH(OH)CH(CH_3)N(C_2H_5)_2\cdot HCl$	243.78	cr (al-ace)		205.6			al	B13^2, 380
9968	Norephedrine, N-ethyl $C_6H_5CH(OH)CH(CH_3)NHC_2H_5$	179.26	cr (lig)	143^{18}	51.5			bz	B9^3, 1728
9969	Normorphine or Desmethylmorphine $C_{16}H_{17}NO_3$	271.32	(w + 3/2)		273 (+ 3/2 w), 263-4 (anh)				B27^2, 117

PHYSICAL CONSTANTS OF ORGANIC COMPOUNDS (Continued)

No.	Name, Synonyms, and Formula	Mol. wt.	Color, crystalline form, specific rotation and λ_{max} (log ε)	b.p. °C	m.p. °C	Density	n_D	Solubility	Ref.
9970	Nornicotine-(l) or *l*-3-(2-Pyrrolidyl) pyridine $C_9H_{12}N_2$	148.21	hyg $[\alpha]^{22}/_D$ −88.8 (undil)	270, $130\text{-}1^{11}$	......	$1.0737^{19.5}$	$1.5378^{18.5}$	w, al, eth, ace, chl	B23[2], 107
9971	Ocimene or 3,7-Dimethyl-1,3,7-octatriene $CH_2{=}C(CH_3)CH_2CH_2CH{=}C(CH_3)CH{=}CH_2$	136.24	....	176-8d, $73\text{-}4^{21}$	......	0.8000^{20}	1.4862^{20}	al, eth, bz, chl, aa	B1[4], 1108
9972	Ocimene, dihydro or 2,6-Dimethyl-2,6-octadiene $CH_3CH{=}C(CH_3)CH_2CH_2CH{=}C(CH_3)_2$	138.25	....	168, 75^{30}	......	$0.775^{20/4}$	1.4498^{20}	al, eth, ace, aa	B1[4], 1058
9973	Octacosane $CH_3(CH_2)_{26}CH_3$	394.77	mcl or rh (bz-al)	431.6, 264^{10}	64.5	$0.8067^{20/}{}_4$, 0.7750^{70}	1.4520^{20}, 1.4330^{70}	ace, bz, chl	B1[4], 588
9974	Octacosanoic acid $CH_3(CH_2)_{26}CO_2H$	424.75	(ace or aa)	....	90.4	0.8191^{100}	1.4313^{100}	....	B2[4], 1318
9975	1-Octacosanol $CH_3(CH_2)_{26}CH_2OH$	410.77	(ace or peth)	sub, $200\text{-}50^{1}$	83.3	....	....	....	B1[4], 1915
9976	9,12-Octadecadienoic acid or -Linoleic acid *(cis,cis)* $CH_3(CH_2)_4CH{=}CHCH_2CH{=}CH(CH_2)_7CO_2H$	280.45	....	229.30^{16}	−5	$0.9022^{20/4}$	1.4699^{20}	al, eth, ace, bz, chl	B2[4], 1754
9977	10,12-Octadecadienoic acid *(trans,trans)* or 10,12-Linoleic acid $CH_3(CH_2)_4CH{=}CH\text{-}CH{=}CH(CH_2)_8CO_2H$	280.45	(bz or al)	....	56-7	$0.8686^{70/4}$	1.4689^{60}	....	B2[4], 1752
9978	7,11-Octadecadiyne $CH_3(CH_2)_5C{\equiv}CCH_2CH_2C{\equiv}C(CH_2)_5CH_3$	246.44	....	$167\text{-}8^{7}$	......	$0.841^{19/4}$	1.4698^{19}	....	B1[3], 1068
9979	Octadecanal or Stearaldehyde $CH_3(CH_2)_{16}CHO$	268.48	nd (peth)	261, $212\text{-}13^{12}$	55	....	....	....	B1[4], 3397
9980	Octadecanal, dimethyl acetal or 1,1-Dimethyoxy octadecane $CH_3(CH_2)_{16}CH(OCH_3)_2$	314.55	....	$167\text{-}70^{3}$	......	....	1.4410^{25}	al, eth	B1[4], 3397
9981	Octadecane $CH_3(CH_2)_{16}CH_3$	254.50	nd (al, eth-MeOH)	316.1, 173.5^{20}	28.2	$0.7768^{20/4}$	1.4390^{20}	eth, ace, lig	B1[4], 553
9982	Octadecane, 1-amino or Stearyl amine $C_{18}H_{37}NH_2$	269.51	(w)	348.8, 199.5^{20}	fr 52.9	$0.8618^{20/4}$	1.4522^{20}	al, eth, bz, chl	B4[4], 825
9983	Octadecane, 1-amino, hydro chloride $C_{18}H_{37}NH_2.HCl$	305.98	orh pl (al)	....	162-3	....	....	....	B4[4], 826
9984	Octadecane, 1-bromo $CH_3(CH_2)_{16}CH_2Br$	333.40	cr (al)	210^{10}	28.2	$0.9848^{20/4}$	1.4631^{20}	al, eth, peth	B1[4], 555
9985	Octadecane, 1-chloro $CH_3(CH_2)_{16}CH_2Cl$	288.96	....	348, 199^{10}	28.6	0.8641^{20}	1.4531^{20}	....	B1[4], 554
9986	Octadecane, 1,18-dibromo $Br(CH_2)_{18}Br$	412.29	nd or lf (al)	$205\text{-}7^{15}$	64	....	....	chl	B1[4], 556
9987	Octadecane, 1-iodo $CH_3(CH_2)_{16}CH_2I$	380.40	lf (lig), nd (ace, al-ace)	383, 223^{10}	34	$1.0994^{20/4}$	1.4810^{20}	....	B1[4], 556
9988	Octadecane, 9-(4-tolyl) $C_8H_{17}CH(4\text{-}C_6H_4CH_3)C_9H_{19}$	344.62	....	185^{10}	......	$0.8549^{20/4}$	1.4811^{20}	....	B5[4], 1221
9989	Octadecandioic acid, diethyl ester or Diethyl eicosanedioate $C_2H_5O_2C(CH_2)_{16}CO_2C_2H_5$	370.57	....	240^{12}	54-5	....	....	al, eth	B2[4], 2176
9990	1,18-Octadecanediol $HO(CH_2)_{18}OH$	286.50	lf (al, bz), nd (bz, diox)	$210\text{-}1^{2}$	97-9	....	....	....	B1[4], 2639
9991	1-Octadecanethiol or *n*-Octadecyl mercaptan $CH_3(CH_2)_{16}CH_2SH$	286.56	....	$188^{1\text{-}2}$	24-8	0.8475^{20}	1.4645^{20}	eth	B1[4], 1894
9992	Octadecanoic acid or Stearic acid $CH_3(CH_2)_{16}CO_2H$	284.48	mcl lf (al)	360d, 232^{15}	71-2	$0.9408^{20/4}$	1.4299^{80}	eth, ace, chl	B1[4], 1206
9993	1-Octadecanol or Stearyl alcohol $CH_3(CH_2)_{16}CH_2OH$	270.50	lf (al)	210.5^{15}	59-60	$0.8124^{59/4}$	....	al, eth, chl	B1[4], 1888
9994	3-Octadecanone or Ethyl pentadecyl ketone $C_{15}H_{31}COC_2H_5$	268.48	....	48-50	......	....	....	....	B1[4], 3398
9995	9,11,13-Octadecatrienoic acid *(cis)* or α-Eleostearic acid $C_4H_9[CH{=}CH]_3(CH_2)_7CO_2H$	278.44	nd (al)	$235^{12}{}_d$, 170^{1}	49	$0.9028^{50/4}$	1.5112^{50}	al, eth	B2[4], 1787
9996	9,12-15-Octadecatrienoic acid *(cis,cis,cis)* or α-Linolenic acid $CH_3[CH_2CH{=}CH]_3(CH_2)_7CO_2H$	278.44	....	$230\text{-}2^{17}$, $125^{0.05}$	−11.3	$0.9164^{20/4}$	1.4800^{20}	al, eth	B2[4], 1781
9997	9-Octadecenal or Olealdehyde $CH_3(CH_2)_7CH{=}CH(CH_2)_7CHO$	266.47	ye nd	$168\text{-}9^{3}$	......	$0.8509^{20/4}$	1.4558^{20}	....	B1[4], 3533
9998	1-Octadecene $C_{16}H_{33}CH{=}CH_2$	252.48	....	179^{15}, 145^{8}	17.5	$0.7891^{20/4}$	1.4448^{20}	bz	B1[4], 930

PHYSICAL CONSTANTS OF ORGANIC COMPOUNDS (Continued)

No.	Name, Synonyms, and Formula	Mol. wt.	Color, crystalline form, specific rotation and λ_{max} (log ε)	b.p. °C	m.p. °C	Density	n_D	Solubility	Ref.
9999	9-Octadecene $C_8H_{17}CH{=}CHC_8H_{17}$	252.48	...	162^{9}	−30.5	$0.7916^{20/4}$	1.4470^{20}	...	$B1^{4}$, 932
10000	9-Octadecenoic acid (trans) or Elaidic acid $CH_3(CH_2)_7CH{=}CH(CH_2)_7CO_2H$	282.47	pl (al)	288^{100}, 234^{15}	45	0.8734^{41}	1.4499^{45}	al, eth, bz, chl	$B2^{4}$, 1647
10001	6-Octadecenoic acid, 6,7-diiodo $CH_3(CH_2)_{10}CI{=}CI(CH_2)_4CO_2H$	534.26	nd (al)	d	48.5	...	...	eth, bz, chl	$B2^{4}$, 1638
10002	9-Octadecenoic acid *(cis)*, 12-hydroxy or Ricinoleic acid $C_6H_{13}CH(OH)CH{=}CH(CH_2)_8CO_2H$	298.47	$[\alpha]^{22/}_{D}$ + 5.05	$226\text{-}8^{10}$	α:7.7 β:16 γ:5.5	$0.9450^{21/4}$	1.4716^{21}	al, eth	$B3^{4}$, 1026
10003	11-Octadecenoic acid *(trans)* or Vaccenic acid $C_6H_{13}CH{=}CH(CH_2)_9CO_2H$	282.47	...	...	44	...	1.4439^{60}	ace	$B2^{4}$, 1640
10004	9-Octadecene, 1-ol *(cis)* or Oleylalcohol $CH_3(CH_2)_7CH{=}CH(CH_2)_7CH_2OH$	268.48	...	$205\text{-}10^{15}$	6-7	$0.8489^{20/4}$	1.4606^{20}	al, eth	$B1^{4}$, 2204
10005	9-Octadecenl-1-ol *(trans)* or Elaidyl alcohol $CH_3(CH_2)_7CH{=}CH(CH_2)_7CH_2OH$	268.48	(al or ace)	198^{10}	36-7	$0.8338^{40/4}$	1.4552^{40}	al, eth	$B1^{4}$, 2204
10006	Octadecyl sulfate $(C_{18}H_{37}O)_2SO_2$	603.04	...	...	70.5	...	...	...	$B1^{4}$, 1892
10007	1-Octadecyne $C_{16}H_{33}C{\equiv}CH$	250.47	(al)	313, 180^{15}	22.5	$0.8025^{20/4}$	1.4774^{20}	...	$B1^{4}$, 1075
10008	2-Octadecyne $C_{15}H_{31}C{\equiv}CCH_3$	250.47	(al)	184^{15}	30	$0.8016^{30/4}$	...	...	$B1^{4}$, 1075
10009	9-Octadecyne $C_8H_{17}C{\equiv}CC_8H_{17}$	250.47	...	$163\text{-}4^{7}$	3	$0.8012^{20/4}$	1.4488^{25}	...	$B1^{4}$, 1076
10010	9-Octadecynoic acid or Stearolic acid $CH_3(CH_2)_7C{\equiv}C(CH_2)_7CO_2H$	280.45	pr (al, peth), nd (dil al)	$189\text{-}90^{1.8}$	48	...	1.4510^{54}	eth	$B2^{4}$, 1751
10011	1,3-Octadiene, 3-chloro $C_4H_9CH{=}CClCH{=}CH_2$	144.64	...	$64\text{-}5^{18}$	...	$0.9366^{20/4}$	1.4794^{20}	eth	$B1^{4}$, 1038
10012	1,6-Octadiene, 7-methyl-3-methylene $(CH_3)_2C{=}CHCH_2CH_2C({=}CH_2)CH{=}CH_2$	136.24	...	$56\text{-}7^{12}$	...	$0.7982^{20/4}$	1.47065^{20}	...	$B1^{4}$, 1108
10013	1,7-Octadiene $CH_2{=}CH(CH_2)_4CH{=}CH_2$	110.20	...	113-8	...	$0.735^{20/20}$	1.424^{20}	...	$B1^{4}$, 1038
10014	2,4-Octadiene, 7-methyl $(CH_3)_2CHCH_2CH{=}CHCH{=}CHCH_3$	124.23	...	149	...	$0.7521^{18/4}$	1.4543^{18}	...	$B1^{2}$, 239
10015	2-6-Octadiene $CH_3CH{=}CHCH_2CH_2CH{=}CHCH_3$	110.20	...	118-20	...	0.748^{17}	1.4292^{17}	...	$B1^{4}$, 1039
10016	2,6-Octadiene, 2,6-dimethyl *(cis,cis)* $CH_3CH{=}C(CH_3)CH_2CH_2CH{=}C(CH_3)_2$	138.25	...	168, 75^{30}	...	$0.775^{21/4}$	1.4498^{20}	al, eth, ace	$B1^{4}$, 1058
10017	2,6-Octadiene, 2,6-dimethyl or Dihydroocimene $CH_3CH{=}C(CH_3)CH_2CH_2CH{=}C(CH_3)_2$	138.25	...	168, 75^{30}	...	$0.775^{21/4}$	1.4498^{20}	al, eth, ace, aa	$B1^{4}$, 1058
10018	2,7-Octadiene, 2,6-dimethyl or Linaloolen $(CH_3)_2C{=}CHCH_2CH_2CH(CH_3)CH{=}CH_2$	138.25	...	168, 58^{12}	...	0.7882^{20}	1.4561^{20}	al	B1, 261
10019	1,5-Octadien, 3-yn-5-propyl $C_2H_5CH{=}C(C_3H_7)C{\equiv}CCH{=}CH_2$	148.25	...	$57\text{-}8^{6}$	...	$0.8047^{20/4}$	1.4949^{20}	eth, ace	$B1^{4}$, 1129
10020	2,6-Octadien-4-yne, 3,6-diethyl $CH_3CH{=}C(C_2H_5)C{\equiv}CC(C_2H_5){=}CHCH_3$	162.27	...	169-71, 99^{12}	...	$0.8196^{20/4}$	1.4965^{20}	eth, ace	$B1^{3}$, 1066
10021	2,6-Octadien-4-yne, 3,6-dimethyl $CH_3CH{=}C(CH_3)C{\equiv}CC(CH_3){=}CHCH_3$	134.22	...	170	−45	$0.8071^{22/4}$	1.4998^{20}	eth, ace	$B1^{4}$, 1129
10022	1,7-Octadiyne $HC{\equiv}C(CH_2)_4C{\equiv}CH$	106.17	...	135-6, $93\text{-}5^{16}$	...	$0.8169^{21/4}$	1.4521^{18}	eth	$B1^{4}$, 1122
10023	2,6-Octadiyne $CH_3C{\equiv}CCH_2CH_2C{\equiv}CCH_3$	106.17	...	62^{19}	27	$0.828^{80/4}$	1.4658^{30}	eth	$B1^{4}$,1123
10024	3,5-Octadiyne $CH_3CH_2C{\equiv}CC{\equiv}CCH_2CH_3$	106.17	...	163-4, 78^{34}	...	$0.826^{0/4}$	1.4968^{0}	eth	$B1^{4}$, 1123
10025	3,5-Octadiyne, 2,7-dimethyl $(CH_3)_2CHC{\equiv}CC{\equiv}CCH(CH_3)_2$	134.22	...	74^{12}	...	$0.8090^{20/4}$	...	eth	$B1^{1}$, 128
10026	Octanal or Caprylaldehyde $CH_3(CH_2)_6CHO$	128.21	...	171, 72^{20}	...	$0.8211^{20/4}$	1.4217^{20}	al, eth, ace, bz	$B1^{4}$, 3337
10027	Octane $CH_3(CH_2)_6CH_3$	114.23	...	125.7, 19.2^{10}	−56.8	$0.7025^{20/4}$	1.3974^{20}	al, eth, ace, bz, chl, peth	$B1^{4}$, 412
10028	Octane, 1-amino or Octyl amine $CH_3(CH_2)_6CH_2NH_2$	129.25	...	179.6, 63.2^{10}	0	$0.7826^{20/4}$	1.4924^{20}	al, eth	$B4^{4}$, 751
10029	Octane, 2-amino (d) $C_6H_{13}CH(NH_2)CH_3$	129.25	$[\alpha]^{17/}_{D}$ + 8.6 (undil)	70^{25}	...	$0.771^{25/4}$	1.4220^{25}	al, eth	$B4^{4}$, 762
10030	Octane, 2-amino (dl) $C_6H_{13}CH(NH_2)CH_3$	129.25	...	163-5, $58\text{-}9^{13}$	...	$0.7745^{20/0}$	1.4232^{25}	al, eth	$B4^{4}$, 763

PHYSICAL CONSTANTS OF ORGANIC COMPOUNDS (Continued)

No.	Name, Synonyms, and Formula	Mol. wt.	Color, crystalline form, specific rotation and λ_{max} (log ε)	b.p. °C	m.p. °C	Density	n_D	Solubility	Ref.
10031	Octane, 1-bromo or Octyl bromide $CH_3(CH_2)_6CH_2Br$	193.13		200.8, 77.3^{10}	−55	$1.1122^{20/4}$	1.4524^{20}	al, eth	$B1^4$, 422
10032	Octane, 1-Bromo-8-fluoro $Br(CH_2)_8F$	211.12		$118\text{-}20^{22.5}$			1.4500^{20}	al, eth	$B1^4$, 424
10033	Octane, 2-bromo (d) $C_6H_{13}CHBrCH_3$	193.13	$[\alpha]^{25}/_D$ + 34.2	71^{14}		$1.0982^{25/4}$	1.4500^{20}	al, eth	$B1^4$, 423
10034	Octane, 2-bromo (dl) $C_6H_{13}CHBrCH_3$	193.13		188-9, 72^{14}		$1.0878^{25/4}$	1.4442^{25}	al, eth	$B1^4$, 423
10035	Octane, 2-bromo (l) $C_6H_{13}CH(Br)CH_3$	193.13	$[\alpha]^{25}/_D$ −37.5	$72\text{-}3^{18}$, 46^1		$1.0920^{20/4}$	1.4475^{25}	al, eth	$B1^4$, 423
10036	Octane, 1-chloro or Octyl chloride $CH_3(CH_2)_6CH_2Cl$	148.68		182, 78^{15}	−57.8	$0.8738^{20/4}$	1.4305^{20}		$B1^4$, 419
10037	Octane, 1-chloro-8-fluoro $F(CH_2)_8Cl$	166.67		87^{10}		$0.978^{20/4}$	1.4266^{25}	al, eth, ace	$B1^4$, 420
10038	Octane, 2-chloro (d) $C_6H_{13}CHClCH_3$	148.68	$[\alpha]^{20}/_D$ + 33.7	171-3, 75^{28}		$0.8658^{17/4}$	1.4273^{21}	al, eth	$B1^4$, 419
10039	Octane, 3-chlor-3-methyl $C_5H_{11}CCl(CH_3)CH_2CH_3$	162.70		$73\text{-}4^{15}$		$0.8680^{25/4}$	1.4351^{20}	eth, ace, bz, chl	$B1^3$, 508
10040	Octane, 4-chloro (d) $C_4H_9CHClC_3H_7$	148.68	$[\alpha]^{25}/_D$ + 0.28 (undil)	92^{50}				al, eth, chl	$B1^3$, 466
10041	Octane, 4-chloro-4-methyl $C_4H_9CCl(CH_3)C_3H_7$	162.70		$71^{14.5}$		$0.8723^{20/4}$	1.4360^{20}	eth, chl	$B1^3$, 509
10042	Octane, 1-cyclohexyl $C_8H_{17}C_6H_{11}$	196.38		263.6	−19.7	0.8138^{20}	1.4504^{20}		$B5^4$, 178
10043	Octane, 1,2-dibromo $C_6H_{13}CHBrCH_2Br$	272.02		240-2, 118.5^{15}		$1.4580^{20/4}$	1.4970^{20}		$B1^3$, 468
10044	Octane, 1,8-dibromo $Br(CH_2)_8Br$	272.02		270-2, $92\text{-}3^{0.45}$	15-6	$1.4594^{25/4}$	1.4971^{25}	eth, chl	$B1^4$,424
10045	Octane, 2,3-dimethyl $C_5H_{11}CH(CH_3)CH(CH_3)_2$	142.28		164.7		$0.7377^{20/4}$	1.4146^{20}		$B1^4$, 476
10046	Octane, 2,6-dimethyl $CH_3CH_2CH(CH_3)(CH_2)_3CH(CH_3)_2$	142.28		160-1, $90\text{-}3^{26}$		$0.7313^{20/4}$	1.4097^{20}		$B1^4$, 477
10047	Octane, 2,7-dimethyl $(CH_3)_2CH(CH_2)_4CH(CH_3)_2$	142.28		159.6	−54.6	$0.7240^{20/4}$	1.4092^{20}	eth	$B1^4$, 478
10048	Octane, 1-fluoro or Octyl fluoride $CH_3(CH_2)_6CH_2F$	132.22		142-3		$0.8103^{20/4}$	1.3935^{20}		$B1^4$, 418
10049	Octane, 1-iodo or Octyl iodide $CH_3(CH_2)_7CH_2I$	240.13		225.5, 86.5^5	−45.7	$1.3297^{20/4}$	1.4889^{20}	al, eth	$B1^4$, 425
10050	Octane, 2-iodo (D-) $C_6H_{13}CHICH_3$	240.13	$[\alpha]^{28}/_D$ −45.5	92^{12}		$1.3219^{20/4}$	1.4863^{25}	al, eth, lig	$B1^4$, 425
10051	Octane, 2-iodo (dl) $C_6H_{13}CHICH_3$	240.13		210, $95\text{-}6^{16}$		$1.3251^{20/4}$	1.4896^{20}	al, eth, lig	$B1^4$, 425
10052	Octane, 2-iodo (l, +) $C_6H_{13}CHICH_3$	240.13	$[\alpha]^{26}/_D$ + 46.3	101^{22}		$1.3314^{17/4}$	1.4877^{22}	al, eth, lig	$B1^4$, 425
10053	Octane, 2-methyl $C_6H_{13}CH(CH_3)_2$	128.26		142.8	−80.1	$0.7107^{20/4}$	1.4029^{20}	al, eth, lig	$B1^4$, 454
10054	Octane, 3-methyl (d) $C_5H_{11}CH(CH_3)C_2H_5$	128.26	$[\alpha]^{17}/_D$ + 9.4	143-4	−107.6	0.7206^{17}	1.4068^{20}	ace, bz	$B1^4$, 455
10055	Octane, 3-methyl (l) $C_5H_{11}CH(CH_3)C_2H_5$	128.26	$[\alpha]^{27}/_D$ −8.5	143		$0.714^{27/4}$	1.4052^{25}	ace, bz	$B1^3$, 509
10056	Octane, 4-methyl (dl) $C_4H_9CH(CH_3)C_3H_7$	128.26		142.4, 32^{10}	−113.2	$0.7199^{20/4}$	1.4061^{20}	al, eth, ace, bz	$B1^4$, 456
10057	Octane, 4-methyl (l) $C_4H_9CH(CH_3)C_3H_7$	128.26	$[\alpha]^{19}/_D$ −1.06	141		$0.717^{19/4}$		al, eth, ace, bz	$B1^3$, 510
10058	Octane, 1-phenyl or Octylbenzene $C_7H_{15}CH_2C_6H_5$	190.33		264-5, $131\text{-}4^{12}$	−7	$0.8582^{20/4}$	1.4851^{20}	bz, eth	$B5^4$, 1157
10059	Octane, 2-phenyl $C_6H_{13}CH(C_6H_5)CH_3$	190.33		$123\text{-}5^{20}$		$0.8611^{20/4}$	1.4837^{20}		$B5^4$, 1157
10060	Octanedial or Suberaldehyde $OHC(CH_2)_6CHO$	142.20		230-40d, $96\text{-}8^3$			1.4439^{20}	w, al	$B1^4$, 3706
10061	Octanedioic acid or Suberic acid $HO_2C(CH_2)_6CO_2H$	174.20	lo nd or pl (w)	300 sub, 219.5^{10}	144			al	$B2^4$, 2028
10062	1,7-Octanediol, 3,7-dimethyl $(CH_3)_2C(OH)(CH_2)_3CH(CH_3)CH_2CH_2OH$	174.28		265		0.937^{20}	1.4599^{20}		$B1^3$, 2233
10063	1,8-Octanediol $HO(CH_2)_8OH$	146.23	nd (bz-lig), pr	172^{20}	63			al, eth, ace	$B1^4$, 2592

No.	Name, Synonyms, and Formula	Mol. wt.	Color, crystalline form, specific rotation and λ_{max} (log ε)	b.p. °C	m.p. °C	Density	n_D	Solubility	Ref.
10064	4,5-Octanediol (dl) or 1,2-Dipropyl ethylene glycol $C_3H_7CHOHCHOHC_3H_7$	146.23		110^{8}	28		1.4419^{25}		$B1^{4}$, 2593
10065	4,5-Octanediol *(meso)* $C_3H_7CHOHCHOHC_3H_7$	146.23			123-4				$B1^{4}$, 2593
10066	2,3-Octanedione $C_5H_{11}COCOCH_3$	142.20		$172\text{-}3^{733}$				al	$B1^{3}$, 3138
10067	2,3-Octanedione dioxime or Methyl pentyl glyoxime $C_5H_{11}C(=NOH)C(=NOH)CH_3$	172.23	nd (dil al)		173			al, ace	$B1^{4}$, 3706
10068	2,3-Octanedione, 3-oxime $C_5H_{11}C(=NOH)COCH_3$	157.21	(lig)	133^{15}	59			eth	B1, 795
10069	2,7-Octanedione $CH_3CO(CH_2)_4COCH_3$	142.20	pl (bz)	114^{10}	44			al	$B1^{4}$, 3707
10070	2,7-Octanedione dioxime $CH_3C(=NOH)(CH_2)_4C(=NOH)CH_3$	172.23	(al)		158			al	B1, 795
10071	3,6-Octanedione $CH_3CH_2COCH_2CH_2COCH_2CH_3$	142.20	pl (al)	98^{14} sub	35-6			al	$B1^{4}$, 3708
10072	3,6-Octanedione, 2,2,7,7-tetramethyl $(CH_3)_3CCOCH_2CH_2COC(CH_3)_3$	198.31		$115\text{-}7^{17}$, $55\text{-}60^{0.5}$	2.5	$0.900^{27/4}$	1.4400^{20}	al	$B1^{4}$, 3734
10073	4,5-Octanedione $C_3H_7COCOC_3H_7$	142.20	ye oil	168, 60^{12}		$0.934^{0/4}$		al, eth, ace	$B1^{4}$, 3708
10074	4,5-Octanedione dioxime $C_3H_7C(=NOH)C(=NOH)C_3H_7$	172.23		sub	186-7			al, eth	$B1^{2}$, 846
10075	1-Octanethiol or Octyl mercaptan $CH_3(CH_2)_6CH_2SH$	146.29		199.1, 86^{15}	−49.2	$0.8433^{20/4}$	1.4540^{20}	al	$B1^{4}$, 1767
10076	2-Octanethiol (dl) $C_6H_{13}CHSHCH_3$	146.29		186.4, 88.9^{30}	−79	$0.8366^{20/4}$	1.4504^{20}	al, eth, bz	$B1^{4}$, 1777
10077	2-Octanethiol (l) $C_6H_{13}CHSHCH_3$	146.29	$[\alpha]^{25/}{}_{546}$ −36.4	$78\text{-}80^{22}$		$0.830^{25/4}$		al, eth, bz	$B1^{4}$, 1777
10078	Octanoic acid or Caprylic acid $CH_3(CH_2)_6CO_2H$	144.21		239.3, 140^{23}	16.5	0.9088^{20}	1.4285^{20}	**al, chl**	$B1^{4}$, 982
10079	1-Octanol or Octyl alcohol $CH_3(CH_2)_6CH_2OH$	130.23		194.4, 98^{19}	−16.7	$0.8270^{20/4}$	1.4295^{20}	**al, eth**	$B1^{4}$, 1756
10080	1-Octanol, 8-chloro $Cl(CH_2)_8OH$	164.68		139^{19}			1.4563^{25}	al, eth	$B1^{4}$, 1766
10081	1-Octanol, 3,7-dimethyl (d) or Tetrahydrogeraniol $(CH_3)_2CH(CH_2)_3CH(CH_3)CH_2CH_2OH$	158.28	$[\alpha]^{20/}{}_{D}$ +4.1	212-3, $105\text{-}6^{10}$		$0.8285^{20/4}$	1.4355^{20}	eth	$B1^{4}$, 1830
10082	1-Octanol, 3,7-dimethyl (l) $(CH_3)_2CH(CH_2)_3CH(CH_3)CH_2CH_2OH$	158.28	$[\alpha]^{11/}{}_{546}$ −3.7	212-3, 109^{15}		0.830^{18}	1.4370^{15}	eth	$B1^{3}$, 1768
10083	1-Octanol, 8-fluoro $F(CH_2)_8OH$	148.22		$106\text{-}7^{10}$		$0.945^{20/4}$	1.4248^{25}	al, eth	$B1^{4}$, 1766
10084	2-Octanol (d) $C_6H_{13}CH(OH)CH_3$	130.23	$[\alpha]^{17/}{}_{D}$ +9.9	86^{20}		$0.8216^{20/4}$	1.4264^{20}	al, eth, ace	$B1^{4}$, 1770
10085	2-Octanol (dl) $C_6H_{13}CH(OH)CH_3$	130.23		180, 87^{20}	−31.6	$0.8193^{20/4}$	1.4203^{20}	al, eth, ace	$B1^{4}$, 1770
10086	2-Octanol (l) $C_6H_{13}CH(OH)CH_3$	130.23	$[\alpha]^{17/}{}_{D}$ −9.9	86^{20}		$0.8201^{20/4}$	1.4264^{20}	al, eth, ace	$B1^{3}$, 1721
10087	2-Octanol, 2-methyl $C_6H_{13}C(OH)(CH_3)_2$	144.26		178, $81\text{-}3^{16}$		$0.8210^{20/4}$	1.4280^{20}	al, eth	$B1^{4}$, 1805
10088	3-Octanol $C_2H_5CH(OH)C_5H_{11}$	130.23		166-72					$B1^{4}$, 1779
10089	3-Octanol, 3,6-dimethyl $C_2H_5CH(CH_3)CH_2CH_2C(OH)(CH_3)C_2H_5$	158.28		202.2	−67.5		1.4370^{20}		$B1^{1}$, 214
10090	3-Octanol, 3,7-dimethyl (dl) or Tetrahydrolinalool $(CH_3)_2CH(CH_2)_3C(OH)(CH_3)CH_2CH_3$	158.28		196-7, $87\text{-}8^{10}$	31-2	0.8280^{20}	1.4335^{20}	al	$B1^{4}$, 1829
10091	3-Octanol, 3-ethyl $C_5H_{11}C(OH)(C_2H_5)_2$	158.28		199, 84^{12}		$0.8361^{25/4}$	1.4390^{20}	al	$B7^{3}$, 1766
10092	4-Octanol (d) $C_3H_7CH(OH)C_4H_9$	130.23	$[\alpha]^{22/}{}_{D}$ 0.74	79^{16}		$0.8159^{25/4}$	1.4275^{25}	al	$B1^{2}$, 452
10093	4-Octanol (dl) $C_3H_7CH(OH)C_4H_9$	130.23		176.3, 81.3^{20}	−40.7	$0.8186^{20/4}$	1.4248^{20}	al	$B1^{4}$, 1779
10094	2-Octanone or Methyl hexyl ketone $C_6H_{13}COCH_3$	128.21		173, $59\text{-}60^{11}$	−16	$0.8202^{20/4}$	1.4151^{20}	**al, eth**	$B1^{4}$, 3339
10095	3-Octanone or Ethyl pentyl ketone $C_5H_{11}COC_2H_5$	128.21		167^{749}		$0.8221^{20/4}$	1.4153^{20}	**al, eth**	$B1^{4}$, 3341
10096	4-Octanone or Butyl propyl ketone $C_4H_9COC_3H_7$	128.21		163, 70^{26}		$0.8146^{25/4}$	1.4173^{14}	**al, eth**	$B1^{4}$, 3342

PHYSICAL CONSTANTS OF ORGANIC COMPOUNDS (Continued)

No.	Name, Synonyms, and Formula	Mol. wt.	Color, crystalline form, specific rotation and λ_{max} (log ε)	b.p. °C	m.p. °C	Density	n_D	Solubility	Ref.
10097	4-Octanone, 5-hydroxy or Butyroin $C_3H_7CH(OH)COC_3H_7$	144.21	...	180-90, 95^{20}	−10	0.9231^{20}	1.4290^{20}	al, eth, ace	$B1^4$, 4042
10098	4-Octanone, 7-methyl or Propyl-*iso*-pentyl ketone $(CH_3)_2CHCH_2CH_2COC_3H_7$	142.24	...	177-9	...	$0.8239^{20/4}$	1.4210^{20}	al, eth	$B1^4$, 3356
10099	Octasiloxane, octadecamethyl $CH_3[-Si(CH_3)_2O.]_7Si(CH_3)_3$	607.31	...	$153^{5.1}$	...	0.913	1.3970^{20}	bz, lig, peth	$B4^3$, 1881
10100	1,3,5,7-Octatetraene $CH_2=CHCH=CHCH=CHCH=CH_2$	106.17	(bz)	sub	ca 50	...	...	aa	$B1^4$, 1124
10101	2,4,6-Octatriene *(trans,trans,trans)* $CH_3CH=CHCH=CHCH=CHCH_3$	108.18	lf	147-8, 43^{10}	52	$0.7961^{23/4}$	1.5131^{27}	al, chl, lig	$B1^4$, 1101
10102	1,3,7-Octatriene, 3,7-dimethyl or Ocimene $CH_2=C(CH_3)CH_2CH_2CH=C(CH_3)CH=CH_2$	136.24	...	176-8d, $73\text{-}4^{21}$	...	0.8000^{20}	1.4862^{20}	al, eth, bz, chl, aa	$B1^4$, 1108
10103	2,4,6-Octatriene, 2,6-dimethyl (4-*trans*,6-*trans*) or Allocimene A $CH_3CH=C(CH_3)CH=CH\text{-}CH=C(CH_3)_2$	136.24	...	188^{750}, 91^{20}	−35.4	$0.8118^{20/4}$	1.5446^{20}	...	$B1^4$, 1106
10104	2,4,6-Octatriene, 2,6-dimethyl (4-*trans*,6-*cis*) or Allocimene B $CH_3CH=C(CH_3)CH=CHCH=C(CH_3)_2$	136.24	...	89^{20}	−20.6	0.8060^{20}	1.5446^{20}	...	$B1^4$, 1106
10105	1-Octene $C_6H_{13}CH=CH_2$	112.22	...	121.3, 15.4^{10}	−101.7	$0.7149^{20/4}$	1.4087^{20}	al, eth, ace, bz, chl	$B1^4$, 874
10106	1-Octene, 2-chloro $C_6H_{13}CCl=CH_2$	146.66	...	168-70	...	$0.9274^{0/0}$	...	eth, ace, bz	$B1^1$, 840
10107	1-Octene, 3,7-dimethyl $(CH_3)_2CH(CH_2)_3CH(CH_3)CH=CH_2$	140.27	...	154	...	$0.7396^{20/4}$	1.4212^{20}	...	$B1^4$, 905
10108	1-Octene, 2-methyl $C_6H_{13}C(CH_3)=CH_2$	126.24	...	114.8	−77.8	$0.7343^{20/4}$	1.4184^{20}	...	$B1^4$, 896
10109	2-Octene *(cis)* $C_5H_{11}CH=CHCH_3$	112.22	...	125.6, 16.5^{10}	−100.2	$0.7243^{20/4}$	1.4150^{20}	al, eth, ace, bz, chl	$B1^4$, 878
10110	2-Octene *(trans)* $C_5H_{11}CH=CHCH_3$	112.22	...	125, 16^{10}	−87.7	$0.7199^{20/4}$	1.4132^{20}	al, eth, ace, bz, chl	$B1^4$, 879
10111	2-Octene, 2-chloro $C_5H_{11}CH=CClCH_3$	146.66	...	167-8	...	$0.8923^{16/16}$	1.4424^{16}	al, eth, ace, bz	$B1^4$, 879
10112	2-Octene, 4-chloro $C_4H_9CHClCH=CHCH_3$	146.66	...	153, $65\text{-}6^{15}$	...	$0.8924^{20/4}$	1.4452^{20}	eth, ace, bz, chl	$B1^1$, 840
10113	2-Octene, 2,6-dimethyl $C_2H_5CH(CH_3)CH_2CH_2CH_2C(CH_3)=CH_2$	140.27	...	163	...	$0.746^{22/4}$	1.425^{22}	...	$B1^4$, 904
10114	3-Octene *(cis)* $C_4H_9CH=CHC_2H_5$	112.22	...	122.9, 14.3^{10}	−126	$0.7189^{20/4}$	1.4135^{20}	al, eth, ace, bz, lig	$B1^4$, 880
10115	3-Octene *(trans)* $C_4H_9CH=CHC_2H_5$	112.22	...	123.3, 14.6^{10}	−110	$0.7152^{20/4}$	1.4126^{20}	al, eth, ace, bz, lig	$B1^4$, 880
10116	4-Octene *(cis)* $C_3H_7CH=CHC_3H_7$	112.22	...	122.5, 14^{10}	−118.7	$0.7212^{20/4}$	1.4148^{20}	al, eth, ace, bz, lig	$B1^4$, 880
10117	4-Octene *(trans)* $C_3H_7CH=CHC_3H_7$	112.22	...	122.3, 13.7^{10}	−93.8	$0.7141^{20/4}$	1.4114^{20}	al, eth, ace, bz, lig	$B1^4$, 880
10118	4-Octene, 4-chloro *(cis)* $C_3H_7CH=CClC_3H_7$	146.66	...	165.3	...	$0.8912^{20/4}$	1.4447^{20}	eth	$B1^4$, 881
10119	1-Octen-3-yne $C_4H_9C\equiv C\text{-}CH=CH_2$	108.18	...	62^{60}	...	$0.7830^{20/4}$	1.4592^{20}	eth	$B1^4$, 1100
10120	1-Octen-3yne-5-ol, 5-methyl $C_3H_7C(OH)(CH_3)C\equiv C\text{-}CH=CH_2$	138.21	...	80^{13}	...	$0.8851^{15/4}$	1.4735^{20}	al	$B1^3$, 2034
10121	Octyl ether or Dioctyl ether $(C_8H_{17})_2O$	242.45	...	286-7	...	$0.8063^{20/4}$	1.4327^{20}	al, eth	$B1^4$, 1760
10122	Octyl nitrate $C_8H_{17}ONO_2$	175.23	...	$110\text{-}2^{20}$	...	$0.8419^{17/17}$	...	al, eth	$B1^4$, 1762
10123	Octyl nitrite $C_8H_{17}ONO$	159.23	ye	174-5, 60^{10}	...	0.862^{17}	1.4127^{20}	al, eth	$B1^4$, 762
10124	Octyl phenyl ether $C_8H_{17}OC_6H_5$	206.33	...	285, $164\text{-}7^{20}$	8	$0.9319^{15/15}$	1.4875^{20}	al, eth	B6, 144
10125	Octyl thiocyanate $C_8H_{17}SCN$	171.30	...	$141\text{-}2^{19}$	105	$0.9149^{25/4}$	1.4649^{20}	al, eth	$B3^3$, 282
10126	1-Octyne or Hexyl acetylene $C_6H_{13}C\equiv CH$	110.20	...	125.2, 19.7^{10}	−79.3	0.7461^{20}	1.4159^{20}	al, eth	$B1^4$, 1034
10127	1-Octyne, 1-chloro $C_6H_{13}C\equiv CCl$	144.64	...	$61\text{-}2^{17}$	...	0.912^{20}	1.445^{20}	al, eth	$B1^3$, 1005
10128	1-Octyne-3-ol, 3-methyl $C_5H_{11}C(OH)(CH_3)C\equiv CH$	140.23	...	75^{10}	...	$0.863^{10/10}$	1.443^{10}	...	$B1^4$, 2268

No.	Name, Synonyms, and Formula	Mol. wt.	Color, crystalline form, specific rotation and λ_{max} (log ε)	b.p. °C	m.p. °C	Density	n_D	Solubility	Ref.
10129	2-Octyne $C_5H_{11}C{\equiv}CCH_3$	110.20		138	−61.6	$0.7596^{20/4}$	1.4278^{20}	al, eth	$B1^4$, 1035
10130	2-Octyne-1-ol $C_5H_{11}C{\equiv}CCH_2OH$	126.20		$98\text{-}9^{15}$	−18	$0.8805^{20/4}$	1.4556^{20}	eth	$B1^4$, 2256
10131	3-Octyne $C_4H_9C{\equiv}CC_2H_5$	110.20		133, 85^{169}	−103.9	$0.7529^{20/4}$	1.4250^{20}	al, eth	$B1^4$, 1036
10132	3-Octyne, 2-chloro-2-methyl $C_4H_9C{\equiv}CCCl(CH_3)_2$	158.67		68^{15}		$0.8929^{20/4}$	1.4480^{20}	eth	$B1^4$, 1048
10133	3-Octyne, 2,2-dimethyl $C_4H_9C{\equiv}CC(CH_3)_3$	138.25		79^{60}		$0.7491^{20/4}$	1.4270^{20}	eth	$B1^3$, 1018
10134	3-Octyne, 7-methyl $(CH_3)_2CHCH_2CH_2C{\equiv}CC_2H_5$	124.23		87^{99}		$0.7599^{20/4}$	1.4280^{20}	eth	$B1^3$, 1014
10135	4-Octyne $C_3H_7C{\equiv}CC_3H_7$	110.20		131.5	−102.5	$0.7509^{20/4}$	1.4248^{20}	al, eth	$B1^4$, 1037
10136	Olealdehyde or 9-Octadecenal $CH_3(CH_2)_7CH{=}CH(CH_2)_7CHO$	266.47	ye nd	$168\text{-}9^3$		$0.8509^{20/4}$	1.4558^{20}		$B1^4$,3533
10137	Oleamide $CH_3(CH_2)_7CH{=}CH(CH_2)_7CONH_2$	281.48			76			eth	$B2^4$, 1668
10138	Oleamide, N-phenyl or Oleanilide $CH_3(CH_2)_7CH{=}CH(CH_2)_7CONHC_6H_5$	357.58	nd	143.5^{10}	41			eth, bz, aa	$B12^2$, 150
10139	Oleic acid or 9,10-Octadecenoic acid (*cis*). $CH_3(CH_2)_7CH{=}CH(CH_2)_7CO_2H$	282.47		286^{100}, $228\text{-}9^{15}$	16.3	$0.8935^{20/4}$	1.4582^{20}	al, eth, ace, bz, chl	$B2^4$, 1641
10140	Oleic acid, benzyl ester or Benzyl oleate $C_{17}H_{33}CO_2CH_2C_6H_5$	372.59		237^1		$0.9330^{25/25}$	1.4875^{25}	al, eth	$B6^4$, 2269
10141	Oleic acid, butyl ester or Butyl oleate $C_{17}H_{33}CO_2C_4H_9$	338.57	ye	$227\text{-}8^{15}$	−26.4	0.8704^{15}	1.4480^{25}	al	$B2^4$, 1653
10142	Oleic acid, ethyl ester or Ethyl oleate. $C_{17}H_{33}CO_2C_2H_5$	310.52		207^{13}	0.8720^{20}	1.4515^{20}		al, eth	$B2^4$, 1651
10143	Oleic acid, methyl ester or Methyl oleate $C_{17}H_{33}CO_2CH_3$	296.49		218.5^{20}	−19.9	0.8739^{20}	1.4522^{20}		$B2^4$, 1649
10144	Oleic acid, isopentyl ester or Isopentyl oleate $C_{17}H_{33}CO_2\text{-i-}C_5H_{11}$	352.60		$223\text{-}4^{10}$		0.897^{15}		al, eth	$B2^3$, 1414
10145	Oleonitrile $C_{17}H_{33}CN$	263.47		330-5d, 204^{12}	−1	$0.848^{17/17}$	1.4566^{20}		$B2^4$, 1668
10146	Oleyl alcohol or 9-Octadecen-1-ol $CH_3(CH_2)_7CH{=}CH(CH_2)_7CH_2OH$	268.48		$205\text{-}10^{15}$	6-7	$0.8489^{20/4}$	1.4606^{20}	al, eth	$B1^4$, 2204
10147	Orcein or Orcin $C_{28}H_{24}N_2O_7$	500.51	br-red pw					al, ace	$B6^3$, 4531
10148	Ornithine (L) or L-2,5-Diamino valeric acid. $H_2N(CH_2)_3CH(NH_2)CO_2H$	132.16	cr (al-eth), $[\alpha]^{25}/_D$ + 11.5 (w, c=6.5)		140			w, al	$B4^3$, 1346
10149	Ornithine monohydrochloride (L) $H_2N(CH_2)_3CH(NH_2)CO_2H.HCl$	168.62	nd $[\alpha]^{25}/_D$ + 11.0 (w, c=5.5)		215 (230)			w	$B4^3$, 1347
10150	Ornithine-sulfate (L) $H_2N(CH_2)_3CH(NH_2)CO_2H.H_2SO_4$	230.24	$[\alpha]^{25}/_D$ + 8.4 (w)		234d			w	$B4^3$, 1347
10151	Orthoacetic acid, triethyl ester or 1,1,1-triethoxyethane, triethyl ortho acetate. $CH_3C(OC_2H_5)_3$	162.23		144-6, 66.5^{41}		$0.8847^{25/4}$	1.3980^{20}	al, eth, chl	$B2^4$, 137
10152	Orthoacetic acid, trimethyl ester or 1,1,1-trimethoxy ethane, trimethyl ortho acetate. $CH_3C(OCH_3)_3$	120.15		107-9		$0.9438^{25/4}$	1.3859^{25}	al, eth	$B2^4$, 127
10153	Orthocarbonic acid, tetraethyl ester or Ethyl ortho carbonate, Tetraethoxy methane. $C(OC_2H_5)_4$	192.26		160-1, 62^{28}		$0.9186^{20/4}$	1.3928^{20}	al, eth	$B3^4$, 6
10154	Orthocarbonic acid, tetrapropyl ester or Propyl ortho carbonate. $C(OC_3H_7)_4$	248.36		224.2		$0.897^{20/4}$	1.4100^{20}	al, eth	$B3^4$, 7
10155	Orthoformic acid, triisobutyl ester or Isobutyl ortho formate $HC(O\text{-i-}C_4H_9)_3$	232.36		224-6		$0.8582^{20/4}$	1.4120^{20}	al, eth	$B2^4$, 29
10156	Orthoformic acid, triethyl ester or Ethyl ortho formate, Triethoxy methane. $HC(OC_2H_5)_3$	148.20		143^{765}, 60^{20}		$0.8909^{20/4}$	1.3922^{20}	al, eth	$B2^4$, 25
10157	Orthoformic acid, trimethyl ester or Methyl ortho formate $HC(OCH_3)_3$	106.12		103-5		$0.9676^{20/4}$	1.3793^{20}	al, eth	$B2^4$, 22

No.	Name, Synonyms, and Formula	Mol. wt.	Color, crystalline form, specific rotation and λ_{max} (log ε)	b.p. °C	m.p. °C	Density	n_D	Solubility	Ref.
10158	Orthoformic acid, triisopentyl ester or isopentyl ortho formate $HC(O\text{-}i\text{-}C_5H_{11})_3$	274.44		267-9d, 166^{25}		$0.8628^{20/4}$	1.4233^{20}		B2^4, 30
10159	Orthoformic acid, triphenyl ester or Phenyl ortho formate $HC(OC_6H_5)_3$	292.33		$269\text{-}70^{50}$ d	76-7			al, eth	B6^4, 611
10160	Orthoformic acid, tripropyl ester or Propyl ortho formate $HC(OC_3H_7)_3$	190.28		$190\text{-}1^{745}$, 93^{30}		$0.8805^{20/4}$	1.4072^{20}	al, eth	B2^4, 27
10161	Orthoformic acid, triisopropyl ester or isopropyl ortho formate $HC[OCH(CH_3)_2]_3$	190.28		166-8		$0.8621^{20/4}$	1.4000^{20}	al, eth	B2^4, 28
10162	Orthoformic acid, trithio, triethyl ester or Ethyl orthothio formate $HC(SC_2H_5)_3$	196.38		235d, $127\text{-}8^{12}$		$1.053^{20/4}$	1.5410^{15}	al, eth	B2^4, 93
10163	Orthopropionic acid, triethyl ester or Ethyl ortho propionate $CH_3CH_2C(OC_2H_5)_3$	176.26		171, 44^{9}			1.4000^{25}	al, eth	B2^4, 707
10164	Orthosilicic acid, tetraethyl ester or Ethyl ortho silicate $Si(OC_2H_5)_4$	208.33	liq	168.8	−82.5	$0.9320^{20/4}$	1.3928^{20}	al, eth	B1^4, 1360
10165	Orthosilicic acid, tetrakis(2-ethylbutyl)ester $Si[OCH_2CH(C_2H_5)_2]_4$	432.76	liq	358		$0.8920^{20/4}$	1.4307^{20}	eth, bz	B1^4, 1725
10166	Orthosilicic acid, tetrakis(2-ethylhexyl)ester $Si[OCH_2CH(C_2H_5)C_4H_9]_4$	544.98		419, 227^{5}	−90	$0.8803^{20/4}$	1.4388^{20}	eth, bz	B1^4, 1787
10167	Orthosilicic acid, tetramethyl ester or Methyl ortho silicate $Si(OCH_3)_4$	152.22	nd	121, $25\text{-}7^{12}$	−2	1.0232^{20}	1.3683^{20}	al	B1^4, 1266
10168	1,3,4-Oxadiazole, 2,5-dimethyl $C_4H_6N_2O$	98.10		178-9				w, al, eth	B27, 565
10169	Oxalic acid or Ethanedoic acid HO_2CCO_2H	90.04	mcl ta or pr (+ 2w, w), orh (anh)	157 sub	a:189.5, β:182 (anh), 101.5 (hyd)	α:$1.900^{17/4}$ β:1.895		w, al	B2^4, 1819
10170	Oxalic acid, diallyl ester or Diallyl oxalate $C_3H_5O_2CCO_2C_3H_5$	170.17		217, 86^{2}		1.1582^{20}	1.4481^{20}	al, ace, bz	B2^4, 1851
10171	Oxalic acid, dibutyl ester or Dibutyl oxalate $C_4H_9O_2CCO_2C_4H_9$	202.25		242^{773}, 96^{2}	−30.5	$0.9873^{20/4}$	1.4234^{20}	al, eth	B2^4, 1850
10172	Oxalic acid, diisobutyl ester $(i\text{-}C_4H_9O_2C\text{-})_2$	202.25		229, 143^{20}		$0.9737^{20/4}$	1.4180^{20}	al, eth, ace	B2^4, 1850
10173	Oxalic acid, di(2-chloroethyl)ester $[ClCH_2CH_2O_2C\text{-}]_2$	215.03	lf (dil al)	132^{3}	45			al, bz	B2^4, 1849
10174	Oxalic acid, dicyclohexyl ester or Dicyclohexyl oxalate $(C_6H_{11}O_2C\text{-})_2$	254.33	(MeOH)	$190\text{-}1^{73}$	42 (47)			al, eth	B6^3, 26
10175	Oxalic acid, diethyl ester or Diethyl oxalate $(C_2H_5O_2I\text{-})_2$	146.14		185.7, 97^{20}	−38.5	$1.0785^{20/4}$	1.4101^{20}	al, eth, ace	B2^4, 1848
10176	Oxalic acid dihydrazide $H_2NNHCOCONHNH_2$	118.10	nd (w)		243d	$1.458^{22.5}$			B2^4, 1868
10177	Oxalic acid, dimethyl ester or Dimethyl oxalate $CH_3O_2CCO_2CH_3$	118.09	mcl ta	164.5	54	1.148^{15}, 1.1716^{60}	1.379^{82}	al, eth, ace	B2^4, 1847
10178	Oxalic acid, di-isopentyl ester or Diisopentyl oxalate $(i\text{-}C_5H_{11}O_2C\text{-})_2$	230.30		267-8, 144^{14}		$0.968^{11/11}$		al, eth	B2^4, 1851
10179	Oxalic acid, dipropyl ester or Dipropyl oxalate $C_3H_7O_2CCO_2C_3H_7$	174.20		211, $78\text{-}80^{3}$	−44.3	$1.0188^{20/4}$	1.4158^{20}	al, eth	B2^4, 1849
10180	Oxalic acid, diisopropyl ester $(i\text{-}C_3H_7O_2C\text{-})_2$	174.20		191^{765}		$1.0010^{20/4}$	1.4100^{20}	al, eth	B2^4, 1850
10181	Oxalic acid, di(2-tolyl)ester $[(2\text{-}CH_3C_6H_4)O_2C\text{-}]_2$	270.28	nd (al)	distb	91			al, eth, ace, bz, chl	B6^2, 330
10182	Oxalic acid, di-(3-tolyl)ester $[(3\text{-}CH_3C_6H_4)O_2C\text{-}]_2$	270.28	nd (al)	distb	106			al, eth, ace, bz, chl	B6^2, 353
10183	Oxalic acid, di-(4-tolyl)ester $[(4\text{-}CH_3C_6H_4)O_2C\text{-}]_2$	270.28	lf or pl, (al-eth)		148			al, eth, ace, bz, chl	B6^4, 2114
10184	Oxalic acid, monochloride, monoethyl ester $ClCOCO_2C_2H_5$	136.53	hyg	137, 30^{10}		$1.2226^{20/4}$		eth, bz	B2^4, 1853
10185	Oxalic acid, monoethyl, monomethyl ester $CH_3O_2CCO_2C_2H_5$	132.12		173.7		$1.5505^{0/0}$		al, eth	B2^1, 232
10186	Oxaluric acid or Oxalic acid ureide $H_2NCONHCOCO_2H$	132.08	cr		208-10d			w	B3^4, 121
10187	Oxalyl chloride $ClCOCOCl$	126.93	nd (eth, peth)	63-4	−16	$1.4785^{20/4}$	1.4316^{20}	eth	B2^4, 1853

No.	Name, Synonyms, and Formula	Mol. wt.	Color, crystalline form, specific rotation and λ_{max} (log ε)	b.p. °C	m.p. °C	Density	n_D	Solubility	Ref.
10188	Oxamic acid or Oxalic acid monoamide HO_2CCONH_2	89.05	cr (w)		210d				$B2^4$, 1857
10189	Oxamic acid, N-acetyl, ethyl ester $CH_3CONHCOCO_2C_2H_5$	159.14	pl (eth)		54-5			al, eth	$B2^2$, 509
10190	Oxamic acid, N-sec-butyl Sec-$C_4H_9NHCOCO_2H$	145.16	cr (eth)		88-9			eth	B4, 162
10191	Oxamide, dithio or Dithiooxamide $H_2NCSCSNH_2$	120.19	og-red	sub					$B2^4$, 1871
10192	Oxamic acid, ethyl ester or Ethyl oxamate $H_2NCOCO_2C_2H_5$	117.10			114-5			w, eth	$B2^4$, 1857
10193	Oxamic acid, N-phenyl or Oxanilic acid $HO_2CCONHC_6H_5$	165.15	nd (bz)		150			al, eth, chl	$B12^3$, 550
10194	Oxamic acid, N-phenyl, ethyl ester $C_6H_5NHCOCO_2C_2H_5$	193.20	pl or pr (al), nd (w)	260-300	65-6			al, eth, ace, bz	$B12^3$, 550
10195	Oxamide $H_2NOCCONH_2$	88.07	nd (w)		419d				$B2^4$, 1860
10196	Oxamide, N,N′-diethyl $C_2H_5NHCOCONHC_2H_5$	144.17	nd (al)		175 (180)	1.169^4		al	$B4^4$, 359
10197	Oxamide, N,N′-dimethyl $CH_3NHCOCONHCH_3$	116.12	pl or nd (al)	sub	217	$1.3^{4/4}$		w, chl	$B4^4$, 246
10198	Oxamide, N,N′-dephenyl or Oxanilide $C_6H_5NHCONHC_6H_5$	240.26	lf (bz)	>360	254			bz	$B12^3$, 551
10199	Oxamide, N,N′-diisopropyl [i-C_3H_7NHCO-]$_2$	172.23	nd (al)		212			al	$B4^4$, 518
10200	1,4-Oxathiane C_4H_8OS	104.17		147^{755}		$1.1174^{20/4}$			$B19^4$, 33
10201	Oxazole C_3H_3NO	69.06		69-70			$1.4285^{17.5}$		$B27^2$, 9
10202	Oxazole, 2,4-dimethyl 2,4$(CH_3)_2(C_3HNO)$	97.12		108		$0.9352^{15/4}$	1.4166^{15}	w, al, eth	$B27^2$, 10
10203	Oxazole, 2,5-dimethyl 2,5-$(CH_3)_2(C_3HNO)$	97.12		117-8		$0.9958^{21/4}$	1.4385^{21}	w	$B27^2$, 10
10204	Oxazole, 2,4-diphenyl 2,4-$(C_6H_5)_2(C_3HNO)$	221.26	lf (al)	338-40	103			al, eth, bz	B27, 78
10205	Oxazole, 2,5-diphenyl 2,5-$(C_6H_5)_2(C_3HNO)$	221.26	nd (lig)	360	74	$1.0940^{100/4}$	1.6231^{100}	al, eth	$B27^2$, 43
10206	Oxazole, 4,5-diphenyl 4,5-$(C_6H_5)_2(C_3HNO)$	221.26	pl or pr (lig)	$192\text{-}5^{15}$	44		1.6283^{100}		B27, 79
10207	Oxazole, 2,4,5-triphenyl or Azobenzil 2,4,5-$(C_6H_5)_3(C_3NO)$	297.36	pr		116			bz	$B27^2$, 56
10208	2,4-Oxazolidenedione, 5,5-dipropyl $C_9H_{15}NO_3$	185.22		$148\text{-}50^3$	42-3				Am67, 522
10209	Oxetane or Trimethylene oxide $CH_2CH_2CH_2$ (O bridging terminal C)	58.08		47.8		$0.8930^{25/4}$	1.3961^{20}	w, al, eth, ace	$B17^4$, 13
10210	Oxetane, 3-chloro or 2-Chloro-1,3-epoxy propane C_3H_5ClO	92.53			132-4			al, eth, chl	$B17^4$, 14
10211	Oximide O=C–C=O (NH bridging)	71.04	pr (al)						B21, 368
10212	Oxindole C_8H_7NO	133.15	nd (w)	227^{23}	127			al, eth	$B21^4$, 3611
10213	Oxindole, 1-ethyl 1-$C_2H_5(C_8H_6NO)$	161.20	nd (ace, w)		97-8			ace	$B21^4$, 3615
10214	Oxindole, 3-hydroxy or Diozindole 3-HO(C_8H_6NO)	149.15	cr (w, al)		180				$B21^4$, 6076
10215	Oxindole, 1-methyl-3-ethyl 1-CH_3-3-$C_2H_5(C_8H_5NO)$	175.23		$280\text{-}5^{745}$, $103\text{-}7^{0.5}$			1.557^{25}	al, eth	$B21^2$, 258
10216	Oxacanthine or Vinetine $C_{37}H_{40}N_2O_6$	608.73	nd (al, eth), $[\alpha]^{20/}_D$ + 131.5 (chl, c=1)		216-7			al, eth, bz, chl	$B27^2$, 892
10217	Oxyacanthine hydrochloride $C_{37}H_{40}N_2O_6.HCl$	645.20	nd, $[\alpha]^{15/}_D$ + 163.8 (w, c=3)		270-1			w	$B27^2$, 893
10218	Oxyacanthine nitrate,dihydrate $C_{37}H_{40}N_2O_6.HNO_3.2H_2O$	707.78	nd		195-200				$B27^2$, 893

No.	Name, Synonyms, and Formula	Mol. wt.	Color, crystalline form, specific rotation and λ_{max} (log ε)	b.p. °C	m.p. °C	Density	n_D	Solubility	Ref.
10219	Oxynarcotine or α-Narcotine-*N*-oxide $C_{22}H_{23}NO_8$	429.43	hyg nd, $[\alpha]_D$ + 135 (chl)					w, al, chl	B27[2], 607
10220	Oxysparteine or Isolupanine $C_{15}H_{24}N_2O$	248.37	ye to col hyg nd (peth), $[\alpha]^{18}/_D$ −10.0 (al, c=18)	209^{12}	111			w, al, eth, chl	B24[2], 56
10221	Oxysparteine monohydrochloride, tetrahydrate $C_{15}H_{24}N_2.HCl.4H_2O$	356.89	wh cr (w)		48-50			w, al	B24[2], 57

No.	Name, Synonyms, and Formula	Mol. wt.	Color, crystalline form, specific rotation and λ_{max} (log ε)	b.p. °C	m.p. °C	Density	n_D	Solubility	Ref.
10222	Palmitaldehyde or Hexadecanal $CH_3(CH_2)_{14}CHO$	240.43	pl (eth), nd (peth)	$200-2^{29}$	34	....	....	al, eth, ace, bz	$B1^4$, 3393
10223	Palmitamide $CH_3(CH_2)_{14}CONH_2$	255.44	lf	236^{12}	107	....	....	....	$B2^4$, 1182
10224	Palmitamide, *N*-phenyl or Palmitanilide $CH_3(CH_2)_{14}CONHC_6H_5$	331.54	nd (al)	$282-4^{17}$	90.5	....	....	al, ace, bz, chl	$B12^3$, 486
10225	Palmitic acid or Hexadecanoic acid $CH_3(CH_2)_{14}CO_2H$	256.43	nd (al)	350, 267^{100}	63	$0.8527^{62/4}$	1.4335^{60}	al, eth, ace, bz, chl	$B2^4$, 1157
10226	Palmitic acid anhydride $(C_{15}H_{31}CO)_2O$	494.84	lf (peth)	....	64	$0.8383^{83/4}$	1.4364^{68}	eth	$B2^4$, 1181
10227	Palmitic acid, benzyl ester or Benzyl palmitate $C_{15}H_{31}CO_2CH_2C_6H_5$	346.55	cr (al)	....	36	$0.9136^{35/25}$	1.4689^{50}	al, eth, bz, chl	$B6^3$, 1481
10228	Palmitic acid, butyl ester or Butyl palmitate $C_{15}H_{31}CO_2C_4H_9$	312.54	cr (dil al)		16.9	....	1.4312^{50}	al, eth	$B2^4$, 1167
10229	Palmitic acid, ethyl ester or Ethyl palmitate $C_{15}H_{31}CO_2C_2H_5$	284.48	nd	191^{10}	α: 24, β: 19.3	$0.8577^{25/4}$	1.4347^{34}	al, eth, ace, bz, chl	$B2^4$, 1165
10230	Palmitic acid, hexadecyl ester or Cetyl palmitate $C_{15}H_{31}CO_2C_{16}H_{33}$	480.86	pl (eth)	360	53-4	$0.8324^{50/4}$	1.4425^{50}	eth, bz, chl	$B2^4$, 1168
10231	Palmitic acid, 16-hydroxy $HO(CH_2)_{15}CO_2H$	272.43	cr (bz-eth)		95			al, ace	$B3^3$, 664
10232	Palmitic acid, (2-hydroxy ethyl) ester or β-hydroxy ethyl palmitate $C_{15}H_{31}CO_2CH_2CH_2OH$	300.48	....	$173-4^3$	51	$0.8768^{60/4}$	....	al	$B2^4$, 1168
10233	Palmitic acid, methyl ester or Methyl palmitate $C_{15}H_{31}CO_2CH_3$	270.46	....	$415-8^{747}$, 148^2	30	....	....	al, eth, ace, bz, chl	$B2^4$, 1165
10234	Palmitic acid, myricyl ester or Myricyl palmitate $C_{15}H_{31}CO_2C_{31}H_{63}$	691.26	cr (eth)	....	72	....	....	eth	B2, 373
10235	Palmitic acid, propyl ester $C_{15}H_{31}CO_2C_3H_7$	298.51	nd	190^{12}	20.4	0.8455^{88}	1.4392^{25}	....	$B2^4$, 1167
10236	Palmitic acid, isopropyl ester $C_{15}H_{31}CO_2CH(CH_3)_2$	298.51	....	160^2	13-4	0.8404^{38}	1.4364^{25}	al, eth, ace, bz, chl	$B2^4$, 1167
10237	Palmitic acid, 9,10,16-trihydroxy or Aleuritic acid $HO(CH_2)_6CH(OH)CH(OH)(CH_2)_7CO_2H$	304.43	lf (dil al), nd (w)	....	102	....	....	....	$B2^4$, 1118
10238	Palmitonitrile $C_{15}H_{31}CN$	237.43	hex	333, 251^{100}	31	$0.8303^{20/4}$	1.4450^{20}	al, eth, ace, bz, chl	$B2^4$, 1183
10239	Palmityl chloride $C_{15}H_{31}COCl$	274.87	....	199^{20}	12	....	1.4514^{20}	eth	$B2^4$, 1182
10240	Palmitone $C_{15}H_{31}COC_{15}H_{31}$	450.83	lf (al)	....	83	$0.7947^{91/4}$	1.4297^{94}	eth	$B1^4$, 3413
10241	Paludrine or Proganil $C_{11}H_{16}ClN_5$	253.73	pl (dil al)	....	130-1	....	....	al	C40, 2931
10242	Paludrine hydrochloride $C_{11}H_{16}ClN_5 \cdot HCl$	290.20	nd (w)	....	245	....	....	al	J1946, 729
10243	Pamelonitrile $NC(CH_2)_5CN$	122.17	....	175^{14}	−31.4	0.949^{18}	1.4472^{20}	al, eth, chl	$B2^4$, 2006
10244	Panthesin $C_{18}H_{32}N_2O_5S$	388.52	pa ye pw (al)	....	157-9	....	....	w, al	$B3^{14}$, 1058
10245	Pantothenic acid *(d)* $HOCH_2C(CH_3)_2CH(OH)CONH(CH_2)_2CO_2H$	219.24	ye visc oil, $[\alpha]^{25}{}_D$ + 37.5 (w)	....	....	....	....	w, eth, bz	$B4^3$, 1283
10246	Pantothenic acid, calcium salt *(d)* or Calcium pantothenate $(C_9H_{16}NO_5)_2Ca$	476.56	wh (MeOH), $[\alpha]^{26}{}_D$ + 28.2 (w)	....	195-6	....	....	w	$B4^3$, 1286
10247	Pantothenic acid, calcium salt-*(l)* or Calcium pantothenate $(C_9H_{16}NO_5)_2Ca$	476.54	cr (MeOH), $[\alpha]^{26}{}_D$ -27.8 (w)	....	187-9	....	....	w	$B4^3$, 1288
10248	Pantothenyl alcohol or Panthenol $HOCH_2C(CH_3)_2CH(OH)CONH(CH_2)_3OH$	205.26	hyg oil	d, $118-20^{0.02}$	....	$1.2^{20/20}$	1.497^{20}	w, al	$B4^4$, 1652
10249	Papaveraldine or Xanthaline $C_{20}H_{19}NO_5$	353.37	nd (al) cr, (bz, peth)	210-1		....	....	bz, chl, aa	$B21^4$, 6738
10250	Papaverine or Papaveroline tetramethyl ether $C_{20}H_{21}NO_4$	339.39	wh pr (al-eth) nd, (chl-peth)	d, sub, $135-40^{11}$	147-8	$1.337^{20/4}$	1.625	al, ace, chl, Py	$B21^4$, 2788
10251	Papaverine hydrochloride $C_{20}H_{21}NO_4.HCl$	375.85	wh mcl pr (w)	....	224-5	....	....	w, al	$B21^4$, 2790

PHYSICAL CONSTANTS OF ORGANIC COMPOUNDS (Continued)

No.	Name, Synonyms, and Formula	Mol. wt.	Color, crystalline form, specific rotation and λ_{max} (log ε)	b.p. °C	m.p. °C	Density	n_D	Solubility	Ref.
10252	Parabanic acid or oxalurea $C_3H_2N_2O_3$	114.06	mcl nd (w)	sub 100	243-5d			w, al	B24^{2}, 263
10253	Parabutyraldehyde or 2,4,6-Tripropyl-1,3,5-trioxane $C_{12}H_{24}O_3$	216.32		98-100^{35}		0.918			B19^{4}, 4725
10254	Para isobutyraldehyde or 2,4,6-Triisopropyl-1,3,5-trioxane $C_{12}H_{24}O_3$	216.32	nd (al)	195 sub	59-60			al, eth	B19^{4}, 4726
10255	Paraconic acid or Hydroxymethyl succinic acid-γ-lactone $C_5H_6O_4$	130.10	dlq		57-8			w	B18^{4}, 5264
10256	Paracyanogen $(CN)_x$		br pw		sub				B2^{4}, 1864
10257	Paraldehyde or 2,4,6-Trimethyl-1,3,5-trioxane $C_6H_{12}O_3$	132.16		128	12.6	0.9943$^{20/4}$	1.4049^{20}	al, eth, chl	B19^{4}, 4715
10258	Paraldol $C_8H_{16}O_4$	176.21	wh tcl pr	90^{15}	89-91	1.116^{20}	1.4610^{20}	w, al, eth	B1^{4}, 3987
10259	Parasorbic acid $C_6H_8O_2$	112.13	oily liq, $[\alpha]^{19/}{}_D$ + 210 (al, c=3)	100^{15}		1.079$^{18/4}$	1.4730^{20}	w, al, eth	B17^{4}, 4305
10260	Parathion or Diethyl-*p*-nitro phenyl mono thiophosphate $C_{10}H_{14}NO_5PS$	291.26	ye liq	375, 157-62$^{0.5}$	6.1	1.2704$^{20/20}$	1.5370^{25}	al, eth, ace, chl	B6^{4}, 1337
10261	Patulin or Clavacin $C_7H_6O_4$	154.12	pl or pr (eth, chl)		111			w, al, eth, ace, bz	B18^{4}, 1184
10262	Patchouli alcohol $C_{15}H_{26}O$	222.37			56	0.9924$^{65/20}$	1.5029^{65}	al, eth	B6^{3}, 426
10263	Paucin, hydrate $C_{27}H_{39}N_5O_5.6½H_2O$	630.74	ye lf		126d				B10^{3}, 1841
10264	Pelargonaldehyde $C_8H_{17}CHO$	142.24		190-2, 93.5^{23}		0.8264$^{22/4}$	1.4275^{20}	eth	B1^{4}, 3352
10265	Pelargonaldehyde oxime or Pelargonaldoxime $C_8H_{17}CH{=}NOH$	157.26	lf (dil al)		64			al, eth, ace	B1^{2}, 761
10266	Pelargonamide $CH_3(CH_2)_7CONH_2$	157.26		sub	99-100	0.8394^{110}	1.4248^{110}		B2^{4}, 1023
10267	Pelargonic acid or Nonanoic acid $CH_3(CH_2)_7CO_2H$	158.24		255, 150^{20}	fp 12.2	0.9057$^{20/4}$	1.4343^{19}	al, eth	B2^{4}, 1018
10268	Pelargonic acid, 9-amino $H_2N(CH_2)_8CO_2H$	173.26		185-7				w, al	B4^{3}, 1479
10269	Pelargonic acid, ethyl ester or Ethyl pelargonate $CH_3(CH_2)_7CO_2C_2H_5$	186.29		227, 96-8^{10}	−36.7	0.8657$^{20/4}$	1.4220^{20}	al, eth, ace	B2^{4}, 1019
10270	Pelargonic acid, 9-fluoro $F(CH_2)_8CO_2H$	176.23		88-90$^{0.2}$	*ca* 18		1.4289^{25}	al, eth	B2^{4}, 1024
10271	Pelargonic acid, methyl ester or Methyl pelargonate $CH_3(CH_2)_7CO_2CH_3$	172.27		213-4^{757}, 104-6^{23}		0.8799^{15}	1.4214^{20}	al, eth	B2^{4}, 1019
10272	Pelargononitrile $CH_3(CH_2)_7CN$	139.24		224.4, 91.9^{10}	−34.2	0.8178$^{20/4}$	1.4255^{20}	al, eth	B2^{4}, 1024
10273	Pelargonyl chloride $CH_3(CH_2)_7COCl$	176.69		215.3, 98^{15}	−60.5	0.9463$^{15/4}$		eth, ace	B2^{4}, 1023
10274	Pelargonidin chloride $C_{15}H_{11}ClO_5$	306.70	red br hyg (anh) pr, or pl (dil HCl)		>350 (anh)			w, al	B18^{4}, 3198
10275	Pellotine or *N*-Methyl anhalonidine $C_{13}H_{19}NO_3$	237.30	pl (al, peth)		111.5			al, eth, ace, chl, peth	B21^{4}, 2525
10276	Penicilic acid $C_8H_{10}O_4$	170.17	rh or hex pl (+1w), nd (peth)		64-5 (+1w) 8-7 (anh)			w, al, eth, ace, bz	B3^{2}, 519
10277	Pentacene or 2,3,6,7-Dibenzanthracene $C_{22}H_{14}$	278.35	deep vt-bl nd or lf (ph NO_2), cr (bz)	290-300 sub (vac)	270-1				B5^{3}, 2551
10278	Pentacene, 6,13-diphenyl $C_{34}H_{22}$	430.55	vt-bl nd	318-20				bz, aa	B5^{3}, 2738
10279	Pentacosane $CH_3(CH_2)_{23}CH_3$	352.69		401.9, 239.9^{10}		0.8012$^{20/4}$	1.4491^{20}	bz, chl	B1^{4}, 582
10280	Pentacosane, 13-phenyl $C_6H_5CH(C_{12}H_{25})_2$	428.78		235^{10}	31.7	0.8537$^{20/4}$	1.4787$^{20/}{}_4$		B5^{4}, 1241
10281	Pentacosane, 13-undecyl $(C_{12}H_{25})_2CHC_{11}H_{23}$	506.98		307^{10}	9.7	0.8168$^{20/4}$	1.4567^{20}		B1^{4}, 600

PHYSICAL CONSTANTS OF ORGANIC COMPOUNDS (Continued)

No.	Name, Synonyms, and Formula	Mol. wt.	Color, crystalline form, specific rotation and λ_{max} (log ε)	b.p. °C	m.p. °C	Density	n_D	Solubility	Ref.
10282	6,9-Pentadecadiyne $H_2C(C\equiv C\text{-}C_5H_{11})_2$	204.36	...	135-6^{4}	...	0.840$^{21/4}$	1.4693^{21}	...	B1^{4}, 1132
10283	Pentadecanal $CH_3(CH_2)_{13}CHO$	226.40	nd	185^{25}	24-5	...	...	al, eth, ace	B1^{4}, 3391
10284	Pentadecanal oxime $CH_3(CH_2)_{13}CH{=}NOH$	241.42	nd (dil al)	...	86	...	...	eth	B1^{2}, 770
10285	Pentadecane $CH_3(CH_2)_{13}CH_3$	212.42	...	270.6, 136.10	10	0.7685$^{20/4}$	1.4315^{20}	al, eth	B1^{4}, 529
10286	Pentadecane, 1-amino or Pentadecylamine $CH_3(CH_2)_{13}CH_2NH_2$	227.43	fl	307.6, 165.8^{10}	37.3	0.8104$^{20/4}$	1.4480^{20}	al, eth	B4^{4}, 817
10287	Pentadecane, 1-bromo or Pentadecyl bromide $CH_3(CH_2)_{13}CH_2Br$	291.32	...	322, 177^{10}	19	1.0675$^{20/4}$	1.4611^{20}	ace, chl	B1^{4}, 531
10288	Pentadecane, 1,15-dibromo $Br(CH_2)_{15}Br$	370.21	lf (al)	215-25^{15}, 192^{2}	27	...	...	chl	B1^{4}, 531
10289	Pentadecane, 1(2,3-dihydroxyphenyl) or Tetra hydra urushiol $(2,3\text{-}(HO)_2C_6H_3)CH_2(CH_2)_{13}CH_3$	320.52	nd (to, eth, peth)	...	59	...	...	al, eth, bz, aa, chl	B6^{3}, 4771
10290	Pentadeconoic acid or Pentadecylic acid $CH_3(CH_2)_{13}CO_2H$	242.40	pl (aq, al, aa), cr (peth)	257^{100}, 158^{1}	53-4	0.8423^{80}	1.4254^{80}	al, eth, ace, bz, chl	B2^{4}, 1147
10291	Pentadecanoic acid, methyl ester or Methyl pentadeconoate $CH_3(CH_2)_{13}CO_2CH_3$	256.43	nd (dil al)	153.5	18.5	0.8618$^{25/4}$	1.4390^{25}	al, eth	B2^{3}, 936
10292	2-Pentadecanol $C_{13}H_{27}CH(OH)CH_3$	228.42	...	299	35	0.8328^{20}	1.4463^{20}	...	B1^{4}, 1871
10293	2-Pentadecanone $C_{13}H_{27}COCH_3$	226.40	...	294	39.5	0.8182^{39}	...	...	B1^{4}, 3391
10294	8-Pentadecanone or Diheptyl ketone $(C_7H_{15})_2CO$	226.40	cr (al)	291, 178^{20}	43	...	...	al, eth, bz, chl	B1^{4}, 3392
10295	1-Pentadecene $CH_3(CH_2)_{12}CH{=}CH_2$	210.40	...	268.2, 133.7^{10}	2-8	0.7764$^{20/4}$	1.4389^{20}	ace	B1^{4}, 926
10296	1-Pentadecyne $CH_3(CH_2)_{12}C\equiv CH$	208.39	...	268, 129.8^{10}	10	0.7928$^{20/4}$	1.4419^{20}	ace	B1^{4}, 1072
10297	2,4-Pentadienal, 5-phenyl or Cinnamylidene acetaldehyde $C_6H_5CH{=}CHCH{=}CHCHO$	158.20	...	155-65^{3}, 92-5$^{0.05}$	42-3	...	...	al, eth, bz	B7^{3}, 1653
10298	1,2-Pentadiene or Ethyl allene $CH_3CH_2CH{=}C{=}CH_2$	68.12	...	44.9	−137.3	0.6926$^{20/4}$	1.4209^{20}	al, eth, ace, bz	B1^{4}, 993
10299	1,2-Pentadiene, 1-chloro-3-ethyl $(C_2H_5)_2C{=}C{=}CHCl$	130.62	...	85-8^{100}	...	0.9297$^{19/4}$	...	eth	B1^{3}, 1002
10300	1,2-Pentadiene, 1-chloro-3-methyl $CH_3CH_2C(CH_3){=}C{=}CHCl$	116.59	...	68-70^{100}	...	0.9562$^{20/4}$	...	eth	B1^{3}, 990
10301	1,3-Pentadiene or Piperylene $CH_3CH{=}CHCH{=}CH_2$	68.12	...	42	−87.5	0.6760$^{20/4}$	1.4301^{20}	al, eth, ace, bz	B1^{4}, 994
10302	1,3-Pentadiene, 1-chloro-3-methyl $CH_3CH{=}C(CH_3)CH{=}CHCl$	116.59	...	62-3^{100}	...	0.9574$^{20/4}$	...	eth, chl	B1^{3}, 990
10303	1,3-Pentadiene, 2-chloro-3-methyl $CH_3CH{=}C(CH_3)CCl{=}CH_2$	116.59	...	57-60^{95}	...	0.9437$^{20/4}$	1.4671^{20}	eth, chl	B1^{3}, 991
10304	1,3-Pentadiene, 3-chloro or Methyl chloroprene $CH_3CH{-}C(Cl)CH{=}CH_2$	102.56	...	99-101	...	0.9576$^{20/4}$	1.4785^{20}	eth, ace, bz, chl	B1^{3}, 962
10305	1,3-Pentadiene, 2,4-dimethyl $(CH_3)_2C{=}CHC(CH_3){=}CH_2$	96.17	...	93^{758}	...	0.7343$^{23/4}$	1.43904^{23}	...	B1^{4}, 1034
10306	1,3-Pentadiene, 4-methyl $(CH_3)_2C{=}CHCH{=}CH_2$	82.15	...	76.5	...	0.7181$^{20/4}$	1.4532^{20}	...	B1^{4}, 1020
10307	1,4-Pentadiene $CH_2{=}CHCH_2CH{=}CH_2$	68.12	...	26	−148.3	0.6608$^{20/4}$	1.3888^{20}	al, eth, ace, bz	B1^{4}, 998
10308	2,3-Pentadiene $CH_3CH{=}C{=}CHCH_3$	68.12	...	48.2	−125.6	0.6950$^{20/4}$	1.4284^{20}	al, eth, ace, bz	B1^{4}, 999
10309	2,4-Pentadienoic acid or β-Vinyl acrylic acid $CH_2{=}CHCH{=}CHCO_2H$	98.10	hyg pr (eth)	d110-5	80	...	...	w, al, eth, bz, peth	B2^{4}, 1694
10310	2,4-Pentadienoic acid, 4-hydroxy-γ-lactone $CH_2{=}C\text{-}CH{=}CHCO$ (C and CO joined through O)	96.09	pa ye oil	73^{11}	...	...	...	chl	B17^{4}, 4495
10311	2,4-Pentadienoic acid, 5-(3,4-methylene dioxy)phenyl or Piperic acid $(3,4\text{-}CH_2O_2)C_6H_3CH{=}CHCH{=}CHCO_2H$	218.21	ye in light nd (al), ye nd (sub)	sub	215	...	...	al	B19^{4}, 3565
10312	2,4-Pentadienoic acid, 5-phenyl or Cinnamylidenacetic acid $C_6H_5CH{=}CHCH{=}CHCO_2H$	174.20	pl (al), pr (bz)	...	166-7	...	...	eth, bz	B9^{3}, 3070

No.	Name, Synonyms, and Formula	Mol. wt.	Color, crystalline form, specific rotation and λ_{max} (log ε)	b.p. °C	m.p. °C	Density	n_D	Solubility	Ref.
10313	2,4-pentadienoic acid, 5-phenyl,ethyl ester $C_6H_5CH=CHCH=CHCO_2C_2H_5$	202.25	ye oil	149-50^{4}	25-7	1.0469$^{20/4}$	1.5768^{80}	al, eth	B9^{2}, 441
10314	2,4-Pentadienoic acid, 5-phenyl,methyl ester $C_6H_5CH=CHCH=CHCO_2CH_3$	188.23	lf or pl	185^{20}	71			al	B9^{3}, 3071
10315	2,4-Pentadienonitrile *(cis)* $CH_2=CHCH=CHCN$	79.10		49.5^{32}, 32.5^{13}	−64	0.8541$^{26/4}$	1.4855^{20}	eth, ace	B2^{4}, 1695
10316	2,4-Pentadienonitrile *(trans)* $CH_2=CHCH=CHCN$	79.10		41^{13}	−43	0.8576$^{20/4}$	1.4986^{20}	eth, ace	B2^{4}, 1696
10317	1,4-Pentadien-3-one, 1,5-bis-(2-ethoxyphenyl) $[(2\text{-}C_2H_5OC_6H_4)CH=CH]_2CO$	322.40	ye lf (dil al)		89			al	B8^{1}, 666
10318	1,4-Pentadiene-3-one, 1,5-bis(2-hydroxyphenyl) $[(2\text{-}HOC_6H_4)CH=CH]_2CO$	266.30	ye nd (dil al)		168d			al, eth , ace, bz, Py	B8^{3}, 2954
10319	1,4-Pentadien-3-one, 1,5-bis-(4-hydroxyphenyl) or Bis-(4-hydroxystyryl)ketone $[(4\text{-}HOC_6H_4)CH=CH]_2CO$	266.30	ye-og nd or lf (dil al)		237-8			ace, al	B8^{3}, 2958
10320	1,4-Pentadien-3-one, 1,5-bis-(2-methoxyphenyl) or Bis (2-methoxystyryl)ketone $[(2\text{-}CH_3OC_6H_4)CH=CH]_2CO$	294.35	ye nd or lf (al)		127				B8^{2}, 405
10321	1,4-Pentadien-3-one, 1,5-bis-(3-methoxyphenyl) or Bis (3-methoxystyryl)ketone $[(3\text{-}CH_3OC_6H_4)CH=CH]_2CO$	294.35	nd (chl-MeOH)		55-6			ace, chl	B81, 666
10322	1,4-Pentadien-3-one, 1,5-bis-(4-methoxyphenyl) or Dianisal acetone $[(4\text{-}CH_3OC_6H_4)CH=CH]_2CO$	294.35	ye lf (aa)		129-30			bz, aa, chl	B8^{3}, 2958
10323	1,4-Pentadien-3-one, 1,5-bis(3,4-methylene dioxyphenyl) or Dipiperonylidene acetone $C_{19}H_{14}O_5$	322.32	ye nd (bz)		185			ace, chl	B19^{4}, 5891
10324	1,4-Pentadien-3-one, 1,5-bis-(4-nitrophenyl) or Bis (2-nitrostyryl)ketone $[(2\text{-}O_2NC_6H_4)CH=CH]_2CO$	324.29	ye nd (aa)		170-1			chl	B7^{2}, 455
10325	1,4-Pentadien-3-one, 1,5-bis-(3-nitrophenyl) $[(3\text{-}O_2NC_6H_4)CH=CH]_2CO$	324.29	ye br (Ac_2O)		238			ace	B7^{2}, 455
10326	1,4-Pentadien-3-one, 1,5-bis-(4-nitrophenyl) or Bis(4-nitrosytyryl) ketone $[(4\text{-}O_2NC_6H_4)CH=CH]_2CO$	324.29	ye (Ac_2O)		254			bz	B7^{2}, 455
10327	1,4-Pentadiene-3-one, 1(2-chlorphenyl)-5-(3-chlorophenyl) $(2\text{-}ClC_6H_4)CH=CH\text{-}CO\text{-}CH=CH(C_6H_4Cl\text{-}3)$	303.19	ye nd (dil al)		67-8				B7^{2}, 454
10328	1,4-Pentadiene-3-one, 1(2-chlorophenyl)-5-(4-chlorophenyl) $(2\text{-}ClC_6H_4)CH=CH\text{-}CO\text{-}CH=CH(C_6H_4Cl\text{-}4)$	303.19	ye nd (al)		109				B7^{2}, 454
10329	1,4-Pentadien-3-one, 1,5-di(2-furyl) $[2\text{-}(C_4H_3O)CH=CH]_2CO$	214.22	dlq pr (peth), ye pr (lig)	d	60-1			al, eth, chl	B19^{4}, 1822
10330	1,4-Pentadien-3-one, 1,5-diphenyl or Dibenzal acetone $[C_6H_5CH=CH]_2CO$	234.30	pl or lf (ace, AcOEt)	d	113d			ace, chl	B7^{3}, 2559
10331	1,3-Pentadiyne $CH_3C\equiv CC\equiv CH$	64.09		55-6		0.7375$^{20/4}$	1.4431^{21}	eth, bz, chl	B1^{4}, 1117
10332	Pentaerythritol or Tetramethylol methane $C(CH_2OH)_4$	136.15	cr (dil HCl)	sub	269		1.548	w	B1^{4}, 2812
10333	Pentaerythritol tetra acetate $(CH_3CO_2CH_2)_4C$	304.30	tetr nd (w, bz)		83-4	1.273$^{18/4}$		w, al, eth	B1^{4}, 264
10334	Pentaerythrityl tetrabromide or 2,2-bis(Bromoethyl)-1,3-dibromo propane $C(CH_2Br)_4$	387.73	cr (ace), nd (lig)	305-6	163	2.596^{15}		al, bz	B1^{4}, 337
10335	Pentaerythrityl tetrachloride or 2,2-bis(chloromethyl)-1,3-dichloropropane $C(CH_2Cl)_4$	209.93		110^{12}	97			eth, chl	B1^{4}, 336
10336	Pentaerythrityl tetraiodide or 2,2-bis(iodomethyl)-1,3-diiodo methane $C(CH_2I)_4$	575.74	nd (to)		225				B1^{4}, 338
10337	Pentaerythritol tetranitrate or PETN $C(CH_2ONO_2)_4$	316.14	tetr (ace), pr (ace-al)		140-1	1.773$^{20/4}$		ace, bz	B1^{4}, 2816
10338	Pentamethylene sulfide or Tetrahydro thiopyran $C_5H_{10}S$	102.19		141.7, 93^{82}	19	0.9861$^{20/4}$	1.5067^{20}	al, eth, ace, bz	B17^{4}, 55
10339	Pentanal or Valeraldehyde $CH_3(CH_2)_3CHO$	86.13		103	−91.5	0.8095$^{20/4}$	1.3944^{20}	al, eth	B1^{4}, 3268

No.	Name, Synonyms, and Formula	Mol. wt.	Color, crystalline form, specific rotation and λ_{max} (log ε)	b.p. °C	m.p. °C	Density	n_D	Solubility	Ref.
10340	Pentane $CH_3(CH_2)_3CH_3$	72.15		36.1	−130	$0.6262^{20/4}$	1.3575^{20}	**al, eth, ace, bz, chl**	$B1^4$, 303
10341	Pentane, 1-amino or Pentylamine $CH_3(CH_2)_3CH_2NH_2$	87.16		104.4, 5.9^{10}	−55	$0.7547^{20/4}$	1.4118^{20}	**al, eth**, ace, bz	$B4^4$, 675
10342	Pentane, 2-amino $C_3H_7CH(NH_2)CH_3$	87.16		91.5^{755}		0.7384^{20}	1.4027^{20}	w, **al, eth**, ace, bz	$B4^4$, 689
10343	Pentane, 3-amino $H_2NCH(C_2H_5)_2$	87.16		91		$0.7487^{20/4}$	1.4063^{20}	al	$B4^4$, 692
10344	Pentane, 3,3-bis(ethyl sulfonyl) $(C_2H_5)_2C(SO_2C_2H_5)_2$	256.38	lf (dil al)		85			al, eth, w	B4, 681
10345	Pentane, 1-bromo or Pentyl bromide $CH_3(CH_2)_3CH_2Br$	151.05		129.6, 21^{10}	−87.9	$1.2182^{20/4}$	1.4447^{20}	al, **eth**, bz, chl	$B1^4$, 312
10346	Pentane, 1-bromo-5-fluoro $F(CH_2)_5Br$	169.04		162		1.3604^{25}	1.4406^{25}	al, eth	$B1^4$, 313
10347	Pentane, 1-bromo-2-methyl $C_3H_7CH(CH_3)CH_2Br$	165.07		142-5, $51\text{-}3^{25}$		$1.1624^{20/4}$	1.4495^{20}	eth, chl	$B1^4$, 361
10348	Pentane, 1-bromo-3-methyl $C_2H_5CH(CH_3)CH_2CH_2Br$	165.07		$148\text{-}9^{766}$		$1.1829^{20/4}$	1.4496^{20}	eth, chl	$B1^4$, 365
10349	Pentane, 1-bromo-4-methyl or Isohexyl bromide $(CH_3)_2CH(CH_2)_3Br$	165.07		147-8		1.1683^{20}	1.4490	eth, chl	$B1^4$, 361
10350	Pentane, 2-bromo $C_3H_7CHBrCH_3$	151.05		117.4, 58.4^{100}	−95.5	$1.2075^{20/4}$	1.4413^{20}	al, eth, bz, chl	$B1^4$, 312
10351	Pentane, 2-bromo-2-methyl $C_3H_7CBr(CH_3)_2$	165.07		142-3, 70^{100}			1.442^{23}	eth, chl	$B1^4$, 361
10352	Pentane, 3-bromo $BrCH(C_2H_5)_2$	151.05		118.6, 10.8^{10}	−126.2	$1.2124^{20/4}$	$1.4441^{20/4}$	al, eth, bz, chl	$B1^4$, 313
10353	Pentane, 3-bromo-3-methyl $(C_2H_5)_2C(Br)CH_3$	165.07		129-31, $82\text{-}3^{145}$		1.1835^{20}	1.4525^{20}	eth, chl	$B1^4$, 365
10354	Pentane, 1-chloro or Pentyl chloride $CH_3(CH_2)_3CH_2Cl$	106.60		107.8, 5^{10}	−99	$0.8818^{20/4}$	1.4127^{20}	**al, eth**, bz, chl	$B1^4$, 309
10355	Pentane, 1-chloro-5-fluoro $F(CH_2)_5Cl$	124.59		143.2		1.0325^{25}	1.4120^{23}	al, eth	$B1^4$, 309
10356	Pentane, 2-chloro $C_3H_7CHClCH_3$	106.60	$[\alpha]_D$ + 34.1	96.9	−137	$0.8698^{20/4}$	1.4069^{20}	al, eth, bz, chl	$B1^4$, 309
10357	Pentane, 2-chloro-2,3-dimethyl $C_2H_5CH(CH_3)CCl(CH_3)_2$	134.65		$38\text{-}9^{20}$			1.4264^{20}	eth, bz	$B1^4$, 406
10358	Pentane, 2-chloro-2,4-dimethyl $(CH_3)_2CHCH_2CCl(CH_3)_2$	134.65		$127\text{-}8^{733}$, $33\text{-}4^{20}$		$0.861^{20/4}$	1.4180^{20}	eth	$B1^4$, 408
10359	Pentane, 2-chloro-3-ethyl $(C_2H_5)_2CHCH(Cl)CH_3$	134.65		83.5^{100}		$0.8951^{25/25}$	1.4318^{20}	eth, chl	$B1^2$, 120
10360	Pentane, 2-chloro-2-methyl $C_3H_7CCl(CH_3)_2$	120.62		$110\text{-}1^{734}{}_d$, $36\text{-}7^{15}$		$0.863^{20/4}$	1.4126^{20}	eth	$B1^4$, 360
10361	Pentane, 2-chloro-4-methyl $(CH_3)_2CHCH_2CHClCH_3$	120.62		$111\text{-}2^{733}$		$0.8610^{20/4}$	1.4113^{20}	eth	$B1^3$, 399
10362	Pentane, 2-chloro-2,4,4-trimethyl $(CH_3)_2CClCH_2C(CH_3)_3$	148.68		145-50d, 44^{16}	−26	0.8746^{20}	1.4308^{20}	al	$B1^4$, 444
10363	Pentane, 3-chloro $ClCH(C_2H_5)_2$	106.60		97.8, $38\text{-}9^{20}$	−105	$0.8731^{20/4}$	1.4082^{20}	al, eth, bz, chl	$B1^4$, 309
10364	Pentane, 3-chloro-2,2-dimethyl-3-ethyl $(C_2H_5)_2CClC(CH_3)_3$	162.70		d $53\text{-}4^{6}$			1.4528^{25}	eth, chl	$B1^4$, 462
10365	Pentane, 3-chloro-2,3-dimethyl $C_2H_5C(CH_3)ClCH(CH_3)_2$	134.65		$135\text{-}8^{757}{}_d$, $41\text{-}2^{20}$		$0.884^{22/22}$	1.4318^{20}	eth, chl	$B1^4$, 406
10366	Pentane, 3-chloro-3-ethyl $(C_2H_5)_3CCl$	134.65		143-4, $43\text{-}4^{20}$		$0.8856^{20/4}$	1.4400^{20}	eth	$B1^4$, 403
10367	Pentane, 3-chloro-3-ethyl-2-methyl $(C_2H_5)_2CClCH(CH_3)_2$	148.68		150-5d, $98\text{-}100^{98}$		1.0325^{25}	1.4120^{23}	al, eth	$B1^4$, 437
10368	Pentane, 3-chloro-2-methyl $C_2H_5CHClCH(CH_3)_2$	120.62		$115\text{-}7^{752}{}_d$				eth, chl	$B1^2$, 111
10369	Pentane, 3-chloro-3-methyl $(C_2H_5)_2C(Cl)CH_3$	120.62		116, 35^{25}		$0.8900^{20/4}$	1.4210^{20}	eth, bz, chl	$B1^4$, 365
10370	Pentane, 3-(chloromethyl) $(C_2H_5)_2CHCH_2Cl$	120.62		125-7		$0.8914^{20/4}$	1.4222^{20}	eth, bz, chl	$B1^3$, 403
10371	Pentane, 4-chloro-2,2-dimethyl $CH_3CHClCH_2C(CH_3)_3$	134.65		93^{250}		$0.855^{20/4}$	1.4180^{20}	eth, chl	$B1^4$, 404
10372	Pentane, 1,5-diamino or Pentamethylendiamine Cadavarine $H_2N(CH_2)_5NH_2$	102.18		178-80	9	0.867^{25}	1.4561^{25}	w, al	$B4^4$, 1310

No.	Name, Synonyms, and Formula	Mol. wt.	Color, crystalline form, specific rotation and λ_{max} (log ε)	b.p. °C	m.p. °C	Density	n_D	Solubility	Ref.
10373	Pentane, 1,5-dibromo $Br(CH_2)_5Br$	229.94		222.3, 98.6^{20}	−39.5	1.7018$^{20/4}$	1.5126^{20}	bz, chl	B1^4, 314
10374	Pentane, 1,2-dichloro $C_3H_7CHClCH_2Cl$	141.04		148-9, 58-9^{20}		1.0872$^{20/4}$	1.4485^{20}	al, chl	B1^3, 341
10375	Pentane, 1,2-dichloro-4,4-dimethyl $(CH_3)_3CCH_2CHClCH_2Cl$	169.09		173-5^{745}, 58-9^{12}		1.0259^{20}	1.4489^{20}	bz, chl	B1^3, 445
10376	Pentane, 1,3-dichloro $C_2H_5CHClCH_2CH_2Cl$	141.04		80.4^{60}		1.0834$^{20/4}$	1.4485^{20}	ace, chl	B1^1, 342
10377	Pentane, 1,4-dichloro $CH_3CHClCH_2CH_2CH_2Cl$	141.04		161-3, 58-60^{15}		1.0840$^{20/4}$	1.4503^{20}	ace, chl	B1^4, 309
10378	Pentane, 1,5-dichoro $Cl(CH_2)_5Cl$	141.04		180, 59^{10}	−72.8	1.1006$^{22/4}$	1.4564^{20}	al, eth, bz, chl	B1^4, 310
10379	Pentane, 1,5-dichloro-3,3-dimethyl $(ClCH_2CH_2)_2C(CH_3)_2$	169.09		135^{80}, 58-9^{8}		1.0563$^{20/4}$	1.4652^{20}	chl	B1^4, 409
10380	Pentane, 2,2-dichloro $C_3H_7CCl_2CH_3$	141.04		128-9, 36-7^{20}		1.040^{20}	1.434^{20}	eth, bz, chl	B1^1, 342
10381	Pentane, 2,4-dichloro $CH_2(CHClCH_3)_2$	141.04		147-50, 62^{12}		1.0634$^{15/4}$	1.447^{18}	eth, bz, chl	B1^1, 343
10382	Pentane, 2,4-dichloro-2,4-dimethyl $CH_2[CCl(CH_3)_2]_2$	169.09		51-7^{8}	23-4	1.0292$^{20/4}$	1.4537^{20}		B1^4, 408
10383	Pentane, 3,3-dichloro $(C_2H_5)_2CCl_2$	141.04		131-2^{750}, 32^{14}		1.053^{20}	1.442^{20}	bz, chl	B1^1, 343
10384	Pentane, 3,3-dichloro-2,4-dimethyl $[(CH_3)_2CH]_2CCl_2$	169.09		118-20d		0.9513^{9}		eth, chl	B1, 158
10385	Pentane, 3,3-diethyl or Tetraethyl methane $C(C_2H_5)_4$	128.26		146.2, 30.7^{10}	−33.1	0.7536$^{20/4}$	1.4206^{20}	eth, bz	B1^4, 462
10386	Pentane, 1,5-diido $I(CH_2)_5I$	323.94		149^{20}, 101-2^{3}	9	2.1903^{15}	1.6046^{15}	eth, chl	B1^4, 317
10387	Pentane, 2,2-dimethyl $C_3H_7C(CH_3)_3$	100.20		79.2	−123.8	0.6739$^{20/4}$	1.3822^{20}	al, eth, **ace**, **bz**, **chl**	B1^4, 403
10388	Pentane, 2,2-dimethyl-3-ethyl $(CH_3)_3CCH(C_2H_5)_2$	128.26		133.83	−99.3	0.74378$^{20/4}$	1.41227^{20}		B1^4, 462
10389	Pentane, 2,3-dimethyl $C_2H_5CH(CH_3)CH(CH_3)_2$	100.20		89.8		0.6951$^{20/4}$	1.3919^{20}	al, eth, **ace**, **bz**, **chl**	B1^4, 405
10390	Pentane, 2,3-dimethyl-3-ethyl $(CH_3)_2CHC(C_2H_5)_2CH_3$	128.26		141.6			1.4186^{20}		B1^4, 462
10391	Pentane, 2,4-dimethyl $CH_2[CH(CH_3)_2]_2$	100.20		80.5	−119.2	0.6727$^{20/4}$	1.3815^{20}	al, eth, **ace**, **bz**, **chl**	B1^4, 406
10392	Pentane, 2,4-dimethyl-3-ethyl $CH_3CH(CH_3)CH(C_2H_5)CH(CH_3)_2$	128.26		136.73	−122.4	0.7365$^{20/4}$	1.4131^{20}		B1^4, 462
10393	Pentane, 2,4-dimethyl-3-phenyl $(CH_3)_2CHCH(C_6H_5)CH(CH_3)_2$	176.29		220-5		0.8822$^{10/0}$	1.512		B5^1, 214
10394	Pentane, 3,3-dimethyl $(C_2H_5)_2C(CH_3)_2$	100.20		86.1	−134.4	0.6936$^{20/4}$	1.3909^{20}	al, eth, **ace**, **bz**, **chl**	B1^4, 409
10395	Pentane, 1,5-dinitro $O_2N(CH_2)_5NO_2$	162.15		134$^{1.2}$			1.461^{20}	bz	B1^4, 319
10396	Pentane, 1,1-diphenyl $C_4H_9CH(C_6H_5)_2$	224.35		307.9	−12	0.9659^{20}	1.5511^{20}		B5^4, 1966
10397	Pentane, 1,5-diphenyl $C_6H_5(CH_2)_5C_6H_5$	224.35		330.6, 187-9^{10}		0.9814$^{19/0}$	1.559^{19}		B5^4, 1963
10398	Pentane, 1,2-epoxy-2,4,4-trimethyl $(CH_3)_3CCH_2\underbrace{C(CH_3)CH_2}_{O}$	128.21		140.9, 20$^{5.4}$	−64	0.8287$^{20/20}$	1.4097^{20}	eth, bz	B17^4, 112
10399	Pentane, 3-ethyl or Triethyl methane $(C_2H_5)_3CH$	100.20		93.5	−118.6	0.6982$^{20/4}$	1.3934^{20}	al, eth, **ace**, **bz**, **chl**	B1^4, 402
10400	Pentane, 3-ethyl-2-methyl $(C_2H_5)_2CHCH(CH_3)_2$	114.23		115.6, 9.5^{10}	−115	0.7193$^{20/4}$	1.4040^{20}	al, eth, ace, bz, chl	B1^4, 437
10401	Pentane, 3-ethyl-3-methyl $(C_2H_5)_3CCH_3$	114.23		118.2, 9.9^{10}	−90.9	0.7274$^{20/4}$	1.4078^{20}	**al**, eth, **ace**, **bz**, **chl**	B1^4, 438
10402	Pentane, 1-fluoro or Pentyl fluoride $C_5H_{11}F$	90.14		62.8	−120	0.7907$^{20/4}$	1.3591^{20}	al, eth	B1^4, 308
10403	Pentane, 1-iodo or Pentyl iodide $C_5H_{11}I$	198.05		157, 39.3^{10}	−85.6	1.5161$^{20/4}$	1.4959^{20}	al, eth	B1^4, 315
10404	Pentane, 2-iodo $C_3H_7CHICH_3$	198.05		144-5		1.5096$^{20/4}$	1.4961^{20}	eth, ace, bz	B1^4, 316
10405	Pentane, 3-iodo $(C_2H_5)_2CHI$	198.05		145-6, 68^{50}		1.5176$^{20/4}$	1.4974^{20}	eth, ace, bz	B1^3, 349

No.	Name, Synonyms, and Formula	Mol. wt.	Color, crystalline form, specific rotation and λ_{max} (log ε)	b.p. °C	m.p. °C	Density	n_D	Solubility	Ref.
10406	Pentane, 2-methyl $C_3H_7CH(CH_3)_2$	86.18		60.3	−153.7	$0.6532^{20/4}$	1.3715^{20}	al, eth, ace, bz, chl	$B1^4$, 358
10407	Pentane, 2-methyl-2-phenyl $C_6H_5C(CH_3)_2CH_2CH_2CH_3$	162.27		205-6		$0.8796^{10/4}$	$1.4955^{16.5}$		$B5^3$, 1019
10408	Pentane, 3-methyl $(C_2H_5)_2CHCH_3$	86.18		63.3		$0.6645^{20/4}$	1.3765^{20}	al, eth, ace, bz	$B1^4$, 363
10409	Pentane, 3-methyl-1-phenyl $C_6H_5CH_2CH_2CH(CH_3)C_2H_5$	162.27		220^{757}		$0.8644^{14.5/4}$	1.4896		$B5^4$, 1116
10410	Pentane, 3-nitro $(C_2H_5)_2CHNO_2$	117.15		153-5		$0.957^{0/4}$		al, eth, ace	$B1^4$, 318
10411	Pentane, perfluoro C_5F_{12}	288.04		57.73, -30.5^{10}		$1.7326^{20/4}$	1.2564^{22}	bz	$B1^4$, 308
10412	Pentane, 3-phenyl $C_6H_5CH(C_2H_5)_2$	148.25		187.5, $83\text{-}5^{22}$		$0.8649^{20/4}$	1.4880^{20}		$B5^4$, 1090
10413	Pentane, 1,1,1,5-tetrachloro $CCl_3(CH_2)_3CH_2Cl$	209.93		112^{24}		1.3416^{25}	1.4859^{25}		$B1^4$, 311
10414	Pentane, 2,2,3,3,-tetramethyl $(CH_3)_3CC(CH_3)_2C_2H_5$	128.26		140.27^{760}			1.4236^{20}		$B1^4$, 463
10415	Pentane, 2,2,3,4-tetramethyl $(CH_3)_3CCH(CH_3)CH(CH_3)_2$	128.26		133, 70.6^{104}		$0.7389^{20/4}$	1.4147^{20}		$B1^4$, 463
10416	Pentane, 2,2,4,4-tetramethyl $CH_2[C(CH_3)_3]_2$	128.26		122.7, 12.5^{10}	−66.5	$0.7195^{20/4}$	1.4069^{20}	al, bz	$B1^4$, 464
10417	Pentane, 2,3,3,4-tetramethyl $(CH_3)_2C[CH(CH_3)_2]_2$	128.26		141.5, 78^{104}	-102.14	$0.7547^{20/4}$	1.4222^{20}		$B1^4$, 464
10418	Pentane, 2,2,3-trimethyl $C_2H_5CH(CH_3)C(CH_3)_3$	114.23		110, 3.9^{10}	−112.3	$0.7161^{20/4}$	$1.4030^{20/4}$	al, eth, ace, bz, chl	$B1^4$, 438
10419	Pentane, 2,2,4-trimethyl or Isooctane $(CH_3)_2CHCH_2C(CH_3)_3$	114.23		99.2, 4.3^{10}	−107.4	$0.6919^{20/4}$	1.3915^{20}	al, eth, ace, bz, chl	$B1^4$, 439
10420	Pentane, 2,3,3-trimethyl $C_2H_5C(CH_3)_2CH(CH_3)_2$	114.23		114.7, 6.9^{10}	−100.7	$0.7262^{20/4}$	1.4075^{20}	al, eth, ace, bz, chl	$B1^4$, 445
10421	Pentane, 2,3,4-trimethyl $CH_3CH[CH(CH_3)_2]_2$	114.23		113.4, 7.1^{10}	−109.2	$0.7191^{20/4}$	1.4042^{20}	al, eth, ace, bz, chl	$B1^4$, 446
10422	Isopentane or 2-Methyl butane $CH_3CH_2CH(CH_3)_2$	72.15		27.8	−159.9	$0.6201^{20/4}$	1.3537^{20}	al, eth	$B1^4$, 320
10423	1,5-Pentanedial or Glutyraldehyde $OHC(CH_2)_3CHO$	100.12		187-9d, $71\text{-}2^{10}$				w, al, bz	$B1^4$, 3659
10424	Pentanedioic acid or Glutaric acid $HO_2C(CH_2)_3CO_2H$	132.12	nd (bz)	302-4d, 200^{20}	99	$1.424^{25/4}$	1.4188^{106}	w, al, eth, chl	$B2^4$, 1934
10425	1,2-Pentanediol *(d)* or 1,2-Pentylene glycol $C_3H_7CHOHCH_2OH$	104.15	[α] +0.95	$210\text{-}12^{751}$, $99\text{-}102^{13}$		$0.9802^{20/20}$	1.4412^{19}		$B1^4$, 2538
10426	1,2-Pentanediol, 2,4,4-trimethyl $(CH_3)_3CCH_2C(OH)(CH_3)CH_2OH$	146.23	pr or pl (bz)		62-3			w, al, eth	$B1^4$, 2604
10427	1,3-Pentanediol, 2,2-dimethyl $C_2H_5CH(OH)C(CH_3)_2CH_2OH$	132.20	(eth)	212-4, 119^{21}	60-3			al	B1, 490
10428	1,3-Pentanediol, 2,2,4-trimethyl $(CH_3)_2CHCHOHC(CH_3)_2CH_2OH$	146.23	pl (bz)	234^{737}, $81\text{-}2^{1}$	51-2	$0.937^{15/15}$	1.4513^{15}	al, eth	$B1^4$, 2604
10429	1,4-Pentanediol or γ-Pentylene glycol $CH_3CHOHCH_2CH_2CH_2OH$	104.15		220^{715}, $124\text{-}6^{10}$		$0.9883^{20/4}$	1.4452^{23}	w, al, chl	$B1^4$, 2539
10430	1,4-Pentanediol, 2,2,4-trimethyl $(CH_3)_2C(OH)CH_2C(CH_3)_2CH_2OH$	146.23	cr (eth)	209.11, $114\text{-}5^{13}$	86			al, eth	B1, 493
10431	1,5-Pentanediol or Pentamethylene glycol $HO(CH_2)_5OH$	104.15		260, $137\text{-}8^{12}$	−18	$0.9939^{20/20}$	1.4494^{20}	w, al	$B1^4$, 2540
10432	1,5-Pentanediol, diacetate or 1,5-Diacetoxy pentane $CH_3CO_2(CH_2)_5O_2CCH_3$	188.22		$122\text{-}3^{3}$		1.0296^{20}	1.4261^{19}		$B2^4$, 226
10433	1,5-Pentanediol, 2,2-dimethyl $HOCH_2CH_2CH_2C(CH_3)_2CH_2OH$	132.20		130^{12}				al, eth	$B1^1$, 254
10434	2,3-Pentanediol $C_2H_5CHOHCHOHCH_3$	104.15		187.5, 97^{17}		$0.9800^{19/0}$	1.4412^{25}	w, al	$B1^4$, 2543
10435	2,3-Pentanediol, 2,4,4-trimethyl $(CH_3)_3CCH(OH)C(CH_3)(OH)CH_3$	146.23	mcl pr (lig)		65-6			al, eth	$B1^4$, 2604
10436	2,4-Pentanediol, 2-methyl $CH_3CH(OH)CH_2C(OH)(CH_3)_2$	118.18		197		$0.9254^{17/4}$	1.4250^{20}	w, al, eth	$B1^4$, 2565
10437	2,4-Pentanediol, 3-methyl $CH_3CH[CH(OH)CH_3]_2$	118.18		211-2, 91^{3}		0.9640^{20}	1.4433^{20}	w, al	$B1^4$, 2572
10438	2,3-Pentanedione $C_2H_5COCOCH_3$	100.12	dk ye liq	108		$0.9565^{19/4}$	1.4014^{19}	w, al, eth, ace	$B1^4$, 3660

PHYSICAL CONSTANTS OF ORGANIC COMPOUNDS (Continued)

No.	Name, Synonyms, and Formula	Mol. wt.	Color, crystalline form, specific rotation and λ_{max} (log ε)	b.p. °C	m.p. °C	Density	n_D	Solubility	Ref.
10439	2,3-Pentanedione, dioxime $C_2H_5C(NOH)C(NOH)CH_3$	130.15	ye nd (al), pl (to,al)	sub	172-3			al	B1^4, 3661
10440	2,3-Pentanedione, 2-oxime $C_2H_5COC)NOH)CH_3$	115.13	lf (dil al)		69-72			al, eth	B1^4, 3661
10441	2,3-Pentanedione, 3-oxime $C_2H_5C(NOH)COCH_3$	115.13	pl (lig)	183-7d	58-9			al, eth, chl	B1^4, 3661
10442	2,4-Pentanedione or Acetylacetone $CH_3COCH_2COCH_3$	100.12		139^{746}	−23	$0.9721^{25/4}$	1.4494^{20}	w, al, eth, ace, chl	B1^4, 3662
10443	2,4-Pentanedione, 3,3-dimethyl $(CH_3CO)_2C(CH_3)_2$	128.17		173, 58^{10}	19	$0.9575^{20/4}$	1.4306^{20}	eth	B1^4, 3705
10444	2,4-Pentanedione, dioxime $CH_2[C(=NOH)CH_3]_2$	130.15	pr (eth)		149-50			al	B1^3, 3123
10445	2,4-Pentanedione, 3-ethyl $(CH_3CO)_2CHC_2H_5$	128.17		177-80, $69\text{-}70^{13}$		$0.9531^{19/4}$	1.4408^{19}	al, eth, chl	B1^4, 3703
10446	2,4-Pentanedione, monoimide $CH_3COCH_2C(NH)CH_3$	99.13		209	43			w, eth	B1^4, 3678
10447	1-Pentanethiol or Pentylmercaptan $CH_3(CH_2)_3CH_2SH$	104.21		126.6^{460}, 99.5^{10}	−75.7	$0.8421^{20/4}$	1.4469^{20}	al, eth	B1^4, 1653
10448	2-Pentanethiol $C_3H_7CH(SH)CH_3$	104.21		112.9, 63.9^{150}	−169	$0.8327^{20/4}$	1.4412^{20}	al, lig	B1^4, 1662
10449	3-Pentanethiol $(C_2H_5)_2CHSH$	104.21		105	−110.8	$0.8410^{20/4}$	1.4447^{20}	al	B1^4, 1665
10450	3-Pentanecarboxylic acid-1,5-dinitrile, 3-acetyl, ethyl ester $CH_3COC(CH_2CN)_2CO_2C_2H_5$	236.27	cr	$190\text{-}200^2$	83				B3^2, 512
10451	1,2,3-Pentanetriol $C_2H_5(CHOH)_2CH_2OH$	120.15	syr	192^{63}		$1.0851^{34/0}$		w, al, eth	B1^4, 2778
10452	1,2,5-Pentanetriol $HOCH_2CHOH(CH_2)_3OH$	120.15		$190\text{-}1^{13}$		$1.136^{20/15}$	1.4730^{20}	w, al	B1^4, 2779
10453	1,3,5-Pentanetriol $(HOCH_2CH_2)_2CHOH$	120.15		$188\text{-}9^{11}$		$1.1291^{20/4}$	1.4785^{20}	w, al, eth, ace	B1^4, 2779
10454	Pentanoic acid or Valeric acid $CH_3(CH_2)_3CO_2H$	102.13		186, 82.7^{10}	−33.8	$0.9391^{20/4}$	1.4085^{20}	w, al, eth	B2^4, 868
10455	1-Pentanol or *n*-Pentyl alcohol $CH_3(CH_2)_3CH_2OH$	88.15		137.3^{748}, 50^{13}	−79	$0.8144^{20/4}$	1.4101^{20}	al, eth, ace	B1^4, 1640
10456	1-Pentanol, 5-amino $HO(CH_2)_5NH_2$	103.16		221-2, $126\text{-}8^{22}$	38-9	$0.9488^{17/4}$	1.4618^{17}	w, al, ace	B4^4, 1750
10457	1-Pentanol, 5-chloro $Cl(CH_2)_5OH$	122.59		112^{12}			1.4518^{20}	al, eth	B1^4, 1650
10458	1-Pentanol, 2,4-dimethyl(*(dl)*) $(CH_3)_2CHCH_2CH(CH_3)CH_2OH$	116.20		160-2, $65\text{-}7^{16}$		$0.793^{20/4}$	1.427^{20}	al, eth	B1^4, 1753
10459	1-Pentanol, 2,4-dimethyl *(l)* $(CH_3)_2CHCH_2CH(CH_3)CH_2OH$	116.20	$[\alpha]^{22}{}_D$ −1.1 (undil)	157		$0.816^{25/4}$		al, eth	B1^4, 1753
10460	1-Pentanol, 5-fluoro $F(CH_2)_5OH$	106.14		$70\text{-}1^{11}$			1.4057^{25}	al, eth	B1^4, 1647
10461	1-Pentanol, 2-methyl $C_3H_7CH(CH_3)CH_2OH$	102.18		148		$0.8263^{20/4}$	1.4182^{20}	al, eth, ace	B1^4, 1713
10462	1-Pentanol, 3-methyl(-*(dl)* $C_2H_5CH(CH_3)CH_2CH_2OH$	102.18		152.4, $51\text{-}3^{8}$		$0.8242^{20/4}$	1.4112^{23}	al, eth	B1^4, 1722
10463	1-Pentanol, 4-methyl or Isohexyl alcohol $(CH_3)_2CH(CH_2)_3OH$	102.18		151.6		$0.8131^{20/4}$	1.4134^{25}	al, eth	B1^4, 1721
10464	1-Pentanol, 1-phenyl $C_4H_9CH(OH)C_6H_5$	164.25	$[\alpha]^{20}{}_D$ + 40.8	$140\text{-}2^{25}$		$0.9672^{20/20}$	1.4086^{25}	al, eth, ace	B6^4, 1952
10465	1-Pentanol, 5-phenyl $C_6H_5(CH_2)_5OH$	164.25		155^{20}		0.9725^{20}	1.5156^{20}	al, eth	B6^4, 1954
10466	2-Pentanol $C_3H_7CH(OH)CH_3$	88.15		118.9, 62^{60}		$0.8103^{20/4}$	1.4053^{20}	w, al, eth	B1^4, 1655
10467	2-Pentanol, 1-chloro $C_3H_7CH(OH)CH_2Cl$	122.59		$157\text{-}60^{735}$, $59\text{-}62^{14}$		$1.037^{20/20}$	1.4404^{25}	al, eth	B1^3, 1613
10468	2-Pentanol, 2,4-dimethyl $(CH_3)_2CHCH_2C(OH)(CH_3)_2$	116.20		133.1, $53\text{-}4^{25}$	>−20	$0.8103^{20/4}$	1.4172^{20}		B1^4, 1753
10469	2-Pentanol, 2-methyl $C_3H_7C(OH)(CH_3)_2$	102.18	col	120-2, $49.5^{27.5}$	−103	$0.8350^{16/4}$	1.4100^{20}	al, eth	B1^4, 1714
10470	2-Pentanol, 3-methyl $C_2H_5CH(CH_3)CH(OH)CH_3$	102.18		134.3, 75.6^{50}		$0.8307^{20/4}$	1.4182^{20}	al, eth	B1^3, 1672
10471	2-Pentanol, 4-methyl $(CH_3)_2CHCH_2CH(OH)CH_3$	102.18		133, $50\text{-}5^{25}$		$0.8075^{20/4}$	1.4100^{20}	al, eth	B1^4, 1717

No.	Name, Synonyms, and Formula	Mol. wt.	Color, crystalline form, specific rotation and λ_{max} (log ε)	b.p. °C	m.p. °C	Density	n_D	Solubility	Ref.
10472	2-Pentanol, 4-methyl,acetate or 4-Methyl-2-pentyl acetate $(CH_3)_2CHCH_2CH(CH_3)O_2CCH_3$	144.21	...	$147\text{-}8, 76.5^{47}$	...	$0.8805^{0/0}$	1.4066^{20}	ace, bz	B2,133
10473	2-Pentanol, 2-phenyl $C_3H_7C(OH)(CH_3)C_6H_5$	164.25	...	$216, 112^{14}$	...	$0.9723^{22/4}$	...	al	$B6^2$, 505
10474	3-Pentanol $(C_2H_5)_2CHOH$	88.15	...	$116.1, 30^{12}$	...	$0.8212^{20/4}$	1.4104^{20}	al, eth, ace	$B1^4$, 1662
10475	2-Pentanol, 2,4,4-trimethyl $(CH_3)_3CCH_2C(OH)(CH_3)_2$	130.23	...	$147.5, 42\text{-}4^{7}$	−20	$0.8225^{20/4}$	1.4284^{20}	eth	$B1^4$, 1796
10476	3-Pentanol, 4-amino-2-methyl $CH_3CH(NH_2)CH(OH)CH(CH_3)_2$	117.19	...	174^{745}	35-6	...	...	al, eth, ace, bz	$B4^4$, 796
10477	3-Pentanol, 1-chloro $C_2H_5CH(OH)CH_2CH_2Cl$	122.59	...	$173, 77^{20}$	...	$1.0327^{25/4}$	1.448^{25}	al, eth	$B1^4$, 1665
10478	3-Pentanol, 2,2-dimethyl $C_2H_5CH(OH)C(CH_3)_3$	116.20	...	$135, 44\text{-}5^{15}$	−5	$0.8253^{20/4}$	1.4223^{20}	al, eth	$B1^4$, 1751
10479	3-Pentanol, 2,3-dimethyl $C_2H_5C(OH)(CH_3)CH(CH_3)_2$	116.20	...	$139.7, 44\text{-}5^{14}$	<−30	$0.833^{20/4}$	1.4287^{20}	al, eth	$B1^4$, 1752
10480	3-Pentanol, 2,4-dimethyl $(i\text{-}C_3H_7)_2CHOH$	116.20	...	$138.7, 87.5^{125}$	<−70	$0.8288^{20/4}$	1.4250^{20}	al, eth	$B1^4$, 1754
10481	3-Pentanol, 2,4-dimethyl-3-phenyl $(i\text{-}C_3H_7)_2C(OH)C_6H_5$	192.30	ye	$229, 157^{60}$	...	$0.9755^{20/4}$	1.5239^{20}	eth	$B6^1$, 273
10482	3-Pentanol, 3-ethyl $(C_2H_5)_3COH$	116.20	...	$143.1, 73^{52}$	...	$0.8407^{22/4}$	1.4294^{20}	al, eth	$B1^4$, 1750
10483	3-Pentanol, 3-ethyl-2-methyl $(C_2H_5)_2C(OH)CH(CH_3)_2$	130.23	...	$159\text{-}61^{750}, 55\text{-}7^{48}$	...	$0.8295^{20/20}$	1.4372^{10}	al, eth	$B1^4$, 1794
10484	3-Pentanol, 2-methyl $C_2H_5CH(OH)CH(CH_3)_2$	102.18	...	126.7	...	$0.8243^{20/4}$	1.4175^{20}	al, eth	$B1^4$,1716
10485	3-Pentanol, 3-methyl $(C_2H_5)_2C(OH)CH_3$	102.18	$[\alpha]^{20}_D$ +40.8	122.4	−23.6	$0.8286^{20/4}$	4186^{20}	al, eth	$B1^4$, 1723
10486	3-Pentanol, 1-phenyl *(d)* $C_2H_5CH(OH)CH_2CH_2C_6H_5$	164.25	$[\alpha]^{20/}_D$ +26.8 (al)	143^{19}	38	$0.9687^{20/4}$	...	al	$B6^4$, 1953
10487	3-Pentanol, 3-phenyl $(C_2H_5)_2C(OH)C_6H_5$	164.25	...	$223\text{-}4^{762}, 110^{12}$	<−17	$0.9831^{20/4}$	1.5165^{20}	vs	$B6^4$, 1963
10488	3-Pentanol, 2,3,4-trimethyl $(CH_3)_2CHC(OH)(CH_3)CH(CH_3)_2$	130.23	...	156.5	...	$0.8507^{20/20}$	1.4353^{20}	...	$B1^4$, 1798
10489	1-Pentanol-2-one $C_3H_7COCH_2OH$	102.13	...	$152, 62\text{-}4^{18}$	...	$0.9860^{20/4}$	1.4234^{12}	w, al, eth	$B1^4$, 4004
10490	1-Pentanol-4-one $HO(CH_2)_3COCH_3$	102.13	...	$208^{730}, 116\text{-}8^{33}$	...	$1.0071^{20/4}$	1.4390^{20}	al, eth, w	$B2^4$, 704
10491	2-Pentanol-4-one, 2-methyl or Diacetone alcohol $(CH_3)_2C(OH)CH_2COCH_3$	116.16	...	$164, 67\text{-}9^{19}$	−44	$0.9387^{20/4}$	1.4213^{20}	w, al, eth	$B1^4$, 4023
10492	3-Pentanol-2-one $C_2H_5CH(OH)COCH_3$	102.13	...	$147\text{-}8, 59^{27}$	...	$0.9500^{20/4}$	1.4350^{10}	al, eth, ace, bz	$B1^3$, 3220
10493	2-Pentanol-3-one $C_2H_5COCH(OH)CH_3$	102.13	...	$152.5, 63^{20}$	...	$0.9742^{20/4}$	1.4128^{20}	al, eth	$B1^4$, 4008
10494	4-Pentanol-2-one $CH_3CH(OH)CH_2COCH_3$	102.13	...	$177, 62\text{-}4^{12}$	...	$1.0071^{20/4}$	1.4265^{20}	al, eth	$B1^4$, 4005
10495	2-Pentanone or Methyl propyl ketone $C_3H_7COCH_3$	86.13	...	102	−77.8	$0.8089^{20/4}$	1.3895^{20}	al, eth	$B1^4$, 3271
10496	2-Pentanone, 4-amino-4-methyl $(CH_3)_2C(NH_2)CH_2COCH_3$	115.18	...	$25^{0.14}$	...	<1	...	al, eth	$B4^3$, 894
10497	2-Pentanone, 3-benzylidene $C_6H_5CH{=}C(C_2H_5)COCH_3$	174.24	...	$136\text{-}8^{12}$	...	$1.0005^{22/4}$	1.5650^{22}	...	$B7^1$, 198
10498	2-Pentanone, 1-chloro $C_3H_7COCH_2Cl$	120.58	...	$154\text{-}6d, 58\text{-}9^{17}$	...	...	...	MeOH	$B1^4$, 3276
10499	2-Pentanone, 5-chloro $Cl(CH_2)_3COCH_3$	120.58	...	76^{34}	...	$1.0523^{20/4}$	1.4375^{20}	eth, bz	$B1^4$, 3277
10500	2-Pentanone, 3-ethyl-4-methyl $(CH_3)_2CHCH(C_2H_5)COCH_3$	128.21	...	154-5	...	$0.812^{20/4}$	1.4105^{20}	al, eth, bz, chl, aa	$B1^3$, 2883
10501	2-Pentanone, 3-methyl (dl) or *sec* butyl methyl ketone $C_2H_5CH(CH_3)COCH_3$	100.16	...	118^{758}	...	$0.8130^{30/4}$	1.4002^{20}	al, eth, chl	$B1^4$, 3309
10502	2-Pentanone, 4-methyl or Methyl isobutyl Ketone $(CH_3)_2CHCH_2COCH_3$	100.16	...	$116.8, 35\text{-}40^{16}$	−84.7	0.7978^{20}	1.3962^{20}	al, eth, ace, bz, chl	$B1^4$, 3305
10503	2-Pentanone oxime or Methyl propyl ketoxime $C_3H_7C(=NOH)CH_3$	101.15	...	167^{748}	...	$0.9095^{20/4}$	1.4450^{20}	w, al, eth	$B1^4$, 3274
10504	3-Pentanone or Diethyl ketone $C_2H_5COC_2H_5$	86.13	...	101.7	−39.8	$0.8138^{20/4}$	1.3924^{20}	w, al, ace	$B1^4$, 3282

No.	Name, Synonyms, and Formula	Mol. wt.	Color, crystalline form, specific rotation and λ_{max} (log ε)	b.p. °C	m.p. °C	Density	n_D	Solubility	Ref.
10505	3-Pentanone, 1-chloro $C_2H_5COCH_2CH_2Cl$	120.58		68^{20}			1.4361^{20}	al, eth	$B1^4$, 3284
10506	3-Pentanone, 2-chloro $C_2H_5COCHClCH_3$	120.58		135				al, eth	$B1^4$, 3284
10507	3-Pentanone, 2,2-dimethyl or Ethyl *tert*-butyl ketone $C_2H_5COC(CH_3)_3$	114.19		124.5^{730}	−45	$0.8125^{20/4}$	1.4065^{20}	al, eth, ace, chl	$B1^4$, 3331
10508	3-Pentanone, 2,4 dimethyl or Diisopropyl ketone $(i\text{-}C_3H_7)_2CO$	114.19		124-5	−69	$0.8108^{20/4}$	1.3999^{20}	al, eth, bz	$B1^4$, 3334
10509	3-Pentanone, 2-methyl or Ethyl isopropyl ketone $(CH_3)_2CHCOC_2H_5$	100.16		$114\text{-}5^{745}$		$0.830^{0/0}$	1.3975^{20}	al, eth, ace, bz, chl	$B1^4$, 3304
10510	3-Pentanone, 2,2,4,4-tetramethyl or Pivalone $[(CH_3)_3C]_2CO$	142.24		$152, 70^{43}$		0.8240^{18}	1.4194^{20}	al, eth, ace, chl, aa	$B1^4$, 3334
10511	Pentaguine $C_{18}H_{27}N_3O$	301.43		$165\text{-}70^{0.02}$			1.5785^{25}	dil HCl	Am68, 1524
10512	Pentasiloxane, dodecamethyl $CH_3[Si(CH_3)_2O\text{-}]_4Si(CH_3)_3$	384.84		$229^{710}, 103\text{-}7^{12}$	−80	0.8755^{20}	1.3925^{20}	bz, lig	Am68, 2284
10513	Pentatriacontane $CH_3(CH_2)_{33}CH_3$	492.96	cr (al)	$490, 311^{10}$	75	$0.8157^{20/4}$	1.4568^{20}	ace	$B1^4$, 598
10514	18-Pentatriacontanone or Stearone $C_{17}H_{35}COC_{17}H_{35}$	506.94	lf (lig)		88.4	$0.793^{95/4}$			$B1^4$, 3413
10515	2-Pentenal, 2-methyl $C_2H_5CH{=}C(CH_3)CHO$	98.14		$136\text{-}7, 38\text{-}9^{18}$		$0.8581^{20/4}$	1.4488^{20}	al, eth, bz	$B1^4$, 3471
10516	4-Pentenal $CH_2{=}CHCH_2CH_2CHO$	84.12		96		$0.852^{20/4}$	1.4191^{20}	eth, ace	$B1^4$, 3459
10517	1-Pentene $CH_3CH_2CH_2CH{=}CH_2$	70.13		30	−138	$0.6405^{20/4}$	1.3715^{20}	al, eth, bz	$B1^4$, 808
10518	1-Pentene, 1-bromo $C_3H_7CH{=}CHBr$	149.03		$121\text{-}2, 43.5^{30}$		$1.2606^{20/4}$	1.4572^{20}	eth, bz, chl	$B1^3$, 775
10519	1-Pentene, 2-bromo $C_3H_7CBr{=}CH_2$	149.03		107-8		1.228^{20}	1.4535^{20}	eth, bz, chl, lig	$B1^3$, 775
10520	1-Pentene, 3-bromo $CH_3CH_2CHBrCH{=}CH_2$	149.03		30.5^{30}		$1.2417^{25/4}$	1.4626^{25}	ace, bz, chl	$B1^3$, 775
10521	1-Pentene, 2-chloro $CH_3CH_2CH_2CCl{=}CH_2$	104.58		95-7		0.872^{5}		al, eth	$B1^3$, 774
10522	1-Pentene, 2-chloro-3-ethyl-3-methyl $(C_2H_5)_2C(CH_3)CCl{=}CH_2$	146.66		$147^{742}, 53^{20}$		$0.9147^{20/4}$	1.4450^{25}	bz, chl	$B1^3$, 847
10523	1-Pentene, 3-chloro $CH_3CH_2CHClCH{=}CH_2$	104.58		$93\text{-}4^{764}$		$0.8978^{20/4}$	1.4254^{20}	al, eth, ace	$B1^3$, 774
10524	1-Pentene, 3-chloro-2-methyl $C_2H_5CHClC(CH_3){=}CH_2$	118.61		121-4			1.4422^{20}	eth, bz, chl	$B1^3$, 809
10525	1-Pentene, 4-chloro $CH_3CHClCH_2CH{=}CH_2$	104.58		97-100		0.934^{15}	1.417^{15}	eth, chl	$B1^3$, 774
10526	1-Pentene, 5-Chloro $ClCH_2CH_2CH_2CH{=}CH_2$	104.58		$103\text{-}4^{773}$		$0.9125^{20/4}$	1.4297^{20}	eth, ace	$B1^4$, 810
10527	1-Pentene, 2,3-dimethyl $C_2H_5CH(CH_3)C(CH_3){=}CH_2$	98.19		84.3	−134.8	$0.7051^{20/4}$	1.4033^{20}	al, eth	$B1^4$, 870
10528	1-Pentene, 2,4-dimethyl $(CH_3)_2CHCH_2C(CH_3){=}CH_2$	98.19		81.6	−123.8	$0.6943^{20/4}$	1.3986^{20}	al, eth, bz, chl	$B1^4$, 870
10529	1-Pentene, 3,3-dimethyl $C_2H_5C(CH_3)_2CH{=}CH_2$	98.19		77.5	−134.3	$0.6974^{20/4}$	1.3984^{20}	al, eth, bz, chl	$B1^4$, 873
10530	1-Pentene, 4,4-dimethyl $(CH_3)_3CCH_2CH{=}CH_2$	98.19		72.5	−136.6	$0.6827^{20/4}$	1.3918^{20}	al, eth, bz	$B1^4$, 869
10531	1-Pentene, 2-ethyl $C_3H_7C(C_2H_5){=}CH_2$	98.19		94		$0.7079^{20/4}$	1.405^{20}	al, eth, bz	$B1^4$, 867
10532	1-Pentene, 3-ethyl $(C_2H_5)_2CHCH{=}CH_2$	98.19		85		0.6948^{22}	1.3966^{23}		$B1^4$, 867
10533	1-Pentene, 2-methyl $C_3H_7C(CH_3){=}CH_2$	84.16		60.7	−135.7	$0.6799^{20/4}$	1.3920^{20}	al, bz, chl, peth	$B1^4$, 841
10534	1-Pentene, 3-methyl $C_2H_5CH(CH_3)CH{=}CH_2$	84.16		51.1	−153	$0.6675^{20/4}$	1.3841^{20}	al, bz, chl, peth	$B1^4$, 847
10535	1-Pentene, 4-methyl $(CH_3)_2CHCH_2CH{=}CH_2$	84.16		53.9	−153.6	$0.6642^{20/4}$	1.3828^{20}	al, bz, chl, peth	$B1^4$, 846
10536	1-Pentene, 2-methyl, perfluoro $F_3C(CF_2)_2C(CF_3){=}CF_2$	300.05		60				bz	$B1^4$, 842
10537	1-Pentene, perfluoro $CF_3CF_2CF_2CF{=}CF_2$	250.04		$29\text{-}30^{740}$			1.2571^{25}	chl	$B1^4$, 810

No.	Name, Synonyms, and Formula	Mol. wt.	Color, crystalline form, specific rotation and λ_{max} (log ε)	b.p. °C	m.p. °C	Density	n_D	Solubility	Ref.
10538	1-Pentene, 2,4,4-trimethyl or Diisobutylene $(CH_3)_3CCH_2C(CH_3)=CH_2$	112.22	....	101.4	−93.5	$0.7150^{20/4}$	1.4086^{20}	eth, bz, lig, chl	$B1^4$, 892
10539	2-Pentene (cis) $CH_3CH_2CH=CHCH_3$	70.13	....	36.9	−151.4	$0.6556^{20/4}$	1.3830^{20}	**al, eth,** bz	$B1^4$, 814
10540	2-Pentene-*(trans)* $CH_3CH_2CHCHCH_3$	70.13	....	36.3	−136	$0.6482^{20/4}$	1.3793^{20}	**al, eth,** bz	$B1^4$, 814
10541	2-Pentene, 1-bromo $CH_3CH_2CH=CHCH_2Br$	149.03	....	123-4, 35^{25}	......	1.2545^{20}	1.4731^{20}	ace, bz, chl	$B1^4$, 816
10542	2-Pentene, 2-bromo $CH_3CH_2CH=CBrCH_3$	149.03	....	110.5^{750}	......	$1.277^{20/20}$	1.4580^{20}	ace, bz, chl	$B1^1$, 784
10543	2-Pentene, 3-bromo $CH_3CH_2CBr=CHCH_3$	149.03	....	115.2^{750}	......	$1.273^{20/20}$	1.4628^{20}	ace, bz, chl	$B1^1$, 784
10544	2-Pentene, 4-bromo $CH_3CHBrCH=CHCH_3$	149.03	....	117d, 22^9	......	1.2312^{21}	1.4752^{21}	ace, bz, chl	$B1^4$, 816
10545	2-Pentene, 5-bromo $BrCH_2CH_2CH=CHCH_3$	149.03	....	121.7^{621}	......	$1.2715^{20/4}$	1.4695^{20}	ace, bz, chl	$B1^4$, 816
10546	2-Pentene, 1-chloro $CH_3CH_2CH=CHCH_2Cl$	104.58	....	109 5, 62^{148}	......	$0.908^{22/4}$	1.4352^{22}	al, eth, ace, chl	$B1^3$, 781
10547	2-Pentene, 2-chloro $CH_3CH_2CH=CClCH_3$	104.58	....	95-7, 45^{130}	......	$0.9067^{20/4}$	1.4261^{20}	eth, ace	$B1^3$, 782
10548	2-Pentene, 3-chloro $CH_3CH_2CCl=CHCH_3$	104.58	....	90-2	......	1.423^{24}	0.9125^{20}	eth, ace	$B1^3$,782
10549	2-Pentene, 3-chloro-2,4-dimethyl $(CH_3)_2CHCCl=C(CH_3)_2$	132.63	....	118-20, 44-5^{30}	......	$0.9513^{9/9}$	....	eth, bz, chl	B1, 221
10550	2-Pentene, 4-chloro $CH_3CHClCH=CHCH_3$	104.58	....	103, 18-20^{12}	......	$0.9004^{20/20}$	1.4322^{20}	eth, ace, chl	$B1^4$, 815
10551	2-Pentene, 5-chloro $ClCH_2CH_2CH=CHCH_3$	104.58	....	107-8^{755}	......	$0.9043^{20/4}$	1.4310^{20}	ace, bz, chl	$B1^4$, 816
10552	2-Pentene, 5-chloro-2-methyl $ClCH_2CH_2CH=C(CH_3)_2$	118.61	....	132-3^{756}	......	0.9135^{20}	....	ace, chl	$B1^4$, 843
10553	2-Pentene, 2,5-dichloro $ClCH_2CH_2CH=CClCH_3$	139.02	....	40-1^8	......	$1.1182^{15/4}$	....	chl	$B1^4$, 816
10554	2-Pentene, 2,3-dimethyl $C_2H_5C(CH_3)=C(CH_3)_2$	98.19	....	97.5	−118.3	0.7277^{20}	1.4208^{20}	al, eth, bz, chl	$B1^4$, 870
10555	2-Pentene, 2,4-dimethyl $(CH_3)_2CHCH=C(CH_3)_2$	98.19	....	83.4	−127.7	$0.6954^{20/4}$	1.4040^{20}	al, eth, bz, chl	$B1^4$, 872
10556	2-Pentene, 3,4-dimethyl $(CH_3)_2CH(CH_3)=CHCH_3$	98.19	....	87	......	$0.7126^{20/4}$	1.4070^{20}	....	$B1^4$, 871
10557	2-Pentene, 4,4-dimethyl (*trans*) $(CH_3)_3CCH=CHCH_3$	98.19	....	76.7	−115.2	$0.6889^{20/4}$	1.3982^{20}	al, eth, bz, chl	$B1^4$, 868
10558	2-Pentene, 3-ethyl $(C_2H_5)_2C=CHCH_3$	98.19	....	94	......	$0.7079^{20/4}$	1.405^{20}	al, eth, bz, chl	$B1^4$, 868
10559	2-Pentene, 2-methyl $C_2H_5CH=C(CH_3)_2$	84.16	....	67.3	−135	$0.6863^{20/4}$	1.4004^{20}	al, bz, chl, peth	$B1^4$, 842
10560	2-Pentene, 3-methyl (cis) $C_2H_5C(CH_3)=CHCH_3$	84.16	....	67.6	−138.4	$0.6986^{20/4}$	1.4045^{20}	al, bz, chl, peth	$B1^4$, 848
10561	2-Pentene, 3-methyl (*trans*) $C_2H_5C(CH_3)=CHCH_3$	84.16	....	70.4	−134.8	$0.6942^{20/4}$	1.4016^{20}	al, bz, chl, peth	$B1^4$, 848
10562	2-Pentene, 4-methyl (*cis*) $(CH_3)_2CHCH=CHCH_3$	84.16	....	56.3	−134.4	$0.6690^{20/4}$	1.3800^{20}	al, bz, chl, peth	$B1^4$, 844
10563	2-Pentene, 4-methyl (*trans*) $(CH_3)_2CHCH=CHCH_3$	84.16	....	58.5	−140.8	$0.6686^{20/4}$	1.3889^{20}	al, bz, chl, peth	$B1^4$, 844
10564	2-Pentene, 2,3-dimethyl. $(CH_3)_2CHC(CH_3)=C(CH_3)_2$	98.19	....	116.5	-113.38	0.74342^{204}	1.4274^{20}	....	$B1^4$, 894
10565	2-Pentene, 2,4,4-trimethyl $(CH_3)_3CCH=C(CH_3)_2$	112.22	....	104.9	−106.3	$0.7218^{20/4}$	1.4160^{20}	eth, bz, lig, chl	$B1^4$, 891
10566	2-Pentenedioic acid, diethyl ester (*trans*) or Ethyl glutaconate $C_2H_5O_2CCH_2CH=CHCO_2C_2H_5$	186.21	....	236-8, 125^{12}	......	$1.0496^{20/4}$	1.4411^{20}	al, eth	$B2^4$, 2227
10567	2-Pentenoic acid, 4-hydroxy, lactone or β-Angelica lactone $CH_3\underbrace{CHCH=CHCO}_{O}$	98.10	....	208-9^{751}, 98^{15}	<−17	$1.0810^{20/4}$	1.4454^{20}	w, al, eth	$B17^4$, 4302
10568	2-Pentenoic acid, 2-methyl (trans) $C_2H_5CH=C(CH_3)CO_2H$	114.14	....	213^{750}, 112^{12}	24.4	0.9751^{20}	1.4513^{20}	eth, chl	$B2^4$, 1568
10569	2-Pentenoic acid, 4-methyl $(CH_3)_2CHCH=CHCO_2H$	114.14	....	217, 115-6^{20}	35	$0.9529^{21/4}$	1.4489^{21}	al, eth, ace	$B2^4$, 1569

No.	Name, Synonyms, and Formula	Mol. wt.	Color, crystalline form, specific rotation and λ_{max} (log ε)	b.p. °C	m.p. °C	Density	n_D	Solubility	Ref.
10570	3-Pentenoic acid, 4-hydroxy, γ-lactone or γ-Angelica lactone $CH_3\text{-}C{=}CHCH_2CO$ (ring O)	98.10	...	$167, 53^{12}$	18	$1.084^{20/4}$	1.4476^{20}	w, al, eth	B17^{4}, 4300
10571	3-Pentenonitrile, 2-hydroxy (*cis*) $CH_3CH{=}CHCH(OH)CN$	97.12	...	139^{70}	...	$0.9675^{15/4}$	1.4460^{21}	al, eth, bz, chl	B3^{4}, 1001
10572	4-Pentenamide, 2,2-diethyl or Novonal $CH_2{=}CHCH_2C(C_2H_5)_2CONH_2$	155.24	wh pw, cr (eth-peth)	155^{10}	75-6	...	...	al, eth	B2^{3}, 1351
10573	4-Pentenoic acid or Allyl acetic acid $CH_2{=}CHCH_2CH_2CO_2H$	100.12	...	$188\text{-}9, 93^{20}$	−22.5	$0.9809^{20/4}$	1.4281^{20}	al, eth	B2^{4}, 1542
10574	4-Pentenoic acid, 2-acetyl ethyl ester $CH_2{=}CHCH_2CH(COCH_3)CO_2C_2H_5$	170.21	...	$211\text{-}2, 102^{12}$	...	$0.9898^{20/4}$	1.4388^{18}	al, eth, bz	B3, 738
10575	4-Pentenonitrile $CH_2{=}CHCH_2CH_2CN$	81.12	...	$140, 60\text{-}1^{40}$	...	0.8239^{24}	1.4213^{14}	al, eth	B2^{4}, 1593
10576	1-Penten-3-ol $C_2H_5CH(OH)CH{=}CH_2$	86.13	...	$114\text{-}6, 37^{20}$	...	$0.8935^{22/4}$	1.4239^{20}	**al, eth**	B1^{4}, 2117
10577	2-Penten-1-ol (cis) $C_2H_5CH{=}CHCH_2OH$	86.13	...	$138, 41.2^{7}$	...	$0.8529^{20/4}$	1.4354^{20}	al, eth, ace	B1^{4}, 2121
10578	2-Penten-1-ol (*trans*) $C_2H_5CH{=}CHCH_2OH$	86.13	...	$139.5, 42^{7}$	...	$0.8471^{20/4}$	1.4341^{20}	al, eth, ace	B1^{4}, 2121
10579	3-Pentene-2-ol (d) $CH_3CH{=}CHCH(OH)CH_3$	86.13	$[\alpha]_D + 0.413$	$118\text{-}21, 33^{10}$	...	$0.8354^{20/4}$	$1.4250^{25}/_{20}$	al, eth, ace	B1^{4}, 2122
10580	3-Pentene-2-ol (dl) $CH_3CH{=}CHCH(OH)CH_3$	86.13	...	$121.6, 60.6^{55}$	...	0.8328^{25}	1.4280^{20}	al, eth, ace	B1^{4}, 2122
10581	3-Penten-2-ol (l) $CH_3CH{=}CHCH(OH)CH_3$	86.13	$[\alpha]_D$ -3.3 (H_2O 1.5%)	119-21	...	$0.8354^{20/4}$	1.4280^{20}	al, eth, ace	B1^{4}, 2124
10582	3-Penten-2-ol, 2-methyl $CH_3CH{=}CHC(OH)(CH_3)_2$	100.16	...	$121\text{-}2^{757}$	...	$0.8347^{20/4}$	1.4302^{20}	**al, eth**	B1^{4}, 2145
10583	4-Pentene-1-ol $CH_2{=}CHCH_2CH_2CH_2OH$	86.13	...	140-2	...	0.8457^{204}	1.4309^{20}	eth	B1^{4}, 2119
10584	4-Penten-2-ol $CH_2{=}CHCH_2CH(OH)CH_3$	86.13	...	$115\text{-}6^{750}$	...	$0.8367^{20/4}$	1.4225^{20}	**al, eth**	B1^{4}, 2118
10585	4-Penten-2-ol, 2-methyl $CH_2{=}CHCH_2C(OH)(CH_3)_2$	100.16	...	119.5	...	$0.8300^{20/4}$	1.4263^{20}	al, eth	B1^{4}, 2146
10586	1-Penten-3-one or Ethyl vinyl ketone $C_2H_5COCH{=}CH_2$	84.12	...	$102^{740}, 44^{90}$	...	$0.8468^{20/4}$	1.4192^{20}	al, eth, ace, bz	B1^{4}, 3457
10587	1-Penten-3-one, 5,5-dimethyl-1-phenyl or Benzalpinacolone $(CH_3)_2CHCH_2COCH{=}CHC_6H_5$	188.27	cr	154^{25}	43	0.9508^{46}	1.5523^{25}	al, bz, chl	B7^{3}, 1487
10588	1-Penten-3-one, 1-(4 methoxy phenyl) or Ethyl-4-methoxy styryl ketone $C_2H_5COCH{=}CH(C_6H_4OCH_3\text{-}4)$	190.24	col-lt ye pl (eth-peth)	...	60	...	...	al, eth	B8^{3}, 832
10589	1-Penten-3-one, 2-methyl or Ethyl isopropenyl ketone $C_2H_5COC(CH_3){=}CH_2$	98.14	...	118.5	−69.5	$1.8530^{20/4}$	1.4289^{20}	al, ace	B1^{4}, 3470
10590	1-Penten-3-one, 1-Phenyl $C_2H_5COCH{=}CHC_6H_5$	160.22	lf (lig)	142^{12}	38-9	$0.8697^{20/4}$	1.5684^{20}	al, eth, bz	B7^{2}, 298
10591	3-Penten-2-one (trans) or Methyl propenyl ketone $CH_3CH{=}CHCOCH_3$	84.12	...	122	...	$0.8624^{20/4}$	1.4350^{20}	eth, ace	B1^{4}, 3460
10592	3-Penten-2-one, 4-methyl or Mesityl oxide $(CH_3)_2C{=}CHCOCH_3$	98.14	...	$129.7, 41^{23}$	−52.8	$0.8653^{20/4}$	1.4440^{20}	w, **al, eth**, ace	B1^{4}, 3471
10593	1-Penten-3-yne or Puryline $CH_3C{\equiv}C\text{-}CH{=}CH_2$	66.10	...	59-60	...	$0.7401^{20/4}$	1.4496^{20}	eth, bz	B1^{4}, 1087
10594	1-Penten-3-yne, 2-methyl $CH_3C{\equiv}CC(CH_3){=}CH_2$	80.13	...	$81\text{-}2^{100}$	...	...	1.4002^{20}	eth, bz	B1^{4}, 1095
10595	1-Penten-4-yne or Allyl acetylene $CH_2{=}CHCH_2C{\equiv}CH$	66.10	...	42-3	...	$0.777^{22/22}$	1.3653^{22}	eth, bz	B1^{4}, 998
10596	3-Penten-1-yne, 3-ethyl $CH_3CH{=}C(C_2H_5)C{\equiv}CH$	94.16	...	$96.5, 41\text{-}3^{100}$	...	0.7886^{25}	1.4338^{25}	eth, bz	B1^{4}, 1099
10597	3-Penten-1-yne, 3-methyl $CH_3CH{=}C(CH_3)C{\equiv}CH$	80.13	...	66-7	...	0.789^{20}	1.4332^{20}	eth, bz	B1^{4}, 1096
10598	Peucedanin $C_{15}H_{14}O_4$	258.27	pr or pl (bz-peth), yesh cr (eth)	$276\text{-}81^{17}$	109	...	...	eth, chl, aa	B19^{4}, 2647
10599	Pentyl amine or 1-Amino pentane $CH_3(CH_2)_3CH_2NH_2$	87.16	...	$104.4, 5.9^{10}$	−55	$0.7547^{20/4}$	1.4118^{20}	**al, eth**, ace, bz	B4^{4}, 674
10600	iso-Pentyl amine $(CH_3)_2CHCH_2CH_2NH_2$	87.16	...	95-7	...	$0.7505^{20/4}$	1.4083^{20}	w, **al, eth**, ace, chl	B4^{4}, 696

No.	Name, Synonyms, and Formula	Mol. wt.	Color, crystalline form, specific rotation and λ_{max} (log ε)	b.p. °C	m.p. °C	Density	n_D	Solubility	Ref.
10601	Isopentyl boric acid $i\text{-}C_5H_{11}B(OH)_2$	115.97	pl (w)		169			w, al, eth, ace	B4[2], 1023
10602	Pentyl ether or Dipentyl ether $C_5H_{11}OC_5H_{11}$	158.28		190, 70^{12}	−69	$0.7833^{20/4}$	1.4119^{20}	**al, eth**	B1[4], 1643
10603	Isopentyl ether or Diisopentyl ether $(i\text{-}C_5H_{11})_2O$	158.28		172-3, 60^{10}		$0.7777^{20/4}$	1.4085^{20}	**al, eth, chl**	B1[4], 1682
10604	Pentyl isocyanide $C_5H_{11}NC$	97.16		155.5, 50^{45}	−51.1	$0.806^{20/4}$		al	B4[3], 331
10605	Pentyl isothiocyanate $C_5H_{11}NCS$	129.22	br ye		193.4			al, eth	B4[4], 685
10606	Isopentyl isothiocyanate $(CH_3)_2CHCH_2CH_2NCS$	129.22		182-4		0.9419^{17}		al, eth	B4[4], 707
10607	Isopentyl α-naphtyl ether $i\text{-}C_5H_{11}O\text{-}\alpha\text{-}C_{10}H_7$	214.31		317^{742}, $148\text{-}53^{3}$		$1.0069^{14/4}$	1.5705^{14}		B6[4], 2925
10608	Isopentyl β-naphthyl ether $i\text{-}C_5H_{11}O\text{-}\beta\text{-}C_{10}H_7$	214.31	lf	323-6d, $155\text{-}60^{6}$	26.5	$1.0155^{12/4}$	1.5768^{12}	al, eth	B6[4], 2974
10609	Isopentyl nitrate $(CH_3)_2CHCH_2CH_2ONO_2$	133.15		147-8		$0.9961^{22/4}$	1.4122^{22}	**al, eth**	B1[4], 1683
10610	Pentyl nitrite $C_5H_{11}ONO$	117.15	ye	104-5, 29^{60}		$0.8817^{20/4}$	1.3851^{20}	**al, eth**	B1[4], 1644
10611	Isopentyl nitrite $(CH_3)_2CHCH_2CH_2ONO$	117.15		99.2, 30^{60}		$0.8828^{20/4}$	1.3918^{20}	**al, eth**	B1[4], 1683
10612	Pentyl sulfide or Dipentyl sulfide $(C_5H_{11})_2S$	174.34		230, 84.5^{4}	−51.3	$0.8409^{20/4}$	1.4556^{20}	eth	B1[4], 1654
10613	Isopentyl sulfide or Disopentyl sulfide $(i\text{-}C_5H_{11})_2S$	174.34	col pa-ye	216, 85.5^{6}		$0.8323^{20/4}$	1.4520^{20}	**al**, eth	B1[4], 1689
10614	Pentyl sulfate $(C_5H_{11}O)_2SO_2$	238.34		$117^{3.5}$	14	$1.029^{20/0}$	1.4290^{20}		B1[2], 418
10615	Isopentyl sulfate $(i\text{-}C_5H_{11}O)_2SO_2$	238.34		$139\text{-}41^{12}$	−20				B1[2], 434
10616	Isopentyl thiocyanate $(CH_3)_2CHCH_2CH_2SCN$	129.22		197				al, eth	B3[3], 282
10617	*Iso*-Pentyl vinyl ether $i\text{-}C_5H_{11}OCH{=}CH_2$	114.19		112-3		$0.7826^{20/4}$	1.4072^{20}	al, eth	B1[4], 2055
10618	1-Pentyne $C_3H_7C{\equiv}CH$	68.12		40.2	−90	$0.6901^{20/4}$	1.3852^{20}	al, **eth**, bz, chl	B1[4], 990
10619	1-Pentyne, 1-bromo $C_3H_7C{\equiv}CBr$	147.01		$44\text{-}6^{52}$		$1.281^{13/4}$	1.4579^{13}		B1[3], 958
10620	1-Pentyne, 3-chloro-3-ethyl $(C_2H_5)_2C(Cl)C{\equiv}CH$	130.62		$73\text{-}6^{100}$		$0.9230^{19/4}$	1.4437^{19}	eth, bz, chl	B1[4], 1032
10621	1-Pentyne, 3-chloro-3-methyl $C_2H_5CCl(CH_3)C{\equiv}CH$	116.59		102.3, 55^{130}		$0.9163^{20/4}$	1.4330^{20}	eth, bz, chl	B1[4], 1021
10622	1-Pentyne, 4,4-dimethyl $(CH_3)_3CCH_2C{\equiv}CH$	96.17		76.1	−75.7	$0.7142^{20/4}$	1.3983^{20}	eth, bz, chl	B1[4], 1033
10623	1-Pentyne, 3-ethyl $(C_2H_5)_2CHC{\equiv}CH$	96.17		87-9		$0.7246^{25/4}$	1.4043^{25}	eth, bz, chl	B1[3], 1002
10624	1-Pentyne, 3-ethyl-3-methyl $(C_2H_5)_2C(CH_3)C{\equiv}CH$	110.20		101-2		$0.7422^{20/4}$	1.4110^{20}	eth, bz, chl	B1[4], 1046
10626	1-Pentyne, 1-iodo $C_3H_7C{\equiv}CI$	194.02		54^{23}		$1.6127^{19/4}$	1.5148^{19}		B1[4], 992
10627	1-Pentyne, 4-methyl $(CH_3)_2CHCH_2C{\equiv}CH$	82.15		61-2	−105.1	$0.7092^{15/4}$	1.3936^{15}	bz, chl	B1[4], 1019
10628	2-Pentyne $C_2H_5C{\equiv}CCH_3$	68.12		56	−101	$0.7107^{20/4}$	1.4039^{20}	al, **eth**, bz, chl	B1[4], 992
10629	2-Pentyne, 4-chloro-4-methyl $(CH_3)_2C(Cl)C{\equiv}CCH_3$	116.59		55^{70}			1.4143^{20}	ace, bz, chl	B1[4], 1018
10630	2-Pentyne, 4,4-dimethyl $(CH_3)_3CC{\equiv}CCH_3$	96.17		83	−82.4	$0.7176^{20/4}$	1.4071^{20}	eth, bz, chl	B1[4], 1032
10631	2-Pentyne, 4-methyl $(CH_3)_2CHC{\equiv}CCH_3$	82.15		72.5	−110.4	$0.716^{19/4}$	1.4078^{19}	bz, chl	B1[4], 1018
10632	2-Pentynoic acid $CH_3CH_2C{\equiv}CCO_2H$	98.10	cr (peth)	122^{10}	50	0.978^{20}	1.4619^{20}	w	B2[4], 1693
10633	4-Pentynoic acid $HC{\equiv}CCH_2CH_2CO_2H$	98.10		102^{17}	57.7			w, al, eth	B2[4], 1693

PHYSICAL CONSTANTS OF ORGANIC COMPOUNDS (Continued)

No.	Name, Synonyms, and Formula	Mol. wt.	Color, crystalline form, specific rotation and λ_{max} (log ε)	b.p. °C	m.p. °C	Density	n_D	Solubility	Ref.
10634	1-Pentyn-3-ol, 3,4-dimethyl ... $(CH_3)_2CHC(OH)(CH_3)C\equiv CH$	112.17	...	133^{735}	...	0.8691$^{20/4}$	1.4372^{20}	w, al, eth	B1^4, 2255
10635	1-Pentyn-3-ol, 3-methyl ... $C_2H_5C(OH)(CH_3)C\equiv CH$	98.14	...	120-1, 61^{70}	30-1	0.8688$^{20/4}$	1.4310^{20}	...	B1^4, 2242
10636	Perbenzoic acid ... C_6H_5COOOH	138.12	mcl pl (peth)	97-110$^{13-5}$ sub	41-3	...	...	al, eth, ace, bz, chl	B9^3, 1049
10637	Pereirine ... $C_{19}H_{26}N_2O$	298.43	pa ye amor pw, [α] + 137.5 (al)	...	135d	...	...	al, eth, chl	C28, 5459
10638	Perimidine or Peri-naphthimidazole ... $C_{11}H_8N_2$	168.20	gr cr (dil al)	...	222	...	...	al, eth, ace, bz	B23^2, 209
10639	Perseitol or α-Mannoheptitol ... $C_7H_{16}O_7$	212.20	nd, [α]^{20} + 4.53 (w)	...	188	...	...	w	B1^4, 2854
10640	Perylene or *peri*-Dinaphthalene ... $C_{20}H_{12}$	252.32	gold-br ye pl (bz, aa)	sub 350-400	277-9	1.35	...	ace, bz, chl	B5^3, 2521
10641	Perylene-3-carboxylic acid ... 3-$C_{20}H_{11}CO_2H$	296.33	og-br nd ($PhNO_2$)	...	330	...	...	...	B9^3, 3664
10642	α-Phellandrene (d) or 5-isopropyl-2-methyl-1,3-cyclohex-adiene ... $C_{10}H_{16}$	136.24	$[\alpha]_D$ + 49.1 (undil)	175-6, 61^{11}	...	0.8463$^{25/4}$	1.4777^{22}	eth	B5^4, 436
10643	β-Phellandrene or 3-isopropyl-6-methylene cyclohexene ... $C_{10}H_{16}$	136.24	$[\alpha]^{20/}{}_D$ + 65.2 (undil)	171-2, 57^{11}	...	0.8520$^{20/4}$	1.4788^{20}	eth	B5^4, 436
10644	Phenacyl acetate or α-Hydroxy acetophene acetate ... $C_6H_5COCH_2O_2CCH_3$	178.19	rh pl (eth, lig)	270, 150-2^{10}	49	1.1169$^{65/4}$	...	al, eth, bz, chl	B8^3, 301
10645	Phenacyl alcohol or α-Hydroxy acetophenone ... $C_6H_5COCH_2OH$	136.15	hex pl (al, eth), pl (w, dil al), pr (lig)	124-6^{12} (sub 56^1)	90.5 (anh)	1.0963$^{99/4}$	...	al, eth, chl	B8^3, 298
10646	Phenacyl bromide or α-Bromo acetophenone ... $C_6H_5COCH_2Br$	199.05	nd (al), rh pr (dil al), pl (peth)	135^{18}	50-1	1.647$^{20/4}$	...	al, eth, bz, chl	B7^3, 979
10647	Phenacyl chloride or α-Chloro acetophenone ... $C_6H_5COCH_2Cl$	154.60	pl (dil al), rh, lf (peth)	247, 139-41^{14}	56-5	1.324$^{15/4}$	...	al, eth, ace, bz	B7^3, 967
10648	Phenacyl chloride-2,4-dimethyl ... 2,4$(CH_3)_2C_6H_3COCH_2Cl$	182.65	nd	...	62	...	...	al, eth, bz, chl	B7^2, 172
10649	Phenanthrene ... $C_{14}H_{10}$	178.23	mcl pl (al), lf (sub)	340, 210-15^{12}	101	0.9800^4	1.5943	al, eth, ace, bz, aa	B5^4, 2297
10650	Phenanthrene, 2-acetyl ... 2-$CH_3COC_{14}H_9$	220.27	nd (MeOH)	...	144-5	...	...	al, bz	B7^3, 2543
10651	Phenanthrene, 3-acetyl ... 3-$CH_3COC_{14}H_9$	220.27	nd (MeOH)	...	72	...	...	al, bz, chl, aa	B7^3, 2544
10652	Phenanthrene, 9-acetyl ... 9-$CH_3COC_{14}H_9$	220.27	nd (MeOH)	...	74.5	...	...	al, eth, bz	B7^3, 2549
10653	Phenanthrene, 2-amino or 2-Phenanthryl amine ... 2-$NH_2C_{14}H_9$	193.25	lt ye (lig)	...	85	...	...	...	B12^3, 3339
10654	Phenanthrene, 3-amino or 3-Phenanthryl amine ... 3-$NH_2C_{14}H_9$	193.25	α:lf (lig) β:cr (lig)	...	α:143, β:87.5	...	...	al	B12^3, 3339
10655	Phenanthrene, 4-amino ... 4-$H_2NC_{14}H_9$	193.25	...	...	104-5	...	...	al, eth, bz, chl	B12^3, 3341
10656	Phenanthrene, 9-amino (a form) or 9-Phenanthryl amine 9-$NH_2C_{14}H_9$	193.25	lt ye cr (al)	sub	137-8	...	...	al, eth, bz, chl	B12^3, 3341
10657	Phenanthrene, 9-amino (b form) ... 9-$NH_2C_{14}H_9$	193.25	lt ye nd (al)	sub	104	...	...	al, eth, bz, chl	B12^1, 555
10658	Phenanthrene, 9-bromo ... 9-$BrC_{14}H_9$	257.13	pr (al)	>360, 190^{12} sub	64-5	1.4093$^{10/4}$	...	al, eth	B5^3, 2145
10659	Phenanthrene, 9,10-diamino ... 9,10-$(NH_2)_2C_{14}H_8$	208.26	pa ye lf	...	166	...	...	...	B13^3, 524
10660	Phenanthrene, 9,10-dihydro ... $C_{14}H_{12}$	180.25	nd (MeOH)	168-9^{15}	34-5	1.0757$^{40/4}$	1.6415$^{20/}{}_4$	al, eth	B5^3, 1989
10661	Phenanthrene, 3,4-dihydroxy or 3,4-Phenanthrenediol ... 3,4-$(HO)_2C_{14}H_8$	210.23	col nd (peth), dk in air	sub 130 (vac)	143	...	...	al, eth	B6^3, 5689
10662	Phenanthrene, 3,4-dimethoxy ... 3,4-$(CH_3O)_2C_{14}H_8$	238.29	bt ye lf (MeOH)	298-303^{112}	45	...	...	al, eth	B6^3, 5689

No.	Name, Synonyms, and Formula	Mol. wt.	Color, crystalline form, specific rotation and λ_{max} (log ε)	b.p. °C	m.p. °C	Density	n_D	Solubility	Ref.
10663	Phenanthrene, 4,5-dimethyl $4,5-(CH_3)_2C_{14}H_8$	206.29	pr (MeOH)		76-7			chl	$B5^{3}$, 2175
10664	Phenanthrene, 9,10-dimethyl $9,10-(CH_3)_2C_{14}H_8$	206.29	lt red pr (aa), nd (MeOH)	sub	144			bz, chl, aa	$B5^{3}$, 2175
10665	Phenanthrene, 9,10-diphenyl $9,10-(C_6H_5)_2C_{14}H_8$	330.43	nd (eth, bz)	sub 270	240			eth, bz	$B5^{3}$, 2624
10666	Phenanthrene, 1,2,3,4,5,6,7,8,9,10,11,12-dodecahydro $C_{14}H_{22}$	190.33		$81-2^{1.5}$		$0.9674^{20/4}$	1.5102^{20}	ace, bz	$B5^{3}$, 1074
10667	Phenanthrene, 9-ethyl $9-C_2H_5C_{14}H_9$	206.29		$199-200^{1}$	62-3 (66)	$1.0603^{78/4}$	1.6582^{78}	al, bz	$B5^{3}$, 2170
10668	Phenanthrene, 1,2,3,4,11,12-hexadydro $C_{14}H_{16}$	184.28		307	−3	1.045^{20}	1.5810^{15}	eth, bz, aa, chl, peth	$B5^{3}$, 1676
10669	Phenanthrene, 1-hydroxy or 1-Phenanthrol $1-HOC_{14}H_9$	194.23	nd (peth, bz-lig, eth)		157				$B6^{3}$, 3557
10670	Phenanthrene, 2-hydroxy or 2-Phenanthrol $2-HOC_{14}H_9$	194.23	pl, lf (al, eth, lig)		168			al, eth, bz	$B6^{3}$, 3557
10671	Phenanthrene, 3-hydroxy or 3-Phenanthrol $3-HOC_{14}H_9$	194.23	nd (al, lig)		122-3			al, eth	$B6^{3}$, 3558
10672	Phenanthrene, 9-hydroxy or 9-Phenanthrol $9-HOC_{14}H_9$	194.23	nd (lig, bz)		158			al, eth, bz, chl, lig	$B6^{3}$, 3560
10673	Phenanthrene, 1-methyl $1-CH_3C_{14}H_9$	192.26	lf, pl (dil al)		123			al	$B5^{3}$, 2151
10674	Phenanthrene, 3-methyl $3-CH_3C_{14}H_9$	192.26	pr or nd (al)	$140-50^{6}$	65			ace	$B5^{3}$, 2154
10675	Phenanthrene, 4-methyl $4-CH_3C_{14}H_9$	192.26		$175-80^{10}$	52.5			al	$B5^{3}$, 2154
10676	Phenanthrene, 9-methyl $9-CH_3C_{14}H_9$	192.26			90-1				$B5^{3}$, 2154
10677	Phenanthrene, 2-nitro $2-O_2NC_{14}H_9$	223.23			119-20			al, eth, ace	$B5^{3}$, 2146
10678	Phenanthrene, 3-nitro $3-O_2NC_{14}H_9$	223.23			172-4			ace, bz, chl	$B5^{3}$, 2146
10679	Phenanthrene, 9-nitro $9-NO_2C_{14}H_9$	223.23	ye nd (al)		116-7			al, eth, bz	$B5^{3}$, 2147
10680	Phenanthrene, 1,2,3,4,5,6,7,8-octahydro or Octathrene $C_{14}H_{18}$	186.30		$295, 169^{15}$	16.7	$1.026^{20/4}$	1.5569^{17}	ace, bz, aa	$B5^{4}$, 1585
10681	Phenanthrene, 1,2,8-trimethyl $1,2,8-(CH_3)_3C_{14}H_7$	220.31		$210-20^{15}$	144-5				$B5^{3}$, 2187
10682	Phenanthrene, 1,4,7-trimethyl $1,4,7-(CH_3)_3C_{14}H_7$	220.31			72.3				$B5^{3}$, 2188
10683	Phenanthrene, 1,2,3,4,9,10,11,12-octahydro (cis) $C_{14}H_{18}$	186.30		$129^{6}, 88-90^{0.1}$		$1.0072^{25/4}$	1.5549^{21}	bz	$B5^{4}$, 1585
10684	Phenanthrene, 1,2,3,4,9,10,11,12-octahydro (*trans*) $C_{14}H_{18}$	186.30	nd	$94-5^{15}$	23-4	$1.0060^{20/4}$	1.5528^{21}	bz	$B5^{3}$, 1402
10685	Phenanthrene, perhydro $C_{14}H_{24}$	192.34		$86-9^{2}$		$0.9447^{20/4}$	1.5011^{20}	eth, ace, bz	$B5^{4}$, 492
10686	Phenanthrene, 7-isopropyl-1-methyl $1-CH_3\ 7\ i\ C_3H_7C_{14}H_8$	234.34	pl (al)	$390, 290^{10}$	100-1	1.035		bz, lig	$B5^{3}$, 2199
10687	Phenanthrene, 1,2,3,4-tetrahydro or Tetranthrene $C_{14}H_{14}$	182.27	lf (MeOH)	173^{11}	33-4	$1.0601^{40/4}$		al, eth, ace, bz, chl, lig	$B5^{4}$, 1909
10688	Phenanthrene, 3,4,5-trihydroxy or 3,4,5-Phenanthrene-triol $3,4,5-(HO)_3C_{14}H_7$	226.23	lf or pl (w)		148			al, eth, chl	$B6^{1}$, 1411
10689	Phenanthrene, 1,6,7-trimethyl $1,6,7-(CH_3)_3C_{14}H_7$	220.31			123-4				$B5^{3}$, 2188
10690	1-Phenanthrene carboxylic acid or 1-Phenanthroic acid $1-C_{14}H_9CO_2H$	222.24	nd (al)		232-3			al, bz	$B9^{3}$, 3496
10691	2-Phenanthrene carboxylic acid or 2-Phenanthroic acid $2-C_{14}H_9CO_2H$	222.24	nd (aa)		258-60			al, bz, aa	$B9^{3}$, 3497
10692	2-Phenanthronitrile $2-C_{14}H_9CN$	203.24	cr (bz-lig, al)		108-10			al, eth, ace	$B9^{3}$, 3498
10693	3-Phenanthrene carboxylic acid or 3-Phenanthroic acid $3-C_{14}H_9CO_2H$	222.24	nd (aa)	sub	270			al, eth, aa	$B9^{3}$, 3498
10694	3-Phenanthronitrile $3-C_{14}H_9CN$	203.24	nd (abs al)		102			al, eth	$B9^{3}$, 3499

No.	Name, Synonyms, and Formula	Mol. wt.	Color, crystalline form, specific rotation and λ_{max} (log ε)	b.p. °C	m.p. °C	Density	n_D	Solubility	Ref.
10695	9-Phenanthrene carboxylic acid or 9-Phenanthroic acid 9-$C_{14}H_9CO_2H$	222.24	nd (aa), lf (sub)	sub	256-7			al, eth, bz, aa	B9[3], 3501
10696	9-Phenanthronitrile 9-$C_{14}H_9CN$	203.24	nd (al)		103			eth, ace	B9[3], 3502
10697	2-Phenanthrene sulfonic acid 2-$C_{14}H_9SO_3H$	258.29	cr (bz), cr (w + 1)		ca 150			w, al, bz	B11[3], 445
10698	3-Phenanthrene sulfonic acid 3-$C_{14}H_9SO_3H$	258.29	lf (bz), cr (w)		88-9 (+2w), 120-1 (+1w), N175-6 (anh)			w	B11[3], 445
10699	9-Phenanthrene sulfonic acid 9-$C_{14}H_9SO_3H$	258.29	lf or nd (bz, w + 2)		134 hyd, 174 sub			w, al, aa	B11[2], 111
10700	Phenanthridine or 3-4-Benzoguinoline $C_{13}H_9N$	179.22	nd (dil al)	349	106-7			al, eth, ace, bz, chl	B20[4], 4016
10701	Phenanthridine, 6-hydroxy $C_{13}H_9NO$	195.22	nd (al, sub)	sub	293-4				B21[4], 1572
10702	1,7-Phenanthrolene or 1,7-Diazaphenanthrene $C_{12}H_8N_2$	180.21	pl (anh), nd (w + 2)	>360	78 (anh) 65.5 (+2w)			al	B23[1], 61
10703	1,10-Phenanthroline or 4,5-Diazaphenanthrene $C_{12}H_8N_2$	180.21	wh nd (bz), cr (w + 1)	>300	117 (anh)			al, ace, bz	B23, 227
10704	4,7-Phenanthroline or 1,8-Dizaphenanthrene $C_{12}H_8N_2$	180.21	nd (w)	sub 100	177			al, chl	B23[1], 61
10705	4,7-Phenanthroline hydrate $C_{12}H_8N_2.H_2O$	198.22	wh nd (w)		100-3				B23[2], 235
10706	1,7-Phenanthroline, 9-nitro $C_{12}H_7N_3O_2$	225.21	nd (dil al)		168				B23[2], 236
10707	9,10-Phenanthroquinone $C_{14}H_8O_2$	208.22	og nd (to), og-red pl (sub)	>360 sub	208-10	$1.405^{22/4}$		eth	B7[3], 4084
10708	9,10-Phenanthroquinone-2-bromo $C_{14}H_7BrO_2$	287.11	red-ye cr (aa)		233-4				B7[3], 4093
10709	9,10-Phenanthroquinone, 3-bromo $C_{14}H_7BrO_2$	287.11	ye nd (aa)		268-9			bz	13, 925
10710	9,10-Phenanthraquinone, 1-chloro $C_{14}H_7O_2Cl$	242.66			229			al, bz	B7[3], 4092
10711	9,10-Phenanthraquinone, 2-Chloro $C_{14}H_7ClO_2$	242.66	ye-red nd (aa)		252-3			al	B7[3], 4093
10712	9,10-Phenanthraquinone, 3-chloro $C_{14}H_7ClO_2$	242.66	og-ye nd (aa, bz-al)		264-5			al, bz, aa	B7[3], 4093
10713	9,10-Phenanthraquinone, 1,2-dihydroxy $C_{14}H_8O_4$	240.22	dk red nd (ace)		d			al, ace, aa	B8[2], 506
10714	9,10-Phenanthraquinone, 2,5-dihydroxy $C_{14}H_8O_4$	240.22	dk red nd (w)		400d				B8[2], 507
10715	9,10-Phenanthraquinone, 2,7-dihydrocy $C_{14}H_8O_4$	240.22	dk red or br nd		>400d			al, eth, ace, bz, aa	B8[2], 507
10716	9,19-Phenanthraquinone, 4,5-dihydroxy $C_{14}H_8O_4$	240.22	dk red (al), nd (w)	d>400				al	B8[2], 508
10717	9,10-Phenanthroquinone, 2,5-dinitro $C_{14}H_6N_2O_6$	298.21	red ye pr (aa)		228			aa	B7[3], 4096
10718	9,10-Phenanthroquinone, 2,7-dinitro $C_{14}H_6N_2O_6$	298.21	gold ye nd (aa)		301-3			aa	B7[2], 730
10719	9,10-Phenanthroquinone, 2-hydroxy $C_{14}H_8O_3$	224.22	br-red or bk-vt nd (aa)	sub	283			w	B8[3], 2929
10720	9,10-Phenanthroquinone, 3-hydroxy $C_{14}H_8O_3$	224.22	ye-red or red nd (aa MeOH)	sub	330d			w, al	B8[3], 2930
10721	9,10-Phenanthroquinone, 2-nitro $C_{14}H_7NO_4$	253.21	ye lf or nd (aa)		260				B7[3], 4095

PHYSICAL CONSTANTS OF ORGANIC COMPOUNDS (Continued)

No.	Name, Synonyms, and Formula	Mol. wt.	Color, crystalline form, specific rotation and λ_{max} (log ε)	b.p. °C	m.p. °C	Density	n_D	Solubility	Ref.
10722	9,10-Phenanthroquinone, 7-isopropyl-1-methyl or Retenoquinone $C_{18}H_{16}O_2$	264.32	og nd (chl-al)	sub	197-8	...	...	al, eth	B7³, 4165
10723	9,10-Phenanthroquinone, 1,2,4-trihydroxy $C_{14}H_8O_5$	256.21	red (al + 1)	...	d	...	...	al	B8², 558
10724	9,10-Phenanthroquinone, 2,3,4-trihydroxy $C_{14}H_8O_5$	256.21	red br pw	...	185d	...	...	w	B8², 559
10725	Phenazine or Azophenylene $C_{12}H_8N_2$	180.21	ye-red nd (aa)	>360 sub	176-7	...	...	bz	B23², 233
10726	Phenazine, 1,4-dihydroxy-di-(N-oxide) $C_{12}H_8N_2O_4$	244.21	purp (chl)	...	236d	...	...	chl	C50, 358
10727	Phenazine, 9,10-dihydro or Hydrazophenylene $C_{12}H_{10}N_2$	182.22	rh lf	...	317 (sealed tube)	...	...	bz	B23², 225
10728	Phenazine, 1-hydroxy or 1-Phenazinol $C_{12}H_8N_2O$	196.21	ye nd (bz, dil MeOH) lf (dil MeOH)	sub	158	...	...	Py	B23², 360
10729	Phenazine, 2-methyl $C_{13}H_{10}N_2$	194.24	lt ye nd or pr	350d	117	...	...	al, eth, chl	B23², 239
10730	β-Phenethyl amine $C_6H_5CH_2CH_2NH_2$	121.18	...	197-8	...	$0.9580^{24/4}$	1.5290^{25}	al, eth	B12³, 2408
10731	β-Phenethyl amine, hydrochloride $C_6H_5CH_2CH_2NH_2.HCl$	157.64	pl or lf (al)	...	218-9	...	...	w, al	B12³, 2409
10732	*o*-Phenetidine or 2-Ethoxy aniline $2\text{-}C_2H_5OC_6H_4NH_2$	137.18	...	232.5, $127\text{-}8^{14}$	<−20	...	1.5560^{20}	al, eth	B13³, 756
10733	o-Phenetidine, 4-nitro or 2-Ethoxy-4-nitro aniline $4\text{-}NO_2\text{-}2\text{-}C_2H_5OC_6H_3NH_2$	182.18	ye nd (dil al)	...	91	...	...	al, eth, ace	B13³, 888
10734	*o*-Phenetidine, 5-nitro or 2-Ethoxy-5-nitro aniline $5\text{-}NO_2\text{-}2\text{-}C_2H_5OC_6H_3NH_2$	182.18	ye nd (dil al)	$205\text{-}6^{14}$	96-7	...	...	al, eth	B13³, 878
10735	*o*-Phenetidine, 6-nitro or 2-Ethoxy-6-nitro aniline $6\text{-}NO_2\text{-}2\text{-}C_2H_5OC_6H_3NH_2$	182.18	ye or og cr (w)	...	60	...	...	w	B13³, 876
10736	*m*-Phenetidine or 3-Ethoxy aniline $3\text{-}C_2H_5OC_6H_4NH_2$	137.18	...	248, $127\text{-}8^{11}$	...	...	...	al, eth	B13³, 932
10737	*m*-Phenetidine, 4-nitro or 3-Ethoxy-4-nitro aniline $4\text{-}NO_2\text{-}3\text{-}C_2H_5OC_6H_3NH_2$	182.18	nd (dil al)	...	122-3	...	...	al, eth, ace, bz	B13³978
10738	*m*-Phenetidine, 5-nitro or 5-Ethoxy-3-nitro aniline $3\text{-}NO_2\text{-}5\text{-}C_2H_5OC_6H_3NH_2$	182.18	ye og red nd (al)	...	115	...	...	al, ace, bz	B13³, 977
10739	*m*-Phenetidine, 6-nitro or 5-Ethoxy-2-nitro aniline $5\text{-}(C_2H_5O)\text{-}2\text{-}(NO_2)C_6H_3NH_2$, $2\text{-}NO_2\text{-}5\text{-}C_2H_5OC_6H_3NH_2$	182.18	ye nd (dil al)	...	105-6	...	...	ace, bz	B13³975
10740	*p*-Phenetidine or 4-Ethoxy aniline $4\text{-}C_2H_5OC_6H_4NH_2$	137.18	...	254, 125^{12}	2.4	$1.0652^{16/4}$	1.5528^{20}	al, eth	B13³, 996
10741	*p*-Phenetidine, 2-nitro or 4-Ethoxy-2-nitro aniline $2\text{-}NO_2\text{-}4\text{-}C_2H_5OC_6H_3NH_2$	182.18	red pr (al)	...	113	...	...	eth, chl	B13³, 1203
10742	*p*-Phenetidine, 3-nitro or 4-Ethoxy-3-nitro aniline $3\text{-}NO_2\text{-}4\text{-}C_2H_5OC_6H_3NH_2$	182.18	og-ye nd (bz, dil al)	...	41	...	...	ace, bz	B13², 284
10743	Phenetole or Ethoxy benzene $C_6H_5OC_2H_5$	122.17	...	170, 60^9	−29.5	$0.9666^{20/4}$	1.5076^{20}	al, eth	B6⁴, 554
10744	Phenetole, 2-bromo or 1-Bromo-2-ethoxybenzene $2\text{-}BrC_6H_4OC_2H_5$	201.06	...	222-6	...	...	...	al, eth	B6⁴, 1038
10745	Phenetole, 3-bromo $3\text{-}BrC_6H_4OC_2H_5$	201.06	...	222	...	...	...	al, eth	B6³, 739
10746	Phenetole, 4-bromo $4\text{-}BrC_6H_4OC_2H_5$	201.06	...	230-2, 109^{17}	4	$1.4071^{25/4}$	1.5517^{20}	al, eth	B6⁴, 1045
10747	Phenetole, 2-chloro or 1-Chloro-2-ethoxy benzene $2\text{-}Cl\text{-}C_6H_4OC_2H_5$	156.61	...	210, $97\text{-}8^{15}$	...	$1.1288^{25/4}$	1.5284^{25}	al, eth, bz	B6⁴, 785
10748	Phenetole, 3-chloro or 1-Chloro-3-ethoxy benzene $3\text{-}ClC_6H_4OC_2H_5$	156.61	...	$204\text{-}5^{717}$	...	$1.1712^{20/4}$	...	al, eth, bz, aa	B6⁴, 811
10749	Phenetole, 4-chloro or 4-Chloro-1-ethoxy benzene $4\text{-}ClC_6H_4OC_2H_5$	156.61	...	212-4, 98^{17}	21	$1.1254^{20/4}$	1.5252^{20}	al, eth, bz, aa	B6⁴, 823
10750	Phenetole, 2,4-dichloro $2,4\text{-}Cl_2\text{-}C_6H_3OC_2H_5$	191.06	...	237	...	...	...	al, eth, bz	B6⁴, 885
10751	Phenetole, 2,4-dinitro $2,4\text{-}(NO_2)_2C_6H_3OC_2H_5$	212.16	nd or lf (al)	...	86-7	...	...	ace	B6⁴, 1373
10752	Phenetole, 2,5-dinitro $2,5\text{-}(NO_2)_2C_6H_3OC_2H_5$	212.16	lf (al)	...	96-8	...	...	...	B6², 245

No.	Name, Synonyms, and Formula	Mol. wt.	Color, crystalline form, specific rotation and λ_{max} (log ε)	b.p. °C	m.p. °C	Density	n_D	Solubility	Ref.
10753	Phenetole, 2,6-dinitro $2,6-(NO_2)_2C_6H_3OC_2H_5$	212.16	nd (eth)	$137-9^{3}$	60-1			eth	$B6^{3}$, 868
10754	Phenetole, 3,5 dinitro $3,5-(NO_2)_2C_6H_3OC_2H_5$	212.16	nd (al)		97.5				$B6^{3}$, 869
10755	Phenetole, 4-ethyl $4-C_2H_5C_6H_4OC_2H_5$	150.22		211, $92-3^{12}$		$0.9385^{17/4}$		al, ace, bz	$B6^{3}$, 1665
10756	Phenetole, 2-fluoro $2-FC_6H_4OC_2H_5$	140.16		171.4, 64^{11}	−16.7	1.0874^{17}	1.4932^{17}	ace, bz	$B6^{4}$, 771
10757	Phenetole, 3-fluoro $3-FC_6H_4OC_2H_5$	140.16		171.4^{755}, 65.2^{15}	−27.5	$1.0716^{16/4}$	1.4847^{17}	bz	$B6^{3}$, 669
10758	Phenetole, 4-fluoro $4-FC_6H_4OC_2H_5$	140.16		173^{766}, 54^{7}	−8.5	1.0715^{18}	1.4826^{18}	bz, chl	$B6^{4}$, 774
10759	Phenetole, 2-iodo $2-IC_6H_4OC_2H_5$	248.06		245^{736}, $121-31^{18}$				al, eth, ace, bz	$B6^{3}$, 769
10760	Phenetole, 4-iodo $4-IC_6H_4OC_2H_5$	248.06	cr (dil MeOH)	$249-50^{729}$	29			al, eth, bz, chl	$B6^{4}$, 1077
10761	Phenetole, 3-mercapto $3-HSC_6H_4OC_2H_5$	154.23		238-9				al, eth, ace	B6, 833
10762	Phenetole, 4-merapto $4-HSC_6H_4OC_2H_5$	154.23		238	1.6			al, eth, ace, bz	$B6^{2}$, 852
10763	Phenetole, 2-nitro $2-NO_2C_6H_4OC_2H_5$	167.16	br ye	267, 149^{15}	2.1	1.1903^{15}	1.5425^{20}	al, eth	$B6^{4}$, 1250
10764	Phenetole, 3-nitro $3-NO_2C_6H_4OC_2H_5$	167.16	br ye	284d, 169^{70} d	36			al, eth	$B6^{4}$, 1271
10765	Phenetole, 4-nitro $4-NO_2C_6H_4OC_2H_5$	167.16	pr (dil al, eth)	283, 168^{15}	60	$1.1176^{100/4}$		eth, ace, bz	$B6^{4}$, 1283
10766	Phenicin or 3,3′-Dihydroxy-5,5′-dimethyl biguinone $C_{14}H_{,}$liO$_6$	274.23	yesh-br (al)		230-1			al, chl, aa	$B8^{3}$, 4251
10767	Phenol C_6H_5OH	94.11		181.7, 70.9^{10}	43	$1.0576^{20/4}$	1.5408^{41}	w, al, eth, ace, bz, chl	$B6^{4}$, 531
10768	Phenol, 2-acetyl or 2-Hydroxy acetophenone $2-CH_3COC_6H_4OH$	136.15		218, 106^{17}	4-6	$1.1307^{20/4}$	1.5584^{20}	al, eth, aa	$B8^{3}$, 261
10769	Phenol, 3-acetyl or 3-Hydroxy acetophenone $3-CH_3COC_6H_4OH$	136.15	nd or lf	296^{756}, 153^{5}	96	1.0992^{109}	1.5348^{109}	al, eth, bz, chl	$B8^{3}$, 272
10770	Phenol, 4-acetyl or 4-Hydroxy acetophenone $4-CH_3COC_6H_4OH$	136.15	nd (eth, dil al)	$147-8^{3}$	109-10	1.1090^{109}	1.5577^{109}		$B8^{3}$, 276
10771	Phenol, 2-allyl $2-(CH_2=CHCH_2)C_6H_4OH$	134.18		220, $93-4^{8}$	−6	$1.0255^{15/15}$	1.5181^{20}	eth	$B6^{3}$, 528
10772	Phenol, 2-allyl-4-chloro $2-(CH_2=CHCH_2)-4-ClC_6H_3OH$	168.62	pr (lig)	$130-2^{15}$	48	1.171^{15}		al, bz	$B6^{1}$, 282
10773	Phenol, 2-allyl-6-methoxy $2-C_3H_4-6-(CH_3O)C_6H_3OH$	164.20		250-1, 115^{9}		$1.2090^{20/4}$	1.5545^{20}	ace, bz	$B6^{3}$, 5013
10774	Phenol, 4-allyl or Chavicol $4-(CH_2=CHCH_2)C_6H_4OH$	134.18		235-6, 120^{12}	16	$1.033^{18/4}$	1.5448^{20}	al, eth, chl, peth	$B6^{3}$, 2415
10775	Phenol, 5-allyl-2-methoxy or Chavibetol $5-(CH_2=CHCH_2)-2-(CH_3O)C_6H_3OH$	164.20		253-4, 111^{8}	8.5	$1.0613^{25/4}$	1.5413^{20}	al, eth	$B6^{3}$, 5024
10776	Phenol, *o*-amino $2-H_2NC_6H_4OH$	109.13	wh rh bipym nd (bz)	sub 153^{11}	174	1.328		al, eth, w	$B13^{3}$, 752
10777	Phenol, 2-amino-3-chloro $3-Cl-2-H_2NC_6H_3OH$	143.57	nd		122			w	$B13^{2}$, 182
10778	Phenol, 2-amino-4-chloro-5-nitro $5-NO_2-4-Cl-2-H_2NC_6H_2OH$	188.57	ye nd		225d (dk at 200)			al	$B13^{2}$, 196
10779	Phenol, 2-amino-4-chloro-6-nitro $6-NO_2-4-Cl-2H_2NC_6H_2OH$	188.57			152			al	$B13^{2}$, 196
10780	Phenol, 2-amino-6-chloro-4-niro $4-NO_2-6-Cl-2-H_2NC_6H_2OH$	188.57	ye nd (w + l)		160				$B13^{3}$, 895
10781	Phenol, 2-amino-5-chloro $5-Cl-2-H_2NC_6H_3OH$	143.57	nd (al), pr (dil al)		154-5			al, eth	$B13^{3}$, 850
10782	Phenol, 2-amino-3,5-dibromo $3,5-Br_2-2-H_2NC_6H_2OH$	266.92	nd (lig)		145			w	$B13^{2}$, 187
10783	Phenol, 2-amino-4,6-dibromo $4,6-Br_2-2-H_2NC_6H_2OH$	266.92	ye nd (dil al)		99			al, eth, bz, chl	$B13^{2}$, 188
10784	Phenol, 2-amino-3,5-dichloro $3,5-Cl_2-2-H_2NC_6H_2OH$	178.02	nd (bz, w)		132-3			al, ace, bz	$B13^{2}$, 185

No.	Name, Synonyms, and Formula	Mol. wt.	Color, crystalline form, specific rotation and λ_{max} (log ε)	b.p. °C	m.p. °C	Density	n_D	Solubility	Ref.
10785	Phenol, 2-amino-4,6-dinitro or Picramic acid $4,6\text{-}(NO_2)_2\text{-}2\text{-}H_2NC_6H_2OH$	199.12	dk red nd (al), pr (chl)		169			al, bz, aa	$B13^3$, 899
10786	Phenol, 2-(2-aminoethyl) $2\text{-}(H_2NCH_2CH_2)C_6H_4OH$	137.18	rh (al-eth)		152-3				$B13^3$, 1624
10787	Phenol, 2-amino-3-nitro $3\text{-}NO_2\text{-}2\text{-}H_2NC_6H_3OH$	154.13	red nd (w)	sub	216-7			w	$B13^3$, 875
10788	Phenol, 2-amino-4-nitro $4\text{-}NO_2\text{-}2\text{-}H_2NC_6H_3OH$	154.13	og pr (+w)		80-90 (+w), 145-7 (anh)			al, eth, aa	$B13^3$, 877
10789	Phenol, 2-amino-5-nitro $5\text{-}NO_2\text{-}2\text{-}NH_2C_6H_3OH$	154.13			207-8			w, al, bz	$B13^3$, 887
10790	Phenol, 2-amino-6-nitro $6\text{-}NO_2\text{-}2\text{-}H_2NC_6H_3OH$	154.13	red nd (dil al)		111-2			al, eth, bz, chl, aa	$B13^2$, 195
10791	Phenol, 3-amino $3\text{-}H_2NC_6H_4OH$	109.13	pr (to)	164^{11}	123			al, eth	$B13^3$, 931
10792	Phenol, *m*-amino,hydrobromide $3\text{-}H_2NC_6H_4OH.HBr$	190.04	pr (w)		224			w	B13, 403
10793	Phenol, *m*-amino,hydrochloride $3\text{-}H_2NC_6H_4OH.HCl$	145.59	pr (w)		229			w	$B13^3$, 932
10794	Phenol, *m*-amino,hydroiodide $3\text{-}H_2NC_6H_4OH.HI$	237.04	pr (w)		209			w	B13, 403
10795	Phenol, 3-(2-aminoethyl), hydrochloride $3\text{-}(H_2NCH_2CH_2)C_6H_4OH.HCl$	173.64	cr (al-eth)		145				$B13^3$, 1630
10796	Phenol, 3-amino-2-chloro $2\text{-}Cl\text{-}3\text{-}H_2NC_6H_3OH$	143.57			85-7			al, eth	B13, 420
10797	Phenol, 3-amino-4,6-dichloro $4,6\text{-}Cl_2\text{-}3\text{-}H_2NC_6H_2OH$	178.02	ye br pr (w)		135-6			al, eth, ace, bz, chl	$B13^3$, 970
10798	Phenol, 3-amino-4-nitro $4\text{-}NO_2\text{-}3\text{-}H_2NC_6H_3OH$	154.13	og nd (w)		185-6			al, eth, bz, chl	$B13^1$, 136
10799	Phenol, *p*-amino $4\text{-}H_2NC_6H_4OH$	109.13	wh pl (w)	$110^{0.3}$	186-7			al	$B13^3$, 991
10800	Phenol, 4-(2 aminoethyl) or Tyramine $4\text{-}(H_2NCH_2CH_2)C_6H_4OH$	137.18	pl or nd (bz), cr (al), nd (w)	$205\text{-}7^{25}$	164-5			al, xyl	$B13^3$, 1637
10801	Phenol, 4-(2 aminopropyl) (l) or L-Paredrine $4\text{-}[CH_3CH(NH_2)CH_2]C_6H_4OH$	151.21	$[\alpha]^{17}{}_D$ −52 (al)		111			al, eth, chl	H34, 2202
10802	Phenol, 4-amino-2-chloro $2\text{-}Cl\text{-}4\text{-}H_2NC_6H_3OH$	143.57	nd (al, eth, w)		153			al, eth	$B13^3$, 1180
10803	Phenol, 4-amino-2-chloro-6-nitro $6\text{-}NO_2\text{-}2\text{-}Cl\text{-}4\text{-}H_2NC_6H_2OH$	188.57			130			al	B13, 524
10804	Phenol, 4-amino-3-chloro $3\text{-}Cl\text{-}4\text{-}H_2NC_6H_3OH$	143.57	nd		160			al, eth	$B13^3$, 1184
10805	Phenol, 4-amino-2,6-dibromo $2,6\text{-}Br_2\text{-}4\text{-}H_2NC_6H_2OH$	266.92	nd (al, bz)		192-3			al, bz	$B13^3$, 1195
10806	Phenol, 4-amino 2,5-dichloro $2,5\text{-}Cl_2\text{-}4\text{-}H_2NC_6H_2OH$	178.02	cr (bz)		178-9			al, eth, aa	$B13^2$, 274
10807	Phenol, 4-amino -2,6-dichloro $2,6\text{-}Cl_2\text{-}4\text{-}H_2NC_6H_2OH$	178.02	nd or lf (w, bz)	sub	167			al, eth	$B13^2$, 274
10808	Phenol, 4-amino-3,5-dichloro $3,5\text{-}Cl_2\text{-}4\text{-}H_2NC_6H_2OH$	178.02	nd (w, bz)		154			al, eth, chl, aa	$B13^3$, 1189
10809	Phenol, 4-amino-2-nitro $2\text{-}NO_2\text{-}4\text{-}H_2NC_6H_3OH$	154.13	dk red pl or nd (w, al)		131			al, eth	$B13^2$, 284
10810	Phenol, 4-amino-3-nitro $3\text{-}NO_2\text{-}4\text{-}H_2NC_6H_3OH$	154.13	dk red pr (eth)		154			w, al, eth, chl	$B13^3$, 1203
10811	Phenol, 4-(2-aminopropyl) $4\text{-}CH_3CH(NH_2)CH_2C_6H_4OH$	151.21	cr (bz)		125-6			w, al, chl	$B13^3$, 1709
10812	Phenol, 2-benzyl $2\text{-}(C_6H_5CH_2)C_6H_4OH$	184.24	cr (peth)	312, $159\text{-}62^{12}$	51-3				$B6^3$, 3349
10813	Phenol, 4-benzyl $4\text{-}(C_6H_5CH_2)C_6H_4OH$	184.24	nd (al)	320-2, $198\text{-}200^{10}$	84			al, eth, bz, chl, aa	$B6^3$, 3357
10814	Phenol, 2-bromo $2\text{-}BrC_6H_4OH$	173.01		194-5, 87.3^{13}	5.6	$1.4924^{20/4}$	1.589^{20}	al, eth	$B6^4$, 1037
10815	Phenol, 2-bromo-3-nitro $2\text{-}Br\text{-}3\text{-}NO_2C_6H_3OH$	218.01	pa ye nd (HCl)	sub	147-8			al, eth	$B6^3$, 844

No.	Name, Synonyms, and Formula	Mol. wt.	Color, crystalline form, specific rotation and λ_{max} (log ε)	b.p. °C	m.p. °C	Density	n_D	Solubility	Ref.
10816	Phenol, 2-bromo-4-nitro 2-Br-4-$NO_2C_6H_3OH$	218.01	cr (to, w), nd (chl, eth, dil al)		114			al, eth, chl	B6[3], 845
10817	Phenol, 2-bromo-5-nitro 2-Br-5-$NO_2C_6H_3OH$	218.01	pa ye nd (w), cr (peth)		129-30			al, eth, ace, bz	B6[4], 1365
10818	Phenol, 2-bromo-6-nitro 2-Br-6-$NO_2C_6H_3OH$	218.01	pa ye nd (chl, al)		68			al, aa	B6[3], 844
10819	Phenol, 3-bromo 3-BrC_6H_4OH	173.01		236.5, 135-40[12]	33			al, eth, chl	B6[4], 1042
10820	Phenol, 3-bromo-5-chloro 3-Br-5-ClC_6H_3OH	207.45	nd (peth)	256-60[756]	70			al, aa	B6[2], 187
10821	Phenol, 3-bromo-2,4-dinitro 2,4-$(NO_2)_2$-3-BrC_6H_2OH	263.02	pa ye nd (w)		175			eth	B6[2], 249
10822	Phenol, 3-bromo-2,6-dinitro 2,6-$(NO_2)_2$-3-BrC_6H_2OH	263.00	nd (peth)		131			al, eth	B6[2], 250
10823	Phenol, 3-bromo-2-nitro 3-Br-2-$NO_2C_6H_3OH$	218.01	ye nd (peth), col nd (+ w)		65-7 (anh)			al	B6[3], 842
10824	Phenol, 3-bromo-4-nitro 3-Br-4-$NO_2C_6H_3OH$	218.01	pa ye nd (bz, peth)		129-30			al, eth, bz	B6[4], 1365
10825	Phenol, 3-bromo-5-nitro 3-Br-5-$NO_2C_6H_3OH$	218.01	cr (w)		145			al, eth	B6[2], 233
10826	Phenol, 4-bromo 4-BrC_6H_4OH	173.01		238, 118.2[11]	66.4	1.840[15]		w, al, eth, chl	B6[4], 1043
10827	Phenol, 4-bromo-2-chloro 4-Br-2-ClC_6H_3OH	207.45	nd (lig, bz, aa)	233-4, 127-30[12]	50-1	1.6170[20/4]	1.5859[20]	al, eth, ace, bz	B6[3], 750
10828	Phenol, 4-bromo-2,6-dinitro 2,6-$(NO_2)_2$-4-BrC_6H_2OH	263.00	pa ye nd (w), nd (al), pr (aa)	sub	78			al, eth, bz, chl	B6[3], 872
10829	Phenol, 4-bromo-2-nitro 4-Br-2-$NO_2C_6H_3OH$	218.01	ye nd or lf (al), pr (eth)	sub	92			al, eth, bz, chl	B6[4], 1363
10830	Phenol, 4-bromo-3-nitro 4-Br-3-$NO_2C_6H_3OH$	218.01	ye nd (w)		147			eth, bz	B6[4], 1364
10831	Phenol, 5-bromo-2,4-dinitro 2,4$(NO_2)_2$-5-BrC_6H_2OH	263.00	pr (al, eth)		92			al, eth	B6[3], 871
10832	Phenol, 5-bromo-2-nitro 5-Br-2-$NO_2C_6H_3OH$	218.01	ye pr or nd (lig)		44			al, eth, lig	B6[3], 844
10833	Phenol, 6-bromo-2,4-dinitro 2,4-$(NO_2)_2$-6-BrC_6H_2OH	263.00	pa ye cr (al), pr (eth)	sub	118-9			eth, bz, lig	B6[3], 871
10834	Phenol, *o*-butyl 2-$C_4H_9C_6H_4OH$	150.22		235, 106.5[10]		0.975[20/4]	1.5180[25.5]	al, eth	B6[3], 1843
10835	Phenol, *o-sec*-butyl 2-$CH_3CH_2CH(CH_3)C_6H_4OH$	150.22		227-8[751], 116[21]	16	0.9804[25]	1.5200[25]		B6[3], 1852
10836	Phenol, *o-tert*-butyl or 2-*tert*-butyl phenol 2-$(CH_3)_3CC_6H_4OH$	150.22		221, 99[10]		0.9783[20/4]	1.5160[20]	al, eth	B6[3], 1861
10837	Phenol, *m*-butyl or 3-Butyl phenol 3-$C_4H_9C_6H_4OH$	150.22		248, 123[10]		0.974[20/4]		al, eth	B6[2], 485
10838	Phenol, *m-tert*-butyl 3-$(CH_3)_3CC_6H_4OH$	150.22	nd (peth)	240, 132.5[20]	41-2			al, eth	B6[3], 1862
10839	Phenol, *p*-butyl or 4-Butyl phenol 4-$C_4H_9C_6H_4OH$	150.22		248, 138-9[18]	22	0.978[20/4]	1.5165[25.5]	al, eth	B6[3], 1844
10840	Phenol, *p*-isobutyl 4-i-$C_4H_9C_6H_4OH$	150.22		235-9	51-2	0.9796[20/20]	1.5319[25]	al, eth, ace	B6[3], 1859
10841	Phenol, *p-sec*-butyl-(d) 4-*sec*$C_4H_9C_6H_4OH$	150.22	nd $[\alpha]^{20}/_d$ + 13.3 (xyl)	240-2	61-2	0.9883[20]	1.5182[21]	al, eth	B6[2], 487
10842	Phenol, *p-tert*-butyl or 4-*tert*-butyl phenol 4-$(CH_3)_3CC_6H_4OH$	150.22	nd (lig)	239.5, 114[10]	101	0.908[80/4]	1.4787[114]	al, eth	B6[3], 1862
10843	Phenol, *o*-chloro 2-ClC_6H_4OH	128.56		174.9, 56.4[10]	9.0	1.2634[20/4]	1.5524[20]	al, eth, bz	B6[4], 782
10844	Phenol, 2-chloro-4-cyclohexyl 2-Cl-4-$C_6H_{11}C_6H_3OH$	210.70		122-3[3]	39				B6[3], 2508

No.	Name, Synonyms, and Formula	Mol. wt.	Color, crystalline form, specific rotation and λ_{max} (log ε)	b.p. °C	m.p. °C	Density	n_D	Solubility	Ref.
10845	Phenol, 2-chloro-3,4-dimethyl 3,4-$(CH_3)_2$-2-Cl-C_6H_2OH	156.61	cr (peth)	187-9	27	1.5538^{20}		peth	B6^{2}, 456
10846	Phenol, 2-chloro-4,6-dinitro 2-Cl-4,6-$(NO_2)_2C_6H_2OH$	218.55			114-6			al, eth, ace, bz	B6^{4}, 1385
10847	Phenol, 2-chloro-4-nitro 2-Cl-4-$NO_2C_6H_3OH$	173.56	wh nd (50% al)		111			al, eth, chl	B6^{4}, 1353
10848	Phenol, 2-chloro-5-nitro 2-Cl-5-$NO_2C_6H_3OH$	173.56	ye nd or pr (w)		121-2				B6^{4}, 1353
10849	Phenol, 2-chloro-6-nitro 2-Cl-6-$NO_2C_6H_3OH$	173.56			70			al, chl	B6^{1}, 837
10850	Phenol, *m*-chloro 3-ClC_6H_4OH	128.56		214	33	1.268^{25}	1.5565^{40}	al, eth, bz	B6^{4}, 810
10851	Phenol, 3-chloro-4-nitro 3-Cl-4-$NO_2C_6H_3OH$	173.56			121-2			al, bz	B6^{4}, 1357
10852	Phenol, 4-Chloro or *p*-Chloro phenol 4-ClC_6H_4OH	128.56		219.7, 125^{18}	43-4	$1.2651^{40/4}$	1.5579^{40}	al, eth, bz	B6^{4}, 820
10853	Phenol, 4-chloro-2-allyl 2$(CH_2{=}CHCH_2)$-4-ClC_6H_3OH	168.62	pr (lig)	130-2^{15}	48	1.171^{15}		al, bz	B6^{1}, 282
10854	Phenol, 4-chloro-2,6-dimethyl 2,6-$(CH_3)_2$-4-ClC_6H_2OH	156.61	nd (w)		83			al, bz, aa	B6^{1}, 1738
10855	Phenol, 4-chloro-2,3-dinitro 4-Cl-2,3$(NO_2)_2C_6H_2OH$	218.55	pr		127			al, eth, bz, chl	B6, 259
10856	Phenol, 4-chloro-2-nitro 4-Cl-2-$NO_2C_6H_3OH$	173.56	ye mcl pr (al)		88-9			al, eth, chl	B6^{4}, 1349
10857	Phenol, 5-chloro-2,4-dinitro 5-Cl-2,4-$(NO_2)_2C_6H_2OH$	218.55	nd (al, peth)		92	1.74^{22}		al, eth, chl, peth	B6^{4}, 1385
10858	Phenol, 5-chloro-2-nitro 5-Cl-2-$NO_2C_6H_3OH$	173.56	ye pr or nd (w)	sub	41			al, eth, aa	B6^{1}, 836
10859	Phenol, *o*-cyclohexyl 2-$C_6H_{11}C_6H_4OH$	176.26	nd (lig)	283, 147^{17}	56-7			al, aa	B6^{3}, 2492
10860	Phenol, *p*-cyclohexyl 4-$C_6H_{11}C_6H_4OH$	176.26	nd (bz)	293-5^{752}, 132-5^{4}	133			al, eth, bz	B6^{3}, 2501
10861	Phenol, 2,4-diamino 2,4-$(H_2N)_2C_6H_3OH$	124.14	lf		78-80d			al, ace, w	B13^{3}, 1338
10862	Phenol, 2,4-diamino, dihydrochloride or Amidol 2,4-$(H_2N)_2C_6H_3OH\cdot 2HCl$	197.06	nd		230-40d			w	B13^{3}, 1338
10863	Phenol, 2,5-diamino 2,5-$(H_2N)_2C_6H_3OH$	124.14	nd		68			w	B13^{2}, 312
10864	Phenol, 3,4-diamino 3,4-$(H_2N)_2C_6H_3OH$	124.14	nd		170-2			w	B13^{1}, 210
10865	Phenol, 3,5-diamino 3,5-$(H_2N)_2C_6H_3OH$	124.14	nd or pr (chl)		168-70 (180)			w	B13^{3}, 1370
10866	Phenol, 2,4-dibromo 2,4-$Br_2C_6H_3OH$	251.91	nd(peth)	238-9, 177^{17}	40			al, eth, bz	B6^{4}, 1061
10867	Phenol, 2,4-dibromo-6-nitro 2,4-Br_2-6-$NO_2C_6H_2OH$	296.90			118			eth, bz, chl	B6^{1}, 848
10868	Phenol, 2,6-dibromo 2,6-$Br_2C_6H_3OH$	251.91	nd (w)	162^{21} (sub)	56-7			al, eth	B6^{4}, 1064
10869	Phenol, 2,6-dibromo-4-nitro 2,6-Br_2-4-$NO_2C_6H_2OH$	296.90	pa ye pr or lf (al)	d>144	145-6			al, eth	B6^{4}, 1366
10870	Phenol, 3,5-dibromo 3,5-$Br_2C_6H_3OH$	251.91		120-2^{3}	81			al, eth	B6^{4}, 1065
10871	Phenol, 2,4-di-*tert*-butyl 2,4-$[(CH_3)_3C]_2$-C_6H_3OH	206.33		263.5, 146^{20}	56.5		1.5080^{20}		B6^{3}, 2062
10872	Phenol, 2,4-di-*tert*-butyl-5-methyl 2,4$(t\text{-}C_4H_9)_2$-5-$CH_3C_6H_2OH$	220.35		282, 167^{20}	62.1	$0.912^{80/4}$		al, eth, ace, bz	B6^{3}, 2072
10873	Phenol, 2,4-di-*tert*-butyl-6-methyl 2,4-$(t\text{-}C_4H_9)_2$-6-$CH_3C_6H_2OH$	220.35		269, 138.5^{10}	51	$0.891^{80/4}$			B6^{3}, 2073
10874	Phenol, 2,6-di-*sec*-butyl 2,6-di-(sec-$C_4H_9)_2C_6H_3OH$	206.33		255-60	−42		1.5080^{20}		ALD 11971, 7
10875	Phenol, 2,6-di-*tert*-butyl 2,6-$[(CH_3)_3C]_2C_6H_3OH$	206.33	pr (al)	133^{20}	39		1.5001^{20}	al	B6^{3}, 2061
10876	Phenol, 2,6-di-*tert*-butyl-4-ethyl 2,6-$(t\text{-}C_4H_9)_2$-4-$C_2H_5C_6H_2OH$	234.38		272, 140^{10}	44				B6^{3}, 2087
10877	Phenol, 3,6-di-*tert*-butyl-4-(2-methyl butyl) 4-$[CH_3CH_2CH(CH_3)CH_2]$-2,6-$(t\text{-}C_4H_9)_2C_6H_2OH$	276.46		135-8^{6}	47				B6^{3}, 2097

No.	Name, Synonyms, and Formula	Mol. wt.	Color, crystalline form, specific rotation and λ_{max} (log ε)	b.p. °C	m.p. °C	Density	n_D	Solubility	Ref.
10878	Phenol, 2,6-di-*tert*-butyl-4-methyl or Ionol $2,6\text{-}(t\text{-}C_4H_9)_2\text{-}4\text{-}CH_3C_6H_2OH$	220.35		$265, 136^{10}$	71	$0.8937^{75/4}$	1.4859^{75}	al, ace, bz, chl	$B6^3$, 2073
10879	Phenol, 2,3-dichloro $2,3\text{-}Cl_2C_6H_3OH$	163.00	cr (lig, bz)		57-9			al, eth	$B6^4$, 883
10880	Phenol, 2,3-dichloro-4,5-dimethyl $2,3\text{-}Cl_2\text{-}4,5\text{-}(CH_3)_2C_6HOH$	191.06	nd (peth)		102.5			eth, ace, bz	$B6^2$, 456
10881	Phenol, 2,3-dichloro-5,6-dimethyl $2,3\text{-}Cl_2\text{-}5,6\text{-}(CH_3)_2C_6HOH$	191.06	cr (peth)		90			eth, ace, bz, chl	$B6^2$, 454
10882	Phenol, 2,4-dichloro $2,4\text{-}Cl_2C_6H_3OH$	163.00	hex nd (bz)	$210, 145\text{-}7^{110}$	45			al, eth, bz, chl	$B6^4$, 885
10883	Phenol, 2,4-dichloro-3,5-dimethyl $2,4\text{-}Cl_2\text{-}3,5\text{-}(CH_3)_2C_6HOH$	191.06			83			eth	$B6^3$, 1760
10884	Phenol, 2,5-dichloro $2,5\text{-}Cl_2C_6H_3OH$	163.00	pr (bz, peth)	211^{764}	59			al, eth, bz	$B6^4$, 942
10885	Phenol, 2,5-dichloro-3,4-dimethyl $2,5\text{-}Cl_2\text{-}3,4\text{-}(CH_3)_2C_6HOH$	191.06	nd (peth)		84			eth	$B6^3$, 1730
10886	Phenol, 2,6-dichloro $2,6\text{-}Cl_2C_6H_3OH$	163.00	nd (peth)	$219\text{-}20^{740}$	68-9			al, eth, bz	$B6^4$, 949
10887	Phenol, 2,6-dichloro-3,4-dimethyl $2,6\text{-}Cl_2\text{-}3,4\text{-}(CH_3)_2C_6HOH$	191.06	cr (peth)		52			eth	$B6^2$, 456
10888	Phenol, 2,6-dichloro-3,5-dimethyl $2,6\text{-}Cl_2\text{-}3,5\text{-}(CH_3)_2C_6HOH$	191.06	cr (peth)	$105\text{-}10^{1}$	87-8			eth, chl	$B6^4$, 3157
10889	Phenol, 3,4-dichloro $3,4\text{-}Cl_2C_6H_3OH$	163.00	nd (bz-peth)	253.5^{767}	68			al, eth, bz	$B6^4$, 952
10890	Phenol, 3,5-dichloro $3,5\text{-}Cl_2C_6H_3OH$	163.00	pr (peth)	233^{757}, $122\text{-}4^{8}$	68			al, eth	$B6^4$, 957
10891	Phenol, 2,6-dichloro-4-nitro $2,6\text{-}Cl_2\text{-}4\text{-}NO_2C_6H_2OH$	208.00	br nd (w)		127d (exp <100)	1.822		eth, chl	$B6^4$, 1361
10892	Phenol, 3-(diethyl amino) $3\text{-}(C_2H_5)_2NC_6H_4OH$	165.24	rh bipym (CS_2-lig)	$276\text{-}80, 170^{15}$	78			al, eth, w	$B13^3$, 942
10893	Phenol, 2,4-diido $2,4\text{-}I_2C_6H_3OH$	345.91	nd (w)	sub 100	72-3			al, eth	$B6^4$, 1082
10894	Phenol, 2,3-dimethoxy or Pyrogallol-1,2-dimethyl ether $2,3(CH_3O)_2C_6H_3OH$	154.17		$232\text{-}4, 124\text{-}5^{17}$			1.5392^{20}		$B6^3$, 6264
10895	Phenol, 3-(dimethyl amino) $3\text{-}(CH_3)_2NC_6H_4OH$	137.18	nd (lig)	$265\text{-}8, 152\text{-}3^{15}$	87		1.5895^{26}	al, eth, ace, bz	$B13^3$, 934
10896	Phenol, 4-(dimethyl amino) $4\text{-}(CH_3)_2NC_6H_4OH$	137.18		165^{30}	76			al, eth	$B13^3$, 1009
10897	Phenol, 2,3-dimethyl or *o*-3-Xylenol $2,3\text{-}(CH_3)_2C_6H_3OH$	122.17	nd (w, dil al)	$218, 95.4^{10}$	75		1.5420^{20}	al, eth	$B6^3$, 1722
10898	Phenol, 2,3,dimethyl-4-*tert*-butyl $2,3\text{-}(CH_3)_2\text{-}4\text{-}(CH_3)_3C\text{-}C_6H_2OH$	178.27		$259, 145^{20}$					$B6^3$, 2019
10899	Phenol, 2,3-dimethyl-4-chloro $2,3\text{-}(CH_3)_2\text{-}4\text{-}Cl\text{-}C_6H_2OH$	156.61	nd (peth)		85			al, eth, ace	$B6^3$, 1724
10900	Phenol, 2,3-dimethyl-6-chloro $2,3\text{-}(CH_3)_2\text{-}6\text{-}Cl\text{-}C_6H_2OH$	156.61		$221\text{-}3$, 100^{17}				al, ace, bz	$B6^3$, 1747

No.	Name, Synonyms, and Formula	Mol. wt.	Color, crystalline form, specific rotation and λ_{max} (log ε)	b.p. °C	m.p. °C	Density	n_D	Solubility	Ref.
10901	Phenol, 2,3 dimethyl-5-nitro 2,3-$(CH_3)_2$-5-$NO_2C_6H_2OH$	167.16	og-ye nd (bz, w)		109 (120)			al, ace, aa	B6², 455
10902	Phenol, 2,3-dimethyl-4,5,6-trichloro 2,3-$(CH_3)_2$-4,5,6-Cl_3C_6OH	225.50	nd (dil al, peth)		180-1			al	B6², 454
10903	Phenol, 2,4-dimethyl 2,4-$(CH_3)_2C_6H_3OH$	122.17	nd (w)	210, 89.3^{10}	27-8	0.9650$^{20/4}$	1.5420^{14}	**al, eth**	B6³, 1741
10904	Phenol, 2,4-dimethyl, acetate or 2,4-Dimethyl phenyl acetate 2,4-$(CH_3)_2C_6H_3O_2CCH_3$	164.20		226, 108^{13}		1.0298$^{15.5/4}$	1.4990^{15}	al, eth	B6², 459
10905	Phenol, 2,4-dimethyl-6-*tert*-butyl 2,4-$(CH_3)_2$-6-$(CH_3)_3C$-C_6H_2OH	178.27		249, 115^{10}	22.3	0.917$^{80/4}$	1.5183^{20}		B6³, 2020
10906	Phenol, 2,4-dimethyl-5-chloro 2,4-$(CH_3)_2$-5-Cl-C_6H_2OH	156.61	nd (lig, w)		90-1			al, eth, ace	B6³, 1747
10907	Phenol, 2,4-dimethyl-6-(hydroxy methyl) 2,4-$(CH_3)_2$-6-$(HOCH_2)C_6H_2OH$	152.19	nd (bz-peth)	57-8				al, eth	B6³, 4653
10908	Phenol, 2,4-dimethyl-6-nitro 2,4$(CH_3)_2$-6-$NO_2C_6H_2OH$	167.16	ye nd (al)		73				B6³, 1750
10909	**Phenol, 2,4-dimethyl-3,5,6-trichloro** **2,4-$(CH_3)_2$-3,5,6-Cl_3C_6OH**	225.50	pa ye nd		174			eth	B6², 460
10910	Phenol, 2,5-dimethyl 2,5-$(CH_3)_2C_6H_3OH$	122.17	nd (w), pr (al-eth)	211.5	75			al, eth	B6³,1769
10911	Phenol, 2,5-dimethyl, acetate or 2,5-Dimethyl phenyl acetate 2,5$(CH_3)_2C_6H_3O_2CCH_3$	164.20		237^{768}	<-20	1.0624^{15}		al, eth	B6², 467
10912	Phenol, 2,5-dimethyl-4-*tert*-butyl 2,5$(CH_3)_2$-4-$(CH_3)_3C$-C_6H_2OH	178.27		264, 136^{10}	71-2	0.939$^{80/4}$, 1.001$^{27/4}$	1.5311^{20}		B6³, 2019
10913	Phenol, 2,5-dimethyl-4-chloro 2,5$(CH_3)_2$-4-Cl-C_6H_2OH	156.61	silv-gr nd (lig)		74-5			al, bz, aa, peth	B6³, 1773
10914	Phenol, 2,5-dimethyl-3-nitro 2,5-$(CH_3)_2$-3-$NO_2C_6H_2OH$	167.16	ye lf (peth)		91			al, eth	B6, 497
10915	Phenol, 2,5-dimethyl-6-nitro 2,5-$(CH_3)_2$-6-$NO_2C_6H_2OH$	167.16	nd (peth)	236d, 150^{15}	34.5			al, eth, ace, bz	B6², 246
10916	Phenol, 2,5-dimethyl-4-nitro 2,5-$(CH_3)_2$-4-$NO_2C_6H_2OH$	167.16	pa ye nd (dil al)		122-3			al, eth, ace	B6³, 1775
10917	Phenol, 2,5-dimethyl-2,4,6-trichloro 2,5-$(CH_3)_2$-2,4,6-Cl_3C_6HOH	225.50	pa gr nd (al, aq MeOH)		175-6			al, eth, bz, chl	B6², 467
10918	Phenol, 2,6-dimethyl 2,6-$(CH_3)_2C_6H_3OH$	122.17	lf or nd (al)	212, 91.2^{10}	49			al, eth	B6³, 1735
10919	Phenol, 2,6-dimethyl-4-*tert*-butyl 2,6-$(CH_3)_2$-4-$(CH_3)_3C$-C_6H_2OH	178.27	pr (peth)	248, 119^{10}	82.4	0.916$^{80/4}$			B6³, 2019
10920	Phenol, 2,6-dimethyl-3-nitro 2,6-$(CH_3)_2$-3-$NO_2C_6H_2OH$	167.16	lf pr (bz), nd (lig)		99-100			al, chl	B6, 485
10921	**Phenol, 2,6-dimethyl-4-nitro** **2,6-$(CH_3)_2$-4-NO_2-C_6H_2OH**	167.16	pr (MeOH)		171			al, ace, chl	B6, 486
10922	**Phenol, 2,6-bis(1,1-dimethylpropyl)-4-methyl** **4-CH_3-2,6-$[CH_3CH_2C(CH_3)_2]_2C_6H_2OH$**	248.41		283, 165^{20}		0.931$^{25/4}$	1.4950^{20}		B6³, 2071
10923	Phenol, 3,4-dimethyl 3,4-$(CH_3)_2C_6H_3OH$	122.17	nd (w)	225, 106-8^{10}	66-8	0.9830$^{20/4}$		**al, eth**	B6³, 1725
10924	Phenol, 3,4-dimethyl, acetate or 3,4-Dimethyl phenyl acetate 3,4-$(CH_3)_2C_6H_3O_2CCH_3$	164.20		235, 140^{80}	22			al, eth, bz	B6³, 1728
10925	Phenol, 3,4-dimethyl-6-*tert*-butyl 3,4-$(CH_3)_2$-6-$(CH_3)_3C$-C_6H_2OH	178.27		145^{20}	46.0	0.920$^{80/4}$, 0.973$^{27/4}$	1.5222^{20}		B6³, 2019

No.	Name, Synonyms, and Formula	Mol. wt.	Color, crystalline form, specific rotation and λ_{max} (log ε)	b.p. °C	m.p. °C	Density	n_D	Solubility	Ref.
10926	Phenol, 3,4 dimethyl-5-chloro $3,4-(CH_3)_2-5-Cl-C_6H_2OH$	156.61	nd (peth)		98			al, ace, chl	B6², 456
10927	Phenol, 3,4-dimethyl-6-chloro $3,4-(CH_3)_2-6-Cl-C_6H_2OH$	156.61	nd (peth)		72			al, bz, aa	B6³, 1729
10928	Phenol, 3,4-dimethyl-6-nitro $3,4-(CH_3)_2-6-NO_2C_6H_2OH$	167.16	ye rh (al)		87-9			al, eth, bz, chl	B6³, 1731
10929	Phenol, 3,4-dimethyl-2,5,6-tribromo $3,4-(CH_3)_2-2,5,6-Br_3-C_6-OH$	358.85	nd (al)		173-4			al	B6³, 1730
10930	Phenol, 3,4-dimethyl-2,5,6-trichloro $3,4-(CH_3)_2-2,5,6-Cl_3C_6OH$	225.50	nd (peth)		182.5			al	B6², 456
10931	Phenol, 3,5-dimethyl $3,5-(CH_3)_2C_6H_3OH$	122.17	nd (w, peth)	219.5 sub, $102\text{-}3^{10}$	68	$0.9680^{20/4}$		al	B6³, 1753
10932	Phenol, 3,5-dimethyl-4-chloro $3,5-(CH_3)_2-4-Cl-C_6H_2OH$	156.61		246	115-8			al, eth, bz	B6³, 1759
10933	Phenol, 3,5-dimethyl-6-chloro $3,5-(CH_3)_2-6-Cl-C_6H_2OH$	156.61	nd		49-50			eth	B6², 464
10934	Phenol, 3,5-dimethyl-2-nitro $3,5-(CH_3)_2-2-NO_2C_6H_2OH$	167.16	ye nd (lig, dil MeOH)		66			al, eth, ace, bz	B6³, 1765
10935	Phenol, 3,5-dimethyl-2,4,6-tribromo $3,5-(CH_3)_2-2,4,6-Br_3C_6OH$	358.85	nd (al)		166			al	B6³, 1763
10936	Phenol, 3,5-dimethyl-2,4,6-trichloro $3,5-(CH_3)_2-2,4,6-Cl_3C_6OH$	225.50	ye nd (peth)		117-8			peth	B6³, 1761
10937	Phenol, 2,3-dinitro $2,3-(NO_2)_2C_6H_3OH$	184.11	ye nd (w)		144-5	1.681^{20}		al, eth, bz	B6⁴, 1369
10938	Phenol, 2,4-dinitro $2,4-(NO_2)_2C_6H_3OH$	184.11	pa ye pl or lf (w)	sub	115-6	1.683^{24}		al, eth, ace, bz, chl	B6⁴, 1369
10939	Phenol, 2,5-dinitro $2,5(NO_2)_2C_6H_3OH$	184.11	ye mcl pr or nd (dil al, w, lig)		108			eth, bz	B6⁴, 1383
10940	Phenol, 2,6-dinitro $2,6-(NO_2)_2C_6H_3OH$	184.11	pa ye rh nd or lf (dil al)		63-4			al, eth, ace, bz, Py	B6⁴, 1383
10941	Phenol, 3,4-dinitro $3,4-(NO_2)_2C_6H_3OH$	184.11	tcl nd (w)		134	1.672		al, eth, bz	B6⁴, 1384
10942	Phenol, 3,5-dinitro $3,5-(NO_2)_2C_6H_3OH$	184.11	lf (w)		126.1	1.702		al, eth, bz, chl	B6⁴, 1385
10943	Phenol, 2,6-dipropyl $2,6(C_3H_7)_2C_6H_3OH$	178.27		256, $114\text{-}6^{5}$	28			eth	B6³, 2013
10944	Phenol, 2-ethyl $2-C_2H_5C_6H_4OH$	122.17		207, 84.1^{10}	<−18	1.0371^{0}	1.5367^{20}	al, eth, ace, bz	B6³, 1655
10945	Phenol, 2-ethylamino $2-C_2H_5NHC_6H_4OH$	137.18	pl (bz)		113-4			al	B13³, 763
10946	Phenol, 2-ethyl-4-*tert*-butyl $2-C_2H_5-4-(CH_3)_3C-C_6H_3OH$	178.27		257, 141^{20}					B6³, 2012

No.	Name, Synonyms, and Formula	Mol. wt.	Color, crystalline form, specific rotation and λ_{max} (log ε)	b.p. °C	m.p. °C	Density	n_D	Solubility	Ref.
10947	Phenol, 3-ethyl $3\text{-}C_2H_5C_6H_4OH$	122.17		214, 99.3^{10}	−4	$1.0283^{20/4}$		al, eth	B6[3], 1660
10948	Phenol, 3-(ethylamino) $3\text{-}C_2H_5NHC_6H_4OH$	137.18	cr (bz-peth)	176^{12}	62			al, eth, bz, chl	B13, 408
10949	Phenol, 4-(ethylamino) $4\text{-}C_2H_5NHC_6H_4OH$	137.18	nd (w)		110-2			al, eth	B13[3], 1012
10950	Phenol, 4-ethyl $4\text{-}C_2H_5C_6H_4OH$	122.17	nd	219, 99.5^{10}	47-8		0.5239^{25}	al, eth, ace, bz	B6[3], 1663
10951	Phenol, 4-ethyl-2-*tert*-butyl $4\text{-}C_2H_5\text{-}2\text{-}(CH_3)_3C\text{-}C_6H_3OH$	178.27		250, 123^{10}	23			al	B6[2], 2012
10952	Phenol, *o*-fluoro $2\text{-}F\text{-}C_6H_4OH$	112.10		151-2	16.1			w	B6[4], 770
10953	Phenol, *p*-fluoro $4\text{-}F\text{-}C_6H_4OH$	112.10		185.5, 87^{23}	48			ace, peth	B6[4], 773
10954	Phenol, 4-fluoro-2-nitro $4\text{-}F\text{-}2\text{-}NO_2C_6H_3OH$	157.10			73.7			al	B6[1], 121
10955	Phenol, 4-[bis-(2-hydroxyethyl) amino $4[(HOCH_2CH_2)_2N]C_6H_4OH$	197.23	cr (w)		140			w	B13[3], 1033
10956	Phenol, 2-hydroxymethyl-4-methyl or Homosaligenine $2\text{-}HOCH_2\text{-}4\text{-}CH_3C_6H_3OH$	138.17	lf (w, chl)		106-7			w, al, eth	B6[3], 4598
10957	Phenol, *o*-iodo $2\text{-}IC_6H_4OH$	220.01	nd	$186\text{-}7^{160}$, $91\text{-}2^{2}$	43	1.8757^{80}		al, eth	B6[4], 1070
10958	Phenol, *m*-iodo $3\text{-}IC_6H_4OH$	220.01	nd (lig)	d	118			al, eth	B6[4], 1073
10959	Phenol, *p*-iodo $4\text{-}IC_6H_4OH$	220.01	nd (w or sub)	$138\text{-}40^{5}$d	93-4	1.8573^{112}		al, eth	B6[4], 1074
10960	Phenol, 2-mercapto $2\text{-}HSC_6H_4OH$	126.17	oil	$216\text{-}7^{751}$, $88\text{-}90^{8}$	5-6	$1.2373^{0/0}$		eth	B6[3], 4276
10961	Phenol, 3-mercapto $3\text{-}HSC_6H_4OH$	126.17	cr	168^{35}	16-7			al	B6[2], 827
10962	Phenol, 4-mercapto $4\text{-}HSC_6H_4OH$	126.17	cr	$166\text{-}8^{45}$, $133\text{-}7^{11}$	29-30	$1.1285^{25/4}$	1.5101^{25}	w, al	B6[3], 4445
10963	Phenol, 2-methoxy-3-nitro or 6-Nitroguiaicol $2\text{-}CH_3O\text{-}3\text{-}NO_2C_6H_3OH$	169.14	yesh rh pr (peth)		102-3			al	B6[3], 4263
10964	Phenol, 2-methoxy-4-nitro or 5-Nitroguiaicol $2\text{-}CH_3O\text{-}4\text{-}NO_2C_6H_3OH$	169.14	ye nd (w)		103-4			al, eth	B6[3], 4264
10965	Phenol, 2-methoxy-5-nitro or 4-Nitroguiaicol $2\text{-}CH_3O\text{-}5\text{-}NO_2C_6H_3OH$	169.14	pa ye nd (w)		105			al, eth	B6[3], 4264
10966	Phenol, 2-methoxy-6-nitro or 3-Nitroguiaicol $2\text{-}CH_3O\text{-}6\text{-}NO_2C_6H_3OH$	169.14	og-ye nd (sub)	sub	62			w, al	B6[2], 789
10967	Phenol, 2-methoxy-4-propenyl (cis) or Isoeugenol (cis) $2\text{-}CH_3O\text{-}4\text{-}(CH_3CH{=}CH)C_6H_3OH$	164.20		$134\text{-}5^{13}$, $80\text{-}1^{0.5}$		$1.0837^{20/4}$	1.5726^{20}	al, eth	B6[3], 4992
10968	Phenol, 3-methoxy-4-nitro $3\text{-}CH_3O\text{-}4\text{-}NO_2C_6H_3OH$	169.14	ye nd (al)		144			al	B6[2], 822
10969	Phenol, 3-methoxy-5-nitro $3\text{-}CH_3O\text{-}5\text{-}NO_2C_6H_3OH$	169.14	ye cr (al)		144			al	B6[3], 4347
10970	Phenol, 4-methoxy-2-nitro $4\text{-}CH_3O\text{-}2\text{-}NO_2C_6H_3OH$	169.14	og ye nd or mcl cr (al, lig)		80 (83)			al	B6[3], 4442
10971	Phenol, 4-methoxy-3-nitro $4\text{-}CH_3O\text{-}3\text{-}NO_2C_6H_3OH$	169.14	pa ye nd (w), cr (bz)		98-100			al	B6[2], 848
10972	Phenol, 5-methoxy-2-nitro $5\text{-}CH_3O\text{-}2\text{-}NO_2C_6H_3OH$	169.14	yesh nd (al)		95			al	B6[3], 4345
10973	Phenol, 2-(methylamino) $2\text{-}CH_3NHC_6H_4OH$	123.15	pl (bz-peth)		96-7			al, bz	B13[3], 761
10974	Phenol, 4-methyl amino,sulfate or Metol,Pictol $(4\text{-}CH_3NHC_6H_4OH)_2H_2SO_4$	344.37	wh nd(w)		250-60d			w, al	B13[3], 1007
10975	Phenol, 4(methyl amino)-2-nitro $2\text{-}NO_2\text{-}4\text{-}(CH_3NH)C_6H_3OH$	168.15	dk red-br nd (al)		113-4			al	B13[1], 186
10976	Phenol, 4-(2-methylaminopropyl) $4\text{-}[CH_3CH(NHCH_3)CH_2]C_6H_4OH$	165.24	cr (MeOH)		161			al, eth	B13[3], 1710
10977	Phenol, 4-(2-methyl-2-butyl) $4\text{-}[H_5C_2\text{-}C(CH_3)_2]\text{-}C_6H_4OH$	164.25	nd	262.5, 138^{15}	94-6				B6[3], 1965
10978	Phenol, 2-methyl-4-*tert*-butyl $2\text{-}CH_3\text{-}4\text{-}(CH_3)_3C\text{-}C_6H_3OH$	164.25	yesh	$235\text{-}7^{740}$, 132^{20}	27-8	$0.965^{20/4}$	1.5230^{20}	eth, ace, bz	B6[3], 1979

No.	Name, Synonyms, and Formula	Mol. wt.	Color, crystalline form, specific rotation and λ_{max} (log ε)	b.p. °C	m.p. °C	Density	n_D	Solubility	Ref.
10979	Phenol, 2-methyl-4(3-methyl-3-pentyl) $2\text{-}CH_3\text{-}4\text{-}[(C_2H_5)_2C(CH_3)]\text{-}C_6H_3OH$	192.30		$145\text{-}6^{11}$	35		1.5200^{25}	ace	C48, 1328
10980	Phenol, 2-methyl-5-isopropyl or Carvacrol $2\text{-}CH_3\text{-}5\text{-}i\text{-}C_3H_7C_6H_3OH$	150.22		237.7, $101\text{-}3^{10}$	1	$0.9772^{20/4}$	1.5230^{20}	al, eth, ace	$B6^3$, 1885
10981	Phenol, 2-methyl-5-isopropyl-3-nitro $2\text{-}CH_3\text{-}5\text{-}i\text{-}C_3H_7\text{-}3\text{-}ONC_6H_2OH$	195.22	yesh pr (bz), nd (dil al)		153			al, eth, bz, chl	$B7^3$,3412
10982	Phenol, 2-methyl-3,4,5-trichloro $2\text{-}CH_3\text{-}3,4,5\text{-}Cl_3C_6HOH$	211.48	nd (lig)		77			ace	$B6^3$, 1268
10983	Phenol, 3-methyl-6-*tert*-butyl $3\text{-}CH_3\text{-}6\text{-}(CH_3)_3C\text{-}C_6H_3OH$	164.25		224, 127^{11}	46-7	$0.922^{80/4}$	1.5250^{20}	al, eth, ace	$B6^3$, 1982
10984	Phenol, 4-methyl-2-*tert*-butyl $4\text{-}CH_3\text{-}2\text{-}(CH_3)_3C\text{-}C_6H_3OH$	164.25	nd (peth)	237, 111^{10}	55	$0.9247^{75/4}$	1.4969^{75}	eth, ace, bz	$B6^3$, 1978
10985	Phenol, 4-(3-methyl-1-butyl) $4\text{-}(CH_3)_2CHCH_2CH_2C_6H_4OH$	164.25	nd (w)	255, 126^{14}	93	$0.9579^{23/20}$	1.5050^{27}	al, eth	$B6^3$, 1960
10986	Phenol, 3-methyl-2,4,6-trichloro $3\text{-}CH_3\text{-}2,4,6\text{-}Cl_3\text{-}C_6HOH$	211.48	nd (w), pl (peth)	265, $162\text{-}3^{28}$	47			al, eth, chl	$B6^4$, 2070
10987	Phenol, 4-methyl-2,3,5-trichloro $4\text{-}CH_3\text{-}2,3,5\text{-}Cl_3C_6HOH$	211.48	nd (aa, lig)		66-7			ace	$B6^1$, 204
10988	Phenol, *o*-nitro $2\text{-}O_2NC_6H_4OH$	139.11	ye nd or pr (eth, al)	216, $96\text{-}7^{10}$	45-6	1.2942^{40}, 1.485^{14}	1.5723^{50}	al, eth, ace, bz, chl	$B6^4$, 1246
10989	Phenol, *m*-nitro $3\text{-}O_2NC_6H_4OH$	139.11	ye mcl (eth, aq HCl)	194^{70}	97			al, eth, ace, bz	$B6^4$, 1269
10990	Phenol, *p*-nitro $4\text{-}O_2NC_6H_4OH$	139.11	ye mcl pr (to)	279d sub	114-6	1.479^{20}		al, eth, ace, Py	$B6^4$, 1279
10991	Phenol, *p*-nitroso $4\text{-}ONC_6H_4OH$	123.11	pa ye rh nd (ace, bz)		d 144			al, eth, ace, bz	$B7^3$, 3367
10992	Phenol, *o*-octyl $4\text{-}C_8H_{17}C_6H_4OH$	206.33		169^{10}	41-2				$B6^3$, 2046
10993	Phenol, penta bromo Br_5C_6OH	488.59	mcl pr (aa), nd (al)	sub	229.5			al, bz	$B6^4$, 1069
10994	Phenol, penta chloro Cl_5C_6OH	266.34	mcl pr (al + 1w), nd (bz)	$309\text{-}10^{754}$d	174 (+ 1w), 191 (anh)	$1.978^{22/4}$		al, eth, bz	$B6^4$, 1025
10995	Phenol, penta fluoro C_6F_5OH	184.07		$72\text{-}3^{48}$			1.4263^{26}		$B6^4$, 782
10996	Phenol, penta methyl $(CH_3)_5C_6OH$	164.25	nd (al, peth, ace)	267	128			al	$B6^3$, 1991
10997	Phenol, *p*-pentyl $4\text{-}C_5H_{11}C_6H_4OH$	164.25		250.5, $119\text{-}20^{3}$	23	$0.960^{20/4}$	1.5272^{25}	al, eth	$B6^3$, 1950
10998	Phenol, *o*-phenyl or 2-Hydroxy biphenyl $2\text{-}HOC_6H_4C_6H_5$	170.21	nd (peth)	286, 145^{14}	58-60	$1.213^{25/4}$		al, eth, ace, bz, lig	$B6^3$, 3281
10999	Phenol, *m*-phenyl or 3-hydroxy biphenyl $3\text{-}HOC_6H_4C_6H_5$	170.21	nd (w or peth)	>300	78			al, eth, bz, peth	$B6^3$, 3311
11000	Phenol, *p*-phenyl or 4-Hydroxy biphenyl $4\text{-}HOC_6H_4C_6H_5$	170.21	nd or pl (dil al)	305-8 sub	165-7			al, eth, chl	$B6^3$, 3319
11001	Phenol, 4-propenyl $4\text{-}(CH_3CH{=}CH)C_6H_4OH$	134.18	pl (w), lf	250d, $140\text{-}5^{15}$	93-4			al, eth	$B6^3$, 2394
11002	Phenol, *o*-propionyl $2(CH_3CH_2CO)C_6H_4OH$	150.18		150^{80}, 115^{15}			1.5501^{20}	al, eth	$B8^3$, 373
11003	Phenol, *p*-propionyl $4\text{-}(CH_3CH_2CO)C_6H_4OH$	150.18	wh nd or pr (w)		149			al, eth	$B8^3$, 379
11004	Phenol, 2-propyl $2\text{-}C_3H_7C_6H_4OH$	136.19		220, 106.7^{10}		$1.015^{20/4}$		al, eth	$B6^3$, 1784
11005	Phenol, 3-propyl $3\text{-}C_3H_7C_6H_4OH$	136.19		228, 111.2^{10}	26	0.987^{20}	1.5223^{20}	al, eth	$B6^3$, 1787
11006	Phenol, 4-propyl $4\text{-}C_3H_7C_6H_4OH$	136.19		232.6, 111.7^{10}	22	$1.009^{20/4}$	1.5379^{25}	al	$B6^3$, 1783
11007	Phenol, 2-isopropyl or *o*-Cumenol $2\text{-}i\text{-}C_3H_7C_6H_4OH$	136.19		213-4	15-6	$1.012^{20/4}$	1.5315^{20}	al, eth, bz	$B6^3$, 1807
11008	Phenol, 2-isopropyl-5-methyl-4-nitroso or 4-Nitroso carvacrol $2\text{-}i\text{-}C_3H_7\text{-}5\text{-}CH_3\text{-}4\text{-}ON\text{-}C_6H_2OH$	179.22	yesh red		175			al, eth, chl	$B7^3$, 3412
11009	Phenol, 3-isopropyl or *m*-Cumenol $3\text{-}i\text{-}C_3H_7C_6H_4OH$	136.19		228	26		1.5261^{20}	eth	$B6^3$, 1810

PHYSICAL CONSTANTS OF ORGANIC COMPOUNDS (Continued)

No.	Name, Synonyms, and Formula	Mol. wt.	Color, crystalline form, specific rotation and λ_{max} (log ε)	b.p. °C	m.p. °C	Density	n_D	Solubility	Ref.
11010	Phenol, 4-isopropyl or *p*-Cumenol 4-*i*-$C_3H_7C_6H_4OH$	136.19	nd (peth)	$228\text{-}30^{745}$, $109\text{-}10^{10}$	62-3	0.990^{20}	1.5228^{20}	al	$B6^3$, 1810
11011	Phenol, seleno or Selenyl benzene C_6H_5SeH	157.07		183.6		1.4865^{15}		al, eth	$B6^4$, 1777
11012	Phenol, 2,3,4,6-tetrabromo 2,3,4,6-Br_4CHOH	409.70	nd (al, aa)	sub	113-4 (120)			al, bz	$B6^3$, 766
11013	Phenol, 2,3,4,5-tetrachloro 2,3,4,5-Cl_4C_6HOH	231.89	nd (peth, sub)	sub	116-7			al	$B6^4$, 1020
11014	Phenol, 2,3,4,6-tetrachloro 2,3,4,6-Cl_4C_6HOH	231.89	nd (lig, aa)	150^{15}	70			al, bz, chl, lig	$B6^4$, 1021
11015	Phenol, 2,3,5,6-tetrachloro 2,3,5,6-Cl_4C_6HOH	231.89	lf (lig)		115			bz	$B6^4$, 1025
11016	Phenol, 2,3,4,5-tetramethyl 2,3,4,5-$(CH_3)_4C_6HOH$	150.22		266	86-7			al, eth	$B6^3$, 1918
11017	Phenol, 2,3,4,6-tetramethyl or Isodurenol 2,3,4,6-$(CH_3)_4C_6HOH$	150.22	cr (peth)	230-50	80-1			al	$B6^3$, 1919
11018	Phenol, 2,3,5,6-tetramethyl or Durenol 2,3,5,6-$(CH_3)_4C_6HOH$	150.22	nd (lig), pr (al)	247	118-9			aa, peth	$B6^3$, 1919
11019	Phenol, 2,3,4,6-tetranitro 2,3,4,6-$(NO_2)_4C_6HOH$	274.10	lt ye nd (chl)	exp	140d			w	$B6^3$, 973
11020	Phenol, 2,4,6-triamino 2,4,6-$(H_2N)_3C_6H_2OH$	139.16		257				w, al, eth	$B13^3$, 1373
11021	Phenol, 2,3,4-tribromo 2,3,4-$Br_3C_6H_2OH$	330.80	nd or pl (w, lig)		94-5			al, eth, ace, lig	$B6^2$, 192
11022	Phenol, 2,4,5-tribromo 2,4,5-$Br_3C_6H_2OH$	330.80	nd (lig)		87			al, lig	$B6^3$, 760
11023	Phenol, 2,4,6-tribromo or Bromol 2,4,6-$Br_3C_6H_2OH$	330.80	nd (al), pr (bz), cr (aa + l)	$282\text{-}90^{764}$ (sub)	95-6	$2.55^{20/20}$		al, eth	$B6^4$, 1067
11024	Phenol, 3,4,5-tribromo 3,4,5-$Br_3C_6H_2OH$	330.80	tcl (bz-lig)		129			al, bz, aa	$B6^2$, 195
11025	Phenol, 2,4,6-tri-*tert*-butyl 2,4,6-$(t\text{-}C_4H_9)_3C_6H_2OH$	262.44	cr (al, peth)	278, 130^{15}	131	$0.864^{27/4}$		ace	$B6^3$, 2094
11026	Phenol, 2,3,4-trichloro 2,3,4-$Cl_3C_6H_2OH$	197.45	nd (bz, lig, sub)	sub	83.5			al, eth, bz, aa	$B6^3$, 716
11027	Phenol, 2,3,5-trichloro 2,3,5-$Cl_3C_6H_2OH$	197.45	nd (al)	$248\text{-}9^{250}$	62			al, eth	$B6^3$, 716
11028	Phenol, 2,3,6-trichloro 2,3,6-$Cl_3C_6H_2OH$	197.45	nd (dil al, lig)		58			al, eth, bz, aa, lig	$B6^4$, 962
11029	Phenol, 2,4,5-trichloro 2,4,5-$Cl_3C_6H_2OH$	197.45	nd (al, peth)	sub	68-70			al, lig	$B6^4$,962
11030	Phenol, 2,4,6-trichloro 2,4,6-$Cl_3C_6H_2OH$	197.45	rh nd (aa)	246	69.5	$1.4901^{75/4}$		al, eth	$B6^4$, 1005
11031	Phenol, 3,4,5-trichloro 3,4,5-$Cl_3C_6H_2OH$	197.45	nd(lig)	$271\text{-}7^{746}$	101			eth	$B6^3$, 729
11032	Phenol, 2,3,5-triido 2,3,5-$I_3C_6H_2OH$	471.80	nd (peth, bz-lig)		114			al, eth, ace, bz	$B6^4$, 1085
11033	Phenol, 2,4,6-triido 2,4,6-$I_3C_6H_2OH$	471.80	nd (dil al)	sub, d	158-9			eth, ace	$B6^4$, 1085
11034	Phenol, 2,4,5-trimethyl or Psuedocumenol 2,4,5-$(CH_3)_3C_6H_2OH$	136.19	nd (lig)	232	72			al, eth	$B6^3$, 1831
11035	Phenol, 2,4,5-trimethyl,acetate 2,4,5-$(CH_3)_3C_6H_2(O_2CCH_3)$	178.23	nd (peth)	245-6	34			al, eth	$B6^2$, 482
11036	Phenol, 2,4,6-trimethyl or Mesitol 2,4,6-$(CH_3)_3C_6H_2OH$	136.19	nd (peth, MeOH)	221 sub	72			al, eth	$B6^3$, 1835
11037	Phenol, 2,3,6-trinitro 2,3,6-$(O_2N)_3C_6H_2OH$	229.11	ye nd (w)		119			al, eth, bz, aa	$B6^2$, 253
11038	Phenol, 2,4,5-trinitro 2,4,5-$(O_2N)_3C_6H_2OH$	229.11	wh nd (w, dil al)		96			al, eth, bz, aa	$B6^4$, 1388
11039	Phenol, 2,4,6-trinitro or Picric acid 2,4,6-$(O_2N)_3C_6H_2OH$	229.11	ye lf (w), pr (eth), pl (al)	sub exp>300	122-3		1.763	al, eth, ace, bz, aa	$B6^4$, 1388
11040	Phenol, *o*-vinyl 2-$(CH_2{=}CH)C_6H_4OH$	120.15	nd	101^{14}	29-30	$1.0609^{18/4}$	1.5851^{20}	al, eth	$B6^3$, 2383
11041	Phenol, *m*-vinyl 3-$(CH_2{=}CH)C_6H_4OH$	120.15		$114\text{-}6^{16}$		$1.0353^{21/4}$	1.5804^{21}		$B6^3$, 2385

No.	Name, Synonyms, and Formula	Mol. wt.	Color, crystalline form, specific rotation and λ_{max} (log ε)	b.p. °C	m.p. °C	Density	n_D	Solubility	Ref.
11042	Phenolphthalein or 2,2-bis(4-hydroxyphenyl)phthalide $C_{20}H_{14}O_4$	318.33	wh rh nd		262-3	$1.277^{32/4}$	$1.277^{32/}{}_4$	al, eth, ace, chl, Py	B18^{4}, 1945
11043	Phenolphthalein-3′,3′′,5′,5′′-tetrabromo $C_{20}H_{10}Br_4O_4$	633.91	nd (al, eth, aa)		295-7			eth	B18^{4}, 1948
11044	Phenolphthalein-3′,3′′,5′,5′′-tetrachloro $C_{20}H_{10}Cl_4O_4$	456.11	cr (bz, aa)		225			al, ace, bz	B18^{2}, 123
11045	Phenolphthalein-3′,3′′,5′,5′′-tetraiodo or Iodophen $C_{20}H_{10}I_4O_4$	821.92	amor		227-9d	$2.0246^{22/22}$			B18^{4}, 1949
11046	Phenolphthalin $C_{20}H_{16}O_4$	320.34	nd (w)		229-32			al	B10^{3}, 2013
11047	Phenolsulfonphthalein or Phenol red $C_{19}H_{14}O_5S$	354.38	dk red nd or pl		>300				B19^{4}, 1128
11048	Phenothiazine $C_{12}H_9NS$	199.27	ye pr (al), ye lf or pl (to)	371, 290^{40} (sub 130^{1})	186-9			al, eth, ace, bz	B27^{2}, 32
11049	Phenothiazine, 2,8-dinitro,5-oxide $C_{12}H_7N_3O_5S$	305.26	ye-red lf (aa)		d				B27^{1}, 229
11050	Phenothiazine, 10-(2-diethylaminoethyl) $C_{18}H_{22}N_2S$	298.45	ye oil	$195\text{-}208^{4-5}$				dil HCl	Am66, 888
11051	Phenothiazine, 10-(2-dimethylaminopropyl) or Promethazine $C_{17}H_{20}N_2S$	284.42		$190\text{-}3^{0.5}$	60				C42, 575
11052	Phenothiazine, 10-(2-dimethylaminopropyl), hydrochloride $C_{17}H_{20}N_2S.HCl$	320.88			230-2			w, al, chl	C42, 575
11053	Phenothiazine, 10-phenyl $C_{18}H_{13}NS$	275.37	ye pr (al)		89-90			al	B27^{1}, 227
11054	Phenothiazine, 10-octadecyl $C_{30}H_{45}NS$	451.75		$290\text{-}300^{0.5}$	$51.0^{-2.5}$				CAS38, 3985
11056	Phenothiazine, N-propyl $C_{15}H_{15}NS$	241.35		$155\text{-}65^{0.8}$	48-9				CAS53, 18045
11057	Phenoxanthin $C_{12}H_8OS$	200.25	nd, cr (MeOH)	311^{745}, $183\text{-}4^{12}$	59-60			al, eth, ace	B19^{4}, 341
11058	Phenoxazine or Dibenzoxazine $C_{12}H_9NO$	183.21	lf (dil al, bz)	d	156			al, eth, bz, aa	B27^{1}, 223
11059	Phenoxy acetamide $C_6H_5OCH_2CONH_2$	151.16	nd (w al)		101.5			al	B6^{4}, 638
11060	Phenoxyacetic acid $C_6H_5OCH_2CO_2H$	152.15	nd or pl (w)	285d	98-9			w, al, eth, bz, aa	B6^{4}, 634
11061	Phenoxyacetic acid, anhydride $(C_6H_5OCH_2CO)_2O$	286.29	lf (eth)		67-9			bz	B6, 162
11062	Phenoxyacetic acid, 2-bromo $2\text{-}BrC_6H_4OCH_2CO_2H$	231.05	nd (dil al), cr (w)		142.5			al, eth	B6^{4}, 1040
11063	Phenoxyacetic acid, 4-bromo $4\text{-}Br\text{-}C_6H_4OCH_2COOH$	231.05	pr (al), cr (w)		161-2			al, eth	B6^{4}, 1052
11064	Phenoxyacetic acid, 2-carboxy or Salicyl acetic acid $(2\text{-}HO_2CC_6H_4O)CH_2CO_2H$	196.16	nd (w)		190-2			al, eth, ace, aa, chl	B10^{3}, 106
11065	Phenoxyacetic acid, 3-carboxy $(3\text{-}HO_2CC_6H_4O)CH_2CO_2H$	196.16			206-7				B10^{1}, 65
11066	Phenoxyacetic acid, 4-carboxy $(4\text{-}HO_2CC_6H_4O)CH_2CO_2H$	196.16	nd (ace, w)		280-2			al, eth, ace, bz, chl	B10^{2}, 94
11067	Phenoxyacetic acid, 2-chloro $2\text{-}ClC_6H_4OCH_2CO_2H$	186.59	nd (w, al)		148-9			w, al	B6^{4}, 796
11068	Phenoxyacetic acid, 3-chloro $3\text{-}ClC_6H_4OCH_2CO_2H$	186.59	cr (w)		110				B6^{4}, 816
11069	Phenoxyacetic acid, 4-chloro $4\text{-}ClC_6H_4OCH_2COOH$	186.59	pr or nd (w)		156-7				B6^{4}, 845
11070	Phenoxyacetic acid, 4-chloro-2-methyl $[4\text{-}Cl\text{-}2(CH_3)C_6H_3O]CH_2CO_2H$	200.62			120			al, eth, bz	B6^{4}, 1991
11071	Phenoxyacetic acid, 2,4-dichloro $(2,4\text{-}Cl_2\text{-}C_6H_3O)CH_2CO_2H$	221.04	cr (bz)	$160^{0.4}$	140-1			al	B6^{4}, 908
11072	Phenoxyacetic acid, 2,4-dinitro $(2,4\text{-}(NO_2)_2C_6H_3O)CH_2CO_2H$	242.14	pa ye pr (w)		147-8			al	B6^{2}, 244
11073	Phenoxyacetic acid, 3,5-dinitro $(3,5\text{-}(NO_2)_2C_6H_3O)CH_2CO_2H$	242.14	pa br cr pw		207			al	B6, 259

No.	Name, Synonyms, and Formula	Mol. wt.	Color, crystalline form, specific rotation and λ_{max} (log ε)	b.p. °C	m.p. °C	Density	n_D	Solubility	Ref.
11074	Phenoxyacetic acid, ethyl ester or Ethyl phenoxy acetate $C_6H_5OCH_2CO_2C_2H_5$	180.20	....	250-1, 136^{19}	......	1.104^{18}	....	al, eth	$B6^4$, 635
11075	Phenoxyacetic acid, 3-hydroxy $(3\text{-}HOC_6H_4O)CH_2CO_2H$	168.15	nd or pr (w, to)	....	158-9	....	....	al	B6, 817
11076	Phenoxy acetic acid, 3-hydroxy, ethyl ester or Ethyl-3-hydroxy phenoxy acetate $(3\text{-}HOC_6H_4O)CH_2CO_2C_2H_5$	196.20	pr (w, bz)	274d, $170\text{-}3^{11}$	55	....	....	....	B6, 817
11077	Phenoxy acetic acid, 4-hydroxy $(4\text{-}HOC_6H_4O)CH_2CO_2H$	168.15	nd (to), pr (+ ½w)	....	154 (hyd)	....	....	....	B6, 847
11078	Phenoxy acetic acid, 2-methoxy $(2\text{-}CH_3OC_6H_4O)CH_2CO_2H$	182.18	nd (w)	....	123-5 (129)	....	....	w, al, eth, bz, aa	$B6^3$, 4234
11079	Phenoxy acetic acid, 3-methoxy $(3\text{-}CH_3OC_6H_4O)CH_2CO_2H$	182.18	nd (w)	....	118	....	....	....	$B6^3$, 4323
11080	Phenoxy acetic acid, methyl ester or Methyl phenoxy acetate $C_6H_5OCH_2CO_2CH_3$	166.18	....	245, 130^{14}	......	$1.1493^{20/4}$	1.5155^{20}	al, eth	$B6^4$, 635
11081	Phenoxy acetic acid, 2-methyl or o-Cresoxyacetic acid $(2\text{-}CH_3C_6H_4O)CH_2CO_2H$	166.18	lf (w)	....	157	....	....	al	$B6^2$, 331
11082	Phenoxy acetic acid, 3-methyl or m-Cresoxyacetic acid $(3\text{-}CH_3C_6H_4O)CH_2CO_2H$	166.18	nd (w)	....	103-4	....	....	al, bz	$B6^2$, 353
11083	Phenoxy acetic acid, 4-methyl or p-Cresoxyacetic acid $(4\text{-}CH_3C_6H_4O)CH_2CO_2H$	166.18	nd (w)	....	136	....	....	al, bz	$B6^2$, 380
11084	Phenoxy acetic acid, 2-nitro $(2\text{-}O_2NC_6H_4O)CH_2CO_2H$	197.15	pr (w)	....	158	....	....	al, eth	$B6^3$, 804
11085	Phenoxy acetic acid, 2-*iso*-propyl-5-methyl or Thymoxy acetic acid $(2\text{-}i\text{-}C_3H_7\text{-}5\text{-}CH_3C_6H_3O)CH_2CO_2H$	208.26	nd (dil al), cr (bz)	....	149-50	....	....	al, eth	$B6^3$, 1902
11086	Phenoxy acetic acid, 2,4,6-tribromo $[2,4,6\text{-}(Br)_3C_6H_2O]CH_2CO_2H$	388.84	nd (dil al)	....	200	....	....	al, eth	$B6^4$, 1068
11087	Phenoxy acetic acid, 2,4,5-trichloro $[2,4,5\text{-}(Cl)_3\text{-}C_6H_2O]CH_2CO_2H$	255.48	cr (bz)	....	157-8	....	....	al	$B6^4$, 973
11088	Phenoxy acetic acid, 2,4,6-trichloro $[2,4,6\text{-}(Cl)_3\text{-}C_6H_2O]CH_2CO_2H$	255.48	cr (al)	....	177	....	....	al	$B6^4$, 1011
11089	Phenoxy acetonitrile $C_6H_5OCH_2CN$	133.15	....	239-40, 128^{17}	......	$1.0991^{20/4}$	1.5246^{20}	al, eth	$B6^4$, 640
11090	Phenoxy acetyl chloride $C_6H_5OCH_2COCl$	170.60	....	225-6, 111^{13}	......	....	....	eth	$B6^3$, 613
11091	β-Phenoxy ethyl ether $(C_6H_5OCH_2CH_2)_2O$	258.32	nd (dil al)	....	66-7	....	....	al	$B6^4$, 573
11092	Phenyl acetaldehyde or α-Tolualdehyde $C_6H_5CH_2CHO$	120.15	(w)	195, 88^{18}	33-4	$1.0272^{20/4}$	1.5255^{20}	al, eth, ace	$B7^3$, 1003
11093	Phenyl acetamide $C_6H_5CH_2CONH_2$	135.17	pl or lf (w)	....	157	....	....	....	$B9^3$, 2193
11094	Phenyl acetamide, 3,4-dimethoxy $3,4\text{-}(CH_3O)_2C_6H_3CH_2CONH_2$	195.22	cr (w)	....	145-7	....	....	al, eth	$B10^3$, 1463
11095	Phenyl acetamide, α-hydroxy (dl) or Mandelamide $C_6H_5CH(OH)CONH_2$	151.16	pl (bz, al)	....	134-5	....	....	al, bz	$B10^3$, 469
11096	Phenyl acetamide, α-hydroxy-N-ethyl (d) or *l*-N-Ethoxy mandelamide $C_6H_5CH(OH)CONHC_2H_5$	179.22	pl (chl-peth), $[\alpha]^{18/}{}_D$ −103.6 (ace)	....	65-6	....	....	al, eth, ace, bz, chl	$B10^2$, 117
11097	Phenyl acetamide, α-hydroxy-N-ethyl (dl) or dl-N-Ethyl mandelamide $C_6H_5CH(OH)CONHC_2H_5$	179.22	pl (bz-peth)	....	53-4	....	....	w, al, eth, ace, bz	$B10^1$, 89
11098	Phenyl acetamide, 2-hydroxy $2\text{-}HOC_6H_4CH_2CONH_2$	151.16	lf (al-chl)	....	118	....	....	....	B10, 188
11099	Phenyl acetamide, N-methyl $C_6H_5CH_2CONHCH_3$	149.20	cr (bz)	....	58	....	....	al, eth, chl	$B9^3$, 2195
11100	Phenyl acetamide, N-phenyl or α-Phenyl acetamide $C_6H_5CH_2CONHC_6H_5$	211.27	pr (al)	....	117-8	....	....	al, eth	$B12^3$, 521
11101	Phenyl acetamide, 4-isopropyl $4\text{-}i\text{-}C_3H_7C_6H_4CH_2CONH_2$	177.25	pl (bz)	....	170	....	....	al	$B9^3$, 2529
11102	Phenyl acetic acid $C_6H_5CH_2CO_2H$	136.15	lf, pl (peth)	265.5, $144\text{-}5^{12}$	77	$1.091^{77/4}$, 1.228^{6}	....	al, eth, ace	$B9^3$, 2169
11103	Phenyl acetic acid, anhydride $(C_6H_5CH_2CO)_2O$	254.29	pr or nd (eth)	$195\text{-}8^{12}$	71-2	....	....	eth, chl	$B9^3$, 2190

PHYSICAL CONSTANTS OF ORGANIC COMPOUNDS (Continued)

No.	Name, Synonyms, and Formula	Mol. wt.	Color, crystalline form, specific rotation and λ_{max} (log ε)	b.p. °C	m.p. °C	Density	n_D	Solubility	Ref.
11104	Phenyl acetic acid, 4-amino or Aminophenyl acetic acid p-$H_2NC_6H_4CH_2CO_2H$	151.16	pl (w)		199-200d				B14^{3}, 1182
11105	Phenyl acetic acid, α-bromo or α-Bromophenyl acetic acid $C_6H_5CH(Br)CO_2H$	215.05			82-3				B9^{3}, 2276
11106	Phenyl acetic acid, α-bromo, ethyl ester or Ethyl-α-bromophenyl acetate $C_6H_5CH(Br)CO_2C_2H_5$	243.10		145-55^{18}					B9^{3}, 2276
11107	Phenyl acetic acid, 2-bromo or 2-Bromotoluic acid 2-$BrC_6H_4CH_2CO_2H$	215.05	cr (aa)		105-6			al, eth, aa	B9^{3}, 2273
11108	Phenyl acetic acid, 3-bromo or 3-Bromotoluic acid 3-$BrC_6H_4CH_2CO_2H$	215.05	nd (w)		100-1				B9^{3}, 2274
11109	Phenyl acetic acid, 4-bromo 4-$BrC_6H_4CH_2CO_2H$	215.05	nd (w)	sub	116			al, eth	B9^{3}, 2275
11110	Phenyl acetic acid, *iso*-butyl ester or *iso*-Butyl phenyl acetate, Eglantine $C_6H_5CH_2CO_2$-i-C_4H_9	192.26		247, 123-5^{14}		0.999^{18}		al, eth	B9^{3}, 2180
11111	Phenyl acetic acid, 2-carboxy 2-$(HO_2C)C_6H_4CH_2CO_2H$	180.16	cr (w, eth)		185-7			al, w	B9^{3}, 4266
11112	Phenyl acetic acid, 3-carboxy 3-$(HO_2C)C_6H_4CH_2CO_2H$	180.16	nd or pl (w)		184-5			al, eth	B9^{3}, 4269
11113	Phenyl acetic acid, 4-carboxy 4-$(HO_2C)C_6H_4CH_2CO_2H$	180.16	cr (dil al)		239-41			al, eth, bz	B9^{3}, 4269
11114	Phenyl acetic acid, α-chloro (d) d-$C_6H_5CHClCOOH$	170.60	cr (peth), $[\alpha]^{20/}{}_D$ + 192 (bz)		60-1			al, eth, bz, chl	B9^{3}, 2265
11115	Phenyl acetic acid, α-chloro (dl) dl-$C_6H_5CHClCOOH$	170.60	lf (peth)		78			al, eth	B9^{3}, 2265
11116	Phenyl acetic acid, α-chloro (l) l-$C_6H_5CHClCOOH$	170.60	nd (peth), $[\alpha]^{18/}{}_D$ −191.3 (bz)		61			al, eth, bz, chl	B9^{2}, 307
11117	Phenyl acetic acid, 2-chloro 2-$ClC_6H_4CH_2CO_2H$	170.60	nd (w)		96			al	B9^{3}, 2262
11118	Phenyl acetic acid, 3-chloro or 3-Chloro phenyl acetic acid 3-$ClC_6H_4CH_2CO_2H$	170.60	pl (dil al), nd (hep)		77-8			eth	B9^{3}, 2263
11119	Phenyl acetic acid, 4-chloro 4-$ClC_6H_4CH_2CO_2H$	170.60	nd (w)		105-6			al, eth, bz	B9^{3}, 2264
11120	Phenyl acetic acid, 2,5-dihydroxy or Homogentisic acid $2,5$-$(HO)_2C_6H_3CH_2CO_2H$	168.15	pr (w + 1), lf (al-chl)		152-4			al, eth, w	B10^{3}, 1456
11121	Phenyl acetic acid, 3,4-dihydroxy $3,4$-$(HO)_2C_6H_3CH_2CO_2H$	168.15			131-2			w, al, eth	B10^{3}, 1458
11122	Phenyl acetic acid, 3,4-dimethoxy or Homoveratric acid $3,4$-$(CH_3O)_2C_6H_3CH_2CO_2H$	196.20	nd (w + 1), cr (bz-peth)		98-9 (anh), 80-2 (+ w)			w, al, eth	B10^{3}, 1459
11123	Phenyl acetic acid, 2,4-dimethyl-α-hydroxy or 2,4-Dimethyl mandelic acid $2,4$-$(CH_3)_2C_6H_3$-$CH(OH)CO_2H$	180.20	rh (w), lf (peth-chl)		119			al, eth, chl, w	B10^{3}, 613
11124	Phenyl acetic acid, 2,5-dimethyl-α-hydroxy or 2,5-Dimethyl mandelic acid $2,5$-$(CH_3)_2C_6H_3CH(OH)CO_2H$	180.20	nd or pr (bz)		116-7			al, eth, chl	B10^{3}, 612
11125	Phenyl acetic acid, 3,4-dimethyl-α-hydroxy or 3,4-Dimethyl mandelic acid $3,4$-$(CH_3)_2C_6H_3CH(OH)CO_2H$	180.20	lf (bz)		135			al, bz	B10^{3}, 611
11126	Phenyl acetic acid, 2,4-dinitro or 2,4-Dinitrophenyl acetic acid $2,4$-$(NO_2)_2C_6H_3CH_2CO_2H$	226.15			179-80d			al, eth	B9^{3}, 2292
11127	Phenyl acetic acid, 2,4-dinitro, ethyl ester $2,4$-$(NO_2)_2C_6H_3CH_2CO_2C_2H_5$	254.20	nd (w)		37			al, eth	B9^{3}, 2292
11128	Phenyl acetic acid, 2,6-dinitro $2,6$-$(NO_2)_2C_6H_3CH_2CO_2H$	226.15	ye lf (aa)		201-2d			al	B9^{1}, 185
11129	Phenyl acetic acid, 2-ethoxy 2-$(C_2H_5O)C_6H_4CH_2CO_2H$	180.20	nd (lig), cr (w)		103-4				B10^{1}, 82
11130	Phenyl acetic acid, ethyl ester or Ethyl phenyl acetate $C_6H_5CH_2CO_2C_2H_5$	164.20		227, 120-1^{20}		1.0333$^{20/4}$	1.4980^{20}	**al, eth**	B9^{3}, 2176
11131	Phenyl acetic acid, α-hydroxy-(D) or *l*-Mandelic acid $C_6H_5CH(OH)CO_2H$	152.15	ta $[\alpha]^{20/}{}_D$ −158 (w, c=2.5)		133-5	1.341		w, al, eth, aa, chl	B10^{3}, 447

PHYSICAL CONSTANTS OF ORGANIC COMPOUNDS (Continued)

No.	Name, Synonyms, and Formula	Mol. wt.	Color, crystalline form, specific rotation and λ_{max} (log ε)	b.p. °C	m.p. °C	Density	n_D	Solubility	Ref.
11132	Phenyl acetic acid, α-hydroxy (DL) or dl-Mandelic acid.. $C_6H_5CH(OH)CO_2H$	152.15	pl (w), rh (bz)	d	121-3	$1.300^{20/4}$	....	al, eth	$B10^3$, 448
11133	Phenyl acetic acid, α-hydroxy (L +) or d-Mandelic acid . $C_6H_5CH(OH)CO_2H$	152.15	pl (w), $[\alpha]^{20/}_{D}$ + 156.6 (w, c=2.9)	....	134-5	....	....	w, al, eth, chl	$B10^3$, 445
11134	Phenyl acetic acid, α-hydroxy,acetate(D) or l-Acetyl mandelic acid .... $C_6H_5CH(O_2CCH_3)CO_2H$	194.19	nd (w + 1) $[\alpha]^{20/}_{D}$ −156−4 (ace)	....	96-8 (anh)	....	....	al, eth, ace, bz, chl	$B10^3$, 453
11135	Phenyl acetic acid, α-hydroxy,acetate (DL) or dl-acetyl mandelic acid........ $C_6H_5CH(CO_2CH_3)COOH$	194.19	cr (bz), amor (chl-peth), cr (w + 1)	....	38-9 (+ 1w), 79-80 (anh)	....	....	al, eth, bz, chl	$B10^3$, 453
11136	Phenyl acetic acid, α-hydroxy-4-bromo or *p*-Bromo mandelic acid .... $4\text{-}Br\text{-}C_6H_4CH(OH)CO_2H$	231.05	nd (bz)	....	118-20	....	....	al, eth, bz, chl	$B10^3$, 480
11137	Phenyl acetic acid, α-hydroxy-4-chloro or 4-Chloro mandelic acid .... $4\text{-}Cl\text{-}C_6H_4CH(OH)CO_2H$	186.59	nd (bz)	....	119-20	....	....	w, al	$B10^1$, 479
11138	Phenyl acetic acid, α-hydroxy, ethyl ester (D-) or l-Ethyl mandelate........ $C_6H_5CH(OH)CO_2C_2H_5$	180.20	(peth), $[\alpha]^{20/}_{D}$ −128.4 (chl, c=6.7)	150^{20}	35	$1.1270^{20/4}$	....	al, eth	$B10^3$, 456
11139	Phenyl acetic acid, α-hydroxy, ethyl ester (DL) or dl-Ethyl mandelate........ $C_6H_5CH(OH)CO_2C_2H_5$	180.20	nd (peth)	253-5, 141^{15}	37	....	....	al, eth, lig	$B10^3$, 457
11140	Phenyl acetic acid, α-hydroxy, ethyl ester (L +) or d-Ethyl mandelate........ $C_6H_5CH(OH)CO_2C_2H_5$	180.20	(peth), $[\alpha]^{20/}_{D}$ + 205 (CS_2, c=0.7)	....	33	$1.1270^{20/4}$	....	al, eth	$B10^3$, 456
11141	Phenyl acetic acid, α-hydroxy, methyl ester (D-) or l-Methyl mandelate .... $C_6H_5CH(OH)CO_2CH_3$	166.18	cr (peth), $[\alpha]^{20/}_{D}$ −131.5 (w)	160^{22}	55	1.1756^{20}	....	w, al, ace, bz, chl	$B10^3$, 454
11142	Phenyl acetic acid, α-hydroxy, methyl ester (DL) or dl-Methyl mandelate .... $C_6H_5CH(OH)CO_2CH_3$	166.18	pl (bz-lig)	250d, 144^{20}	58	1.1756^{20}	....	al, chl	$B10^3$, 454
11143	Phenyl acetic acid, α-hydroxy, methyl ester (L +) or d-Methyl mandelate .... $C_6H_5CH(OH)CO_2CH_3$	166.18	cr (peth), $[\alpha]^{20/}_{D}$ −133.6 (w)	160^{32}	55.5	1.1756^{20}	....	w, al, ace, bz, chl	$B10^3$, 454
11144	Phenyl acetic acid, α-hydroxy-4-*iso*-propyl (d) or d-4-*iso*-Propyl mandelic acid........ $4\text{-}i\text{-}C_3H_7C_6H_4CH(OH)CO_2H$	194.23	lf (w), $[\alpha]^{17/}_{D}$ + 135 (abs al, c = 4)	....	153-4	....	....	al, eth	B10, 279
11145	Phenyl acetic acid, α-hydroxy-4-*iso*-propyl (dl) or dl-4-iso-Propyl mandelic acid .... $4\text{-}i\text{-}C_3H_7C_6H_4CH(OH)CO_2H$	194.23	nd (w)	....	159-60	....	....	al, eth	$B10^3$, 629
11146	Phenyl acetic acid, α-hydroxy-4-*iso*propyl (l) or l-4-*iso*-propyl mandelic acid........ $4\text{-}i\text{-}C_3H_7C_6H_4CH(OH)CO_2H$	194.23	ta (20% al), $[\alpha]^{17/}_{D}$ −135 (abs al, c=4)	....	153-4	....	....	al, eth	B10, 279
11147	Phenyl acetic acid, 2-hydroxy .... $2\text{-}HOC_6H_4CH_2CO_2H$	152.15	....	240-3d	147-9	....	....	eth	$B10^3$, 422
11148	Phenyl acetic acid, 2-hydroxy-5-nitro .... $2\text{-}(HO)\text{-}5(NO_2)C_6H_3CH_2CO_2H$	197.15	nd	....	160-2	....	....	w, al, eth	B10, 189
11149	Phenyl acetic acid, 2-hydroxy-5-nitro, ethyl ester or Ethyl-2-hydroxy-2-nitro phenyl acetate .... $5\text{-}NO_2\text{-}2\text{-}HOC_6H_3CH_2CO_2C_2H_5$	225.20	pr or pl (al)	....	154-5	....	....	bz, chl	B10, 189
11150	Phenyl acetic acid, 2-hydroxy, hydrazide .... $2\text{-}HO\text{-}C_6H_4\text{-}CH_2CONHNH_2$	166.18	lf (chl, bz), nd (al)	....	154	....	....	bz, al	$B10^2$, 112
11151	Phenyl acetic acid, 3-hydroxy .... $3\text{-}HOC_6H_4CH_2CO_2H$	152.15	nd (bz-lig)	190^{11}	131-4	....	....	w, al, eth, bz, chl	$B10^3$, 428
11152	Phenyl acetic acid, 4-hydroxy .... $4\text{-}HOC_6H_4CH_2CO_2H$	152.15	nd (w)	sub	149-51	....	....	al, eth	$B10^3$, 430
11153	Phenyl acetic acid, 3-hydroxy-4-methoxy or Homoisovanillic acid .... $3\text{-}(HO)\text{-}4\text{-}(CH_3O)C_6H_3CH_2CO_2H$	182.18	....	....	130-1	....	....	w, al, eth	$B10^3$, 1458

PHYSICAL CONSTANTS OF ORGANIC COMPOUNDS (Continued)

No.	Name, Synonyms, and Formula	Mol. wt.	Color, crystalline form, specific rotation and λ_{max} (log ε)	b.p. °C	m.p. °C	Density	n_D	Solubility	Ref.
11155	Phenyl acetic acid, 2-mercapto $2\text{-}HSC_6H_4CH_2CO_2H$	168.21	pl (w, bz-lig)		96-7			al, eth, bz	$B10^1$, 1782
11156	Phenyl acetic acid, α-methoxy (D-) $C_6H_5CH(OCH_3)CO_2H$	166.18	nd (peth), $[\alpha]^{13/}{}_D$ -150 (al)		63-4			w, al, ace	$B10^3$, 451
11157	Phenyl acetic acid, α-methoxy (DL) $C_6H_5CH(OCH_3)CO_2H$	166.18	pl (lig)		71-2			al, eth	$B10^3$, 452
11158	Phenyl acetic acid, 2-methoxy $2\text{-}CH_3OC_6H_4CH_2CO_2H$	166.18	nd (w)	$100\text{-}1^2$	123			al, eth, ace, bz, chl	$B10^3$, 422
11159	Phenyl acetic acid, 4-methoxy or Homoanisic acid $4\text{-}CH_3OC_6H_4CH_2CO_2H$	166.18	pl (w)	$138\text{-}40^{2\text{-}3}$	86			al, eth	$B10^3$, 431
11160	Phenyl acetic acid, 4-methoxy-2-nitro $4\text{-}(CH_3O)\text{-}2\text{-}(NO_2)C_6H_3CH_2CO_2H$	211.17	ye nd (50% al)		157-8d			al, aa	$B10^2$, 113
11161	Phenyl acetic acid, methyl ester or Methyl phenyl acetate $C_6H_5CH_2CO_2CH_3$	150.18		$218, 131\text{-}2^{50}$		$1.0633^{16/16}$	1.5075^{20}	al, eth, ace	$B9^3$, 2175
11162	Phenyl acetic acid, 2-methyl or 2-Tolylacetic acid $(2\text{-}CH_3C_6H_4)CH_2CO_2H$	150.18	nd (w)		88-90				$B9^3$, 2426
11163	Phenyl acetic acid, 3-methyl or 3-Methyl-α-toluic acid $3\text{-}CH_3C_6H_4CH_2CO_2H$	150.18	nd (w)	$120\text{-}3^{26}$	62				$B9^3$, 2429
11164	Phenyl acetic acid, 4-methyl or 4-Methyl phenyl acetic acid $4\text{-}CH_3C_6H_4CH_2CO_2H$	150.18	nd or pl (al, w)	265-7 sub	91-3			al, eth, bz, chl	$B9^3$, 2432
11165	Phenyl acetic acid, 2-nitro $2\text{-}NO_2C_6H_4CH_2CO_2H$	181.15	nd (w), pl (dil al)		141-2			al	$B9^3$, 2282
11166	Phenyl acetic acid, 3-nitro $3\text{-}NO_2C_6H_4CH_2CO_2H$	181.15	nd (w)		122			al	$B9^3$, 2283
11167	Phenyl acetic acid, 4-nitro $4\text{-}NO_2C_6H_4CH_2CO_2H$	181.15	pa ye nd (w)		153-5			al, eth, bz	$B9^3$, 2284
11168	Phenyl acetic acid, phenethyl ester $C_6H_5CH_2CO_2CH_2CH_2C_6H_5$	240.30		$177\text{-}8^{4.5}$	26.5	$1.080^{25/25}$		al	$B9^3$, 2183
11169	Phenyl acetonitrile $C_6H_5CH_2CN$	117.15		$234, 107^{12}$	−23.8	$1.0157^{20/4}$	1.5230^{20}	al, eth, ace	$B9^3$, 2252
11170	Phenyl acetonitrile, α-amino (dl) $C_6H_5CH(NH_2)CN$	132.16	hyg lf (lig)		55				$B14^3$, 1189
11171	Phenyl acetonitrile, 2-amino $2\text{-}H_2NC_6H_4CH_2CN$	132.16	lf (dil al)		72			al, bz	$B14^3$, 1181
11172	Phenyl acetonitrile, 4-amino or 4-Amino phenyl acetonitrile $4\text{-}H_2NC_6H_4CH_2CN$	132.16	lf (w)	$312, 177^{11}$	46			al	$B14^3$, 1182
11173	Phenyl acetonitrile, α-bromo $C_6H_5CH(Br)CN$	196.05	yesh cr (dil al)	$242d, 132\text{-}4^{12}$	29	$1.539^{29/4}$		al, eth, ace, bz	$B9^3$, 2278
11174	Phenyl acetonitrile, 2-bromo or 2-Bromo phenyl acetonitrile $2\text{-}Br\text{-}C_6H_4CH_2CN$	196.05		$145\text{-}7^{14}$	1			al	$B9^3$, 2274
11175	Phenyl acetonitrile, 4-bromo or 4-Bromo phenyl acetonitrile $4\text{-}BrC_6H_4CH_2CN$	196.05	pa ye cr (al)		47			al, bz	$B9^3$, 2275
11176	Phenyl acetonitrile, 2-carboxy $2\text{-}(HO_2C)C_6H_4CH_2CN$	161.16	cr (aa), nd (w, bz)		116d			al, eth, bz	$B9^3$, 4267
11177	Phenyl acetonitrile, 2-chloro $2\text{-}ClC_6H_4CH_2CN$	151.60	grsh-ye nd	$251, 170^{120}$	24	$1.1737^{18/4}$	1.534^{18}		$B9^3$, 2263
11178	Phenyl acetonitrile, α-hydroxy (D+) or d-Mandelonitrile $C_6H_5CH(OH)CN$	133.15	nd, $[\alpha]^{25/}{}_{546.1}$ +46.9 (bz)		28-9			al, eth	$B10^3$, 473
11179	Phenyl acetomitrile, α-hydroxy (DL) or dl-Mandelonitrile $C_6H_5CH(OH)CN$	133.15	ye pr	170d	22	$1.1165^{20/4}$	1.5201^{20}	al, eth	$B10^3$, 474
11180	Phenyl acetonitrile, α-hydroxy (L-) or l-Mandelonitrile $C_6H_5CH(OH)CN$	133.15						al, eth	$B10^1$, 86
11181	Phenyl acetonitrile, 2-hydroxy $2\text{-}HOC_6H_4CH_2CN$	133.15	nd (bz-lig)		117-9			w, eth, ace, bz	$B10^3$, 425
11182	Phenyl acetonitrile, 3-hydroxy $3\text{-}HOC_6H_4CH_2CN$	133.15	pl (w)		52-3			w, al, eth	B10, 189
11183	Phenyl acetonitrile, 4-hydroxy $4\text{-}HOC_6H_4CH_2CN$	133.15	pl (w), mcl pr	330^{756}, 210^{10}	69-70			al, eth	$B10^3$, 438
11184	Phenyl acetonitrile, 2-methoxy or 2-Methoxy phenyl acetonitrile $2\text{-}CH_3OC_6H_4CH_2CN$	147.18	pr (bz-lig)	$141\text{-}3^{11}$	68.5			bz	$B10^3$, 425

No.	Name, Synonyms, and Formula	Mol. wt.	Color, crystalline form, specific rotation and λ_{max} (log ε)	b.p. °C	m.p. °C	Density	n_D	Solubility	Ref.
11185	Phenyl acetonitrile, 4-methoxy $4\text{-}CH_3OC_6H_4CH_2CN$	147.18		286-7, 152^{16}		$1.0845^{20/4}$	1.5309^{20}	al, eth	B10³, 439
11186	Phenyl acetonitrile, 2-methyl or 2-Totyl acetonitrile $2\text{-}CH_3C_6H_4CH_2CN$	131.18		244		1.0156^{22}	1.5252^{20}	al, eth, bz	B9³, 2427
11187	Phenyl acetonitrile, 3-methyl or 3-Totyl acetonitrile $3\text{-}CH_3C_6H_4CH_2CN$	131.18		$245\text{-}7^{745}$d, 133^{15}		1.0022^{22}	1.5233^{20}	al, eth, bz	B9², 349
11188	Phenyl acetonitrile, 4-methyl or 4-Totyl acetonitrile $4\text{-}CH_3C_6H_4CH_2CN$	131.18		242-3, 122^{13}	18	0.9922^{22}	1.5167^{20}	al, eth, bz	B9³, 2433
11189	Phenyl acetonitrile, α-nitro $C_6H_5CH(NO_2)CN$	162.15			39-40			al	B9³, 2291
11190	Phenyl acetonitrile, 2-nitro $2\text{-}NO_2C_6H_4CH_2CN$	162.15	nd (al-w), pr (aa, al)	178^{12}	84			al, eth, ace, bz, chl	B9³, 2283
11191	Phenyl acetonitrile, 3-nitro $3\text{-}NO_2C_6H_4CH_2CN$	162.15	(eth-lig)	$180^{1.5}$	63			al, eth, bz, chl	B9², 312
11192	Phenyl acetonitrile, 4-nitro $4\text{-}NO_2C_6H_4CH_2CN$	162.15	pl	$195\text{-}7^{12}$	116-7			al, eth, bz, chl	B9³, 2291
11193	Phenyl acetyl chloride $C_6H_5CH_2COCl$	154.60		170^{250}, $104\text{-}5^{24}$		$1.1682^{20/4}$	1.5325^{20}	eth	B9³, 2192
11194	Phenyl acetyl chloride, α-acetyl (DL) or dl-Acetyl mandelyl chloride $C_6H_5CH(CO_2CH_3)COCl$	196.63		$150\text{-}5^{33}$				eth, bz, chl	B10¹, 89
11195	Phenyl acetyl chloride, 2,4-dinitro $2,4\text{-}(NO_2)_2C_6H_3CH_2COCl$	244.59	ye lf (CS_2)		77			eth, bz, chl	B9¹, 185
11196	Phenyl acetylene $C_6H_5C{\equiv}CH$	102.14		142-4, 44^{18}	−44.8	$0.9281^{20/4}$	1.5485^{20}	al, eth, ace	B5⁴, 1525
11197	α-Phenyl acrylic acid or Atropic acid $CH_2{=}C(C_6H_5)CO_2H$	148.16	lf (al), nd (w)	267d	106-7			al, eth, bz, chl	B9³, 2751
11198	α-Phenyl acrylic acid, 3-(2-nitro phenyl)(*trans*) $(3\text{-}NO_2C_6H_4)CH{=}C(C_6H_5)CO_2H$	269.26	ye cr (al)		196-7			eth, bz	B9³, 3423
11199	α-Phenyl acrylic acid, 3-(3-nitro phenyl)-(*cis*) $(3\text{-}NO_2C_6H_4)CH{=}C(C_6H_5)CO_2H$	269.26	nd (al, aa)		195-6				B9³, 3423
11200	α-Phenyl acrylic acid, 3-(3-nitrophenyl)-(*trans*) $(3\text{-}NO_2C_6H_4)CH{=}C(C_6H_5)CO_2H$	269.26	ye pr (eth), nd (dil al)		182			al, ace, bz, chl	B9³, 3424
11201	α-Phenyl acrylic acid, 3-(4-nitrophenyl)-cis $(4\text{-}NO_2C_6H_4)CH{=}C(C_6H_5)CO_2H$	269.26	ye cr pr (dil al + 1w), lf (bz + ½)		144			al, eth	B9³, 3424
11202	α-Phenyl acrylic acid, 3-(4-nitrophenyl)-trans $(4\text{-}NO_2C_6H_4)CH{=}C(C_6H_5)CO_2H$	269.26	ye pr or nd (al)		213-4			al	B9³, 3425
11203	Phenylalanine $C_6H_5CH_2CH(NH_2)CO_2H$	165.19	pr (w), $[\alpha]^{18/}{}_D$ + 70 (w)	sub at 295	283-4d			w	B14³, 1228
11204	Phenylalanine (D) $C_6H_5CH_2CH(NH_2)CO_2H$	165.19	nd or pr (w), $[\alpha]^{20/}$ -69.5 (w)	sub at 295	283d		1.600	w	B14³, 1228
11205	Phenylalanine (DL) $C_6H_5CH_2CH(NH_2)CO_2H$	165.19	red-br lf (dil al), pr or nd, (w)	sub d	284-8d			w	B14³, 1229
11206	Phenylalanine, N acetyl (d,t) $C_6H_5CH_2CH(NHCOCH_3)CO_2H$	207.23	cr (w), $[\alpha]^{20/}{}_D$ -50.9		172			al	B14², 297
11207	Phenylalanine, N-acetyl (dl) $C_6H_5CH_2CH(NHCOCH_3)CO_2H$	207.23	nd or pl (w), hex tab (ace)		152-3			w, al	B14³, 1238
11208	Phenylalanine, N-acetyl (l) $C_6H_5CH_2CH(NHCOCH_3)CO_2H$	207.23	$[\alpha]^{26/}{}_D$ + 35.1		172			al	B14², 298
11209	Phenylalanine, ethyl ester $C_6H_5CH_2CH(NH_2)CO_2C_2H_5$	193.25		148^{15}		1.065^{15}			B14³, 1232
11210	Phenylboric acid or Benzeneboronic acid $C_6H_5B(OH)_2$	121.93	nd (w)		218-20			al, eth, bz	B16², 638
11211	*o*-Phenylenediamine or 1,2-Diamino benzene $1,2\text{-}(H_2N)_2C_6H_4$	108.14	brsh-ye lf (w), pl (chl)	256-8	102-3			al, eth, bz, chl	B13³, 28
11212	*o*-Phenylenediamine, 3,5-dichloro $1,2\text{-}(H_2N)_2\text{-}3,5\text{-}Cl_2C_6H_2$	177.03	nd (al)		60.5				B13, 27
11213	*o*-Phenylenediamine, 3,6-dichloro $1,2\text{-}(H_2N)_2\text{-}3,6\text{-}Cl_2C_6H_2$	177.03	nd (50% al)		100			w, al, eth, ace	B13², 20

No.	Name, Synonyms, and Formula	Mol. wt.	Color, crystalline form, specific rotation and λ_{max} (log ε)	b.p. °C	m.p. °C	Density	n_D	Solubility	Ref.
11214	*o*-Phenylenediamine, 4,5-dimethoxy $1,2\text{-}(H_2N)_2\text{-}4,5\text{-}(CH_3O)_2C_6H_2$	168.20	bl pr		131-2			w, al	$B13^3$, 2127
11215	*o*-Phenylenediamine, 3-methoxy-6-methyl $1,2\text{-}(H_2N)_2\text{-}3\text{-}CH_3O\text{-}6\text{-}CH_3\text{-}C_6H_2$	152.20	pr (eth-bz)		75-6			al, eth	$B13^2$, 349
11216	*o*-Phenylenediamine, 4-methoxy $1,2\text{-}(H_2N)_2C_6H_3OCH_3\text{-}4$	138.17	gr pl	$167\text{-}70^{11}$	50-2			eth	$B13^3$, 1362
11217	*o*-Phenylenediamine, 4-methyl $1,2\text{-}(H_2N)_2\text{-}4\text{-}CH_3C_6H_3$	122.17	pl (lig)	265, 92^1	89-90			w	$B13^3$, 292
11218	*o*-Phenylene diamine, 3-nitro $1,2\text{-}(H_2N)_2\text{-}3\text{-}NO_2\text{-}C_6H_3$	153.14	dk red nd (dil al)		158-9				$B13^3$, 61
11219	*o*-Phenylenediamine, 4-nitro $1,2\text{-}(H_2N)_2\text{-}4\text{-}NO_2\text{-}C_6H_3$	153.14	dk red nd		199-200				$B13^3$, 62
11220	*o*-Phenylenediamine, N-phenyl $2\text{-}H_2NC_6H_4NHC_6H_5$	184.24	nd (w)	312.5^{744}	80-1			ace, bz, chl	$B13^3$, 34
11221	*m*-Phenylenediamine or 1,3-Diamino benzene $1,3\text{-}(H_2N)_2C_6H_4$	108.14	rh (al)	282-4	63-4	$1.0696^{58/4}$	1.6339^{58}	w, al, eth, bz	$B13^3$, 65
11222	Phenylenediamine, N-acetyl $3\text{-}(CH_3CONH)C_6H_4NH_2$	150.18	lf (ace-bz), nd or pl (bz)	d at 100	87-9			w, al, eth, ace	$B13^3$, 75
11223	*m*-Phenylenediamine, 2,5-dichloro $2,5\text{-}Cl_2\text{-}1,3\text{-}(H_2N)_2C_6H_2$	177.03	nd (w)		100			w, al, bz	$B13^2$, 29
11224	*m*-Phenylenediamine, 4,6-dichloro $4,6\text{-}Cl_2\text{-}1,3\text{-}(H_2N)_2C_6H_2$	177.03	nd (dil al)		136-7			al	$B13^3$, 98
11225	*m*-Phenylenediamine, 6-methoxy $6\text{-}CH_3O\text{-}1,3\text{-}(NH_2)_2C_6H_3$	138.17	nd (eth)		67-8			al, eth	$B13^2$, 308
11226	*p*-Phenylenediamine or 1,4-Diamino benzene $1,4\text{-}(H_2N)_2C_6H_4$	108.14	wh pl (bz, eth)	267	140			eth, chl	$B13^3$,104
11227	*p*-Phenylenediamine, N-acetyl,sulfate $4\text{-}(CH_3CONH)C_6H_4NH_2.H_2SO_4$	248.25	nd (eth-al)		285 d			w, al	B13, 95
11228	*p*-Phenylenediamine, 2,5-dichloro $2,5\text{-}Cl_2\text{-}1,4\text{-}(H_2N)_2C_6H_2$	177.03	pr (w)		170				B13, 118
11229	*p*-Phenylenediamine, 2,6-dichloro $2,6\text{-}Cl_2\text{-}1,4\text{-}(H_2N)_2C_6H_2$	177.03	nd, pr (dil al)		124-6			al, eth, ace, bz	$B13^3$, 269
11230	*p*-Phenylenediamine, 2-methoxy-5-methyl $2\text{-}CH_3O\text{-}5\text{-}CH_3\text{-}1,4\text{-}(H_2N)_2C_6H_2$	152.20			166			al, eth	$B13^2$, 349
11231	*p*-Phenylenediamine, N-phenyl $4\text{-}H_2NC_6H_4NHC_6H_5$	184.24	nd (al), cr (lig)	354, $155^{0.026}$	66 (al), 75 (lig)			al, eth, lig	$B13^3$, 115
11232	*p*-Phenylenediamine, 2,3,5,6-tetramethyl or Diamino durene $1,4\text{-}(H_2N)_2\text{-}2,3,5,6\text{-}(CH_3)_4C_6$	164.25	nd (w)		148			al, eth, chl	$B13^3$, 360
11233	bis-(1-phenylethyl) amine $[C_6H_5CH(CH_3)]_2NH$	225.13	ye	295-8, 190^{10}		1.018^{15}	1.573		$B12^2$, 589
11234	Phenyl-2-aminophenyl sulfide $2\text{-}H_2NC_6H_4SC_6H_5$	201.29	pl (al)	257.5^{100}	35-6			al	$B13^3$, 904
11235	Phenyl-2-aminophenyl sulfone $(2\text{-}H_2NC_6H_4)SO_2C_6H_5$	233.28	lf (dil al)		122-4			al, bz, aa	$B3^3$, 905
11236	Phenyl-4-aminophenyl sulfide $(4\text{-}H_2NC_6H_4)SC_6H_5$	201.22	nd (dil al), cr (lig)	242.5^{29}	95.8			al, eth	$B13^3$, 1224
11237	Phenyl-4-aminophenyl sulfone $(4\text{-}H_2NC_6H_4)SO_2C_6H_5$	233.28	nd (al)		176			al, bz, aa	$B13^3$, 1226
11238	Phenyl-4-aminophenyl sulfoxide $(4\text{-}H_2NC_6H_4)SOC_6H_5$	217.29	nd (w)		152			al, eth	$B13^3$, 1226
11239	Phenyl-4-bromophenyl ether $4\text{-}BrC_6H_4OC_6H_5$	249.11		310.1, 163^{10}	18.7	$1.4208^{20/4}$	1.6084^{20}	eth	$B6^4$, 1047
11240	Phenyl *tert*-butyl ketone or Pivalophenone $C_6H_5COC(CH_3)_3$	162.23		219-21, $97\text{-}8^{16}$		0.963^{26}	1.5086^{19}	ace	$B7^3$, 1125
11241	Phenyl-α-chlorobenzyl ketone or Desylchloride $C_6H_5CO(CHClC_6H_5)$	230.69	nd (al)	d	68.5			al	$B7^3$, 2106
11242	Phenyl-4-chlorophenyl sulfone or Sulphenone $(4\text{-}ClC_6H_4)SO_2C_6H_5$	252.72	cr (al)		98			eth, ace, bz	$B6^4$, 1587
11243	Phenyl-2,4-diaminophenyl sulfone $[2,4\text{-}(NH_2)_2C_6H_3]SO_2C_6H_5$	248.30	nd (al)		188			al	B13, 553
11244	Phenyl-2,5-dihydroxyphenyl sulfone $[2,5\text{-}(HO)_2C_6H_3]SO_2C_6H_5$	250.27	pr (w), nd (dil al)		196			al	$B6^2$, 1072
11245	Phenyl-2,4-dinitrophenyl ether $[2,4\text{-}(NO_2)_2C_6H_3]OC_6H_5$	260.21	pl (al), nd (al-ace)	$230\text{-}50^{27}$	71			al, eth	$B6^4$, 1375

No.	Name, Synonyms, and Formula	Mol. wt.	Color, crystalline form, specific rotation and λ_{max} (log ε)	b.p. °C	m.p. °C	Density	n_D	Solubility	Ref.
11246	Phenyl-2,4-dinitrophenyl sulfide $[2,4\text{-}(O_2N)_2C_6H_3]SC_6H_5$	276.27	pa ye nd, (ace, bz-al)		121			ace, bz, aa	$B6^4$, 1746
11247	Phenyl-2,6-dinitrophenyl ether $[2,6\text{-}(NO_2)_2C_6H_3]OC_6H_5$	260.21	lf (al)		99-100				$B6^2$, 245
11248	Phenyl-3,4-dinitrophenyl ether $[3,4\text{-}(NO_2)_2C_6H_3]OC_6H_5$	260.21	lf (al)		99-100				$B6^1$, 127
11249	Phenyl ether or Diphenyl ether $(C_6H_5)_2O$	170.21		257.9, 121^{10}		1.0748^{20}	1.5787^{25}	al, eth, bz, aa	$B6^4$, 568
11250	α-Phenyl ethyl alcohol (d) $C_6H_5CH(OH)CH_3$	122.17	$[\alpha]^{19/}{}_D$ +42.9 (undil)	203, 100^{18}		$1.0129^{20/4}$	1.5272^{20}	al, chl	$B6^3$, 1671
11252	α-Phenyl ethyl alcohol (dl) $C_6H_5CH(OH)CH_3$	122.17	glassy	203.4, 87.2^{10}	20	$1.0135^{20/4}$	1.5275^{20}	al, eth	$B6^3$, 1673
11253	α-Phenyl ethyl alcohol (l) $C_6H_5CH(OH)CH_3$	122.17	$[\alpha]^{20/}{}_D$ -45.5 (MeOH, c=5)	202-4, 93^{14}		$1.0129^{20/4}$	1.5272^{20}	al, eth	$B6^3$, 1672
11254	α-Phenyl ethyl alcohol, 3-methyl $[3\text{-}CH_3C_6H_4]CH(OH)CH_3$	136.19		112^{12}		$0.9974^{15/4}$	1.5240^{20}	al, eth	$B6^3$, 1823
11255	α-Phenyl ethyl alcohol, 4-methyl $(4\text{-}CH_3C_6H_4)CH(OH)CH_3$	136.19		219^{756}, 120^{19}		$0.9944^{20/4}$	1.5246^{20}	al, eth	$B6^3$, 1826
11256	β-Phenyl ethyl alcohol $C_6H_5CH_2CH_2OH$	122.17	glass	218.2, 97.4^{10}	fr−27	$1.0202^{20/4}$	1.5325^{20}	al, eth	$B6^3$, 1703
11257	β-Phenyl ethyl alcohol, 2-amino $(2\text{-}H_2NC_6H_4)CH_2CH_2OH$	137.18	ye in air	$152\text{-}3^6$			1.5849^{19}	w	$B13^3$, 1679
11258	β-Phenyl ethyl alcohol, 4-amino $(4\text{-}H_2NC_6H_4)CH_2CH_2OH$	137.18	nd (al)		108				$B13^3$, 1679
11259	β-Phenyl ethyl alcohol, 2-methoxy or 2-Anisyl methyl carbinol $(2\text{-}CH_3OC_6H_4)CH_2CH_2OH$	152.19		128^{17}		$1.0862^{15/4}$	1.5312^{25}	al, eth	$B6^3$, 4569
11260	β-Phenyl ethyl alcohol, 3-methoxy or 3-Anisyl methyl carbinol $(3\text{-}CH_3OC_6H_4)CH_2CH_2OH$	152.19		133^{15}		$1.0781^{19/4}$	1.5325^{20}	al, eth	$B6^3$, 4570
11261	β-Phenyl ethyl alcohol, 4-methoxy or 4-Anisyl methyl carbinol $(4\text{-}CH_3OC_6H_4)CH_2CH_2OH$	152.19		$140\text{-}1^{17}$		$1.0794^{20/4}$	1.5310^{25}	al, eth	$B6^3$, 4571
11262	bis-(2-Phenyl ethyl)amine $[C_6H_5CH_2CH_2]_2NH$	225.33		$335\text{-}7^{603}$, 190^{15}	28-30		1.5550^{25}	al, eth	$b12^3$, 2415
11263	bis-(α-Phenyl ethyl) ether (dl) $[C_6H_5CH(CH_3)]_2O$	226.32		280.2, $167\text{-}8^{23}$		$1.0058^{15/4}$	1.5454^{21}	eth, chl	$B6^3$, 1677
11264	bis-(β-Phenyl ethyl) ether $(C_6H_5CH_2CH_2)_2O$	226.32	vt-bl flr	317-20, 194.5^{20}		$1.0141^{18/4}$	1.5488^{18}	eth, chl	$B6^3$, 1707
11265	Phenyl ethynyl ether or Phenoxy acetylene $C_6H_5OC{\equiv}CH$	118.14		$61\text{-}2^{25}$	−36	$1.0614^{20/4}$	1.5125^{20}	al, eth	$B6^4$, 565
11266	Phenyl glycidol $C_6H_5CHCHCH_2OH$ (epoxide O bridging CHCH)	150.18		138^3	26.5	1.512^{27}	1.5432^{27}	al, eth	$B17^4$, 1349
11267	Phenyl glyoxal C_6H_5COCHO	134.13	nd (+w)	142^{125}, $95\text{-}7^{25}$	91 (+w)			al, eth, ace, bz, chl	$B7^3$, 3443
11268	Phenyl glyoxime $C_6H_5C(NOH)C(NOH)H$	164.16	nd (chl)		180			w, al, eth	$B7^3$, 3448
11269	Phenyl glyoxylic acid, 2-nitro $(2\text{-}NO_2C_6H_4)COCO_2H$	195.13	pr (w + 1)		123			w, al, ace, aa	$B10^1$, 315
11270	Phenyl glyxylonitrile C_6H_5COCN	131.13	ta	206-8, 99^{19}	32-3			al, eth	$B10^3$, 2976
11271	Phenylhydrazine, 1-Benzoyl $C_6H_5NHNHCOC_6H_5$	212.25	pr (al), nd (w), lf (dil al)	314	168			bz, chl	$B15^3$, 163
11272	Phenylhydrazine, o-bromo $2\text{-}BrC_6H_4NHNH_2$	187.04	nd		48				$B15^1$, 117
11273	Phenyl hydrazine, p-bromo $4\text{-}BrC_6H_4NHNH_2$	187.04	nd (w), lf (lig), cr (al)		108			al, eth	$B15^3$, 289
11274	Phenyl hydrazine, 1-butyl $H_2NN(C_4H_9)C_6H_5$	164.25		250^{763}					$B15^1$, 28
11275	Phenylhydrazine, 2,4,6-tribromo $(2,4,6\text{-}Br_3C_6H_2)NHNH_2$	344.83	nd (peth), cr (lig)		146			bz, chl, lig	$B15^1$, 126

No.	Name, Synonyms, and Formula	Mol. wt.	Color, crystalline form, specific rotation and λ_{max} (log ε)	b.p. °C	m.p. °C	Density	n_D	Solubility	Ref.
11276	Phenylhydrazine, 2,4,6-trichloro $(2,4,6\text{-}Cl_3C_6H_2)NHNH_2$	211.48	cr (bz)	...	143	...	...	...	B15^3, 281
11277	Phenylhydroxylamine C_6H_5NHOH	109.13	nd (w, bz, peth)	...	83-4	...	...	al, eth, bz, chl	B15^3, 5
11278	Phenylhydroxylamine, N-nitroso $C_6H_5N(NO)OH$	138.13	nd (lig)	...	59	...	...	al, eth	B16^3, 638
11279	Phenyl, 2-hydroxyphenyl amine $(2HOC_6H_4)NHC_6H_5$	185.23	pr (w)	180-9^{20}	69-70	...	...	al, eth, aa	B13^3, 764
11280	Phenyl, 3-hydroxyphenyl amine $(3\text{-}HOC_6H_4)NHC_6H_5$	185.23	lf (w)	340	81-2	...	...	al, eth, ace, bz	B13^3, 931
11281	Phenyl, 4-hydroxyphenyl amine $(4\text{-}HOC_6H_4)NHC_6H_5$	185.23	lf (w)	330, 215-6^{12}	73	...	...	al, eth, bz, chl	B13^3, 991
11282	Phenyl isocyanate C_6H_5NCO	119.12	...	162-3^{750}, 55^{13}	...	1.0956$^{20/4}$	1.5368^{20}	eth	12^3, 903
11283	Phenyl isothiocyanate or Phenyl mustard oil C_6H_5NCS	135.18	...	221, 95^{12}	−21	1.1303$^{20/4}$	1.6492^{23}	al, eth	12^3, 908
11284	Phenyl, 2-methoxyphenyl ether $(2\text{-}CH_3OC_6H_4)OC_6H_5$	200.24	cr (MeOH), nd (lig)	288^{745}, 91-2^{7}	79	...	...	al, eth, bz	B6^3, 4215
11285	Phenyl nitromethane or α-Nitrotoluene $C_6H_5CH_2NO_2$	137.14	ye liq	225-7, 118-9^{16}	...	1.1598$^{20/0}$	1.5323^{20}	eth, ace	B5^4, 850
11286	Phenyl 2-nitrophenyl amine $(2\text{-}NO_2C_6H_4)NHC_6H_5$	214.22	og pl (dil al), rh bi-pym	...	75.5	...	...	al	B12^3, 1517
11287	Phenyl 3-Nitrophenyl amine $(3\text{-}NO_2C_6H_4)NHC_6H_5$	214.22	red nd or pl (dil al)	...	114	...	...	al, eth, bz	B9^3, 3585
11288	Phenyl 4-nitrophenyl amine $(4\text{-}NO_2C_6H_4)NHC_6H_5$	214.22	ye nd, tab (CCl_4)	211^{20}	133-4	...	...	al, aa	B12^3, 1586
11289	Phenyl 4-nitrosophenyl amine $(4\text{-}ONC_6H_4)NHC_6H_5$	198.22	ye pw pl (bz)	...	143	...	...	al, eth, bz, chl	B12^3, 347
11290	Phenyl 2-nitrophenyl ether $(2\text{-}O_2NC_6H_4)OC_6H_5$	215.21	ye liq	235^{60}, 183-5^{8}	<−20	1.2539^{22}	1.575^{20}	al, eth, bz, chl, aa	B6^4, 1252
11291	Phenyl 4-nitrophenyl ether $(4\text{-}O_2NC_6H_4)OC_6H_5$	215.21	pl (peth, MeOH)	188-90^{8}	61	...	...	eth, bz	B6^4, 1287
11292	Phenyl 2-nitrophenyl sulfide $(2\text{-}NO_2C_6H_4)SC_6H_5$	231.27	ye-og nd (lig, al-eth)	210^{15}	82	...	...	al, eth	B6^4, 1663
11293	Phenyl 4-nitrophenyl sulfide $(4\text{-}NO_2C_6H_4)SC_6H_5$	231.27	pa ye mcl pr (lig)	240^{25}	55	...	...	al, eth	B6^4, 1694
11294	Phenyl isopentyl ketone or Isocaprophenone $i\text{-}C_5H_{11}COC_6H_5$	176.26	...	255-6, 145-7^{30}	−2	0.9623$^{15/4}$	1.533^{20}	al, eth, ace, bz, chl	B7^3, 1158
11295	Phenyl phosphinic acid $C_6H_5P(OH)_2$	142.09	...	...	82-3	...	...	w, al	B16^3, 874
11296	2-Phenyl propionic acid, 2-hydroxy (D) or 2-Hydroxy hydratropic acid $C_6H_5C(OH)(CH_3)CO_2H$	166.18	$[\alpha]^{10.5/}_{D}$ +37.7 (al, c=3.5)	...	116-7	...	...	ace, bz	B10^3, 560
11297	2-Phenyl propionic acid, 2-hydroxy (DL) or 2-Hydroxy hydratropic acid $C_6H_5C(OH)(CH_3)CO_2H$	166.18	nd pl (lig)	...	93-5	...	...	ace, bz	B10^3, 560
11298	2-Phenyl propionic acid, 3-hydroxy (*d*) or 3-Hydroxy hydratropic acid $HOCH_2CH(C_6H_5)CO_2H$	166.18	nd (w, bz), pr (eth, w)	...	130	...	...	al, eth	B10^2, 158
11299	2-Phenyl propionic acid, 3-hydroxy (*dl*) or 3-Hydroxy hadratropic acid $HOCH_2CH(C_6H_5)CO_2H$	166.18	nd, pl (al, bz, w)	d	118	...	...	w, al, eth	B10^3, 564
11300	2-Phenyl propionic acid, 3-hydroxy-(*l*) or 3-Hydroxy hydratropic acid $HOCH_2CH(C_6H_5)CO_2H$	166.18	pl (AcOEt), nd (w), $[\alpha]^{15/}_{D}$ -81.2 (w, c=1.5)	...	...	...	...	al, eth, AcOEt	B10^3, 564
11301	2-Phenyl propionic acid, 2-hydroxy (L) or 2-Hydroxy hydratropic acid $C_6H_5C(OH)(CH_3)CO_2H$	166.18	nd (bz, w), $[\alpha]^{13.8/}_{D}$ -37.7 (al, c=3.4)	...	116-7	...	...	w, al, ace, bz	B10^3, 560
11302	3-Phenyl propionic acid, 2-hydroxy (d) or 2-Hydroxy hydrocinnamic acid $C_6H_5CH_2CH(OH)CO_2H$	166.18	nd (w), $[\alpha]^{20/}_{D}$ +22.2 (w, c=2.2)	...	124-6	...	...	w, al, ace	B10^3, 554

No.	Name, Synonyms, and Formula	Mol. wt.	Color, crystalline form, specific rotation and λ_{max} (log ε)	b.p. °C	m.p. °C	Density	n_D	Solubility	Ref.
11303	3-Phenyl propionic acid, 2-hydroxy-(dl) or 2-Hydroxy hydrocinnamic acid $C_6H_5CH_2CH(OH)CO_2H$	166.18	cr (chl, bz) pr (w)	148-50[15]	98			al, eth, ace	B10[3], 554
11304	3-Phenyl propionic acid, 2-hydroxy (l) or 2-Hydroxy hydrocinnamic acid $C_6H_5CH_2CH(OH)CO_2H$	166.18	nd (w), $[\alpha]^{20/}_{D}$ -19.9 (w, c=3.2)		124.5			al, eth, ace	B10[3], 554
11305	3-Phenyl propionic acid, 3-hydroxy-(d) or 3-Hydroxy hydrocinnamic acid $C_6H_5CH(OH)CH_2CO_2H$	166.18	cr (bz), $[\alpha]^{18/}_{D}$ +20.6 (MeOH, c=5)		116			al, eth	B10[3], 545
11306	3-Phenyl propionic acid, 3-hydroxy-(dl) or 3-Hydroxy hydrocinnamic acid $C_6H_5CH(OH)CH_2CO_2H$	166.18	pr (w)		96			al, ace, chl	B10[3], 546
11307	3-Phenyl propionic acid, 3-hydroxy (l) or 3-Hydroxy hydrocinnamic acid $C_6H_5CH(OH)CH_2CO_2H$	166.18	nd (bz), $[\alpha]^{18/}_{D}$ -19.8 (al, c=4.7)		115-6				B10[3], 546
11308	Phenyl sulfide or Diphenyl sulfide $(C_6H_5)_2S$	186.27		296, 145[8]	−25.9	1.1136[20/4]	1.6334[20]	eth, bz	B6[4], 1488
11309	Phenyl sulfide, 2,4′-diamino $(2-H_2NC_6H_4)S(C_6H_4NH_2-4)$	216.30	nd (w, al), pr (dil al)		62.5			al, eth, bz	B13[3], 1245
11310	Phenyl sulfide, 4,4′-diamino or bis-(4-Amino phenyl)sulfide $(4-H_2NC_6H_4)_2S$	216.30	nd (w)		108-9			al, eth, bz	B13[3], 1246
11311	Phenyl sulfide, 4,4′-dibromo $(4-BrC_6H_4)_2S$	344.06		268.5[40]	115	1.84		eth, chl	B6[4], 1651
11312	Phenyl sulfide, 4,4′-dichloro-2,2′-dinitro $(4-Cl_2-2-NO_2C_6H_3)_2S$	345.16	br-ye nd (90% aa)		149-50			bz	B6[2], 312
11313	Phenyl sulfide, 4,4′-dihydroxy or bis-(4-Hydroxy phenyl)sulfide $(4-HOC_6H_4)_2S$	218.27	mcl pr or lf (al)		151			al, eth	B6[3], 4455
11314	Phenyl sulfide, 2,2′-dimethoxy or *o*-Anisyl disulfide $(2-CH_3OC_6H_4)_2S$	246.32	lf (al)	252-3[10]	73			al, eth, bz	B6, 794
11315	Phenyl sulfide, 4,4′-dinitro or bis-(4-Nitro phenyl)sulfide $(4-O_2NC_6H_4)_2S$	276.27	og pl (aa)		160-1				B6[4], 1696
11316	Phenyl sulfide, 2,2′-dinitro or bis-(2-Nitro phenyl)sulfide $(2-O_2NC_6H_4)_2S$	276.27	gold-ye pl (aa)	sub d	122-3				B6[4], 1665
11317	Phenyl sulfide, 2,2′,4,4′-tetranitro or bis-(2,4-Dinitro phenyl)sulfide $[2,4-(NO_2)_2C_6H_3]_2S$	366.26	ye nd or pl (aa)		197				B6[4], 1748
11318	Phenyl sulfone $(C_6H_5)_2SO_2$	218.27	mcl pr (bz), pl (al), nd (w)	379, 232[18]	128-9	1.252[20/4]		eth, bz	B6[4], 1490
11319	Diphenyl sulfone, 4,4′-diacetamide $(4-CH_3CONHC_6H_4)_2SO_2$	332.37	pa ye nd (eth, dil aa), lf (dil al)		282-5			al	B13[3], 1286
11320	Phenyl sulfone, 3,3′-diamino or bis-(3-Amino phenyl)sulfone $(3-H_2NC_6H_4)_2SO_2$	248.30	pr		168			w, al	B13[3], 984
11321	Phenyl sulfone, 4,4′-diamino or bis-(4-Amino phenyl)sulfone $(4-H_2NC_6H_4)_2SO_2$	248.30	lf (dil al)		178			al	B13[3], 1246
11322	Phenyl sulfone, 4,4′-dichloro or bis-(4-Chloro phenyl)sulfone $(4-ClC_6H_4)_2SO_2$	287.16	mcl	sub	148-9				B6[4], 1587
11323	Phenyl sulfone, 4,4′-diethoxy or bis-(4-Ethoxy phenyl)sulfone $(4-C_2H_5OC_6H_4)_2SO_2$	306.38	pl (al, aa)		163			al, eth	B6[3], 4459
11324	Phenyl sulfone, 4,4′-diethyl $(4-C_2H_5C_6H_4)_2SO_2$	274.38			102			eth, bz	B6, 475
11325	Phenyl sulfone, 2,2′-dihydroxy or bis-(2-Hydroxy phenyl)sulfone $(2-HOC_6H_4)_2SO_2$	250.27	nd (bz)		164-5 (179)			w, al, eth, aa	B6[3], 4278
11326	Phenyl sulfone, 3,3′-dihydroxy or bis-(3-Hydroxy phenyl)sulfone $(3-HOC_6H_4)_2SO_2$	250.27	190-1					al, eth	B6[3], 4365

No.	Name, Synonyms, and Formula	Mol. wt.	Color, crystalline form, specific rotation and λ_{max} (log ε)	b.p. °C	m.p. °C	Density	n_D	Solubility	Ref.
11327	Phenyl sulfone, 4,4′-dihydroxy or bis-(4-Hydroxy phenyl)sulfone ... $(4\text{-}HOC_6H_4)_2SO_2$	250.27	nd (w), rh bipym	...	240-1	1.3663^{15}	...	al, eth	B6^{3}, 4456
11328	Phenyl sulfone, 2,2′-dimethoxy or bis-(2-Methoxy phenyl)sulfone ... $(2\text{-}CH_3OC_6H_4)_2SO_2$	278.32	nd (bz)	...	157-8	...	...	al, aa	B6^{3}, 4278
11329	Phenyl sulfone, 4,4′-dimethoxy or bis-(4-Methoxy phenyl)sulfone ... $(4\text{-}CH_3OC_6H_4)_2SO_2$	278.32	lf or pr (al), nd (al-eth)	sub	130	...	...	al	B6^{3}, 4458
11330	Phenyl-o-tolyl sulfide ... $(2\text{-}CH_3C_6H_4)SC_6H_5$	200.30	...	304.5^{224}, 164^{12}	...	$1.0893^{20/4}$	...	ace, bz	B6^{4}, 2018
11331	Phenyl-m-tolyl sulfide ... $(3\text{-}CH_3H_4)SC_6H_5$	200.30	...	309.5, 164.5^{11}	−6.5	$1.0937^{15/4}$	...	ace, bz	B6^{4}, 2080
11332	Phenyl-*p*-tolyl sulfide ... $(4\text{-}CH_3C_6H_4)SC_6H_5$	200.30	...	317, 167.5^{11}	15.7	$1.0986^{25/4}$	1.6225^{25}	ace, bz	B6^{4}, 2169
11333	Phenyl-2-tolyl sulfone ... $(2\text{-}CH_3C_6H_4)SO_2C_6H_5$	232.30	pl (al)	...	81	...	...	al, eth, bz	B6^{4}, 2018
11334	Phenyl-4-tolyl sulfone ... $(4\text{-}CH_3C_6H_4)SO_2C_6H_5$	232.30	pl (al)	...	127-8	...	...	...	B6^{4}, 2171
11335	Phenyl sulfoxide or Diphenyl sulfoxide ... $(C_6H_5)_2SO$	202.27	pr (lig)	340d, 210^{15}	70.5	...	...	al, eth, bz, aa	B6^{4}, 1489
11336	Phenyl sulfoxide, 4-amino-4′-nitro ... $(4\text{-}H_2NC_6H_4)SO(C_6H_4NO_2\text{-}4)$	262.28	ye (al)	...	132	...	...	al	B13^{3}, 1226
11337	Phenyl sulfoxide, 4,4′-diamino or bis(4-Amino phenyl)sulfoxide ... $(4\text{-}H_2NC_6H_4)_2SO$	232.30	pr (w, al)	...	175d	...	...	...	B13^{3}, 1246
11338	Phenyl sulfoxide, 4,4′-dibromo ... $(4\text{-}BrC_6H_4)_2SO$	360.06	...	...	153-4	...	...	al, bz, chl	B6^{4}, 1651
11339	Phenyl sulfoxide, 4,4′-dichloro or bis-(4-Chloro phenyl)sulfoxide ... $(4\text{-}ClC_6H_4)_2SO$	271.16	...	...	143-4	...	...	chl	B6^{4}, 1587
11340	Phenyl sulfoxide, 4,4′-dihydroxy or bis-(4-Hydroxy phenyl)sulfone ... $(4\text{-}HOC_6H_4)_2SO$	234.27	nd (ace)	...	195	...	...	al, ace	B6^{3}, 4456
11341	Phenyl thiocyanate ... C_6H_5SCN	135.18	...	232-3, $71\text{-}3^{1.5}$	...	$1.155^{18/18}$	...	al, eth	B6^{4}, 1536
11342	Phenyl thiocyanate, 4-amino ... $(4\text{-}H_2NC_6H_4)SCN$	150.20	nd (w), cr (dil al)	...	57-8	...	...	al, eth, bz	B13^{3}, 1239
11343	Phenyl thiocyanate, 4-chloro ... $(4\text{-}ClC_6H_4)SCN$	169.63	nd (al)	...	35-6	...	...	al	B6^{4}, 1601
11344	Phenyl thiocyanate, 4-(dimethyl amino) ... $[4\text{-}(CH_3)_2NC_6H_4]SCN$	178.25	nd (lig, w, al)	...	73-4	...	...	eth	B13^{3}, 1251
11345	Phenyl trimethyl ammonium bromide ... $C_6H_5N^+(CH_3)_3Br^-$	216.12	hygm pr (al, al-eth)	...	213-4	...	...	w	B12^{3}, 254
11346	Phenyl trimethyl ammonium iodide ... $C_6H_5N^+(CH_3)_3I^-$	263.12	lf (al)	...	224	...	...	w, al, aa	B12^{3}, 254
11347	Phenyl vinyl ether ... $C_6H_5OCH{=}CH_2$	120.15	...	155-6	...	$0.9770^{20/4}$	1.5224^{20}	eth	B6^{4}, 561
11348	Phloretin or Dihydronaringenin ... $C_{15}H_{14}O_5$	274.27	nd (dil al), cr (dil ace)	...	262-4d	...	...	al, bz	B8^{3}, 4076
11349	Phlorhizin, dihydrate or Asebotin (a β-glucoside) ... $C_{21}H_{24}O_{10}\cdot 2H_2O$	472.45	nd (w), $[\alpha]^{25}{}_D$ -52 (96% al)	...	108 (+2w), 170d (anh)	1.4298	...	al, ace, Py	B17^{4}, 3042
11350	Phloroglucinol or 1,3,5-Trihydroxybenzene ... $1,3,5\text{-}(HO)_3C_6H_3$	126.11	lf or pl (w+2)	sub	218-9 (anh)	1.46	...	al, eth, bz, Py	B6^{3}, 6301
11351	Phloroglucinol, 2-acetyl or 2,4,6-Trihydroxy acetophenone ... $2,4,6(HO)_3C_6H_2COCH_3$	168.15	nd (w+1)	...	222-4 (anh)	...	...	al, eth, ace, aa	B8^{3}, 3386
11352	Phloroglucinol diacetate ... $C_{10}H_{10}O_5$	210.19	pl (w)	...	104	...	...	al, eth, ace	B6^{1}, 547
11353	Phloroglucinol dimethyl ether or 3,5-Dimethoxy phenol ... $3,5\text{-}(CH_3O)_2C_6H_3OH$	154.17	cr (bz-lig)	$172\text{-}5^{17}$	36-8	...	...	eth, bz	B6^{3}, 6305
11354	Phloroglucinol triacetate ... $1,3,5(CH_3CO_2)_3C_6H_3$	252.22	pr (w), nd (dil al)	...	105-6	...	...	al	B6^{3}, 6306
11355	Phloroglucinol, 2,4,6-trichloro or 2,4,6-Trichloro-1,3,5-trihydroxy benzene ... $C_6H_3Cl_3O_3$	229.45	cr (al)	sub	136	...	...	al	B6, 1104

No.	Name, Synonyms, and Formula	Mol. wt.	Color, crystalline form, specific rotation and λ_{max} (log ε)	b.p. °C	m.p. °C	Density	n_D	Solubility	Ref.
11356	Phloroglucinol, triethyl ether or 1,3,5-Triethoxy benzene 1,3,5-$(C_2H_5O)_3C_6H_3$	210.27	cr (al, dil al)	175^{24}	43.5			al, eth	$B6^3$, 6306
11357	Phloroglucinol-trimethyl ether or 1,3,5-Trimethoxy benzene 1,3,5-$(CH_3O)_3C_6H_3$	168.19	pr (al), lf (peth)	255.5	54-5			al, eth, bz	$B6^3$,6305
11358	Phorone or 2,6-Dimethyl, 2,5-heptadiene-4-one $CH_3C(CH_3)=CHCOCH=C(CH_3)_2$	138.21	ye gr pr	197.8	28	$0.8850^{20/4}$	1.4998^{20}	al, eth, ace	B^4, 3564
11359	Phosgene or Carbonyl chloride $COCl_2$	98.92		7.6	−118	$1.381^{20/4}$		bz, chl, aa	$B3^4$, 31
11360	Phosphine, bis trifluoromethyl $(F_3C)_2PH$	169.99	spont flam	1	−137				$B3^4$, 255
11361	Phosphine, bis(trifluoromethyl) chloro $(F_3C)_2PCl$	204.44	spont flamm	21					$B3^4$, 257
11362	Phosphine, bis(trifluoromethyl) cyano $(F_3C)_2PCN$	195.00	spont. flamm	48			1.3248^{20}		$B3^4$, 256
11363	Phosphine, bis(trifluoromethyl) iodo $(F_3C)_2PI$	295.89		73			1.403^{15}		$B3^4$, 257
11364	Phosphine, dichloro-2,4-(dimethyl phenyl) $[2,4-(CH_3)_2C_6H_3]PCl_2$	207.04		256-8					B16, 773
11365	Phosphine, dichloro-(2,5-dimethyl phenyl) $[2,5-(CH_3)_2C_6H_3]PCl_2$	207.04		253-4	−30	$1.25^{18/18}$			$B16^3$, 848
11366	Phosphine, dichloro(4-ethyl phenyl) $[4-C_2H_5C_6H_4]PCl_2$	207.04		250-2, $85^{0.4}$		$1.237^{20/4}$	1.584^{20}		$B16^3$, 848
11367	Phosphine, dichloro-phenyl $C_6H_5PCl_2$	178.99		224-6, $99\text{-}101^{11}$		$1.356^{20/4}$	1.6030^{20}	bz	$B16^3$, 847
11368	Phosphine, dichloro-trifluoromethyl F_3CPCl_2	170.89		37					$B3^4$, 258
11369	Phosphine, diethyl $(C_2H_5)_2PH$	90.11		85					$B4^3$, 1761
11370	Phosphine, diido-trifluoromethyl F_3CPI_2	353.79	ye fum	d 760, 73^{37}			1.630^{20}		$B3^4$, 258
11371	Phosphine, dimethyl $(CH_3)_2PH$	62.05		25	<1			al, eth	$B4^3$, 1759
11372	Phosphine, diphenyl-ethyl $(C_6H_5)_2PC_2H_5$	214.25		293					$B16^3$, 833
11373	Phosphine, ethyl $C_2H_5PH_2$	62.05		25	<1				$B4^3$, 1761
11374	Phosphine, methyl CH_3PH_2	48.02	gas	−14				eth	$B4^3$, 1759
11375	Phosphine, phenyl $C_6H_5PH_2$	110.10		160-1, 40^{10}		1.001^{15}	1.5796^{20}		$B16^3$, 831
11376	Phosphine, triethyl $(C_2H_5)_3P$	118.16		129^{762}	−88	$0.8006^{19/4}$	1.458^{15}	al, eth	$B4^3$, 1761
11377	Phosphine, triethyl, oxide $(C_2H_5)_3PO$	134.16	wh hyg nd	243	50			al, eth, w	$B4^3$, 1775
11378	Phosphine, triethyl, sulfide $(C_2H_5)_3PS$	150.22	cr (al)		94			w	$B4^3$, 1775
11379	Phosphine, trifluoromethyl F_3CPH_2	102.00	spont. flamm	−26.5					$B3^4$, 255
11380	Phosphine, trimethyl $(CH_3)_3P$	76.08		37.8	−85	<1		eth	$B4^3$, 1759
11381	Phosphine, triphenyl $(C_6H_5)_3P$	262.29		188^1	80	$1.0749^{80/4}$	1.6358^{80}	al, eth, bz, chl	$B16^3$, 833
11382	Phosphine, triphenyl, oxide $(C_6H_5)_3PO$	278.29	pr	>360	156-7	$1.2124^{23/4}$		al, bz	$B16^3$, 864
11383	Phosphine, tris (chloromethyl) $(ClCH_2)_3P$	179.41		100^7		1.414^{20}		al, eth, ace, bz	$B1^3$, 2608
11384	Phosphine, tris (trichloromethyl) $(Cl_3C)_3P$	386.08			53				$B3^4$, 259
11385	Phosphine, tris(trifluoromethyl) $(F_3C)_3P$	237.99	spont, flamm.	17.3	−112				$B3^4$, 255
11386	Phosphine, tris(trifluoromethyl) oxide $(F_3C)_3PO$	253.99		23.6					J1955, 574
11387	Phosphinic acid, diethyl $(C_2H_5)_2PO(OH)$	122.10		320, $134^{0.7}$	18.5			w, al, eth	$B4^3$, 1776
11388	Phosphinic acid, diethyl, anhydride $[(C_2H_5)_2PO]_2O$	226.19		188^{14}		$1.1053^{20/4}$	1.4647^{20}		Am73, 5466

No.	Name, Synonyms, and Formula	Mol. wt.	Color, crystalline form, specific rotation and λ_{max} (log ε)	b.p. °C	m.p. °C	Density	n_D	Solubility	Ref.
11389	Phosphinic acid, diethyl, ethyl ester $(C_2H_5)_2PO(OC_2H_5)$	150.16	...	95^{14}	...	$0.9908^{20/4}$	1.4337^{20}	al, eth	AM73, 5466
11390	Phosphinic acid, dimethyl $(CH_3)_2PO(OH)$	94.05	cr (bz)	377	92	...	...	w, al, eth	$B4^3$, 1776
11391	Phosphinic acid, dimethyl, anhydride $[(CH_3)_2PO]_2O$	170.09	nd (bz)	192^{15}	119-21	...	...	...	Am73, 5466
11392	Phosphinic acid, bis(trifluoromethyl) $(F_3C)_2PO(OH)$	201.99	visc liq	182, $137\text{-}8^{238}$	...	<1.9	...	w	J1955, 563
11393	Phosphinyl chloride dimethyl $(CH_3)_2POCl$	112.50	hyg nd (bz, peth)	204	67-8	...	...	...	Am73, 5466
11394	Phosphinyl chloride, diethyl $(C_2H_5)_2POCl$	140.55	...	104^{15}	...	$1.1394^{20/4}$	1.4647^{20}	...	$B4^3$, 1776
11395	Phosphonic acid, acetyl,diethyl ester $CH_3COPO(OC_2H_5)_2$	180.14	...	$114\text{-}5^{20}$	...	...	1.4200^{26}	...	$B2^4$, 440
11396	Phosphonic acid, benzyl,diethyl ester $C_6H_5CH_2PO(OC_2H_5)_2$	228.23	...	110^2	...	...	1.4930^{20}	...	$B16^3$, 889
11397	Phosphonic acid, carboxy methyl $HO_2CCH_2PO(OH)_2$	140.03	...	...	142-3	...	...	...	$B4^2$, 975
11398	Phosphonic acid, diethyl ester $(C_2H_5O)_2P(O)H$	138.10	...	$53.0\text{-}55^6$	...	...	...	...	$B1^4$, 1329
11399	Phosphonic acid, diisobutyl ester $(i\text{-}C_4H_9O)_2P(O)H$	194.21	...	235-6	...	$0.9759^{20/4}$	...	...	$B1^4$, 1596
11400	Phosphonic acid, diisopropyl ester $(i\text{-}C_3H_7O)_2\text{-}P(O)H$	166.16	...	$76\text{-}7^{10}$	...	$0.9972^{18/0}$	...	...	$B1^4$, 1475
11401	Phosphonic acid, diphenyl ester $(C_6H_5O)_2P(O)H$	234.19	...	$218\text{-}9^{26}$	12	...	1.5564^{25}	...	$B6^4$, 703
11402	Phosphonic acid, methyl, diisopropyl ester $CH_3PO(O\text{-}i\text{-}C_3H_7)_2$	180.18	...	66^3	...	...	$1.4120^{16.5}$	...	$B4^3$, 1778
11403	Phosphonic acid, methyl, diphenyl ester $CH_3P(O)(OC_6H_5)_2$	248.22	...	205^{13}	35	$1.2051^{20/0}$	...	...	$B6^2$, 164
11404	Phosphonic acid, methyl,ethyl ester $CH_3P(O)(OC_2H_5)(OH)$	124.08	...	$106\text{-}7^{0.1}$	...	1.1800^{20}	1.4258^{20}	...	CAS58, 5720
11405	Phosphoric acid, diethyl ester or Diethyl phosphate $(C_2H_5O)_2PO(OH)$	154.10	syr	203d	...	$1.186^{25/4}$	1.4170^{20}	eth	$B1^4$, 1339
11406	Phosphoric acid, dimethyl ester or Dimethyl phosphate $(CH_3O)_2PO(OH)$	126.05	...	172-6d	...	1.335^{25}	1.408^{25}	w, al, ace	$B1^4$, 1259
11407	Phosphoric acid, monoethyl ester $C_2H_5OPO(OH)_2$	126.05	hyg cr	d	...	$1.430^{25/4}$	1.427	w, al, eth, ace	$B1^4$, 1338
11408	Phosphoric acid, phenyl ester or Phenyl phosphate $C_6H_5OPO(OH)_2$	174.09	pl (chl), nd (w)	...	99.5	...	...	w, al, eth, bz, chl	$B6^4$, 708
11409	Phosphoric acid triamide, hexamethyl $[(CH_3)_2N]_3PO$	179.20	...	$98\text{-}100^6$	...	$1.024^{25/25}$	1.4579^{20}	al, eth	$B4^4$, 284
11410	Phosphoric acid, tributyl ester or Tributyl phosphate $(C_4H_9O)_3PO$	266.32	...	289, $160\text{-}2^{15}$	...	$0.9727^{25/4}$	1.4224^{25}	w, al, eth, bz	$B1^4$, 1531
11411	Phosphoric acid, tris (2,4-dimethylphenyl) ester $[2,4\text{-}(CH_3)_2C_6H_3O]_3PO$	410.46	glassy	232-5	...	$1.142^{38/4}$	1.5550^{20}	bz, hx	B6, 488
11412	Phosphoric acid, tris (-2,5-dimethylphenyl) ester $[2,5\text{-}(CH_3)_2C_6H_3O]_3PO$	410.46	cr (dil al)	$260\text{-}5^8$	78-81	1.197^{25}	...	eth, bz	$B6^3$, 1773
11413	Phosphoric acid, tris(2,6-dimethylphenyl) ester $[2,6\text{-}(CH_3)_2C_6H_3O]_3PO$	410.46	wax	$262\text{-}4^6$	136-8	...	...	bz	J1956, 3043
11414	Phosphoric acid, tris (3,4-dimethylphenyl) ester $[3,4\text{-}(CH_3)_2C_6H_3O]_3PO$	410.46	wax	$260\text{-}3^7$	71-2	...	...	bz	B6, 482
11415	Phosphoric acid, tris (3,5-dimethylphenyl) ester $[3,5\text{-}(CH_3)_2C_6H_3O]_3PO$	410.46	wax	290^{10}	46	...	...	aa	C31, 187
11416	Phosphoric acid, triisobutyl ester $(i\text{-}C_4H_9O)_3PO$	266.32	...	264, 138^{10}	...	$0.9681^{20/4}$	1.4193^{20}	w, al, eth, bz	$B1^4$, 1598
11417	Phosphoric acid, triethyl ester or Triethyl phosphate $(C_2H_5O)_3PO$	182.16	...	215-6, 103^{25}	−56.4	$1.0695^{20/4}$	1.4053^{20}	al, eth, bz	$B1^4$, 1339
11418	Phosphoric acid, trimethyl ester or Trimethyl phosphate $(CH_3O)_3PO$	140.08	...	197.2, 85^{26}	α −46 (st), β-62	$1.2144^{20/4}$	1.3967^{20}	w, eth	$B1^4$, 1259
11419	Phosphoric acid, tripentyl ester or Tripentyl phosphate $(C_5H_{11}O)_3PO$	308.40	...	225^{50}, 167^5	...	$0.9608^{20/4}$	1.4319^{20}	al, eth, to	$B1^4$, 1645
11420	Phosphoric acid, triphenyl ester or Triphenyl phosphate $(C_6H_5O)_3PO$	326.29	cr (abs al-lig), pr, (al), nd (eth-lig)	245^{11}	50-1	$1.2055^{50/4}$	...	al, eth, bz, chl	$B6^4$, 720

No.	Name, Synonyms, and Formula	Mol. wt.	Color, crystalline form, specific rotation and λ_{max} (log ε)	b.p. °C	m.p. °C	Density	n_D	Solubility	Ref.
11421	Phosphoric acid, tripropyl ester or Tripropyl phosphate .. $(C_3H_7O)_3PO$	224.24	....	252, 107.5[5]	......	1.0121[20/4]	1.4165[20]	al, eth, to	B1[4], 1428
11422	Phosphoric acid, triisopropyl ester or Triisopropyl phosphate .... $(i\text{-}C_3H_7O)_3PO$	224.24	....	218-20, 95-6[8]	......	0.9867[20/4]	1.4057[20]	al	B1[4], 1478
11423	Phosphoric acid, tri(2-tolyl) ester or Tri-*o*-cresyl phosphate........ $(2\text{-}CH_3C_6H_4O)_3PO$	368.37	col or pa ye	410, 283-5[20]	11	1.955[20/4]	1.5575[20]	al, eth, aa, to	B6[4], 1979
11424	Phosphoric acid, tri(3-tolyl) ester or Tri-*m*-cresyl phosphate........ $(3\text{-}CH_3C_6H_4O)_3PO$	368.37	wax	260[15]	25-6	1.150[25]	1.5575[20]	eth	B6[4], 2057
11425	Phosphoric acid, tri (4-tolyl) ester or Tri-*p*-cresyl phosphate T.P.C. .... $(4\text{-}CH_3C_6H_4O)_3PO$	368.37	nd (al), ta (eth)	244[3.5]	77-8	1.247[25]	....	al, eth, bz, chl, aa	B6[4], 2130
11426	Phosphoric acid, thiono, tri(2-tolyl) ester .... $(2\text{-}CH_3C_6H_4O)_3PS$	384.43	nd (al)	260-5[1]	45-6	....	....	al, aa	B6[4], 1980
11427	Phosphonic acid, thiono, tri-3-tolyl ester .... $(3\text{-}CH_3C_6H_4O)_3PS$	384.43	nd (al)	270-2[1]	40-1	....	....	aa	B6[4], 2058
11428	Phosphonic acid, thiono, tri-4-tolyl ester .... $(4\text{-}CH_3C_6H_4O)_3PS$	384.43	nd (al)	....	93-4	....	....	....	B6[4], 2132
11430	Phosphorous acid, diethyl ester or Diethyl phosphite .... $(C_2H_5O)_2P(OH)$	138.10	......	87[20]	......	1.0720[20/4]	1.4101[20]	al, eth	B1[4], 1329
11431	Phosphorous acid, dimethyl ester or Dimethyl phosphite $(CH_3O)_2P(OH)$	110.05	....	170-1, 70[25]	......	1.2004[20/0]	1.4036[20]	al, Py	B1[4], 1255
11432	Phosphorous acid, tributyl ester or Tributyl phosphite... $(C_4H_9O)_3P$	250.32	....	122[12]	......	0.9259[20/4]	1.4321[19]	al, eth	B1[4], 1527
11433	Phosphorous acid, triethyl ester or Triethyl phosphite ... $(C_2H_5O)_3P$	166.16	....	157.9, 49[12]	......	0.9629[20/4]	1.4127[20]	....	B1[4], 1333
11434	Phosphorous acid, trimethyl ester or Trimethyl phosphite $(CH_3O)_3P$	124.08	....	111-2, 22[23]	......	1.0520[20/0]	1.4095[20]	al, eth	B1[4], 1256
11435	Phosphorous acid, triphenyl ester or Triphenyl phosphite $(C_6H_5O)_3P$	310.29	....	360, 200-1[5]	*ca* 25	1.1844[20/0]	1.5900[20]	al	B6[4], 695
11436	Phosphorous acid, tripropyl ester or Tripropyl phosphite $(C_3H_7O)_3P$	208.24	....	206-7, 92[14]	......	0.9417[20]	1.4282[20]	al, eth	B1[4], 1426
11437	Phosphorous acid, trisopropyl ester or Trisopropyl phosphite .... $(i\text{-}C_3H_7O)_3P$	208.24	....	60-1[10]	......	0.9687[18.5/0]	1.4085[25]	al, eth	B1[4], 1476
11438	Phosphorous acid, tris (2,2,2-trichloro ethyl) ester ...... $(Cl_3CCH_2O)_3P$	476.16	....	263, 127-31[0.1]	......	....	1.5174[20]	....	B1[4], 1384
11439	Phosphorous acid, tri(2-tolyl) ester or Tri-*o*-cresyl phosphite .... $(2\text{-}CH_3C_6H_4O)_3P$	352.38	ye	238[11]	......	1.1423[20/4]	1.5740[28]	eth	B6[4], 1977
11440	Phosphorous acid, tri(4-tolyl) ester or Tri-*p*-cresyl phosphite .... $(4\text{-}CH_3C_6H_4O)_3P$	352.37	pa ye	250-5[10]	......	1.1313[25/25]	1.5703[28]	eth	B6[4], 2128
11441	Phthaladehyde or *o*-Phthalic aldehyde........ $1,2\text{-}C_6H_4(CHO)_2$	134.13	ye cr or nd (lig)	....	53-6	....	....	al, eth	B7[3], 3457
11442	Phthalamic acid or Phthalic acid monoamide .... $2\text{-}HO_2CC_6H_4CONH_2$	165.15	pr	....	148-9	....	....	al	B9[3], 4191
11443	Phthalamic acid, *N*-(1-naphthyl) or Alanap-1 .... $2\text{-}HO_2CC_6H_4CONH\text{-}\alpha\text{-}C_{10}H_7$	291.31	....	....	185	....	....	....	B12[3], 2876
11444	Phthalamide .... $1,2\ C_6H_4(CONH_2)_2$	164.16	cr	d	222	....	....	....	B9[3], 4197
11445	Phthalmide, *N,N,N′N′*-tetraethyl .... $1,2\text{-}C_6H_4[CON(C_2H_5)_2]_2$	276.38	....	204[16]	36	....	....	....	B9[3], 4198
11446	Phthalazine .... $C_8H_6N_2$	130.15	....	175[17]	90-1	....	....	w, al, bz	B23, 174
11447	Phthalazine, 1,2-dihydro-1-oxo or Phthalazone .... $C_8H_6N_2O$	146.15	nd (w), pr (sub)	337[755], sub	184-5	....	....	al, bz	B24[2], 70
11448	Phthalhydrazide or Phthalic acid hydrazide........ $C_8H_6N_2O_2$	162.15	mcl nd (w, dil al, aa)	....	342-4	....	....	aa	B24[2], 194
11449	Phthalic acid or 1,2 Benzene dicarboxylic acid........ $1,2\ C_6H_4(CO_2H)_2$	166.13	pl (w)	d	210-11d, 191 (sealed tube)	1.593	....	al	B9[3], 4094
11450	Phthalic acid , 3-amino .... $3\text{-}H_2NC_6H_3(CO_2H)_2\text{-}1,2$	181.15	nd	....	231-2	....	....	....	B14[3], 1393

No.	Name, Synonyms, and Formula	Mol. wt.	Color, crystalline form, specific rotation and λ_{max} (log ε)	b.p. °C	m.p. °C	Density	n_D	Solubility	Ref.
11451	Phthalic acid, 4-amino, dimethyl ester $4\text{-}H_2NC_6H_3(CO_2CH_3)_2\text{-}1,2$	209.20	pl (al, bz), pr (w)		54			al, chl, Py	B14^3, 1397
11452	Phthalic acid, 3-benzoyl or 2,3-Benzophenone dicarboxylic acid $3\text{-}C_6H_5COC_6H_3(CO_2H)_2\text{-}1,2$	270.26	pl, nd (w + 1)	d	140-1			al, bz	B10^3, 4007
11453	Phthalic acid, 4-benzoyl or 3,4-Benzophenone dicarboxylic acid $4\text{-}C_6H_5COC_6H_3(CO_2H)_2\text{-}1,2$	270.26	lf (xyl)		177			al	B10^3, 4010
11454	Phthalic acid, 3-bromo $3\text{-}BrC_6H_3(CO_2H)_2\text{-}1,2$	245.03	nd		188d			al, eth	B9^3, 4212
11455	Phthalic acid, 4-bromo $4\text{-}BrC_6H_3(CO_2H)_2\text{-}1,2$	245.03				173-5		al, eth	B9^3, 4213
11456	Phthalic acid, butyl ester *(mono)* or *mono*-Butyl phthalate $2\text{-}HO_2CC_6H_4CO_2C_4H_9$	222.24	pl (ace, al)		73-4			al, chl	B9^3, 4101
11457	Phthalic acid, *sec*-butyl ester *(mono) (d)* or *mono-sec*-Butyl phthalate $2\text{-}sec\text{-}C_4H_9O_2CC_6H_4CO_2H$	222.24	$[\alpha]^{20}{}_D$ + 38.4 (al)		48			al, chl	B9^3, 4104
11458	Phthalic acid, *sec*-butyl ester *(mono) (dl)* or mono-*sec*-butyl phthalate $2\text{-}HO_2CC_6H_4CO_2\text{-}sec\text{-}C_4H_9$	222.24	cr (peth)		63			al, chl	B9^3, 4104
11459	Phthalic acid, 3-chloro $3\text{-}ClC_6H_3(CO_2H)_2\text{-}1,2$	200.58	nd (w)		186-7			al, eth	B9^3, 4202
11460	Pathalic acid, 4-chloro $4\text{-}ClC_6H_3(CO_2H)_2\text{-}1,2$	200.58	nd (dil al)		157			al, eth	B9^3, 4202
11461	Phthalic acid, dibenzyl ester or Dibenzyl phthalate $1,2\text{-}C_6H_4(CO_2CH_2C_6H_5)_2$	346.39	pr (al)	277^{15}	42-3			al, eth	B9^3, 4158
11462	Phthalic acid, dibutyl ester or Dibutyl phthalate $1,2\text{-}C_6H_4(CO_2C_4H_9)_2$	278.35		340, 206^{20}		$1.047^{20/20}$	1.4911^{20}	al, eth, bz	B9^3, 4102
11463	Phthalic acid, diisobutyl ester or Diisobutyl phthalate $1,2\text{-}C_6H_4(CO_2\text{-}i\text{-}C_4H_9)_2$	278.35		295-8, $182\text{-}4^{10}$		1.0490^{15}			B9^3, 4105
11464	Phthalic acid, 3,4-dichloro $3,4\text{-}Cl_2C_6H_2(CO_2H)_2\text{-}1,2$	235.02	pl (w)		195			al, eth	B9, 817
11465	Phthalic acid, 3,5-dichloro $3,5\text{-}Cl_2C_6H_2(CO_2H)_2\text{-}1,2$	235.02	nd, ta (aq HCl)	sub	164d			al, eth, ace	B9^3, 4204
11466	Phthalic acid, 3,6-dichloro $3,6\text{-}Cl_2C_6H_2(CO_2H)_2\text{-}1,2$	235.02	pl (w)		d 100			al, eth	B9^3, 4204
11467	Phthalic acid, 4,5-dichloro $4,5\text{-}Cl_2C_6H_2(CO_2H)_2\text{-}1,2$	235.02	nd (w)		*ca* 200d			eth	B9^3, 4205
11468	Phthalic acid, dicyclohexyl ester or Dicyclohexyl phthalate $1,2\text{-}C_6H_4(CO_2C_6H_{11})_2$	330.42	pr (al)		66	$1.383^{20/4}$	1.451^{20}	al, eth	B9^3, 4123
11469	Phthalic acid, di-(2-ethoxyethyl) ester $1,2\text{-}C_6H_4(CO_2CH_2CH_2OC_2H_5)_2$	310.35		345, $233\text{-}5^{23}$	34	1.1229^{21}			B9^2, 597
11470	Phthalic acid, diethyl ester or Diethylphthalate $1,2\text{-}C_6H_4(CO_2C_2H_5)_2$	222.24		298, 172^{12}		$1.1175^{20/4}$	1.5000^{21}	**al, eth**, ace, bz	B9^3, 4099
11471	Phthalic acid, 4,5-dimethoxy or *m*-Hemipic acid $4,5\text{-}(CH_3O)_2C_6H_2(CO_2H)_2\text{-}1,2$	226.19	nd (w), pr (w + 2)		174-5				B10^3, 2431
11472	Phthalic acid, di-(2-methoxyethyl) ester $1,2\text{-}C_6H_4(CO_2CH_2CH_2OCH_3)_2$	282.29		230^{10}		1.1708^{15}			B9^3, 4173
11473	Phthalic acid, dimethyl ester or Dimethyl phthalate $1,2\text{-}C_6H_4(CO_2CH_3)_2$	194.19	pa ye	283.8	0-2	$1.1905^{20/4}$	1.5138^{20}	**al, eth**, bz	B9^3, 4098
11474	Phthalic acid, diisopentyl ester or Diisopentylphthalate $1,2\text{-}C_6H_4(CO_2\text{-}i\text{-}C_5H_{11})_2$	306.40		330-8d, 225^{40}		$1.0220^{16/16}$	1.4871^{20}	al	B9^3, 4107
11475	Phthalic acid, diphenyl ester or Diphenyl phthalate $1,2\text{-}C_6H_4(CO_2C_6H_5)_2$	318.33	pr (al, lig)	$250\text{-}7^{14}$, sub	73				B9^3, 4157
11476	Phthalic acid, dipropyl ester or Dipropyl phthalate $1,2\text{-}C_6H_4(CO_2C_3H_7)_2$	250.29		304-5				al, eth	B9^3, 4101
11477	Phthalic acid, ethyl ester or Mono-ethylphthalate $2\text{-}HO_2CC_6H_4CO_2C_2H_5$	194.19		d	2	$1.1877^{22/4}$	1.509^{22}	al, eth	B9^3, 4099
11478	Phthalic acid, 3-hydroxy $3\text{-}HOC_6H_3(CO_2H)_2\text{-}1,2$	182.13	nd, pr (eth-peth, w)	sub	150d			al, eth	B10^3, 2189
11479	Phthalic acid, 4-hydroxy $4\text{-}HOC_6H_3(CO_2H)_2\text{-}1,2$	182.13	rosettes (w)		204-5d			al, eth	B10^3, 2190
11480	Phthalic acid mono amide or Phthalamic acid $2\text{-}HO_2CC_6H_4\text{-}CONH_2$	165.15	pr		148-9			al	B9^3, 4191
11481	Phthalic acid mononitrile $2\text{-}NCC_6H_4CO_2H$	147.13	nd (al)		187d			ace, al	B9^3, 4199

No.	Name, Synonyms, and Formula	Mol. wt.	Color, crystalline form, specific rotation and λ_{max} (log ε)	b.p. °C	m.p. °C	Density	n_D	Solubility	Ref.
11482	Phthalic acid mononitrile, monoamide or Phthalamonitrile $2\text{-}NCC_6H_4CONH_2$	146.15	nd (MeOH), cr (aa)	...	175	...	...	al, ace	B9[3], 4199
11483	Phthalic acid, 3-nitro $3\text{-}NO_2C_6H_3(CO_2H)_2\text{-}1,2$	211.13	pa ye pr (w)	...	218	...	...	al	B9[3], 4215
11484	Phthalic acid, 3-nitro, diethyl ester $3\text{-}NO_2C_6H_3(CO_2C_2H_5)_2\text{-}1,2$	267.24	pr (al), nd (peth)	...	46	...	...	al, eth	B9[3], 4216
11485	Phthalic acid, 4-nitro $4\text{-}NO_2 1,2\text{-}C_6H_3(CO_2H)_2$	211.13	pa ye nd (w, eth)	...	165-6	...	...	al, w	B9[3], 4234
11486	Phthalic acid, 4-nitro, dimethyl ester $4\text{-}NO_2C_6H_3(CO_2CH_3)_2\text{-}1,2$	239.19	cr (dil al)	...	69-71	...	...	al, MeOH	B9[2], 607
11487	Phthalic acid, 2-octyl ester *(D)* or 2-Octyl phthalate (mono) $2\text{-}HO_2CC_6H_4CO_2[CH(CH_3)C_6H_{13}]$	278.35	pr (peth), $[\alpha]^{20/}{}_D$ + 48.7 (al)	...	75	$1.027^{72/4}$	...	al, bz, chl	B9[3], 4112
11488	Phthalic acid, 2-octyl ester-mono *(dl)* $2\text{-}HO_2CC_6H_4CO_2[CH(CH_3)C_6H_{13}]$	278.35	pr (peth)	...	55	...	...	al, ace, bz, chl	B9[1], 358
11489	Phthalic acid, 2-octyl ester-mono *(l)* $2\text{-}HO_2CC_6H_4CO_2[CH(CH_3)C_6H_{13}]$	278.35	pr (peth)	...	75	...	...	al, bz, chl	B9[3], 4112
11490	Phthalic acid, tetrabromo $C_6Br_4(CO_2H)_2\text{-}1,2$	481.72	nd (w)	...	266d (anh)	...	...	...	B9[1], 367
11491	Phthalic acid, tetrachloro $C_6Cl_4(CO_2H)_2\text{-}1,2$	303.91	pl (w)	2	250d (anh)	...	...	ace	B9[3], 4205
11492	Phthalic acid, tetrachloro, monoethyl ester $HO_2CC_6Cl_4(CO_2C_2H_5)$	331.97	pr (dil al)	d 250	94-5	...	...	al, eth	B9[3], 4206
11493	Phthalic anhydride $1,2\text{-}C_6H_4(CO)_2O$	148.12	wh nd (al, bz)	295 sub	131.6	...	...	al	B17[4], 6135
11494	Phthalic anhydride, 3-chloro $3\text{-}Cl\text{-}1,2\text{-}C_6H_3(CO)_2O$	182.56	nd (sub)	sub	124-5	...	...	...	B17[4], 6142
11495	Phthalic anhydride, 4-chloro $4\text{-}Cl\text{-}1,2\text{-}C_6H_3(CO)_2O$	182.56	pr	$294\text{-}5^{720}$	98.5	...	...	al, eth	B17[4], 6142
11496	Phthalic anhydride, 3,4-dichloro $3,4\text{-}Cl_2\text{-}1,2\text{-}C_6H_2(CO)_2O$	217.01	pl	329	121	...	...	al, chl	B17[4], 6142
11497	Phthalic anhydride, 3,5-dichloro $3,5\text{-}Cl_2\text{-}1,2\text{-}C_6H_2(CO)_2O$	217.01	nd	...	89	...	...	bz, chl	B17[4], 6143
11498	Phthalic anhydride, 3,6-dichloro $3,6\text{-}Cl_2\text{-}1,2\text{-}C_6H_2(CO)_2O$	217.01	nd	339	194.5	...	...	...	B17[4], 6143
11499	Phthalic anhydride, 4,5-dichloro $4,5\text{-}Cl_2\text{-}1,2\text{-}C_6H_2(CO)_2O$	217.01	ta or pr (to)	313	187-9	...	...	al, eth, to	B17[4], 6143
11500	Phthalic anhydride, 3-nitro $3\text{-}(NO_2)\text{-}1\text{-},2\text{-}C_6H_3(CO)_2O$	193.12	nd (aa, ace, al)	...	164	...	...	ace, al	B17[4], 6149
11501	Phthalic anhydride, tetra bromo $1,2\text{-}C_6Br_4(CO)_2O$	463.70	nd (aa-xyl)	...	279-81	...	...	...	B17[4], 6147
11502	Phthalic anhydride, tetrachloro $1,2\text{-}C_6Cl_4(CO)_2O$	285.90	pr, nd (sub)	sub	255-7	...	...	...	B17[4], 6144
11503	Phthalic anhydride, tetra iodo $1,2\text{-}C_6I_4(CO)_2O$	651.70	ye pr, nd (aa), nd (sub)	sub	327-8	...	...	al	B17[4], 6149
11504	Phthalimide $1,2\ C_6H_4(CO)_2NH$	147.13	nd (w), pr (aa), lf (sub)	...	238	...	...	bz	B21[4], 5017
11505	Phthalimide, *N*-acetyl $1,2\text{-}C_6H_4(CO)_2NCOCH_3$	189.17	nd (bz), cr (al, aa)	...	135-6	...	...	eth, chl	B21[4], 5171
11506	Phthalimide, *N*-benzyl $1,2\text{-}C_6H_4(CO)_2NCH_2C_6H_5$	237.26	ye nd (al), cr (aa)	...	116	1.343^{18}	...	al	B21[4], 5053
11507	Phthalimide, *N*(2-bromo-isobutyl) $1,2\text{-}C_6H_4(CO)_2N[CH_2CBr(CH_3)_2]$	282.14	nd (al), lf (chl)	...	97	...	...	al, bz, chl	B21[2], 349
11508	Phthalimide, *N*(2-bromo ethyl) $1,2\text{-}C_6H_4(CO)_2NCH_2CH_2Br$	254.08	nd (w)	...	82-4	...	...	eth	B21[4], 5033
11509	Phthalimide, *N*-bromo methyl $1,2\text{-}C_6H_4(CO)_2NCH_2Br$	240.06	pr (chl, bz, aa)	...	151.5	...	...	ace	B21[4], 5110
11510	Phthalimide, *N*-(2-bromo propyl) $1,2\text{-}C_6H_4(CO)_2N(CH_2CHBrCH_3)$	268.11	nd (al, MeOH)	...	110-1	...	...	al, eth	B21[2], 349
11511	Phthalimide, *N*-(3-bromo propyl) $1,2\text{-}C_6H_4(CO)_2N[(CH_2)_3Br]$	268.11	nd (lig)	...	72-3	...	...	al, eth	B21[4], 5033
11512	Phthalimide, *N*-isobutyl $1,2\text{-}C_6H_4(CO)_2N\text{-}i\text{-}C_4H_9$	203.24	...	293-5	93	...	...	...	B21[4], 5036
11513	Phthalimide, 3,6-dihydroxy $3,6\text{-}(HO)_2\text{-}1,2\text{-}C_6H_2(CO)_2NH$	179.13	gr-ye nd (w + 3)	...	273-4	...	...	w	B21[1], 478

PHYSICAL CONSTANTS OF ORGANIC COMPOUNDS (Continued)

No.	Name, Synonyms, and Formula	Mol. wt.	Color, crystalline form, specific rotation and λ_{max} (log ε)	b.p. °C	m.p. °C	Density	n_D	Solubility	Ref.
11514	Phthalimide, *N*-ethyl $1,2\text{-}C_6H_4(CO)_2NC_2H_5$	175.19	nd (al)	285.6	79			eth	B21^4, 5032
11515	Phthalimide, *N*-(2-hydroxy ethyl) $1,2\text{-}C_6H_4(CO)_2NCH_2CH_2OH$	191.19	nd (al), lf (w)		129.5				B21^4, 5063
11516	Phthalimide, *N*-(hydroxy methyl) $1,2\text{-}C_6H_4(CO)_2NCH_2OH$	177.16	lf, pr (to)		141-2			to	B21^4, 5108
11517	Phthalimide, *N*-methyl $1,2\text{-}C_6H_4(CO)_2NCH_3$	161.16	nd (al), lf (sub)	285-7	134				B21^4, 5030
11518	Phthalimide, *N*-α-naphthyl $1,2\text{-}C_6H_4(CO)_2N\text{-}\alpha\text{-}C_{10}H_7$	273.29	pr, pl (al-aa)		180-1				B21^4, 5059
11519	Phthalimide, *N*-β-naphthyl $1,2\text{-}C_6H_4(CO)_2N\text{-}\beta\text{-}C_{10}H_7$	273.29	nd (aa)		216			al, aa	B21^4, 5060
11520	Phthalimide, 4-nitro $4\text{-}NO_2\text{-}1,2\text{-}C_6H_3(CO)_2NH$	192.13	col nd (w), ye lf (al-ace)		202			ace, aa	B21^2, 373
11521	Phthalimide, *N*-(4-nitrophenyl) $1,2\text{-}C_6H_4(CO)_2N\text{-}(C_6H_4NO_2\text{-}4)$	268.23	cr (aa)		271-2				B21^4, 5049
11522	Phthalimide, *N*-phenyl $1,2\text{-}C_6H_4(CO)_2NC_6H_5$	223.23	wh nd (al)	sub	210			chl	B21^4, 5047
11523	Phthalonitrile $1,2\text{-}C_6H_4(CN)_2$	128.13	nd (w, lig)		141			al, eth, ace, bz	B9^3, 4199
11524	Phthalonitrile, 3,6-dihydroxy $3,6\text{-}(HO)_2C_6H_2(CN)_2\text{-}1,2$	160.13	yesh lf (w + 2)		230			al, eth	B10^3, 2430
11525	Phthalyl alcohol or *o*-Xylylene glycol $1,2\text{-}C_6H_4(CH_2OH)_2$	138.17	pl (eth, peth)		65-6			w, al, eth	B6^3, 4587
11526	Phthalyl chloride $1,2\text{-}C_6H_4(COCl)_2$	203.02		281.1, 131-3^9	15-6	1.4089$^{20/4}$	1.5684^{20}		B9^3, 4190
11527	Phthalyl fluoride $1,2\text{-}C_6H_4(COF)_2$	170.12		227-8, 84^{13}	42-3			peth	B9^3, 4190
11528	Phthalide or α-Hydroxy-*o*-toluic acid lactone $C_8H_6O_2$	134.13	nd or pl (w)	290	75	1.1636$^{99/4}$	1.536^{99}	al, eth	B17^4, 4948
11529	Phthalide, 3-benzylidene *(trans)* $C_{15}H_{10}O_2$	222.24	mcl pr		108			al	B17^4, 5433
11530	Phthalide, 3,3-diphenyl or Phthalophenone $C_{20}H_{14}O_2$	286.33	lf (al)	235^{15}	120				B17^4, 5561
11531	Phthalide, 6-nitro $C_8H_5NO_4$	179.13	ye nd (al, aa)		145			al, eth, bz, aa	B17^4, 4953
11532	Phthalocyanine or Tetrabenzoporphyrazine $C_{32}H_{18}N_8$	514.55	grsh-bl mcl (guino-line)	sub 55d (vac)					J1938, 1151
11533	Physostigmine or Eserine $C_{15}H_{21}N_3O_2$	275.35	orth pr (eth, bz), $[\alpha]^{17}_D$ -82 (chl, c=1.3)		(i) 105-6 (st), (ii) 86-7 (unst)			al, eth, bz, chl	B23^2, 330
11534	Physostigmine, 2-hydroxybenzoate or Eserin salicylate $C_{22}H_{27}N_3O_5$	413.47	pr (al)		185-7			al, chl	B23^2, 332
11535	Physostigmine sulfate $(C_{15}H_{21}N_3O_2)_2.H_2SO_4$	648.77	dlq sc (ace-eth), $[\alpha]_D$ -130 (w)		140-2 (anh)			w, al, ace	B23^2, 332
11536	Phytadiene-*(d)* or 2,6,10,14-Tetramethyl-13,15-hexadeca-diene $C_{20}H_{38}$	278.52	$[\alpha]_D$ + 0.89 (undil)	186-8^{14}		0.826$^{0/4}$		peth, aa, MeOH	B1^4, 1078
11537	Phytol *(dl)* or 3,7,11,15-Tetramethyl-2-hexadecene-1-ol $C_{20}H_{40}O$	296.54		202-4^{10}, 140-1$^{0.03}$		0.8497$^{25/4}$	1.4595^{25}		B1^4, 2208
11538	Picein or *p*-Hydroxyacetophenone-*D*-glucoside $C_{14}H_{18}O_7$	298.29	nd (w + 1), nd (MeOH), $[\alpha]_D$ -86.5		195-6			al, eth, al	B17^4, 3013
11539	Picene or 1,2-Benzochrysene $C_{22}H_{14}$	278.35	lf, pl (xyl, Py, sub)	518-20, sub 300^2	367-9				B5^3, 2555
11540	α-Picoline or 2-Methyl pyridine $2\text{-}CH_3(C_5H_4N)$	93.13		128.8	−66.8	0.9443$^{20/4}$	1.4957^{20}	w, al, eth, ace	B20^4, 2679
11541	α-Picoline, 6-amino or 6-Amino-2-methylpyridine $6\text{-}NH_2\text{-}2\text{-}CH_3(C_5H_3N)$	108.14	hyg (lig)	208-9	41			w, al, eth, ace, bz	B22^4, 4133
11542	α-Picoline, 6-dimethyl amino or 6-Dimethyl amino-2-methyl pyridine $6\text{-}(CH_3)_2N\text{-}2\text{-}CH_3(C_5H_3N)$	136.20		198-200. 88^{15}				al, eth	B22^4, 4134

PHYSICAL CONSTANTS OF ORGANIC COMPOUNDS (Continued)

No.	Name, Synonyms, and Formula	Mol. wt.	Color, crystalline form, specific rotation and λ_{max} (log ε)	b.p. °C	m.p. °C	Density	n_D	Solubility	Ref.
11543	α-Picoline, 4-ethyl $4\text{-}C_2H_5\text{-}2\text{-}CH_3(C_5H_3N)$	121.18		179		$0.9130^{25/4}$		w, al, eth, ace, bz	$B20^4$, 2798
11544	α-Picoline, 5-ethyl or 5-ethyl-2-methyl pyridine $5\text{-}C_2H_5\text{-}2\text{-}CH_3(C_5H_3N)$	121.18		$178.3, 65\text{-}6^{17}$		$0.9219^{20/20}$	1.4971^{20}	al, eth, ace, bz	$B20^4$, 2798
11545	α-Picoline, 6-ethyl $6\text{-}C_2H_5\text{-}2\text{-}CH_3(C_5H_3N)$	121.18		$160\text{-}1, 73\text{-}6^{12}$		$0.9207^{25/4}$	1.4920^{25}	al, eth, ace	$B20^4$, 2803
11546	β-Picoline or 3-Methylpyridine $3\text{-}CH_3(C_5H_4N)$	93.13		144.1	−18.3	$0.9566^{20/4}$	1.5040^{20}	w, al, eth, ace	$B20^4$, 2710
11547	β-Picolene, 2-amino or 2, Amino-3-methyl pyridine $2\text{-}NH_2\text{-}3\text{-}CH_3(C_5H_3N)$	108.14	hyg	$221.5^{748}, 95^{8}$	33.5			w, al, eth, ace, bz	$B22^4$, 4154
11548	γ-Picoline or 4-Methylpyridine $4\text{-}CH_3(C_5H_4N)$	93.13		144.9	3.6	$0.9548^{20/4}$	1.5037^{20}	w, al, eth, ace	$B20^4$, 2732
11549	γ-Picoline, 2-amino or 2-Amino-4-methyl pyridine $2\text{-}NH_2\text{-}4\text{-}CH_3(C_5H_3N)$	108.14	lf or pl (lig)	$115\text{-}7^{11}$	100			w, al, eth, ace, bz	$B22^4$, 4172
11550	γ-Picoline, 3-Amino or 3-Amino-4-methyl pyridine $3\text{-}NH_2\text{-}4\text{-}CH_3(C_5H_3N)$	108.14	pr (bz-peth)	254^{735}	106			w, al, eth, ace, bz	$B22^4$, 4181
11551	γ-Picoline-2-ethyl or 2-Ethyl-4-methylpyridine $2\text{-}C_2H_5\text{-}4\text{-}CH_3(C_5H_3N)$	121.18		$173\text{-}5^{748}$		$0.9239^{20/0}$		al, eth, ace	$B20^4$, 2798
11552	γ-Picoline-3-ethyl or β-Collidine $3\text{-}C_2H_5\text{-}4\text{-}CH_3(C_5H_3N)$	121.18		$198, 76^{12}$		$0.9286^{17/4}$		al, eth, ace, chl	$B20^4$, 2804
11553	Picolinamide or 2-Pyridine carboxamide $2\text{-}(C_5H_4N)CONH_2$	122.13	mcl pr (w)		107-8			al, bz	$B22^4$, 311
11554	Picolinic acid or 2-Pyridine carboxylic acid $2\text{-}(C_5H_4N)CO_2H$	123.11	nd (w, al, bz)	sub	136-7			al, aa	$B22^4$, 303
11555	Picolinic acid, ethyl ester or Ethyl picolinate $2\text{-}(C_5H_4N)CO_2C_2H_5$	151.16	ye in air	$243, 122^{13}$	0-2	1.1194^{20}	1.5104^{20}	w, al, eth	$B22^4$, 308
11556	Picolinic acid, *N*-Methyl or Trigonelline $C_7H_7NO_2$	137.14	pr (aq al + 1 w)		218d (anh), 130 (+1w)			w	$B22^4$, 462
11557	Picolinic acid, 2,3,6-tetrahydro or Baikiain $2\text{-}(C_5H_8N)CO_2H$	127.14	pr (MeOH), $[\alpha]^{20}_D$ −201.6		274d			w	$B22^4$, 182
11558	Picolino nitrile $2\text{-}(C_5H_4N)CN$	104.11	nd or pr (eth)	222-7	29	$1.0810^{25/4}$	1.5242^{25}	w, al, eth, bz chl	$B22^4$, 320
11559	Picric acid or 2,4,6-Trinitro phenol $2,4,6\text{-}(NO_2)_3C_6H_2OH$	229.11	ye lf (w), pr (eth), pl (al)	sub exp >300	122-3		1.763	al, eth, ace, bz, aa, Py	$B6^4$, 1388
11560	Picrolonic acid $C_{10}H_8N_4O_5$	264.20	ye nd (al)		116-7, (d 125)				$B24^2$, 25
11561	Picropodophyllin $C_{22}H_{22}O_8$	414.41	col nd (al, bz)		228			al, eth, ace, bz, chl	$B19^4$, 5298
11562	Picrotoxin or Cocculin $C_{30}H_{34}O_{13}$	602.59	rh lf, $[\alpha]^{16}_D$ −29.3 (abs al, c=4)		203-4			al, Py	$B19^4$, 5245
11563	Pilocarpidine *(d)* $C_{10}H_{14}N_2O_2$	194.23	syr, $[\alpha]^{20}_D$ +81.3 (w, al)					w, al	$B27^2$, 694
11564	Pilocarpidine nitrate $C_{10}H_{14}N_2O_2.HNO_3$	257.25	pr (w), $[\alpha]^{20}_D$ +73.2 (w)		137			w, al	$B27^2$, 694
11565	Pilocarpine $C_{11}H_{16}N_2O_2$	208.26	nd $[\alpha]^{20}_D$ +100.5 (w)	260^5	34			w, al, chl	$B27^2$, 694
11566	Pilocarpine hydrochloride $C_{11}H_{16}N_2O_2.HCl$	244.72	hyg cr		204-5			w, al	$B27^2$, 695
11567	Pilocarpine, 2-hydroxybenzoate or Pilocarpine salicylate $C_{18}H_{22}N_2O_5$	346.38	nd or lf (al), $[\alpha]_D$ +63		120			w, al, eth	$B27^2$, 696
11568	Pilocarpine nitrate $C_{11}H_{16}N_2O_2.HNO_3$	271.27	wh pw or cr (al), $[\alpha]_D$ + 80 (w, c=4)		178			w	$B27^2$, 695

PHYSICAL CONSTANTS OF ORGANIC COMPOUNDS (Continued)

No.	Name, Synonyms, and Formula	Mol. wt.	Color, crystalline form, specific rotation and λ_{max} (log ε)	b.p. °C	m.p. °C	Density	n_D	Solubility	Ref.
11569	Pilocarpine, sulfate $2(C_{11}H_{16}N_2O_2)_2H_2SO_4$	514.59	hyg cr (al-eth), $[\alpha]^{20/}{}_D$ +85 (w, c=7)		132			w, al	B27, 635
11570	Pimaric acid *(d)* or Dextropimaric acid $C_{20}H_{30}O_2$	302.46	orh (ace), pr (al) $[\alpha]^{20/}{}_D$ +87.3 (chl)	282^{18}	218-9			al, eth, Py	B9³, 2911
11571	Pimelic acid or Heptanedioic acid $HO_2C(CH_2)_5CO_2H$	160.17	pr (w)	272^{100} (sub), 212^{10}	106	1.329^{15}		w, al, eth	B2⁴, 2003
11572	Pimelic acid, diethyl ester or Diethyl pimelate $C_2H_5O_2C(CH_2)_5CO_2C_2H_5$	216.28		$252\text{-}5^{745}$, $139\text{-}41^{15}$	−24	0.9945^{20}	1.4305^{20}	al, eth	B2⁴, 2004
11573	Pimelic acid, dimethyl ester or Pimethyl pamelate $CH_3O_2C(CH_2)_5CO_2CH_3$	188.22		120^{10}, 80^{1}	−21	$1.0625^{20/4}$	1.4309^{20}	al, eth	B2⁴, 2004
11574	Pimelic acid, ethyl ester, mono $HO_2C(CH_2)_5CO_2C_2H_5$	188.22	cr (eth)	182^{18}, 160^{4}	10		1.4415^{20}		B2³, 1742
11575	Pimelic acid, 4-oxo $OC(CH_2CH_2CO_2H)_2$	174.15	rh pl (w)		143			al, w	B3³, 1380
11576	Pimelonitrile or 1,5-Dicyanopentane $NC(CH_2)_5CN$	122.17		175^{14}	−31.4	0.949^{18}	1.4472^{20}	al, eth, chl	B2⁴, 2006
11577	Pimelyl chloride or Heptanedioyl chloride $ClOC(CH_2)_5COCl$	197.06		137^{15}					B2⁴, 2005
11578	Pinacolone or 3,3 Dimethyl-2-butanone $(CH_3)_3CCOCH_3$	100.16		106	−49.8	$0.8012^{25/4}$	1.3952^{20}	al, eth, ace	B1³, 354
11580	Pinacolyl alcohol or 3,3-Dimethyl-2-butanol $(CH_3)_3CCH(OH)CH_3$	102.18		120.14	5.6	0.8122^{25}	1.4148^{20}	al, eth	B1⁴, 1727
11581	Pinane *(d-cis)* $C_{10}H_{18}$	138.25	$[\alpha]^{20/}{}_D$ +23.3 (undil)	169, 60.1^{18}	−53	$0.8560^{20/4}$	1.4629^{20}	eth, bz	B5⁴, 318
11582	Pinane *(dl)* or Pincocamphane $C_{10}H_{18}$	138.25		164-5		$0.8551^{20/4}$	1.4609^{20}	eth, bz	B5¹, 48
11583	Pinane *(l, cis)* $C_{10}H_{18}$	138.25	$[\alpha]^{20}{}_D$ −47 (undil)	$167\text{-}8^{757}$		0.8556^{21}	1.4645^{20}	eth, bz	B5⁴, 318
11584	α-Pinene *(dl)* $C_{10}H_{16}$	136.24	oil	156.2, 51.4^{20}	−55	$0.8582^{20/4}$	1.4658^{20}	al, eth, chl	B5⁴, 456
11585	β-Pinene *(d)* or Nopinene $C_{10}H_{16}$	136.24	$[\alpha]_D$ +28.6	164-6, 59.7^{20}		$0.8654^{20/4}$	1.4789^{20}	al, eth, bz, chl	B5⁴, 456
11586	β-Pinene *(l)* $C_{10}H_{16}$	136.24	$[\alpha]^{25/}{}_D$ −21.5	164, 59.7^{20}		$0.8694^{20/4}$	1.4762^{20}	al, eth, bz, chl	B5⁴, 457
11587	Pinic acid *(dl)*(α-form) or 3-Carboxy-2,2-dimethyl cyclobutyl acetic acid $C_9H_{14}O_4$	186.21	wh pr (w)	$214\text{-}6^{9}$	101-2	$1.0925^{109/4}$	1.4458^{109}		B9³, 3852
11588	Pinic acid-*(dl)*(β form) $C_9H_{14}O_4$	186.21		$214\text{-}6^{9}$	68				B9³, 3852
11589	Pinic acid *(dl)*(c form) $C_9H_{14}O_4$	186.21	pl	$214\text{-}6^{9}$	58				B9³, 3852
11590	Pinic acid *(l)* $C_9H_{14}O_4$	186.21	nd (w)		135-6				B9³, 3853
11591	Pinol *(dl)* or *(dl)*-Sobrerone-6,8-epoxy-1-*p*-menthene $C_{10}H_{16}O$	152.24	$[\alpha]_D$ −7.1 (ace)	183-4, $76\text{-}7^{14}$		$0.9515^{20/4}$	1.4695^{20}	al, eth	B17⁴, 327
11592	Pinol hydrate *(trans, dl)* $C_{10}H_{16}O.H_2O$	170.25	pl or nd (w)	270-1, $157\text{-}8^{12}$	132			w, al, eth	B6³, 4136
11593	Piperitenone or 3-Terpinolenone $C_{10}H_{14}O$	150.22	$[\alpha]_{546}$ −0.1	$120\text{-}2^{14}$		0.9774^{20}	1.5294^{20}	al, eth	B7³, 559
11594	Piperazine or Hexahydropyrazine $C_4H_{10}N_2$	86.14	hyg pl or lf (al)	146	106		1.446^{113}	w, al	B23², 3
11595	Piperazine, *N*-benzyl $C_{11}H_{16}N_2$	176.26		$145\text{-}7^{12}$			1.5430^{28}	w, al, eth	Am72, 753
11596	Piperazine, *N,N′*-bis-(4 methoxybenzoyl) or *N,N′*-Dianisoylpiperazine $C_{20}H_{22}N_2O_4$	354.41	wh		192-4				Am56, 150
11597	Piperazine, *N,N′*-bis-(phenylacetyl) or *N,N′*-Di-(α-toluyl) piperazine $N,N'\text{-}(C_6H_5CH_2CO)_2(C_4H_8N_2)$	322.41	wh		150-1				Am56, 150

No.	Name, Synonyms, and Formula	Mol. wt.	Color, crystalline form, specific rotation and λ_{max} (log ε)	b.p. °C	m.p. °C	Density	n_D	Solubility	Ref.
11598	Piperazine, *N,N'*-bis-(3-phenyl propionyl) or *N,N'*-Bis-(hydrocinnanmyl)-piperazine *N,N*-$(C_6H_5CH_2CH_2CO)_2(C_4H_8N_2)$	350.46	wh		122-3				Am56, 150
11599	Piperazine, dihydrobromide $C_4H_{10}N_2.2HBr$	247.96	wh nd		d			w	J1957, 1881
11600	Piperazine, dihydrochloride, monohydrate $C_4H_{10}N_2 \cdot 2HCl \cdot H_2O$	177.07	nd (dil al)		82-3			w	B23, 5
11601	Piperazine, *N,N'*-dimethyl or *N,N'*-Dimethylpiperazine $C_6H_{14}N_2$	114.19		$131\text{-}2^{764}$		$0.8600^{20/4}$	1.4474^{20}	w, al, eth	B23², 5
11602	Piperazine, 2,5-dimethyl *(cis)* $2,5\text{-}(CH_3)_2C_4H_8N_2$	114.19	rh bipym nd or pr (chl)	162 sub	114		1.4720^{20}	w, al, chl	B23², 21
11603	Piperazine, 2,5-dimethyl *(trans)* $2,5\text{-}(CH_3)_2(C_4H_8N_2)$	114.19	mcl pl or pr (bz, chl)	162 sub	118-9			w, al, chl	B23², 19
11604	Piperazine, 2,6-dimethyl *(cis)* $2,6\text{-}(CH_3)_2(C_4H_8N_2)$	114.19	lf or pl (bz)	162	111-3			w, al, peth, chl	B23¹, 8
11605	Piperazine, *N,N'*-dinitroso or *N,N'*-Dinitrosopiperazine $C_4H_8N_4O_2$	144.13	pa ye pl (w)		158			al	B23¹, 7
11606	Piperazine, *N,N'*-diphenyl $C_{16}H_{18}N_2$	238.33		$230\text{-}5^{112}$, dec 300	164				B23², 5
11607	Piperazine, *N*-ethyl $C_6H_{14}N_2$	114.19		154^{753}				w, al, eth	B23², 5
11608	Piperazine hexahydrate $C_4H_{10}N_2 \cdot 6H_2O$	194.23		125-3	44-5			w, al	CAS47, 10014
11609	Piperazine, *N*-methyl *N*-$CH_3(C_4H_9N_2)$	100.16		138			1.4378^{20}	w, al, eth	C51, 10538
11610	Piperazine, 2-methyl $2\text{-}CH_3(C_4H_9N_2)$	100.16	hyg lt (al)	155^{763}	62			w, al, eth, bz, chl	B23¹, 16
11611	Piperazine, *N*-phenyl *N*-$C_6H_5(C_4H_9N_2)$	162.23	pa ye oil	286.5, $156\text{-}7^{10}$		$1.0621^{20/4}$	1.5875^{20}	**al, eth**	C49, 11662
11612	Piperazine, 1,2,4-trimethyl $1,2,4\text{-}(CH_3)_3C_4H_7N_2$	128.22		149-50			1.4433^{20}		
11613	*N*-Piperazine carboxaldehyde *N*-$OHC(C_4H_9N_2)$	114.15		$94\text{-}7^{0.5}$			$1.5094^{20/}{}_D$		
11614	*N*-Piperazine carboxylic acid, ethyl ester $C_7H_{14}N_2O_2$	158.20		237, $116\text{-}7^{12}$			1.4760^{25}	w, al, eth	B23², 9
11615	*N,N'*-Piperazine dicarboxylic acid-diethyl ester *N,N'*-$(C_2H_5O_2C)_2C_4H_8N_2$	230.26	nd (hex)	315, $131\text{-}3^{3}$	49			al, eth	
11616	2,5-Piperazinedione $C_4H_6N_2O_2$	114.10	ta or pl (w)		318-20d, sub 260			al	B24², 141
11617	Piperidine or hexahydropyridine $C_5H_{10}NH$	85.15		106, 17.7^{20}	−9	$0.8606^{20/4}$	1.4530^{20}	**w, al,** eth, ace, bz, chl	B20⁴, 287
11618	Piperidine, *N*-acetyl *N*-$CH_3CO(NC_5H_{10})$	127.19		226-7, 109^{18}		1.011^{9}	1.4790^{25}	w, al	B20⁴, 965
11619	Piperidine, 2-allyl $2\text{-}CH_2{=}CHCH_2(C_5H_9NH)$	125.21			170-1	$0.8823^{15/4}$			B20, 147
11620	Piperidine, *N*-benzoyl $C_6H_5CO(NC_5H_{10})$	189.26	tcl	320-1, 180^{15}	49			al, eth	B20⁴, 972
11621	Piperidine, 4-benzoyl-*N*-methyl $C_{13}H_{17}NO$	203.28	nd	$160\text{-}3^{13}$, $130\text{-}7^{2}$	35-7		1.5430^{23}	al, eth, ace, bz	B21⁴, 3689
11622	Piperidine, 4-benzyl $4\text{-}C_6H_5CH_2(C_5H_9NH)$	175.27		279, $150\text{-}2^{17}$	6-7	$0.9972^{20/0}$	1.5337^{25}	al, eth	B20⁴, 3055
11623	Piperidine, *N*-butyl $C_4H_9(NC_5H_{10})$	141.26		175-7, $47\text{-}8^{20}$		$0.8245^{20/4}$	1.4467^{20}		B20⁴, 311
11624	Piperidine, 2-butyl-*(d)* $2\text{-}C_4H_9(C_5H_9NH)$	141.26	$[\alpha]^{t}{}_D$ + 15.7			0.8512			B20, 127
11625	Piperidine, 2-butyl *(dl)* $2\text{-}C_4H_9(C_5H_9NH)$	141.26		191-3, 75^{14}		$0.8529^{15/4}$			B20⁴, 1633
11626	Piperidine, 2-butyl *(l)* $2\text{-}C_4H_9(C_5H_9NH)$	141.26	$[\alpha]^{16}{}_D$ −18.7			0.8533			B20, 127
11627	Piperidine, 3-butyl $3\text{-}C_4H_9(C_5H_9NH)$	141.26		196-7				al, eth, chl	B20⁴, 1635
11628	Piperidine, *N*-isobutyl *i*-$C_4H_9(NC_5H_{10})$	141.26		160-1, 87^{43}		0.8161^{25}	1.4382^{25}	w	B20⁴, 314
11629	Piperidine, 2-isobutyl $2\text{-}i\text{-}C_4H_9(C_5H_9NH)$	141.26		181-2		$0.8510^{22/4}$	1.4553^{22}	al, eth	B20⁴, 1636

No.	Name, Synonyms, and Formula	Mol. wt.	Color, crystalline form, specific rotation and λ_{max} (log ε)	b.p. °C	m.p. °C	Density	n_D	Solubility	Ref.
11630	Piperidine, *N-sec*-butyl *(dl)* $CH_3CH_2CH(CH_3)(NC_5H_{10})$	141.26		175-6		$0.8378^{20/4}$	1.4506^{20}	al, eth	B20[4], 313
11631	Piperidine, *N-sec*-butyl-*(l)* $CH_3CH_2CH(CH_3)(NC_5H_{10})$	141.26	$[\alpha]^{25}_D$ −54.6	175		$0.835^{25/4}$	1.4486^{21}	al, eth	B20[4], 313
11632	Piperidine, *N-tert*-butyl $(CH_3)_3C(NC_5H_{10})$	141.26		166		$0.8465^{20/4}$	1.4532^{20}	al, eth	B20[4], 315
11633	Piperidine, 2,4-diethyl 2,4-$(C_2H_5)_2(C_5H_8NH)$	141.26		174-9		0.8722^{0}			B20, 128
11634	Piperidine, 2,5-diethyl 2,5$(C_2H_5)_2(C_5H_8NH)$	141.26		190, $100\text{-}5^{22}$		0.8722^{0}		eth, chl	B20, 128
11635	Piperidine, 3,4-diethyl *(cis)* 3,4-$(C_2H_5)_2(C_5H_8NH)$	141.26	$[\alpha]^{22}_D$ +26.0 (90% al, c=4.35)	70^{12}					B20[4], 1637
11636	Piperidine, 3,4-diethyl *(trans)* 3,4-$(C_2H_5)_2(C_5H_8NH)$	141.26		193^{720}					B20[4], 1637
11637	Piperidine, *N*, 2-dimethyl *(d)* or *N*-Methyl-α-pipecoline *N*-2-$(CH_3)_2(NC_5H_9)$	113.20	$[\alpha]^{15}_D$ +68.8 (undil)	127		0.825^{16}	1.4395^{20}		B20[4], 1444
11638	Piperidine, *N*,2-dimethyl-*(dl)* *N*,2-$(CH_3)_2(NC_5H_9)$	113.20		127.5		$0.824^{15/4}$	1.4395^{20}	w, al, eth	B20[4], 1444
11639	Piperidine, *N*, 3-dimethyl-*(dl)* or *N*-Methyl)-β-pipecoline *N*, 3-$(CH_3)_2(NC_5H_9)$	113.20		124-6		0.818^{15}			B20[4], 1499
11640	Piperidine, 2,3-dimethyl or α, β-Lupetidine 2,3-$(CH_3)_2(C_5H_8NH)$	113.20		$138\text{-}40^{720}$				w, al	B20[4], 1573
11641	Piperidine, 2,4-dimethyl *(d)* or α, γ-Lupetidine 2,4-$(CH_3)_2(C_5H_8NH)$	113.20	$[\alpha]_D$ +23.2	140-2		0.845			B20, 108
11642	Piperidine, 2,4-dimethyl *(dl)*. 2,4-$(CH_3)_2(C_5H_8NH)$	113.20		140-2		0.8615^{0}	1.4366^{25}	al, eth	B20[4], 1574
11643	Piperidine, 2,5-dimethyl or α, β'-Lupetidine 2,5-$(CH_3)_2(C_5H_8NH)$	113.20		138-40			1.4452^{25}	al	B20[4], 1574
11644	Piperidine, 2,6-dimethyl or α, α'-Lupetidine 2,6-$(CH_3)_2(C_5H_8NH)$	113.20		$127\text{-}8^{768}$		$0.8158^{25/4}$	1.4377^{20}	w, al, eth	B20[4], 1579
11645	Piperidine, 3,3-dimethyl or β, β'-Lupetidine 3,3-$(CH_3)_2(C_5H_8NH)$	113.20		137, $45\text{-}6^{20}$			1.4452^{25}	al	B20[4], 1595
11646	Piperidine, 4,4-dimethyl or γ, γ-Lupetidine 4,4-$(CH_3)_2(C_5H_8NH)$	113.20		145-6, $30\text{-}2^{12}$			1.4489^{25}	w	B20[4], 1596
11647	Piperidine, *N*-(2,2-dimethyl propyl) or *N*-Neopentyl piperdine $CH_3C(CH_3)_2CH_2(NC_5H_{10})$	155.28		188		$0.8608^{20/4}$	1.4593^{20}	al, eth	B20[2], 13
11648	Piperidine, *N*-dodecyl $C_{12}H_{25}(NC_5H_{10})$	253.47	pa ye	161^{5}, $114\text{-}6^{0.6}$		$0.8378^{20/4}$	1.4588^{20}		B20[4], 321
11649	Piperidine, *N*-ethyl $C_2H_5(NC_5H_{10})$	113.20		130.8		$0.8237^{20/4}$	1.4480^{20}		B20[4], 307
11650	Piperidine, 2-ethyl *(d)* 2-$C_2H_5(C_5H_9NH)$	113.20	$[\alpha]_D$ +17.1		142-4	0.8680^{4}			B20[4], 1564
11651	Piperidine, 2-ethyl *(dl)* 2-$C_2H_5(C_5H_9NH)$	113.20		142-3, $73\text{-}5^{52}$		$0.8650^{0/0}$	1.4494^{21}		B20[1], 28
11652	Piperidine, 2-ethyl *(l)* 2-$C_2H_5(C_5H_9NH)$	113.20	$[\alpha]_D$ −14.9	143^{720}		0.8680^{4}	1.4544^{20}		B20[4], 1564
11653	Piperidine, 3-ethyl *(dl)* 3-$C_2H_5(C_5H_9NH)$	113.20	fum in air	152.6		$0.8565^{23/4}$	1.4531^{20}	ace	B20[4], 1567
11654	Piperidine, 3-ethyl *(l)* 3-$C_2H_5(C_5H_9NH)$	113.20	$[\alpha]^{15}_D$ −4.5	155				ace	B20[4], 1567
11655	Piperidine, 4-ethyl 4-$C_2H_5(C_5H_9NH)$	113.20		156-8		0.8759^{0}	1.4503^{25}		B20[4], 1569
11656	Piperidine, *N*-heptyl $C_7H_{15}(NC_5H_{10})$	183.34		259.5, $100\text{-}3^{9}$		0.8316^{20}	1.4531^{20}		B20[4], 317
11657	Piperidine, *N*-hexyl $C_6H_{13}(NC_5H_{10})$	169.31		219.2, $103\text{-}4^{20}$		$0.8292^{20/4}$	1.4522^{20}		B20[4], 317
11658	Piperidine, 2-(1-hydroxyethyl) 2-$[CH_3CH(OH)](C_5H_9NH)$	129.20		$106\text{-}10^{18}$					B21[4], 85
11659	Piperidine, *N*-(2-hydroxyethyl) $HOCH_2CH_2(NC_5H_{10})$	129.20		$200\text{-}2^{742}$, 90^{12}		$0.9732^{25/25}$	1.4749^{20}	w, al	B20[4], 387
11660	Piperidine, 2-(2-hydroxyethyl) 2-$(HOCH_2CH_2)(C_5H_9NH)$	129.20		234.5, $145\text{-}6^{36}$	39-40	1.01^{27}		w, al, eth	B21[4], 85
11661	Piperidine, 3-(2-hydroxyethyl) 3-$(HOCH_2CH_2)(C_5H_9NH)$	129.20		$121\text{-}3^{6}$		$1.0106^{25/4}$	1.4888^{25}	w, al, eth	B21[4], 92

No.	Name, Synonyms, and Formula	Mol. wt.	Color, crystalline form, specific rotation and λ_{max} (log ε)	b.p. °C	m.p. °C	Density	n_D	Solubility	Ref.
11662	Piperidine, 4-(2-hydroxyethyl) $4\text{-}(HOCH_2CH_2)(C_5H_9NH)$	129.20	syr	227-8, $120\text{-}5^{15}$	132-3	$1.0059^{15/4}$	1.4907^{20}	w, al, eth	B21[4], 93
11663	Piperidine, *N*-methyl $CH_3(NC_5H_{10})$	99.18	...	107	...	$0.8159^{20/4}$	1.4355^{20}	w, al, eth	B20[4], 305
11664	Piperidine, 2-methyl *(d)* or *d*-α-Pipecoline $2\text{-}CH_3(C_5H_9NH)$	99.18	$[\alpha]^{22/}{}_D$ +5.6 (al)	117^{745}	...	...	1.4459^{20}	...	B20[4], 1444
11665	Piperidine, 2-methyl *(dl)* or α-Pipecoline $2\text{-}CH_3(C_5H_9NH)$	99.18	...	$117\text{-}8^{747}$	−4.9	$0.8436^{24/4}$	1.4459^{20}	w, al, eth	B20[4], 1444
11666	Piperidine, 3-methyl *(dl)* or β-Pipecoline $3\text{-}CH_3(C_5H_9NH)$	99.18	...	$125\text{-}6^{763}$	0-5	$0.8446^{26/4}$	1.4470^{20}	w	B20[4],1499
11667	Piperidine, 3-methyl *(l)* $3\text{-}CH_3(C_5H_9NH)$	99.18	$[\alpha]^{25/}{}_D$ −4	124	...	...	...	w	B20[4], 1498
11668	Piperidine, 4-methyl or γ-Pipecoline $4\text{-}CH_3(C_5H_9NH)$	99.18	...	132-4	...	0.8674^{0}	1.4458^{20}	w	B20[4], 1511
11669	Piperidine, *N*-methyl-4-benzoyl $C_{13}H_{17}NO$	203.28	...	$160\text{-}3^{13}$	35-7	...	1.5430^{25}	al, eth, ace, bz	B21[4], 3689
11670	Piperidine, *N*-(2-methyl-2-pentyl) $C_{11}H_{23}N$	169.31	...	205-7	...	$0.8517^{20/4}$	1.4592^{20}	ace	B20[2], 14
11671	Piperidine, *N*-(3-methyl-3-pentyl) $C_{11}H_{23}N$	169.31	...	214^{752}	...	$0.8614^{20/4}$	1.4637^{20}	...	B20[2], 14
11672	Piperidine, *N*-nitroso $N\text{-}ON(C_5H_{10}N)$	114.15	pa ye	217^{721}, 109^{20}	...	$1.0631^{18.5/4}$	$1.4933^{18.5}$	w	B20[4], 1371
11673	Piperidine, *N*-nonyl $C_9H_{19}(NC_5H_{10})$	211.39	...	$135\text{-}7^{11}$	...	0.8313^{25}	1.4538^{25}	...	B20[4], 319
11674	Piperidine, *N*-octyl $C_8H_{17}(NC_5H_{10})$	197.36	...	$136\text{-}8^{13}$, 89^{1}	...	$0.8324^{20/4}$	1.4544^{20}	...	B20[4], 318
11675	Piperidine, *N*-pentyl $C_5H_{11}(NC_5H_{10})$	155.28	...	198.2, 80^{8}	...	0.8282^{20}	1.4498^{20}	...	B20[4], 315
11676	Piperidine, *N*-isopentyl $i\text{-}C_5H_{11}(NC_5H_{10})$	155.28	...	188-9, $76\text{-}9^{20}$	...	...	...	ace	B20[4], 316
11677	Piperidine, *N*-phenyl $C_6H_5(NC_5H_{10})$	161.25	...	$257\text{-}8^{752}$, 126.5^{15}	...	...	...	al, eth, bz, chl	B20[4], 339
11678	Piperidine, *N*-phenyl-4-dimethyl amino or Irenal $4\text{-}(CH_3)_2N(C_5H_9NC_6H_5)$	204.32	lf (bz)	$123\text{-}6^{0.5}$	47-8	...	...	bz	B22[4], 3751
11679	Piperidine, 2-propenyl *(d)* or β-Coniceine $2\text{-}(CH_2{=}CHCH_2)C_5H_9NH$	125.21	nd, $[\alpha]^{45/}{}_D$ +49.9	168-9	39	...	...	al, eth	B20, 146
11680	Piperidine, 2-propenyl-*(dl)* $2\text{-}(CH_2{=}CHCH_2)C_5H_9NH$	125.21	nd	$168\text{-}70^{753}$	8	$0.8716^{25/4}$	...	al, eth	B20, 146
11681	Piperidine, 2-propenyl *(l)* $2\text{-}(CH_2{=}CHCH_2)C_5H_9NH$	125.21	nd, $[\alpha]^{15/}{}_D$ -50.5	168-9	39-40	$0.8672^{15/4}$	...	...	B20, 146
11682	Piperidine, *N*-propyl $C_3H_7(NC_5H_{10})$	127.23	...	151.2	...	0.8231^{20}	1.4446^{20}	w, al, eth	B20[4], 309
11683	Piperidine, 3-propyl *(d)* $3\text{-}C_3H_7(C_5H_9NH)$	127.23	$[\alpha]^{16/}{}_D$ +5.9	174^{752}	...	$0.8517^{19/4}$	...	w, al	B20, 120
11684	Piperidine, 3-propyl *(dl)* $3\text{-}C_3H_7(C_5H_9NH)$	127.23	...	174^{758}	...	$0.8475^{26/4}$	...	al, w	B20, 119
11685	Piperidine, 3-propyl *(l)* $3\text{-}C_3H_7(C_5H_9NH)$	127.23	oil, $[\alpha]^{16/}{}_D$ -6.6	174^{752}	...	$0.8517^{19/4}$	...	w, al	B20, 120
11686	Piperidine, 4-propyl $4\text{-}C_3H_7(C_5H_9NH)$	127.23	...	172^{748}	...	0.864^{22}	1.4465^{23}	al	B20[4], 1616
11687	Piperidine, *N*-isopropyl $(CH_3)_2CH(NC_5H_{10})$	127.23	...	$149\text{-}50^{757}$	...	$0.8389^{20/4}$	1.4491^{20}	w	B20[4], 311
11688	Piperidine, 2-isopropyl *(dl)* $2\text{-}(CH_3)_2CH(C_5H_9NH)$	127.23	...	162	...	0.8668^{0}	...	...	B20[4], 1620
11689	Piperidine, 2-isopropyl *(l)* $2\text{-}(CH_3)_2CH(C_5H_9NH)$	127.23	$[\alpha]_D$ -13.1	161.5	...	0.8503^{19}	...	...	B20, 121
11690	Piperidine, 4-isopropyl $4\text{-}(CH_3)_2CH(C_5H_9NH)$	127.23	...	168-71, $66\text{-}70^{15}$	...	...	...	...	B20[4], 1620
11691	Piperidine, 2,2,6,6-tetramethyl $2,2,6,6\text{-}(CH_3)_4(C_5H_6NH)$	141.26	...	155-7	...	$0.8367^{16/4}$	1.4455^{20}	eth	B20[4], 1639
11692	Piperidine, 2,2,4-trimethyl $2,2,4\text{-}(CH_3)_3(C_5H_7NH)$	127.23	...	148	...	0.832^{15}	1.4458^{20}	al, eth	B20[4], 1624
11693	Piperidine, 2,3,6-trimethyl $2,3,6\text{-}(CH_3)_3(C_5H_7NH)$	127.23	...	36^{5}	...	$0.8302^{20/4}$	1.4434^{20}	...	B20[4], 1625
11694	Piperidine, 2,4,6-trimethyl $2,4,6\text{-}(CH_3)_3(C_5H_7NH)$	127.23	...	165-6	...	$0.8315^{19/4}$	1.4412^{20}	al, eth	B20[4], 1625
11695	*N*-Piperidine carboxaldehyde or *N*-Formylpiperidine $(C_5H_{10}N)CHO$	113.16	pa ye	222, $106\text{-}10^{17}$	...	$1.0205^{25/4}$	1.4700^{20}	w, al, eth, bz, chl, lig	B20[4], 964

No.	Name, Synonyms, and Formula	Mol. wt.	Color, crystalline form, specific rotation and λ_{max} (log ε)	b.p. °C	m.p. °C	Density	n_D	Solubility	Ref.
11696	4-Piperidine carboxylic acid or Hexahydroisonicotinic acid 4-$(HNC_5H_9)CO_2H$	129.16	nd (w)	...	*ca* 326	...	...	w	B22^{4}, 128
11697	4-Piperidine carboxylic acid, ethyl ester 4-$(HNC_5H_9)CO_2C_2H_5$	157.21	col oil	100-1^{10}	...	...	1.4591^{20}	w, al, eth, bz	B22', 128
11698	4-Piperidine carboxylic acid, methyl ester 4-$(HNC_5H_9)CO_2CH_3$	143.19	col oil	107-10^{22}	...	...	1.4635^{25}	w, al, eth, bz	B22^{4}, 128
11699	4-Piperidine carboxylic acid, *N*-methyl, methyl ester *N*-$CH_3(NC_5H_9)CO_2CH_3$-4	157.21	...	96-100^{20}	...	...	1.4539^{24}	w, al, eth, bz	B22^{4}, 131
11700	4-Piperidine carboxylic acid, *N*, methyl-4-phenyl,ethyl ester hydrochlo or Demerol hydrochloride *N*-$CH_3(NC_5H_8)(C_6H_5)(CO_2C_2H_5)$-4,4	283.80	cr (al)	...	186-9	...	...	w, ace	B22^{4}, 1004
11701	2,4-Piperidine dione, 3,3-diethyl or Piperidone $C_9H_{15}NO_2$	169.22	nd (w)	...	103-5	...	...	w, al	B21^{4}, 4614
11702	2-Piperidone or 5-Amino pentanoic acid lactone C_5H_9NO	99.13	hyg	256, 137^{14}	39-40	...	...	w, al, eth	B21^{4}, 3170
11703	2-Piperidone, 3-hydroxy *(dl)* $C_5H_9NO_2$	115.13	nd or pr (AcOEt)	...	141-2	...	...	w, eth	B21^{4}, 6022
11704	2-Piperidone, 5-hydroxy $C_5H_9NO_2$	115.13	(al, al-AcOEt)	...	145-6	...	...	w, al	B21^{4}, 6023
11705	4-Piperidone hydrochloride $C_5H_9NO.HCl$	135.59	cr (+ $^{3}/_{2}$ al, al-eth)	...	147-9	...	...	w	B21^{4}, 3183
11706	4-Piperidone, 2,3,6,6-tetramethyl or Triacetone amine $C_9H_{17}NO$	155.24	rh pl (moist eth), nd (eth)	205	58	...	...	w, al, eth	B21^{4}, 3278
11707	Piperine or 1-Piperylpiperidine $C_{17}H_{19}NO_3$	285.34	pr (AcOEt), pl or mcl pr (al), cr (bz-lig)	...	130-3	...	...	al, bz, chl, aa, Py	B20^{4}, 1341
11708	Piperitone *(d)* or 1-Methyl-4-isopropyl-3-cyclohexenone $C_{10}H_{16}O$	152.24	yesh in air, $[\alpha]^{20/}{}_{D}$ + 49.1 (undil)	222-30, 116-8^{20}	...	0.9344$^{20/4}$	1.4843^{20}	eth	B7^{3}, 326
11709	Piperitone- *(dl)* $C_{10}H_{16}O$	152.24	...	232-3, 113^{18}	...	0.9331$^{20/4}$	1.4845^{20}	...	B7^{3}, 326
11710	Piperitone *(l)* $C_{10}H_{16}O$	152.24	$[\alpha]^{20/}{}_{D}$ -51.5 (undil)	235, 109-10^{15}	...	0.9324$^{20/4}$	1.4848^{20}	...	B7^{3}, 324
11711	Piperoin $C_{16}H_{12}O_6$	300.27	...	...	120	...	...	al, chl	B19^{4}, 5958
11712	Piperolidine *(dl)* or δ -Coniceine $C_8H_{15}N$	125.21	...	161.5^{750}, 65-7^{10}	...	0.9012$^{15/4}$	1.4748	al, eth	B20^{4}, 1989
11713	Piperolidine *(l)* $C_8H_{15}N$	125.21	$[\alpha]^{27/}{}_{D}$ -7.9	158^{729}, 65-7^{18}	...	0.8976$^{20/4}$	1.4748^{20}	al, eth	B20^{4}, 1989
11714	Piperonal or 3,4-Methylenedioxy benzaldehyde $(3,4\text{-}CH_2O_2)C_6H_3CHO$	150.13	wh-ye (w)	263, 140^{15}	37	...	...	al, **eth**, ace	B19^{4}, 1649
11715	Piperonal oxime $(3,4\text{-}CH_2O_2)C_6H_3CH{=}NOH$	165.15	...	...	146	...	...	eth	B19^{4}, 1666
11716	Piperonyl alcohol or 3,4-Methylenedioxy benzyl alcohol $(3,4\text{-}CH_2O_2)C_6H_3CH_2OH$	152.15	nd (peth)	157^{16}	58	...	...	al, eth, bz, chl	B19^{4}, 734
11717	Piperonylic acid or 3,4-Methylene dioxy benzoic acid $(3,4\text{-}CH_2O_2)C_6H_3CO_2H$	166.13	nd (al), pr (sub)	sub	229-31	...	...	...	B19^{4}, 3493
11718	Piperonylic acid, ethyl ester or Ethyl piperonylate $(3,4\text{-}CH_2O_2)C_6H_3CO_2C_2H_5$	194.19	pr	285-6, 164-5^{11}	18.5	...	...	al, eth, peth	B19^{2}, 293
11719	Piperonylic acid, methyl ester or Methyl piperonylate $(3,4\text{-}CH_2O_2)C_6H_3CO_2CH_3$	180.16	nd or lf (peth)	273-4d	53	...	...	al, eth	B19^{4}, 3493
11720	Piperonyl chloride $(3,4\text{-}CH_2O_2)C_6H_3COCl$	184.58	cr	155^{25}	80	...	...	...	B19^{4}, 3497
11721	Piperylene or 1,3,-Pentadiene $CH_3CH{=}CHCH{=}CH_2$	68.12	...	42	−87.5	0.6760$^{20/4}$	1.4301^{20}	**al, eth, ace, bz**	B1^{4}, 994
11722	Pivaldehyde or 2,2-Dimethyl propionaldehyde $(CH_3)_3CCHO$	86.13	...	77-8	6	0.7923^{17}	1.3791^{20}	al, eth	B1^{4}, 3295
11723	Pivaldehyde oxime or Pivaldoxime $(CH_3)_3CCH{=}NOH$	101.15	...	65^{20}	48	...	...	...	B1^{3}, 2824
11724	Pivalamide, *N,N*-diethyl $(CH_3)_3CCON(C_2H_5)_2$	157.26	...	203	...	0.891^{15}	...	al, eth	B4^{3}, 211
11725	Pivalic acid or 2,2-Dimethyl propionic acid $(CH_3)_3CCO_2H$	102.13	nd	164, 70^{14}	35	0.905^{50}	1.3931$^{30\ 5}$	al, eth	B2^{4}, 908

No.	Name, Synonyms, and Formula	Mol. wt.	Color, crystalline form, specific rotation and λ_{max} (log ε)	b.p. °C	m.p. °C	Density	n_D	Solubility	Ref.
11726	Pivalic acid, chloro or 3-Chloro-2,2-dimethyl propionic acid $ClCH_2C(CH_3)_2CO_2H$	136.58		108-12[10]	41-2				B2[4], 914
11728	Pivalic acid, ethyl ester or Ethyl pivalate $(CH_3)_3CCO_2C_2H_5$	130.19		118	−89.5	$0.856^{20/4}$	1.3906^{20}	al, eth	B2[4], 910
11729	Pivalic acid, methyl ester or Methyl pivalate $(CH_3)_3CCO_2CH_3$	116.16		101		$0.891^{0/4}$	1.3880^{20}	al, eth	B2[4], 909
11730	Pivalonitrile or *tert*-Butyl cyanide $(CH_3)_3CCN$	83.13		105-6	15-6	$0.7586^{25/4}$	1.3774^{20}		B2[4], 913
11731	Pivalophenone or *tert*-Butyl phenyl ketone $C_6H_5COC(CH_3)_3$	162.23		219-21, 97-8[16]		0.963^{26}	1.5086^{19}	ace	B7[3], 1125
11732	Pivalyl chloride or 2,2-Dimethyl propionyl chloride $(CH_3)_3CCOCl$	120.58		107, 48[100]		1.003^{20}	1.4139^{20}	eth	B2[4], 912
11733	Pivalyl chloride-chloro $ClCH_2C(CH_3)_2COCl$	155.02		85-6[60]			1.4539^{20}		B2[4], 914
11734	Podophyllotoxin $C_{22}H_{22}O_8$	414.41	$[\alpha]^{20/}{}_D$ -132.7 ($CHCl_3$, 2%)		114-8			al, ace, bz, chl	B19[4], 5299
11735	Polyglycolid $[O\text{-}CH_2CO]_n$				223				B19[1], 679
11736	Polyporic acid $C_{18}H_{12}O_4$	292.29			310-2				B8[1], 3854
11737	Popalin or Salicinbenzoate (an α-glucoside) $C_{20}H_{22}O_8$	390.39	nd (w + 2), pr (al), $[\alpha]_D$ 2.0 (Py, c=5)		180			eth, aa	B17[4], 3298
11738	Porphin or Tetramethene tetrapyrrole $C_{20}H_{14}N_4$	310.36	red or og lf (chl-MeOH)	sub 300[12]	darkens 360	1.336		bz, diox	B26[2], 228
11739	5-α-Pregnane $C_{21}H_{36}$	288.52	$[\alpha]^{19/}{}_D$ ($CHCl_3$, c=1.69)		84-5				B5[4], 1215
11740	5α-Pregnane-20β -ol-3 one $C_{21}H_{34}O_2$	318.50	$[\alpha]_D$ (20 $CHCl_3$, c=1.2)		185				B8[3], 614
11741	5β -Pregnane or 17β -Ethyletiocholane $C_{21}H_{36}$	288.52	mcl sc or pl (MeOH), $[\alpha]^{19/}{}_D$ + 21.2 (chl, c=2)		83.5	$1.032^{15/4}$		chl	B5[4], 1215
11742	5β -Pregnane-3α, 20 α -diol $C_{21}H_{36}O_2$	320.52	pl (ace), $[\alpha]^{20/}{}_D$ + 27.4 (al, c=0.7)		243-4	1.15			B6[1], 4779
11743	5 β -Pregnane 3 α -20 β-diol $C_{21}H_{36}O_2$	320.52	cr (al), $[\alpha]^{20/}{}_D$ + 10		244-6			al	B6[3], 4779
11744	5 β -Pregnane-3 β -20 α -diol $C_{21}H_{36}O_2$	320.52	cr (al, ace)		182			al	B6[3], 4778
11745	5 β -Pregnane-3 β -20 β -diol $C_{21}H_{36}O_2$	320.52	cr (AcOEt-peth, dil al)		174-6			AcOEt	B6[3], 4778
11746	5 β -Pregnane-3,20-dione $C_{21}H_{32}O_2$	316.48	nd (dil al), cr (dil ace)		123			al, eth, ace	B7[3], 3568
11747	5 β -Pregnan-3 α -ol-20-one $C_{21}H_{34}O_2$	318.50	nd (bz), cr (dil al)		149.5			al	B8[3], 618
11748	5 β -Pregnan-3 β -ol-20-one $C_{21}H_{34}O_2$	318.50	cr (dil al)		149			al	B8[3], 617
11749	5 β -Pregnan-20 α -ol-3-one $C_{21}H_{34}O_2$	318.50	pr (ace)		152				B8[3], 614
11750	5 β-Pregnan-20 β -ol-3-one $C_{21}H_{34}O_2$	318.50	cr (dil MeOH)		172			MeOH	B8[3], 614
11751	5-Pregnen-3 β -ol-20-one $C_{21}H_{32}O_2$	316.48	nd (dil al), $[\alpha]^{20/}{}_D$ + 28 (al)		192				B8[3], 949

PHYSICAL CONSTANTS OF ORGANIC COMPOUNDS (Continued)

No.	Name, Synonyms, and Formula	Mol. wt.	Color, crystalline form, specific rotation and λ_{max} (log ε)	b.p. °C	m.p. °C	Density	n_D	Solubility	Ref.
11752	4-Pregnen-11 β, 17 α, 20 β, 21-tetrol-3-one or Reichstein's substance E $C_{21}H_{32}O_5$	364.48	cr (aq, ace)		*ca* 125d			al, ace	B8^{3}, 4027
11753	4-Pregnen-17 α, 20 β, 21-triol-3-one $C_{21}H_{32}O_4$	348.48	cr (MeOH), $[\alpha]_D$ + 63 (diox, c=1)		190			diox, chl, MeOH	B8^{3}, 3533
11754	Prehnitene or 1,2,3,4-Tetramethyl benzene 1,2,3,4-$(CH_3)_4C_6H_2$	134.22	cr (peth)	205, 79.4^{10}	−6.2	0.9052$^{20/4}$	1.5203^{20}	**al, eth, ace, bz**	B5^{4}, 1072
11755	Primeverose or 6-*O*-β-D-xylopyranosyl-α-D-glucose $C_{11}H_{20}O_{10}$	312.27	cr (MeOH), $[\alpha]^{20}{}_D$ + 23→-3.2 (w, c=5) (mut)		210			w, MeOH	B17^{4}, 2447
11756	α-Progesterone or 17 α-Progesterone $C_{21}H_{30}O_2$	314.47	orh pr (dil al), $[\alpha]_D$ + 192		129-31	1.166^{21}			B7^{3}, 3648
11757	β-Progesterone or 17 β-Progesterone $C_{21}H_{30}O_2$	314.47	nd (peth), $[\alpha]^{20}{}_D$ + 172 (diox, c=2)		121-2	1.171^{20}		al, ace, diox	B7^{3}, 3648
11758	Progesterone dioxime $C_{21}H_{32}N_2O_2$	344.50	pl (dil al)		243			al	B7^{3}, 3654
11759	Proline *(D)* or *d*-2-Pyrrolidene carboxylic acid $C_5H_9NO_2$	115.13	hyg pr (al-eth), $[\alpha]^{20}{}_D$ + 81.9 (w)		215-20d			w, al	B22^{4}, 8
11760	Proline-*(DL)* $C_5H_9NO_2$	115.13	hyg nd (al-eth), cr (+ w)		205d (anh)			w, al	B22^{4}, 12
11761	Proline *(L)* $C_5H_9NO_2$	115.13	nd (al-eth), pr (w), $[\alpha]^{20}{}_D$ -80.9 (w, c=1)		220-2d			w	B22^{4}, 8
11762	Proline, 4-hydroxy *(D-cis)* or *Allo*-4-hydroxyproline $C_5H_9NO_3$	131.13	nd (w + 1), $[\alpha]^{18}{}_D$ + 58.6 (w, p=5)		237-41			w	B22^{4}, 2045
11763	Proline-4-hydroxy *(DL, cis)* or *Allo*-4-hydroxyproline $C_5H_9NO_3$	131.13	cr (w, dil al)		250			w	B22^{4}, 2046
11764	Proline-4-hydroxy *(L, cis)* or *Allo*-4-hydroxyproline $C_5H_9NO_3$	131.13	nd (w + 1), $[\alpha]^{18}{}_D$ -58.1 (w, p=5.2)		238-41			w, al	B22^{4}, 2046
11765	Proline-4-hydroxy *(D-trans)* or α-4-Hydroxyproline $C_5H_9NO_3$	131.13	lf (dil al), $[\alpha]^{21}{}_D$ + 75.2 (w)		274			w	B22^{4}, 2047
11766	Proline-4-hydroxy *(DL,trans)* or α-4-Hydroxyproline $C_5H_9NO_3$	131.13	pl (MeOH)		270			w	B22^{4}, 2049
11767	Proline-4-hydroxy-*(L,trans)* or α-4-Hydroxyproline $C_5H_9NO_3$	131.13	lf (dil al), pr (w), $[\alpha]_D$ -76.5 (w, c=2.5)		274			w	B22^{4}, 2047
11768	Proline, 4-hydroxy, betaine *(d)* $C_7H_{13}NO_3$	159.19	pr (w + 1), $[\alpha]^{21}{}_D$ + 36		249d				B22^{4}, 2053
11769	Prontosil $C_{12}H_{14}N_5O_2ClS$	327.79	og-red pw		248-51			al, ace, oils	B16^{3}, 439
11770	Propadiene or Allene $CH_2{=}C{=}CH_2$	40.06	gas	−34.5	−136	0.787	1.4168	bz, peth	B1^{4}, 966
11771	Propanal or Propionaldehyde CH_3CH_2CHO	58.08		48.8	−81	0.8058$^{20/4}$	1.3636^{20}	w, **al, eth**	B1^{4}, 3165
11772	Propane $CH_3CH_2CH_3$	44.10		−42.1	−189.7	0.5853$^{-45/4}$	1.2898^{20}	al, eth, bz, chl	B1^{4}, 176
11773	Propane, 1-amino or *n*-Propyl amine $CH_3CH_2CH_2NH_2$	59.11		47.8	−83	0.7173$^{20/4}$	1.3870^{20}	w, al, eth, ace, bz, chl	B4^{4}, 464

PHYSICAL CONSTANTS OF ORGANIC COMPOUNDS (Continued)

No.	Name, Synonyms, and Formula	Mol. wt.	Color, crystalline form, specific rotation and λ_{max} (log ε)	b.p. °C	m.p. °C	Density	n_D	Solubility	Ref.
11774	Propane, 1-amino-2,2-dimethyl or Neopentyl amine $(CH_3)_3CCH_2NH_2$	87.16		81-2^{741}		0.7455$^{20/4}$	1.4023^{20}	eth	B4^4, 707
11775	Propane, 1-amino-3-dodecyloxy $C_{12}H_{25}O(CH_2)_3NH_2$	243.43		140^1		0.8439$^{20/4}$	1.4487^{20}	ace, bz, chl, MeOH	C55, 22136
11776	Propane, 1-amino-3-methoxy $CH_3O(CH_2)_3NH_2$	89.14		116-9		0.8727$^{20/4}$	1.4391^{20}	w, al, bz, MeOH, chl	B4^4, 1623
11777	Propane, 2-amino or Isopropyl amine $(CH_3)_2CHNH_2$	59.11		32.4	−95.2	0.6891^{20}	1.3742^{20}	w, **al**, **eth**, ace, bz, chl	B4^4, 504
11778	Propane, 2,2,bis(4-aminophenyl) $(4\text{-}H_2NC_6H_4)_2C(CH_3)_2$	226.32		200-31$^{5.5}$	128.2-9.6				B13^3, 495
11779	Propane, 1,3-bis (dimethylamino) $CH_2[CH_2N(CH_3)_2]_2$	130.23		144		0.7837$^{18/4}$		w, al, eth	B4^4, 1259
11780	Propane, 1,1-bis(4-hydroxyphenyl) or 4,4′-Propylidenediphenol $CH_3CH_2CH(C_6H_4OH\text{-}4)_2$	228.29	nd (w)	275^{20}	130			al, eth, aa	B6^3, 5457
11781	Propane, 2,2-bis(4-hydroxyphenyl) or Bis-phenol A -4,4′-isopropylidene diphenol $(4\text{-}HOC_6H_4)_2C(CH_3)_2$	228.29	pr (dil aa), nd (w)	250-2^{13}	152-3			al, eth, bz, aa	B6^3, 5459
11782	Propane, 1-bromo or *n*-Propylbromide $CH_3CH_2CH_2Br$	122.99		71	−110	1.3537$^{20/4}$	1.4343^{20}	al, eth, ace, bz	B1^4, 205
11783	Propane, 1-bromo-2-chloro $CH_3CHClCH_2Br$	157.44		118		1.531$^{20/4}$	1.4745^{20}	al, eth, ace, bz	B1^4, 212
11784	Propane, 1-bromo-3-chloro $Cl(CH_2)_3Br$	157.44		143.3, 32.4^{10}	−58.9	1.5969$^{20/4}$	1.4864^{20}	al, eth, chl	B1^4, 212
11785	Propane, 1-bromo 2,3-dimethyl or Neopentyl bromide $(CH_3)_3CCH_2Br$	151.05		106, 34.6^{100}		1.1997$^{20/4}$	1.4370^{20}	al, eth, ace, bz, chl	B1^4, 337
11786	Propane, 1-bromo-3-fluoro $F(CH_2)_3Br$	140.98		101.4		1.542$^{25/4}$	1.4290^{25}	al, eth, bz, chl	B1^4, 210
11787	Propane, 1-bromo-1-nitro $CH_3CH_2CH(Br)NO_2$	167.99		160-5, 82.5^{50}				al, eth	B1^3, 260
11788	Propane, 2-bromo or Isopropyl bromide $(CH_3)_2CHBr$	122.99		59.4	−89	1.3140$^{20/4}$	1.4251^{20}	**al**, **eth**, ace, bz, chl	B1^4, 208
11789	Propane, 2-bromo-1-chloro $CH_3CHBrCH_2Cl$	157.44		118^{756}		1.537$^{20/4}$	1.4795^{20}	al, eth, ace, bz, chl	B1^4, 212
11790	Propane, 2-bromo-2-chloro $(CH_3)_2CClBr$	157.44		93-5^{745}		1.474^{22}	1.4575^{20}	al, eth, ace, bz, chl	B1^4, 213
11791	Propane, 3-bromo-1,2-epoxy *(d)* or Epibromohydrin $BrCH_2CHCH_2$ (O bridging)	136.98	$[\alpha]^{16}{}'_D$ +45.4 (undil)	134-6^{50}				eth, bz, chl	B17^4, 22
11792	Propane, 3-bromo-1,2-epoxy *(dl)* or Epibromohydrin $BrCH_2CHCH_2$	136.98		138-40, 61-2^{50}		1.615^{14}	1.4841^{20}	eth, bz, chl	B17^4 22
11793	Propane, 2-bromo-2-nitro $(CH_3)_2CBrNO_2$	167.99		152^{745}, 73-5^{50}		1.6562^0		al, eth	B1^4, 233
11794	Propane, 2-(bromomethyl)-1,2,3-tribromo $(BrCH_2)_3CBr$	373.71		150-1^{14}	25	2.5595$^{20/4}$	1.6246^{20}	eth, ace	B1^2, 91
11795	Propane, 1-chloro or *n*-Propyl chloride $CH_3CH_2CH_2Cl$	78.54		46.6	−122.8	0.8909$^{20/4}$	1.3879^{20}	**al**, **eth**, bz, chl	B1^4, 189
11796	Propane, 1-chloro-2,2, difluoro $CH_3CF_2CH_2Cl$	114.52		55	−56.2	1.2001$^{20/4}$	1.3520^{20}	eth, bz, chl	B1^4, 193
11797	Propane, 1-chloro-2,2-dimethyl or Neopentyl chloride $(CH_3)_3CCH_2Cl$	106.60		84.3	−20	0.8660$^{20/4}$	1.4044^{20}	al, eth, bz, chl	B1^4, 336
11798	Propane, 1-chloro-3-fluoro $F(CH_2)_3Cl$	96.53		79.5^{740}			1.3871^{25}	al, eth, bz, chl	B1^4, 193
11799	Propane, 1-chloro-3-iodo $ICH_2CH_2CH_2Cl$	204.44		170-2, 57.10		1.904^{20}	1.5472^{20}	eth, bz, chl	B1^4, 226
11800	Propane, 1-chloro-1-nitro $CH_3CH_2CH(Cl)NO_2$	123.54		141-3, 67^{56}		1.209$^{20/20}$	1.4251^{20}	al, eth	B1^4, 232
11801	Propane, 1-chloro-2-nitro $CH_3CH(NO_2)CH_2Cl$	123.54		172-3, 94^{46}		1.245^{22}	1.4432^{25}	al, eth, chl	B1^4, 233
11802	Propane, 1-chloro-3-nitro $ClCH_2CH_2CH_2NO_2$	123.54		197d, 115-6^{40}		1.267^{20}		al, eth	B1^3, 259
11803	Propane, 1-chloro-2-phenyl *(dl)* $CH_3CH(C_6H_5)CH_2Cl$	154.64		85^{13}		1.0484^{70}	1.5245^{20}		B5^4, 992
11804	Propane, 1-chloro-3-phenyl $C_6H_5(CH_2)_3Cl$	154.64		219-20, 110^{21}		1.056$^{21/4}$	1.5160^{25}		B5^4, 980
11805	Propane, 2-chloro or Isopropyl chloride $(CH_3)_2CHCl$	78.54		35.7	−117.2	0.8617$^{20/4}$	1.3777^{20}	**al**, **eth**, bz, chl	B1^4, 191

PHYSICAL CONSTANTS OF ORGANIC COMPOUNDS (Continued)

No.	Name, Synonyms, and Formula	Mol. wt.	Color, crystalline form, specific rotation and λ_{max} (log ε)	b.p. °C	m.p. °C	Density	n_D	Solubility	Ref.
11806	Propane, 2-chloro-1-nitro $CH_3CHClCH_2NO_2$	123.54	...	172, 75^{15}	...	1.2361^{15}	1.4447^{20}	al, eth	$B1^4$, 232
11807	Propane, 2-chloro-2-nitro $(CH_3)_2CClNO_2$	123.54	...	134d, 57^{50}	...	1.230^{19}	1.4378^{19}	al, eth	$B1^4$, 233
11808	Propane, 2-chloro-1-phenyl *(d)* $CH_3CHClCH_2C_6H_5$	154.64	$[\alpha]_{4359}$, +21.2	94^{17}	...	$1.038^{19/4}$	1.5198^{20}	ace, bz, chl	$B5^4$, 980
11809	Propane, 2-chloro-1-phenyl *(dl)* $CH_3CHClCH_2C_6H_5$	154.64	...	79^{10}	...	$1.0367^{17/4}$	1.5134^{20}	ace, bz, chl	$B5^4$, 980
11810	Propane, 2-chloro-1-Phenyl *(l)* $CH_3CHClCH_2C_6H_5$	154.64	$[\alpha]_{5461}$ -24.9	94^{17}	...	$1.038^{19/4}$	1.5198^{22}	ace, bz, chl	$B5^4$, 980
11811	Propane, 2-(chloromethyl)-1,1,2,3-tetrachloro $(ClCH_2)_2CClCHCl_2$	230.35	...	226^{737}, 95^9	...	$1.5686^{25/4}$	1.5165^{25}	al, bz, chl	$B1^3$, 321
11812	Propane, 2-(chloro methyl)-1,2,3-trichloro $(ClCH_2)_3CCl$	195.90	...	$209\text{-}10^{737}$ 87^9	...	$1.5036^{25/4}$	1.508^{20}	chl	$B1^3$, 321
11813	Propane, 3-chloro-1,2-epoxy *(dl)* or α- Epichlorohydrin $ClCH_2CHCH_2$ (O)	92.53	...	116.5, $60\text{-}1^{100}$	−48	$1.1801^{20/4}$	1.4361^{20}	al, eth, bz	$B17^4$, 22
11814	Propane, 3-chloro-1,2,-epoxy *(l)* or α -Epichlorohydrin $ClCH_2CHCH_2$ (O)	92.53	$[\alpha]^{18}/_D$ -25.6	$92\text{-}3^{360}$	...	1.2007	...	al, eth, bz	$B17^2$, 12
11815	Propane, 3-chloro-2-methyl-1,2-epoxy $ClCH_2C(CH_3)CH_2$ (O)	106.55	...	122, 51^{55}	...	$1.1011^{20/4}$	1.4340^{20}	w, eth	$B17^4$, 47
11816	Propane, 1,2-diamino *(d)* or 1,2-Propanediamine $CH_3CH(NH_2)CH_2NH_2$	74.13	$[\alpha]^{25}/_D$ +29.8	120.5	...	$0.8584^{25/4}$	...	w, chl	$B4^4$, 1255
11817	Propane, 1,2-diamino-*N,N*′-diacetyl or 1,2-Diacetamido propane $CH_3CH(NHCOCH_3)CH_2NHCOCH_3$	158.20	nd (bz)	190^{18}	138-9	...	...	w, al, chl	$B4^3$, 552
11818	Propane, 1,3-diamino or Trimethylene diamine $H_2N(CH_2)_3NH_2$	74.13	...	135.5	...	$0.884^{25/4}$	1.4600^{20}	w, al, eth	$B4^4$, 1258
11819	Propane, 1,1-dibromo or Propylidine bromide $CH_3CH_2CHBr_2$	201.89	...	133.5, 28.4^{10}	...	$1.982^{20/4}$	1.5100^{20}	al, eth, chl	$B1^4$, 215
11820	Propane, 1,1-dibromo-2,2-dimethyl or Neopentylidene bromide $(CH_3)_3CCHBr_2$	229.94	...	180, 66^{20}	14	$1.6695^{20/4}$	1.5047^{20}	eth, bz, chl	$B1^3$, 371
11821	Propane , 1,2-dibromo $CH_3CHBrCH_2Br$	201.89	...	140, 35.7^{10}	−55.2	$1.9324^{20/4}$	1.5201^{20}	al, eth, chl	$B1^4$, 215
11822	Propane, 1,2-dibromo-2-methyl $(CH_3)_2CBrCH_2Br$	215.92	...	149-50, 61^{40}	9-12	$1.759^{20/4}$	1.509	al, eth, bz, chl	$B1^4$, 298
11823	Propane, 1,3,-dibromo $Br(CH_2)_3Br$	201.89	...	167.3, 56.6^{20}	fr−34.2	$1.9822^{20/4}$	1.5232^{20}	al, eth	$B1^4$, 216
11824	Propane, 1,3-dibromo-2,2-dimethyl $(CH_3)_2C(CH_2Br)_2$	229.94	...	185-90d, 72^{14}	...	$1.6934^{20/4}$	1.5050^{20}	al, eth, bz, chl	$B1^4$, 337
11825	Propane, 2,2-dibromo $(CH_3)_2CBr_2$	201.89	...	$114\text{-}5^{740}$	...	$1.7825^{20/4}$	...	al, eth, chl	$B1^4$, 217
11826	Propane, 1,1-dichloro or Propylidene dichloride $CH_3CH_2CHCl_2$	112.99	...	88.1	...	$1.1321^{20/4}$	1.4289^{20}	al, eth, bz, chl	$B1^4$, 195
11827	Propane, 1,1-dichloro-2-methyl or Isobutylidene chloride $(CH_3)_2CHCHCl_2$	127.01	...	105-6	...	$1.0111^{12/12}$	1.4330^{15}	al, eth, bz, chl	$B1^4$, 292
11828	Propane, 1,2-dichloro $CH_3CHClCH_2Cl$	112.99	...	96.4, -3.7^{10}	−100.4	$1.1560^{20/4}$	1.4394^{20}	al, eth, bz, chl	$B1^4$, 195
11829	Propane, 1,2-dichloro-2-fluoro $CH_3CClFCH_2Cl$	130.98	...	88.6	−91.7	$1.2624^{20/4}$	1.4099^{20}	ace, bz	$B1^4$, 197
11830	Propane, 1,2-dichloro-2-methyl or Isobutylene chloride $(CH_3)_2CClCH_2Cl$	127.01	...	108, $38\text{-}9^{70}$	...	$1.093^{20/4}$	1.4370^{20}	al, eth, ace, bz	$B1^4$, 292
11831	Propane, 1,3-dichloro $Cl(CH_2)_3Cl$	112.99	...	120.4, 14^{10}	−99.5	$1.1876^{20/4}$	1.4487^{20}	al, eth, bz, chl	$B1^4$, 196
11832	Propane, 1,3-dichloro-2-methyl $CH_3CH(CH_2Cl)_2$	127.01	...	134.6, 60^{49}	...	$1.1325^{25/4}$	1.4488^{25}	al, eth, bz	$B1^4$, 293
11833	Propane, 2,2-dichloro $(CH_3)_2CCl_2$	112.99	...	69.3	−33.8	$1.1136^{20/4}$	1.4148^{20}	al, eth, bz, chl	$B1^4$, 196
11834	Propane, 1,1-dicyclohexyl $CH_3CH_2CH(C_6H_{11})_2$	208.39	...	270-1	...	$0.8887^{23/0}$	1.485^{23}	...	$B5^4$, 350
11835	Propane, 1,2-dicyclohexyl $CH_3CH(C_6H_{11})CH_2C_6H_{11}$	208.39	...	272-3	...	$0.8725^{21/0}$	1.479^{21}	...	$B5^4$, 350
11836	Propane, 1,3-dicyclohexyl $CH_2(CH_2C_6H_{11})_2$	208.39	...	291-2	−17	$0.8752^{24/24}$	1.4736^{24}	...	$B5^4$, 349
11837	Propane, 1,3-dicyclohexyl-2-ethyl $C_2H_5CH(CH_2C_6H_{11})_2$	236.44	...	296	...	$0.8846^{21/0}$	1.483^{21}	...	$B5^4$, 363
11838	Propane, 1,3-dicylohexyl-2-methyl $CH_3CH(CH_2C_6H_{11})_2$	222.41	...	295.2	0.6	0.8715^{20}	1.4756^{20}	...	$B5^4$, 358

PHYSICAL CONSTANTS OF ORGANIC COMPOUNDS (Continued)

No.	Name, Synonyms, and Formula	Mol. wt.	Color, crystalline form, specific rotation and λ_{max} (log ε)	b.p. °C	m.p. °C	Density	n_D	Solubility	Ref.
11839	Propane, 2,2-dicyclohexyl $(C_6H_{11})_2C(CH_3)_2$	308.39		273-4		$0.9002^{23/0}$	1.490^{23}		B5^4, 350
11840	Propane, 2,2-di(ethyl sulfonyl) or Sulfonal $(CH_3)_2C(SO_2C_2H_5)_2$	228.32	mcl (w), pr (al)	300d	125.8			al, bz, chl	B1^3, 2754
11841	Propane, 1,1-difluoro-1,2,2,3,3-pentachloro $CClF_2CCl_2CCl_2H$	252.30		168.4		$1.73162^{20/4}$	1.46241^{20}		B1^4, 204
11842	Propane, 1,3-difluoro $F(CH_2)_3F$	80.08		41.6		$1.0057^{25/4}$	1.3190^{26}	bz	B1^3, 218
11843	Propane, 2,2-difluoro $(CH_3)_2CF_2$	80.08		−0.4	−104.8	$0.9205^{20/4}$	1.2904^{20}		B1^4, 187
11844	Propane, 1,2-diido CH_3CHICH_2I	295.89				$2.490^{18.5}$		al, eth	B1, 115
11845	Propane, 1,3-diido $I(CH_2)_3I$	295.89		227d, 110^{19}	fr−20	$2.5755^{20/4}$	1.6423^{20}	eth, chl	B1^4, 228
11846	Propane, 2,2-diido $(CH_3)_2CI_2$	295.89		173, 53^{10}		$2.5755^{20/4}$	1.651^{20}	eth, chl	B1^3, 255
11847	Propane, 2,2-dimethyl or Neopentane $(CH_3)_4C$	72.15	gas	9.5	−16.5	0.6135^{20}	1.3476^{0}	al, eth	B1^4, 333
11848	Propane, 1,1-dinitro $CH_3CH_2CH(NO_2)_2$	134.09		184	−42	1.2610^{25}	1.4339^{20}		B1^4, 234
11849	Propane, 1,3-dinitro $O_2N(CH_2)_3NO_2$	134.09		103^{1}	−21.4	$1.353^{26/4}$	1.4654^{20}	eth	B1^4, 234
11850	Propane, 2,2-dinitro $(CH_3)_2C(NO_2)_2$	134.09		185.5, $48\text{-}50^{2}$	53	1.30^{25}			B1^4, 234
11851	Propane, 1,2-diphenoxy $CH_3CH(OC_6H_5)CH_2OC_6H_5$	228.29	rh (MeOH)	$175\text{-}8^{12}$	32	$1.0748^{33.3/4}$	$1.5542^{33.3}$	al, eth, ace, bz, chl	B6^2, 151
11852	Propane, 1,3-diphenoxy $C_6H_5O(CH_2)_3OC_6H_5$	228.29	lf (al)	338-40, 160^{25}	61			al, eth	B6^4, 577
11853	Propane, 1,1-diphenyl $CH_3CH_2CH(C_6H_5)_2$	196.29		280		$0.9951^{14/4}$	1.5681^{14}		B5^4, 1920
11854	Propane, 1,2-diphenyl $CH_3CH(C_6H_5)CH_2C_6H_5$	196.29		$280\text{-}1^{758}$, 109^{2}	52	$0.9807^{20/4}$	1.5700^{20}		B5^4, 1914
11855	Propane, 1,3-diphenyl $C_6H_5(CH_2)_3C_6H_5$	196.29		300.3, 124^{2}	6	$1.007^{20/4}$	1.5760^{20}		B5^4, 1913
11856	Propane, 2,2-diphenyl $(C_6H_5)_2C(CH_3)_2$	196.29		282-3	29				B5^4, 1925
11857	Propane, 1,2-epoxy or Propylene oxide CH_3CHCH_2	58.08		34.3		$0.859^{0/4}$	1.3670^{20}	**w, al, eth**	B17^4, 16
11858	Propane, 1,2-epoxy-3-fluoro FCH_2CHCH_2	76.07		85.0-6.5		$1.090^{20/20}$	1.3730^{20}		B17^4, 20
11859	Propane, 1,2-epoxy-3 iodo or Epiiodohydrin ICH_2CHCH_2	183.98		160-2		1.982^{24}		al, eth	B17^4, 23
11860	Propane, 1,2-epoxy,3,3,3-trichloro or 3,3,3-Trichloro propylene oxide Cl_3CCHCH_2	161.42		149, $44\text{-}5^{13}$		$1.495^{20/4}$	1.4737^{25}	eth	B17^4, 22
11861	Propane, 1-ethoxy-3-phenoxy $C_6H_5O(CH_2)_3OC_2H_5$	180.25	oil	328-30				al, eth	B6, 147
11862	Propane, 1-fluoro or Propyl fluoride $CH_3CH_2CH_2F$	62.09		2.5	−159	$0.7956^{20/4}$	1.3115^{20}	al, eth	B1^4, 187
11863	Propane, 1,1,1,2,2,3,3-heptachloro $Cl_2CHCCl_2CCl_3$	285.21	amor	247-8, 132^{30}	29.4	$1.8048^{24/4}$		chl	B1^4, 205
11864	Propane, 1,1,1,2,3,3,3,-heptachloro $(Cl_3C)_2CHCl$	258.21		249, 93^{2}	11	$1.7921^{34/4}$	1.5427^{21}	chl	B1^4, 205
11865	Propane, heptafluoro-1-nitro $CF_3CF_2CF_2NO_2$	215.03		25					B1^4, 232
11866	Propane, heptafluoro-1-nitroso $CF_3CF_2CF_2NO$	199.03	deep bl	−12	−150				B1^4, 229
11867	Propane, 1,1,1,2,2,3-hexafluoro $CH_2FCF_2CF_3$	152.04		1.2					B1^4, 188
11868	Propane, 1-iodo or *n*-Propyl iodide $CH_3CH_2CH_2I$	169.99		102.4	−101	$1.7489^{20/4}$	1.5058^{20}	**al, eth, bz, chl**	B1^4, 222
11869	Propane, 1-iodo-2,2-dimethyl or Neopentyl iodide $(CH_3)_3CCH_2I$	198.05		127-9d, $42\text{-}4^{20}$		1.4940^{20}	1.4890^{20}	al, eth	B1^4, 338
11870	Propane, 2-iodo or Isopropyl iodide $(CH_3)_2CHI$	169.99		89.4	−90.1	$1.7033^{20/4}$	1.5028^{20}	**al, eth, bz, chl**	B1^4, 223
11871	Propane, 1-methoxy-3-phenoxy $C_6H_5O(CH_2)_3OCH_3$	166.22	oil	230-1					B6, 147

No.	Name, Synonyms, and Formula	Mol. wt.	Color, crystalline form, specific rotation and λ_{max} (log ε)	b.p. °C	m.p. °C	Density	n_D	Solubility	Ref.
11872	Propane, 2-methyl or Isobutane $(CH_3)_2CHCH_3$	58.12		−11.633	−159.4	0.549^{20}		al, eth, chl	$B1^4$, 282
11873	Propane, 1-nitro $CH_3CH_2CH_2NO_2$	89.09		130-1	−108	$1.0081^{24/4}$	1.4016^{20}	**al, eth** chl	$B1^4$, 229
11874	Propane, 2-nitro $(CH_3)_2CHNO_2$	89.09		120	−93	$0.9876^{20/4}$	1.3944^{20}	chl	$B1^4$, 230
11875	Propane, 3-nitro-1,1,1-trifluoro $F_3CCH_2CH_2NO_2$	143.07		135		$1.4203^{20/20}$	1.3549^{20}	eth	$B1^4$, 232
11876	Propane, 1,1,1,2,3-pentachloro $CH_2ClCHClCCl_3$	216.32		179-80			1.5130^{20}	al, eth	$B1^4$, 203
11877	Propane, 1,1,2,3,3-pentachloro $Cl_2CHCHClCHCl_2$	216.32		198-200, $78\text{-}100^{20}$		$1.6086^{34/4}$	1.5131^{17}	al, eth, chl	$B1^4$, 204
11878	Propane, 1,1,2,3,3-pentafluoro-1,2,3-trichloro $CF_2ClCClFC.ClF_2$	237.38		73.7	−72	$1.6631^{20/4}$	1.3512^{20}		$B1^4$, 200
11879	Propane, perchloro C_3Cl_8	319.66		$268\text{-}9^{734}$	160			al, eth	$B1^4$, 205
11880	Propane, perfluoro C_3F_8	188.02		−36	−183				$B1^4$, 189
11881	Propane, 1,1,1,2-tetrachloro $CH_3CHClCCl_3$	181.88		150, 37^{10}		$1.473^{20/4}$	1.4867^{20}	al, eth, chl	$B1^4$, 201
11882	Propane, 1,1,1,3-tetrachloro $CH_2ClCH_2CCl_3$	181.88		159, 59^{24}		$1.4510^{20/6}$	1.4825^{20}	al, eth, bz, chl	$B1^4$, 201
11883	Propane, 1,1,2,2-tetrachloro $CH_3CCl_2CHCl_2$	181.88		153		1.47^{12}	1.4850^{25}	**al, eth**, chl	$B1^4$, 201
11884	Propane, 1,1,2,3-tetrachloro $CH_2ClCHClCHCl_2$	181.88		179-80		1.513^{17}	1.5037^{17}	al, eth	$B1^4$, 202
11885	Propane, 1,2,2,3-tetrachloro $CH_2ClCCl_2CH_2Cl$	181.88		165, 51^{12}		1.500^{18}	1.4940^{18}	al, eth, chl	$B1^4$, 202
11886	Propane, 1,2,2,3-tetrachloro,-1,1,3,3,-tetrafluoro $CF_2ClCCl_2CF_2Cl$	253.84		112	−42.9	$1.7199^{20/4}$	1.39584^{20}		$B1^4$, 203
11887	Propane, 1,1,2-tribromo $CH_3CHBrCHBr_2$	280.78		200-1, 83^{6}		$2.3548^{20/4}$	1.5790^{20}	al, eth, chl, aa	$B1^3$, 251
11888	Propane, 1,2,2-tribromo $CH_3CBr_2CH_2Br$	280.78		190-1, 81^{20}		$2.2985^{20/4}$	1.5670^{20}	al, eth, chl, aa	$B1^2$, 77
11889	Propane, 1,2,3-tribromo $CH_2BrCHBrCH_2Br$	280.78		222.1, 98.5^{10}	16.9	$2.4209^{20/4}$	1.5862^{20}	al, eth	$B1^4$, 221
11890	Propane, 1,1,1-trichloro $CH_3CH_2CCl_3$	147.43		107-9		$1.287^{23/4}$		al, eth, chl	$B1^4$, 198
11891	Propane, 1,1,2- trichloro $CH_3CHClCHCl_2$	147.43		140		1.372^{15}		al, eth, chl	$B1^4$, 199
11892	Propane, 1,1,3- trichloro $CH_2ClCH_2CHCl_2$	147.43		145.5, 33.4^{10}	−59	$1.3557^{20/4}$	1.4718^{20}	al, eth, chl, aa	$B1^4$, 199
11893	Propane, 1,2,2-trichloro $CH_3CCl_2CH_2Cl$	147.43		123-5		1.318^{25}	1.4609^{20}	al, eth, chl	$B1^4$, 199
11894	Propane, 1,2,3- trichloro $CH_2ClCHClCH_2Cl$	147.43		156.8, 41.9^{10}	−14.7	$1.3889^{20/4}$	1.4852^{20}	al, eth, chl	$B1^4$, 199
11895	Propane, 1,1,1-triphenyl $CH_3CH_2C(C_6H_5)_3$	272.39	pr		51				$B5^3$, 2340
11896	1-Propanearsonic acid $CH_3CH_2CH_2AsO_3H_2$	168.02	nd (al), pl (w)		134-5			w, al	$B4^3$, 1824
11897	1-Propane boronic acid or *n*-Propylboric acid $CH_3CH_2CH_2B(OH)_2$	87.91	wh nd	d	107			w, al, eth	$B4^3$, 1964
11898	1,2-Propanediol or Propylene glycol $CH_3CH(OH)CH_2OH$	76.10		189, $96\text{-}8^{21}$		$1.0361^{20/4}$	1.4324^{20}	**w, al,** eth, bz	$B1^4$, 2468
11899	1,2-Propanediol, 3-amino or 3-Amino-1,2-propylene glycol $H_2NCH_2CH(OH)CH_2OH$	91.11		265d, 145^{9}		$1.1752^{20/4}$	1.4910^{25}	w, al	$B4^4$, 1865
11900	1,2-Propanediol carbonate or 1,2-Propylene glycol carbonate $C_4H_6O_3$	102.09		240, 110^{10}	−48.8	$1.2041^{20/4}$	1.4189^{20}	w, al, ace, bz	$B19^4$, 1564
11901	1,2-Propanediol, 3-chloro $ClCH_2CH(OH)CH_2OH$	110.54	yesh liq	213d, 116^{11}		$1.326^{18/15}$	1.4809^{20}	w, al, eth	$B1^4$, 2484
11902	1,2-Propanediol, 3-chloro, diacetate $ClCH_2CH(OCOCH_3)CH_2OCOCH_3$	194.61		245, 116^{22}		$1.199^{25/4}$	1.4407^{20}	al, eth	$B2^3$, 313
11903	1,2, Propanediol, 3-chloro-2-methyl $ClCH_2C(OH)(CH_3)CH_2OH$	124.57		$114\text{-}7^{20}$		$1.2362^{20/4}$	1.4748^{20}	**w, al, eth**	$B1^4$, 2536
11904	1,2-Propanediol diacetate $CH_3CH(O_2CCH_3)CH_2O_2CCH_3$	160.17		$190\text{-}1^{762}$		$1.059^{20/4}$	1.4173^{20}	w, al, eth	$B2^4$, 220

PHYSICAL CONSTANTS OF ORGANIC COMPOUNDS (Continued)

No.	Name, Synonyms, and Formula	Mol. wt.	Color, crystalline form, specific rotation and λ_{max} (log ε)	b.p. °C	m.p. °C	Density	n_D	Solubility	Ref.
11905	1,2-Propane diol, 3-(diethyl amino) $(C_2H_5)_2NCH_2CH(OH)CH_2OH$	147.22	syr	233-5				w, al, eth, chl	$B4^3$, 840
11906	1,2-Propanediol, 3-(dimethylamino) $(CH_3)_2NCH_2CH(OH)CH_2OH$	119.16		220^{749}				w, al, eth, chl	$B4^3$, 840
11907	1,2-Propanediol, 3-mercapto or 1-Thioglycerol $HSCH_2CH(OH)CH_2OH$	108.16	visc	$100\text{-}1^1$		1.2455^{20}	1.5268^{20}	w, al, ace	$B1^3$, 2339
11908	1,2-Propanediol, 2-methyl or Isobutylene glycol $(CH_3)_2C(OH)CH_2OH$	90.12		176, $79\text{-}80^{12}$		$1.0024^{20/4}$	1.5268^{20}	w, al, eth	$B1^4$, 2533
11909	1,2-Propanediol-sulfite $C_3H_6O_3S$	122.14		175, 85^{25}	<−60	$1.2960^{20/4}$	1.4370^{20}	w, al, eth, ace, bz	$B1^4$, 2476
11910	1,3-Propanediol or Trimethylene glycol $HOCH_2CH_2CH_2OH$	76.10		213.5, 110^{12}		$1.0597^{20/4}$	1.4398^{20}	w, al, eth	$B1^4$, 2493
11911	1,3-Propanedioll, 2-amino-2-ethyl $C_2H_5C(CH_2OH)_2NH_2$	119.16	ye	$143\text{-}5^{10}$	37-8	$1.099^{20/4}$	1.490^{20}	w	$B4^4$, 1883
11912	1,3-Propanediol, 2-amino-2-(hydroxymethyl) $(HOCH_2)_3CNH_2$	121.14	nd or fl (MeOH)	$219\text{-}20^{18}$	170-1			w	$B4^4$, 1903
11913	1,3,-Propanediol, 2-amino-2-methyl $(HOCH_2)_2C(CH_3)NH_2$	105.14		151.2^{10}	109-11			w, al	$B4^4$, 1881
11914	1,3-Propanediol, 2-butyl-2-ethyl or 3,3-Di(hydroxymethyl) heptane $(HOCH_2)_2C(C_2H_5)(C_4H_9)$	160.26	wh	262, 123^{15}	43.8	$0.929^{20/20}$	1.4587^{25}	al	$B1^4$, 2611
11915	1,3-Propanediol, 2-chloro $(HOCH_2)_2CHCl$	110.54		146^{18}		$1.3219^{20/4}$	1.4831^{20}	w, al, ace	$B1^4$, 2499
11916	1,3 Propanediol diacetate $CH_3CO_2CH_2CH_2CH_2O_2CCH_3$	160.17		209-10, 84.5^{10}		1.070^{14}	1.4192	w, al	$B2^4$, 221
11917	1,3-Propanediol, 2,2-diethyl $(HOCH_2)_2C(C_2H_5)_2$	132.20	wh	240-1, 131^{13}	61-2	$1.052^{20/20}$		w, al, eth	$B1^4$, 2589
11918	1,3-Propanediol, 2,2-dimethyl or Neopentylene glycol $(HOCH_2)_2C(CH_3)_2$	104.15	nd (bz)	206^{747}, $120\text{-}30^{15}$	130			w, al, eth	$B1^4$, 2551
11919	1,3-Propanediol, 2,2-dinitro $(HOCH_2)_2C(NO_2)_2$	166.09	wh pl (bz)		142				$B1^4$, 2501
11920	1,3-Propanediol, 2-ethyl-2-hydroxymethyl or TMP Trimethylol propane $CH_3CH_2C(CH_2OH)_3$	134.18	wh pw or pl	160^5	58			w, al	$B1^4$, 2786
11921	1,3-Propanediol, 2-ethyl-2-nitro $(HOCH_2)_2C(NO_2)C_2H_5$	149.15	nd (w)	d	57-8			w, al, eth	$B1^4$, 2550
11922	1,3-Propanediol, 2-hydroxymethyl-2-nitro $(HOCH_2)_3CNO_2$	151.12	nd or pr	d	165			w, al, eth	$B1^4$, 2777
11923	1,3,Propanediol, 2-methyl-2-(hydroxy methyl) or Trimethylol ethane $CH_3C(CH_2OH)_3$	120.15	wh pw or nd (al)	$135\text{-}7^{15}$	204			w, al	$B1^4$, 2780
11924	1,3-Propanediol, 2-methyl-2-nitro $(HOCH_2)_2C(CH_3)NO_2$	135.12	mcl	d	149-50			w, al	$B1^4$, 2537
11925	1,3-Propane diol, 2-methyl-2-propyl or 2,2-Bis-(hydroxymethyl) pentane $(HOCH_2)_2C(CH_3)C_3H_7$	132.20	cr (hx)	234	62-3			w	$B1^4$, 2585
11926	1,2 Propanedithiol $CH_3CH(SH)CH_2SH$	108.22		$41\text{-}3^{11}$		$1.08^{20/4}$	1.532^{20}	chl	$B1^4$, 2492
11927	1,3-Propanedithol $HSCH_2CH_2CH_2SH$	108.22		172.9, 63^{15}	−79	$1.0783^{20/4}$	1.5392^{20}	al, eth, bz, chl	$B1^4$, 2503
11928	1-Propane phosphonic acid or *n*-Propylphosphonic acid $CH_3CH_2CH_2PO(OH)_2$	124.08	pl (bz)	d	23			w, al, eth	$B4^3$, 1781
11929	2-Propane phosphonic acid or Isopropyl phosphonic acid $(CH_3)_2CHPO_3H_2$	124.08	pl (bz)	d	74-5			w, al, eth	$B4^3$, 1781
11930	1-Propane sulfonamide $C_3H_7SO_2NH_2$	123.17	pr (eth), cr (bz)		53.5			w, al	$B4^4$, 39
11931	1-Propanesulfonyl chloride $C_3H_7SO_2Cl$	142.60		180d, 77^{12}		$1.2826^{15/4}$	1.452^{20}		$B4^4$, 39
11932	2-Proponesulfonamide or Isopropyl sulfonamide $(CH_3)_2CHSO_2NH_2$	123.17	cr (eth-peth)		67.5			w, al, eth	$B4^4$, 42
11933	2-Propanesulfonic acid or Isopropyl sulfonic acid $(CH_3)_2CHSO_3H$	124.15		$159^{1.4}$	−37	1.187^{25}	1.4332^{20}	w	$B4^4$, 42
11934	1,1,2,3-Propane tetra carboxylic acid, tetra ethyl ester $C_2H_5O_2CCH_2CH(CO_2C_2H_5)CH(CO_2C_2H_5)_2$	332.35		$203\text{-}4^{18}$		$1.1184^{20/4}$	1.4395^{20}	al	$B2^3$, 2077
11935	1,1,3,3-Propane tetra carboxylic acid, tetra ethyl ester or Ethyl methane dimalonate $[(C_2H_5O_2C)_2CH]_2CH_2$	332.35		300-10d, 195^8	−30	1.116^{20}	1.4398^{20}	al	$B2^4$, 2417

PHYSICAL CONSTANTS OF ORGANIC COMPOUNDS (Continued)

No.	Name, Synonyms, and Formula	Mol. wt.	Color, crystalline form, specific rotation and λ_{max} (log ε)	b.p. °C	m.p. °C	Density	n_D	Solubility	Ref.
11936	1-Propanethiol or *n*-Propyl mercaptan $CH_3CH_2CH_2SH$	76.16	...	67-8	−113.3	0.8411^{20}	1.4380^{20}	al, eth, ace, bz	$B1^4$, 1449
11937	2- Propane thiol or Isopropyl mercaptan $(CH_3)_2CHSH$	76.16	...	52.5	−130.5	$0.8143^{20/4}$	1.4255^{20}	**al, eth**, ace	$B1^4$, 1498
11938	1,2,3-Propane tricarboxylic acid or Tricarballic acid $(HO_2CCH_2)_2CHCO_2H$	176.13	orh (w, eth)	...	166	...	...	w, al	$B2^4$, 2366
11939	1,2,3-Propanetricarboxylic acid-1,2-dihydroxy *(l)* $HO_2CCH_2C(OH)(CO_2H)CH(OH)(CO_2H)$	208.12	nd, $[\alpha]_D$ 17.7 (acid)	...	159-60	1.39^{25}	...	w, eth	$B3^4$, 1298
11940	1,2,3-Propanetricarboylic acid, 1-hydroxy or Isocitric acid $HO_2CCH_2CH(CO_2H)CH(OH)CO_2H$	192.13	yesh syr	...	105	...	...	...	$B3^4$, 1270
11941	Propanoic acid or Propionic acid $CH_3CH_2CO_2H$	74.08	...	141, 41.6^{10}	−20.8	0.9930^{20}	1.3869^{20}	**w, al**, eth	$B2^4$, 695
11942	1-Propanol or *n*-Propyl alcohol $CH_3CH_2CH_2OH$	60.10	...	97.4	−126.5	$0.8035^{20/4}$	1.3850^{20}	**w, al, eth**, ace, bz	$B1^4$, 1413
11943	1-Propanol, 2-amino *(dl)* $CH_3CH(NH_2)CH_2OH$	75.11	...	173-6, 80^{18}	...	...	1.4502^{20}	w, al, eth	$B4^4$, 1615
11944	1-Propanol, 3-amino or Propanolamine $H_2NCH_2CH_2CH_2OH$	75.11	...	187-8	...	$0.9824^{26/4}$	1.4617^{20}	w, al, eth	$B4^4$, 1623
11945	1-Propanol, 3-benzyloxy $C_6H_5CH_2OCH_2CH_2CH_2OH$	166.22	...	172^{43}		$1.0474^{20/4}$	...	al, eth	$B6^4$, 2243
11946	1-Propanol, 3-bromo $BrCH_2CH_2CH_2OH$	138.99	...	$98\text{-}112^{185}$, 62^5	...	$1.5374^{20/4}$	1.4834^{25}	w, **al, eth**	$B1^4$, 1446
11947	1-Propanol, 2-chloro or Propylene chlorohydrin $CH_3CHClCH_2OH$	94.54	...	133-4	...	1.103^{20}	1.4390^{20}	w, al, eth	$B1^4$, 1440
11948	1-Propanol, 2-chloro-2-methyl or 2-Chloro isobutyl alcohol $(CH_3)_2CClCH_2OH$	108.57	visc	132-3d, $59\text{-}61^{50}$	...	$1.0477^{20/4}$	1.4388^{20}	...	$B1^4$, 1603
11949	1-Propanol, 3-chloro or Trimethylene chlorohydrin $ClCH_2CH_2CH_2OH$	94.54	...	165, 53^6	...	$1.1309^{20/4}$	1.4459^{20}	w, al, eth	$B1^4$, 1441
11950	1-Propanol, 2,3-dibromo *(d)* $CH_2BrCHBrCH_2OH$	217.89	$[\alpha]_D$ + 7.3	219d	...	2.11	...	**al, eth, ace, bz**	$B1^2$, 371
11951	1-Propanol, 2,3-dibromo *(dl)* $CH_2BrCHBrCH_2OH$	217.89	...	219d, 118^{17}	...	$2.0739^{20/4}$	1.5466^{20}	**al, eth, ace, bz**	$B1^4$, 1446
11952	1-Propanol, 2,3-dichloro $CH_2ClCHClCH_2OH$	128.99	visc	183-5, $70\text{-}80^{17}$	...	$1.3607^{20/4}$	1.4819^{20}	**al, eth, ace, bz**	$B1^4$, 1442
11953	1-Propanol, 2-diethylamino $(C_2H_5)_2NCH(CH_3)CH_2OH$	131.22	...	$166\text{-}9^{749}$, 78^{12}	...	$0.8665^{27/20}$	1.4332^{20}	al, eth, ace, bz	$B4^4$, 1618
11954	1-Propanol, 3-diethylamino $(C_2H_5)_2NCH_2CH_2CH_2OH$	131.22	...	189.5, 87^{16}	...	$0.8600^{20/4}$	1.4439^{20}	al, eth, ace, bz	$B4^4$, 1633
11955	1-Propanol, 2,3-dimercapto or 1,2 Dithioglycerol $HSCH_2CH(SH)CH_2OH$	124.22	visc liq	120^{15}	...	$1.2463^{20/4}$	1.5733^{20}	al, eth, oils	$B1^4$, 2770
11956	1-Propanol, 3-(3,5-dimethoxy-4-hydroxyphenyl) or 2-Syringyl ethanol. Hydrosinapyl alcohol $(3,5\text{-}(CH_3O)_2\text{-}4\text{-}HOC_6H_2)CH_2CH_2CH_2OH$	212.25	wh nd (eth-peth)	...	75-6	...	...	w	$B6^3$, 6669
11957	1-Propanol, 2,2-dimethyl or Neopentyl alcohol $(CH_3)_3CCH_2OH$	88.15	...	113-4	52-3	0.812	...	al, eth	$B1^4$, 1690
11958	1-Propanol, 2,3-epoxy or Glycidol CH_2CHCH_2OH (with O bridge)	74.08	...	166-7d, $65\text{-}6^{2.5}$	...	1.1143^{25}	1.4287^{20}	**w, al**, eth, ace. bz, chl	$B17^4$, 985
11959	1-Propanol, 2-ethoxy $CH_3CH(OC_2H_5)CH_2OH$	104.15	...	140-1	...	$0.9044^{20/4}$	1.4122^{20}	w, al, eth	$B1^3$, 2147
11960	1-Propanol, 3-fluoro $FCH_2CH_2CH_2OH$	78.09	...	127.8	...	$1.0390^{25/4}$	1.3771^{25}	w, al, eth	$B1^4$, 1437
11961	1-Propanol, 3-(4-hydroxy-3-methoxyphenyl) or Hydroconiferyl alcohol $(3\text{-}CH_3O\text{-}4\text{-}HOC_6H_3)CH_2CH_2CH_2OH$	182.22	...	197^{15}	65	...	1.5545^{25}	al, eth	$B6^3$, 6347
11962	1-Propanol, 2-methoxy $CH_3CH(OCH_3)CH_2OH$	90.12	...	130^{758}	...	$0.938^{20/4}$	1.4070^{20}	...	$B1^4$, 2471
11963	1-Propanol, 2-methyl or Isobutyl alcohol $(CH_3)_2CHCH_2OH$	74.12	...	108	...	$0.8018^{20/4}$	1.3955^{20}	w, al, eth, ace	$B1^4$, 1588
11964	1-Propanol, 2-methyl-2-nitro $(CH_3)_2C(NO_2)CH_2OH$	119.12	nd or pl (MeOH)	$94\text{-}5^{10}$	89-90	...	...	al, eth	$B1^4$, 1604
11965	1-Propanol, 2-nitro $CH_3CH(NO_2)CH_2OH$	105.09	...	100^{12}	...	$1.1841^{25/4}$	1.4379^{20}	al, eth	$B1^4$, 1448
11966	1-Propanol, 2-phenoxy $CH_3CH(OC_6H_5)CH_2OH$	152.19	...	244, $124\text{-}6^{20}$	...	$0.9830^{25/25}$	1.4760^{25}	al, eth	$B6^4$, 582
11967	1-Propanol, 3-phenoxy $C_6H_5OCH_2CH_2CH_2OH$	152.19	oil	$249\text{-}50^{764}$, 170^{60}	...	...	1.491^{20}	...	$B6^4$, 584

No.	Name, Synonyms, and Formula	Mol. wt.	Color, crystalline form, specific rotation and λ_{max} (log ε)	b.p. °C	m.p. °C	Density	n_D	Solubility	Ref.
11968	1-Propanol, 1-phenyl- *(dl)* $CH_3CH_2CH(OH)C_6H_5$	136.19	...	213-5^{740}, 98^{10}	...	0.9938$^{23/4}$	1.5210^{23}	al, eth	B6^3, 1793
11969	1-Propanol, 3-phenyl $C_6H_5CH_2CH_2CH_2OH$	136.19	...	236-7^{750}, 132^{21}	<−18	1.008$^{20/4}$	1.5278^{20}	w, al, eth	B6^3, 1880
11970	1-Propanol, 2,2,3,3-tetrafluoro $CHF_2CF_2CH_2OH$	132.06	...	109-1	−15	1.4853$^{20/4}$	1.3197^{20}	al, ace, chl	B1^4, 1438
11971	2-Propanol or Isopropyl alcohol $(CH_3)_2CHOH$	60.10	...	82.4	−89.5	0.7855$^{20/4}$	1.3776^{20}	w, al, eth, ace, bz	B1^4, 1461
11972	2-Propanol, 1-amino *(dl)* or 150 Propanol amine $CH_3CH(OH)CH_2NH_2$	75.11	...	159.4, 59.5^{10}	1.7	0.9611$^{20/4}$	1.4479^{20}	w, al, eth, ace, bz	B4^4, 1665
11973	2-Propanol, 1-amino *(l)* $CH_3CH(OH)CH_2NH_2$	75.11	...	156-8^{758}	...	0.973^{18}	...	w, al, eth, ace, bz	B4^4, 1664
11974	2-Propanol, 1-amino-3-(diethylamino) $(C_2H_5)_2NCH_2CH(OH)CH_2NH_2$	146.23	...	223, 116-8^{25}	...	1.937$^{20/4}$	1.465^{20}	w	B4^3, 767
11975	2-Propanol, 2-benzyl $C_6H_5CH_2C(OH)(CH_3)_2$	150.22	...	214-6, 104-5^{10}	24	0.9790$^{20/25}$	1.5174^{20}	al	B6^3, 1860
11976	2-Propanol, 1,3-bis (dimethylamino) $[(CH_3)_2NCH_2]_2CHOH$	146.23	...	178-85, 79-81^{18}	...	0.8788$^{20/4}$	1.4418^{20}	w	B4^4, 1695
11977	2-Propanol, 1-bromo $CH_3CH(OH)CH_2Br$	138.99	...	145-8, 49.6^{12}	...	1.5585^{30}	1.4801^{20}	w, al, eth	B1^4, 1495
11978	2-Propanol, 1-butoxy or 1,2-Propylene glycol-1-monobutyl ether $CH_3CH(OH)CH_2OC_4H_9$	132.20	...	168.75	...	1.0035$^{20/4}$	1.4168^{20}	al, eth, bz	B1^4, 2471
11979	2-Propanol, 1-chloro $CH_3CH(OH)CH_2Cl$	94.54	...	126-7^{750}	...	1.115$^{20/20}$	1.4392^{20}	w, al, eth	B1^4, 1490
11980	2-Propanol, 1-chloro-2-methyl $(CH_3)_2C(OH)CH_2Cl$	108.57	...	128-9, 71^{100}	−20	1.0628$^{20/4}$	1.4380^{24}	w, al	B1^4, 1628
11981	2-Propanol, 1-chloro-3-propoxy $C_3H_7OCH_2CH(OH)CH_2Cl$	152.62	...	92-5^{15}	...	1.0526$^{25/4}$	1.4378^{25}	eth, ace	B1^3, 2153
11982	2-Propanol, 1-chloro-3-isopropoxy $(CH_3)_2CHOCH_2CH(OH)CH_2Cl$	152.62	...	87^{20}		1.0530^{25}	1.4370^{25}	al, eth	B1^4, 2485
11983	2-Propanol, 1,3-diamino $(H_2NCH_2)_2CHOH$	90.13	cr	235, 93-5^{2}	42-5	...	...	w	B4^4, 1694
11984	2-Propanol, 1-3 diamino, dihydrochloride $(H_2NCH_2)_2CHOH.2HCl$	163.05	hyg pr or nd (dil al)	...	184.5	...	...	w	B4^2, 739
11985	2-Propanol, 1,3-dibromo $(BrCH_2)_2CHOH$	217.89	yesh liq	219d, 105^{16}	...	2.1202$^{25/4}$	1.5495^{25}	al, eth, ace	B1^4, 1496
11986	2-Propanol, 1-1,dichloro $CH_3CH(OH)CHCl_2$	128.99	...	146-8^{765}	...	1.3334$^{22/4}$	...	al, eth, ace	B1^2, 383
11987	2-Propanol, 1,1-dichloro-2-methyl $(CH_3)_2C(OH)CHCl_2$	143.01	...	150-5, 52^{10}	8	1.2363$^{19/4}$	1.4598^{19}	al, bz	B1^4, 1629
11988	2-Propanol, 1,3-dichloro $(ClCH_2)_2CHOH$	128.99	...	176, 69^{12}	...	1.3506$^{17/4}$	1.4837^{20}	w, al, eth, ace	B1^4, 1491
11989	2-Propanol, 1,3-dichloro-2-methyl $(ClCH_2)_2C(OH)CH_3$	143.01	...	174-5, 55-6^{10}	...	1.2745$^{20/4}$	1.4744^{21}	al, ace	B1^4, 1629
11990	2-Propanol, 1-(diethylamino) $CH_3CH(OH)CH_2N(C_2H_5)_2$	131.22	...	158-9^{756}, 63^{22}	...	0.8511$^{20/0}$	1.4255^{20}	w, al, ace	B4^3, 759
11991	2-Propanol, 1-ethoxy $CH_3CH(OH)CH_2OC_2H_5$	104.15	...	131	...	0.9028$^{20/4}$	1.4075^{20}	w, al, eth	B1^4, 2471
11992	2-Propanol, 1-methoxy $CH_3CH(OH)CH_2OCH_3$	90.12	...	118	...	0.9620$^{20/4}$	1.4034^{20}	...	B1^4, 2471
11993	2-Propanol, 1-methoxy-2-methyl $CH_3OCH_2C(OH)(CH_3)_2$	104.15	...	116.6^{747}	...	0.9021$^{15/15}$	...	...	B1^4, 2533
11994	2-Propanol, 2-methyl or *tert*-Butyl alchol $(CH_3)_3COH$	74.12	...	82.3, 20^{31}	25.5	0.7887$^{20/4}$	1.3878^{20}	w, al, eth	B1^4, 1609
11995	2-Propanol, 1-nitro $CH_3CH(OH)CH_2NO_2$	105.09	...	112^{13}, 68^{1}	...	1.1906^{20}	1.4383^{20}	w, al, eth, bz	B1^4, 1496
11996	2-Propanol, 1-nitro-3,3,3-trichloro $O_2NCH_2CH(OH)CCl_3$	208.43	pr or pl (chl)	105-6$^{1.5}$	45-6	...	...	al, eth	B1^4, 1497
11997	2-Propanol, 1-phenoxy $CH_3CH(OH)CH_2OC_6H_5$	152.19	...	134.5^{20}	...	1.0622$^{20/4}$	1.5210^{22}	al, eth	B6^4, 582
11998	2-Propanol, 2-phenyl $(CH_3)_2C(OH)C_6H_5$	136.19	pr	202, 93^{13}	35-7	0.9735$^{20/4}$	1.5325^{20}	al, eth, bz, aa	B6^3, 1813
11999	2-Propanol, 1-propoxy $CH_3CH(OH)CH_2OC_3H_7$	118.18	...	148-9^{730}	...	0.8886$^{20/4}$	1.4130^{20}	...	B1^3, 2147
12000	2-Propanol, 1-isopropoxy $CH_3CH(OH)CH_2OCH(CH_3)_2$	118.18	...	137-8	...	0.879$^{20/4}$	1.4070^{20}	...	B1^4, 2471

No.	Name, Synonyms, and Formula	Mol. wt.	Color, crystalline form, specific rotation and λ_{max} (log ε)	b.p. °C	m.p. °C	Density	n_D	Solubility	Ref.
12001	2-Propanol, 1,1,1,3-tetrachloro $CH_2ClCH(OH)CCl_3$	197.88	....	$95\text{-}6^{17}$	....	$1.610^{20/4}$	1.5145^{20}	eth	$B1^3$, 1474
12002	2-Propanol, 1,1,3,3-tetrachloro $(Cl_2CH)_2CHOH$	197.88	....	$80\text{-}90^{14}$	....	$1.612^{20/4}$	1.5133^{20}	eth	$B1^3$, 1474
12003	2-Propanol, 1,1,1-trichloro $CH_3CH(OH)CCl_3$	163.43	....	161.8^{771}, $53\text{-}5^{12}$	50-1	....	....	al, eth, ace, bz	$B1^3$, 1474
12004	2-Propanol, 1,1,1-trifluoro (dl) $F_3CCH(OH)CH_3$	114.07	....	77.7^{754}	−52	$1.2799^{15/4}$	1.3172^{15}	al, eth, ace, bz	$B1^4$, 1489
12005	2-Propanone or Acetone, Dimethyl Ketone CH_3COCH_3	58.08	....	56.2	−95.3	$0.7899^{20/4}$	1.3588^{20}	**w, al, eth, bz, chl**	$B1^4$, 3180
12006	Propargyl aldehyde or Propynal $HC{\equiv}CCHO$	54.05	....	59-61	....	....	1.4033^{25}	w, al, eth, ace, bz	$B1^4$, 3537
12007	Propargyl aldehyde, diethyl acetyl $HC{\equiv}CCH(OC_2H_5)_2$	128.17	....	130.4, 37^{11}	....	$0.8942^{22/4}$	1.4140^{20}	al, eth, ace, chl	$B1^4$, 3538
12008	Propargyl aldehyde, phenyl $C_6H_5C{\equiv}CCHO$	130.15	....	$127\text{-}8^{28}$ $65^{0.1}$	....	$1.0639^{16/4}$	1.6079^{18}	....	$B7^3$, 1644
12009	Propargyl bromide or 3-Bromo propyne $BrCH_2C{\equiv}CH$	118.96	....	88-90, 33^{130}	....	1.579^{19}	1.4922^{20}	al, eth, bz, chl	$B1^4$, 964
12010	Propargyl alcohol or 2-Propyn-1-ol $HC{\equiv}CCH_2OH$	**56.06**	....	113.6, 30^{21}	−48	0.9485^{20}	1.4322^{20}	w, **al, eth**	$B1^4$, 2214
12011	Propargyl alcohol, acetate or Propargyl acetate $CH_3CO_2CH_2C{\equiv}CH$	98.10	....	124-5	....	$1.0082^{20/4}$	1.4205^{20}	al, eth	$B2^4$, 197
12012	Propargyl alcohol, 3-phenyl $C_6H_5C{\equiv}CCH_2OH$	132.16	....	$137\text{-}8^{15}$	....	$1.07^{18/18}$	1.5873^{18}	eth, ace, bz	$B6^3$, 2736
12013	Propargylic acid or Propynoic acid $HC{\equiv}CCO_2H$	**70.05**	cr (CS_2)	144d, $83\text{-}4^{50}$	18 (anh)	$1.1380^{20/4}$	1.4306^{20}	**w, al, eth**, ace, **chl**	$B2^4$, 1687
12014	Propargylic acid, ethyl ester or Ethyl propargylate $HC{\equiv}CCO_2C_2H_5$	98.10	....	119^{745}	....	$0.9583^{25/25}$	1.4105^{20}	al, eth, chl	$B2^4$, 1688
12015	Propargylic acid, (2-chlorophenyl) $(2\text{-}ClC_6H_4)C{\equiv}CCO_2H$	180.59	cr (bz, 50% aa)	....	133-4	....	....	aa	$B9^3$, 3066
12016	Propargylic acid, 3-chlorophenyl $(3\text{-}ClC_6H_4)C{\equiv}CCO_2H$	180.59	cr (aa, bz-peth)	....	144-5	....	....	aa	$B9^3$, 3066
12017	Propargylic acid, (4-chlorophenyl) $(4\text{-}ClC_6H_4)C{\equiv}CCO_2H$	180.59	cr (aa), pl (bz)	....	192-3	....	....	aa	$B9^3$, 3066
12018	Propargylic acid, (2-nitrophenyl) $(2\text{-}O_2NC_6H_4)C{\equiv}CCO_2H$	**191.14**	....	exp	157d	....	....	al, eth	$B9^3$, 3067
12019	Propargylic acid, (4-nitrophenyl) $(4\text{-}O_2NC_6H_4)C{\equiv}CCO_2H$	**191.14**	nd (eth, al)	....	181	....	....	eth, aa, chl	$B9^3$, 3067
12020	**Propargylic acid, phenyl** $C_6H_5C{\equiv}CCO_2H$	146.15	....	sub	137	....	....	al, eth	$B9^3$, 3061
12021	Propargylic acid, phenyl, ethyl ester $C_6H_5C{\equiv}C\text{-}CO_2C_2H_5$	174.20	....	260-70d, 144^{11}	....	$1.0550^{25/4}$	1.5535^{20}	eth	$B9^3$, 3063
12022	Propargylonitrile or Propynonitrile $HC{\equiv}CCN$	51.05	....	42.5	5	$0.8167^{17/4}$	1.3868^{25}	al	$B2^4$, 1689
12023	**Propenal or Acrolein** $CH_2{=}CHCHO$	**56.06**	....	52-3	−87	$0.8410^{20/4}$	1.4017^{20}	w, al, eth, ace	$B1^4$, 3435
12024	2-Propen-1-arsonic acid or Allylarsonic acid $CH_2{=}CHCH_2AsO_3H_2$	166.01	....	....	130-1	....	....	w, al	$B4^3$, 1826
12025	Propene or Propylene $CH_3CH{=}CH_2$	42.08	gas	−47.4	−185.2	$0.5193^{20/4}$ liq	1.3567^{-70}	al, aa	$B1^4$, 725
12026	**1-Propene-3,3, diol, diacetate** $(CH_3CO_2)_2CHCH{=}CH_2$	**158.15**	....	180, 76^{13}	**−37.6**	$1.0760^{20/4}$	1.4193^{20}	**al, eth, ace, bz, lig**	$B2^4$, 291
12027	**1,2,3-Propene tricarboxylic acid (*cis*) or *cis*-Aconitic acid** $HO_2CCH_2C(CO_2H){=}CHCO_2H$	**174.11**	**nd (w)**	....	**130**	....	....	**w**	$B2^4$, 2405
12028	**1,2,3-Propene tricarboxylic acid (*trans*) or *trans*-Aconitic acid** $HO_2CCH_2C(CO_2H){=}CHCO_2H$	**174.11**	**lf (w), nd (w, eth)**	....	**198-9**	....	....	**w, al**	$B2^4$, 2405
12029	**Propenyl benzene (*cis*)** $(CH_3CH{=}CH)C_6H_5$	**118.18**	....	69^{28}	**−60.5**	$0.9088^{20/4}$	1.5420^{20}	**al, eth, ace, bz**	$B5^4$, 1359
12030	**Propenyl benzene-(*trans*)** $(CH_3CH{=}CH)C_6H_5$	**118.18**	....	**175-6**	**−27.1**	0.9019^{25}	1.5508^{20}	**al, eth, ace, bz**	$B5^4$, 1359
12031	**Propenyl benzene, α-chloro** $CH_3CH{=}CClC_6H_5$	**152.62**	....	$90.5^{0.9}$	......	$1.085^{20/4}$	1.5635^{15}	**ace, bz**	$B5^2$, 372
12032	**Propenyl benzene, β-chloro** $C_6H_5CH{=}CClCH_3$	**152.62**	....	$118\text{-}23^{20}$, $61\text{-}2^{1}$	......	$1.0738^{19/4}$	1.5565^{19}	**ace, bz, chl**	$B5^3$, 1186
12033	**Propenyl benzene, γ-chloro-(*trans*) or *trans*-Cinnamyl chloride** $C_6H_5CH{=}CHCH_2Cl$	**152.62**	....	$106\text{-}7^{13}$	**8-9**	$1.0926^{20/4}$	1.5851^{20}	**al, eth**, ace, bz	$B5^4$, 1360

No.	Name, Synonyms, and Formula	Mol. wt.	Color, crystalline form, specific rotation and λ_{max} (log ε)	b.p. °C	m.p. °C	Density	n_D	Solubility	Ref.
12034	Propenyl benzene, γ-ethoxy $C_2H_5OCH_2CH{=}CHC_6H_5$	162.23		127-8^{22}		0.970$^{15/4}$	1.547^{15}		B6^{3}, 2404
12035	Propenyl benzene, β-nitro $C_6H_5CH{=}C(NO_2)CH_3$	163.18	ye nd (peth)		65-6			al	B5^{4}, 1360
12036	Isopropenyl methyl ketone $CH_2{=}C(CH_3)COCH_3$	84.12		98	−54	0.8527$^{20/4}$	1.4220^{20}	al	B1^{4}, 3462
12037	Propenoic acid or Aerylic acid $CH_2{=}CHCO_2H$	72.06		141.6, 48.5^{15}	13	1.0511$^{20/4}$	1.4224^{20}	w, al, eth, ace, bz	B2^{4}, 1455
12038	Propenoic acid, 2-methyl or Methacrylic acid $CH_2{=}C(CH_3)CO_2H$	86.09	pr	162-3^{757}, 60^{12}	16	1.0153$^{20/4}$	1.4314^{20}	w, al, eth	B2^{4}, 1518
12039	2-Propene-1-ol, 2-methyl or Methallyl alcohol $CH_2{=}C(CH_3)CH_2OH$	72.11		114.5		0.8515^{20}	1.4255^{20}	w, al, eth	B1^{4}, 2114
12040	2-Propene-1-ol, 1-phenyl or α-Vinyl benzyl alcohol $CH_2{=}CH\text{-}CH(OH)C_6H_5$	134.18		215-6, 111^{18}		1.0251$^{21/0}$	1.5406^{20}	al, eth, bz, chl	B6^{3}, 2417
12041	β-Propiolactone CH_2CH_2CO (ring closed through O)	72.06		162d, 51^{10}	−33.4	1.1460$^{20/5}$	1.4105^{20}	eth, chl	B17^{4}, 4157
12042	Propionaldehyde or Propanal CH_3CH_2CHO	58.08		48.8	−81	0.8058$^{20/4}$	1.3636^{20}	w, al, eth	B1^{4}, 3165
12043	Propionaldehyde, 2-Bromo $CH_3CHBrCHO$	136.98		109-10, 52-4^{80}		1.592^{20}	1.4813^{20}	eth	B1^{4}, 3177
12044	Propionaldehyde, 2-chloro $CH_3CHClCHO$	92.53		86		1.182$^{15/4}$	1.431^{17}	eth, bz	B1^{3}, 2691
12045	Propionaldehyde, 3-chloro $ClCH_2CH_2CHO$	92.53		130-1, 40^{19}		1.268^{15}	1.475^{25}	al, eth	B1^{4}, 3174
12046	Propionaldehyde, 3-chloro, diethyl acetal $ClCH_2CH_2CH(OC_2H_5)_2$	166.65		84^{25}		0.9951$^{19/4}$	1.4268^{20}	ace, bz	B1^{3}, 2692
12047	Propionaldehyde, 2,3-dibromo $BrCH_2CHBrCHO$	215.87	pa ye tum liq	73-5^{10}		2.198^{13}	1.5082^{10}	eth	B1^{4}, 3178
12048	Propionaldehyde, 2,2-dichloro CH_3CCl_2CHO	126.97	(peth)	80	38-9				B1^{4}, 3174
12049	Propionaldehyde, 2,3-dichloro $ClCH_2CHClCHO$	126.97	hyd (w)	73^{50}		1.400^{20}	1.4762^{20}		B1^{4}, 3174
12050	Propionaldehyde diethyl acetal or 1,1-Diethoxypropane $CH_3CH_2CH(OC_2H_5)_2$	132.20		122.8^{744}		0.8232$^{20/4}$	1.3924^{19}	w, al, eth, ace, bz	B1^{4}, 3168
12051	Propionaldehyde, 2,3-epoxy CH_2CHCHO (epoxy ring through O)	72.06		112-3	−62		1.4265^{20}		B17^{4}, 4159
12052	Propionaldehyde, 3-ethoxy, diethyl acetal or 1,1,3-Triethoxy propane $C_2H_5OCH_2CH_2CH(OC_2H_5)_2$	176.26		184-6d, 78^{14}		0.898$^{19/4}$	1.4067^{20}	al	B1^{4}, 3971
12053	Propionaldehyde, 3-hydroxy-2-oxo (enol form) or Reductone $HOCH{=}C(OH)CHO$	88.06	ye nd (w)		200-20d			w, al	B1^{4}, 4145
12054	Propionaldehyde oxime or Propionaldoxime $CH_3CH_2CH{=}NOH$	73.09		131.5	40	0.9258$^{20/4}$	1.4287^{20}		B1^{4}, 3170
12055	Propionaldehyde, 2-phenoxy $CH_3CH(OC_6H_5)CHO$	150.18		229-30, 99-101^{19}				al, eth, bz	B6, 151
12056	Propionaldehyde, 2-phenoxy, oxime $CH_3CH(OC_6H_5)CH{=}NOH$	165.19	nd (dil al)		110			al, eth	B6, 151
12057	Propionaldehyde, 2-phenyl or Hydratropic aldehyde $CH_3CH(C_6H_5)CHO$	134.18		202-5, 92.5^{10}		1.0089$^{20/4}$	1.5176^{20}	al	B7^{3}, 1050
12058	Propionaldehyde, 3-phenyl or Hydrocinnamaldehyde $C_6H_5CH_2CH_2CHO$	134.18	mcl	223^{745}, 104-5^{13}	47			al, eth	B7^{3}, 1046
12059	Propionaldehyde, 3-(4-tolyl) $(4\text{-}CH_3C_6H_4)CH_2CH_2CHO$	148.20		122^{15}		0.999$^{14/4}$	1.525^{14}		B7^{2}, 247
12060	Propionaldehyde, 2,2,3-trichloro $ClCH_2CCl_2CHO$	161.42		63-5^{45}		1.470^{25}	1.473^{25}	eth	B1^{3}, 2693
12061	Propionamide $CH_3CH_2CONH_2$	73.09	rh, pl (bz)	213	81.3	0.9262^{110}	1.4180^{110}	w, al, ace, chl	B2^{4}, 725
12062	Propionamide, 3-bromo $CH_2BrCH_2CONH_2$	151.99	cr (w)		111			al, eth, ace	B2^{4}, 766
12063	Propionamide, 3-chloro-N-(2-tolyl) $ClCH_2CH_2CONH(C_6H_4CH_3\text{-}2)$	197.66	(w, dil al)		78			al	B12^{2}, 441
12064	Propionamide, N,N-diethyl $CH_3CH_2CON(C_2H_5)_2$	129.20		191, 81-5^{20}			1.4425^{20}	al	B4^{4}, 353
12065	Propionamide, 2-hydroxy $CH_3CH(OH)CONH_2$	89.09			75.5	1.1381$^{80/4}$		w, al	B3^{4}, 674
12066	Propionamide, 2-phenoxy $CH_3CH(OC_6H_5)CONH_2$	165.19	nd or pl (to, w)		132-3			al, eth, aa	B6^{4}, 643

No.	Name, Synonyms, and Formula	Mol. wt.	Color, crystalline form, specific rotation and λ_{max} (log ε)	b.p. °C	m.p. °C	Density	n_D	Solubility	Ref.
12067	Propionamide, 3-phenoxy $C_6H_5OCH_2CH_2CONH_2$	165.19	nd (w)		119			al, eth	B6^3, 616
12068	Propionamide, N-phenyl or Propionanilide $CH_3CH_2CONHC_6H_5$	149.19	pl (eth, al, bz)	222.2	105-6	1.175		al, eth	B12^3, 472
12069	Propionamide, N-propionyl or Dipropionamide $(CH_3CH_2CO)_2NH$	129.16	nd (w, eth)	210-20 sub	154				B2^4, 727
12070	Propionamide, N-(2-tolyl) $CH_3CH_2CONH(C_6H_4CH_3\text{-}2)$	163.22	nd (bz)	298-9	89.5			al, eth, chl, aa	B12^2, 440
12071	Propionamide, N-(3-tolyl) $CH_3CH_2CONH(C_6H_4CH_3\text{-}3)$	163.22	nd (eth)		81			al, eth, lig	B12, 861
12072	Propionic acid or Propanoic acid $CH_3CH_2CO_2H$	74.08		141, 41.6^{10}	−20.8	0.9930^{20}	1.3809^{20}	**w, al, eth**	B2^4, 695
12073	Propionic acid, allyl ester or Allyl propionate $CH_3CH_2CO_2CH_2CH{=}CH_2$	114.14		124-5^{774}		0.9140^{20}	1.4105^{20}	al, eth, ace	B2^4, 711
12074	Propionic acid anhydride or Propionic anhydride $(CH_3CH_2CO)_2O$	130.14		168$^{71.2}$, 67.5^{18}	−45	1.0110$^{20/4}$	1.4038^{20}	**eth**	B2^4, 722
12075	Propionic acid, β-benzoyl or β-Benzoyl propionic acid $C_6H_5COCH_2CH_2CO_2H$	178.19	lf (dil al)		116			al, eth, bz, chl	B10^3, 3035
12076	Propionic acid, 2-bromo-(dl) $CH_3CHBrCO_2H$	152.98	pr	203.5, 96^{10}	25.7	1.7000$^{20/4}$	1.4753^{20}	w, al, eth	B2^4, 761
12077	Propionamide, 2-bromo-N-(2-tolyl) $CH_3CHBrCONH(C_6H_4CH_3\text{-}2)$	242.12	nd		131			al, eth, bz, chl	B12, 794
12078	Propionic acid, 2-bromo,ethyl ester $CH_3CHBrCO_2C_2H_5$	181.03		159-61d, 71^{26}		1.4135$^{20/4}$	1.4490^{20}	**al, eth**, chl	B2^4, 762
12079	Propionic acid, 2-bromo,methyl ester(d) $CH_3CHBrCO_2CH_3$	167.00	$[\alpha]^{17/}{}_D$ +42.65	144, 61-2^{26}		1.482^{17}		al, eth, chl	B2, 253
12080	Propionic acid, 2-bromo,methyl ester (l) $CH_3CHBrCO_2CH_3$	167.00	$[\alpha]^{20/}{}_{578}$ -55.5	61-3^{32}		1.484^{20}		al, eth	B2^2, 229
12080a	Propionic acid, 2-bromo,methyl ester-(dl) $CH_3CHBrCO_2CH_3$	167.00		143-5, 51.5^{19}		1.4966$^{25/4}$	1.4451^{22}	al, eth	B9^3, 2423
12081	Propionic acid, 2-bromo-2-phenyl (dl) $CH_3CBr(C_6H_5)CO_2H$	229.07	pl (CS_2)		93-4			bz	B2^4, 764
12082	Propionic acid, 3-bromo $BrCH_2CH_2CO_2H$	152.98	pl (CCl_4)	140-2^{45}	62.5	1.48		al, eth, bz, chl	B2^4, 765
12083	Propionic acid, 3-bromo, butyl ester $CH_2BrCH_2CO_2C_4H_9$	209.09		130^{26}		1.4549^{20}	1.3051^{20}	al, eth, ace	B2^4, 765
12084	Propionic acid, 3-bromo, ethyl ester $CH_2BrCH_2CO_2C_2H_5$	181.03		179, 70^{12}			1.4516^{20}	al, eth, ace	B2^4, 765
12085	Propionic acid, 3-bromo,methyl ester $CH_2BrCH_2CO_2CH_3$	167.00		105.5^{60}		1.4897^{15}	1.4542^{20}	al, eth, ace	B2^4, 762
12087	Propionic acid, 3-bromo,isopentyl ester $CH_2BrCH_2CO_2\text{-i-}C_5H_{11}$	223.11		110-1^{11}		1.2217$^{15/4}$	1.4556^{9}	al, eth	B2^2, 231
12088	Propionic acid, 3-bromo-2-phenyl $CH_2BrCH(C_6H_5)CO_2H$	229.07	pr (CS_2)		93-4			al, eth, bz	B9, 526
12089	Propionic acid, 3-bromo-3-phenyl $C_6H_5CHBrCH_2CO_2H$	229.07	mcl pr (al)		137			al	B9^3, 2401
12090	Propionic acid, 3-bromo-3-phenyl,methyl ester $C_6H_5CHBrCH_2CO_2CH_3$	243.10	pr		37-8			al	B9^1, 201
12091	Propionic acid, butyl ester or Butyl propionate $CH_3CH_2CO_2C_4H_9$	130.19		145.5	−89.5	0.8754$^{20/4}$	1.4014^{20}	**al, eth**	B2^4, 708
12092	Propionic acid, isobutyl ester $CH_3CH_2CO_2\text{-i-}C_4H_9$	130.19		136.8, 66.5^{60}	−71.4	0.8687$^{20/4}$	1.3973^{20}	**al, eth**	B2^4, 709
12093	Propionic acid, *sec*-butyl ester or *sec*-Butyl propionate $C_2H_5CO_2CH(CH_3)C_2H_5$	130.19		132-1		0.8657^{20}	1.3952^{20}	al, eth	B2^4, 709
12094	Propionic acid, 2-chloro $CH_3CHClCO_2H$	108.52		186, 84^{12}		1.2585$^{20/4}$	1.4380^{20}	w, **al, eth**, ace	B2^4, 745
12095	Propionyl chloride, 2-chloro-(d) $CH_3CHClCOCl$	126.97	$[\alpha]^{18/}{}_D$ +0.2	110^{744}		1.2394$^{7.5}$			B2^4, 747
12096	Propionic acid, 2-chloro, butyl ester $CH_3CHClCO_2C_4H_9$	164.63		183-5, 72-3^{10}		1.0253$^{20/4}$	1.4263^{20}	eth	B2^4, 746
12097	Propionic acid, 2-chloro, ethyl ester $CH_3CHClCO_2C_2H_5$	136.58		147-8, 52-4^{18}		1.0793$^{20/4}$	1.4178^{20}	**al, eth**	B2^4, 746
12098	Propionic acid, 2-chloro, isobutyl ester (d) $CH_3CHClCO_2\text{-i-}C_4H_9$	164.63	$[\alpha]_D$ +5.2	175-7		1.0312$^{20/4}$	1.4247^{20}	al, eth, ace	B2, 248
12099	Propionic acid, 2-chloro, isopropyl ester $CH_3CHClCO_2CH(CH_3)_2$	150.61		151-2, 46-7^{12}		1.0315$^{20/4}$	1.4149^{20}	al, eth	B2^4, 746

PHYSICAL CONSTANTS OF ORGANIC COMPOUNDS (Continued)

No.	Name, Synonyms, and Formula	Mol. wt.	Color, crystalline form, specific rotation and λ_{max} (log ε)	b.p. °C	m.p. °C	Density	n_D	Solubility	Ref.
12100	Propionic acid, 2-chloro, methyl ester (d) $CH_3CHClCO_2CH_3$	122.55	$[\alpha]_D$ + 19.9 (undil)	133-4, 50^{35}		$1.1815^{20/4}$		al	B2, 248
12101	Propionic acid, 2-chloro, methyl ester-(dl) $CH_3CHClCO_2CH_3$	122.55		132.5		$1.209^{20/4}$	1.4182^{20}	al	$B2^4$, 746
12102	Propionic acid, 2-chloro, methyl ester-(l) $CH_3CHClCO_2CH_3$	122.55	$[\alpha]_D$ -26.9 (undil)	$79\text{-}80^{120}$		$1.158^{5/4}$		al	$B2^3$, 553
12103	Propionamide, 2-chloro-N-(2-tolyl) $CH_3CHClCONH(C_6H_4CH_3\text{-}2)$	197.66	nd (abs al)		111			al	B12, 794
12104	Propionic acid, 3-chloro $CH_2ClCH_2CO_2H$	108.52	lf (w), hyg cr (lig)	204d	41 (61)			w, al, eth	$B2^4$, 748
12105	Propionic acid, 3-chloro, butyl ester $CH_2ClCH_2CO_2C_4H_9$	164.63		104^{22}		$1.0370^{20/4}$	1.4321^{20}	w, eth	$B2^4$, 749
12106	Propionic acid, 3-chloro, isobutyl ester $CH_2ClCH_2CO_2\text{-i-}C_4H_9$	164.63		191.3		$1.0323^{20/4}$	1.4295^{20}	al, eth	$B2^4$, 749
12107	Propionic acid, 3-chloro, ethyl ester $CH_2ClCH_2CO_2C_2H_5$	136.58		162, 56^{11}		$1.1086^{20/4}$	1.4254^{20}	al, eth	$B2^4$, 749
12108	Propionic acid, 3-chloro, methyl ester $CH_2ClCH_2CO_2CH_3$	122.55		155-7, $40\text{-}2^{10}$		$1.1861^{15/4}$	1.4263^{20}	al	$B2^4$, 748
12109	Propionic acid, 3-chloro, isopentyl ester $CH_2ClCH_2CO_2\text{-i-}C_5H_{11}$	178.66		$207\text{-}8^{740}$, 87^{12}		$1.0171^{20/4}$	1.4343^{20}	al, eth	$B2^4$, 749
12110	Propionic acid, 3-chloro, propyl ester $CH_2ClCH_2CO_2C_3H_7$	150.61		180, $77\text{-}8^{12}$		$1.0656^{20/4}$	1.4290^{20}	al, eth	$B2^4$, 749
12111	Propionic acid, 3-cyclohexyl $C_6H_{11}CH_2CH_2CO_2H$	156.22		275-8, 143^{11}	16	$0.9966^{20/4}$	1.4634^{20}	eth	$B9^3$, 64
12112	Propionic acid, cyclohexyl ester or Cyclohexyl propionate $CH_3CH_2CO_2C_6H_{11}$	156.22		193^{750}, 93^{35}		$0.9359^{20/4}$	1.4403^{20}	al, eth, ace	$B6^4$, 37
12113	Propionic acid, 2,3-diamino $H_2NCH_2CH(NH_2)CO_2H$	104.11	hyg rosettes		ca 110-20				$B4^3$, 1292
12114	Propionic acid, 2,3-dibromo $CH_2BrCHBrCO_2H$	231.87		220-40d, 160^{20}	66-7			al, eth, bz	$B2^4$, 767
12115	Propionic acid, 2,3-dibromo, ethyl ester $CH_2BrCHBrCO_2C_2H_5$	259.93		214-5, 112^{23}		$1.7966^{20/4}$	1.5007^{20}	al, eth	$B2^4$, 767
12116	Propionic acid, 2,3-dibromo, methyl ester $CH_2BrCHBrCO_2CH_3$	245.90		206, 115^{25}		$1.9333^{20/4}$	1.5127^{20}	al	$B2^4$, 767
12117	Propionic acid, 2,3-dibromo-3-phenyl (d) or *d*-Cinnamic acid dibromide $C_6H_5CHBrCHBrCO_2H$	307.97	pr (chl), $[\alpha]^{15}_D$ + 45.8 (abs al)	182				al, eth	$B9^3$, 2404
12118	Propionic acid, 2,3-dibromo-3-phenyl (dl) $C_6H_5CHBrCHBrCO_2H$	307.97	mcl pr (chl)	sub	240d			al, eth	$B9^3$, 2405
12119	Propionic acid, 2,3-dibromo-3-phenyl (*meso*) $C_6H_5CHBrCHBrCO_2H$	307.97	nd		91.3			bz	$B9^3$, 2406
12120	Propionic acid, 2,3-dibromo-3-phenyl, ethyl ester-(d) $C_6H_5CHBrCHBrCO_2C_2H_5$	336.02	cr (CS_2)		71			al	B9, 518
12121	Propionic acid, 2,3-dibromo-3-phenyl, ethyl ester-(dl) $C_6H_5CHBrCHBrCO_2C_2H_5$	336.02	mcl pr or pl		15-6				$B9^3$, 2406
12122	Propionic acid, 2,2-dichloro $CH_3CCl_2CO_2H$	142.97		185-90, $90\text{-}2^{14}$		$1.389^{12/4}$		al, eth	$B2^4$, 753
12123	Propionic acid, 2,2-dichloro-3,3,3-trifluoro, methyl ester $CF_3CCl_2CO_2CH_3$	210.97		$116\text{-}7^{625}$		1.5092^{20}	1.3806^{20}		$B2^4$, 758
12124	Propionic acid, 2,2-dichloro-3,3,3-trifluoro, propyl ester $CF_3CCl_2CO_2C_3H_7$	239.02		144.5^{625}		1.3531^{20}	1.3888^{20}		$B2^4$, 758
12125	Propionic acid, 2,3-dichloro $CH_2ClCHClCO_2H$	142.97	hyg nd (peth)	210, 113^{12}	50		1.4650^{20}	al, eth, w	$B2^4$, 756
12126	Propionic acid, 2,3-dichloro, ethyl ester $CH_2ClCHClCO_2C_2H_5$	171.02		183-4, $76\text{-}7^{15}$		$1.2461^{20/4}$	1.4482^{20}	al, eth	$B2^4$, 756
12127	Propionic acid, 2,3-dichloro, methyl ester-(d) $CH_2ClCHClCO_2CH_3$	157.00	$[\alpha]^{20}_D$ + 1.7	92^{50}		$1.3282^{20/4}$		al, eth, ace	$B2^1$, 111
12128	Propionic acid, 2,3-dichloro, isopropyl ester $CH_2ClCHClCO_2CH(CH_3)_2$	185.05		$61\text{-}2^5$		1.2010	1.4470		$B2^4$, 756
12129	Propionic acid, 3,3-dichloro $CHCl_2CH_2CO_2H$	142.97	pr		56			al, ace, bz, chl	$B2^4$, 758
12130	Propionic acid, 2,3-dihydroxy or Glyceric acid $HOCH_2CH(OH)CO_2H$	106.08	syr	d				w, al, ace	$B3^4$, 1050

No.	Name, Synonyms, and Formula	Mol. wt.	Color, crystalline form, specific rotation and λ_{max} (log ε)	b.p. °C	m.p. °C	Density	n_D	Solubility	Ref.
12131	Propionic acid, 2,3-dihydroxy, propyl ester $HOCH_2CH(OH)CO_2C_3H_7$	148.16		$132\text{-}4^{3}$		$1.1537^{20/4}$	1.4503^{20}		B3[3], 854
12132	Propionic acid, 2,2-dimethyl or Pivalic acid $(CH_3)_3CCO_2H$	102.13	nd	$164, 70^{14}$	35	0.905^{50}	$1.3931^{36.5}$	al, eth	B2[4], 908
12133	Propionic acid, 2,2-diphenyl $CH_3C(C_6H_5)_2CO_2H$	226.27	pl (bz-peth), nd (w), lf (dil al)	sub >300	175-7			al, eth, bz, chl	B9[3], 3342
12134	Propionic acid, 2,3-diphenyl (d) $C_6H_5CH_2CH(C_6H_5)CO_2H$	226.27	cr (dil al), $[\alpha]^{20}{}_D$ +94 (bz)		83-9			al, eth, bz	B9[3], 3333
12135	Propionic acid, 2,3-diphenyl (dl) $C_6H_5CH_2CH(C_6H_5)CO_2H$	226.27	(i) pr (chl) (ii) pl (chl) (iii) cr (MeOH)	330-40	(i)88-9 (ii)95-6 (iii)82			al, eth, bz	B9[3], 3333
12136	Propionic acid, 2,3-diphenyl (l) $C_6H_5CH_2CH(C_6H_5)CO_2H$	226.27	nd (dil al), $[\alpha]^{20}{}_D$ -85.1 (bz)		83-9			al, eth, bz	B9[3], 3333
12137	Propionic acid, 2,3-diphenyl-2-hydroxy or α-Benzylmandelic acid $C_6H_5CH_2C(C_6H_5)(OH)CO_2H$	242.27	nd (bz), cr (dil al)		165.6			al, eth, aa	B10[3] 1193
12138	Propionic acid, 3,3-diphenyl or Benzhydryl acetic acid $(C_6H_5)_2CHCH_2CO_2H$	226.27	nd (dil al)		155			al, eth	B9[3], 3338
12139	Propionic acid, 3,3-diphenyl-2-hydroxy or β,β-Diphenyl lactic acid $(C_6H_5)_2CHCH(OH)CO_2H$	242.27	nd (w)		159d			al, eth	B10[2], 228
12140	Propionic acid, 3,3-diphenyl-3-hydroxy $(C_6H_5)_2C(OH)CH_2CO_2H$	242.27	nd (dil al)		212			al, ace, aa	B10[3], 1196
12141	Propionic acid, 3,3-diphenyl-3-hydroxy, ethyl ester $(C_6H_5)_2C(OH)CH_2CO_2C_2H_5$	270.33	pr (dil al)		87			al	B10[2], 228
12142	Propionic acid, 2,3-epoxy or Glycidic acid CH_2CHCO_2H (O ring)	88.06						w, al, eth	B18[1], 435
12143	Propionic acid, ethyl ester or Ethyl propionate $CH_3CH_2CO_2C_2H_5$	102.13		99.1	−73.9	0.8917^{20}	1.3839^{20}	al, eth, ace	B2[4], 705
12144	Propionic acid, 3-fluoro $FCH_2CH_2CO_2H$	92.07		$83\text{-}4^{14}$			1.3889^{25}	w, al, eth	B2[4], 734
12145	Propionic acid, 3-(2-furyl) or Furfuryl acetic acid $(2\text{-}C_4H_3O)CH_2CH_2CO_2H$	140.14	cr (chl-lig, w, peth)	$229, 108\text{-}10^{10}$	58			w, eth, chl	B18[4], 4090
12146	Propionic acid, 3-(2-furyl), ethyl ester $(2\text{-}C_4H_3O)CH_2CH_2CO_2C_2H_5$	168.19		$212, 108\text{-}10^{10}$			1.4812^{25}		B18[4], 4090
12147	Propionic acid, furfuryl ester or Furfuryl propionate $CH_3CH_2CO_2CH_2(C_4H_3O)$	154.17		$195\text{-}6^{762}$		$1.1085^{20/4}$		al, eth, ace	B17[2], 115
12148	Propionic acid, heptyl ester or Heptyl propionate $CH_3CH_2CO_2C_7H_{15}$	172.27		$210, 124\text{-}5^{16}$	−50.9	$0.8679^{20/4}$	1.4201^{15}	al, eth, ace	B2[4], 710
12149	Propionic acid, hexyl ester or Hexyl propionate $CH_3CH_2CO_2C_6H_{13}$	158.24		$190, 73\text{-}4^{10}$	−57.5	0.8698^{20}	1.4162^{15}	al, eth, ace	B2[4], 709
12150	Propionic acid, 2-hydroxy (D) or D-Lactic acid $CH_3CH(OH)CO_2H$	90.08	pl (chl aa), $[\alpha]_D$ -2.3 (w, c=1.24)	103^{2}	53			w, al	B3[4], 633
12151	Propionic acid, 3-hydroxy or Hydracrylic acid $HOCH_2CH_2CO_2H$	90.08	syr	d			1.4489^{20}	w, al, eth	B3[4], 689
12152	Propionic acid, 3-hydroxy, lactone or β-Propiolactone CH_2CH_2CO (O ring)	72.06		$162d, 51^{10}$	−33.4	$1.1460^{20/5}$	1.4105^{20}	eth, chl	B17[4], 4157
12153	Propionic acid, 3-Hydroxy, methyl ester $HOCH_2CH_2CO_2CH_3$	104.11		121^{94}		1.118^{25}	1.4306^{23}	w, al, eth	B3[3], 525
12154	Propionic acid, 3-(2-hydroxyphenyl) or o-Hydrocoumaric acid $(2\text{-}HOC_6H_4)CH_2CH_2CO_2H$	166.18	pr (w)		82-4			w, al, eth	B10[3], 534
12155	Propionic acid, 3-(4-hydroxy phenyl) $(4\text{-}HOC_6H_4)CH_2CH_2CO_2H$	166.18		$208\text{-}10^{14}$	129-30			w, al, eth, bz	B10[3], 539
12156	Propionic acid, 3-(4-hydroxy-3-methoxy phenyl) or Hydroferulic acid $(3\text{-}CH_3O\text{-}4\text{-}HOC_6H_3)CH_2CH_2CO_2H$	196.20	pl (w)		89-90			al, eth	B10[3], 1517
12157	Propionic acid, 2-hydroxy-3,3,3-trichloro $Cl_3CCH(OH)CO_2H$	193.41	pr (eth)	$140\text{-}7^{49}$	125			al, eth, chl	B3[4], 680

No.	Name, Synonyms, and Formula	Mol. wt.	Color, crystalline form, specific rotation and λ_{max} (log ε)	b.p. °C	m.p. °C	Density	n_D	Solubility	Ref.
12158	Propionic acid, 2,2′-imino-di-(dl) $HN[CH(CH_3)CO_2H]_2$	161.16	(i) nd (w), pr (eth) (ii) cr		(i) 234-5d (ii) 254-5			w	B4^{3}, 1251/1252
12159	Propionic acid, 3,3-imino-di, N-methyl, diethyl ester $CH_3N[CH_2CH_2CO_2C_2H_5]_2$	231.29		136-8^{4}		1.0190$^{20/20}$	1.4421^{20}	al, eth	B4^{2}, 829
12160	Propionic acid, β-(3-indolyl) $C_{11}H_{11}NO_2$	189.21	pl (w)		134			al, eth, ace, bz, chl	B22^{4}, 1112
12161	Propionic acid, β-(3-indolyl), methyl ester $C_{12}H_{13}NO_2$	203.24	pr (MeOH)		79-80			al	B22^{2}, 53
12162	Propionic acid, 2-iodo (dl) CH_3CHICO_2H	199.98	nd (bz, w, al)	93-6$^{0.2}$	45-7	2.073$^{18/4}$		al, eth	B2^{3}, 573
12163	Propionic acid, 3-iodo $ICH_2CH_2CO_2H$	199.98	lf (w)		85			al, eth, ace	B2^{4}, 770
12164	Propionic acid, 3-iodo, methyl ester $ICH_2CH_2CO_2CH_3$	214.00		188^{756}		1.8408^{7}		al	B2^{3}, 574
12165	Propionic acid, 2-mercapto $CH_3CH(SH)CO_2H$	106.14		106-7^{15}	10-4	1.1938$^{20/4}$	1.4810^{20}	w, al, eth	B3^{4}, 682
12166	Propionic acid, 3-mercapto $HSCH_2CH_2CO_2H$	106.14	amor	110-12^{15}	17-9	1.218^{21}	1.4911^{20}	w, al, eth	B3^{4}, 726
12167	Propionitrile, 2-methoxy $CH_3\text{-}CH(OCH_3)CN$	85.11		118^{740}		0.8928$^{20/4}$	1.3818^{20}	al	B3^{4}, 675
12168	Propionitrile, 3-methoxy $CH_3OCH_2CH_2CN$	85.11		163.5^{763}, 85.5^{49}		0.9379$^{20/4}$	1.4043^{20}	al, eth	B3^{4}, 708
12169	Propionic acid, methyl ester or Methyl propionate $CH_3CH_2CO_2CH_3$	88.11		79.9	−87.5	0.9150$^{20/4}$	1.3775^{20}		B2^{4}, 704
12170	Propionic acid, 2 methyl or Isobuturic acid $(CH_3)_2CHCO_2H$	88.11		153.2, 53.7^{10}	−46.1	0.9681$^{20/4}$	1.3930^{20}	al, eth, w	B2^{4}, 843
12171	Propionic acid, 3-(α-naphthyl) $(\alpha\text{-}C_{10}H_7)CH_2CH_2CO_2H$	200.24	cr (bz), nd (al)	179^{11}	156-7			al	B9^{3}, 3219
12172	Propionic acid, 3-(β-naphthyl) $(\beta\text{-}C_{10}H_7)CH_2CH_2CO_2H$	200.24	lf or nd (w, al)		135			al	B9^{3}, 3221
12173	Propionic acid, octyl ester or Octyl propionate $CH_3CH_2CO_2C_8H_{17}$	186.29		228	−42.6	0.8663^{20}	1.4221$^{15/D}$	al, eth, bz	B2^{4}, 710
12174	Propionic acid, 2-oxo or Pyruvic acid CH_3COCO_2H	88.06		165d, 54^{10}	13.8	1.2272$^{20/4}$	1.4280^{20}	w, al, eth, ace	B3^{4}, 1505
12175	Propionic acid, 3-oxo-2-phenyl, ethyl ester $C_6H_5CH(CHO)CO_2C_2H_5$	192.21	pl (chl)	136^{16}	70-1	1.1204$^{20/20}$	1.532^{21}	al, eth	B10^{3}, 3023
12176	Propionic acid, pentachloro $CCl_3CCl_2CO_2H$	246.30	cr (CCl_4)		200-15d				B2^{2}, 228
12177	Propionic acid, pentyl ester or Pentyl propionate $CH_3CH_2CO_2C_5H_{11}$	144.21		168.6	−73.1	0.8761$^{25/4}$	1.4096^{15}	al, eth, bz	B2^{4}, 709
12178	Propionic acid, isopentyl ester or Isopentyl propionate $CH_3CH_2CO_2\text{-i-}C_5H_{11}$	144.21		160.7		0.8697$^{20/4}$	1.4069^{20}		B2^{4}, 709
12179	Propionic acid, 2-phenoxy (D) $CH_3CH(OC_6H_5)CO_2H$	166.18	nd (w), $[\alpha]^{1'}_{D}$ + 39.3 (al,c=1.2)	265-6^{758}	87			al, eth	B6^{4}, 642
12180	Propionic acid, 2-phenoxy (DL) $CH_3CH(OC_6H_5)CO_2H$	166.18	nd (w)	265-6, 105-6^{5}	115-6	1.1865$^{20/4}$	1.5184^{20}	al, eth	B6^{4}, 642
12181	Propionic acid, 2-phenoxy, ethyl ester $CH_3CH(OC_6H_5)CO_2C_2H_5$	194.23		243-4, 120-5^{6}		1.360$^{17/4}$			B6^{4}, 643
12182	Propionic acid, 3-phenoxy $C_6H_5OCH_2CH_2CO_2H$	166.18	nd (w), lf (lig)	234-45^{771}, 188-9^{24}	97-8				B6^{4}, 643
12183	Propionic acid, 3-phenoxy, ethyl ester $C_6H_5OCH_2CH_2CO_2C_2H_5$	194.23	nd (peth)	170^{40}	24	1.0821$^{25/25}$	1.5007^{18}	al, eth	B6^{4}, 644
12184	Propionic acid, phenyl ester or Phenyl propionate $CH_3CH_2CO_2C_6H_5$	150.18	pr	211, 100^{16}	20	1.0467$^{25/25}$	1.4980^{20}	al, eth, bz	B6^{4}, 615
12185	Propionic acid, 2-phenyl (d) or Hydratropic acid $CH_3CH(C_6H_5)CO_2H$	150.18	$[\alpha]^{20/}_{D}$ + 81.1 (al,c=3)	152^{16}					B9^{3}, 2417
12186	Propionic acid, 3-phenyl or Hydrocinnamic acid $C_6H_5CH_2CH_2CO_2H$	150.18	nd (w)		106-8			al, eth	B9^{3}, 2382
12187	Propionic acid, 3-(3-pyrenyl) $(3\text{-}C_{16}H_9)CH_2CH_2CO_2H$	274.32	pl (aa)		180			ace, bz	E14, 441
12188	Propionic acid, propyl ester or Propyl propionate $CH_3CH_2CO_2C_3H_7$	116.16		122.3	−75.9	0.8809$^{20/4}$	1.3935^{20}	al, eth, ace, chl	B2^{4}, 707

PHYSICAL CONSTANTS OF ORGANIC COMPOUNDS (Continued)

No.	Name, Synonyms, and Formula	Mol. wt.	Color, crystalline form, specific rotation and λ_{max} (log ε)	b.p. °C	m.p. °C	Density	n_D	Solubility	Ref.
12189	Propionic acid, isopropyl ester or Isopropyl propionate $CH_3CH_2CO_2CH(CH_3)_2$	116.16	...	109-10	...	$0.8660^{20/4}$	1.3872^{20}	**al, eth**	B2^4, 708
12190	Propionic acid, 2,2,3,3-tetrachloro $Cl_2CHCCl_2CO_2H$	211.86	cr (CS_2-chl)	...	76	...	...	w	B2^4, 760
12191	Propionic acid, 2,3,3,3 tetrafluoro, ethyl ester $CF_3CHFCO_2C_2H_5$	174.10	...	108-9	...	$1.289^{20/4}$	1.3260^{20}	...	B2^4, 737
12192	Propionic acid, tetrahydrofurfuryl ester $CH_3CH_2CO_2CH_2(C_4H_7O)$	158.20	...	$204\text{-}7^{756}$, $85\text{-}7^{3}$	...	$1.044^{20/0}$	...	**al, eth, chl**	B17^4, 1104
12193	Propionic acid, 3-(2-tetrahydrofuryl) $(2\text{-}C_4H_7O)CH_2CH_2CO_2H$	144.17	...	263, $118\text{-}20^{2}$	...	$1.1155^{20/20}$	1.4578^{25}	al, ace	B18^4, 3846
12194	Propionic acid, 3-(2-tetrahydrofuryl), ethyl ester $(2\text{-}C_4H_7O)CH_2CH_2CO_2C_2H_5$	172.22	...	$221\text{-}2^{750}$, 73^{2}	...	$1.024^{7/15}$	1.440^{20}	al, ace	B18^4, 3846
12195	Propionic acid, 2,2,3-trichloro $CH_2ClCCl_2CO_2H$	177.41	hyg pr (CS_2)	140^{40}	65-6	...	...	al, bz	B2^4, 759
12196	Propionic acid, 2,3,3-triphenyl $(C_6H_5)_2CHCH(C_6H_5)CO_2H$	302.37	nd (dil al, peth)	...	222-3	...	...	al, eth	B9^3, 3598
12197	Propionic acid, 3,3,3-triphenyl $(C_6H_5)_3CCH_2CO_2H$	302.37	pr (al)	...	179-80	...	...	al, eth	B9^3, 3603
12198	Propionitrile CH_3CH_2CN	55.08	...	97.3	−92.9	$0.7818^{20/4}$	1.3655^{20}	w, al, eth, ace, bz	B2^4, 728
12199	Propionitrile, 2-bromo $CH_3CHBrCN$	133.98	...	59^{24}	...	$1.5505^{20/4}$	1.4585^{20}	eth, ace	B2^4, 764
12200	Propionitrile, 3-bromo $BrCH_2CH_2CN$	133.98	...	92^{25}, 69^{7}	...	$1.6152^{20/4}$	1.4800^{20}	al, eth	B2^4, 766
12201	Propionitrile, 2-chloro $CH_3CHClCN$	89.52	...	123-4, 73^{144}	...	1.0792^{10}	...	...	B2^4, 748
12202	Propionitrile, 3-chloro $ClCH_2CH_2CN$	89.52	...	175-6, $85\text{-}7^{20}$	...	1.1573^{20}	1.4360^{20}	...	B2^4, 751
12203	Propionitrile, 2,3-dichloro $CH_2ClCHClCN$	123.97	...	$58\text{-}9^{8}$	...	1.3500^{20}	1.4640^{20}	...	B2^4, 757
12204	Propionitrile, 2,2-dimethyl $(CH_3)_3CCN$	83.13	...	105-6	15-6	$0.7586^{25/4}$	1.3774^{20}	...	B2^4, 913
12205	Propionitrile, 2-(dimethylamino) $[(CH_3)_2N]CH(CH_3)CN$	98.15	...	144	...	...	...	al, eth	B4^3, 1236
12206	Propionitrile, 2-ethoxy $CH_3CH(OC_2H_5)CN$	99.13	...	136^{765}	...	$0.8743^{20/4}$	1.3890^{22}	al, eth, aa	B3^4, 675
12207	Propionitrile, 3-ethoxy $C_2H_5OCH_2CH_2CN$	99.13	...	171, 65^{15}	...	$0.9285^{15/4}$	1.4068^{20}	al, eth	B3^4, 708
12208	Propionitrile, 2-hydroxy or Acetaldehyde cyanohydrin $CH_3CH(OH)CN$	71.08	...	182-4, 102^{30}	−40	$0.9877^{20/4}$	1.4058^{18}	**w, al**, eth, chl	B3^4, 675
12209	Propionitrile, 2-hydroxy-3,3,3-trichloro or Chloral cyanohydrin $Cl_3CCH(OH)CN$	174.41	pl (w)	215-20d	61	...	...	w, al, eth	B3^4, 680
12210	Propionitrile, 3-hydroxy or Hydracrylonitrile $HOCH_2CH_2CN$	71.08	...	230, 110^{15}	...	1.0588^{0}	1.4240^{20}	**w, al**	B3^4, 708
12211	Propionitrile, 2,2′-imino-di- or 2,2′-imino dipropionitrile $HN[CH(CH_3)CN]_2$	123.16	nd (eth)	...	68	...	...	al, eth	B4^3, 1251
12212	Propionitrile, 3-(2-oxocyclohexyl) $C_9H_{13}NO$	151.21	...	$138\text{-}42^{10}$	...	$1.0181^{20/4}$	1.4755^{20}	...	B10^3, 2835
12213	Propionitrile, 2-propoxy $CH_3CH(OC_3H_7)CN$	113.16	...	150^{727}	...	$0.866^{20/4}$	1.398^{20}	...	B3, 285
12214	Propionitrile, 3-phenyl $C_6H_5CH_2CH_2CN$	131.18	...	261, $125\text{-}6^{15}$	...	1.0016^{20}	1.5266^{20}	al, eth	B9^3, 2395
12215	Propionyl bromide CH_3CH_2COBr	136.98	...	$103\text{-}1^{770}$	...	$1.5210^{16/4}$	1.4578^{16}	eth	B2^4, 724
12216	Propionyl bromide, 2-bromo $CH_3CHBrCOBr$	215.87	...	152-4	...	$2.0612^{16/4}$	...	...	B2^4, 764
12217	Propionyl chloride CH_3CH_2COCl	92.53	...	80	−94	1.0646^{20}	1.4032^{20}	eth	B2^4, 724
12218	Propionyl chloride, 2-bromo $CH_3CHBrCOCl$	171.42	...	131-3	...	1.697^{11}	...	eth, chl	B2^4, 764
12219	Propionyl chloride, 3-chloro CH_2ClCH_2COCl	126.97	yesh	$143\text{-}5^{763}$, 82^{102}	...	1.3307^{13}	1.4549^{20}	al, eth, chl	B2^4, 750
12220	Propionyl chloride, 2,2-dichloro CH_3CCl_2COCl	161.42	...	$117\text{-}8^{750}$, 68-72	...	$1.4062^{20/4}$	1.4524^{20}	...	B2^4, 755
12221	Propionyl chloride, 2,3-dichloro $CH_2ClCHClCOCl$	161.42	...	$52\text{-}4^{16}$	...	$1.4757^{20/4}$	1.4764^{20}	eth	B2^4, 757

No.	Name, Synonyms, and Formula	Mol. wt.	Color, crystalline form, specific rotation and λ_{max} (log ε)	b.p. °C	m.p. °C	Density	n_D	Solubility	Ref.
12222	Propionyl chloride, 3,3-dichloro $CHCl_2CH_2COCl$	161.42		$43-4^{10}$		$1.4557^{20/4}$	1.4738^{20}	eth, diox	$B2^3$, 562
12223	Propionyl chloride, pentachloro CCl_3CCl_2COCl	264.75	nd		42			bz	$B2^4$, 761
12224	Propionyl chloride, 2-phenoxy $CH_3CH(OC_6H_5)COCl$	184.62		$146-7^{55}$, $115-7^{10}$		$1.1865^{20/4}$	1.5178^{20}	eth	$B6^4$, 643
12225	Propionyl fluoride CH_3CH_2COF	76.07		44		$0.972^{15/4}$	1.329^{13}		$B2^4$, 724
12226	Propionyl iodide CH_3CH_2COI	183.98		127-8					$B2^3$, 542
12227	Propioryl urea $CH_3CH_2CONHCONH_2$	116.12	cr (w)		210-11			w, al	$B3^3$, 121
12228	Propiophenone or Ethyl phenyl ketone $C_6H_5COC_2H_5$	134.18		217.5, 91.6^{10}	18.6	1.0096^{20}	1.5269^{20}	al, eth	$B7^3$, 1022
12229	Propiophenone, α-amino, hydrochloride $C_6H_5COCH(NH_2)CH_3.HCl$	185.65	nd (al-eth)		187			w	$B14^3$, 147
12230	Propiophenone, o-amino $(2\text{-}H_2NC_6H_4)COCH_2CH_3$	149.19	pa ye lf (peth) pl (dil al)	$93^{0.8}$	46-7			w, al, eth, ace	$B14^3$, 143
12231	Propiophenone, [p] -amino $(4\text{-}H_2NC_6H_4)COCH_2CH_3$	149.19	pl (al, w), nd (w)		140			w, al, chl	$B14^3$, 146
12232	Propiophenone, α-bromo $C_6H_5COCHBrCH_3$	213.07		245-50, $134-5^{18}$		$1.4298^{20/4}$	1.5720^{20}	al, eth, ace, bz, chl	$B7^3$, 1033
12233	Propiophenone, [p] -bromo $(4\text{-}BrC_6H_4)COC_2H_5$	213.07	nd	169^{15}	48			al, eth, ace	$B7^3$, 1032
12234	Propiophenone, β-chloro $C_6H_5COCH_2CH_2Cl$	168.62	lf (eth), cr (peth al)		49-50				$B7^3$, 1030
12235	Propiophenone, [p]- chloro $(4\text{-}ClC_6H_4)COC_2H_5$	168.62		$134-7^{31}$	36-7			al	$B7^3$, 1029
12236	Propiophenone, [p] -chloro, oxime $4\text{-}ClC_6H_4C(=NOH)C_2H_5$	183.64	pl (al)		62-3			al	B7, 301
12237	Propiophenone, α,β-dibromo-β-phenyl or [threo] -Chalone dibromide $C_6H_5COCHBrCHBrC_6H_5$	368.07	nd (al)		122-3			al	$B7^3$, 2155
12238	Propiophenone, [m] -methoxy $(3\text{-}CH_3OC_6H_4)COCH_2CH_3$	164.20		258-60, $95-7^{0.7}$		1.0812^0	1.5230^{25}	al, eth	B8, 106
12239	Propiophenone, β,β-diphenyl $C_6H_5COCH_2CH(C_6H_5)_2$	286.37	nd (al)		96			ace, bz, chl	$B7^3$, 2756
12240	Propiophenone, [p] -methoxy $(4\text{-}CH_3OC_6H_4)COCH_2CH_3$	164.20		267-9, 142^{14}	<-15	$1.0670^{18/4}$	1.5253^{20}	al, eth	$B8^3$, 381
12241	Propiophenone, [m] -methyl or Ethyl-3-tolyl ketone $(3\text{-}CH_3C_6H_4)COC_2H_5$	148.20		234^{745}, $130-5^{33}$		$1.0059^{0/4}$		al, eth, ace, bz	$B7^3$, 1092
12242	Propiophenone, [p] -methyl or Ethyl-4-tolyl ketone $(4\text{-}CH_3C_6H_4)COC_2H_5$	148.20		238-9, 120^{18}		$0.9926^{20/4}$	1.5278^{20}	al, eth, ace, bz	$B7^3$, 1093
12243	Propiophenone oxime or Propiophenoxime $C_6H_5C(=NOH)C_2H_5$	149.19	pl (peth)	245-6d, 165^{38}	53-5			al, eth	$B7^3$, 1025
12244	n-Propyl alcohol or 1-Propcenol $CH_3CH_2CH_2OH$	60.10		97.4	−126.5	$0.8035^{20/4}$	1.3850^{20}	w, **al**, **eth**, ace, bz	$B1^4$, 1413
12245	Isopropyl alcohol or 2-Propanol $(CH_3)_2CHOH$	60.10		82.4	−89.5	$0.7855^{20/4}$	1.3776^{20}	**w**, **al**, **eth**, ace, bz	$B1^4$, 1461
12246	n-Propyl amine or l-Amino propane $CH_3CH_2CH_2NH_2$	59.11		47.8	−83	$0.7173^{20/4}$	1.3870^{20}	w, al, eth, ace, bz, chl	$B4^4$, 464
12247	Propyl amine, 3-(diethyl amino) $(C_2H_5)_2NCH_2CH_2CH_2NH_2$	130.23		165-72	$0.825^{20/20}$	$1.443^{20}{}_D$			$B4^4$, 1260
12248	n-Propyl amine, N,N-dimethyl or -l-(dimethyl amino)propane $CH_3CH_2CH_2N(CH_3)_2$	87.16		65.5^{752}		$0.7152^{20/4}$	1.3860^{20}	al, eth, bz, chl	$B4^4$, 467
12249	Propyl amine, N-nitro $CH_3CH_2CH_2NHNO_2$	104.11		$128-9^{40}$	−21	1.1046^{15}	1.4610^{20}	**al**, **eth**	$B4^1$, 569
12250	Isopropyl amine or 2-Amino propane $(CH_3)_2CHNH_2$	59.11		32.4	−95.2	0.6891^{20}	1.3742^{20}	**w**, **al**, **eth**, ace, bz, chl	$B4^4$, 504
12251	n-Propyl bromide or l-Bromo propane $CH_3CH_2CH_2Br$	122.99		71	−110	$1.3537^{20/4}$	1.4343^{20}	al, eth, ace, bz	$B1^4$, 205
12252	Isopropyl bromide or 2-Bromo propane $(CH_3)_2CHBr$	122.99		59.4	−89	$1.3140^{20/4}$	1.4251^{20}	**al**, **eth**, ace, bz, chl	$B1^4$, 208
12253	n-Propyl chloride or l-Chloro propane $CH_3CH_2CH_2Cl$	78.54		46.6	−122.8	$0.8909^{20/4}$	1.3879^{20}	**al**, **eth**, bz, chl	$B1^4$, 189

No.	Name, Synonyms, and Formula	Mol. wt.	Color, crystalline form, specific rotation and λ_{max} (log ε)	b.p. °C	m.p. °C	Density	n_D	Solubility	Ref.
12254	Isopropyl chloride or 2-Chloro propane $(CH_3)_2CHCl$	78.54		35.7	−117.2	$0.8617^{20/4}$	1.3777^{20}	al, eth, bz, chl	$B1^{4}$, 191
12255	Propyl, 1,2-dichloro propyl ether $C_3H_7O(CHClCHClCH_3)$	171.07		176		$1.129^{15/4}$	1.447^{16}	eth	$B1^{3}$, 2691
12256	Propyl, 1,3-dichloro propyl ether $C_3H_7O(CHClCH_2CH_2Cl)$	171.07		65^{12}		1.112^{20}	1.4476^{20}	al, eth	$B1^{2}$, 690
12257	Propylene or Propene $CH_3CH{=}CH_2$	42.08	gas	−47.4	−185.2	$0.5193^{20/4}$ liq	1.3567^{-70}	al, aa	$B1^{4}$, 725
12258	Propylene, 3-amino or Allyl amine $H_2NCH_2CH{=}CH_2$	57.10		58		$0.7621^{20/4}$	1.4205^{20}	w, al, eth, chl	$B4^{4}$, 1057
12259	Propylene, 1-bromo (cis) or 1-Bromopropene, Propenyl bromide $CH_3CH{=}CHBr$	120.98		57.8	−113	$1.4291^{20/4}$	1.4560^{20}	eth, ace, chl	$B1^{4}$, 754
12260	Propylene, 2-bromo or Isopropenyl bromide $CH_3CBr{=}CH_2$	120.98		48.4^{748}	−124.8	$1.362^{20/4}$	1.4440^{20}	eth, ace, chl	$B1^{4}$, 754
12261	Propylene, 2-bromo-3-cyclohexyl $C_6H_{11}CH_2CBr{=}CH_2$	203.12		90^{15}		1.215^{17}	1.495^{17}	eth	$B5^{3}$, 224
12262	Propylene, 3-bromo or Allyl bromide $BrCH_2CH{=}CH_2$	120.98		70^{752}	−119.4	$1.398^{20/4}$	1.4697^{20}	al, eth	$B1^{4}$, 754
12263	Propylene, 3-bromo-3,3-difluoro $BrCF_2CH{=}CH_2$	156.96		42		$1.543^{25/4}$	1.3773^{25}		$B1^{4}$, 756
12264	Propylene, 1-chloro or [cis] -Propenyl chloride $CH_3CH{=}CHCl$	76.53		32.8	−134.8	$0.9347^{20/4}$	1.4055^{20}	eth, ace, bz, chl	$B1^{4}$, 737
12265	Propylene, 1-chloro-([trans]) or [trans] -Propenyl chloride $CH_3CH{=}CHCl$	76.53		37.4	−99	$0.9350^{20/7}$	1.4054^{20}	eth, ace, bz, chl	$B1^{4}$, 737
12266	Propylene, 2-chloro or Isopropenyl chloride $CH_3CCl{=}CH_2$	76.53		22.6	−137.4	$0.9014^{20/4}$	1.3973^{20}	eth, ace, bz, chl	$B1^{4}$, 737
12267	Propylene, 3-chloro or Allyl chloride $ClCH_2CH{=}CH_2$	76.53		45	−134.5	$0.9376^{20}{}_{D}$	1.4157^{20}	al, eth, ace, bz, lig	$B1^{4}$, 738
12268	Propylene, 3-chloro-2-(chloro methyl) $(ClCH_2)_2C{=}CH_2$	125.00		138, $30\text{-}1^{9}$	−15	$1.1782^{20/3}$	1.4754^{20}	al, chl	$B1^{4}$, 805
12269	Propylene, 2-cyclopropyl $CH_2{=}C(C_3H_5)CH_3$	82.15		$69.5\text{-}70^{751}$		$0.7500^{20/4}$	1.4252^{20}		$B5^{4}$, 243
12270	Propylene, 1,1-dibromo $CH_3CH{=}CBr_2$	199.87		127.4, 41.5^{30}		$1.9803^{20/20}$	1.5260^{20}	chl, bz	$B1^{4}$, 759
12271	Propylene, 2,3-dibromo or 2,3-Dibromo propene $CH_2BrCBr{=}CH_2$	199.87		141, $37\text{-}7^{11}$		$2.0346^{25/4}$	1.5416^{25}	eth, ace, chl	$B1^{4}$, 760
12272	Propylene, 1,1-dichloro or 1,1-Dichloro propene $CH_3CH{=}CCl_2$	110.97		76-7		$1.1864^{25/4}$	1.4430^{25}	eth, ace, chl	$B1^{4}$, 742
12273	Propylene, 1,1-dichloro-2-fluoro $CH_3CF{=}CCl_2$	128.96		77.7^{745}		$1.3026^{25/4}$	1.4196^{25}		$B1^{4}$, 745
12274	Propylene, 1,1-dichloro-3-phenyl or Cinnamylidene chloride $C_6H_5CH_2CH{=}CCl_2$	187.07	cr (eth, chl), pl (peth)	$142\text{-}3^{30}$	59			eth, bz, chl	$B5^{4}$, 1363
12275	Propylene, 1,2-dichloro ([cis]) $CH_3CCl{=}CHCl$	110.97		92.5^{742}			1.4549^{20}	ace, bz, chl	$B1^{4}$, 742
12276	Propylene, 1,2-dichloro ([trans]) or 1,2-Dichloro propene $CH_3CCl{=}CHCl$	110.97		77^{757}		$1.1818^{20/4}$	1.4471^{20}	al	$B1^{4}$, 742
12277	Propylene, 1,2-dichloro-1,3,3,3-tetrafluoro $CF_3CCl{=}CFCl$	182.93		47.3	−137	1.5468	1.3511^{20}		$B1^{4}$, 747
12278	Propylene, 1,2-dichloro-3,3,3-trifluoro $CF_3CCl{=}CHCl$	164.94		53.7	−109.2	$1.4653^{20/4}$	1.3670^{20}		$B1^{4}$, 746
12279	Propylene, 1,3-dichloro ([cis]) or 1,3-Dichloro propene $ClCH_2CH{=}CHCl$	110.97		104.3		$1.217^{20/4}$	1.4730^{20}	eth, bz, chl	$B1^{4}$, 743
12280	Propylene, 1,3-dichloro-([trans]) $ClCH_2CH{=}CHCl$	110.97		112		$1.224^{20/4}$	1.4682^{20}	eth, bz, chl	$B1^{4}$, 744
12281	Propylene, 2,3-dichloro $CH_2ClCCl{=}CH_2$	110.97		94		$1.211^{20/4}$	1.4603^{20}	al, eth, bz, chl	$B1^{4}$, 744
12282	Propylene, 3,3-dichloro or 3,3-Dichloro propene $CHCl_2CH{=}CH_2$	110.97		84.4		$1.175^{20/4}$	1.4510^{20}	al, eth, bz, chl	$B1^{4}$, 745
12283	Propylene, 1,1-diphenyl or 1,1-Diphenyl propene $CH_3CH{=}C(C_6H_5)_2$	194.28	lf (al)	280-1, 149^{11}	52	$1.0250^{20/4}$	1.5880^{20}	al, bz	$B5^{3}$, 1998
12284	Propylene, 2,3-diphenyl $C_6H_5CH_2C(C_6H_5){=}CH_2$	194.28		289^{757}	48	$1.10143^{20/4}$	1.5903^{20}		$B5^{2}$, 553
12285	Propylene, 1,2-epoxy or Methyl oxirene $CH_3C{=}CH$ (with O bridging the two C)	56.06		63				al, eth	B17, 20
12286	Propylene, 3-fluoro or Allyl fluoride $FCH_2CH{=}CH_2$	60.07	gas	−3				al, eth, chl	$B1^{4}$, 733

No.	Name, Synonyms, and Formula	Mol. wt.	Color, crystalline form, specific rotation and λ_{max} (log ε)	b.p. °C	m.p. °C	Density	n_D	Solubility	Ref.
12287	Propylene glycol or 1,2-Propanediol $CH_3CH(OH)CH_2OH$	76.10	......	189, 96-8[21]	......	$1.0361^{20/4}$	1.4324^{20}	**w, al, eth, bz**	B1[4], 2468
12288	Propylene, 3-iodo or Allyl iodide $CH_2{=}CHCH_2I$	167.98	......	102	−99.3	$1.8494^{20/4}$	1.5530^{20}	al, eth, chl	B1[4], 761
12289	Propylene, 1-nitro $CH_3CH{=}CHNO_2$	87.08	......	59-60[34]	......	$1.0661^{20/4}$	1.4527^{20}	eth, ace, chl	B1[4], 763
12290	Propylene, 2-nitro $CH_2{=}C(NO_2)CH_3$	87.08	......	57[100], 32[30]	......	$1.0643^{20/4}$	1.4358^{20}	eth, ace, chl	B1[4], 764
12291	Propylene, 1,1,2,3,3-pentachloro $CHCl_2CCl{=}CCl_2$	214.31	......	185, 116[9]	......	$1.6317^{34/4}$	1.5313^{20}	eth	B1[4], 753
12292	Propylene, perchloro $CCl_3CCl{=}CCl_2$	248.75	......	209-10, 140[100]	......	$1.7652^{20/4}$	1.5455^{20}	chl	B1[4], 753
12293	Propylene, perfluoro $CF_3CF{=}CF_2$	150.02	gas	−29.4	−156.2	$1.583^{-40/4}$	......	......	B1[4], 735
12294	Propylene, 2-phenyl or α-Methyl styrene $C_6H_5C(CH_3){=}CH_2$	118.18	......	163-4, 60[17]	24.3	$0.9082^{20/4}$	1.5303^{20}	eth, bz, chl	B5[4], 364
12295	Propylene, 1,2,3,3-tetrachloro $CHCl_2CCl{=}CHCl$	179.86	......	165, 50.6[12]	......	$1.537^{19/4}$	1.5121^{19}	bz, chl	B1[4], 752
12296	Propylene, 1,1,2-trichloro $CH_3CCl{=}CCl_2$	145.42	......	118, 41[52]	......	$1.382^{20/4}$	1.4827^{20}	al, eth, bz, chl	B1[4], 747
12297	Propylene, 1,1,2-trichloro-3,3,3-trifluoro $CF_3CCl{=}CCl_2$	199.39	......	88.1	−114.6	$1.617^{20/4}$	......	......	B1[4], 751
12298	Propylene, 1,2,3-trichloro $CH_2ClCCl{=}CHCl$	145.42	......	142, 32-3[14]	......	$1.414^{20/20}$	1.5030^{20}	al, eth, bz, chl	B1[4], 748
12299	Propylene, 3,3,3-trichloro $CCl_3CH{=}CH_2$	145.42	......	114-5, 57[102]	−30	$1.369^{20/20}$	1.4827^{20}	al, eth, bz, chl	B1[4], 749
12300	Propylene oxide or 1,2-Epoxy propane $CH_3\underbrace{CHCH_2}_{O}$	58.08	......	34.3	......	$0.859^{0/4}$	1.3670^{20}	**w, al, eth**	B17[4], 16
12301	Propyl ether or Dipropyl ether $C_3H_7OC_3H_7$	102.18	......	91	fr−122	$0.7360^{20/4}$	1.3809^{20}	**al, eth**	B1[4], 1422
12302	Isopropyl ether or Diisopropyl ether $(CH_3)_2CHOCH(CH_3)_2$	102.18	......	68	−85.9	$0.7241^{20/4}$	1.3679^{20}	**al, eth,** ace	B1[4], 1471
12303	Propyl ethynyl ether $C_3H_7OC{\equiv}CH$	84.12	......	75	......	$0.8080^{20/4}$	1.3935^{20}	al, eth	B1[4], 2213
12304	*n*-Propyl fluoride or 1-Fluoro propane $CH_3CH_2CH_2F$	62.09	......	250	−159	$0.7956^{20/4}$	1.3115^{20}	al, eth	B1[4], 187
12305	Propyl hexedrine or 1-Cyclohexyl-2-methyl amino propane $C_{10}H_{21}N$	135.28	......	205, 92-3[20]	......	$0.8501^{20/4}$	1.4600^{20}	al	B12[3], 108
12306	*n*-Propyl iodide or 1-Iodo propane $CH_3CH_2CH_2I$	169.99	......	102.4	−10.3	$1.7489^{20/4}$	1.5058^{20}	al, eth, bz, chl	B1[4], 222
12307	Isopropyl iodide or 2-Iodo propane $(CH_3)_2CHI$	169.99	......	891.4	−90.1	$1.7033^{20/4}$	1.5028^{20}	**al, eth, bz, chl**	B1[4], 223
12308	Propyl isocyanide $CH_3CH_2CH_2NC$	69.11	......	99.5	......	......	......	**al, eth**	B4[4], 474
12309	Isopropyl isocyanide $i\text{-}C_3H_7NC$	69.11	......	87	......	0.7596^{0}	......	**al, eth**	B4, 154
12310	Propyl isothiocyanate C_3H_7NCS	101.17	......	153	......	$0.9781^{16/4}$	1.5085^{16}	**al, eth**	B4[4], 491
12311	Isopropyl isothiocyanate or Isopropyl mustard oil $(CH_3)_2CHNCS$	101.17	......	138, 29-30[10]	......	......	......	al, eth	B4[4], 532
12312	Isopropyl methyl ketone or 3-Methyl-2-butanone $i\text{-}C_3H_7COCH_3$	86.13	......	94-5	−92	$0.8051^{20/4}$	1.3880^{20}	**al, eth,** ace	B1[4], 3287
12313	Propyl-α-naphthyl ether $C_3H_7O\text{-}\alpha\text{-}C_{10}H_7$	186.25	......	293.5, 167[18]	......	$1.0447^{18/4}$	1.5928^{18}	......	B6[3], 2924
12314	Propyl-β-naphthyl ether $C_3H_7O\text{-}\beta\text{-}C_{10}H_7$	186.25	nd (al)	305, 144[10]	41	......	......	al	B6[3], 2973
12315	Propyl nitrate $C_3H_7ONO_2$	105.09	......	110[762]	......	$1.0538^{20/4}$	1.3973^{20}	al, eth	B1[4], 1424
12316	Isopropyl nitrate $(CH_3)_2CHONO_2$	105.09	......	100-1	......	$1.036^{19/19}$	1.3912^{16}	al, eth	B1[4], 1475
12317	Propyl nitrite C_3H_7ONO	89.09	......	79	......	0.935^{21}	1.3604^{20}	al, eth	B1[4], 1424
12318	Isopropyl nitrite $(CH_3)_2CHONO$	89.09	......	45	......	$0.8684^{15/4}$	......	al, eth	B1[4], 1474
12319	Propyl isopropyl ether $i\text{-}C_3H_7OC_3H_7$	102.18	......	83	......	$0.7370^{20/4}$	1.376^{21}	al, eth, ace	B1[4], 1471

No.	Name, Synonyms, and Formula	Mol. wt.	Color, crystalline form, specific rotation and λ_{max} (log ε)	b.p. °C	m.p. °C	Density	n_D	Solubility	Ref.
12320	Propyl red or 4′-Dipropyl aminoazobenzene-2-carboxylic acid $C_{19}H_{23}N_3O_2$	325.41	vt-bl or purp-red cr (al)					w, al	$B16^3$, 368
12321	Propyl sulfide or Dipropyl sulfide $(C_3H_7)_2S$	118.24		142.4, 32.3^{10}	−102.5	$0.8377^{20/4}$	1.4487^{20}	al, eth	$B1^4$, 1452
12322	Propyl sulfide, 2,2′-dichloro or bis(2-chloropropyl)sulfide $(CH_3CHClCH_2)_2S$	187.13		122^{23}	−40	$1.1569^{25/4}$	1.5020^{20}	al	$B1^3$, 1437
12323	Propyl sulfide, 3,3′-dichloro or bis(3-chloropropyl)sulfide $(ClCH_2CH_2CH_2)_2S$	187.13		162^{43}		$1.1774^{25/4}$	1.5075^{20}	al, eth, to	$B1^3$, 1438
12324	Isopropyl sulfide or Diisopropyl sulfide $[(CH_3)_2CH]_2S$	118.24		120	−78.1	$0.8142^{20/4}$	1.4438^{20}	al, eth	$B1^4$, 1502
12325	Propyl sulfone or Dipropyl sulfone $(C_3H_7)_2SO_2$	150.24	sc		29-30	$1.0278^{50/4}$	1.4456^{30}	al, eth	$B1^4$, 1453
12326	Isopropyl sulfone or Diisopropyl sulfone $(i\text{-}C_3H_7)_2SO_2$	150.24	(eth)		36			w	$B1^4$, 1502
12327	Propyl sulfate or Dipropyl sulfate $(C_3H_7O)_2SO_2$	182.23		121^{20}		$1.1064^{20/4}$	1.4135^{20}	peth	$B1^4$, 1423
12328	Propyl sulfoxide or Dipropyl sulfoxide $(C_3H_7)_2SO$	134.24	nd	80^2	22-3	$0.9654^{20/4}$	1.4663^{20}	al, eth	$B1^4$, 1453
12329	Propyl thiocyanate C_3H_7SCN	101.17		163				al, eth	$B3^4$, 329
12330	Isopropyl thiocyanate $(CH_3)_2CHSCN$	101.17		$152\text{-}3^{754}$		0.9784^{20}		**al, eth**	$B3^4$, 329
12331	Isopropyl vinyl ether $i\text{-}C_3H_7OCH{=}CH_2$	86.13		55-6		$0.7534^{20/4}$	1.3840^{20}	al, eth, ace, bz	$B1^4$, 2052
12332	Propynal or Propargyl aldehyde $HC{\equiv}CCHO$	54.05		59-61			1.4033^{25}	**w**, al, eth, ace, bz	$B1^4$, 3537
12333	Propyne or Methyl acetylene $CH_3C{\equiv}CH$	40.06	gas	−23.2	−101.5	0.7062^{-50}	1.3863^{-40}	al, bz, chl	$B1^4$, 958
12334	Propyne, 3-bromo or Propargyl bromide $BrCH_2C{\equiv}CH$	118.96		88-90, 33^{130}		1.579^{18}	1.4922^{20}	al, eth, bz, chl	$B1^4$ 964
12335	Propyne, 3-chloro or Propargyl chloride $ClCH_2C{\equiv}CH$	74.51		65		$1.0297^{20/4}$	1.4320^{20}	**al, eth**, bz, chl	$B1^4$, 963
12336	Propyne, 3-cyclohexyl $C_6H_{11}CH_2C{\equiv}CH$	122.21		157-8, 48^{11}		$0.8449^{20/4}$	1.4605^{20}	eth	$B5^4$, 419
12337	Propyne, 1,3-dibromo $BrCH_2C{\equiv}CBr$	197.86		$73\text{-}4^{30}$		2.1894^{20}	1.5690^{20}	eth, chl	$B1^4$, 965
12338	Propyne, 1-iodo $CH_3C{\equiv}CI$	165.96		110		2.08^{22}		eth, al	$B1^4$, 965
12339	Propyne, 3-iodo or Propargyl iodide $ICH_2C{\equiv}CH$	165.96		115		2.0177^0		eth	$B1^4$, 965
12340	Propyne, 3-methoxy or Methyl propargyl ether $CH_3OCH_2C{\equiv}CH$	70.09		63		0.83^{12}	1.5035^{20}	**al, eth**	$B1^4$, 2215
12341	Propyne, 1-phenyl $CH_3C{\equiv}CC_6H_5$	116.16		183, 77^{17}		$0.9388^{20/4}$	1.5650^{20}	eth	$B5^4$, 1530
12342	Propyne, 3,3,3-trifluoro $F_3CC{\equiv}CH$	94.04	gas	−48.3					$B1^4$, 962
12343	Propynoic acid or Propargylic acid $HC{\equiv}CCO_2H$	70.05	cr (CS_2)	144d, $83\text{-}4^{50}$	18 (anh)	$1.1380^{20/4}$	1.4306^{20}	**w, al, eth**, ace, **chl**	$B2^4$, 1687
12344	2-Propyn-1-ol or Propargyl alcohol $HC{\equiv}CCH_2OH$	56.06		113.6, 30^{21}	−48	0.9485^{20}	1.4322^{20}	w, **al, eth**	$B1^4$, 2214
12345	Prostigmine bromide or Neostigmine bromide $C_{12}H_{19}O_2N_2Br$	303.20	(al-eth)		167d			w, al	$B13^3$, 939
12346	Protopine or Fumarine $C_{20}H_{19}NO_5$	353.37	mcl pr (al-chl)		208			chl	$B27^2$, 625
12347	Protoveratridine $C_{32}H_{51}NO_9$	593.76	cr (al-chl), $[\alpha]^{20/}{}_D$ −14 (Py, c=1)		272-3			al	$B21^4$, 6801
12348	Protoveratrine $C_{39}H_{61}NO_{12}$	751.91	pl (ace), $[\alpha]^{24/}{}_D$ −8.3 (chl)		225d				$B21^4$, 6845
12349	Protoveratrine A $C_{41}H_{63}NO_{14}$	793.95	lf (al), $[\alpha]^{20/}{}_D$ −44.1 (Py, c=1.12)		305d				$B21^4$, 6845

No.	Name, Synonyms, and Formula	Mol. wt.	Color, crystalline form, specific rotation and λ_{max} (log ε)	b.p. °C	m.p. °C	Density	n_D	Solubility	Ref.
12350	Protoveratrine B or Neoprotoveratrine $C_{41}H_{63}NO_{15}$	809.95	lf (chl-al), $[\alpha]^{20/}{}_D$ −39.8 (Py, c=1.24)		285-90d				B21[4], 6847
12351	Protoverine $C_{27}H_{43}NO_9$	525.64	nd (MeOH)		220-2			al, bz	B21[4], 6841
12352	Prulaurasin or dl-Mandelonitrile glucoside $C_{14}H_{17}NO_6$	295.29	wh nd or pl (al), $[\alpha]_D$ −54		122			w, al	B31, 240
12353	Prunasin or d-Mandelonitrile-β-d-glucoside $C_{14}H_{17}NO_6$	295.29	$[\alpha]^{22/}{}_D$ −27.0		149-50			w, al, ace	B17[4], 3356
12354	Pseudoaconitine $C_{36}H_{49}NO_{12}$	687.78	tcl (MeOH), $[\alpha]^{15/}{}_D$ +18.4 (al)		214			al, eth	B21[4], 2890
12355	Pseudocodeine or Neoisocodeine $C_{18}H_{21}NO_3$	299.37	wh nd, $[\alpha]_D$ −96.6 (al)		181-2	1.290^{80}	1.574	al	B27[2], 112
12356	Pseudoconhydrine or 5-Hydroxyconine $C_8H_{17}NO$	143.23	hyg nd (eth), $[\alpha]^{20/}{}_D$ +11.0 (al, c=10)	236	106 (anh), 60 (+1 w)			w, al, eth	B21[4], 121
12357	Psuedoconiceine (L) or γ-Coniceine $C_8H_{17}N$	127.23	hyg, $[\alpha]^{15/}{}_D$ +122.6	171-2		$0.8776^{15/4}$	1.4607^{10}	al, eth, chl	B20, 146
12358	Pseudocumene or 1,2,4-Trimethyl benzene 1,2,4-$(CH_3)_3C_6H_3$	120.19		169.3, 51.6^{10}	−43.8	$0.8758^{20/4}$	1.5048^{20}	al, eth, ace, bz, peth	B5[4], 1010
12359	Pseudocumene-5-acetyl or 2,4,5-Trimethyl acetophenone 2,4,5-$(CH_3)_3C_6H_2COCH_3$	162.23		246-7, $137\text{-}8^{20}$	10-1	$1.0039^{15/4}$	1.541^{15}	al, eth, bz, aa	B7[1], 1145
12360	Pseudoephedrine(d) or Isoephedrine $C_{10}H_{15}NO$	165.24	pr or lf (eth), $[\alpha]^{20/}{}_D$ +51.9 (abs al, c=0.6)		118.7			al, eth, bz	B13[1], 1719
12361	Pseudoephedrine (dl) $C_{10}H_{15}NO$	165.24	nd (eth)	130^{16}	118.2			al, eth, bz	B13[1], 1720
12362	Pseudoephedrine (l) $C_{10}H_{15}NO$	165.24	lf or pr (eth), $[\alpha]^{20/}{}_D$ −51.9 (abs al, c=0.6)		118.7			al, eth, bz	B13[1], 1719
12363	Pseudoephedrine hydrochloride (d) $C_{10}H_{15}NO \cdot HCl$	201.70	rh pl or nd (al), $[\alpha]^{20/}{}_D$ +62 (w, c=0.8)		181-2				B13[1], 1719
12364	Pseudoephedrine hydrochloride (dl) $C_{10}H_{15}NO \cdot HCl$	201.70	nd (abs al)		164			w, al, eth	B13[1], 1720
12365	Pseudoephedrine hydrochloride (l) $C_{10}H_{15}NO.HCl$	201.70	nd (abs al or AcOEt), $[\alpha]^{20/}{}_D$ −62.1 (w, c=1.8)		182			w	B13[2], 377
12366	Pseudohyoscyamine or Norhyoscyanine $C_{16}H_{21}NO_3$	275.35	nd, $[\alpha]_D$ −22 (al)		140.5			al, chl	B21[4], 167
12367	Pseudoionone or Citrylidenacetone $(CH_3)_2CCH_2CH_2C(CH_3)COCH_3$	140.23	pa ye oil	$143\text{-}5^{12}$		0.8984^{20}	1.5335^{20}	al, eth, chl	B1[4], 3598
12368	Pseudojervine $C_{33}H_{49}NO_8$	587.75	wh nd or hex pl, $[\alpha]_D$ −133 (al-chl, ⅓)		304-5d				C39, 1413
12369	Pseudomorphine or 2,2′-Bimorphine $C_{34}H_{36}N_2O_6$	568.67	cr (aq NH_3 + 3w), $[\alpha]^{24/}{}_D$ +44.8 (in HCl, c=0.86)		282-3			Py	B27[2], 886

No.	Name, Synonyms, and Formula	Mol. wt.	Color, crystalline form, specific rotation and λ_{max} (log ε)	b.p. °C	m.p. °C	Density	n_D	Solubility	Ref.
12370	Pseudomorphine hydrochloride $C_{24}H_{36}N_2O_6.2HCl.2H_2O$	677.62	pw, $[\alpha]_D$ −114.76						$B27^{2}$, 886
12371	Pseudopelletierine or Pseudoplenicine $C_9H_{15}NO$	153.22	orh pr (peth)	246	54	1.001^{100}	1.4760^{100}	w, al, eth, chl	$B21^{4}$, 3315
12372	Pseudoreserpine $C_{32}H_{38}N_2O_9$	594.66	$[\alpha]^{24/}_D$ -N−65 (chl)		257-8			ace	C52, 2876
12373	Pseudothiohydantiol or 2-Imino-4-thiazolidine $C_3H_4N_2OS$	116.14	pr or nd (w)		255-8d				$B27^{2}$, 284
12374	Pseudotropine or Pseudotropanol $C_8H_{15}NO$	141.21	rh ta or pr (eth), rh bipym (peth-bz)		109			w, al, bz, chl	$B21^{4}$, 169
12375	D-Psicose or D-Allulose $C_6H_{12}O_6$	180.16	$[\alpha]^{25/}_D$ +4.7 (w, c=4.3)		58			w, al	$B1^{4}$, 4400
12376	Pteridine or Pyrimido[(4,5)]pyrazine $C_6H_4N_4$	132.12	ye pl (bz, sub)	sub $125\text{-}30^{20}$	139-40			w, al	J1951, 474
12377	Pteridine, 2-amino-4,6-dihydroxy or Uropterin $C_6H_5N_5O_2$	179.14	og-ye (w + 1)		>410d darkens 360				$B26^{2}$, 313
12378	Pteridine, 2-amino-4-hydroxy or 2-Amino-4-pteridol $C_6H_5N_5O$	163.14	ye		>360				C47, 5945
12379	Pukateine (l) or 4-Hydroxy-5,6-methylenedioxyaporphin $C_{18}H_{17}NO_3$	295.34	cr (al, eth), $[\alpha]^{15/}_D$ −200 (al, c=6)	$210\text{-}5^{2}$	200				$B27^{1}$, 461
12380	Pulegenone or 4-Methyl-1-isopropyl cyclopentinen-5-one $C_9H_{14}O$	138.21		188-9		$0.9144^{20/0}$	1.4660^{20}	al, eth, ace	$B7^{3}$, 291
12381	Pulegone or 4(8)p-Menthen-3-one $C_{10}H_{16}O$	152.24	$[\alpha]^{20/}_D$ +23.4 (undil)	224, 103^{17}		0.9346^{45}	1.4894^{20}	al, eth, chl	$B7^{3}$, 334
12382	Purine or 7-Imidazo[4,5]pyrimidine $C_5H_4N_4$	120.11	nd (to al)	sub	216-7			w, al, ace	B26, 354
12383	Purine, 6-amino or Adenine $C_5H_5N_5$	135.13	nd or lf (w + 3)	sub at 220	360-5d (anh)				$B26^{2}$, 252
12384	Purine, 6-mercapto or 6-Purinethiol $C_5H_4N_4S$	152.17	ye pr (w + 1)		313-4d				Am74, 411
12385	Purine, 2,6,8-trichloro $C_5HCl_3N_4$	223.45	nd (al)		159-61			al, eth, ace, bz	B26, 356
12386	Pyocyanine $C_{13}H_{10}N_2O$	210.24	dk bl nd (w + 1), (chl-peth)	sub	133d			ace, chl, Py	$B23^{2}$, 361
12387	Pyraconitine $C_{32}H_{43}O_9N$	585.69	nd		171			al	$B21^{4}$, 6794
12388	γ-Pyran C_5H_6O	82.10	col oil	80			1.4559^{20}	al, eth, bz	B17, 36
12389	γ-Pyran, 2,3-dihydro C_5H_8O	84.12		86-7		$0.922^{19/15}$	1.4402^{19}	w, al	$B17^{4}$, 148
12390	γ-Pyran, tetrahydro or Pentamethylene oxide $C_5H_{10}O$	86.13		88		$0.8810^{20/4}$	1.4200^{20}	al, eth, bz	$B17^{4}$, 51
12391	Pyranthrone or 8,16-Pyranthenedione $C_{30}H_{14}O_2$	406.44	red-ye or red-br nd ($PhNO_2$)	sub, vac	d				$B7^{3}$, 4514
12392	Pyrazine or 1,4-Diazine $C_4H_4N_2$	80.09	pr (w)	$115\text{-}6^{768}$	54	$1.0311^{61/4}$	1.4953^{61}	w, al, eth, ace	$B23^{2}$, 80
12393	Pyrazine, 2,3-dimethyl $2,3(CH_3)_2C_4H_2N_2$	108.14		156		$1.0281^{0/4}$		w, al, eth	$B23^{2}$, 80
12394	Pyrazine, 2,5-dimethyl-3-ethyl $2,5\text{-}(CH_3)_2\text{-}3\text{-}C_2H_5(C_4HN_2)$	136.20		180-1		$0.9657^{24/4}$	1.5014^{24}		B23, 99
12395	Pyrazine, 2,5-dimethyl or Ketine $2,5\text{-}(CH_3)_2(C_4H_2N_2)$	108.15		155	15	$0.9887^{20/4}$	1.4980^{20}	w, al, eth, ace	$B23^{2}$, 80
12396	Pyrazine, 2,6-dimethyl $2,6\text{-}(CH_3)_2(C_4H_2N_2)$	108.14		155.6	47-8	$0.9647^{50/4}$		w, al, eth	B23, 97
12397	Pyrazine, 2-methyl $2\text{-}CH_3(C_4H_3N_2)$	94.12		136-7		$1.0290^{20/4}$	1.5067^{19}	w, al, eth, ace	B23, 94
12398	Pyrazine carboxamide $(C_4H_3N_2)CONH_2$	123.11	wh nd (w, al)	sub	191-3				C48, 2074

No.	Name, Synonyms, and Formula	Mol. wt.	Color, crystalline form, specific rotation and λ_{max} (log ε)	b.p. °C	m.p. °C	Density	n_D	Solubility	Ref.
12399	2,3-Pyrazine dicarboxylic acid $2,3-(C_4H_2N_2)(CO_2H)_2$	168.11	pr (w + 2)		193d			w, ace, MeOH	$B25^2$, 164
12400	2,3-Pyrazine dicarboxylic acid, 5-methyl $5-CH_3-2,3-(C_4HN_2)(CO_2H)_2$	182.14	(w, ace, dil al-eth)		174-5			w, al	$B25^2$, 165
12401	2,5-Pyrazine dicarboxylic acid $2,5-(C_4H_2N_2)(CO_2H)_2$	168.11	nd (w + 2)		255-6d (sealed tube)				$B25^2$, 164
12402	2,6-Pyrazine dicarboxylic acid $2,6-(C_4H_2N_2)(CO_2H)_2$	168.11	mcl nd (w + 2)		217-8d			al	B25, 168
12403	Pyrazole or 1,2-Diazole $C_3H_4N_2$	68.08	nd or pr (lig)	186-8	69-70		1.4203	w, al, eth, bz	$B23^2$, 33
12404	Pyrazole, 4-bromo-1,3-dimethyl $4-Br-1,3-(CH_3)_2(C_3HN_2)$	175.03		$76-7^{10}$		$1.4975^{15/4}$	1.5214^{15}	al, eth, ace	$B23^2$, 54
12405	Pyrazole, 4-bromo-1,5-dimethyl $4-Br-1,5-(CH_3)_2(C_3HN_2)$	175.03	cr	85^{10}	38-9				$B23^2$, 54
12406	Pyrazole, 4-bromo-3,5-dimethyl $4-Br-3,5-(CH_3)_2(C_3HN_2)$	175.03	nd (dil al)		123			eth, ace	$B23^2$, 68
12407	Pyrazole, 4-bromo-3-methyl $4-Br-3-CH_3(C_3H_2N_2)$	161.00	cr (dil al)		76-7	$1.5638^{100/4}$	1.5182^{100}	al, eth, ace	$B23^2$, 54
12408	Pyrazole, 3-chloro-1,5-dimethyl $3-Cl-1,5-(CH_3)_2(C_3HN_2)$	130.58	pl (w)	210-2, 138^{72}	47-8	$1.0823^{100/4}$	1.4648^{100}	al, eth, ace, bz	$B23^2$, 49
12409	Pyrazole, 4-chloro-3,5-dimethyl $4-Cl-3,5-(CH_3)_2(C_3HN_2)$	130.58	pr (w), cr (al)	220-2	117-8			al, eth, ace, bz	$B23^2$, 67
12410	Pyrazole, 5-chloro-1,3-dimethyl $5-Cl-1,3-(CH_3)_2(C_3HN_2)$	130.58		157-8		$1.1367^{18/4}$	1.4877^{18}	w	$B23^2$, 49
12411	Pyrazole, 5-chloro-3-methyl $5-Cl-3-CH_3(C_3H_2N_2)$	116.55	cr (eth, lig)	258, 138^{15}	118-9			w	$B23^2$, 49
12412	Pyrazole, 1,3-dimethyl $1,3-(CH_3)_2(C_3H_2N_2)$	96.13		136-8, 31^1		$0.9561^{17/4}$	1.4734^{15}	w	$B23^2$, 44
12413	Pyrazole, 1,5-dimethyl $1,5-(CH_3)_2(C_3H_2N_2)$	96.13		153		$0.9813^{17/4}$	1.4782^{16}	w, al, eth	$B23^2$, 44
12414	Pyrazole, 3,4-dimethyl $3,4-(CH_3)_2(C_3H_2N_2)$	96.13	(peth)	111^{10}	58			al, eth, ace, bz	$B23^2$, 64
12415	Pyrazole, 3,5-dimethyl $3,5-(CH_3)_2(C_3H_2N_2)$	96.13	cr (peth, al)	218^{758}	107-8	$0.8839^{16/4}$		al, eth, ace, bz, chl	$B23^2$, 65
12416	Pyrazole, 1-ethyl $1-C_2H_5(C_3H_3N_2)$	96.13		139		$0.9537^{20/4}$	1.4700^{20}	al, eth, ace, bz, Py	C55, 22291
12417	Pyrazole, 3-methyl $3-CH_3(C_3H_3N_2)$	82.11		204^{752}, 108^{25}	36-7	$1.0203^{16/4}$	1.4915^{20}	w, al, eth	$B23^2$, 44
12418	Pyrazole, 1-isopropyl $1-i-C_3H_7(C_3H_3N_2)$	110.16		143				w, al	C53, 21043
12419	2-Pyrazoline or 4,5-Dihydropyrazole $C_3H_6N_2$	70.09		144		$1.0200^{17/4}$	1.4796^{17}	w, al, eth	$B23^2$, 24
12420	2-Pyrazoline, 1-phenyl $1-C_6H_5(C_3H_5N_2)$	146.19	pl (lig)	273^{754}, 151^{17}	52	$1.0689^{58/4}$	1.6015^{58}	al, eth, bz	$B23^2$, 25
12421	5-Pyrazolinone, 3-methyl $C_4H_6N_2O$	98.10			215			w	$B24^2$, 8
12422	5-Pyrazolinone, 1-methyl-3-phenyl $C_{10}H_{10}N_2O$	174.20		330-40, 235^{68}	213-7			al, eth	B24, 148
12423	5-Pyrazolinone, 3-methyl-1-phenyl $C_{10}H_{10}N_2O$	174.20	[α] + 1.637	287^{105}, 191^{17}	127			w, al	$B24^2$, 9
12424	5-Pyrazolone or 2-Pyrazolin-5-on $C_3H_4N_2O$	84.08	nd (to, w)	sub	165			w, al	$B24^2$, 6
12425	5-Pyrazolone, 3-methyl $C_4H_6N_2O$	98.10	pr (w), nd (al), lf (sub)	sub	215 (219)			w	$B24^2$, 8
12426	5-Pyrazolone, 3-methyl-1(nitrophenyl) $C_{10}H_9N_3O_3$	219.20	ye amor (aa)		185				$B24^1$, 191
12427	5-Pyrazolone, 3-methyl-1-phenyl $C_{10}H_{10}N_2O$	174.20	mcl pr (w)	287^{105}, 191^{17}	127		1.637		$B24^2$, 9
12428	5-Pyrazolone, 3-methyl-1-(4-sulfophenyl) $C_{10}H_{10}N_2O_4S$	254.26	nd (w + 1)		290-320d				$B24^2$, 20
12429	5-Pyrazolone-3-carboxylic acid, 1-phenyl $C_{10}H_8N_2O_3$	204.19	nd (w, al)		261			al	$B25^2$, 219
12430	Pyrene or Benzo[d,e,f]phenanthrene $C_{16}H_{10}$	202.26	pa ye pl (to, sub)	393, 260^{60}	156	$1.271^{23/4}$		al, eth, bz, lig	$B5^4$, 2467

No.	Name, Synonyms, and Formula	Mol. wt.	Color, crystalline form, specific rotation and λ_{max} (log ε)	b.p. °C	m.p. °C	Density	n_D	Solubility	Ref.
12431	Pyrene, 1-acetyl 1-$(CH_3CO)C_{16}H_9$	244.29	gr-ye lf (al, MeOH)		89-90			eth, bz	B7[3], 2726
12432	Pyrene, 1-amino 1-$NH_2C_{16}H_9$	217.27	ye nd (hx), lf (dil al)		117-8			al, ace	B12[3], 3368
12433	Pyrene, 1,2-dihydro $C_{16}H_{12}$	204.27			132				BCHE, 4239
12434	Pyrene, 2,7-dimethyl $C_{18}H_{14}$	230.31			228-32				B5[3], 2307
12435	Pyrene perhydro (3a,8a-*cis*) $C_{16}H_{26}$	218.38		318.5	90				B5[4], 1199
12436	3-Pyrene carboxylic acid 3-$(C_{16}H_9)CO_2H$	246.27	ye nd (eth-al, sub)	sub	274			eth, chl	B9[3], 3575
12437	4-Pyrene carboxylic acid 4-$(C_{16}H_9)CO_2H$	246.27	gr lf or nd (Ph NO_2)		327-8			chl	C5[2], 11081
12438	Pyrethrin I or Chrysatnthemum monocarboxylic acid $C_{21}H_{28}O_3$	328.45	visc liq, $[\alpha]^{25/}{}_D$ −32.3 (eth, c=5.66)	$170^{0.1}$d			1.5192^{18}	al, eth, peth	B9[3], 215
12439	Pyrethrin II or Chrysanthemum dicarboxylic acid, methyl ester $C_{22}H_{28}O_5$	372.46	visc liq, $[\alpha]^{20/}{}_D$ −6 (eth, c=5)	$200^{0.1}$d			1.5258^{20}	al, eth, peth	B9[3], 3988
12440	Pyridazine or 1,2-Diazine $C_4H_4N_2$	80.09		208, $47\text{-}8^{1}$	−8	$1.1035^{20/4}$	1.5218^{20}	**w, al**, eth, ace, bz	B23[1], 28
12441	Pyridine C_5H_5N	79.10		115.5	−42	0.9819^{20}	1.5095^{20}	**w, al, eth, ace, bz**	B20[4], 2205
12442	Pyridine, 3-acetamido 3-$CH_3CONH(C_5H_4N)$	136.15		326.7	133			w, al	B22[4], 4073
12443	Pyridine, 2-acetyl or Methyl 2-pyridyl ketone 2-$CH_3CO(C_5H_4N)$	121.14	ye in air	192, 78^{12}			1.5203^{20}	al, eth	B21[4], 3544
12444	Pyridine, 3-acetyl or Methyl-3-pyridyl ketone 3-$CH_3CO(C_5H_4N)$	121.14	ye in air	220, 106^{12}	13-4		1.5341^{20}	w, al, eth	B21[4], 3548
12445	Pyridine, 3-acetyl,oxime 3-$CH_3C(=NOH)(C_5H_4N)$	136.15	cr (al, bz)		113 (130.5)				B21[4], 3549
12446	Pyridine, 2-allyl 2-$(CH_2=CHCH_2)(C_5H_4N)$	119.17		190, 58^{10}		0.959^{20}		**al, eth**	B20[4], 2890
12447	Pyridine, 2-amino or α-Pyridyl amine 2-$H_2N(C_5H_4N)$	94.12	lf (liq)	204 (sub), $104\text{-}6^{20}$	57-8			al, eth, ace, bz	B22[4], 3840
12448	Pyridine, 2-amino-5-bromo 5-Br-2-$H_2N(C_5H_3N)$	173.01	cr (bz)		137			al, bz	B22[4], 4031
12449	Pyridine, 2-amino-5-chloro 5-Cl-2-$H_2N(C_5H_3N)$	128.56	pl	$127\text{-}8^{11}$	136-8			w, al	B22[4], 4019
12450	Pyridine, 2-amino-3,5-dibromo 3,5-Br_2-2-$H_2N(C_5H_2N)$	251.92	nd (dil al, peth)		104			al, eth	B22[4], 4041
12451	Pyridine, 2-amino-3,5-dichloro 3,5-Cl_2-2-$H_2N(C_5H_2N)$	163.01	nd or pr (dil al)		84-5			al, ace, lig	B22[4], 4028
12452	Pyridine, 2-amino-5-iodo 5-I-2-$H_2N(C_5H_3N)$	220.01	nd or lf (dil al)		129			al, eth	B22[4], 4045
12453	Pyridine, 2-amino-3-nitro 3-NO_2-2-$H_2N(C_5H_3N)$	139.11	ye nd (dil al)		165-7			al	B22[4], 4053
12454	Pyridine, 2-amino-5-nitro 5-NO_2-2-$H_2N(C_5H_3N)$	139.11	ye lf (dil al)		188			al	B22[4], 4054
12455	Pyridine, 3-amino or β-Pyridyl amine 3-$H_2N(C_5H_4N)$	94.12	lf (bz-lig)	252, $131\text{-}2^{12}$	64-5			w, al, eth	B22[4], 4067
12456	Pyridine, 4-amino or γ-Pyridyl amine 4-$H_2N(C_5H_4N)$	94.12	nd (bz)	180^{13}	158-9			w, al, eth, bz	B22[4], 4098
12457	Pyridine, 4-amino-2,6-dichloro 2-6-Cl_2-4-$H_2N(C_5H_2N)$	163.01	nd (dil al)		176			al, eth, ace, bz	B22[4], 4119
12458	Pyridine, 4-amino-3,5-dinitro 3,5-$(NO_2)_2$-4-$H_2N(C_5H_2N)$	184.11	ye pl (dil al)		170-1				B22[4], 4128
12459	Pyridine, 4-amino-3-nitro 3-NO_2-4-$H_2N(C_5H_3N)$	139.11	ye nd (w)		200			al	B22[4], 4122
12460	Pyridine, 5-amino-2-butoxy 5-NH_2-2-$C_4H_9O(C_5H_3N)$	166.22		$148\text{-}50^{12}$		1.037^{25}	1.5373^{20}	al, eth	B22[4], 5574
12461	Pyridine, 2-benzoyl or Phenyl-2-pyridyl ketone 2-$C_6H_5CO(C_5H_4N)$	183.21		$114\text{-}5^{0.01}$	41-3			chl	B21[4], 4119

No.	Name, Synonyms, and Formula	Mol. wt.	Color, crystalline form, specific rotation and λ_{max} (log ε)	b.p. °C	m.p. °C	Density	n_D	Solubility	Ref.
12462	Pyridine, 4-benzoyl or Phenyl-4-pyridyl ketone 4-$C_6H_5CO(C_5H_4N)$	183.21	nd (peth), pl (w)	314^{742}, $170\text{-}2^{10}$	71-3	....	....	al, eth, bz	B21⁴, 4125
12463	Pyridine, 2-benzyl 2-$C_6H_5CH_2(C_5H_4N)$	169.23	nd	276^{742}, 149^{16}	11-4	$1.067^{0/0}$	1.5785^{20}	al, eth	B20⁴, 3649
12464	Pyridine, 3-benzyl 3-$C_6H_5CH_2(C_5H_4N)$	169.23	nd	286^{740}	34	1.061	....	al, eth	B20⁴, 3651
12465	Pyridine, 4-benzyl 4-$C_6H_5CH_2(C_5H_4N)$	169.23	....	287^{742}, $180\text{-}1^{31}$	......	$1.0614^{20/0}$	1.5818^{20}	al, eth	B20⁴, 3652
12466	Pyridine, 2-bromo 2-$Br(C_5H_4N)$	158.00	....	$193\text{-}4^{764}$, $74\text{-}5^{13}$	......	1.657^{15}	1.5734^{20}	al, eth	B20⁴, 2503
12467	Pyridine, 3-bromo 3-$Br(C_5H_4N)$	158.00	nd (al)	$172\text{-}3^{751}$, $68\text{-}70^{18}$	142-3	$1.645^{0/4}$	1.5694^{20}	w, al, eth	B20⁴, 2505
12468	Pyridine, 3-bromo-6-hydroxy 3-Br-6-$HO(C_5H_3N)$	174.00	pr (w, al)	....	177-8	....	....	al	B21⁴, 362
12469	Pyridine, 4-bromo 4-$Br(C_5H_4N)$	158.00	....	$172\text{-}3^{752}$, $68\text{-}70^{18}$	0-1	$1.6450^{0/4}$	1.5694^{20}	w, al, eth	B20⁴, 2512
12470	Pyridine, 5-bromo-2-hydroxy 2-HO-5-$Br(C_5H_3N)$	174.00	pr (w, al)	....	177-8	....	....	al	B21⁴, 2512
12471	Pyridine, 2-chloro 2-$Cl(C_5H_4N)$	113.55	oil	170, $54\text{-}8^{10}$	......	1.205^{15}	1.5320^{20}	al, eth	B20⁴, 2493
12472	Pyridine, 2-chloro-5-nitro 2-Cl-5-$NO_2(C_5H_3N)$	158.54	....	....	108-10	....	....	....	B20⁴, 2533
12473	Pyridine, 3-chloro 3-$Cl(C_5H_4N)$	113.55	....	148^{764}, $85\text{-}7^{100}$	......	....	1.5304^{20}	....	B20⁴, 2497
12474	Pyridine, 4-chloro 4-$Cl(C_5H_4N)$	113.55	....	147-8, $85\text{-}7^{100}$	−42.5	....	....	w, al	B20⁴, 2499
12475	Pyridine, 5-chloro-2-trichloro methyl 5-Cl-2-$Cl_3C(C_5H_3N)$	230.91	....	$139\text{-}42^{25}$	45.5-6.5	....	....	....	B20⁴, 2704
12476	Pyridine, 2,3-diamino-(one form) 2,3-$(NH_2)_2(C_5H_3N)$	109.13	lf or pl (dil al)	$148\text{-}50^{5}$	122	....	....	al	B22², 395
12477	Pyridine, 2,3-diamino (one form) 2,3-$(NH_2)_2(C_5H_3N)$	109.13	nd (bz)	sub	116	....	....	w, al, bz	B22², 394
12478	Pyridine, 2,4-diamino 2,4-$(NH_2)_2(C_5H_3N)$	109.13	hyg lf or nd	....	107	....	....	w	B22², 394
12479	Pyridine, 2,5-diamino 2,5-$(NH_2)_2(C_5H_3N)$	109.13	nd	$180\text{-}5^{12}$	109-10	....	....	w, al	B22², 394
12480	Pyridine, 3,4-diamino 3,4-$(NH_2)_2(C_5H_3N)$	109.13	nd or lf	....	218-9	....	....	....	B22², 395
12481	Pyridine, 3,5-diamino 3,5-$(NH_2)_2(C_5H_3N)$	109.13	hyg nd or lf	....	119-20	....	....	w	B22¹, 648
12482	Pyridine, 2,6-dibromo 2,6-$Br_2(C_5H_3N)$	236.89	....	255	118	....	....	....	B20⁴, 2515
12483	Pyridine, 3,5-dibromo 3,5-$Br_2(C_5H_3N)$	236.89	nd (al)	222, sub 100	112	....	....	eth	B20⁴, 2516
12484	Pyridine, 1,2-dihydro-1-methyl-2-oxo C_6H_7NO	109.13	nd	250, 121^{10}	7	....	....	w	B21⁴, 3348
12485	Pyridine, 2,4-dihydroxy or 2,4-Pyridinediol 2,4-$(HO)_2(C_5H_3N)$	111.10	rh bipym (al, w)	....	260-5d	....	....	w, al	B21⁴, 2058
12486	Pyridine, 2,5-dihydroxy 2,5-$(HO)_2(C_5H_3N)$	111.10	....	....	248	....	....	w, al	B21⁴, 2062
12487	Pyridine, 3,5-diido-2-hydroxy 3,5-I_2-2-$HO(C_5H_2N)$	346.89	br nd (al)	....	261-2	....	....	....	B21⁴, 365
12488	Pyridine, 2,3-dimethyl or 2,3-Lutidine 2,3-$(CH_3)_2(C_5H_3N)$	107.16	....	163-4	......	$0.9319^{25/4}$	1.5057^{20}	w, al, eth	B20⁴, 2765
12489	Byridine, 2,6-dimethyl-3-ethyl 2,6-$(CH_3)_2$-3-$C_2H_5(C_5H_2N)$	135.21	....	$83\text{-}5^{23}$	......	0.9120^{18}	....	....	B20⁴, 2828
12490	Pyridine, 2,6-dimethyl-4-ethyl 2,6-$(CH_3)_2$-4-$C_2H_5(C_5H_2N)$	135.21	....	$187.5\text{-}8^{758}$	......	$0.9089^{25/4}$	1.4964^{25}	....	B20⁴, 2828
12491	Pyridine, 3,5-dimethyl-2-ethyl or α-Parvoline 3,5-$(CH_3)_2$-2-$C_2H_5(C_5H_2N)$	135.21	....	$198\text{-}9^{764}$, $85\text{-}7^{15}$	......	0.9338^{0}	....	al, eth, ace	B20⁴, 2828
12492	Pyridine, 2-(dimethylamino) 2-$(CH_3)_2N(C_5H_4N)$	122.17	....	196, 88^{15}	......	$1.0157^{14/14}$	1.5663^{20}	al, eth, bz	B22⁴, 3847
12493	Pyridine, 4-(dimethylamino) 4-$(CH_3)_2N(C_5H_4N)$	122.17	pl (eth)	....	114	....	....	w, al, bz, chl	B22⁴, 4101

No.	Name, Synonyms, and Formula	Mol. wt.	Color, crystalline form, specific rotation and λ_{max} (log ε)	b.p. °C	m.p. °C	Density	n_D	Solubility	Ref.
12495	Pyridine, 2-ethyl $2\text{-}C_2H_5(C_5H_4N)$	107.16		148.6	−63.1	0.9502^{0}	1.4964^{20}	al, eth, ace, w	$B20^4$, 2755
12496	Pyridine, 2-ethyl-4-methyl or 2-Ethyl-γ-picoline $2\text{-}C_2H_5\text{-}4\text{-}CH_3(C_5H_3N)$	121.18		$173\text{-}5^{748}$		$0.9239^{20/0}$		al, eth, ace	$B20^4$, 2798
12497	Pyridine, 3-ethyl $3\text{-}C_2H_5(C_5H_4N)$	107.16		165	−76.9	0.9539^{0}	1.5021^{20}	w, al, eth, ace	$B20^4$, 2758
12498	Pyridine, 4-ethyl $4\text{-}C_2H_5(C_5H_4N)$	107.16		167.7	−90.5	0.9417^{20}	1.5009^{20}	w, al, eth, ace	$B20^4$, 2761
12499	Pyridine, 2-fluoro $2\text{-}F(C_5H_4N)$	97.09		125^{758}			1.4574^{20}		$B20^4$, 2491
12500	Pyridine hydrochloride $C_5H_5N.HCl$	115.56	hyg pl or sc (al)	281-9	82			w, al, chl	$B20^4$, 2230
12501	Pyridine, 2-hydroxy or 2-Pyridol $2\text{-}HO(C_5H_4N)$	95.10	nd (bz)	280-1	106-7			w, al, bz, chl	$B21^4$, 344
12502	Pyridine, 2-(β-hydroxyethyl) $2\text{-}(HOCH_2CH_2)(C_5H_4N)$	123.15	hyg	$118\text{-}21^{15}$		$1.1111^{4/0}$	1.5368^{20}	w, al, chl	$B21^4$, 512
12503	Pyridine, 2-hydroxy-4-methyl $4\text{-}CH_3\text{-}2\text{-}HO(C_5H_3N)$	109.13		307-9	130			w, al, chl	$B21^4$, 505
12504	Pyridine, 2-hydroxymethyl $2\text{-}HOCH_2(C_5H_4N)$	109.13		$112\text{-}3^{16}$, 102.5^{8}		$1.1317^{20/4}$	1.5444^{20}	w, al, eth, ace, bz	$B21^4$, 487
12505	Pyridine, 2-(1-hydroxypropyl) $2\text{-}[CH_3CH_2CH(OH)](C_5H_4N)$	137.18	pa ye	213-6, $112\text{-}3^{12}$		$1.0501^{20/4}$	1.5197^{20}	w, al	$B21^4$, 544
12506	Pyridine, 2-(β-hydroxypropyl) $2\text{-}[CH_3CH(OH)CH_2](C_5H_4N)$	137.18	pr	123.5^{20}	32			w, al, chl	$B21^4$, 545
12507	Pyridine, 2-(β-hydroxyisopropyl) $2\text{-}[HOCH_2CH(CH_3)](C_5H_4N)$	137.18		$128\text{-}31^{17}$				w, al	B21, 57
12508	Pyridine, 3-hydroxy or 3-Pyridol $3\text{-}HOC_5H_4N$	95.10	nd (bz)		129			w, al	$B21^4$, 402
12509	Pyridine, 3-(α-hydroxy isopropyl) $3\text{-}[(CH_3)_2C(OH)](C_5H_4N)$	137.18	cr	$140\text{-}1^{12}$	58			w, al, chl	$B21^4$, 552
12510	Pyridine, 4-hydroxy or 4-Pyridol $4\text{-}HOC_5H_4N$	95.10	pr or nd (w + 1)	>350, $257\text{-}60^{10}$	148.5 (anh), 65 (hyd)			w, al	$B21^4$, 446
12511	Pyridine, 4-(hydroxymethyl) $4\text{-}HOCH_2(C_5H_4N)$	109.13		$140\text{-}2^{12}$	51-5			chl	$B21^4$, 507
12512	Pyridine, 2-iodo $2\text{-}I\text{-}C_5H_4N$	205.00		93^{13}		1.9735^{20}	1.6366^{20}	al, eth, ace, bz	$B20^4$, 2522
12513	Pyridine, 2-mercapto $2\text{-}HS(C_5H_4N)$	111.16			128			w, al, bz, chl	$B21^4$, 373
12514	Pyridine, 2-methoxy $2\text{-}CH_3O(C_5H_4N)$	109.13		142-3			$1.5042.^{20}$		$B21^4$, 345
12515	Pyridine, 4-methoxy $4\text{-}CH_3O(C_5H_4N)$	109.13		191^{738}, 95^{45}				w	$B21^4$, 447
12516	Pyridine, 2-methyl or α-Picoline $2\text{-}CH_3(C_5H_4N)$	93.13		128.8	−66.8	$0.9443^{20/4}$	1.4957^{20}	w, al, eth, ace	$B20^4$, 2679
12517	Pyridine, 2-methylamino $2\text{-}CH_3NH(C_5H_4N)$	108.14		200-1, 90^{9}	15	$1.052^{29/29}$		w, al, eth, bz, aa	$B22^4$, 3847
12518	Pyridine, 4-methylamino $4\text{-}CH_3NH(C_5H_4N)$	108.14	pl (eth)		117-8			w, al, eth, ace, bz	$B22^4$, 4100
12519	Pyridine-N-oxide C_5H_5NO	95.10		$146\text{-}7^{13}$	65-6				$B20^4$, 2305
12520	Pyridine, pentachloro C_5Cl_5N	251.33		sub 280	125-6			al, bz, lig	$B20^4$, 2503
12521	Pyridine, 2-phenyl or α-Pyridyl benzene $2\text{-}C_6H_5(C_5H_4N)$	155.20		270-2, 146^{15}		1.0833^{25}	1.6210^{20}	al, eth	$B20^4$, 3639
12522	Pyridine, 3-phenyl or β-Pyridyl benzene $3\text{-}C_6H_5(C_5H_4N)$	155.20	pa ye oil	273-4			1.6123^{25}	al, eth	$B20^4$, 3643
12523	Pyridine, 4-phenyl $4\text{-}C_6H_5(C_5H_4N)$	155.20	pl (w)	280-2	77-8			al, eth	$B20^4$, 3645
12524	Pyridine picrate $C_5H_5N.C_6H_3N_3O_7$	308.21	ye nd (al)		167-8				$B20^4$, 2287
12525	Pyridine, 2-propyl or Conyrine $2\text{-}C_3H_7(C_5H_4N)$	121.18		166-8, 60^{11}	2	$0.9119^{20/4}$	1.4925^{20}	al, eth, ace	$B20^4$, 2790
12526	Pyridine, 4-propyl $4\text{-}C_3H_7(C_5H_4N)$	121.18		184-6, 80^{2}		0.9381^{15}	1.4966^{20}	al, eth	$B20^4$, 2792

No.	Name, Synonyms, and Formula	Mol. wt.	Color, crystalline form, specific rotation and λ_{max} (log ε)	b.p. °C	m.p. °C	Density	n_D	Solubility	Ref.
12527	Pyridine, 2-isopropyl $2\text{-i-}C_3H_7(C_5H_4N)$	121.18	...	159.8	...	0.9342^{0}	1.4915^{20}	al, eth, ace	B20⁴, 2794
12528	Pyridine, 4-isopropyl $4\text{-i-}C_3H_7(C_5H_4N)$	121.18	...	178	−54.9	$0.9382^{25/4}$	1.4962^{20}	al, eth, ace	B20⁴, 2795
12529	Pyridine, 2,3,4,6-tetramethyl or β-Parvoline $2,3,4,6\text{-}(CH_3)_4(C_5HN)$	135.21	...	203^{750}	...	$0.9322^{25/4}$	1.5087^{25}	al, eth	B20⁴, 2830
12530	Pyridine, 2-(trichloromethyl) $2\text{-}Cl_3C(C_5H_4N)$	196.46	...	$125\text{-}6^{25}$	−10	1.4526^{25}	1.5596^{25}	...	B20⁴, 2703
12531	Pyridine, 2,4,6-trihydroxy $2,4,6(HO)_3(C_5H_2N)$	127.10	ye nd or pw	...	230d	...	...	al, eth, ace	B21⁴, 2504
12532	Pyridine, 2,3,4-trimethyl $2,3,4\text{-}(CH_3)_3(C_5H_2N)$	121.18	...	192-3, $79\text{-}82^{14}$	...	$0.9127^{15/4}$	1.5150^{20}	w, al, eth	B20⁴, 2806
12533	Pyridine, 2,3,5-trimethyl $2,3,5\text{-}(CH_3)_3(C_5H_2N)$	121.18	...	186.7	...	$0.9352^{19/4}$	1.5057^{25}	al, eth, ace, bz	B20⁴, 2807
12534	Pyridine, 2,3,6-trimethyl $2,3,6\text{-}(CH_3)_3(C_5H_2N)$	121.18	...	176-8	...	$0.9220^{25/4}$	1.5053^{20}	al, eth, ace, bz	B20⁴, 2808
12535	Pyridine, 2,4,5-trimethyl $2,4,5\text{-}(CH_3)_3(C_5H_2N)$	121.18	...	188	...	$0.9330^{25/4}$	1.5054^{25}	al, eth, ace, bz	B20⁴, 2809
12536	Pyridine, 2,4,6-trimethyl or γ-Collidine $2,4,6\text{-}(CH_3)_3(C_5H_2N)$	121.18	...	170.5^{762}	−44.5	$0.9166^{22/4}$	1.4959^{25}	w, al, eth, ace	B20⁴, 2810
12537	Pyridine, 2-vinyl $2\text{-}CH_2{=}CH(C_5H_4N)$	105.14	...	159-60, $50\text{-}5^{4}$	...	$0.9985^{20/0}$	1.5495^{20}	al, eth, ace, chl	B20⁴, 2884
12538	Pyridine, 3-vinyl $3\text{-}CH_2{=}CH(C_5H_4N)$	105.14	...	160, $67\text{-}8^{18}$	...	...	1.5530^{20}	al, chl	B20⁴, 2886
12539	Pyridine, 4-vinyl $4\text{-}CH_2{=}CH(C_5H_4N)$	105.14	red to dk-br	65^{15}	...	$0.9800^{20/4}$	1.5449^{20}	...	B20⁴, 2887
12540	2-Pyridine carboxaldehyde $2\text{-}(C_5H_4N)CHO$	107.11	...	$180^{7.50}$, 81.5^{25}	...	$1.1201^{20/4}$	1.53653^{20}	w, al, eth	B21⁴, 3495
12541	2-Pyridine carboxylic acid or Picolinic acid $2\text{-}(C_5H_4N)CO_2H$	123.11	nd (w, al, bz)	sub	136-7	...	...	al, aa	B22⁴, 303
12542	3-Pyridine carboxaldehyde $3\text{-}(C_5H_4N)CHO$	107.11	...	89.5^{14}	...	$1.415^{20/4}$	...	w, al, ace, chl	B21⁴, 3517
12543	3-Pyridine carboxylic acid or Nicotinic acid, Niacin $3\text{-}(C_5H_4N)CO_2H$	123.11	nd (w, al)	sub	236-7	1.473	...	...	B22⁴, 348
12544	4-Pyridine carboxaldehyde or Isonicotinaldehyde $4\text{-}OHC(C_5H_4N)$	107.11	...	$77\text{-}8^{12}$	...	...	1.5423^{20}	w, eth	B21⁴, 3529
12545	4-Pyridine carboxylic acid or Isonicotinic acid $4\text{-}(C_5H_4N)CO_2H$	123.11	nd (w)	sub at 260^{15}	319	...	...	...	B22⁴, 518
12546	4-Pyridine carboxylic acid anhydride $C_{12}H_8N_2O_3$	228.21	nd	...	103-4	...	...	...	B22⁴, 526
12547	4-Pyridine carboxylic acid, ethyl betaine $C_8H_9O_2N$	151.16	nd	...	241d	...	...	w, al	B22, 47
12548	2,3-Pyridine dicarboxylic acid or Quinolinic acid $2,3\text{-}(C_5H_3N)(CO_2H)_2$	167.12	mcl pr (w)	...	228-9	...	...	...	B22⁴, 1618
12549	2,4-Pyridine dicarboxylic acid or Lutidinic acid $2,4\text{-}(C_5H_3N)(CO_2H)_2$	167.12	lf (w + 1)	...	248-50 (anh)	0.942	...	...	B22⁴, 1630
12550	2,4-Pyridine dicarboxylic acid, 6-methyl or Uvitonic acid $6\text{-}CH_3\text{-}2,4\text{-}(C_5H_2N)(CO_2H)_2$	181.15	cr (w)	...	282d	...	...	...	B22², 107
12551	2,5-Pyridine dicarboxylic acid or Isocinchomeronic acid $2,5\text{-}(C_5H_3N)(CO_2H)_2$	167.12	lf (w + 1)	sub	256-8d	...	...	...	B22⁴, 1632
12552	2,6-Pyridine dicarboxylic acid or Dipicolinic acid $2,6\text{-}(C_5H_3N)(CO_2H)_2$	167.12	nd (w + 3/2)	...	252 (anh)	...	...	...	B22⁴, 1635
12553	3,4-Pyridine dicarboxylic acid or Cinchomeronic acid $3,4\text{-}(C_5H_3N)(CO_2H)_2$	167.12	pr, nd or lf (w)	sub	262d	...	...	...	B22⁴, 1641
12554	3,4-Pyridine dicarboxylic acid, 4,5-dihydro-2,6-dimethyl,diethyl ester $C_{13}H_{19}NO_4$	253.30	bl flr pl (eth)	...	85	...	...	al, ace	B22², 98
12555	3,4-Pyridine dicarboxylic acid, 2,6-dimethyl,diethyl ester $2,6\text{-}(CH_3)_2(C_5HN)(CO_2C_2H_5)_2\text{-}3,4$	251.28	...	270d, 163^{13}	16	...	...	...	B22², 109
12556	3,4-Pyridine dicarboxylic acid, 5-methoxy-6-methyl $6\text{-}CH_3\text{-}5\text{-}CH_3O\text{-}(C_5HN)(CO_2H)_2\text{-}3,4$	211.17	cr (w-ace)	...	213-5	...	...	...	B22⁴, 2591
12557	3,5-Pyridine dicarboxylic acid or Dinicotinic acid $3,5\text{-}(C_5H_3N)(CO_2H)_2$	167.12	cr (w)	sub	325d	...	...	...	B22⁴, 1643
12558	3,5-Pyridine dicarboxylic acid, 1,4-dihydro-2,6-dimethyl,diethyl ester $C_{13}H_{19}NO_4$	253.30	ye nd or lf (al)	...	184-5	...	...	chl	B22⁴, 1582

No.	Name, Synonyms, and Formula	Mol. wt.	Color, crystalline form, specific rotation and λ_{max} (log ε)	b.p. °C	m.p. °C	Density	n_D	Solubility	Ref.
12559	3,5-Pyridine dicarboxylic acid, 1,4-dihydro-2,4,6-trimethyl,diethyl est $C_{14}H_{21}NO_4$	267.33	lt bl flr pl (al)		131			chl	B22^4, 1594
12560	3,5-Pyridine dicarboxylic acid, 2,6-dimethyl,diethyl ester $2,6\text{-}(CH_3)_2(C_5HN)(CO_2C_2H_5)_2\text{-}3,5$	251.28	nd (al), pr (eth)	301-2, 180^{16}	75-6			al, eth, bz, chl, lig	B22^4, 1656
12561	2,6-Pyridine diol hydrate $C_5H_5NO_2.H_2O$	129.12	ye pr		202-3			w, al	B21^4, 2063
12562	Pyridine sulfate $C_5H_5NH_2SO_4$	177.17	cr (al)					w, al	B20^4, 2233
12563	4-Pyridine sulfonic acid $4\text{-}C_5H_4NSO_3H$	159.16			333-4			w, al	B22^4, 3465
12564	Pyridine pentacarboxylic acid-dihydrate $C_5N(CO_2H)_5 \cdot 2H_2O$	335.18	cr (eth, w)		220d			w	B22^2, 142
12565	3-Pyridine sulfonic acid $3\text{-}C_5H_4NSO_3H$	159.16	orh		375d	1.718$^{25/25}$		w	B22^4, 3458
12566	2,3,4,5-Pyridine tetracarboxylic acid $2,3,4,5\text{-}(C_5HN)(CO_2H)_4$	255.14	cr		160d			w	B22, 188
12567	2,3,4,6-Pyridine tetracarboxylic acid $2,3,4,6\text{-}(C_5HN)(CO_2H)_4$	255.14	nd (w + 3)		236d			w, aa	B22^2, 142
12568	2,3,5,6-Pyridine tetracarboxylic acid $2,3,5,6\text{-}(C_5HN)(CO_2H)_4$	255.14	cr (w + 2)		200d			w	B22^1, 544
12569	2,3,4-Pyridine tricarboxylic acid or γ-Carbocinchomeronic acid $2,3,4\text{-}(C_5H_2N)(CO_2H)_3$	211.13	lf (w + $^3/_2$)		250 (anh)				B22^4, 1777
12570	2,3,5-Pyridine tricarboxylic acid or Carbodinicotinic acid $2,3,5\text{-}(C_5H_2N)(CO_2H)_3$	211.13	pl, lf or nd (w, dil al)		323 (anh)			w	B22^2, 136
12571	2,4,5-Pyridine tricarboxylic acid or Berberonic acid $2,4,5(C_5H_2N)(CO_2H)_3$	211.13	tcl pr (dil HCl + 2w)		243 (anh)				B22^4, 1778
12572	2,4,6-Pyridine tricarboxylic acid or Trimesic acid $2,4,6\text{-}(C_5H_2N)(CO_2H)_3$	211.13	nd (w + 2)	sub	227d				B22^4, 1778
12573	3,4,5-Pyridine tricarboxylic acid or β-Carbocinchomeronic acid $3,4,5\text{-}(C_5H_2N)(CO_2H)_3$	211.13	lf or pl (w + 3), ta (w)		261d −w115			al	B22^4, 1779
12574	2,4,6-Pyridine triol $2,4,6\text{-}(HO)_3(C_5H_2N)$	127.10	ye nd or pw		230d			al, eth, ace	B21^4, 2504
12575	Pyridoxal hydrochloride $C_8H_9NO_3 \cdot HCl$	203.63	rh		165d			w	B21^4, 6419
12576	Pyridoxal oxime $C_8H_{10}N_2O_3$	182.18	cr (al)		205-6d			al	B21^4, 6427
12577	Pyridoxamine dihydrochloride $C_8H_{12}N_2O_2 \cdot 2HCl$	241.12	pr (al)		226-7d			w	C51, 14833
12578	Pyrimidine or 1,3-Diazin $C_4H_4N_2$	80.09		123-4	22		1.4998^{20}	w, al	B23, 89
12579	Pyrimidine, 2-amino $2\text{-}H_2N(C_4H_3N_2)$	95.11	nd (AcOEt)	sub	127-8			w	B24, 80
12580	Pyrimidine, 2-amino-4,6-dihydroxy $4,6\text{-}(HO)_2\text{-}2\text{-}H_2N(C_4HN_2)$	127.10	pr (w + 1)		>330				B24, 468
12581	Pyrimidine, 2-amino-4,5-dimethyl $4,5\text{-}(CH_3)_2\text{-}2\text{-}H_2N(C_4HN_2)$	123.16	nd (w)	sub	214-5			al, ace, bz, chl	B24, 91
12582	Pyrimidine, 2-amino-4,6-dimethyl $4,6\text{-}(CH_3)_2\text{-}2\text{-}H_2N(C_4HN_2)$	123.16			153-4			w, al, ace, bz, chl	B24^1, 234
12583	Pyrimidine, 2-amino-4-methyl $4\text{-}CH_3\text{-}2\text{-}H_2N(C_4H_2N_2)$	109.13	pl (w), nd (sub)	sub	159-60			al	B24, 84
12584	Pyrimidine, 2-amino-5-methyl $5\text{-}CH_3\text{-}2\text{-}H_2N(C_4H_2N_2)$	109.13	pl (sub), pr (w)	sub	193.5			al	B24, 87
12585	Pyrimidine, 2-amino-5-nitro $5\text{-}NO_2\text{-}2\text{-}H_2N(C_4H_2N_2)$	140.10	nd (al)		236-7			al, ace	B24^1, 231
12586	Pyrimidine, 4-amino $4\text{-}H_2N(C_4H_3N_2)$	95.10	pl (AcOEt)		151-2			w, al	B24, 81
12587	Pyrimidine, 4-amino-2,6-dimethyl or Kyanmethin $2,6\text{-}(CH_3)_2\text{-}4\text{-}H_2N(C_4HN_2)$	123.16	nd (al), pl (bz)	sub	183 (192)				B24^2, 46
12588	Pyrimidine, 4-amino-2-methyl $2\text{-}CH_3\text{-}4\text{-}H_2N(C_4H_2N_2)$	109.13	rh (ace)		205			w, al, ace, bz	B24, 84
12589	Pyrimidine, 4-amino-5-methyl $5\text{-}CH_3\text{-}4\text{-}H_2N(C_4H_2N_2)$	109.13	pl (al, AcOEt)		176			al, w	B24, 87
12590	Pyrimidine, 4-amino-6-methyl $6\text{-}CH_3\text{-}4\text{-}H_2N(C_4H_2N_2)$	109.13	pr (w), nd, lf (sub)	sub	197			w, al	B24, 85

No.	Name, Synonyms, and Formula	Mol. wt.	Color, crystalline form, specific rotation and λ_{max} (log ε)	b.p. °C	m.p. °C	Density	n_D	Solubility	Ref.
12591	Pyrimidine, 6-amino-2,4-dihydroxy or 4-Amino uracil $2,4\text{-}(HO)_2\text{-}6\text{-}H_2N(C_4HN_2)$	127.10	cr (w)		d			w	B24, 469
12592	Pyrimidine, 6-amino-4,5-dimethyl $4,5\text{-}(CH_3)_2\text{-}6\text{-}NH_2(C_4HN_2)$	123.16	nd (w)		230			bz, ace, chl	B24, 92
12593	Pyrimidine, 2-chloro-4-dimethylamino-6-methyl $4\text{-}[(CH_3)_2N]\text{-}6\text{-}CH_3\text{-}2\text{-}Cl(C_4HN_2)$	171.63	br wax	140-7^{4}	87			al	C36, 911
12594	Pyrimidine, 2,4-dichloro-5-methyl $5\text{-}CH_3\text{-}2,4\text{-}Cl_2(C_4HN_2)$	163.01	pl (al)	235	25-7			al, eth, bz, chl	B23, 93
12595	Pyrimidine, 2,4-dichloro-6-methyl $6\text{-}CH_3\text{-}2,4\text{-}Cl_2(C_4HN_2)$	163.01	nd (lig)	219	46-7			al, eth, bz, chl	B23, 92
12596	Pyrimidine, 2,4-dichloro-5-nitro $2,4\text{-}Cl_2\text{-}5\text{-}NO_2(C_4HN_2)$	193.98		153-5^{58}	29.3				B23, 90
12597	Pyrimidine, 2-hydroxy,hydrochloride $2\text{-}HO(C_4H_3N_2).HCl$	132.55	rods (al)		205			w	B24^{1}, 231
12598	Pyrimidine, 2-methyl $2\text{-}CH_3(C_4H_3N_2)$	94.12		138^{758}	−4			w	B23, 92
12599	Pyrimidine, 4-methyl $4\text{-}CH_3(C_4H_3N_2)$	94.12		142	32	1.031$^{16/16}$	1.5000^{20}	w	B23, 92
12600	Pyrimidine, 5-methyl $5\text{-}CH_3(C_4H_3N_2)$	94.12		152-4	30.5			w	B23, 93
12601	Pyrimidine, 4-methylamino $4\text{-}CH_3NH(C_4H_3N_2)$	109.13	lf (al), nd (peth)	142-4^{16}	74-5			w	B24^{2}, 38
12602	Pyrocalciferol or 9α-Lumisterol $C_{28}H_{44}O$	396.66	nd (MeOH), $[\alpha]^{20/}{}_D$ + 512 (al)		93-5			al	B6^{3}, 3098
12603	Pyrocoll $C_{10}H_6N_2O_2$	186.17	ye mcl pl (aa)	sub	268			al, eth, bz, chl	B24^{1}, 360
12604	Pyrogallol or 1,2,3-Trihydroxy benzene $1,2,3\text{-}(HO)_3C_6H_3$	126.11	lf or nd (bz)	309, 171^{12}	133-4	1.453$^{4/4}$	1.561^{134}	w, al, eth	B6^{3}, 6260
12605	Pyrogallol, 4-acetyl or 2,3,4-Trihydroxy acetophenone $2,3,4\text{-}(HO)_3C_6H_2(COCH_3\text{-}1)$	168.15	pa ye nd or lf (w)		173			al, eth, ace, aa	B8^{3}, 3376
12606	Pyrogallol, 1,2-dimethyl ether or 2,3-Dimethoxy phenol $2,3(CH_3O)_2C_6H_3OH$	154.17		232-4, 124-5^{17}		1.5392^{20}			B6^{3}, 6264
12607	Pyrogallol, 1,3-dimethyl ether or 2,6-Dimethoxy phenol $2,6\text{-}(CH_3O)_2C_6H_3OH$	154.17	mcl pr (w)	262-7	56-7			al, eth	B6^{3}, 6264
12608	Pyrogallol triacetate $1,2,3\text{-}(CH_3CO_2)_3C_6H_3$	252.22	pr (al)		165			al	B6^{3}, 6269
12609	Pyrogallol, 4,5,6-trichloro or 4,5,6-Trichloro-1,2,3-hydroxy benzene $C_6H_3Cl_3O_3$	229.45	nd (al, bz)		185			w, al, eth	B6, 1084
12610	Pyrogallol trimethyl ether or 1,2,3-Trimethoxy benzene $1,2,3\text{-}(CH_3O)_3C_6H_3$	168.20	rh nd (al)	235, 140^{12}	48-9	1.1118$^{45/45}$		al, eth, bz	B6^{3}, 6265
12611	2-Pyrone, 4,6-dimethyl $4,6\text{-}(CH_3)_2(C_5H_2O_2)$	124.14	lf (eth)	245, 126^{11}	51.5			w, al, eth	B17^{4}, 4531
12612	2-Pyrone, 3-hydroxy $3\text{-}HO\text{-}(C_5H_3O_2)$	112.08	nd (w + 2)	112^{20}	95 (anh), 80-5 (hyd)			w, al, eth, chl	B17^{4}, 5908
12613	4-Pyrone or 4H-Pyran-4-one $C_5H_4O_2$	96.09							
12614	4-Pyrone, 2,6-dimethyl $2,6\text{-}(CH_3)_2(C_5H_2O_2)$	124.14	hyg cr	215-7^{742}, 105^{23}	32.5	1.190	1.5238	w, al, eth, bz, chl	B17^{4}, 4399
12615	4-Pyrone, 3-Hydroxy-2-methyl or Larixinic acid, Maltol $2\text{-}CH_3\text{-}3\text{-}HO(C_5H_2O_2)$	126.11	pl, nd (sub)	248-9^{713}, 139-40^{25} sub	132			w, al, eth, ace	B17^{4}, 4532
12616	4-Pyrone, 5-hydroxy-2-(hydroxymethyl) or Kojic acid $2\text{-}HOCH_2\text{-}5\text{-}HO(C_5H_2O_2)$	142.11	red mcl pr (chl)	sub at 93	162.4			chl	B17^{4}, 5916
12617	4-Pyrone-2-carboxylic acid-5,6-dihydro-6,6-dimethyl, butyl ester $6,6\text{-}(CH_3)_2\text{-}2\text{-}C_5H_3O_2CO_2C_4H_9$	226.27	ye or pa red-br	256-70, 113-4^{14}		1.052^{25}	1.4745^{29}	**al, eth, bz, chl**	B18^{4}, 5345
12618	4-Pyrone-2-carboxylic acid-5-hydroxy or Comenic acid $C_6H_4O_5$	156.09	ye cr		>270d				B18^{4}, 5985
12619	Pyrophosphoramide, octamethyl $[(CH_3)_2N]_4P_2O_3$	286.25		118-22$^{0.3}$		1.1343^{25}	1.462^{25}	w, al, chl	B4^{4}, 288
12620	Pyrophosphoric acid, tetraethyl ester or Tetraethyl pyrophosphate $(C_2H_5O)_4P_2O_3$	290.19		155^{3}		1.1847$^{20/4}$	1.4180^{20}	**al, eth, ace, chl**	B1^{4}, 1340

No.	Name, Synonyms, and Formula	Mol. wt.	Color, crystalline form, specific rotation and λ_{max} (log ε)	b.p. °C	m.p. °C	Density	n_D	Solubility	Ref.
12621	Pyrophosphoric acid, dithiono, tetraethyl ester or Sulfotep $(C_2H_5O)_4P_2OS_2$	322.31	col oil	136-9^{2}		1.196$^{25/4}$	1.4753^{25}	al	B1^{4}, 1351
12622	Pyrotartaric acid (dl) or Methyl succinic acid $HO_2CCH(CH_3)CH_2CO_2H$	132.12	pr	d	115		1.4303	w, al, eth	B2^{4}, 1948
12623	Pyrrole C_4H_4NH	67.09		130-1		0.9691$^{20/4}$	1.5085^{20}	al, eth, ace, bz	B20^{4}, 2072
12624	Pyrrole, N-acetyl $C_4H_4N(COCH_3)$	109.13		181-2					B20^{4}, 2089
12625	Pyrrole, 2-acetyl $2\text{-}CH_3CO(C_4H_3NH)$	109.13	mcl nd (w)	220	90			w, al, eth	B21^{4}, 3437
12626	Pyrrole, 2-acetyl-N-methyl $2\text{-}CH_3CO(C_4H_3N)CH_3$	123.15		200-2^{252}, 92-4^{22}		1.0445$^{15/4}$	1.5403^{15}	al, bz, chl	B21^{4}, 3439
12627	Pyrrole, N-benzyl $(C_4H_4N)CH_2C_6H_5$	157.22		247, 138^{27}	15		1.5655^{24}	al, eth	B20^{4}, 2086
12628	Pyrrole, N-butyl $(C_4H_4N)C_4H_9$	123.20	oil	170-1, 53-4^{11}			1.4727^{20}		B20^{4}, 2082
12629	Pyrrole, 2,3-dimethyl-4-ethyl or Hemopyrrole $2,3\text{-}(CH_3)_2\text{-}4\text{-}C_2H_5(C_4HNH)$	123.20		198^{725}, 113^{16}	16-7	0.915$^{20/4}$		w	B20^{4}, 2152
12630	Pyrrole, 2,4-dimethyl $2,4\text{-}(CH_3)_2(C_4H_2NH)$	95.14	pa bl flr	171, 62-3^{10}		0.9236$^{20/4}$	1.5048^{20}	al, eth, bz	B20^{4}, 2107
12631	Pyrrole, 2,4-dimethyl-3-ethyl or Krypto pyrrole $2,4\text{-}(CH_3)_2\text{-}3\text{-}C_2H_5(C_4HNH)$	123.20	pr	197^{710}, 96^{16}	0	0.913$^{20/4}$	1.4961^{20}	al, eth, bz, chl	B20^{4}, 2153
12632	Pyrrole, 2,5-dimethyl $2,5\text{-}(CH_3)_2(C_4H_2NH)$	95.14		170-2^{765}, 50-3^{8}		0.9353$^{20/4}$	1.5036^{20}	al, eth	B20^{4}, 2152
12633	Pyrrole, 1,3-diphenyl $3\text{-}C_6H_5(C_4H_3N)C_6H_5$	219.29	pl (al)		122-3			eth, bz, chl	B20^{1}, 148
12634	Pyrrole, 2,5-diphenyl $2,5\text{-}(C_6H_5)_2(C_4H_2NH)$	219.29	lf (aa, dil al)		143.5			al, eth, bz, aa	B20^{4}, 4162
12635	Pyrrole, N-ethyl $(C_4H_4N)C_2H_5$	95.14		129-30^{762}		0.9009$^{20/4}$	1.4841^{20}	al	B20^{4}, 2082
12636	Pyrrole, 2-ethyl $2\text{-}C_2H_5(C_4H_3NH)$	95.14		163-5, 59-60^{15}		0.9042$^{20/4}$	1.4942^{20}	al	B20^{4}, 2106
12637	Pyrrole, 3-ethyl-4-methyl or Opsopyrrole $3\text{-}C_2H_5\text{-}4\text{-}CH_3(C_4H_2NH)$	109.17	ye oil	70^{11}	3	0.9059$^{20/4}$	1.4913^{20}	al, eth	B20^{4}, 2145
12638	Pyrrole, 3-ethyl-2,4,5-trimethyl or Phyllopyrrole $3\text{-}C_2H_5\text{-}2,4,5\text{-}(CH_3)_3(C_4NH)$	137.22	pl (sub), lf (eth), wh lf (peth)	213^{725}, 92-3^{12}	67-8			al, eth	B20^{4}, 2174
12639	Pyrrole, N-methyl $(C_4H_4N)CH_3$	81.12		114-5^{747}		0.9145$^{15/4}$	1.4875^{20}	al, eth	B20^{4}, 2080
12640	Pyrrole, 2-methyl $2\text{-}CH_3(C_4H_3NH)$	81.12		147-8^{750}		0.9446$^{15/4}$	1.5035^{16}	al, eth	B20^{4}, 2103
12641	Pyrrole, 3-methyl $3\text{-}CH_3(C_4H_3NH)$	81.12		142-3^{743}, 45^{11}			1.4970^{20}	al, eth	B20^{4}, 2105
12642	Pyrrole, N-Phenyl $(C_4H_4N)C_6H_5$	143.19	pl (sub), red in air	234, 140^{38} sub	62			al, eth, ace, bz, peth	B20^{4}, 2084
12643	Pyrrole, 2-phenyl $2\text{-}C_6H_5(C_4H_3NH)$	143.19	pl (al, sub)	271-2^{726}	129			al, eth, bz, chl	B20^{4}, 3452
12644	Pyrrole, N-Propyl $(C_4H_4N)C_3H_7$	109.17		145-6		0.8833$^{20/4}$		al, eth	B20^{4}, 3082
12645	Pyrrole, 3-Propyl $3\text{-}C_3H_7(C_4H_3NH)$	109.17		90^{30}, 49^{2}			1.4900^{25}	eth, ace	B20^{4}, 2143
12646	Pyrrole, 2-isopropyl $2\text{-}i\text{-}C_3H_7(C_4H_3NH)$	109.17		171-2^{741}		0.908$^{25/4}$	1.491^{25}	al, eth	B20^{4}, 2143
12647	Pyrrole, tetraiodo or Iodol C_4I_4NH	570.68	ye nd (al)		150d			eth, ace, chl, aa	B20^{4}, 2098
12648	Pyrrole, tetramethyl $C_4(CH_3)_4NH$	123.20	lf (dil al, peth)	130^{7}	111			al, eth, ace, bz	B20^{4}, 2156
12649	2-Pyrrole carboxaldehyde $(HNC_4H_3)CHO\text{-}(2)$	95.10	rh pr (peth)	217-9, 114^{15}	46-7		1.5939^{16}		B21^{4}, 3419
12650	2-Pyrrole carboxylic acid $(HNC_4H_3)CO_2H\text{-}2$	111.10	lf (w)		208d			w, al, eth	B22^{4}, 225
12651	2-Pyrrole carboxylic acid, 3,5-dimethyl, ethyl ester $3,5\text{-}(CH_3)_2\ (C_4HNH)CO_2C_2H_5\text{-}2$	167.21	cr (al)		125			al, ace	B22^{4}, 260
12652	3-Pyrrole carboxylic acid $(HNC_4H_3)CO_2H\text{-}3$	111.10	nd (lig)		161-2				B22^{4}, 235

No.	Name, Synonyms, and Formula	Mol. wt.	Color, crystalline form, specific rotation and λ_{max} (log ε)	b.p. °C	m.p. °C	Density	n_D	Solubility	Ref.
12653	3-Pyrrole carboxylic acid, 2,4-dimethyl, ethyl ester $2,4\text{-}(CH_3)_2\text{-}3\text{-}(C_4HNH)CO_2C_2H_5$	167.21	cr (eth-lig, peth)	291, 181-2^{35}	78-9			al, eth	B22^4, 255
12654	3-Pyrrole carboxylic acid, 2,5-dimethyl, ethyl ester $2,5\text{-}(CH_3)_2\text{-}3\text{-}(C_4HNH)CO_2C_2H_5$	167.21	rh (al)	290^{731}, 130^{15}	117-8			al	B22^4, 267
12655	3-Pyrrole carboxylic acid, 4,5-dimethyl, ethyl ester $4,5\text{-}(CH_3)_2\text{-}3\text{-}(C_4HNH)CO_2C_2H_5$	167.21	cr (dil al)		110-1			al, eth, chl	B22^4, 260
12656	2,4-Pyrrole dicarboxylic acid $2,4\text{-}(C_4H_2NH)(CO_2H)_2$	155.11	cr (w)		295d			w	B22^4, 1542
12657	2,4-Pyrrole dicarboxylic acid, 3,5-dimethyl, diethyl ester $3,5\text{-}(CH_3)_2\text{-}2,4\text{-}(C_4NH)(CO_2C_2H_5)_2$	239.27	nd (dil al)		136-7			ace, bz, chl	B22^4, 1568
12658	2,4-Pyrrole dicarboxylic acid, 3,5-dimethyl-N-ethyl, diethyl ester $C_{14}H_{21}NO_4$	267.33	cr (dil al)		40-1			al	B22^4, 1574
12659	2,4-Pyrrole dicarboxylic acid, 5-formyl-3-methyl, diethyl ester $3\text{-}CH_3\text{-}5\text{-}HCO\text{-}2,4\text{-}(C_4NH)(CO_2C_2H_5)_2$	253.25	nd		124-5			al, to	B22^4, 3265
12660	2-Pyrone $C_5H_4O_2$	96.09		206-9, 120^{30}	8-9	1.2000$^{20/4}$	1.5270^{25}	w, ace	B17^4, 4399
12661	2-Pyrone, tetrahydro $C_5H_8O_2$	100.12		218-20, 113-4^{14}	−12.5	1.0794^{20}	1.4503^{20}	w, al, eth	B17^4, 4169
12662	Pyrrolidine or Tetrahydropyrrol C_4H_8NH	71.12		88-9		0.8520$^{22/4}$	1.4431^{20}	al, eth, chl	B20^4, 61
12663	Pyrrolidine, N-butyl $(C_4H_8N)C_4H_9$	127.23		154-5		0.816^{25}	1.4373^{25}	al	B20^4, 67
12664	Pyrrolidine, 2-butyl $2\text{-}C_4H_9(C_4H_7NH)$	127.23		173-4^{741}, 67^{18}		0.8277$^{20/4}$	1.4490^{20}	w, al	B20^4, 1627
12665	Pyrrolidine, N-(chloroacetyl) $(C_4H_8N)COCH_2Cl$	147.60		112$^{0.5}$	44-6				B20^4, 186
12666	Pyrrolidine, N,2-dimethyl $2\text{-}CH_3(C_4H_7N)CH_3$	99.18		96		0.7994$^{20/4}$	1.4252^{20}	w, al, eth	B20^4, 1384
12667	Pyrrolidine, 2,4-dimethyl $2,4\text{-}(CH_3)_2(C_4H_6NH)$	99.18		115-7		0.8297$^{20/4}$	1.4325^{20}	w, al, eth	B20^4, 1537
12668	Pyrrolidine, 2,5-dimethyl (*cis*) $2,5\text{-}(CH_3)_2(C_4H_6NH)$	99.18			106-7	0.8205$^{20/4}$	1.4299^{20}	w, al, eth	B20^4, 1538
12669	Pyrrolidine, N-methyl $(C_4H_8N)CH_3$	85.15		81.3		0.8188$^{20/4}$	1.4247^{20}	w, eth	B20^4, 64
12670	Pyrrolidine, N-phenyl $(C_4H_8N)C_6H_5$	147.22		119-20^{12}		1.0260^{25}	1.5813^{20}	eth	B20^4, 78
12671	2-Pyrrolidone or γ-Butyrolactam C_4H_7NO	85.11	cr (peth)	250.5^{742}, 133^{12}	24.6	1.120$^{20/4}$	1.4806^{30}	w, al, eth, bz	B21^4, 3142
12672	2-Pyrrolidone, N,5-dimethyl $C_6H_{11}NO$	113.16		215-7^{743}, 87.5^{10}			1.4650^{20}	w, eth	B21^4, 3192
12673	2-Pyrrolidone, 3,3-dimethyl $C_6H_{11}NO$	113.16	lf (bz)	237	65-7			w, al, eth, ace	B21, 242
12674	2-Pyrrolidone, N-methyl C_5H_9NO	99.13		202, 84-5^{14}	−23	1.0260$^{25/25}$	1.4684^{20}	w, eth, ace	B21^4, 3145
12675	3-Pyrroline or Dihydropyrrole C_4H_6NH	69.11		90-1		0.9097$^{20/4}$	1.4664^{20}	al, w, eth, ace	B20^4, 1906
12676	Pyruvic acid or 2-oxo-Propionic acid CH_3COCO_2H	88.06		165d, 54^{10}	13.8	1.2272$^{20/4}$	1.4280^{20}	w, al, eth, ace	B3^4, 1505
12677	Pyruvic acid-acetyl $CH_3COCH_2COCO_2H$	130.10	nd (bz-chl), pr (bz)	130^{37} sub	101 (anh) 55-63 (hyd)			w, al, eth, ace, bz, chl	B3^3, 1331
12678	Pyruvic acid, acetyl, ethyl ester $CH_3COCH_2COCO_2C_2H_5$	158.15		213-5, 111-2^{16}	18	1.1251^{20}	1.4757^{17}	al, eth	B3^4, 1777
12679	Pyruvic acid, 3-(2,5-dimethoxyphenyl) $[2,5\text{-}(CH_3O)_2C_6H_3]CH_2COCO_2H$	224.21	yesh cr (aa)		166-70d			ace, MeOH	B10^2, 723
12680	Pyruvic acid, 3-(3,4-dimethoxy phenyl) $[3,4\text{-}(CH_3O)_2C_6H_3]CH_2COCO_2H$	224.21	lf (aa)		ca 187d			al, ace	B10^3, 4521
12681	Pyruvic acid, ethyl ester or Ethyl pyruvate $CH_3COCO_2C_2H_5$	116.12		155, 69-71^{42}	−50	1.0596$^{15.6/4}$	1.4052^{20}	al, eth, ace	B3^4, 1513
12682	Pyruvic acid, methyl ester or Methyl pyruvate $CH_3COCO_2CH_3$	102.09		134-7, 53^{15}		1.154$^{0/4}$	1.4046^{25}	al, eth, ace	B3^3, 1160
12683	Pyruvic acid, 3-phenyl $C_6H_5CH_2COCO_2H$	164.16	lf (chl, bz)	157-8				al, eth	B10^3, 3000
12684	Pyruvonitrile CH_3COCN	69.06	rh	92-3		0.9745$^{20/4}$	1.3764^{20}	eth, ace	B3^4, 1515
12685	Pyruvic acid, 3-(2-nitro phenyl) $(2\text{-}NO_2C_6H_4)CH_2COCO_2H$	209.16	ye nd or lf (w, al, bz)		130			al, eth, aa	B10^3, 3017

PHYSICAL CONSTANTS OF ORGANIC COMPOUNDS (Continued)

No.	Name, Synonyms, and Formula	Mol. wt.	Color, crystalline form, specific rotation and λ_{max} (log ε)	b.p. °C	m.p. °C	Density	n_D	Solubility	Ref.
12686	o,o′-Quaterphenyl or o,o′-Diphenyl biphenyl $C_{24}H_{18}$	306.41	pr (al)	420	118-9			al, eth, ace, bz, chl	B5^{3}, 2561
12687	p,p′-Quaterphenyl or Tetraphenyl $C_{24}H_{18}$	306.41	lf (bz)	428^{18}	320				B5^{3}, 2562
12688	Quercitrin or 3-D-α-L-rhamnopyranosylquercitrin $C_{21}H_{20}O_{11}$	448.38	pa ye nd or pl (+2w dil al), $[\alpha]_{578}$ −73.5 (al, c=4)		182-5 (hyd), 250-2 (anh)			al	B18^{4}, 3491
12689	Quillaic acid or Quillaja sapogenin $C_{30}H_{46}O_5$	486.69	nd (dil al), $[\alpha]^{20/}{}_D$ +56.1 (py, c=2.91)		294			al, eth, ace, aa, Py	B10^{3}, 4652
12690	Quinacrine dihydrochloride (dl) or Atebrine $C_{25}H_{30}ClN_3O.2HCl.2H_2O$	532.94	yesh nd (w), ye cr pw		248-50d			w, MeOH	B21^{4}, 6247
12691	Quinamine $C_{19}H_{24}N_2O_2$	312.41	pr (bz), nd (80%al), $[\alpha]^{15/}{}_D$ +93 (chl, c=2)		185-6			eth, ace, bz, lig	B27^{2}, 667
12692	Quinazoline or 1,3-Benzodiazine $C_8H_6N_2$	130.15	ye pl (peth)	241.5, $117\text{-}20^{15}$	48			w, al, eth, ace, bz	B23^{2}, 177
12693	Quinazoline, 3,4-dihydro-4-oxo or Quinazolinone $C_8H_6N_2O$	146.15	nd (dil aa)	360	216-8				B24^{2}, 71
12694	Quinazoline, 3,4-dihydro-2-phenyl $C_{14}H_{12}N_2$	208.26	lf (dil al)		142			al, eth, chl, aa	B23, 239
12695	Quinazoline, 3,4-dihydro-3-phenyl or Phenzoline $C_{14}H_{12}N_2$	208.26	pl (eth-lig)	d	96-7	1.290^{4}		al, eth, bz, chl	B23^{2}, 155
12696	Quinazoline, 3,4-dihydro-4-phenyl $C_{14}H_{12}N_2$	208.26	pl (al, AcOEt)		166-7			al, eth	B23, 239
12697	Quinazoline, 2,4-dihydroxy-6,8-dinitro $C_8H_4N_4O_6$	252.14	ye gr pr (aa)		274-5d				B24^{1}, 344
12698	2,4-Quinazolinedione or Benzoylene urea $C_8H_6N_2O_2$	162.15	nd (w, al), lf (aa)		356				B24^{2}, 197
12699	Quinhydrone or Benzoquinhydrone $C_{12}H_{10}O_4$	218.21	red br nd	sub	171	$1.401^{20/4}$		al, eth	B7^{3}, 3363
12700	Quinicine or Quinotoxin $C_{20}H_{24}N_2O_2$	324.42	red ye amor, $[\alpha]^{15/}{}_D$ +44.1 (chl)		ca 60			al, eth, chl	B25^{2}, 20
12701	Quinicine oxalate (d) $(C_{20}H_{24}N_2O_2)_2H_2C_2O_4.9H_2O$	901.02	pr (chl), nd (al), $[\alpha]^{15/}{}_D$ +19.5 (al-chl), (+9.5w)		149			w, al, chl	B25^{2}, 20
12702	Quinidine $C_{20}H_{24}N_2O_2$	324.42	cr (+2.5w, dil al)		174-5 (anh)			al, bz, chl	B23^{2}, 414
12703	Quinidine hydrate or Conquinine $C_{20}H_{24}N_2O_2.2½H_2O$	369.46	cr (dil al)		171d			al, eth, bz, chl	B23^{2}, 414
12704	Quinidine hydrochloride $C_{20}H_{24}N_2O_2.HCl.H_2O$	378.90	pr (w), $[\alpha]^{20/}{}_D$ +200 (w, c=1)		258-9d (anh)			chl	B23^{2}, 415
12705	Quinidine sulfate (d) $(C_{20}H_{24}N_2O_2)_2.H_2SO_4.2H_2O$	782.95	pr, nd (w), $[\alpha]^{25/}{}_D$ +212 (al)					w, al, chl	B23^{2}, 415
12706	Quinine $C_{20}H_{24}N_2O_2.3H_2O$	378.47	cr (+3w, eth), nd (+3w, al), rh nd (abs al), $[\alpha]^{15/}{}_D$ −145.2 (al)		177 (anh), 57 (hyd)		1.625	al, eth, chl, Py	B23^{2}, 416

No.	Name, Synonyms, and Formula	Mol. wt.	Color, crystalline form, specific rotation and λ_{max} (log ε)	b.p. °C	m.p. °C	Density	n_D	Solubility	Ref.
12707	Quinine bisulfate $C_{20}H_{24}N_2O_2.H_2SO_4.7H_2O$	548.60	pr (+w, w), pr (+5w, al), $[\alpha]^{15/}_D$ +168.4 (al)		160d			chl	B23[2], 416
12708	Quinine, o-ethyl carbonate $C_{23}H_{28}N_2O_4$	396.49	nd (w), cr (dil al)		95			al, eth, chl	B23[2], 424
12709	Quinine formate or Quinoform $C_{20}H_{24}N_2O_2.HCO_2H$	370.45	nd $[\alpha]^{20/}_D$ −144.2 (w, c=1)		149-50 (anh), 126 (+1w)			w, al, chl	B23[1], 169
12710	Quinine hydrobromide $C_{20}H_{24}N_2O_2.HBr.H_2O$	423.35	silky efflor nd (w)		ca 208 softens 152			w, al, chl	B23[1], 168
12711	Quinine hydrochloride $C_{20}H_{24}N_2O_2.HCl$	360.88	silky efflor nd (w), $[\alpha]^{15/}_D$ −145 (al)		158-60 (anh)			w, al, chl	B23[2], 420
12712	Quinine hydrochloride, hydrate $C_{20}H_{24}N_2O_2.HCl.2H_2O$	396.91	silky efflor nd (w), $[\alpha]^{20/}_D$ −149.8 (w, c=1.3)		156-90			w, al, chl	B23[2], 420
12713	Quinine salicylate or Quinine-2-hydroxy benzoate $C_{20}H_{24}N_2O_2.C_7H_6O_3.H_2O$	480.56	wh pr (al), cr (+2w, w)		195			al, chl	B23[2], 422
12714	Quinine sulfate $2(C_{20}H_{24}N_2O_2).H_2SO_4$	746.92	silky nd (w)		235.2			al	B23[2], 420
12715	Quinine sulfate, dihydrate $2(C_{20}H_{24}N_2O_2)_2.H_2SO_4.2H_2O$	782.95	silky nd (w), $[\alpha]^{15/}_D$ −220 (0.5N HCl)		205			al	B23[2], 420
12716	Quinine valerate $C_{20}H_{24}N_2O_2.C_4H_9CO_2H.H_2O$	444.57	wh		ca 96			al, eth	M, 892
12717	Quininone $C_{20}H_{22}N_2O_2$	322.41	nd lf (eth), $[\alpha]^{20/}_D$ +75.5 (al, c=2)		108 (rapid htng)			al, eth, bz, chl	B25[2], 23
12718	Quinizarin or 1,4-Dihydroxy anthraquinone $1,4-(HO)_2C_{14}H_6O_2$	240.22	ye red lf (eth), dk red nd (al), red cr (to, aa)		200-2 (aa), 194 (to)				B8[1], 3775
12719	Quinoline or Benzo[b]-pyridine C_9H_7N	129.16		238, 114^{17}	fr−15.6	$1.0929^{20/4}$	1.6268^{20}	al, eth, ace, bz	B20[4], 3334
12720	Quinoline, 3-acetamido $3-(CH_3CONH)(C_9H_6N)$	186.21	cr (w)		166-7			al	B22[4], 4607
12721	Quinoline, 4-acetamide $4-(CH_3CONH)(C_9H_6N)$	186.21	nd (w + 1)	sub	176			al	B22[4], 4614
12722	Quinoline, 5-acetamido $5-(CH_3CONH)(C_9H_6N)$	186.21	pl (w), pr (dil al)		178			al	B22[4], 4671
12723	Quinoline, 6-acetamido $6-(CH_3CONH)(C_9H_6N)$	186.21	nd (w)		138			w, al	B22[2], 355
12724	Quinoline, 8-acetamido $8-(CH_3CONH)(C_9H_6N)$	186.21	cr (w, al)		167.5				B22[2], 356
12725	Quinoline, 3-acetamido $8-(CH_3CONH)(C_9H_6N)$	186.21	nd (al)		103			al	B22[1], 640
12726	Quinoline, 2-amino or α-Quinolyl amine $2-H_2N(C_9H_6N)$	144.18	lf (w)	sub	131.5			al, eth, ace, chl, aa	B22[4], 4587
12727	Quinoline, 2-amino-4-hydroxy $4-HO-2-H_2N(C_9H_5N)$	160.18	nd (w + 1), rh (al)		303-4				B22[4], 5718
12728	Quinoline, 2-amino-4-methyl or 2-Aminolepidine $4-CH_3-2-H_2N(C_9H_5N)$	158.20	cr pw (bz)	320	133			al, eth, chl, aa	B22[4], 4801
12729	Quinoline, 3-amino or β-Quinolyl amine $3-H_2N(C_9H_6N)$	144.18	rh (w, dil al)		94			al, eth, chl	B22[4], 4605

No.	Name, Synonyms, and Formula	Mol. wt.	Color, crystalline form, specific rotation and λ_{max} (log ε)	b.p. °C	m.p. °C	Density	n_D	Solubility	Ref.
12730	Quinoline, 3-amino-2-methyl or 3-Amino quinaldine 2-CH_3-3-$H_2N(C_9H_5N)$	158.20	pa ye nd (eth, peth)	278d, 198^{16}	160			al, eth, bz, chl	$B22^4$, 4756
12731	Quinoline, 4-amino or γ-Quinolyl amine 4-$H_2N(C_9H_6N)$	144.18	nd (w + 1), nd (bz, dil al)	180^{12}	154 (anh), 70 (+ w)			w, al, eth, chl	$B22^4$, 4611
12732	Quinoline, 4-amino-2-methyl or 4-Amino quinaldine 2-CH_3-4-$H_2N(C_9H_5N)$	158.20	nd (bz-lig), pr (eth-bz)	333	168			al, eth, ace, bz	$B22^4$, 4559
12733	Quinoline, 5-amino or 5-Quinalyl amine 5-$H_2N(C_9H_6N)$	144.18	ye nd (al), lf (eth)	310 sub, 184^{10}	110			al, eth, bz	$B22^4$, 4669
12734	Quinoline, 5-amino-6-hydroxy 6-HO-5-$H_2N(C_9H_5N)$	160.18	gr nd (w + 2)		185			al	B22, 501
12735	Quinoline, 5-amino-8-hydroxy 8-HO-5-$H_2N(C_9H_5N)$	160.18	nd (bz)		143				B22, 5865
12736	Quinoline, 5-amino-2-methyl or 5-Amino quinaldine 2-CH_3-5-$H_2N(C_9H_5N)$	158.20	grsh pl or nd (w + 1)		117-8			w, al, bz, lig	$B22^2$, 360
12737	Quinoline, 5-amino-8-methyl 8-CH_3-5-$H_2N(C_9H_5N)$	158.20	yesh nd (w, dil al)		143			al	$B22^4$, 4828
12738	Quinoline, 6-amino or 6-Quinolyl amine 6-$H_2N(C_9H_6N)$	144.18	cr (w + 2), pr (eth)	187^{12}	114 (anh)			al	$B22^4$, 4681
12739	Quinoline, 6-amino-2-methyl or 6-Amino quinaldine 2-CH_3-6-$H_2N(C_9H_5N)$	158.20	pa br (w, dil al)		187.5			w, al, chl	$B22^4$, 4780
12740	Quinoline, 6-amino-4-methyl 4-CH_3-6-$H_2N(C_9H_5N)$	158.20	nd (w)		169-70			al, eth, chl	$B22^4$, 4812
12741	Quinoline, 7-amino or 7-Quinolyl amine 7-$H_2N(C_9H_6N)$	144.18	ye nd (+ 1w)		93-4 (anh), 74-5 (+ w)			al	$B22^4$, 4704
12742	Quinoline, 7-amino-8-hydroxy 8-HO-7-$H_2N(C_9H_5N)$	160.18	br pr (eth, dil al)		124			al, eth, bz, chl	$B22^4$, 5873
12743	Quinoline, 7-amino-2-methyl 2-CH_3-7-$H_2N(C_9H_5N)$	158.20	pr (dil al)	304	129			al, eth, ace, bz	$B22^4$, 4784
12744	Quinoline, 7-amino-8-methyl 8-CH_3-7-$H_2N(C_9H_5N)$	158.20	pr (dil al)	304	125			al, eth, ace, bz	B22, 456
12745	Quinoline, 8-amino or 8-Quinolyl amine 8-$H_2N(C_9H_6N)$	144.18	pa ye nd (sub), cr (al, lig)	157-8^{19}	70			w, al	$B22^4$, 4708
12746	Quinoline, 8-amino-6-methoxy 6-CH_3O-8-$H_2N(C_9H_5N)$	174.20	cr	137-8^{1}	51				$B22^4$, 5747
12747	Quinoline, 8-amino-2-methyl or 8-Amino quinaldine 2-CH_3-8-$H_2N(C_9H_5N)$	158.20	pr (lig)		57-8			al, eth, ace, bz	$B22^4$, 4785
12748	Quiniline, 8-amino-6-methyl 6-CH_3-8-$H_2N(C_9H_5N)$	158.20	nd	sub	73			w, al, eth, ace, bz	$B22^4$, 4822
12749	Quinoline, 2-bromo 2-$Br(C_9H_6N)$	208.06	nd (al)		49			al, eth, bz, chl	$B20^4$, 3387
12750	Quinoline, 3-bromo 3-$Br(C_9H_6N)$	208.06	ye oil	274-6, $98^{0.5}$	13-5		1.6641^{20}	aa	$B20^4$, 3388
12751	Quinoline, 3-bromo-2-hydroxy or β-Bromo carbostyril 3-Br-2-$HO(C_9H_5N)$	224.06	pr (al)	sub	253				$B21^4$, 1063
12752	Quiniline, 4-bromo 4-$Br(C_9H_6N)$	208.06	cr	270d	29-30				$B20^4$, 3389
12753	Quinoline, 4-bromo-2-hydroxy or γ-Bromo carbostyril 4-Br-2-$HO(C_9H_5N)$	224.06	nd (al)	sub	266-7			al	$B21^4$, 1063
12754	Quinoline, 5-bromo 5-$Br(C_9H_6N)$	208.06	nd	280^{756}, 105-$7^{1.2}$	52				$B20^4$, 3389
12755	Quinoline, 5-bromo-2-hydroxy or 5-Bromo carbostyril 5-Br-2-$HO(C_9H_5N)$	224.06	nd (al)		300				$B21^4$, 1063
12756	Quinoline, 5-bromo-6-hydroxy 5-Br-6-$HO(C_9H_5N)$	224.06	nd (dil al)		186				$B21^4$, 1119
12757	Quinoline, 5-bromo-8-hydroxy 5-Br-8-$HO(C_9H_5N)$	224.06	nd (al), nd or lf (sub)	sub	124			w, al, bz, chl	$B21^4$, 1184
12758	Quinoline, 6-bromo 6-$Br(C_9H_6N)$	208.06		278, 155-6^{15}	24			al, eth	$B20^4$, 3390
12759	Quinoline, 6-bromo-2-hydroxy 6-Br-2-$HO(C_9H_5N)$	224.06	ye nd (al)		269			w, al, eth, chl	$B21^4$, 1063
12760	Quinoline, 7-bromo 7-$Br(C_9H_6N)$	208.06	nd	290	34 (52)			al, eth	$B20^4$, 3390

No.	Name, Synonyms, and Formula	Mol. wt.	Color, crystalline form, specific rotation and λ_{max} (log ε)	b.p. °C	m.p. °C	Density	n_D	Solubility	Ref.
12761	Quinoline, 7-bromo-2-hydroxy or 7-Bromo carbostyril 7-Br-2-HO(C_9H_5N)	224.06	nd (aa), pl (al)	sub	228	...	...	al, eth, chl	B21[4], 1064
12762	Quinoline, 8-Bromo 8-Br(C_9H_6N)	208.06	...	302-4, 165-6^{20}	<−10 (80)	...	...	al	B20[4], 3391
12763	Quinoline, 8-bromo-5-hydroxy 8-Br-5-HO(C_9H_5N)	224.06	nd (al)	...	190d	...	...	al, chl	B21, 85
12764	Quinoline, 2-chloro 2-Cl(C_9H_6N)	163.61	nd (aq al)	265-6^{750}, 153-4^{22}	38	1.2464$^{25/4}$	1.6342^{25}	al, eth, bz, lig	B20[4], 3376
12765	Quinoline, 2-chloro-4-methyl or 2-Chloro lepidine 4-CH_3-2-Cl(C_9H_5N)	177.63	nd (dil al)	296	59	...	...	al, eth, chl	B20[4], 3482
12766	Quinoline, 2-chloro-6-methyl 6-CH_3-2-Cl(C_9H_5N)	177.63	nd (dil al)	...	112 (116)	...	...	al, eth, bz, chl	B20[4], 3492
12767	Quinoline, 2-chloro-8-methyl 8-CH_3-2-Cl(C_9H_5N)	177.63	nd (eth)	286^{734}	61	...	...	al, eth, bz, chl	B20[1], 152
12768	Quinoline, 3-chloro 3-Cl(C_9H_6N)	163.61	hyg cr	255, 141^{15}	...	...	...	...	B20[4], 3377
12769	Quinoline, 3-chloro-2-methyl or 3-Chloro quinaldine 2-CH_3-3-Cl(C_9H_5N)	177.63	nd (dil al)	...	71 2	...	...	al, eth	B20[4], 3461
12770	Quinoline, 3-chloro-4-methyl or 3-Chloro lepidine 4-CH_3-3-Cl(C_9H_5N)	177.63	nd (dil al)	...	55	...	...	al	B20[1], 150
12771	Quinoline, 3-chloro-6-methyl 6-CH_3-3-Cl(C_9H_5N)	177.63	nd (dil MeOH)	...	85.5	...	...	al, MeOH	B20[2], 246
12772	Quinoline, 4-chloro 4-Cl(C_9H_6N)	163.61	cr	261^{764}, 130^{15}	34-5	1.251	...	al, eth	B20[4], 3377
12773	Quinoline, 4 chloro 2 methyl or 4 Chloro quinaldine 2-CH_3-4-Cl(C_9H_5N)	177.63	nd (+1w)	269-70	42-3	...	...	al, eth, bz, chl	B20[4], 3461
12774	Quinoline, 5-chloro 5-Cl(C_9H_6N)	163.61	cr (al)	256-7^{756}	45	...	...	al	B20[4], 3379
12775	Quinoline, 5-chloro-7-iodo-8-hydroxy or Vioform 5-Cl-7-I-8-HO(C_9H_4N)	305.50	ye br nd (al, aa)	...	178-9	...	...	...	B21[4], 1190
12776	Quinoline, 6-chloro 6-Cl(C_9H_6N)	163.61	pr (eth), nd (al)	262-4	44-5	...	1.6110^{56}	...	B20[4], 3379
12777	Quinoline, 6-chloro-2-methyl or 6-Chloro quinaldine 2-CH_3-6-Cl(C_9H_5N)	177.63	lf or nd (dil al)	...	91	...	...	eth	B20[4], 3462
12778	Quinoline, 6-chloro-4-methyl or 6-Chloro lepidine 4-CH_3-6-Cl(C_9H_5N)	177.63	nd (al)	...	70-2	...	...	al, eth, ace, bz	B20[4], 3482
12779	Quinoline, 7-chloro 7-Cl(C_9H_6N)	163.61	nd or pr	267-8	31-2	1.2158$^{58/4}$	1.6108$^{58/}{}_4$	al, eth, ace, bz, chl	B20[4], 3381
12780	Quinoline, 7-chloro-4-hydroxy 7-Cl-4-HO(C_9H_5N)	179.61	nd (al-w)	...	276-80	...	...	...	B21[4], 1084
12781	Quinoline, 7-chloro-2-methyl or 7-Chloro quinaldine 2-CH_3-7-Cl(C_9H_5N)	177.63	nd (eth), cr (lig)	87$^{0.5}$	75-6	...	...	...	B20[4], 3462
12782	Quinoline, 8-chloro 8-Cl(C_9H_6N)	163.61	...	288-9	fr−20	1.2834$^{14/4}$	1.6408$^{14.3}$	al, eth, ace, bz, chl	B20[4], 3377
12783	Quinoline, 8-chloro-5-methyl 5-CH_3-8-Cl(C_9H_5N)	177.63	nd (w)	...	49	...	...	al, eth, bz	B20[2], 246
12784	Quinoline, 8-(chloromethyl) 8-$ClCH_2$(C_9H_6N)	177.63	nd (w)	...	49	...	...	al, eth, bz	B20[4], 3502
12785	Quinoline, decahydro (*cis*) $C_9H_{17}N$	139.24	...	205^{735}, 90^{20}	−40	0.9426$^{20/4}$	1.4926^{20}	al, eth	B20[4], 2017
12786	Quinoline, decahydro (*trans,d*) $C_9H_{17}N$	139.24	$[\alpha]^{25/}{}_D$ +4.8 (al, c=3)	200-2	75	...	...	al, eth, ace, bz	B20[4], 2017
12787	Quinoline, decahydro (*trans,dl*) $C_9H_{17}N$	139.24	pr (lig, sub)	203^{735} sub	48	0.9610^{22}	1.4692^{56}	al, eth, ace, bz	B20[4], 2017
12788	Quinoline, decahydro (*trans,l*) $C_9H_{17}N$	139.24	$[\alpha]^{25/}{}_D$ −4.5 (al, c=3)	200-1	74-5	...	...	...	B20[4], 2017
12789	Quinoline, 5,7-dibromo-8-hydroxy 5,7-Br_2-8-HO(C_9H_4N)	302.95	nd (al)	sub	196	...	...	al, ace, bz, aa, chl	B21[4], 1180
12790	Quinoline, 6,8-dibromo-2-hydroxy or 6,8-Dibromo carbostyril 6,8-Br_2-2-HO(C_9H_4N)	302.95	nd (dil al)	...	230	...	...	al	B21[4], 1064
12791	Quinoline, 2,3-dichloro 2,3-Cl_2(C_9H_5N)	198.05	(dil al)	...	104-5	...	...	al, eth, bz	B20[4], 3382
12792	Quinoline, 2,4-dichloro 2,4-Cl_2(C_9H_5N)	198.05	nd (dil al)	280-2	67-8	...	...	al, eth, bz, chl	B20[4], 3382

No.	Name, Synonyms, and Formula	Mol. wt.	Color, crystalline form, specific rotation and λ_{max} (log ε)	b.p. °C	m.p. °C	Density	n_D	Solubility	Ref.
12793	Quinoline, 2,6-dichloro 2,6-$Cl_2(C_9H_5N)$	198.05	nd (eth)		156 (161)			al, eth	B20^4, 3383
12794	Quinoline, 2,7-dichloro 2,7-$Cl_2(C_9H_5N)$	198.05	nd (al)	sub 100^2	120			al, eth	B20^4, 3383
12795	Quinoline, 3,4-dichloro 3,4-$Cl_2(C_9H_5N)$	198.05	peth		69-70			al, eth	B20^4, 3383
12796	Quinoline, 4,5-dichloro 4,5-$Cl_2(C_9H_5N)$	198.05		134$^{5.5}$	118			al, eth	B20^4, 3383
12797	Quinoline, 4,6-dichloro 4,6-$Cl_2(C_9H_5N)$	198.05	(peth)		104			al, eth	B20^4, 3383
12798	Quinoline, 4,7-dichloro 4,7-$Cl_2(C_9H_5N)$	198.05	cr (MeOH), nd (80% al)	148^{10}	93				B20^4, 3384
12799	Quinoline, 4,8-dichloro 4,8-$Cl_2(C_9H_5N)$	198.05			155-6			al, eth	B20^4, 3385
12800	Quinoline, 5,6-dichloro 5,6-$Cl_2(C_9H_5N)$	198.05	nd (al)		85			al, eth, peth	B20, 361
12801	Quinoline, 5,7-dichloro 5,7-$Cl_2(C_9H_5N)$	198.05	nd (al)		117			al, eth	B20^4, 3385
12802	Quinoline, 5,7-dichloro-8-hydroxy 5,7-Cl_2-8-$HO(C_9H_4N)$	214.05	nd (al)		183			bz, peth	B21^4, 1180
12803	Quinoline, 5,8-dichloro 5,8-$Cl_2(C_9H_5N)$	198.05	nd (al), pl (eth)	sub	97-8			al, eth	B20^4, 3385
12804	Quinoline, 6,8-dichloro 6,8-$Cl_2(C_9H_5N)$	198.05	nd (al)		104-5			eth	B20^4, 3385
12805	Quinoline, 7,8-dichloro 7,8-$Cl_2(C_9H_5N)$	198.05	nd		85.5			al, eth	Prak 48, Prak 279
12806	Quinoline, 1,2-dihydro-1-methyl-2-oxo or N-Methyl carbostyril $C_{10}H_9NO$	159.19	nd (lig)	324^{728}	74			al, eth, ace, bz, chl	B21^4, 3737
12807	Quinoline, 2,4-dihydroxy 2,4-$(HO)_2(C_9H_5N)$	161.16			355				B21^4, 2219
12808	Quinoline, 5,7-diido-8-hydroxy or Diidoquin 5,7-I_2-8-$HO(C_9H_4N)$	396.95	yesh nd (aa, xyl)		210d			al	B21^4, 1191
12809	Quinoline, 5,8-diido-6-hydroxy 5,8-I_2-6-$HO(C_9H_4N)$	396.95	yesh		191			al, eth	B21^2, 54
12810	Quinoline, 2,3-dimethyl or 3-Methyl quinaldine 2,3-$(CH_3)_2(C_9H_5N)$	157.22	ye rh (eth)	261^{729}	68-9	1.1013		al, eth, lig	B20^4, 3521
12811	Quinoline, 2,4-dimethyl or 4-Methyl quinaldine 2,4-$(CH_3)_2(C_9H_5N)$	157.22	rh pr (eth)	264-6, 143^{15}		1.0611^{15}	1.6075^{20}	al, eth	B20^4, 3522
12812	Quinoline, 2,4-dimethyl-6-hydroxy 2,4-$(CH_3)_2$-6-$HO(C_9H_4N)$	173.21	pr or pl (al)	360d	214			al, ace	B21^2, 68
12813	Quinoline, 2,4-dimethyl-7-hydroxy 2,4-$(CH_3)_2$-7-$HO(C_9H_4N)$	173.21	nd (al)		218			al	B2^1, 116
12814	Quinoline, 2,4-dimethyl-8-hydroxy 2,4-$(CH_3)_2$-8-$HO(C_9H_4N)$	173.21	pr (eth)	281 sub	65			al, eth, ace, bz, chl	B21^4, 1292
12815	Quinoline,2,3-dimethyl-4-hydroxy 2,3-$(CH_3)_2$-4-$HO(C_9H_4N)$	173.21	pr (w + 1)	sub	319-20				B21^4, 1293
12816	Quinoline, 2,6-dimethyl or 6-Methyl quinaldine 2,6-$(CH_3)_2(C_9H_5N)$	157.22	rh pr (eth)	266-7, 152-5^{13}	60			bz	B20^4, 3525
12817	Quinoline, 2,6-dimethyl-4-hydroxy 2,6-$(CH_3)_2$-4-$HO(C_9H_4N)$	173.21	nd (w + 1)		279			w	B21^4, 1294
12818	Quinoline, 2,7-dimethyl 2,7-$(CH_3)_2(C_9H_5N)$	157.22		264-5, 115-6^7	61			al, eth, chl	B20^4, 3528
12819	Quinoline, 2,8-dimethyl or *o*-Toluquinaldine 2,8-$(CH_3)_2(C_9H_5N)$	157.22		255.3, 103-4^5	27	1.0394$^{20/4}$	1.6022^{20}	al, eth	B20^4, 3529
12820	Quinoline, 2,8-dimethyl-4-hydroxy 2,8-$(CH_3)_2$-4-$HO(C_9H_4N)$	173.21	lf or pl (w + 1)	sub	260-1			al	B21^4, 1296
12821	Quinoline, 3,4-dimethyl or 3-Methyl lepidine 3,4-$(CH_3)_2(C_9H_5N)$	157.22	cr (eth)	290^{737}	73-4			al, eth	B20^4, 3530
12822	Quinoline, 4,6-dimethyl-2-hydroxy 4,6-$(CH_3)_2$-2-$HO(C_9H_4N)$	173.21	pr (al)		249-50				B21^4, 1301
12823	Quinoline, 4,7-dimethyl-2-hydroxy 4,7-$(CH_3)_2$-2-$HO(C_9H_4N)$	173.21	cr (aa)		220			al, aa	B21^4, 1303
12824	Quinoline, 4,8-dimethyl-2-hydroxy 4,8-$(CH_3)_2$-2-$HO(C_9H_4N)$	173.21	pl (aq aa)		217-8			al	B21^4, 1304
12825	Quinoline, 5,8-dimethyl 5,8-$(CH_3)_2(C_9H_5N)$	157.22	nd	265^{736}	4-5	1.070^{21}		al, eth	B20^4, 3538

No.	Name, Synonyms, and Formula	Mol. wt.	Color, crystalline form, specific rotation and λ_{max} (log ε)	b.p. °C	m.p. °C	Density	n_D	Solubility	Ref.
12826	Quinoline, 6,8-dimethyl or β-Cytisolidine $6,8\text{-}(CH_3)_2(C_9H_5N)$	157.22	...	268-9, $133\text{-}4^{14}$	...	1.0665^{4}	...	al, eth	$B20^{4}$, 3539
12827	Quinoline, 6,8-dimethyl-2-hydroxy $6,8\text{-}(CH_3)_2\text{-}2\text{-}HO(C_9H_4N)$	173.21	nd (al)	...	201-2	...	...	al	$B21^{1}$, 225
12828	Quinoline, 6,8-dimethyl-5-hydroxy or Cytisoline $6,8\text{-}(CH_3)_2\text{-}5\text{-}HO(C_9H_4N)$	173.21	pl (chl), cr (al)	sub	197-8	...	...	al, bz, chl	B21, 117
12829	Quinoline, 6-(dimethylamino)-2-methyl or 6-Dimethyl amino quinaldine $2\text{-}CH_3\text{-}6\text{-}[(CH_3)_2N](C_9H_5N)$	186.26	ye pr (aa, AcOEt)	319	101	...	...	al, eth, bz	$B22^{2}$, 361
12830	Quinoline, 3-ethyl-2-hydroxy or 3-Ethyl carbostyril $3\text{-}C_2H_5\text{-}2\text{-}HO(C_9H_5N)$	173.21	(dil HCl)	...	168	...	...	bz, chl	B21, 115
12831	Quinoline, 2-hydrazino $2\text{-}H_2NNH(C_9H_6N)$	159.19	(bz)	...	142-3	...	...	al	$B22^{1}$, 690
12832	Quinoline, 5-hydrazino $5\text{-}H_2NNH(C_9H_6N)$	159.19	ye nd (w)	...	150-1	...	...	al	B22, 565
12833	Quinoline hydrochloride or Quinilirium chloride $C_9H_7N.HCl$	165.62	pr (w)	...	134.5	...	...	w, al, chl	$B20^{2}$, 226
12834	Quinoline hydrogen sulfate $C_9H_7N.H_2SO_4$	227.23	cr (al, aa)	...	164	...	...	w, aa	$B20^{2}$, 226
12835	Quinoline, 2-hydroxy or 2-Quinolinol $2\text{-}HO(C_9H_6N)$	145.16	pr (al, dil al + 1w), nd (sub)	sub	199-200 (anh)	...	...	al, eth	$B21^{4}$, 1057
12836	Quinoline, 2-hydroxy-3-methyl or 3-Methyl carbostyril $3\text{-}CH_3\text{-}2\text{-}HO(C_9H_5N)$	159.19	yesh nd (ace, dil al)	sub	234-5	...	...	al	$B21^{4}$, 1246
12837	Quinoline, 2-hydroxy-4-methyl or 2-Hydroxy lepidine $4\text{-}CH_3\text{-}2\text{-}HO(C_9H_5N)$	159.19	nd (w)	>360, 270^{17}	245	...	...	al	$B21^{4}$, 1252
12838	Quinoline, 2-hydroxy-6-methyl or 6-Methyl carbostyril $6\text{-}CH_3\text{-}2\text{-}HO(C_9H_5N)$	159.19	nd (al)	$240\text{-}1^{12}$	237	...	...	al, eth, ace, bz	$B21^{4}$, 1271
12839	Quinoline, 3-hydroxy or 3-Quinolinol $3\text{-}HO(C_9H_6N)$	145.16	cr (bz, dil al)	...	200-1	...	...	al, bz	$B21^{4}$, 1075
12840	Quinoline, 3-hydroxy-2-methyl or 3-Hydroxy quinaldine $2\text{-}CH_3\text{-}3\text{-}HO(C_9H_5N)$	159.19	nd (al)	...	203-5	...	...	al, eth, chl	$B21^{4}$, 1216
12841	Quinoline, 4-hydroxy or 4-Quinolinol $4\text{-}HO(C_9H_6N)$	145.16	nd (w + 3)	...	210 (anh), 100 (hyd)	...	...	al	$B21^{4}$, 1079
12842	Quinoline, 4-hydroxy-2-methyl or 4-Hydroxy quinaldine $2\text{-}CH_3\text{-}4\text{-}HO(C_9H_5N)$	159.19	pr (w + 2)	300d	232	...	...	al	$B21^{4}$, 1218
12843	Quinoline, 4-hydroxy-2-phenyl $2\text{-}C_6H_5\text{-}4\text{-}HO(C_9H_5N)$	221.26	pl or pr (al), nd (aa)	...	256-7	...	...	al	$B21^{4}$, 1625
12844	Quinoline, 5-hydroxy or 5-Quinolinol $5\text{-}HO(C_9H_6N)$	145.16	nd (al), pl	sub	224d	...	...	...	$B21^{4}$, 1103
12845	Quinoline, 5-hydroxy-2-methyl or 5-Hydroxy quinaldine $2\text{-}CH_3\text{-}5\text{-}HO(C_9H_5N)$	159.19	pl (al)	...	246-7	...	...	eth	$B21^{2}$, 63
12846	Quinoline, 5-hydroxy-6-methyl $6\text{-}CH_3\text{-}5\text{-}HO(C_9H_5N)$	159.19	nd (al, sub)	sub	230	...	...	al, eth, ace	B21, 111
12847	Quinoline, 5-hydroxy-8-methyl $8\text{-}CH_3\text{-}5\text{-}HO(C_9H_5N)$	159.19	nd (dil al, sub)	sub	762-3	...	...	...	B21, 112
12848	Quinoline, 6-hydroxy or 6-Quinolinol $6\text{-}HO(C_9H_6N)$	145.16	pr (al, eth)	>360	193	...	...	...	$B21^{2}$, 53
12849	Quinoline, 6-hydroxy-2-methyl or 6-Hydroxy quinaldine $2\text{-}CH_3\text{-}6\text{-}HO(C_9H_5N)$	159.19	(w)	304-5, 186^{35}	213	1.1665^{0}	...	al, eth	B21, 106
12850	Quinoline, 6-hydroxy-4-methyl or 6-Hydroxy lepidine $4\text{-}CH_3\text{-}6\text{-}HO(C_9H_5N)$	159.19	nd (w, dil al)	...	222-4	...	...	al, ace, chl	$B21^{4}$, 1265
12851	Quinoline, 6-hydroxy-8-methyl $8\text{-}CH_3\text{-}6\text{-}HO(C_9H_5N)$	159.19	nd (dil al)	...	200	...	...	al	B21, 113
12852	Quinoline, 7-hydroxy or 7-Quinolinol $7\text{-}HO(C_9H_6N)$	145.16	pr (al), nd (dil al-eth)	sub	238-40	...	...	al	$B21^{4}$, 1130
12853	Quinoline, 7-hydroxy-6-methyl $6\text{-}CH_3\text{-}7\text{-}HO(C_9H_5N)$	159.19	nd (al)	240^{22} sub	244	...	...	al, eth, bz	B21, 111
12854	Quinoline, 8-hydroxy or 8-Quinolinol $8\text{-}HO(C_9H_6N)$	145.16	nd (dil al)	266.6^{752} sub	75-6	1.034^{209}	...	al, ace, bz, chl	$B21^{4}$, 1135
12855	Quinoline, 8-hydroxy-2-methyl or 8-Hydroxy quinaldine $2\text{-}CH_3\text{-}8\text{-}HO(C_9H_5N)$	159.19	pr (dil al)	267, $145\text{-}60^{22}$	74-5	...	...	eth, bz	$B21^{4}$, 1232
12856	Quinoline, 8-hydroxy-4-methyl or 8-Hydroxy lepidine $4\text{-}CH_3\text{-}8\text{-}HO(C_9H_5N)$	159.19	nd (lig)	...	141	...	...	ace, bz, chl, aa	$B21^{4}$, 1266

No.	Name, Synonyms, and Formula	Mol. wt.	Color, crystalline form, specific rotation and λ_{max} (log ε)	b.p. °C	m.p. °C	Density	n_D	Solubility	Ref.
12857	Quinoline, 8-hydroxy-5-methyl 5-CH_3-8-HO(C_9H_5N)	159.19	nd (dil al)	sub 100	122-4				B21[4], 1269
12858	Quinoline, 8-hydroxy-6-methyl 6-CH_3-8-HO(C_9H_5N)	159.19	nd (chl, bz)	sub	95-6			al	B21[4], 1274
12859	Quinoline, 8-hydroxy-7-methyl 7-CH_3-8-HO(C_9H_5N)	159.19	nd (dil al)	sub 100	72-4			al	B21[4], 1279
12860	Quinoline, 8-hydroxy-5-nitroso 5-NO-8-HO(C_9H_5N)	174.16	nd (al)		245d				B21[1], 405
12861	Quinoline, 8-hydroxy, sulfate or Chinosol $(8\text{-}HOC_9H_6N)_2 \cdot H_2SO_4$	388.39			177.5			w, al	B21, 92
12862	Quinoline, 2-iodo 2-I(C_9H_6N)	255.06	nd (dil al)		52-3			al, eth, ace	B20[4], 3393
12863	Quinoline, 4-iodo 4-I(C_9H_6N)	255.06	nd or pr	sub	97 (100)			al, eth	B20, 370
12864	Quinoline, 5-iodo 5-I(C_9H_6N)	255.06	nd (al, eth)	sub	101-2				B20[1], 141
12865	Quinoline, 6-iodo 6-I(C_9H_6N)	255.06	lf (w), nd (sub)	sub	91			al, eth	B20, 370
12866	Quinoline, 8-iodo 8-I(C_9H_6N)	255.06	lo nd (al)		36			al, eth, ace, bz, lig	B20[1], 141
12867	Quinoline, 2-methoxy-6-nitro 6-NO_2-2-CH_3O(C_9H_5N)	204.19	nd (dil aa, bz, sub)		189-90			al, eth, bz, chl, lig	B21, 81
12868	Quinoline, 2-methoxy-8-nitro 8-NO_2-2-CH_3O(C_9H_5N)	204.19	nd (dil al)		124-5			al, bz	B21, 82
12869	Quinoline, 4-methoxy 4-CH_3O(C_9H_6N)	159.19		245, 167^{20}	41			al, eth, bz	B21[4], 1080
12870	Quinoline, 5-methoxy 5-CH_3O(C_9H_6N)	159.19		282^{750}				al, eth	B21[4], 1103
12871	Quinoline, 6-methoxy or *p*-Quinanisole 6-CH_3O(C_9H_6N)	159.19	hyg lf	305^{740}, 153^{12}	26.5	$1.154^{20/20}$, 1.000^{209}		al, eth	B21[4], 1108
12872	Quinoline, 6-methoxy-5-nitro 5-NO_2-6-CH_3O(C_9H_5N)	204.19	cr (al)		104-5			ace	B21[4], 1121
12873	Quinoline, 6-methoxy-8-nitro 8-NO_2-6-CH_3O(C_9H_5N)	204.19	yesh nd (al)		159-60			chl	B21[4], 1121
12874	Quinoline, 6-methoxy-1,2,3,4-tetrahydro or Thalline 6-CH_3O($C_9H_{10}N$)	163.22	pr (peth, al), rh pym (w)	283^{735}, $127\text{-}30^{1}$	42-3		1.5718^{20}	al, eth, bz	B22, 61
12875	Quinoline, 8-methoxy or *o*-Quinanisole 8-CH_3O(C_9H_6N)	159.19	nd (peth)	282^{750}, 164^{14}	49-50	1.034^{29}		al, eth, bz, peth	B21[4], 1154
12876	Quinoline, 8-methoxy-5-nitro 5-NO_2-8-CH_3O(C_9H_5N)	204.19	cr (al)		151.5			al	B21[4], 1193
12877	Quinoline, 1-methyl-1,2,3,4-tetrahydro or Kairoline ($C_9H_{10}N$)CH_3	147.22		247-50, $123\text{-}6^{14}$		$1.022^{20/4}$	1.5802^{23}	al	B20[2], 174
12878	Quinoline, 2-methyl or Quinaldine 2-CH_3(C_9H_6N)	143.19		247.6, 118^{10}	−2	$1.0585^{20/4}$	1.8116^{20}	al, eth, ace, chl	B20[4], 3454
12879	Quinoline, 2-methyl-5-nitro or 5-Nitro quinaldine 5-NO_2-2-CH_3(C_9H_5N)	188.19	nd (dil al)		82			al, eth	B20[4], 3468
12880	Quinoline, 2-methyl-6-nitro or 6-Nitro quinaldine 6-NO_2-2-CH_3(C_9H_5N)	188.19	cr (80% MeOH)		165 (173)			al	B20[4], 3468
12881	Quinoline, 2-methyl-8-nitro or 8-Nitro quinaldine 8-NO_2-2-CH_3(C_9H_5N)	188.19	pa ye nd (dil al)		137			al, eth, bz	B20[4], 3468
12882	Quinoline, 3-methyl or β-Methyl quinoline 3-CH_3(C_9H_6N)	143.19	pr	259.8, $140\text{-}2^{25}$	16-7	$1.0673^{20/4}$	1.6171^{20}	al, eth, ace	B20[4], 3472
12883	Quinoline, 4-methyl or Lepidine 4-CH_3(C_9H_6N)	143.19	red br	264.2, 133^{15}	9-10	$1.0862^{20/4}$	1.6206^{20}	al, eth, ace, bz, lig	B20[4], 3477
12884	Quinoline, 4-methyl-3-nitro or 3-Nitro lepidine 3-NO_2-4-CH_3(C_9H_5N)	188.19	pr (w)		118			al	B20[4], 3485
12885	Quinoline, 4-methyl-8-nitro 8-NO_2-4-CH_3(C_9H_5N)	188.19	lf (abs al)		126-7			eth	B20[4], 3485
12886	Quinoline, 5-methyl 5-CH_3(C_9H_6N)	143.19		262.7	19	$1.0832^{20/4}$	1.6219^{20}	al, eth, ace	B20[4], 3488
12887	Quinoline, 6-methyl or *p*-Toluquinoline 6-CH_3(C_9H_6N)	143.19		258.6, 130^{15}	ca -22	$1.0654^{20/4}$	1.6157^{20}	al, eth, ace	B20[4], 3489
12888	Quinoline, 6-methyl-5-nitro 5-NO_2-6-CH_3(C_9H_5N)	188.19	pa ye nd (al)		116-7			al, eth, ace, bz	B20[4], 3495
12889	Quinoline, 6-methyl-8-nitro 8-NO_2-6-CH_3(C_9H_5N)	188.19	pa ye nd (w)		122			al, eth, ace, bz	B20, 400

PHYSICAL CONSTANTS OF ORGANIC COMPOUNDS (Continued)

No.	Name, Synonyms, and Formula	Mol. wt.	Color, crystalline form, specific rotation and λ_{max} (log ε)	b.p. °C	m.p. °C	Density	n_D	Solubility	Ref.
12890	Quinoline, 7-methyl or *m*-Toluquinoline 7-$CH_3(C_9H_6N)$	143.19	ye	257.6, 144^{18}	39	$1.0609^{20/4}$	1.6150^{20}	al, eth, ace	B20^4, 3497
12891	Quinoline, 8-methyl or *o*-Toluquinoline 8-$CH_3(C_9H_6N)$	143.19		247^8, 143^{34}		$1.0719^{20/4}$	1.6164^{20}	al, eth, ace	B20^4, 3500
12892	Quinoline, 8-methyl-5-nitro 5-NO_2-8-$CH_3(C_9H_5N)$	188.19	pa ye nd (al)		93			al, eth, ace, bz	B20^4, 3504
12893	Quinoline, 8-methyl-6-nitro 6-NO_2-8-$CH_3(C_9H_5N)$	188.19	cr (al)		129			al	B20^4, 3504
12894	Quinoline, 3-nitro 3-$NO_2(C_9H_6N)$	174.16	nd (dil al)		128			al, ace	B20^4, 3395
12895	Quinoline, 4-nitro, oxide 4-$NO_2(C_9H_6NO)$	190.16	ye nd pl (ace)		154				B20^4, 3396
12896	Quinoline, 5-nitro 5-$NO_2(C_9H_6N)$	174.16	pl (w or al), nd (+w, w)	sub	73-5 (sub)			bz	B20^4, 3397
12897	Quinoline, 6-nitro 6-$NO_2(C_9H_6N)$	174.16	ye pl (HCl-aa)	sub	153-4			bz	B20^4, 3397
12898	Quinoline, 7-nitro 7-$NO_2(C_9H_6N)$	174.16	nd or lf, w al pl (sub)	sub	132-3			eth, chl	B20^4, 3398
12899	Quinoline, 8-nitro 8-$NO_2(C_9H_6N)$	174.16	mcl pr (al)		91-2			al, eth	B20^4, 3399
12900	Quinoline, 2-oxo-1,2,3,4-tetrahydro C_9H_9NO	147.18	pr (al, eth)	201^{45}	163-4			al, eth	B21^4, 3638
12901	Quinoline, 2-phenyl 2-$C_6H_5(C_9H_6N)$	205.26	nd (dil al)	363, 310^{187}	86			al, eth, ace, bz	B20^4, 4137
12902	Quinoline, 3-phenyl 3-$C_6H_5(C_9H_6N)$	205.26	pl (eth)	$205\text{-}7^{12}$	52			al, eth, ace, bz, chl	B20^4, 4145
12903	Quinoline, 4-phenyl 4-$C_6H_5(C_9H_6N)$	205.26		260^{77}	61-2			al, ace, bz, chl	B20^4, 4149
12904	Quinoline, 5-phenyl 5-$C_6H_5(C_9H_6N)$	205.26	nd (dil al)		83			al, eth, ace	B20^4, 4150
12905	Quinoline, 6-phenyl 6-$C_6H_5(C_9H_6N)$	205.26	pl (al, bz)	260^{77}	110	1.945^{20}		al, ace, bz, chl	B20^4, 4151
12906	Quinoline, 8-phenyl 8-$C_6H_5(C_9H_6N)$	205.26	ye gr oil	283^{187}				al, ace, eth, bz, chl	B20^4, 4153
12907	Quinoline, 1,2,3,4-tetrahydro $C_9H_{11}N$	133.19	nd	251	20	$1.0588^{20/4}$	1.6062^{19}	al, eth	B20^4, 2923
12908	Quinoline, 5,6,7,8-tetrahydro $C_9H_{11}N$	133.19		222, $92\text{-}5^{12}$		1.0304^{13}	1.5435^{20}	al, eth, ace, bz	B20^4, 2922
12909	Quinoline, 2,3,4-trimethyl or 3,4-Dimethyl quinaldine 2,3,4-$(CH_3)_3(C_9H_4N)$	171.24		285, $156\text{-}8^{12}$	92				B20^4, 3557
12910	Quinoline, 2,3,8-trimethyl 2,3,8-$(CH_3)_3(C_9H_4N)$	171.24		285	86-7			al, eth, chl	B20^4, 3557
12911	Quinoline, 2,4,6-trimethyl or 4,6-Dimethyl quinaldine 2,4,6-$(CH_3)_3(C_9H_4N)$	171.24	nd (w, dil al + 1w)	281-2, $146\text{-}8^{13}$	65.5 (anh), 40 (+w)			al, eth, ace, bz, chl	B20^4, 3558
12912	Quinoline, 2,4,7-trimethyl or 4,7-Dimethyl quinaldine 2,4,7-$(CH_3)_3(C_9H_4N)$	171.24	nd (w)	200-1	63-4	$1.0337^{20/4}$	1.5973^{24}	al, eth, ace, bz	B20^4, 3558
12913	Quinoline, 2,4,8-trimethyl or 4,8-Dimethyl quinaldine 2,4,8-$(CH_3)_3(C_9H_4N)$	171.24		287	50-1		1.5855^{50}		B20^4, 3559
12914	Quinoline, 2,5,6-trimethyl 2,5,6$(CH_3)_3(C_9H_4N)$	171.24	nd		69-70			al, eth, bz	B20, 415
12915	Quinoline, 2,5,7-trimethyl or Tetracoline 2,5,7-$(CH_3)_3(C_9H_4N)$	171.24	pr	286.6^{746}, $107\text{-}8^{719}$				al, eth, bz, peth	B20^4, 3559
12916	Quinoline, 2,6,8-trimethyl 2,6,8-$(CH_3)_3(C_9H_4N)$	171.24	pr (peth), lf (dil al)	260^{719}	46			al, eth, bz, peth	B20^4, 3560
12917	Quinoline, 4,5,8-trimethyl 4,5,8-$(CH_3)_3(C_9H_4N)$	171.24		155^{13}	73-4			al, eth, bz	B20^4, 3562
12918	2-Quinoline carbonitrite 2-$NC(C_9H_6N)$	154.17	nd (lig, chl)	$100\text{-}70^{20\text{-}23}$	94			al, eth, bz, chl, lig	B22^4, 1157
12919	2-Quinoline carboximide 2-$H_2NCO(C_9H_6N)$	172.19	nd (dil al, bz-lig)		133			al, bz, chl	B22^4, 1154
12920	2-Quinoline carboxylic acid or Quinaldinic acid 2-$HO_2C(C_9H_6N)$	173.17	nd (+3w), (bz)		157 (anh)			w, bz	B22^4, 1149
12921	2-Quinoline carboxylic acid, 4,8-dihydroxy or Xanthurenic acid 2-HO_2C-4,8-$(HO)_2(C_9H_4N)$	205.17	ye micr cr (w)		289 (297)			al	B22^4, 2513

No.	Name, Synonyms, and Formula	Mol. wt.	Color, crystalline form, specific rotation and λ_{max} (log ε)	b.p. °C	m.p. °C	Density	n_D	Solubility	Ref.
12922	2-Quinoline carboxylic acid, 4-hydroxy or Kynurenic acid $4\text{-HO-2-HO}_2\text{C}(\text{C}_9\text{H}_5\text{N})$	189.17	ye nd (+w dil aa)		282-3				B22[4], 2245
12923	2-Quinoline carboxylic acid, methyl ester $2\text{-CH}_3\text{O}_2\text{C}(\text{C}_9\text{H}_6\text{N})$	187.20	nd (lig)		86			al, lig	B22[4], 1152
12924	2-Quinoline carboxylyl chloride $2\text{-ClCO}(\text{C}_9\text{H}_6\text{N})$	191.62	nd (eth, lig)		(i)97-8, (ii)175-6			eth, bz	B22[4], 509
12925	3-Quinoline carbonitrile or 3-Cyanoquinoline $3\text{-NC}(\text{C}_9\text{H}_6\text{N})$	154.17	(al or sub)		108			eth, ace, bz	B22[4], 1171
12926	3-Quinoline carboxylic acid $3\text{-HO}_2\text{C}(\text{C}_9\text{H}_6\text{N})$	173.17	pl (al, dil al)		275d			al	B22[4], 1167
12927	3-Quinoline carboxylic acid, 2-phenyl $2\text{-C}_6\text{H}_5\text{-3-HO}_2\text{C}(\text{C}_9\text{H}_5\text{N})$	249.27	nd (al)		230d			al, ace, aa	B22[4], 1355
12928	4-Quinoline carbonitrile or Cinchoninonitrile $4\text{-NC}(\text{C}_9\text{H}_6\text{N})$	154.17	cr (chl, lig, eth) nd(sub)	240-5	103-4			al, eth, ace, bz	B22[4], 1181
12929	4-Quinoline carboxylic acid or Cinchoninic acid $4\text{-HO}_2\text{C}(\text{C}_9\text{H}_6\text{N})$	173.17	mcl pr (w), nd (+1w), mcl or tcl (+2w)		257-8 (anh)				B22[4], 1177
12930	4-Quinoline carboxylic acid, 2-(3-carboxy-4-hydroxy phenyl) or Hexophan $\text{C}_{17}\text{H}_{11}\text{NO}_5$	309.28	yesh pw		283-4d			chl	B22[2], 206
12931	4-Quinoline carboxylic acid, 6-methoxy or Quininic acid $6\text{-CH}_3\text{O-4-HO}_2\text{C}(\text{C}_9\text{H}_5\text{N})$	203.20	pa ye pr (dil al)	sub	285d				B22[4], 2297
12932	4-Quinoline carboxylic acid, 8-methoxy-2-phenyl or Isatophan $8\text{-CH}_3\text{O-2-C}_6\text{H}_5\text{-4-HO}_2\text{C}(\text{C}_9\text{H}_4\text{N})$	279.30	ye nd (al)		216			al, chl	B22[1], 559
12933	4-Quinoline carboxylic acid, 6-methyl-2-phenyl,ethyl ester or Novatophan $6\text{-CH}_3\text{-2-C}_6\text{H}_5\text{-4-}(\text{C}_2\text{H}_5\text{O}_2\text{C})(\text{C}_9\text{H}_4\text{N})$	291.35	ye cr (al)		75-6			eth, ace, bz, chl	B22[4], 1395
12934	4-Quinoline carboxylic acid, 2-phenyl or Cinchophene $2\text{-C}_6\text{H}_5\text{-4-HO}_2\text{C}(\text{C}_9\text{H}_5\text{N})$	249.27	nd (MeOH, dil al), ye in air		218			eth	B22[4], 1358
12935	4-Quinoline carboxylic acid, 2-phenyl,allyl ester or Atoquinol $2\text{-C}_6\text{H}_5\text{-4-}(\text{C}_3\text{H}_5\text{O}_2\text{C})(\text{C}_9\text{H}_5\text{N})$	289.33	nd (dil al)	265^{15}	30			al, eth, ace, bz	B22[4], 1361
12936	5-Quinoline carbonitrile $5\text{-NC}(\text{C}_9\text{H}_6\text{N})$	154.17	nd (lig), nd(+3/2 w, dil al)		89, 70 (+w)			al, bz	B22[4], 1195
12937	5-Quinoline carboxylic acid $5\text{-HO}_2\text{C}(\text{C}_9\text{H}_6\text{N})$	173.17	cr (aa, sub)	sub<338	342				B22[4], 1195
12938	6-Quinoline carboxylic acid $6\text{-HO}_2\text{C}(\text{C}_9\text{H}_6\text{N})$	173.17	nd, pr or pl (sub)	sub<290	291-2				B22[4], 1196
12939	7-Quinoline carboxylic acid $7\text{-HO}_2\text{C}(\text{C}_9\text{H}_6\text{N})$	173.17	nd (w al)	sub	249-50			al	B22[4], 1199
12940	8-Quinoline carbonitrile or 8-Cyano quinoline $8\text{-NC}(\text{C}_9\text{H}_6\text{N})$	154.17	nd (dil al)		84			al	B22[4], 1203
12941	8-Quinoline carboxylic acid $8\text{-HO}_2\text{C}(\text{C}_9\text{H}_6\text{N})$	173.17	nd (w)	sub	187			al	B22[4], 1200
12942	5-Quinoline sulfonic acid, 8-hydroxy-7-iodo $\text{C}_9\text{H}_6\text{NO}_4\text{IS}$	351.12			260d				B22[4], 3497
12943	5-Quinoline sulfonic acid, 6-hydroxy $6\text{-HO-5-}(\text{HO}_3\text{S})(\text{C}_9\text{H}_5\text{N})$	225.22	ye nd (+½ w, w, al)		270d			w	B22, 407
12944	5-Quinoline sulfonic acid, 8-hydroxy $8\text{-HO-5-HO}_3\text{S}(\text{C}_9\text{H}_5\text{N})$	225.22	ye lf, nd (+1w, dil HCl)		322-3				B22[4], 3493
12945	5-Quinoline sulfonic acid, 8-hydroxy-7-iodo or Loretin $8\text{-HO-7-I-5-HO}_3\text{S}(\text{C}_9\text{H}_4\text{N})$	351.12	ye pr or lf (al)		260d				B22[4], 3497
12946	7-Quinoline sulfonic acid, 8-hydroxy $8\text{-HO-7-HO}_3\text{S}(\text{C}_9\text{H}_5\text{N})$	225.22	ye nd (al)		314-5			w, al	B22, 408
12947	8-Quinoline sulfonic acid, 5-hydroxy-6-iodo or Lorenite $5\text{-HO-6-I-8-HO}_3\text{S}(\text{C}_9\text{H}_4\text{N})$	351.12	ye nd or lf		210-30d				B22, 406
12948	2-Quinolinone, 1-methyl $\text{C}_{10}\text{H}_9\text{NO}$	159.19		324^{728}	74			al, eth, ace, bz, chl	B20[4], 3505
12949	Quinovic acid or Quinovaic acid $\text{C}_{30}\text{H}_{46}\text{O}_5$	486.69	pl or nd $[\alpha]^{16}{}_D$ +87 (aq KOH)		298d				B10[3], 2307

No.	Name, Synonyms, and Formula	Mol. wt.	Color, crystalline form, specific rotation and λ_{max} (log ε)	b.p. °C	m.p. °C	Density	n_D	Solubility	Ref.
12950	Quinoxaline or 1,4-Benxodiazine $C_8H_6N_2$	130.15	cr (peth)	229.5, $108\text{-}11^{12}$	28	$1.1334^{48/4}$	1.6231^{48}	w, al, **eth**, **ace**, bz	$B23^2$, 177
12951	Quinoxaline, 6-amino $6\text{-}H_2N(C_8H_5N_2)$	145.16	ye nd (eth)	sub	159			w, al, eth, chl	B25, 326
12952	Quinoxaline, 6-chloro $6\text{-}Cl(C_8H_5N_2)$	164.59	nd (w)	$117\text{-}9^{10}$ sub	64				$B23^2$, 177
12953	Quinoxaline, 2,3-dichloro $2,3\text{-}Cl_2(C_8H_4N_2)$	199.04	cr (al, bz)		151-3			al, bz, chl, aa	$B23^2$, 177
12954	Quinoxaline, 2,3-dihydroxy $2,3(HO)_2(C_8H_4N_2)$	162.15	nd (w)		410			w	$B24^2$, 200
12955	Quinoxaline, 2,3-dimethyl $2,3\text{-}(CH_3)_2(C_8H_4N_2)$	158.20	nd (w + 3) (ace)		106 (anh), 85 (+ w)			al, eth, ace, bz	$B23^2$, 197
12956	Quinoxaline, 2,6-dimethyl $2,6\text{-}(CH_3)_2(C_8H_4N_2)$	158.20		267-9	54			w, al, ace, bz	B23, 192
12957	Quinoxaline, 2-hydroxy or 2-Quinoxalinol $2\text{-}HO(C_8H_5N_2)$	146.15	lf (al)	sub $200^{0.5}$	271				$B24^2$, 72
12958	Quinoxaline, 6-methoxy $6\text{-}CH_3O(C_8H_5N_2)$	160.18	nd (w)	128^7 sub	57.5			eth, ace	B23, 387
12959	Quinoxaline, 2-methyl $2\text{-}CH_3(C_8H_5N_2)$	144.18	ye	245-7, 118^{16}	180-1			w, al, **eth**, **ace**, bz	$B23^2$, 190
12960	Quinoxaline, 6-methyl $6\text{-}CH_3(C_8H_5N_2)$	144.18		248^{748}, 141.5^{29}	218-9	$1.1164^{20/4}$	$1.6211^{18.4}$	**w**, **al**, **eth**, **ace**, **bz**	B23, 184
12961	Quinoxaline, 1,2,3,4-tetrahydro $C_8H_{10}N_2$	134.18	lf (w, eth, peth)	289, $153\text{-}4^{14}$	99			al, eth, bz, chl	$B23^2$, 106
12962	Quinoxaline, 5-hydroxy or 5-Quinoxalol $5\text{-}HO(C_8H_5N_2)$	146.15	(eth)	184^7, sub $90^{2.5}$	101-2				C48, 8232
12963	Quinuclidine or 1,4-Ethylene piperidine $C_7H_{13}N$	111.19	(eth)		158 (sealed tube)			w, al, eth, ace, bz	$B20^4$, 1966
12964	Quinuclidine, 3-hydroxy $C_7H_{13}NO$	127.19	cr (bz)		221-1			ace	$B21^4$, 237
12965	2-Quinuclidine carboxylic acid $C_8H_{13}NO_2$	155.20	col hyg cr (al-ace)		280d				$B22^4$, 200

No.	Name, Synonyms, and Formula	Mol. wt.	Color, crystalline form, specific rotation and λ_{max} (log ε)	b.p. °C	m.p. °C	Density	n_D	Solubility	Ref.
12966	Raffinose or Melitriose $C_{18}H_{32}O_{16}\cdot 5H_2O$	562.52	wh pw, pr or nd (+5w, dil al) $[\alpha]^{20}{}_D$ +101, +123	d 130	118-9 (anh), 80 (hyd)	1.465^{0}		w, MeOH, Py	B17^{4}, 3801
12967	Raunescine hydrate $C_{11}H_{16}H_2O_8\cdot H_2O$	582.65	wh hex pr (90% MeOH), $[\alpha]^{25}{}_D$ −74 (chl)		160-70			chl, aa	Am79, 250
12968	Reductone or 3-Hydroxy-2-oxo-propionaldehyde (enol-form) HOCH=C(OH)CHO	88.06	ye nd (w)	200-20d				w, al	B1^{4}, 4145
12969	Resazurin $C_{12}H_7NO_4$	229.19	dk red to grsh pr or pl (aa, AcOEt)	sub vac	d				B27, 128
12970	Rescinnamine or Moderil $C_{35}H_{42}N_2O_9$	634.73	nd (bz) $[\alpha]^{24}{}_D$ −98 (chl, c=0.1)		238-9			ace, chl	Am77, 2241
12971	Reserpic acid or Reserpinolic acid $C_{22}H_{28}N_2O_5$	400.47	(MeOH), $[\alpha]^{23}{}_D$ −81 (of hydro-chloride)		241-3				Am78, 2023
12972	Reserpine or Rivasin-Serparsin $C_{33}H_{40}N_2O_9$	608.69	lo pr (dil ace), $[\alpha]_D$ −117.7 (chl, c=1)		264-5, 277 (sealed tube)			bz, chl	H37, 67
12973	Reserpinine or Raubasinine $C_{22}H_{26}N_2O_4$	382.46	pa ye pl (aq ace), $[\alpha]^{20}{}_D$ −131 (chl, c=1.18)		243-4d			al, chl	C49, 11672
12974	Resorcinol or 1,3-Dihydroxy benzene 1,3-$(HO)_2C_6H_4$	110.11	nd (bz), pl (w)	178^{16}	111	1.2717		w, al, eth, aa	B6^{3}, 4292
12975	Resorcinol monoacetate 3-$(CH_3CO_2)C_6H_4OH$	152.18	ye	283d				al	B6^{3}, 4319
12976	Resorcinol, 4-acetyl or 2,4-Dihydroxy acetophenone. Resacetophenone 2,4-$(HO)_2C_6H_3COCH_3$	152.15	nd or lf		147	1.1800^{141}		aa, Py	B8^{3}, 2082
12977	Resorcinol, 5-acetyl or 3,5-Dihydroxy acetophenone 3,5-$(HO)_2C_6H_3COCH_3$	152.15	cr (w)		147-8			w, al, eth, ace	B8^{2}, 301
12978	Resorcinol benzoate (mono) 3-$C_6H_5CO_2C_6H_4OH$	214.22	pr (dil al)		135-6			chl, aa	B9^{3}, 555
12979	Resorcinol, 4-(2-aminoethyl), hydrochloride 4-$(H_2NCH_2CH_2)$-1,3-$(HO)_2C_6H_3$.HCl	189.64	nd (w)		237d			w, al	B13^{2}, 486
12980	Resorcinol acetate (mono) or 3-Hydroxy phenyl acetate 3-HO-$C_6H_4O_2CCH_3$	152.15	ye	283d				al	B6^{3}, 4319
12981	Resorcinol, monobenzyl ether 3-$C_6H_5CH_2OC_6H_4OH$	200.24	cr (CCl_4)	200^{5}	69.2			w	B6^{3}, 4314
12982	Resorcinol, 2-bromo or 1,3-Dihydroxy-2-bromo benzene 1,3-$(HO)_2$-2-BrC_6H_3	189.01	nd (chl), cr (peth)		102-3			w, al, eth, chl	B6^{3}, 4336
12983	Resorcinol, 4-bromo or 1,3-Dihydroxy-4-bromo benzene 1,3-$(HO)_2$-4-BrC_6H_3	189.01		$138\text{-}41^{12}$	103			w, eth	B6^{3}, 4336
12984	Resorcinol, 5-bromo or 1,3-Dihydroxy-5-bromo benzene 1,3-$(HO)_2$-5-BrC_6H_3	189.01	nd (bz), pr(+1)		87, 79 (+w)			w, eth, bz	B6^{2}, 820
12985	Resorcinol, 4-butyl 4-C_4H_9-1,3-$(HO)_2C_6H_3$	166.22		$164\text{-}6^{6-7}$	47-8				B6^{3}, 4657
12986	Resorcinol, 4-isobutyl 4-i-C_4H_9-1,3-$(HO)_2C_6H_3$	166.22		$166\text{-}8^{6-7}$	62-3			al, eth	B6^{2}, 899
12987	Resorcinol, 2-chloro or 2-Chloro-1,3-dihydroxy benzene 2-Cl-1,3-$(HO)_2C_6H_3$	144.56			97-8				B6^{3}, 4333
12988	Resorcinol, 4-chloro or 4-Chloro-1,3-dihydroxy benzene 4-Cl-1,3-$(HO)_2C_6H_3$	144.56		259, 147^{18}	(i)89 (ii)105			w, al, eth, ace, bz	B6^{3}, 4333
12989	Resorcinol, 5-chloro or 5-Chloro-1,3-dihydroxy benzene 5-Cl-1,3-$(HO)_2C_6H_3$	144.56	hyg nd (sub)	sub	117			w, eth, ace	B6^{2}, 819
12990	Resorcinol, 4-cyclohexyl or 4-Cyclohexyl-1,3-dihydroxy benzene 4-$C_6H_{11}C_6H_3(OH)_2$(1,3)	192.26	nd (bz chl), cr (bz-peth)		127-8				B6^{3}, 5057

PHYSICAL CONSTANTS OF ORGANIC COMPOUNDS (Continued)

No.	Name, Synonyms, and Formula	Mol. wt.	Color, crystalline form, specific rotation and λ_{max} (log ε)	b.p. °C	m.p. °C	Density	n_D	Solubility	Ref.
12991	Resorcinol diacetate $1,3\text{-}(CH_3CO_2)_2C_6H_4$	194.19		$278, 153\text{-}4^{12}$				al, eth	$B6^3$, 4320
12992	Resorcinol, 4,6-diacetyl or Resodiacetophenone $4,6\text{-}(CH_3CO)_2C_6H_2(OH)_2\text{-}(1,3)$	194.19	nd (al)		185			eth, ace, bz	$B8^3$, 3511
12993	Resorcinol dibenzoate $3\text{-}(C_6H_5CO_2)C_6H_4(O_2CC_6H_5)$	318.33	pl (dil al)		117			al, eth	$B9^3$, 555
12994	Resorcinol, 4,6-dichloro $4,6\text{-}Cl_2\text{-}1,3\text{-}(HO)_2C_6H_2$	179.00		254 sub	113			w, al, eth, ace	$B6^3$, 4335
12995	Resorcinol diethyl ether or 1,3-Diethoxy benzene $1,3\text{-}(C_2H_5O)_2C_6H_4$	166.22	pr	235^{756}	12.4			al, eth	$B6^3$, 4307
12996	Resorcinol, 2,5-dimethyl $2,5\text{-}(CH_3)_2\text{-}1,3\text{-}(HO)_2C_6H_2$	138.17	nd (bz), pr (w)	277-80	163			al, eth	$B6^3$, 4606
12997	Resorcinol, 4,5-dimethyl $4,5\text{-}(CH_3)_2\text{-}1,3\text{-}(HO)_2C_6H_2$	138.17	nd (bz), pr (w + 1)	sub	136-7, 115-7 (+ w)			w, al, eth, aa	$B6^3$, 4581
12998	Resorcinol, 4,6-dimethyl $4,6\text{-}(CH_3)_2\text{-}1,3\text{-}(HO)_2C_6H_2$	138.17	mcl pr (w + 1), nd (sub)	276-9 sub	125			w, al, eth	$B6^3$, 4595
12999	Resorcinol dimethyl ether or 1,3-Dimethoxy benzene $1,3\text{-}(CH_3O)_2C_6H_4$	138.17		217-8	−52	$1.0552^{25/25}$	1.5231^{20}	al, eth, bz	$B6^3$, 4305
13000	Resorcinol, 2,4-dinitro $2,4\text{-}(NO_2)_2\text{-}1,3\text{-}(HO)_2C_6H_2$	200.11	ye lf (al)		147-8				$B6^3$, 4351
13001	Resorcinol, 2,4-dinitroso $2,4\text{-}(ON)_2\text{-}1,3\text{-}(HO)_2C_6H_2$	168.11	ye rh pl (50% al), lf (aq MeOH)		168			w, al	$B7^3$, 4732
13002	Resorcinol dipropyl ether or 1,3-Dipropoxy benzene $1,3\text{-}(C_3H_7O)_2C_6H_4$	194.27		$251, 127\text{-}8^{12}$		$1.035^{20/21}$	1.5138^{83}		$B6^3$, 4308
13003	Resorcinol, dithio or 1,3-Dimercapto benzene $1,3\text{-}(HS)_2C_6H_4$	142.23	lf (dil al)	$245, 123^{17}$	27			al, eth, ace, bz	$B6^3$, 4366
13004	Resorcinol, 4-ethyl $4\text{-}C_2H_5\text{-}1,3\text{-}(HO)_2C_6H_3$	138.17	pr (chl, bz)	131^{15}	98-9				$B6^3$, 4554
13005	Resorcinol ethyl ether(mono) or 3-Ethoxy phenol $3\text{-}C_2H_5OC_6H_4OH$	138.17	ye	246-7, $117^{5.5}$		$1.0705^{4/4}$		al, eth, bz	$B6^3$, 4307
13006	Resorcinol, 4-hexyl $4\text{-}C_6H_{13}\text{-}1,3\text{-}(HO)_2C_6H_3$	194.27	pa ye nd (lig)	$333, 178^{8}$	68-9			al, eth, ace, bz, chl	$B6^3$, 4712
13007	Resorcinol, 4-iodo $4\text{-}I\text{-}1,3\text{-}(HO)_2C_6H_3$	236.01	pr, nd (+ 1w, w)		67			al, eth, ace, chl	$B6^3$, 4341
13008	Resorcinol, 5-iodo $5\text{-}I\text{-}1,3\text{-}(HO)_2C_6H_3$	236.01	nd (bz, sub)	sub	92.3			al	$B6^2$, 821
13009	Resorcinol, 2-methoxy $2\text{-}CH_3O\text{-}1,3\text{-}(HO)_2C_6H_3$	140.14	cr (bz)	$154\text{-}5^{24}$	85-7				$B6^3$, 6264
13010	Resorcinol, 5-methoxy $5\text{-}CH_3O\text{-}1,3\text{-}(HO)_2C_6H_3$	140.14	pl (bz)	213^{16}	78-81			al, eth	$B6^3$, 6304
13011	Resorcinol, 2-methyl $2\text{-}CH_3\text{-}1,3\text{-}(HO)_2C_6H_3$	124.14	pr (bz)	$264, 168^{16}$	119-21			w, al, eth, bz	$B6^3$, 4512
13012	Resorcinol methyl ether(mono) or 3-Methoxy phenol $3\text{-}CH_3OC_6H_4OH$	124.14		244.3, 144^{25}	<−17		1.5520^{20}	al, eth	$B6^3$, 4303
13013	Resorcinol, 4-(4-methyl pentyl) $4[(CH_3)_2CH(CH_2)_3]\text{-}1,3\text{-}(HO)_2C_6H_3$	194, 27	lf (peth)	190^{12}	71		1.5292^{15}	al, eth	$B6^3$, 4717
13014	Resorcinol, 2-nitro $2\text{-}NO_2\text{-}1,3\text{-}(HO)_2C_6H_3$	155.11	og-red pr (al)		87-8			al	$B6^3$, 4343
13015	Resorcinol, 4-nitro $4\text{-}NO_2\text{-}1,3\text{-}(HO)_2C_6H_3$	155.11			122			al, eth, bz	$B6^3$, 4344
13016	Resorcinol, 4-nitroso or 2-Hydroxy-4-benzoquinone oxime $4\text{-}ON\text{-}1,3\text{-}(HO)_2C_6H_3$	139.11	ye nd (+ w), br nd (chl)		150d			w, al, eth, ace, chl	$B8^3$, 1965
13017	Resorcinol, 4-pentyl $4\text{-}C_5H_{11}\text{-}1,3\text{-}(HO)_2C_6H_3$	180.25		$168\text{-}70^{6-7}$	72-3			al, eth, bz	$B6^3$, 4693
13018	Resorcinol, 4-isopentyl $4\text{-}i\text{-}C_5H_{11}\text{-}1,3\text{-}(HO)_2C_6H_3$	180.25		$177\text{-}8^{8}$	68-70			al, eth, bz	$B6^3$, 4700
13019	Resorcinol, 5-pentyl or Olivetol $5\text{-}C_5H_{11}\text{-}1,3\text{-}(HO)_2C_6H_3$	180.25	nd (+ w), pr (bz-lig)	164^{5}	49			w, al, eth, ace, bz	$B6^3$, 4695
13020	Resorcinol, 4-propionyl $4\text{-}(CH_3CH_2CO)C_6H_3(OH)_2\text{-}1,3$	166.18	ye nd (al)	$176\text{-}8^{6}$	97 (anh)			al, eth, bz, aa	$B8^3$, 2144
13021	Resorcinol monopropyl ether or 3-Propoxy phenol $3\text{-}C_3H_7OC_6H_4OH$	152.19	cr (w, al)	$256\text{-}7, 120^{5}$	55				$B6^3$, 4308

No.	Name, Synonyms, and Formula	Mol. wt.	Color, crystalline form, specific rotation and λ_{max} (log ε)	b.p. °C	m.p. °C	Density	n_D	Solubility	Ref.
13022	Resorcinol, 4-propyl $4\text{-}C_3H_7\text{-}1,3\text{-}(HO)_2C_6H_3$	152.19	pr (bz), nd (peth)	172^{14}	82-3			w, al, eth	$B6^{3}$, 4611
13023	Resorcinol, 4-iso-propyl $4\text{-}(CH_3)_2CH\text{-}1,3\text{-}(HO)_2C_6H_3$	152.19	cr (aq aa)	265-81, $114^{0.2}$	105			al, eth	$B6^{2}$, 896
13024	Resorcinol, 5-propyl or Divarinol $5\text{-}C_3H_7\text{-}1,3\text{-}(HO)_2C_6H_3$	152.19	lf (+1w), lf (bz)	$148\text{-}9^{3}$	83-4, 51 (+w)			al, eth, ace, bz, aa	$B6^{3}$, 4622
13025	Resorcinol, tetrachloro $1,3\text{-}(HO)_2C_6Cl_4$	247.89	nd (w)		141			al, eth, bz, aa	$B6^{2}$, 819
13026	Resorcinol, 2,4,6-tribromo $2,4,6\text{-}Br_3C_6H(OH)_2\text{-}1,3$	346.80	nd (w)		112			al, eth	$B6^{3}$, 4340
13027	Resorcinol, 2,4,6-trichloro $2,4,6\text{-}Cl_3C_6H(OH)_2\text{-}1,3$	213.45	nd (w)		83			al, eth	$B6^{3}$, 4336
13028	Resorcinol, 2,4,5-trimethyl $2,4,5\text{-}(CH_3)_3C_6H(OH)_2\text{-}1,3$	152.19	cr		156				$B6^{3}$, 4643
13029	Resorcinol, 2,4,6-trimethyl or Mesorcinol $2,4,6\text{-}(CH_3)_3\text{-}C_6H(OH)_2\text{-}1,3$	152.19	pl (al), lf	275	149-50			al, eth, bz	$B6^{3}$, 4653
13030	Resorcinol, 4,5,6-trimethyl $4,5,6\text{-}(CH_3)_3C_6H(OH)_2\text{-}1,3$	152.19	nd or lf		163-4			al, eth, ace, bz, chl	$B6^{3}$, 4643
13031	Resorcinol, 2,4,6-trinitro or Styphnic acid $2,4,6\text{-}(NO_2)_3C_6H(OH)_2\text{-}1,3$	245.11	hex (al), ye cr (aa)	sub	179-80			al, eth	$B6^{3}$, 4354
13032	Resorufin or 7-Hydroxy-2-phenoxazone $C_{12}H_7NO_3$	213.19	br nd ($PhNH_2$), pr (HCl)						$B27^{2}$, 108
13033	Retene or 7-isopropyl-1-methyl phenanthrene $C_{18}H_{18}$	234.34	pl (al)	390, 200^{10}	100-1	1.035		bz, lig	$B5^{3}$, 2199
13034	Retromecine or Senecifolinene $C_8H_{13}NO_2$	155.20	pr (ace) $[\alpha]'_D$ +27.4 (w), $^{26}/_D$ +50.2 (al)	$80^{0.01}$	121-2 (130)			w, al, ace	$B21^{4}$, 2048
13035	Rhamnetin $C_{16}H_{12}O_7$	316.27	ye nd (al)		294-6			ace	$B18^{4}$, 3474
13036	D-Rhamnitol or Rhamnite $C_6H_{14}O_5$	166.17	pr (ace), $[\alpha]^{20}/_D$ −12.4 (w,c =0.5)		123			w, al, Py	$B1^{4}$, 2837
13037	D-Rhamnose, hydrate (α-anomer) $C_6H_{12}O_5.H_2O$	182.17	(wh), $[\alpha]^{18.5}/_D$ −8.25 (w)		90-1			w	$B1^{4}$, 4260
13038	DL-Rhamnose $C_6H_{12}O_5$	164.16	(w)		151-3 (anh)			w, al	$B1^{4}$, 4261
13039	L-Rhamnose, hydrate $C_6H_{12}O_5.H_2O$	182.17	mcl pl (al w+1) $[\alpha]^{20}/_D$ −77→ +8.9 (mut)	105^{2} sub	92	1.4708^{20}		w, al	$B1^{4}$, 4261
13040	L-Rhamnose (β-anomer) $C_6H_{12}O_5$	164.16	nd (ace), $[\alpha]^{20}/_D$ +38.4 +8.9 (w,mut)		122-6			w, al	$B1^{4}$, 4262
13041	Rheadin or Rhoedine $C_{21}H_{21}NO_6$	383.40	nd (chl, eth, al), $[\alpha]^{17.5}/_D$ +232 (al)	sub	256-8				C38, 6060
13042	Rhizopterin or 12-Formylpteroic acid $C_{15}H_{12}N_6O_4$	340.30	lt ye pl (w)		>300 darkens 285				Am69, 2751
13043	Rhodamine β or Tetraethyl rhodamine $C_{28}H_{30}N_2O_3$	442.56	gr lf (w+4), col pr (al, xyl)		165 (anh)			w, al, eth, bz	$B19^{4}$, 4301
13044	Rhodamine, β-hydrochloride $C_{28}H_{30}N_2O_3.HCl$	479.02	gr or red vt pw (w), lf (dil HCl)					w, al, bz	$B19^{4}$, 4301
13045	Rhodamine or 4-Thioxo-4-thiazolidone $C_3H_3NOS_2$	133.18	lt ye pr (al, w, aa)		170	0.868		al, eth	$B27^{2}$, 288

No.	Name, Synonyms, and Formula	Mol. wt.	Color, crystalline form, specific rotation and λ_{max} (log ε)	b.p. °C	m.p. °C	Density	n_D	Solubility	Ref.
13046	Rhodanine, 3-amino $C_3H_4N_2OS_2$	148.20			100				B27^{2}, 289
13047	Rhodanine, 3-amino-5-benzylidene $C_{10}H_8N_2OS_2$	236.31			194-7				C61, 7190
13048	Rhodanine, 5-(dimethylamino) amino-benzylidene $C_{12}H_{12}N_2OS_2$	264.36	red or og nd (al, Py-w)		296			eth, ace	B27^{2}, 484
13049	Rhodanine, 3-ethyl $C_5H_7NOS_2$	161.24			37-9				B27^{1}, 309
13050	Rhodanind, 5-ethyl $C_5H_7NOS_2$	161.24	ye amor (dil al)		105			al, eth, ace, aa	B27^{1}, 313
13051	Rhodanine, 3-methyl $C_4H_5NOS_2$	147.21			69-71				B27^{2}, 288
13052	Rhodanine, 3-phenyl $C_9H_7NOS_2$	209.28	ye pl (aa), nd or pr (al)		194-5				B27^{2}, 288
13053	Rhodanine, 5-isopropylidene $C_6H_7NOS_2$	173.25			197-8				C44, 2979
13054	Rhodizonic acid or 5,6-Dihydroxy-5-cyclohexene-1,2,3,4-tetrone $C_6H_2O_6$	170.08	dk og nd (sub)		155-60			al	B8^{3}, 4214
13055	Riboflavin or Lactoflavin Vitamin B2 $C_{17}H_{20}N_4O_6$	376.37	ye or og-ye nd (w, dil aa), $[\alpha]^{25}/_D$ −112→ −122 (0.02N NaOH, c = 0.5), (mut)		280d				C28, 2036
13056	D-Ribose $C_5H_{10}O_5$	150.13	pl (abs al), $[\alpha]_D$ −21.5 (w)		95			w	B1^{4}, 4211
13057	Ricinidine or 1-Methyl, 2-pyridone-3-carbonitrile $C_7H_6N_2O$	134.14	nd (sub), (chl, al)	243^{28}	140			al	B22^{2}, 222
13058	Ricinine or Ricidine $C_8H_8N_2O_2$	164.16	pr or lf (w al)	sub $170\text{-}80^{20}$	201.5			Py	B22^{4}, 3354
13059	Ricinoleic acid or 12-Hydroxy-9-octadecenoic acid (*cis*) $C_6H_{13}CH(OH)CH_2CH{=}CH(CH_2)_7CO_2H$	298.47	$[\alpha]^{12}/_D$ + 5.05	226.8^{10}	α:7.7 β:16 γ:5.5	$0.9450^{21/4}$	1.4716^{21}	al, eth	B3^{4}, 1026
13060	Ricinoleic acid, butyl ester (*cis*) or Butyl ricinoleate $C_6H_{13}CH(OH)CH_2CH{=}CH(CH_2)_7CO_2C_4H_9$	354.57	$[\alpha]^{25}/_D$ + 3.7	278^{12}		0.9058^{22}	1.4566^{22}	eth	B3^{3}, 711
13061	Ricinoleic acid, isobutyl ester or Isobutyl ricinoleate $C_6H_{13}CHOHCH_2CH{=}CH(CH_2)_7CO_2\text{-i-}C_4H_9$	354.57	[α] + 4.01	282^{9}		0.9078^{22}	1.4538^{22}	eth	B3^{1}, 138
13062	Ricinoleic acid, ethyl ester (*cis*) or Ethyl ricinoleate $C_6H_{13}CH(OH)CH_2CH{=}CH(CH_2)_7CO_2C_2H_5$	326.52	$[\alpha]^{22}/_D$ + 5.3	258^{13}		0.9045^{22}	1.4618^{22}		B3^{4}, 1029
13063	Rosaniline $C_{20}H_{21}N_3O$	319.41	nd (w)		186d			al	B13^{3}, 2078
13064	Rosinduline, anhydro base $C_{22}H_{15}N_3$	321.38	red-br lf (eth-bz)		198-9			al, eth, bz	B25^{2}, 322
13065	Rotenone or Tubotoxine $C_{23}H_{22}O_6$	394.42	nd or lf (al, aq ace), $[\alpha]^{29.5}/_D$ −225.2 (bz)	$210\text{-}20^{0.5}$	(i)163 (ii) 176			al, ace, bz, chl, aa	B19^{4}, 5227
13066	Rubicene $C_{26}H_{14}$	326.40	red nd (xyl)		306			CS_2	B5^{3}, 2673
13067	Rubignol $C_5H_4O_4$	128.08			203.5			w, al, eth, ace	B17^{4}, 6679

No.	Name, Synonyms, and Formula	Mol. wt.	Color, crystalline form, specific rotation and λ_{max} (log ε)	b.p. °C	m.p. °C	Density	n_D	Solubility	Ref.
13068	Rubijervine $C_{27}H_{43}NO_2$	413.64	nd (+1w dil al), $[\alpha]^{25/}{}_D$ +19 (al, c=1)		242			al, bz, chl	B21[4], 2310
13069	Rubixanthin or 3-Hydroxy-γ-carotene $C_{40}H_{56}O$	552.88	dk red nd (bz-MeOH), og-red (bz-peth)		160			bz, chl	B6[3], 3772
13070	Rubrene or 9,10,11,12-Tetraphenyl naphthacene $C_{42}H_{28}$	532.68	og red (bz-lig)		331			bz	B5[3], 2803
13071	Rufol or 1,5-Dihydroxy anthracene 1,5-$(HO)_2C_{14}H_8$	210.23	ye nd		265d			al, eth, bz	B6, 1032
13072	Rutaecarpine $C_{18}H_{13}N_3O$	287.32	yesh nd (al, AcOEt)		259-60				B26[2], 104
13073	Rutinose or 6-D-β-L-rhamnopyransoyl-D-glucose $C_{12}H_{22}O_{10}$	326.30	hyg pw (al, eth), $[\alpha]^{20/}{}_D$ −10 (al)		189-92d			w, al	B17[4], 2536
13074	Sabadine $C_{29}H_4$li9NO_8	541.70	nd (eth), $[\alpha]_D$ −11 (al)		256-60			al, ace	B21[4], 2883
13075	Sabinane or d-Thujane $C_{10}H_{18}$	138.25	$[\alpha]_D$ +73.1	157^{758}		$0.8139^{20/4}$	1.4376^{20}	al, eth, ace, bz	B5[4], 317

PHYSICAL CONSTANTS OF ORGANIC COMPOUNDS (Continued)

No.	Name, Synonyms, and Formula	Mol. wt.	Color, crystalline form, specific rotation and λ_{max} (log ε)	b.p. °C	m.p. °C	Density	n_D	Solubility	Ref.
13076	Sabinene (d) or 4(10)-Thujene $C_{10}H_{16}$	136.24	$[\alpha]^{20/}{}_D$ + 101.4	163-5^{758}, 49^{13}		0.8437$^{20/4}$	1.4676^{20}	**al, eth**, bz, chl	B5^{3}, 365
13077	Sabinene (l) $C_{10}H_{16}$	136.24	$[\alpha]^{15/}{}_D$ −42.5	162-6		0.8468^{20}	1.4674^{17}	**al, eth**, bz	B5^{3}, 365
13078	Sabinol (d) or 4(10)Thujene-3-01 $C_{10}H_{16}O$	152.24	$[\alpha]^{18/}{}_D$ + 3.94	208, 90^{11}		0.9488$^{19/4}$	1.4871^{25}	eth	B6^{4}, 382
13079	Sabinol acetate(d) $C_{12}H_{18}O_2$	194.27	$[\alpha]_D$ + 79	81-2^{3}		0.972^{15}			B6^{3}, 384
13080	D-Saccaharic acid $C_6H_{10}O_8$	210.14	nd (95%al), $[\alpha]_D$ + 6.9→ + 20.6 (w, c=2.5, mut)		125-6			w, al	B3^{4}, 1291
13081	Saccharin or 2-Sulfobenzoic acid imide $C_7H_5NO_3S$	183.18	mcl (ace), pr (al), lf (w)	sub vac	229d	0.828		al, ace	B27^{2}, 214
13082	Safrole or 1-Allyl-3,4-(methylene dioxy) benzene $C_{10}H_{10}O_2$	162.19	mcl	234.5, 104-5^{6}	11.2	1.000$^{20/4}$	1.5381^{20}	al, **eth, chl**	B19^{4}, 275
13083	Salicylaldehyde or 2-Hydroxybenzaldehyde 2-HOC_6H_4CHO	122.12		197, 93^{25}	−7	1.1674$^{20/4}$	1.5740^{20}	**al, eth**, ace, bz	B8^{3}, 135
13084	Salicylaldehyde azine or Salazine $C_{14}H_{12}N_2O_2$	240.26	ye nd or lf (al)		214			al, bz, chl	B8^{2}, 43
13085	Salicylaldehyde oxime or Salicylaldoxime 2-$HOC_6H_4CH{=}NOH$	137.14	pr (bz-peth)		63			al, eth, bz	B8^{3}, 168
13086	Salicylamide or 2-Hydroxy benzamide 2-$HOC_6H_4CONH_2$	137.14	ye lf (dil al)	181.5^{14}	142	1.175$^{140/4}$		al	B10^{3}, 152
13087	Salicylamide, N-phenyl or N-Phenyl salicylamide 2-$HOC_6H_4CONHC_6H_5$	213.24	pr (w, al)		136-7			al	B12^{3}, 944
13088	Salicylamide, N-(2-tolyl) or N-(2-Tolyl) salicylamide 2-$HOC_6H_4CONH(C_6H_4CH_3$-2)	227.26	nd (al)		144			al	B12^{3}, 1888
13089	Salicylic acid or 2-Hydroxybenzoic acid 2-$HOC_6H_4CO_2H$	138.12	nd (w), mcl pr (al)	211^{20} sub	159	1.443$^{20/4}$	1.565	al, eth, ace	B10^{3}, 87
13090	Salicylic acid, allyl ester or Allyl salicylate 2-$HOC_6H_4CO_2CH_2CH{=}CH_2$	178.19		247-50, 105-6^{5}		1.1000^{15}			B10^{3}, 125
13091	Salicylic acid, 4-acetamido, phenyl ester or Salophene 4-CH_3CONH-2-$HOC_6H_3CO_2C_6H_5$	271.27	pl (w), lf (al)		187-8			al, eth, bz	B13^{3}, 1066
13092	Salicylic acid acetate or Aspirin 2-$CH_3CO_2C_6H_4CO_2H$	180.16	nd (w), mcl ta (w)		135			al, eth, chl	B10^{3}, 102
13093	Salicylic acid, 3-amino or 3-Amino-2-hydroxybenzoic acid 3-NH_2-2-$HOC_6H_3CO_2H$	153.14			235d				B14^{3}, 1434
13094	Salicylic acid, 4-amino or 4-Amino-2-hydroxybenzoic acid 4-H_2N-2-$HOC_6H_3CO_2H$	153.14			150-1d			w, al, eth, ace	B14^{3}, 1436
13095	Salicylic acid, 5-amino or 5-Amino-2-hydroxybenzoic acid 5-H_2N-2-$HOC_6H_3CO_2H$	153.14			283				B14^{3}, 1456
13096	Salicylic acid anhydride (2-$HOC_6H_4CO)_2O$	258.23	amor		255d			al, eth	B19^{4}, 2059
13097	Salicylic acid, benzyl ester or Benzyl salicylate 2-$HOC_6H_4CO_2CH_2C_6H_5$	228.25		320, 170^{5}		1.1799$^{20/4}$	1.5805^{20}	al, eth	B10^{3}, 132
13098	Salicylic acid, benzyl ester, acetate 2-$(CH_3CO_2)C_6H_4CO_2CH_2C_6H_5$	270.28		197-200^{7}	26			al, eth, ace, bz	B10^{2}, 52
13099	Salicylic acid, 3-bromo or Hydroxy-3-bromobenzoic acid 3-Br-2-$HOC_6H_3CO_2H$	217.02	nd (dil al)		184-5			al, eth, ace	B10^{3}, 174
13100	Salicylic acid, 4-bromo 4-Br-2-HO-$C_6H_3CO_2H$	217.02	pl, nd (w)		214			al	B10^{2}, 63
13101	Salicylic acid, 5-bromo 5-Br-2-$HOC_6H_3CO_2H$	217.02	nd (w, dil al)	sub>100	168-9			w, al, eth	B10^{3}, 176
13102	Salicylic acid, 5-bromo,acetate 5-Br-2$(CH_3CO_2)C_6H_3CO_2H$	259.06	nd (al)		60			al, eth	B10^{2}, 64
13103	Salicylic acid, butyl ester or Butyl salicylate 2-$HOC_6H_4CO_2C_4H_9$	194.23		270-2, 136-8^{10}	−5.9	1.0728$^{20/4}$	1.5115^{20}		B10^{3}, 121
13104	Salicylic acid, *iso* butyl ester or *Iso* Butyl salicylate 2-$HOC_6H_4CO_2$-i-C_4H_9	194.23		260-2, 136-8^{10}	5.9	1.0639$^{20/4}$	1.5087^{20}	al, eth	B10^{3}, 121
13105	Salicylic acid, 3-chloro 3-Cl-2-$HOC_6H_3CO_2H$	172.57	nd(w)		180			al, chl, aa	B10^{3}, 163
13106	Salicylic acid, 3-chloro,ethyl ester or Ethyl 3-chloro salicylate 3-Cl-2-$HOC_6H_3CO_2C_2H_5$	200.62	nd	269-70, 147^{12}	21			al	B10, 101

No.	Name, Synonyms, and Formula	Mol. wt.	Color, crystalline form, specific rotation and λ_{max} (log ε)	b.p. °C	m.p. °C	Density	n_D	Solubility	Ref.
13107	Salicylic acid, 3-chloro,methyl ester or Methyl-3-chloro salicylate $3\text{-Cl-2-HOC}_6H_3CO_2CH_3$	186.59	nd (MeOH or al)	260d	38				$B10^{2}$, 61
13108	Salicylic acid, 4-chloro or 4-Chloro-2-hydroxy benzoic acid $4\text{-Cl-2-HOC}_6H_3CO_2H$	172.57	nd (w)	sub (d)	207			al, bz, chl	$B10^{3}$, 164
13109	Salicylic acid, 5-chloro $5\text{-Cl-2-HOC}_6H_3CO_2H$	172.57	nd (w, or al)		173-4			al, eth, bz, chl	$B10^{3}$, 165
13110	Salicylic acid, 5-chloro,ethyl ester $5\text{-Cl-2-HOC}_6H_3CO_2C_2H_5$	200.62	nd (al)		25			al	B10, 103
13111	Salicylic acid, 5-chloro, methyl ester $5\text{-Cl-2-HOC}_6H_3CO_2CH_3$	186.59	nd (al)	249d	50			al	$B10^{3}$, 166
13112	Salicylic acid, 6-chloro $6\text{-Cl-2-HOC}_6H_3CO_2H$	172.57	nd (w)		166			al, eth, ace, bz	B10, 104
13113	Salicylic acid, 3-diazo or 3-Diazo-2-hydroxy benzoic acid $3\text{-N}_2\text{-2-HOC}_6H_3CO_2H$	164.12	ye nd (ace)		155d			ace	B16, 553
13114	Salicylic acid, 3,5-dichloro $3,5\text{-Cl}_2\text{-2-HOC}_6H_2CO_2H$	207.01	nd (dil al), rh pr	sub d	220-1			al, eth	$B10^{3}$, 169
13115	Salicylic acid, 3,5-diido $3,5\text{-I}_2\text{-2-HOC}_6H_2CO_2H$	389.92	nd (al, aa)		235-6			al, eth	$B10^{3}$, 189
13116	Salicylic acid, 3,5-diido, ethyl ester $3,5\text{-I}_2\text{-2-HOC}_6H_2CO_2C_2H_5$	417.97	lf (al)		133			al, eth, bz	B10, 114
13117	Salicylic acid, 3,4-dimethoxy $3,4\text{-(CH}_3\text{O)}_2\text{-2-HOC}_6H_2CO_2H$	198.18	nd (w)		169-72			chl	$B10^{3}$, 2058
13118	Slaicylic acid, 4,5-dimethoxy $4,5\text{(CH}_3\text{O)}_2\text{-2-HOC}_6H_2CO_2H$	198.18	nd (w)		202d			al	$B10^{3}$, 2065
13119	Salicylic acid, 4,5-dimethoxy,methyl ester $4,5\text{-(CH}_3\text{O)}_2\text{-2-HOC}_6H_2CO_2CH_3$	212.20	nd (w)		95				$B10^{3}$, 2067
13120	Salicylic acid, 4,6-dimethoxy $4,6\text{-(CH}_3\text{O)}_2\text{-2-HOC}_6H_2CO_2H$	198.18	nd (eth-bz)		152-4d			al, eth, ace	$B10^{3}$, 2069
13121	Salicylic acid, 2,4-dinitrobenzyl ester $2\text{-HOC}_6H_4CO_2[CH_2C_6H_3(NO_2)_2\text{-2,4}]$	318.24	ye pl (aa)		168				$B10^{2}$, 52
13122	Salicylic acid, 3,5-dinitro $3,5(\text{NO}_2)_2\text{-2-HOC}_6H_2CO_2H$	228.12	ye nd or pl (+1w)		182			al, eth, bz	$B10^{3}$, 207
13123	Salicylic acid, dithio or Dithiosalicylic acid $2\text{-HOC}_6H_4CS_2H$	170.24	og-ye nd (peth)		48-50			w, al, eth, bz	$B10^{2}$, 78
13124	Salicylic acid, 4-ethoxy $4\text{-C}_2H_5O\text{-2-HOC}_6H_3CO_2H$	182.18	nd (w, bz)		154			al, eth, bz	B10, 379
13125	Salicylic acid, ethyl ester or Ethyl salicylate $2\text{-HOC}_6H_4CO_2C_2H_5$	166.18		234, 132.8^{27}	2-3	$1.1326^{20/4}$	1.5296^{20}	al, eth	$B10^{3}$, 115
13126	Salicylic acid, ethyl ester, acetate or Ethyl aspirin $2(\text{CH}_3CO_2)C_6H_4CO_2C_2H_5$	208.21		272, $148\text{-}50^{15}$		1.1566^{15}		al, eth, ace, bz	$B10^{2}$, 48
13127	Salicylic acid, 3-formyl $3\text{-HCO-2-HOC}_6H_3CO_2H$	166.13	nd (w + 1)		179			al	$B10^{3}$, 4207
13128	Salicylic acid, 5-formyl $5\text{-HCO-2-HOC}_6H_3CO_2H$	166.13	nd (w)		250d			eth	$B10^{3}$, 4209
13129	Salicylic acid hydrazide $2\text{-HOC}_6H_4CONHNH_2$	152.15	pl (al), pr (w)		147			al, bz	$B10^{3}$, 161
13130	Salicylic acid, β-hydroxyethyl ester $2\text{-HOC}_6H_4CO_2CH_2CH_2OH$	182.18		173^{15}	37	$1.2537^{15/15}$		al, eth, bz, chl	$B10^{3}$, 138
13131	Salicylic acid, 3-iodo $3\text{-I-2-HOC}_6H_3CO_2H$	264.02	nd (w)		199			al, eth	$B10^{3}$, 186
13132	Salicylic acid, 4-iodo $4\text{-I-2-HOC}_6H_3CO_2H$	264.02	pr or nd (al)		230d			al, eth	$B10^{2}$, 65
13133	Salicylic acid, 5-iodo $5\text{-I-2-HOC}_6H_3CO_2H$	264.02	nd (w)		197			al, eth	$B10^{3}$, 188
13134	Salicylic acid, menthyl ester or Menthyl salicylate $C_{17}H_{24}O_3$	276.38		190^{15}, $143^{0.5}$		1.0467^{20}	1.5198^{26}	al, eth, ace, bz	$B10^{3}$, 126
13135	Salicylic acid, 4-methoxy-6-methyl or Everninic acid $4\text{-CH}_3O\text{-6-CH}_3\text{-2-HOC}_6H_2CO_2H$	182.18	nd (peth, w)		171-2			al, ace, aa	$B10^{2}$, 273
13136	Salicylic acid, 4-methoxy-6-methyl,methyl ester or Sparassol $4\text{-CH}_3O\text{-6-CH}_3\text{-2-HOC}_6H_2CO_2CH_3$	196.20	pr (w), lf (MeOH)		67-8			al, eth, ace, chl, peth	$B10^{2}$, 273
13137	Salicylic acid, (methoxymethyl) ester or Mesotan $2\text{-HOC}_6H_4CO_2CH_2OCH_3$	182.18		162^{742}		1.2^{15}		al, eth, ace, bz, chl	$B10^{3}$, 143
13138	Salicylic acid, (2-methoxyphenyl) ester $2\text{-HOC}_6H_4CO_2(C_6H_4OCH_3\text{-2})$	244.25			70			al, eth, chl	B10, 81

PHYSICAL CONSTANTS OF ORGANIC COMPOUNDS (Continued)

No.	Name, Synonyms, and Formula	Mol. wt.	Color, crystalline form, specific rotation and λ_{max} (log ε)	b.p. °C	m.p. °C	Density	n_D	Solubility	Ref.
13139	Salicylic acid, methyl ester $2\text{-}HOC_6H_4CO_2CH_3$	152.15	....	223.3	−8	$1.1738^{20/4}$	1.5360^{20}	al, eth	B10[3], 107
13140	Salicylic acid, methyl ester, benzoate $2\text{-}(C_6H_5CO_2)C_6H_4CO_2CH_3$	256.26	pr (al, eth)	385d, $270\text{-}80^{120}$	92	....	....	al, eth, bz, chl	B10[3], 112
13141	Salicylic acid, 3-methyl or 2,3-Cresotic acid $3\text{-}CH_3\text{-}2\text{-}HOC_6H_3CO_2H$	152.15	nd (w, dil al)	....	169-70	....	....	al, eth, chl	B10[3], 505
13142	Salicylic acid, 4-methyl or 2,4-Cresotic acid $4\text{-}CH_3\text{-}2\text{-}HOC_6H_3CO_2H$	152.15	nd (w), pr (al), pl (chl)	sub	177.8	....	....	al, eth, chl	B10[3], 521
13143	Salicylic acid, 5-methyl or 2,5-Cresotic acid $5\text{-}CH_3\text{-}2\text{-}HOC_6H_3CO_2H$	152.15	nd (w, peth)	....	153	....	....	al, eth, bz, chl	B10[3], 516
13144	Salicylic acid, 6-methyl or 2,6-Cresotic acid $6\text{-}CH_3\text{-}2\text{-}HOC_6H_3CO_2H$	152.15	nd (chl)	....	173 (184)	....	....	al, eth, chl	B10[3], 496
13145	Salicylic acid, α-naphthyl ester or Alphol $2\text{-}HOC_6H_4CO_2\text{-}\alpha\text{-}C_{10}H_7$	264.28	....	....	83	....	....	eth	B10[2], 52
13146	Salicylic acid, β-naphthyl ester or Betol $2HOC_6H_4CO_2\text{-}\beta\text{-}C_{10}H_7$	264.28	cr (al)	....	95.5	1.11^{116}	....	eth, bz	B10[3], 136
13147	Salicylic acid, 3-nitro $3\text{-}NO_2\text{-}2\text{-}HOC_6H_3CO_2H$	183.12	yesh nd (aa, w + 1)	....	148-9 (anh), 128-9 (+ w)	....	....	al, eth, ace, bz, chl	B10[3], 190
13148	Salicylic acid, 4-nitro $4\text{-}NO_2\text{-}2\text{-}HOC_6H_3CO_2H$	183.12	ye nd (w, dil al)	....	235	....	....	al, ace, chl, aa	B10[3], 194
13149	Salicylic acid, 5-nitro $5\text{-}NO_2\text{-}2\text{-}HOC_6H_3CO_2H$	183.12	nd (w)	....	229-30	1.650^{20}	....	al, eth, ace, bz	B10[3], 197
13150	Salicylic acid, 6-nitro $6\text{-}NO_2\text{-}2\text{-}HOC_6H_3CO_2H$	183.12	pa ye nd	....	....	....	....	w, al	B10[3], 205
13151	Salicylic acid, 4-nitrobenzyl ester $2\text{-}HOC_6H_4CO_2(CH_2C_6H_4NO_2\text{-}4)$	273.25	cr (dil al)	....	97-8	....	....	al, ace	B10[2], 52
13152	Salicylic acid, pentyl ester $2\text{-}HOC_6H_4CO_2C_5H_{11}$	208.26	....	148-54	....	$1.065^{15/15}$	1.506^{20}	**al, eth**	B10[3], 122
13153	Salicylic acid, isopentyl ester or Isopentyl salicylate $2\text{-}HOC_6H_4CO_2\text{-}i\text{-}C_5H_{11}$	208.26	....	$276\text{-}7^{741}$, $151\text{-}2^{15}$	....	$1.0535^{20/4}$	1.5080^{20}	al, eth, chl	B10[3], 123
13154	Salicylic acid, phenyl ester or Phenyl salicylate Salol $2\text{-}HOC_6H_4CO_2C_6H_5$	214.22	pl (MeOH)	173.12	43	$1.2614^{30/4}$	....	al, eth, ace, bz, aa	B10[3], 127
13155	Salicylic acid, propyl ester $2\text{-}HOC_6H_4CO_2C_3H_7$	180.20	....	239	96-8	$1.0979^{20/4}$	1.5161^{20}	**al, eth**	B10[3], 119
13156	Salicylic acid, *iso*-propyl ester $2\text{-}HOC_6H_4CO_2CH(CH_3)_2$	180.20	....	240-2, 118^{17}	....	$1.0729^{20/4}$	1.5065^{20}	**al, eth**	B10[3], 120
13157	Salicylic acid, 3-*iso*-propyl-6-methyl or *o*-Thymotinic acid $3\text{-}i\text{-}C_3H_7\text{-}6\text{-}CH_3\text{-}2\text{-}HOC_6H_2CO_2H$	194.23	nd (w, bz, lig)	sub	127	....	....	al, eth, bz, aa	B10[3], 629
13158	Salicylic acid, 5-*iso*-propyl-3-methyl $5\text{-}i\text{-}C_3H_7\text{-}3\text{-}CH_3\text{-}2\text{-}HOC_6H_2CO_2H$	194.23	nd (w)	....	147	....	....	al	B10, 282
13159	Salicylic acid, 5-*iso*-propyl-4-methyl $5\text{-}i\text{-}C_3H_7\text{-}4\text{-}CH_3\text{-}2\text{-}HOC_6H_2CO_2H$	194.23	nd (bz)	....	189-90	....	....	....	B10[2], 171
13160	Salicylic acid, 6-*iso*-propyl-3-methyl or *o*-Carvacrotinic acid $6\text{-}i\text{-}C_3H_7\text{-}3\text{-}CH_3\text{-}2\text{-}HOC_6H_2CO_2H$	194.23	nd (w)	sub	136	....	....	al, eth	B10, 282
13161	Salicylic acid, 3-sulfo $3\text{-}SO_3H\text{-}2\text{-}HOC_6H_3CO_2H$	218.18	pr (w + 2)	....	120	....	....	w, al, eth, aa	B11[3], 701
13162	Salicylic acid, 5-sulfo $5\text{-}SO_3H\text{-}2\text{-}HOC_6H_3CO_2H$	218.28	hyg nd (w + 2)	....	198d 120 (+ 2w)	....	....	**w, al**, eth	B11[3], 704
13163	Salicylonitrile $2\text{-}HOC_6H_4CN$	119.12	pr (bz)	149^{14}	98	$1.052^{100/4}$	1.5372^{100}	al, eth, bz, chl	B10[3], 159
13164	Salicyl chloride $2\text{-}HOC_6H_4COCl$	156.57	....	92^{15}	19	1.3112^{20}	1.5812^{20}	eth	B10[3], 150
13165	Salicyl chloride, 3-methyl or 3-Methyl salicyl chloride $3\text{-}CH_3\text{-}2\text{-}HOC_6H_3COCl$	170.60	....	$87\text{-}9^{16}$	27-8	....	....	....	B10[2], 132
13166	Salsoline $C_{11}H_{15}NO_2$	193.25	pw or cr (al), $[\alpha]^{20/}{}_D$ + 34.5 (0.1 NHCl, c=1)	....	221-2	....	....	al, chl	B21[4], 2121

No.	Name, Synonyms, and Formula	Mol. wt.	Color, crystalline form, specific rotation and λ_{max} (log ε)	b.p. °C	m.p. °C	Density	n_D	Solubility	Ref.
13167	Salvarsan or 606. Arspenamine $C_{12}H_{12}N_2As_2O_2 \cdot 2HCl \cdot 2H_2O$	475.04	pw	....	185-95d	....	....	w	$B16^2$, 560
13168	Sambunigrin or d-Mandelonitrile glucoside $C_6H_5CH(CN)OC_6H_{11}O_5$	295.29	nd (bz-peth), $[\alpha]^{18/}_D$ −75.1	....	151-2	....	....	w, al	$B17^4$, 3356
13169	Samidin $C_{21}H_{22}O_7$	386.40	cubic, $[\alpha]^{21/}_D$ +49.1 (chl, c=1.59)	....	138-9	....	....	al, eth	$B19^4$, 2788
13170	Sanquinarine $C_{20}H_{15}NO_5$	349.34	cr (eth al)	....	266	....	....	al, eth, ace, bz	$B27^2$, 614
13171	Santalic acid or Guerbet's acid $C_{15}H_{22}O_2$	234.34	red	....	β:202 γ:189	....	β:1.5136^{20} γ:1.5055^{2}	....	$B9^3$, 2619
13172	α-Santalol or Arheol $C_{15}H_{24}O$	220.35	$[\alpha]^{20/}_D$ +17	301-2, 167^{14}	......	$0.9679^{20/4}$	1.5023^{20}	al	$B6^3$, 2083
13173	β-Santalol $C_{15}H_{24}O$	230.35	$[\alpha]^{20/}_D$ −90.5	$167\text{-}8^{10}$	......	$0.9750^{20/4}$	1.5115^{20}	....	$B6^3$, 2080
13174	Santene or 2,3-Dimethyl-2-norbornene C_9H_{14}	122.21	....	140-1, 35^{15}	......	$0.8698^{17/4}$	1.4688^{17}	eth, ace, bz	$B5^3$, 333
13175	Santenic acid (*cis,d*) or π-Norcamphenic acid $C_9H_{14}O_4$	186.21	pl (w), $[\alpha]^{33/}_D$ +38.3 (al)	....	170-1	....	....	al	$B9^3$, 3847
13176	β-Santenol (*cis,exo*) or 2,3-Dimethyl-2-norbornano $C_9H_{16}O$	140.23	nd (al), tab (lig)	19^2	101-2	....	....	eth	$B6^4$, 243
13177	α-Santenone (*cis*) or 1,7-Dimethyl-1,2-norbarnanone $C_9H_{14}O$	138.21	pl $[\alpha]^{22/}_D$ +11.4 (al)	191	55	....	....	al, ace	$B7^3$, 305
13178	Santonin or Santonic acid $C_{15}H_{18}O_3$	246.31	orh (w, aa, eth) $[\alpha]^{18/}_D$ −173 (al, c=2)	....	174-6	$1.187^{66/4}$	1.590	bz, chl, Py	$B17^4$, 6232
13179	Sarmentogenin $C_{23}H_{34}O_5$	390.52	pr (95% al, MeOH-eth)	270-5	......	....	....	al, Py	$B18^4$, 2443
13180	Sarpagine or Raupine $C_{19}H_{22}N_2O_2$	310.40	nd $[\alpha]^{20/}_D$ +54 (Py)	....	320	....	....	al	Am84, 622
13181	Sarsaspogenin or Parigenin $C_{27}H_{44}O_3$	416.64	lo pr nd (ace), $[\alpha]^{25/}_D$ −75 (chl, c=0.5)	....	200-1	....	....	al, ace, bz, chl	$B19^4$, 824
13182	Scarlet red $C_{24}H_{20}N_4O$	380.45	dr br pw or nd	d 260	186d	....	....	chl, peth	B16, 172
13183	Scilliroside (a β-glucoside) $C_{32}H_{44}O_{12}$	620.69	lo pr (dil MeOH) $[\alpha]^{20/}_D$ -60 (MeOH)	d	168-70	....	....	al, diox	$B18^4$, 3178
13184	Scopoline or Oscin $C_8H_{13}NO_2$	155.20	hyg nd (lig, eth, chl, peth)	248	108-9	$1.0891^{134/4}$	....	w, al, ace	$B27^2$, 61
13185	Scyllitol $C_6H_{12}O_6$	180.16	pr (+3w)	....	353d	$1.659^{19/4}$	....	....	$B6^3$, 6926
13186	Sebacamide or Decandiamide $H_2NOC(CH_2)_8CONH_2$	200.28	pr or pl (aa)	....	210	....	....	aa	$B2^4$, 2088
13187	Sebacic acid or Decanedioic acid $HO_2C(CH_2)_8CO_2H$	202.25	lf	295^{100}	134.5	$1.2705^{20/4}$	1.422^{133}	al, eth	$B2^4$, 2078
13188	Sebacic acid, dibutyl ester or di-Butyl sebacate $C_4H_9O_2C(CH_2)_8CO_2C_4H_9$	314.47	....	344-5, 227^{17}	−10	0.9405^{15}	1.4433^{15}	eth	$B2^4$, 2081
13189	Sebacic acid, diethyl ester or Diethyl sebacate $C_2H_5O_2C(CH_2)_8CO_2C_2H_5$	258.36	....	306^{773}, 188^{19}	5	$0.9646^{20/4}$	1.4366^{20}	al, ace	$B2^4$, 2080
13190	Sebacic acid, di(2-ethyl butyl) ester $(C_2H_5)_2CHCH_2O_2C(CH_2)_8CO_2CH_2CH(C_2H_5)_2$	370.57	....	344-6	−22	$0.920^{25/4}$	....	al, ace, bz	$B2^4$, 2082
13191	Sebacic acid, di(2-ethyl hexyl) ester $C_4H_9CH(C_2H_5)CH_2O_2C(CH_2)_8CO_2CH_2CH(C_2H_5)C_4H_9$	426.68	....	256^5	−48	$0.912^{25/4}$	1.451^{25}	al, ace, bz	$B2^4$, 2083
13192	Sebacic acid, dimethyl ester or Dimethyl sebacate $CH_3O_2C(CH_2)_8CO_2CH_3$	230.30	lo pr	175^{20}, 144^5	38	0.9882^{28}	1.4355^{28}	al, eth, ace	$B2^4$, 2080
13193	Sebacic acid, monomethyl ester, mononitrile $NC(CH_2)_8CO_2CH_3$	197.28	....	178^{16}	3-4	0.934^{20}	1.4398^{25}	eth	$B2^4$, 2079

PHYSICAL CONSTANTS OF ORGANIC COMPOUNDS (Continued)

No.	Name, Synonyms, and Formula	Mol. wt.	Color, crystalline form, specific rotation and λ_{max} (log ε)	b.p. °C	m.p. °C	Density	n_D	Solubility	Ref.
13194	Sebaconitrile $NC(CH_2)_8CN$	164.25		204^{16}		$0.9313^{20/4}$	1.4474^{20}		$B2^4$, 2089
13195	Sebacyl chloride $ClOC(CH_2)_8COCl$	239.14		220^{75}, 165^{11}	−2.5	$1.1212^{20/4}$	1.4684^{18}		$B2^4$, 2088
13196	Sedarmide $H_2NCONHCOCH(CH_2CH{=}CH_2)CH(CH_3)_2$	184.24	nd (al)		194			al, chl	$B3^2$, 53
13197	Selanthrone $C_{12}H_8Se_2$	310.12	pr (al), nd (aa)	223^{11}	181			CS_2	$B19^4$, 352
13198	Seleno urea $H_2NCSeNH_2$	123.02	pr or nd (w)	200d				w	$B3^4$, 435
13199	Seleno urea, 1-ethyl $C_2H_5NHCSeNH_2$	151.07	nd (al, peth, w)		ca 125			al	$B4^2$, 610
13200	α-Selinene $C_{15}H_{24}$	204.36	$[\alpha]'_D$ + 49.5	268-72, $128\text{-}32^{11}$		$0.9196^{20/15}$	1.5048^{20}		$B5^3$, 1090
13201	B-Selinene $C_{15}H_{24}$	204.36	$[\alpha]^{20}_D$ + 38.2	$121\text{-}2^6$		$0.9170^{18/15}$	1.4956^{21}		$B5^3$, 1090
13202	Semicarbazide or Aminourea $H_2NCONHNH_2$	75.07	pr (al)		96			w, al	$B3^4$, 177
13203	Semicarbazide, 1,1-diphenyl $(C_6H_5)_2NNHCONH_2$	227.27	nd (al, bz)		195			al, bz	B15, 304
13204	Semicarbazide, 1,1-diphenyl-3-thio $(C_6H_5)_2NNHCSNH_2$	243.33	cr (al, bz)		202			al, ace, bz, chl	$B15^3$, 193
13205	Semicarbazide, 1,4-diphenyl $C_6H_5NHNHCONHC_6H_5$	227.27	nd or lf (al,bz)		177			al	$B15^3$, 184
13206	Semicarbazide, 1,4-diphenyl-3-thio $C_6H_5NHCSNHNHC_6H_5$	243.33			176-7			al, ace, chl	$B15^3$, 190
13207	Semicarbazide, 2,4-diphenyl $H_2NN(C_6H_5)CONHC_6H_5$	227.27	lf (al)		165.5			al, eth, bz, chl	B15, 277
13208	Semicarbazide, 2,4-diphenyl-3-thio $H_2NN(C_6H_5)CSNHC_6H_5$	243.33	lf (al)		139			ace, bz	$B15^3$, 182
13209	Semicarbazide, 4,4-diphenyl $(C_6H_5)_2NCONHNH_2$	227.27			154			al, eth, chl	$B12^3$, 896
13210	Semicarbazide hydrochloride $H_2NCONHNH_2.HCl$	111.52	pr (dil al)		175-7d			w	$B3^4$, 177
13211	Semicarbazide, 4-methyl-3-thio $CH_3NHCSNHNH_2$	105.16			135-8			w, al	$B4^4$, 220
13212	Semicarbazide, 1-phenyl $H_2NCONHNHC_6H_5$	151.17			172			al, ace, MeOH	$B15^3$, 184
13213	Semicarbazide, 1-phenyl-3-thio $C_6H_5NHNHCSNH_2$	167.23	pr (al)		200-1d				$B15^3$, 190
13214	Semicarbazide, 2-phenyl $H_2NN(C_6H_5)CONH_2$	151.17	nd (bz, al)		120			w, al, chl	$B15^2$, 103
13215	Semicarbazide, 2-phenyl-3-thio $H_2NN(C_6H_5)CSNH_2$	167.23	w		153			al	$B15^1$, 70
13216	Semicarbazide, 4-phenyl $H_2NNHCONHC_6H_5$	151.17	nd (bz), pl (w)		128			al, chl	$B12^3$, 822
13217	Semicarbazide, 4-phenyl-3-thio $H_2NNHCSNHC_6H_5$	167.23	pl (al)		140d				$B12^2$, 232
13218	Semicarbazide, 3-thio or Thiosemicarbazide $H_2NNHCSNH_2$	91.13	lo nd (w)		183			w, al	$B3^4$, 374
13219	Semicarbazide, 1-(3-tolyl) or Maretine $(3\text{-}CH_3C_6H_4)NHNHCONH_2$	165.19	lf (w, dil al)		183-4			al	B15, 508
13220	Semicarbazide, 1-(4-tolyl) $(4\text{-}CH_3C_6H_4)NHNHCONH_2$	165.19			190-1			w	$B15^2$, 239
13221	Semicarbazide, 1,1,4-triphenyl $(C_6H_5)_2NNHCONHC_6H_5$	303.36	nd (al)		206-7				$B15^2$, 115
13222	Semicarbazide, 1,4,4,-triphenyl $C_6H_5NHNHCON(C_6H_5)_2$	303.36	pl (al)		151-2			al	$B15^3$, 185
13223	Semicarbazide, 2,4,4-triphenyl $H_2NN(C_6H_5)CON(C_6H_5)_2$	303.36	cr (dil al)		128			al	$B15^3$, 181
13224	Serine *(D)* or *D*-2-Amino-3-hydroxy propionic acid $HOCH_2CH(NH_2)CO_2H$	105.09	nd or hex pr (w), $[\alpha]^{20/}_D$ + 6.9 (w, c=10)	d	228d			w	$B4^3$, 1572
13225	Serine *(DL)* or *DL*-2-Amino-3-hydroxy propionic acid $HOCH_2CH(NH_2)CO_2H$	105.10	mcl pr or lf (w)		246d (sealed tube)	$1.603^{22.5}$		w	$B4^3$, 1573

PHYSICAL CONSTANTS OF ORGANIC COMPOUNDS (Continued)

No.	Name, Synonyms, and Formula	Mol. wt.	Color, crystalline form, specific rotation and λ_{max} (log ε)	b.p. °C	m.p. °C	Density	n_D	Solubility	Ref.
13226	Serine *(L)* or *L*-2-Amino-3-hydroxy propionic acid $HO\text{-}CH_2CH(NH_2)CO_2H$	105.09	hex pl or pr (w), $[\alpha]^{20}_D$ −6.8 (w, c=10)	sub$150^{10^{-4}}$	228d	....	....	w	$B4^3$, 1568
13227	Serpentine $C_{21}H_{20}N_2O_3$	348.40	ye rods or lf (al), $[\alpha]^{25}_D$ + 292 (MeOH)	....	175	....	....	al, eth, ace, MeOH	Am76, 2843
13228	**Sesamin or Asaranin** $C_{20}H_{18}O_6$	354.36	nd (al), $[\alpha]^{20}_D$ + 68.4 (chl, c=24)	....	123-4	....	....	al, ace, bz, chl	$B19^4$, 6236
13229	**Shikonine or *d*-Alkanin** $C_{16}H_{16}O_5$	288.30	br-red nd (bz), $[\alpha]^{20}_{644}$ + 135 (bz, c=1.3)	....	147	....	....	al, eth, ace, bz	$B8^3$, 4088
13230	Silane, allyl-trimethyl $(CH_3)_3SiCH_2CH{=}CH_2$	114.26	....	85.4^{752}	......	$0.7193^{20/4}$	1.4074^{20}	....	$B4^3$, 1854
13231	**Silane, benzyl-trimethyl** $(CH_3)_3SiCH_2C_6H_5$	164.32	....	$191\text{-}2^{748}$, 93^{35}	......	$0.8933^{20/4}$	1.5042^{20}	....	$B16^3$, 1201
13232	**Silane, 4-bromophenoxy-trimethyl** $4\text{-}BrC_6H_4OSi(CH_3)_3$	245.19	....	126^{25}	......	$1.2619^{20/4}$	1.5145^{20}	....	$B6^4$, 1056
13233	**Silane, butyl-chloro-dimethyl** $C_4H_9Si(Cl)(CH_3)_2$	245.19	....	138.4^{747}	......	$0.8751^{20/4}$	1.5145^{20}	....	BOSC2^1, 180
13234	Silane, butyl-trichloro $C_4H_9SiCl_3$	191.56	....	148.9	......	$1.1606^{20/4}$	1.4363^{20}	bz, eth	$B4^1$, 582
13235	Silane, isobutyl-trichloro $i\text{-}C_4H_9SiCl_3$	191.56	....	....	141-6	$1.154^{20/4}$	......	....	$B4^1$, 582
13236	Silane, butyl-trimethyl $C_4H_9Si(CH_3)_3$	130.31	....	115	......	$0.7353^{0/4}$	......	....	$B4^3$, 1851
13237	**Silane, isobutyl-trimethyl** $i\text{-}C_4H_9Si(CH_3)_3$	130.31	....	....	108-9	$0.7330^{0/4}$	......	....	$B4^1$, 580
13238	Silane, chloro-dimethyl-ethyl $(CH_3)_2Si(Cl)C_2H_5$	122.67	....	89-90	......	0.8675^{20}	1.4105^{20}	....	$B4^3$, 1865
13239	Silane, chloro-dimethyl-phenyl $(CH_3)_2Si(Cl)C_6H_5$	170.71	....	196	......	1.0646^{20}	1.5184^{20}	....	$B16^3$, 1206
13240	Silane, chloro-dimethyl-vinyl $(CH_3)_2Si(Cl)CH{=}CH_2$	120.65	....	83-4	......	$0.8744^{20/4}$	1.4141^{20}	....	KHOC, 240
13241	Silane, chloro-ethyl-methyl-phenyl $C_6H_5SiCl(CH_3)C_2H_5$	184.74	....	....	......	....	....	CS_2	....
13242	Silane, chloro-phenyl $C_6H_5SiH_2Cl$	142.66	....	162-3	......	$1.0683^{20/4}$	1.5340^{20}	....	BOSC. 2., 147
13243	Silane, chloro-triethoxy $(C_2H_5O)_3SiCl$	198.72	....	156, 69^{12}	−51	$1.032^{20/20}$	1.3999^{20}	al	$B1^3$, 1336
13244	Silane, chloro-triethyl $(C_2H_5)_3SiCl$	150.72	....	144-5	......	$0.8967^{20/4}$	1.4314^{20}	....	$B4^3$, 1867
13245	Silane, chloro-trimethyl $(CH_3)_3SiCl$	108.64	....	57.7, $0^{61.5}$	−57.7	$0.8580^{20/4}$	1.3885^{20}	....	$B4^3$, 1857 334, BOSC2^1.7
13246	Silane, chloro-triphenyl $(C_6H_5)_3SiCl$	294.86	....	$240\text{-}3^{35}$	......	....	......	....	$B16^3$, 1208
13247	Silane, chloromethyl-dimethyl-phenyl $(CH_3)_2Si(C_6H_5)CH_2Cl$	184.74	....	225	......	....	......	CS_2	$B16^3$, 1200
13248	Silane, (4-chlorophenoxy)-trimethyl $4\text{-}ClC_6H_4OSi(CH_3)_3$	200.74	....	214^{758}	......	$1.0320^{20/4}$	1.4930^{20}	....	$B6^4$, 878
13249	Silane, (2-chlorophenoxy)-trimethyl $2\text{-}ClC_6H_4OSi(CH_3)_3$	200.74	....	$95\text{-}6^{15}$	......	....	......	....	$B6^4$, 809
13250	Silane, 3-chlorophenyl-trimethyl $3\text{-}ClC_6H_4Si(CH_3)_3$	200.74	....	206-7, $84\text{-}5^{9}$	......	$1.0071^{20/4}$	1.5108^{20}	....	$B16^3$, 1200
13251	Silane, 3-chloropropyl-trimethyl $Cl(CH_2)_3Si(CH_3)_3$	150.72	....	148	......	0.8825^{20}	$1.4310^{20}{}_{D}$	....	KHOC, 263
13252	**Silane, dibenzyl-diethoxy** $(C_6H_5CH_2)_2Si(OC_2H_5)_2$	300.47	....	$139^{0.5}$	......	....	1.5250^{20}	....	BOSC 2^1, 674
13253	Silane, dichloro-diisobutyl $(i\text{-}C_4H_9)_2SiCl_2$	213.22	....	93^{16}	......	$1.00^{20/4}$	......	....	$B4^3$, 1893
13254	Silane, dichloro-diethoxy $(C_2H_5O)_2SiCl_2$	189.11	....	135.9, 51.6^{12}	−130	$1.1290^{20/4}$	......	....	$B1^4$, 1364

No.	Name, Synonyms, and Formula	Mol. wt.	Color, crystalline form, specific rotation and λ_{max} (log ε)	b.p. °C	m.p. °C	Density	n_D	Solubility	Ref.
13255	Silane, dichloro-diethyl $(C_2H_5)_2SiCl_2$	157.11		d128-30	−96.5	1.0504^{20}	1.4309^{20}		B4³, 1890
13256	Silane, dichloro-dimethyl $(CH_3)_2SiCl_2$	129.06		70.5	−76	$1.070^{25/25}$	$1.405^{25/D}$		B4³, 1877
13257	Silane, dichloro-diphenyl $(C_6H_5)_2SiCl_2$	253.20		302-5, $163\text{-}5^{10}$		$1.2216^{20/4}$	1.5819^{20}	al, eth, ace, bz	B16³, 1214
13258	Silane, dichloro-ethyl-methyl $C_2H_5SiCl_2CH_3$	143.09		100^{747}		$1.0630^{20/4}$	1.4197^{20}		B4³, 1889
13259	Silane, dichloro-methyl-phenyl $C_6H_5SiCl_2CH_3$	191.13		206-7		1.1866^{20}	1.5180^{20}		B16³, 1211
13260	Silane, dichloro-methyl-isopropyl $i\text{-}C_3H_7SiCl_2CH_3$	157.11		121-2		1.0385^{20}	1.4270^{20}		C52, 6159
13261	Silane, dichloro methyl vinyl $CH_2{=}CHSiCl_2CH_3$	141.07		93		$1.085^{27/27}$	$1.4295^{20/D}$		B4³, 1894
13262	Silane, dichloro-phenyl $C_6H_5SiHCl_2$	177.11		180-2		$1.225^{25/25}$			B16³, 1210
13263	Silane, dichloro-isopropyl-methyl $CH_3SiCl_2CH(CH_3)_2$	157.11		121-2		1.0385^{20}	1.4270^{20}		C52, 6159
13264	Silane, diethoxy-difluoro $(C_2H_5O)_2SiF_2$	156.20		83	−122				B1³, 1336
13265	Silane, diethoxy-dimethyl $(C_2H_5O)_2Si(CH_3)_2$	148.28		113.8, 69.5^{179}	−87	$0.8395^{20/4}$	$1.3805^{20/D}$		B4³, 1874
13266	Silane, diethoxy-diphenyl $(C_2H_5O)_2Si(C_6H_5)_2$	272.42		$302\text{-}4^{767}$, $151\text{-}3^{6}$		$1.0329^{20/4}$	$1.52695^{20/D}$		B16², 608
13267	Silane, diethoxy-methyl $(C_2H_5O)_2SiHCH_3$	134.25		$135\text{-}8^{731}$		0.829^{25}	$1.3724^{25/D}$		KHOC, 358
13268	Silane, diethoxy-methyl-vinyl $(C_2H_5O)_2Si(CH_3)CH{=}CH_2$	160.29		133		0.8620^{20}	$1.4001^{20/D}$		KHOC, 388
13269	Silane, diethyl $(C_2H_5)_2SiH_2$	88.22		56	−134.4	$0.6832^{20/4}$	1.3920^{20}		B4³, 1846
13270	Silane, diethyl-difluoro $(C_2H_5)_2SiF_2$	124.21		60.9^{755}	−78.7	$0.9348^{20/4}$	1.3385^{20}		B4³, 1890
13271	Silane, difluoro-diphenyl $(C_6H_5)_2SiF_2$	220.29		252, 158^{50}		$1.145^{17/4}$	1.5221^{25}	bz	B16³, 1213
13272	Silane, dimethoxy-dimethyl $(CH_3O)_2Si(CH_3)_2$	120.22		80.5-82		0.8535^{20}	$1.3679^{20/D}$		KHOC, 346
13273	Silane, dimethoxy-diphenyl $(CH_3O)_2Si(C_6H_5)_2$	244.37		$90\text{-}5^{13}$		1.0771^{20}	$1.5447^{20/D}$		KHOC, 489
13274	Silane, dimethyl $(CH_3)_2SiH_2$	60.17		−20.1	−150.2	0.68^{-80}			B4¹, 579
13275	Silane, dimethyl-diphenoxy $(C_6H_5O)_2Si(CH_3)_2$	244.37		$130\text{-}2^{5}$	−23	1.0599^{25}	1.5330^{20}		B6⁴, 764
13276	Silane, dimethyl-diphenyl $(C_6H_5)_2Si(CH_3)_2$	212.37		276^{740}, 173^{45}		$0.9867^{20/4}$	1.5644^{20}		B16², 605
13277	Silane, dimethylphenyl $C_6H_5SiH(CH_3)_2$	136.27		157		0.8891^{20}	1.4995^{20}		KHOC, 396
13278	Silane, dimethylpropyl $C_3H_7SiH(CH_3)_2$	102.25		73-4					B4³, 1850
13279	Silane, dimethylamino-trimethyl $(CH_3)_3SiN(CH_3)_2$	117.27		86, $0^{22.5}$			1.4379^{24}		BOSC 2¹, 138
13280	Silane, diphenyl $(C_6H_5)_2SiH_2$	184.31		75				CS_2	
13281	Silane, diphenyl-methyl $(C_6H_5)_2SiHCH_3$	198.34		266.8, 93.5^{1}		$0.9973^{20/4}$	1.5694^{20}		BOSC 2², 522
13282	Silane, diphenyl-divinyl $(C_6H_5)_2Si(CH{=}CH_2)_2$	236.39		$130\text{-}1^{0.05}$		$1.0092^{25/4}$	$1.5350^{25/4}$		BOSC 2¹, 614
13283	Silane, ethoxy-trichloro $C_2H_5OSiCl_3$	179.51		101.9	−135	1.2274^{20}	1.4045^{20}	al	B1⁴, 1364
13284	Silane, ethoxy-triethyl $(C_2H_5)_3SiOC_2H_5$	160.33		154-5		0.8160^{20}	1.4140^{20}	al, eth	B4³, 1866
13285	Silane, ethoxy-trifluoro $C_2H_5OSiF_3$	130.14	gas	−7	−122				B1³, 1336
13286	Silane, ethoxy-trimethyl $C_2H_5OSi(CH_3)_3$	118.25		75^{745}		$0.7573^{20/4}$	$1.3741^{20/D}$	al, eth, ace	B4³, 1856
13287	Silane, ethoxy-triphenyl $(C_6H_5)_3SiOC_2H_5$	304.46		344, 207^{3}	65			chl	B16³, 1207

No.	Name, Synonyms, and Formula	Mol. wt.	Color, crystalline form, specific rotation and λ_{max} (log ε)	b.p. °C	m.p. °C	Density	n_D	Solubility	Ref.
13288	Silane, ethyl trichloro $C_2H_5SiCl_3$	163.51		97.9	−105.6	$1.2381^{20/4}$	1.4257^{20}		$B4^3$, 1900
13289	Silane, ethyl-triethoxy $(C_2H_5O)_3SiC_2H_5$	192.33		158.9		$0.8594^{20/4}$	1.3955^{20}	al, eth	$B4^3$, 1899
13290	Silane, ethyl-trifluoro $C_2H_5SiF_3$	114.14		−4.4	−105	1.227^{-76}			$B4^3$, 1900
13291	Silane, ethyl-trimethoxy $(CH_3O)_3SiC_2H_5$	150.25		124.3		$0.9488^{20/4}$	1.3838^{20}	al	$B4^3$, 1899
13292	Silane, ethyl-triphenyl $(C_6H_5)_3SiC_2H_5$	288.46			76			chl	
13293	Silane, fluoro-triethoxy $(C_2H_5O)_3SiF$	182.27		134.6					$B1^3$, 1336
13294	Silane, fluoro-triethyl $(C_2H_5)_3SiF$	134.27		110		$0.8354^{25/4}$	1.3900^{25}	peth	$B4^3$, 1866
13295	Silane, fluoro-triphenyl $(C_6H_5)_3SiF$	278.40		$245\text{-}52^{30}$, $160\text{-}70^{0.5}$	64				$B16^3$, 1208
13296	Silane, methyl CH_3SiH_3	46.14		−57	−156.5				$B4^1$, 579
13297	Silane, methyl-phenyl $C_6H_5SiH_2CH_3$	122.24		140, $46\text{-}7^{20}$		$0.8895^{20/4}$	1.5058^{20}		BOSC 2^1, 209
13298	Silane, methyl-tribromo CH_3SiBr_3	282.83		131.3^{744}	−28.4	2.2130^{25}	1.5152^{25}		$B4^3$, 1897
13299	Silane, methyl-trichloro CH_3SiCl_3	149.48		66.4	−77.8	$1.273^{25/23}$			$B4^3$, 1896
13300	Silane, methyl-triethoxy $(C_2H_5O)_3SiCH_3$	178.30		143		0.8923^{20}	1.3835^{20}	al	$B4^3$, 1895
13301	Silane, methyl-trimethoxy $(CH_3O)_3SiCH_3$	136.22		102		$0.951^{25/4}$	1.3687^{25}	chl	BOSC 2^1, 109
13302	Silane, methyl-triphenoxy $(C_6H_5O)_3SiCH_3$	322.44		$267\text{-}71^{100}$, $210\text{-}1^{12}$		1.135^{20}	1.5599^{20}		$B6^4$, 765
13303	Silane, methyl-triphenyl $(C_6H_5)_3SiCH_3$	274.44		$196\text{-}200^{9}$	69.5			CS_2	$B16^3$, 1199
13304	Silane, 2-methylphenoxy-trimethyl $2\text{-}CH_3C_6H_4OSi(CH_3)_3$	180.32		192		$0.9287^{20/4}$	1.4830^{20}		BOSC 2^1, 386
13305	Silane, 3-methylphenoxy-trimethyl $3\text{-}CH_3C_6H_4OSi(CH_3)_3$	180.32		198^{3}		$0.9186^{20/4}$	1.4791^{20}		BOSC 2^1, 386
13306	Silane, 4-methylphenoxy-trimethyl $4\text{-}CH_3C_6H_4OSi(CH_3)_3$	180.32		199		$0.9183^{20/4}$	1.4790^{20}		BOSC 2^1, 386
13307	Silane, pentyl-trichloro $C_5H_{11}SiCl_3$	205.59		171^{742}, 60.5^{15}		$1.1330^{20/4}$	1.4503^{20}		$B4^3$, 1905
13308	Silane, pentyl-triethoxy $C_5H_{11}Si(OC_2H_5)_3$	234.41		95^{13}		0.8862^{20}	1.4059^{20}		$B4^3$, 1905
13309	Silane, phenyl $C_6H_5SiH_3$	108.22		118-20		$0.8681^{20/4}$	1.5125^{20}		$B16^3$, 1198
13310	Silane, phenyl-tri-butyl $(C_4H_9)_3SiC_6H_5$	276.54		$140\text{-}2^{4}$		0.8753^{20}	1.4915^{20}		$B16^3$, 1199
13311	Silane, phenyl-trichloro $C_6H_5SiCl_3$	211.55		201				chl	$B16^3$, 1216
13312	Silane, phenyl-triethoxy $C_6H_5Si(OC_6H_5)_3$	240.37		235-8		$0.9961^{20/4}$	1.4718^{20}		$B16^3$, 1215
13313	Silane, phenyl-triethyl $C_6H_5Si(C_2H_5)_3$	192.38		283		$0.8915^{20/4}$	1.4999^{20}		$B16^3$, 1198
13314	Silane, phenyl-trifluoro $C_6H_5SiF_3$	162.19		101.5, 16.5^{24}		$1.2169^{20/4}$	1.4110^{20}	al, bz	$B16^3$, 1216
13315	Silane, phenyl-trimethyl $C_6H_5Si(CH_3)_3$	150.30		170				CS_2	$B16^3$, 1198
13316	Silane, phenyl-tripropyl. $(C_3H_7)_3SiC_6H_5$	234.46		$146\text{-}7^{18}$		$0.8799^{20/4}$	1.4950^{20}		BOSC 2^1, 601
13317	Silane, propyl-trichloro $C_3H_7SiCl_3$	177.53		122.2^{740}		$1.1851^{20/4}$	1.4290^{20}		$B4^3$, 1902
13318	Silane, propyl-trichloro $(CH_3)_2CHSiCl_3$	177.53		120.3^{748}	−87.7	$1.1934^{20/4}$	$1.4319^{20}/_D$		$B4^3$, 1903
13319	Silane, tetra(allyloxy) $(C_3H_5O)_4Si$	256.37		$134\text{-}5^{34}$		$0.9824^{20/4}$	1.4349^{20}		$B1^3$, 1886
13320	Silane, tetraethoxy $(C_2H_5O)_4Si$	208.33		165.5-8	−77	$0.9356^{20/20}$	1.3832^{20}		$B1^4$, 1360
13321	Silane, tetraethyl $(C_2H_5)_4Si$	144.33		153		0.7658^{20}	1.4268^{20}		$B4^3$, 1847

No.	Name, Synonyms, and Formula	Mol. wt.	Color, crystalline form, specific rotation and λ_{max} (log ε)	b.p. °C	m.p. °C	Density	n_D	Solubility	Ref.
13322	Silane, tetramethoxy $(CH_3O)_4Si$	152.22		121-2		1.032^{20}	1.3683^{20}		B1[4], 1266
13323	Silane, tetramethyl $(CH_3)_4Si$	88.22		26.5	−102.2	$0.648^{19/4}$	1.3587^{20}	al, eth	B4[3], 1843
13324	Silane, tetraphenyl $(C_6H_5)_4Si$	336.51		228^{5}	236-7	$1.078^{20/4}$		CS_2	B16[3], 1199
13325	Silane, tetrapropoxy $(C_3H_7O)_4Si$	264.44		$225\text{-}7^{757}$, 94^{5}		$0.9158^{20/4}$	1.4012^{20}		B1[4], 1435
13326	Silane, tetravinyl $(CH_2{=}CH)_4Si$	136.27		130.2		0.7999^{20}	1.4625^{20}		KHOC, 278
13327	Silane, 2-tolyl-trichloro $(2\text{-}CH_3C_6H_4)SiCl_3$	225.58		226.7	−26	$1.306^{25/4}$	1.5336^{25}		BOSC 2[1], 204
13328	Silane, tributyl $(c_4H_9)_3SiH$	200.44		215-20 $88\text{-}9^{5}$		$0.7794^{20/4}$	$1.4380^{20}{}_D$		BOSC 2[1], 509
13329	Silane, triethoxy $(C_2H_5O)_3SiH$	164.28		132-5		$0.8745^{20/4}$			B1[4], 1359
13330	Silane, triethoxy-vinyl $(C_2H_5O)_3SiCH{=}CH_2$	190.31							BOSC 2[1], 294
13331	Silane, triethyl $(C_2H_5)_3SiH$	116.28		109^{755}		0.7302^{20}	1.4117^{20}		B4[3], 1847
13332	Silane, trimethyl vinyl $(CH_3)_3SiCH{=}CH_2$	100.24		55.5^{767}		0.6910^{20}	$1.3914^{20}{}_D$		KHOC, 354
13333	Silane, trioctyl $(C_8H_{17})_3SiH$	368.76		$163\text{-}5^{0.15}$		$0.8207^{20/20}$	1.4545^{20}		BOSC 2[2], 81
13334	Silane, triphenyl $(C_6H_5)_3SiH$	260.41		152-61	36-7				B16[3], 1199
13335	Silane, tripropyl $(C_3H_7)_3SiH$	158.36		171 3		$0.7723^{0/4}$	1.4280^{20}		B4[3], 1850
13336	Silane, vinyl-trichloro $CH_2{=}CHSiCl_3$	161.49		91-2	−95	$1.2426^{20/4}$	1.4295^{20}	chl	B4[3], 1908
13337	Silanol, dimethyl-ethyl $(CH_3)_2Si(OH)C_2H_5$	104.22		120^{774}, 58^{50}		$0.8332^{20/4}$	1.4070^{20}		BOSC 2[1], 108
13338	Silanol, diphenyl-methyl $(C_6H_5)_5Si(OH)CH_3$	214.34		148^{3}	167			CS_2	
13339	Silanol, triethyl $(C_2H_5)_3Si(OH)$	132.28		154, 46.6^{9}		$0.8647^{20/4}$	1.4329^{20}	al, eth	B4[3], 1866
13340	Silanediol, dimethyl, diacetate $(CH_3CO_2)_2Si(CH_3)_2$	176.24		165^{750}, $44\text{-}5^{3}$		$1.0540^{20/4}$	1.4030^{20}		B4[3], 1876
13341	Silanetriol, methyl-triacetate $(CH_3CO_2)_3SiCH_3$	220.25		$110\text{-}2^{17}$	40.5	$1.1677^{25/4}$	1.4083^{20}		B4[3], 1896
13342	Silin, 1,1-diphenyl-perhydro $C_{17}H_{20}Si$	252.43		$193\text{-}8^{5}$		1.0319^{25}	1.5779^{20}		BOSC2[1], 647
13343	Sinapine hydrogen sulfate $C_{16}H_{23}NO_5.H_2SO_4$	407.44	lf (al)		186-7d			w	B10[2], 354
13344	Sinapine, thiocyanate, monohydrate $C_{16}H_{23}NO_5.HSCN.H_2O$	386.46	ye nd		180-1				B10[2], 354
13345	Sinomenine or Cuculine $C_{19}H_{23}NO_4$	329.40	nd (bz), $[\alpha]^{26}{}_D$ −71 (al, c=2.1)		162			al, ace, chl	B21[4], 6670
13346	α-Sitosterol $C_{29}H_{48}O$	412.70	nd (al), $[\alpha]^{28}{}_D$ −1.7 (chl,c=2)		166			al, chl	B6[3], 2876
13347	α_2-Sitosterol $C_{30}H_{50}O$	426.73	cr (al-peth), $[\alpha]^{25}{}_D$ +3.5 (chl,c=2)		156			al	B6[3], 2697
13348	α_3-Sitosterol $C_{29}H_{48}O$	412.70	pl (al), $[\alpha]^{20}{}_D$ +5.2 (chl)		142-3			al	B6[3], 2707
13349	β-Sitosterol or Verosterol $C_{29}H_{50}O$	414.72	pl (al), nd (MeOH), $[\alpha]^{25}{}_D$ −37 (chl,c=2)		140			al, eth, aa	B6[3], 2696
13350	Skimmin or 7-Hydroxycoumarin-β-D-glucose $C_{15}H_{16}O_8$	324.29	cr (w + 1), $[\alpha]^{18}{}_D$ −80		219-21			al	B18[4], 301

PHYSICAL CONSTANTS OF ORGANIC COMPOUNDS (Continued)

No.	Name, Synonyms, and Formula	Mol. wt.	Color, crystalline form, specific rotation and λ_{max} (log ε)	b.p. °C	m.p. °C	Density	n_D	Solubility	Ref.
13351	Smilagenin or Isosapogenin $C_{27}H_{44}O_3$	416.64	silky nd (ace), $[\alpha]^{25/}_{D}$ −69 (chl,c=0.5)		185			al, ace, bz, chl	B19[4], 826
13352	Solanidine-T or Solatubine $C_{27}H_{43}NO$	397.64	lo nd (chl-MeOH)	sub d	218-9			bz, chl	B21[4], 1398
13353	Solanine-S or Purapurine $C_{45}H_{73}NO_{16}$	884.07	nd (al), fl (diox) $[\alpha]^{20/}_{D}$ −69 (al)		*ca* 190				J1963, 745
13354	Solasodine or Purapuridine $C_{27}H_{43}NO_2$	413.64	hex pl (sub), $[\alpha]^{20/}_{D}$ −92.4 (bz)		202			ace, bz, chl, MeOH, Py	J1942, 13
13355	Sorbaldehyde or 2,4-Hexadienal $CH_3CH{=}CHCH{=}CHCHO$	96.13		$173\text{-}4^{756}$, 76^{30}		0.898^{20}	1.5384^{20}		B1[4], 3545
13356	Sorbamide $CH_3CH{=}CHCH{=}CHCONH_2$	111.14	nd (w)		171-2			w, al	B2[4], 1705
13357	Sorbic Acid or 2,4-Hexadienoic acid $CH_3CH{=}CHCH{=}CHCO_2H$	112.13	nd (dil al)	228d, 153^{50}	134.5	$1.204^{19/4}$		al, eth	B2[4], 1701
13358	Sorbic acid, ethyl ester or Ethyl sorbate $CH_3CH{=}CHCH{=}CHCO_2C_2H_5$	140.18		195.5, 85^{20}		$0.9506^{20/4}$	1.4951^{20}	al, eth, chl	B2[4], 1703
13359	Sorbic acid, methyl ester or Methyl sorbate $CH_3CH{=}CHCH{=}CHCO_2CH_3$	126.16	lf	180, 70^{20}	15	$0.9777^{20/4}$	1.5025^{22}	al, eth	B2[4], 1703
13360	Sorbylalcohol or 2,4-Hexadien-1-ol $CH_3CH{=}CHCH{=}CHCH_2OH$	98.14	nd	76^{12}	30-1	$0.8967^{20/4}$	1.4981^{20}	al, eth	B1[4], 2240
13361	Sorbyl chloride $CH_3CH{=}CHCH{=}CHCOCl$	130.57		78^{15}		$1.0666^{19/4}$	1.5545^{20}	ace	B2[3], 1460
13362	Sorbierite or L-Iditol $HOCH_2(CHOH)_4CH_2OH$	182.17	pr (al), $[\alpha]^{\prime}_{D}$ −3.5 (w,p=2)		73-4			w	B1[4], 2843
13363	D-Sorbitol or D-Gulcitol $C_6H_{14}O_6$, $HOCH_2(CHOH)_4CH_2OH$	182.17	nd (w + 0.5), $[\alpha]^{25/}_{D}$ −1.98 (w)	$295^{3.5}$	110-2 (anh), 75 (hyd)	$1.489^{20/4}$	1.3330^{20}	w, ace, aa	B1[4], 2839
13364	L-Sorbitol or D-Gulitol $C_6H_{14}O_6$,$HOCH_2(CH_2OH)_4CH_2OH$	182.17	nd (w + 0.5), $[\alpha]$ + 1.7 (w)		77			w, ace, aa	B1[3], 2405
13365	D-Sorbitol, hexaacetate $C_{18}H_{26}O_{12}$	434.40	pr (w), $[\alpha]^{18/}_{D}$ + 6.8 (aa)		99.5 (120)			al	B2[4], 275
13366	D-Sorbose $C_6H_{12}O_6$	180.16	rh (al), $[\alpha]^{20/}_{D}$ + 42.9 (w)		165	1.612^{17}		w	B1[4], 4411
13367	DL-Sorbose or β-Acrose $C_6H_{12}O_6$	180.16	rh (dil al)		162-3	1.638^{17}		w, MeOH	B31, 348
13368	L-Sorbose $C_6H_{12}O_6$	180.16	rh (al), $[\alpha]^{20/}_{D}$ −43.2 (w,c=5)		165	1.612^{17}		w	B1[4], 4412
13369	Sparteine *(D)* or Lupinidine $C_{15}H_{26}N_2$	234.38	$[\alpha]^{20/}_{D}$ −19.5 (al)	325^{754}, 173^{8}	30-1	$1.0196^{20/4}$	1.5312^{20}	al, eth, chl	B23[2], 97
13370	Sparteine sulfate, pentahydrate $C_{15}H_{26}N_2\cdot H_2SO_4\cdot 5H_2O$	422.53	pr $[\alpha]^{17/}_{D}$ −15.3		242		1.5289	al	B23[2], 99
13371	Spiro [5,5]-hendecane-2,4,8,10-tetraoxa or Pentacrythritol dimethylene ether $C_7H_{12}O_4$	160.17		147^{53}	46.5			w, al, eth, ace, bz	B19[4], 5650
13372	Spiropentane C_5H_8	68.12		$39\text{-}40^{746}$		$0.7266^{20/4}$	1.4120^{20}		B5[4], 218
13373	Squalene or Spinacene $C_{30}H_{50}$	410.73		280^{17}	<−20	$0.8584^{20/4}$	1.4990^{20}	eth, ace	B1[4], 1146
13374	Squalene, perhydro or 2,6,10,15,19,23-Hexamethyltetracosane $C_{30}H_{62}$	422.82	oil	*ca* 350, 263^{10}	−38	$0.8125^{15/4}$	1.4525^{20}	eth, **bz**, peth, chl	B1[3], 585

PHYSICAL CONSTANTS OF ORGANIC COMPOUNDS (Continued)

No.	Name, Synonyms, and Formula	Mol. wt.	Color, crystalline form, specific rotation and λ_{max} (log ε)	b.p. °C	m.p. °C	Density	n_D	Solubility	Ref.
13375	Stachydrine *(L)* or Hygric acid methylbetaine $C_7H_{13}NO_2$	143.19	cr (w + 1), $[\alpha]^{25/}_D$ −40.2 (w,c=4)		235d (anh), 116-8 (+ w)			w, al	$B22^4$, 26
13376	Stachydrine-*(L)*-oxalate $C_7H_{13}NO_2.H_2C_2O_4$	233.22	nd		105-7				$B22^1$, 484
13377	Starch or Amylum $(C_6H_{10}O_5)n$		amor pw		d				C51, 11746
13378	Starch triacetate $C_{216}H_{288}O_{144}$	5188.6	pw $[\alpha]^{25/}_D$ + 72.5		180d			ace, bz	C46, 8401
13379	Stearamide $CH_3(CH_2)_{16}CONH_2$	283.50	lf (al)	$250\text{-}1^{12}$	109			eth, chl	$B2^4$, 1240
13380	Stearamide, *N*-phenyl or Stearanilide $C_{17}H_{35}CONHC_6H_5$	359.60	nd (al)	153.5^{10}	94			al, eth, ace, bz, chl	$B12^3$, 486
13381	Stearic acid or Octadecanoic acid $CH_3(CH_2)_{16}CO_2H$	284.48	mcl lf (al)	360d, 232^{15}	71.2	$0.9408^{20/4}$	1.4299^{20}	eth, ace, chl	$B2^4$, 1206
13382	Stearic acid anhydride or Stearic anhydride $[C_{17}H_{35}CO]_2O$	550.95			72	$0.8365^{82/4}$	1.4362^{80}		$B2^4$, 1239
13383	Stearic acid, benzyl ester or Benzyl stearate $C_{17}H_{35}CO_2CH_2C_6H_5$	374.61	pa ye		28 (45)	$0.9075^{50/25}$	1.4663^{50}		$B6^3$, 1481
13384	Stearic acid, butyl ester $C_{17}H_{35}CO_2C_4H_9$	340.59		$343,223^{15}$	27.5	$0.855^{20/4}$	1.4328^{50}	al, ace	$B2^4$, 1219
13385	Stearic acid, isobutyl ester $C_{17}H_{35}CO_2\text{-}i\text{-}C_4H_9$	340.59	wax	199^5	(i)22.5 (ii)28.9	$0.8498^{20/4}$		eth	$B2^3$, 1017
13386	Stearic acid, cyclohexyl ester or Cyclohexyl stearate $C_{17}H_{35}CO_2C_6H_{11}$	366.63			44	$0.890^{15/15}$		eth	$B6^4$, 38
13387	Stearic acid, 9,10-dibromo *(trans)* $CH_3(CH_2)_7CHBrCHBr(CH_2)_7CO_2H$	442.27	col or ye		29-30	$1.2458^{30/4}$	1.4893^{42}	eth	$B2^3$, 1048
13388	Stearic acid, 2,3-dihydroxy $C_{15}H_{31}CHOHCHOHCO_2H$	316.48	nd (aa)		α: 107 β: 126				$B3^4$, 1089
11388a	Stearic acid, 9,10-dihydroxy $C_8H_{17}CH(OH)CH(OH)(CH_2)_7CO_2H$	316.48	lf		95				$B2^4$, 1218
13389	Stearic acid, ethyl ester or Ethyl stearate $C_{17}H_{35}CO_2C_2H_5$	312.54		199^{10}	31-3	$1.057^{20/4}$	1.4349^{40}	al, eth, ace	$B2^4$, 1249
13390	Stearic acid, 9,10,12,13,15,16-hexabromo $CH_3[CH_2CHBrCHBr]_3(CH_2)_7CO_2H$	757.86	cr (diox)		182				$B2^4$, 1220
13391	Stearic acid, hexadecyl ester or Cetyl stearate $C_{17}H_{35}CO_2C_{16}H_{33}$	508.91	lf or pl (eth, aa)		57		1.4410^{70}	eth, ace, chl	$B3^4$, 934
13392	Stearic acid, 2-hydroxy $C_{16}H_{33}CH(OH)CO_2H$	300.48	flat nd (chl, AcOEt)		93			al, eth, ace, bz, chl	$B3^4$, 935
13393	Stearic acid, 3-hydroxy $C_{15}H_{31}CH(OH)CH_2CO_2H$	300.48	pl (chl)		89-90			al, eth, chl	$B3^4$, 940
13394	Stearic acid, 10-hydroxy $C_8H_{17}CHOH(CH_2)_8CO_2H$	300.48	pl		84			al	$B3^4$, 941
13395	Stearic acid, 11-hydroxy $C_7H_{15}CH(OH)(CH_2)_9CO_2H$	300.48	ta or pl (al)		81-2				$B3^4$, 942
13396	Stearic Acid, 12-hydroxy $C_6H_{13}CH(OH)(CH_2)_{10}CO_2H$	300.48	(al)		82			al, eth, chl	$B2^4$, 1222
13397	Stearic acid, 2-hydroxyethyl ester or Glycolmonostearate $C_{17}H_{35}CO_2CH_2CH_2OH$	328.54	(peth)	$189\text{-}91^3$	60-1	$0.8780^{60/4}$	1.4310^{60}		$B2^4$, 1216
13398	Stearic acid, methyl ester or Methylstearate $C_{17}H_{35}CO_2CH_3$	298.51		$442\text{-}3^{747}$, 215^{15}	39.1	$0.8498^{40/4}$	1.4367^{40}	al, eth, ace, chl	$B2^4$, 1272
13399	Stearic acid, 10-methyl *(D)* or Tuberculostearic acid $C_8H_{17}CH(CH_3)(CH_2)_8CO_2H$	298.51	$[\alpha]^{22/}_D$ −0.11 (ace)	$175\text{-}8^{0.7}$	12-3	$0.8771^{25/4}$	1.4512		$B3^4$, 1093
13401	Stearic acid, 9,10-dioxo or Stearoxylic acid $CH_3(CH_2)_7COCO(CH_2)_7CO_2H$	312.45	ye pl (al)		86			al, eth, ace, lig	$B3^4$, 1796
13402	Stearic acid, 9,10,12,13,15,16-hexabromo or α-Linolenic acid hexabromide $CH_3[CH_2CHBrCHBr]_3(CH_2)_7CO_2H$	757.86	cr (diox)		182				$B2^4$, 1249
13403	Stearic acid, 3-oxo or Palmitoyl acetic acid $C_{15}H_{31}COCH_2CO_2H$	298.47			102-3			al	$B3^4$, 1687
13404	Stearic acid, 3-oxo, ethyl ester $C_{15}H_{31}COCH_2CO_2C_2H_5$	326.52	(al or peth)		37-8			al, peth	$B3^4$, 1687
13405	Stearic acid, 6-oxo or Lactaric acid $C_{12}H_{25}CO(CH_2)_4CO_2H$	298.47	pl (al, peth)		87			eth, chl	$B3^4$, 1688

PHYSICAL CONSTANTS OF ORGANIC COMPOUNDS (Continued)

No.	Name, Synonyms, and Formula	Mol. wt.	Color, crystalline form, specific rotation and λ_{max} (log ε)	b.p. °C	m.p. °C	Density	n_D	Solubility	Ref.
13406	Stearic acid, 6-oxo, ethyl ester $C_{12}H_{25}CO(CH_2)_4CO_2C_2H_5$	326.52	cr		47			al, eth	$B3^{1}$, 253
13407	Stearic acid, 10-oxo $C_8H_{17}CO(CH_2)_8CO_2H$	298.47	pl (al)		76 (82)				$B3^{4}$, 1689
13408	Stearic acid, 10-oxo, ethyl ester $C_8H_{17}CO(CH_2)_8CO_2C_2H_5$	326.53	pl (al)		41			al	B3, 725
13409	Stearic acid, 12-oxo $C_6H_{13}CO(CH_2)_{10}CO_2H$	298.47	lf (aa), cr (lig)		82			al	$B3^{4}$, 1690
13410	Stearic acid, 12-oxo, ethyl ester $C_6H_{13}CO(CH_2)_{10}CO_2C_2H_5$	326.52	lf (al)	$199\text{-}200^{3}$	38			al	$B3^{3}$, 1294
13411	Stearic acid, pentyl ester or Pentylo, stearate Amylstearate $C_{17}H_{35}CO_2C_5H_{11}$	354.62	pl		30		1.4342^{50}	al, eth	$B2^{4}$, 1220
13412	Stearic acid, isopentyl ester or Isoamyl stearate $C_{17}H_{35}CO_2CH_2CH_2CH(CH_3)_2$	354.62		192^{2}	25.5	$0.855^{20/4}$	1.433^{50}	eth, ace	$B2^{3}$, 1017
13413	Stearic acid, phenyl ester or Phenyl stearate $C_{17}H_{35}CO_2C_6H_5$	360.59		267^{15}	51-3			al, eth	$B6^{4}$, 618
13414	Stearic acid, propyl ester or Propyl stearate $C_{17}H_{35}CO_2C_3H_7$	326.56	cr (eth), pr (peth)	186.8^{2}	fr 28.9	0.8452^{38}	1.4400^{30}	al, eth, ace	$B2^{4}$, 1219
13415	Stearic acid, isopropyl ester or Isopropyl stearate $C_{17}H_{35}CO_2CH(CH_3)_2$	326.56		207^{6}	28	0.8403^{38}		al, eth, ace, chl	$B2^{4}$, 1219
13416	Stearic acid, 9,10,11,12-tetrabromo or Linoleic acid tetrabromide $C_5H_{11}[CHBrCHBrCH_2]_2(CH_2)_6CO_2H$	600.07	pl or lf (aa)		114-5			al, eth, bz, chl, aa	$B2^{4}$, 1247
13417	Stearic acid, 9,10,12,13-tetrabromo, ethyl ester $C_5H_{11}[CHBrCHBrCH_2]_2(CH_2)_6CO_2C_2H_5$	628.12	nd		63				$B2^{3}$, 1049
13418	Stearic acid, 9,10,12,13-tetrabromo, methyl ester $C_5H_{11}[CHBrCHBrCH_2]_2(CH_2)_6CO_2CH_3$	614.09	lf	215^{15}	63		1.4346^{45}	al, eth, chl	$B2^{4}$, 1248
13419	Stearic acid, tetrahydrofurfuryl ester $C_{17}H_{35}CO_2CH_2(C_4H_70)$	368.60			22	$0.917^{25/25}$		al, eth	Am50, 134
13420	Stearolic acid or 9-Octadecynoic acid $CH_3(CH_2)_7C{\equiv}C(CH_2)_7CO_2H$	280.45	pr (al, peth), nd (dil al)	260, $189\text{-}90^{1.8}$	48	1.4510^{54}		eth	$B2^{4}$, 1751
13421	Stearolic acid, ethyl ester or Ethylstearolate $CH_3(CH_2)_7C{\equiv}C(CH_2)_7CO_2C_2H_5$	308.50		$180^{2.5}$			1.4555^{20}		$B2^{4}$, 1752
13422	Stearolic acid, methyl ester or Methylstearolate $CH_3(CH_2)_7C{\equiv}C(CH_2)_7CO_2CH_3$	294.48		175^{3}			1.4562^{20}		$B2^{4}$, 1751
13423	Stearone or 18-Pentatriacontanone $(C_{17}H_{35})_2CO$	506.94	lf (lig)		88.4	$0.793^{95/4}$			$B1^{4}$, 3413
13424	Stearonitrile $C_{17}H_{35}CN$	265.48		362, 193^{10}	41	$0.8325^{20/4}$	1.4389^{45}	al, eth, ace, chl	$B2^{4}$, 1242
13425	Stearophenone $C_{17}H_{35}COC_6H_5$	344.58			56			ace	$B7^{3}$, 1317
13426	Stearoxylic acid or 9,10-Dioxostearic acid $C_8H_{17}COCO(CH_2)_7CO_2H$	312.45	yr pl (al)		86			al, eth, ace, lig	$B3^{4}$, 1796
13427	Stearoyl chloride $C_{17}H_{35}COCl$	302.93		215^{15}	23		1.4523^{24}	al	$B2^{4}$, 1240
13428	Stearyl alcohol or 1-Octadecanol $CH_3(CH_2)_{16}CH_2OH$	270.50	lf (al)	210.5^{15}	59-60	$0.8124^{59/4}$		al, eth, chl	$B1^{4}$, 1888
13429	Stearyl sulfate or Octadecyl sulfate $(C_{18}H_{37}O)_2SO_2$	603.04			70.5				$B1^{4}$, 1892
13430	Stibine, bis (trifluoromethyl)-bromo $(F_3C)_2SbBr$	339.67		113					J1957, 3708
13431	Stibene, bis (trifluoromethyl)-chloro $(F_3C)_2SbCl$	295.22		*ca* 88, 17^{20}					J1957, 3708
13432	Stibene, bis(trifluoromethyl)-iodo $(F_3C)_2SbI$	386.67		*ca* 129, 16^{8}	−42				J1957, 3708
13433	Stibene, dibromo-trifluoromethyl CF_3SbBr_2	350.56		*ca* 157, $34^{2.5}$					J1957, 3708
13434	Stibene, dichloro-phenyl $C_6H_5SbCl_2$	269.76	nd or ta	$110\text{-}5^{10}$	69-70			al, ace, aa	$B16^{3}$, 1167
13435	Stibene, diido-trifluoromethyl F_3CSbI_2	444.57	bt ye	>200d	4-8				J1957, 3708
13436	Stibene, triethyl $(C_2H_5)_3Sb$	208.94		161.4	−98	1.3224^{15}		al, eth	$B4^{3}$, 1834
13437	Stibene, trimethyl $(CH_3)_3Sb$	166.85		80.6	−62	1.523^{15}	1.42^{15}	al, eth	$B4^{3}$, 1834

PHYSICAL CONSTANTS OF ORGANIC COMPOUNDS (Continued)

No.	Name, Synonyms, and Formula	Mol. wt.	Color, crystalline form, specific rotation and λ_{max} (log ε)	b.p. °C	m.p. °C	Density	n_D	Solubility	Ref.
13438	Stibene, triphenyl $(C_6H_5)_3Sb$	353.07	pr (peth)	>360, $>220^{1}$	53.5	$1.4343^{25/4}$	1.6948^{42}	eth, ace, bz, chl, aa	$B16^3$, 1159
13439	Stibene, tris(trifluoromethyl) $(CF_3)_3Sb$	328.77	....	73	−58	....	....	....	J1957, 3708
13440	Stibene, tri-(2-tolyl) $(2\text{-}CH_3C_6H_4)_3Sb$	395.15	(al)	....	102	....	....	eth, bz, chl, peth	B16, 892
13441	Stibene, tri-(3-tolyl) $(3\text{-}CH_3C_6H_4)_3Sb$	395.15	(peth)	....	72	$1.3957^{16/4}$	....	al, eth, bz, chl, aa	B16, 892
13442	Stibene, tri(4-tolyl) $(4\text{-}CH_3C_6H_4)_3Sb$	395.15	rh (eth, MeOH)	....	127-8	1.3595^{16}	....	eth, bz, chl	$B16^3$, 1161
13443	Stigmastanol or β-Sitostanol $C_{20}H_{52}O$	416.74	pl (al), cr (+1 w), $[\alpha]^{20'}_D$ +24.8 (chl, c=1.1)	....	144.5	....	....	chl	$B6^3$, 2172
13444	Stigmasterol $C_{29}H_{48}O$	412.70	cr (al, +1 w), $[\alpha]^{22'}_D$ −51 (chl,c=2)	....	170	....	....	eth, ace, bz, chl	$B6^3$, 2857
13445	2-Stilbazole *(cis)* $C_{13}H_{11}N$	181.24	....	141^{10}	−50	....	....	al, eth, bz	J1958, 2202
13446	2-Stilbazole *(trans)* $C_{13}H_{11}N$	181.24	....	$324\text{-}5^{750}$, 194^{14}	91-2	....	....	al, eth, bz, lig	$B20^4$, 3874
13447	4-Stilbazole -*(trans)* $C_{13}H_{11}N$	181.24	....	208^{35}	131	....	....	al, eth, chl	$B20^4$, 3880
13448	Stilbene - *(cis)* $C_6H_5CH{=}CHC_6H_5$	180.25	....	141^{13}	5-6	$1.0143^{20/4}$	1.6130^{20}	al, eth, ace, bz, chl	$B5^3$, 1958
13449	Stilbene *(trans)* or *trans*-1,2-Diphenylthylene $C_6H_5CH{=}CHC_6H_5$	180.25	cr (al)	305^{720}, $166\text{-}7^{12}$	124-5	0.9707	1.6264^{17}	eth, bz	$B5^3$, 1953
13450	Stilbene, 2-bromo *(cis)* $2\text{-}BrC_6H_4CH{=}CHC_6H_5$	259.15	....	$121^{0.5}$	....	....	1.6404^{25}	al, lig	Am78, 475
13451	Stilbene, 2-bromo *(trans)* $2\text{-}BrC_6H_4CH{=}CHC_6H_5$	259.15	....	$145^{0.5}$	34	....	1.6822^{25}	al, lig	$B5^3$, 1964
13452	Stilbene, α-chloro *(cis)* $C_6H_5CCl{=}CHC_6H_5$	214.69	....	$160\text{-}2^{12}$	....	....	1.6281^{19}	al, eth, bz, chl	$B5^3$, 1961
13453	Stilbene, α-chloro-*(trans)* $C_6H_5CCl{=}CHC_6H_5$	214.69	....	320-4	53-4	....	....	al, eth, bz, chl	$B5^3$, 1961
13454	Stilbene, 2-chloro *(trans)* $2\text{-}ClC_6H_4CH{=}CHC_6H_5$	214.69	....	$208\text{-}10^{30}$, $138\text{-}40^{2}$	39-40	....	....	al, chl	$B5^3$, 1961
13455	Stilbene, 3-chloro *(trans)* $3\text{-}ClC_6H_4CH{=}CHC_6H_5$	214.69	....	$175\text{-}80^{0.2}$	73-4	....	....	al, chl	$B5^3$, 1961
13456	Stilbene, 2,2′-diamino *(cis)* $2\text{-}H_2NC_6H_4CH{=}CHC_6H_4NH_2\text{-}2$	210.28	red nd (w)	....	123	....	....	ace	$B13^3$, 510
13457	Stilbene, 2,2′-diamino *(trans)* $2\text{-}H_2NC_6H_4CH{=}CH(C_6H_4NH_2\text{-}2)$	210.28	gold-ye pr (al)	....	176	....	....	ace	$B13^3$, 510
13458	Stilbene, 4,4′-diamino (trans) $4\text{-}H_2NC_6H_4CH{=}CH(C_6H_4NH_2\text{-}4)$	210.28	ye nd or lf (al)	sub	231	....	....	MeOH	$B13^3$, 513
13459	Stilbene, 2,2′-dibromo *(trans)* $2\text{-}BrC_6H_4CH{=}CH(C_6H_4Br\text{-}2)$	338.04	....	....	215-6	....	....	al, bz	$B5^3$, 1965
13460	Stilbene, α, β-dichloro *(cis)* $C_6H_5CCl{=}CClC_6H_5$	249.14	....	180^{18}	67.5-8	....	....	al, eth	$B5^3$, 1963
13461	Stilbene, α β-dichloro-*(trans)* $C_6H_5CCl{=}CClC_6H_5$	249.14	....	180^{18}	143-4	....	....	al, eth	$B5^3$, 1963
13462	Stilbene, 2,2′-dihydroxy (α-form) $2\text{-}HOC_6H_4CH{=}CH(C_6H_4OH\text{-}2)$	212.25	nd (al)	....	95	....	....	al, eth	B6, 1022
13463	Stilbene, 2,2′-dihydroxy-(β form) $2\text{-}HOC_6H_4CH{=}CH(C_6H_4OH\text{-}2)$	212.25	flat nd (al)	....	197	....	....	eth, bz	$B6^3$, 5574
13464	Stilbene, 3,5-dihydroxy *(trans)* or Pinosylvin $3,5\text{-}(HO)_2C_6H_3CH{=}CHC_6H_5$	212.25	nd (aa)	....	156	....	....	ace, bz, chl, aa	$B6^3$, 5577
13465	Stilbene, 4,4′-dihydroxy *(trans)* or Stilbestrol $4\text{-}HOC_6H_4CH{=}CH(C_6H_4OH\text{-}4)$	212.25	nd (aa), ta (al)	....	284	....	....	ace, bz	$B6^3$, 5581
13466	Stilbene, 4,4′-dimethoxy or Bianisal $4\text{-}CH_3OC_6H_4CH{=}CH(C_6H_4OCH_3\text{-}4)$	240.30	lf (bz, aa)	sub	214-5	....	....	ace	$B6^3$, 5582
13467	Stilbene, α, β-dimethyl *(cis)* $C_6H_5C(CH_3){=}C(CH_3)C_6H_5$	208.30	....	....	67-8	$0.9537^{78/4}$	1.5612^{78}	al	$B5^3$, 2009

PHYSICAL CONSTANTS OF ORGANIC COMPOUNDS (Continued)

No.	Name, Synonyms, and Formula	Mol. wt.	Color, crystalline form, specific rotation and λ_{max} (log ε)	b.p. °C	m.p. °C	Density	n_D	Solubility	Ref.
13468	Stilbene, α, β-dimethyl *(trans)* $C_6H_5C(CH_3){=}C(CH_3)C_6H_5$	208.30			107	0.987^{20}	1.6173^{20}	eth	B5^3, 2010
13469	Stilbene, α, β-dinitro *(cis)* $C_6H_5C(NO_2){=}C(NO_2)C_6H_5$	270.24	ye pym pr (al)	d	108-9			al, eth, ace, bz, chl	B5^3, 1973
13470	Stilbene, α, β dinitro-*(trans)* $C_6H_5C(NO_2){=}C(NO_2)C_6H_5$	270.24	pa ye nd or pr (al)		187-8			ace, bz, chl, aa	B5^3, 1973
13471	Stilbene, 2,2′-dinitro *(cis)* $2\text{-}O_2NC_6H_4CH{=}CH(C_6H_4NO_2\text{-}2)$	270.24	ye nd (aa)		126				B5, 637
13472	Stilbene, 2,2′-dinitro *(trans)* $2\text{-}O_2NC_6H_4CH{=}CH(C_6H_4NO_2\text{-}2)$	270.24	pa ye nd (chl)	420 exp	199				B5^3, 1970
13473	Stilbene, 2,4-dinitro *(cis)* $2,4\text{-}(NO_2)_2C_6H_3CH{=}CHC_6H_5$	270.24	ye pl (aa)		127			bz, chl, aa	B5^2, 541
13474	Stilbene, 2,4-dinitro *(trans)* $2,4\text{-}(NO_2)_2C_6H_3CH{=}CHC_6H_5$	270.24	pa ye (aa)	412 exp	143-5			chl, xyl	B5^3, 1970
13475	Stilbene, 2,6-dinitro *(trans)* $2,6\text{-}(NO_2)_2C_6H_3CH{=}CHC_6H_5$	270.24	ye nd (bz, aa)		114			chl	B5^3, 1970
13476	Stilbene, 3,4′-dinitro *(cis)* $3\text{-}O_2NC_6H_4CH{=}CH(C_6H_4NO_2\text{-}4)$	270.24	ye nd (aa)		155			ace, bz, chl	B5^2, 541
13477	Stilbene, 3,4′-dinitro *(trans)* $3\text{-}O_2NC_6H_4CH{=}CH(C_6H_4NO_2\text{-}4)$	270.24	ye nd (aa, Py)		220-2				B5^3, 1971
13478	Stilbene, 4,4′-dinitro *(cis)* $4\text{-}O_2NC_6H_4CH{=}CH(C_6H_4NO_2\text{-}4)$	270.25	ye nd (aa, chl)		186			bz, chl	B5^3, 1972
13479	Stilbene, 4,4′-dinitro *(trans)* $4\text{-}O_2NC_6H_4CH{=}CH(C_6H_4NO_2\text{-}4)$	270.25	pa ye lf (aa), nd (aa)		303-4			al, eth, ace, bz, chl	B5^3, 1971
13480	Stilbene, α-ethyl $C_6H_5C(C_2H_5){=}CHC_6H_5$	208.30		296-7	57				B5^3, 2009
13481	Stilbene, 2-methoxy *(trans)* $2\text{-}CH_3OC_6H_4CH{=}CHC_6H_5$	210.28			59				B6^3, 3494
13482	Stilbene, 3-methoxy *(trans)* $3\text{-}CH_3OC_6H_4CH{=}CHC_6H_5$	210.28		$173\text{-}4^4$	34			al, eth, ace	B6^3, 3497
13483	Stilbene, 4-methoxy *(trans)* $4\text{-}CH_3OC_6H_4CH{=}CHC_6H_5$	210.28		142.5^{15}	136-6.5			al, eth, ace, bz	B6^3, 3498
13484	Stilbene, α-methyl $C_6H_5CH{=}C(CH_3)C_6H_5$	194.28		$195\text{-}200^{45}$	82-3	$0.9565^{100/4}$	1.5635^{17}	al, eth, bz	B5^3, 1995
13485	Stilbene, 2-nitro *(cis)* $2\text{-}O_2NC_6H_4CH{=}CHC_6H_5$	225.25		187^{11}	62.5-3.5			al, eth	B5^3, 1967
13486	Stilbene, 2-nitro *(trans)* $2\text{-}O_2NC_6H_4CH{=}CHC_6H_5$	225.25		209^{11}	73			al, eth	B5^3, 1966
13487	Strophanthiden or Corchorgenin $C_{23}H_{32}O_6$	404.50	orh ta (MeOH-w), lf (w + 2)		235 (anh), 171-5 (+ w)			al, ace, bz, chl, aa	B18^4, 3127
13488	g-Strophanthin or Quabain $C_{29}H_{44}O_{12}$	584.66	hyg pl (+ aw), $[\alpha]^{25}{}_D$ −34		200			al	B18^4, 3554
13489	Strychnine $C_{21}H_{22}N_2O_2$	334.42	orh pr (al), $[\alpha]^{18}{}_D$ −139.3 (chl, c=1)	270^5	286-8	$1.36^{20/4}$		chl	B27^2, 723
13490	Strychnine hydrochloride $C_{21}H_{22}N_2O_2.HCl.2H_2O$	406.91	nd (w), $[\alpha]_D$ −28.3 (w, c=0.7)					w	B27^2, 730
13491	Strychnine nitrate $C_{21}H_{22}N_2O_2.HNO_3$	397.44	nd (w)		280-310	1.627		w, MeOH	B27^2, 732
13492	Strychnine sulfate $(C_{21}H_{22}N_2O_2)_2.H_2SO_4.5H_2O$	856.99	wh mcl $[\alpha]_D$ + 13.2		200d			w, MeOH	B27^2, 731
13493	Styracin or Cinnamyl cinnamate $C_6H_5CH{=}CHCO_2(CH_2CH{=}CHC_6H_5)$	264.32	nd		44	1.1565^4		al, eth	B9^3, 2693
13494	Styracitol or 1,5-anhydro-D-mannitol $C_6H_{12}O_5$	164.16	pr (90% al), $[\alpha]^{20}{}_D$−49.9, −71.7 (w)		157			w	B17^4, 2579
13495	Styrene or Vinyl benzene $C_6H_5CH{=}CH_2$	104.15		145.2, 33.6^{10}	−30.6	$0.9060^{20/4}$	1.5468^{20}	al, eth, ace, bz, peth	B5^4, 1334

No.	Name, Synonyms, and Formula	Mol. wt.	Color, crystalline form, specific rotation and λ_{max} (log ε)	b.p. °C	m.p. °C	Density	n_D	Solubility	Ref.
13496	Styrene, 2-amino or 2-Vinyl aniline $2\text{-}H_2NC_6H_4CH{=}CH_2$	119.17		112^{20}		$1.0181^{20/4}$	1.6124^{20}	ace, bz	$B12^{3}$, 2785
13497	Styrene, 3-amino or 3-Vinyl aniline $3\text{-}H_2NC_6H_4CH{=}CH_2$	119.17		$112\text{-}5^{12}$		$1.0216^{20/20}$	1.6069^{26}	ace, bz	$B12^{3}$, 2786
13498	Styrene, 4-amino or 4-Vinyl aniline $4\text{-}H_2NC_6H_4CH{=}CH_2$	119.17		116^{9}	23.5	$1.012^{20/20}$	1.6250^{22}	ace, bz	$B12^{3}$, 2786
13499	Styrene, α-bromo $C_6H_5CBr{=}CH_2$	183.05		$86\text{-}7^{14}$	−44	1.4025^{23}	1.5881^{20}		$B5^{4}$, 1349
13500	Styrone, β-bromo-(*trans*) $C_6H_5CH{=}CHBr$	183.05		219d, 108^{20}	7	1.4269^{16}	1.6093^{20}	al, eth	$B5^{4}$, 1349
13501	Styrene, 2-bromo $2\text{-}BrC_6H_4CH{=}CH_2$	183.05		206.2, 98^{20}	fr−52.8	$1.4160^{20/4}$	1.5927^{20}		$B5^{4}$, 1349
13502	Styrene, 3-bromo $3\text{-}BrC_6H_4CH{=}CH_2$	183.05		$90\text{-}4^{20}$		$1.4059^{20/4}$	1.5933^{20}		$B5^{3}$, 1176
13503	Styrone, 4-bromo $4\text{-}BrC_6H_4CH{=}CH_2$	183.05		103^{20}, $50^{2.5}$	4.5	$1.3984^{20/4}$	1.5947^{20}	chl, aa	$B5^{4}$, 1349
13504	Styrene, α-chloro $C_6H_5CCl{=}CH_2$	138.60		199, 73^{16}	−23	$1.1016^{18/4}$	1.5612^{20}	al, eth	$B5^{4}$, 1345
13505	Styrene, β-chloro-(*trans*) $C_6H_5CH{=}CHCl$	138.60		199, 90^{18}		$1.1095^{18/4}$	1.5648^{20}	al, eth, ace	$B5^{4}$, 1346
13506	Styrene, 2-chloro $2\text{-}ClC_6H_4CH{=}CH_2$	138.60		188.7, 64.6^{10}	−63.1	1.1000^{20}	1.5649^{20}	al, eth, ace, peth, aa	$B5^{4}$, 1345
13507	Styrene, 3-chloro $3\text{-}ClC_6H_4CH{=}CH_2$	138.60		$62\text{-}3^{6}$		$1.1168^{20/4}$	1.5625^{20}	al, eth	$B5^{4}$, 1345
13508	Styrene, 4-chloro $4\text{-}ClC_6H_4CH{=}CH_2$	138.60		192, 66.3^{10}	−15.9	$1.0868^{20/4}$	1.5660^{20}	al, eth, ace, bz, peth	$B5^{4}$, 1345
13509	Styrene, ββ-dichloro $C_6H_5CH{=}CCl_2$	173.04		225, $85\text{-}6^{5}$		$1.2531^{20/4}$	1.5852^{20}	ace, chl	$B5^{4}$, 1348
13510	Styrene, 2,4-dimethyl $2,4\text{-}(CH_3)_2C_6H_3CH{=}CH_2$	132.20		$79\text{-}80^{12}$		$0.9022^{21.5}$	$1.5214^{21.5}$		$B5^{4}$, 1386
13511	Styrene, 3-ethyl $3\text{-}C_2H_5C_6H_4CH{=}CH_2$	132.20		190	−101	0.8945^{20}	1.5351^{20}		$B5^{3}$, 1217
13512	Styrene, 4-ethyl $4\text{-}C_2H_5C_6H_4CH{=}CH_2$	132.20		192.3	−49.7	0.8925^{20}	1.5376^{70}		$B5^{4}$, 1384
13513	Styrene, 2-fluoro $2\text{-}F\text{-}C_6H_4CH{=}CH_2$	122.14		46^{32}		1.0282^{20}	1.5200^{20}	al, eth, bz	$B5^{4}$, 1342
13514	Styrene, 3-fluoro $3\text{-}FC_6H_4CH{=}CH_2$	122.14		$30\text{-}1^{4}$		1.0177^{20}	1.5170^{20}	al, eth, bz	$B5^{4}$, 1343
13515	Styrene, 4-fluoro $4\text{-}FC_6H_4CH{=}CH_2$	122.14		67.4^{50}, $29\text{-}30^{4}$	−34.5	$1.0220^{20/4}$	1.5150^{20}	al, eth, bz	$B5^{4}$, 1343
13516	Styrene, α-methyl $C_6H_5C(CH_3){=}CH_2$	118.18		163-4, 60^{17}	24.3	$0.9082^{20/4}$	1.5303^{20}	eth, bz, chl	$B5^{3}$, 1192
13517	Styrene, 2-methyl or 2-Vinyl toluene $2\text{-}CH_3C_6H_4CH{=}CH_2$	118.18		171		$0.9106^{20/4}$	1.5450^{20}	bz, chl	$B5^{4}$, 1367
13518	Styrene, 3-methyl or 3-Vinyl toluene $3\text{-}CH_3C_6H_4CH{=}CH_2$	118.18		168, $61\text{-}2^{18}$	−70	$0.9028^{20/20}$	1.5410^{20}	al, eth, bz	$B5^{4}$, 1367
13519	Styrene, 4-methyl or 4-Vinyl toluene $4\text{-}CH_3C_6H_4CH{=}CH_2$	118.18		169, 63^{15}	−37.8	$0.8760^{20/4}$	1.5428^{20}	bz	$B5^{4}$, 1369
13520	Styrene, β-nitro-(*trans*) $C_6H_5CH{=}CHNO_2$	149.15	ye pr (peth, al)	250-60, 150^{14}	60			al, eth, ace, chl, peth	$B5^{4}$, 1352
13521	Styrene, 3-nitro $3\text{-}O_2NC_6H_4CH{=}CH_2$	149.15		$120\text{-}1^{11}$	−1	$1.1552^{32/4}$	1.5836^{20}	al, eth, bz, chl, lig	$B5^{4}$, 1351
13522	Styrene, 4-nitro $4\text{-}O_2NC_6H_4CH{=}CH_2$	149.15	pr (lig)	d	29			al, eth, chl, lig, aa	$B5^{4}$, 1351
13523	Styrene oxide or (1,2-Epoxy ethyl) benzene $C_6H_5\text{-}CH\text{-}CH_2$ (epoxide, O bridging)	120.15		194.1, $84\text{-}5^{25}$	−35.6	$1.0523^{16/4}$	1.5342^{20}	al, eth, bz	$B17^{4}$, 398
13524	Styrene, 2,4,6-trimethyl or Vinyl mesitylene $2,4,6\text{-}(CH_3)_3C_6H_2CH{=}CH_2$	146.23		208-10, 83^{12}		$0.9057^{20/4}$	1.5296^{20}		$B5^{4}$, 1408
13525	β-Styrene sulfonyl chloride $C_6H_5CH{=}CHSO_2Cl$	202.66	fl (bz)		88				$B11^{3}$, 370
13526	Subboric acid, tetramethyl ester $(CH_3O)_2BB(OCH_3)_2$	145.76		93	−24.3				$B1^{4}$, 1273
13527	Suberaldehyde or Octanedial $OHC(CH_2)_6CHO$	142.20		230-40d, $96\text{-}8^{3}$			1.4439^{20}	w, al	$B1^{4}$, 3706
13528	Suberaldehyde oxime or Suberaldoxime $HON{=}C(CH_2)_6C{=}NOH$	172.23	pr (al)		155-6			al	$B1^{4}$, 3706
13529	Suberic acid or Octanedioic acid $HO_2C(CH_2)_6CO_2H$	174.20	lo nd or pl (w)	300 sub, 219.5^{10}	144			al	$B2^{4}$, 2028

No.	Name, Synonyms, and Formula	Mol. wt.	Color, crystalline form, specific rotation and λ_{max} (log ε)	b.p. °C	m.p. °C	Density	n_D	Solubility	Ref.
13530	Suberic acid, diethyl ester or Diethyl suberate $C_2H_5O_2C(CH_2)_6CO_2C_2H_5$	230.30	....	282-6, $140\text{-}1^{8}$	5.9	$0.9811^{20/4}$	1.4328^{20}	al, eth	$B2^{4}$, 2029
13531	Suberic acid, dimethyl ester $CH_3O_2C(CH_2)_6CO_2CH_3$	202.25	....	268, 120^{6}	−3.1	$1.0217^{20/4}$	1.4341^{20}	al, eth, ace	$B2^{4}$, 2029
13532	Succinaldehyde or Butanedial OHCCH$_2$CH$_2$CHO	86.09		169-70d, 58.5^{9}	......	$1.064^{20/4}$	1.4262^{18}	w, al, eth, ace, bz	$B1^{4}$, 3642
13533	Succinamicacid or Succinic acid monoamide $HO_2CCH_2CH_2CONH_2$	117.10	nd (w, ace)	....	156-8	....	....	w	$B2^{4}$, 1922
13534	Succinamic acid, N-phenyl or Succinanilic acid $HO_2CCH_2CH_2CONHC_6H_5$	194.19	nd (w)	....	148.5	....	....	al, eth	B12, 295
13535	Sucinamide $H_2NCOCH_2CH_2CONH_2$	116.12	orh nd (w)	125.5 (sub)	268-70d	....	....	....	$B2^{4}$, 1922
13536	Succinic acid or Butanedioic acid $HO_2CCH_2CH_2CO_2H$	118.09	tcl or mcl pr	235d	188	$1.572^{25/4}$	1.450	al, eth, ace	$B2^{4}$, 1908
13537	Succinic acid, acetyl,diethyl ester or Diethyl acetyl succinate $C_2H_5O_2CCH(COCH_3)CH_2CO_2C_2H_5$	216.23	....	254-6, 139^{12}	......	$1.081^{20/4}$	1.438^{16}	al, eth, bz	$B3^{4}$, 1826
13538	Succinic acid anhydride or Succinic anhydride $C_2H_4(CO)_2O$	100.07	nd (al), rh pym (chl)	261, 139^{15}	119.6	$1.2340^{20/4}$	....	al, chl	$B17^{4}$, 5820
13539	Succinic acid, benzyl ester-(mono) $HO_2CCH_2CH_2CO_2CH_2C_6H_5$	208.21	cr (bz)	....	59	....	....	....	$B6^{4}$, 2271
13540	Succinic acid, benzyl-(dl) $C_6H_5CH_2CH(CO_2H)CH_2CO_2H$	208.21	nd or lf	....	163-4	....	....	al, eth	$B9^{3}$, 4300
13541	Succinic acid, bromo $HO_2CCHBrCH_2CO_2H$	196.99	(w)	....	161	2.073	....	w, al	$B2^{4}$, 1929
13542	Succinic acid, bromo,dimethyl ester $CH_3O_2CCHBrCH_2CO_2CH_3$	225.04	....	$132\text{-}6^{30}$	......	1.5094^{15}	....	....	$B2^{4}$, 1930
13543	Succinic acid, dibenzyl ester $C_6H_5CH_2O_2CCH_2CH_2CO_2CH_2C_6H_5$	298.34	....	245^{15}	49-50	1.256	1.596	al, eth, bz, chl	$B6^{4}$, 2271
13544	Succinic acid, 2,3-dibromo (d) $HO_2CCHBrCHBrCO_2H$	275.88	pl (aa-CCl_4), $[\alpha]^{24/}_{D}$ +147.8 (AcOEt)	....	157-8	....	....	w, al, ace	$B2^{3}$, 1679
13545	Succinic acid, 2,3-dibromo(dl) $HO_2CCHBrCHBrCO_2H$	275.88	(w or AcOEt)	....	171	....	....	w, al, bz	$B2^{4}$, 1930
13546	Succinic acid, 2,3-dibromo(l) $HO_2CCHBrCHBrCO_2H$	275.88	nd (bz), $[\alpha]^{18/}_{D}$ −148 (AcOEt, c=5.8)	....	157-8d	....	....	w, al, ace	$B2^{3}$, 1679
13547	Succinic acid, 2,3-dibromo (*meso*) $HO_2CCHBrCHBrCO_2H$	275.88	....	sub 275	255 (sealed tube)	....	....	al, eth	$B2^{4}$, 1930
13548	Succinic acid, dibutyl ester or Dibutyl succinate $C_4H_9O_2CCH_2CH_2CO_2C_4H_9$	230.30	....	274.5, 145^{4}	−29.2	$0.9752^{20/4}$	1.4299^{20}	al, eth, bz	$B2^{4}$, 1916
13549	Succinic acid, di-*sec*-butyl ester sec-$C_4H_9O_2CCH_2CH_2CO_2$-*sec*-C_4H_9	230.30	....	256	......	$0.9735^{20/4}$	1.4238^{25}	al, eth, bz	$B2^{3}$, 1665
13550	Succinic acid, di(2-carboxy phenyl) ester or Diaspirin. Succinyl salicylic acid	358.30	nd (aa, al)	....	176-8	....	....	....	$B10^{2}$, 43
13551	Succinic acid, 2,3-dichloro(d) $HO_2CCHClCHClCO_2H$	186.98	mcl, $[\alpha]^{19/}_{D}$ +3.6 (w, c=3)	....	168	1.820^{15}	....	w, eth, ace, chl	$B2^{4}$, 1928
13552	Succinic acid, 2,3-dichloro (dl) $HO_2CCHClCHClCO_2H$	186.98	pr (w, eth-peth)	....	175d	1.844^{15}	....	eth, ace, chl	$B2^{4}$, 1929
13553	Succinic acid, 2,3-dichloro-(meso) $HO_2CCHClCHClCO_2H$	186.98	hex pr (w)	....	221d	....	....	w, al, eth, ace, chl	$B2^{4}$, 1928
13554	Succinic acid, 2,3-dichloro,diethyl ester(dl) $C_2H_5O_2CCHClCHClCO_2C_2H_5$	243.09	....	132^{15}	......	$1.1963^{77/4}$	1.4512^{20}	eth	$B2^{2}$, 558
13555	Succinic acid, 2,3-dichloro,diethyl ester(*meso*) $C_2H_5O_2CCHClCHClCO_2C_2H_5$	243.09	nd (dil al)	$125.5^{12.5}$	63	$1.490^{99/4}$	1.4266^{65}	al, eth	$B2^{2}$, 558
13556	Succinic acid, diethyl ester or Diethyl succinate $C_2H_5O_2CCH_2CH_2CO_2C_2H_5$	174.20	....	216.5, 105^{15}	−20.6	$1.0402^{20/4}$	1.4198^{20}	al, eth, ace	$B2^{4}$, 1914
13557	Succinic acid, 2,3-dimethoxy,dimethyl ester $CH_3O_2CCH(OCH_3)CH(OCH_3)CO_2CH_3$	206.20	pr, $[\alpha]^{18/}_{D}$ +79.9 (MeOH, c=2)	$130\text{-}2^{12}$	53-4	$1.1317^{60/4}$	1.4340^{20}	MeOH	$B3^{3}$, 1019
13558	Succinic acid, dimethyl ester or Dimethyl succinate $CH_3O_2CCH_2CH_2CO_2CH_3$	146.14	....	196.4, 80^{11}	19	$1.1198^{20/4}$	1.4197^{20}	al, eth, ace	$B2^{4}$, 1913

No.	Name, Synonyms, and Formula	Mol. wt.	Color, crystalline form, specific rotation and λ_{max} (log ε)	b.p. °C	m.p. °C	Density	n_D	Solubility	Ref.
13559	Succinic acid, diphenyl ester or Diphenyl succinate $C_6H_5O_2CCH_2CH_2CO_2C_6H_5$	270.28	lf (al)	222.5^{15}	121			eth, ace, bz	$B6^3$, 605
13560	Succinic acid, dipropyl ester $C_3H_7O_2CCH_2CH_2CO_2C_3H_7$	202.25		250.8, 101.5^3	−5.9	$1.0020^{20/4}$	1.4250^{20}	eth, ace, bz	$B2^4$, 1916
13561	Succinic acid, di-(4-tolyl)ester $(4\text{-}CH_3C_6H_4)O_2CCH_2CH_2CO_2(C_6H_4CH_3\text{-}4)$	298.34	nd or lf (al)		121			al, eth, ace, bz, aa	$B6^2$, 379
13562	Succinic acid, ethyl ester(mono) *dl* $HO_2CCH_2CH_2CO_2C_2H_5$	146.14	pr or nd	172^{42}, 119^3	8	$1.1466^{20/4}$	1.4327^{20}	w, al, eth	$B2^4$, 1915
13563	Succinic acid, ethyl,methyl ester $C_2H_5O_2CCH_2CH_2CO_2CH_3$	160.17		208.2, $90\text{-}5^3$	<−20	$1.076^{20/4}$		al, eth	B2, 609
13564	Succinic acid, ethyl-(d) $HO_2CCH(C_2H_5)CH_2CO_2H$	146.14	pr or nd, $[\alpha]^{18/}{}_D$ + 20.6 (aa, c=3.7)	180-3	96	$1.0017^{20/4}$		w, al, eth	$B2^4$, 1995
13565	Succinic acid, ethyl (l) $HO_2CCH(C_2H_5)CH_2CO_2H$	146.14	pr or nd, $[\alpha]^{24/}{}_D$ −20.8 (ace, c=4.6)	180-3	96	1.0018^{20}		w, al, eth	$B2^4$, 1995
13566	Succinic acid, formyl,diethyl ester $C_2H_5O_2CCH(CHO)CH_2CO_2C_2H_5$	202.21		$130\text{-}4^{15}$			1.4486^{25}	al, eth	$B3^4$, 1819
13567	Succinic acid, 2-hydroxy-2-methyl (*dl*) $HO_2C(CH_3)(OH)CH_2CO_2H$	148.12	mcl pr	sub	123			w, al, ace	$B3^4$, 1149
13568	Succinic acid, 2-hydroxy-3-methyl (d) or d-Citramalic acid $HO_2CC(CH_3)CH(OH)CH_2CO_2H$	148.12	$[\alpha]^{14/}{}_D$ + 34.7 (w)		108-9			w, al, ace	$B3^4$, 1151
13569	Succinic acid, mercapto (d) or d-Thiomalic acid $HO_2CCH(SH)CH_2CO_2H$	150.15	cr (AeOEt-bz), $[\alpha]^{17/}{}_D$ + 64.4 (al)		154			w, al, ace	$B3^4$, 1130
13570	Succinic acid, mercapto (dl) $HO_2CCH(SH)CH_2CO_2H$	150.15	cr (eth)		151			w, al, eth, ace	$B3^4$, 1130
13571	Succinic acid, mercapto (l) $HO_2CCH()SH)CH_2CO_2H$	150.15	$[\alpha]^{17/}{}_D$ −64.8 (al)		152-3			w, al, ace	$B3^3$, 287
13572	Succinic acid, methyl ester (mono) $HO_2CCH_2CH_2CO_2CH_3$	132.12		$121\text{-}3^4$	56			w	$B2^4$, 1913
13573	Succinic acid, methyl (dl) or Pyrotartaric acid $HO_2CCH(CH_3)CH_2CO_2H$	132.12	pr	d	115		1.4303	w, al, eth	$B2^4$, 1948
13574	Succinic acid, monochloride,monomethyl ester $CH_3O_2CCH_2CH_2COCl$	150.56		102^{35}, $53\text{-}4^1$			1.4412^{20}		$B2^4$, 1921
13575	Succinic acid, oxo-diethyl ester or Diethyl oxaloacetate $C_2H_5O_2CCOCH_2CO_2C_2H_5$	188.18		$131\text{-}2^{24}$		$1.131^{20/4}$	1.4561^{17}	al, eth, ace, bz	$B3^4$, 1809
13576	Succinic acid, 2-oxo-3-methyl,diethyl ester or Diethyl methyl oxaloacetate $C_2H_5O_2CCH(CH_3)COCO_2C_2H_5$	202.21		$137\text{-}8^{23}$, $75\text{-}8^2$		$1.0970^{20/4}$	1.4313^{20}	al, eth	$B3^4$, 1818
13577	Succinic acid, 2-oxo-3-phenyl-1-ethyl ester-4-nitrile or Ethyl phenyl cyanopyruvate $C_2H_5O_2CCOCH(C_6H_5)CN$	217.23	(eth-lig)	206^{20}	120			al, chl	$B10^3$, 3957
13578	Succinic acid, phenyl (d) $HO_2CCH(C_6H_5)CH_2CO_2H$	194.19	pr (w), $[\alpha]^{16.5/}{}_D$ + 148.3 (al, c=1.5)		173-4			al, eth, ace	$B9^3$, 4276
13579	Succinic acid, phenyl (dl) $HO_2CCH(C_6H_5)CH_2CO_2H$	194.19	lf or nd (w)	d	168			al, eth, ace	$B9^3$, 4276
13580	Succinic acid, phenyl (l) $HO_2CCH(C_6H_5)CH_2CO_2H$	194.19	$[\alpha]^{15/}{}_D$ −173.3 (ace)		173-4			al, eth, ace	$B9^3$, 4276
13581	Succinic acid, phenyl,anhydride (d) $C_{10}H_8O_3$	176.17	nd (bz-peth), $[\alpha]^{15/}{}_D$ + 100.9 (bz)		83-4			al, bz, chl	$B17^1$, 259
13582	Succinic acid, phenyl,anhydride (dl) $C_{10}H_8O_3$	176.17	mcl pr or nd (eth)	$204\text{-}6^{22}$	54			al, eth, ace, bz	$B17^4$, 6173
13583	Succinic acid, phenyl,anhydride (l) $C_{10}H_8O_3$	176.17	$[\alpha]^{14/}{}_D$ −100.9 (bz)		83-4			bz, chl	$B17^4$, 6173
13584	Succinic acid, (3-phenyl propenyl) $(C_6H_5CH_2CH{=}CH)CH(CO_2H)CH_2CO_2H$	234.25	lf (eth, bz)		112			eth, ace, bz	B9, 909
13585	Succinic acid, isopropylidene or Tetraconic acid $(CH_3)_2C{=}C(CO_2H)CH_2CO_2H$	158.15	tcl nd (eth)		160-1			al, eth	$B2^4$, 2251

No.	Name, Synonyms, and Formula	Mol. wt.	Color, crystalline form, specific rotation and λ_{max} (log ε)	b.p. °C	m.p. °C	Density	n_D	Solubility	Ref.
13586	Succinic acid, tetrahydroxy $HO_2CC(OH)_2C(OH)_2CO_2H$	182.09			114-5				$B3^4$, 1883
13587	Succinic acid, tetramethyl $HO_2CC(CH_3)_2C(CH_3)_2CO_2H$	174.20	tcl (60% MeOH, lig), mcl (eth, ace)	sub	200	1.30		al, bz, chl	$B2^4$, 2054
13588	Succinimide $C_4H_5NO_2$	99.09	pl (+1w, al), rh (ace)	287-8d	126-7	1.418		w	$B21^4$, 4539
13589	Succinimide, N-benzyl $C_{11}H_{11}NO_2$	189.21	nd (w), pr (al)	390-400	103			al, eth, bz, chl	$B21^4$, 4550
13590	Succinimide, N-bromo $C_4H_4BrNO_2$	177.99	cr (bz)	173.5d		2.098		ace, AcOEt	$B21^4$, 4575
13591	Succinimide, N-(2-bromophenyl) $C_{10}H_8BrNO_2$	254.08	(dil al)		91			al	$B21^2$, 304
13592	Succinimide, N-(3-bromophenyl) $C_{10}H_8BrNO_2$	254.08	nd (al)		118			al	$B21^2$, 304
13593	Succinimide, N-(4-bromophenyl) $C_{10}H_8BrNO_2$	254.08	pr (al)		172			ace, bz	$B21^4$, 4547
13594	Succinimide, N-chloro $C_4H_4ClNO_2$	133.53	plCCl_4)		150	1.65		ace, aa	$B21^4$, 4575
13595	Succinimide, N-chloromethyl $C_5H_6ClNO_2$	147.56		158-60^{12}	58			w, ace, bz	$B21^2$, 305
13596	Succinimide, N-(4-chlorophenyl) $C_{10}H_8ClNO_2$	209.63	nd (dil al)		170			al, ace, bz	$B21^4$, 4547
13597	Succinimide, N-(ethoxymethyl) $C_7H_{11}NO_3$	157.17	nd	262, 151-2^{14}	31-2			w, al	$B21^2$, 304
13598	Succinimide, N-(4-ethoxyphenyl) $C_{12}H_{13}NO_3$	219.24	pr(al)		155				$B21^4$, 4554
13599	Succinimide, N-ethyl $C_6H_9NO_2$	127.14	(eth)	236	26			w, al, eth	$B21^4$, 4544
13600	Succinimide, N-(hydroxymethyl) $C_5H_7NO_3$	129.12	lf (bz)		66			w	$B21^2$, 304
13601	Succinimide, N-iodo $C_4H_4INO_2$	224.99	(ace)	234	135			w, al, eth	$B21^4$, 4576
13602	Succinimide, N-methyl $C_5H_7NO_2$	113.12	nd (eth-peth, al, ace)	234	71			w, al, eth	$B21^4$, 4544
13603	Succinimide, N-α-naphthyl $C_{14}H_{11}NO_2$	225.25	nd (dil al)		153			al	B21, 376
13604	Succinimide, N-β-naphthyl $C_{14}H_{11}NO_2$	225.25	nd (al)		183			al, bz	$B21^4$, 4551
13605	Succinimide, N-phenyl or Succinanil $C_{10}H_9NO_2$	175.19	mcl pr or nd (w, al)	ca 400	156	1.356		eth	$B21^4$, 4547
13606	Succinonitrile $NCCH_2CH_2CN$	80.09		265-7, 124^5	57	0.9867$^{60/4}$	1.4173^{60}	al, ace, bz, chl	$B2^4$, 1923
13607	Succinonitrile, tetramethyl $NCC(CH_3)_2C(CH_3)_2CN$	136.20	mcl pl, lf and pr (dil al)		170-1	1.070		al	$B2^4$, 2054
13608	Succinyl chloride $ClCOCH_2CH_2COCl$	154.98	pl or lf	193.3, 88.5^{19}	20	1.3748$^{20/4}$	1.4683^{20}	eth, ace, bz	$B2^4$, 1921
13609	Succinyl chloride, 2,3-dichloro-dl) ClCOCHClCHClCOCl	223.87		78.5^7	39			eth	$B2^2$, 558
13610	Succinyl chloride, 2,3-dichloride (*meso*) ClCOCHClCHClCOCl	223.87		105-6^{45}				CCl_4	$B2^2$, 558
13611	Sucrose or Saccharose $C_{12}H_{22}O_{11}$	342.30	mcl, $[\alpha]^{20/}{}_D$ + 66.37 (w)		185-6	1.5805$^{17.5}$	1.5376	w, Py	B31, 424
13612	Sucrose octaacetate $C_{28}H_{38}O_{19}$	678.60	nd, (al), $[\alpha]^{20/}{}_D$ + 59.6 (chl)	d 285, 260^1	86-87	1.27^{16}	1.4660	eth, ace, bz, chl	B31, 453
13613	Sudan III or Tetrazobenzene-β-naphthol $C_{22}H_{16}N_4O$	352.40	br lf with gr lustre(aa)		195			al, eth, ace, bz	$B16^3$, 148
13614	Sulfadiazine or 2-Sulfanilamido pyrimidine $C_{10}H_{10}N_4O_2S$	250.28	cr (w), wh pw		255-6d				C55, 25956
13615	Sulfaguanidine $C_7H_{10}N_4O_2S$	214.24	nd (w)		190-3				$B14^3$, 1970

No.	Name, Synonyms, and Formula	Mol. wt.	Color, crystalline form, specific rotation and λ_{max} (log ε)	b.p. °C	m.p. °C	Density	n_D	Solubility	Ref.
13616	Sulfaguanidine, monohydrate $C_7H_{10}N_4O_2S.H_2O$	232.26	nd (w)		143 (sealed tube)				B14[3], 1971
13617	Sulfamerazine or Sulfamethyldiazine $C_{11}H_{12}N_4O_2S$	264.30	cr		234-8				C55, 5501
13618	Sulfomethazine $C_{12}H_{14}N_4O_2S$	278.33	pa ye(w + 1/2) cr (diox-w)		198-9 (205-7)			w	Am64, 567
13619	Sulfamethylthiazole $C_{10}H_{11}N_3O_2S_2$	269.34			236-8			al	Am64, 2905
13620	Sulfamic acid, N-cyclohexyl $C_6H_{11}NHSO_2OH$	179.23	(al)		169-70				B12[3], 1594
13621	Sulfamide, tetramethyl $[(CH_3)_2N]_2SO_2$	152.21	pl or nd (dil al)	225	73			al	B4[4], 270
13622	Sulfanilamide or 4-Amino sulfonamide $4\text{-}H_2NC_6H_4SO_2NH_2$	172.20	lf (aq al)		165-6	1.08		al, eth, ace	B14[3], 1920
13623	Sulfanilamide, N-acetyl $4\text{-}(CH_3CO)NHC_6H_4SO_2NH_2$	214.24	pr (w)		182-4			al, ace	B14[3], 2042
13624	Sulfanilic acid or 4-Aminobenzene sulfonic acid $4\text{-}H_2NC_6H_4SO_3H$	173.19	rh pl or mcl (w + 2)		288	$1.485^{25/4}$			B14[3], 1916
13625	Sulfapyrazine or N-(2-Pyrazinyl)sulfanilamide $C_{10}H_{10}N_4O_2S$	250.28	nd ($PhNO_2$)		251			Py	Am63, 3153
13626	Sulfapyridine $C_{11}H_{11}N_3O_2S$	249.29	ye og (al)		191-3				C34, 1814
13627	Sulfaquinoxaline $C_{14}H_{12}N_4O_2S$	300.33			247-8				C49, 5525
13628	Sulfathiadiazole $C_8H_8N_4S_2O_2$	256.30			218			al, Py	C40, 5411
13629	Sulfathiazole $C_9H_9N_3O_2S_2$	255.31	br pl, rods or pw (45 al)		(i) 202.5 (ii) 125				C40, 7518
13630	Sulfathiazole, 4-nitro $C_9H_8N_4O_4S_2$	300.31	pa ye pw		258-62				C53, 12281
13631	Sulfathiazole, phthalyl, $C_{17}H_{13}N_3O_5S_2$	403.43			272-4				C51, 9689
13632	Sulfathiazole, succinyl $C_{13}H_{13}N_3O_5S_2$	355.38	cr		192-5				C51, 12148
13633	Sulfoacetic acid or Sulfoethanoic acid $HO_3SCH_2CO_2H$	140.11	hyg ta (w + 1)	245d	84-6			w, al, ace	B4[4], 102
13634	Sylvestrene (d) or Carvestrene $C_{10}H_{16}$	136.24	$[\alpha]^{18}{}_D$ + 83.2 (chl, c=4.3), (undil), + 66.3	175^{751}		$0.8479^{15/4}$	1.4760^{18}	al, eth	B14[3], 1916
13635	Syringenin or Sinapyl alcohol, 3-(3,5-dimethoxy-4-hydroxyphenyl)-2-propen-1-ol. $[3,5\text{-}(CH_3O)_2\text{-}4\text{-}HOC_6H_2]CH{=}CHCH_2OH$	210.23	nd (eth-peth)		66-7			eth	B6[3], 6690
13636	Syringin or Methoxy coniferine $C_{17}H_{24}O_9$	372.38	cr (w), nd (al), $[\alpha]_D$ −17.1		192			al	B31, 222
13637	Syringoyl methyl ketone $C_{11}H_{12}O_5$	224.21	ye nd		80-1			al, bz	Am62, 986

No.	Name, Synonyms, and Formula	Mol. wt.	Color, crystalline form, specific rotation and λ_{max} (log ε)	b.p. °C	m.p. °C	Density	n_D	Solubility	Ref.
13638	Tachysterol $C_{28}H_{44}O$	396.66	$[\alpha]^{18/}_{D}$ −70 (bz)	220 vac				al, eth, ace, bz	B6³, 3087
13639	D-Tagatose $C_6H_{12}O_6$	180.16	cr (dil al), $[\alpha]^{20/}_{D}$ −5 (w, c=1)		134-5			w	B31, 348
13640	D-Talitol or D-Altritol, D-Talaite $C_6H_{14}O_6$	182.17	pr (al)$[\alpha]^{20}_{D}$ + 3.1		87-8			w, al	B1⁴, 2839
13641	D-Talonic acid, hemihydrate $C_6H_{12}O_6 \cdot 1/2H_2O$	205.16	cr (aq al + 1/2 w), $[\alpha]^{25/}_{D}$ + 16.7→ −21.6 (w, c=4, mut)		138-9			w	B3³, 1068
13642	D-Talonic acid, γ-lactone $C_6H_{10}O_6$	178.14	pr (al), $[\alpha]^{25/}_{D}$ −34.6→ −28.4 (w), (mut)		135-7				B18⁴, 30273
13643	D-Talose $C_6H_{12}O_6$ $HOCH(CHOH)_4CHO$	180.16	[α]cr (al) β:cr (MeOH), $[\alpha]^{27/}_{D}$ + 29→ + 19.7 (w, c=1, mut)		α:130-5 β:120-1			w	B31, 283
13644	Tannic acid or Gallotannic acid Tannin $C_{76}H_{52}O_{46}$	1701.22	pa ye br amor or fl		210-5d			w, al, ace	B31, 133
13645	Taraxanthin $C_{40}H_{56}O_4$	600.88	ye pr (MeOH), $[\alpha]^{20/}_{Cd}$ + 200 (AcOEt)		185-6			al, ace, peth	B17⁴, 2239
13646	L-Tartaric acid or L-Threoic acid or L-2,3-Dihydroxy butanedoic acid $HO_2CCH(OH)CH(OH)CO_2H$	150.09	mcl (anh), rh pr (w + 1), $[\alpha]^{20/}_{D}$ + 12.7 (w, c=17.4)		171-4	1.7598^{20}	1.4955	w, al, ace	B3⁴, 1229
13647	DL-Tartaric acid or Racemic acid $HO_2CCH(OH)CH(OH)CO_2H$	150.09	mcl pr (w, al + 1w)		206	1.788		w	B3⁴, 1229
13648	Tartaric acid (*meso*) $HO_2CCH(OH)CH(OH)CO_2H$	150.09	tcl pl (w)		146-8	$1.666^{20/4}$	1.5	w, al	B3⁴, 1218
13649	Tartaric acid anhydride,diacetate (d) or α,α-Diacetoxy succinic anhydride $C_8H_8O_7$	216.16	nd (bz)		135			al, eth, ace	B18⁴, 2296
13650	Tartaric acid dibenzyl ester(d) or d-Dibenzyl tartrate $C_6H_5CH_2O_2C(CHOH)_2CO_2CH_2C_6H_5$	330.34	$[\alpha]^{15/}_{D}$ + 19.3	$250\text{-}70^{4}$	50	1.2036^{72}		al, Py	B6³, 1537
13651	Tartaric acid, dibutyl ester (*d*) or d-Dibutyl tartrate $C_4H_9O_2C(CHOH)_2CO_2C_4H_9$	262.30	pr $[\alpha]^{20/}_{D}$ + 11.3 (al)	320, 178^{12}	22	$1.0909^{20/4}$	1.4451^{20}	w, al, ace	B3⁴, 1232
13652	DL-Tartaric acid, dibutyl ester $C_4H_9O_2C(CHOH)_2CO_2C_4H_9$	262.30		320^{765}, 185^{12}		$1.0879^{25/4}$	1.4474^{15}	w, al, ace	B3³, 1032
13653	Tartaric acid, diisobutyl ester-(d) or d-Diisobutyl tartrate $i\text{-}C_4H_9O_2C(CHOH)_2CO_2\text{-}i\text{-}C_4H_9$	262.30	$[\alpha]_D$ + 11.8 (al)	323-5, 183^{11}	73	$1.0265^{81/4}$		al	B3³, 1021
13654	DL-Tartaric acid, diiosbutyl ester or DL-Diisobutyl tartrate $i\text{-}C_4H_9O_2C(CHOH)_2CO_2\text{-}i\text{-}C_4H_9$	262.30	cr (bz)	311^{768}, 195^{13}	63	$1.0386^{68/4}$		al	B3⁴, 1232
13655	Tartaric acid, diethyl ester (d) or d-Diethyl tartrate $C_2H_5O_2C(CHOH)_2CO_2C_2H_5$	206.20	$[\alpha]^{16/}_{D}$ + 7.9 (undil)	280, 142^{8}	18.7	$1.2036^{20/4}$	1.4468^{20}	w, al, eth, ace, aa	B3³, 1025
13656	DL-Tartaric acid, diethyl ester or DL-Diethyl racemate $C_2H_5O_2C(CHOH)_2CO_2C_2H_5$	206.20		281^{765}, 158^{14}	18.7	$1.2046^{20/4}$	1.4438^{20}	**al, eth**, ace	B3⁴, 1232
13657	Tartaric acid, diethyl ester-(*l*) $C_2H_5O_2C(CHOH)_2CO_2C_2H_5$	206.20	$[\alpha]^{20/}_{D}$ −7.55 (undil)	280, 162^{15}		$1.2054^{20/4}$	1.4468^{20}	al, eth, ace	B3³, 1020
13658	Tartaric acid, diethyl ester-(*meso*) $C_2H_5O_2C(CHOH)_2CO_2C_2H_5$	206.20		157.5^{14}	60	$1.1350^{99/4}$	1.4315^{65}	eth, ace	B3³, 1031
13659	Tartaric acid, diethyl ester, diacetate-(*d*) $C_2H_5O_2C(CHO_2CCH_3)_2CO_2C_2H_5$	290.27	mcl cr (lig), $[\alpha]^{100/}_{D}$ + 6.3 (undil)	296^{764}, 163^{10}	67	$1.1149^{66/4}$		al, eth	B3³, 1021

No.	Name, Synonyms, and Formula	Mol. wt.	Color, crystalline form, specific rotation and λ_{max} (log ε)	b.p. °C	m.p. °C	Density	n_D	Solubility	Ref.
13660	Tartaric acid, dimethyl ester (d) $CH_3O_2C(CHOH)_2CO_2CH_3$	178.14	(i)cr (bz), (ii)cr (bz), (iii)cr (w), $[\alpha]^{50}{}'_D$ +6.7 (MeOH, c=16)	280, 166^{12}	(i)48(ii) 50(iii) 61	1.306^{45}		w, al, eth, ace, chl	$B3^3$, 1018
13661	DL-Tartaric acid, dimethyl ester or DL-Dimethyl tartrate $CH_3O_2C(CHOH)_2CO_2CH_3$	178.14	orh nd (bz), ta (chl)	280, 169^{20}	90	1.2604^{90}		w, al, eth, ace	$B3^4$, 1232
13662	Tartaric acid, dimethyl ester-(*meso*) or *meso*-Dimethyl tartrate $CH_3O_2C(CHOH)_2CO_2CH_3$	178.14	nd (chl), cr (MeOH)	$98^{0.01}$ sub	114			w, al, ace	$B3^3$, 1031
13663	Tartaric acid, di(2-methyl butyl)ester-(*d*) $C_5H_{11}O_2C(CHOH)_2CO_2C_5H_{11}$	290.36	$[\alpha]^{20}{}'_D$ +14.1	208^{20}		$1.0636^{20/4}$		al, ace	B3, 519
13664	Tartaric acid, dinitrate (d) $HO_2C(CHONO_2)_2CO_2H$	240.08	nd (eth-bz), $[\alpha]^{20}{}'_D$ +13.7 (MeOH, p=9)		d			al, eth, ace	$B3^3$, 1018
13665	Tartaric acid, dipropyl ester-(d) or d-Dipropyl tartrate $C_3H_7O_2C(CHOH)_2CO_2C_3H_7$	234.25	$[\alpha]^{20}{}'_D$ +12.4 (w)	303, 181^{23}		$1.1390^{20/4}$		w, al, eth, ace	$B3^2$, 331
13666	DL-Tartaric acid, dipropyl ester $C_3H_7O_2C(CHOH)_2CO_2C_3H_7$	234.25	pr (al-eth)	286^{765}, 167^{11}		$1.1256^{20/4}$		w, al, eth, ace	$B3^2$, 337
13667	Tartaric acid, diisopropyl ester-(d) or d-Diisopropyl tartrate i-$C_3H_7O_2C(CHOH)_2CO_2$-i-C_3H_7	234.25	$[\alpha]^{20}{}'_D$ +14.9	275^{765}, 152^{12}		$1.1300^{20/4}$		al, eth, ace	$B3^2$, 331
13668	DL-Tartaric acid, diisopropyl ester or DL-Diisopropyl tartrate i-$C_3H_7O_2C(CHOH)_2CO_2$-i-C_3H_7	234.25		275^{765}, 154^{12}	34	$1.1166^{20/4}$		al, eth, ace	$B3^2$, 337
13669	Tartaric acid, ethyl ester-(mono) $HO_2C(CHOH)_2CO_2C_2H_5$	178.14	$[\alpha]'_D$ +21.8 (w)		90			w, al	$B3^3$, 1020
13670	Taurine or 2-Amino ethane sulfonic acid $H_2NCH_2CH_2SO_3H$	125.14	mcl pr (w)		328			w	$B4^3$, 1697
13671	Taurine, N,N-dimethyl or 2-(N,N-dimethylamino)ethane sulfonic acid $(CH_3)_2NCH_2CH_2SO_2H$	153.20	pr (MeOH), pl (w + 1)		315-6 (anh), 270-80 d (+ w)			w, aa	$B4^3$, 1699
13672	Taruine, N-methyl or 2-(N-Methyl)ethane sulfonic acid $CH_3NHCH_2CH_2SO_3H$	139.17	pr		241-2			w	$B4^3$, 1699
13673	Taurine, N-methyl-N-phenyl $C_6H_5NC(CH_3)CH_2CH_2SO_3H$	215.27	pa vt (al)		239-40			al	$B12^2$, 285
13674	Taurine, N-phenyl $C_6H_5NHCH_2CH_2SO_3H$	201.24	lf (w), pr (al)		277-80d			w	$B12^2$, 284
13675	Taurocholic acid or Cholaic acid. Cholyl taurine $C_{26}H_{45}NO_7S$	515.71	pr (al-eth), $[\alpha]^{18}{}'_D$ +38.8 (al, c=2)		ca 125d			w, al	E14, 195
13676	Tephrosin or Hydroxy deguelin $C_{23}H_{22}O_7$	410.42	pr (chl-MeOH)		198 (218)			eth, ace, chl	$B19^4$, 5271
13677	Terebic acid-(dl) or 3.3-Dimethyl paraconic acid $C_7H_{10}O_4$	158.15	mcl pr (al)		176	$0.8155^{24/4}$			$B18^4$, 5287
13678	Terephthaldehyde 1,4-$(OHC)_2C_6H_4$	134.13	nd (w)	247^{771}	116			al, eth	$B7^3$, 3460
13679	Terephthalamide 1,4-$C_6H_4(CONH_2)_2$	164.16	nd (w), pl (aa)		>250				$B9^3$, 4253
13680	Terephthalamide, N,N,N′,N′-tetraethyl 1,4-$C_6H_4[CON(C_2H_5)_2]_2$	276.38	cr (eth-al)		127			al, bz	$B9^3$, 4253
13681	Terephthalic acid or 1,4-Benzene dicarboxylic acid 1,4-$C_6H_4(CO_2H)_2$	166.13	nd (sub)	>300 (sub without melting)	sub				$B9^3$, 4249
13682	Terephthalic acid, 2-amino 2-$H_2NC_6H_3(CO_2H)_2$-1,4	181.15	ye cr (w)		324-5d				$B14^1$, 637
13683	Terephthalic acid, 2-amino, dimethyl ester 2-$H_2NC_6H_3(CO_2CH_3)_2$-1,4	209.20	nd (bz), cr (al)		134			ace, chl	$B14^2$, 338
13684	Terephthalic acid, 2-benzyl or Benzophenone-2,5-dicarboxylic acid 2-$(C_6H_5CO)C_6H_3(CO_2H)_2$-1,4	270.24	nd		291-2			al, eth	$B10^3$, 4008
13685	Terephthalic acid, 2-bromo 2-$BrC_6H_3(CO_2H)_2$-1,4	245.03	nd (w, al)		299			al	$B9^3$, 4258

No.	Name, Synonyms, and Formula	Mol. wt.	Color, crystalline form, specific rotation and λ_{max} (log ε)	b.p. °C	m.p. °C	Density	n_D	Solubility	Ref.
13686	Terephthalic acid, 2-chloro $2\text{-ClC}_6H_3(CO_2H)_2\text{-(1,4)}$	200.58	cr (w)	....	320	....	....	al, eth	$B9^{3}$, 4256
13687	Terephthalic acid, 2,5-dichloro $2,5\text{-Cl}_2C_6H_2(CO_2H)_2\text{-1,4}$	235.02	nd (w)	sub	306	....	....	al, eth	$B9^{3}$, 4257
13688	Terephthalic acid, diethyl ester or Diethyl terephthalate $1,4\text{-C}_6H_4(CO_2C_2H_5)_2$	222.24	nd pr (al, peth)	302, 142^{2}	44	$1.1098^{45/45}$	....	al, eth	$B9^{3}$, 4250
13689	Terephthalic acid, 2,5-dihydroxy $2,5\text{-(HO)}_2C_6H_2(CO_2H)_2\text{-1,4}$	198.13	ye cr (al, w)	....	d	....	....	....	$B10^{3}$, 2438
13690	Terephthalic acid, dimethyl ester $1,4\text{-C}_6H_4(CO_2CH_3)_2$	194.19	nd (eth)	sub	141-2	....	....	eth	$B9^{3}$, 4250
13691	Terephthalic acid, monoamide or Terephthalamic acid $4\text{-HO}_2CC_6H_4CONH_2$	165.15	....	sub 250	>300	....	....	....	$B9^{3}$, 4253
13692	Terephthalic acid, mononitrile or 4-Cyanobenzoic acid $4\text{-NCC}_6H_4CO_2H$	147.13	pl or lf (w)	....	219	....	....	al, eth	$B9^{3}$, 4254
13693	Terephthalic acid, 2-nitro $2\text{-NO}_2C_6H_3(CO_2H)_2\text{-1,4}$	211.13	nd (w)	....	270-5	....	....	al	$B9^{3}$, 4258
13694	Terephthalic acid, tetrabromo $Br_4C_6(CO_2H)_2\text{-1,4}$	481.72	nd (w)	....	266 (anh)	....	....	....	$B9^{1}$, 367
13695	Terephthalonitrile $1,4\text{-C}_6H_4(CN)_2$	128.13	nd (w, MeOH)	sub	222	....	....	bz, aa	$B9^{3}$, 4255
13696	Terephthalyl chloride $1,4\text{-C}_6H_4(COCl)_2$	203.02	nd or pl (lig)	$125\text{-}7^{9}$	83-4	....	....	eth	$B9^{3}$, 4252
13697	Terephthalyl alcohol or *p*-Xylylene glycol $1,4\text{-(HOCH}_2)_2C_6H_4$	138.17	nd (w)	....	115-6	....	....	w, al, eth, ace	$B6^{3}$, 4608
13698	Terpenolic acid or Terpenylic acid $C_8H_{12}O_4$	172.18	lf or pr (w + 1)	sub 130-40	90 (anh), 57 (+ w)	....	....	w	$B18^{4}$, 5302
13699	*o*-Terphenyl or 1,2-Diphenyl benzene $1,2\text{-(C}_6H_5)_2C_6H_4$	230.31	mcl pr (MeOH)	332, $160\text{-}70^{2}$	58	....	....	ace, bz, chl, MeOH	$B5^{3}$, 2292
13700	*m*-Terphenyl or 1-3-Diphenyl benzene $1,3\text{-(C}_6H_5)_2C_6H_4$	230.31	ye nd (al)	365	89	....	....	al, eth, bz, aa	$B5^{3}$, 2294
13701	*p*-Terphenyl $1,4\text{-(C}_6H_5)_2C_6H_4$	230.31	....	sub 250^{45}	213	....	....	eth, bz	$B5^{3}$, 2296
13702	α-Terpinene or p-Mentha-1,3-diene $C_{10}H_{16}$	136.24	....	177.2, $68\text{-}70^{12}$	......	$0.8502^{20/4}$	1.4784^{10}	al, eth	$B5^{4}$, 435
13703	β-Terpinene $C_{10}H_{16}$	136.24	....	173-4	......	0.838^{23}	1.4754^{22}	....	$B5^{4}$, 437
13704	γ-Terpinene $C_{10}H_{16}$	136.24	....	183	......	$0.849^{20/4}$	$1.4765^{14.5}$	....	$B5^{4}$, 436
13705	α-Terpineol (dl) or *dl*-p-Menth-1-en-8-ol $C_{10}H_{18}O$	154.25	cr (peth)	220, 85^{2}	40-1	$0.9337^{20/4}$	1.4831^{20}	al, eth, ace, bz, chl	$B6^{3}$, 251
13706	α-Terpineol acetate $C_{12}H_{20}O_2$	196.29	*d*: $[\alpha]_D$ + 52.5 (undil), *l*: $[\alpha]_D$ −73 (undil)	140^{40}, $104\text{-}6^{11}$	......	$0.9659^{20/4}$	1.4689^{21}	al, eth, bz	$B6^{4}$, 252
13707	Terpin hydrate (*cis*) or *cis*-Terpino hydrate $C_{10}H_{20}O_2.H_2O$	214.30	rh cr	sub 100	123d	....	1.51	al	$B6^{3}$, 4113
13708	Terpinolene $C_{10}H_{16}$	136.24	....	185, 76^{10}	......	$0.8623^{20/4}$	1.4883^{20}	al, eth, bz	$B5^{4}$, 437
13709	3-Terpinolenone or Piperitenone $C_{10}H_{14}O$	150.22	$[\alpha]_{546}$ −0.1	$120\text{-}2^{14}$	......	$0.9774^{20/4}$	1.5294^{20}	al, eth	$B7^{3}$, 559
13710	Terramycin or Oxytetracycline $C_{22}H_{22}N_2O_9$	460.44	....	....	184-5	1.634^{20}	....	....	AM87, 134
13711	Testosterone or 17β-Hydroxy-4-androsten-3-one $C_{19}H_{28}O_2$	288.43	nd (dil ace), $[\alpha]^{24}{}_D$ + 109 (al, c=4)	....	155	....	....	al, eth, ace	$B8^{3}$, 892
13712	Testosterone, 4,5-dihydro-17-methyl $C_{20}H_{32}O_2$	304.47	cr (AcOEt)	....	192-3	....	....	....	$B8^{3}$, 609
13713	Testosterone, 17-ethenyl or 17-Vinyl testosterone $C_{21}H_{30}O_2$	314.47	pr nd (peth-eth), $[\alpha]_D$ + 87.6	....	140-1	....	....	eth, ace, chl, MeOH	$B8^{3}$, 1067
13714	Testosterone, 17-ethyl $C_{21}H_{32}O_2$	316.48	nd (AcOEt)	....	143-4	....	....	....	$B8^{3}$, 946

PHYSICAL CONSTANTS OF ORGANIC COMPOUNDS (Continued)

No.	Name, Synonyms, and Formula	Mol. wt.	Color, crystalline form, specific rotation and λ_{max} (log ε)	b.p. °C	m.p. °C	Density	n_D	Solubility	Ref.
13715	Testosterone, 17-ethynyl $C_{21}H_{28}O_2$	312.45	cr (chl-MeOH, AcOEt), $[\alpha]^{20}/_D$ + 22.5 (diox)	(sub, vac)	270-2			diox, Py	$B8^{3}$, 1206
13716	Testosterone, 17-methyl $C_{20}H_{30}O_2$	302.46	nd (bz), $[\alpha]^{20}/_D$ + 82 (al)		165-6			al, eth	$B8^{3}$, 939
13717	Testosterone propionate $C_{22}H_{32}O_3$	344.49	$[\alpha]^{25}/_D$ + 83-90 (diox, c=1)		118-22			al, eth, Py	$B8^{3}$, 897
13718	Tetrabenzotriazaporphyrin $C_{32}H_{18}N_8$	514.55	purple nd and pl (quinoline)					Py	J1939, 1809
13719	Tetrabutyl ammonium iodide $(C_4H_9)_4N^+I^-$	369.37	lf(w bz)		148			al, w	$B4^{4}$, 558
13720	Tetracosane $CH_3(CH_2)_{22}CH_3$	338.66	cr (eth)	391.3, 231.3^{10}	54	$0.7665^{70/4}$, 0.7991^{20}	1.4283^{70}, 1.4480^{20}	eth	$B1^{4}$, 578
13721	Tetracoscene, 11-decyl $(C_{10}H_{21})_2CHC_{13}H_{27}$	478.93		301^{10}	10.8	$0.8161^{20/4}$	1.4556^{20}		$B1^{4}$, 598
13722	Tetracosane, 3-ethyl $(C_2H_5)_2CHC_{21}H_{43}$	366.71		255.5^{10}	30.1	$0.7949^{40/4}$	1.4436^{40}		$B1^{4}$, 584
13723	Tetracyclone or Tetraphenyl cyclopentadienone $C_{29}H_{20}O$	384.48	bk-vt lf, cr (aa, xyl)		220-1			al, bz	$B7^{3}$, 2997
13724	6,8-Tetradecadiyne $C_5H_{11}C{\equiv}CC{\equiv}CC_5H_{11}$	190.33		$118\text{-}9^{4}$	2	$0.8699^{16/4}$		eth	$B1^{3}$, 1067
13725	Tetradecanal or Myristaldehyde $C_{13}H_{27}CHO$	212.38	lf	166^{24}	30			al, eth, ace	$B1^{4}$, 3389
13726	Tetradecane $CH_3(CH_2)_{12}CH_3$	198.39		253.7, 121.9^{10}	5.9	$0.7628^{20/4}$	1.4290^{20}	al, eth	$B1^{4}$, 520
13727	Tetradecane, 1-amino or Myristyl amine $C_{13}H_{27}CH_2NH_2$	213.41		291.2, 162^{15}	83.1	$0.8079^{20/4}$	1.4463^{20}	al, eth, ace, bz, chl	$B4^{4}$, 812
13728	Tetradecane, bromo or Myristyl bromide $C_{13}H_{27}CH_2Br$	277.29		307, 181^{21}	5.6	$1.0170^{20/4}$	1.4603^{20}	al, ace, bz, chl	$B1^{4}$, 523
13729	Tetradecane, 1-chloro or Myristyl chloride $C_{13}H_{27}CH_2Br$	232.84		292, 153^{10}	4.9	$0.8665^{20/4}$	1.4473^{20}	al, ace, bz, chl	$B1^{3}$, 550
13730	Tetradecane, 1,14-dibromo $Br(CH_2)_{14}Br$	356.18	lf (al-eth), cr (al)	$190\text{-}2^{8}$	50.4			al, eth, chl	$B1^{4}$, 523
13731	Tetradecane, 1,1-dicyclohexyl $C_{13}H_{27}CH(C_6H_{11})_2$	362.68		406	37.6	0.8735^{20}	1.4799^{20}		$B5^{3}$, 300
13732	Tetradecane, 1-phenyl or Tetradecyl benzene $C_{13}H_{27}CH_2C_6H_5$	274.49		358.9, 210^{12}	16.1	$0.8559^{20/4}$	1.4818^{20}		$B5^{4}$, 1212
13733	Tetradecanedioic acid, dimethyl ester $CH_3O_2C(CH_2)_{12}CO_2CH_3$	286.41	nd (MeOH)	$191\text{-}2^{10}$	43-5				$B2^{4}$, 2149
13734	1,14-Tetradecanediol or Tetradecamethylene glycol $HO(CH_2)_{14}OH$	230.39	nd (bz)	200^{9}	84.5			al, eth	$B1^{4}$, 2631
13735	1-Tetradecane sulfonic acid $C_{13}H_{27}CH_2SO_3H$	278.45			65.5 (anh), 55-6 (+1w)	$0.9996^{25/4}$		w	$B4^{4}$, 66
13736	1-Tetradecanethiol or Myristyl mercaptan $C_{13}H_{27}CH_2SH$	230.45		$176\text{-}80^{22}$		$0.8484^{20/20}$	1.4597^{20}	al, eth	$B1^{4}$, 1867
13737	Tetradecanoic acid or Myristic acid $C_{13}H_{27}CO_2H$	228.38	lf (eth 80% aa)	250.5^{100}, 149.3^{2}	58	$0.8439^{80/4}$	1.4305^{80}	al, ace, bz, chl	$B2^{4}$, 1126
13738	1-Tetradecanol or Myristyl alcohol $C_{13}H_{27}CH_2OH$	214.40	lf	263.2, 167^{11}	39-40	$0.8236^{38/4}$		al, eth, ace, bz, chl	$B1^{4}$, 1864
13739	2-Tetradecanone or n-Dodecyl methyl ketone $C_{12}H_{25}COCH_3$	214.39	cr (dil al)	$205\text{-}6^{100}$, 134^{13}	33-4			al, ace	$B1^{4}$, 3389
13740	3-Tetradecanone or Ethyl undecyl ketone $C_{11}H_{23}COC_2H_5$	212.38	cr (MeOH)	152^{16}	34			al, ace	$B1^{4}$, 3389
13741	5,9-Tetradecadien-7-yne, 6,9-dimethyl $C_4H_9CH{=}C(CH_3)C{\equiv}CC(CH_3){=}CHC_4H_9$	218.38		$95\text{-}8^{0.3}$		$0.8241^{20/4}$	1.4866^{20}	eth, bz	$B1^{4}$, 1133
13742	1-Tetradecene $C_{12}H_{25}CH{=}CH_2$	196.38		232-4, 125^{15}	−12	$0.7745^{20/4}$	1.4351^{20}	al, eth, bz	$B1^{4}$, 924
13743	2-Tetradecyne $C_{11}H_{23}C{\equiv}CCH_3$	194.36		252.5, 134^{15}	6.5	$0.8000^{20/4}$		al, eth	$B1^{4}$, 1070

No.	Name, Synonyms, and Formula	Mol. wt.	Color, crystalline form, specific rotation and λ_{max} (log ε)	b.p. °C	m.p. °C	Density	n_D	Solubility	Ref.
13744	7-Tetradecyne $C_6H_{13}C\equiv CC_6H_{13}$	194.36		144^{30}		$0.7991^{20/4}$	1.4330^{25}	al, eth	$B1^3$, 1067
13745	Tetraethyl ammonium bromide $(C_2H_5)_4N^+Br^-$	210.16	dlq (al)			$1.3970^{20/4}$		w, al, chl	$B4^4$, 332
13746	Tetraethyl ammonium hydroxide, hydrate $(C_2H_5)_4N^+OH^-.4H_2O$	219.32	nd (w + 4)	d (vac)	49-50			w, al	$B4^4$, 331
13747	Tetraethylene glycol $HOCH_2(CH_2OCH_2)_3CH_2OH$ $C_8H_{18}O_5$	194.23		328, 198^{14}	−6.2	$1.1285^{15/4}$	1.4577^{20}	w, al, eth	$B1^4$, 2403
13748	Tetraethylene glycol, dimethyl ether $CH_3OCH_2(CH_2OCH_2)_3CH_2OCH_3$	222.29		275.8		$1.0132^{20/20}$		w, al, eth	$B1^4$, 2404
13749	Tetraethylene glycol, mono stearate $C_{17}H_{35}CO_2CH_2(CH_2OCH_2)_3CH_2OH$	476.69		328	40	$1.1285^{15/4}$	1.4593^{20}		C37, 3202
13750	Tetraethylene pentamine $H_2NCH_2(CH_2NHCH_2)_3CH_2NH_2$	189.31		340.3, $186\text{-}92^{14}$			1.5042^{20}		$B4^1$, 1244
13751	Tetraethyl hypophosphate $(C_2H_5O)_2POOP(OC_2H_5)_2$	274.19		$116\text{-}7^{2}$		1.1283^{18}	1.4284^{20}	bz	$B1^4$, 1358
13752	Tetralin or 1,2,3,4-Tetrahydronaphthalene $C_{10}H_{12}$	132.21		207.6, 79.4^{10}	−35.8	$0.9702^{20/4}$	1.5413^{20}	al, eth	$B5^4$, 1388
13753	Tetralin, 6-acetyl 6-$(CH_3CO)C_{10}H_{11}$	174.24		289-91d, 182^{20}					$B7^3$, 1473
13754	Tetralin, 2-amino-(*dl*) 2-$H_2NC_{10}H_{11}$	147.22		249^{710}, 140^{20}	38	$1.0295^{22/4}$	1.5604^{22}	al, eth, ace, bz	$B12^3$, 2811
13755	Tetralin, 5-amino 5-$H_2NC_{10}H_{11}$	147.22		276.8, 155^{22}		1.0625^{16}	1.6050^{20}	al, eth	$B12^3$, 2805
13756	Tetralin, 1,1-dimethyl 1,1-$(CH_3)_2.C_{10}H_{10}$	160.26		221		0.950^{20}	1.5292^{20}		$B5^4$, 1430 (35.5214)
13757	Tetralin, 1-ethyl 1-$C_2H_5C_{10}H_{11}$	160.26		239-46		0.95285^{20}	1.5318^{20}		$B5^4$, 1429
13758	Tetralin, 2-ethyl 2-$C_2H_5C_{10}H_{11}$	160.26		237, $63\text{-}5^{0.5}$		$0.9401^{15.5}$	$1.5250^{15.5}$		$B5^3$, 1261
13759	Tetralin, 6-ethyl 6-$C_2H_5C_{10}H_{11}$	160.26		244		$0.9632^{17/4}$	1.5414^{16}		$B5^4$, 1429
13761	Tetralin, 1-hydroxy 1-$HOC_{10}H_{11}$	148.20		$102\text{-}4^{2}$			1.5658^{20}		$B6^3$, 2457
13762	Tetralin, 1-methyl 1-$CH_3C_{10}H_{11}$	146.23		220.59		0.9583^{20}	1.5353^{20}		$B5^4$, 1412 (35.5210)
13763	Tetralin, 2-methyl 2-$CH_3C_{10}H_{11}$	146.23		220-2, $99\text{-}101^{13}$					$B5^4$, 1414
13764	Tetralin, 6-methyl 6-$CH_3C_{10}H_{11}$	146.23		220-2		$0.9541^{15/4}$	1.5332^{15}		$B5^4$, 1413
13765	Tetralin, 1,2,3,4-tetrabromo or Naphthalene tetrabromide 1,2,3,4-$Br_4C_{10}H_8$	447.79	mcl pr (chl)		111d			CS_2, bz	$B5^3$, 1229
13766	Tetralin, 5,6,7,8-tetrachloro 5,6,7,8-$Cl_4C_{10}H_8$	269.99		180^{26}	172				$B5^3$, 1227
13767	1,4-Tetralin dicarboxylic acid, 1-phenyl-(*d,α*) or d-α-Isatropic acid $C_{18}H_{16}O_4$	296.33	pr $[\alpha]^{20}{}_D$ + 9.4 (al, c=12.6)		239d				$B9^3$, 416
13768	5-Tetralin sulfonamide 5-$C_{10}H_{11}SO_2NH_2$	211.28	lf (al, 30% aa)		139-40			al	$B11^2$, 87
13769	5-Tetralin sulfonic acid 5-$C_{10}H_{11}SO_3H$	212.26	cr (chl + 1w)		105-10 (+ 1w)			w	$B11^2$, 87
13770	5-Tetralin sulfonyl chloride 5-$C_{10}H_{11}SO_2Cl$	230.71	pl (peth)		70.5			eth	$B11^2$, 87
13771	6-Tetralin sulfonic acid 6-$C_{10}H_{11}SO_3H$	212.26	cr (chl, dil sulf)		75			w, eth, chl	$B11^2$, 88
13772	6-Tetralin sulfonyl chloride 6-$C_{10}H_{11}SO_2Cl$	230.71	pl (eth)	$197\text{-}200^{18}$	58			eth	$B11^2$, 88
13773	1-Tetralone or 1-oxo-1,2,3,4-tetrahydronaphthalene 1-$C_{10}H_{10}O$	146.19		255-7, 129^{12}	8	$1.0989^{16/4}$	1.5672^{20}		$B7^3$, 1416
13773a	1-Tetralone, 7-ethyl 7-C_2H_5-(1-$C_{10}H_9O$)	174.24		$152\text{-}3^{12}$		$1.0556^{17/4}$	1.5599^{17}	al, eth, bz	$B7^3$, 1473
13774	1-Tetralone, 4-methyl 4-$CH_3(C_{10}H_9O)$	160.22		$133\text{-}4^{12}$		$1.0779^{19/4}$	1.5620^{19}		$B7^3$, 1444
13775	2-Tetralone 2-$C_{10}H_{10}O$	146.19		234-40, 138^{16}	18	$1.1055^{27/4}$	1.5598^{20}	eth, bz	$B7^3$, 1424

PHYSICAL CONSTANTS OF ORGANIC COMPOUNDS (Continued)

No.	Name, Synonyms, and Formula	Mol. wt.	Color, crystalline form, specific rotation and λ_{max} (log ε)	b.p. °C	m.p. °C	Density	n_D	Solubility	Ref.
13776	Tetramethyl ammonium bromide $(CH_3)_4N^+Br^-$	154.05	dlq ditetr bipym	(360 sub,vac)	230d	1.56		w, MeOH	B4^4, 145
13777	Tetramethyl ammonium chloride $(CH_3)_4N^+Cl^-$	109.60	dlq ditetr bipym (dil)		420 (sealed tube)	$1.169^{20/4}$		w, MeOH	B4^4, 145
13778	Tetramethyl ammonium hydroxide, monohydrate $(CH_3)_4N^+OH^-.H_2O$	109.17			d 130-5			w	B4, 50
13779	Tetramethyl ammonium hydroxide, trihydrate $(CH_3)_4N^+OH^-.3H_2O$	145.20			60			w	B4, 50
13780	Tetramethyl ammonium hydroxide, pentahydrate $(CH_3)_4N^+OH^-.5H_2O$	181.23		d	62-3			w, al	B4^2, 557
13781	Tetramethyl ammonium iodide $(CH_3)_4NI$	201.05			>230d	1.829			B4^4, 146
13782	Tetraphenylene or Tetrabenzocyclooctatetraene $C_{24}H_{16}$	304.39	cr (al, AcOEt)	$200^{0.2}$ sub	233			al	B5^3, 2595
13783	Tetrapropyl ammonium bromide $(C_3H_7)_4N^+Br^-$	266.27			252			w, chl	B4^4, 471
13784	Tetrapropyl ammonium iodide $(C_3H_7)_4N^+I^-$	313.27	rh bipym		280d	$1.3138^{25/4}$		w, al, chl, aa	B4^4, 472
13785	Tetrasiloxane, decamethyl $(CH_3)_3Si(OSi(CH_3)_2)_3CH_3$	310.69		194, 88^{20}	−76	$0.8536^{20/4}$	1.3895^{20}	bz, peth	B4^3, 1879
13786	Tetrasiloxane, 1,1,1,3,5,7,7,7,-octamethyl $[(CH_3)_3SiOSi(CH_3)_2]_2O$	310.69		170		$0.8559^{20/4}$	1.3854^{20}		B4^3, 1874
13787	1-Tetratriacontanol or *n*-Carnatyl alcohol $CH_3(CH_2)_{33}OH$	494.93	nd (ace)		913			ace	B1^4, 1921
13788	1,2,4,5-Tetrazine or *s*-Tetrazine $C_2H_2N_4$	82.06	dk red pr	sub	99			w, al, eth	B26^2, 212
13789	1,2,3,4-Tetrazole CH_2N_4	70.05	pl (al)	sub	156			w, al, ace, aa	B26^2, 196
13790	Thebaine or Paramorphine $C_{19}H_{21}NO_3$	311.38	pl (eth), pr (dil al), $[\alpha]^{25}{}_D$ −218.5 (al, p=2)	sub $91^{0.01}$	193	$1.305^{20/4}$		al, bz, chl, Py	B27^2, 177
13791	Thebaine, dihydro $C_{19}H_{23}NO_3$	313.40	$[\alpha]^{20}{}_D$ −267 (bz, c=1.02)		162-3			al, bz, AcOEt	B27^2, 110
13792	Thebaine, hydrochloride monohydrate $C_{19}H_{21}NO_3.HCl.H_2O$	365.86	orh pr (al), $[\alpha]^{20}{}_D$ −157 (al)					w, al, eth	B27^2, 181
13793	Thebainone A $C_{18}H_{21}NO_3$	299.37	nd (dil al + ½ w), nd or pr (al, aa)		151-2			ace, bz, chl	B21^4, 6548
13794	Theobromine or 3,7-Dimethylxanthine $C_7H_8N_4O_2$	180.17	rh or mcl nd (w)	sub 290	351 (357)				B26^2, 264
13795	Theophylline or 1,3-Dimethylxanthine $C_7H_8N_4O_2$	180.17	nd or pl (w + 1)		272-4			w	B26^2, 263
13796	Thevetin $C_{42}H_{66}O_{18}$	858.98	nd (al), pl or nd (i-proh)		210			al, Py, MeOH	B18^4, 1493
13797	Thiacyclobutane or Trimethylene sulfide C_3H_6S	74.14		94.7, 14^{30}	−73.2	$1.0200^{20/4}$	1.5102^{20}	al, ace, bz	B17^4, 14
13798	1,3,4-Thiadiazole, 2.5-dimethyl $C_4H_6N_2S$	114.17		202-3	64				B27, 565
13799	Thialdine or Thioacetaldehyde ammonia $C_6H_{13}NS_2$	163.30	pl (al-eth)	d	46	$1.0632^{50/20}$		al, eth	B27^2, 525
13800	Thiane $C_5H_{10}S$	102.19		141.8, 93^{82}	19	$0.9861^{20/4}$	1.5067^{20}	al, eth, ace, bz, chl	B17^4, 55
13801	Thianthrene or Dephenylene disulfide $C_{12}H_8S_2$	216.32	mcl pr or pl (al)	336, 204^{11}	158-9			eth, bz	B19^4, 347
13802	Thiazole C_3H_3NS	85.12		116.8		$1.998^{17/4}$	1.5969^{20}	al, eth, ace	B27^2, 9
13803	Thiazole, 2-acetamido 2-$CH_3CONH(C_3H_2NS)$	142.18			208				C35, 5110
13804	Thiazole, 2-amino or Abadol 2-$H_2N(C_3H_2NS)$	100.14	ye pl (al)	140^{11}	93				B27^2, 205
13805	Thiazole, 2-amino-4-methyl 4-CH_3-2-$H_2N(C_3HNS)$	114.17	hyg cr	281-2d, 136^{30-40}	45-6			w, al, eth	B27^2, 206

No.	Name, Synonyms, and Formula	Mol. wt.	Color, crystalline form, specific rotation and λ_{max} (log ε)	b.p. °C	m.p. °C	Density	n_D	Solubility	Ref.
13806	Thiazole, 2-amino-5-methyl $5\text{-}CH_3\text{-}2\text{-}H_2N(C_3HNS)$	114.17	pl (w)		96			al, eth	B27, 162
13807	Thiazole, 2-amino-5-sulfanilyl $C_9H_9N_3O_2S_2$	255.31	nd (al)		219-21d			al, eth, ace, diox	C41, 447
13808	Thiazole, 2,4-dimethyl $2,4\text{-}(CH_3)_2(C_3HNS)$	113.18		$144\text{-}5^{719}$, $70\text{-}3^{50}$		$1.0562^{15/4}$	1.5091^{20}	al, eth	$B27^2$, 10
13809	Thiazole, 4,5-dimethyl $4,5\text{-}(CH_3)_2(C_3HNS)$	113.18		158, $81\text{-}3^{59}$	83-4			al, eth	AM74, 5778
13810	Thiazole, 2,4-diphenyl $2,4\text{-}(C_6H_5)_2(C_3HNS)$	237.32	lf (al)	>360	92-3	$1.1554^{98/4}$		al, eth	$B27^2$, 43
13811	Thiazole, 5-(2-hydroxyethyl)-4-methyl $4\text{-}CH_3\text{-}5\text{-}(HOCH_2CH_2)(C_3HNS)$	143.20	col to pa ye	135^7		$1.196^{24/4}$		w, al, eth, bz, chl	AM71, 2931
13812	Thiazole, 2-mercapto-4-methyl $4\text{-}CH_3\text{-}2\text{-}HS(C_3HNS)$	131.21	ye (dil al)	188^3	88-9			al	$B27^2$, 208
13813	Thiazole, 2-methyl $2\text{-}CH_3(C_3H_2NS)$	99.15		128-9, $65\text{-}70^{80}$			1.510	w, al, ace	B27, 16
13814	Thiazole, 4-methyl $4\text{-}CH_3(C_3H_2NS)$	99.15		132^{745}, 70^{90}		1.112^{25}		w, al, eth	$B27^2$, 9
13815	5-Thiazole carboxylic acid $5\text{-}(C_3H_2NS)\text{-}CO_2H$	129.13	lf ye nd (dil HCl)		217-8			eth	C48, 2688
13816	5-Thiazole carboxylic acid, 2-amino,ethyl ester $2\text{-}H_2N\text{-}(C_3HNS)CO_2C_2H_5\text{-}5$	172.20		213-5d	163-4				C47, 7453
13817	5-Thiazole carboxylic acid, 4-methyl $4\text{-}CH_3\text{-}(C_3HNS)CO_2H\text{-}5$	143.16	pr or pl (w), nd (al)	sub >250	280d				B27, 316
13818	5-Thiazole carboxylic acid, 4-methyl,ethyl ester $4\text{-}CH_3\text{-}(C_3HNS)CO_2C_2H_5\text{-}5$	171.21	pr	$232\text{-}3^{735}$, $110\text{-}5^{15}$	28			al, eth, ace	B27, 316
13819	5-Thiazole carboxylic acid, 4-methyl,ethyl ester hydrochloride $C_7H_9NO_2S.HCl$	207.67	nd (al)		155d			w, al	J1939, 443
13820	5-Thiazole carboxylic acid, 4-methyl-2-sulfanilamide $C_{11}H_{11}N_3O_4S_2$	313.35	cr		241-2				C38, 2250
13821	Thiazolidine or Tetrahydrothiazole C_3H_7NS	89.16		164-5		$1.131^{25/4}$	1.551^{20}	w, al, eth, ace	AM59, 200
13822	2,4-Thiazolidinedione or 2,4-Dioxothiazolidine $C_3H_3O_2NS$	117.12	pl (w), pr (al)	179^{19}	128			eth	$B27^2$, 284
13823	2-Thiazolidine thione or 2-thiothiazolidone $C_3H_5NS_2$	119.20	nd (w, MeOH)		106-7				$B27^2$, 198
13824	2-Thiazoline or 4,5-Dihydrothiazole C_3H_5NS	87.14		138^{750}				eth, ace, bz	$B27^1$, 206
13825	2-Thiazoline, 2-amino $C_3H_6N_2S$	102.15	nd or lf (bz)	d	84-5			w, al, bz, chl	$B27^2$, 194
13826	2-Thiazoline-2,5-dimethvl C_5H_9NS	115.19		152					B27, 14
13827	2-Thiazoline, 5-ethyl-2-methyl $5\text{-}C_2H_5\text{-}2\text{-}CH_3(2\text{-}C_3H_3NS)$	129.22		$62\text{-}3^{16}$					
13828	2-Thiazoline, 4-hexyl-2-methyl $4\text{-}C_6H_{13}\text{-}2\text{-}CH_3(2\text{-}C_3H_3NS)$	185.33		$92\text{-}3^2$					
13829	2-Thiazoline, 5-hexyl-2-methyl $5\text{-}C_6H_{13}\text{-}2\text{-}CH_3(2\text{-}C_3H_3NS)$	185.33		$82\text{-}4^1$					
13830	2-Thiazoline, 2-mercapto or 2-Thiazolinethiol $2HS\text{-}(2\text{-}C_3H_4NS)$	119.20	nd (w, MeOH)		106-7			w, al	$B27^2$, 198
13831	2-Thiazoline, 2,4,4-trimethyl $2,4,4\text{-}(CH_3)_3(2\text{-}C_3H_2NS)$	129.22		146-8		0.969^{25}	1.4825^{25}		C53, 11368
13832	Thienone or Di-(2-thienyl)ketone $(C_4H_3S)_2CO$	194.27	nd (al)	326	90			eth, ace	$B19^4$, 1745
13833	Thiepane $C_6H_{12}S$	116.22		$169\text{-}71^{747}$		$0.9883^{20/4}$	1.5138^{20}	eth, ace, chl	$B17^4$, 81
13834	Thietane C_3H_6S	74.14		94.7, 14^{30}	−73.25	$1.0200^{20/4}$	1.5102^{20}	al, ace, bz	$B17^4$, 14
13835	Thietane, 1,1-dioxide $C_3H_6SO_2$	106.14		91.2^{14}	75.5-6		1.5156^{20}	w, al, eth, lig	$B17^4$, 16
13836	Thietane, 2-methyl $2\text{-}CH_3(C_3H_5S)$	88.17		$105.5\text{-}7.5^{747}$		$0.9571^{20/4}$	1.4852^{20}	al, eth, ace, bz, chl	$B17^4$, 44
13837	Thiirane or Ethylene sulfide C_2H_4S	60.11		55-6d		$1.0368^{0/4}$	1.4935^{20}	ace, chl	$B17^4$, 11
13838	Thioacetamide CH_3CSNH_2	75.13	cr (al), pl (eth)		115-6			w, al	$B2^4$, 565

PHYSICAL CONSTANTS OF ORGANIC COMPOUNDS (Continued)

No.	Name, Synonyms, and Formula	Mol. wt.	Color, crystalline form, specific rotation and λ_{max} (log ε)	b.p. °C	m.p. °C	Density	n_D	Solubility	Ref.
13839	Thioacetamide, *N*-phenyl or Thioacetanilide $CH_3CSNHC_6H_5$	151.23	nd (w)	d	75-6				B12^{2}, 142
13840	Thioacetic acid CH_3COSH	76.11	ye	87, 26-7^{35}	<−17	1.064$^{20/4}$	1.4648^{20}	w, al, eth, ace	B2^{4}, 542
13841	Thioacetic acid, ethyl ester or Ethyl thioacetate $CH_3COSC_2H_5$	104.17		116.4		0.9792$^{20/4}$	1.4583^{21}	al, eth	B2^{4}, 543
13842	Thioanisole or Methyl phenyl sulfide $CH_3SC_6H_5$	124.20		193, 74^{10}		1.0579$^{20/4}$	1.5868^{20}	al, bz	B6^{4}, 1466
13843	Thiobenzophenone $(C_6H_5)_2CS$	198.28	bl nd (peth)	174^{14}	53-4			bz, chl	B7^{3}, 2087
13844	Thiobenzophenone, 4,4′ bis(dimethyl amino) $[4\text{-}(CH_3)_2NC_6H_4]_2CS$	284.42	pl		204			bz, chl, aa	B14^{3}, 233
13845	Thiocarbamoyl chloride, *N,N*-diethyl $(C_2H_5)_2NCSCl$	151.65	pr	108^{10}	48-51				B4^{4}, 389
13846	Thiocarbamoylchloride, *N,N*-dimethyl $(CH_3)_2NCSCl$	123.60	pr	98^{10}	42-3			eth, chl	B4^{3}, 147
13847	Thiochrome $C_{12}H_{14}N_4OS$	262.33	ye pr (chl)	sub, vac	227-8			w, MeOH	C29, 6242
13848	Thiocyanuric acid or Trithiocyanuric acid $C_3S_3N_3H_3$	177.26	ye pr		200d				B26, 256
13849	Thiodiglycolic acid $S(CH_2CO_2H)_2$	150.15	cr (AcOEt-bz, w)		129			w, al	B3^{4}, 612
13850	Thioglycolamide, *N*-β-naphthyl $HSCH_2CONH\text{-}(\beta\text{-}C_{10}H_7)$	217.29	nd		111-2			al	C29, 3330
13851	Thioglycolic acid $HSCH_2CO_2H$	92.11		120^{20}	−165	1.3253^{20}	1.5030^{20}	**w, al, eth**	B3^{4}, 600
13852	Thioglycolic acid, acetate or Acetylthioglycolic acid $CH_3COSCH_2CO_2H$	134.15	ye	158-9^{17}				w	B3^{4}, 610
13853	Thioindigo $C_{16}H_8O_2S_2$	296.36	br-red nd (xyl), red mcl nd (bz)	sub	359			xyl	B19^{4}, 2091
13854	Thiolane or Tetrahydrothiophene C_4H_8S	88.17		121.1, 14.5^{10}	−96.2	0.9987$^{20/4}$	1.5048^{20}	**al, eth, ace, bz, chl**	B17^{4}, 34
13855	Thiolane, 1,1-dioxide $C_4H_8SO_2$	120.17		285	27		1.4840^{20}	chl	B17^{4}, 37
13856	Thiolane, 2-methyl $2\text{-}CH_3(C_4H_7S)$	102.19		132.5^{750}		0.9541$^{20/4}$	1.4900^{20}	al, eth, ace, bz, chl	B17^{4}, 62
13857	Thiolane, 3-methyl $3\text{-}CH_3(C_4H_7S)$	102.19		137^{742}		0.9625$^{20/4}$	1.4917^{20}	al, eth, ace, bz, chl	B17^{4}, 64
13858	Thiomorpholine or 1,4-Thiazan $C_{14}H_9NS$	103.18		174^{746}, 110^{100}		1.0882$^{20/4}$	1.5386^{20}	w, al, eth, ace, bz	B27^{2}, 4
13859	Thionamide, tetramethyl $[(CH_3)_2N]_2SO$	136.21		209, 70^{16}	31			eth	B4^{4}, 269
13860	Thiomin hydrochloride or 7-Amino-3-imino-3H-2-phenothiazine hydrochloride Lauth's violet $C_{12}H_9N_3S\text{-}HCl$	263.74	dk br or gr pl or nd					bz, chl	B27^{2}, 447
13861	Thiophene or Thiofuran C_4H_4S	84.14		84.2	−38.2	1.0649$^{20/4}$	1.5289^{20}	**al, eth, ace, bz, Py**	B17^{4}, 234
13862	Thiophene, acetamido $2\text{-}CH_3CONH(C_4H_3S)$	141.19	lf (w)		161-2			al, ace	B17^{1}, 136
13863	Thiophene, 2-acetyl or Methyl-2-thienyl ketone $2\text{-}CH_3CO(C_4H_3S)$	126.17		213.5, 94-6^{13}	10-11	1.1679$^{20/4}$	1.5667^{20}	**al, eth**	B17^{4}, 4507
13864	Thiophene, 2-acetyl-5-bromo $5\text{-}Br\text{-}2\text{-}CH_3CO(C_4H_2S)$	205.07	nd (al)	105-7$^{4.5}$	94-5				B17^{4}, 4512
13865	Thiophene, 2-acetyl-5-chloro $5\text{-}Cl\text{-}2\text{-}CH_3CO(C_4H_2S)$	160.62	ta (al, eth)	88-9$^{4.5}$	52			w, al, eth	B17^{4}, 4510
13866	Thiophene, 2-acetyl-3-hydroxy $2\text{-}CH_3CO\text{-}3\text{-}HO(C_4H_2S)$	142.17		47-9$^{0.2}$	51.5-2.5		1.5795^{20}	al	TETRA21, 3331
13867	Thiophene, 2-acetyl-5-methyl $2\text{-}CH_3CO\text{-}5\text{-}CH_3(C_4H_2S)$	140.20		232-3, 98-100^{8}	27-8	1.1185$^{25/4}$	1.5604	eth, ace, bz	B17^{4}, 4550
13868	Thiophene, 2-amino or Thiophenine $2\text{-}H_2N(C_4H_3S)$	99.15	pa ye tab (al)	77-9^{11}				w, al	B17^{2}, 296
13869	Thiophene, 2-benzoyl or Phenyl 2-thienyl ketone $(C_4H_3S)COC_6H_5$	188.24	nd (dil al)	300	56-7	1.1890$^{54/4}$	1.6181^{54}	al, eth, ace	B17^{4}, 5187
13870	Thiophene, 3-benzoyl or Phenyl 3-thienyl ketone $3\text{-}C_6H_5CO(C_4H_3S)$	188.24		129-30^{3}	63-4			al, eth	B17^{4}, 5193

PHYSICAL CONSTANTS OF ORGANIC COMPOUNDS (Continued)

No.	Name, Synonyms, and Formula	Mol. wt.	Color, crystalline form, specific rotation and λ_{max} (log ε)	b.p. °C	m.p. °C	Density	n_D	Solubility	Ref.
13871	Thiophene, 2-bromo 2-Br-(C_4H_3S)	163.03		149.51, 42-6^{13}		1.684$^{20/4}$	1.5868^{20}	eth, ace	B17^4, 245
13872	Thiophene, 2-bromo-5-chloro 5-Cl-2-Br-(C_4H_2S)	197.48		70^{18}	−20	1.803$^{25/25}$	1.5925^{25}	eth, ace	B17^4, 246
13873	Thiophene, 2-bromo-5-iodo 5-I-2-Br-(C_4H_2O)	288.93		116^{13}					B17^4, 252
13874	Thiophene, 2-bromo-3-methyl 2-Br-3-CH_3(C_4H_2S)	177.06		175^{729}, 27$^{1.8}$		1.5844$^{18/4}$	1.5714^{20}	eth, bz	B17^4, 279
13875	Thiophene, 2-bromo-5-methyl 2-Br-5-CH_3(C_4H_2S)	177.06	col to pa ye	177^{740}, 29$^{1.8}$		1.5529$^{20/4}$	1.5673^{20}	eth, bz	B17^4, 272
13876	Thiophene, 3-bromo 3-Br(C_4H_3S)	163.03		159-60, 66-8^{31}		1.735$^{20/4}$	1.5919^{20}	ace, bz	B17^4, 245
13877	Thiophene, 2-butyl 2-C_4H_9(C_4H_3S)	140.24		181-2		0.9537$^{20/4}$	1.5090^{20}		B17^4, 305
13878	Thiophene, 2-chloro 2-Cl(C_4H_3S)	118.58		128.3	−71.9	1.2863$^{20/4}$	1.5487^{20}	**al, eth**	B17^4, 241
13879	Thiophene, 2-chloro-5-butyl 2-Cl-5-C_4H_9(C_4H_2S)	174.69		117-8^{88}		1.0842$^{17/4}$	1.5162^{20}	bz	B17, 44
13880	Thiophene, 2-chloro-5-iodo 2-Cl-5-I(C_4H_2S)	244.48		95-6^{14}	−25				B17^4, 252
13881	Thiophene, 2-chloro-5-methyl 2-Cl-5-CH_3(C_4H_2S)	132.61		154-5^{742}, 55^{19}		1.2147$^{25/4}$	1.5372^{20}	al, eth, ace, bz	B17^4, 271
13882	Thiophene, 2,5-dibromo 2,5-Br_2(C_4H_2S)	241.93		210.3, 76-80^{10}	−6	2.147$^{23/23}$	1.6288^{20}	al, eth	B17^4, 248
13883	Thiophene, 2,5-dibromo 3,4-dinitro 3,4-$(NO_2)_2$-2,5-Br_2(C_4S)	331.92	cr (al)		139-40				B17^4, 261
13884	Thiophene, 2,3-dichloro 2,3-Cl_2(C_4H_2S)	153.03		173-4	−26.2	1.4605$^{20/4}$	1.5651^{20}	**al**, eth, ace, bz	B17^4, 242
13885	Thiophene, 2,4-dichloro 2,4-Cl_2(C_4H_2S)	153.03		167.6	−34	1.4553$^{20/4}$	1.5660^{20}	al, eth, ace, bz	B17^4, 242
13886	Thiophene, 2,5-dichloro 2,5-Cl_2(C_4H_2S)	153.03		162	−40.5	1.4422$^{20/4}$	1.5626^{20}	**al, eth**	B17^4, 243
13887	Thiophene, 2,5-diido 2,5-I_2(C_4H_2S)	335.93	lf (al)	139-40^{15}	41-2			al	B17^4, 253
13888	Thiophene, 2,3-dimethyl or 2,3-Thioxene 2,3-$(CH_3)_2$(C_4H_2S)	112.19		141.6, 29.3^{10}	−49	1.0021$^{20/4}$	1.5192^{20}	al, eth, bz	B17^4, 287
13889	Thiophene, 2,4-dimethyl or 2,4-Thioxene 2,4$(CH_3)_2$(C_4H_2S)	112.19		140.7, 29.9^{10}		0.9956$^{20/20}$	1.5104^{20}	al, eth, bz	B17^4, 288
13890	Thiophene, 2,5-dimethyl or α,α′-Thioxene 2,5-$(CH_3)_2$(C_4H_2S)	112.19		136.7, 26.2^{10}	−62.6	0.985$^{20/4}$	1.5129^{20}	al, eth, bz	B17^4, 290
13891	Thiophene, 3,4-dimethyl 3,4-$(CH_3)_2$(C_4H_2S)	112.19		144-6^{762}, 70-1^{55}		0.994$^{25/15}$	1.5206^{20}	al, eth	B17^4, 293
13892	Thiophene, 2,5-dinitro 2,5-$(NO_2)_2C_4H_2S$)	174.13	(i) ye lf (al) (ii) ye nd (al, w)	290	(i)52 (ii)80-2			al, eth	B17^4, 260
13893	Thiophene, 2,4-diphenyl 2,4-$(C_6H_5)_2$(C_4H_2S)	236.33			119-20			al, ace, chl	B17^4, 680
13894	Thiophene, 2-ethyl 2-C_2H_5(C_4H_3S)	112.19		104, 24^{10}		0.9930$^{20/4}$	1.5122^{20}	al, eth	B17^4, 285
13895	Thiophene, 3-ethyl 3-C_2H_5(C_4H_3S)	112.19		136, 26^{10}	−89.1	0.9980$^{20/4}$	1.5146^{20}	al, eth	B17^4, 286
13896	Thiophene, 2-hydroxy 2-HO(C_4H_3S)	100.14		217-9, 91-3^{13}		1.255$^{20/4}$	1.5644^{20}	w, ace, chl	B17^4, 4286
13897	Thiophene, 2-hydroxy-5-methyl or 2,5-Thiotenol 5-CH_3-2-HO(C_4H_2S)	114.16		85^{40}				al, eth	B17^4, 4301
13898	Thiophene, 2-hydroxymethyl 2-$HOCH_2$(C_4H_3S)	114.16		207, 96^{12}		1.2053$^{16/4}$	1.5280^{20}	al, eth, ace	B17^4, 1242
13899	Thiophene, 2-iodo 2-I-(C_4H_3S)	210.03		180-2, 73^{15}	−40		1.6465^{25}	al, eth	B17^4, 250
13900	Thiophene, 2-iodo-5-nitro 5-NO_2-2-I(C_4H_2S)	255.03	ye pr (al)		74			al	B17^4, 259
13901	Thiophene, 2-methyl or α-Thiololene 2-CH_3(C_4H_3S)	98.16		112.6, 9.2^{10}	−63.4	1.0193$^{20/4}$	1.5203^{20}	**al, eth, ace, bz**	B17^4, 269
13902	Thiophene, 3-methyl or β-Thiololene 3-CH_3(C_4H_3S)	98.16		115.4, 11^{10}	−69	1.0218$^{20/4}$	1.5204^{20}	**al, eth, ace, bz,** chl	B17^4, 277
13903	Thiophene, 2-methyl-5-phenyl 2-CH_3-5-C_6H_5(C_4H_2S)	174.26	nd	270-2	51			al, eth, lig	B17^4, 551

No.	Name, Synonyms, and Formula	Mol. wt.	Color, crystalline form, specific rotation and λ_{max} (log ε)	b.p. °C	m.p. °C	Density	n_D	Solubility	Ref.
13904	Thiophene, 2-methylamino or *N*-Methyl-2-thiophenine $2\text{-}CH_3NH(C_4H_3S)$	113.18		$88\text{-}92^{15}$				eth, ace	B17^1, 136
13905	Thiophene, 2-nitro $2\text{-}O_2N(C_4H_3S)$	129.13	lf ye mcl nd (peth)	224-5	46.5	$1.3644^{43/4}$		al	B17^4, 255
13906	Thiophene, 3-nitro $3\text{-}3NO_2(C_4H_3S)$	129.13		$225, 95^{12}$	78-9			al, lig	B17^4, 256
13907	Thiophene, 3-nitro-2,4,5-trichloro $2,4,5\text{-}Cl_3\text{-}3\text{-}O_2N(C_4S)$	232.47	red-ye nd (al)		86			al, eth, bz	B17, 35
13908	Thiophene, 4-nitro-2,3,5-tribromo $2,3,5\text{-}Br_3\text{-}4\text{-}O_2N(C_4S)$	365.82	red-ye nd (al)		106			eth	B17, 35
13909	Thiophene, 2-octyl $2\text{-}C_8H_{17}(C_4H_3S)$	196.35		257-9		$0.8118^{20.5}_{20.5}$		eth	B17^4, 335
13910	Thiophene, 2-isopentyl $2\text{-}i\text{-}C_5H_{11}(C_4H_3S)$	154.27		$74\text{-}5^{11}$		$0.9481^{12/4}$	$1.5014^{12/}_{587.6}$		B17^4, 316
13911	Thiophene, 2-propyl $2\text{-}C_3H_7(C_4H_3S)$	126.22		157-9, 55.5^{20}		$0.9683^{20/4}$	1.50757^{20}	al, eth, bz	B17^4, 297
13912	Thiophene, 3-propyl $3\text{-}C_3H_7(C_4H_3S)$	126.22		163.2		$0.97377^{20/4}$	1.5057^{20}	al, eth, bz	B17^2, 42
13913	Thiophene, tetrabromo C_4Br_4S	399.72	nd (al)	326, $170\text{-}3^{14}$	117-8			eth	B17^4, 250
13914	Thiophene, tetrachloro C_4Cl_4S	221.92	nd (dil al)	233.4, $75\text{-}7^{2}$	30-1	$1.7036^{30/4}$	1.5915^{30}	al, eth	B17^4, 244
13915	Thiophene, tetrahydro or Thiolane C_4H_8S	88.17		121.1, $14\text{-}5^{10}$	−96.6	$0.9987^{20/4}$	1.5018^{20}	al, eth, ace, bz, chl	B17^4, 34
13916	Thiophene, tetraphenyl $(C_6H_5)_4(C_4S)$	388.53		400	189-90			eth, bz	B17^4, 810
13917	Thiophene, 2,3,5-tribromo $2,3,5\text{-}Br_3(C_4HS)$	320.82	nd (al)	260	29			al, eth	B17^4, 249
13918	Thiophene, 2,3,5-trichloro $2,3,5\text{-}Cl_3(C_4HS)$	187.47		198.7	−16.1	$1.5856^{20/4}$	1.5791^{20}	al, eth	B17^4, 244
13919	Thiophene, 2,3,5-trimethyl $2,3,5\text{-}(CH_3)_3(C_4HS)$	126.22		164.5, 46.8^{10}		$0.9753^{20/4}$	1.5112^{20}	al, eth, ace, bz	B17^4, 301
13920	2-Thiophenacetic acid $2\text{-}HO_2CH_2C(C_4H_8S)$	142.17	cr (w)		76			al, eth	B18^4, 4062
13921	2-Thiophene carboxaldehyde or 2-Thiophenealdehyde $2\text{-}OHC(C_4H_3S)$	112.15	pa ye liq	$197, 85\text{-}6^{16}$		$1.215^{21/21}$	1.5920^{20}	al, eth	B17^4, 4477
13922	2-Thiophene carboxaldehyde oxime or 2-Thiophenaldoxime $2\text{-}(C_4H_3S)CH{=}NOH$	127.16	nd		133			eth	B17^4, 4482
13923	2-Thiophene carboxaldehyde phenylhydrazone $2\text{-}(C_4H_3S)CH{=}NNHC_6H_5$	202.27	ye nd (al)		134.5			al	B17^4, 4482
13924	2-Thiophene carboxaldehyde, 5-bromo $5\text{-}Br\text{-}(C_4H_2S)\text{-}CHO\text{-}2$	191.04		$105\text{-}7^{11}$			1.6378^{20}		B17^4, 4487
13925	2-Thiophene carboxaldehyde, 5-chloro $5\text{-}Cl\text{-}(C_4H_2S)CHO\text{-}2$	146.59		77.5^{5}			1.6036^{25}		B17^4, 4485
13926	2-Thiophene carboxaldehyde, 5-methyl $5\text{-}CH_3\text{-}(C_4H_2S)CHO\text{-}2$	126.17		114^{25}			1.5825^{20}	chl	B17^4, 4529
13927	3-Thiophene carboxaldehyde $3\text{-}(C_4H_3S)CHO$	112.15		$86\text{-}7^{20}$			1.5855^{20}	al, eth	B17^4, 4497
13928	2-Thiophene carboxylic acid or α-Thiophenic acid $2\text{-}(C_4H_3S)CO_2H$	128.15	nd (w)	260d	129-30			w, al, eth, chl	B18^4, 4011
13929	2-Thiophene carboxylic acid, ethyl ester or Ethyl-α-thiophene carboxylate $2\text{-}(C_4H_3S)CO_2C_2H_5$	156.20		218, 94^{10}		$1.1623^{16/4}$	1.5248^{20}	al, ace	B18^4, 4012
13930	2-Thiophene carboxylic acid hydrazide $2\text{-}(H_2NNHCO)(C_4H_3S)$	142.17			136				B18^4, 4024
13931	2-Thiophene carboxylic acid, 3-methyl $3\text{-}CH_3(C_4H_2S)CO_2H\text{-}2$	142.17	nd (dil al, w)		144			al, eth, ace	B18^4, 4068
13932	2-Thiophene carboxylic acid, 5-methyl $5\text{-}CH_3(C_4H_2S)CO_2H\text{-}2$	142.17			138-9			al, eth	B18^4, 4086
13933	2-Thiophene (carboxylyl) chloride $2\text{-}(C_4H_3S)COCl$	146.59		$201, 77^{10}$					B18^4, 4016
13934	3-Thiophene carboxylic acid $3\text{-}HO_2C(C_4H_3S)$	128.15			138.4			w	B18^4, 4054
13935	2,3-Thiophene dicarboxylic acid $2,3\text{-}(C_4H_2S)(CO_2H)_2$	172.16	pr or nd (w)		272-4			eth	B18^4, 4478
13936	2,4-Thiophene dicarboxylic acid $2,4\text{-}(C_4H_2S)(CO_2H)_2$	172.16	cr	sub>200	280d				B18^4, 4480

No.	Name, Synonyms, and Formula	Mol. wt.	Color, crystalline form, specific rotation and λ_{max} (log ε)	b.p. °C	m.p. °C	Density	n_D	Solubility	Ref.
13937	2,5-Thiophene dicarboxylic acid $2,5-(C_4H_2S)(CO_2H)_2$	172.16		sub 150-300	359 (sealed tube)			al, eth	$B18^4$, 4495
13938	2,5-Thiophene dicarboxylic acid, diethyl ester or Diethyl-2,5-thiophene dicarboxylate $2,5-(C_4H_2S)(CO_2C_2H_5)_2$	228.26	nd (al)		51.5			al	B18, 331
13939	2-Thiophene sulfonamide $2-(C_4H_3S)SO_2NH_2)$	163.21	nd (w)		147				$B18^4$, 6706
13940	2-Thiophene sulfonyl chloride $2-(C_4H_3S)(SO_2Cl)$	182.64		$99-101^6$, sub	28			eth	$B18^4$, 6706
13941	3-Thiophene sulfonamide $3-(C_4H_3S)(SO_2NH_2)$	163.21	pl (w)		152-3				$B18^4$, 6714
13942	3-Thiophene sulfonyl chloride $3-(C_4H_3S)(SO_2Cl)$	182.64	cr (eth)	$98-9^{0.5}$	43			eth	$B18^4$, 6713
13943	Thiophenol or Mercaptobenzene C_6H_5SH	110.17		168.7, 46.4^{10}	−14.8	$1.0766^{20/4}$	1.5893^{20}	al, eth, bz	$B6^4$, 1463
13944	Thiophenol, *o*-amino $2H_2NC_6H_4SH$	125.19	nd	234, $125-7^6$	26		1.4606^{20}	al, eth	$B13^3$, 902
13945	Thiophenol, *m*-amino $3-H_2N-C_6H_4SH$	125.19	pa ye oil	$180-90^{16}$				al, aa, eth	$B13^3$, 982
13946	Thiophenol, *m*-amino, hydrochloride $3-H_2NC_6H_4SH.HCl$	161.65		sub	232			w, al	B13, 425
13947	Thiophenol, *p*-amino $4H_2NC_6H_4SH$	125.19	cr	$140-5^{15}$	46			w, al	$B13^3$, 1221
13948	Thiophenol, *p*-bromo $4-BrC_6H_4SH$	189.07	lf (al)	230-1	73			eth, chl	$B6^4$, 1650
13949	Thiophenol, *o*-chloro $2-Cl_6H_4SH$	144.62		205-6, 117^{15}		1.2752^{10}			$B6^4$, 1570
13950	Thiophenol, *m*-chloro $3-ClC_6H_4SH$	144.62		205-7		1.2637^{13}		al, eth, peth, chl	$B6^4$, 1576
13951	Thiophenol, *p*-chloro $4-ClC_6H_4SH$	144.62		198-9	61	$1.1911^{20/4}$	1.5480^{20}	al, eth	$B6^4$, 1581
13952	Thiophenyl, *m*-ethoxy $3-C_2H_5OC_6H_4SH$	154.23		238-9				al, eth, ace, bz	B6, 833
13953	Thiophenol, *p*-ethoxy $4-C_2H_5OC_6H_4SH$	154.23		238	1.6			al, eth, ace, bz	$B6^3$, 4446
13954	Thiophenol, *m*-methoxy $3-CH_3OC_6H_4SH$	140.20		223-6			1.5874^{20}	chl	$B6^3$, 4363
13955	Thiophenol, *p*-methoxy or 4-Mercaptoanisole $4-CH_3OC_6H_4SH$	140.20		227-9, $89-90^5$		$1.1313^{25/4}$	1.5801^{25}	al, eth, bz	$B6^3$, 4445
13956	Thiophenol, *o*-nitro $2-NO_2C_6H_4SH$	155.17			58.5			al, eth, bz	$B6^4$, 1661
13957	Thiophenol, *p*-nitro $4-NO_2C_6H_4SH$	155.18	(eth, chl, ace)		79			al, eth, ace	$B6^4$, 1687
13958	Thiophenol, 2,4,6-trinitro or Thiopicric acid $2,4,6-(NO_2)_3C_6H_2SH$	245.17	ye nd		114			al, eth, ace, bz, chl	$B6^2$, 316
13959	Thiophosgene or Thiocarbonyl chloride $CSCl_2$	114.98	red	73		1.508^{15}	1.5442^{20}	al, eth	$B3^4$, 281
13960	Thiophthene (solid) or Thieno[3,2-b]thiophene $C_6H_4S_2$	140.22	bipym orh (lig)	221-4	56			eth	$B19^4$, 189
13961	Thiophthene (liquid) or Thieno[3,2-b]thiophene $C_6H_4S_2$	140.22		224-6, 106^{16}	6.5			eth	$B19^4$, 189
13962	Thiopyran, tetrahydro or Pentamethylene sulfide $C_5H_{10}S$	102.19		141.7, 93^{82}	19	$0.9861^{20/4}$	1.5067^{20}	al, eth, ace, bz	$B17^4$, 55
13963	Thiopyrine or 1,5-Dimethyl-2-phenyl-3-thio-3-pyrazolone $C_{11}H_{12}N_2S$	204.29	cr (w)		166			al, eth	$B24^2$, 28
13964	Thiosalicylic acid or 2-Mercaptobenzoic acid $2-HSC_6H_4CO_2H$	154.18	lf or nd (al, w, aa)	sub	168-9			al, eth, aa	$B10^3$, 265
13965	Thiosemicarbazide or Semicarbazide-3-thio $H_2NNHCSNH_2$	91.13	lo nd (w)		183			w, al	$B3^4$, 374
13966	2-Thiouracil $C_4H_4N_2OS$	128.15	pr (w, al)		>340d				$B24^2$, 171
13967	2-Thiouracil, 6-methyl $C_5H_6N_2OS$	142.18	pl (w)		299-303d				$B24^2$, 183
13968	2-Thiouracil, 6-propyl $C_7H_{10}N_2OS$	170.23	pw (w)		219				C43, 674

No.	Name, Synonyms, and Formula	Mol. wt.	Color, crystalline form, specific rotation and λ_{max} (log ϵ)	b.p. °C	m.p. °C	Density	n_D	Solubility	Ref.
13969	4-Thiouracil $C_4H_4N_2OS$	128.15	yesh nd or pr (w)		328d				B24, 323
13970	4-Thiouracil, 6-methyl $C_5H_6N_2OS$	142.18	ye pr (w)		>250d				B24, 352
13971	Thiourea H_2NCSNH_2	76.12	rh (al)		182	1.405		w, al	$B3^4$, 342
13972	Thiourea, 1-acetyl $CH_3CONHCSNH_2$	118.15	pr (w), rh (al)		165			al	$B3^4$, 354
13973	Thiourea, *S*-acetyl, hydrochloride $H_2NC(=NH)SCOCH_3.HCl$	154.61	pl		109d			w	$B3^3$, 314
13974	Thiourea, 1-allyl $(CH_2=CHCH_2)NHCSNH_2$	116.18	mcl or rh, pr (w)		(i)71, (ii)78.4	$1.219^{20/20}$	1.5936^{78}	w, al	$B4^4$, 1072
13975	Thiourea, 1-allyl-3-phenyl $(CH_2=CHCH_2)NHCSNHC_6H_5$	192.28			98				$B12^3$, 855
13976	Thiourea, 1-(4-aminobenzene sulfonyl) or 1-Sulfanyl thiourea $(4-H_2NC_6H_4SO_2)NHCSNH_2$	231.29	pw, pl (w)		182				$B14^3$, 1975
13977	Thiourea, 1-benzoyl $C_6H_5CONHCSNH_2$	180.22	pr (dil al)		171			al	$B9^3$, 1120
13978	Thiourea, 1-benzyl $C_6H_5CH_2NHCSNH_2$	166.24	pr (w)		164-5				$B12^3$, 2277
13979	Thiourea, 1-benzyl-3-phenyl $C_6H_5CH_2NHCSNHC_6H_5$	242.34			153-4			al, eth	$B12^2$, 564
13980	Thiourea, 1-butyl-3-phenyl $C_4H_9NHCONHC_6H_5$	208.32	pr		85			bz	$B12^3$, 853
13981	Thiourea, 1-(2-chlorophenyl) $2-ClC_6H_4NHCSNH_2$	186.66	nd or pl		146			al, bz	$B12^3$, 1296
13982	Thiourea, 1-(4-chlorophenyl) $4-ClC_6H_4NHCSNH_2$	186.66	pl or nd (al)		178			al, bz	$B12^3$, 1368
13983	Thiourea, 1,3-dibenzyl $(C_6H_5CH_2NH)_2CS$	256.37			147-8			al, eth	$B12^3$, 2278
13984	Thiourea, 1,3-diethyl $(C_2H_5NH)_2CS$	132.22		d	144			w, al, eth	$B4^4$, 375
13985	Thiourea, 1,3-diethyl-1,3-diphenyl $[C_6H_5N(C_2H_5)]_2CS$	284.42	rh pl (lig), pr (al)		75.5			al	$B12^3$, 884
13986	Thiourea, 1,3-dimethyl $(CH_3NH)_2CS$	104.17	dlq pl		62			w, al, ace, chl	$B12^4$, 217
13987	Thiourea, 1,3-di(α-naphthyl) $(\alpha-C_{10}H_7NH)_2CS$	328.43	lf (to)		207.5			to	$B12^2$, 696
13988	Thiourea, 1,3-di(β-naphthyl) $(\beta-C_{10}H_7NH)_2CS$	328.43	lf (aa)		203				$B12^3$, 3048
13989	Thiourea, 1,1-diphenyl $(C_6H_5)_2NCSNH_2$	228.31	pr (al)		210d, (218)			al	$B12^3$, 898
13990	Thiourea, 1,3-diphenyl $(C_6H_5NH)_2CS$	228.31	lf (al)		154, (189)			al, eth, chl	$B12^3$, 858
13991	Thiourea, 1,1-dipropyl $(C_3H_7)_2NCSNH_2$	160.28			67				B4, 144
13992	Thiourea, 1,3-dipropyl $(C_3H_7NH)_2CS$	160.28	lf (w)		71			eth	B4, 143
13993	Thiourea, 1,3-diisopropyl $(i-C_3H_7NH)_2CO$	160.28			141			chl	$B4^4$, 526
13994	Thiourea, 1,3-di(2-tolyl) $(2-CH_3C_6H_4NH)_2CS$	256.37	nd (al, sub)	216-8 (sub)	165-6			al, bz, chl, aa	$B12^3$, 1881
13995	Thiourea, 1,3-di(4-tolyl) $(4-CH_3C_6H_4NH)_2CS$	256.37	rh bipym pr		176			eth	$B12^3$, 2099
13996	Thiourea, 1-ethyl-3-phenyl $C_2H_5NHCSNHC_6H_5$	180.27			107			al, bz	$B12^3$, 853
13997	Thiourea, ethylidene $CH_3CH=NCSNH_2$	102.15	(al)		*ca* 212d				$B3^1$, 76
13998	Thiourea, *S*-methyl $HN=C(SCH_3)NH_2$	90.14	lf (ace)		79			w, al, ace	$B3^4$, 358
13999	Thiourea, *S*-methyl, hydroiodide or *S*-Methyl isothiouronium iodide $HN=C(SCH_3)NH_2.HI$	218.06	pr		117			w, al	$B3^1$, 78
14000	Thiourea, *S*-methyl, nitrate or *S*-Methyl isothiouronium nitrate $HN=C(SCH_3)NH_2.HNO_3$	153.16			109-10			w, al	$B3^1$, 78

No.	Name, Synonyms, and Formula	Mol. wt.	Color, crystalline form, specific rotation and λ_{max} (log ε)	b.p. °C	m.p. °C	Density	n_D	Solubility	Ref.
14001	Thiourea, *S*-methyl, sulfate or *S*-Methyl isothiouronium sulfate $[HN{=}C(SCH_3)NH_2]_2.H_2SO_4$	278.36	nd (al, w)		244d				B3[4], 358
14002	Thiourea, *S*-methyl-1,3-diphenyl $C_6H_5N{=}C(SCH_3)NHC_6H_5$	242.34	nd (al)		109-10			al	B12[2], 911
14003	Thiourea, 1-methyl $CH_3NHCSNH_2$	90.14	pr		120-1			w, al, ace	B4[4], 216
14004	Thiourea, 1-methyl-3-(α-naphthyl) α-$C_{10}H_7NHCSNHCH_3$	216.30	pl (al)		198			al	B12[2], 696
14005	Thiourea, 1-methyl-3-phenyl $CH_3NHCSNHC_6H_5$	166.24	ta, pl		112-3			al	B12[3], 853
14006	Thiourea, 1-(α-naphthyl) α-$C_{10}H_7NHCSNH_2$	202.27	pr (al)		198				B12[3], 2941
14007	Thiourea, 1-(α-naphthyl)-3-phenyl α-$C_{10}H_7NHCSNHC_6H_5$	278.37	lf		162-3				B12, 1241
14008	Thiourea, 1-β-naphthyl β-$C_{10}H_7NHCSNH_2$	202.27	lf (al)		186			al	B12[3], 3045
14009	Thiourea, *S*-phenyl $HN{=}C(SC_6H_5)NH_2$	152.22	nd (bz)		96-7d			bz	B6[1], 146
14010	Thiourea, 1-phenyl $C_6H_5NHCSNH_2$	152.22	nd (w), pr (al)		154			al	B12[3], 852
14011	Thiourea, 1,1,3,3,-tetramethyl $[(CH_3)_2N]_2CS$	132.22		245	78-9			w, al	B4[4], 232
14012	Thiourea, 1,1,3,3-tetraphenyl $[(C_6H_5)_2N]_2CS$	380.51	nd (al, MeOH)		194-5			eth, bz	B12[3], 898
14013	Thiourea, 1-(2-tolyl) $(2\text{-}CH_3C_6H_4)NHCSNH_2$	166.24	nd (dil al, w)		162			w, al	B12[3], 1878
14014	Thiourea, 1-(3-tolyl) $(3\text{-}CH_3C_6H_4)NHCSNH_2$	166.24	pr (al)		110-1			w	B12[3], 1975
14015	Thiourea, 1-(4-tolyl) $(4\text{-}CH_3C_6H_4)NHCSNH_2$	166.24	pl (al)		188-9			al	B12[3], 2097
14016	Thiourea, 1,1,3-trimethyl $(CH_3)_2NCSNHCH_3$	118.20	pr (bz-lig)		87-8			w, al, bz, chl	B4[4], 232
14017	Thiourea, 1,1,3-triphenyl $(C_6H_5)_2NCSNHC_6H_5$	304.41	nd (al)		152				B12, 432
14018	Thioxanthene or Dibenzthiopyran $C_{13}H_{10}S$	198.28	nd (al-chl)	340^{730}, sub	128-9			chl, peth	B17[4], 615
14019	Thioxanthone or 9-Oxothioxanthene $C_{13}H_8OS$	212.27	ye nd (chl)	$371\text{-}3^{715}$, sub	209 (212)			bz, aa	B17[4], 5303
14020	D-Threonic acid or 2,3,4-Trihydroxy butyric acid $HOCH_2(CHOH)_2CO_2H$	136.10	nd (al), $[\alpha]_D$ −30 (w)		197-8			w	B3[4], 1112
14021	DL-Threonic acid $HOCH_2(CHOH)_2CO_2H$	136.10	cr (al, ace)		99			w	B3[3], 895
14022	L-Threonic acid $HOCH_2(CHOH)_2CO_2H$	136.10	nd (al-eth), $[\alpha]_D$ +9.5 (w)		169-70			w, ace	B3[4], 1112
14023	Threonine *(D)* or *D*-2-Amino-3-hydroxy butyric acid $CH_3CH(OH)CH(NH_2)CO_2H$	119.12	cr (80% al), $[\alpha]^{26/}{}_D$ −28.3 (w, c=1.1)		255-7d			w	B4[3], 1625
14024	Threonine *(DL)* or *DL*-2-Amino-3-hydroxy butyric acid $CH_3CH(OH)CH(NH_2)CO_2H$	119.12	cr (dil al)		234-5d, 229-30 (+ ½ w)			w	B4[3], 1625
14025	D-Threose or D-Trihydroxy butyraldehyde $HOCH_2(CHOH)_2CHO$	120.11	hyg syr or nd(w), $[\alpha]^{22/}{}_D$ +29.1→ 19.6 (w, mut)		126-32			w	B1[4], 4173
14026	Thujane *(d)* or Sabinane $C_{10}H_{18}$	138.25	$[\alpha]_D$ +73.1	157^{758}		$0.8139^{20/4}$	1.4376^{20}	al, eth, ace, bz	B5[4], 317
14027	3-Thujene $C_{10}H_{16}$	136.24	$[\alpha]^{/}{}_D$ −37.20	151		$0.8301^{20/4}$	1.4515^{20}		B5[4], 451
14028	4(10)-Thujene *(d)* or Sabinene $C_{10}H_{16}$	136.24	$[\alpha]_D$ +80.2	163-5, 66^{30}		0.842^{20}	1.4678^{20}	al, eth, ace, bz, chl	B5[4], 451
14029	4(10)Thujene *(l)* or *l*-Sabinene $C_{10}H_{16}$	136.24	$[\alpha]^{15/}{}_D$ −42.5	163-5		$0.8464^{20/4}$	1.4515^{20}	al, eth, ace, bz, chl	B5[4], 452

PHYSICAL CONSTANTS OF ORGANIC COMPOUNDS (Continued)

No.	Name, Synonyms, and Formula	Mol. wt.	Color, crystalline form, specific rotation and λ_{max} (log ε)	b.p. °C	m.p. °C	Density	n_D	Solubility	Ref.
14030	3-Thujone $C_{10}H_{16}O$	152.24	$[\alpha]^{18/}_{D}$ −19.9	200-1, 75^{9}		0.9152^{20}	1.4490^{25}	al, eth, ace	B7[3], 379
14031	Thymidine or Thymine-β-D-2-desoxy-riboside $C_{10}H_{14}N_2O_5$	242.23	nd (AcOEt), $[\alpha]^{25/}_{D}$ +30.6 (w, c=1.03)		186-7			w, MeOH, aa, Py	C48, 226
14032	Thymol or 2-Methyl-5-isopropyl phenol $2\text{-}CH_3\text{-}5\text{-}i\text{-}C_3H_7C_6H_3OH$	150.22	pl (aa, ace)	233, 92^{2}	52	$0.925^{80/4}$	1.5227^{20}	al, eth	B6[3], 1893
14033	Thymol acetate $5\text{-}CH_3\text{-}2\text{-}(CH_3)_2CHC_6H_3O_2CCH_3$	192.26		245, 131^{21}		1.009^{9}		al, eth, bz, chl	B6[3], 1901
14034	Thymol, 4-amino or 2-Amino-3-methyl-6-*iso*-propyl phenol $2\text{-}H_2N\text{-}3\text{-}CH_3\text{-}6\text{-}i\text{-}C_3H_7\text{-}C_6H_2OH$	165.24	nd, sc (bz)		178-9			al, bz	B13[3], 1803
14035	Thymol carbonate or Thymatol Tyranol $2\text{-}CH_3\text{-}5\text{-}i\text{-}C_3H_7C_6H_3OH.H_2CO_3$	212.25	nd or pr		49			eth, chl	B6[2], 499
14036	Thymol, 6-chloro or 4-Chloro-2-*iso*-propyl-5-methyl phenol $4\text{-}Cl\text{-}2\text{-}i\text{-}C_3H_7\text{-}5\text{-}CH_3C_6H_2OH$	184.67	nd or pl (lig)	259-63	59-61			w, al, eth, bz, peth	B6[3], 1906
14037	Thymol, 2,6-dinitro or 2,4-Dinitro-3-methyl-6-*iso*-propyl phenol $2,4(NO_2)_2\text{-}3\text{-}CH_3\text{-}6\text{-}[(CH_3)_2CH]C_6HOH$	240.22	pr (peth)		55.5			al, eth	B6[3], 1911
14038	α-Thymolsulfonic acid or 4-Hydroxy-5-*iso*-propyl-2 methyl benzene sulfonic acid $4\text{-}HO\text{-}5(CH_3)_2CH\text{-}2\text{-}CH_3C_6H_2SO_3H$	230.28	pl (+1w)		91-2			w	B11[3], 537
14039	Thymol blue or Thymolsulfonephthalein $C_{27}H_{30}O_5S$	466.59	gr-red (al, eth, aa)		221-4d			al, aa	B19[4], 1135
14040	Thymolphthalein $C_{28}H_{30}O_4$	430.56	pr or nd (al)		253			eth, ace	B18[4], 1955
14041	Thyroxine *(d)* $C_{15}H_{11}I_4NO_4$	776.87	nd $[\alpha]^{25/}_{548}$ +3 (al-NaOH)		237d				B14[3], 1566
14042	Thyroxine-*(l)* $C_{15}H_{11}I_4NO_4$	776.87	nd $[\alpha]^{20/}_{D}$ −4.4 (al-NaOH)		235-6				B14[3], 1566
14043	Tiglaldehyde or 2-Methyl-2-butenal $CH_3CH{=}C(CH_3)CHO$	84.12		$116\text{-}7^{738}$, $63\text{-}5^{119}$		$0.8710^{20/4}$	1.4475^{20}	w, al, eth	B1[4], 3464
14044	Tiglic acid or *trans*-2-Methyl-2-butenoic acid $CH_3CH{=}C(CH_3)CO_2H$	100.12	ta (w)	198.5	64-5	$0.9641^{20/4}$	1.4330^{76}	al, eth	B2[4], 1552
14045	Tiglyl chloride or *trans*-2-Methyl-2-butenyl chloride $CH_3CH{=}C(CH_3)COCl$	118.56		146-8, 45^{12}				eth	B2[4], 1554
14046	Tigogenin or 5-α,22α-Spirostan-3β-ol $C_{27}H_{44}O_3$	416.64	lf (al + 1w), pr (ace), $[\alpha]^{25/}_{D}$ −49 (Py, c=1)		205-6			eth, ace, MeOH, peth	B19[4], 828
14047	α-Tocopherol or Vitamin E. 5,7,8-Trimethyl tocol $C_{29}H_{50}O_2$	430.71	pa ye visc oil $[\alpha]^{25/}_{D}$ +0.65 (al)	350d, $140^{10\text{-}6}$	2-3			al, eth, ace	B17[4], 1436
14048	α-Tocopherol acetate $C_{31}H_{52}O_3$	472.75		$184^{0.01}$	−27.5	$0.9533^{21.5}$	1.495-1.497^{20}	eth, ace, chl	B17[4], 1439
14049	α-Tocopherolquinone or α-tocoquinone $C_{29}H_{50}O_3$	446.71	ye oil	$120^{0.002}$				eth, peth	Am73, 5148
14050	β-Tocopherol or Vitamin E. 5,8-Dimethyl tocol $C_{28}H_{48}O_2$	416.69	pa ye visc oil $[\alpha]^{20/}_{D}$ +6.4	$200\text{-}10^{0.1}$				al, eth, ace, chl	B17[4], 1427
14051	γ-Tocopherol or Vitamin E. 7,8-Dimethyltocol $C_{28}H_{48}O_2$	416.69	pa ye visc oil $[\alpha]^{20/}_{D}$ −2.4 (al)	$200\text{-}10^{0.1}$	−3			al, eth, ace, chl	B17[4], 1429
14052	γ-Tocopherol or γ-Methyltocol $C_{27}H_{46}O_2$	402.66	pa ye visc oil $[\alpha]^{25/}_{546}$ +3.4 (al, c=1.5)	$150^{00.001}$				al, eth, ace, chl	B17[4], 1411

No.	Name, Synonyms, and Formula	Mol. wt.	Color, crystalline form, specific rotation and λ_{max} (log ε)	b.p. °C	m.p. °C	Density	n_D	Solubility	Ref.
14053	Tolbutamide or 1-Butyl-3(*p*-tolylsulfonyl)urea $C_{12}H_{18}N_2O_3S$	270.35	orth cr	....	128-9	1.245^{25}	....	al, eth, chl	C53, 13084
14054	*o*-Tolualdehyde or 2-Methylbenzaldehyde 2-$CH_3C_6H_4CHO$	120.15	....	200, 94^{10}	....	$1.0386^{19/4}$	1.5481^{20}	al, eth, ace, bz, chl	$B7^3$, 1011
14055	*m*-Tolualdehyde or 3-Methylbenzaldehyde 3-$CH_3C_6H_4CHO$	120.15	....	199, $93\text{-}4^{17}$	....	1.0189^{21}	....	al, eth, ace, bz, chl	$B7^3$, 1013
14056	*p*-Tolualdehyde or 4-Methylbenzaldehyde 4-$CH_3C_6H_4CHO$	120.15	....	204-5, 106^{10}	....	$1.0194^{17/4}$	1.5454^{20}	al, eth, ace, chl	$B7^3$, 1016
14057	*o*-Toluamide or 2-Methylbenzamide 2-$CH_3C_6H_4CONH_2$	135.17	pl or nd (w)	....	147	....	....	al	$B9^3$, 2304
14058	*m*-Toluamide or 3-Methylbanzamide 3-$CH_3C_6H_4CONH_2$	135.17	nd (eth)	....	97	....	....	al	$B9^3$, 2322
14059	*p*-Toluamide or 4-Methylbenzamide 4-$CH_3C_6H_4CONH_2$	135.17	nd or pl (w)	....	160	....	....	al, eth	$B9^3$, 2343
14060	Toluene or Methyl benzene $C_6H_5CH_3$	92.14	....	110.6, $14.5^{14.5}$	−95	$0.8669^{20/4}$	1.4961^{20}	al, eth, ace, bz, lig	$B5^4$, 766
14061	Toluene, 2-allyl 2-CH_2=$CHCH_2C_6H_4CH_3$	132.21	....	$182\text{-}3^{757}$, $88\text{-}90^{25}$	....	$0.9005^{20/4}$	1.5187^{20}	....	$B5^3$, 1212
14062	Toluene, 2-amino or *o*-Toluidine 2-$CH_3C_6H_4NH_2$	107.16	....	200.2, 80.1^{10}	−14.7	$0.9984^{20/4}$	1.5725^{20}	al, eth	$B12^3$, 1837
14063	Toluene, 3-amino or *m*-Toluidine 3-CH_3-$C_6H_4NH_2$	107.16	....	203.3, 82.3^{10}	−30.4	$0.9889^{20/4}$	1.5681^{20}	al, eth, ace, bz	$B12^3$, 1949
14064	Toluene, 4-amino or *p*-Toluidine 4-$CH_3C_6H_4NH_2$	107.16	lf (w + l)	200.5, 79.6^{10}	44-5	$0.9619^{20/4}$	1.5534^{45}, 1.5636^{20}	al, eth, ace, Py	$B12^3$, 2017
14065	Toluene, 2-bromo or *o*-Tolylbromide 2-Br-$C_6H_4CH_3$	171.04	....	181.7	−27.8	$1.4232^{20/4}$	1.5565^{20}	al, eth, bz	$B5^4$, 825
14066	Toluene, 2-bromo-3-nitro 2-Br-3-$NO_2C_6H_3CH_3$	216.03	ye pr (al)	157^{22}	41^2	....	....	al	$B5^4$, 860
14067	Toluene, 2-bromo-4-nitro 2-Br-4-$NO_2C_6H_3CH_3$	216.03	nd (al)	$150\text{-}1^{20}$	78	....	....	eth	$B5^4$, 861
14068	Toluene, 2-bromo-5-nitro 2-Br-5-$NO_2C_6H_3CH_3$	216.03	cr (al)	$140\text{-}3^{17}$	78	....	....	eth	$B5^4$, 860
14069	Toluene, 2-bromo-6-nitro 2-Br-6-$NO_2C_6H_3CH_3$	216.03	pe ye nd (dil al)	143^{22}	42	....	....	al	B53, 751
14070	Toluene, 3-bromo or *m*-Tolylbromide 3-Br$C_6H_4CH_3$	171.04	....	183.7	−39.8	$1.4099^{20/4}$	1.5510^{20}	al, eth, ace, chl	$B5^4$, 827
14071	Toluene, 3-bromo-2-nitro 3-Br-2-$NO_2C_6H_3CH_3$	216.03	pa ye nd	$129\text{-}30^{10}$	28	....	....	al, eth	$B5^4$, 860
14072	Toluene, 3-bromo-4-nitro 3-Br-4-$NO_2C_6H_3CH_3$	216.03	pa ye pr or nd (MeOH)	$154\text{-}5^{20}$	37	....	....	al, eth	$B5^4$, 861
14073	Toluene, 3-bromo-5-nitro 3-Br-5-$NO_2C_6H_3CH_3$	216.03	pa ye nd or pr (MeOH)	269-70	84	....	....	al, eth	$B5^3$, 752
14074	Toluene, 4-bromo or *p*-Tolylbromide 4-Br-$C_6H_4CH_3$	171.04	cr (al)	184.3, 61.9^{10}	28.5	$1.3995^{20/4}$	1.5477^{20}	al, eth, ace, bz, chl	$B5^4$, 827
14075	Toluene, 4-bromo-2-nitro 4-Br-2-$NO_2C_6H_3CH_3$	216.03	pa ye nd (dil al)	256-7, 130-2	47	....	....	al, eth	$B5^4$, 860
14076	Toluene, 4-bromo-3-nitro 4-Br-3-$NO_2C_6H_3CH_3$	216.03	pa ye nd (MeOH)	....	35	....	1.5682^{20}	....	$B5^4$, 860
14077	Toluene, 5-bromo-2-nitro 5-Br-2-$NO_2C_6H_3CH_3$	216.03	cr (al)	267, 143^{10}	56	....	....	....	$B5^4$, 860
14078	Toluene, 2-chloro or *o*-Tolylchloride 2-Cl$C_6H_4CH_3$	126.59	....	159.1,42.6^{10}	-35.1	$1.0825^{20/4}$	1.5268^{20}	al, eth, ace bz chl	$B5^4$, 805
14079	Toluene, 2-chloro-4-nitro 2-Cl-4-$NO_2C_6H_3CH_3$	171.58	nd (al)	260	68	....	1.5470^{69}	al, eth	$B5^4$, 855
14080	Toluene, 2-chloro-6-nitro 2-Cl-6-$NO_2C_6H_3CH_3$	171.58	nd (dil al)	238	37-40	....	1.5377^{69}	al	$B5^4$, 854
14081	Toluene, 3-chloro or *m*-Tolyl chloride 3-Cl-$C_6H_4CH_3$	126.59	....	162	−47.8	$1.0722^{20/4}$	1.5214^{19}	al, eth, bz, chl	$B5^4$, 806
14082	Toluene, 3-chloro-4-nitro 3-Cl-4-$NO_2C_6H_3CH_3$	171.58	pa ye nd	146^{19}	24	....	....	....	$B5^4$, 856
14083	Toluene, 3-chloro-5-nitro 3-Cl-5-$NO_2C_6H_3CH_3$	171.58	ye nd (al)	....	61	....	1.5404^{69}	al	$B5^4$, 855
14084	Toluene, 4-chloro or *p*-Tolyl chloride 4-Cl$C_6H_4CH_3$	126.59	....	162, 44^{10}	7.5	$1.0697^{20/4}$	1.5150^{20}	al, eth, chl, aa	$B5^4$, 806
14085	Toluene, 4-chloro-3-nitro 4-Cl-2-$NO_2C_6H_3CH_3$	171.58	mcl nd	240^{720}, 115.5^{11}	38	1.2559^{80}	....	eth	$B5^4$, 853

PHYSICAL CONSTANTS OF ORGANIC COMPOUNDS (Continued)

No.	Name, Synonyms, and Formula	Mol. wt.	Color, crystalline form, specific rotation and λ_{max} (log ε)	b.p. °C	m.p. °C	Density	n_D	Solubility	Ref.
14086	Toluene, 4-chloro-3-nitro 4-Cl-3-$NO_2C_6H_3CH_3$	171.58		260^{745}, 118^{11}	7		1.5572^{70}		$B5^4$, 855
14087	Toluene, 5-chloro-2-nitro 5-Cl-2-$NO_2C_6H_3CH_3$	171.58	ye		24.9		1.5496^{65}		$B5^4$, 854
14088	Toluene, 2,3-diamino 2,3-$(NH_2)_2C_6H_3CH_3$	122.17	cr	255	63-4			w, al, eth	$B13^3$, 277
14089	Toluene, 2,4-diamino 2,4-$(NH_2)_2C_6H_3CH_3$	122.17	nd (w), cr (al)	29^2, $148\text{-}50^8$	99			w, al, eth, bz	$B13^3$, 278
14090	Toluene, 2,5-diamino 2,5-$(NH_2)_2C_6H_3CH_3$	122.17	pl (bz)	273-4	64			w, al, eth	$B13^3$, 284
14091	Toluene, 2,6-diamino 2,6-$(NH_2)_2C_6H_3CH_3$	122.17	pr (bz, w)		106			w, al	$B13^3$, 297
14092	Toluene, 3,4-diamino 3,4-$(NH_2)_2C_6H_3CH_3$	122.17	lf (lig)	265 sub	89-90			w	$B13^3$, 292
14093	Toluene, 3,5-diamino 3,5-$(NH_2)_2C_6H_3CH_3$	122.17		283-5	<0			w, al, eth	$B13^3$, 299
14094	Toluene, 3,5-diamino, hydrochloride 3,5-$(NH_2)_2C_6H_3CH_3$.2HCl	195.09	nd (dil al)		255-60d			w	B13, 164
14095	Toluene, 2,5-dibromo 2,5-$Br_2C_6H_3CH_3$	249.93		236, $135\text{-}6^{35}$	fr 5.6	1.8127^{19}	1.5982^{18}		$B5^4$, 835
14096	Toluene, 3,5-dibromo 3,5-$Br_2C_6H_3CH_3$	249.93	nd	246	39				$B5^4$, 836
14097	Toluene, 2,3-dichloro 2,3-$Cl_2C_6H_3CH_3$	161.03		207-8, $61\text{-}2^3$			1.5511^{20}	bz	$B5^4$, 815
14098	Toluene, 2,4-dichloro 2,4-$Cl_2C_6H_3CH_3$	161.03		196-7	−13.5	$1.2498^{20/20}$	1.5511^{20}		$B5^4$, 815
14099	Toluene, 2,5-dichloro 2,5-$Cl_2C_6H_3CH_3$	161.03		200^{770}	5	1.2535^{20}	1.5449^{20}	bz	$B5^3$, 694
14100	Toluene, 2,6-dichloro 2,6-$Cl_2C_6H_3CH_3$	161.03		198			1.5507^{20}	chl	$B5^4$, 815
14101	Toluene, 3,4-dichloro 3,4-$Cl_2C_6H_3CH_3$	161.03		208.9, 81.8^{10}	−15.2	$1.2564^{20/4}$	1.5471^{20}	al, eth, ace, bz	$B5^4$, 815
14102	Toluene, 3,5-dichloro 3,5-$Cl_2C_6H_3CH_3$	161.03		201-2	26				$B5^4$, 816
14103	Toluene, 2,5-diethoxy 2,5-$(C_2H_5O)_2C_6H_3CH_3$	180.25	nd (lig)	247-9	24-5	1.0134^{15}		al, eth, bz, chl	$B6^3$, 4499
14104	Toluene, 2,3-dihydroxy or Isohomocatechol 2,3-$(HO)_2C_6H_3CH_3$	124.14	lf (bz)	241, 127^{12}	68			w, al, eth, bz, chl	$B6^3$, 4492
14105	Toluene, 2,4-dihydroxy or Cresorcinol 2,4-$(HO)_2C_6H_3CH_3$	124.14	cr (bz-peth)	267-70	105-7			w, al, eth	$B6^3$, 4495
14106	Toluene, 2,5-dihydroxy or Toluhydroquinone 2,5-$(HO)_2C_6H_3CH_3$	124.14	rh pl (bz)	163^{11} sub	126-7			w, al, eth	$B6^3$, 4498
14107	Toluene, 2,5-dihydroxy, diacetate or 2.5-Diacetoxy toluene 2.5-$(CH_3CO_2)_2C_6H_3CH_3$	208.21	nd (aa), pr (lig), nd or pr (w)		52			eth, al, aa	$B6^3$, 4501
14108	Toluene, 3,4-dihydroxy or 4-Homocatechol 3,4-$(HO)_2C_6H_3CH_3$	124.14	lf (bz-lig), pr (bz)	251, $143\text{-}6^{26}$ sub	65	$1.1287^{74/4}$	1.5425^{74}	w, al, eth, ace	$B6^3$, 4514
14109	Toluene, 3,5-dihydroxy or Orcinol 3.5-$(HO)_2C_6H_3CH_3$	124.14	pr (w + l), lf (chl)	289-90, 147^5	107-8 (anh), 58 (hyd)	1.290^4		w, al, eth, bz	$B6^3$, 4531
14110	Toluene, 2,4-dimercapto or Dithiocresorcinol 2,4-$(HS)_2C_6H_3CH_3$	156.26		263	36.7				B6, 873
14111	Toluene, 3,4-dimercapto 3,4-$(HS)_2C_6H_3CH_3$	156.26			28-30			chl	$B6^3$, 4530
14112	Toluene, 2,3-dimethoxy or Isohomoveratrol 2,3-$(CH_3O)_2C_6H_3CH_3$	152.19		202-3, $92\text{-}3^{18}$		$1.0335^{20/4}$	1.5121^{25}	al, eth	$B6^3$, 4493
14113	Toluene, 2,4-dimethoxy 2,4-$(CH_3O)_2C_6H_3CH_3$	152.19		211, $110\text{-}20^{30}$				al, eth	$B6^3$, 4496
14114	Toluene, 3,4-dimethoxy or Homoveratrol 3,4-$(CH_3O)_2C_6H_3CH_3$	152.19	pr (eth)	219-21, $122\text{-}4^{27}$	24	$1.0509^{25/4}$	1.5257^{25}		$B6^3$, 4516
14115	Toluene, 3,5-dimethoxy 3,5-$(CH_3O)_2C_6H_3CH_3$	152.19		244, 102^8		1.0478^{15}	1.5234^{20}	al, eth, bz, aa	$B6^3$, 4533
14116	Toluene, 2,4-dinitro 2,4-$(NO_2)_2C_6H_3CH_3$	182.14	ye nd or mcl pr	300d	71	1.3208^{71}	1.442	al, eth, ace, bz, Py	$B5^4$, 865
14117	Toluene, 2,5-dinitro 2,5-$(NO_2)_2C_6H_3CH_3$	182.14	nd (al)		52.5	1.282^{111}		al, bz	$B5^4$, 866

No.	Name, Synonyms, and Formula	Mol. wt.	Color, crystalline form, specific rotation and λ_{max} (log ε)	b.p. °C	m.p. °C	Density	n_D	Solubility	Ref.
14118	Toluene, 2,6-dinitro 2,6-$(NO_2)_2C_6H_3CH_3$	182.14	rh nd (al)		66	1.2833^{111}	1.479	al	$B5^4$, 866
14119	Toluene, 3,4-dinitro 3,4-$(NO_2)_2C_6H_3CH_3$	182.14	ye nd (CS_2)		58.3	1.2594^{111}		al	$B5^4$, 866
14120	Toluene, 3,5-dinitro 3,5-$(NO_2)_2C_6H_3CH_3$	182.14	ye rh nd (aa)	sub	93	1.2772^{111}		al, eth, bz, chl	$B5^4$, 867
14121	Toluene, 4,6-Dinitro-3-Fluoro 4,6-$(NO_2)_2$-3-$FC_6H_2CH_3$	200.13		$129\text{-}30^{0.5}$	78				$B5^4$, 867
14122	Toluene, 2,4-diisopropyl 2,4-$(i\text{-}C_3H_9)_2C_6H_3CH_3$	176.30		82^7		0.8636^{20}	1.4912^{20}	CS_2	$B5^4$, 1152
14123	Toluene, 3,5-diisopropyl 3,5-$(i\text{-}C_3H_7)_2C_6H_3CH_3$	176.30		215-8		0.8668^{20}	1.4950^{20}		$B5^4$, 1153
14124	Toluene, 2-ethoxy or Ethyl *o*-tolyl ether $(2\text{-}CH_3C_6H_4)OC_2H_5$	136.19		$184, 70^{12}$		$0.9592^{13/4}$	1.508^{13}	al, eth	$B6^4$, 1944
14125	Toluene, 3-ethoxy or Ethyl *m*-tolyl ether $(3\text{-}CH_3C_6H_4)OC_2H_5$	136.19		192		0.949^{20}	1.513^{20}	al, eth	$B6^4$, 2039
14126	Toluene, 4-ethoxy or Ethyl *p*-tolyl ether $(4\text{-}CH_3C_6H_4)OC_2H_5$	136.19		188-9		$0.9509^{18/4}$	1.5058^{18}	al, eth	$B6^4$, 2099
14127	Toluene, 2-fluoro or *o*-Tolyl fluoride 2-$FC_6H_4CH_3$	110.13		$114, 30^{26}$	−62	$1.0041^{13/4}$	1.4704^{20}	al, eth	$B5^4$, 799
14128	Toluene, 3-fluoro or *m*-Tolyl fluoride 3-$FC_6H_4CH_3$	110.13		116	−87.7	0.9986^{20}	1.4691^{20}	al, eth	$B5^4$, 799
14129	Toluene, 4-fluoro or 4-Tolyl fluoride 4-$FC_6H_4CH_3$	110.13		116.6	−56.8	$1.0007^{16/4}$	1.4699^{20}	al, eth	$B5^4$, 800
14130	Toluene, 3-(α-hydroxy ethyl) 3-$[CH_3CH(OH)]C_6H_4CH_3$	136.19		112^{12}		$0.9974^{15/4}$	1.5240^{20}	al, eth	$B6^3$, 1823
14131	Toluene, 4-(α-hydroxy ethyl) 4-$[CH_3CH(OH)]C_6H_4CH_3$	136.19		$219^{756}, 120^{19}$		$0.9944^{20/4}$	1.5246^{20}	al, eth	$B6^3$, 1826
14132	Toluene, 2-hydroxylamino or *o*-Toyl hydroxyl amine 2-$HONHC_6H_4CH_3$	123.15	nd (eth-bz)		44			al, eth, bz	$B15^3$, 15
14133	Toluene, 2-hydroxylamino-6-nitro 2-HONH-6-$NO_2C_6H_3CH_3$	168.15	col or ye (bz)		120-1			al, eth	$B15^2$, 14
14134	Toluene, 3-hydroxylamino or *m*-Tolyl hydroxyl amine 3-$(HONH)C_6H_4CH_3$	123.15	lf (bz-peth)		68.5			al, eth, chl	$B15^3$, 17
14135	Toluene, 4-hydroxylamino or *p*-Tolyl hydroxyl amine 4-$(HONH)C_6H_4CH_3$	123.15	lf (bz)	115-20d	98			al, eth, chl	$B15^3$, 18
14136	Toluene, 4-hydroxylamino-2-nitro 4-(HONH)-2-$NO_2C_6H_3CH_3$	168.15	ye (bz)		108-9			bz	$B15^2$, 16
14137	Toluene, 2-iodo or *o*-Tolyl iodide 2-$IC_6H_4CH_3$	218.04		211-2, $73\text{-}5^7$		$1.713^{20/4}$	1.6079^{20}	al, eth	$B5^4$, 838
14138	Toluene, 3-iodo or *m*-Tolyl iodide 3-$IC_6H_4CH_3$	218.04		$213, 80\text{-}2^{10}$	−27.2	1.705^{20}	1.6053^{20}	al, eth	$B5^4$, 839
14139	Toluene, 4-iodo or *p*-Tolyl iodide 4-$IC_6H_4CH_3$	218.04	lf (al)	211 sub	36-7	$1.678^{20/4}$		al, eth	$B5^4$, 840
14140	Toluene, 2-mercapto or *o*-Thiocresol 2-$CH_3C_6H_4SH$	124.20	pl or lf	194.2, 67.1^{10}	15	$1.041^{20/4}$	1.570^{20}	al, eth	$B6^4$, 2014
14141	Toluene, 3-mercapto or *m*-Thiocresol 3-$CH_3C_6H_4SH$	124.20		195.1, 67.8^{10}	<−20	$1.044^{20/4}$	1.572^{20}	al, eth	$B6^4$, 2079
14142	Toluene, 4-mercapto or *p*-Thiocresol 4-$CH_3C_6H_4SH$	124.20	lf (eth, dil al)	$195, 67.6^{10}$	44			al, eth	$B6^4$, 2153
14143	Toluene, 2-methoxy or Methyl-*o*-tolyl ether 2-$CH_3OC_6H_4CH_3$	122.17		171.3		$0.9851^{15/15}$	1.5161^{20}	al, eth, ace	$B6^4$, 1943
14144	Toluene, 3-methoxy or Methyl-*m*-tolyl ether 3-$CH_3OC_6H_4CH_3$	122.17		177.2		$0.9697^{25/25}$	1.5164^{13}	al, eth, ace, bz	$B6^4$, 2039
14145	Toluene, 4-methoxy or Methyl-*p*-tolyl ether 4-$CH_3OC_6H_4CH_3$	122.17		176.5		$0.9689^{25/25}$	1.5124^{19}	al, eth	$B6^4$, 2098
14146	Toluene, 2-nitro 2-$NO_2C_6H_4CH_3$	137.14	(i) nd, (ii) cr	221.7, 118^{16}	(i)-9.5 (ii)-2.9	1.1629^{20}	1.5450^{20}	al, eth	$B5^4$, 845
14147	Toluene, 3-nitro 3-$NO_2C_6H_4CH_3$	137.14	pa ye	232.6	16	$1.1571^{20/4}$	1.5466^{20}	al, eth, bz	$B5^4$, 847
14148	Toluene, 3-nitro-4-triazo 4-N_3-3-NO_2-$C_6H_3CH_3$	178.15	ye nd or pl (lig, dil al)		38			al	$B5^3$, 773
14149	Toluene, 4-nitro 4-$NO_2C_6H_4CH_3$	137.14	orth cr (al, eth)	238.3, 105^9	54.5	$1.1038^{75/4}$		al, eth, ace, bz	$B5^4$, 848

No.	Name, Synonyms, and Formula	Mol. wt.	Color, crystalline form, specific rotation and λ_{max} (log ε)	b.p. °C	m.p. °C	Density	n_D	Solubility	Ref.
14150	Toluene, 2-nitroso 2-$ONC_6H_4CH_3$	121.14	nd or pr		72.5			al, eth, chl	B5^3, 728
14151	Toluene, 3-nitroso 3-$ONC_6H_4CH_3$	121.14	nd		53.5			eth, chl	B5^4, 844
14152	Toluene, 4-nitroso 4-$ONC_6H_4CH_3$	121.14	nd (lig)		48.5			bz, chl	B5^4, 845
14153	Toluene, 4-isopropenyl 4-$[CH_2{=}C(CH_3)]C_6H_4CH_3$	132.21		185.3		$0.8936^{23/4}$	1.52832^{23}		B5^4, 1383
14154	Toluene, 4-isopropyl-2-nitro 4-i-C_3H_7-2-$NO_2C_6H_3CH_3$	179.22					1.5280^{20}		B5^4, 1064
14155	Toluene, 2,3,4,5,6-pentabromo $C_6Br_5CH_3$	486.62			288	2.97^{17}		bz	B5^3, 720
14156	Toluene, 2,3,4,5,6-pentachloro $C_6Cl_5CH_3$	264.37	nd (bz, peth)	301	224			to	B5^4, 823
14157	Toluene, 2-phenyl or 2-Methyl biphenyl 2-C_6H_5-$C_6H_4CH_3$	168.24		255.3, $130\text{-}6^{27}$	−0.2	$1.010^{22/4}$	1.5914^{20}	al, eth	B5^4, 1855
14158	Toluene, 3-phenyl or 3-Methyl biphenyl 3-C_6H_5-$C_6H_4CH_3$	168.24		272.7, $148\text{-}50^{20}$	4.5	$1.0182^{17/4}$	1.5972^{20}	al, eth	B5^4, 1858
14159	Toluene, 4-phenyl 4-$C_6H_5C_6H_4CH_3$	168.24	pl (lig, MeOH)	267-8, $134\text{-}6^{15}$	49-50	1.015^{27}		al, eth	B5^4, 1860
14160	Toluene, 2-*iso*-propenyl 2-$[CH_2{=}C(CH_3)]C_6H_4CH_3$	132.21		175, $59\text{-}62^{11}$		$0.9181^{15/0}$	1.5112^{30}		B5^3, 1214
14161	Toluene, 2-propoxy or Propyl-*o*-tolyl ether 2-$C_3H_7OC_6H_4CH_3$	150.22		204.1		$0.9517^{0/0}$			B6^3, 1246
14162	Toluene, 3-propoxy or Propyl-*m*-tolyl ether 3-$C_3H_7OC_6H_4CH_3$	150.22		210.6		$0.9484^{0/0}$			B6^4, 2040
14163	Toluene, 4-propoxy or Propyl-*p*-tolyl ether 4-$C_3H_7OC_6H_4CH_3$	150.22		210.4		$0.9497^{0/0}$			B6^3, 1354
14164	Toluene, 2-propyl or 1-Methyl-2-propyl benzene 2-$C_3H_7C_6H_4CH_3$	134.22		185, 63.5^{10}	−60.2	0.8744^{20}	1.4998^{20}		B5^4, 1056
14165	Toluene, 3-propyl or 1-Methyl-3-propyl benzene 3-$C_3H_7C_6H_4CH_3$	134.22		182, 61.4^{10}		0.8610^{20}	1.4936^{20}		B5^4, 1056
14166	Toluene, 4-propyl or 1-Methyl-4-propyl benzene 4-$C_3H_7C_6H_4CH_3$	134.22		183, 61.9^{10}	−63.6	0.8584^{20}	1.4919^{20}	al, eth	B5^4, 1057
14167	Toluene, 2,3,5,6-tetrabromo 2,3,5,6-$Br_4C_6HCH_3$	407.72	nd		116-7			ace, bz	B5, 310
14168	Toluene, 2,3,4,5-tetrachloro 2,3,4,5-$Cl_4C_6HCH_3$	229.92	nd (MeOH, dil al)		98.1			al, eth, ace	B5^2, 233
14169	Toluene, 2,3,4,6-tetrachloro 2,3,4,6-$Cl_4C_6HCH_3$	229.92	nd (al, eth)	276.5	96			al, bz	B5^3, 702
14170	Toluene, 2,3,5,6-tetrachloro 2,3,5,6--$Cl_4C_6HCH_3$	229.92	nd (MeOH)	sub	93-4			al, eth	B5^3, 702
14171	Toluene, 2,4,5-triacetoxy 2,4,5-$(CH_3CO_2)_3C_6H_2CH_3$	266.25	cr (al)		114-5			w, al, bz	B6, 1109
14172	Toluene, 2,4,6-triacetoxy 2,4,6-$(CH_3CO_2)_3C_6H_2CH_3$	266.25	(i) cr (eth-peth) (ii) nd (peth)		(i) 76 (ii) 58			al, bz	B6^3, 6320
14173	Toluene, 2,4,6-triamino 2,4,6-$(NH_2)_3C_6H_2CH_3$	137.18	col nd (bz), red in air		105			w, al, bz	B13^3, 557
14174	Toluene, 2-triazo or *o*-Tolyl azide 2-$N_3C_6H_4CH_3$	133.15	pa ye liq	90.5^{30}	<−10			eth	B5^4, 876
14175	Toluene, 3-triazo or *m*-Tolyl azide 3-$N_3C_6H_4CH_3$	133.15		92.5^{31}				eth	B5^4, 876
14176	Toluene, 4-triazo or *p*-Tolyl azide 4-$N_3C_6H_4CH_3$	133.15		d 180, 80^{10}		$1.0527^{23/4}$		al, eth	B5^4, 876
14177	Toluene, 2,3,4-tribromo 2,3,4-$Br_3C_6H_2CH_3$	328.83	rh pl (aa, lig-CS_2)		45-6	2.456^{20}			B5^1, 156
14178	Toluene, 2,3,5-tribromo 2,3,5-$Br_3C_6H_2CH_3$	328.83	mcl pr (eth-to)		53-4	2.467^{17}		eth	B5^3, 719
14179	Toluene, 2,3,6-tribromo 2,3,6-$Br_3C_6H_2CH_3$	328.83	mcl pr or lf (lig, chl)		60.5	2.471^{17}			B5^1, 156
14180	Toluene, 2,4,5-tribromo 2,4,5-$Br_3C_6H_2CH_3$	328.83	mcl pr (eth-al), nd (al)		113.5	2.472^{17}			B5^3, 719
14181	Toluene, 2,4,6-tribromo 2,4,6-$Br_3C_6H_2CH_3$	328.83	lo nd or mcl pr (eth-AcOEt)	290	70	2.479^{17}			B5^3, 719

PHYSICAL CONSTANTS OF ORGANIC COMPOUNDS (Continued)

No.	Name, Synonyms, and Formula	Mol. wt.	Color, crystalline form, specific rotation and λ_{max} (log ε)	b.p. °C	m.p. °C	Density	n_D	Solubility	Ref.
14182	Toluene, 2,3,4-trichloro $2,3,4\text{-}Cl_3C_6H_2CH_3$	195.48	nd (al, MeOH)	244	43-4	....	....	al, eth, ace	B5[4], 819
14183	Toluene, 2,3,5-trichloro $2,3,5\text{-}Cl_3C_6H_2CH_3$	195.48	nd(al)	229-31[757]	45-6	....	....	al, ace	B5, 299
14184	Toluene, 2,3,6-trichloro $2,3,6\text{-}Cl_3C_6H_2CH_3$	195.48	nd (al)	....	45-6	....	....	al, ace	B5[4], 819
14185	Toluene, 2,4,5-trichloro $2,4,5\text{-}Cl_3C_6H_2CH_3$	195.48	nd or lf (al)	229-30[716]	82.4	....	....	al, ace	B5[4], 819
14186	Toluene, 2,4,6-trichloro $2,4,6\text{-}Cl_3C_6H_2CH_3$	195.48	nd (al)	....	38	....	....	al, ace	B5[4], 819
14187	Toluene, 3,4,5-trichloro $3,4,5\text{-}Cl_3C_6H_2CH_3$	195.48	....	246-7[768]	45-6	....	....	al, ace	B[5], 299
14188	Toluene, 2,4,6-trihydroxy $2,4,6\text{-}(HO)_3C_6H_2CH_3$	140.14	....	....	222-3	....	....	w, al, eth	B6[3], 6318
14189	Toluene, 3,4,5-trihydroxy or 5-Methyl pyrogallol $3,4,5\text{-}(HO)_3C_6H_2CH_3$	140.14	pa br nd (bz)	sub	129	....	....	bz	B6[3], 6320
14190	Toluene, 2,3,4-triido $2,3,4\text{-}I_3C_6H_2CH_3$	469.83	pa br nd (al)	....	92	....	....	al, bz	B5[1], 157
14191	Toluene, 2,3,5-triido $2,3,5\text{-}I_3C_6H_2CH_3$	469.83	og pl (al)	....	72-3	....	....	....	B5[1], 157
14192	Toluene, 2,3,6-triido $2,3,6\text{-}I_3C_6H_2CH_3$	469.83	nd (al)	....	80.5	....	....	....	B5[1], 158
14193	Toluene, 2,4,5-triido $2,4,5\text{-}I_3C_6H_2CH_3$	469.83	pa br nd (al)	....	118-20	....	....	....	B5[1], 158
14194	Toluene, 2,4,6-triido $2,4,6\text{-}I_3C_6H_2CH_3$	469.83	nd (al, bz)	300d	118-9	....	....	al	B5[1], 158
14195	Toluene, 3,4,5-triido $3,4,5\text{-}I_3C_6H_2CH_3$	469.83	nd(al)	....	122-3	....	....	al	B5, 317
14196	Toluene, 2,3,4-trinitro $2,3,4\text{-}(NO_2)_3C_6H_2CH_3$	227.13	tcl lf (al), pr (ace)	....	112	1.62	....	eth, ace, bz	B5[4], 872
14197	Toluene, 2,4,5-trinitro $2,4,5\text{-}(NO_2)_3C_6H_2CH_3$	227.13	yesh pl (ace), pa ye rh bipym (al)	exp 290-310	104	....	....	eth, ace, bz	B5[4], 872
14198	Toluene, 2,4,6-trinitro or TNT $2,4,6\text{-}(NO_2)_3C_6H_2CH_3$	227.13	orh (al)	240 exp	82	1.654	....	eth, ace, bz, Py	B5[4], 873
14199	Toluene, 2-vinyl $2\text{-}(CH_2{=}CH)C_6H_4CH_3$	118.18	....	171	......	$0.9106^{20/4}$	1.5450^{20}	bz, chl	B5[4], 1367
14200	Toluene, 3-vinyl $3\text{-}(CH_2{=}CH)C_6H_4CH_3$	118.18	....	168, 61-2[18]	−70	$0.9028^{20/20}$	1.5410^{20}	al, eth, bz	B5[4], 1367
14201	Toluene, 4-vinyl $4\text{-}(CH_2{=}CH)C_6H_4CH_3$	118.18	....	169, 63[15]	−37.8	$0.8760^{20/4}$	1.5428^{20}	bz	B5[4], 1369
14202	*o*-Toluene arsonic acid or *o*-Tolyl arsinic acid $2\text{-}CH_3C_6H_4AsO(OH)_2$	216.07	nd (w)	....	163-4	....	....	al	B16[3], 1060
14203	*m*-Toluene arsonic acid or *m*-Tolyl arsinic acid $3\text{-}CH_3C_6H_4AsO(OH)_2$	216.07	nd (w)	....	150	....	....	al	B16[3], 1061
14204	*p*-Toluene arsonic acid or *p*-Tolyl arsinic acid $4\text{-}CH_3C_6H_4AsO(OH)_2$	216.07	nd (w)	....	3 (−w, 105-10)	....	....	al	B16[3], 1061
14205	*o*-Toluene boronic acid or *o*-Tolyl boronic acid $2\text{-}CH_3C_6H_4B(OH)_2$	135.96	nd or pl (w)	d	165-8	....	....	al, eth, bz	B16[3], 1277
14206	*m*-Toluene boronic acid or *m*-Tolyl boronic acid $3\text{-}CH_3C_6H_4B(OH)_2$	135.96	cr (w)	....	137-40	....	....	al, eth	B16[3], 1277
14207	*p*-Toluene boronic acid or *p*-Tolyl boronic acid $4\text{-}CH_3C_6H_4B(OH)_2$	135.96	nd (w)	....	245	....	....	eth	B16[3], 1277
14208	*o*-Toluene sulfinic acid or *o*-Tolyl sulfinic acid $2\text{-}CH_3C_6H_4SO_2H$	156.20	nd (w)	d	80	....	....	w, al, eth, ace	B11[3], 6
14209	*p*-Toluene sulfinic acid or *p*-Tolyl sulfinic acid $4\text{-}CH_3C_6H_4SO_2H$	156.20	rh pl or lo nd (w)	....	86-7	....	....	al, eth	B11[3], 8
14210	*p*-Toluene sulfinyl chloride $4\text{-}CH_3C_6H_4SOCl$	174.64	nd	115-20[4]	54-8	....	....	chl	B11[3], 11
14211	*o*-Toluene sulfonamide or 2-Toluene sulfonamide $2\text{-}CH_3C_6H_4SO_2NH_2$	171.21	oct (al), pr (w)	....	156.3	....	....	al	B11[3], 167
14212	*o*-Toluene sulfonamide, 4-amino $4\text{-}NH_2\text{-}2\text{-}CH_3C_6H_3SO_2NH_2$	186.23	nd or pl (w)	....	164	....	....	al	B14[3], 2208
14213	*o*-Toluene sulfonamide, N-methyl $2\text{-}CH_3C_6H_4SO_2NHCH_3$	185.24	pl (bz-lig)	....	74-5	....	....	al, ace, chl	B11, 87

PHYSICAL CONSTANTS OF ORGANIC COMPOUNDS (Continued)

No.	Name, Synonyms, and Formula	Mol. wt.	Color, crystalline form, specific rotation and λ_{max} (log ε)	b.p. °C	m.p. °C	Density	n_D	Solubility	Ref.
14214	*o*-Toluene sulfonamide, N-phenyl $2-CH_3C_6H_4SO_2NHC_6H_5$	247.31	pr (dil al)		136			al	B12, 566
14215	*o*-Toluene sulfonamide, 4-propyl $4-C_3H_7-2-CH_3C_6H_3SO_2NH_2$	213.29	pl (bz, dil al)		101-2			w, al	B11, 138
14216	*o*-Toluene sulfonamide, 5-isopropyl $5-i-C_3H_7-2-CH_3C_6H_3SO_2NH_2$	213.29	pl (al), pr or lf (dil al)		75			lig	B11, 140
14217	*o*-Toluene sulfonic acid or 2-Toluene sulfonic acid $2-CH_3C_6H_4SO_3H$	172.21	hyg pl (w + 2)	128.8^{25}	67.5			w, al	$B11^3$, 167
14218	*o*-Toluene sulfonic acid, 4-amino $4-H_2N-2-CH_3C_6H_3SO_3H$	187.21	mcl pr (w + l)		d				$B14^3$, 2208
14219	*o*-Toluene sulfonic acid, 5-amino $5-H_2N-2-CH_3C_6H_3SO_3H$	187.21	pl (w + l)		>275				$B14^3$, 2210
14220	*o*-Toluene sulfonic acid, 6-amino $6-H_2N-2-CH_3C_6H_3SO_3H$	187.21	nd (+ ½w)		130d				$B14^1$, 723
14221	*o*-Toluene sulfonic acid, 4-nitro, dihydrate $4-NO_2-2-CH_3C_6H_3SO_3H.2H_2O$	253.23	pl (w)		133.5			w, al, eth, chl	$B11^3$, 173
14222	*o*-Toluene sulfonic acid, 5-nitro $5-NO_2-2-CH_3C_6H_3SO_3H$	217.20	pr or pl (w + 2)		130-3			w, al, eth, chl	$B11^3$, 173
14223	*o*-Toluene sulfonic acid, 2-tolyl ester or (2-Tolyl)-*o*-toluene sulfonate $2-CH_3C_6H_4SO_2(OC_6H_4CH_3-2)$	262.32	cr (al)		50-1			bz, al	B11, 85
14224	2-Toluene sulfonic acid, 3-tolyl ester $2-CH_3C_6H_4SO_2(OC_6H_4CH_3-3)$	262.32	cr (al)		60			al, bz	B11, 85
14225	2-Toluene sulfonic acid, 4-tolyl ester $2-CH_3C_6H_4SO_2(OC_6H_4CH_3-4)$	262.32	cr (al)		70-1			al, bz	B11, 85
14226	*o*-Toluene sulfonyl bromide or 2-Toluene sulfonyl bromide $2-CH_3C_6H_4SO_2Br$	235.10	cr	138^{10}	13				B11, 86
14227	*o*-Toluene sulfonyl chloride or 2-Toluene sulfonyl chloride $2-CH_3C_6H_4SO_2Cl$	190.65		154^{36}	10.2	$1.3383^{20/4}$	1.5565^{20}	eth, bz	$B11^3$, 167
14228	*m*-Toluene sulfonamide or 3-Toluene sulfonamide $3-CH_3C_6H_4SO_2NH_2$	171.21	pl (w), mcl pr		108			al	$B11^1$, 23
14229	*m*-Toluene sulfonamide, N-phenyl $3-CH_3C_6H_4SO_2NHC_6H_5$	247.31	pr (al)		96			al, eth	$B11^1$, 23
14230	*m*-Toluene sulfonamide, 4-isopropyl $4-i-C_3H_7-3-CH_3C_6H_3SO_2NH_2$	213.29	fl (dil al)		149.9			al, eth	$B11^3$, 350
14231	*m*-Toluene sulfonic acid or 3-Toluene sulfonic acid $3-CH_3C_6H_4SO_3H$	172.21	oil					w, al, eth	$B11^3$, 176
14232	*m*-Toluene sulfonic acid, 4-amino $4-NH_2-3-CH_3C_6H_3SO_3H$	187.21	lf ye nd (w + ½)		132d (hyd)			w	$B14^3$, 2213
14233	*m*-Toluene sulfonyl chloride-6-acetamido $6-(CH_3CONH)-3-CH_3C_6H_4SO_2Cl$	247.70	nd (bz)		159				$B14^2$, 448
14234	*p*-Toluene sulfonamide or 4-Toluene sulfonamide $(4-CH_3C_6H_4)SO_2NH_2$	171.21	mcl pl (w + 2)		138-9 (anh), 105 (hyd)			al	$B11^3$, 266
14235	*p*-Toluene sulfonamide, 2-amino $2-NH_2-4-CH_3C_6H_3SO_2NH_2$	186.23	pr (w)		176				$B14^3$, 2220
14236	*p*-Toluene sulfonamide, N,N-dichloro or Dichloramine T $4-CH_3C_6H_4SO_2NCl_2$	240.10	pa ye pr (peth-chl)		83			al, eth, bz, aa, chl	$B11^3$, 301
14237	*p*-Toluene sulfonamide, N-ethyl $4-CH_3C_6H_4SO_2NHC_2H_5$	199.27	pl (dil al, lig)		64			al	$B11^3$, 268
14238	*p*-Toluene sulfonamide, N-methyl $4-CH_3C_6H_4SO_2NHCH_3$	185.24	pl (dil al)		78-9	1.340		al, eth	$B11^3$, 267
14239	*p*-Toluene sulfonamide, N-methyl-N-nitroso $4-CH_3C_6H_4SO_2N(NO)CH_3$	214.24	cr		60			al, eth	$B11^1$, 29
14240	*p*-Toluene sulfonamide, N-methyl-N-phenyl $4-CH_3C_6H_4SO_2N(CH_3)C_6H_5$	261.34	pl or mcl pr (AcOEt)		95			al, eth	$B12^3$, 1102
14241	*p*-Toluene sulfonamide, N-methyl-N-(2-tolyl) $4-CH_3C_6H_4SO_2N(CH_3)(C_6H_4CH_3-2)$	275.37	pr (al)		119-20			al	$B12^1$, 388
14242	*p*-Toluene sulfonamide, N-methyl-N-(4-tolyl) $4-CH_3C_6H_4SO_2N(CH_3)(C_6H_4CH_3-4)$	275.37	pr		60				$B12^3$, 2145
14243	*p*-Toluene sulfonamide, N-phenyl or 4-Toluene sulfoanilide $4-CH_3C_6H_4SO_2NHC_6H_5$	247.33	dimorphic α: tcl, β: mcl pr (dil al, bz)		103-4			al, aa	$B12^3$, 1081

No.	Name, Synonyms, and Formula	Mol. wt.	Color, crystalline form, specific rotation and λ_{max} (log ε)	b.p. °C	m.p. °C	Density	n_D	Solubility	Ref.
14244	*p*-Toluene sulfonamide, 3-isopropyl $3\text{-}i\text{-}C_3H_7\text{-}4\text{-}CH_3C_6H_3SO_2NH_2$	213.29	nd (w)		162			al	B11[3], 349
14245	*p*-Toluene sulfonamide, N-(2-tolyl) $4\text{-}CH_3C_6H_4SO_2NH(C_6H_4CH_3\text{-}2)$	261.34	rh bipym (al), nd (dil aa)		110			al, eth, ace, bz, chl	B12[2], 452
14246	*p*-Toluene sulfonamide, N-(4-tolyl) $4\text{-}CH_3C_6H_4SO_2NH(C_6H_4CH_3\text{-}4)$	261.34	tcl pr or nd (aa)		118-9			al, bz, chl	B12[3], 2143
14247	*p*-Toluene sulfonic acid or 4-Toluene sulfonic acid $4\text{-}CH_3C_6H_4SO_3H$	172.20	hyg pl (w + 1), mcl lf or pl	140^{20}	104-5			w, al, eth	B11[3], 183
14248	*p*-Toluene sulfonic acid, 2-amino $2\text{-}NH_2\text{-}4\text{-}CH_3C_6H_3SO_3H$	187.21	pl, nd or pr (w)						B14[3], 2219
14249	*p*-Toluene sulfonic acid, butyl ester $4\text{-}CH_3C_6H_4SO_2(OC_4H_9)$	228.31		$164\text{-}6^{6}$		$1.1319^{20/4}$	1.5050^{20}	eth	B11[3], 189
14250	*p*-Toluene sulfonic acid, 2-chloro ethyl ester or 2-Chloro ethyl-*p*-toluene sulfonate $4\text{-}CH_3C_6H_4SO_2(OCH_2CH_2Cl)$	234.70		210^{21}					B11[3], 188
14251	*p*-Toluene sulfonic acid, 2-chloro propyl ester or 2-Chloro propyl-*p*-toluene sulfonate $4\text{-}CH_3C_6H_4SO_2(OCH_2CHClCH_3)$	248.72		$216\text{-}7^{17}$		$1.2674^{20/4}$	1.5225^{21}		B11[2], 45
14252	*P*-Toluene sulfonic acid, ethyl ester or Ethyl-*p*-toluene sulfonate $4\text{-}CH_3C_6H_4SO_2(OC_2H_5)$	200.25	mcl pr (aa, AcOEt)	173^{15}	34-5	$1.166^{48/4}$		al, eth	B11[3], 188
14253	4-Toluene sulfonic acid, ethylene ester or Ethylene-*p*-toluene sulfonate $(4\text{-}CH_3C_6H_4SO_2)CH_2CH_2(O_2SC_6H_4CH_3\text{-}4)$	370.44	cr (bz)		128				B11[3], 225
14254	*p*-Toluene sulfonic acid, methyl ester or Methyl-*p*-toluene sulfonate $(4\text{-}CH_3C_6H_4)SO_2(OCH_3)$	186.23	mcl lf or pr (eth-lig)	292, $168\text{-}70^{13}$	28-9			al, eth, bz, chl	B11[3], 187
14255	*p*-Toluene sulfonic acid, propyl ester or Propyl-*p*-toluene sulfonate $(4\text{-}CH_3C_6H_4)SO_2(OC_3H_7)$	214.28		189^{9}	<−20	$1.144^{20/4}$	1.4998^{20}		B11[3], 188
14256	*p*-Toluene sulfonic acid, isopropyl ester $(4\text{-}CH_3C_6H_4)SO_2(O\text{-}i\text{-}C_3H_7)$	214.28			21	1.5065^{20}			B11[3], 189
14257	*p*-Toluene sulfonic acid, 2-tolyl ester or 2-Tolyl-*p*-toluene sulfonate $4\text{-}CH_3C_6H_4SO_2(OC_6H_4CH_3\text{-}2)$	262.32	nd		54-5			al	B11[3], 204
14258	*p*-Toluene sulfonic acid, (2,4,6-tribromo phenyl)ester $(2,4,6\text{-}Br_3C_6H_2O)SO_2C_6H_4CH_3\text{-}4$	484.98	cr (al)		113			al	B11[2], 47
14259	*p*-Toluene sulfonyl chloride $4\text{-}CH_3C_6H_4SO_2Cl$	190.64	tcl (eth, peth)	$145\text{-}6^{15}$	71			al, eth, bz	B11[3], 265
14260	*o*-Toluic acid or 2-Methyl benzoic acid $2\text{-}CH_3C_6H_4CO_2H$	136.15	pr or nd (w)	$258\text{-}9^{751}$	107-8	1.062^{115}	1.512^{115}	al, eth, chl	B9[3], 2298
14261	*o*-Toluic acid anhydride $(2\text{-}CH_3C_6H_4CO)_2O$	254.29		>325	38-9			eth	B9[3], 2303
14262	*o*-Toluic acid, 3-chloro $3\text{-}Cl\text{-}2\text{-}CH_3C_6H_3CO_2H$	170.60	nd (al)		159			al, eth	B9[3], 2309
14263	*o*-Toluic acid, 4-chloro or 4-Chloro-2-methyl benzoic acid $4\text{-}Cl\text{-}2\text{-}CH_3C_6H_3CO_2H$	170.60	nd (w, al, bz)		173			w, al, aa	B9[2], 320
14264	*o*-Toluic acid, 5-chloro or 5-Chloro-2-methyl benzoic acid $5\text{-}Cl\text{-}2\text{-}CH_3C_6H_3CO_2H$	170.60	nd (al)		168-9			al	B9[2], 320
14265	*o*-Toluic acid, 4,6-dihydroxy or *o*-Orsellinic acid $4,6\text{-}(HO)_2\text{-}2\text{-}CH_3C_6H_2CO_2H$	168.15	nd (dil aa, + 1w)		176d			al, eth	B10[3], 1479
14266	*o*-Toluic acid, 4,6-dihydroxy, ethyl ester $4,6\text{-}(HO)_2\text{-}2CH_3C_6H_2CO_2C_2H_5$	196.20	wh nd		147			al, eth	B10[3], 1482
14267	*o*-Toluic acid, ethyl ester or Ethyl-*o*-toluate $2\text{-}CH_3C_6H_4CO_2C_2H_5$	164.21		227, 102.5^{12}	<−10	$1.0325^{21/4}$	1.507^{22}	al, eth	B9[3], 2301
14268	*o*-Toluic acid, methyl ester or Methyl-*o*-toluate $2\text{-}CH_3C_6H_4CO_2CH_3$	150.18		215, 97^{15}	<−15	$1.068^{20/4}$		al, eth	B9[3], 2301
14269	*o*-Toluic acid, 6-methyl $6\text{-}CH_3\text{-}2\text{-}CH_3C_6H_3CO_2H$	150.18	nd (w)		102				B9[3], 2310
14270	*o*-Toluonitrile, or 2-methyl benzonitrile $2\text{-}CH_3C_6H_4CN$	117.15		205, 90^{15}	−14	$0.9955^{20/4}$	1.5279^{20}	al, eth	B9[3], 2307
14271	*o*-Toluonitrile, 4-nitro $4\text{-}NO_2\text{-}2\text{-}CH_3C_6H_3CN$	162.15	lf (sub)	sub	100 (105)			ace, bz, chl	B9[1], 188
14272	*o*-Toluonitrile, 5-nitro $5\text{-}NO_2\text{-}2\text{-}CH_3C_6H_3CN$	162.15	nd (95% al)	$174\text{-}5^{18}$	105			al, eth, ace, bz, chl	B9[3], 2314

No.	Name, Synonyms, and Formula	Mol. wt.	Color, crystalline form, specific rotation and λ_{max} (log ε)	b.p. °C	m.p. °C	Density	n_D	Solubility	Ref.
14273	*o*-Toluonitrile, 6-nitro or 6-Nitro-*o*-toluonitrile $6\text{-}NO_2\text{-}2\text{-}CH_3C_6H_3CN$	162.15	pl (bz)	....	109-10	....	....	al, ace, bz	B9², 323
14274	*o*-Toluyl chloride or 2-Methyl benzoyl chloride $2\text{-}CH_3C_6H_4COCl$	154.60	....	213-4, 88-90^{12}	....	....	1.5549^{20}	eth	B9³, 2304
14275	*m*-Toluic acid or 3-Methyl benzoic acid $3\text{-}CH_3C_6H_4CO_2H$	136.15	pr (w, al)	263 sub	111-3	1.054^{112}	1.509	al, eth	B9³, 2318
14276	*m*-Toluic acid, 2-amino or 2-Amino-3-methyl benzoic acid $2\text{-}H_2N\text{-}3\text{-}CH_3C_6H_3CO_2H$	151.16	nd (al), pr (w)	....	172	....	....	al, eth	B14², 291
14277	*m*-Toluic acid, 4-amino or 4-Amino-3-methyl benzoic acid $4\text{-}H_2N\text{-}3\text{-}CH_3C_6H_3CO_2H$	151.16	nd (w)	....	170	....	....	w	B14², 290
14278	*m*-Toluic acid anhydride $(3\text{-}CH_3C_6H_4CO)_2O$	254.29	cr (peth)	230^{17}	71	....	....	al, eth, ace, bz, chl	B9³, 2321
14279	*m*-Toluic acid, 4-chloro or 4-Chloro-3-methyl benzoic acid $4\text{-}Cl\text{-}3\text{-}CH_3C_6H_3CO_2H$	170.60	nd (w)	....	209-10	....	....		B9³, 2327
14280	*m*-Toluic acid, 5-chloro or 5-Chloro-3-methyl benzoic acid $5\text{-}Cl\text{-}3\text{-}CH_3C_6H_3CO_2H$	170.60	nd (dil al)	....	178	....	....	al	B9, 479
14281	*m*-Toluic acid, 6-chloro $6\text{-}Cl\text{-}3\text{-}CH_3C_6H_3CO_2H$	170.60	nd (w al)	....	167	....	....	w, al	B9³, 2327
14282	*m*-Toluic acid, ethyl ester or Ethyl-*m*-toluate $3\text{-}CH_3C_6H_4CO_2C_2H_5$	164.20	....	234, 103-5^{10}	....	1.0265$^{21/4}$	1.5052^{22}	al, eth	B9³, 2320
14283	*m*-Toluic acid, methyl ester or Methyl-*m*-toluate $3\text{-}CH_3C_6H_4CO_2CH_3$	150.18	....	221^{758}	....	1.061$^{20/4}$	....	al	B9³, 2320
14284	*m*-Toluonitrile $3\text{-}CH_3C_6H_4CN$	117.15	....	213, 845^{10}	−23	1.0316$^{20/4}$	1.5232^{20}	al, eth	B9³, 2324
14285	*m*-Toluonitrile, 2-nitro $2\text{-}NO_2\text{-}3\text{-}CH_3C_6H_3CN$	162.15	nd (al)	....	84	....	....	ace, bz, chl	B9³, 2330
14286	*m*-Toluonitrile, 4-nitro $4\text{-}NO_2\text{-}3\text{-}CH_3C_6H_3CN$	162.15	pr (al), nd	....	80	....	....	al	B9³, 2331
14287	*m*-Toluonitrile, 5-nitro $5\text{-}NO_2\text{-}3\text{-}CH_3C_6H_3CN$	162.15	nd (lig)	....	104.5	....	....	al	B9³, 2331
14288	*m*-Toluonitrile, 6-nitro $6\text{-}NO_2\text{-}3\text{-}CH_3C_6H_3CN$	162.15	nd (al)	....	93-4	....	....	al, bz, aa	B9³, 2332
14289	*m*-Toluyl chloride or 3-Methyl benzoyl chloride $3\text{-}CH_3C_6H_4COCl$	154.60	....	219.20, 105^{20}	−23	1.0265$^{21/4}$	1.505^{22}	al, eth	B9³, 2321
14290	*p*-Toluic acid or 4-Methyl benzoic acid $4\text{-}CH_3C_6H_4CO_2H$	136.15	nd (w)	275 (sub)	182	....	....	al, eth	B9³, 2334
14291	*p*-Toluic acid anhydride $(4\text{-}CH_3C_6H_4CO)_2O$	254.29	pl (MeOH), nd (al)	....	95	....	....	eth, ace, bz, chl	B9³, 2341
14292	*p*-Toluic acid, 2-chloro or 2-Chloro-4-methyl benzoic acid $2\text{-}Cl\text{-}4\text{-}CH_3C_6H_3CO_2H$	170.60	nd (al)	....	155.6	....	....	al, eth, bz, chl	B9, 497
14293	*p*-Toluic acid, 3-chloro $3\text{-}Cl\text{-}4\text{-}CH_3C_6H_3CO_2H$	170.60	nd or lf (dil al)	....	200-2	....	....	al	B9³, 2355
14294	*p*-Toluic acid, ethyl ester or Ethyl-*p*-toluate $4\text{-}CH_3C_6H_4CO_2C_2H_5$	164.20	....	235.7, 110^{12}	....	1.0269$^{18/4}$	1.5089^{18}	al, eth	B9³, 2337
14295	*p*-Toluic acid, methyl ester or Methyl-*p*-toluate $4\text{-}CH_3C_6H_4CO_2CH_3$	150.18	cr (aq MeOH, peth)	222.5	33.2	....	....	al, eth	B9³, 2337
14296	*p*-Toluonitrile or 4-Methyl benzonitrile $4\text{-}CH_3C_6H_4CN$	117.15	nd (al)	217.6, 91^{11}	29.5	0.9805$^{30/30}$	....	al, eth	B9³, 2348
14297	*p*-Toluonitrile, 2-nitro $2\text{-}NO_2\text{-}4\text{-}CH_3C_6H_3CN$	162.15	nd (w)	....	101	....	....	al, bz, chl	B9², 334
14298	*p*-Toluonitrile, 3-nitro $3\text{-}NO_2\text{-}4\text{-}CH_3C_6H_3CN$	162.15	pa ye nd (w)	171^{12}	107-8	....	....	al, eth, ace, bz, chl	B9³, 2359
14299	*p*-Toluyl chloride or 4-Methyl benzoyl chloride $4\text{-}CH_3C_6H_4COCl$	154.60	....	225-7, 102^{15}	−2	1.1686$^{20/4}$	1.5547^{20}		B9³, 2342
14300	*o*-Toluidine or 2-Amino toluene $2\text{-}H_2NC_6H_4CH_3$	107.16	....	200.2, 80.1^{10}	−14.7	0.9984$^{20/4}$	1.5725^{20}	al, eth	B12³, 1837
14301	*o*-Toluidine, N-acetyl-4-bromo $4\text{-}Br\text{-}2\text{-}CH_3C_6H_3(NHCOCH_3)$	228.09	nd (dil al, lig)	....	159-60	....	....	al	B12², 456
14302	*o*-Toluidine, N-acetyl-5-bromo $5\text{-}Br\text{-}2\text{-}CH_3 \cdot C_6H_3(NHCOCH_3)$	228.09	nd (bz)	....	165.5	....	....	al, bz	B12³, 1923
14303	*o*-Toluidine, hydrochloride $2\text{-}CH_3C_6H_4NH_2.HCl$	143.62	mcl pr (w)	242	215	....	....	w, al	B12³, 1840

No.	Name, Synonyms, and Formula	Mol. wt.	Color, crystalline form, specific rotation and λ_{max} (log ε)	b.p. °C	m.p. °C	Density	n_D	Solubility	Ref.
14304	*o*-Toluidine, N-acetyl-6-bromo 6-Br-2-$CH_3C_6H_3(NHCOCH_3)$	228.09	nd (bz)		166			al	B12^{3}, 1922
14305	*o*-Toluidine, N-benzyl 2-$CH_3C_6H_4NHCH_2C_6H_5$	197.28	cr (al, eth)	300-5, 176^{10}	60	1.0142$^{65/4}$	1.5861^{65}	al, ace, chl	B12^{3}, 2220
14306	*o*-Toluidine, N-benzyl-N-ethyl 2-$CH_3C_6H_4N(C_2H_5)CH_2C_6H_5$	225.33	ye	230$^{20-5}$					B12, 1033
14307	*o*-Toluidine, N-benzyl-N-methyl 2-$CH_3C_6H_4N(CH_3)CH_2C_6H_5$	211.31	ye	167^{13}				al	B12, 1033
14308	*o*-Toluidine, 4-bromo 4-Br-2-CH_3-$C_6H_3NH_2$	186.05	cr (al)	240	59.5			al, eth, aa	B12^{2}, 456
14309	*o*-Toluidine, 5-bromo 5-Br-2-$CH_3C_6H_3NH_2$	186.05	lf	253-7d, 139^{17}	33			al, eth	B12^{3}, 1923
14310	*o*-Toluidine, 6-bromo 6-Br-2-$CH_3C_6H_3NH_2$	186.05		130^{16}				eth	B12^{3}, 1922
14311	*o*-Toluidine, 3-chloro 3-Cl-2-$CH_3C_6H_3NH_2$	141.60		245, 96-9^{10}	0-2		1.5880^{20}	al	B12^{3}, 1919
14312	*o*-Toluidine, 4-chloro 4-Cl-2-$CH_3C_6H_3NH_2$	141.60	lf (al)	241	29-30			al	B12^{3}, 1914
14313	*o*-Toluidine, 5-chloro 5-Cl-2-$CH_3C_6H_3NH_2$	141.60		237^{722}, 140^{38}	26			al	B12^{3}, 1910
14314	*o*-Toluidine, N,N-diethyl 2-$CH_3C_6H_4N(C_2H_5)_2$	163.26		208-9^{755}				al, eth	B12^{3}, 1844
14315	*o*-Toluidine, N,N-dimethyl 2-$CH_3C_6H_4N(CH_3)_2$	135.21		185.3, 70-2^{15}	−60	0.9286$^{20/4}$	1.5152^{20}	**al, eth**	B12^{3}, 1843
14316	*o*-Toluidine, N-ethyl 2-$CH_3C_6H_4NHC_2H_5$	135.21		218, 95.5^{10}	<−15	0.948$^{25/4}$	1.5456^{20}	al, eth	B12^{3}, 1843
14317	*o*-Toluidine, 4-iodo 4-I-2-$CH_3C_6H_3NH_2$	233.05	nd (dil al), pr (lig)		91-2			al, eth, bz, aa, lig	B12^{3}, 1927
14318	*o*-Toluidine, 5-iodo 5-I-2-$CH_3C_6H_3NH_2$	233.05	nd (aq al)	273d	48-9			al, aa	B12^{1}, 391
14319	*o*-Toluidine, 4-methoxy or 2-Methyl-*p*-anisidine 4-CH_3O-2-$CH_3C_6H_3NH_2$	137.18	cr (lig)	248-9, 146-7^{23}	29-30		1.5647^{20}	al	B13^{3}, 1560
14320	*o*-Toluidine, 5-methoxy or 6-Methyl-*m*-anisidine 5-CH_3O-2-$CH_3C_6H_3NH_2$	137.18	nd (w)	253, 140^{20}	47			eth	B13^{3}, 1573
14321	*o*-Toluidine, 6-methoxy or 6-Methyl-*o*-anisidine 6-CH_3O-2-$CH_3C_6H_3NH_2$	137.18	nd (w)	119-21^{16}	31			al	B13^{3}, 1538
14322	*o*-Toluidine, N-Methyl 2-$CH_3C_6H_4NHCH_3$	121.18		207-8, 99^{17}		0.9769$^{20/4}$	1.5649^{20}	**al, eth**, ace	B12^{3}, 1842
14323	*o*-Toluidine, 3-nitro 3-NO_2-2-$CH_3C_6H_3NH_2$	152.15	ye rh nd (w), ye lf (al)	305d	92 (97)			al, eth, bz, chl	B12^{3}, 1944
14324	*o*-Toluidine, 4-nitro 4-NO_2-2-$CH_2C_6H_3NH_2$	152.15	ye mcl pr or nd (w, al, lig)		134-5	1.1586$^{140/4}$		al, bz, aa	B12^{3}, 1938
14325	*o*-Toluidine, 5-nitro 5-NO_2-2-$CH_3C_6H_3NH_2$	152.15	ye mcl pr (al)		107-8			al, eth, ace, bz, chl	B12^{3}, 1932
14326	*o*-Toluidine, 6-nitro 6-NO_2-2-$CH_3C_6H_3NH_2$	152.15	og-ye pr (dil al)		97	1.1900$^{100/4}$		al, eth, bz, chl	B12^{3}, 1929
14327	*m*-Toluidine or 3-Amino toluene 3-$CH_3C_6H_4NH_2$	107.16		203.3, 82.3^{10}	−30.4	0.9889$^{20/4}$	1.5681^{20}	**al, eth, ace, bz**	B12^{3}, 1949
14328	*m*-Toluidine, N-benzyl 3-$CH_3C_6H_4NHCH_2C_6H_5$	197.28	pa ye oil	312, 180^{10}		1.0083$^{65/4}$	1.5845^{65}	eth, ace, bz, chl	B12^{2}, 552
14329	*m*-Toluidine, 4-bromo 4-Br-3-$CH_3C_6H_3NH_2$	186.05	pl (50% al), cr (al)	240	81			al	B12^{3}, 1999
14330	*m*-Toluidine, 5-bromo 5-Br-3-$CH_3C_6H_3NH_2$	186.05		255-60, 150-1^{15}	37-8	1.1422^{19}		al	B12^{2}, 474
14331	*m*-Toluidine, 6-bromo 6-Br-3-CH_3-$C_6H_3NH_2$	186.05	pr	129-30^{15}	46	1.474$^{25/25}$	1.5990^{25}	al, eth	B12^{3}, 1998
14332	*m*-Toluidine, 4-chloro 4-Cl-3-$CH_3C_6H_3NH_2$	141.60	nd (peth)	241	83-4			al, ace, bz	B12^{3}, 1993
14333	*m*-Toluidine, 6-chloro 6-Cl-3-$CH_3C_6H_3NH_2$	141.60	pl	228-30	29-30			al	B12^{3}, 1992
14334	*m*-Toluidine, N,N-dimethyl 3-$CH_3C_6H_4N(CH_3)_2$	135.21		212		0.9410$^{20/4}$	1.5492^{20}	**al, eth**	B12^{3}, 1953
14335	*m*-Toluidine, N-ethyl 3-$CH_3C_6H_4NHC_2H_5$	135.21		221, 111-2^{20}			1.5451^{20}	al, eth	B12^{3}, 1954
14336	*m*-Toluidine hydrochloride 3-$CH_3C_6H_4NH_2$.HCl	143.62	lf (w)	250	228			w, al	B12^{3}, 1951

No.	Name, Synonyms, and Formula	Mol. wt.	Color, crystalline form, specific rotation and λ_{max} (log ε)	b.p. °C	m.p. °C	Density	n_D	Solubility	Ref.
14337	*m*-Toluidine, 2-iodo $2\text{-I-3-}CH_3C_6H_3NH_2$	233.05	pr		41-2			al, eth, ace	$B12^3$, 2003
14338	*m*-Toluidine, 4-iodo $4\text{-I-3-}CH_3C_6H_3NH_2$	233.05	lf or pl (al, peth)		46			al, eth, bz, aa, lig	$B12^2$, 475
14339	*m*-Toluidine, 5-iodo $5\text{-I-3-}CH_3C_6H_3NH_2$	233.05	nd (peth)		78.5			al, eth, ace	$B12^1$, 406
14340	*m*-Toluidine, 6-iodo $6\text{-I-3-}CH_3C_6H_3NH_2$	233.05	nd (dil al), br in air		48			al, eth, chl	$B12^2$, 475
14341	*m*-Toluidine, 5-methoxy or 3-Methyl-*p*-anisidine $5\text{-}CH_3O\text{-3-}CH_3C_6H_3NH_2$	137.18	cr (dil al)		59-60			al, eth, ace, bz	$B13^2$, 320
14342	*m*-Toluidine, 6-methoxy or 5-Methyl-*o*-anisidine $6\text{-}CH_3O\text{-3-}CH_3C_6H_3NH_2$	137.18	nd or lf (al, lig, peth)	235	93-4			al, eth, bz	$B13^3$, 1577
14343	*m*-Toluidine, N-methyl $3\text{-}CH_3C_6H_4NHCH_3$	121.18		206-7, $120\text{-}1^{40}$			1.5557^{25}	al, eth, ace	$B12^3$, 1953
14344	*m*-Toluidine, 2-nitro $2\text{-}NO_2\text{-3-}CH_3C_6H_3NH_2$	152.15	ye-og pr or nd (bz-peth)		108			al, eth	$B12^3$, 2004
14345	*m*-Toluidine, 4-nitro $4\text{-}NO_2\text{-3-}CH_3C_6H_3NH_2$	152.15	lt ye nd (w, dil al)		138			al, eth	$B12^3$, 2008
14346	*m*-Toluidine, 5-nitro $5\text{-}NO_2\text{-3-}CH_3C_6H_3NH_2$	152.15	ye-red or red-br nd (al)		98			al, eth, bz	$B12^3$, 2007
14347	*m*-Toluidine, 6-nitro $6\text{-}NO_2\text{-3-}CH_3C_6H_3NH_2$	152.15	ye lf (w), pl (dil al)		112			al, eth, bz, chl	$B12^3$, 2004
14348	*p*-Toluidine or 4-Amino toluene $4\text{-}CH_3C_6H_4NH_2$	107.16	lf (w + 1)	200.5, 79.6^{10}	44-5	$0.9619^{20/4}$	1.5534^{45}, 1.5636^{20}	al, eth, ace, Py	$B12^3$, 2017
14349	*p*-Toluidine, N-acetyl-2-bromo $2\text{-Br-4-}CH_3C_6H_3NH(COCH_3)$	228.09	nd (al)		118			al	$B12^3$, 2159
14350	*p*-Toluidine, N-acetyl-3-bromo $3\text{-Br-4-}CH_3C_6H_3NH(COCH_3)$	228.09	nd (bz, dil al)		113			w, al, bz	$B12^3$, 2157
14351	*p*-Toluidine, N-benzyl $4\text{-}CH_3C_6H_4NHCH_2C_6H_5$	197.28	ye lf	319^{765}, 181^{10}	19-20	$1.0064^{65/4}$	1.5832^{65}	al, eth, bz, chl	$B12^3$, 2220
14352	*p*-Toluidine, 2-bromo $2\text{-Br-4-}CH_3C_6H_3NH_2$	186.05	lf	240, $120\text{-}2^{30}$	26	1.510^{20}	1.5999^{20}	al, eth	$B12^3$, 2158
14353	*p*-Toluidine, 2-bromo-5-nitro $2\text{-Br-5-}NO_2\text{-4-}CH_3C_6H_2NH_2$	231.05	br to pa ye nd (al, aa)		121				$B12^1$, 441
14354	*p*-Toluidine, 3-bromo $3\text{-Br-4-}CH_3C_6H_3NH_2$	186.05		254-7	26			eth	$B12^3$, 2157
14355	*p*-Toluidine, 2-chloro $2\text{-Cl-4-}CH_3C_6H_3NH_2$	141.60		219^{732}	7	1.151^{20}	1.5748^{22}		$B12^3$, 2152
14356	*p*-Toluidine, 3-chloro $3\text{-Cl-4-}CH_3C_6H_3NH_2$	141.60		242-4, $112\text{-}3^{3}$	26			al	$B12^3$, 2151
14357	*p*-Toluidine, N,N-diethyl $4\text{-}[(C_2H_5)_2N]C_6H_4CH_3$	163.26		229^{770}		0.9242^{16}		**al, eth**	$B12^3$, 2028
14358	*p*-Toluidine, N,N-dimethyl $4\text{-}CH_3C_6H_4N(CH_3)_2$	135.21		211		$0.9366^{20/4}$	1.5366^{20}	**al, eth**	$B12^3$, 2026
14359	*p*-Toluidine, 3,5-dinitro $3,5\text{-}(NO_2)_2\text{-4-}CH_3C_6H_2NH_2$	197.15	ye nd (w, aa)		171			al, eth, ace, bz, chl	$B12^3$, 2184
14360	*p*-Toluidine, N-ethyl $4\text{-}CH_3C_6H_4NHC_2H_5$	135.21		217		0.9391^{16}		al, eth	$B12^3$, 2027
14361	*p*-Toluidine hydrochloride $4\text{-}CH_3C_6H_4NH_2.HCl$	143.62	mcl nd (aa-eth)	258	243			w, al, aa	$B12^3$, 2021
14362	*p*-Toluidine, 2-iodo $2\text{-I-4-}CH_3C_6H_3NH_2$	233.05	pr	d	40			al, eth, ace, bz, chl, peth	$B12^2$, 533
14363	*p*-Toluidine, 3-iodo $3\text{-I-4-}CH_3C_6H_3NH_2$	233.05	nd (dil al, peth)		39-40			al, eth, ace, aa	$B12^2$, 533
14364	*p*-Toluidine, 2-methoxy or 4-Methyl-*o*-anisidine $2\text{-}CH_3O\text{-4-}CH_3C_6H_3NH_2$	137.18	pa ye	237-9, $179\text{-}80^{46}$				al, eth, ace	$B13^3$, 1552
14365	*p*-Toluidine, 3-methoxy or 4-Methyl-*m*-anisidine $3\text{-}CH_3O\text{-4-}CH_3C_3H_3NH_2$	137.18		250-2	58			al, eth, bz, lig	$B13^1$, 213
14366	*p*-Toluidine, N-methyl $4\text{-}CH_3C_6H_4NHCH_3$	121.18		209-11, 102^{20}		$0.9348^{55/4}$	1.5568^{20}	**al, eth**, ace	$B12^3$, 2025
14367	*p*-Toluidine, 2-nitro $2\text{-}NO_2\text{-4-}CH_3C_6H_3NH_2$	152.15	red lf (dil al), mcl pr (al)		117	$1.164^{121/4}$		al	$B12^3$, 2174
14368	*p*-Toluidine, 3-nitro $3\text{-}NO_2\text{-4-}CH_3C_6H_3NH_2$	152.15	ye nd (w)		78-9			eth, bz	$B12^3$, 2165

No.	Name, Synonyms, and Formula	Mol. wt.	Color, crystalline form, specific rotation and λ_{max} (log ε)	b.p. °C	m.p. °C	Density	n_D	Solubility	Ref.
14369	*o*-Tolyl acetic acid or *o*-Methyl-α-toluic acid $2\text{-}CH_3C_6H_4CH_2CO_2H$	150.18	nd (w)	...	88-90	...	...	w	B9[3], 2426
14370	2-Tolyl acetonitrile or *o*-Methyl-α-tolunitrile $2\text{-}CH_3C_6H_4CH_2CN$	131.18	...	244	...	1.0156^{22}	1.5252^{20}	al, eth, bz	B9[3], 2427
14371	3-Tolyl acetic acid or *m*-Methyl-α-toluic acid $3\text{-}CH_3C_6H_4CH_2CO_2H$	150.18	nd (w)	$120\text{-}3^{26}$	62	...	...	...	B9[3], 2429
14372	3-Tolyl acetonitrile or *m*-Methyl-α-tolunitrile $3\text{-}CH_3C_6H_4CH_2CN$	131.18	...	$245\text{-}7^{745}/_d$, 133^{15}	...	1.0022^{22}	1.5233^{20}	al, eth, bz	B9[2], 349
14373	4-Tolyl acetic acid or *p*-Methyl-α-toluic acid $4\text{-}CH_3C_6H_4CH_2CO_2H$	150.18	nd or pl (al, w)	265-7 sub	91-3	...	...	al, eth, bz, chl	B9[3]
14374	2-Tolyl ether or Di-2-Tolyl ether $(2\text{-}CH_3C_6H_4)_2O$	198.26	...	$146\text{-}7^{17}$	...	1.047^{24}	...	al, eth, bz	B6[4], 1947
14375	3-Tolyl ether or Di-3-Tolyl ether $(3\text{-}CH_3C_6H_4)_2O$	198.26	...	284, $135\text{-}7^{14}$	...	1.0323^{21}	...	al, eth, bz	B6[4], 2043
14376	4-Tolyl ether or Di-4-Tolyl ether $(4\text{-}CH_3C_6H_4)_2O$	198.26	...	285	51	...	...	al, eth, bz	B6[4], 2103
14377	2-Tolyl hydrazine $2\text{-}CH_3C_6H_4NHNH_2$	122.17	nd (dil al)	...	59	...	...	al, eth	B15[3], 654
14378	2-Tolyl isocyanate $2\text{-}CH_3C_6H_4NCO$	133.15	...	184-6	...	...	1.5282^{20}	eth, bz	B12[3], 1886
14379	3-Tolyl isocyanate $3\text{-}CH_3C_6H_4NCO$	133.15	...	195-8	...	...	...	eth, bz	B12[3], 1979
14380	4-Tolyl isocyanate $4\text{-}CH_3C_6H_4NCO$	133.15	...	187^{751}	...	...	...	eth, bz	B12[3], 2110
14381	2-Tolyl selenide or Di-(2-Tolyl)selenide $(2\text{-}CH_3C_6H_4)_2Se$	261.23	pr or lf (al)	186^{16}	65	...	...	...	B6[2], 343
14382	4-Tolyl selenide or Di-(4-Tolyl)selenide $(4\text{-}CH_3C_6H_4)_2Se$	261.23	rods or nd (al)	196^{16}	69-70	...	...	...	B6[4], 2218
14383	2-Tolyl sulfone or 2,2′-Ditolyl sulfone $(2\text{-}CH_3C_6H_4)_2SO_2$	246.32	nd (al)	...	134-5	...	...	al, eth, bz, chl	B6[4], 2020
14384	4-Tolyl sulfone or Di-4-Tolyl sulfone $(4\text{-}CH_3C_6H_4)_2SO_2$	246.32	pr (bz), nd (w, al), pl (al)	405^{714}	159	...	...	bz, chl	B6[4], 2174
14385	4-Tolyl sulfoxide or Di-4-Tolyl sulfoxide $(4\text{-}CH_3C_6H_4)_2SO$	230.32	cr (lig-peth)	...	94	...	...	al, eth, bz, chl, aa	B6[4], 2173
14386	4-Tolyl thiocyanate $4\text{-}CH_3C_6H_4SCN$	149.21	...	240-5, $116\text{-}8^{10}$	...	...	...	al, bz, chl	B6[3], 1421
14387	Tomatidine $C_{27}H_{45}NO_2$	415.66	pl, $[\alpha]^{20}/_D$ + 5 (MeOH)	...	210-1	...	...	eth	Am7[3], 4018
14388	Tomatine or Lycopersicin $C_{50}N_{83}NO_{21}$	1034.20	nd (MeOH), $[\alpha]^{20}/_D$ −30 (Py)	...	270	...	...	al, diox	C41, 3502
14389	Torularhodin or Torulene $C_{37}H_{48}O_2$	524.79	red nd (MeOH-eth), vt-bk (bz-MeOH)	...	201-3d	...	...	ace, chl, Py	B9[3], 3661
14390	α-Toxicarol (dl) or Hydroxydequelin $C_{23}H_{22}O_7$	410.42	gr ye pl (al)	...	219-23 (231)	...	1.580	chl	B19[4], 5270
14391	α-Toxicarol (l) $C_{23}H_{22}O_7$	410.42	gr-yesh pl or nd (AcOEt-al)	...	125-7	...	...	ace, AcOEt	B19[4], 5270
14392	α-Toxicarol, dihydro - (l) $C_{23}H_{24}O_7$	412.44	pa ye pl or rods, $[\alpha]^{20}/_D$ −30 (bz, c=5)	...	179 (206)	...	...	...	B19[4], 5262
14393	β-Toxicarol-(dl) $C_{23}H_{22}O_7$	410.42	pa ye pl (al)	...	169-70	...	...	al	B19[4], 5270
14394	Trasentin hydrochloride or Adephenine. 2-(Diethyl amino)ethyl diphenyl acetate hydrochloride $(C_6H_5)_2CHCO_2CH_2CH_2N(C_2H_5)_2.HCl$	347.88	nd	...	114-5	...	...	w	B9[3], 3297
14395	α,α-Trehalose or Mycose $C_{12}H_{22}O_{11}$	342.30	orh cr, $[\alpha]^{20}/_D$ + 199 (w, c=6)	...	214-6 (anh), 97 (hyd)	...	$1.58^{24/24}$	w	B17[4], 3505

No.	Name, Synonyms, and Formula	Mol. wt.	Color, crystalline form, specific rotation and λ_{max} (log ε)	b.p. °C	m.p. °C	Density	n_D	Solubility	Ref.
14396	α,α-Trehalose, dihydrate $C_{12}H_{22}O_{11}\cdot 2H_2O$	378.33	cr (dil al), $[\alpha]^{20/}{}_D$ +178.3 (w, c=7)		103			w	$B17^4$, 3505
14397	Triacetamide or N,N-Diacetyl acetamide $(CH_3CO)_3N$	143.14	nd (eth)		79			eth	$B2^4$, 416
14398	Triacetin or Glycerol triacetate $CH_3CO_2CH(CH_2O_2CCH_3)_2$	218.21	cr (al)	258-60, 130.5^7	4.1	$1.1596^{20/4}$	1.4301^{20}	al, eth, ace, bz, chl	$B2^4$, 253
14399	Triacontane $CH_3(CH_2)_{28}CH_3$	422.82	orh (eth bz)	449.7, 304^{15}	65.8	$0.7750^{78/4}$	1.4352^{70}, 1.4536^{20}	eth, bz	$B1^4$, 592
14400	1-Triacontanol or Myricyl alcohol $CH_3(CH_2)_{28}CH_2OH$	438.82	nd (eth), pl (bz)		88	0.777^{95}		al, eth, bz	$B1^4$, 1918
14401	Triallylamine $(CH_2{=}CHCH_2)_3N$	137.22		155-6		$0.809^{20/4}$	1.4502^{20}	al, eth, ace, bz	$B4^4$, 1061
14402	1,3,5-Triazine $C_3H_3N_3$	81.08		114	86	1.38		al, eth	Am76, 5646
14403	1,3,5-Triazine, 2,4-diamino or Guanamine 2,4-$(NH_2)_2C_3HN_3$	111.11	nd (w)		329d			w	$B26^1$, 65
14404	1,3,5-Triazine, 2,4-diamino-6-phenyl or Benzoguanamine 2,4-$(NH_2)_2$-6-$C_6H_5C_3N_3$	187.20	nd or pr (al)		226-8			al, eth	$B26^1$, 69
14405	1,3,5-Triazine, perhydro, 1,3,5-tributyl 1,3,5-$(C_4H_9)_3C_3H_6N_3$ $C_{15}H_{33}N_3$	255.45		$130\text{-}2^3$			1.4602^{25}		B26, 3
14406	1,3,5-Triazine, perhydro, 1,3,5-triethyl 1,3,5-$(C_2H_5)_3C_3H_6N_3$	171.29		$78\text{-}9^6$			1.4580^{25}		$B26^2$, 3
14407	1,3,5-Triazine, perhydro, 1,3,5-trinitro or Cyclonite. Hexogen 1,3,5-$(NO_2)_3C_3H_6N_3$	222.12	orh cr (ace)		205-6	1.82^{20}		ace, aa	$B26^2$, 5
14408	1,3,5-Triazine-perhydro-1,3,5-triphenyl 1,3,5-$(C_6H_5)_3C_3H_6N_3$	315.42	nd (lig), pr (eth, chl-al)	185	143			eth, ace, bz, chl	$B26^2$, 3
14409	1,3,5-Triazino-2,4,6-tricarboxylic acid or Cyanuric tricarboxylic acid 2,4,6 $C_3N_3(CO_2H)_3$	213.11	pw		>250d				$B26^2$, 168
14410	1,3,5-Triazino-2,4,6-tricarboxylic acid, triethyl ester 2,4,6-$C_3N_3(CO_2C_2H_5)_3$	297.27	nd (al)	d	168				$B26^2$, 168
14411	1,3,5-Triazino-2,4,6-tricarbonitrile 2,4,6-$C_3N_3(CN)_3$	156.11	mcl pr (bz)	262^{771}, 119^1	119			bz	$B26^1$, 91
14412	1,2,3-Triazole or Osotriazole $C_2N_3H_3$	69.07	hyg cr	203^{739}	23	$1.1861^{25/4}$	1.4854^{25}	w, eth, ace	$B26^1$, 5
14413	1,2,4-Triazole or Pyrrodiazole $C_2N_3H_3$	69.07	pr (w), nd (al, eth, chl, bz)	260	120-1	1.132^{153}	1.4854^{25}	w, al	$B26^2$, 7
14414	1,2,4-Triazole, 3-amino or Amizol. ATA 3-$NH_2C_2N_3H_2$	84.08	cr (w, al, AcOEt)		159			w, al, chl	$B26^2$, 76
14415	1,2,4-Triazole, 4-amino 4-$NH_2C_2N_3H_2$	84.08	hyg nd (al, chl)		82-3			w, al	$B26^2$, 7
14416	1,2,4-Triazole, 3,5-diamino $C_2H_5N_5$	99.10			211-2			w, al	$B26^1$, 57
14417	Tribenzylamine $(C_6H_5CH_2)_3N$	287.40	pl, (eth), mcl (al)	380-90, 230^{13}	91-2	$0.9912^{95/4}$		eth	$B12^3$, 2226
14418	Triabromo acetamide Br_3CCONH_2	295.77	mcl pr (al)	sub	121-2			w, al, eth	$B2^3$, 485
14419	Tribromoacetic acid Br_3CCOOH	296.74	mcl	245	135			w, al, eth	$B2^4$, 534
14420	Tribromoacetic acid, ethyl ester or Ethyl tribromo acetate $Br_3CCOOC_2H_5$	324.79		225, 148^{73}		$2.2300^{20/20}$	1.5438^{13}	al, eth	$B2^3$, 485
14421	Tribromoacetyl bromide Br_3CCOBr	359.64		210-5				eth, bz, chl	$B2^3$, 485
14422	Bis-(tribromomethyl) trisulfide $(Br_3C)_2S_3$	599.63	rh pr (eth)	d	125d			bz, chl, peth	$B3^2$, 107
14423	Tributylamine $(C_4H_9)_3N$	185.35	hyg	213, $91\text{-}2^9$		$0.7771^{20/0}$	1.4297^{20}	al, eth, ace, bz	$B4^4$, 554
14424	Tributylamine, perfluoro $(C_4F_9)_3N$	671.10		179		$1.873^{25/4}$	1.291^{25}	ace	$B2^4$, 819
14425	Tri-*iso*-butyl amine $(i\text{-}C_4H_9)_3N$	185.35		191.5, 84^{15}	−21.8	$0.7684^{20/4}$	1.4252^{17}	al, eth	$B4^4$, 631
14426	Tributyrin or Glycerol tributyrate $C_4H_9CO_2CH(CH_2O_2CC_4H_9)_2$	302.37		305-10, 190^{15}	−75	$1.0350^{20/4}$	1.4359^{20}	al, eth, ace, bz	$B2^4$, 799

No.	Name, Synonyms, and Formula	Mol. wt.	Color, crystalline form, specific rotation and λ_{max} (log ε)	b.p. °C	m.p. °C	Density	n_D	Solubility	Ref.
14427	Trichloro acetamide ..... Cl_3CCONH_2	162.40	mcl pr (w)	238-9^{746}	142	....	....	al, eth	B2^{4}, 520
14428	Trichloro acetamide, *N,N*-diethyl ..... $Cl_3CCON(C_2H_5)_2$	218.52	pr	109^{9}	27	....	1.4900^{24}	....	B4^{4}, 351
14429	Trichloro acetamide, *N,N*-dimethyl or *N,N*-Dimethyl trichloroacetamide ..... $Cl_3CON(CH_3)_2$	190.46	....	230-3d, 84^{4}	12	1.390^{20}	1.5017^{25}	bz, chl	B4^{4}, 182
14430	Trichloro acetamide, *N*-phenyl or α-Trichloroacetanilide $Cl_3CCONHC_6H_5$	238.50	lf (dil al)	168-70	95-7	....	....	al	B12^{3}, 464
14431	Trichloroacetic acid ..... Cl_3CCO_2H	163.39	dlq cr	197.5, 141-2^{25}	α:58 β:49.6	1.62$^{25/4}$, 1.6218$^{64/1}$	1.4603^{61}	al, eth, w	B2^{4}, 508
14432	Trichloroacetic acid anhydride or Trichloroacetic anhydride ..... $(Cl_3CCO)_2O$	308.76	....	222-4d, 98-100^{12}	......	1.6908^{20}	....	eth, aa	B2^{4}, 518
14433	Trichloroacetic acid butyl ester or Butyl trichloroacetate .. $Cl_3CCO_2C_4H_9$	219.50	....	203-5, 97-9^{19}	......	1.2778$^{20/4}$	1.4525^{25}	al, eth, ace, bz	B2^{4}, 515
14434	Trichloroacetic acid, *iso*-butyl ester or *iso*-Butyl trichloroacetate..... $Cl_3CCO_2\text{-}i\text{-}C_4H_9$	219.50	....	187-9, 93-4^{20}	......	1.2636$^{20/4}$	1.4483^{20}	al, eth, bz	B2^{4}, 515
14435	Trichloroacetic acid, *sec*-butyl ester or *sec*-Butyl trichloroacetate..... Cl_3CCO_2 *sec*-C_4H_9	219.50	....	93-4^{24}	......	1.2636$^{20/4}$	1.4483^{20}	al, eth, bz	B2^{3}, 472
14436	Trichloro acetic acid, *tert*-butyl ester or *tert*-Butyl trichloroacetate ..... $Cl_3CCO_2C(CH_3)_3$	219.50	cr (MeOH)	54-5^{7}	25.5	1.2363$^{25/4}$	1.4398^{25}	al, eth	B2^{4}, 516
14437	Trichloroacetic acid, 2-chloroethyl ester or 2-Chloroethyl trichloro acetate ..... $Cl_3CCO_2CH_2CH_2Cl$	225.89	....	217, 100^{14}	......	1.5357$^{20/4}$	1.4813^{20}	eth	B2^{4}, 515
14438	Trichloroacetic acid, ethyl ester or Ethyl trichlorro acetate $Cl_3CCO_2C_2H_5$	191.44	....	167-8, 62^{12}	......	1.3836$^{20/4}$	1.4505^{20}	al, eth, bz	B2^{4}, 514
14439	Trichloroacetic acid, 2-hydroxyethyl ester ..... $Cl_3CCO_2CH_2CH_2OH$	207.44	....	130-4^{12}	......	1.532$^{20/4}$	1.4775^{20}	al	B2^{3}, 474
14440	Trichloroacetic acid, 2-methoxyethyl ester or 2-Methoxyethyl trichloro acetate..... $Cl_3CCO_2CH_2CH_2OCH_3$	221.47	....	98-9^{17}	14.5	1.3826$^{20/4}$	1.4563^{20}	al, eth, bz	B2^{3}, 474
14441	Trichloroacetic acid, methyl ester or Methyl trichloro acetate ..... $Cl_3CCO_2CH_3$	177.41	....	153.8, 44.5^{12}	−17.5	1.4874$^{20/4}$	1.4572^{20}	al, eth	B2^{4}, 513
14442	Trichloroacetic acid, 2-(2-methyl butyl) ester or 2-(2-Methylbutyl) trichloroacetate ..... $Cl_3CCO_2CH_2CH(CH_3)CH_2CH_3$	233.52	....	217, 92-5^{12}	......	1.2314$^{20/4}$	1.4521^{20}	al, eth	B2^{3}, 473
14443	Trichloroacetic acid, pentyl ester or Pentyltrichloroacetate $Cl_3CCO_2C_5H_{11}$	233.52	....	220-2, 118^{30}	......	1.2475$^{20/20}$	....	al, eth	B2^{3}, 473
14444	Trichloroacetic acid, isopentyl ester or Isopentyl trichloro acetate ..... $Cl_3CO_2\text{-}i\text{-}C_5H_{11}$	233.52	....	217, 92-5^{11}	......	1.2314$^{20/4}$	1.4521^{20}	al, eth	B2^{4}, 516
14445	Trichloroacetic acid, propyl ester or Propyl trichloro acetate ..... $Cl_3CCO_2C_3H_7$	205.47	....	187, 69^{10}	......	1.3221$^{20/4}$	1.4501^{20}	al, eth	B2^{4}, 515
14446	Trichloroacetic acid, *iso*-propyl ester or *iso*-Propyl trichloroacetate ..... $Cl_3CCO_2CH(CH_3)_2$	205.47	....	173.5^{747}, 65-7^{15}	......	1.3034$^{20/4}$	1.4428^{20}	al, eth, bz	B2^{4}, 515
14447	Trichloroacetic acid, trichlomethyl ester or Trichloromethyl trichloro acetate..... $Cl_3CCO_2CCl_3$	280.75	....	191-2, 73-4^{10}	34	....	....	eth, bz, chl, lig	B3^{3}, 35
14448	Trichloroacetonitrile ..... Cl_3CCN	144.39	....	84.6^{741}	−42	1.4403$^{25/4}$	1.4409^{20}	....	B2^{4}, 524
14449	Trichloroacetyl bromide ..... Cl_3CCOBr	226.28	....	143	......	1.900$^{15/15}$	....	eth, bz	B2^{3}, 476
14450	Trichloroacetyl chloride ..... Cl_3CCOCl	181.83	....	118	......	1.6202$^{20/4}$	1.4695^{20}	eth	B2^{4}, 519
14451	Trichloromethyl perchlorate ..... Cl_3COClO_3	217.82	....	exp d 40	−55	....	....	eth	B3^{3}, 37
14452	Tricosane ..... $CH_3(CH_2)_{21}CH_3$	324.63	lf (eth-al)	380, 243^{15}	47.6	0.7785$^{48/4}$, 0.7969$^{20/4}$	1.4468^{20}	eth	B1^{4}, 576
14453	Tricosane, 2-methyl ..... $(CH_3)_2CH(CH_2)_{20}CH_3$	338.66	....	205^{3}	37.6	0.7539$^{90/4}$	1.4201^{90}	....	B1^{3}, 576
14454	Tricosanoic acid, methyl ester or Methyltricosonate..... $CH_3(CH_2)_{21}CO_2CH_3$	368.64	....	....	55.6	....	....	....	B1^{3}, 1078

PHYSICAL CONSTANTS OF ORGANIC COMPOUNDS (Continued)

No.	Name, Synonyms, and Formula	Mol. wt.	Color, crystalline form, specific rotation and λ_{max} (log ε)	b.p. °C	m.p. °C	Density	n_D	Solubility	Ref.
14455	1-Tricosanol $C_{22}H_{43}CH_2OH$	340.63		$191\text{-}3^{0.7}$	74				B1^{4}, 1908
14456	Tricyclene $C_{10}H_{16}$	136.24		153.5	67.5	$0.8268^{80/4}$	1.4389^{20}		B5^{4}, 468
14457	1,12-Tridecadiyne $HC{\equiv}C(CH_2)_9C{\equiv}CH$	176.30		115.5^{12}	−3	$0.8262^{21/4}$	1.454^{20}		B1^{2}, 248
14458	Tridecanal $C_{12}H_{25}CHO$	198.35		156^{13}	14	$0.8356^{18/4}$	1.4384^{18}	al	B1^{4}, 3386
14459	Tridecanal oxime $C_{12}H_{25}CH{=}NOH$	213.36	nd (dil al)		80.5			eth, chl	B1^{2}, 769
14460	Tridecane $CH_3(CH_2)_{11}CH_3$	184.37		235.4, 107^{10}	−5.5	$0.7564^{20/4}$	1.4256^{20}	al, eth	B1^{4}, 512
14461	Tridecane, 1-amino or Tridecyl amine $C_{12}H_{25}CH_2NH_2$	199.38		275.8, 140.1^{10}	27.4	$0.8049^{20/4}$	1.4443^{20}	al, eth	B4^{4}, 810
14462	Tridecane, 1-bromo or Tridecyl bromide $C_{12}H_{25}CH_2Br$	263.26		296, 162^{16}	6.2	$1.0177^{20/4}$	1.4593^{20}	chl	B1^{4}, 514
14463	Tridecane, 1,13-dibromo $Br(CH_2)_{13}Br$	342.16		188.92^{12}	8-10	1.276^{15}	1.4880^{27}	eth, chl	B1^{3}, 548
14464	Tridecane, 7-hexyl $(C_6H_{13})_3CH$	268.53		170.5^{10}	−28.3	$0.7877^{20/4}$	1.4409^{20}		B1^{4}, 562
14465	Tridecane, 7-methyl $(C_6H_{13})_2CHCH_3$	198.39		115.5^{10}	−37.2	$0.7634^{20/4}$	1.4291^{20}		B1^{4}, 525
14466	Tridecane, 7-phenyl $(C_6H_{13})_2CHC_6H_5$	260.46		$183\text{-}4^{20}$		$0.8723^{20/4}$	1.49307^{18}		B5^{4}, 1210
14467	Tridecanedioic acid $HO_2C(CH_2)_{11}CO_2H$	244.33			114			al, eth, chl	B2^{4}, 2141
14468	1,12-Tridecanediol $HO(CH_2)_{11}CH(OH)CH_3$	216.36	cr (dil al)	$188\text{-}90^{8}$	60-1			al	B1^{4}, 563
14469	1,13-Tridecanediol or Tridecamethylene glycol $HO(CH_2)_{13}OH$	216.36	cr (bz)	$195\text{-}7^{10}$	76.5			al, aa	B1^{4}, 2630
14470	Tridecan amide $C_{12}H_{25}CONH_2$	213.36	lf (al)		100			al, eth	B2^{4}, 1119
14471	Tridecanoic acid $C_{12}H_{25}CO_2H$	214.35	cr (peth, ace)	236^{100}, 140.5^{1}	44-5			al, eth, ace, aa	B2^{4}, 1117
14472	Tridecanoic acid, methyl ester or Methyl tridecanoate $C_{12}H_{25}CO_2CH_3$	228.38		$90\text{-}5^{1}$	fr 6.5		1.4405^{20}	al	B2^{4}, 1118
14473	Tridecanonitrile $C_{12}H_{25}CN$	195.35		293, 142^{10}	9.7	$0.8257^{20/4}$	1.4378^{20}	al, eth	B2^{3}, 906
14474	1-Tridecanol or Tridecyl alcohol $C_{12}H_{25}CH_2OH$	200.36	cr (al)	152^{14}	32-3	$0.8223^{31/4}$		al, eth	B1^{4}, 1860
14475	2-Tridecanol $C_{11}H_{23}CH(OH)CH_3$	200.36		161^{30}, $95\text{-}6^{0.5}$	23		1.4188^{70}	al, eth, bz	B1^{4}, 1861
14476	2-Tridecanone or Undecyl methyl ketone $C_{11}H_{23}COCH_3$	198.35		263, 160^{16}	30.5	$0.8217^{30/4}$	1.4318^{20}	al, eth, ace, bz	B1^{4}, 3386
14477	3-Tridecanone or Ethyl decyl ketone $C_{10}H_{21}COC_2H_5$	198.35	pl	140^{17}	31			ace	B1^{4}, 3387
14478	7-Tridecanone or Dihexyl ketone $(C_6H_{13})_2CO$	198.35	lf (al)	261, 138^{12}	33	0.825^{30}		al, eth, chl, lig	B1^{4}, 3387
14479	1-Tridecene $C_{11}H_{23}CH{=}CH_2$	182.35		232.8, 104^{11}	−13	$0.7658^{20/4}$	1.4340^{20}	al, eth, bz	B1^{4}, 921
14480	1-Tridecyne $C_{11}H_{23}C{\equiv}CH$	180.33		94.5^{25}		$0.7729^{20/4}$	1.4309^{20}	eth, bz	B1^{4}, 1069
14481	Triethanol amine or Tris (2-hydroxyethyl) amine $(HOCH_2CH_2)_3N$	149.19	hyg cr	277^{150}	21-2	$1.1242^{20/4}$	1.4852^{20}	w, al, chl	B4^{4}, 1524
14482	Triethanol amine hydrochloride $(HOCH_2CH_2)_3N.HCl$	185.65	cr (al)		179-80				B4^{4}, 1525
14483	Triethylamine $(C_2H_5)_3N$	101.19		89.3	−114.7	$0.7275^{20/4}$	1.4010^{20}	w, al, eth, ace, bz	B4^{4}, 322
14484	Triethylamine hydrochloride $(C_2H_5)_3N.HCl$	137.65	hex (al)	sub 245	260d	$1.0689^{21/4}$		w, al, chl	B4^{4}, 327
14485	Triethylamine hydroiodide $(C_2H_5)_3N \cdot HI$	229.10	pr (95% al)		181	1.924		w, al, chl	B4^{4}, 328
14486	Triethylamine, perfluoro $(C_2F_5)_3N$	371.05		70.3		$1.736^{20/4}$	1.262^{25}		B2^{4}, 471
14487	Triethylene glycol $HO(CH_2CH_2O)_2CH_2CH_2OH$	150.17	hyg liq	278.3, 165^{14}	−5	$1.1274^{15/4}$	1.4531^{20}	w, al, bz	B1^{4}, 2400
14488	Triethylene glycol, 3-amino propyl ether $HO(CH_2CH_2O)_2CH_2CH_2OCH_2CH_2CH_2NH_2$	207.27	glassy	184^{10}	−50	$1.0682^{20/20}$	1.4668^{20}	w, al	C55, 5935

No.	Name, Synonyms, and Formula	Mol. wt.	Color, crystalline form, specific rotation and λ_{max} (log ε)	b.p. °C	m.p. °C	Density	n_D	Solubility	Ref.
14489	Triethylene glycol diacetate $CH_3CO_2(CH_2CH_2O)_2CH_2CH_2O_2CCH_3$	234.25		300				w, al, eth	$B2^4$, 215
14490	Triethylene glycol, monobutyl ether $HO(CH_2CH_2O)_2CH_2CH_2OC_4H_9$	206.28		278		$0.9890^{20/4}$	1.4389^{20}	w, al, MeOH	$B1^4$, 2402
14491	Triethylene tetramine $H_2NCH_2(CH_2NHCH_2)_2CH_2NH_2$	146.24		266-7, 157^{20}	12		1.4971^{20}	w, al	$B4^4$, 1242
14492	Trifluoro acetamide, *N*-phenyl or Trifluoro acetanilide $F_3CCONHC_6H_5$	189.14	(60% al)		87.6			al	$B12^2$, 141
14493	Trifluoroacetic acid F_3CCO_2H	114.02		72.4	−15.2	1.5351^0		w, al, eth, ace	$B2^4$, 458
14494	Trifluoroacetic acid anhydride $(F_3CCO)_2O$	210.03		39-40	−65	$1.490^{25/4}$	1.269^{25}	eth, aa	$B2^4$, 469
14495	Trifluoro acetic acid, butyl ester or Butyltrifluoro acetate $F_3CCO_2C_4H_9$	170.13		100.2		$1.0268^{22/4}$	1.353^{20}	chl	$B2^4$, 464
14496	Trifluoro acetic acid, *tert*-butyl ester $F_3CCO_2C(CH_3)_3$	170.13		83			1.3300^{25}	chl	$B2^4$, 465
14497	Trifluoroacetic acid, ethyl ester or Ethyltrifluoroacetate $F_3CCO_2C_2H_5$	142.08		60-2		$1.19^{20/4}$	1.308^{20}		$B2^4$, 463
14498	Trifluoroacetic acid, methyl ester or Methyl trifluoro acetate $F_3CCO_2CH_3$	128.05				$1.28^{20/4}$			$B2^4$, 463
14499	Trifluoroacetic acid, propyl ester or Propyltrifluoro acetate $F_3CCO_2C_3H_7$	156.10		82.5		$1.1285^{25/4}$	$1.3233^{22.5}$	chl	$B2^4$, 464
14500	Trifluoroacetonitrile F_3CCN	95.02		−64					$B2^4$, 472
14501	Trifloromethyl peroxide CF_3-OO-CF_3	170.01		−32					$B3^4$, 22
14502	Trifluoromethyl sulfide $(CF_3)_2S$	170.07		−22.2					$B3^4$, 278
14503	Trifurfuryl amine $(C_5H_5O)_3N$	257.29		$136\text{-}8^1$				eth	$B18^4$, 7094
14504	Triglycine $H_2N(CH_2CONH)_2CH_2CO_2H$	189.17	nd (dil al)		246d			w	$B4^3$, 1198
14505	Triglycine-*N*-phthalyl $C_{14}H_{13}N_3O_6$	319.27	nd (al)		234-5d			w, al	$B21^4$, 5186
14506	Tri-heptyl amine $(C_7H_{15})_3N$	311.60		330^{762}, $151\text{-}4^1$				al, eth	$B4^4$, 736
14507	Tri-hexyl amine $(C_6H_{13})_3N$	269.51		263-4, 119^{15}				al, eth	$B4^4$, 711
14508	Triiodoacetic acid I_3CCO_2H	437.74	ye lf		150d			w, al, eth	$B2^4$, 537
14509	Trilaurin or Glycerol trilaurate $C_{11}H_{23}CO_2CH(CH_2O_2CC_{11}H_{23})_2$	639.01	nd (al)		46.4	0.8986^{55}	1.4404^{60}	al, eth, ace, bz, chl	$B2^4$, 1098
14510	Trilauryl amine or Tridodecylamine $(C_{12}H_{25})_3N$	522.00		$220\text{-}8^{0.03}$	fr 15.7		1.4567^{25}	eth, bz, chl	Am74, 428
14511	Trimellitic acid or 1,2,4-Benzene tricarboxylic acid 1,2,4-$C_6H_3(CO_2H)_3$	210.14	nd (w), cr (aa, al)		238d			w, al, eth	$B9^3$, 4792
14512	Trimesic acid or 1,3,5-Benzene tricarboxylic acid 1,3,5-$C_6H_3(CO_2H)_3$	210.14	pr or nd (w + 1)		380 (anh)			al, eth	$B9^3$, 4793
14513	Trimethadione or 3,5,5-trimethyl 2,4-oxazoldinedione $C_6H_9NO_3$	143.13	cr (50% MeOH)	$78\text{-}80^5$	46			al, eth, ace, bz, chl	$B4^4$, 8013
14514	Trimethyl amine $(CH_3)_3N$	59.11		2.9	−117.2	$0.6356^{20/4}$	1.3631^0	w, al, eth, bz, chl	$B4^4$, 134
14515	Trimethyl amine hydrochloride $(CH_3)_3N.HCl$	95.57	mcl dlq nd (al)	sub at 200	277-8			w, al, chl	$B4^4$, 138
14516	Trimethylamine oxide $(CH_3)_3NO$	75.11	hyg nd (w + 2)		255-7			w, al	$B4^4$, 144
14517	Trimethylene oxide, hydrochloride $(CH_3)_3NO.HCl$	111.57	nd (al)		218-20d			w, al	$B4^4$, 144
14518	Trimethylamine oxide, hydroiodide $(CH_3)_3NO.HI$	203.02	pr (al)		130d			w, al	B4, 50
14519	Trimethylamine, perfluoro $(CF_3)_3N$	221.03			−7				$B3^4$, 79
14520	Tris-(2-methylbutyl) amine $[CH_3CH_2CH(CH_3)CH_2]_3N$	227.43		230-7, 94^4		0.7964^{13}	1.4330^{20}	al, eth, ace, bz	B4, 179
14521	Trimethylene sulfide or Thiacyclobutane C_3H_6S	74.14		94.7, 14^{30}	−73.2	$1.0200^{20/4}$	1.5102^{20}	al, ace, bz	$B17^4$, 14

PHYSICAL CONSTANTS OF ORGANIC COMPOUNDS (Continued)

No.	Name, Synonyms, and Formula	Mol. wt.	Color, crystalline form, specific rotation and λ_{max} (log ε)	b.p. °C	m.p. °C	Density	n_D	Solubility	Ref.
14522	Trimethylol ethane $CH_3C(CH_2OH)_3$	120.15	wh pw or nd (al)	$135\text{-}7^{15}$	204			w, al	$B1^4$, 2780
14523	1,1,1-Trimethylol propane or TMP $CH_3CH_2C(CH_2OH)$	134.18	wh pw or pl	160^5	58			w, al	$B1^4$, 2786
14524	Trimethyloxonium fluoborate $[(CH_3)_3O]^+BF_4^-$	147.91	hyg nd		148d			ace, chl	$B1^4$, 1248
14525	Bis-(trimethylsilyl) amine $[(CH_3)_3Si]_2NH$	161.39		126.2		0.7741^{25}	1.4090^{20}		$B4^3$, 1861
14526	Trinitro acetonitrile $(NO_2)_3CCN$	176.05	wax	220 exp	41.5			eth	B2, 229
14527	Tri-octylamine $(C_8H_{17})_3N$	353.68		$365, 182^5$			1.4510^{19}		$B4^4$, 754
14528	Triolein or Glycerol trioleate $C_{17}H_{33}CO_2CH(CH_2O_2CC_{17}H_{33})_2$	885.45	poly-morphic	$235\text{-}40^{18}$	−5.5	0.8988^{40}	1.4621^{40}	eth, chl, peth	$B2^4$, 1664
14529	1,3,5-Trioxane or Metaformaldehyde $C_3H_6O_3$	90.08	rh nd (eth)	114.5, sub 46^1	64	1.17^{65}		w, al, eth, bz, chl	$B19^4$, 4710
14530	Tripalmitin or Glycerol tripalmitate $C_{15}H_{31}CO_2CH(CH_2O_2CC_{15}H_{31})_2$	807.34	nd (eth)	310-20	66.4	$0.8752^{70/4}$	1.4381^{80}	eth, bz, chl	$B2^3$, 971
14531	Tripentylamine or Triamylamine $(C_5H_{11})_3N$	227.43		240-5, 130^{14}		$0.7907^{20/4}$	1.4366^{20}	al, eth	$B4^4$, 676
14532	Tri-isopentylamine $(i\text{-}C_5H_{11})_3N$	227.43		$235, 94^4$		$0.7848^{20/4}$	1.4331^{20}	al, **eth, bz**	$B4^4$, 700
14533	Triphenyl acetic acid $(C_6H_5)_3CCO_2H$	288.35	mcl pr (al), lf (aa)		271			al, aa, lig	$B9^3$, 3585
14534	Triphenyl amine $(C_6H_5)_3N$	245.32	mcl (MeOH, AcOEt, bz)	365	127	$0.774^{0/0}$	1.353^{16}	eth, bz	$B12^3$, 292
14535	Triphenylene $C_{18}H_{12}$	228.29		425	199			al, bz, chl	$B5^3$, 2384
14536	Triphenylethylene, chloro $(C_6H_5)_2C{=}C(Cl)C_6H_5$	290.79			117			al, eth, ace, bz, chl	$B5^3$, 2400
14537	Triphenyl methane or Tritan $(C_6H_5)_3CH$	244.34	rh (al)	$358\text{-}9^{754}$, $190\text{-}215^{10}$	94	$1.014^{99/4}$	1.5839^{99}	eth, bz, chl	$B5^3$, 2307
14538	Triphenyl methane, 3-amino or 3-Amino tritan $3\text{-}H_2NC_6H_4CH(C_6H_5)_2$	259.35	nd (eth)		120			al, eth	$B12^2$, 790
14539	Triphenylmethane, 4-amino or 4-Aminotritan $4\text{-}H_2NC_6H_4CH(C_6H_5)_2$	259.35	pr or lf (eth, lig), ta (Peth)	ca 248^{12}	84-5			eth, bz, lig	$B12^2$, 790
14540	Triphenyl methane, α-chloro or Trityl chloride $(C_6H_5)_3CCl$	278.78	nd or pr (bz-peth)	310, $230\text{—}5^{20}$	113-4			eth, ace, bz, chl	$B5^3$, 2315
14541	Triphenylmethane, 4,4′-diamino $(4\text{-}H_2NC_6H_4)_2CHC_6H_5$	274.37	pr (bz, eth)		139-40			al, eth, chl, lig	$B13^3$, 529
14542	Triphenyl methane, 4,4′-dimethyl $(4\text{-}CH_3C_6H_4)_2CHC_6H_5$	272.39	nd (MeOH)	$218\text{-}20^{12}$	56			al, eth, bz, chl, lig	$B13^3$, 2342
14543	Triphenylmethane, 2,4-bis (dimethylamino) $2,4\text{-}[(CH_3)_2N]_2C_6H_3CH(C_6H_5)_2$	330.47	pl (peth)		122-3			al, eth, ace, bz	B13, 273
14544	Triphenyl methane, 4,4′-bis(dimethyl amino) $(4\text{-}(CH_3)_2NC_6H_4)_2CHC_6H_5$	330.47	nd or lf (al, bz)		102			eth, bz	$B13^3$, 529
14545	Triphenyl methane, 4,4′-bis(dimethyl amino)-4″-amino $[4\text{-}(CH_3)_2NC_6H_4]_2CHC_6H_4NH_2\text{-}4$	345.49	(al)		151-2				$B13^3$, 566
14546	Triphenyl methane, 4,4′-bis(dimethyl amino)-2″-hydroxy $[4\text{-}(CH_3)_2NC_6H_4]_2CH(C_6H_4OH\text{-}2)$	346.47	nd (al)		127-8			bz, lig	$B13^3$, 2063
14547	Triphenyl methane, 4,4′-bis(dimethyl amino)-3″-hydroxy $[4\text{-}(CH_3)_2NC_6H_4]_2CH(C_6H_4OH\text{-}3)$	346.47	cr (al)		149			bz	$B13^2$, 440
14548	Triphenyl methane, 4,4′-bis(dimethyl amino)-4″-hydroxy $[4\text{-}(CH_3)_2NC_6H_4]_2CH(C_6H_4OH\text{-}4)$	346.47	cr (al)		165			al, bz	$B13^2$, 440
14549	Triphenylmethane, 3-methyl or 3-Methyltritan $(3\text{-}CH_3C_6H_4)CH(C_6H_5)_2$	258.36	pr (al)	354^{706}	62			eth, bz, chl, aa, lig	$B5^3$, 2332
14550	Triphenylmethane, 2,4′,4″-triamino $(4\text{-}H_2NC_6H_4)_2CH(C_6H_4NH_2\text{-}2)$	289.38	cr (al)		165			al	$B13^3$, 565
14551	Triphenylmethane, 3,4′,4″-triamino $(4\text{-}H_2NC_6H_4)_2CH(C_6H_4NH_2\text{-}3)$	289.38	nd (eth, eth-lig)		150			al	$B13^3$, 565
14552	Triphenyl methane, 4,4′,4″-triamino or *p*-Leucaniline $(4\text{-}H_2NC_6H_4)_3CH$	289.38	lf (w, al, bz)		208			al, eth	$B13^3$, 566
14553	Triphenyl methane, 4,4′,4″-tris(dimethylamino) or Leuco crystal violet $[(CH_3)_2NC_6H_4]_3CH$	373.54	lf (al), nd (bz, lig)		175			eth, bz, chl, aa	$B13^3$, 566

No.	Name, Synonyms, and Formula	Mol. wt.	Color, crystalline form, specific rotation and λ_{max} (log ε)	b.p. °C	m.p. °C	Density	n_D	Solubility	Ref.
14554	Triphenyl methane, 4,4′,4″-trihydroxy or Leucoaurin $(4\text{-}HOC_6H_4)_3CH$	292.33	pr (aa, dil al)		240			al, chl, aa	B6^3, 6578
14555	Triphenylmethane, 2,2′,2″-trimethyl or Tri-2-tolylmethane $[2\text{-}CH_3C_6H_4]_3CH$	286.42	nd (al)		130-1			al, eth	B5^3, 2347
14556	Triphenylmethane, 4,4′,4″-trinitro $(4\text{-}O_2NC_6H_4)_3CH$	379.33	sc (bz)		212.5				B5^4, 2501
14557	Triphenylmethanol or Tritanol $(C_6H_5)_3COH$	260.34	pl (al), trg (bz)	380	164.2	1.199$^{0/4}$		al, eth, ace, bz, aa	B6^3, 3640
14558	Triphenylmethanol, 2,2′-bis-(dimethylamino) $[2\text{-}(CH_3)_2NC_6H_4]_2C(OH)C_6H_5$	346.47	pr (lig)		105				B13, 741
14559	Triphenylmethanol, 2,3′-bis(dimethylamino) 2,3′$[(CH_3)_2NC_6H_4]_2C(OH)C_6H_5$	346.47	pl (bz)		183-4			bz, HCl	B13, 742
14560	Triphenylmethanol, 2,4′-bis (dimethylamino) $C_{23}H_{26}N_2O$	346.47	(al)		169-70			bz	B13, 742
14561	Triphenyl methanol, 3,3′-bis(dimethylamino) $[3\text{-}(CH_3)_2NC_6H_4]_2$	346.47	cr (eth)		128-9			eth, aa	B13, 742
14562	Triphenylmethanol, 3,4′-bis(dimethyl amino) $C_{23}H_{26}N_2O$	346.47	pr (bz-al)		140			bz	B13, 742
14563	Triphenylmethanol, 4,4′-bis(dimethylamino) or Michler's hydrol $[4\text{-}(CH_3)_2NC_6H_4]_2C(OH)C_6H_5$	346.47	cr (eth, bz, lig, MeOH)		121-3			eth, bz, lig	B13^3, 2068
14564	Triphenyl methanol, 4,4′-dihydroxy or Benzaurin $[4\text{-}HOC_6H_4]_2C(OH)C_6H_5$	292.33	ye-red pw		110-20				B6^3, 6582
14565	Triphenyl methanol, 4,4,4″-triamino or Pararosaniline $(4\text{-}H_2NC_6H_4)_3COH$	305.38	col to red lf		189-205			al	B13^3, 2072
14566	Triphenylmethanol, 3,3′,3″-trinitro $(3\text{-}NO_2C_6H_4)_3COH$	395.33	rh (MeOH, chl)		167			bz, aa	B6^1, 352
14567	Triphenylmethanol, 4,4′,4″-trinitro $[\text{-}O_2NC_6H_4]_3COH$	395.33	mcl pr (bz, aa)		(i) 190, (ii) 167			bz, aa	B6^3, 3673
14568	Triphenylselenonium chloride $(C_6H_5)_3SeCl$	345.73	orh (AcOEt)		230d			w, al, chl	B6^4, 1780
14569	Triphenylselenonium fluoride $(C_6H_5)_3SeF$	329.28	oct deliq		145d			w, al, ace, chl	B6^3, 1108
14570	Bis-(2 nitrophenyl) trisulfide $(2\text{-}NO_2C_6H_4)_2S_3$	340.39	ye nd (al)		175-6				B6^3, 1062
14571	2,3,5-Triphenyl,1,2,3,4-tetrazolium chloride or Tetrazolium salt . T.T.C. $C_{19}H_{15}N_4Cl$	334.81	nd (al, chl)		243d			w, al, ace	B26, 363
14572	Triisopropanolamine $[CH_3CH(OH)CH_2]_3N$	191.27		170-80^{10}	45	1.0$^{20/4}$		w, al	B4^4, 1680
14573	Tripropylamine $(C_3H_7)_3N$	143.27		156	−93.5	0.7558$^{20/4}$	1.4181^{20}	al, eth	B4^4, 470
14574	Tripropylamine, perfluoro $(C_3F_7)_3N$	521.07		130		1.822$^{4/4}$	1.279^{25}		B2^4, 742
14575	Triptane or 2,2,3-Trimethyl butane $(CH_3)_2CHC(CH_3)_3$	100.20		80.9	−24.2	0.6901$^{20/4}$	1.3894^{20}	al, eth, ace, bz	B1^4, 410
14576	Trisiloxane, 1,5-dichloro-1,1,3,3,5,5-hexamethyl $ClSi(CH_3)_2OSi(CH_3)_2OSi(CH_3)_2Cl$	277.37		184	−53	1.018$^{20/4}$			B4^3, 1884
14577	Trisiloxane, 1,5-dihydroxy-1,1,3,3,5,5-hexamethyl $HOSi(CH_3)_2OSi(CH_3)_2OSi(CH_3)_2OH$	240.48			−23	0.9950^{20}	1.4090$^{20/D}$		KHOC, 653
14578	Trisiloxane, 1,1,1,3,5,5,5-heptamethyl $(CH_3)_3SiOSiH(CH_3)OSi(CH_3)_3$	222.51				0.8194$^{20/4}$	1.3818^{20}		B4^3, 1874
14579	Trisiloxane, octamethyl $(CH_3)_3SiOSi(CH_3)_2OSi(CH_3)_3$	236.53		153^{774}, 50-2^{17}	−80	0.8200$^{20/4}$	1.3840^{20}	bz, peth	B4^3, 1879
14580	Trisiloxane, 1,1,3,5,5-pentamethyl-1,3,5-triphenyl $C_6H_5Si(CH_3)_2OSi(CH_3)(C_6H_5)OSi(CH_3)_2C_6H_5$	422.75		169$^{0.7}$		1.0227$^{20/4}$	1.5280^{20}	bz, lig	Am70, 1116
14581	Tristearin or Glycerol tristearate $C_{17}H_{35}CO_2CH(CH_2O_2CC_{17}H_{35})_2$	891.50	cr (eth, peth)		73	0.8559$^{90/4}$	1.4399^{80}	ace, chl, CS_2	B2^4, 1233
14582	Tristearylamine or Trioctadecylamine $(C_{18}H_{37})_3N$	774.48			54.6			eth, bz, chl	B4^4, 829
14583	Tritetracontane $C_{43}H_{88}$ $(CH_3(CH_2)_{41}CH_3$	605.17		332^3	85.5	0.7812$^{90/4}$	1.4340^{90}		B1^3, 593
14584	Tris-(4-tolyl) amine $(4\text{-}CH_3C_6H_4)_3N$	287.40	cr (aa)		117			eth, ace, bz, chl	B12^3, 2034
14585	3,3′,4″-Tritolylamine $(3\text{-}CH_3C_6H_4)_2N(C_6H_4CH_3\text{-}4)$	287.40	nd (al)		89.90			al	B12^1, 415

No.	Name, Synonyms, and Formula	Mol. wt.	Color, crystalline form, specific rotation and λ_{max} (log ϵ)	b.p. °C	m.p. °C	Density	n_D	Solubility	Ref.
14586	Bis-(tribromomethyl) trisulfide $(CBr_3)_2S_3$	599.63	rh pr (eth)	d	125d	....	....	bz, chl, peth	$B3^2$, 107
14587	1,3,5-Trithiane or Trithioformaldehyde $C_3H_6S_3$	138.26	hex (bz), pr (w), nd (al)	sub	220	$1.6374^{24/4}$	....	bz	$B19^4$, 4711
14588	1,3,5-Trithiane, 2,4,6-trimethyl-(α-form) or Trithioacetaldehyde $2,4,6\text{-}(CH_3)_3(C_3H_3S_3)$	180.34	mcl (al, ace)	245-8	101	....	....	al, eth, ace, bz, chl	$B19^4$, 4719
14589	1,3,5-Trithiane, 2,4,6-trimethyl-(β-form) $1,3,5\text{-}(CH_3)_3(C_3H_3S_3)$	180.34	rh nd (ace)	245-8	126-7	....	....	al, eth, ace, bz, chl	$B19^4$, 4718
14590	1,3,5-Trithiane-2,4,6-trimethyl-2,4-6-triphenyl $C_{24}H_{24}S_3$	408.63	nd (al)	....	122	....	....	eth, ace, chl	$B19^4$, 4775
14591	1,3,5-Trithiane, 2,4,6-triphenyl (α-form) $2,4,6(C_6H_5)_3C_3H_3S_3$	366.55	nd (bz-al)	d	167	....	....	bz, chl	$B19^4$, 4774
14592	1,3,5-Trithiane, 2,4,6-triphenyl-(β-form) $2,4,6\text{-}(C_6H_5)_3C_3H_3S_3$	366.55	cr (bz)	....	229-30	....	....	bz	$B19^4$, 4773
14593	α-Tritisterol $C_{30}H_{50}O$	426.73	nd (MeOH-ace), $[\alpha]^{20}{}_D$ + 54.3 (al)	....	114-5	....	....	al, peth, chl	$B6^3$, 2911
14594	β-Tritisterol $C_{30}H_{50}O$	426.73	nd (MeOH), $[\alpha]_D$ + 49.2 (al)	....	97	....	....	al	H20, 424
14595	Tropacocaine or Benzoyl-ψ-tropeine $C_{15}H_{19}NO_2$	245.32	pl or tab	d	49	$1.0426^{100/4}$	1.5080^{100}	al, eth, bz, chl, peth	$B21^4$, 174
14596	Tropacocaine hydrochloride $C_{15}H_{19}NO_2 . HCl$	281.78	pl (al)	....	283d	....	....	w	$B21^4$, 174
14597	Tropane or 2,3-Dihydro-8-methylnortropidine $C_8H_{15}N$	125.21	....	167	....	$0.9259^{15/15}$	1.4732^{20}	....	$B20^4$, 1963
14598	3-Tropanol $C_8H_{15}NO$	141.21	....	229	64	....	....	w, al, eth, chl	$B21^4$, 168
14599	3-Tropanone $C_8H_{13}NO$	139.20	....	$224\text{-}5^{714}$, 113^{25}	42-4	$1.9872^{100/4}$	1.4598^{100}	al, eth, ace, bz	$B21^4$, 3299
14600	Tropeine, benzoyl $C_{15}H_{19}NO_2$	245.32	cr (eth)	175-70	41-2	....	....	w, al, eth	$B21^4$, 173
14601	Tropeine, benzoyl, dihydrate $C_{15}H_{19}NO_2 . 2H_2O$	281.35	pl (w)	....	58	....	....	w, al, eth	B21, 19
14602	Tropine or 3-Tropanol $C_8H_{15}NO$	141.21	hyg pl (eth)	229	64	....	....	w, al, eth, chl	$B21^2$, 17
14603	Tropinic acid [(d)] $C_8H_{13}NO_4$	187.20	cr (w, dil al), $[\alpha]'_D$ + 14.8 (w)	....	253d	....	....	w	B22, 123
14604	Tropinic acid [(dl)] $C_8H_{13}NO_4$	187.20	nd (dil al)	....	251d	....	....	w	B22, 124
14605	Tropinic acid [(l)] $C_8H_{13}NO_4$	187.20	cr (a), $[\alpha]^{20}{}_D$ −14.8 (w)	....	243	....	....	w	B22, 124
14606	Tropinone or 3-Tropanone $C_8H_{13}NO$	139.20	lo nd (peth)	$224\text{-}5^{714}$, 113^{25}	42-4	$1.9872^{100/4}$	1.4598^{100}	al, eth, ace, bz	$B21^2$, 225
14607	Truxane or Di-indene $C_{18}H_{16}$	232.33	pl (peth)	....	116	....	....	al, eth, aa	$B5^3$, 2253
14608	Truxane, tetraphenyl or Triphenyl allene dimer $(C_6H_5)_4C_{18}H_{12}$	536.72	lf (bz-al)	....	210	....	....	chl	$B5^3$, 2790
14609	α-Truxilline or Cocamine $C_{36}H_{46}N_2O_8$	602.77	amor pw	....	[ca] 80	....	....	al, eth, bz, chl	B22, 202
14610	Tryptophan [(D)] or α-Amino-β-indolyl propionic acid $C_{11}H_{12}N_2O_2$	204.23	pl (dil al) $[\alpha]^{20}{}_D$ + 33 (w)	....	281-2	....	....	al	$B22^4$, 6765
14611	Tryptophan [(DL)] $C_{11}H_{12}N_2O_2$	204.23	pl (50% al)	....	282, (293)	....	....	....	$B22^4$, 6768
14612	Tryptophan [(L)] $C_{11}H_{12}N_2O_2$	204.23	lf or pl (dil al), $[\alpha]^{20}{}_D$ + 6.1 (1N NaOH, p=11)	....	290-2d	....	....	w	$B22^4$, 6765
14613	Tryptophan, [N]-acetyl-[(DL)] $C_{13}H_{14}N_2O_3$	246.27	pl (dil al)	....	206-7	....	....	al, eth	$B22^4$, 6782

No.	Name, Synonyms, and Formula	Mol. wt.	Color, crystalline form, specific rotation and λ_{max} (log ε)	b.p. °C	m.p. °C	Density	n_D	Solubility	Ref.
14614	Tryptophan, [N] -acetyl-[(L)] $C_{13}H_{14}N_2O_3$	246.27	nd (dil MeOH), $[\alpha]^{15}{}_D$ +25 (95% al, c=1)		189-90			w, al	B22[4], 6781
14615	Tryptophan, [N] -methyl [(DL)] $C_{12}H_{14}N_2O_2$	218.26	nd (dil al)		297d			al	B22[4], 6776
14616	Tryptophan, [N] -methyl [(L)] or Abrin $C_{12}H_{14}N_2O_2$	218.26	pr (w) $[\alpha]^{21}{}_D$ +44.4 (dil HCl)		295d			al	B22[4], 6776
14617	Tuduranine $C_{18}H_{19}NO_3$	297.35	nd (eth), $[\alpha]^{20}{}_D$ −127.5 (al)		204			al, eth, ace	B21[4], 2643
14618	Turanose or 3-*O*-α-D-glucoside-D-fructose $C_{12}H_{22}O_{11}$	342.30	pr (w-al, MeOH) $[\alpha]^{20}{}_D$ +27.2→ +75.8 (w,c=4,mut)		168			w, al	B17[4], 3092
14619	Tyramine or [p] -(β -aminoethyl)phenol 4-$(H_2NCH_2CH_2)C_6H_4OH$	137.18	pl or nd (bz), cr (al) nd (w)	205-7[25], 165-7[2]	164-5			al, xyl	B13[3], 1637
14620	Tyrosine [(D)] or 2-Amino-3-(4-hydroxyphenyl)propionic acid $(4\text{-}HOC_6H_4)CH_2CH(NH_2)CO_2H$	181.19	cr (w), $[\alpha]^{25}{}_D$ +10.3 (1N HCl,c=4)		310-4d			dil HCl	B14[3], 1504
14621	Tyrosine [(DL)] $(4\text{-}HOC_6H_4)CH_2CH(NH_2)CO_2H$	181.19	nd		340d				B14[3], 1506
14622	Tyrosine [(L)] $(4\text{-}HOC_6H_4)CH_2CH(NH_2)CO_2H$	181.19	nd (w), $[\alpha]^{22}{}_D$ −10.6 (1N HCl,c=4)	sub	342-4d (sealed tube)				B14[3], 1504
14623	Tyrosine amide [(L)] $(4\text{-}HOC_6H_4)CH_2CH(NH_2)CONH_2$	180.21	pl or pr (al), $[\alpha]^{20}{}_D$ +19.5 (w)		153-4			w, al, MeOH	B14[3], 1511
14624	Tyrosine-3-bromo [(L)] $(3\text{-}Br\text{-}4\text{-}HOC_6H_3)CH_2CH(NH_2)CO_2H$	260.09	cr (w + 1), nd (w + 2)		246-9d				B14[2], 377
14625	Tyrosine-3,5-dibromo [(L)] $(3,5\text{-}Br_2\text{-}4\text{-}HOC_6H_2)CH_2CH(NH_2)CO_2H$	338.98	nd or pl (w + 2), $[\alpha]^{20}{}_D$ −5.5 (1N HCl, c=5)		245 (anh)				B14[3], 1561
14626	Tyrosine, ethyl ester [(L)] $(4\text{-}HOC_6H_4)CH_2CH(NH_2)CO_2C_2H_5$	209.25	pr (AcOEt), $[\alpha]^{20}{}_D$ +20.4 (MeOH)		108-9			al, bz, AcOEt, MeOH	B14[3], 1510
14627	Tyrosine, [N] -methyl-[(L)] $(4\text{-}HOC_6H_4)CH_2CH(NHCH_3)CO_2H$	195.22	nd $[\alpha]^{21}{}_D$ +19.8 (dil HCl)		257, (d 280)				B14[3], 1513
14628	Tyrosine, methyl ester $(4\text{-}HOC_6H_4)CH_2CH(NH_2)CO_2CH_3$	195.22	pr (AcOEt), $[\alpha]^{20}{}_D$ +25.7 (MeOH)		136-7			al, MeOH	B14[3], 1510

No.	Name, Synonyms, and Formula	Mol. wt.	Color, crystalline form, specific rotation and λ_{max} (log ε)	b.p. °C	m.p. °C	Density	n_D	Solubility	Ref.
14629	Umbellatine $C_{21}H_{21}NO_8$	415.40	ye nd		206-7			al	C48, 10034
14630	1,10-Undecadiyne $HC\equiv C(CH_2)_7C\equiv CH$	148.25		83^{12}	−17	$0.8182^{21/4}$	1.453^{21}	ace, bz	$B1^2$, 248
14631	Undecanal or Undecylaldehyde $CH_3(CH_2)_9CHO$	170.30		117^{18}	−4	$0.8251^{23/4}$	1.4520^{20}	al, eth	$B1^4$, 3374
14632	Undecanal oxime $CH_3(CH_2)_9CH{=}NOH$	185.31	wh nd (dil MeOH)		72			w, al, eth	$B1^4$, 3374
14633	Undecanal, 2-methyl $CH_3(CH_2)_8CH(CH_3)CHO$	184.32		114^{10}		$0.830^{15/4}$	1.4321^{20}	al, eth	$B1^4$, 3383
14634	Undecane $C_{11}H_{24}$	156.31		196, 75^{10}	−25.6	$0.7402^{20/4}$	1.4398^{20}	al	$B1^4$, 487
14635	Undecane, 1-amino or Undecylamine $CH_3(CH_2)_9CH_2NH_2$	171.33	cr (eth, al)	242, 112.3^{10}	17	$0.7979^{20/4}$	1.4398^{20}	al	$B4^4$, 792
14636	Undecane, 1-bromo-11-fluoro $F(CH_2)_{11}Br$	253.20		$95^{0.6}$			1.4518^{25}	al, eth	$B1^4$, 490
14637	Undecanoic acid or Undecylic acid $CH_3(CH_2)_9CO_2H$	186.29	cr (ace)	280, 164^{15}	28.6	$0.8907^{20/4}$	1.4294^{45}	al, eth, ace, bz, chl	$B2^4$, 1068
14638	Undecanoic acid, 2-bromo $CH_3(CH_2)_8CHBrCO_2H$	265.19		$178\text{-}83^{14}$	10				$B2^4$, 1073
14639	Undecanoic acid, 10-bromo $CH_3CHBr(CH_2)_8CO_2H$	265.19	cr (peth-eth, bz, ace)		35.7			al	$B2^3$, 861
14640	Undecanoic acid, 11-bromo $BrCH_2(CH_2)CO_2H$	265.19	nd (lig)	188^{18}	57			al, eth, ace, bz	$B2^4$, 1073
14641	Undecanoic acid, 10,11-dibromo $BrCH_2CH(Br)(CH_2)_8CO_2H$	344.09	cr		38			al	$B2^4$, 1075
14642	Undecanoic acid, ethyl ester or Ethyl undecanoate $CH_3(CH_2)_9CO_2C_2H_5$	214.35		131^{14}	−15	$0.8633^{20/4}$	1.4285^{20}	al, eth, ace, bz	$B2^3$, 858
14643	Undecanoic acid, 11-fluoro $F(CH_2)_{10}CO_2H$	204.28		$113\text{-}5^{0.25}$	36			al, eth, lig	$B2^4$, 1071
14644	Undecanoic acid, 4-hydroxy-8-lactone $C_7H_{15}CHCH_2CH_2CO$ (—O—)	184.28		286, 162^{13}		$0.9494^{20/4}$	1.4512^{20}	al	$B17^4$, 4260
14645	Undecanoic acid -α-lactone or 2-Undecalactone $C_{11}H_{20}O_2$	184.28		286, 162^{13}		$0.9494^{20/4}$	1.4512^{20}	al	$B17^2$, 284
14646	1-Undecanol or 1-Undecylalcohol $CH_3(CH_2)_9CH_2OH$	172.31		243, 131^{15}	19	$0.8298^{20/4}$	1.4392^{20}	al, eth	$B1^4$, 1835
14647	1-Undecanol, 2-methyl $CH_3(CH_2)_8CH(CH_3)CH_2OH$	186.34		$129\text{-}31^{12}$		$0.8300^{15/4}$	1.4382^{20}	al, eth	$B1^4$, 1855
14648	2-Undecanol [(d)] $CH_3(CH_2)_8CHOHCH_3$	172.31	(bz), $[\alpha]^{20}_{D}$ + 10.3 (bz)	128^{20}	12	$0.8270^{20/4}$	1.4369^{20}	al, eth	$B1^3$, 1774
14649	2-Undecanol [(dl)] $CH_3(CH_2)_8CHOHCH_3$	172.31		228, 119^{12}	0	0.8268^{19}	1.4369^{20}	al, eth	$B1^4$, 1838
14650	2-Undecanol -[(l)] $CH_3(CH_2)_8CHOHCH_3$	172.31	$[\alpha]_D$ −0.02	231-3		0.8302^{20}	1.4381^{20}	al, eth	$B1^3$, 1774
14651	3-Undecanol *(l)* $CH_3(CH_2)_7CHOHCH_2CH_3$	172.31	$[\alpha]^{20}_{D}$ −6.2 (al)	229	17	$0.8295^{20/4}$	1.4367^{20}	al, eth, bz	$B1^2$, 462
14652	5-Undecanol $C_6H_{13}CHOH(C_4H_9)$	172.31		229, 107^{12}	−3.5	$0.8292^{20/4}$	1.4354^{24}	al, eth	$B1^4$, 1839
14653	6-Undecanol $C_5H_{11}CH(OH(C_5H_{11})$	172.31	cr (ace)	228, $117\text{-}8^{16}$	25	$0.8334^{20/4}$	1.4374^{20}	al, ace	$B1^4$, 1839
14654	1-Undecanethiol or Undecyl mercaptan $CH_3(CH_2)_9CH_2SH$	188.37		257.4	−3	0.8448^{20}	1.4585^{20}		$B1^4$, 1838 2(1.1000)
14655	2-Undecanone $CH_3(CH_2)_8COCH_3$	170.30		231-2, 106^{12}	15	$0.8250^{20/4}$	1.4291^{20}	al, eth, ace, bz, chl	$B1^4$, 3374
14656	3-Undecanone $CH_3(CH_2)_7COCH_2CH_3$	170.30		227, $104\text{-}6^{11}$	12	$0.8272^{20/4}$	1.4296^{20}	al, eth	$B1^4$, 3375
14657	4-Undecanone $CH_3(CH_2)_6CO(CH_2)_2CH_3$	170.30		106^{13}	4-5	$0.8274^{25/4}$	1.4248^{24}	al, eth	$B1^4$, 3375
14658	4-Undecanone, 7-ethyl,2-methyl $C_4H_9CH(C_2H_5)CH_2CH_2COCH_2CH(CH_3)_2$	212.38		252-3, $101\text{-}3^{4}$		$0.8362^{20/4}$	1.4370^{20}	al, eth, ace, bz	$B1^3$, 2920
14659	5-Undecanone $CH_3(CH_2)_5CO(CH_2)_3CH_3$	170.30		227, $105\text{-}6^{12}$	2	$0.8278^{19/4}$	1.4275^{18}	al, eth	$B1^4$, 3375
14660	6-Undecanone or *n*-Caprone $(C_5H_{11})_2CO$	170.30		228, 145.7^{29}	14-5	$0.8308^{20/4}$	1.4270^{20}	al, eth	$B1^4$, 3376
14661	Undecanonitrile $CH_3(CH_2)_9CN$	167.29		253, 143^{22}		$0.8254^{30/4}$	1.4293^{30}	al, eth	$B3^2$, 860

PHYSICAL CONSTANTS OF ORGANIC COMPOUNDS (Continued)

No.	Name, Synonyms, and Formula	Mol. wt.	Color, crystalline form, specific rotation and λ_{max} (log ε)	b.p. °C	m.p. °C	Density	n_D	Solubility	Ref.
14662	Undecasiloxane, tetra cosamethyl $(CH_3)_3SiO[Si(CH_3)_2O]_9Si(CH_3)_3$	829.77		322.8, $202^{4.7}$		0.930^{20}	1.3994^{20}	bz	$B4^3$, 1881
14663	1-Undecene $CH_3(CH_2)_8CH{=}CH_2$	154.30		192.7, 72^{10}	−49.2	$0.7503^{20/4}$	1.4261^{20}	eth, chl, lig	$B1^4$, 910
14664	2-Undecene *(trans)* $CH_3(CH_2)_7CH{=}CHCH_3$	154.30		192-3, 75^{10}	−48.3	$0.7528^{20/4}$	1.4292^{20}		$B1^4$, 911
14665	5-Undecene *(trans)* $CH_3(CH_2)_4CH{=}CH(CH_2)_3CH_3$	154.30		192, 73^{10}	−61.1	$0.7497^{20/4}$	1.4285^{20}	eth, chl, lig	$B1^4$, 911
14666	9-Undecenoic acid *(trans)* $CH_3CH{=}CH(CH_2)_7CO_2H$	184.28		273, $121\text{-}3^{0.7}$	19	$0.9119^{25/0}$	1.4519^{20}	eth	$B2^4$, 1616
14667	10-Undecenoic acid $CH_2{=}CH(CH_2)_8CO_2H$	184.28	cr	275, 165^{13}	24-5	$0.9072^{24/4}$	1.4486^{24}	al, eth	$B2^4$, 1612
14668	10- Undecenoic acid, ethyl ester or Ethyl-10-undecenoate $CH_2{=}CH(CH_2)_8CO_2C_2H_5$	212.33		264-5, 131.5^{16}	−38	0.8827^{15}	1.4449^{23}	al, eth	$B2^4$, 1614
14669	10- Undecenoic acid, methyl ester $H_2C{=}CH(CH_2)_8CO_2CH_3$	198.31		248, 124^{10}	−27.5	0.889^{15}	1.4393^{20}	al, eth, aa	$B2^4$, 1613
14670	10 -Undecenoyl chloride $CH_2{=}CH(CH_2)_8COCl$	202.72		$120\text{-}2^{10}$		$0.944^{20/4}$	1.454^{20}		$B2^4$, 1615
14671	10-Undecen-1-ol or Undecenyl alchohol $CH_2{=}CH(CH_2)_8CH_2OH$	170.30		250, $132\text{-}3^{15}$	−2	$0.8495^{15/4}$	1.4500^{20}	al, eth	$B1^4$, 2194
14672	1-Undecene-3-yne $CH_3(CH_2)_6C{\equiv}CCH{=}CH_2$	150.26		74^{9}		$0.7962^{20/4}$	1.4606^{20}		$B1^3$, 1054
14673	1- Undecyne $CH_3(CH_2)_8C{\equiv}CH$	152.28		196, 73.2^{10}	−25	$0.7728^{20/4}$	1.4306^{20}	al, eth, ace, bz	$B1^4$, 1064
14674	5-Undecyne $CH_3(CH_2)_4C{\equiv}C(CH_2)_3CH_3$	152.28		198, 78.3^{10}	−74.1	$0.7753^{20/4}$	1.4369^{20}		$B1^4$, 1064
14675	9-Undecynoic acid $CH_3C{\equiv}C(CH_2)_8CO_2H$	182.26	pl (dil al)	170^{15}	61			w, al, eth	$B2^3$, 1471
14676	Uneicosane $CH_3(CH_2)_{19}CH_3$	296.58	cr (w)	358.5, 203^{10}	40.5	$0.7917^{20/4}$	1.4441^{20}	peth	$B1^4$, 569
14677	Uneicosane, 1-cyclopentyl $C_5H_9C_{21}H_{43}$	364.70		420	42	0.8286^{20}	1.4602^{20}		$B5^4$, 200 (3.1031)
14678	Uneicosane, 11-decyl $(C_{10}H_{21})_2CH$	436.85		282^{10}	10.0	$0.8116^{20/4}$	1.4540^{20}		$B1^4$, 595
14679	Uneicosane, 11-(2,2-dimethyl propyl) $(CH_3)_3CCH_2CH(C_{10}H_{21})_2$	366.71		238^{10}	−21	$0.8031^{20/4}$	1.4491^{20}		$B1^4$, 585
14680	Uneicosane, 11-phenyl $C_6H_5CH(C_{10}H_{21})_2$	372.68		$205^{1.0}$	20.8	$0.8531^{20/4}$	1.4788^{20}		$B5^4$, 1226
14681	Uneicosanoic acid $CH_3(CH_2)_{19}CO_2H$	326.56	nd (ace)		82			eth, bz, chl	$B2^4$, 1285
14682	10-Uneicosene, 11-Phenyl $CH_3(CH_2)_8CH{=}C(C_6H_5)(CH_2)_9CH_3$	370.66		203^{1}	48.2	$0.8638^{20/4}$	1.4922^{20}		$B5^4$, 2057
14683	Untriacontane $CH_3(CH_2)_{29}CH_3$	436.85	lf (AcOEt)	458, 302^{15}	67.9	$0.781^{68/4}$	1.4278^{90}	peth	$B1^4$, 594
14684	Untriacontanoic acid or Melissic acid $CH_3(CH_2)_{29}CO_2H$	466.83	sc or nd (al or ace)		93			bz	$B2^2$, 382
14685	16-Untriacontanone or Palmitone $(C_{15}H_{31})_2CO$	450.83	lf (al)		83	$0.7947^{91/4}$	1.4297^{94}	eth	$B1^4$, 3413
14686	Uracil $C_4H_4N_2O_2$	112.09	nd (w)		338			al, eth	$B24^2$, 168
14687	Uracil, 5-amino $C_4H_5N_3O_2$	127.10	nd (w)		d				$B24^2$, 266
14688	Uracil, 5,6 dihydro or β-Lactylurea $C_4H_6N_2O_2$	114.10	nd (w)		275-6			w, al, chl	$B24^2$, 140
14689	Uracil, 5,6-dihydro-5-methyl or Dihydrothymine $C_5H_8N_2O_2$	128.13	(w or al)		264-5			w	$B24^1$, 306
14690	Urcil, 5-hydroxy or Isobarbituric acid $C_4H_4N_2O_3$	128.09	pr (w)		d>300			w	$B24^1$, 408
14691	Uracil, 1-methyl $C_5H_6N_2O_2$	126.11	pr (w)		179			w, al	$B24^2$, 170
14692	Uracil, 3-methyl $C_5H_6N_2O_2$	126.11	pr (al) nd (w)		232			w, al	$B24^2$, 170
14693	Uracil, 5-methyl or Thymine $C_5H_6N_2O_2$	126.11	nd (al), pl (w)	sub	326				$B24^2$, 183
14694	Uracil, 6-methyl $C_5H_6N_2O_2$	126.11	oct, pr or nd (w, al)		270-80d			w, al	$B24^2$, 182

PHYSICAL CONSTANTS OF ORGANIC COMPOUNDS (Continued)

No.	Name, Synonyms, and Formula	Mol. wt.	Color, crystalline form, specific rotation and λ_{max} (log ε)	b.p. °C	m.p. °C	Density	n_D	Solubility	Ref.
14695	Uracil, 5-nitro $C_4H_3N_3O_4$	157.09	gold nd (al)		exp> 300			al	$B24^{2}$, 171
14696	Uracil, 2-thio or 2-Thiouracil $C_4H_4N_2OS$	128.15	pr (w, al)		>340d				$B24^{2}$, 171
14697	6-Uracil carboxylic acid or Orotic acid $C_5H_4N_2O_4$	156.10	pr (w + 1)		347d				$B25^{2}$, 249
14698	6-Uracil carboxylic acid, monohydrate $C_5H_4N_2O_4 \cdot H_2O$	174.11	pr (w)		125-30d				$B25^{2}$, 249
14699	Uranin or Fluorescein, sodium salt $C_{20}H_{10}O_5Na_2$	376.28	ye pw					w, al	$B19^{4}$, 2909
14700	Urea or Carbamide H_2NCONH_2	60.06	tetr pr (al)	d	135	$1.3230^{20/4}$	1.484	w, al, Py	$B3^{4}$, 94
14701	Urea, 1-acetyl $CH_3CONHCONH_2$	102.09		sub 180-190	218			al	$B3^{3}$, 119
14702	Urea, 1-acetyl-3-methyl $CH_3CONHCONHCH_3$	116.12	tcl (w, al), pr (w)	d	180-1			w	$B4^{4}$, 207
14703	Urea, 1-allyl $CH_2{=}CHCH_2NHCONH_2$	100.12	nd (al)		85			w, al	$B4^{4}$, 1070
14704	Urea, 1-allyl-3-phenyl $CH_2{=}CHCH_2NHCONHC_6H_5$	176.22	nd (bz)		115-6			al, bz	$B12^{3}$, 765
14705	Urea, 1-(4-amino benzene sulfonyl) or 1-Sulfonyl urea $4\text{-}H_2NC_6H_4SO_2NHCONH_2$	215.23	(w)	sub 320	146-8d				$B14^{3}$, 1968
14706	Urea, 1-benzoyl $C_6H_5CONHCONH_2$	164.16	fl or lf (al)		214-5				$B9^{3}$, 1115
14707	Urea, 1-benzyl or Phthalamic acid $C_6H_5CH_2NHCONH_2$	150.18	nd (al)	d 200	149			al, ace	$B12^{3}$, 2271
14708	Urea, 1-(2-bromo-2-ethyl-butanoyl) or Adalin. Uradal $H_2NCONH[COCBr(C_2H_5)_2]$	237.10	rh (dil al)		118	1.544^{25}		ace, bz	$B3^{4}$, 117
14710	Urea, 1-(2-bromo-3-methyl butanoyl) or Adabine Bromural $H_2NCONH[COCHBrCH(CH_3)_2]$	223.07	nd or lf (to)	sub	154 (160)	1.56^{15}		al, eth, ace, bz	$B3^{3}$, 123
14711	Urea, 1-(2-bromo phenyl) $2\text{-}BrC_6H_4NHCONH_2$	215.05	nd (al)		202			bz, chl	$B12^{3}$, 1419
14712	Urea, 1-(3-bromophenyl) $3\text{-}BrC_6H_4NHCONH_2$	215.05	nd (bz, al)		164-5			al, eth	B12, 634
14713	Urea, 1-(4-bromophenyl) $4\text{-}BrC_6H_4NHCONH_2$	215.05	nd (bz, al)	d 260	225-7			al, eth, bz, aa	$B12^{3}$, 1452
14714	Urea, 1-butyl $C_4H_9NHCONH_2$	116.16	ta (w), nd (bz)		96			al, eth	$B4^{4}$, 578
14715	Urea, 1-isobutyl $H_2NCONH\text{-}i\text{-}C_4H_9$	116.16	pr (w), nd (ace)		141				$B4^{4}$, 648
14716	Urea, 1-*sec*-butyl $C_2H_5CH(CH_3)NHCONH_2$	116.16	nd $[\alpha]^{20}{}_{D}$ + 24.1 (al, c=1.5)		166-(d) 169-70-(dl)			al, eth	$B4^{4}$, 622
14717	Urea, 1-*tert*-butyl $(CH_3)_3CNHCONH_2$	116.16	nd (w dil al)	sub >100	191d			w, al	$B4^{4}$, 665
14718	Urea, 1-(2-chlorophenyl) $(2\text{-}ClC_6H_4)NHCONH_2$	170.60	pr (w)		152			w, al, ace	$B12^{3}$, 1293
14719	Urea, 3-(4-chlorophenyl)-1,1-dimethyl or CNV Weed Killer $3\text{-}(4\text{-}ClC_6H_4)NHCON(CH_3)_2$	198.65	wh pl (MeOH)		170-1				C51, 16534
14720	Urea, 1,3-diacetyl $(CH_3CONH)_2CO$	144.13	nd (aa, 50% al)	sub 179-80d	154-5			eth, ace, bz	$B3^{3}$, 121
14721	Urea, 1,3-dibutyl $(C_4H_9NH)_2CO$	172.27			72-4				$B4^{4}$, 578
14722	Urea, 1,3-bis-(2,4-dinitro phenyl) $(2,4(NO_2)_2C_6H_3NH)_2CO$	392.24	lf or pr (al)		204				$B12^{3}$, 1690
14723	Urea, 1,3-bis-(2-ethoxyphenyl) $(2\text{-}C_2H_5OC_6H_4NH)_2CO$	300.36	pr (dil al)		125			al	$B13^{3}$, 816
14724	Urea, 1,3-bis-(4-ethoxyphenyl) $(4\text{-}C_2H_5OC_6H_4NH)_2CO$	300.36	nd (aa), pr (abs al)		255-6			al	$B13^{3}$, 1112
14725	Urea, 1,1-diethyl $H_2NCON(C_2H_5)_2$	116.16	pl nd (eth)	$94\text{-}6^{0.02}$	75			w, al, eth, bz, lig	$B4^{4}$, 380
14726	Urea, 1,3-diethyl $(C_2H_5NH)_2CO$	116.16	ta (lig), dlq nd (al)	263	112.5	1.0415	1.4616^{40}	w, al, eth	$B4^{4}$, 370
14727	Urea, 1,3-diethyl-1,3-diphenyl $[C_6H_5N(C_2H_5)]_2CO$	268.36	(al, w)		79			al	$B12^{3}$, 882

No.	Name, Synonyms, and Formula	Mol. wt.	Color, crystalline form, specific rotation and λ_{max} (log ε)	b.p. °C	m.p. °C	Density	n_D	Solubility	Ref.
14728	Urea, 1,3-bis-(hydroxymethyl) or N,N′-Dimethylol urea $(HOCH_2NH)_2CO$	120.11	pr (abs al), pl (w-al)	d 260	126	1.49^{25}	...	w, al	B3^4, 107
14729	Urea, 1,3-bis-(1-hydroxy-2,2,2-trichloro ethyl) or Dichloral urea. Cragherbicide $[CCl_3CH(OH)NH]_2CO$	354.83	...	...	196	...	...	al, ace	B3^3, 116
14730	Urea, 1,1-dimethyl $H_2NCON(CH_3)_2$	88.11	mcl pr (al, chl)	...	182	1.255	...	w	B4^4, 224
14731	Urea, 1,3-dimethyl $(CH_3NH)_2CO$	88.11	rh bipym (chl-eth)	268-70	108	1.142	...	w, al	B4^4, 207
14732	Urea, 1,3-dimethyl-1,3-diphenyl $[C_6H_5N(CH_3)]_2CO$	240.30	pl (al)	350	121	...	...	w, al, ace, chl	B12^3, 875
14733	Urea, 1,1-di(β-naphthyl) $(\beta\text{-}C_{10}H_7)_2NCONH_2$	312.37	nd (al)	...	192-3	...	...	...	B12, 1297
14734	Urea, 1,3-di(α-naphthyl) $(\alpha\text{-}C_{10}H_7NH)_2CO$	312.37	nd (aa, Py)	sub	296	...	...	Py	B12^3, 2926
14735	Urea, 1,3-di(β-naphthyl) $(\beta\text{-}C_{10}H_7NH)_2CO$	312.37	nd (ace, aa)	...	310	...	...	aa	B12^3, 3033
14736	Urea, 1,1-diphenyl or Acardite $(C_6H_5)_2NCONH_2$	212.25	ta (al)	d	189	1.276	...	al, eth, chl	B12^3, 893
14737	Urea, 1,3-diphenyl or Carbanilide $(C_6H_5NH)_2CO$	212.25	rh bipym, pr (al)	262d	238	1.239	...	eth, aa	B12^3, 767
14738	Urea, 1,3-diphenyl-1-methyl $C_6H_5NHCON(CH_3)C_6H_5$	226.28	nd (al, xyl), cr (lig)	203-5d	106	...	...	eth, bz, chl	B12^3 878
14739	Urea, 1-(3-ethoxyphenyl) $(3\text{-}C_2H_5OC_6H_4)NHCONH_2$	180.21	nd	...	112	...	...	...	B13^3, 960
14740	Urea, 1-(4-ethoxyphenyl) or Dulcin $(4\text{-}C_2H_5OC_6H_4)NHCONH_2$	180.21	lf (dil al), pl (w)	d	173-4	...	...	al, AcOEt	B13^3, 1109
14741	Urea, 1-ethyl $C_2H_5NHCONH_2$	88.11	nd (bz, al-eth)	d	92-3	...	...	w, al, eth, bz, chl	B4^4, 369
14742	Urea, 1-ethyl-1-phenyl $C_6H_5N(C_2H_5)CONH_2$	164.21	ta (peth), cr (aa)	...	62-3	...	...	w, eth, ace	B12^3, 882
14743	Urea, 1-ethyl-3-phenyl $C_2H_5NHCONHC_6H_5$	164.21	nd (dil al)	...	104	...	...	al	B12^3, 761
14744	Urea hydrochloride $(H_2N)_2CO.HCl$	96.52	...	...	145d	...	...	...	B3^4, 102
14745	Urea, 1-hydroxy $HONHCONH_2$	76.06	nd (al)	d	141	...	...	w	B3^4, 170
14746	Urea, 1-hydroxymethyl or Methylol urea $HOCH_2NHCONH_2$	90.08	pr (al)	...	111	...	...	w, al, aa, MeOH	B3^4, 105
14747	Urea, 1-(2-iodo-3-methyl butanoyl) or α-Iodo isovaleryl urea $H_2NCONHCOCHICH(CH_3)_2$	270.07	lf (al)	...	180-1	...	...	al	B3^1, 29
14748	Urea, 1-(4-methoxyphenyl) $(4\text{-}CH_3OC_6H_4)NHCONH_2$	166.18	pl (w)	...	168	...	...	eth	B13^3, 1096
14749	Urea, 1-methyl $CH_3NHCONH_2$	74.08	rh pr (w, al)	d	103	...	...	w, al	B4^4, 205
14750	Urea, 1-methyl-1-nitroso $H_2NCON(NO)CH_3$	103.08	col or yesh pl (eth)	...	123-4d	...	...	al, eth, ace, bz, chl	B4^4, 272
14751	Urea, 1-methyl-3-phenyl $C_6H_5NHCONHCH_3$	150.18	...	...	151	...	...	w, al, bz	B12^3, 761
14752	Urea, 1-(2-methyl-2-butyl) $H_2NCONHC(CH_3)_2C_2H_5$	130.19	mcl (w)	...	162	...	...	...	B4^4, 695
14753	Urea, 1-α-naphthyl $\alpha\text{-}C_{10}H_7NHCONH_2$	186.21	nd (al)	...	219-20	...	...	al, eth	B12, 1253
14754	Urea, 1-(β-naphthyl) $\beta\text{-}C_{10}H_7NHCONH_2$	186.21	nd (al)	...	219	...	...	al	B12^3, 3029
14755	Urea nitrate $H_2NCONH_2.HNO_3$	123.07	mcl lf (w)	...	157d	$1.690^{20/4}$	...	al	B3^4, 102
14756	Urea, 1-nitro $H_2NCONHNO_2$	105.05	nd (al-peth), lf or pr(al)	exp	158-9	...	...	al, eth, ace, aa	B3^4, 248
14757	Urea oxalate $2(H_2NCONH_2).HO_2CCO_2H$	210.15	mcl (w + l)	...	173d	1.585	...	w	B3^2, 48
14758	Urea, 1-isopentyl $H_2NCONH\text{-}(i\text{-}C_5H_{11})$	130.19	pl (dil al)	...	96 (150)	...	...	al	B4^4, 695
14759	Urea, 1-(2-phenoxy ethyl) $(C_6H_5OCH_2CH_2)NHCONH_2$	180.21	nd (w + 2), cr (50% al)	...	120-1	...	...	al	B6^1, 91

No.	Name, Synonyms, and Formula	Mol. wt.	Color, crystalline form, specific rotation and λ_{max} (log ε)	b.p. °C	m.p. °C	Density	n_D	Solubility	Ref.
14760	Urea, 1-phenyl or Phenyl carbamide $C_6H_5NHCONH_2$	136.15	nd or pl (w), tab (al)	238	147			al, AcOEt	B12³, 760
14761	Urea, 1-phenylacetyl $C_6H_5CH_2CONHCONH_2$	178.19	(al)		209			al	B9³, 2208
14762	Urea, propionyl or Propionyl urea $CH_3CH_2CONHCONH_2$	116.12	cr (w)		210-11			w	B3³, 121
14763	Urea, 1-propyl $C_3H_7NHCONH_2$	102.14	pr (al)		110			al	B4⁴, 482
14764	Urea, 1,1,3,3-tetraethyl $[(C_2H_5)_2N]_2CO$	172.27		209, $94\text{-}5^{12}$		$0.919^{20/4}$	1.4474^{20}		B4⁴, 380
14765	Urea, 1,1,3,3-tetramethyl $[(CH_3)_2N]_2CO$	116.16		166-7, $63\text{-}4^{12}$	−1.2	0.9687^{20}	1.4496^{23}		B4⁴, 225
14766	Urea, 1,1,3,3-tetraphenyl $[(C_6H_5)_2N]_2CO$	364.45	rh (bz)		183	1.222			B12³, 894
14767	Urea, 1-(2-tolyl) $(2\text{-}CH_3C_6H_4)NHCONH_2$	150.18	lf (al, w)		195-6			al, eth	B12³, 1867
14768	Urea, 1-(3-tolyl) $(3\text{-}CH_3C_6H_4)NHCONH_2$	150.18	lf (w)		142			w, al	B12³, 1968
14769	Urea, 1-(4-tolyl) $(4\text{-}CH_3C_6H_4)NHCONH_2$	150.18	nd (w), pl (w-aa)		183			al, ace	B12³, 2082
14770	Urea, 1,1,3-trimethyl $(CH_3)_2NCONHCH_3$	102.14	pr (eth)	232.5^{764}	75.5			w, al	B4⁴, 224
14771	1-Urea carboxylic acid, ethyl ester or Allophanic acid, ethyl ester $H_2NCONHCO_2C_2H_5$	132.12	nd (w, bz)	d	195				B3⁴, 127
14772	Urethane, N-amino $H_2NNHCO_2C_2H_5$	104.11	cr	198d, 93^{9}	46			al, eth	B3⁴, 174
14773	Uric acid or 2,6,8-Purine trione $C_5H_4N_4O_3$	168.11	rh pr or pl	d	d	1.89		glycerol	B26², 293
14774	Uric acid, 1-methyl $C_6H_6N_4O_3$	182.14	nd		400				B26², 299
14775	Uric acid, 3-methyl $C_6H_6N_4O_3$	182.14	pr (w + 1)		>350	1.6104^{25}			B26², 299
14776	Uric acid, 7-methyl $C_6H_6N_4O_3$	182.14	pl (w)		370-80d	$1.706^{25/4}$			B26², 299
14777	Uric acid, 9-methyl $C_6H_6N_4O_3$	182.14			380-400d			w	B26², 299
14778	Uric acid, 1,3,7-trimethyl $C_8H_{10}N_4O_3$	210.19			373-5				B26², 301
14779	Uridine or 1-β-D-Ribofuranosyluracil $C_9H_{12}N_2O_6$	244.20	nd (aq al), $[\alpha]^{20/}{}_D$ + 4 (w)		165			w, Py	J1947, 358
14780	Uridylic acid or Uridine-3′-phosphate $C_9H_{13}N_2O_9P$	324.18	pr (MeOH), $[\alpha]^{20/}{}_D$ + 10.5 (w)		202d			w, MeOH	J1948, 746
14781	Urocanic acid or 4-Imidazolyl acrylic acid $C_6H_6N_2O_2$	138.13	pr(w + 2)		*cis* 175-6, *trans* 218-24			w, ace	B25², 121
14782	Urochloralic acid or 2,2,2-trichloroethyl-β-D-glucuronide $C_8H_{11}O_7Cl_3$	325.23	nd		142			w, al	B18⁴, 5113
14783	Ursodesoxycholic acid or 3α-7β-Dihydroxy-5β-cholanic acid $C_{24}H_{40}O_4$	325.53	pl (al), $[\alpha]^{20/}{}_D$ + 57 (abs al, c=2)		203			al	B10³, 1635
14784	Ursolic acid or Malol. Urson $C_{30}H_{48}O_3$	456.71	pr (al), $[\alpha]^{21/}{}_D$ + 72.4 (MeOH)		284			eth, ace, chl, MeOH	B10³, 1038
14785	Usnic acid (d) or Usninic acid $C_{18}H_{16}O_7$	344.32	ye orh pr (ace), $[\alpha]^{20/}{}_D$ + 469 (chl, c=0.7)		204			al, eth	B18⁴, 3522
14786	Usnic acid (dl) $C_{18}H_{16}O_7$	344.32			193-4		1.710	al, eth, chl	B18⁴, 3523

No.	Name, Synonyms, and Formula	Mol. wt.	Color, crystalline form, specific rotation and λ_{max} (log ε)	b.p. °C	m.p. °C	Density	n_D	Solubility	Ref.
14787	Usnic acid (l) $C_{18}H_{16}O_7$	344.32	$[\alpha]^{20/}{}_D$ −480		203			al, eth, chl	B18[4], 3522
14788	Uzarin $C_{35}H_{54}O_{14}$	698.81	pr, $[\alpha]_D$ −27 (Py)		268-70			Py	B18[4], 1497

PHYSICAL CONSTANTS OF ORGANIC COMPOUNDS (Continued)

No.	Name, Synonyms, and Formula	Mol. wt.	Color, crystalline form, specific rotation and λ_{max} (log ε)	b.p. °C	m.p. °C	Density	n_D	Solubility	Ref.
14789	Vacciniin or 6-*O*-Benzoyl-D-glucose $C_{13}H_{16}O_7$	284.27	amor (aq ace + 1w), $[\alpha]^{21/}_D$ + 48 (al)		104-6			w, al, ace	B31, 123
14790	Valeraldehyde or Pentanal $CH_3(CH_2)_3CHO$	86.13		103	−91.5	$0.8095^{20/4}$	1.3944^{20}	al, eth	$B1^4$, 3268
14791	Valeraldehyde, diethyl acetyl or Pentanal diethyl acetal $CH_3(CH_2)_3CH(OC_2H_5)_2$	160.26		59^{12}		0.829^{22}	1.4029^{22}		$B1^4$, 3269
14792	Valeraldehyde, 3-hydroxy-2-methyl or Propionaldol $C_2H_5CH(OH)CH(CH_3)CHO$	116.16		$94\text{-}6^{23}$		$0.986^{25/4}$	1.4502^{20}	w, al, eth, ace	$B1^4$, 4022
14793	Valeraldehyde, 3-hydroxy-2,2,4-trimethyl $(CH_3)_2CHCH(OH)C(CH_3)_2CHO$	144.21		$118\text{-}20^{14}$		$0.9482^{20/4}$	1.4501^{20}		$B1^4$, 4049
14794	Valeraldehyde, 4-hydroxy (dl) $CH_3CH(OH)CH_2CH_2CHO$	102.13		$63\text{-}5^{10}$		$1.019^{20/4}$	1.4359^{17}	w, ace	$B1^4$, 4002
14795	Valeraldehyde, 4-hydroxy (l) $CH_3CH(OH)CH_2CH_2CHO$	102.13	$[\alpha]^{23/}_D$ −7.8	$43\text{-}6^{1}$				w, ace	$B1^2$, 872
14796	Valeraldehyde, 2-methyl $C_3H_7CH(CH_3)CHO$	100.16		116^{737}				w, eth, ace	$B1^4$, 3304
14797	Valeraldehyde, 4-methyl-2-oxo or Formyl isobutyl ketone $(CH_3)_2CHCH_2COCHO$	114.14	ye-gr	$45\text{-}6^{12}$					$B1^1$, 406
14798	Valeraldehyde oxime or Valeraldoxime $CH_3(CH_2)_3CH{=}NOH$	101.15	cr		52			eth	$B1^3$, 2798
14799	Valeraldehyde, 2-oxo or Butyryl formaldehyde C_3H_7COCHO	100.12		112, 36^{16}			1.4043^{25}	w, al, eth	$B1^4$, 3658
14800	Valeraldehyde, 4-oxo or Levalinaldehyde $CH_3COCH_2CH_2CHO$	100.12		186-8d, 70^{12}	<−21	$1.0184^{21/4}$	1.4257^{22}	w, al, eth, ace, bz	$B1^4$, 3659
14801	Valeramide $CH_3(CH_2)_3CONH_2$	101.15	micalike mcl pl (peth, al)		106 (114)	0.8735^{110}	1.4183^{110}	w, al, eth	$B2^4$, 874
14802	Valeramide, 2-bromo-4-methyl (d) $(CH_3)_2CHCH_2CHBrCONH_2$	194.07	(w), $[\alpha]^{20/}_D$ −48.3 (al, c=5.9)		118			al, eth, chl	$B2^2$, 291
14803	Valeramide, N,N-dimethyl $CH_3(CH_2)_3CON(CH_3)_2$	129.20		141^{100}	−51	$0.8962^{25/4}$	1.4419^{25}	w, al, eth	$B4^3$, 127
14804	Valeramide, 4-methyl $(CH_3)_2CHCH_2CH_2CONH_2$	115.18	nd (al)		120-1			w, al	$B2^4$, 946
14805	Valeramide, 4-methyl-N-phenyl or Isocaproanilide $(CH_3)_2CHCH_2CH_2CONHC_6H_5$	191.27	nd (bz, dil al)		112			eth	$B12^3$, 478
14806	Valeramide, N-phenyl $C_4H_9CONHC_6H_5$	177.25	mcl pr (al), cr (peth)		63			eth	$B12^3$, 476
14807	Valeramide, 2,2,4-trimethyl $(CH_3)_2CHCH_2C(CH_3)_2CONH_2$	143.23	lf (lig)		71			al	$B2^2$, 305
14808	Valeric acid or Pentanoic acid $CH_3(CH_2)_3CO_2H$	102.13		186, 82.7^{10}	−33.8	$0.9391^{20/4}$	$1.4085^{20/4}$	w, al, eth	$B2^4$, 868
14810	Valeric acid, 2-acetyl, ethyl ester $CH_3COCH(C_3H_7)CO_2C_2H_5$	172.22		$210\text{-}2^{749}$, 90^{10}		$0.9682^{15/4}$	1.4271^{20}	al, eth	$B3^4$, 1602
14811	Valeric acid, 2-amino (D+) $C_3H_7CH(NH_2)CO_2H$	117.15	lf (w), $[\alpha]^{23/}_D$ + 32 (6NHCl, c=2)	sub	ca 307			w	$B4^3$, 1331
14812	Valeric acid, 2-amino (DL) or Norvaline $C_3H_7CH(NH_2)CO_2H$	117.15	lf (al, w)	sub	303 (sealed tube)			w	$B4^3$, 1333
14813	Valeric acid, 2-amino (L-) $C_3H_7CH(NH_2)CO_2H$	117.15	cr (dil al, w), $[\alpha]^{20/}_D$ −23 (20% HCl, c=10)	sub	ca−305			w	$B4^3$, 1332
14814	Valeric acid, 4-amino (D) $CH_3CH(NH_2)CH_2CH_2CO_2H$	117.15	cr (dil al), $[\alpha]^{20/}_D$ + 12 (w, p=10)		214			w	$B4^3$, 1342
14815	Valeric acid, 4-amino (DL) $CH_3CH(NH_2)CH_2CH_2CO_2H$	117.15	cr (w)	d	199 (214)			w	$B4^3$, 1342
14816	Valeric acid, 5-amino $H_2N(CH_2)_4CO_2H$	117.15	lf (dil al)	d	157-8d			w	$B4^3$, 1343
14817	Valeric acid anhydride $(C_4H_9CO)_2O$	186.25		218^{754}, 111^{15}	−56.1	$0.924^{20/4}$	1.4171^{26}	al, eth	$B2^4$, 874

No.	Name, Synonyms, and Formula	Mol. wt.	Color, crystalline form, specific rotation and λ_{max} (log ε)	b.p. °C	m.p. °C	Density	n_D	Solubility	Ref.
14818	Valeric acid, 2-bromo (dl) $C_3H_7CHBrCO_2H$	181.03	...	118^{12}	...	1.381^{20}	...	al, eth	B2^{4}, 883
14819	Valeric acid, 2-bromo, ethyl ester $C_3H_7CHBrCO_2C_2H_5$	209.08	...	190-2, $92\text{-}4^{18}$	...	$1.226^{18/4}$	1.4496^{20}	al, eth	B2^{3}, 681
14820	Valeric acid, 2-bromo-4-methyl (d) $(CH_3)_2CHCH_2CHBrCO_2H$	195.06	oil, $[\alpha]^{20/}_{D}$ + 29.8 (eth, c=6)	$131\text{-}2^{18}$	...	...	...	al, eth	B2^{3}, 747
14821	Valeric acid, 2-bromo-4-methyl (l) $(CH_3)_2CHCH_2CHBrCO_2H$	195.06	oil, $[\alpha]^{20/}_{D}$ −12.1	$94^{0.3}$	...	...	...	al, eth	B2^{3}, 747
14822	Valeric acid, 2-bromo-4-methyl, ethyl ester $(CH_3)_2CHCH_2CHBrCO_2C_2H_5$	223.11	pa ye	202-4, $86\text{-}7^{11}$	...	...	...	...	B2^{3}, 747
14823	Valeric acid, butyl ester or Butyl valarate $C_4H_9CO_2C_4H_9$	158.24	...	185.8, $84\text{-}5^{8}$	−92.8	$0.8710^{15/4}$	1.4128^{20}	al, eth	B2^{3}, 671
14824	Valeric acid, *iso*-butyl ester or Isobutyl valarate $C_4H_9CO_2CH_2CH(CH_3)_2$	158.24	...	179	...	$0.8625^{25/4}$	1.4046^{20}	al, eth, ace	B2^{3}, 671
14825	Valeric acid, *sec*-butyl ester (d) or *sec*-Butyl valarate $C_4H_9CO_2CH(CH_3)C_2H_5$	158.24	$[\alpha]^{20/}_{D}$ + 20.7	174.5, 67^{18}	...	$0.8605^{20/4}$	1.4070^{20}	al, eth, bz, Py	B2^{3}, 671
14826	Valeric acid, 2-chloro (dl) $C_3H_7CHClCO_2H$	136.58	...	222^{763}, $133\text{-}5^{39}$	−15	1.141^{13}	1.4481^{30}	al, eth	B2^{4}, 878
14827	Valeric acid, 2-chloro, ethyl ester $C_3H_7CHClCO_2C_2H_5$	164.63	...	185^{752}	...	1.040^{12}	1.4307^{11}	eth	B2^{3}, 678
14828	Valeric acid, 3-chloro $C_2H_5CHClCH_2CO_2H$	136.58	cr (bz)	112^{10}	33	$1.1484^{20/4}$	1.4462^{20}	al, eth	B2^{4}, 879
14829	Valeric acid, 3-chloro, ethyl ester $C_2H_5CHClCH_2CO_2C_2H_5$	164.63	...	189, $66\text{-}7^{10}$	...	$1.0330^{20/4}$	1.4278^{20}	al, eth	B2^{3}, 678
14830	Valeric acid, 4-chloro $CH_3CHClCH_2CH_2CO_2H$	136.58	...	116^{10}	...	1.1514^{20}	1.4458^{20}	al, eth	B2^{4}, 879
14831	Valeric acid, 4-chloro, ethyl ester $CH_3CHClCH_2CH_2CO_2C_2H_5$	164.63	...	196, 70.5^{9}	...	$1.0393^{20/4}$	1.4310^{20}	al, eth	B2^{4}, 879
14832	Valeric acid, 5-chloro $Cl(CH_2)_4CO_2H$	136.58	...	230, $141\text{-}9^{12}$	18	$1.3416^{25/4}$	1.4555^{20}	al, eth	B2^{4}, 880
14833	Valeric acid, 5-chloro, ethyl ester $Cl(CH_2)_4CO_2C_2H_5$	164.63	...	205-6, 93^{16}	...	1.0561^{20}	1.4355^{20}	al, eth	B2^{4}, 880
14834	Valeric acid, 2,4-dimethyl (dl) $(CH_3)_2CHCH_2CH(CH_3)CO_2H$	130.19	...	$111^{0.9}$	143-5	1.0818^{27}	...	al, eth	B2^{4}, 979
14835	Valeric acid, 2,4-dioxo or Acetyl pyruvic acid $CH_3COCH_2COCO_2H$	130.10	nd (bz-chl), pr (bz)	130^{37} sub	55-63 (+1w) 101 (anh)	...	...	w, al, eth, ace, bz	B3^{3}, 1331
14836	Valeric acid, 2,4-dioxo,ethyl ester or Ethyl acetone oxalate $CH_3COCH_2COCO_2C_2H_5$	158.15	...	213-5, $111\text{-}2^{16}$	18	1.1251^{20}	1.4757^{17}	al, eth	B3^{4}, 1777
14837	Valeric acid, ethyl ester or Ethyl valerate $C_4H_9CO_2C_2H_5$	130.19	...	144.6^{736}, 50.5^{29}	−91.2	$0.8770^{20/4}$	1.4120^{20}	al, eth	B2^{4}, 872
14838	Valeric acid, 2-ethyl (dl) or 3-Hexane carboxylic acid $C_3H_7CH(C_2H_5)CO_2H$	130.19	...	209.2, $105\text{-}7^{18}$	...	$0.9361^{33/33}$	...	al, eth	B2^{4}, 975
14839	Valeric acid, 3-ethyl $(C_2H_5)_2CHCH_2CO_2H$	130.19	...	212, $104\text{-}5^{13}$	...	...	1.4250^{20}	...	B2^{4}, 976
14840	Valeric acid, 5-fluoro $F(CH_2)_4CO_2H$	120.12	...	83^{2}	...	...	1.4080^{25}	al, eth	B2^{4}, 876
14841	Valeric acid, furfuryl ester or Furfuryl valerate $C_4H_9CO_2CH_2(C_4H_3O)$	182.22	...	$228\text{-}9^{764}$, $82\text{-}3^{1}$	...	$1.0284^{20/4}$	...	al, eth	B17^{2}, 115
14842	Valeric acid, heptyl ester or Heptyl valerate $C_4H_9CO_2C_7H_{15}$	200.32	...	245.2	−46.4	0.8623^{20}	$1.4254^{15/}_{He}$	al, eth, ace	B2^{4}, 872
14843	Valeric acid, hexyl ester or Hexyl valerate $C_4H_9CO_2C_6H_{13}$	186.29	...	226.3	−63.1	0.8635^{20}	1.4228^{15}	al, eth, ace	B2^{3}, 671
14844	Valeric acid, 2-hydroxy or Valerolactic acid $C_3H_7CH(OH)CO_2H$	118.13	hyg pl	sub	34	...	...	w, al, eth	B3^{4}, 807
14845	Valeric acid, 4-hydroxy, lactone (d) or γ-Valerolactone $CH_3CHCH_2CH_2CO$ (ring closed through O)	100.12	$[b\alpha]^{20/}_{D}$ + 13.5 (undil)	$86\text{-}90^{14}$	...	...	...	w, al, ace	B17^{2}, 288
14846	Valeric acid, 4-hydroxy-lactone (dl) or γ-Valerolactone $CH_3CHCH_2CH_2CO$ (ring closed through O)	100.12	...	206, $83\text{-}4^{13}$	−31	1.0465^{25}	1.4328^{20}	w, al, ace	B17^{4}, 4176
14847	Valeric acid, 4-hydroxy-lactone (l) $CH_3CHCH_2CH_2CO$ (ring closed through O)	100.12	$[\alpha]^{20/}_{D}$ + 4.6 (eth, c=10)	$78\text{-}80^{8}$	...	...	1.4322^{20}	w, al, ace	B17^{4}, 4176
14848	Valeric acid, 5-hydroxy-lactone or γ-Valerolactone $CH_2(CH_2)_3CO$ (ring closed through O)	100.12	...	218-20, $113\text{-}4^{14}$	−12.5	1.0794^{20}	1.4503^{20}	w, al, eth	B17^{4}, 4169
14849	Valeric acid, methyl ester or Methyl valerate $C_4H_9CO_2CH_3$	116.16	...	126.5	...	$0.8947^{20/4}$	1.4003^{20}	al, eth, ace	B2^{4}, 871

No.	Name, Synonyms, and Formula	Mol. wt.	Color, crystalline form, specific rotation and λ_{max} (log ε)	b.p. °C	m.p. °C	Density	n_D	Solubility	Ref.
14850	Valeric acid, 2-methyl (d) $C_3H_7CH(CH_3)CO_2H$	116.16	$[\alpha]^{25}{}_D$ + 18.5 (undil)	96^{15}		$0.927^{16/4}$	1.4112^{25}	w, al, eth	$B2^4$, 942
14851	Valeric acid, 2-methyl (dl) $C_3H_7CH(CH_3)CO_2H$	116.16		195-6, $102\text{-}5^{12}$		$0.9230^{20/4}$	1.413^{20}	w, al, eth	$B2^4$, 942
14852	Valeric acid, 2-methyl (l) $C_3H_7CH(CH_3)CO_2H$	116.16	$[\alpha]^{25}{}_D$ −7.1 (eth)	190-3, 96^{15}		$0.9781^{20/4}$	1.4117^{15}	w, al, eth	$B2^3$, 740
14853	Valeric acid, 3-methyl (d) $C_2H_5CH(CH_3)CH_2CO_2H$	116.16	$[\alpha]^{20}{}_D$ + 8.5 (undil)	199, $92\text{-}3^{10}$		$0.9276^{20/4}$	1.4158^{20}	al, eth	$B2^4$, 948
14854	Valeric acid, 3-methyl (dl) $C_2H_5CH(CH_3)CH_2CO_2H$	116.16		197-8	−41.6	0.9262^{20}	1.4159^{20}	al, eth	$B2^4$, 948
14855	Valeric acid, 3-methyl (l) $C_2H_5CH(CH_3)CH_2CO_2H$	116.16	$[\alpha]^{20}{}_D$ −8.9 (al)	196-7, 105^{30}		$0.923^{25/4}$	1.4152^{20}	al, eth	$B2^4$, 948
14856	Valeric acid, 3-methyl-3-phenyl $C_2H_5C(CH_3)(C_6H_5)CH_2CO_2H$	192.26		174^{14}		1.050^{25}	1.5197^{25}		$B9^3$, 2548
14857	Valeric acid, 4-methyl or Isocaproic acid $(CH_3)_2CHCH_2CH_2CO_2H$	116.16		200-1, $86\text{-}8^{11}$	−33	$0.9225^{20/4}$	1.4144^{20}	al, eth	$B2^4$, 944
14858	Valeric acid, 4-methyl-2-phenyl or α-Phenyl isocaproic acid $(CH_3)_2CHCH_2CH(C_6H_5)CO_2H$	192.24	pr (peth)	$178\text{-}80^{20}$	78-9			al	$B9^3$, 2546
14859	Valeric acid, octyl ester or Octyl valerate $C_4H_9CO_2C_8H_{17}$	214.35		261.6	−42.3	0.8615^{20}	1.4273^{15}	al, eth, ace	$B2^3$, 672
14860	Valeric acid, 2-oxo $C_3H_7COCO_2H$	116.12		179, 81^{12}	6-7			eth, bz, chl	$B3^4$, 1558
14861	Valeric acid, 3-oxo, ethyl ester or Ethyl propionyl acetate $C_2H_5COCH_2CO_2C_2H_5$	144.17		191, 90^{13}			1.4230^{20}	al, eth, bz	$B3^4$, 1559
14862	Valeric acid, 4-oxo or Levulinic acid $CH_3COCH_2CH_2CO_2H$	116.12	lf or pl	245-6d, $139\text{-}40^{8}$	37.2	$1.1335^{20/4}$	1.4396^{20}	w, al, eth	$B3^4$, 1560
14863	Valeric acid, pentyl ester or Pentyl valerate $C_4H_9CO_2C_5H_{11}$	172.27		203.7, 103^{23}	−78.8	$0.8638^{20/4}$	1.4164^{20}	al, eth	$B2^4$, 872
14864	Valeric acid, 2-phenyl (d) $C_3H_7CH(C_6H_5)CO_2H$	178.23	$[\alpha]_D$ + 58.8 (chl)	165^{14}		1.047^{20}		al, eth, chl	$B9^3$, 2506
14865	Valeric acid, 2-phenyl (dl) $C_3H_7CH(C_6H_5)CO_2H$	178.23	nd (lig)	280	58			al, eth	$B9^3$, 2506
14866	Valeric acid, 4-phenyl $C_6H_5CH(CH_3)CH_2CH_2CO_2H$	178.23		210^{85}, 165^{12}	ca 13	$1.0554^{15/4}$	1.5167^{20}	al, eth	$B9^3$, 2505
14867	Valeric acid, 5-phenyl $C_6H_5(CH_2)_4CO_2H$	178.23	pl (w), pr (peth)	$190\text{-}3^{30}$	57-8			al	$B9^3$, 2502
14868	Valeric acid, propyl ester or Propyl valerate $C_4H_9CO_2C_3H_7$	144.21		167.5		$0.8699^{20/4}$	1.4065^{20}	al, eth, chl	$B2^4$, 872
14869	Valeric acid, isopropyl ester or Isopropyl valerate $C_4H_9CO_2CH(CH_3)_2$	144.21		153.5		$0.8579^{20/4}$	1.4061^{20}	al, eth, ace	$B2^3$, 671
14870	Valeronitrile $CH_3(CH_2)_3CN$	83.13		141.3, 30.9^{10}	−96	$0.8008^{20/4}$	1.3971^{20}	eth, ace, bz	$B2^4$, 875
14871	Valeronitrile, 2-hydroxy-2-propyl $(C_3H_7)_2C(OH)CN$	141.21		$119\text{-}20^{21}$		0.9077^{18}	1.4337^{18}	al	$B3^4$, 880
14872	Valeronitrile, 4-methyl or Isocapronitrile $(CH_3)_2CHCH_2CH_2CN$	97.16		156-7, 50^{13}	−51	$0.8030^{20/4}$	1.4059^{20}	al, eth	$B2^4$, 946
14873	Valeronitrile, 4-methyl-2-phenyl or Phenyl isocapronitrile $(CH_3)_2CHCH_2CH(C_6H_5)CN$	173.26		$263\text{-}6^{765}$, $136\text{-}8^{15}$		0.942^{16}		al, eth, bz, aa	$B9^1$, 220
14874	Valeronitrile, 2-phenyl $C_3H_7CH(C_6H_5)CN$	159.23		$254\text{-}5^{750}$, $125\text{-}8^{13}$		0.9425	1.5000^{20}	al, bz	$B9^3$, 2508
14875	Valerophenone or Butyl phenyl ketone $C_4H_9COC_6H_5$	162.23		248.5, $131\text{-}3^{13}$		$0.988^{20/20}$	1.5158^{20}	al, eth	$B7^3$, 1114
14876	Valerophenone, γ-oxo or Phenacyl acetone $C_6H_5COCH_2CH_2COCH_3$	176.22	ye oil	162^{12}			1.5250^{30}	ace	$B7^3$, 3509
14877	Valeryl chloride C_4H_9COCl	120.58		$107\text{-}10^{756}$	−110	1.0155^{15}	1.4200^{20}		$B2^4$, 874
14878	Valeryl chloride, 2-chloro $C_3H_7CHClCOCl$	155.02		$155\text{-}7^{763}$, $61\text{-}2^{28}$		1.1765^{20}	1.4465^{20}	eth	$B2^4$, 879
14879	Valeryl chloride, 4-chloro $CH_3CHClCH_2CH_2COCl$	155.02		61^{8}				eth	$B2^3$, 679
14880	Valeryl chloride, 5-chloro $Cl(CH_2)_4COCl$	155.02		83^{12}		1.210^{18}	1.4639^{20}	eth	$B2^4$, 881
14881	Valeryl chloride, 4,4-dimethyl or Neopentyl acetyl chloride $(CH_3)_3CCH_2CH_2COCl$	148.63		150-2, 58^{2}			1.4294^{20}	eth	$B2^3$, 780

No.	Name, Synonyms, and Formula	Mol. wt.	Color, crystalline form, specific rotation and λ_{max} (log ε)	b.p. °C	m.p. °C	Density	n_D	Solubility	Ref.
14882	Valeryl chloride, 2-ethyl $C_3H_7CH(C_2H_5)COCl$	148.63		158-60, 50^{11}				eth	$B2^4$, 975
14883	Valeryl chloride, 2-methyl (dl) $C_3H_7CH(CH_3)COCl$	134.61		$140\text{-}1^{745}$		$0.9781^{20/4}$	1.4330^{27}	eth	$B2^3$, 741
14884	Valeryl chloride, 3-methyl (dl) $C_2H_5CH(CH_3)CH_2COCl$	134.61		142-3		$0.9781^{20/4}$		eth	$B2^4$, 949
14885	Valeryl chloride, 4-methyl $(CH_3)_2CHCH_2CH_2COCl$	134.61		$144\text{-}5^{745}$		0.9725^{20}			$B2^4$, 945
14886	Valine (D) or *l*-α-Amino isovaleric acid $(CH_3)_2CHCH(NH_2)CO_2H$	117.15	pl (aq al), $[\alpha]^{20/}{}_D$ −29 (20% al)		156-7 (hyd) 293d (sealed tube)			w	$B4^3$, 1369
14887	Valine (DL) $(CH_3)_2CHCH(NH_2)CO_2H$	117.15		sub	298d (sealed tube)	1.316		w	$B4^3$, 1370
14888	Valine (L) or *d*-α-Amino isovaleric acid $(CH_3)_2CHCH(NH_2)CO_2H$	117.15	lf (w-al), $[\alpha]^{23/}{}_D$ + 22.9 (20%al, c=0.8)		93-6, 315 (sealed tube)	1.230		w	$B4^3$, 1365
14889	Valine, β-hydroxy-dl) $(CH_3)_2C(OH)CH(NH_2)CO_2H$	133.15	pl (dil al)		240d			w	$B4^3$, 1660
14890	Vanillan or 4-Hydroxy-3-methoxy benzaldehyde $4\text{-}HO\text{-}3\text{-}CH_3O\text{-}C_6H_3CHO$	152.15	(i) nd (w, lig), (ii) tetr (w, lig)	285, 170^{15}	(i) 77-9, (ii) 81-2	1.056		al, eth, ace, bz	$B8^3$, 2011
14891	Vanillan acetate or 4-Acetoxy-3-methoxy benzaldehyde $4\text{-}CH_3CO_2\text{-}3\text{-}CH_3OC_6H_3CHO$	194.19	nd (eth)		102-3			al, eth	$B8^3$, 2030
14892	Vanillan, 5-bromo $5\text{-}Br\text{-}4\text{-}HO\text{-}3\text{-}CH_3OC_6H_2CHO$	231.05	pl (aa), nd, pl (al)		164-6				$B8^3$, 2050
14893	Vanillan, 5-chloro $5\text{-}Cl\text{-}4\text{-}HO\text{-}3\text{-}CH_3OC_6H_2CHO$	186.59	tetr		165				$B8^3$, 2043
14894	Vanillan, 2-iodo $4\text{-}HO\text{-}2\text{-}I\text{-}3\text{-}CH_3OC_6H_2CHO$	278.05			155-6			al	$B8^3$, 2058
14895	Vanillan, 5-Iodo $4\text{-}HO\text{-}5\text{-}I\text{-}3\text{-}CH_3OC_6H_2CHO$	278.05	pa ye		180				$B8^3$, 2059
14896	*o*-Vanillan or 2-Hydroxy-3-methoxy benzaldehyde $2\text{-}HO\text{-}3\text{-}CH_3OC_6H_3CHO$	152.15	lt ye-lt gr nd (w, lig)	265-6, 128^{10}	44-5			al, eth, lig	$B8^3$, 1979
14897	Vasicine (dl) or Peganine $C_{11}H_{12}N_2O$	188.23	nd (al)		209-10			al, ace, chl	$B23^2$, 342
14898	Vasicine (l) $C_{11}H_{12}N_2O$	188.23	nd (al), $[\alpha]^{14/}{}_D$ −62 (al, c=2.4)		211-2			al, chl	$B23^2$, 342
14899	Veraridine or 3-Veratroyl veracerine $C_{36}H_{51}NO_{11}$	673.80	yesh amor, $[\alpha]^{22/}{}_D$ + 8 (al)		180				$B21^4$, 6824
14900	Veratramide or 3,4-Dimethoxy amino benzaldehyde $3,4\text{-}(CH_3O)_2C_6H_3CONH_2$	181.19	cr (w)		164			w, eth, bz	$B10^3$, 1427
14901	Veratramine $C_{27}H_{39}NO_2$	409.61	nd, $[\alpha]^{19/}{}_D$ −70 (MeOH)		209-10 (hyd)			bz, chl	$B21^4$, 2385
14902	Veratric acid or 3,4-Dimethoxy benzoic acid $3,4\text{-}(CH_3O)_2C_6H_3CO_2H$	182.18	nd (w, aa), rh (sub)	sub	181-2			al, eth	$B10^3$, 1404
14903	Veratrine or Cevadine $C_{32}H_{49}NO_9$	591.74	rh (+ 2 al), $[\alpha]^{17/}{}_D$ + 12.5 (al)		205d			al, chl, Py	$B21^4$, 6820
14904	Veratronitrile or 3,4-Dimethoxy benzonitrile $3,4\text{-}(CH_3O)_2C_6H_3CN$	163.18	nd (w)		67-8			bz	$B10^2$, 264
14905	Veratrosine or Veratramine-β-D-glucoside $C_{33}H_{49}NO_7$	571.75	nd (aq MeOH), $[\alpha]^{24/}{}_D$ −55 (al-chl, c=0.94)		242-3d				$B21^4$, 2386
14906	Verbenone (d) or d-4-Piperone $C_{10}H_{14}O$	150.22	$[\alpha]^{18/}{}_D$ + 249.6	227-8, $103\text{-}4^{16}$	9.8	0.9978^{20}	1.4993^{18}	w, al, ace, bz	$B7^3$, 583
14907	Verbenone (l) $C_{10}H_{14}O$	150.22	$[\alpha]_D$−144	253-5, 100^{16}	6.5	0.9731^{20}	1.4961^{20}	w, al, bz	$B7^3$, 583

PHYSICAL CONSTANTS OF ORGANIC COMPOUNDS (Continued)

No.	Name, Synonyms, and Formula	Mol. wt.	Color, crystalline form, specific rotation and λ_{max} (log ε)	b.p. °C	m.p. °C	Density	n_D	Solubility	Ref.
14908	α-Vetivone or α-Vetiverone $C_{15}H_{22}O$	218.34	(peth), $[\alpha]_D$ +238.2	144^{2}	51.5	$1.0035^{20/4}$	1.5370^{20}	ace	$B7^{3}$, 1265
14909	β-Vetivone or β-Vetiverone $C_{15}H_{22}O$	218.34	(pentane), $[\alpha]^{20/}_D$ −38.9 (al, c=10.6)	$141\text{-}2^{2}$	44.5	$1.0001^{20/4}$	1.5309^{20}	ace	$B7^{3}$, 1256
14910	Vicianose or 6-*O*-(α-L-arabinosido)-β-D-glucose $C_{11}H_{20}O_{10}$	312.27	nd (dil al), $[\alpha]^{20/}_D$ +56.5→ +39.7 (w, mut)	....	ca 210d	....	....	w	$B17^{4}$, 2447
14911	Vicine (a β-D-glucoside) $C_{10}H_{16}N_4O_7$	304.26	nd (w, dil al +1w), $[\alpha]^{25/}_D$ −12 (w, c=10)	....	239-42d	....	....	....	B31, 163
14912	Vincamine $C_{21}H_{26}N_2O_3$	354.45	$[\alpha]^{23/}_D$ +41 (Py)	....	231-2	....	....	....	M, 1107
14913	Violaxanthin or Zeaxanthin diepoxide $C_{40}H_{56}O_4$	600.88	red pr (MeOH, al-eth), $[\alpha]^{20/}_d$ +35 (chl, c=0.08)	....	208	....	....	al, eth	$B19^{4}$, 1139
14914	Violuric acid or Alloxan-5-oxime $C_4H_3N_3O_4$	157.09	pa ye rh	....	203-4d	....	....	al	$B24^{2}$, 304
14915	Visnadin $C_{21}H_{24}O_7$	388.42	nd, $[\alpha]_D$ +9 (al)	....	85-6	....	....	al, eth	$B19^{4}$, 2787
14916	Visnagin or 5-Methoxy-2-methyl furanochromone $C_{13}H_{10}O_4$	230.22	nd (w, MeOH)	....	144-5	....	....	chl	$B19^{4}$, 2640
14917	Vitamin A_1 or Axerophytol $C_{20}H_{30}O$	286.46	ye pr (peth)	$137\text{-}8^{10.6}$	63-4	....	....	al, eth, ace, bz	$B6^{3}$, 2787
14918	Vitamin B_1 or Thiamine hydrochloride $C_{12}H_{18}Cl_2N_4OS$	337.27	(i) mcl pr (MeOH-al), (ii) pl (MeOH-al, w-al)	....	(i) 233-4, (ii) 250	....	....	w	Am74, 2409
14919	Vitamin B_6 or Pyroxidin $C_8H_{11}NO_3$	169.18	nd (ace)	sub	160	....	....	w, al, ace	$B21^{4}$, 2509
14920	Vitamin B_6 hydrochloride or Pyridoxin hydrochloride $C_8H_{11}NO_3.HCl$	205.64	pl (al, ace)	sub	206-8	....	....	w	$B21^{4}$, 2511
14921	Vitamin K_1 $C_{31}H_{46}O_2$	450.71	$[\alpha]^{20/}_D$ −0.4 (bz 57.5%)	$140\text{-}5^{0.001}$ dec>120	−20	$0.967^{25/25}$	1.5250^{25}	al, eth, ace, bz, peth	$B7^{3}$, 3792
14922	Vitamin B_{12} $C_{63}H_{88}N_{14}O_{14}CoP$	1355.42	....	>300	....	....	....	....	....
14923	Vitamin B_1-hydrochloride $C_{12}H_{17}N_4OSCl.HCl$	337.27	....	....	233-4	....	....	w	Am74, 2409
14924	Vitamin D_2 $C_{28}H_{44}O$	396.66	$[\alpha]^{25/}_D$ +82.6 (ace, c=3)	....	115-8	....	....	....	$B6^{3}$, 3089
14925	Vitamin D_3 $C_{27}H_{44}O$	384.65	$[\alpha]^{20/}_D$ +84.8 (ace, c=1.6)	....	84-5	....	....	....	$B6^{3}$, 2811
14926	Vomicine $C_{22}H_{24}N_2O_4$	380.44	nd (80% al), pr (ace), $[\alpha]^{22/}_D$ +80.4 (al, p=0.4)		282	....	....	chl	$B27^{2}$, 795
14927	Vinyl acetate $CH_3CO_2CH=CH_2$	86.09	....	72.2	−93.2	$0.9317^{20/4}$	1.3959^{20}	al, eth, ace, bz, chl	$B2^{4}$, 176
14928	Vinyl acetylene or 1-Buten-3-yne $CH_2=CH\text{-}C\equiv CH$	52.08	....	5.1	......	$0.7095^{0/0}$	1.4161^{1}	bz	$B1^{4}$, 1083
14929	Vinyl amine $CH_2=CHNH_2$	43.07	....	$55\text{-}6^{750}$	......	0.8321^{24}	....	w, al, eth	$B20^{4}$, 3
14930	Vinyl bromide $CH_2=CHBr$	106.95	....	15.8	−139.5	1.4933	1.4410^{20}	al, eth, ace, bz, chl	$B1^{4}$, 718

No.	Name, Synonyms, and Formula	Mol. wt.	Color, crystalline form, specific rotation and λ_{max} (log ε)	b.p. °C	m.p. °C	Density	n_D	Solubility	Ref.
14931	Vinyl chloride or Chloro ethylene $CH_2{=}CHCl$	62.50	gas	−13.4	−1538	$0.9106^{20/4}$	1.3700^{20}	al, eth	$B1^4$, 700
14932	Vinyl ether or Divinyl ether $(CH_2{=}CH)_2O$	70.09		28	−101	$0.773^{20/4}$	1.3989^{20}	al, eth, ace	$B1^4$, 2058
14933	Vinyl ether, hexachloro $(Cl_2C{=}CCl)_2O$	276.76		210		1.654^{21}			B1, 725
14934	Vinyl fluoride $CH_2{=}CHF$	46.04	gas	−72.2	−160.5			al, ace	$B1^4$, 694
14935	Vinyl iodide $CH_2{=}CHI$	153.95		56		2.037^{20}	1.5385^{20}	al, eth	$B1^4$, 722
14936	Vinyl propionate $CH_3CH_2CO_2CH{=}CH_2$	100.12		91-2					$B2^4$, 711
14937	2-Vinyl pyridine 2-$(CH_2{=}CH)C_5H_4N$	105.14		159-60, $50\text{-}5^4$		$0.9985^{20/0}$	1.5495^{20}	al, eth, ace, chl	$B20^4$, 2884
14938	3-Vinyl pyridine 3-$(CH_2{=}CH)C_5H_4N$	105.14		162, $67\text{-}8^{18}$			1.5530^{20}	al, eth	$B20^4$, 2886
14939	4-Vinyl pyridine 4-$(CH_2{=}CH)C_5H_4N$	105.14		65^{15}		$0.9800^{20/4}$	1.5449^{20}	w, al, chl	$B20^4$, 2887
14940	Vinyl sulfide $(CH_2{=}CH)_2S$	86.15		84	$0.9174^{15/4}$			al, eth, ace	$B1^4$, 2068
14941	2-Vinyl toluene or 2-Methyl styrene 2-$CH_3C_6H_4CH{=}CH_2$	118.18		171, 51^9		$0.9106^{20/4}$	1.5450^{20}	bz, chl	$B5^4$, 1367
14942	*m*-Vinyl toluene or 3-Methyl styrene 3-$CH_3C_6H_4CH{=}CH_2$	118.18		168, $61\text{-}2^{18}$	−70	$0.9028^{20/20}$	1.5428^{20}	bz	$B5^4$, 1367
14943	*p*-Vinyl toluene or 4-Methyl styrene 4-$CH_3C_6H_4CH{=}CH_2$	118.18		169, 63^{15}	−37.8	$0.8760^{20/4}$	1.5428^{20}	bz	$B5^4$, 1369

PHYSICAL CONSTANTS OF ORGANIC COMPOUNDS (Continued)

No.	Name, Synonyms, and Formula	Mol. wt.	Color, crystalline form, specific rotation and λ_{max} (log ε)	b.p. °C	m.p. °C	Density	n_D	Solubility	Ref.
14944	Xanthene or Dibenzo-1,4-Puran $C_{13}H_{10}O$	182.22	wh ye lf (al)	310-2	100.5			eth, bz, chl, lig	$B17^4$, 614
14945	Xanthene, 9-hydroxy or Dibenz-γ-pyranol $C_{13}H_{10}O_2$	198.22	nd (aq al)		ca 125			al, eth, chl	$B17^4$, 1602
14946	Xanthene, 9-phenyl $C_{19}H_{14}O$	258.32	(al)		145-6			bz, aa	$B17^4$, 733
14947	9-Xanthene carboxylic acid or Xanthanoic acid $C_{14}H_{10}O_3$	226.23	nd (dil al, MeOH)		223.4			eth	$B18^4$, 4351
14948	Peri-xanthenoxanthene or 1,1-Binaphthalene-2:8′,2:8-dioxide $C_{20}H_{10}O_2$	282.30	ye pr (chl)	$400^{20.25}$	242			bz	$B19^4$, 448
14949	Xanthine or 2,6-Purinedione $C_5H_4N_4O_2$	152.11	yesh pl (w)	sub	d				$B26^2$, 260
14950	Xanthine, 2-methyl or Heteroxanthine $C_6H_6N_4O_2$	166.14	nd (w)		380d				$B26^2$, 263
14951	Xanthione or 9-Xanethenethione $C_{13}H_8OS$	212.27	red nd (al)		156			eth, bz	$B17^4$, 5301
14952	Xanthogen-diethyl or Auligen $C_2H_5OC(:S)SSC(:S)OC_2H_5$	242.30	ye nd or pl (al)	$107\text{-}9^{0.05}$	31-2	$1.2604^{25/4}$		eth, ace, bz, peth, chl	$B3^3$, 349
14953	Xanthogenic acid or Ethoxy dithioformic acid $C_2H_5OCS_2H$	122.20	unst	25d	fr−53				$B3^4$, 401
14953a	Xanthogenic acid, ethyl ester $CH_3CH_2OCS_2C_2H_5$	150.25		200		1.085^{19}	$1.5370^{18.2}$		$B3^2$, 153
14954	Xanthone or Dibenzopyrone $C_{13}H_8O_2$	196.21	nd (al)	349-50, 146^3	174			al, eth, bz, chl	$B17^4$, 5292
14955	Xanthone, 1,2-dihydroxy $C_{13}H_8O_4$	228.20	pa ye nd (aq al + 3 w)		166-7			al, Py	$B18^4$, 1678
14956	Xanthone, 1,3-dihydroxy $C_{13}H_8O_4$	228.20	nd (aq al + 1 w)	sub	259			al	$B18^4$, 1680
14957	Xanthone, 1,6-dihydroxy or Isoeuxanthone $C_{13}H_8O_4$	228.20	ye nd (dil al)		245-6			al, eth	$B18^4$, 1682
14958	Xanthone, 1,7-dihydroxy or Euxanthone $C_{13}H_8O_4$	228.20	ye nd (to), pl (al)	sub d	240				$B18^4$, 1682
14959	Xanthone, 1,8-dihydroxy $C_{13}H_8O_4$	228.20	ye lf (bz)		187				$B18^4$, 1683
14960	Xanthone, 2,3-dihydroxy $C_{13}H_8O_4$	228.20	ye nd (al)		294			al, ace, bz, aa	$B18^4$, 1683
14961	Xanthone, 2,7-dihydroxy or β-Isoeuxanthone $C_{13}H_8O_4$	228.20	ye nd (al, eth)	sub	>330			al, eth	$B18^2$, 357
14962	Xanthone, 3,4-dihydroxy $C_{13}H_8O_4$	228.20	pa red-ye nd (dil al + 3w)		240			al	$B18^4$, 1684
14963	Xanthone, 3,6-dihydroxy $C_{13}H_8O_4$	228.20	nd (dil al), pr (sub)	sub	300-50d			al, aa	$B18^4$, 1685
14964	Xanthone, 1,7-dihydroxy-3-methoxy or Gentisin $C_{14}H_{10}O_5$	258.23	ye rh	400 sub	266-7			al, Py	$B18^4$, 2602
14966	Xanthophyll or Lutein. Lateol $C_{40}H_{56}O_2$	568.88	ye or vt pr (eth-MeOH), $[\alpha]_{Cd}$ + 160 (chl)		196			al, eth, bz, peth	$B6^3$, 5870
14967	Xanthopterin or 2-Amino-4,6-pteridenediol. Uropteim $C_6H_5N_5O_2$	179.14	hyg ye amor or og (aa)	$98\text{-}100^{18}$	>410d	1.559			$B26^2$, 313
14968	Xanthotoxin or Ammoidin $C_{12}H_8O_4$	216.19	pr (dil al), nd (peth)		148			al	$B19^4$, 2633
14969	Xanthotoxol or 8-Hydroxy-4′:5′,6:7-furocoumarin $C_{11}H_6O_4$	202.17			251-2				$B19^4$, 2633
14970	Xanthoxyletin or Alloxanthylethin $C_{15}H_{14}O_4$	258.27	pr (MeOH, peth)		133			ace, bz, chl	$B19^4$, 2645
14971	Xanthyletin or 2,2-Dimethyl chromeno coumarin $C_{14}H_{12}O_3$	228.25	pr	$140\text{-}5^{0.1}$	131.5			al	$B19^4$, 1828
14972	o-Xylene or 1,2-Dimethyl benzene $1,2\text{-}(CH_3)_2C_6H_4$	106.17		144.4, 32^{10}	−25.2	$0.8802^{10/4}$	1.5055^{20}	al, eth, ace, bz	$B5^4$, 917
14973	o-Xylene, 4-chloro or 4-Chloro-1,2-dimethyl benzene $1,2\text{-}(CH_3)_2C_6H_3Cl\text{-}(4)$	140.61		194^{755}	−6	$1.0692^{15/15}$		ace, bz	$B5^3$, 816
14974	o-Xylene, 3,4-dinitro or 1,2-Dimethyl-3,4-dinitro benzene $1,2\text{-}(CH_3)_2\text{-}3,4\text{-}(NO_2)_2C_6H_2$	196.16	nd (al)		82			eth, bz	$B5^3$, 822
14975	o-Xylene, 3,5-dinitro $1,2\text{-}(CH_3)_2\text{-}3,5\text{-}(NO_2)_2C_6H_2$	196.16	ye nd (al, peth)		77			al, ace, bz, chl	$B5^3$, 823

No.	Name, Synonyms, and Formula	Mol. wt.	Color, crystalline form, specific rotation and λ_{max} (log ε)	b.p. °C	m.p. °C	Density	n_D	Solubility	Ref.
14976	o-Xylene, 3,6-dinitro $1,2-(CH_3)_2-3,6-(NO_2)_2C_6H_2$	196.16	nd (al)	....	89-90	....	....	al, eth, ace, bz, peth	B5^{1}, 181
14977	o-Xylene, 4,5-dinitro $1,2-(CH_3)_2-4,5-(NO_2)_2C_6H_2$	196.16	nd (al, bz, aa)	....	118	....	....	eth, ace, bz	B5^{3}, 823
14978	o-Xylene, 4-ethyl $1,2-(CH_3)_2-4-C_2H_5C_6H_3$	134.22	....	189.7, 67.8^{10}	−67	0.8745$^{20/4}$	1.5031^{20}	al, eth, ace, bz, peth	B5^{4}, 1069
14979	o-Xylene, 3-iodo or 3-Iodo-1,2-dimethyl benzene $3-I-1,2-(CH_3)_2C_6H_3$	232.06	....	228-30, 125-6^{15}	....	1.6395$^{20/4}$	1.6074^{20}	ace	B5^{3}, 820
14980	o-Xylene, 4-iodo $4-I-1,2-(CH_30_2C_6H_3$	232.06	....	231-2, 111^{11}	....	1.6334$^{18/4}$	1.6049^{18}	ace	B5^{3}, 821
14981	o-Xylene, 3-methoxy or 2,3-Dimethyl anisole $2,3-(CH_3)_2C_6H_3OCH_3$	136.19	....	199, 85^{18}	29	0.9596^{40}	1.5120^{40}	al, eth, ace, bz	B6^{3}, 1723
14982	o-Xylene, 4-methoxy $4-CH_3O-1,2-(CH_3)_2C_6H_3$	136.19	....	204-5, 96-7^{17}	....	0.9744$^{14/4}$	1.5198^{14}	al, eth, bz	B6^{3}, 1727
14983	o-Xylene, 3-nitro $3-NO_2-1,2-(CH_3)_2C_6H_3$	151.16	nd (al)	240, 131^{20}	15	1.1402$^{/20}$	1.5441^{20}	al	B5^{4}, 930
14984	o-Xylene, 4-nitro $4-NO_2-1,2-(CH_3)_2C_6H_3$	151.16	ye pr (al)	254^{750}, 143^{21}	30-1	1.112^{15}	1.5202^{20}	al	B5^{4}, 930
14985	o-Xylene, 3,4,5,6-tetrabromo $3,4,5,6-Br_4-1,2-(CH_3)_2C_6$	421.75	nd (bz)	374-5	262	....	....	bz	B5^{3}, 820
14986	*m*-Xylene or 1,3-Dimethyl benzene $1,3-(CH_3)_2C_6H_4$	106.17	....	139.1, 28.1^{10}	−47.9	0.8642$^{20/4}$	1.4972^{10}	**al, eth, ace, bz**	B5^{4}, 932
14987	m-Xylene 4-chloro $1,3-(CH_3)_2C_6H_3Cl-(4)$	140.61	....	187-8, 89^{24}	....	1.0598$^{20/20}$	1.5230^{25}	ace, bz	B5^{3}, 834
14988	*m*-Xylene, 5-chloro $1,3-(CH_3)_2-C_6H_3-Cl-(5)$	140.61	....	187-8, 66^{12}	....	....	....	ace, bz	B5^{3}, 834
14989	*m*-Xylene, 2,5-dinitro $1,3-(CH_3)_2-2,5-(NO_2)_2C_6H_2$	196.16	ye (al)	....	101	....	....	al, eth, ace, bz, chl	B5^{2}, 295
14990	*m*-Xylene, 4-ethyl $1,3-(CH_3)_2-4-C_2H_5C_6H_3$	134.22	....	188.4, 66.5^{10}	−62.9	0.8763$^{20/4}$	1.5038^{20}	**al, eth, ace, bz,** peth	B5^{4}, 1070
14991	*m*-Xylene, 5-ethyl $1,3-(CH_3)_2-5-C_2H_5C_6H_3$	134.22	....	183.7, 63^{10}	−84.3	0.8648$^{20/4}$	1.4981^{20}	**al, eth, ace**, bz, peth	B5^{4}, 1071
14992	*m*-Xylene, 2-iodo $2-I-1,3-(CH_3)_2C_6H_3$	232.06	oil	229-30, 102-3^{14}	11.2	1.6158$^{20/4}$	1.6035^{20}	ace, bz	B5^{4}, 947
14993	*m*- Xylene, 4-iodo $4-I-1,3-(CH_3)_2C_6H_3$	232.06	....	231d, 111^{14}	....	1.6282$^{16/4}$	1.6008^{16}	ace, bz	B5^{4}, 947
14994	*m*-Xylene, 5-iodo $5-I-1,3-(CH_3)_2C_6H_3$	232.06	oil	230-1, 117^{27}	....	1.6085$^{18.5/4}$	1.5967$^{18.5}$	ace	B5^{3}, 840
14995	*m*-Xylene, 2-methoxy $2-CH_3O-1,3-(CH_3)_2C_6H_3$	136.19	....	182-3	....	0.9619$^{14/4}$	1.5053^{14}	al, eth, bz	B6^{3}, 1737
14996	*m*-Xylene, 4-methoxy $4-CH_3O-1,3-(CH_3)_2C_6H_3$	136.19	....	192, 83-4^{15}	....	0.9740$^{16/4}$	1.5190^{16}	al, eth, bz	B6^{3}, 1744
14997	*m*-Xylene, 5-methoxy $5-CH_3O-1,3-(CH_3)_2C_6H_3$	136.19	....	194.5, 89^{15}	....	0.9627$^{15/4}$	1.5110^{20}	al, eth, bz, aa	B6^{3}, 1756
14998	*m*-Xylene, 2-nitro $2-NO_2-1,3-(CH_3)_2C_6H_3$	151.16	....	222, 84.6$^{0.05}$	13	1.112^{15}	1.5202^{20}	al	B5^{4}, 948
14999	*m*-Xylene, 4-nitro $4-NO_2-1,3-(CH_3)_2C_6H_3$	151.16	....	246^{744}, 122^{18}	9	1.135^{15}	1.5473^{25}	eth, ace, bz	B5^{4}, 948
15000	*m*-Xylene, 5-nitro $5-NO_2-1,3-(CH_3)_2C_6H_3$	151.17	nd (al)	273^{739}	75	....	....	al, eth	B5^{4}, 948
15001	*m*-Xylene, 2,4,5,6-tetrabromo $2,4,5,6-Br_4-1,3-(CH_3)C_6$	421.75	nd (xyl, al)	....	248 (252)	....	....	bz	B5^{3}, 839
15002	*m*-Xylene, 2,4,6-trinitro $2,4,6-(NO_2)_3-1,3-(CH_3)_2-C_6H$	241.16	pa ye pr or lf (al-bz)	....	184	1.604^{19}	....	bz, chl, Py	B5^{4}, 950
15003	*p*-Xylene or 1,4-Dimethyl benzene $1,4-(CH_3)_2C_6H_4$	106.17	mcl pr (al)	138.3, 27.2^{10}	13.3	0.8611$^{20/4}$	1.4958^{20}	**al, eth, ace, bz**	B5^{4}, 951
15004	*p*-Xylene, 2-chloro $1,4-(CH_3)_2C_6H_3Cl-(2)$	140.61	....	187	1.6	1.0589$^{15/4}$	....	ace, bz	B5^{4}, 965
15005	*p*-Xylene, 2,3-dinitro $1,4-(CH_3)_2-2,3-(NO_2)_2C_6H_2$	196.16	mcl pr (al)	....	93	....	....	al, eth, ace, bz, chl	B5^{4}, 972
15006	*p*-Xylene, 2,5-dinitro $1,4-(CH_3)_2-2,5-(NO_2)_2C_6H_2$	196.16	ye nd (al)	....	147-8	....	....	ace, bz, chl	B5^{4}, 973
15007	*p*-Xylene, 2,6-dinitro $1,4-(CH_3)_2-2,6(NO_2)_2C_6H_2$	196.16	nd (al)	....	123-4	....	....	al, ace, bz, chl	B5^{4}, 973
15008	*p*-Xylene, 2-ethyl $1,4-(CH_3)_2-2-C_2H_5C_6H_3$	134.22	....	186.9, 64.9^{10}	−53.7	0.8772$^{20/4}$	1.5043^{20}	**al, eth, ace, bz,** peth	B5^{4}, 1070
15009	*p*-Xylene, 2-iodo $2-I-1,4-(CH_3)_2C_6H_3$	232.06	....	227-8d, 106-8^{13}	....	1.6168$^{17/4}$	1.5992^{17}	ace, bz	B5^{4}, 970

No.	Name, Synonyms, and Formula	Mol. wt.	Color, crystalline form, specific rotation and λ_{max} (log ε)	b.p. °C	m.p. °C	Density	n_D	Solubility	Ref.
15010	*p*-Xylene, 2-methoxy 2-CH_3O-1,4-$(CH_3)_2C_6H_3$	136.19	....	194^{772}	....	$0.9693^{13/4}$	1.5182^{13}	al, eth, bz, peth	$B6^3$, 1772
15011	*p*-Xylene, 2-nitro 2-NO_2-1,4-$(CH_3)_2C_6H_3$	151.16	pa ye	240-1, $64\text{-}5^{0.35}$	....	1.132^{15}	1.5413^{20}	al	$B5^4$, 971
15012	*p*-Xylene, 2,3,5-trinitro 2,4,5-$(NO_2)_3$-1,4-$(CH_3)_2C_6H$	241.16	mcl nd (al), lf (al-bz)	410 exp	139-40	1.59^{19}	....	al	$B5^3$, 864
15013	2,3-Xylidine or 2,3-Dimethyl aniline 2,3-$(CH_3)_2C_6H_3NH_2$	121.18	....	221-2, 106^{15}	<−15	0.9931^{20}	1.5684^{20}	al, eth	$B12^3$, 2438
15014	2,4-Xylidine or 2,4-Dimethyl aniline 2,4-$(CH_3)_2C_6H_3NH_2$	121.18	....	214, 91^{10}	−14.3	$0.9723^{20/4}$	1.5569^{20}	al, eth, bz	$B12^3$, 2469
15015	2,5-Xylidine 2,5-$(CH_3)_2C_6H_3NH_2$	121.18	ye lf (lig)	214, $92\text{-}100^{10}$	15.5	$0.9790^{21/4}$	1.5591^{21}	eth	$B12^3$, 2503
15016	2,6-Xylidine or 2,6-Dimethyl aniline 2,6-$(CH_3)_2C_6H_3NH_2$	121.18	....	214^{739}	11.2	0.9842^{20}	1.5610^{20}	al, eth	$B12^3$, 2462
15017	3,4-Xylidine or 3,4-Dimethyl aniline 3,4-$(CH_3)_2C_6H_3NH_2$	121.18	pl or pr (lig)	228	51	1.076^{18}	....	eth, lig	$B12^3$, 2443
15018	3,5-Xylidine or 3,5-Dimethyl aniline 3,5-$(CH_3)_2C_6H_3NH_2$	121.18	....	220-1, $99\text{-}100^{20}$	9.8	$0.9706^{20/4}$	1.5581^{20}	eth	$B12^3$, 2495
15019	Xylitol $C_5H_{12}O_5$, $HOCH(CHOH)_3CH_2OH$	152.15	(i) rh (al), (ii) mcl (al) (st)	(ii) 215-7	(i) 61, (ii) 93-4	....	....	w, al, Py	$B1^4$, 2832
15020	D-Xylose or Wood sugar $C_5H_{10}O_5$	150.13	mcl nd, $[\alpha]^{20/}{}_D$ +22.5 (chl)	....	145	$1.525^{20/4}$	....	w	$B1^4$, 4223
15021	D-Xylose osazone $C_{17}H_{20}N_4O_3$, $HOCH_2(CHOH)_2C(=NNHC_6H_5)CH=NNHC_6H_5$	328.37	pa ye nd, $[\alpha]_D$ −40.9 (al)	....	159(167 d)	....	....	al, eth, ace	B31, 61

No.	Name, Synonyms, and Formula	Mol. wt.	Color, crystalline form, specific rotation and λ_{max} (log ε)	b.p. °C	m.p. °C	Density	n_D	Solubility	Ref.
15022	Yamogenin $C_{27}H_{42}O_3$	414.63	pl, $[\alpha]^{25}{}_D$ −123		201				B19[4], 865
15023	Yobyrine $C_{19}H_{16}N_2$	272.35	nd(dil al)	$150^{0.01}$	218-9			al, chl	B23[2], 263
15024	Yohimbine or Corynine $C_{21}H_{26}N_2O_3$	354.45	nd (dil al)	$159^{0.01}$ sub	241			al, eth, chl	B25[2], 201
15025	Yohimbine hydrochloride $C_{21}H_{26}N_2O_3.HCl$	390.91	orh nd or pl (w, dil HCl)		302			w	B25[2], 204
15025	Yohimbine-hydrochloride $C_{21}H_{26}N_2O_3.HCl$	390.91	orh nd or pl (w, dil HCl)		302			w	B25[2], 201
15026	Yohimbine-nitrate $C_{21}H_{26}N_2O_3.HNO_3$	417.46	pr (w)		276			w	B25[2], 204
15027	Yuccagenin $C_{27}H_{42}O_4$	430.63	nd (al), $[\alpha]^{25}{}_D$ −113		252			al, eth, diox	B19[4], 1083

PHYSICAL CONSTANTS OF ORGANIC COMPOUNDS (Continued)

No.	Name, Synonyms, and Formula	Mol. wt.	Color, crystalline form, specific rotation and λ_{max} (log ε)	b.p. °C	m.p. °C	Density	n_D	Solubility	Ref.
15028	Zagadinine $C_{27}H_{43}NO_7$	493.64	orh (al), nd (bz), [α] −45 (chl)	....	201-4	....	....	al, chl	Am35, 258
15029	Zeaxanthin or Zeaxanthol $C_{40}H_{56}O_2$	568.88	ye pr (MeOH), rh (chl-eth)	$226\text{-}9^{0.06}$	215.5	....	....	eth, ace, bz, chl, Py	B6[3], 5865
15030	Zymosterol or 8(14).24(25′) cholestadienol $C_{27}H_{44}O$	384.65	pl (MeOH), nd, $[\alpha]^{20}/_D$ + 149 (chl)	$160^{0.001}$ sub	110	....	....	ace, chl	B6[3], 2828
15031	Zeaxanthin diacetate $C_{44}H_{60}O_4$	652.96	....	....	154-5	....	....	....	B6[3], 5868

STRUCTURAL FORMULAS OF ORGANIC COMPOUNDS

In Numeric as they occur in Organic Compounds Table

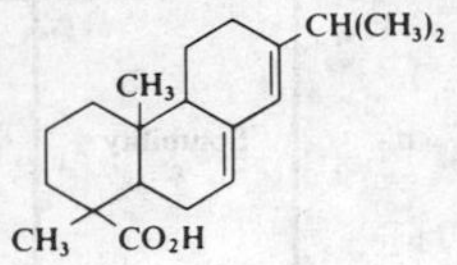

1

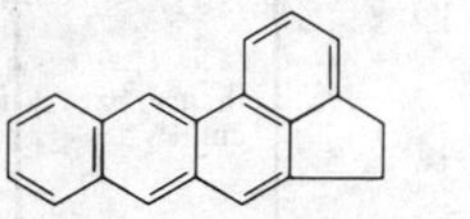
3

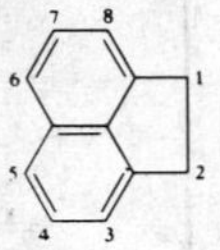

4

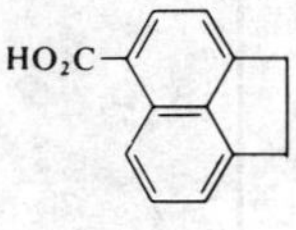

16

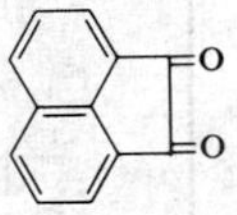

17

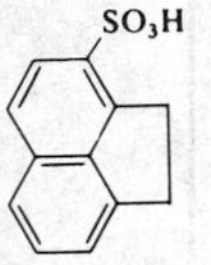

18

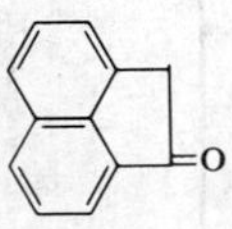

19

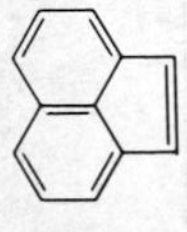
20

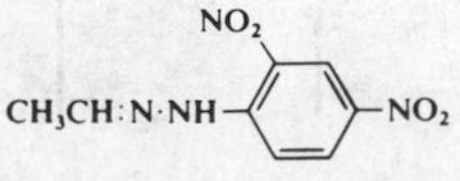

22

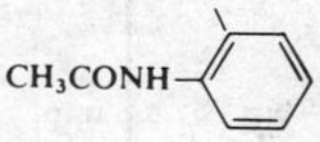

90

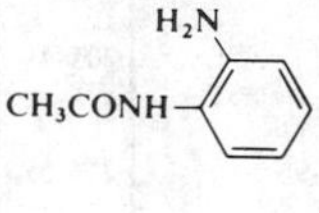

92

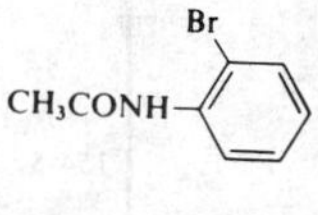

96

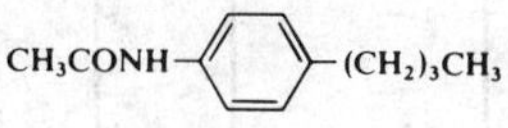

100

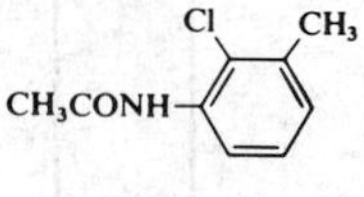

103

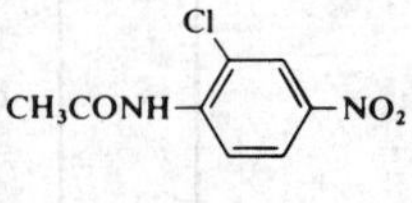

104

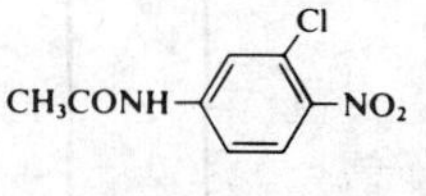

111

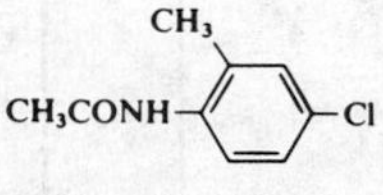

113

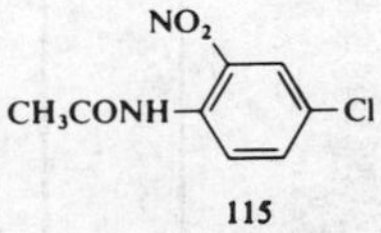

115

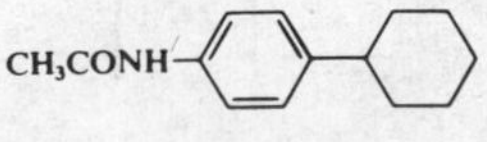

118

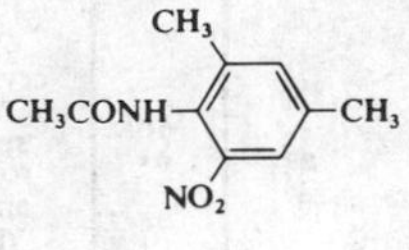

120

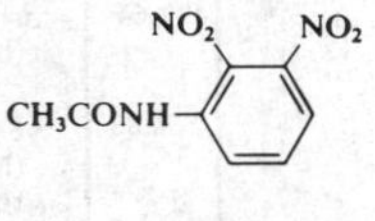

121

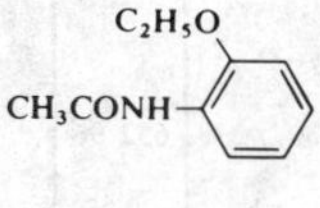

128

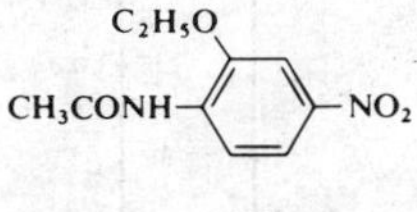

129

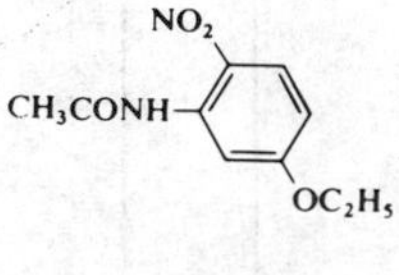

135

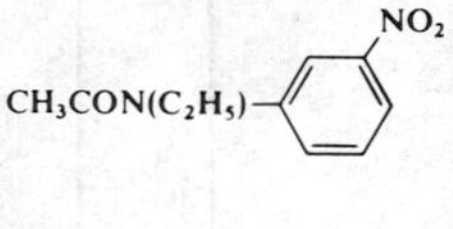

137

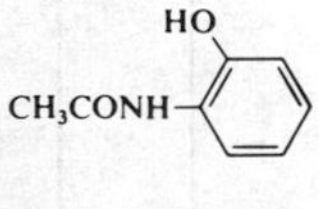

139

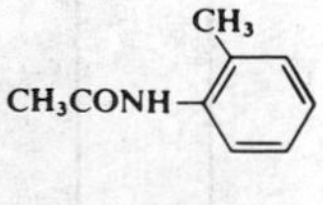

143

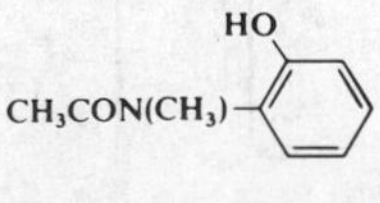

146

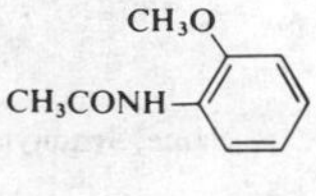

148

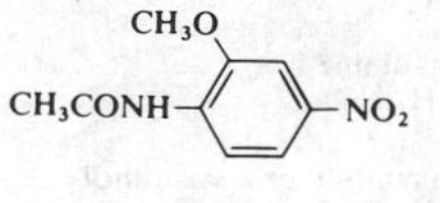

150

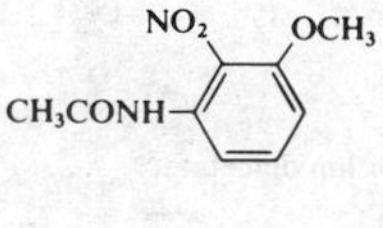

154

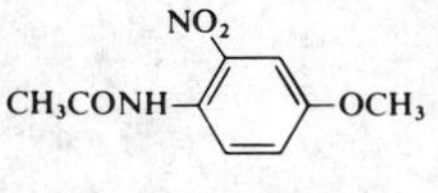

158

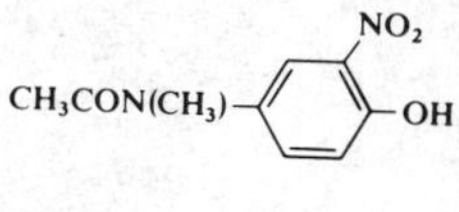

162

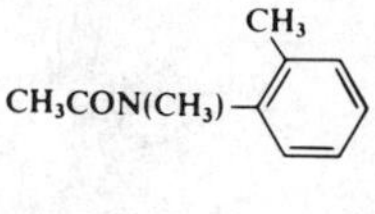

164

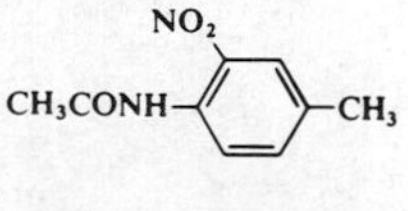

167

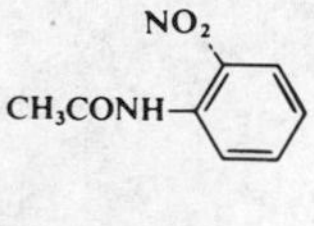

168

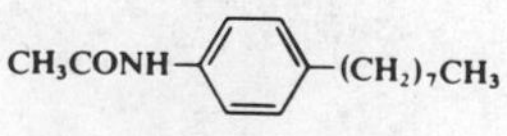

171

STRUCTURAL FORMULAS OF ORGANIC COMPOUNDS (Continued)

In Numeric as they occur in Organic Compounds Table

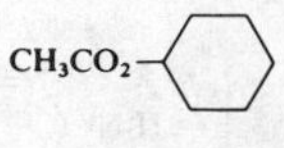

196

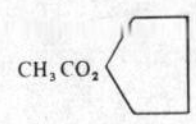

197

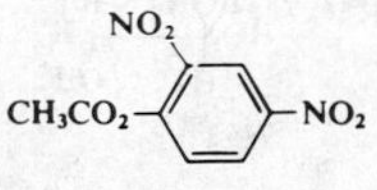

201

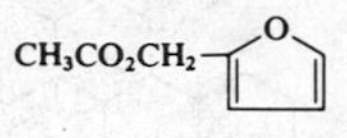

210

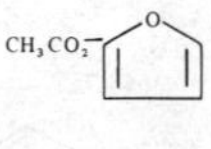

211

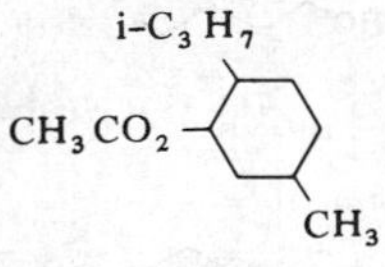

221

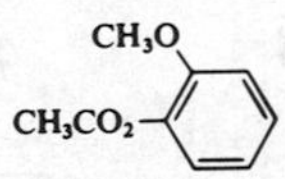

224

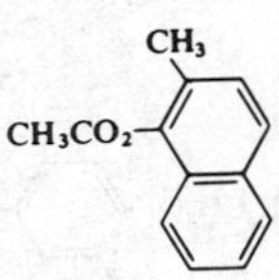

233

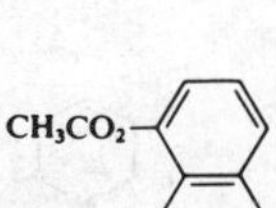

241

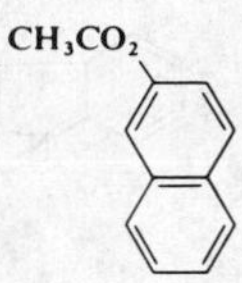

242

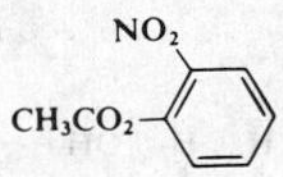

243

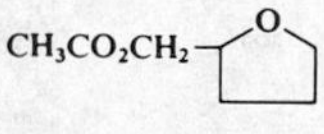

264

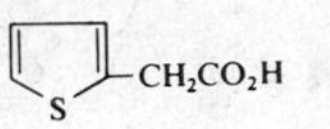

265

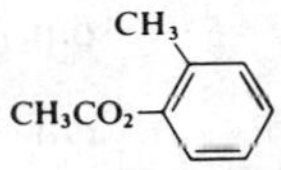

266

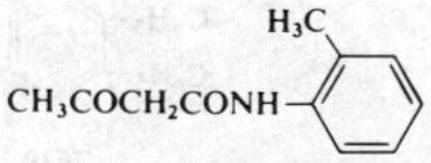

272

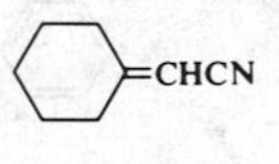

343

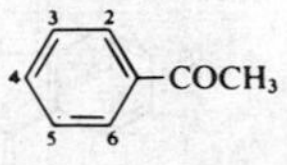

351

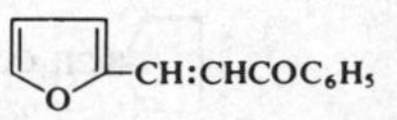

395

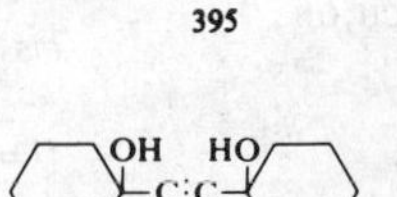

450

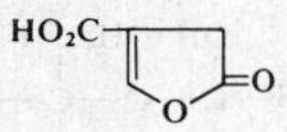

452

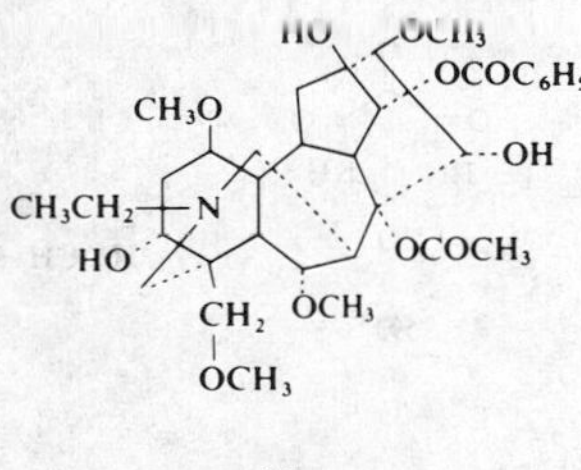

460

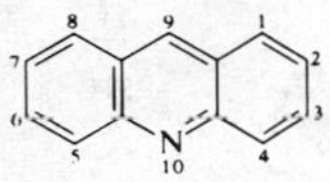

464

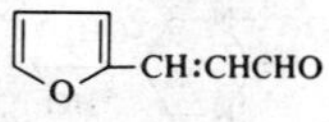

483

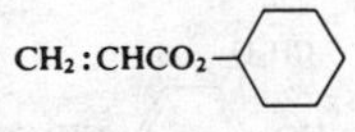

497

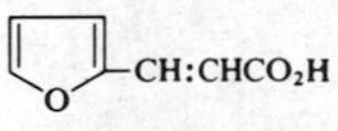

502

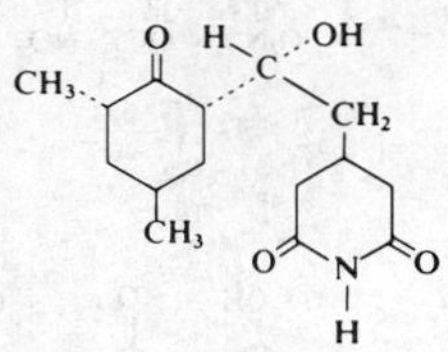

521

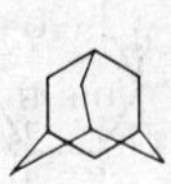

522

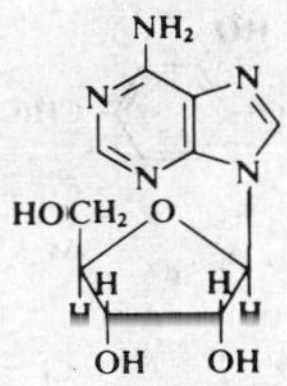

523

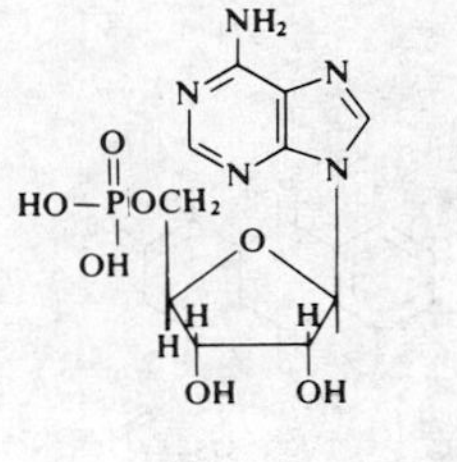

524

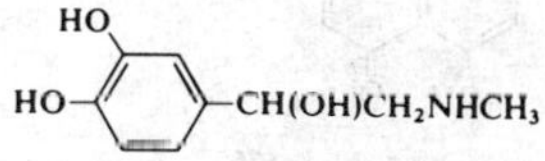

548

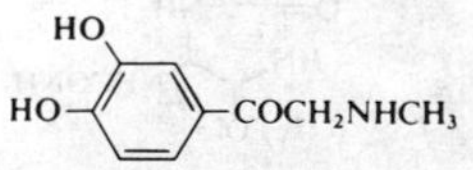

550

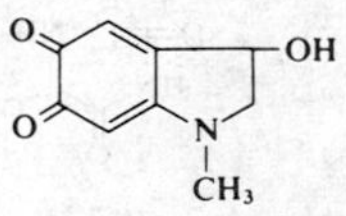

551

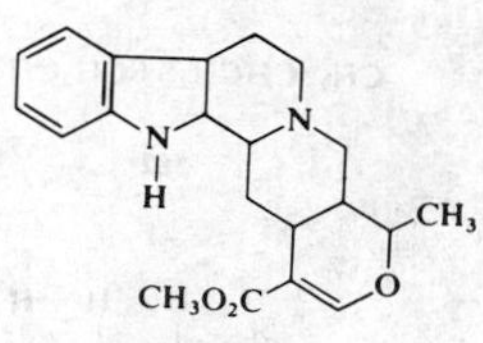

552

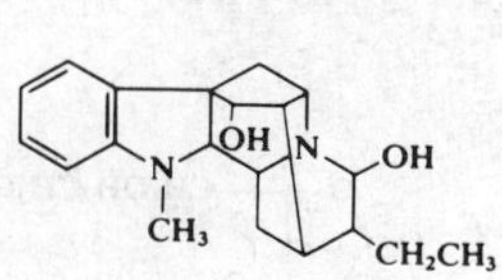

553

STRUCTURAL FORMULAS OF ORGANIC COMPOUNDS (Continued)

In Numeric as they occur in Organic Compounds Table

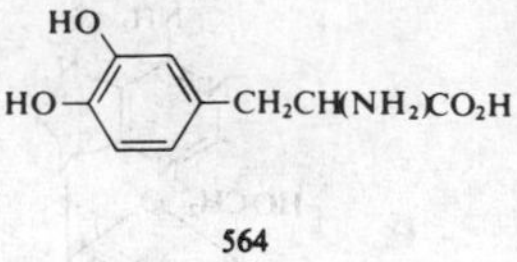

564

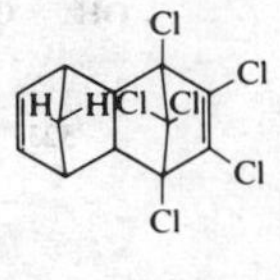

574

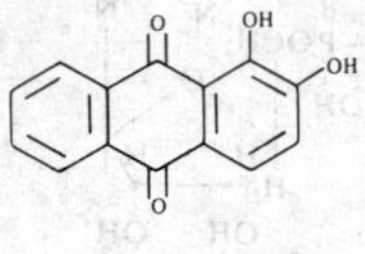

575

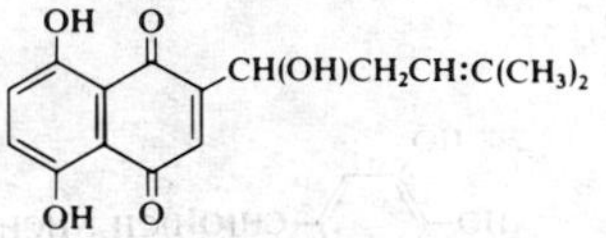

576

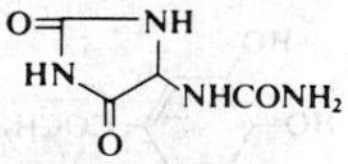

578

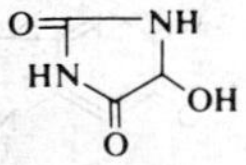

579

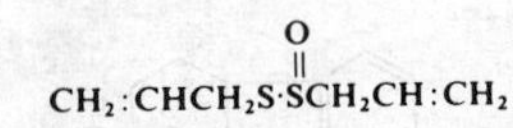

582

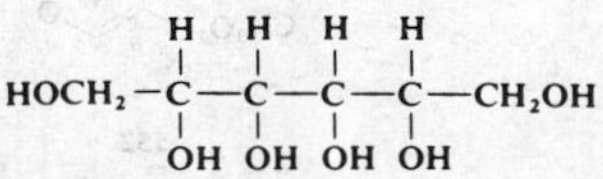

583

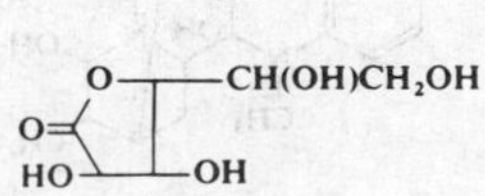

586

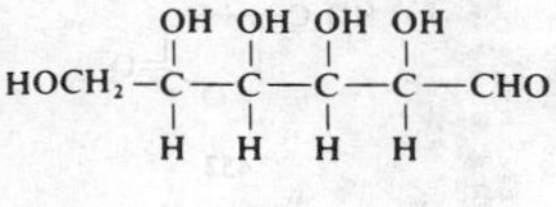

592

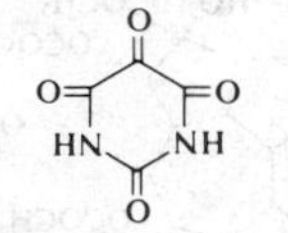

593

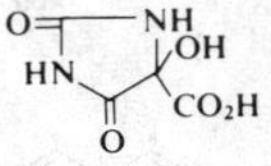

594

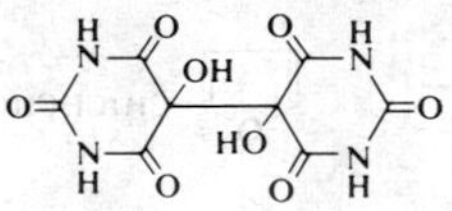

595

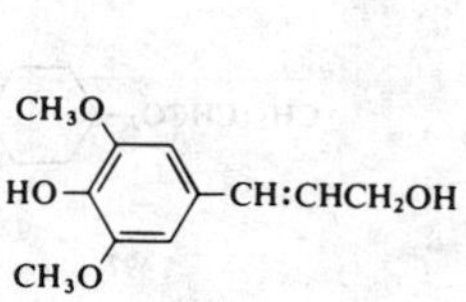

600

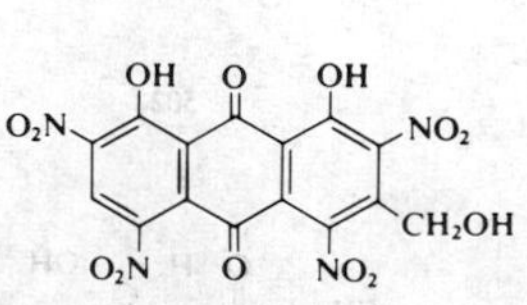

631

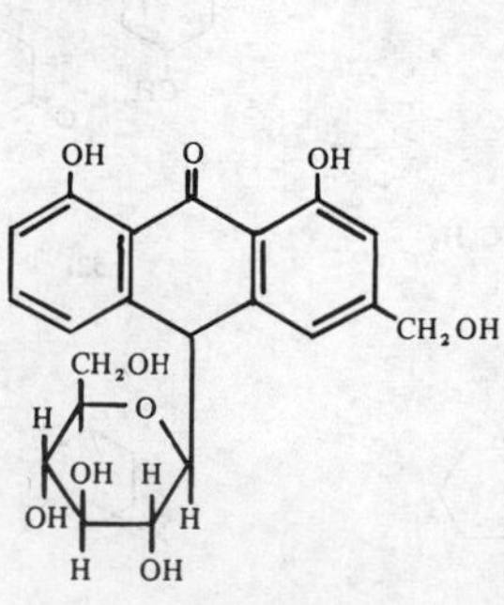

632

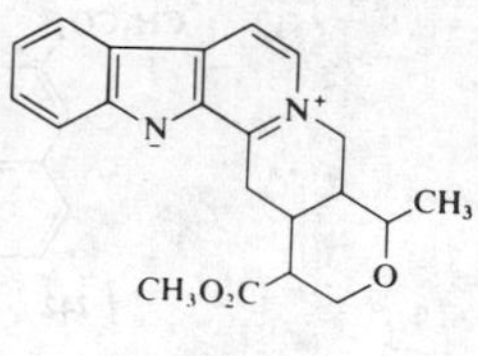

633

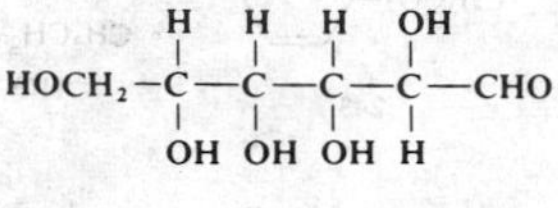

634

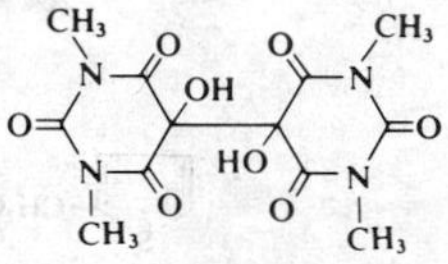

636

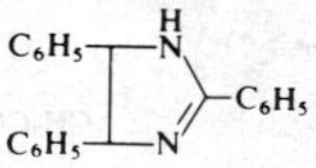

637

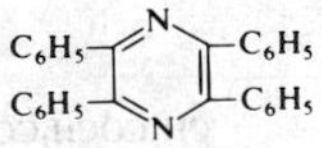

639

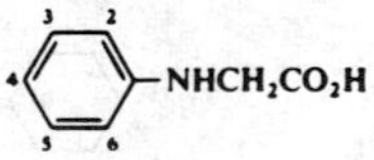

670

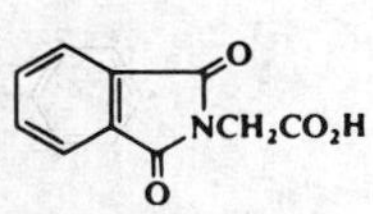

673

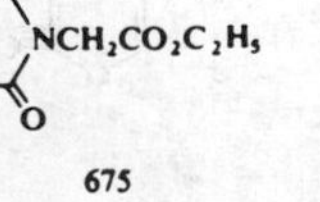

675

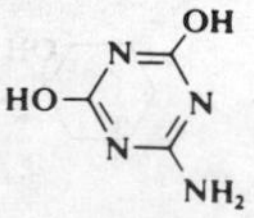

738

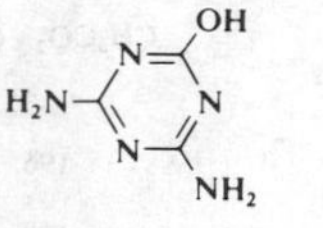

739

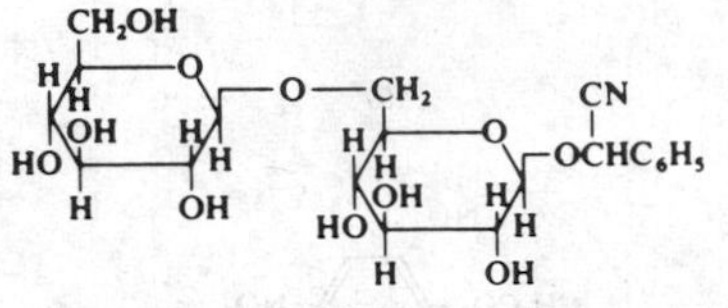

744

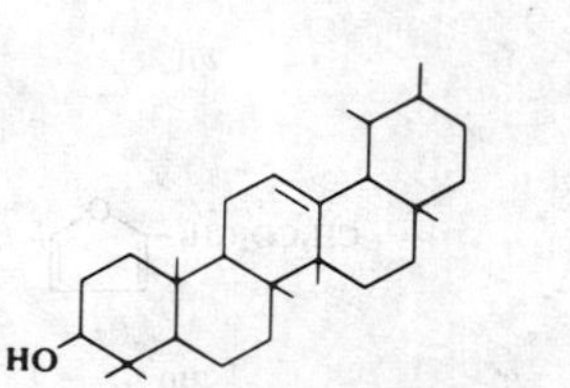

746

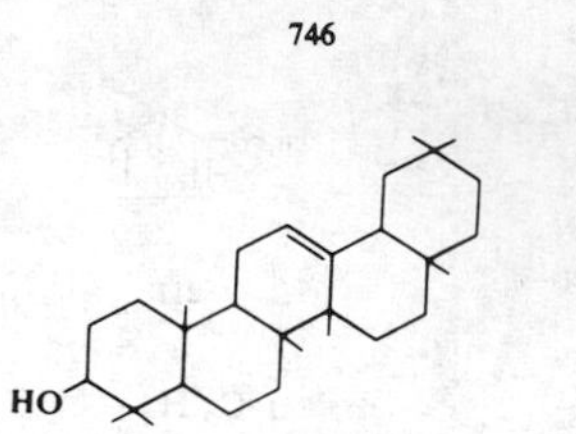

747

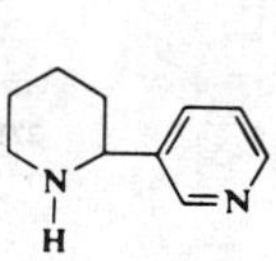

748

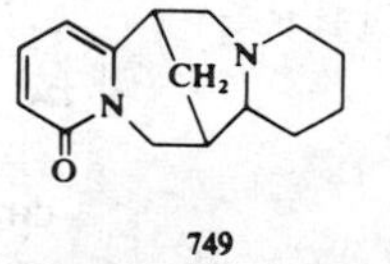

749

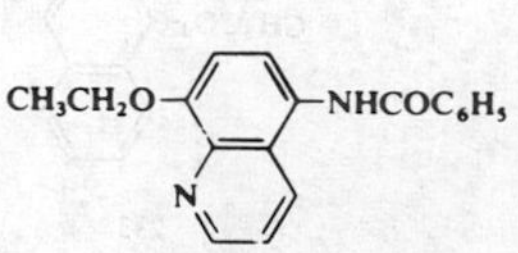

750

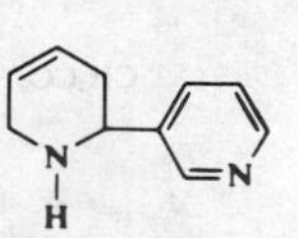

751

STRUCTURAL FORMULAS OF ORGANIC COMPOUNDS (Continued)

In Numeric as they occur in Organic Compounds Table

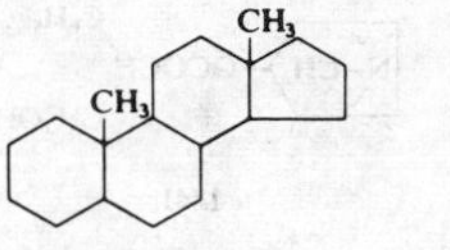

752

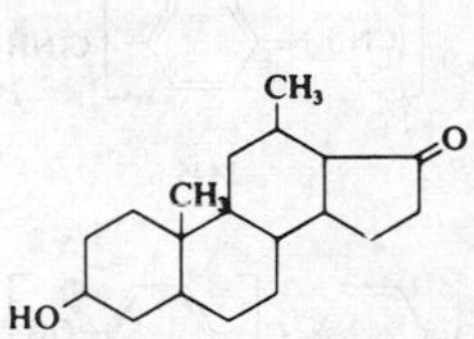

753

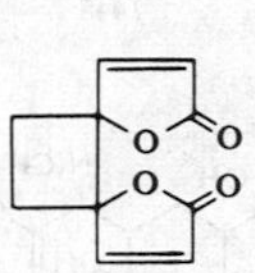

754

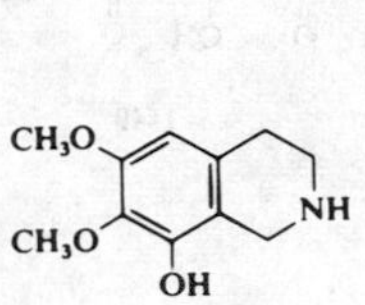

757

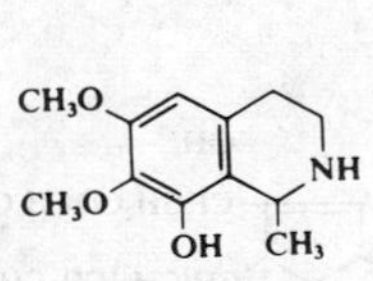

758

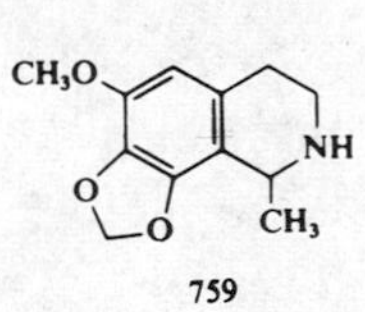

759

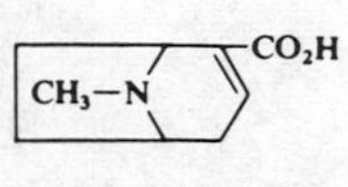

763

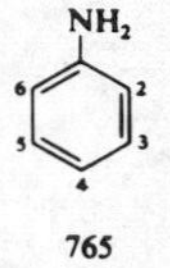

765

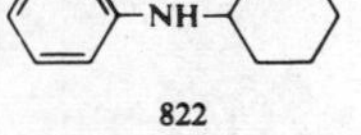

822

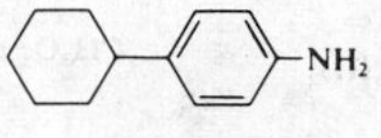

823

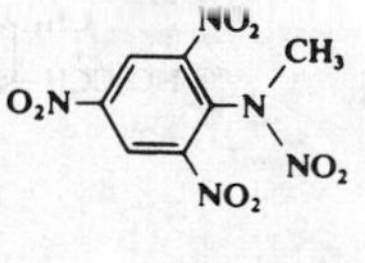

946

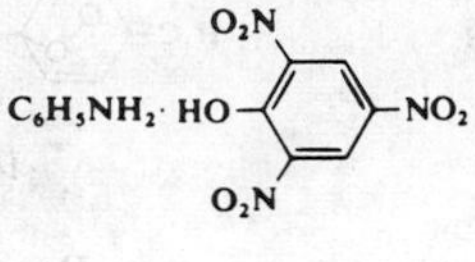

970

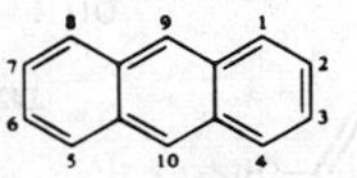

1096

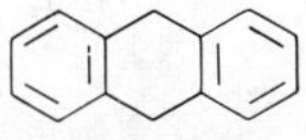

1111

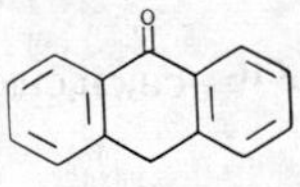

1117

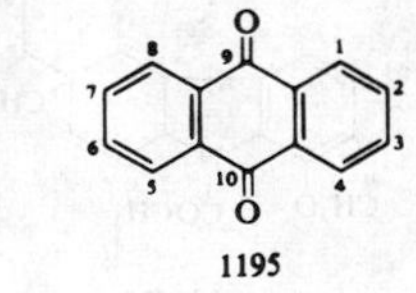

1195

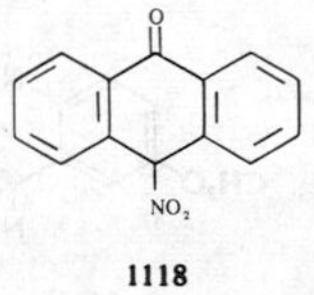

1118

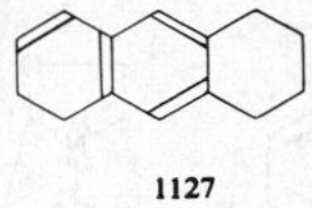

1127

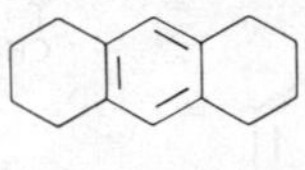

1135

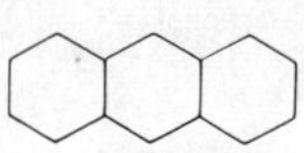

1138

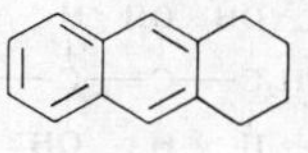

1139

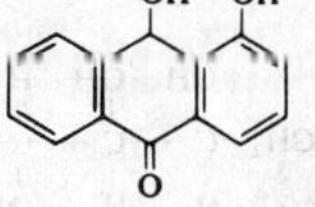

1146

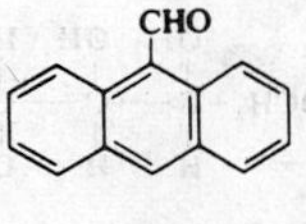

1150

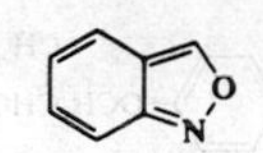

1151

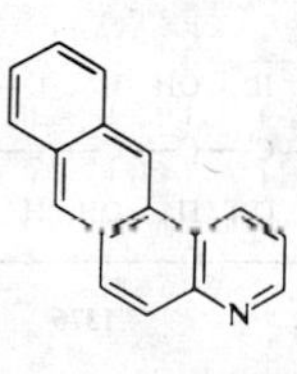

1194

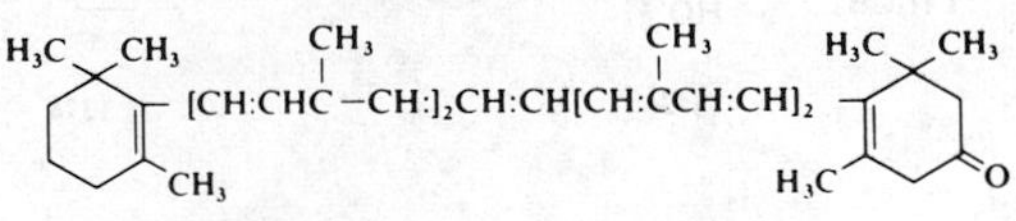

1336

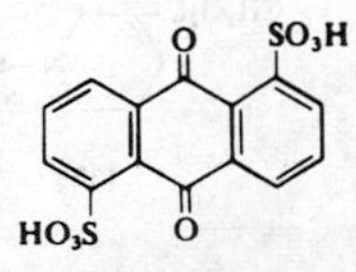

1273

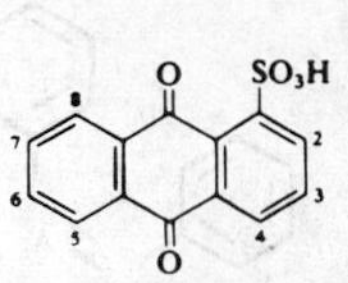

1292

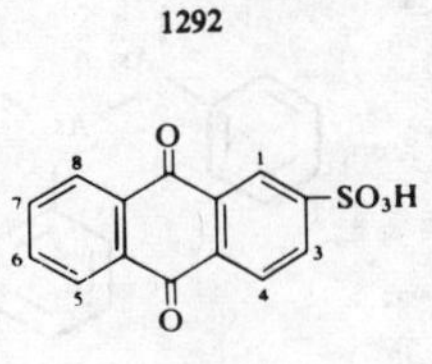

1294

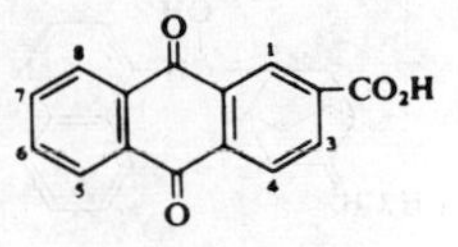

1313

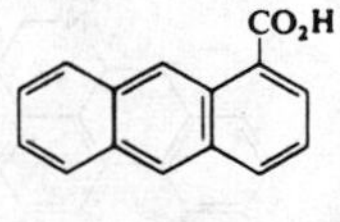

1321

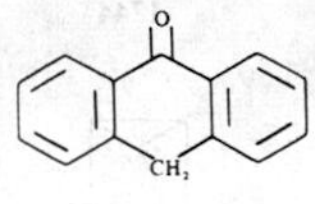

1324

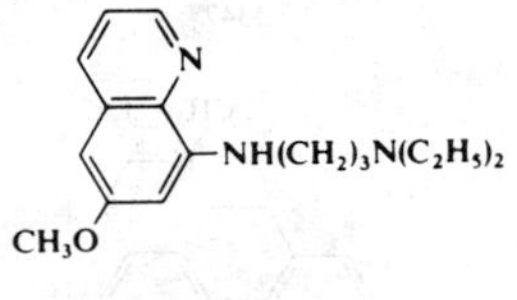

1326

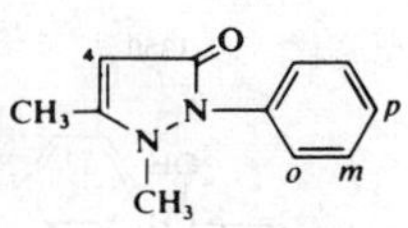

1327

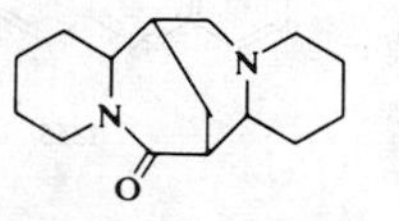

1337

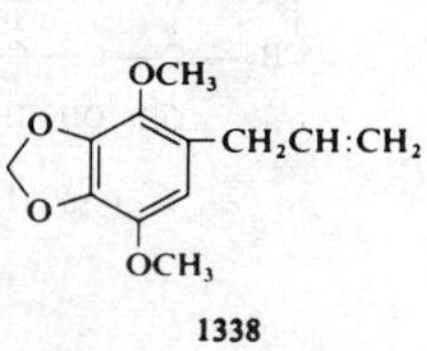

1338

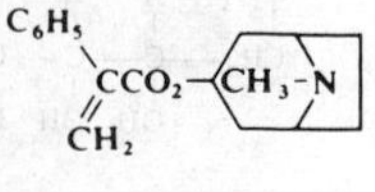

1339

STRUCTURAL FORMULAS OF ORGANIC COMPOUNDS (Continued)

In Numeric as they occur in Organic Compounds Table

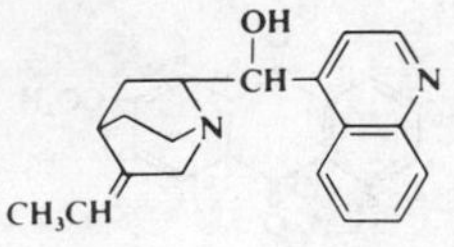

1342

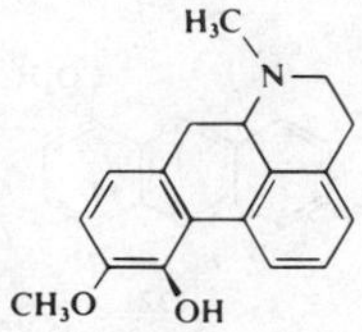

1344

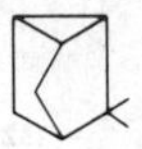

1347

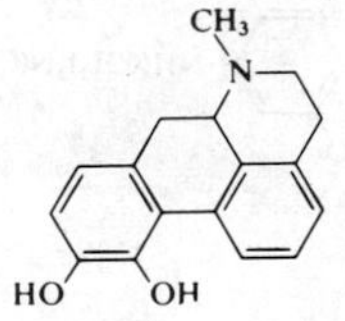

1350

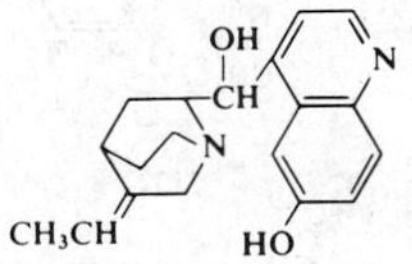

1353

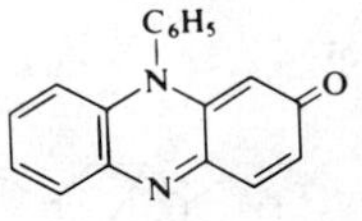

1355

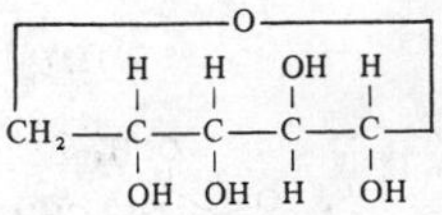

1356

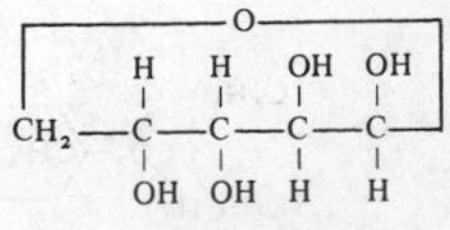

1359

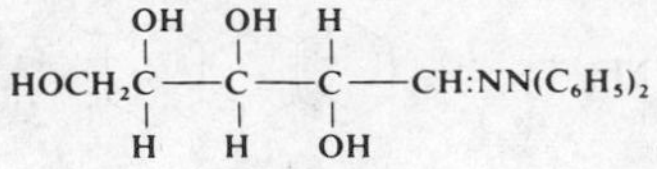

1363

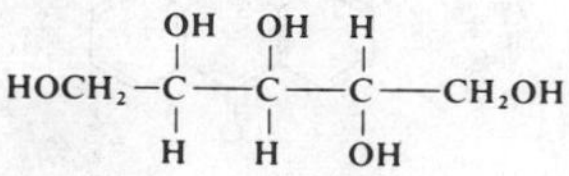

1366

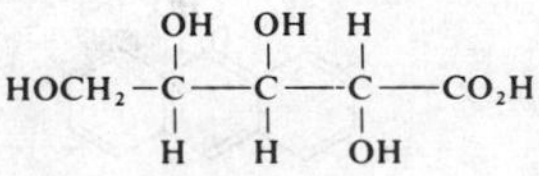

1369

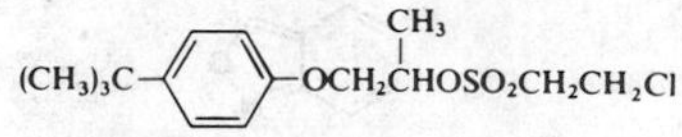

1375

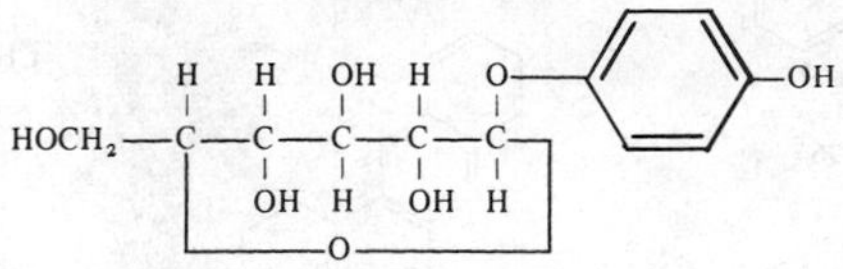

1376

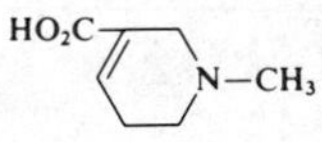

1378

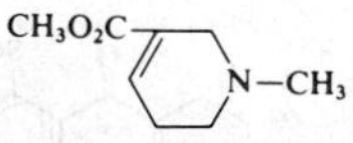

1379

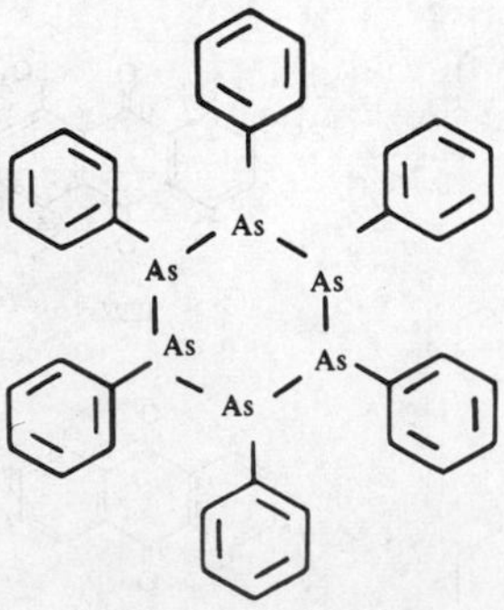

1394

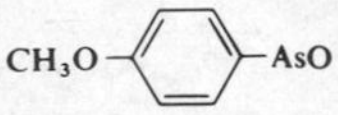

1410

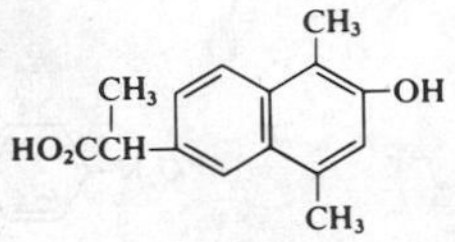

1422

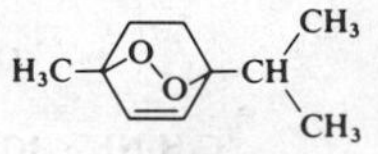

1423

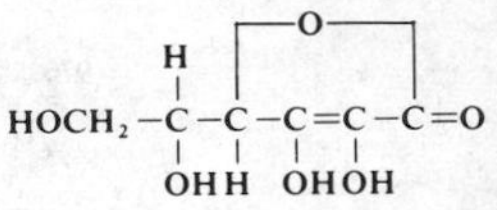

1424

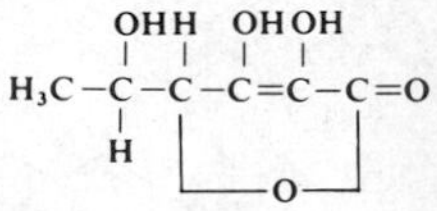

1427

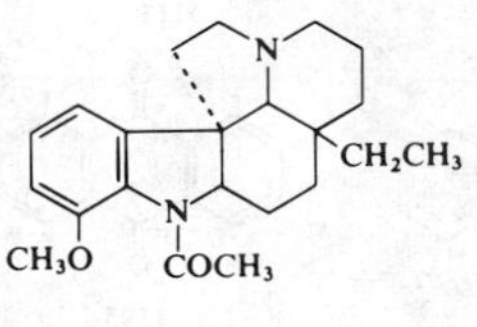

1435

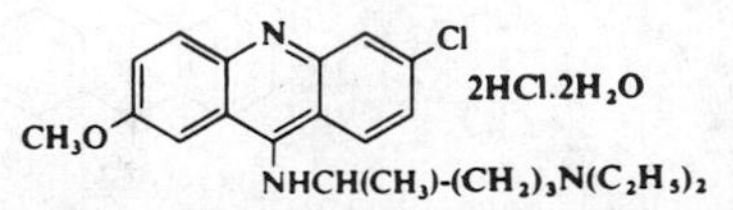

1436

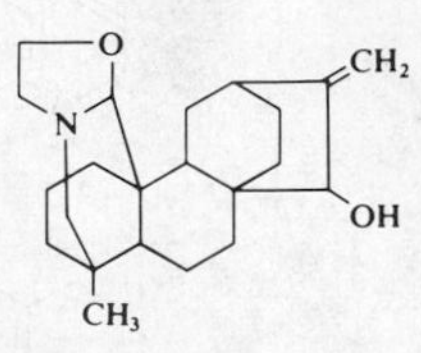

1437

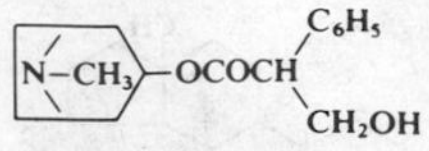

1441

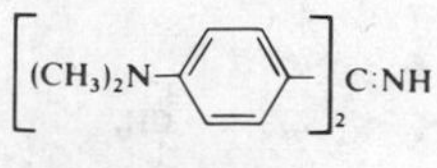

1445

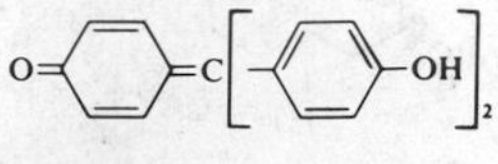

1448

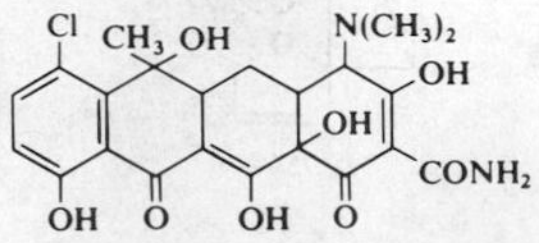

1449

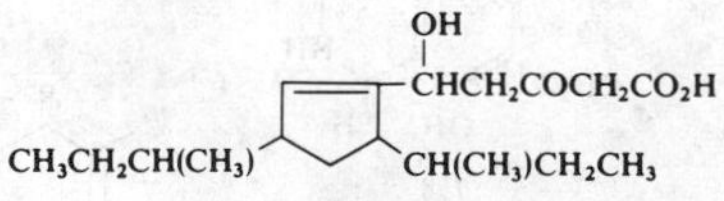

1451

STRUCTURAL FORMULAS OF ORGANIC COMPOUNDS (Continued)

In Numeric as they occur in Organic Compounds Table

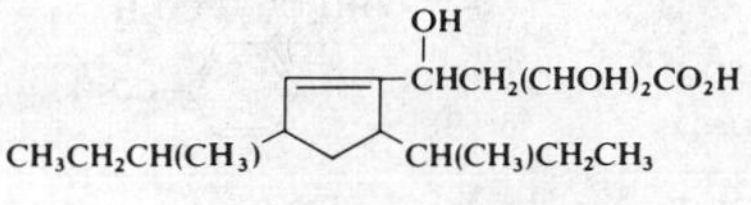

1452

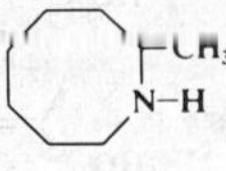

1461

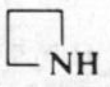

1462

1463

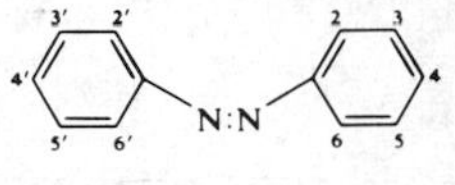

1464

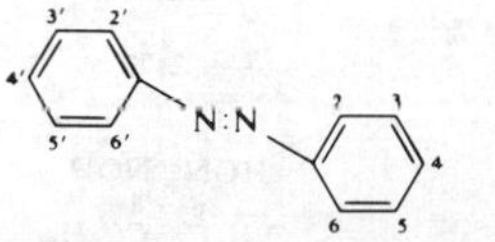

1465

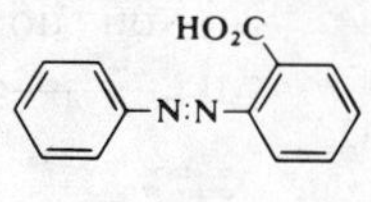

1559

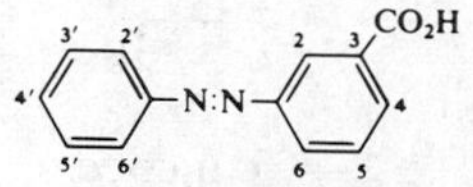

1561

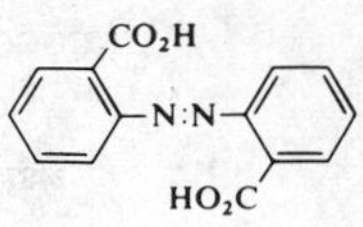

1568

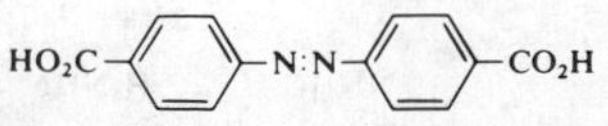

1574

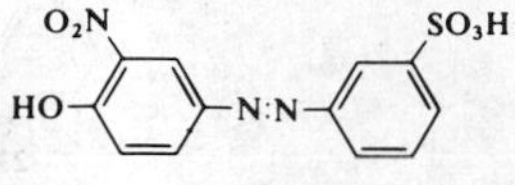

1580

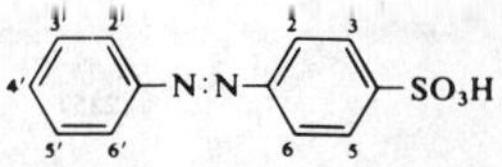

1581

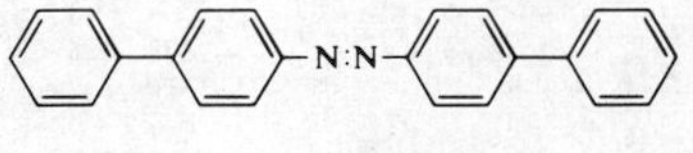

1583

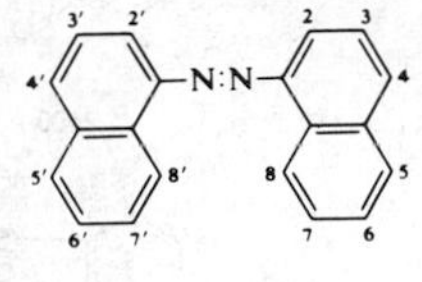

1586

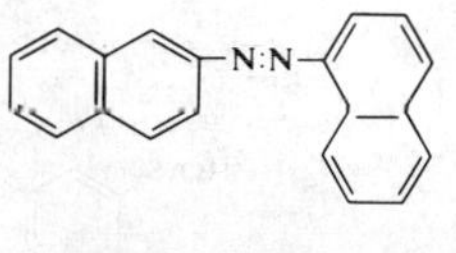

1588

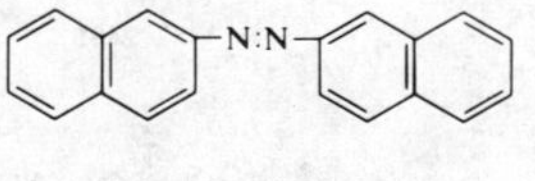

1589

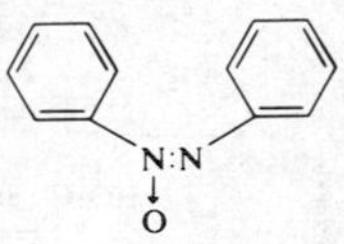

1590

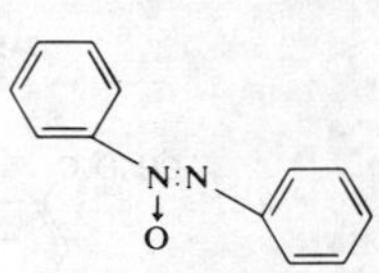

1591

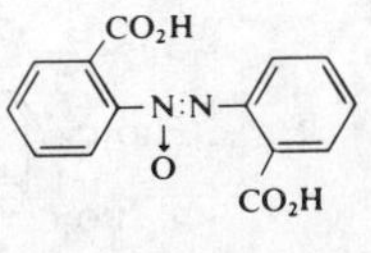

1598

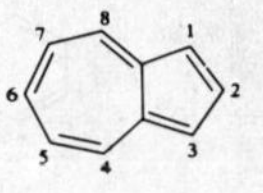

1603

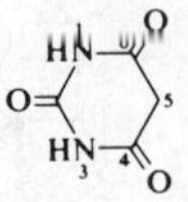

1608

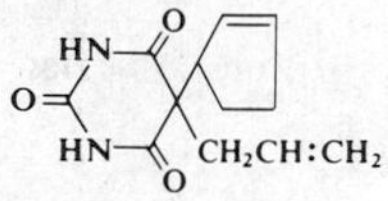

1610

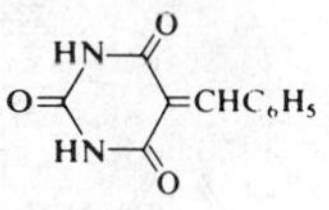

1616

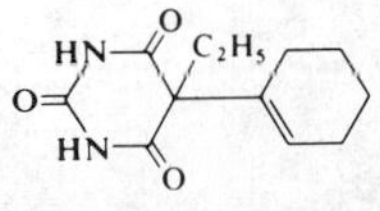

1619

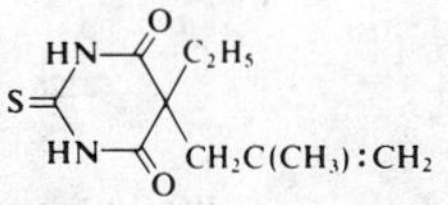

1626

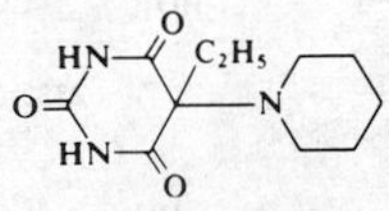

1635

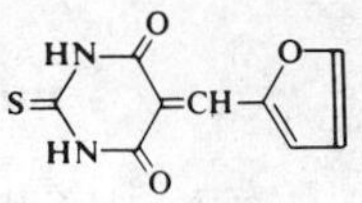

1638

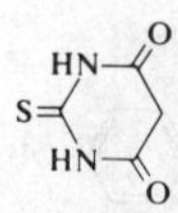

1644

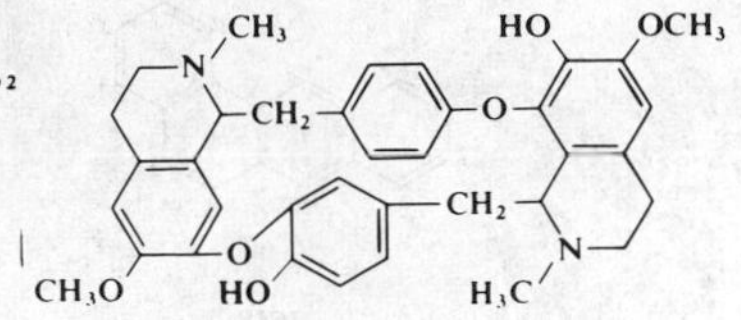

1645

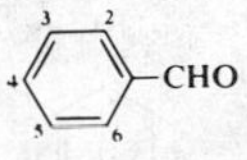

1654

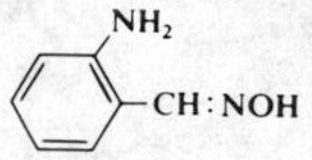

1660

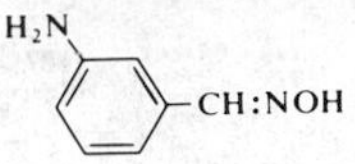

1662

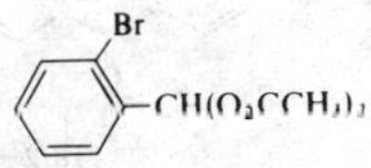

1667

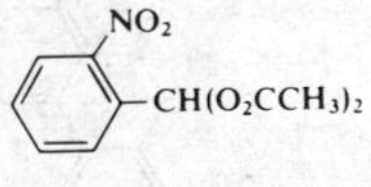

1784

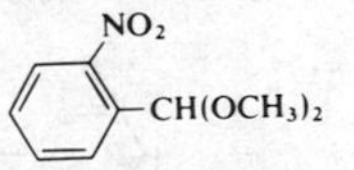

1785

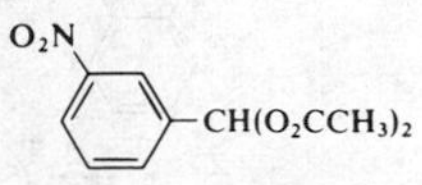

1787

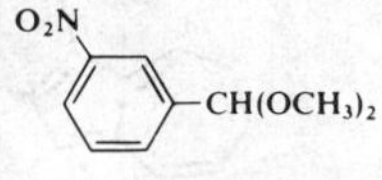

1788

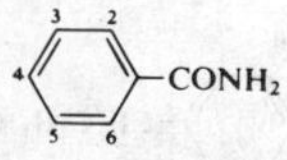

1809

STRUCTURAL FORMULAS OF ORGANIC COMPOUNDS (Continued)

In Numeric as they occur in Organic Compounds Table

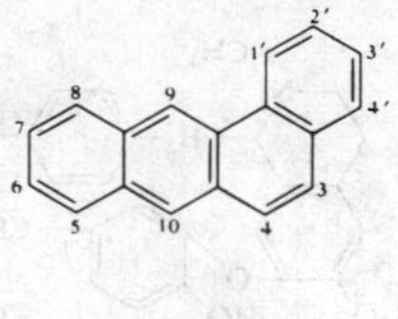

1858

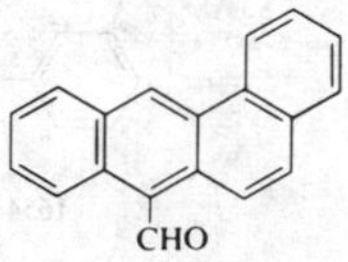

1870

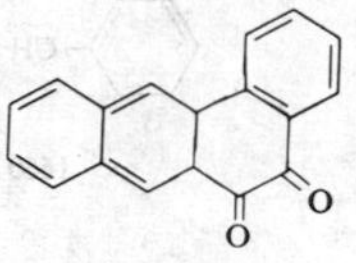

1871

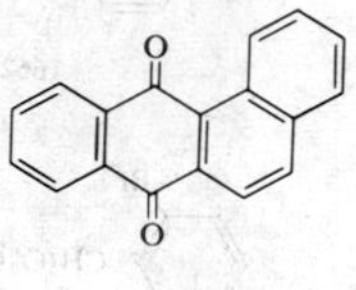

1872

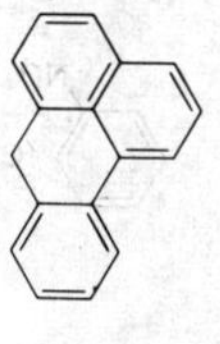

1873

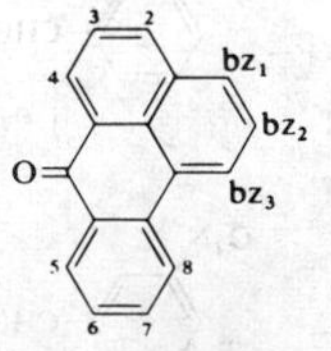

1874

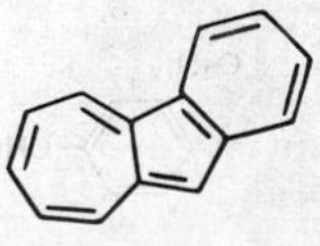

1879

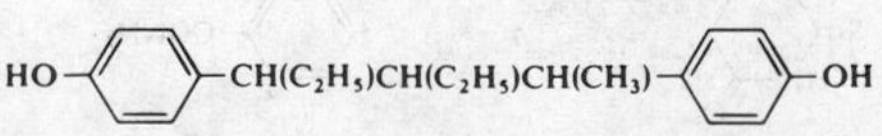

1881

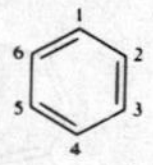

1882

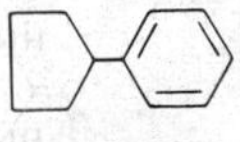

2058

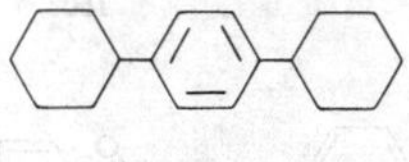

2106

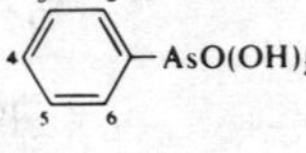

2298

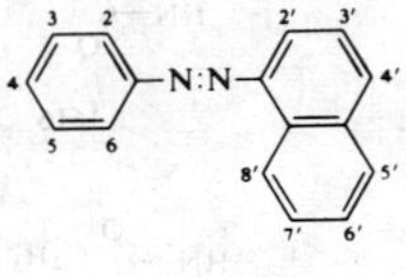

2314

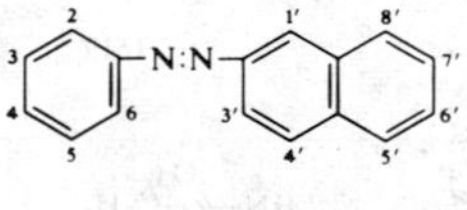

2326

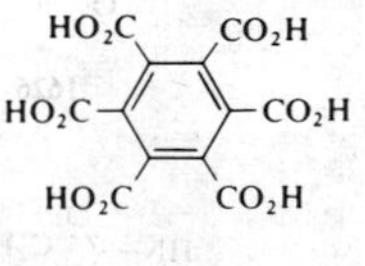

2333

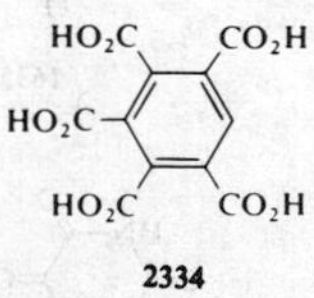

2334

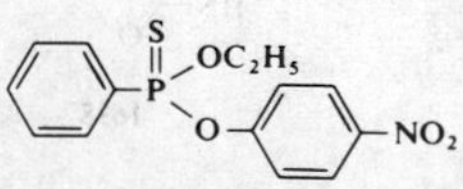

2340

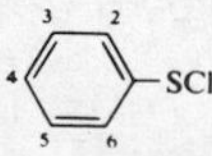

2344

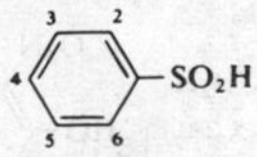

2348

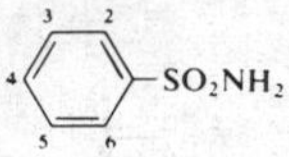

2357

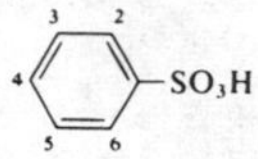

2378

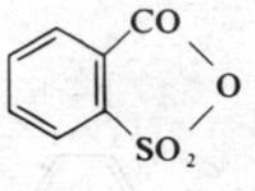

2400

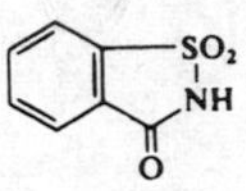

2401

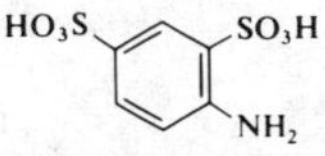

2451

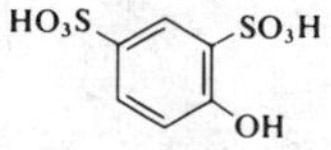

2452

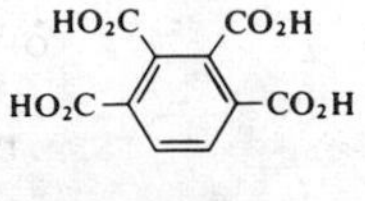

2453

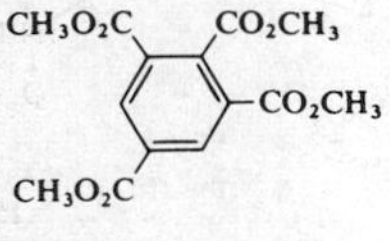

2455

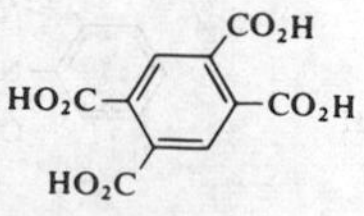

2456

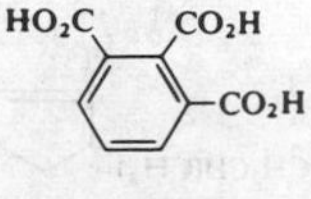

2459

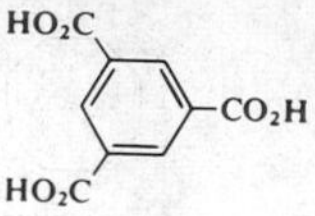

2462

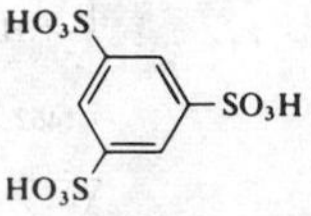

2467

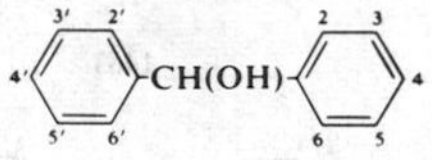

2468

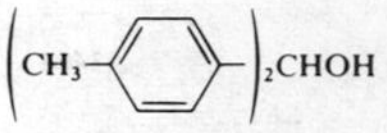

2470

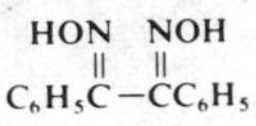

2483

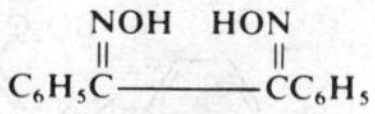

2484

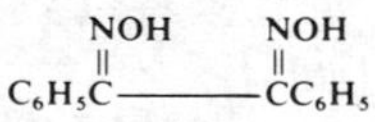

2485

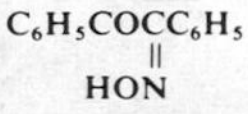

2486

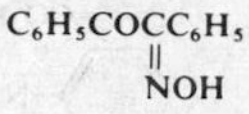

2487

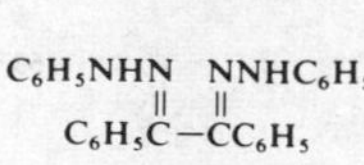

2488

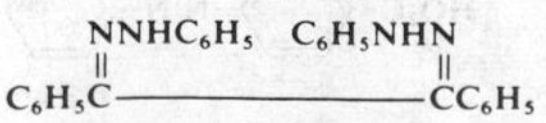

2489

STRUCTURAL FORMULAS OF ORGANIC COMPOUNDS (Continued)

In Numeric as they occur in Organic Compounds Table

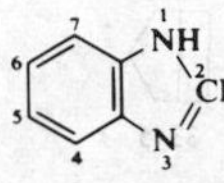

2493

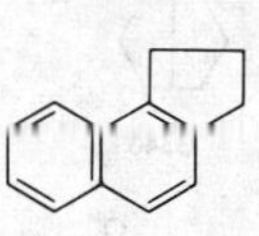

2506

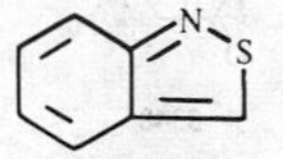

2507

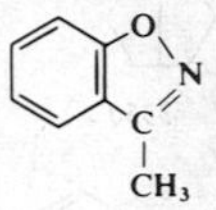

2509

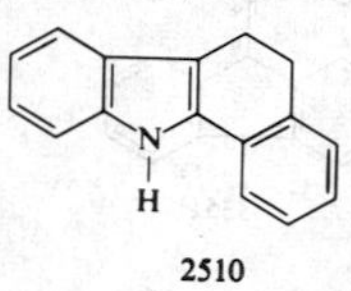

2510

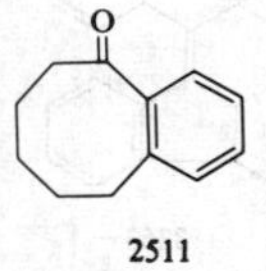

2511

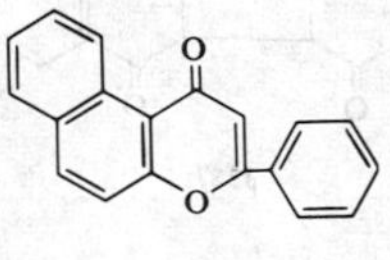

2513

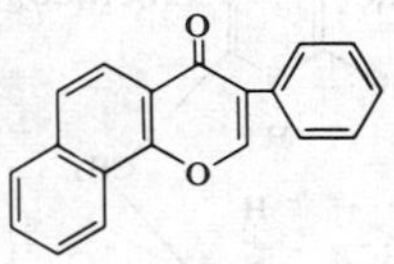

2514

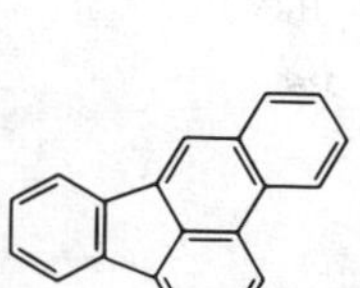

2515

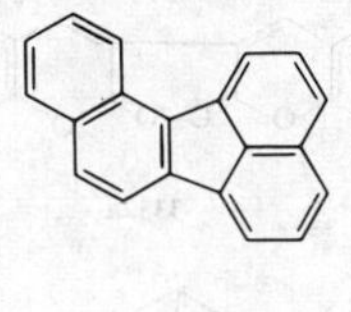

2516

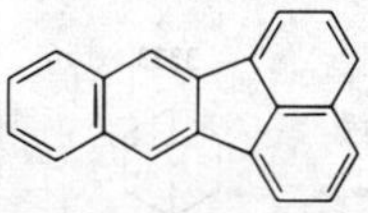

2517

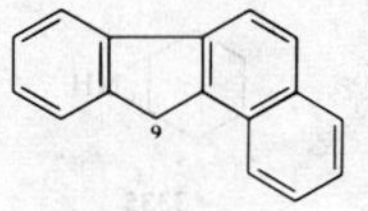

2518

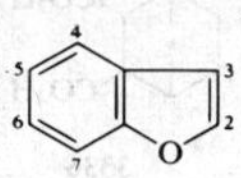

2520

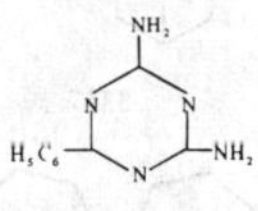

2528

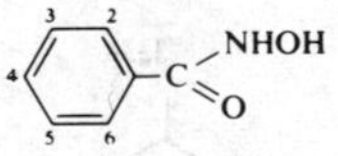

2529

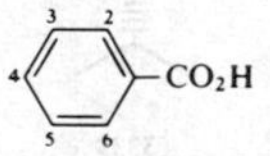

2531

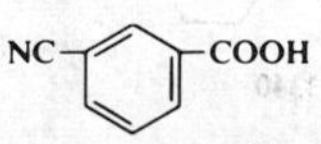

2625

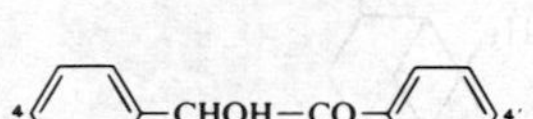

2837

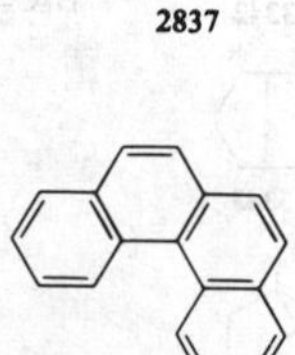

2879

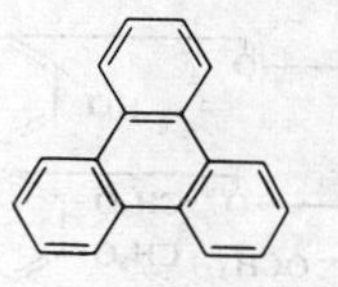

2880

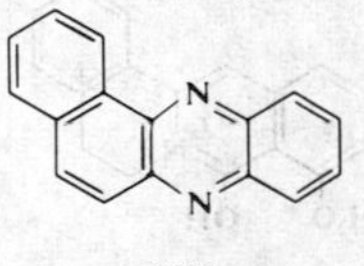

2881

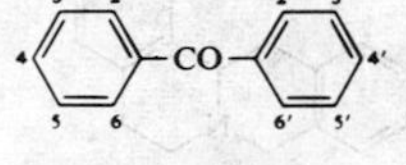

2882

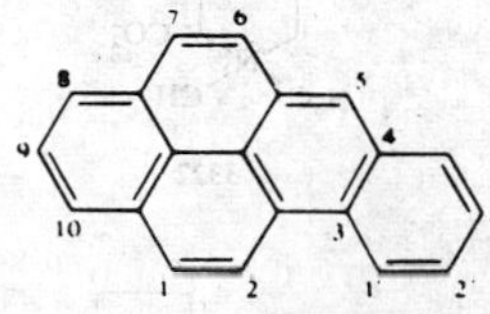

3008

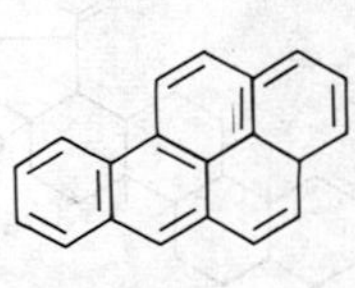

3009

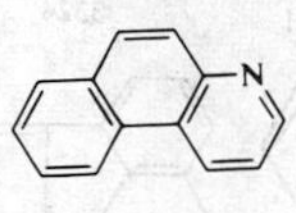

3013

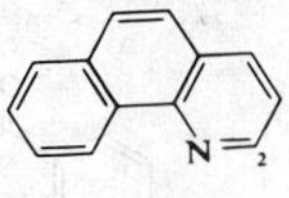

3014

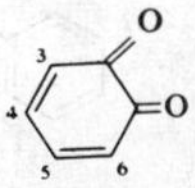

3016

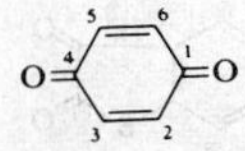

3023

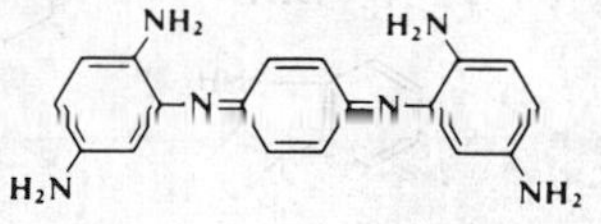

3031

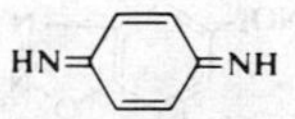

3046

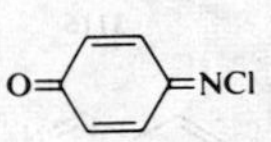

3048

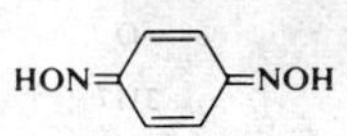

3053

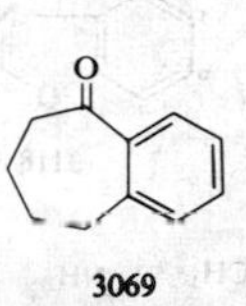

3069

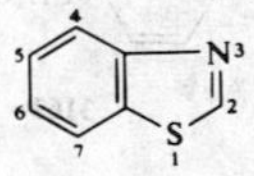

3078

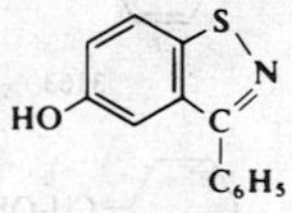

3094

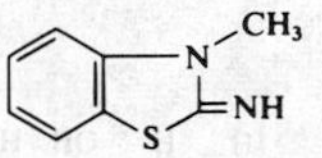

3108

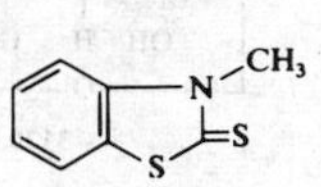

3109

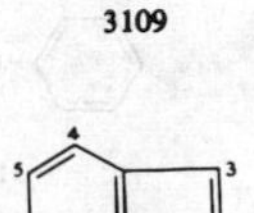

3110

In Numeric as they occur in Organic Compounds Table

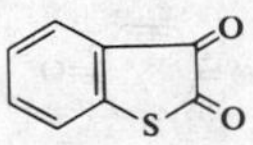

3114

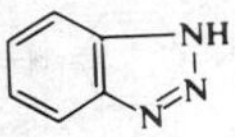

3115

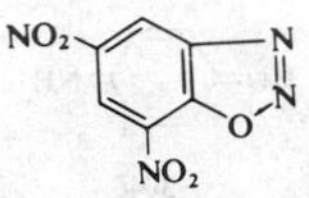

3116

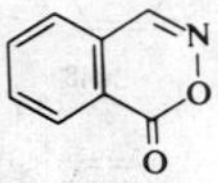

3117

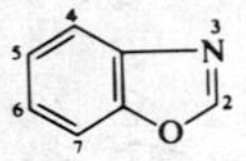

3118

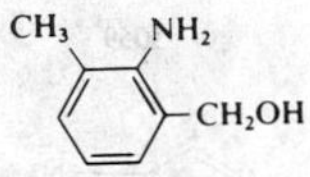

3162

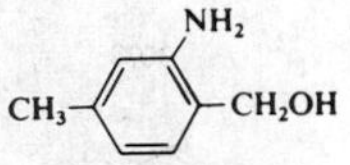

3163

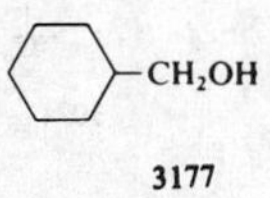

3177

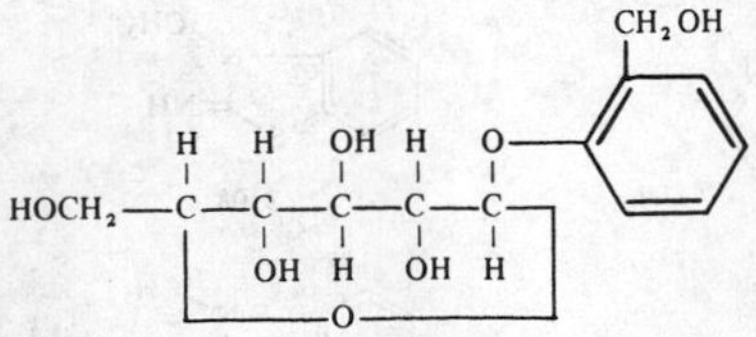

3179

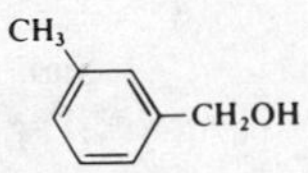

3186

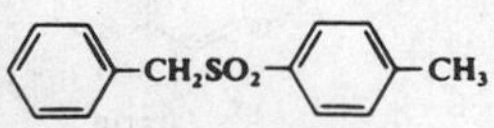

3289

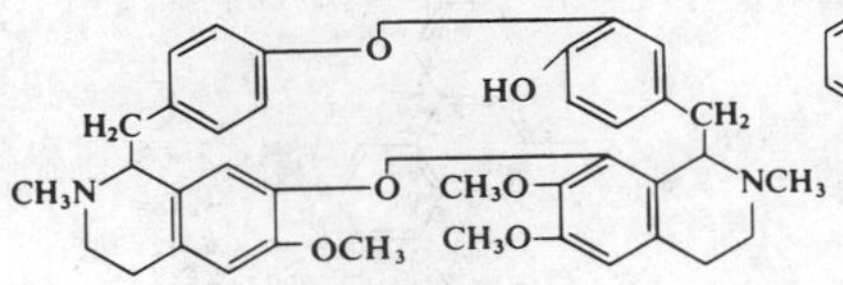

3314

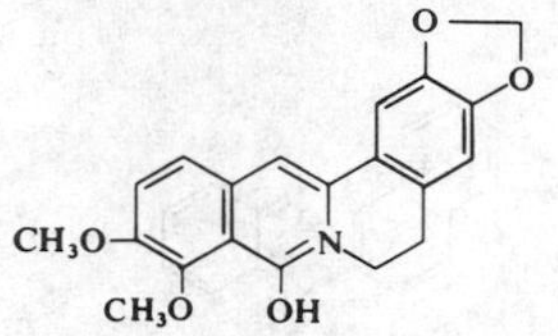

3315

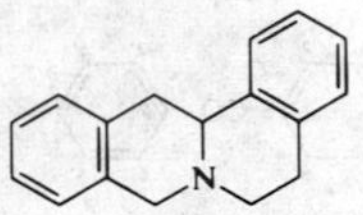

3320

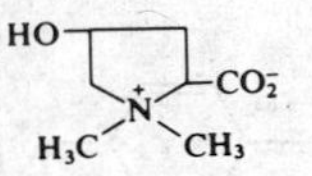

3322

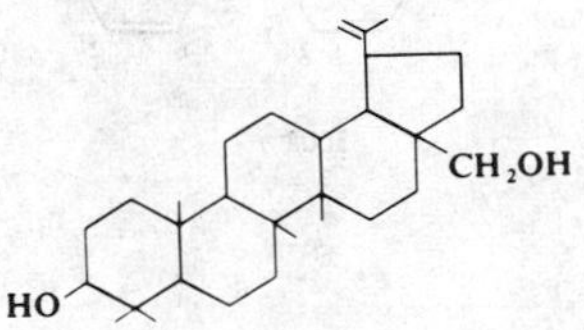

3323

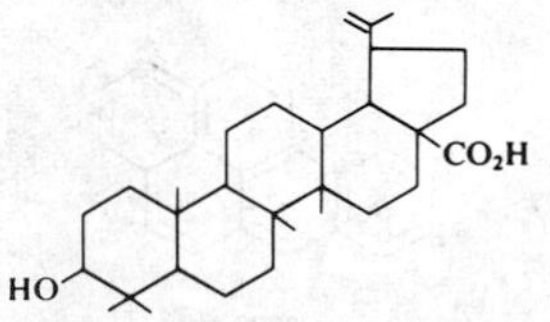

3324

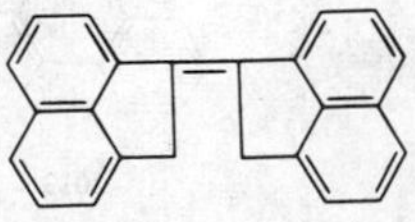

3325

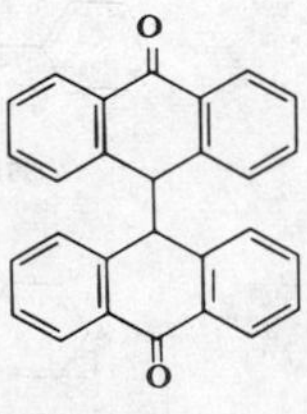

3329

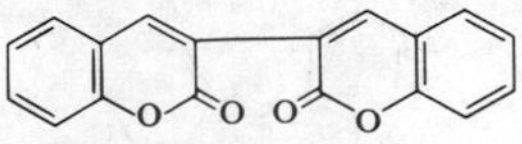

3332a

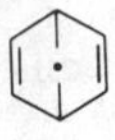

3333

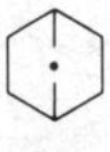

3334

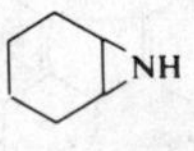

3335

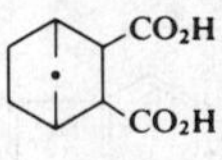

3336

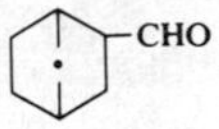

3337

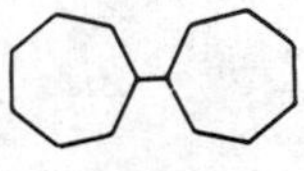

3338

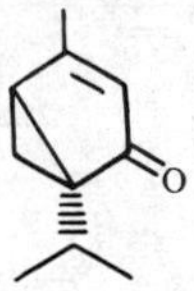

3339

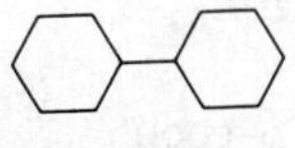

3340

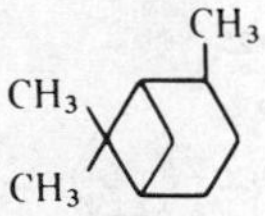

3342

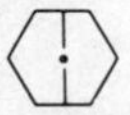

3343

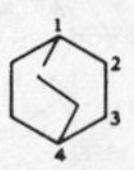

3344

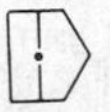

3345

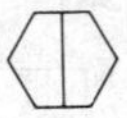

3346

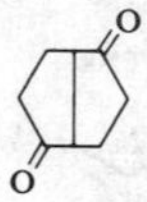

3348

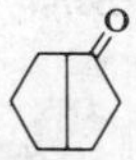

3350

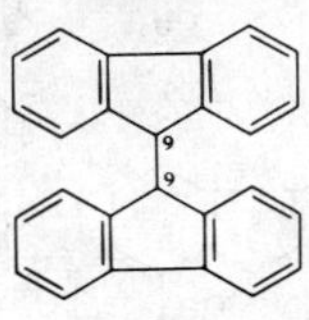

3351

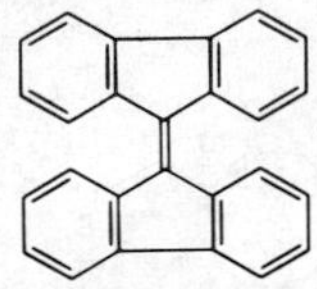

3353

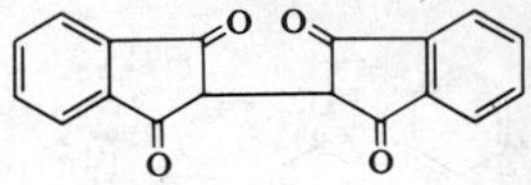

3357

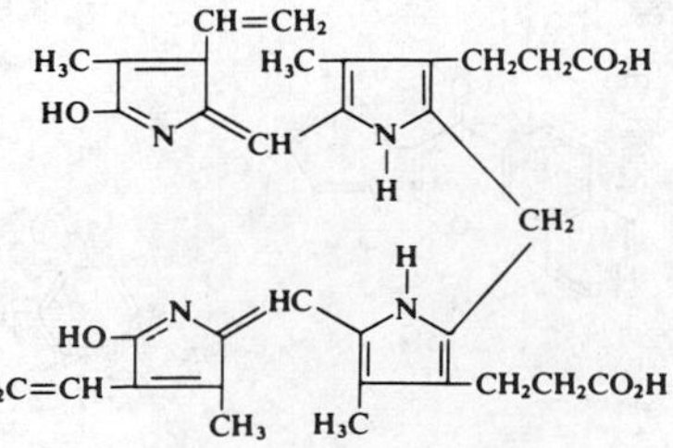

3360

STRUCTURAL FORMULAS OF ORGANIC COMPOUNDS (Continued)

In Numeric as they occur in Organic Compounds Table

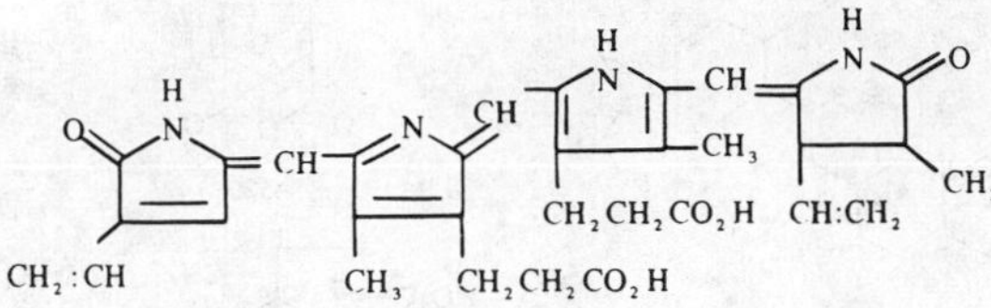

3361

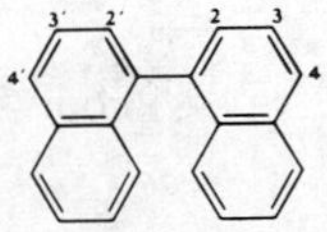

3362

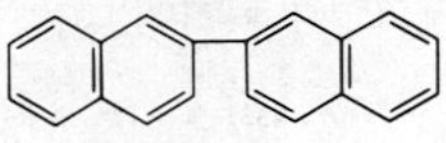

3366

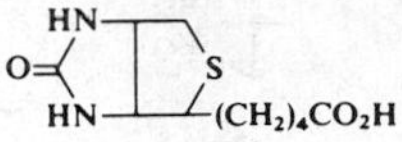

3367

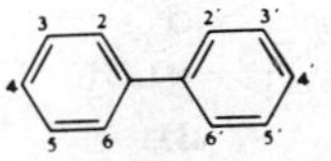

3369

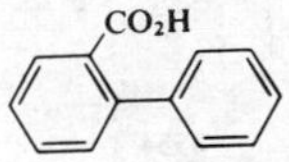

3500

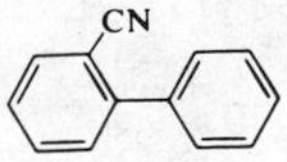

3501

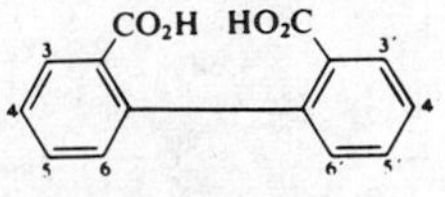

3505

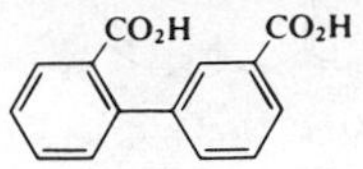

3511

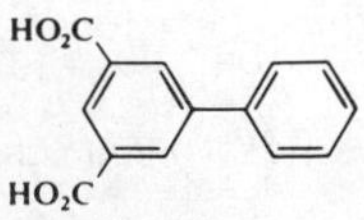

3517

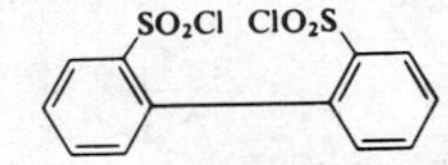

3523

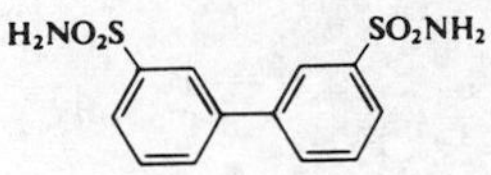

3524

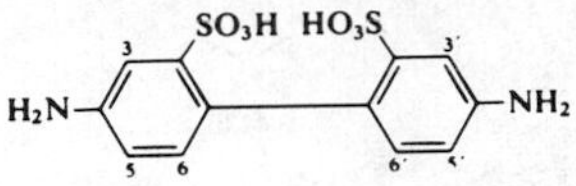

3529

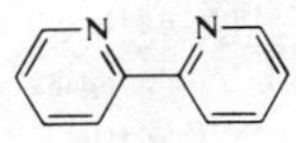

3530

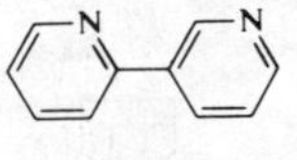

3531

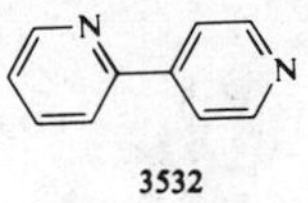

3532

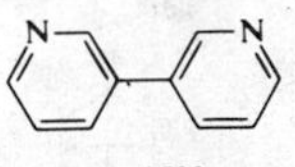

3533

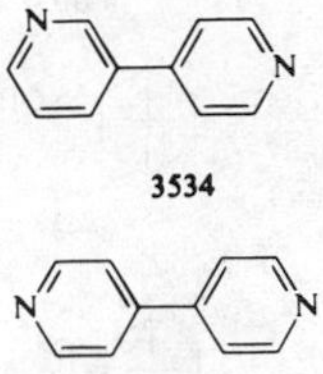

3534

3534a

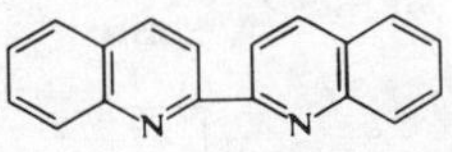

3535

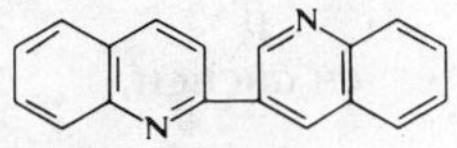

3536

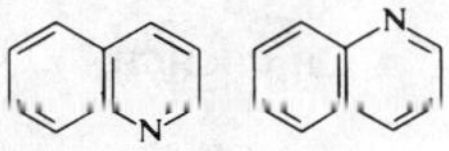

3537

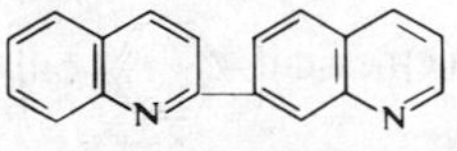

3538

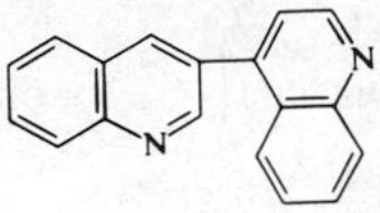

3540

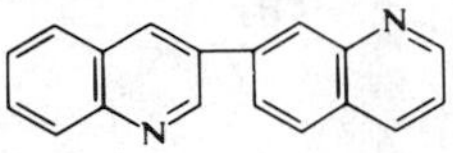

3541

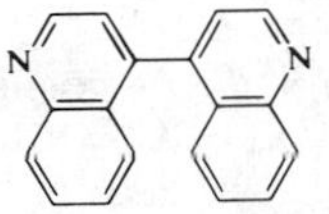

3542

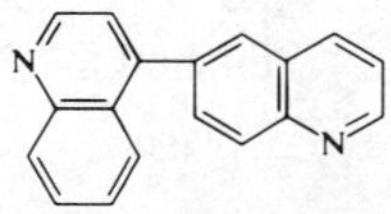

3543

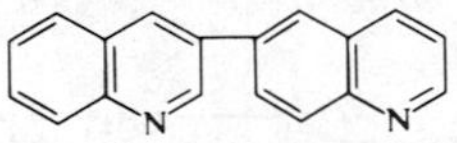

3544

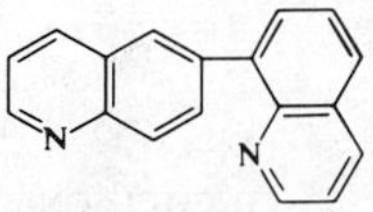

3545

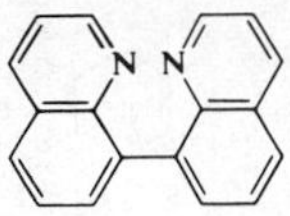

3546

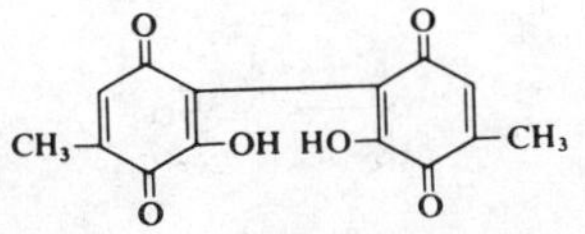

3548

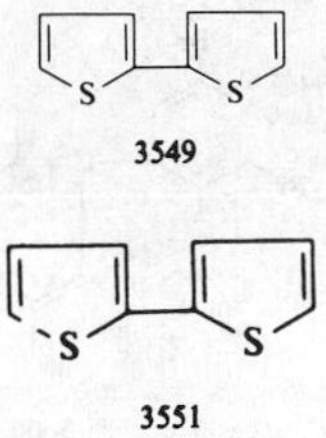

3549

3551

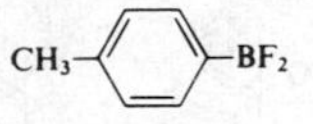

3570

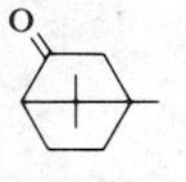

3581

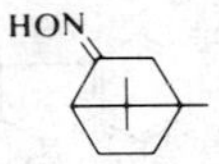

3583

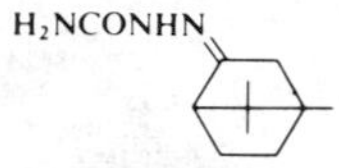

3586

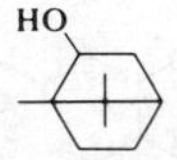

3588

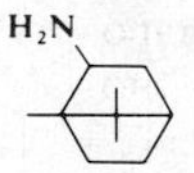

3595

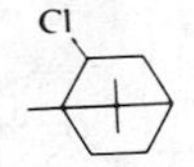

3596

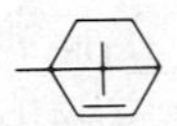

3597

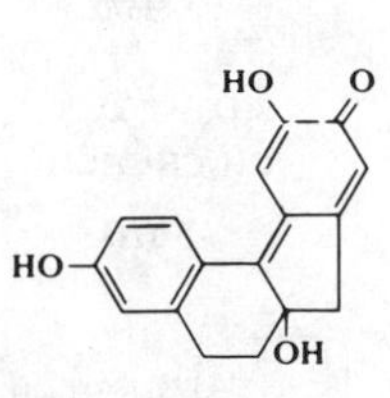

3599

In Numeric as they occur in Organic Compounds Table

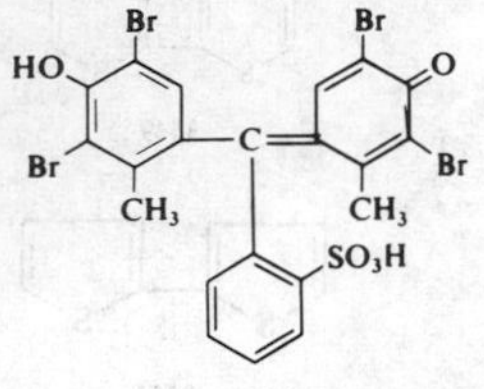

3600

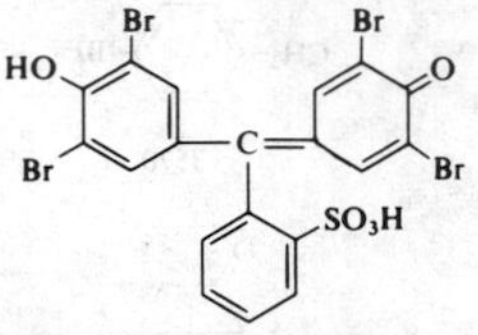

3620

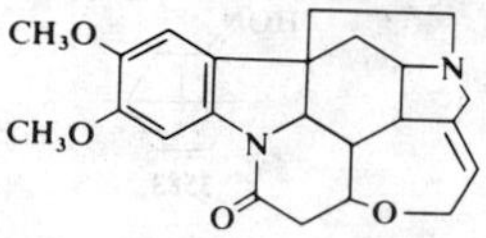

3624

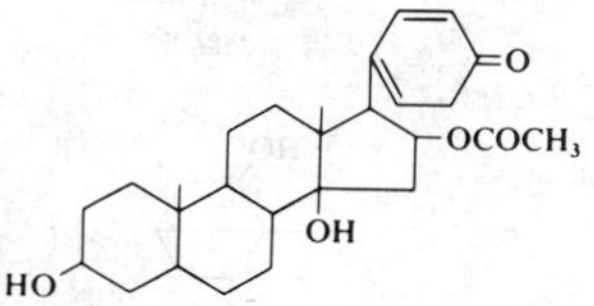

3628

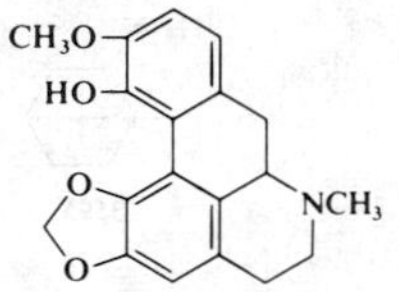

3629

O O

$CH_2CHCHCH_2$

3650

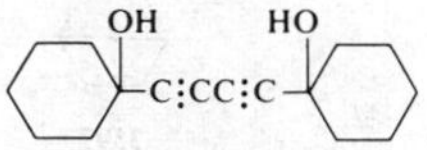

3670

O O

$CH_2CHCHCH_2$

3738

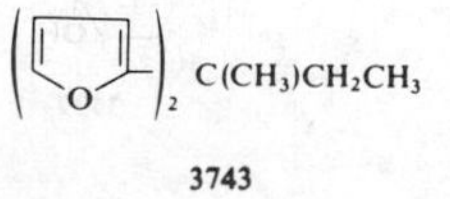

3743

O

$CH_3CHCHCH_3$

3757

O

$CH_3CHCHCH_3$

3758

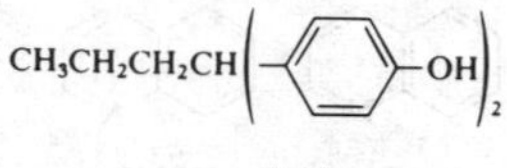

3765

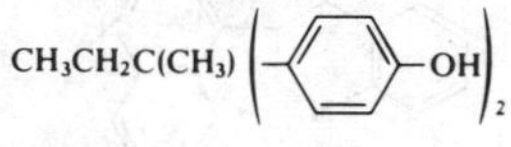

3766

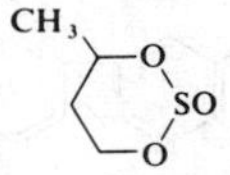

3832

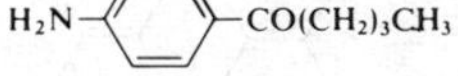

4093

O

$CH_2C(CH_3)_2$

4194

O

$CH_3C(C_6H_5)CHCO_2C_2H_5$

4327

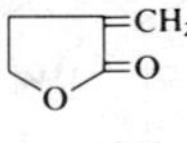

4350

O

CH_2CHCH_2CN

4424

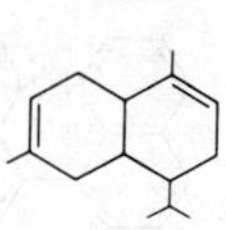

4466

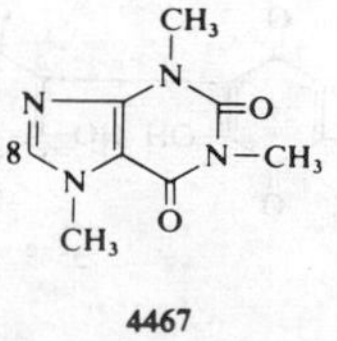

4467

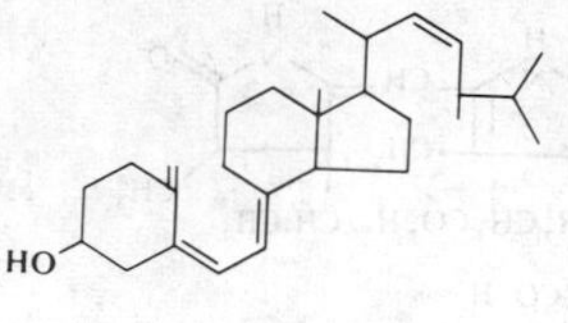

4477

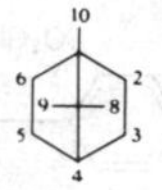

4478

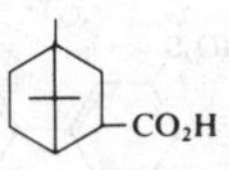

4479

4480

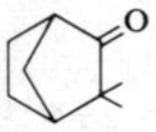

4483

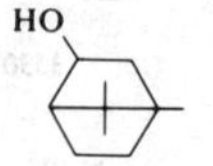

4484

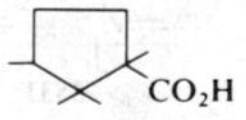

4486

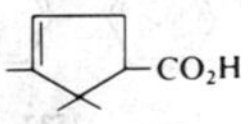

4489

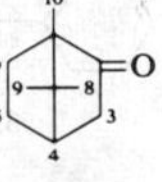

4490

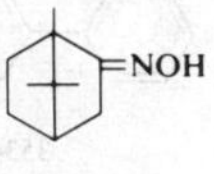

4493

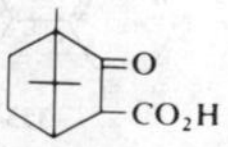

4514

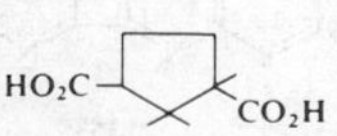

4518

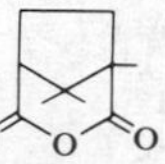

4520

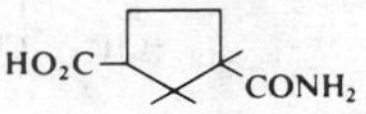

4525

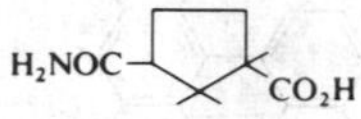

4526

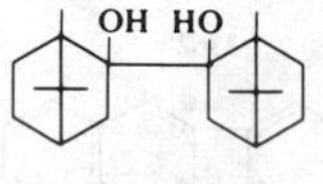

4531

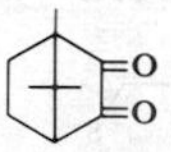

4532

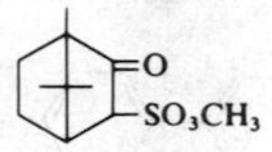

4533

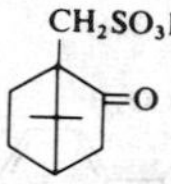

4534

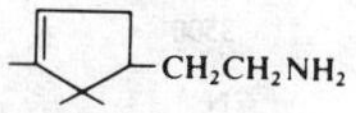

4537

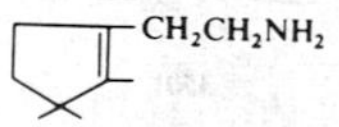

4538

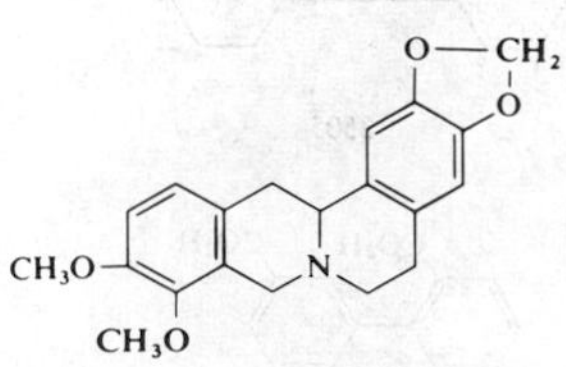

4539

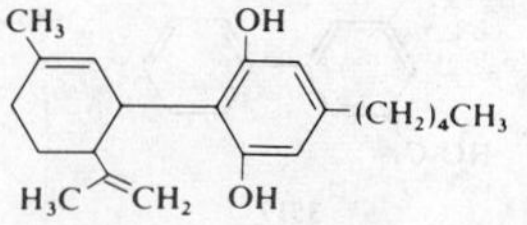

4544

STRUCTURAL FORMULAS OF ORGANIC COMPOUNDS (Continued)

In Numeric as they occur in Organic Compounds Table

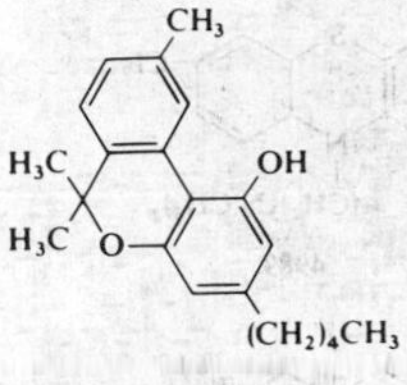

4545

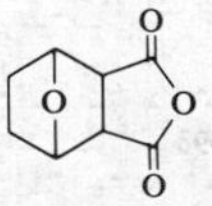

4546

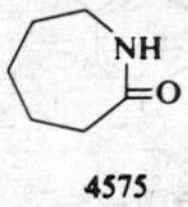

4575

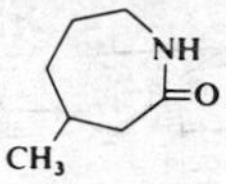

4576

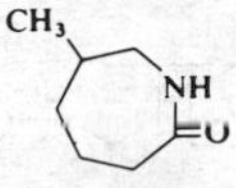

4577

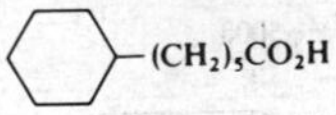

4588

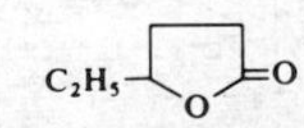

4597

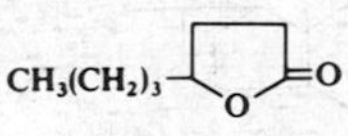

4637

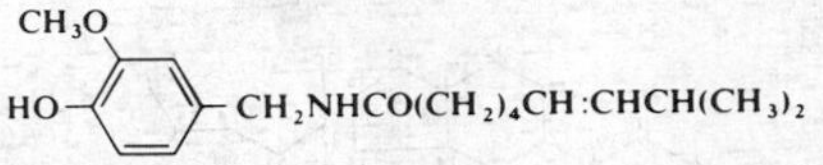

4649

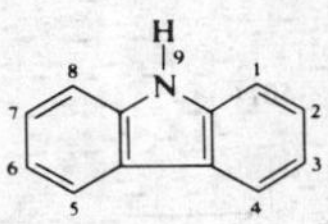

4697

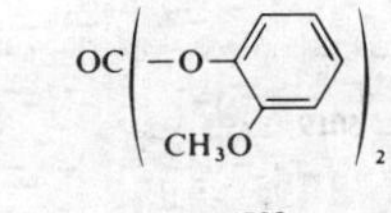

4722

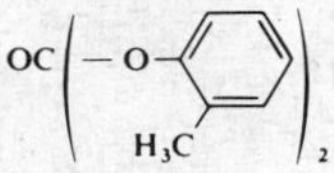

4734

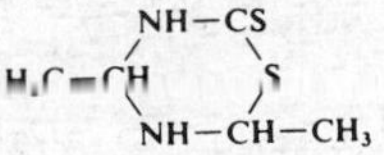

4743

4744

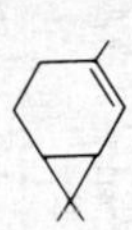

4746

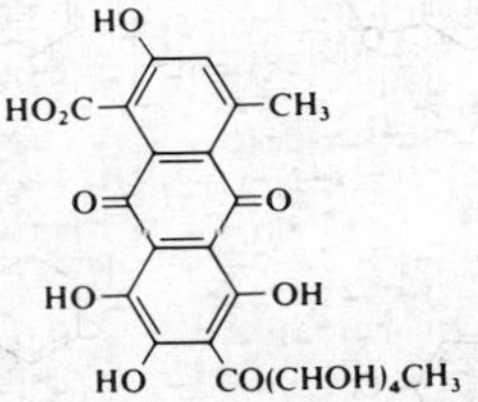

4747

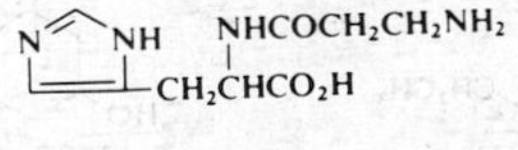

4749

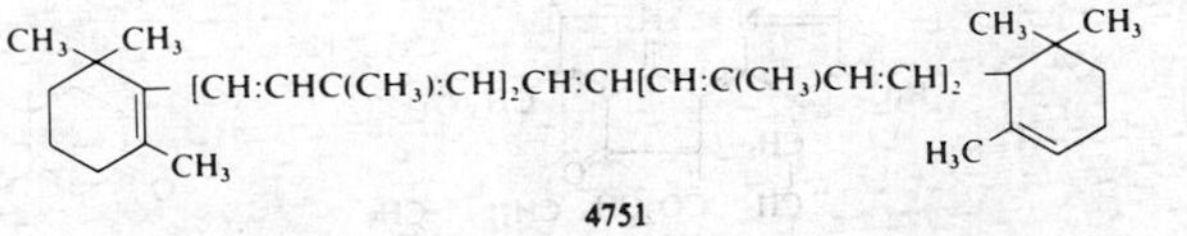

4751

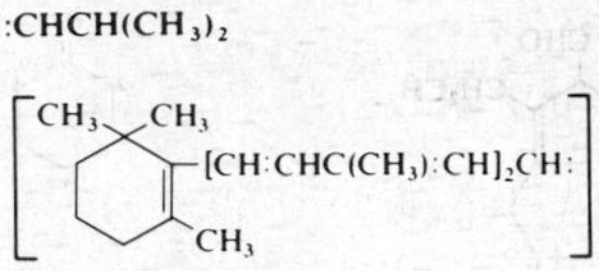

4752

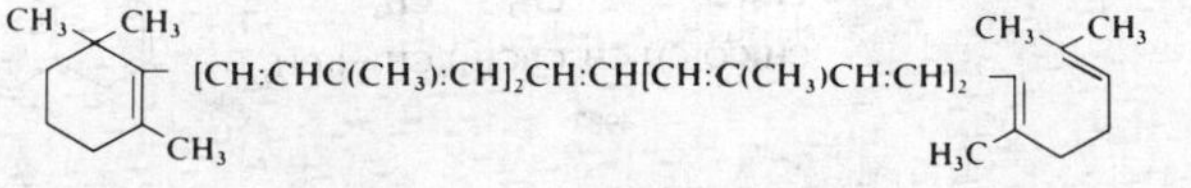

4753

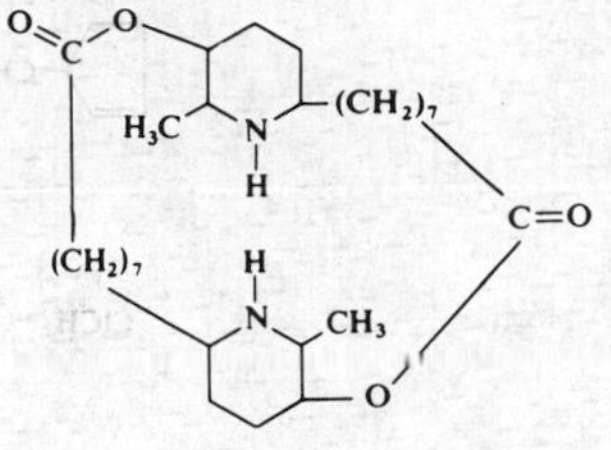

4754

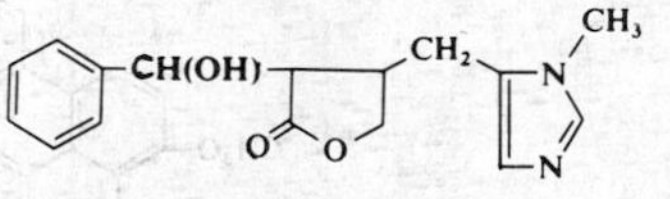

4756

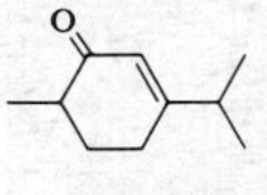

4760

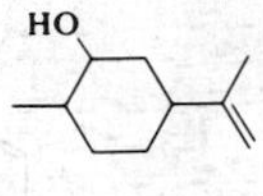

4762

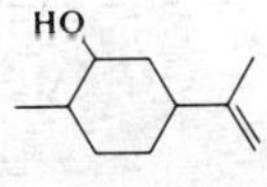

4764

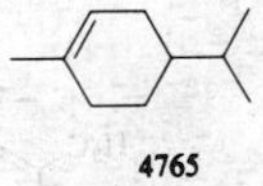

4765

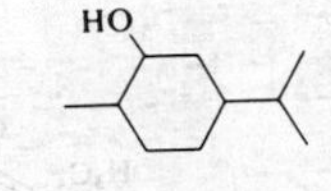

4766

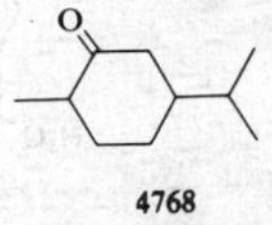

4768

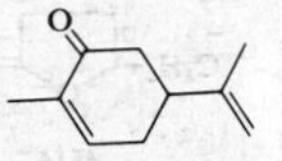

4769

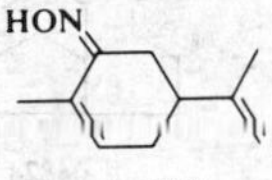

4772

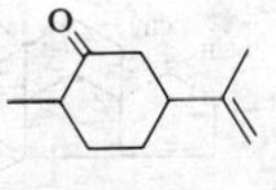

4777

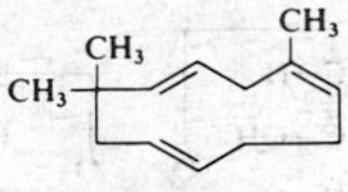

4779

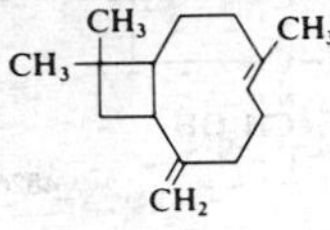

4780

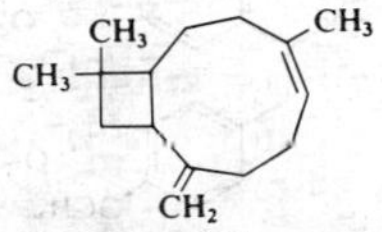

4781

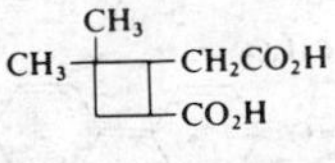

4782

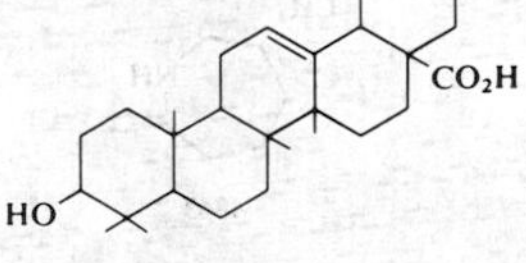

4784

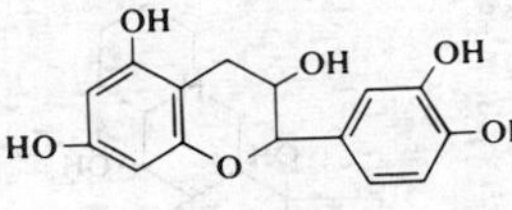

4785

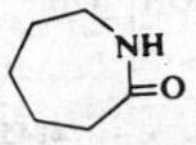

4811

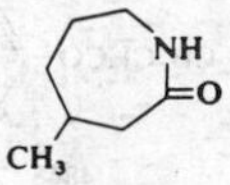

4812

STRUCTURAL FORMULAS OF ORGANIC COMPOUNDS (Continued)

In Numeric as they occur in Organic Compounds Table

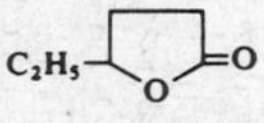

4814

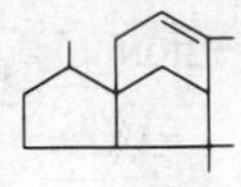

04824

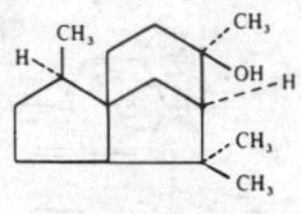

4825

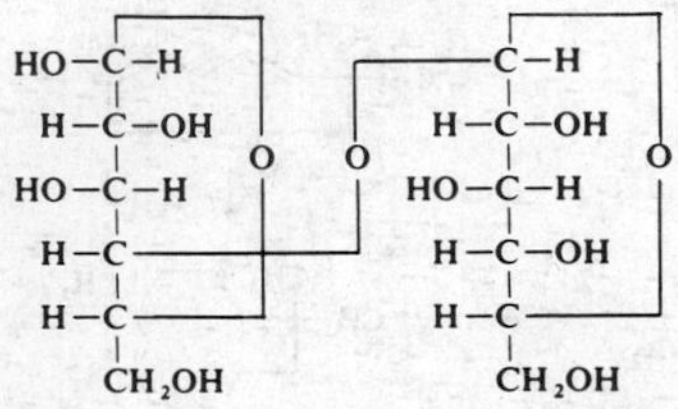

4826

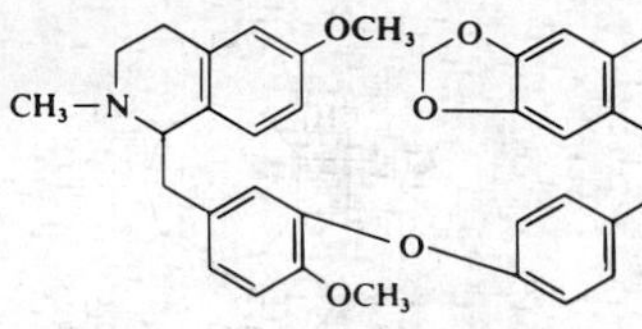

4836

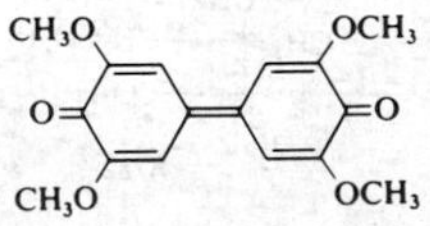

4838

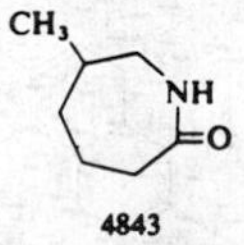

4843

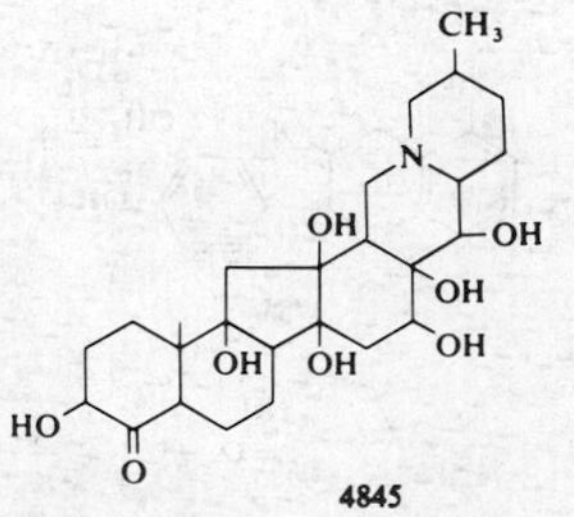

4845

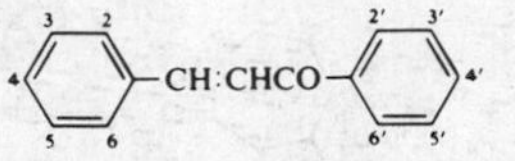

4848

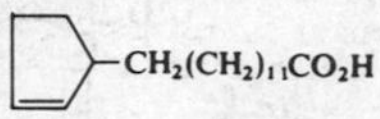

4860

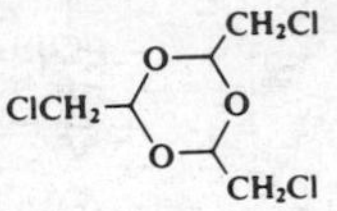

4862

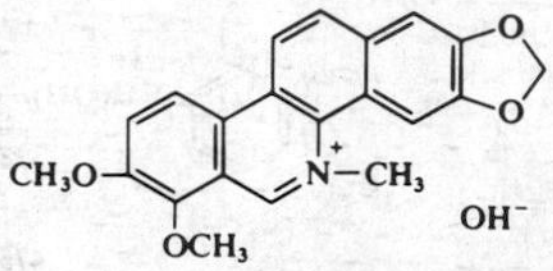

4863

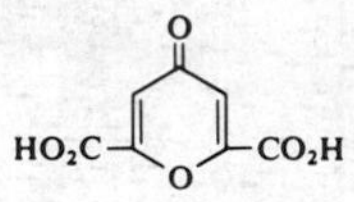

4864

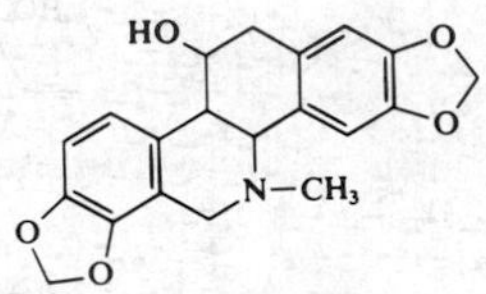

4866

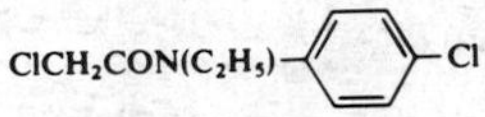

4889

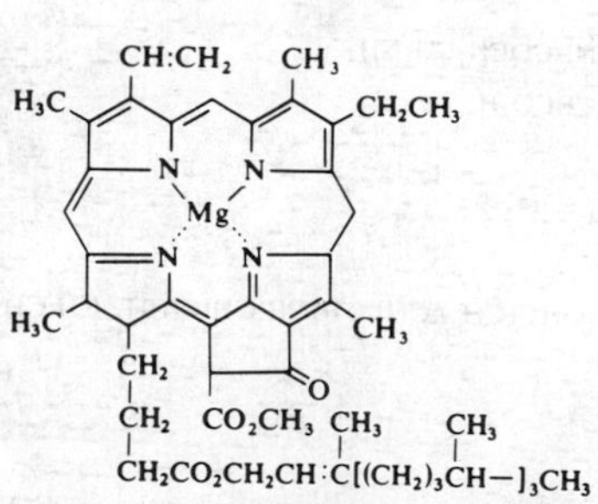

4986

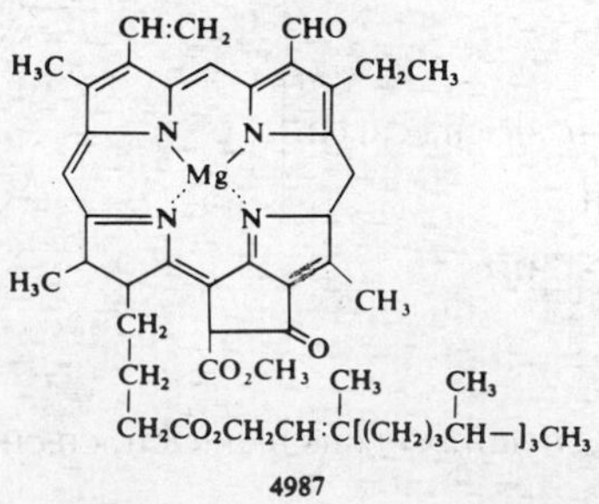

4987

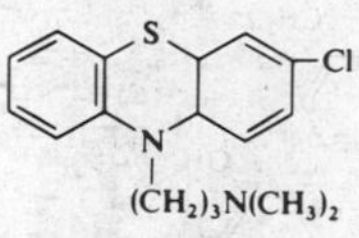

4989

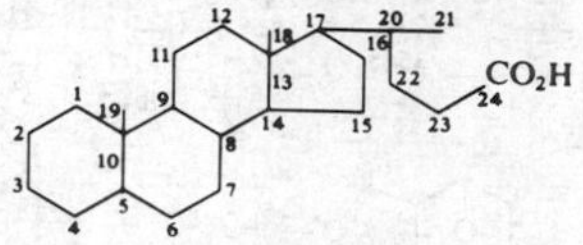

4998

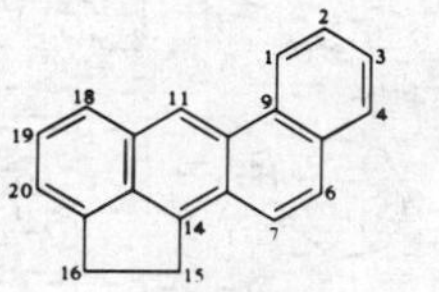

5003

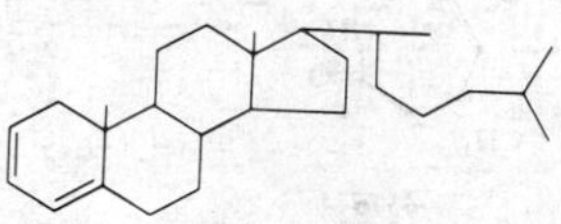

5007

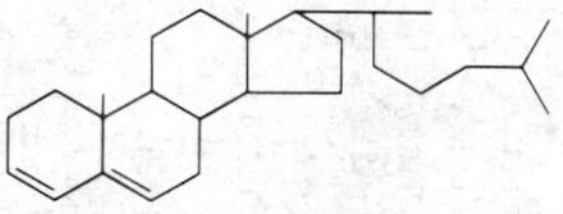

5008

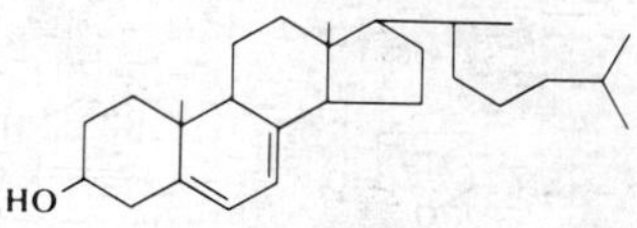

5009

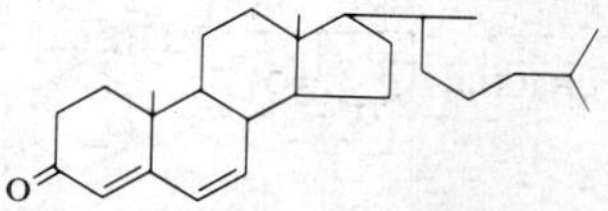

5010

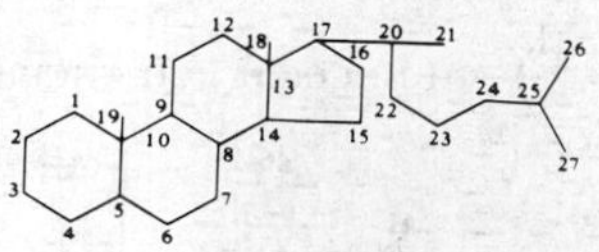

5011

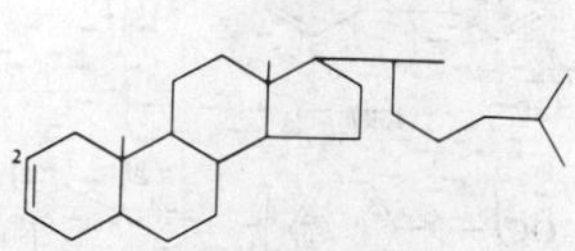

5019

STRUCTURAL FORMULAS OF ORGANIC COMPOUNDS (Continued)

In Numeric as they occur in Organic Compounds Table

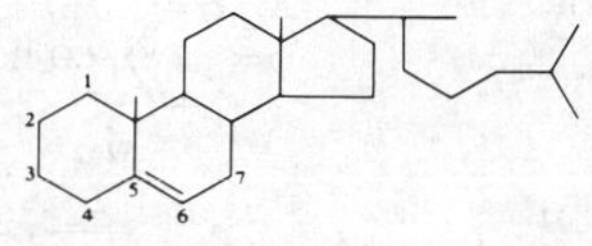

5020

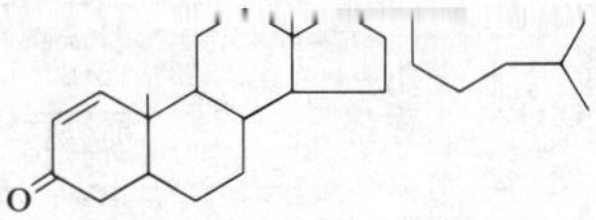

5025

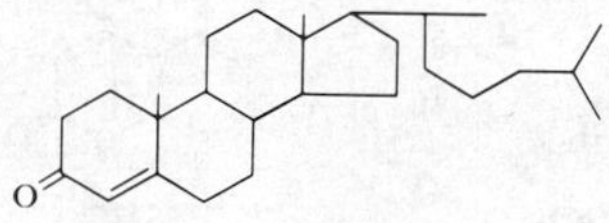

5026

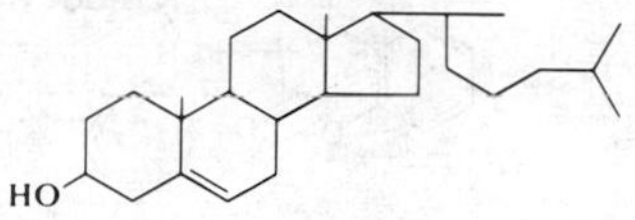

5028

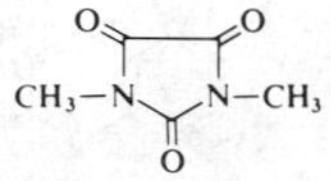

5032

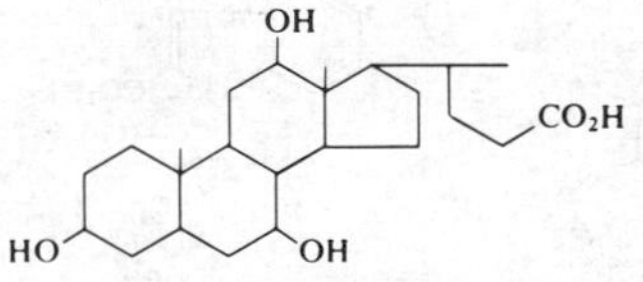

5033

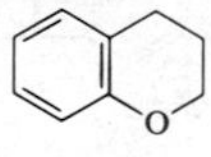

5039

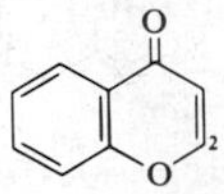

5043

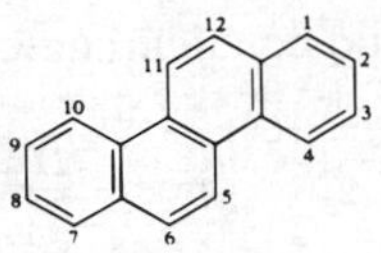

5047

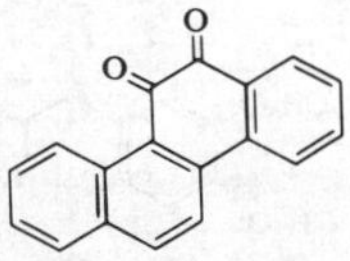

5052

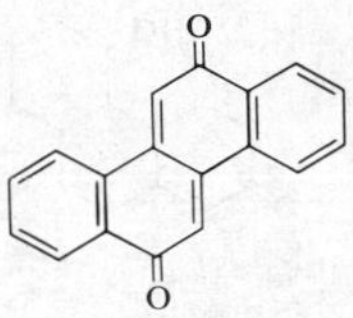

5053

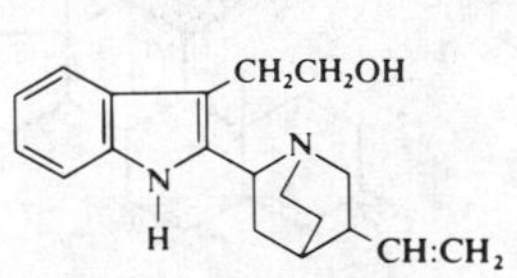

5054

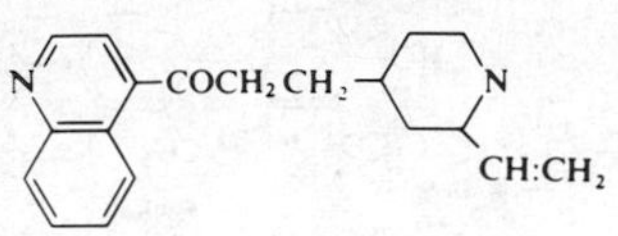

5055

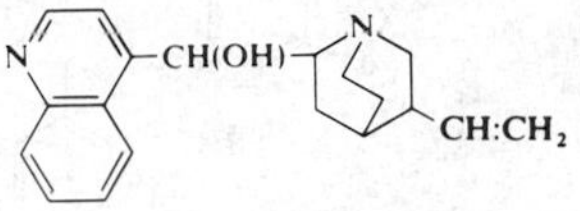

5056

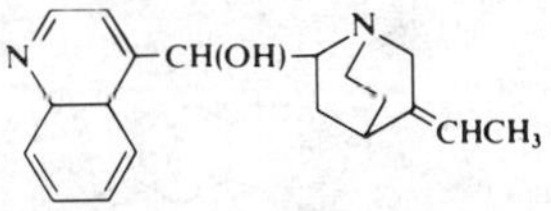

5059

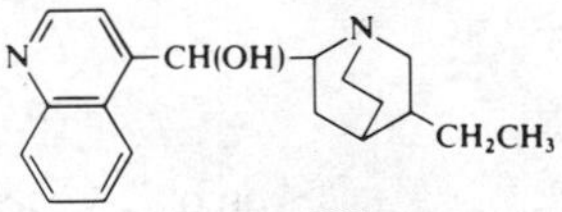

5064

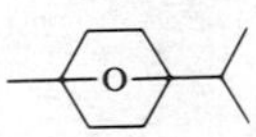

5065

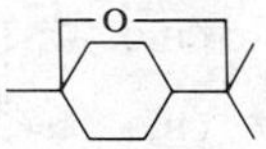

5066

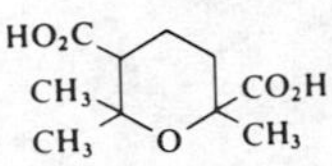

5067

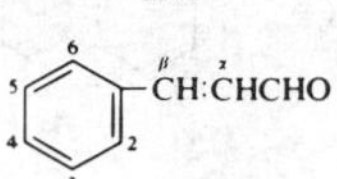

5070

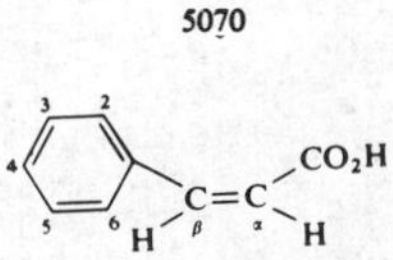

5081

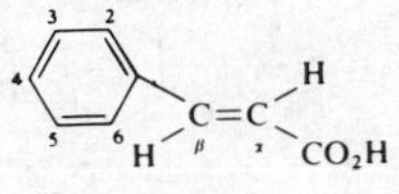

5084

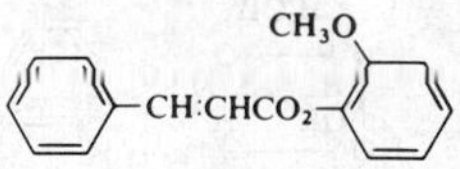

5128

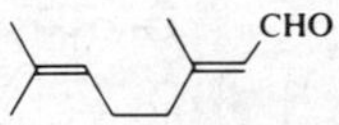

5167

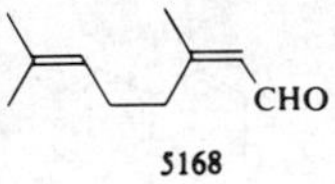

5168

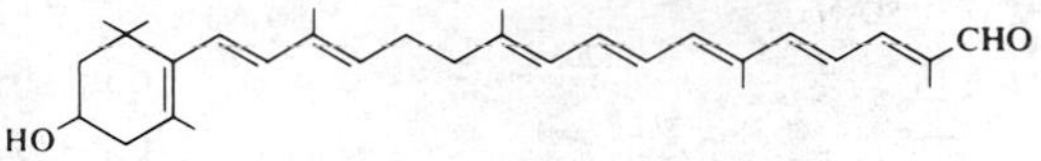

5169

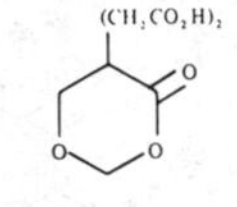

5171

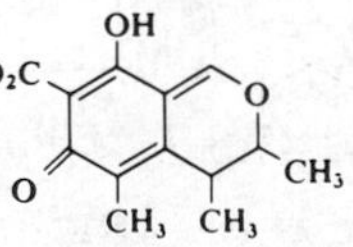

5179

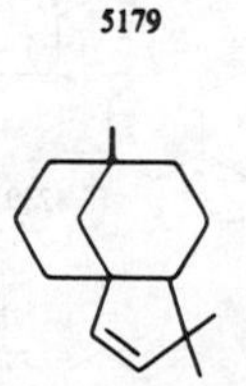

5187

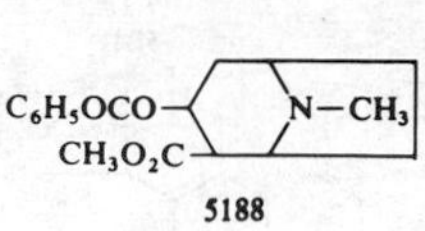

5188

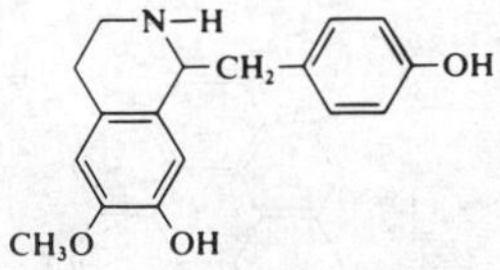

5193

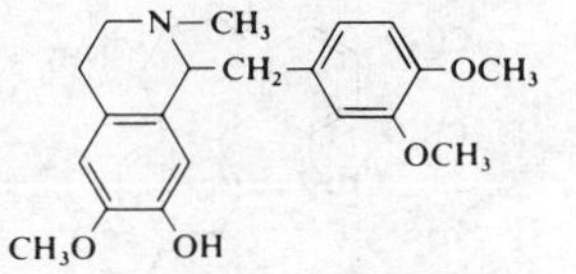

5194

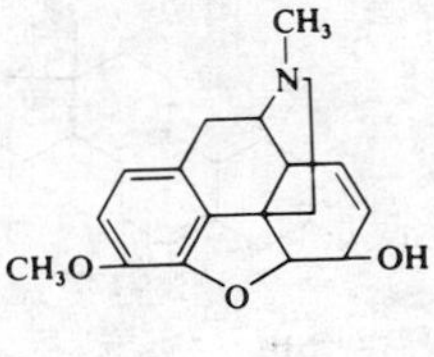

5195

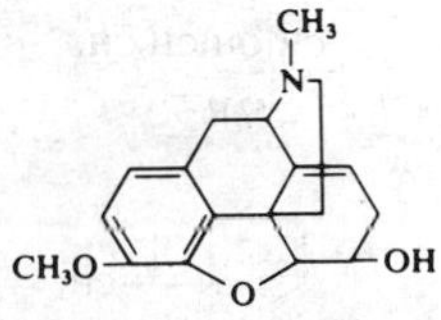

5200

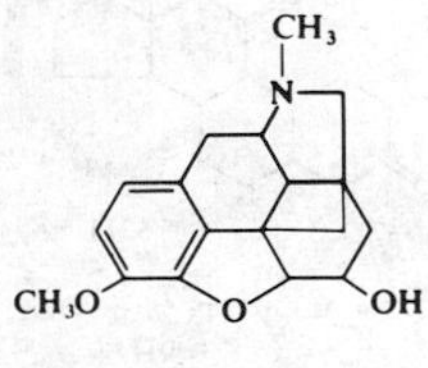

5201

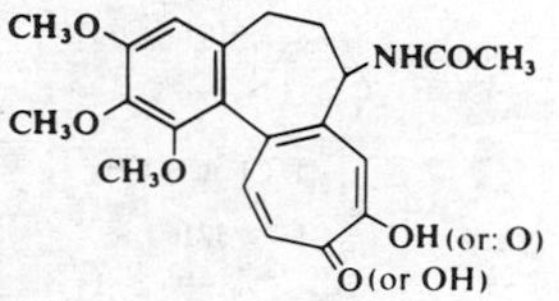

5202

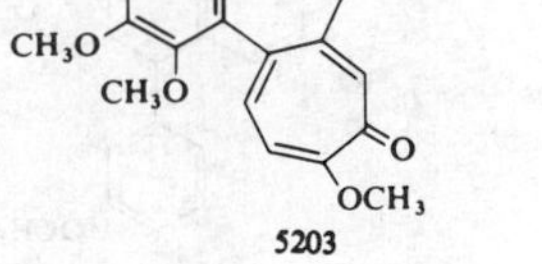

5203

STRUCTURAL FORMULAS OF ORGANIC COMPOUNDS (Continued)

In Numeric as they occur in Organic Compounds Table

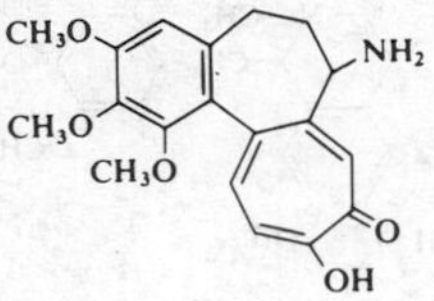

5204

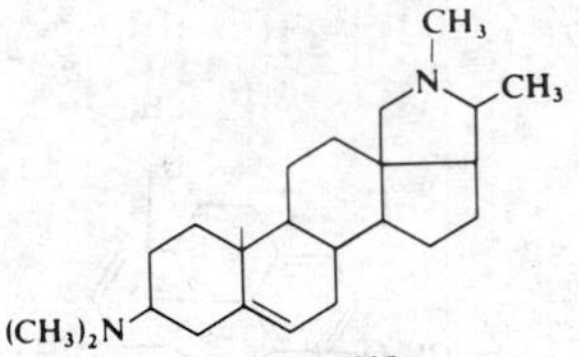

5205

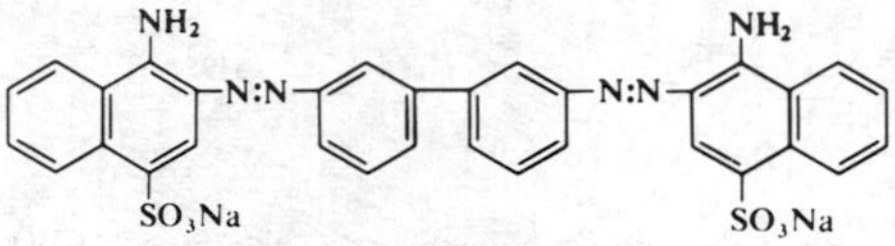

5206

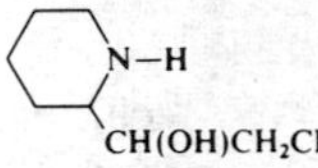

5207

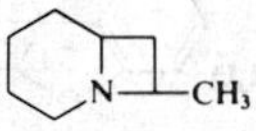

5210

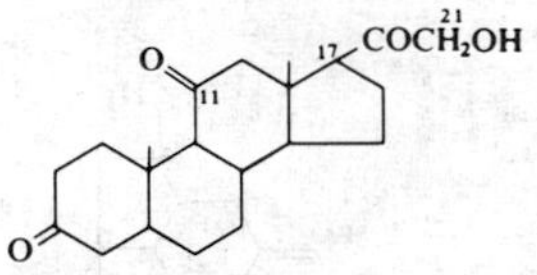

5211

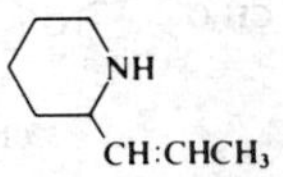

5213

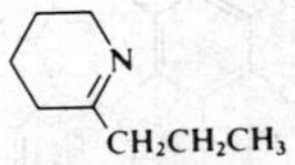

5216

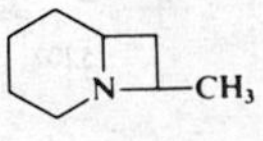

5217

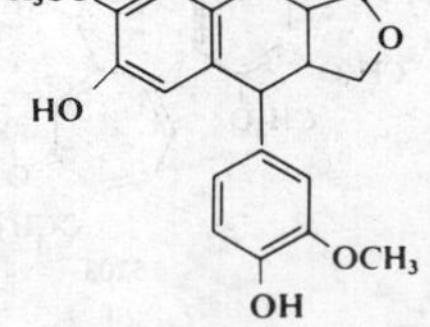

5220

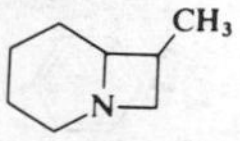

5221

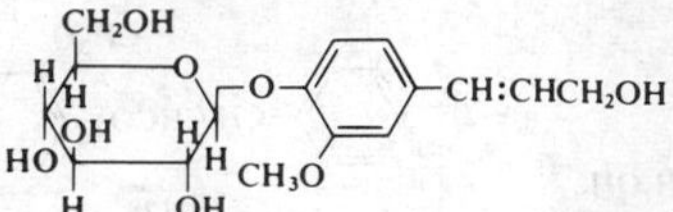

5224

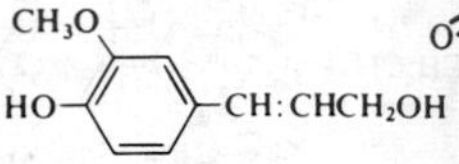

5225

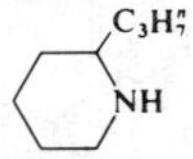

5226

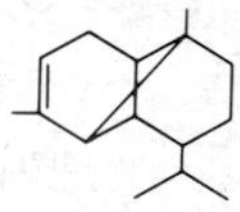

5237

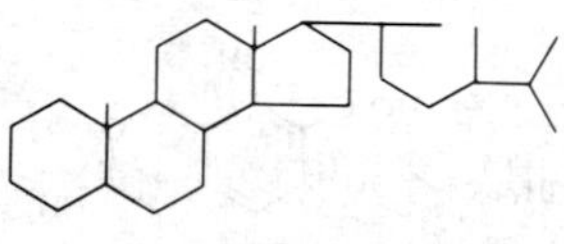

5238

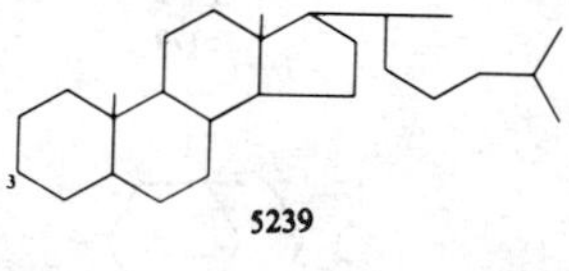

5239

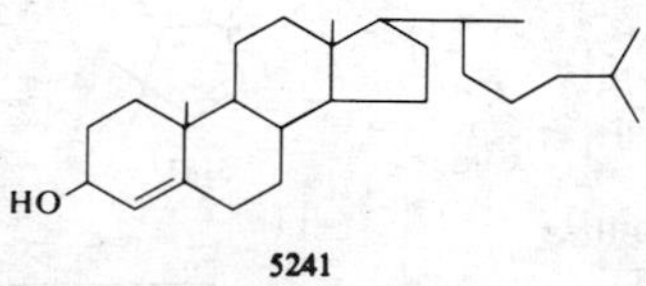

5241

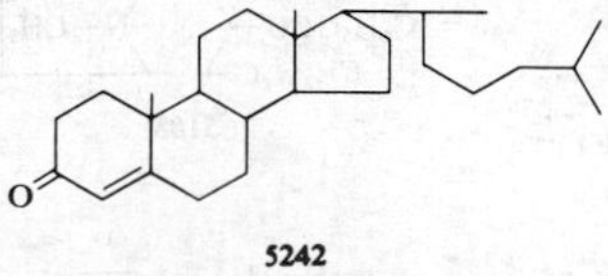

5242

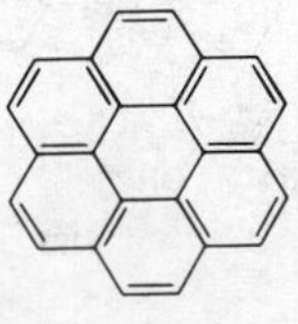

5243

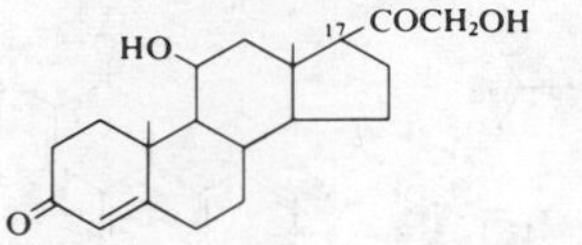

5244

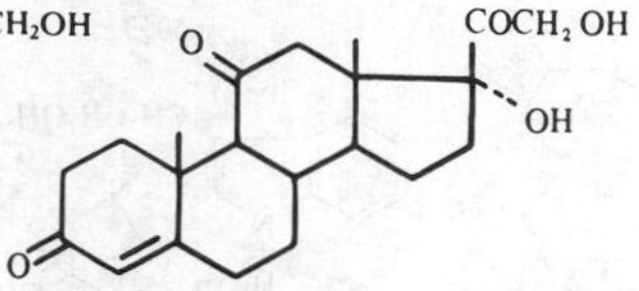

5246

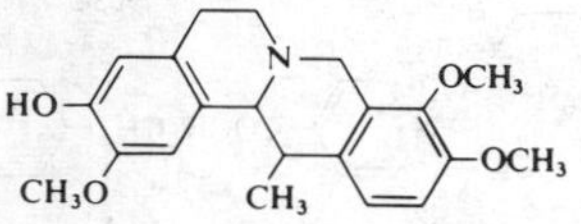

5248

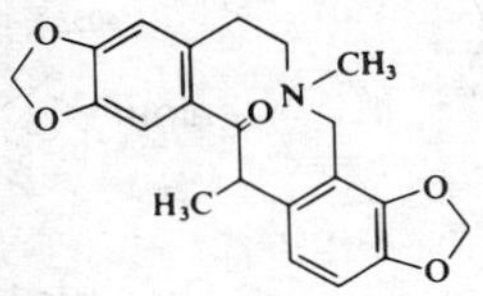

5250

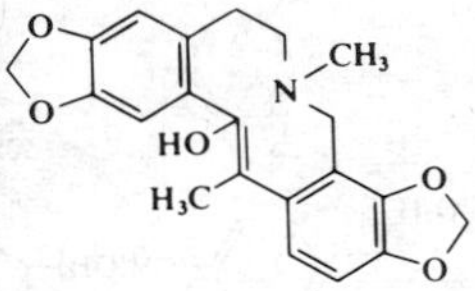

5251

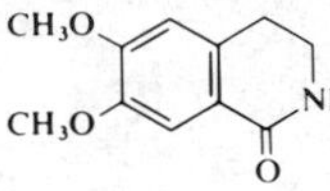

5252

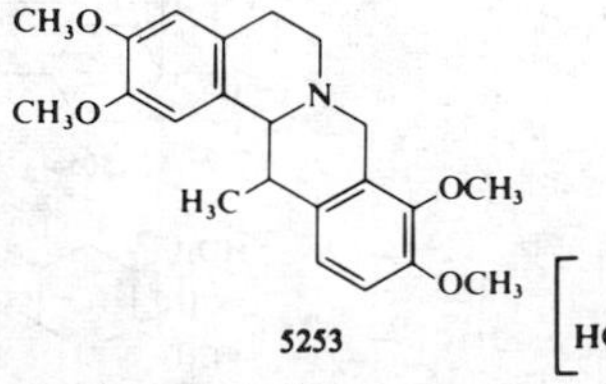

5253

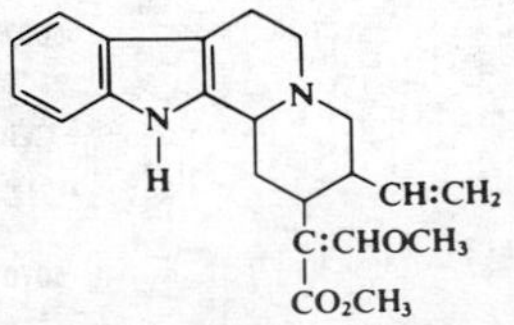

5258

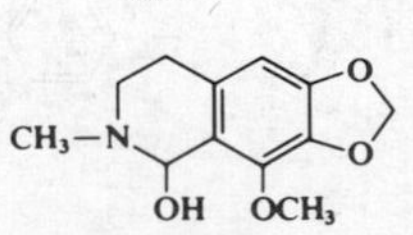

5259

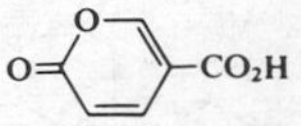

5262

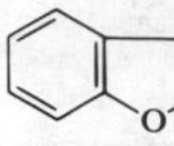

5263

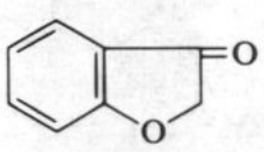

5264

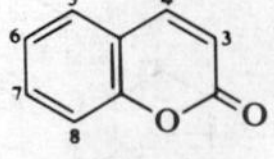

5265

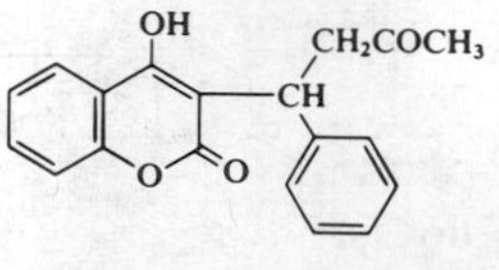

5279

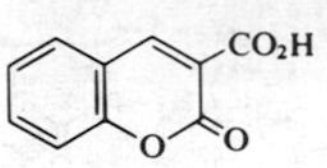

5295

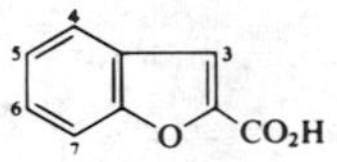

5296

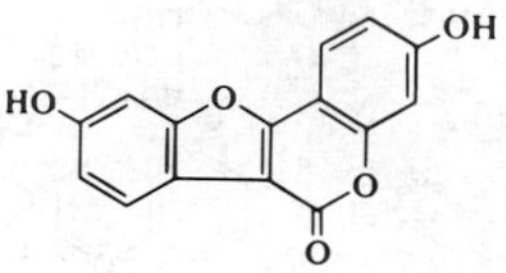

5300

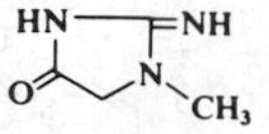

5303

$[HO_2C\overset{CH_3}{C}{:}CHCH{:}CH\overset{CH_3}{C}{:}CHCH=]_2$

5379

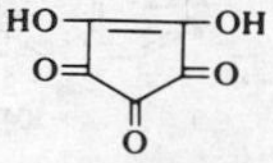

5380

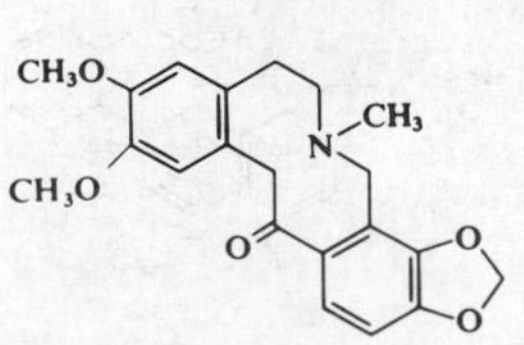

5392

In Numeric as they occur in Organic Compounds Table

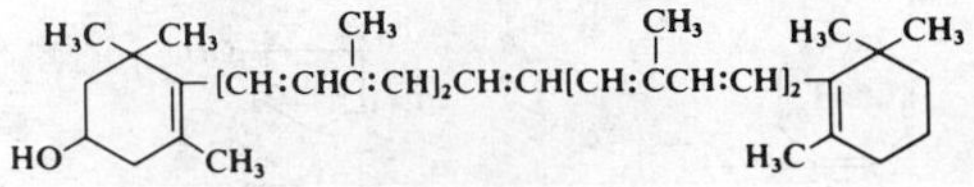
5393

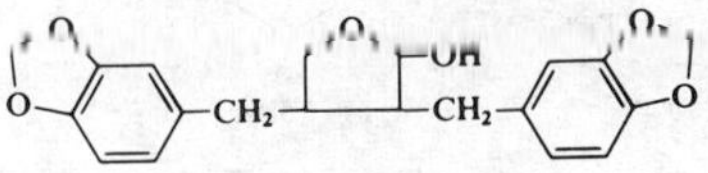
5397

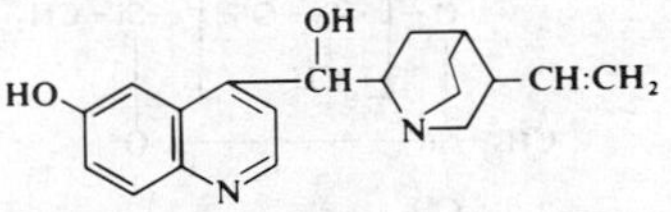
5399

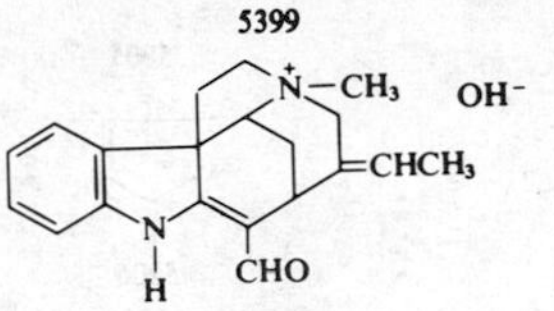
5400

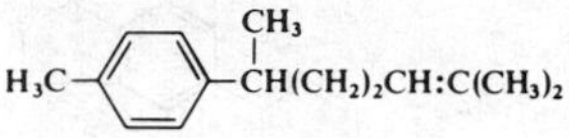
5401

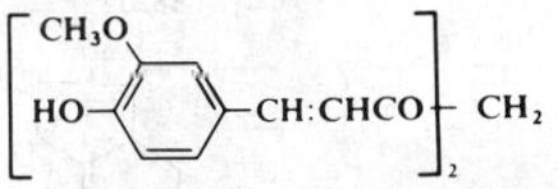
5402

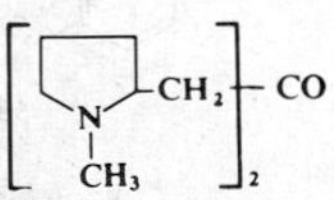
5403

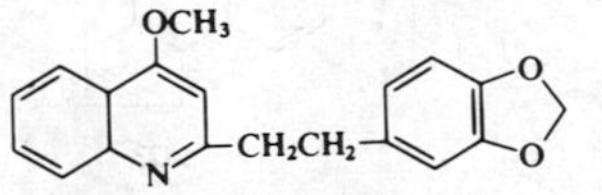
5405

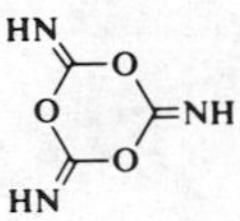
5406

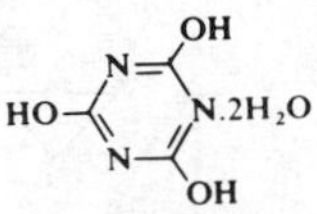
5434

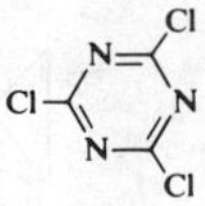
5436

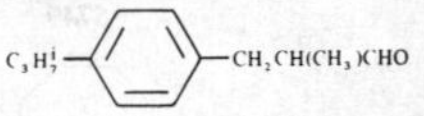
5437

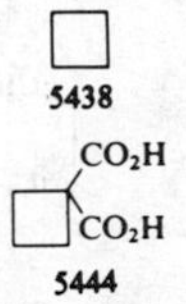
5438

5444

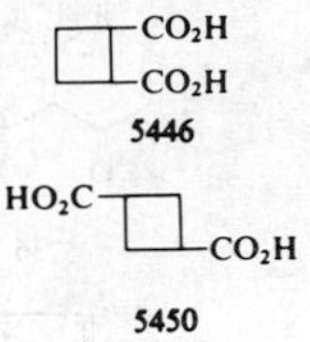
5446

5450

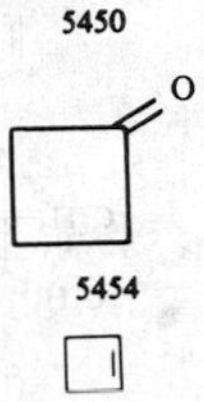
5454

5455

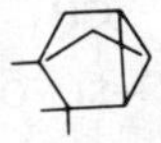
5458

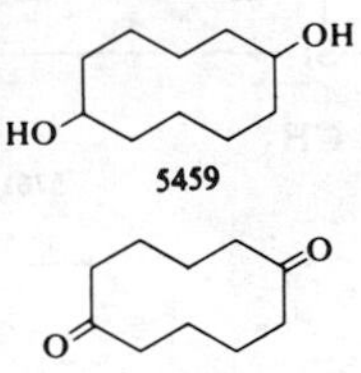
5459

5460

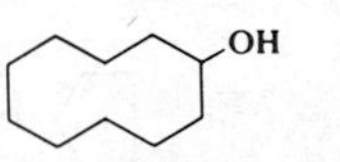
5461

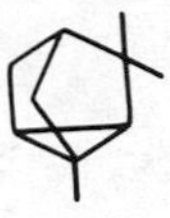
5463

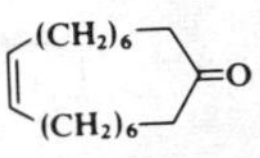
5464

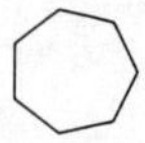
5465

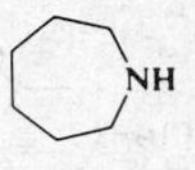
5466

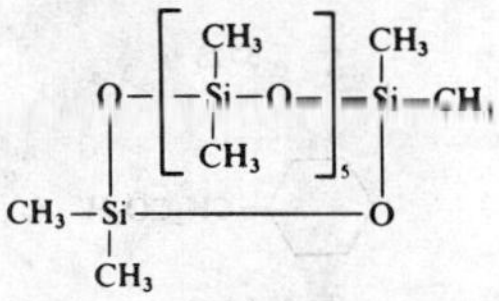
5473

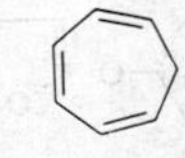
5474

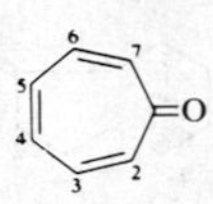
5475

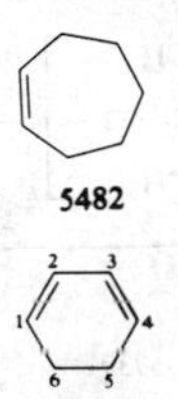
5482

5483

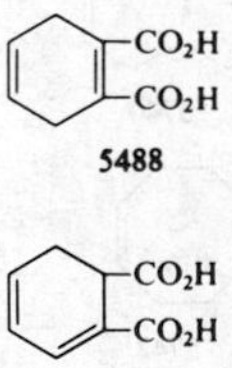
5488

5489

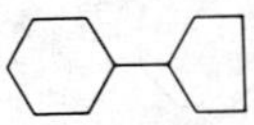
5507

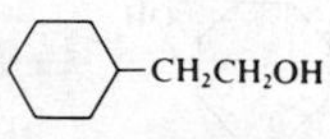
5542

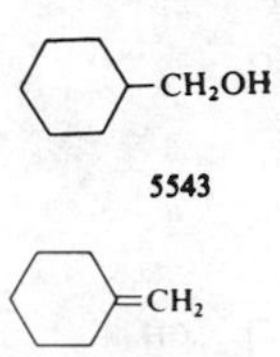
5543

5551

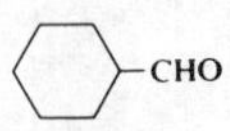
5563

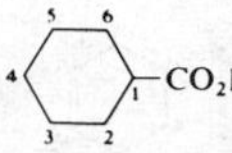
5564

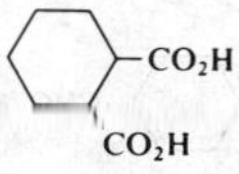
5574

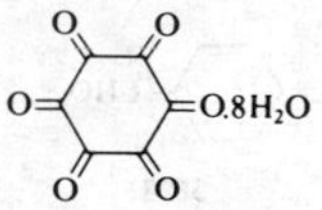
5576

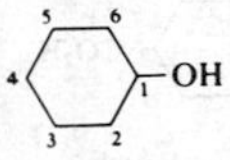
5577

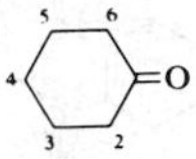
5621

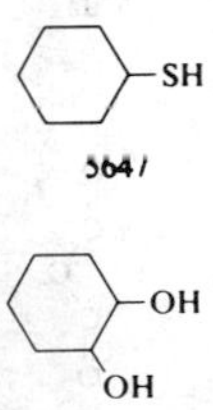
5647

5663

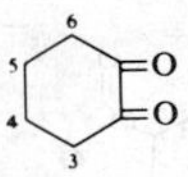
5667

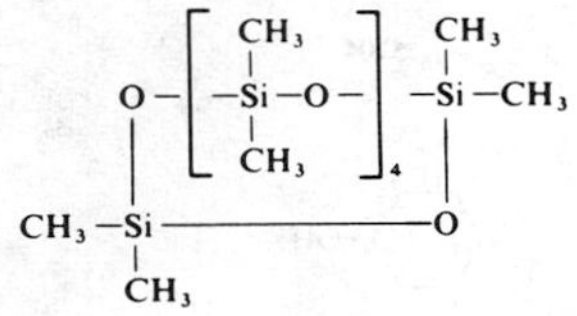
5676

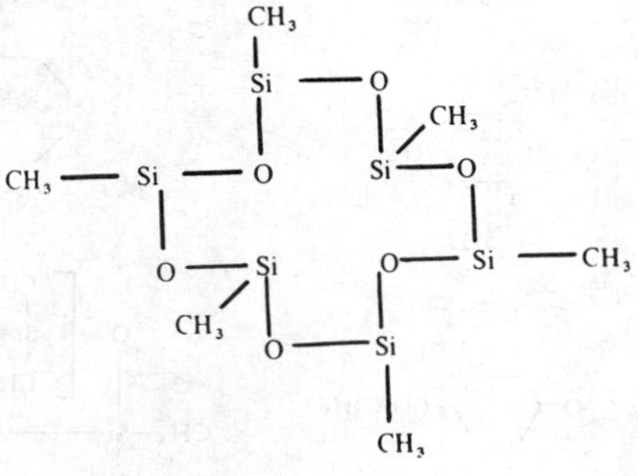
5677

In Numeric as they occur in Organic Compounds Table

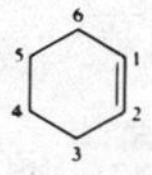
5678

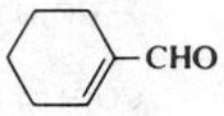
5697

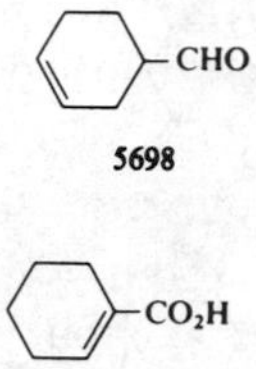
5698

5699

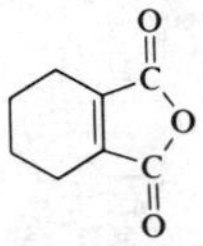
5702

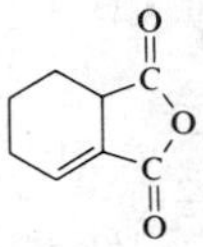
5703

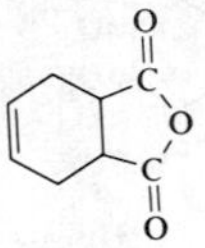
5704

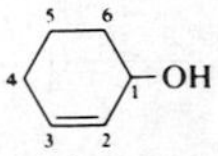
5707

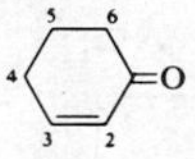
5713

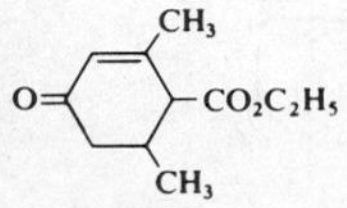
5717

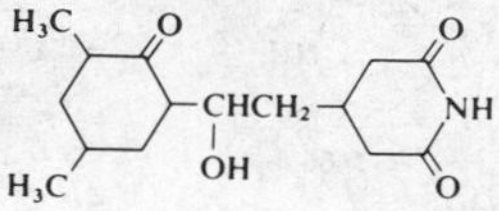
5725

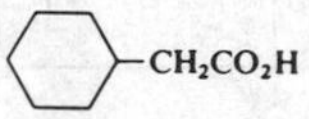
5726

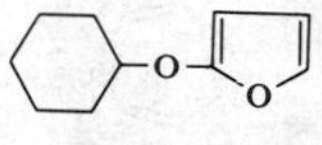
5732

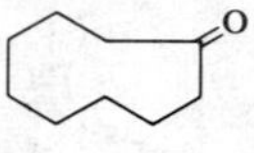
5738

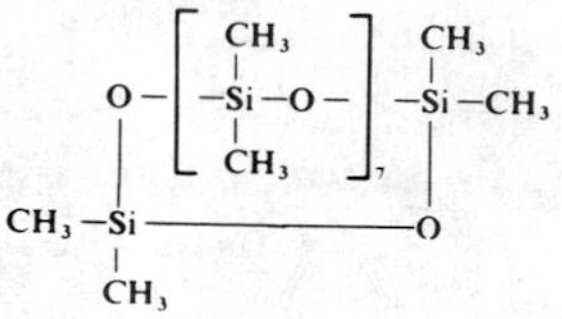
5739

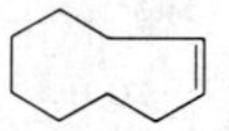
5740

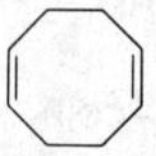
5742

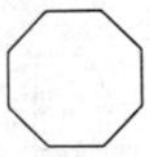
5744

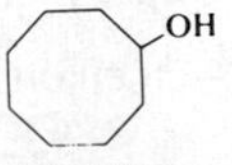
5745

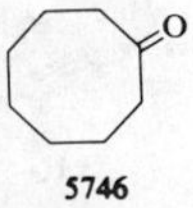
5746

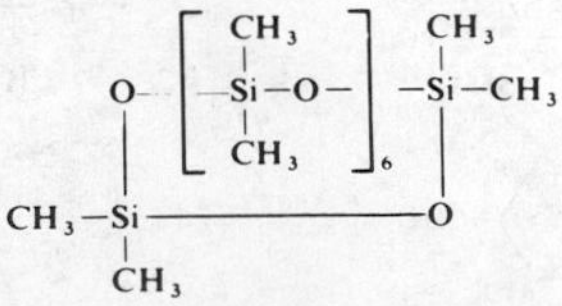
5748

5749

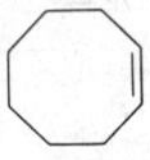
5752

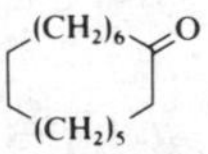
5754

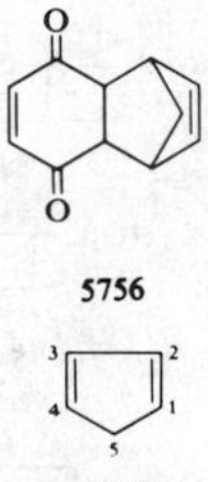
5756

5756a

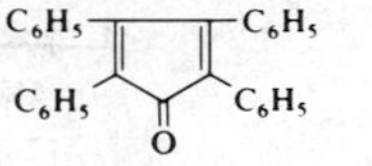
5760

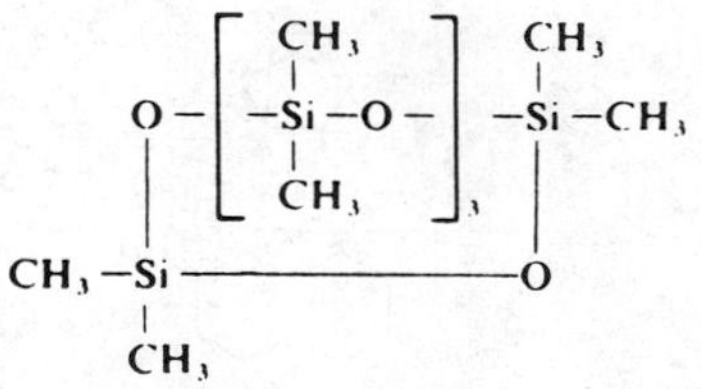
5761

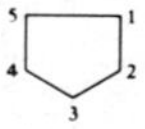
5762

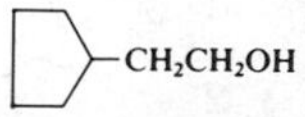
5774

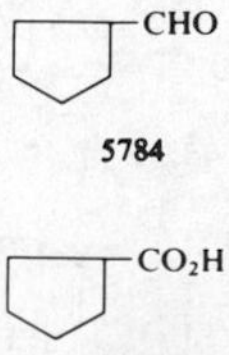
5784

5785

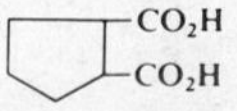
5789

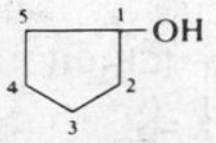
5798

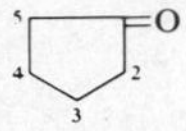
5801

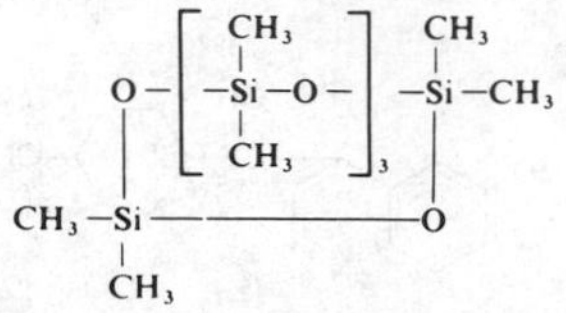
5805

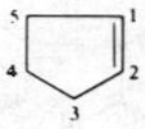
5806

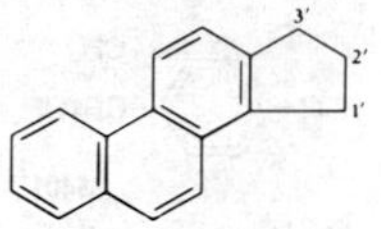
5820

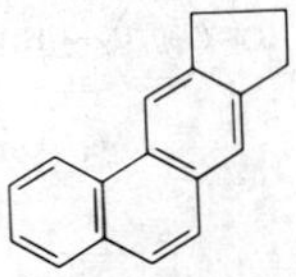
5822

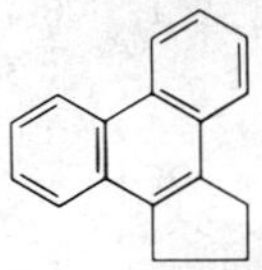
5823

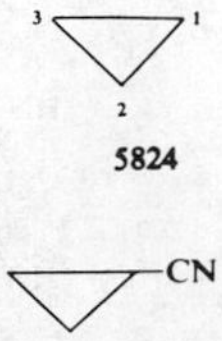
5824

5838

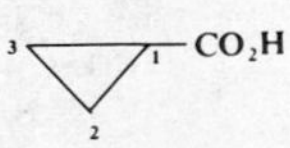
5839

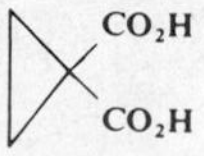
5847

STRUCTURAL FORMULAS OF ORGANIC COMPOUNDS (Continued)

In Numeric as they occur in Organic Compounds Table

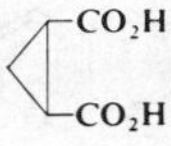
5849

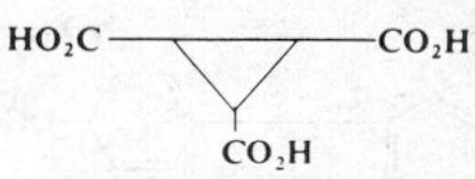
5856

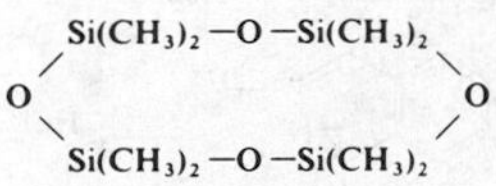
5858

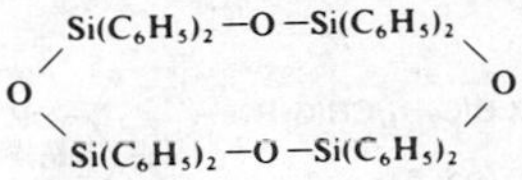
5859

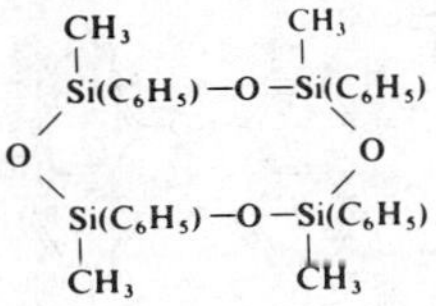
5860

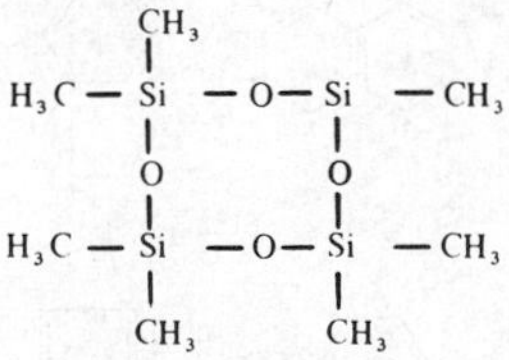
5861

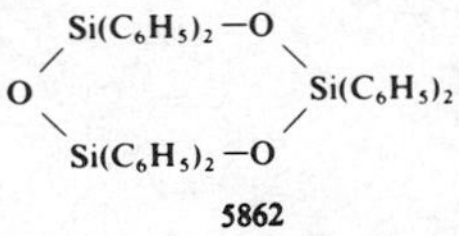
5862

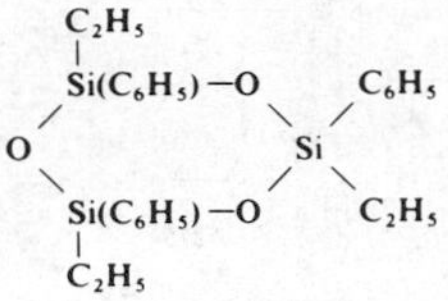
5863

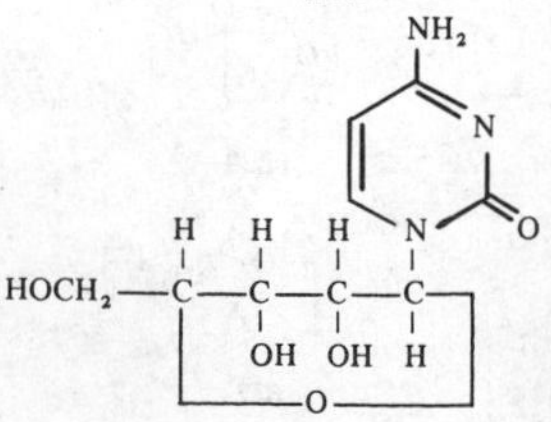
5877

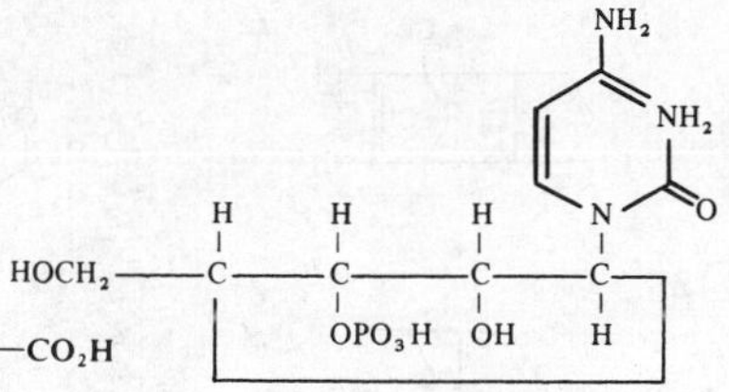
5878

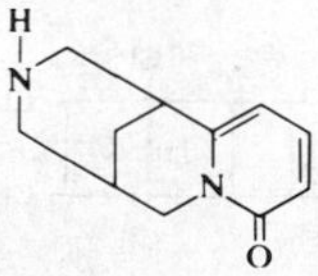
5879

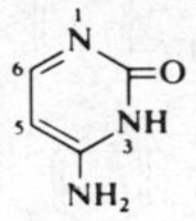
5881

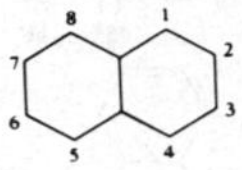
5887

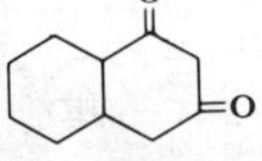
5893

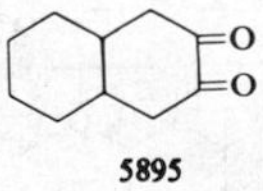
5895

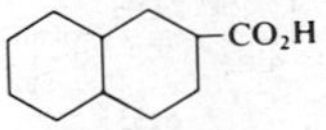
5897

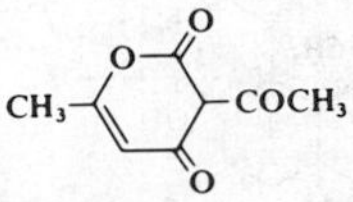
5947

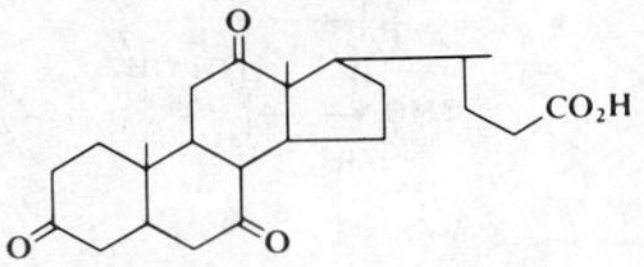
5948

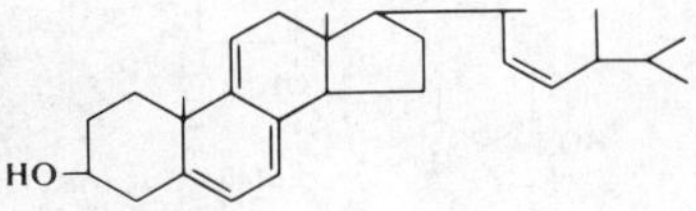
5949

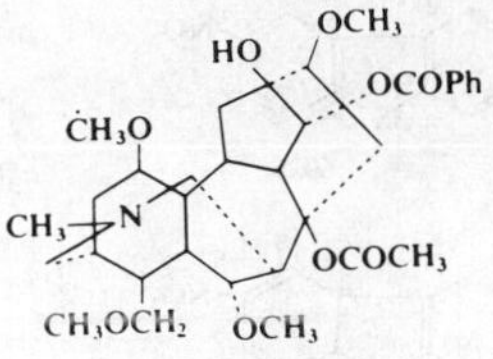
5950

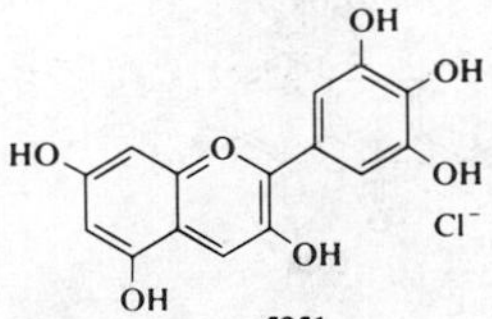
5951

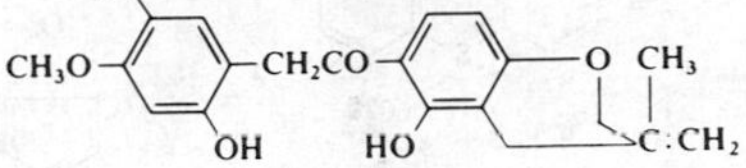
5953

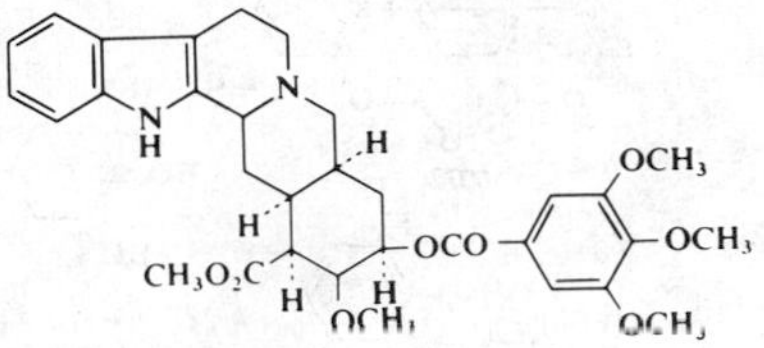
5954

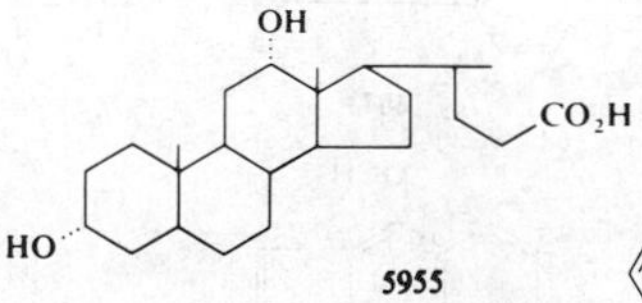
5955

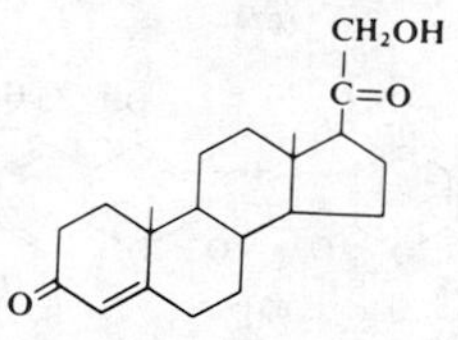
5956

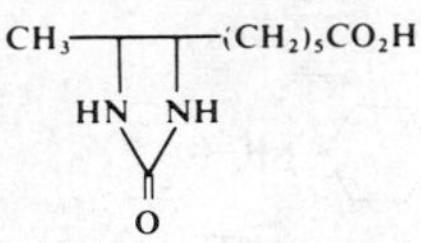
5957

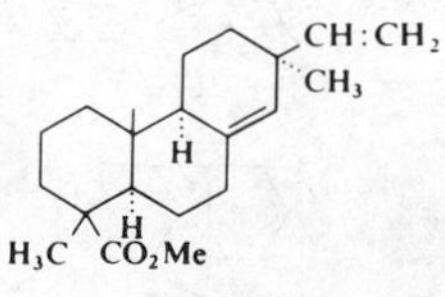
5960

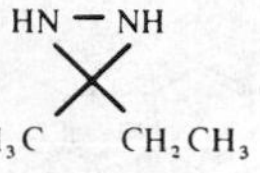
5966

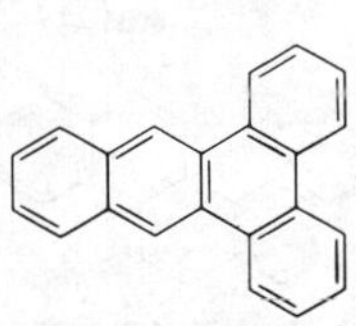
5982

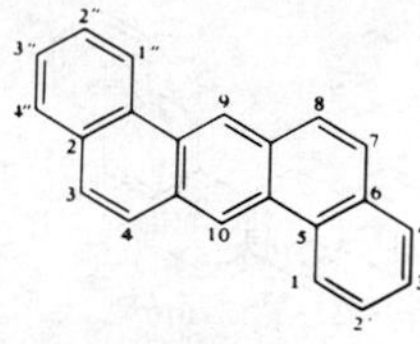
5983

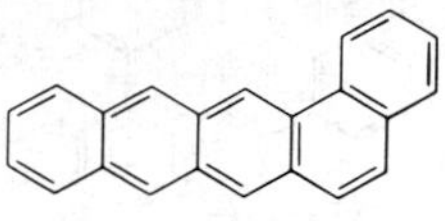
5985

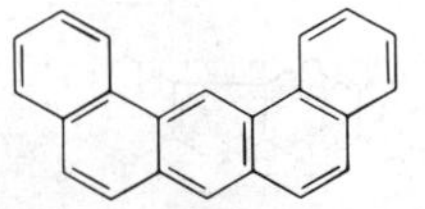
5986

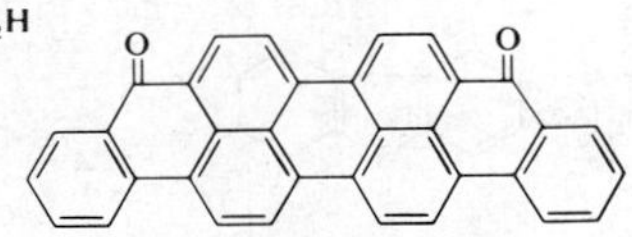
5987

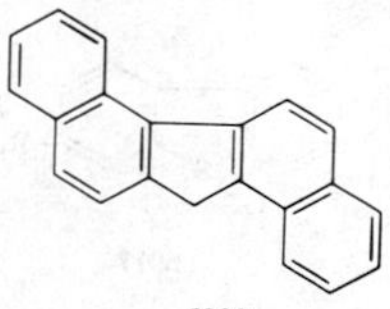
5989

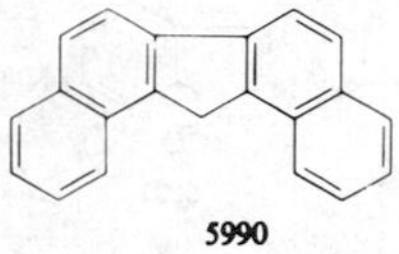
5990

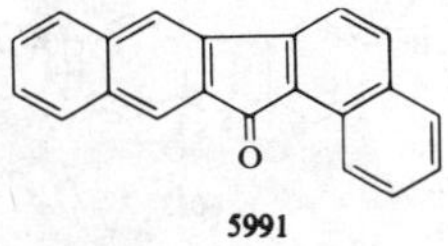
5991

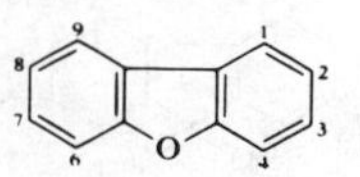
5992

STRUCTURAL FORMULAS OF ORGANIC COMPOUNDS (Continued)

In Numeric as they occur in Organic Compounds Table

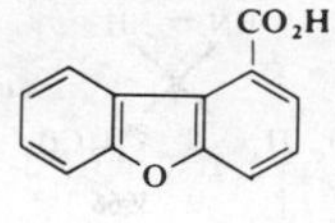

6006

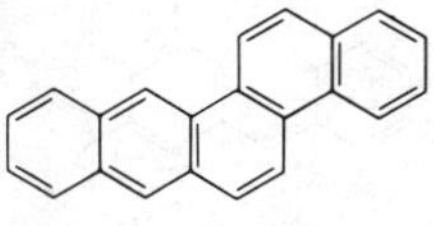

6009

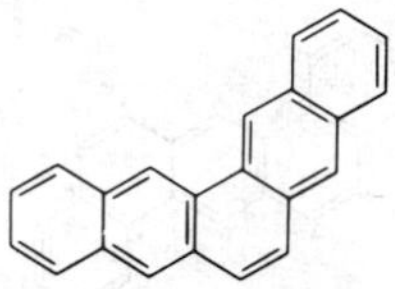

6010

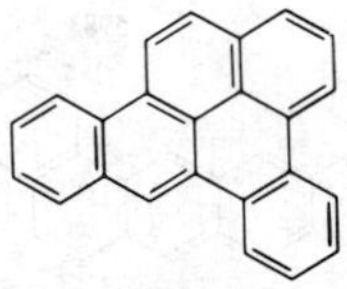

6011

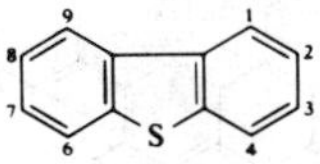

6012

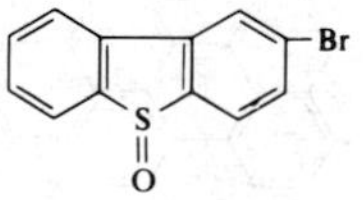

6016

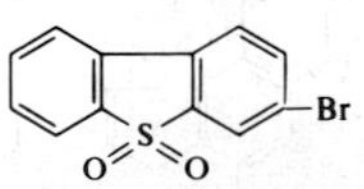

6017

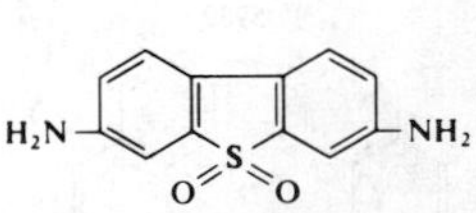

6021

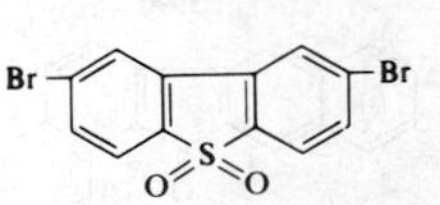

6023

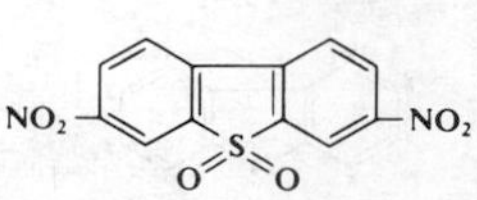

6024

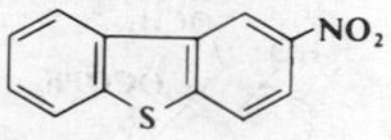

6025

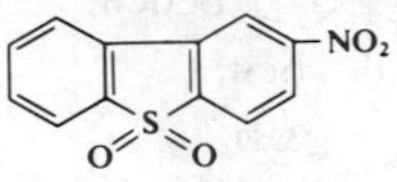

6026

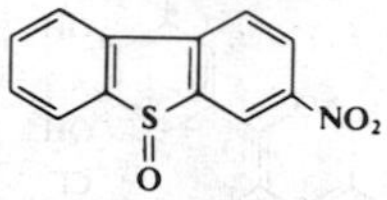

6028

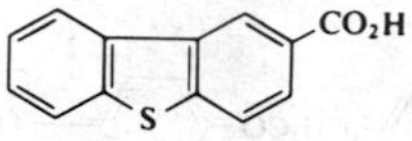

6029

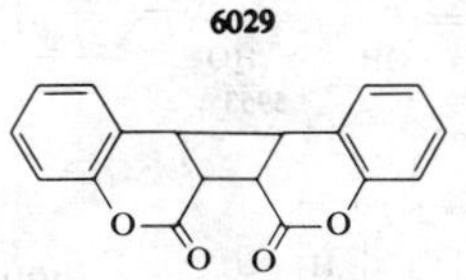

6072

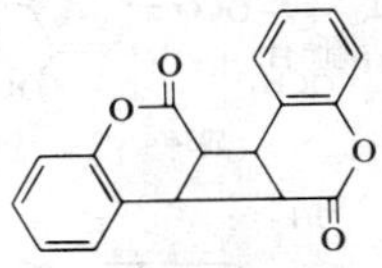

6073

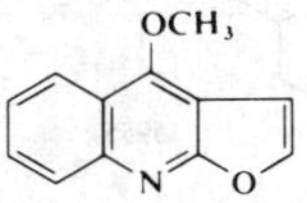

6074

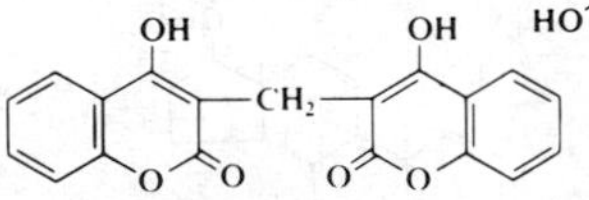

6075

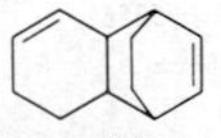

6076

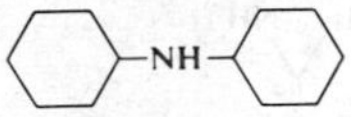

6077

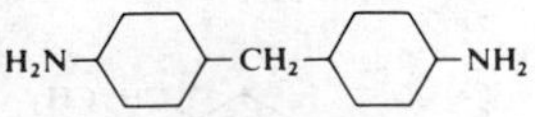

6079

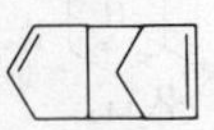

6082

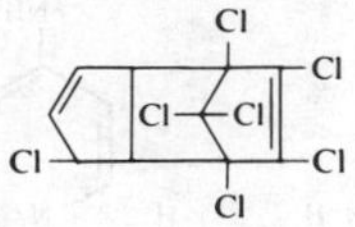

6083

6084

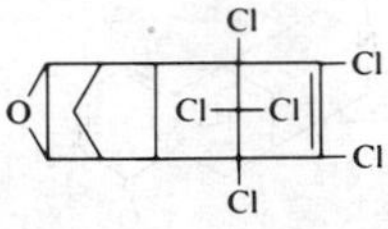

6086

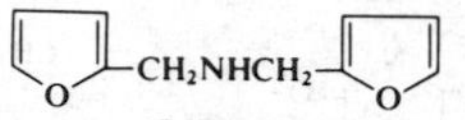

6129

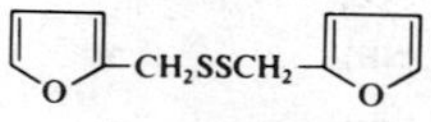

6130

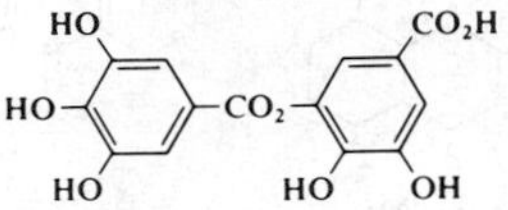

6131

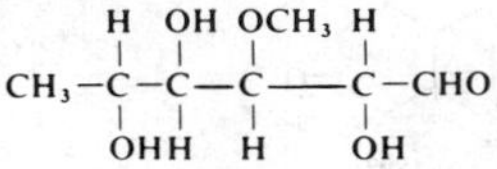

6132

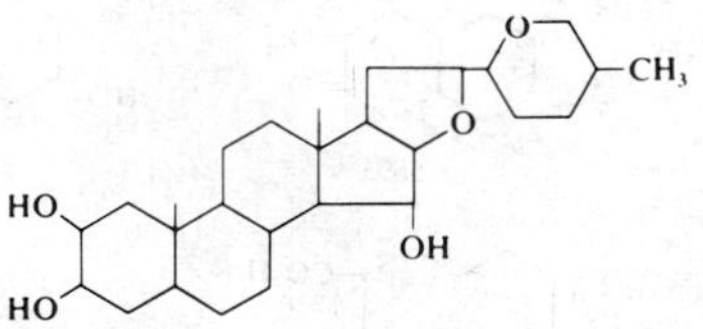

6133

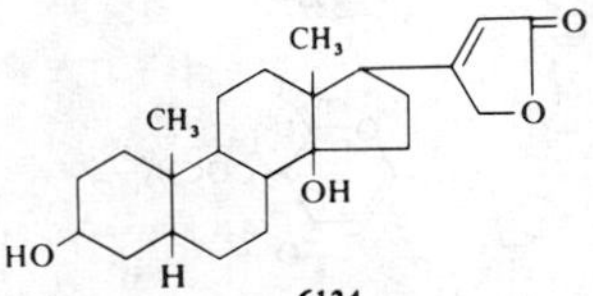

6134

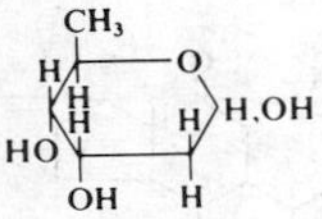

6136

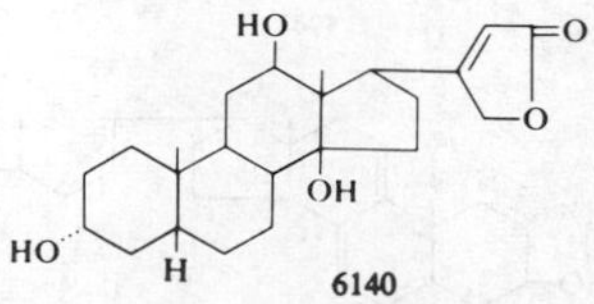

6140

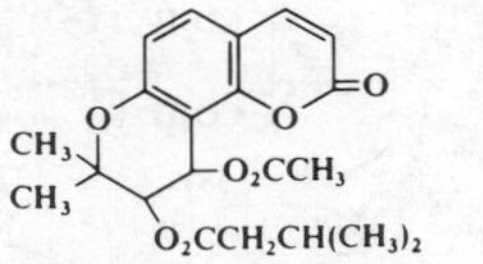

6145

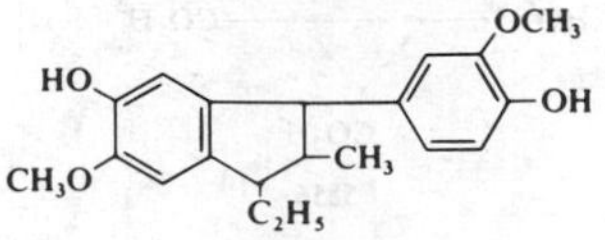

6152

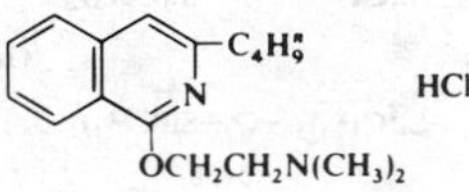

6155

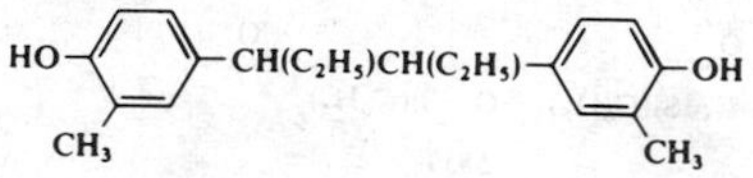

6168

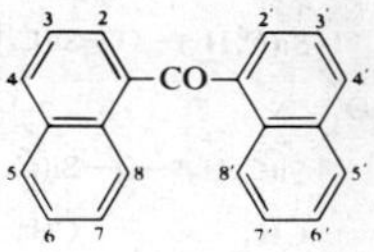

6181

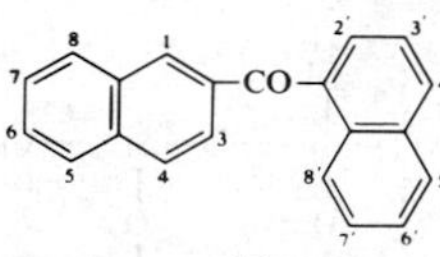

6182

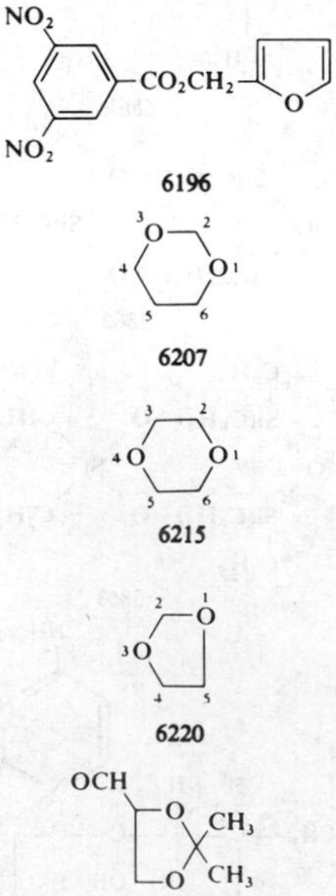

6196

6207

6215

6220

6223

STRUCTURAL FORMULAS OF ORGANIC COMPOUNDS (Continued)

In Numeric as they occur in Organic Compounds Table

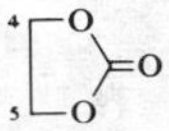
6224

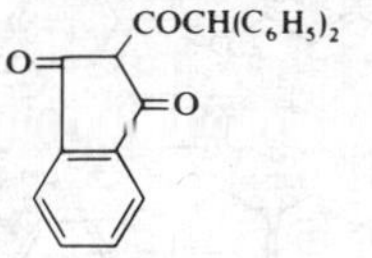
6231

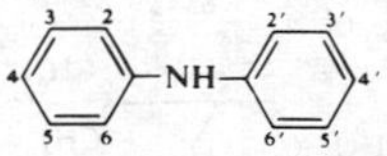
6249

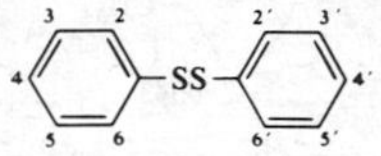
6267

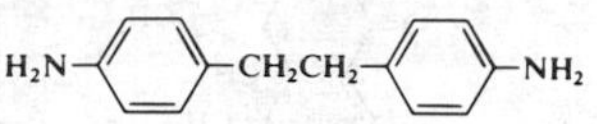
6269

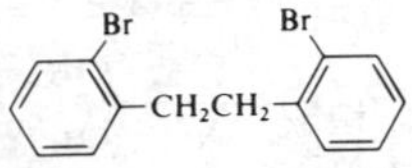
6270

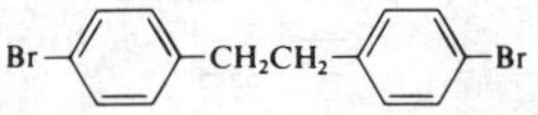
6271

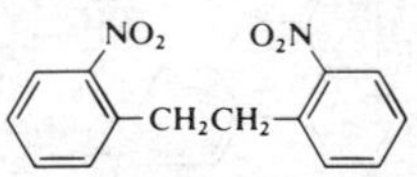
6274

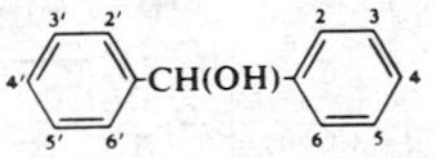
6314

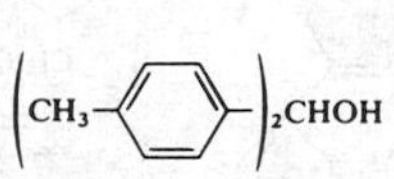
6315

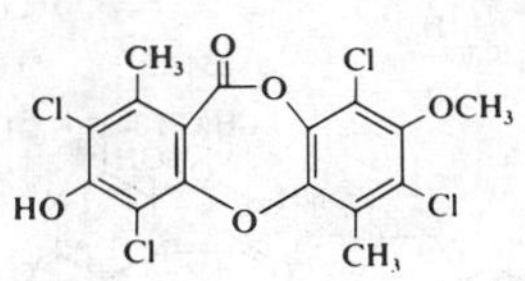
6327

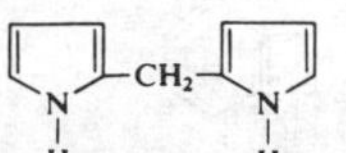
6342

6357

6358

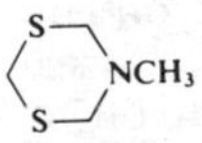
6359

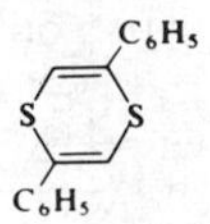
6360

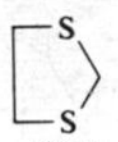
6361

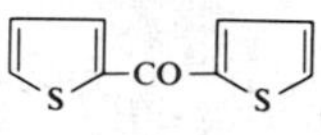
6362

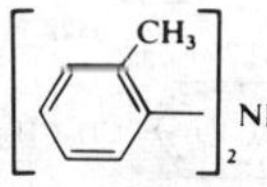
6365

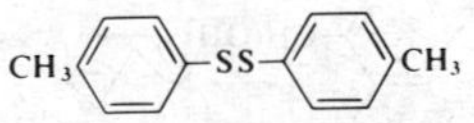
6370

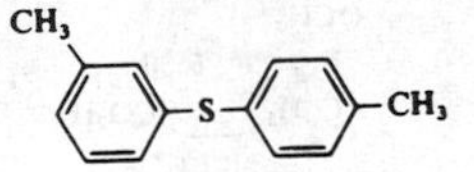
6372

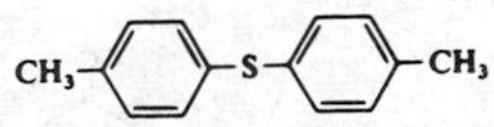
6373

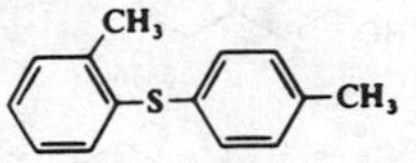
6374

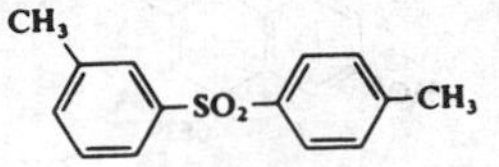
6376

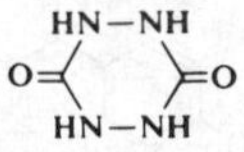
6379

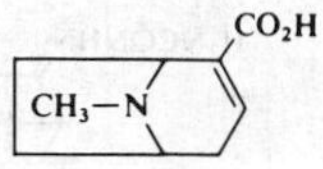
6423

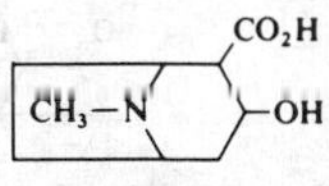
6424

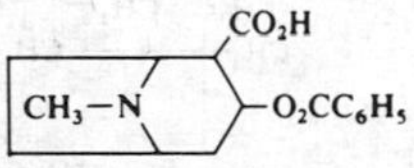
6428

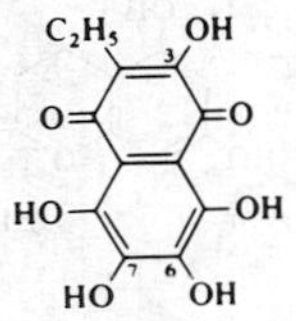
6431

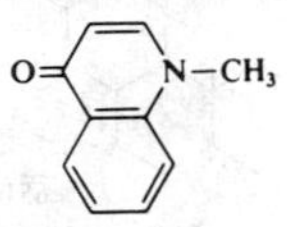
6433

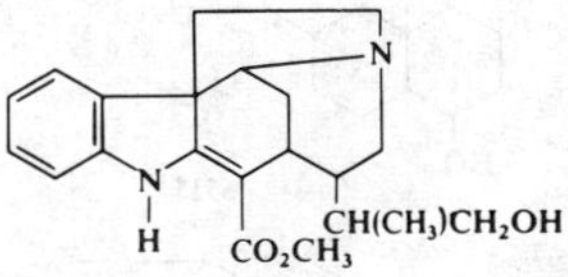
6434

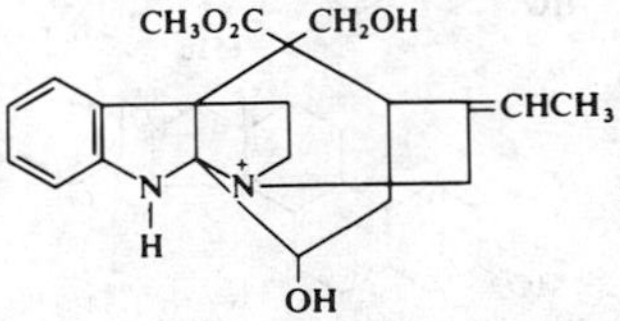
6435

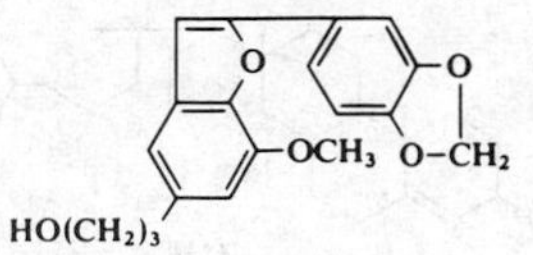
6438

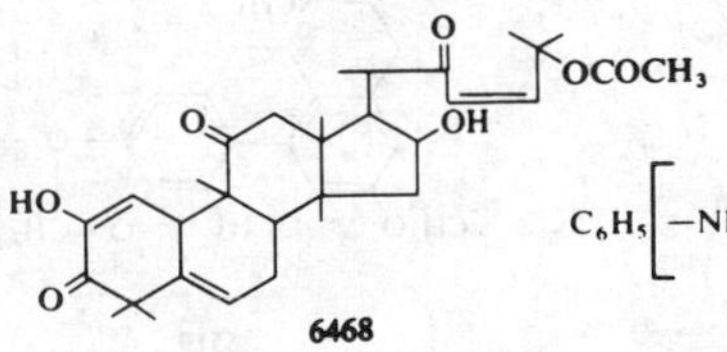
6468

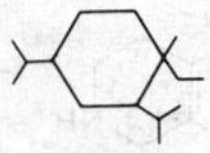
6470

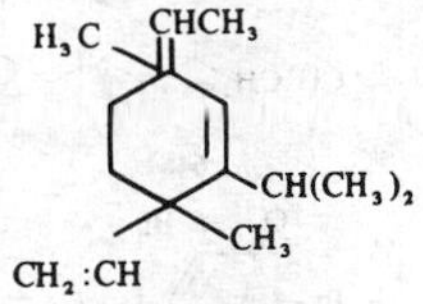
6471

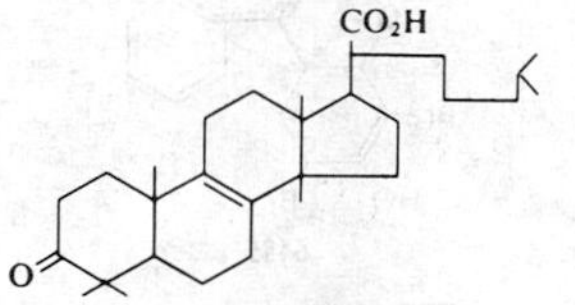
6472

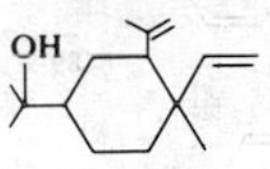
6473

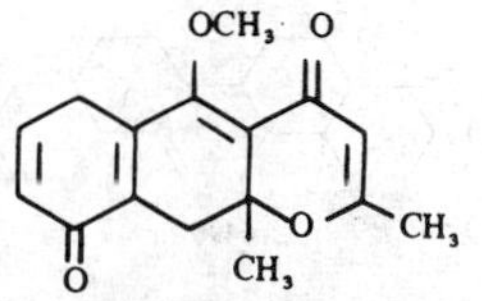
6476

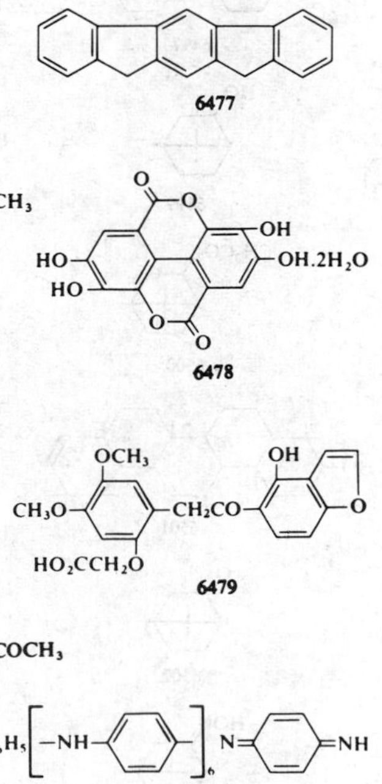
6477

6478

6479

6480

In Numeric as they occur in Organic Compounds Table

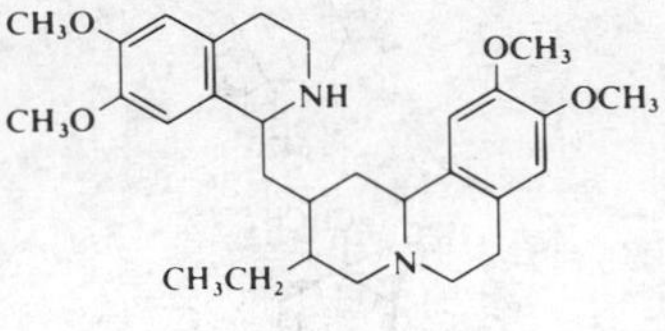

6481

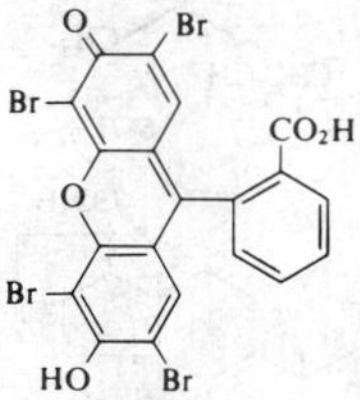

6485

$H_2N-C_6H_4-CH(OH)CH(NH_2)CH_3 . 2HCl$

6489

$NO_2-C_6H_4-CON(CH_3)CH(CH_3)CH(OH)C_6H_5$

6494

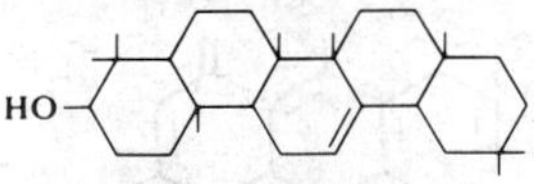

6496

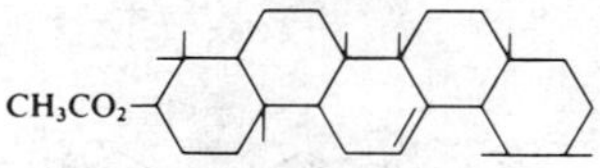

6497

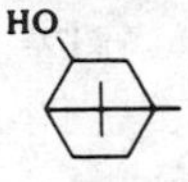

6499

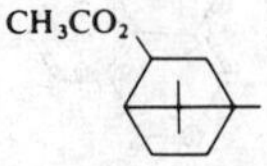

6500

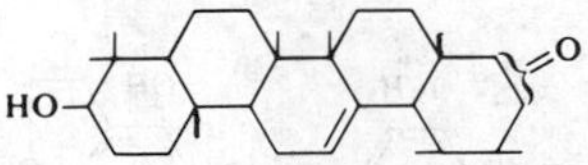

6501

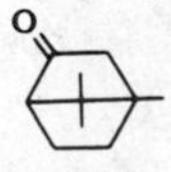

6502

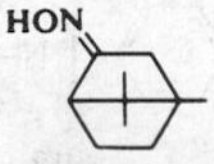

6503

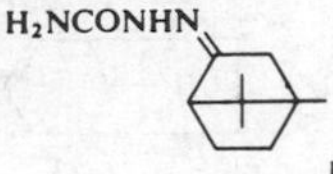

6504

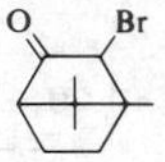

6508

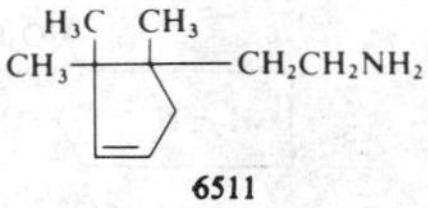

6511

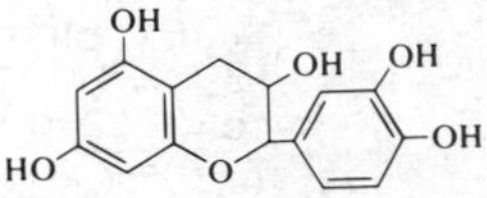

6512

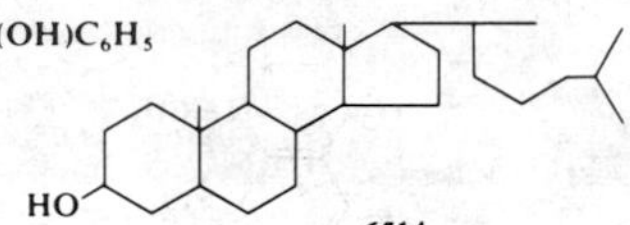

6514

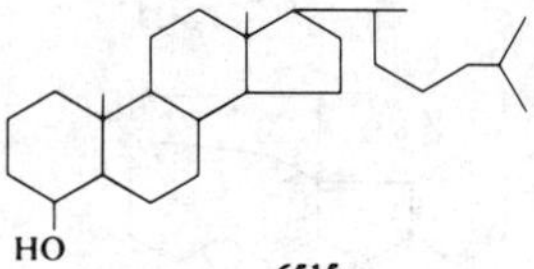

6515

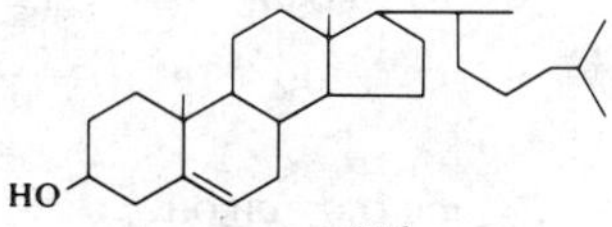

6516

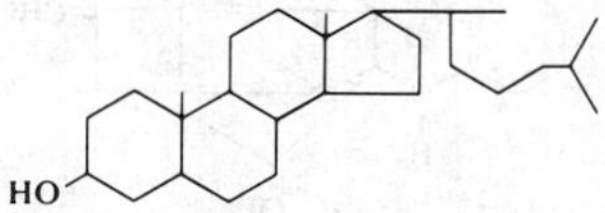

6517

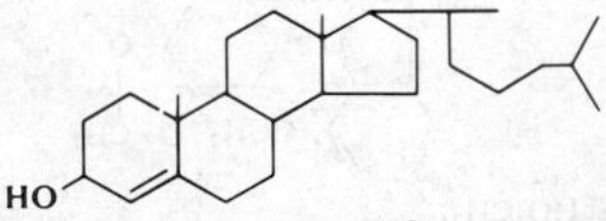

6518

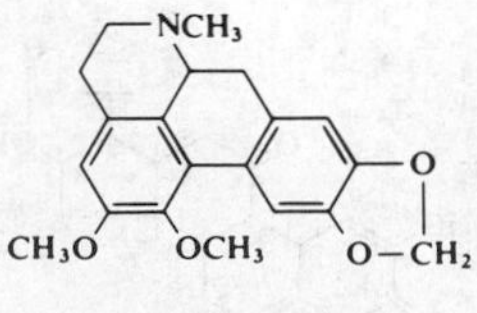

6519

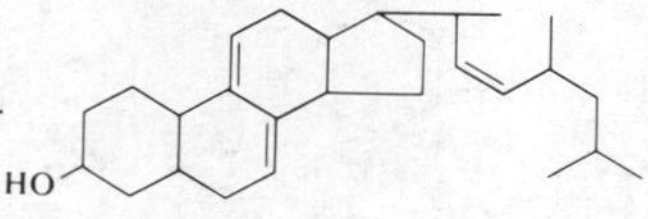

6521

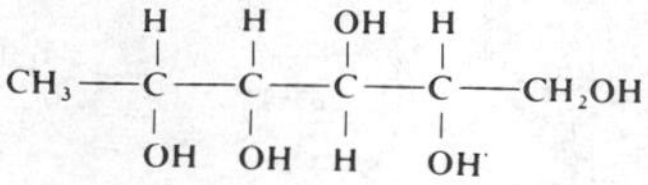

6522

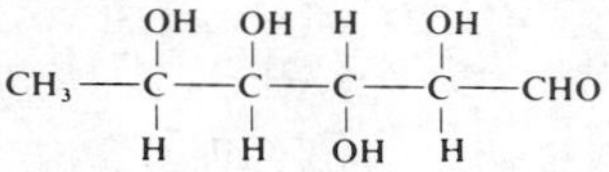

6524

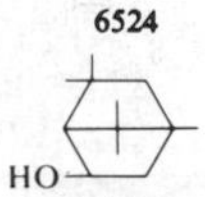

6525

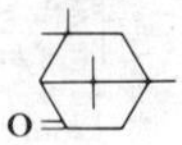

6526

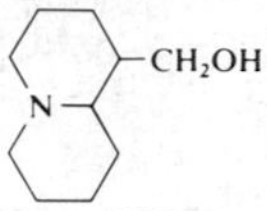

6528

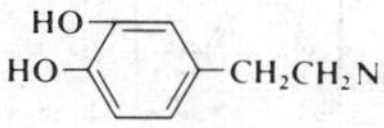

6529

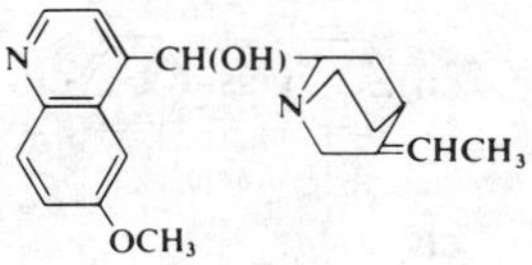

6530

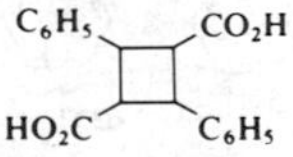

6533

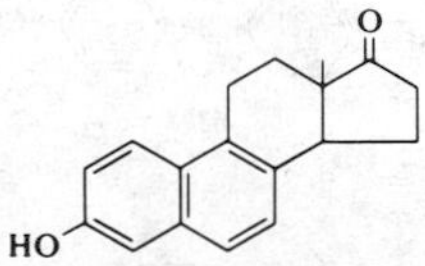

6536

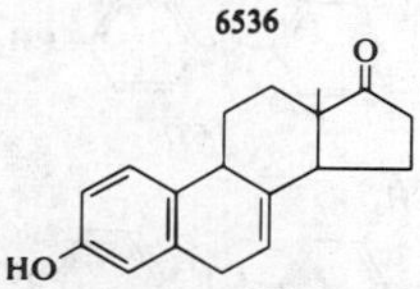

6539

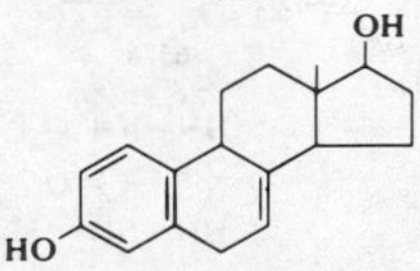

6540

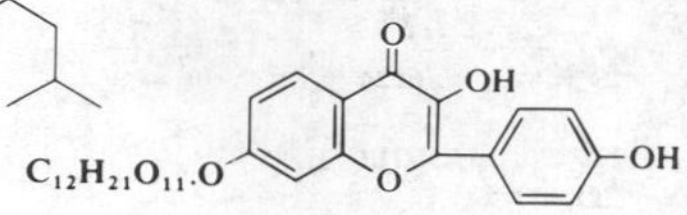

6541

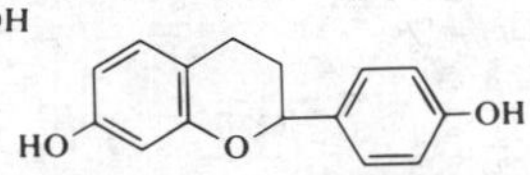

6542

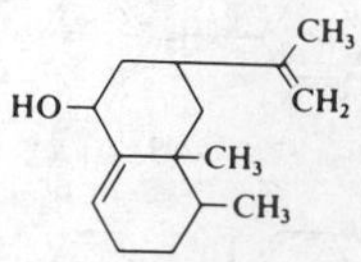

6543

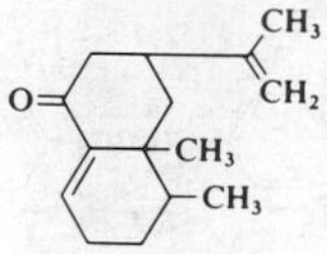

6544

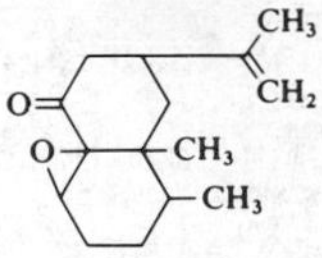

6545

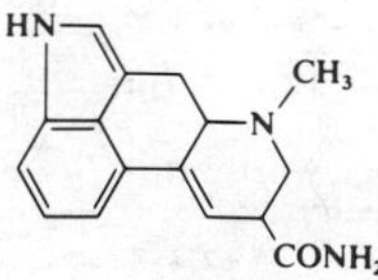

6546

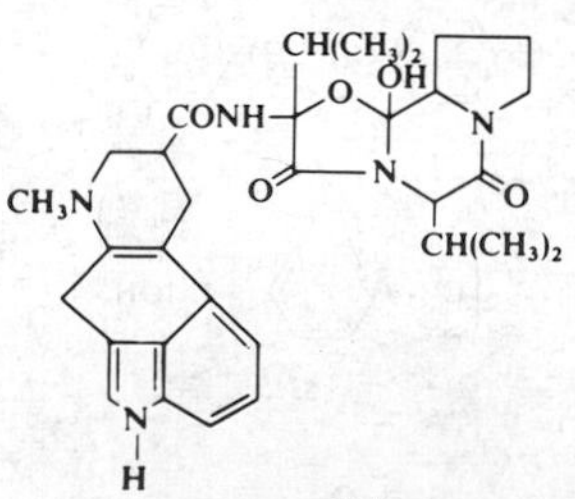

6547

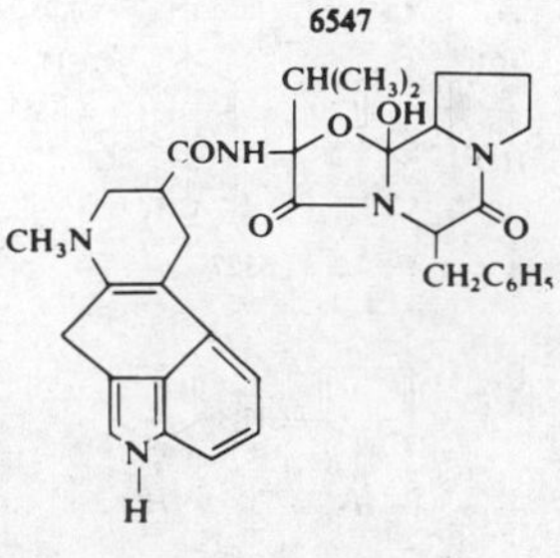

6549

In Numeric as they occur in Organic Compounds Table

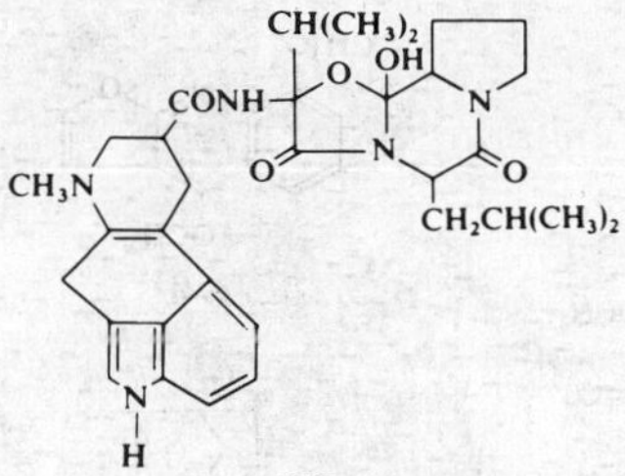
6551

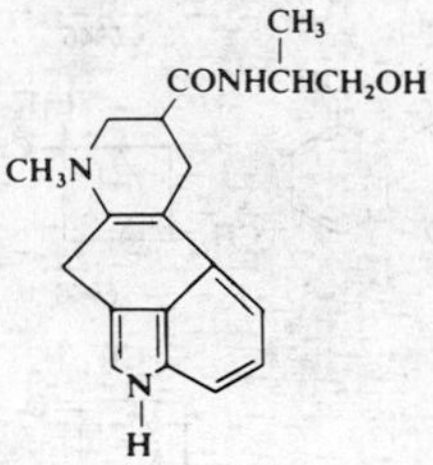
6553

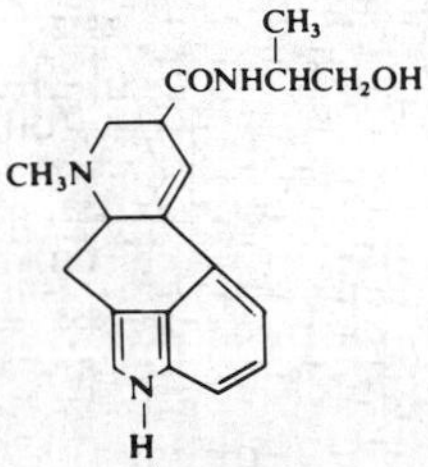
6554

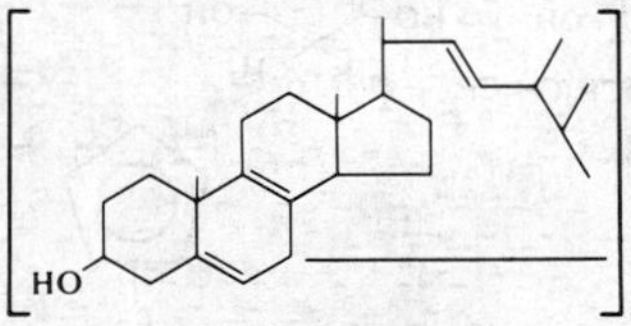
6556

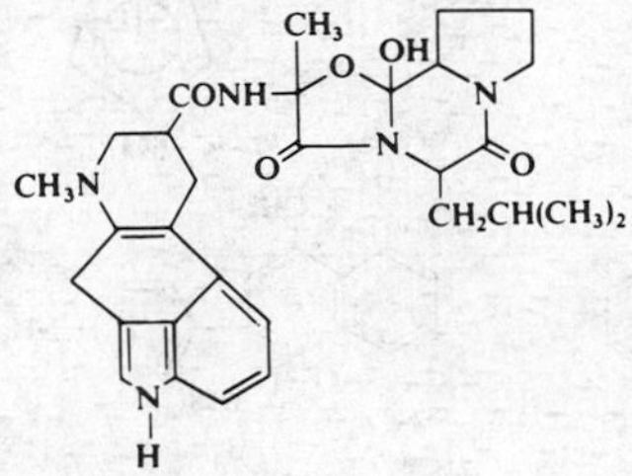
6557

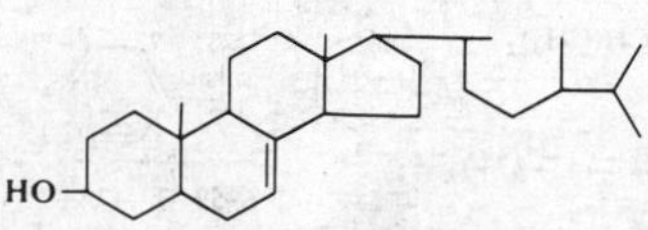
6559

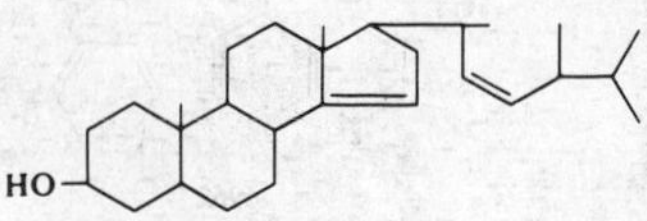
6560

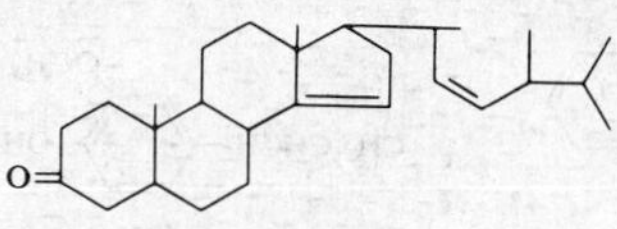
06561

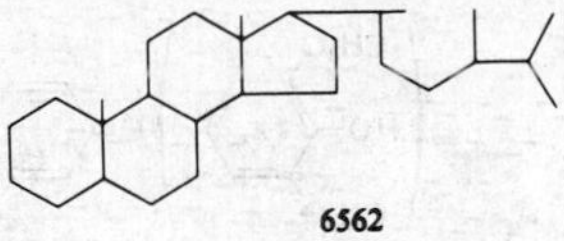
6562

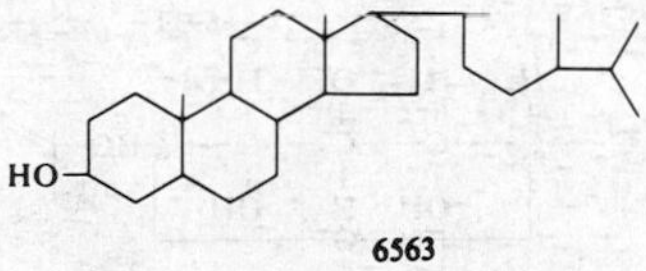
6563

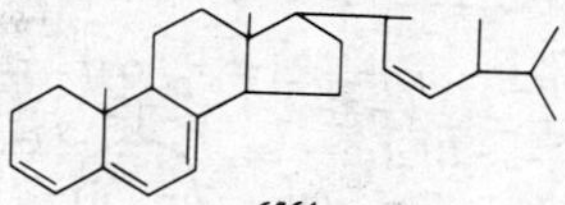
6564

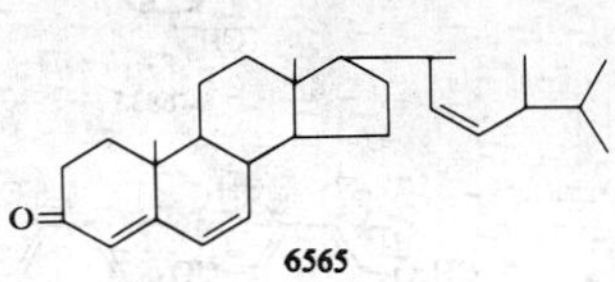
6565

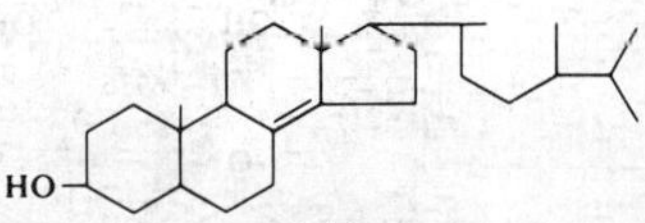
6566

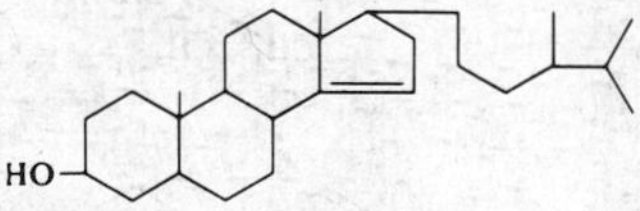
6567

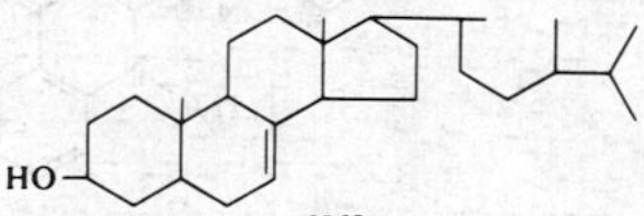
6568

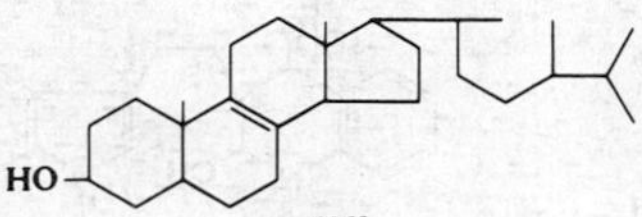
6569

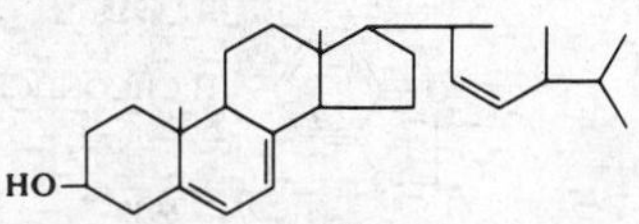
6570

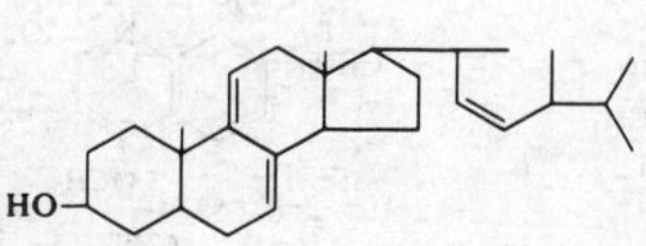
6571

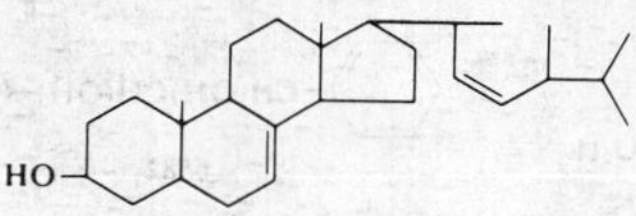
6572

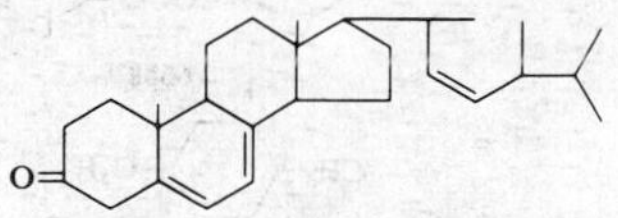
6573

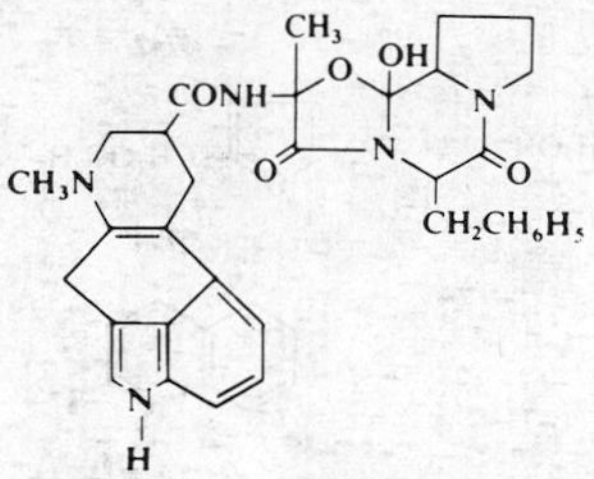
6574

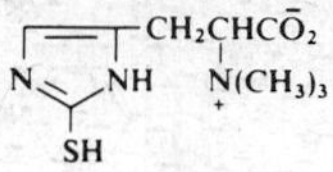
6576

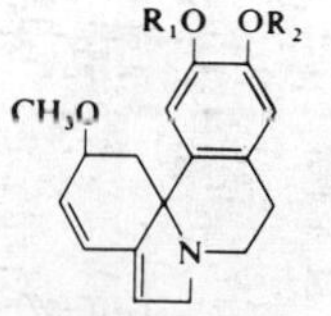
$R_1 = CH_3$ or H
$R_2 = H$ or CH_3
6579

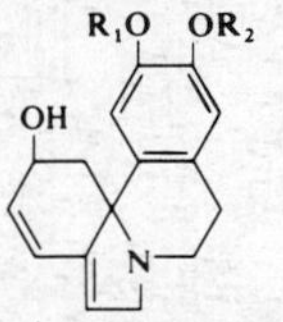
$R_1 = CH_3$ or H
$R_2 = H$ or CH_3
6580

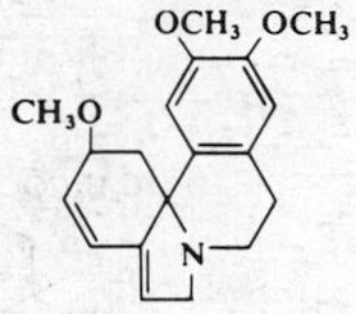
6581

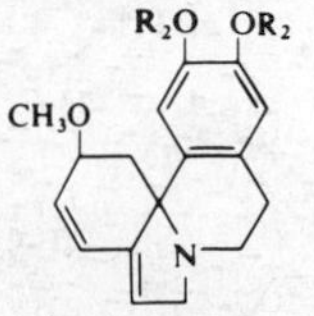
$R_1 = CH_3$ or H
$R_2 = H$ or CH_3
6582

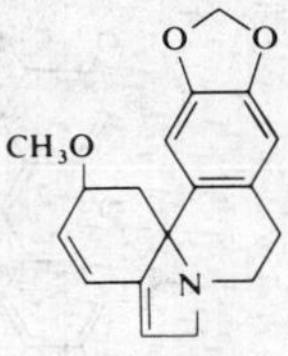
6584

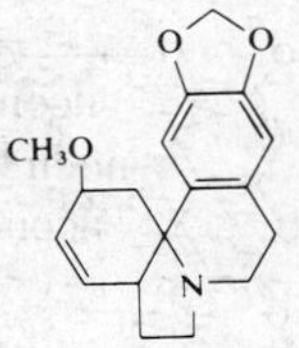
6585

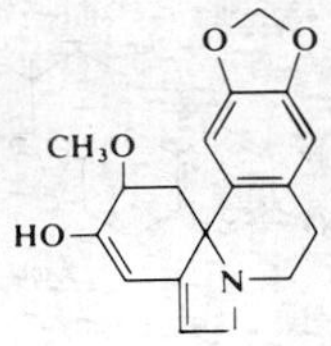
6586

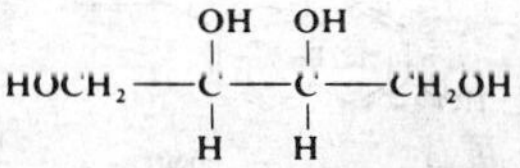
6589

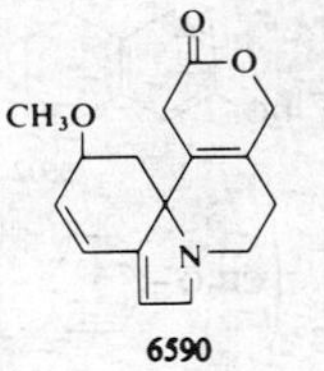
6590

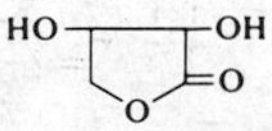
6591

6593

In Numeric as they occur in Organic Compounds Table

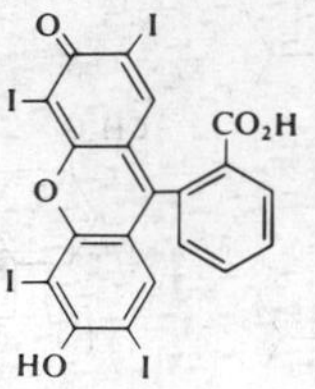
6595

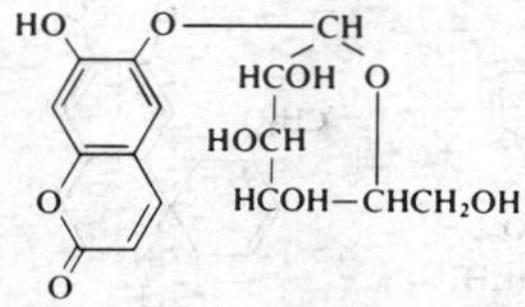
6598

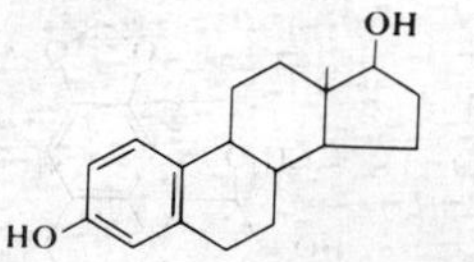
6599

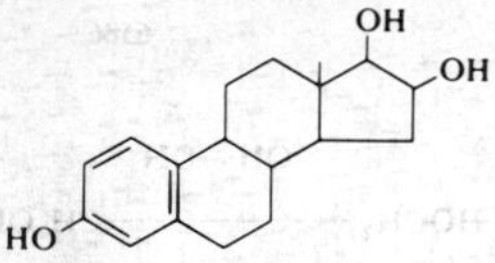
6601

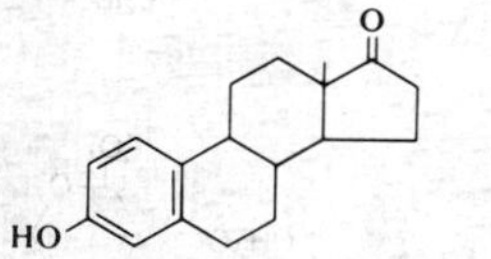
6602

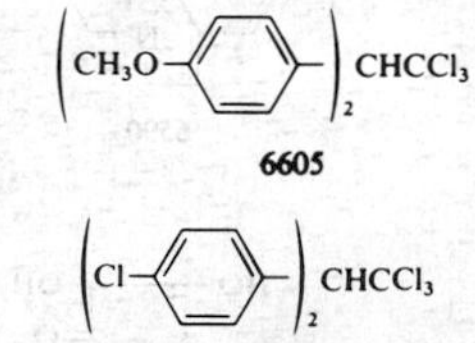
6605

6609

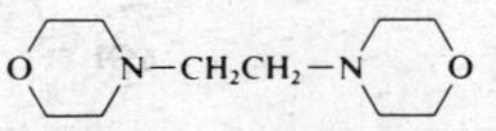
6638

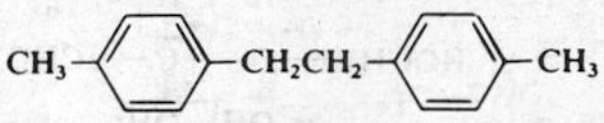
6643

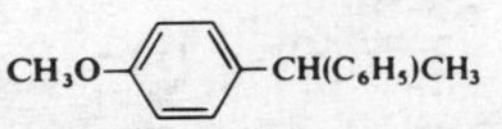
6655

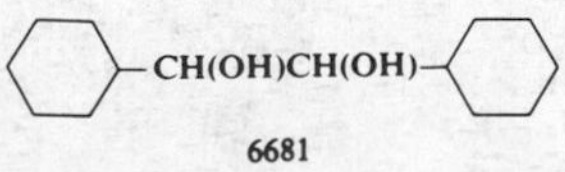
6681

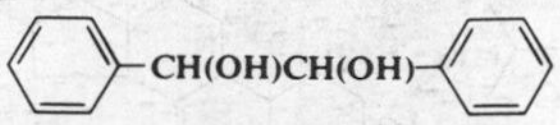
6682

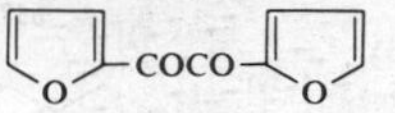
6688

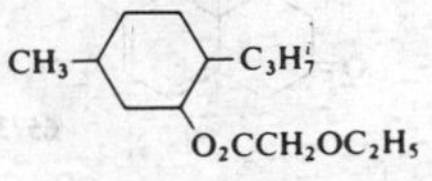
6742

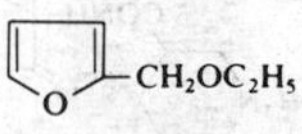
6794

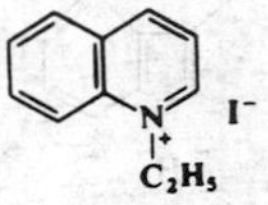
6830

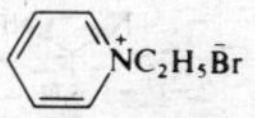
6831

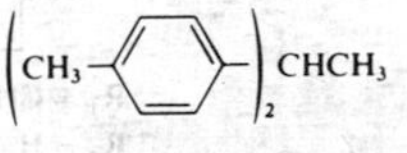
6889

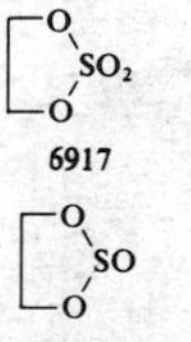
6917

6918

6919

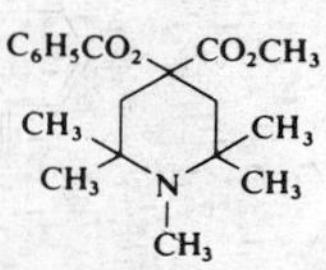
6925

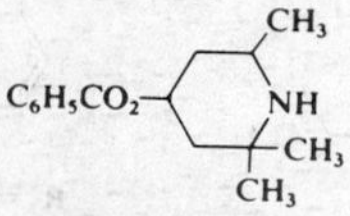
6927

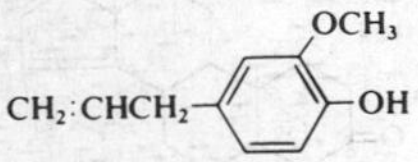
6931

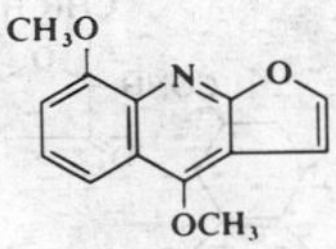
6941

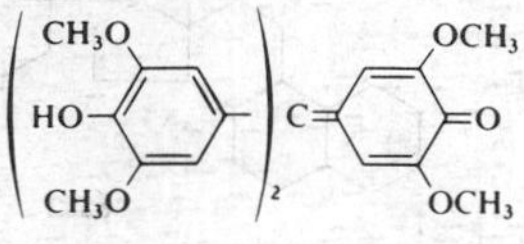
6933

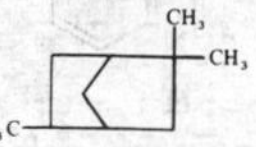
6946

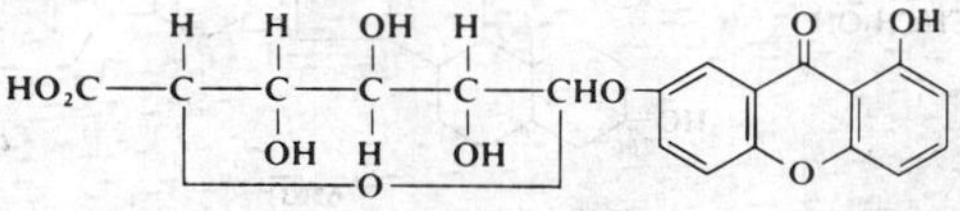
6934

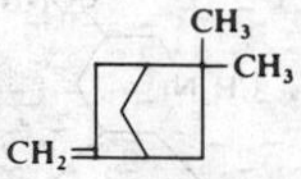
6948

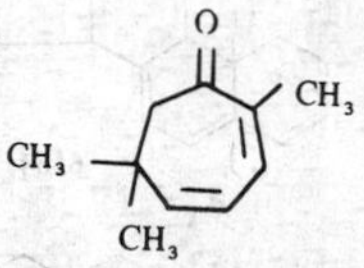
6935

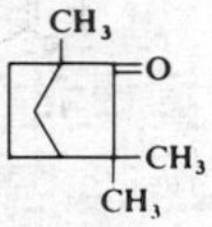
6952

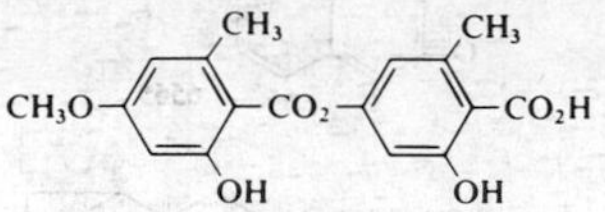
6936

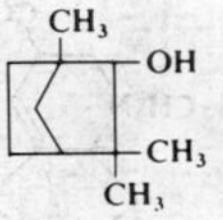
6955

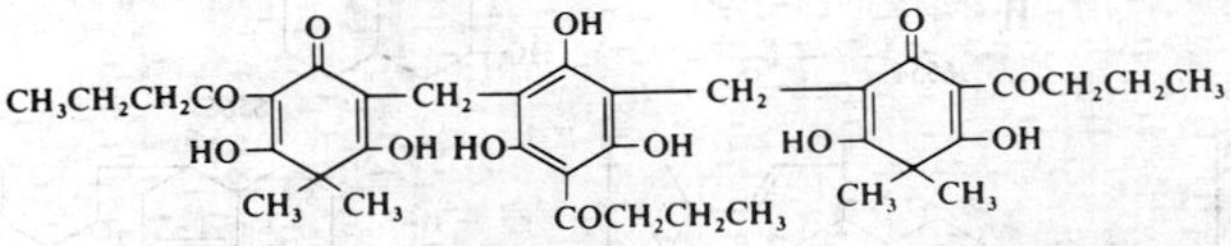
6956

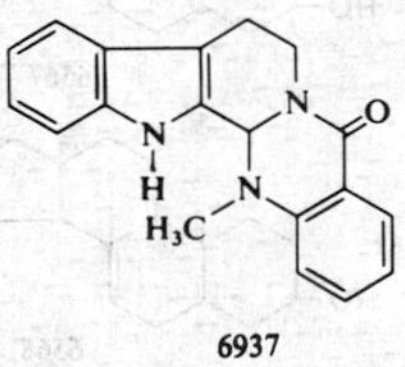
6937

6957

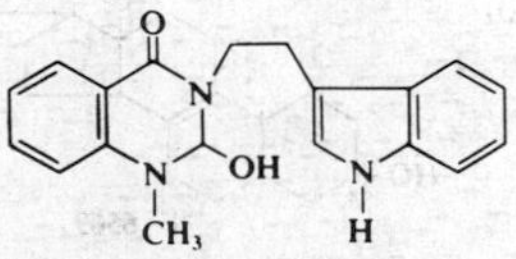
6938

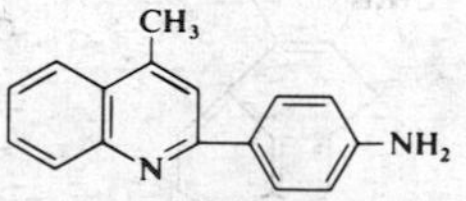
6958

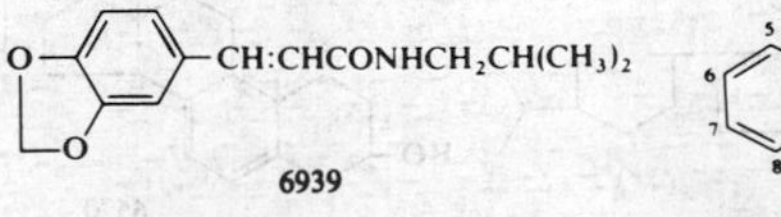
6939

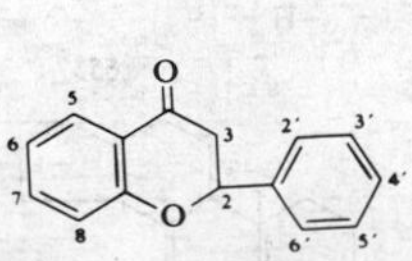
6959

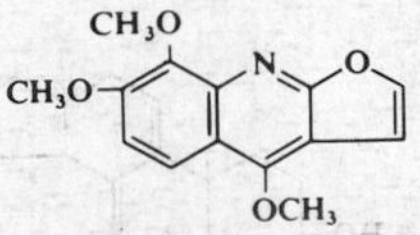
6940

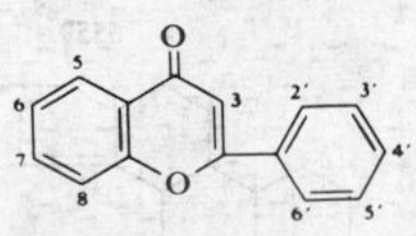
6962

STRUCTURAL FORMULAS OF ORGANIC COMPOUNDS (Continued)

In Numeric as they occur in Organic Compounds Table

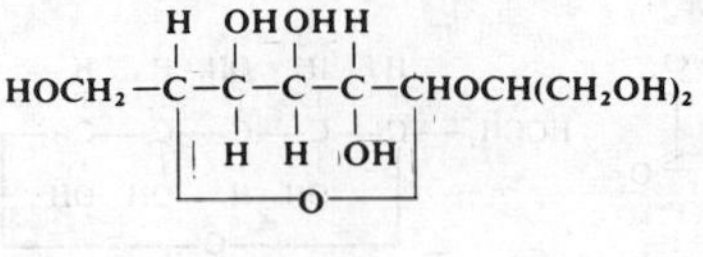
6981

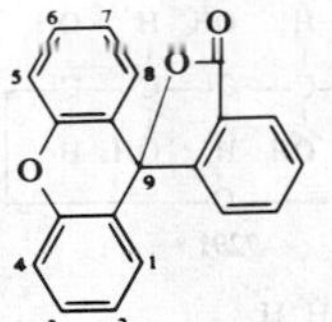
6982

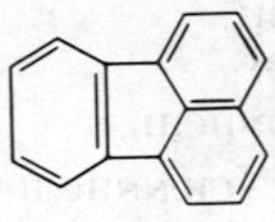
6986

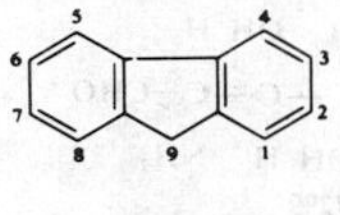
6987

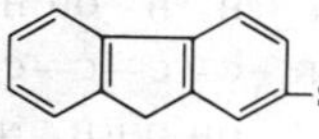
7020

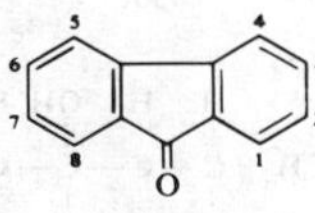
7021

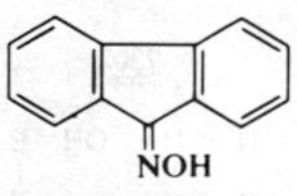
7030

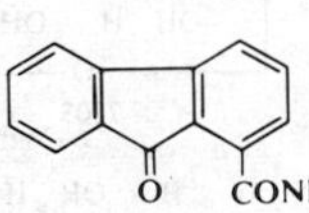
7033

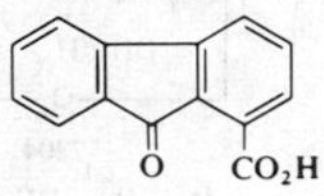
7034

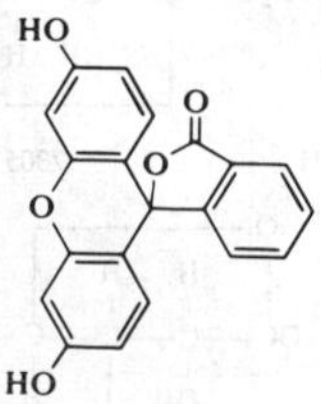
7041

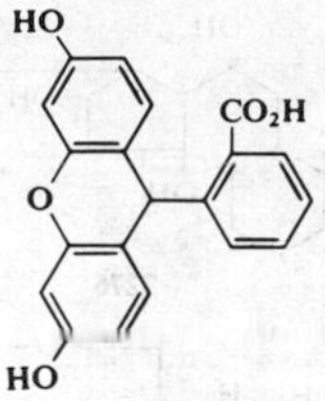
7042

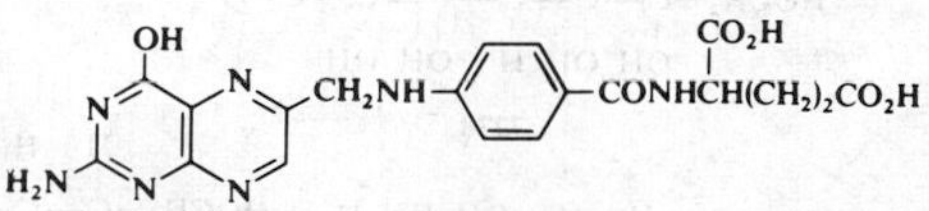
7045

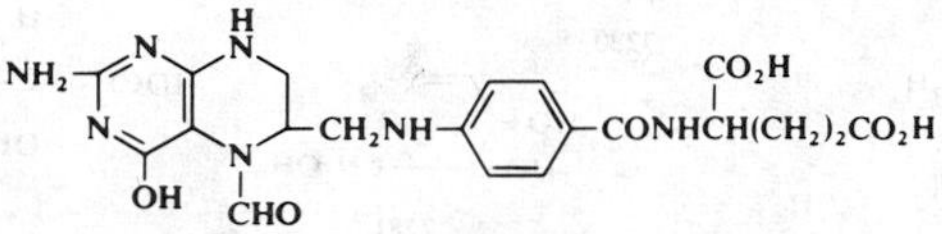
7046

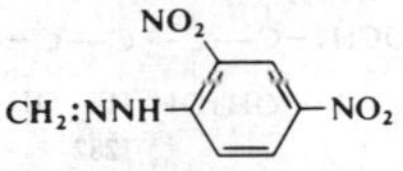
7053

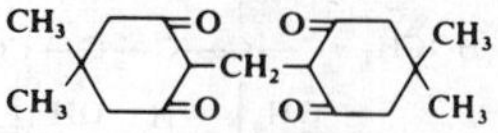
7057

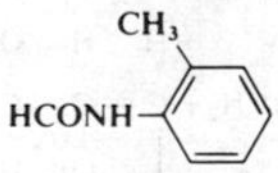
7066

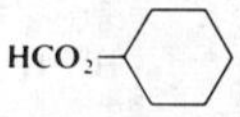
7078

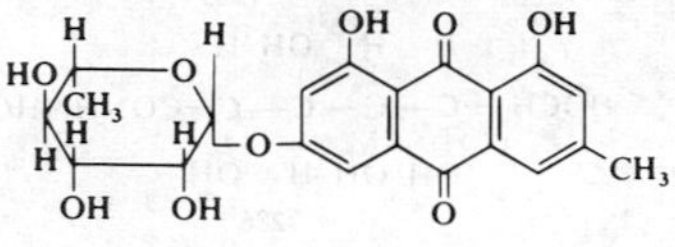
7091

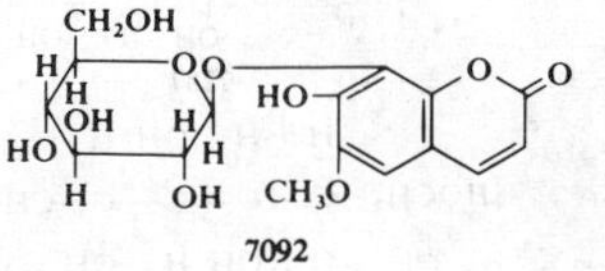
7092

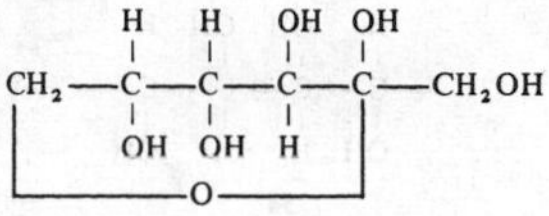
7093

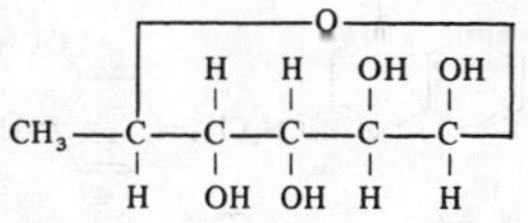
7094

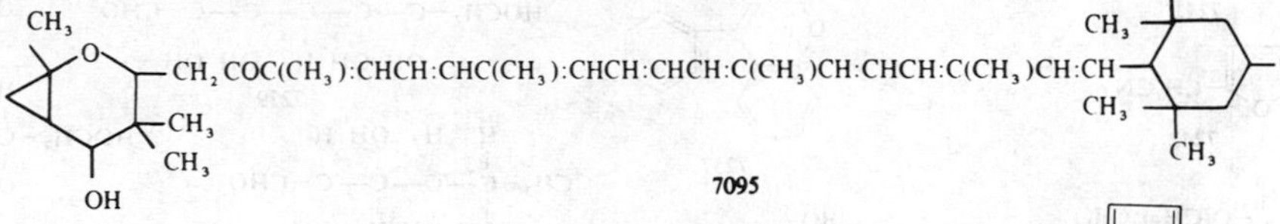
7095

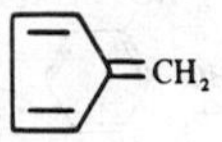
7096

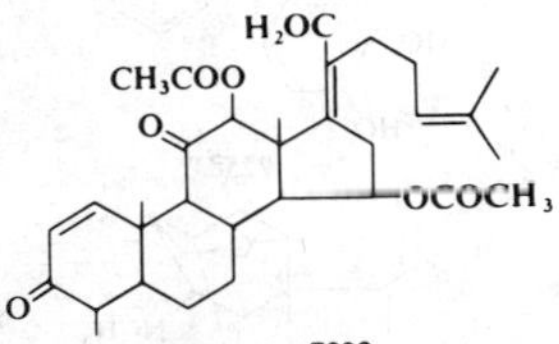
7098

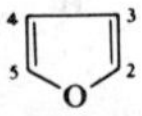
7116

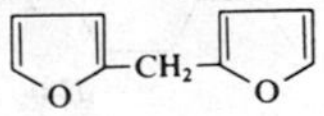
7133

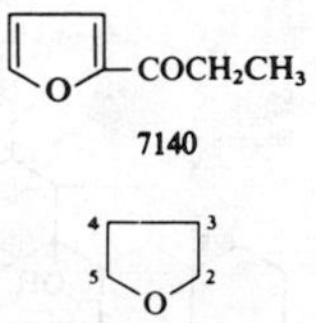
7140

7142

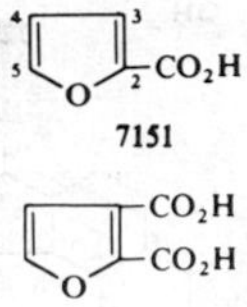
7151

7158

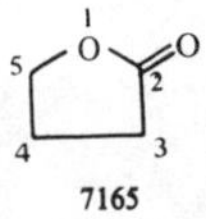
7165

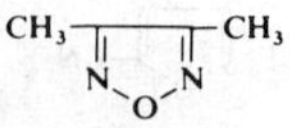
7169

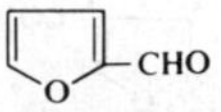
7170

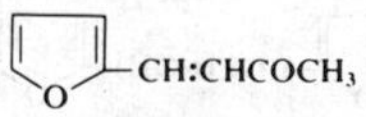
7171

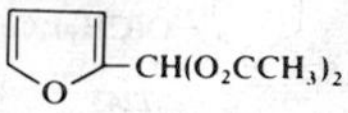
7174

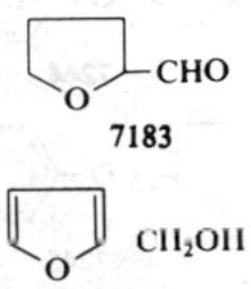
7183

7184

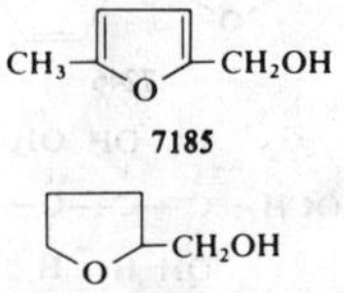
7185

7186

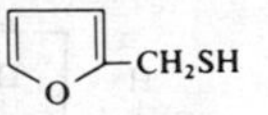
7192

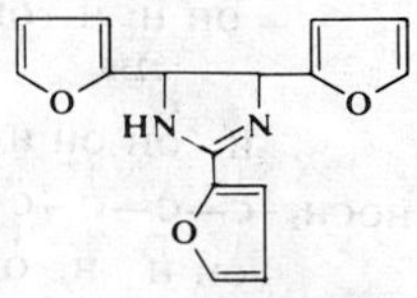
7193

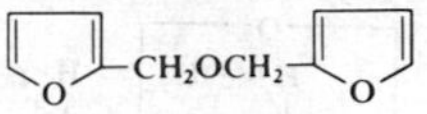
7194

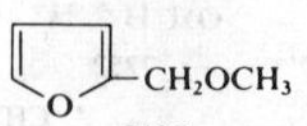
7196

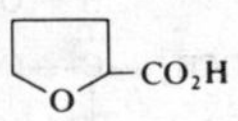
7224

In Numeric as they occur in Organic Compounds Table

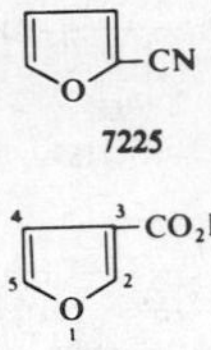
7225
7227

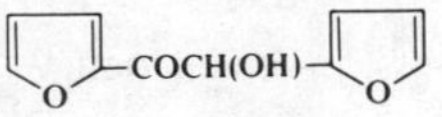
7240

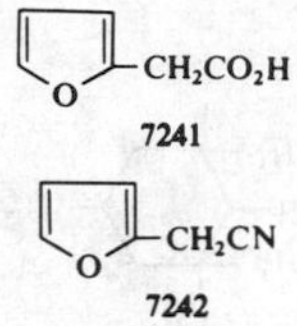
7241
7242

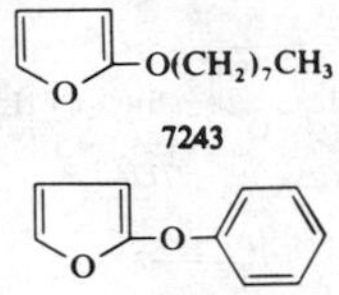
7243
7244

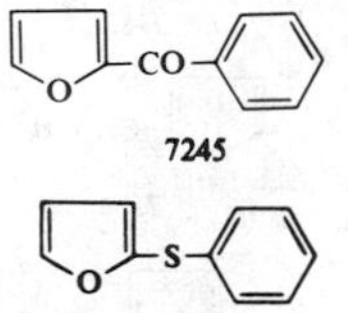
7245
7246

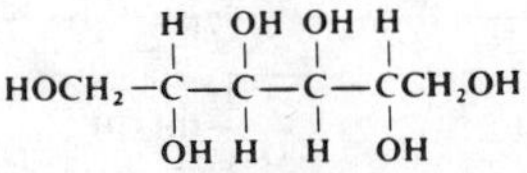
7249

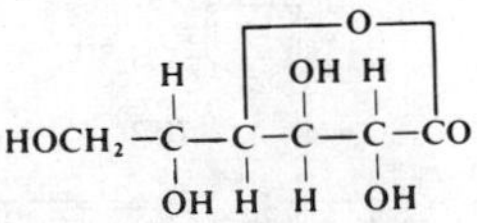
7250

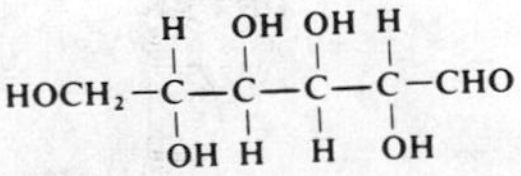
7251

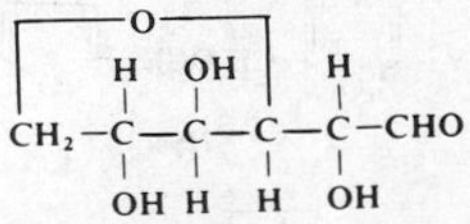
7252

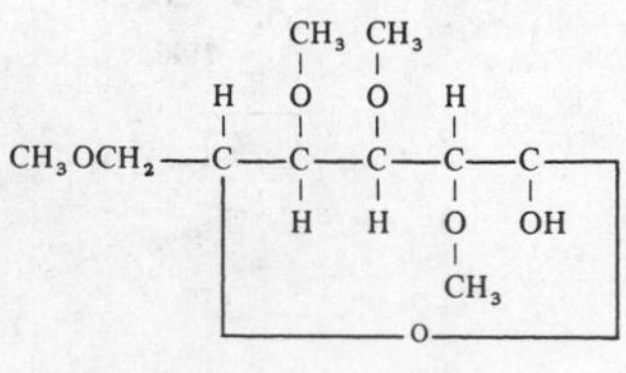
7253

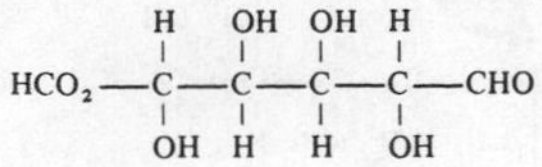
7254

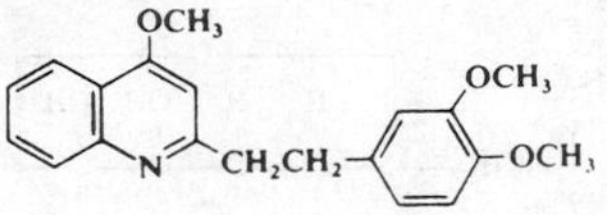
7256

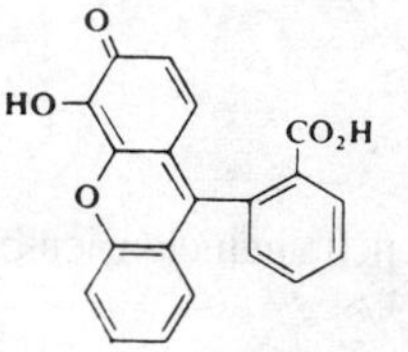
7257

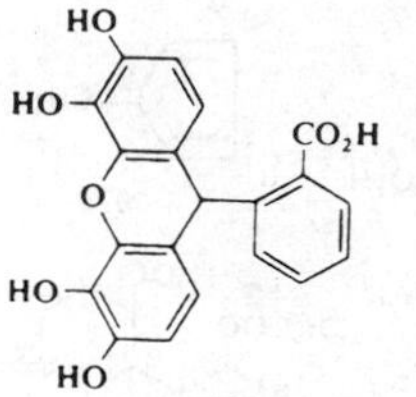
7258

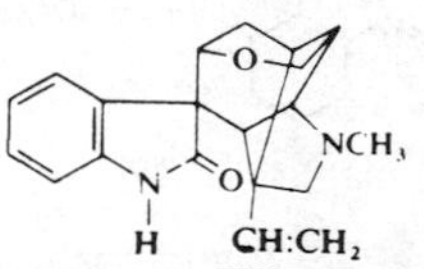
7259

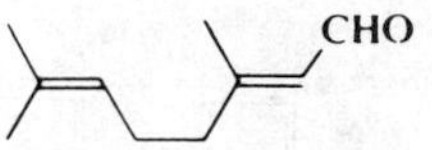
7261

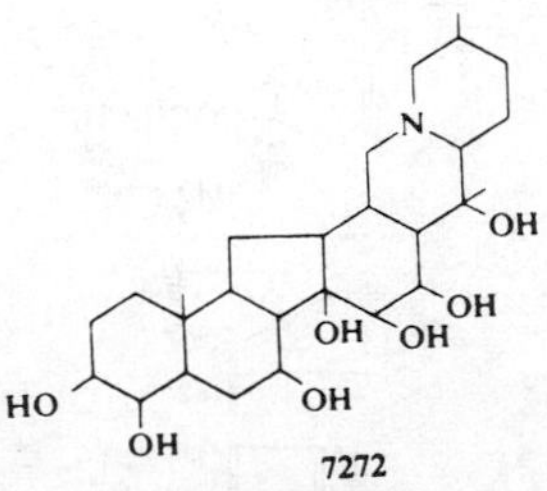
7272

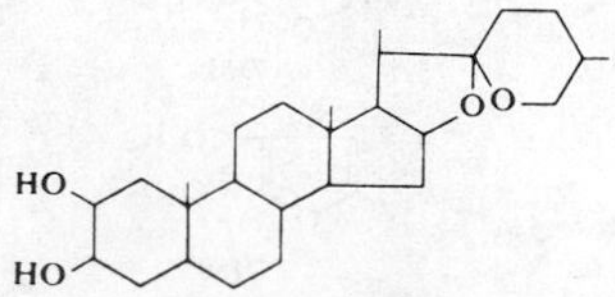
7275

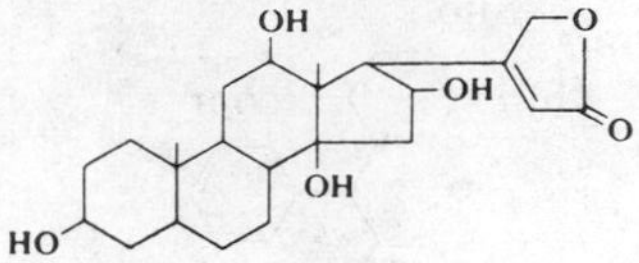
7276

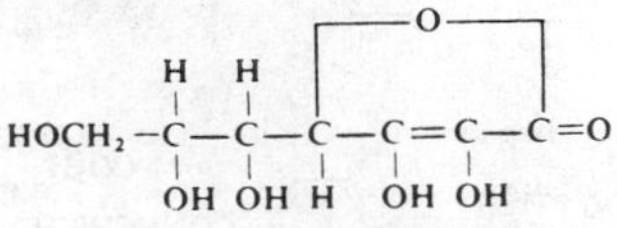
7278

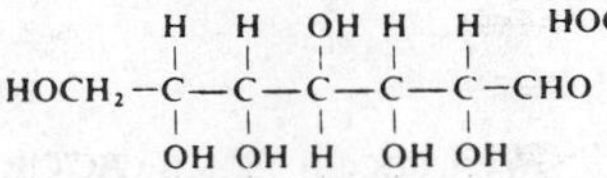
7279

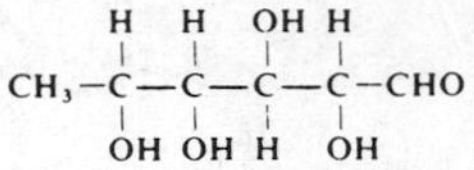
7280

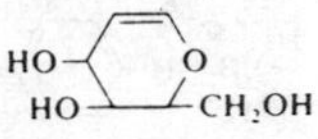
7281

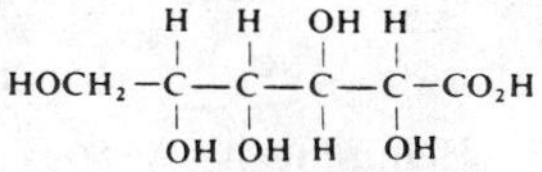
7282

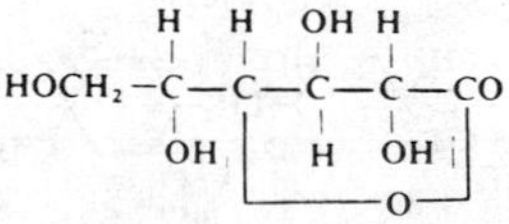
7283

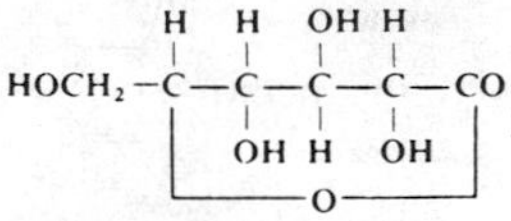
7284

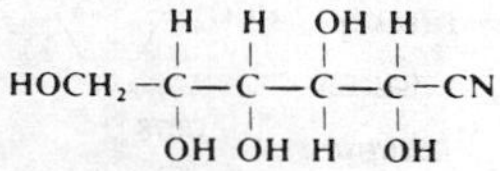
7285

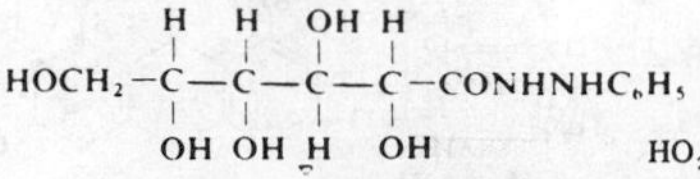
7286

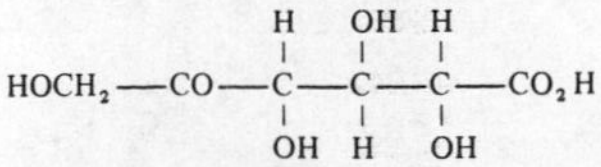
7287

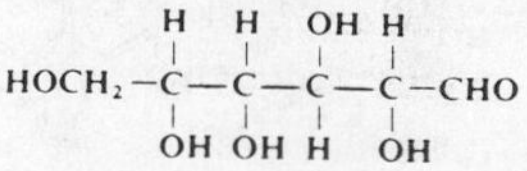
7288

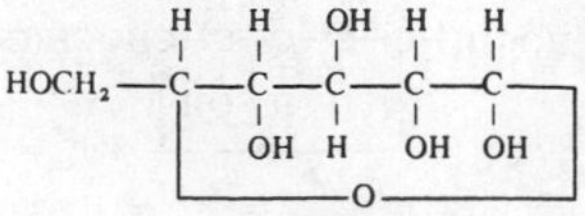
7289

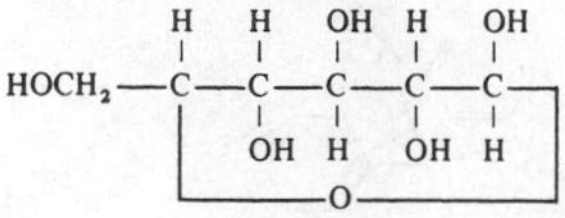
7291

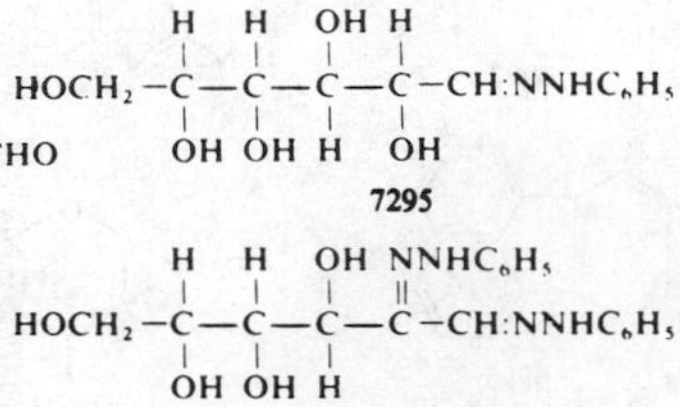
7295
7297

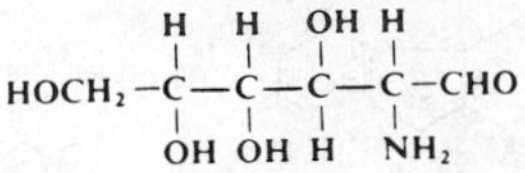
7299

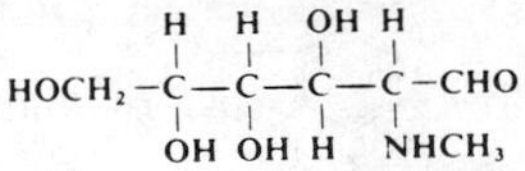
7301

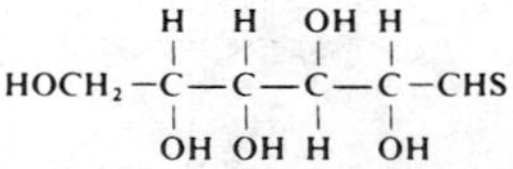
7302

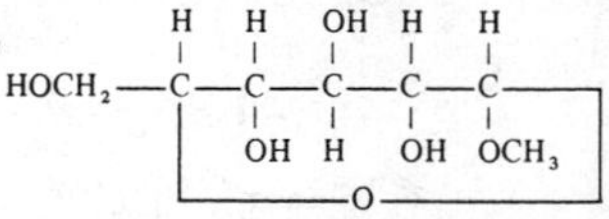
7303

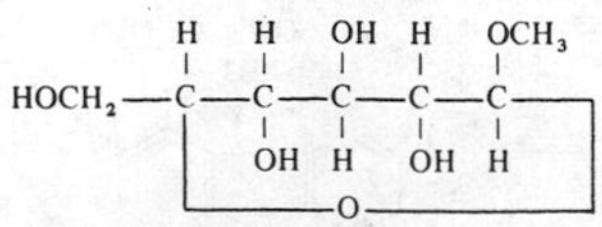
7304

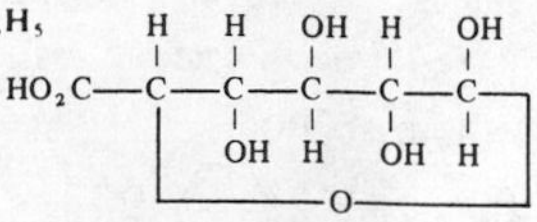
7305

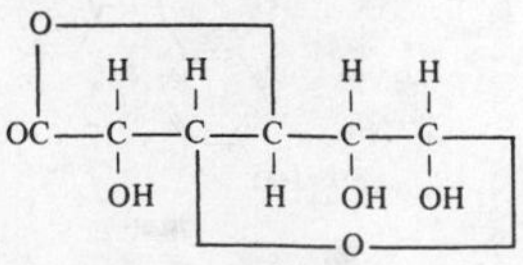
7306

STRUCTURAL FORMULAS OF ORGANIC COMPOUNDS (Continued)

In Numeric as they occur in Organic Compounds Table

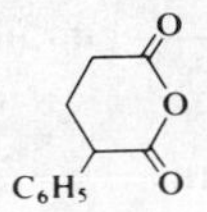

7320

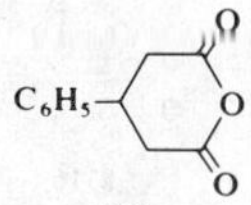

7321

7358

7371

7375

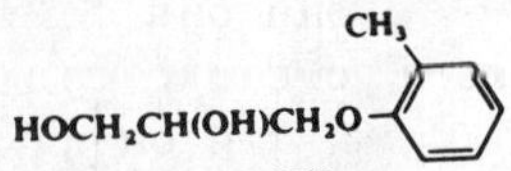

7389

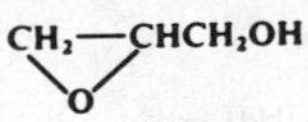

7406

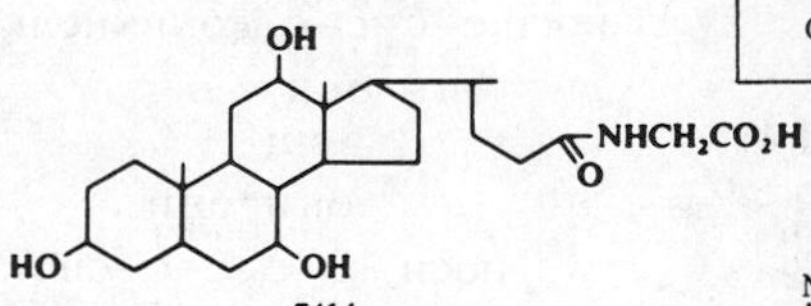

7414

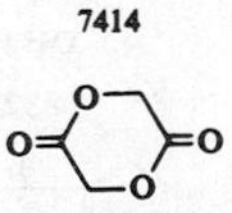

7419

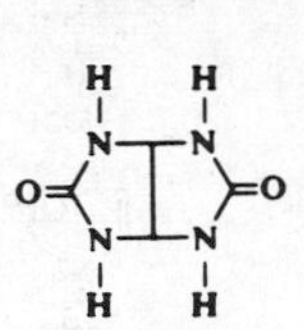

7420

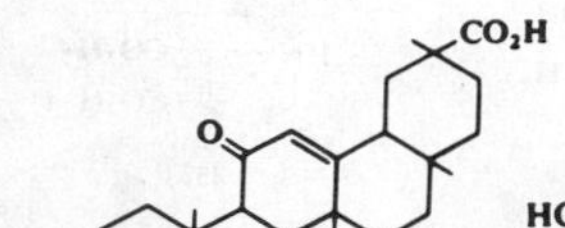

7421

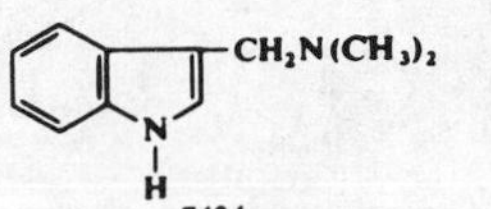

7434

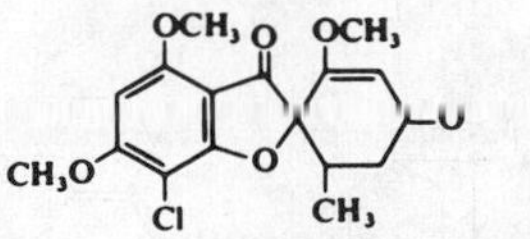

7435

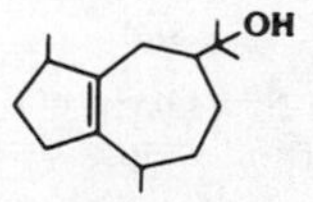

7435a

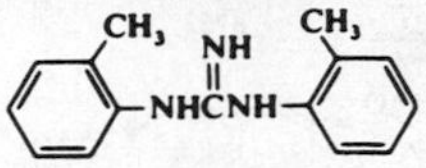

7443

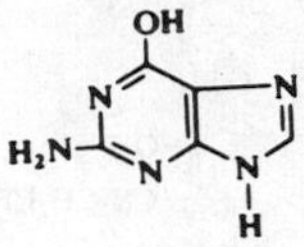

7453

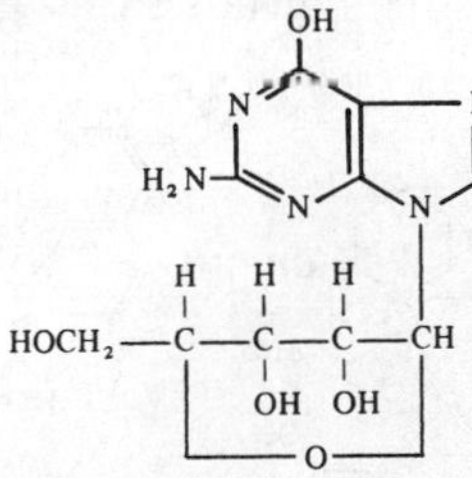

7454

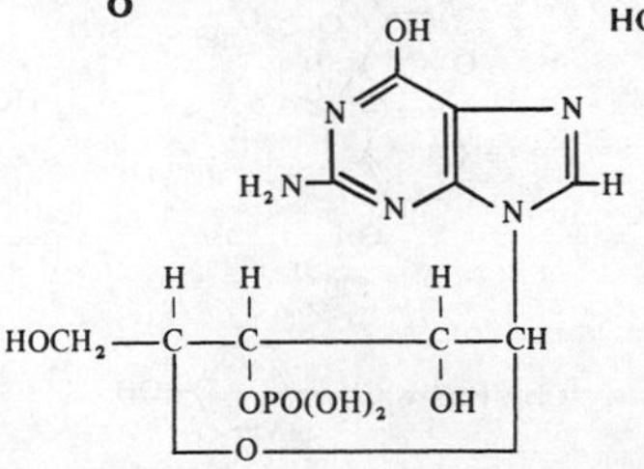

7455

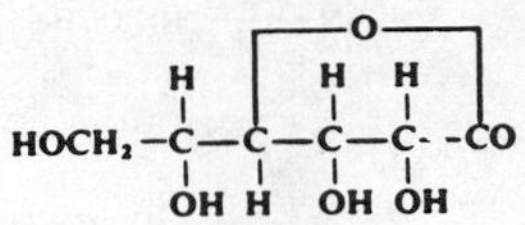

7462

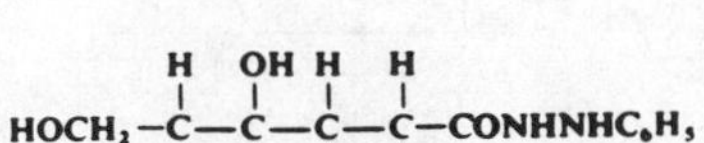

7464

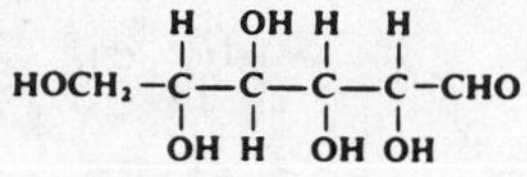

7465

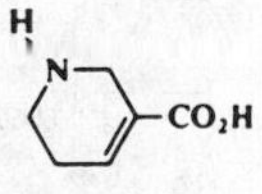

7467

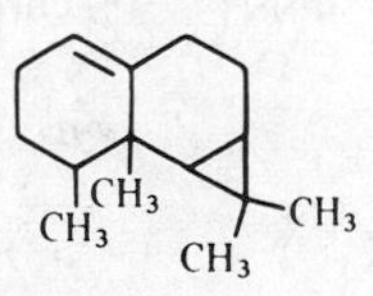

7468

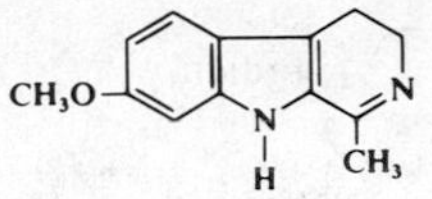

7470

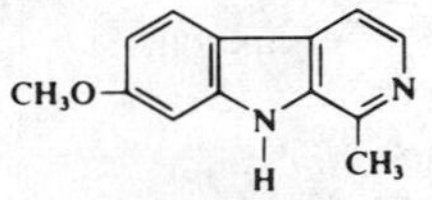

7471

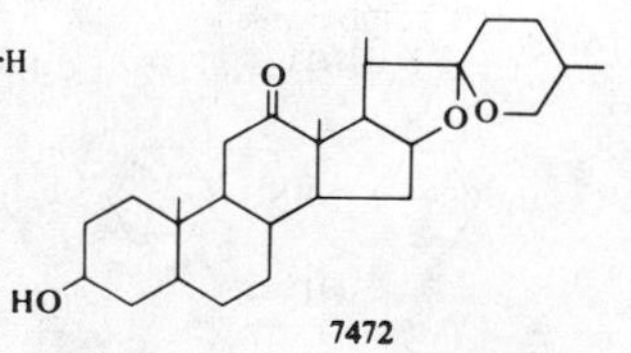

7472

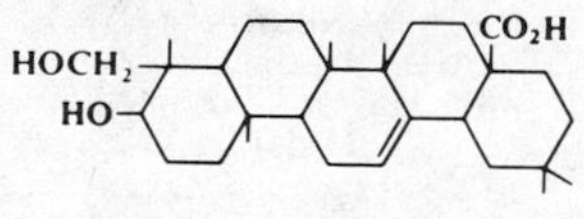

7474

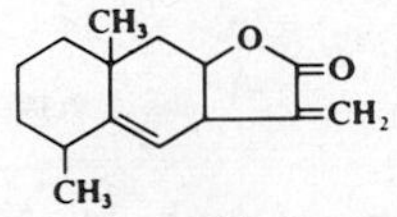

7475

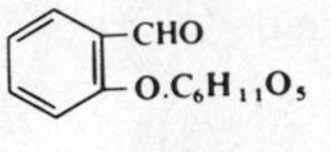

7476

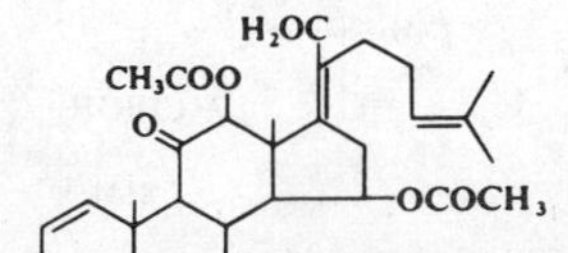

7476a

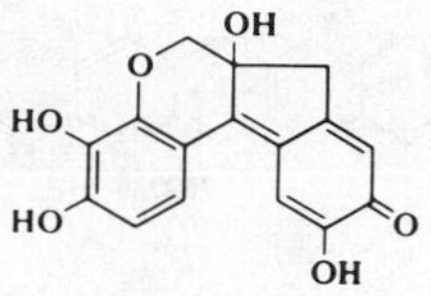

7477

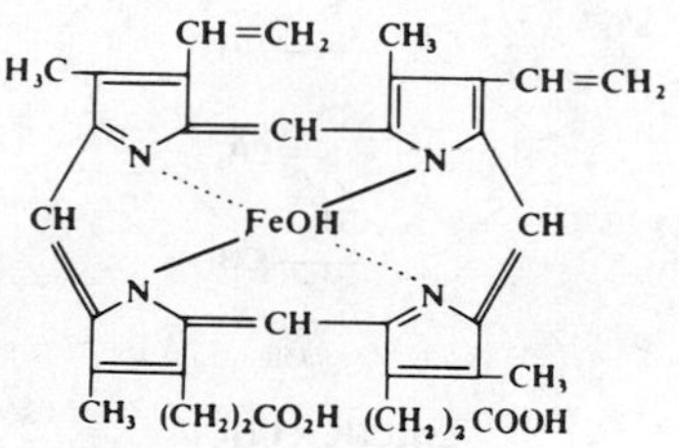

7478

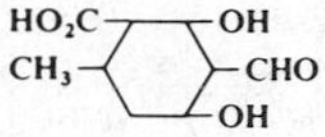

7479

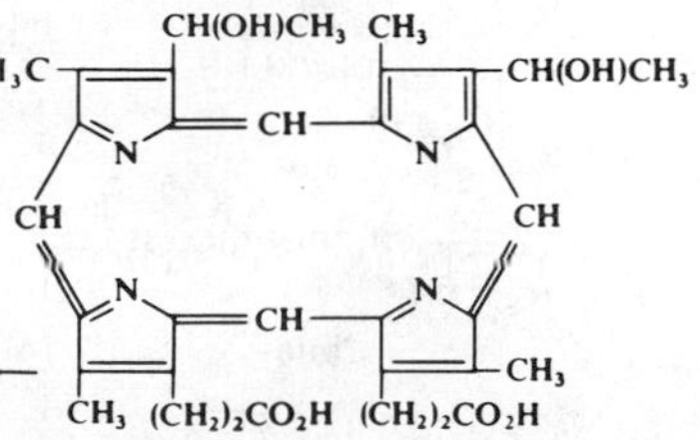

7480

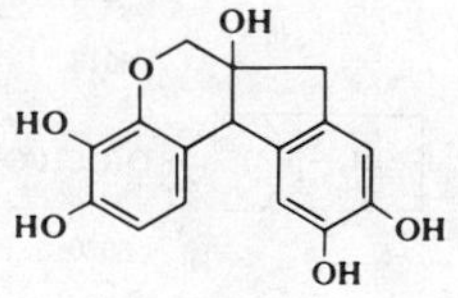

7481

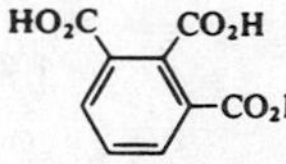

7482

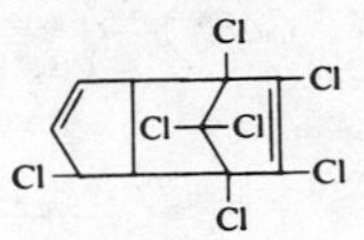

7484

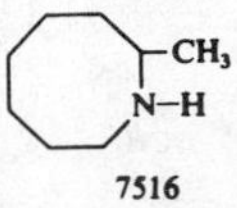

7516

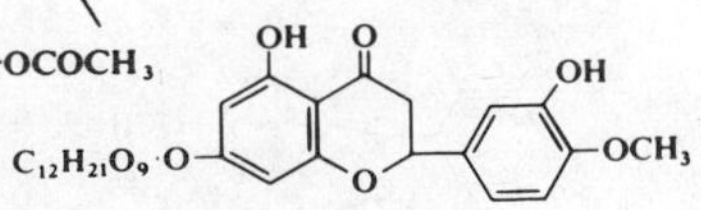

7689

STRUCTURAL FORMULAS OF ORGANIC COMPOUNDS (Continued)

In Numeric as they occur in Organic Compounds Table

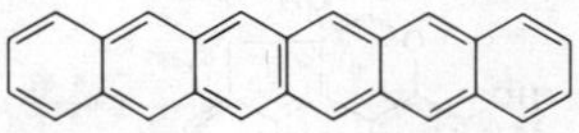
7690

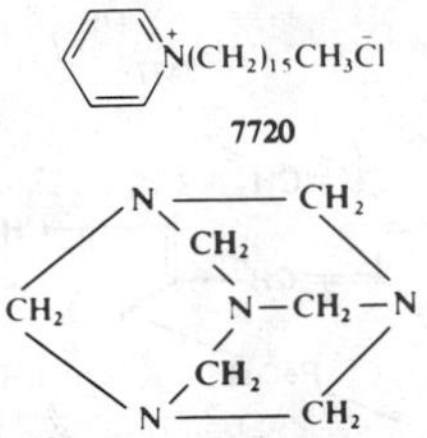
7720

7758

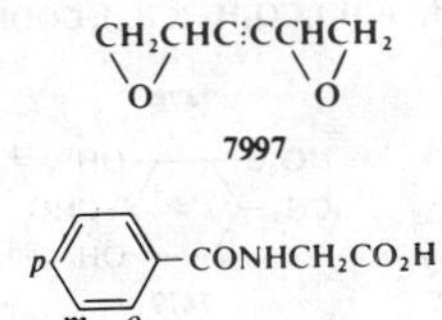
7997

8003

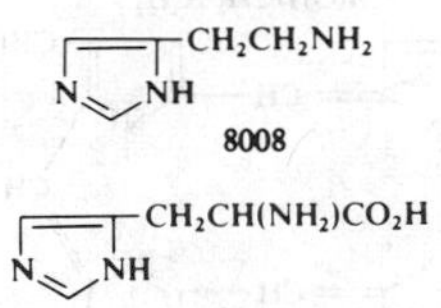
8008

8010

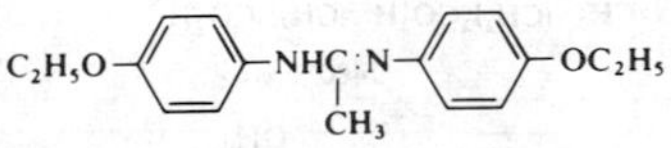
8018

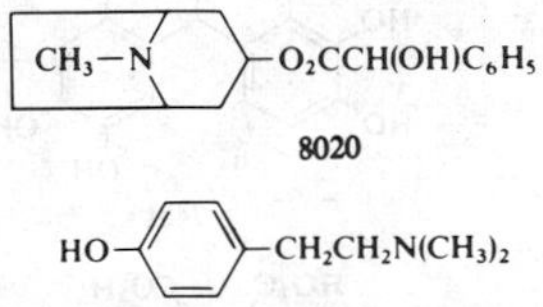
8020

8027

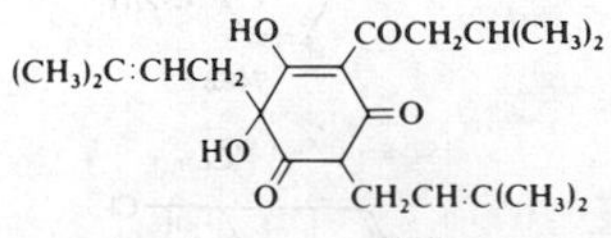
8030

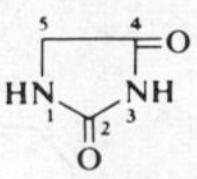
8034

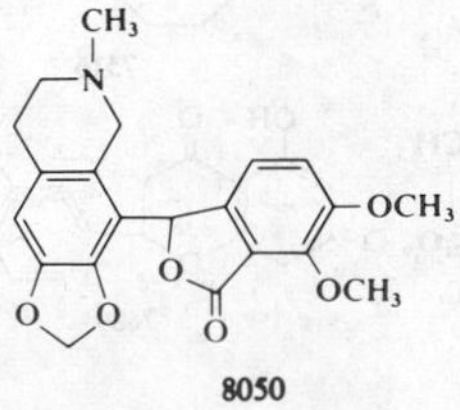
8050

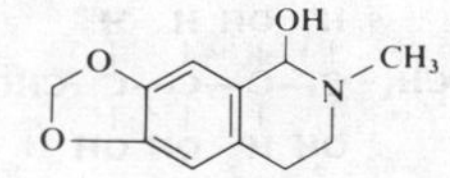
8052

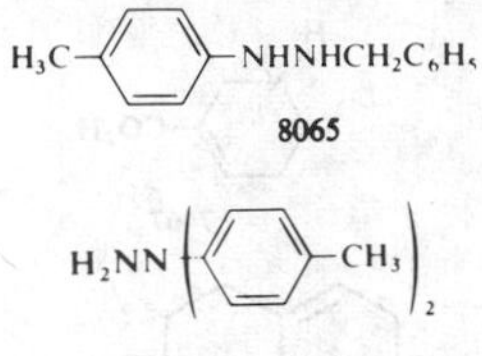
8065

8093

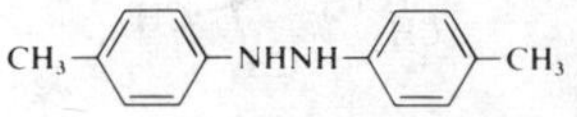
8096

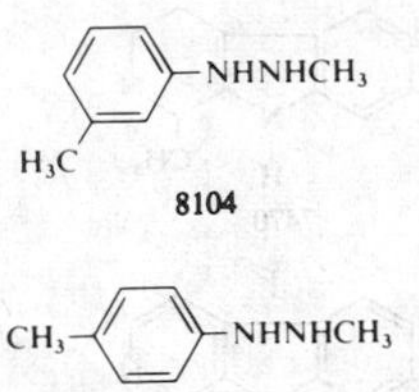
8104

8105

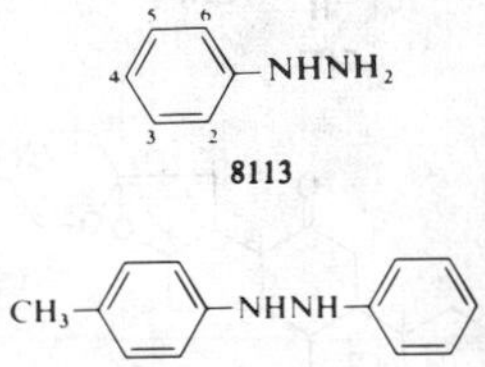
8113

8118

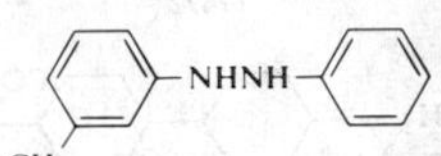
8134

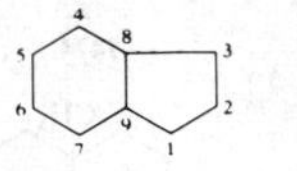
8138

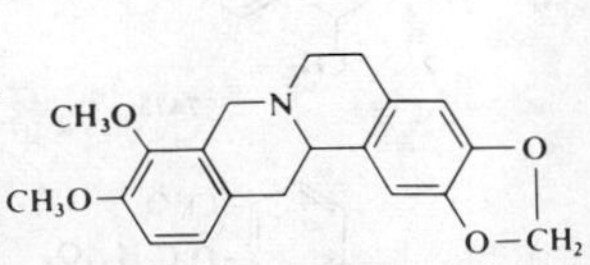
8142

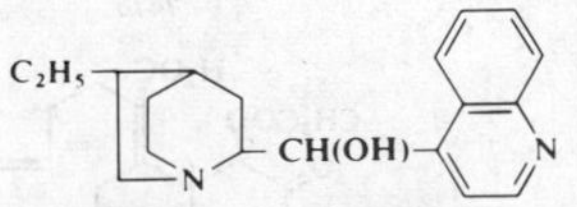
8144

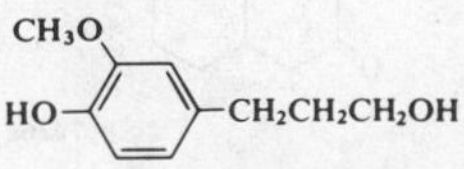
8158

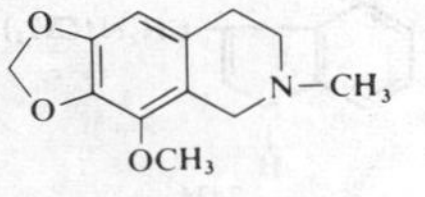
8159

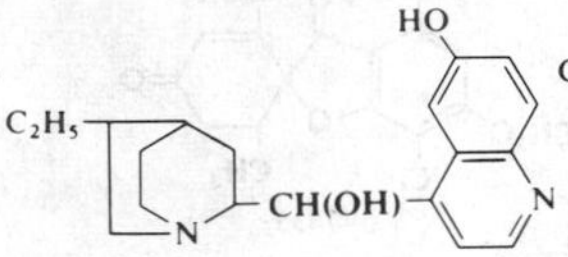
8160

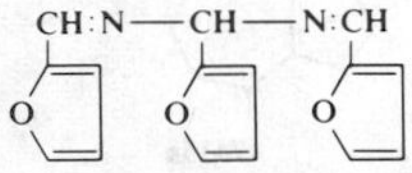
8162

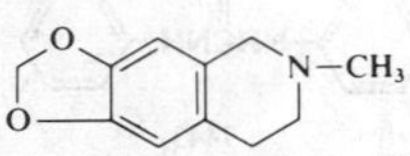
8163

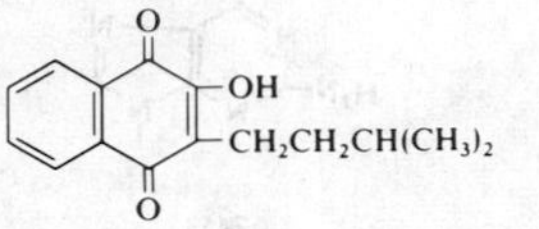
8164

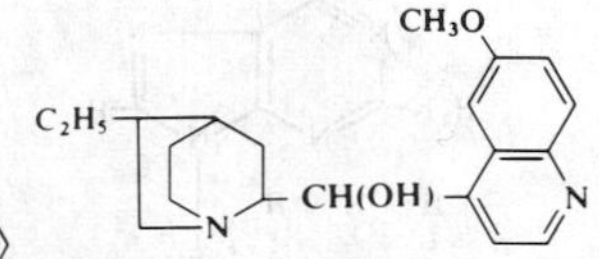
8165

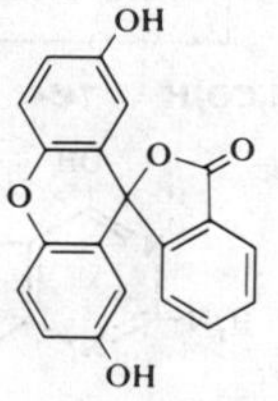
8207

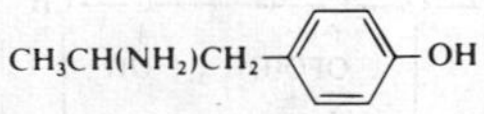
8217

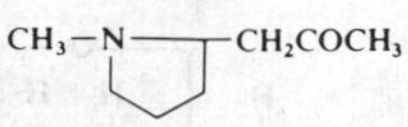
8297

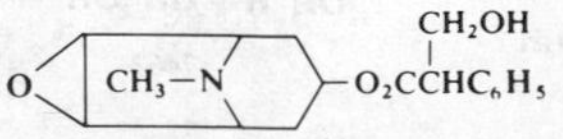
8298

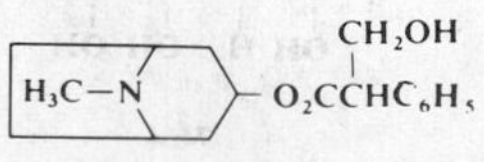
8309

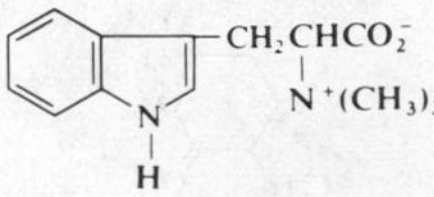
8315

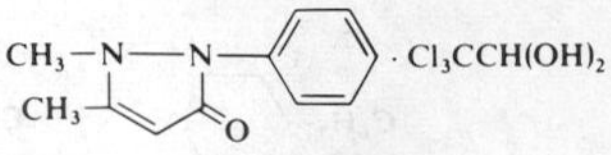
8316

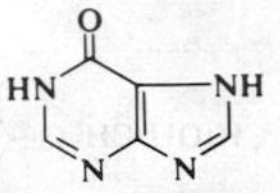
8317

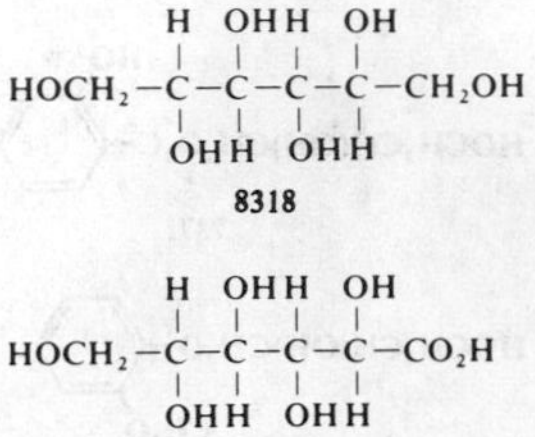
8318

8319

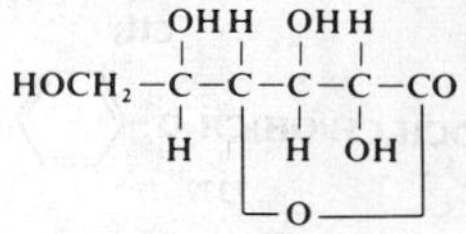
8320

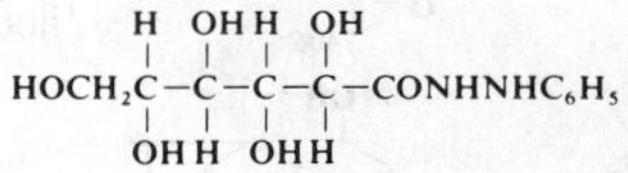
8321

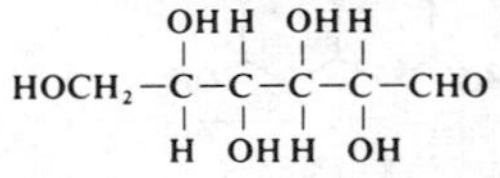
8323

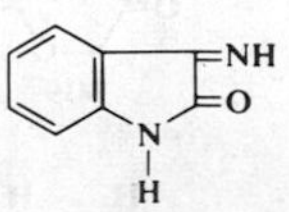
8324

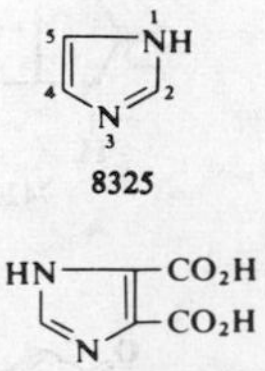
8325

8331

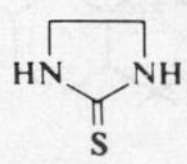
8332

In Numeric as they occur in Organic Compounds Table

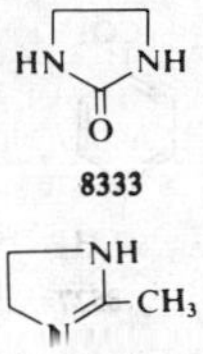

8333

8334

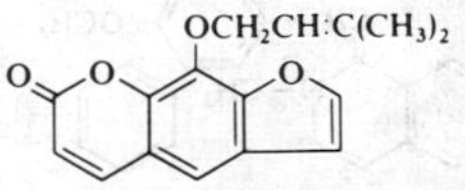

8336

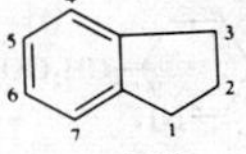

8338

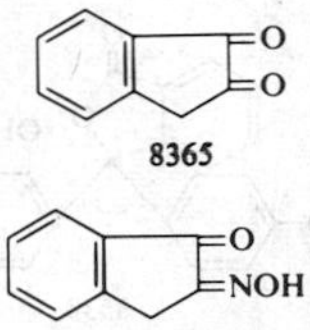

8365

8366

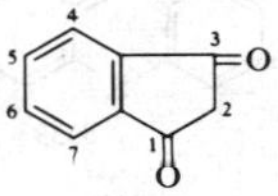

8367

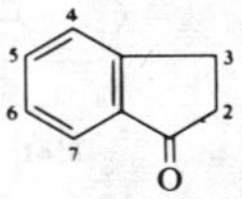

8375

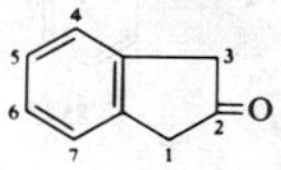

8379

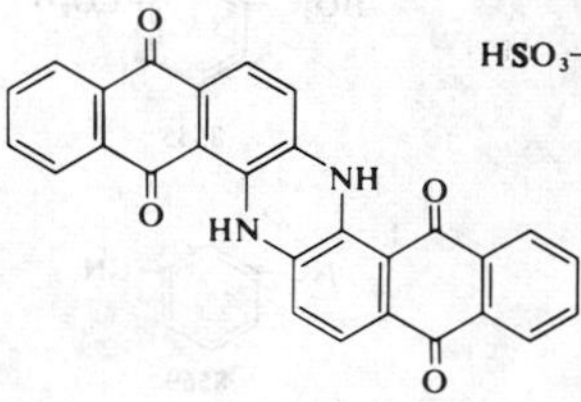

8381

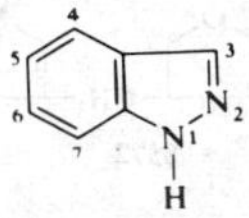

8382

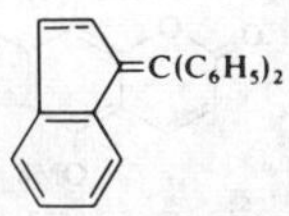

8383

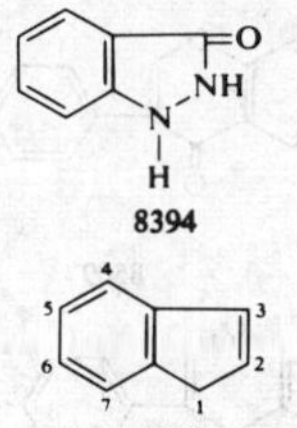

8394

8395

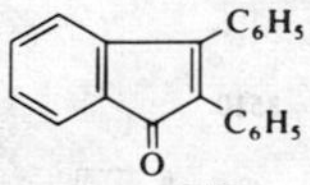

8403

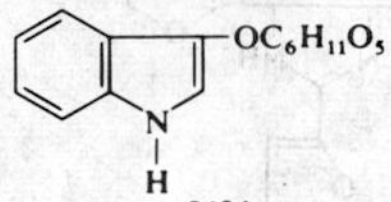

8404

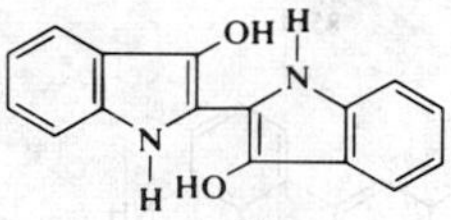

8405

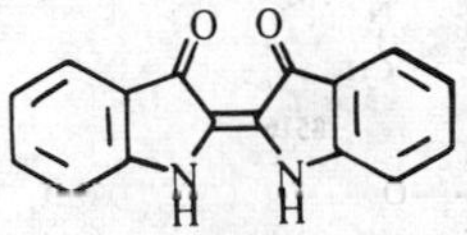

8406

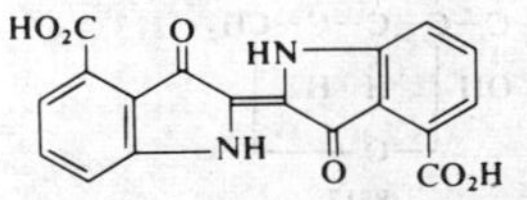

8407

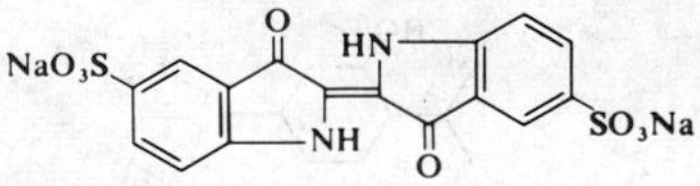

8408

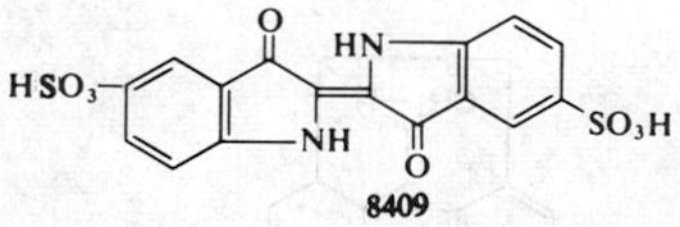

8409

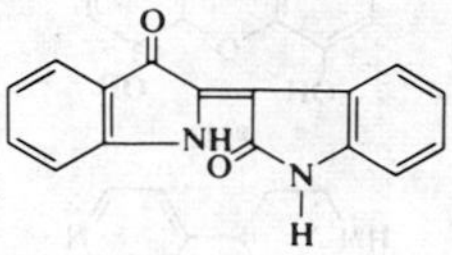

8410

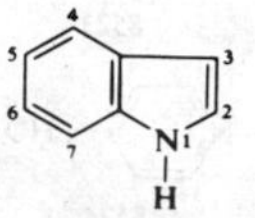

8411

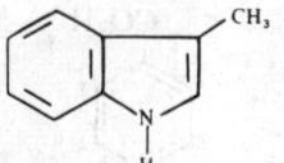

8421

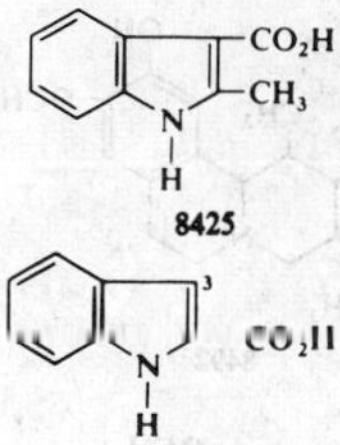

8425

8426

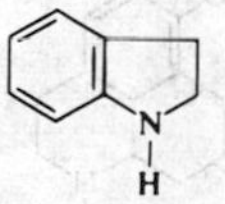

8428

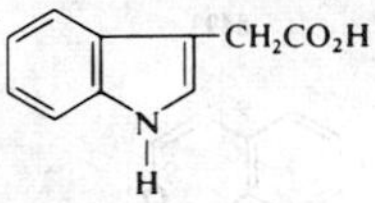

8429

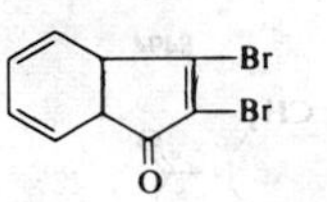

8430

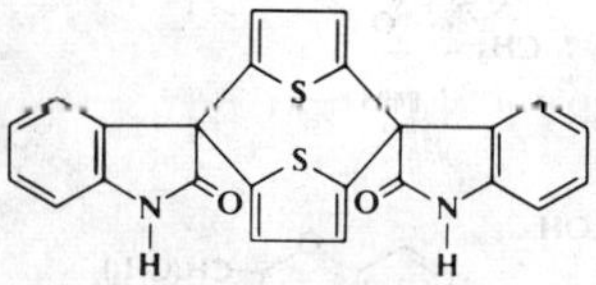

8431

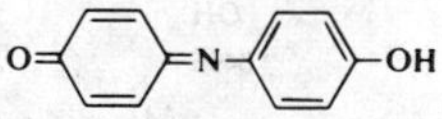

8432

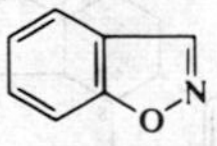

8433

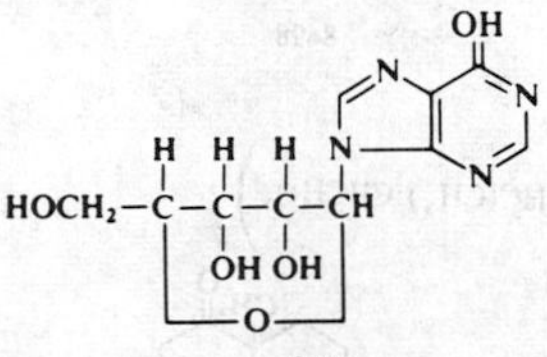

8434

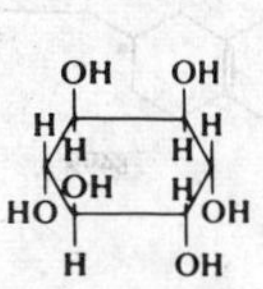

8435

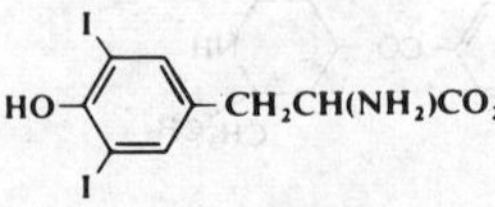

8442

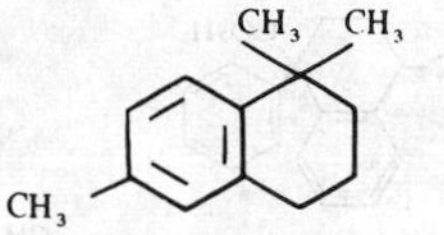

8445

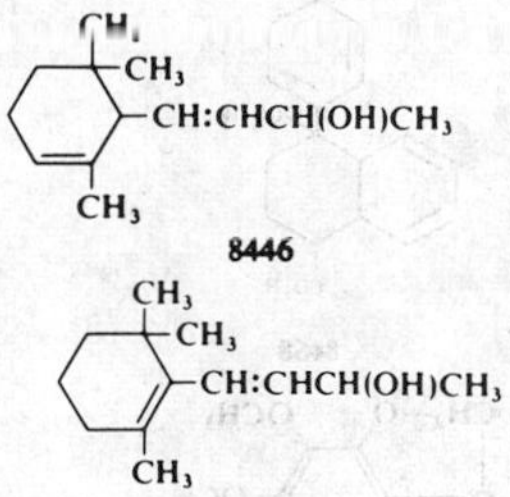

8446

8447

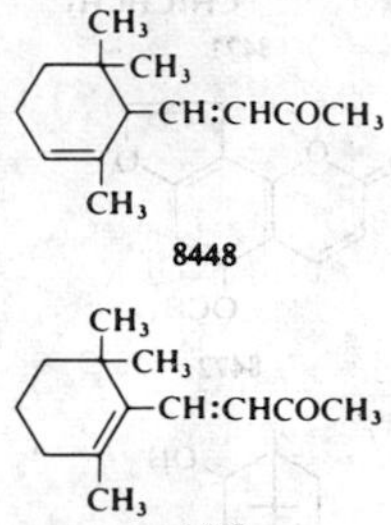

8448

8452

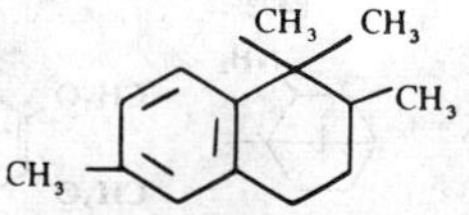

8454

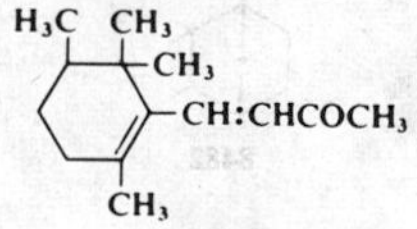

8455

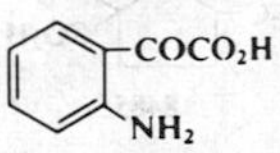

8456

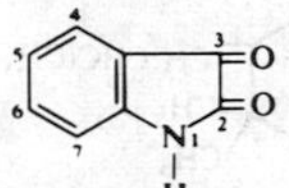

8457

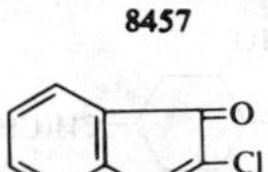

8459

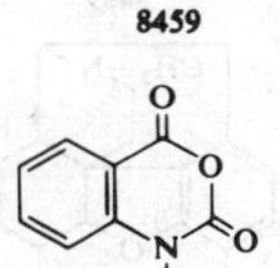

8466

STRUCTURAL FORMULAS OF ORGANIC COMPOUNDS (Continued)

In Numeric as they occur in Organic Compounds Table

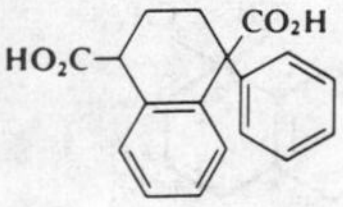

8467

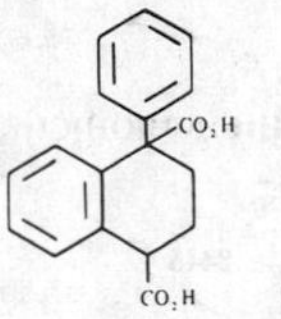

8468

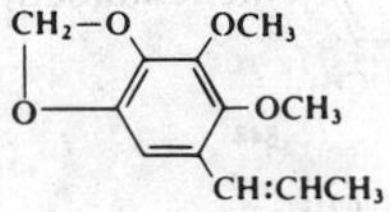

8471

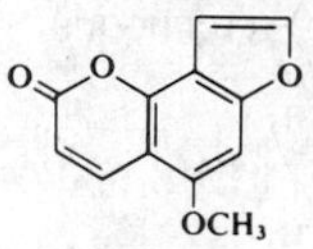

8472

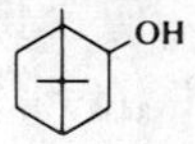

8473

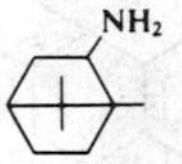

8480

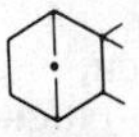

8482

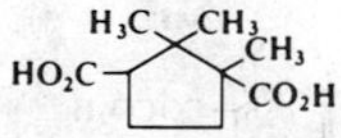

8485

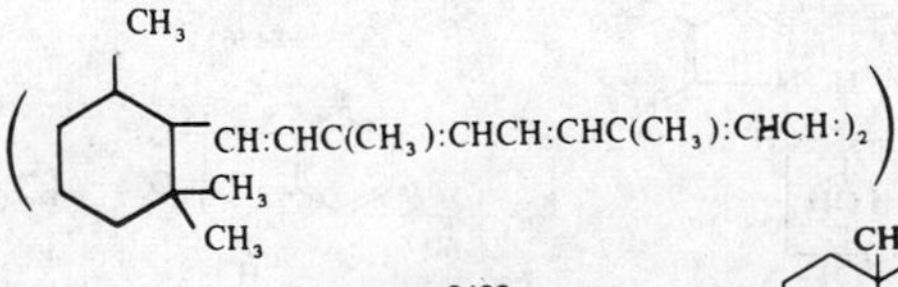

8488

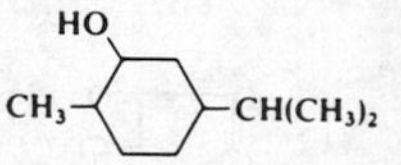

8489

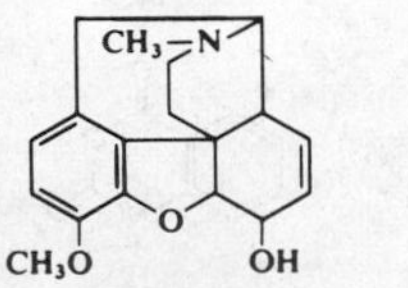

8491

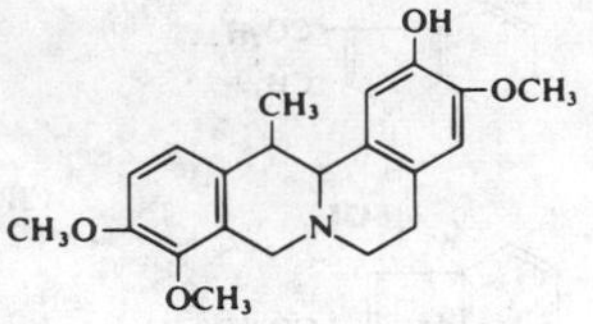

8492

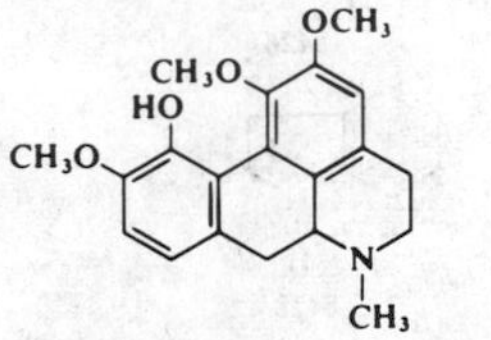

8493

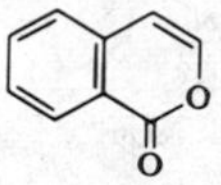

8494

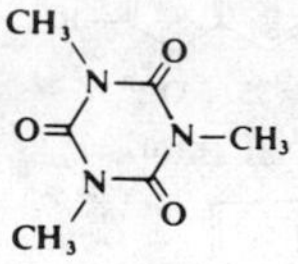

8495

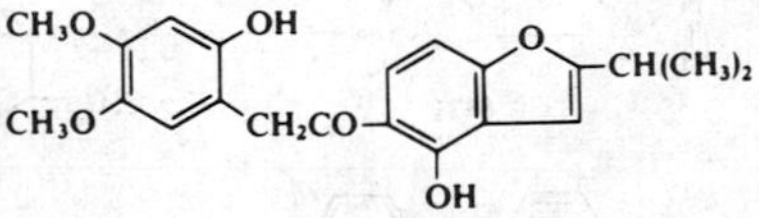

8496

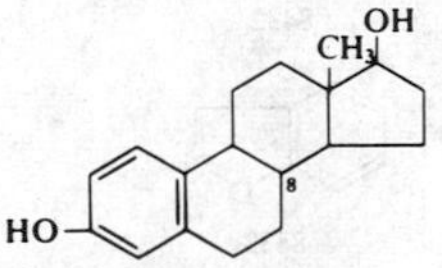

8498

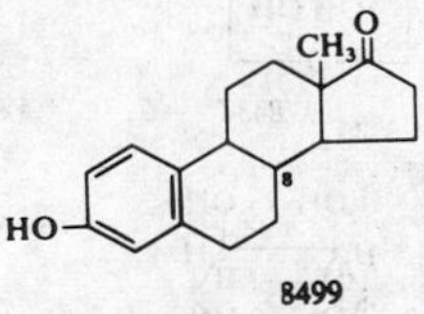

8499

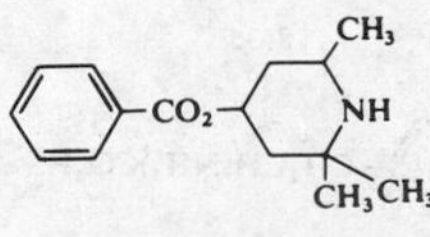

8500

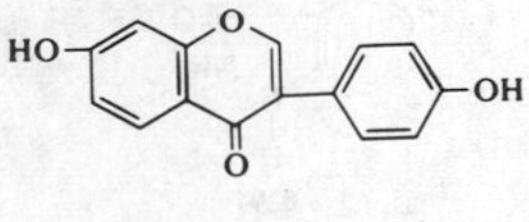

8509

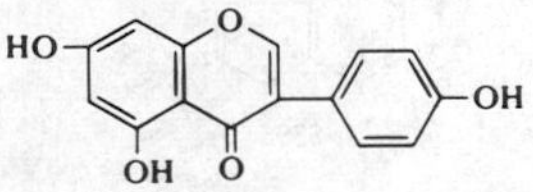

8510

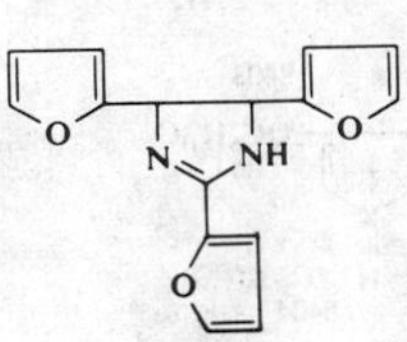

8511

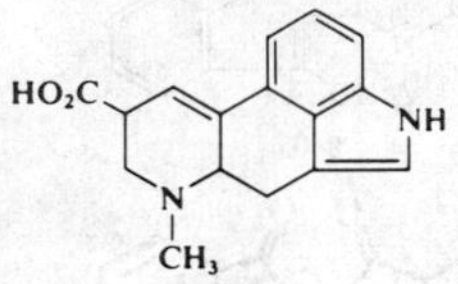

8516

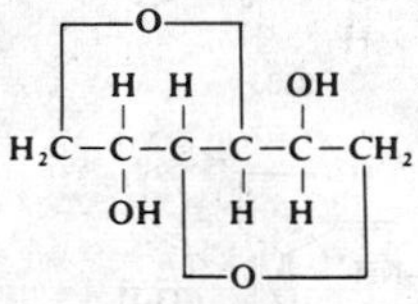

8517

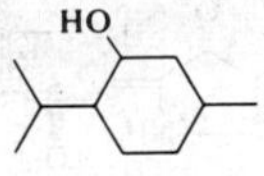

8518

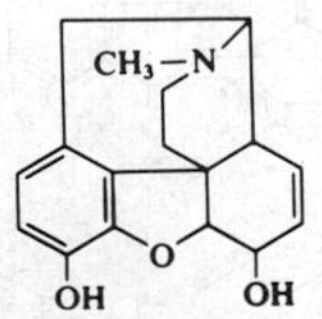

8522

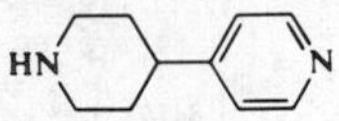

8523

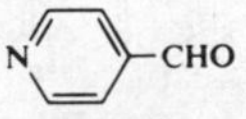

8524

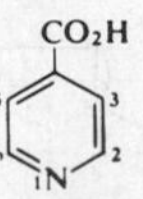

8525

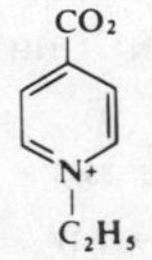

8527

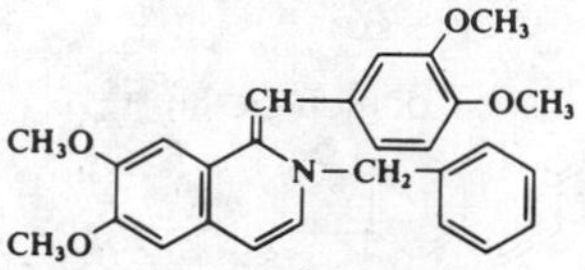

8533

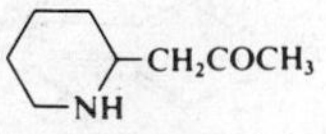

8536

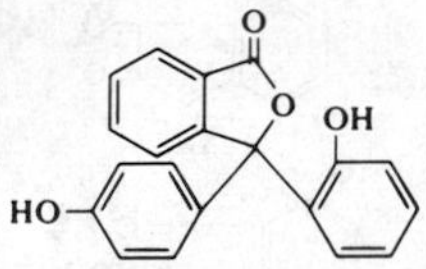

8538

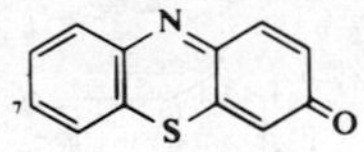

8539

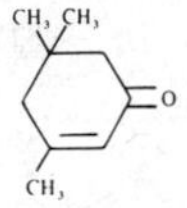

8541

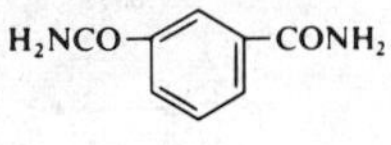

8543

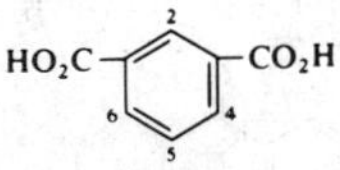

8545

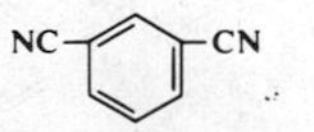

8569

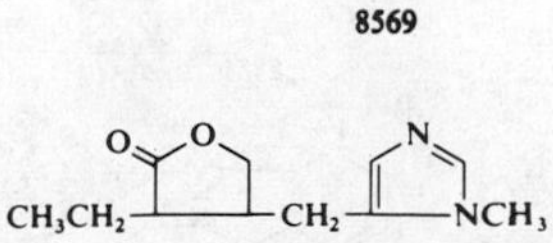

8572

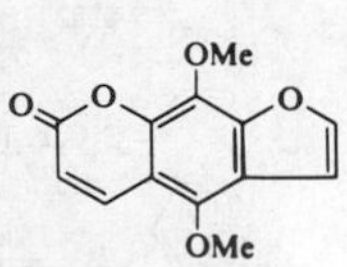

8573

STRUCTURAL FORMULAS OF ORGANIC COMPOUNDS (Continued)

In Numeric as they occur in Organic Compounds Table

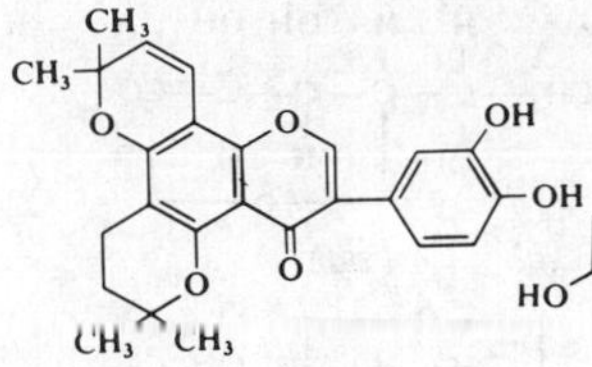
8574

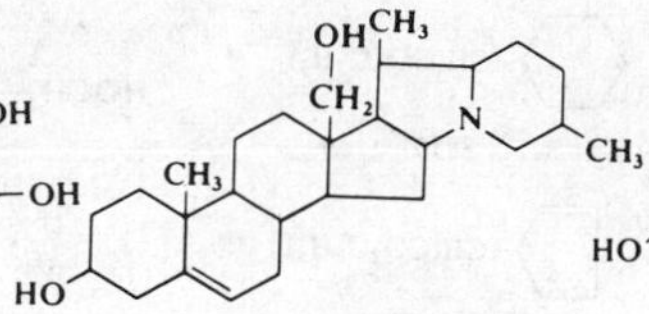
8605

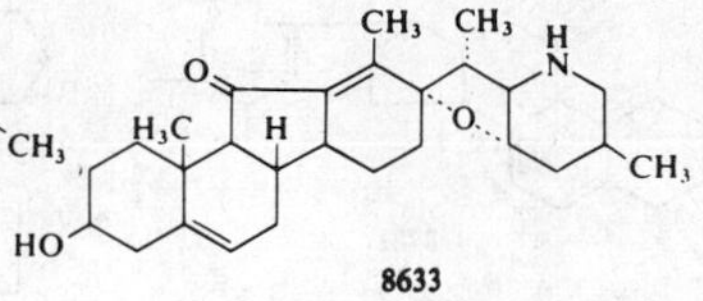
8633

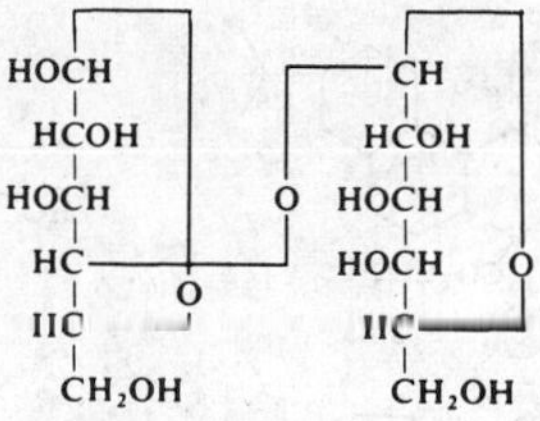
8681

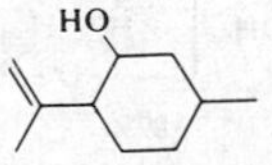
8577

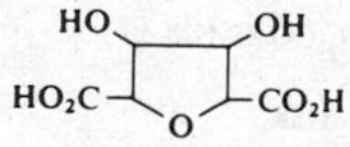
8607

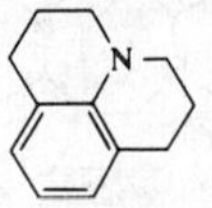
8634

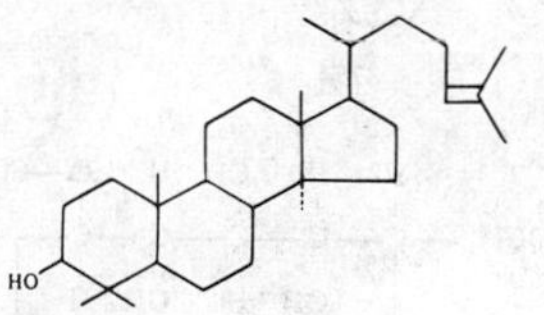
8685

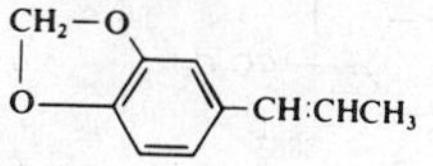
8608

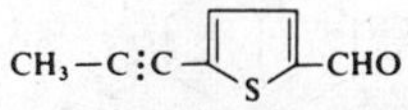
8636

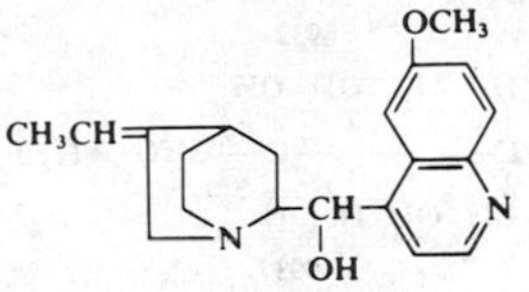
8579

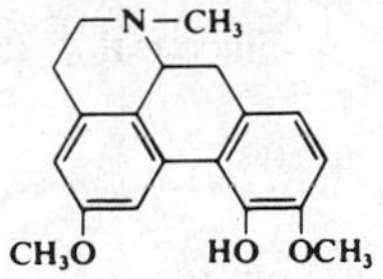
8610

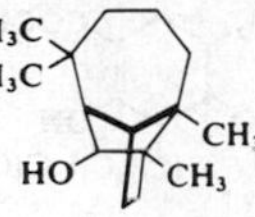
8637

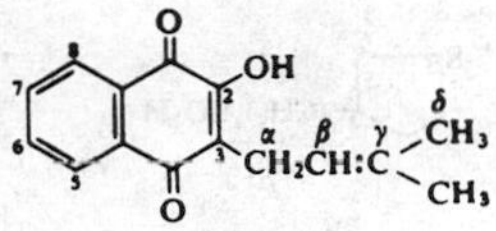
8692

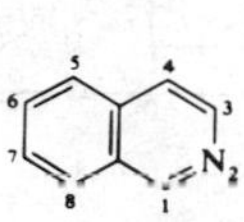
8581

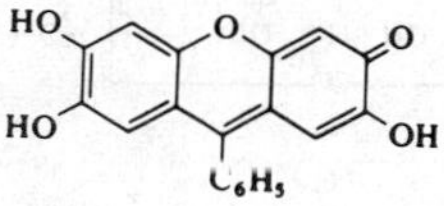
8617

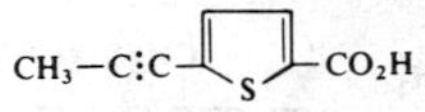
8638

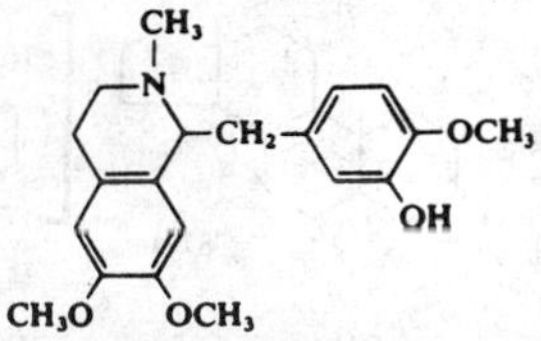
8695

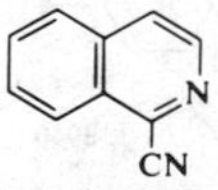
8601

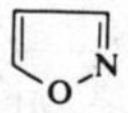
8618

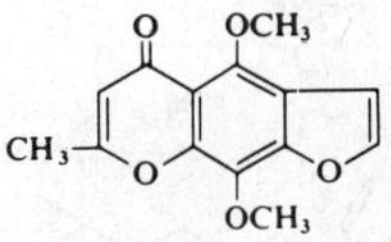
8645

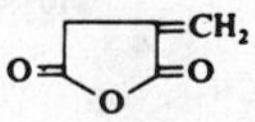
8621

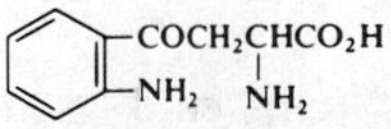
8646

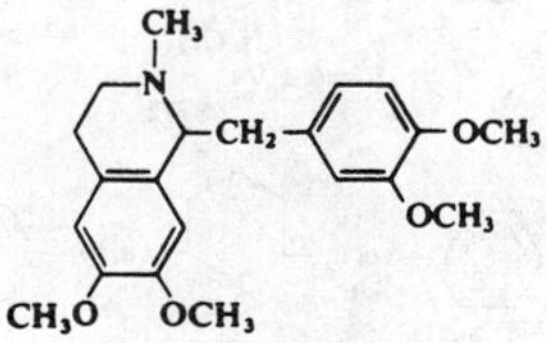
8698

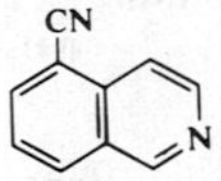
8602

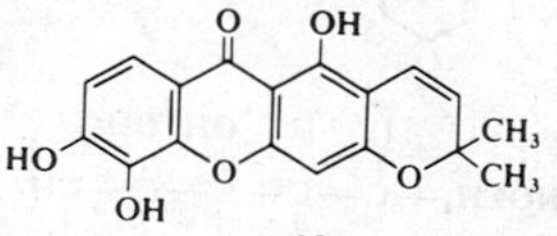
8625

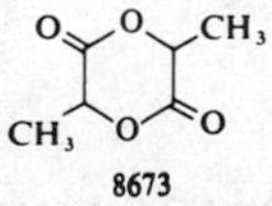
8673

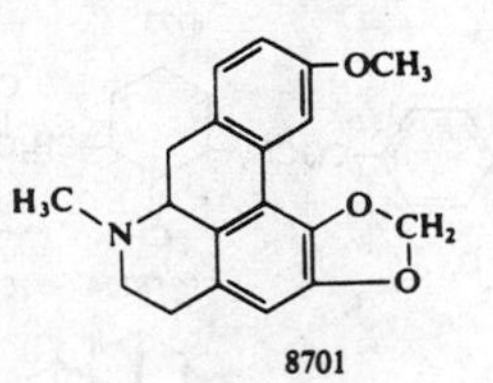
8701

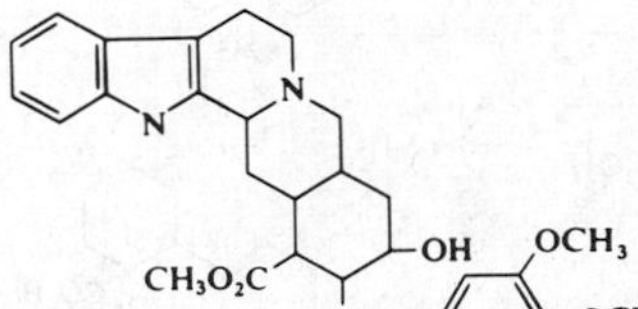
8603

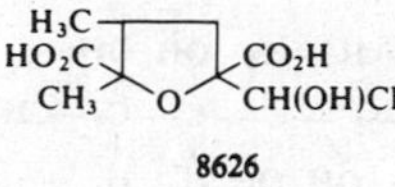
8626

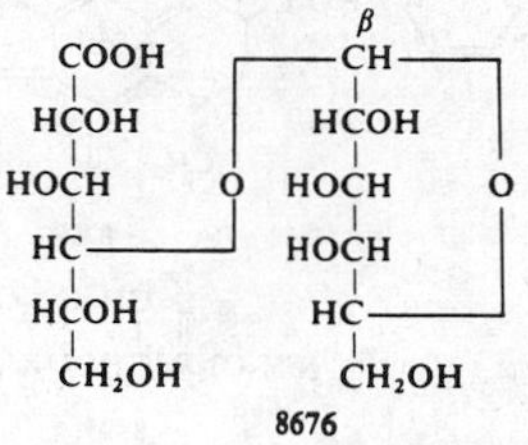
8676

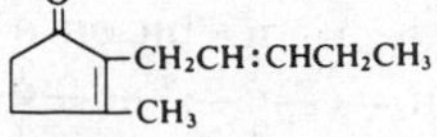
8631

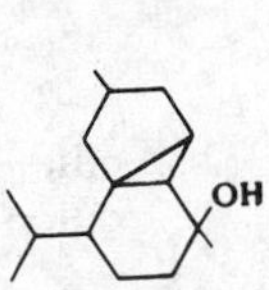
8729

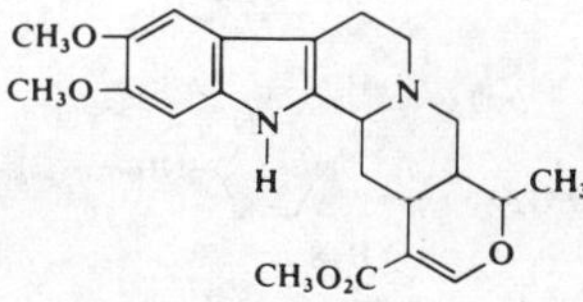
8604

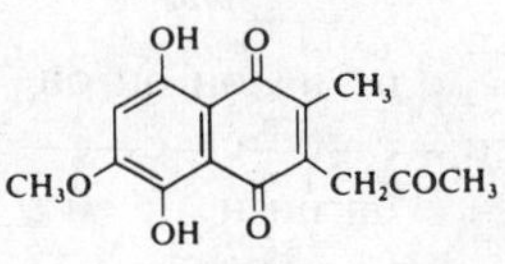
8632

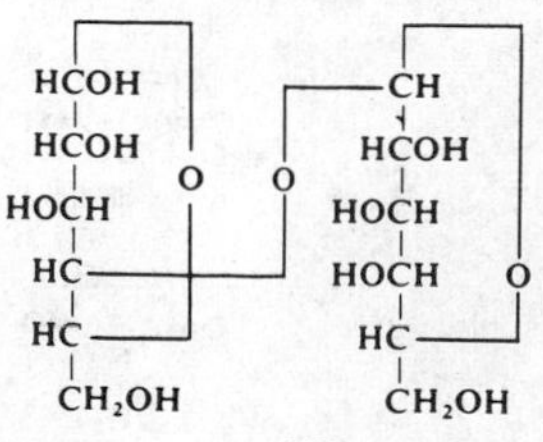
8680

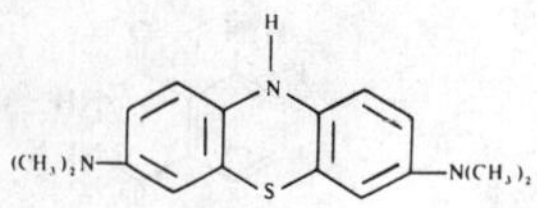
8742

STRUCTURAL FORMULAS OF ORGANIC COMPOUNDS (Continued)

In Numeric as they occur in Organic Compounds Table

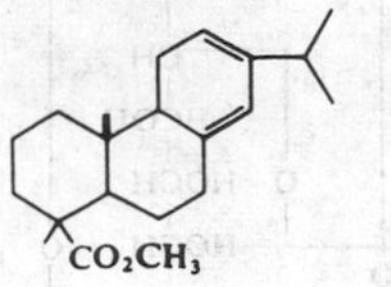

8743

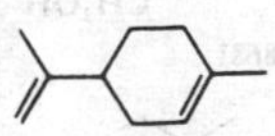

8754

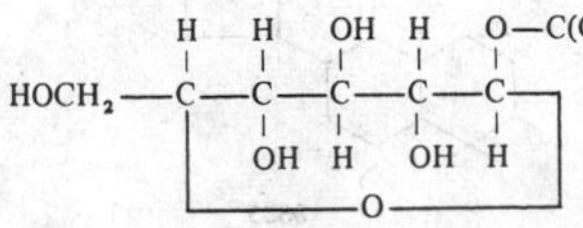

8763

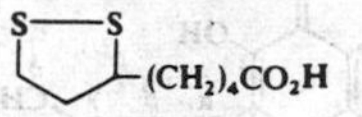

8770

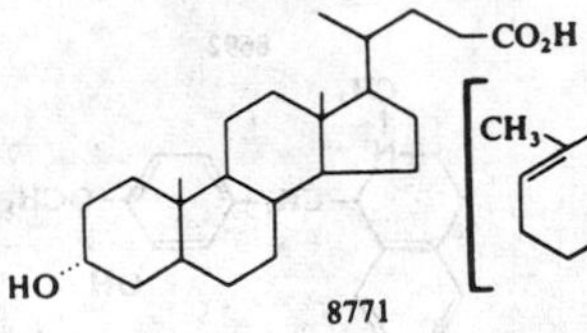

8771

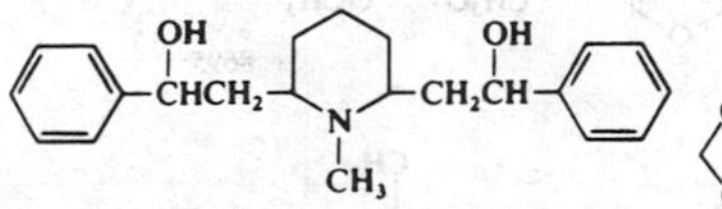

8772

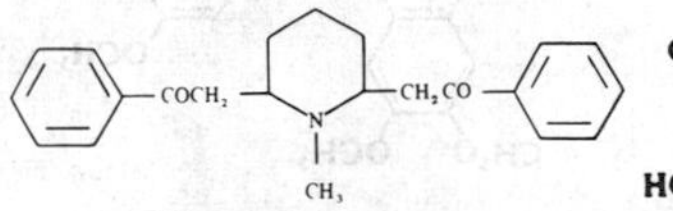

8773

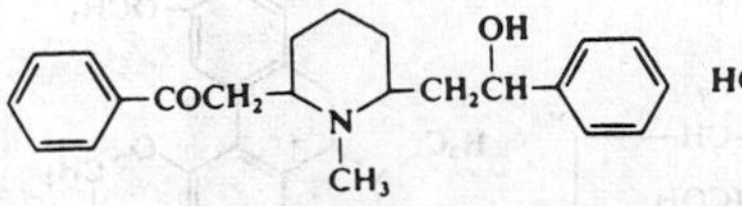

8774

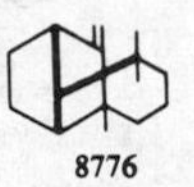

8776

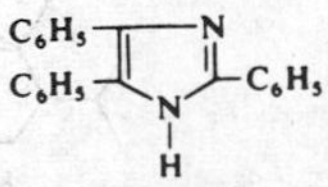

8777

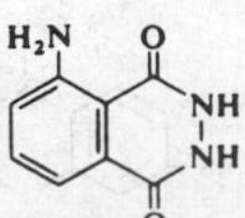

8780

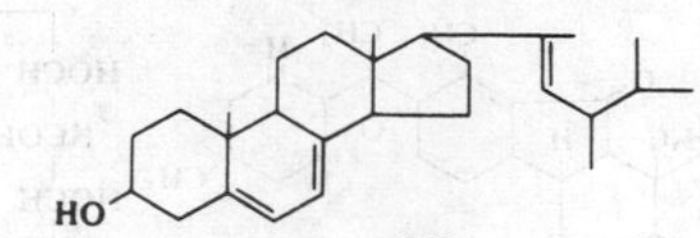

8781

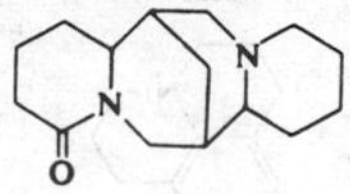

8782

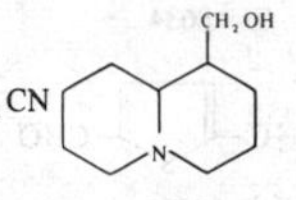

8785

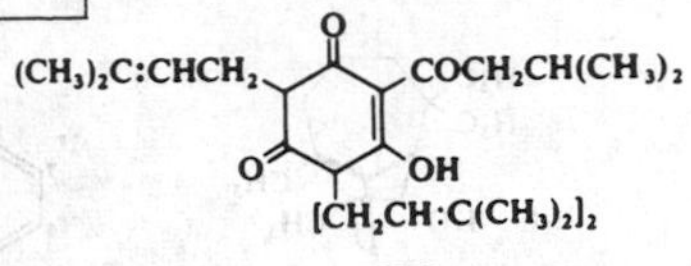

8787

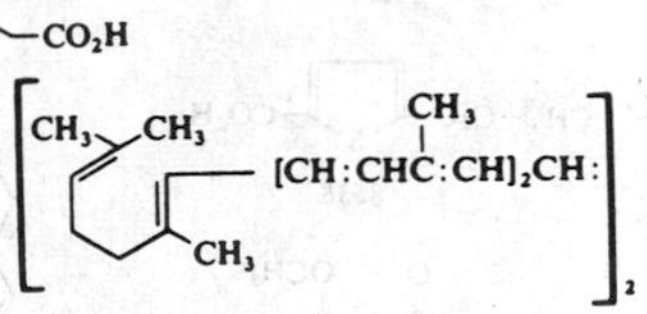

8796

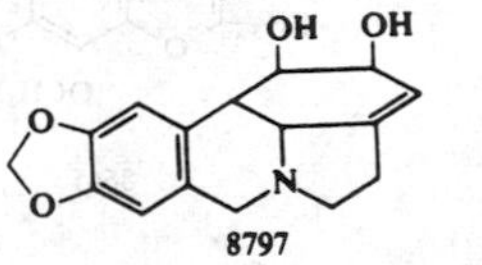

8797

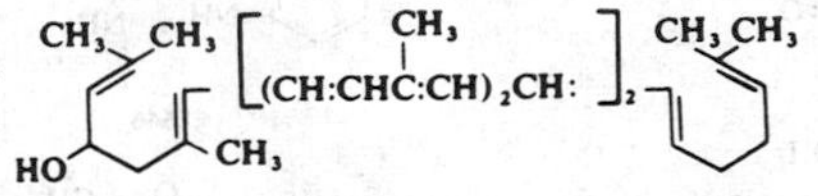

8798

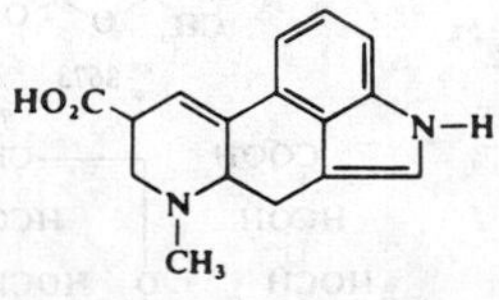

8799

$$(CH_3O)_2\overset{S}{\overset{\|}{P}}S\overset{CH_2CO_2C_2H_5}{CH}CO_2C_2H_5$$

8805

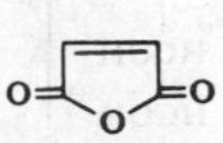

8820

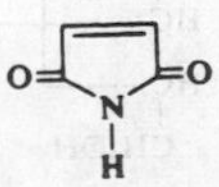

8824

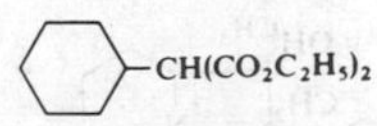

8856

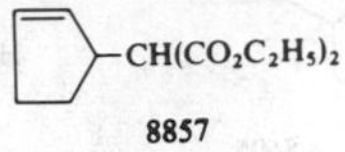

8857

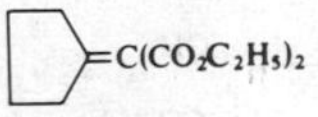

8858

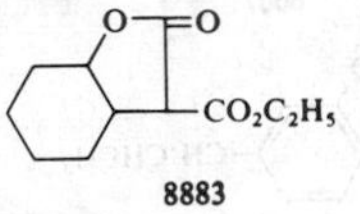

8883

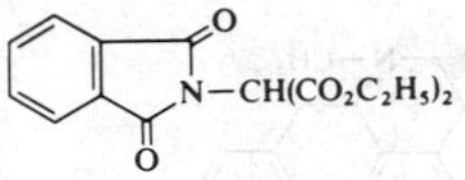

8898

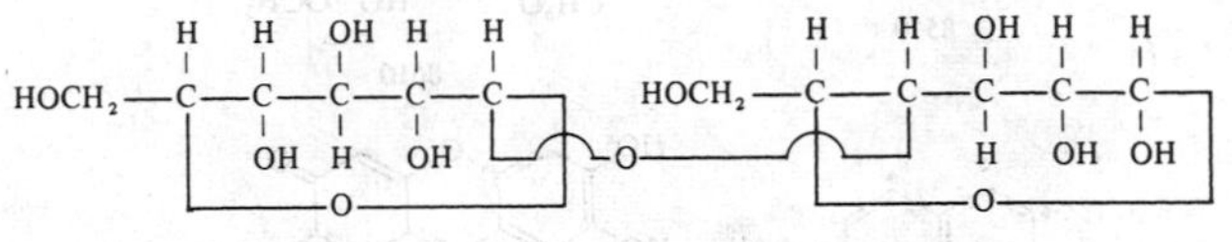

8908

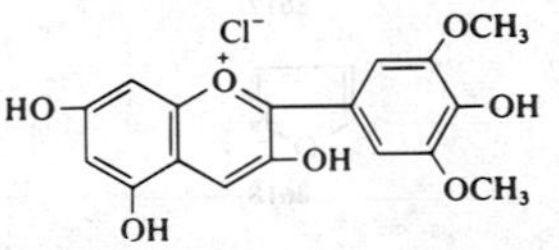

8910

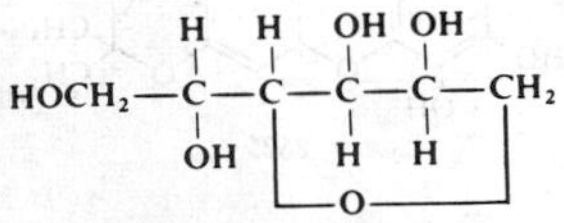

8921

$$HOCH_2-\underset{OH}{\overset{H}{C}}-\underset{OH}{\overset{H}{C}}-\underset{H}{\overset{OH}{C}}-\underset{H}{\overset{OH}{C}}-CH_2OH$$

8922

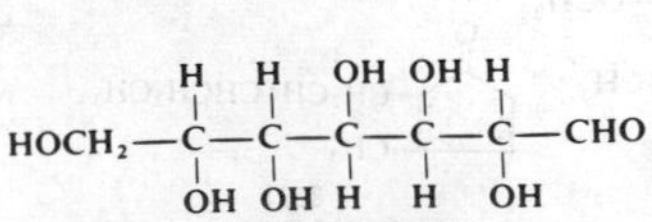

8926

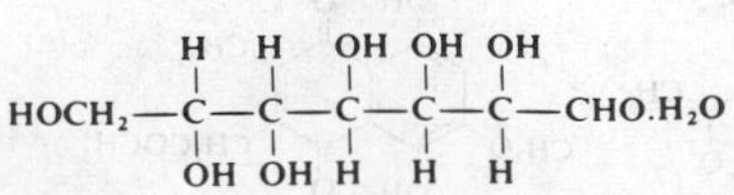

8927

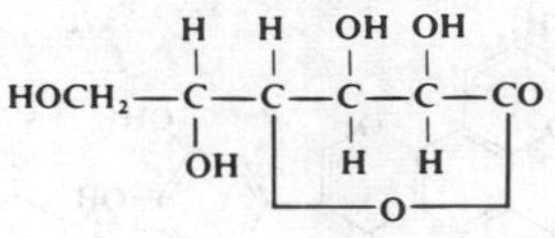

8928

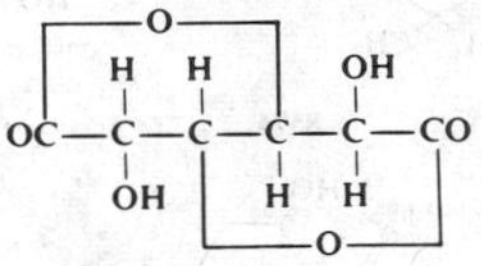

8930

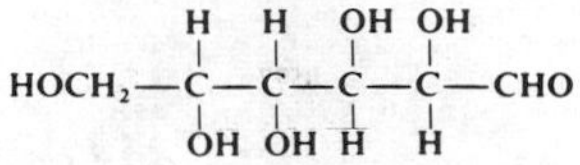

8932

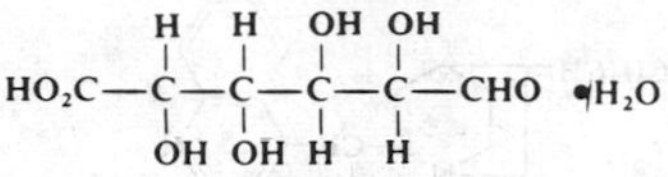

8937

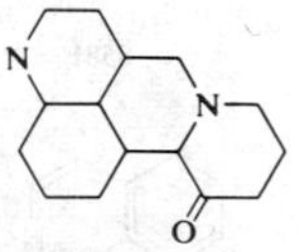

8940

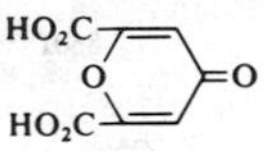

8941

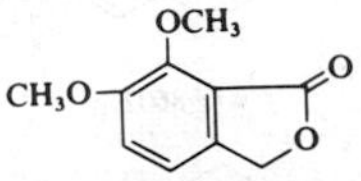

8943

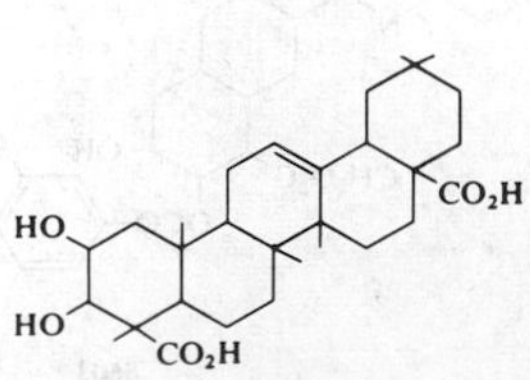

8944

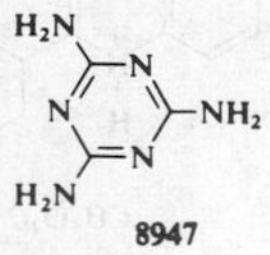

8947

STRUCTURAL FORMULAS OF ORGANIC COMPOUNDS (Continued)

In Numeric as they occur in Organic Compounds Table

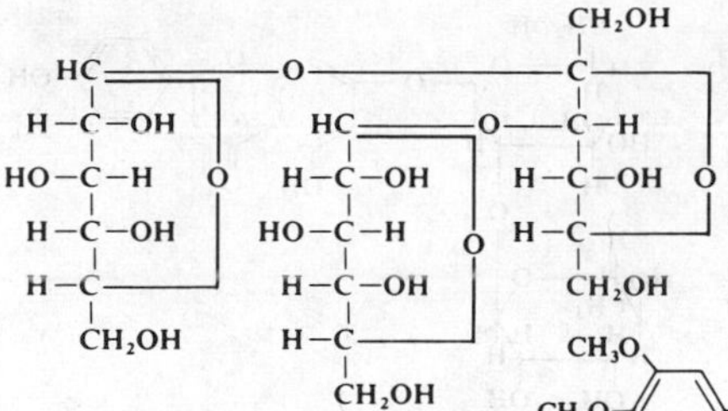
8950

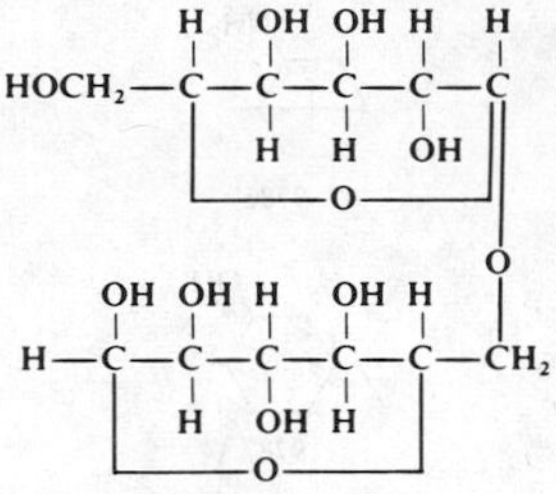
8951

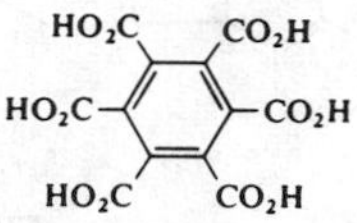
8953

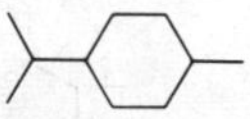
8954

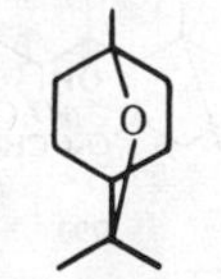
8956

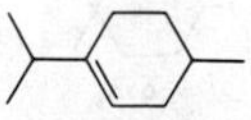
8957

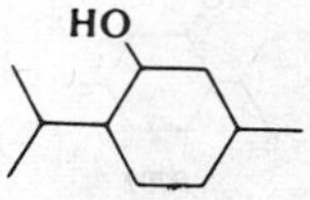
8961

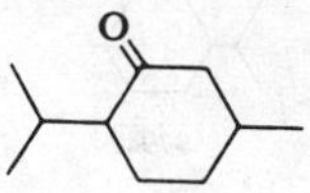
8966

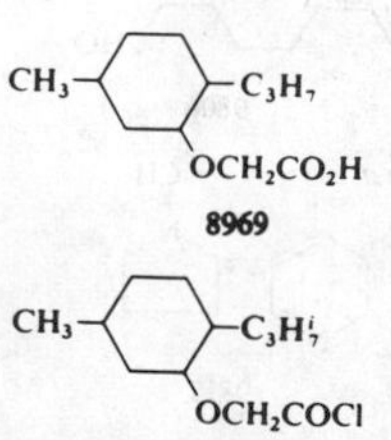
8969

8970

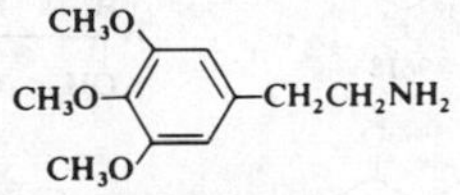
8986

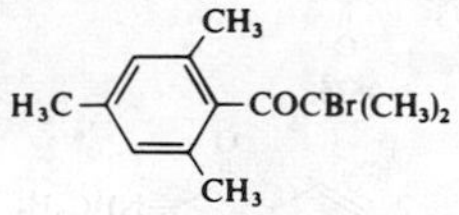
8989

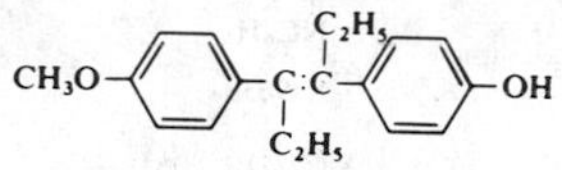
8997

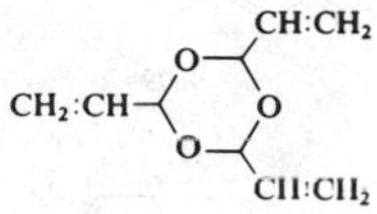
8998

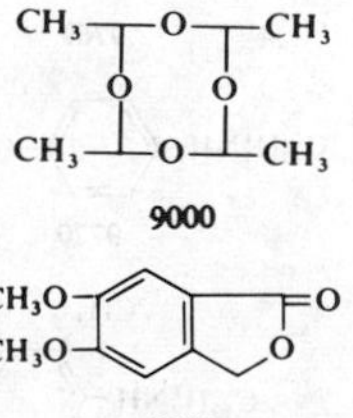
9000

9001

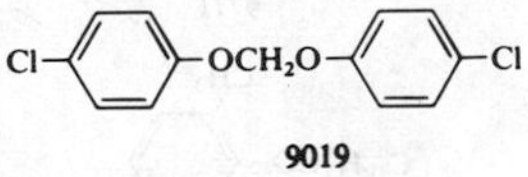
9019

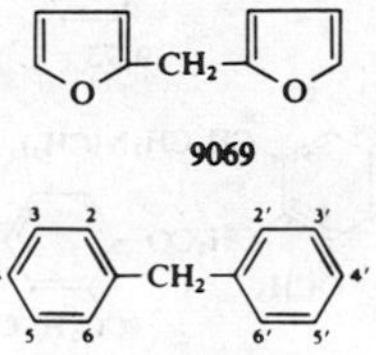
9069

9074

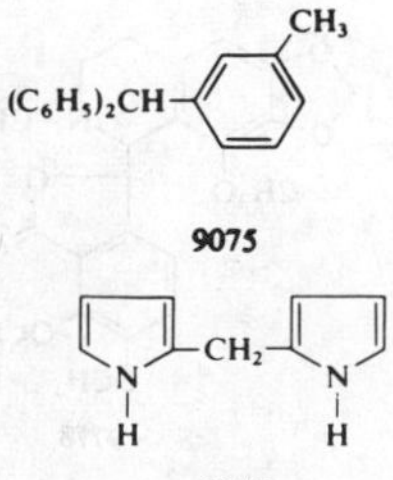
9075

9076

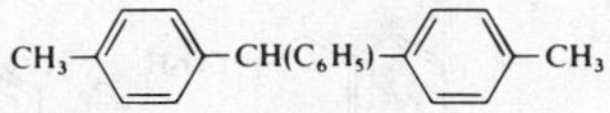
9078

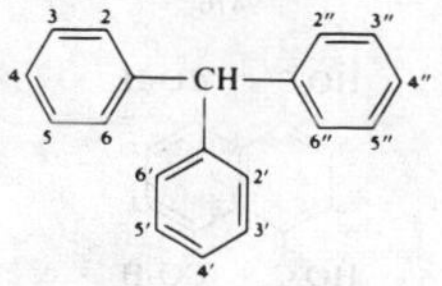
9103

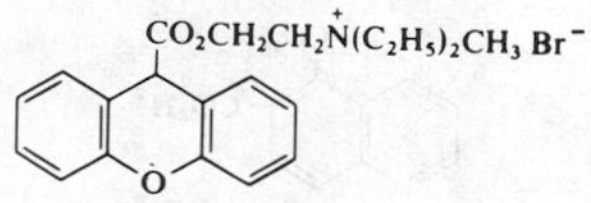
9126

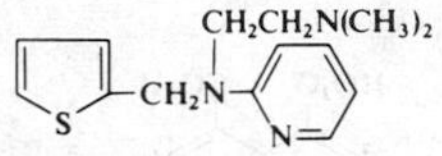
9129

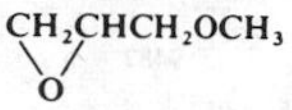
9159

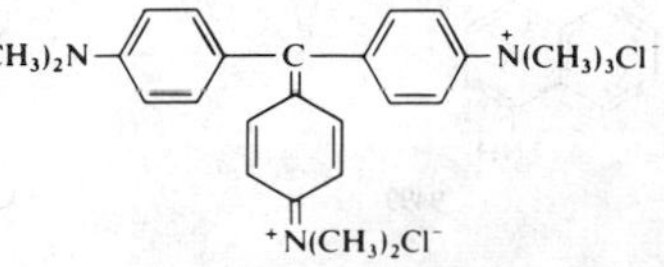
9166

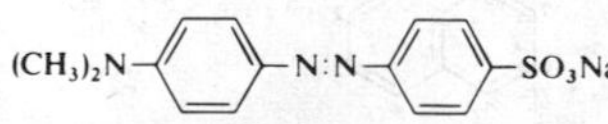
9179

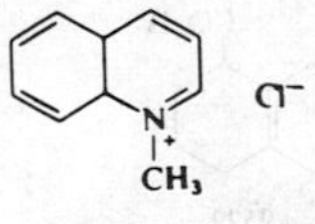
9195

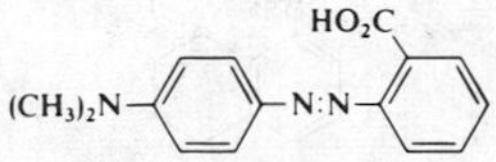
9196

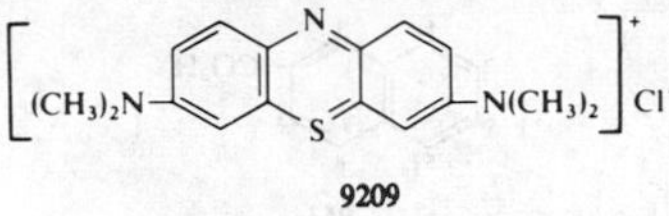
9209

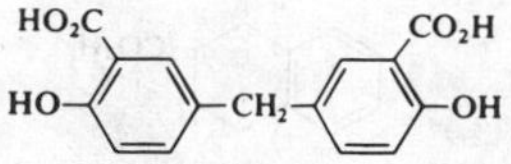
9210

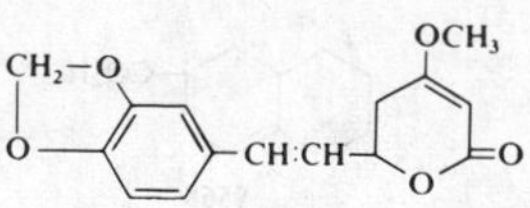
9211

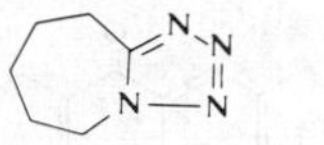
9212

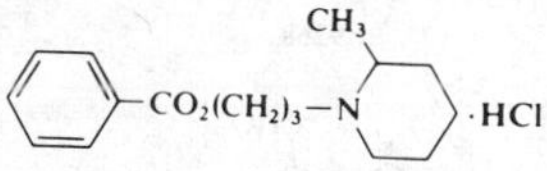
9213

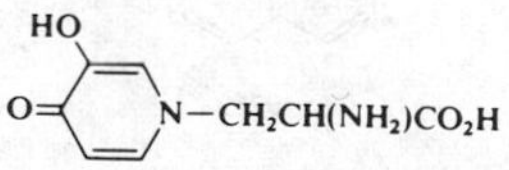
9215

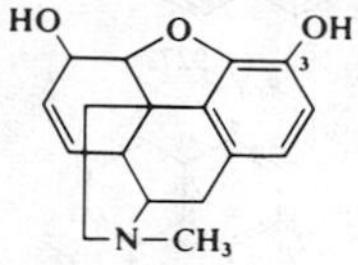
9216

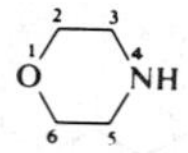
9225

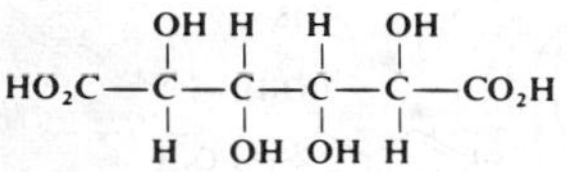
9239

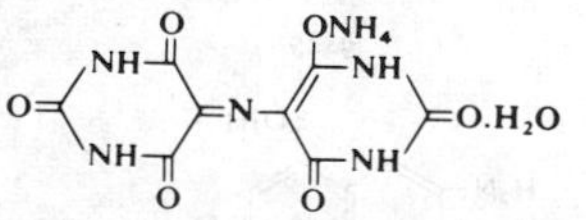
9242

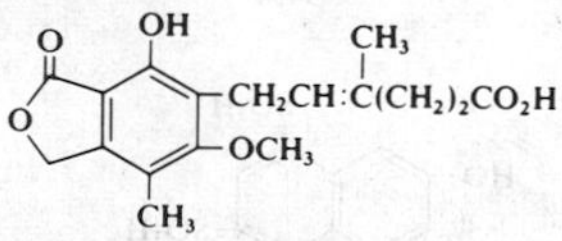
9243

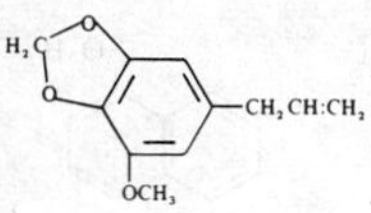
9258

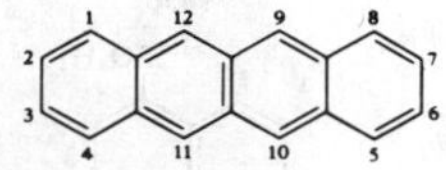
9262

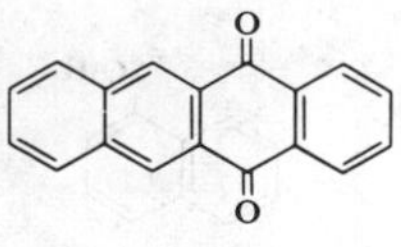
9267

STRUCTURAL FORMULAS OF ORGANIC COMPOUNDS (Continued)

In Numeric as they occur in Organic Compounds Table

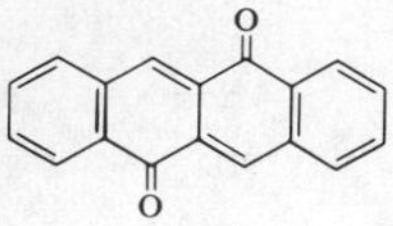
9268

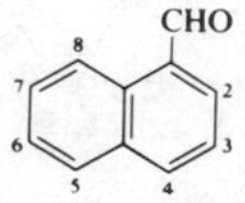
9269

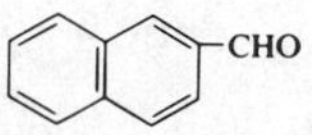
9273

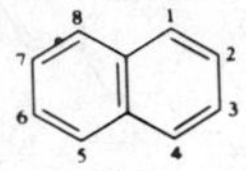
9275

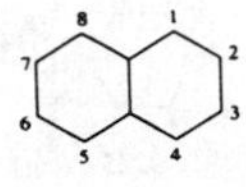
9316

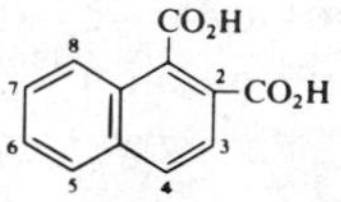
9445

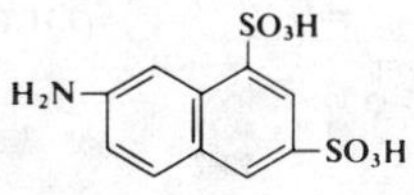
9454

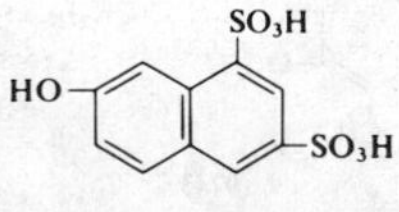
9455

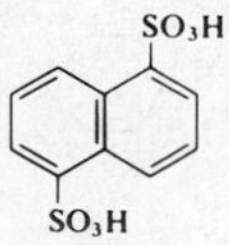
9456

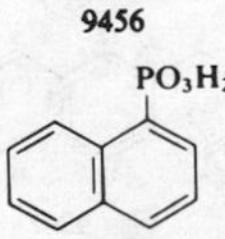
9462

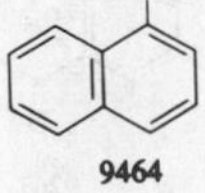
9464

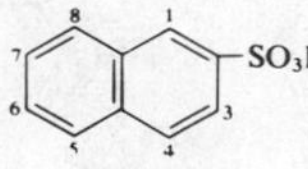
9476

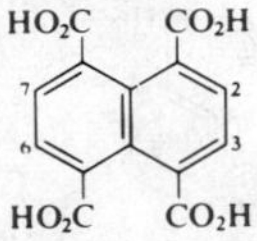
9483

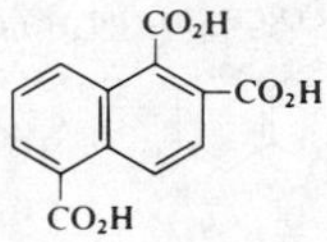
9485

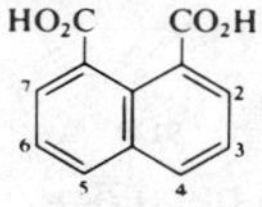
9487

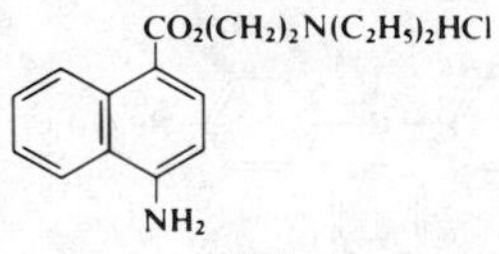
9499

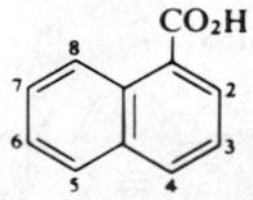
9500

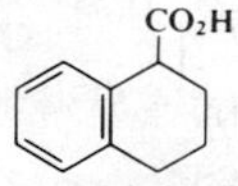
9529

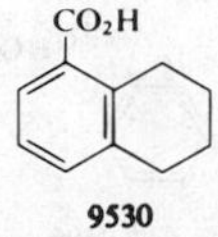
9530

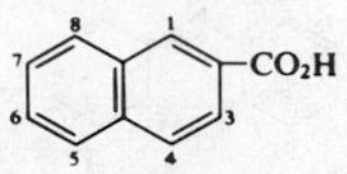
9534

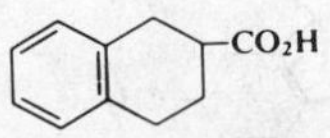
9567

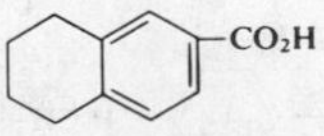
9568

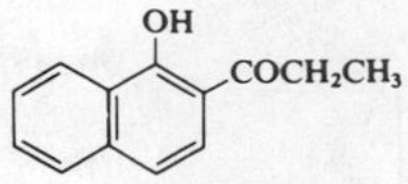
9586

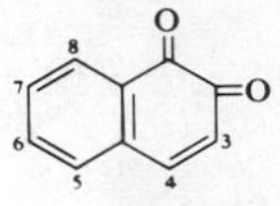
9618

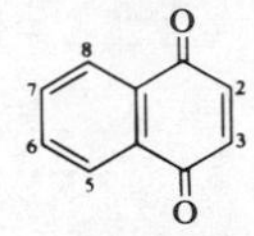
9636

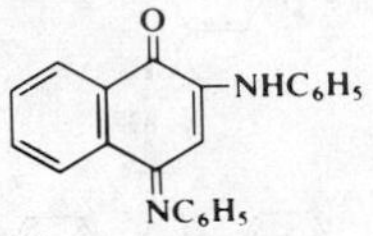
9638

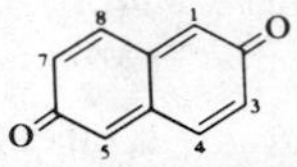
9679

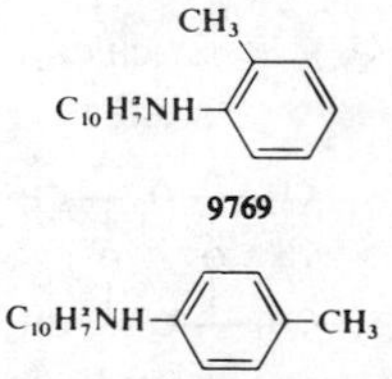
9769

9770

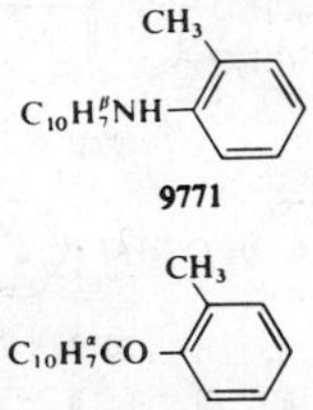
9771

9773

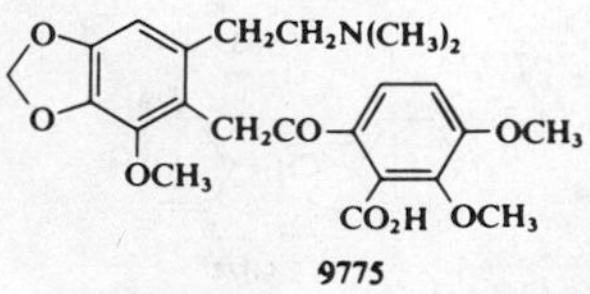
9775

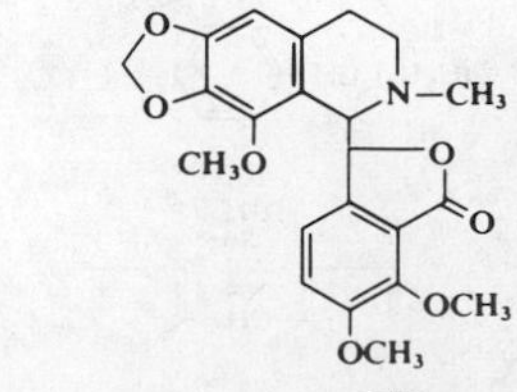
9778

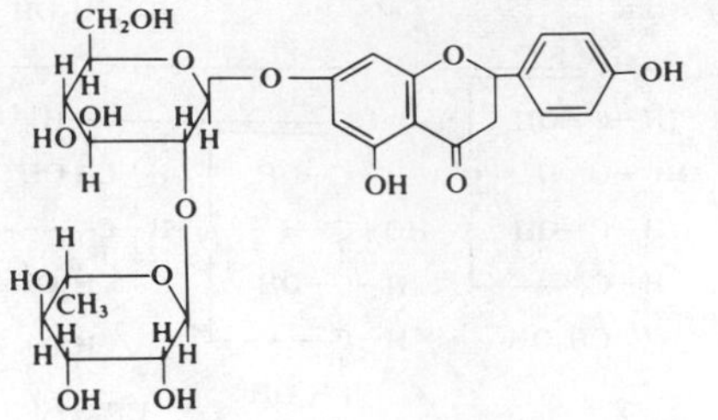
9783

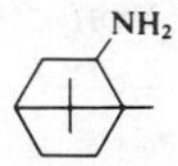
9786

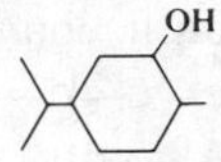
9787

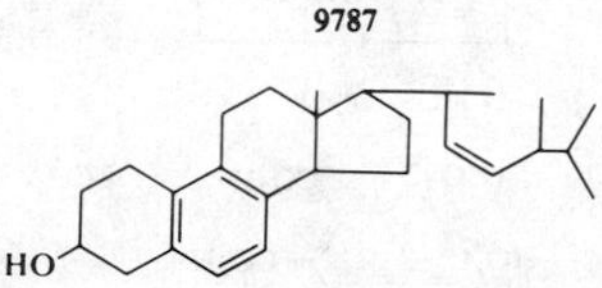
9789

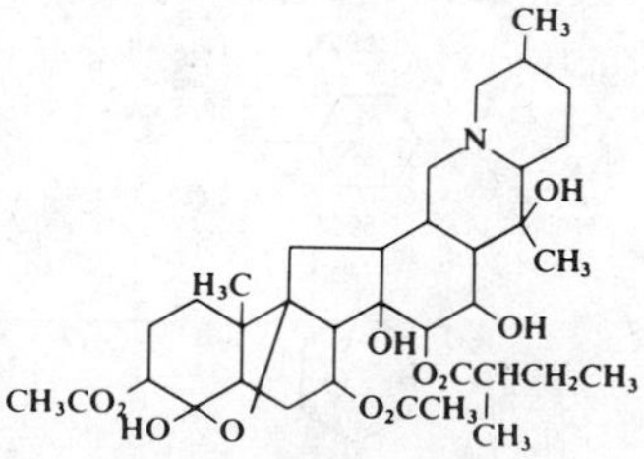
9790

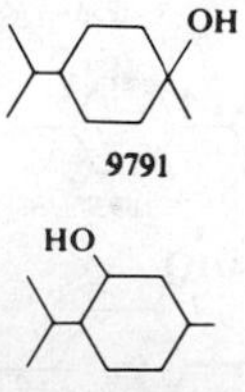
9791

9792

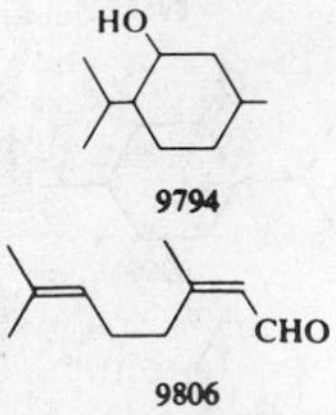
9794

9806

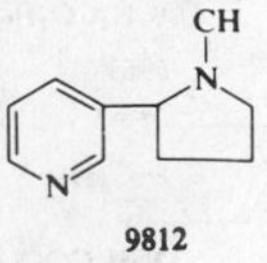
9812

STRUCTURAL FORMULAS OF ORGANIC COMPOUNDS (Continued)

In Numeric as they occur in Organic Compounds Table

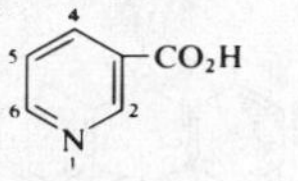

9818

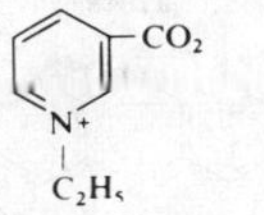

9820

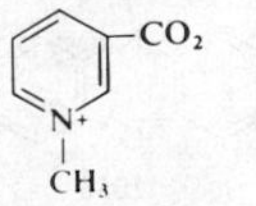

9827

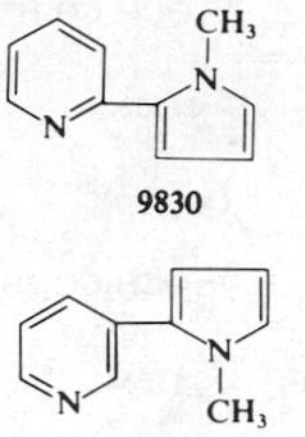

9830

9832

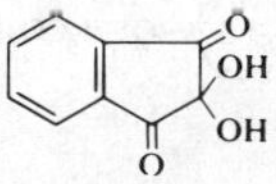

9833

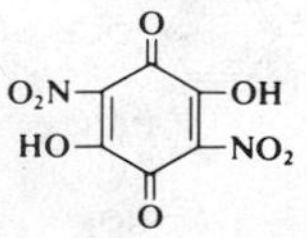

9834

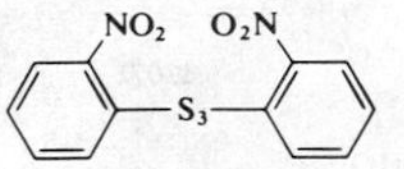

9892

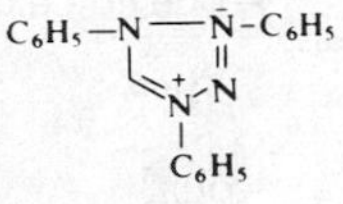

9899

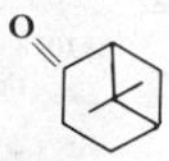

9958

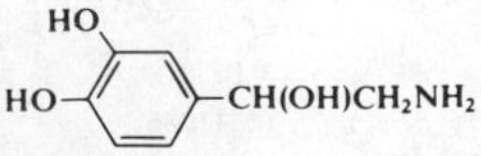

9959

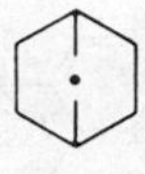

9960

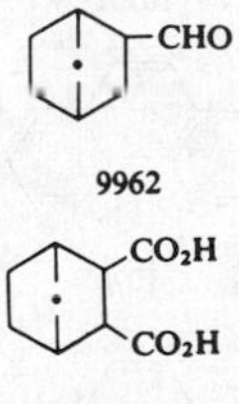

9962

9963

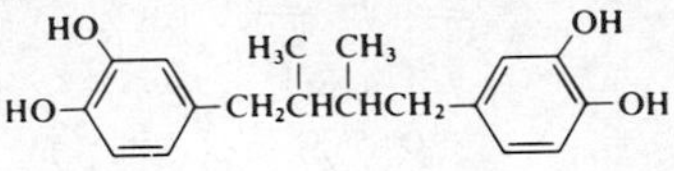

9965

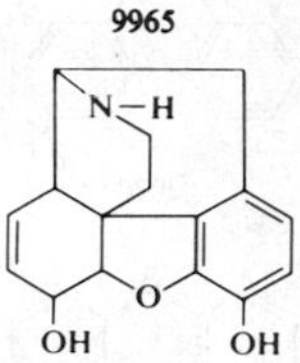

9969

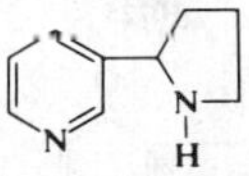

9970

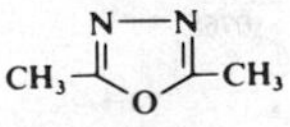

10168

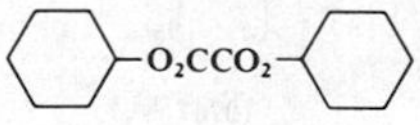

10174

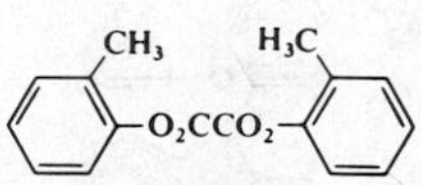

10181

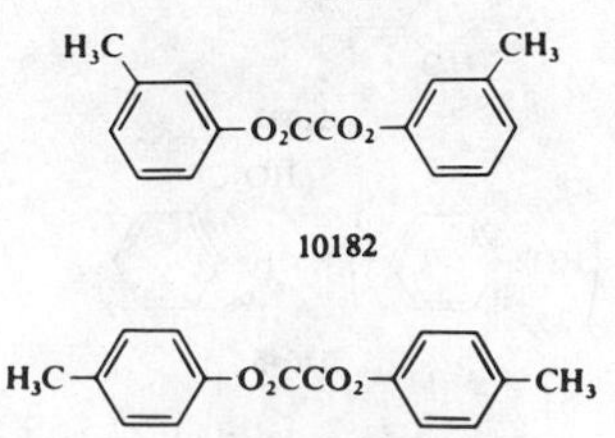

10182

10183

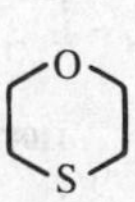

10200

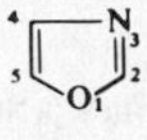

10201

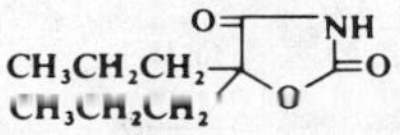

10208

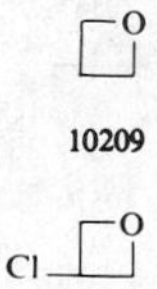

10209

10210

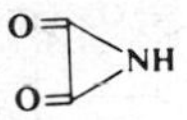

10211

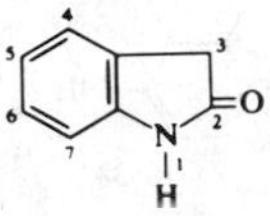

10212

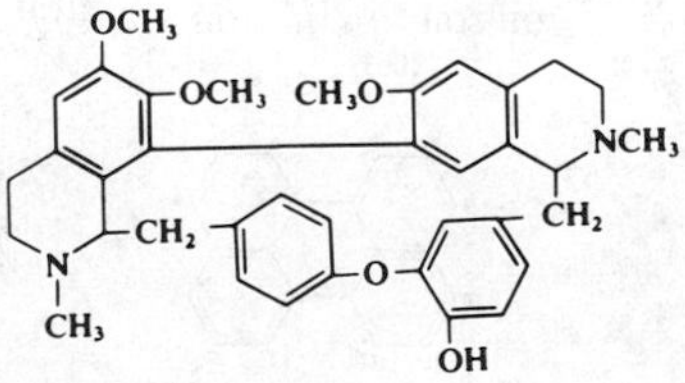

10216

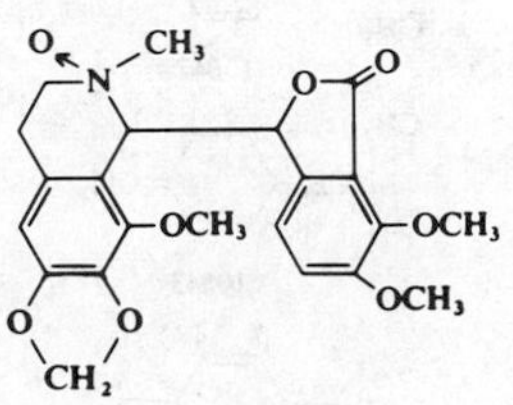

10219

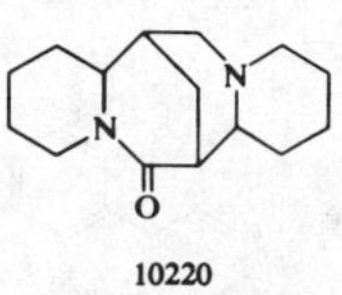

10220

10241

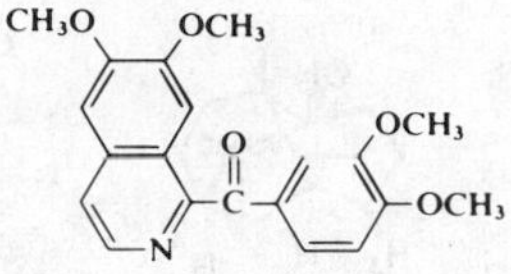

10249

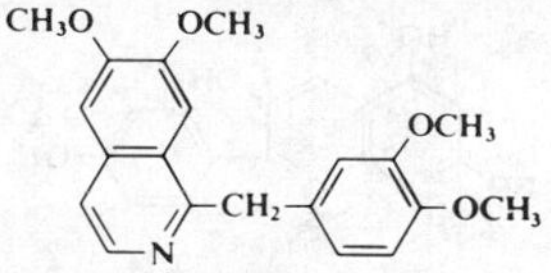

10250

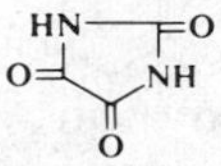

10252

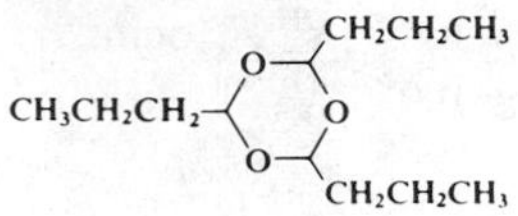

10253

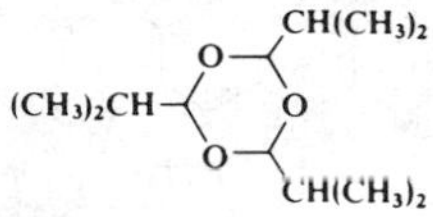

10254

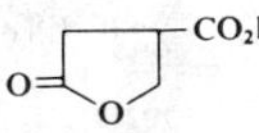

10255

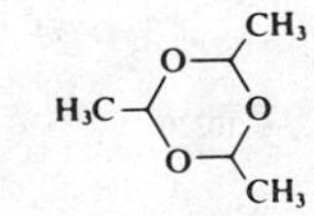

10257

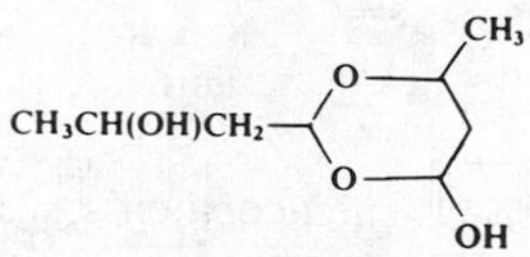

10258

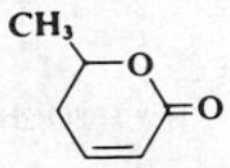

10259

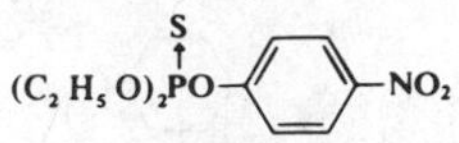

10260

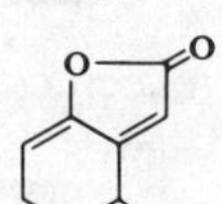

10261

In Numeric as they occur in Organic Compounds Table

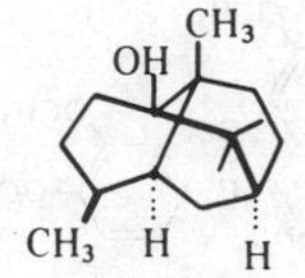
10262

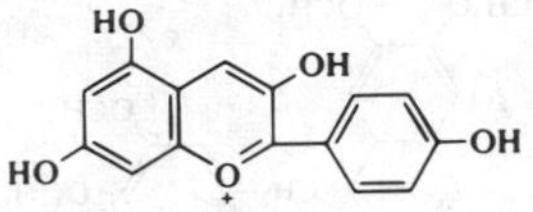
10274

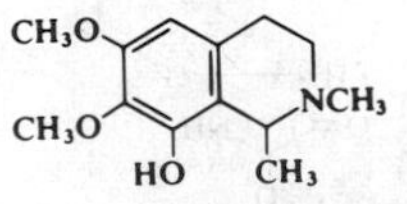
10275

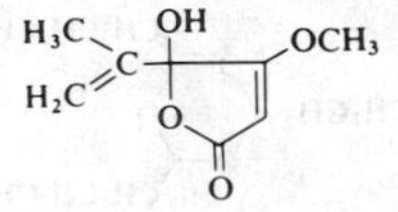
10276

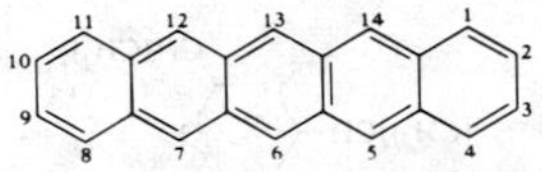
10277

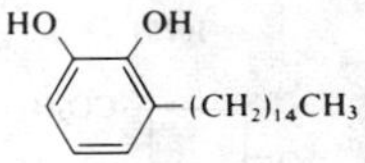
10289

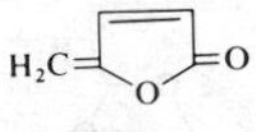
10310

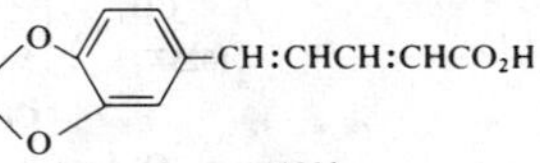
10311

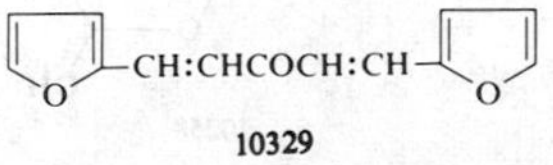
10329

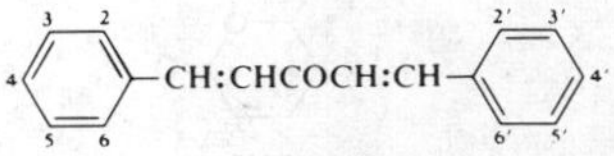
10330

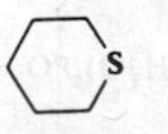
10338

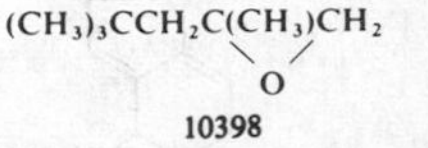
10398

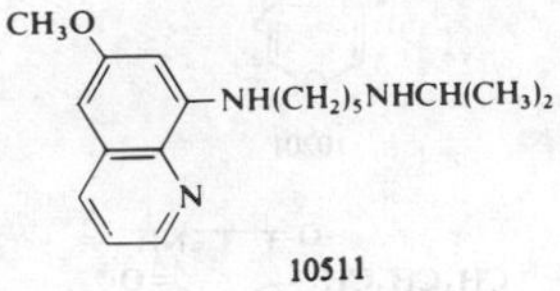
10511

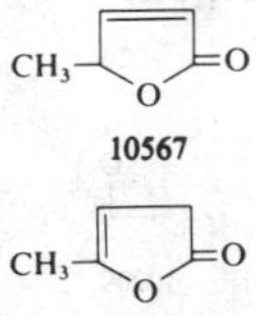
10567

10570

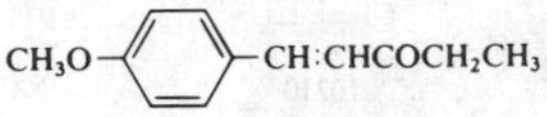
10588

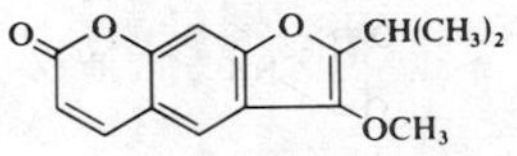
10598

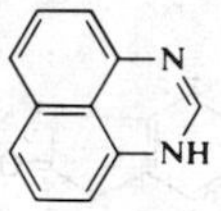
10638

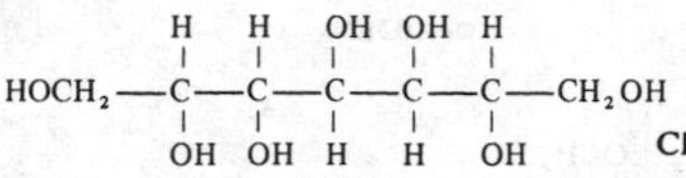
10639

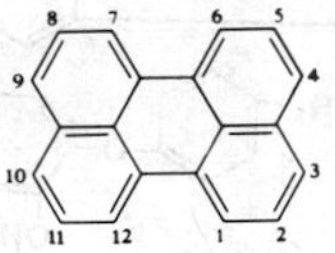
10640

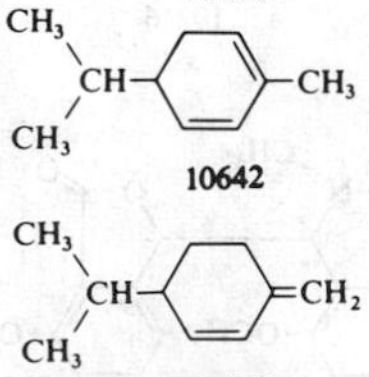
10642

10643

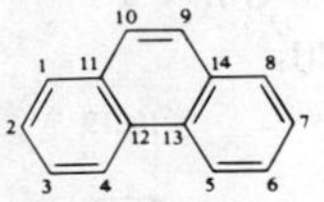
10649

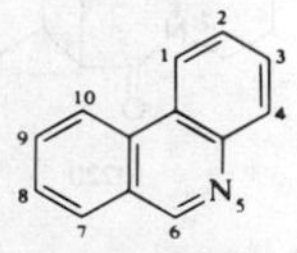
10700

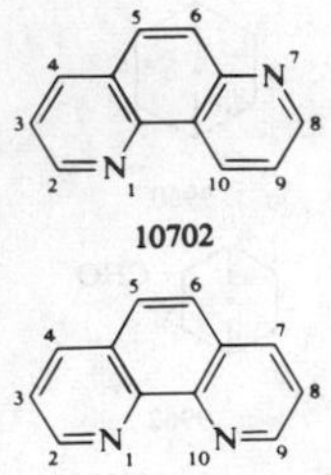
10702

10703

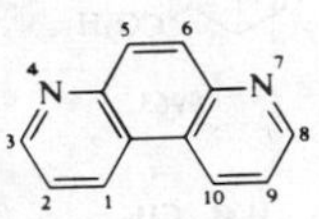
10704

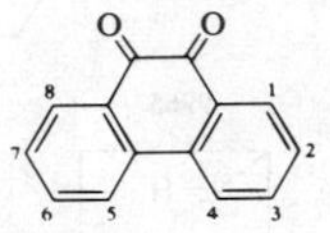
10707

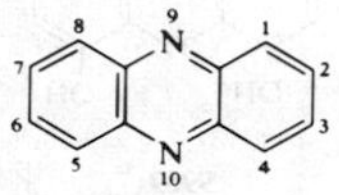
10725

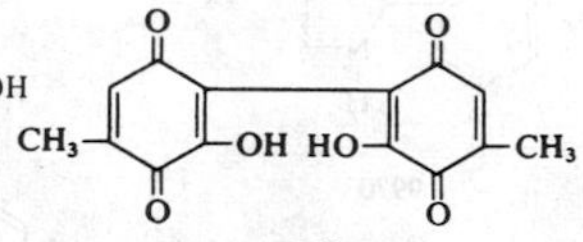
10766

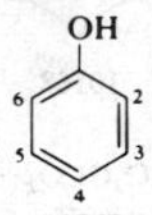
10767

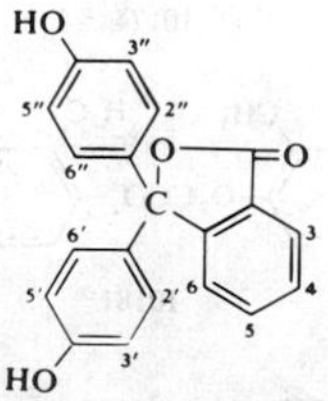
11042

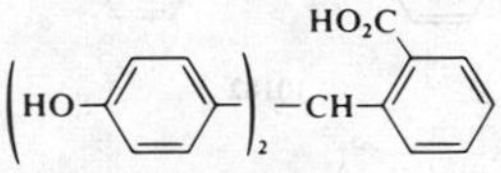
11046

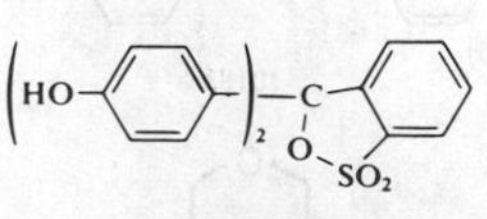
11047

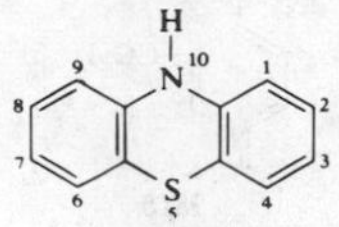
11048

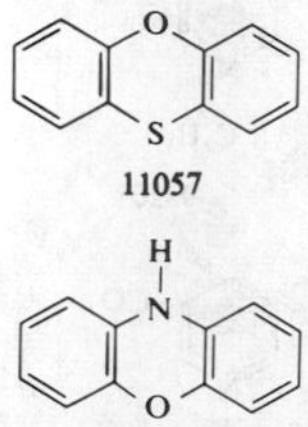
11057

11058

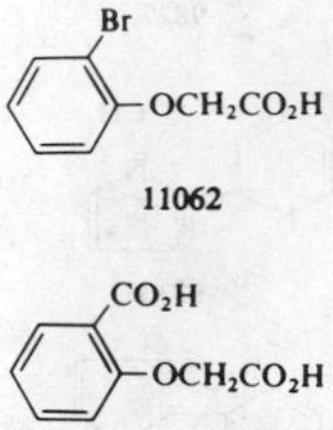
11062

11064

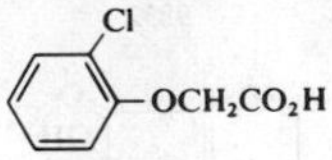
11067

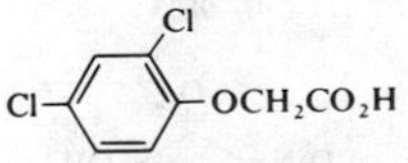
11071

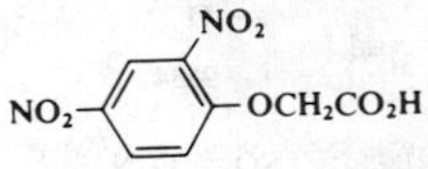
11072

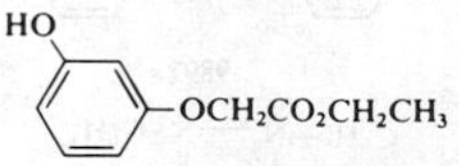
11076

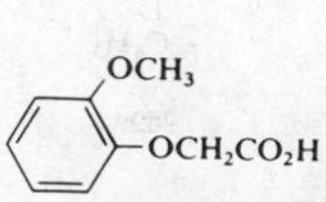
11078

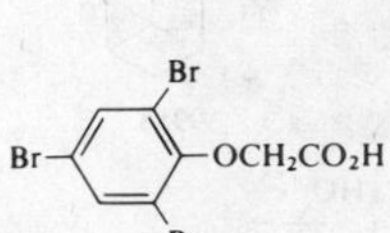
11086

In Numeric as they occur in Organic Compounds Table

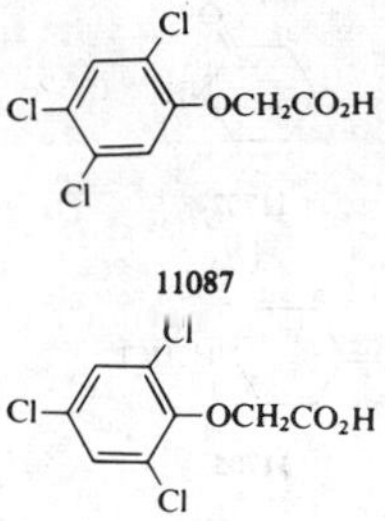

11087

11088

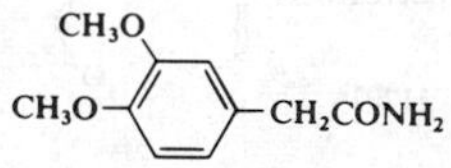

11094

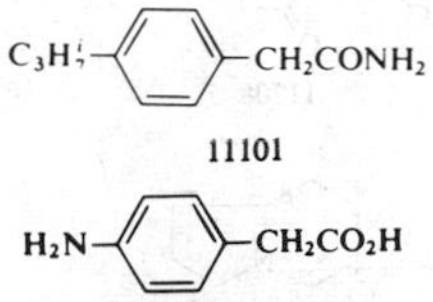

11101

11104

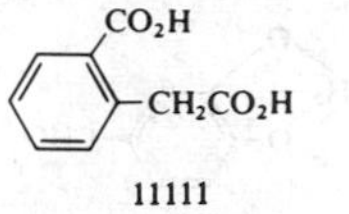

11111

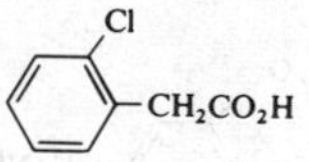

11117

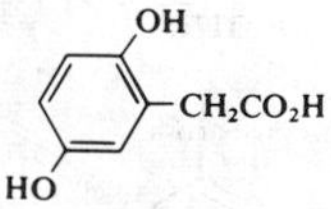

11120

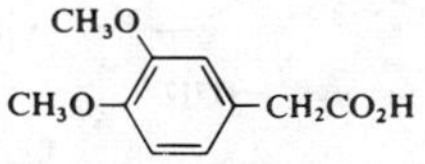

11122

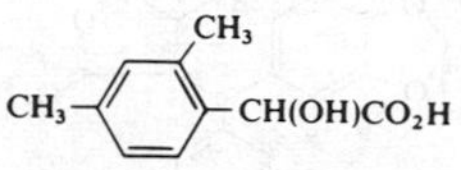

11123

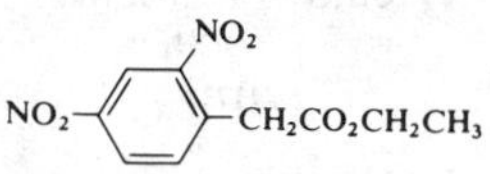

11127

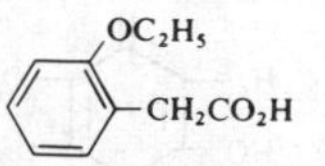

11129

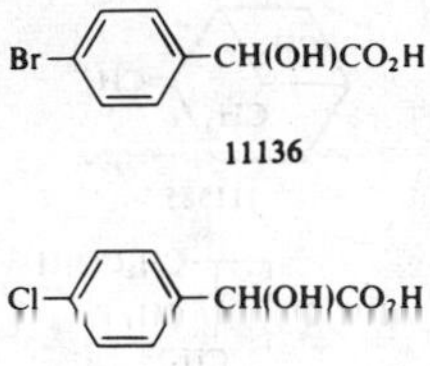

11136

11137

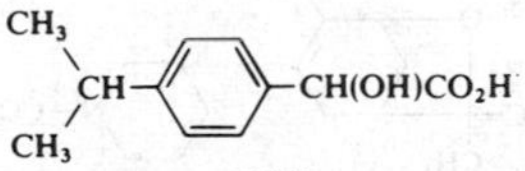

11144

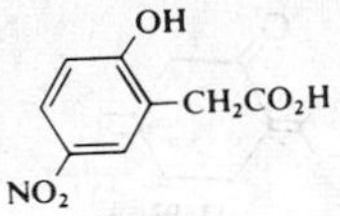

11148

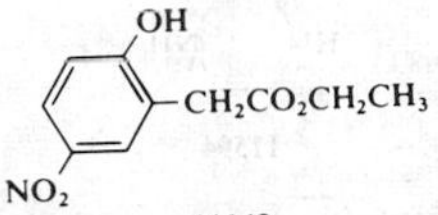

11149

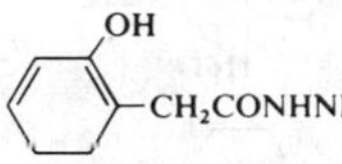

11150

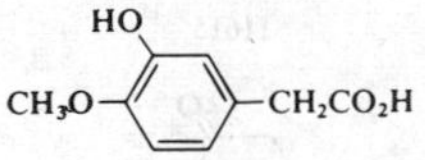

11153

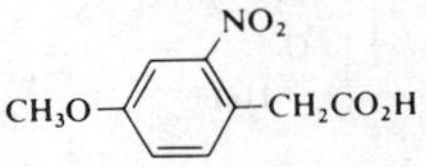

11160

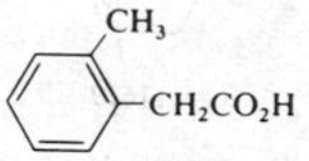

11162

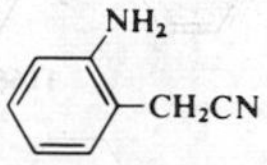

11171

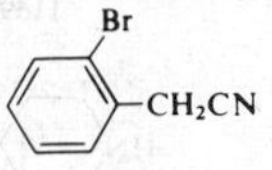

11174

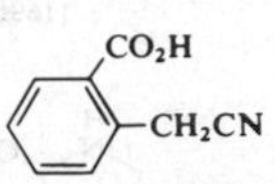

11176

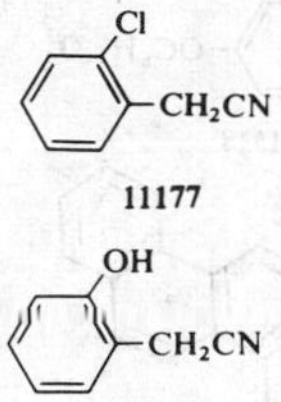

11177

11181

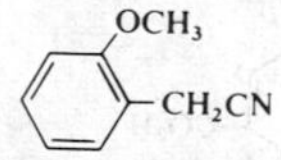

11184

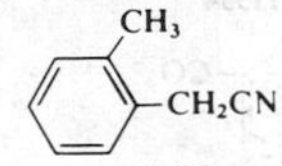

11186

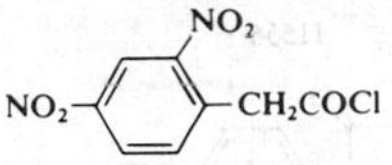

11195

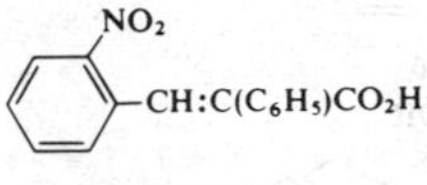

11198

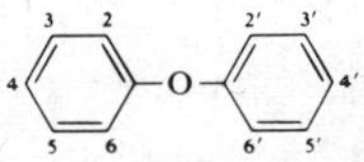

11249

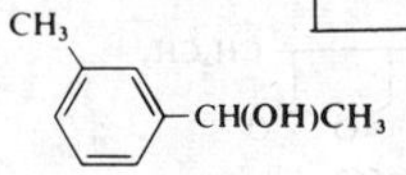

11254

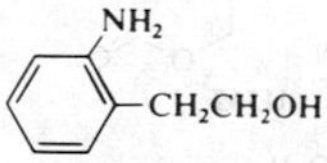

11257

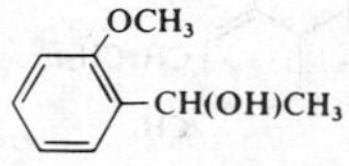

11259

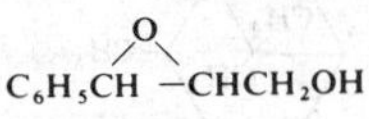

11266

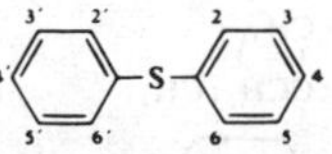

11308

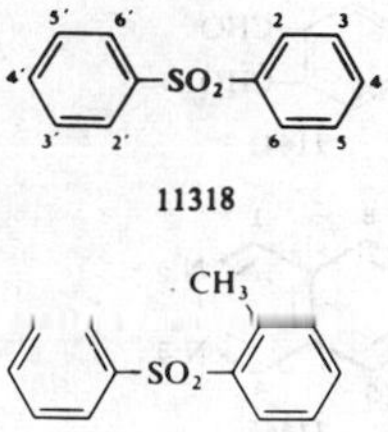

11318

11333

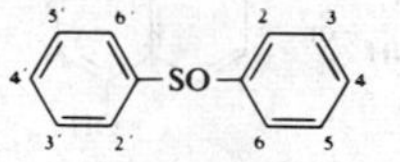

11335

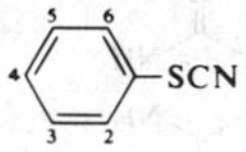

11341

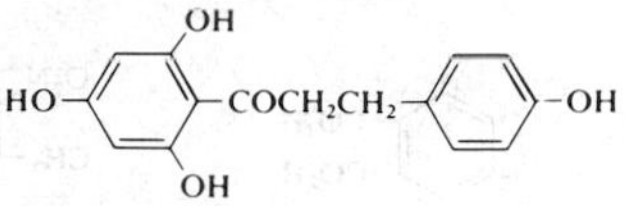

11348

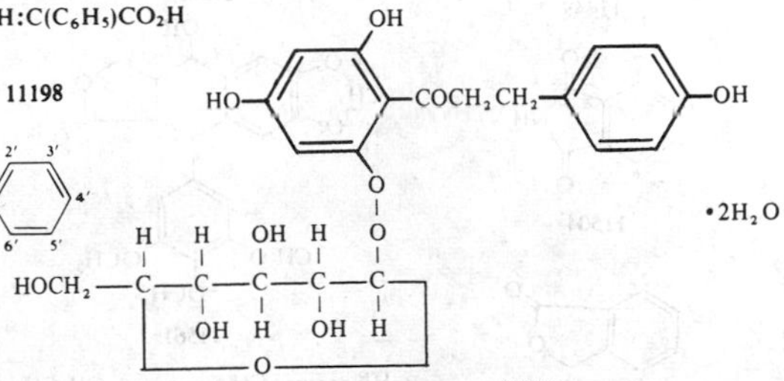

11349

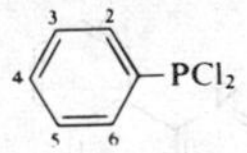

11367

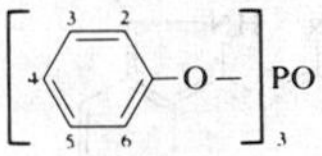

11420

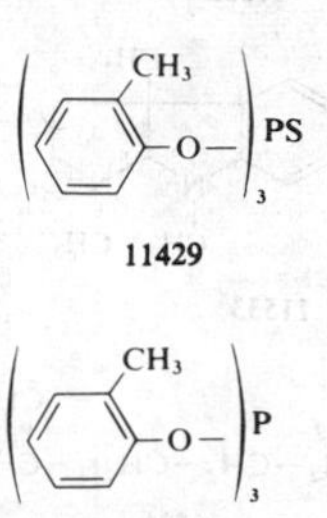

11429

11439

STRUCTURAL FORMULAS OF ORGANIC COMPOUNDS (Continued)

In Numeric as they occur in Organic Compounds Table

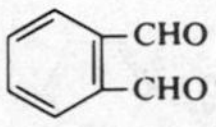
11441

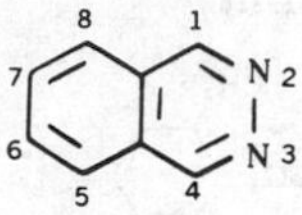
11446

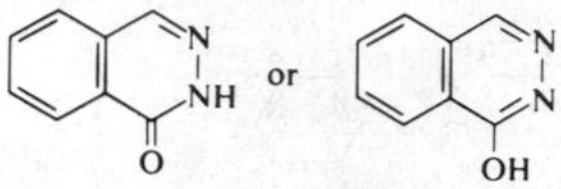
11447

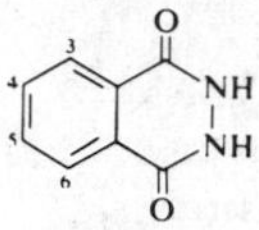
11448

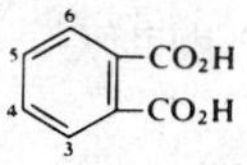
11449

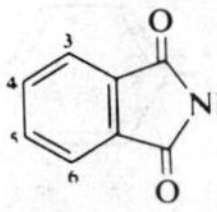
11504

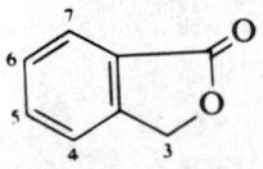
11528

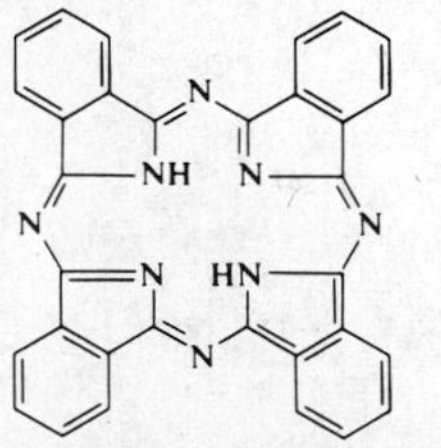
11532

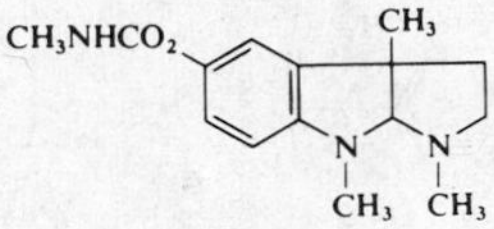
11533

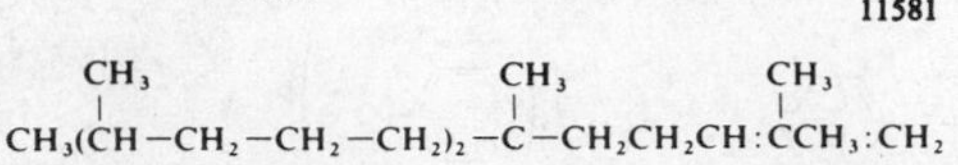
11536

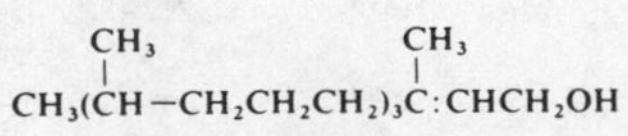
11537

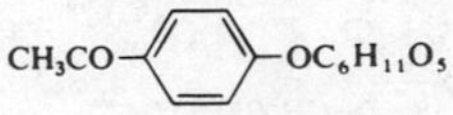
11538

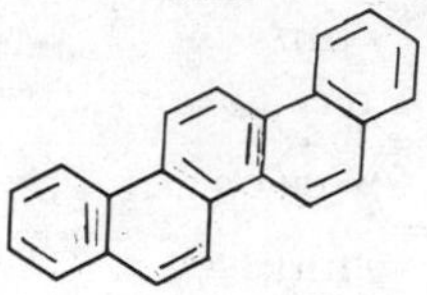
11539

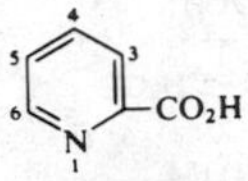
11554

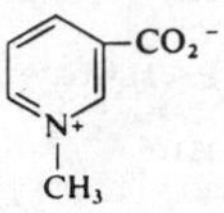
11556

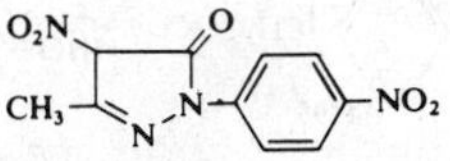
11560

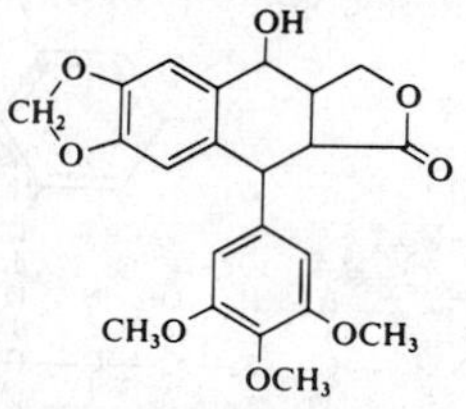
11561

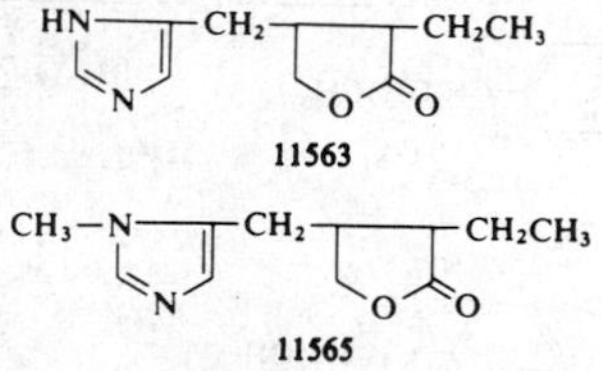
11563 11565

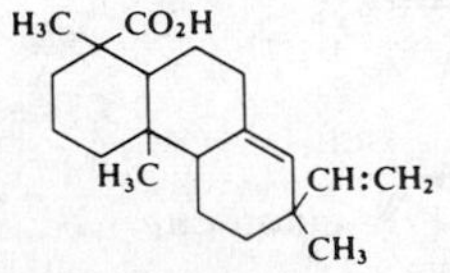
11570

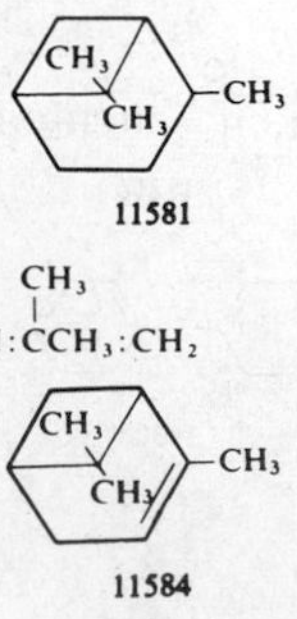
11581 11584

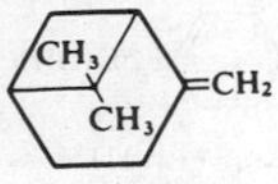
11585

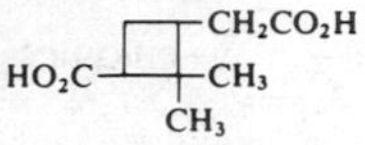
11587

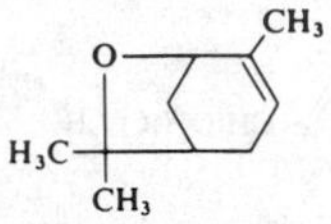
11591

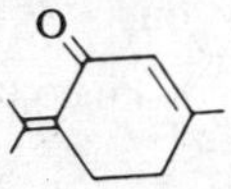
11593

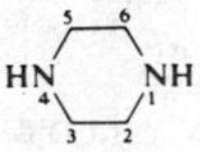
11594

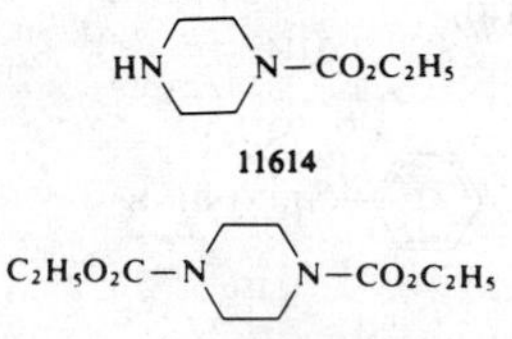
11614 11615

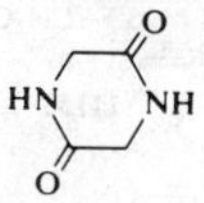
11616

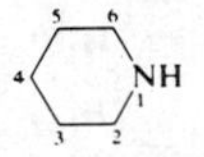
11617

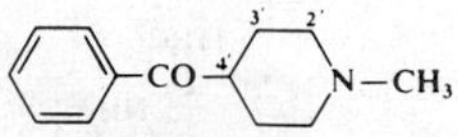
11669

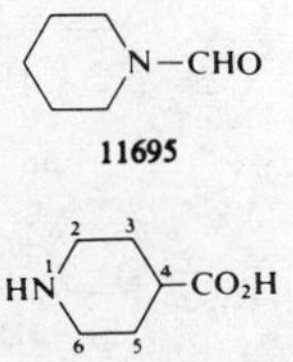
11695 11696

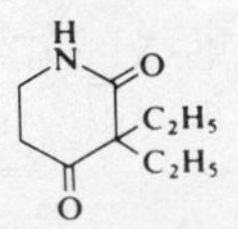
11701

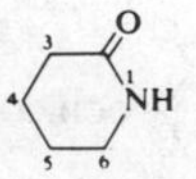
11702

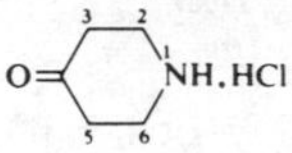
11705

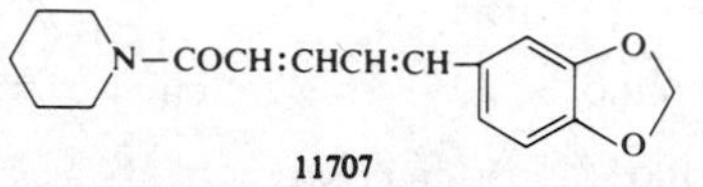
11707

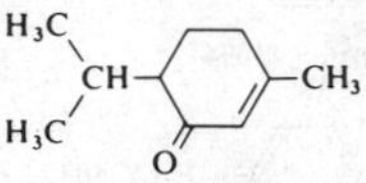
11708

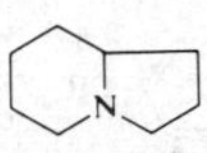
11712

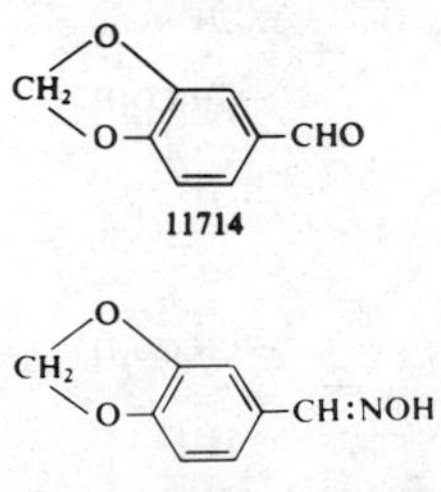
11714 11715

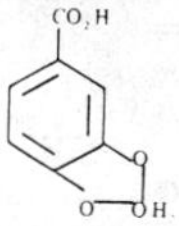
11717

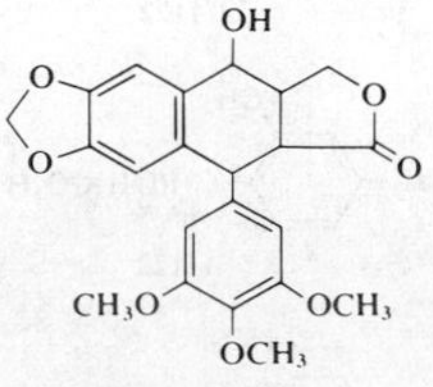
11734

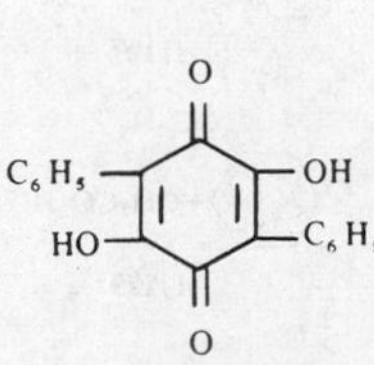
11736

STRUCTURAL FORMULAS OF ORGANIC COMPOUNDS (Continued)

In Numeric as they occur in Organic Compounds Table

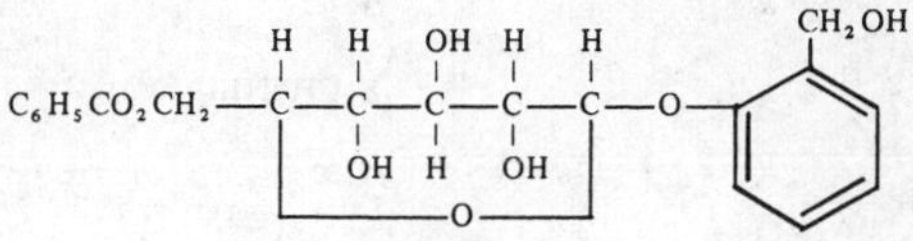
11737

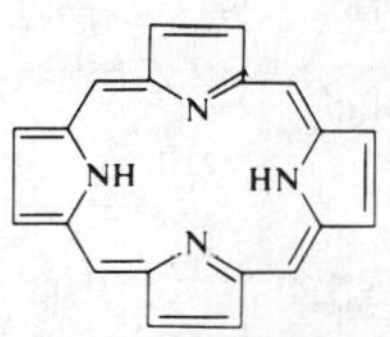
11738

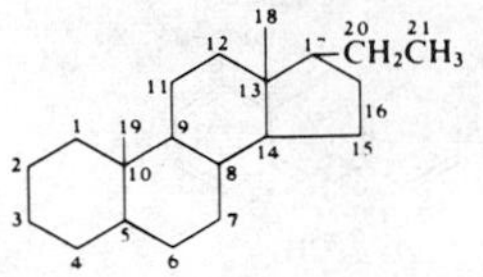
11741

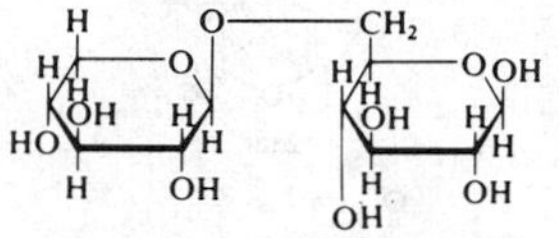
11755

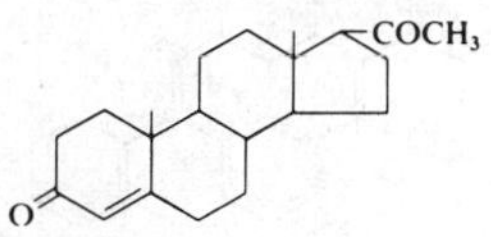
11756

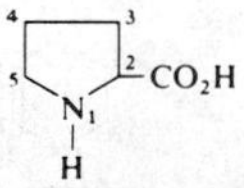
11759

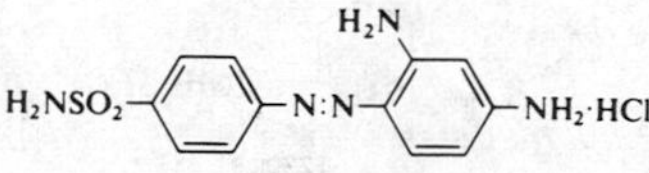
11769

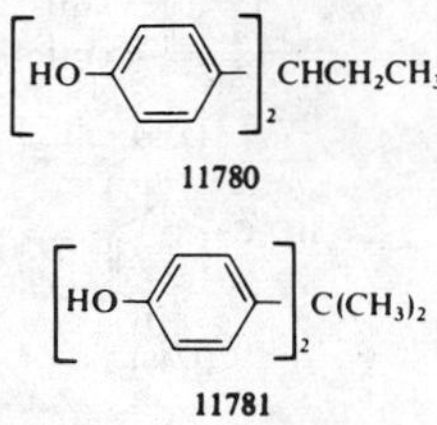
11780

11781

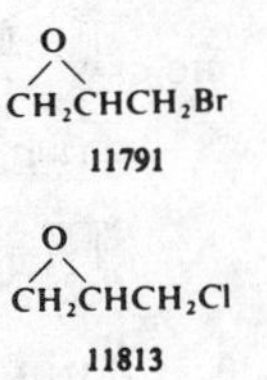
11791

11813

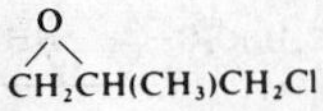
11815

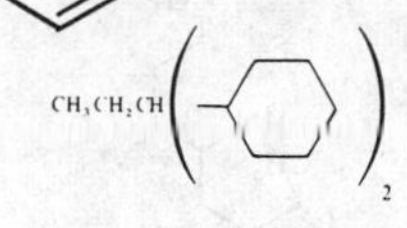
11834

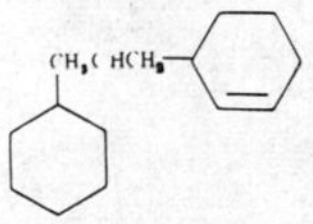
11835

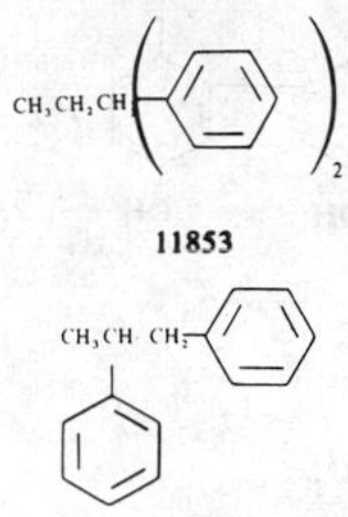
11853

11854

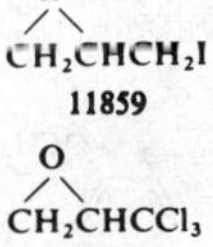
11859

11860

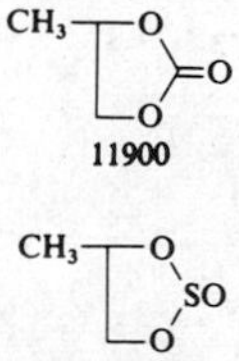
11900

11909

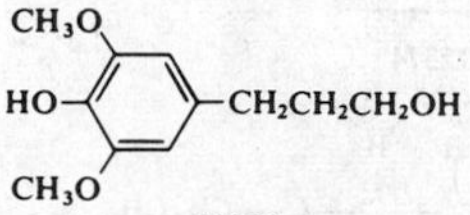
11956

11958

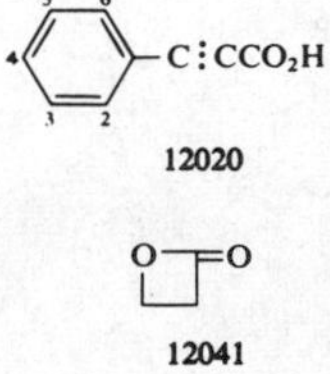
12020

12041

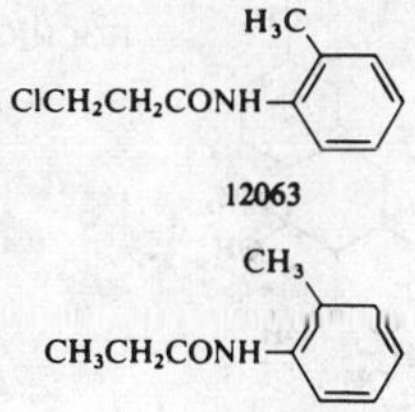
12063

12070

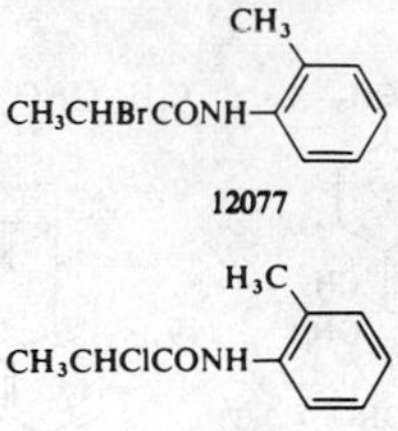
12077

12103

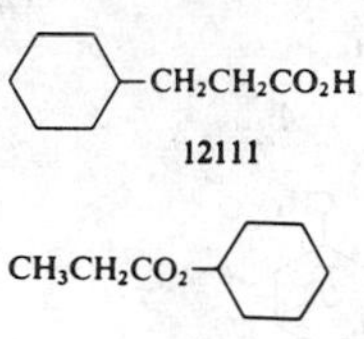
12111

12112

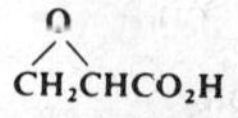
12142

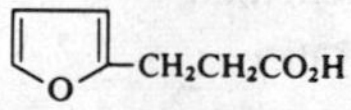
12145

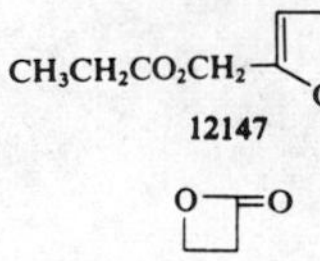
12147

12152

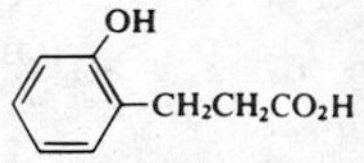
12154

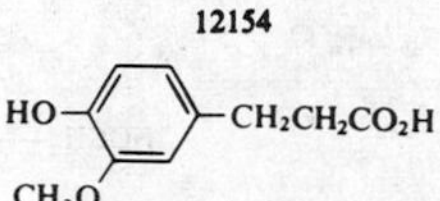
12156

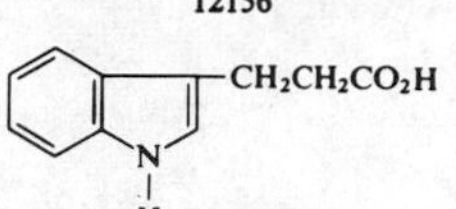
12160

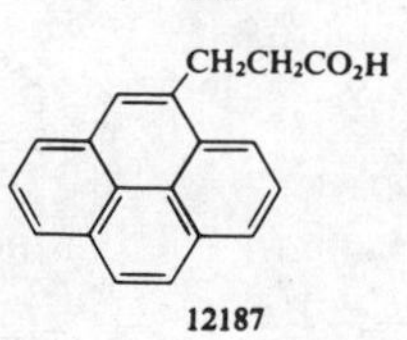
12187

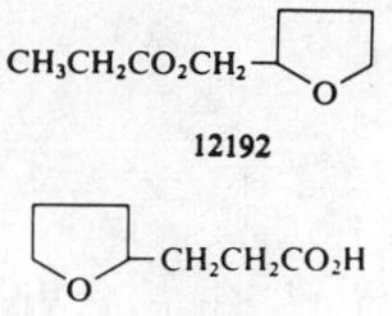
12192

12193

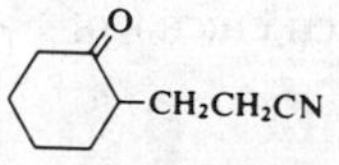
12212

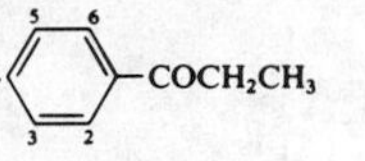
12228

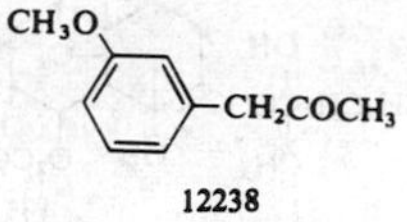
12238

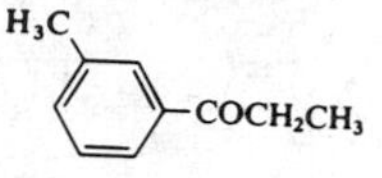
12241

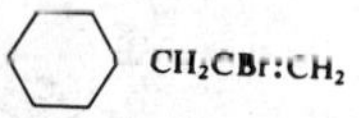
12261

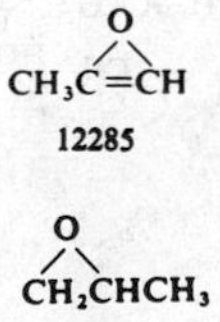
12285

12300

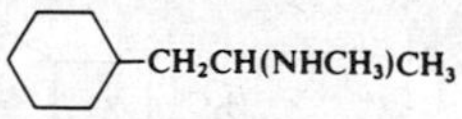
12305

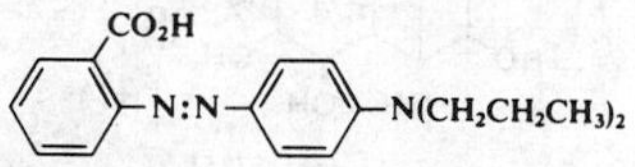
12320

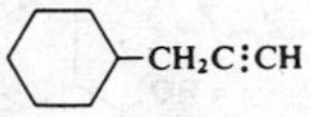
12336

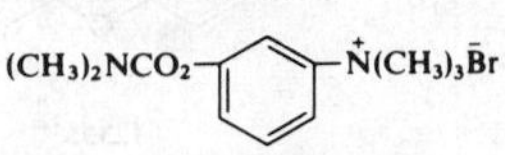
12345

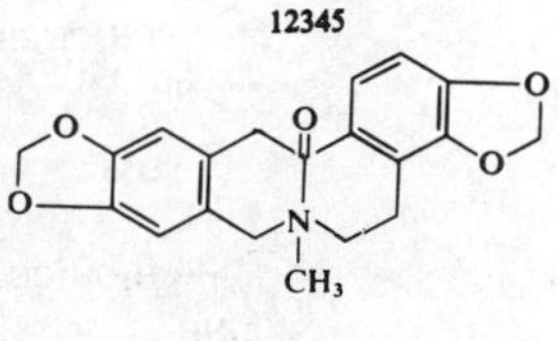
12346

STRUCTURAL FORMULAS OF ORGANIC COMPOUNDS (Continued)

In Numeric as they occur in Organic Compounds Table

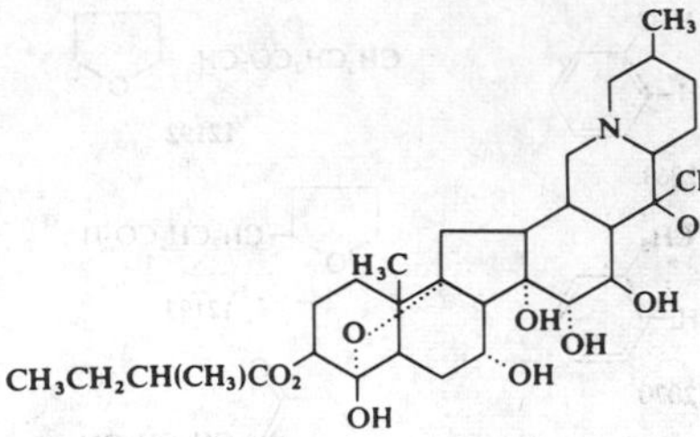

12347

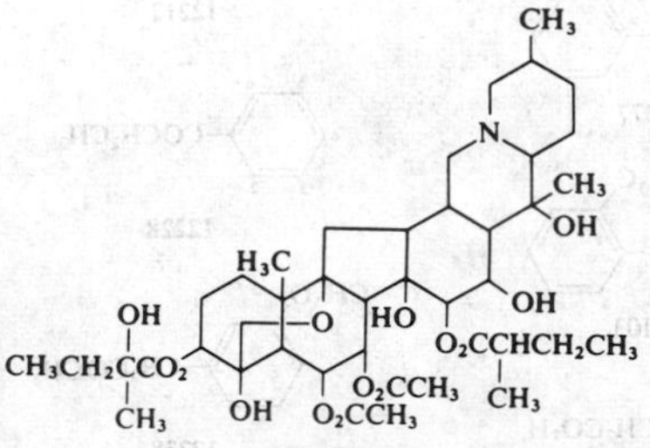

12349

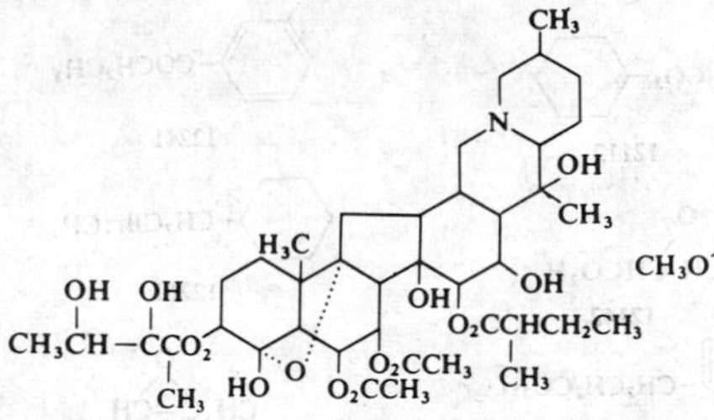

12350

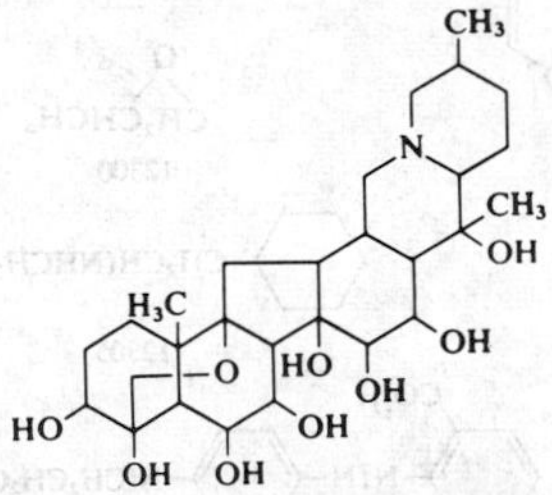

12351

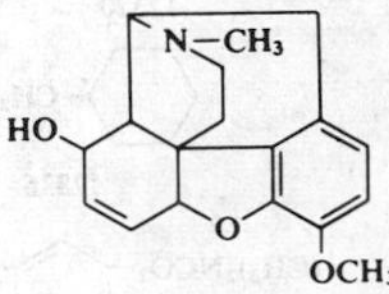

12355

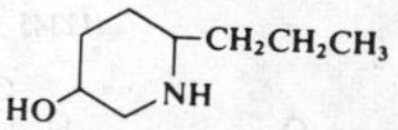

12356

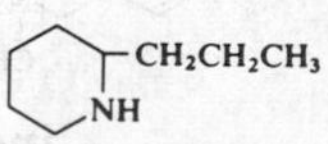

12357

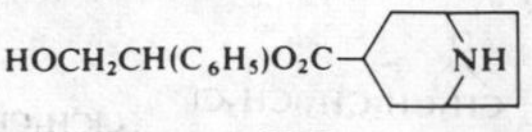

12366

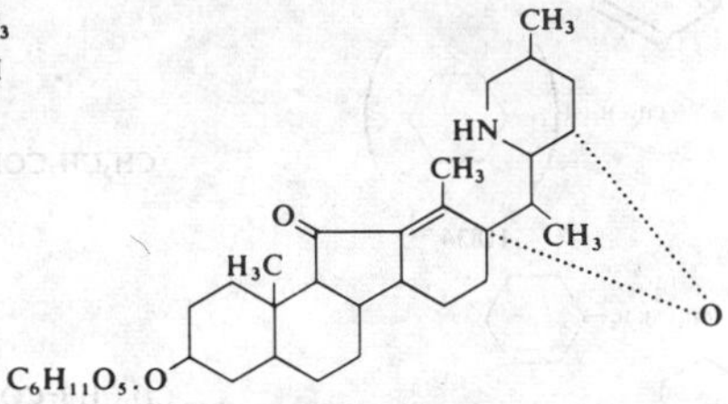

12368

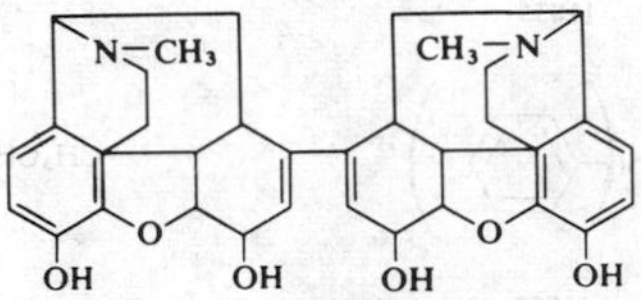

12369

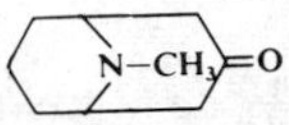

12371

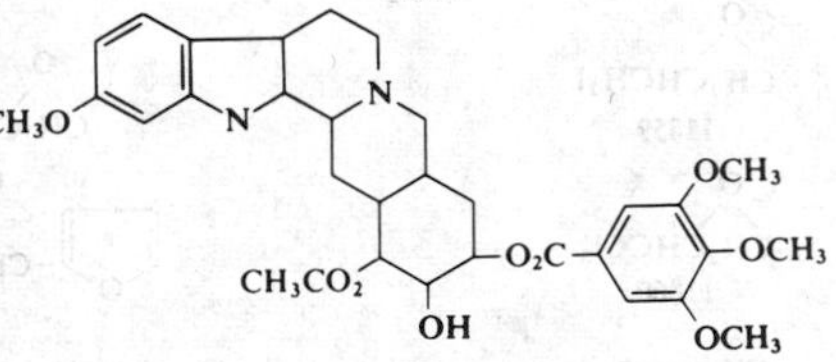

12372

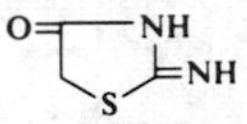

12373

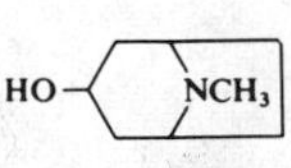

12374

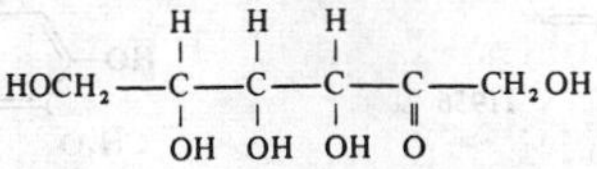

12375

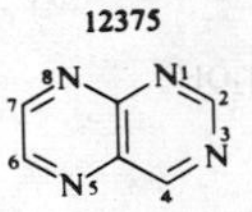

12376

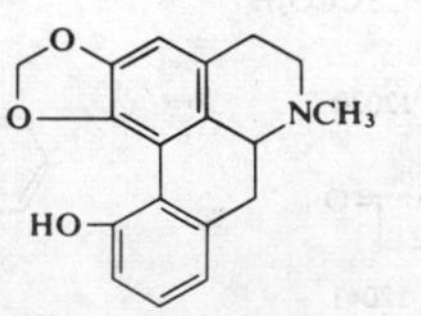

12379

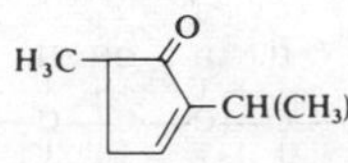

12380

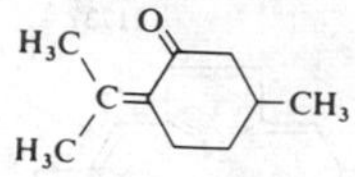

12381

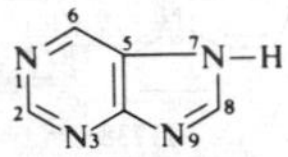

12382

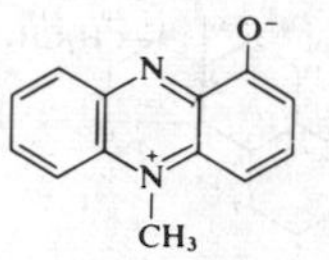

12386

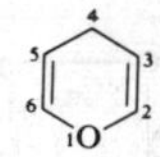

12388

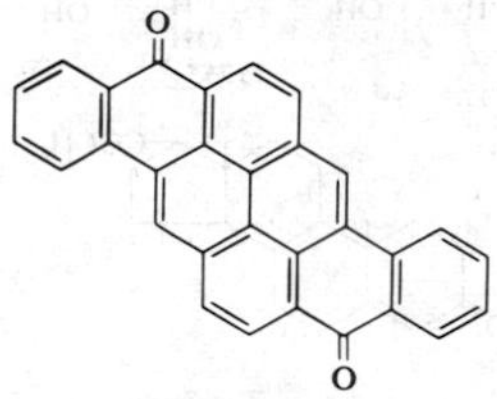

12391

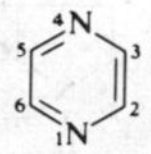

12392

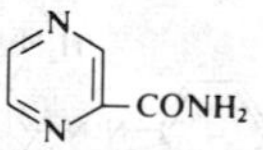

12398

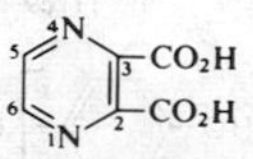

12399

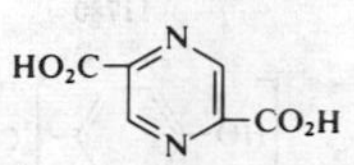

12401

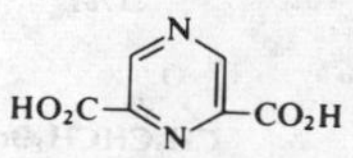

12402

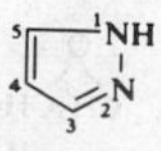

12403

STRUCTURAL FORMULAS OF ORGANIC COMPOUNDS (Continued)

In Numeric as they occur in Organic Compounds Table

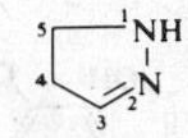
12419

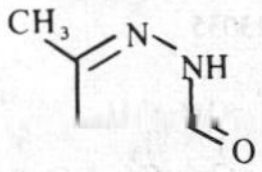
12421

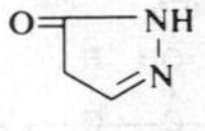
12424

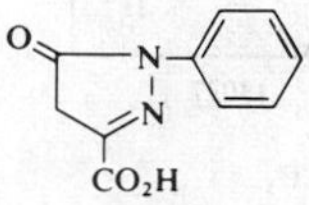
12429

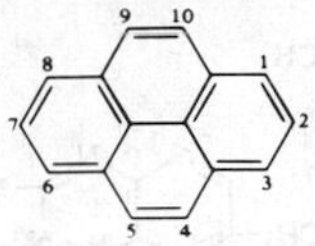
12430

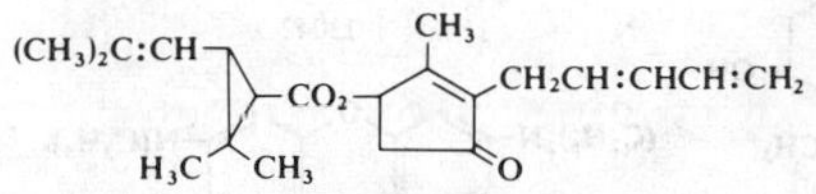
12438

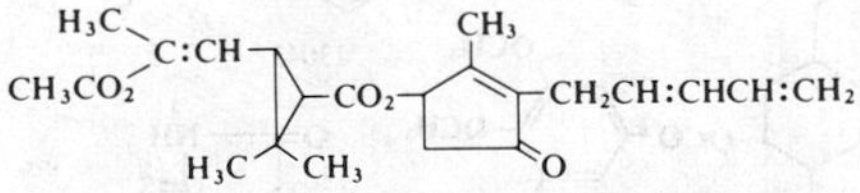
12439

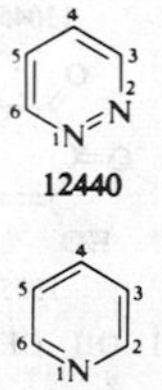
12440

12441

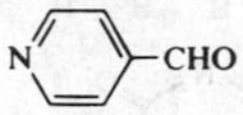
12544

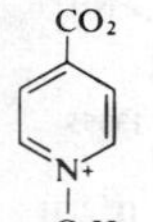
12547

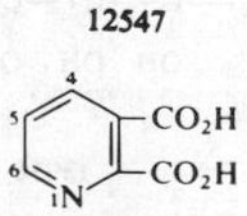
12548

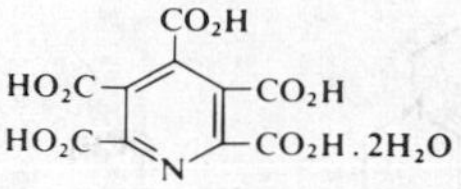
12564

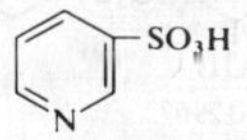
12565

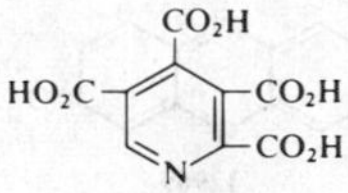
12566

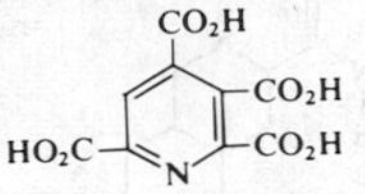
12567

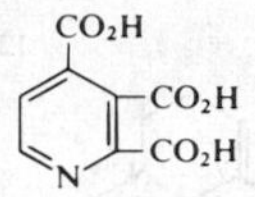
12569

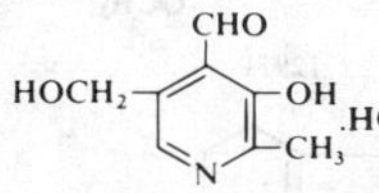
12575

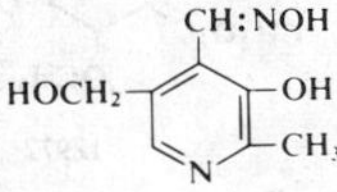
12576

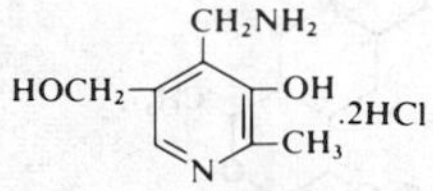
12577

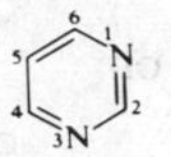
12578

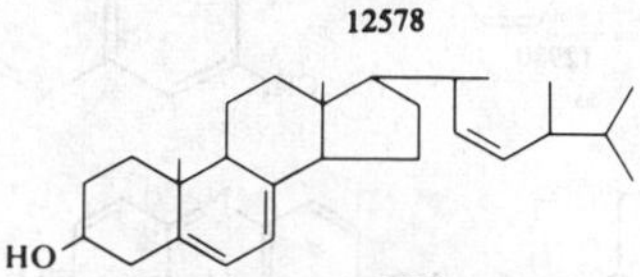
12602

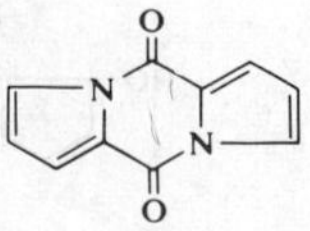
12603

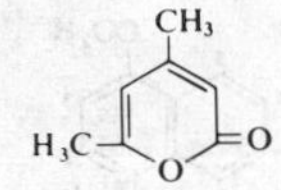
12611

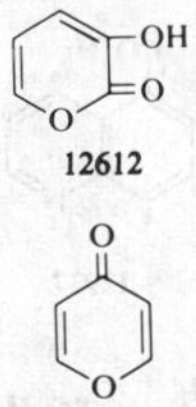
12612

12613

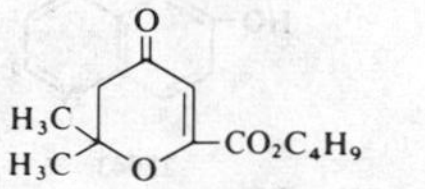
12617

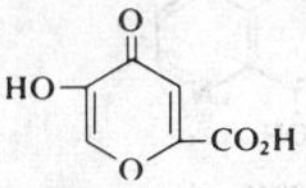
12618

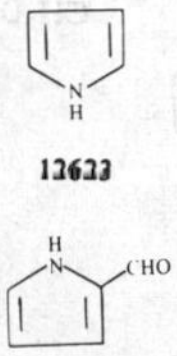
12623

12649

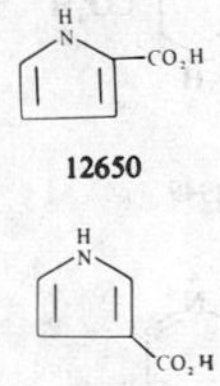
12650

12652

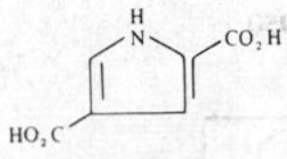
12656

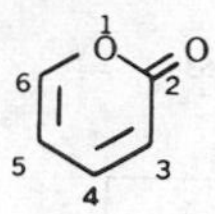
12660

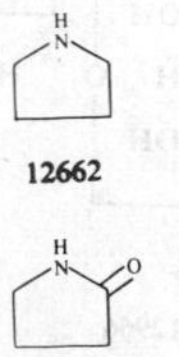
12662

12671

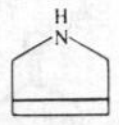
12675

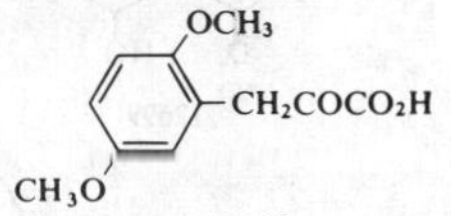
12679

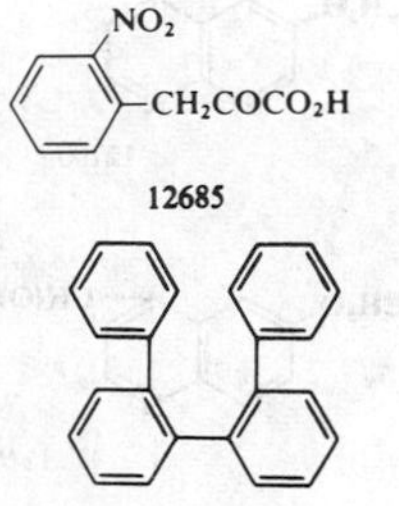
12685

12686

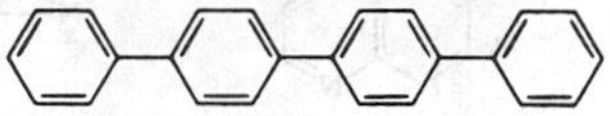
12687

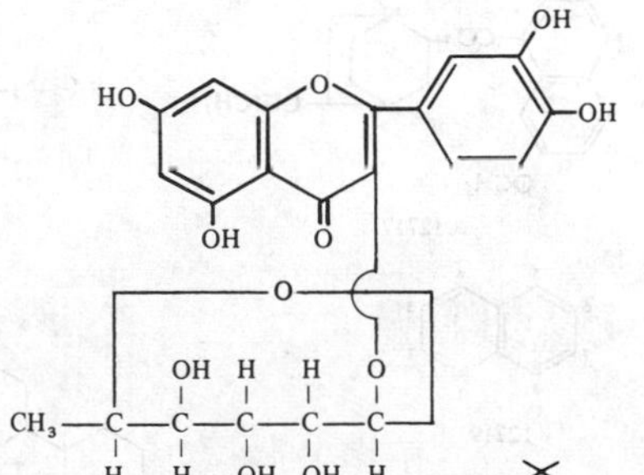
12688

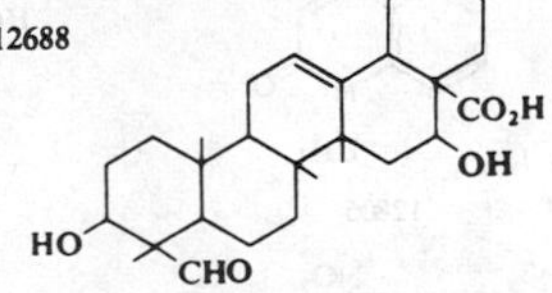
12689

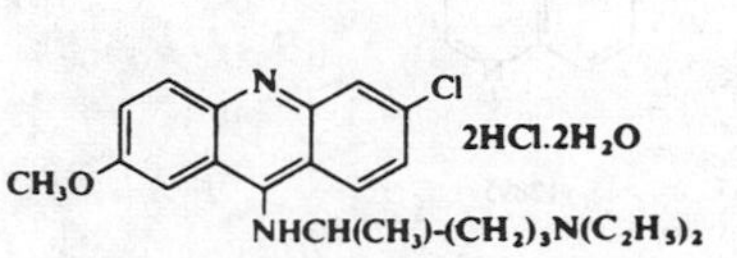
12690

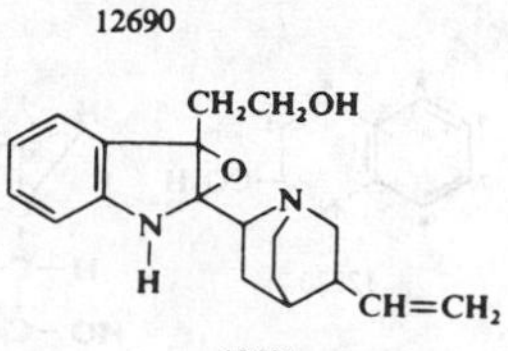
12691

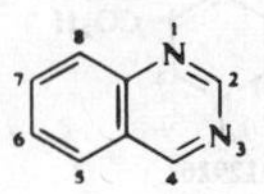
12692

STRUCTURAL FORMULAS OF ORGANIC COMPOUNDS (Continued)

In Numeric as they occur in Organic Compounds Table

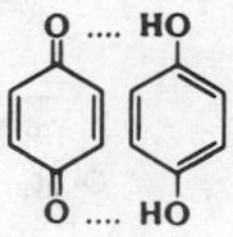
12699

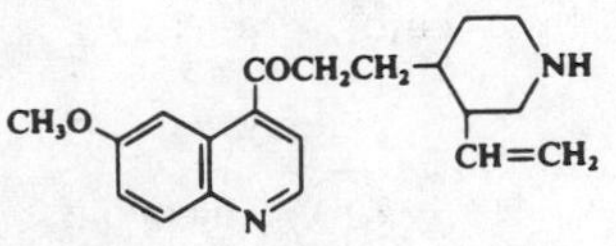
12700

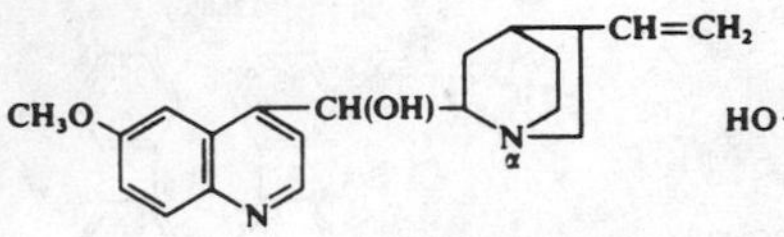
12702

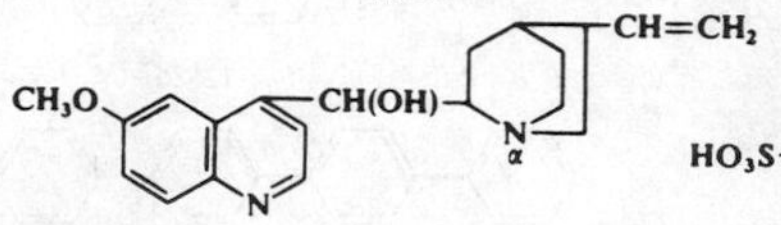
12706

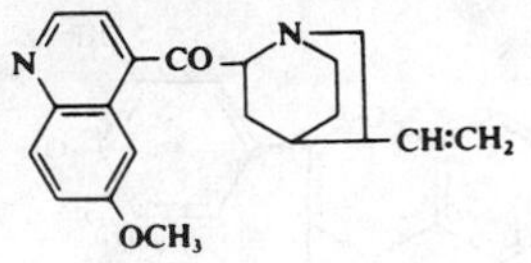
12717

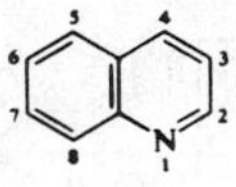
12719

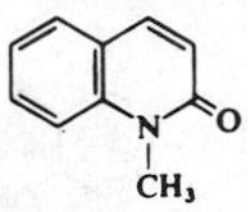
12806

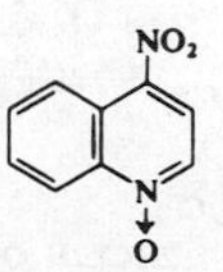
12895

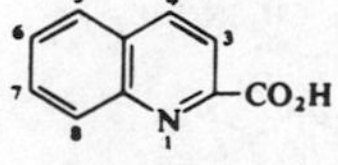
12920

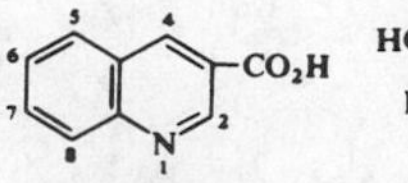
12926

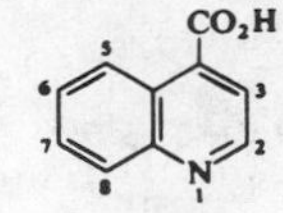
12929

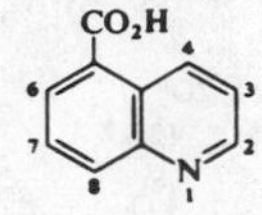
12937

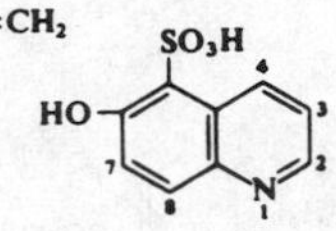
12943

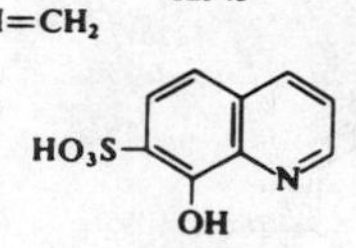
12946

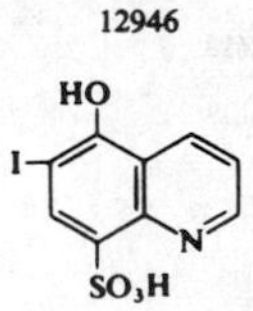
12947

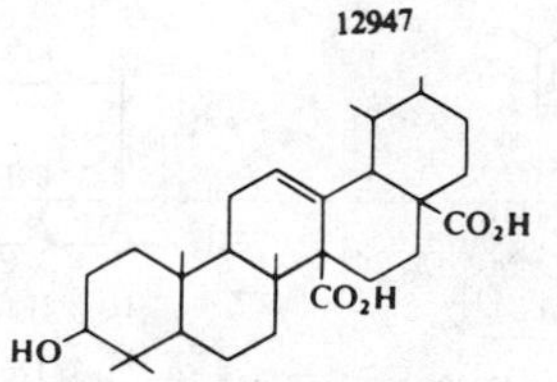
12949

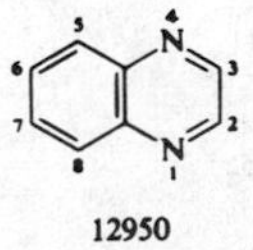
12950

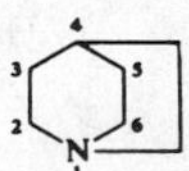
12963

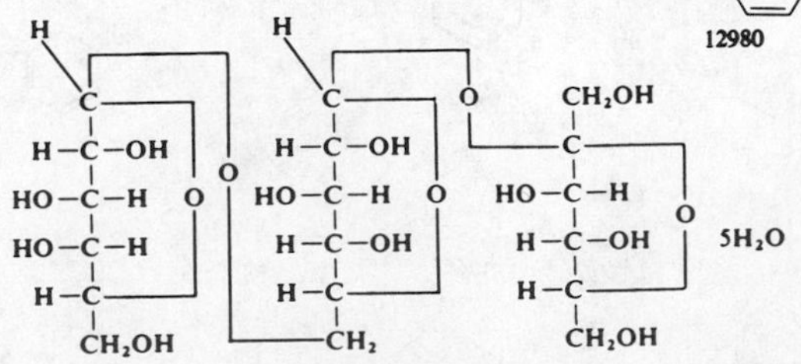
12966

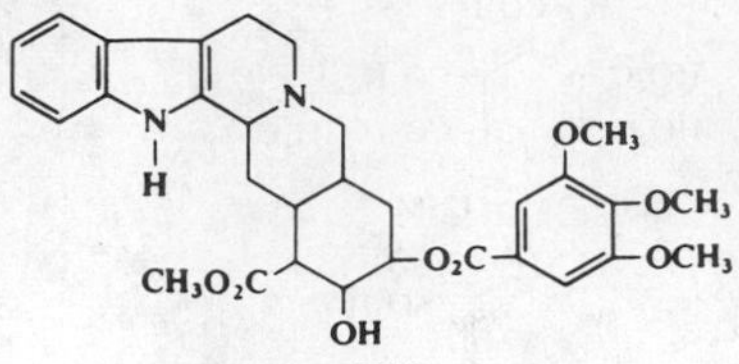
12967

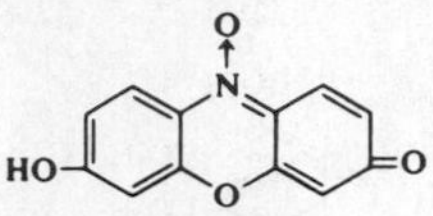
12969

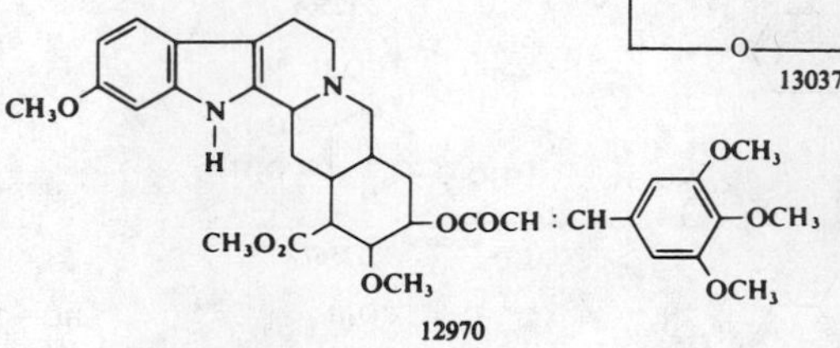
12970

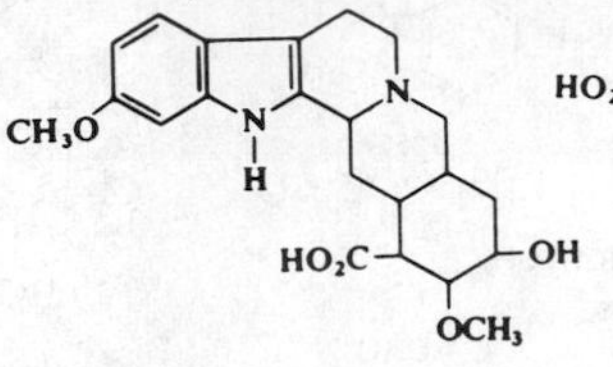
12971

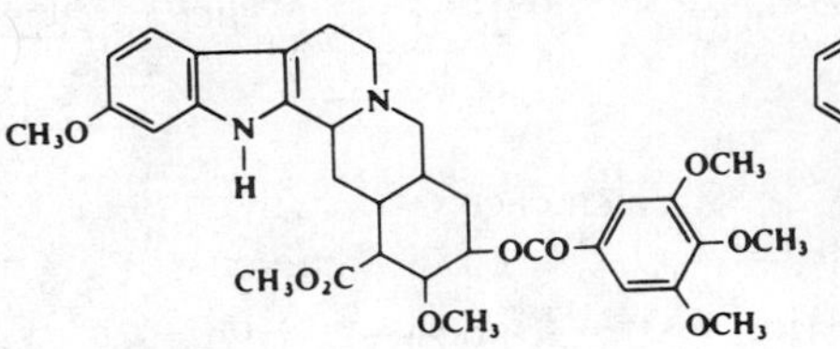
12972

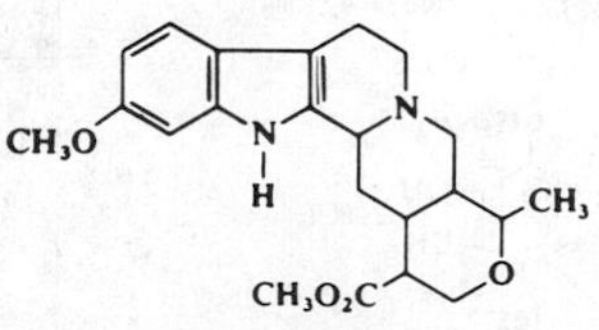
12973

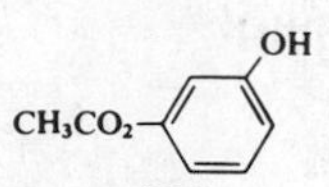
12980

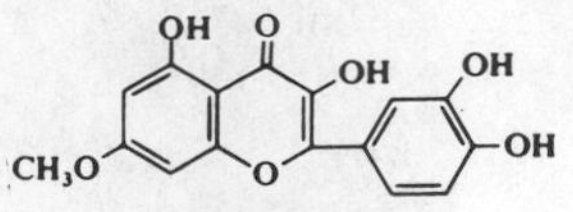
13035

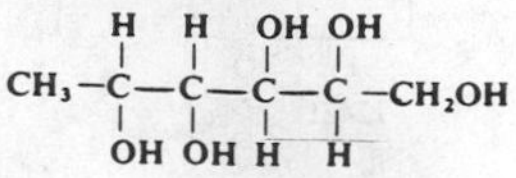
13036

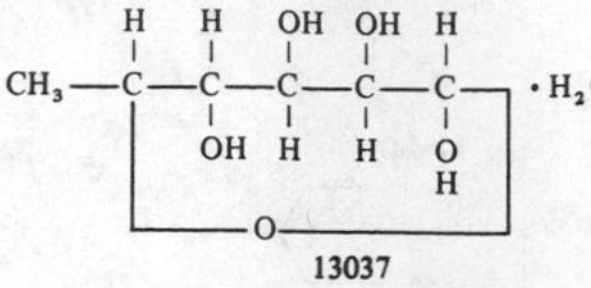
13037

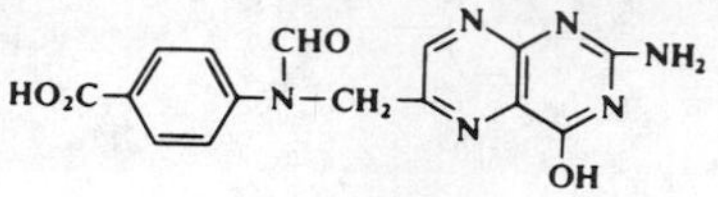
13042

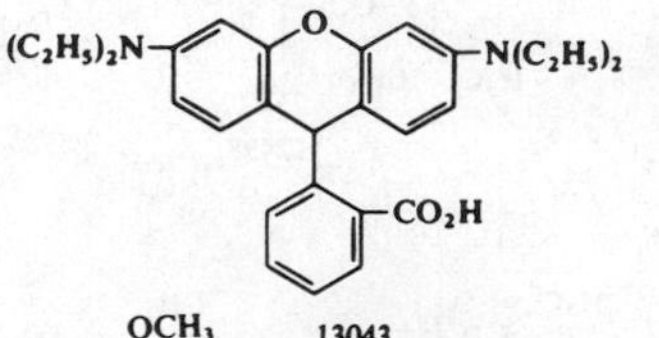
13043

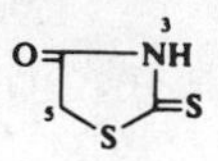
13045

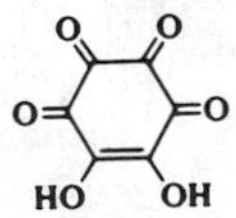
13054

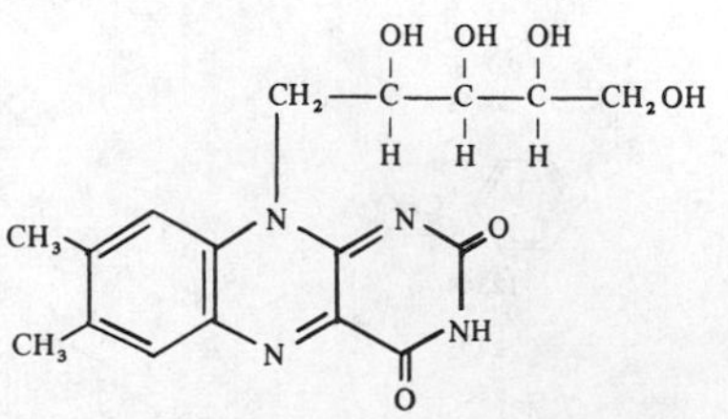
13055

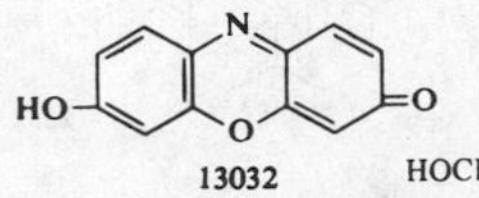
13032

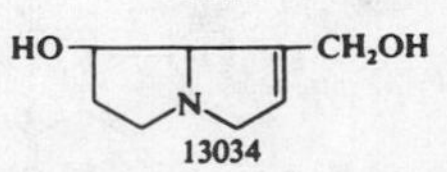
13034

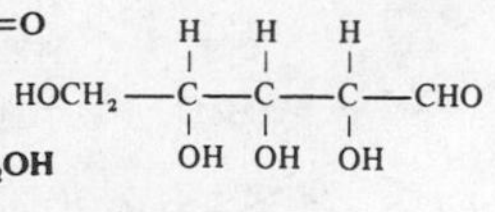
13056

In Numeric as they occur in Organic Compounds Table

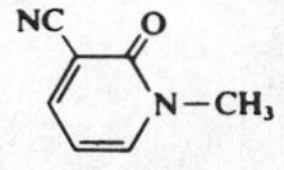
13057

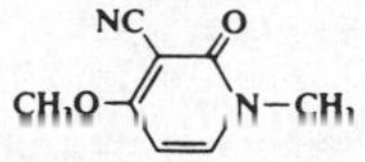
13058

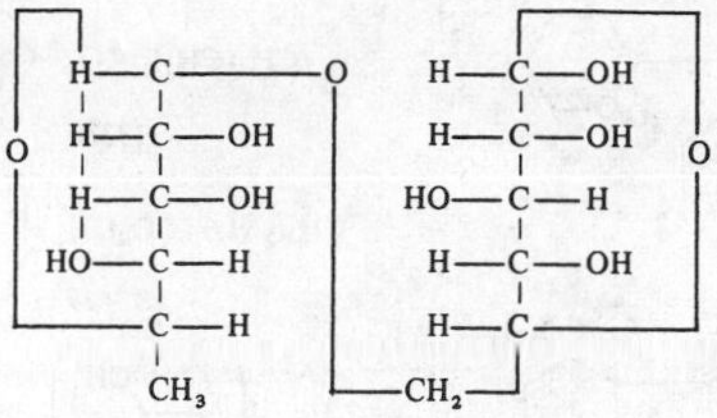
13073

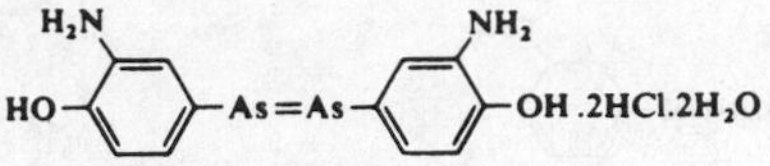
13167

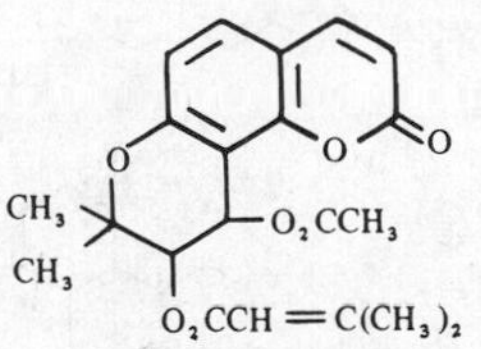
13169

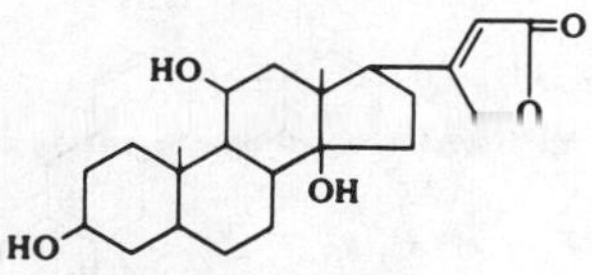
13179

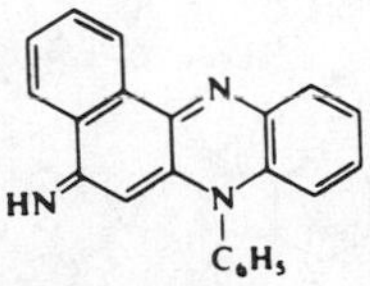
13064

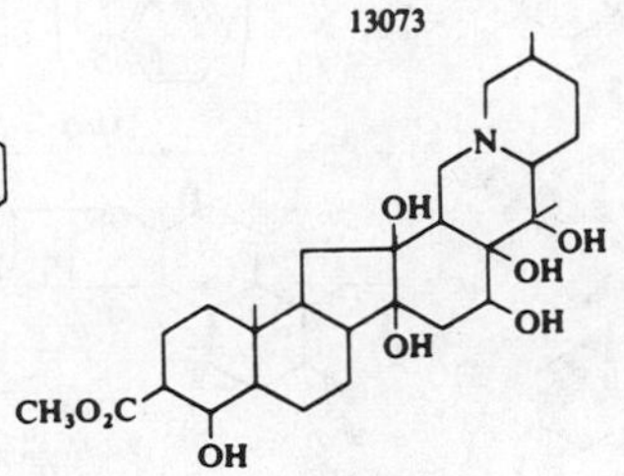
13074

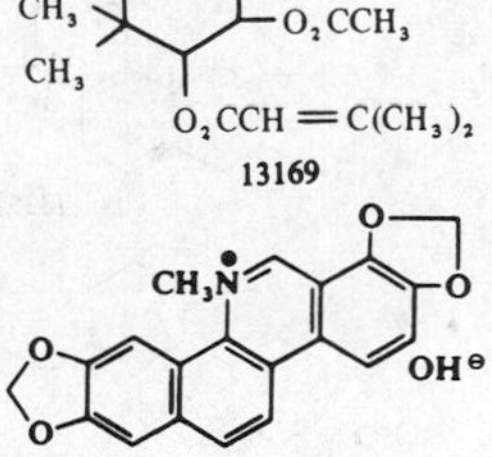
13170

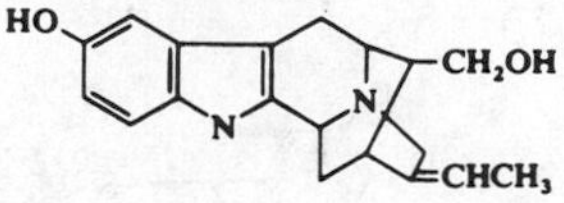
13180

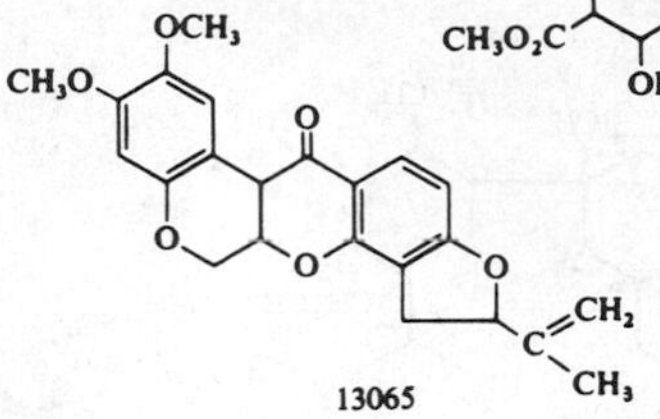
13065

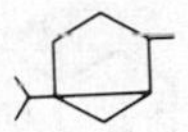
13075

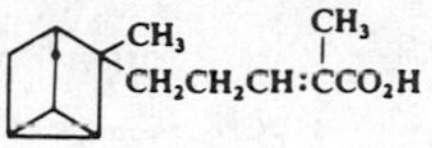
13171

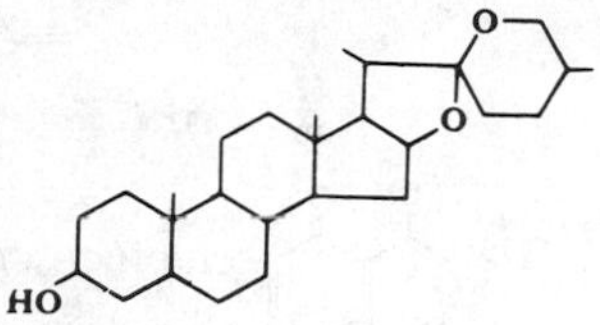
13181

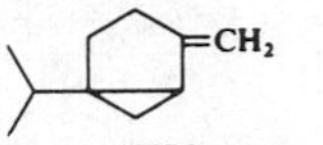
13076

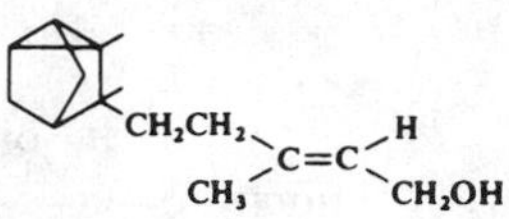
13172

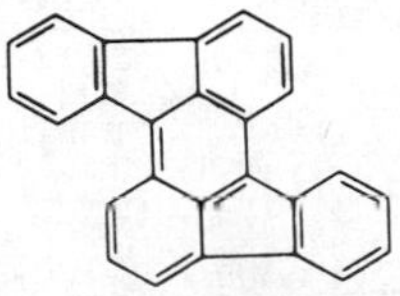
13066

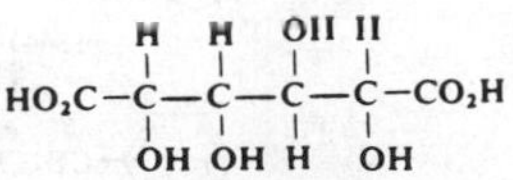
13080

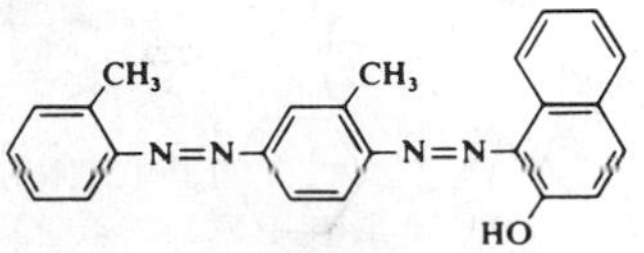
13182

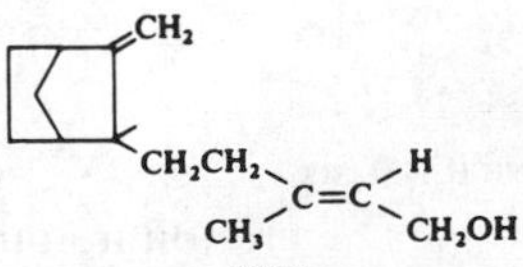
13173

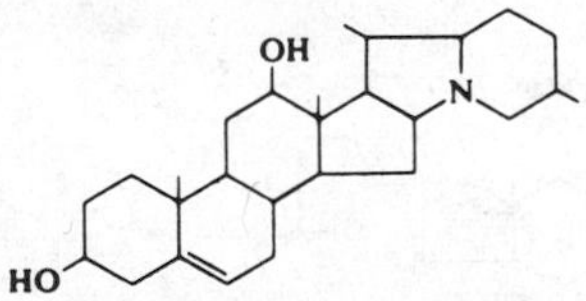
13068

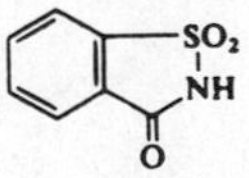
13081

13174

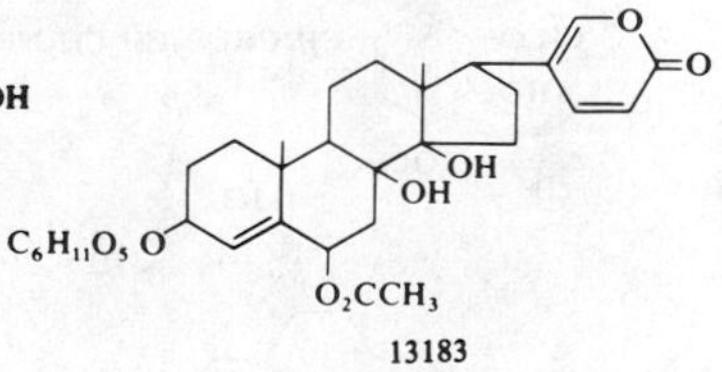
13183

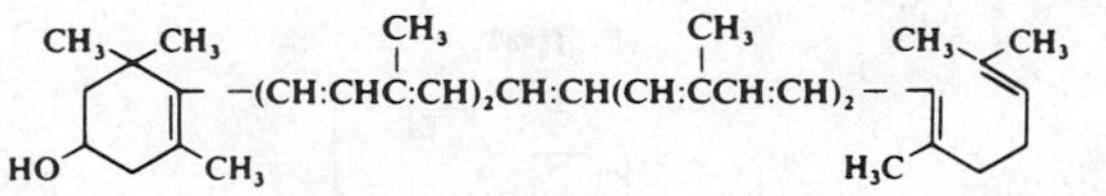
13069

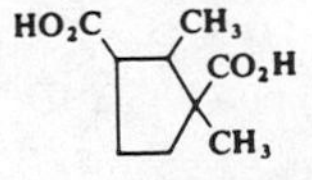
13175

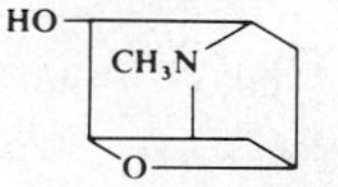
13184

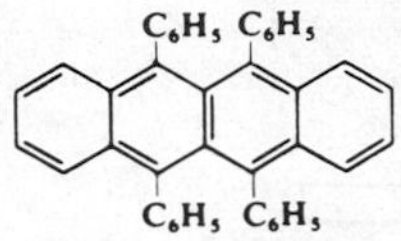
13070

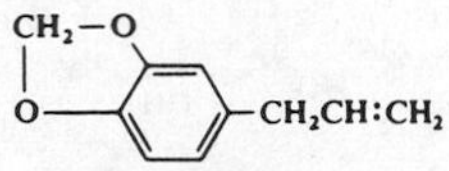
13082

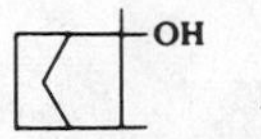
13176

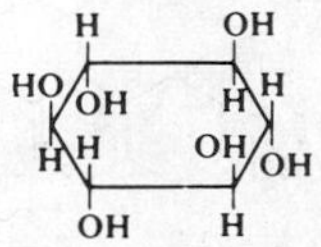
13185

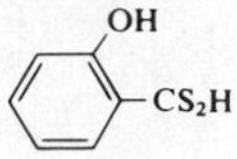
13123

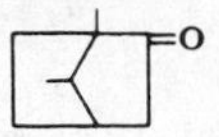
13177

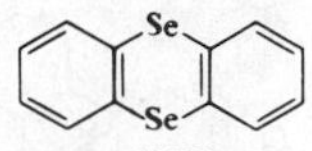
13197

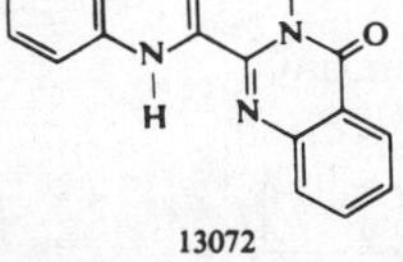
13072

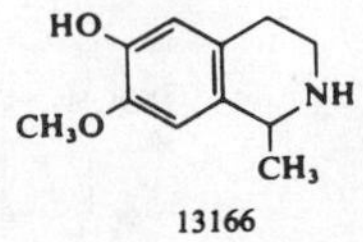
13166

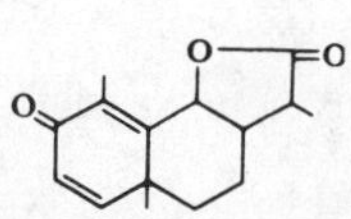
13178

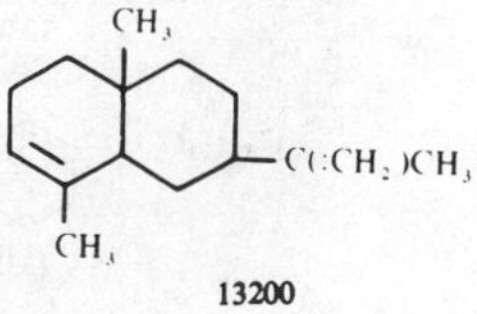
13200

STRUCTURAL FORMULAS OF ORGANIC COMPOUNDS (Continued)

In Numeric as they occur in Organic Compounds Table

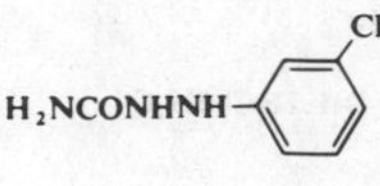

13219

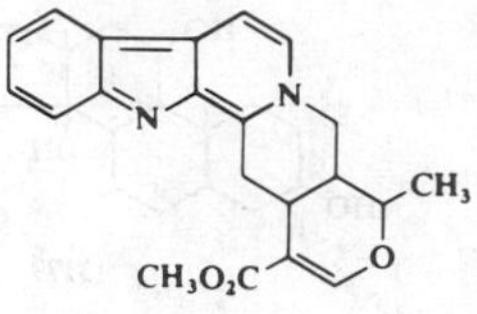

13227

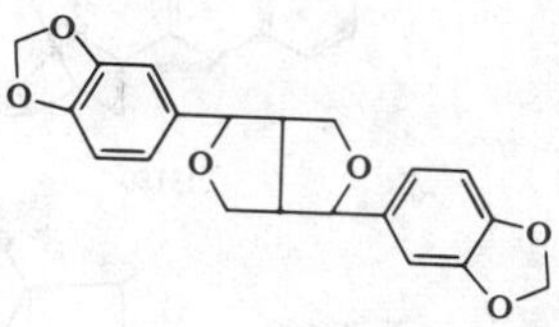

13228

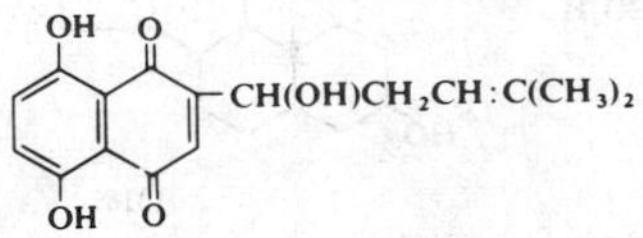

13229

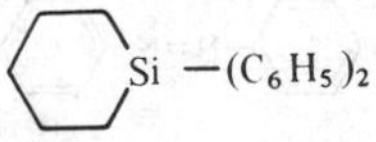

13342

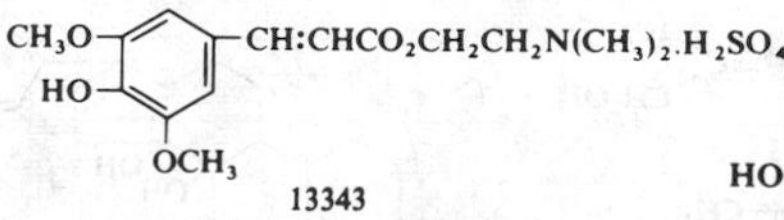

13343

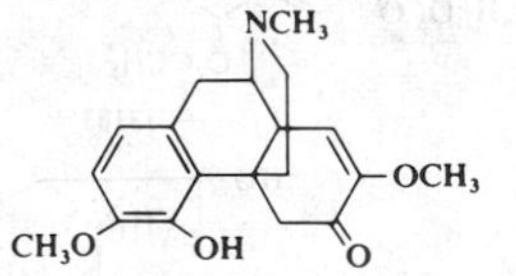

13345

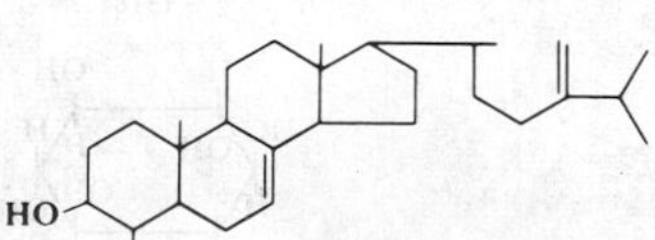

13346

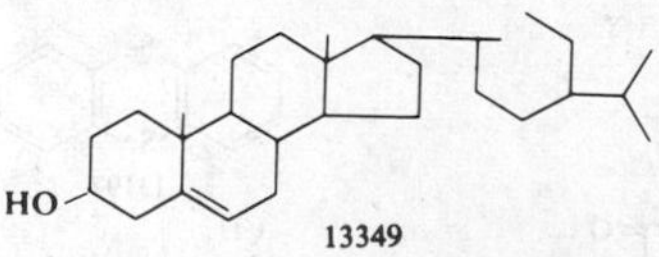

13349

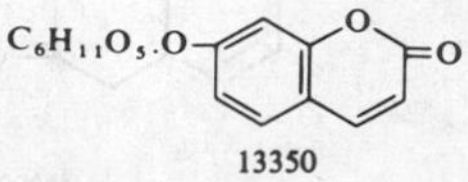

13350

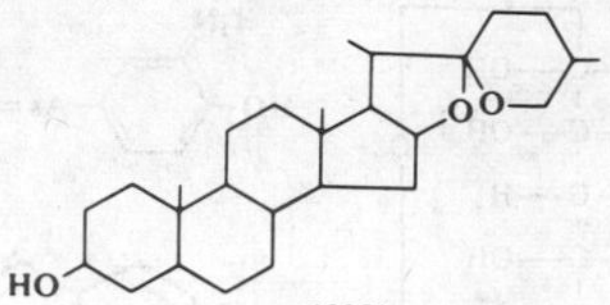

13351

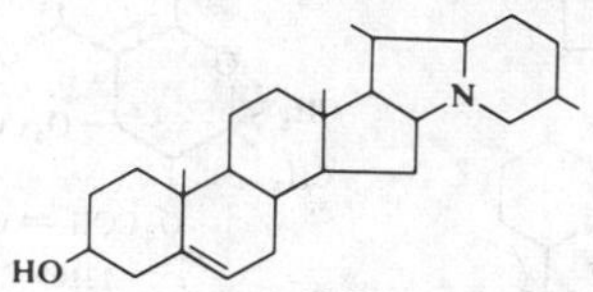

13352

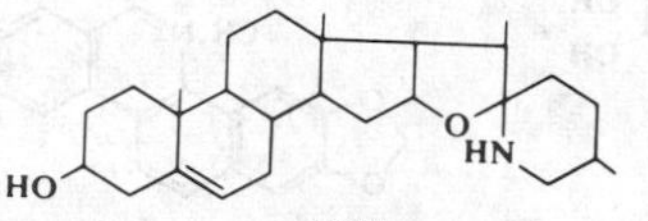

13354

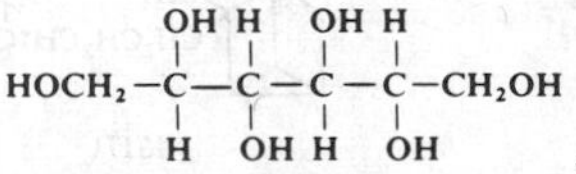

13362

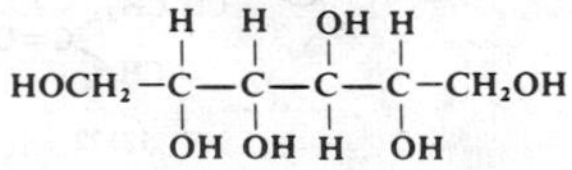

13363

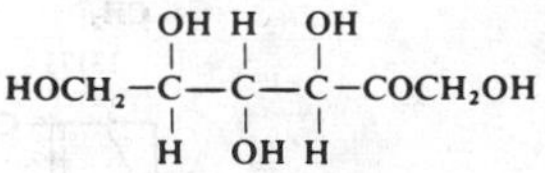

13368

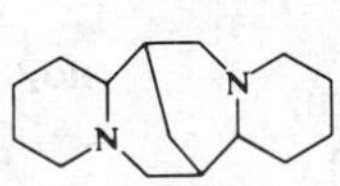

13369

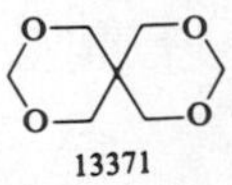

13371

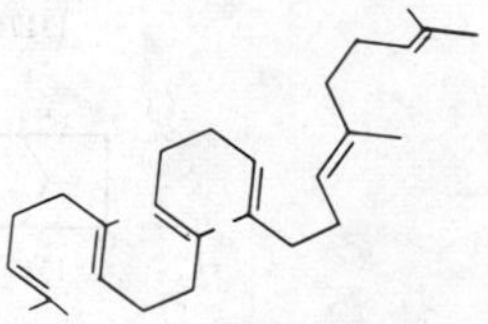

13373

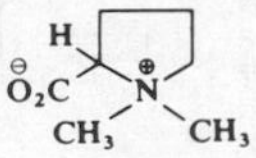

13375

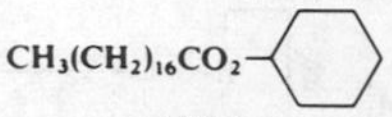

13386

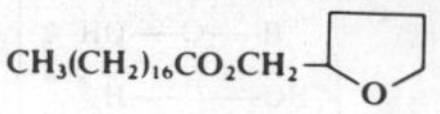

13419

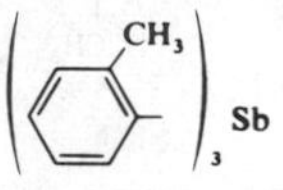

13440

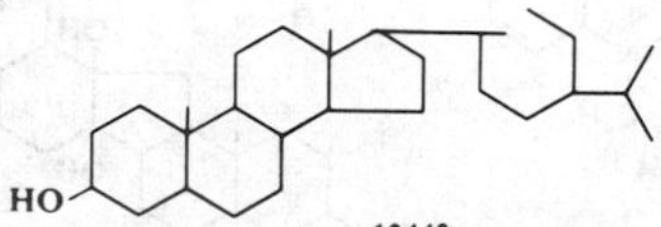

13443

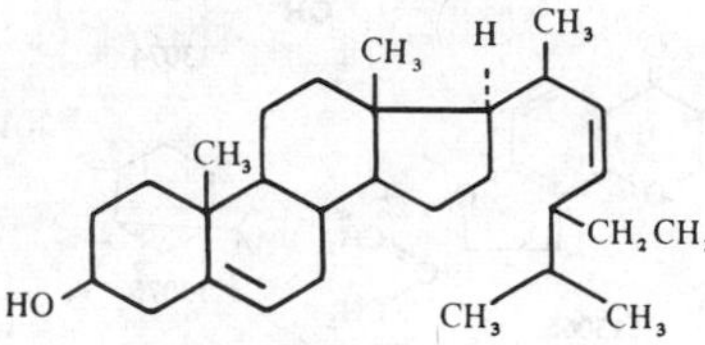

13444

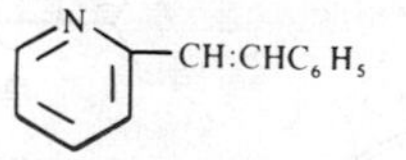

13445

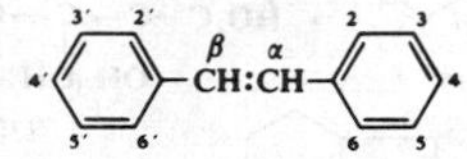

13449

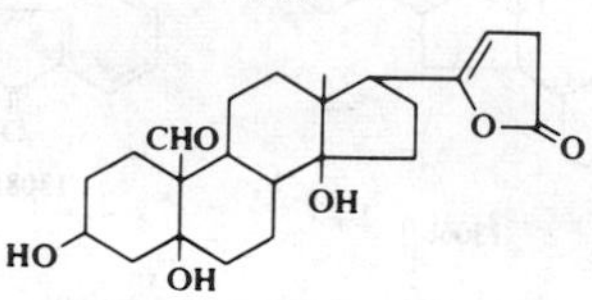

13487

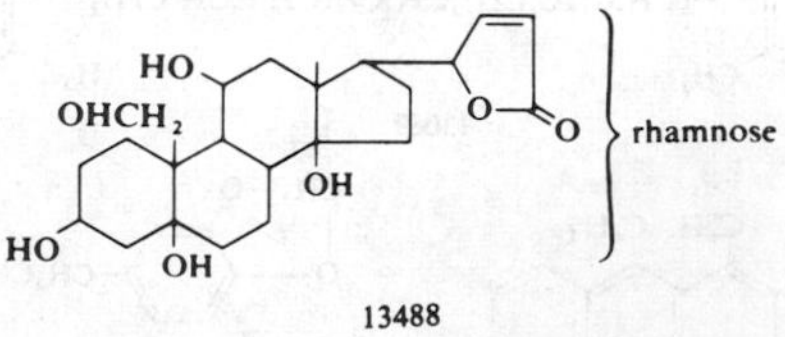

13488

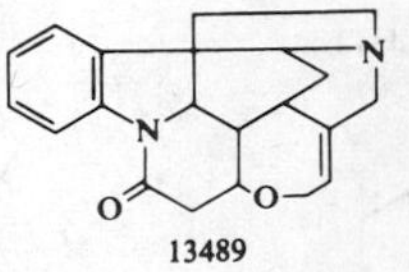

13489

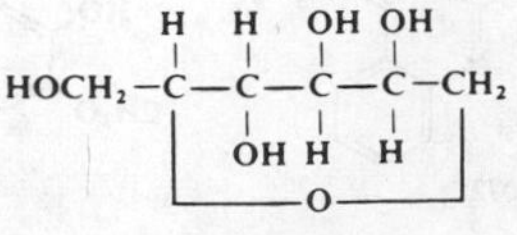

13494

In Numeric as they occur in Organic Compounds Table

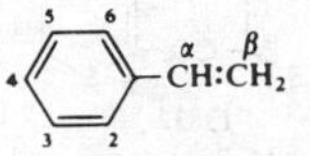
13495

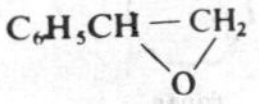
13523

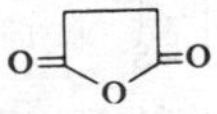
13538

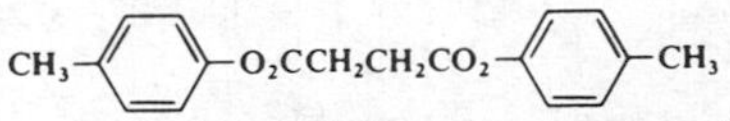
13561

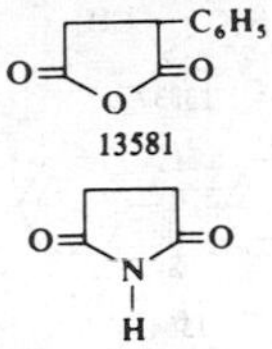
13581

13588

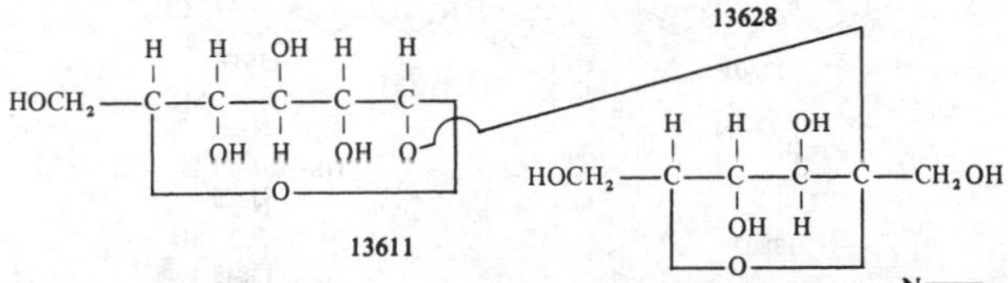
13611

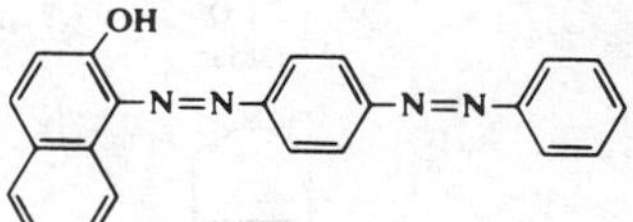
13613

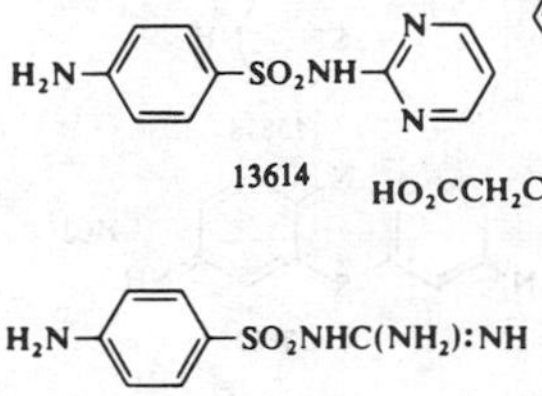
13614

13615

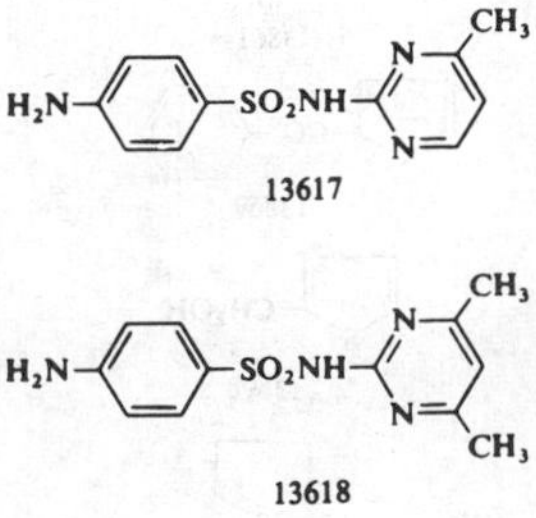
13617

13618

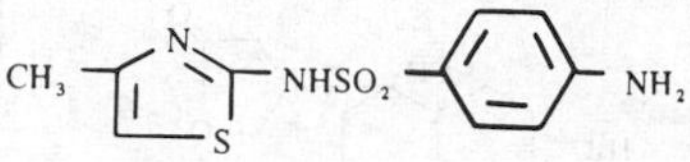
13619

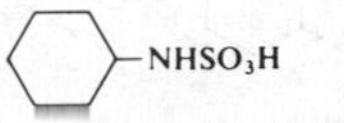
13620

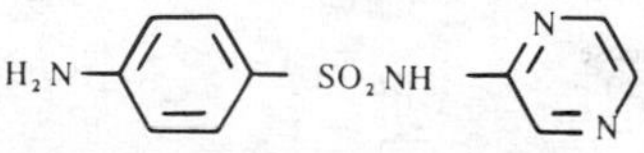
13625

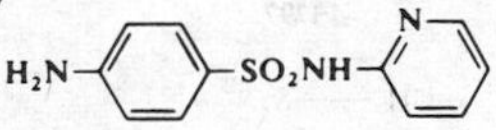
13626

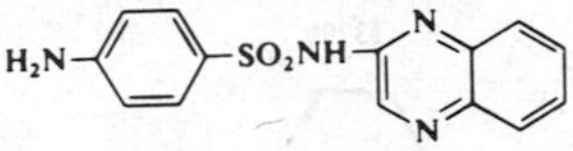
13627

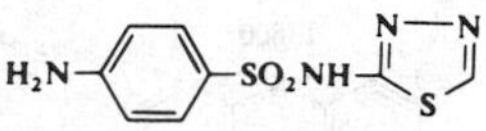
13628

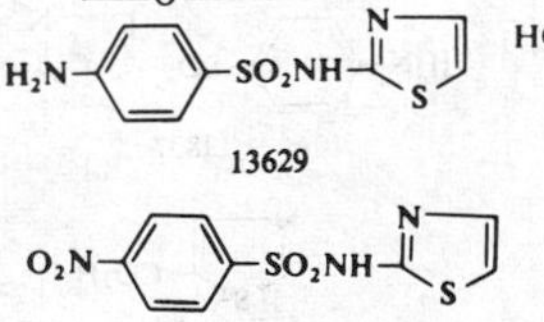
13629

13630

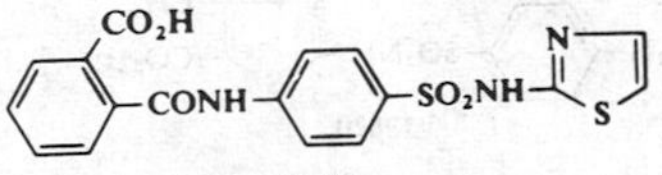
13631

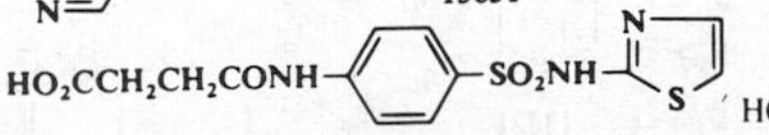
13632

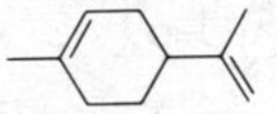
13634

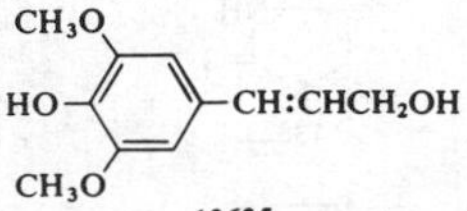
13635

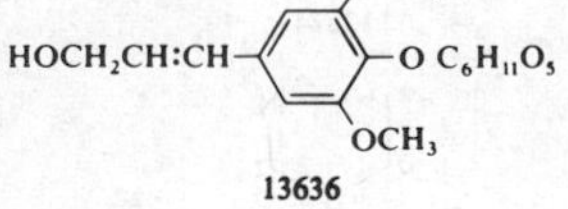
13636

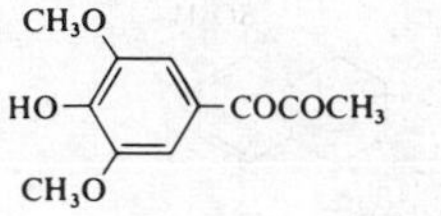
13637

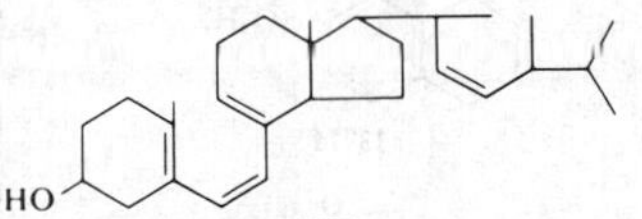
13638

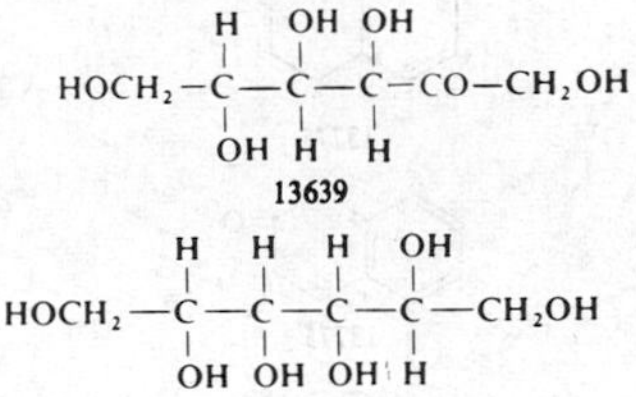
13639

13640

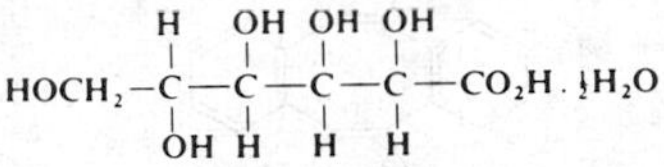
13641

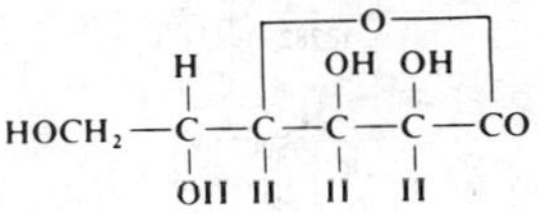
13642

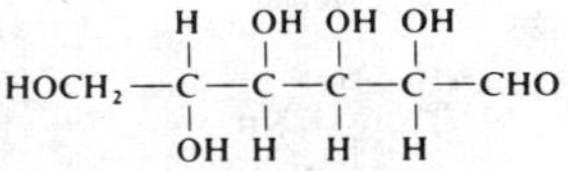
13643

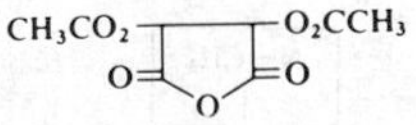
13649

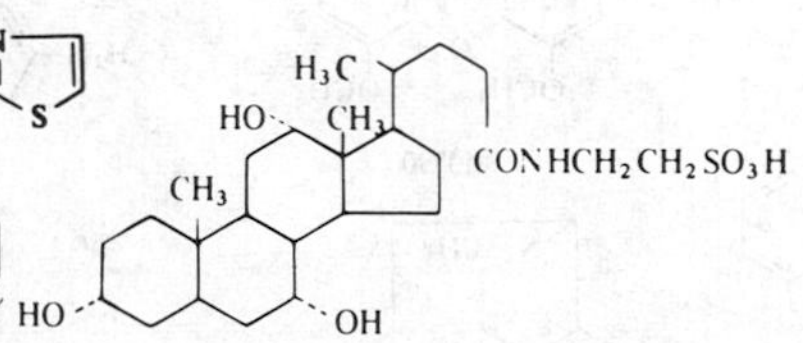
13675

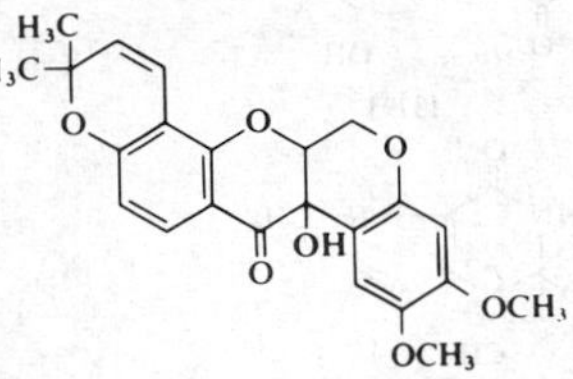
13676

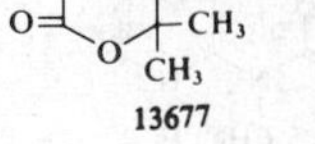
13677

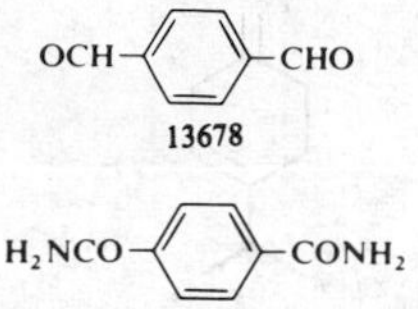
13678

13679

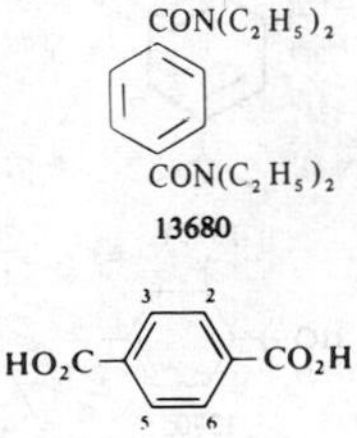
13680

13681

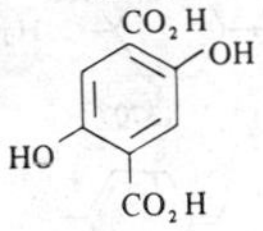
13689

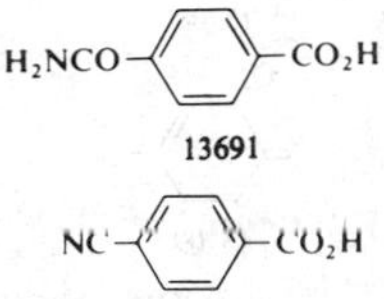
13691

13692

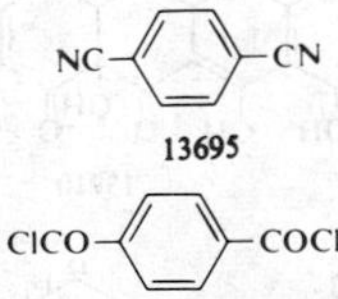
13695

13696

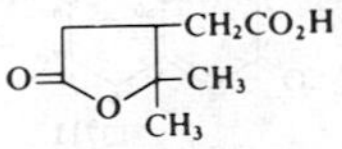
13698

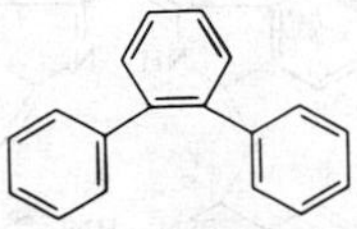
13699

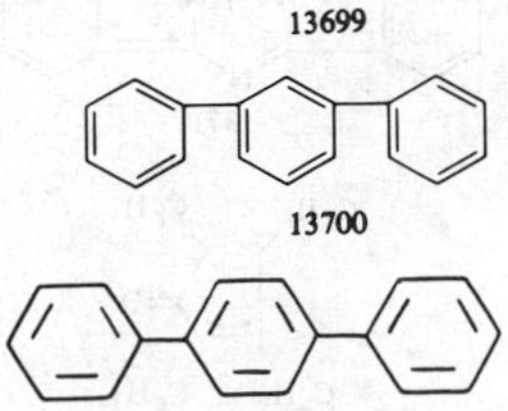
13700

13701

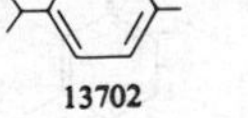
13702

In Numeric as they occur in Organic Compounds Table

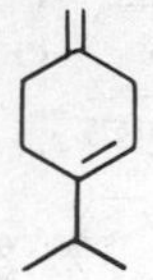
13703

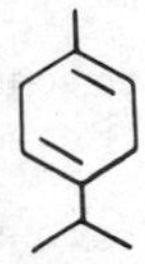
13704

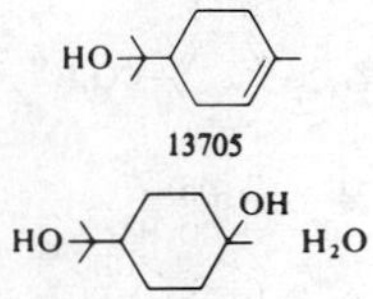
13705

13707

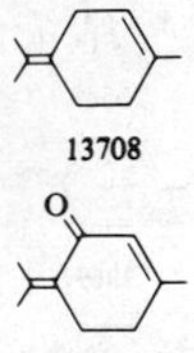
13708

13709

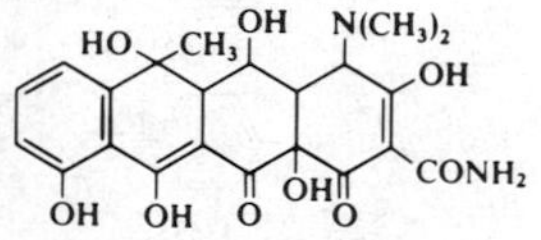
13710

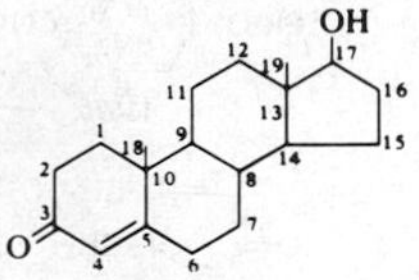
13711

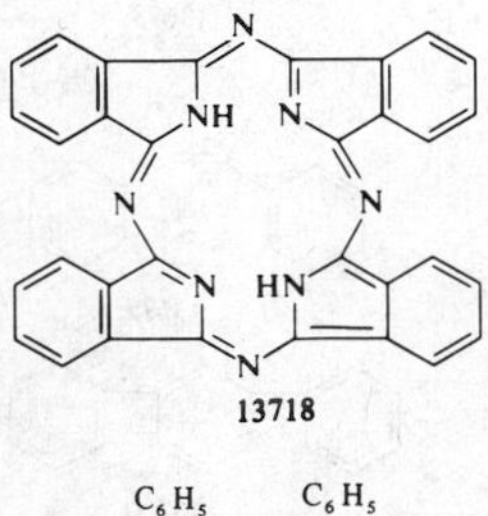
13718

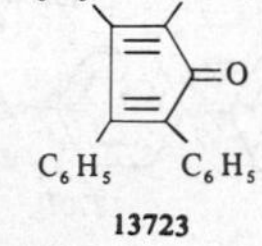
13723

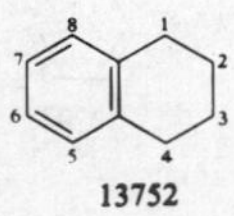
13752

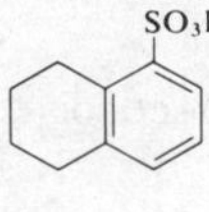
13769

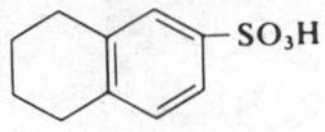
13771

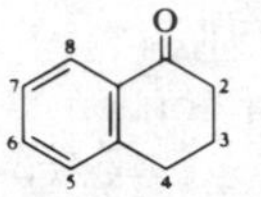
13773

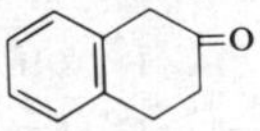
13775

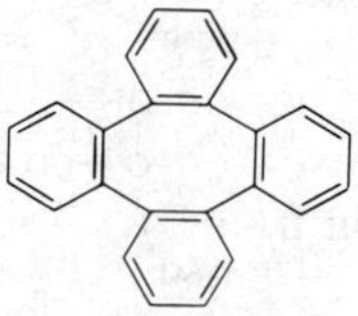
13782

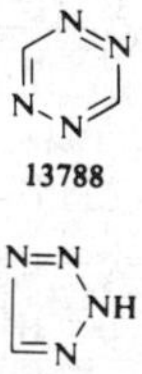
13788

13789

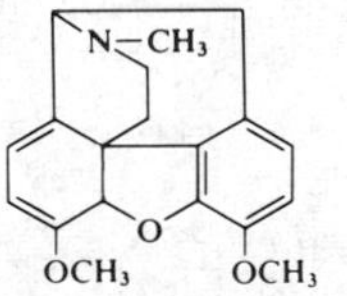
13790

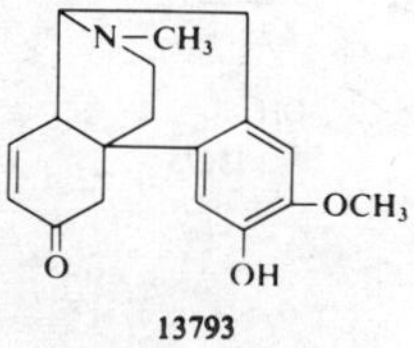
13793

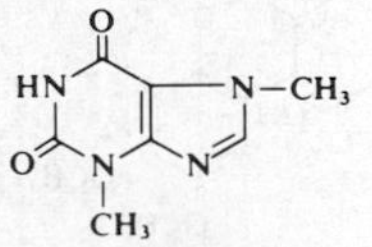
13794

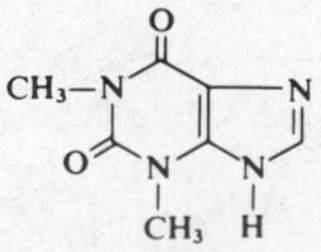
13795

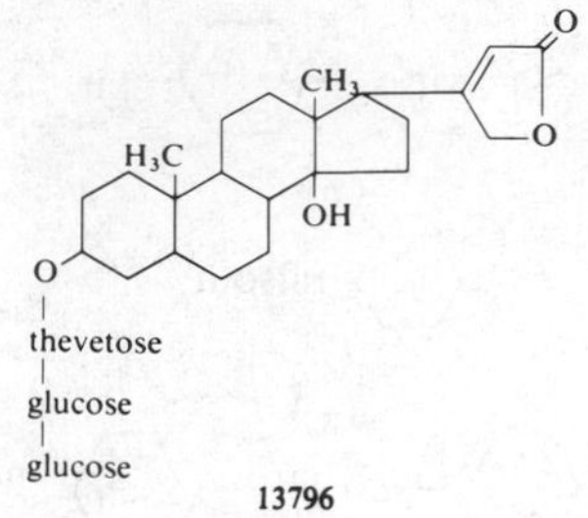
13796

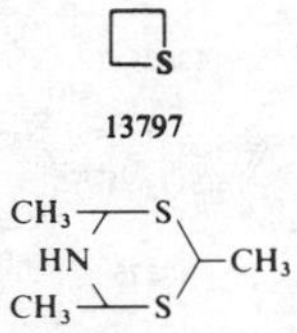
13797

13799

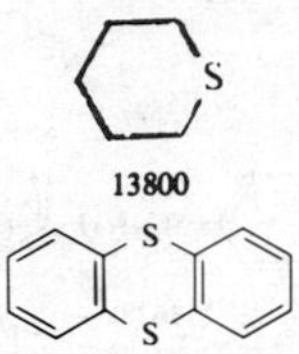
13800

13801

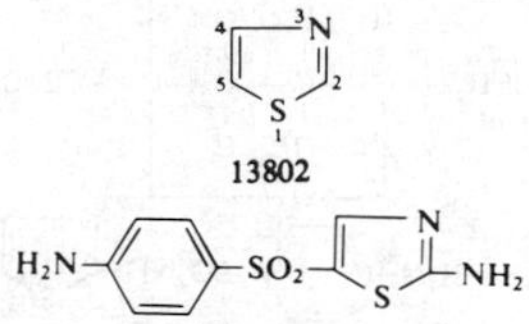
13802

13807

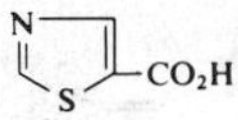
13815

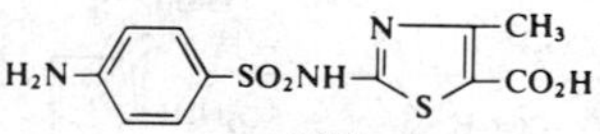
13820

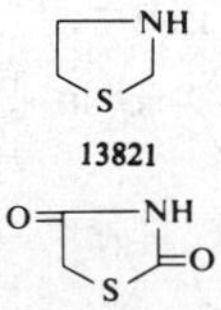
13821

13822

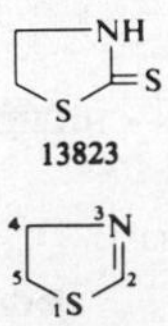
13823

13824

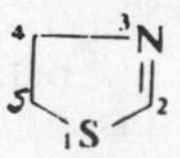
13825

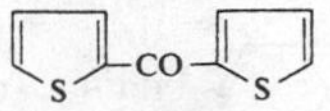
13832

13833

13834

13835

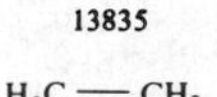
13837

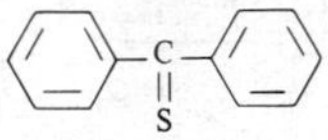
13843

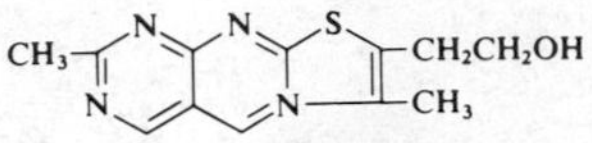
13847

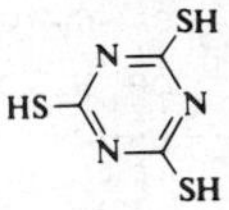
13848

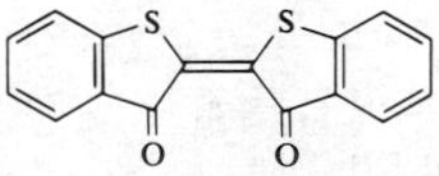
13853

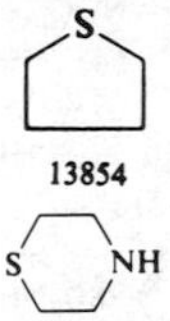
13854

13858

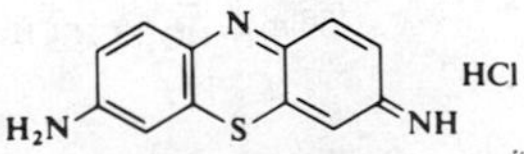
13860

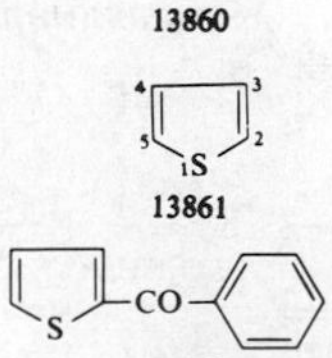
13861

13869

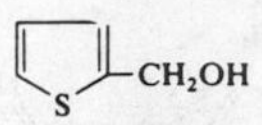
13898

13915

STRUCTURAL FORMULAS OF ORGANIC COMPOUNDS (Continued)

In Numeric as they occur in Organic Compounds Table

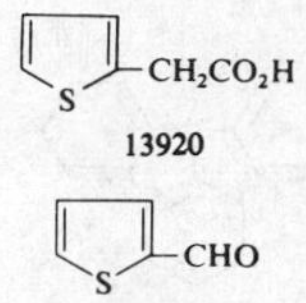
13920
13921

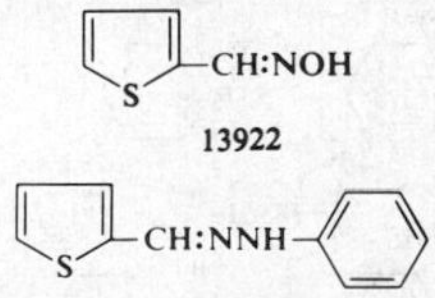
13922
13923

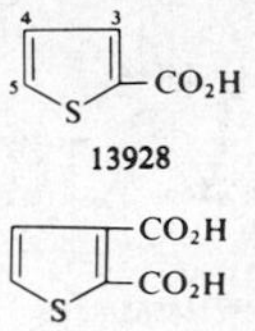
13928
13935

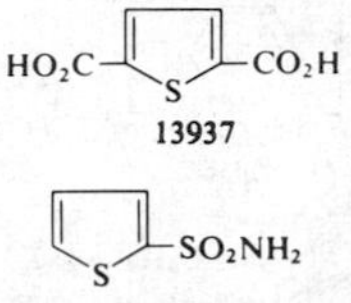
13937
13939

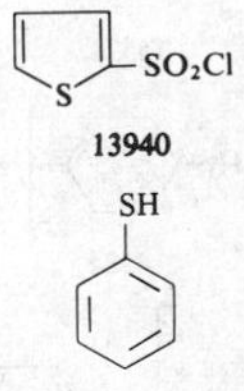
13940
13943

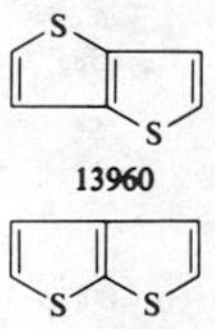
13960
13961

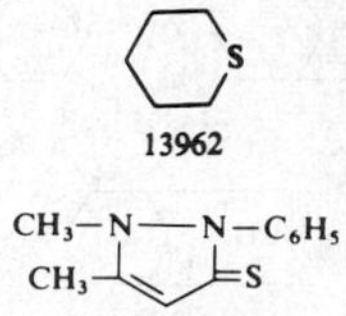
13962
13963

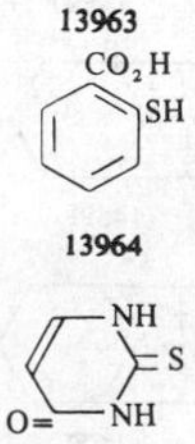
13964
13966

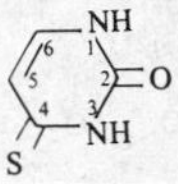
13969

13994

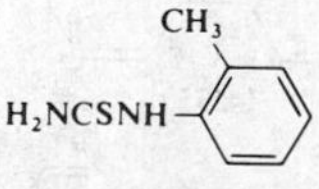
14013

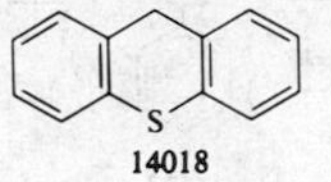
14018

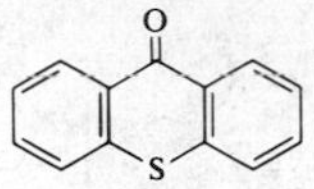
14019

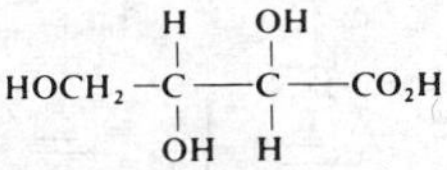
14020

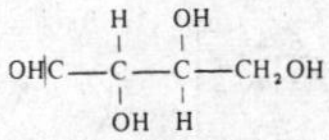
14025

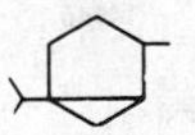
14026

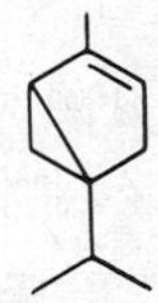
14027

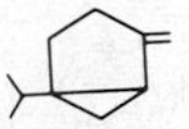
14028

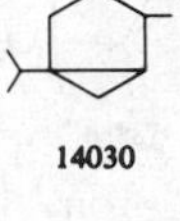
14030

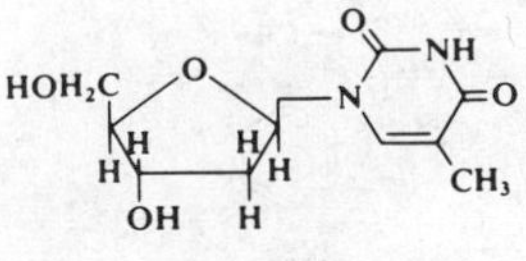
14031

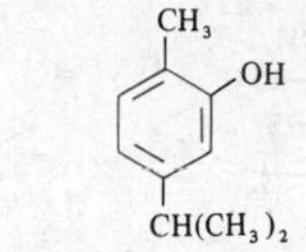
14032

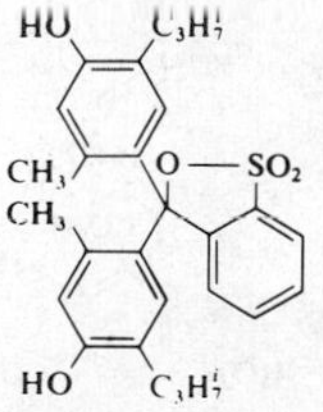
14039

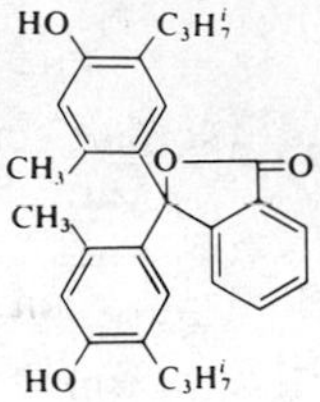
14040

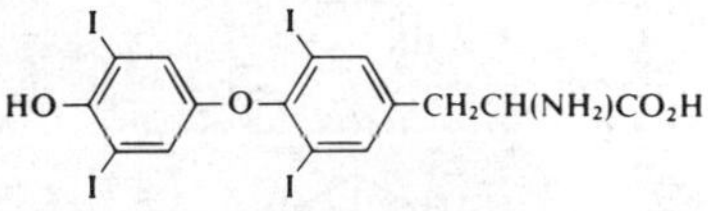
14041

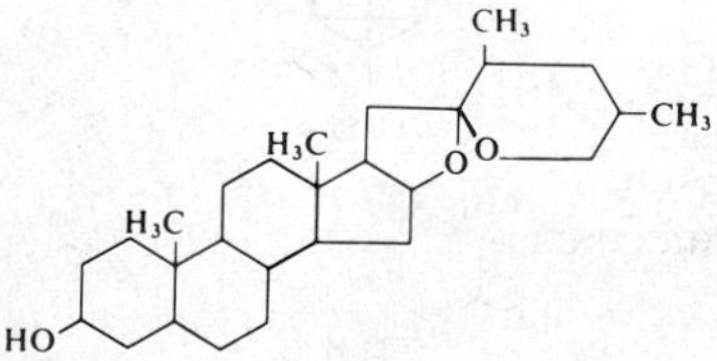
14046

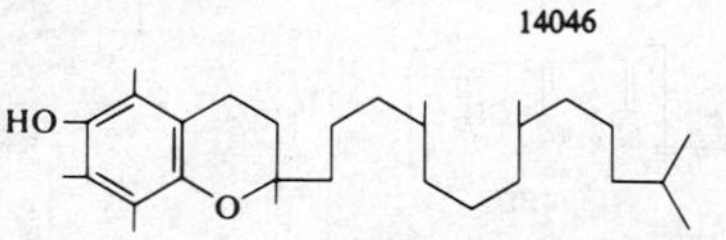
14047

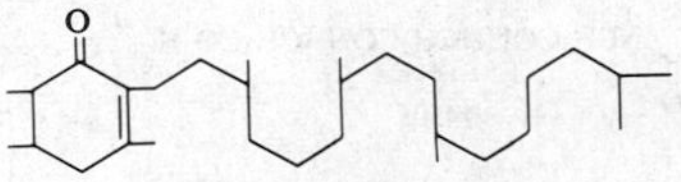
14049

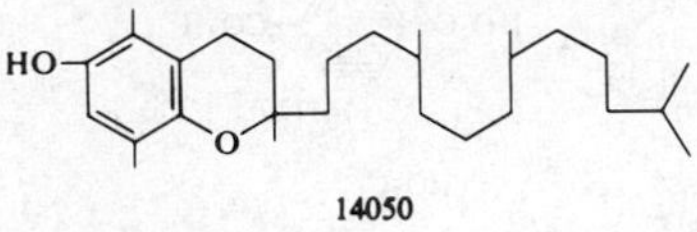
14050

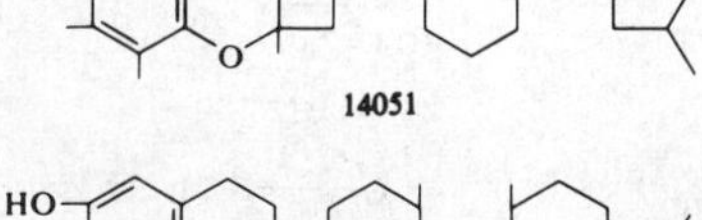
14051

14052

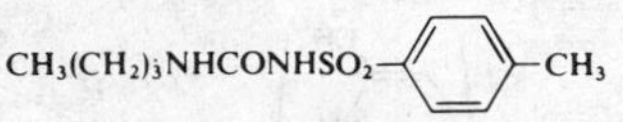
14053

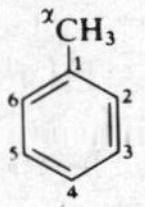
14060

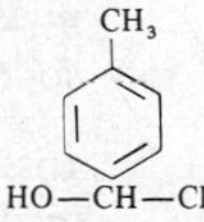
14131

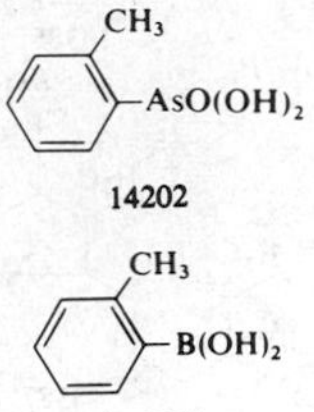
14202
14205

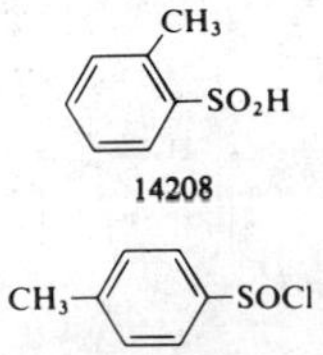
14208
14210

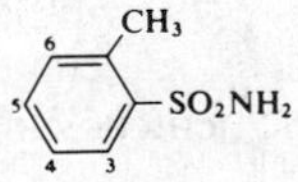
14211

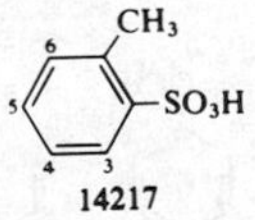
14217

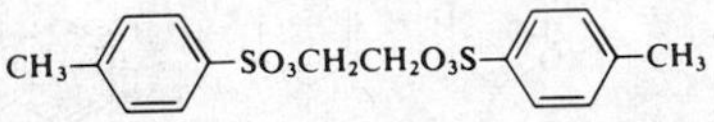
14253

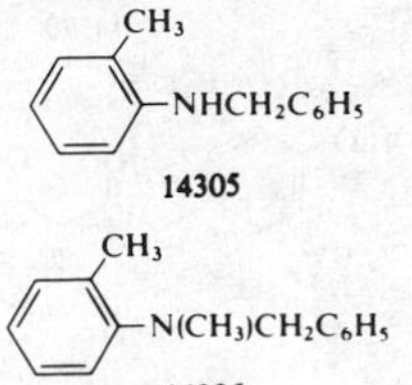
14305
14306

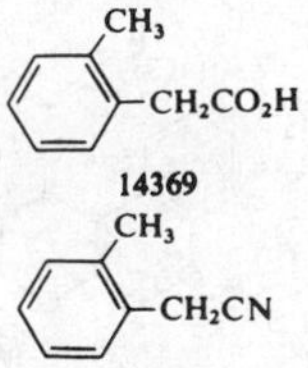
14369
14370

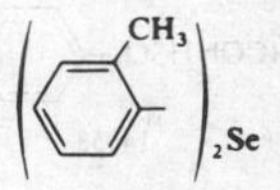
14381

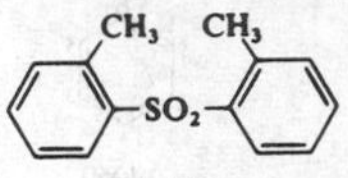
14383

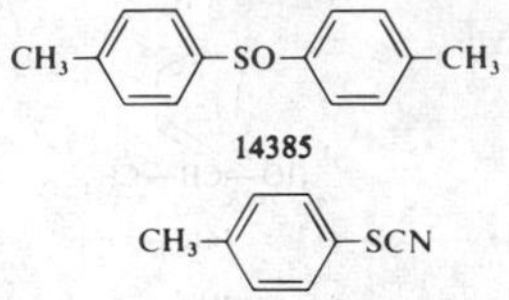
14385

14386

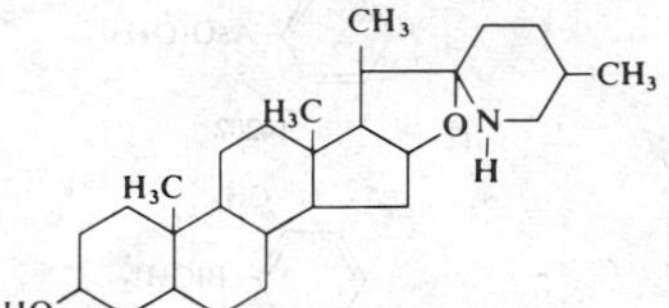
14387

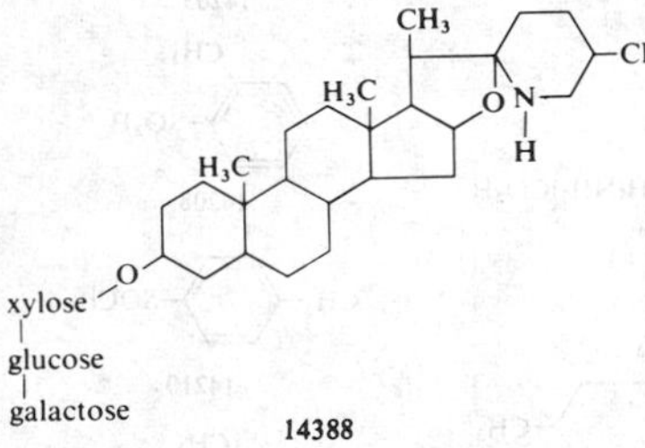
14388

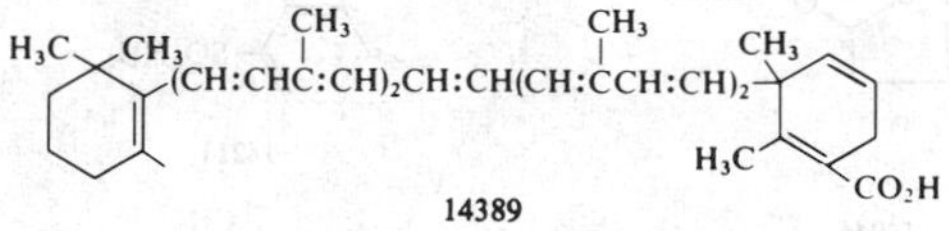
14389

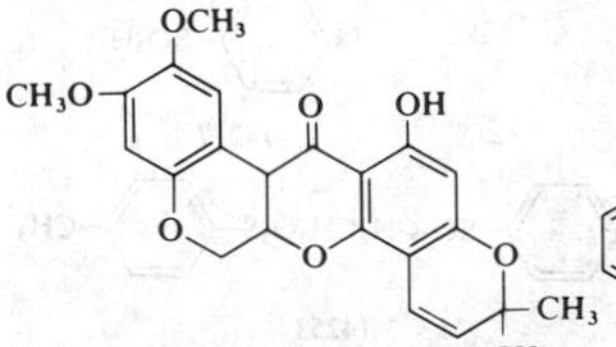
14390

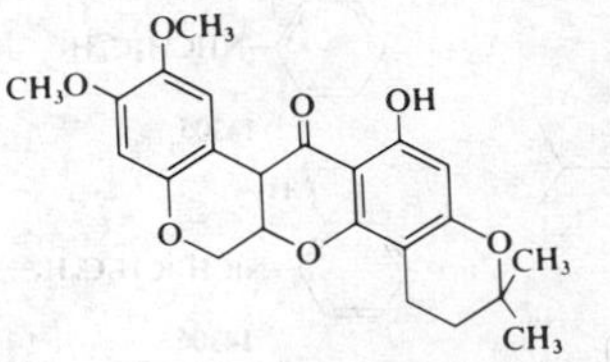
14392

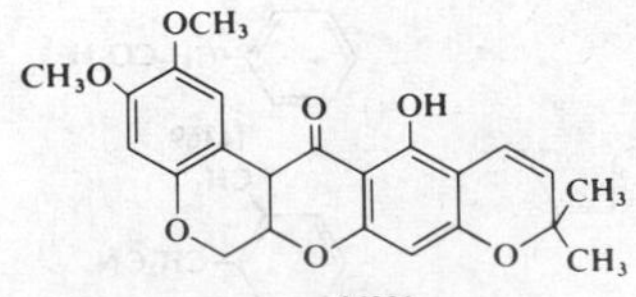
14393

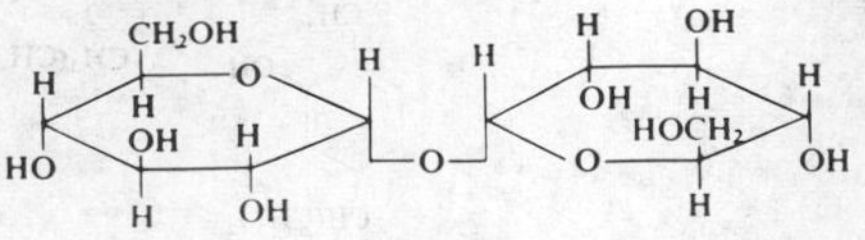
14395

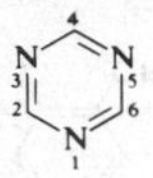
14402

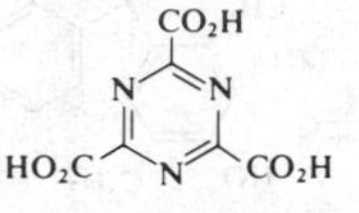
14409

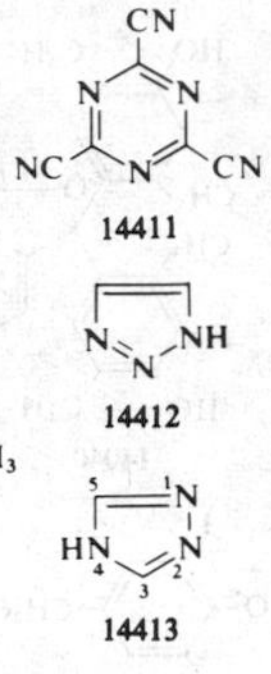
14411

14412

14413

14456

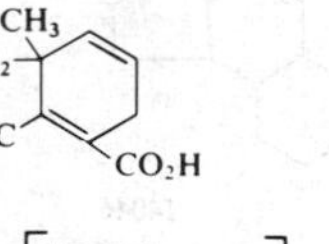
14503

14505

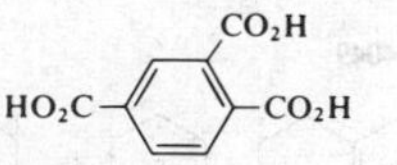
14511

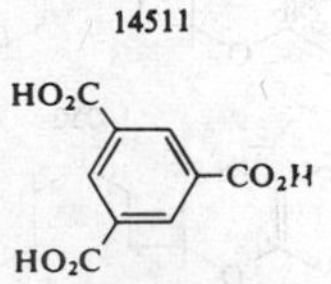
14512

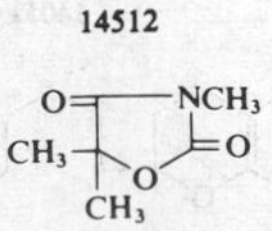
14513

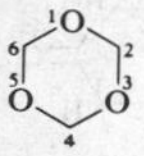
14529

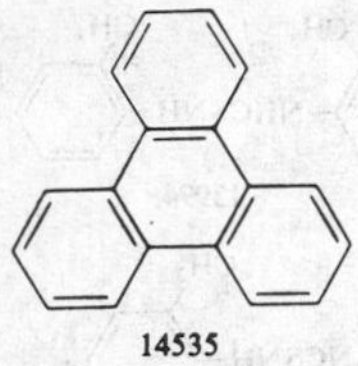
14535

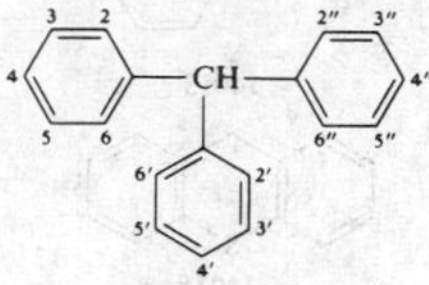
14537

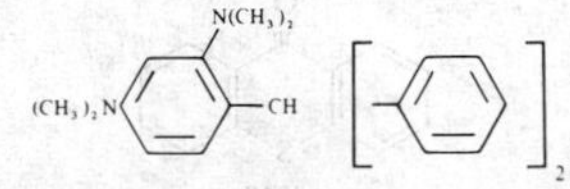
14543

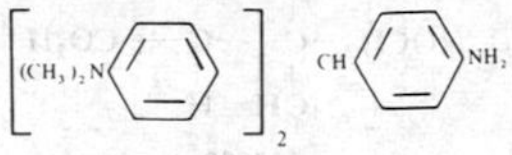
14545

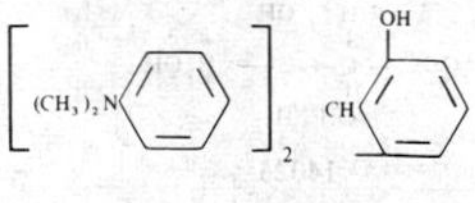
14546

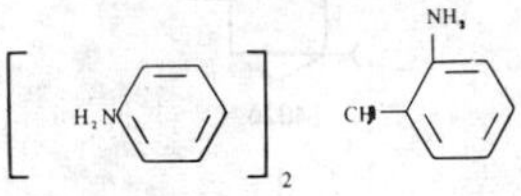
14550

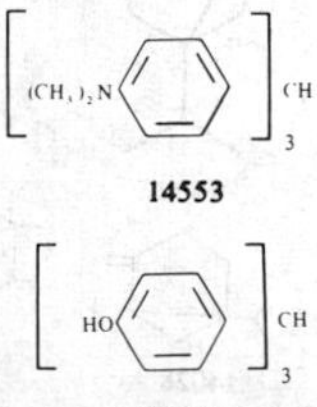
14553

14554

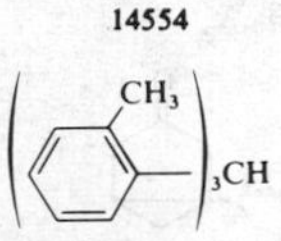
14555

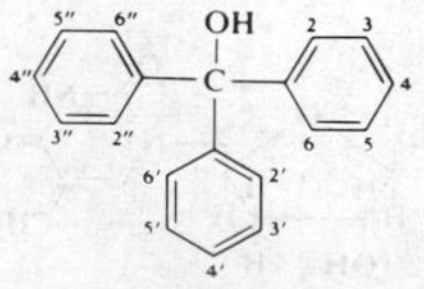
14557

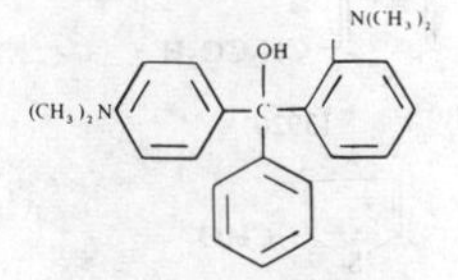
14560

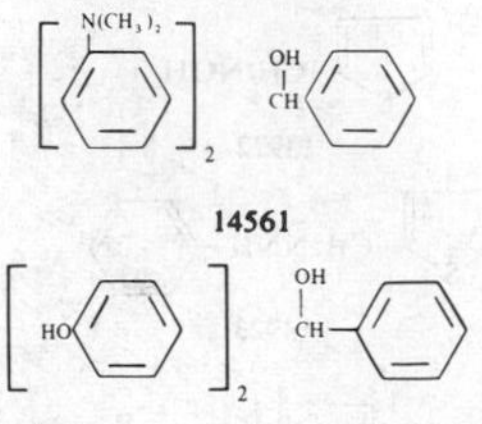
14561

14564

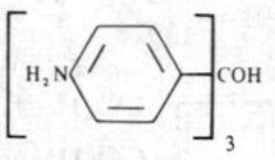
14565

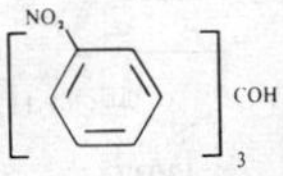
14566

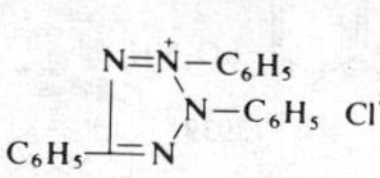
14571

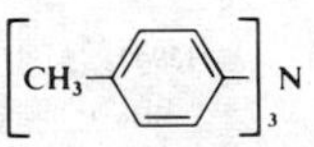
14584

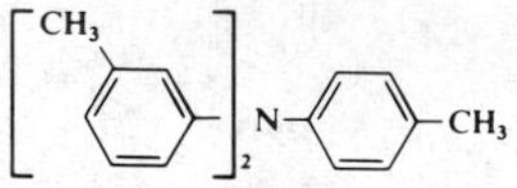
14585

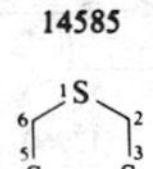
14587

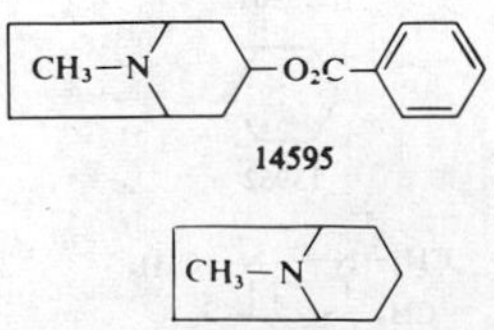
14595

14597

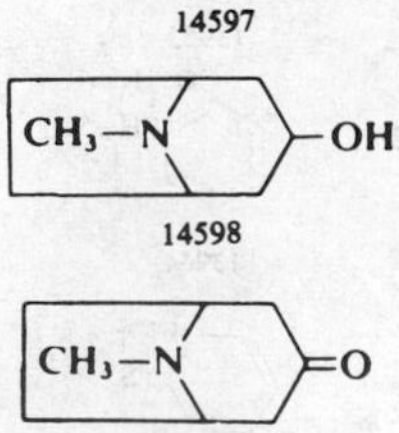
14598

14599

STRUCTURAL FORMULAS OF ORGANIC COMPOUNDS (Continued)

In Numeric as they occur in Organic Compounds Table

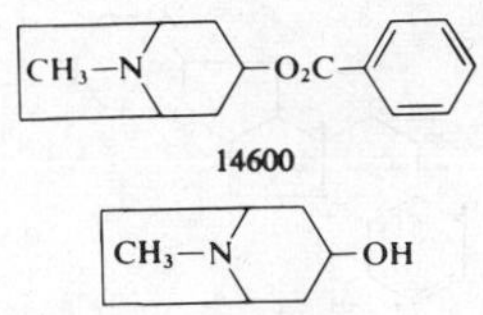

14600

14602

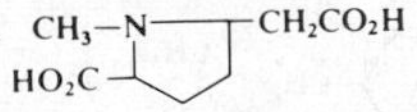

14603

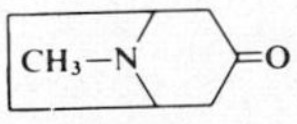

14606

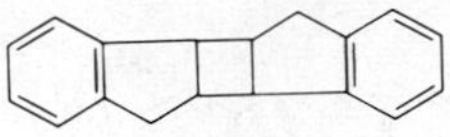

14607

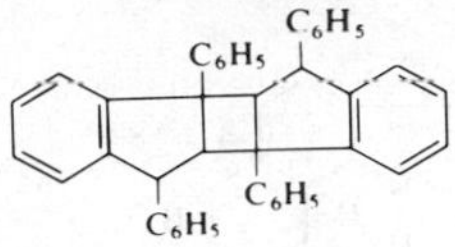

14608

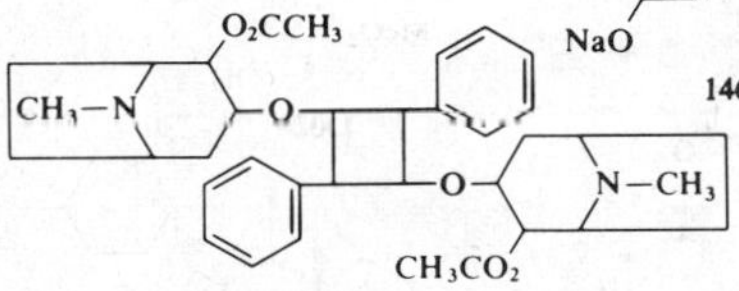

14609

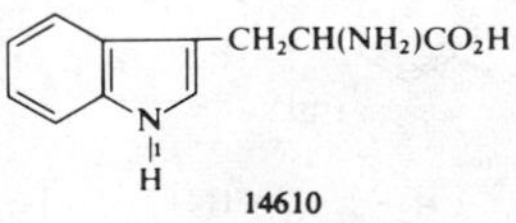

14610

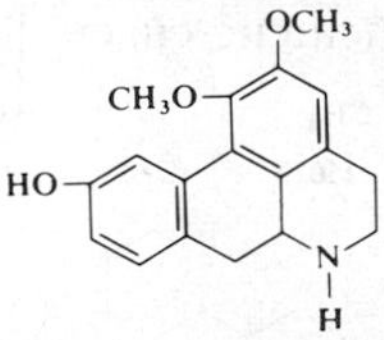

14617

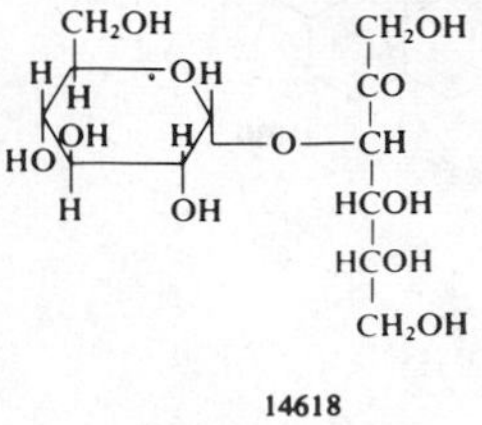

14618

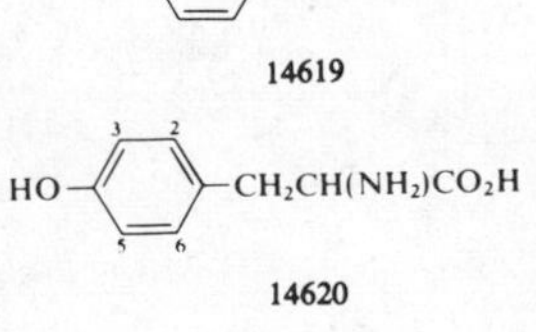

14619

14620

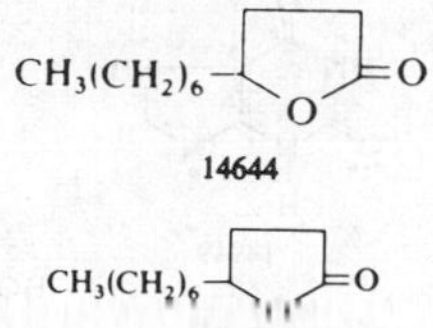

14644

14645

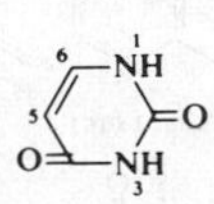

14686

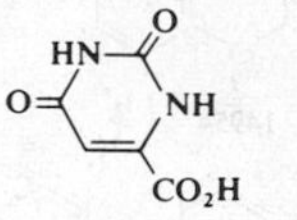

14697

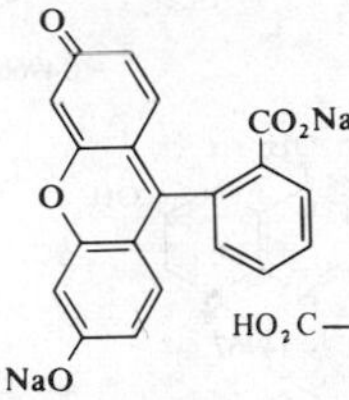

14699

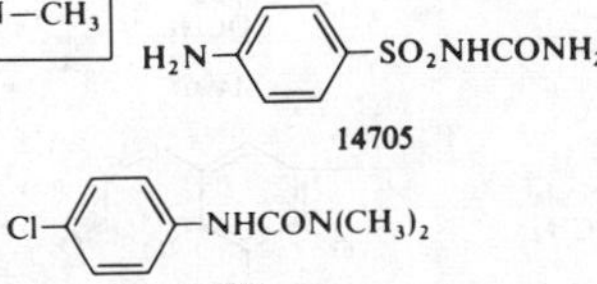

14705

14719

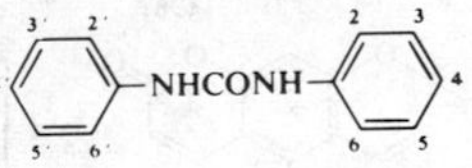

14737

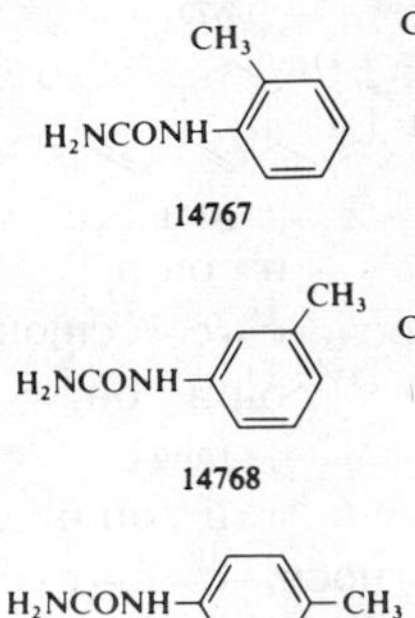

14767

14768

14769

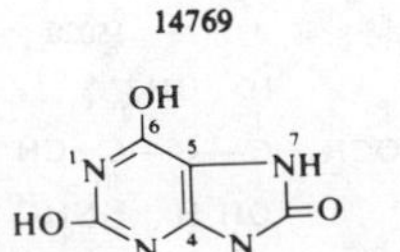

14773

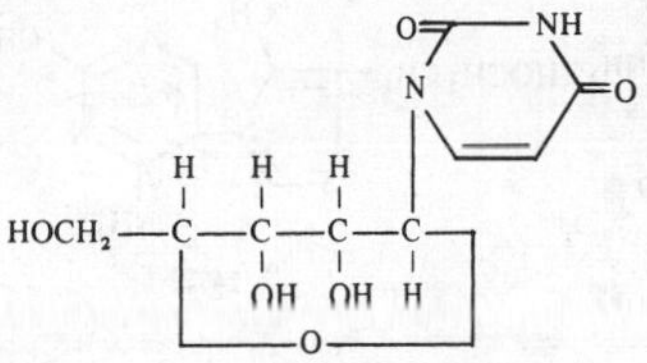

14779

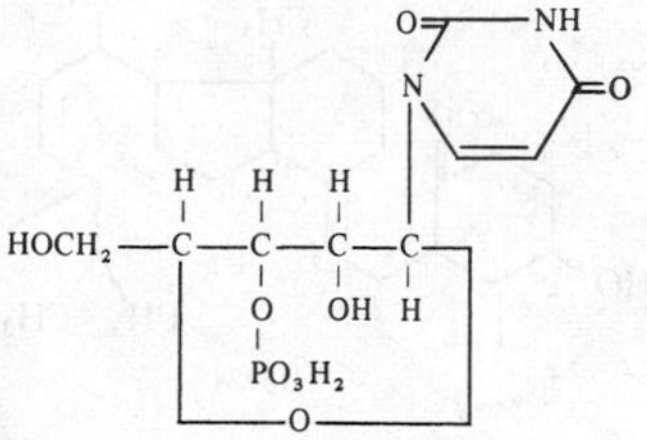

14780

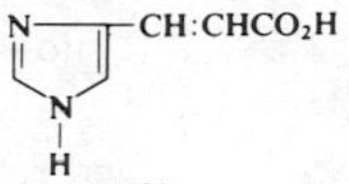

14781

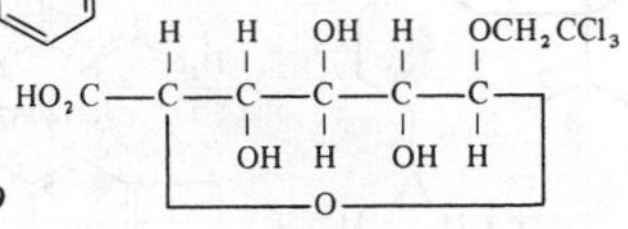

14782

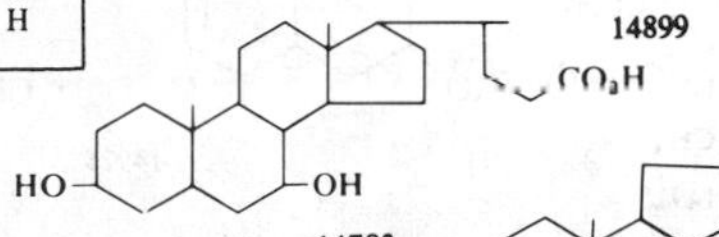

14783

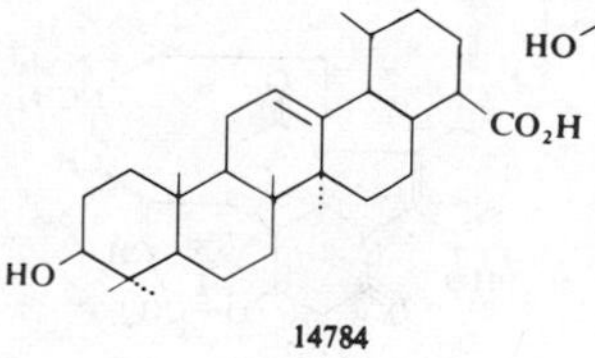

14784

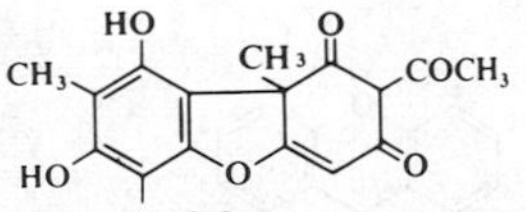

14785

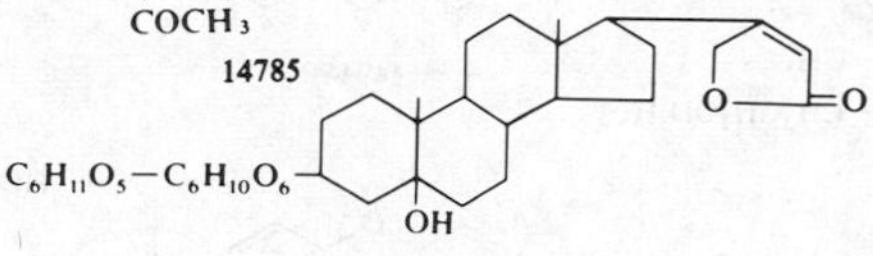

14788

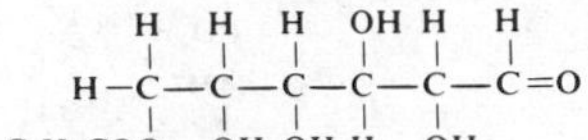

14789

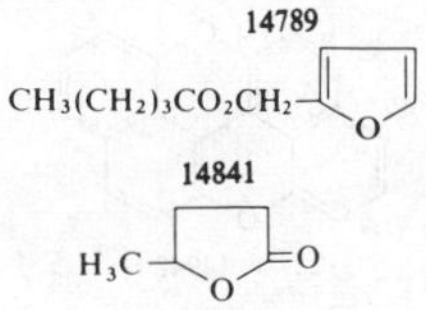

14841

14845

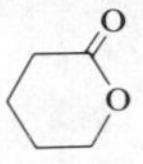

14848

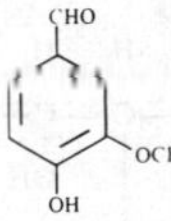

14890

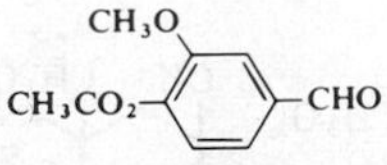

14891

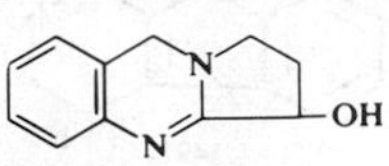

14897

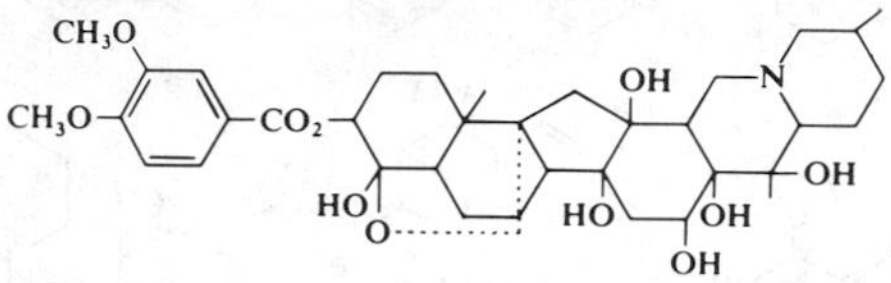

14899

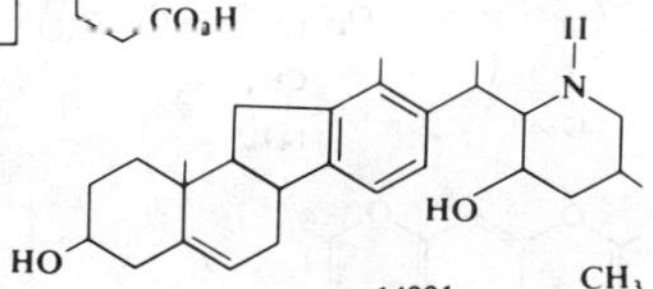

14901

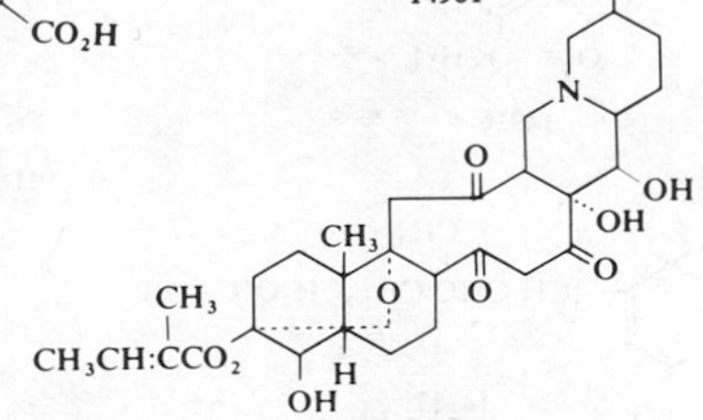

14903

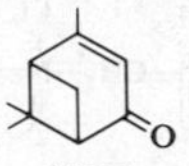

14906

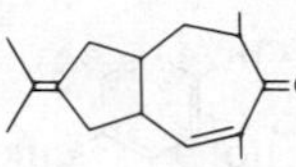

14908

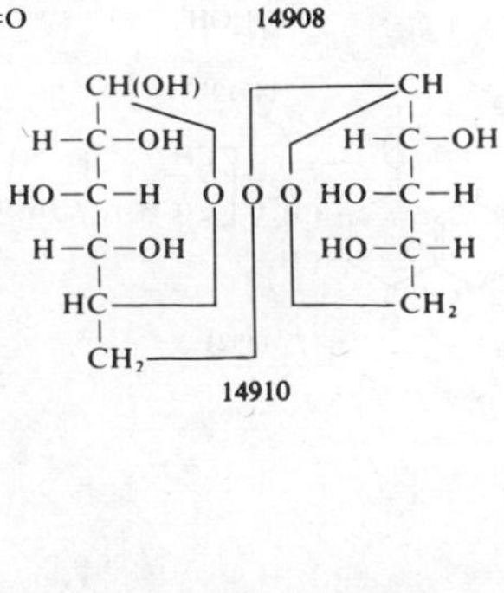

14910

STRUCTURAL FORMULAS OF ORGANIC COMPOUNDS (Continued)

In Numeric as they occur in Organic Compounds Table

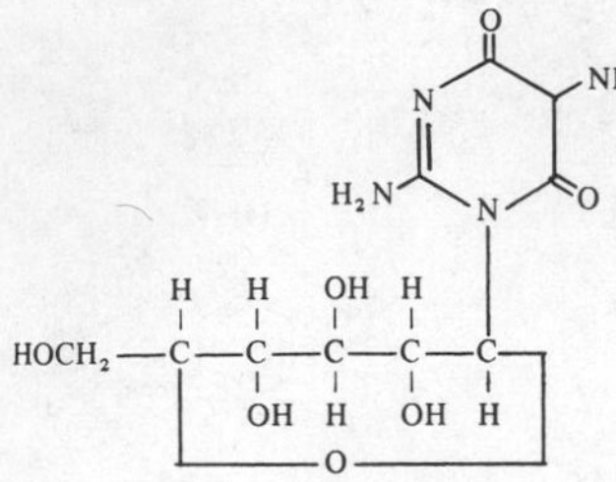

14911

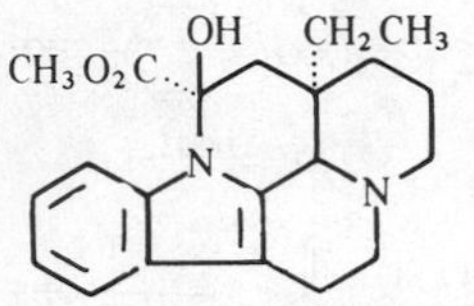

14912

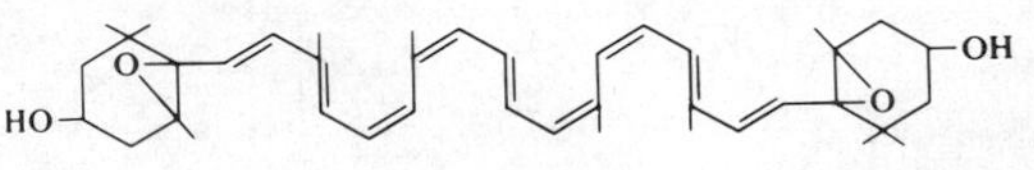

14913

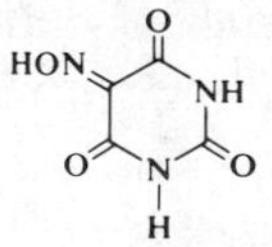

14914

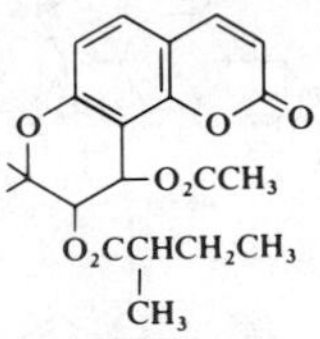

14915

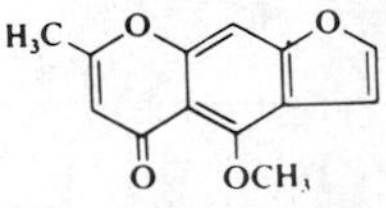

14916

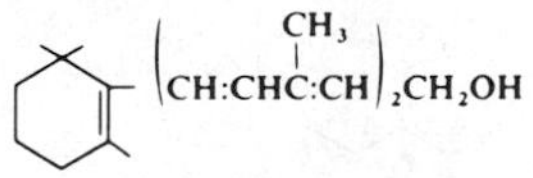

14917

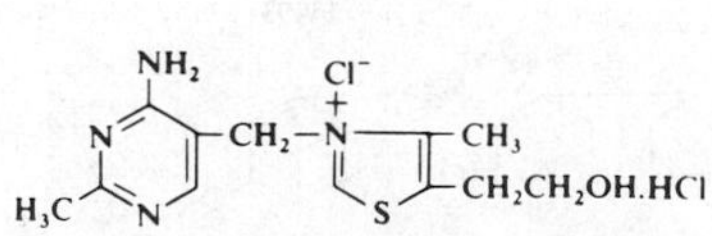

14918

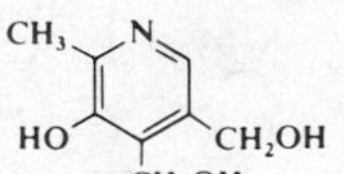

14919

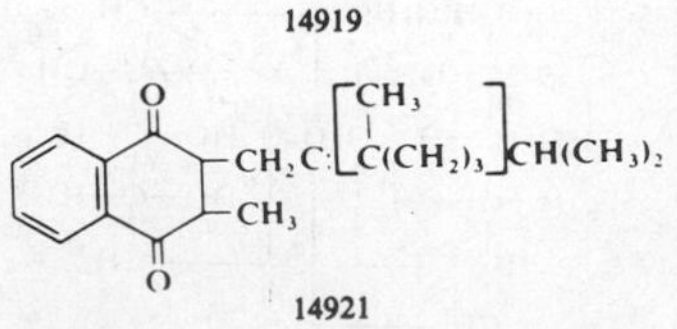

14921

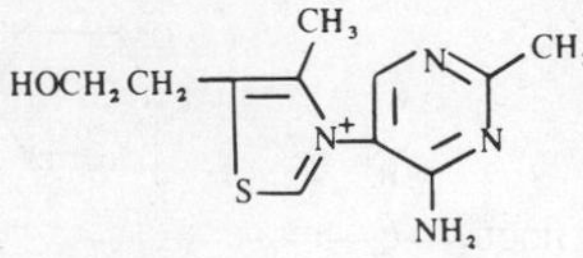

14923

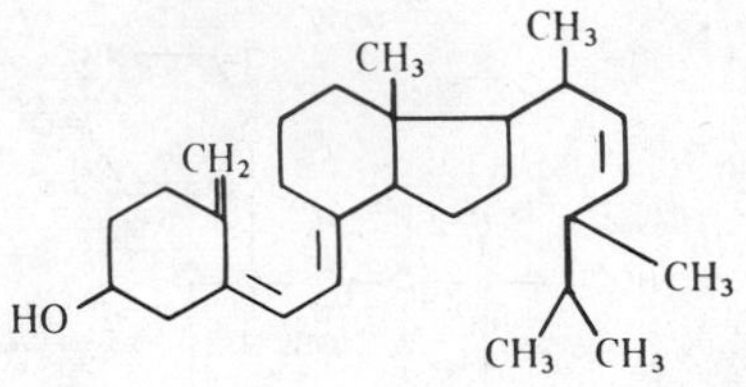

14924

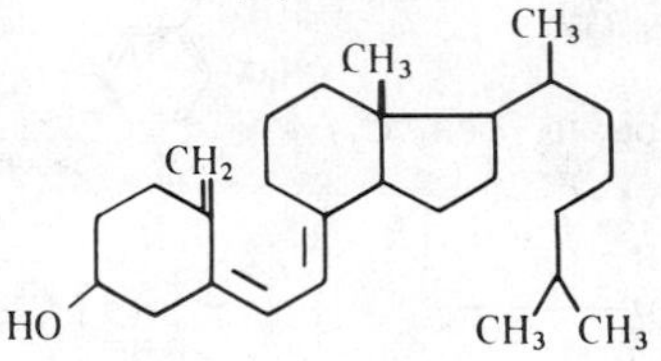

14925

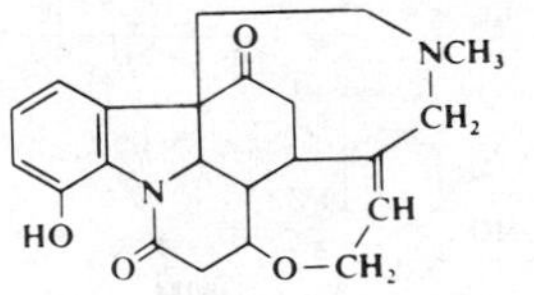

14926

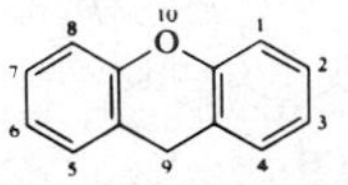

14944

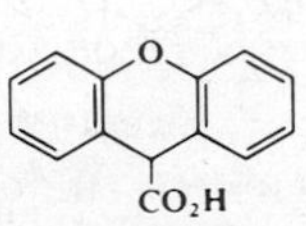

14947

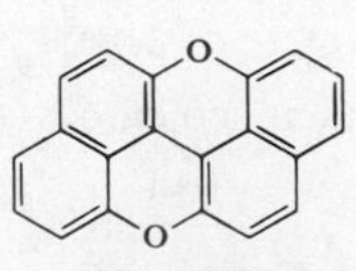

14948

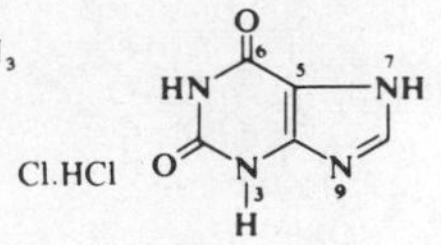

14949

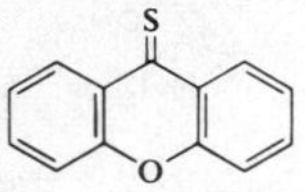

14951

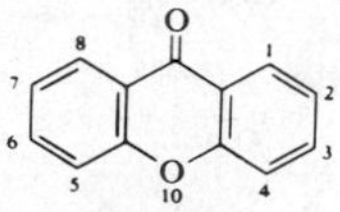

14954

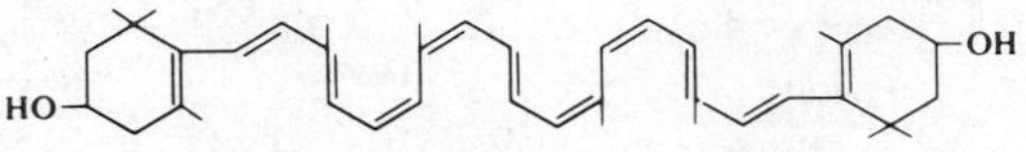

14966

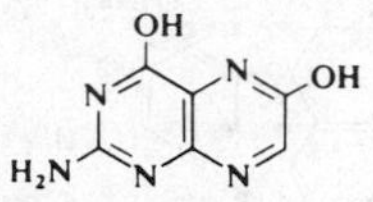

14967

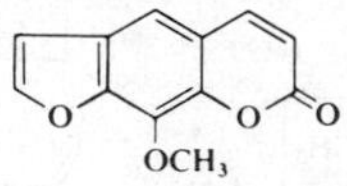

14968

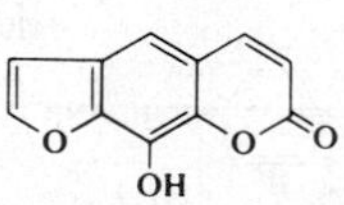

14969

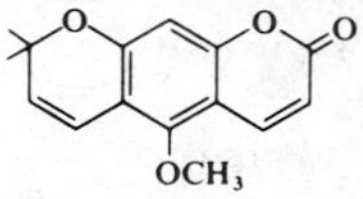

14970

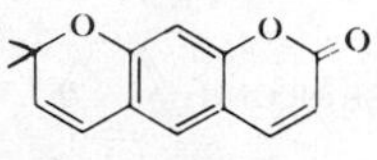

14971

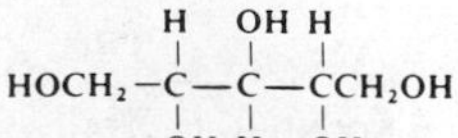

15019

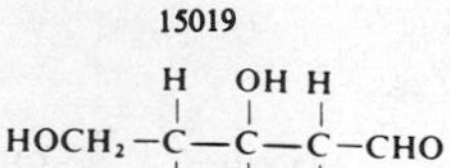

15020

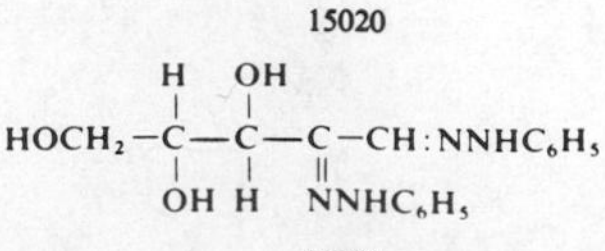

15021

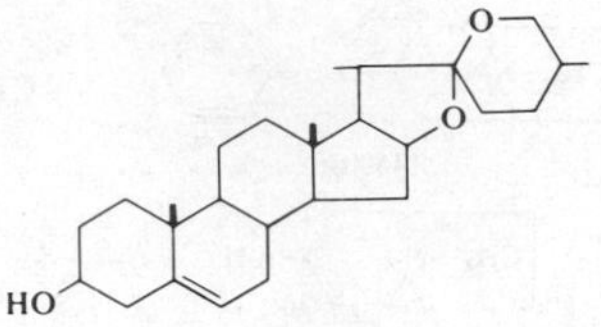

15022

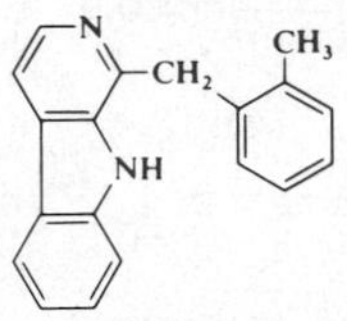

15023

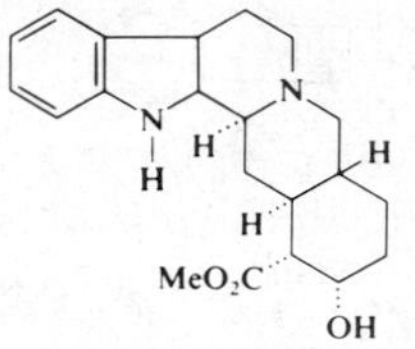

15024

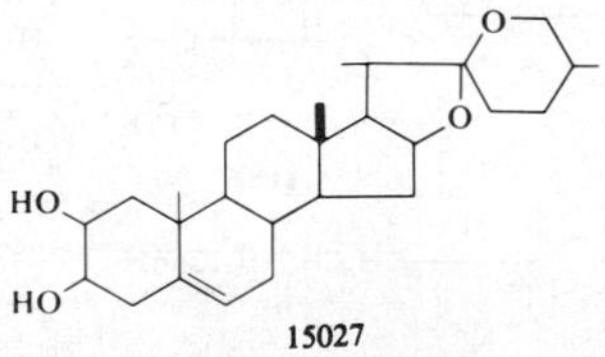

15027

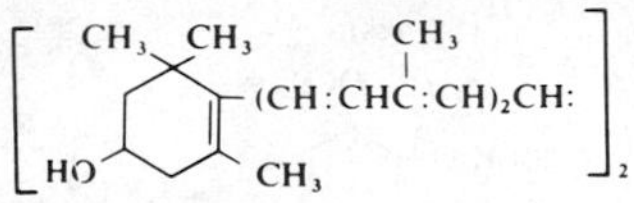

15029

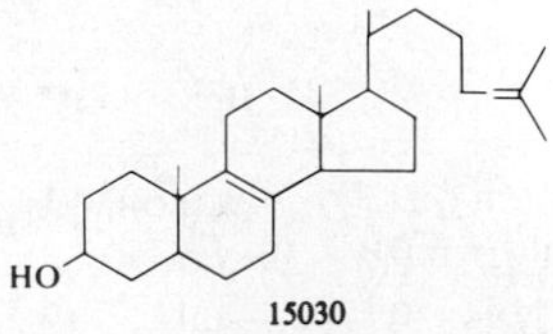

15030

MELTING POINT INDEX OF ORGANIC COMPOUNDS

Temperatures in °C; where values are not precisely known, or where there is a range of melting points, the compound is listed according to the lower temperature.

−197: 9089
−189: 11772
−185: 3942, 12025, 12257
−183: 6603, 11880
−182: 9015
101. 9047
−169: 6840, 10448
−168: 3971
−165: 13851
−161: 3576
−160: 9100, 14934
−159: 3777, 3808, 10422, 11862, 11872, 12304
−158: 9061
−157: 3963, 6847
−156: 12293, 13296
−153: 10406, 10534, 10535, 14931
−151: 8640, 10539
−150: 9093, 11866, 13274
−148: 10307
−147: 7853
−146: 3661, 8575, 9037
−145: 9051
−144: 6704
−143: 1411, 6646
−142: 5440, 5776, 6876
−141: 7731, 7954, 9079
−140: 3707, 4104, 4189, 10563
−139: 7941, 14930
−138: 3672, 3980, 4742, 5773, 5837, 9155, 10517, 10560
−137: 3970, 7960, 10298, 10356, 11360, 12266, 12277
−136: 580, 3631, 6614, 6775, 7654, 10530, 10540, 11770
−135: 5806, 9064, 10533, 10559, 13283
−134: 3761, 10394, 10527, 10529, 10561, 10562, 12264, 12267, 13269
−133: 1585, 3945, 4013, 7955
−132: 3657, 3782
−131: 3706, 3968, 4101, 7991
−130: 4100, 6158, 6867, 10340, 11937, 13254
−129: 324, 596, 4014
−128: 3715, 3746, 7733
−127: 5815, 5824, 10555
−126: 445, 5545, 7805, 10114, 10352, 11942, 12244
−125: 4195, 7992, 10308
−124: 12260
−123: 3698, 4099, 5812, 7833, 9124, 10387, 10528
−122: 6863, 9068, 9205, 10392, 11795, 12253, 13264, 13285
−121: 21, 5685, 7571, 7798
−120: 3636, 5772, 7569, 10402
−119: 5832, 7643, 7819, 10391, 12262
−118: 6610, 6770, 7818, 10116, 10399, 10554, 11359
−117: 3993, 4096, 4211, 5780, 5833, 6632, 6708, 6827, 11805, 12254, 14514
−116: 7553, 7832, 7999
−115: 3851, 3964, 4139, 4145, 5687, 5688, 6839, 9023, 10400, 10557
−114: 4741, 12297, 14483
−113: 37, 7552, 7961, 9193, 10056, 10564, 11936, 12259
−112: 435, 3683, 3691, 4095, 4186, 4421, 5937, 6625, 10418, 11385
−111: 3692, 3985, 4097, 4717, 5531, 5781, 6676, 6920
−110: 10115, 10449, 10631, 11782, 12251, 14877
−109: 3979, 4148, 5827, 7566, 7648, 10421, 12278
−108: 3638, 5761, 5767, 6653, 6805, 6871, 11873
−107: 4740, 5562, 10054, 10419
−106: 5551, 6618, 7606, 10565
−105: 3679, 3981, 3994, 4171, 5814, 6619, 6810, 7831, 10363, 10627, 13288, 13290
−104: 3675, 3704, 3771, 4085, 4089, 5777, 7052, 11843
−103: 3767, 5678, 6102, 6627, 7813, 7995, 9157, 10131, 10469
−102: 7556, 10135, 10417, 12321, 13323
−101: 6046, 6098, 6630, 7836, 9194, 9928, 10105, 10628, 12333, 13511, 14932
−100: 3944, 4328, 4427, 6663, 6799, 9128, 9232, 10109, 10420, 11828
−99: 2236, 3671, 3745, 4222, 4364, 6349, 7083, 10354, 10388, 11831, 12265, 12288
−98: 186, 225, 434, 4106, 4191, 9199, 13436
−97: 4149, 4383, 5756, 6170, 6626, 7054, 8642, 9035
−96: 2237, 5394, 13255, 13854, 13915, 14870
−95: 262, 306, 2140, 2145, 4117, 4125, 5502, 5769, 6220, 6764, 7076, 7760, 8755, 9060, 10350, 11777, 12005, 12250, 13336, 14060
−94: 4123, 5558, 6628, 6651, 6813, 7769, 9183, 12217
−93: 206, 270, 5762, 6156, 7086, 9020, 9127, 9148, 10117, 10538, 11874, 14573, 14927
−92: 261, 3571, 3574, 3930, 4185, 5912, 7047, 7087, 7994, 9192, 12198, 12312, 14823
−91: 4294, 4295, 7075, 7804, 10339, 11829, 14790, 14837
−90: 4460, 4462, 5524, 7522, 7646, 10166, 10401, 10618, 11870, 12307, 12498
−89: 3860, 4069, 4199, 4445, 5483, 5521, 7993, 8954, 11728, 11788, 11971, 12091, 12245, 12252, 13895
−88: 1982, 4407, 7749, 8812, 11376
−87: 1416, 4617, 5525, 7824, 10110, 10301, 10345, 11721, 12023, 12169, 13265, 13318, 14128
−86: 480, 3918, 6843, 9021, 9161, 9929
−85: 279, 626, 1058, 1985, 3759, 5765, 6837, 6912, 7116, 7515, 10403, 11380, 12302
−84: 3573, 4047, 4351, 4412, 6164, 7765, 9150, 9927, 10502, 14991
−83: 204, 2099, 7601, 7620, 8862, 11773, 12246
−82: 2027, 7625, 9173, 10164, 10630
−81: 4198, 5416, 6754, 7673, 7852, 11771, 12042
−80: 215, 443, 625, 2144, 3733, 3758, 4399, 4604, 4615, 4967, 6720, 6865, 7079, 7865, 8644, 10053, 10512, 14579
−79: 4168, 5474, 5677, 7738, 9814, 10076, 10126, 10455, 11927
−78: 227, 1456, 1584, 4337, 7563, 7642, 12324, 13270, 14863
−77: 185, 8869, 10108, 10495, 13299, 13320
−76: 3649, 4105, 4190, 12497, 13256, 13785
−75: 1987, 1988, 2228, 4278, 5523, 5802, 7392, 7732, 7816, 10447, 10622, 12188, 14426
−74: 2100, 3731, 3935, 4008, 5381, 5501, 5734, 8754, 9926, 14674
−73: 177, 263, 3709, 4372, 5642, 5938, 5946, 6052, 6881, 7085, 7562, 12143, 12177, 13797,

−15: 13508, 14101, 14493, 14642, 14826

−14: 791, 803, 891, 3903, 4084, 4321, 4738, 5331, 6869, 9025, 9156, 11894, 13943, 14062, 14270, 14300, 15014

−13: 182, 2107, 2848, 2874, 3914, 5639, 6169, 7493, 8161, 9363, 9377, 9850, 14098, 14479

−12: 307, 1924, 2052, 2149, 2740, 5465, 5752, 7585, 7935, 7936, 8140, 10396, 12661, 13742, 14848

−11: 5692, 5925, 6680, 6883, 6918, 8768, 9996

−10: 807, 3708, 4634, 6105, 6111, 10097, 11868, 12306, 12530, 13188

−9: 1726, 1955, 5599, 5757, 5928, 6393, 6395, 6417, 7937, 9043, 9379, 11617

−8: 286, 525, 1952, 3905, 6836, 7837, 9340, 9424, 9792, 10758, 12440, 13139

−7: 325, 1742, 2072, 2137, 2295, 5070, 5475, 5785, 6319, 6931, 7169, 7583, 9378, 9937, 10058, 13083, 14519

−6: 765, 2011, 2156, 2253, 4463, 5431, 5516, 6862, 7753, 8328, 8791, 9358, 10771, 11331, 11754, 13747, 13882, 14973

−5: 350, 996, 3464, 4547, 6838, 7846, 8764, 9032, 9297, 9933, 9976, 10478, 13103, 13560, 14460, 14487, 14528

−4: 921, 1885, 3070, 3140, 3665, 4263, 4283, 4948, 5158, 5509, 5595, 5597, 5749, 9040, 9225, 9298, 9939, 10947, 11665, 12598, 14631

−3: 3226, 5676, 10668, 13531, 14051, 14457, 14652, 14654

−2: 323, 3077, 3326, 3788, 3848, 3862, 4217, 4571, 5226, 5443, 5924, 7226, 7786, 7849, 7859, 8881, 9299, 10167, 11294, 12878, 13195, 14299, 14671

−1: 537, 804, 931, 1977, 4324, 4598, 4629, 5515, 5598, 8395, 9260, 9357, 9919, 9957, 10145, 13521, 14765

0: 924, 1000, 1605, 1777, 1934, 1937, 3461, 3487, 3645, 3714, 4560, 5330, 6077, 6127, 6399, 6441, 6664, 6665, 6845, 9958, 10028, 10142, 11473, 11555, 11666, 11838, 12247, 12469, 12631, 14157, 14311, 14649, 14940

1: 545, 1402, 1658, 2009, 2697, 2720, 2991, 3206, 3397, 3669, 3802, 4004, 4005, 4144, 4167, 4213, 4757, 4831, 5065, 5066, 5511, 5730, 5927, 7106, 8956, 8960, 10762, 10980, 11174, 11972, 13953, 15004

2: 333, 639, 872, 913, 1030, 3078, 4728, 5455, 5470, 6321, 6801, 7056, 7058, 8805, 10072, 10295, 10740, 10763, 11477, 12525, 13125, 13724, 14047, 14659

3: 1423, 3155, 3255, 8977, 10009, 11548, 12637, 13193, 14204

4: 398, 1745, 1946, 2247, 3340, 3341, 3488, 3650, 3663, 3738, 3742, 4009, 4553, 4582, 4840, 7390, 7715, 9048, 9385, 10746, 10768, 12825, 13435, 13503, 13729, 14158, 14398, 14657

5: 1882, 2163, 2194, 2864, 3567, 3744, 3832, 3904, 5371, 6465, 6811, 6954, 9057, 9375, 10814, 10960, 11580, 12022, 13104, 13189, 13448, 13530, 13726, 13728, 14099

6: 57, 929, 1013, 3405, 3824, 4127, 4325, 5491, 6308, 6952, 9070, 10004, 10146, 10260, 11622, 11722, 11855, 13743, 13961, 14462, 14860, 14907

7: 360, 1011, 1967, 2057, 2068, 3119, 3839, 5164, 5557, 5593, 6288, 8822, 9144, 9359, 12484, 13500, 14084, 14086, 14355

8: 3128, 3432, 4718, 5214, 5329, 6849, 6870, 7072, 8531, 8708, 9097, 9821, 10124, 10775, 11680, 11987, 12033, 12660, 13562, 13773, 14463

9: 539, 748, 902, 1729, 1935, 1954, 2108, 3465, 3809, 4554, 5508, 6402, 6622, 9022, 10281, 10372, 10386, 10843, 11822, 12883, 14473, 14906, 14999, 15018

10: 426, 535, 1065, 1844, 2245, 4138, 4181, 6130, 6727, 6854, 7394, 7699, 7797, 8139, 8593, 9086, 10285, 10296, 11574, 12165, 12359, 13721, 13863, 14227, 14638, 14678

11: 877, 896, 2188, 2244, 3138, 3409, 4437, 5335, 5336, 6215, 7504, 7603, 8557, 11423, 11864, 12463, 13082, 14992, 15016

12: 15, 1679, 2050, 3834, 4438, 5122, 5922, 6346, 6815, 7971, 9254, 9432, 10239, 10257, 11401, 12995, 13399, 14429, 14491, 14648, 14656

13: 486, 1032, 1169, 1737, 5533, 6060, 6104, 7399, 7671, 7693, 7867, 8330, 10236, 12037, 12174, 12444, 12676, 12750, 14226, 14998, 15003

14: 941, 1807, 2431, 2878, 3103, 3468, 5744, 5926, 6108, 6673, 6834, 7374, 7741, 8860, 9095, 9226, 9437, 9793, 10614, 11820, 14440, 14458, 14660

15: 895, 3143, 4017, 7423, 7721, 9932, 10044, 11007, 11332, 11526, 11730, 12121, 12204, 12395, 12517, 12627, 13359, 14140, 14655, 14983, 15015

16: 175, 822, 909, 1741, 3142, 3404, 3886, 4306, 4309, 4624, 7641, 8895, 9004, 10078, 10139, 10228, 10680, 10774, 10835, 10952, 10961, 11889, 12038, 12111, 12555, 12629, 12882, 13732, 14147

17: 801, 1684, 1962, 2266, 5700, 5845, 5858, 5900, 6121, 6447, 7230, 7703, 7704, 9034, 9998, 12166, 14635, 14651

18: 31, 777, 2746, 3891, 4839, 5045, 5513, 5839, 5843, 6175, 6921, 7701, 7707, 7714, 7737, 8424, 8651, 9202, 10291, 10570, 11188, 11239, 11387, 11718, 12013, 12228, 12343, 12678, 13655, 13656, 13775, 14832, 14836

19: 1177, 1923, 3113, 3458, 3840, 6410, 6723, 7793, 8113, 9102, 9255, 9259, 10287, 10338, 10443, 12886, 13164, 13558, 13800, 13962, 14351, 14646, 14666

20: 351, 352, 371, 376, 958, 1733, 1974, 2053, 2216, 2682, 3198, 3335, 3833, 4207, 4305, 5962, 6287, 6744, 7354, 7722,

42: 181, 337, 504, 816, 1347, 1405, 1443, 1899, 2446, 2545, 2896, 3139, 3277, 3300, 3459, 3508, 4064, 5081, 5407, 6039, 6106, 6442, 6544, 8224, 8375, 8961, 9081, 9828, 10174, 10208, 10297, 11461, 11527, 11983, 12223, 12773, 12874, 13846, 14069, 14599, 14606, 14677

43: 390, 392, 824, 951, 1522, 1783, 1961, 2085, 2271, 2855, 3288, 3842, 4288, 4800, 4807, 4945, 5408, 5662, 6237, 7469, 7549, 7551, 7840, 7842, 7978, 8571, 9830, 10294, 10446, 10587, 10767, 10852, 10957, 11356, 11914, 13154, 13733, 13942, 14182

44: 966, 1397, 1524, 1702, 1722, 1743, 2218, 2262, 2615, 2954, 3617, 4244a, 4286, 4287, 4338, 4934, 5140, 6329, 6381, 6392, 6407, 7695, 7713, 8281, 8354, 8702, 8706, 8963, 10003, 10069, 10206, 10832, 10876, 11608, 12665, 12776, 13386, 13493, 13688, 14064, 14132, 14142, 14348, 14471, 14896, 14909

45: 314, 756, 850, 1081, 1724, 2379, 2721, 3063, 3251, 3348, 4024, 4038, 4430, 4882, 4985, 5347, 5799, 6286, 6463, 6884, 6905, 8825, 8882, 8897, 9028, 9799, 9931, 10000, 10173, 10662, 10882, 10988, 11426, 11996, 12162, 12475, 12774, 13805, 14177, 14183, 14184, 14187, 14572

46: 319, 339, 767, 839, 1509, 1512, 1694, 2172, 2677, 3233, 3304, 3310, 4381, 4792, 4842, 4923, 6055, 6199, 6284, 7012, 7396, 7702, 8084, 8130, 8718, 9125, 10925, 10983, 11172, 11415, 11484, 12230, 12595, 12649, 12916, 13371, 13799, 13905, 13947, 14331, 14338, 14509, 14513, 14772

47: 46, 1687, 1768, 1917, 2241, 2441, 2692, 2936, 3843, 4379, 5415, 6370, 6766, 7212, 7723, 8073, 8145, 8494, 8770, 9000, 9383, 9867, 10877, 10950, 10986, 11175, 11678, 12058, 12396, 12408, 12985, 13406, 14075, 14320, 14452

48: 672, 680, 2283, 2426, 2623, 3618, 3750, 4162, 4175, 4233, 4343, 4397, 4661, 4818, 4907, 6048, 6335, 6414, 6440, 7502, 8066, 8147, 8191, 8726, 8877, 9099, 9315, 9709, 9846, 10001, 10010, 10221, 10772, 10853, 10953, 11056, 11272, 11457, 11723, 12233, 12284, 12610, 12692, 12787, 13123, 13420, 13845, 14152, 14318, 14340, 14682

49: 173, 241, 257, 397, 956, 1012, 1671, 2032, 2844, 3280, 3489, 3785, 4344, 4556, 4651, 5918, 6474, 7489, 7506, 7581, 7594, 7717, 7734, 8090, 8592, 8647, 8719, 9145, 9276, 9333, 9403, 9995, 10644, 10918, 10933, 11615, 11620, 12234, 12749, 12783, 12784, 12875, 13019, 13543, 13746, 14035, 14159, 14595

50: 358, 361, 752, 837, 1371, 1648, 3173, 3241, 3250, 3496, 3602, 4244, 4711, 4735, 4841, 4946, 5479, 6248, 6383, 6451, 7065, 7712, 8998, 9688, 9829, 9912, 10632, 10646, 10827, 11216, 11377, 11420, 12003, 12125, 12913, 13111, 13650, 13730, 14223

51: 63, 365, 385, 717, 842, 899, 1189, 1527, 1594, 1740, 1912, 2434, 2442, 2849, 2930, 3375, 3392, 3873, 4481, 4499, 4565, 5172, 5314, 5478, 5614, 5969, 8896, 9238, 9570, 9714, 9715, 9745, 9890, 9968, 10232, 10428, 10812, 10840, 10873, 11054, 11895, 12511, 12611, 12746, 13413, 13866, 13903, 13938, 14376, 14908, 15017

52: 9, 403, 404, 813, 833, 1337, 1675, 1688, 1719, 1911, 2347, 2552, 2902, 2903, 3014, 3105, 3666, 3705, 3755, 4382, 4390, 4480, 4482, 4887, 4922, 5430, 5604, 5972, 6268, 6365, 6458, 6473, 6642, 6736, 7174, 7904, 8356, 8411, 9737, 9795, 9798, 10101, 10675, 10887, 11182, 11854, 11957, 12283, 12420, 12754, 12862, 12902, 13865, 14032, 14107, 14117, 14798

53: 52, 402, 703, 827, 875, 1063, 1528, 2089, 2198, 2265, 2850, 2992, 3382, 3427, 3996, 4206, 4643, 4673, 6366, 7015, 7068, 7503, 8519, 8650, 8969, 9252, 10230, 10290, 11097, 11384, 11441, 11719, 11850, 11930, 12150, 12243, 13438, 13453, 13557, 13843, 14151, 14178

54: 483, 769, 1067, 1372, 1649, 1770, 2094, 2173, 2226, 2242, 2284, 2457, 2727, 3389, 3414, 4653, 4668, 5844, 6119, 6217, 6249, 6279, 6384, 6449, 6452, 7082, 7326, 7373, 7379, 7499, 7500, 8372, 9386, 9414, 9989, 10177, 10189, 11357, 11451, 12371, 12392, 12956, 13582, 13720, 14149, 14210, 14257, 14582

55: 136, 164, 244, 823, 878, 1499, 1685, 1900, 2207, 2374, 2438, 2788, 2852, 2998, 3028, 3050, 5321, 5368, 5534, 5620, 6085, 6154, 6292, 7719, 8302, 8303, 9046, 9435, 9676, 9757, 9767, 9979, 10321, 10984, 11076, 11141, 11143, 11170, 11293, 11488, 12770, 13021, 13177, 14037, 14454, 14835

56: 34, 55, 368, 903, 1034, 1554, 1613, 1800, 2091, 2199, 2445, 2755, 3133, 4921, 5343, 5353, 5365, 5366, 5603, 6247, 6660, 6738, 6873, 6892, 7398, 7692, 7705, 8159, 8198, 8306, 8329, 8374, 8386, 8471, 8767, 9078, 9283, 9293, 9295, 9977, 10262, 10647, 10859, 10868, 10871, 12129, 12607, 13425, 13572, 13869, 13960, 14077, 14542

57: 56, 273, 372, 831, 932, 1021, 1024, 1039, 1437, 1914, 2036, 2208, 3026, 3260, 4675, 4809, 4870, 5487, 6373, 6679, 6927, 6928, 6929, 7461, 7519, 8085, 8194, 8404, 8570, 8653, 8994, 9237, 9882, 9891, 9910, 10255, 10633, 10879, 11342, 11921, 12447, 12747, 12958, 13391, 13480, 13606, 14640, 14867

89: 2209, 3256, 3320, 4073, 4722, 5268, 5287, 5308, 6038, 6040, 6421, 6725, 6820, 8542, 8698, 8700, 9335, 10258, 10317, 11053, 11217, 11497, 11964, 12070, 12156, 12431, 12936, 13393, 13700, 14092, 14585, 14976

90: 303, 396, 704, 736, 1486, 1516, 1757, 1784, 1804, 2125, 2332, 3030, 3055, 3109, 3486, 4326, 4479, 4676, 4710, 4793, 4859, 5352, 5559, 5644, 6362, 6588, 6901, 7002, 8108, 8325, 8434, 8826, 9091, 9466, 9974, 10224, 10645, 10676, 10881, 10906, 11446, 12435, 12625, 3413037, 13661, 13669, 13698, 13832, 15020

91: 101, 114, 781, 1078, 1799, 2003, 2281, 2522, 3395, 3400, 3601, 4650, 5288, 6179, 6297, 7131, 7181, 7427, 7435a, 8105, 8118, 8422, 8536, 8905, 9516, 10181, 10733, 10914, 11164, 11267, 12119, 12777, 12865, 12899, 13446, 13591, 14038, 14317, 14373, 14417

92: 20, 356, 573, 722, 886, 1335, 1539, 1956, 2048, 2342, 2550, 2937, 4504, 4678, 4806, 6195, 6289, 6429, 6644, 6999, 7844, 9835, 10829, 10831, 10857, 11390, 12909, 13008, 13039, 13140, 13810, 14190, 14323, 14741

93: 540, 1138, 1533, 1592, 1665, 2480, 2750, 3059, 3286, 3415, 3473, 4503, 4776, 4931, 5001, 5020, 5163, 5794, 5796, 6185, 6462, 8093, 8109, 8666, 8787, 8818, 8824, 8949, 10959, 10985, 11001, 11297, 11428, 11512, 12081, 12088, 12602, 12741, 12798, 12892, 13392, 13804, 14120, 14170, 14288, 14342, 14684, 14888, 15005

94: 168, 171, 541, 1050, 1114, 1190, 1652, 1674, 2443, 2763, 2857, 3013, 3019, 3433, 3507, 4033, 4507, 5995, 6378, 6605, 7970, 8075, 8164, 9103, 9305, 9308, 9769, 10977, 11021, 11378, 11492, 12729, 12918, 13380, 13864, 14385, 14537

95: 80, 81, 135, 274, 414, 728, 737, 858, 1027, 1412, 1704, 1894, 1965, 2204, 2259, 2427, 2482, 2540, 2603, 2894, 2935, 2968, 3024, 3210, 4570, 4709, 5310, 6035, 6083, 6280, 6677, 6678, 7320, 7484, 7998, 8439, 8673, 8675, 8831, 9657, 9771, 10231, 10972, 11023, 11236, 12612, 12708, 12858, 13056, 13119, 13146, 13388a, 13462, 14240, 14291, 14430

96: 4, 167, 363, 399, 989, 1997, 2718, 2860, 2885, 3193, 3232, 3399, 4785, 4787, 4897, 5022, 5325, 7113, 7582, 8287, 8821, 8878, 8899, 9556, 9571, 9721, 9885, 10734, 10752, 10769, 10973, 11038, 11117, 11134, 11155, 11306, 12239, 12695, 13155, 13202, 13564, 13565, 13806, 14009, 14169, 14229, 14714, 14758

97: 24, 42, 131, 380, 542, 1051, 1559, 1999, 2277, 3000, 3004, 3165, 3440, 4028, 4060, 4254, 4789, 5973, 6958, 7014, 7182, 8197, 8294, 8420, 8421, 9567, 9990, 10213, 10335, 10754, 10989, 11507, 12182, 12803, 12863, 12987, 13020, 13151, 14058, 14326, 14594

98: 50, 353, 1037, 1083, 1471, 1535, 1840, 1905, 1964, 2115, 2354, 2504, 2701, 2869, 2944, 3003, 3516, 3584, 4079, 4374, 4699, 5188, 5190, 5209, 5317, 5968, 6234, 6506, 6898, 7430, 8188, 8669, 8723, 8783, 9214, 9317, 9397, 9465, 9651a, 10926, 10971, 11060, 11122, 11242, 11303, 11495, 13004, 13163, 13975, 14135, 14168, 14346

99: 72, 96, 488, 1029, 1603, 1691, 1713, 2197, 2239, 2345, 2352, 2742, 2847, 3235, 3269, 3385, 5025, 5663, 5860, 6012, 6059, 6590, 6705, 6819, 6917, 7318, 7633, 8020, 8773, 9307, 9701, 10266, 10424, 10783, 10920, 11247, 11248, 11408, 12961, 13365, 13788, 14021, 14089

100: 641, 1902, 2813, 2824, 2884, 2929, 3038, 3115, 3172, 3796, 4030, 5021, 5107, 5460, 5576, 5864, 5896, 6962, 8055, 9313, 9439, 9478, 9834, 9852, 10686, 10705, 11108, 11213, 11223, 11549, 13033, 13046, 14271, 14470, 14944

101: 82, 377, 1003, 1164, 1477, 1939, 2758, 2868, 2985, 4568, 4899, 5320, 6648, 8116, 9135, 10649, 10842, 11031, 11059, 11587, 12677, 12829, 12864, 12962, 13176, 14215, 14297, 14588, 14989

102: 161, 357, 1366, 2061, 2315, 2397, 2599, 2810, 3037, 4569, 5240, 5264, 5334, 6295, 7130, 7236, 7460, 8336, 8909, 8943, 9697, 10237, 10694, 10880, 10963, 11211, 11324, 12982, 13403, 13440, 14269, 14544, 14891

103: 14, 149, 172, 502, 634, 815, 2258, 2931, 3130, 3416, 3481, 3583, 3585, 5261, 5294, 5395, 5704, 6304, 6503, 6509, 6585, 7007, 7093, 7108, 7325, 7459, 8547, 9345, 9413, 9420, 9728, 9772, 10204, 10696, 10964, 11082, 11129, 11701, 12546, 12725, 12928, 12983, 13589, 14243, 14396, 14749

104: 115, 133, 401, 547, 574, 775, 1150, 1345, 1518, 3514, 6181, 6564, 6591, 6925, 6998, 7754, 8845, 9464, 9490, 9611, 9750, 10655, 10657, 11352, 12450, 12791, 12797, 12804, 12872, 14197, 14247, 14287, 14743, 14789

105: 100, 108, 391, 843, 1052, 1077, 1677, 2007, 2103, 2279, 2471, 2967, 3029, 3281, 4576, 4612, 4788, 4898, 5447, 5449, 5664, 5671, 6522, 6523, 6592, 7321, 7452, 7458, 8415, 8729, 8833, 9250, 9312, 9365, 9372, 9406, 9549, 9582, 9603, 9754, 10125, 10739, 10965, 11107, 11119, 11354, 11940, 12068, 13023, 13050, 13376, 13769, 14105, 14173, 14272, 14558

106: 354, 514, 638, 1074, 1085, 1086, 1106, 1139, 1365, 1440, 1453, 1505, 1789,

106: 1798, 2074, 2875, 3025, 3156, 4160, 4256, 4262, 4486, 4488, 5024, 5121, 5476, 5623, 6132, 6584, 7017, 7575, 8146, 8309, 8734, 9427, 9473, 9600, 9735, 9930, 10182, 10700, 10956, 11197, 11550, 11571, 11594, 12186, 12356, 12501, 12668, 12955, 13823, 13830, 13908, 14091, 14738, 14801

107: 272, 635, 1097, 1546, 1650, 1907, 1942, 2743, 2756, 5477, 6254, 7328, 8334, 9292, 9332, 9673, 10223, 11553, 11897, 12415, 12478, 13468, 13996, 14109, 14260, 14298, 14325

108: 8, 65, 468, 806, 981, 1536, 1749, 3129, 3282, 3419, 4549, 4910, 5372, 6609, 6709, 7011, 8067, 8173, 8310, 8368, 8398, 8808, 9109, 9599, 9764, 10692, 10939, 11258, 11273, 11310, 11349, 11529, 12472, 12717, 12925, 13184, 13237, 13469, 13568, 14136, 14228, 14344, 14626, 14731

109: 400, 1328, 1332, 1545, 1746, 2889, 2981, 3597, 3764, 4487, 4684, 5293, 5358, 5423, 5947, 6184, 6428, 6756, 7013, 7161, 7215, 9645, 10328, 10598, 10770, 10901, 11913, 12374, 12479, 13379, 13973, 14000, 14002, 14273

110: 143, 375, 865, 1009, 1152, 1193, 1548, 1555, 1682, 1810, 1992, 2360, 2474, 2829, 2883, 3434, 4496, 4822, 5997, 6151, 6174, 6565, 6726, 7299, 8141, 8172, 8599, 8704, 8774, 8971, 9003, 9139, 9201, 9753, 9765, 10949, 11068, 11510, 12056, 12655, 12733, 12905, 13363, 14014, 14245, 14564, 14763, 15030

111: 464, 785, 889, 1111, 2104, 2455, 2890, 2964, 3067, 4231, 4947, 5102, 5566, 6070, 6358, 6689, 6700, 6831, 6858, 8081, 8794, 8972, 9303, 9674, 9851, 10220, 10261, 10275, 10790, 10801, 10847, 11604, 12062, 12103, 12648, 12974, 13765, 13850, 14275, 14746

112: 450, 674, 808, 1523, 1625, 1802, 1852, 1859, 2206, 2670, 2676, 2695, 2892, 2914, 3154, 3557, 4241, 5201, 5510, 5539, 6136, 6153, 6191, 6197, 7163, 7250, 7292, 7418, 8637, 8835, 8925, 9440, 9592, 9610, 9635, 12483, 12766, 13026, 13584, 14005, 14196, 14347, 14726, 14739, 14805

113: 1070, 1501, 1725, 2693, 2704, 2705, 2771, 2841, 2871, 3169, 3285, 3435, 3484, 3500, 3598, 5637, 5665, 6530, 6767, 6997, 8290, 8388, 8843, 9049, 9320, 9347, 9722, 10330, 10741, 10945, 10975, 11012, 12445, 12994, 14180, 14258, 14350, 14540

114: 84, 90, 562, 726, 888, 954, 1327, 1367, 1410, 1507, 1805, 2060, 2351, 2502, 2536, 2729, 2854, 2990, 3062, 3106, 3492, 3534a, 4061, 4240, 4502, 4903, 4904, 5276, 7071, 8365, 8383, 8701, 8871, 9338, 10192, 10816, 10846, 10990, 11032, 11287, 11602, 11734, 12493, 12738, 13416, 13475, 13586, 13662, 13958, 14171, 14394, 14467, 14593

115: 85, 383, 408, 1348, 1513, 1915, 3001, 3021, 3023, 3182, 3466, 3539, 4477, 4493, 4730, 4790, 4943, 5029, 5841, 6177, 6232a, 6271, 7256, 7304, 7335, 8192, 8321, 8699, 8847, 9188, 9270, 9481, 9600, 10738, 10932, 10938, 11015, 11307, 11311, 12180, 12622, 13573, 13697, 13838, 14704, 14924

116: 158, 364, 814, 991, 992, 1439, 1475, 1537, 1751, 1906, 2055, 2280, 2661, 2716, 2916, 2941, 3056, 3159, 3380, 4799, 4871, 4900, 5077, 6560, 6844, 6987, 7193, 8051, 8052, 8276, 8665, 8667, 8891, 8915, 9311, 9318, 9632, 10207, 10679, 11013, 11109, 11124, 11176, 11192, 11296, 11301, 11305, 11506, 11560, 12075, 12477, 12888, 13678, 13802, 14167, 14607

117: 477, 1488, 1497, 1758, 2278, 2782, 3035, 5286, 5458, 5632, 6145, 6517, 6848, 7967, 8018, 8027, 8106, 8162, 8377, 8389, 8997, 9325, 9402, 9643, 10703, 10729, 10936, 11100, 11181, 12409, 12432, 12518, 12654, 12736, 12801, 12989, 12993, 13913, 13999, 14367, 14536, 14584

118: 138, 890, 942, 980, 1015, 1049, 1441, 1484, 1867, 2124, 2256, 2477, 2553, 2557, 2593, 2862, 3358, 3431, 3472, 3791, 4494, 4495, 4508, 4700, 4888, 5307, 5326, 5974, 6236, 6301, 6360, 6438, 7022, 7448, 8212, 8649, 8781, 9428, 9702, 9719, 9873, 10833, 10867, 10958, 11018, 11079, 11098, 11136, 11299, 11603, 12360, 12361, 12302, 12411, 12482, 12686, 12796, 12884, 12966, 13592, 13717, 14193, 14194, 14246, 14349, 14708, 14802, 14977

119: 609, 1047, 1595, 1808, 2202, 3305, 3499, 4161, 5363, 5725, 6250, 6294, 6939, 7157, 7239, 8119, 8811, 8914, 9115, 9356, 9405, 10677, 11037, 11123, 11137, 11391, 12067, 12481, 13011, 13538, 13893, 14241, 14411

120: 725, 1020, 2410, 2451, 3041, 3117, 3394, 3428, 4712, 4743, 4987, 5095, 5971, 5998, 6003, 7830, 8208, 8258, 8413, 8611, 8937, 8992, 9319, 9336, 9471, 9551, 11070, 11530, 11567, 11711, 12794, 13161, 13214, 13577, 14003, 14133, 14413, 14538, 14759, 14804

121: 19, 68, 123, 821, 1531, 1689, 2263, 2469, 2473, 2529, 3114, 3370, 3374, 3844, 4754, 4872, 5207, 5418, 5793, 6589, 8838, 8850, 8912, 9411, 9736, 10848, 10851, 11132, 11246, 11496, 11757, 13034, 13559, 13561, 14353, 14418, 14563, 14732

122: 386, 471, 861, 985, 1333, 1534, 1860, 2531, 2680, 2965, 2987, 3543, 4846, 6013, 6202, 6683, 7155, 7221, 7227, 8176, 8823, 8873, 9441, 10671, 10737, 10777, 10916, 11039, 11166, 11235, 11316,

122: 11559, 11598, 12237, 12352, 12476, 12633, 12857, 12889, 13015, 13040, 14195, 14543, 14590

123: 134, 473, 544, 893, 945, 987, 1344, 1578, 1732, 2820, 3122, 3164, 3442, 4252, 5145, 5236, 7252, 8174, 8271, 8430, 8567, 8584, 9101, 9224, 9408, 9588, 9848, 10065, 10673, 10689, 10791, 11078, 11158, 11269, 11746, 12406, 13036, 13228, 13456, 13567, 13707, 14750, 15007

124: 601, 856, 1018, 1479, 1664, 1672, 1714, 2212, 2341, 2698, 2805, 2891, 2913, 2949, 3046, 3181, 3923, 5175, 5299, 5893, 6260, 8107, 8668, 8670, 8674, 9476, 11229, 11302, 11304, 11494, 12659, 12742, 12757, 12868, 13449

125: 122, 160, 551, 590, 859, 1148, 1181, 1846, 2205, 2316, 2318, 2922, 2986, 3051, 3377, 3407, 3417, 3470, 4855, 5205, 6015, 6448, 7027, 7042, 7287, 8217, 8672, 9457, 9517, 9578, 10811, 11840, 12157, 12520, 12651, 12744, 12998, 13080, 14391, 14422, 14586, 14698, 14723

126: 202, 818, 1863, 2359, 2375, 2918, 2961, 2983, 3047, 4667, 5157, 5701, 5821, 6808, 7710, 9195, 9446, 9501, 9573, 9724, 9747, 10263, 10942, 12885, 13471, 13588, 14025, 14106, 14589, 14728

127: 543, 643, 670, 694, 1474, 1478, 1581, 1601, 1790, 1963, 2291, 2549, 2590, 2659, 2747, 2828, 2900, 3401, 3471, 4259, 4516, 4849, 5027, 5076, 5194, 5200, 6274, 8693, 8864, 9604, 9615, 9656, 10212, 10320, 10855, 10891, 11334, 12423, 12427, 12579, 12990, 13157, 13442, 13473, 13680, 14534, 14546

128: 119, 318, 591, 592, 826, 1192, 1480, 1483, 1547, 1609, 1618, 1692, 1747, 1755, 2472, 2481, 2722, 3108, 3525, 4280, 4514, 4854, 4856, 5016, 5048, 5292, 5705, 5994, 6323, 6404, 7006, 7340, 8193, 8211, 8586, 9579, 9636, 9733, 10996, 11318, 11778, 12513, 12894, 13216, 13223, 13822, 14018, 14053, 14253, 14561

129: 71, 118, 782, 1001, 1023, 1115, 1832, 2162, 2400, 2814, 4581, 5305, 5351, 5819, 6014, 6037, 6681, 7228, 7429, 8535, 9301, 9581, 9605, 9741, 9746, 9816, 10322, 10817, 10824, 11024, 11515, 11756, 12155, 12452, 12508, 12643, 12743, 12893, 13849, 13928, 14189

130: 157, 406, 455, 1100, 1627, 1633, 1707, 1752, 1792, 2112, 2412, 2424, 2790, 2819, 2923, 2963, 2980, 4247, 4810, 5128, 6934, 7301, 7449, 8060, 8177, 8199, 8775, 9106, 9585, 9804, 10241, 10803, 11153, 11298, 11329, 11707, 11780, 11918, 12024, 12027, 12503, 12685, 14220, 14222, 14518, 14555

131: 88, 946, 953, 1490, 1676, 2111, 2117, 2326, 2717, 3420, 3430, 5096, 5397, 5448, 5605, 6566, 6991, 7214, 7282, 8091, 8134, 8283, 8333, 8367, 8549, 9304, 9409, 9665, 10809, 10822, 11025, 11121, 11151, 11214, 11493, 12077, 12559, 12726, 13447, 14971

132: 92, 453, 653, 786, 1504, 1640, 1809, 1824, 1993, 2648, 2660, 3079, 3093, 3097, 3373, 3551, 3596, 4116, 4511, 4539, 5259, 6573, 6701, 8050, 8142, 8888, 8932, 8933, 8934, 9606, 9639, 10210, 10784, 11336, 11569, 11592, 11662, 12066, 12433, 12614, 12898, 14232

133: 103, 819, 904, 1447, 2274, 2436, 2454, 2465, 2837, 2839, 2993, 3614, 3766, 4243, 5104, 6074, 6235, 6432, 6508, 7151, 7197, 8195, 8852, 8911, 9497, 9613, 9684, 10860, 11131, 11288, 12015, 12386, 12442, 12604, 12728, 12919, 13116, 13922, 14221, 14970

134: 367, 478, 610, 668, 1334, 1487, 2906, 2959, 3287, 3294, 3421, 3498, 4248, 4541, 4759, 5152, 5328, 5654, 5656, 5914, 6045, 6375, 6857, 7211, 7283, 7293, 7450, 7745, 8460, 8926, 9623, 9723, 10699, 10941, 11095, 11133, 11517, 11896, 12160, 12833, 13187, 13357, 13639, 13683, 13923, 14324, 14383

135: 710, 1146, 1422, 1508, 1556, 1632, 1660, 1669, 1818, 1994, 2541, 2942, 3332, 3447, 5084, 5098, 5254, 5820, 6269, 6497, 6524, 6546, 7234, 7313, 7336, 8032, 8096, 8133, 8564, 8602, 8830, 8885, 8916, 8917, 9543, 9659, 9679, 10637, 10797, 11125, 11505, 11590, 12172, 12978, 13092, 13211, 13601, 13642, 13649, 14419, 14700

136: 637, 1445, 1588, 2250, 2270, 2476, 3354, 3713, 4515, 4866, 5253, 6031, 6182, 7154, 7209, 8205, 8807, 9310, 9323, 9658, 11083, 11224, 11355, 11413, 11554, 12449, 12541, 12657, 12997, 13087, 13160, 13483, 13930, 14214, 14628

137: 132, 276, 1166, 2213, 2486, 2685, 2838, 2971, 3381, 3418, 3438, 3765, 4045, 4250, 4472, 5361, 5880, 6746, 7847, 8739, 9211, 9580, 9710, 9717, 9743, 10656, 11564, 12020, 12089, 12448, 12881, 14206

138: 209, 275, 407, 1542, 1611, 1868, 2329, 2969, 3092, 3123, 4510, 5005, 5079, 5090, 5113, 5446, 6005, 6520, 6703, 7156, 7238, 7240, 7434, 9350, 9515, 11817, 12723, 13169, 13641, 13932, 13934, 14234, 14345

139: 104, 110, 1014, 1610, 1680, 1686, 1825, 1918, 2243, 2343, 2635, 2767, 3157, 3345, 4509, 5339, 5849, 6180, 6258, 6685, 7280, 8533, 8692, 8806, 12376, 13208, 13768, 13883, 14541, 15012

140: 113, 905, 1118, 1264, 1325, 1612, 2063, 2276, 2586, 2600, 2627, 2760, 2994, 3225, 3425, 4059, 5789, 5847, 7036, 7152, 7203, 7296, 7426, 7481, 8087, 8137, 8293, 8685, 8744, 9122, 9337, 9843, 9855, 9881,

MELTING POINT INDEX OF ORGANIC COMPOUNDS (Continued)

140: 10148, 10337, 10955, 11019, 11071, 11226, 11452, 11535, 12231, 12366, 13057, 13217, 13349, 13713, 14562

141: 503, 776, 908, 1698, 1820, 1856, 1869, 2398, 2419, 2752, 2781, 2909, 3121, 3163, 5015, 5078, 5147, 5540, 5706, 5956, 6516, 6567, 8380, 8414, 9243, 9662, 11165, 11516, 11523, 11703, 12856, 13025, 13235, 13690, 13993, 14715, 14745

142: 86, 1377, 1701, 1817, 1838, 2364, 2594, 2618, 2881, 3084, 3523, 3578, 5570, 6519, 6941, 6990, 7070, 8129, 8326, 8884, 9845, 9853, 9905, 11062, 11397, 11650, 11919, 12467, 12694, 12831, 13086, 13348, 14427, 14768, 14782

143: 1558, 2458, 3045, 3355, 4475, 5000, 5035, 5105, 5450, 5666, 6669, 8127, 8511, 9748, 9752, 10661, 11276, 11289, 11339, 11575, 12634, 12735, 12737, 13461, 13474, 13616, 13714, 14408, 14834

144: 125, 521, 657, 1349, 1413, 1467, 1519, 1549, 1765, 2196, 2466, 2642, 2655, 2825, 2919, 2970, 3279, 3289, 3356, 3537, 5797, 6563, 6759, 8458, 9114, 9352, 9716, 9941, 10061, 10650, 10664, 10681, 10937, 10968, 10969, 11201, 12016, 13088, 13443, 13529, 13931, 13984, 14916

145: 111, 862, 1102, 1845, 1896, 2349, 2417, 2501, 2678, 2772, 2966, 2989, 3057, 3082, 3315, 3343, 3443, 3457, 4836, 4857, 6168, 6200, 6652, 7025, 7346,

145: 7814, 7929, 8045, 8063, 8089, 8249, 8635, 8763, 9507, 9775, 10782, 10795, 10825, 10869, 11094, 11531, 11704, 14569, 14744, 14946

146: 570, 682, 857, 1134, 1133, 1624, 2321, 2544, 2637, 2725, 2792, 2988, 3278, 3408, 5138, 5432, 5949, 6155, 6231, 6938, 7288, 7289, 7451, 8006, 8126, 8180, 8921, 9618, 9884, 11275, 11715, 13648, 13981, 14705

147: 381, 384, 812, 1681, 2610, 2649, 2757, 2915, 5169, 5324, 7000, 7464, 8382, 8840, 9367, 9612, 9667, 9839, 9865, 10250, 10815, 10830, 11072, 11147, 11705, 12976, 12977, 13000, 13129, 13158, 13229, 13939, 13983, 14057, 14266, 14760, 15006

148: 140, 145, 355, 632, 720, 955, 1103, 1173, 1331, 1822, 1870, 1887, 2251, 2373, 2903a, 2933, 3424, 3445, 3545, 4260, 5028, 5080, 5115, 5275, 6137, 6139, 6233, 6568, 6682, 6684, 7019, 7397, 8200, 8384, 8778, 9649, 10183, 10688, 11067, 11232, 11322, 11442, 11480, 12510, 13147, 13534, 13719, 14524, 14968

149: 576, 867, 1140, 1495, 2433, 2631, 2972, 3371, 5250, 7026, 7231, 8122, 8313, 8371, 8453, 9887, 10444, 11003, 11085, 11152, 11312, 11747, 11748, 11924, 12353, 12701, 12709, 13029, 14230, 14547, 14707

150: 116, 583, 959, 1302, 1517, 1521, 1623, 1865, 2561, 2613, 2634, 2921, 4979, 4986, 5009, 5017,

150: 5337, 5655, 5669, 7291, 7442, 8317, 8496, 8772, 9530, 9593, 9751, 10193, 11478, 11597, 12647, 12832, 13016, 13094, 13594, 14203, 14508, 14551

151: 13, 109, 141, 146, 560, 687, 788, 2490, 2657, 2811, 2846, 3293, 3732, 4694, 5459, 6239, 7003, 8062, 8264, 8573, 8730, 8928, 8929, 9640, 9766, 11313, 11509, 12586, 12876, 12953, 13038, 13168, 13222, 13570, 13793, 14545, 14751

152: 330, 940, 964, 1550, 1753, 2363, 2830, 3036, 3652, 5030, 5894, 6559, 8025, 8311, 9421, 10779, 10786, 11120, 11207, 11238, 11749, 11781, 13120, 13571, 13941, 14017, 14718

153: 150, 159, 163, 528, 651, 774, 1715, 1762, 2299, 2320, 2564, 3436, 5036, 5170, 5488, 6027, 7838, 8086, 8136, 8186, 8403, 8919, 8920, 8950, 9511, 9626, 9854, 9868, 10802, 10981, 11144, 11146, 11167, 11338, 12582, 12596, 12616, 12897, 13143, 13215, 13603, 13979, 14623

154: 169, 476, 688, 866, 1168, 2367, 2641, 2689, 2694, 4692, 5277, 5436, 5879, 6243, 6272, 7010, 7341, 7611, 8143, 8272, 8645, 9443, 9474, 9496, 9568, 9620, 9670, 10781, 10808, 10810, 11077, 11149, 11150, 12069, 12731, 12895, 13124, 13209, 13569, 13990, 14010, 14710, 14720, 15031

155: 1084, 1117, 1160, 1324, 1356, 1357, 1540, 1760, 1815,

155: 1837, 2571, 2574, 2597, 2831, 2962, 3317, 3452, 5004, 5032, 5203, 5204, 5255, 5257, 5823, 6044, 6176, 6569, 7004, 7020, 8047, 8813, 9001, 9789, 9030, 12130, 12799, 13054, 13113, 13476, 13528, 13598, 13711, 13819, 14292, 14894

156: 105, 511, 698, 718, 810, 1142, 1262, 1461, 1614, 1630, 1657, 1795, 1829, 1862, 2357, 3158, 3429, 5364, 5444, 5673, 5957, 5979, 7516, 8231, 8385, 8880, 9369, 9682, 11058, 11069, 11382, 12171, 12430, 12712, 12793, 13028, 13347, 13464, 13533, 13605, 13789, 14211, 14886, 14951

157: 83, 107, 1381, 1466, 1629, 1716, 1841, 2514, 2558, 2559, 2633, 2939, 3931, 4713, 5123, 5195, 6252, 8044, 8236, 8288, 9164, 9614, 10244, 10669, 11081, 11087, 11093, 11160, 11328, 11460, 12018, 12920, 13494, 13544, 13546, 14755, 14816

158: 152, 553, 754, 789, 1128, 1628, 1706, 1847, 2298, 2602, 2710, 3020, 3116, 3441, 4478, 4531, 5087, 5142, 5965, 6830, 7016, 7024, 8110, 9634, 10070, 10672, 10728, 11033, 11075, 11084, 11218, 11605, 12456, 12711, 12963, 13801, 14756

159: 1359, 1360, 1668, 2355, 2508, 2579, 2606, 3094, 3376, 4847, 5097, 5435, 5537, 6244, 6377, 6553, 6755, 8033, 8229, 8267, 8918, 9574, 9859, 11145, 11939, 12139, 12385, 12583, 12873, 12951, 13089, 14233, 14262,

180: 6152, 6275, 6607, 7031, 7462, 8031, 8253, 8459, 8638, 9641, 9683, 9781, 9860, 10214, 10902, 11268, 11518, 11737, 12187, 12959, 13105, 13344, 13378, 14702, 14717, 14895, 14899

181: 655, 772, 970, 1204, 1617, 2546, 2619, 2654, 3100, 3383, 3544, 4484, 5790, 6004, 6313, 6499, 7018, 7038, 8392, 8498, 9502, 9650, 9653, 9862, 12019, 12355, 12363, 13197, 14485, 14902

182: 1198, 1282, 1429, 1876, 2240, 2674, 3006, 3581, 6502, 6547, 6561, 6662, 6687, 6982, 9638, 10930, 11200, 11744, 12365, 12688, 13122, 13390, 13402, 13623, 13971, 13976, 14290, 14730

183: 708, 1452, 1560, 1587, 1812, 1866, 3359, 4525, 5116, 5211, 5402, 5421, 6186, 6322, 6993, 8626, 8931, 9196, 12587, 12802, 13218, 13219, 13604, 13965, 14559, 14766, 14769

184: 142, 479, 1185, 1186, 1289, 2282, 3027, 3101, 4517, 4752, 6507, 7217, 8048, 8480, 8695, 8696, 9627, 9786, 9961, 11112, 11447, 11984, 12558, 13099, 13710, 15002

185: 659, 753, 993, 1154, 2059, 2203, 2268, 2289, 2532, 2699, 5284, 6514, 7100, 7315, 7463, 8235, 8252, 8300, 8493, 8607, 8742, 9142, 9534, 9861, 9965, 10323, 10724, 10798, 11111, 11443, 11534, 11740, 12426, 12609, 12691, 12734, 12992, 13167, 13351, 13611, 13645

186: 746, 773, 783, 1468, 2297, 2302, 2327, 2366, 2708, 2893, 3098, 3504, 5054, 5224, 6025, 6650, 8128, 8240, 8286, 8771, 9725, 10074, 10799, 11048, 11459, 11700, 12756, 13063, 13182, 13343, 13478, 14008, 14031

187: 121, 757, 1233, 2726, 2803, 3366, 4705, 4751, 4756, 5114, 5191, 6147, 6264, 6515, 8461, 8492, 10247, 11481, 11499, 12229, 12739, 12941, 13091, 13470, 14959

188: 686, 711, 1212, 2372, 2644, 2938, 4518, 5014, 5298, 5675, 6305, 6529, 6757, 8393, 10639, 11243, 11454, 12454, 13536, 14015

189: 1766, 2518, 5356, 6490, 6542, 7057, 7249, 7850, 8218, 8423, 8538, 9462, 12867, 13073, 13159, 13916, 14565, 14614, 14736

190: 338, 648, 685, 1098, 1197, 1353, 1354, 1586, 2305, 2626, 2951, 3448, 3541, 3552, 5295, 5333, 5377, 5862, 6479, 7229, 8003, 8019, 8580, 8930, 9322, 9355, 9595, 9681, 11064, 11326, 11753, 12763, 13220, 13615

191: 126, 251, 840, 2402, 2568, 3126, 5088, 5091, 5670, 6259, 6963, 7278, 8646, 8686, 9544, 9694, 9857, 12106, 12398, 12809, 13626, 14717

192: 669, 997, 1418, 1424, 1426, 1712, 1816, 2730, 3336, 4163, 4821, 5119, 5296, 5574, 6489, 6877, 7034, 7103, 8005, 8488, 9346, 9596, 9664, 9692, 9777, 9778, 9849, 9963, 10805, 11596, 11751, 12017, 13632, 13636, 13712, 14733

193: 662, 673, 1252, 2296, 2399, 2612, 2799, 3553, 4271, 5046, 5085, 5125, 5453, 7270, 7279, 9675, 9780, 10605, 12399, 12504, 12840, 13790, 14786

194: 646, 1213, 1279, 1444, 2707, 2783, 2834, 3295, 3353, 4536, 5355, 6019, 6988, 8088, 9240, 9430, 9535, 9631, 9966, 11498, 13047, 13052, 13196, 14012

195: 524, 589, 986, 1350, 1705, 1898, 2519, 2800, 4534, 6257, 6427, 6469, 6541, 6554, 7030, 7344, 7584, 8299, 8841, 8936, 9533, 9554, 9609, 9648, 9677, 9718, 9727, 10218, 10246, 11199, 11340, 11464, 11538, 12713, 13203, 13613, 14767, 14771

196: 697, 1107, 1254, 1281, 1386, 1451, 2002, 2973, 3127, 3535, 6555, 8579, 9546, 11198, 11244, 12789, 14729, 14966

197: 124, 747, 2459, 2578, 2798, 3314, 5192, 5260, 5986, 7028, 7153, 7208, 7274, 7482, 8029, 8239, 8470, 8486, 9416, 9879, 10722, 11317, 12590, 12828, 13053, 13133, 13463, 14020

198: 456, 585, 649, 925, 2555, 2687, 2801, 3393, 3555, 4527, 4999, 5033, 5376, 5399, 5951, 6426, 6748, 6875, 7314, 7440, 8004, 8246, 8464, 8779, 9576, 9678, 12028, 13064, 13162, 13618, 13676, 14004, 14006

199: 117, 130, 667, 1155, 1376, 1842, 2880, 3080, 3554, 4532, 5362, 5419, 6000, 7309, 7312, 8042, 8609, 8935, 9327, 9458, 11104, 11219, 12835, 13131, 13472, 14535, 14815

200: 661, 745, 1143, 1247, 1351, 1387, 1390, 1494, 2309, 2607, 2652, 2974, 2975, 2995, 3474, 4273, 5037, 5859, 5876, 6933, 6992, 7938, 8308, 8332, 8409, 8443, 8691, 9217, 9219, 9691, 9698, 11086, 12053, 12176, 12379, 12459, 12568, 12718, 12839, 12851, 13181, 13213, 13488, 13492, 13587, 13848, 14293

201: 156, 1329, 1711, 1831, 1839, 3388, 4258, 5340, 6265, 8232, 8262, 8682, 8801, 9597, 9875, 11128, 12827, 13058, 14389, 15022, 15028

202: 129, 1237, 1385, 1794, 2667, 2711, 2818, 2976, 3044, 3411, 4535, 4828, 6148, 6188, 6368, 8337, 8627, 9668, 9686, 11520, 12561, 13118, 13204, 13354, 14711, 14780

203: 642, 1235, 1636, 2585, 3387, 4275, 5134, 6424, 6521, 8457, 8610, 9858, 11562, 12840, 13067, 13988, 14783, 14787, 14914

204: 382, 460, 1145, 1214, 1363, 1384, 2645, 2665, 2982, 5143, 5283, 6501, 6532, 6579, 7112, 7286, 8170, 8175, 8241, 8274, 8983, 9498, 11479, 11566, 11923, 13844, 14522, 14617, 14722, 14785

205: 633, 1203, 2317, 2323, 2495, 2650, 2669, 3179, 3528, 3546, 5058, 5262, 5982, 6190, 6300, 6425, 6556, 6598, 7092, 8319, 8390, 8426, 9505, 9536, 9598,

241: 13672, 13820, 15024
242: 721, 1266, 2658, 2762, 4695, 5057, 5139, 7271, 8740, 9707, 9877, 13068, 13370, 14905, 14948
243: 1199, 2611, 3291, 3792, 7411, 7473, 8115, 8466, 8633, 9210, 9541, 10176, 10252, 11742, 11758, 12571, 12973, 14361, 14571, 14605
244: 1244, 1383, 1848, 2908, 3396, 6261, 6326, 6434, 8201, 8277, 8560, 11743, 12853, 14001
245: 147, 636, 663, 863, 1108, 1243, 1568, 3327, 5111, 6001, 6167, 6495, 6513, 6552, 7635, 8556, 9513, 10242, 12837, 12860, 14207, 14625, 14957
246: 1268, 2370, 2977, 4750, 4845, 6007, 6415, 6430, 8999, 9451, 12845, 13225, 14504, 14624
247: 3351, 4697, 5873, 6850, 8335, 8437, 8499, 13627
248: 684, 1231, 1436, 1608, 2533, 2675, 2822, 3520, 7046, 7412, 7783, 8043, 8275, 8522, 9633, 9739, 11769, 12486, 12549, 12690, 15001
249: 1315, 1567, 6472, 8009, 8435, 11768, 12822, 12939
250: 1489, 2577, 2706, 3599, 5281, 7045, 7470, 7477, 8394, 9841, 10974, 11491, 11763, 12569, 13128
251: 1321, 3323, 8014, 13625, 14604, 14969
252: 1125, 1174, 1234, 1341, 2547, 3322, 6575, 8016, 9493, 10711, 12552, 13783, 15027
253: 595, 1129, 1196, 2816, 3214, 5535, 6134, 8436, 8681, 12751, 14040, 14603
254: 762, 1598, 2252, 2580, 8463, 9216, 9218, 9742, 9825, 10198, 10326
255: 1295, 3102, 3213, 5047, 5060, 5220, 6029, 6135, 8247, 8315, 9239, 11502, 12373, 12401, 13096, 13547, 13614, 14023, 14094, 14516, 14724
256: 593, 712, 1261, 1287, 1583, 1616, 2534, 2703, 3049, 3329, 3352, 5049, 5136, 8625, 8741, 9522, 10695, 12551, 12843, 13041, 13074
257: 1270, 1565, 2325, 3550, 6010, 6026, 8132, 12372, 12929, 14627
258: 552, 1388, 2415, 6536, 6538, 8037, 8203, 10691, 12704, 13630
259: 1305, 2403, 8017, 13072, 14956
260: 466, 1216, 2385, 4749, 4829, 5108, 5870, 5874, 5875, 6602, 8565, 9487, 10721, 12485, 12820, 12942, 12945, 14484
261: 17, 1291, 5272, 6030, 6965, 7689, 9509, 12429, 12487, 12573
262: 644, 1871, 4864, 6072, 6989, 7410, 11042, 11348, 12553, 14985
263: 3087, 5985
264: 6980, 10712, 12972, 14689
265: 154, 961, 1120, 7472, 8574, 9348, 13071
266: 7160, 8559, 9486, 11490, 12753, 13170, 13694, 14964
267: 1446, 3519, 6961, 8462, 9518
268: 522, 1222, 1246, 5064, 6131, 8803, 10709, 12603, 13535, 14788
269: 1183, 1431, 5983, 6482, 8269, 8502, 9563, 10332, 12759
270: 1248, 1285, 2728, 4519a, 4626, 5882, 6379, 8023, 9485, 9545, 9548, 10217, 10277, 10693, 11766, 12943, 13693, 13715, 14388, 14694
271: 1238, 1311, 7275, 8501, 8503, 9525, 11521, 12957, 14533
272: 3512, 5871, 7471, 12347, 13631, 13795, 13935
273: 1485, 1511, 1520, 6024, 9454, 9969, 11513
274: 3444, 5093, 5273, 9220, 9488, 9823, 11557, 11765, 11767, 12436, 12697
275: 3384, 5100, 6964, 8777, 12926, 14688
276: 1250, 2456, 4627, 5271, 5952, 6537, 6978, 9654, 12780, 15026
277: 3325, 6930, 6975, 10640, 13674, 14515
278: 1306, 2461, 2548, 5199, 6937, 8554, 9373
279: 8606, 11501, 12817
280: 567, 584, 743, 1208, 1249, 2382, 6133, 8013, 8543, 8555, 8797, 11066, 12965, 13055, 13491, 13784, 13817, 13936
281: 1322, 8024, 9130, 14610
282: 1286, 5270, 8688, 9652, 11319, 12369, 12550, 12922, 14611, 14926
283: 1230, 2304, 3040, 8513, 9131, 10719, 11203, 11204, 12930, 13095, 14596
284: 472 732, 11205, 13465, 14784
285: 564, 631, 2463, 3524, 5379, 6533, 7277, 7758, 8011, 8514, 11227, 12350, 12931
286: 1195, 2333, 2420, 5146, 8039, 8953, 13489
287: 1310, 5197, 8010, 8012, 8552
288: 1147, 1318, 1572, 2391, 5053, 6075, 6601, 8568, 13624, 14155
289: 575, 1241, 1245, 12921
290: 1221, 1225, 1313, 2371, 2395, 3065, 6576, 12428, 14612
291: 716, 1239, 9539, 12938, 13684
292: 1251, 2823, 4267, 4269
293: 1276, 2505, 3321, 8562, 8687, 8689, 8735, 8736, 8737, 10701
294: 1877, 2324, 2394, 2414, 6009, 9267, 12689, 13035, 14960
295: 555, 1002, 1122, 1271, 6436, 6484, 6485, 7467, 8553, 9565, 9566, 11043, 12656, 14616
296: 13048, 14734
297: 1224, 3357, 4573, 14615
298: 557, 699, 1205, 2497, 5171, 8563, 12949, 14887
299: 7039, 13685, 13967
300: 1317, 2383, 3365, 3527, 3682, 6380, 6855, 6968, 7099, 7420, 9321, 9492, 12755, 14963
301: 8510, 9265, 10718
302: 15025
303: 1206, 1220, 5302, 6969, 12727, 13479, 14812
304: 4268, 8690, 9826, 12368
305: 1571, 2392, 7744, 9241, 12349
306: 1280, 2464, 6970, 13066, 13687
308: 1448, 6972, 9450
309: 9447
310: 1273, 2381, 3494, 4784, 8561, 9449, 11736, 14620, 14735
312: 1255, 1272, 5660, 9819
313: 1304, 12384
314: 554, 556, 5538, 7041, 12946
315: 1637, 2389, 3717, 8615, 13671
316: 6971
317: 10727
318: 1301, 2496, 11616
319: 1223, 8525, 12545, 12815
320: 1228, 1599, 5881,

BOILING POINT INDEX OF ORGANIC COMPOUNDS

Temperatures in °C; where there is a range of values, the compound is listed according to the lower value.

−164: 9015

−129: 9093

−103: 6840

−88: 6603

−84: 443, 9089

−83: 4741

−82: 9100

−81: 9047

−79: 6651, 6663

−78: 4715, 9079

−77: 4187

−76: 6876

−72: 14934

−64: 14500

−59: 9033

−57: 13296

−56: 8640

−51: 9067

−50: 4742

−48: 12342

−47: 6676, 12025, 12257

−42: 6659, 11772

−40: 9037

−38: 581, 6618

−37: 6159, 6646

−36: 11880

−35: 6762

−34: 580, 11770

−32: 5824, 14501

−31: 1585, 9088

−30: 445

−29: 9061, 12293

−28: 324

−26: 6668, 6847, 11379

−25: 9155

−24: 4211, 6632, 7055, 9035

−23: 12333

−22: 9084, 14502

−21: 5429, 7047, 9123

−20: 9090, 13274

−14: 11374

−13: 14931

−12: 9026, 9174, 11866

−11: 1415, 3808, 11872

−7: 13285

−6: 3942, 4104, 4189, 6158, 9148

−4: 3638, 5442, 13290

−3: 12286

−2: 6170

0: 3672, 3981, 4014, 6657, 9113, 11843

1: 1584, 1863, 11360, 11867

2: 1411, 11491, 14514

3: 3715, 3980, 5456, 6628, 9020

4: 444, 3972, 5833

5: 3151, 4065, 14928

6: 3657, 4740, 6619, 9124

7: 6156, 7096, 11359

8: 4195

9: 9064, 9797, 11847

10: 3631, 3668, 6809

12: 3655, 3814, 5389, 5431, 5438, 6614, 6775, 9205

13: 5980, 6920

14: 9077

15: 14930

16: 6754, 6814

17: 11385

18: 9029

19: 13176

20: 21, 439, 3576, 3971, 5827

21: 3572, 6867, 9068, 11361

22: 12266

23: 4769, 5416, 11386

24: 6225, 9056

25: 5807, 8161, 9038, 10496, 11371, 11373, 11865, 14953

26: 4064, 7514, 10307, 13323

27: 3777, 4202, 9947, 10422

28: 5830, 14932

29: 1465, 4199, 5829, 10537, 14089

30: 6634, 10517, 10520, 13514

31: 3970, 7083, 7116, 9119

32: 448, 3761, 5966, 6627, 9191, 11777, 12250, 12264

33: 1421, 3667, 4225, 6173, 7189

34: 3661, 4068, 5831, 6165, 6791, 8575, 8642, 9041, 11857, 12300

35: 6704, 6839, 11805, 12254

36: 1407, 1409, 5441, 6166, 6804, 6807, 9023, 9906, 10340, 10539, 10540, 11693

37: 5828, 6863, 6864, 9199, 11368, 11380, 12265

38: 4013, 6610, 6770, 7932, 9168, 9190, 10357

39: 178, 3662, 4198, 6349, 13372, 14494

40: 481, 3636, 4450, 5756, 8661, 9060, 10553, 10618

41: 3964, 3977, 7778, 10316, 11842, 11926

42: 3639, 3646, 5733, 7940, 8981, 9082, 10301, 10595, 11721, 12022, 12263

43: 7134, 12222, 14795

44: 4091, 5806, 5832, 10298, 10619, 12225

45: 610, 6673, 7052, 7097, 7934, 7964, 12267, 12318, 14797

46: 616, 4717, 6625, 6630, 7532, 7537, 11795, 12253, 13513

47: 6675, 6866, 10209, 11773, 12246, 12277, 13866

48: 3335, 7982, 9994, 10308, 11362, 11771, 12042, 12260

49: 3745, 5758, 5762, 6343, 7191, 7968, 10315

50: 435, 4227, 5750, 5826, 6793, 6919, 6922, 7423, 7533, 7542, 7933, 9162

51: 197, 493, 8684, 10382, 10534

52: 480, 494, 1416, 3813, 4194, 4995, 5691, 5837, 6780, 8345, 8359, 11937, 12023, 12221

53: 4056, 5800, 6046, 9080, 10535, 11398, 12278

54: 5743, 7079, 7536, 9197, 10626, 14436

55: 3913, 3963, 4066, 4148, 5310, 5936, 6841, 8361, 9118, 10629, 11796, 12331, 13332, 14929

56: 306, 1463, 3759, 4658, 6092, 7406, 7408, 8654, 9034, 10012, 10562, 10628, 12005, 13269, 14935

57: 225, 2219, 4739, 6626, 7739, 7824, 7854, 7889, 7952, 10019, 10303, 10411, 10907, 12259, 12290, 13245

58: 604, 676, 3746, 3875, 3953, 4146, 4228, 4968, 6640, 8658, 8663, 8907, 9948, 10563, 12203, 12258

59: 3643, 3758, 4988, 5640, 6615, 7731, 8364, 9157, 9169, 10593, 11788, 12199, 12252, 12289, 12332, 14791

60: 3247, 3647, 3664, 3978, 4147, 4971, 5538, 5540, 6806, 6865, 7650, 10406, 10533, 10536, 11437, 13270, 14497

61: 1989, 3805, 5430, 6823, 6923, 7897, 7962, 9050, 9098, 10127, 10627, 11265, 12080, 12128, 14879

62: 326, 3994, 4687, 7044, 7528, 7661, 7665, 7848, 7910, 8292, 9081, 10023, 10119, 10302, 10402, 13507, 13827

63: 442, 1462, 3675, 3676, 3677, 3757, 3902, 3949, 4085, 4086, 4087, 4103a, 4154, 4953, 6825, 6826, 7136, 7543, 7941, 7947, 8078, 10187, 10408, 12060, 12285, 12340, 14794

64: 37, 3743, 3955, 3968, 4035, 4050, 4144, 4210, 4236, 7539, 7781, 9058, 9173, 10011

65: 605, 624, 3571, 6821, 7137, 7407, 7728, 8996, 9127, 9128, 11723, 12248, 12256, 12335, 12539, 14939

66: 613, 1673, 3219, 4052, 4662, 4680, 6100, 6364, 6810, 7960, 10597, 11402, 13299

67: 615, 3560, 3899, 4016, 4153, 4966, 6090, 7142, 7202, 7541,

67: 7766, 7961, 10559, 10561, 11936, 13515

68: 484, 3640, 3649, 3706, 3707, 3754, 3950, 3954, 4100, 4101, 4107, 4113, 4188, 4192, 4196, 4882, 5710, 6719, 7088, 7547, 7727, 7790, 7954, 7955, 8641, 9002, 9021, 10132, 10300, 10505, 12302

69: 2230, 5816, 5836, 7685, 7760, 8660, 9207, 10201, 11833, 12029, 12269

70: 321, 609, 2260, 3089, 3337, 3967, 3993, 4206, 4226, 4969, 5440, 5624, 6356, 6824, 7173, 7531, 7983, 9949, 9962, 10029, 10460, 10560, 11635, 12262, 12637, 13256, 13872, 14486

71: 41, 1401, 2356, 3573, 4108, 4145, 4193, 5776, 6612, 6871, 7991, 8890, 10033, 10041, 11782, 12251

72: 270, 3201, 3350, 4666, 4683, 4952, 5697, 6653, 6805, 6869, 7213, 7617, 7906, 9163, 10035, 10530, 10631, 10995, 14493, 14927

73: 1502, 2344, 3199, 3604, 4008, 4098, 4123, 5934, 6631, 7726, 7885, 8450, 10039, 10310, 10620, 11363, 11878, 12047, 12049, 12337, 13278, 13439, 13959

74: 3866, 5648, 5944, 6219, 6223, 6672, 6724, 7534, 7610, 10025, 13910, 14672

75: 322, 519, 752, 3660, 3671, 3958, 4019, 4176, 4386, 5518, 5718, 5815, 6721, 7242, 9022, 9951, 10128, 12303, 13280, 13286

76: 434, 446, 1404, 2134, 3663, 3751, 4074, 4207, 4331, 4346, 5500, 5550, 5933, 7725, 7746, 9092, 10306, 10499, 10557, 10622, 11400, 12272, 12404, 13360

77: 57, 204, 516, 3673, 3679, 3979, 4080, 4128, 4197, 5604, 7122, 7548, 7747, 7877, 7908, 8524, 8656, 9178, 10529, 11722, 12004, 12273, 12276, 12544, 13868, 13925

78: 3698, 4015, 4099, 5607, 5735, 6220, 6412, 6708, 7168, 7752, 7896, 7935, 7936, 10077, 13361, 13609, 14406, 14513, 14847

79: 3711, 3918, 4664, 4682, 5589, 5610, 6654, 7123, 7887, 8344, 9161, 10092, 10133, 10387, 11798, 11809, 12102, 12169, 12317, 13510

80: 509, 571, 1175, 1778, 1882, 2163, 2744, 3204, 3705, 3807, 3861, 4335, 4370, 5483, 5491, 5513, 5681, 6163, 6216, 6698, 6785, 7145, 7738, 7740, 7925, 8980, 9054, 9175, 10120, 10376, 10391, 12002, 12048, 12217, 12328, 12388, 13034, 13272, 13437, 14575

81: 342, 563, 2023, 4057, 4122, 5606, 6064, 6773, 7087, 7901, 7995, 8079, 9206, 9801, 10528, 10594, 10666, 11774, 12669, 13079

82: 3864, 4076, 4435, 4915, 5626, 5709, 5711, 6694, 6829, 7563, 7631, 9016, 9783, 9955, 11971, 11994, 12245, 13829, 14122, 14499

83: 620, 2123, 2138, 3856, 4143, 4186, 4230, 5678, 5708, 6629, 6779, 6893, 7073, 7628, 7751, 8657, 10359, 10555, 10630, 12144, 12319, 12489, 13240, 13264, 14496, 14630, 14840, 14880

84: 3680, 3868, 3925, 3988, 3989, 5751, 6325, 6334, 6828, 6843, 7529, 7742, 7950, 7993, 9194, 9803, 10527, 11797, 12046, 12282, 13861, 14448, 14940

85: 27, 433, 2022, 2024, 2153, 2786, 3709, 3786, 3854, 3855, 4131, 4141, 4142, 4264, 4350, 4373, 4408, 4420, 4932, 5486, 5611, 6396, 7237, 7526, 7749, 7770, 7872, 7981, 8077, 8455, 8976, 9925, 10299, 10532, 11369, 11733, 11803, 11858, 12405, 13230, 13897

86: 3208, 3348, 3937, 3943, 3952, 4907, 5913, 5945, 7146, 7166, 7550, 7753, 7774, 8360, 9182, 10084, 10086, 10394, 10685, 12044, 12389, 13279, 13499, 13927, 14845

87: 2511, 3635, 4964, 5583, 5614, 5770, 6813, 6881, 7619, 7951, 8101, 8342, 9791, 10037, 10134, 10556, 10623, 11430, 11982, 12309, 12781, 13165, 13840

88: 3326, 3633, 3823, 3848, 3945, 3976, 4026, 4140, 4654, 4890, 4912, 5385, 5886, 7140, 7513, 7686, 7946, 9920, 10270, 11826, 11829, 12009, 12297, 12334, 12390, 12662, 13865, 13904

89: 588, 2107, 2247, 3202, 3333, 3969, 6172, 6357, 7051, 8521, 8624, 9005, 10104, 10389, 11870, 12542, 13238, 14483

90: 32, 263, 619, 2176, 3594, 3693, 3747, 3795, 4237, 4728, 4884, 5294, 5497, 5778, 5814, 6057, 7676, 7733, 7736, 7817, 7818, 7953, 9012, 9024, 10258, 10548, 12031, 12261, 12645, 12675, 13273, 13502, 14174, 14472

91: 507, 2102, 3659, 3692, 4096, 4097, 5042, 5627, 5772, 5834, 6633, 7813, 7949, 9059, 9181, 10342, 10343, 12301, 13336, 13835, 14936

92: 44, 261, 525, 1396, 3710, 4204, 4231, 4232, 4412, 4460, 4462, 4892, 4895, 4956, 5635a, 5636, 6102, 6344, 6722, 6776, 6859, 7132, 7819, 7820, 7821, 7837, 7992, 9235, 10040, 10050, 11814, 11981, 12127, 12200, 12275, 12684, 13164, 13828, 14175

93: 2246, 3261, 3644, 3985, 4005, 4185, 4646, 5582, 5584, 5714, 5811, 6635, 6799, 7129, 7428, 7507, 7549, 7612, 7643, 7903, 9203, 10305, 10371, 10399, 10523, 11790, 12162, 12230, 12512, 13253, 13261, 13526, 14435

94: 496, 611, 1403, 3654, 3828, 3930, 3944, 3957, 3995, 4073, 5579, 5741, 6218, 7128, 7133, 7508, 7687, 8346, 8347, 8479, 9069, 9192, 10531, 10558, 10684, 11613, 11808, 11810, 11964, 12281, 12312, 13797, 13834, 14480, 14521, 14792, 14821

95: 2038, 2136, 2444, 3574, 3674, 3784, 3876, 4081, 4334, 4910, 4967, 5044, 5628, 5840, 5842, 6418, 6761, 7219, 7653, 7654, 7729, 7959, 7999, 8362, 8618, 9193, 10521, 10547, 10600, 11389, 12001, 13249, 13308, 13741, 13880, 14636

138: 3735, 3739, 3774, 3787, 3895, 4591, 4886, 4959, 5369, 5466, 5560, 5561, 5629, 5752, 6270, 6347, 7409, 7645, 7776, 7809, 7882, 7980, 8131, 9030, 9233, 10129, 10480, 10577, 10959, 11159, 11266, 11609, 11640, 11643, 11792, 12212, 12268, 12311, 12598, 12983, 13233, 13824, 14226, 15003

139: 177, 410, 432, 626, 2069, 2216, 3259, 3313, 4434, 4575, 4611, 4793, 5802, 6052, 6350, 7147, 7655, 7833, 8354, 8621, 10080, 10442, 10479, 10571, 10578, 10615, 12416, 12475, 13252, 13887, 14986

140: 759, 761, 1124, 1606, 1741, 1805, 2495, 2540, 3163, 3438, 3463, 3607, 3776, 3782, 3815, 4138, 4181, 4381, 4456, 4553, 4630, 4672, 4875, 5401, 5407, 5409, 5562, 5591, 5602, 5749, 5798, 5960, 5967, 6312, 6616, 7054, 7346, 7500, 7553, 7558, 7649, 7810, 7811, 7836, 7883, 7976, 8452, 8672, 8858, 9152, 9436, 10398, 10414, 10464, 10575, 10583, 10674, 11261, 11641, 11642, 11775, 11821, 11891, 11959, 12082, 12157, 12195, 12509, 12511, 12593, 13174, 13297, 13310, 13706, 13804, 13889, 13947, 14068, 14247, 14477, 14883, 14921, 14971

141: 269, 486, 2374, 2438, 3217, 3459, 3893, 3898, 6079, 7561, 7675, 7768, 7797, 8505, 8507, 8857, 9010, 10057, 10125, 10338, 10390, 10417, 11184, 11800, 11941, 12037, 12072, 12271, 13445, 13448, 13800, 13888, 13962, 14803, 14870, 14909

142: 227, 255, 848, 3870, 4022, 4117, 4266, 4361, 4432, 4567, 4794, 4862, 5682, 6065, 6406, 6473, 6797, 6884, 7183, 7427, 7524, 7853, 7914, 7915, 8725, 9083, 9133, 10048, 10053, 10056, 10347, 10351, 10590, 11196, 11267, 11651, 12274, 12298, 12321, 12514, 12599, 12601, 12641, 13483, 14884

143: 305, 3152, 3740, 3806, 3872, 3929, 4217, 4383, 4442, 4948, 5506, 5730, 5753, 5803, 7886, 7900, 8512, 9968, 10054, 10055, 10138, 10156, 10355, 10366, 10482, 10486, 11652, 11784, 11911, 12080a, 12086, 12219, 12367, 12418, 13300, 14069, 14449

144: 45, 223, 428, 3606, 3650, 3738, 3769, 3871, 4055, 4125, 4448, 4734, 4929, 5071, 5463, 5804, 6851, 6915, 7535, 7540, 7638, 7699, 7767, 7916, 8210, 8341, 8662, 10151, 10404, 11546, 11548, 11779, 12013, 12079, 12124, 12205, 12343, 12419, 13244, 13744, 13808, 13891, 14837, 14885, 14908, 14972

145: 192, 325, 751, 3164, 3307, 3613, 3728, 3824, 3873, 4260, 4298, 4301, 4425, 5161, 5509, 5546, 5695, 5844, 6786, 6837, 6842, 7127, 7187, 7224, 7919, 7937, 8002, 8069, 8724, 8791, 8896, 9183, 9687, 9809, 10362, 10405, 10925, 10979, 11106, 11174, 11595, 11646, 11892, 11977, 12091, 12644, 13451, 13495, 14259

146: 491, 870, 883, 1719, 1854, 3597, 3598, 3694, 3874, 6333, 6667, 6690, 6713, 7143, 7225, 7671, 7939, 8449, 9231, 9435, 9437, 9935, 9942, 10385, 11594, 11915, 11986, 12224, 12519, 13316, 13831, 14045, 14082, 14374

147: 239, 400, 505, 2404, 2960, 3770, 3936, 3938, 4640, 4732, 4924, 5382, 6743, 7150, 7431, 7573, 7632, 7902, 7921, 7979, 8854, 9112, 9256, 9438, 9943, 10101, 10200, 10349, 10381, 10472, 10475, 10492, 10522, 10609, 10770, 12097, 12474, 12640, 13371

148: 26, 237, 238, 390, 656, 1910, 3768, 4120, 4399, 4402, 4449, 4459, 5738, 5744, 7365, 8669, 9031, 10208, 10348, 10374, 10461, 11209, 11303, 11692, 11999, 12460, 12473, 12476, 12495, 12798, 13024, 13152, 13234, 13251, 13338

149: 193, 218, 252, 341, 1686, 2869, 3809, 4316, 6620, 6710, 6794, 7060, 7912, 9097, 9146, 10014, 10313, 10386, 11612, 11687, 11822, 11860, 13163, 13871

150: 2188, 2313, 2721, 3143, 3609, 3894, 4058, 4288, 4410, 4446, 4465, 4936, 5074, 5218, 5358, 5458, 5547, 5742, 5897, 6735, 6905, 6914, 7101, 7486, 7603, 7772, 7888, 7977, 8673, 8675, 8683, 9055, 9428, 9816, 9918, 9950, 9954, 10367, 11002, 11014, 11138, 11194, 11794, 11881, 11987, 12213, 14052, 14067, 14881, 15023

151: 536, 1724, 3961, 4002, 4295, 4599, 4996, 5219, 5661, 6723, 6781, 6946, 6947, 6951, 7188, 7627, 7668, 7852, 7922, 8214, 10463, 10952, 11682, 11913, 12099, 14027

152: 200, 425, 558, 625, 658, 693, 1952, 2237, 3742, 3803, 3932, 4003, 4224, 4279, 4297, 4819, 5217, 5264, 5394, 5603, 5777, 5981, 6311, 7517, 7660, 7695, 8056, 8058, 8412, 8906, 9226, 9399, 10462, 10489, 10493, 10510, 11257, 11653, 11793, 12185, 12216, 12330, 12600, 13334, 13740, 13760, 13826, 14474

153: 412, 597, 963, 1783, 1922, 2435, 3234, 3729, 3781, 4392, 4479, 4617, 5769, 8863, 9953, 10099, 10112, 10291, 10410, 11883, 12170, 12310, 12413, 13321, 13380, 14441, 14456, 14579, 14869

154: 1056, 1057, 1914, 3153, 3206, 3619, 3892, 4103, 4877, 4970, 5075, 5473, 5559, 6098, 6161, 6949, 7012, 7163, 7978, 7986, 8570, 8659, 10107, 10498, 10500, 10587, 11607, 12663, 13009, 13284, 13339, 13881, 14072, 14227

155: 316, 365, 490, 1025, 2180, 2397, 3741, 3956, 4004, 4308, 4367, 5592, 5621, 6063, 6330, 6742, 6948, 7081, 7192, 7378, 7379, 7509, 7567, 7574, 7640, 7644, 7669, 7765, 7779, 7867, 8943, 9007, 9316, 10297, 10465, 10572, 10604, 11056, 11347, 11610, 11654, 11691, 11720,

155: 12108, 12395, 12396, 12620, 12681, 12917, 14401, 14878

156: 207, 298, 825, 1157, 1457, 1883, 1921, 2416, 3296, 4041, 4115, 4961, 5212, 5496, 5558, 5767, 6778, 7169, 7523, 7613, 7666, 7974, 7975, 8805, 8975, 10488, 11584, 11655, 11894, 11973, 12393, 13243, 14458, 14573, 14872

157: 28, 40, 64, 217, 3486, 3722, 4296, 5072, 5073, 5601, 5717, 6408, 6718, 7615, 7697, 7985, 8709, 8790, 10169, 10403, 10459, 10467, 11433, 11716, 12336, 12410, 12683, 12745, 13075, 13277, 13658, 13911, 14026, 14066

158: 216, 403, 409, 419, 515, 520, 539, 3309, 3349, 3716, 4481, 4482, 4644, 4726, 5210, 5221, 5222, 5223, 5647, 5763, 6734, 6741, 6950, 7114, 7355, 7520, 7636, 7868, 7913, 8244, 8974, 9046, 9057, 9952, 11713, 11990, 13289, 13595, 13809, 13852, 14882

159: 311, 374, 2236, 3580, 3891, 4049, 4134, 4135, 4136, 4368, 5563, 6051, 6078, 6904, 6908, 7530, 7658, 8789, 10047, 10483, 11882, 11933, 11972, 12078, 12527, 12537, 13876, 14078, 14937, 15024

160: 208, 510, 1127, 1862, 1927, 2780, 2913, 3134, 3877, 3928, 4093, 4137, 4480, 4636, 5041, 5906, 5907, 6638, 6738, 6944, 7235, 7608, 7609, 7656, 7722, 7898, 8213, 8506, 8871, 8989, 9006, 9273, 9880, 10046, 10153, 10236, 10458, 11071, 11141, 11143, 11375, 11545, 11621, 11628, 11669, 11787, 11859, 11920, 12178, 12538, 13452, 14523, 15030

161: 628, 1605, 2145, 3721, 4021, 4032, 4233, 4874, 4935, 5343, 5577, 7170, 7621, 7625, 7876, 8138, 9121, 9944, 10377, 11648, 11689, 11712, 12003, 13436, 14475

162: 182, 205, 240, 334, 1461, 1788, 2146, 2179, 2429, 2479, 3413, 3825, 4347, 4444, 4937, 5417, 6118, 6661, 7078, 7516, 7682, 7791, 7923, 8826, 9004, 9117, 9685, 9709, 9915, 9999, 10346, 10868, 11282, 11602, 11603, 11604, 11688, 12038, 12041, 12107, 12152, 12323, 13077, 13137, 13242, 13886, 14081, 14084, 14876, 14938

163: 23, 195, 337, 1892, 2113, 4216, 4234, 4300, 4566, 4615, 4618, 4927, 5225, 5412, 6067, 6342, 6731, 6880, 7620, 7637, 7862, 8788, 8792, 8856, 9076, 10009, 10024, 10030, 10096, 10113, 12168, 12294, 12329, 12488, 12636, 13076, 13333, 13516, 13912, 14028, 14029, 14106

164: 228, 1786, 2287, 3563, 4358, 4398, 5532, 5680, 5698, 5707, 5712, 5810, 5930, 5961, 6105, 6328, 6543, 6836, 7124, 7624, 7674, 7683, 7873, 7874, 8484, 8987, 10045, 10177, 10491, 10791, 11582, 11585, 11586, 11725, 12132, 12985, 13019, 13821, 13919, 14249

165: 74, 530, 1395, 1718, 2144, 2233, 3804, 4071, 4170, 4172, 4205, 4263, 4322, 4339, 4463, 4586, 5205, 5593, 5639, 6227, 7043, 7099, 7334, 7435a, 7527, 7796, 7909, 8355, 8483, 8762, 8778, 9244, 9584, 9928, 9929, 10118, 10511, 10896, 11694, 11885, 11949, 12174, 12247, 12295, 12497, 12676, 13320, 13340, 14864

166: 78, 679, 1460, 3718, 3924, 4294, 4357, 5226, 5227, 5228, 5495, 5587, 5594, 5596, 5729, 5775, 5863, 6728, 6796, 7614, 7680, 7904, 8482, 8671, 8743, 9247, 9383, 9909, 9926, 10088, 10161, 10962, 11632, 11953, 11958, 12525, 12986, 13725, 14765

167: 43, 1604, 3272, 4079, 4132, 4340, 4619, 4744, 4746, 5595, 5638, 5740, 6278, 6405, 7618, 7629, 7630, 7679, 7771, 7795, 8653, 8958, 8960, 9116, 9238, 9245, 9524, 9927, 9978, 9980, 10095, 10111, 10503, 10570, 11216, 11583, 11823, 12498, 13173, 13885, 14307, 14438, 14597, 14868

168: 2159, 3456, 3605, 3730, 3748, 4133, 4414, 4589, 4620, 4731, 4745, 5213, 5214, 5215, 5642, 5883, 7190, 7639, 7849, 7875, 8761, 8957, 8959, 9567, 9972, 9997, 10016, 10017, 10018, 10073, 10106, 10136, 10164, 10660, 10961, 11679, 11680, 11681, 11690, 11841, 11978, 12074, 12177, 13017, 13518, 13943, 14200, 14430, 14942

169: 31, 1708, 1991, 2119, 2286, 3342, 3343, 3916, 4017, 4042, 5530, 5641, 5713, 5721, 5761, 6733, 6921, 7364, 7696, 10020, 10992, 11581, 12233, 12358, 13519, 13532, 13833, 14201, 14580, 14943

170: 38, 119, 213, 349, 677, 1009, 1996, 2139, 3265, 3323, 3501, 3858, 3867, 4171, 4285, 4366, 4377, 4667, 4671, 4926, 5643, 5931, 5937, 5938, 6082, 6539, 6727, 6860, 7102, 7369, 7748, 7879, 7880, 7905, 7907, 8074, 8215, 8589, 8874, 8884, 8889, 8954, 8955, 8998, 9122, 9139, 9315, 10021, 10743, 11179, 11193, 11431, 11799, 12183, 12438, 12471, 12536, 12628, 12632, 13315, 13786, 14464, 14572, 14675

171: 215, 295, 4363, 4621, 4659, 5292, 5502, 5504, 5635, 6544, 6696, 6907, 7184, 7890, 7895, 8793, 8969, 10026, 10038, 10163, 10643, 10756, 10757, 12207, 12357, 12630, 12646, 13307, 13335, 13517, 14143, 14199, 14298, 14941

172: 229, 391, 487, 1986, 3859, 4441, 5634, 6739, 7253, 7595, 7788, 8678, 8995, 10063, 10066, 10603, 11353, 11406, 11686, 11801, 11806, 11927, 11945, 12467, 12469, 13022, 13562

173: 196, 314, 713, 1987, 1988, 2086, 3232, 4044, 4451, 4811, 5065, 5216, 5233, 5290, 5599, 5600, 5693, 5835, 6287, 6374, 6918, 7226, 7662, 7724, 7878, 7963, 8027, 8111, 8877, 9042, 9129, 9931, 10094, 10185, 10232, 10375, 10443, 10477, 10687, 10758,

173: 11551, 11846, 11943, 12496, 12664, 13130, 13154, 13355, 13482, 13590, 13703, 13884, 14252, 14446

174: 51, 922, 1408, 1710, 2089, 2520, 2875, 2992, 3612, 3670, 3681, 3719, 3842, 4105, 4817, 5590, 5597, 5598, 5694, 5899, 5942, 6891, 7501, 7518, 7538, 7578, 7657, 8630, 8905, 9260, 9286, 10123, 10476, 10843, 11633, 11683, 11684, 11685, 11989, 13843, 13858, 14272, 14825, 14856

175: 650, 2189, 2191, 2231, 2271, 2312, 3093, 3192, 3251, 3847, 3966, 4180, 4303, 4311, 4765, 5858, 5866, 5915, 5943, 6361, 7117, 7402, 7616, 7784, 7996, 8033, 8903, 9305, 10243, 10642, 10675, 11356, 11446, 11576, 11623, 11630, 11631, 11851, 11909, 12030, 12098, 12202, 13192, 13399, 13422, 13455, 13634, 13874, 14160, 14600

176: 1613, 1730, 2235, 2285, 3608, 3749, 4031, 4110, 4290, 4352, 4354, 4359, 4663, 5066, 6066, 6095, 6210, 6717, 7602, 7622, 7652, 8612, 8887, 8956, 9715, 9971, 10093, 10102, 10948, 11908, 11988, 12255, 12534, 13020, 13736, 14145

177: 210, 297, 1723, 2192, 2205, 3273, 3975, 4284, 4353, 4668, 4963, 5867, 5946, 6277, 6337, 7059, 7186, 7572, 7579, 7659, 7792, 8756, 10098, 10445, 10494, 11168, 13018, 13702, 13875, 14144

178: 1654, 2027, 2104, 2190, 2512, 3840, 3862, 4304, 4362, 4955, 5293, 5471, 5633, 5719, 5865, 6116, 6247, 6623, 6853, 7080, 7525, 7793, 8338, 8441, 8754, 8755, 9382, 10087, 10168, 10372, 11190, 11344, 11970, 12528, 12974, 13193, 14638, 14858

179: 429, 1751, 3238, 3562, 4190, 4355, 5503, 5722, 6372, 6636, 7049, 7199, 9998, 10028, 11543, 11876, 11884, 12084, 12171, 13822, 14424, 14824, 14860

180: 25, 290, 340, 924, 2082, 3207, 3791, 3838, 4251, 4588, 4946, 5492, 5544, 5573, 5588, 6004, 6248, 6385, 6655, 6751, 6898, 6909, 7001, 7064, 7294, 7363, 8645, 8986, 9288, 9309, 9610, 10085, 10097, 10378, 11191, 11279, 11820, 11931, 12026, 12110, 12394, 12456, 12479, 12540, 12731, 13262, 13359, 13421, 13460, 13461, 13564, 13565, 13766, 13899, 13945

181: 346, 451, 2099, 3841, 4394, 4678, 5498, 5501, 5609, 6245, 7210, 7816, 7857, 8867, 9111, 10767, 11629, 12624, 13086, 13877, 14065

182: 294, 450, 497, 1044, 1156, 1326, 2155, 3118, 3780, 3794, 3839, 4941, 5839, 6050, 6606, 7623, 8395, 8592, 8677, 9018, 9070, 9285, 9394, 10036, 10606, 11392, 11574, 12117, 12208, 14061, 14165, 14995

183: 33, 36, 339, 468, 1010, 1406, 1982, 2098, 2101, 3177, 3618, 3779, 4607, 4925, 5543, 5567, 5774, 5929, 6058, 7600, 7605, 7970, 8216, 9143, 9293, 10441, 11011, 11591, 11952, 12096, 12126, 12341, 13704, 14070, 14166, 14466, 14991

184: 232, 289, 765, 1890, 2028, 2400, 4034, 4213, 4323, 4345, 6715, 7115, 7497, 10008, 11848, 12052, 12526, 12962, 14048, 14074, 14124, 14378, 14488, 14576

185: 73, 230, 299, 756, 1769, 2010, 2897, 3196, 3257, 3312, 3328, 3330, 3832, 4017a, 4024, 4158, 4168, 4179, 4246, 4255, 4545, 4661, 4945, 4978, 5384, 5403, 5414, 5470, 5632, 5922, 6237, 6359, 6392, 6639, 6997, 7496, 7871, 7967, 8702, 8831, 8997, 9158, 9270, 9988, 10175, 10268, 10283, 10314, 10953, 11824, 11850, 12122, 12291, 13708, 14153, 14164, 14315, 14408, 14823, 14827

186: 531, 1053, 3778, 4289, 4293, 4310, 4321, 4372, 4688, 5325, 5613, 5909, 6292, 6650, 6815, 7113, 7372, 8745, 9882, 10076, 10454, 10957, 11536, 12094, 12403, 12533, 13414, 14381, 14800, 14808, 15008

187: 220, 231, 296, 923, 968, 1888, 1889, 2008, 2009, 2012, 2492, 3075, 4544, 4610, 4643, 4853, 4920, 4921, 4922, 4993, 5141, 5190, 5410, 5612, 5888, 6160, 6221, 6241, 6903, 7177, 7316, 7546, 7590, 8351, 8399, 9938, 10412, 10423, 10434, 10845, 11944, 12490, 12738, 13485, 14380, 14434, 14445, 14987, 14988, 15004

188: 236, 587, 1971, 2154, 2169, 2844, 2866, 3455, 3575, 3603, 3665, 4991, 5263, 5316, 5411, 5739, 5912, 6228, 6475, 6671, 6892, 7606, 8350, 8500, 9198, 9279, 9401, 9939, 9991, 10034, 10103, 10453, 10573, 11291, 11381, 11388, 11647, 11676, 12164, 12380, 12535, 13506, 13812, 14126, 14463, 14468, 14640, 14990

189: 2234, 2775, 2878, 3483, 4126, 4177, 4254, 4299, 4326, 4423, 4809, 5516, 5541, 5620, 5715, 5911, 6117, 6175, 6916, 7048, 7494, 9019, 9091, 9202, 9899, 10010, 11898, 11954, 12287, 13397, 14255, 14829, 14978

190: 293, 542, 1720, 1731, 2527, 2848, 2991, 3821, 3827, 3939, 4169, 4200, 4223, 4325, 4698, 4703, 4725, 4893, 4902, 5436, 5443, 6005, 6886, 7368, 7626, 7856, 8635, 8782, 9916, 9924, 10160, 10174, 10235, 10264, 10450, 10452, 10602, 11051, 11151, 11634, 11817, 11888, 11904, 12149, 12446, 13013, 13134, 13511, 13730, 14819, 14852, 14867

191: 618, 4324, 5304, 5617, 7175, 8343, 10180, 10229, 11625, 12064, 12515, 13177, 13231, 13733, 14425, 14447, 14455, 14861

192: 3, 212, 333, 1042, 1725, 2151, 2185, 3820, 3826, 4119, 4376, 4638, 4679, 5517, 6143, 6954, 9257, 9272, 9897, 10206, 10451, 11391, 12443, 12532, 13304, 13412, 13508, 13512, 14125, 14663, 14664, 14665, 14996

193: 49, 1035, 1553, 1998, 2183, 2193, 4483, 5616, 5667, 6062, 6084, 6125, 6286, 6336, 6952, 6953, 7108, 7822, 8297, 8401, 9028, 9189, 9894, 9898, 11636, 12112, 12466, 13342, 13608, 13842

194: 30, 61, 249, 279, 350, 872, 1043, 1046, 1981, 2011, 2842, 2953, 3211, 3264, 3268, 4157, 4329, 4537, 4990, 5724, 5746, 5958, 6049, 6060, 6097, 6233, 6335, 6720, 7185, 7743, 7846, 9110, 9212, 10079, 10814, 10989, 13523, 13785, 14140, 14973, 14997, 15010

195: 47, 60, 259, 459, 1055, 1088, 1375, 1552, 1767, 2040, 3267, 3311, 3756, 3849, 4020, 4214, 4648, 4704, 5367, 5645, 5887, 5989, 6048, 6122, 6238, 6526, 6527, 6624, 6714, 7519, 7917, 8328, 8334, 8425, 9475, 9495, 9896, 9937, 10254, 11050, 11092, 11103, 11192, 12147, 13358, 13484, 14141, 14142, 14379, 14469, 14851

196: 248, 250, 394, 933, 1054, 1133, 1925, 1926, 1979, 2255, 2525, 3907, 3911, 3965, 3973, 4309, 4677, 5347, 5556, 5565, 5618, 6203, 6209, 6422, 6447, 6845, 7204, 8715, 8751, 8760, 10090, 11627, 12492, 13239, 13303, 13558, 14098, 14382, 14634, 14673, 14831, 14855

197: 285, 1036, 1742, 1891, 1990, 2025, 2039, 2524, 2526, 3137, 3277, 3720, 4733, 5615, 6446, 6736, 6895, 7062, 7512, 8158, 8820, 8868,

197: 10436, 10616, 10730, 11358, 11418, 11802, 11961, 12631, 13083, 13098, 13772, 13921, 14431, 14854

198: 284, 637, 675, 1033, 1459, 2052, 2157, 2182, 2229, 2254, 3262, 3974, 4040, 4397, 5176, 6680, 6693, 6883, 7084, 7223, 7735, 8130, 8400, 8497, 8757, 9769, 10005, 11542, 11552, 11675, 11877, 12491, 12494, 12629, 13305, 13918, 14044, 14100, 14674, 14772

199: 206, 1041, 1131, 1781, 1935, 1950, 2122, 2186, 2740, 3887, 4278, 6089, 6358, 6445, 6707, 6729, 6730, 8704, 8866, 8883, 10075, 10091, 10239, 10667, 13306, 13389, 13410, 13504, 13505, 14055, 14853, 14981

200: 62, 533, 1003, 1337, 1780, 1793, 2041, 2227, 2486, 3128, 3595, 3801, 3802, 4302, 4843, 4989, 5048, 5275, 5426, 5720, 5892, 6348, 6637, 6699, 7303, 7385, 7698, 9040, 9135, 9381, 10031, 10222, 11659, 11778, 11887, 12439, 12517, 12626, 12786, 12788, 12912, 12968, 12981, 13198, 13734, 13782, 14030, 14050, 14051, 14054, 14062, 14064, 14099, 14300, 14348, 14857, 14953a

201: 300, 504, 2491, 3119, 3226, 4385, 5360, 5619, 5679, 6240, 7181, 8886, 9482, 12900, 13311, 13933, 14102

202: 199, 323, 351, 420, 1948, 3077, 3175, 3846, 4574, 4928, 5336, 5555, 6226, 6444, 7074, 7265, 8664, 8814, 9888, 9919, 9957, 10089,

202: 11253, 11537, 11998, 12057, 12674, 13798, 14112, 14822

203: 740, 741, 879, 917, 974, 1880, 1936, 1947, 2132, 2450, 3996, 4215, 4723, 4923, 5927, 5988, 7789, 7972, 8106, 8985, 9132, 9922, 11251, 11252, 11405, 11724, 11934, 12076, 12529, 12787, 14063, 14327, 14412, 14433, 14682, 14738, 14863

204: 79, 264, 914, 1045, 1090, 1782, 1924, 1938, 1951, 2019, 2054, 2131, 3115, 3509, 3829, 3912, 4405, 4492, 4653, 6443, 7487, 7564, 7705, 7757, 7782, 7841, 7969, 8358, 8966, 9828, 10748, 11393, 11445, 12104, 12192, 12417, 12447, 13194, 13582, 14056, 14161, 14982

205: 54, 76, 288, 621, 1934, 1984, 2100, 2150, 2228, 2253, 2480, 2874, 3160, 3298, 3415, 3482, 4422, 4571, 4584, 4647, 4814, 5153, 5182, 5425, 5552, 6678, 6782, 6822, 7212, 7216, 7456, 7859, 7998, 8352, 8714, 8895, 9326, 9561, 9986, 10004, 10146, 10407, 10734, 10800, 11403, 11670, 11706, 11754, 12305, 12785, 12902, 13739, 13949, 13950, 13951, 14270, 14453, 14619, 14680, 14833

206: 1050, 2158, 3392, 3845, 3865, 4033, 4245, 4349, 4419, 4538, 4801, 5344, 5515, 5644, 5928, 6094, 6332, 6740, 7165, 7167, 7198, 7433, 9232, 9340, 9804, 11270, 11436, 11918, 12116, 12660, 13250, 13259, 13501, 13577, 14343, 14846

207: 829, 1985, 2108, 3596, 3830, 4656, 4724, 4837, 4863, 5180, 5181, 5505, 5542, 6126, 7599, 8872, 9732, 10142, 10944, 12109, 13415, 13752, 13898, 14097, 14322

208: 266, 803, 804, 1525, 1983, 2267, 3602, 3822, 4330, 4337, 4547, 4587, 4632, 5716, 5898, 6391, 6832, 6911, 7491, 7589, 7826, 7971, 8623, 8990, 8993, 10490, 10567, 11541, 12155, 12440, 13078, 13447, 13454, 13524, 13563, 13663, 14101, 14314

209: 75, 127, 278, 540, 919, 1379, 2143, 2634, 3630, 4038, 4174, 4265, 6419, 6819, 7827, 8008, 8327, 8531, 8708, 8753, 8968, 9958, 10220, 10430, 10446, 11812, 11916, 12292, 13486, 13859, 14366, 14764, 14838

210: 247, 281, 1031, 1069, 1978, 2018, 2133, 2194, 2504, 2867, 3287, 3590, 4027, 4156, 4305, 4649, 5166, 5805, 5851, 5925, 5926, 6045, 6370, 6458, 6660, 7125, 7222, 8713, 8967, 9984, 9990, 9993, 10051, 10249, 10425, 10681, 10747, 10882, 10903, 11292, 12069, 12125, 12148, 12379, 12408, 13065, 13428, 13882, 14162, 14163, 14250, 14421, 14810, 14866, 14933

211: 622, 1679, 2141, 2682, 2710, 2777, 3237, 3915, 5151, 5924, 6744, 7881, 8229, 9343, 9380, 9504, 9794, 9795, 9796, 9912, 10179, 10437, 10574, 10755, 10884, 10910, 11288, 12184, 13089, 14113, 14137, 14139, 14358

BOILING POINT INDEX OF ORGANIC COMPOUNDS (Continued)

295: 280, 408, 545, 1348, 1371, 1833, 2712, 3470, 3531, 5914, 6316, 6318, 6451, 8242, 8922, 9254, 9255, 9329, 9588, 9766, 10680, 11233, 11463, 11493, 11838, 13187, 13363
296: 143, 399, 2071, 2128, 2232, 3397, 3481, 6643, 9282, 9867, 10769, 11308, 11837, 12765, 13480, 13659, 14462
297: 1076, 1893, 2468, 2716, 3534, 6205, 6306, 6314, 8276, 9532, 9908
298: 2126, 2162, 3255, 6142, 6288, 6399, 6889, 10662, 11470, 12070
299: 3375, 3398, 9531, 10292
300: 449, 674, 768, 985, 1078, 1152, 1193, 1333, 1810, 2784, 2825, 3180, 3222, 5084, 5660, 6041, 6285, 7504, 8839, 9688, 10061, 11840, 11855, 11935, 12842, 13529, 13869, 14116, 14194, 14305, 14489
301: 2641, 5265, 6321, 7488, 9283, 9384, 12560, 13172, 13721, 14156
302: 3382, 4046, 6249, 7318, 8557, 9385, 10424, 12762, 13257, 13266, 13688
303: 144, 148, 5291, 8163, 8471, 9708, 13665
304: 84, 90, 2478, 6283, 9410, 9570, 11330, 11476, 12743, 12744, 12849
305: 954, 1075, 1111, 1904, 2882, 3477, 3534a, 6291, 7392, 7694, 9063, 9423, 9517, 9737, 10334, 11000, 12314, 12871, 13449, 14323, 14426
306: 319, 1077, 1647, 3221, 4641, 4729, 6382, 6784, 9569, 9722, 13189
307: 145, 966, 2787, 3452, 8356, 9940, 10281, 10286, 10396, 10668, 12503, 13728
308: 7499, 8842, 9386, 9553
309: 1373, 2274, 2956, 6453, 10994, 11331, 12604
310: 722, 769, 958, 1060, 1770, 1838, 2862, 2934, 3400, 3927, 5296, 6177, 6267, 7401, 8151, 9049, 9519, 9853, 11239, 12733, 14530, 14540, 14944
311: 7398, 11057, 13654
312: 845, 4980, 6365, 9412, 10812, 11172, 11220, 14328
313: 10007, 11499
314: 1771, 2774, 2788, 2957, 7713, 11271, 12462
315: 967, 1772, 2014, 2015, 2016, 2293, 3424, 11615
316: 2789, 9740, 9763, 9981
317: 395, 1836, 3405, 3755, 9417, 9744, 9837, 10607, 11264, 11332
318: 1773, 10278, 12435
319: 10, 1327, 2124, 3409, 6366, 8436, 12829, 14351
320: 1113, 1939, 3270, 3275, 3484, 3490, 6093, 7502, 7744, 9241, 9695, 10813, 11387, 11620, 12728, 13097, 13453, 13651, 13652
322: 2161, 3423, 4842, 7702, 7704, 9099, 9760, 10287, 14662
323: 2556, 5539, 9418, 10608, 13653
324: 3015, 12806, 12948, 13446
325: 3374, 3434, 4712, 6307, 6879, 9572, 13369
326: 6279, 6362, 12442, 13832, 13913
327: 2958, 5755, 9914
328: 1370, 3453, 6450, 11861, 13747, 13749
329: 2239, 6589, 9904, 11496
330: 1095a, 2903, 2917, 5859, 6367, 7404, 10145, 10397, 11183, 11281, 11474, 12135, 12422, 14506
331: 320, 955, 2137, 4714
332: 2905, 6012, 13699, 14583
333: 2935, 6467, 9738, 10238, 12732, 13006
334: 3305, 9419
335: 9, 3109, 6317, 9761, 9762, 11262
336: 2258, 7489, 7506, 7703, 13801
337: 1397, 7061, 11447
338: 3014, 3122, 3621, 10204, 11852
339: 1526, 11498
340: 1070, 1096, 1554, 2259, 3492, 6324, 6461, 9151, 10649, 11280, 11335, 11462, 13750, 14018
341: 6460, 7021
342: 2954, 3439, 5464
343: 3500, 6439, 7131
344: 2838, 4841, 7712, 13188, 13190, 13287
345: 464, 1653, 2896, 4848, 9420, 11469
346: 2482, 6280
348: 6679, 9982, 9985
349: 7490, 7498, 10700, 14954
350: 2552, 2898, 2902, 3013, 3652, 5094, 7711, 9289, 9290, 10225, 10729, 14047, 14732
353: 2915
354: 2955, 9075, 11231, 14549
355: 2771, 3370, 3422, 4697
356: 7483
357: 6454, 7708
358: 6670, 9103, 10165, 13732, 14537, 14676
360: 2522, 2545, 3092, 4660, 5028, 9770, 9992, 10205, 10230, 11435, 12693, 12812, 13381
362: 13424
363: 3414, 12901
364: 9441
365: 9773, 13700, 14527, 14534
368: 6381
371: 11048, 14019
374: 14985
375: 6986, 10260
377: 11390
379: 1195, 6323, 11318
380: 8948, 14417, 14452, 14557
383: 9987
385: 13140
390: 6297, 10686, 13033, 13589
391: 13720
393: 12430
395: 2910, 9764
398: 6289
400: 2481, 3417, 9771, 13916, 14948, 14964
401: 479, 10279
405: 6377, 14384
406: 13731
410: 5335, 11423, 15012
412: 7692, 13474
413: 2518
415: 2290, 6878, 10233
418: 2129
419: 10166
420: 12686, 13472, 14677
422: 6440
425: 2880, 14535
428: 12687
430: 575, 1241
431: 9096, 9973
435: 1858
440: 9416, 9900
442: 7485, 13398
445: 9370
446: 1194
448: 5047
449: 14399
452: 3366
458: 14683
459: 1307, 2294
462: 1308
467: 6420
471: 6178
480: 2517
490: 10513
518: 11539
525: 5243
727: 4520
891: 12307

FORMULA INDEX FOR ORGANIC COMPOUNDS (Continued)

$C_6H_{12}Cl_3N$: 04948
$C_6H_{12}I_2$: 07797
$C_6H_{12}N_2$: 00307, 00677, 08069
$C_6H_{12}N_2O_2$: 00527, 07847, 08405, 10196
$C_6H_{12}N_2O_4$: 04680, 04683, 07807, 08131, 08133
$C_6H_{12}N_2O_4S$: 08687, 08688, 08689
$C_6H_{12}N_2O_4S_2$: 05873, 05874, 05875, 05876
$C_6H_{12}N_2S_4$: 06176
$C_6H_{12}N_4$: 07758
$C_6H_{12}O$: 00619, 00620, 03747, 03926, 04185, 04186, 04229, 04564, 05577, 05800, 07144, 07759, 07911, 07918, 07973, 07974, 07975, 09150, 10501, 10502, 10509, 10582, 10585, 11578, 11579, 14796
$C_6H_{12}O_2$: 00185, 00186, 00187, 00188, 00189, 00190, 04321, 04324, 04325, 04328, 04329, 04365, 04407, 04571, 05663, 05664, 05665, 05666, 05733, 05961, 06208, 07085, 07086, 07859, 07924, 08641, 10491, 11729, 12188, 12189, 14792, 14849, 14850, 14851, 14852, 14853, 14854, 14855, 14857
$C_6H_{12}O_2S$: 08976
$C_6H_{12}O_3$: 04339, 04340, 04345, 04410, 04594, 04595, 04596, 06741, 08671, 08731, 08732, 08733, 10257
$C_6H_{12}O_4$: 06136, 08812, 12131
$C_6H_{12}O_4Si$: 13340
$C_6H_{12}O_5$: 05553, 05554, 06524, 07094, 07280, 08921, 13038, 13040, 13494
$C_6H_{12}O_5S$: 07302
$C_6H_{12}O_6$: 00592, 00634, 00635, 07093, 07251, 07288, 07289, 07290, 07291, 07465, 07466, 08322, 08323, 08435, 08436, 08437, 08932, 08933, 08934, 12375, 13185, 13366, 13367, 13368, 13639, 13643
$C_6H_{12}O_7$: 07282, 08319, 13641
$C_6H_{12}S$: 05647, 13833
$C_6H_{12}S_3$: 14588, 14589
$C_6H_{13}Br$: 03685, 07765, 07767, 07768, 10347, 10348, 10349, 10351, 10353
$C_6H_{13}BrO_2$: 00025
$C_6H_{13}Cl$: 03699, 03700, 03708, 03714, 07769, 07773, 07777, 10360, 10361, 10368, 10369, 10370
$C_6H_{13}ClO$: 04103, 07869, 07885, 07894, 07895, 07896
$C_6H_{13}ClO_2$: 00028, 11981, 11982
$C_6H_{13}F$: 07813
$C_6H_{13}FO$: 07872
$C_6H_{13}I$: 07816, 07817
$C_6H_{13}N$: 05466, 05494, 05727, 11663, 11664, 11665, 11666, 11667, 11668, 12666, 12667, 12668
$C_6H_{13}NO$: 00066, 00073, 04246, 04247, 04248, 04565, 04568, 05580, 09233, 10496, 14804
$C_6H_{13}NO_2$: 00571, 00584, 01352, 04573, 04574, 04669, 04670, 04679, 04682, 07822, 07988, 08513, 08514, 08515, 08735, 08736, 08737, 09234
$C_6H_{13}NO_3S$: 13620
$C_6H_{13}NO_4$: 00662
$C_6H_{13}NO_5$: 07299
$C_6H_{13}NS_2$: 13799
$C_6H_{13}N_3$: 07255
$C_6H_{13}N_3O_3$: 05186
$C_6H_{13}O_4P$: 11395
C_6H_{14}: 03745, 03746, 07760, 10406, 10408, 10556
$C_6H_{14}Br$: 03695
$C_6H_{14}ClN$: 05728
$C_6H_{14}ClNO_5$: 07300
$C_6H_{14}Cl_2O_2Si_3$: 14576
$C_6H_{14}FO_3P$: 07044
$C_6H_{14}N$: 04088
$C_6H_{14}N_2$: 05453, 05981, 11601, 11602, 11603, 11604, 11607
$C_6H_{14}N_2O$: 06332, 06335, 08734, 09227, 14752, 14758
$C_6H_{14}N_2O_2$: 08800
$C_6H_{14}N_4O_2$: 01382, 01383
$C_6H_{14}O$: 03869, 03870, 03871, 03872, 03874, 03904, 04121, 04122, 04123, 06773, 07868, 07882, 07883, 07884, 07891, 07892, 07893, 09180, 09181, 09182, 10461, 10462, 10463, 10469, 10470, 10471, 10484, 10485, 11580, 12301, 12302, 12319
$C_6H_{14}OS$: 12328
$C_6H_{14}O_2$: 00035, 03842, 03843, 06890, 06907, 06908, 07840, 07841, 07842, 10436, 10437, 11999, 12000
$C_6H_{14}O_2S$: 06339, 06340, 12325, 12326
$C_6H_{14}O_3$: 00043, 06118, 06122, 06338, 07364, 07397, 07855, 07856, 07857, 07858, 11920
$C_6H_{14}O_4$: 07851, 14487
$C_6H_{14}O_4S$: 12327
$C_6H_{14}O_4Si_3$: 14577
$C_6H_{14}O_5$: 06522, 06523, 13036
$C_6H_{14}O_6$: 00583, 07249, 07850, 08318, 08922, 08923, 08924, 10639, 13037, 13039, 13362, 13363, 13364
$C_6H_{14}S$: 04125, 06774, 07852, 07853, 07854, 09183, 12321, 12324
$C_6H_{14}S_2$: 00036, 06336, 06337, 06604
$C_6H_{14}Si$: 13230
$C_6H_{15}As$: 01414
$C_6H_{15}AsO_3$: 01395
$C_6H_{15}AsO_4$: 01393
$C_6H_{15}B$: 03574
$C_6H_{15}BO_3$: 03559
$C_6H_{15}Br_2N$: 06096
$C_6H_{15}ClN_2O_2$: 05038, 08801
$C_6H_{15}ClO_3Si$: 13243
$C_6H_{15}ClSi$: 13233, 13244, 13251
$C_6H_{15}Cl_2N$: 04951
$C_6H_{15}FO_3Si$: 13293
$C_6H_{15}FSi$: 13294
$C_6H_{15}N$: 03678, 03681, 04081, 04082, 04089, 06163, 06331, 06334, 07761, 07762, 07763, 07987, 14483
$C_6H_{15}NO$: 06099, 06729, 06730, 06731, 10476
$C_6H_{15}NO_2$: 00023, 06329, 06800, 08223, 08225, 08226
$C_6H_{15}NO_3$: 08228, 14481
$C_6H_{15}OP$: 11377
$C_6H_{15}O_2P$: 11389
$C_6H_{15}O_3P$: 06693, 11400, 11433
$C_6H_{15}O_4P$: 11417
$C_6H_{15}P$: 11376
$C_6H_{15}PS$: 11378
$C_6H_{15}Sb$: 13436
$C_6H_{16}ClN$: 03682, 14484
$C_6H_{16}ClNO$: 08227, 09154
$C_6H_{16}ClNO_3$: 14482
$C_6H_{16}Cl_2OSi_2$: 06348
$C_6H_{16}IN$: 14485
$C_6H_{16}N_2$: 06851, 07757, 07782, 07784, 08092
$C_6H_{16}OSi$: 13339
$C_6H_{16}O_2Si$: 13265
$C_6H_{16}O_3Si$: 13329
$C_6H_{16}O_{14}$: 05576
$C_6H_{16}Si$: 13331
$C_6H_{17}NO_3$: 04258
$C_6H_{18}Cl_2N_2$: 07783
$C_6H_{18}N_3OP$: 11409
$C_6H_{18}N_4$: 14491
$C_6H_{18}OSi_2$: 06355
$C_6H_{18}O_3Si_2$: 06350
$C_6H_{18}Si_2$: 06346
$C_6H_{19}NSi_2$: 14525
$C_6H_{19}O_2Si_3$: 14578
C_6H_{20}: 13640
$C_6H_{24}O_6Si_6$: 05677
C_6I_6: 02166
C_6N_4: 06875
C_6N_6: 14411
C_7Cl_6O: 03151
C_7F_{14}: 05550
C_7F_{16}: 07563
C_7HCl_5O: 01794
$C_7HCl_5O_2$: 02768
C_7HCl_7: 03305
$C_7H_2Cl_4O$: 01798
$C_7H_2Cl_4O_2$: 02783
$C_7H_3BrN_2O_6$: 02565
$C_7H_3Br_3O_2$: 02798, 02799, 02800, 02801, 02802, 03060
$C_7H_3Br_4$: 15001
$C_7H_3Br_5$: 14155
$C_7H_3ClN_2O_2$: 09852
$C_7H_3ClN_2O_5$: 03145, 06203
$C_7H_3Cl_3O$: 01799, 01800, 01801, 01802, 01803, 01804, 03143, 03144
$C_7H_3Cl_3O_2$: 02803, 02805,

$C_8H_{17}NO_2$: 04627, 04628, 09232, 10123
$C_8H_{17}NO_3$: 10122
$C_8H_{17}O$: 07926
$C_8H_{17}OSi_2$: 06349
$C_8H_{17}O_2$: 04417
C_8H_{18}: 03796, 07566, 07568, 07569, 07570, 07571, 07798, 07799, 07800, 07801, 07804, 07805, 07806, 07808, 07835, 10027, 10400, 10401, 10418, 10419, 10420, 10421
$C_8H_{18}BrN$: 05229
$C_8H_{18}ClN$: 05230, 05231, 05232
$C_8H_{18}ClNO_2$: 00430, 00431
$C_8H_{18}Cl_2Si$: 13253
$C_8H_{18}O$: 04117, 04118, 04120, 06797, 07605, 07606, 07613, 07614, 07617, 07618, 07619, 07620, 07624, 07625, 07668, 07871, 07897, 07898, 10079, 10084, 10085, 10086, 10088, 10092, 10093, 10475, 10483, 10488
$C_8H_{18}OS$: 04178, 06056
$C_8H_{18}O_2$: 03740, 03741, 04155, 06911, 07577, 07839, 07844, 10063, 10064, 10065, 10426, 10428, 10430, 10435
$C_8H_{18}O_2S$: 06055
$C_8H_{18}O_3$: 06112, 06117, 07870, 10151
$C_8H_{18}O_3S$: 04173, 04174
$C_8H_{18}O_3Si$: 13330
$C_8H_{18}O_4S$: 04167
$C_8H_{18}O_5$: 13747
$C_8H_{18}O_6S_2$: 03760
$C_8H_{18}S$: 04168, 04169, 04171, 04172, 10075, 10076, 10077
$C_8H_{18}S_2$: 06054
$C_8H_{19}N$: 06051, 06052, 06053, 06789, 07567, 10028, 10029, 10030
$C_8H_{19}NO_2$: 00038, 04083
$C_8H_{19}NO_3$: 06330
$C_8H_{19}O_3P$: 11399
$C_8H_{20}As_2$: 03330
$C_8H_{20}BrN$: 13745
$C_8H_{20}ClNO$: 07611
$C_8H_{20}N_2$: 03748, 08074
$C_8H_{20}OSi$: 13284
$C_8H_{20}O_3P_2$: 11388
$C_8H_{20}O_3Si$: 13289
$C_8H_{20}O_4Si$: 10164, 13320
$C_8H_{20}O_5P_2S_2$: 12621
$C_8H_{20}O_6P_2$: 13751
$C_8H_{20}O_7P_2$: 12620
$C_8H_{20}Si$: 13321
$C_8H_{20}Si_2$: 06842
$C_8H_{23}N_5$: 13750
$C_8H_{24}N_4O_3P_2$: 12619
$C_8H_{24}O_2Si_3$: 14579
$C_8H_{24}O_4Si_4$: 05858
$C_8I_4O_3$: 11503
$C_9F_{21}N$: 14574
$C_9H_4O_2$: 04489
$C_9H_5Br_2NO$: 12789, 12790
C_9H_5ClINO: 12775
$C_9H_5ClO_2$: 05267, 12015, 12016, 12017
$C_9H_5ClO_6$: 02463
$C_9H_5Cl_2N$: 12791, 12792, 12793, 12794, 12795, 12796, 12797, 12798, 12799, 12800, 12801, 12803, 12804, 12805
$C_9H_5Cl_2NO$: 12802
$C_9H_5I_2NO$: 12808, 12809
$C_9H_5NO_4$: 12018, 12019
$C_9H_5NO_8$: 12566, 12567, 12568
C_9H_6BrN: 08587, 12749, 12750, 12752, 12754, 12758, 12760, 12762
C_9H_6BrNO: 12751, 12753, 12755, 12756, 12757, 12759, 12761, 12763
$C_9H_6BrNO_2$: 11509
$C_9H_6Br_2O$: 08430
$C_9H_6Br_2O_2$: 05107
C_9H_6ClN: 08588, 12764, 12768, 12772, 12774, 12776, 12779, 12782
C_9H_6ClNO: 12780
$C_9H_6Cl_3NO_4$: 09886
C_9H_6IN: 12862, 12863, 12864, 12865, 12866
$C_9H_6INO_4S$: 12942, 12945, 12947
$C_9H_6NO_3$: 05076
$C_9H_6N_2$: 08906
$C_9H_6N_2O_2$: 08599, 12860, 12894, 12896, 12897, 12898, 12899
$C_9H_6N_2O_3$: 12895
$C_9H_6N_2O_3S$: 01638
$C_9H_6N_2O_4$: 01637
C_9H_6O: 12008
$C_9H_6OS_2$: 13832
$C_9H_6O_2$: 05043, 05265, 08365, 08367, 08494, 12020
$C_9H_6O_3$: 05277, 05278, 05280, 05281, 05282, 05285, 05296
$C_9H_6O_4$: 05271, 05272, 09833
$C_9H_6O_6$: 02459, 02460, 02462, 14511, 14512
$C_9H_6O_7$: 02461, 02464
C_9H_7BrO: 05071
$C_9H_7BrO_2$: 05095, 05096, 05097, 05098
$C_9H_7BrO_4$: 13102
$C_9H_7Br_3O$: 00629
C_9H_7ClO: 02520, 05162
$C_9H_7ClO_2$: 05102, 05104, 05105
$C_9H_7FO_2$: 05123
C_9H_7N: 05158, 05159, 08581, 12719
C_9H_7NO: 03134, 08590, 08591, 12835, 12839, 12841, 12844, 12848, 12852, 12854
$C_9H_7NOS_2$: 13052
$C_9H_7NO_2$: 05266, 08366, 08426, 08460, 08461, 08462, 11176, 11517, 12807
$C_9H_7NO_3$: 05077, 05078, 08377, 08378, 08380, 08427, 11516
$C_9H_7NO_4$: 05138, 05139, 05142, 05143, 05146
$C_9H_7NO_4S$: 12943, 12944, 12946
$C_9H_7NO_5$: 12685
$C_9H_7N_2$: 08905
C_9H_8: 00451, 08395, 12341
$C_9H_8BrNO_3$: 08005, 08006, 08007
$C_9H_8Br_2$: 08341
$C_9H_8Br_2O_2$: 12117, 12118, 12119
C_9H_8ClN: 08582, 12833
$C_9H_8ClNO_4$: 02620, 09863
$C_9H_8Cl_2$: 08342, 12274
$C_9H_8Cl_3NO$: 04897, 04898
$C_9H_8I_2O_3$: 08271, 13116
$C_9H_8N_2$: 08330, 08584, 08585, 08586, 12726, 12729, 12731, 12733, 12738, 12741, 12745, 12959, 12960
$C_9H_8N_2O$: 00117, 05419, 12727, 12734, 12735, 12742, 12958
$C_9H_8N_2O_2$: 08048, 08369
$C_9H_8N_2O_2S$: 08033
$C_9H_8N_2O_6$: 02673, 06195
$C_9H_8N_2O_7$: 14266
$C_9H_8N_4O_4S_2$: 13630
C_9H_8O: 02524, 02525, 02526, 02527, 05042, 05070, 08375, 08379, 12012
$C_9H_8O_2$: 00514, 01440, 05041, 05081, 05082, 05083, 05084, 05269, 07133, 07425, 09069, 09185, 11197
$C_9H_8O_2S$: 08639
$C_9H_8O_3$: 02533, 02536, 02538, 02539, 03130, 05124, 05125, 05126, 07432, 12683
$C_9H_8O_4$: 00433, 05134, 05108, 05109, 05110, 05111, 07418, 08563, 08872, 11111, 11112, 11113, 11719, 13092
$C_9H_8O_5$: 11064, 11065, 11066
$C_9H_8O_6$: 07482
C_9H_8S: 03113
C_9H_9Br: 01884, 01966
C_9H_9BrO: 00365, 12232, 12233
$C_9H_9BrO_2$: 02566, 02575, 02582, 12081, 12088, 12089
$C_9H_9Br_2NO_3$: 14625
$C_9H_9Br_3$: 02264
C_9H_9Cl: 00608, 12031, 12032, 12033
$C_9H_9ClN_2$: 00562
C_9H_9ClO: 00311, 00372, 00612, 08157, 10772, 10853, 12234, 12235
$C_9H_9ClO_2$: 02592, 02593, 02595, 02604, 02614, 04924, 04937, 12224
$C_9H_9ClO_3$: 11070, 13106, 13110
$C_9H_9I_2NO_2$: 00720, 01164, 01166
$C_9H_9I_2NO_3$: 08442, 08443, 08444
C_9H_9N: 08059, 08156, 08418, 08419, 08420, 08421, 11186, 11187, 11188, 12214, 14370, 14372
C_9H_9NO: 02864, 02865, 03120, 05079, 05080, 11184, 11185, 12900
$C_9H_9NO_2$: 01655, 01656, 01657, 02860, 02861,
$C_9H_9NO_2$: 02862, 02863, 05087, 05088, 05089, 08354, 09187, 09188, 12035, 14904
$C_9H_9NO_3$: 00551, 00648, 00684, 00712, 01154, 02532, 02534, 08003, 08888
$C_9H_9NO_4$: 00652, 01029, 02759, 09847, 09867, 09882

FORMULA INDEX FOR ORGANIC COMPOUNDS (Continued)

$C_{15}H_{11}NO_4$: 11199, 11200, 11201, 11202
$C_{15}H_{11}NS$: 13810
$C_{15}H_{11}O_4$: 02742
$C_{15}H_{12}$: 01131, 01132, 01133, 10673, 10674, 10675, 10676
$C_{15}H_{12}Br_2O$: 04846, 04847, 12237
$C_{15}H_{12}N_2OS$: 08040
$C_{15}H_{12}N_2O_2$: 08039
$C_{15}H_{12}N_2O_3$: 07193, 08162, 08511
$C_{15}H_{12}N_6O_4$: 13042
$C_{15}H_{12}O$: 01653, 04848, 07028
$C_{15}H_{12}O_2$: 00209, 00500, 05152, 06032, 06033, 06034, 06959
$C_{15}H_{12}O_3$: 02537, 02541, 02552, 02553, 02902
$C_{15}H_{12}O_4$: 13140
$C_{15}H_{12}O_6$: 06961, 09210
$C_{15}H_{13}N$: 08422
$C_{15}H_{13}NO$: 04710, 06988, 06989
$C_{15}H_{13}NO_2$: 03129
$C_{15}H_{13}NO_3$: 02535
$C_{15}H_{13}NO_4$: 13091
$C_{15}H_{14}$: 12283, 12284, 13484
$C_{15}H_{14}N_2O$: 07470
$C_{15}H_{14}N_2O_2$: 01469, 01470, 01471, 08829
$C_{15}H_{14}O$: 00319, 00320, 02934, 02935, 13481, 13482, 13483
$C_{15}H_{14}O_2$: 02775, 02793, 02844, 06242, 09211, 12133, 12134, 12135, 12136, 12138
$C_{15}H_{14}O_3$: 02492, 02928, 02929, 02931, 02932, 02933, 04734, 04735, 04736, 06241, 06542, 08692, 12137, 12139, 12140
$C_{15}H_{14}O_4$: 08693, 09347, 10598, 14970
$C_{15}H_{14}O_5$: 04722, 06513, 11348
$C_{15}H_{14}O_6$: 04786, 04787, 06512, 08632
$C_{15}H_{14}O_8$: 04785
$C_{15}H_{15}Cl_3O_2$: 06605
$C_{15}H_{15}N$: 04711
$C_{15}H_{15}NO$: 02936, 02937
$C_{15}H_{15}NO_2$: 04660
$C_{15}H_{15}NO_3$: 14503
$C_{15}H_{15}NO_6$: 08898
$C_{15}H_{15}NS$: 11056
$C_{15}H_{15}N_2O$: 00473
$C_{15}H_{15}N_3O_2$: 01560, 09196
$C_{15}H_{16}$: 01893, 06306, 11853, 11854, 11855, 11856
$C_{15}H_{16}NO$: 01531
$C_{15}H_{16}N_2O$: 01528, 01529, 01530, 14732
$C_{15}H_{16}N_2S$: 13983, 13994, 13995
$C_{15}H_{16}N_4O_4$: 04468
$C_{15}H_{16}N_4O_5$: 04472
$C_{15}H_{16}O$: 02470, 06315, 06655
$C_{15}H_{16}O_2$: 03482, 11780, 11781, 11851, 11852
$C_{15}H_{16}O_3$: 01422, 07367, 08164
$C_{15}H_{16}O_7$: 06432
$C_{15}H_{16}O_8$: 13350
$C_{15}H_{16}O_9$: 06598
$C_{15}H_{17}N$: 03212, 14307
$C_{15}H_{17}NO$: 02471, 02472, 02473
$C_{15}H_{17}NO_2S$: 14241, 14242
$C_{15}H_{17}N_2O$: 01527
$C_{15}H_{17}N_2O_2$: 08315
$C_{15}H_{18}$: 01604, 01605, 01606, 09362
$C_{15}H_{18}ClN$: 03294
$C_{15}H_{18}N_2$: 11778
$C_{15}H_{18}N_4O_4$: 04473
$C_{15}H_{18}O$: 09417, 09418, 09760, 09761, 10607, 10608
$C_{15}H_{18}O_3$: 13178
$C_{15}H_{18}O_6$: 02465
$C_{15}H_{19}NO_2$: 14595, 14600, 14601
$C_{15}H_{19}Si_2$: 06345
$C_{15}H_{20}ClNO_2$: 14596
$C_{15}H_{20}N_2O$: 00749
$C_{15}H_{20}O_2$: 07475
$C_{15}H_{20}O_4$: 08874
$C_{15}H_{21}NO_2$: 06927, 06928, 06929, 08500, 11700
$C_{15}H_{21}N_3O_2$: 11533
$C_{15}H_{22}$: 05401, 08349
$C_{15}H_{22}ClNO_2$: 06930, 08501, 08502, 08503
$C_{15}H_{22}O$: 05988, 06544, 14908, 14909
$C_{15}H_{22}O_2$: 02767, 06545
$C_{15}H_{23}$: 04466
$C_{15}H_{23}ClO_4S$: 01375
$C_{15}H_{24}$: 02220, 02295, 04779, 04780, 04781, 04824, 05187, 05237, 06471, 06942, 06943, 07468, 08776, 10282, 13200, 13201
$C_{15}H_{24}N_2O$: 01337, 08940, 10220
$C_{15}H_{24}O$: 06543, 08729, 10872, 10873, 10878, 13172, 13173
$C_{15}H_{24}O_6$: 00459
$C_{15}H_{24}O_8$: 11934, 11935
$C_{15}H_{26}ClN_4O$: 00709
$C_{15}H_{26}N_2$: 13369
$C_{15}H_{26}O$: 06473, 06944, 06945, 08637, 09808, 09809, 09810, 10262
$C_{15}H_{26}O_6$: 07392
$C_{15}H_{26}O_7$: 05176
$C_{15}H_{26}Si$: 13316
$C_{15}H_{28}$: 10296, 11834, 11835, 11836, 11839
$C_{15}H_{28}N_2O_4S$: 13370
$C_{15}H_{28}O$: 05754
$C_{15}H_{28}O_2$: 08964
$C_{15}H_{28}O_4$: 08863
$C_{15}H_{30}$: 06470, 10295
$C_{15}H_{30}Br_2$: 10288
$C_{15}H_{30}O$: 10283, 10293, 10294
$C_{15}H_{30}O_2$: 04634, 07596, 08714, 08715, 09255, 10290
$C_{15}H_{30}O_4$: 07372, 07373
$C_{15}H_{31}Br$: 10287
$C_{15}H_{31}NO$: 10284
$C_{15}H_{32}$: 10285
$C_{15}H_{32}O$: 10292
$C_{15}H_{33}B$: 03577, 03579
$C_{15}H_{33}BO_3$: 03561, 03564
$C_{15}H_{33}ClN_2O_5$: 10221
$C_{15}H_{33}N$: 10286, 14520, 14531, 14532
$C_{15}H_{33}NO$: 11775
$C_{15}H_{33}N_3$: 14405
$C_{15}H_{33}O_4P$: 11419
$C_{15}H_{34}ClN$: 06415
$C_{16}H_8O_2S_2$: 13853
$C_{16}H_{10}$: 06986, 12430
$C_{16}H_{10}Cl_4O_5$: 06327
$C_{16}H_{10}N_2$: 02881
$C_{16}H_{10}N_2O_2$: 08406, 08410
$C_{16}H_{10}N_2O_8S_2$: 08409
$C_{16}H_{10}O_2$: 09674
$C_{16}H_{10}O_3$: 09667
$C_{16}H_{10}O_7$: 01317
$C_{16}H_{11}N$: 12432
$C_{16}H_{11}NO_2$: 09675, 12927, 12934
$C_{16}H_{11}NO_4$: 01277
$C_{16}H_{11}N_3O_3$: 02324, 02325
$C_{16}H_{11}N_3O_7$: 09421
$C_{16}H_{12}$: 01148, 01149, 09419, 09420, 12433
$C_{16}H_{12}N_2$: 02314, 02326
$C_{16}H_{12}N_2O$: 02322, 02323, 02329
$C_{16}H_{12}N_2O_6$: 06243
$C_{16}H_{12}O$: 01097, 01098, 01099, 07131, 10650, 10651, 10652
$C_{16}H_{12}O_2$: 01262, 01263, 01265, 01266, 01267, 06857, 06858
$C_{16}H_{12}O_3$: 04854, 07035
$C_{16}H_{12}O_4$: 01260, 01261, 07109, 08815
$C_{16}H_{12}O_5$: 03599, 06965, 06966
$C_{16}H_{12}O_6$: 07477, 11711
$C_{16}H_{12}O_7$: 13035
$C_{16}H_{12}S$: 13893
$C_{16}H_{12}S_2$: 06360
$C_{16}H_{13}N$: 09762, 09764, 12633, 12634
$C_{16}H_{13}NO_3$: 05091, 05092, 05093
$C_{16}H_{13}N_3$: 02315, 02316
$C_{16}H_{14}$: 01124, 01125, 01126, 03651, 03652, 10663, 10664, 10667
$C_{16}H_{14}ClN_3$: 02317
$C_{16}H_{14}O$: 09151
$C_{16}H_{14}O_2$: 04851, 04852, 04853, 05094, 10662
$C_{16}H_{14}O_3$: 02551, 02840, 02901, 05128, 11103, 14261, 14278, 14291
$C_{16}H_{14}O_4$: 03509, 03514, 03518, 03522, 06887, 08336, 10181, 10182, 10183, 13098, 13559
$C_{16}H_{14}O_5$: 11061
$C_{16}H_{14}O_6$: 06960, 07481
$C_{16}H_{15}ClO_2$: 04945, 06237
$C_{16}H_{15}Cl_3$: 06644
$C_{16}H_{15}N$: 00344, 06040, 06205
$C_{16}H_{16}$: 01113, 03966, 13467, 13468, 13480
$C_{16}H_{16}ClNO$: 04893
$C_{16}H_{16}N_2O_2$: 03410, 03411, 03412, 06850, 08799
$C_{16}H_{16}N_2O_4S$: 11319
$C_{16}H_{16}O_2$: 06038, 06238, 08151, 11168, 13466
$C_{16}H_{16}O_3$: 02491, 06240
$C_{16}H_{16}O_4$: 02841, 02998, 02999, 03000, 03001, 03002, 03003, 03004, 06476, 09489
$C_{16}H_{16}O_5$: 00576, 13229
$C_{16}H_{16}O_6$: 04838
$C_{16}H_{16}Si$: 13282
$C_{16}H_{17}Br$: 03686, 03687
$C_{16}H_{17}N$: 04701
$C_{16}H_{17}NO$: 00071, 06037

SUBLIMATION DATA FOR ORGANIC COMPOUNDS

Compiled by Mansel Davies

The tables quote the parameters from what appear to be the best data in the literature expressed in the form

$$\log_{10} p(\text{mm}) = A - B/T$$

and the temperature range for which they apply. The corresponding heats and entropies (taking the standard state of the vapor to be

$$\Delta H\ (\text{sublimation}) = 2.303\ \text{R.B. cal/mol}$$
$$\Delta S\ (\text{sublimation}) = 2.303\text{R}(A - 2.881)\ \text{cal/mol K.}$$

Compound	Temp. range °C	A	B	Ref.
Acenaphthene	18 to 37	11.758	4290.5	1
Acetamide	25 to 77	11.8468	4050.1	2
Acetic acid	−35 to + 10	8.502	2177.4	3
Acetic Acid. *m*-cresyl ester	2 to 44	9.759	3170	14
Acetophenone, 1-chloro-	5 to 50	13.779	4740	14
Acetophenone, 1-chloro-*o*-nitro	23 to 54	14.24	5413	14
Acetophenone, 1-chloro-*m*-nitro	26 to 70	14.080	5700	14
Acetophenome p-methoxy	3 to 27	11.367	4056	1
Acetone, benzoyl	5 to 26	12.317	4375	1
Adipic acid	86 to 133	15.463	6757	4
Anthracene	65 to 80	12.638	5320	5
Anthracene	105 to 125	12.002	5102	6
Anthracene, 9,10 diphenyl	208 to 229	16.058	8213	22
Anthraquinone	224 to 286	12.305	5747	3
Arachidic acid	63 to 73	25.453	10,424	23
Arsine, Diphenylcyano	23 to 53	10.724	4420	14
Azobenzene (*cis*)	30 to 60	9.652	3914	7
Azobenzene (*trans*)	30 to 60	9.721	3911	7
Behenic acid	71 to 79	23.604	10,100	23
Benzanthrone	—	13.416	6030	8
Benzene	−30 to 5	9.846	2309	3
Benzene	−58 to −30	9.556	2241	3
Benzene, *p*-chloroiodo	30 to 50	9.819	3200	15
Benzene, -dichloro	10 to 50	11.985	3570	17
Benzene, *α*-hexachloro	51 to 71	11.950	4850	14
Benzene, *β*-hexachloro	95 to 117	11.790	5375	14
Benzene, *γ*-hexchloro	60 to 92	15.515	6022	14
Benzene,*λ*-hexachloro	55 to 75	12.635	5100	14
Benzene, -hexamethyl		11.070	4215	37
Benzene, 1, 2, 3,-trichloro	16 to 30	10.662	3440	6
Benzene, 1, 2, 4-trichloro	6 to 25	10.445	3254	6
Benzene, 1,3,5-trichloro	9 to 28	9.176	2956	6
Benzil	45 to 67	12.708	5140	1
Benzoic Acid	70 to 114	12.870	4776	9
Benzoic acid, *p*-hydroxy	125 to 160	13.623	6063	9
Benzoic acid, *o*-methoxy	80 to 95	11.871	4746	9
Benzophenone (stable)	16 to 42	17.46	4966	10
Benzophenone, (meta stable)	11 to 25	17.19	4818	10
Benzoquinone	—	10.00	3280	18
Benzoquinone, 2.6-dichloro	1 to 42	9.85	3670	18
Benzoquinone, trichloro	28 to 54	12.03	4630	18
Benzoquinone, tetrachloro	60 to 83	12.06	5170	18
Benzoquinone, *p*-xylo	0 to 20	11.53	4030	18
Bibenzyl	13 to 34	12.194	4386	1
Biphenyl	6 to 26	ll.168	3959	1
Butyramide	25 to 68	12.739	4546	2
Butyramide	63 to 109	12.594	4513	2
Camphor	0 to 180	8.799	2797	3
Capramide	80 to 97	16.471	6577	2
Capramide, *N*-methyl	30 to 52	14.594	5371	ll
Capric acid	16 to 28	17.130	6119	23
Caproamide	65 to 95	13.328	4968	2
*ε*Caprolactam	21 to 41	11.839	4339	34
Caprylamide	52 to 101	14.920	5783	2
Carbamic acid, *n*-butyl ester	19 to 43	14.582	4919	11
Carbamic acid, ethyl ester	19 to 43	14.090	4646	11
Carbamic acid, *n*-hexyl ester	18 to 41	14.748	5018	11
Carbamic acid, methyl ester	14 to 32	11.966	3883	11
Carbon tetrabromide (monoclinic)	22 to 46	9.3867	2841	12
(cubic)	48 to 56	8.5670	2579	12
Carbon tetrachloride	−64 to −48	9.089	2027	13
o-Cresol, 3, 5-dinitro	17 to 51	14.140	5400	14
Cyclohexane	−5 to + 5	8.594	1953	3
Cyclo-trimethylene-trinitramine	110 to 138	11.870	5850	16
Diphenylamine	25 to 51	12.434	4654	21
Dodecanedioic acid	102 to 123	17.728	8006	4
Eicosanedioic acid	107 to 122	18.185	8644	4
Enanthamide	72 to 93	13.617	5182	2
Ethane, 1.1 p dichloro diphenyl tri-chloro	66 to 100	14.191	6160	14
Ethane, hexachloro (cubic)	13 to 174	8.731	2677	26
Ethane, hexachloro (triclinic)	13 to 174	9.890	3077	2b

SUBLIMATION DATA FOR ORGANIC COMPOUNDS (continued)

Compound	Temp. range C	A	B	Ref.
Ethylene dibromide	−21 to +8	9.884	2606	13
Ethylene, *trans* di-iodo	−8 to 20	5.86	2130	19
Fluorene	33 to 49	11.325	4324	5
Formic acid	−5 to +8	12.486	3160	36
2 Fuoric acid	44 to 55	14.62	5667	25
Hendecanoic acid	20 to 28	16.432	6037	24
Heneicosanoic acid	68 to 73	22.602	9642	24
Heptadecanoic acid	48 to 58	21.836	8769	24
Hydroquinone tetrachloro	77 to 86	10.08	4650	18
Hydroquinone tetrachloro, *p*-xylo	59 to 88	12.36	5280	18
Lauramide	76 to 95	19.169	7980	2
Lauric acid	22 to 41	19.897	7322	23
Methane	−194 to −184	7.651	5169	3
Methane, triphenyl	52 to 76	12.661	5228	1
Myristamide	85 to 100	20.940	8746	2
Myristic acid	38 to 52	18.740	7291	23
Naphthalene	6 to 21	11.597	3783	5
1-Naphthol	25 to 39	13.074	4873	1
	39 to 50	11.526	4389	1
2-Naphthol	25 to 39	13.356	5109	1
	39 to 58	11.660	4579	1
Nonadecanoic acid	58 to 64	35.916	13,815	24
n-Octadecane	15 to 25	22.83	7995	27
Oxalic acid, anhyd.	60 to 105	12.223	4727	29
Oxalic acid, anhyd. (α)	38 to 52	13.17	5130	28
Oxalic acid, anhyd. (β)	38 to 50	12.57	4875	28
Oxamic acid	82 to 90	12.58	5639	30
Oxamide	80 to 96	12.57	5893	30
Palmitamide	91 to 105	22.690	9489	2
Palmitic acid	46 to 60	20.217	8069	23
Pelargonamide	80 to 97	15.249	5997	2
Pentadecanoic acid	38 to 48	23.110	8813	24
Pentaerythritol (tetrag)	106 to 135	16.17	7528	28
Pentaerythritol, tetranitrate	97 to 138	17.73	7750	16
Phenanthrene	37 to 50	11.388	4519	5
Phenol	5 to 32	11.421	3540	14
Phenol, *p*-acetyl	47 to 75	12.216	5003	31
Phenol, *p*-benzyl	40 to 62	12.600	5072	31
Phenol, *p*-*tert* butyl	8 to 30	12.332	4402	31
Phenol, 2-*tert* butyl-4- methyl	2 to 20	11.685	4036	31
Phenol, 4-*tert* butyl-2-methyl	3 to 24	11.199	3952	31
Phenol, *p*- formyl	39 to 63	11.795	4762	31
Phenol, *p*-methoxy	5 to 27	13.132	4624	31
Phenol, *o*-phenyl	19 to 40	11.754	4331	31
Phenol, *p*-phenyl	54 to 74	12.056	5068	31
Phenol, 2:4:6-tri*tert* butyl	18 to 40	11.507	4383	31
Phthalic anhydride	30 to 60	12.249	4632	32
Propionamide	45 to 73	12.041	4139	2
Pyrene	72 to 85	11.270	4904	5
Pyrrole 2-carboxylic acid	77 to 81	16.60	6633	25
Rubeanic acid	87 to 105	12.713	5515	30
Salicylic acid	95 to 134	12.859	4969	9
Sebacic acid	102 to 130	18.911	8395	4
Stearamide	94 to 106	24.449	10,230	2
Steraric acid	57 to 67	21.180	8696	23
Suberic acid	106 to 134	16.937	7472	4
Succinic acid	99 to 128	14.068	6132	4
d-Tartaric acid, dimethyl ester	35 to 44	16.610	5903	20
dl-Tartaric acid, dimenthyl ester	42 to 85	16.127	5941	20
Thapsic acid	104 to 125	17.165	7885	4
2-Thenoic acid	42 to 50	13.53	5065	25
Thymol	0 to 40	14.201	4766	14
Toluene, 2, 4, 6-trinitro	50 to 143	15.34	6180	33
Tridecanoic acid	31 to 39	20.939	7764	24
Valeramide	60 to 101	12.846	4666	2

References

1. Aihara, *Bull. Chem. Soc.*, Japan
2. Davies, Jones, and Thomas, *Trans. Faraday Soc.*, 55, 1100, 1959.
3. *CRC Handbook of Chemistry and Physics*, 41st Ed. p. 2428 et seq.
4. Davies and Thomas, *Trans. Faraday, Soc.*, 56, 185, 1960.
5. Bradley and Cleasby, *J. Chem. Soc.*, 1690, 1953.
6. Sears and Hopke, *J. Am. Chem. Soc.*, 71, 1632, 1949.
7. Bright et al., *Research*, 3, 185, 1950.
8. Inokuchi et al., *Bull. Chem. Soc.*, Japan, 25, 299, 1952.
9. Davies and Jones, *Trans. Soc.*, 50, 1042, 1954.
10. Neumann and Volker, *Z. Physik. Chem.*, 161A, 33, 1932.
11. Davies and Jones, *Trans. Soc.*, 55, 1329, 1959.
12. Bradley and Drury, ibid., 55, 1844, 1959.
13. Nitta and Seki, *J. Chem. Soc.*, Japan, 69, 85, 1948.
14. Balson, *Trans. Faraday Soc.*, 43, 54, 1947.
15. Ewald, ibid., 49, 1401, 1953.
16. Edwards, ibid., 49, 152, 1953.
17. Darkis et al., *Ind. Eng. Chem.*, 32, 946, 1940.
18. Coolidge and Coolidge, *J. Am. Chem. Soc.* 49, 100, 1927.
19. Broadway and Fraser, *J. Chem. Soc.*, 429, 1933.
20. Crowell and Jones, *J. Phys. Chem.*, 58, 666, 1954.
21. Aihara, *J. Chem. Soc.*, Japan, 74, 437, 1953.
22. Stevens, *J. Chem. Soc.*, 2973, 1953.
23. Davies, Malpass, and Stenhagan, *Ark. Kemi.*
24. Thomas, M. Sc. thesis, Univ. of Wales, 1959.
25. Bradley and Care, *J. Chem. Soc.*, 1688, 1953.
26. Ivin and Dainton, *Trans. Soc.*, 43, 32, 1947.
27. Bradley and Shellard, *Proc. Roy. Soc.*, 198A, 239, 1949.
28. Bradley and Cotson, *J. Chem. Soc.*, 1684, 1953.
29. Noyes and Wobbe, *J. Am. Chem. Soc.*, 48, 1882, 1926.
30. Bradley and Cleasby, *J. Chem. Soc.*, 1681, 1953.
31. Aihara, *Bull Chem. Soc.*, Japan.
32. Crooks and Feetham, *J. Chem. Soc.*, 899, 1946.
33. Edwards, *Trans. Faraday Soc.*, 46, 423, 1950.
34. Aihara, *J. Chem. Soc.*, Japan, 74, 631, 1953.
35. Seki and Suzuki, *Bull. Chem. Soc.*, Japan, 70, 387, 1949.
36. Coolidge, *J. Am. Chem. Soc.*, 53, 1874, 1930.
37. Nitta et al., *J. Chem. Soc.*, Japan, 70, 387, 1949.

HEATS OF FUSION OF SOME ORGANIC COMPOUNDS

William E. Acree, Jr.

Compounds in this table are listed in order of increasing number of carbon and hydrogen atoms in the molecules. Melting point temperatures are listed in degrees Celcius and heats of fusion in calories per gram, joules per gram, and joules per gram molecular weight.

				Heat of fusion (H_f)		
Formula	**Compound**	**Mol. wt**	**M.P. (°C)**	**cal/g**	**J/g**	**J/mol**
$CHCl_3$	Trichloromethane	119.38	−63.6	17.62	73.72	8,800
CHN	Hydrogen cyanide	27.03	−13.4	74.38	311.21	8,412
CH_2Cl_2	Dichloromethane	84.93	−95.14	16.89	70.67	6,002
CH_2N_2	Cyanamide	42.04	44.0	49.81	208.41	8,761
CH_2O_2	Formic acid	46.03	8.3	66.05	276.35	12,720
CH_3D	Monodeuteromethane	17.05	−182.7	12.76	62.97	910
CH_3Br	Bromomethane	94.94	−93.7	15.05	62.97	5,978
CH_4	Methane	16.04	−182.5	13.96	58.41	936
CD_4	Deuteromethane	20.07	−183.4	10.75	44.98	902
CH_4O	Methanol	32.04	−97.9	23.70	99.16	3,177
CH_4S	Methyl mercaptan	48.10	−121.0	29.35	122.8	5,906
CH_5N	Methylamine	31.06	−93.5	47.20	197.48	6,133
$CBrCl_3$	Bromotrichloromethane	198.27	−5.7	3.05	12.76	2,539
CCl_2O	Phosgene	98.92	−127.9	13.86	57.99	5,736
CCl_3NO_2	Chloropicrin	164.38	−64.0	48.16	201.50	33,122
CCl_4	Tetrachloromethane	153.82	−23.0	5.09	21.30	3,276
CS_2	Carbon disulfide	76.13	−111.5	13.80	57.74	4,395
$C_2HCl_3O_2$	Trichloroacetic acid	163.39	57.5	8.60	35.98	5,878
$C_2H_2Br_2Cl_2$	1,2-Dibromo-1,1-dichloromethane	256.75	−66.9	7.73	32.34	8,303
$C_2H_2Cl_2O_2$	Dichloroacetic acid	128.94	10.8	14.21	59.45	7,665
$C_2H_3Br_3$	1,1,2-Tribromomethane	266.76	−29.2	8.16	34.14	9,107
$C_2H_3ClO_2$	α-Chloroacetic acid	94.50	61.2	31.06	129.96	12,281
$C_2H_3ClO_2$	β-Chloroacetic acid	94.50	56.0	35.12	146.94	13,885
$C_2H_3Cl_3$	1,1,1-Trichloroethane	133.40	−30.4	4.90	20.50	2,734
$C_2H_3Cl_3$	1,1,2-Trichloroethane	133.40	−36.6	20.68	86.53	11,543
$C_2H_3F_3$	1,1,1-Trifluoroethane	84.04	−111.3	17.61	73.68	6,192
$C_2H_4Br_2$	1,2-Dibromoethane	187.86	9.93	13.97	57.70	10,839
$C_2H_4Cl_2$	1,2-Dichlororoethane	99.96	−35.5	21.12	88.37	8,833
$C_2H_4O_2$	Acetic acid	60.05	16.6	45.91	192.09	11,535
C_2H_5Cl	Chloroethane	64.51	−138.3	16.49	68.99	4,450
C_2H_6	Ethane	30.07	−183.3	22.73	95.10	2,859
C_2H_6O	Dimethyl ether	46.07	−141.5	25.62	107.19	4,938
C_2H_6O	Ethanol	46.07	−114.5	26.05	108.99	5,021
$C_2H_6O_2$	Ethylene glycol	62.07	−11.5	43.26	181.00	11,234
C_2H_6S	Dimethyl sulfide	62.13	−98.3	30.73	128.57	7,988
C_2H_6S	Ethanethiol	62.13	121.0	19.14	80.08	4,975
$C_2H_6S_2$	Methyl disulfide	94.19	−120.5	23.32	97.57	9,190
C_2H_7N	Dimethylamine	45.08	−92.2	31.51	131.84	5,943
$C_2H_8N_2$	Ethylene diamine	60.10	11.1	89.81	35.77	22,583
C_3H_3N	Acrylonitrile	53.06	−83.5	28.06	117.40	6,229
$C_3H_4O_2$	Acrylic acid	72.06	12.3	37.03	154.93	11,164
$C_3H_5Br_3$	1,2,3-Tribromopropane	280.78	16.19	20.24	84.68	23,776
$C_3H_5N_3O_3$	Trinitroglycerol	131.09	12.3	23.02	96.32	12,627
C_3H_6	Cyclopropane	42.08	−127.4	30.92	129.37	5,444
C_3H_6	Propene	42.08	−185.3	17.06	71.38	3,004
$C_3H_6Br_2$	1,3-Dibromopropane	201.89	−34.2	16.10	67.36	13,599
$C_3H_6Cl_2$	1,2-Dichloropropane	112.99	−100.5	13.53	56.61	6,396
C_3H_6O	Acetone	58.08	−94.8	23.42	97.99	5,691
C_3H_7Cl	2-Chloropropane	78.54	−117.2	22.48	94.06	7,387
C_3H_7N	Cyclopropylamine	57.10	−35.39	55.18	230.87	13,183
$C_3H_7NO_2$	Ethyl carbamate	89.09	48.7	40.85	170.92	15,227
C_3H_8	Propane	44.10	−181.7	19.11	79.96	3,526
C_3H_8O	Propanol	60.10	−126.1	20.66	86.44	5,195
C_3H_8O	Isopropanol	60.10	−89.5	21.37	89.41	5,373
$C_3H_8O_3$	Glycerol	92.09	18.2	47.95	200.62	8,475
C_3H_9N	Triethylamine	59.11	−117.1	26.47	110.75	6,546
$C_4H_4N_2$	Succinonitrile	80.09	54.5	11.71	48.99	3,924
$C_4H_4O_3$	Succinic anhydride	100.07	119.0	48.47	203.93	20,407
C_4H_4S	Thiophene	84.14	−39.4	14.11	59.04	4,968
C_4H_5N	Pyrrole	67.09	−23.41	28.17	117.86	7,907
C_4H_6	1,3-Butadiene	54.09	−108.9	35.28	147.61	7,984
C_4H_6	2-Butyne	54.09	−32.36	40.80	170.71	9,234
$C_4H_6O_2$	Crotonic acid	86.09	72.0	25.32	105.94	9,120
$C_4H_6O_2$	*cis*-Crotonic acid	86.09	71.2	34.90	146.02	12,571
$C_4H_6O_2$	γ-Butyrolactone	86.09	−43.37	26.57	111.17	9,571
$C_4H_6O_4$	Dimethyl oxalate	118.09	54.35	42.64	178.41	21,068
$C_4H_6O_4$	Succinic acid	118.09	183.8	66.68	278.99	32,946

Formula	Compound	Mol. wt	M.P. (°C)	Heat of fusion (H_f) cal/g	J/g	J/mol
C_4H_8	Isobutene	56.11	−140.4	25.25	105.65	5,928
C_4H_8	*cis*-2-Butene	56.11	−138.9	32.30	135.14	7,583
$C_4H_8N_2S$	Allyl thiourea	116.18	77.0	33.45	139.95	16,259
C_4H_8O	Tetrahydrofuran	72.11	−108.39	28.31	118.45	8,541
C_4H_8O	2-Butanone	72.11	−86.67	27.97	117.03	8,439
$C_4H_8O_2$	Ethyl acetate	88.11	−83.6	28.43	118.95	10,481
$C_4H_8O_2$	*n*-Butyric acid	88.11	−5.7	30.04	125.69	11,075
$C_4H_8O_2$	*p*-Dioxane	88.11	11.0	34.85	145.81	12,847
C_4H_9Br	2-Bromobutane	137.02	−112.7	12.01	50.25	6,885
C_4H_{10}	*n*-Butane	58.12	−138.3	19.18	80.25	4,664
C_4H_{10}	Isobutane	58.12	−159.42	18.96	79.33	4,611
$C_4H_{10}O$	*n*-Butanol	74.12	−89.8	29.93	125.23	9,282
$C_4H_{10}O$	*tert*-Butanol	74.12	25.4	21.88	91.55	6,786
$C_4H_{10}O$	Ethyl ether	74.12	−116.3	23.45	98.11	7,272
$C_4H_{12}Si$	Tetramethylsilane	88.19	−99.04	18.64	77.99	6,878
C_5H_8	Cyclopentane	68.12	−135.1	11.80	49.37	3,363
C_5H_8	Isoprene	68.12	−145.9	16.8	70.29	4,788
C_5H_8	Methylene cyclobutane	68.12	−134.6	20.22	84.60	5,763
C_5H_8	1,4-Pentadiene	68.12	−148.8	21.55	90.17	6,142
$C_5H_8O_2$	δ-Valerolactone	100.12	−10.33	25.14	105.19	10,532
$C_5H_8O_3$	Levulinic acid	116.12	33.0	18.97	79.37	9,216
$C_5H_8O_4$	Glutaric acid	132.12	97.8	37.81	158.20	20,901
C_5H_{10}	1-Pentene	70.13	−166.2	19.81	82.89	5,813
C_5H_{10}	*cis*-2-Pentene	70.13	−151.4	24.25	101.46	7,115
C_5H_{10}	*trans*-2-Pentene	70.13	−140.2	28.48	119.16	8,357
C_5H_{10}	Cyclopentane	70.13	−93.8	2.07	8.66	607
$C_5H_{10}O_2$	Valeric acid	102.13	−59.0	27.83	116.44	11,892
$C_5H_{11}N$	Cyclopentylamine	85.15	−82.7	23.33	97.61	8,311
C_5H_{12}	*n*-Pentane	72.15	−129.7	27.89	116.69	8,419
C_5H_{12}	Isopentane	72.15	−159.9	17.05	71.34	5,147
C_5H_{12}	2,2-Dimethylpropane	72.15	−16.6	10.79	45.15	3,258
$C_5H_{12}O$	1-Pentanol	88.15	−78.9	26.65	111.50	9,829
C_6HCl_5O	Pentachlorophenol	266.34	189.3	15.39	64.39	17,150
$C_6H_3Br_3O$	2,4,6-Tribromophenol	330.80	93.0	13.38	55.98	18,518
$C_6H_3Cl_3$	1,2,3-Trichlorobenzene	181.45	53.7	27.00	112.97	20,498
$C_6H_3Cl_3$	1,3,5-Trichlorobenzene	181.45	63.5	23.97	100.29	18,198
$C_6H_3Cl_3O$	2,4,5-Trichlorophenol	197.45	68.5	28.87	120.79	23,850
C_6H_4BrI	*o*-Bromoiodobenzene	282.91	21.0	12.18	50.96	14,417
C_6H_4BrI	*m*-Bromoiodobenzene	282.91	9.3	10.27	42.97	12,157
C_6H_4BrI	*p*-Bromoiodobenzene	282.91	90.1	16.16	67.61	19,128
$C_6H_4Br_2$	*o*-Dibromobenzene	235.91	1.8	12.78	53.47	12,614
$C_6H_4Br_2$	*m*-Dibromobenzene	235.91	−6.9	13.38	55.98	13,206
$C_6H_4Br_2$	*p*-Dibromobenzene	235.91	86.0	20.55	85.98	20,284
$C_6H_4Br_2O$	2,4-Dibromophenol	251.91	12.0	13.97	58.45	14,724
$C_6H_4ClNO_2$	*m*-Chloronitrobenzene	157.56	44.4	29.38	122.93	19,369
$C_6H_4ClNO_2$	*p*-Chloronitrobenzene	157.56	83.5	31.51	131.84	20,773
$C_6H_4Cl_2$	*o*-Dichlorobenzene	147.00	−16.7	21.02	87.95	12,929
$C_6H_4Cl_2$	*m*-Dichlorobenzene	147.00	−24.8	20.55	85.98	12,639
$C_6H_4Cl_2$	*p*-Dichlorobenzene	147.00	52.7	27.89	116.69	17,153
$C_5H_4Cl_2O$	2,3-Dichlorophenol	163.00	56.8	31.32	131.04	21,360
$C_6H_4Cl_2O$	2,4-Dichlorophenol	163.00	44.8	29.46	123.26	20,091
$C_6H_4Cl_2O$	2,5-Dichlorophenol	163.00	57.8	32.89	137.61	22,430
$C_6H_4Cl_2O$	2,6-Dichlorophenol	163.00	66.8	32.47	135.85	22,144
$C_6H_4Cl_2O$	3,4-Dichlorophenol	163.00	67.8	30.69	128.41	20.931
$C_6H_4Cl_2O$	3,5-Dichlorophenol	163.00	67.8	30.07	125.81	20,507
$C_6H_4I_2$	*o*-Diiodobenzene	329.91	23.4	10.15	42.47	14,011
$C_6H_4I_2$	*m*-Diiodobenzene	329.91	34.2	11.54	48.28	15,928
$C_6H_4I_2$	*p*-Diiodobenzene	329.91	129.0	16.20	67.78	22,361
$C_6H_4N_2O_4$	*o*-Dinitrobenzene	168.11	116.93	32.25	134.93	22,683
$C_6H_4N_2O_4$	*m*-Dinitrobenzene	168.11	89.7	24.70	103.34	17,372
$C_6H_4N_2O_4$	*p*-Dinitrobenzene	168.11	173.5	39.99	167.32	28,128
$C_6H_4N_2O_5$	2,3-Dinitrophenol	184.11	143.8	34.06	142.51	26,238
$C_6H_4N_2O_5$	2,4-Dinitrophenol	184.11	114.8	31.38	131.29	24,172
$C_6H_4N_2O_5$	2,5-Dinitrophenol	184.11	107.8	30.80	128.87	23,726
$C_6H_4N_2O_5$	2,6-Dinitrophenol	184.11	62.8	25.41	106.32	19,575
$C_6H_4N_2O_5$	3,4-Dinitrophenol	184.11	133.8	32.94	137.82	25,374
$C_6H_4O_2$	*p*-Benzoquinone	108.10	112.9	40.97	171.42	18,531
C_6H_5Br	Bromobenzene	157.01	−30.6	16.17	67.66	10,623
C_6H_5BrO	4-Bromophenol	173.01	63.5	20.50	85.77	14,839
C_6H_5Cl	Chlorobenzene	112.56	−45.2	20.40	85.35	9,607
C_6H_5ClO	2-Chlorophenol	128.56	9.8	23.28	97.40	12,522
C_6H_5ClO	3-Chlorophenol	128.56	32.6	27.71	115.94	14,905
C_6H_5ClO	4-Chlorophenol	128.56	42.7	26.15	109.41	14,066

HEATS OF FUSION OF SOME ORGANIC COMPOUNDS (continued)

Formula	Compound	Mol. wt	M.P. (°C)	Heat of fusion (H_f)		
				cal/g	J/g	J/mol
C_6H_5F	Fluorobenzene	96.10	−42.21	28.12	117.65	11,306
C_6H_5I	Iodobenzene	204.01	−31.3	11.43	47.82	9,756
$C_6H_5NO_2$	Nitrobenzene	123.11	5.7	22.50	94.14	11,590
$C_6H_5NO_3$	2-Nitrophenol	139.11	44.8	29.97	125.39	17,443
$C_6H_5NO_3$	3-Nitrophenol	139.11	96.8	32.98	137.99	19,196
$C_6H_5NO_3$	4-Nitrophenol	139.11	113.8	31.36	131.21	18,253
C_6H_6	Benzene	78.11	5.53	30.45	127.40	9,951
$C_6H_6N_2O_2$	*o*-Nitroaniline	138.13	71.2	27.88	116.65	16,113
$C_6H_6N_2O_2$	*m*-Nitroaniline	138.13	147.0	36.50	152.72	21,095
$C_6H_6N_2O_2$	*p*-Nitroaniline	138.13	114.0	40.97	171.42	23,678
C_6H_6O	Phenol	94.11	40.9	28.67	119.96	11,289
$C_6H_6O_2$	1,2-Dihydroxybenzene	110.11	105.0	49.40	206.69	22,759
$C_6H_6O_2$	1,3-Dihydroxybenzene	110.11	110.0	46.22	193.38	21,293
$C_6H_6O_2$	1,4-Dihydroxybenzene	110.11	172.3	58.84	246.19	27,108
C_6H_6S	Thiophene	110.17	−14.9	24.90	104.18	11,478
C_6H_7N	Aniline	93.13	−6.3	27.09	113.34	10,555
$C_6H_8N_2$	Phenylhydrazine	108.14	19.6	36.31	151.92	16,429
$C_6H_8O_4$	Methyl fumarate	144.13	102.0	57.93	242.38	34,934
C_6H_{10}	Cyclohexene	82.15	−103.5	9.58	40.08	3,293
$C_6H_{10}O_4$	Adipic acid	146.14	153.2	57.00	238.49	34,853
C_6H_{12}	Methylcyclopentane	84.16	−142.5	19.68	82.34	6,930
C_6H_{12}	Cyclohexane	84.16	6.6	7.47	31.25	2,630
C_6H_{12}	Tetramethylethylene	84.16	−74.6	15.51	64.89	5,461
$C_6H_{12}O$	Cyclohexanol	100.16	25.46	4.19	17.53	1,756
$C_6H_{12}O_2$	ϵ-Caprolactone	116.16	−1.02	28.94	121.08	14,065
C_6H_{14}	2,2-Dimethylbutane	86.18	−99.0	1.61	6.74	581
C_6H_{14}	2,3-Dimethylbutane	86.18	−128.8	2.22	9.29	801
C_6H_{14}	*n*-Hexane	86.18	−95.3	36.27	151.75	13,078
C_6H_{14}	2-Methylpentane	86.18	−153.7	17.38	72.72	6,267
$C_6H_{14}O$	Isopropyl ether	102.18	−86.8	25.79	17.91	11,026
$C_6H_{14}O$	*n*-Propyl ether	12.18	−126.1	20.66	86.44	8,832
$C_6Cl_5NO_2$	Pentachloronitrobenzene	295.34	144.8	14.90	62.34	18,412
C_6Cl_6	Hexachlorobenzene	284.78	231.8	20.02	83.76	23,853
$C_7H_3Cl_3O_2$	2,3,6-Trichlorobenzoic acid	225.46	129.5	25.28	105.77	23,847
$C_7H_3I_3O_2$	2,3,5-Triiodobenzoic acid	499.81	230.6	15.41	64.48	32,228
$C_7H_5ClO_2$	2-Chlorobenzoic acid	156.57	140.2	39.27	164.31	25,726
$C_7H_5ClO_2$	3-Chlorobenzoic acid	156.57	154.2	36.39	152.26	23,839
$C_7H_5ClO_2$	4-Chlorobenzoic acid	156.57	239.7	49.23	205.98	32,250
$C_7H_5NO_4$	2-Nitrobenzoic acid	167.12	145.8	40.06	167.61	28,011
$C_7H_5NO_4$	3-Nitrobenzoic acid	167.12	141.1	27.59	115.44	19,292
$C_7H_5NO_4$	4-Nitrobenzoic acid	167.12	239.2	52.80	220.92	36,920
$C_7H_5N_3O_6$	2,4,6-Trinitrotoluene	227.13	80.83	22.34	93.47	21,230
$C_7H_6N_2O_4$	2,4-Dinitrotoluene	182.14	70.14	26.40	110.46	20,119
$C_7H_6O_2$	Benzoic acid	122.12	122.4	33.89	141.80	17,317
C_7H_7Br	4-Bromotoluene	171.04	28.0	20.86	87.28	14,928
$C_7H_7NO_2$	2-Aminobenzoic acid	137.14	145.0	35.98	150.54	20,645
$C_7H_7NO_2$	3-Aminobenzoic acid	137.14	179.5	38.03	159.12	21,822
$C_7H_7NO_2$	4-Aminobenzoic acid	137.11	188.5	36.46	152.55	20,921
$C_7H_7NO_3$	4-Nitro-5-methylphenol	153.14	127.8	42.77	178.95	27,404
$C_7H_7NO_3$	2-Nitro-5-methylphenol	153.14	29.6	32.45	135.77	20,792
C_7H_8	Toluene	92.14	−94.99	17.77	74.35	6,851
C_7H_8	1,3,5-Cycloheptatriene	92.14	−75.23	3.01	12.59	1,160
C_7H_8O	Benzyl alcohol	108.14	−15.2	19.83	82.97	8,972
C_7H_8O	2-Methylphenol	108.14	29.8	30.81	128.91	13,940
C_7H_8O	3-Methylphenol	108.14	11.8	20.80	87.03	9,411
C_7H_8O	4-Methylphenol	108.14	35.8	26.27	109.91	11,886
C_7H_9O	*p*-Toluidine	109.15	43.3	39.90	166.94	18,222
$C_7H_{12}O_4$	Pimelic acid	160.17	104.3	41.22	172.46	27,623
C_7H_{14}	1-Heptene	98.19	−119.7	30.82	128.95	12,662
C_7H_{14}	Methylcyclohexane	98.19	−126.6	16.43	68.74	6,750
C_7H_{14}	Cycloheptane	98.19	−8.03	4.58	19.16	1,881
C_7H_{16}	*n*-Heptane	100.20	−90.6	33.78	141.34	14,162
C_7H_{16}	2-Methylhexane	100.20	−118.2	21.16	88.53	8,871
C_7H_{16}	2,2-Dimethylpentane	100.20	−123.8	13.98	58.49	5,861
C_7H_{16}	2,4-Dimethylpentane	100.20	−119.9	15.95	66.73	6,686
C_7H_{16}	3,3-Dimethylpentane	100.20	−134.9	16.86	70.54	7,068
C_7H_{16}	3-Ethylpentane	100.20	−118.6	22.78	95.31	9,550
C_7H_{16}	2,2,3-Trimethylbutane	100.20	−25.0	5.25	21.97	2,201
$C_8H_6Cl_4$	Tetrachloro-*o*-xylene	243.95	86.0	21.02	87.95	21,455
$C_8H_6Cl_4$	Tetrachloro-*p*-xylene	243.95	95.0	22.10	92.47	22,588
$C_8H_8Br_2$	α-α′-Dibromo-*o*-xylene	263.96	95.0	24.25	101.46	26,781
$C_8H_8Br_2$	α-α′-Dibromo-*m*-xylene	263.96	77.0	21.45	89.75	23,690
$C_8H_8Cl_2$	α-α′-Dichloro-*o*-xylene	175.06	55.0	29.03	121.46	21,263

HEATS OF FUSION OF SOME ORGANIC COMPOUNDS (continued)

Formula	Compound	Mol. wt	M.P. (°C)	Heat of fusion (H_f)		
				cal/g	J/g	J/mol
$C_8H_8Cl_2$	α-α′-Dichloro-*m*-xylene	175.06	34.0	26.64	111.46	19,512
$C_8H_8Cl_2$	α-α′-Dichloro-*p*-xylene	175.06	100.0	32.73	136.94	23,973
$C_8H_8O_2$	Phenylacetic acid	136.15	76.7	25.44	106.44	14,492
$C_8H_8O_2$	*o*-Toluic acid	136.15	103.7	35.40	148.11	20,165
$C_8H_8O_2$	*m*-Toluic acid	136.15	108.75	27.59	115.44	15,717
$C_8H_8O_2$	*p*-Toluic acid	136.15	179.6	39.90	166.94	22,729
C_8H_{10}	*o*-Xylene	106.17	−25.2	30.64	128.20	13,611
C_8H_{10}	*m*-Xylene	106.17	−47.8	26.01	108.83	11,554
C_8H_{10}	*p*-Xylene	106.17	13.2	37.83	158.28	16,805
$C_8H_{10}O$	2,3-Dimethylphenol	122.17	72.8	41.13	172.09	21,024
$C_8H_{10}O$	2,5-Dimethylphenol	122.17	74.8	45.73	191.33	23.375
$C_8H_{10}O$	2,6-Dimethylphenol	122.17	45.7	36.97	154.68	18,897
$C_8H_{10}O$	3,4-Dimethylphenol	122.17	60.8	35.46	148.36	18,125
$C_8H_{10}O$	3,5-Dimethylphenol	122.17	63.6	35.21	147.32	17,998
C_8H_{12}	1,5-Cyclooctadiene	108.18	−69.17	21.71	90.83	9,826
C_8H_{14}	Bicyclooctane	108.18	174.63	18.10	75.73	8,192
$C_8H_{14}O_4$	Suberic acid	172.18	142.13	40.01	167.40	28,823
C_8H_{16}	Cyclooctane	112.21	14.83	5.13	21.46	2,408
C_8H_{16}	Ethylcyclohexane	112.21	−33.3	17.75	74.27	8,334
C_8H_{16}	*trans*-1,1-Dimethylcyclohexane	112.21	33.3	4.38	18.33	2,057
C_8H_{16}	*cis*-1,2-Dimethylcyclohexane	112.21	−49.9	3.50	14.64	1,643
C_8H_{16}	*trans*-1,2-Dimethylcyclohexane	112.21	−88.2	22.35	93.51	10,493
C_8H_{16}	*cis*-1,3-Dimethylcyclohexane	112.21	−75.6	23.05	96.44	10,822
C_8H_{16}	*trans*-1,3-Dimethylcyclohexane	112.21	−90.1	21.01	87.91	9,864
C_8H_{16}	*cis*-1,4-Dimethylcyclohexane	112.21	−87.4	19.82	82.93	9,306
C_8H_{16}	*trans*-1,4-Dimethylcyclohexane	112.21	−36.9	26.27	109.91	12,333
$C_8H_{16}O_2$	Caprylic acid	144.21	16.3	35.40	148.11	21,359
C_8H_{18}	*n*-Octane	114.23	−56.8	43.21	180.79	20,652
C_8H_{18}	3-Methylheptane	114.23	−120.5	23.81	99.62	11,380
C_8H_{18}	4-Methylheptane	114.23	121.0	22.68	94.89	10,839
C_8H_{18}	2,2,4-Trimethylpentane	114.23	−107.3	18.92	79.16	9,042
$C_8H_{18}N_2$	1,1-Dimethylazoethane	142.24	−14.6	17.27	72.26	10,278
$C_8H_{18}N_2O$	1,1-Dimethylazoxyethane	158.24	15.2	17.40	72.84	11,526
C_9H_7N	Quinoline	129.16	15.6	19.98	83.60	10,798
$C_9H_8O_2$	Cinnamic acid	148.16	133.0	36.50	152.72	22,627
$C_9H_8O_2$	Allocinnamic acid	148.16	68.0	27.35	114.43	16,954
$C_9H_{10}O_2$	Hydrocinnamic acid	150.18	48.0	28.14	117.74	17,682
C_9H_{12}	1,2,4-Trimethylbenzene	120.19	−43.8	7.47	31.25	3,756
C_9H_{12}	1,2,3-Trimethylbenzene	120.19	−25.4	16.65	69.66	8,372
C_9H_{12}	1,3,5-Trimethylbenzene	120.19	−44.7	18.90	79.08	9,505
C_9H_{12}	*n*-Propylbenzene	120.19	−99.56	18.43	77.11	9,268
$C_9H_{16}O_4$	Azelaic acid	188.22	106.8	41.49	173.59	32,673
C_9H_{18}	*n*-Propylcyclohexane	126.24	−94.9	19.64	82.17	10,373
$C_9H_{18}O_2$	Pelargonic acid	158.24	12.35	30.63	128.16	20,280
C_9H_{20}	*n*-Nonane	128.26	−53.5	28.83	120.62	15,471
C_9H_{20}	2,2,3,3-Tetramethylpentane	128.26	−9.8	4.35	18.20	2,334
C_9H_{20}	2,2,4,4-Tetramethylpentane	128.26	−66.54	18.16	75.98	9,745
C_9H_{20}	3,3-Diethylepentane	128.26	−33.1	18.80	78.66	10,089
$C_{10}H_7Br$	α-Bromonaphthalene	207.07	−1.8	17.50	73.22	15,162
$C_{10}H_7Br$	β-Bromonaphthalene	207.07	58.8	13.82	57.82	11,973
$C_{10}H_7Cl$	α-Chloronaphthalene	162.62	−2.5	18.96	79.33	12,901
$C_{10}H_7Cl$	β-Chloronaphthalene	162.62	58.8	21.60	90.37	14,696
$C_{10}H_7$	α-Iodonaphthalene	254.07	6.8	14.97	62.63	15,912
$C_{10}H_7I$	β-Iodonaphthalene	254.07	54.35	15.09	63.14	16,042
$C_{10}H_7NO_2$	α-Nitronaphthalene	173.17	56.7	25.44	106.44	18,432
$C_{10}H_8$	Naphthalene	128.17	78.2	35.66	149.20	19,123
$C_{10}H_8O$	α-Naphthol	144.17	94.0	38.68	161.84	23,332
$C_{10}H_8O$	β-Naphthol	144.17	123.0	29.03	121.46	17,511
$C_{10}H_9N$	α-Aminonaphthalene	143.19	50.0	24.19	101.21	14,492
$C_{10}H_9N$	β-Aminonaphthalene	143.19	113.0	39.41	164.89	23,611
$C_{10}H_{14}$	*n*-Butylbenzene	134.22	−87.9	19.98	83.60	11,221
$C_{10}H_{14}$	1,2,4,5-Tetramethylbenzene	134.22	79.3	37.40	156.48	21,003
$C_{10}H_{14}$	1,2,3,4-Tetramethylbenzene	134.22	−7.7	20.0	83.68	11,232
$C_{10}H_{14}$	1-Methyl-4-isopropylbenzene	134.22	−68.9	17.10	71.55	9,603
$C_{10}H_{14}O$	Thymol	150.22	51.5	27.47	114.93	17,265
$C_{10}H_{18}O_4$	Sebacic acid	202.25	130.8	48.22	201.75	40,804
$C_{10}H_{20}$	*n*-Butylcyclohexane	140.27	−74.73	24.12	100.92	14,156
$C_{10}H_{20}O_2$	*n*-Capric acid	172.27	31.99	38.87	162.63	28,016
$C_{10}H_{22}$	*n*-Decane	142.28	−29.7	48.34	202.25	28,776
$C_{11}H_8O_2$	α-Naphthoic acid	172.18	161.0	27.61	115.52	19,890
$C_{11}H_8O_2$	β-Naphthoic acid	172.18	185.0	32.68	136.73	23,542
$C_{11}H_{10}$	2-Methylnaphthalene	142.20	34.4	20.11	89.14	11,965
$C_{11}H_{14}$	1,1-Dimethylindan	142.20	−45.8	19.60	82.01	11,662

Formula	Compound	Mol. wt	M.P. (°C)	Heat of fusion (H_f) cal/g	J/g	J/mol
$C_{11}H_{14}$	4,6-Dimethylindan	142.20	−16.70	21.05	88.07	12,524
$C_{11}H_{14}$	4,7-Dimethylindan	142.20	−0.52	22.09	92.42	13,142
$C_{11}H_{20}O_4$	Undecanedioic acid	216.28	111.8	43.82	183.34	39,653
$C_{11}H_{22}O_2$	*n*-Undecilic acid	186.29	28.25	32.20	134.72	25,097
$C_{11}H_{24}$	*n*-Undecane	156.31	−25.6	34.12	142.76	22,315
$C_{12}H_8S$	Dibenzothiophene	184.26	97.8	19.85	83.05	15,303
$C_{12}H_9N$	Carbazole	167.21	243.0	42.05	175.94	29,419
$C_{12}H_{10}$	Acenaphthene	154.21	93.4	33.38	139.66	21,537
$C_{12}H_{10}$	Biphenyl	154.21	69.0	28.83	120.62	18,601
$C_{12}H_{10}N_2$	Azobenzene	182.22	67.1	28.91	120.96	22,041
$C_{12}H_{10}N_2O$	Azoxybenzene	198.22	36.0	21.62	90.46	17,931
$C_{12}H_{11}N$	Diphenylamine	168.23	52.98	25.23	105.56	17,864
$C_{12}H_{12}$	1,4-Dimethylnaphthalene	156.23	6.8	24.32	101.75	15,896
$C_{12}H_{12}$	2,3-Dimethylnaphthalene	156.23	104.8	38.40	160.67	25,101
$C_{12}H_{12}$	2,6-Dimethylnaphthalene	156.23	110.17	38.33	160.37	25,055
$C_{12}H_{12}$	2,7-Dimethylnaphthalene	156.23	95.66	35.72	149.45	23,349
$C_{12}H_{12}$	1,8-Dimethylnaphthalene	156.23	63.18	24.12	100.92	15,767
$C_{12}H_{12}N_2$	Hydrazobenzene	184.24	134.0	22.89	95.77	17,645
$C_{12}H_{16}$	Cyclohexylbenzene	160.26	7.3	22.82	95.48	15,302
$C_{12}H_{22}O_4$	Dodecanedioic acid	230.30	129.3	52.48	219.58	50,569
$C_{12}H_{24}O_2$	*n*-Lauric acid	200.32	43.22	43.72	182.92	36,643
$C_{12}H_{26}$	*n*-Dodecane	170.34	−9.6	51.33	214.76	36,582
$C_{12}Cl_{10}$	Decachlorobiphenyl	498.66	304.5	18.90	79.08	39,434
$C_{13}H_8Cl_2O$	*p,p′*-Dichlorobenzophenone	251.11	146.8	28.67	119.96	30,123
$C_{13}H_{10}$	Fluorene	166.22	114.8	28.15	117.78	19,577
$C_{13}H_{10}O$	Benzophenone	182.22	47.88	23.86	99.83	18,191
$C_{13}H_{13}N$	Benzylaniline	183.25	32.37	21.86	91.46	16,760
$C_{13}H_{18}$	1,1,4,6-Tetramethylindan	174.29	0.36	21.59	90.33	15,743
$C_{13}H_{18}$	1,1,4,7-Tetramethylindan	174.29	27.6	−15.47	64.73	11,282
$C_{13}H_{24}O_4$	Tridecanedioic acid	244.33	114.3	44.31	185.39	45,296
$C_{14}H_8O_2$	Anthroquinone	208.22	284.8	37.48	156.82	32,653
$C_{14}H_{10}$	Anthracene	178.23	219.5	38.66	161.75	28,829
$C_{14}H_{10}$	Phenanthrene	178.23	99.24	22.08	92.38	16,465
$C_{14}H_{10}O_2$	Benzil	213.23	95.2	22.15	92.68	19,762
$C_{14}H_{28}O_2$	Myristic acid	228.37	53.96	47.49	198.70	45,377
$C_{16}H_{10}$	Pyrene	202.26	151.2	20.22	84.60	17,111
$C_{16}H_{10}$	Fluoranthene	202.26	107.8	22.30	93.30	18,871
$C_{16}H_{32}$	*n*-Decylcyclohexane	224.43	−1.72	41.10	171.96	38,593
$C_{16}H_{32}O_2$	Palmitic acid	256.43	61.82	39.18	163.93	42,037
$C_{16}H_{34}O$	1-Hexanedecanol	242.44	49.27	33.80	141.42	34,286
$C_{18}H_{12}$	Chrysene	228.29	258.2	27.38	114.56	26,153
$C_{18}H_{12}$	Triphenylene	228.29	200.3	26.28	109.96	25,103
$C_{18}H_{12}$	1,2-Benzanthracene	228.29	161.1	22.38	93.64	21,377
$C_{18}H_{12}$	3,4-Benzophenanthrene	228.29	61.5	17.08	71.46	16,314
$C_{18}H_{14}$	*p*-Terphenyl	230.31	210.1	36.84	154.14	35,500
$C_{18}H_{14}O_3$	Cinnamic anhydride	278.31	48.0	28.14	117.74	32,768
$C_{18}H_{34}O_2$	Elaidic acid	282.47	44.4	52.08	217.90	61,550
$C_{18}H_{36}O_2$	Stearic acid	284.48	68.82	47.54	198.91	56,586
$C_{18}H_{38}$	*n*-Octadecane	254.5	28.2	57.65	241.21	61,388
$C_{19}H_{40}$	*n*-Nonadecane	268.53	32.1	40.78	170.62	45,817
$C_{20}H_{12}$	Perylene	252.31	280.7	30.08	125.85	31,753
$C_{20}H_{12}$	1,2-Benzopyrene	252.31	181.3	15.69	65.65	16,564
$C_{20}H_{12}$	3,4-Benzopyrene	252.31	181.0	16.41	68.66	17,324
$C_{20}H_{42}$	*n*-Eicosane	282.55	36.8	59.11	247.32	69,880
$C_{21}H_{16}$	1,2′-Dinaphthylmethane	268.36	96.4	27.21	113.85	30,553
$C_{21}H_{44}$	*n*-Heneicosane	296.58	40.5	38.44	160.83	47,700
$C_{22}H_{12}$	1,12-Benzoperylene	276.34	281.0	15.02	62.84	17,365
$C_{22}H_{12}$	*o*-Phenylenepyrene	276.34	162.0	18.60	77.82	21,505
$C_{22}H_{14}$	1,2:3,4-Dibenzanthracene	278.35	280.3	22.17	92.76	25,820
$C_{22}H_{14}$	1,2:5,6-Dibenzanthracene	278.35	271.0	26.76	119.96	31,164
$C_{22}H_{46}$	*n*-Docosane	310.61	44.4	37.67	157.61	49,955
$C_{23}H_{46}$	*n*-Tricosane	322.62	47.6	30.74	128.62	41,495
$C_{24}H_{12}$	Coronene	300.36	437.3	15.28	63.93	19,202
$C_{24}H_{14}$	3,4:9,10-Dibenzopyrene	302.37	283.6	22.03	92.17	27,869
$C_{24}H_{14}$	1,2:3,4-Dibenzopyrene	302.37	228.0	19.51	81.63	24,682
$C_{24}H_{14}$	1,2:4,5-Dibenzopyrene	302.37	247.0	24.11	100.88	30,503
$C_{24}H_{18}$	*p*-Quaterphenyl	306.41	314.0	29.48	123.34	37,793
$C_{24}H_{50}$	*n*-Tetracosane	338.66	50.9	38.74	162.09	54,893
$C_{25}H_{52}$	*n*-Pentacosane	352.69	53.7	39.13	163.72	57,742
$C_{26}H_{14}$	1,12-Phenyleneperylene	326.40	268.3	12.65	52.93	17,276
$C_{27}H_{56}$	*n*-Heptacosane	380.76	59.0	37.93	158.70	60,427
$C_{28}H_{58}$	*n*-Octacosane	394.77	61.4	39.14	163.76	64,648

HEATS OF VAPORIZATION OF ORGANIC COMPOUNDS

Numerical values in the following table are in the units of gram calories per gram mole. To convert to joules per gram mole, multiply the listed value by 4.184.

Formula	Name	ΔH_v
CBrN	Cyanogen bromide	10,882.8
CBr_4	Carbon tetrabromide	10,771.4
$CBrF_3$	Bromotrifluoromethane	–
CBr_2F_2	Dibromodifluoromethane	–
$CClF_3$	Chlorotrifluoromethane	3,996.3
CClN	Cyanogen chloride	5,243.4
CCl_2F_2	Dichlorodifluoromethane	8,363 1
CCl_2O	Phosgene	6,224.3
CCl_3F	Trichlorofluoromethane	[illegible]
CCl_3NO_2	Trichloronitromethane	9,109.7
CCl_4	Carbon tetrachloride	8,271.5
		7,628.8
CFN	Cyanogen fluoride	5,875.3
CF_4	Carbon tetrafluoride	3,016.5
$CHBr_3$	Tribromomethane	9,673.3
$CHClF_2$	Chlorodifluoromethane	5,212.9
$CHCl_2F$	Dichlorofluoromethane	6,286.8
$CHCl_3$	Chloroform	7,500.5
CHF_3	Trifluoromethane	–
CHN	Hydrogen cyanide	7,338.8
CH_2Br_2	Dibromomethane	8,722.0
CH_2Cl_2	Dichloromethane	7,572.3
CH_2O	Formaldehyde	5,917.9
CH_2O_2	Formic acid	9,896.5
CH_3AsCl_2	Dichloromethylarsine	9,636.8
CH_3BO	Borine carbonyl	4,867.6
CH_3Br	Methyl bromide	5,925.9
CH_3Cl	Methyl chloride	5,375.3
CH_3Cl_3Si	Trichloromethylsilane	7,450.0
CH_3F	Methyl fluoride	3,986.4
CH_3I	Methyl iodide	6,616.5
CH_3NO	Formamide	15,556.6
CH_3NO_2	Nitromethane	9,210.9
CH_4	Methane	2,128.8
CH_4Cl_2Si	Dichloromethylsilane	7,011.0
CH_4O	Methanol	9,377.2
		8,978.8
CH_4S	Methanethiol	6,331.9
CH_5ClSi	Chloromethylsilane	6,349.5
CH_5N	Methylamine	6,469.5
CH_6Si	Methylsilane	4,683.6
CH_9NSi_2	2-Methyldisilazane	7,185.6
CIN	Cyanogen iodide	14,065.4
CN_4O_8	Tetranitromethane	9,848.7
CO	Carbon monoxide	1,613.3
COS	Carbonyl sulfide	4,992.2
COSe	Carbonyl selenide	5,366.5
CO_2	Carbon dioxide	5,539.0
CSSe	Carbon selenosulfide	8,003.0
CS_2	Carbon disulfide	6,786.8
C_2BrCl_2O	Trichloroacetyl bromide	9,673.9
C_2HClF_2	1-Chloro-2,2-difluoroethylene	–
C_2ClF_3	1-Chloro-1,2,2-trifluoroethylene	5,421.5
$C_2Cl_2F_2$	1,2-Dichloro-1,2-difluoroethylene	7,185.6
		–
C_2F_4	Tetrafluoroethylene	
$C_2Cl_2F_4$	1,1-Dichloro-1,2,2,2-tetrafluoroethane	–
$C_2Cl_2F_4$	1,2-Dichloro-1,1,2,2-tetrafluoroethane	6,134.6
$C_2Cl_3F_3$	1,1,2-Trichloro-1,2,2-trifluoroethane	7,115.4
C_2Cl_4	Tetrachloroethylene	9,240.5
CCl_4F_2	Tetrachlorodifluoroethane	–
$C_2Cl_4F_2$	1,1,2,2-Tetrachloro-1,2-difluoroethane	8,746.2
$C_2Cl_3F_3$	1,1,2-Trifluoro-1,2,2-trichloroethane	–
C_2Cl_6	Hexachloroethane	11,711.3
C_2ClF_5	Chloropentafluoroethane	–
C_2F_6	Hexafluoroethane	–
C_2HBr_3O	Tribromoacetaldehyde	11,057.8
C_2HCl_3	Trichloroethylene	8,314.7
C_2HCl_3O	Trichloroacetaldehyde	8,469.2
$C_2HCl_3O_2$	Trichloroacetic acid	13,817.9
C_2HCl_5	Pentachloroethane	9,800.1
C_2H_2	Acetylene	4,665.8
$C_2H_2Br_4$	1,1,1,2-Tetrabromoethane	14,517.3
$C_2H_2Br_4$	1,1,2,2-Tetrabromoethane	12,911.5
$C_2H_2O_4$	Oxalic acid	21,630.6
$C_2H_2Cl_2$	*cis*-1,2-Dichloroethylene	7,420.6
$C_2H_2Cl_2$	*trans*-1,2-Dichloroethylene	7,243.1
$C_2H_2Cl_2$	1,1-Dichloroethylene	7,211.8
$C_2H_2F_2$	1,1-Difluoroethylene	
$C_2H_2Cl_2O_2$	Dichloroacetic acid	12,952.9
$C_2H_2Cl_4$	1,1,1,2-Tetrachloroethane	9,296.5
		8,725.6
$C_2H_2Cl_4$	1,1,2,2-Tetrachloroethane	9,917.1
C_2H_3Br	1-Bromoethylene	6,076.9
$C_2H_3BrO_2$	Bromoacetic acid	13,537.8
$C_2H_3Br_3$	1,1,2-Tribromoethane	11,874.1
C_2H_3Cl	1-Chloroethylene	6,263.0
$C_2H_3ClO_2$	Chloroacetic acid	13,134.5
$C_2H_3Cl_3$	1,1,1-Trichloroethane	8,012.7
$C_2H_3Cl_3$	1,1,2-Trichloroethane	9,163.2
$C_2H_3ClF_2$	1-Chloro-1,1-difluoroethane	–
$C_2H_3F_3$	1,1,1-Trifluoroethane	–
$C_2H_3Cl_3O_2$	Trichloroacetaldehyde-hydrate	12,141.5
C_2H_3F	1-Fluoroethylene	4,198.1
C_2H_3N	Acetonitrile	8,173.2
C_2H_3NS	Methyl thiocyanate	9,424.1
C_2H_3NS	Methyl isothiocyanate	7,990.1
C_2H_4	Ethylene	3,453.7
C_2H_4BrCl	1-Bromo-2-chloroethane	9,314.9
C_2H_4BrCl	1-Bromo-2-chloroethane	8,995.6
$C_2H_4Br_2$	1,2-Dibromoethane	9,229.4
$C_2H_4Cl_2$	1,1-Dichloroethane	7,288.0
$C_2H_4Cl_2$	1,2-Dichloroethane	7,950.7
$C_2H_4F_2$	1,1-Difluoroethane	6,068.8
C_2H_4O	Acetaldehyde	7,267.8
		6,622.1
C_2H_4O	Ethylene oxide	6,823.3
$C_2H_4O_2$	Acetic acid	9,963.9
		9,486.6
$C_2H_4O_2$	Methyl formate	7,027.8
$C_2H_4O_2S$	Mercaptoacetic acid	13,790.7
C_2H_5Br	Ethyl bromide	6,843.1
C_2H_5Cl	Ethyl chloride	6,310.6
C_2H_5ClO	2-Chloroethanol	10,740.6
$C_2H_5Cl_3Si$	Trichloroethylsilane	9,457.8
$C_2H_5Cl_3OSi$	Trichloroethoxysilane	8,811.4
C_2H_5F	Ethyl fluoride	5,519.5
$C_2H_5F_3Si$	Ethyltrifluorosilane	6,945.7
C_2H_5I	Ethyl iodide	7,851.8
C_2H_5NO	Acetamide	14,025.3
C_2H_5NO	Acetaldoxime	11,317.8
$C_2H_5NO_2$	Nitroethane	9,531.1
$C_2H_5N_3O_2$	Di(nitrosomethyl)amine	10,326.7
C_2H_6	Ethane	3,739.5
$C_2H_6Cl_2Si$	Dichlorodimethylsilane	7,995.7
C_2H_6O	Ethyl alcohol	9,673.9
C_2H_6O	Dimethyl ether	5,409.8
$C_2H_6O_2$	1,2-Ethanediol	14,032.4
C_2H_6S	Dimethyl sulfide	6,742.3
C_2H_6S	Ethane thiol	6,728.7
C_2H_6Sb	Dimethylantimony	12,075.7
C_2H_7N	Ethylamine	6,845.1
C_2H_7N	Dimethylamine	6,660.0
$C_2H_8N_2$	1,2-Ethanediamine	10,510.5
C_2H_8Si	Dimethylsilane	5,497.8
$C_2H_{10}B_2$	Dimethyldiborane	5,696.7
$C_2H_{11}NSi_2$	2-Ethyldisilazane	7,348.3
C_2N_2	Cyanogen	6,597.3
C_3H_3N	Acrylonitrile	7,941.4
C_3H_4	Propadiene	5,141.2
C_3H_4	Propyne	5,632.4
$C_3H_4Br_2$	2,3-Dibromopropene	9,886.2
$C_3H_4Cl_2O_2$	Methyl dichloroacetate	10,820.5
C_3H_4O	2-Propenal	7,628.8
$C_3H_4O_2$	Acrylic acid	10,955.1
$C_3H_4O_3$	Pyruvic acid	11,815.7
$C_3H_5Br_3$	1,2,3-Tribromopropane	12,047.1
C_3H_5Cl	1-Chloropropene	6,594.3
C_3H_5Cl	Allyl chloride	7,386.8
C_3H_5ClO	Epichlorohydrine	9,815.4
$C_3H_5ClO_2$	Methyl chloroacetate	10,815.0
$C_3H_5Cl_3$	1,1,1-Trichloropropane	8,933.9
$C_3H_5Cl_3$	1,2,3-Trichloropropane	10,714.3
$C_3H_5Cl_3Si$	Allyltrichlorosilane	9,386.1
C_3H_5N	Propionitrile	8,769.0
C_3H_5NO	3-Hydroxypropionitrile	13,287.2
C_3H_5NS	Ethylisothiocyanate	9,574.7
$C_3H_5N_3O_9$	Nitroglycerine	13,753.1
C_3H_6	Propene	4,697.4
C_3H_6	Cyclopropane	5,897.7
C_3H_6BrNO	2-Bromo-2-nitrosopropane	9,619.6
$C_3H_6Br_2$	1,2-Dibromopropane	9,801.9
$C_3H_6Br_2$	1,3-Dibromopropane	10,374.4
$C_3H_6Br_2O$	2,3-Dibromo-1-propanol	13,190.0
$C_3H_6Cl_2$	1,2-Dichloropropane	8,428.5
$C_3H_6Cl_2O$	1,3-Dichloro-2-propanol	12,067.6
C_3H_6O	Acetone	7,641.5
C_3H_6O	Allyl alcohol	10,577.7
C_3H_6O	Propylene oxide	7,295.8
$C_3H_6O_2$	Propanoic acid	12,454.4
$C_3H_6O_2$	Methyl acetate	7,732.8
$C_3H_6O_2$	Ethyl formate	7,511.7
$C_3H_6O_3$	Methyl glycolate	11,105.0
$C_3H_6O_3$	Methoxyacetic acid	13,451.0
C_3H_7Br	*n*-Propyl bromide	8,029.8
C_3H_7Br	2-Bromopropane	7,391.7
C_3H_7Cl	*n*-Propyl chloride	7,485.7
		6,905.8
C_3H_7Cl	2-Chloropropane	6,855.2
$C_3H_7Cl_3Si$	Trichloroisopropylsilane	8,973.3
C_3H_7I	*n*-Propyl iodide	8,467.1
C_3H_7I	2-Iodopropane	8,243.4
	Propionamide	14,554.0
C_3H_7NO	1-Nitropropane	9,949.9
$C_3H_7NO_2$	2-Nitropropane	9,476.9
$C_3H_7NO_2$	Ethyl carbamate	13,078.6
$C_3H_7NO_2$	Propane	4,550.0
C_3H_8		4,811.8
C_3H_8O	*n*-Propanol	11,298.8
		10,421.1
C_3H_8O	Isopropanol	10,063.5
C_3H_8O	Ethyl methyl ether	6,388.3
$C_3H_8O_2$	1,2-Propanediol	13,575.2
$C_3H_8O_2$	1,3-Propanediol	13,782.3
$C_3H_8O_2$	2-Methoxyethanol	9,893.8
$C_3H_8O_2$	Glycerol	18,188.9
C_3H_8S	Methyl ethyl sulfide	–
C_3H_8S	Propanethiol	7,855.3
C_3H_9B	Trimethylborine	5,375.4
C_3H_9ClSi	Chlorotrimethylsilane	7,589.1
C_3H_9Ga	Trimethylgallium	7,758.8
C_3H_9N	*n*-Propylamine	7,408.0
C_3H_9N	Trimethylamine	6,361.7
$C_3H_9O_4P$	Trimethyl phosphate	11,019.7
$C_3H_{12}B_2$	Trimethyldiborane	6,981.8
C_3O_2	Carbon suboxide	6,446.3
C_3S_2	Carbon subsulfide	10,466.0
$C_4Cl_6O_3$	Trichloroacetic anhydride	12,929.0
C_3F_8	Octafluoropropane	–
C_4H_4O	Furan	–
C_4H_2	1,3-Butadiyne	7,761.0
$C_4H_2Br_2O_3$	α,β-Dibromomaleic anhydride	12,579.2
$C_4H_2Cl_2O_2$	*trans*-Fumaryl chloride	11,251.0
$C_4H_2O_3$	Maleic anhydride	12,122.3
$C_4H_3NO_2S$	2-Nitrothiophene	11,926.2
C_4H_4	Butenyne	6,677.2
$C_4H_4Cl_2O_2$	Succinyl chloride	12,466.1
$C_4H_4Cl_2O_3$	Chloroacetic anhydride	14,645.1
$C_4H_4O_3$	Succinic anhydride	14,726.0
$C_4H_4O_4$	1,4-Dioxane-2,6-dione	14,013.6
C_4H_4S	Thiophene	8,748.3
C_4H_4Se	Selenophene	7,766.1
$C_4H_5ClO_2$	α-Chlorocrotonic acid	15,440.1
$C_4H_5ClO_3$	Ethyl chloroglyoxylate	10,268.4
$C_4H_5Cl_3O_2$	Ethyl trichloroacetate	11,625.1
C_4H_5N	3-Butenenitrile	9,447.8
C_4H_5N	Methacrylonitrile	8,083.8
C_4H_5N	*cis*-Crotononitrile	8,905.4
C_4H_5N	*trans*-Crotononitrile	9,277.1
$C_4H_5NO_2$	Succinimide	16,422.0
C_4H_5NS	Allylisothiocyanate	9,967.8
C_4H_6	1,2-Butadiene	6,539.1
C_4H_6	1,3-Butadiene	5,688.2
C_4H_6	Cyclobutene	6,167.5
C_4H_6	1-Butyne	6,596.9
C_4H_6	2-Butyne	7,868.5
$C_4H_6Cl_2O_2$	Ethyl dichloroacetate	10,842.8
$C_4H_6Cl_2O_2$	2-Chloroethyl chloroacetate	12,588.7
$C_4H_6O_2$	*cis*-Crotonic acid	12,964.7
$C_4H_6O_2$	*trans*-Crotonic acid	13,252.2
$C_4H_6O_2$	Methyl acrylate	8,598.0
$C_4H_6O_2$	Methacrylic acid	12,526.6
$C_4H_6O_2$	Vinyl acetate	8,470.4
$C_4H_6O_2$	Acetic anhydride	10,930.4
$C_4H_6O_4$	Dimethyl oxalate	11,519.4
C_4H_7Br	*cis*-1-Bromo-1-butene	8,300.2
C_4H_7Br	*trans*-1-Bromo-1-butene	8,515.7
C_4H_7Br	2-Bromo-1-butene	8,389.7
C_4H_7Br	*cis*-2-Bromo-2-butene	8,486.3
C_4H_7Br	*trans*-2-Bromo-2-butene	8,238.1
C_4H_7BrO	1-Bromo-2-butanone	10,980.7
C_4H_7BrO	2-Methylpropionyl bromide	10,974.6
$C_4H_7Br_3$	1,1,2-Tribromobutane	11,936.5
$C_4H_7Br_3$	1,2,2-Tribromobutane	11,622.3

HEATS OF VAPORIZATION OF ORGANIC COMPOUNDS (continued)

Formula	Name	Δ H_v
$C_4H_7Br_3$	1,2,2,-Tribromobutane	11.622.3
$C_4H_7Br_3$	2,2,3-Tribromobutane	11,664.2
$C_4H_7ClO_2$	Ethyl chloroacetate	10,522.6
$C_4H_8Cl_2$	1,1-Dichloro-2-methylpropane	9,111.1
$C_4H_7Cl_3$	1,2,3-Trichlorobutane	9,447.0
C_4H_7N	Butyronitrile	9,462.9
$C_4H_7NO_2$	Diacetamide	14,508.1
C_4H_8	1-Butene	5,996.7
C_4H_8	*cis*-2-Butene	6,401.0
C_4H_8	*trans*-2-Butene	6,221.6
C_4H_8	2-Methylpropene	5,742.9
C_4H_8	Cyclobutane	6,464.8
C_4H_8BrClO	2-Bromoethyl-2-chloroethyl-ether	12,010.5
$C_4H_8Br_2$	1,2-Dibromobutane	10,182.1
$C_4H_8Br_2$	*dl*-2,3-Dibromobutane	10,136.1
$C_4H_8Br_2$	*meso*-2,3-Dibromobutane	9,966.9
$C_4H_8Br_2$	1,4-Dibromobutane	11,369.3
$C_4H_8Br_2O$	Di(2-bromoethyl)ether	12,454.4
$C_4H_8Cl_2$	1,2-Dichlorobutane	8;850.6
$C_4H_8Cl_2$	2,3-Dichlorobutane	8,975.3
$C_4H_8Cl_2$	1,1-Dichloro-2-methylpropane	8,795.6
$C_4H_8Cl_2$	1,2-Dichloro-2-methylpropane	9,260.1
$C_4H_8Cl_2$	1,3-Dichloro-2-methylpropane	10,519.7
$C_4H_8Cl_2O$	Di(chloroethyl)ether	11,376.8
C_4H_8O	1,2-Epoxy-2-methylpropane	7,066.6
C_4H_8O	Methyl ethyl ketone	8,149.5
$C_4H_8O_2$	Dioxane	8,546.2
$C_4H_8O_2$	*n*-Butyric acid	11,881.2
$C_4H_8O_2$	Isobutyric acid	11,182.8
$C_4H_8O_2$	Ethyl acetate	8,301.1
$C_4H_8O_2$	Methyl propanoate	8,356.2
$C_4H_8O_2$	*n*-Propyl formate	8,208.1
$C_4H_8O_2$	Isopropyl formate	8,230.2
$C_4H_8O_3$	α-Hydroxyisobutyric acid	15.967.0
$C_4H_8O_3$	Ethyl glycolate	11,318.1
C_4H_9Br	*n*-Butyl bromide	8,789.1
C_4H_9BrO	1-Bromo-2-butanol	13,473.7
C_4H_9Cl	*n*-Butyl chloride	8,144.8
C_4H_9Cl	*sec*-Butyl chloride	7,407.9
C_4H_9Cl	Isobutyl chloride	8,045.1
C_4H_9Cl	*tert*-Butyl chloride	6,876.0
$C_4H_9ClO_2$	2-(2-Chloroethoxy)ethanol	14,082.1
C_4H_9I	*n*-Butyl iodide	–
C_4H_9I	1-Iodo-2-methylpropane	9,650.7
$C_4H_9NO_2$	Ethyl methylcarbamate	12,161.2
$C_4H_9NO_2$	Propyl carbamate	14,071.8
$C_4H_9N_2O_2$	Di(nitrosoethyl)amine	10,894.8
C_4H_{10}	*n*-Butane	5,801.2
C_4H_{10}	2-Methylpropane	5,084.4
		5,416.2
$C_4H_{10}Cl_2Si$	Dichlorodiethylsilane	10,038.6
$C_4H_{10}F_2Si$	Diethyldifluorosilane	8,214.9
$C_4H_{10}O$	*n*-Butyl alcohol	10,970.5
$C_4H_{10}O$	*sec*-Butyl alcohol	10,712.3
$C_4H_{10}O$	Isobutyl alcohol	10,936.0
$C_4H_{10}O$	*tert*-Butyl alcohol	10,413.2
$C_4H_{10}O$	Diethyl ether	6,946.2
$C_4H_{10}O$	Methyl propyl ether	7,409.7
$C_4H_{10}O_2$	1,3-Butanediol	10,479.1
$C_4H_{10}O_2$	2,3-Butanediol	13,708.6
$C_4H_{10}O_2$	1,2-Dimethoxyethane	7,681.0
$C_4H_{10}O_2S$	2,2-Thiodiethanol	6,597.0
$C_4H_{10}O_3$	Diethylene glycol	16,146.7
$C_4H_{10}O_3$	1,2,3-Butanetriol	16,345.8
$C_4H_{10}O_3S$	Diethyl sulfite	10,783.0
$C_4H_{10}O_4S$	Diethyl sulfate	12,518.2
$C_4H_{10}S$	*n*-Butanethiol	–
$C_4H_{10}S$	Diethyl sulfide	8,210.8
$C_4H_{10}Se$	Diethyl selenide	9,274.7
$C_4H_{10}Zn$	Diethyl zinc	9,162.3
$C_4H_{11}N$	Diethyl amine	7 307.5
$C_4H_{11}N$	Isobutylamine	478.3
$C_4H_{12}Cl_2Si_2$	1,3-Diethoxytetramethyl-disiloxane	9,881.6
$C_4H_{12}Pb$	Tetramethyllead	8,843.8
$C_4H_{12}Si$	Tetramethylsilane	6,439.2
$C_4H_{12}Sn$	Tetramethyl tin	7,897.8
$C_4H_{14}B_2$	Tetramethyldiborane	7.517.1
C_4F_8	Octafluorocyclobutane	–
C_4F_{10}	Perfluoro-*n*-butane	–
C_5H_4BrN	3-Bromopyridine	10,863.7
C_5H_4Cln	2-Chloropyridine	10,614.5
$C_5H_4O_2$	2-Furaldehyde	11,614.6
$C_5H_4O_3$	Citraconic anhydride	12,307.8
C_5H_5N	Pyridine	9,649.4
$C_5H_6Cl_2O_2$	Glutaryl chloride	13,192.1
$C_5H_6N_2$	Glutaronitrile	13,767.5
$C_5H_6O_2$	Furfuryl alcohol	12,815.8
$C_5H_6O_3$	Glutaric anhydride	14,814.1
$C_5H_6O_3$	Pyrotartaric anhydride	13,251.2

Formula	Name	Δ H_v
C_5H_6S	2-Methylthiophene	8,884.2
C_5H_6S	3-Methylthiophene	9,084.1
$C_5H_7ClO_3$	Propyl chloroglyoxylate	11,430.0
C_5H_7N	Tiglonitrile	8,704.6
C_5H_7N	Angelonitrile	9,707.5
C_5H_7N	α-Ethylacrylonitrile	8,679.1
$C_5H_7NO_2$	Ethyl cyanoacetate	15,615.6
C_5H_8	Cyclopentene	–
C_5H_8	Isoprene	6,901.8
C_5H_8	1,3-Pentadiene	7,313.9
C_5H_8	1,4-Pentadiene	6,826.6
C_5H_8O	Tiglaldehyde	9,009.2
$C_5H_8O_2$	Levulinaldehyde	11,483.8
$C_5H_8O_2$	Tiglic acid	13,756.5
$C_5H_8O_2$	α-Valerolactone	11,537.0
$C_5H_8O_2$	α-Ethylacrylic acid	14,417.8
$C_5H_8O_2$	Ethyl acrylate	9,259.4
$C_5H_8O_2$	Methyl methacrylate	8,974.9
$C_5H_8O_3$	Levulinic acid	17,795.0
$C_5H_8O_4$	Glutaric acid	22,085.2
$C_5H_8O_4$	Dimethyl malonate	12,608.1
$C_5H_9ClO_2$	Ethyl α-chloropropionate	11,032.8
$C_5H_9ClO_2$	Isopropyl chloroacetate	10,575.7
C_5H_9N	Valeronitrile	9,931.3
C_5H_9NO	α-Hydroxybutyronitrile	13,577.0
C_5H_{10}	1-Pentene	6,931.2
C_5H_{10}	2-Pentene	–
C_5H_{10}	3-Methyl-2-butene	7,112.8
C_5H_{10}	2-Methyl-1-butene	6,474.6
C_5H_{10}	3-Methyl-1-butene	–
C_5H_{10}	Cyclopentane	7,411.1
C_5H_{10}	Methylcyclobutane	6,413.2
$C_5H_{10}Br_2$	1,2-Dibromopentane	11,130.0
$C_5H_{10}Br_2$	1,2-Dibromo-2-methylbutane	7,616.9
$C_5H_{10}Br_2$	1,3-Dibromo-3-methylbutane	10,639.6
$C_5H_{10}Cl_2Si$	Allyldichloroethylsilane	9,833.9
$C_5H_{10}O$	Diethyl ketone	11,183.0
$C_5H_{10}O$	Methyl *n*-propyl ketone	11:240.6
$C_5H_{10}O$	Methyl isopropyl ketone	11,073.2
$C_5H_{10}Cl_2O$	2-Chloroethyl 2-chloroisopropyl ether	11,420.8
$C_5H_{10}Cl_2O$	2-Chloroethyl 2-chloropropyl ether	11,316.9
	Di(2-chloroethoxy)methane	12,908.0
$C_5H_{10}Cl_2O_2$	4-Hydroxy-3-methyl-2-butanone	13,639.4
$C_5H_{10}O_2$	Valeric acid	13,370.3
$C_5H_{10}O_2$	Isovaleric acid	12,951.1
$C_5H_{10}O_2$	Ethyl propanoate	8,877.8
$C_5H_{10}O_2$	*n*-Propyl acetate	8,921.1
$C_5H_{10}O_2$	Isopropyl acetate	8,794.8
$C_5H_{10}O_2$	Methyl butyrate	8,886.0
$C_5H_{10}O_2$	Methyl isobutyrate	8,593.3
$C_5H_{10}O_2$	*n*-Butyl formate	9,285.9
$C_5H_{10}O_2$	Isobutyl formate	8,678.8
$C_5H_{10}O_2$	*sec*-Butyl formate	8,975.7
$C_5H_{10}O_2$	*tert*-Butyl formate	8,955.3
$C_5H_{10}O_2$	Diethyl carbonate	10,159.0
$C_5H_{10}O_3$	1-Bromopentane	–
$C_5H_{11}Br$	1-Bromo-3-methylbutane	9,282.7
$C_5H_{11}Br$	1-Chloropentane	–
$C_5H_{11}Br$	1-Iodopentane	–
$C_5H_{11}I$	1-Iodo-3-methylbutane	9,951.6
$C_5H_{11}I$	Piperidine	8,911.8
$C_5H_{11}N$	Pentanoic acid	–
$C_5H_{11}NO_2$	Isobutyl carbamate	13,897.1
$C_5H_{11}NO_3$	Isoamyl nitrate	10,817.2
C_5H_{12}	*n*-Pentane	6,595.1
C_5H_{12}	2-Methylbutane	6,470.8
C_5H_{12}	2,2-Dimethylpropane	5,648.6
$C_5H_{12}O$	Amyl alcohol	12,495.5
$C_5H_{12}O$	Isoamyl alcohol	12,497.9
$C_5H_{12}O$	2-Pentanol	12,086.2
$C_5H_{12}O$	*tert*-Amyl alcohol	11,239.2
$C_5H_{12}O$	Ethyl propyl ether	7,092.7
$C_5H_{12}O$	Methyl *n*-butyl ether	–
$C_5H_{12}O_3$	2,3,4-Pentanetriol	19,694.4
$C_5H_{12}S$	1-Pentanethiol	–
$C_5H_{14}OSi$	Ethoxytrimethylsilane	8,030.6
$C_5H_{14}Si$	Ethyltrimethylsilane	7,633.4
$C_5H_{14}Sn$	Ethyltrimethyltin	8,820.9
$C_6Cl_4O_2$	Chloranil	21,514.3
C_6Cl_6	Hexachlorobenzene	15,199.1
C_6HCl_5	Pentachlorobenzene	15,124.2
C_6HCl_5O	Pentachlorophenol	16,742.6
$C_6H_2BrCl_3O$	3-Bromo-2,4,6-trichlorophenol	15,231.9
$C_6H_2Cl_4$	1,2,3,4-Tetrachlorobenzene	12,872.5
$C_6H_2Cl_4$	1,2,3,5-Tetrachlorobenzene	11,982.1
$C_6H_2Cl_4$	1,2,4,5-Tetrachlorobenzene	12,828.8
$C_6H_2Cl_4O$	2,3,4,6-Tetrachlorophenol	15,362.7
$C_6H_3BrCl_2O$	2-Bromo-4,6-dichlorophenol	13,829.1
$C_6H_3Cl_3$	1,2,3-Trichlorobenzene	11,349.5

Formula	Name	Δ H_v
$C_6H_3Cl_3$	1,2,4-Trichlorobenzene	11,425.1
$C_6H_3Cl_3$	1,3,5-Trichlorobenzene	11,211.0
$C_6H_3Cl_3O$	2,4,5-Trichlorophenol	13,237.0
$C_6H_3Cl_3O$	2,4,6-Trichlorophenol	14,092.8
$C_6H_4Br_2$	1,4-Dibromobenzene	13,047.8
C_6H_4BrCl	1,4-Bromochlorobenzene	16,671.8
		11,451.1
$C_6H_4Cl_2$	1,2-Dichlorobenzene	10,943.0
$C_6H_4Cl_2$	1,3-Dichlorobenzene	10,446.8
$C_6H_4Cl_2$	1,4-Dichlorobenzene	17,260.5
		10.611.0
$C_6H_4Cl_2O$	2,4-Dichlorophenol	13,230.4
$C_6H_4Cl_2O$	2,6-Dichlorophenol	13,472.0
$C_6H_4Cl_3N$	2,4,6-Trichloroaniline	22,297.3
$C_6H_5AsCl_2$	Dichlorophenylarsine	12,229.5
C_6H_5Br	Bromobenzene	10,157.7
C_6H_5Cl	Chlorobenzene	10,098.0
		9,067.3
C_6H_5ClO	2-Chlorophenol	10,341.1
C_6H_5ClO	3-Chlorophenol	11,979.7
C_6H_5ClO	4-Chlorophenol	12,281.6
$C_6H_5ClO_2S$	Benzenesulfonylchloride	12,621.0
$C_6H_5Cl_2O_2P$	Phenyl dichlorophosphate	13,319.6
$C_6H_5Cl_3Si$	Trichlorophenylsilane	11,385.9
C_6H_5F	Fluorobenzene	7,980.4
$C_6H_5F_3Si$	Trifluorophenylsilane	9,171.6
C_6H_5I	Iodobenzene	10,277.2
		10,377.8
$C_6H_5NO_2$	Nitrobenzene	12,168.2
$C_6H_5NO_3$	2-Nitrophenol	12,497.3
C_6H_6	1,5-Hexadiene-3-yne	8,288.0
C_6H_6	Benzene	10,254.3
		8,146.5
C_6H_6ClN	2-Chloroaniline	12,441.0
C_6H_6ClN	3-Chloroaniline	13,385.6
C_6H_6ClN	4-Chloroaniline	12,832.8
C_6H_6ClO	4-Chlorophenol	12,964.7
$C_6H_6N_2O_2$	2-Nitroaniline	15,284.0
$C_6H_6N_2O_2$	3-Nitroaniline	15,996.3
$C_6H_6N_2O_2$	4-Nitroaniline	17,220.2
C_6H_6O	Phenol	11,891.5
$C_6H_6O_2$	Pyrocatechol	13,779.7
$C_6H_6O_2$	Resorcinol	16,400.8
$C_6H_6O_2$	Hydroquinone	18,734.0
$C_6H_6O_3$	Pyrogallol	15,731.8
C_6H_6S	Benzenethiol	11,320.1
C_6H_7N	Aniline	11,307.6
C_6H_7N	2-Picoline	9,933.2
C_6H_7N	3-Methylpyridine (β-picoline)	–
C_6H_8	1,3-Cyclohexadiene	–
$C_6H_8Cl_2O_4$	Ethylene-bis-chloroacetate	16,499.1
$C_6H_8N_2$	1,3-Phenylenediamine	14,761.1
$C_6H_8N_2$	Phenylhydrazine	13,711.9
$C_6H_8O_3$	α-Methylglutaric anhydride	14,204.9
$C_6H_8O_3$	α,α-Dimethylsuccinic anhydride	13,683.1
$C_6H_8O_4$	Dimethyl maleate	12,615.7
C_6H_{10}	Cyclohexene	–
C_6H_{10}	1,5-Hexadiene	–
$C_6H_{10}Cl_2O_2$	Isobutyl dichloroacetate	11,733.1
$C_6H_{10}Cl_2Si$	Diallyldichlorosilane	10,462.8
$C_6H_{10}O$	Cyclohexanone	10,037.6
$C_6H_{10}O$	Mesityl oxide	10,109.4
$C_6H_{10}O_2$	Isocaprolactone	11,685.0
$C_6H_{10}O_3$	Propionic anhydride	11,572.6
$C_6H_{10}O_3$	Ethyl acetoacetate	11,842.0
$C_6H_{10}O_3$	Methyl levulinate	12,249.8
$C_6H_{10}O_4$	Adipic acid	19,570.2
$C_6H_{10}O_4$	Diethyl oxalate	14,016.9
$C_6H_{10}O_4$	Glycol diacetate	12,496.1
$C_6H_{10}O_5$	Dimethyl-*l*-malate	14,127.6
$C_6H_{10}O_6$	Dimethyl-*d*-tartrate	15,372.6
$C_6H_{10}O_6$	Dimethyl-*dl*-tartrate	14,999.1
$C_6H_{10}S$	Diallyl sulfide	9,652.6
$C_6H_{11}BrO_2$	Ethyl α-bromoisobutyrate	10,635.8
$C_6H_{11}ClO_2$	*sec*-Butylchloroacetate	11,152.0
$C_6H_{11}N$	Capronitrile	10,492.3
C_6H_{12}	1-Hexene	7,787.6
C_6H_{12}	2-Hexene	–
C_6H_{12}	Cyclohexane	7,830.9
C_6H_{12}	Methylcyclopentane	7,940.0
$C_6H_{12}Cl_2O$	Dichlorodiisopropyl ether	11,881.1
$C_6H_{12}Cl_2O_2$	bis(2-Chloroethyl)acetal	13,497.1
$C_6H_{12}O$	2-Hexanone	12,358.3
$C_6H_{12}O$	4-Methyl-2-pentanone	11,669.6
$C_6H_{12}O$	Allyl propyl ether	8,621.5
$C_6H_{12}O$	Allyl isopropyl ether	8,637.5
$C_6H_{12}O$	Cyclohexanol	11,935.8
$C_6H_{12}O_2$	Caproic acid	16,189.4
$C_6H_{12}O_2$	Isocaproic acid	14.874.8
$C_6H_{12}O_2$	4-Hydroxy-4-methyl-2-pentanone	11,718.8

Formula	Name	ΔH_v
$C_6H_{12}O_2$	Methyl pentanoate	–
$C_6H_{12}O_2$	Methyl isovalerate	9,567.5
$C_6H_{12}O_2$	Ethyl *n*-butyrate	9,468.5
$C_6H_{12}O_2$	Ethyl isobutyrate	8,945.7
$C_6H_{12}O_2$	*n*-Propyl propanoate	9,857.2
$C_6H_{12}O_2$	*n*-Butyl acetate	–
$C_6H_{12}O_2$	Isobutyl acetate	9,300.8
$C_6H_{12}O_2$	*n*-Amyl formate	–
$C_6H_{12}O_2$	Isoamyl formate	9,438.2
$C_6H_{12}O_3$	Paraformaldehyde	10,348.2
C_6H_{14}	Hexane	[illegible]
C_6H_{14}	2-Methylpentane	7,676.6
C_6H_{14}	3-Methylpentane	7,743.9
C_6H_{14}	2,2-Dimethylbutane	7,271.0
C_6H_{14}	2,3-Dimethylbutane	7,120.0
$C_6H_{14}O$	1-Hexanol	12,708.5
$C_6H_{14}O$	2-Hexanol	12,386.5
$C_6H_{14}O$	3-Hexanol	11,157.9
$C_6H_{14}O$	2-Methyl-1-pentanol	12,036.6
$C_6H_{14}O$	2-Methyl-2-pentanol	11,132.0
$C_6H_{14}O$	2-Methyl-4-pentanol	10,985.5
$C_6H_{14}O$	Ethyl butyl ether	–
$C_6H_{14}O$	Di-*n*-propyl ether	8,229.6
$C_6H_{14}O$	Diisopropyl ether	7,777.3
$C_6H_{14}O_2$	Acetal	9,853.9
$C_6H_{14}O_2$	1,2-Diethoxyethane	8,102.6
$C_6H_{14}O_3$	Di(2-methoxyethyl)ether	11,105.2
$C_6H_{14}O_3$	Diethyleneglycol-diethyl ether	12,669.0
$C_6H_{14}O_3$	Dipropyleneglycol	14,610.4
$C_6H_{14}O_4$	Triethyleneglycol	17,097.1
$C_6H_{15}N$	Dipropyl sulfide	–
$C_6H_{15}N$	Di-*n*-Propylamine	–
$C_6H_{15}N$	Triethylamine	–
$C_6H_{15}B$	Triethylboron	2,535.0
$C_6H_{15}ClSi$	Chlorotriethylsilane	9,806.9
$C_6H_{15}O_4P$	Triethyl phosphate	11,549.9
$C_6H_{15}Tl$	Triethylthallium	9,458.6
$C_6H_{16}O_2Si$	Diethoxydimethylsilane	9,758.2
$C_6H_{16}Si$	Trimethylpropylsilane	7,964.6
$C_6H_{16}Sn$	Trimethylpropyltin	9,659.6
$C_6H_{18}Cl_2O_2Si_3$	1,5 Dichlorohexamethyltrisiloxane	11,391.5
$C_6H_{18}O_3Si_3$	Hexamethylcyclotrisiloxane	10,503.3
$C_7H_3Cl_2F_3$	3,4-Dichloro-α,α,α-trifluorotoluene	10,253.5
$C_7H_4ClF_3$	2-Chloro-α,α,α-trifluorotoluene	10,016.9
$C_7H_4Cl_4$	2-α,α,α-Tetrachlorotoluene	12,501.3
C_7H_5BrO	Benzoyl bromide	12,070.8
C_7H_5ClO	Benzoyl chloride	11,438.0
$C_7H_5Cl_3$	α,α,α-Trichlorotoluene	12,168.6
$C_7H_5F_3$	α,α,α-Trifluorotoluene	8,869.7
C_7H_5N	Benzonitrile	11,341.0
C_7H_5N	Phenyl isocyanide	10,736.7
C_7H_5NO	Phenyl isocyanate	10,556.7
$C_7H_5NO_3$	2-Nitrobenzaldehyde	13,773.6
$C_7H_5NO_3$	3-Nitrobenzaldehyde	14,726.9
C_7H_5NS	Phenyl isothiocyanate	12,132.7
$C_7H_6Cl_2$	α,α-Dichlorotoluene	11,075.9
C_7H_6O	Benzaldehyde	11,657.8
$C_7H_6O_2$	Benzoic acid	15,253.3
		16,295.1
$C_7H_6O_2$	Salicylaldehyde	11,536.5
$C_7H_6O_2$	4-Hydroxybenzaldehyde	16,043.4
$C_7H_6O_3$	Salicylic acid	18,920.7
C_7H_7Br	α-Bromotoluene	11,360.4
C_7H_7Br	2-Bromotoluene	11,365.0
C_7H_7Br	3-Bromotoluene	10,537.1
C_7H_7Br	4-Bromotoluene	10,076.2
C_7H_7BrO	4-Bromoanisole	12,075.4
C_7H_7Cl	α-Chlorotoluene	11,158.7
C_7H_7Cl	2-Chlorotoluene	10,279.3
C_7H_7Cl	3-Chlorotoluene	10,081.1
C_7H_7Cl	4-Chlorotoluene	10,151.7
C_7H_7F	2-Fluorotoluene	9,164.8
C_7H_7F	3-Fluorotoluene	9,251.8
C_7H_7F	4-Fluorotoluene	9,281.0
C_7H_7I	2-Iodotoluene	11,380.7
$C_7H_7NO_2$	2-Nitrotoluene	12,239.1
$C_7H_7NO_2$	3-Nitrotoluene	11,831.1
$C_7H_7NO_2$	4-Nitrotoluene	11,915.0
C_7H_8	Toluene	9,368.5
		8,580.5
$C_7H_8Cl_2Si$	Benzyldichlorosilane	13,128.7
$C_7H_8Cl_2Si$	Dichloromethylphenylsilane	11,464.7
$C_7H_8Cl_2Si$	Dichloro-4-tolysilane	13,125.7
C_7H_8O	Anisole	10,440.9
C_7H_8O	Benzyl alcohol	14,093.2
C_7H_8O	*o*-Cresol	12,487.3
C_7H_8O	*m*-Cresol	13,483.8

Formula	Name	ΔH_v
C_7H_8O	*p*-Cresol	13,611.7
$C_7H_8O_2$	3,5-Dimethyl-1,2-pyrone	14,470.6
$C_7H_8O_2$	2-Methoxyphenol	13,425.8
$C_7H_8O_3$	Ethyl 2-furoate	12,144.0
C_7H_9N	2,6-Dimethylpyridine	–
C_7H_9N	Benzylamine	11,703.2
C_7H_9N	*N*-Methylaniline	11,982.3
C_7H_9N	2-Toluidine	12,663.4
C_7H_9N	3-Toluidine	12,104.1
$C_7H_{10}O_4$	Dimethyl citraconate	12,917.3
$C_7H_{10}O_4$	Dimethyl itaconate	16,612.7
$C_7H_{10}O_4$	*trans*-Dimethyl mesaconate	12,688.1
$C_7H_{11}NO_2$	2-Cyano-2-butyl acetate	12,720.8
$C_7H_{12}O_2$	Butyl acrylate	10,194.0
C_7H_9N	4-Toluidine	12,428.6
C_7H_9NO	2-Methoxyaniline	13,684.6
$C_7H_{10}N_2$	Toluene-2,4-diamine	15,928.1
$C_7H_{10}N_2$	4-Tolylhydrazine	15,063.1
$C_7H_{10}O_3$	Trimethylsuccinic anhydride	12,196.7
$C_7H_{12}O_3$	Ethyl levulinate	12,733.6
$C_7H_{12}O_4$	Pimelic acid	19,840.8
$C_7H_{12}O_4$	Diethyl malonate	12,227.7
$C_7H_{13}ClO$	Enanthyl chloride	15,242.7
$C_7H_{13}N$	Heptanonitrile	10,830.5
C_7H_{14}	Ethylcyclopentane	8,797.7
C_7H_{14}	2-Heptene	8,643.2
C_7H_{14}	Methylcyclohexane	8,549.2
$C_7H_{14}O$	Enanthaldehyde	11,413.4
$C_7H_{14}O$	2-Heptanone	12,478.9
$C_7H_{14}O$	4-Heptanone	13,451.9
$C_7H_{14}O$	2,5-Dimethyl-3-pentanone	12,266.9
$C_7H_{14}O_2$	Enanthic acid	15,893.8
$C_7H_{14}O_2$	Methyl caproate	10,676.8
$C_7H_{14}O_2$	Ethyl isovalerate	10,183.9
$C_7H_{14}O_2$	Propyl butyrate	10,283.7
$C_7H_{14}O_2$	Propyl isobutyrate	10,259.7
$C_7H_{14}O_2$	Isopropyl isobutyrate	9,717.6
$C_7H_{14}O_2$	Isobutyl propionate	10,495.8
$C_7H_{14}O_2$	Isoamyl acetate	10,494.9
C_7H_{16}	Perfluoro-*n*-Heptane	–
C_7H_{16}	*n*-Heptane	8,928.8
		8,409.6
C_7H_{16}	2-Methylhexane	8,538.7
C_7H_{16}	3-Methylhexane	8,596.3
C_7H_{16}	3-Ethylpentane	8,642.8
C_7H_{16}	2,2-Dimethylpentane	8,106.7
C_7H_{16}	2,3-Dimethylpentane	8,390.9
C_7H_{16}	2,4-Dimethylpentane	8,167.1
C_7H_{16}	3,3-Dimethylpentane	8,145.4
C_7H_{16}	2,2,3-Trimethylbutane	7,767.1
$C_7H_{16}O$	*n*-Heptanol	13,920.9
$C_7H_{16}O_3$	Triethyl orthoformate	10,935.0
$C_7H_{18}O_3Si$	Triethoxymethylsilane	10,306.7
$C_7H_{18}Si$	Butyltrimethylsilane	9,206.0
$C_7H_{18}Si$	Triethylmethylsilane	9,232.5
$C_8H_4Cl_2O_2$	Phthaloyl chloride	13,716.0
$C_8H_4O_3$	Phthalic anhydride	13,919.0
$C_8H_5Cl_2N$	α,α-Dichlorophenylacetonitrile	12,829.9
$C_8H_5Cl_5$	Pentachloroethylbenzene	13,728.7
$C_8H_6Cl_2$	2,3-Dichlorostyrene	12,827.2
$C_8H_6Cl_2$	2,4-Dichlorostyrene	12,511.7
$C_8H_6Cl_2$	2,5-Dichlorostyrene	12,592.5
$C_8H_6Cl_2$	2,6-Dichlorostyrene	12,186.0
$C_8H_6Cl_2$	3,4-Dichlorostyrene	12,626.5
$C_8H_6Cl_2$	3,5-Dichlorostyrene	12,511.7
$C_8H_6Cl_4$	3,4,5,6-Tetrachloro-1,2-xylene	14,763.1
$C_8H_6Cl_4$	1,2,3,5-Tetrachloro-4-ethylbenzene	12,980.3
$C_8H_6O_2$	Phenylglyoxal	13,731.6
$C_8H_6O_2$	Phthalide	14,021.6
$C_8H_6O_3$	Piperonal	14,425.5
C_8H_7Cl	3-Chlorostyrene	10,990.2
C_8H_7ClO	Phenylacetyl chloride	12,627.1
C_8H_7N	2-Tolunitrile	11,557.7
C_8H_7N	4-Tolunitrile	11,562.8
C_8H_7N	Phenylacetonitrile	12,796.2
C_8H_7N	2-Tolyl isocyanide	11,303.3
$C_8H_7NO_4$	2-Nitrophenyl acetate	16,875.3
C_8H_7NS	2-Methylbenzothiazole	14,492.3
C_8H_8	Styrene	9,634.7
$C_8H_8Br_2$	(1,2-Dibromoethyl)benzene	14,874.7
$C_8H_8Cl_2$	1,2-Dichloro-3-ethylbenzene	11,784.3
$C_8H_8Cl_2$	1,2-Dichloro-4-ethylbenzene	11,711.5
$C_8H_8Cl_2$	1,4-Dichloro-2-ethylbenzene	11,262.7
C_8H_8O	Acetophenone	11,731.5
$C_8H_8O_2$	Phenylacetate	12,174.9
$C_8H_8O_2$	Phenylacetic acid	15,568.7
$C_8H_8O_2$	Anisaldehyde	13,581.8
$C_8H_8O_2$	Methyl benzoate	12,077.2
$C_8H_8O_2$	Methyl salicylate	12,658.8
$C_8H_8O_3$	Vanillin	15,703.2

Formula	Name	ΔH_v
$C_8H_8O_4$	Dihydroacetic acid	14,663.8
C_8H_9Br	2-Bromo-1,4-xylene	11,603.7
C_8H_9Br	1-Bromo-4-ethylbenzene	10,170.0
C_8H_9Br	(2-Bromoethyl)benzene	12,152.5
C_8H_9Cl	1-Chloro-2-ethylbenzene	10,749.7
C_8H_9Cl	1-Chloro-3-ethylbenzene	10,724.1
C_8H_9Cl	1-Chloro-4-ethylbenzene	10,659.9
C_8H_9ClO	1-Chloro-2-ethoxybenzene	12,411.1
C_8H_9ClO	4-Chlorophenylethyl alcohol	14,298.5
$C_8H_{10}Cl_2Si$	Dichlorophenylethylsilane	11,895.1
C_8H_9NO	Acetanilide	15,474.1
$C_8H_9NO_2$	Methyl anthranilate	13,186.3
$C_8H_9NO_2$	4-Nitro-1,3-xylene	12,948.0
C_8H_{10}	Ethylbenzene	9,301.3
C_8H_{10}	*o*-Xylene	9,998.5
C_8H_{10}	*m*-Xylene	9,904.2
C_8H_{10}	*p*-Xylene	9,809.9
$C_8H_{10}Cl_2OSi$	Dichloroethoxyphenylsilane	12,516.5
$C_8H_{10}Cl_2Si$	Dichloroethylphenylsilane	11,721.2
$C_8H_{10}O$	2-Ethylphenol	12,516.7
$C_8H_{10}O$	3-Ethylphenol	13,856.4
$C_8H_{10}O$	4-Ethylphenol	13,437.9
$C_8H_{10}O$	Xylenol	–
$C_8H_{10}O$	2,3-Xylenol	13,106.9
$C_8H_{10}O$	2,4-Xylenol	13,130.2
$C_8H_{10}O$	2,5-Xylenol	13,130.2
$C_8H_{10}O$	3,4-Xylenol	13,991.0
$C_8H_{10}O$	3,5-Xylenol	13,767.7
$C_8H_{10}O$	Phenetole	11,075.8
$C_8H_{10}O$	α-Methyl benzyl alcohol	13,087.4
$C_8H_{10}O$	Phenylethylalcohol	13,307.4
$C_8H_{10}O_2$	4,6-Dimethylresorcinol	12,433.1
$C_8H_{10}O_2$	2-Phenoxyethanol	14,368.3
$C_8H_{10}O_6$	Diethyl dioxosuccinate	13,973.3
$C_8H_{11}ClSi$	Chlorodimethylphenylsilane	11,382.2
$C_8H_{11}N$	*N*-Ethylaniline	11,817.0
$C_8H_{11}N$	*N,N*-Dimethylaniline	11,320.4
$C_8H_{11}N$	4-Ethylaniline	12,679.9
$C_8H_{11}N$	2,4-Xylidine	13,099.2
$C_8H_{11}N$	2,6-Xylidine	11,742.6
$C_8H_{11}NO$	2-Phenetidine	13,877.8
$C_8H_{11}NO$	2-Anilinoethanol	15,643.2
$C_8H_{12}AsNO_2$	Dimethyl arsanilate	11,277.7
$C_8H_{12}Cl_2O_5$	Diethyleneglycol-bis-chloroacetate	19,830.5
$C_8H_{12}O_4$	Diethyl maleate	12,908.0
$C_8H_{12}O_4$	Diethyl fumarate	12,747.4
$C_8H_{12}Si$	Dimethylphenylsilane	10,274.2
$C_8H_{14}O_3$	Ethyl-α-ethylacetoacetate	12,344.2
$C_8H_{14}O_4$	Propyl levulinate	13,354.4
$C_8H_{14}O_4$	Isopropyl levulinate	12,689.6
$C_8H_{14}O_4$	Dipropyl oxalate	13,056.4
$C_8H_{14}O_4$	Diisopropyl oxalate	12,949.3
$C_8H_{14}O_4$	Diethyl succinate	13,076.1
$C_8H_{14}O_4$	Diethyl isosuccinate	12,087.6
$C_8H_{14}O_4$	Suberic acid	21,089.8
$C_8H_{14}O_5$	Diethyl malate	14,202.9
$C_8H_{14}O_6$	Diethyl-*dl*-tartrate	15,150.4
$C_8H_{14}O_6$	Diethyl-*d*-tartrate	15,517.8
$C_8H_{15}Br$	(2-Bromoethyl)cyclohexane	11,462.7
$C_8H_{15}N$	*n*-Caprylonitrile	12,221.8
$C_8H_{15}NO_3$	Ethyl-*N,N*-diethyloxamate	13,758.4
C_8H_{16}	1-Octene	–
C_8H_{16}	2-Octene	–
C_8H_{16}	2-Methyl-2-heptene	9,643.8
C_8H_{16}	1,1-Dimethylcyclohexane	8,949.1
C_8H_{16}	*cis*-1,2-Dimethylcyclohexane	9,364.9
C_8H_{16}	*trans*-1,2-Dimethylcyclohexane	9,097.1
C_8H_{16}	*cis*-1,3-Dimethylcyclohexane	9,232.6
C_8H_{16}	*trans*-1,3-Dimethylcyclohexane	9,080.3
C_8H_{16}	*cis*-1,4-Dimethylcyclohexane	9,188.9
C_8H_{16}	*trans*-1,4-Dimethylcyclohexane	8,951.2
C_8H_{16}	Ethylcyclohexane	9,441.2
$C_8H_{16}O$	Caprylaldehyde	21,201.0
$C_8H_{16}O$	Cyclohexaneethanol	13,152.4
$C_8H_{16}O$	6-Methyl-3-hepten-2-ol	13,864.1
$C_8H_{15}O$	6-Methyl-5-hepten-2-ol	13,999.1
$C_8H_{16}O$	2-Octanone	11,649.2
$C_8H_{16}O$	2,2,4-Trimethyl-3-pentanone	12,854.6
$C_8H_{16}O_2$	Caprylic acid	16,745.7
$C_8H_{16}O_2$	Ethyl isocaproate	10,826.7
$C_8H_{16}O_2$	Propyl isovalerate	10,715.7
$C_8H_{16}O_2$	Isobutyl butyrate	10,283.9
$C_8H_{16}O_2$	Isobutyl isobutyrate	10,706.3
$C_8H_{16}O_2$	Amylisopropionate	10,567.2
$C_8H_{16}ClO_4$	Tetraethyleneglycol-chlorohydrin	16,371.2
$C_8H_{17}I$	1-Iodooctane	11,625.1
$C_8H_{17}NO_2$	Ethyl-*l*-leucinate	11,383.5
C_8H_{18}	Octane	9,221.0
C_8H_{18}	2-Methylheptane	9,362.0

Formula	Name	Δ H_v
C_8H_{18}	3-Methylheptane	9,432.0
C_8H_{18}	4-Methylheptane	9,404.8
C_8H_{18}	2,2-Dimethylhexane	8,927.8
C_8H_{18}	2,3-Dimethylhexane	9,224.9
C_8H_{18}	2,4-Dimethylhexane	9,086.6
C_8H_{18}	2,5-Dimethylhexane	9,110.2
C_8H_{18}	3,3-Dimethylhexane	9,065.2
C_8H_{18}	3,4-Dimethylhexane	9,239.4
C_8H_{18}	3-Ethylhexane	9,416.3
C_8H_{18}	2,2,3-Trimethylpentane	8,861.1
C_8H_{18}	2,2,4-Trimethylpentane	8,548.0
C_8H_{18}	2,3,3-Trimethylpentane	8,960.9
C_8H_{18}	2,3,4-Trimethylpentane	8,988.2
C_8H_{18}	2-Methyl-3-ethylpentane	9,134.3
C_8H_{18}	3-Methyl-3-ethylpentane	9,028.7
C_8H_{18}	2,2,3,3-Tetramethylbutane	10,351.5
$C_8H_{18}N_2$	Tetramethylpiperazine	11,187.5
C_8H_{18}	*n*-Octanol	14,262.4
C_8H_{18}	2-Octanol	12,468.4
$C_8H_{18}O$	Di *n*-butyl ether	–
$C_8H_{18}O$	Methyl heptyl ether	–
$C_8H_{18}O_2$	1,2-Dipropoxy ethane	6,370.7
$C_8H_{18}O_3$	Diethylene glycol butyl ether	14,127.0
$C_8H_{18}O_5$	Tetraethylene glycol	21,296.6
$C_8H_{18}S$	Di *n*-butyl sulfide	11,183.6
$C_8H_{18}S_2$	Dibutyl disulfide	8,254.1
$C_8H_{19}N$	Diisobutylamine	10,058.3
$C_8H_{20}O_4Si$	Tetraethoxysilane	10,968.6
$C_8H_{20}Pb$	Tetraethyllead	12,959.7
$C_8H_{20}Si$	Amyltrimethylsilane	9,659.6
$C_8H_{20}Si$	Tetraethylsilane	9,893.0
$C_8H_{20}Sb_2$	Tetraethylbistibine	12,975.4
$C_8H_{22}O_3Si_2$	1,3-Dichlorotetramethyl-disiloxane	11,261.9
$C_8H_{24}Cl_2O_3Si_4$	1,7-Dichlorooctamethyl-tetrasiloxane	12,602.9
$C_8H_{24}O_2Si_3$	Octamethyltrisiloxane	10,956.0
$C_8H_{24}O_4Si_4$	Octamethylcyclotetra siloxane	11,515.0
$C_9H_6O_2$	Coumarin	15,202.7
C_9H_7N	Quinoline	12,575.4
C_9H_7N	Isoquinoline	12,847.6
C_9H_8	Indene	10,496.7
C_9H_8O	Cinnamylaldehyde	14,048.4
$C_9H_8O_2$	*trans*-Cinnamic acid	17,492.9
C_9H_9N	Skatole	15,232.7
$C_9H_9NO_4$	Ethyl 3-nitrobenzoate	15,056.1
C_9H_{10}	α-Methyl styrene	10,214.6
C_9H_{10}	β-Methyl styrene	10,701.3
C_9H_{10}	2-Methyl styrene	–
C_9H_{10}	3-Methyl styrene	–
C_9H_{10}	4-Methyl styrene	10,724.2
$C_9H_{10}O$	2,4-Xylaldehyde	13,618.4
$C_9H_{10}O$	Cinnamyl alcohol	13,421.6
$C_9H_{10}O$	Propiophenone	12,407.6
$C_9H_{10}O$	3-Vinylanisole	12,756.4
$C_9H_{10}O$	3-Vinylanisole	12,735.8
$C_9H_{10}O$	4-Vinylanisole	12,554.7
$C_9H_{10}O_2$	Benzyl acetate	12,107.2
$C_9H_{10}O_2$	Ethyl benzoate	11,981.5
$C_9H_{10}O_2$	Hydrocinnamic acid	15,411.9
$C_9H_{10}O_2$	Ethyl salicylate	13,030.1
$C_9H_{11}NO$	*N*-Methylacetanilide	13,235.2
$C_9H_{11}NO_2$	Ethyl carbanilate	19,791.8
C_9H_{12}	1,2,3-Trimethylbenzene	10,781.9
C_9H_{12}	1,2,4-Trimethylbenzene	10,710.2
C_9H_{12}	1,3,5-Trimethylbenzene	10,516.8
C_9H_{12}	*o*-Ethyl toluene	10,488.8
C_9H_{12}	*m*-Ethyl toluene	10,416.6
C_9H_{12}	*p*-Ethyl toluene	10,461.1
C_9H_{12}	Isopropylbenzene	10,335.3
C_9H_{12}	*N*-Propylbenzene	10,424.1
$C_9H_{12}O$	2-Ethylanisole	11,642.8
$C_9H_{12}O$	3-Ethylanisole	11,616.7
$C_9H_{12}O$	4-Ethylanisole	11,625.7
$C_9H_{12}O$	3-Phenyl-1-propanol	14,493.9
$C_9H_{12}O$	2-Isopropylphenol	13,402.3
$C_9H_{12}O$	3-Isopropylphenol	13,292.2
$C_9H_{12}O$	4-Isopropylphenol	13,878.7
$C_9H_{12}O$	Benzyl ethyl ether	11,315.5
$C_9H_{13}ClOSi$	Chloroethoxymethyl-phenylsilane	12,270.3
$C_9H_{13}N$	2,4,5-Trimethylaniline	13,975.0
$C_9H_{13}N$	*N,N*-Dimethyl-*O*-toluidine	11,648.3
$C_9H_{13}N$	*N,N*-Dimethyl-4-toluidine	12,738.4
$C_9H_{13}N$	4-Cumidine	13,127.9
$C_9H_{14}O$	Phorone	12,557.2
$C_9H_{14}O$	Isophorone	11,277.6
$C_9H_{14}O_4$	*cis*-Diethyl citraconate	12,913.2
$C_9H_{14}O_4$	Diethyl itaconate	12,075.8
$C_9H_{14}O_4$	Diethyl mesaconate	13,326.1
$C_9H_{14}O_7$	Trimethyl citrate	15,807.7
$C_9H_{16}O_3$	Isobutyl levulinate	13,571.2
$C_9H_{16}O_4$	Azelaic acid	20,944.2
$C_9H_{16}O_4$	Diethyl ethylmalonate	12,842.0
$C_9H_{16}O_4$	Diethyl glutarate	13,261.5
$C_9H_{18}O$	2-Nonanone	11,529.5
$C_9H_{18}O$	Di-isobutyl ketone	–
$C_9H_{18}O$	Azelaldehyde	12,143.4
$C_9H_{18}O_2$	Pelargonic acid	17,807.8
$C_9H_{18}O_2$	Methyl caprylate	11,914.9
$C_9H_{18}O_2$	Isobutyl isovalerate	10,999.7
$C_9H_{18}O_2$	Isoamyl butyrate	11,104.5
$C_9H_{18}O_2$	Isoamyl isobutyrate	10,870.6
$C_9H_{19}I$	1-Iodononane	14,853.0
C_9H_{20}	*n*-Nonane	10,456.9
C_9H_{20}	2,6-Dimethylheptane	–
C_9H_{20}	2-Methyloctane	–
C_9H_{20}	3-Methyloctane	–
$C_9H_{20}O$	1-Nonanol	13,849.2
$C_9H_{20}O$	Diisobutyl carbinol	–
$C_9H_{20}O_3$	Dipropyleneglycol isopropyl ether	12,583.8
$C_9H_{20}O_4$	Tripropyleneglycol	15,291.4
$C_9H_{22}Si$	Hexyltrimethylsilane	10,264.9
$C_9H_{22}Si$	Triethylpropylsilane	10,709.3
$C_{10}H_7Br$	1-Bromonaphthalene	13,274.9
$C_{10}H_7Cl$	1-Chloronaphthalene	13,570.5
$C_{10}H_{12}$	Dicyclopentadiene	10,165.9
$C_{10}H_8$	Naphthalene	17,065.2
		12,311.6
$C_{10}H_8Cl_2Si$	Dichloro-1-naphthysilane	16,325.3
$C_{10}H_8O$	1-Naphthol	14,205.6
$C_{10}H_8O$	2-Naphthol	14,138.5
$C_{10}H_9N$	1-Naphthylamine	14,529.5
$C_{10}H_9N$	2-Naphthylamine	14,679.6
$C_{10}H_9N$	2-Methylquinoline	14,154.0
$C_{10}H_{10}$	1,3-Divinylbenzene	11,384.7
$C_{10}H_{10}O$	4-Phenyl-3-buten-2-one	13,913.9
$C_{10}H_{10}O_2$	α-Methylcinnamic acid	18,149.4
$C_{10}H_{10}O_2$	Methyl cinnamate	13,325.5
$C_{10}H_{10}O_2$	Safrole	13,255.8
$C_{10}H_{10}O_4$	1,2-Phenylene diacetate	14,986.0
$C_{10}H_{10}O_4$	Dimethylphthalate	14,922.2
$C_{10}H_{12}$	2,4-Dimethylstyrene	11,454.0
$C_{10}H_{12}$	2,5-Dimethylstyrene	11,283.5
$C_{10}H_{12}$	3-Ethylstyrene	11,285.7
$C_{10}H_{12}$	4-Ethylstyrene	11,146.6
$C_{10}H_{12}$	Tetralin	11,613.0
$C_{10}H_{12}O$	Anethole	13,006.8
$C_{10}H_{12}O$	4-Methylpropiophenone	12,505.0
$C_{10}H_{12}O$	Estragole	12,879.3
$C_{10}H_{12}O$	Cuminal	12,668.0
$C_{10}H_{12}O$	4-Vinylphenetole	13,728.7
$C_{10}H_{12}O_2$	Eugenol	13,907.8
$C_{10}H_{12}O_2$	Isoeugenol	14,084.2
$C_{10}H_{12}O_2$	Chavibetol	14,527.7
$C_{10}H_{12}O_2$	Propyl benzoate	12,318.7
$C_{10}H_{12}O_3$	2-Phenoxyethyl acetate	14,070.3
$C_{10}H_{13}ClO$	2-Chloroethyl-α-methylbenzyl ether	12,969.2
$C_{10}H_{13}Cl_2O_2P$	4-*tert*-Butylphenyl dichlorophosphate	13,711.0
$C_{10}H_{14}$	1,2,3,4-Tetramethylbenzene	12,258.0
$C_{10}H_{14}$	1,2,3,5-Tetramethylbenzene	12,358.4
$C_{10}H_{14}$	1,2,4,5-Tetramethylbenzene	12,583.6
$C_{10}H_{14}$	4-Ethyl-1,3-xylene	11,070.4
$C_{10}H_{14}$	5-Ethyl-1,3-xylene	11,045.5
$C_{10}H_{14}$	2-Ethyl-1,4-xylene	11,144.6
$C_{10}H_{14}$	1,2-Diethylbenzene	11,695.5
$C_{10}H_{14}$	1,3-Diethylbenzene	10,993.9
$C_{10}H_{14}$	1,4-Diethylbenzene	10,746.3
$C_{10}H_{14}$	1-Methyl-2-isopropylbenzene	–
$C_{10}H_{14}$	1-Methyl-4-isopropylbenzene	11,038.7
$C_{10}H_{14}$	*N*-Butylbenzene	11,052.1
$C_{10}H_{14}$	Isobutylbenzene	8,567.8
$C_{10}H_{14}$	*sec*-Butylbenzene	11,069.3
$C_{10}H_{14}$	*tert*-Butylbenzene	10,705.5
$C_{10}H_{14}O$	Carvacrol	13,765.7
$C_{10}H_{14}O$	Carbone	12,796.2
$C_{10}H_{14}O$	Cuminyl alcohol	13,799.2
$C_{10}H_{14}O$	4-Ethylphenetole	12,766.4
$C_{10}H_{14}O$	2-Isopropyl-5-methylphenol	13,352.8
$C_{10}H_{14}O$	4-Isobutylphenol	14,053.5
$C_{10}H_{14}O$	4-*sec*-Butylphenol	13,690.2
$C_{10}H_{14}O$	2-*sec*-Butylphenol	12,781.3
$C_{10}H_{14}O$	2-*tert*-Butylphenol	13,112.3
$C_{10}H_{14}O$	4-*tert*-Butylphenol	13,787.7
$C_{10}H_{14}N_2$	Nicotine	12,337.1
$C_{10}H_{15}N$	*N*-Diethylaniline	12,539.2
$C_{10}H_{15}NO_2$	*N*-Phenyliminodiethanol	17,482.1
$C_{10}H_{16}$	Camphene	10,505.4
$C_{10}H_{16}$	Dipentene	10,538.3
$C_{10}H_{16}$	*d*-Limonene	10,508.4
$C_{10}H_{16}$	Myrcene	10,704.8
$C_{10}H_{16}$	α-Phellandrene	11,139.5
$C_{10}H_{16}$	α-Pinene	9,813.6
$C_{10}H_{16}$	β-Pinene	10,235.8
$C_{10}H_{16}$	Terpenoline	12,030.8
$C_{10}H_{16}AsNO_3$	Diethyl arsanilinate	12,973.9
$C_{10}H_{16}O$	*d*-Camphor	12,800.9
		11,978.0
$C_{10}H_{16}O$	*l*-dihydrocarvone	11,825.9
$C_{10}H_{16}O$	α-Citral	13,255.5
$C_{10}H_{16}O$	*d*-Fenchone	11,273.4
$C_{10}H_{16}O$	Pulegone	13,395.4
$C_{10}H_{16}O$	α-Thujone	11,950.8
$C_{10}H_{16}OSi$	Ethoxydimethylphenylsilane	11,718.6
$C_{10}H_{16}O_2$	Campholenic acid	16,324.1
$C_{10}H_{16}O_2$	Diosphenol	13,644.0
$C_{10}H_{16}O_2$	Fencholic acid	16,442.8
$C_{10}H_{18}$	*cis*-Decalin	10,515.4
$C_{10}H_{18}$	*trans*-Decalin	8,749.1
$C_{10}H_{18}O$	*d*-Citronellal	12,305.1
$C_{10}H_{18}O$	Cineol	10,570.8
$C_{10}H_{18}O$	Dihydrocarveol	13,698.5
$C_{10}H_{18}O$	*dl*-Fenchyl alcohol	12,955.9
$C_{10}H_{18}O$	Geraniol	14,060.7
$C_{10}H_{18}O$	*d*-Linalool	12,269.7
$C_{10}H_{18}O$	Nerol	13,366.1
$C_{10}H_{18}O$	α-Terpineol	12,754.5
$C_{10}H_{18}O_2$	Citronellic acid	16,455.4
$C_{10}H_{18}O_3$	Amyl levulinate	14,321.7
$C_{10}H_{18}O_3$	Isoamyl levulinate	13,867.9
$C_{10}H_{18}O_4$	Diethyl ethylmethylmalonate	12,345.6
$C_{10}H_{18}O_4$	Diethyl adipate	14,240.6
$C_{10}H_{18}O_4$	Diisobutyl oxalate	13,343.1
$C_{10}H_{18}O_4$	Dipropyl succinate	13,975.7
$C_{10}H_{18}O_4$	Sebacic acid	21,978.3
$C_{10}H_{18}O_6$	Dipropyl-*d*-tartrate	15,754.0
$C_{10}H_{18}O_6$	Diisopropyl-*d*-tartrate	15,836.6
$C_{10}H_{19}N$	Camphylamine	13,224.1
$C_{10}H_{20}$	Menthane	10,293.1
$C_{10}H_{20}$	1-Decene	10,233.3
$C_{10}H_{20}$	*n*-Decane	10,912.0
$C_{10}H_{20}Br_2$	1,2-Dibromodecane	16,407.7
$C_{10}H_{20}O$	Decanol	14,065.1
$C_{10}H_{20}O$	Citronellol	14,214.1
$C_{10}H_{20}O$	Capraldehyde	13,154.9
$C_{10}H_{20}O$	*l*-Menthol	13,475.3
$C_{10}H_{20}O$	Decan-2-one	12,114.7
$C_{10}H_{20}O_2$	Capric acid	19,372.6
$C_{10}H_{20}O_2$	Isoamyl isovalerate	11,040.8
$C_{10}H_{22}$	2,7-Dimethyloctane	10,339.3
$C_{10}H_{22}O$	Diisoamyl ether	11,072.2
$C_{10}H_{22}O_2$	2-Butyl-2-ethylbutane-1,3-diol	15,833.7
$C_{10}H_{22}O$	Dihydrocitronellol	16,769.8
$C_{10}H_{22}O_3$	Dipropylene glycol monobutyl ether	13,721.1
$C_{10}H_{22}S$	Diisoamyl sulfide	11,829.9
$C_{10}H_{24}Si$	Heptyltrimethylsilane	10,987.3
$C_{10}H_{24}Si$	Butyltriethylsilane	11,124.0
$C_{10}H_{28}O_4Si_3$	1,5-Diethoxyhexamethyl trisiloxane	12,586.4
$C_{10}H_{30}O_3Si_4$	Decamethyltetrasiloxane	11,981.2
$C_{10}H_{30}O_5Si_5$	Decamethylcyclopenta siloxane	12,272.1
$C_{11}H_8O_2$	1-Naphthoic acid	22,581.4
$C_{11}H_8O_2$	2-Naphthoic acid	22,630.8
$C_{11}H_{10}$	1-Methylnaphthalene	–
$C_{11}H_{12}O_2$	Ethyl-*trans*-cinnamate	13,639.9
$C_{11}H_{12}O_2$	1 Phenyl-1,3-pentanedione	15,033.9
$C_{11}H_{12}O_3$	Ethyl benzoylacetate	17,115.4
$C_{11}H_{12}O_3$	Myristicine	14,471.4
$C_{11}H_{14}$	1-Phenylpentane	–
$C_{11}H_{14}$	2,4,5-Trimethylstyrene	12,076.1
$C_{11}H_{14}$	2,4,6-Trimethylstyrene	11,588.8
$C_{11}H_{14}$	4-Isopropylstryene	11,471.0
$C_{11}H_{14}O$	Isobutyrophenone	12,878.8
$C_{11}H_{14}O$	Pivalophenone	13,221.3
$C_{11}H_{14}O$	2,3,5-Trimethylacetophenone	14,283.6
$C_{11}H_{14}O_2$	Isobutyl benzoate	13,105.8
$C_{11}H_{14}O_2$	4-Allylveratrole	15,027.1
$C_{11}H_{16}$	Pentamethylbenzene	–
$C_{11}H_{16}$	3,5-Diethyltoluene	11,167.4
$C_{11}H_{16}$	1,2,4-Trimethyl-5-ethylbenzene	12,145.3
$C_{11}H_{16}$	1,3,5-Trimethyl-2-ethylbenzene	11,677.3
$C_{11}H_{16}$	3-Ethylcumene	11,233.5
$C_{11}H_{16}$	4-Ethylcumene	11,425.6
$C_{11}H_{16}$	*sec*-Amylbenzene	11,886.0
$C_{11}H_{16}O$	4-*tert*-Butyl-2-cresol	13,798.1
$C_{11}H_{16}O$	2-*tert*-Butyl-4-cresol	14,037.9
$C_{11}H_{16}O$	4-*tert*-Amylphenol	13,154.3
$C_{11}H_{16}O_5$	Ethylcamphoronic anhydride	16,373.6

HEATS OF VAPORIZATION OF ORGANIC COMPOUNDS (continued)

Formula	Name	Δ H_v
$C_{11}H_{18}O_2$	Bornyl formate	12,276.0
$C_{11}H_{18}O_2$	Geranyl formate	13,189.7
$C_{11}H_{18}O_2$	Neryl formate	12,959.3
$C_{11}H_{18}O_2Si$	Diethoxymethylphenyl silane	13,267.3
$C_{11}H_{18}O_5$	Diethyl-gamma-oxoazelate	17,543.6
$C_{11}H_{20}O_2$	10-Hendecenoic acid	17,247.5
$C_{11}H_{20}O_2$	Menthyl formate	12,077.7
$C_{11}H_{20}O_2$	2-Ethylhexyl acrylate	12,522.5
$C_{11}H_{20}O_2$	Octyl acrylate	12,957.5
$C_{11}H_{20}O_3$	Hexyl levulinate	14,626.2
$C_{11}H_{22}O$	Hendecan-2-one	14,353.5
$C_{11}H_{22}O_2$	Methyl [illegible]	13,831.7
$C_{11}H_{22}O_2$	Hendecanoic acid	14,689.9
$C_{11}H_{24}$	Undecane	11,481.7
$C_{11}H_{24}O$	Hendecan-2-ol	14,216.2
$C_{11}H_{26}Si$	Trimethyloctylsilane	12,285.8
$C_{11}H_{26}Si$	Amyltriethylsilane	11,859.7
$C_{12}H_9Br$	4-Bromobiphenyl	13,493.4
$C_{12}H_9BrO$	2-Bromo-4-phenylphenol	13,589.9
$C_{12}H_9Cl$	2-Chlorobiphenyl	13,925.7
$C_{12}H_9Cl$	4-Chlorobiphenyl	14,017.4
$C_{12}H_9ClO$	2-Chloro-3-phenylphenol	15,258.0
$C_{12}H_9ClO$	2-Chloro-6-phenylphenol	15,508.4
$C_{12}H_9Cl_2PO$	2-Xenyl dichlorophosphate	17,127.6
$C_{12}H_9N$	Carbazole	15,421.6
$C_{12}H_{10}$	Acenaphthene	13,078.5
$C_{12}H_{10}$	Diphenyl	12,910.0
$C_{12}H_{10}ClPO_3$	Diphenyl chlorophosphate	13,191.2
$C_{12}H_{10}Cl_2Si$	Dichlorodiphenylsilane	14,968.5
$C_{12}H_{10}F_2Si$	Difluorodiphenylsilane	12,913.3
$C_{12}H_{10}N_2$	Azobenzene	14,786.7
$C_{12}H_{10}O$	1-Acetonaphthone	16,095.5
$C_{12}H_{10}O$	2-Acetonaphthone	16,496.7
$C_{12}H_{10}O$	Diphenyl ether	12,325.5
$C_{12}H_{10}O$	2-Phenylphenol	15,397.8
$C_{12}H_{10}O$	4-Phenylphenol	16,974.3
$C_{12}H_{10}S$	Diphenyl sulfide	13,974.8
$C_{12}H_{10}S$	Diphenyl disulfide	17,452.0
$C_{12}H_{10}Se$	Diphenyl selenide	14,603.4
$C_{12}H_{11}N$	Diphenylamine	14,920.3
$C_{12}H_{12}$	1-Ethylnaphthalene	12,751.3
$C_{12}H_{12}N_2$	1,1-Diphenylhydrazine	15,940.4
$C_{12}H_{14}N_2O_5$	2-Cyclohexyl-4,6-dinitrophenol	19,100.0
$C_{12}H_{14}O_3$	Eugenyl acetate	15,120.7
$C_{12}H_{14}O_4$	Apiole	16,881.7
$C_{12}H_{14}O_4$	Diethyl phthalate	15,383.0
$C_{12}H_{16}$	2,5-Diethylstyrene	12,150.3
$C_{12}H_{16}$	Phenylcyclohexane	13,345.6
$C_{12}H_{16}O_2$	Isoamyl benzoate	12,782.9
$C_{12}H_{18}$	Hexamethylbenzene	
$C_{12}H_{18}$	1,2,4-Triethylbenzene	11,957.9
$C_{12}H_{18}$	1,3,4-Triethylbenzene	12,215.0
$C_{12}H_{18}$	1,3,5-Triethylbenzene	
$C_{12}H_{18}$	1,2-Diisopropylbenzene	11,751.4
$C_{12}H_{18}$	1,3-Diisopropylbenzene	11,498.9
$C_{12}H_{18}O$	2-*tert*-Butyl-4-ethylphenol	13,994.0
$C_{12}H_{18}O$	4-*tert*-Butyl-2,5-xylenol	14,477.9
$C_{12}H_{18}O$	4-*tert*-Butyl-2,6-xylenol	14,142.5
$C_{12}H_{18}O$	6-*tert*-Butyl-2,4-xylenol	13,882.4
$C_{12}H_{18}O$	6-*tert*-Butyl-3,4-xylenol	14,848.3
$C_{12}H_{20}O_2$	*d*-Bornyl acetate	11,838.7
$C_{12}H_{20}O_2$	Geranyl acetate	13,879.9
$C_{12}H_{20}O_2$	Linalyl acetate	12,910.6
$C_{12}H_{20}O_3Si$	Triethoxyphenylsilane	14,117.9
$C_{12}H_{20}O_7$	Triethyl citrate	14,818.4
$C_{12}H_{21}PO_4$	Trimethallyl phosphate	12,566.1
$C_{12}H_{22}O_2$	Citronellyl acetate	15,781.3
$C_{12}H_{22}O_2$	Menthyl acetate	12,819.2
$C_{12}H_{22}O_4$	Dimethyl sebacate	14,861.3
$C_{12}H_{22}O_4$	Diisoamyl oxalate	14,123.7
$C_{12}H_{22}O_6$	Diisobutyl-*d*-tartrate	14,874.9
$C_{12}H_{24}$	1-Dodecene	12,587.8
$C_{12}H_{24}$	Triisobutylene	10,790.4
$C_{12}H_{24}O$	Dodecan-2-one	14,138.7
$C_{12}H_{24}O$	Lauraldehyde	13,644.2
$C_{12}H_{24}O_2$	Lauric acid	16,585.3
$C_{12}H_{26}$	*n*-Dodecane	11,857.7
$C_{12}H_{26}O$	Dodecyl alcohol	15,16[illegible]
$C_{12}H_{26}O_4$	Tripropylene glycol monoisopropyl ether	14,171.5
$C_{12}H_{27}N$	Triisobutylamine	12,390.6
$C_{12}H_{27}N$	Dodecylamine	14,836.4
$C_{12}H_{28}Si$	Triethylhexylsilane	12,119.4
$C_{12}H_{34}O_5Si_4$	1,7-Diethoxyoctamethyltetrasiloxane	14,095.9
$C_{12}H_{36}O_4Si_5$	Dodecamethylpentasiloxane	12,942.6
$C_{12}H_{36}O_6Si_6$	Dodecamethylcyclohexasiloxane	13,760.6
$C_{13}H_9N$	Acridine	15,174.6
$C_{13}H_{10}$	Fluorene	13,682.8
$C_{13}H_{10}O$	Benzophenone	14,725.4
$C_{13}H_{10}O_2$	Phenyl benzoate	14,181.7
$C_{13}H_{10}O_3$	Salol	15,441.6
$C_{13}H_{12}$	Diphenylmethane	13,089.4
$C_{13}H_{12}O$	Benzhydrol	15,220.2
$C_{13}H_{12}O$	Benzyl phenyl ether	14,156.7
$C_{13}H_{12}O$	1-Proprionaphthone	16,630.8
$C_{13}H_{13}ClSi$	Chloromethyldiphenylsilane	14,924.6
$C_{13}H_{14}Si$	Methyldiphenylsilane	15,396.8
$C_{13}H_{14}$	2-Isopropylnaphthalene	13,036.9
$C_{13}H_{18}O$	Enanthophenone	15,597.7
$C_{13}H_{20}$	Heptylbenzene	13,535.4
$C_{13}H_{20}O$	α-Ionone	14,253.4
$C_{13}H_{22}O_2$	Bornyl propionate	13,245.0
$C_{13}H_{26}O$	2-Tridecanone	14,416.1
$C_{13}H_{26}O_2$	Methyl laurate	14,853.5
$C_{13}H_{26}O_2$	Tridecanoic acid	19,214.8
$C_{13}H_{28}$	Tridecane	12,991.3
$C_{13}H_{28}O_4$	Tripropyleneglycol monobutyl ether	15,937.6
$C_{13}H_{30}Si$	Decyltrimethylsilane	13,311.1
$C_{13}H_{30}Si$	Triethylheptylsilane	13,298.3
$C_{14}H_8O_2$	Anthraquinone	21,163.1
$C_{14}H_8O_4$	1,4-Dihydroxyanthraquinone	17,677.9
$C_{14}H_{10}$	Anthracene	16,823.6
$C_{14}H_{10}$	Phenanthrene	14,184.0
$C_{14}H_{10}O_2$	Benzil	15,046.4
$C_{14}H_{10}O_3$	Benzoic anhydride	16,060.9
$C_{14}H_{12}$	1,1-Diphenylethylene	13,778.1
$C_{14}H_{12}$	*trans*-Diphenylethylene	15,010.1
$C_{14}H_{12}O$	Desoxybenzoin	15,642.1
$C_{14}H_{12}O$	Benzoin	15,952.5
$C_{14}H_{14}$	Dibenzyl	13,387.6
$C_{14}H_{14}O$	2-Isobutyronaphthone	17,133.8
$C_{14}H_{15}O$	Dibenzylamine	16,260.1
$C_{14}H_{15}N$	Ethyldiphenylamine	14,569.4
$C_{14}H_{20}Cl_2$	1,2-Dichlorotetraethylbenzene	14,629.0
$C_{14}H_{20}Cl_2$	1,4-Dichlorotetraethylbenzene	13,397.5
$C_{14}H_{20}O_3$	2-(4-*tert*-Butylphenoxy) ethyl acetate	16,017.6
$C_{14}H_{22}$	1,2,3,4-Tetraethylbenzene	12,763.5
$C_{14}H_{22}O$	2,4-Di-*tert*-butylphenol	14,237.7
$C_{14}H_{24}O_2$	Bornyl butyrate	13,746.1
$C_{14}H_{24}O_2$	Bornyl isobutyrate	13,501.8
$C_{14}H_{24}O_2$	Geranyl butyrate	16,086.4
$C_{14}H_{24}O_2$	Geranyl isobutyrate	15,699.5
$C_{14}H_{26}O_4$	Diethyl sebacate	16,819.6
$C_{14}H_{28}O$	2-Tetradecanone	15,102.7
$C_{14}H_{28}O$	Myristaldehyde	14,088.9
$C_{14}H_{28}O_2$	Myristic acid	18,380.1
$C_{14}H_{29}Cl$	1-Chlorotetradecane	14,083.5
$C_{14}H_{30}$	Tetradecane	13,750.0
$C_{14}H_{31}N$	Tetradecylamine	14,840.8
$C_{14}H_{32}Si$	Triethyloctylsilane	12,954.8
$C_{14}H_{40}O_6Si_5$	1,9-Diethoxydecamethylpentasiloxane	15,296.9
$C_{14}H_{42}O_5Si_6$	Tetradecamethylhexasiloxane	13,800.0
$C_{14}H_{42}O_7Si_7$	Tetradecamethylcycloheptasiloxane	14,263.8
$C_{15}H_{14}O$	1,3-Diphenyl-2-propanone	15,429.8
$C_{15}H_{14}O_2$	1-Biphenyloxy-2,3-epoxypropane	16,160.6
$C_{15}H_{16}O_2$	4,4-Isopropylidenebisphenol	23,254.0
$C_{15}H_{18}O$	Isocapronaphthone	17,360.3
$C_{15}H_{18}OSi$	Ethoxymethyldiphenylsilane	16,106.4
$C_{15}H_{20}O_2$	Helenin	26,532.7
$C_{15}H_{24}$	Cadinene	15,518.3
$C_{15}H_{24}O$	2,6-Di-*tert*-butyl-4-cresol	14,338.6
$C_{15}H_{24}O$	4,6-Di-*tert*-butyl-2-cresol	14,006.9
$C_{15}H_{24}O$	4,6-Di-*tert*-butyl-3-cresol	15,464.6
$C_{15}H_{26}O$	Champacol	14,655.9
$C_{15}H_{26}O_6$	Triethyl camphoronate	16,112.2
$C_{15}H_{30}O_2$	Methyl myristate	16,051.0
$C_{15}H_{32}$	Pentadecane	14,635.9
$C_{15}H_{32}O_5$	Tetrapropylene glycol monoisopropyl ether	16,494.6
$C_{15}H_{34}Si$	Dodecyltrimethylsilane	14,374.6
$C_{16}H_{14}O_2$	Benzyl cinnamate	20,840.6
$C_{16}H_{18}O$	Di(α-methylbenzyl)ether	14,628.1
$C_{16}H_{20}O_2Si$	Diethoxydiphenylsilane	15,828.8
$C_{16}H_{22}O_4$	Dibutyl phthalate	17,747.0
$C_{16}H_{25}Cl$	Pentaethylchlorobenzene	13,707.3
$C_{16}H_{26}$	Pentaethylbenzene	13,670.1
$C_{16}H_{26}O$	2,6-Di-*tert*-butyl-4-ethylphenol	14,438.0
$C_{16}H_{26}O$	4,6-Di-*tert*-butyl-3-ethylphenol	15,954.8
$C_{16}H_{30}O$	Muscone	14,722.5
$C_{16}H_{31}N$	Palmitonitrile	16,433.7
$C_{16}H_{32}$	1-Hexadecene	15,634.7
$C_{16}H_{32}$	Tetraisobutylene	12,937.2
$C_{16}H_{32}O$	2-Hexadecanone	15,194.4
$C_{16}H_{32}O$	Palmitaldehyde	15,454.2
$C_{16}H_{32}O_2$	Palmitic acid	17,603.6
$C_{16}H_{34}$	Hexadecane	15,405.5
$C_{16}H_{34}O$	Cetyl alcohol	14,483.4
$C_{16}H_{35}N$	Cetylamine	15,238.0
$C_{16}H_{36}Si$	Decyltriethylsilane	15,393.7
$C_{16}H_{46}O_7Si_6$	1,1,1-Diethoxydodecamethylhexasiloxane	15,945.3
$C_{16}H_{48}O_6Si_7$	Hexadecamethylheptasiloxane	14,841.5
$C_{16}H_{48}O_8Si_8$	Hexadecamethylcyclooctasiloxane	14,986.3
$C_{17}H_{10}O$	Benzanthrone	18,309.6
$C_{17}H_{18}O_3$	4-*tert*-Butylphenyl salicylate	16,455.6
$C_{17}H_{24}O_2$	Menthyl benzoate	16,804.5
$C_{17}H_{34}O$	2-Heptadecanone	16,559.8
$C_{17}H_{34}O_2$	Methyl palmitate	17,003.5
$C_{17}H_{36}$	Heptadecane	15,608.5
$C_{17}H_{38}Si$	Tetradecyltrimethylsilane	16,439.7
$C_{18}H_{12}Cl_3O_3PS$	Tri-2-chlorophenylthiophosphate	24,386.1
$C_{18}H_{15}O_4P$	Triphenyl phosphate	19,272.3
$C_{18}H_{30}$	Hexaethylbenzene	14,184.9
$C_{18}H_{30}O$	2,4,6-Tri-*tert*-butylphenol	14,703.7
$C_{18}H_{34}O_2$	Oleic acid	20,326.7
$C_{18}H_{34}O_2$	Elaidic acid	19,538.0
$C_{18}H_{36}O$	Stearaldehyde	16,555.6
$C_{18}H_{36}O_2$	Stearic acid	19,306.6
$C_{18}H_{38}$	Octadecane	15,447.0
$C_{18}H_{38}$	2-Methylheptadecane	16,095.9
$C_{18}H_{38}O$	1-Octadecanol	17,508.0
$C_{18}H_{39}N$	Ethylcetylamine	15,718.3
$C_{18}H_{52}O_8Si_7$	1,1,3-Diethoxytetradecamethylheptasiloxane	16,765.8
$C_{18}H_{54}O_7Si_8$	Octadecamethyloctasiloxane	15,270.3
$C_{19}H_{16}$	Triphenylmethane	34,470.8
$C_{19}H_{40}$	Nonadecane	16,497.3
$C_{20}H_{20}OSi$	Ethoxytriphenylsilane	20,214.2
$C_{20}H_{43}N$	Diethylhexadecylamine	15,871.3
$C_{20}H_{58}O_9Si_8$	1,1,5-Diethoxyhexadecamethyloctasiloxane	17,626.6
$C_{20}H_{60}O_8Si_9$	Eicosamethylnonasiloxane	19,522.9
$C_{21}H_{21}O_4P$	Tritolyl phosphate	20,835.9
$C_{21}H_{44}$	Heneicosane	17,702.2
$C_{22}H_{42}O_2$	Erucic acid	23,655.2
$C_{22}H_{42}O_2$	Brassidic acid	24,085.7
$C_{22}H_{46}$	Docosane	16,941.1
$C_{22}H_{66}O_9Si_{10}$	Docosamethyldecasiloxane	21,878.6
$C_{23}H_{48}$	Tricosane	19,082.1
$C_{24}H_{50}$	Tetracosane	19,642.5
$C_{24}H_{72}O_{10}Si_{11}$	Tetracosamethylhendecasiloxane	23,941.2
$C_{25}H_{52}$	Pentacosane	20,815.9
$C_{26}H_{54}$	Hexacosane	21,605.7
$C_{27}H_{33}O_4P$	Dicarvacryl-2-tolyl phosphate	24,233.3
$C_{27}H_{56}$	Heptacosane	21,958.1
$C_{28}H_{58}$	Octacosane	24,144.2
$C_{29}H_{60}$	Nonacosane	24,816.8
$C_{32}H_{34}ClO_4P$	Dicarvacryl-mono-(6-chloro-2-xenyl)-phosphate	25,299.5

SOLUBILITY PARAMETERS OF ORGANIC COMPOUNDS

While the solubility of organic compounds in selected solvents may be estimated from the "like dissolves like" rule of thumb, more useful information is available under the "Solubility" heading of the table "Physical Constants of Organic Compounds" in Section C of this book. This table provides qualitative data on solubility in water (w), ethanol (al), ethyl ether (eth), acetone (ace), and benzene (bz).

More quantitative solubility data for nonpolar organic compounds may be calculated from the Hildebrand expression for the square root of the cohesive energy density which is defined as the solubility parameter (δ). As shown by the following expression, δ values may be calculated if information for ΔH_v, Kelvin temperature (T), molecular weight (M), and density (D) is available:

$$\delta = \left(\frac{\Delta E_v}{V}\right)^{1/2} = \left(\frac{D(\Delta H_v - RT)}{M}\right)^{1/2}$$

The value for ΔH_v may be found in the table "Heats of Vaporization of Organic Compounds" immediately following this table. Values for D and M are listed under physical constants in Section C. Thus, the δ value for heptane may be calculated as follows:

$$\delta = \left[\frac{0.684\,(8670 - 1.99(298))}{100}\right]^{1/2} = 7.4\ \mathrm{H}$$

The dimensions for δ are $(\mathrm{cal\ cm^{-3}})^{1/2}$ but the Hildebrand unit (H) is used for convenience. It is important to note that the law of mixtures applies for δ values of mixed nonpolar solvents. Thus, the δ value for an equimolar mixture of heptane and carbon disulfide (δ = 10.0 H) is 8.7 H.

When ΔH_v values are not available, Small's molar attraction constants shown in Table I may be used.[a] As illustrated in the following expression which is solved for heptane, the summation of these constants (ΣG) may be used to estimate δ values at 298 K:

$$\delta = \frac{D\Sigma G}{M} = \frac{(2 \times 214) + 5(133)0.684}{100} = 7.5\ \mathrm{H}$$

Solvents such as acetone (δ = 9.9 H) and water (δ = 23.4 H) are completely miscible when a large difference in δ values exists. The critical δ range for solubility of solid solutes and liquid solutes is less than that for liquids, and the critical δ range for nonpolar polymers in nonpolar liquids is less than 2 H at temperatures below 50°C.

These δ values are most useful for nonpolar solvents that are listed in Section A of Table II. Some consideration must be given to the dipole-dipole interactions in more polar solvents that are listed in Section B of Table II. The values shown for hydrogen-bonded solvents that are shown in Section C of Table II must be used with more discretion because of the stronger intermolecular forces present. However, these values are much more useful than the "like dissolves like" rule of thumb and can be used to predict the solubility of most solutes in most solvents.

[a]Small, P. A., *J. Appl. Chem.*, 3, 71, 1953.

Table I
MOLAR-ATTRACTION CONSTANTS AT 298 K

Group	G	Group	G
$-CH_3$	214	CO ketones	275
$-CH_2-$ single bonded	133	COO esters	310
$-CH<$	28	CN	410
$>C<$	−93	Cl (mean)	260
$CH_2=$	190	Cl single	270
$-CH=$ double bonded	111	Cl twinned as in $>CCl_2$	260
$>C=$	19	Cl triple as in $-CCl_3$	250
$CH\equiv C-$	285	Br single	340
$-C\equiv C-$	222	I single	425
Phenyl	735	CF_2 (n-fluorocarbons only)	150
Phenylene (*o,m,p*)	658	CF_3 (n-fluorocarbons only)	274
Naphthyl	1146	S sulfides	225
Ring, 5 membered	105–115	SH thiols	315
Ring, 6 membered	95–105	ONO_2 nitrates	~440
Conjugation	20–30	NO_2 (aliphatic nitro-compounds)	~440
H (variable)	80–100	PO_4 (organic phosphates)	~500
O ether	70	Si (in silicones)	−38

Table compiled by R. B. Seymour.

Table II
SOLUBILITY PARAMETER VALUES

A. Nonpolar Solvents

Name	δ (H)	Name	δ (H)
Acetic acid nitrile (acetonitrile)	11.9	Benzene, isopropyl (cumene)	8.5
Anthracene	9.9	Benzene, 1-isopropyl-4-methyl (*p*-cymene)	8.2
Benzene	9.2	Benzene, nitro	10.0
Benzene, chloro	9.5	Benzene, propyl	8.6
Benzene, 1,2-dichloro	10.0	Benzene, 1,3,5-trimethyl (mesitylene)	8.8
Benzene, ethyl	8.8	Benzoic acid nitrile (benzonitrile)	8.4

SOLUBILITY PARAMETER VALUES

A. Nonpolar Solvents

Name	δ (H)
Biphenyl, perchloro	8.8
1,3-Butadiene	7.1
1,3-Butadiene, 2-methyl (isoprene)	7.4
Butane	6.8
Butanoic acid nitrile	10.5
Carbon disulfide	10.0
Carbon tetrachloride	8.6
Chloroform	9.3
Cyclohexane	[illegible]
Cyclohexane, methyl	7.8
Cyclohexane, perfluoro	6.0
Cyclopentane	8.7
Decalin	8.8
Decane	8.0
Dimethyl sulfide	9.4
Ethane	6.0
Ethane, bromo (ethyl bromide)	9.6
Ethane, chloro (ethyl chloride)	9.2
Ethane, 1,2-dibromo	10.4
Ethane, 1,1-dichloro (ethylidene chloride)	8.9
Ethane, difluoro-tetrachloro (Freon 112®)	7.8
Ethane, nitro	11.1
Ethane, pentachloro	9.4
Ethane, 1,1,2,2-tetrachloro	9.7
Ethanethiol (ethyl mercaptan)	9.2
Ethane, 1,1,2-trichloro	9.6
Ethane trichloro-trifluoro (Freon 113®)	7.3
Ethene, (ethylene)	6.1
Ethene, tetrachloro (perchloroethylene)	9.3
Ethene, trichloro	9.2
Heptane	7.4
Heptane, perfluoro	5.8
Hexane	7.3
Hexene-1	7.4
Malonic acid dinitrile (malononitrile)	15.1
Methane	5.4
Methane, bromo	9.6
Methane, dichloro (methylene chloride)	9.7
Methane, dichloro-difluoro (Freon 12®)	5.5
Methane, dichloro, manofluoro (Freon 21®)	8.3
Methane, nitro	12.7
Methane, tetrachloro difluoro (Freon 112®)	7.8
Methane, trichloro-monofluoro (Freon 11®)	7.6
Naphthalene	9.9
Nonane	7.8
Octane	7.6
Pentane	7.0
Pentane, 1-bromo	7.6
Pentane, 1-chloro	8.3
Pentanoic acid, nitrile (valeronitrile)	9.6
Pentene-1	6.9
Phenanthrene	9.8
Propane	6.4
Propane, 1-bromo	8.9
Propane, 2,2-dimethyl (neopentane)	6.3
Propane, 1-nitro	16.3
Propane-2-nitro	9.9
Propene (propylene)	6.5
Propene, 2-methyl (isobutylene)	6.7
Propenoic acid nitrile (acrylonitrile)	10.5
Propionic acid nitrile	10.8
Styrene	9.3
Terphenyl, hydrogenated	9.0
Tetralin	9.5
Toluene	8.9
Xylene, *m*-	8.8

B. Moderately Polar Solvents

Name	δ (H)
Acetic acid, butyl ester	8.5
Acetic acid, ethyl ester	9.1
Acetic acid, methyl ester	9.6
Acetic acid, pentyl ester	8.0
Acetic acid, propyl ester	8.8
Acetic acid amide, *N,N*-diethyl	9.9
Acetic acid amide, *N,N*-dimethyl	10.8
Acrylic acid, butyl ester	8.4
Acrylic acid, ethyl ester	8.6
Acrylic acid, methyl ester	8.9
Adipic acid, dioctyl ester	8.7
Aniline, *N,N*-dimethyl	9.7
Benzene, 1-methoxy-4-propenyl (anethole)	8.4
Benzoic acid, ethyl ester	8.2
Benzoic acid, methyl ester	10.5
Butanal	9.0
Butane, 1-iodo	8.6
Butanoic acid, 4-hydroxylactone (butyrolactone)	12.6
2-Butanone	9.3
Carbonic acid, diethyl ester	8.8
Carbonic acid, dimethyl ester	9.9
Cyclohexanone	9.9
Cyclopentanone	10.4
2-Decanone	7.8
Diethylene glycol, monobutyl ether (butyl carbitol)	9.5
Diethylene glycol, monoethyl ether (ethyl carbitol)	10.2
Dimethyl sulfoxide	12.0
1,4-Dioxane	10.0
Ethene, chloro (vinyl chloride)	7.8
Ether, 1,1-dichloroethyl	10.0
Ether, diethyl	7.4
Ether, dimethyl	8.8
Ether, dipropyl	7.8
Ethylene glycol, monobutyl ether (butyl Cellosolve®)	9.5
Ethylene glycol, monoethyl ether (ethyl Cellosolve)	10.5
Ethylene glycol, monomethyl ether (methyl Cellosolve)	11.4
Formic acid amide, *N,N*-diethyl	10.6
Formic acid amide, *N,N*-dimethyl	12.1
Formic acid, ethyl ester	9.4
Formic acid, methyl ester	10.2
Formic acid, 2-methylbutyl ester	8.0
Formic acid, propyl ester	9.2
Furan	9.4
Furan, tetrahydro	9.1
Furfural	11.2
2-Heptanone	8.5
Hexanoic acid, 6-aminolactam (ε-caprolactam)	12.7
Hexanoic acid, 6-hydroxylactone (caprolactone)	10.1
Isophorone	9.1
Lactic acid, butyl ester	9.4
Lactic acid, ethyl ester	10.0
Methacrylic acid, butyl ester	8.3
Methacrylic acid, ethyl ester	8.5
Methacrylic acid, methyl ester	8.8
Oxalic acid, diethyl ester	8.6
Oxalic acid, dimethyl ester	11.0
Oxirane (ethylene oxide)	11.1
Pentane, 1-iodo	8.4
2-Pentanone	8.7
Pentanone-2,4-hydroxy,4-methyl (diacetone alcohol)	9.2
Pentanone-2,4-methyl (mesityl oxide)	9.0
Phosphoric acid, triphenyl ester	8.6
Phosphoric acid, tri-2-tolyl ester	8.4
Phthalic acid, dibutyl ester	9.3
Phthalic acid, diethyl ester	10.0
Phthalic acid, dihexyl ester	8.9
Phthalic acid, dimethyl ester	10.7
Phthalic acid, di-2-methylnonyl ester	7.2
Phthalic acid dioctyl ester	7.9
Phthalic acid, dipentyl ester	9.1
Phthalic acid, dipropyl ester	9.7
Propane, 1,2-epoxy (propylene oxide)	9.2

Table II (continued)

SOLUBILITY PARAMETER VALUES

B. Moderately Polar Solvents

Name	δ (H)	Name	δ (H)
Sebacic acid, dioctyl ester	8.6	Propionic acid, ethyl ester	8.4
Stearic acid, butyl ester	7.5	Propionic acid, methyl ester	8.9
Sulfone, diethyl	12.4	4-Pyrone	13.4
Sulfone, dimethyl	14.5	2-Pyrrolidone, 1-methyl	11.3
Sulfone, dipropyl	11.3	Sebacic acid, dibutyl ester	9.2

C. Hydrogen-bonded Solvents

Acetic acid	10.1	1-Hexanol	10.7
Acetic acid amide, *N*-ethyl	12.3	1-Hexanol-2-ethyl	9.5
Acetic acid, dichloro	11.0	Maleic acid anhydride	13.6
Acetic acid, anhydride	10.3	Methacrylic acid	11.2
Acrylic acid	12.0	Methacrylic acid amide, *N*-Methyl	14.6
Amine, diethyl	8.0	Methanol	14.5
Amine, ethyl	10.0	Methanol, 2-furil (furfuryl alcohol)	12.5
Amine, methyl	11.2	1-Nonanol	8.4
Ammonia	16.3	Pentane, 1-amino	8.7
Aniline	10.3	1,3-Pentanediol, 2-methyl	10.3
1,3-Butanediol	10.9	1-Pentanol	11.6
1,4-Butanediol	10.0	2-Pentanol	12.1
2,3-Butanediol	8.7	Piperidine	11.1
1-Butanol	13.6	2-Piperidone	11.4
2-Butanol	12.6	1,2-Propanediol	10.8
1-Butanol, 2-ethyl	11.9	1-Propanol	10.5
1-Butanol, 2-methyl	11.5	2-Propanol	10.0
Butyric acid	10.5	1-Propanol, 2-methyl	10.5
Cyclohexanol	10.6	2-Propanol, 2-methyl	11.4
Diethylene glycol	11.8	2-Propenol (allyl alcohol)	12.1
1-Dodecanol	9.9	Propionic acid	8.1
Ethanol	10.0	Propionic acid anhydride	12.7
Ethanol, 2-chloro (ethylene chlorohydrin)	12.6	1,2-Propanediol	12.2
Ethylene glycol	10.7	Pyridine	14.6
Formic acid	14.7	2-Pyrrolidone	12.1
Formic acid amide, *N*-ethyl	10.8	Quinoline	13.9
Formic acid amide, *N*-methyl	15.4	Succinic acid anhydride	16.1
Glycerol	9.9	Tetraethylene glycol	16.5
2,3-Hexanediol	10.2	Toluene, 3-hydroxy (meta cresol)	10.3
1,3-Hexanediol-2-ethyl	23.4	Water	9.4

Table compiled by R. B. Seymour.

MISCIBILITY OF ORGANIC SOLVENT PAIRS

Table A

Doctor J. S. Drury

Industrial and Engineering Chemistry Vol. 44, No. 11, Nov. 1952

(Reprinted by permission)

The classifications were made by shaking together 5 ml. of each of the solvents listed in a test tube for 1 minute, then allowing the mixture to settle. If no interfacial meniscus was observed, the solvent pair was considered miscible. If such a meniscus was present, the solvent pair was regarded as immiscible. The classification of immiscible is a qualitative one since solvent pairs may exhibit some degree of partial miscibility while existing as separate phases. Solvent pairs possessing a pronounced degree of partial miscibility are designated by the symbol Is.

Compound number	Compounds	Acetone	Acetyl acetone	2-Amino-2-methyl-1-propanol	Aniline	Benzaldehyde	Benzene	Benzin	Benzyl alcohol	Butyl acetate	Butyl alcohol	*n*-Butyl ether	Capryl alcohol	Carbon tetrachloride	Diacetone alcohol	Diethanolamine	Diethyl cellosolve	Diethyl ether	Dimethylaniline	Ethyl alcohol	Ethyl benzoate	Ethylene glycol	2-Ethylhexanol	Formamide	Furfuryl alcohol	Glycerol	Hydroxyethyl-ethylenediamine	Isoamyl alcohol	Methyl isobutyl ketone	Nitromethane	Dibutoxytetra-ethylene glycol	Pyridine	Triethanolamine	Trimethylene glycol
1	Acetone	..	M	M	..	M	M	M	M	M	M	M	M	M	M	M	M	M	M	M	M	M	M	M	M	I	M	M	M	M	M	M	..	M
2	Acetyl acetone	M	..	R	..	M	M	M	M	M	M	M	M	M	M	R	M	M	M	M	M	M	M	M	M	I	R	M	M	M	M	M	..	M
3	Adiponitrile	M	M	M	M	..	M	..	M	M	M	I	..	I	M	..	M	I	M	M	M	I	I	M	M	I	M	I	M	..	M	M	M	M
4	2-Amino-2-methyl-1-propanol	M	R	..	..	M	M	I	M	M	M	Is	M	M	R	M	M	M	M	M	M	M	M	M	M	M	M	M	M	M	M	M	..	M
5	Benzaldehyde	M	M	M	..	..	M	M	M	M	M	M	M	M	M	I	M	M	M	M	M	Is	M	M	M	Is	R	M	M	M	M	M	..	M
6	Benzene	M	M	M	..	M	..	M	M	M	M	M	M	M	Is	I	M	M	M	M	M	I	M	I	M	I	I	M	M	I	M	M	..	I
7	Benzin	M	M	I	..	M	M	..	I	M	M	M	M	M	I	I	M	M	M	M	M	M	M	I	I	I	Is	M	M	M	M	M	..	I
8	Benzonitrile	M	M	M	M	..	M	..	M	M	M	M	..	M	M	..	M	M	M	M	M	I	M	I	M	I	M	M	M	..	M	M	M	I
9	Benzothiazole	M	M	M	M	..	M	..	M	M	M	M	..	M	M	..	M	M	M	M	M	M	M	I	M	I	M	M	M	..	M	M	M	M
10	Benzyl alcohol	M	M	M	..	M	M	I	..	M	M	M	M	M	M	M	M	M	M	M	M	M	M	M	M	M	M	M	M	M	M	M	..	M
11	Benzyl mercaptan	M	M	I	M	..	M	..	M	M	M	M	..	M	M	..	M	M	M	M	M	I	M	I	M	I	M	M	M	..	M	M	R	I
12	Butyl acetate	M	M	M	..	M	M	M	M	..	M	M	M	M	M	I	M	M	M	M	M	Is	M	I	M	I	I	M	M	M	M	M	..	Is
13	Butyl alcohol	M	M	M	..	M	M	M	M	M	..	M	M	M	M	M	M	M	M	M	M	M	M	M	M	M	M	M	M	M	M	M	..	M
14	*n*-Butyl ether	M	M	Is	..	M	M	M	M	M	M	..	M	M	M	I	M	M	M	M	M	I	M	I	M	I	I	M	M	I	M	M	..	I
15	Capryl alcohol	M	M	M	..	M	M	M	M	M	M	M	..	M	M	M	M	M	M	M	M	M	M	I	M	I	M	M	M	Is	M	M	..	M
16	Carbon tetrachloride	M	M	M	..	M	M	M	M	M	M	M	M	..	Is	I	M	M	M	M	M	I	M	I	M	I	I	M	M	M	M	M	..	I
17	Diacetone alcohol	M	M	R	..	M	Is	I	M	M	M	M	M	Is	..	M	M	M	M	M	M	M	M	M	M	I	R	M	M	M	M	M	..	M
18	Diethanolamine	M	R	M	..	I	I	I	M	I	M	I	M	I	M	..	I	I	Is	M	M	M	M	M	M	M	M	M	I	I	I	M	..	M
19	Diethyl Cellosolve	M	M	M	..	M	M	M	M	M	M	M	M	M	M	I	..	M	M	M	M	M	M	M	M	I	I	M	M	M	M	M	..	M
20	Diethyl ether	M	M	M	..	M	M	M	M	M	M	M	M	M	M	I	M	..	M	M	M	I	M	I	M	I	I	M	M	M	M	M	..	I
21	Dimethylaniline	M	M	M	..	M	M	M	M	M	M	M	M	M	M	Is	M	M	..	M	M	I	M	I	M	I	I	M	M	M	M	M	..	I
22	Di-*N*-propylaniline	M	M	I	M	..	M	..	M	M	M	M	..	M	I	..	M	M	M	M	M	I	M	I	M	I	I	M	M	..	M	M	I	I
23	Ethyl alcohol	M	M	M	..	M	M	M	M	M	M	M	M	M	M	M	M	M	M	..	M	M	M	M	M	M	M	M	M	M	M	M	..	M
24	Ethyl benzoate	M	M	M	..	M	M	M	M	M	M	M	M	M	M	M	M	M	M	M	..	I	M	I	M	I	M	M	M	M	M	M	..	Is
25	Ethyl isothiocyanate	M	M	R	M	..	M	..	M	M	M	M	..	M	M	..	M	M	M	M	M	I	M	I	M	I	R	M	M	..	M	M	M	I
26	Ethyl thiocyanate	M	M	M	M	..	M	..	M	M	M	M	..	M	M	..	M	M	M	M	M	I	M	I	M	I	M	M	M	..	M	M	M	I
27	Ethylene glycol	M	M	M	..	Is	I	M	M	Is	M	I	M	I	M	M	M	I	I	M	I	..	M	M	M	M	M	M	I	I	M	M	..	M
28	2-Ethylhexanol	M	M	M	..	M	M	M	M	M	M	M	M	M	M	M	M	M	M	M	M	M	..	I	M	I	M	M	M	I	M	M	..	M
29	Formamide	M	M	M	..	M	I	I	M	I	M	I	i	I	M	M	M	I	I	M	I	M	I	..	M	M	M	M	Is	M	M	M	..	M
30	Furfuryl alcohol	M	M	M	..	M	M	I	M	M	M	M	M	M	M	M	M	M	M	M	M	M	M	M	..	M	M	M	M	M	M	M	..	M
31	Glycerol	I	I	M	..	I	I	I	M	I	M	I	I	I	I	M	I	I	I	M	I	M	I	M	M	..	M	I	I	I	I	M	..	M
32	Hydroxyethyl-ethylenediamine	M	R	M	..	R	I	Is	M	I	M	I	M	I	R	M	I	I	I	M	M	M	M	M	M	M	..	M	M	M	M	M	..	M
33	Isoamyl alcohol	M	M	M	..	M	M	M	M	M	M	M	M	M	M	M	M	M	M	M	M	M	M	M	M	I	M	..	M	M	M	M	..	M
34	Isoamyl sulfide	M	M	I	M	..	M	..	M	M	M	M	..	M	I	..	M	M	M	M	M	I	M	I	I	I	I	M	M	..	M	M	I	I
35	Isobutyl mercaptan	M	M	M	M	..	M	..	M	M	M	M	..	M	M	..	M	M	M	M	M	I	M	I	M	I	I	M	M	..	M	M	R	R
36	Methyl disulfide	M	M	M	M	..	M	..	M	M	M	M	..	M	M	..	M	M	M	M	M	I	M	I	M	I	I	M	M	.	M	M	I	R
37	Methyl isobutyl ketone	M	M	M	..	M	M	M	M	M	M	M	M	M	M	I	M	M	M	M	M	I	M	Is	M	I	M	M	..	M	M	M	..	I
38	Nitromethane	M	M	M	..	M	I	M	M	M	M	I	Is	M	M	I	M	M	M	M	M	I	I	M	M	I	M	M	M	..	M	M	..	I
39	Dibutoxytetra-ethylene glycol	M	M	M	..	M	M	M	M	M	M	M	M	M	M	I	M	M	M	M	M	M	M	M	M	I	M	M	M	M	..	M	..	M
40	Pyridine	M	M	M	..	M	M	M	M	M	M	M	M	M	M	M	M	M	M	M	M	M	M	M	M	M	M	M	M	M	M	..	..	M
41	Tri-*n*-butylamine	M	M	I	I	..	M	..	M	M	M	M	..	M	I	..	M	M	M	M	M	I	M	I	M	I	I	M	M	..	M	M	I	I
42	Trimethylene glycol	M	M	M	..	M	I	I	M	Is	M	I	M	I	M	M	M	I	I	M	Is	M	M	M	M	M	M	M	I	I	M	M	..	..

MISCIBILITY OF ORGANIC SOLVENT PAIRS (Continued)

Tables B and C

W. M. Jackson and J. S. Drury

The classifications were made at 20°C in the following manner. One-milliliter portions of each solvent comprising a pair were shaken together for approximately a minute. If no interfacial meniscus was observed after the contents of the tube were allowed to settle, the solvent pair was considered to be miscible, M. If a meniscus was observed without apparent change in the volume of either solvent, the pair was regarded as immiscible, I. This classification is a qualitative one, since solvent pairs may exhibit various degrees of partial miscibility while existing as separate phases. If an obvious change occurred in the volume of each solvent, but a meniscus was present, the pair was classified as partially miscible, S. The designation R indicates that the two solvents reacted.

Table B

Compound number	Compounds	Acetone	Isoamyl acetate	*n*-Amyl cyanide	Benzene	Benzyl ether	2-Bromoethyl acetate	Chloroform	Cinnamaldehyde	Di-*n*-amylamine	Di-*n*-butyl carbonate	Diethylacetic acid	Diethylenetriamine	Diethyl formamide	Diisobutyl ketone	Diisopropylamine	Di-*n*-propyl aniline	Ethyl alcohol	Ethyl benzoate	Ethyl ether	Ethyl phenylacetate	Heptadecanol[a]	3-Heptanol	*n*-Heptyl acetate	*n*-Hexyl ether	Methyl isopropyl ketone	4-Methyl-*n*-valeric acid	*o*-Phenetidine	Sulfuric acid (concd.)	Tetradecanol[a]	Tri-*n*-butyl phosphate	Triethylene glycol	Triethylenetetramine	2,6,8-Trimethyl 4-nonanone	Compound number
1	Acetone	..	M	M	M	M	M	M	M	M	M	M	M	M	M	M	M	M	M	M	M	M	M	M	M	M	M	M	R	M	M	M	M	M	1
2	Isoamyl acetate	M	..	M	M	M	M	M	M	M	M	M	M	M	M	M	M	M	M	M	M	M	M	M	M	M	M	M	R	M	M	I	M	M	2
3	*n*-Amyl cyanide	M	M	..	M	M	M	M	M	M	M	M	M	M	M	M	M	M	M	M	M	M	M	M	M	M	M	M	R	M	M	M	M	M	3
4	Benzene	M	M	M	..	M	S	M	M	M	M	M	M	M	M	M	M	M	M	M	M	M	M	M	M	M	M	M	I	M	M	S	M	M	4
5	Benzyl ether	M	M	M	M	..	M	M	M	M	M	M	M	M	M	M	M	M	M	M	M	M	M	M	M	M	M	M	R	M	M	I	M	M	5
6	2-Bromoethyl acetate	M	M	M	S	M	..	M	M	R	M	M	R	R	M	R	M	M	M	M	M	M	M	M	S	M	M	M	R	M	M	M	R	M	6
7	Chloroform	M	M	M	M	M	M	..	M	M	M	M	M	M	M	M	M	M	M	M	M	M	M	M	M	M	M	M	I	M	M	M	M	M	7
8	Cinnamaldehyde	M	M	M	M	M	M	M	..	M	M	M	R	M	M	M	M	M	M	M	M	M	M	M	M	M	M	R	R	M	M	M	R	M	8
9	Di-*n*-amylamine	M	M	M	M	M	R	M	M	..	M	R	I	R	M	M	M	M	M	M	M	M	M	M	M	M	M	M	R	M	M	S	I	M	9
10	Di-*n*-butyl carbonate	M	M	M	M	M	M	M	M	M	..	M	M	M	M	M	M	M	M	M	M	M	M	M	M	M	M	M	R	M	M	I	I	M	10
11	Diethylacetic acid	M	M	M	M	M	M	M	M	R	M	..	R	M	M	R	M	M	M	M	M	M	M	M	M	M	M	M	R	M	M	M	R	M	11
12	Diethylenetriamine	M	M	M	M	M	R	M	R	I	M	R	..	R	M	M	M	M	M	M	M	M	M	R	I	R	R	M	R	M	M	M	M	I	12
13	Diethyl formamide	M	M	M	M	M	R	M	M	R	M	M	R	..	M	R	M	M	M	M	M	M	M	M	I	M	M	M	R	M	M	M	R	M	13
14	Diisobutyl ketone	M	M	M	M	M	M	M	M	M	M	M	M	M	..	M	M	M	M	M	M	M	M	M	M	M	M	M	R	M	M	I	M	M	14
15	Diisopropylamine	M	M	M	M	M	R	M	M	M	M	R	M	R	M	..	M	M	M	M	M	M	M	M	M	M	R	M	R	M	M	M	M	M	15
16	Di-*n*-propylaniline	M	M	M	M	M	M	M	M	M	M	M	M	M	M	M	..	M	M	M	M	M	M	M	M	M	M	M	R	M	M	I	M	M	16
17	Ethyl alcohol	M	M	M	M	M	M	M	M	M	M	M	M	M	M	M	M	..	M	M	M	M	M	M	M	M	M	M	M	M	M	M	M	M	17
18	Ethyl benzoate	M	M	M	M	M	M	M	M	M	M	M	M	M	M	M	M	M	..	M	M	M	M	M	M	M	M	M	M	M	M	M	M	M	18
19	Ethyl ether	M	M	M	M	M	M	M	M	M	M	M	M	M	M	M	M	M	M	..	M	M	M	M	M	M	M	M	R	M	M	I	M	M	19
20	Ethyl phenylacetate	M	M	M	M	M	M	M	M	M	M	M	M	M	M	M	M	M	M	M	..	M	M	M	M	M	M	M	R	M	M	M	M	M	20
21	Heptadecanol[a]	M	M	M	M	M	M	M	M	M	M	M	M	M	M	M	M	M	M	M	M	..	M	M	M	M	M	M	R	M	M	I	M	M	21
22	3-Heptanol	M	M	M	M	M	M	M	M	M	M	M	M	M	M	M	M	M	M	M	M	M	..	M	M	M	M	M	R	M	M	M	M	M	22
23	*n*-Heptyl acetate	M	M	M	M	M	M	M	M	M	M	M	R	M	M	M	M	M	M	M	M	M	M	..	M	M	M	M	R	M	M	I	R	M	23
24	*n*-Hexyl ether	M	M	M	M	M	S	M	M	M	M	M	I	I	M	M	M	M	M	M	M	M	M	M	..	M	M	M	R	M	M	I	I	M	24
25	Methyl isopropyl ketone	M	M	M	M	M	M	M	M	M	M	M	R	M	M	M	M	M	M	M	M	M	M	M	M	..	M	M	R	M	M	M	R	M	25
26	4-Methyl-*n*-valeric acid	M	M	M	M	M	M	M	M	M	M	M	R	M	M	R	M	M	M	M	M	M	M	M	M	M	..	M	R	M	M	M	R	M	26
27	*o*-Phenetidine	M	M	M	M	M	M	M	R	M	M	M	M	M	M	M	M	M	M	M	M	M	M	M	M	M	M	..	R	M	M	M	M	M	27
28	Sulfuric acid (concd.)	R	R	R	I	R	R	I	R	R	R	R	R	R	R	R	R	M	M	R	R	R	R	R	R	R	R	R	..	R	R	R	R	R	28
29	Tetradecanol[a]	M	M	M	M	M	M	M	M	M	M	M	M	M	M	M	M	M	M	M	M	M	M	M	M	M	M	M	R	..	M	I	M	M	29
30	Tri-*n*-butyl phosphate	M	M	M	M	M	M	M	M	M	M	M	M	M	M	M	M	M	M	M	M	M	M	M	M	M	M	M	R	M	..	M	M	M	30
31	Triethylene glycol	M	I	M	S	I	M	M	M	S	I	M	M	M	I	M	I	M	M	I	M	I	M	I	I	M	M	M	R	I	M	..	M	I	31
32	Triethylenetetramine	M	M	M	M	M	R	M	R	I	I	R	M	R	M	M	M	M	M	M	M	M	M	R	I	R	R	M	R	M	M	M	..	I	32
33	2,6,8-Trimethyl 4-nonanone	M	M	M	M	M	M	M	M	M	M	M	I	M	M	M	M	M	M	M	M	M	M	M	M	M	M	M	R	M	M	I	I	..	33

[a] Union Carbide name.

MISCIBILITY OF ORGANIC SOLVENT PAIRS (Continued)

Table C

Compound number	Compounds	Acetone	Isoamyl acetate	n-Amyl cyanide	Anisaldehyde	Benzene	Benzyl ether	Chloroform	o-Cresol	Diisobutyl ketone	Diethylacetic acid	Diethyl formamide	Di-n-propyl aniline	Ethyl alcohol	Ethyl ether	3-Heptanol	n-Heptyl acetate	n-Hexyl ether	α-Methylbenzylamine	α-Methylbenzyldiethanolamine	α-Methylbenzyldimethylamine	α-Methylbenzylethanolamine	2-Methyl-5-ethylpyridine	Methyl isopropyl ketone	4-Methyl-n-valeric acid	o-Phenetidine	2-Phenylethylamine	Isopropanolamine	Pyridine	Salicylaldehyde	Tetradecanol[a]	Tri-n-butyl phosphate	Triethylenetetramine	2,6,8-Trimethyl 4-nonanone
1	1,3-Butylene glycol	M	I	M	I	I	I	M	M	I	M	M	I	M	S	M	I	I	M	M	M	M	M	M	M	M	M	M	M	M	M	M	M	I
2	2,3-Butylene glycol	M	M	M	M	S	I	M	M	M	M	M	I	M	M	M	M	I	M	M	M	M	M	M	M	M	M	M	M	M	M	M	M	I
3	2-Chloroethanol	M	M	M	M	M	M	M	M	M	M	M	M	M	M	M	M	M	R	M	M	M	R	M	M	M	R	R	M	M	M	M	M	M
4	3-Chloro-1,2-propanediol	M	M	M	M	I	M	M	M	M	M	M	I	M	M	M	M	I	R	M	M	M	M	M	M	M	R	R	M	M	S	M	R	S
5	Dibutyl hydrogen phosphite	M	M	M	M	M	M	M	M	M	M	M	M	M	M	M	M	M	M	M	M	M	M	M	M	M	M	M	M	M	M	M	M	M
6	Diethylene glycol dibutyl ether	M	M	M	M	M	M	M	M	M	M	M	M	M	M	M	M	M	R	M	S	M	M	M	M	M	R	R	M	M	M	M	R	M
7	Diethylene glycol diethyl ether	M	M	M	M	M	M	M	M	M	M	M	M	M	M	M	M	M	M	M	M	M	M	M	M	M	M	M	M	M	M	M	M	M
8	Diethylene glycol monobutyl ether	M	M	M	M	M	M	M	M	M	M	M	M	M	M	M	M	M	M	M	M	M	M	M	M	M	M	M	M	M	M	M	M	M
9	Diethylene glycol monoethyl ether	M	M	M	M	M	M	M	M	M	M	M	M	M	M	M	M	I	M	M	M	M	M	M	M	M	M	M	M	M	M	M	M	M
10	Diethylene glycol monomethyl ether	M	M	M	M	M	M	M	M	M	M	M	M	M	M	M	M	I	M	M	M	M	M	M	M	M	M	M	M	M	M	M	M	M
11	Dipropylene glycol	M	M	M	M	M	M	M	M	M	M	M	M	M	M	M	M	I	M	M	M	M	M	M	M	M	M	M	M	M	M	M	M	M
12	Ethylene diacetate	M	M	M	M	M	M	M	M	M	M	M	M	M	M	M	M	I	M	M	M	M	M	M	M	M	M	M	M	M	M	M	M	M
13	Ethylene glycol	M	I	I	I	I	I	S	M	I	M	M	I	M	I	M	I	I	M	M	M	M	M	I	M	M	M	M	M	I	I	S	M	I
14	Ethyl glycol ethylbutyl ether	M	M	M	M	M	M	M	M	M	M	M	M	M	M	M	M	M	M	M	M	M	M	M	M	M	M	M	M	M	M	M	M	M
15	Ethylene glycol monobutyl ether	M	M	M	M	M	M	M	M	M	M	M	M	M	M	M	M	M	M	M	M	M	M	M	M	M	M	M	M	M	M	M	M	M
16	Ethylene glycol monoethyl ether	M	M	M	M	M	M	M	M	M	M	M	M	M	M	M	M	M	M	M	M	M	M	M	M	M	M	M	M	M	M	M	M	M
17	Ethylene glycol monomethyl ether	M	M	M	M	M	M	M	M	M	M	M	M	M	M	M	M	M	M	M	M	M	M	M	M	M	M	M	M	M	M	M	M	M
18	Ethylene glycol monophenyl ether	M	M	M	M	M	M	M	M	M	M	M	M	M	M	M	M	M	M	M	M	M	M	M	M	M	M	M	M	M	M	M	M	M
19	Glycerol	I	I	I	I	I	I	I	M	I	I	M	I	M	I	I	I	I	M	M	I	M	M	I	I	I	M	M	M	I	I	I	M	I
20	1,2-Propanediol	M	M	M	M	I	I	M	M	I	M	M	I	M	S	M	I	I	M	M	M	M	M	M	M	M	M	M	M	M	M	M	M	I
21	1,3-Propanediol	M	I	I	I	I	I	M	M	I	M	M	I	M	I	M	I	I	M	M	M	M	M	M	M	M	M	M	M	I	S	M	M	I
22	Triethylene glycol	M	I	M	M	S	I	M	M	I	M	M	I	M	I	M	I	I	M	M	M	M	M	M	M	M	M	M	M	M	I	M	M	I
23	Triethyl phosphate	M	M	M	M	M	M	M	M	M	M	M	M	M	M	M	M	M	M	M	M	M	M	M	M	M	M	M	M	M	M	M	M	M
24	Trimethylene chlorohydrin	M	M	M	M	M	M	M	M	M	M	M	M	M	M	M	M	M	R	M	M	M	R	M	M	M	R	R	M	M	M	M	R	M

[a] Union Carbide name.

STEROID HORMONES AND OTHER STEROIDAL SYNTHETICS

Compiled by Erwin Di Cyan, Ph.D.

The field of steroids has expanded considerably and rapidly in degree and in kind, because synthetic steroids have been synthesized which though resembling the hormones in the body have no natural counterpart, but exert an effect comparable to those of the natural hormones.

In fact, the term *steroid hormone* thus becomes a misnomer when applied to the newer synthetically prepared steroids which do not have a counterpart in the body of man or other animals—as prednisone. (A hormone, by definition, is a material with certain functions and characteristics, *secreted by the ductless glands*. That part of the definition cannot be met by prednisone or by similar steroids as these are not secreted by the ductless, or endocrine glands.)

All the hormones as well as the synthetic analogues have in common the cyclopentanophenanthrene nucleus. Although chemically very similar, a comparatively slight structural change is in many instances productive of substances which have physiologically dissimilar effects, often acting upon different physiologic systems. But in many cases a small change in structure will result merely in an accentuation of certain effects.

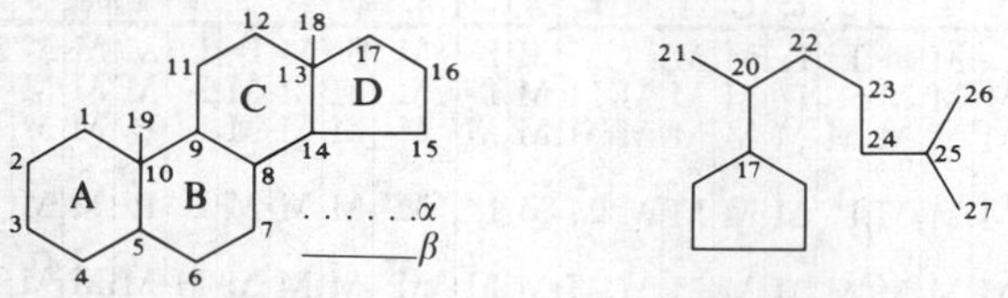

The Cyclopentanophenanthrene Nucleus

Classification. Classification becomes a bizarre problem by reason of the (a) overlapping uses to which these substances are put, and (b) the multiple purposes for which the hormones or synthetic substances are used. Indeed, the steroids may be classified by structure; that however would be uninformative to the student as to their use. Classification by origin, as adrenal, would also be unsuitable because, for example, a number of the adrenal corticosteroids are not found in the adrenal cortex at all, but merely resemble the natural hormones found in the adrenal cortex.

For those reasons the hormonal or hormonelike entries in the tables are classified by-and-large, by their predominant pharmacologic effects. Even that classification has its disparities as for example, the use of male sex hormones, i.e. the androgens, is neither limited to men, nor to uses which entail their effect upon male sex characteristics.

Uses. Originally, the use of steroid hormones was largely based upon one or more of the following predicates:

(a) To supplement the progressively declining secretion of a specific hormone due to natural biologic aging of the organism; in the menopause as an example of such declining secretion, a female sex hormone is used for such supplementation;

(b) To make available to the body a specific hormone, the natural secretion of which is inhibited because of a congenital or developmental anomaly; the underdevelopment of male secondary sex characteristics is an example of such an inhibited secretion, in which a male sex hormone is used—and correspondingly, female sex hormones in underdevelopment in females;

(c) To cause a reversal of hormonal balance in the treatment of diseases peculiar to a sex; for example, in the case of cancer of the female breast, a male sex hormone is administered, and in cancer of the prostate, a female sex hormone is used;

(d) To mimic a natural function, as menstruation, by the administration of estrogens—on withdrawal of which bleeding occurs; or by the alternate use of estrogenic and progestational—both female sex hormones.

(e) To delay a function, as ovulation, as in oral contraceptives, or *birth control pills*.

Since the finding that cortisone ameliorates the symptoms of rheumatoid arthritis (1949) the adrenal corticosteroid hormones and especially the synthetically prepared steroid analogues which have no natural counterpart in the body, have been successfully employed in the treatment of diseases not related to sex or sex function.

Androgens and Anabolic Agents. The agents listed in the tables under this classification have the effect of male sex hormones (androgens) i.e., to stimulate sexual maturation, in the "male climacteric," etc. But all androgens have in greater or lesser degree the ability to stimulate muscle development, i.e., an anabolic effect. Among the synthetically prepared agents which have no counterpart in the body (Methandrostenolone or Oxymetholone) are those which have a lessened androgenic, but a heightened anabolic effect. These qualities are determined by biological tests on animals but principally confirmed by clinical use in man. The anabolic effect includes remineralization of bone, which may be partially demineralized (osteoporosis) by age, or by certain drugs, as the adrenal corticosteroids (q.v.).

Anabolic agents are used for muscle and bone nutrition in men as well as women. The reason for the high interest in synthetic steroidal substances for anabolic use, is based on the need for materials, which within a given effective dose have a greater anabolic-to-androgenic ratio than such androgens as methyl testosterone. Otherwise, the administration of androgens to women produces manifestations of virilism, such as growth of hair on the face, a deepening of the voice, etc. Androgens are also used in the female in the suppression of excessive bleeding and in the treatment of cancer of the breast and cervix. (For other androgen-like agents, see also Progestogens and Progestins.)

Estrogens. Estrogenic agents hasten sexual maturation in the female. Therefore, they are used in underdevelopment in the female. The widest use of estrogens is in the treatment of the menopause, in which they supplement from without, the secretion of natural estrogens by the ovary, which begins to decline at about the 40th year. The menopause is usually a slow process, and the declining secretion gives rise to various symptoms during the time that the secretion declines, until adjustment to the new status takes place. The menopause, a period of physical and psychological stress, is made less precipitous by estrogens.

Frequently, a menopause must be quickly induced, as in cancer of the ovary or in uterine hemorrhage. This is done by radiation or by the removal of the uterus. Severe vasomotor symptoms occur when the menopause is thus suddenly induced. Estrogens—among other drugs—are used in the amelioration of these symptoms.

Estrogens (especially diethylstilbestrol which though not a hormone has an estrogenic effect) are also used in the control of cancer of the prostate in the male. Note the inverse correspondence to the use of male sex hormones in cancer of the breast in the female.

Progestogens and Progestins (Including 19-Norsteroid Compounds). The agents under that listing include progesterone, a female sex hormone, as well as progestins, i.e., synthetic progesterone-like compounds which have no natural counterpart in the body. Their use includes a variety of conditions: functional uterine bleeding, absence of menstruation (amenorrhea) used at times with estrogens, painful menstruation (dysmenorrhea), infertility, habitual abortion in order to maintain pregnancy, and in fact, to suppress ovulation hence their use as antifertility drugs. Certain progestins—as norethindrone combined with an estrogen, are the principal components of birth control pills—suppressing ovulation, there is no egg to fertilize, hence conception does not take place.

Adrenal Corticosteroids, Including Antiinflammatory, Antiallergic and Antirheumatic Agents. The adrenal cortex secretes a large number of hormones. They usually differ from each other in the accentuation of some phases of their properties. Virtually all of the cortical hormones are catabolic, thus having an effect in this respect, diametrically opposed to the androgens which are anabolic. Nearly all the cortical hormones—differing in degree from each other—cause retention of sodium and water by the body and hasten the excretion of potassium. These effects are utilized in the treatment of adrenal insufficiency or Addison's disease, in which conversely, there is an undue excretion of sodium and a strong retention of potassium. Desoxycorticosterone is used in Addison's disease because it has a particularly strong sodium retaining and potassium excreting effect.

Since the finding in 1949 of the usefulness of cortisone in profoundly reducing the symptoms of rheumatoid arthritis, the adrenal corticosteroids, including hydrocortisone, a natural hormone secreted by the adrenal cortex, and particularly the synthetic analogues not found in the body, as prednisone, have been used in the treatment of a wide variety of inflammatory diseases—especially diseases of collagen tissue. The same antiinflammatory effect is also brought into use in the reduction of inflammations associated with diseases of the skin, allergy, asthma, and in such systematic diseases as disseminated lupus erythematosus, also a collagen disease.

The drawbacks of cortisone, also shared in lesser measure by hydrocortisone, gave the impetus to the synthesis of steroidal substances not native to the body but differing somewhat from cortisone and hydrocortisone, in order to reduce the drawbacks attendant to the use of the latter. The sideeffects—especially those of cortisone—are retention of water and sodium, excretion of potassium, loss of mineral from bone leading to osteoporosis and fractures, hypertension, at times diabetes, personality changes or gastric ulcer. Prednisone and prednisolone among others (see tables) are two such steroidal synthetics which have the effects of cortisone, but fewer or less severe sideeffects. Whereas the synthetic steroidal substances are superior to cortisone with respect to lessened sideeffects, it cannot be said that the sideeffects are absent—they vary in degree from substance to substance.

Diuretic, Antidiuretic and Local Anesthetic Agents. Aldosterone, a natural hormone of the adrenal cortex promotes retention in the body of sodium and water, and facilitates excretion of potassium. Hence its effect is almost diametrically opposed to diuretics—especially the thiazide diuretics. Aldosterone is much more active in this respect than desoxycorticosterone, and is used in the treatment of Addison's disease, a hypofunction of the adrenal glands.

Spironolactone is an antagonist to aldosterone—the latter when elaborated in the body in excessive amounts gives rise to a syndrome called aldosteronism. Spironolactone, a synthetically produced steroid does not have a natural counterpart in the body, is diuretic when mercurial or thiazide diuretics are ineffective; it prevents sodium retention and potassium excretion—effects opposite to aldosterone. Hence spironolactone is used in aldosteronism, against edema, in the treatment of congestive heart failure and in other conditions in which an accumulation of water, and water-retaining salt, is to be corrected.

Doses. The amount of substance which comprises a dose of steroid hormones, or of the steroidal synthetics varies from substance to substance—from 0.1 mg for an estradiol ester, to 50 mg for a 19-norsteroid compound. The dose is conditioned upon the order of activity of the substance, the purpose for which it is administered, as well as the patient's response. However, as additional steroids for hormonal use are synthesized—especially those with adrenocortical activity, their average dose is usually smaller than the previously available steroid. The smaller effective dose of the more recent steroid is cited as an advantage over the previously available steroid.

However, a smaller dose cannot be claimed as an inherent advantage of a new steroid in comparison with an existing one, unless the lower dosage exhibits either greater or more prolonged activity or lesser sideeffects. One cannot meaningfully compare a dose, milligram for milligram, without taking into consideration if a heightened effect of the smaller dose produces fewer sideeffects. For example, it does not make any difference if a given effect and the same accompanying sideeffects are produced by a 50 mg or a 5 mg dose.

ADRENAL CORTICOSTEROIDS, INCLUDING ANTIINFLAMMATORY, ANTIALLERGIC AND ANTIRHEUMATIC AGENTS

Names & synonyms:	BETAMETHASONE; 9α-fluoro-16β-methylprednisolone; 16β-methyl-11β,17α,21-trihydroxy-9α-fluoro-1,4-pregnadiene-3,20-dione.	BETAMETHASONE ACETATE; 9α-fluoro-16β-methylprednisolone-21-acetate.	BETAMETHASONE DISODIUM PHOSPHATE; 9α-fluoro-16β-methylprednisolone-21-disodium phosphate.
Formulae:	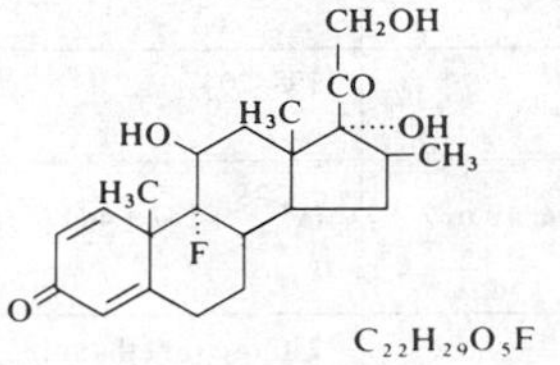 $C_{22}H_{29}O_5F$	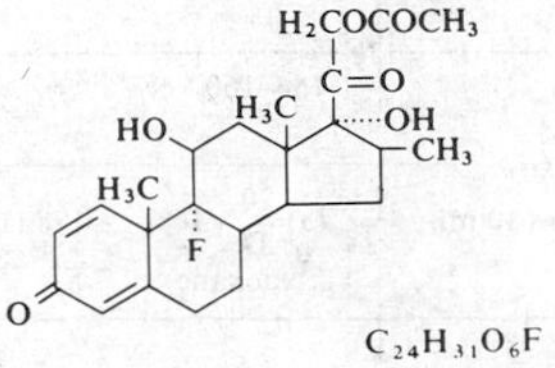 $C_{24}H_{31}O_6F$	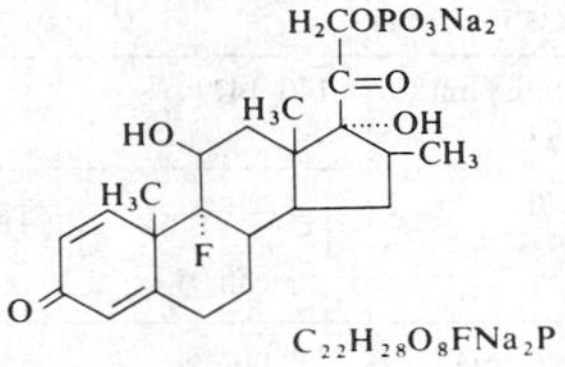 $C_{22}H_{28}O_8FNa_2P$
Molecular weight	392.5	434.5	516.4
Melting point (°C)	240 (dec.)	200 to 220 (dec.)	decomposes
Specific rotation	$(\alpha)\frac{25}{D}$ +112 to +120 (100 mg. in 10 ml. dioxane)	$(\alpha)\frac{25}{D}$ +120 to +128 (100 mg. in 10 ml. dioxane)	$(\alpha)\frac{25}{D}$ +99 to +105 (100 mg. in 10 ml. water)
Absorption max.	239 mμ, E(1 %, 1 cm) 390, methanol	239 mμ, methanol	241 mμ, water

STEROID HORMONES AND OTHER STEROIDAL SYNTHETICS (Continued)

Names & synonyms:	CHLOROPREDNISONE ACETATE; 6α-chloroprednisone acetate; 6α-chloro-$\Delta^{1,4}$-pregnadien-17β,21-diol-3,11,20-trione 21-acetate.	CORTICOSTERONE; 11,21-dihydroxyprogesterone; Δ^4-pregnene-11β,21-diol-3,20-dione; 11β,21-dihydroxy-4-pregnene-3,20-dione; Kendall compound B; Reichstein substance H.	CORTISONE; 17-hydroxy-11-dehydrocorticosterone; 17α,21-dihydroxy-4-pregnene-3,11,20-trione; Δ^4-pregnene-17α,21-diol-3,11,20-trione; Kendall compound E; Wintersteiner compound F.
Formulae:	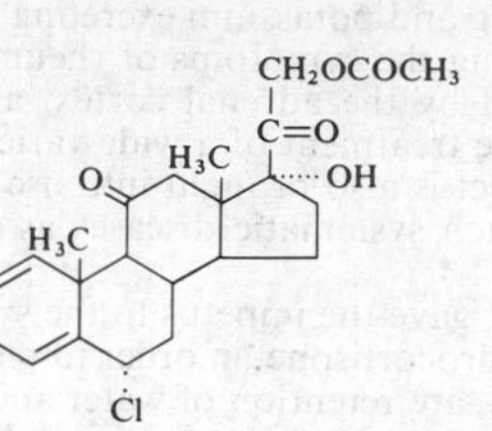	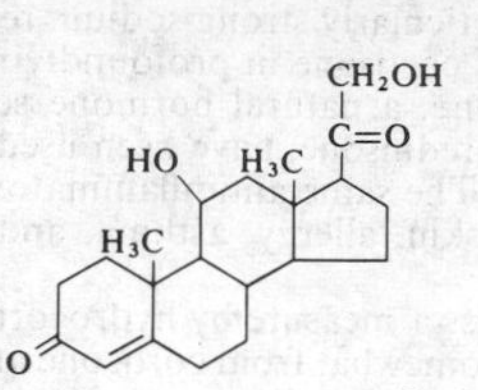 $C_{21}H_{30}O_4$	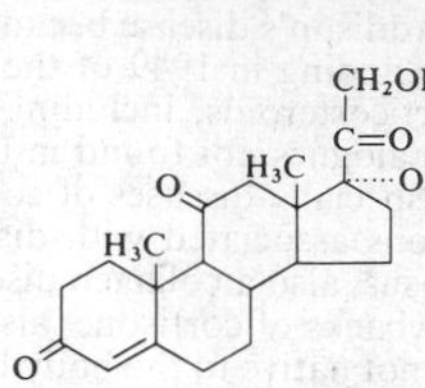 $C_{21}H_{28}O_5$
Molecular weight	436.6	346.40	360.4
Melting point (°C)	207–213	180–182	220–224
Specific rotation	$(\alpha)^{25}_{D}$ +137 to +142 (100 mg. in 10 ml. chloroform)	$(\alpha)^{15}_{D}$ +222 (110 mg. in 10 ml. alcohol)	$(\alpha)^{25}_{D}$ +209 (120 mg. in 10 ml. alcohol)
Absorption max.		240 mμ	237 mμ

Names & synonyms:	DESOXYCORTICOSTERONE; deoxycorticosterone; 11-desoxycorticosterone; 21-hydroxyprogesterone; 4-pregnen-21-ol-3,20-dione; Kendall desoxy compound B; Reichstein substance Q.	DESOXYCORTICOSTERONE ACETATE; DCA; 11-desoxycorticosterone acetate.	DESOXYCORTICOSTERONE PIVALATE; desoxycorticosterone trimethylacetate; 21-hydroxy-4-pregnene-3,20-dione pivalate.
Formulae:	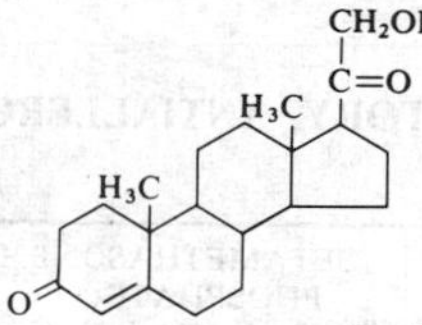 $C_{21}H_{30}O_3$	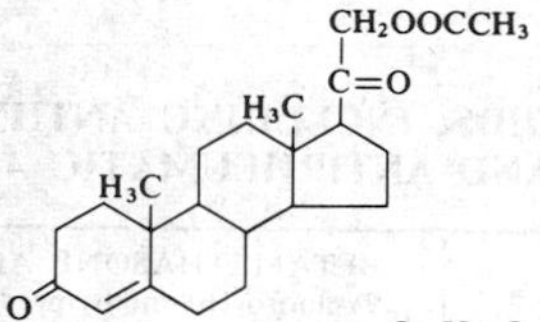 $C_{23}H_{32}O_4$	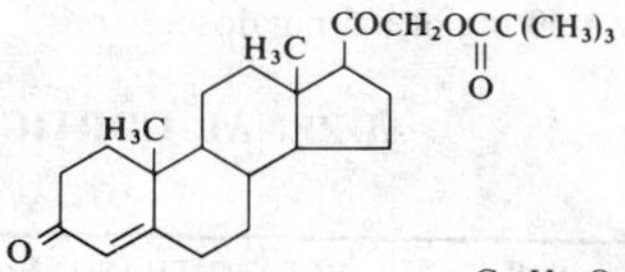 $C_{26}H_{38}O_4$
Molecular weight	330.2	372.4	414.6
Melting point (°C)	140–142	154–160	198–204
Specific rotation	$(\alpha)^{22}_{D}$ +176 – +178 (100 mg. in 10 ml. alcohol)	$(\alpha)^{20}_{D}$ +168 – +178 (100 mg. in 10 ml. dioxane)	$(\alpha)^{25}_{D}$ +157 ±4 (1 % in dioxane)
Absorption max.	240 mμ		240 mμ (in ethanol)

Names & synonyms:	DEXAMETHASONE; hexadecadrol; 9α-fluoro-16α-methyl prednisolone; 9α-fluoro-11β,17α-21-trihydroxy-16α-methyl-1,4-pregnadiene-3,20-dione; 16α-methyl-9α-fluoro-1,4-pregnadiene-11β,17α-21-triol-3,20-dione; 16α-methyl-9α-fluoro-Δ[1]-hydrocortisone; 1-dehydro-16α-methyl-9α-fluorohydrocortisone.	DICHLORISONE ACETATE; 9α-11β-dichloro-1,4-pregnadiene-17α,21-diol-3,20-dione-21-acetate	FLUOCINOLONE ACETONIDE; 6α,9α-difluoro-16α hydroxyprednisolone-16,17-acetonide; 6α,9α-difluoro-16α,17α-isopropylidenediosy-1,4-pregnadiene-3,20-dione.
Formulae:	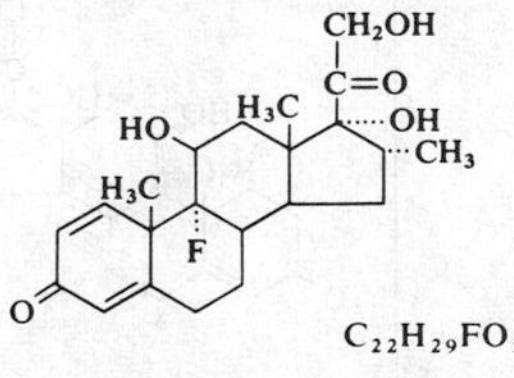 $C_{22}H_{29}FO_5$	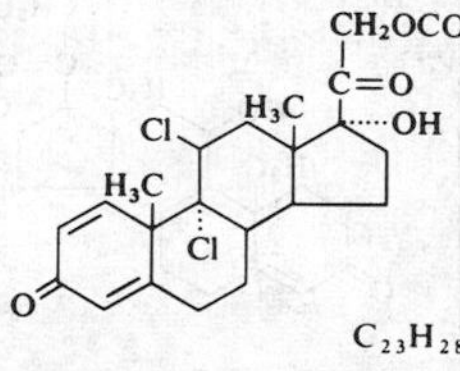 $C_{23}H_{28}O_5Cl$	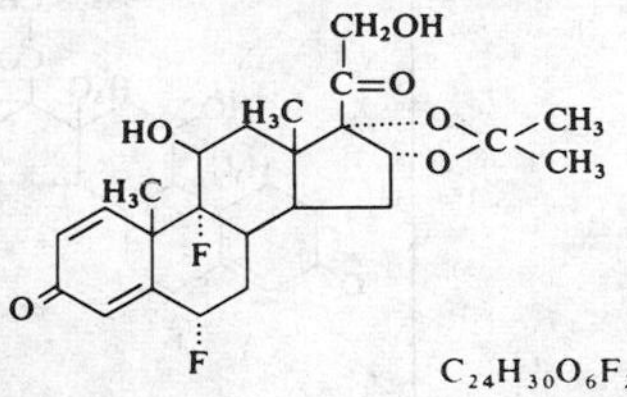 $C_{24}H_{30}O_6F_2$
Molecular weight	392.4	455.3	452.50
Melting point (°C)	262–264	235 (dec.)	255–266
Specific rotation	$(\alpha)\frac{25}{D}$ +78 (100 mg. in 10 ml. dioxane)	$(\alpha)\frac{25}{D}$ +160 – 168 (100 mg. in 10 ml. dioxane)	not less than +95° and not more than +105°C at 25°C.
Absorption max.		237 mμ – 316 – 337 (ε_1^1)	237 mμ ± 1 mμ

Names & synonyms:	FLUOROHYDROCORTISONE; fludrocortisone; 9α-fluorohydrocortisone; 9α-fluorocortisol; fluohydrisone; 9α-fluoro-11β,17α,21-trihydroxy-4-pregnene-3,20-dione; 9α-fluoro-17-hydroxycorticosterone.	FLUOROMETHOLONE; 9α-fluoro-11β,17α-dihydroxy-6α-methyl-1,4-pregnadiene-3,20-dione; 21-desoxy-9α-fluoro-6α-methyl-prednisolone.	FLUPREDNISOLONE; 6α-fluoroprednisolone; 6α-fluoro-1-dehydrohydrocortisone; 6α-fluoro-11β,17α,21-trihydroxy-1,4-pregnadiene-3,20-dione.
Formulae:	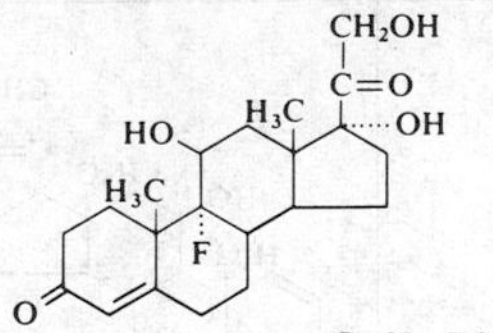 $C_{21}H_{29}FO_5$	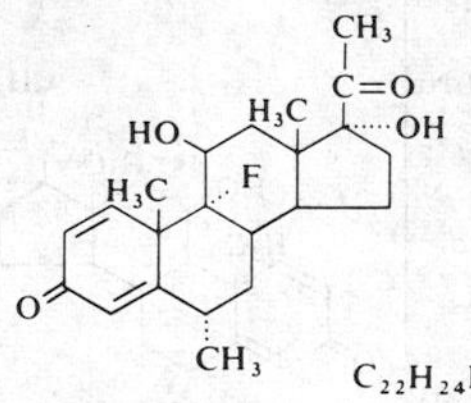 $C_{22}H_{24}FO_4$	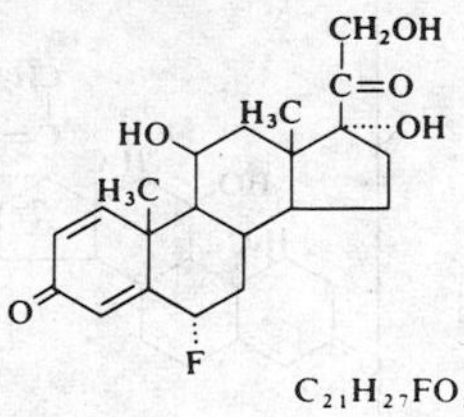 $C_{21}H_{27}FO_5$
Molecular weight	380.4	376.4	378.4
Melting point (°C)	260–262 (dec.)	290 (dec.)	205–210
Specific rotation	$(\alpha)\frac{23}{D}$ +139 (55 mg. in 10 ml. alcohol)	$(\alpha)\frac{25}{D}$ +56 (pyridine)	$(\alpha)_D$ +88 (dioxane)
Absorption max.		239 mu (a_M = 15,050) methanol	λ_{max} 241.5 mu (ε 16,000)

Names & synonyms:	FLURANDRENOLONE; 6-fluoro-16α-hydroxyhydrocortisone-16,17-acetonide; 6α-fluoro-11β,21-dehydroxy-16α,17α-isopropylidenedioxy-pregna-4-ene-3, 20-dione.	HYDROCORTISONE; cortisol; 17-hydroxycorticosterone; hydrocortisone free alcohol; 11β,17α,21-trihydroxy-4-pregnene-3,20-dione; 4-pregnene-11β,17α,21-triol-3,20-dione; Kendall compound F; Reichstein substance M.	HYDROCORTISONE ACETATE; cortisol acetate; hydrocortisone-21-acetate; 17-hydroxycorticosterone-21-acetate.
Formulae:	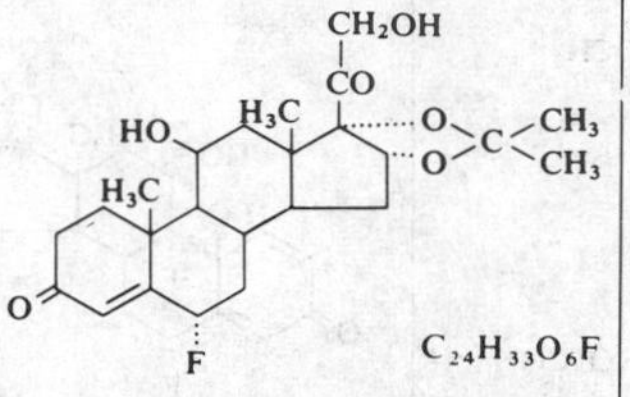 $C_{24}H_{33}O_6F$	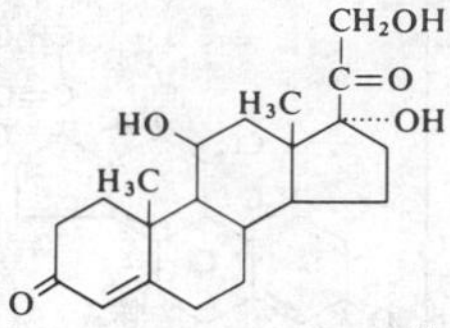 $C_{21}H_{30}O_5$	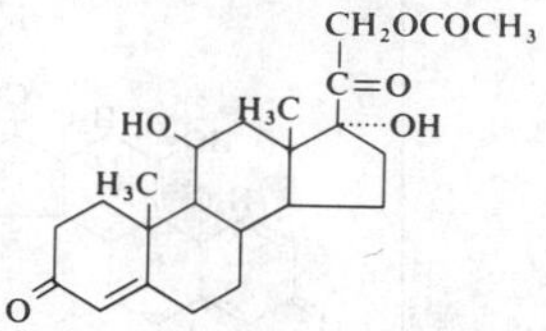 $C_{23}H_{32}O_6$
Molecular weight	436.5	362.5	404.5
Melting point (°C)	240–250	215–220 (dec.)	223 (dec.)
Specific rotation	$(\alpha)\frac{25}{D} = +145$ (1 % in $CHCl_3$)	$(\alpha)\frac{25}{D}$ +150 – +156 (100 mg. in 10 ml. dioxane)	$(\alpha)\frac{25}{D}$ +158 – +165 (100 mg. in 10 ml. dioxane)
Absorption max.	236 mμ (methanol)	242 mμ	242 mμ (methanol)

Names & synonyms:	HYDROCORTISONE SODIUM SUCCINATE; 11β, 17α, 21-trihydroxy-4-pregnene-3, 20-dione, 21 hydrogen succinate, sodium salt; hydrocortisone, 21 hydrogen succinate, sodium salt.	METHYLPREDNISOLONE; 6α-methylprednisolone; Δ^1-6α-methylhydrocortisone; 1-dehydro-6α-methylhydrocortisone; 11β, 17α,21-trihydroxy-6α-methyl-1,4-pregnadiene-3,20-dione.	METHYLPREDNISOLONE SODIUM SUCCINATE; 1-dehydro-6α-methylhydrocortisone, 21-hydrogen succinate, sodium salt; 6α-methylprednisolone 21-hydrogen succinate, sodium salt; 11β, 17α, 21-trihydroxy-6α-methyl-1, 4-pregnadiene-3, 20-dione, 21-hydrogen succinate, sodium salt.
Formulae:	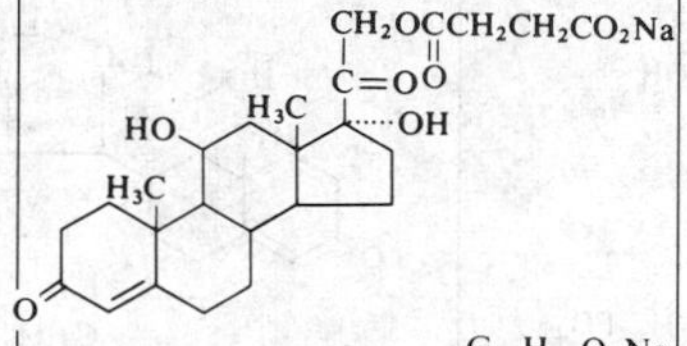 $C_{25}H_{33}O_8Na$	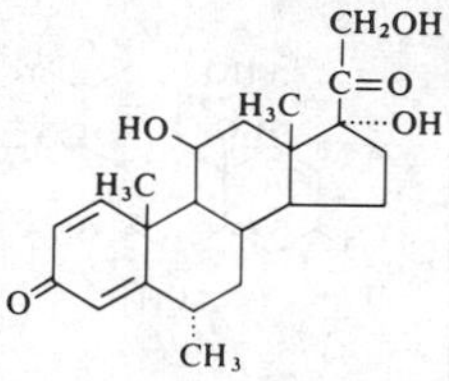 $C_{22}H_{30}O_5$	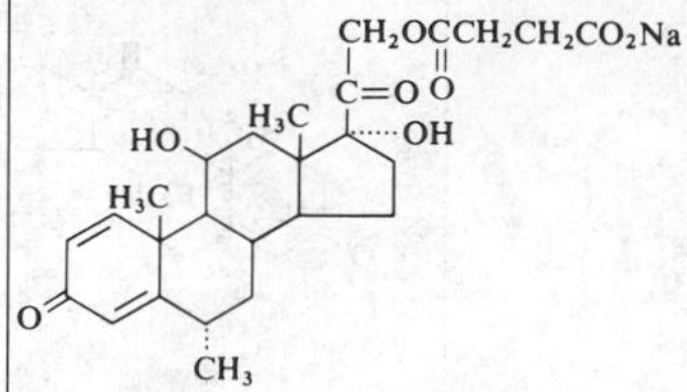 $C_{26}H_{33}O_8Na$
Molecular weight	484.5	374.5	496.5
Melting point (°C)	decomposes	230–240 (dec.)	decomposes
Specific rotation	$(\alpha)_D + 140 \pm 5$ (alcohol)	$(\alpha)\frac{25}{D}$ +85 (dioxane)	$(\alpha)_D + 100 \pm 4$ (alcohol)
Absorption max.	λ 242 mμ (ε 15,700)	243 mμ	λ_{max} 242 mμ (ε 14,500)

Names & synonyms:	PARAMETHASONE; 6α-fluoro-16α-methylprednisolone; 6α-fluoro-11β-17α,21-trihydroxy-16α-methyl-1,4-pregnadiene-3,20-dione.	PARAMETHASONE ACETATE; 6α-fluoro-16α-methylprednisolone-21-acetate; 6α-fluoro-16α-methylpregna-1,4-diene-11β,21-diol-3,20-dione-21-acetate; 6α-fluoro-17β,17α,21-trihydroxy-16α-methyl-1,4-pregnadiene-3,20-dione-21-acetate.	PREDNISOLONE; metacortandralone; Δ^1-dehydrocortisol; delta F; Δ^1-hydrocortisone; Δ^1-dehydrohydrocortisone; 1,4-pregnadiene-3,20-dione-11β,17α,21-triol; 11β,17α,21-trihydroxy-1,4-pregnadiene-3,20-dione.
Formulae:	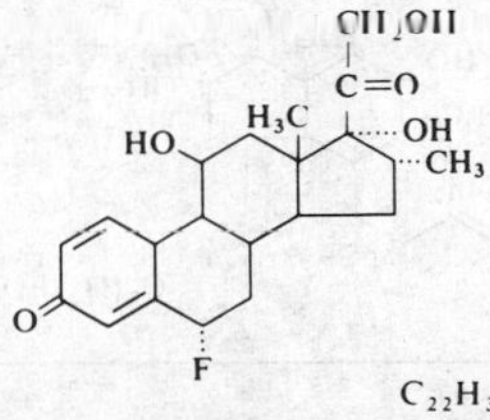 $C_{22}H_{30}O_5$	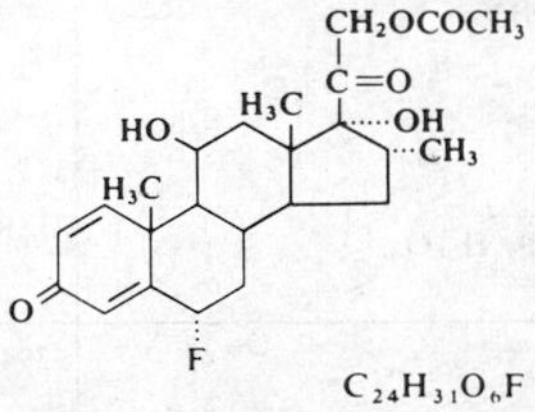 $C_{24}H_{31}O_6F$	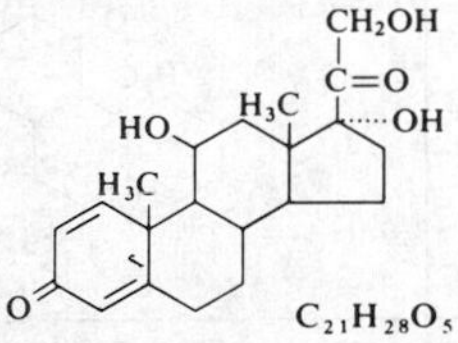 $C_{21}H_{28}O_5$
Molecular weight	392.45	434.5	360.4
Melting point (°C)	228–241	233–246	240 (dec.)
Specific rotation	+59 to +69 at 25°C	$(\alpha)\frac{25}{D}$ +72 (1 % in $CHCl_3$)	$(\alpha)\frac{25}{D}$ +97 – +103 (100 mg. in 10 ml. dioxane)
Absorption max.	242 mμ	242 mμ (methanol)	242 mμ (ε = 15,000) methanol

Names & synonyms:	PREDNISOLONE PHOSPHATE SODIUM; disodium prednisolone 21-phosphate.	PREDNISOLONE PIVALATE; prednisolone trimethylacetate; 11β,17α,21-trihydroxy-1,4-pregnadiene-3,20-dione 21-pivalate.
Formulae:	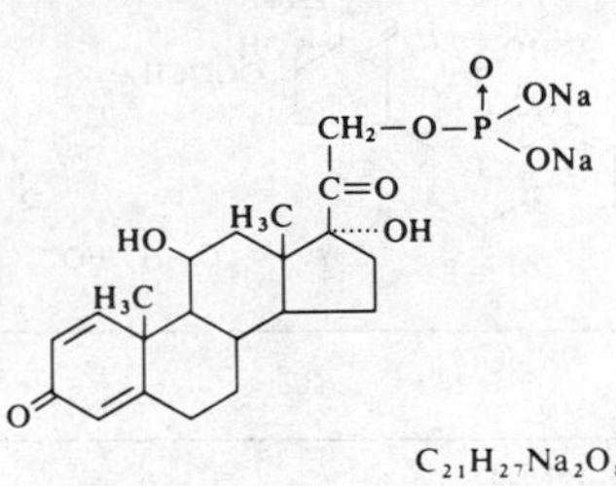 $C_{21}H_{27}Na_2O_8P$	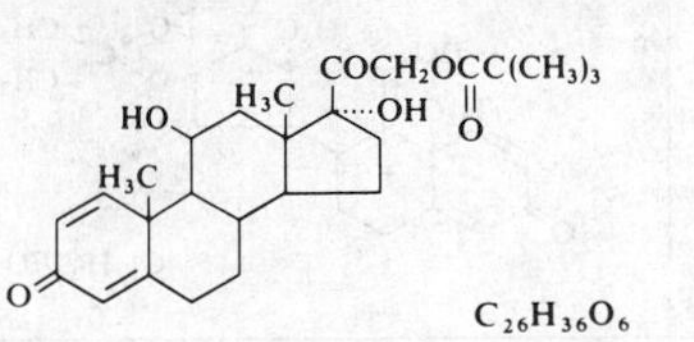 $C_{26}H_{36}O_6$
Molecular weight	484.4	444.6
Melting point (°C)		229
Specific rotation	$(\alpha)\frac{25}{D}$ +102.5 (100 mg. in 10 ml. H_2O)	+108 ±4 (1 % in dioxane)
Absorption max.	243 mμ	240 and 263 mμ (in absolute ethanol)

Names & synonyms:	PREDNISONE; metacortandricin; Δ^1-dehydrocortisone; delta E; Δ^1-cortisone; 1,4-pregnadiene-17α,21-diol-3,11,20-trione; 17α,21-dihydroxy-1,4-pregnadiene-3,11,20-trione.	TRIAMCINOLONE; 9α-fluoro-16α-hydroxyprednisolone; 9α-fluoro-11β,16α,17α,21-tetrahydroxy-1,4-pregnadiene-3,20-dione.
Formulae:	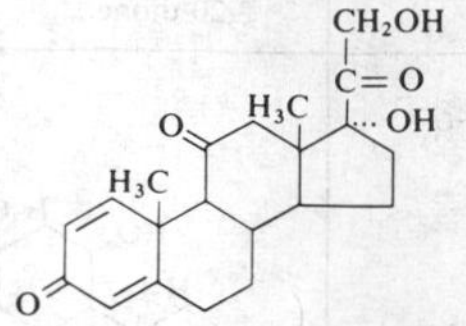 $C_{21}H_{26}O_5$	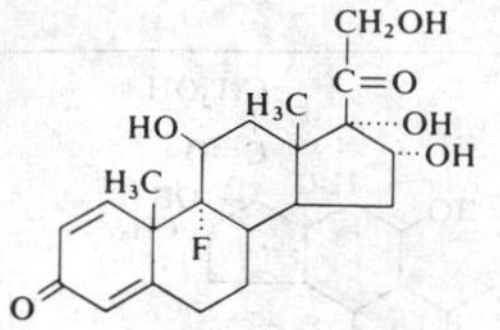 $C_{21}H_{27}FO_6$
Molecular weight	358.4	394.4
Melting point (°C)	225 (dec.)	260–262.5 (dec.)
Specific rotation	$(\alpha)^{25}_{D}$ +167 – +175 (100 mg. in 10 ml. dioxane)	$(\alpha)^{25}_{D}$ +75 (200 mg. in 100 ml. acetone)
Absorption max.	239 mμ (ε = 15,500) methanol	238 mμ (ε = 15,800)

Names & synonyms:	TRIAMCINOLONE ACETONIDE; 9α-fluoro-11β,21-dihydroxy-16α,17α-isopropylidene-dioxy-1,4-pregnadiene-3,20-dione; 9α-fluoro-16α-hydroxyprednisolone 16,17-acetonide.	TRIAMCINOLONE DIACETATE; 16α,21-diacetoxy-9α-fluoro-11β,17α-dihydroxy-1,4-pregnadiene-3,20-dione; 9α-fluoro-16α-hydroxyprednisolone 16,21-diacetate.
Formulae:	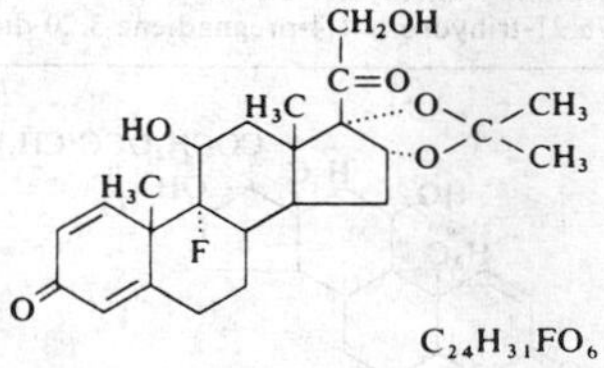 $C_{24}H_{31}FO_6$	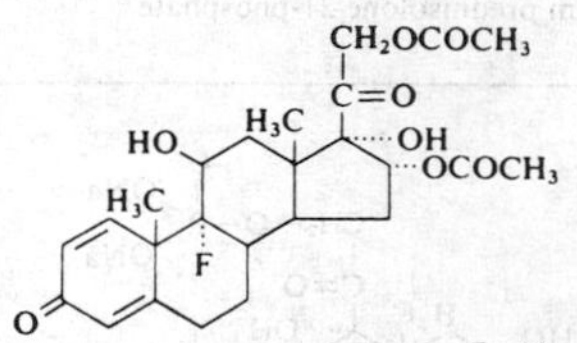 $C_{25}H_{31}FO_8$
Molecular weight	434.4	478.49
Melting point (°C)	274–278 (dec.); 292–294	variable: 158–235
Specific rotation	$(\alpha)^{25}_{D}$ +109 – +112 (53.7 mg. in 10 ml. chloroform)	$(\alpha)^{25}_{D}$ +22 (78.8 mg. in 10 ml. chloroform)
Absorption max.	238–239 mμ (ε = 14,600)	239 mμ (ε = 15,200)

ANDROGENS AND ANABOLIC AGENTS

Names & synonyms:	ANDROSTERONE; cis-androsterone; 3α-hydroxy-17-androstanone; androstane-3α-ol-17-one.	FLUOXYMESTERONE; 9α-fluoro-11β-hydroxy-17α-methyltestosterone 9α-fluoro-11β,17β-dihydroxy-17α-methyl-4-androsten-3-one.	ALDOSTERONE; electrocortin; 18-oxocorticosterone; 18-formyl-11β,21-dihydroxy-4-pregnene-3,20-dione.
Formulae:	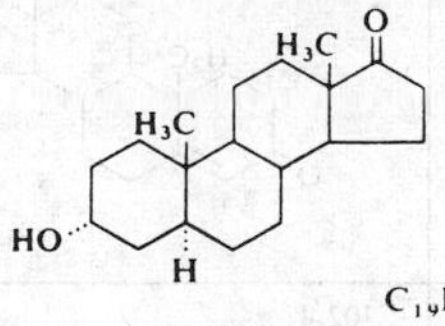 $C_{19}H_{30}O_2$	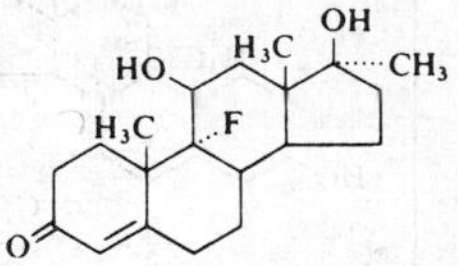 $C_{20}H_{29}FO_3$	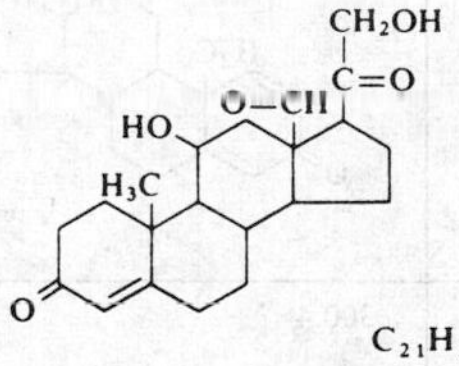 $C_{21}H_{28}O_5$
Molecular weight	290.4	336.4	360.4
Melting point (°C)	185–185.5	270 (dec.)	108–112 (hydrate); 164 (anhydrous)
Specific rotation	$(\alpha)\frac{15}{D}$ +85 – +90 (150 mg. in 10 ml. dioxane)	$(\alpha)\frac{25}{D}$ +107 – +109 (alcohol)	$(\alpha)\frac{25}{D}$ +161 (10 mg. in 10 ml. chloroform)
Absorption max.		240 mμ (ε = 16,700) alcohol	240 mμ (log ε = 4.20 monohydr.; ε mol. 15,000 anhydr.)

Names & synonyms:	HYDROXYDIONE SODIUM; 21-hydroxypregnane-3,20-dione-21-sodium hemisuccinate.	SPIRONOLACTONE; 3-(3-oxo-7α-acetylthio-17β-hydroxy-4-androsten-17α-yl)-propionic acid γ lactone.
Formulae:	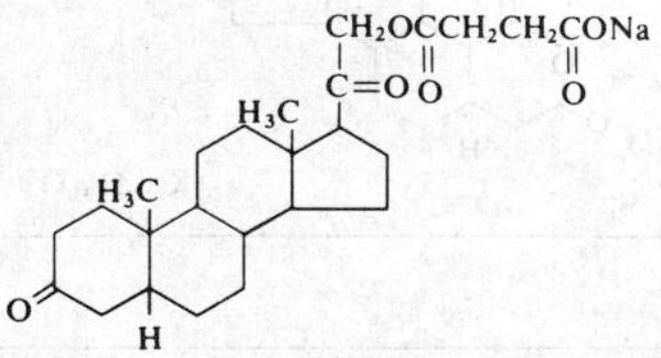 $C_{25}H_{35}O_6Na$	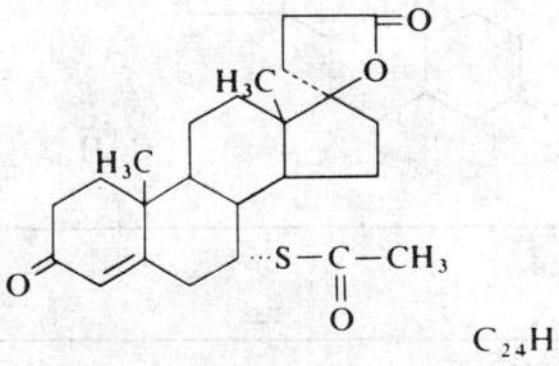 $C_{24}H_{32}O_4S$
Molecular weight	454.5	416.5
Melting point (°C)	193–203 (dec.)	135 (preliminary)–202 (dec.)
Specific rotation	$(\alpha)\frac{25}{D}$ +95 (chloroform) for free acid.	$(\alpha)\frac{25}{D}$ – 34 (chloroform)
Absorption max.	280 mμ (ε = 93.2)	ε^{238} = 20,200

Names & synonyms:	METHANDROSTENOLONE; 17α-methyl-17β-hydroxy-1,4-androstadien-3-one.	METHYLANDROSTENEDIOL; MAD; methandriol; 17α-methyl-5-androsten-3β,17β-diol.	METHYL TESTOSTERONE; 17-methyl testosterone; 17α-methyl-Δ^4-androsten-17-β-ol-3-one; 17(β)-hydroxy-17(α-methyl-4-androsten-3-one.
Formulae:	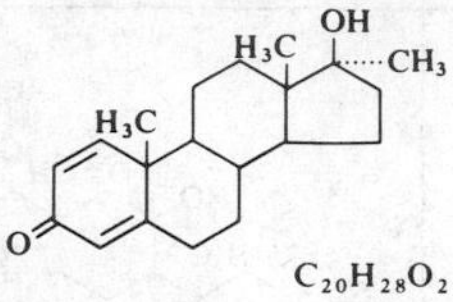 $C_{20}H_{28}O_2$	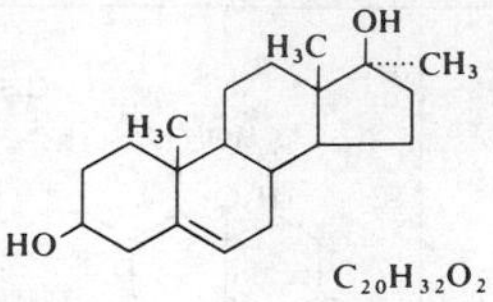 $C_{20}H_{32}O_2$	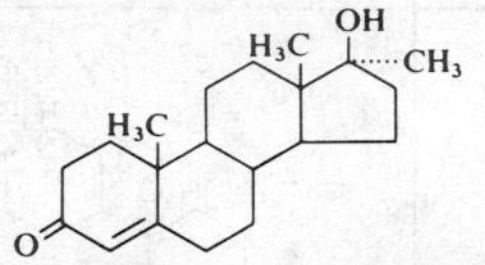 $C_{20}H_{30}O_2$
Molecular weight	300.4	304.4	302.4
Melting point (°C)	166–167	205–207	161–166
Specific rotation	$(\alpha)^{20}_{D}$ +9 – +17 (100 mg. in 10 ml. alcohol)	$(\alpha)^{20}_{D}$ – 73 (100 mg. in 10 ml. alcohol)	$(\alpha)^{25}_{D}$ +69 – +75 (100 mg. in 10 ml. dioxane)
Absorption max.			

Names & synonyms:	NORETHANDROLONE; 17α-ethyl-19-nortestosterone; 17α-ethyl-17-hydroxy-4-norandrosten-3-one; 17α-ethyl-17-hydroxy-19-norandrost-4-en-3-one.	OXANDROLONE; 17β-hydroxy-17α-methyl-2-oxa-5α-androstane-3-one.
Formulae:	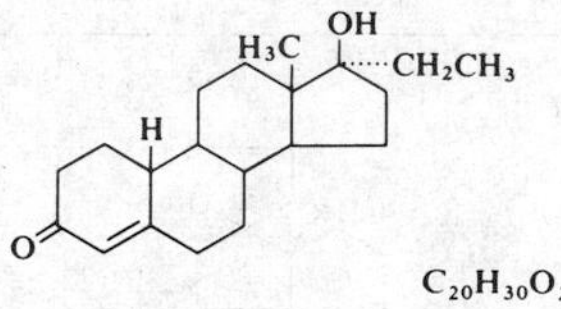 $C_{20}H_{30}O_2$	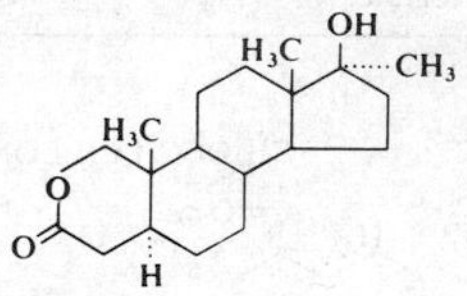 $C_{19}H_{30}O_3$
Molecular weight	302.4	306.4
Melting point (°C)	130–136	230–233
Specific rotation	$(\alpha)^{25}_{D}$ +21 (dioxane)	$(\alpha)^{25}_{D}$ – 21 (1 % in chloroform)
Absorption max.	240 mμ (ε = 16,500)	None

Names & synonyms:	OXYMETHOLONE: 17β-hydroxy-2-hydroxymethylene-17α-methyl-3-androstanone; 2-hydroxymethylene-17-α-methyl dihydrotestosterone.	PROMETHOLONE: 2α-methyl-dihydro-testosterone propionate; 2α-methyl-5α-androstane-17β-ol-3-one-propionate.
Formulae:	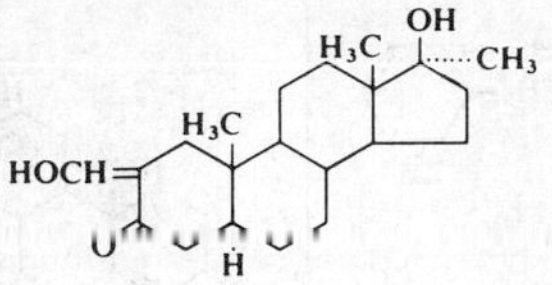 $C_{21}H_{32}O_3$	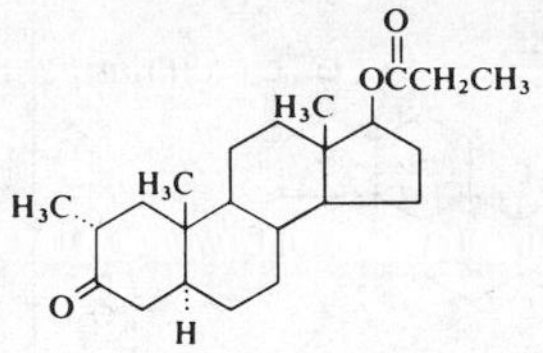
Molecular weight	332.4	360.5
Melting point (°C)	182	124–130
Specific rotation	$(\alpha)\frac{25}{D} = +36$ (200 mg. in 10 ml. dioxane)	$(\alpha)\frac{25}{D} +22 - +29$ (200 mg. in 10 ml. chloroform)
Absorption max.	$E^1_1 = 547$ at 315 mμ (in alkaline methanol made 0.01 N with NaOH)	without significant absorption from 220–300 mμ (methanol)

Names & synonyms:	TESTOSTERONE: trans-testosterone; Δ^4-androsten-17-β-ol-3-one; 17β-hydroxy-4-androsten-3-one.	TESTOSTERONE CYPIONATE: testosterone cyclopentylpropionate; 17β-hydroxy-4-androsten-3-one, cyclopentanepropionate.
Formulae:	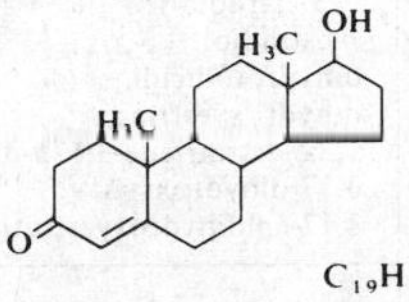 $C_{19}H_{28}O_2$	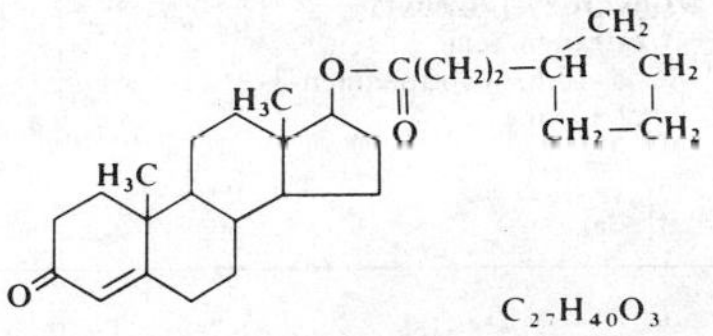 $C_{27}H_{40}O_3$
Molecular weight	288.4	412.6
Melting point (°C)	151–156	100–102
Specific rotation	$(\alpha)\frac{24}{D} +109$ (400 mg. in 10 ml. alcohol)	$(\alpha)_D +88.5 \pm 3.5$ ($CHCl_3$)
Absorption max.	238 mμ	λ_{max} 241 mμ (ε 16,125)

Names & synonyms:	TESTOSTERONE ENANTHATE; testosterone heptanoate; 17β-hydroxyandrost-4-en-3-one-17-enanthate.	TESTOSTERONE PHENYLACETATE; 17β-hydroxy-4-androsten-3-one phenyl-acetate; testosterone α-toluate.	TESTOSTERONE PROPIONATE; Δ^4-androstene-17-β-propionate-3-one.
Formulae:	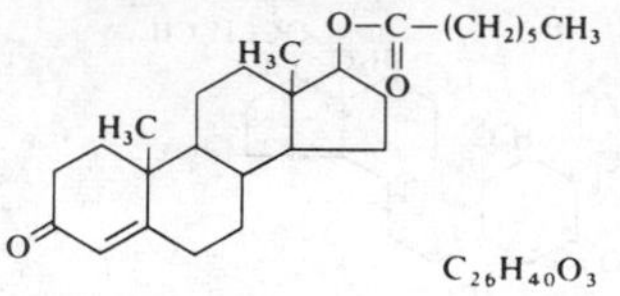 $C_{26}H_{40}O_3$	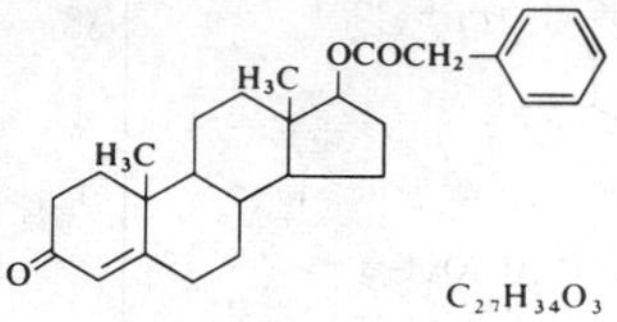 $C_{27}H_{34}O_3$	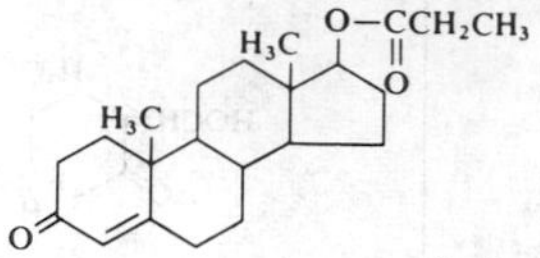 $C_{22}H_{32}O_3$
Molecular weight	400.6	406.5	344.4
Melting point (°C)	34–39	129–131	118–122
Specific rotation	$(\alpha)\frac{25}{D}$ +77 – +82 (2 % in dioxane)	$(\alpha)\frac{25}{D}$ +101 ±3 (1 % in chloroform)	$(\alpha)\frac{25}{D}$ +83 – +90 (100 mg. in 10 ml. dioxane)
Absorption max.	241 mμ (in ethanol)	241 mμ (in ethanol)	

ESTROGENS

Names & synonyms:	EQUILENIN; 3-hydroxy-17-keto-$\Delta^{1,3,5-10,6,8}$ estrapentaene; 1,3,5–10,6,8-estrapentaen-3-ol-17-one.	EQUILIN; 3-hydroxy-17-keto-$\Delta^{1,3,5-10,7}$ estratetraene; 1,3,5,7-estratetraen-3-ol-17-one.	ESTRADIOL (formerly called α-estradiol); β-estradiol; dihydrofolliculin; dihydroxyestrin; 1,3,5-estratriene-3,17β-diol; 3,17-dihydroxy-$\Delta^{1,3,5-10}$-estratriene; 3,17-epidihydroxyestratriene.
Formulae:	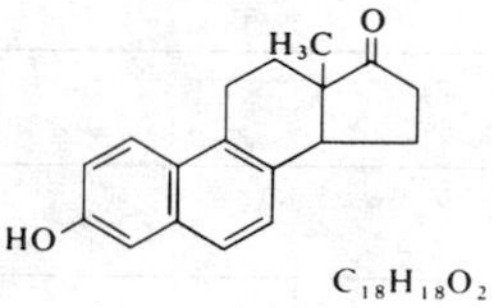 $C_{18}H_{18}O_2$	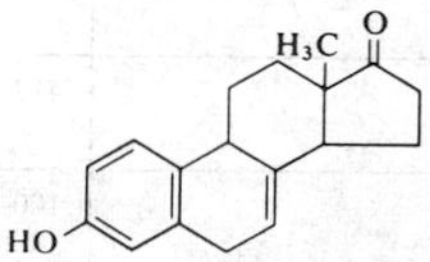 $C_{18}H_{20}O_2$	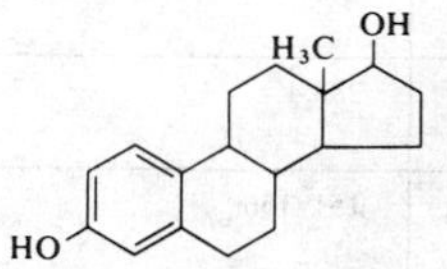 $C_{18}H_{24}O_2$
Molecular weight	266.3	268.3	272.3
Melting point (°C)	258–259	236–240	173–179
Specific rotation	$(\alpha)\frac{25}{D}$ +89 (dioxane)	$(\alpha)\frac{25}{D}$ +308 (200 mg. in 10 ml. dioxane); +325 (200 mg. in 10 ml. alcohol).	$(\alpha)\frac{25}{D}$ +76 – +83 (100 mg. in 10 ml. dioxane)
Absorption max.	231, 270, 282, 292, 325, 340 mμ	283–285 mμ	225, 280 mμ

Names & synonyms:	ESTRADIOL BENZOATE; β-estradiol-3-benzoate; estradiol monobenzoate.	ESTRADIOL CYPIONATE; estradiol cyclopentylpropionate; β-estradiol 17-cyclopentanepropionate; 1,3,5(10)-estratriene-3,17β-diol,17-cyclopentanepropionate.
Formulae:	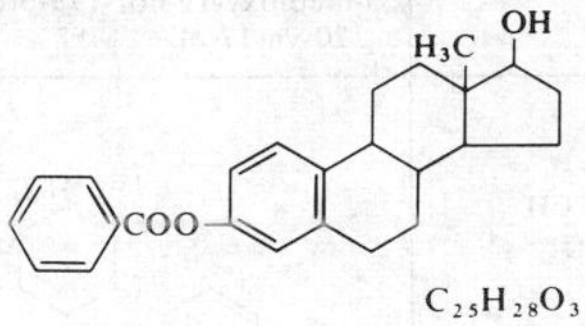 $C_{25}H_{28}O_3$	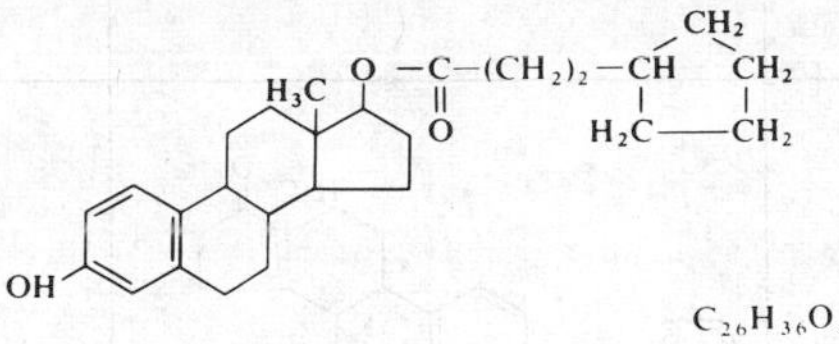 $C_{26}H_{36}O_3$
Molecular weight	376.4	396.6
Melting point (°C)	191–196	151–154
Specific rotation	$(\alpha)\frac{25}{D}$ +58 – +63 (200 mg. in 10 ml. dioxane)	$(\alpha)_D$ +41.5 ±3.5 (dioxane)
Absorption max.		223 mμ

Names & synonyms:	ESTRADIOL DIPROPIONATE; α-estradiol dipropionate; 17β-estradiol dipropionate.	ESTRIOL; trihydroxyestrin; $\Delta^{1,3,5-10}$-estratriene-3-16-cis-17-trans-diol; 1,3,5-estratriene-3,16α,17β-triol.	ESTRONE; folliculin; ketohydroxyestrin; 1,3,5-estratrien-3-ol-17-one.
Formulae:	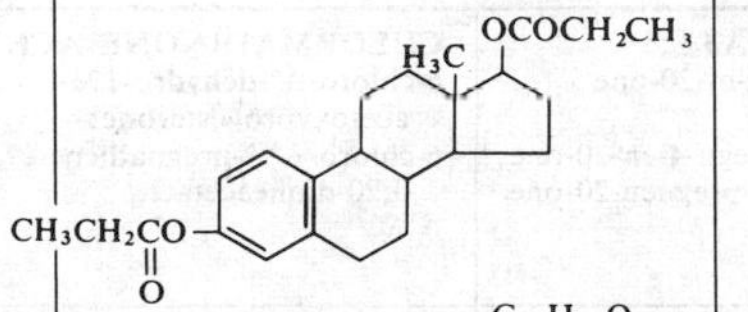 $C_{24}H_{32}O_4$	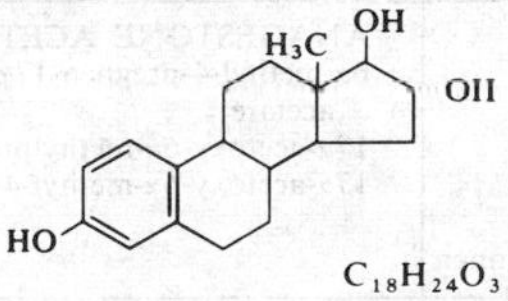 $C_{18}H_{24}O_3$	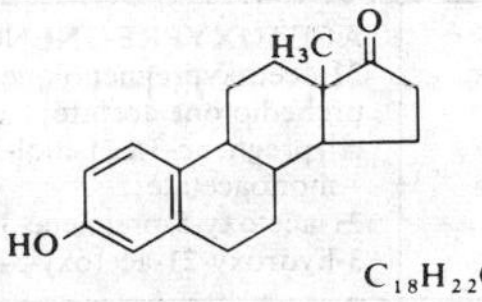 $C_{18}H_{22}O_2$
Molecular weight	384.5	288.3	270.3
Melting point (°C)	104–109	282	258–262
Specific rotation	$(\alpha)\frac{25}{D}$ +39 ±2 (1 % in dioxane)	$(\alpha)\frac{25}{D}$ +53 – +63 (40 mg. in 1 ml. dioxane)	$(\alpha)\frac{25}{D}$ +158 – +168 (100 mg. in 10 ml. dioxane)
Absorption max.	268 mμ	280 mμ	283–285 mμ

STEROID HORMONES AND OTHER STEROIDAL SYNTHETICS (Continued)

Names & synonyms:	ESTRONE BENZOATE	ETHYNYL ESTRADIOL; 17-ethinyl estradiol; 17α-ethynyl-1,3,5-estratriene-3,17β-diol.	MESTRANOL; ethynylestradiol 3-methyl ether; 3-methoxy-17α-ethynyl-1,3,5(10)-estratriene-17β-ol; 17α-ethynyl-estradiol-3-methyl ether; 3-methoxy-19-nor-17α-pregna-1,3,5,trien-20-yn-17-ol.
Formulae:	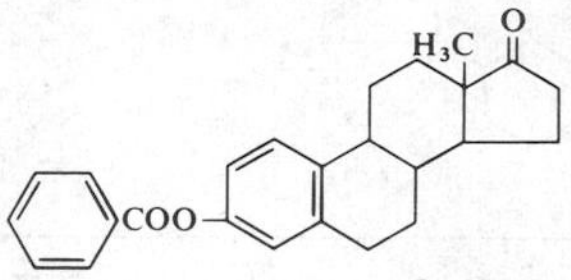 $C_{25}H_{26}O_3$	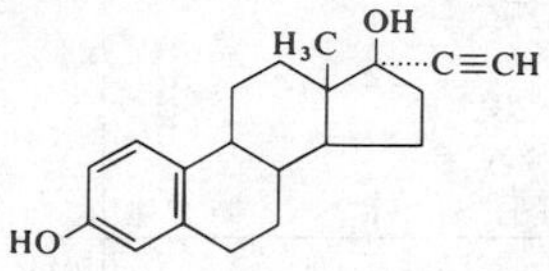 $C_{20}H_{24}O_2$	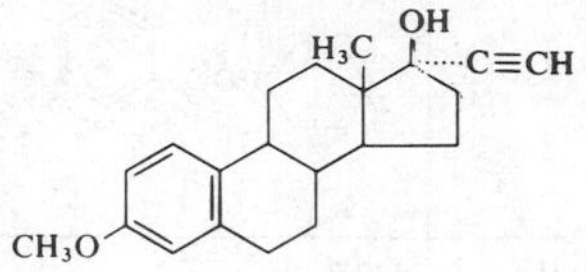 $C_{21}H_{26}O_2$
Molecular weight	374.4	296.4	310.4
Melting point (°C)	220	141–146	148–154
Specific rotation	$(\alpha)^{25}_{D}$ + 120 (dioxane)	$(\alpha)^{25}_{D}$ +1 – +10 (100 mg. in 10 ml. dioxane)	$(\alpha)^{25}_{D}$ +2 to +8 (200 mg. in 10 ml. dioxane)
Absorption max.		248 mμ	278 to 287 mμ (methanol)

PROGESTOGENS AND PROGESTINS (INCLUDING 19-NORSTEROID COMPOUNDS)

Names & synonyms:	ACETOXYPREGNENOLONE; 21-acetoxypregnenolone; prebediolone acetate; Δ^5-pregnene-3β,21-diol-20-one-21-monoacetate; 21-acetoxy-5-pregnene-3-ol-20-one; 3-hydroxy-21-acetoxy-5-pregnen-20-one.	ANAGESTONE ACETATE; 6α-methyl-4-pregnen-17α-ol-20-one acetate; 17α-acetoxy-6α-methylpregn-4-en-20-one; 17α-acetoxy-6α-methyl-4-pregnen-20-one.	CHLORMADINONE ACETATE; 6-chloro-Δ^6-dehydro-17α-acetoxyprogesterone; 6-chloro-$\Delta^{4,6}$-pregnadiene-17α-ol-3,20-dioneacetate.
Formulae:	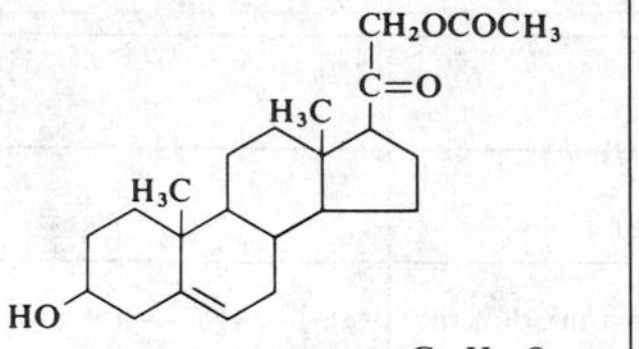 $C_{23}H_{34}O_4$	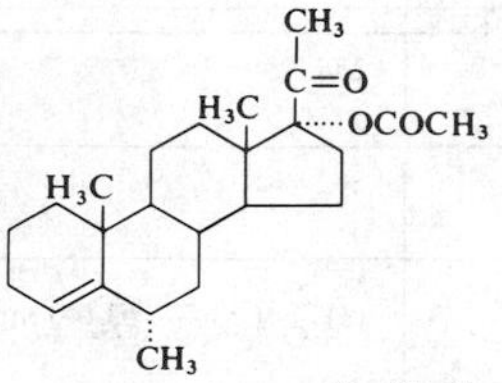 $C_{24}H_{36}O_3$	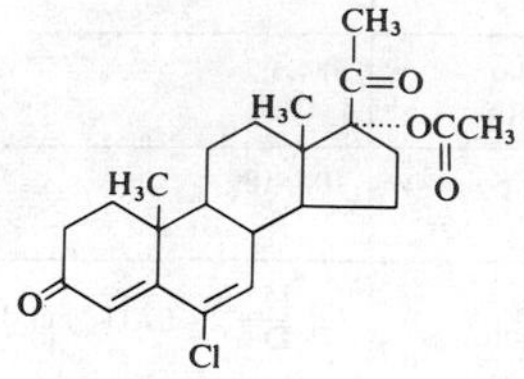 $C_{23}H_{29}ClO_4$
Molecular weight	374.5	372.6	404.9
Melting point (°C)	184–185	172–178	204–212
Specific rotation	$(\alpha)^{20}_{D}$ +37 – +43 (dioxane)	$(\alpha)^{25}_{D}$ +40 to +45 (10 mg. in 10 ml. chloroform)	$(\alpha)^{25}_{D}$ 0 to −6 (200 mg. in 10 ml. chloroform)
Absorption max.			284 mμ (methanol) Log ε = 4.34 ±0.02

Names & synonyms:	DIMETHISTERONE; 6α,21-dimethylethisterone; 6α,21-dimethyl-17β-hydroxy-17α-pregn-4-en-20-yn-3-one; 6α-methyl-17α-propynylandrost-4-en-17β-ol-3-one; 17β-hydroxy-6α-methyl-17α-(prop-1-ynyl)-androst-4-ene-3-one.	ETHISTERONE; anhydrohydroxyprogesterone; ethinyl testosterone; pregneninolone; 17α ethynyl testosterone; 17α-ethynyl-17β-hydroxy-4-androsten-3-one.	ETHYNODIOL DIACETATE; 17α-ethynyl-4-estrene-3β,17β-diol-17-diacetate; 19-nor-17α-pregn-4-en-20-yne-3β,17-diol diacetate.
Formulae:	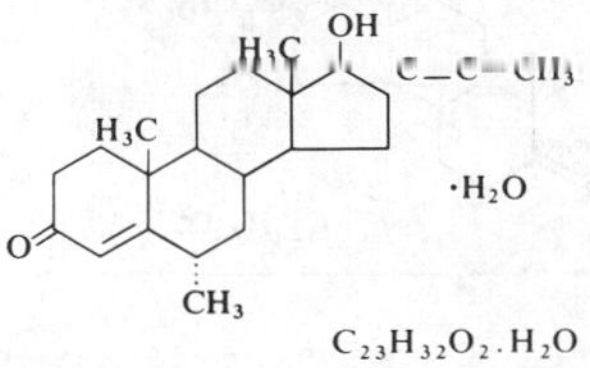 $C_{23}H_{32}O_2.H_2O$	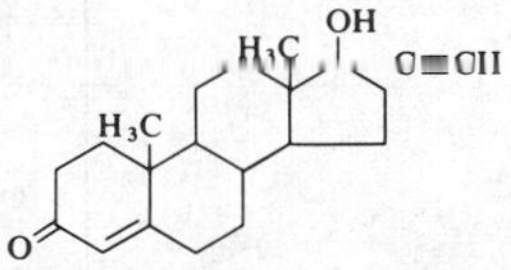 $C_{21}H_{28}O_2$	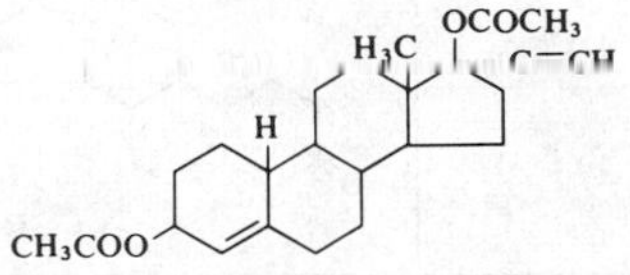 $C_{24}H_{32}O_4$
Molecular weight	358.5	312.4	384.5
Melting point (°C)	App. 100 (dec.)	266–273	126–132
Specific rotation	$(\alpha)\frac{20}{D}$ +16.5 to +18.5 (2% solution in chloroform) (calculated to the anhydrous basis)	$(\alpha)\frac{25}{D}$ −32° (100 mg. in 10 ml. pyridine)	$(\alpha)\frac{25}{D}$ −74 (1% in chloroform)
Absorption max.	App. 240 mμ (anhydrous ethanol) $E_1^{1\%}$ cm = 443	241 mμ (methanol)	None

Names & synonyms:	FLUROGESTONE ACETATE; 17α-acetoxy-9α-fluoro-11β-hydroxy-4-pregnene-3,20-dione.	HYDROXYMETHYLPROGESTERONE; medroxyprogesterone; 17α-hydroxy-6α-methylprogesterone; 17α-hydroxy-6α-methyl-4-pregnene-3,20-dione.	HYDROXYMETHYLPROGESTERONE ACETATE; medroxyprogesterone acetate; 17α-hydroxy-6α-methylprogesterone acetate; 17α-hydroxy-6α-methyl-4-pregnene-3,20-dione acetate.
Formulae:	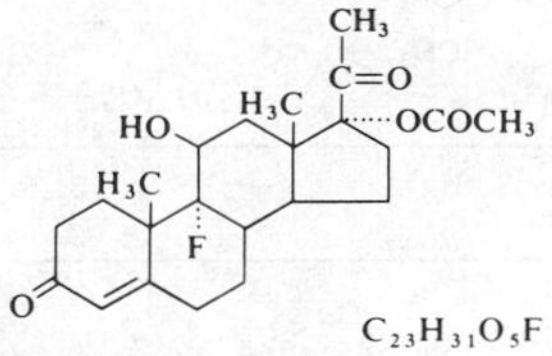 $C_{23}H_{31}O_5F$	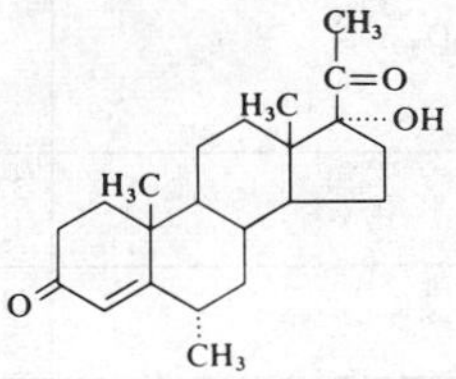 $C_{22}H_{32}O_3$	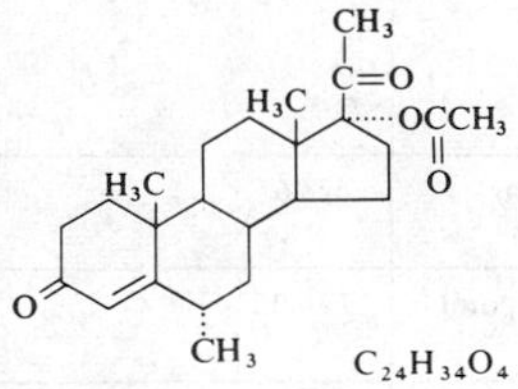 $C_{24}H_{34}O_4$
Molecular weight	406.5	344.5	386.5
Melting point (°C)	250–251	220–223.5	202–207
Specific rotation	$(\alpha)\frac{25}{D}$ +78	$(\alpha)\frac{25}{D}$ +75	$(\alpha)\frac{25}{D}$ +51 (dioxane)
Absorption max.	238 mμ (ε = 17,100)	241 mμ (ε = 16,150)	241 mμ (α_M = 16,500) ethanol

Names & synonyms:	HYDROXYPROGESTERONE; 17α-hydroxyprogesterone; 17α-hydroxy-4-pregene-3,20 dione; 4-pregnen-17α-ol-3,20-dione.	HYDROXYPROGESTERONE ACETATE; 17α-acetoxyprogesterone; 17α-hydroxyprogesterone acetate; 17α-hydroxy-4-pregnene-3,20 dione acetate.
Formulae:	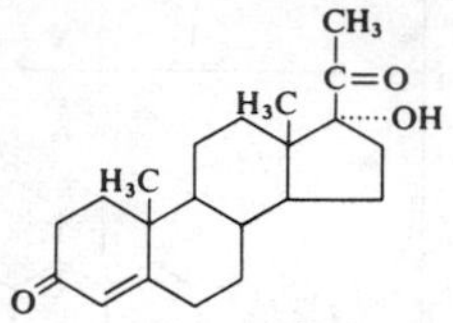 $C_{21}H_{30}O_3$	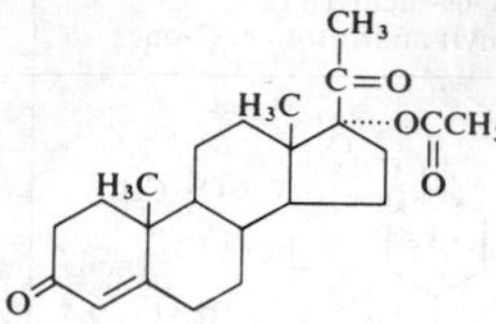 $C_{23}H_{32}O_4$
Molecular weight	330.4	372.5
Melting point (°C)	276	249–250
Specific rotation	$(\alpha)\frac{17}{D}$ +105 (104 mg. in 10 ml. chloroform)	$(\alpha)\frac{25}{D}$ +72 (chloroform)
Absorption max.		240 mμ (a_M = 16,875) ethanol

Names & synonyms:	HYDROXYPROGESTERONE CAPROATE; 17α-hydroxyprogesterone caproate; 17α-hydroxy-4-pregnene-3,20-dione caproate.	MELENGESTROL ACETATE: MGA; 17α-hydroxy-6-methyl-16-methylene-4,6-pregnadiene-3, 20-dione acetate; 6-dehydro-17-hydroxy-6-methyl-16-methylene-progesterone acetate.
Formulae:	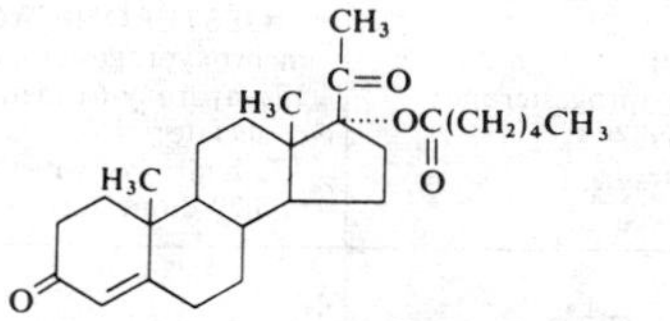 $C_{27}H_{40}O_4$	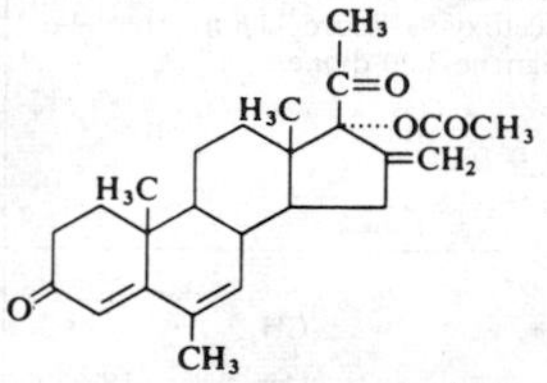 $C_{25}H_{32}O_4$
Molecular weight	428.6	396.51
Melting point (°C)	121–123	215–227
Specific rotation	$(\alpha)\frac{25}{D}$ +57 (chloroform)	$(\alpha)_D$ −127 to −135 (in $CHCl_3$)
Absorption max.		288 mμ (ε_1^1 = 24,000) (ethanol)

Names & synonyms:	NORETHINDRONE: Norethisterone; 17α-ethynyl-19-nortestosterone; 17α-ethynyl-17-hydroxy-19-nor-17α-4-en-20-yn-3-one.	NORETHINDRONE ACETATE: 17α-ethinyl-19-nortestosterone acetate.
Formulae:	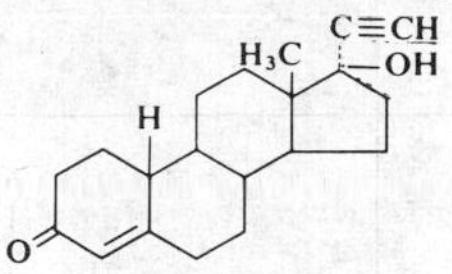 $C_{20}H_{26}O_2$	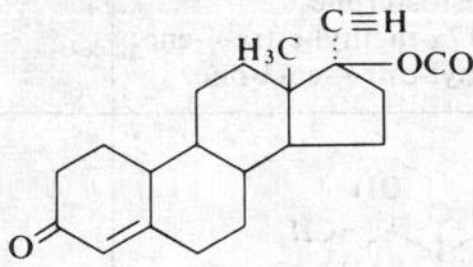 $C_{22}H_{28}O_3$
Molecular weight	298.4	340.4
Melting point (°C)	202 and 208	157–163
Specific rotation	$(\alpha)_{D}^{25}$ −30 – −35 (200 mg. in 10 ml. dioxane)	$(\alpha)_{D}^{25}$ −32 – −35 (200 mg. in 10 ml. dioxane)
Absorption max.	α (1 %, 1 cm) λ 240 = 535 ± 15	α (1 %, 1 cm.) λ 240 = 490 to 520 (505 ± 15) (ethanol)

Names & synonyms:	NORETHISTERONE; norethindrone; 19-norethisterone; 17α-ethynyl-19-nor-Δ^4-androstan-17β-ol-3-one; 17α-ethynyl-19-nor-testosterone; 17-hydroxy-3-oxo-19-nor-17α-pregn-4-ene-20-yne; 17-hydroxy-19-nor-17α-pregn-4-en-20-yn-3-one.	NORETHYNODREL: 17α-ethynyl-17β-hydroxy-5(10)-estren-3-one.
Formulae:	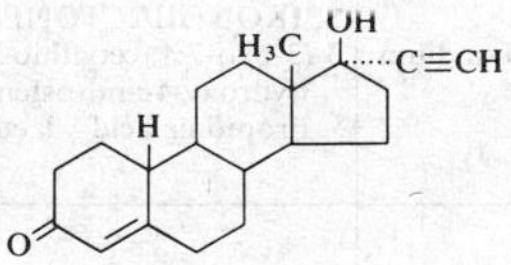 $C_{20}H_{26}O_2$	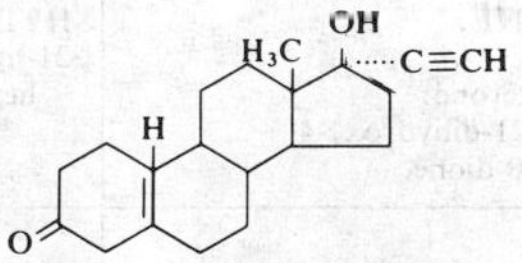 $C_{20}H_{26}O_2$
Molecular weight	298.4	298.4
Melting point (°C)	200–207	174–184
Specific rotation	$(\alpha)_{D}^{25}$ −30 – −38 (200 mg. in 10 ml. dioxane)	$(\alpha)_{D}^{25}$ +125 (dioxane)
Absorption max.	240 mμ (ε_1^1 = 576)	

STEROID HORMONES AND OTHER STEROIDAL SYNTHETICS (Continued)

Names & synonyms:	NORMETHISTERONE; 19-normethisterone; normethandrolone; metalutin; normetandrone; 17α-methyl-19-nor-Δ^4-androsten-17β-ol-3-one; 17α-methyl-19-nor-testosterone; 17β-hydroxy-3-oxo-17α-methyl-estra-4-ene; 17β-hydroxy-17-methyl-estr-4-en-3-one.	PREGNENOLONE; Δ^5-pregnenolone; Δ^5-pregnen-3β-ol-20-one; 17β(1-ketoethyl)-Δ^5-androstene-3β-ol.	PROGESTERONE; progestin; progestone; pregnendione; Δ^4-pregnene-3,20-dione.
Formulae:	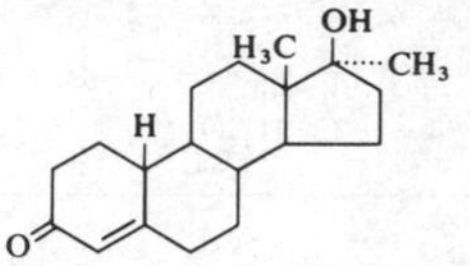 $C_{19}H_{28}O_2$	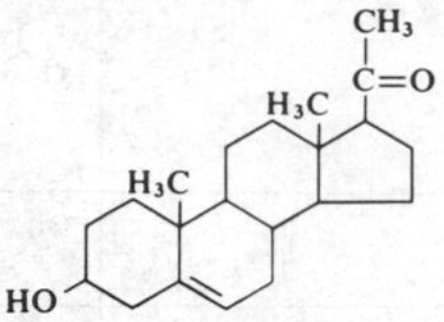 $C_{21}H_{32}O_2$	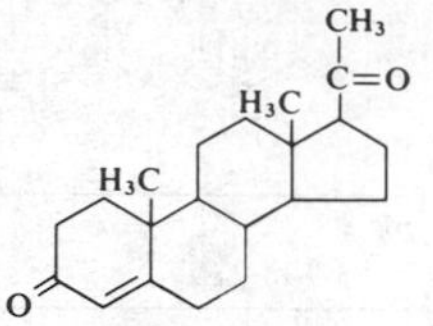 $C_{21}H_{30}O_2$
Molecular weight	288.4	308.4	314.4
Melting point (°C)	153–158	193	(β) isomer 121; (α) isomer 127–131
Specific rotation	$(\alpha)\frac{25}{D}$ +25 to +29 (200 mg. in 10 ml. chloroform)	$(\alpha)\frac{20}{D}$ +28 – +30 (alcohol)	$(\alpha)\frac{20}{D}$ +172 – +182 (200 mg. in 10 ml. dioxane)
Absorption max.	241 mμ – 565 ± 15		240 mμ

DIURETIC, ANTIDIURETIC AND LOCAL ANESTHETIC AGENTS

Names & synonyms:	ALDOSTERONE; electrocortin; 18-oxocorticosterone; 18-formyl-11β,21-dihydroxy-4-pregnene-3,20-dione.	HYDROXYDIONE SODIUM; 21-hydroxypregnane-3,20-dione-21-sodium hemisuccinate.	SPIRONOLACTONE; 3-(3-oxo-7α-acetylthio-17β-hydroxy-4-androsten-17α-yl)-propionic acid γ lactone.
Formulae:	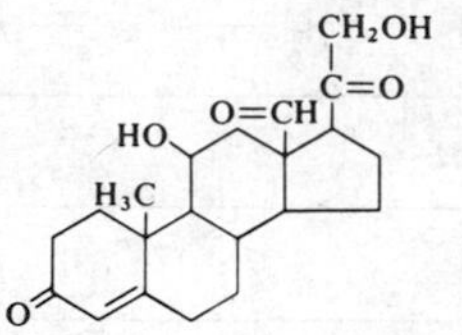 $C_{21}H_{28}O_5$	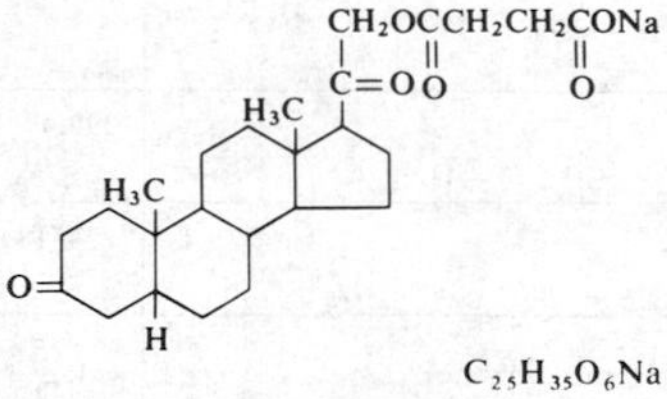 $C_{25}H_{35}O_6Na$	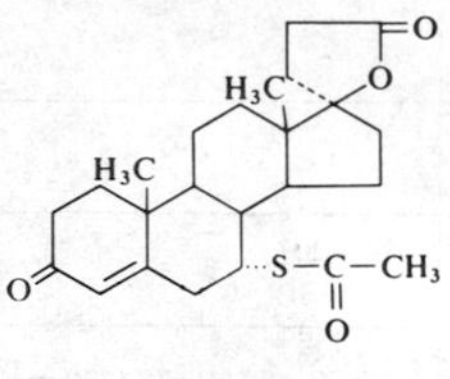 $C_{24}H_{32}O_4S$
Molecular weight	360.4	454.5	416.5
Melting point (°C)	108–112 (hydrate); 164 (anhydrous)	193–203 (dec.)	135; 202 (dec.)
Specific rotation	$(\alpha)\frac{25}{D}$ +161 (10 mg. in 10 ml. chloroform)	$(\alpha)\frac{25}{D}$ +95 (chloroform) for free acid.	$(\alpha)\frac{25}{D}$ – 34 (chloroform)
Absorption max.	240 mμ (log ε = 4.20 monohydr.; ε mol. 15,000 anhydr.)	280 mμ (ε = 93.2)	ε^{238} = 20,200

PROPERTIES OF THE AMINO ACIDS

IONIZATION CONSTANTS AND pH VALUES AT THE ISOELECTRIC POINTS OF THE AMINO ACIDS IN WATER AT 25°C

The majority of the recorded values are true thermodynamic constants calculated from electrometric force measurements of cells without liquid junctions. The values for the constants given in the table were derived from the classical, the zwitterionic (Bjerrum), and the acidic (Bronsted) formulations of ionization and the corresponding mass law expressions. pH values at the isoelectric points were calculated from the expression, $pI = \frac{1}{2}(pk_{a1} + pk_w - pk_{b1})$. The error is approximately 0.5% when this expression is used to calculate pI values for cystine, tyrosine, and diiodotyrosine.

	Classical				Zwitterionic				Acidic					
Amino acid	pk_{a1}	pk_{a2}	pk_{b1}	pk_{b2}	pK_{A1}	pK_{A2}	pK_{B1}	pP_{B2}	pK_1	pK_2	pK_3	pK_4	pI	Ref.
DL-Alanine	9.866	—	11.649	—	2.348	—	4.131	—	2.348	9.866	—	—	6.107	1
L-Arginine	12.48	—	4.96	11.99	2.01	—	1.52	4.96	2.01	9.04	12.48	—	10.76	2
L-Aspartic acid	3.86	9.82	11.93	—	2.10	3.86	4.18	—	2.10	3.86	9.82	—	2.98	3
L-Cystine	8.00	10.25	11.95	12.96	1.04	2.05	3.75	6.00	1.04	2.05	8.00	10.25	5.02	4
Diiodo-L-tyrosine	6.48	7.82	11.88	—	2.12	6.48	6.18	—	2.12	6.48	7.82	—	4.29	5,6
L-Glutamic acid	4.07	9.47	11.90	—	2.10	4.07	4.53	—	2.10	4.07	9.47	—	3.08	7
Glycine	9.778	—	11.647	—	2.350	—	4.219	—	2.350	9.778	—	—	6.064	8
L-Histidine	9.18	—	7.90	12.23	1.77	—	4.82	7.90	1.77	6.10	9.18	—	7.64	3
Hydroxy-L-proline	9.73	—	12.08	—	1.92	—	4.27	—	1.92	9.73	—	—	5.82	9
DL-Isoleucine	9.758	—	11.679	—	2.318	—	4.239	—	2.318	9.758	—	—	6.038	1
DL-Leucine	9.744	—	11.669	—	2.328	—	4.253	—	2.328	9.744	—	—	6.036	1
L-Lysine	10.53	—	5.05	11.82	2.18	—	3.47	5.05	2.18	8.95	10.53	—	9.47	2
DL-Methionine	9.21	—	11.72	—	2.28	—	4.79	—	2.28	9.21	—	—	5.74	10
DL-Phenylalanine	9.24	—	11.42	—	2.58	—	4.76	—	2.58	9.24	—	—	5.91	11
L-Proline	10.60	—	12.0	—	2.00	—	3.40	—	2.00	10.60	—	—	6.3	12
DL-Serine	9.15	—	11.79	—	2.21	—	4.85	—	2.21	9.15	—	—	5.68	9
L-Tryptophan	9.39	—	11.62	—	2.38	—	4.61	—	2.38	9.39	—	—	5.88	13
L-Tyrosine	9.11	10.07	11.80	—	2.20	9.11	3.93	—	2.20	9.11	10.07	—	5.63	6
DL-Valine	9.719	—	11.711	—	2.286	—	4.278	—	2.286	9.719	—	—	6.002	1

REFERENCES

1. Smith, P. K., Taylor, A. C., and Smith, E. R. B., *J. Biol. Chem.*, 122, 109, 1937—1938.
2. Schmidt, C. L. A., Kirk, P. L., and Appleman, W. K., *J. Biol. Chem.*, 88, 285, 1930.
3. Greenstein, J. P., *J. Biol. Chem.*, 93, 479, 1931.
4. Borsook, H., Ellis, E. L., and Huffman, H. M., *J. Biol. Chem.*, 117, 281, 1937.
5. Dalton, J. B., Kirk, P. L., and Schmidt, C. L. A., *J. Biol. Chem.*, 88, 589, 1930.
6. Winnek, P. S. and Schmidt, C. L. A., *J. Gen. Physiol.*, 18, 889, 1935.
7. Simms, H. S., *J. Gen. Physiol.*, 11, 629, 1928; 12, 231, 1928.
8. Owen, B. B., *J. Am. Chem. Soc.*, 56, 24, 1934.
9. Kirk, P. L. and Schmidt, C. L. A., *J. Biol. Chem.*, 81, 237, 1929.
10. Emerson, O. H., Kirk, P. L., and Schmidt, C. L. A., *J. Biol. Chem.*, 92, 449, 1931.
11. Miyamoto, S. and Schmidt, C. L. A., *J. Biol. Chem.*, 90, 165, 1931.
12. McCay, C. M. and Schmidt, C. L. A., *J. Gen. Physiol.*, 9, 333, 1926.
13. Schmidt, C. L. A., Appleman, W. K., and Kirk, P. L., *J. Biol. Chem.*, 85, 137, 1929—1930.

IONIZATION CONSTANTS OF THE AMINO ACIDS IN AQUEOUS ETHANOL SOLUTIONS

Amino acid	pK_1	pK_2	pK_3	Volume % ethanol	Temperature (°C)	Ref.
Alanine	3.55	10.02	—	72	25	1
Arginine	3.34	9.40	14.1	72	25	1
Aspartic acid	2.85	5.20	10.51	72	25	1
Glutamic acid	3.16	5.63	10.75	72	25	2
Glycine	2.66	9.82	—	10	19.5	2
	2.96	9.76	—	40	19.5	2
	3.46	9.82	—	72	25	1
	3.79	9.99	—	90	19.5	2
Histidine	3.00	5.85	9.45	72	25	1
Isoleucine	3.69	9.81	—	72	25	1
Lysine	2.75	8.95	10.53	48	25	1
	3.56	8.95	10.49	84	25	1
Proline	3.04	10.55	—	72	25	1
Valine	3.60	9.73	—	72	25	1

REFERENCES

1. Jukes, T. H. and Schmidt, C. L. A., *J. Biol. Chem.*, 105, 359, 1934.
2. Michaelis, L. and Mizutani, M., *Z. Physik. Chem.*, 116, 135, 1925.

IONIZATION CONSTANTS OF THE AMINO ACIDS IN AQUEOUS FORMALDEHYDE SOLUTION[a]

Amino acid	0.99	3.95	5.60	10.0	17.9
	Mole % formaldehyde				
DL-Alanine	8.36	7.42	6.96[b]	6.56	6.10
L-Arginine	—	3.45[c]	3.40[d]	—	—
L-Aspartic acid	—	—	7.21[d]	≤3.8[e]	—
				6.85[f]	
L-Glutamic acid	—	—	6.91[d]	≤4.2[e]	—
				6.8[f]	
Glycine	7.16	6.08	5.92[b]	5.34	5.04
L-Histidine	—	7.90[c]	7.90[d]	—	—
Hydroxy-L-proline	—	—	7.19[d]	—	—
L-Leucine	8.44	7.50	6.92[d]	6.62	6.20

Amino acid	0.99	3.95	5.60	10.0	17.9
	Mole % formaldehyde				
DL-Leucine	8.44	7.48	—	6.60	6.20
L-Lysine	—	7.35[c]	7.15[d]	—	—
L-Phenylalanine	—	—	6.62[d]	5.9[e]	—
DL-Phenylalanine	8.09	7.16	6.80[b]	6.35	6.13
L-Proline	—	—	7.78[d]	—	—
DL-Serine	6.66	5.74	5.63[b]	—	4.94
L-Tryptophan	—	—	6.88[d]	—	—
L-Tyrosine	—	—	7.50[d]	6.2[e]	—
				>9[f]	
DL-Valine	8.52	7.65	7.47[b]	—	6.52

[a] pK_2 at 22°.[1]
[b] pK_2 at 22°.[2]
[c] pK_2 at 30° for arginine and pK_3 at 30° for histidine and lysine.[3]
[d] pK_2 at 30°, pK_3 at 30° for histidine and lysine.[4]
[e] pK_2 at 25° for aspartic acid, glutamic acid, phenylalanine, and tyrosine.[5]
[f] pK_3 at 30° for aspartic acid, glutamic acid, and tyrosine.[5]

REFERENCES

1. Dunn, M. S. and Weiner, J. G., *J. Biol. Chem.*, 117, 381, 1937.
2. Dunn, M. S. and Loshakoff, A., *J. Biol. Chem.*, 113, 691, 1936.
3. Levy, M., *J. Biol. Chem.*, 109, 365, 1935.
4. Levy, M. and Silberman, D. E., *J. Biol. Chem.*, 118, 723, 1937.
5. Harris, L. J., *Proc. R. Soc. London, Series B*, 95, 440, 1923—1924.

SPECIFIC ROTATIONS OF THE AMINO ACIDS USING SODIUM LIGHT (5893 Å)

Abbreviations

c —grams of solute per 100 ml of solution.
d —density of the solution.
p —grams of solute per 100 g of solution.
l —length of the tube in decimeters.
α —observed rotation in angular degrees.
[α] —specific rotation in angular degrees calculated from
$[\alpha]_\lambda^t = \alpha \times 100/c \times l = \alpha \times 100/p \times d \times l$ where t is temperature in °C and λ is wave length of the incident light in Ångstroms.

A—prepared from a protein or other naturally occurring material.
B—prepared by resolution of the inactive synthetic form.
C—prepared by resolution of the inactive racemized form.
D—prepared from the inactive synthetic form by a biological method.
E—prepared from the inactive racemized form by a biological method.
?—source not given.

Source	c	Solvent	d	p	Moles acid or base per mole amino acid	l	Temp. (°C)	α	[α]	Ref.
L-Alanine										
A	5.790	0.97 *N*HCl	1.033	5.605	1.5	2	15	+ 1.70	+ 14.7	1
A	10.3	Water	1.03	1.00	0	2	22	+ 0.55	+ 2.7	2
A	1.781	3 *N*NaOH	—	—	15	2	20	—	+ 3.0	3
D-Alanine										
B	71.344	6 *N*HCl	—	—	39.4	2	30.4	−0.392	−14.6	4
L-Arginine										
A	1.653	6.0 *N*HCl	—	—	63	4.001	23.4	+ 1.777	+ 26.9	5
A	3.48	Water	—	—	0	2	20	—	+ 12.5	6
A	0.87	0.50 *N*NaOH	—	—	10	2	20	—	+ 11.8	6
L-Aspartic acid										
A	2.002	6.0 *N*HCl	—	—	39	4.001	24.0	+ 1.972	+ 24.6	7
A	1.3300	Water	—	—	0	3	18	—	4.7	3
A	1.3300	3 *N*NaOH	—	—	30	3	18	—	−1.7	3
D-Aspartic acid										
C	4.289	0.97 *N*HCl	1.032	4.156	3	1	20	−1.09	−25.5	8
L-Cystine										
A	0.9974	1.02 *N*HCl	1.0181	0.9797	24.6	2	24.35	−4.277	−214.40	9
A	0.400	0.20 *N*NaOH	—	—	12	2	18.5	—	−70.0	3
D-Cystine										
C	—	1 *N*HCl	—	1	24	—	20	—	+ 223	10
Diiodo-D-tyrosine										
A	5.08	1.1 *N*HCl	1.05	4.84	9.4	1	20	+ 0.15	+ 2.89	11
A	4.41	13.4 *N*NH_4OH	0.9779	4.51	132	1	20	+ 0.10	+ 2.27	11
L-Glutamic acid										
A	1.002	6.0 *N*HCl	—	—	87	4.001	22.4	+ 1.25	+ 31.2	12
A	1.471	Water	—	—	0	2	18	—	+ 11.5	3
A	1.471	1 *N*NaOH	—	—	10	2	18	—	+ 10.96	3
D-Glutamic acid										
C	5.425	0.37 *N*HCl	1.0233	5.3011	1	1	20	−1.63	−30.05	8
L-Histidine										
A	1.480	6.0 *N*HCl	—	—	63	4.001	22.7	+ 0.766	+ 13.0	7
A	1.128	Water	1.0012	1.127	0	4	25.00	−1.714	−39.01	13
A	0.775	0.50 *N*NaOH	—	—	10	2	20	—	−10.9	6
D-Histidine										
?	4.000	1.0 *N*HCl	—	—	4	1	20	−0.407	−10.2	14
B	2.66	Water	—	—	0	2	23	+ 2.11	+ 39.8	14
Hydroxy-L-proline										
A	1.31	1.0 *N*HCl	—	—	10	2	20	—	−47.3	6
A	1.001	Water	—	—	0	4.001	22.5	−3.009	−75.2	7
A	0.655	0.50 *N*NaOH	—	—	10	2	20	—	−70.6	6
Hydroxy-D-proline										
B	4.48	Water	1.03	4.35	0	1	21	+ 3.37	+ 75.2	16

Source	c	Solvent	d	p	Moles acid or base per mole amino acid	l	Temp. (°C)	α	[α]	Ref.
Allo-Hydroxy-L-proline										
B	2.617	Water	1.014	2.581	0	1	18	−1.52	−58.1	16
Allo-Hydroxy-D-proline										
B	2.530	Water	1.013	2.998	0	1	17	+1.48	+58.5	16
L-Isoleucine										
B	5.09	6.1 *N*HCl	1.098	4.64	15	1	20	+2.07	+40.61	17
B	3.10	Water	1.008	3.08	0	2	20	+0.70	+11.29	17
A	3.34	0.33 *N*NaOH	1.017	3.28	1.3	2	20	+0.74	+11.09	18
D-Isoleucine										
B	4.53	6.1 *N*HCl	1.083	4.18	17	1	20	−1.85	−40.86	17
B	3.12	Water	1.006	3.10	0	2	20	−0.66	−10.55	17
D-allo-Isoleucine										
D	5.14	6.0 *N*HCl	1.094	4.70	15.0	2	20	−3.80	−36.95	19
B	2.00	Water	—	—	0	1	20	−0.285	−14.2	20
L-allo-Isoleucine										
B	3.97	6.0 *N*HCl	—	—	20	1	20	+1.50	−38.1	20
B	2.00	Water	—	—	0	1	20	+0.28	+14.0	20
L-Leucine										
A	1.999	6.0 *N*HCl	—	—	38	4.001	25.9	+1.212	+15.1	5
A	2.001	Water	—	—	0	4.001	24.7	−0.863	−10.8	5
A	1.31	3.00 *N*NaOH	—	—	30	2	20	—	+7.6	3
D-Leucine										
?	4.0	6.0 *N*HCl	1.1	3.664	19	2	20	+1.26	−15.6	21
?	—	Water	—	2.08	0	2	20	+0.43	+10.34	38
L-Lysine										
A	2.00	6.0 *N*HCl	—	—	43	4	22.9	+1.652	+25.9	5
A	6.496	Water	—	—	0	2	20	+1.90	+14.6	22
D-Lysine										
B	2.00	0.27 *N*HCl	—	—	2	2	20	−0.939	−23.48	23
L-Methionine										
B	0.80	Water	—	—	0	2	25	−0.13	−8.11	24
D-Methionine										
B	0.80	0.2001 *N*HCl	—	—	4	2	25	−0.34	−21.18	24
B	0.80	Water	—	—	0	2	25	+0.13	+8.12	24
B	0.80	0.6 *N*$NaHCO_3$	—	—	11	2	25	−0.12	−7.47	24
L-Phenylalanine										
B	1.936	Water	1.0040	1.928	0	2	20	−1.36	−35.14	27
D-Phenylalanine										
B	3.814	5.4 *N*HCl	1.0895	3.501	23	2	20	+0.54	+7.07	28
B	2.043	Water	1.0045	2.034	0	2	20	+1.43	+35.0	27
L-Proline										
A	0.575	0.50 *N*HCl	—	—	10	2	20	—	−52.6	6
A	1.001	Water	—	—	0	4.001	23.4	−3.402	−85.0	7
B	2.42	0.6 *N*KOH	1.031	2.35	3	1	20	−2.25	−93.0	29
D-Proline										
B	3.90	Water	1.01	3.865	0	1	20	+3.18	+81.5	29
L-Serine										
B	9.344	1 *N*HCl	1.0465	8.929	1	1	25	+1.35	+14.45	30
B	10.414	Water	1.0414	9.997	0	2	20	−1.42	−6.83	30
D-Serine										
B	9.359	1 *N*HCl	1.0465	8.943	1	1	25	−1.34	−14.32	30
B	10.412	Water	1.0414	9.998	0	2	20	+1.43	+6.87	30
D-Threonine										
B	—	Water	—	1.092	0	2	26	−0.625	−28.3	31
L-Threonine										
B	—	Water	—	1.331	0	2	26	+0.780	+28.4	31
D-allo-Threonine[a]										
B	—	Water	—	1.634	0	2	26	−0.302	−9.1	31
L-allo-Threonine										
B	—	Water	—	1.643	0	2	26	+0.320	+9.6	31
L-Thyroxine										
A	—	0.13 *N*NaOH in 70% EtOH by weight	—	3	3	1	—	−0.147	−4.4	32
L-Tryptophan										
A	1.02	0.50 *N*HCl	—	—	10	2	20	—	+2.4	6
A	1.004	Water	—	—	0	4.001	22.7	−1.266	−31.5	7
A	2.426	0.5 *N*NaOH	1.0243	2.368	4.2	1	20	+0.15	+6.17	33
D-Tryptophan										
C	0.5024	Water	—	—	0	2	25	+0.326	+32.45	34
L-Tyrosine										
B	4.40	6.3 *N*HCl	1.116	3.94	28	2	20	−0.76	−8.64	35
A	0.906	3.0 *N*NaOH	—	—	60	3	18	—	−13.2	3
D-Tyrosine										
B	5.1484	6.3 *N*HCl	1.1175	4.6071	24	2	20	+0.89	+8.64	35
L-Valine										
B	3.4	6.0 *N*HCl	1.1	3.05	20	2	20	+1.93	+28.8	36
B	3.58	Water	1.007	3.56	0	2	20	+0.46	+6.42	36
D-Valine										
B	3.2	6.0 *N*HCl	1.1	2.91	21	2	20	−1.86	−29.04	36
E	6.24	Water	1.00	6.24	0	1	20	−0.37	−6.06	37

[a] The levorotatory allothreonine probably belongs to the D family and its enantiomorph to the L family.

REFERENCES

1. Clough, G. W., *J. Chem. Soc.*, 113, 526, 1918.
2. Fischer, E. and Raske, K., *Ber*, 40, 3717, 1907.
3. Lutz, O. and Jirgensons, B., *Ber.*, 63, 448, 1930.
4. Dunn, M. S., Butler, A. W., and Naiditch, M. J., unpublished data.
5. Dunn, M. S. and Courtney, G., unpublished data.
6. Lutz, O. and Jirgensons, B., *Ber.*, 64, 1221, 1931.
7. Dunn, M. S. and Stoddard, M. P., unpublished data.
8. Fischer, E., *Ber.*, 32, 2451, 1899.
9. Toennies, G. and Lavine, T. F., *J. Biol. Chem.*, 89, 153, 1930.
10. Loring, H. S. and du Vigneaud, V., *J. Biol. Chem.*, 107, 267, 1934.
11. Abderhalden, E. and Guggenheim, M., *Ber.*, 41, 1237, 1908.
12. Dunn, M. S. and Sexton, E. L., unpublished data.
13. Dunn, M. S. and Frieden, E. H., unpublished data.
14. Cox, G. J. and Berg, C. P., *J. Biol. Chem.*, 107, 497, 1934.
15. Dakin, H. D., *Biochem. J.*, 13, 398, 1919.
16. Leuchs, H. and Bormann, K., *Ber.*, 52, 2086, 1919.
17. Locquin, R., *Bull. Soc. Chim.*, (4)1, 601, 1907.
18. Ehrlich, F., *Ber.*, 37, 1809, 1904.
19. Ehrlich, F., *Ber.*, 40, 2538, 1907.
20. Abderhalden, E. and Zeisset, W., *Z. Physiol. Chem.*, 196, 121, 1931.
21. Fischer, E., and Warburg, O., *Ber.*, 38, 3997, 1905.
22. Vickery, H. B., private communication, April, 1940.
23. Berg, C. P., *J. Biol. Chem.*, 115, 9, 1936; private communication, June, 1940.
24. Windus, W. and Marvel, C. S., *J. Am. Chem. Soc.*, 53, 3490, 1931.
27. Fischer, E. and Schoeller, W., *Ann.*, 357, 1, 1907.
28. Fischer, E. and Mouneyrat, A., *Ber.*, 33, 2383, 1900.
29. Fischer, E. and Zemplén, G., *Ber.*, 42, 2989, 1909.
30. Fischer, E. and Jacobs, W. A., *Ber.*, 39, 2942, 1906.
31. West, H. D. and Carter, H. E., *J. Biol. Chem.*, 119, 109, 1937; private communication from H. E. Carter, July, 1940.
32. Foster, G. L., Palmer, W. W., and Leland, J. P., *J. Biol. Chem.*, 115, 467, 1936.
33. Abderhalden, E. and Baumann, L., *Z. Physiol. Chem.*, 55, 412, 1908.
34. Berg, C. P., J. Biol. Chem., 100, 79 1933; private communication, July, 1940.
35. Fischer, E., *Ber.*, 32, 3638, 1899.
36. Fischer, E., *Ber.*, 39, 2320, 1906.
37. Ehrlich, F. and Wendel, A., *Biochem. Z.*, 8, 399, 1908.
38. Ehrlich, F., *Biochem. Z.*, 1, 8, 1906.

SOLUBILITIES OF THE AMINO ACIDS IN GRAMS PER 100 GRAMS OF WATER

	Temperature (°C)					
Amino acid	0°	25°	50°	75°	100°	Ref.
DL-Alanine	12.11	16.72	23.09	31.89	44.04	1
L-Alanine	12.73	16.65	21.79	28.51	37.30	1
DL-Aspartic acid	0.262	0.778	2.000	4.456	8.594	1
L-Aspartic acid	0.209	0.500	1.199	2.875	6.893	1
L-Cystine[c] × 10^2	0.502	1.096	2.394	5.229	11.42	2
Diiodo-DL-tyrosine × 10	0.149	0.340	0.773	—	—	3
Diiodo-L-tyrosine × 10	0.204	0.617	1.862	5.62	17.00	1
DL-Glutamic acid	0.855	2.054	4.934	11.86	28.49	1
L-Glutamic acid	0.341	0.864	2.186	5.532	14.00	1
Glycine	14.18	24.99	39.10	54.39	67.17	1
L-Histidine	—	4.19	—	—	—	4
Hydroxy-L-proline	28.86	36.11	45.18	51.67[a]	—	5
DL-Isoleucine	1.826	2.229	3.034	4.607	7.802	1
L-Isoleucine	3.791	4.117	4.818	6.076	8.255	2
DL-Leucine	0.797	0.991	1.406	2.276	4.206	1
L-Leucine	2.270	2.426[c]	2.887[b]	3.823	5.638	1
DL-Methionine	1.818	3.381	6.070	10.52	17.60	2
DL-Phenylalanine	0.997	1.411	2.187	3.708	6.886	1
L-Phenylalanine	1.983	2.965	4.431	6.624	9.900	2
L-Proline × 10^{-1}	12.74	16.23	20.67	23.90[a]	—	3
DL-Serine	2.204	5.023	10.34	19.21	32.24	2
L-Tryptophan	0.823	1.136	1.706	2.795	4.987	2
DL-Tyrosine × 10	0.147	0.351	0.836	—	—	3
L-Tyrosine × 10	0.196	0.453	1.052	2.438	5.650	1
D-Tyrosine × 10	0.196	0.453	1.052	—	—	3
DL-Valine	5.98	7.09	9.11	12.61	18.81	1
L-Valine	8.34	8.85	9.62	10.24[a]	—	6

[a] Value at 65°.

[b] Dunn and Stoddard[7] report 2.19 g at 25° for L-leucine rendered methionine-free by repeated recrystallization from 6 *N* HCl. Hlynka[8] found 2.20 g at 25° and 2.66 g at 50° for L-leucine rendered methionine-free (by S. W. Fox[9]) by fractional crystallization of the formyl derivative and identical values for D-leucine obtained by resolution of the DL form.

[c] The following values were found by Loring and du Vigneaud[10]: DL-cystine (0.0049 g), D-cystine (0.0108 g), and *meso*-cystine (0.0056 g) at 25°.

REFERENCES

1. Dalton, J. B. and Schmidt, C. L. A., *J. Biol. Chem.*, 103, 549, 1933.
2. Dalton, J. B. and Schmidt, C. L. A., *J. Biol. Chem.*, 109, 241, 1935.
3. Winnek, P. S. and Schmidt, C. L. A., *J. Gen. Physiol.*, 18, 889, 1934—1935.
4. Dunn, M. S., Frieden, E. H., and Brown, H. V., unpublished data.
5. Tomiyama, T. and Schmidt, C. L. A., *J. Gen. Physiol.*, 19, 379, 1935—1936.
6. Dalton, J. B. and Schmidt, C. L. A., *J. Gen. Physiol.*, 19, 767, 1935—1936.
7. Dunn, M. S. and Stoddard, M. P., unpublished data.
8. Hlynka, I., Thesis (1939), California Institute of Technology, Pasadena, California.
9. Fox, S. W., *Science*, 84, 163, 1936.
10. Loring, H. S. and du Vigneaud, V., *J. Biol. Chem.*, 107, 270, 1934.

SOLUBILITIES OF THE AMINO ACIDS IN GRAMS PER 100 GRAMS OF WATER-ETHANOL MIXTURES

% ethanol by volume	Temp. °C	Grams amino acid per 100 grams solvent	Ref.
DL-Alanine			
24.93	0.00	3.84	1
50.10	0.00	1.16	1
74.50	0.00	0.305	1
95.14	0.00	0.0167	1
10	25	12.25	2
24.93	24.97	7.09	1
50.10	24.97	2.52	1
74.20	24.97	0.573	1
95.14	25.09	0.0329	1
25.28	45.16	10.6	1
50.10	44.96	4.25	1
74.20	44.98	0.949	1
95.14	45.19	0.0545	1
24.93	64.96	15.9	1
50.10	64.94	6.68	1
74.20	64.94	1.48	1
95.09	65.15	0.0851	1
DL-Aspartic acid			
24.93	0.03	0.0703	1
50.10	0.03	0.0267	1
74.20	0.02	0.0111	1
24.55	25.06	0.266	1
50.25	25.06	0.0992	1
74.28	25.14	0.0317	1
95.14	25.07	0.0020	1
24.74	45.25	0.680	1
50.18	45.25	0.255	1
74.28	45.27	0.0608	1
95.14	45.21	0.0042	1
24.93	64.91	1.53	1
50.10	64.91	0.588	1
74.20	65.07	0.132	1
95.14	65.00	0.0129	1
L-Aspartic acid			
20	25	0.204	3
50	25	0.0633	3
70	25	0.0224	3
90	25	0.0034	3
L-Glutamic acid			
24.74	0.01	0.0855	1
50.18	0.01	0.0371	1
74.28	0.03	0.0163	1
24.56	25.05	0.292	1
50.25	25.08	0.131	1
74.35	25.07	0.0370	1
95.14	25.04	0.0044	1
24.55	45.01	0.811	1
50.18	45.27	0.378	1
74.35	44.93	0.0885	1
95.14	45.20	0.0127	1
Glycine			
24.93	0.02	3.95	1
50.10	0.02	1.03	1
74.50	0.02	0.200	1
95.09	0.01	0.0080	1
10	25	17.13	2
24.93	24.97	8.72	1
50.10	24.97	2.47	1
74.20	24.97	0.448	1
95.14	25.09	0.0172	1
24.93	44.98	15.0	1
50.10	44.98	4.62	1
74.20	44.97	0.756	1
95.14	45.19	0.0294	1
24.93	65.11	24.5	1
50.10	65.10	8.03	1
74.20	65.07	1.23	1
95.14	65.00	0.0488	1
L-Isoleucine			
80	20	0.46	4
80	78—80	1.16	4
L-allo-Isoleucine			
80	20	0.81	4
80	78—80	1.97	4
DL-Leucine			
24.93	0.00	0.251	1
50.10	0.00	0.118	1
74.50	0.00	0.0693	1
95.14	0.00	0.0116	1
10	25	0.771	2
24.93	24.97	0.493	1
50.10	24.97	0.318	1
74.20	24.97	0.175	1
95.14	25.09	0.0258	1
24.93	45.24	0.833	1
50.10	45.24	0.633	1
74.50	45.18	0.323	1
95.14	45.18	0.0471	1
24.93	65.16	1.45	1
50.10	65.20	1.16	1
74.20	65.15	0.584	1
95.09	65.07	0.0844	1
L-Leucine			
20	25	1.33	2
60	25	0.641	2
90	25	0.123	2
L-Proline			
100	19	1.5	5
DL-Serine			
24.93	0.00	0.1530	1
50.10	0.00	0.146	1
74.50	0.00	0.0304	1
95.14	0.00	0.0008	1
24.93	25.14	1.54	1
50.10	25.14	0.461	1
74.50	25.10	0.0840	1
95.14	25.09	0.0028	1
24.93	45.15	3.14	1
50.10	45.04	0.985	1
74.20	45.04	0.185	1
95.14	45.18	0.0058	1
24.93	65.26	5.99	1
50.10	65.25	1.88	1
74.50	65.24	0.318	1
95.14	65.01	0.0152	1
DL-Threonine			
95	25	0.07[a]	6
DL-allo-Threonine			
95	25	0.03[a]	6
L-Tyrosine			
95	17	0.10	7
DL-Tyrosine			
95.09	0.00	0.0031	8
25.28	24.85	0.0285	8
50.99	24.75	0.0226	8
74.63	24.75	0.0117	8
95.09	25.24	0.0032	8
25.28	45.15	0.0630	8
50.99	45.16	0.0513	8
74.63	44.93	0.0230	8
95.09	44.98	0.0035	8
95.09	65.06	0.0067	8
DL-Valine			
24.93	0.02	2.10	1
50.10	0.02	0.769	1
74.20	0.02	0.269	1
95.14	0.01	0.0277	1
10	25	5.50	2
25.28	24.85	3.30	1
50.99	24.85	1.53	1
74.35	24.93	0.570	1
95.14	25.04	0.0569	1
24.55	44.91	5.10	1
50.25	44.92	2.74	1
74.35	44.92	0.999	1
95.14	45.21	0.0979	1
24.55	65.07	7.44	1
50.10	64.94	4.49	1
74.20	64.34	1.62	1
95.09	65.15	0.167	1
L-Valine			
20	25	5.11	2
40	25	2.93	2
60	25	1.61	2
80	25	0.52	2

[a] Grams per 100 mℓ of solution.

REFERENCES

1. Dunn, M. S. and Ross, F. J., *J. Biol. Chem.*, 125, 309, 1938.
2. Cohn, E. J., McMeekin, T. L., Edsall, J. T., and Weare, J. H., *J. Am. Chem. Soc.*, 56, 2270, 1934.
3. McMeekin, T. L., Cohn, E. J., and Weare, J. H., *J. Am. Chem. Soc.*, 57, 626, 1935.
4. Abderhalden, E. and Zeisset, W., *Z. Physiol. Chem.*, 196, 121, 1931.
5. Kapfhammer, J. and Eck, R., *Z. Physiol. Chem.*, 170, 294, 1927.
6. West, H. D. and Carter, H. E., *J. Biol. Chem.*, 119, 109, 1937.
7. Stutzer, A., *Z. Anal. Chem.*, 31, 501, 1892.
8. Dunn, M. S. and Ross, F. J., unpublished data.

SOLUBILITIES OF THE AMINO ACIDS IN GRAMS PER 100 GRAMS OF ORGANIC SOLVENT

Solvent	Grams amino acid per 100 grams solvent	Temp. (°C)	Ref.
DL-Alanine			
Ethanol	0.0087	25	1
L-Aspartic acid			
Ethanol	0.000196	25	2
L-Glutamic acid			
Ethanol	0.000347	25	2
Ethanol	0.0056	44.93	3
Glycine			
Acetone	0.000291	25	4
Butanol	0.000892	25	4
Ethanol	0.0037	25	1
Formamide	0.558	25	4
Methanol	0.0407	25	4
L-Isoleucine			
Ethanol	0.09	20	5
Ethanol	0.13	78—80	5
L-allo-Isoleucine			
Ethanol	0.13	20	5
Ethanol	0.19	78—80	5
L-Leucine			
Ethanol	0.0217	25	1
L-Proline			
Ethanol	1.5	19	6
DL-Valine			
Ethanol	0.0136	0.03	3
Ethanol	0.019	25	1

REFERENCES

1. Cohn, E. J., McMeekin, T. L., Edsall, J. T., and Weare, J. H., *J. Am. Chem. Soc.*, 56, 2270, 1934.
2. McMeekin, T. L., Cohn, E. J., and Weare, J. H., *J. Am. Chem. Soc.*, 57, 626, 1935.
3. Dunn, M. S. and Ross, F. J., *J. Biol. Chem.*, 125, 309, 1938.
4. McMeekin, T. L., Cohn, E. J., and Weare, J. H., *J. Am. Chem. Soc.*, 58, 2173, 1936.
5. Abderhalden, E. and Zeisset, W., *Z. Physiol. Chem.*, 196, 121, 1931.
6. Kapfhammer, J. and Eck, R., *Z. Physiol. Chem.*, 170, 294, 1927.

DENSITIES OF CRYSTALLINE AMINO ACIDS

Amino acid	Density	Ref.
DL-Alanine	1.424	1
L-Alanine	1.401	2
β-Alanine	1.404	1
DL-α-Amino-*n*-butyric acid	1.231	1
α-Aminoisobutyric acid	1.278	1
L-Arginine	1.1	3
L-Aspartic acid	1.66	3
DL-Glutamic acid	1.460	4
L-Glutamic acid	1.538	4
Glycine[a]	1.601	3
	1.607	1
DL-Leucine	1.191	1
L-Leucine	1.165	1
DL-Methionine	1.340	5
DL-Serine	1.537	5
L-Tyrosine	1.456	1
DL-Valine	1.316	1
L-Valine	1.230	1

[a] The density of glycine at 50° is 1.5753 according to Houck[6] who concluded that the figure 1.1607, reported by Curtius[7] and reproduced in chemical handbooks, is a typographical error.

REFERENCES

1. Cohn, E. J., McMeekin, T. L., Edsall, J. T., and Weare, J. H., *J. Am. Chem. Soc.*, 56, 2270, 1934.
2. Dalton, J. B. and Schmidt, C. L. A., *J. Biol. Chem.*, 103, 549, 1933.
3. Huffman, H. M., Ellis, E. L., and Fox, S. W., *J. Am. Chem. Soc.*, 58, 1728, 1936; Huffman, H. M., Fox, S. W., and Ellis, E. L., *J. Am. Chem. Soc.*, 59, 2144, 1937.
4. Schmidt, C. L. A., *Chemistry of the Amino Acids and Proteins*, Charles C Thomas, Springfield, 1938, p. 900.
5. Albrecht, G. and Dunn, M. S., unpublished data.
6. Houck, R. C., *J. Am. Chem. Soc.*, 52, 2420, 1930.
7. Curtius, T., *J. Prakt. Chem.*, 26, 145, 1882.

CARBOHYDRATES

These data for carbohydrates were compiled originally for the Biology Data Book by M. L. Wolfram, G. G. Maher and R. G. Pagnucco (1964). Data are reproduced here by permission of the copyright owners of the above publication, the Federation of American Societies for Experimental Biology, Washington, D.C. pp. 351–359.

All data are for crystalline substances, unless otherwise specified. Selection of substances was restricted to natural carbohydrates found free (or in chemical combination and released on hydrolysis) and to biological oxidation products of the natural carbohydrates. The nomenclature conforms with that of the British-American report as published in the *Journal of Organic Chemistry*, 28:281 (1963). Substances have been arranged alphabetically under the name of the parent sugar within groups formulated according to increasing carbon content (excluding carbon in substituents), with synonymous common names in parentheses. **Melting Point:** b.p. = boiling point; d. = decomposes; s. = sinters. **Specific Rotation** was determined in water at concentrations of 1–5 g per 100 ml. of solution and at 20°–25°C, unless otherwise specified; other temperatures or wavelengths are shown in brackets; *c* = grams solute per 100 ml of solution.

Part I. NATURAL MONOSACCHARIDES: ALDOSES AND KETOSES

	Substance (Synonym)	Chemical Formula	Melting Point °C	Specific Rotation $[\alpha]_D$
	(A)	(B)	(C)	(D)
	Aldoses			
1	D-Glyceraldehyde	$C_3H_6O_3$	....	+13.5 ± 0.5 (syrup)
2	D-Glyceraldehyde, 3-deoxy-3,3-*C*-bis-(hydroxymethyl)- (Cordycepose)	$C_5H_{10}O_4$	....	−26 (*c* 0.6, C_2H_5OH
3	D-Glyceraldehyde, 3,3-bis(*C*-hydroxy-methyl)- (Apiose)	$C_5H_{10}O_5$	....	+5.6 (*c* 10) [15°] syrup
4	β-D-Arabinose	$C_5H_{10}O_5$	155	−175 → −103
5	D-Arabinose, 2-*O*-methyl-	$C_6H_{12}O_5$	Syrup	−102
6	α-L-Arabinose	$C_5H_{10}O_5$	158 amorphous	+55.4 → +105
7	β-L-Arabinose	$C_5H_{10}O_5$	160	+190.6 → +104.5
8	DL-Arabinose	$C_5H_{10}O_5$	163.5–164.5	None
9	α-L-Lyxose	$C_5H_{10}O_5$	105	+5.8 → +13.5
10	L-Lyxose, 5-deoxy-3-*C*-formyl-(Streptose)	$C_6H_{10}O_5$	....	....
11	L-Lyxose, 3-*C*-formyl- (Hydroxy-streptose)	$C_6H_{10}O_6$	....	....
12	Pentose, 4,5-anhydro-5-deoxy-D-*erythro*-	$C_5H_8O_3$	....	....
13	Pentose, 2-deoxy-D-*erythro*-	$C_5H_{10}O_4$	96–98	−91 → −58
14	D-Ribose	$C_5H_{10}O_5$	87	−23.1 → −23.7
15	D-Ribose, 2-*C*-hydroxymethyl-(Hamamelose)	$C_6H_{12}O_6$	....	−7.1 [λ578]
16	α-D-Xylose	$C_5H_{10}O_5$	145	+93.6 → +18.8
17	D-Xylose, 5-deoxy-	$C_5H_{10}O_4$	....	+16
18	β-D-Xylose, 2-*O*-methyl-	$C_6H_{12}O_5$	137–138	−21 → +34
19	α-D-Xylose, 3-*O*-methyl-	$C_6H_{12}O_5$	95	+45 → +19
20	D-Allose, 6-deoxy-	$C_6H_{12}O_5$	140–143 146–148	+1.6 [18°] (*c* 0.6) −4.7 → 0
21	D-Allose, 6-deoxy-2,3-di-*O*-methyl-(Mycinose)	$C_8H_{16}O_5$	102–106	−46 → −29
22	Amicetose (a trideoxy hexose)	$C_6H_{12}O_3$	Oil, b.p. 65–70	+28.6 ($CHCl_3$)
23	Antiarose	$C_6H_{12}O_5$	....	Levo
24	α-D-Galactose	$C_6H_{12}O_6$	167	+150.7 → +80.2
25	β-D-Galactose	$C_6H_{12}O_6$	143–145	+52.8 → +80.2
26	D-Galactose, 3,6-anhydro-	$C_6H_{10}O_5$	....	+21.3 [10°]
27	α-D-Galactose, 6-deoxy- (D-Fucose; Rhodeose)	$C_6H_{12}O_5$	140–145	+127 → +76.3 (*c* 10)
28	D-Galactose, 6-deoxy-3-*O*-methyl-(Digitalose)	$C_7H_{14}O_5$	106[1], 119[2]	+106
29	D-Galactose, 6-deoxy-4-*O*-methyl-	$C_7H_{14}O_5$	131–132	+82
30	D-Galactose, 6-deoxy-2,3-di-*O*-methyl-	$C_8H_{16}O_5$	....	+73
31	α-D-Galactose, 3-*O*-methyl-	$C_7H_{14}O_6$	144–147	+150.6 → +108.6
32	α-D-Galactose, 6-*O*-methyl-	$C_7H_{14}O_6$	122–123	+117 → +77.3
33	L-Galactose	$C_6H_{12}O_6$	....	*See* D-Galactose
34	α-L-Galactose, 3,6-anhydro-	$C_6H_{10}O_5$	....	−39.4 → −25.2
35	α-L-Galactose, 6-deoxy- (L-Fucose)	$C_6H_{12}O_5$	145	−124.1 → −76.4
36	L-Galactose, 6-deoxy-2-*O*-methyl-	$C_7H_{14}O_5$	149–150	−75 ± 4 (*c* 0.5)
37	L-Galactose, 6-sulfate	$C_6H_{12}O_9S$	....	−47 (*c* 0.2) (Na salt)
38	DL-Galactose	$C_6H_{12}O_6$	143–144, 163	None (racemic)
39	α-D-Glucose	$C_6H_{12}O_6$	146, 83 (H_2O)	+112 → +52.7
40	β-D-Glucose	$C_6H_{12}O_6$	148–150	+18.7 → +52.7
41	D-Glucose, 6-acetate	$C_7H_{14}O_7$	135	+48
42	D-Glucose, 2,3-di-*O*-methyl-	$C_8H_{16}O_6$	85–86, 121	+50
43	D-Glucose, 6-*O*-benzoyl- (Vaccinin)	$C_{13}H_{16}O_7$	Amorphous	+48 (C_2H_5OH)
44	α-D-Glucose, 6-deoxy- (Chinovose; Epirhamnose; Glucomethylose; Isorhamnose; Isorhodeose; Quinovose)	$C_6H_{12}O_5$	139–140	+73.3 → +29.7 (*c* 8)
45	α-D-Glucose, 6-deoxy-3-*O*-methyl-(D-Thevetose)	$C_7H_{14}O_5$	116	+84 → +33

	Substance (Synonym)	Chemical Formula	Melting Point °C	Specific Rotation $[\alpha]_D$
	(A)	(B)	(C)	(D)
	Aldoses (Con't)			
46	D-Glucose, 6-sulfonic acid, 6-deoxy- (6-Sulfoquinovose)	$C_6H_{12}O_8S$	173–174	+87[3]
47	D-Glucose, 3-*O*-methyl-	$C_7H_{14}O_6$	162–167	+98 → +59.5
48	α-L-Glucose	$C_6H_{12}O_6$	141–143	−95.5 → −51.4
49	L-Glucose, 6-deoxy-3-*O*-methyl- (L-Thevetose)	$C_7H_{14}O_5$	126–129	−36.9 ± 2
50	D-Gulose, 6-deoxy-	$C_6H_{12}O_5$	...	...
51	Hexose, 2-deoxy-D-*arabino*-[4]	$C_6H_{12}O_5$	148	+46.6 [18°]
52	Hexose, 2,6-dideoxy-3-*O*-methyl-D-*arabino*- (D-Oleandrose)	$C_7H_{14}O_4$	...	−11
53	Hexose, 3,6-dideoxy-D-*arabino*- (Tyvelose)	$C_6H_{12}O_4$	...	+24 ± 2
54	Hexose, 2,6-dideoxy-3-*O*-methyl-L-*arabino*- (L-Oleandrose)	$C_7H_{14}O_4$	62–63	+11.9 ± 2.5
55	Hexose, 3,6-dideoxy-L-*arabino*- (Ascarylose)	$C_6H_{12}O_4$	...	−24 ± 2
56	Hexose, 2,6-dideoxy-3-*O*-methyl-D-*lyxo*- (Diginose)	$C_7H_{14}O_4$	90–92	+56 ± 4
57	Hexose, 2,6-dideoxy-L-*lyxo*- (L-Fucose, 2-deoxy-)	$C_6H_{12}O_4$	103–106	−61.6
58	Hexose, 2,6-dideoxy-3-*O*-methyl-L-*lyxo*-	$C_7H_{14}O_4$	78–85	−65
59	Hexose, 2,6-dideoxy-D-*ribo*- (Digitoxose; D-Altrose, 2,6-dideoxy-)	$C_6H_{12}O_4$	110	+46.4
60	Hexose, 2,6-dideoxy-3-*O*-methyl-D-*ribo*- (Cymarose)	$C_7H_{14}O_4$	93	+52
61	Hexose, 3,6-dideoxy-D-*ribo*- (Paratose)	$C_6H_{12}O_4$	...	+10 ± 2 (*c* 0.9)
62	Hexose, 4,6-dideoxy-3-*O*-methyl-D-*ribo*- (D-Gulose, 4,6-dideoxy-3-*O*-methyl-; Chalcose)	$C_7H_{14}O_4$	96–99	+120 → +76
63	Hexose, 2,6-dideoxy-D-*xylo*- (Boivinose)	$C_6H_{12}O_4$	96–98	−3.9 → +3.9
64	Hexose, 2,6-dideoxy-3-*O*-methyl-D-*xylo*- (Sarmentose)	$C_7H_{14}O_4$	78–79	+12 → +15.8
65	Hexose, 3,6-dideoxy-D-*xylo*- (Abequose)	$C_6H_{12}O_4$	...	−3.2 ± 0.6
66	Hexose, 2,6-dideoxy-3-*C*-methyl-L-*xylo*- (Mycarose)	$C_7H_{14}O_4$	129–129	−31.1
67	Hexose, 2,6-dideoxy-3-*C*-methyl-3-*O*-methyl-L-*xylo*-(Cladinose)	$C_8H_{16}O_4$	oil, b.p. 120–132 (0.25 mm)	−23.1
68	Hexose, 3,6-dideoxy-L-*xylo*- (Colitose)	$C_6H_{12}O_4$	...	+4 (H_2O); −51 ± 2 (CH_3OH)
69	D-Idose[5]	$C_6H_{12}O_6$	...	...
70	L-Idose, 1,6-anhydro-	$C_6H_{10}O_5$	...	...
71	α-D-Mannose	$C_6H_{12}O_6$	133	+29.3 → +14.5
72	β-D-Mannose	$C_6H_{12}O_6$	132	−16.3 → +14.5
73	D-Mannose, 6-deoxy- (D-Rhamnose)	$C_6H_{12}O_5$	86–90	−7.0
74	α-L-Mannose, 6-deoxy-monohydrate (L-Rhamnose)	$C_6H_{14}O_6$	93–94	−8.6 → +8.2
75	β-L-Mannose, 6-deoxy-	$C_6H_{12}O_5$	123–125	+38.4 → +8.9
76	L-Mannose, 6-deoxy-2-*O*-methyl-	$C_7H_{14}O_5$	...	...
77	L-Mannose, 6-deoxy-3-*O*-methyl- (L-Acofriose)	$C_7H_{14}O_5$	114–115	+30 [18°]
78	L-Mannose, 6-deoxy-2,4-di-*O*-methyl-	$C_8H_{16}O_5$	82	−19 [16°]
79	L-Mannose, 6-deoxy-5-*C*-methyl-4-*O*-methyl-(Noviose)	$C_8H_{16}O_5$	128–130	+19.9 (50% C_2H_5OH)
80	Rhodinose (a 2,3,6-trideoxyhexose)	$C_6H_{12}O_3$	...	−11 ± 1.6
81	D-Talose	$C_6H_{12}O_6$	128–132	+16.9
82	D-Talose, 6-deoxy- (D-Talomethylose)	$C_6H_{12}O_5$	129–131	+20.6
83	L-Talose, 6-deoxy- (L-Talomethylose)	$C_6H_{12}O_5$	116–118	−19.5 ± 2 [18°]
84	L-Talose, 6-deoxy-2-*O*-methyl- (L-Acovenose)	$C_7H_{14}O_5$	...	−19.4
85	Heptose, D-*glycero*-D-*galacto*-	$C_7H_{14}O_7$	139–140	+47 → +64 (*c* 0.5)
86	Heptose, D-*glycero*-D-*manno*-	$C_7H_{14}O_7$	...	...
87	Heptose, D-*glycero*-L-*manno*-	$C_7H_{14}O_7$	...	...
	Ketoses			
88	Dihydroxyacetone	$C_3H_6O_3$	80 (dimer)	None
89	Tetrulose, L-*glycero*-[8] (L-Erythrulose; Ketoerythritol; L-Threulose)	$C_4H_8O_4$	Syrup	+12
90	Pentulose, D-*erythro*- (Adonose; D-Ribulose)	$C_5H_{10}O_5$	Syrup	+16.6 [27°]
91	Pentulose, L-*erythro*- (L-Ribulose)	$C_5H_{10}O_5$	...	−16.6

CARBOHYDRATES (Continued)

Part I. NATURAL MONOSACCHARIDES: ALDOSES AND KETOSES (Continued)

	Substance (Synonym)	Chemical Formula	Melting Point °C	Specific Rotation $[\alpha]_D$
	(A)	(B)	(C)	(D)
	Ketoses (Con't)			
92	Pentulose, D-*threo*- (D-Xylulose)	$C_5H_{10}O_5$	....	−33
93	Pentulose, 5-deoxy-D-*threo*-	$C_5H_{10}O_4$	....	−5 ± 1 (CH_3OH)
94	Pentulose, L-*threo*- (L-Xylulose; L-Lyxulose; Xyloketose)	$C_5H_{10}O_5$	Syrup	+33.1
95	Hexulose, β-D-*arabino*-(β-D-Fructose; Levulose)	$C_6H_{12}O_6$	102–104[7]	−133.5 → −92
96	Hexulose, 6-deoxy-D-*arabino*- (D-Rhamnulose)	$C_6H_{12}O_5$	....	−13 ± 2
97	Hexulose, D-*lyxo*- (D-Tagatose)	$C_6H_{12}O_6$	131–132	+2.7 → −4, −5
98	5-Hexulose, D-*lyxo*	$C_6H_{12}O_6$	158	−86.6
99	Hexulose, 6-deoxy-L-*lyxo*- (L-Fuculose)	$C_6H_{12}O_5$	....	....
100	Hexulose, D-*ribo*- (D-Psicose)	$C_6H_{12}O_6$	Amorphous	+4.7
101	Hexulose, L-*xylo*- (L-Sorbose)	$C_6H_{12}O_6$	159–161	−43.1
102	Hexulose, 6-deoxy-L-*xylo*-	$C_6H_{12}O_5$	88	−25 ± 2 (*c* 0.7)
103	Heptulose, D-*altro*- (Sedoheptulose; Sedoheptose)	$C_7H_{14}O_7$	Amorphous	+2.5 (*c* 10)
104	Heptulose·hemihydrate, L-*galacto*- (Perseulose)	$C_7H_{14}O_7 \cdot \frac{1}{2}H_2O$	110–115	−90 → −80
105	Heptulose, L-*gulo*-	$C_7H_{14}O_7$	....	−28
106	Heptulose, D-*ido*-	$C_7H_{14}O_7$	172	−34 ± 8 (*c* 0.3)
107	Heptulose, D-*manno*- (Mannoketoheptose; D-Mannotagatoheptose)	$C_7H_{14}O_7$	152	+29.4
108	Heptulose, D-*talo*-	$C_7H_{14}O_7$	....	....
109	Octulose, D-*glycero*-L-*galacto*-	$C_8H_{16}O_8$	....	−57, −43.4 → −13.4
110	Octulose, D-*glycero*-D-*manno*-	$C_8H_{16}O_8$	....	+20 (CH_3OH)

[1] Original melting point. [2] Melting point after four-months' storage. [3] As a methyl glycoside cyclohexylamine salt. [4] Included because of speculations concerning it in biological processes. [5] Either D-idose or L-altrose is in the polysaccharide varianose. [6] Early literature refers to this as D-erythrose. [7] The $\cdot\frac{1}{2}H_2O$ and $\cdot 2H_2O$ forms also exist.

Part II. NATURAL MONOSACCHARIDES: AMINO SUGARS

	Substance (Synonym)	Chemical Formula	Melting Point °C	Specific Rotation $[\alpha]_D$
	(A)	(B)	(C)	(D)
	Aldosamines			
1	D-Ribose, 3-amino-3-deoxy-	$C_5H_{11}NO_4$	158–158.5 d.	−24.6 (hydrochloride)
2	D-Galactose, 2-amino-2-deoxy- (Galactosamine; Chondrosamine)	$C_6H_{13}NO_5$	185	+121 → +80 (hydrochloride)
3	α-L-Galactose, 2-amino-2,6-dideoxy- (L-Fucosamine)	$C_6H_{13}NO_4$	192–193 d.	−119 → −92 [27°] (hydrochloride)
4	α-D-Glucose, 2-amino-2-deoxy- (Glucosamine; Chitosamine)	$C_6H_{13}NO_5$	88	+100 → +47.5
5	β-D-Glucose, 2-amino-2-deoxy-	$C_6H_{13}NO_5$	110–111	+28 → +47.5
6	D-Glucose, 3-amino-3-deoxy- (Kanosamine)	$C_6H_{13}NO_5$	128 d.	+19 [14°]
7	D-Glucose, 6-amino-6-deoxy-	$C_6H_{13}NO_5$	161–162 d.	+23 → +50.1 (hydrochloride)
8	D-Glucose, 2,6-diamino-2,6-dideoxy- (Neosamine C)	$C_6H_{14}N_2O_4$	>230	+61.5 (dihydrochloride)
9	D-Glucose, 3,6-dideoxy-3-dimethylamino- (Mycaminose)	$C_8H_{17}NO_4$	115–116	+31 (hydrochloride)
10	D-Glucose, 4,6-dideoxy-4-dimethylamino-	$C_8H_{17}NO_4$	192–193	+45.5 (hydrochloride)
11	L-Glucose, 2-deoxy-2-methylamino-	$C_7H_{15}NO_5$	130–132	−64
12	D-Gulose, 2-amino-1,6-anhydro-2-deoxy-	$C_6H_{11}NO_4$	250–260 d.	+41 ± 2 (hydrochloride)
13	D-Gulose, 2-amino-2-deoxy-	$C_6H_{13}NO_5$	152–162 d.	+5.6 → −18.7 (hydrochloride)
14	Hexose, 3,4,6-trideoxy-3-dimethylamino-D-*xylo*- (Desosamine; Picrocine)	$C_8H_{17}NO_3$	189–191 d.	+49.5 (*c* 10) (hydrochloride)
15	Hexose, a 4-acetamido-2-amino-2,4,6-trideoxy-	$C_8H_{16}N_2O_4$	216–219	+115 → +94 [26°] (*c* 0.05)
16	Hexose, an amino-deoxy-3-*O*-carboxyethyl-	$C_9H_{17}NO_7$	....	....
17	Hexose, a 2,6-diamino-2,6-dideoxy- (Neosamine B; Paramose)	$C_6H_{14}N_2O_4$	135–150 d.	+17.5 (*c* 0.9 (hydrochloride)

Part II. NATURAL MONOSACCHARIDES: AMINO SUGARS (Continued)

	Substance (Synonym)	Chemical Formula	Melting Point °C	Specific Rotation $[\alpha]_D$
	(A)	(B)	(C)	(D)
	Aldosamines (Con't)			
18	Hexose, a 3-dimethylamino-2,3,6-trideoxy- (Rhodosamine)	$C_8H_{17}NO_3$	....	....
19	D-Mannose, 2-amino-2-deoxy- (Mannosamine)	$C_6H_{13}NO_5$	142 d.	−4.3 (*c* 9) (hydrochloride)
20	D-Mannose, 3-amino-3,6-dideoxy- (Mycosamine)	$C_6H_{13}NO_4$	162	−11.5 (hydrochloride)
21	D-Talose, 2-amino-2-deoxy- (Talosamine)	$C_6H_{13}NO_5$	151–153	+3.4 → −5.7 (*c* 0.9) (hydrochloride)
22	L-Talose, 2-amino-2,6-dideoxy- (Pneumosamine)	$C_6H_{13}NO_4$	162–163	+6.9 → +10.4 (hydrochloride)
	Ketosamines			
23	Pentulose, 1-(*o*-carboxyanilino)-1-deoxy-D-*erythro*-	$C_{12}H_{14}NO_6$	....	....
24	Hexulose, 1-(*o*-carboxyanilino)-1-deoxy-D-*arabino*-	$C_{13}H_{16}NO_7$	....	....
25	Hexulose, 5-amino-5-deoxy-L-*xylo*-	$C_6H_{13}NO_5$	174–176	−62
26	Hexulose, 6-deoxy-6-(*N*-methylacetamido)-L-*xylo*-	$C_9H_{17}NO_6$	....	....

Part III. NATURAL ALDITOLS AND INOSITOLS (with Inososes and Inosamines)

	Substance (Synonym)	Chemical Formula	Melting Point °C	Specific Rotation $[\alpha]_D$
	(A)	(B)	(C)	(D)
	Alditols			
1	Glycerol	$C_3H_8O_3$	20	None
2	Glycerol, 1-deoxy- (1,2-Propane-diol)[1]	$C_3H_8O_2$	Oil, b.p. 188–189	None (racemic)
3	Erythritol	$C_4H_{10}O_4$	118–120	None (meso)
4	Erythritol, 1,4-dideoxy- (2,3-Butyleneglycol)	$C_4H_{10}O_2$	25, 34	None (meso)
5	D-Threitol, 1,4-dideoxy-	$C_4H_{10}O_2$	19	−13.0
6	L-Threitol, 1,4-dideoxy-	$C_4H_{10}O_2$	....	+10.2
7	DL-Threitol, 1,4-dideoxy-	$C_4H_{10}O_2$	7.6	None (racemic)
8	D-Arabinitol	$C_5H_{12}O_5$	103	+7.82 (*c* 8, borax solution)
9	L-Arabinitol	$C_5H_{12}O_5$	101–102	−32 (*c* 0.4, 5% molybdate)
10	Ribitol (Adomitol)	$C_5H_{12}O_5$	102	None (meso)
11	Galactitol (Dulcitol)	$C_6H_{14}O_6$	186–188	None (meso)
12	D-Glucitol (Sorbitol)	$C_6H_{14}O_6$	112	−1.8 [15°]
13	D-Glucitol, 1,5-anhydro- (Polygalitol)	$C_6H_{12}O_5$	140–141	+42.4
14	L-Iditol	$C_6H_{14}O_6$	73.5	−3.5 (*c* 10)
15	D-Mannitol	$C_6H_{14}O_6$	166	−0.21
16	D-Mannitol, 1,5-anhydro- (Styracitol)	$C_6H_{12}O_5$	157	−49.9
17	Heptitol, D-*glycero*-D-*galacto*- (Heptitol, L-*glycero*-D-*manno*-; Perseitol)	$C_7H_{16}O_7$	183–185, 188	−1.1
18	Heptitol, D-*glycero*-D-*gluco*- (Heptitol, L-*glycero*-D-*talo*-; β-Sedoheptitol)	$C_7H_{16}O_7$	131–132	+46 (5% NH_4 molybdate)
19	Heptitol, D-*glycero*-D-*manno*- (Heptitol, D-*glycero*-D-*talo*-; Volemitol)	$C_7H_{16}O_7$	153	+2.65
20	Octitol, D-*erythro*-D-*galacto*-	$C_8H_{18}O_8 \cdot H_2O$	169–170	−11 (5% NH_4 molybdate)
	Inositols			
21	Betitol (a dideoxy inositol)	$C_6H_{12}O_4$	224	....
22	Bioinosose (*scyllo*-Inosose; *myo*-Inosose-2; a deoxy keto inositol)	$C_6H_{10}O_6$	198–200	None (meso)
23	*h*-Bornesitol (a *myo*-inositol monomethyl ether)	$C_7H_{14}O_6$	200	+31.6
24	*l*-Bornesitol (a *myo*-inositol monomethyl ether)	$C_7H_{14}O_6$	205–206	−32.1
25	Conduritol (a 2,3-dehydro-2,3-dideoxyinositol)	$C_6H_{10}O_4$	142–143	None (meso)
26	Cordycepic acid (a tetrahydroxycyclohexanecarboxylic acid)[2]	$C_7H_{12}O_6$	....	....
27	Dambonitol (a *myo*-inositol dimethyl ether)	$C_8H_{16}O_6$	206	None (meso)
28	DL-Inositol	$C_6H_{12}O_6$	253	None (racemic)

CARBOHYDRATES (Continued)
Part III. NATURAL ALDITOLS AND INOSITOLS
(with Inososes and Inosamines) (Continued)

	Substance (Synonym)	Chemical Formula	Melting Point °C	Specific Rotation $[\alpha]_D$
	(A)	(B)	(C)	(D)
	Inositols (Con't)			
29	*d*-Inositol	$C_6H_{12}O_6$	....	+60
30	*l*-Inositol	$C_6H_{12}O_6$	240	−65
31	Laminitol (a *C*-methyl *myo*-inositol)	$C_7H_{14}O_6$	266–269	3
32	Liriodendritol (a *myo*-inositol dimethyl ether)	$C_8H_{16}O_6$	224	−25
33	*muco*-Inositol monomethyl ether	$C_7H_{14}O_6$	322–325	....
34	*myo*-Inositol (*meso*-Inositol)	$C_6H_{12}O_6$	217–218	None (meso)
35	*d*-*myo*-Inosose-1 (a deoxy keto inositol)	$C_6H_{10}O_6$	138–139	+19.6
36	Mytilitol (a *C*-methyl *scyllo*-inositol)	$C_7H_{14}O_6$	259	None (meso)
37	*neo*-Inosamine-2 (a deoxy amino inositol)	$C_6H_{13}O_5N$	239–241 d.	None (meso)
38	*d*-Ononitol (a *myo*-inositol monomethyl ether)	$C_7H_{14}O_6$	172	+6.6
39	*h*-Pinitol (a *dextro*-inositol monomethyl ether)	$C_7H_{14}O_6$	186	+65.5
40	*l*-Pinitol (a *levo*-inositol monomethyl ether)	$C_7H_{14}O_6$	186	−65
41	*l*-Quebrachitol (a *levo*-inositol monomethyl ether)	$C_7H_{14}O_6$	190–191	−80.2 [28°]
42	*d*-Quercitol (a deoxy *dextro*-inositol)	$C_6H_{12}O_5$	235	+24.2
43	*d*-Quinic acid (a trideoxy carboxy *dextro*-inositol)	$C_7H_{12}O_6$	164	+44 (*c* 10)
44	*l*-Quinic acid (a trideoxy carboxy *levo*-inositol)	$C_7H_{12}O_6$	162	−42.1
45	Quinic acid, 5-dehydro-	$C_7H_{10}O_6$	140–142 (138 s.)	−82.4 [28°]
46	Scyllitol (*scyllo*-Inositol; Cocositol)	$C_6H_{12}O_6$	352–353	None (meso)
47	Sequoyitol (a *myo*-inositol monomethyl ether)	$C_7H_{14}O_6$	234–235	None (meso)
48	Shikimic acid (a 3,4-anhydro-quinic acid)	$C_7H_{10}O_5$	183–184	−200 [16°]
49	Shikimic acid, 5-dehydro-	$C_7H_8O_5$	150–152	−57.5 [28°] (EtOH)
50	Streptamine (2,4-diaminodideoxy-scyllitol)	$C_6H_{14}O_4N_2$	88, 210–250 d.	None (meso)
51	Streptamine, 2-deoxy-	$C_6H_{14}O_3N_2$	....	None (meso)
52	Streptadine (1,3-Dideoxy-1,3-diguanidino-scyllitol)	$C_8H_{18}N_6O_4$	....	None (meso)
53	Viburnitol (a deoxy *levo*-inositol)[3]	$C_6H_{12}O_5$	174	−73.9

[1] The 1-phosphate ester of this diol is said to occur in brain tissue and sea-urchin eggs. [2] Strong evidence that cordycepic acid is really D-mannitol. [3] Not an enantiomorph of *d*-quercitol; other isomeric relationship is involved.

Part IV. NATURAL ALDONIC, URONIC, AND ALDARIC ACIDS

	Substance (Synonym)	Chemical Formula	Melting Point °C	Specific Rotation $[\alpha]_D$
	(A)	(B)	(C)	(D)
	Aldonic Acids			
1	D-Glyceric acid	$C_3H_6O_4$	Gum	Dextro
2	L-Glyceric acid	$C_3H_6O_4$	Gum	Levo
3	D-Arabinonic acid	$C_5H_{10}O_6$	114–116	+10.5 (*c* 6)
4	L-Arabinonic acid	$C_5H_{10}O_6$	118–119	−9.6 → −41.7[1]
5	L-Arabinonic-1,4-lactone	$C_5H_8O_5$	97–99	−72
6	D-Ribonic acid	$C_5H_{10}O_6$	112–113	−17.0
7	D-Xylonic acid	$C_5H_{10}O_6$	....	−2.9 → +20.1[1]
8	L-Xylonic acid	$C_5H_{10}O_6$	....	−91.8[1]
9	D-Altronic acid	$C_6H_{12}O_7$	....	+11.5 → +24.8[1] (Ca salt, *N* HCl)
10	D-Galactonic acid	$C_6H_{12}O_7$	122	−11.2 → +57.6[1]
11	D-Gluconic acid	$C_6H_{12}O_7$	130–132 (110–112 s.)	−6.7 → +11.9[1]
12	L-Gulonic acid	$C_6H_{12}O_7$	Exists only in soln.	[ca. 0°]
13	Hexsonic acid, 2-deoxy-D-*arabino*-	$C_6H_{12}O_6$	93–95	+68 (lactone)
14	2-Hexulosonic acid, D-*arabino*-	$C_6H_{10}O_7$	....	−81.7 (Na salt)
15	2-Hexulosonic acid, 3-deoxy-D-*erythro*-	$C_6H_{10}O_6$	....	−29.2 (*c* 6, Ca salt)
16	2-Hexulosonic acid, D-*lyxo*-	$C_6H_{10}O_7$	169	−5
17	5-Hexulosonic acid, D-*arabino*-	$C_6H_{10}O_7$	108–109	....
18	5-Hexulosonic acid, D-*xylo*-	$C_6H_{10}O_7$	....	−14.5
19	D-Mannonic acid	$C_6H_{12}O_7$	....	−15.6
20	D-Gluconic acid, *O*-β-D-galactopyranosyl- (1 → 4)- (Lactobionic acid)	$C_{12}H_{22}O_{12}$	....	+25.1 (Ca salt)

Part IV. NATURAL ALDONIC, URONIC, AND ALDARIC ACIDS (Continued)

	Substance (Synonym)	Chemical Formula	Melting Point °C	Specific Rotation $[\alpha]_D$
	(A)	(B)	(C)	(D)
	Uronic Acids			
21	L-Lyxuronic acid	$C_5H_8O_6$	....	....
22	β-D-Galacturonic acid	$C_6H_{10}O_7$	160	+27 → +55.6
23	α-D-Galacturonic acid·monohydrate	$C_6H_{12}O_8$	159–160 (110–115 s.)	+97.9 → +50.9
24	D-Galacturonic acid, 2-amino-2-deoxy-	$C_6H_{11}O_6N$	160 d.	+84.5 (pH 2 HCl)
25	β-D-Glucuronic acid	$C_6H_{10}O_7$	156	+11.7 → +36.3
26	D-Glucuronic acid, 2-amino-2-deoxy-	$C_6H_{11}O_6N$	120–172 d.	+55
27	D-Glucuronic acid, 3-*O*-methyl-	$C_7H_{12}O_7$	Syrup	+6
28	L-Guluronic acid	$C_6H_{10}O_7$	....	....
29	L-Iduronic acid	$C_6H_{10}O_7$	....	+30
30	β-D-Mannuronic acid	$C_6H_{10}O_7$	165–167	−47.9 → −23.9
31	α-D-Mannuronic acid·monohydrate	$C_6H_{12}O_8$	110 s., 120–130 d.	+16 → −6.1 (*c* 6.8)
	Aldaric Acids			
32	D-Tartaric acid	$C_4H_6O_6$	170	−15
33	L-Tartaric acid	$C_4H_6O_6$	170	+15 [15°]
34	L-Malic acid	$C_4H_6O_5$	100	−2.3 (*c* 8.4)

[1] Equilibrates with the lactone.

WAXES

These data for waxes were compiled originally for the Biology Data Book by A. H. Warth. Data are reproduced here by permission of the copyright owners of the above publication, the American Societies for Experimental Biology, Washington, D.C. p. 382.

Specific Gravity (column C) was calculated at the specified temperature, degrees centigrade, and referred to water at the same temperature. **Density.** shown in parentheses (column C), and **Refractive Index** (column D) were measured at the specified temperature, degrees centigrade.

	Wax	Melting Point °C	Specific Gravity or (Density)	Refractive Index $n\frac{°C}{D}$	Iodine Value	Acid Value	Saponification Value
	(A)	(B)	(C)	(D)	(E)	(F)	(G)
1	Bamboo leaf	79–80	(0.961$^{25°}$)	....	7.8[1]	14.5	43.4
2	Bayberry (myrtle)	46.7–48.8	(0.985$^{15°}$)	1.436$^{80°}$	2.9[2]–3.9[3]	3.5	20.5–21.7
3	Beeswax, crude	62–66	(0.927–0.970$^{15°}$)	1.439–1.483$^{40°}$	6.8–16.4[2]	16.8–35.8	89.3–149.0
4	Beeswax, white, U.S.P.	61–69	(0.959–0.975$^{15°}$)	1.447–1.465$^{65°}$	7–11[3]	17–24	90–96
5	Beeswax, yellow	62–65	(0.960–0.964$^{15°}$)	1.443–1.449$^{65°}$	6–11	18–24	90–97
6	Candelilla, refined	67–69	(0.982–0.986$^{15°}$)	1.454–1.463$^{85°}$	14.4–20.4	12.7–18.1	35–86
7	Cape berry[4]	40.5–45.0	(1.004–1.007$^{15°}$)	1.450$^{45°}$	0.6–2.4	2.5–3.7	211–215
8	Carandá	79.7–84.5	(0.990$^{25°}$)	....	8.0–8.9	5.0–9.5	64.5–78.5
9	Carnauba	83–86	0.990–1.001$^{15°}$	1.467–1.472$^{40°}$	7.2–13.5	2.9–9.7	78–95
10	Castor oil, hydrogenated	83–88	(0.980–0.990$^{20°}$)	....	2.5–8.5	1.0–5.0	177–181
11	Chinese insect	81.5–84.0	0.950–0.970$^{15°}$	1.457$^{40°}$	1.4	0.2–1.5	73–93
12	Cotton	68–71	0.959$^{15°}$	....	24.5	32	70.6
13	Cranberry	207–218	(0.970–0.975$^{15°}$)	....	44.2–53.2[2]	42.2–59.1	131–134
14	Douglas-fir bark	59.0–72.8	(1.030$^{25°}$)	1.468$^{80°}$	25.8–62.5	58.6–80.7	112–200
15	Esparto	67.5–78.1	0.988$^{15°}$	....	22–23	22.7–23.9	69.8–79.3
16	Flax	61.5–69.8	0.908–0.985$^{15°}$	....	21.6–28.8	17.5–48.3	77.5–101.5
17	Ghedda, E. Indian beeswax	60.5–66.4	0.956–0.973$^{15°}$	1.440$^{50°}$	5.6–12.6	5.8–7.9	84.5–118.3
18	Indian corn	80–81	....	....	4.2[2]	1.9	120.3
19	Japan wax	48–53	0.975–0.993$^{15°}$	....	4.5–12.5	6–20	206.5–237.5
20	Jojoba	11.2–11.8	0.864–0.899$^{25°}$	1.465$^{25°}$	81.7–88.4[2]	0.2–0.6	92.2–95.0
21	Madagascar	88	....	....	3.2–5.3	17.7–28.0	140.0–159.6
22	Microcrystalline, amber	64–91	0.913–0.943$^{15°}$	1.424–1.452$^{80°}$	0	0	0
23	Microcrystalline, white	71–89	0.928–0.941$^{15°}$	1.441$^{80°}$	0	0	0
24	Montan, crude	76–86	(1.010–1.020$^{25°}$)	....	13.9–17.6	22.7–31.0	59.4–92.0
25	Montan, refined	77–84	(1.010–1.030$^{25°}$)	....	10–14	24–43	72–103
26	Orange peel	44.0–46.5	0.985$^{15°}$	1.502$^{20°}$	115.7[2]	48.3	120.9
27	Ouricury, refined	79.0–83.8	1.053$^{15°}$	....	6.9–7.8[2]	3.4–21.1	61.8–85.8
28	Ozocerite, refined	74.4–75.0	0.907–0.920$^{15°}$	....	0	0	0
29	Palm	74–86	(0.991–1.045$^{15°}$)	....	8.9–16.9[2]	5.0–10.6	64.5–104.0
30	Paraffin, American	49–63	0.896–0.925$^{15°}$	1.442–1.448$^{80°}$	0	0	0
31	Peat wax, natural	73–76	0.980$^{15°}$	....	16–40	60.0–73.3	73.9–136.0
32	Rice bran, refined	75.3–79.9	....	1.469$^{30°}$	11.1–19.4	15–17	56.9–104.4

WAXES (Continued)

	Wax	Melting Point °C	Specific Gravity or (Density)	Refractive Index $n\frac{°C}{D}$	Iodine Value	Acid Value	Saponification Value
	(A)	(B)	(C)	(D)	(E)	(F)	(G)
33	Shellac wax	79–82	0.971–0.980$^{15°}$	....	6.0–8.8[3]	12.1–24.3	63.8–83.0
34	Sisal hemp	74–81	1.007–1.010$^{15°}$	....	28–29[2]	16–19[2]	56–58
35	Sorghum grain	77–82	....	....	15.7–20.9	10.1–16.2	16–44
36	Spanish moss	79–80	....	....	33.0	25.0	120.4
37	Spermaceti	42–50	0.905–0.945$^{15°}$	1.440$^{70°}$	4.8–5.9	2.0–5.2	108–134
38	Sugarcane, crude	52–67	0.988–0.998$^{25°}$	....	32–84	24–57	128–177
39	Sugarcane, double-refined	77–82	0.961–0.979$^{25°}$	1.510$^{25°}$	13–29	8–23	55–95
40	Wool wax, refined	36–43	0.932–0.945$^{15°}$	1.478–1.482$^{40°}$	15.0–46.9	5.6–22.0	80–127

[1] Wijs test. [2] Hanus test. [3] Hubl test. [4] *Myrica cordifolia*.

TRADE NAMES OF DYESTUFF INTERMEDIATES

Trade name	Chemical name
A acid	1,7-Hydroxynaphthalene-3,6-disulfonic acid
Acetyl H acid	N-Acetyl-1-amino-8-naphthol-3,6-disulfonic acid
Alen's acid	1-Naphthylamine-3,6-disulfonic acid (also Freund's ac.)
Alizarin	1,2-Dihydroxyanthraquinone
Amido acid	2-Amino-7-hydroxynaphthalene-5-sulfonic acid
Amido J acid	2-Naphthylamine-5,7-disulfonic acid
Amino G acid	7-Amino-1,3-naphthalene disulfonic acid
	2-Naphthylamine-6,8-disulfonic acid
Amino R acid	3-Amino-2,7-naphthalene disulfonic acid
	2-Naphthylamine-3,6-disulfonic acid
Aminophenolic acid V	1-Amino-3-oxybenzene-5-sulfonic acid
Aminophenol sulfonic acid III	1-Amino-3-oxybenzene-6-sulfonic acid
Andresen's acid	1-Naphthol-3,8-disulfonic acid
Anisidine	o-Aminophenol methylether
Anthrachrysone	1,3,5,7-Tetrahydroanthraquinone
Anthraflavic acid	2,6-Dihydroxyanthraquinone
Anthranilic acid	o-Aminobenzoic acid
Anthrarufin	1,5-Dihydroxyanthraquinone
Anthranol	9-Hydroxyanthracene
α Anthrol	1-Hydroxyanthracene
Armstrong's acid	Naphthalene-1,5-disulfonic acid
Armstrong & Wynne acid	1-Naphthol-3-sulfonic acid
Armstrong & Wynne acid II	2-Naphthylamino-5,7-disulfonic acid
B acid	8-Amino-1-naphthol-4,6-disulfonic acid
Badische acid	2-Naphthylamino-8-sulfonic acid
Bayer's acid	2-Naphthol-8-sulfonic acid
Benzidine	p,p′-Diaminodiphenyl
Bronner's acid	2-Naphthylamino-6-sulfonic acid
β acid	Anthraquinone-2-sulfonic acid
C acid } CLT acid }	6-Chloro-m-toluidine-4-sulfonic acid
Casella's acid	2-Naphthol-7-sulfonic acid (F acid)
Chicago acid	1-Amino-8-naphthyl-2,4-disulfonic acid
Chloro H acid	8-Chloro-1-naphthol-3,6-disulfonic acid
Chromogene I	4,5-Dihydroxy-2,7-naphthalene disulfonic acid
Chromotrope acid	1,8-Dihydronaphthalene-3,6-disulfonic acid
Chromotropic acid	4,5-Dihydroxy-2,7-naphthalene disulfonic acid
Chrysazine	1,8-Dihydroxyanthraquinone
Cleve's acid	1-Naphthylamine-3-sulfonic acid
Cleve's acid	1-Naphthylamine-5-sulfonic acid
Cleve's acid	1-Naphthylamine-6-sulfonic acid
Cleve's acid	1-Naphthylamine-7-sulfonic acid
α Coccinic acid	1-Methyl-5-oxybenzyl-2,4-dicarbonic acid
Cresidine	3-Amino-4-methoxytoluene
Cresotic acid	Cresol carboxylic acid
Croceine acid	2-Naphthol-1-sulfonic acid
DS	4,4′-Diamino-2,2′-stilbene disulfuric acid
DTS	Dihydrothio-p-toluidine sulfonic acid
Dahl's acid	2-Naphthylamine-5-sulfonic acid
Dahl's acid II	1-Naphthylamine-4,6-disulfonic acid
Dahl's acid III	1-Naphthylamine-4,7-disulfonic acid
Dimethyl-γ-acid	7-Dimethylamino-1-naphthol sulfonic acid
Dioxy G acid	1,7-Dihydroxynaphthalene-3-sulfonic acid
Dioxy J acid	1,6-Dihydroxynaphthalene-3-sulfonic acid
Dioxy S acid	1,8-Dihydroxynaphthalene-4-sulfonic acid
Diphenylblack base	p-Aminodiphenylamine

TRADE NAMES OF DYESTUFF INTERMEDIATES (Continued)

Trade name	Chemical name
Disulfo acid S	1-Naphthylamine-4,8-disulfonic acid
δ acid	1-Naphthol-4,8-disulfonic acid 1-Naphthylamine-4,8-disulfonic acid
Ebert & Merz acid	Naphthalene-2,7-disulfonic acid Naphthalene-2,6-disulfonic acid
Ethyl-γ-acid	7-Ethylamino-1-naphthol-3-sulfonic acid
Ethyl F acid	7-Ethylamino-2-naphthalene sulfonic acid
Ewer & Pick's acid	Naphthalene-1,6-disulfonic acid
ε acid	1-Naphthol-3,8-disulfonic acid 1-Naphthylamine-3,8-disulfonic acid
F acid	2-Naphthol-7-sulfonic acid (Casella's acid)
Fast Black B base	4,4′-Diamino diphenylamine
Fast Blue base	Dianisidine
Fast Blue Red O base	3-Nitro-p-phenetidine
Fast Bordeaux GP	3-Nitro-p-anisidine
Fast Orange GR	o-Nitroaniline
Fast Orange R	m-Nitroaniline
Fast Red base AL	α-Aminoanthraquinone
Fast Red B base	5-Nitro-o-anisidine
Fast Red GG base	p-Nitroaniline
Fast Red GL base	3-Nitro-p-toluidine
Fast Red 3 GL base	4-Chloro-2-nitroaniline
Fast Red RL base	5-Nitro-o-toluidine
Fast Scarlet G base	4-Nitro-α-toluidine
Fast Scarlet R base	4-Nitro-O-anisidine
Forsling's acid I	2-Naphthylamine-8-sulfonic acid
Forsling's acid II	2-Naphthylamine-5-sulfonic acid
Freund's acid	1-Naphthylamine-3,6-disulfonic acid
G acid	2-Naphthol-6,8-disulfonic acid
GR acid	α-Naphthol-3,6-disulfonic acid
Gallic acid	3,4,5-Trihydroxybenzoic acid
γ acid	2-Amino-8-naphthol-6-sulfonic acid
H acid	1-Amino-8-naphthol-3,6-disulfonic acid
Histazarin	2,3-Dihydroxyanthraquinone
Isoanthraflavic acid	2,7-Dihydroxyanthraquinone
J acid	2-Amino-5-naphthol-7-sulfonic acid
K acid	1-Amino-8-naphthol-4,6-disulfonic acid
Kalle's acid	1-Naphthylamine-2,7-disulfonic acid
Ketone base	Tetramethyl aminobenzophenone
Koch's acid	1-Naphthylamine-3,6,8-trisulfonic acid
L acid	1-Naphthol-5-sulfonic acid
Laurent's acid	1-Naphthylamine-5-sulfonic acid
Lepidine	4-Methylquinoline
Leucotrop	Phenyldimethyl benzylammonium chloride
M acid	1-Amino-5-naphthol-7-sulfonic acid
Mesidine	2,4,6-Thimethylaniline
Metanilic acid	Aniline-m-sulfonic acid
Methyl-γ-acid	7-Methyl-8-naphthol disulfonic acid
Michler's hydrol	Tetramethyl diaminobenzohydrol
Michler's ketone	Tetramethyl diaminobenzophenone
Myrbane oil	Nitrobenzene
Naphthacetol	1-Acetylamino-4-naphthol
Naphthazarin	5,8-Dihydroxy-1,4-naphthoquinone
Naphthionic acid	1-Naphthylamine-4-sulfonic acid
o-Naphthionic acid	1-Naphthylamine-2-sulfonic acid
Naphthol AS	Anilide of hydronaphthoic acid
Naphthoresorcine	1,3-Dihydroxynaphthalene
Nekal BX	Na-salt of 1,4-bis,sec-butylnaphthalene-6-sulfonic acid
Nevile and Winther's acid	1-Naphthol-4-sulfonic acid
Nigrotic acid	1,3,6,7-Dihydroxysulfonaphthoic acid
Nitron 1,2,4-acid	1-Amino-8-nitro-7-naphthol-4-sulfonic acid
Nitroso base	p-Nitrodimethyl aniline
NW acid	Nevile and Winther s acid
Oxy L acid	1-Naphthol-5-sulfonic acid
Oxy Tobias acid	β-Naphthol-1-sulfonic acid
Peri acid	1-Naphthylamine-8-sulfonic acid
p-Phenetidine	p-Aminophenol ethylether
Phenyl gamma acid	2-Phenylamine-8-naphthol-6-sulfonic acid
Phenyl Peri acid	Phenyl-1-naphthylamine-8-sulfonic acid
Phosxgene	Carbonyl chloride
Phthalic acid	O-Benzenedicarbolic acid
Picramic acid	2-Amino-4,6-dinitrophenol
Picric acid	2,4,6-Trinitrophenol
Pirio's acid	4-Amino-1-naphthalene sulfonic acid
Primuline base	p-Toluidine heated with sulfur
Purpurine	1,2,4-Trihydroxyanthraquinone
Pyrogallol	1,2,3-Trihydroxybenzene
Quinaldine	2-Methylquinoline
Quinazarin	1,4-Dihydroxyanthraquinone

TRADE NAMES OF DYESTUFF INTERMEDIATES (Continued)

Trade name	Chemical name
R acid	2-Naphthol-3,6-disulfonic acid
2 R acid	2-Amino-8-naphthol-3,6-disulfonic acid
Red acid	1,5-Dihydroxynaphthalene-3,7-disulfonic acid
RG acid	1-Naphthol-3,6-disulfonic acid
Resorcinol	1,3-Dihydroxybenzene
Rumpff acid	2-Naphthol-8-sulfonic acid (Croceine acid)
S acid	1-Amino-8-naphthol-4-sulfonic acid
2 S acid	1-Amino-8-naphthol-2,4-disulfonic acid
Salicylic acid	o-Hydroxybenzoic acid
Schäffer's acid	2-Naphthol-6-sulfonic acid
Schäffer and Baum acid	α-Naphthol-2-sulfonic acid
Schollkopf's acid	1-Naphthol-4,8-disulfonic acid 1-Naphthylamine-4,8-disulfonic acid 1-Naphthylamine-8-sulfonic acid
Sulfanilic acid	Aniline-p-sulfonic acid
Thiocarbanilide	Diphenylthiourea
Tobias acid	2-Naphthylamine-1-sulfonic acid
Tolidine	Di-p-aminoditolyl
Toluidine	Aminotoluene
Violet acid	α-Naphthol-3,6-disulfuric acid
Xylidine	Aminoxylene
Y acid	2-Naphthol-6,8-disulfuric acid
Yellow acid	1,3-Dihydroxynaphthalene-5,7-disulfuric acid
1:2:4 acid	1-Amino-2-naphthol-4-sulfonic acid

NOMENCLATURE OF SOME MONOMERS AND POLYMERS

Calvin E. Schildknecht

The small molecules from which industrial polymers are formed are of two general types. The first are the unsaturated or ethylenic compounds. They are often called vinyl-type monomers. Table 1 gives formulas, common names and newer indexing names used by *Chemical Abstracts.* The second type of monomer useful in polymer synthesis are better called intermediates. These are shown in Table 2. Recent names as 2,5-Furandione for Maleic anhydride and 1,3-isobenzofurandione for Phthalic anhydride are employed in *Chemical Abstracts* for indexing, but older names are generally retained in the abstracts. Recent books[1-3] have used capital letters for abbreviations of polymer types, e.g., PVA for polyvinyl alcohols and VC-VAC for vinyl chloride-vinyl acetate copolymers. Preceding small letters may show proved structures, for example, i-PP for substantially isotactic or stereoregular polypropylenes. Experimentally determined physical properties may be added using small numbers and letters following for example, x-i-PP-mp 150-170C indicates a crosslinked isotactic polypropylene of crystal melting range 150-170C. Abbreviations used to indicate the type of polymerization processes are presented in Table 3.

Table 1
NAMES OF SOME ETHYLENIC MONOMERS

Formula	Common Name	New Name
$CH_2{=}CH_2$	Ethylene	Ethene
$CH_2{=}CH$	Vinyl (group)	Ethenyl
$CH_2{=}CHC_6H_5$	Styrene	Ethylbenzene
$CH_2{=}CHOOCCH_3$	Vinyl acetate	Ethenyl acetate
$CH_2{=}CHCl$	Vinyl chloride	Chloroethene
$CH_2{=}CHOCH_3$	Methyl vinyl ether	Methoxyethene
$CH_2{=}CHOCH{=}CH_2$	Divinyl ether	1,1-Oxybis (ethene)
$CH_2{=}CCl_2$	Vinylidene chloride	1,1-Dichloroethene
$CH_2{=}CHCOOH$	Acrylic acid	2-Propenoic acid
$CH_2{=}CHCN$	Acrylonitrile	2-Propenenitrile
$CH_2{=}CHCONH_2$	Acrylamide	2-Propenamide
$CH_2{=}CHCOOC_2H_5$	Ethyl acrylate	Ethyl 2-Propenoate
$CH_2{=}C(CH_3)COOH$	Methacrylic acid	2-Methyl-2-propenoic acid
$CH_2{=}C(CH_3)COOCH_3$	Methyl methacrylate	Ethyl 2-propenoate
$CH_2{=}CHCH_3$	Propylene	Propene
$CH_2{=}CHCH_2$	Allyl (group)	2-Propenyl
$CH_2{=}CHCH_2OH$	Allyl alcohol	2-Propen-1-ol
$CH_2{=}CHCH_2OOCCH_3$	Allyl acetate	2-Propenyl ethanoate

Table 2

SOME INTERMEDIATES FOR POLYMERS

Formula	Common name	New name
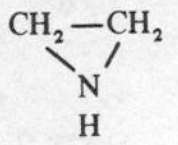	Ethylene imine	Aziridine
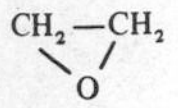	Ethylene oxide	Oxirane
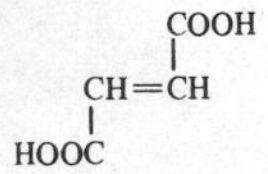	Fumaric acid	2-Butenedioic acid (Z)
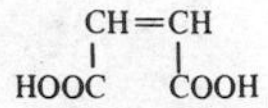	Maleic acid	2-Butenedioic acid (E)
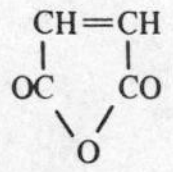	Maleic anhydride	2,5-Furandione
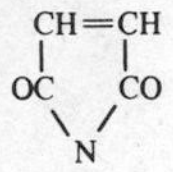	Maleimide	1 H-Pyrrole-2,5-dione
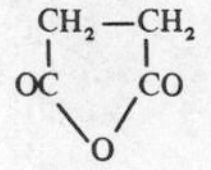	Succinic anhydride	Dihydro-2,5-furandione
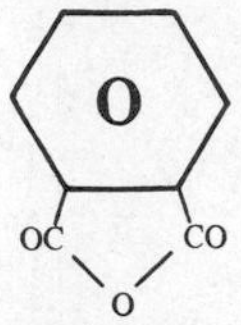	Phthalic anhydride	1,3-Isobenzofurandione
$HOOC(CH_2)_4COOH$	Adipic acid	Hexanedioic acid
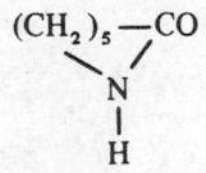	Caprolactam	Hexahydro-2H-azepin-2-one

Table 3

ABBREVIATIONS USED IN POLYMERIZATION PROCESSES

PE	An ethylene high polymer
hd-PE	High density polyethylene
S-AN	Copolymer of styrene and acrylonitrile
PIB	An isobutene high polymer
PMP	Poly-4-methyl-1-pentene
HD	1,4-Hexadiene
EP	Copolymer of ethylene + propylene
EPDM	Ethylene-propylene-difunctional monomer copolymer
VAC	Vinyl acetate
VC	Vinyl chloride
AN	Acrylonitrile
MMA	Methyl methacrylate
S	Styrene
ECH	Epichlorohydrin
PF	Phenol-formaldehyde polymer
MF	Melamine-formaldehyde polymer
DMP	2,6-dimethylphenol
DEB	m-Diethynylbenzene
DMT	Dimethyl terephthalate
BHET	bis(2-Hydroxyethyl)terephtholate
BCMO	3,3-bis (Chloromethyl)oxelane
PMA	Pyromellitic dionhydride
BPDA	Benzophenone tetracarboxylate dianhydride
TMA	Trimellitic anhydride
TDI	Toluene diisocyanate
PP	A propylene high polymer
i-PP	Substantially isotactic PP
syn-PP	Substantially syndiotactic PP (if it exists)
PB	Poly-1-butene
AA	Acrylamide
DAA	Diacetone acrylamide
PVAC	Vinyl acetate polymer
PVC	Vinyl chloride polymer
PVA	Polyvinyl alcohol
PMMA	Methyl methacrylate polymer
PS	A polystyrene
DGEBA	Diglycidyl ether of bisphenol A
UF	Ureo-formaldehyde polymer
THF	Tetrahydrofuran
PET	Polyethylene terephthalate
PBT	Polybutylene terephthalate
PO	1,2-Propylene oxide
PBI	Polybenzimidazole
MDI	Methylene bis (4-phenyl isocyanate)
MOCA	Methylene bis (o-chloroaniline)
PABM	Polyaminobismaleimide
TAHT	1,3,5-Triacryloyl hexahydrotriazine
HMM	Hexamethylalmelamine
HMMM	Hexamethoxymethylolmelamine

REFERENCES

1. **Schildknecht, C. E.,** *Mod. Paint Coat.*, 69 (6), 41—45, 1979.
2. **Schildknecht, C. E.,** *Vinyl and Related Polymers*, (1952); *Polymer Processes*, (editor, 1956); *Allyl Compounds and Their Polymers*, (1973); *Polymerization Processes*, (editor, 1977), Wiley-Interscience, New York.
3. **Billmeyer, F. W.,** *Textbook of Polymer Science*, 2nd ed., Wiley-Interscience, New York, 1971.

IONIC EXCHANGE RESINS

ANION EXCHANGE RESINS

The following table is divided into two parts; the first lists properties of some anionic resins and the second, properties of some cationic resins.

Character S = strong W = weak	Trade name	Manu-facturer*	Active group	Matrix	Effective pH	Selectivity	Order of selectivity	Total exchange capacity; meq/ml	Total exchange capacity; meq/gm	Maximum thermal stability; °C	Physical form; s = sphere b = beads	Standard mesh range	Ionic form as shipped	Shipping density; lb./cu. ft.
S	Dowex 1	1	Trimethyl benzyl ammonium	Polystyrene	0–14	Cl/H approx. 25	I, NO_3, Br, Cl, Acetate, OH, F	1.33	3.5	OH^- 50 Cl^- 150	s	20–50 (wet)	Cl^-	44
S	Dowex 21 K	1	Trimethyl benzyl ammonium	Polystyrene	0–14	Cl/H approx. 15	I, NO_3, Br, Cl, Acetate, OH, F	1.25	4.5	OH^- 50 Cl^- 150	s	20–50 (wet)	Cl^-	43
S	Duolite A-101 D	2	Quaternary ammonium	Polystyrene	0–14	—	—	1.4	4.2	OH^- 60 Cl^- 100	b	16–50	Cl^-	—
S	Ionac A-540	3	Quaternary ammonium	Polystyrene	0–14	—	—	1.0	3.6	salt 100 OH^- 60	b	16–50	salt	43–66
S	Dowex 2	1	Dimethyl ethanol benzyl ammonium	Polystyrene	0–14	Cl/H approx. 1.5	I, NO_3, Br, Cl, Acetate, OH, F	1.33	3.5	OH^- 30 Cl^- 150	s	20–50 (wet)	Cl^-	44
S	Duolite A-102 D	2	Quaternary ammonium	Polystyrene	0–14	—	—	1.4	4.2	OH^- 40 Cl^- 100	b	16–50	Cl^-	—
S	Ionac A-550	3	Dimethyl ethanol benzyl ammonium	Polystyrene	0–14	—	—	1.3	3.5	salt 100 OH^- 40	b	16–50	salt	43–46
W	Duolite A-30 B	2	Tertiary amine; Quaternary ammonium	Epoxy polyamines	0–9	—	—	2.6	8.7	80	b	16–50	salt	—
W	Ionac A-300	3	Tertiary amine; Quaternary ammonium	Epoxy amine	0–12	—	—	1.8	5.5	40	g	16–50	salt	19–21
W	Duolite A-6	2	Tertiary amine	Phenolic	0–5	—	—	2.4	7.6	60	g	16–50	salt	—
W	Duolite A-7	2	Secondary amine	Phenolic	0–4	—	—	2.4	9.1	40	g	16–50	salt	—

CATION EXCHANGE RESINS

Character S = strong W = weak	Trade name	Manu-facturer*	Active group	Matrix	Effective pH	Selectivity	Order of selectivity	Total exchange capacity; meq/ml	Total exchange capacity; meq/mg	Maximum thermal stability; °C	Physical form; s = sphere b = beads	Standard mesh range	Ionic form as shipped	Shipping density; lb./cu. ft.
S	Dowex 50	1	Nuclear sulfonic acid	Polystyrene	0–14	Na/H approx. 1.2	Ag, Cs, Rb, K, NH_4, Na, H, Li, Ba, Sr, Ca, Mg, Be	Na^+ 1.9 H^+ 1.7	Na^+ 4.8 H^+ 5.0	150	s	20–50 (wet)	H^+ or Na^+	H^+ 50 Na^+ 53
S	Dowex MPC-1	4	Nuclear sulfonic acid	Polystyrene	0–14	—	—	1.6–1.8 H^+ form	4.5–4.9 H^+ form	150	b	20–40 (wet)	Na^+	50
S	Duolite C-20	2	Nuclear sulfonic acid	Polystyrene	0–14	—	—	2.2	5.1	150	b	16–50	Na^+	—
S	Ionac 240	3	Nuclear sulfonic acid	Polystyrene	0–14	—	—	1.9	4.6	140 (Na^+) 130 (H^+)	b	16–50	Na^+	50–55
S	Duolite C-3	2	Methylene sulfonic	Phenolic	0–9	—	—	1.1	2.9	60	g	16–50	H^+	—
W	Dowex CCR-1	4	Carboxylic	Phenolic	0–9	—	—	—	—	38	g	20–50 (wet)	H^+ (dry)	21
W	Duolite ES-63	2	Phosphonic	Polystyrene	4–14	—	—	3.3	6.5	100	b	16–50	H^+	—
W	Duolite ES-80	2	Aliphatic	Acrylic	6–14	—	—	3.5	10.2	100	b	16–50	H^+	—

* 1. Dow
2. Diamond Shamrock
3. Ionac
4. Nalco

LIMITS OF SUPERHEAT OF PURE LIQUIDS

From the *Journal of Physical and Chemical Reference Data*, Volume 14, No. 3, 695, 1985. Reproduced by permission of the copyright owners, the American Chemical Society and the American Institute of Physics, and the author, C. T. Avedisian. One should refer to the original publication for discussions of the significance of the limit of superheat, limiting liquid superheat as a physical property, experimental methods for measuring limits of superheat of liquids, nucleation rates commensurate with experimental conditions, and criteria for selection of data in the table. In addition to the data in the table below, the original table contains data on limits of superheat for 27 binary mixtures and one ternary (Ethane, n-Propane, and n-Butane).

The homogeneous nucleation limit, or limit of superheat, represents the deepest penetration of a liquid into the domain on metastable states. At constant pressure and composition, it is the highest temperature below the critical point a liquid can sustain without undergoing a phase transition; at constant temperature, it is the lowest pressure. The practical significance of this limit resides in the consequences of the phase transition that eventually occurs when this limit is reached.

NOTE: In the column, *J [nuclei/(cm³·s)]*, 1E+2 means 1×10^2, 1E−2 means 1×10^{-2}, etc.

Substance	P [MPa]	T [K]	J [nuclei/(cm³-s)]
Ar	-1.220	85.0	1E+02
Argon	0.101	130.8	1E+02
	0.190	131.2	1E+02
	0.260	131.5	1E+05
	0.360	131.8	1E+01
	0.410	131.9	1E+05
	0.600	132.8	1E+05
	0.810	133.5	1E+03
	1.100	134.3	1E+01
	1.150	135.1	1E+05
	1.400	135.3	1E+01
	1.420	136.0	1E+05
	1.720	137.1	1E+05
	2.140	138.6	1E+05
	2.450	139.5	1E+05
	2.710	141.3	1E+05
H_2	.076	27.8	1E-02
Hydrogen	.149	27.9	1E-02
	.381	29.4	1E-02
	.751	30.6	1E-02
	.834	30.8	1E-02
H_2O	-27.70	283.2	1E+03
Water	0.101	553.0	1E+06
	0.101	575.2	1E+15
	1.293	580.4	1E+21
	2.519	584.9	1E+21
	2.710	588.3	1E+21
	5.000	593.6	1E+21
	6.808	600.4	1E+21
	8.500	606.5	1E+21
	9.731	607.2	1E+21
	10.746	610.3	1E+21
	11.978	615.6	1E+21
	12.873	616.7	1E+21
	13.731	620.2	1E+21
	15.789	627.0	1E+21
	17.556	632.3	1E+21
	20.113	642.2	1E+21
He I	0.012	4.05	1E+07
Helium I	0.017	4.12	1E+07
	0.037	4.22	1E+07
	0.054	4.31	1E+07
	0.066	4.37	1E+07
	0.081	4.45	1E+07
	0.100	4.55	1E+07
	0.112	4.62	1E+07
	0.129	4.70	1E+07
	0.143	4.76	1E+07
He II Helium II	-0.06	2.09	1E+05
Kr	0.400	182.5	1E+05
Krypton	0.820	184.3	1E+05
	1.200	187.0	1E+05
	1.410	187.6	1E+05
	1.630	189.1	1E+05
	1.900	189.9	1E+05
	2.200	192.1	1E+05
	2.430	192.9	1E+05
	2.800	194.8	1E+05
	3.140	196.6	1E+05
	3.460	198.0	1E+05
	3.800	199.4	1E+05
N_2	-1.010	75.0	1E+02
Nitrogen	0.101	110.0	1E+00
	0.410	111.4	1E+05
	0.520	112.0	1E+05
	0.610	112.1	1E+05
	0.700	112.7	1E+05
	0.820	113.2	1E+05
	0.940	113.8	1E+05
	1.060	114.2	1E+05
	1.210	114.8	1E+05
	1.240	115.2	1E+05
	1.330	115.5	1E+05
	1.360	115.6	1E+05
	1.460	116.2	1E+05
	1.590	116.8	1E+05
	1.620	117.0	1E+05
	1.730	117.6	1E+05
	1.770	117.7	1E+05
	1.870	118.3	1E+05
	1.920	118.4	1E+05
	2.070	119.1	1E+05
N_2O_3	0.154	395.6	1E+02
Nitrogen-	0.554	396.2	1E+02
tetroxide	0.980	398.2	1E+01
	2.000	401.5	1E+01
	3.040	405.2	1E+01
	3.920	408.1	1E+01
	4.500	410.2	1E+01
	5.000	412.5	1E+01
	5.500	414.5	1E+01
	6.000	416.4	1E+01
O_2	-1.520	75.0	1E+02
Oxygen	0.101	134.1	1E+00
	0.400	135.4	1E+05
	0.500	136.2	1E+05
	0.680	136.5	1E+05
	0.920	137.4	1E+05
	1.060	137.5	1E+05
	1.180	138.3	1E+05
	1.350	138.9	1E+05
	1.480	139.3	1E+05
	1.740	140.7	1E+05
	2.030	141.9	1E+05
	2.260	142.8	1E+05
	2.500	143.6	1E+05
	2.700	144.5	1E+05
	2.970	145.9	1E+05
SO_2 Sulphur-dioxide	0.101	323.2	1E+02
Xe	0.500	254.1	1E+05
Xenon	0.830	256.3	1E+05
	1.070	257.2	1E+05
	1.260	258.2	1E+05
	1.470	259.6	1E+05
	1.550	260.3	1E+05
	1.680	261.0	1E+05
	1.750	261.6	1E+05
	1.860	261.9	1E+05
	1.970	262.8	1E+05
	2.070	263.4	1E+05
	2.170	263.8	1E+05
	2.370	265.2	1E+05
	2.480	266.1	1E+05
	2.630	266.9	1E+05
	2.750	267.5	1E+05
	2.850	267.8	1E+05
	2.970	269.1	1E+05
	3.050	269.7	1E+05
	3.130	270.0	1E+05
	3.450	272.0	1E+05
	3.630	273.0	1E+05
$ClCHF_2$	0.101	327.8	1E+04
Chloro-	0.236	328.2	1E+04
difluoro-	0.280	329.4	1E+04
methane	0.510	330.8	1E+04
	0.560	331.5	1E+04
	0.710	332.4	1E+04
	0.810	332.9	1E+04
	0.910	334.2	1E+04
CCl_4 Carbontetra-chloride	-27.60	268.2	1E+03
$CHCl_3$	-31.70	258.2	1E+03
Chloroform	0.101	466.2	1E+02
CH_2Cl_2 Methylene-chloride	0.101	394.8	1E+01
CH_3Cl Chloro-methane	0.101	366.2	1E+05
CH_4	0.400	167.6	1E+05
Methane	0.620	168.3	1E+05
	0.820	169.3	1E+05
	1.030	170.5	1E+05
	1.230	171.4	1E+05
	1.430	172.1	1E+05
	1.630	173.1	1E+05
	1.830	174.0	1E+05
	2.030	175.2	1E+05
	2.220	176.4	1E+05
	2.430	177.6	1E+05
	2.630	178.6	1E+05
	2.820	180.0	1E+05
CH_4O	0.101	458.4	1E+01
Methanol	0.101	461.2	1E+05
	0.101	466.2	1E+18
	0.600	469.2	1E+19
	1.050	471.2	1E+20
	2.030	476.7	1E+16
	2.030	478.2	1E+20
	3.000	482.2	1E+21
	4.000	488.7	1E+22
	4.980	494.7	1E+22
	5.970	501.2	1E+23
	6.960	507.7	1E+23
$C_2Cl_2H_2F_2$	0.221	342.5	1E+06
Dichloro-	0.427	344.3	1E+06
difluoro-	0.462	344.7	1E+06
ethane	0.655	346.6	1E+06
	0.896	348.8	1E+06
	0.931	349.0	1E+06
	1.227	351.7	1E+06
	1.489	354.4	1E+06
	1.917	358.8	1E+06
	2.399	363.7	1E+06
	2.910	369.0	1E+06
	3.289	373.0	1E+06
	3.323	373.4	1E+06
	3.585	376.2	1E+06
	3.634	376.9	1E+06
C_2H_3Cl Chloroethane	0.101	374.1	1E+05
C_2H_3F Fluoroethene	0.101	290.1	1E+05
C_2H_3N Acetonitrile	0.101	497.0	1E+06
$C_2H_4F_2$ 1,1-Difluoro-ethane	0.101	343.6	1E+05
$C_2H_4O_2$	-28.80	292.7	1E+03
Acetic acid	0.101	526.2	1E+06
$C_2H_4O_2$ Methyl-formate	0.101	423.2	1E+01
C_2H_5Br Ethylbromide	0.101	422.2	1E+01
C_2H_5Cl Ethylchloride	0.101	399.2	1E+01
C_2H_6 Ethane	0.101	269.2	1E+05

Substance	P [MPa]	T [K]	J [nuclei/(cm³-s)]
C_2H_6O	0.101	464.1	1E+01
Ethanol	0.101	466.0	1E+04
	0.101	471.5	1E+17
	0.580	474.2	1E+19
	0.980	471.0	1E+02
	1.070	477.2	1E+20
	1.540	481.7	1E+20
	2.030	484.2	1E+21
	2.520	486.7	1E+21
	3.010	490.2	1E+21
	3.500	494.2	1E+22
C_3H_3N Acrylonitrile	0.101	489.0	1E+05
C_3H_4 Propadiene	0.101	346.2	1E+05
C_3H_4 Propyne	0.101	356.8	1E+05
C_3H_6 Cyclopropane	0.101	350.7	1E+05
C_3H_6 Propene	0.101	325.6	1E+05
C_3H_6O	0.101	454.5	1E+01
Acetone	0.101	456.4	1E+02
	0.101	458.7	1E+13
	0.101	462.7	1E+18
	0.980	462.6	1E+01
$C_3H_6O_2$ Ethylformate	0.101	428.5	1E+01
$C_3H_6O_2$ Methylacetate	0.101	416.6	1E+01
C_3H_8	0.101	326.4	1E+06
N-Propane	0.302	332.8	1E+04
	0.491	336.8	1E+04
	0.715	339.1	1E+04
	0.907	343.2	1E+04
C_3H_8O	0.101	487.4	1E+04
N-Propanol	0.101	493.0	1E+15
	0.101	495.7	1E+18
C_3H_8O Isopropanol	0.101	473.0	1E+06
C_4H_6 1,3-Butadene	0.101	377.3	1E+05
C_4H_8 1-Butene	0.101	371.0	1E+05
C_4H_8 Butylene	0.101	510.2	1E+16
C_4H_8 Cis-2-Butene	0.101	385.4	1E+05
C_4H_8 Trans-2-Butene	0.101	379.7	1E+05
C_4H_8 2-Methylpropene	0.101	369.6	1E+05
C_4H_{10} N-Butane	0.101	377.6	1E+05
C_4H_{10} 2-Methylpropane	0.101	361.0	1E+05
$C_4H_{10}O$	0.101	509.6	1E+02
N-Butanol	0.101	511.9	1E+04
	0.101	513.2	1E+13
	0.101	516.2	1E+16
	0.101	518.2	1E+18
	0.980	519.4	1E+02
$C_4H_{10}O$	-1.75	293.	1E+02
Ether	-1.52	402.7	1E+04
	-1.22	407.6	1E+04
	-1.12	409.2	1E+04
	-1.00	410.2	1E+04
	-0.74	413.4	1E+04
	0.101	417.5	1E+02
	0.101	425.7	1E+19
	0.211	419.4	1E+01
	0.415	420.1	1E+01
	0.480	427.7	1E+18
	0.500	421.1	1E+02
	0.641	424.3	1E+02
	0.777	426.3	1E+01
	0.880	432.7	1E+18
	1.000	428.4	1E+01
	1.280	436.7	1E+01
	1.366	433.6	1E+01
	1.442	435.1	1E+02
	1.575	437.2	1E+01
	1.660	440.7	1E+19
	1.865	441.2	1E+01
	2.089	443.3	1E+01
	2.450	450.7	1E+21
	2.850	455.7	1E+20
$C_4H_{10}O$ Isobutanol	0.101	437.2	1E+01
$C_4H_{11}N$ Diethylamine	0.101	408.5	1E+01
C_5F_{12}	0.101	381.5	1E+06
Perfluoro-pentane	0.300	385.4	1E+06
	0.500	388.7	1E+06
	0.700	392.2	1E+06
	0.890	396.4	1E+06
	1.090	399.0	1E+06
	1.280	403.1	1E+06
	1.480	407.4	1E+06
C_5H_8 Cyclopentene	0.101	451.4	1E+06
C_5H_{10} Cyclopentane	0.101	455.1	1E+06
C_5H_{10} 1-Pentene	0.101	417.2	1E+06
C_5H_{12} 2,2-Dimethyl-propane	0.101	386.1	1E+06
C_5H_{12}	0.101	409.2	1E+01
Isopentane	0.101	411.7	1E+07
C_5H_{12}	0.101	418.8	1E+04
N-Pentane	0.101	426.2	1E+18
	0.490	423.7	1E+02
	0.880	429.1	1E+02
	1.280	435.3	1E+02
	2.600	451.2	1E+06
$C_5H_{12}O$ N-Pentanol	0.101	532.2	1E+17
C_6F_6	0.101	464.8	1E+01
Hexafluoro-benzene	0.101	467.9	1E+06
	0.500	469.9	1E+02
	0.570	474.1	1E+06
	1.000	477.1	1E+02
	1.050	480.4	1E+06
	1.540	486.0	1E+06
	2.030	494.2	1E+06
C_6F_{14}	0.101	409.8	1E+06
Perfluoro-hexane	0.300	414.4	1E+06
	0.500	418.5	1E+06
	0.700	422.3	1E+06
	0.880	425.6	1E+06
	1.050	430.3	1E+06
	1.240	434.6	1E+06
C_6H_5Br Bromobenzene	0.101	534.2	1E+02
C_6H_5Cl Chlorobenzene	0.101	523.2	1E+02
C_6H_6	-15.0	291.2	1E+03
Benzene	0.101	498.9	1E+02
	0.101	510.2	1E+18
	0.490	502.2	1E+02
	0.580	514.2	1E+19
	0.980	509.2	1E+02
	1.070	516.7	1E+18
	1.470	513.8	1E+02
	1.540	520.7	1E+19
	2.030	525.7	1E+19
	2.520	532.2	1E+20
	3.010	537.7	1E+17
	3.500	544.7	1E+18
C_6H_7N	-30.0	272.2	1E+03
Aniline	0.101	535.2	1E+02
C_6H_{12}	0.101	490.8	1E+02
Cyclohexane	0.300	493.1	1E+02
	0.420	495.2	1E+06
	0.720	499.7	1E+06
	0.950	501.7	1E+06
	0.980	502.1	1E+02
	1.110	504.2	1E+06
	1.350	506.2	1E+06
	1.700	512.2	1E+06
	2.100	518.2	1E+06
	2.370	519.2	1E+06
	2.550	523.2	1E+06
C_6H_{12} 1-Hexyne	0.101	465.2	1E+06
C_6H_{12} Methylcyclo-pentane	0.101	476.1	1E+06
C_6H_{14} 2,3-Dimethyl-butane	0.101	446.4	1E+06
C_6H_{14}	0.101	453.5	1E+02
N-Hexane	0.101	454.9	1E+05
	0.101	459.2	1E+13
	0.101	463.7	1E+20
	0.290	465.2	1E+15
	0.420	461.7	1E+06
	0.490	459.3	1E+02
	0.490	468.2	1E+22
	0.760	466.7	1E+06
	0.980	467.0	1E+02
	0.980	475.2	1E+23
	1.080	471.7	1E+06
	1.120	478.2	1E+15
	1.280	474.7	1E+06
	1.420	475.7	1E+06
	1.590	479.7	1E+06
	1.600	486.2	1E+16
	1.720	481.7	1E+06
	1.960	487.7	1E+17
	2.060	493.2	1E+16
	2.390	496.7	1E+06
	2.570	501.2	1E+16
$C_6H_{14}O$ N-Hexanol	0.101	555.7	1E+04
C_7F_8	0.101	485.3	1E+01
Octafluoro-toluene	0.490	489.7	1E+01
	0.980	499.8	1E+01
C_7F_{16}	0.101	434.8	1E+06
Perfluoro-heptane	0.230	436.9	1E+06
	0.400	440.5	1E+06
	0.570	444.4	1E+06
	0.770	448.3	1E+06
	0.920	452.7	1E+06
	1.070	456.1	1E+06
	1.150	459.0	1E+06
	1.280	461.3	1E+06
C_7H_8 Toluene	0.101	526.7	1E+02
C_7H_{14} Methylcyclo-hexane	0.101	510.4	1E+06
C_7H_{16}	0.101	486.9	1E+06
N-Heptane	0.101	493.7	1E+18
	0.294	489.2	1E+06
	0.392	490.7	1E+06
	0.490	493.7	1E+06
	0.589	494.2	1E+06
	0.736	498.7	1E+06
	0.952	500.7	1E+06
	1.275	505.2	1E+06
	1.373	509.7	1E+06
	1.570	512.7	1E+06
	1.736	515.2	1E+06
	1.805	516.7	1E+06
	2.001	519.7	1E+06
$C_7H_{16}O$ N-Heptanol	0.101	566.3	1E+04
C_8F_{18}	0.101	457.0	1E+06
Perfluoro-octane	0.300	461.1	1E+06
	0.500	467.1	1E+06

Substance	P [MPa]	T [K]	J [nuclei/(cm³-s)]
	0.700	471.2	1E+06
	0.890	476.9	1E+06
	1.090	482.8	1E+06
	1.190	484.1	1E+06
C_8H_{10} Cyclo-octane	0.101	560.7	1E+06
C_8H_{10} 2,3-dimethyl-benzene	0.101	508.2	1E+02
C_8H_{16} 1-Octene	0.101	510.3	1E+06
C_8H_{18} N-Octane	0.101	513.8	1E+06
	0.377	519.3	1E+04
	0.653	525.2	1E+04
	0.929	528.6	1E+04
	1.204	532.4	1E+04
C_8H_{18} 2,2,4-Trimethyl-pentane	0.101	488.5	1E+06
$C_8H_{18}O$ N-Octanol	0.101	586.0	1E+04
C_9F_{20} Perfluoro-nonane	0.101	478.5	1E+06
	0.300	484.4	1E+06
	0.500	489.3	1E+06
	0.700	493.3	1E+06
	0.890	499.7	1E+06
	1.090	505.7	1E+06
C_9H_{20} N-Nonane	0.101	538.5	1E+06
$C_{10}F_{22}$ Perfluoro-decane	0.101	497.1	1E+06
	0.300	503.2	1E+06
	0.500	508.6	1E+06
	0.700	515.6	1E+06
	0.890	521.2	1E+06
	1.090	527.7	1E+06
$C_{10}H_{22}$ N-Decane	0.101	558.3	1E+06
$C_{12}H_{10}O$ Diphenylether	0.101	703.2	1E+17

AZEOTROPES

Zdzislaw M. Kurtyka

GENERAL CLASSIFICATION OF AZEOTROPES AND THEIR SYMBOLISM

Although the first reported observations of the appearance of minimum vapor pressure in binary mixtures were in 1802 by Dalton,[1] the term "azeotrope" was not introduced until 1911 by Wade and Merriman.[2]

Azeotrope from the Greek "not to boil with change" or "to boil unchanged" means literally the same as the English "constant-boiling", i.e., the vapor boiling from a liquid has the same composition as the liquid.

A liquid mixture of two or more components which can be separated by distillation was termed a "zeotrope" by Swietoslawski.[3] Such a mixture is also called a "nonazeotrope" or a "non-azeotropic" system in many countries. However, some objection may be placed against the use of the terms "nonazeotrope" and "non-azeotropic" system due to the use of two negatives in the same word. In Greek "a" means "non". Therefore, accuracy indicates that the word nonazeotrope should be replaced by non-non-azeotrope. It is obvious the latter is awkward and that it is reasonable to replace it with Swietoslawski's suggestion, zeotrope.

Azeotropic systems may be classified broadly in relation to the character of the extremum (maximum or minimum), the number of components in the system and whether they form one or more liquid phases.

Positive and Negative Azeotropes

In 1926 Lecat[4] proposed to divide azeotropes into positive and negative. Positive azeotropes are characterized by a minimum boiling temperature at constant preassure, i.e. a maximum in the vapor pressure at constant temperature. Negative azeotropes, on the other hand, have a maximum boiling temperature and a minimum vapor pressure.

In Anglo-Saxon literature a positive azeotrope is equivalent to a pressure-maximum azeotrope, while a negative one — to a pressure-minimum azeotrope. This description of azeotropes cannot be extended to systems exhibiting neither a minimum nor a maximum in either boiling temperature or vapor pressure (saddle or positive-negative azeotropes).

One may say at this point that neither azeotrope nor zeotrope, thus described, gives any indication whether the vapor phase consists of one, two or more liquid phases. To make this perfectly clear, the terms homo- and heteroazeotrope, and homo- and heterozeotrope were introduced. For practical purposes negative azeotropes are divided into three groups.[5]

In view of the fact that the number of different types of azeotropes continue to increase, symbols are given below for positive and negative azeotropes and zeotropes.

Type of homoazeotrope	Type of zeotrope
1. (A,B) binary positive	$(A,B)_z$ binary positive
2. A,B,C) ternary positive	$(A,B,C)_z$ ternary positive
3. (A,B,C,D)quaternary positive	$(A,B,C,D)_z$quaternary positive
4. [(−)A,B]binary negative	$[(-)A,B]_z$ binary negative

Symbols are also used for heteroazeotropes. For example, there is a combination of letters, dots, and dashes. There is one letter for each of the components in the system, one dash for each of the phases in the azeotrope and one less dot than there are dashes. The system may be illustrated by the following two examples. The system benzene-ethanol-water forms a two-phase heteroazeotrope. Symbolically this is written as (B,E,W, -·-). The system nitromethane-water-n-paraffin forms three liquid phases at the boiling temperature. This applies for n-paraffins from heptane to tridecane. If the symbol H_i is assigned to the n-parafffin, the azeotrope will be designated by the symbols (N,W,H_i,-·-·-).

Saddle or Positive-Negative Azeotropes

In spite of the fact that a ternary saddle azeotrope was predicted by Ostwald at the end of the last century, the first azeotrope of this type was found in 1945 by Ewell and Welch[7] in the system acetone-chloroform-methanol.

Saddle azeotropic systems, also called positive-negative systems, exhibit a hyperbolic point which is neither a minimum nor a maximum in either boiling temperature or vapor pressure, and are characterized by the presence of a "top-ridge" line. They also exhibit some peculiar properties called distillation anomalies.[7-10]

In general, ternary saddle azeotropes are classified according to the number of binary negative systems forming such an azeotrope. From this point of view, ternary saddle azeotropes may be divided into bipositive-negative and binegative-positive azeotropes. All possible types of ternary bipositive-negative azeotropes were found.[11-14] The bipositive-negative azeotropes may be designated by the symbols [(−)A,B(±)H], where, A, B, and H are the components forming these azeotropes. For the ternary binegative-positive azeotropes with B, E, and C as the components, the symbols are [(+)B,E(−)C] (two binary negative azeotropes [(−)C,B] and [(−)C,E] occur in this system). This type of a ternary saddle azeotrope is rather a rare phenomenon.[15,16]

Although the terms "saddle azeotrope" and "positive-negative" azeotrope are equivalent in the context of ternary systems, the latter is more preferable when discussing multicomponent systems.

THE COMMONNESS OF THE PHENOMENON OF AZEOTROPY

Azeotropes occur in organic and inorganic systems, although most of the known azeotropes are formed by organic compounds. This is because the very large number of organic compounds boil without decomposition in the easily accessible ranges of temperature and pressure.

In the last 2 decades of the 19th century the appearance of an extremum vapor pressure and boiling temperature was believed to be a rare phenomenon. For this reason Ostwald[17] used the term "ausgezeichnete Lösungen" to emphasize the phenomenon was not often encountered. This view survived in certain circles for many years despite convincing evidence the phenomenon of azeotropy is definitely a common one.

The first investigator who provided evidence of the commonness of azeotropy was Ryland.[18] His word dealt with 80 systems, 45 of which were azeotropic; 80 further new azeotropes were described in Lecats doctoral dissertation, published in 1908 to 1909. Lecat was able to list 1000 azeotropes in his monograph 10 years later.[19] Among them were numerous binary and ternary heteroazeotropes. This monograph showed that the appearance of maximum or miminum vapor pressures of binary and ternary mixtures should not be regarded as a rare phenomenon. From then on many physical chemists and technologists began to take an interest in the theoretical and practical application of azeotropy.

In 1949 Lecat[20] published the azeotropic data for 13,290 binary systems. The number of azeotropes reached 6287 or 47% of the systems examined.

In 1973 Horsley[21] published his Azeotropic Data-III. In this volume 15,823 binary, 725 ternary, 21 quaternary, and 2 quinary systems are reported. The number of the azeotropic systems is as follows: binary 7945 (52%), ternary 371 (51%), quaternary 9 (43%), and quinary 1.

It is interesting to see that 119 binary azeotropes (47% of the systems examined), including 32 negative, occur in the systems composed either of two inorganic compounds (elements) or an inorganic-organic compound system; 768 binary systems contain water as one component (among them 665 [86%] are azeotropic). The ternary positive-negative (saddle) azeotropes occur in 40 systems; 267 (72%) ternary azeotropic systems contain water as one component. There are also 4 ternary negative azeotropes. As far as the quaternary systems are concerned, 8 systems form positive azeotropes and one, a positive-negative.

In the last decade, studies of vapor-liquid equilibria were developed rapidly, mostly at a constant temperature, with the aim of correlating and predicting the equilibrium parameters. Unfortunately, a large number of investigators did not report whether an azeotrope or a zeotrope is formed in a particular system. In view of the fact that the number of the systems examined in that period of time is small, compared to that of the systems examined previously, it becomes clear that the current situation regarding the commonness of the phenomenon of azeotropy remains essentially unchanged.

From the existing experimental work one might conclude that the frequency of occurence of azeotropes diminishes as the number of components in the system increases, and that multicomponent azeotropes should be expected to be very rare. Such a point of view, however, would be wrong. Although the conditions of the formation of multicomponent azeotropes are complex, they are not difficult to fulfil in practice.[22-24]

THE AZEOTROPIC RANGE

The idea of relating a certain "azeotropic ability" to the chemical character of a substance is already apparent in the first works of Lecat,[19] where he arranged the experimental data on azeotropes according to the chemical character of the components.

The term "azeotropic range" is due to Swietoslawski.[25] There is also another closely related term known, namely, the "relative azeotropic effect",[26] but its use is very limited. Malesinski[27] developed the concept of Swietoslawski's azeotropic range on a general assumption that the components of the system form a regular solution.

The symmetrical azeotropic range, Z, is given by the formula

$$(\delta_1^{1/2} + \delta_2^{1/2}) = Z_{12} = 0.5Z \qquad (1)$$

where z_{12} is the half-value of the symmetrical azeotropic range and the quantities δ_1 and δ_2, known otherwise as azeotropic deviations, for positive azeotropes are defined by $\delta_1 = T_1 - T^{Az}$ and $\delta_2 = T_2 - T^{Az}$; T_1, T_2 are the boiling temperatures of the pure components and T^{Az} is the boiling temperature of the azeotrope. For negative azeotropes we have $\delta_1 = T^{Az} - T_1$ and $\delta_2 = T^{Az} - T_2$. The symmetrical azeotropic range means that both parts of the range (the upper and the lower part) are equal.

Equation 1 which is also valid at high pressures[28] enables us to compute z_{12} of any binary azeotropic system, provided that the boiing temperature of the pure components and of the azeotrope are known. It should be remembered that by convention the z_{12} values are positive for positive azeotropic systems, and negative for negative ones.

The knowledge of the azeotropic ranges usually leads to a better understanding of the distillation course of complex azeotropic mixtures, e.g. high- and low-temperature coal tars, petroleum and synthetic gasoline. In addition, the azeotropic ranges appear in the equations for calculating the boiling temperatures and compositions of ternary homoazeotropes. These equations were found useful in predicting the existence of a large number of ternary azeotropes.[24]

THE PREDICTION OF AZEOTROPIC DATA IN BINARY SYSTEMS

Empirical Correlations

Lecat[19] first observed that the composition of a binary azeotrope is related to the difference between the boiling temperatures of the components. He used a power series to relate the above quantities for the systems formed by a common substance with members of a homologous series.

His relation may be written in the form

$$x_1 = A_0 + A_1\Delta + A_2\Delta^2 + A_3\Delta^3 + \cdots \quad (2)$$

where x_1 is the weight fraction of component 1 in the azeotrope and Δ is the absolute difference between the boiling temperatures of the components.

Lecat[19] proposed also a relation between Δ and the azeotropic deviation, δ, i.e. the absolute difference between the boiling temperatures of the azeotrope and that of the more volatile component, for positive azeotropes, and the less volatile component, for negative azeotropes. The Δ—δ relation, otherwise known as Lecat's rule is

$$\delta = C_0 + C_1\Delta + C_2\Delta^2 + C_3\Delta^3 + \cdots \quad (3)$$

For most cases the terms with Δ higher than Δ^2 may be neglected to give

$$\delta = C_0 + C_1\Delta + C_2\Delta^2 \quad (4)$$

The existence of an approximate relation of this kind was expected for a series of binary regular solutions.[30] For instance, the constants of Equation 4 for the azeotropes of ethanol with aliphatic halogen derivatives are $C_o = 12$, $C_1 = -0.5$ and $C_2 = 0.00526$. The general agreement with Equation 4 for that series is satisfactory, despite that these alcohol solutions show large deviations from regularity.

In the 1940s several graphical correlations of empirical nature for the composition of azeotropes within a series of organic compounds were reported, notably by Mair et al.,[31] Horsley,[32] Meissner and Greenfield,[33] Skolnik,[26] and by Seymour.[34] A decade later Johnson and Madonis[35] described another correlation that included a number of other series of compounds.

The empirical treatment suffered mainly due to many variations in the form of the equations and the number of constants required for their evaluation.

A further attempt in improving the situation in this field was made by Seymour et al.[36] They proposed a correlation between x and Δt in the form of a master equation to correlate, among other things, the data from other series already reported. This equation is

$$\log(10\,\frac{x_1}{x_2} = mf(\tau)\,\Delta t + b \quad (5)$$

where x_1 and x_2 are the mole fractions of component 1 and 2 in the azeotrope, respectively, Δt is the difference between the boiling temperatures of the two components, m and b are constants and $f(\tau) = \tau_1/\tau_2$. The quantity τ is defined as the ratio of the boiling temperatures (°K) of a compound and a hypothetical n-paraffin of the same molecular weight. The differences between the calculated and observed azeotropic compositions are reported for 15 series of organic compounds.

An approximate linear relation between the logarithm of the mole fraction, $\log x_i$, of the main and the secondary azeotropic agent, and the average condensation temperature, and was found for several ternary homo- and heteropolyazeotropic mixtures.[37] Empirical correlations for the azeotropic composition apply to both homo- and heteroazeotropes.

Composition of a Binary Homoazeotrope

The theory of regular solutions offers methods for calculating the composition of a binary (positive or negative) azeotrope from certain properties of the pure components and of the azeotrope.

Generally speaking there are three known methods serving this purpose. Two of them involve the boiling temperatures of the pure components, T_1, T_2, and that of the azeotrope, T^{Az}, and were developed by Prigogine[30] and Malesinski.[23] The third method is based on the activity coefficients of the components at the azeotropic point, and is due to Kireev.[38]

The Prigogine equation for equal molar vaporization entropies of the components, $\Delta S_1^\circ = \Delta S_2^\circ$, may be written in the form

$$x_2 = \alpha(1 + \alpha)^{-1} \quad (6)$$

were α is the square root of the ratio of azeotropic deviations, i.e., $(\delta_1/\delta_2)^{1/2}$, and x_2 is themole fraction of components 2 in the azeotrope.

When $\Delta S_1^\circ \neq \Delta S_2^\circ$, Equation 6 takes the form

$$x_2 = \alpha'(1 + \alpha')^{-1} \quad (7)$$

In this case $\alpha = c\alpha$ and $c = (\Delta S_1^\circ/\Delta S_2^\circ)^{1/2}$.

The Malesinski equation for the components having equal molar vaporization entropies is given by

$$x_2 = 0.5 + \frac{T_1 - T_2}{2z_{12}} \quad (8)$$

where z_{12} is the half-value of the symmetrical azeotropic range. For systems with unequal molar vaporization entropies, Malesinski's equation requires evaluation of z_u (the upper part of the azeotropic range), the quantity which is not easily available.

Recently, Equations 6, 7, and 8 were evaluated with regard to their usefulness for the calculation of the composition of binary azeotropes.[39]

The simplest and most suitable form of Kireev's equation for the case in which $\Delta S_1^o = \Delta S_2^o$, is the expression

$$x_2 = (1 + b)^{-1} \quad (9)$$

where b is the square root of the ratio of the logarithms of the activity coefficients of the components γ_2 and γ_1, i.e. $(\ln \gamma_2/\ln\gamma_2)^{1/2}$. At the azeotropic point the composition of the liquid and the vapor are equal, $x_2 = y_2$, and in the case of an ideal vapor phase the expressions for γ_1 and γ_2 are given by

$$\gamma_1 = P/p_1^0 \text{ and } \alpha_2 = P/p_2^0$$

In these relations p_1^o and p_2^o are the vapor pressures of the pure components at the boiling temperature of the aceotrope, and P is the total pressure of the mixture at equilibrium.

Equation 9 for $\Delta S_1^o \neq \Delta S_2^o$ becomes

$$x_2 = (1 + b')^{-1} \quad (10)$$

where $b' = cb$ and $c = (\Delta S_2^o/\Delta S_1^o)^{1/2}$.

The possible error in the composition of the binary azeotrope under isobaric conditions is due not only to the deviations of the system from regularity but also to the change of the regular solutions constant, A_{12}, with temperature, and to the differences in the vaporization entropies of the pure components. One of the causes of the deviations of the system from regularity is the nonideality of the vapor phase.

The effect of the nonideality of the vapor phase and the differences in the vaporization entropies of the components on the azeotropic composition were studied by Kurtyka and Kurtyka[40] on the systems of acetic acid with n-paraffins. Acetic acid is associated (dimerized) in the vapor phase and its vaporization entropy is 14.85 cal/g-mole, while those of the hydrocarbons are ~20 cal/g-mole.

To get a general idea about those effects on the azeotropic composition, the results obtained for some systems of acetic acid with n-paraffins are reproduced in part from this paper. The system, $\Delta x_2 = x_{2(calcd.)} - x_{2(obsd.)}$, with x_2 computed by Equation 9; Δx_2, with x_2 computed with the corrections for the dimerization of the acid in the vapor phase, and Δx_2, with x_2 computed by Equation 10 are: acetic acid — n-heptane, 12.3, 1.8, 9.3; acetic acid — n-octane, 12.8, 3.8, 9.2, and acetic acid — n-nonane, 9.8 mole %, 4.3 mole %, and 7.0 mole %, respectively.

These results show that the Δx_2 values are reduced to a reasonable magnitude when the corrections for the dimerization of the acid are taken into account. This simply means that the dimerization of acetic acid in the vapor phase is the dominant factor contributing to the large differences in Δx_2. The effect of the differences in the vaporization entropies of the components on the azeotropic composition as exemplified by the systems containing acetic acid and n-paraffins is small compared to the deviations from regularity caused, among other things, by the dimerization of acetic acid in the vapor phase.

Equation 9 has a built-in unfavorable factor because it involves the vapor pressure, p_i^o, of each pure component at the boiling temperature of the azeotrope. p_i^o values are almost exclusively computed by means of the Antoine equation. Therefore, the accuracy of the computed p_i^o is related to the constants of that equation. However, the effect of the differences in p_i^o values on the azeotropic composition is usually small, except in cases where the boiling temperature of the azeotrope is close to the boiling temperature of one component of the system.

The Azeotropic Data of a Binary Heteroazeotrope

Many systems of two or more components exhibit a limited solubility and most of them form positive heteroazeotropes. In binary systems it is easy to predict from the data on critical solution temperatures (CST) whether an azeotrope that occurs at a certain temperature is a homoazeotrope or heteroazeotrope.[41]

To describe the vapor-liquid equilibrium in a heterogeneous system, it is necessary to introduce certain simplifying assumptions.

The simplest case is obtained upon the assumption that liquids are completely inmiscible in each other and that the vapor phase is ideal.

Then the composition of a binary heteroazeotrope, y_1, is given by the formula

$$x_1 = y_1 = \frac{p_1^0\,(T^{Az})}{p_1^0\,(T^{Az}) + p_2^0\,(T^{Az})} \quad (11)$$

where p_1^o and p_2^o are the vapor pressures of pure components 1 and 2, respectively, expressed as functions of the boiling temperature of the azeotrope, T^{Az}, or simply, at the boiling temperature of the azeotrope.

And for the boiling temperature of the azeotrope, T^{Az} we have the relation

$$P = p_1^0 (T^{Az}) + p_2^0 (T^{Az}) \qquad (12)$$

where P is the total pressure at equilibrium.

In the cases in which the condition of complete immiscibility is not satisfied, we introduce the correction factors in terms of the activities of the components, α_1 and α_2 in the liquid phase.

Accordingly, the expression for the composition of an azeotrope takes the form

$$y_1 = \frac{p_1^0 (T^{Az}) \alpha_1}{p_1^0 (T^{Az}) \alpha_1 + p_2^0 (T^{Az}) \alpha_2} \qquad (13)$$

And the boiling temperature of the azeotrope may be calculated from

$$P = p_1^0 (T^{Az}) \alpha_1 + p_2^0 (T^{Az}) \alpha_2 \qquad (14)$$

There is also another possibility, namely to assume that the activities of both components in the liquid phase are equal, i.e., $\alpha_1 = \alpha_2 = \alpha$. The systems behavior of which is well described by a common activity, are those of nitromethane with n-paraffins.[42] The computation of the activities, α_1 and α_2 of the components is a straightforward procedure involving the solution of Equations 13 and 14.

Equations 11 and 12 were found to be satisfied for the systems of aromatic and n-paraffin hydrocarbons with water. Examples of the systems in which considerable miscibility of the components occurs are: n-butanol-water, aniline-water, and acetonitrile-n-paraffins. For instance, the solubility of aniline in water at 90°C is 6.4 g/100 mℓ.

For calculating p_1^o and p_2^o at the boiling temperatures of the respective azeotropes, the Antoine equation is recommended.[43] The constants of the Antoine equation, viz. $\log p^o = A - B/(C + t)$, where p^o is expressed in millimeters Hg and t is in this case the boiling temperature of the azeotrope (°C), are compiled for quite a large number of organic compounds.[44,45]

For example, the activities of the components, α_1 and α_2 in the systems aniline-water[2] and n-butanol-water were found to be: 0.315, 1.00, and 0.649 and 0.9824, respectively.

The relations, thus described, can be easily extended to ternary and multicomponent heteroazeotropes.

THE PREDICTION OF AZEOTROPIC DATA IN TERNARY SYSTEMS

To separate nonideal liquid mixtures by fractional distillation, it is important to establish, among other things, whether the mixtures to be separated form azeotropes. In view that the experimental methods for determining the azeotropic composition are rather difficult and time-consuming operations, expecially in the case of azeotropes containing three and more components, several computational methods were developed to predict the composition of ternary and multicomponent azeotropes.

In general, two approaches in this area may be distinguished. The first one, originated by Haase[46,47] and developed by Malesinski,[48] is based on the theory of regular solutions and is restricted to homoazeotropes. The Malesinski method makes it possible to predict the appearance of ternary azeotropes of various types and to calculate their composition and boiling temperatures. The second approach, which is not limited only to the azeotropic points, is based on the use of various equations of empirical and semiempirical nautre, that relate the liquid-phase activity coefficients to the composition of the liquid phase.

The Azeotropic Data of a Ternary Homoazeotrope from the Binary Azeotropic Data

The composition of a ternary homoazeotrope (1,2,3) under isobaric conditions, when the vaporization entropies of the components are equal, is related to the two pairs of binary azeotropes by the equations.

Pairs (1,2) and (2,3):

$$x_1 = \frac{x_1^{(1,2)} + a x_3^{(2,3)}}{1 - ab} \qquad (15)$$

$$x_3 = \frac{x_3^{(2,3)} + b x_1^{(1,2)}}{1 - ab} \qquad (16)$$

where

$$a = \frac{z_{13} - z_{23} - z_{12}}{2z_{12}} \quad \text{and} \quad b = \frac{z_{13} - z_{23} - z_{12}}{2z_{23}}$$

$x_1^{(1,2)}$ and $x_3^{(2,3)}$ are the mole fractions of components 1 and 3 in the binary azeotropes (1,2) and (2,3), respectively; x_1 and x_3 are the mole fractions of the above components in the ternary azeotrope and z_{12}, z_{13}, and z_{23} are the half-values of the symmetrical azeotropic range.

When the vaporization entropies of the components are not equal, the regular solution constants, A_{12}, A_{13}, and A_{23} take the place of z_{12}, z_{13}z, and z_{23}, respectively.

Pairs (1,2), and (2,3):

$$x_1 = \frac{x_1^{(1,3)} + c x_2^{(2,3)}}{1 - cd} \quad (17)$$

$$x_2 = \frac{x_2^{(2,3)} + d x_1^{(1,3)}}{1 - cd} \quad (18)$$

where

$$c = \frac{z_{12} - z_{23} - z_{13}}{2z_{13}} \quad \text{and} \quad d = \frac{z_{12} - z_{23} - z_{13}}{2z_{23}}$$

Pairs (1,2) and (1,3):

$$x_2 = \frac{x_2^{(1,2)} + e x_3^{(1,3)}}{1 - ef} \quad (19)$$

$$x_3 = \frac{x_3^{(1,3)} + f x_2^{(1,2)}}{1 - ef} \quad (20)$$

where

$$e = \frac{z_{23} - z_{13} - z_{12}}{2z_{12}} \quad \text{and} \quad f = \frac{z_{23} - z_{13} - z_{12}}{2z_{13}}$$

The values of z_{12}, z_{13}, and z_{23} should be computed from Equation 1.

The ternary system is zeotropic if, for example, the composition of one component of the system is zero or takes a negative value.

In the case when one binary system constituting the ternary system is zeotropic, the respective z_{ij} value may be estimated from that of any close member of a homologous series, its isomers or closely related substances.

The sources of error in the calculated composition of the ternary azeotrope are similar to those of the binary azeotrope.

The boiling temperature of a ternary homoazeotrope, T, with component 2 as the reference component, can be computed from the equation

$$T = T_2 - \frac{\delta_2^{(1,2)} + \left(\dfrac{\delta_2^{(1,2)}}{z_{12}} \cdot \dfrac{\delta_2^{(2,3)}}{z_{23}}\right)^{1/2} (z_{13} - z_{23} - z_{12}) + \delta_2^{(2,3)}}{1 - \dfrac{(z_{13} - z_{23} - z_{12})^2}{4z_{12}z_{23}}} \quad (21)$$

where $\delta_2^{(1,2)}$ and $\delta_2^{(2,3)}$ are the azeotropic depressions or elevations in the azeotropes (1,2) and (2,3) in relation to component 2, e.g. $\delta_2^{(1,2)} = T_2 - T^{(1,2)}$: $T^{(1,2)}$ is the boiling temperature of the azeotrope (1,2).

By convention, the $\delta_i^{(i,j)}$ and z_{ij} are positive for a positive azeotrope and negative for a negative one.

By interchanging the components, the boiling temperature of the ternary homoazeotrope may be computed from the two remaining sets of the pairs of the components.

For mixtures which exactly fulfil the requirements for regular solutions, the result is independent of the choice of the reference component. If there are deviations from regularity, then the values of z_{ij} or $\delta_i^{(i,j)}$ depend on which experimentally determined quantity was used in the calculations.

For positive azeotropes Equations 15 to 21 usually give good results for the calculated azeotropic data.[23] In the case of positive-negative (saddle) azeotropes the agreement between the calculated and observed azeotropic data is less satisfactory. But this is understandable in view that these systems are complex mixtures, which often contain polar and associated components, e.g. alcohols and low-molecular fatty acids.

For the series of ternary saddle systems acetic acid-pyridine(2-picoline)-n-paraffins, the results obtained for the calculated azeotropic compositions improved considerably, when the corrections for the association (dimerization) of acetic acid in the vapor phase were taken into account.[49]

Equations 15 to 21 made it possible to predict the existence of a large number of ternary saddle (positive-negative) azeotropes that may appear in the course of fractional distillation of certain fractions of coal tar.[24]

It has been found, for instance, that ternary saddle azeotropes are formed in the series of ternary systems aniline-phenol-n-paraffins (ranging from nonane to tetradecane). Only one saddle azeotrope of this type was examined in the system aniline(1)-phenol(2)-tridecane(3).[50] For this system the differences between the calculated and observed azeotropic compositions, Δx_1, Δx_2, and Δx_3, and that of the boiling temperatures of the azeotrope, ΔT, are: 1.0, −2.1, and 1.0 mole %, and −0.11°C, respectively.

In general, good results for ternary homoazeotropes are obtained for systems which show moderate deviations from regularity, any of the binary azeotropes is close to what is called a tangent azeotrope,[23] and the differences between the boiling temperature of the ternary azeotrope and those of the components are not large.[24]

It should be also mentioned that two empirical correlations for the azeotropic composition in the series of ternary saddle azeotropes were proposed by Zeiborak.[51]

Composition of a Ternary Homoazeotrope from Vapor-Liquid Equilibrium Data of the Binary Systems

A general method for predicting the vapor-liquid equilibrium data in ternary and multicomponent systems can be restricted to the azeotropic point by the use of the relative volatility, α_{ij}, defined as

$$\alpha_{ij} = \frac{y_i x_j}{x_i y_j} = \frac{p_i^0 \gamma_i}{p_j^0 \gamma_j} = 1 \qquad (22)$$

In relation (Equation 22) p_i^0 and p_j^0 are the vapor pressures of components i and j at the boiling temperature of the azeotrope, and γ_i and γ_j are the liquid-phase activity coefficients of these components.

The method involves minimization of the function, f, which for a ternary azeotrope becomes

$$f = |\alpha_{13} - 1| + |\alpha_{23} - 1| \qquad (23)$$

α_{13} and α_{23} are the relative volatilities of the pairs of components (1,3) and (2,3), respectively.

The value of the function, f, sufficiently close to zero, corresponds to the azeotropic composition.

This method was used by Aristovicz and Stepanova[52] for the calculation of the azeotropic composition in 19 ternary systems and 1 quaternary system. The results obtained were good. In this procedure the liquid-phase activity coefficients were correlated by the Wilson equation.[53,54]

For ternary and multicomponent heteroazeotropes the above procedure remains essentially unchanged but requires the use of an equation that is applicable to partially miscible systems, e.g. the NRTL (Non-Random, Two-Liquid) equation.[55]

The Azeotropic Data of a Ternary Heteroazeotrope

On the basis of the arguments similar to those described previously, which are applicable to heteroazeotropic systems of any number of components, we can obtain the expressions for the composition and the boiling temperature of a ternary heteroazeotrope.

For the case in which the components are immiscible in each other, the expressions for y_1, y_2, and T^{az} of a ternary heteroazeotrope are

$$y_1 = \frac{p_1^0}{p_1^0 + p_2^0 + p_3^0} \qquad (24)$$

$$y_2 = \frac{p_2^0}{p_1^0 + p_2^0 + p_3^0} \qquad (25)$$

and

$$P = p_1^0 + p_2^0 + p_3^0 \qquad (26)$$

where p_1^0, p_2^0 and p_3^0 are the vapor pressures of the pure components at the boiling temperature of the azeotrope, P is the total pressure of the mixture at equilibrium, and y_1 and y_2 are the mole fractions of component 1 and 2 in the vapor of the azeotrope, respectively.

Other cases, in which the condition of complete immiscibility of the components is not fulfilled, may be described by introducing the activities of the components, α_1, α_2, and α_3.

The procedure, which is straightforward, and involves the multiplication of each p°_i by α_i in expressions (24-26), will not be reproduced here.

The activities, α_1, α_2, and α_3 are related to P, p_1^0, p_2^0, and p_3^0 by

$$\alpha_1 = \frac{P - p_2^0 \alpha_2 - p_3^0 \alpha_3}{p_1^0} \qquad (27)$$

$$\alpha_2 = \frac{P y_2}{p_2^0} \qquad (28)$$

and

$$\alpha_3 = \frac{P y_3}{p_3^0} \qquad (29)$$

Only one series of ternary heteroazeotropes with three liquid phases has been investigated to date.[6] The heteroazeotropes occur in the systems nitromethane-water-n-paraffins. For this series satisfactory agreement was obtained between the calculated (using Equations 24 and 25) and observed azeotropic compositions, but the calculated boiling temperatures of the azeotropes by Equation 26 were found to be much lower than those observed. Low calculated boiling temperatures are due to the relatively high miscibility of nitromethane in water at the respective boiling temperatures of the azeotropes. The difference is lower for the systems with higher-boiling hydrocarbons and tends to the difference between the calculated and observed boiling temperatures for the nitromethane-water system.

It is interesting to note that in some cases the behavior of a series of ternary heteroazeotropes, in which two components remain unchanged for the series, may be described by the activities α_1, α_2, and α_3, which are common for each component within the whole series. This case is exemplified by the series of ternary heteroazeotropes water-pyridine-n-paraffins.[56]

Studies of ternary heteroazeotropes based on the theory of regular solutions were made by Malesinska and Malesinski[57] and by Stecki.[58]

All the examined heteroazeotropes were found to be positive. The existence of negative heteroazeotropes is rather doubtful.

REFERENCES

1. **Dalton, J.,** *Mem. Manchester Phil. Soc.*, 5, 585, 1802; *Ann. Phil.*, 9, 186, 1817.
2. **Wade, J. and Merriman, R. W.,** *J. Chem. Soc. Trans.*, 99, 997, 1911.
3. **Swietoslawski, W.,** *Ebulliometric Measurements,* Reinhold, New York, 1945.
4. **Lecat, M.,** *Compt. Rend.*, 183, 880, 1926.
5. **Swietoslawski, W.,** *Rocz. Chem.* 26, 632, 1952; *Bull. Acad. Polon. Sci. Cl. III,* 1, 63, 1953.
6. **Malesinska, B. and Malesinski, W.,** *Bull. Acad. Polon. Sci. Ser. Sci. Chim.*, 11, 475, 1963.
7. **Ewell, R. L. and Welch, L. M.,** *Ind. Eng. Chem.*, 37, 1244, 1945.
8. **Lang, H.,** *Z. Physik. Chem.*, 196, 278, 1950.
9. **Swietoslawski, W. and Trabczynski, W.,** *Bull. Acad. Polon. Sci. Cl. III,* 3, 333, 1955.
10. **Galska-Krajewska, A.,** *Bull. Acad. Polon. Sci. Ser. Sci. Chim.*, 10, 45, 51, 1962.
11. **Zieborak, K.,** *Bull. Acad. Polon. Sci. Cl. III,* 3, 53, 1955.
12. **Kurtyka, Z. M.,** *J. Chem. Eng. Data,* 16, 310, 1971.
13. **Kurtyka, Z.,** *Bull. Acad. Polon. Sci. Ser. Sci. Chim.*, 9, 741, 1961.
14. **Zieborak, K. and Wyrzykowska-Stankiewicz, D.,** *Bull. Acad. Polon. Sci. Ser. Sci. Chim.*, 8, 137, 1960.
15. **Orszagh, A., Lelakowska, J., and Beldowicz, M.,** *Bull. Acad. Polon. Sci. Ser. Sci. Chim. Geol. et Geogr.*, 6, 419, 1958.
16. **Orszagh, A., Lelakowska, J., and Radecki, A.,** *Bull. Acad. Polon. Sci. Ser. Sci. Chim. Geol. et Geogr.*, 6, 605, 1958.
17. **Ostwald, W.,** *Lehrbuch der Allgemeinen Chimie,* Vol. 2, Engelman, Leipzig, 1899.
18. **Ryland, G.,** *Am. Chem. J.*, 22, 384, 1899; *Chem. News,* 81, 15, 42, 50, 1900.
19. **Lecat, M.,** *L'Azeotropisme,* Lamartin, Bruxelles, 1918.
20. **Lecat, M.,** *Tables Azeotropiques,* Vol. 1, L'Auteur, Bruxelles, 1949.
21. **Horsley, L. H.,** *Azeotropic Data-III,* American Chemical Society, Washington, D.C., 1973.
22. **Swietoslawski, W.,** *Azeotropy and Polyazeotrophy,* Pergamon Press, Oxford, London, 1963.
23. **Malesinski, W.,** *Azeotropy and Other Theoretical Problems of Vapour-Liquid Equilibrium,* Wiley-Interscience, New York, 1965.
24. **Kurtyka, Z. M.,** *Azeotropy and Its Applications,* to be published.
25. **Swietoslawski, W.,** *Bull. Acad. Sci. Polon. Ser. A,* 19, 29, 1950, *Przem. Chem.*, 7, 363, 1951.
26. **Skolnik, H.,** *Ind. Eng. Chem.*, 40, 442, 1948.
27. **Malesinski, W.,** *Bull. Acad. Polon. Sci. Cl. III,* 3, 601, 1955; 4, 295, 1956.
28. **Zawisza, A.,** *Bull. Acad. Polon. Sci. Ser. Sci. Chim.*, 9, 141, 1961.
29. **Lecat, M.,** *Azeotropisme et Distillation, Traite de Chimie Organique, Vol. I,* V. Grignard, Ed., Mason et Cie., Paris, 1935.
30. **Prigogine, I., and Defay, R.,** *Chemical Thermodynamics,* translated by D. H. Everett, Longmans, Green, London, 1954.
31. **Mair, B. J., Glasgow, A. R., and Rossini, F. D.,** *J. Res. Bur. Std.*, 27, 39, 1941.
32. **Horsley, L. H.,** *Anal. Chem.*, 19, 508, 1947.
33. **Meissner, H. P. and Greenfield, S. H.,** *Ind. Eng. Chem.*, 40, 438, 1948.
34. **Seymour, K. M.,** Abstracts 50-I, 110th National Meeting of the American Chemical Sciety, Chicago, September 1946.
35. **Johnson, A. I. and Madonis, J. A.,** *Can. J. Chem. Eng.*, 37, 71, 1959.
36. **Seymour, K. M., Carmichael, R. H., Carter, J., Ely, J., Isaacs, E., King, J., Taylor, R., and Northern, T.,** *Ind. Eng. Chem. Fundam.*, 16, 200, 1977.
37. **Orszagh, A.,** *Rocz. Chem.*, 29, 623, 636, 1955.
38. **Kireev, V. A.** *Acta Physicochim. URSS,* 14, 371, 1941.
39. **Kurtyka, Z. M. and Kurtyka, A.,** *Ind. Eng. Chem. Fundam.*, 19, 225, 1980.
40. **Kurtyka, Z. M. and Kurtyka, Z.,** *Ind. Eng. Chem. Fundam.*, in press.
41. **Francis, A. W.,** *Critical Solution Temperatures,* American Chemical Society, Washington, D.C. 1961.

42. **Malesinska, B . and Malesinski, W.,** *Bull. Acad. Polon. Sci. Ser. Sci. Chim.*, 11, 469, 1963.
43. **Antoine, C.,** *Compt. Rend.*, 107, 681, 836, 1143, 1888.
44. **Hala, E., Wichterle, I., Polak J., and Boublik, T.,** *Vapour-Liquid Equilibrium Data at Normal Pressures*, Pergamon Press, Oxford, London, 1968.
45. **Hirata, M., Ohe, S., and Nagahama, K.,** *Computer-Aided Data Book of Vapour-Liquid Equilibria*, Kodansha-Elsevier, Tokyo, Amsterdam, 1975.
46. **Haase, R.,** *Z. Physik. Chem.*, 195, 362, 1950.
47. **Haase, R.,** *Termodynamik des Mischphasen*, Springer-Verlag, Berlin, 1956.
48. **Malesinski, W.,** *Bull. Acad. Polon. Sci. Cl. III*, 4, 701, 709, 1956; 5, 177, 183, 1957.
49. **Zeiborak, K. and Wyrzykowska-Stankiewicz, D.,** *Bull. Acad. Polon. Sci. Ser. Sci. Chim. Geol. et Geogr.*, 6, 755, 1958.
50. **Stadnicki, J. S.,** *Bull. Acad. Polon. Sci. Ser. Sci. Chim.*, 10, 357, 1962.
51. **Zieborak, K.,** *Bull. Acad. Polon. Sci. Cl. III*, 3, 531, 1955.
52. **Aristovicz, V. Y. and Stepanova, E. I.,** *Zhur. Prikl. Khim.*, 43, 2192, 1970.
53. **Wilson, G. M.,** *J. Am. Chem. Soc.*, 86, 127, 1964.
54. **Prausnitz, J. M.,** *Molecular Thermodynamics of Fluid-Phase Equilibria*, Prentice Hall, Englewood Cliffs, New Jersey, 1969.
55. **Renon, H. and Prausnitz, J. M.,** *A. I. Ch. E. Journal*, 14, 135, 1968.
56. **Trabczynski, W.,** *Bull. Acad. Polon. Sci. Ser. Sci. Chim. Geol. et Geogr.*, 6, 269, 1958.
57. **Malesinska, B. and Malesinski, W.,** *Bull. Acad. Polon. Sci. Ser. Sci. Chim.*, 12, 861, 867, 1964.
58. **Stecki, J.,** *Bull. Acad. Polon. Sci. Cl. III*, 5, 421, 1957; *Ser. Sci. Chim. Geol. et Geogr.*, 6, 47, 1958.

TABLES OF AZEOTROPES AND ZEOTROPES

In Tables 1, 2, and 3 the different types of azeotropes are identified as:

1. Homoazeotrope, positive; no marking
2. Homoazeotrope, negative; N
3. Homoazeotrope, saddle or positive-negative; S
4. Heteroazeotrope, positive, with two phases; H
5. Heteroazeotrope, positive, with three phases; H-3

Throughout Tables 1,2, and 3 the azeotropic composition is expressed as weight percent (wt %). In Table 1 only the weight percent of component 2, x_2, the variable component, is listed. However, in Tables 2, and 3 the weight percents of all of the components are listed.

Compounds are listed in the following tables according to the empirical formula convention employed by Chemical Abstracts. As further assistance in locating particular systems the following index has been arranged in alphabetical order with respect to component X_1 of each system. The entry number for the particular system is in the second column of the list.

Table 1
BINARY SYSTEMS

Component X_1	Entry No.
Acetal	1364
Acetaldehyde	377
Acetic acid	383
Acetone	544
Acetonitrile	369
Acetophenone	1529
Acrylonitrile	525
Allyl alcohol	554
Aluminum chloride	1
n-Amyl alcohol	1143
p-tert-Amyl alcohol	1729
Aniline	1255
o-Anisidine	1511
Benzaldehyde	1408
Benzene	1180
Benzoic acid	1415
Benzonitrile	1400
Benzyl alcohol	1448
Benzyl phenyl ether	1743
Borneol	1717
Boron fluoride	2
Boron hydride	10
Bromoacetic acid	349
Bromodichloromethane	217
Bromoform	224
Bromomethane	271
1-Butanethiol	1032
1-Butanol	908
2-Butanol	939
2-Butanone	770
2-Butoxyethanol	1368
Butyl acetate	1335
Butyl alcohol	908
Butyl formate	1100
Butyl nitrite	893
Butyraldehyde	782

Table 1 (Continued)
BINARY SYSTEMS

Component X_1	Entry No.
Butyronotile	764
Camphene	1715
Capric acid	1722
Caproic acid	1338
Capronitrile	1325
Caprylic acid	1622
Carbon disulfide	204
Carbon tetrachloride	185
Carvacol	1696
Carvone	1703
Chloroacetic acid	354
Chloroform	232
m-Cresol	1473
o-Cresol	1462
p-Cresol	1485
Cumene	1655
Cyclohexanol	1327
Cyclohexanone	1297
Diethylene glycol	1008
Doxane	811
Dipropylene glycol	1386
Enanthic acid	1514
Ethanol	460
2-Ethoxyethanol	994
Ethyl	
acetate	821
acetoacetate	1301
alcohol	460
aniline	1592
benzene	1551
bromoacetate	750
carbamate	631
chloroacetate	756
formate	563
fumarate	1613
lactate	1136
maleate	1616
methyl sulfide	712
nitrate	455
oxalate	1319
phenol, para	1554
pyruvate	1071
salicylate	1651
succinate	1619
Ethylene glycol	476
Ethylene sulfide	411
Ethylenediamine	522
Ethylidine diacetate	1313
Formic acid	242
2-Furaldehyde	1044
Glycol diacetate	1324
Glycol monoacetate	865
Glycerol	687
Guaiacol	1497
1-Heptanol	1521
n-Heptyl alcohol	1521
Hexachloroethane	337
1-Hexanol	1357
n-Hexyl alcohol	1357

Component X_1	Entry No.
Hydrogen chloride	14
Hydrogen cyanide	11
Hydrogen fluoride	15
Indole	1527
Iodobenzene	1168
Iodoethane	420
Iodomethane	272
Isoamyl alcohol	1148
Isoamyl benzoate	1735
Isoamyl formate	1345
Isoamyl oxalate	1739
Isobutyl nitrate	902
Isobutyronitrile	767
Isobutryic acid	829
Isopropyl alcohol	663
Isopropyl lactate	1354
Isopropyl methyl sulfide	1038
Isovaleric acid	1105
Levulinic acid	1074
2,4-Lutidine	1501
Menthol	301
Mesitol	1662
Mesitylene	1656
Methanol	301
Methyl	
acetate	567
acetoacetate	1085
acetophenone, para	1643
alcohol	301
aniline	1503
anisole, para	1562
butyrate	1120
chloroacetate	534
disulfide	520
formate	414
fumarate	1284
lactate	876
maleate	1290
1-Methylnaphthalene	1724
2-Methylnaphthalene	1728
2-Methyl-2-propanol	949
2-Methylthiophene	1063
3-Methylthiophene	1067
2-Octanol	1630
sec-Octyl alcohol	1630
Naphthalene	1679
Nitrobenzene	1159
Nitroethane	453
Nitromethane	276
1-Nitropropane	654
o-Nitrotoluene	1426
p-Nitrotoluene	1437
Pelargonic acid	1669
2-Pentanone	1096
Perfluorobutyric acid	724
Phenethyl alcohol	1568
o-Phenetidine	1600
p-Phenetidine	1607
Phenol	1190

Table 1 (Continued)
BINARY SYSTEMS

Component X_1	Entry No.
Phenyl acetate	1538
Phenyl benzoate	1741
Phenyl ether	1732
o-Phenylenediamine	1278
3-Phenylpropanol	1664
Phosphorus oxychloride	16
Phosphorus trichloride	17
2-Picoline	1273
3-Picoline	1276
Pinacol	1377
1-Propanethiol	716
1-Propanol	673
Propioamide	593
Propionic acid	574
Propionitrile	541
Propiophenone	1648
Propyl	
acetate	1123
alcohol	673
benzene	1659
benzoate	1690
formate	858
isovalerate	1626
lactate	1350
nitrite	661
propionate	1348
Pseudocumene	1660
Pyridine	1056
Pyrogallol	1252
Pyrrol	734
Pyruvic acid	527
Qunialdine	1683
Quinoline	1635
Resorcinol	1238
Silicon tetrachloride	20
Tetrachlorothylene	327
Tetrahydrothiophene	888
Thioacetic acid	380
Thiophene	727
Thymol	1707
Tin chloride	26
Toluene	1442
o-Toluidine	1505
Trichloroacetic acid	346
Trichloroethylene	342
Trichlorofluoromethane	173
Trinitromethane	176
Triethylene glycol	1391
Valeric acid	1128
Water	28
2,4-Xylenol	1576
3,4-Xylenol	1582
2,4-Xylidine	1595

Table 2
TERNARY SYSTEMS

Component X_1	Entry No.
Acetic acid	108
Acetone	141
Acetonitrile	106
Aniline	167
Argon	1
1-Butanol	158
2-Butanone	154
Butyric acid	156
Carbon tetrachloride	92
Chlorine trifluoride	4
Chloroform	93
m-Cresol	175
m,p-Cresol (mixture)	172
p-Dioxane	157
Ethanol	130
Ethyl benzene	166, 177
Ethylene glycol	135
Hydrogen bromide	2
Hydrogen chloride	5
Hydrogen cyanide	3
Hydrogen fluoride	7
Isobutyl alcohol	161
Isobutyl lactate	176
Isopropyl alcohol	151
Methanol	101
Methyl formate	
Nitromethane	129
Phenol	163
1-Propanol	152
1-Propanol, 2-methyl	161
2-Propanol	151
Propionic acid	145
Propyl lactate	171
Pyridine	162
Silicon tetrafluoride	8
Trichloroethylene	107

Table 3
QUATERNARY AND QUINARY SYSTEMS

Component X_1	Entry No.
Acetic acid	18
Acetone	20
Chloroform	16
Hydrogen cyanide	1
Isopropyl alcohol	21
Water	3

BINARY SYSTEMS

No.	System	B.P. (°C)	Azeotropic data	
			Compn. X_2 (wt. %)	B.P. (°C)
	Aluminum chloride	183		
1	-tantalum chloride (2)	242	90.4	235
	Boron fluoride	−100		
2	-boron hydride	−92	22.8	−106
3	-acetonitrile, N	81.6	38	101
4	-methyl formate, N	31.9	47	91
5	-methyl ether, N	−21	40	127
6	-ethyl formate, N	54.1	52	102
7	-methyl acetate, N	57.1	52	110
8	-ethyl ether, N	34.5	52	125
9	-ethyl propionate, N	99.15	60	116
	Boron hydride	−92.5		
10	-hydrogen chloride	−85	36	−94
	Hydrogen cyanide	26		
11	-methyl alcohol	64.7	Zeotropic	
12	-methyl formate	31.7	48	24
13	-ethyl nitrite	17.4	85	16.5
	Hydrogen chloride	−85		
14	-methyl ether, N	−22	62	−2
	Hydrogen fluoride	19.4		
15	-ethyl ether, N	34.5	60	74
	Phosphorus oxychloride	107.2		
16	-titanium tetrachloride, N	136.5	53.4	143.2
	Phosphorus trichloride	76		
17	-cyclohexane	80.75	Zeotropic	
18	-2,3-dimethylpentane	80.5	27	74.2
19	-2,2,3-trimethylbutane	80.9	23	74.5
	Silicon tetrachloride	56.9		
20	-carbon tetrachloride	76.75	Zeotropic	
21	-chloroform	61.0	30	55.6
22	-nitromethane	101	6	53.8
23	-acetonitrile	82	9.4	49.0
24	-acrylonitrile	79	11	51.2
25	-propionitrile	97	8	55.6
	Tin chloride	113.85		
26	-toluene	110.7	48	109.15
27	-2,5-dimethylhexane	109.4	60	107.5
	Water	100		
28	-hydrogen chloride, N	−85	20.22	108.58
29	-hydrogen bromide, N	−73	47.5	126
30	-hydrogen iodide, N	−34	57	127
31	-hydrogen fluoride, N	19.4	35.6	111.35
32	-nitric acid, N	86	67.4	120.7
33	-hydrogen peroxide	152.1	Zeotropic	
34	-hydrazine, N	113.8	67.7	120
35	-carbon tetrachloride, H	76.75	95.9	66
36	-carbon disulfide, H	46.25	97.2	42.6
37	-chloroform, H	61	97.2	56.1
38	-formic acid, N	100.75	77.4	107.2
39	-nitromethane, H	101.2	76.4	83.59
40	-tetrachloroethylene, H	121	82.8	88.5
41	-trichloroethylene	86.2	83	73.4
42	-acetonitrile, H	80.1	83.7	76.5
43	-acetic acid	118.1	Zeotropic	
44	-acetamide, H	221.2	Zeotropic	
45	-nitroethane, H	114.07	71.5	87.22
46	-ethyl nitrate	87.68	78	74.35
47	-ethyl alcohol	78.32	96	78.17
48	-methyl sulfate	189.1	27	98.6
49	-acrylonitrile, H	77.2	85.7	70.6
50	-acrolein, H	52.8	97.4	52.4
51	-acetone	56.1	Zeotropic	
52	-allyl alcohol	96.9	72.3	88.9
53	-propionaldehyde, H	47.9	98	47.5
54	-ethyl formate	54.2	95	52.6
55	-propionic acid	141.1	17.7	99.9
56	-trioxane	114.5	70	91.4
57	-l-chloropropane	46.6	97.8	44
58	-isopropyl alcohol	82.3	87.4	80.3
59	-propyl alcohol, 740 mm	97.3	71.7	87
60	-perfluorobutyric acid	122.0	29	97
61	-crotonic acid	189	2.2	99.9
62	-methyl acrylate	80	92.8	71
63	-ethyl chloroacetate	143.5	54.9	95.2
64	-butyronitrile, H	117.6	67.5	88.7
65	-isobutyronitrile, H	103	77	82.5
66	-ethyl vinyl ether, H	35.5	98.5	34.6
67	-butyric acid	163.5	3	99.4
68	-ethyl acetate	77.15	91.53	70.38
69	-isopropyl formate	68.8	97	65.0
70	-propyl formate	80.9	97.7	71.6
71	-methyl lactate	144.8	20	99
72	-butyl alcohol, H	117.4	57.5	92.7
73	-*sec*-butyl alcohol	99.5	73.2	87.0
74	-pyridine	115.5	58.7	93.6
75	-furfuryl alcohol	169.35	20	98.5
76	-furfurylamine	144	26	99
77	-isoprene	34.1	99.86	32.4
78	-cyclopentanone	130.8	57.6	94.6
79	-allyl acetate	104.1	85.3	83
80	-cyclopentanol	140.85	42	96.25
81	-valeraldehyde, H	103.3	81	83
82	-butyl formate	106.6	85.5	83.8
83	-isopropyl acetate	88.6	89.4	76.6
84	-isovaleric acid	176.5	18.4	99.5
85	-methyl butyrate	102.65	88.5	82.7
86	-methyl isobutyrate	92.3	93.2	77.7
87	-valeric acid	188.5	11	99.8
88	-piperidine	105.8	65	92.8
89	-n-pentane, H	36.1	98.6	34.6
90	-n-amyl alcohol, H	137.8	45.6	95.8
91	-*tert*-amyl alcohol	102.25	72.5	87.35
92	-2-pentanol	119.3	63.5	91.7
93	-N-methylbutylamine	91.1	85	82.7
94	-chlorobenzene	131.8	71.6	90.2
95	-nitrobenzene, H	210.85	—	98.6
96	-benzene, H	80.1	91.17	69.25
97	-phenol	182	9.2	99.52
98	-aniline, H, 742 mm	184.3	19.2	98.6
99	-2-picoline	129.5	52	93.5
100	-3-picoline	144.1	40	97
101	-4-picoline	144.3	37.2	97.35
102	-2,5-dimethylfuran	93.3	88.3	77.0
103	-cyclohexene, H	82.75	91.07	70.8
104	-ethyl crotonate	137.8	62	93.5
105	-ethylene glycol diacetate	190.8	15.4	99.7
106	-butyl chloroacetate	181.9	24.5	98.12
107	-cyclohexane, H	80.8	91.6	69.5
108	-amyl formate	132	71.6	91.6
109	-butyl acetate	126.2	71.3	90.2
110	-ethyl butyrate	120.1	78.5	87.9
111	-isoamyl formate	124.2	79	90.2
112	-isobutyl acetate	117.2	83.5	87.4
113	-isopropyl propionate	110.3	80.1	85.2
114	-propyl propionate	122.1	77	88.9
115	-paraldehyde, H	124	71.5	90
116	-n-hexane, H	68.7	94.4	61.6
117	-butyl ethyl ether, H	92.2	88.1	76.6
118	-n-hexyl alcohol, H	157.1	32.8	97.8
119	-acetal	103.6	85.5	82.6
120	-pinacol	174.35	Zeotropic	
121	-toluene, H	110.7	86.5	84.1
122	-anisole	153.85	59.5	95.5
123	-benzyl alcohol, H	205.2	9	99.9
124	-guaiacol	205.0	12.5	99.5
125	-2,6-lutidine, H	144.0	48.2	96.02
126	-o-toluidine	199.7	84.6	—
127	-p-toluidine, H	200.4	86.2	—
128	-2-heptanone	149	52	95
129	-3-heptanone	147.6	57.8	94.6
130	-4-heptanone	143.7	59.5	94.3
131	-ethyl valerate	145.45	60	94.5
132	-isoamyl acetate	142	63.7	93.8
133	-isobutyl propionate	136.85	47.8	92.75
134	-n-heptane, H	98.4	87.1	79.2
135	-benzyl formate	202.3	20	99.2
136	-methyl benzoate	199.45	20.8	99.08
137	-phenyl acetate	195.7	24.9	98.9
138	ethylbcnzene,H	136.2	67.0	92.0
139	-m-xylene, H	139	64.2	92
140	-N-ethylaniline	204.8	16.1	99.2
141	-1-octene, H	121.28	71.3	88.0
142	-hexyl acetate	171.0	39	97.4
143	-isoamyl propionate	160.3	51.5	96.55
144	-isobutyl butyrate	156.8	54	96.3
145	-isobutyl isobutyrate	147.3	60.6	95.5
146	-propyl isovalerate	155.8	54.8	96.2
147	-n-octane, H	125.7	75.5	89.6
148	-isooctane, H	99.3	88.9	78.8
149	-butyl ether, H	142.6	67	92.9
150	-n-octyl alcohol, H	195.15	10	99.4
151	-dibutylamine, H	159.6	49.5	97

No.	System	B.P. (°C)	Azeotropic data: Compn. X_2 (wt. %)	Azeotropic data: B.P. (°C)
152	-quinoline, H	237.3	3.4	—
153	-ethyl benzoate	212.4	16.0	99.4
154	-cumene, H	152.4	56.2	95
155	-mesitylene, H	164.6	—	96.5
156	-triallylamine, H	151.1	62	95
157	-isoamyl butyrate	178.5	36.5	98.05
158	-isobutyl carbonate	190.3	20	98.6
159	-n-nonane, H	150.8	18	94.8
160	-naphthalene, H	218.0	16	98.8
161	-methyl phthalate, H	283.2	2.5	99.95
162	-nicotine, H	—	2.5	99.85
163	-camphene, H	159.6	—	96.0
164	-n-decane, H	173.3	—	97.2
165	-n-undecane, H	194.5	4.0	98.85
166	-o-phenyl phenol	—	1.25	99.95
167	-phenyl ether, H	259.3	3.25	99.33
168	-ethyl phthalate, H	298.5	2.0	99.98
169	-isoamyl benzoate, H	262.3	4.4	99.9
170	-n-dodecane, H	214.5	2	99.45
171	-dihexylamine, H	239.8	7.2	99.8
172	-tributylamine, H	213.9	20.3	99.65
	Trichlorofluoromethane	24.9		
173	-acetaldehyde	20.2	45	15.6
174	-methyl formate	32	18	20.0
175	-2-methylbutane	27	8	23.16
	Trichloronitromethane	111.9		
176	-acetic acid	118.1	19.5	107.65
177	-ethyl alcohol	78.3	66	77.5
178	-isopropyl alcohol	82.4	65	81.95
179	-propyl alcohol	97.2	41.5	94.05
180	-isoamyl alcohol	131.9	7	111.15
181	-n-pentanol	119.8	17	108.0
182	-toluene	110.75	Zeotropic	
183	-methylcyclohexane	101.15	73	100.8
184	-n-heptane	98.4	93	98.32
	Carbon tetrachloride	76.75		
185	-carbon disulfide	46.25	Zeotropic	
186	-chloroform	62.1	Zeotropic	
187	-formic acid	100.7	18.5	66.65
188	-nitromethane	101.2	17	71.3
189	-methyl alcohol	64.7	20.56	55.7
190	-acetonitrile	81.6	17	65.1
191	-acetic acid	118.1	1.54	76
192	-ethyl alcohol	78.3	15.8	65.04
193	-acrylonitrile	77.3	21	66.2
194	-acetone	56.15	88.5	56.08
195	-propyl alcohol	97.25	7.9	73.4
196	-thiophene	84	Zeotropic	
197	-butyl nitrite	78.2	30	75.3
198	-butyl alcohol	117.75	2.4	76.55
199	-ethyl ether	34.6	Zeotropic	
200	-pyridine	115.5	Zeotropic	
201	-benzene	80.1	Zeotropic	
202	-n-heptane	98.4	Zeotropic	
203	-o-xylene	143.6	Zeotropic	
	Carbon disulfide	46.25		
204	-chloroform	61.2	Zeotropic	
205	-formic acid	100.75	17	42.55
206	-nitromethane	101.2	81.4	41.2
207	-methyl alcohol	64.7	29	39.8
208	-acetic acid	118.1	Zeotropic	
209	-propyl nitrite	47.75	38	40.15
210	-ethyl alcohol	78.3	9	42.6
211	-acetone	56.15	33	39.25
212	-propyl alcohol	97.1	5.5	45.65
213	-ethyl acetate	76.7	3	46.1
214	-n-pentane	36.15	89	35.7
215	-n-hexane	68.95	Zeotropic	
216	-toluene	110.7	Zeotropic	
	Bromodichloromethane	90.2		
217	-nitromethane	101.2	25	87.3
218	-methyl alchol	64.7	40	63.8
219	-ethyl alcohol	78.3	28	75.5
220	-ethyl acetate	77.1	12	90.55
221	-benzene	80.2	Zeotropic	
222	-cyclohexane	80.75	Zeotropic	
223	-n-hexane	68.8	Zeotropic	
	Bromoform	149.5		
224	-formic acid	100.75	48	97.4
225	-acetamide	221.15	Zeotropic	
226	-butyric acid	162.45	6.8	146.8
227	-phenol	182.2	Zeotropic	
228	-aniline	184.35	Zeotropic	
229	-toluene	110.65	Zeotropic	
230	-o-cresol	191.1	Zeotropic	
231	-α-pinene	155.8	25	146.5
	Chloroform	61.2		
232	-formic acid	100.75	15	59.15
233	-methyl alcohol	64.7	12.6	53.43
234	-ethyl alcohol	78.3	7	59.35
235	-acetone, N	56.5	21.9	64.4
236	-propyl alcohol	97.2	Zeotropic	
	Chloroform	61.2		
237	-p-dioxane	101	Zeotropic	
238	-cyclohexane	80.75	Zeotropic	
239	-methylcyclopentane	72.0	20	60.5
240	-n-hexane	68.7	16.5	60.4
241	-toluene	110.65	Zeotropic	
	Formic acid	100.75		
242	-nitromethane	101.22	54.5	97.07
243	-trichloroethylene	86.95	75	74.1
244	-tetrachloroethylene	121.1	50	88.15
245	-acetic acid	118.1	Zeotropic	
246	-nitroethane	114.2	Zeotropic	
247	-ethyl ether	34.6	Zeotropic	
248	-ethyl sulfide	92.2	65	82.2
249	-pyridine, N	115.5	38.6	127.43
250	-2-methylbutane	27.95	96	27.2
251	-n-pentane	36.15	80	34.2
252	-bromobenzene	156.1	32	98.1
253	-chlorobenzene	131.75	41	93.7
254	-fluorobenzene	84.9	73	73.0
255	-benzene	80.2	69	71.05
256	-aniline	184.35	Zeotropic	
257	-2-picoline, N	129	75	158.0
258	-cyclohexane	80,75	30	70.7
259	-methylcyclopentane	72.0	71	63.3
260	-n-hexane	68.95	72	60.6
261	-propyl sulfide	141.5	17	98.0
262	-isopropyl sulfide	120.5	38	93.5
263	-toluene	110.7	50	85.8
264	-o-chlorotoluene	159.3	17	100.2
265	-methylcyclohexane	101.1	53.5	80.2
266	-n-heptane	98.45	43.5	78.2
267	-styrene	145.8	27	97.75
268	-o-xylene	143.6	26	95.5
269	-m-xylene	139.0	28.2	92.8
270	-n-octane	125.8	37	90.5
	Bromomethane	3.65		
271	-methyl alcohol	64.7	0.45	3.55
	Iodomethane	42.5		
272	-methyl alcohol	64.7	4.5	37.8
273	-ethyl alcohol	78.3	3.2	41.2
274	-acetone	56.15	5	42.4
275	-n-hexane	68.85	Zeotropic	
	Nitromethane	101.2		
276	-methyl alcohol	64.7	90.9	64.4
277	-acetic acid	118.1	4	101.2
278	-ethyl alcohol	78.3	71	76.05
279	-propyl alcohol	97.15	51.6	89.09
280	-p-dioxane	101.35	43.5	100.55
281	-n-butyl alcohol	117.73	28.6	98.0
282	-n-pentane, H	36.07	99	35
283	-cyclohexane	80.75	73.5	69.5
284	-methylcyclopentane	72.0	77	64.2
285	-n-hexane, H	68.74	81.5	61.7
286	-toluene	110.75	45	96.5
287	-n-heptane, 748 mm Hg, H	98.4	64.4	79.7
288	-styrene	145.8	Zeotropic	
289	-o-xylene	144.3	Zeotropic	
290	-n-octane, 748 mm; H	125.75	44.8	90.23
291	-cumene	152.8	Zeotropic	
292	-mesitylene	164.6	Zeotropic	
293	-n-nonane, 748 mm; H	150.85	28.4	96.14
294	-n-decane, 748 mm; H	174.12	16.1	98.81
295	-n-undecane, 748 mm; H	194.5	9.3	100.01
296	-n-dodecane, 748 mm; H	216.0	4.2	100.60
	Methyl nitrate	64.8		
297	-methyl alcohol	64.65	27	52.5
298	-cyclohexane	80.75	23	61.0
299	-n-hexane	68.8	44	56.0
300	-n-heptane	98.4	Zeotropic	

BINARY SYSTEMS (Continued)

No.	System	B.P. (°C)	Azeotropic data: Compn. X_2 (wt. %)	Azeotropic data: B.P. (°C)
	Methyl alcohol	64.7		
301	-trichloroethylene	87	62	59.3
302	-bromoethane	38	94.7	34.9
303	-acetic acid	118.1	Zeotropic	
304	-acetone	56.15	88	55.5
305	-methyl acetate	57.1	81	53.5
306	-thiophene	84	83.6	59.71
307	-methyl acrylate	80	46	62.5
308	-p-dioxane	101.05	Zeotropic	
309	-ethyl sulfide	92.2	38	61.2
310	-pyridine	115.4	Zeotropic	
311	-cyclopentane	49.4	86	38.8
312	-isobutyl formate	97.9	5	64.6
313	-piperidine	106.4	Zeotropic	
314	-n-pentane	36.15	93	30.85
315	-chlorobenzene	132.0	Zeotropic	
316	-fluorobenzene	85.15	68	59.7
317	-benzene	80.1	60.9	57.5
318	-cyclohexane	80.7	63.6	53.9
319	-toluene	110.6	27.5	63.5
320	-methylcyclohexane	100.8	46	59.2
321	-n-heptane, H	98.45	48.5	59.1
322	-o-xylene	143.6	Zeotropic	
323	-n-octane	125.75	32.5	62.75
324	-n-nonane	150.7	16.6	64.1
325	-n-decane	173.8	Zeotropic	
326	-methyl *tert*-butyl ether	55.06	85.7	51.27
	Tetrachloroethylene	121.1		
327	-acetic acid	118.1	38.5	107.35
328	-acetamide	221.2	2.6	120.45
329	-ethylene glycol	197.4	6	119.1
330	-acetone	56.1	Zeotropic	
331	-propionic acid	140.9	8.5	119.1
332	-propyl alcohol	97.25	48	94.05
333	-n-butyl alcohol	117.7	32	110.0
334	-pyridine	115.4	48.5	112.85
335	-n-amyl alcohol	138.2	15	117.0
336	-toluene	110.75	Zeotropic	
	Hexachloroethane	185		
337	-trichloroacetic acid	196	15	181
338	-phenol	182.2	30	173.7
339	-aniline	184.35	34	176.75
340	-benzyl alcohol	205.15	12	182.0
341	-p-cresol	201.7	10	183.0
	Trichloroethylene	86.9		
342	-acetic acid	118.1	3.8	86.5
343	-benzene	80.2	Zeotropic	
344	-cyclohexane	80.7	83.4	80.5
345	-n-heptane	98.45	Zeotropic	
	Trichloroacetic acid	197.55		
346	-pentachloroethane	161.95	96.5	161.8
347	-naphthalene	218.05	Zeotropic	
348	-butylbenzene	183.1	80	181.3
	Bromoacetic acid	205.1		
349	-o-dichlorobenzene	179.5	84	177.0
350	-o-bromotoluene	181.5	82	179.0
351	-acetophenone	202.0	30	206.5
352	-butylbenzene	183.1	75	179.5
353	-cymene	176.7	85	174.7
	Chloroacetic acid	189.35		
354	-bromobenzene	156.1	89	154.3
355	-phenol	181.5	Zeotropic	
356	-m-bromotoluene	183.8	70	174
357	-p-bromotoluene	185.0	66	174.1
358	-styrene	145.8	86	144.8
359	-o-xylene	144.3	88	143.5
360	-m-xylene	139.2	93	139.05
361	-n-octane	125.75	Zeotropic	
362	-cumene	152.8	79	150.8
363	-mesitylene	164.6	83	162.0
364	-pseudocumene	168.2	66	162.8
365	-naphthalene	218.05	22	187.1
366	-cymene	176.7	58	169.0
367	-n-decane	173.3	58	165.2
368	-1,3,5-triethylbenzene	215.5	25	185.5
	Acetonitrile	81.6		
369	-acetic acid	118.1	Zeotropic	
370	-ethyl alcohol	78.3	56	72.5
371	-pyridine	115.5	Zeotropic	
372	-isoprene	34.1	97.6	33.7
373	-isopropyl acetate	89.5	40.0	79.5
374	-toluene	110.7	20	81.4
375	-ethylbenzene	136.2	Zeotropic	
376	-n-undecane	195.4	Zeotropic	
	Acetaldehyde	20.4		
377	-acetone	56.15	Zeotropic	
378	-ethyl ether	34.5	23.5	18.9
379	-benzene	80.1	Zeotropic	
	Thioacetic acid	89.5		
380	-benzene	80.15	Zeotropic	
381	-cyclohexane	80.75	Zeotropic	
382	-methylcyclopentane	72.0	Zeotropic	
	Acetic acid	118.1		
383	-nitroethane	114.2	70	112.4
384	-dioxane	101.35	23	119.5
385	-acetone	56.1	Zeotropic	
386	-pyridine, N	115.5	48.9	138.1
387	-2-picoline, N	129.3	59.6	144.12
388	-3-picoline, N	144	69.6	152.5
389	-4-picoline, N	144.3	69.7	154.3
390	-benzene	80.2	98.0	80.05
391	-cyclohexane	80.75	90.4	78.8
392	-n-hexane	68.6	94.0	68.25
393	-isopropyl sulfide	120	52	111.5
394	-toluene	110.7	71.9	100.6
395	-triethylamine, N	89	33	163
396	-2,6-lutidine, N	144.0	77.1	148.1
397	-methylcyclohexane	101.1	69	96.3
398	-n-heptane	98.25	67	91.72
399	-styrene	145.2	14.3	116.8
400	-ethylbenzene	136.15	34	114.65
401	-o-xylene	143.6	22	116.6
402	-m-xylene	139.0	27.5	115.35
403	-p-xylene	138.4	28	115.25
404	-ethylcyclohexane	131.8	—	107.9
405	-n-octane	125.75	46.3	105.7
406	-cumene	152.8	16	116.0
407	-mesitylene	164.6	Zeotropic	
408	-n-nonane	150.8	31	112.9
409	-n-decane	173.3	20.5	116.75
410	-n-undecane	194.5	5	117.87
	Ethylene sulfide	55.7		
411	-acetone	56.15	43	51.5
412	-n-hexane	68.8	Zeotropic	
413	-2,3-dimethylbutane	58.0	35	54.0
	Methyl formate	31.7		
414	-ethyl ether	34.6	45	28.4
415	-isoprene	34.1	50	22.5
416	-2-methylbutane	27.95	53	17.05
417	-n-pentane	36.15	47	21.8
418	-n-hexane	69.0	Zeotropic	
419	-2,3-dimethylbutane	58.0	15	30.5
	Iodoethane	72.3		
420	-ethyl alcohol	78.3	14	63
421	-propyl alcohol	97.2	7	70
422	-n-hexane	68.85	24	68.0
	Acetamide	221.2		
423	-benzaldehyde	179.2	93.5	178.6
424	-methylaniline	196.25	86	193.8
425	-m-cresol	202.1	Zeotropic	
426	-styrene	145.8	88	144
427	-o-xylene	144.3	89	142.6
428	-m-xylene	139.0	90	138.4
429	-p-xylene	138.2	92	137.75
430	-2,4-xylenol	210.5	Zeotropic	
431	-3,4-xylenol	226.8	4	221.1
432	-ethylaniline	205.5	82	199.0
433	-quinoline	237.3	Zeotropic	
434	-indene	183.0	82.5	177.2
435	-naphthalene	218.05	73	199.55
436	-safrol	235.9	68	208.8
437	-eugenol	255.0	12	220.8
438	-p-cymene	176.7	81	170.5
439	-diethylaniline	217.05	76	198.05
440	-camphene	159.6	88	155.5
441	-dipentene	177.7	82	169.15
442	-camphor	209.1	77	199.8
443	-isoamyl valerate	192.7	84	184.85
444	-isoamyl sulfide	214.8	83	199.5
445	-1-methylnaphthalene	245.1	56.2	209.8
446	-2-methylnaphthalene	241.15	60	208.25
447	-acenaphthene	277.9	35.8	217.1
448	-biphenyl	255.9	49.5	212.95
449	-phenyl ether	259.3	48	214.55

No.	System	B.P. (°C)	Azeotropic data: Compn. X_2 (wt. %)	Azeotropic data: B.P. (°C)
450	-diphenylmethane	265.6	43.5	215.15
451	-1,2-diphenylethane	284	32	218.2
452	-benzyl ether	297	Zeotropic	
	Nitroethane	114.2		
453	-n-hexane, H	68.74	89.4	59.4
454	-toluene	110.75	75	106.2
	Ethyl nitrate	[illegible]		
455	-thiophene	84.7	Zeotropic	
456	-benzene	80.15	88	80.03
457	-cyclohexane	80.75	64	74.5
458	-n-hexane	68.8	76	66.25
459	-n-heptane	98.4	37	82.6
	Ethyl alcohol	78.3		
460	-acrylonitrile	77.3	59	70.8
461	-acetone	56.1	Zeotropic	
462	-ethyl sulfide	92.2	44	72.6
463	-pyridine	115.4	Zeotropic	
464	-cyclopentane	49.4	92.5	44.7
465	-n-pentane	36.15	95	34.3
466	-fluorobenzene	85.15	25	70.0
467	-benzene	80.1	68.3	67.9
468	-cyclohexane	80.8	70.8	64.8
469	-n-hexane	68.95	79	58.68
470	-propyl ether	90.4	56	74.4
471	-toluene	110.7	32	76.7
472	-ethylbenzene	136.15	Zeotropic	
473	-p-xylene	138.3	Zeotropic	
474	-n-octane	125.6	22	77.0
	Ethylene glycol	197.4		
475	-pyridine	115.5	Zeotropic	
476	-benzene	80.2	Zeotropic	
477	-phenol	182.2	Zeotropic	
478	-aniline	184.35	76	180.55
479	-o-bromotoluene	181.75	75	166.8
480	-o-nitrotoluene	221.75	51.5	188.55
481	-toluene	110.6	97.7	110.1
482	-m-toluidine	200.3	58	188.55
483	-o-cresol	191.1	73	189.6
484	-m-cresol	202.1	40	195.2
485	-2,6-lutidine	144.0	Zeotropic	
486	-n-heptane	98.45	97	97.9
487	-styrene	145.8	83.5	139.5
488	-m-xylene	139.1	93.45	135.1
489	-p-xylene	138.4	93.6	134.5
490	-3,4-xylenol	226.8	11	197.2
491	-2,4,6-collidine	171.3	90.3	170.5
492	-2,4-xylidine	214.0	53	188.6
493	-butyl ether	142.1	93.6	139.5
494	-quinoline	237.3	20.5	196.35
495	-indene	183.0	74	168.4
496	-cumene	152.8	82	147.0
497	-mesitylene	164.6	87	156
498	-propylbenzene	158.8	81	152
499	-cymene	176.7	74.5	163.2
500	-camphene	159.5	80	152.5
501	-camphor	209.1	60	186.15
502	-menthol	216.3	48.5	188.55
503	-n-decane	173.3	77	161.0
504	-naphthalene	218.05	49	183.9
505	-1-methylnaphthalene	245.1	40.0	190.25
506	-2-methylnaphthalene	241.15	42.8	189.1
507	-acenaphthene	277.9	25.8	194.65
508	-biphenyl	256.1	33.5	192.25
509	-fluorene	296.4	18	196.0
510	-diphenylmethane	265.6	31.5	193.3
511	-benzyl phenyl ether	286.5	13	195.5
512	-n-tridecane	234.0	45	188.0
513	-anthracene	340	1.7	197
514	-stilbene	306.4	13	196.8
	Methyl sulfide	37.3		
515	-acetone	56.15	Zeotropic	
516	-isoprene	34.3	65	32.5
517	-cyclopentane	49.35	12.5	37.1
518	-n-pentane	36.15	53.4	31.8
519	-2,2-dimethylbutane	49.7	20.2	36.5
	Methyl disulfide	109.44		
520	-n-heptane	98.4	73.7	96.44
521	-2,3-dimethylhexane	109.15	51.8	102.84
	Ethylenediamine	116.5		
522	-n-butyl alcohol	117.7	64.3	124.7
523	-benzene	80.1	Zeotropic	
524	-toluene	110.7	69.2	104
	Acrylonitrile	77.3		
525	-isopropyl alcohol	82.55	44	71.7
526	-benzene	80.2	53	73.3
	Pyruvic acid	166.8		
527	-propionic acid	141.3	Zeotropic	
528	-benzene	80.15	Zeotropic	
529	-toluene	110.75	92.5	110.05
530	-o-xylene	144.3	72	137.0
531	-ethylbenzene	136.15	78	130.5
532	-mesitylene	164.6	60	151.2
533	-propylbenzene	159.3	63	147.6
	Methyl chloroacetate	129.95		
534	-isobutyl alcohol	107.85	88	107.55
535	-cyclopentanol	140.85	23	127.5
536	-amyl alcohol	138.2	30	126.8
537	-isoamyl alcohol	131.3	39.5	124.9
538	-ethylbenzene	136.15	37.5	127.2
539	-m-xylene	139.2	10	128.25
540	-p-xylene	138.45	15	128.3
	Propionitrile	97.2		
541	-propyl alcohol	97.2	50	90.5
542	-n-hexane	68.8	91	63.5
543	-ethylbenzene	136.15	Zeotropic	
	Acetone	56.15		
544	-methyl acetate	57	51.7	55.8
545	-diethylamine	55.5	61.8	51.4
546	-pyridine	115.4	Zeotropic	
547	-cyclopentane	49.3	64	41.0
548	-n-pentane	36.15	80	32.5
549	-benzene	80.1	Zeotropic	
550	-cyclohexane	80.75	32.5	53.0
551	-n-hexane	68.95	41	49.8
552	-isopropyl ether	69.0	39	54.2
553	-n-heptane	98.4	10.5	55.85
	Allyl alcohol	96.95		
554	-ethyl sulfide	92.1	55	85.1
555	-pyridine	115.4	Zeotropic	
556	-benzene	80.2	82.64	76.75
557	-cyclohexane	80.8	42	74.0
558	-n-hexane	68.95	95.5	65.5
559	-methylcyclohexane	101.1	58	85.0
560	-m-xylene	139.0	Zeotropic	
561	-2,5-dimethylhexane	109.4	50	89.3
562	-n-octane	125.75	32	93.4
	Ethyl formate	54.1		
563	-n-pentane	36.2	70	32.5
564	-benzene	80.2	Zeotropic	
565	-methylcyclopentane	72.0	25	51.2
566	-n-hexane	68.95	33	49.0
	Methyl acetate	56.95		
567	-cyclopentane	49.3	62.1	43.2
568	-benzene	80.2	0.3	56.7
569	-cyclohexane	80.7	22.0	55.5
570	-n-hexane	68.95	39.3	51.75
571	-n-heptane	98.45	3.55	56.65
572	-2-methylhexane	90.0	11.4	56.0
573	-2,2,3-trimethylbutane	80.9	25.8	55.1
	Propionic acid	141.0		
574	-pyridine, N	115.5	32.8	148.6
575	-2-picoline, N	129.3	45.0	154.5
576	-chlorobenzene	132.0	82	128.9
577	-benzene	80.15	Zeotropic	
578	-o-xylene	143.6	57	135.4
579	-p-xylene	138.2	66	132.5
580	-n-hexane	68.85	Zeotropic	
581	-n-heptane	98.15	98	97.82
582	-n-octane	125.12	78.5	120.89
583	-n-nonane	150.67	46.0	134.27
584	-n-decane	174.06	19.5	139.76
585	-n-undecane	193.85	Zeotropic	
586	-propyl sulfide	141.5	55	136.5
587	-quinoline	237.5	Zeotropic	
588	-cumene	152.8	35	139.0
589	-mesitylene	164.0	23	139.3
590	-propylbenzene	158.0	25	139.5
591	-camphene	159.6	35	138.0
592	-α-pinene	155.8	41.5	136.4
	Propionamide	222.2		
593	-p-bromochlorobenzene	196.4	84	189.5
594	-p-dibromobenzene	220.25	78	204.9

No.	System	B.P. (°C)	Azeotropic data: Compn. X_2 (wt. %)	Azeotropic data: B.P. (°C)
595	-iodobenzene	188.45	90	183.5
596	-nitrobenzene	210.75	76	205.4
597	-o-nitrophenol	217.25	75.2	211.15
598	-phenol	182.2	Zeotropic	
599	-p-bromotoluene	185.0	90	181.0
600	-m-nitrotoluene	230.8	56	214.5
601	-toluene	110.75	Zeotropic	
602	-o-cresol	191.1	Zeotropic	
603	-m-cresol	202.2	Zeotropic	
604	-o-toluidine	200.35	97.5	200.25
605	-m-toluidine	203.1	Zeotropic	
606	-acetophenone	202	85	200.35
607	-methyl salicylate	222.35	66	210.55
608	-o-xylene	144.3	98	144.0
609	-dimethylaniline	194.15	84.5	190.5
610	-3,4-xylidine	225.5	72	217.2
611	-quinoline	237.3	Zeotropic	
612	-indene	182.6	88	179.5
613	-ethyl benzoate	212.6	75	205.0
614	-cumene	152.8	96	151.8
615	-mesitylene	164.6	90	162.3
616	-naphthalene	218.05	68.5	204.65
617	-cymene	176.7	85	172.8
618	-carvone	231.0	52	214.5
619	-camphene	159.6	87	156.5
620	-camphor	209.1	83	203.5
621	-borneol	213.4	78	209.2
622	-n-decane	173.3	88.2	168
623	-1-methylnaphthalene	245.1	48	213.8
624	-2-methylnaphthalene	241.15	50	213.0
625	-n-undecane	194.5	79	183
626	-acenaphthene	277.9	25	220.8
627	-biphenyl	256.1	45	216.0
628	-n-dodecane	216.0	68.4	193.0
629	-fluorene	295	10	221.5
630	-diphenylmethane	265.6	40	218.2
	Ethyl carbamate	185.25		
631	-bromobenzene	156.1	90.2	153.95
632	-iodobenzene	188.45	67	174.5
633	-nitrobenzene	210.75	12	184.95
634	-phenol	182.2	46.5	190.75
635	-benzonitrile	191.1	43	182.1
636	-anisole	153.85	95	153.5
637	-2,4-xylenol	210.5	Zeotropic	
638	-n-octyl alcohol	195.2	27.5	183.5
639	-isobutyl sulfide	172.0	77	166.5
640	-indene	182.6	65	172.65
641	-cumene	152.8	94	151.5
642	-mesitylene	164.6	78	159.0
643	-propylbenzene	159.3	85	157.0
644	-pseudocumene	168.2	75	161.4
645	-naphthalene	218.0	23	184.05
646	-butylbenzene	183.1	63	172.0
645	-camphene	159.6	85	157.0
648	-limonene	177.6	68	168.07
649	-camphor	209.1	16	184.85
650	-2-methylnaphthalene	241.15	Zeotropic	
651	-amyl ether	187.4	63	171.0
652	-isoamyl ether	173.35	73	163.15
653	-methyl pelargonate	213.8	15	184.3
	1-Nitropropane	131		
654	-propyl alcohol	97.15	91.2	96.95
655	-n-butyl alcohol	117.73	67.8	115.3
656	-isobutyl alcohol	107.89	84.8	105.28
657	-n-heptane	98.43	86.5	96.6
658	-ethylbenzene	136.19	44.0	129.0
659	-n-octane	125.66	65.8	115.8
660	-n-nonane	150.8	38.4	126.6
	Propyl nitrite	47.75		
661	-n-pentane	36.15	91	35.8
662	-cyclopentane	49.3	46	45.5
	Isopropyl alcohol	82.45		
663	-butylamine	77.8	40	74.7
664	-n-pentane	36.15	94	35.5
665	-fluorobenzene	85.15	70	74.5
666	-benzene	80.2	66.3	71.74
667	-cyclohexane	80.7	68	69.4
668	-n-hexane	68.85	77	62.7
669	-toluene	110.6	31	80.6
670	-n-heptane	98.45	49.5	76.4
671	-o-xylene	144.3	Zeotropic	
672	-n-octane	124.75	16	81.6
	Propyl alcohol	97.2		
673	-dioxane	101.35	45	95.3
674	-butyl formate	106.8	36	95.5
675	-chlorobenzene	132	20	96.5
676	-fluorobenzene	85.15	82	80.2
677	-benzene	80.2	83.1	77.12
678	-cyclohexane	80.75	81.5	74.69
679	-toluene	110.6	48.8	92.5
680	-methylcyclohexane	100.8	65.2	87.0
681	-n-heptane	98.4	65.3	84.6
682	-styrene	145.8	92	97.0
683	-o-xylene	143.6	Zeotropic	
684	-m-xylene	139.2	6	97.08
685	-p-xylene	138.4	7.8	96.88
686	-n-octane	125.6	30	93.9
	Glycerol	290.5		
687	-p-chloronitrobenzene	239.1	87	235.6
688	-triethylene glycol	288.7	63	285.1
689	-m-nitrotoluene	230.8	87	228.8
690	-p-cresol	201.7	Zeotropic	
691	-methyl salicylate	222.35	92.5	221.4
269	-3,4-xylenol	226.8	Zeotropic	
693	-o-xylene	143.6	Zeotropic	
694	-quinoline	237.3	Zeotropic	
695	-ethyl salicylate	233.7	89.7	230.5
696	-naphthalene	218.05	90	215.2
697	-safrol	235.9	85.5	231.3
698	-methyl phthalate	283.2	69	271.5
699	-estragol	215.6	92.5	213.5
700	-eugenol	254.5	96	251.3
701	-propyl benzoate	230.85	92	228.8
702	-carvone	231.0	97	230.85
703	-2-methylnaphthalene	241.15	83.5	233.7
704	-acenaphthene	277.9	71	259.1
705	-biphenyl	254.9	75	246.1
706	-phenyl ether	259.3	78	247.9
707	-1,3,5-triethylbenzene	215.5	92	212.9
708	-bornyl acetate	227.7	90	226.0
709	-diphenylmethane	265.6	73	250.8
710	-benzyl phenyl ether	286.5	70	264.5
711	-benzyl ether	297.0	64	269.5
	Ethyl methyl sulfide	66.61		
712	-cyclohexane	80,75	Zeotropic	
713	-methylcyclopentane	71.85	35.9	65.6
714	-n-hexane	68.75	43.4	63.94
715	-2,2-dimethylpentane	79.2	11.8	66.37
	1-Propanethiol	67.3		
716	-thiophene	84.7	Zeotropic	
717	-cyclohexane	80.75	2.4	67.77
718	-2,3-dimethylbutane	58.0	83.7	57.54
719	-n-hexane	68.75	47.4	64.35
720	-2-methylpentane	60.27	76.1	59.2
721	-isopropyl ether	68.3	35	66.0
722	-2,2-dimethylpentane	79.2	18.7	67.2
723	-2,2,3-trimethylbutane	80.97	12.6	67.57
	Perfluorobutyric acid	122.0		
724	-ethyl-benzene	136.15	20	115.4
725	-m-xylene	139.0	17	117.5
726	-p-xylene	138.4	18	117.6
	Thiophene	84.7		
727	-benzene	80.15	Zeotropic	
728	-cyclohexane	80.8	58.8	77.9
729	-methylcyclopentane	71.85	86	71.47
730	-n-hexane	68.75	88.8	68.46
731	-2,3-dimethylpentane	89.9	36	80.9
732	-2,4-dimethylpentane	80.55	57.3	76.58
733	-n-heptane	98.4	16.8	83.09
	Pyrrol	129.2		
734	-chlorobenzene	131.75	57	124.5
735	-isopropyl sulfide	120.5	80	117.5
736	-propyl sulfide	140.8	35	127.5
737	-toluene	110.75	Zeotropic	
	Methyl pyruvate	137.5		
738	-isoamyl acetate	142.1	35	135.0
739	-m-xylene	139.2	50	130.0
	Methyl oxalate	163.3		
740	-p-dichlorobenzene	174.35	35	162.05
741	-pinacol	174.35	19	163.15
742	-o-bromotoluene	181.5	2	164.1
743	-butyl butyrate	166.4	42	160.5
744	-ethyl caproate	167.7	40	161.0
745	-indene	182.6	17	163.6

BINARY SYSTEMS (Continued)

			Azeotropic data	
No.	System	B.P. (°C)	Compn. X_2 (wt. %)	B.P. (°C)
746	-mesitylene	164.0	50.2	154.8
747	-naphthalene	218.0	Zeotropic	
748	-2,7-dimethyloctane	160.6	55	147.0
749	-1,3,5-triethylbenzene	215.5	Zeotropic	
	Ethyl bromoacetate	158.8		
750	-butyric acid	164.0	16	157.4
751	-isobutyric acid	154.6	60	153.0
752	-bromobenzene	156.1	72	155.3
753	-cyclohexanol	160.8	35	155.5
754	-o-chlorotoluene	159.3	48	156.2
755	-propylbenzene	159.3	50	155.8
	Ethyl chloroacetate	143.55		
756	-isoamyl acetate	142.1	60	141.7
757	-isoamyl alcohol	131.3	77	131.0
758	-allyl sulfide	139.35	78	138.5
759	-propyl butyrate	142.8	53	141.7
760	-ethylbenzene	136.15	82	135.3
761	-o-xylene	144.3	42	140.2
762	-m-xylene	139.0	68	137.45
763	-butyl ether	142.4	55	139.8
	Butyronitrile	117.9		
764	-n-butyl alcohol	117.8	50	113.0
765	-toluene	110.75	73	107.0
766	-methylcyclohexane	101.15	80	90.5
	Isobutyronitrile	103.85		
767	-benzene	80.15	Zeotropic	
768	-methylcyclohexane	101.15	60	85.5
769	-n-heptane	98.4	62	80.5
	2-Butanone	79.6		
770	-methyl propionate	79.85	40	79.0
771	-ethyl acetate	77.1	88.2	77.05
772	-1-chlorobutane	78.5	62	77.0
773	-butyl nitrite	78.2	70	76.7
774	-*tert*-butyl alcohol	82.45	31	78.7
775	-butylamine	77.8	65	74.0
776	-fluorobenzene	84.9	25	79.3
777	-benzene	80.1	56	78.33
778	-cyclohexane	80.75	60	71.8
779	-n-hexane	68.8	71.4	64.2
780	-n-heptane	98.5	30	77.0
781	-2,5-dimethylhexane	109.4	5	79.0
	Butyraldehyde	74.8		
782	-benzene	80.1	Zeotropic	
783	-n-hexane	68.7	74	60.0
	Butyric acid	164.0		
784	-iodobutane	130.4	97.5	129.8
785	-2-furaldehyde	161.45	57.5	159.4
786	-pyridine	115.5	8.0	163.2
787	-propyl chloroacetate	162.5	60	160.5
788	-isoamyl nitrate	149.75	88	147.85
789	-p-dichlorobenzene	174.4	43	162.0
790	-chlorobenzene	132.0	97.2	131.75
791	-o-bromotoluene	181.5	28	163.0
792	-m-bromotoluene	184.3	20.5	163.62
793	-p-bromotoluene	185.0	25	161.5
794	-anisole	153.85	88	152.85
795	-n-heptane	98.4	Zeotropic	
796	-styrene	145.8	85	143.5
797	-ethylbenzene	136.15	96	135.8
798	-o-xylene	144.3	90	143.0
799	-m-xylene	139.0	94	138.5
800	-p-xylene	138.45	94.5	137.8
801	-indene	182.6	16	163.65
802	-cumene	152.8	80	149.5
803	-mesitylene	164.8	62	158.0
804	-propylbenzene	158.9	72	154.5
805	-pseudocumene	169	55	159.5
806	-naphthalene	218.1	Zeotropic	
807	-butylbenzene	183.1	25	162.5
808	-cymene	176.7	40	161.0
809	-camphene	159.6	97.2	152.3
810	-n-undecane	194.5	15.5	162.4
	Dioxane	101.35		
811	-ethyl acetate	77.1	Zeotropic	
812	-1-bromobutane	101.5	53	98.0
813	-pyridine	115.5	Zeotropic	
814	-piperidine	106.4	Zeotropic	
815	-*tert*-amyl alcohol	102.35	20	100.65
816	-benzene	80.15	Zeotropic	
817	-cyclohexane	80.75	75.4	79.5
818	-ethyl borate	118.6	8	100.7
819	-toluene	110.75	Zeotropic	
820	-n-heptane	98.4	56	91.85

			Azeotropic data	
No.	System	B.P. (°C)	Compn. X_2 (wt. %)	B.P. (°C)
	Ethyl acetate	77.1		
821	-butyl nitrite	78.2	29	76.3
822	-isobutyl nitrite	67.1	Zeotropic	
823	-*tert*-butyl alcohol	82.45	27	76.0
824	-benzene	80.15	Zeotropic	
825	-cyclohexane	80.75	44	71.6
826	-methylcyclopentane	72.0	62	67.2
827	-n-hexane	68.7	60.1	65.15
828	-methylcyclohexane	101.1	Zeotropic	
	Isobutyric acid	154.6		
829	-iodobutane	130.4	93	128.8
830	-ethyl pyruvate	155.5	40	153.0
831	-bromobenzene	156.15	65	148.6
832	-chlorobenzene	132.0	92	131.2
833	-phenol	182.2	Zeotropic	
834	-1-bromohexane	156.5	65	148.0
835	-o-bromotoluene	181.5	15	153.9
836	-toluene	110.75	Zeotropic	
837	-anisole	153.85	58	149.0
838	-styrene	145.8	73	142.0
839	-o-xylene	144.3	78	141.0
840	-m-xylene	139.0	85	136.9
841	-p-xylene	138.4	87	136.4
842	-cumene	152.8	65	146.8
843	-propylbenzene	158.9	51	149.3
844	-pseudocumene	168.2	37	152.3
845	-cymene	176.7	20	153.4
846	-camphene	159.6	55	148.1
847	-d-limonene	177.8	22	152.5
848	-2,7-dimethyloctane	160.2	52	148.55
849	-isoamyl ether	173.2	7	154.2
	Methyl propionate	79.85		
850	-1-chlorobutane	78.05	62	76.8
851	-butyl nitrite	78.2	88	77.7
852	-n-butyl alcohol	117.8	Zeotropic	
853	-benzene	80.2	48	79.45
854	-cyclohexane	80.75	48	75.0
855	-methylcyclopentane	72.0	72	69.5
856	-propyl ether	90.5	Zeotropic	
857	-methylcyclohexane	101.1	11.5	79.3
	Propyl formate	80.85		
858	-1-chlorobutane	78.5	62	76.1
859	-butyl nitrite	78.2	65	76.8
860	-*tert*-butyl alcohol	82.6	60	78.0
861	-benzene	80.2	53	78.5
862	-cyclohexane	80.75	52	75.0
863	-n-hexane	68.95	70.5	63.6
864	-n-heptane	98.5	29	78.2
	Glycol monoacetate	190.9		
865	-phenol, N	182.2	35	197.5
866	-m-bromotoluene	184.3	68	182.0
867	-o-cresol, N	191.1	49	199.45
868	-m-cresol, N	202.2	69	206.5
869	-p-cresol, N	201.7	67	206.0
870	-n-octyl alcohol	195.2	29	189.5
871	-indene	182.6	80	180.0
872	-naphthalene	218.0	Zeotropic	
873	-amyl ether	187.5	58	180.8
874	-isoamyl ether	173.2	72	170.2
875	-1,3,5-triethylbenzene	215.5	Zeotropic	
	Methyl lactate	143.8		
876	-phenol	182.2	Zeotropic	
877	-anisole	153.85	18	142.8
878	-4-heptanone	143.55	53	142.7
879	-ethyl valerate	145.45	42	140.0
880	-methyl caproate	149.8	30	141.7
881	-m-xylene	139.0	57.5	131.2
882	-p-xylene	138.2	60	130.8
883	-n-octane	125.8	70	120.3
884	-butyl ether	142.8	58	137.0
885	-cumene	152.8	38	137.8
886	-camphene	159.6	15	140.0
887	-2,7-dimethyloctane	160.1	32	137.8
	Tetrahydrothiophene	118.8		
888	-pyridine	115.4	55	113.5
889	-1-methylpyrrol	112.8	82	111.5
890	-ethylcyclohexane	131.85	19.3	117.46
891	-2-methylheptane	117.70	61.8	113.96
892	-n-octane	125.7	39.7	117.79
	Butyl nitrite	78.2		
893	-benzene	80.15	25	77.95
894	-cyclohexane	80.75	37	76.5

No.	System	B.P. (°C)	Azeotropic data: Compn. X_2 (wt. %)	Azeotropic data: B.P. (°C)
895	-n-hexane	68.8	82	68.5
896	-methylcyclohexane	101.15	Zeotropic	
897	-n-heptane	98.4	Zeotropic	
	Isobutyl nitrite	67.1		
898	-benzene	80.15	Zeotropic	
899	-cyclohexane	80.75	Zeotropic	
900	-methylcyclopentane	72.0	32	65.9
901	-n-hexane	68.8	46	65.0
	Isobutyl nitrate	123.5		
902	-n-butyl alcohol	117.8	55	112.8
903	-isobutyl alcohol	107.85	64	105.6
904	-chlorobenzene	131.75	Zeotropic	
905	-propyl sulfide	141.5	Zeotropic	
906	-toluene	110.75	Zeotropic	
907	-ethylbenzene	136.15	Zeotropic	
	n-Butyl alcohol	117.75		
908	-pyridine	115.5	31	118.6
909	-butyl formate	106.6	76.4	105.8
910	-ethyl carbonate	125.9	37	116.5
911	-chlorobenzene	132.0	44	115.3
912	-fluorobenzene	84.9	Zeotropic	
913	-benzene	80.1	Zeotropic	
914	-2-picoline	129.4	Zeotropic	
915	-cyclohexene	82.7	95	82.0
916	-cyclohexane	80.75	90.5	79.8
917	-hexaldehyde	128.3	22.9	116.8
918	-ethyl isobutyrate	110.1	83	109.2
919	-isoamyl formate	123.8	31	115.9
920	-isobutyl acetate	117.2	50	114.5
921	-methyl isoyalerate	116.3	60	113.5
922	-paraldehyde	123.9	48	115.75
923	-n-hexane	68.95	96.8	68.2
924	-acetal	103.55	87	101.0
925	-isopropyl sulfide	120.5	55	112.0
926	-ethyl borate	118.6	48	113.0
927	-toluene	110.7	72.2	105.5
928	-methylcyclohexane	100.8	80	95.3
929	-n-heptane	98.4	82	93.85
930	-ethylbenzene	136.15	34.9	115.85
931	-o-xylene	143.6	25	116.8
932	-m-xylene	139.0	28.5	116.5
933	-p-xylene	138.3	32	115.7
934	-n-octane	125.75	54.8	108.45
935	-butyl ether	142.1	17.5	117.65
936	-isobutyl ether	122.3	52	113.5
937	-n-nonane	150.7	28.5	115.9
938	-2,7-dimethyloctane	160.2	Zeotropic	
	***sec*-Butyl alcohol**	99.5		
939	-butyl formate	106.8	32	98.0
940	-ethyl propionate	99.15	53	95.7
941	-benzene	80.15	84.6	78.5
942	-cyclohexane	80.75	82	76.0
943	-methylcyclopentane	72.0	88.5	69.7
944	-propyl ether	90.4	78	87.0
945	-toluene	110.7	45	95.3
946	-methylcyclohexane	101.5	61.8	89.7
947	-n-heptane	98.4	63.3	88.1
948	-isooctane	99.3	66.2	88.0
	***tert*-Butyl alcohol**	82.9		
949	-ethyl sulfide	92.1	30	79.8
950	-flurobenzene	85.15	69	76.0
951	-benzene	80.2	63.4	73.95
952	-cyclohexane	80.7	65.8	71.2
953	-methylcyclopentane	72.0	74	66.6
954	-n-hexane	68.85	78	63.7
955	-isopropyl ether	68.3	92.1	67.3
956	-propyl ether	90.4	48	79.0
957	-toluene	110.7	Zeotropic	
958	-methylcyclohexane	100.8	34	78.8
959	-n-heptane	98.45	38	78.0
960	-p-xylene	138.45	Zeotropic	
961	-2,5-dimethylhexane	109.2	23	81.5
	Ethyl ether	34.6		
962	-isoprene	34.3	52	33.2
963	-2-methyl-2-butene	37.1	15	34.2
964	-n-pentane	36.16	44	33.7
965	-benzene	80.2	Zeotropic	
966	-n-hexane	68.85	Zeotropic	
	Isobutyl alcohol	108.0		
967	-2-pentanone	102.35	81	101.8
968	-3-pentanone	102.05	80	101.7
969	-butyl formate	106.8	60	103.0
970	-methyl butyrate	102.65	75	101.3
971	-propyl acetate	101.6	83	101.0
972	-n-pentane	36.15	Zeotropic	
973	-chlorobenzene	132.0	37	107.1
974	-fluorobenzene	84.9	91	84.0
975	-benzene	80.1	92.6	79.3
976	-cyclohexene	82.7	85.8	80.5
977	-cyclohexane	80.75	86	78.3
978	-methylcyclopentane	72.0	95	71.0
979	-isobutyl vinyl ether	83.0	93.8	82.7
980	-ethyl isobutyrate	110.1	48	105.5
981	-n-hexane	68.9	97.5	68.3
982	-propyl ether	90.55	90	89.5
983	-acetal	103.55	80	98.2
984	-isopropyl sulfide	100.5	27	105.8
985	-toluene	110.7	55	101.2
986	-methylcyclohexane	100.8	68	92.6
987	-n-heptane	98.45	73	90.8
988	-ethylbenzene	136.15	20	107.2
989	-p-xylene	138.4	11.4	107.1
990	-1,3-dimethylcyclohexane	120.7	44	102.2
991	-2,5-dimethylhexane	109.2	58	98.7
992	-2,2,4-trimethylpentane	99.3	73	92.0
993	-butyl ether	142.4	Zeotropic	
	2-Ethoxyethanol	135.3		
994	-toluene	110.75	89.2	110.15
995	-methylcyclohexane	101.15	85	98.6
996	-propyl butyrate	143.7	28	133.5
997	-n-heptane	98.4	86	96.5
998	-styrene	145.8	45	130.0
999	-ethylbenzene	136.15	52	127.8
1000	-p-xylene	138.45	50	128.6
1001	-n-octane	125.75	62	116.0
1002	-cumene	152.8	33	133.2
1003	-propylbenzene	159.3	20	134.6
1004	-camphene	159.6	35	131.0
	Methyl propyl ether	38.95		
1005	-2-methyl-2-butene	37.15	75	36.3
1006	-n-pentane	36.2	78	35.6
1007	-isoprene	34.3	Zeotropic	
	Diethylene glycol	245.5		
1008	-p-dibromobenzene	220.25	87	212.85
1009	-nitrobenzene	210.75	90	210.0
1010	-o-nitrophenol	217.2	89.5	216.0
1011	-pyrocatechol, N	245.9	54	259.5
1012	-m-nitrotoluene	230.8	75	224.2
1013	-methyl salicylate	222.95	85	220.55
1014	-p-cresol	202.0	Zeotropic	
1015	-ethyl fumarate	217.85	90	217.1
1016	-quinoline	237.3	71	233.6
1017	-benzyl acetate	215.0	93	214.85
1018	-naphthalane	218.0	78	212.6
1019	-isosafrol	252.0	54	233.5
1020	-safrol	235.9	67	225.5
1021	-methyl phthalate	283.7	3.7	245.4
1022	-thymol	232.9	87	232.25
1023	-1-methylnaphthalene	244.6	55	277.0
1024	-2-methylnaphthalene	241.15	61	225.45
1025	-biphenyl	256.1	52	232.65
1026	-acenaphthene	277.9	38	239.6
1027	-1,3,5-triethylbenzene	215.5	78	210.0
1028	-bornyl acetate	227.6	82	223.0
1029	-fluorene	295.0	20	243.0
1030	-diphenylmethane	265.4	48	236.0
1031	-benzyl phenyl ether	286.5	20	241.5
	1-Butanethiol	97.8		
1032	-benzene	80.15	Zeotropic	
1033	-pyridine	115.4	Zeotropic	
1034	-n-heptane	98.42	50.6	95.45
1035	-2-methylhexane	90.05	84.6	89.74
1036	-3-methylhexane	91.95	77.2	91.2
1037	-2,5-dimethylhexane	109.1	12	98.22
	Isopropyl methyl sulfide	84.76		
1038	-cyclohexane	80.85	70	79.76
1039	-3-methylhexane	91.6	17.6	84.38
1040	-2,4-dimethylpentane	80.55	70.3	79.39
	Methyl propyl sulfide	95.47		
1041	-ethylcyclopentane	103.45	9.3	95.41
1042	-methylcyclohexane	101.05	22.0	95.06
1043	-3-methylhexane	91.6	67.05	90.53
	2-Furaldehyde	161.45		
1044	-n-heptane	98.4	94.7	98.3

No.	System	B.P. (°C)	Azeotropic data: Compn. X_2 (wt. %)	Azeotropic data: B.P. (°C)
1045	-ethylbenzene	136.15	Zeotropic	
1046	-o-xylene	143.6	87	140.5
1047	-m-xylene	139.0	88	138.4
1048	-p-xylene	138.4	80	138.0
1049	-cumene	152.8	73	148.5
1050	-mesitylene	164.6	40	155.2
1051	-pseudocumene	168.2	33	157.0
1052	-propylbenzene	159.2	58	151.4
1053	-cymene	176.7	32	157.8
1054	-camphene	159.5	60	146.75
1055	-cineol	176.35	41	157.25
	Pyridine	115.4		
1056	-piperidine	105.8	92	106.1
1057	-phenol, N	181.4	86.9	183.1
1058	-toluene	110.75	77.8	110.1
1059	-n-heptane	98.4	74.7	95.6
1060	-n-octane	125.75	43.9	109.5
1061	-n-nonane	150.7	10.1	115.1
1062	-n-decane	173.3	Zeotropic	
	2-Methylthiophene	111.92		
1063	-n-heptane	98.4	97.8	97.77
1064	-2-methylheptane	117.7	32.2	109.77
1065	-2,2-dimethylhexane	106.85	66.8	104.62
1066	-2,5-dimethylhexane	109.15	60.4	106.12
	3-Methylthiophene	114.96		
1067	-ethylcyclopentane	103.45	96.1	102.82
1068	-n-octane	125.75	18.0	114.15
1069	-2-methylheptane	117.7	41.2	111.86
1070	-2,5-dimethylhexane	109.15	68.3	107.12
	Ethyl pyruvate	155.1		
1071	-bromobenzene	156.1	52	149.5
1072	-m-xylene	139.2	70	137.2
1073	-cumene	152.8	55	146.2
	Levulinic acid	252.0		
1074	-m-nitrotoluene	230.8	85	229.5
1075	-p-nitrotoluene	238.9	78	236.4
1076	-methyl salicylate	222.95	94	222.75
1077	-3,4-xylenol	226.8	Zeotropic	
1078	-ethyl salicylate	233.8	82	230.5
1079	-naphthalene	218.0	89	216.7
1080	-safrol	235.9	83	232.5
1081	-1-methylnaphthalene	244.6	64	237.0
1082	-2-methylnaphthalene	241.15	71	234.55
1083	-isobutyl benzoate	241.9	75	238.6
1084	-1,3,5-triethylbenzene	215.5	89	214.0
	Methyl acetoacetate	169.5		
1085	-siobutyl sulfide	172.0	42	166.0
1086	-mesitylene	164.6	57	159.5
1086	-cymene	176.7	44	165.0
1088	-camphene	159.6	60	152.8
1089	-isoamyl ether	173.2	40	160.5
	Methyl malonate	181.4		
1090	-acetophenone	202.0	61	201.0
1091	-naphthalene	218.0	Zeotropic	
1092	-butylbenzene	183.2	48	173.0
1093	-cymene	176.7	60	169.0
1094	-camphene	159.6	74	154.6
1095	-d-limonene	177.8	52	167.3
	2-Pentanone	102.25		
1096	-methyl butyrate	102.65	50	101.9
1097	-toluene	110.7	Zeotropic	
1098	-methylcyclohexane	101.15	60	95.2
1099	-n-heptane	98.4	66	93.2
	Butyl formate	106.8		
1100	-*tert*-amyl alcohol	102.35	65	101.0
1101	-benzene	80.15	Zeotropic	
1102	-pinacolone	106.2	62	106.0
1103	-methylcyclohexane	101.15	65	96.0
1104	-n-heptane	98.45	60	90.7
	Isovaleric acid	176.5		
1105	-ethyl acetoacetate	180.4	23	176.1
1106	-ethyl oxalate	185.65	16	176.3
1107	-o-xylene	144.3	95	143.8
1108	-butyl sulfide	185.0	27	175.0
1109	-indene	183.0	40	173.0
1110	-cumene	152.8	88	152.0
1111	-mesitylene	164.6	81	162.5
1112	-pseudocumene	168.2	77	165.7
1113	-naphthalene	218.05	Zeotropic	
1114	-butylbenzene	183.1	50	173.0
1115	-cymene	175.3	62	170.8
1116	-camphene	159.6	83	156.5
1117	-cineol	176.3	57.5	175.0
1118	-n-decane	173.3	67	167.0
1119	-n-tridecane	234.0	Zeotropic	
	Methyl butyrate	102.65		
1120	-methylcyclohexane	101.1	55	97.0
1121	-n-heptane	98.45	65	95.1
1122	-n-octane	125.8	Zeotropic	
	Propyl acetate	101.6		
1123	-*tert*-amyl alcohol	102.0	42	99.5
1124	-benzene	80.2	Zeotropic	
1125	-cyclohexane	80.75	Zeotropic	
1126	-n-hexane	69.0	Zeotropic	
1127	-acetal	103.55	32	101.25
	Valeric acid	186.35		
1128	-phenol	182.2	Zeotropic	
1129	-indene	182.6	70	178.5
1130	-mesitylene	164.6	90	164.0
1131	-naphthalene	218.0	4	186.0
1132	-cymene	176.7	78	176.5
1133	-camphene	159.6	92	158.5
1134	-amyl ether	187.5	55	181.5
1135	-isoamyl ether	173.2	87.5	171.8
	Ethyl lactate	154.1		
1136	-toluene	110.75	Zeotropic	
1137	-o-xylene	144.3	70	140.2
1138	-p-xylene	138.45	83	136.6
1139	-cumene	152.8	52	143.5
1140	-mesitylene	164.9	27	150.05
1141	-pseudocumene	168.2	27	152.4
1142	-camphene	159.5	45	144.95
	n-Amyl alcohol	138.2		
1143	-benzene	80.2	Zeotropic	
1144	-phenol	182.2	Zeotropic	
1145	-amyl formate	132.0	57	131.4
1146	-ethylbenzene	136.15	60	129.8
1147	-p-xylene	138.45	50.1	130.9
	Isoamyl alcohol	131.9		
1148	-bromobenzene	156.15	15	131.65
1149	-butyl acetate	126.0	83.5	125.85
1150	-paraldehyde	124.0	78.0	123.5
1151	-o-fluorotoluene	114.0	86.0	112.1
1152	-toluene	110.7	90	109.7
1153	-n-heptane	98.45	93	97.7
1154	-ethylbenzene	136.15	51	125.7
1155	-n-octane	125.8	70	117.0
1156	-butyl ether	142.1	35	129.8
1157	-cumene	152.8	6	131.6
1158	-camphene	159.6	76	130.9
	Nitrobenzene	210.75		
1159	-aniline	184.35	Zeotropic	
1160	-methyl maleate	204.05	93	203.9
1161	-benzyl alcohol	205.25	62	204.2
1162	-3,4-xylenol	226.8	Zeotropic	
1163	-ethyl benzoate	212.5	19	210.6
1164	-camphor	208.9	65	208.4
1165	-borneol	215.0	42	207.8
1166	-1,3,5-triethylbenzene	215.5	Zeotropic	
1167	-ethyl bornyl ether	204.9	70	203.0
	Iodobenzene	188.55		
1168	-nitrobenzene	210.75	Zeotropic	
1169	-phenol	181.5	47	177.7
1170	-ethyl oxalate	185.65	52	181.0
1171	-caproic acid	205.15	12	186.8
1172	-isocaproic acid	199.5	15	185.5
1173	-benzyl alcohol	205.2	12	187.75
1174	-p-cresol	201.7	10	188.1
1175	-o-toluidine	200.35	Zeotropic	
1176	-isobutyl lactate	182.15	70	180.5
1177	-indene	182.6	Zeotropic	
1178	-isoamyl butyrate	178.5	Zeotropic	
1179	-butylbenzene	183.1	Zeotropic	
	Benzene	80.15		
1180	-aniline	184.35	Zeotropic	
1181	-cyclohexene	82.1	35.3	78.9
1182	-cyclohexane	80.75	48.1	77.56
1183	-methylcyclopentane	71.85	84	71.7
1184	-n-hexane	69.0	95.3	68.5
1185	-2,2-dimethylpentane	79.1	53.7	75.85
1186	-2,3-dimethylpentane	89.79	21.2	79.4
1187	-2,4-dimethylpentane	80.8	51.7	75.2
1188	-n-heptane	98.4	0.7	80.1
1189	-2,2,4-trimethylpentane	99.2	2.3	80.1

BINARY SYSTEMS (Continued)

			Azeotropic data	
No.	System	B.P. (°C)	Compn. X_2 (wt. %)	B.P. (°C)
	Phenol	182.2		
1190	-aniline, N	183.91	58.1	185.84
1191	-2-picoline, N	129.2	24.6	185.5
1192	-3-picoline, N	143.5	29.8	188.93
1194	-4-picoline, N	144.8	32.5	190.0
1195	-ethylene diacetate, N	189.86	60.8	195.53
1196	-benzaldehyde	179.2	49.0	175.6
1197	-o-cresol	191.1	Zeotropic	
1198	-2,4-lutidine, N	159.0	43.0	193.4
1199	-2,6-lutidine, N	144.0	27.5	185.5
1200	-o-toluidine	200.35	Zeotorpic	
1201	-2,4,6-collidine, N	171.0	47.7	195.23
1202	-n-octyl alcohol	195.15	87	195.4
1203	-*sec*-octyl alcohol	179.0	50	184.5
1204	-indene	182.2	53	177.8
1205	-mesitylene	164.5	79	163.5
1206	-pseudocumene	168.2	75	166.0
1207	-naphthalene	218.1	Zeotropic	
1208	-butylbenzene	183.1	54	175.0
1209	-camphene	159.6	78	156.1
1210	-n-decane	173.3	65	168.0
1211	-2,7-dimethyloctane	160.25	94	159.5
1212	-amyl ether	187.5	22	180.2
1213	-isoamyl ether	173.2	85	172.2
1214	-isoamyl sulfide	214.8	Zeotropic	
1215	-1,3,5-triethylbenzene	215.5	Zeotropic	
1216	-n-tridecane	235.42	16.9	180.56
	Pyrocatechol	245.9		
1217	-indole	253.5	85.0	255.0
1218	-o-phenetidine, N	232.5	8	246.0
1219	-p-phenetidine, N	249.9	66	253.8
1220	-quinoline, N	237.4	39	257.9
1221	-naphthalene	218.05	88.5	217.45
1222	-quinaldine, N	246.5	52.0	252.5
1223	-safrole	235.9	77.0	233.55
1224	-isosafrole	252.0	30	243.0
1225	-eugenol	254.8	1.5	245.85
1226	-carvone	231.0	29	248.3
1227	-thymol	232.9	83	232.2
1228	-1-methylnapthalene	244.9	60	235.1
1229	-2-methylnaphthalene	241.15	63	233.25
1230	-acenaphthene	277.9	16	245.25
1231	-biphenyl	255.9	43.5	239.85
1232	-phenyl ether	259.3	40.7	242.0
1233	-1,3,5-triethylbenzene	215.5	91.1	214.7
1234	-fluorene	295.0	Zeotropic	
1235	-diphenyl methane	265.6	35.0	243.05
1236	-n-tridecane	234.0	70.0	229.7
1237	-1,2-diphenylethane	284.9	Zeotropic	
	Resorcinol	281.4		
1238	-naphthalene	218.05	Zeotropic	
1239	-1-naphthol	288.0	30	280.2
1240	-2-naphthol	295.0	15	280.8
1241	-methyl phthalate	283.7	62	287.5
1242	-1-methylnaphthalene	244.6	85.5	243.1
1243	-2-methylnaphthalene	241.15	89.5	240.05
	Resorcinol	281.4		
1244	-p-*tert*-amylphenol	266.5	85	265.8
1245	-acenaphthene	277.9	59	266.2
1246	-biphenyl	255.9	79	252.15
1247	-phenyl ether	259.3	77	255.65
1248	-fluorene	295.0	52	274.0
1249	-n-tridecane	234.0	88	233.25
1250	-stilbene	306.5	44	277.5
1251	-1,2-diphenylethane	284.9	53	269.7
	Pyrogallol	309.0		
1252	-2-naphthol	295.0	22	293.5
1253	-acenaphthene	277.9	80	272.8
1254	-biphenyl	256.1	90	253.5
	Aniline	184.35		
1255	-o-cresol, N	191.1	92	191.25
1256	-n-octyl alcohol	195.2	17	183.95
1257	-o-xylene	144.3	Zeotropic	
1258	-indene	182.6	58.5	179.75
1259	-mesitylene	164.7	88.0	164.35
1260	-pseudocumene	169.35	86.5	168.64
1261	-naphthalene	218.0	Zeotropic	
1262	-butylbenzene	183.1	54	177.8
1263	-n-nonane	150.7	86.5	149.2
1264	-n-decane	174.6	64	167.28
1265	-amyl ether	187.5	45	177.5
1266	-isoamyl ether	173.2	72	169.35
1267	-2-methylnaphthalene	241.15	Zeotropic	
1268	-n-undecane	194.5	42.5	175.31
1269	-1,3,5-triethylbenzene	215.5	Zeotropic	
1270	-n-dodecane	216.5	28.5	180.37
1271	-n-tridecane	235.4	13.8	182.94
1272	-n-tetradecane	252.5	4.8	183.90
	2-Picoline	129.3		
1273	-n-octane	125.75	58.0	121.12
1274	-n-nonane	150.7	15.9	129.2
1275	-n-decane	174.6	Zeotropic	
	3-Picoline	144.0		
1276	-allyl sulfide	139.35	70	135.5
1277	-2,6-lutidine	144.06	27.3	143.5
	o-Phenylenediamine	258.6		
1278	-isosafrole	252.0	70	249.2
1279	-isafrole	235.9	Zeotropic	
1280	-biphenyl	256.1	63	249.7
1281	-phenyl ether	259.0	54	251.2
1282	-diphenylmethane	265.4	30	254.0
1283	-1,2-diphenylethane	284.5	Zeotropic	
	Methyl fumarate	193.25		
1284	-m-bromotoluene	184.3	84	183.65
1285	-o-cresol, N	191.1	40	197.8
1286	-m-cresol, N	202.2	28	204.3
1287	-benzyl ethyl ether	185.0	68	183.5
1288	-naphthalene	218.0	Zeotropic	
1289	-dipentene	177.7	30	172.5
	Methyl maleate	204.05		
1290	-caproic acid	205.15	37	201.5
1291	-o-cresol, N	191.1	22	204.65
1292	-m-cresol, N	202.2	45	208.75
1293	-p-cresol, N	201.7	44	208.6
1294	-naphthalene	218.0	13	203.7
1295	-borneol	215.0	22	202.95
1296	-isoamyl sulfide	214.8	18	203.0
	Cyclohexanone	155.7		
1297	-n-hexyl alcohol	157.85	6	155.65
1298	-cumene	152.8	35	152.0
1299	-camphene	159.6	42.5	150.55
1300	-2,7-dimethyloctane	160.1	45	151.5
	Ethyl acetoacetate	180.4		
1301	-phenetole	170.45	76	169.8
1302	-isobutyl sulfide	172.0	90	171.0
1303	-indene	182.6	32	177.15
1304	-propylbenzene	159.3	76	158.3
1305	-pseudocumene	168.2	63	165.2
1306	-butylbenzene	183.1	48	172.0
1307	-cymene	176.7	59	170.5
1308	-camphene	159.6	70	156.15
1309	-dipentene	177.7	57	169.05
1310	-d-limonene	177.8	57	169.05
1311	-2,7-dimethyloctane	160.1	76	156.0
1312	-amyl ether	187.5	30	174.5
	Ethylidene diacetate	168.5		
1313	-phenetole	170.45	44	164.5
1314	-butyl butyrate	166.4	63	163.5
1315	-ethyl caproate	167.7	55	164.0
1316	-*sec*-octyl alcohol	180.4	6.5	168.3
1317	-cineole	176.35	34	164.95
1318	-isoamyl ether	173.2	43	161.5
	Ethyl oxalate	185.65		
1319	-o-cresol, N	191.1	64	194.1
1320	-camphene	159.6	84	158.5
1321	-2,7-dimethyloctane	160.1	78	159.5
1322	-amyl ether	187.5	46	177.7
1323	-isoamyl ether	173.2	71	170.15
	Glycol diacetate	186.3		
1324	-o-cresol, N	191.1	65	194.5
	Capronitrile	163.9		
1325	-cumene	152.8	82	150.8
1326	-camphene	159.6	65	143.0
	Cyclohexanol	160.8		
1327	-o-xylene	143.6	86	143.0
1328	-m-xylene	139.0	95	138.9
1329	-indene	181.7	25	160.0
1330	-propylbenzene	158.8	60	153.8
1331	-naphthalene	218.05	Zeotropic	
1332	-cymene	176.7	28	159.5
1333	-camphene	159.5	59	151.9
1334	-cineole	176.35	8	160.55
	Butyl acetate	126.0		
1335	-paraldehyde	124.35	91	124.25

No.	System	B.P. (°C)	Azeotropic data: Compn. X_2 (wt. %)	Azeotropic data: B.P. (°C)
1336	-n-octane	125.8	48	119.0
1337	-butyl ether	142.1	5	125.9
	Caproic acid	205.3		
1338	-m-cresol	202.2	87	201.9
1339	-guaiacol	205.05	58	200.8
1340	-acetophenone	202.0	68	200.5
1341	-naphthalene	218.05	39	203.75
1342	-1-methylnaphthalene	244.6	Zeotropic	
1343	-2-methylnaphthalene	241.15	Zeotropic	
1344	-1,3,5-triethylbenzene	215.5	37	202.0
	Isoamyl formate	123.8		
1345	-paraldehyde	124.1	44	123.0
1346	-ethylbenzene	136.15	Zeotropic	
1347	-isobutyl ether	122.3	35	121.5
	Propyl propionate	122.5		
1348	-toluene	110.75	Zeotropic	
1349	-n-octane	125.8	40	118.2
	Propyl lactate	171.7		
1350	-o-cresol	191.1	Zeotropic	
1351	-isobutyl sulfide	172.0	52	169.0
1352	-mesitylene	164.6	72	160.5
1353	-isoamyl ether	173.2	47	167.5
	Isopropyl lactate	166.8		
1354	-o-cresol	191.1	Zeotropic	
1355	-mesitylene	164.6	40	159.5
1356	-camphene	159.6	70	154.2
	n-Hexyl alcohol	157.8		
1357	-o-cresol	191.1	zeotropic	
1358	-anisole	153.85	63.5	151.0
1359	-m-xylene	139.0	85	138.3
1360	-cumene	152.8	65	149.5
1361	-mesitylene	164.6	45	153.5
1362	-pesudocumene	168.2	32	156.3
1363	-propylbenzene	158.8	45	152.5
	Acetal	103.55		
1364	-methylcyclohexane	101.15	60	99.65
1365	-n-heptane	98.45	72	97.75
1366	-2,5-dimethylhexane	109.3	25	103.0
1367	-n-octane	125.75	Zeotropic	
	2-Butoxyethanol	171.15		
1368	-benzaldehyde	179.2	9	170.95
1369	-o-cresol, N	191.1	85	191.55
1370	-phenetole	170.45	48	167.1
1371	-isobutyl sulfide	172.0	58	163.8
1372	-mesitylene	164.6	68	162.0
1373	-butylbenzene	183.4	26.6	169.6
1374	-camphene	159.6	70	154.5
1375	-dipentene	177.7	47	164.0
1376	-cineole	176.35	41.5	168.9
	Pinacol	174.35		
1377	-o-cresol, N	191.1	92	191.5
1378	-p-cresol	201.7	Zeotropic	
1379	-n-octane	125.75	Zeotropic	
1380	-pseudocumene	168.2	62	162.9
1381	-propylbenzene	159.3	72	156.3
1382	-naphthalene	218.05	Zeotropic	
1383	-p-cymene	176.7	50	167.7
1384	-cineole	176.35	55	168.5
1385	-isoamyl ether	173.4	60	167.2
	Dipropylene glycol	229.2		
1386	-p-cresol	201.7	Zeotropic	
1387	-methyl salicylate	222.95	65	213.0
1388	-isosafrole	252.0	40	225.5
1389	-safrole	235.9	50	222.0
1390	-2-methylnaphthalene	241.1	Zeotropic	
	Triethylene glycol	288.7		
1391	-methyl phthalate	283.2	67	277.0
1392	-1-methylnaphthalene	244.6	Zeotropic	
1393	-acenaphthene	277.9	65	271.5
1394	-biphenyl	256.1	90	255.3
1395	-fluorene	294.0	Zeotropic	
1396	-phenyl benzoate	315.0	20	286.0
1397	-diphenylmethane	265.4	80	263.0
1398	-stilbene	306.5	40	284.5
1399	-1,2-diphenylethane	284.5	58	275.5
	Benzonitrile	191.1		
1400	-o-cresol, N	191.1	51	195.95
1401	-m-cresol, N	202.2	89	202.5
1402	-p-cresol, N	201.7	86	202.1
1403	-o-toluidine	200.35	Zeotropic	
1404	-isoamyl butyrate	181.05	92	180.85
1405	-cineole	176.35	86	175.6
1406	-amyl ether	187.5	58	180.5
1407	-isoamyl ether	173.2	84	171.4
	Benzaldehyde	179.2		
1408	-o-cresol, N	191.1	77	192.0
1409	-p-cresol	2017	Zeotropic	
1410	-naphthalene	218.0	Zeotropic	
1411	-p-cymene	175.3	72	171.0
1412	-d-limonene	177.8	57	171.2
1413	-camphene	159.6	84.5	158.45
1414	-isoamyl ether	173.2	62.5	168.6
	Benzoic acid	250.8		
1415	-p-nitrotoluene	238.9	89	237.4
1416	-3,4-xylenol	226.8	Zeotropic	
1417	-propyl succinate	250.5	57	248.0
1418	-naphthalene	218.05	95	217.7
1419	-1-methylnaphthalene	244.6	73	239.6
1420	-2-methylnaphthalene	241.15	75	237.25
1421	-biphenyl	277.9	49.5	246.05
1422	-phenyl ether	259.3	41	247.3
1423	-fluorene	295.0	Zeotropic	
1424	-diphenylmethane	265.6	18	248.95
1425	-1,2-diphenylethane	284.0	Zeotropic	
	o-Nitrotoluene	221.75		
1426	-benzyl alcohol	205.2	91	204.75
1427	-methyl salicylate	222.95	14	221.65
1428	-3,4-xylenol	226.8	Zeotropic	
1429	-2,4-xylidine	214.0	Zeotropic	
1430	-naphthalene	218.0	Zeotropic	
1431	-diethylaniline	217.05	88	216.85
1432	-geraniol	229.6	19	220.7
1433	-menthol	216.3	66	214.65
1434	-n-decyl alcohol	232.8	15	221.0
1435	-2-methylnaphthalene	241.15	Zeotropic	
1436	-bornyl acetate	227.6	27	221.15
	p-Nitrotoluene	238.9		
1437	-quinoline	237.3	92	237.2
1438	-safrole	235.9	82	234.5
1439	-geraniol	229.6	75	228.8
1440	-n-decyl alcohol	232.8	67	231.5
1441	-bornyl acetate	227.6	90	227.45
	Toluene	110.7		
1442	-2,6-lutidine	144.0	Zeotropic	
1443	-ethylcyclopentane	103.5	93	103.0
1444	-n-heptane	98.4	Zeotropic	
1445	-2,5-dimethylhexane	109.4	65	107.0
1446	-2-methylheptane	117.6	18	110.3
1447	-2,3,4-trimethylpentane	113.5	40	109.5
	Benzyl alcohol	205.2		
1448	-o-cresol	191.1	Zeotropic	
1449	-m-cresol, N	202.2	39	207.1
1450	-p-cresol, N	201.7	38	206.8
1451	-methylaniline	196.25	70	195.8
1452	-o-toluidine	200.35	Zeotropic	
1453	-3,4-xylenol	226.8	Zeotropic	
1454	-dimethylaniline	194.05	93.5	193.9
1455	-ethylaniline	205.5	50.0	202.8
1456	-2,4-xylidine	214.0	Zeotropic	
1457	-naphthalene	218.05	40	204.1
1458	-diethylaniline	217.05	28	20.42
1459	-d-limonene	177.8	89	176.4
1460	-borneol	215.0	14.2	205.07
1461	-1,3,5-triethylbenzene	215.5	43	203.2
	o-Cresol	191.1		
1462	-benzylamine, N	185.0	33	201.45
1463	-phenyl acetate, N	195.7	64	198.5
1464	-2,4,6-collidine, N	171.3	37	197.2
1465	-n-octyl alcohol	195.15	62	196.9
1466	-butyl sulfide	185.0	75	183.8
1467	-indene	183.0	91	182.9
1468	-naphthalene	218.05	Zeotropic	
1469	-terpinene	181.5	72	177.8
1470	-terpinolene	184.6	66	179.5
1471	-thymene	179.7	27	176.6
1472	-camphor, N	209.1	85	209.85
	m-Cresol	202.2		
1473	-o-toluidine, N	200.35	38.5	203.65
1474	-m-toluidine, N	203.1	47	205.5
1475	-p-toluidine, N	200.55	38	204.3
1476	-phenyl acetate, N	195.7	30	204.4
1477	-2,4,6-collidine, N	171.3	27	206.2
1478	-isoamyl lactate, N	202.4	50	207.6
1479	-n-octyl alcohol, N	195.15	38	203.3

No.	System	B.P. (°C)	Azeotropic data: Compn. X_2 (wt. %)	Azeotropic data: B.P. (°C)
1480	-propiophenone, N	217.7	83	218.6
1481	-phorone, N	197.8	45	206.5
1482	-naphthalene	218.05	Zeotropic	
1483	-camphor, N	209.1	63.5	213.35
1484	-1,3,5-triethylbenzene	215.5	Zeotropic	
	p-Cresol	201.6		
1485	-o-toluidine, N	200.35	43	203.5
1486	-m-toluidine, N	203.1	53	204.9
1487	-p-toluidine, N	200.55	43	204.05
1488	-o-anisidine	219.0	Zeotropic	
1489	-acetophenone, N	202.0	53.5	208.4
1490	-benzyl formate, N	202.4	58.0	207.0
1491	-methyl benzoate, N	199.4	60.0	204.35
1492	-phenyl acetate, N	195.7	32	204.3
1493	-isoamyl lactate, N	202.4	52	207.25
1494	-n-octyl alcohol, N	195.2	30	202.25
1495	-camphor, N	209.1	69.5	213.5
1496	-ethyl caprylate, N	208.35	75	209.5
	Guaiacol	205.05		
1497	-acetophenone	202.0	32.5	205.25
1498	-m-toluidine	203.1	Zeotropic	
1499	-ethylaniline	205.5	45	204.4
1500	-ethyl caprylate, N	208.35	85	208.9
	2,4-Lutidine	159.0		
1501	-n-nonane	150.7	67.75	148.3
1502	-n-undecane	195.4	Zeotropic	
	Methylaniline	196.25		
1503	-n-octyl alcohol	195.2	55	193.0
1504	-d-limonene	177.8	97	174.5
	o-Toluidine	200.35		
1505	-acetophenone, N	202.0	68	203.65
1506	-n-octyl alcohol	195.2	77	194.7
1507	-n-decane	174.6	87	173.76
1508	-n-undecane	195.5	60.3	188.25
1509	-n-dodecane	216.5	37	195.75
1510	-n-tridecane	234.6	14.5	199.45
	o-Anisidine	219.0		
1511	-naphthalene	218.0	50	217.0
1512	-2-methylnaphthalene	241.15	Zeotropic	
1513	-1,3,5-triethylbenzene	215.5	65	214.5
	Enanthic acid	222.0		
1514	-ethyl fumarate	217.85	78	216.4
1515	-ethyl maleate	223.3	50	220.0
1516	-ethyl succinate	217.85	80	216.0
1517	-propiophenone	217.7	80	216.5
1518	-naphthalene	218.0	70	214.2
1519	-biphenyl	256.1	Zeotropic	
1520	1,3,5-triethylbenzene	215.5	73	211.0
	n-Heptyl alcohol	176.15		
1521	-benzyl methyl ether	167.8	80	167.0
1522	-p-methylanisole	177.05	48	173.3
1523	-phenetole	170.45	72	169.0
1524	-p-cymene	176.0	52	173.0
1526	-isoamyl ether	173.35	63	170.35
	Indole	253.5		
1527	-carvacrol	237.85	12	254.5
1528	-p-*tert*-amylphenol	266.5	88	268.0
	Acetophenone	202.		
1529	-p-ethylphenol N	218.8	85	219.5
1530	-2,4-xylenol, N	210.5	70	213.0
1531	-3,4-xylenol	226.8	Zeotropic	
1532	-dimethylaniline	194.15	Zeotropic	
1533	-ethylaniline	205.5	Zeotropic	
1534	-2,4-xylidine	214.0	Zeotropic	
1535	-n-octyl alcohol	195.2	87.5	194.95
1536	-naphthalene	218.0	Zeotropic	
1537	-1,3,5-triethylbenzene	215.5	Zeotropic	
	Phenyl acetate	195.7		
1538	-2,4-xylenol	210.5	Zeotropic	
1539	-n-octyl alcohol	195.15	47	192.4
1540	-indene	182.6	Zeotropic	
1541	-naphthalene	218.05	Zeotropic	
1542	-thymene	179.7	82	179.3
1543	-linalool	198.6	39	193.5
	Methyl salicylate	222.95		
1544	-phenethyl alcohol	219.4	57	218.0
1545	-3,4-xylenol	226.8	Zeotropic	
1546	-ethyl maleate	223.3	40	221.95
1547	-quinoline	237.3	Zeotropic	
1548	-geraniol	229.7	3	222.2
1549	-menthol	216.4	85	216.25
1550	-n-tridecane	234.0	Zeotropic	
	Ethylbenzene	136.15		
1551	-ethylcyclohexane	131.8	85	131.2
1552	-n-octane	125.75	Zeotropic	
1553	-n-nonane	150.7	Zeotropic	
	p-Ethylphenol	218.8		
1554	-ethyl fumarate, N	217.85	52	223.0
1555	-ethyl maleate, N	223.3	62	226.3
1556	-p-methylacetophenone, N	226.35	70	229.5
1557	-benzyl acetate, N	215.0	40	221.0
1558	-ethyl benzoate, N	212.5	20	219.8
1559	-naphthalene	218.0	55	215.0
1560	-diethylaniline	217.05	40	214.0
1561	-1,3,5-triethylbenzene	215.5	60	212.0
	p-Methylanisole	177.05		
1562	-*sec*-octyl alcohol	180.4	21	176.3
1563	-pseudocumene	169.0	Zeotropic	
1564	-butyl isovalerate	177.6	42	176.4
1565	-butylbenzene	183.2	Zeotropic	
1566	-cineole	176.35	65	175.35
1567	-isoamyl ether	173.2	70.5	172.5
	Phenethyl alcohol	219.4		
1568	-3,4-xylenol	226.8	Zeotropic	
1569	-2,4-xylidine	214.0	Zeotropic	
1570	-naphthalene	218.05	56	214.2
1571	-diethylaniline	217.05	60	213.95
1572	-borneol	213.4	80	213.0
1573	-menthol	216.3	70	215.05
1574	-1-methylnaphthalene	244.9	Zeotropic	
1575	-biphenyl	256.1	Zeotropic	
	2,4-Xylenol	210.5		
1576	-ethyl fumarate, N	217.85	68	219.65
1577	-quinoline, N	237.3	92	239.0
1578	-p-methylacetophenone, N	226.35	15	227.0
1579	-propiophenone, N	217.7	35	221.0
1580	-benzyl acetate, N	215.0	64	216.8
1581	-camphor, N	209.1	50	217.0
	3,4-Xylenol	226.8		
1582	-o-phenetidine, N	232.5	92	232.65
1583	-ethyl fumarate, N	217.85	35	228.2
1584	-ethyl maleate, N	223.3	45	230.0
1585	-quinoline, N	237.3	65	241.95
1586	-p-methylacetophenone, N	226.35	49	231.35
1587	-propiophenone, N	217.7	33	228.5
1588	-naphthalene	218.0	84	217.6
1589	-diethylaniline	217.05	92	217.0
1590	-camphor, N	209.1	27	227.55
1591	-n-tridecane	234.0	42	223.5
	Ethylaniline	205.5		
1592	-n-octyl alcohol	195.2	85	194.9
1593	-naphthalene	218.0	Zeotropic	
1594	-camphor	209.1	Zeotropic	
	2,4-Xylidine	217.4		
1595	-menthol	216.3	30	213.5
1596	-n-undecane	195.5	88	194.98
1597	-n-dodecane	216.5	63	209.8
1598	-n-tridecane	234.6	29	215.28
1599	-n-tetradecane	252.0	2.5	217.38
	o-Phenetidine	232.5		
1600	-ethyl salicylate	233.8	18	232.2
1601	-naphthalene	218.0	Zeotropic	
1602	-safrole	235.9	14	232.38
1603	-anethole	235.7	25	232.25
1604	-carvacrol	237.85	87	238.0
1605	-thymol, N	232.9	54.9	234.3
1606	-2-methylnaphthalene	241.15	Zeotropic	
	p-Phenetidine	249.9		
1607	-safrole	235.9	Zeotropic	
1608	-isosafrole	252.0	36	248.8
1609	-1-methylnaphthalene	244.6	73	243.95
1610	-2-methylnaphthalene	241.15	85	240.85
1611	-biphenyl	256.1	10	249.5
1612	-phenyl ether	259.0	15	249.75
	Ethyl fumarate	217.85		
1613	-naphthalene	218.0	42	216.7
1614	-thymol, N	232.9	87.5	233.35
1615	-menthol	216.3	70	216.0
	Ethyl maleate	223.3		
1616	-p-methylacetophenone	226.35	12	223.15
1617	-naphthalene	218.0	77	217.65
1618	-thymol, N	232.9	73	234.9
	Ethyl succinate	217.25		

No.	System	B.P. (°C)	Azeotropic data: Compn. X_2 (wt. %)	Azeotropic data: B.P. (°C)
1619	-propiophenone	217.7	33	216.7
1620	-naphthalene	218.05	38.5	216.3
1621	-2-methylnaphthalene	241.15	Zeotropic	
	Caprylic acid	238.5		
1622	-naphthalene	218.05	94	216.2
1623	-carvacrol	237.85	75	237.6
1624	-1-methylnaphthalene	244.6	48	233.5
1625	-2-methylnaphthalene	241.15	52	235.0
	Propyl isovalerate	155.7		
1626	-cumene	152.8	Zeotropic	
1627	-propylbenzene	158.9	Zeotropic	
1628	-camphene	159.6	35	145.0
1629	-nopinene	163.8	25	155.0
	***sec*-Octyl alcohol**	179.0		
1630	butylbenzene	183.1	50	178.2
1631	-p-cymene	176.7	56	174.0
1632	-thymene	179.7	48	176.0
1633	-cineole	176.35	73.5	175.85
1634	-amyl ether	187.5	14	178.8
	Quinoline	237.3		
1635	-mesitol, N	220.5	15	240.4
1636	-safrole	235.9	73	235.15
1637	-carvacrol, N	237.85	52	244.3
1638	-thymol, N	232.9	45	243.1
1639	-1-methylnaphthalene	244.6	Zeotropic	
1640	2-methylnaphththalene	241.15	7	237.25
1641	-p-*tert*-amylphenol, N	266.5	94	267.5
1642	-biphenyl	256.1	Zeotropic	
	p-Methylacetophenone	226.35		
1643	-thymol, N	232.9	68	234.9
1644	-geraniol	229.6	5	226.25
1645	-citronellol	224.4	68	223.7
1646	-2-methylnaphthalene	241.15	Zeotropic	
1647	-bornyl acetate	227.6	40	225.8
	Propiophenone	217.7		
1648	-benzyl acetate	215.0	Zeotropic	
1649	-borneol	215.0	Zeotropic	
1650	-1,3,5-triethylbenzene	215.5	75	215.4
	Ethylsalicylate	233.8		
1651	-safrole	235.9	12	233.65
1652	-geraniol	229.7	60	228.5
1653	-n-decyl alcohol	232.9	52	230.5
1654	-2-methylnaphthalene	241.15	Zeotropic	
	Cumene	152.8		
1655	-n-nonane	150.75	77	148.0
	Mesitylene	164.6		
1656	-propylbenzene	159.3	Zeotropic	
1657	-camphene	159.6		
1658	2,7-dimethyloctane	160.1	72	158.6
	Propylbenzene	159.3		
1659	-camphene	159.6	53	158.0
	Pseudocumene	168.2		
1660	-p-cymene	176.7	Zeotropic	
1661	-n-decane	173.3	25	166.5
	Mesitol	230.5		
1662	-naphthalene	218.0	63	215.5
1663	-1,3,5-triethylbenzene	215.5	70	213.0
	3-Phenylpropanol	235.6		
1664	-naphthalene	218.05	~80	217.8
1665	-safrole	235.9	53	233.8
1666	-anethole	235.7	52	234.0
1667	-thymol, N	232.9	38	237.5
1668	-biphenyl	254.9	—	235.4
	Pelargonic acid	254.0		
1669	-naphthalene	218.0	Zeotropic	
1670	-isosafrole	252.0	65	249.5
1671	-eugenol	254.8	48	250.5
1672	-thymol	232.9	Zeotropic	
1673	-1-methylnaphthalene	244.6	82	243.0
1674	-2-methylnaphthalene	241.15	90	240.2
1675	-biphenyl	256.1	55	250.0
1676	-phenyl ether	259.0	45	250.5
1677	-1,3,5-triethylbenzene	215.5	Zeotropic	
1678	-diphenylmethane	265.4	25	252.7
	Naphthalene	218.0		
1679	-borneol	213.4	65	213.0
1680	-citronellol	224.5	30	217.8
1681	-menthol	216.4	74.5	215.5
1682	-n-tridecane	234.0	Zeotropic	
	Quinaldine	246.5		

No.	System	B.P. (°C)	Azeotropic data: Compn. X_2 (wt. %)	Azeotropic data: B.P. (°C)
1683	-safrole	235.9	Zeotropic	
1684	-carvacrol, N	237.85	33	250.8
1685	-thymol, N	232.9	20	250.0
	Methyl phthalate	283.2		
1686	-acenaphthene	277.9	66.5	276.35
1687	-biphenyl	255.9	Zeotropic	
1688	-diphenylmethane	265.6	Zeotropic	
1689	-1,2-diphenylethane	284.0	47	280.5
	Propyl benzoate	230.85		
1690	-carvacrol, N	237.85	82	238.85
1691	-thymol, N	232.8	55	235.5
1692	-2-methylnaphthalene	241.15	Zeotropic	
	p-Cymene	176.7		
1693	-dipentene	177.7	40	175.8
1694	-d-limonene	177.8	25	174.5
1695	-cineoloe	176.35	55	176.2
	Carvacrol	237.85		
1696	-carvenone, N	234.5	45	243.0
1697	-menthenone, N	222.5	25	239.5
1698	-propyl succinate, N	250.5	75	251.5
1699	-n-decyl alcohol	232.8	Zeotropic	
1700	-isobutyl benzoate, N	241.9	67	243.85
1701	-biphenyl	256.1	Zeotropic	
1702	-bornyl acetate, N	227.6	25	238.2
	Carvone	230.95		
1703	-thymol, N	232.9	52	238.65
1704	-geraniol	229.6	60	229.2
1705	-n-decyl alcohol	232.8	19	230.85
1706	-2-methylnaphthalene	241.15	Zeotropic	
	Thymol	232.9		
1707	-carvenone, N	234.5	50	241.0
1708	-pulegone, N	223.8	35	235.3
1709	-geraniol	229.6	42.5	225.6
1710	-menthone, N	209.5	8	233.2
1711	-2-methylnaphthalene	241.15	Zeotropic	
1712	-isobutyl benzoate, N	242.15	80	243.2
1713	-1,3,5-triethylbenzene	215.5	Zeotropic	
1714	-bornyl acetate, N	227.7	40	235.6
	Camphene	159.6		
1715	-dipentene	177.7	Zeotropic	
1716	-2,7-dimethyloctane	160.25	38	158.0
	Borneol	211.8		
1717	-methol	216.4	Zeotropic	
1718	-1,3,5-triethylbenzene	215.5	38	212.2
	Menthol	216.3		
1719	-2-methylnaphthalene	241.15	Zeotropic	
1720	-terpineol methyl ether	216.2	50	215.3
1721	-1,3,5-triethylbenzene	215.5	~45	214.0
	Capric acid	268.8		
1722	-1-methylnaphthalene	244.6	Zeotropic	
1723	-diphenylmethane	265.4	72	262.5
	1-Methylnaphthalene	244.6		
1724	-2-methylnaphthalene	241.15	Zeotropic	
1725	-biphenyl	256.1	Zeotropic	
1726	-phenyl ether	259.0	Zeotropic	
1727	-diphenylmethane	265.4	Zeotropic	
	2-Methylnaphthalene	241.15		
1728	-isobutyl benzoate	241.9	40	240.8
	p-*tert*-Amylphenol	266.5		
1729	-acenaphthene	277.9	Zeotropic	
1730	-fluorene	295.0	Zeotropic	
1731	-diphenylmethane	265.4	60	263.0
	Phenyl ether	259.0		
1732	-isoamyl benzoate	262.05	10	258.9
1733	-isoamyl oxalate	268.0	Zeotropic	
1734	-diphenylmethane	265.6	Zeotropic	
	Isoamyl benzoate	262.0		
1735	-isoamyl oxalate	268.0	Zeotropic	
1736	-diphenylmethane	265.6		
	1,3,5-Triethylbenzene	215.5		
1737	-bornyl acetate	227.2	Zeotropic	
1738	-bornyl ethyl ether	204.9	Zeotropic	
	Isoamyl oxalate	268.0		
1739	-diphenylmethane	265.4	86	265.25
1740	-1,2-diphenylethane	284.5	Zeotropic	
	Phenyl benzoate	315.0		
1741	-stilbene	306.5	Zeotropic	
1742	-benzyl ether	297.0	Zeotropic	
	Benzyl phenyl ether	286.5		
1743	-1,2-diphenylethane	284.5	Zeotropic	

TERNARY SYSTEMS

No.	System	B.P. (°C)	Azeotropic data Compn. (wt. %)	B.P. (°C)	Type of azeotrope	Ref.
1	Argon	−186			—	1
	Nitrogen	−195	Zeotropic			
	Oxygen	−183	90—120°K			
2	Hydrogen bromide	−67	10.4		H	2
	Water	100	11.0	105		
	Chlorobenzene	131.8	78.6			
3	Hydrogen cyanide	26			—	3
	Acetonitrile	81.6	Zeotropic			
	Acrolein	52.45				
4	Chlorine trifluoride	—			—	4
	Hydrogen fluoride	19.4	Zeotropic			
	Uranium hexafluoride	56				
5	Hydrogen chloride	−80	15.8		H	5
	Water	100	64.8	107.33		
	Phenol	182	19.4			
6	Hydrogen chloride	−80	5.3		H	5
	Water	100	20.2	96.9		
	Chlorobenzene	131.8	74.5			
7	Hydrogen fluoride	19.4	10		N	6
	Fluosilicic acid	—	36	116.1		
	Water	100	54			
8	Silicon tetrafluoride	—	24.6		—	7
	Hexafluoroethane	−78	32.7	−104		
	Ethane	−88	42.7			
9	Water	100	4.5		H	8
	Carbon tetrachloride	76.75	85.5	62		
	Ethyl alcohol	78.3	10			
10	Water	100	5		H	8
	Carbon tetrachloride	76.75	84	65.15		
	Allyl alcohol	96.95	11			
11	Water	100	4.05		H	9
	Carbon tetrachloride	76.7	91.0	65		
	sec-Butyl alcohol	99	4.95			
12	Water	100	1.6		H	10
	Carbon disulfide	46.25	93.4	41.3		
	Ethyl alcohol	78.3	5.0			
13	Water	100			—	11
	Carbon disulfide	46.25	Zeotropic			
	Dioxane	101.4				
14	Water	100			—	12
	Chloroform	61	Zeotropic			
	Formic acid	100.75				
15	Water	100	1.3		—	13
	Chloroform	61	90.5	52.3		
	Methyl alcohol	64.7	8.2			
16	Water	100	2.3		—	14
	Chloroform	61	94.2	55.3		
	Ethyl alcohol	78.3	3.5			
17	Water	100	18.6		S	15
	Formic acid	100.8	71.9	107.2		
	Propionic acid	140.7	9.5			
18	Water	100	19.5		S	16
	Formic acid	100.8	75.9	107.62		
	Butyric acid	162.4	4.6			
19	Water	100	15.5		S	16
	Formic acid	100.8	66.8	107.02		
	Isobutyric acid	154	17.7			
20	Water	100	21.3		S	16
	Formic acid	100.8	76.3	107.64		
	Isovaleric acid	176.5	2.4			
21	Water	100			—	16
	Formic acid	100.8	Zeotropic			
	Valeric acid	186				
22	Water	100	2.1		H-3	17
	Nitromethane	101.2	6.5	33.1		
	n-Pentane	36.07	91.4			
23	Water	100	7.88		H-3; 748 mm Hg	18
	Nitromethane	101.2	29.73	71.43		
	n-Heptane	98.43	62.39			
24	Water	100	12.4		H-3; 748 mm Hg	18
	Nitromethane	101.2	44.25	77.35		
	n-Octane	125.7	43.35			
25	Water	100	17.4		H-3; 748 mm Hg	18
	Nitromethane	101.2	58.3	80.72		
	n-Nonane	150.8	24.3			
26	Water	100	19.1		H-3; 748 mm Hg	18
	Nitromethane	101.2	68.1	82.35		
	n-Decane	174.12	12.8			

TERNARY SYSTEMS (Continued)

No.	System	B.P. (°C)	Azeotropic data: Compn. (wt. %)	Azeotropic data: B.P. (°C)	Type of azeotrope	Ref.
27	Water	100	20.6		H-3; 748 mm Hg	18
	Nitromethane	101.2	73.3	82.82		
	n-Undecane	194.5	6.1			
28	Water	100	21.5		H-3; 748 mm Hg	18
	Nitromethane	101.2	75.3	83.13		
	n-Dodecane	214.5	3.2			
29	Water	100	22.8		H-3; 748 mm Hg	18
	Nitromethane	101.2	75.4	83.21		
	n-Tridecane	234	1.8			
30	Water	100			—	19
	Methyl alcohol	64.7	Zeotropic			
	Ethyl alcohol	78.3				
31	Water	100	5.26		—	20
	Methyl alcohol	64.7	81.20	67.85		
	Methyl chloroacetate	131.4	13.54			
32	Water	100			—	21
	Methyl alcohol	64.7	Zeotropic			
	Methyl acetate	57.1				
33	Water	100	0.6		—	22
	Methyl alcohol	64.7	5.4	30.2		
	Isoprene	34.0	94.0			
34	Water	100	12.45		—	23
	Tetrachloroethylene	120.8	66.75	81.18		
	n-Propyl alcohol	97.2	20.8			
35	Water	100	6.4		—	24
	Trichloroethylene	86.95	73.1	67		
	Acetonitrile	81.6	20.5			
36	Water	100	5.5		—	25
	Trichloroethylene	86.95	78.4	67.0		
	Ethyl alcohol	78.3	16.1			
37	Water	100	7		—	26
	Trichloroethylene	86.95	73	69.4		
	Isopropyl alcohol	82.45	20			
38	Water	100			—	24
	Acetonitrile	81.6	Zeotropic			
	Acetone	56.4				
39	Water	100	1		—	26
	Acetonitrile	81.6	44	72.9		
	Ethyl alcohol	78.3	55			
40	Water	100			—	27
	Acetonitrile	81.6	Zeotropic			
	Diethylamine	55.5				
41	Water	100	8.2		—	24
	Acetonitrile	81.6	23.3	66		
	Benzene	80.2	68.5			
42	Water	100	3.5		—	13
	Acetonitrile	81.6	9.6	68.6		
	Triethylamine	89.7	86.9			
43	Water	100			—	8
	Acetic acid	118.1	Zeotropic			
	Toluene	110.7				
44	Water	100	8.4		H	17
	Nitroethane	114.07	9.3	59.5		
	n-Hexane	68.74				
45	Water	100	11.5		H	17
	Nitroethane	114.07	24.5	75.1		
	n-Heptane	98.43	64.0			
46	Water	100	8.7		—	26
	Ethyl alcohol	78.3	20.3	69.5		
	Acrylonitrile	77.2	71.0			
47	Water	100	4.8		—	26
	Ethyl alcohol	78.3	87.9	78.0		
	Crotonaldehyde	102.4	7.3			
48	Water	100	9.0		—	28
	Ethyl alcohol	78.3	8.4	70.23		
	Ethyl acetate	77.05	82.6			
49	Water	100	7.5		—	26
	Ethyl alcohol	78.3	42.5	81.8		
	Butylamine	77.8	50.0			
50	Water	100	6.3		—	26
	Ethyl alcohol	78.3	8.6	62		
	Butyl methyl ether	70.3	85.1			
51	Water	100	7.4		H	29
	Ethyl alcohol	78.3	18.5	64.86		
	Benzene	80.2	74.1			
52	Water	100	4.8		H	30
	Ethyl alcohol	78.3	19.7	62.60		
	Cyclohexane	80.75	75.5			

TERNARY SYSTEMS (Continued)

No.	System	B.P. (°C)	Azeotropic data Compn. (wt. %)	B.P. (°C)	Type of azeotrope	Ref.
53	Water	100	9		—	31
	Ethyl alcohol	78.3	13	74.7		
	Triethylamine	89.4	78			
54	Water	100	12		H	26
	Ethyl alcohol	78.3	37	74.4		
	Toluene	110.6	51			
55	Water	100	3		H	26
	Ethyl alcohol	78.3	12	56.0		
	n-Hexane	68.7	85			
56	Water	100	6.1		H	26
	Ethyl alcohol	78.3	33.0	68.8		
	n-Heptane	98.45	60.9			
57	Water	100	0.4		—	32
	Acetone	56.7	7.6	32.5		
	Isoprene	34.7	92.0			
58	Water	100	8.5		H	33
	Allyl alcohol	96.95	5.1	59.7		
	n-Hexane	68.95	86.4			
59	Water	100	12.5		H	26
	Isopropyl alcohol	82.3	40.5	83		
	Butylamine	77.8	47.0			
60	Water	100	7.5		H	34
	Isopropyl alcohol	82.45	19.0	66.3		
	Benzene	80.2				
61	Water	100	7.5		H	8
	Isopropyl alcohol	82.45	18.5	64.3		
	Cyclohexane	80.75	74.0			
62	Water	100	13.1		H	26
	Isopropyl alcohol	82.3	38.2	76.3		
	Toluene	110.6	48.7			
63	Water	100	17.0		—	35
	Propyl alcohol	97.3	10.0	82.45		
	Propyl acetate	101.6	73.0			
64	Water	100	7.6		740mm Hg; H	26
	Propyl alcohol	97.2	10.1	67		
	Benzene	80.1	82.3			
65	Water	100	8.5		H	8
	Propyl alcohol	97.2	10.0	66.55		
	Cyclohexane	80.75	81.5			
66	Water	100	19.2		H	36
	n-Butyl alcohol	117.75	2.9	61.5		
	n-Hexane	68.95	77.9			
67	Water	100	41.4		H	36
	n-Butyl alcohol	117.75	7.6	78.1		
	n-Heptane	98.4	51.0			
68	Water	100	60.0		H	36
	n-Butyl alcohol	117.75	14.6	86.1		
	n-Octane	125.75	25.4			
69	Water	100	69.9		H	36
	n-Butyl alcohol	117.75	18.3	90.0		
	n-Nonane	150.7	11.8			
70	Water	100	29.9		H	26
	n-Butyl alcohol	117.75	34.6	90.6		
	Butyl ether	142.1	35.5			
71	Water	100	5		H	37
	2-Butanone	79.6	35	63.6		
	Cyclohexane	80.7	60			
72	Water	100	4		H	37
	Butyraldehyde	74.8	21	55.0		
	n-Hexane	68.7	75			
73	Water	100	8.9		H	14
	sec-Butyl alcohol	99.6	10.8	69.7		
	Cyclohexane	80.75	80.3			
74	Water	100	9		H	13
	sec-Butyl alcohol	99.4	19	76.3		
	Isooctane	99.0	72			
75	Water	100	8.1		H	38
	tert-Butyl alcohol	82.55	21.4	67.3		
	Benzene	80.2	70.5			
76	Water	100			—	22
	tert-Butyl alcohol	82.55	Zeotropic			
	Isoprene	34.0				
77	Water	100	8		H	8
	tert-Butyl alcohol	82.55	21	65.0		
	Cyclohexane	80.75	71			
78	Water	100			H	38
	Isobutyl alcohol	108	Zeotropic			
	Benzene	80.2				
79	Water	100	17.9		H	39
	Isobutyl alcohol	108	16.4	81.3		
	Toluene	110.7	65.7			

TERNARY SYSTEMS (Continued)

No.	System	B.P. (°C)	Azeotropic data Compn. (wt. %)	B.P. (°C)	Type of azeotrope	Ref.
80	Water	100			—	37
	Ethyl acrylate	99.3	Zeotropic			
	Isopropyl ether	68.3				
81	Water	100			—	40
	Toluene	110.7	Zeotropic			
	Benzyl alcohol	204.7				
82	Water	100			—	2
	Pyridine	115.5	Zeotropic			
	Benzene	80.1				
83	Water	100	14.0		H	41
	Pyridine	115.5	15.5	78.6		
	n-Heptane	98.4	70.5			
84	Water	100	22.5		H	41
	Pyridine	115.5	25.5	86.7		
	n-Octane	125.75	52.0			
85	Water	100	30.5		H	41
	Pyridine	115.5	37.0	90.5		
	n-Nonane	150.7	32.5			
86	Water	100	35.5		H	41
	Pyridine	115.5	45.5	92.3		
	N- Decane	173.3	19.0			
87	Water	100	38.5		H	41
	Pyridine	115.5	51.0	93.1		
	n-Undecane	194.5	10.5			
88	Water	100	40.5		H	41
	Pyridine	115.5	54.5	93.5		
	n-Dodecane	216.0	5.0			
89	Water	100	32.4		H	42
	Isoamyl alcohol	131.5	19.6	89.8		
	Isoamyl formate	124.2	48.0			
90	Water	100	44.8		—	43
	Isoamyl alcohol	131.5	31.2	93.6		
	Isoamyl acetate	142.0	24.0			
91	Water	100			—	37
	2-Picoline	129.2	Zeotropic			
	Paraldehyde	124.5				
92	Carbon tetrachloride	76.8				44
	Methyl alcohol	64.7	Zeotropic			
	Benzene	80.1				
93	Chloroform	61			—	45
	Formic Acid	100.75	Zeotropic			
	Acetic acid	118.1				
94	Chloroform	61	47		S	46
	Methyl alcohol	64.7	23	57.5		
	Acetone	56.1	30			
95	Chloroform	61	65.3		S	47
	Ethyl alcohol	78.3	10.4	63.2		
	Acetone	56.1	24.3			
96	Chloroform	61.0	56.1		—	48
	Ethyl alcohol	78.3	9.5	57.3		
	n-Hexane	68.7	34.4			
97	Chloroform	61.2	68.8		—	48
	Acetone	56.5	3.6	60.79		
	n-Hexane	68.7	27.6			
98	Chloroform	61.0			—	49
	Acetone	56.4	Zeotropic			
	Toluene	110.7				
99	Chloroform	61.2	79.70		S	50
	Ethyl formate	54.1	5.3	61.97		
	Isopropyl bromide	59.4	15.7			
100	Chloroform	61.2			—	51
	2-Bromopropane	59.4	Zeotropic			
	Isopropyl formate	68.8				
101	Methyl alcohol	64.7	17.4		—	26
	Acetone	56.1	5.8	53.7		
	Methyl acetate	56.3	76.8			
102	Methyl alcohol	64.7	14.6		—	52
	Acetone	56.25	30.8	47		
	n-Hexane	68.95	59.6			
103	Methyl alcohol	64.7	17.8		—	53
	Methyl acetate	57	48.6	50.8		
	Cyclohexane	80.75	33.6			
104	Methyl alcohol	64.7	14.6		—	14
	Methyl acetate	56.3	36.8	47.4		
	n-Hexane	68.7	48.6			
105	Methyl alcohol	64.7			—	8
	Benzene	80.1	Zeotropic			
	Cyclohexane	80.75				
106	Acetonitrile	81.6	34		—	26
	Ethyl alcohol	78.3	8	70.1		
	Triethylamine	89.7	58			

TERNARY SYSTEMS (Continued)

No.	System	B.P. (°C)	Azeotropic data: Compn. (wt. %)	Azeotropic data: B.P. (°C)	Type of azeotrope	Ref.
107	Trichloroethylene	87.2			—	54
	Benzene	80.1	Zeotropic			
	Cyclohexane	80.7				
108	Acetic acid	118.1	23		S	55
	Acetic anhydride	139.6	55	134.4		
	Pyridine	115.5	22			
109	Acetic acid	118.1	3.4		S	56
	Pyridine	115.5	10.6	98.5		
	n-Heptane	98.4	86.0			
110	Acetic acid	118.1	10.4		S	56
	Pyridine	115.5	20.1	115.7		
	n-Octane	125.75	69.5			
111	Acetic acid	118.1	20.7		S	57
	Pyridine	115.5	29.4	128.0		
	n-Nonane	150.7	49.9			
112	Acetic acid	118.1	31.4		S	56
	Pyridine	115.5	38.2	134.1		
	n-Decane	173.3	30.4			
113	Acetic acid	118.1	37.5		S	58
	Pyridine	115.5	43.5	137.1		
	n-Undecane	194.5	19.0			
114	Acetic acid	118.1	13.5		S	59
	Pyridine	115.5	25.2	129.08		
	Ethylbenzene	136.5	61.3			
115	Acetic acid	118.1	17.7		S	57
	Pyridine	115.5	30.5	132.2		
	o-Xylene	143.6	51.8			
116	Acetic acid	118.1	10.2		S	60
	Pyridine	115.5	22.5	129.22		
	p-Xylene	138.4	67.3			
117	Acetic acid	118.1	15		—	61
	Isoamyl alcohol	132	54	132		
	Isoamyl acetate	142	31			
118	Acetic acid	118.1	7.6		—	62
	Benzene	80.1	34.4	77.2		
	Cyclohexane	80.75	58.0			
119	Acetic acid	118.1	3.6		S	63
	2-Picoline	129.45	24.8	121.3		
	n-Octane	125.75	71.6			
120	Acetic acid	118.1	12.8		S	63
	2-Picoline	129.45	38.4	135.0		
	n-Nonane	150.7	48.8			
121	Acetic acid	118.1	19.9		—	63
	2-Picoline	129.45	46.8	141.3		
	n-Decane	173.3	33.3			
122	Acetic acid	118.1	30.5		—	63
	2-Picoline	129.45	55.2	143.4		
	n-Undecane	194.5	14.3			
123	Acetic acid	118.1			—	64
	2,6-Lutidine	144.0	Zeotropic			
	n-Octane	125.75				
124	Acetic acid	118.1	12.6		S	64
	2,6-Lutidine	144.0	74.3	147.0		
	n-Decane	173.3	13.1			
125	Acetic acid	118.1	75.0		S	58
	2,6-Lutidine	144.0	13.8	163.0		
	n-Undecane	194.5	11.2			
126	Acetic acid	118.1			—	65
	Ethylbenzene	136.15	Zeotropic			
	n-Nonane	150.7				
127	Acetic acid	118.1			—	66
	Acetic anhydride	139.6	Zeotropic			
	Methylene diacetate	164.0				
128	Methyl formate	31.9			—	14
	Ethyl ether	34.6	Zeotropic			
	n-Pentane	36.15				
129	Nitroethane	114.2	31.7		S	67
	p-Dioxane	101.3	17.7	102.87		
	Isobutyl alcohol	108	50.6			
130	Ethyl alcohol	78.3	29.6		—	68
	Benzene	80.1	12.8	64.7		
	Cyclohexane	80.75	57.6			
131	Ethyl alcohol	78.3			—	69
	Benzene	80.1	Zeotropic			
	n-Hexane	68.7				
132	Ethyl alcohol	78.3			—	70
	Aniline	184.35	Zeotropic			
	Toluene	110.7				
133	Ethyl alcohol	78.3			—	70
	Aniline	184.35	Zeotropic			
	n-Heptane	98.4				

TERNARY SYSTEMS (Continued)

No.	System	B.P. (°C)	Azeotropic data Compn. (wt. %)	B.P. (°C)	Type of azeotrope	Ref.
134	Ethyl alcohol	78.3			—	70
	Toluene	110.7	Zeotropic			
	n-Heptane	98.4				
135	Ethylene glycol	197.4			—	71
	Pyridine	115.5	Zeotropic			
	Phenol	181.4				
136	Ethylene glycol	197.4	5.9		S	71
	Phenol	181.4	79.1	185.01		
	2-Picoline	128.8	15.0			
137	Ethylene glycol	197.4	15.9		S	71
	Phenol	181.4	67.7	186.41		
	3-Picoline	143.5	16.4			
138	Ethylene glycol	197.4	8.7		S	71
	Phenol	181.4	74.6	185.04		
	2,6-Lutidine	144.0	16.7			
139	Ethylene glycol	197.4	29.5		S	71
	Phenol	181.4	54.8	188.55		
	2,4,6-Collidine	171.0	15.7			
140	Ethylene glycol	197.45	33.6		S	72
	o-Cresol	191.0	62.4	189.65		
	2,4,6-Collidine	171.3	4.0			
141	Acetone	56.1	51.1		—	14
	Methyl acetate	56.3	5.6	49.7		
	n-Hexane	68.7	43.3			
142	Acetone	56.1			—	73
	2-Butanone	79.6	Zeotropic			
	Ethyl acetate	77.0				
143	Acetone	56.4			—	74
	Benzene	80.1	Zeotropic			
	Cyclohexane	80.75				
144	Acetone	56.1			—	75
	Benzene	80.1	Zeotropic			
	Toluene	110.7				
145	Propionic acid	140.7	55.5		S	58
	Pyridine	115.5	26.4	147.1		
	n-Undecane	194.5	18.1			
146	Propionic acid	141.05	4.5		S	76
	2-Picoline	129.3	10.5	123.7		
	n-Octane	125.4	85.0			
147	Propionic acid	141.05	16.5		S	76
	2-Picoline	129.3	21.5	140.1		
	n-Nonane	150.6				
148	Propionic acid	141.05	29.5		S	76
	2-Picoline	129.3	32.0	149.33		
	n-Decane	174.0	38.5			
149	Propionic acid	141.05	43.0		S	76
	2-Picoline	129.3	40.0	153.4		
	n-Undecane	194.8	17.0			
151	Propionic acid	141.05			S	76
	2-Picoline	129.3	Zeotropic			
	n-Dodecane	216.1				
151	Isopropyl alcohol	82.3	31.1		—	76
	Benzene	80.1	15.0	69.1		
	Cyclohexane	80.75	53.9			
152	Propyl alcohol	97.2	15.5		—	77
	Benzene	80.1	30.4	73.81		
	Cyclohexane	80.75	54.2			
153	Propyl alcohol	97.2			—	78
	Benzene	80.1	Zeotropic			
	n-Heptane	98.4				
154	2-Butanone	79.6			—	79
	Ethyl acetate	77.1	Zeotropic			
	n-Hexane	68.7				
155	2-Butanone	79.6			—	80
	Benzene	80.1	Zeotropic			
	Cyclohexane	80.75				
156	Butyric acid	162.45			—	58
	Pyridine	115.5	Zeotropic			
	n-Undecane	194.5				
157	p-Dioxane	101.1	44.3		S	81
	Isobutyl alcohol	107.0	26.7	101.8		
	Toluene	110.7				
158	n-Butyl alcohol	117.75	11.9		—	82
	Pyridine	115.5	20.7	108.7		
	Toluene	110.7	67.4			
159	n-Butyl alcohol	117.75	4		—	83
	Benzene	80.1	48	77.42		
	Cyclohexane	80.75	48			
160	n-Butyl alcohol	117.7			—	84
	Benzene	80.1	Zeotropic			
	n-Heptane	98.4				

TERNARY SYSTEMS (Continued)

			Azeotropic data			
No.	System	B.P. (°C)	Compn. (wt. %)	B.P. (°C)	Type of azeotrope	Ref.
161	Isobutyl alcohol	107.0	43.2		—	85
	Benzene	80.1	47.0	77.2		
	Cyclohexane	80.75				
162	Pyridine	115.5	8.6		S	86
	Isoamyl alcohol	131.0	4.1	110.79		
	Toluene	110.7				
163	Phenol	181.4	33.5		—	87
	Aniline	183.95	48.5	184.45		
	n-Tridecane	234.0	18.0			
164	Phenol	182.0	26.4		S	88
	Ethylene diacetate	186.0	34.4	194.45		
	Phenyl acetate	195.7	39.2			
165	Phenol	181.4	19.88		S	89
	2,4-Lutidine	159.0	21.52	181.78		
	n-Undecane	194.5	58.60			
166	Ethylbenzene	136.15			—	59
	Pyridine	115.5	Zeotropic			
	n-Nonane	150.7				
167	Aniline	184.35			—	70
	Toluene	110.7	Zeotropic			
	n-Heptane	98.4				
168	Aniline	184.35			—	8
	Benzyl alcohol	205.5	Zeotropic			
	d-Limonene	177.8				
169	Aniline	184.35			—	8
	sec-Octyl alcohol	178.7	Zeotropic			
	d-Limonene	177.8				
170	Aniline	184.35			—	8
	o-Bromotoluene	181.75	Zeotropic			
	sec-Octyl alcohol	178.7				
171	Propyl lactate	171.7	31		—	8
	Phenetole	171.5	33	163.0		
	Menthene	170.8	36			
172	m-,p-Cresol (mixt.)	202	81		S	90
	Pyridine bases (mixt.)	143	9	202.81		
	Naphthalene	218.1	10			
173	m-, p-Cresol (mixt.)	202	65.5		S	90
	Pyridine bases (mixt.)	157	16.5	202.03		
	Naphthalene	218.1				
174	m-, p-Cresol (mixt.)	202	62		S	90
	Pyridine bases (mixt.)	163	17	202.39		
	Naphthalene	218.1	21			
175	m-Cresol	202.8	61.5		S	91
	2,4,6-Collidine	171.3	20.8	205.82		
	Naphthalene	217.9	17.7			
176	Isobutyl lactate	182.15			—	8
	sec-Octyl alcohol	178.7	Zeotropic			
	Terpinene	180.5				
177	Ethylbenzene	136.1			—	92
	Isopropylbenzene	152.8	Zeotropic			
	Butylbenzene	183.1				

REFERENCES

1. **Narinskii,** *Tr. Vses. Nauch. Issled. Inst. Kisloror. Mashinostr.*, 11, 3, 1967.
2. Dow Chemical Co., unpublished data.
3. **Sokolov, Sevryogova, Zhavoronskov,** *Rev. Chim. (Bucharest)*, 20, 169, 1969.
4. **Ellis, Johnson,** *J. Inorg. Nucl. Chem.*, 6, 194, 199, 1958.
5. **Prahl, Mathes,** *Angew. Chem.*, 47, 11, 1934.
6. **Munter, Aepli, Kossatz,** *Ind. Eng. Chem.*, 39, 427, 1947.
7. **Calfee, Fukuhara, Bigelow,** *J. Amer. Chem. Soc.*, 61, 3552, 1939.
8. **Lecat,** *L'Azeotropisme*, Lamartin, Bruxelles, 1918.
9. **Marinichev, Susarev,** *Zhur. Fiz. Khim.*, 43, 1132, 1969.
10. **Ghysels,** *Bull. Soc. Chim. Belges*, 33, 57, 1924.
11. **De Mol,** *Ingr. Chim.*, 22, 262, 1938.
12. **Conti, Othmer, Gilmont,** *J. Chem. Eng. Data*, 5, 301, 1960.
13. **Kudryavtseva, Susarev, Eisen,** *Zhur. Fiz. Khim.*, 43, 437, 1969.
14. **Kudryavtseva, Eisen, Susarev,** *Zhur. Fiz. Khim.* 40, 1285, 1652, 1966.
15. **Kushner, Tatsievskaya, Serafimov,** *Zhur. Fiz. Khim.*, 41, 237, 1967.
16. **Kushner, Tatsievskaya, Serafimov,** *Zhur. Fiz. Khim.*, 42, 2248, 1968.
17. **Riddick,** Commercial Solvents Corp., unpublished data.
18. **Malesinska, Malesinski,** *Bull. Acad. Polon. Sci. Ser. Sci. Chim.*, 11, 475, 1963.

19. Delzenne, *J. Chem. Eng. Data*, 3, 224, 1958.
20. Calices, Hannotte, *Ingr. Chim.*, 20, 1, 1936.
21. Balashov, Serafimov, Bessonova, *Zhur. Fiz. Khim.*, 40, 2294, 1966.
22. Lesteva, Kachalova, Morozova, Ogorodnikov, Trenke, *Zhur. Prikl. Khim.*, 40, 1808, 1967.
23. Malesinska, *Bull. Acad. Polon. Sci. Ser. Sci. Chim.*, 12, 853, 1964.
24. Pratt, Preprint, Trans. Inst. Chem. Engrs. (London), March 1947.
25. Licht, Denzler, *Chem. Eng. Progr.*, 44, 627, 1948.
26. Union Carbide Chemicals, *Alcohols*, 1961.
27. Union Carbide Chemicals, unpublished data.
28. Merriman, *J. Chem. Soc. Trans.*, 103, 1790, 1801, 1913.
29. Young, Fortey, *J. Chem. Soc. Trans.*, 81, 717, 1902.
30. Zieborak, Galska, *Bull. Acad. Polon. Sci. Cl. III*, 3, 383, 1955.
31. Tyerman, Br. Pat. 590.713, 1947.
32. Patterson, U.S. Pat. 2,407.997, 1946.
33. Kogan, Tolstova, *Zhur. Fiz. Khim.*, 33, 276, 1959.
34. Yorizane, Yoshimura, *Hiroshima Daigaku Kogakuba Kenkya Hokoku*, 13, 41, 1965.
35. Smirnova, Moraczevskii, Storonkin, *Vest. Leningrad Univ.*, 14, 70, 1959.
36. Kogan, Fridman, Deizenrot, *Zhur. Prikl. Khim.* 30, 1339, 1957.
37. Union Carbide Chemicals Co., unpublished data.
38. Young, Fortey, *J. Chem. Soc. Trans.*, 81, 739, 1902.
39. Frolov, Loginova, Nazarova, *Zhur. Fiz. Khim.* 43, 2632, 1969.
40. Susarev, Gorbunov, *Zhur. Prikl. Khim.*, 36, 459, 1963.
41. Trabczynski, *Bull. Acad. Polon. Sci. Ser. Sci. Chim. Geol. Geogr.*, 6, 269, 1958.
42. Hannotte, *Bull. Soc. Chim. Belges*, 35, 85, 1926.
43. Hyatt, U.S. Patent 2,176.500, 1939.
44. Hirata, Hirose, *Mem. Fac. Technol. Tokyo Metrop. Univ.*, No. 11, 876, 1961.
45. Conti, Othmer, Gilmont, *J. Chem. Eng. Data*, 5, 301, 1960.
46. Ewell, Welch, *Ind. Eng. Chem.*, 37, 1224, 1945.
47. Morachevskii, Leontev, *Zhur. Fiz. Khim.*, 34, 2347, 1960.
48. Kidryavtseva, Susarev, *Zhur. Prikl. Khim.*, 36, 1231, 1471, 1710, 2025, 1963.
49. Satapathy et al., *J. Appl. Chem. (London)*, 6, 261, 1956.
50. Orszagh, Lelakowska, Beldowicz, *Bull. Acad. Polon. Sci. Ser. Sci. Chim. Geol. Geogr.*, 6, 419, 1958.
51. Lelakowska, *Bull. Acad. Polon. Sci. Ser. Sci. Chim. Geol. Geogr.*, 6, 645, 1958.
52. Forman, U.S. Patent 2,581.789, 1952.
53. Fisher, U.S. Patent 2,341.433, 1944.
54. Rao, Dakshinamurty, Rao, *J. Sci. Ind. Res. (India)*, 20B, 218, 1961.
55. Jones, *J. Chem. Eng. Data*, 7, 13, 1962.
56. Zieborak, *Bull. Acad. Polon. Sci. Cl. III*, 3, 531, 1955.
57. Zieborak, Wyrzykowska-Stankiewicz, *Bull. Acad. Polon. Sci. Ser. Sci. Chim. Geol. Geogr.*, 7, 247, 1959.
58. Zieborak, Wyrzykowska-Stankiewicz, *Bull. Acad. Polon. Sci. Ser. Sci. Chim. Geol. Geogr.*, 6, 517, 1958.
59. Galska-Krajewska, Zieborak, *Rocz. Chem.*, 36, 119, 1962.
60. Galska-Krajewska, *Bull. Acad. Polon. Sci. Ser. Sci. Chim.*, 9, 455, 1961.
61. Krokhin, *Zhur. Fiz. Khim.*, 43, 442, 1969.
62. Baradarajan, Satyanarayana, *J. Chem. Eng. Data*, 13, 148, 1968.
63. Zieborak, Wyrzykowska-Stankiewicz, *Bull. Acad. Polon. Sci. Ser. Sci. Chim. Geol. Geogr.*, 6, 377, 1958.
64. Zieborak, Kaczorowana-Badyoczek, Maczynska, *Rocz. Chem.*, 29, 783, 1955.
65. Zieborak, Galska-Krajewska, *Bull. Acad. Polon. Sci. Ser. Sci. Chim. Geol. Geogr.*, 7, 253, 1959.
66. Tatscheff et al., *Z. Phys. Chem. (Leipzig)*, 237, 52, 1968.
67. Malesinska, Malinsinski, *Bull. Acad. Polon. Sci. Ser. Sci. Chim.*, 8, 191, 1960.
68. Morachevskii, Zharov, *Zhur. Prikl. Khim.*, 36, 2771, 1963.
69. Yuan, Ho, Keshpande, Lu, *J. Chem. Eng. Data*, 8, 549, 1963.
70. Hollo, Ember, Lengyel, Weig, *Acta Chim. Acad. Sci. Hung.*, 13, 307, 1957.
71. Razniewska, *Rocz. Cham.*, 38, 851, 1964.
72. Kurtyka, *Bull. Acad. Polon. Sci. Cl. III*, 4, 49, 1956.
73. Babicz, Ivanchikova, Serafimov, *Zhur. Prikl. Khim.*, 42, 1354, 1969.
74. Kurmanadharao, Krishnamurty, Rao, *Rec. Trav. Chim.*, 76, 769, 1957.
75. Vitman, Zharov, *Zhur. Prikl. Khim.*, 42, 2858, 1969.
76. Trabczynski, *Bull. Acad. Polon. Sci. Ser. Sci. Chim.*, 12, 335, 1965.
76. Nagata, *Can. J. Chem. Eng.*, 42, 82, 1964.
77. Moraczevskii, Cheng, *Zhur. Fiz. Khim.*, 35, 2535, 1961.
78. Fu, Lu, *J. Chem. Eng. Data*, 13, 6, 1968.
79. Gorbunova, Lutigina, Malenko, *Zhur. Prikl. Khim.*, 38, 374, 622, 1965.
80. Donald, Ridgway, *J. Appl. Chem. (London)*, 8, 403, 408, 1958.

81. **Wyrzykowska-Stankiewicz, Zieborak,** *Bull. Acad. Polon. Sci. Ser. Sci. Chim.*, 8, 655, 1960.
82. **Hollo, Lengyel,** *Ind. Eng. Chem.*, 51, 957, 1959.
83. **Zieborak, Galska-Krajewska,** *Bull. Acad. Polon. Sci. Ser. Sci. Chim. Geol. Geogr.*, 6, 763, 1958.
84. **Vijayaraghavan, Deshpande, Kuloor,** *J. Chem. Eng. Data*, 12, 13, 1967.
85. **Nataraj, Rao,** *Trans. Indian Inst. Chem. Eng.*, 95, 1968.
86. **Zieborak, Wyrzykowska-Stankiewicz,** *Bull. Acad. Polon. Sci. Ser. Sci. Chim.*, 8, 137, 1960.
87. **Stadnicki,** *Bull. Acad. Polon. Sci. Ser. Sci. Chim.*, 10, 357, 1962.
88. **Orszagh, Lelakowska, Radecki,** *Bull. Acad. Polon. Sci. Ser. Sci. Chim. Geol. Geogr.*, 6, 605, 1958.
89. **Fahmy, Assal,** *Bull. Acad. Polon. Sci. Ser. Sci. Chim.*, 14, 773, 1966.
90. **Zieborak, Markowska-Majewska,** *Bull. Acad. Polon. Sci. Cl. III*, 2, 341, 1954.
91. **Kurtyka,** *Bull. Acad. Polon. Sci. Ser. Sci. Chim.*, 9, 741, 1961.
92. **Linek, Fried, Pick,** *Coll. Czech. Chem. Commun.*, 30, 1358, 1965.

QUATERNARY AND QUINARY SYSTEMS

			Azeotropic data			
No.	System	B.P. (°C)	Compn. (wt. %)	B.P. (°C)	Type	Ref.
1	Hydrocyanic acid	26			—	1
	Water	100.0	Zeotropic			
	Acrylonitrile	77.3				
	Acrolein	52.4				
2	Hydrocyanic acid	26			—	1
	Acetonitrile	81.6	Zeotropic			
	Acrylonitrile	77.3				
	Acrolein	52.4				
3	Water	100.0			—	2
	Formic acid	100.8	Zeotropic			
	Acetic acid	118.1				
	Butyric acid	162.4				
4	Water	100.0	7.38		H	3
	Nitromethane	101.2	20.65	76.88		
	Tetrachloroethylene	120.8	59.45			
	n-Propyl alcohol	97.2	12.52			
5	Water	100.0	9.86		H	3
	Nitromethane	101.2	34.40	77.06		
	Tetrachloroethylene	120.8	32.60			
	n-Octane	125.75	23.14			
6	Water	100.0	—		H	3
	Tetrachloroethylene	120.8	—	80.98		
	n-Propyl alcohol	97.2	—			
	n-Octane	125.75	—			
7	Water	100.0	9.98		H	3
	Nitromethane	101.2	41.00	76.34		
	n-Propyl alcohol	97.2	12.42			
	n-Octane	125.75	36.60			
8	Water	100.0			—	4
	Acetonitrile	81.6	Zeotropic			
	Ethyl alcohol	78.3				
	Triethylamine	89.7				
9	Water	100.0	8.7		—	5
	Ethyl alcohol	78.3	11.1			
	Crotonaldehyde	102.2	0.1	70		
	Ethyl acetate	77.1	80.1			
10	Water	100.0	6.1		H	6
	Ethyl alcohol	78.3	19.2			
	Benzene	80.1	20.4	62.14		
	Cyclohexane	80.75	54.3			
11	Water	100.0			—	7
	Ethyl alcohol	78.3	Zeotropic			
	Benzene	80.1				
	n-Hexane	68.95				
12	Water	100.0			—	7
	Ethyl alcohol	78.3	Zeotropic			
	Benzene	80.1				
	Methylcyclohexane	100.88				
13	Water	100.0	6.8		H	8
	Ethyl alcohol	78.3	18.7	64.97		
	Benzene	80.1	62.4			
	n-Heptane	98.4	12.1			
14	Water	100.0	6.7		H	8
	Ethyl alcohol	78.3	17.7	64.69		
	Benzene	80.1	61.4			
	Isooctane	99.3	14.1			
15	Water	100.0			—	9
	1-Chlorobutane	78.44	Zeotropic			
	n-Butyl alcohol	117.73				
	Butyl ether	—				

No.	System	B.P. (°C)	Azeotropic data		Type	Ref.
			Compn. (wt. %)	B.P. (°C)		
16	Chloroform	61.2			—	10
	Methyl alcohol	64.6	Zeotropic			
	Methyl acetate	56.9				
	Benzene	80.1				
17	Acetic acid	118.1	17		S	11
	Pyridine	115.4	27	127.9		
	Ethylbenzene	136.4	18			
	n-Nonane	150.8	38			
18	Acetic acid	118.1			—	12
	Pyridine	115.4	Zeotropic			
	p-Xylene	138.4				
	n-Nonane	150.8				
19	Acetone	56.1			—	13
	Isopropyl alcohol	82.3	Zeotropic			
	Benzene	80.1				
	Toluene	110.7				
20	Acetone	56.1			—	14
	Benzene	80.1	Zeotropic			
	Cyclohexane	80.9				
	Toluene	110.7				
21	Isopropyl alcohol	82.3			—	15
	2-Butanone	79.6	Zeotropic			
	Benzene	80.1				
	Cyclohexane	80.9				
22	Water	100.0	9.45		H	3
	Nitromethane	101.2	37.30			
	Tetrachloroethylene	120.8	21.15	76.5		
	n-Propyl alcohol	97.2	10.58			
	n-Octane	125.75	21.52			
23	Chloroform	61.2			—	10
	Methyl alcohol	64.6				
	Acetone	56.15	Zeotropic			
	Methyl acetate	56.9				
	Benzene	80.1				

REFERENCES

1. **Sokolov, Sevryoguva, Zhavoronkov,** *Teor. Osn. Khim. Tekhnol.*, 3, 288, 1969.
2. **Kushner, Lebedeva, Tatsievskaya, Serafimov,** *Shur. Prikl. Khim.*, 42, 1104, 1968.
3. **Malesinska,** *Bull. Acad. Polon. Sci. Ser. Sci. Chim.*, 12, 853, 1964.
4. Union Carbide Chemicals, unpublished data.
5. Eastman Chemical Products, unpublished data.
6. **Zieborak, Galska,** *Bull. Acad. Polon. Sci. Cl. III*, 3, 383, 1955.
7. **Swietoslawski, Zieborak, Galska-Krajewska,** *Bull. Acad. Polon. Sci. Ser. Sci. Chim. Geol. Geogr.*, 7, 43, 1959.
8. **Swietoslawski, Zieborak,** *Bull. Acad. Polon. Sci. Ser. A*, 9, 13, 1950.
9. **Riddick,** Commercial Solvents, unpublished data.
10. **Hudson, van Winkle,** *J. Chem. Eng. Data*, 14, 310, 1969.
11. **Zieborak, Galska-Krajewska,** *Bull. Acad. Polon. Sci. Ser. Sci. Chim. Geol. Geogr.*, 7, 253, 1959.
12. **Galska-Krajewska,** *Bull. Acad. Polon. Sci. Ser. Sci. Chim.*, 9, 455, 1961.
13. **Vitman, Zharov,** *Zhur. Prikl. Khim.*, 42, 2858, 1969.
14. **Vitman, Markova,** *Zhur. Prikl. Khim.*, 42, 2360, 1969.
15. **Lutugina, Kolbina,** *Zhur. Prikl. Khim.*, 41, 2766, 1968.

HEAT OF FORMATION OF INORGANIC OXIDES

$lb/in^2 \times 6894.8 = Pa$ $\quad$ $lb/in^2 \times 0.070307 = Kg/cm^2$

$lb/in^2 \times 6894.8 = N/m^2$ $\quad$ $lb/in^2 \times 51.7149 = Torr$

$lb/in^2 \times 0.068046 = $ Std. atm. $\quad$ $lb/in^2 \times 51.7149 = $ mmHg at 0°C

The ΔH_o values are given in gram calories per mole. The a, b, and I values listed here make it possible for one to calculate the ΔF and ΔS values by use of the following equations:

$$\Delta F_t = \Delta H_o + 2.303aT \log T + b \times 10^{-3}T^2 + c \times 10^5 T^{-1} + IT$$
$$\Delta S_t = -a - 2.303a \log T - 2b \times 10^{-3}T + c \times 10^5 T^{-2} - I$$

Ref: Bulletin 542, U. S. Bureau of Mines, 1954.

Coefficients in Free-Energy Equations

Reaction and temperature range of validity		ΔH_o	2.303a	b	c	I
2 Ac(*c*) + 3/2 O_2(*g*) = Ac_2O_3(*c*)	(298.16°–1,000° K.)	−446,090	−16.12	......	.....	+109.89
2 Al(*c*) + 1/2 O_2(*g*) = Al_2O(*g*)	(298.16°–931.7° K.)	−31,660	+14.97	......	.....	−72.74
2 Al(*l*) + 1/2 O_2(*g*) = Al_2O(*g*)	(931.7°–2,000° K.)	−38,670	+10.36	......	.....	−51.53
Al(*c*) + 1/2 O_2(*g*) = AlO(*g*)	(298.16°–931.7° K.)	+10,740	+5.76	......	.....	−37.61
Al(*l*) + 1/2 O_2(*g*) = AlO(*g*)	(931.7°–2,000° K.)	+8,170	+5.76	......	.....	−34.85
2 Al(*c*) + 3/2 O_2(*g*) = Al_2O_3 (*corundum*).	(298.16°–931.7° K.)	−404,080	−15.68	+2.18	+3.935	+123.64

HEAT OF FORMATION OF INORGANIC OXIDES (Continued)
Coefficients in Free-Energy Equations (Continued)

Reaction	Temperature range of validity	ΔH_0	2.303a	b	c	I
$2Al(l) + 3/2O_2(g) = Al_2O_3$ (corundum).	(931.7°–2,000° K.)	−407,950	−6.19	−.78	+3.935	+102.37
$2Am(c) + 3/2O_2(g) = Am_2O_3(c)$	(298.16°–1,000° K.)	−422,090	−16.12	......	.....	+107.89
$Am(c) + O_2(g) = AmO_2(c)$	(298.16°–1,000° K.)	−240,600	−4.61	......	.....	+55.91
$2Sb(c) + 3/2O_2(g) = Sb_2O_3$ (cubic)	(298.16°–842° K.)	−169,450	+6.12	−6.01	−.30	+52.21
$2Sb(c) + 3/2O_2(g) = Sb_2O_3$ (orthorhombic).	(298.16°–903° K.)	−168,060	+6.12	−6.01	−.30	+50.56
$2Sb(l) + 3/2O_2(g) = Sb_2O_3$ (orthorhombic).	(903°–928° K.)	−175,370	+15.29	−7.75	−.30	+33.12
$2Sb(l) + 3/2O_2(g) = Sb_2O_3(l)$	(928°–1,698° K.)	−173,940	−32.84	+.75	−.30	+166.52
$2Sb(l) + 3/2O_2(g) = 1/2Sb_4O_6(g)$	(1,698°–1,713° K.)	−132,760	+10.91	+.75	−.30	+.96
$2Sb(g) + 3/2O_2(g) = 1/2Sb_4O_6(g)$	(1,713°–2,000° K.)	−234,760	−.74	+.75	−.30	+98.17
$2Sb(c) + 2O_2(g) = Sb_2O_4(c)$	(298.16°–903° K.)	−208,310	+6.31	−5.36	−.40	+73.02
$2Sb(l) + 2O_2(g) = Sb_2O_4(c)$	(903°–1,500° K.)	−215,610	+15.47	−7.10	−.40	+55.61
$6Sb(c) + 13/2O_2(g) = Sb_6O_{13}(c)$	(298.16°–903° K.)	−649,160	+38.46	−25.13	−1.30	+192.54
$6Sb(c) + 13/2O_2(g) = Sb_6O_{13}(c)$	(903°–1,500° K.)	−691,370	+14.13	−30.35	−1.30	+315.93
$2Sb(c) + 5/2O_2(g) = Sb_2O_5(c)$	(298.16°–903° K.)	−226,060	+37.12	−22.66	−.50	+18.61
$2Sb(l) + 5/2O_2(g) = Sb_2O_5(c)$	(903°–1,500° K.)	−240,130	+29.01	−24.40	−.50	+59.74
$2As(c) + 3/2O_2(g) = As_2O_3$ (orthorhombic).	(298.16°–542° K.)	−154,870	+29.54	−21.33	−.30	−8.83
$2As(c) + 3/2O_2(g) = As_2O_3$ (monoclinic).	(298.16°–586° K.)	−150,760	+29.54	−21.33	−.30	−16.95
$2As(c) + 3/2O_2(g) = As_2O_3(l)$	(542°–730.3° K.)	−156,260	−43.29	+2.97	−.30	+180.95
$2As(c) + 3/2O_2(g) = 1/2As_4O_6(g)$	(730.3°–883° K.)	−135,930	+.46	+2.97	−.30	+26.88
$1/2As_4(g) + 3/2O_2(g) = 1/2As_4O_6(g)$	(883°–2,000° K.)	−154,450	−2.90	+.75	−.30	+59.71
$2As(c) + 2O_2(g) = As_2O_4(c)$	(298.16°–883° K.)	−173,690	+21.52	−13.42	−.40	+34.38
$1/2As_4(g) + 2O_2(g) = As_2O_4(c)$	(883°–1,500° K.)	−192,210	+18.15	−15.64	−.40	+67.22
$2As(c) + 5/2O_2(g) = As_2O_5(c)$	(298.16°–883° K.)	−217,080	+12.32	−4.65	−.50	+80.50
$1/2As_4(g) + 5/2O_2(g) = As_2O_5(c)$	(883°–2,000° K.)	−235,600	+8.96	−6.87	−.50	+113.33
$Ba(\alpha) + 1/2O_2(g) = BaO(c)$	(298.16°–648° K.)	−134,590	−7.60	+.87	+.42	+45.76
$Ba(\beta) + 1/2O_2(g) = BaO(c)$	(648°–977° K.)	−134,140	−3.34	−.56	+.42	+34.01
$Ba(l) + 1/2O_2(g) = BaO(c)$	(977°–1,911° K.)	−135,900	−2.19	−.56	+.42	+32.37
$Ba(g) + 1/2O_2(g) = BaO(c)$	(1,911°–2,000° K.)	−176,400	−8.01	−0.56	+0.42	+72.66
$Ba(\alpha) + O_2(g) = BaO_2(c)$	(298.16°–648° K.)	−154,830	−11.05	+.87	+.42	+74.48
$Ba(\beta) + O_2(g) = BaO_2(c)$	(648°–977° K.)	−154,380	−6.79	−.56	+.42	+62.73
$Ba(l) + O_2(g) = BaO_2(c)$	(977°–1,500° K.)	−156,140	−5.64	−.56	+.42	+61.09
$Be(c) + 1/2O_2(g) = BeO(c)$	(298.16°–1,556° K.)	−144,220	−1.91	−.46	+1.24	+30.64
$Be(l) + 1/2O_2(g) = BeO(c)$	(1,556°–2,000° K.)	−144,300	+6.06	−1.75	+1.485	+7.25
$Bi(c) + 1/2O_2(g) = BiO(c)$	(298.16°–544° K.)	−50,450	−4.61	......	.....	+35.51
$Bi(l) + 1/2O_2(g) = BiO(c)$	(544°–1,600° K.)	−52,920	−4.61	......	.....	+40.05
$2Bi(c) + 3/2O_2(g) = Bi_2O_3(c)$	(298.16°–544° K.)	−139,000	−11.56	+2.15	−.30	+96.52
$2Bi(l) + 3/2O_2(g) = Bi_2O_3(c)$	(544°–1,090° K.)	−142,270	+2.30	−3.25	−.30	+67.55
$2Bi(l) + 3/2O_2(g) = Bi_2O_3(l)$	(1,090°–1,600° K.)	−147,350	−32.84	+.75	−.30	+174.59
$2B(c) + 3/2O_2(g) = B_2O(c)$	(298.16°–723° K.)	−304,690	+11.72	−7.55	+.355	+34.25
$2B(c) + 3/2O_2(g) = B_2O_3(gl)$	(298.16°–723° K.)	−298,670	+26.57	−15.90	−.30	−10.40
$2B(c) + 3/2O_2(g) = B_2O_3(l)$	(723°–2,000° K.)	−308,100	−38.41	+5.15	−.30	+173.24
$Cd(c) + 1/2O_2(g) = CdO(c)$	(298.16°–594° K.)	−62,330	−2.05	+.71	−.10	+29.17
$Cd(l) + 1/2O_2(g) = CdO(c)$	(594°–1,038° K.)	−63,240	+2.07	−.76	−.10	+20.14
$Cd(g) + 1/2O_2(g) = CdO(c)$	(1,038°–2,000° K.)	−89,320	−2.83	−.76	−.10	+60.05
$Ca(\alpha) + 1/2O_2(g) = CaO(c)$	(298.16°–673° K.)	−151,850	−6.56	+1.46	+.68	+43.93
$Ca(\beta) + 1/2O_2(g) = CaO(c)$	(673°–1,124° K.)	−151,730	−4.14	+.41	+.68	+37.63
$Ca(l) + 1/2O_2(g) = CaO(c)$	(1,124°–1,760° K.)	−153,480	−1.36	−.29	+.68	+31.49
$Ca(g) + 1/2O_2(g) = CaO(c)$	(1,760°–2,000° K.)	−194,670	−7.18	−.29	+.68	+73.84
$Ca(\alpha) + O_2(g) = CaO_2(c)$	(298.16°–500° K.)	−158,230	−12.32	+1.46	+.68	+78.28
$C(graphite) + 1/2O_2(g) = CO(g)$	(298.16°–2,000° K.)	−25,400	+2.05	+.27	−1.095	−28.79
$C(graphite) + O_2(g) = CO_2(g)$	(298.16°–2,000° K.)	−93,690	+1.63	−.07	−.23	−5.64
$2Ce(c) + 3/2O_2(g) = Ce_2O_3(c)$	(298.16°–1,048° K.)	−435,600	−4.60	......	.....	+92.84
$2Ce(l) + 3/2O_2(g) = Ce_2O_3(c)$	(1,048°–1,900° K.)	−440,400	−4.60	......	.....	+97.42
$Ce(c) + O_2(g) = CeO_2(c)$	(298.16°–1,048° K.)	−245,490	−6.42	+2.34	−.20	+67.79
$Ce(l) + O_2(g) = CeO_2(c)$	(1,048°–2,000° K.)	−247,930	+.71	−.66	−.20	+51.73
$2Cs(c) + 1/2O_2(g) = Cs_2O(c)$	(298.16°–301.5° K.)	−75,900	.....	......	.....	+36.60
$2Cs(l) + 1/2O_2(g) = Cs_2O(c)$	(301.5°–763° K.)	−76,900	.....	......	.....	+39.92
$2Cs(l) + 1/2O_2(g) = Cs_2O(l)$	(763°–963° K.)	−75,370	−9.21	......	.....	+64.47
$2Cs(g) + 1/2O_2(g) = Cs_2O(l)$	(963°–1,500° K.)	−113,790	−23.03	......	.....	+145.60
$2Cs(c) + O_2(g) = Cs_2O_2(c)$	(298.16°–301.5° K.)	−96,500	−2.30	......	.....	+62.30
$2Cs(l) + O_2(g) = Cs_2O_2(c)$	(301.5°–870° K.)	−97,800	−4.61	......	.....	+72.34
$2Cs(l) + O_2(g) = Cs_2O_2(l)$	(870°–963° K.)	−96,060	−18.42	......	.....	+110.94
$2Cs(g) + O_2(g) = Cs_2O_2(l)$	(963°–1,500° K.)	−134,000	−31.08	......	.....	+188.11
$2Cs(c) + 3/2O_2(g) = Cs_2O_3(c)$	(298.16°–301.5° K.)	−112,690	−11.51	......	.....	+110.10
$2Cs(l) + 3/2O_2(g) = Cs_2O_3(c)$	(301.5°–775° K.)	−113,840	−12.66	......	.....	+116.77
$2Cs(l) + 3/2O_2(g) = Cs_2O_3(l)$	(775°–963° K.)	−110,740	−26.48	......	.....	+152.70
$2Cs(g) + 3/2O_2(g) = Cs_2O_3(l)$	(963°–1,500° K.)	−148,680	−39.14	......	.....	+229.87
$Cs(c) + O_2(g) = CsO_2(c)$	(298.16°–301.5° K.)	−63,590	−11.51	......	.....	+72.29
$Cs(l) + O_2(g) = CsO_2(c)$	(301.5°–705° K.)	−64,240	−12.66	......	.....	+77.30
$Cs(l) + O_2(g) = CsO_2(l)$	(705°–963° K.)	−61,770	−18.42	......	.....	+90.20
$Cs(g) + O_2(g) = CsO_2(l)$	(963°–1,500° K.)	−80,500	−24.18	......	.....	+126.83
$Cl_2(g) + 1/2O_2(g) = Cl_2O(g)$	(298.16°–2,000° K.)	+17,770	−.71	−.12	+.49	+16.81
$1/2Cl_2(g) + 1/2O_2(g) = ClO(g)$	(298.16°–1,000° K.)	+33,000	.....	......	.....	−.24
$1/2Cl_2(g) + O_2(g) = ClO_2(g)$	(298.16°–2,000° K.)	+24,150	−.76	−.105	−.665	+19.08
$1/2Cl_2(g) + 3/2O_2(g) = ClO_3(g)$	(298.16°–500° K.)	+37,740	+5.76	......	.....	+21.42
$Cl_2(g) + 7/2O_2(g) = Cl_2O_7(g)$	(298.16°–500° K.)	+65,040	+12.66	......	.....	+78.01
$2Cr(c) + 3/2O_2(g) = Cr_2O_3(\beta)$	(298.16°–1,823° K.)	−274,670	−14.07	+2.01	+.69	+105.65
$2Cr(l) + 3/2O_2(g) = Cr_2O_3(\beta)$	(1,823°–2,000° K.)	−278,030	+2.33	−.35	+1.57	+58.29
$Cr(c) + O_2(g) = CrO_2(c)$	(298.16°–1,000° K.)	−142,500	.....	......	.....	+42.00
$Cr(c) + 3/2O_2(g) = CrO_3(c)$	(298.16°–471° K.)	−141,590	−13.82	......	.....	+103.90
$Cr(c) + 3/2O_2(g) = CrO_3(l)$	(471°–600° K.)	−141,580	−32.24	......	.....	+153.14
$Co(\alpha, \beta) + 1/2O_2(g) = CoO(c)$	(298.16°–1,400° K.)	−56,910	+.69	......	.....	+16.03
$Co(\gamma) + 1/2O_2(g) = CoO(c)$	(1,400°–1,763° K.)	−58,160	−1.15	......	.....	+22.71
$Co(l) + 1/2O_2(g) = CoO(c)$	(1,763°–2,000° K.)	−65,680	−6.22	......	.....	+43.43
$3Co(\alpha, \beta, \gamma) + 2O_2(g) = Co_3O_4(c)$	(298.16°–1,500° K.)	−207,300	−2.30	......	.....	+90.56
$2Cu(c) + 1/2O_2(g) = Cu_2O(c)$	(298.16°–1,357° K.)	−40,550	−1.15	−1.10	−.10	+21.92
$2Cu(l) + 1/2O_2(g) = Cu_2O(c)$	(1,357°–1,502° K.)	−43,880	+8.47	−2.60	−.10	−3.72
$2Cu(l) + 1/2O_2(g) = Cu_2O(l)$	(1,502°–2,000° K.)	−37,710	−12.48	+.25	−.10	+54.44
$Cu(c) + 1/2O_2(g) = CuO(c)$	(298.16°–1,357° K.)	−37,740	−.64	−1.40	−.10	+24.87
$Cu(l) + 1/2O_2(g) = CuO(c)$	(1,357°–1,720° K.)	−39,410	+4.17	−2.15	−.10	+12.05
$Cu(l) + 1/2O_2(g) = CuO(l)$	(1,720°–2,000° K.)	−41,060	−11.35	+.25	−.10	+59.09
$F_2(g) + 1/2O_2(g) = F_2O(g)$	(298.16°–2,000° K.)	+5,070	−.41	−.15	+.535	+16.04
$2Ga(c) + 1/2O_2(g) = Ga_2O(c)$	(298.16°–302.7° K.)	−81,110	+10.32	−5.75	−.10	−3.66

Coefficients in Free-Energy Equations (Continued)

Reaction and temperature range of validity		ΔH_0	2.303a	b	c	I
$2\,Ga(l) + 1/2\,O_2(g) = Ga_2O(c)$	(302.7°–1,000° K.)	−83,360	+13.49	−5.75	−.10	−4.08
$2\,Ga(c) + 3/2\,O_2(g) = Ga_2O_3(c)$	(298.16°–302.7° K.)	−256,240	+14.64	−3.75	−.30	+32.23
$2\,Ga(l) + 3/2\,O_2(g) = Ga_2O_3(c)$	(302.7°–2,000° K.)	−258,490	+17.82	−3.75	−.30	+31.79
$Ge(c) + 1/2\,O_2(g) = GeO(c)$	(298.16°–1,200° K.)	−60,900	+1.27	−1.49	−.10	+17.19
$Ge(c) + 1/2\,O_2(g) = GeO(g)$	(298.16°–1,200° K.)	−21,870	+6.72	−.075	−.10	−41.25
$Ge(c) + O_2(g) = GeO_2(gl)$	(298.16°–1,200° K.)	−127,830	+4.28	−2.52	−0.20	+30.54
$2\,Au(c) + 3/2\,O_2(g) = Au_2O_3(c)$	(298.16°–500° K.)	−2,160	−10.36	.....	.....	+95.14
$Hf(c) + O_2(g) = HfO_2(monocl.)$	(298.16°–2,000° K.)	−268,380	−9.74	−.28	+1.54	+78.16
$H_2(g) + 1/2\,O_2(g) = H_2O(l)$	(298.16°–373.16° K.)	−70,600	−18.26	+.64	−.04	+91.67
$H_2(g) + 1/2\,O_2(g) = H_2O(g)$	(298.16°–2,000° K.)	−56,930	+6.75	−.64	−.08	−8.74
$D_2(g) + 1/2\,O_2(g) = D_2O(l)$	(298.16°–374.5° K.)	−72,760	−18.10	.....	.....	+93.59
$D_2(g) + 1/2\,O_2(g) = D_2O(g)$	(298.16°–2,000° K.)	−58,970	+5.50	−.75	+.085	−3.74
$1/2\,H_2(g) + 1/2\,O_2(g) = OH(g)$	(298.16°–2,000° K.)	+10,350	+.90	+.005	−.26	−6.69
$H_2(g) + O_2(g) = H_2O_2(l)$	(298.16°–425° K.)	−47,140	−13.52	−7.13	.....	+99.39
$H_2(g) + O_2(g) = H_2O_2(g)$	(298.16°–1,500° K.)	−32,570	+4.77	−.96	+.97	+13.84
$2\,In(c) + 3/2\,O_2(g) = In_2O_3(c)$	(298.16°–429.6° K.)	−220,410	+5.43	−.50	−.30	+59.49
$2\,In(l) + 3/2\,O_2(g) = In_2O_3(c)$	(429.6°–2,000° K.)	−220,970	+13.22	−3.00	−.30	+41.36
$I_2(c) + 5/2\,O_2(g) = I_2O_5(c)$	(298.16°–386.8° K.)	−42,040	+2.30	.....	.....	+113.71
$I_2(l) + 5/2\,O_2(g) = I_2O_5(c)$	(386.8°–456° K.)	−43,490	+16.12	.....	.....	+81.70
$I_2(g) + 5/2\,O_2(g) = I_2O_5(c)$	(456°–500° K.)	−58,020	−6.91	.....	.....	+174.79
$Ir(c) + O_2(g) = IrO_2(c)$	(298.16°–1,300° K.)	−39,480	+8.17	−6.39	−.20	+20.33
$0.947\,Fe(\alpha) + 1/2\,O_2(g) = Fe_{0.947}O(c)$	(298.16°–1,033° K.)	−65,320	−11.26	+2.61	+0.44	+48.60
$0.947\,Fe(\beta) + 1/2\,O_2(g) = Fe_{0.947}O(c)$	(1,033°–1,179° K.)	−62,380	+4.08	−.75	+.235	+3.00
$0.947\,Fe(\gamma) + 1/2\,O_2(g) = Fe_{0.947}O(c)$	(1,179°–1,650° K.)	−66,750	−8.04	+.67	−.10	+42.28
$0.947\,Fe(\gamma) + 1/2\,O_2(g) = Fe_{0.947}O(l)$	(1,650°–1,674° K.)	−64,200	−18.72	+1.67	−.10	+73.45
$0.947\,Fe(\delta) + 1/2\,O_2(g) = Fe_{0.947}O(l)$	(1,674°–1,803° K.)	−59,650	−6.84	+.25	−.10	+34.81
$0.947\,Fe(l) + 1/2\,O_2(g) = Fe_{0.947}O(l)$	(1,803°–2,000° K.)	−63,660	−7.48	+.25	−.10	+39.12
$3\,Fe(\alpha) + 2\,O_2(g) = Fe_3O_4\,(magnetite)$	(298.16°–900° K.)	−268,310	+5.87	−12.45	+.245	+73.11
$3\,Fe(\alpha) + 2\,O_2(g) = Fe_3O_4(\beta)$	(900°–1,033° K.)	−272,300	−54.27	+11.65	+.245	+233.52
$3\,Fe(\beta) + 2\,O_2(g) = Fe_3O_4(\beta)$	(1,033°–1,179° K.)	−262,990	−5.71	+1.00	−.40	+89.19
$3\,Fe(\gamma) + 2\,O_2(g) = Fe_3O_4(\beta)$	(1,179°–1,674° K.)	−276,990	−44.05	+5.50	−.40	+213.52
$3\,Fe(\delta) + 2\,O_2(g) = Fe_3O_4(\beta)$	(1,674°–1,803° K.)	−262,560	−6.40	+1.00	−.40	+91.05
$3\,Fe(l) + 2\,O_2(g) = Fe_3O_4(\beta)$	(1,803°–1,874° K.)	−275,280	−8.74	+1.00	−.40	+104.84
$3\,Fe(l) + 2\,O_2(g) = Fe_3O_4(l)$	(1,874°–2,000° K.)	−257,240	−26.89	+1.00	−.40	+155.46
$2\,Fe(\alpha) + 3/2\,O_2(g) = Fe_2O_3\,(hematite)$	(298.16°–950° K.)	−200,000	−13.84	−1.45	+1.905	+108.26
$2\,Fe(\alpha) + 3/2\,O_2(g) = Fe_2O_3(\beta)$	(950°–1,033° K.)	−202,960	−42.64	+7.85	+.13	+188.48
$2\,Fe(\beta) + 3/2\,O_2(g) = Fe_2O_3(\beta)$	(1,033°–1,050° K.)	−196,740	−10.27	+.75	−.30	+92.26
$2\,Fe(\beta) + 3/2\,O_2(g) = Fe_2O_3(\gamma)$	(1,050°–1,179° K.)	−193,200	−.39	−.13	−.30	+59.96
$2\,Fe(\gamma) + 3/2\,O_2(g) = Fe_2O_3(\gamma)$	(1,179°–1,674° K.)	−202,540	−25.95	+2.87	−.30	+142.85
$2\,Fe(\delta) + 3/2\,O_2(g) = Fe_2O_3(\gamma)$	(1,674°–1,800° K.)	−192,920	−.85	−.13	−.30	+61.21
$2\,La(c) + 3/2\,O_2(g) = La_2O_3(c)$	(298.16°–1,153° K.)	−431,120	−13.31	+.80	+1.34	+112.36
$2\,La(l) + 3/2\,O_2(g) = La_2O_3(c)$	(1,153°–2,000° K.)	−434,330	−4.88	−.80	+1.34	+91.17
$Pb(c) + 1/2\,O_2(g) = PbO\,(red)$	(298.16°–600.5° K.)	−52,800	−2.76	−.80	−.10	+32.49
$Pb(l) + 1/2\,O_2(g) = PbO\,(red)$	(600.5°–762° K.)	−53,780	−.51	−1.75	−.10	+28.44
$Pb(c) + 1/2\,O_2(g) = PbO\,(yellow)$	(298.16°–600.5° K.)	−52,040	+.81	−2.00	−.10	+22.13
$Pb(l) + 1/2\,O_2(g) = PbO\,(yellow)$	(600.5°–1,159° K.)	−53,020	+3.06	−2.95	−.10	+18.08
$Pb(l) + 1/2\,O_2(g) = PbO(l)$	(1,159°–1,745° K.)	53,060	19.01	.95	.10	61.22
$Pb(c) + 1/2\,O_2(g) = PbO(g)$	(298.16°–600.5° K.)	+10,270	+1.91	+1.08	+.295	−23.21
$Pb(l) + 1/2\,O_2(g) = PbO(g)$	(600.5°–2,000° K.)	+9,300	+4.17	+.13	+.295	−27.29
$3\,Pb(c) + 2\,O_2(g) = Pb_3O_4(c)$	(298.16°–600.5° K.)	−174,920	+8.82	−8.20	−.40	+72.78
$3\,Pb(l) + 2\,O_2(g) = Pb_3O_4(c)$	(600.5°–1,000° K.)	−177,860	+15.59	−11.05	−.40	+60.57
$Pb(c) + O_2(g) = PbO_2(c)$	(298.16°–600.5° K.)	−66,120	+.64	−2.45	−.20	+45.58
$Pb(l) + O_2(g) = PbO_2(c)$	(600.5°–1,100° K.)	−67,100	+2.90	−3.40	−.20	+41.50
$2\,Li(c) + 1/2\,O_2(g) = Li_2O(c)$	(298.16°–452° K.)	−142,220	−3.06	+5.77	−.10	+34.19
$2\,Li(l) + 1/2\,O_2(g) = Li_2O(c)$	(452°–1,600° K.)	−141,380	+16.97	−2.63	−.10	−17.05
$2\,Li(c) + O_2(g) = Li_2O_2(c)$	(298.16°–452° K.)	−151,880	−1.38	.....	.....	+54.83
$2\,Li(l) + O_2(g) = Li_2O_2(c)$	(452°–500° K.)	−153,260	−1.38	.....	.....	+57.88
$Mg(c) + 1/2\,O_2(g) = MgO\,(periclase)$	(298.16°–923° K.)	−144,090	−1.06	+.13	+.25	+29.16
$Mg(l) + 1/2\,O_2(g) = MgO\,(periclase)$	(923°–1,393° K.)	−145,810	+1.84	−.62	+.64	+23.07
$Mg(g) + 1/2\,O_2(g) = MgO\,(periclase)$	(1,393°–2,000° K.)	−180,700	−3.75	−.62	+.64	+65.69
$Mg(c) + O_2(g) = MgO_2(c)$	(298.16°–500° K.)	−150,230	−9.12	+.13	+.25	+70.84
$Mn(\alpha) + 1/2\,O_2(g) = MnO(c)$	(298.16°–1,000° K.)	−92,600	−4.21	+.97	+.155	+29.66
$Mn(\beta) + 1/2\,O_2(g) = MnO(c)$	(1,000°–1,374° K.)	−91,900	+1.84	−.39	+.34	+12.15
$Mn(\gamma) + 1/2\,O_2(g) = MnO(c)$	(1,374°–1,410° K.)	−89,810	+7.30	−.72	+.34	−6.05
$Mn(\delta) + 1/2\,O_2(g) = MnO(c)$	(1,410°–1,517° K.)	−89,390	+8.68	−.72	+.34	−10.70
$Mn(l) + 1/2\,O_2(g) = MnO(c)$	(1,517°–2,000° K.)	−93,350	+7.99	−.72	+.34	−5.90
$3\,Mn(\alpha) + 2\,O_2(g) = Mn_3O_4(\alpha)$	(298.16°–1,000° K.)	−332,400	−7.41	+.66	+.145	+106.62
$3\,Mn(\beta) + 2\,O_2(g) = Mn_3O_4(\alpha)$	(1,000°–1,374° K.)	−330,310	+10.75	−3.42	+.70	+54.07
$3\,Mn(\gamma) + 2\,O_2(g) = Mn_3O_4(\alpha)$	(1,374°–1,410° K.)	−324,050	+27.12	−4.41	+.70	−.50
$3\,Mn(\delta) + 2\,O_2(g) = Mn_3O_4(\alpha)$	(1,410°–1,445° K.)	−322,800	+31.27	−4.41	+.70	−14.46
$3\,Mn(\delta) + 2\,O_2(g) = Mn_3O_4(\beta)$	(1,445°–1,517° K.)	−328,870	−4.56	+1.00	−.40	+95.20
$3\,Mn(l) + 2\,O_2(g) = Mn_3O_4(\beta)$	(1,517°–1,800° K.)	−340,730	−6.63	+1.00	−.40	+109.60
$2\,Mn(\alpha) + 3/2\,O_2(g) = Mn_2O_3(c)$	(298.16°–1,000° K.)	−230,610	−5.96	−.06	+.945	+80.74
$2\,Mn(\beta) + 3/2\,O_2(g) = Mn_2O_3(c)$	(1,000°–1,374° K.)	−229,210	+6.15	−2.78	+1.315	+45.70
$2\,Mn(\gamma) + 3/2\,O_2(g) = Mn_2O_3(c)$	(1,374°–1,410° K.)	−225,030	+17.06	−3.44	+1.315	+9.33
$2\,Mn(\delta) + 3/2\,O_2(g) = Mn_2O_3(c)$	(1,410°–1,517° K.)	−224,200	+19.82	−3.44	+1.315	+0.05
$2\,Mn(l) + 3/2\,O_2(g) = Mn_2O_3(c)$	(1,517°–1,700° K.)	−232,110	+18.44	−3.44	+1.315	+9.65
$Mn(\alpha) + O_2(g) = MnO_2(c)$	(298.16°–1,000° K.)	−126,400	−8.61	+.97	+1.555	+70.14
$2\,Hg(l) + 1/2\,O_2(g) = Hg_2O(c)$	(298.16°–629.88° K.)	−22,400	−4.61	.....	.....	+43.29
$2\,Hg(g) + 1/2\,O_2(g) = Hg_2O(c)$	(629.88°–1,000° K.)	−53,800	−16.12	.....	.....	+125.36
$Hg(l) + 1/2\,O_2(g) = HgO\,(red)$	(298.16°–629.88° K.)	−21,760	+.85	−2.47	−.10	+24.81
$Hg(g) + 1/2\,O_2(g) = HgO\,(red)$	(629.88°–1,500° K.)	−36,920	−2.92	−2.47	−.10	+59.42
$Mo(c) + O_2(g) = MoO_2(c)$	(298.16°–2,000° K.)	−132,910	−3.91	.....	.....	+47.42
$Mo(c) + 3/2\,O_2(g) = MoO_3(c)$	(298.16°–1,068° K.)	−182,650	−8.86	−1.55	+1.54	+90.07
$Mo(c) + 3/2\,O_2(g) = MoO_3(l)$	(1,068°–1,500° K.)	−179,770	−36.34	+1.40	−.30	+167.61
$2\,Nd(c) + 3/2\,O_2(g) = Nd_2O_3\,(hexagonal)$.	(298.16°–1,113° K.)	−435,150	−16.19	+3.21	+1.78	+125.68
$2\,Nd(l) + 3/2\,O_2(g) = Nd_2O_3\,(hexagonal)$.	(1,113°–1,500° K.)	−437,090	+4.03	−2.13	+1.78	+71.77
$Np(c) + O_2(g) = NpO_2(c)$	(298.16°–913° K.)	−246,450	−3.45	.....	.....	+52.44
$Np(l) + O_2(g) = NpO_2(c)$	(913°–1,500° K.)	−249,010	−4.61	.....	.....	+58.68
$Ni(\alpha) + 1/2\,O_2(g) = NiO(c)$	(298.16°–633° K.)	−57,640	−4.61	+2.16	−.10	+34.41
$Ni(\beta) + 1/2\,O_2(g) = NiO(c)$	(633°–1,725° K.)	−57,460	−.14	−.46	−.10	+23.27
$Ni(l) + 1/2\,O_2(g) = NiO(c)$	(1,725°–2,000° K.)	−58,830	+7.23	−1.36	−.10	+1.76
$2\,Nb(c) + 2\,O_2(g) = Nb_2O_4(c)$	(298.16°–2,000° K.)	−382,050	−9.67	.....	.....	+116.23
$2\,Nb(c) + 5/2\,O_2(g) = Nb_2O_5(c)$	(298.16°–1,785° K.)	−458,640	−16.14	−.56	+1.94	+157.66
$2\,Nb(c) + 5/2\,O_2(g) = Nb_2O_5(l)$	(1,785°–2,000° K.)	−463,630	−66.04	+2.21	−.50	+317.84
$N_2(g) + 1/2\,O_2(g) = N_2O(g)$	(298.16°–2,000° K.)	+18,650	−1.57	−.27	+.92	+23.47

Coefficients in Free-Energy Equations (Continued)

Reaction and temperature range of validity		ΔH_0	2.303a	b	c	I
1/2 $N_2(g)$ + 1/2 $O_2(g)$ = $NO(g)$	(298.16°–2,000° K.)	+21,590	−.28	+.45	−.03	−2.20
$N_2(g)$ + 3/2 $O_2(g)$ = $N_2O_3(g)$	(298.16°–500° K.)	+17,390	−.35	.....	.....	+54.30
1/2 $N_2(g)$ + $O_2(g)$ = $NO_2(g)$	(298.16°–2,000° K.)	+7,730	+.53	−.265	+.605	+13.74
$N_2(g)$ + 2 $O_2(g)$ = $N_2O_4(g)$	(298.16°–1,000° K.)	+1,370	+2.14	−3.24	+1.38	+68.34
$N_2(g)$ + 5/2 $O_2(g)$ = $N_2O_5(c)$	(298.16°–305° K.)	−10,200	.....	.....	.....	+131.70
$N_2(g)$ + 5/2 $O_2(g)$ = $N_2O_5(g)$	(298.16°–500° K.)	+3,600	.....	.....	.....	+86.50
$Os(c)$ + 2 $O_2(g)$ = OsO_4 (*yellow*)	(298.16°–329° K.)	−92,260	+14.97	.....	.....	+32.45
$Os(c)$ + 2 $O_2(g)$ = OsO_4 (*white*)	(298.16°–315° K.)	−90,560	+14.97	.....	.....	+27.60
$Os(c)$ + 2 $O_2(g)$ = $OsO_4(l)$	(315°–403° K.)	−88,970	+9.67	.....	.....	+35.78
$Os(c)$ + 2 $O_2(g)$ = $OsO_4(g)$	(298.16°–1,000° K.)	−81,200	+8.17	−2.86	+1.94	+20.34
3/2 $O_2(g)$ = $O_3(g)$	(298.16°–2,000° K.)	+33,980	+2.03	−.48	+.36	+11.45
P (*white*) + 1/2 $O_2(g)$ = $PO(g)$	(298.16°–317.4° K.)	−9,370	+2.53	.....	.....	−25.40
$P(l)$ + 1/2 $O_2(g)$ = $PO(g)$	(317.4°–553° K.)	−9,390	+3.45	.....	.....	−27.63
1/4 $P_4(g)$ + 1/2 $O_2(g)$ = $PO(g)$	(553°–1,500° K.)	−12,640	+2.30	.....	.....	−18.61
4 P (*white*) + 5 $O_2(g)$ = P_4O_{10} (*hexagonal*).	(298.16°–317.4° K.)	−711,520	+95.67	−51.50	−1.00	−28.24
4 $P(l)$ + 5 $O_2(g)$ = P_4O_{10} (*hex.*)	(317.4°–553° K.)	−711,800	+97.98	−51.50	−1.00	−33.13
$P_4(g)$ + 5 $O_2(g)$ = P_4O_{10} (*hex.*)	(553°–631° K.)	−725,560	+87.45	−51.07	−2.405	+20.87
$P_4(g)$ + 5 $O_2(g)$ = $P_4O_{10}(g)$	(631°–1,500° K.)	−722,330	−43.45	+2.93	−2.405	+348.20
$Pu(c)$ + $O_2(g)$ = $PuO_2(c)$	(298.16°–1,500° K.)	−246,450	−3.45	.....	.....	+52.48
$Po(c)$ + $O_2(g)$ = $PoO_2(c)$	(298.16°–900° K.)	−61,510	−9.21	.....	.....	+72.80
2 $K(c)$ + 1/2 $O_2(g)$ = $K_2O(c)$	(298.16°–336.4° K.)	−86,400	.....	.....	.....	+33.90
2 $K(l)$ + 1/2 $O_2(g)$ = $K_2O(c)$	(336.4°–1,049° K.)	−87,380	+1.15	.....	.....	+33.90
2 $K(g)$ + 1/2 $O_2(g)$ = $K_2O(c)$	(1,049°–1,500° K.)	−133,090	−16.12	.....	.....	+129.64
2 $K(c)$ + $O_2(g)$ = $K_2O_2(c)$	(298.16°–336.4° K.)	−118,300	−2.30	.....	.....	+59.60
2 $K(l)$ + $O_2(g)$ = $K_2O_2(c)$	(336.4°–763° K.)	−119,780	−4.61	.....	.....	+69.85
2 $K(l)$ + $O_2(g)$ = $K_2O_2(l)$	(763°–1,049° K.)	−118,250	−18.42	.....	.....	+107.66
2 $K(g)$ + $O_2(g)$ = $K_2O_2(l)$	(1,049°–1,500° K.)	−161,870	−31.08	.....	.....	+187.49
2 $K(c)$ + 3/2 $O_2(g)$ = $K_2O_3(c)$	(298.16°–336.4° K.)	−126,640	−12.66	.....	.....	+111.75
2 $K(l)$ + 3/2 $O_2(g)$ = $K_2O_3(c)$	(336.4°–703° K.)	−127,790	−12.66	.....	.....	+115.16
2 $K(l)$ + 3/2 $O_2(g)$ = $K_2O_3(l)$	(703°–1,000° K.)	−125,330	−27.63	.....	.....	+154.28
$K(c)$ + $O_2(g)$ = $KO_2(c)$	(298.16°–336.4° K.)	−68,940	−10.36	.....	.....	+66.45
$K(l)$ + $O_2(g)$ = $KO_2(c)$	(336.4°–653° K.)	−69,510	−10.36	.....	.....	+68.15
$K(l)$ + $O_2(g)$ = $KO_2(l)$	(653°–1,000° K.)	−67,880	−18.42	.....	.....	+88.34
$K(c)$ + 3/2 $O_2(g)$ = $KO_3(c)$	(298.16°–336.4° K.)	−63,340	−10.36	.....	.....	+85.85
$K(l)$ + 3/2 $O_2(g)$ = $KO_3(c)$	(336.4°–500° K.)	−63,910	−10.36	.....	.....	+87.55
2 $Pr(c)$ + 3/2 $O_2(g)$ = Pr_2O_3(*c*, *C-type*)	(298.16°–1,205° K.)	−440,600	−4.60	.....	.....	+78.38
2 $Pr(l)$ + 3/2 $O_2(g)$ = Pr_2O_3(*c*, *C-type*)	(1,205°–2,000° K)	−446,100	−4.60	.....	.....	+82.94
6 $Pr(c)$ + 11/2 $O_2(g)$ = $Pr_6O_{11}(c)$	(298.16°–1,205° K.)	−1,374,000	.....	.....	.....	+241.04
6 $Pr(l)$ + 11/2 $O_2(g)$ = $Pr_6O_{11}(c)$	(1,205°–1,500° K.)	−1,390,500	.....	.....	.....	+254.73
$Pr(c)$ + $O_2(g)$ = $PrO_2(c)$	(298.16°–1,200° K.)	−230,990	−6.42	+2.34	−.20	+61.07
$Ra(c)$ + 1/2 $O_2(g)$ = $RaO(c)$	(298.16°–1,000° K.)	−130,000	.....	.....	.....	+23.50
$Re(c)$ + 3/2 $O_2(g)$ = $ReO_3(c)$	(298.16°–433° K.)	−149,090	−16.12	.....	.....	+110.49
$Re(c)$ + 3/2 $O_2(g)$ = $ReO_3(l)$	(433°–1,000° K.)	−146,750	−31.32	.....	.....	+145.16
2$Re(c)$ + 7/2 $O_7(g)$ = $Re_2O_7(c)$	(298.16°–569° K.)	−301,470	−34.54	.....	.....	+250.57
2 $Re(c)$ + 7/2 $O_7(g)$ = $Re_2O_7(l)$	(569°–635.5° K.)	−295,810	−73.68	.....	.....	+348.45
2 $Re(c)$ + 7/2 $O_2(g)$ = $Re_2O_7(g)$	(635.5°–1,500° K.)	−256,460	+3.45	.....	.....	+70.33
2 $Re(c)$ + 4 $O_2(g)$ = $Re_2O_8(c)$	(298.16°–420° K.)	−313,870	−41.45	.....	.....	+293.57
2 $Re(c)$ + 4 $O_2(g)$ = $Re_2O_8(l)$	(420°–600° K.)	−318,470	−87.50	.....	.....	+425.32
2 $Rh(c)$ + 1/2 $O_2(g)$ = $Rh_2O(c)$	(298.16°–2,000° K.)	−23,740	−8.06	.....	.....	+35.64
$Rh(c)$ + 1/2 $O_2(g)$ = $RhO(c)$	(298.16°–1,500° K.)	−22,650	−7.37	.....	.....	+40.54
2 $Rh(c)$ + 3/2 $O_2(g)$ = $Rh_2O_3(c)$	(298.16°–1,500° K.)	−70,060	−13.58	.....	.....	+101.72
2 $Rb(c)$ + 1/2 $O_2(g)$ = $Rb_2O(c)$	(298.16°–312.2° K.)	−78,900	.....	.....	.....	+32.20
2 $Rb(l)$ + 1/2 $O_2(g)$ = $Rb_2O(c)$	(312.2°–750° K.)	−79,950	.....	.....	.....	+35.56
2 $Rb(l)$ + 1/2 $O_2(g)$ = $Rb_2O(l)$	(750°–952° K.)	−78,830	−10.36	.....	.....	+63.85
2 $Rb(g)$ + 1/2 $O_2(g)$ = $Rb_2O(l)$	(952°–1,500° K.)	−120,290	−23.03	.....	.....	+145.14
2 $Rb(c)$ + $O_2(g)$ = $Rb_2O_2(c)$	(298.16°–312.2° K.)	−102,000	−2.30	.....	.....	+57.40
2 $Rb(l)$ + $O_2(g)$ = $Rb_2O_2(c)$	(312.2°–840° K.)	−103,360	−4.61	.....	.....	+67.52
2 $Rb(l)$ + $O_2(g)$ = $Rb_2O_2(l)$	(840°–952° K.)	−101,680	−18.42	.....	.....	+105.91
2 $Rb(g)$ + $O_2(g)$ = $Rb_2O_2(l)$	(952°–1,500 K.)	−143,130	−31.08	.....	.....	+187.16
2 $Rb(c)$ + 3/2 $O_2(g)$ = $Rb_2O_3(c)$	(298.16°–312.2° K.)	−118,190	−11.51	.....	.....	+104.70
2 $Rb(l)$ + 3/2 $O_2(g)$ = $Rb_2O_3(c)$	(312.2°–760° K.)	−119,400	−12.66	.....	.....	+111.43
2 $Rb(l)$ + 3/2 $O_2(g)$ = $Rb_2O_3(l)$	(760°–952° K.)	−116,740	−27.63	.....	.....	+151.06
2 $Rb(g)$ + 3/2 $O_2(g)$ = $Rb_2O_3(l)$	(952°–1,500° K.)	−157,720	−39.14	.....	.....	+228.39
$Rb(c)$ + $O_2(g)$ = $RbO_2(c)$	(298.16°–312.2° K.)	−52,330	−10.36	.....	.....	+66.25
$Rb(l)$ + $O_2(g)$ = $RbO_2(c)$	(312.2°–685° K.)	−65,120	−11.51	.....	.....	+71.30
$Rb(l)$ + $O_2(g)$ = $RbO_2(l)$	(685°–952° K.)	−63,070	−18.42	.....	.....	+87.89
$Rb(g)$ + $O_2(g)$ = $RbO_2(l)$	(952°–1,500° K.)	−83,560	−24.18	.....	.....	+126.57
Ru(α, β, γ) + $O_2(g)$ = $RuO_2(c)$	(298.16°–1,500° K.)	−57,290	−6.91	.....	.....	+62.01
2 $Sm(c)$ + 3/2 $O_2(g)$ = $Sm_2O_3(c)$	(298.16°–1,623° K.)	−430,600	−4.60	.....	.....	+78.38
2 $Sm(l)$ + 3/2 $O_2(g)$ = $Sm_2O_3(c)$	(1,623°–2,000° K.)	−438,000	−4.60	.....	.....	+82.94
2 $Sc(c)$ + 3/2 $O_2(g)$ = $Sc_2O_3(c)$	(298.16°–1,673° K.)	−409,960	+7.78	−1.84	−0.30	+52.73
2 $Sc(l)$ + 3/2 $O_2(g)$ = $Sc_2O_3(c)$	(1,673°–2,000° K.)	−412,950	+18.47	−2.93	−.30	+21.88
$Se(c)$ + 1/2 $O_2(g)$ = $SeO(g)$	(298.16°–490° K.)	+9,280	−3.04	+4.40	+.30	−14.78
$Se(l)$ + 1/2 $O_2(g)$ = $SeO(g)$	(490°–1,027° K.)	+9,420	+8.70	.....	+.30	−44.50
1/2 $Se_2(g)$ + 1/2 $O_2(g)$ = $SeO(g)$	(1,027°–2,000° K.)	−7,400	−.37	.....	+.19	−.80
$Se(c)$ + $O_2(g)$ = $SeO_2(c)$	(298.16°–490° K.)	−53,770	+14.94	−9.41	−.20	+6.94
$Se(l)$ + $O_2(g)$ = $SeO_2(c)$	(490°–595° K.)	−53,640	+27.59	−14.31	.....	−25.05
$Se(l)$ + $O_2(g)$ = $SeO_2(g)$	(595°–1,027° K.)	−32,840	+6.79	.....	.....	−10.80
1/2 $Se_2(g)$ + $O_2(g)$ = $SeO_2(g)$	(1,027°–2,000° K.)	−49,000	−.74	.....	.....	+27.61
$Si(c)$ + 1/2 $O_2(g)$ = $SiO(g)$	(298.16°–1,683° K.)	−21,090	+3.84	+.16	−.295	−33.14
$Si(l)$ + 1/2 $O_2(g)$ = $SiO(g)$	(1,683°–2,000° K.)	−30,170	−7.78	−.12	+.25	−40.01
$Si(c)$ + $O_2(g)$ = SiO_2(α-*quartz*)	(298.16°–848° K.)	−210,070	+3.98	−3.32	+.605	+34.59
$Si(c)$ + $O_2(g)$ = SiO_2(β-*quartz*)	(848°–1,683° K.)	−209,920	−3.36	−.19	−.745	+53.44
$Si(l)$ + $O_2(g)$ − SiO_2(β-*quartz*)	(1,683°–1,883° K.)	−219,000	+.58	−.47	−.20	+46.58
$Si(l)$ + $O_2(g)$ = $SiO_2(l)$	(1,883°–2,000° K.)	−228,590	−15.66	.....	.....	+103.97
$Si(c)$ + $O_2(g)$ = SiO_2(α-*crist.*)	(298.16°–523° K.)	−207,330	+19.96	−9.75	−.745	−9.78
$Si(c)$ + $O_2(g)$ = SiO_2(β-*crist.*)	(523°–1,683° K.)	−209,820	−3.34	−.24	−.745	+53.35
$Si(l)$ + $O_2(g)$ = SiO_2(β-*crist.*)	(1,683°–2,000° K.)	−218,900	+.60	−.52	−.20	+46.49
$Si(c)$ + $O_2(g)$ = SiO_2(α-*trid.*)	(298.16°–390° K.)	−207,030	+22.29	−11.62	−.745	−15.64
$Si(c)$ + $O_2(g)$ = SiO_2(β-*trid.*)	(390°–1,683° K.)	−209,350	−1.59	−.54	−.745	+47.86
$Si(l)$ + $O_2(g)$ = SiO_2(β-*trid.*)	(1,683°–1,953° K.)	−218,430	+2.35	−.82	−.20	+41.00
2 $Ag(c)$ + 1/2 $O_2(g)$ = $Ag_2O(c)$	(298.16°–1,000° K.)	−7,740	−4.14	.....	.....	+27.84
2 $Ag(c)$ + $O_2(g)$ = $Ag_2O_2(c)$	(298.16°–500° K.)	−6,620	−3.22	.....	.....	+52.17
2 $Na(c)$ + 1/2 $O_2(g)$ = $Na_2O(c)$	(298.16°–371° K.)	−99,820	−7.51	+5.47	−.10	+50.43
2 $Na(l)$ + 1/2 $O_2(g)$ = $Na_2O(c)$	(371°–1,187° K.)	−100,150	+4.97	−2.45	−.10	+22.19
2 $Na(g)$ + 1/2 $O_2(g)$ = $Na_2O(c)$	(1,187°–1,190° K.)	−156,200	−20.72	.....	.....	+145.48
2 $Na(g)$ + 1/2 $O_2(g)$ = $Na_2O(l)$	(1,190°–2,000° K.)	−150,250	−23.03	.....	.....	+147.58
2 $Na(c)$ + $O_2(g)$ = $Na_2O_2(c)$	(298.16°–371° K.)	−122,500	−2.30	.....	.....	+57.51
2 $Na(l)$ + $O_2(g)$ = $Na_2O_2(c)$	(371°–733° K.)	−124,320	−5.76	.....	.....	+71.30

Coefficients in Free-Energy Equations (Continued)

Reaction and temperature range of validity	ΔH_0	2.303*a*	*b*	*c*	*I*
$2\,Na(l) + O_2(g) = Na_2O_2(l)$ (733°–1,187° K.)	−123,220	−20.72	.....	.....	+112.66
$2\,Na(g) + O_2(g) = Na_2O_2(l)$ (1,187°–1,500° K.)	−174,800	−31.08	.....	.....	+187.97
$Na(c) + O_2(g) = NaO_2(c)$ (298.16°–371° K.)	−63,040	−8.06	.....	.....	+56.98
$Na(l) + O_2(g) = NaO_2(c)$ (371°–1,000° K.)	−64,220	−11.51	.....	.....	+69.04
$Sr(c) + 1/2\,O_2(g) = SrO(c)$ (298.16°–1,043° K.)	−142,410	−6.79	+.305	+.675	+44.33
$Sr(l) + 1/2\,O_2(g) = SrO(c)$ (1,043°–1,657° K.)	−143,370	−2.42	−.38	+.675	+32.77
$Sr(g) + 1/2\,O_2(g) = SrO(c)$ (1,657°–2,000° K.)	−181,180	−8.24	−.38	+.675	+74.32
$Sr(c) + O_2(g) = SrO_2(c)$ (298.16°–1,000° K.)	−155,510	−11.40	+.305	+.675	+75.44
$S(rh) + 1/2\,O_2(g) = SO(g)$ (298.16°–368.6° K.)	+19,250	−1.24	+2.95	+.225	−18.84
$S(mon) + 1/2\,O_2(g) = SO(g)$ (368.6°–392° K.)	+19,200	−1.29	+3.31	+.225	−18.72
$S(l\lambda,\mu) + 1/2\,O_2(g) = SO(g)$ (392°–718° K.)	+20,320	+10.22	−.17	+.225	−50.05
$1/2\,S_2(g) + 1/2\,O_2(g) = SO(g)$ (298.16°–2,000° K.)	+3,890	+.07	.....	.....	−1.50
$S(rh) + O_2(g) = SO_2(g)$ (298.16°–368.6° K.)	−70,980	+.83	+2.35	+.51	−5.85
$S(mon) + O_2(g) = SO_2(g)$ (368.6°–392° K.)	−71,020	+.78	+2.71	+.51	−5.74
$S(l\lambda,\mu) + O_2(g) = SO_2(g)$ (392°–718° K.)	−69,900	+12.30	−.77	+.51	−37.10
$1/2\,S_2(g) + O_2(g) = SO_2(g)$ (298.16°–2,000° K.)	−86,330	+2.42	−.70	+.31	+10.71
$S(rh) + 3/2\,O_2(g) = SO_3(c—I)$ (298.16°–335.4° K.)	−111,370	−6.45	.....	.....	+88.32
$S(rh) + 3/2\,O_2(g) = SO_3(c—II)$ (298.16°–305.7° K.)	−108,680	−11.97	.....	.....	+94.95
$S(rh) + 3/2\,O_2(g) = SO_3(l)$ (298.16°–335.4° K.)	−107,430	−21.18	.....	.....	+113.76
$S(rh) + 3/2\,O_2(g) = SO_3(g)$ (298.16°–368.6° K.)	−95,070	+1.43	+0.66	+1.26	+16.81
$S(mon) + 3/2\,O_2(g) = SO_3(g)$ (368.6°–392° K.)	−95,120	+1.38	+1.02	+1.26	+16.93
$S(l\lambda,\mu) + 3/2\,O_2(g) = SO_3(g)$ (392°–718° K.)	−94,010	+12.89	−2.46	+1.26	−14.40
$1/2\,S_2(g) + 3/2\,O_2(g) = SO_3(g)$ (298.16°–1,500° K.)	−110,420	+3.02	−2.39	+1.06	+33.41
$2\,Ta(c) + 5/2\,O_2(g) = Ta_2O_5(c)$ (298.16°–2,000° K.)	−492,790	−17.18	−1.25	+2.46	+161.68
$Tc(c) + O_2(g) = TcO_2(c)$ (298.16°–500° K.)	−103,400	.....	.....	.....	+41.00
$Tc(c) + 3/2\,O_2(g) = TcO_3(c)$ (298.16°–500° K.)	−129,000	.....	.....	.....	+64.50
$2\,Tc(c) + 7/2\,O_2(g) = Tc_2O_7(c)$ (298.16°–392.7° K.)	−266,000	.....	.....	.....	+147.00
$2\,Tc(c) + 7/2\,O_2(g) = Tc_2O_7(l)$ (392.7°–500° K.)	−258,930	.....	.....	.....	+129.00
$Te(c) + 1/2\,O_2(g) = TeO(g)$ (298.16°–723° K.)	+43,110	+1.91	+.84	+.315	−27.22
$Te(l) + 1/2\,O_2(g) = TeO(g)$ (723°–1,360° K.)	+39,750	+6.08	+.09	+.315	−33.94
$1/2\,Te_2(g) + 1/2\,O_2(g) = TeO(g)$ (1,360°–2,000° K.)	+23,730	−.90	+.09	+.315	−.29
$Te(c) + O_2(g) = TeO_2(c)$ (298.16°–723° K.)	−78,090	−2.10	−2.35	−.20	+51.27
$Te(l) + O_2(g) = TeO_2(c)$ (723°–1,006° K.)	−81,530	+1.84	−3.10	−.20	+45.30
$Te(l) + O_2(g) = TeO_2(l)$ (1,006°–1,300° K.)	−82,090	−21.74	+.50	−.20	+113.04
$2\,Tl(\alpha) + O_2(g) = Tl_2O(c)$ (298.16°–505.5° K.)	−44,110	−6.91	.....	.....	+42.30
$2\,Tl(\beta) + O_2(g) = Tl_2O(c)$ (505.5°–573° K.)	−44,260	−6.91	.....	.....	+42.60
$2\,Tl(\beta) + 1/2\,O_2(g) = Tl_2O(l)$ (573°–576° K.)	−40,880	−13.82	.....	.....	+55.76
$2\,Tl(l) + 1/2\,O_2(g) = Tl_2O(l)$ (576°–773° K.)	−42,320	−11.51	.....	.....	+51.89
$2\,Tl(l) + 1/2\,O_2(g) = Tl_2O(g)$ (773°–1,730° K.)	−18,400	+11.51	.....	.....	−45.55
$2\,Tl(g) + 1/2\,O_2(g) = Tl_2O(g)$ (1,730°–2,000° K.)	−104,670	.....	.....	.....	+41.59
$2\,Tl(\alpha) + 3/2\,O_2(g) = Tl_2O_3(c)$ (298.16°–505.5° K.)	−99,410	−16.12	.....	.....	+119.09
$2\,Tl(\beta) + 3/2\,O_2(g) = Tl_2O_3(c)$ (505.5°–576° K.)	−99,560	−16.12	.....	.....	+119.39
$2\,Tl(l) + 3/2\,O_2(g) = Tl_2O_3(c)$ (576°–990° K.)	−101,010	−13.82	.....	.....	+115.55
$2\,Tl(l) + 3/2\,O_2(g) = Tl_2O_3(l)$ (990°–1,500° K.)	−94,550	−27.63	.....	.....	+150.39
$2\,Tl(\alpha) + 2\,O_2(g) = Tl_2O_4(c)$ (298.16°–505.5° K.)	−117,680	−23.03	.....	.....	+161.19
$2\,Tl(\beta) + 2\,O_2(g) = Tl_2O_4(c)$ (505.5°–576° K.)	−117,830	−23.03	.....	.....	+161.49
$2\,Tl(l) + 2\,O_2(g) = Tl_2O_4(c)$ (576°–1,000° K.)	−119,270	−20.72	.....	.....	+157.63
$Th(c) + O_2(g) = ThO_2(c)$ (298.16°–2,000° K.)	−294,350	−5.25	+.59	+.775	+62.81
$Sn(c) + 1/2\,O_2(g) = SnO(c)$ (298.16°–505° K.)	−68,600	−3.57	+1.65	−.10	+32.59
$Sn(l) + 1/2\,O_2(g) = SnO(c)$ (505°–1,300° K.)	−69,670	+3.06	−1.50	−.10	+18.39
$Sn(c) + 1/2\,O_2(g) = SnO(g)$ (298.16°–505° K.)	−1,000	−.97	+3.24	+.32	−17.41
$Sn(l) + 1/2\,O_2(g) = SnO(g)$ (505°–2,000° K.)	−2,070	+5.66	+.09	+.32	−31.62
$Sn(c) + O_2(g) = SnO_2(c)$ (298.16°–505° K.)	−142,010	−14.00	+2.45	+2.38	+90.74
$Sn(l) + O_2(g) = SnO_2(c)$ (505°–1,898° K.)	−143,080	−7.37	−.70	+2.38	+76.53
$Sn(l) + O_2(g) = SnO_2(l)$ (1,898°–2,000° K.)	−139,130	−21.97	+.50	−.20	+120.11
$Ti(\alpha) + 1/2\,O_2(g) = TiO(\alpha)$ (298.16°–1,150° K.)	−125,010	−4.01	−.29	+.83	+36.28
$Ti(\beta) + 1/2\,O_2(g) = TiO(\alpha)$ (1,150°–1,264° K.)	−125,040	+1.17	−1.55	+.83	+21.90
$Ti(\beta) + 1/2\,O_2(g) = TiO(\beta)$ (1,264°–2,000° K.)	−125,210	−1.77	−1.25	−.10	+30.83
$Ti(\alpha) + 1/2\,O_2(g) = TiO(g)$ (298.16°–1,150° K.)	+11,710	+3.71	+1.07	−.10	−35.50
$Ti(\beta) + 1/2\,O_2(g) = TiO(g)$ (1,150°–2,000° K.)	+11,680	+8.89	−.19	−.10	−49.88
$2\,Ti(\alpha) + 3/2\,O_2(g) = Ti_2O_3(\alpha)$ (298.16°–473° K.)	−360,660	+32.08	−23.49	−.30	−10.66
$2\,Ti(\alpha) + 3/2\,O_2(g) = Ti_2O_3(\beta)$ (473°–1,150° K.)	−369,710	−30.95	+2.62	+4.80	+162.79
$2\,Ti(\beta) + 3/2\,O_2(g) = Ti_2O_3(\beta)$ (1,150°–2,000° K.)	−369,760	−20.59	+.10	+4.80	+134.03
$3\,Ti(\alpha) + 5/2\,O_2(g) = Ti_3O_5(\alpha)$ (298.16°–450° K.)	−587,980	−4.19	−9.72	−.50	+131.05
$3\,Ti(\alpha) + 5/2\,O_2(g) = Ti_3O_5(\beta)$ (450°–1,150° K.)	−586,330	−18.31	+1.03	−.50	+159.98
$3\,Ti(\beta) + 5/2\,O_2(g) = Ti_3O_5(\beta)$ (1,150°–2,000° K.)	−586,420	−2.76	−2.75	−.50	+116.81
$Ti(\alpha) + O_2(g) = TiO_2$ (*rutile*) (298.16°–1,150° K.)	−228,360	−12.80	+1.62	+1.975	+82.81
$Ti(\beta) + O_2(g) = TiO_2$ (*rutile*) (1,150°–2,000° K.)	−228,380	−7.62	+.36	+1.975	+68.43
$W(c) + O_2(g) = WO_2(c)$ (298.16°–1,500° K.)	−137,180	−1.38	.....	.....	+45.56
$4\,W(c) + 11/2\,O_2(g) = W_4O_{11}(c)$ (298,16°–1,700° K.)	−745,730	−32.70	.....	.....	+321.84
$W(c) + 3/2\,O_2(g) = WO_3(c)$ (298.16°–1,743° K.)	−201,180	−2.92	−1.81	−.30	+70.89
$W(c) + 3/2\,O_2(g) = WO_3(l)$ (1,743°–2,000° K.)	−203,140	−35.74	+1.13	−.30	+173.27
$U(\alpha) + O_2(g) = UO_2(c)$ (298.16°–935° K.)	−262,880	−19.92	+3.70	+2.13	+100.54
$U(\beta) + O_2(g) = UO_2(c)$ (935°–1,045° K.)	−260,660	−4.28	−.31	+1.78	+55.50
$U(\gamma) + O_2(g) = UO_2(c)$ (1,045°–1,405° K.)	−262,830	−6.54	−.31	+1.78	+64.41
$U(l) + O_2(g) = UO_2(c)$ (1,405°–1,500° K.)	−264,790	−5.92	.....	.....	+63.50
$3\,U(\alpha) + 4\,O_2(g) = U_3O_8(c)$ (298.16°–935° K.)	−863,370	−56.57	+10.68	+5.20	+330.19
$3\,U(\beta) + 4\,O_2(g) = U_3O_8(c)$ (935°–1,045° K.)	−856,720	−9.67	−1.35	+4.15	+195.12
$3\,U(\gamma) + 4\,O_2(g) = U_3O_8(c)$ (1,045°–1,405° K.)	−863,230	−16.44	−1.35	+4.15	+221.79
$3\,U(l) + 4\,O_2(g) = U_3O_8(c)$ (1,405°–1,500° K.)	−869,460	−10.91	−1.35	+4.15	+208.82
$U(\alpha) + 3/2\,O_2(g) = UO_3$ (*hex*) (298.16°–935° K.)	−294,090	−18.33	+3.49	+1.535	+114.94
$U(\beta) + 3/2\,O_2(g) = UO_3$ (*hex*) (935°–1,045° K.)	−291,870	−2.69	−.52	+1.185	+69.90
$U(\gamma) + 3/2\,O_2(g) = UO_3$ (*hex*) (1,045°–1,400° K.)	−294,040	−4.95	−.52	+1.185	+78.80
$V(c) + 1/2\,O_2(g) = VO(c)$ (298.16°–2,000° K.)	−101,090	−5.39	−.36	+.53	+38.69
$V(c) + 1/2\,O_2(g) = VO(g)$ (298.16°–2,000° K.)	+52,090	+1.80	+1.04	+.35	−28.42
$2\,V(c) + 3/2\,O_2(g) = V_2O_3(c)$ (298.16°–2,000° K.)	−299,910	−17.98	+.37	+2.41	+118.83
$2\,V(c) + 2\,O_2(g) = V_2O_4(\alpha)$ (298.16°–345° K.)	−342,890	−11.03	+3.00	−.40	+117.38
$2\,V(c) + 2\,O_2(g) = V_2O_4(\beta)$ (345°–1,818° K.)	−345,330	−24.36	+1.30	+3.545	+155.55
$2\,V(c) + 2\,O_2(g) = V_2O_4(l)$ (1,818°–2,000° K.)	−339,880	−59.59	+3.00	−0.40	+264.42
$6\,V(c) + 13/2\,O_2(g) = V_6O_{13}(c)$ (298.16°–1,000° K.)	−1,076,340	−95.33	.....	.....	+557.61
$2\,V(c) + 5/2\,O_2(g) = V_2O_5(c)$ (298.16°–943° K.)	−381,960	−41.08	+5.20	+6.11	+228.50
$2\,V(c) + 5/2\,O_2(g) = V_2O_5(l)$ (943°–2,000° K.)	−365,840	−38.91	+3.25	−.50	+207.54
$2\,Y(c) + 3/2\,O_2(g) = Y_2O_3(c)$ (298.16°–1,773° K.)	−419,600	+2.76	−1.73	−.30	+66.36
$2\,Y(l) + 3/2\,O_2(g) = Y_2O_3(c)$ (1,773°–2,000° K.)	−422,850	+13.36	−2.75	−.30	+35.56
$Zn(c) + 1/2\,O_2(g) = ZnO(c)$ (298.16°–692.7° K.)	−84,670	−6.40	+.84	+.99	+43.25
$Zn(l) + 1/2\,O_2(g) = ZnO(c)$ (692.7°–1,180° K.)	−85,520	−1.45	−.36	+.99	+31.25
$Zn(g) + 1/2\,O_2(g) = ZnO(c)$ (1,180°–2,000° K.)	−115,940	−7.28	−.36	+.99	+74.94
$Zr(\alpha) + O_2(g) = ZrO_2(\alpha)$ (298.16°–1,135° K.)	−262,980	−6.10	+.16	+1.045	+65.00
$Zr(\beta) + O_2(g) = ZrO_2(\alpha)$ (1,135°–1,478° K.)	−264,190	−5.09	−.40	+1.48	+63.58
$Zr(\beta) + O_2(g) = ZrO_2(\beta)$ (1,478°–2,000° K.)	−262,290	−7.76	+.50	−.20	+69.50

REFRACTORY MATERIALS

Borides

Name	Formula	Molecular weight	Melting point, °C	Crystalline form	Lattice parameter, Å	X-ray density, g/cm³
Chromium boride	CrB_2	73.65	1,850[29][a]	Hexagonal AlB_2 type [C 32]	a = 2.969 c = 3.066[39]	5.16
Hafnium boride	HfB_2	200.14	3,100[31]	Hexagonal AlB_2 type [C 32]	a = 3.14[6] c = 3.47	10.5
Molybdenum boride	MoB	106.77	2,180[40]	Tetragonal	a = 3.110[41] c = 16.95	8.77
	MoB_2	117.59	2,100[40]	Hexagonal AlB_2 type [C 32]	a = 3.05 c = 3.113[42]	7.78
	Mo_2B	202.72	2,000[40] (decomposes)	Tetragonal $CuAl_2$ type [C 16]	a = 5.543[41] c = 4.735	9.31
Niobium boride	NbB	103.73	>0.2000[3]	Orthorhombic	a = 3.298 b = 8.724 c = 3.137[35]	
	NbB_2	114.55	2,900[36] (decomposes)	Hexagonal AlB_2 type [C 32]	a = 3.086[28] c = 3.306	7.21
Tantalum boride	TaB	191.77	>0.2000[3]	Orthorhombic	a = 3.276 b = 8.669 c = 3.157[37]	14.29
	TaB_2	202.59	3,000[6]	Hexagonal AlB_2 type [C 32]	a = 3.088[28] c = 3.241	12.60
Thorium boride	ThB_4	275.53	>0.2500[6]	Tetragonal D^5_{4k}-P4/mbm	a = 7.256[43] c = 4.113	8.45
Titanium boride	TiB_2	69.54	2,980[6]	Hexagonal AlB_2 type [C 32]	a = 3.028[28] c = 3.228	4.52
Tungsten boride	WB	194.68	2,860[29]	Tetragonal	a = 3.115[6] c = 16.92	16.0
	W_2B	378.54	2,770[29]	Tetragonal $CuAl_2$ type [C 16]	a = 5.564[41] c = 4.740	16.72
Uranium boride	UB_{12}	367.91	>0.1500[6] (decomposes)	Face-centered cubic	a = 7.473[44]	5.82
Vanadium boride	VB_2	72.59	2,100[29]	Hexagonal AlB_2 type [C 32]	a = 2.998[28] c = 3.057	5.10
Zirconium boride	ZrB_2	112.86	3,040 ± 50[30]	Hexagonal AlB_2 type [C 32]	a = 3.169[28] c = 3.530	6.09
	ZrB_{12}	221.06	2,680[30]	Face-centered cubic	a = 7.408[33]	

Name	Thermal conductivity, $cal\text{-}sec^{-1}\text{-}cm^{-2}\text{-}cm\text{-}°C^{-1}$	Electrical resistivity, microhm-cm	Ductility relative scale[b]	Resistance to oxidation[c]	Hardness[d]
Chromium boride		21 at room temperature[38]			1,800 kg/mm^2[29]
Hafnium boride		10[31]	3[3]		
Molybdenum boride		α-MoB = 45 at room temperature[40]	3[3]	3[3]	8 Mohs[20]
		β-MoB = 25 at room temperature			1,570 kg/mm^2[29]
		45 at room temperature[40]			1,280 kg/mm^2[29]
		40 at room temperature[40]			8–9 Mohs[20]
Niobium boride		6.45 at room temperature[36]			
	0.040 at 20°C[36]	32.0 at room temperature[36]			>0.8 Mohs[36]
Tantalum boride		100 at room temperature[38]	3[3]	3[3]	
	0.026 at 20°C[6]	68 at room temperature[38]			
Thorium boride					
Titanium boride	0.0624 at 200°C[28] (15% porosity)	28.4[28]	3[3]	2–3[3]	3,400 kg/mm^2[29]
Tungsten boride			3[3]	3[3]	9 Mohs[6]
Uranium boride					
Vanadium boride	16 at 20°C[34]				8–9 Mohs[6]
Zirconium boride	0.0550 at 200°C[28] 0.029 at room temperature[33]	9.2 at 20°C[31] 60–80 at room temperature[30]	2[3]	2–3[3]	8 Mohs[32]

Carbides

Name	Formula	Molecular weight	Melting point, °C	Crystalline form	Lattice parameter, Å	X-ray density, g/cm³
Boron carbide	B_4C	55.29	2,450	Hexagonal	a_0 = 5.60[6] c_0 = 12.12	2.52
Chromium carbide	Cr_3C_2	180.05	1,895[15]	Orthorhombic ($D5_{10}$)	a = 2.82 b = 5.53 c = 11.47[16]	6.7
	Cr_7Cr_3	400.01	1,780[15]	Hexagonal [C_{3v}^4]	a = 14.01 c = 4.532[17]	6.92
Graphite	C	12.01	3,700 ± 100[3]	Hexagonal	Orthohexagonal axes a_0 = 2.46 b_0 = 4.28 c_0 = 6.71	2.25
Hafnium carbide	HfC	190.51	3,890[9]	Cubic NaCl type (B1)	a = 4.46[10]	12.7

REFRACTORY MATERIALS (Continued)

Carbides (Continued)

Name	Formula	Molecular weight	Melting point, °C	Crystalline form	Lattice parameter, Å	X-ray density, g/cm³
	MoC	107.96	2,695[9]	Face-centered cubic	a = 4.28[19]	
Molybdenum carbide	Mo_2C	203.91	2,690[6]	Hexagonal (L'3)	a = 3.002 c = 4.724[18]	9.2
Niobium carbide	NbC	104.92	3,500[9]	Cubic NaCl type (B1)	a = 4.461	7.85
Silicon carbide						
β	SiC	40.07	Trans. to α at 2,100	Face-centered cubic	a_o = 4.3590[6]	3.22
α	SiC		2,700[6]	Hexagonal (Wurtzite)	a_o = 3.081[6] c_o = 5.0394	
Tantalum carbide	TaC	192.96	3,880[9], 4,730[1]	Cubic NaCl type (B1)	a = 4.455[14]	[illegible]
Thorium carbide	ThC	244.06	2,625[23]	Cubic NaCl type (B1)	a = 5.34[9]	10.67
	ThC_2	256.07	2,655[29]	Monoclinic [C 2/e]	a = 6.53[24] b = 4.24	9.6
Titanium carbide	TiC	59.91	3,160[4]	Cubic NaCl type (B1)	a = 4.32[6]	4.938
Tungsten carbide	W_2C	379.73	2,730[21]	Hexagonal (L'3)	a = 2.98[6] c = 4.71	17.34
	WC	195.87	2,630[21] (decomposes)	Hexagonal (L'3)	a = 2.900[22] c = 2.831	15.77
Uranium carbide	UC	250.08	2,450–2,500[25]	Cubic NaCl type (B1)	a = 4.955[26]	13.6
	UC_2	262.09	2,350–2,400[25]	Body-centered tetragonal $CaCl_2$ type	a = 3.517[27] c = 5.987	11.68
Vanadium carbide	VC	62.96	2,830[1]	Cubic NaCl type (B1)	a = 4.16[6]	5.8
Zirconium carbide	ZrC	103.23	3,030[4]	Cubic NaCl type (B1)	a = 4.689[5]	6.44

Name	Thermal conductivity, cal-sec⁻¹-cm⁻²-cm-°C⁻¹	Electrical resistivity, microhm-cm	Ductility relative scale[b]	Resistance to oxidation[c]	Hardness[d]
Boron carbide	0.05 at 20–425°C[1]	0.30–0.80[1]	3[3]	3[3]	9.3 Mohs
			3[3]	3[3]	1,300 kg/mm²[8]
Chromium carbide			3[3]	3[3]	
			3[3]	4[3]	
Graphite	0.268–0.451 at room temperature	65 at room temperature[71]	3[3]	3[3]	
Hafnium carbide		109 at room temperature[1]	3[3]	5[3]	1,800 kg/mm²[6]
Molybdenum carbide		97[6]	3[3]	5[3]	>0.7 Mo[illegible][20]
Niobium carbide	0.034 at room temperature[7]	74[6]	2–3[3]	3[3]	2,470 kg/mm²[13]
Silicon carbide			3[3]	2[3]	9.2 Mohs
β	0.10 at 20–425°C[1]	107–200 ohm cm[1] at room temperature			
Tantalum carbide	0.053 at room temperature[6]	30[6]	2–3[3]	3[3]	1,800 kg/mm²[8]
Thorium carbide			3[3]	3[3]	
Titanium carbide	0.049 at room temperature[6]	180–250[4]	3[3]	3[4]	3,200 kg/mm²[8]
Tungsten carbide		80[6]	3[3]	5[3]	3,000 kg/mm²[6]
		53[6]	3[3]	5[3]	2,400 kg/mm²[6]
			3[3]		
Uranium carbide				5[3]	
Vanadium carbide		150[6]	3[3]		2,800 kg/mm²[11]
					2,100 kg/mm²[12]
Zirconium carbide	0.049 at room temperature[6]	70 at room temperature[4]	3[3]	3[3]	2,600 kg/mm²[6]

Nitrides

Name	Formula	Molecular weight	Melting point, °C	Crystalline form	Lattice parameter, Å	X-ray density, g/cm³
Boron nitride	BN	24.83	2,730[6]	Hexagonal (B 12)	a = 2.51 ± .02 c = 6.70 ± .04[52]	2.25
Chromium nitride	CrN	66.02	1,500 (decomposes)	Cubic NaCl type (B1)	a = 4.140[6]	6.14
Hafnium nitride	HfN	192.60	3,310[6]			
Niobium nitride	NbN	106.92	2,050[46]	Cubic NaCl type (B1)	a = 4.41–4.375[6]	7.28
Tantalum nitride	Ta_2N	375.77	3,090[46]	Hexagonal	a = 3.05[48] c = 4.95[53]	14.1
Thorium nitride	ThN	246.13	2,630[48]	Cubic NaCl type (B1)	a = 5.2[48]	
Titanium nitride	TiN	61.91	2,930 N_2 liberated on melting[46]	Cubic NaCl type (B1)	a = 4.23[53]	5.43
Uranium nitride	UN		2,650	Cubic NaCl type (B1)	a = 4.880[58]	14.32
Vanadium nitride	VN	64.96	2,050[46]	Cubic NaCl type (B1)	a = 4.129[6]	6.102
Zirconium nitride	ZrN	105.22	2,980[47]	Cubic NaCl type (B1)	a = 4.567[53]	7.349

Nitrides (Continued)

Name	Thermal conductivity, cal-sec^{-1}-cm^{-2}-cm-°C^{-1}	Electrical resistivity, microhm-cm	Ductility relative scale[b]	Resistance to oxidation[c]	Hardness[d]
Boron nitride		1,900 at 2,000°C(1)	3(3)	2(3)	2.0 Mohs
Chromium nitride					
Hafnium nitride					
Niobium nitride		200 at room temperature(46)	5(3)	3(3)	+8 Mohs(46)
Tantalum nitride		135 at room temperature(47)	3(3)	5(3)	+8
Thorium nitride					
Titanium nitride		21.7 at room temperature	3(3)	3(3)	Between 9 and 10 Mohs(46)
Uranium nitride					
Vanadium nitride		200 at room temperature(46)	3(3)		
Zirconium nitride		13.6 at room temperature(46)	3(3)	3(3)	+8 Mohs(6)

Oxides

Name	Formula	Molecular weight	Melting point, °C	Crystalline form	Lattice parameter, Å	X-ray density, g/cm³
Aluminum oxide	Al_2O_3	101.92	2,015(1)	Rhombohedral	a = 5.13 axial angle = 55° 6′	3.965
Beryllium oxide	BeO	25.02	2,550(1)	Hexagonal	a = 2.70 c = 4.39	3.03
Cerium oxide	CeO_2	172.3	2,600(1)	Face-centered cubic	a = 5.41	7.13
Chromic oxide	Cr_2O_3	152.02	2,265(1)	Rhombohedral	a = 5.38 axial angle = 54° 50′	5.21
Hafnium oxide	HfO_2	210.6	2,777(1)	Face-centered cubic	a = 5.11	9.68
Magnesium oxide	MgO	40.32	2,800(1)	Face-centered cubic	a = 4.20	3.58
Silicon oxide	SiO_2	60.06	1,728(1)	Hexagonal	a = 4.90 c = 5.39	2.32 (low cristobalite)
Thorium oxide	ThO_2	264.12	3,300(1)	Face-centered cubic	a = 5.59	9.69
Titanium oxide	TiO_2	79.90	1,840(1)	Tetragonal	a = 4.58 c = 2.98	4.24
Uranium oxide	UO_2	270.07	2,280(1)	Face-centered cubic	a = 5.47	10.96
Zirconium oxide	ZrO_2	123.22	2,677(1)	Monoclinic	a = 5.21 c = 5.37	5.56

Name	Thermal conductivity, cal-sec^{-1}-cm^{-2}-cm-°C^{-1}	Electrical resistivity, microhm-cm	Ductility relative scale[b]	Resistance to oxidation[c]	Hardness[d]
Aluminum oxide	0.0723 at 100°C(2)	1×10^{22} at 14°C(1) 3×10^{19} at 300°C(1) 3.5×10^{14} at 800°C(1)	3(3)	1(3)	9 Mohs(1)
Beryllium oxide	0.525 at 100°C(2)	4×10^{14} at 600°C(1) 5×10^{12} at 1,100°C(1) 8×10^{8} at 2,100°C(1)			9 Mohs(1)
Cerium oxide		6.5×10^{10} at 800°C(1) 3.4×10^{8} at 1,200°C(1)			6 Mohs(1)
Chromium oxide		1.3×10^{9} at 350°C(1) 2.3×10^{7} at 1,200°C(1) 6.8×10^{3} at 600°C(1) 4.5×10^{3} at 1,100°C(1)	3(3)	1(3)	
Hafnium oxide		5×10^{15} at 400°C(1) 1×10^{3} at 1,500°C(1)			
Magnesium oxide	0.0860 at 100°C(2)	2×10^{14} at 850°C(1) 3×10^{13} at 980°C(1) 4.5×10^{8} at 2,100°C(1)			6 Mohs(1)
Silicon oxide		1×10^{21} at 20°C(1) 7×10^{12} at 600°C(1) 4×10^{5} at 1,300°C(1) (vitreous)	3(4) (vitreous)	1(4) (vitreous)	6–7 Mohs(1) (cristobalite)
Thorium oxide	0.0245 at 100°C(2)	2.6×10^{13} at 550°C(1) 8×10^{11} at 800°C(1) 1.5×10^{10} at 1,200°C(1)			6.5 Mohs(1)
Titanium oxide	0.0156 at 100°C(2)	1.2×10^{10} at 800°C(1) 8.5×10^{6} at 1,200°C(1)			5.5–6.0 Mohs(1)
Uranium oxide	0.0234 at 100°C(2)	3.8×10^{10} at 20°C(1) 5×10^{8} at 500°C(1)			
Zirconium oxide	0.00466 at 100°C(2)	1×10^{12} at 385°C(1) 2.2×10^{10} at 700°C(1) 3.6×10^{8} at 1,200°C(1)	3(3)	1(3)	6.5 Mohs(1)

REFRACTORY MATERIALS (Continued)

Silicides

Name	Formula	Molecular weight	Melting point, °C	Crystalline form	Lattice parameter, Å	X-ray density g/cm³
Chromium silicide	$CrSi_2$	108.13	1,570[51]	Hexagonal	a = 4.42 c = 6.35[51]	
	CrSi	80.07	1,870 (decomposes in presence of C)	Tetragonal (C 11b)	a = 3.20 c = 7.86[6]	
Molybdenum silicide	$MoSi_2$	152.07	1,870 (decomposes in presence of C)	Tetragonal (C 11b)	a = 3.20 c = 7.86[64][51]	6.24
Niobium silicide	$NbSi_2$		1,950[50]	Hexagonal $CrSi_2$ type	a = 4.785 c = 6.576[51]	5.29
Tantalum silicide	$TaSi_2$	237.00	2,400[50]	Hexagonal $CrSi_2$ type	a = 4.773 c = 6.552[51]	8.83
Titanium silicide	$TiSi_2$	104.02	1,540[49]	Orthorhombic	a = 8.24 b = 4.79 c = 8.52[59]	4.13
	Ti_5Si_3	323.68	2,120[49]	Hexagonal	a = 7.465 c = 5.162[60]	4.32
Tungsten silicide	WSi_2	240.04	2,050[50]	Tetragonal $MoSi_2$ Structure (C 11b)	a = 3.21 c = 7.83[66]	9.3
Uranium silicide	βUSi_2	294.19	1,700[6]	Hexagonal	a = 3.85[6] c = 4.06	9.25
	U_3Si_2	770.33	1,665[6]	Tetragonal	a = 7.3151[6] c = 3.8925	12.20
Vanadium silicide	VSi_2	107.07	1,750[50]	Hexagonal $CrSi_2$ type	a = 4.562 c = 6.359[51]	4.71

Name	Thermal conductivity, $cal\cdot sec^{-1}\cdot cm^{-2}\cdot cm\cdot °C^{-1}$	Electrical resistivity, microhm-cm	Ductility relative scale[b]	Resistance to oxidation[c]	Hardness[d]
Chromium silicide					1,150 kg/mm^2 [50]
Molybdenum silicide	0.075 at room temperature to 200°C[6]	21.5 at room temperature 18.9 at −80°C[69]	2[3]	1[3]	1,290 kg/mm^2 [50]
Niobium silicide		6.3[68]	2[3]	4[3]	1,050 kg/mm^2 [50]
Tantalum silicide		8.5[68]	2[3]	3[3]	1,560 kg/mm^2 [50]
Titanium silicide		123 (hot pressed)[1]	3[3] 3[3]	4[3] 4[3]	870 kg/mm^2 [70] 986 kg/mm^2 [49]
Tungsten silicide		33.4[68]		1–2[3]	1,310 kg/mm^2 [68] 1,090 kg/mm^2 [68]
Uranium silicide					
Vanadium silicide		9.5[68]			1,090 kg/mm^2 [50]

[a]Numbers in parentheses refer to references at end of table.
[b]Ductility – 1: capable of being severely drawn, rolled, or otherwise worked without failure; 2: capable of withstanding slight deformation, or consisting of individually ductile crystals fragilely bound together; 3: incapable of being worked, of glasslike brittleness.
[c]Resistance to oxidation – classed according to the temperature range in which the rate of attack by air would cause severe erosion or failure of the coated specimen within a few hours. 1: above 1,700°C; 2: 1,400 to 1,700°C; 1,100 to 1,400°C; 4: 800 to 1,000°C; 5: 500 to 800°C.
[d]Microhardness values taken with 100-g load.

REFERENCES

1. **Campbell, I. E.,** *High Temperature Technology,* Electrochemical Society, John Wiley & Sons New York, 1956.
2. **Kingery, W. D., Franch, J., Coble, R. L., and Vasilos, T.,** Thermal conductivity: X, data for several pure oxide materials corrected to zero porosity, *J. Am. Ceram. Soc.,* 37, 2, 107, 1954.
3. **Powel, C. F., Campbell, I. E., and Gonser, B. W.,** *Vapor Plating,* John Wiley & Sons, New York, 1955.
4. **Friederich, E. and Sittig, L.,** *Z. Anorg. Allg. Chem.,* 144, 169, 1925.
5. **Norton, J. T. and Morory, A. L.,** *Trans. Am. Inst. Min. Metall. Pet. Eng.,* 185, 133, 1949.
6. **Schwarzkopf, P. and Kieffer, R.,** *Refractory Hard Metals,* Macmillan, New York, 1953.
7. **Schwarzkopf, P. and Sindeband, S. J.,** Electrochemical Society, 97th Meeting, Cleveland, Ohio, 1950.
8. **Kieffer, R. and Kölbl, F.,** *Powder Metall. Bull.,* 4, 4, 1949.
9. **Agte, C. and Alterthum, H.,** *Z. Tech. Phys.,* 11, 182, 1930.
10. **Becker, K. and Ebert, F.,** *Z. Phys.,* 31, 268, 1925.
11. **Ruff, O. and Martin, W.,** *Z. Angew. Chem.,* 25, 49, 1912.
12. **Hinnuber, J.,** *Z. VDI,* 92, 111, 1950.

13. Foster, L. S., Forbes, L. W., Jr., Friar, L. B., Moody, L. S., and Smith, W. H., *J. Am. Ceram. Soc.*, 33, 27, 1950.
14. Ellinger, F. H., *Trans. Am. Soc. Met.*, 31, 89, 1943.
15. Bloom, D. S. and Grant, N. J., *Trans. Am. Inst. Min. Metall. Pet. Eng.*, 188, 41, 1950.
16. Hellstrom, K. and Westgren, A., *Kem. Tidskr.*, 45, 141, 1933.
17. Westgren, A., *Jernkontorets Ann.*, 119, 231, 1935.
18. Kuo, K. and Hagg, G., *Nature*, 170, 245, 1952.
19. Nowotny, H. and Kieffer, R., *Z. Anorg. Allg. Chem.*, 267, 261, 1952.
20. Weiss, G., *Ann. Chem.*, 1, 446, 1946.
21. Brewer, L., Bromely, L. A., Gilles, P. W., and Lofgren, N. L., *The Chemistry and Metallurgy of Miscellaneous Materials: Thermodynamics*, McGraw-Hill, New York, 1950.
22. Becker, K., *Z. Phys.*, 51, 481, 1928.
23. Wilhelm, H. A. and Chiotti, P., *Trans. Am. Soc. Met.*, 42, 1295, 1950.
24. Hunt, E. B. and Rundle, R. E., *J. Am. Chem. Soc.*, 73, 4777, 1951.
25. Mallet, W., Gerds, A. F., and Nelson, H. R., *J. Electrochem. Soc.*, 99, 197, 1952.
26. Litz, L. M., Gurrett, A. B., and Croxton, F. C., *J. Am. Chem. Soc.*, 70, 1718, 1948.
27. Rundle, R. E., Baenziger, N. C., Wilson, A. S., and McDonald, R. A., *J. Am. Chem. Soc.*, 70, 99, 1948.
28. Norton, J. T., Blumenthal, H., and Sindeband, S. J., *Trans. Am. Inst. Min. Metall. Pet. Eng.*, 185, 749, 1949.
29. Honak, E. R., Thesis, Tech. Hochsch. Graz, 1951.
30. Glaser, F. W. and Post, B., *J. Met.*, 1953.
31. Moers, K., *Z. Anorg. Allg. Chem.*, 198, 262, 1931.
32. Andrieux, L., *Rev. Mét.*, 45, 49, 1948.
33. Post, B. and Glaser, F. W., *J. Met.*, 4, 631, 1952.
34. Moers, K., *Z. Anorg. Allg. Chem.*, 198, 243, 1931.
35. Anderson, L. H. and Kiessling, R., *Acta Chem. Scand.*, 4, 160, 209, 1950.
36. Glaser, F. W., *J. Met.*, 4, 391, 1952.
37. Kiessling, R., *Acta Chem. Scand.*, 3, 603, 1949.
38. Moers, K., *Z. Anorg. Allg. Chem.*, 198, 262, 1931.
39. Kiessling, R., *Acta Chem. Scand.*, 3, 595, 1949.
40. Steintz, R., *J. Met.*, 4, 148, 1952.
41. Kiessling, R., *Acta Chem. Scand.*, 1, 893, 1947.
42. Bertaut, F. and Blum, P., *Acta Crystallogr.*, 4, 72, 1951.
43. Zalkin, A. and Templeton, D. H., *J. Chem. Phys.*, 18, 391, 1950.
44. Bertaut, F. and Blum, P., *Comptes rendus*, 229, 666, 1949.
45. Baumann, H. N., Jr., Electrochemical Society, 99th Meeting, Washington, D.C., April 1951.
46. Friederich, E. and Sittig, L., *Z. Anorg. Allg. Chem.*, 143, 293, 1925.
47. Agte, C. and Moers, K., *Z. Anorg. Chem.*, 198, 233, 1931.
48. Chiotti, P., *J. Am. Ceram. Soc.*, 35, 123, 1952.
49. Hansen, M., Klasler, H. D., and McPherson, D. J., *Trans. Am. Soc. Met.*, 44, 518, 1952.
50. Cerwenka, E., Thesis, Tech. Hochsch. Graz, 1951.
51. Wallbaum, H. J., *Z. Metallkd.*, 33, 378, 1941.
52. Pease, R. S., *Acta Crystallogr.*, 5, 356, 1952.
53. van Arkel, A. E., *Physica*, 4, 296, 1924.
54. Horn, F. H. and Ziegler, W. T., *J. Am. Chem. Soc.*, 69, 2762, 1947.
55. Pauling, L., Killeffer, D. H., and Linz, A., *Molybdenum Compounds*, Interscience, New York, 1952.
56. Hagg, G., *Z. Phys. Chem. Abt. B*, 7, 339, 1930.
57. Kiessling, R. and Lier, Y. H., *J. Met.*, 3, 639, 1951.
58. Rundle, R. E., Baenziger, N. C., Wilson, A. S., and McDonald, R. A., *J. Am. Chem. Soc.*, 70, 99, 1941.
59. Laues, F. and Wallbaum, H. J., *Z. Kristallogr. A*, 101, 78, 1939.
60. Pietrokowsky, P. and Duwez, P., *J. Met.*, 3, 772, 1951.
61. Naray Szako, S. V., *Z. Kristallogr. A*, 97, 223, 1937.
62. Lundin, C. E., McPherson, D. J., and Hansen, M., American Society for Metals, 34th Ann. Convention, Preprint No. 41, 1952.
63. Wallbaum, H. J., *Z. Metallkd.*, 31, 362, 1939.
64. Zachariasen, W. H., *Z. Phys. Chem.*, 128, 39, 1927.
65. Templeton, D. H. and Dauben, C. H., *Acta Crystallogr.*, 3, 261, 1950.
66. Nowotny, H., Kieffer, R., and Schachner, H., *Mh. Chem.*, 83, 1243, 1952.
67. Brauer, G. and Mitices, A., *Z. Anorg. Allg. Chem.*, 249, 325, 1942.
68. Gallistl, E., Thesis, Tech. Hochsch. Graz, 1951.
69. Glaser, F. W., *J. Appl. Phys.*, 22, 103, 1951.
70. Cerwenka, E., Thesis, Tech. Hochsch. Graz, 1951.
71. Currie, L. M., Hamister, V. C., and MacPherson, H. G., paper presented at the United Nations International Conference on The Peaceful Uses of Atomic Energy, Geneva, 1955.

THERMODYNAMIC PROPERTIES OF ELEMENTS AND OXIDES

Thermodynamic calculations over a wide range of temperatures are generally made with the aid of algebraic equations representing the characteristic properties of the substances being considered. The necessary integrations and differentiations, or other mathematical manipulations, are then most easily effected.

The most convenient starting point in making such calculations for a given substance is the heat capacity at constant pressure. From this quantity and a knowledge of the properties of any phase transitions, the other thermodynamic properties may be computed by the well-known equations given in standard texts on thermodynamics.

Users of the following equations and tables are cautioned that the units for a, b, c, and d are cal/g mole, whereas those for A are Kcal/g mole. The necessary adjustment must be made when the data are substituted into the equations.

Empirical heat capacity equations are generally of the form of a power series with the absolute temperature T as the independent variable:

$$C_p = a' + (b' \times 10^{-3})T + (c' \times 10^{-6})T^2$$

or

$$C_p = a'' + (b'' \times 10^{-3})T + \frac{d \times 10^5}{T^2}$$

Since both forms are used in the ensuing, let

(1) $$C_p = a + (b \times 10^{-3})T + (c \times 10^{-6})T^2 + \frac{d \times 10^5}{T^2}$$

The constants a, b, c, and d are to be determined either experimentally or by some theoretical or semi-empirical approach.

The heat content or enthalpy H is determined from the heat capacity by a simple integration over the range of temperatures for which (1) is applicable. Thus, if 298°K is taken as a reference temperature,

(2) $$H_T - H_{298} = \int_{298}^{T} C_p dT$$

$$= a(T - 298) + \frac{1}{2}(b \times 10^{-3})(T^2 - 298^2) + \frac{1}{3}(c \times 10^{-6})(T^3 - 298^3) - (d \times 10^5)\left(\frac{1}{T} - \frac{1}{298}\right)$$

$$= aT + \frac{1}{2}(b \times 10^{-3})T^2 + \frac{1}{3}(c \times 10^{-6})T^3 - \frac{d \times 10^5}{T} - A$$

where all the constants on the right hand side of the equation have been incorporated in the term −A.

In general, the enthalpy is given by a sum of terms such as (2) for each phase of the substance involved in the temperature range considered plus terms which represent the heats of transitions:

$$H_T - H_{298} = \sum \int_{T_1}^{T_2} C_p dT + \sum \Delta H_{tr}$$

In a similar manner, the entropy Si is obtained from (1) by performing the integration

(3) $$S_T - S_{298} = \int_{298}^{T} (C_p/T)dt$$

$$= a\ln(T/298) + (b \times 10^{-3})(T - 298) + \frac{1}{2}(c \times 10^{-6})(T^2 - 298^2) - \frac{1}{2}(d \times 10^5)\left(\frac{1}{T^2} - \frac{1}{298^2}\right)$$

$$= a\ln T + (b \times 10^{-3})T + \frac{1}{2}(c \times 10^{-6})T^2 - \left(\frac{\frac{1}{2}(d \times 10^5)}{T^2}\right) - B'$$

or

(4) $$S_T = 2.303\, a \log T + (b \times 10^{-3})T + \frac{1}{2}(c \times 10^{-6})T^2 - \frac{1}{2}\left(\frac{d \times 10^{-5}}{T^2}\right) - B$$

where

(5) $$B = B' - S_{298}$$

From the definition of free energy F:

$$F = H - TS$$

the quantity

$$F_T - H_{298} = (H - H_{298}) - TS_T$$

is obtained from (2) and (4):

(6) $$F_T - H_{298} = -2.303aT \log T - \frac{1}{2}(b \times 10^{-3})T^2 - \frac{1}{6}(c \times 10^{-6})T^3 - \frac{\frac{1}{2}(d \times 10^5)}{T} + (B + a)T - A$$

and also the free energy function

$$\frac{F_T - H_{298}}{T} = -2.303a \log T - \frac{1}{2}(b \times 10^{-3})T - \frac{1}{6}(c \times 10^{-6})T^2 - \frac{\tfrac{1}{2}(d \times 10^5)}{T^2} + (B + a) - \frac{A}{T} \quad (7)$$

In the following 2 tables there has been collected the values of the constants. The first column lists the element or the oxide. The second column gives the phase to which they are applicable. The third, fourth, and fifth columns specify the thermodynamic properties for the transition to the succeeding phase. In column 6, the value of the entropy at 298.15°K, the reference temperature, is given. The remaining columns, except for the last, give the values of the constants a, b, c, d, A, and B required in the thermodynamic equations.

All values throughout the table which represent estimates have been enclosed in parentheses.

The heat capacities at temperatures beyond the range of experimental determination were estimated by extrapolation. Where no experimental values were found, use of analogy with compounds of neighboring elements in the Periodic Table was employed.

THERMODYNAMIC PROPERTIES OF THE ELEMENTS*

Element	Phase	Temperature of Transition (°K)	Heat of Transition (kcal/g mole)	Entropy of Transition (e.u.)	Entropy at 298°K (e.u.)	cal/g mole a	cal/g mole b	cal/g mole c	cal/g mole d	A (kcal/g mole)	B(e.u.)
Ac	solid	(1090)	(2.5)	(2.3)	(13)	(5.4)	(3.0)	—	—	(1.743)	(18.7)
	liquid	(2750)	(70)	(25)	—	(8)	—	—	—	(0.295)	(31.3)
Ag	solid	1234	2.855	2.313	10.20	5.09	1.02	—	0.36	1.488	19.21
	liquid	2485	60.72	24.43	—	7.30	—	—	—	0.164	30.12
	gas	—	—	—	—	(4.97)	—	—	—	(−66.34)	(−12.52)
Al	solid	931.7	2.57	2.76	6.769	4.94	2.96	—	—	1.604	22.26
	liquid	2600	67.9	26	—	7.0	—	—	—	0.33	30.83
Am	solid	(1200)	(2.4)	(2.0)	(13)	(4.9)	(4.4)	—	—	(1.657)	(16.2)
	liquid	2733	51.7	18.9	—	(8.5)	—	—	—	(0.409)	(34.5)
As	solid	883	3¼	35.¼	8.4	5.17	2.34	—	—	1.646	21.8
Au	solid	1336.16	3.03	2.27	11.32	6.14	−0.175	0.92	—	1.831	23.65
	liquid	2933	74.21	25.30	—	7.00	—	—	—	−0.631	26.99
B	solid	2313	(3.8)	(1.6)	1.42	1.54	4.40	—	—	0.655	8.67
	liquid	2800	75	27	—	(6.0)	—	—	—	(−4.599)	(31.4)
Ba	solid, α	648	0.14	0.22	16	5.55	1.50	—	—	1.722	16.1
	solid, β	977	1.83	1.87	—	5.55	1.50	—	—	1.582	15.9
	liquid	1911	35.665	18.63	—	(7.4)	—	—	—	(0.843)	(25.3)
	gas	—	—	—	—	(4.97)	—	—	—	(−39.65)	(−11.7)
Be	solid	1556	2.919	1.501	2.28	5.07	1.21	—	−1.15	1.951	27.62
	liquid	—	—	—	—	5.27	—	—	—	−1.611	25.68
Bi	solid	544.2	2.63	4.83	13.6	5.38	2.60	—	—	1.720	17.8
	liquid	1900	41.1	21.6	—	7.60	—	—	—	−0.087	25.6
	gas	—	—	—	—	(4.97)	—	—	—	(−46.19)	(−15.9)
C	solid	—	—	—	1.3609	4.10	1.02	—	−2.10	1.972	23.484
Ca	solid, α	723	0.24	0.33	9.95	5.24	3.50	—	—	1.718	20.95
	solid, β	1123	2.2	1.96	—	6.29	1.40	—	—	1.689	26.01
	liquid	1755	38.6	22.0	—	7.4	—	—	—	−0.147	30.28
	gas	—	—	—	—	(4.97)	—	—	—	(−43.015)	(−9.88)
Cd	solid	594.1	1.46	2.46	12.3	5.31	2.94	—	—	1.714	18.8
	liquid	1040	23.86	22.94	—	7.10	—	—	—	0.798	26.1
	gas	—	—	—	—	(4.97)	—	—	—	(−25.28)	(−11.7)
Ce	solid	1048	2.1	2.0	13.8	4.40	6.0	—	—	1.579	13.1
	liquid	2800	73	26	—	(7.9)	—	—	—	(−0.148)	(29.1)
Cl_2	gas	—	—	—	53.286	8.76	0.27	—	−0.65	2.845	−2.929
Co	solid, α	723	0.005	0.007	6.8	4.72	4.30	—	—	1.598	21.4
	solid, β	1398	0.095	0.068	—	3.30	5.86	—	—	0.974	13.1
	solid, γ	1766	3.7	2.1	—	9.60	—	—	—	3.961	50.5
	liquid	3370	93	28	—	8.30	—	—	—	−2.034	38.7
Cr	solid	2173	3.5	1.6	5.68	5.35	2.36	—	−0.44	1.848	25.75
	liquid	2495	72.97	29.25	—	9.40	—	—	—	1.556	50.13
	gas	—	—	—	—	(4.97)	—	—	—	(−82.47)	(−13.8)
Cs	solid	301.9	0.50	1.7	19.8	7.42	—	—	—	2.212	22.5
	liquid	963	16.32	17.0	—	8.00	—	—	—	1.887	24.1
	gas	—	—	—	—	(4.97)	—	—	—	(−17.35)	(−13.6)
Cu	solid	1356.2	3.11	2.29	7.97	5.41	1.50	—	—	1.680	23.30
	liquid	2868	72.8	25.4	—	7.50	—	—	—	0.024	34.05
F_2	gas	—	—	—	48.58	8.29	0.44	—	−0.80	2.760	−0.76
Fe	solid, α	1033	0.410	0.397	6.491	3.37	7.10	—	0.43	1.176	14.59
	solid, β	1180	0.217	0.184	—	10.40	—	—	—	4.281	55.66
	solid, γ	1673	0.15	0.084	—	4.85	3.00	—	—	0.396	19.76
	solid, δ	1808	3.86	2.14	—	10.30	—	—	—	4.382	55.11
	liquid	3008	84.62	28.1	—	10.00	—	—	—	−0.021	50.73
Ga	solid	302.94	1.335	4.407	9.82	5.237	3.03	—	—	1.710	21.01
	liquid	2700	—	—	—	(6.645)	—	—	—	(0.648)	(23.64)
Ge	solid	1232	8.3	6.7	10.1	5.90	1.13	—	—	1.764	23.8
	liquid	2980	68	23	—	(7.3)	—	—	—	(−5.668)	(25.7)
H_2	gas	—	—	—	31.211	6.62	0.81	—	—	2.010	6.75
Hf	solid	(2600)	(6.0)	(2.3)	13.1	(6.00)	(0.52)	—	—	(1.812)	(21.2)
Hg	liquid	629.73	13.985	22.208	18.46	6.61	—	—	—	1.971	19.20
	gas	—	—	—	—	4.969	—	—	—	−13.048	−13.54
In	solid	430	0.775	1.80	13.88	5.81	2.50	—	—	1.844	19.97
	liquid	2440	53.8	22.0	—	7.50	—	—	—	1.564	27.34
	gas	—	—	—	—	(4.97)	—	—	—	(−58.42)	(−14.46)
Ir	solid	2727	6.6	2.4	8.7	5.56	1.42	—	—	1.721	23.4
K	solid	336.4	0.5575	1.657	15.2	1.3264	19.405	—	—	1.258	−1.86
	liquid	1052	18.88	17.95	—	8.8825	−4.565	2.9369	—	1.923	32.55
	gas	—	—	—	—	(4.97)	—	—	—	(−19.689)	(−9.46)

* From the U.S. Atomic Energy Commission Report ANL-5750.

THERMODYNAMIC PROPERTIES OF THE ELEMENTS (Continued)

Element	Phase	Temperature of Transition (°K)	Heat of Transition (kcal/g mole)	Entropy of Transition (e.u.)	Entropy at 298°K (e.u.)	cal/g mole a	cal/g mole b	cal/g mole c	cal/g mole d	A (kcal/g mole)	B (e.u.)
La	solid	1153	(2.3)	(2.0)	13.7	6.17	1.60	—	—	1.911	21.9
	liquid	3000	80	27	—	(7.3)	—	—	—	(−0.15)	(26.0)
Li	solid	459	0.69	1.5	6.70	3.05	8.60	—	—	1.292	12.92
	liquid	1640	32.48	19.81	—	7.0	—	—	—	1.509	32.00
	gas	—	—	—	—	(4.97)	—	—	—	(−34.30)	(−2.84)
Mg	solid	923	2.2	2.4	7.77	5.33	2.45	—	−0.103	1.733	23.39
	liquid	1393	31.5	22.6	—	(8.0)	—	—	—	0.942	30.97
	gas					(4.97)				(−31.78)	(−7.60)
Mn	solid, α	1000	0.535	0.535	7.59	5.70	3.38	—	−0.37	1.974	26.11
	solid, β	1374	0.545	0.397	—	8.33	0.66	—	—	2.672	41.02
	solid, γ	1410	0.430	0.305	—	10.70	—	—	—	4.760	56.84
	solid, δ	1517	3.5	2.31	—	11.30	—	—	—	5.176	60.88
	liquid	2368	53.7	22.7	—	11.00	—	—	—	1.221	56.38
	gas	—	—	—	—	6.26	—	—	—	−63.704	−3.13
Mo	solid	2883	(5.8)	(2.0)	6.83	5.48	1.30	—	—	1.692	24.78
N_2	gas	—	—	—	45.767	6.76	0.606	0.13	—	2.044	−7.064
Na	solid	371	0.63	1.7	12.31	5.657	3.252	0.5785	—	1.836	20.92
	liquid	1187	23.4	20.1	—	8.954	−4.577	2.540	—	1.924	36.0
	gas	—	—	—	—	(4.97)	—	—	—	(−24.40)	(−8.7)
Nb	solid	2760	(5.8)	(2.1)	8.3	5.66	0.96	—	—	1.730	24.24
Nd	solid	1297	(2.55)	(1.97)	13.9	5.61	5.34	—	—	1.910	19.7
	liquid	(2750)	(61)	(22)	—	(9.1)	—	—	—	−0.606	35.8
Ni	solid, α	626	0.092	0.15	7.137	4.06	7.04	—	—	1.523	18.095
	solid, β	1728	4.21	2.44	—	6.00	1.80	—	—	1.619	27.16
	liquid	3110	90.48	29.0	—	9.20	—	—	—	0.251	45.47
Np	solid	913	(2.3)	(2.5)	(14)	(5.3)	(3.4)	—	—	(1.731)	(17.9)
	liquid	(2525)	(55)	(22)	—	(9.0)	—	—	—	(1.392)	(37.5)
O_2	gas	—	—	—	49.003	8.27	0.258	—	−1.877	3.007	−0.750
Os	solid	2970	(6.4)	(2.2)	7.8	5.69	0.88	—	—	1.736	24.9
P_4	solid, white	317.4	0.601	1.89	42.4	13.62	28.72	—	—	5.338	43.8
	liquid	553	11.9	21.5	—	19.23	0.51	—	−2.98	6.035	66.7
	gas	—	—	—	—	(19.5)	(−0.4)	(1.3)	—	(−6.32)	(46.1)
Pa	solid	(1825)	(4.0)	(2.2)	(13.5)	(5.2)	(4.0)	—	—	(1.728)	(17.3)
	liquid	(4500)	(115)	(26)	—	(8.0)	—	—	—	(−3.823)	(28.8)
Pb	solid	600.6	1.141	1.900	15.49	5.64	2.30	—	—	1.784	17.33
	liquid	2023	42.5	21.0	—	7.75	−0.73	—	—	1.362	27.11
	gas	—	—	—	—	(4.97)	—	—	—	(−45.25)	(−13.6)
Pd	solid	1828	4.12	2.25	8.9	5.80	1.38	—	—	1.791	24.6
	liquid	3440	89	26	—	(9.0)	—	—	—	(1.215)	(43.8)
Po	solid	525	(2.4)	(4.6)	13	(5.2)	(3.2)	—	—	(1.693)	(17.6)
	liquid	(1235)	(24.6)	(19.9)	—	(9.0)	—	—	—	(0.847)	(35.2)
	gas	—	—	—	—	(4.97)	—	—	—	(−28.73)	(−13.5)
Pr	solid	1205	(2.5)	(2.1)	(13.5)	(5.0)	(4.6)	—	—	(1.705)	(16.4)
	liquid	3563	—	—	—	(8.0)	—	—	—	(−0.519)	(30.0)
Pt	solid	2042.5	5.2	2.5	10.0	5.74	1.34	—	0.10	1.737	23.0
	liquid	4100	122	29.8	—	(9.0)	—	—	—	(0.406)	(42.6)
Pu	solid	913	(2.26)	(2.48)	(13.0)	(5.2)	(3.6)	—	—	(1.710)	(17.7)
	liquid	—	—	—	—	(8.0)	—	—	—	(0.506)	(31.0)
Ra	solid	1233	(2.3)	(1.9)	(17)	(5.8)	(1.2)	—	—	(1.783)	(16.4)
	liquid	(1700)	(35)	(21)	—	(8.0)	—	—	—	(1.284)	(28.6)
	gas	—	—	—	—	(4.97)	—	—	—	(−38.87)	(−14.5)
Rb	solid	312.0	0.525	1.68	16.6	3.27	13.1	—	—	1.557	5.9
	liquid	952	18.11	19.0	—	7.85	—	—	—	1.814	26.5
	gas	—	—	—	—	(4.97)	—	—	—	(−19.04)	(−12.3)
Re	solid	3440	(7.9)	(2.3)	(8.89)	(5.85)	(0.8)	—	—	(1.780)	(24.7)
Rh	solid	2240	(5.2)	(2.3)	7.6	5.40	2.19	—	—	1.707	23.8
	liquid	4150	127	30.7	—	(9.0)	—	—	—	(−0.923)	(44.4)
Ru	solid, α	1308	0.034	0.026	6.9	5.25	1.50	—	—	1.632	23.5
	solid, β	1473	0	0	—	7.20	—	—	—	2.867	35.5
	solid, γ	1773	0.23	0.13	—	7.20	—	—	—	2.867	35.5
	solid, δ	2700	(6.1)	(2.3)	—	7.50	—	—	—	3.169	37.6
S	solid, α	368.6	0.088	0.24	7.62	3.58	6.24	—	—	1.345	14.64
	solid, β	392	0.293	0.747	—	3.56	6.95	—	—	1.298	14.54
	liquid	717.76	2.5	3.5	—	5.4	5.0	—	—	1.576	24.02
½S_2	gas	—	—	—	—	(4.25)	(0.15)	—	(−1.0)	(−2.859)	(9.57)
Sb	solid (α,β,γ)	903.7	4.8	5.3	10.5	5.51	1.74	—	—	1.720	21.4
	liquid	1713	46.665	27.3	—	7.50	—	—	—	−1.992	28.1
½Sb_2	gas	—	—	—	—	4.47	—	—	−0.11	−53.876	−21.7
Sc	solid	1670	(4.0)	(2.4)	(9.0)	(5.13)	(3.0)	—	—	1.663	21.1
	liquid	3000	80	27	—	(7.50)	—	—	—	(−2.563)	31.3
Se	solid	490.6	1.25	2.55	10.144	3.30	8.80	—	—	1.375	11.28
	liquid	1000	14.27	14.27	—	7.0	—	—	—	0.881	27.34
Si	solid	1683	11.1	6.60	4.50	5.70	1.02	—	−1.06	2.100	28.88
	liquid	2750	71	26	—	7.4	—	—	—	−7.646	33.17
Sm	solid	1623	3.7	2.3	(15)	(6.7)	(3.4)	—	—	(2.149)	(24.2)
	liquid	(2800)	(70)	(25)	—	(9.0)	—	—	—	(−2.296)	(33.4)
Sn	solid (α,β)	505.1	1.69	3.35	12.3	4.42	6.30	—	—	1.598	14.8
	liquid	2473	(55)	(22)	—	7.30	—	—	—	0.559	26.2
	gas	—	—	—	—	(4.97)	—	—	—	(−60.21)	(−14.3)

THERMODYNAMIC PROPERTIES OF THE ELEMENTS (Continued)

Element	Phase	Temperature of Transition (°K)	Heat of Transition (kcal/g mole)	Entropy of Transition (e.u.)	Entropy at 298°K (e.u.)	cal/g mole a	cal/g mole b	cal/g mole c	cal/g mole d	A (kcal/g mole)	B (e.u.)
Sr	solid	1043	2.2	2.1	13.0	(5.60)	(1.37)	—	—	(1.731)	(19.3)
	liquid	1657	33.61	20.28	—	(7.7)	—	—	—	(0.976)	(30.4)
	gas	—	—	—	—	(4.97)	—	—	—	(−37.16)	(−10.2)
Ta	solid	3250	7.5	2.3	9.9	5.82	0.78	—	—	1.770	23.4
Te	solid	(2400)	(5.5)	(2.3)	(8.0)	(5.6)	(2.0)	—	—	(1.759)	(24.5)
	liquid	(3800)	(120)	(32)	—	(11)	—	—	—	(3.459)	(59.4)
Te	solid, α	621	0.13	0.21	11.88	4.58	5.25	—	—	1.599	15.78
	solid, β	723	4.28	5.92	—	4.58	5.25	—	—	1.469	15.57
	liquid	1360	11.9	8.75	—	9.0	—	—	—	−0.988	34.96
½ Te_2	gas	—	—	—	—	4.47	—	—	−0.10	−19.048	−6.47
Th	solid	2173	(4.6)	(2.1)	12.76	8.2	−0.77	2.04	—	2.591	33.64
	liquid	4500	(130)	(29)	—	(8.0)	—	—	—	(−7.602)	(26.84)
Ti	solid, α	1155	0.950	0.822	7.334	5.25	2.52	—	—	1.677	23.33
	solid, β	2000	(4.6)	(2.3)	—	7.50	—	—	—	1.645	35.46
	liquid	3550	(101)	(28)	—	(7.8)	—	—	—	(−2.355)	(35.45)
Tl	solid, α	508.3	0.082	0.16	15.4	5.26	3.46	—	—	1.722	15.6
	solid, β	576.8	1.03	1.79	—	7.30	—	—	—	2.230	26.4
	liquid	1730	38.81	22.4	—	7.50	—	—	—	1.315	25.9
	gas	—	—	—	—	(4.97)	—	—	—	(−41.88)	(−45.4)
U	solid, α	938	0.665	0.709	12.03	3.25	8.15	—	0.80	1.063	8.47
	solid, β	1049	1.165	1.111	—	10.28	—	—	—	3.493	48.27
	solid, γ	1405	(3.0)	(2.1)	—	9.12	—	—	—	1.110	39.09
	liquid	3800	—	—	—	(8.99)	—	—	—	(−2.073)	36.01
V	solid	2003	(4.0)	(2.0)	7.05	5.57	0.97	—	—	1.704	24.97
	liquid	3800	—	—	—	(8.6)	—	—	—	1.827	44.06
W	solid	3650	8.42	2.3	8.0	5.74	0.76	—	—	1.745	24.9
Y	solid	1750	(4.0)	(2.3)	(11)	(5.6)	(2.2)	—	—	(1.767)	(21.6)
	liquid	3500	(90)	(26)	—	(7.5)	—	—	—	(−2.277)	(29.6)
Zn	solid	692.7	1.595	2.303	9.95	5.35	2.40	—	—	1.702	21.25
	liquid	1180	27.43	23.24	—	7.50	—	—	—	1.020	31.35
	gas	—	—	—	—	—	—	—	—	(−29.407)	(−9.81)
Zr	solid, α	1135	0.920	0.811	9.29	6.83	1.12	—	−0.87	2.378	30.45
	solid, β	2125	(4.9)	(2.3)	—	7.27	—	—	—	1.159	31.43
	liquid	(3900)	(100)	(26)	—	(8.0)	—	—	—	(−2.190)	(34.7)

THERMODYNAMIC PROPERTIES OF THE OXIDES*,**

Oxide	Phase	Temperature of Transition (°K)	Heat of Transition kcal/mole	Entropy of Transition (e.u.)	Entropy at 298°K (e.u.)	cal/g mole a	cal/g mole b	cal/g mole c	cal/g mole d	A (kcal/mole)	B (e.u.)
Ac_2O_3	solid	(2250)	(20)	(8.9)	(36.5)	(20.0)	(20.4)	—	—	(6.870)	(80.9)
	liquid	—	—	—	—	(40)	—	—	—	(−19.767)	(180.5)
Ag_2O	solid	dec. 460	—	—	29.09	13.26	7.04	—	—	4.266	48.56
Ag_2O_2	solid	dec	—	—	(20.4)	(16.4)	(12.2)	—	—	(5.432)	(76.7)
Al_2O_3	solid	2300	26	11	12.186	26.12	4.388	—	−7.269	10.422	142.03
	liquid	dec.	—	—	—	(33)	—	—	—	(−11.655)	(174.1)
Am_2O_3	solid	(2225)	(17)	(7.6)	(37)	(20.0)	(15.6)	—	—	(6.657)	(81.6)
	liquid	(3400)	(85)	(25)	—	(38.5)	—	—	—	(−7.796)	(181.8)
AmO_2	solid	dec.	—	—	(20)	(14.0)	(6.8)	—	—	(4.477)	(61.8)
As_2O_3	solid, α	503	4.1	8.2	25.6	8.37	48.6	—	—	4.656	36.6
	solid, β	586	4.4	7.5	—	8.37	48.6	—	—	0.556	28.4
	liquid	730	7.15	9.79	—	(39)	—	—	—	(5.760)	(187.6)
	gas	—	—	—	—	(21.5)	—	—	—	(−14.164)	(62.5)
AsO_2	solid	(1200)	(9.0)	(7.5)	(13)	(8.5)	(9.4)	—	—	(2.952)	(38.2)
	liquid	(dec.)	—	—	—	(21)	—	—	—	(2.184)	(108.0)
As_2O_5	solid	dec. >1100	—	—	25.2	(31.1)	(16.4)	—	(−5.4)	(11.813)	(159.9)
Au_2O_3	solid	dec.	—	—	30	(23.5)	(4.8)	—	—	(7.220)	(105.3)
B_2O_3	solid	723	5.27	7.29	12.91	8.73	25.40	—	−1.31	4.171	45.04
	liquid	2520	(55)	(22)	—	30.50	—	—	—	7.822	161.59
Ba_2O	solid	(880)	(5.2)	(5.9)	(23.5)	(20.0)	(2.2)	—	—	(6.061)	(91.1)
	liquid	(1040)	(20)	(19)	—	(22)	—	—	—	(1.769)	(96.8)
	gas	—	—	—	—	(15)	—	—	—	(−25.51)	(29.0)
BaO	solid	2196	13.8	6.28	16.8	12.74	1.040	—	−1.984	4.510	57.2
	liquid	3000	(62)	(21)	—	(13.9)	—	—	—	(−9.341)	(57.5)
BaO_2	solid	723	(5.7)	(7.9)	(18.5)	(13.6)	(2.0)	—	—	(4.144)	(59.6)
	liquid	dec.1110	—	—	—	(21)	—	—	—	(3.241)	(99.0)
BeO	solid	dec.	—	—	3.37	8.69	3.65	—	−3.13	3.803	48.99
BiO	solid	(1175)	(3.7)	(3.1)	(15)	(9.7)	(3.0)	—	—	(3.025)	(41.2)
	liquid	(1920)	(54)	(28)	—	(14)	—	—	—	(2.306)	(64.9)
	gas	—	—	—	—	(8.9)	—	—	—	(−61.49)	(−1.8)
Bi_2O_3	solid	1090	6.8	6.2	36.2	23.27	11.05	—	—	7.429	99.7
	liquid	(dec.)	—	—	—	(35.7)	—	—	—	(7.614)	(168.3)
CO	gas	—	—	—	47.30	6.60	1.2	—	—	2.021	−9.34
CO_2	gas	—	—	—	51.06	7.70	5.3	−0.83	—	2.490	−5.64
CaO	solid	2860	(18)	(6.3)	9.5	10.00	4.84	—	−1.08	3.559	49.5
CdO	solid	dec.	—	—	13.1	9.65	2.08	—	—	2.970	42.5
Ce_2O_3	solid	1960	(20)	(10)	(33.5)	(23.0)	(9.0)	—	—	(7.258)	(100.2)
	liquid	(3500)	(80)	(23)	—	(37)	—	—	—	(−2.591)	(178.5)
CeO_2	solid	3000	(19)	(6.3)	17.7	15.0	2.5	—	—	4.579	68.5

*For description of headings and symbols see preceding table.
**From U.S. Atomic Energy Report ANL-5750.

THERMODYNAMIC PROPERTIES OF THE OXIDES (Continued)

Oxide	Phase	Temperature of Transition (°K)	Heat of Transition kcal/mole	Entropy of Transition (e.u.)	Entropy at 298°K (e.u.)	cal/g mole a	cal/g mole b	cal/g mole c	cal/g mole d	A (kcal/mole)	B (e.u.)
CoO	solid	2078	(12)	(5.8)	10.5	(9.8)	(2.2)	—	—	(3.020)	(46.0)
	liquid	(2900)	(61)	(21)	—	(15.5)	—	—	—	(−1.886)	(79.2)
Co_3O_4	solid	dec., 1240	—	—	(35.5)	(29.5)	(17.0)	—	—	(9.551)	(137.6)
Cr_2O_3	solid	2538	(25)	(10)	19.4	28.53	2.20	—	−3.736	9.857	145.9
CrO_2	solid	dec. 700	—	—	(11.5)	(16.1)	(3.0)	—	(−3.0)	(5.946)	(82.8)
CrO_3	solid	460	(6.1)	(13)	(17.5)	(18.1)	(4.0)	—	(−2.0)	(6.245)	(87.9)
	liquid	(1000)	(25)	(25)	—	(27)	—	—	—	(3.381)	(127.0)
	gas	—	—	—	—	(20)	—	—	—	(−28.62)	(53.6)
Cs_2O	solid	763	(4.58)	(6.0)	(23)	(16.5)	(5.4)	—	—	(5.160)	(72.6)
	liquid	dec.	—	—	—	(22)	—	—	—	(3.205)	(99.0)
Cs_2O_2	solid	867	(5.5)	(6.3)	(40)	(21.4)	(11.4)	—	—	(6.887)	(85.3)
	liquid	dec.	—	—	—	(29.5)	—	—	—	(4.125)	(123.8)
Cs_2O_3	solid	775	(7.75)	(10)	(47)	(24.0)	(22.6)	—	—	(8.160)	(96.5)
	liquid	dec.	—	—	—	(35)	—	—	—	(2.148)	(142.2)
Cu_2O	solid	1503	13.4	8.92	22.44	(13.4)	(8.6)	—	—	(4.378)	(54.9)
	liquid	dec.	—	—	—	(21.5)	—	—	—	(3.721)	(96.0)
CuO	solid	1609	(8.9)	(5.5)	10.4	14.34	6.2	—	—	4.551	61.11
	liquid	dec.	—	—	—	(22)	—	—	—	(−4.339)	(98.91)
FeO	solid	1641	7.5	4.6	12.9	9.27	4.80	—	—	(2.977)	(43.8)
	liquid	(2700)	(55)	(20)	—	(14.5)	—	—	—	(−3.721)	(69.2)
Fe_3O_4	solid, α	900	(0)	(0)	35.0	12.38	1.62	—	−0.38	3.826	58.3
	solid, β	dec.	—	—	—	(14.5)	—	—	—	(−2.399)	(66.7)
Fe_2O_3	solid, α	950	0.16	0.17	21.5	21.88	48.20	—	—	8.666	104.0
	solid, β	1050	0	0	—	48.00	—	—	—	12.652	238.3
	solid, γ	dec.	—	—	—	23.49	18.6	—	−3.55	9.021	119.9
Ga_2O	solid	(925)	(8.5)	(9.2)	(22.5)	36.00	—	—	—	11.979	187.6
	liquid	(1000)	(20)	(20.)	—	31.71	1.8	—	—	8.467	159.7
	gas	—	—	—		(13.8)	(8.6)	—	—	(4.497)	(58.7)
Ga_2O_3	solid	2013	(22)	(11.)	20.23	(21.5)	—	—	—	(−0.559)	(94.1)
	liquid	(2900)	(75)	(26.)	—	(14.)	—	—	—	(−28.06)	(22.3)
GeO	solid	983	(50)	(51)	(12.5)	11.77	25.2	—	—	(4.630)	(54.35)
	gas	—	—	—	—	(35.5)	—	—	—	(−20.66)	173.2
GeO_2	solid, (α,β)	1389	10.5	7.56	(12.5)	(10.4)	(2.6)	—	(−0.5)	(3.384)	(47.8)
	liquid	(2625)	(61)	(23.)	—	(8.2)	(0.4)	—	(−0.2)	(−49.67)	(−20.2)
H_2O	liquid	373.16	9.770	26.18	16.716	11.2	7.17	—	—	(3.658)	(53.5)
	gas	—	—			(21.7)				(1.149)	(111.9)
HfO_2	solid	3063	(17)	(5.6)	14.18	18.03	—	—	—	5.376	86.01
Hg_2O	solid	dec.	—	—	(30.)	7.17	2.56	—	(−0.08)	−8.290	−3.56
HgO	solid	dec.	—	—	16.839	17.39	2.08	—	−3.48	6.445	87.48
In_2O	solid	(600)	(4.5)	(7.5)	(28.)	(14.7)	(7.8)	—	—	(4.730)	(58.1)
	liquid	(800)	(16)	(20.)	—	(22.)	—	—	—	(3.206)	(92.6)
	gas	—	—	—	—	(15.)	—	—	—	(−18.39)	(25.8)
InO	solid	(1325)	(4.0)	(3.0)	(14.5)	(10.0)	(3.2)	—	—	(3.124)	(43.4)
	liquid	(2000)	(60.)	(30)	—	(14.)	—	—	—	(1.615)	(64.9)
	gas	—	—	—	—	(9.0)	—	—	—	(−68.38)	(−3.1)
In_2O_3	solid	(2000)	(20.)	(10.)	30.1	(22.6)	(6.0)	—	—	7.005	(100.5)
	liquid	(3600)	(85)	(24)	—	(35)	—	—	—	(−0.195)	(172.8)
Ir_2O_3	solid	(1450)	(10)	(6.8)	(26.5)	(21.8)	(14.4)	—	—	(7.140)	(102.0)
	liquid	(2250)	(50)	(22)	—	(35)	—	—	—	(0.706)	(170.3)
	gas	—	—	—	—	(20)	(10)	—	—	(−57.73)	(54.8)
IrO_2	solid	dec. 1373	—	—	(15.9)	9.17	15.20	—	—	3.410	40.9
K_2O	solid	(980)	(6.8)	(6.9)	(23)	(15.9)	(6.4)	—	—	(5.025)	(69.5)
	liquid	dec.	—	—	—	(22)	—	—	—	(1.130)	(98.3)
K_2O_2	solid	763	(7.0)	(9.2)	(27)	(20.8)	(5.4)	—	—	(6.442)	(93.1)
	liquid	(1800)	(45)	(25)	—	(29)	—	—	—	(4.127)	(134.2)
	gas	—	—	—	—	(20)	—	—	—	(−57.07)	(41.7)
K_2O_3	solid	703	(6.1)	(8.7)	(33.5)	(19.1)	(23.2)	—	—	(6.750)	(82.2)
	liquid	(975)	(25)	(26)	—	(35.5)	—	—	—	(6.447)	(164.7)
	gas	—	—	—	—	(20)	(5.0)	—	—	(−31.29)	(37.3)
KO_2	solid	653	(4.9)	(7.5)	27.9	(15.0)	(12.0)	—	—	(5.006)	(61.1)
	liquid	dec.	—	—	—	(24)	—	—	—	(3.424)	(105.5)
La_2O_3	solid	2590	(18)	(7)	(36.5)	28.86	3.076	—	−3.275	9.840	(130.7)
Li_2O	solid	2000	(14)	(7)	9.06	(11.4)	(5.4)	—	—	(3.639)	(57.5)
	liquid	2600	(56)	(22)	—	(21)	—	—	—	(−1.961)	(112.7)
Li_2O_2	solid	dec.470	—	—	(16.5)	(17.0)	(5.4)	—	—	(5.309)	(82.0)
MgO	solid	3075	18.5	5.8	6.4	10.86	1.197	—	−2.087	3.991	57.0
MgO_2	solid	dec. 361	—	—	(20.5)	(12.1)	(2.4)	—	—	(3.714)	(49.2)
MnO	solid	2058	13.0	6.32	14.27	11.11	1.94	—	−0.88	3.689	50.10
	liquid	dec.	—	—	—	(13.5)	—	—	—	(−8.543)	(58.02)
Mn_3O_4	solid, α	1445	4.97	3.44	35.5	34.64	10.82	—	−2.20	11.312	166.3
	solid, β	1863	(33)	(18)	—	50.20	—	—	—	17,376	260.4
	liquid	(2900)	(75)	(26)	—	(49)	—	—	—	(−17.86)	(233.4)
Mn_2O_3	solid	dec. 1620	—	—	26.4	24.73	8.38	—	−3.23	8.829	118.8
MnO_2	solid	dec. 1120	—	—	12.7	16.60	2.44	—	−3.88	6.359	84.8
MoO_2	solid	(2200)	(16)	(7.3)	(14.5)	(16.2)	(3.0)	—	(−3.0)	(5.973)	(80.4)
	liquid	dec. 2250	—	—	—	(23)	—	—	—	(−2.463)	(118.4)
MoO_3	solid	1068	12.54	11.74	18.68	13.6	13.5	—	—	4.655	62.83
	liquid	1530	33	22	—	(28.4)	—	—	—	(0.222)	(139.88)
	gas	—	—	—	—	(18.1)	—	—	—	(−48.54)	(42.8)

Oxide	Phase	Temperature of Transition (°K)	Heat of Transition kcal/mole	Entropy of Transition (e.u.)	Entropy at 298°K (e.u.)	cal/g mole a	cal/g mole b	cal/g mole c	cal/g mole d	A (kcal/mole)	B (e.u.)
N_2O	gas	—	—	—	52.58	10.92	2.06	—	−2.04	4.032	11.40
Na_2O	solid	1193	(7.1)	(6.0)	17.4	15.70	5.40	—	—	4.921	73.7
	liquid	dec.	—	—	—	(22)	—	—	—	(1.494)	(105.9)
Na_2O_2	solid	dec. 919	—	—	22.6	(20.2)	(3.8)	—	—	(6.192)	(93.6)
NaO_2	solid	(825)	(6.2)	(7.5)	27.7	(16.2)	(3.6)	—	—	(4.990)	(65.7)
	liquid	(1300)	(28)	(22)	—	(23)	—	—	—	(3.175)	(100.9)
	gas	—	—	—	—	(15)	—	—	—	(−35.22)	(22.0)
NbO	solid	(2650)	(16)	(6.0)	(12)	(9.6)	(4.4)	—	—	(3.058)	(44.0)
NbO_2	solid	(2275)	(16)	(7.0)	(12.7)	(17.1)	(1.6)	—	(−2.8)	(6.109)	(84.6)
	liquid	(3800)	(85)	(22)	—	(24)	—	—	—	(1.033)	(127.2)
Nb_2O_5	solid	1733	(28)	(16)	32.8	21.88	28.2	—	—	7.776	100.3
	liquid	(3200)	(80)	(25)	—	(44.2)	—	—	—	(−24.09)	(201.6)
Nd_2O_3	solid	2545	(22)	(8.8)	(35.3)	28.99	5.760	—	(−4.159)	10.295	(133.9)
NiO	solid	2230	(12.1)	(5.43)	9.22	13.69	0.83	—	−2.915	5.097	70.67
	liquid	dec.	—	—	—	(14.3)	—	—	—	(−7.861)	(67.91)
NpO_2	solid	(2600)	(15)	(5.7)	19.19	(17.7)	(3.2)	—	(−2.6)	(6.292)	(84.08)
Np_2O_5	solid	dec.	—	—	(43)	(32.4)	(12.6)	—	—	(10.22)	(145.4)
		800—900°K									
OsO_2	solid	dec. 923	—	—	(14.5)	(11.5)	(6.0)	—	—	(3.696)	(52.8)
OsO_4	solid	813.3	3.41	10.9	34.7	(16.4)	(23.1)	—	(−2.4)	(6.726)	(67.0)
	liquid	403	9.45	23.4	—	(33)	—	—	—	(6.612)	(143.0)
	gas	—	—	—	—	16.46	8.60	—	−4.6	(−7.644)	(25.3)
P_2O_3	liquid	448.5	4.5	10	(34)	(34.5)	—	—	—	(10.287)	(162.6)
	gas	—	—	—	—	(15)	(10)	—	—	(−1.953)	(38.0)
PO_2	solid	(350)	(2.7)	(7.7)	(11.5)	(11.3)	(5.0)	—	—	(3.591)	(54.4)
	liquid	(dec)	—	—	—	(20)	—	—	—	(3.640)	(95.9)
P_2O_5	solid	631	8.8	13.9	33.5	8.375	5.40	—	—	4.897	30.3
	gas	—	—	—	—	36.80	—	—	—	3.284	165.6
PaO_2	solid	(2560)	(20)	(7.8)	(17.8)	(14.4)	(2.6)	—	—	(4.409)	(65.0)
Pa_2O_5	solid	(2050)	(26)	(13)	(37.5)	(28.4)	(11.4)	—	—	(8.975)	(127.7)
	liquid	(3350)	(95)	(28)	—	(48)	—	—	—	(−0.800)	(241.1)
PbO	solid, red	762	(0.4)	(0.5)	16.2	10.60	4.00	—	—	3.338	45.4
	solid, yellow	1159	2.8	2.4	—	9.05	6.40	—	—	2.454	36.4
	liquid	1745	51	29	—	(14.6)	—	—	—	1.788	65.7
	gas	—	—	—	—	(8.1)	(0.4)	—	—	(−59.94)	(−11.0)
Pb_3O_4	solid	dec.	—	—	50.5	(31.1)	(17.6)	—	—	(10.055)	(132.0)
PbO_2	solid	dec.	—	—	18.3	12.7	7.80	—	—	4.133	56.4
PdO	solid	dec. 1150	—	—	(9.1)	3.30	14.2	—	—	1.615	(13.9)
PoO_2	solid	(825)	(5.5)	(6.7)	(17)	(14.3)	(5.6)	—	—	(4.513)	(66.1)
	liquid	(dec.)	—	—	—	(22)	—	—	—	(3.460)	(106.5)
Pr_2O_3	solid	(2200)	(22)	(10)	(35.5)	(29.0)	(4.0)	—	(−4.0)	(10.166)	(133.2)
	liquid	(4000)	(90)	(23)	—	(36)	—	—	—	(−6.298)	(168.3)
PrO_2	solid	dec. 700	—	—	(17)	(17.6)	(3.4)	—	(−2.8)	(6.338)	(85.9)
PtO	solid	dec. 780	—	—	(13.5)	(9.0)	(6.4)	—	—	(2.968)	(39.7)
Pt_3O_4	solid	(dec.)	—	—	(41)	(30.8)	17.4	—	—	(9.957)	(139.7)
PtO_2	solid	723	(4.6)	(6.4)	(16.5)	(11.1)	(9.6)	—	—	(3.736)	(49.6)
	liquid	dec. 750	—	—	—	(21)	—	—	—	(3.785)	(101.5)
PuO	solid	(1290)	(7.2)	(5.6)	(20)	(12.0)	(2.4)	—	—	(3.685)	(49.1)
	liquid	(2325)	(47)	(20)	—	(14.5)	—	—	—	(−2.287)	(58.3)
	gas	—	—	—	—	(8.9)	—	—	—	(−62.307)	(−5.3)
Pu_2O_3	solid	(1880)	(16)	(8.5)	(38)	(21.2)	18.2	—	—	(7.130)	(88.2)
	liquid	(3250)	(75)	(23)	—	(40)	—	—	—	(−5.691)	(187.2)
PuO_2	solid	(2400)	(15)	(6.2)	19.7	(17.1)	(3.4)	—	(−2.6)	(6.122)	(80.2)
	liquid	(3500)	(90)	(26)	—	(20.5)	—	—	—	(−10.62)	(92.2)
RaO	solid	(>2500)	—	—	(17)	(10.5)	(2.0)	—	—	(3.220)	(43.4)
Rb_2O	solid	(910)	(5.7)	(6.3)	(27)	(15.4)	(5.8)	—	—	(4.850)	(62.5)
	liquid	dec.	—	—	—	(22)	—	—	—	(2.754)	(95.9)
Rb_2O_2	solid	843	(7.3)	(8.7)	(27.5)	(10.9)	(8.0)	—	—	(6.587)	(94.0)
	liquid	(dec.)	—	—	—	(29)	—	—	—	(3.273)	(133.2)
Rb_2O_3	solid	762	(7.6)	(10)	(32.5)	(20.5)	(13.0)	—	—	(6.690)	(88.2)
	liquid	dec.	—	—	—	(34)	—	—	—	(5.603	(157.8)
RbO_2	solid	685	(4.1)	(6.0)	(21.5)	(13.8)	(6.4)	—	—	(4.399)	(59.0)
	liquid	dec.	—	—	—	(21)	—	—	—	(3.720)	(95.7)
ReO_2	solid	(1475)	(12)	(8.1)	(15)	(10.8)	(9.8)	—	—	(3.656)	(49.5)
ReO_2	liquid	(3250)	(80)	(25)	—	(24.5)	—	—	—	(1.204)	(127.0)
ReO_3	solid	433	5.2	12	19.8	(18.0)	(5.8)	—	—	(5.625)	(84.5)
	liquid	dec.	—	—	—	29	—	—	—	(4.644)	(136.8)
Re_2O_7	solid	569	15.8	27.8	44	(41.8)	(14.8)	—	(−3.0)	(14.127)	(200.3)
	liquid	635.5	17.7	27.9	—	(65.7)	—	—	—	(9.203)	(314.7)
	gas	—	—	—	—	(38.2)	—	—	—	(−25.97)	(109.3)
ReO_4	solid	420	(4.2)	(10)	(34.5)	(21.4)	(10.8)	—	(−2.0)	(7.531)	(91.8)
	liquid	(460)	(9.3)	(20)	—	(33)	—	—	—	(6.775)	(146.7)
	gas	—	—	—	—	(16.5)	(8.6)	—	(−5.0)	(−8.118)	(30.6)
Rh_2O	solid	dec. 1400	—	—	(25.5)	15.59	6.47	—	—	4.936	(65.3)
RhO	solid	dec. 1394	—	—	(12)	(9.84)	(5.53)	—	—	(3.179)	(45.7)
Rh_2O_3	solid	dec. 1388	—	—	(23)	20.73	13.80	—	—	6.794	(99.2)
RuO_2	solid	dec. 1400	—	—	(12.5)	(11.4)	(6.0)	—	—	3.666	(54.2)
RuO_4	solid	300	(3.2)	(11)	(32.5)	(20)	—	—	—	(5.963)	(81.5)
	liquid	dec.	—	—	—	(33)	—	—	—	(6.663)	(144.9)
SO_2	gas	—	—	—	59.40	11.4	1.414	—	−2.045	4.148	7.12
Sb_2O_3	solid	928	14.74	15.88	29.4	19.10	17.1	—	—	6.455	84.5
	liquid	1698	8.92	5.25	—	(36)	—	—	—	(0.035)	(168.2)
	gas	—	—	—	—	(20.8)	—	—	—	(−34.70)	(49.9)

THERMODYNAMIC PROPERTIES OF THE OXIDES (Continued)

Oxide	Phase	Temperature of Transition (°K)	Heat of Transition kcal/mole	Entropy of Transition (e.u.)	Entropy at 298°K (e.u.)	cal/g mole a	cal/g mole b	cal/g mole c	cal/g mole d	A (kcal/mole)	B (e.u.)
SbO_3	solid	dec.	—	—	15.2	11.30	8.1	—	—	3.725	51.6
Sb_2O_3	solid	dec.	—	—	29.9	(22.4)	(23.6)	—	—	(7.728)	(104.8)
Sc_2O_3	solid	(2500)	(23)	(9.3)	24.8	23.17	5.64	—	—	7.159	108.9
SeO	solid	(1375)	(7.6)	(5.5)	(11)	(9.1)	(3.8)	—	—	(2.882)	(42.0)
	liquid	(2075)	(45)	(22)	—	(15.5)	—	—	—	(0.490)	(77.5)
	gas	—	—	—	—	8.20	0.50	—	−0.80	(−58.54)	(0.7)
SeO_2	solid	603	(24.5)	(40.6)	(15)	(12.8)	(6.1)	—	(−0.2)	(4.150)	(59.9)
	gas	—	—	—	—	(14.3)	—	—	—	(−20.43)	(20.4)
SiO	solid	(2550)	(12)	(4.7)	(6.5)	(7.3)	(2.4)	—	—	(2.283)	(35.8)
SiO_2	solid, β	856	0.15	0.18	10.06	11.22	8.20	—	−2.70	4.615	57.83
	solid, α	1883	2.04	1.08	—	14.41	1.94	—	—	4.602	73.67
	liquid	dec. 2250	—	—	—	(20)	—	—	—	(9.649)	(111.08)
Sm_2O_3	solid	(2150)	(20)	(9.3)	(36.5)	(25.9)	(7.0)	—	—	(8.033)	(113.2)
	liquid	(3800)	(80)	(21)	—	(36)	—	—	—	(−6.431)	(166.3)
SnO	solid	(1315)	(6.4)	(4.9)	13.5	9.40	3.62	—	—	2.964	41.1
	liquid	(1800)	(60)	(33)	—	(14.5)	—	—	—	(0.141)	(68.1)
	gas	—	—	—	—	(9.0)	—	—	—	(−69.76)	(−6.4)
SnO_2	solid	1898	(11.39)	(5.95)	12.5	17.66	2.40	—	−5.16	7.103	91.7
	liquid	(3200)	(75)	(23)	—	(22.5)	—	—	—	(0.304)	(117.7)
SrO	solid	2703	16.7	6.2	13.0	12.34	1.120	—	−1.806	4.335	58.7
SrO_2	solid	dec. 488	—	—	(14.8)	(16.8)	(2.2)	—	(−3.0)	(6.113)	(83.3)
Ta_2O_5	solid	2150	(16)	(7.4)	34.2	29.2	10.0	—	—	9.151	135.2
	liquid	—	—	—	—	(46)	—	—	—	(6.158)	(235.1)
TcO_2	solid	(2400)	(18)	(7.5)	(13.5)	(10.4)	(9.2)	—	—	(3.510)	(48.6)
	liquid	(4000)	(105)	(26)	—	(25)	—	—	—	(−5.946)	(132.7)
TcO_3	solid	(dec. <1200)	—	—	(19.5)	(19.4)	(5.2)	—	(−2.0)	(6.686)	(93.7)
Tc_2O_7	solid	392.7	(11)	(28)	(42.5)	(39.1)	(18.6)	—	(−2.4)	(13.29)	(187.2)
	liquid	583.8	(14)	(24)	—	(64)	—	—	—	(10.02)	(299.8)
	gas	—	—	—	—	(25)	(28)	—	—	(−21.98)	(43.8)
TeO	solid	(1020)	(7.1)	(7.0)	(13)	(8.6)	(6.2)	—	—	(2.840)	(37.8)
	liquid	(1775)	(50)	(28)	—	(15.5)	—	—	—	(−0.448)	(72.3)
	gas	—	—	—	—	(8.9)	—	—	—	(−62.16)	(−5.2)
TeO_2	solid	1006	3.2	3.2	16.99	13.85	6.87	—	—	4.435	63.97
	liquid	dec.	—	—	—	(20)	—	—	—	(3.940)	(96.4)
ThO	solid	(2150)	(13)	(6.0)	(16)	(11.0)	(2.4)	—	—	(3.386)	(47.4)
	liquid	(3250)	(65)	(20)	—	(15)	—	—	—	(−6.561)	(66.9)
ThO_2	solid	3225	(18)	(5.6)	15.59	16.45	2.346	—	−2.124	5.721	80.03
TiO	solid, α	1264	0.82	0.65	8.31	10.57	3.60	—	−1.86	3.935	54.03
	solid, β	dec. 2010	—	—	—	11.85	3.00	—	—	4.108	61.71
Ti_2O_2	solid, α	473	0.215	0.455	18.83	7.31	53.52	—	—	4.559	38.78
	solid, β	2400	(24)	(10)	—	34.68	1.30	—	−10.20	13.605	184.48
	liquid	3300	—	—	—	(37.5)	—	—	—	(−7.796)	(193.2)
Ti_5O_5	solid, α	450	2.24	4.98	30.92	35.47	29.50	—	—	11.887	179.98
	solid, β	(2450)	(50)	(20)	—	41.60	8.00	—	—	10.230	202.80
	liquid	(3600)	(85)	(24)	—	(60)	—	—	—	(−18.701)	(306.4)
TiO_2	solid	2128	(16)	(7.5)	12.01	17.97	0.28	—	−4.35	6.829	92.92
	liquid	dec.3200	—	—	—	(21.4)	—	—	—	(−2.610)	(111.08)
Ti_2O	solid	573	(5.0)	(8.7)	23.8	(15.8)	(6.0)	—	(−0.3)	(5.078)	(68.2)
	liquid	773	(17)	(22)	—	(22.1)	—	—	—	(2.651)	(96.0)
	gas	—	—	—	—	(13.7)	—	—	—	(−20.94)	(18.0)
Ti_2O_3	solid	990	(12.4)	(13)	(33.5)	(23.0)	(5.0)	—	—	(7.080)	(99.0)
	liquid	(dec)	—	—	—	(35.5)	—	—	—	(4.604)	(167.8)
UO	solid	(2750)	(14)	(5.1)	(16)	(10.6)	(2.0)	—	—	(3.249)	(45.0)
UO_2	solid	3000	—	—	18.63	19.20	1.62	—	−3.957	7.124	93.37
U_3O_3	solid	dec.	—	—	(66)	(65)	(7.5)	—	(−10.9)	(23.37)	(312.7)
UO_2	solid	dec. 925	—	—	23.57	22.09	2.54	—	−2.973	7.696	104.72
VO	solid	(2350)	(15)	(6.4)	9.3	11.32	1.61	—	−1.26	3.869	56.4
	liquid	(3400)	(70)	(21)	—	(14.5)	—	—	—	(−8.157)	(70.9)
V_2O_3	solid	2240	(24)	(11)	23.58	29.35	4.76	—	−5.42	10.780	148.12
	liquid	dec. 3300	—	—	—	(38)	—	—	—	(−6.028)	(193.4)
V_2O_5	solid	(2100)	(42)	(20)	(32)	(36)	(30)	—	—	(12.07)	(182.1)
	liquid	(dec.)	—	—	—	(55.6)	—	—	—	(−54.72)	(249.1)
VO_2	solid, α	345	1.02	2.96	12.32	14.96	—	—	—	4.460	72.92
	solid, β	1818	13.60	7.48	—	17.85	1.70	—	−3.94	5.680	89.09
	liquid	dec. 3300	—	—	—	25.50	—	—	—	2.962	135.87
V_2O_5	solid	943	15.56	16.50	313	46.54	−3.90	—	−13.22	18.136	240.2
	liquid	(2325)	(63)	(27)	—	45.60	—	—	—	2.122	220.1
	gas	—	—	—	—	(40)	—	—	—	(−73.90)	(149.6)
VO_2	solid	(1543)	(11.5)	(7.45)	(15)	(17.6)	(4.2)	—	(−4.0)	(6.772)	(88.8)
	liquid	dec. 2125	—	—	—	(24)	—	—	—	(−0.112)	(121.8)
WO_3	solid	1743	(17)	(9.8)	19.90	17.33	7.74	—	—	5.511	81.15
	liquid	(2100)	(43)	(20)	—	(30)	—	—	—	(−1.162)	(152.5)
	gas	—	—	—	—	(18)	—	—	—	(−69.36)	(40.2)
Y_2O_3	solid	(2500)	(25)	(10)	(29.5)	(26.0)	(8.2)	—	(−2.2)	(8.846)	(122.3)
ZnO	solid	dec.	—	—	10.4	11.71	1.22	—	−2.18	4.277	57.88
ZrO_2	solid, α	1478	1.420	0.961	12.03	16.64	1.80	—	−3.36	6.168	85.21
	solid, β	2950	20.8	7.0	—	17.80	—	—	—	4.270	89.96

SELECTED VALUES OF CHEMICAL THERMODYNAMIC PROPERTIES

(From National Bureau of Standards Technical Notes 270-3, 270-4, 270-5, 270-6, 270-7 and 270-8

D. D. Wagman, W. H. Evans, V. B. Parker, R. H. Schumm, S. M. Bailey, I. Halow, K. L. Churney, and R. L. Nuttall

The compounds listed in the table represent only a small fraction of those for which data are given in the six Technical Notes referenced above. Copies of these Technical Notes may be purchased from the Superintendent of Documents, U.S. Government Printing Office, Washington, D.C. 20402. Conversion of units used in the table to SI units or other units may be accomplished by employing the following factors.

CONVERSION FACTORS FOR UNITS OF MOLECULAR ENERGY

	$J\ mol^{-1}$	$cal\ mol^{-1}$	$cm^3\ atm\ mol^{-1}$	$kWh\ mol^{-1}$	$Btu\ lb^{-1}\ mol^{-1}$	$cm^{-1}\ molecule^{-1}$	$eV\ molecule^{-1}$
$J\ mol^{-1}$ =	1	2.390057×10^{-1}	9.86923	2.77778×10^{-7}	0.429923	8.35940×10^{-2}	1.036409×10^{-5}
$cal\ mol^{-1}$ =	4.18400[a]	1	41.2929	1.162222×10^{-6}	1.798796	3.49757×10^{-1}	4.33634×10^{-5}
$cm^3\ atm\ mol^{-1}$ =	0.1013250[a]	2.42173×10^{-2}	1	2.81458×10^{-8}	4.35619×10^{-2}	8.47016×10^{-3}	1.050141×10^{-6}
$kWh\ mol^{-1}$ =	3,600,000	860,421	3.55292×10^{7}	1	1,547,721	300,938	37.3107
$Btu\ lb^{-1}\ mol^{-1}$ =	2.32600[a]	5.55927×10^{-1}	22.9558	6.46111×10^{-7}	1	1.944396×10^{-1}	2.41069×10^{-5}
$cm^{-1}\ molecule^{-1}$ =	11.96258	2.85912	118.0614	3.32294×10^{-6}	5.14299	1	1.239812×10^{-4}
$eV\ molecule^{-1}$ =	96487.0[a]	23060.9	952,252	2.68019×10^{-2}	41482.0	8065.73	1

[a] These numbers represent the fundamental values used in deriving the data in the table. The remaining factors were obtained by applying the relationships:

$$n_{ij} = n_{ik} \cdot n_{kj} \qquad n_{ii} = n_{ik} \cdot n_{ki} = 1$$

INTRODUCTION

Substances and Properties Included in the Tables

The tables contain values of the enthalpy and Gibbs energy of formation, enthalpy, entropy and heat capacity at 298.15 K (25°C), and the enthalpy of formation at 0 K, for inorganic substances and organic molecules contianing not more than two carbon atoms.

No values are given in these tables for metal alloys or other solid solutions, fused salts, or for substances of undefined chemical composition.

Physical States

The physical state of each substance is indicated in the column headed "State" as crystalline solid (c), liquid (l), glassy or amorphous (amorp), or gaseous (g). For solutions, the physical state is that normal for the indicated solvent at 298.15 K. Isomeric substances or various crystalline modifications of a given substance are designated by a number following the letter designation, as c2, g2, etc.

Definition of Symbols

The symbols used in these tables are defined as follows: P = pressure; V = volume; T = absolute temperature; S = entropy; H = enthalpy (heat content); G = H − TS = Gibbs energy (formerly the free energy); $C_p = (dH/dT)_p$ = heat capacity at constant pressure.

Conventions Regarding Pure Substances

The values of the thermodynamic properties of the pure substances given in these tables are for the substances in their standard states (indicated by the superscript ° on the thermodynamic symbol). These standard states are defined as follows:

1. For a pure solid or liquid, the standard state is the substance in the condensed phase under a pressure of one atmosphere.*
2. For a gas the standard state is the hypothetical ideal gas at unit fugacity, in which state the enthalpy is that of the real gas at the same temperature and zero pressure.

The values of ΔHf° and ΔGf° given in the tables represent the change in the appropriate thermodynamic quantity when one gram-formula weight of the substance in its standard state is formed isothermally at the indicated temperature from the elements, each in its appropriate standard reference state. The standard reference state at 298.15 K for each element except phosphorus has been chosen to be the standard state that is thermodynamically stable at that temperature and one atmosphere pressure. For phosphorus the standard reference state is the crystalline white form; the more stable forms have not been well characterized thermochemically. The same reference states have been maintained for the elements at 0 K except for the liquid elements bromine and mercury, for which the reference states have been chosen as the stable crystalline forms.

The value of $H°_{298} - H°_0$ represents the enthalpy difference for the given substance between 298.15 K and 0 K. If the indicated standard state at 298.15 K is the gas, the corresponding state at 0 K is the hypothetical ideal gas; if the state at 298.15 K is solid or liquid, the corresponding state at 0 K is the thermodynamically stable crystalline solid, unless otherwise specifically indicated.

The values of S° represent the virtual or "thermal" entropy of the substance in the standard state at 298.15 K, omitting contributions from nuclear spins. Isotope mixing effects, etc., are also excluded except in the case of the hydrogen-deuterium system. Where data have been available only for a particular isotope, they have been corrected when possible to the normal isotopic composition.

The values of the enthalpies of formation of gaseous ionic species are computed on the convention that the value of ΔHf° for the electron is zero. Conversions between 0 and 298.15 K are claculated using the value of $H°_{298} - H°_0$ = 1.481 kcal per mole of electrons, and assuming that the values of $H°_{298} - H°_0$ for the ionized and nonionized molecules are the same.

Conventions Regarding Solutions

For all dissolved substances the composition of the solvent is indicated following the chemical formula of the solute. In most instances the number of moles of solvent associated with 1 mole of solute is stated explicitly. In some cases the concentration of the solute cannot be specified. For aqueous solutions this is indicated in the State column by "au" (aqueous, unspecified). Such solutions may be assumed to be "dilute".

* One standard atmosphere equals 101325 pascal.

The standard state for a strong electrolyte in aqueous solution is the ideal solution at unit mean molality (unit activity). For a non-dissociating solute in aqueous solution the standard state is the ideal solution at unit molality.

The value of ΔHf° for a solute in its standard state is equal to the apparent molal enthalpy of formation of the substance in the infinitely dilute solution, since the enthalpy of dilution of an ideal solution is zero. At this dilution the partial molal enthalpy is equal to the apparent molal quantity. At concentrations other than the standard state, the value of ΔHf° represents the apparent enthalpy of the reaction of formation of the solution from the elements comprising the solute, each in its standard reference state, and the appropriate total number of moles of solvent. In this representation the value of ΔHf° for the solvent is not required. The experimental value for an enthalpy of dilution is obtained directly as the difference between the two values of ΔHf° at the corresponding concentrations. At finite concentrations the partial molal enthalpy of formation differs from the apparent enthalpy. In some instances the partial molal enthalpy of formation is given in the Tables. In this case the concentration designation is preceded by "D:".

The values for the thermodynamic properties for an individual ion in aqueous solution are for that undissociated ion in the standard state and are based on the convention that ΔHf°, ΔGf°, S°, and Cp° for H^+(a) are zero. The properties of the neutral strong electrolyte in aqueous solution in the standard state are equal to the sum of these values for the appropriate number of ions assumed to constitute the molecule of the given electrolyte. By adopting the above convention with respect to H^+(a), it follows that for an individual ionic species the G=H=S relation becomes

$$\Delta Gf^\circ = \Delta Hf^\circ - T\{\Delta Sf^\circ + (n/2)S^\circ[H_2(gs)]\}$$

with n = the algebraic value of the ionic charge. For neutral electrolytes and gaseous ions the normal consistency relation holds (see below).

Unit of Energy and Fundamental Constants

All of the energy values given in these tables are expressed in terms of the thermochemical calorie. This unit, defined as equal to 4.1840 joules, has been generally accepted for presentation of chemical thermodyanmic data. Values reported in other units have been converted to calories by means of the conversion factors for molecular energy given in the table which precedes this discussion.

The following values of the fundamental physical constants have been used in these calculations:

- R = gas constant = 8.3143 ± 0.0012 J/deg mol = 1.98717 ± 0.00029 cal/deg mol
- F = Faraday constant = 96487.0 ± 1.6 coulombs/mol = 23060.9 ± 0.4 cal/V equivalent
- Z = Nhc = 11.96258 ± 0.00107 J/cm^{-1} mol = 2.85912 ± 0.00026 cal/cm^{-1} mol
- c_2 = second radiation constant = hc/k = 1.43879 ± 0.00015 cm deg 0°C = 273.15 K

These constants are consistent with those given in the Table of Physical Constants, recommended by the National Academy of Sciences — National Research Council.* The formula weights listed in the tables have been calculated for the empirical molecular formula given in the Formula and Description column.

Internal Consistency of the Tables

The various aspects of internal consistency are specified below:

1. Subsidiary and auxiliary quantities used. All of the values given in these tables have been calculated from the original articles, using consistent values for all subsidiary and auxiliary quantities. The original data were corrected where possible for differences in energy units, molecular weights, temperature scales, etc. Thus we have sought to maintain a uniform scale of energies for all substances in the tables.

2. Physical and thermodynamic relationships for the tabulated properties of a substance. The tabulated values of the properties of a substance satisfy all the known physical and thermodynamic relationships among these properties. The quantities ΔHf°, ΔGf°, and S° at 298.15 K satisfy the relation (within the assumed uncertainty)

$$\Delta Gf^\circ = \Delta Hf^\circ - T\,\Delta Sf^\circ$$

Substance			0 K	298.15 K (25°C)				
Formula and Description	State	Formula weight	ΔHf°_0 kcal/mol	ΔHf°	ΔGf° kcal/mol	$H^\circ_{298} - H^\circ_0$	S° cal/deg mol	C°_p
ACTINIUM								
Ac	c	227.0280	0	0	0	—	13.5	6.5
	g	227.0280	—	97.0	87.6	1.481	44.92	4.98
ALUMINUM								
Al	c	26.9815	0	0	0	1.094	6.77	5.82
	g	—	77.44	78.0	68.3	1.654	39.30	5.11
Al^+	g	—	215.476	217.517	—	—	—	—
Al^{2+}	g	—	649.663	653.185	—	—	—	—
Al^{3+}	g	—	1305.70	1310.70	—	—	—	—
std. state, m = 1	aq	—	—	−127.	−116.	—	−76.9	—
Al_2	g	53.9630	116.	116.14	103.57	2.33	55.7	8.7
AlO_2^-								
std. state, m = 1	aq	58.9803	—	−219.6	−196.7	—	−5.	—
Al_2O_3								
α, corundum	c	101.9612	−397.59	−400.5	−378.2	2.394	12.17	18.89
δ	c	—	—	−398.	—	—	—	—
ϱ	c	—	—	−391.	—	—	—	—
κ	c	—	—	−397.	—	—	—	—
γ	c	—	—	−395.	—	—	—	—
	amorp	—	—	−390.	—	—	—	—
$Al_2O_3 \cdot H_2O$								
boehmite	c	119.9765	—	−472.0	−436.3	—	23.15	31.37
diaspore	c	—	—	−478.	−440.	—	16.86	25.22
$Al_2O_3 \cdot 3H_2O$								
gibbsite	c	156.0072	—	−612.5	−546.7	—	33.51	44.49
bayerite	c	—	—	−610.1	—	—	—	—
AlH_3	c	30.0054	—	−11.	—	—	—	—
$AlOH^{2+}$								
std. state, m = 1	aq	43.9889	—	—	−165.9	—	—	—
$Al(OH)_3$	amorp	78.0036	—	−305.	—	—	—	—
AlF_3	c	83.9767	−358.02	−359.5	−340.6	2.778	15.88	17.95
AlF_3	g	—	−287.01	−287.9	−284.0	3.37	66.2	14.97
un-ionized; std. state, m = 1	aq	—	—	−363.	−338.	—	−6.	—
ionized; std. state, m = 1	aq	—	—	−366.	−316.	—	−86.8	—

* NBS Technical News Bulletin, October 1963. See also Report of the CODATA Task Group on Fundamental Constants, CODATA Bulletin 11, December 1973.

Substance			0 K	298.15 K (25°C)				
Formula and Description	State	Formula weight	ΔHf_0° kcal/mol	ΔHf°	ΔGf° kcal/mol	$H_{298}^\circ - H_0^\circ$	S° cal/deg mol	C_p°
$AlCl_3$	c	133.3405	−168.02	−168.3	−150.3	4.104	26.45	21.95
	g	—	—	−139.4	—	—	—	—
std. state, m = 1	aq	—	—	−247.	−210.	—	−36.4	—
$AlCl_3 \cdot 6H_2O$	c	241.4325	—	−643.3	—	—	—	—
Al_2Cl_6	g	266.6810	—	−308.5	−291.7	—	117.	—
$AlBr_3$	c	266.7085	—	−126.0	—	—	—	24.3
	g	—	—	−101.6	—	—	—	—
std. state, m = 1	aq	—	—	−214.	−191.	—	−17.8	—
Al_2Br_6	g	533.4170	—	−232.0	—	—	—	—
AlI_3	c	407.6947	—	−75.0	−71.9	—	38.	23.6
	g	—	—	−49.6	—	—	—	—
	aq	—	—	−165.8	—	—	—	—
Al_2I_6	g	815.3894	—	−123.5	—	—	—	—
Al_2S_3	c	150.1550	—	−173.	—	—	—	—
$Al_2(SO_4)_3$	c	342.1478	—	−822.38	−740.95	—	57.2	62.00
std. state, m = 1	aq	—	—	−906.	−766.	—	−139.4	—
$Al_2(SO_4)_3 \cdot 6H_2O$	c	450.2398	—	−1269.53	−1104.82	—	112.1	117.8
Al_2Se_3	c	290.843	—	−135.	—	—	—	—
Al_2Te_3	c	436.763	—	−78.	—	—	—	—
AlN	c	40.9882	−74.80	−76.0	−68.6	0.925	4.82	7.20
$Al(NO_3)_3$								
std. state, m = 1	aq	212.9962	—	−276.	−196.	—	28.1	—
$Al(NO_3)_3 \cdot 6H_2O$	c	321.0882	—	−681.28	−526.74	—	111.8	103.5
$AlCl_3 \cdot NH_3$	c	150.3711	—	−212.6	—	—	—	—
$AlCl_3 \cdot 3NH_3$	c	184.4323	—	−283.0	—	—	—	—
$AlCl_3 \cdot 6NH_3$	c	235.5242	—	−363.4	—	—	—	94.
$AlBr_3 \cdot NH_3$	c	283.7391	—	−177.6	—	—	—	—
$AlBr_3 \cdot 3NH_3$	c	317.8003	—	−252.7	—	—	—	—
$AlBr_3 \cdot 6NH_3$	c	368.8922	—	−343.0	—	—	—	—
$AlI_3 \cdot NH_3$	c	424.7253	—	−119.0	—	—	—	—
$AlI_3 \cdot 3NH_3$	c	458.7865	—	−205.9	—	—	—	—
$AlI_3 \cdot 6NH_3$	c	509.8784	—	−312.9	—	—	—	—
$NH_4Al(SO_4)_2$	c	237.1433	—	−562.2	−487.2	—	51.7	54.12
std. state, m = 1	aq	—	—	−593.	−491.	—	−40.2	—
$NH_4Al(SO_4)_2 \cdot 12H_2O$	c	453.3274	—	−1420.26	−1180.21	—	166.6	163.3
AlP	c	57.9553	—	−39.8	—	—	—	—
$AlPO_4$								
berlinite	c	121.9529	−411.40	−414.4	−386.7	3.528	21.70	22.27
$AlAs$	c	101.9031	—	−27.8	—	—	—	—
Al_4C_3	c	143.9594	−48.71	−49.9	−46.9	3.936	21.26	27.91
$Al(CH_3)_3$	c	72.0867	−29.76	—	—	—	—	—
	liq	—	—	−32.6	−2.4	8.114	50.05	37.19
	g	—	—	−17.7	—	—	—	—
$Al_2(CH_3)_6$	g	144.1734	—	−55.19	−2.34	—	125.4	—
$Al_2Si_2O_7 \cdot 2H_2O$								
kaolinite	c	258.1615	—	−979.6	−903.0	—	48.5	58.62
halloysite	c	—	—	−975.0	−898.5	—	48.6	58.86
$Al_6Si_2O_{13}$								
mullite	c	426.0532	—	−1632.8	−1541.2	—	61.	77.94
ANTIMONY								
Sb								
III	c	121.75	0	0	0	1.410	10.92	6.03
IV, explosive	amorp	—	—	2.54	—	—	—	—
	g	—	62.63	62.7	53.1	1.481	43.06	4.97
Sb^+	g	—	261.91	263.46	—	—	—	—
Sb^{2+}	g	—	643.1	646.1	—	—	—	—
Sb^{3+}	g	—	1227.1	1231.6	—	—	—	—
Sb^{4+}	g	—	2245.4	2251.4	—	—	—	—
SbO	g	137.749	48.	47.67	—	2.122	—	—
SbO^+								
std. state, m = 1	aq	—	—	—	−42.33	—	—	—
SbO_2^-								
std. state, m = 1	aq	153.749	—	—	−81.32	—	—	—
Sb_2O_3	aq	291.498	—	−164.9	—	—	—	—
Sb_2O_4	c	307.498	—	−216.9	−190.2	—	30.4	27.39
Sb_2O_5	c	323.497	—	−232.3	−198.2	—	29.9	—
SbH_3	g	124.774	36.625	34.681	35.31	2.502	55.61	9.81
$HSbO_2$								
undissoc.; std. state, m = 1	aq	154.757	—	−116.6	−97.4	—	11.1	—
$Sb(OH)_3$	c	172.772	—	—	−163.8	—	—	—
undissoc.; std. state, m = 1	aq	—	—	−184.9	−154.1	—	27.8	—
H_3SbO_4	aq	188.772	—	−216.8	—	—	—	—
SbF_3	c	178.745	—	−218.8	—	—	—	—
$SbCl_3$	c	228.109	—	−91.34	−77.37	—	44.0	25.8
$SbCl_3$	g	—	−74.57	−75.0	−72.0	4.269	80.71	18.33
$SbCl_5$	liq	299.015	—	−105.2	−83.7	—	72.	—
$SbOCl$	c	173.202	—	−89.4	—	—	—	—
	g	—	−25.	−25.5	—	—	—	—
$SbBr_3$	c	361.477	—	−62.0	−57.2	—	49.5	—
	g	—	−41.03	−46.5	−53.5	4.727	89.09	19.17
in CS_2	—	—	—	−58.4	—	—	—	—
SbI_3	c	502.463	—	−24.0	—	—	—	—
	aq	—	—	−23.6	—	—	—	—
Sb_2S_3								
black	c	339.692	—	−41.8	−41.5	—	43.5	28.65
orange	amorp	—	—	−35.2	—	—	—	—
$Sb_2(SO_4)_3$	c	531.685	—	−574.2	—	—	—	—
ARGON								
Ar	g	39.948	0	0	0	1.481	36.9822	4.9679
std. state, m = 1	aq	—	—	−2.9	3.9	—	14.2	—
Ar^+	g	—	363.42	364.90	—	—	—	—
Ar^{2+}	g	—	1000.5	1003.5	—	—	—	—
Ar^{3+}	g	—	1943.9	1948.3	—	—	—	—
ARSENIC								
As								
α, gray, metallic	c	74.9216	0	0	0	1.226	8.4	5.89
γ, yellow, cubic	c	—	—	3.5	—	—	—	—
β	amorp	—	—	1.0	—	—	—	—
	g	—	72.04	72.3	62.4	1.481	41.61	4.968
As^+	g	—	298.38	300.12	—	—	—	—
As^{2+}	g	—	764.42	767.64	—	—	—	—

SELECTED VALUES OF CHEMICAL THERMODYNAMIC PROPERTIES (Continued)

Substance			0 K	298.15 K (25°C)				
Formula and Description	State	Formula weight	ΔHf_0° kcal/mol	ΔHf°	ΔGf° kcal/mol	$H_{298}^\circ - H_0^\circ$	S° cal/deg mol	C_p°
As^{3+}	g	—	1417.44	1422.14	—	—	—	—
As^{4+}	g	—	2573.58	2579.76	—	—	—	—
As_2	g	149.8432	53.30	53.1	41.1	2.251	57.2	8.366
AsO	g	90.9210	16.88	16.72	—	2.101	—	—
AsO^+								
undissoc.; std. state, m = 1	aq	—	—	—	−39.15	—	—	—
$AsO_2^-{}_{3-}$								
std. state, m = 1	aq	106.9204	—	−102.54	−83.66	—	9.7	—
AsO_4								
std. state, m = 1	aq	138.9192	—	−212.27	−155.00	—	−38.9	—
As_2O_5	c	229.8402		−221.05	−187.0		25.2	27.85
$As_2O_5 \cdot 4H_2O$	c	301.9016	—	−503.0	—	—	—	—
AsH_3	g	77.9455	17.70	15.88	16.47	2.438	53.22	9.10
$HAsO_2$								
undissoc.; std. state, m = 1	aq	107.9284	—	−109.1	−96.25	—	30.1	—
$HAsO_4^{2-}$								
undissoc.; std. state, m = 1	aq	139.9272	—	−216.62	−170.82	—	−0.4	--
$H_2AsO_3^-$								
undissoc.; std. state, m = 1	aq	124.9357	—	−170.84	−140.35	—	26.4	—
$H_2AsO_4^-$								
undissoc.; std. state, m = 1	aq	140.9351	—	−217.39	−180.04	—	28.	—
H_3AsO_4	c	141.9431	—	−216.6	—	—	—	—
	aq	—	—	−216.2	—	—	—	—
AsF_3	liq	131.9168	—	−196.3	−185.04	—	43.31	30.25
	g	—	−186.82	−187.80	−184.22	3.413	69.07	15.68
$AsCl_3$	liq	181.2806	—	−72.9	−62.0	—	51.7	—
	g	—	−62.12	−62.5	−59.5	4.137	78.17	18.10
$AsBr_3$	c	314.6486	—	−47.2	—	—	—	—
	g	—	−25.55	−31.	−38.	4.569	86.94	18.92
AsI_3	c	455.6348	−13.91	−13.9	−14.2	5.964	50.92	25.28
	g	—	—	—	—	4.834	92.79	19.27
As_2S_2	c	213.9712	—	−34.1	—	—	—	—
As_2S_3	c	246.0352	—	−40.4	−40.3	—	39.1	27.8
$NH_4H_2AsO_3$								
std. state, m = 1, (NH_4^+ + $H_2AsO_3^-$)	aq	143.9743	—	−202.51	−159.32	—	53.5	—
$(NH_4)_2HAsO_4$	c	176.0043	—	−282.4	—	—	—	—
std. state, m = 1, ($2NH_4^+$ + $HAsO_4^{2-}$)	aq	—	—	−279.96	−208.76	—	53.8	—
BARIUM								
Ba	c	137.34	0	0	0	1.65	15.0	6.71
	g	—	43.2	43.	35.	1.481	40.663	4.968
Ba^+	g	—	163.39	164.67	—	—	—	—
Ba^{2+}	g	—	394.09	396.86	—	—	—	—
std. state, m = 1	aq	—	—	−128.50	−134.02	—	+2.3	—
BaO_2	c	169.339	—	−151.6	—	—	—	16.0
$BaO_2 \cdot H_2O$	c	187.354	—	−222.3	—	—	—	—
$BaO_2 \cdot 8H_2O$	c	313.462	—	−718.6	—	—	—	—
BaH	g	138.348	53.6	53.	47.	2.082	52.29	7.19
BaH_2	c	139.356	—	−42.7	—	—	—	—
$Ba(OH)_2$	c	171.355	—	−225.8	—	—	—	—
$Ba(OH)_2 \cdot H_2O$	c	189.370	—	−298.4	—	—	—	—
$Ba(OH)_2 \cdot 3H_2O$	c	225.401	—	−442.0	—	—	—	—
$Ba(OH)_2 \cdot 8H_2O$	c	315.477	—	−798.8	−667.6	—	102.	—
BaF_2	c	175.337	−288.19	−288.5	−276.5	3.452	23.03	17.02
	g	—	−195.0	−195.5	−198.0	3.262	71.91	12.85
std. state, m = 1	aq	—	—	−287.50	−267.30	—	−4.3	—
$BaCl_2$	c	208.246	−205.35	−205.2	−193.7	3.993	29.56	17.96
	g	—	−125.37	−125.7	−128.5	3.511	77.72	13.43
std. state, m = 1	aq	—	—	−208.40	−196.76	—	29.3	—
$BaCl_2 \cdot H_2O$	c	226.261	—	−277.4	−252.32	—	39.9	—
$BaCl_2 \cdot 2H_2O$	c	244.277	—	−348.98	−309.86	—	48.5	38.71
$Ba(ClO_2)_2$	c	272.244	−162.6	−127.0	—	—	47.	—
$Ba(ClO_3)_2$	c	304.242	—	−184.4	—	—	—	—
$Ba(ClO_4)_2$	c	336.241	—	−191.2	—	—	—	—
$Ba(ClO_4)_2 \cdot 3H_2O$	c	390.287	—	−404.3	−303.7	—	94.	—
$BaBr_2$	c	297.158	—	−181.0	−176.1	—	35.	—
	g	—	−101.4	−105.	−113.	3.9	79.	14.7
$BaBr_2$								
std. state, m = 1	aq	—	—	−186.60	−183.72	—	41.7	—
∞ H_2O	aq	—	—	−186.60	—	—	—	—
$BaBr_2 \cdot H_2O$	c	315.173	—	−255.3	—	—	—	—
$BaBr_2 \cdot 2H_2O$	c	333.189	—	−326.5	−294.1	—	54.	—
$Ba(BrO_3)_2$	c	393.154	—	−171.65	−130.1	—	59.	—
$Ba(BrO_3)_2$								
std. state, m = 1	aq	—	—	−160.56	−125.16	—	79.6	—
$Ba(BrO_3)_2 \cdot H_2O$	c	411.170	−236.99	−243.84	−188.6	9.94	68.9	53.5
BaI_2	c	391.149	—	−143.9	—	—	—	—
std. state, m = 1	aq	—	—	−154.88	−158.68	—	55.5	—
$BaI_2 \cdot H_2O$	c	409.164	—	−219.7	—	—	—	—
$BaI_2 \cdot 2H_2O$	c	427.179	—	−290.8	—	—	—	—
$BaI_2 \cdot 7H_2O$	c	517.256	—	−639.7	—	—	—	—
$Ba(IO_3)_2$	c	487.145	−243.31	−245.5	−206.7	8.84	59.6	44.8
std. state, m = 1	aq	—	—	−234.3	−195.2	—	58.9	—
$Ba(IO_3)_2 \cdot H_2O$	c	505.161	—	−316.0	−263.9	—	71.	—
BaS	c	169.404	—	−110.	−109.	—	18.7	11.80
Ba_2S_2	g	338.808	—	−90.	—	—	—	—
$BaSO_3$	c	217.402	—	−281.9	—	—	—	—
$BaSO_4$	c	233.402	—	−352.1	−325.6	—	31.6	24.32
std. state, m = 1	aq	—	—	−345.82	−311.99	—	7.1	—
BaS_2O_3	c	249.466	—	—	—	—	—	40.7
$BaSl_2O_6$	aq	297.464	—	−415.4	—	—	—	—
$Ba(HSO_3)_2$	aq	299.480	—	−430.0	—	—	—	—
$BaI_2 \cdot 4SO_2$	c	647.400	—	−470.2	—	—	—	—
$BaSe$	c	216.30	—	−89.	—	—	—	—
$BaSeO_3$	c	264.298	—	−248.7	−231.4	—	40.	—
$BaSeO_4$	c	280.298	—	−274.0	−249.7	—	42.	—
BaN_2	c	165.353	—	−41.	—	—	—	—
$Ba(N_3)_2 \cdot H_2O$	c	239.396	—	−73.7	−25.1	—	45.	—
$Ba(NO_2)_2$	c	229.351	—	−183.6	—	—	—	—

Substance			0 K	298.15 K (25°C)				
Formula and Description	State	Formula weight	ΔHf°_0 kcal/mol	ΔHf°	ΔGf° kcal/mol	$H^\circ_{298} - H^\circ_0$	S° cal/deg mol	C°_p
	aq	—	—	−178.5	—	—	—	—
$Ba(NO_3)_2$	c	261.350	—	−237.11	−190.42	—	51.1	36.18
std. state, m = 1	aq	—	—	−227.62	−187.24	—	72.3	—
$BaHPO_4$	c	233.319	—	−433.7	—	—	—	—
$Ba(H_2PO_2)_2$	c	267.317	—	−421.2	—	—	—	—
$Ba(H_2PO_2)_2 \cdot H_2O$	c	285.332	—	−490.5	—	—	—	—
$Ba(H_2PO_4)_2$	c	331.315	—	−747.	—	—	—	—
$BaHAsO_4 \cdot H_2O$	c	295.283	—	−412.6	—	—	—	—
$Ba(H_2AsO_4)_2 \cdot 2H_2O$	c	455.241	—	−696.9	—	—	—	—
BaC_2	c	161.362	—	−18.				
$BaCO_3$								
witherite	c	197.349	—	−290.7	−271.9	—	26.8	20.40
std. state, m = 1	aq	—	—	−290.34	−260.19	—	−11.3	—
BaC_2O_4	c	225.360	—	−327.1	—	—	—	—
$BaC_2O_4 \cdot 2H_2O$	c	261.391	—	−471.1	—	—	—	—
$Ba(HCO_3)_2$								
std. state, m = 1	aq	259.375	—	−459.28	−414.54	—	45.9	—
$Ba(C_2H_3O_2)_2$	c	255.430	—	−354.8	—	—	—	—
std. state, m = 1	aq	—	—	−360.82	−310.60	—	43.7	—
$BaCN_2$	c	177.365	—	−63.6	—	—	—	—
$Ba(CN)_2$	c	189.376	—	−52.2	—	—	—	—
	aq	—	—	−55.0	—	—	—	—
$BaO \cdot SiO_2$	c	213.424	—	−388.05	−368.13	—	26.2	21.51
glassy	amorp	—	—	−376.	—	—	—	—
$BaO \cdot 2SiO_2$	c	273.509	—	−609.0	−576.2	—	36.6	32.05
$2BaO \cdot SiO_2$	c	366.764	—	−546.8	−519.8	—	42.1	32.24
$2BaO \cdot 3SiO_2$	c	486.933	—	−1000.2	−947.2	—	61.7	53.68
$Ba_2Fe(CN)_6$	aq	486.634	—	−145.7	—	—	—	—
$Ba_2Fe(CN)_6 \cdot 6H_2O$	c	594.726	—	−567.0	—	—	—	—
$Ba_3[Fe(CN)_6]_2$								
std. state, m = 1	aq	835.928	—	−116.9	−53.5	—	136.1	—
$BaMnO_4$								
barium manganate	c	256.276	—	—	−267.5	—	—	—
$BaCrO_4$	c	253.334	—	−345.6	−321.53	—	379	—
$BaMoO_4$	c	297.278	—	−370.	−344.1	—	33.	33.6
$BaWO_4$	c	385.188	—	−407.	—	—	—	—
$BaTiO_3$	c	233.238	—	−396.7	−375.8	—	25.8	24.49
$BaZrO_3$	c	276.558	—	−425.3	−405.0	—	29.8	24.31
BERYLLIUM								
Be	c	9.0122	0	0	0	0.466	2.27	3.93
	g	—	76.49	77.5	68.5	1.481	32.543	4.968
Be^+	g	—	291.474	293.965	—	—	—	—
Be^{2+}	g	—	711.426	715.398	—	—	—	—
Std. state, m = 1	aq	—	—	−91.5	−90.75	—	−31.0	—
BeO	c	25.0116	−144.86	−145.7	−138.7	0.669	3.38	6.10
	g	—	—	28.	—	—	—	—
BeO_2^{2-}								
std. state, m = 1	aq	41.0110	—	−189.0	−153.0	—	−38.	—
BeH	g	10.0202	75.	75.6	68.3	2.062	42.21	6.95
BeH_2	c	11.0282	—	−4.60	—	—	—	—
$Be(OH)_2$								
α	c	43.0269	—	−215.7	−194.8	—	12.4	—
β	c	—	—	−216.5	−195.4	—	12.	—
fresh precipitated	amorp	—	—	−214.6	—	—	—	—
$Be_3(OH)_3^{3+}$								
std. state, m = 1	aq	78.0587	—	—	−430.6	—	—	—
BeF_2								
α, quartz	c	47.0090	−244.85	−245.4	−234.1	2.024	12.75	12.39
β, cristobalite	c	—	—	−244.7	—	—	—	—
glassy	amorp	—	—	−244.3	—	—	—	—
$BeCl_2$								
α	c	79.9182	−117.40	−117.2	−106.5	2.863	19.76	15.50
β	c	—	−118.57	−118.5	−107.3	2.729	18.12	14.92
$BeCl_2 \cdot 4H_2O$	c	151.9796	—	−432.2	—	—	—	—
Be_2Cl_4	g	159.8364	—	−202.	—	—	—	—
$BeBr_2$	c	168.8302	—	−84.5	—	—	—	—
BeI_2	c	262.8210	—	−46.0	—	—	—	—
BeS	c	41.0762	—	−56.0	—	—	—	—
$BeSO_4$								
α, tetragonal	c	105.0738	−285.505	−288.05	−261.44	3.125	18.62	20.48
std. state, m = 1	aq	—	—	−308.8	−268.72	—	−26.2	—
$BeSO_4 \cdot H_2O$	c	123.0891	—	−364.2	—	—	—	—
$BeSO_4 \cdot 2H_2O$	c	141.1045	−429.814	−435.74	−381.99	5.865	39.01	36.63
$BeSO_4 \cdot 4H_2O$								
tetragonal	c	177.1352	−569.682	−579.29	−497.29	8.306	55.68	51.77
$BeSeO_4$	c	151.970	—	−212.8	—	—	—	—
std. state, m = 1	aq	—	—	−234.7	−196.2	—	−18.1	—
$BeSeO_4 \cdot 2H_2O$	c	188.000	—	−360.3	—	—	—	—
$BeSeO_4 \cdot 4H_2O$	c	224.031	—	−505.0	—	—	—	—
Be_3N_2								
α, cubic	c	55.0500	—	−140.6	—	—	—	—
β, hexagonal	c	—	—	−136.5	—	—	—	—
$Be(NO_3)_2$	aq	133.0220	—	−191.0	—	—	—	—
Be_2C	c	30.0356	—	−28.0	—	—	—	—
$BeCO_3$	c	69.0216	—	−245.	—	—	—	—
Be_2SiO_4	c	110.1080	−510.77	−513.7	−485.8	2.922	15.37	22.84
$Be(BO_2)_2$	g	94.6318	—	−325.	—	—	—	—
$BeO \cdot Al_2O_3$								
chrysoberyl	c	126.9728	−546.32	−549.9	−520.7	3.128	15.84	25.19
$BeO \cdot 3Al_2O_3$	c	330.8952	−1335.65	−1344.9	−1271.6	8.153	42.0	63.38
$BeMoO_4$	c	168.950	—	−328.	—	—	—	—
$PuBe_{13}$	c	356.21	—	−36.	—	—	—	—
BISMUTH								
Bi	c	208.980	0	0	0	1.536	13.56	6.10
	g	—	49.56	49.5	40.2	1.481	44.669	4.968
Bi^+	g	—	217.65	219.07	—	—	—	—
Bi^{2+}	g	—	602.5	605.4	—	—	—	—
Bi^{3+}	g	—	1192.0	1196.4	—	—	—	—
std. state, m = 1	aq	—	—	—	19.8	—	—	—
Bi_2	g	417.960	53.12	52.5	—	2.454	—	8.83

Substance			0 K	298.15 K (25°C)				
Formula and Description	State	Formula weight	ΔHf°_0 kcal/mol	ΔHf°	ΔGf° kcal/mol	$H^\circ_{298} - H^\circ_0$	S° cal/deg mol	C°_p
BiO^+								
std. state, m = 1	aq	—	—	—	−35.0	—	—	—
Bi_2O_3	c	465.9582	—	−137.16	−118.0	—	36.2	27.13
$BiO \cdot OH$	c	241.9868	—	—	−88.0	—	—	—
$Bi(OH)_3$	c	260.0021	—	−170.0	—	—	—	—
$BiCl_3$	c	315.339	—	−90.6	−75.3	—	42.3	25.
$BiCl_4^-$								
std. state, m = 1	aq	350.792	—	—	−115.1	—	—	—
$BiCl_6^{3-}$								
std. state, m = 1	aq	421.698	—	—	−178.51	—	—	—
$BiOCl$	c	260.4324	—	−87.7	−77.0	—	28.8	—
$Bi(OH)_2Cl$	c	278.4477	—	—	−128.71	—	—	—
$BiOBr$	c	304.8884	—	—	−71.0	—	—	—
BiI_3	c	589.6932	—	—	−41.9	—	—	—
BiI_4								
std. state, m = 1	aq	716.5976	—	—	−49.9	—	—	—
BiS	g	241.044	—	43.	29.	—	68.	—
Bi_2S_3	c	514.152	—	−34.2	−33.6	—	47.9	29.2
$Bi_2(SO_4)_3$	c	706.1448	—	−608.1	—	—	—	—
Bi_2Te_3	c	800.760	−18.43	−18.5	−18.4	7.387	62.36	28.8
$BiAsO_4$	c	347.8992	—	—	−148.	—	—	—
BORON								
B								
β	c	10.811	0	0	0	0.290	1.40	2.65
	amorp	—	—	0.9	—	0.315	1.56 + X*	2.86
	g	—	133.28	134.5	124.0	1.511	36.65	4.971
B^+	g	—	324.64	327.34	—	—	—	—
B^{++}	g	—	904.74	908.92	—	—	—	—
B^{+++}	g	—	1779.43	1785.09	—	—	—	—
B^{+++++}	g	—	15606.2	15614.9	—	—	—	—
BO_2^-								
std. state, m = 1	aq	—	—	−184.60	−162.27	—	−8.9	—
B_2O_3	c	69.6202	−302.731	−304.20	−285.30	2.223	12.90	15.04
	amorp	—	—	−299.84	−282.6	—	18.6	14.6
	g	—	−201.4	−201.67	−198.85	3.426	66.85	15.98
$B_4O_7^{--}$								
std. state, m = 1	aq	155.2398	—	—	−622.6	—	—	—
BH	g	11.8190	106.7	107.46	100.29	2.065	41.05	6.97
BH_3	g	13.8349	—	24.	—	—	—	—
BH_4^-								
std. state, m = 1	aq	14.8429	—	11.51	27.31	—	26.4	—
B_2H_6	g	27.6698	12.29	8.5	20.7	2.857	55.45	13.60
		—	—	6.0	23.0	—	39.4	—
B_4H_{10}	g	53.3237	—	15.8	—	—	—	—
H_3BO_3	c	61.8331	−258.312	−261.55	−231.60	—	21.23	19.45
	g	—	—	−237.6	—	—	—	—
un-ionized; std. state, m = 1	aq	—	—	−256.29	−231.56	—	38.8	—
$B(OH)_4^-$								
std. state, m = 1	aq	78.8405	—	−321.23	−275.65	—	24.5	—
$H_2B_4O_7$								
std. state, m = 1, (undissoc.)	aq	157.2557	—	—	−650.1	—	—	—
BF_3	g	67.8062	−271.082	−271.75	−267.77	2.784	60.71	12.06
BF_4^-								
std. state, m = 1	aq	86.8046	—	−376.4	−355.4	—	43.	—
B_2F_4	g	97.6156	−343.48	−344.2	−337.1	4.08	75.8	18.90
BCl_3	c	117.170	−105.40	—	—	—	—	—
BCl_3	liq	117.170	—	−102.1	−92.6	6.88	49.3	25.5
	g	—	−96.28	−96.50	−92.91	3.362	69.31	14.99
$BClF_2$	g	84.2608	—	−212.8	−209.4	—	65.	—
BCl_2F	g	100.7154	—	−154.2	−150.9	—	68.	—
BBr_3	liq	250.538	—	−57.3	−57.0	—	54.9	—
	g	—	−43.83	−49.15	−55.56	3.755	77.47	16.20
$BBrF_2$	g	128.7168	—	—	—	3.054	68.42	13.49
$BFBr_2$	g	189.6274	—	—	—	3.386	74.06	14.89
BCl_2Br	g	161.626	—	—	—	3.477	74.16	15.39
$BClBr_2$	g	206.082	—	—	—	3.609	76.90	15.78
BI_3	g	391.5242	18.	17.00	4.96	4.024	83.43	16.92
BS	g	42.875	81.	81.74	69.02	2.085	51.65	7.18
BN	c	24.8177	−60.10	−60.8	−54.6	0.628	3.54	4.71
NH_4BO_2								
std. state, m = 1	aq	60.8484	—	−216.27	−181.24	—	18.2	—
NH_4BO_3	aq	76.8478	—	−195.4	—	—	—	—
B_2C	g	33.6332	181.	—	—	—	—	—
B_4C	c	55.2552	−16.93	−17.	−17.	1.343	6.48	12.62
$B(CH_3)_3$	liq	55.9162	—	−34.2	−7.7	—	57.1	—
	g	—	−23.38	−29.7	−8.6	3.831	75.2	21.15
$B(C_2H_5)_3$	liq	97.9974	−42.54	−46.5	2.2	13.02	80.47	57.65
$BH(OCH_3)_2$								
dimethoxyborane	liq	73.8879	—	−144.7	−113.2	—	57.	—
$B(OCH_3)_3$								
trimethoxyborane	liq	103.9144	—	−223.2	−178.0	11.45	67.8	45.89
$B(OC_2H_5)_3$								
triethoxyborane	liq	145.9958	—	−250.8	—	—	—	—
BSi_2	g	66.983	175.	—	—	—	—	—
$BSiC$	g	50.9082	164.	—	—	—	—	—
BROMINE								
Br	g	79.909	28.189	26.741	19.701	1.481	41.805	4.968
Br^+	g	—	301.38	301.41	—	—	—	—
Br^{2+}	g	—	799.2	800.7	—	—	—	—
Br^{3+}	g	—	1627.0	1630.0	—	—	—	—
Br^-	g	—	−53.0	−55.9	—	—	—	—
std. state, m = 1	aq	—	—	−29.05	−24.85	—	19.7	−33.9
Br_2	c	159.818	0	—	—	—	—	—
	liq	—	—	0	0	5.859	36.384	18.090
	g	—	10.923	7.387	0.751	2.323	58.641	8.61
std. state, m = 1	aq	—	—	−0.62	0.94	—	31.2	—
Br_3^-								
std. state, m = 1	aq	239.727	—	−31.17	−25.59	—	51.5	—
Br_5^-								
std. state, m = 1	aq	399.545	—	−34.0	−24.8	—	75.7	—

* X = undetermined residual entropy.

Substance			0 K	298.15 K (25°C)				
Formula and Description	State	Formula weight	ΔHf°_0 kcal/mol	ΔHf°	ΔGf° kcal/mol	$H^\circ_{298} - H^\circ_0$	S° cal/deg mol	C°_p
BrO^-								
std. state, m = 1	aq	95.9084	—	−22.5	−8.0	—	10.	—
BrO_2	c	111.9078	—	11.6	—	—	—	—
BrO_3^-								
std. state, m = 1	aq	127.9072	—	−16.03	4.43	—	38.65	—
BrO_4^-								
std. state, m = 1	aq	—	—	3.1	28.2	—	47.7	—
HBr	g	80.9170	−6.826	−8.70	−12.77	2.067	47.463	6.965
std. state, m = 1	aq	—	—	−29.05	−24.85	—	19.7	−33.9
$HBrO_3$								
std. state, m = 1	aq	128.9152	—	−16.03	4.43	—	38.65	—
BrF_3	liq	136.9042	—	−71.9	−57.5	—	42.6	29.78
	g	—	−58.41	−61.09	−54.84	3.416	69.89	15.92
BrCl	g	115.362	5.28	3.50	−0.23	2.245	57.36	8.36
CADMIUM								
Cd								
γ	c	112.40	0	0	0	1.491	12.37	6.21
α	c	—	—	−0.14	−0.14	—	12.37	—
	g	—	26.78	26.77	18.51	1.481	40.066	4.968
in Hg; two-phase amalgam	—	—	—	−5.078	−2.328	—	3.145	—
Cd^+	g	—	234.18	235.65	—	—	—	—
Cd^{2+}	g	—	624.09	627.04	—	—	—	—
std. state, m = 1	aq	—	—	−18.14	−18.542	—	−17.5	—
CdO	c	128.399	—	−61.7	−54.6	—	13.1	10.38
$Cd(OH)_2$								
precipitated	c	146.415	—	−134.0	−113.2	—	23.	—
std. state, m = 1	aq	—	—	−128.08	−93.73	—	−22.6	—
undissoc.;	aq	—	—	—	−105.8	—	—	—
CdF_2	c	150.397	—	−167.4	−154.8	—	18.5	—
std. state, m = 1	aq	—	—	−177.14	−151.82	—	−24.1	—
$CdCl_2$	c	183.306	−93.677	−93.57	−82.21	3.791	27.55	17.85
std. state, m = 1	aq	—	—	−98.04	−81.286	—	9.5	—
undissoc.; std. state, m = 1	aq	—	—	−96.8	−85.88	—	29.1	—
$CdCl_2 \cdot H_2O$	c	201.321	—	−164.54	−140.310	—	40.1	—
$CdCl_2 \cdot 5/2H_2O$	c	228.344	—	−270.54	−225.644	—	54.3	—
$Cd(ClO_4)_2$								
std. state, m = 1	aq	311.301	—	−79.96	−22.66	—	69.5	—
$Cd(ClO_4)_2 \cdot 6H_2O$	c	419.393	—	−490.6	—	—	—	—
$CdCl_2 \cdot 2HCl \cdot 7H_2O$	c	382.335	—	−654.8	—	—	—	—
$CdBr_2$	c	272.218	−72.455	−75.57	−70.82	4.235	32.8	18.32
std. state, m = 1 1	aq	—	—	−76.24	−68.24	—	21.9	—
$CdBr_2 \cdot 4H_2O$	c	344.279	—	−356.73	−298.287	—	75.6	—
CdI_2	c	366.209	−48.52	−48.6	−48.13	4.565	38.5	19.11
std. state m = 1	aq	—	—	−44.52	−43.20	—	35.7	—
CdS	c	144.464	—	−38.7	−37.4	—	15.5	—
$CdSO_4$	c	208.462	−220.720	−223.06	−196.65	4.354	29.407	23.80
std. state, m = 1	aq	—	—	−235.46	−196.51	—	−12.7	—
$CdSO_4 \cdot H_2O$	c	226.477	−292.087	−296.26	−255.46	5.582	36.814	32.16
$CdSO_4 \cdot 8/3H_2O$	c	256.502	−406.960	−413.33	−350.224	8.497	54.883	50.97
$CdSO_4 \cdot 2\frac{1}{2}H_2SO_4$	c	453.655	—	−769.6	—	—	—	—
CdSe	c	191.36	—	—	—	—	—	—
$CdSeO_3$	c	239.358	—	−137.5	−119.0	—	34.0	—
std. state, m = 1	aq	—	—	−139.8	−106.9	—	−14.4	—
$CdSeO_4$	c	255.358	—	−151.3	−127.1	—	39.3	—
std. state, m = 1	aq	—	—	−161.3	−124.0	—	−4.6	—
$CdSeO_4 \cdot H_2O$	c	273.373	—	−225.2	—	—	—	—
CdTe	c	240.00	—	−22.1	−22.0	—	24.	—
$Cd(NO_3)_2$	c	236.410	—	−109.06	—	—	—	—
std. state, m = 1	aq	—	—	−117.26	−71.76	—	52.5	—
$Cd(NO_3)_2 \cdot 2H_2O$	c	272.440	—	−252.30	—	—	—	—
$Cd(NO_3)_2 \cdot 4H_2O$	c	308.471	—	−394.11	—	—	—	—
$Cd(NH_3)^{2+}$								
std. state, m = 1	aq	129.431	—	—	−28.4	—	—	—
$Cd(NH_3)_2^{2+}$								
std. state, m = 1	aq	146.461	—	−63.6	−38.0	—	34.6	—
$Cd(NH_3)_4^{2+}$								
std. state, m = 1	aq	180.522	—	−107.6	−54.1	—	80.4	—
$CdCl_2 \cdot 2NH_3$	c	217.367	—	−152.0	−106.1	—	51.	—
$CdCl_2 \cdot 4NH_3$	c	251.428	—	−195.4	−116.2	—	79.	—
$CdCl_2 \cdot 6NH_3$	c	285.490	—	−237.9	−126.2	—	109.2	—
$CdBr_2 \cdot NH_3$	c	289.249	—	−103.3	—	—	—	—
$CdBr_2 \cdot 2NH_3$	c	306.279	—	−131.5	—	—	—	—
$CdBr_2 \cdot 6NH_3$	c	374.402	—	−218.4	—	—	—	—
$CdI_2 \cdot 2NH_3$	c	400.270	—	−104.0	—	—	—	—
$CdI_2 \cdot 6NH_3$	c	468.392	—	−173.0	—	—	—	—
$Cd_3(PO_4)_2$	c	527.143	—	—	−587.1	—	—	—
$CdCO_3$	c	172.409	—	−179.4	−160.0	—	22.1	—
CdC_2O_4	c	200.420	—	−218.1	—	—	—	—
std. state, m = 1	aq	—	—	−215.3	−179.6	—	−6.6	—
undissoc.; std. state, m = 1	aq	200.420	—	—	−185.1	—	—	—
$CdC_2O_4 \cdot 3H_2O$	c	254.466	—	—	−360.4	—	—	—
$CdCN^+$								
std. state, m = 1	aq	138.418	—	—	15.4	—	—	—
$Cd(CN)_2$	c	164.436	—	38.8	—	—	—	—
std. state, m = 1	aq	—	—	53.9	63.9	—	27.5	—
undissoc.;	aq	—	—	—	49.7	—	—	—
$Cd(CN)_3^-$								
std. state, m = 1	aq	190.454	—	—	84.8	—	—	—
$Cd(CN)_4^{2-}$								
std. state, m = 1	aq	216.471	—	102.3	121.3	—	77.	—
$Cd(CNS)_2$	c	228.564	—	12.43	—	—	—	—
std. state, m = 1	aq	—	—	18.40	25.76	—	51.4	—
undissoc.; std. state, m = 1	aq	—	—	—	23.2	—	—	—
$CdSiO_3$	c	188.484	—	−284.20	−264.0	—	23.3	21.17
$Cd(BO_2)_2$	c	198.020	—	—	−354.87	—	—	—
CALCIUM								
Ca	c	40.08	0	0	0	1.364	9.90	6.05
	g	—	42.48	42.6	34.5	1.481	36.992	4.968
Ca^+	g	—	183.45	185.05	—	—	—	—
Ca^{2+}	g	—	457.21	460.29	—	—	—	—

Substance			0 K	298.15 K (25°C)				
Formula and Description	State	Formula weight	ΔHf°_0 kcal/mol	ΔHf° kcal/mol	ΔGf° kcal/mol	$H^\circ_{298} - H^\circ_0$	S° cal/deg mol	C°_p cal/deg mol
std. state, m = 1	aq	—	—	−129.74	−132.30	—	−12.7	—
CaO	c	56.079	—	−151.79	−144.37	—	9.50	10.23
	g	—	+11.	—	—	—	—	—
CaO_2	c	72.079	—	−156.0	—	—	—	—
CaH	g	41.088	55.	54.7	47.9	2.076	48.19	7.11
CaH_2	c	42.096	—	−44.5	−35.2	—	10.	—
$CaOH^+$								
std. state, m = 1	aq	—	—	—	−171.7	—	—	—
$Ca(OH)_2$	c	74.095	—	−235.68	−214.76	—	19.93	20.91
	g	—	—	−130.	—	—	—	—
std. state, m = 1	aq	—	—	−239.68	−207.49	—	−17.8	—
CaF	g	59.078	−64.76	−65.0	−71.2	2.181	54.8	8.03
CaF_2	c	78.077	—	−291.5	−279.0	—	16.46	16.02
	g	—	−186.35	−186.8	−188.9	3.025	65.55	12.25
std. state, m = 1	aq	—	—	−288.74	−265.58	—	−19.3	—
$CaCl$	g	75.533	−23.23	−23.4	−29.7	2.292	57.70	8.58
$CaCl_2$	c	110.986	—	−190.2	−178.8	—	25.0	17.35
	g	—	−112.76	−112.7	−114.54	3.613	69.35	14.18
std. state, m = 1	aq	—	—	−209.64	−195.04	—	14.3	—
$CaCl_2 \cdot H_2O$	c	129.001	—	−265.1	—	—	—	—
$CaCl_2 \cdot 2H_2O$	c	147.017	—	−335.3	—	—	—	—
$CaCl_2 \cdot 4H_2O$	c	183.047	—	−480.3	—	—	—	—
$CaCl_2 \cdot 6H_2O$	c	219.078	—	−623.3	—	—	—	—
$CaOCl_2$	c	126.985	—	−178.4	—	—	—	—
$CaOCl_2 \cdot H_2O$	c	145.001	—	−249.1	—	—	—	—
$Ca(OCl)_2$	aq	142.985	—	−180.3	—	—	—	—
$Ca(ClO_2)_2$	c	174.984	—	−162.1	—	—	—	—
$Ca(ClO_4)_2$	c	238.981	—	−176.09	—	—	—	—
std. state, m = 1	aq	—	—	−191.56	−136.42	—	74.3	—
8 H_2O	aq	—	—	−188.36	—	—	—	—
10 H_2O	aq	—	—	−189.16	—	—	—	—
$Ca(ClO_4)_2 \cdot 4H_2O$	c	311.043	—	−465.8	−352.97	—	103.6	—
$CaBr_2$	c	199.898	—	−163.2	−158.6	—	31.	—
	g	—	—	−95.2	—	—	—	—
std. state, m = 1	aq	—	—	−187.84	−182.00	—	26.7	—
$CaBr_2 \cdot 6H_2O$	c	307.990	—	−599.0	−514.6	—	98.	—
$Ca(BrO_3)_2$	c	295.894	—	−163.9	—	—	—	—
CaI_2	c	293.889	—	−127.5	−126.4	—	34.	—
	g	—	—	−65.	—	—	—	—
std. state, m = 1	aq	—	—	−156.12	−156.96	—	40.5	—
$CaI_2 \cdot 8H_2O$	c	438.012	—	−700.2	—	—	—	—
$Ca(IO_3)_2$	c	389.885	—	−239.6	−200.6	—	55.	—
$Ca(IO_3)_2 \cdot H_2O$	c	407.901	—	−309.1	—	—	—	—
$Ca(IO_3)_2 \cdot 6H_2O$	c	497.977	—	−664.6	−542.0	—	108.	—
CaS	c	72.144	—	−115.3	−114.1	—	13.5	11.33
	g	—	+32.	—	—	—	—	—
$CaSO_3$	c	120.142	—	—	—	—	24.23	21.92
$CaSO_3 \cdot 2H_2O$	c	156.173	—	−418.9	−371.7	—	44.	42.7
$CaSO_4$								
insol., anhydrite	c	136.142	—	−342.76	−315.93	—	25.5	23.82
sol., α	c	—	—	−340.64	−313.93	—	25.9	23.95
sol., β	c	—	—	−339.58	−312.87	—	25.9	23.67
std. state, m = 1	aq	—	—	−347.06	−310.27	—	−7.9	—
$CaSO_4 \cdot 1/2H_2O$	c	145.149	—	−376.85	−343.41	—	31.2	28.54
macro; α								
micro; β	c	—	—	−376.35	−343.18	—	32.1	29.69
$CaSO_4 \cdot 2H_2O$								
selenite	c	172.172	—	−483.42	−429.60	—	46.4	44.46
$CaSe$	c	119.04	—	−88.0	−86.8	—	16.	—
$CaSeO_4$	c	183.038	—	−265.25	—	—	—	—
$CaSeO_4 \cdot 2H_2O$	c	219.068	—	−407.9	−355.4	—	53.	—
$Ca(N_3)_2$	c	124.120	—	+3.5	—	—	—	—
Ca_3N_2	c	148.253	—	−103.	—	—	—	—
$Ca(NO_2)_2$	c	132.091	—	−177.2	—	—	—	—
in 800 H_2O	aq	—	—	−179.5	—	—	—	—
$Ca(NO_2)_2 \cdot 4H_2O$	c	204.152	—	−450.7	—	—	—	—
$Ca(NO_3)_2$	c	164.090	—	−224.28	−177.63	—	46.2	35.70
std. state, m = 1	aq	—	—	−228.86	−185.52	—	57.3	—
∞ H_2O	aq	—	—	−228.86	—	—	—	—
$Ca(NO_3)_2 \cdot 2H_2O$	c	200.120	—	−368.25	−293.82	—	64.4	—
$Ca(NO_3)_2 \cdot 3H_2O$	c	218.136	—	−439.3	−351.8	—	76.3	—
$Ca(NO_3)_2 \cdot 4H_2O$	c	236.151	—	−509.64	−409.53	—	89.7	—
$Ca(NH_2)_2$								
Ca_3P_2	c	182.188	—	−121.	—	—	—	—
$Ca(PO_3)_2$								
β	c	198.024	—	—	—	5.715	35.05	34.68
glassy	amorp	—	—	−587.0	—	—	—	—
$Ca_2P_2O_7$								
β	c	254.103	−792.88	−798.0	−748.6	7.430	45.23	44.89
$Ca_3(PO_4)_2$								
β, low temp. form	c	310.183	—	−984.9	−928.5	—	56.4	54.45
α, high temp form	c	—	—	−982.3	−926.3	—	57.58	55.35
std. state, m = 1	aq	—	—	−999.8	−883.9	—	−144.	—
$CaHPO_4$	c	136.059	−430.299	−433.65	−401.83	4.455	26.62	26.30
std. state, m = 1	aq	—	—	−438.57	−392.64	—	−20.7	—
$CaHPO_4 \cdot 2H_2O$	c	172.090	−568.032	−574.47	−515.00	7.490	45.28	47.10
$Ca(H_2PO_2)_2$	c	170.057	—	−418.9	—	—	—	—
std. state, m = 1	aq	—	—	−423.1	—	—	—	—
$Ca(H_2PO_4)_2$	c	234.055	—	−742.04	—	—	—	—
std. state, m = 1	aq	—	—	−749.38	−672.64	—	30.5	
$Ca(H_2PO_4)_2 \cdot H_2O$	c	252.070	−805.547	−814.93	−730.98	9.950	62.1	61.86
$Ca_3(AsO_4)_2$	c	398.078	—	−788.4	−732.1	—	54.	—
hydrated precipitate	—	—	—	−799.	—	—	—	—
$CaHAsO_4$	aq	180.007	—	−345.6	—	—	—	—
$Ca(H_2AsO_4)_2$	aq	321.950	—	−563.6	—	—	—	—
CaC_2	c	64.102	−15.14	−14.3	−15.5	2.711	16.72	14.99
$CaCO_3$								
calcite	c	100.089	—	−288.46	−269.80	—	22.2	19.57
aragonite	c	—	—	−288.51	−269.55	—	21.2	19.42
std. state, m = 1	aq	—	—	−291.58	−258.47	—	−26.3	—

SELECTED VALUES OF CHEMICAL THERMODYNAMIC PROPERTIES (Continued)

Substance			0 K	298.15 K (25°C)				
Formula and Description	State	Formula weight	ΔHf_0° kcal/mol	ΔHf° kcal/mol	ΔGf° kcal/mol	$H_{298}^\circ - H_0^\circ$ kcal/mol	S° cal/deg mol	C_p° cal/deg mol
CaC_2O_4	c	128.100	—	−325.2	—	—	—	—
std. state, m = 1	aq	—	—	−326.9	−293.37	—	−1.8	—
$CaC_2O_4 \cdot H_2O$	c	146.115	—	−400.30	−361.85	—	37.4	36.52
$CaCN_2$	c	80.105	—	−83.8	—	—	—	—
$Ca(CN)_2$	c	92.116	—	−44.1	—	—	—	—
	aq	—	—	−56.9	—	—	—	—
$CaO \cdot SiO_2$								
wollastonite	c	116.164	—	−390.76	−370.39	—	19.58	20.38
pseudowollastonite	c	—	—	−389.2	−369.2	—	20.88	20.67
glassy	amorp	—	—	−382.65	—	—	—	—
$2CaO \cdot SiO_2$								
β	c	172.244	−548.95	−551.5	−524.1	5.098	30.53	30.78
γ	c	—	−551.25	−554.0	−526.1	4.898	28.87	30.27
$3CaO \cdot SiO_2$	c	228.323	—	−700.1	−665.4	—	40.3	41.08
$3CaO \cdot 2SiO_2$								
rankinite	c	288.408	−942.23	−946.7	−899.0	8.424	50.38	51.24
$CaO \cdot Al_2O_3$	c	158.041	−552.87	−556.0	−527.9	4.569	27.30	28.87
$CaO \cdot 2Al_2O_3$	c	260.002	−944.98	−950.7	−901.2	7.286	42.50	48.00
$2CaO \cdot Al_2O_3$	c	214.120	—	−707	—	—	—	—
$3CaO \cdot Al_2O_3$	c	270.199	−853.21	−857.5	−815.4	8.216	49.2	50.16
$CaO \cdot Al_2O_3 \cdot 2SiO_2$								
anorthite, triclinic	c	278.210	—	−1009.2	−955.5	—	48.4	50.46
anorthite, hexagonal	c	—	—	−1004.3	−949.8	—	45.8	49.76
glassy	amorp	—	—	−991.8	—	—	—	—
$CaO \cdot Fe_2O_3$	c	215.772	−361.787	−363.37	−337.67	6.076	34.74	36.71
$2CaO \cdot Fe_2O_3$	c	271.851	−508.804	−511.30	−478.44	7.564	45.12	46.19
$CaCrO_4$	aq	156.074	—	−340.9	—	—	—	—
$CaMoO_3$	c	184.018	—	−296.	—	—	—	—
$CaMoO_4$	c	200.018	—	−368.4	−342.9	—	29.3	27.32
	g	—	−197.	—	—	—	—	—
std. state, m = 1	aq	—	—	−368.2	−332.2	—	−6.2	—
$CaWO_4$	c	287.928	−391.272	−393.20	−367.71	4.775	30.21	27.28
std. state, m = 1	aq	—	—	−386.8	—	—	—	—
$CaZrO_3$	c	179.298	—	−422.3	−401.8	—	23.92	23.88
$CaHfO_3$	c	266.568	—	−433.0	—	—	—	—
$CaMgC_2O_6$								
dolomite	c	184.411	−552.93	−556.0	−517.1	6.210	37.09	37.65
$2CaO \cdot 5MgO \cdot 8SiO_2 \cdot H_2O$								
tremolite	c	812.410	—	−2954.	−2780.	23.34	131.2	156.7
$CaUO_4$	c	342.107	—	−478.4	—	—	—	—
CARBON								
C								
graphite, Acheson spectroscopic	c	12.0112	0	0	0	0.251	1.372	2.038
diamond	c	—	0.5797	0.4533	0.6930	0.125[a]	0.568	1.4615
	g	—	169.98	171.291	160.442	1.562	37.7597	4.9805
C^+	g	—	429.628	432.420	—	—	—	—
C^{2+}	g	—	991.900	996.173	—	—	—	—
C^{3+}	g	—	2095.98	2101.73	—	—	—	—
CO	g	28.0106	−27.199	−26.416	−32.780	2.0716	47.219	6.959
std. state, m = 1	aq	—	—	−28.91	−28.66	—	25.0	—
in CH_3COOH	—	—	—	−26.416	−28.15	—	31.7	—
CO^+	g	—	295.9	298.16	—	—	—	—
CO^{2+}	g	—	942.	945.5	—	—	—	—
CO_2	g	44.0100	−93.963	−94.051	−94.254	2.2378	51.06	8.87
undissoc.; std. state, m = 1	aq	—	—	−98.90	−92.26	—	28.1	—
CO_2^+	g	—	223.8	225.23	—	—	—	—
CO_3^{2-}								
std. state, m = 1	aq	60.0094	—	−161.84	−126.17	—	−13.6	—
CH	g	13.0191	141.6	142.4	—	—	—	—
CH^+	g	—	398.1	400.4	—	—	—	—
CH_2	g	14.0271	93.9	93.7	—	—	—	—
CH_2^+	g	—	333.6	334.9	—	—	—	—
CH_3	g	15.0351	34.0	33.2	—	—	—	—
CH_3^+	g	—	261.0	261.7	—	—	—	—
CH_4	g	16.0430	−15.970	−17.88	−12.13	2.388	44.492	8.439
std. state, m = 1	aq	—	—	−21.28	−8.22	—	20.0	—
CH_4^+	g	—	277.1	276.7	—	—	—	—
HCO	g	29.0185	−4.2	−4.12	−7.76	2.386	53.68	8.26
$HCOO^-$								
std. state, m = 1	aq	45.0180	—	−101.71	−83.9	—	22.	−21.0
HCO_3^-								
std. state, m = 1	aq	61.0174	—	−165.39	−140.26	—	21.8	—
HCHO	g	30.0265	−27.1	−28.	−27.	2.394	52.26	8.46
unhydrolyzed	aq	—	—	−35.9	—	—	—	—
HCOOH	liq	—	—	−101.51	−86.38	—	30.82	23.67
	g	—	—	−90.48	—	—	—	—
un-ionized; std. state, m = 1	aq	—	—	−101.68	−89.0	—	39.	—
ionized; std. state, m = 1	aq	—	—	−101.71	−83.9	—	22.	−21.0
in ∞ H_2O	aq	—	—	−101.71	—	—	—	—
H_2CO_3								
std. state, m = 1, (undissoc.)	aq	62.0253	—	−167.22	−148.94	—	44.8	—
CH_3OH	liq	32.0424	—	−57.04	−39.76	—	30.3	19.5
	g	−45.355	−47.96	−38.72	2.731	57.29	10.49	
std. state, m = 1	aq	—	—	−58.779	−41.92	—	31.8	—
CF_3	g	69.0064	−113.	−114.	—	—	—	—
CF_3^+	g	—	119.9	120.4	—	—	—	—
CF_4	g	88.0048	−219.6	−221.	−210.	3.043	62.50	14.60
COF_2	g	66.0074	−150.95	−151.7	−148.0	2.642	61.78	11.19
CH_3F	g	34.0335	—	—	—	2.422	53.25	8.96
CH_2F_2	g	52.0239	−104.97	−106.8	−100.2	2.555	58.94	10.25
CHF_3	g	70.0143	−162.84	−164.5	−156.3	2.764	62.04	12.20
CCl_3	g	118.3702	14.	14.	—	—	—	—
CCl_4	liq	153.8232	—	−32.37	−15.60	—	51.72	31.49
	g	—	−24.08	−24.6	−14.49	4.117	74.03	19.91
$COCl_2$	g	98.9166	−51.89	−52.3	−48.9	3.067	67.74	13.78
CH_3Cl	g	50.4881	−17.426	−19.32	−13.72	2.489	56.04	9.74
std. state, m = 1	aq	—	—	−24.3	−12.3	—	34.6	—

[a] Relative to C, diamond

Substance			0 K	298.15 K (25°C)				
Formula and Description	State	Formula weight	ΔHf_0° kcal/mol	ΔHf°	ΔGf° kcal/mol	$H_{298}^\circ - H_0^\circ$	S° cal/deg mol	C_p°
CH_2Cl_2	liq	84.9331	—	−29.03	−16.09	—	42.5	23.9
	g	—	−20.462	−22.10	−15.75	2.830	64.56	12.18
CF_3Cl	g	104.4594	−164.8	−166.	−156.	3.293	68.16	15.98
CF_2Cl_2	g	120.9140	−113.0	−114.	−105.	3.543	71.86	17.27
$CFCl_3$	liq	137.3686	—	−72.02	−56.61	—	53.86	29.05
	g	—	−65.2	−66.	−57.	3.843	74.05	18.66
COFCl	g	86.4620	—	—	—	2.845	66.11	12.52
CH_2ClF	g	68.4785	—	—	—	2.689	63.17	11.24
$CHClF_2$	g	86.4689	—	—	—	2.955	67.11	13.35
$CHCl_2F$	g	102.9235	—	—	—	3.170	70.02	14.56
CBr_4								
monoclinic	c	331.6472	—	4.5	11.4	—	50.8	34.5
	g	26.10	19.	16.	4.873	85.55	21.79	
$COBr_2$	liq	187.8286	—	−30.4	—	—	—	—
	g	—	−19.19	−23.0	−26.5	3.340	73.85	14.78
CH_3Br	g	94.9441	−4.72	−8.4	−6.2	2.536	58.86	10.14
in C_2H_5OH	—	—	—	−13.08	—	—	—	—
CH_2Br_2	g	173.8451	—	—	—	3.020	70.06	13.07
$CHBr_3$	liq	252.7461	—	−6.8	−1.2	—	52.8	31.
	g	—	10.24	4.	2.	3.811	79.07	17.02
CF_3Br	g	148.9154	−150.72	−153.6	−147.3	3.457	71.14	16.57
CCl_3Br	g	198.2792	−8.81	−11.0	−5.1	4.285	79.55	20.38
CH_3I	liq	141.9395	—	−3.7	3.2	—	39.0	30.
	g	—	5.38	3.1	3.5	2.585	60.71	10.54
CHI_3	c	393.7323	—	33.7	—	—	—	—
	g	—	—	—	—	4.106	85.1	17.92
CF_3I	g	195.9108	—	—	—	3.579	73.44	16.94
CH_2ClI	g	176.3845	—	—	—	3.002	70.7	13.02
CH_2IBr	g	220.8405	—	—	—	3.102	73.5	13.46
CS_2	liq	76.1392	—	21.44	15.60	—	36.17	18.1
	g	—	27.86	28.05	16.05	2.547	56.82	10.85
	aq	—	—	21.3	—	—		
COS	g	60.0746	−33.991	−33.96	−40.47	2.373	55.32	9.92
CH_3SH	liq	48.1070	—	−11.08	−1.85	—	40.44	21.64
	g	—	−2.885	−5.34	−2.23	2.898	60.96	12.01
CN	g	26.0178	108.	109.	102.	2.07	48.4	6.97
CNO^-								
std. state, m = 1	aq	42.0172	—	−34.9	−23.3	—	25.5	—
HCN	liq	27.0258	—	26.02	29.86	—	26.97	16.88
	g	—	32.39	32.3	29.8	2.208	48.20	8.57
std. state, m = 1	aq	—	—	36.0	41.2	—	22.5	—
HCN^+								
CH_3NH_2								
methylamine	liq	31.0577	—	−11.3	8.5	—	35.90	—
	g	—	—	5.49	7.67		58.15	12.7
std. state, m = 1	aq	—	—	−16.77	4.94	—	29.5	—
CH_2N_2								
diazomethane	g	—	—	—	—	2.887	58.02	12.55
NH_2CN								
cyanamide	c	—	—	14.1	—	—	—	—
NH_4CN	c	44.0564	—	0.10	—	—	—	32.
std. state, m = 1	aq	—	—	4.3	22.2	—	49.6	—
	aq	—	—	7.7	—	—	—	—
$C{=}NH(NH_2)_2$								
guanidine	c	59.0711	—	−18.1	—	—	—	—
HNCO								
isocyanic acid	g	43.0252	—	—	—	2.615	56.85	10.72
HCNO								
cyanic acid, std. state, m = 1, ionized	aq	—	—	−34.9	−23.3	—	25.5	—
un-ionized, std. state, m = 1	aq	—	—	−36.90	−28.0	—	34.6	—
CH_3NO_2								
nitromethane	liq	61.0406	—	−27.03	−3.47	—	41.05	25.33
	g	—	−14.546	−17.86	−1.65	3.083	65.69	13.70
CH_3ONO								
methyl nitrite	g	61.0406	—	−16.5	—	—	—	—
CH_3NO_3	liq	77.0400	−38.82	−38.0	−10.4	8.26	51.9	37.6
	g	—	—	−29.8	−9.4	—	76.1	—
NH_4HCO_3	c	79.0559	—	−203.0	−159.2	—	28.9	
std. state, m = 1	aq	—	—	−197.06	−159.23	—	48.9	
NH_4CNO	c	60.0558	—	−72.75	—	—	—	—
std. state, m = 1	aq	—	—	−66.6	−42.3	—	52.6	—
$CO(NH_2)_2$								
urea	c	60.0558	—	−79.71	−47.19	—	25.00	22.26
NH_2COONH_4								
ammonium carbamate	c	78.0712	—	−154.17	−107.09	—	31.9	—
$(NH_4)_2CO_3$								
std. state, m = 1	aq	96.0865	—	−225.18	−164.11	—	40.6	—
CNBr	c	105.9268	—	33.58	—	—	—	—
	g	—	46.07	44.5	39.5	2.648	59.32	11.22
CNI	c	152.9222	—	39.71	44.22	—	23.0	—
	g	—	54.04	53.9	47.0	2.724	61.35	11.54
std. state, m = 1	aq	—	—	42.5	44.95	—	29.9	—
CNS^-								
thiocyanate ion std. state, m = 1	aq	58.0818	—	18.27	22.15	—	34.5	−9.6
HCNS								
thiocyanic acid, undissoc. std. state, m = 1	aq	—	—	—	23.31	—	—	—
ionized; std. state, m = 1	aq	—	—	18.27	22.15	—	34.5	−9.6
NH_4CNS								
ammonium thiocyanate	c	76.1204	—	−18.8	—	—	—	—
std. state, m = 1	aq	—	—	−13.40	3.18	—	61.6	9.5
in 200 H_2O	aq	—	—	−13.4	—	—	—	—
$CS(NH_2)_2$								
thiourea	c	76.1204	—	−21.1	—	—	—	—
$C_2O_4^{2-}$								
std. state, m = 1	aq	88.0199	—	−197.2	−161.1	—	10.9	—
C_2H_4	g	28.0542	14.515	12.49	16.28	2.525	52.45	10.41
std. state, m = 1	aq	—	—	8.69	19.43	—	29.2	—

SELECTED VALUES OF CHEMICAL THERMODYNAMIC PROPERTIES (Continued)

Substance			0 K	298.15 K (25°C)				
Formula and Description	State	Formula weight	ΔHf_0° kcal/mol	ΔHf°	ΔGf° kcal/mol	$H_{298}^\circ - H_0^\circ$	S° cal/deg mol	C_p°
C_2H_5								
ethyl radical	g	29.0622	28.	25.	31.	—	59.2	—
$C_2H_5^+$	g	—	222.	220.	—	—	—	—
C_2H_6	g	30.0701	−16.523	−20.24	−7.86	2.856	54.85	12.58
std. state, m = 1	aq	—	—	−24.40	−4.09	—	28.3	—
$C_2H_6^+$	g	—	252.2	250.0	—	—	—	—
$HC_2O_4^-$								
std. state, m = 1	aq	89.0279	—	−195.6	−166.93	—	35.7	—
CH_2CO								
ketene	g	42.0376	−13.86	−14.6	−14.8	2.819	59.16	12.37
$(COOH)_2$								
oxalic acid	c	90.0358	—	−197.7	—	—	—	28.
std. state, m = 1	aq	—	—	−197.2	−161.1	—	10.9	—
CH_3COO^-								
std. state, m = 1	aq	59.0450	—	−116.16	−88.29	—	20.7	−1.5
$C_2H_3O_3^-$								
glycolate ion, std. state m = 1	aq	75.0444	—	−155.9	—	—	—	—
C_2H_4O								
ethylene oxide	liq	44.0536	—	−18.60	−2.83	—	36.77	21.02
	g	—	−9.589	−12.58	−3.12	2.596	57.94	11.45
CH_3CHO								
acetaldehyde	liq	—	—	−45.96	−30.64	—	38.3	—
1/3 $(CH_3CHO)_3$								
paraldehyde	liq	—	—	−54.73	—	—	—	—
1/4 $(CH_3CHO)_4$								
metaldehyde	c	—	—	−56.2	—	—	—	—
$HCOOCH_3$								
methyl formate	liq	60.0530	—	−90.60	—	—	—	29.
	g	—	—	−83.7	—	—	—	—
CH_3COOH								
acetic acid	liq	60.0530	—	−115.8	−93.2	—	38.2	29.7
	g	—	−99.972	−103.31	−89.4	3.286	67.5	15.9
ionized; std. state, m = 1	aq	—	—	−116.16	−88.29	—	20.7	−1.5
un-ionized; std. state, m = 1	aq	—	—	−116.10	−94.8	—	42.7	—
$CH_2OHCOOH$								
hydroxyacetic acid (glycolic acid)	c	76.0524	—	−158.7	—	—	—	—
$CH(OH)_2COOH$								
dihydroxyacetic acid (glyoxylic acid)	c	92.0518	—	−199.7	—	—	—	—
$CH_3CH_2O^-$								
std. state, m = 1	aq	45.0616	—	—	−24.5	—	—	—
C_2H_5OH								
ethanol	liq	46.0695	—	−66.37	−41.80	—	38.4	26.64
	g	—	−51.969	−56.19	−40.29	3.390	67.54	15.64
std. state, m = 1	aq	—	—	−68.9	−43.44	—	35.5	—
C_2H_5OH								
in ∞ H_2O	aq	—	—	−68.9	—	—	—	—
$(CH_2OH)_2$								
ethylene glycol	liq	—	—	−108.70	−77.25	—	39.9	35.8
C_2F_4								
tetrafluoroethylene	g	100.0159	−154.68	−155.5	−147.2	3.903	71.69	19.23
C_2F_6								
hexafluoroethane	g	138.0127	−308.0	−310.	−290.	4.87	79.4	25.5
$CH{\equiv}CF$	g	44.0287	—	—	—	2.739	55.34	12.52
CH_3CH_2F								
ethyl fluoride	g	48.0606	—	—	—	3.06	63.2	14.0
$CH_2{=}CF_2$	g	—	−76.95	−78.6	−73.0	2.980	63.6	14.36
CH_3CHF_2	g	66.0510	−110.98	−114.3	−100.6	3.34	67.5	16.2
CH_3CF_3								
1,1,1-trifluoroethane	g	84.0414	−172.93	−176.0	−159.5	3.631	66.87	18.69
CF_3CH_2F	g	102.0318	—	—	—	4.079	75.85	20.82
C_2Cl_4								
tetrachloroethylene	liq	165.8343	—	−12.5	1.1	—	63.8	33.7
	g	—	−2.70	−2.9	5.4	4.686	81.5	22.69
C_2Cl_6								
I, cubic	c	236.7403	—	−46.0	—	—	—	—
II, monoclinic	c	—	—	−47.9	—	—	—	—
III, triclinic	c	—	—	−48.5	—	—	—	—
$CH_2{=}CHCl$								
vinyl chloride	liq	62.4992	—	3.5	—	—	—	—
	g	—	10.31	8.5	12.4	2.825	63.07	12.84
$1/n(CH_2{=}CHCl)_n$								
polyvinyl chloride	c	—	—	−22.5	—	—	—	14.2
C_2H_5Cl								
ethyl chloride	liq	64.5152	—	−32.63	−14.20	—	45.60	24.94
	g	—	−23.331	−26.81	−14.45	3.179	65.94	15.01
$CHCl{=}CHCl$								
cis-1,2-dichloroethylene	liq	—	—	−6.6	5.27	—	47.42	27.
trans-1,2-dichloroethylene	liq	—	—	−5.53	6.52	—	46.81	27.
CH_2ClCH_2Cl								
1,2-dichloroethane	liq	—	—	−39.49	−19.03	—	49.84	30.9
	g	—	−28.357	−31.02	−17.67	4.08	73.68	18.8
$CHCl{=}CCl_2$								
trichloroethylene	liq	131.3893	—	−10.1	2.9	—	54.6	28.8
	g	—	−1.032	−1.86	4.31	3.975	77.6	19.18
CH_3COCl								
acetyl chloride	liq	78.4986	—	−65.44	−49.73	—	48.0	28.
	g	—	−56.054	−58.20	−49.20	3.53	70.5	16.2
CH_2ClCH_2OH								
ethylene chlorohydrin	liq	80.5146	—	−70.6	—	—	—	—
$CHCl_2COOH$								
dichloracetic acid	liq	128.9430	—	−119.0	—	—	—	44.
ionized	aq	—	—	−122.4	—	—	—	—
un-ionized	aq	—	—	−120.4	—	—	—	—
CCl_3CHO								
chloral (trichloroacetaldehyde)	liq	147.3887	—	−56.45	—	—	—	36.
CCl_3COOH								
trichloroacetic acid	c	163.3881	—	−120.7	—	—	—	—
ionized	aq	—	—	−123.4	—	—	—	—

SELECTED VALUES OF CHEMICAL THERMODYNAMIC PROPERTIES (Continued)

Substance			0 K	298.15 K (25°C)				
Formula and Description	State	Formula weight	ΔHf_0° kcal/mol	ΔHf°	ΔGf° kcal/mol	$H_{298}^\circ - H_0^\circ$	S° cal/deg mol	C_p°
$CCl_3CH(OH)_2$								
chloral hydrate	c	165.4040	—	−137.7	—	—	—	34.
	g	—	—	−107.2	—	—	—	—
in 150 $CHCl_3$	—	—	—	−131.2	—	—	—	—
$CF_2{=}CFCl$	g	116.4705	−132.04	−132.7	−125.2	4.096	76.96	20.06
CF_3CCl_3	g	187.3765	—	—	—	5.61	88.6	28.8
$CF_2ClCFCl_2$	liq	—	—	−188.37	—	—	—	41.5
	g	—	—	−181.5	—	—	—	—
CCl_3CF_2Cl	g	203.8311	−115.75	−117.1	−97.3	5.65	91.5	29.5
$CF_2{=}CHCl$	g	98.4801	−74.25	−75.4	−69.1	3.570	72.39	17.23
CH_3CF_2Cl	liq	100.4960	—	—	—	—	—	31.4
CBr_3CBr_3								
hexabromoethane	g	503.4763	—	—	—	7.108	105.6	33.30
$CH_2{=}CHBr$								
vinyl bromide	g	106.9552	22.26	18.7	19.3	2.905	65.90	13.27
C_2H_5Br								
ethyl bromide	liq	108.9712	—	−21.99	−6.64	—	47.5	24.1
	g	—	−10.188	−15.42	−6.34	3.259	68.50	15.42
in 2000 CH_3OH	—	—	−21.71	—	—	—	—	
$CHBr{=}CHBr$								
cis 1,2-dibromoethylene	g	185.8562	—	—	—	3.491	74.38	16.44
trans, 1,2-dibromoethylene	g	—	—	—	—	3.682	74.90	16.79
CH_3CHBr_2	g	187.8722	—	—	—	3.94	78.3	19.3
CH_2BrCH_2Br								
ethylene bromide	liq	—	—	−19.4	−5.0	—	53.37	32.51
	g	—	—	−9.16	−2.47	—	79.1	20.7
CH_3COBr								
acetyl bromide	liq	122.9546	—	−53.39	—	—	—	—
$CF_2{=}CFBr$	g	160.9265	—	—	—	4.238	80.0	20.51
CF_3CF_2Br	g	198.9233	—	—	—	4.98	85.7	26.1
$CF_2{=}CBr_2$	g	221.8371	—	—	—	4.537	83.5	21.58
CF_2BrCF_2Br	g	259.8339	—	−186.5	—	—	—	—
$CF_2{=}CHBr$	g	142.9361	—	—	—	3.682	75.1	17.63
CF_3CH_2Br	g	162.9424	—	—	—	4.299	80.59	21.67
CHF_2CF_2Br	g	180.9329	—	−197.0	—	—	—	—
CH_2ClCH_2Br	liq	143.4162	—	—	—	—	—	31.1
$CHClBrCHClBr$	g	256.7622	—	−8.8	—	—	—	—
$CF_2{=}CBrCl$	g	177.3811	—	—	—	4.385	82.0	21.16
$CF_2BrCHCl_2$	g	213.8421	—	−107.9	—	—	—	—
$CI{\equiv}CI$	g	277.8311	—	—	—	3.901	74.80	16.81
C_2I_4	c	531.6399	—	73.	—	—	—	—
$CH_2{=}CHI$								
vinyl iodide	g	153.9506	—	—	—	3.027	68.1	13.84
C_2H_5I								
ethyl iodide	liq	155.9666	—	−9.6	3.5	—	50.6	27.5
CH_2ICH_2I								
ethylene iodide	c	281.8630	—	0.1	13.8	—	47.	—
	g	—	—	15.9	18.8	—	83.2	19.2
C_2H_5SH								
ethanethiol	liq	—	—	−17.53	−1.28	—	49.48	28.17
	g	—	−6.940	−10.95	−1.05	3.617	70.77	17.37
$(CH_3)_2SO$								
dimethyl sulfoxide	liq	78.1335	—	−48.6	−23.7	—	45.0	35.2
	g	—	−31.427	−35.96	−19.48	4.132	73.20	21.26
$(CH_3)_2SO_2$								
dimethyl sulfone	c	94.1329	—	−107.8	−72.3	—	34.	—
	g	—	−83.3	−88.7	−65.2	4.3	74.2	23.9
$C_2H_5HSO_4$								
ethyl sulfuric acid	aq	126.1317	—	−209.3	—	—	—	—
$N{\equiv}C{-}C{\equiv}N$								
cyanogen	g	52.0357	73.386	73.84	71.07	3.028	57.79	13.58
CH_3CN								
acetonitrile	liq	—	—	12.8	23.7	—	35.76	21.86
	g	—	22.58	20.9	25.0	2.892	58.67	12.48
$C_2H_5NH_2$								
ethylamine	liq	45.0848	—	−17.7	—	—	—	31.
	g	—	—	−11.27	—	—	—	16.7
$(CH_3)_2NH$								
dimethylamine	liq	—	—	−10.5	16.7	—	43.58	32.9
	g	—	—	−4.41	16.35	—	65.24	16.9
std. state, m = 1	aq	—	—	−16.88	13.85	—	31.8	—
$NH_2CH_2CH_2NH_2$								
ethylenediamine	liq	—	—	−5.82	—	—	—	50.
CH_3CONH_2								
acetamide	c	—	—	−76.0	—	—	—	16.
$C_2H_5NO_2$								
nitroethane	liq	75.0676	—	−33.5	—	—	—	33.
	g	—	—	−23.56	—	—	—	—
CH_3CH_2ONO								
ethyl nitrite	liq	—	—	−30.8	—	—	—	—
NH_2CH_2COOH								
glycine	c	—	−121.415	−126.22	−88.09	3.867	24.74	23.71
ionized; std. state, m = 1	aq	—	—	−112.280	−75.278	—	28.54	—
un-ionized, std. state, m = 1	aq	—	—	−122.846	−88.618	—	37.84	—
$CH_3CH_2ONO_2$								
ethyl nitrate	c	—	−45.023	—	—	—	—	—
	liq	—	—	−45.49	−10.29	9.242	59.08	40.7
CH_3COONH_4								
ammonium acetate	c	77.0836	—	−147.26	—	—	—	—
std. state, m = 1	aq	—	—	−147.83	−107.26	—	47.8	17.6
$CH_2OHCOONH_4$								
ammonium glycolate	c	93.0830	—	−190.6	—	—	—	—
$(CH{=}NOH)_2$								
glyoxime	c	—	—	−21.2	—	—	—	—
$(CH_3)_2NH_2NO_3$								
dimethylammonium nitrate	c	108.0977	—	−83.7	—	—	—	—
std. state, m = 1	aq	—	—	−78.30	−27.41	—	76.2	—
$(NH_4)_2C_2O_4$								
ammonium oxalate	c	—	—	−268.4	—	—	—	54.
std. state, m = 1	aq	—	—	−260.5	−199.0	—	65.1	—
$(NH_4)_2C_2O_4 \cdot H_2O$	c	142.1124	—	−340.7	—	—	—	—

Substance			0 K	298.15 K (25°C)				
Formula and Description	State	Formula weight	ΔHf°_0 kcal/mol	ΔHf° kcal/mol	ΔGf° kcal/mol	$H^\circ_{298} - H^\circ_0$	S° cal/deg mol	C°_p cal/deg mol
CH_3NCS								
methyl isothiocyanate	c	73.1169	—	19.0	—	—	—	—
	g	—	33.46	31.3	34.5	3.464	69.29	15.65
CH_3SCN								
methyl thiocyanate	liq	—	—	28.4	—	—	—	—
	g	—	—	38.3	—	—	—	—
CH_3COOCH_3	liq	74.0801	—	−106.42	—	—	—	—
$(CH_3)_3N$								
trimethylamine	liq	59.1119	—	−11.0	24.1	—	49.82	32.31
	g	—	—	−5.81	23.65	—	68.6	—
std. state, m = 1	aq	—	—	−18.17	22.22	—	31.9	—
$(C_2H_5)_2O$								
diethyl ether	liq	74.1237	—	−66.82	—	—	—	—
	g	—	—	−60.26	—	—	—	—
$(C_2H_5)_2S$								
diethyl sulfide	liq	90.1883	—	−28.43	2.81	—	64.36	40.97
	g	—	−13.15	−19.86	4.34	5.467	87.96	27.97
$(C_2H_5)_2NH$								
diethylamine	liq	73.1390	—	−24.7	—	—	—	—
	g	—	—	−17.07	—	—	—	—
CERIUM								
Ce								
γ	c	140.12	0	0	0	1.8	17.2	6.44
	g	—	101.2	101.	92	1.594	45.807	5.515
Ce^{+}	g	—	227.3	228.5	—	—	—	—
Ce^{2+}	g	—	478.	480.	—	—	—	—
Ce^{3+}	g	—	942.7	947.4	—	—	—	—
std. state, m = 1	aq	—	—	−166.4	−160.6	—	−49.	—
CeO_2	c	172.119	−256.69	−258.80	−244.9	2.478	14.89	14.73
Ce_2O_3	c	328.238	−427.64	−429.3	−407.8	5.13	36.0	27.4
CeH_2	c	142.136	−47.83	−49.	−39.	1.776	13.3	9.78
CeF_3	c	197.115	—	—	—	4.237	27.5	22.3
$CeF_3 \cdot H_2O$	c	215.131	—	−472.4	—	—	—	—
$CeCl_3$	c	246.479	—	−251.8	−233.7	—	36.	20.9
	g	—	—	−174.	—	—	—	—
std. state, m = 1	aq	—	—	−286.2	−254.7	—	−9.	—
$CeCl_3 \cdot 7H_2O$	c	372.587	—	−757.5	—	—	—	—
$CeOCl$	c	191.572	—	−239.	—	—	—	—
$CeClO_4^{2+}$								
std. state, m = 1	aq	239.571	—	−209.1	−165.2	—	−37.	—
$CeBr^{2+}$								
std. state, m = 1	aq	220.029	—	—	−186.3	—	—	—
CeI_3	c	520.833	—	−155.3	—	—	—	—
$Ce(I_3)_3$	c	664.828	—	−332.	—	—	—	—
$Ce(IO_3)_3 \cdot 2H_2O$	c	700.858	—	—	−378.3	—	—	—
CeS_2	c	204.248	—	−146.3	—	—	—	—
Ce_2S	g	312.304	73.	71.6	60.0	3.25	81.	12.9
Ce_2S_3	c	376.432	—	−284.	—	—	—	—
$CeSO_4^{+}$								
std. state	aq	236.182	—	−380.2	−343.3	—	−17.	—
$Ce(SO_4)_2^{-}$								
std. state	aq	332.243	—	−595.9	−523.6	—	2.	—
$Ce_2(SO_4)_3$	c	568.425	—	−945.1	—	—	—	—
$Ce_2(SO_4)_3 \cdot 5H_2O$	c	658.502	—	—	—	—	—	132.
$Ce_2(SO_4)_3 \cdot 8H_2O$	c	712.548	—	—	−1320.6	—	—	—
$Ce(NO_3)_3$	c	326.135	—	−293.0	—	—	—	—
$Ce(NO_3)_3 \cdot 3H_2O$	c	380.181	—	−516.	—	—	—	—
$Ce(NO_3)_3 \cdot 4H_2O$	c	398.196	—	−588.9	—	—	—	—
$Ce(NO_3)_3 \cdot 6H_2O$	c	434.227	—	−729.14	—	—	—	—
CeC_2	c	164.142	—	−15.	15.2	—	20.	—
	g	—	136.	136.2	122.9	2.47	64.	10.5
CeC_4	g	188.165	167.1	168.	152.	3.68	73.	17.3
$CeCrO_3$	c	240.114	—	−368.	−347.	—	25.	—
CESIUM								
Cs	c	132.9054	0	0	0	1.843	20.37	7.69
	g	132.9054	18.542	18.180	11.748	1.481	41.942	4.968
Cs^{+}	g	132.9054	108.337	109.456	—	—	—	—
Cs^{2+}	g	132.9054	686.63	689.23	—	—	—	—
Cs^{+}	a	132.9054	—	−61.73	−69.79	—	31.80	−2.5
CsO_2	c	164.9042	—	−68.4	—	—	—	—
Cs_2O	c	281.8102	−82.142	−82.64	−73.65	4.225	35.10	18.16
	g	281.8102	—	−37.	—	—	—	—
CsH	c	133.9134	—	−12.950	—	—	—	—
	g	133.9134	28.4	27.7	23.1	2.114	51.40	7.54
$CsOH$	c	149.9128	—	−99.72	—	—	—	—
	g	149.9128	−57.9	−59.	−59.1	2.828	60.88	11.88
	a	149.9128	—	−116.70	−107.38	—	29.23	—
$CsOH \cdot H_2O$	c	167.9282	—	−180.22	—	—	—	—
CsF	c	151.9038	−132.205	−132.3	−125.6	2.802	22.18	12.21
	g	151.9038	−85.21	−85.8	−89.8	2.306	58.11	8.57
$CsCl$	c	168.3584	−105.926	−105.89	−99.08	2.976	24.18	12.54
	g	168.3584	−56.89	−57.41	−61.62	2.42	61.15	8.83
$CsClO$	a	184.3578	—	−87.3	−78.6	—	42.	—
$CsClO_2$	a	200.3572	—	−77.6	−65.7	—	56.0	—
$CsClO_3$	c	216.3566	—	−98.4	−73.6	—	37.3	—
	a	216.3566	—	−86.58	−71.71	—	70.6	—
$CsClO_4$	c	232.3560	−104.136	−105.90	−75.13	5.325	41.84	25.88
	a	232.3560	—	−92.64	−71.85	—	75.3	—
	c	212.8144	−95.358	−96.99	−93.55	3.140	27.02	12.65
	g	212.8144	−47.69	−50.0	−57.6	2.46	63.89	8.86
$CsBrO$	a	228.8138	—	−84.2	−77.8	—	42.	—
$CsBrO_3$	c	260.8126	—	−89.82	−68.11	—	39.1	—
	a	260.8126	—	−77.76	−65.36	—	70.45	—
CsI	c	259.8098	−82.652	−82.84	−81.40	3.232	29.41	12.62
	g	259.8098	−35.40	−36.3	−45.7	2.52	65.77	8.95
$CsIO$	a	275.8092	—	−87.4	−79.0	—	30.5	—
$CsIO_3$	c	307.8080	—	—	−103.7	—	—	—
$CsIO_4$	a	307.8080	—	−114.6	−100.4	—	60.1	—
$CsIO_4$	c	323.8074	—	—	−91.0	—	—	—
Cs_2S	c	297.8748	—	−86.0	—	—	—	—
	a	297.8748	—	−115.6	−119.1	—	60.1	—

SELECTED VALUES OF CHEMICAL THERMODYNAMIC PROPERTIES (Continued)

Substance			0 K	298.15 K (25°C)				
Formula and Description	State	Formula weight	ΔHf°_0 kcal/mol	ΔHf° kcal/mol	ΔGf° kcal/mol	$H^\circ_{298} - H^\circ_0$ kcal/mol	S° cal/deg mol	C°_p cal/deg mol
Cs_2SO_3	c	345.8730	—	−271.2	—	—	—	—
	a	345.8730	—	−275.4	−255.9	—	56.6	—
Cs_2SO_4	c	361.8724	−342.63	−344.89	−316.36	6.63	50.65	32.24
	g	361.8724	—	−263.6	—	—	—	—
	a	361.8724	—	−340.78	−317.55	—	68.4	—
$CsHSO_3$								
from HSO_3	a	213.9756	—	−211.40	−195.94	—	65.2	—
$CsHSO_4$	c	229.9750	—	−276.8	—	—	—	—
from HSO_4^-	a	229.9750	—	−273.81	−250.48	—	63.3	—
Cs_2Se	a	344.7708	—	—	−108.7	—	—	—
[illegible]	a	[illegible]	—	[illegible]	[illegible]	—	67	—
Cs_2SeO_4	c	408.7684	—	−272.34	—	—	—	—
	a	408.7684	—	−266.7	−245.1	—	76.5	—
CsN_3	c	174.9255	—	−4.7	—	—	—	—
$CsNO_2$	a	178.9109	—	−86.7	−77.5	—	61.2	—
$CsNO_3$	c	194.9103	—	−120.93	−97.18	—	37.1	—
	g	194.9103	—	−89.4	—	—	—	—
	a	194.9103	—	−111.29	−96.40	—	66.8	−23.7
$CsPO_3$	c	211.8774	—	−296.7	—	—	—	—
Cs_3PO_4	a	493.6876	—	−490.5	−452.9	—	42.	—
$Cs_4P_2O_7$	a	705.5650	—	−789.7	−737.9	—	99.2	—
CsH_2PO_4	c	229.8928	—	−374.0	—	—	—	—
$CsAsO_2$	a	239.8258	—	−164.27	−153.45	—	41.5	—
Cs_3AsO_4	a	537.6354	—	−397.46	−364.37	—	56.5	—
$Cs_2C_2O_4$								
oxalate	c	353.8308	—	—		7.45	56.92	—
	a	353.8308	—	−320.7	−300.7	—	74.5	—
$CsHCO_3$	c	193.9228	—	−230.9	—	—	—	—
from HCO_3^-	a	193.9228	—	−227.12	−210.05	—	53.6	—
$CsHC_2O_4$								
from $HC_2O_4^-$	a	221.9334	—	−257.3	−236.72	—	67.5	—
CH_3COOCs								
acetate	a	191.9506	—	−177.89	158.08	—	52.5	
$CsCN$	c	158.9233	—	—	—	4.330	33.40	15.70
	a	158.9233	—	−25.7	−28.6	—	54.3	—
$CsCNS$								
thiocyanate	a	190.9873	—	−43.46	−47.64	—	66.3	—
$CsMnO_4$	a	251.8410	—	−191.1	−176.7	—	77.5	—
	a	664.7288	—	−305.7	−280.5	—	125.0	—
Cs_2CrO_4	c	381.8044	—	−341.57	—	—	—	—
$Cs_2Cr_2O_7$	c	481.7986	−497.023	−499.24	−456.09	10.671	78.89	55.34
	a	481.7986	—	−479.7	−450.6	—	126.2	—
Cs_2WO_4	a	513.6584	—	−380.6	—	—	—	—
$CsVO_3$	a	231.8456	—	−274.0	−257.1	—	44.	—
Cs_3VO_4	a	513.6558	—	—	−424.3	—	—	—
Cs_2UO_4	c	367.8374	−439.01	−461.0	−431.7	7.365	52.50	36.51
CHLORINE								
Cl	g	35.453	28.68	29.082	25.262	1.499	39.457	5.220
Cl^+	g	—	328.86	330.74	—	—	—	—
Cl^{2+}	g	—	877.81	881.17	—	—	—	—
Cl^{3+}	g	—	1798.26	1803.10	—	—	—	—
Cl^-	g	—	−57.7	−58.8	—	—	—	—
std. state, m = 1	aq	—	—	−39.952	−31.372	—	13.5	−32.6
Cl_2	g	70.906	0	0	0	2.193	53.288	8.104
std. state, m = 1	aq	—	—	−5.6	1.65	—	29.	—
Cl_3^-								
std. state, m = 1	aq	106.359	—	—	−28.8	—	—	—
ClO	g	51.4524	24.36	24.34	23.45	2.114	54.14	7.52
ClO^-								
std. state, m = 1	aq	—	—	−25.6	−8.8	—	10.	—
ClO_2	g	67.4518	25.09	24.5	28.8	2.580	61.36	10.03
std. state, m = 1	aq	—	—	17.9	28.7	—	39.4	—
ClO_2^-								
std. state, m = 1	aq	—	—	−15.9	4.1	—	24.2	
ClO_3	g	83.4512	—	37.	—	—	—	
ClO_3^-								
std. state, m = 1	aq	—	—	−23.7	−0.8	—	38.8	—
ClO_4								
std. state, m = 1	aq	99.4506	—	−30.91	−2.06	—	43.5	—
Cl_2O	g	86.9054	19.71	19.2	23.4	2.719	63.60	10.85
HCl	g	36.4610	−22.020	−22.062	−22.777	2.066	44.646	6.96
std state, m = 1	aq	—	—	−39.952	−31.372	—	13.5	−32.6
HCl								
∞ H_2O	aq	—	—	−39.952	—	—	—	—
$HClO$	g	52.4604	—	—	—	2.440	56.54	8.88
undissoc.; std. state, m = 1	aq	—	—	−28.9	−19.1	—	34.	—
$HClO_2$								
undissoc.; std. state, m = 1	aq	68.4598	—	−12.4	1.4	—	45.0	—
$HClO_3$								
std. state, m = 1	aq	84.4592	—	24.85	−1.92	—	38.8	—
$HClO_4$	liq	100.4586	—	−9.70	—	—	—	—
std. state, m = 1	aq	—	—	−30.91	−2.06	—	43.5	—
$HClO_4$								
∞ H_2O	aq	—	—	−30.91	—	—	—	—
ClF	g	54.4514	−13.0	−13.02	−13.37	2.127	52.05	7.66
ClF_3	liq	92.4482	—	−45.3	—	—	—	—
CHROMIUM								
Cr	c	51.996	0	0	0	0.970	5.68	5.58
	g	—	94.29	94.8	84.1	1.481	41.68	4.97
Cr^+	g	—	250.30	252.29	—	—	—	—
Cr^{2+}	g	—	630.7	634.2	—	—	—	—
std. state, m = 1	aq	—	—	−34.3	—	—	—	—
Cr^{3+}	g	—	1345.	1350.	—	—	—	—
CrO_3	c	99.9942	—	−140.9	—	—	—	—
	g	—	—	−92.2	—	—	—	—
CrO_4^{2-}								
std. state, m = 1	aq	115.9936	—	−210.60	−173.96	—	12.00	—
Cr_2O_3	c	151.9902	—	−272.4	−252.9	—	19.4	28.38
$Cr_2O_7^{2-}$								
std. state, m = 1	aq	215.9878	—	−356.2	−311.0	—	62.6	—

SELECTED VALUES OF CHEMICAL THERMODYNAMIC PROPERTIES (Continued)

Substance			0 K	298.15 K (25°C)				
Formula and Description	State	Formula weight	ΔHf°_0 kcal/mol	ΔHf°	ΔGf° kcal/mol	$H^\circ_{298} - H^\circ_0$	S° cal/deg mol	C°_p
ErI_3	c	547.973	—	−146.5	—	—	—	—
$Er(IO_3)_3$	c	691.968	—	−330.	—	—	—	—
ErC_2	g	191.282	138.	138.2	125.4	2.47	63.	10.5
EUROPIUM								
Eu	c	151.96	0	0	0	1.913	18.59	6.61
	g	—	42.232	41.9	34.0	1.481	45.097	4.968
Eu^+	g	—	172.93	174.08	—	—	—	—
Eu^{2+}	g	—	432.	435.	—	—	—	—
std. state, m = 1	aq	—	—	−126.	−129.1	—	−2.	—
Eu^{3+}								
std. state, m = 1	aq	—	—	−144.6	−137.2	—	−53.	2.
EuO	c	167.959	—	−141.5	−133.1	—	15.	—
Eu_2O_3								
cubic	c	351.918	—	−397.4	—	—	—	29.6
monoclinic	c	—	—	−394.7	−372.1	—	35.	29.2
Eu_3O_4	c	519.878	—	−543.	−512.	—	49.	—
$Eu(OH)_3$	c	202.982	—	—	−285.5	—	—	—
EuF	g	170.958	—	−70.	—	—	—	—
EuF_3	c	208.955	—	—	—	—	—	—
$EuCl_2$	c	222.866	—	−197.	—	—	—	—
	g	—	—	−110.	—	—	—	—
$EuCl_3$	c	258.319	—	−223.7	—	—	—	—
std. state, m = 1	aq	—	—	−264.4	−231.3	—	−13.	−96.
$EuCl_3 \cdot 6H_2O$	c	366.411	—	−665.6	−565.5	—	97.3	87.7
$Eu(BrO_3)_3 \cdot 9H_2O$	c	697.820	—	−823.4	—	—	—	—
$Eu(IO_3)_3$	c	676.668	—	−308.4	—	—	—	—
EuS	g	184.024	27.	—	—	—	—	—
$Eu_2(SO_4)_3 \cdot 8H_2O$	c	736.228	—	—	—	—	160.6	146.0
EuC_2	c	175.982	—	−15.	−16.	—	24.	—
FLUORINE								
F	g	18.9984	18.38	18.88	14.80	1.558	37.917	5.436
F^+	g	—	420.16	422.14	—	—	—	—
F^{2+}	g	—	1226.98	1230.44	—	—	—	—
F^{3+}	g	—	2672.0	2676.9	—	—	—	—
F^{4+}	g	—	4684.2	4690.6	—	—	—	—
F^-	g	—	−63.7	−64.7	—	—	—	—
std. state, m = 1	aq	—	—	−79.50	−66.64	—	−3.3	−25.5
F_2	g	37.9968	0	0	0	2.108	48.44	7.48
F_2^+	g	—	365.1	366.6	—	—	—	—
FO	g	34.9978	41.	41.	—	—	—	—
F_2O	g	53.9962	−4.7	−5.2	−1.1	2.604	59.11	10.35
HF	liq	20.0064	—	−71.65	—	—	$18.02 + x^a$	12.35
	g	—	−64.789	−64.8	−65.3	2.055	41.508	6.963
ionized; std. state, m = 1	aq	—	—	−79.50	−66.64	—	−3.3	−25.5
HF								
∞H_2O	aq	—	—	−79.50	—	—	—	—
HF_2^-								
std. state, m = 1	aq	39.0048	—	−155.34	−138.18	—	22.1	—
XeF_4	c	207.294	—	−62.5	—	—	—	—
FRANCIUM								
Fr	cs	223.0000	0	0	0	—	22.8	—
FrF	c	241.9984	—	—	—	2.80	26.0	12.6
FrCl	c	258.4530	—	—	—	3.10	27.0	12.80
FrBr	c	302.9040	—	—	—	3.30	31.0	12.90
FrI	c	349.9045	—	—	—	3.40	33.0	12.90
GADOLINIUM								
Gd	c	157.25	0	0	0	2.178	16.27	8.85
	g	—	95.353	95.0	86.0	1.825	46.416	6.584
Gd^+	g	—	237.0	238.1	—	—	—	—
Gd^{2+}	g	—	517.	519.	—	—	—	—
Gd^{3+}								
std. state, m = 1	aq	—	—	−164.	−158.	—	−49.2	0.
GdO	g	173.249	−17.	—	—	—	—	—
Gd_2O_3								
monoclinic	c	362.498	—	−434.9	—	—	—	25.5
cubic	c	—	—	—	—	4.45	36.0	25.22
GdH_2	c	159.266	—	−45.5	—	—	—	—
GdF	g	176.248	—	−41.	—	—	—	—
GdF_2	g	195.247	—	−169.	—	—	—	—
GdF_3	g	214.245	—	−310.	—	—	—	—
$GdCl_3$	c	263.609	—	−241.	—	—	—	21.
std. state, m = 1	aq	—	—	−284.	−253.	—	−8.8	−98.
$GdCl_3 \cdot 6H_2O$	c	371.701	−675.66	−685.	−586.	14.43	97.56	83.0
$Gd(BrO_3)_3 \cdot 9H_2O$	c	703.110	—	−842.9	—	—	—	—
GdI_3	c	537.963	—	−142.	—	—	—	—
$Gd(IO_3)_3$	c	681.958	—	−327.	—	—	—	—
GdS	g	189.314	38.	—	—	—	—	—
$Gd_2(SO_4)_3 \cdot 8H_2O$	c	746.808	—	−1513.	−1322.	—	155.8	140.5
$Gd(NO_3)_3 \cdot 6H_2O$	c	451.357	—	—	—	19.16	133.2	106.2
$GdPO_4 \cdot H_2O$	c	270.237	—	—	−490.	—	—	—
GdC_2	c	181.272	—	−25.	—	—	—	—
	g	—	128.2	128.2	—	2.5	—	—
GALLIUM								
Ga	c	69.72	—	—	—	1.331	9.77	6.18
	liq	—	—	1.33	—	—	—	—
	g	—	66.	66.2	57.1	1.566	40.38	6.06
Ga^+	g	—	204.32	206.00	—	—	—	—
Ga^{2+}	g	—	677.38	680.54	—	—	—	—
std. state, m = 1	aq	—	—	—	−21.	—	—	—
Ga^{3+}	g	—	1385.	1390.	—	—	—	—
std. state, m = 1	aq	—	—	−50.6	−38.0	—	−79.	—
Ga^{4+}	g	—	2865.	2871.	—	—	—	—
GaO	g	85.719	67.	66.8	60.6	2.127	55.2	7.66
Ga_2O	c	155.439	—	−85.	—	—	—	—
Ga_2O_3								
β, rhombic	c	187.438	—	−260.3	−238.6	—	20.31	22.00
GaH	g	70.728	53.	52.7	46.3	2.07	46.69	7.00
$Ga(OH)_3$	c	120.742	—	−230.5	−198.7	—	24.	—
GaF	g	88.718	−60.	−60.2	—	2.167	—	7.95
GaCl	g	105.173	−19.	−19.1	−25.4	2.29	57.4	8.50

[a] x = Undetermined residual entropy

Substance			0 K	298.15 K (25°C)				
Formula and Description	State	Formula weight	ΔHf_0° kcal/mol	ΔHf°	ΔGf° kcal/mol	$H_{298}^\circ - H_0^\circ$	S° cal/deg mol	C_p°
Cr_3O_4	c	219.9856	—	−366.	—	—	—	—
$HCrO_4^-$								
std. state, m = 1	aq	117.0016	—	−209.9	−182.8	—	44.0	—
$Cr(OH)_3$								
precipitated	c	103.0181	—	−254.3	—	—	—	—
CrF_2	c	89.9928	—	−186.	—	—	—	—
	g	—	—	−99.	—	—	—	—
CrF_3	c	108.9912	−276.22	−277.	−260.	3.357	22.44	18.82
$CrCl_2$	c	122.902	−94.93	−94.5	−85.1	3.593	27.56	17.01
	g	—	—	−30.7	—	—	—	—
	aq	—	—	−114.2	—	—	—	—
$CrCl_2 \cdot 2H_2O$								
light green	c	158.9327	—	−237.1	—	—	—	—
$CrCl_2 \cdot 3H_2O$								
pale blue	c	176.9480	—	−308.9	—	—	—	—
$CrCl_2 \cdot 4H_2O$								
dark blue	c	194.9634	—	−384.4	—	—	—	—
$CrCl_3$	c	158.355	−132.96	−133.0	−116.2	4.22	29.4	21.94
CrO_2Cl_2	liq	154.9008	—	−138.5	−122.1	—	53.0	—
	g	—	−127.68	−128.6	−119.9	4.32	78.8	20.2
$CrBr_2$	c	211.814	—	−72.2	—	—	—	—
	g	—	—	−17.	—	—	—	—
$(CrBr_2)_2$	g	423.628	—	−84.	—	—	—	—
CrI_2	c	305.8048	—	−37.5	—	—	—	—
	g	—	—	24.	—	—	—	—
	aq	—	—	−60.1	—	—	—	—
CrI_3	c	432.7092	—	−49.0	—	—	—	—
$CrICl_2$	c	249.8064	—	−100.	—	—	—	—
$CrIBr_2$	c	338.7184	—	−79.	—	—	—	—
$Cr_2(SO_4)_3$	c	392.1768	—	—	—	—	—	67.5
CrN	c	66.0027	—	−29.8	—	—	—	11.0
Cr_2N	c	117.9987	—	−30.5	—	—	—	—
NH_4HCrO_4								
std. state, m = 1	aq	135.0402	—	−241.6	−201.8	—	71.1	—
$(NH_4)_2CrO_4$	c	152.0708	—	−279.0	—	—	—	—
std. state, m = 1	aq	—	—	−273.94	−211.90	—	66.2	—
in 300 H_2O	aq	—	—	−273.5	—	—	—	—
$(NH_4)_2Cr_2O_7$	c	252.0650	—	−431.8	—	—	—	—
std. state, m = 1	aq	—	—	−419.5	−348.9	—	116.8	—
Cr_3C_2	c	180.0103	−19.51	−19.3	−19.5	3.621	20.42	23.53
Cr_7C_3	c	400.0054	—	−38.7	−39.9	—	48.0	49.92
$Cr_{23}C_6$	c	1267.9749	—	−87.2	−89.3	—	145.8	149.2
$Cr(CO)_6$	c	220.0593	—	−257.4	—	—	—	—
	g	—	—	−240.4	—	—	—	—
$PbCrO_4$	c	323.184	—	−222.5	—	—	—	—
Tl_2CrO_4	c	524.734	—	−225.8	−205.9	—	67.5	—
Ag_2CrO_4	c	331.7336	—	−174.89	−153.40	—	52.0	34.00
std. state, m = 1	aq	—	—	−160.13	−137.09	—	46.8	—
$FeCr_2O_4$	c	223.8366	—	−345.3	−321.2	—	34.9	31.94
COBALT								
Co								
α, hexagonal	c	58.9332	0	0	0	1.139	7.18	5.93
β, f. c. cubic	c	—	—	0.11	0.06	—	7.34	—
	g	—	101.119	101.5	90.9	1.520	42.879	5.502
Co^+	g	—	282.486	284.348	—	—	—	—
Co^{2+}	g	—	675.82	679.17	—	—	—	—
std. state, m = 1	aq	—	—	−13.9	−13.0	—	−27.	—
Co^{3+}	g	—	1448.35	1453.18	—	—	—	—
std. state, m = 1	aq	—	—	22.	32.	—	−73.	—
Co_2CoO	c	74.9326	—	−56.87	−51.20	—	12.66	13.20
Co_3O_4	c	240.7972	—	−213.	−185.	—	24.5	29.5
$Co(OH)_2$								
blue, precipitated	c	92.9479	—	—	−107.6	—	—	—
pink, precipitated	c	—	—	−129.0	−108.6	—	19.	—
pink, precipitated, aged	c	—	—	—	−109.5	—	—	—
std. state, m = 1	aq	—	—	−123.8	−88.2	—	−32.	—
undissoc.; std. state, m = 1	aq	—	—	—	−100.8	—	—	—
$Co(OH)_3$								
precipitated	c	109.9553	—	−171.3	—	—	—	—
CoF_3	c	115.9284	—	−193.8	—	—	—	—
$CoCl_2$	c	129.8392	−74.74	−74.7	−64.5	3.375	26.09	18.76
std. state, m = 1	aq	—	—	−93.8	−75.7	—	0.	—
$CoCl_2$								
in ∞ H_2O	aq	—	—	−93.8	—	—	—	—
$CoCl_2 \cdot H_2O$	c	147.8545	—	−147.	—	—	—	—
$CoCl_2 \cdot 2H_2O$	c	165.8699	—	−220.6	−182.8	—	45.	—
$CoCl_2 \cdot 6H_2O$	c	237.9312	—	−505.6	−412.4	—	82.	—
$Co(ClO_4)_2$								
std. state, m = 1	aq	257.8344	—	−75.7	−17.1	—	60.	—
in ∞ H_2O	aq	—	—	−75.7	—	—	—	—
$CoBr_2$	c	218.7512	—	−52.8	—	—	—	19.0
std. state, m = 1	aq	—	—	−72.0	−62.7	—	12.	—
$CoBr_2 \cdot 6H_2O$	c	326.8432	—	−482.8	—	—	—	—
CoI_2	c	312.7420	—	−21.2	—	—	—	—
std. state, m = 1	aq	—	—	−40.3	−37.7	—	26.	—
$Co(IO_3)_2$								
std. state, m = 1	aq	408.7384	—	−119.7	−74.2	—	30.	—
$Co(IO_3)_2 \cdot 2H_2O$	c	444.7691	—	−258.6	−190.2	—	64.	—
CoS	c	90.9972	—	−19.8	—	—	—	—
Co_2S_3								
precipitated	c	214.0584	—	−35.2	—	—	—	—
$CoSO_4$	c	154.9948	—	−212.3	−187.0	—	28.2	—
std. state, m = 1	aq	—	—	−231.2	−191.0	—	−22.	—
$CoSO_4 \cdot 6H_2O$	c	263.0868	−630.257	−641.4	−534.35	13.525	87.86	84.46
$CoSO_4 \cdot 7H_2O$	c	281.1022	−699.547	−712.22	−591.26	15.097	97.05	93.33
$CoSe$	c	137.893	—	−14.6	—	—	—	—
$CoTe_2$	c	314.133	—	−31.	—	—	—	—
$Co(NO_3)_2$	c	182.9430	—	−100.5	—	—	—	—
std. state, m = 1	aq	—	—	−113.0	−66.2	—	43.	—
$[Co(NH_3)]^{2+}$								
std. state, m = 1	aq	75.9638	—	−34.7	−22.1	—	3.	—

Substance			0 K	298.15 K (25°C)				
Formula and Description	State	Formula weight	ΔHf°_0 kcal/mol	ΔHf° kcal/mol	ΔGf° kcal/mol	$H^\circ_{298} - H^\circ_0$	S° cal/deg mol	C°_p cal/deg mol
$[Co(NH_3)_2]^{2+}$ std. state, m = 1	aq	92.9944	—	—	−30.5	—	—	—
$[Co(NH_3)_3]^{2+}$ std. state, m = 1	aq	110.0250	—	—	−38.1	—	—	—
$[Co(NH_3)_4]^{2+}$ std. state, m = 1	aq	127.0556	—	—	−45.3	—	—	—
$[Co(NH_3)_6]^{3+}$ std. state, m = 1	aq	161.1169	—	−139.8	−38.9	—	40.0	—
$[Co(NH_3)_6]N_3{}^{3+}$ std. state, m = 1	aq	203.1370	—	—	42.9	—	60	—
$[Co(NH_3)_6](NO_3)_3$	c	347.1316	—	−306.4	−125.5	—	107.	—
std. state, m = 1	aq	—	—	−288.5	−117.4	—	140.	—
$CoBr_2 \cdot NH_3$	c	235.7818	—	−85.1	—	—	—	—
$CoBr_2 \cdot 2NH_3$ rose	c	252.8124	—	−116.0	—	—	—	—
$[Co(NH_3)_6]Br^{2+}$ std. state, m = 1	aq	241.0259	—	−171.7	−60.5	—	39.	—
$[Co(NH_3)_6]Br_2$	c	320.9349	—	−216.4	—	—	—	—
$[Co(NH_3)_6]Br_3$	c	400.8439	−217.52	−239.7	−119.8	12.18	77.7	78.1
std. state. m = 1	aq	—	—	−227.0	−112.2	—	94.	—
Co_2P	c	148.8402	—	−45.	—	—	—	—
$Co_3(PO_4)_2$	c	366.7424	—	—	−573.3	—	—	—
$CoHPO_4$	c	154.9126	—	—	−282.5	—	—	—
$CoAs$	c	133.8548	—	−9.7	—	—	—	—
$CoAs_2$	c	208.7764	—	−14.7	—	—	—	—
Co_2As	c	192.7880	—	−9.5	—	—	—	—
Co_2As_3	c	342.6312	—	−23.3	—	—	—	—
Co_3As_2	c	326.6428	—	−19.4	—	—	—	—
Co_5As_2	c	444.5092	—	−19.0	—	—	—	—
$Co_3(AsO_4)_2$	c	454.6380	—	—	−387.4	—	—	—
$CoSb$	c	180.683	—	−10.	—	—	—	—
$CoSb_2$	c	302.433	—	−13.	—	—	—	—
$CoSb_3$	c	424.183	—	−16.	—	—	—	—
Co_2C	c	129.8776	—	−10.	—	—	—	—
$CoCo_3$	c	118.9426	—	−170.4	—	—	—	—
CoC_2O_4	c	146.9531	—	−203.5	—	—	—	—
std. state, m = 1	aq	—	—	−211.1	−174.1	—	−16.	—
$Co(C_2O_4)_2{}^{2-}$ std. state, m = 1	aq	234.9730	—	−408.5	−344.8	—	26.	—
$CoSi$	c	87.0192	−23.87	−24.0	−23.6	1.78	10.3	10.6
$CoSi_2$	c	115.1052	—	−24.6	—	—	—	—
$CoSi_3$	c	143.1912	—	−25.6	—	—	—	—
Co_2SiO_4	c	209.9500	—	−353.	—	—	—	—
COPPER								
Cu	c	63.54	0	0	0	1.196	7.923	5.840
	g	—	80.58	80.86	71.37	1.481	39.74	4.968
Cu^+	g	—	258.752	260.513	—	—	—	—
std. state, m = 1	aq	—	—	17.13	11.95	—	9.7	—
Cu^{2+}	g	—	726.69	729.93	—	—	—	—
std. state, m = 1	aq	—	—	15.48	15.66	—	−23.8	—
Cu^{3+}	g	—	1576.1	1580.8	—	—	—	—
Cu_2	g	127.08	115.7	115.72	103.24	2.370	57.71	8.75
CuO	c	79.539	—	−37.6	−31.0	—	10.19	10.11
Cu_2O	c	143.079	—	−40.3	−34.9	—	22.26	15.21
CuH	c	64.548	—	5.1	—	—	—	—
$Cu(OH)_2$	c	97.555	—	−107.5	—	—	—	—
std. state, m = 1	aq	—	—	−94.46	−59.53	—	−28.9	—
CuF^+ std. state, m = 1	aq	82.538	—	−62.4	−52.7	—	−16.	—
CuF_2	c	101.537	—	−129.7	—	—	—	—
$CuF_2 \cdot 2H_2O$	c	137.567	—	—	−234.6	—	—	—
$CuCl$	c	98.993	—	−32.8	−28.65	—	20.6	11.6
$CuCl^+$ std. state, m = 1	aq	—	—	—	−16.3	—	—	—
$CuCl_2$	c	134.446	−52.79	−52.6	−42.0	3.581	25.83	17.18
$CuCl_2 \cdot 2\ H_2O$	c	170.477	—	−196.3	−156.8	—	40.	—
$CuCl_2$ std. state. m = 1	aq	134.446	—	—	−57.4	—	—	—
$Cu(ClO_4)_2$ std. state, m = 1	aq	262.441	—	−46.34	11.54	—	63.2	—
$Cu(ClO_4)_2 \cdot 6H_2O$	c	370.533	—	−460.9	—	—	—	—
Cu_2OCl_2	c	213.985	—	−90.	—	—	—	—
$CuBr$	c	143.449	−23.76	−25.0	−24.1	2.893	22.97	13.08
$CuBr^+$ std. state, m = 1	aq	—	—	—	−11.9	—	—	—
$CuBr_2$	c	223.358	—	−33.9	—	—	—	—
$CuBr_2 \cdot 4H_2O$	c	295.419	—	−317.0	—	—	—	—
$CuBr_2 \cdot 3Cu(OH)_2$	c	516.022	—	−378.1	−306.2	—	68.	—
$Cu(BrO_3)_2 \cdot 3Cu(OH)_2$	c	612.019	—	—	−244.2	—	—	—
CuI	c	190.444	—	−16.2	−16.6	—	23.1	12.92
$Cu(IO_3)_2$ std. state, m = 1	aq	413.345	—	−90.3	−45.5	—	32.8	—
$Cu(IO_3)_2 \cdot H_2O$	c	431.361	—	−165.4	−112.0	—	59.1	—
$Cu(IO_3)_2 \cdot 3Cu(OH)_2$	c	706.009	—	—	−160.0	—	—	—
CuS	c	95.604	—	−12.7	−12.8	—	15.9	11.43
Cu_2S α	c	159.144	—	−19.0	−20.6	—	28.9	18.24
$CuSO_4$	c	159.602	—	−184.36	−158.2	—	26.	23.9
std. state, m = 1	aq	—	—	−201.84	−162.31	—	−19.0	—
$CuSO_4$ in ∞ H_2O	aq	—	—	−201.84	—	—	—	—
$CuSO_4 \cdot H_2O$	c	177.617	—	−259.52	−219.46	—	34.9	32.
$CuSO_4 \cdot 3H_2O$	c	213.648	—	−402.56	−334.65	—	52.9	49.
$CuSO_4 \cdot 5H_2O$	c	249.678	—	−544.85	−449.344	—	71.8	67.
Cu_2SO_4	c	223.142	—	−179.6	—	—	—	—
$CuO \cdot CuSO_4$	c	239.141	—	−223.8	—	—	—	—
$CuSO_4 \cdot 2Cu(OH)_2$ antlerite	c	354.711	—	—	−345.8	—	—	—
$CuSO_4 \cdot 3Cu(OH)_2$ brochantite	c	452.266	—	—	−434.5	—	—	—

SELECTED VALUES OF CHEMICAL THERMODYNAMIC PROPERTIES (Continued)

Substance			0 K	298.15 K (25°C)				
Formula and Description	State	Formula weight	ΔHf°_0 kcal/mol	ΔHf° kcal/mol	ΔGf° kcal/mol	$H^\circ_{298} - H^\circ_0$	S° cal/deg mol	C°_p cal/deg mol
$CuSO_4 \cdot 3Cu(OH)_2 \cdot H_2O$								
langite	c	470.281	—	−594.	−488.6	—	80.	—
$CuSe$	c	142.50	—	−9.45	—	—	—	—
$CuSe_2$	c	221.46	—	−10.3	—	—	—	—
Cu_2Se	c	206.04	—	−14.2	—	—	—	—
Cu_2Te	c	254.68	—	5.	—	—	—	—
CuN_3	c	105.560	—	66.7	82.4	—	24.	—
$Cu(N_3)_2$	c	147.580	—	143.0	—	—	—	—
Cu_3N	c	204.627	—	17.8	—	—	—	22.
$Cu(NO_3)_2$	c	187.550	—	−72.4	—	—	—	—
std. state, m = 1	aq	—	—	−83.64	−37.56	—	46.2	—
$Cu(NO_3)_2$								
in ∞ H_2O	aq	—	—	−83.64	—	—	—	—
$Cu(NO_3)_2 \cdot 3H_2O$	c	241.596	—	−290.9	—	—	—	—
$Cu(NO_3)_2 \cdot 6H_2O$	c	295.642	—	−504.5	—	—	—	—
$Cu(NH_3)^{2+}$								
std. state, m = 1	aq	80.571	—	−9.3	3.72	—	2.9	—
$Cu(NH_3)_2{}^{2+}$								
std. state, m = 1	aq	97.601	—	−34.0	−7.28	—	26.6	—
$Cu(NH_3)_3{}^{2+}$								
std. state, m = 1	aq	114.632	—	−58.7	−17.48	—	47.7	—
$Cu(NH_3)_4{}^{2+}$								
std. state, m = 1	aq	131.662	—	−83.3	−26.60	—	65.4	—
$Cu(NH_3)_5{}^{2+}$								
std. state, m = 1	aq	148.693	—	—	−32.13	—	—	—
$CuSO_4 \cdot NH_3$	c	176.632	—	−218.0	—	—	—	—
$CuSO_4 \cdot 2NH_3$	c	193.663	—	−248.2	—	—	—	—
$CuSO_4 \cdot 5NH_3$	c	244.755	—	−328.1	—	—	—	—
CuP_2	c	125.488	—	−29.	—	—	—	—
$Cu_3(PO_4)_2$	c	380.563	—	—	−490.3	—	—	—
Cu_3As	c	265.542	—	−2.8	—	—	—	—
$Cu_3(AsO_4)_2$	c	468.458	—	—	−310.9	—	—	—
std. state, m = 1	aq	—	—	−378.10	−263.02	—	−149.2	—
CuC_2O_4	c	151.560	—	—	−158.2	—	—	—
std. state, m = 1	aq	—	—	−181.7	−145.5	—	−12.9	—
undissoc.; std. state, m = 1	aq	—	—	—	−153.1	—	—	—
$CuCO_3 \cdot Cu(OH)_2$								
malachite	c	221.104	—	−251.3	−213.6	—	44.5	—
$2CuCO_3 \cdot Cu(OH)_2$								
azurite	c	344.653	—	−390.1	—	—	—	—
$CuCN$	c	89.558	—	23.0	26.6	—	20.2	—
$Cu(CN)_2{}^-$								
std. state, m = 1	aq	115.576	—	—	61.6	—	—	—
$Cu(CN)_3{}^{2-}$								
std. state, m = 1	aq	141.594	—	—	96.5	—	—	—
$Cu(CN)_4{}^{3-}$								
std. state, m = 1	aq	167.611	—	—	135.4	—	—	—
$CuONC$								
cuprous fulminate	c	105.557	—	26.3	—	—	—	—
$CuCNS$	c	121.622	—	—	16.7	—	—	—
std. state, m = 1	aq	—	—	35.40	34.10	—	44.2	—
$(Cu(CNS)_2$								
std. state, m = 1	aq	179.704	—	52.02	59.96	—	45.2	—
$CuAl$	c	90.522	—	−9.8	—	—	—	—
$CuAl_2$	c	117.503	—	−9.75	—	—	—	—
Cu_2Al	c	154.062	—	−16.5	—	—	—	—
Cu_3Al	c	217.602	—	−16.8	—	—	—	—
Cu_3Al_2	c	244.583	—	−26.2	—	—	—	—
DYSPROSIUM								
Dy	c	162.50	0	0	0	2.116	17.87	6.73
	g	—	70.04	69.4	60.8	1.48	46.97	4.97
Dy^+	g	—	206.8	207.8	—	—	—	—
Dy^{2+}	g	—	476.	477.0	—	—	—	—
Dy^{3+}								
std. state, m = 1	aq	—	—	−167.	−159.	—	−55.2	5.
DyO	g	178.499	−19.	—	—	—	—	—
Dy_2O_3	c	372.998	−443.02	−445.3	−423.4	5.04	35.8	27.79
DyF	g	181.498	—	−43.	—	—	—	—
DyF^{2+}								
std. state, m = 1	aq	—	—	—	−231.7	—	—	—
$DyCl_3$								
β	c	268.859	—	−239.	—	—	—	—
γ	c	—	—	−236.	—	—	—	—
std. state, m = 1	aq	—	—	−286.	−253.	—	−14.8	−93.
$DyCl_3$								
in ∞ H_2O	aq	—	—	−286.	—	—	—	—
$DyCl_3 \cdot 6H_2O$	c	376.951	−676.83	−686.	−586.	14.42	96.00	82.7
DyI_3	c	543.213	—	−145.	—	—	—	—
$Dy(IO_3)_3$	c	687.208	—	−329.	—	—	—	—
DyC_2	g	186.522	206.	206.1	193.2	2.47	64.	10.5
ERBIUM								
Er	c	167.26	0	0	0	1.765	17.49	6.72
	g	—	76.08	75.8	67.1	1.48	46.72	4.97
Er^+	g	—	216.8	218.0	—	—	—	—
Er^{2+}	g	—	492.	495.	—	—	—	—
Er^{3+}								
std. state, m = 1	aq	—	—	−168.6	−159.9	—	−58.4	5.
ErO	g	183.259	−13.	—	—	—	—	—
Er_2O_3	c	382.518	−451.69	−453.6	−432.3	4.78	37.2	25.93
ErH_2	c	169.276	—	−49.0	—	—	—	—
Er^2H_2	c	171.288	—	−49.3	—	—	—	—
ErH_3	c	170.284	—	−58.	—	—	—	—
ErF	g	186.258	—	−45.	—	—	—	—
ErF_2	g	205.257	—	−164.	—	—	—	—
ErF_3	c	224.255	—	−409.	—	—	—	—
	g	—	—	−294.	—	—	—	—
$ErCl_3$	c	273.619	—	−238.7	—	—	—	24.
$ErCl_3$	g	—	—	−167.	—	—	—	—
$ErCl_3 \cdot 6H_2O$	c	381.711	−677.92	−687.0	−586.6	14.39	95.3	82.0
$Er(BrO_3)_3 \cdot 9H_2O)$	c	713.120	—	−844.9	—	—	—	—

Substance			0 K	298.15 K (25°C)				
Formula and Description	State	Formula weight	ΔHf°_0 kcal/mol	ΔHf° kcal/mol	ΔGf° kcal/mol	$H^\circ_{298} - H^\circ_0$ kcal/mol	S° cal/deg mol	C°_p cal/deg mol
$GaCl_3$	c	176.079	—	−125.4	−108.7	—	34.	—
	g	—	—	−107.0	—	—	—	—
GaBr	g	149.629	−10.	−11.9	−21.5	2.37	60.2	8.70
$GaBr_3$	c	309.447	—	−92.4	−86.0	—	43.	—
GaI	g	196.624	7.4	6.9	—	2.41	—	8.76
GaI_3	c	450.433	—	−57.1	—	—	—	—
	g	—	—	−34.0	—	—	—	—
$Ga_2(SO_4)_3$	c	427.625	—	—	—	—	—	62.4
GaN	c	83.727	—	−26.4	—	—	—	—
GaP	c	100.694	—	−21.	—	—	—	—
$GaPO_4$	c	164.691	—	—	−310.1	—	—	—
GaAs	c	144.642	—	−17.	−16.2	—	15.34	11.05
GaSb	c	191.47	—	−10.0	−9.3	—	18.18	11.60
Ga_2C_2	g	163.462	—	134.	—	—	—	—
GERMANIUM								
Ge	c	72.59	0	0	0	1.105	7.43	5.580
	g	—	89.34	90.0	80.3	1.768	40.103	7.345
Ge^+	g	—	271.18	273.32	—	—	—	—
Ge^{2+}	g	—	638.63	642.25	—	—	—	—
Ge^{3+}	g	—	1427.85	1432.95	—	—	—	—
GeO								
brown	c	88.589	—	−62.6	−56.7	—	12.	—
yellow	c	—	—	—	−49.5	—	—	—
	g	—	−11.	−11.04	−17.49	2.102	53.58	7.39
GeO_2								
hexagonal	c	104.589	—	−131.7	−118.8	—	13.21	12.45
	amorp	—	—	−128.4	—	—	—	—
Ge_2O_2	g	177.179	—	−112.	—	—	—	—
Ge_3O_3	g	265.768	—	−212.	—	—	—	—
GeH_4	g	76.622	24.29	21.7	27.1	2.567	51.87	10.76
Ge_2H_6	liq	151.228	—	32.82	—	—	—	—
	g	—	—	38.8	—	—	—	—
Ge_3H_8	liq	225.834	—	46.3	—	—	—	—
	g	—	—	54.2	—	—	—	—
H_2GeO_3	aq	122.604	—	−195.73	—	—	—	—
GeF	g	91.588	−8.	−7.97	—	2.185	—	8.30
GeF_2	g	110.587	−121.	—	—	—	—	—
GeF_4	g	148.584	—	—	—	4.163	72.36	19.56
GeCl	g	108.043	37.	37.09	29.7	2.290	59.	8.81
$GeCl_4$	liq	214.402	—	−127.1	−110.6	—	58.7	—
GeH_3Cl	g	111.067	—	—	—	2.865	63.00	13.08
$GeHCl_3$	g	—	—	—	—	4.192	79.06	19.40
GeBr	g	152.499	58.	56.32	—	2.355	—	8.87
$GeBr_2$	g	232.408	—	−15.0	−25.5	—	79.1	—
$GeBr_4$	liq	392.226	—	−83.1	−79.2	—	67.1	—
	g	—	−64.61	−71.7	−76.0	5.736	94.66	24.34
GeI_2	c	326.399	—	−21.	−20.	—	32.	—
	g	—	—	11.2	−1.0	—	76.	—
GeI_4	c	580.208	—	−33.9	−34.5	—	64.8	—
	g	—	−12.31	−13.6	−25.4	6.12	102.49	24.89
GeS	c	104.654	—	−16.5	−17.1	—	17.	—
	g	—	22.0	22.	10.	2.185	56.	8.05
GeS_2	c	136.718	—	−45.3	—	—	—	—
GeSe	c	151.55	—	−22.0	—	—	—	—
GeTe	c	200.19	—	−6.	—	—	—	—
Ge_3N_4	c	273.797	—	−15.1	—	—	—	—
GeP	c	103.564	—	−5.	−4.	—	15.	—
GeC	g	84.601	150.	151.	—	—	—	—
GeC_2	g	96.612	142.	143.	—	—	—	—
Ge_2C	g	157.191	130.	131.	—	—	—	—
GeSi	g	100.676	126.	127.	—	—	—	—
Ge_2Si	g	173.266	123.	124.	—	—	—	—
Ge_3Si	g	245.856	121.	122.	—	—	—	—
GOLD								
Au	c	196.967	—	—	—	1.436	11.33	6.075
	g	—	87.46	87.5	78.0	1.481	43.115	4.968
Au^+	g	—	300.20	301.73	—	—	—	—
Au^{2+}	g	—	773.0	776.0	—	—	—	—
AuO_3^{3-}								
std. state, m = 1	aq	244.9652	—	—	−12.4	—	—	—
AuH	g	197.9750	70.9	70.5	63.5	2.068	50.441	6.968
$HAuO_3^{2-}$								
std. state, m = 1	aq	245.9732	—	—	−34.0	—	—	—
$H_2AuO_3^-$								
std. state, m = 1	aq	246.9811	—	—	−52.2	—	—	—
$Au(OH)_3$								
precipitated	c	247.9891	—	−101.5	−75.77	—	45.3	—
AuF_3	c	253.9622	—	−86.9	—	—	—	—
AuCl	c	232.420	—	−8.3	—	—	—	—
$AuCl_2^-$								
std. state, m = 1	aq	267.873	—	—	−36.13	—	—	—
$AuCl_3$	c	303.326	—	−28.1	—	—	—	—
$AuCl_3 \cdot 2H_2O$	c	339.357	—	−170.9	—	—	—	—
$AuCl_4^-$								
std. state, m = 1	aq	338.779	—	−77.0	−56.22	—	63.8	—
$HAuCl_4$								
std. state, m = 1	aq	339.787	—	−77.0	−56.22	—	63.8	—
$HAuCl_4 \cdot 3H_2O$	c	393.833	—	−284.9	—	—	—	—
$HAuCl_4 \cdot 4H_2O$	c	411.848	—	−355.9	—	—	—	—
AuBr	c	276.876	—	−3.34	—	—	—	—
$AuBr_2^-$								
std. state, m = 1	aq	356.785	—	−30.7	−27.49	—	52.5	—
$AuBr_3$	c	436.695	—	−12.73	—	—	—	—
$AuBr_4^-$								
std. state, m = 1	aq	516.603	—	−45.8	−40.0	—	80.3	—
$HAuBr_4 \cdot 5H_2O$	c	607.688	—	−398.7	—	—	—	—
AuI	c	323.8714	—	0.	—	—	—	—
$AuSb_2$	c	440.467	—	−2.6	—	—	—	—
$Au(CN)_2^-$								
std. state, m = 1	aq	249.003	—	57.9	68.3	—	41.	—

Substance			0 K	298.15 K (25°C)				
Formula and Description	State	Formula weight	ΔHf°_0 kcal/mol	ΔHf°	ΔGf° kcal/mol	$H^\circ_{298} - H^\circ_0$	S° cal/deg mol	C°_p
$Au(SCN)_2^-$								
std. state, m = 1	aq	313.131	—	—	60.2	—	—	—
$Au(SCN)_4^-$								
std. state, m = 1	aq	429.294	—	—	134.2	—	—	—
$Au(SCN)_5^{2-}$								
std. state, m = 1	aq	487.376	—	—	156.4	—	—	—
$Au(SCN)_6^{3-}$								
std. state, m = 1	aq	545.458	—	—	178.5	—	—	—
AuSn	c	315.657	—	−7.28	−7.15	—	23.22	12.06
$AuSn_2$	c	434.347	—	−10.14	−9.07	—	32.4	—
$AuSn_4$	c	671.727	—	−9.25	−9.04	—	59.9	—
$AuPb_2$	c	611.347	—	−1.5	—	—	—	—
AuIn	c	311.787	—	−10.8	—	—	—	—
$AuIn_2$	c	426.607	—	−18.0	—	—	—	—
AuCd	c	309.367	—	−9.28	−9.34	—	23.9	—
AuCu	c	260.507	—	−4.46	−4.40	—	19.1	11.9
$AuCu_3$	c	387.587	—	−6.84	−6.97	—	35.7	24.0
HAFNIUM								
Hf								
α, hexagonal	c	178.49	0	0	0	1.397	10.41	6.15
	g	—	147.92	148.0	137.8	1.481	44.642	4.972
Hf^+	g	—	309.	311.	—	—	—	—
Hf^{2+}	g	—	653.	656.	—	—	—	—
Hf^{3+}	g	—	1190.	1195.	—	—	—	—
HfO	g	194.489	—	12.	—	—	—	—
HfO_2	c	210.489	—	−273.6	−260.1	—	14.18	14.40
HfF_4								
monoclinic	c	254.484	—	−461.4	−437.5	—	27.	—
$HfCl_4$	c	320.302	—	−236.70	−215.42	—	45.6	28.80
HfN	c	192.497	—	−88.3	—	—	—	—
HfC	c	190,501	—	−60.1	—	—	—	—
HfB	c	189.301	—	−47.	—	—	—	—
HfB_2	c	200.112	−80.09	−80.3	−79.4	1.77	10.2	11.89
HELIUM								
He	g	4.0026	0	0	0	1.481	30.1244	4.9679
std. state, m = 1	aq	—	—	−0.4	4.6	—	13.3	—
He^+	g	—	566.978	568.459	—	—	—	—
HOLMIUM								
Ho	c	164.930	0	0	0	1.91	18.0	6.49
	g	—	72.33	71.9	63.3	1.48	46.72	4.97
Ho^+	g	—	211.1	212.2	—	—	—	—
Ho^{2+}	g	—	483.	486.	—	—	—	—
Ho^{3+}								
std. state, m = 1	aq	—	—	−168.5	− 161.0	—	−54.2	4.
HoO	g	180.929	−22.	—	—	—	—	—
Ho_2O_3	c	377.858	−447.59	−449.5	−428.1	5.02	37.8	27.48
HoH_2	c	166.946	—	−31.7	—	—	—	—
HoF	g	183.928	—	−43.	—	—	—	—
HoF_2	g	202.927	—	−163.	—	—	—	—
HoF_3	c	221.925	—	−408.	—	—	—	—
	g	—	—	−294.	—	—	—	—
$HoCl_3$	c	271.289	—	−240.3	—	—	—	21.
	g	—	—	−168.	—	—	—	—
$HoCl_3$								
∞ H_2O	aq	—	—	−288.4	—	—	—	—
$HoCl_3 \cdot 6H_2O$	c	379.381	−678.80	−687.9	−588.0	14.49	97.08	83.0
HoOCl	c	216.382	—	−239.2	—	—	—	—
HoI_3	c	545.643	—	−149.0	—	—	—	—
HoS	g	196.994	43.	—	—	—	—	—
HoAs	c	239.852	—	−72	—	—	—	—
HoC_2	c	188.952	—	−26.	−26.7	—	23.	—
	g	—	135.	135.1	—	2.47	—	10.5
HoC_4	g	212.975	165.	—	—	—	—	—
Ho_2C_3	c	365.893	—	−56.	—	—	—	—
HoAu	g	361.897	100.	99.06	—	2.41	—	8.77
HYDROGEN								
H	g	1.0080	51.626	52.095	48.581	1.481	27.391	4.9679
^1H	g	1.0078	51.626	52.095	48.581	1.481	27.391	4.9679
^2H	g	2.0141	52.524	52.981	49.360	1.481	29.455	4.9679
H^+	g	1.0080	365.211	367.161	—	—	—	—
std. state, m = 1	aq	—	—	0	0	—	0	0
H^-	g	—	34.40	33.39	—	—	—	—
H_2	g	2.0159	0	0	0	2.0238	31.208	6.889
$^1H_2$	g	2.0156	0	0	0	2.0238	31.208	6.889
$^2H_2$	g	4.0282	0	0	0	2.0481	34.620	6.978
$^1H^2H$	g	3.0219	0.079	0.076	−0.350	2.0328	34.343	6.978
H_2								
std. state, m = 1	aq	2.0159	—	−1.0	4.2	—	13.8	—
H_2^+	g	—	355.74	357.23	—	—	—	—
OH	g	17.0074	9.25	9.31	8.18	2.1070	43.890	7.143
OH^-	g	—	−32.3	−33.67	—	—	—	—
std. state, m = 1	aq	—	—	−54.970	−37.594	—	−2.57	−35.5
HO_2	g	33.0068	6.	5.	—	—	—	—
H_2O	liq	18.0153	—	−68.315	−56.687	—	16.71	17.995
^2H_2O	liq	20.0276	—	−70.411	−58.195	—	18.15	20.16
$^1H^2HO$	liq	19.0213	—	−69.285	−57.817	—	18.95	—
H_2O	g	18.0153	−57.102	−57.796	−54.634	2.3667	45.104	8.025
^1H_2O	g	18.0150	−57.102	−57.796	−54.634	2.3667	45.103	8.025
^2H_2O	g	20.0276	−58.855	−59.560	−56.059	2.3801	47.378	8.19
$^1H^2HO$	g	19.0213	−57.927	−58.628	−55.719	2.3721	47.658	8.08
H_2O_2	liq	34.0147	—	−44.88	−28.78	−26.2	—	21.3
	g	—	−31.08	−32.58	−25.24	2.594	55.6	10.3
undissoc.; std. state, m = 1	aq	—	—	−45.69	−32.05	—	34.4	—
H_2O_2								
∞ H_2O	aq	—	—	−45.69	—	—	—	—
INDIUM								
In	c	114.82	0	0	0	1.578	13.82	6.39
	g	—	58.25	58.15	49.89	1.48	41.51	4.98
In^+	g	—	191.68	193.06	—	—	—	—
std. state, m = 1	aq	—	—	—	−2.9	—	—	—

Substance			0 K	298.15 K (25°C)				
Formula and Description	State	Formula weight	ΔHf°_0 kcal/mol	ΔHf° kcal/mol	ΔGf° kcal/mol	$H^\circ_{298} - H^\circ_0$	S° cal/deg mol	C°_p cal/deg mol
In^{2+}	g	—	626.82	629.68	—	—	—	—
std. state, m = 1	aq	—	—	—	−12.1	—	—	—
In^{3+}	g	—	1273.2	1277.56	—	—	—	—
std. state, m = 1	aq	—	—	−25.	−23.4	—	−36.	—
InO	g	130.819	92.	93.	87.1	2.14	56.5	7.78
In_2O_3	c	277.638	—	−221.27	−198.55	—	24.9	22.
InH	g	115.828	52.	51.5	45.49	2.075	49.60	7.07
InOH	g	131.827	−18.	−19.	—	—	—	—
InF	g	133.818	−48.2	−48.61	—	2.198	—	—
$InCl_3$	c	221.179	—	−128.4	—	—	—	—
In_2Cl_3	g	335.999	—	−103.6	—	—	—	—
InBr	c	194.729	—	−41.9	−40.4	—	27.	—
	g	—	−11.5	−13.6	−22.54	2.406	61.99	8.76
$InBr_3$	c	354.547	—	−102.5	—	—	—	—
	g	—	—	−67.4	—	—	—	—
InI	c	241.724	—	−27.8	−28.8	—	31.	—
	g	—	2.52	1.8	−9.0	2.437	63.87	8.80
InI_3	c	495.533	—	−57.	—	—	—	—
	g	—	—	−28.8	—	—	—	—
InS	c	146.884	—	−33.0	−31.5	—	16.	—
	g	—	—	57.	—	—	—	—
In_2S	g	261.704	15.	2.9	—	76.	—	—
In_2S_3	c	325.832	—	−102.	−98.6	—	39.1	28.20
$In_2(SO_4)_3$	c	517.825	—	−666.	−583.	—	65.	67.
InN	c	128.827	—	−4.2	—	—	—	—
InP	c	145.794	—	−21.2	−18.4	—	14.3	10.86
InAs	c	189.742	—	−14.0	−12.8	—	18.1	11.42
InSb	c	236.57	—	−7.3	−6.1	—	20.6	11.82
$InSb_2$	g	358.32	—	75.	—	—	—	—
IODINE								
I	g	126.9044	25.631	25.535	16.798	1.481	43.184	4.968
I^+	g	—	266.77	268.16	—	—	—	—
I^{2+}	g	—	707.22	710.09	—	—	—	—
I^-	g	—	−45.4	−47.0	—	—	—	—
std. state, m = 1	aq	—	—	−13.19	−12.33	—	26.6	−34.0
I_2	c	253.8088	0	0	0	3.154	27.757	13.011
	g	—	15.659	14.923	4.627	2.418	62.28	8.82
std. state, m = 1	aq	—	—	5.4	3.92	—	32.8	—
I_3^-								
std. state, m = 1	aq	380.7132	—	−12.3	−12.3	—	57.2	—
IO^-								
std. state, m = 1	aq	—	—	−25.7	−9.2	—	−1.3	—
IO_3^-								
std. state, m = 1	aq	174.9026	—	−52.9	−30.6	—	28.3	—
IO_4^-								
std. state, m = 1	aq	190.9020	—	−36.2	−14.0	—	−53.	—
I_2O_5	c	333.8058	—	−37.78	—	—	—	—
HI	g	127.9124	6.850	6.33	0.41	2.069	49.351	6.969
std. state, m = 1	aq	—	—	−13.19	−12.33	—	26.6	−34.0
HI								
in ∞ H_2O	aq	—	—	−13.19	—	—	—	—
HIO								
undissoc.; std. state, m = 1	aq	143.9118	—	−33.0	−23.7	—	22.8	—
HIO_3	c	175.9106	—	−55.0	—	—	—	—
undissoc.; std. state, m = 1	aq	—	—	−50.5	−31.7	—	39.9	—
IF	g	145.9028	−22.40	−22.86	−28.32	2.174	56.42	7.99
IF_5	liq	221.8964	—	−206.7	—	—	—	—
ICl								
α	c	162.3574	—	−8.4	—	—	—	—
	liq	—	—	−5.71	−3.25	—	32.3	—
	g	—	4.64	4.25	−1.30	2.282	59.140	8.50
std. state, m = 1	aq	—	—	—	−4.1	—	—	—
ICl_3	c	233.2634	—	−21.4	−5.34	—	40.0	—
IBr	c	206.8134	—	−2.5	—	—	—	—
	g	—	11.90	9.76	0.89	2.367	61.822	8.71
std. state, m = 1	aq	—	—	—	−1.0	—	—	—
IRIDIUM								
Ir	c	192.2	0	0	0	1.260	8.48	6.00
	g	—	158.78	159.0	147.7	1.481	46.240	4.968
Ir^+	g	—	373.	375.	—	—	—	—
IrO_2	c	224.20	—	−65.5	—	—	—	13.7
IrO_3	g	240.20	—	1.9	—	—	—	—
IrF_6	c	306.19	—	−138.54	−110.34	—	59.2	—
IrCl	c	227.65	—	−19.5	—	—	—	—
$IrCl_3$	c	298.56	—	−58.7	—	—	—	—
IrS_2	c	256.33	—	−33.	—	—	—	—
Ir_2S_3	c	480.59	—	−56.	—	—	—	—
IRON								
Fe								
α	c	55.847	0	0	0	1.073	6.52	6.00
	g	—	98.94	99.5	88.6	1.6374	43.112	6.137
Fe^+	g	—	281.12	283.16	—	—	—	—
Fe^{2+}	g	—	654.2	657.8	—	—	—	—
std. state, m = 1	aq	—	—	−21.3	−18.85	—	−32.9	—
Fe^{3+}	g	—	1360.8	1365.9	—	—	—	—
std. state, m = 1	aq	—	—	−11.6	−1.1	—	−75.5	—
$Fe_{0.947}O$								
wustite	c	68.8865	−63.85	−63.64	−58.89	2.26	13.74	11.50
FeO	c	71.8464	—	−65.0	—	—	—	—
Fe_2O_3								
hematite	c	159.6922	−195.46	−197.0	−177.4	3.719	20.89	24.82
Fe_3O_4								
magnetite	c	231.5386	−265.80	−267.3	−242.7	5.87	35.0	34.28
FeO(OH)								
goethite	c	88.8538	—	−133.6	—	—	—	—
$Fe(OH)_2$								
precipitated	c	89.8617	—	−136.0	−116.3	—	21.	—
$Fe(OH)_3$								
precipitated	c	106.8691	—	−196.7	−166.5	—	25.5	—
undissoc.; std. state, m = 1	aq	—	—	—	−157.6	—	—	—

SELECTED VALUES OF CHEMICAL THERMODYNAMIC PROPERTIES (Continued)

Substance			0 K	298.15 K (25°C)				
Formula and Description	State	Formula weight	ΔHf°_0 kcal/mol	ΔHf°	ΔGf° kcal/mol	$H^\circ_{298} - H^\circ_0$	S° cal/deg mol	C°_p
FeF_2	c	93.8438	—	—	—	3.049	20.79	16.28
FeF_2^+								
std. state, m = 1	aq	—	—	−166.4	−150.2	—	−15.	—
FeF_3	aq	112.8422	—	−242.9	—	—	—	—
$FeCl_2$	c	126.753	−82.313	−81.69	−72.26	3.889	28.19	18.32
std. state, m = 1	aq	—	—	−101.2	−81.59	—	−5.9	—
$FeCl_2 \cdot 2H_2O$	c	162.7837	—	−227.8	—	—	—	—
$FeCl_2 \cdot 4H_2O$	c	198.8143	—	−370.3	—	—	—	—
$FeCl_3$	c	162.206	−95.828	−95.48	−79.84	4.710	34.0	23.10
std. state, m = 1	aq	—	—	−131.5	−95.2	—	−35.0	—
$FeCl_3 \cdot 6H_2O$	c	270.2980	—	−531.5	—	—	—	—
FeOCl	c	107.2994		−99.1			—	
$Fe(ClO_4)_2$								
std. state, m = 1	aq	254.7482	—	−83.1	−22.97	—	54.1	—
$FeBr_2$	c	215.665	—	−59.7	—	—	—	—
std. state, m = 1	aq	—	—	−79.4	−68.55	—	6.5	—
$FeBr_3$	c	295.574	—	−64.1	—	—	—	—
std. state, m = 1	aq	—	—	−98.8	−75.7	—	−16.4	—
FeI_2	c	309.6558	—	−27.0	—	—	—	—
std. state, m = 1	aq	—	—	−47.7	−43.51	—	20.3	—
FeI_3	g	436.5602	—	17.	—	—	—	—
std. state, m = 1	aq	—	—	−51.2	−38.1	—	4.3	—
$Fe_{1.000}S$								
Iron-rich pyrrhotite, α	c	87.911	−24.01	−23.9	−24.0	2.235	14.41	12.08
FeS_2								
pyrite	c	119.975	−41.72	−42.6	−39.9	2.302	12.65	14.86
markasite	c	—	—	−37.0	—	—	—	—
$FeSO_4$	c	151.9086	—	−221.9	−196.2	—	25.7	24.04
std. state, m = 1	aq	—	—	−238.6	−196.82	—	−28.1	—
$FeSO_4 \cdot H_2O$	c	169.9239	—	−297.25	—	—	—	—
$FeSO_4 \cdot 4H_2O$	c	223.9700	—	−508.9	—	—	—	—
$FeSO_4 \cdot 7H_2O$	c	278.0160	—	−720.50	−599.97	—	97.8	94.28
FeSe	c	134.807	—	−18.0	—	—	—	—
FeTe	c	183.447	—	−15.0	—	—	—	—
Fe_4N	c	237.3947	—	−2.5	0.9	—	37.	—
$Fe(NO_3)_3$	aq	241.8617	—	−161.3	—	—	—	—
std. state, m = 1	aq	—	—	−160.3	−80.9	—	29.5	—
FeP	c	86.8208	—	−30.	—	—	—	—
FeP_2	c	117.7946	—	−46.	—	—	—	—
Fe_2P	c	142.6678	—	−39.	—	—	—	—
Fe_3P	c	198.5148	—	−39.	—	—	—	—
$FePO_4$	c	150.8184	—	−310.1	—	—	—	—
$FePO_4 \cdot 2H_2O$								
strengite	c	186.8491	−445.28	−451.3	−396.2	6.607	40.93	43.15
Fe_3C								
α, cementite	c	179.5522	—	6.0	4.8	—	25.0	25.3
$FeCO_3$								
siderite	c	115.8564	—	−177.00	−159.35	—	22.2	19.63
$Fe(CO)_5$	liq	195.8998	—	−185.0	−168.6	—	80.8	57.5
	g	—	—	−175.4	−166.65	—	106.4	—
$HFe(CN)_6^{3-}$								
std. state, m = 1	aq	212.9621	—	108.9	160.40	—	42.	—
$H_2Fe(CN)_6^{2-}$								
std.state, m = 1	aq	213.9700	—	108.9	157.37	—	52.	
$H_3Fe(CN)_6^-$								
std. state, m = 1	aq	—	—	108.9	—	—	—	—
$Fe(SCN)^{2+}$								
thiocyanate; std. state, m = 1	aq	113.9288	—	5.6	17.0	—	−31.	—
FeSi	c	83.933	−17.67	−17.6	−17.6	1.91	11.0	11.4
$FeSi_2$								
β-lebeanite	c	112.019	−19.19	−19.4	−18.7	2.40	13.3	15.79
$FeSi_{2.33}$								
α-lebeanite	c	121.287	−14.03	−14.	−14.	2.89	16.6	17.62
Fe_3Si	c	195.627	−22.55	−22.4	−22.6	4.14	24.8	23.50
Fe_2SiO_4								
fayalite	c	203.7776	—	−353.7	−329.6	—	34.7	31.76
$FeAl_2O_4$	c	173.8076	—	−470.	−442.	—	25.4	29.53
KRYPTON								
Kr	g	83.80	0	0	0	1.481	39.1905	4.9679
std. state, m = 1	aq	—	—	−3.7	3.6	—	14.7	—
Kr^+	g	—	322.84	324.32	—	—	—	—
Kr^{2+}	g	—	889.47	892.43	—	—	—	—
Kr^{3+}	g	—	1741.6	1746.0	—	—	—	—
LANTHANUM								
La								
α	c	138.91	0	0	0	1.593	13.6	6.48
	g	—	103.084	103.0	94.07	1.509	43.563	5.438
La^+	g	—	231.690	233.087	—	—	—	—
La^{2+}	g	—	487.	490.	—	—	—	—
La^{3+}	g	—	929.2	933.2	—	—	—	—
std. state, m = 1	aq	—	—	−169.0	−163.4	—	−52.0	−3.
LaO	g	154.909	−28.5	−29.01	−34.72	2.121	57.27	7.59
La_2O	g	293.819	−2.	−3.2	—	3.00	—	12.0
La_2O_2	g	309.819	−145.	−146.6	—	3.67	—	16.2
La_2O_3	c	325.818	−427.19	−428.7	−407.7	4.742	30.43	26.00
LaH_2	c	140.926	—	−48.3	—	—	—	—
$La(OH)_3$	c	189.932	—	−337.0	—	—	—	—
LaF	g	157.908	—	—	—	2.173	56.9	7.98
LaF_3	g	195.905	—	—	—	4.14	78.7	17.4
$LaCl_3$	c	245.269	—	−256.0	—	—	—	26.0
std. state, m = 1	aq	—	—	−288.9	−257.5	—	−12.	−101.
$LaCl_3$								
in ∞ H_2O	aq	—	—	−288.9	—	—	—	—
$LaCl_3 \cdot 7H_2O$	c	371.376	—	−759.7	−648.5	17.12	110.6	103.0
$La(BrO_3)_3 \cdot 9H_2O$	c	684.770	—	−846.0	—	—	—	—
LaI_3	c	519.623	—	−159.4	—	—	—	—
$La(IO_3)_3$	c	663.618	—	−334.	−270.4	—	62.	—
LaS	c	170.974	−108.9	−109.	−107.9	2.6	17.5	14.
La_2S_3	c	374.012	—	−289	—	—	—	—
$La_2(SO_4)_3$	c	566.005	—	−942.0	—	—	—	67.
$La_2(SO_4)_3 \cdot 9H_2O$	c	728.143	—	−1589.	—	—	—	152.

SELECTED VALUES OF CHEMICAL THERMODYNAMIC PROPERTIES (Continued)

Substance			0 K	298.15 K (25°C)				
Formula and Description	State	Formula weight	ΔHf°_0 kcal/mol	ΔHf° kcal/mol	ΔGf° kcal/mol	$H^\circ_{298} - H^\circ_0$ kcal/mol	S° cal/deg mol	C°_p cal/deg mol
LaN	c	152.917	—	−72.5	—	—	—	—
$La(NO_3)_3$	c	324.925	—	−299.8	—	—	—	—
$La(NO_3)_3$								
in ∞ H_2O	aq	—	—	−317.7	—	—	—	—
$La(NO_3)_3 \cdot 3H_2O$	c	378.971	—	−520.0	—	—	—	—
$La(NO_3)_3 \cdot 4H_2O$	c	396.986	—	−592.3	—	—	—	—
$La(NO_3)_3 \cdot 6H_2O$	c	433.017	—	−732.23	—	—	—	—
LaC_2	c	162.932	—	−17.	−17.3	—	17.	—
$La_2(CO_3)_3$	c	457.848	—	—	−750.9	—	—	—
LaB_6	c	203.776	—	−31.	—	—	—	—
LaAu	g	335.877	111.4	110.8	98.2	2.46	67.	8.8
LEAD								
Pb	c	207.19	0	0	0	1.644	15.49	6.32
	g	—	46.76	46.6	38.7	1.481	41.889	4.968
Pb^+	g	—	217.795	219.116	—	—	—	—
Pb^{2+}	g	—	564.44	567.25	—	—	—	—
std. state, m = 1	aq	—	—	−0.4	−5.83	—	2.5	—
Pb^{3+}	g	—	1300.93	1305.22	—	—	—	—
Pb^{4+}	g	—	2276.9	2282.7	—	—	—	—
PbO								
yellow	c	223.189	−51.766	−51.466	−44.91	2.207	16.42	10.94
red	c	—	—	−52.34	−45.16	—	15.9	10.95
PbO_2	c	239.189	—	−66.3	−51.95	—	16.4	15.45
Pb_2O_3	c	462.378	—	—	—	—	36.3	25.74
Pb_3O_4	c	685.568	—	−171.7	−143.7	—	50.5	35.1
$Pb(OH)_2$	c	241.205	—	—	−108.1	—	—	—
precipitated	c	—	—	−123.3	—	—	—	—
PbF_2	c	245.187	—	−158.7	−147.5	—	26.4	—
std. state, m = 1	aq	—	—	−159.4	−139.11	—	−4.1	—
PbF_4	c	283.184	—	−225.1	—	—	—	—
$PbCl_2$	c	278.096	—	−85.90	−75.98	—	32.5	—
ionized; std. state, m = 1	aq	—	—	−80.3	−68.57	—	29.5	—
$PbCl_4$	liq	349.002	—	−78.7	—	—	—	—
$PbBr_2$	c	367.008	—	−66.6	−62.60	—	38.6	19.15
ionized; std. state, m = 1	aq	—	—	−58.5	−55.53	—	41.9	—
PbI_2	c	460.999	−41.81	−41.94	−41.50	4.666	41.79	18.49
ionized; std. state, m = 1	aq	—	—	−26.8	−30.49	—	55.7	—
PbS	c	239.254	—	−24.0	−23.6	—	21.8	11.83
$PbSO_3$	c	287.252	—	−160.1	—	—	—	—
$PbSO_4$	c	303.252	−217.82	−219.87	−194.36	4.795	35.51	24.667
PbS_2O_3	c	319.316	—	−161.1	—	—	—	—
PbSe	c	286.15	—	−24.6	−24.3	—	24.5	12.0
$PbSeO_4$	c	350.148	—	−145.6	−120.7	—	40.1	—
$PbTeO_3 \cdot 0.667H_2O$ (amorp)	—	394.804	—	−185.5	—	—	—	—
$Pb_3(PO_4)_2$	c	811.513	—	—	—	—	84.4	61.25
$PbCO_3$	c	267.199	—	−167.1	−149.5	—	31.3	20.89
PbC_2O_4								
std. state, m = 1	c	295.210	—	−197.6	−166.9	—	34.9	25.2
$Pb(CH_3)_4$	liq	267.330	—	23.4	—	—	—	—
$Pb(C_2H_5)_4$	liq	323.439	—	12.6	—	—	—	—
	g	—	—	26.19	—	—	—	—
$PSiO_3$	c	283.274	—	−273.83	−253.86	—	26.2	21.52
Pb_2SiO_4	c	506.464	—	−325.8	−299.4	—	44.6	32.78
LITHIUM								
Li	c	6.941	0	0	0	1.106	6.96	5.92
	g	6.941	37.715	38.09	30.28	1.481	33.14	4.968
Li^+	g	6.941	162.045	162.42	—	—	—	—
Li^{2+}	g	6.941	—	1936.7	—	—	—	—
Li^{3+}	g	6.941	—	4760.5	—	—	—	—
Li^+	a	6.941	—	−66.56	−70.10	—	3.2	16.4
Li								
in D:99Hg		6.941	—	−19.60	—	—	—	—
LI								
Hg:x		6.941	—	—	−19.5	—	—	—
LiO	g	22.9404	18.10	18.1	12.5	2.14	50.40	7.75
LiO_2	g	38.9398	—	—	—	2.60	58.27	10.34
Li_2O	c	29.8814	−141.393	−142.91	−134.13	1.732	8.98	12.93
	g	29.8814	−38.18	−38.4	−43.4	3.03	55.30	11.91
Li_2O_2	c	45.8808	—	−151.6	—	—	—	—
	a	45.8808	—	−158.8	—	—	—	—
LiH	c	7.9490	−20.425	−21.64	−16.34	0.903	4.782	6.66
LiD	c	8.9551	−20.684	−21.73	−16.18	1.085	5.640	8.20
LiOH	c	23.9484	−114.517	−115.90	−104.92	1.772	10.23	11.87
LiOD	c	24.9545	—	−116.8	—	—	—	—
LiF	c	25.9394	−146.657	−147.22	−140.47	1.547	8.52	9.94
Li_2F_2	g	51.8788	−223.67	−224.8	−224.7	3.19	61.79	15.09
Li_3F_3	g	77.8182	−360.30	−361.9	−357.	4.88	76.	24.5
$LiHF_2$	c	45.9458	−223.66	−225.22	−209.11	2.826	17.0	16.77
LiCl	c	42.394	−97.68	−97.66	−91.87	2.224	14.18	11.47
	g	42.394	−46.663	−46.7	−51.8	2.165	50.840	7.94
	a	42.394	—	−106.51	−101.48	—	16.7	−16.2
$LiCl \cdot H_2O$	c	60.4094	—	−170.31	−151.01	—	24.58	—
$LiCl \cdot 2H_2O$	c	78.4248	—	−242.03	—	—	—	—
$LiCl \cdot 3H_2O$	c	96.4402	—	−313.4	—	—	—	—
Li_2Cl_2	g	84.788	−141.20	−141.9	−142.5	3.70	69.0	17.26
$LiClO_3$	c	90.3922	—	−88.2	—	—	—	—
$LiClO_4$	c	106.3916	—	−91.06	—	—	—	—
	a	106.3916	—	−97.472	−72.2	—	46.7	−1.8
$LiClO_4 \cdot H_2O$	c	124.4070	—	−166.6	−121.8	—	37.1	—
$LiClO_4 \cdot 3H_2O$	c	160.4378	—	−310.22	−239.30	—	60.9	—
LiBr	c	86.850	—	−83.942	−81.74	—	17.75	—
	a	86.850	—	−95.612	−94.95	—	22.9	−17.5
$LiBr \cdot H_2O$	c	104.8654	—	−158.36	−142.05	—	26.2	—
$LiBr \cdot 2H_2O$	c	122.8808	—	−230.1	−200.9	—	38.8	—
$LiBrO_3$	c	134.8482	—	−82.93	—	—	—	—
LiI	c	133.8454	−64.663	−64.63	−64.60	2.716	20.74	12.20
	a	133.8454	—	−79.75	−82.4	—	29.8	−17.6
$LiI \cdot H_2O$	c	151.8608	—	−141.09	−127.0	—	29.4	—
$LiI \cdot 2H_2O$	c	169.8762	—	−212.81	−186.5	—	44.	—
$LiI \cdot 3H_2O$	c	187.8916	—	−284.93	—	—	—	—
$LiIO_3$	c	181.8436	—	−120.31	—	—	—	—

Substance			0 K	298.15 K (25°C)				
Formula and Description	State	Formula weight	ΔHf_0° kcal/mol	ΔHf°	ΔGf° kcal/mol	$H_{298}^\circ - H_0^\circ$	S° cal/deg mol	C_p°
	a	181.8436	—	−119.46	−100.70	—	31.4	−13.2
Li_2S	c	45.946	—	−105.5	—	—	—	—
Li_2S_2	c	78.010	—	−104.7	—	—	—	—
Li_2SO_3	c	93.9442	—	−281.3	—	—	—	—
Li_2SO_4	c	109.9436	−340.367	−343.33	−315.91	4.452	27.5	28.10
	a	109.9436	—	−350.44	−318.18	—	11.3	−37.2
$LiSO_4 \cdot H_2O$	c	127.9590	−410.02	−414.8	−374.2	5.697	39.1	36.11
$Li_2SO_4 \cdot D_2O$	c	129.9712	—	−416.9	−375.6	—	40.	—
LiHS	c	40.0130	—	−60.1	—	—	—	—
$LiNO_2$	c	52.9465	—	−89.0	−72.2	—	23.	—
	a	52.9465	—	−91.56	−77.8	—	32.7	−6.9
$LiNO_3$	c	68.9459	—	−115.47	−91.1	—	21.5	—
	a	68.9459	—	−116.12	−96.7	—	38.3	−4.3
$LiNO_3 \cdot 3H_2O$	c	122.9921	—	−328.5	−263.8	—	53.4	—
Li_3PO_4	c	115.7944	—	−500.9	—	—	—	—
Li_2C_2	c	37.9044	—	−14.2	—	—	—	—
Li_2CO_3	c	73.8914	−288.652	−290.6	−270.58	3.627	21.60	23.69
Li_2SiO_3	c	89.9662	−391.26	−393.9	−372.2	3.453	19.08	23.68
Li_2SiF_6	c	155.9584	—	−704.4	—	—	—	—
$LiBO_2$	c	49.7508	−245.37	−246.7	−233.3	2.144	12.3	14.3
$LiBH_4$	c	21.7840	−43.20	−45.6	−29.9	3.049	18.13	19.73
	a	21.7840	—	−55.05	−42.8	—	29.7	—
$LiAlSi_2O_6$								
α spodumene	c	186.0909	—	−730.1	−688.71	—	30.90	38.0
β spodumene	c	186.0909	—	−723.4	−683.79	—	36.90	38.9
Li_2CrO_4	c	129.8756	—	−331.9	—	—	—	—
	a	129.8756	—	−343.7	−314.2	—	18.5	—
Li_2MoO_4	c	173.8196	—	−363.36	−336.9	—	30.	—
Li_2WO_4	c	261.7296	—	−335.	—	—	—	—
Li_2TiO_3	c	109.7802	−396.780	−399.3	−377.6	3.953	21.93	26.54
Li_2ZrO_3	c	153.1002	—	−420.7	—	—	—	—
Li_4ZrO_4	c	182.9816	—	−567.	—	—	—	—
Li_4PuF_8	c	418.8012	—	−1033.	—	—	—	—
LUTETIUM								
Lu	c	174.97	0	0	0	1.524	12.18	6.42
	g	—	102.242	102.2	96.7	1.482	44.142	4.986
Lu^+	g	—	227.36	228.80	—	—	—	—
Lu^{2+}	g	—	547.	549.9	—	—	—	—
Lu^{3+}								
std. state, m = 1	aq	—	—	−159.	−150.	—	−63.	6.
Lu_2O_3	c	397.938	−446.93	−448.9	−427.6	4.192	26.28	24.32
$LuCl_3$	c	281.329	—	−226.0	—	—	—	—
std. state, m = 1	aq	—	—	−279.	−244.	—	−23.	−92.
$LuCl_3 \cdot 6H_2O$	c	389.421	−667.41	−676.6	−576.3	13.96	89.9	82.0
LuOCl	c	226.422	—	−227	—	—	—	—
$Lu(BrO_3)_3 \cdot 9H_2O$	c	720.830	—	−833.8	—	—	—	—
LuI_3	c	333.683	—	−131.	—	—	—	—
$Lu(IO_3)_3$	c	699.678	—	320.	—	—	—	—
LuS	g	207.034	48.	—	—	—	—	—
MAGNESIUM								
Mg	c	24.312	0	0	0	1.195	7.81	5.95
	g	—	35.014	35.30	27.04	1.481	35.502	4.968
Mg^+	g	—	211.333	213.100	—	—	—	—
Mg^{2+}	g	—	558.052	561.299	—	—	—	—
std. state, m = 1	aq	—	—	−111.58	−108.7	—	−33.0	—
Mg^{3+}	g	—	2406.08	2410.81	—	—	—	—
Mg^{4+}	g	—	4927.14	4933.35	—	—	—	—
MgO								
macrocrystal (periclase)	c	40.3114	−142.813	−143.81	−136.10	1.235	6.44	8.88
microcrystal	c	—	−141.954	−142.92	−135.27	1.266	6.67	9.00
MgH_2	c	26.3279	−16.05	−18.0	−8.6	1.270	7.43	8.45
$Mg(OH)_2$	c	58.3267	−218.402	−220.97	−199.23	2.725	15.10	18.41
precipitate	amorp	—	—	−220.0	—	—	—	—
std. state, m = 1	aq	—	—	−221.52	−183.9	—	−38.1	—
MgF	g	43.3104	−52.89	−53.0	−59.2	2.143	52.79	7.78
MgF_2	c	62.3088	−267.57	−268.5	−255.8	2.370	13.68	14.72
$MgCl_2$	c	95.218	−153.180	−153.28	−141.45	3.288	21.42	17.06
$MgCl_2$								
in ∞ H_2O	aq	—	—	−191.48	—	—	—	—
$MgCl_2 \cdot H_2O$	c	113.2333	—	−231.03	−205.98	—	32.8	27.55
$MgCl_2 \cdot 2H_2O$	c	131.2487	—	−305.86	−267.24	—	43.0	38.05
$MgCl_2 \cdot 4H_2O$	c	167.2793	—	−453.87	−388.03	—	63.1	57.70
$MgCl_2 \cdot 6H_2O$	c	203.3100	—	−597.28	−505.49	—	87.5	75.30
$Mg(ClO_4)_2$	c	223.2132	—	−135.97	—	—	—	—
std. state, m = 1	aq	—	—	−173.40	−112.8	—	54.0	—
$Mg(ClO_4)_2$								
in ∞ H_2O	aq	—	—	−173.40	—	—	—	—
$Mg(ClO_4)_2 \cdot 2H_2O$	c	259.2439	—	−291.3	—	—	—	—
$Mg(ClO_4)_2 \cdot 4H_2O$	c	295.2746	—	−439.1	—	—	—	—
$Mg(ClO_4)_2 \cdot 6H_2O$	c	331.3052	—	−584.5	−445.3	—	124.5	—
$MgBr_2$	c	184.130	—	−125.3	−120.4	—	28.0	—
	g	—	—	−74.0	—	—	—	—
std. state, m = 1	aq	—	—	−169.68	−158.4	—	6.4	—
$MgBr_2$								
in ∞ H_2O	aq	—	—	−169.68	—	—	—	—
MgI_2	c	278.1208	—	−87.0	−85.6	—	31.0	—
	g	—	—	−41.	—	—	—	—
std. state, m = 1	aq	—	—	−137.96	−133.4	—	20.2	—
in ∞ H_2O	aq	—	—	−137.96	—	—	—	—
MgS	c	56.376	−82.44	−82.7	−81.7	1.992	12.03	10.89
$MgSO_3$	c	104.3742	—	−241.0	—	—	—	—
$MgSO_4$	c	120.3736	—	−307.1	−279.8	—	21.9	23.06
undissociated; std. state,	aq	—	—	−324.1	−289.74	—	−1.7	—
m = 1	aq	—	—	−328.90	−286.7	—	−28.2	—
$MgSO_4$								
in ∞ H_2O	aq	—	—	−328.90	—	—	—	—
$MgSO_4 \cdot 2H_2O$	c	156.4043	—	−453.2	—	—	—	—
$MgSO_4 \cdot 4H_2O$	c	192.4350	—	−596.7	—	—	—	—
$MgSO_4 \cdot 6H_2O$	c	228.4656	—	−737.8	−629.1	—	83.2	83.20
$MgSO_4 \cdot 7H_2O$	c	246.4810	—	−809.92	−686.4	—	89.	—

Substance			0 K	298.15 K (25°C)				
Formula and Description	State	Formula weight	ΔHf_0° kcal/mol	ΔHf° kcal/mol	ΔGf°	$H_{298}^\circ - H_0^\circ$	S° cal/deg mol	C_p°
$Mg(NO_3)_2$	c	148.3218	—	−188.97	−140.9	—	39.2	33.92
std. state, m = 1	aq	—	—	−210.70	−161.9	—	37.0	—
$Mg(NO_3)_2$								
in ∞ H_2O	aq	—	—	−210.70	—	—	—	—
$Mg(NO_3)_2 \cdot 2H_2O$	c	184.3525	—	−336.8	—	—	—	—
$Mg(NO_3)_2 \cdot 6H_2O$	c	256.4138	—	−624.59	−497.3	—	108.	—
$Mg_2P_2O_7$	c	222.5674	—	—	—	6.467	37.02	42. 53
$Mg_3(PO_4)_2$	c	262.8788	−896.98	−903.6	−845.8	7.825	45.22	51.02
MgC_2	c	48.3343	—	+ 20.	—	—	—	—
Mg_2C_3	c	84.6574	—	+ 17	—	—	—	—
$MgCO_3$								
magnesite	c	84.3214	—	−261.9	−241.9	—	15.7	18.05
$MgCO_3 \cdot 3H_2O$								
nesquehonite	c	138.3674	—	—	−412.6	—	—	—
$MgCO_3 \cdot 5H_2O$								
lansfordite	c	174.3980	—	—	−525.7	—	—	—
MgC_2O_4	c	112.3319	—	−303.3	—	—	—	—
std. state, m = 1	aq	—	—	−308.8	−269.8	—	−22.1	—
Mg_2Si	c	76.710	—	−18.6	−18.0	—	18.	17.6
$MgSiO_3$								
clinoenstatite	c	100.3962	−368.039	−370.22	−349.46	2.895	16.19	19.45
Mg_2SiO_4								
forsterite	c	140.7076	−516.42	−519.6	−491.2	4.129	22.74	28.32
$Mg_3Si_2O_5(OH)_4$								
chrysotile	c	277.1345	—	−1043.4	−965.1	—	52.9	65.41
antigorite	c	—	—	—	—	—	53.2	65.47
$Mg_3Si_4O_{10}(OH)_2$								
talc	c	379.2887	−1405.57	−1415.5	−1324.8	11.20	62.3	76.9
$Mg_2Al_4Si_5O_{18}$								
cordierite	c	584.9692	—	−2177.	−2055.	—	97.3	108.1
$MgCrO_4$	c	140.3056	—	−321.1	—	—	—	—
$MgCr_2O_4$	c	192.3016	—	−426.3	−398.9	—	25.34	30.30
$MgMoO_4$	c	184.250	—	−334.81	−309.69	—	28.4	26.57
$MgWO_4$	c	272.160	−363.88	−366.3	−339.6	4.115	24.18	26.14
MgV_2O_6								
metavanadate	c	222.1924		−526.19	−487.43	—	38.4	39.47
$Mg_2V_2O_7$								
pyrovanadate	c	262.5038		−677.80	−632.24	—	47.9	48.63
$MgTiO_3$								
metatitanate	c	120.210	−373.69	−375.9	−354.8	3.240	17.82	21.96
Mg_2TiO_4								
orthotitanate	c	160.522	−514.32	−517.5	−489.2	4.502	26.13	30.75
$MgUO_4$	c	326.339	—	−443.9	−418.2	—	31.5	30.6
MgU_3O_{10}	c	898.393	—	—	—	—	80.9	73.
MANGANESE								
Mn								
α	c	54.9380	0	0	0	1.194	7.65	6.29
β	c	—	—	—	—	1.234	8.22	6.34
γ	c	—	.343	.37	.34	1.221	7.75	6.59
	g	—	66.77	67.1	57.0	1.481	41. 49	4.97
Mn^+	g	—	238.20	240.0	—	—	—	—
Mn^{++}	g	—	598.87	602.1	—	—	—	—
std. state, m = 1	aq	—	—	−52.76	−54.5	—	−17.6	12.
MnO	c	70.9374	—	−92.07	−86.74	—	14.27	10.86
MnO_2	c	86.9368	—	−124.29	−111.18	—	12.68	12.94
precipitated	amorp	—	—	−120.1	—	—	—	—
Mn_2O_3	c	157.8742	—	−229.2	−210.6	—	26.4	25.73
Mn_3O_4	c	228.8116	—	−331.7	−306.7	—	37.2	33.38
MnH	g	55.9460	—	—	—	2.077	51.03	7.035
$Mn(OH)_2$								
precipitated	amorp	88.9527	—	−166.2	−147.0	—	23.7	—
MnF	g	73.9364	−5.	−5.2	—	—	—	—
MnF_2	c	92.9348	—	—	—	3.11	22.05	15.96
$MnCl_2$	c	125.8440	−115.245	−115.03	−105.29	3.602	28.26	17.43
std. state, m = 1	aq	—	—	−132.66	−117.3	—	9.3	−53.
$MnCl_2 \cdot H_2O$	c	143.8593	—	−188.8	−166.4	—	41.6	
$MnCl_2 \cdot 2H_2O$	c	161.8747	—	−261.0	−225.2	—	52.3	—
$MnCl_2 \cdot 4H_2O$	c	197.9054	—	−403.3	−340.3	—	72.5	
$MnBr$	g	134.8470	20.8	19.1	—	—	—	—
$MnBr_2$	c	214.7560	—	−92.0	—	—	—	—
std. state, m = 1	aq	—	—	−110.9	—	—	—	—
$MnBr_2 \cdot H_2O$	c	232.7713	—	−168.5	—	—	—	—
$MnBr_2 \cdot 4H_2O$	c	286.8174	—	−380.1	—	—	—	—
MnI	g	181.8424	25.7	25.5	—	—	—	—
MnI_2								
std. state, m = 1	aq	308.7468	—	−79.1	—	—	—	—
$MnI_2 \cdot 2H_2O$	c	344.7748	—	−201.4	—	—	—	—
$MnI_2 \cdot 4H_2O$	c	380.8082	—	−343.9	—	—	—	—
$Mn(IO_3)_2$	c	404.7432	—	−160.	−124.4	—	63.	—
MnS								
green	c	87.0020	—	−51.2	−52.2	—	18.7	11.94
precipitated, pink	amorp	—	—	−51.1	—	—	—	—
$MnSO_4$	c	150.9996	—	−254.60	−228.83	—	26.8	24.02
std. state, m = 1	aq	—	—	−270.1	−232.5	—	−12.8	−58.
$MnSO_4$								
in ∞ H_2O	aq	—	—	−270.1	—	—	—	—
$MnSO_4 \cdot H_2O$								
α	c	169.0149	—	−329.0	—	—	—	—
β	c	—	—	−322.2	—	—	—	—
$MnSO_4 \cdot 4H_2O$	c	223.0610	—	−539.7	—	—	—	—
$MnSO_4 \cdot 5H_2O$	c	241.0763	—	−610.2	—	—	—	78.
$MnSO_4 \cdot 7H_2O$	c	277.1070	—	−750.3	—	—	—	—
$Mn(N_3)_2$								
manganese azide	c	138.9782	—	92.2	—	—	—	—
Mn_5N_2	c	302.7034	—	−48.8	—	—	—	—
$Mn(NO_3)_2$	c	178.9478	—	−137.73	—	—	—	—
std. state, m = 1	aq	—	—	−151.9	−107.8	—	52.	−29.
MnP	c	85.9118	—	−27.	—	—	—	—
MnP_3	c	147.8594	—	−51.	—	—	—	—
$Mn_3(PO_4)_2$	c	354.7568	—	−744.9	—	—	—	—
$MnHPO_4$	c	150.9174	—	—	−332.5	—	—	—

Substance			0 K	298.15 K (25°C)				
Formula and Description	State	Formula weight	ΔHf°_0 kcal/mol	ΔHf° kcal/mol	ΔGf° kcal/mol	$H^\circ_{298} - H^\circ_0$ kcal/mol	S° cal/deg mol	C°_p cal/deg mol
Mn_3C	c	176.8252	—	1.1	1.3	—	23.6	22.33
Mn_7C_3	c	420.5995	—	−10.	—	—	—	—
$MnCO_3$								
natural	c	114.9474	—	−213.7	−195.2	—	20.5	19.48
precipitated	c	—	—	−211.1	—	—	—	—
MnC_2O_4	c	142.9579	—	−245.9	—	—	—	—
undissoc.; std. state, m = 1	aq	—	—	−248.5	−221.0	—	16.1	—
$MnC_2O_4 \cdot 2H_2O$	c	178.9886	—	−389.2	−338.2	—	48.	—
$MnC_2O_4 \cdot 3H_2O$	c	197.0039	—	−459.1	—	—	—	—
$MnSiO_3$	c	131.0222	—	−315.7	−296.5	—	21.3	20.66
glassy	amorp	—	—	−307.2	—	—	—	—
Mn_2SiO_4	c	201.9590	—	−413.6	−390.1	—	39.0	31.04
MERCURY								
Hg	c	200.59	0	—	—	—	—	—
	liq	—	—	0	0	2.233	18.17	6.688
	g	—	15.407	14.655	7.613	1.481	41.79	4.968
std. state, m = 1	aq	—	—	9.0	9.4	—	17.	—
Hg^+	g	—	256.10	256.82	—	—	—	—
Hg^{2+}	g	—	688.63	690.83	—	—	—	—
std. state, m = 1	aq	—	—	40.9	39.30	—	−7.7	—
Hg^{3+}	g	—	1478.	1480.	—	—	—	—
Hg_2^{2+}								
std. state, m = 1	aq	—	—	41.2	36.70	—	20.2	—
HgO								
red, orthorhombic	c	216.589	—	−21.71	−13.995	—	16.80	10.53
yellow	c	—	—	−21.62	−13.964	—	17.0	—
hexagonal	c	—	—	−21.4	−13.92	—	17.6	—
HgH	g	201.598	58.36	57.20	51.63	2.078	52.46	7.16
$Hg(OH)_2$								
undissoc.; std. state, m = 1	aq	234.605	—	−84.9	−65.7	—	34.	—
Hg_2F_2	c	439.177	—	—	−104.1	—	—	—
$HgCl^+$								
std. state, m = 1	aq	—	—	−4.5	−1.3	—	18.	—
$HgCl_2$	c	271.496	—	−53.6	−42.7	—	34.9	—
undissoc.; std. state, m = 1	aq	—	—	−51.7	−41.4	—	37.	—
Hg_2Cl_2	c	472.086	—	−63.39	−50.377	—	46.0	—
$HgBr_2$	c	360.408	—	−40.8	−36.6	—	41.	—
undissoc.; std. state, m = 1	aq	—	—	−38.4	−34.2	—	41.	—
Hg_2Br_2	c	560.998	—	−49.45	−43.278	—	52.	—
HgI	g	327.494	32.9	31.64	21.14	2.546	67.26	8.99
HgI_2								
red	c	454.399	—	−25.2	−24.3	—	43.	—
yellow	c	—	—	−24.6	—	—	—	—
Hg_2I_2	c	654.989	—	−29.00	−26.53	—	55.8	—
HgS								
red	c	232.654	—	−13.9	−12.1	—	19.7	11.57
black	c	—	—	−12.8	−11.4	—	21.1	—
Hg_2SO_4	c	497.242	—	−177.61	−149.589	—	47.96	31.54
$Hg_2(N_3)_2$	c	485.220	—	142.0	178.4	—	49.	—
HgC_2O_4	c	288.610	—	−162.1	—	—	—	—
Hg_2CO_3	c	461.189	—	−132.3	−111.9	—	43.	—
$Hg_2C_2O_4$	c	489.200	—	—	−141.8	—	—	—
$HgCH_3$	g	215.625	—	40.	—	—	—	—
$Hg(CH_3)_2$	liq	230.660	—	14.3	33.5	—	50.	—
$Hg(C_2H_5)_2$	liq	258.715	—	7.2	—	—	—	—
	g	—	—	18.0	—	—	—	—
$Hg(CN)_2$	c	252.626	—	63.0	—	—	—	—
undissoc.; std. state, m = 1	aq	—	—	66.5	74.6	—	39.5	—
$Hg(CN)_3^-$								
std. state, m = 1	aq	278.644	—	94.9	110.7	—	53.8	—
$Hg(CN)_4^{2-}$								
std. state, m = 1	aq	304.662	—	125.8	147.8	—	73.	—
$Hg(ONC)_2$								
mercuric fulminate	c	284.624	—	64.	—	—	—	—
MOLYBDENUM								
Mo	c	95.94	0	0	0	1.098	6.85	5.75
	g	—	156.92	157.3	146.4	1.4812	43.461	4.968
Mo^+	g	—	320.6	322.5	—	—	—	—
Mo^{2+}	g	—	693.2	696.5	—	—	—	—
Mo^{3+}	g	—	1319.0	1323.8	—	—	—	—
Mo^{4+}	g	—	2388.9	2395.2	—	—	—	—
Mo^{5+}	g	—	3799.4	3807.2	—	—	—	—
MoO	g	111.939	—	101.	—	—	—	—
MoO_2	c	127.939	—	−140.76	−127.40	—	11.06	13.38
	g	—	—	3.	—	—	—	—
MoO_3	c	143.939	—	−178.08	−159.66	—	18.58	17.92
	g	—	—	−78.	—	—	—	—
	aq	—	—	−172.5	—	—	—	—
MoO_4	aq	159.938	—	−158.0	—	—	—	—
MoO_4^{2-}								
std. state, m = 1	aq	—	—	−238.5	−199.9	—	6.5	—
MoO_5	aq	175.937	—	−139.6	—	—	—	—
H_2MoO_4								
white	c	161.954	—	−250.0	—	—	—	—
$H_2MoO_4 \cdot H_2O$								
yellow	c	179.969	—	−325.	—	—	—	—
MoF_6	liq	209.930	−381.733	−378.95	−352.08	10.205	62.06	40.58
	g	—	−370.608	−372.29	−351.88	5.74	83.75	28.82
$MoCl_2$	c	166.846	—	−67.4	—	—	—	—
$MoCl_3$	c	202.299	—	−92.5	—	—	—	—
$MoCl_4$	c	237.752	—	−114.8	—	—	—	—
$MoCl_5$	c	273.205	—	−126.0	—	—	—	—
MoO_2Cl_2	c	198.845	—	−171.4	—	—	—	—
$MoO_2Cl_2 \cdot H_2O$	c	216.860	—	−245.4	—	—	—	—
$MoOCl_4$	c	253.751	—	−153.0	—	—	—	—
$MoBr_2$	c	255.758	—	−62.4	—	—	—	18.3
$MoBr_4$	c	415.576	—	−76.8	—	—	—	—
MoO_2Br_2	c	287.757	—	−150.4	—	—	—	—
MoS_2	c	160.068	−55.52	−56.2	−54.0	2.528	14.96	15.19
Mo_2S_3	c	288.072	—	−87.	—	—	—	—

Substance			0 K	298.15 K (25°C)				
Formula and Description	State	Formula weight	ΔHf°_0 kcal/mol	ΔHf°	ΔGf° kcal/mol	$H^\circ_{298} - H^\circ_0$	S° cal/deg mol	C°_p
Mo_2N	c	205.887	—	−19.50	—	—	—	—
MoC	c	107.951	—	−2.4	—	—	—	—
Mo_2C	c	203.891	—	−10.9	—	—	—	—
$Mo(CO)_6$	c	264.003	—	−234.9	−209.8	—	77.9	57.90
$FeMoO_4$	c	215.785	—	−257.	−233.	—	30.9	28.31
NEODYMIUM								
Nd	c	144.24	0	0	0	1.73	17.1	6.56
	g	—	78.53	78.3	69.9	1.498	45.243	5.280
Nd^+	g	—	205.1	206.3	—	—	—	—
Nd^{2+}	g	—	452.	455.	—	—	—	—
Nd^{3+}								
std. state, m = 1	aq	—	—	−166.4	−160.5	—	−49.4	−5.
NdO	g	160.239	−30.2	—	—	—	—	—
Nd_2O_3								
hexagonal	c	336.478	−430.50	−432.1	−411.3	5.00	37.9	26.60
NdH_2	c	146.256	—	−46.	—	—	—	—
NdF	g	163.238	—	−38.	—	—	—	—
NdF_2	g	182.237	—	−165.	—	—	—	—
NdF_3	c	201.235	—	−396.	—	—	—	—
$NdCl_2$	c	215.146	—	−163.	—	—	—	—
$NdCl_3$	c	250.599	—	−248.8	—	—	—	27.
$NdCl_3$								
∞H_2O	aq	—	—	−286.3	—	—	—	—
$NdCl_3 \cdot 6H_2O$	c	358.691	−678.72	−687.0	−588.1	15.14	99.7	86.25
NdI_3	c	524.953	—	−152.8	—	—	—	—
$Nd(IO_3)_3$	c	668.948	—	−332.	—	—	—	—
NdS	g	176.304	33.	—	—	—	—	—
Nd_2S_3	c	384.672	—	−284.	−280.2	6.160	44.28	29.28
$Nd_2(SO_4)_3 \cdot 8H_2O$	c	720.788	—	−1513.1	—	—	160.9	144.9
Nd_2Se_3	c	525.36	—	—	—	7.13	53.6	31.1
$Nd_2(SeO_3)_3 \cdot 8H_2O$								
amorp.	c	813.477	—	−1230.3	—	—	—	—
$Nd_2(SeO_4)_3 \cdot 5H_2O$	c	807.430	—	−1100.4	—	—	—	—
$Nd(NO_3)_3$	c	330.255	—	−294.2	—	—	—	—
$Nd(NO_3)_3 \cdot 3H_2O$	c	384.301	—	−515.	—	—	—	—
$Nd(NO_3)_3 \cdot 4H_2O$	c	402.316	—	−588.6	—	—	—	—
$Nd(NO_3)_3 \cdot 6H_2O$	c	438.347	—	−728.39	—	—	—	—
NdC_2	g	168.262	130.5	130.75	117.9	2.48	6.3	10.6
$Nd_2(CO_3)_3$	c	468.508	—	—	−744.5	—	—	—
NEON								
Ne	g	20.183	0	0	0	1.481	34.9471	4.9679
std. state, m = 1	aq	—	—	−1.1	4.6	—	15.8	—
Ne^+	g	—	497.29	498.77	—	—	—	—
Ne^{2+}	g	1444.7	1447.6	—	—	—	—	—
Ne^{3+}	g	—	2915.	2919.	—	—	—	—
Ne^{4+}	g	—	5156	5162	—	—	—	—
Ne^{5+}	g	—	8072.	8079.	—	—	—	—
NICKEL								
Ni	c	58.71	0	0	0	1.144	7.14	6.23
	g	—	102.213	102.7	91.9	1.631	43.519	5.583
Ni^+	g	—	278.275	280.243	—	—	—	—
Ni^{2+}	g	—	696.87	700.32	—	—	—	—
std. state, m = 1	aq	—	—	−12.9	−10.9	—	−30.8	—
Ni^{3+}	g	—	1508.0	1512.9	—	—	—	—
NiO	c	74.709	−56.7	−57.3	−50.6	1.6	9.08	10.59
Ni_2O_3	c	165.418	—	−117.0	—	—	—	—
$Ni(OH)_2$	c	92.725	—	−126.6	−106.9	—	21.	—
std. state, m = 1	aq	—	—	−122.8	−86.1	—	−35.9	—
$Ni(OH)_3$								
precipitated	c	109.732	—	−160.	—	—	—	—
NiF_2	c	96.707	−155.18	−155.7	−144.4	2.729	17.59	15.31
$NiCl_2$	c	129.616	−73.077	−72.976	−61.918	3.438	23.34	17.13
$NiCl_2$								
in ∞H_2O	aq	—	—	−92.8	—	—	—	—
$NiCl_2 \cdot 2H_2O$	c	165.647	—	−220.4	−181.7	—	42.	—
$NiCl_2 \cdot 4H_2O$	c	201.677	—	−362.5	−295.2	—	58.	—
$NiCl_2 \cdot 6H_2O$	c	237.708	—	−502.67	−409.54	—	82.3	—
$Ni(ClO_4)_2$								
std. state, m = 1	aq	257.611	—	−74.7	−15.0	—	56.2	—
$Ni(ClO_4)_2 \cdot 6H_2O$	c	365.703	—	−486.6	—	—	—	—
$NiBr_2$	c	218.528	—	−50.7	—	—	—	—
std. state, m = 1	aq	—	—	−71.0	−60.6	—	8.6	—
$NiBr_2 \cdot 3H_2O$	c	272.574	—	−274.0	—	—	—	—
NiI_2	c	312.519	—	−18.7	—	—	—	—
std. state, m = 1	aq	—	—	−39.3	−35.6	—	22.4	—
$Ni(IO_3)_2$	c	408.515	—	−116.9	−78.0	—	51.	—
NiS	c	90.774	—	−19.6	−19.0	—	12.66	11.26
precipitated	c	—	—	−18.5	—	—	—	—
Ni_3S_2	c	240.258	—	−48.5	−47.1	—	32.0	28.12
$NiSO_4$	c	154.772	—	−208.63	−181.6	—	22.	33.
std. state, m = 1	aq	—	—	−230.2	−188.9	—	−26.0	—
$NiSO_4$								
∞H_2O	aq	—	—	−230.2	—	—	—	—
$NiSO_4 \cdot 4H_2O$	c	226.833	—	−502.9	—	—	—	—
$NiSO_4 \cdot 6H_2O$								
α, tetragonal, green	c	262.864	−628.887	−641.21	−531.78	12.391	79.94	78.36
β, monoclinic, blue	c	—	—	−638.7	—	—	—	—
$NiSO_4 \cdot 7H_2O$	c	280.879	−697.670	−711.36	−588.49	14.085	90.57	87.14
Ni_3N	c	190.137	—	0.2	—	—	—	—
$Ni(NO_3)_2$	c	182.720	—	−99.2	—	—	—	—
std. state, m = 1	aq	—	—	−112.0	−64.2	—	39.2	—
$Ni(NO_3)_2 \cdot 3H_2O$	c	236.766	—	−317.0	—	—	—	—
$Ni(NO_3)_2 \cdot 6H_2O$	c	290.812	—	−528.6	—	—	—	111.
Ni_3C	c	188.141	—	16.1	—	—	—	—
$NiCO_3$	c	118.719	—	—	−146.4	—	—	—
$Ni(CO)_4$	liq	170.752	—	−151.3	−140.6	—	74.9	48.9
	g	—	−144.877	−144.10	−140.36	7.074	98.1	34.70
$Ni(HCO_2)_2$	c	148.746	—	−208.4	—	—	—	—
$Ni(CN)_2$								
precipitated	c	110.746	—	30.5	—	—	—	—
std. state, m = 1	aq	—	—	59.1	71.5	—	14.2	—

Substance			0 K	298.15 K (25°C)				
Formula and Description	State	Formula weight	ΔHf_0° kcal/mol	ΔHf°	ΔGf° kcal/mol	$H_{298}^\circ - H_0^\circ$	S° cal/deg mol	C_p°
$Ni(CNS)_2$	c	174.874	—	22.8	—	—	—	—
NIOBIUM								
Nb	c	92.906	0	0	0	1.255	8.70	5.88
	g	—	172.758	173.5	162.8	1.997	44.490	7.208
Nb^+	g	—	330.32	332.54	—	—	—	—
Nb^{2+}	g	—	660.55	664.25	—	—	—	—
Nb^{3+}	g	—	1238.0	1243.1	—	—	—	—
Nb^{4+}	g	—	2121.3	2127.9	—	—	—	—
NbO	c	108.9054	—	−97.0	−90.5	—	11.5	9.86
	g	—	51.2	51.	44.	2.099	57.09	7.36
NbO_2	c	124.9048	−189.19	−190.3	−177.0	2.222	13.03	13.74
	g		[illegible]	−31.3	−32.3	—	61.0	—
Nb_2O_5								
(high temp. form)	c	265.809	−451.63	−454.0	−422.1	5.325	32.80	31.57
NbF_5	c	187.898	−432.68	−433.5	−406.1	5.707	38.3	32.2
$NbCl_5$	c	270.171	—	−190.6	−163.3	—	50.3	35.4
$NbOCl_2$	c	179.8114	—	−185.1	—	—	—	—
$NbOCl_3$	c	215.2644	—	−210.2	−187.	—	34.	—
$NbBr_5$	c	492.451	—	−132.9	—	—	—	—
$NbOBr_3$	c	348.6174	—	−179.3	—	—	—	—
NbI_5	c	727.428	—	−64.2	—	—	—	—
NbN	c	106.9127	−55.35	−56.2	−49.2	1.439	8.25	9.32
Nb_2N	c	199.8187	—	−59.9	—	—	—	—
NbC	c	104.9172	−33.12	−33.2	−32.7	1.422	8.46	8.81
Nb_2C	c	197.8232	—	−45.4	−44.4	—	15.3	14.48
NITROGEN								
N	g	14.0067	112.534	112.979	108.886	1.481	36.613	4.968
N^+	g	—	447.663	449.589	—	—	—	—
N^{2+}	g	—	1130.55	1133.96	—	—	—	—
N^{3+}	g	—	2224.52	2229.41	—	—	—	—
N^{4+}	g	—	4011.04	4017.41	—	—	—	—
N^{5+}	g	—	6268.41	6276.26	—	—	—	—
N_2	g	28.0134	0	0	0	2.072	45.77	6.961
N_3	g	—	45.	43.2	—	—	—	—
std. state, m = 1	aq	—	—	65.76	83.2	—	25.8	—
NO	g	30.0061	21.45	21.57	20.69	—	50.347	7.133
NO_2	g	46.0055	8.60	7.93	12.26	2.438	57.35	8.89
NO_2^-								
std. state, m = 1	aq	—	—	−25.0	−7.7	—	29.4	−23.3
NO_3^-								
nitrate; std. state, m = 1	aq	62.0049	—	−49.56	−26.61	—	35.0	−20.7
N_2O	g	44.0128	20.435	19.61	24.90	2.284	52.52	9.19
N_2O_3	liq	76.0116	—	12.02	—	—	—	—
	g	—	21.628	20.01	33.32	3.566	74.61	15.68
N_2O_4	liq	92.0110	—	−4.66	23.29	—	50.0	34.1
	g	—	4.49	2.19	23.38	3.918	72.70	18.47
N_2O_5	c	108.0104	—	−10.3	27.2	—	42.6	34.2
	g	—	5.7	2.7	27.5	4.237	85.0	20.2
NH_3	g	17.0306	−9.34	−11.02	−3.94	2.388	45.97	8.38
undissoc.; std. state, m = 1	aq	—	—	−19.19	−6.35	—	26.6	—
NH_4^+								
std. state, m = 1	aq	18.0386	—	−31.67	−18.97	—	27.1	19.1
N_2H_4	liq	32.0453	—	12.10	35.67	—	28.97	23.63
	g	—	26.18	22.80	38.07	2.743	56.97	11.85
undissoc.; std. state, m = 1	aq	—	—	8.20	30.6	—	33.	—
HN_3	liq	43.0281	—	63.1	78.2	—	33.6	—
	g	—	71.82	70.3	78.4	2.599	57.09	10.44
undissoc.; std. state, m = 1	aq	—	—	62.16	76.9	—	34.9	—
HNO_2								
cis	g	47.0135	−17.12	−18.64	−10.27	2.608	59.43	10.70
trans	g	—	−17.68	−19.15	−10.82	2.652	59.54	11.01
cis-trans mixture, equil.	g	—	—	−19.0	−11.0	—	60.7	10.9
undissoc.; std. state, m = 1	aq	—	—	−28.5	−12.1	—	32.4	—
HNO_3	liq	63.0129	—	−41.61	−19.31	—	37.19	26.26
	g	—	−29.94	−32.28	−17.87	2.815	63.64	12.75
std. state, m = 1	aq	—	—	−49.56	−26.61	—	35.0	−20.7
HNO_3								
∞ H_2O	aq	—	—	−49.56	—	—	—	—
NH_4OH	liq	35.0460	—	−86.33	−60.74	—	39.57	37.02
undissoc.; std. state, m = 1	aq	—	—	−87.505	−63.04	—	43.3	—
ionized; std. state, m = 1	aq	—	—	−86.64	−56.56	—	24.5	−16.4
NH_4OH								
∞ H_2O	aq	—	—	−86.64	—	—	—	—
NH_4NO_2	c	64.0441	—	−61.3	—	—	—	—
std. state, m = 1	aq	—	—	−56.7	−26.7	—	56.5	−4.2
NH_4NO_3	c	80.0435	—	−87.37	−43.98	—	36.11	33.3
std. state, m = 1	aq	—	—	−81.23	−45.58	—	62.1	−1.6
NH_4NO_3								
∞ H_2O	aq	—	—	−81.23	—	—	—	—
$(NH_4)_2O$	liq	52.0766	—	−102.94	−63.84	—	63.94	59.08
NF_3	g	71.0019	−28.43	−29.8	−19.9	2.827	62.29	12.7
NH_4F	c	37.0370	−107.41	−110.89	−83.36	2.655	17.20	15.60
std. state, m = 1	aq	—	—	−111.17	−85.61	—	23.8	−6.4
NCl_3	liq	120.3657	—	55.	—	—	—	—
NOCl	g	65.4591	12.81	12.36	15.79	2.716	62.52	10.68
NH_4Cl	c	53.4916	—	−75.15	−48.51	—	22.6	20.1
std. state, m = 1	aq	—	—	−71.62	−50.34	—	40.6	−13.5
NH_4Cl								
∞ H_2O	aq	—	—	−71.62	—	—	—	—
$NH_2OH \cdot HCl$	c	69.4910	—	−75.9	—	—	—	22.2
NH_4ClO								
std. state, m = 1	aq	69.4910	—	−57.3	−27.8	—	37.	—
NH_4ClO_2								
std. state, m = 1	aq	85.4904	—	−47.6	−14.9	—	51.3	—
std. state, m = 1	aq	101.4898	—	−56.52	−20.89	—	65.9	—
NH_4ClO_4	c	117.4892	—	−70.58	−21.25	—	44.5	—
std. state, m = 1	aq	—	—	−62.58	−21.03	—	70.6	—
NH_4Br	c	94.9477	—	−64.73	−41.9	—	27.	23.
std. state, m = 1	aq	—	—	−60.72	−43.82	—	46.8	−14.8
NH_4Br								
∞ H_2O	aq	—	—	−60.72	—	—	—	—

Substance			0 K	298.15 K (25°C)				
Formula and Description	State	Formula weight	ΔHf_0° kcal/mol	ΔHf°	ΔGf° kcal/mol	$H_{298}^\circ - H_0^\circ$	S°	C_p° cal/deg mol
NH_4BrO								
std. state, m = 1	aq	113.9470	—	−54.2	−27.0	—	37.	—
NH_4BrO_3								
std. state, m = 1	aq	145.9458	—	−47.70	−14.54	—	65.75	—
NH_4Br_2Cl								
std. state, m = −72.4	aq	213.3096	—	−72.4	−49.7	—	72.2	—
NH_4I	c	144.9430	—	−48.14	−26.9	—	28.	—
std. state, m = 1	aq	—	—	−44.86	−31.30	—	53.7	−14.9
NH_4I								
∞ H_2O	aq	—	—	−44.86	—	—	—	—
NH_4IO								
std. state, m = 1	aq	160.9424	—	−57.4	−28.2	—	25.8	—
NH_4IO_3	c	192.9412	—	−92.2	—	—	—	—
std. state, m = 1	aq	—	—	−84.6	−49.6	—	55.4	—
NH_4IO_4	aq	208.9406	—	−67.9	—	—	—	—
NH_4HS	c	51.1106	—	−37.5	−12.1	—	23.3	—
std. state, m = 1 (NH_4^+ + HS^-)	aq	—	—	−35.9	−16.09	—	42.1	—
$(NH_4)_2S$								
std. state, m = 1	aq	68.1412	—	−55.4	−17.4	—	50.7	—
$(NH_4)_2S_2$								
std. state, m = 1	aq	100.2052	—	−56.1	−18.9	—	61.0	—
$(NH_4)_2S_3$								
std. state, m = 1	aq	132.2692	—	−57.1	−20.3	—	70.0	—
H_2NSO_3H								
sulfamic acid	c	97.0928	—	−161.3	—	—	—	—
	aq	—	—	−156.3	—	—	—	—
NH_4HSO_3	c	99.1088	—	−183.7	—	—	—	—
NH_4HSO_4	c	115.1082	—	−245.45	—	—	—	—
$SO_2(NH_2)_2$								
sulfamide	c	96.1081	—	−129.3	—	—	—	—
$(NH_4)_2SO_3$	c	116.1394	—	−211.6	—	—	—	—
std. state, m = 1	aq	—	—	−215.2	−154.2	—	47.2	—
$(NH_4)_2SO_4$	c	132.1388	−282.23	−215.56	—	52.6	44.81	
std. state, m = 1, ($2NH_4^+$ + SO_4^{2-})	aq	—	—	−280.66	−215.91	—	59.0	−31.8
OSMIUM								
Os	c	190.2	0	0	0	—	7.8	5.9
	g	—	—	189.	178.	1.481	46.000	4.968
Os^+	g	—	—	391.91	—	—	—	—
Os^{2+}	g	—	—	785.	—	—	—	—
OsO_3	g	238.20	—	−67.8	—	—	—	—
OsO_4								
yellow	c	254.20	—	−94.2	−72.9	—	34.4	—
white	c	—	—	−92.2	−72.6	—	40.1	—
$Os(OH)_4$	amorp	258.23	—	—	−161.0	—	—	—
$OsCl_3$	c	296.56	—	−45.5	—	—	—	—
$OsCl_4$	c	332.01	—	−60.9	—	—	—	—
OsS_2	c	254.33	—	−34.9	—	—	—	—
O	g	15.9994	58.983	59.553	55.389	1.607	38.467	5.237
O^+	g	—	373.019	375.070	—	—	—	—
O^{2+}	g	—	1183.73	1187.26	—	—	—	—
O^{3+}	g	—	2450.87	2455.88	—	—	—	—
O^{4+}	g	—	4236.1	4242.6	—	—	—	—
O^{5+}	g	—	6862.8	6870.8	—	—	—	—
O_2	g	31.9988	0	0	0	2.0746	49.003	7.016
std. state, m = 1	aq	—	—	−2.8	3.9	—	26.5	—
O_3	g	47.9982	34.74	34.1	39.0	2.4736	57.08	9.37
	aq	—	—	30.1	—	—	—	—
PALLADIUM								
Pd	c	106.4	0	0	0	1.299	8.98	6.21
	g	—	90.2	90.4	81.2	1.481	39.90	4.968
Pd^+	g	—	282.436	284.117	—	—	—	—
Pd^{2+}	g	—	730.5	733.6	—	—	—	—
std. state, m = 1	aq	—	—	35.6	42.2	—	−44.	—
Pd^{3+}	g	—	1490.	1494.	—	—	—	—
PdO	c	122.40	—	−20.4	—	—	—	7.5
Pd_2H	c	213.81	—	−4.7	—	—	—	—
$Pd(OH)_2$								
precipitated	c	140.41	—	−94.4	—	—	—	—
$Pd(OH)_4$								
precipitated	c	174.43	—	−171.1	—	—	—	—
$PdCl_2$	c	177.31	—	−47.5	—	—	—	—
$PdBr_2$	c	266.22	—	−24.9	—	—	—	—
PdI_2	c	360.21	—	−15.1	−17.1	—	43.	—
PdS	c	138.46	—	−18.	−16.	—	11.	—
PdS_2	c	170.53	—	−19.4	−17.8	—	19.	—
Pd_4S	c	457.66	—	−16.	−16.	—	43.	—
$Pd(CN)_2$	c	158.44	—	49.1	—	—	—	—
$Pd(CNS)_2$	c	222.56	—	—	56.0	—	—	—
PHOSPHORUS								
P								
α, white	c	30.9738	0	0	0	1.281[a]	9.82	5.698
red, triclinic	c	—	−3.78	−4.2	−2.9	0.862	5.45	5.07
black	c	—	—	−9.4	—	—	—	—
red	amorp	—	—	−1.8	—	—	—	—
	g	—	75.	75.20	66.51	1.481	38.978	4.968
in CS_2	—	—	—	0.5	—	—	—	—
P^+	g	—	328.20	329.88	—	—	—	—
P^{2+}	g	—	781.51	784.67	—	—	—	—
P^{3+}	g	—	1477.11	1481.75	—	—	—	—
P^{4+}	g	—	2661.67	2667.79	—	—	—	—
P^{5+}	g	—	4161.2	4168.8	—	—	—	—
P_2	g	61.9476	34.94	34.5	24.8	2.126	52.108	7.66
P_4	g	123.8952	15.83	14.08	5.85	3.378	66.89	16.05
PO	g	46.9732	−6.7	−6.8	−12.4	2.245	53.22	7.59
P_4O_6	c	219.8916	—	−392.0	—	—	—	—
P_4O_{10}								
hexagonal	c	283.8892	−705.82	−713.2	−644.8	8.117	54.70	50.60
	amorp	—	—	−727.	—	—	—	—
PH_3	g	33.9977	3.20	1.3	3.2	2.420	50.22	8.87

Substance			0 K	298.15 K (25°C)				
Formula and Description	State	Formula weight	ΔHf°_0 kcal/mol	ΔHf°	ΔGf° kcal/mol	$H^\circ_{298} - H^\circ_0$	S° cal/deg mol	C°_p
std. state, m = 1	aq	—	—	−2.27	6.05	—	28.7	—
HPO_3	c	79.9800	—	−226.7	—	—	—	—
H_3PO_3	c	81.9959	—	−230.5	—	—	—	—
H_3PO_4	c	97.9953	−301.29	−305.7	−267.5	4.059	26.41	25.35
ionized, std. state, m = 1	aq	—	—	−305.3	−243.5	—	−53.	—
PF	g	49.9722	—	—	—	2.117	53.74	7.56
PF_3	g	87.9690	−218.25	−219.6	−214.5	3.092	65.28	14.03
PF_5	g	125.9658	—	−381.4	—	—	—	—
PCl_3	liq	137.3328	—	−76.4	−65.1	—	51.9	—
	g	—	−67.85	−68.6	−64.0	3.817	74.49	17.17
PCl_5	c	208.2388	—	−106.0	—	—	—	—
$POCl_3$	c	153.3322	−145.81	—	—	—	—	—
PBr_3	liq	270.7008	—	−44.1	−42.0	—	57.4	—
	g	—	−27.47	−33.3	−38.9	4.240	83.17	18.16
PBr_5	c	430.5188	—	−64.5	—	—	—	—
$POBr_3$	c	286.7002	—	−109.6	—	—	—	—
PI_3	c	411.6870	—	−10.9	—	—	—	—
	g	—	—	—	—	4.542	89.45	18.73
PH_4I	c	161.9101	—	−16.7	0.2	—	29.4	26.2
P_2S_3	c	158.1396	—	−19.2	—	—	—	—
PN	g	44.9805	26.5	26.26	20.97	2.080	50.45	7.10
P_3N_5	c	162.9549	—	−71.4	—	—	—	36.
$NH_4H_2PO_4$	c	115.0259	—	−345.38	−289.33	—	36.32	34.00
$(NH_4)_2HPO_3$	aq	116.0571	—	−294.9	—	—	—	—
$(NH_4)_2HPO_4$	c	132.0565	—	−374.50	—	—	—	45.
$(NH_4)_3PO_4$	c	149.0871	—	−399.6	—	—	—	—
std. state, m = 1	aq	—	—	−400.3	−300.4	—	28.	—
PLATINUM								
Pt	c	195.09	0	0	0	1.372	9.95	6.18
	g	—	134.90	135.1	124.4	1.572	45.960	6.102
Pt^+	g	—	341.6	343.3	—	—	—	—
std. state, m = 1	aq	—	—	—	60.9	—	—	—
PtO_2	g	227.089	—	41.0	40.1	62.	—	—
Pt_3O_4	c	649.268	—	−39.	—	—	—	—
$Pt(OH)_2$	c	229.105	—	−84.1	—	—	—	—
PtCl	c	230.543	—	−13.5	—	—	—	—
$PtCl_2$	c	265.996	—	−29.5	—	—	—	—
std. state, m = 1	aq	—	—	—	—	—	—	—
$PtCl_3$	c	301.449	—	−43.5	—	—	—	—
$PtCl_4$	c	336.902	—	−55.4	—	—	—	—
$HPtCl_5 \cdot 2H_2O$	c	409.394	—	−242.0	—	—	—	—
$H_2PtCl_6 \cdot 6H_2O$	c	517.916	—	566.7	—	—	—	—
PtBr	c	274.999	—	−9.2	—	—	—	—
$PtBr_2$	c	354.908	—	−19.6	—	—	—	—
$PtBr_3$	c	434.817	—	−28.9	—	—	—	—
$PtBr_4$	c	514.726	—	−37.4	—	—	—	—
PtI_4	c	702.708	—	−17.1				
PtS	c	227.154	−19.020	−19.5	−18.2	1.946	13.16	10.37
PtS_2	c	259.218	−25.333	−26.0	−23.8	2.813	17.85	15.75
POLONIUM								
Po	c	210.	0	0	0	—	—	—
Po^{2+}								
std. state, m = 1	aq	—	—	—	17.	—	—	—
Po^{4+}								
std. state, m = 1	aq	—	—	—	70.	—	—	—
$Po(OH)_4$	c	278.0	—	—	−130.	—	—	—
PoS	c	242.1	—	—	−1.	—	—	—
POTASSIUM								
K	c	39.1020	0	0	0	1.695	15.34	7.07
	g	39.1020	21.544	21.33	14.49	1.481	38.295	4.968
K^+	g	39.1020	—	122.92	—	—	—	—
K^{2+}	g	39.1020	—	853.70	—	—	—	—
K^{3+}	g	39.1020	—	1909.5	—	—	—	—
K^{4+}	g	39.1020	—	3315.6	—	—	—	—
K^+	a	39.1020	—	−60.32	−67.70	—	24.5	5.2
K								
in 88.81 Hg		39.1020	—	—	−24.41	—	—	—
KO_2	c	71.1008	—	−68.10	−57.23	—	27.9	18.53
KO_3	c	87.1002	—	−62.2	—	—	—	—
K_2O	c	94.2034	—	−86.4	—	—	—	—
	g	94.2034	—	−15.	—	—	—	—
K_2O_2	c	110.2028	—	−118.1	−101.6	—	24.4	—
	g	110.2028	—	−38.	—	—	—	—
KH	c	40.1100	—	−13.80	—	—	—	—
KD	c	41.1160	—	−13.21	—	—	—	—
KOH	c	56.1094	−100.681	−101.521	−90.61	−2.904	18.85	15.51
	a	56.1094	—	−115.29	−105.29	—	21.9	−30.3
in ∞ H_2O	a	56.1094	—	−115.29	—	—	—	—
KF	c	58.1004	−135.223	−135.58	−128.53	2.392	15.91	11.72
KF								
in ∞ H_2O	a	58.1004	—	−139.82	—	—	—	—
$KF \cdot 2H_2O$	c	94.1312	—	−278.112	−244.17	—	37.1	—
KHF_2	cα	78.1068	−220.560	−221.72	−205.48	3.655	24.92	18.39
KCl	c	74.5550	−104.310	−104.385	−97.79	2.717	19.74	12.26
	a	74.5550	—	−100.27	−99.07	—	38.0	−27.4
KClO	a	90.5544	—	−85.9	−76.5	—	35.	—
$KClO_2$	a	106.5538	—	−76.2	−63.6	—	48.7	—
$KClO_3$	c	122.5532	—	−95.06	−70.82	—	34.2	23.96
	a	122.5532	—	−85.17	−69.62	—	63.3	—
$KClO_4$	c	138.5526	−101.525	−103.43	−72.46	5.036	36.1	26.86
	a	138.5526	—	−91.23	−69.76	—	68.0	—
KBr	c	119.0110	−92.414	−94.120	−90.98	2.919	22.92	12.50
	g	119.0110	−40.832	−43.04	−50.89	2.416	59.85	8.824
	a	119.0110	—	−89.37	−92.55	—	44.2	−28.7
KBr_3	a	278.8290	—	−91.49	−93.29	—	76.0	—
KBrO	a	135.0104	—	−82.8	−75.7	—	34.	—
$KBrO_3$	c	167.0092	−83.957	−86.10	−64.82	5.593	35.65	28.72
	a	167.0092	—	−76.35	−63.27	—	63.15	—
$KBrO_4$	c	183.0086	−65.619	−68.80	−41.70	5.593	40.65	28.72
	a	183.0086	—	−57.2	−39.5	—	72.2	—
KI	c	166.0065	−78.137	−78.370	−77.651	3.039	25.41	12.65

Substance		0 K		298.15 K (25°C)				
Formula and Description	State	Formula weight	ΔHf°_0 kcal/mol	ΔHf°	ΔGf° kcal/mol	$H^\circ_{298} - H^\circ_0$	S° cal/deg mol	C°_p
KI_3	c	419.8152	—	−78.4	—	—	—	—
	a	419.8152	—	−72.6	−80.0	—	81.7	—
KIO	a	182.0059	—	−86.0	−76.9	—	23.2	—
KIO_3	c	214.0047	−118.54	−119.83	−100.00	5.09	36.20	25.45
	a	214.0047	—	−113.2	−98.3	—	52.8	—
KIO_4	c	230.0041	—	−111.67	−86.38	—	42.	—
	a	230.0041	—	−96.5	−81.7	—	77.	—
K_2S	c	110.2680	—	−91.0	−87.0	—	25.0	
	a	110.2680	—	−112.7	−114.9	—	45.5	
K_2SO_3	c	158.2662	—	−269.0	—	—	—	—
K_2SO_3	a	158.2662	—	−272.5	−251.7	—	42.	—
K_2SO_4	c	174.2656	−341.126	−343.64	−315.83	6.079	41.96	31.42
	a	174.2656	—	−337.96	−313.37	—	53.8	−60.
$K_2S_2O_3$	c	190.3302	—	−280.5	—	—	—	—
	a	190.3302	—	−276.5	−260.3	—	65.	—
$K_2S_2O_7$	c	254.3278	—	−474.8	−428.2	—	61.	—
	a	254.3278	—	−455.5	—	—	—	—
$KHSO_4$	c	136.1716	—	−277.4	−246.5	—	33.0	—
	a	136.1716	—	272.40	−248.39	—	56.0	−15.
KNO_2								
rhombic	c	85.1075	−88.45	−88.39	−73.28	4.871	36.35	25.67
	a	85.1075	—	−85.3	−75.4	—	53.9	—
KNO_3	c	101.1069	−116.86	−118.22	−94.39	4.488	31.80	23.04
KNO_3	a	101.1069	—	−109.88	−94.31	—	59.5	−15.5
KPO_3	c	118.0740	—	—	—	3.886	25.93	21.56
	a	118.0740	—	−293.8	—	—	—	—
K_3PO_4	c	212.2774	—	−466.1	—	—	—	—
	a	212.2774	—	−486.3	−446.6	—	21.0	—
K_3AsO_4	a	256.2252	—	−393.23	−358.10	—	34.6	—
KH_2AsO_4	c	180.0372	−278.60	−282.2	−247.6	5.490	37.05	30.29
K_2CO_3	c	138.2134	−273.76	−275.1	−254.2	5.417	37.17	27.35
$K_2C_2O_4$								
oxalate	c	166.2240	—	−321.9	—	—	—	—
HCOOK								
formate	c	84.1200	—	−162.46	—	—	—	—
$KHCO_3$	c	100.1194	—	−230.2	−206.4	—	27.6	—
CH_3COOK								
acetate	c	98.1472	—	−172.8	—	—	—	—
	a	98.1472	—	−176.48	−155.99	—	45.2	3.7
KCN	c	65.1199	−28.174	−27.0	−24.35	4.157	30.71	15.84
	g	65.1199	21.494	21.7	15.34	3.188	62.57	12.51
KCNO								
cyanate	c	81.1193	—	−100.06	—	—	—	—
	a	81.1193	—	−95.2	−91.0	—	50.0	—
KCNS	c	97.1839	−47.980	−47.84	−42.62	4.176	29.70	21.16
	a	97.1839	—	−42.05	−45.55	—	59.0	−4.4
K_2SiO_3	c	154.2882	—	—	—	5.230	34.9	28.3
K_2SiF_6	c	220.2804	—	−706.5	−668.9	—	54.0	—
$KAl(SO_4)_2$	c	258.2067	—	−590.4	−535.4	—	48.9	46.12
$KAl(SO_4)_2 \cdot 3H_2O$	c	312.2529	—	−808.1	−711.0	—	75.0	—
$KAl(SO_4)_2 \cdot 12H_2O$	c	474.3915	—	−1448.8	−1228.9	—	164.3	155.6
$KAl_3Si_3O_{10}(OH)_2$								
muscovite	c	398.3133	—	−1430.3	−1340.5	—	73.2	—
$K_3Fe(CN)_6$	c	329.2604	—	−59.7	−31.0	—	101.83	—
	a	329.2604	—	−46.7	−28.8	—	138.1	
$K_4Fe(CN)_6$	c	368.3624	−141.30	−142.0	−108.3	14.871	100.1	79.40
$K_4Fe(CN)_6$	a	368.3624	—	−132.4	−104.71	—	120.7	—
$KMnO_4$	c	158.0376	—	−200.1	−176.3	—	41.04	28.10
	a	158.0376	—	—	—	—	—	−14.4
K_2CrO_4	c	194.1976	−333.796	−335.5	−309.7	6.805	47.83	34.89
	a	194.1976	—	−331.24	−309.36	—	61.0	—
$C_2Cr_2O_7$	c	294.1918	—	−492.7	−449.8	—	69	52.4
	a	294.1918	—	−476.8	−446.4	—	111.6	—
NaK	l	62.0918	—	1.5	—	—	—	—
Na_2K	l	85.0816	—	2.0	—	—	—	—
NaK_2	l	101.1938	—	2.5	—	—	—	—
PRASEODYMIUM								
Pr	c	140.907	0	0	0	1.74	17.5	6.50
	g	—	85.25	85.0	76.7	1.487	45.339	5.105
Pr^+	g	—	210.3	211.5	—	—	—	—
Pr^{2+}	g	—	453.6	456.3	—	—	—	—
Pr^{3+}	g	—	952.3	956.5	—	—	—	—
std. state, m = 1	aq	—	—	−168.4	−162.3	—	−50.	−7.
Pr^{4+}	g	—	1851.	1856.	—	—	—	—
Pr^{5+}	g	—	3176.	3183.	—	—	—	—
PrO	g	156.906	−38.	—	—	—	—	—
PrO_2	c	172.906	—	−226.9	—	—	—	28.06
Pr_2O_3								
hexagonal	c	329.812	—	−432.5	—	—	—	28.06
cubic	c	—	—	−432.5	—	—	—	—
PrH_2	c	142.923	−45.47	−47.4	−36.9	1.83	13.6	9.8
$PrOH^{2+}$								
std. state, m = 1	aq	157.914	—	—	−206.	—	—	—
$Pr(OH)_2^+$								
std. state, m = 1	aq	174.922	—	—	−257.	—	—	—
$Pr(OH)_3$	c	191.929	—	—	−307.1	—	—	—
$PrCl_3$	c	247.266	—	−252.6	—	—	—	24.
	g	—	—	−187.	—	—	—	—
std. state, m = 1	aq	—	—	−288.3	−256.4	—	−10.	−105.
PrOCl	c	192.359	—	−242.	—	—	—	—
PrI_3	c	521.620	—	−156.4	—	—	—	—
$Pr(IO_3)_3$	c	665.615	—	−333.8	—	—	—	—
$PrSO_4^+$								
std. state, m = 1	aq	236.969	—	−382.3	−345.1	—	−17.	—
$Pr(SO_4)_2^-$								
std. state, m = 1	aq	333.030	—	−598.4	−525.6	—	0.	—
$Pr(NO_3)_3$	c	326.922	—	−293.8	—	—	—	—
in HNO_3(aq)	aq	—	—	−316.7	—	—	—	—

Substance			0 K	298.15 K (25°C)				
Formula and Description	State	Formula weight	ΔHf°_0 kcal/mol	ΔHf°	ΔGf° kcal/mol	$H^\circ_{298} - H^\circ_0$	S° cal/deg mol	C°_p
$Pr(NO_3)_3 \cdot 6H_2O$	c	435.014	—	−731.05	—	—	—	—
PrC	c	152.918	—	−13.0	—	—	—	—
PrC_2	g	164.929	131.	131.3	118.7	2.48	62.6	10.6
$Pr_2(CO_3)_3$	c	461.842	—	−768.	—	—	—	—
PROMETHIUM								
147_{Pm}	g	146.915	—	—	—	1.545	44.692	5.797
PROTACTINIUM								
Pa	c	231.0359	0	0	0	—	12.4	—
	g	231.0359	—	145.	134.6	1.518	47.31	5.48
Pa^{4+}	a	231.0359	—	−148.0	—	—	—	—
in HCl + 3.43 H_2O:A		231.0359	—	−159.6	—	—	—	—
Pa^{5+}								
in HCl + 3.43 H_2O:A		231.0359	—	−161.8	—	—	—	—
Pa^{4+}								
in HCl + 8.16 H_2O:A		231.0359	—	−144.8	—	—	—	—
in HCl + 54.4 H_2O:A		231.0359	—	−147.7	—	—	—	—
PaO_2	c	263.0347	—	—	—	—	17.8	—
$PaCl_4$	c	372.8479	—	−249.3	−227.7	—	46.0	—
in HCl + 3.43 H_2O:Au		372.8479	—	−290.1	—	—	—	—
in HCl + 8.16 H_2O:Au		372.8479	—	−291.6	—	—	—	—
in HCl + 54.4 H_2O:Au		372.8479	—	−304.9	—	—	—	—
$PaCl_5$	c	408.3009	—	−273.6	−247.2	—	57.	—
	g	408.3009	—	−251.	−236.	—	94.	—
in HCl + 3.43 H_2O:Au		408.3009	—	−322.2	—	—	—	—
$PaBr_4$	c	550.6719	—	−197.0	−188.3	—	56.0	—
in HCl + 8.16 H_2O:Au		550.6719	—	−248.70	—	—	—	—
in HCl + 54.4 H_2O:Au		550.6719	—	−262.44	—	—	—	—
$PaBr_5$	c	630.5809	—	−206.	−196.	—	69.	—
	g	630.5809	—	−180.	−182.	—	111.	—
in HCl + 3.43 H_2O:Au		630.5809	—	−264.8	—	—	—	—
$PaOBr_2$	c	406.8533	—	−239.	—	—	—	—
PaI_4	c	738.6535	—	−123.2	—	—	—	—
in HCl + 54.4 H_2O:Au		738.6535	—	−199.1	—	—	—	—
RADIUM								
Ra	c	226.025	0	0	0	—	17.	—
	g	—	—	38.	31.	1.481	42.15	4.97
Ra^+	g	—	—	161.22	—	—	—	—
Ra^{2+}	g	—	—	396.70	—	—	—	—
std. state, m = 1	aq	—	—	−126.1	−134.2	—	13.	—
RaO	c	242.0244	—	−125.	—	—	—	—
$RaCl_2$	c	296.931	—	—	—	—	32.	—
std. state, m = 1	aq	—	—	−206.0	−196.9	—	40.	—
$RaCl_2 \cdot 2H_2O$	c	332.9617	—	−350.	−311.4	—	51.	—
$Ra(IO_3)_2$	c	575.8302	—	−245.4	−207.6	—	65.	—
$RaSO_4$	c	322.0866	—	−351.6	−326.4	—	33.	—
std. state, m = 1	aq	—	—	−343.4	−312.2	—	18.	—
$Ra(NO_3)_2$	c	350.0348	—	−237.	−190.3	—	53.	—
std. state, m = 1	aq	—	—	−225.2	−187.4	—	83.	—
RADON								
Rn	g	222.	0	0	0	1.481	42.09	4.968
Rn^+	g	—	247.86	249.34	—	—	—	—
RHENIUM								
Re	c	186.2	0	0	0	1.296	8.81	6.09
	g	—	183.8	184.0	173.2	1.481	45.131	4.968
Re^+	g	—	365.5	367.1	—	—	—	—
std. state, m = 1	aq	—	—	—	−8.	—	—	—
ReO_2	c	218.20	—	—	−88.	—	—	—
ReO_3	c	234.20	—	−144.6	—	—	—	—
$HReO_4$	c	251.21	—	−182.2	−156.9	—	37.8	—
std. state, m = 1	aq	—	—	−188.2	−166.0	—	48.1	−3.2
$HReO_4$								
∞ H_2O	aq	—	—	−188.2	—	—	—	—
$ReCl_3$	c	292.56	−62.73	−63.	−45.	4.319	29.6	22.08
$ReCl_5$	c	363.47	—	−89.	—	—	—	—
H_2ReCl_4	c	330.03	—	−152.	—	—	—	—
$ReBr_3$	c	425.93	—	−40.	—	—	—	—
ReS_2	c	250.33	—	−43.	—	—	—	—
RHODIUM								
Rh	c	102.905	0	0	0	1.174	7.53	5.97
	g	—	132.79	133.1	122.1	1.483	44.383	5.022
Rh^+	g	—	304.90	306.69	—	—	—	—
Rh^{2+}	g	—	721.8	725.0	—	—	—	—
Rh^{3+}	g	—	1438.	1443.	—	—	—	—
RhO	g	118.9044	—	92.	—	—	—	—
RhO^+	g	—	—	308.	—	—	—	—
RhO_2	g	134.9038	—	44.	—	—	—	—
Rh_2O_3	c	253.8082	—	−82.	—	—	—	24.8
$RhCl_2$	g	173.811	—	30.3	—	—	—	—
$RhCl_3$	c	209.264	—	−71.5	—	—	—	—
$RhCl_6^{3-}$	aq	315.623	—	−202.8	—	—	—	—
$RhCl_3 \cdot 3(C_2H_5)_2S$	c	479.8289	—	−166.	—	—	—	—
RUBIDIUM								
Rb	c	85.4678	0	0	0	1.790	18.35	7.424
	g	85.4678	19.639	19.330	12.690	1.481	40.626	4.968
Rb^+	g	85.4678	115.965	117.137	—	—	—	—
Rb^{2+}	g	85.4678	745.10	747.76	—	—	—	—
Rb^{3+}	g	85.4678	1660.0	1664.1	—	—	—	—
Rb^+	a	85.4678	—	−60.03	−67.87	—	29.04	—
Rb								
in 185 Hg		85.4678	—	—	−25.25	—	—	—
RbO_2	c	117.4666	—	−66.6	—	—	—	—
Rb_2O	c	186.9350	—	−81.	—	—	—	—
Rb_2O_2	c	202.9344	—	−112.8	—	—	—	—
RbH	c	86.4758	—	−12.5	—	—	—	—
RbOH	c	102.4752	—	−99.95	—	—	—	—
$RbOH \cdot H_2O$	c	120.4906	—	−178.98	—	—	—	—
$RbOH \cdot 2H_2O$	c	138.5060	—	−251.73	—	—	—	—
RbF	c	104.4662	—	−133.3	—	—	—	—
	a	104.4662	—	−139.53	−134.510	—	25.70	—
$RbHF_2$	c	124.4726	−219.52	−220.5	−204.5	3.932	28.70	18.97

SELECTED VALUES OF CHEMICAL THERMODYNAMIC PROPERTIES (Continued)

Substance			0 K	298.15 K (25°C)				
Formula and Description	State	Formula weight	ΔHf°_0 kcal/mol	ΔHf° kcal/mol	ΔGf° kcal/mol	$H^\circ_{298} - H^\circ_0$ kcal/mol	S° cal/deg mol	C°_p cal/deg mol
RbCl	c	120.9208	−104.080	−104.05	−97.47	2.917	22.92	12.52
$RbCl_3$	a	191.8268	—	—	−96.7	—	—	—
RbClO	a	136.9202	—	−85.6	−76.7	—	39.	—
$RbClO_2$	a	152.9196	—	−75.9	−63.8	—	53.2	—
$RbClO_3$	c	168.9190	—	−96.3	−71.8	—	36.3	24.66
	a	168.9190	—	−84.88	−69.77	—	67.80	—
$RbClO_4$	c	184.9184	—	−104.50	−73.54	—	39.2	—
	a	184.9184	—	−90.94	−69.93	—	72.5	—
RbBr	c	165.3768	−92.714	−94.31	−91.25	3.124	26.28	12.63
	a	165.3768	—	−89.08	−92.72	—	48.74	—
$RbBr_3$	c	325.1948	—	−100.0	—	—	—	—
	a	325.1948	—	−91.20	−93.46	—	80.5	—
$RbBr_5$	a	485.0128	—	−94.0	−92.7	—	104.7	—
$RbBrO_3$	c	213.3750	—	−87.78	−66.47	—	38.5	—
$RbBrO_3$	a	213.3750	—	−76.06	−63.44	—	67.69	—
$RbBrO_4$	a	229.3744	—	−56.9	−39.7	—	76.7	—
$RbBrCl_2$	c	236.2828	—	−116.2	—	—	—	—
RbI	c	212.3722	−79.603	−79.78	−78.60	3.190	28.30	12.71
	a	212.3722	—	−73.22	−80.20	—	55.6	—
RbI_3	c	466.1810	—	−82.8	−81.0	—	53.9	—
	a	466.1810	—	−72.3	−80.2	—	86.2	—
$RbIO_3$	c	260.3704	—	—	−101.90	—	—	—
	a	260.3704	—	−112.90	−98.50	—	57.30	—
Rb_2S	c	202.9996	—	−86.2	—	—	—	—
	a	202.9996	—	−112.2	−115.2	—	54.6	—
Rb_2S_2	a	235.0636	—	−112.9	−116.7	—	64.9	—
Rb_2SO_3	a	250.9978	—	−272.0	−252.0	—	51.	—
Rb_2SO_4	c	266.9972	−340.795	−343.12	−314.76	6.458	47.19	32.04
Rb_2SO_4	a	266.9972	—	−337.38	−313.71	—	62.9	—
$RbNO_3$	c	147.4727	—	−118.32	−94.61	—	35.2	24.4
	a	147.4727	—	−109.59	−94.48	—	64.0	—
Rb_3PO_4	a	351.3748	—	−485.4	−447.1	—	34.	—
$Rb_4P_2O_7$	a	515.8146	—	−782.9	−730.2	—	88.	—
$RbAsO_2$	a	192.3882	—	−162.57	−151.53	—	38.7	—
Rb_3AsO_4	a	395.3226	—	−392.36	−358.61	—	48.2	—
Rb_2CO_3	c	230.9450	−270.41	−271.5	−251.2	5.851	43.34	28.11
	a	230.9450	—	−281.90	−261.91	—	44.5	—
$RbHCO_3$	c	146.4852	—	−230.2	−206.4	—	29.0	—
	a	146.4852	—	−225.42	−208.13	—	50.84	—
RbCN	c	111.4857	—	—	—	4.159	33.67	16.20
	a	111.4857	—	−24.0	−26.7	—	51.5	—
RbCNO cyanate	a	127.4851	—	−94.9	−91.2	—	54.5	—
RbCNS thiocyanate	a	143.5497	—	−41.76	−45.72	—	63.5	—
$RbBO_2$	c	128.2776	−231.0	−232.0	−218.2	3.181	22.54	17.7
$RbBH_4$	a	100.3108	—	−48.52	−40.56	—	55.4	—
$RbMnO_4$	a	204.4034	—	−189.4	−174.8	—	74.7	—
Rb_2MnO_4	a	289.8712	—	−276.	−255.4	—	72.	—
Rb_2CrO_4	c	286.9292	—	−338.0	—	—	—	—
	a	286.9292	—	−330.66	−309.70	—	70.08	—
$Rb_2Cr_2O_7$	a	386.9234	—	−476.3	−446.7	—	120.7	—
RUTHENIUM								
Ru	c	101.07	0	0	0	1.100	6.82	5.75
	g	—	153.210	153.6	142.4	1.490	44.550	5.144
Ru^+	g	—	323.07	324.94	—	—	—	—
Ru^{2+}	g	—	709.6	713.0	—	—	—	—
Ru^{3+}	g	—	1366.0	1370.9	—	—	—	—
RuO_4	c	133.069	—	−72.9	—	—	—	—
$RuCl_3$ black	c	207.429	—	−49.	—	—	—	—
$RuCl_4$	g	242.882	—	−12.4	—	—	—	—
$RuBr_3$	c	340.797	—	−33.	—	—	—	—
RuI_3	c	481.783	—	−15.7	—	—	—	—
RuS_2	c	165.198	—	−47.	—	—	—	—
SAMARIUM								
Sm	c	150.35	0	0	0	1.81	16.63	7.06
	g	—	49.26	49.4	41.3	1.953	43.722	7.255
Sm^+	g	—	179.1	180.7	—	—	—	—
Sm^{2+}	g	—	434.	438.	—	—	—	—
std. state, m = 1	aq	—	—	—	−118.9	—	—	—
Sm^{3+} std. state, m = 1	aq	—	—	−165.3	−159.3	—	−50.6	−5.
SmO	g	166.349	−31.	—	—	—	—	—
Sm_2O_3 monoclinic	c	348.698	−433.89	−435.7	−414.6	5.02	36.1	27.37
cubic	c	—	—	—	—	5.0	—	26.86
SmF	g	169.348	—	−63.	—	—	—	—
SmF_3	c	207.345	—	−425	—	—	—	—
$SmF_3 \cdot H_2O$	c	216.353	—	−436.2	—	—	—	—
$SmCl_3$	c	256.709	—	−245.2	—	—	—	—
∞H_2O	aq	—	—	−285.2	—	—	—	—
$SmCl_3 \cdot 6H_2O$	c	364.801	—	−686.0	−587.1	—	99.	86.4
SmI_3	c	531.063	—	−148.2	—	—	—	—
$Sm(IO_3)_3$	c	675.058	—	−330.	—	—	—	—
$Sm_2(SO_3)_3$	c	540.887	—	—	−697.4	—	—	—
$Sm_2(SO_4)_3$	c	588.885	—	−931.9	—	—	—	—
$Sm(NO_3)_3$	c	336.365	—	−289.7	—	—	—	—
SmC_2	c	174.372	—	−17.	−18.1	—	23.	—
$Sm_2(CO_3)_3$	c	480.728	—	—	−741.4	—	—	—
SCANDIUM								
Sc	c	44.956	0	0	0	1.247	8.28	6.10
	g	—	89.87	90.3	80.32	1.674	41.75	5.28
Sc^+	g	—	240.69	—	—	—	—	—
Sc^{2+}	g	—	535.87	—	—	—	—	—
Sc^{3+}	g	—	1106.86	—	—	—	—	—
std. state, m = 1	aq	—	—	−146.8	−140.2	—	−61.	—
Sc^{4+}	g	—	2801.1	—	—	—	—	—

SELECTED VALUES OF CHEMICAL THERMODYNAMIC PROPERTIES (Continued)

Substance			0 K	298.15 K (25°C)				
Formula and Description	State	Formula weight	ΔHf°_0 kcal/mol	ΔHf°	ΔGf° kcal/mol	$H^\circ_{298} - H^\circ_0$	S° cal/deg mol	C°_p
Sc^{5+}	g	—	4914.9	—	—	—	—	—
ScO	g	60.955	−13.5	−13.68	−19.90	2.100	53.65	7.38
Sc_2O	g	105.911	−6.	−6.9	—	2.7	—	11.2
Sc_2O_3	c	137.910	−453.95	−456.22	−434.85	3.34	18.4	22.52
$Sc(OH)_3$	c	95.978	—	−325.9	−294.8	—	24.	—
ScF	g	63.9544	−33.	−33.2	−39.3	2.138	53.11	7.74
ScF_2	g	82.9528	−153.	−153.5	−156.6	2.84	67.0	11.5
ScF_3	c	101.9512	—	−389.4	−371.8	—	22.	—
$ScCl_2$	g	115.862	—	—	—	3.1	72.5	12.6
$ScCl_3$	c	151.315	—	−221.1	—	—	—	—
in 5500 H_2O	aq	—	—	−268.8	—	—	—	—
$ScCl_3 \cdot 6H_2O$	c	259.411	—	−666.6	—	—	—	—
$Sc(OH)_2Cl$	c	114.424	—	−303.	−278.3	—	20.	—
$ScBr_3$	c	284.683	—	−177.6	—	—	—	—
ScI_2	g	298.765	—	—	—	3.4	81.6	13.3
ScS	g	77.020	41.9	41.8	29.7	2.2	56.3	8.0
$Sc_2(SO_4)_3$	c	378.097	—	—	—	—	—	62.0
ScC_2	g	68.978	143.	143.6	—	—	—	—
$Sc(HCO_2)_3$	c	180.009	—	—	—	—	—	54.
$Sc_2(C_2O_4)_3$	c	353.971	—	—	—	—	—	124.
SELENIUM								
Se								
hexagonal, black	c	78.96	0	0	0	1.319	10.144	6.062
monoclinic, red	c	—	—	1.6	—	—	—	—
	g	—	54.11	54.27	44.71	1.4815	42.21	4.978
glassy	amorp	—	—	1.2	—	—	—	—
Se^{2-}								
std. state, m = 1	aq	—	—	—	30.9	—	—	—
Se_2	g	157.92	35.26	34.9	23.0	2.275	60.2	8.46
Se_6	g	473.76	—	39.2	—	—	—	—
SeO	g	94.959	13.0	12.75	6.41	2.108	55.9	7.47
SeO_2	c	110.959	—	−53.86	—	—	—	—
	aq	—	—	−52.97	—	—	—	—
SeO_3	c	126.958	—	−39.9	—	—	—	—
Se_2O_5	c	237.917	—	−97.6	—	—	—	—
H_2SeO_3	c	128.974	—	125.35	—	—	—	—
undissoc.; std. state, m = 1	aq	—	—	−121.29	−101.87	—	49.7	—
	aq	—	—	−121.24	—	—	—	—
H_2SeO_4	c	144.974	—	−126.7	—	—	—	—
$H_2SeO_4 \cdot H_2O$	c	162.989	—	−200.9	—	—	—	—
	liq	—	—	−196.1	—	—	—	—
SeF_6	g	192.950	−264.1	−267.	−243.	4.740	74.99	26.4
$SeCl_2$	g	149.866	—	−7.6	—	—	—	—
$SeCl_4$	c	220.772	—	−43.8	—	—	—	—
SILICON								
Si	c	28.086	0	0	0	0.769	4.50	4.78
	amorp	—	—	1.0	—	—	—	—
	g	—	107.86	108.9	98.3	1.805	40.12	5.318
Si^{+}	g	—	295.83	298.35	—	—	—	—
Si^{2+}	g	—	672.71	676.71	—	—	—	—
Si^{3+}	g	—	1444.50	1449.98	—	—	—	—
Si^{4+}	g	—	2485.50	2492.46	—	—	—	—
Si^{5+}	g	—	6331.3	6339.8	—	—	—	—
Si^{6+}	g	—	11062.6	11072.6	—	—	—	—
Si_2	g	56.172	141.32	142.	128.	2.22	54.92	8.22
Si_3	g	84.258	146.4	147.	—	2.9	—	12.9
SiO	g	44.0854	−24.08	−23.8	−30.2	2.082	50.55	7.15
SiO_2								
α, quartz	c	60.0848	—	−217.72	−204.75	1.657	10.00	10.62
α, cristobalite	c	—	—	−217.37	−204.46	1.671	10.20	10.56
SiO_2								
α, tridymite	c	—	—	−217.27	−204.42	1.693	10.4	10.66
	amorp	—	—	−215.94	−203.33	—	11.2	10.6
	g	—	—	−77.	—	—	—	—
	aq	—	—	−214.4	—	—	—	—
SiH	g	29.0940	86.	86.28	—	2.069	—	—
SiH_4	g	32.1179	10.30	8.2	13.6	2.517	48.88	10.24
Si_2H_6	g	62.2198	23.04	19.2	30.4	3.768	65.14	19.31
Si_3H_8	liq	92.3218	—	22.1	—	—	—	—
	g	—	—	28.9	—	—	—	—
H_2SiO_3	c	78.1001	—	−284.1	−261.1	—	32.	—
undissoc.; std. state, m = 1	aq	—	−282.7	−258.0	—	26.	—	
H_4SiO_4	c	96.1155	—	−354.0	−318.6	—	46.	—
SiF	g	47.0844	1.	1.7	−5.8	2.260	53.94	7.80
SiF_2	g	66.0828	−147.75	−148.	−150.	2.630	60.38	10.49
SiF_4	g	104.0796	−384.66	−385.98	−375.88	3.663	67.49	17.60
std. state, m = 1	aq	—	—	—	−384.2	—	—	—
SiF_6^{2-}								
std. state, m = 1	aq	142.0764	—	−571.0	−525.7	—	29.2	—
H_2SiF_6	aq	144.0923	—	—	—	—	—	—
SiCl	g	63.539	45.	45.39	—	2.267	—	8.81
$SiCl_2$	g	98.992	−39.61	−39.59	−42.35	2.98	67.0	12.16
$SiCl_4$	liq	169.898	—	−164.2	−148.16	—	57.3	34.73
	g	—	−156.508	−157.03	−147.47	4.633	79.02	21.57
SiBr	g	107.995	51.30	50.	—	2.40	—	9.23
$SiBr_4$	liq	347.722	—	−109.3	−106.1	—	66.4	—
	g	—	—	−99.3	−103.2	5.317	90.29	23.21
SiI_4	c	535.7036	—	−45.3	—	—	—	—
SiH_3I	g	158.0143	—	—	—	2.886	64.73	13.00
SiS	g	60.150	26.6	26.88	14.56	2.135	53.43	7.71
SiS_2	c	92.214	—	−49.5	—	—	—	—
Si_3N_4								
α	c	140.2848	—	−177.7	−153.6	—	24.2	—
SiC								
β, cubic	c	40.0972	−15.36	−15.6	−15.0	0.781	3.97	6.42
α, hexagonal	c	—	—	−15.0	−14.4	—	3.94	6.38
	g	—	175.6	177.	—	2.4	—	—
SiC_2	g	52.1083	145.7	147.	—	2.5	—	—
Si_2C	g	68.1832	131.1	132.	—	2.7	—	—
Si_2C_2	g	80.1943	167.	—	—	—	—	—

Substance			0 K	298.15 K (25°C)				
Formula and Description	State	Formula weight	ΔHf_0° kcal/mol	ΔHf°	ΔGf° kcal/mol	$H_{298}^\circ - H_0^\circ$	S° cal/deg mol	C_p°
Si_2C_3	g	92.2054	176.	—	—	—	—	—
Si_3C	g	96.2692	161.	—	—	—	—	—
SILVER								
Ag	c	107.870	0	0	0	1.373	10.17	6.059
	g	—	67.90	68.01	58.72	1.481	41.321	4.9679
Ag^+	g	—	242.00	243.59	—	—	—	—
std. state, m = 1	aq	—	—	25.234	18.433	—	17.37	5.2
Ag_2O	c	231.7394	−7.034	−7.42	−2.68	3.397	29.0	15.74
Ag_2O_2	c	247.7388	—	−5.8	6.6	—	28.	21.
Ag_2O_3	c	263.7382	—	8.1	29.0	—	24.	—
AgF	c	126.8684	—	−48.9	—	—	—	—
std. state, m = 1	aq	—	—	−54.27	−48.21	—	14.1	−20.3
in ∞ H_2O	aq	—	—	−54.27	—	—	—	—
AgF_2	c	145.8668	—	−86.	—	—	—	—
AgCl	c	143.323	—	−30.370	−26.244	—	23.0	12.14
std. state, m = 1	aq	—	—	−14.718	−12.939	—	30.9	−27.4
$AgClO_2$	c	175.3218	—	2.10	18.1	—	32.16	20.87
$AgClO_3$	c	191.3212	—	−7.24	15.4	—	34	—
std. state, m = 1	aq	—	—	0.38	16.51	—	56.2	—
$AgClO_4$	c	207.3206	—	−7.44	—	—	—	—
std. state, m = 1	aq	—	−5.68	16.37	—	60.9	—	
AgBr	c	187.779	—	−23.99	−23.16	—	25.6	12.52
std. state, m = 1	aq	—	—	−3.82	−6.42	—	37.1	−28.7
$AgBrO_3$	c	235.7772	—	−2.5	−17.04	—	36.3	—
std. state, m = 1	aq	—	—	9.20	22.86	—	56.02	—
AgI	c	234.7744	—	−14.78	−15.82	—	27.6	13.58
std. state, m = 1	aq	—	—	12.04	6.10	—	44.0	−28.8
AgI_2^-								
std. state, m = 1	aq	361.6788	—	—	−20.8	—	—	—
AgI_3^{2-}								
std. state, m = 1	aq	488.5832	—	−43.5	−36.8	—	60.5	—
AgI_4^{3-}								
std. state, m = 1	aq	615.4876	—	—	−50.1	—	—	—
$AgIO_3$	c	282.7726	—	−40.9	−22.4	—	35.7	24.60
std. state, m = 1	aq	—	—	−27.7	−12.2	—	45.7	—
Ag_2S								
α, orthorhombic	c	247.804	−8.126	−7.79	−9.72	4.136	34.42	18.29
β	c	—	—	−7.03	−9.43	—	36.0	—
Ag_2SO_3	c	295.8022	—	−117.3	−98.3	—	37.8	—
std. state, m = 1	aq	—	—	−101.4	−79.4	—	27.8	—
Ag_2SO_4	c	311.8016	—	−171.10	−147.82	—	47.9	31.40
std. state, m = 1	aq	—	—	−166.85	−141.10	—	39.6	−60.
Ag_2Se	c	294.70	−9.42	−9.	−10.6	4.48	36.02	19.54
AgN_3	c	149.8901	—	73.8	89.9	—	24.9	—
std. state, m = 1	aq	—	—	90.99	101.6	—	43.2	
Ag_3N	c	337.6167	—	47.6	—	—	—	—
$AgNO_2$	c	153.8755	—	−10.77	4.56	—	30.64	19.17
std. state, m = 1	aq	—	—	10.7	9.5	—	46.8	−18.1
$AgNO_3$	c	169.8749	—	−29.73	−8.00	—	33.68	22.24
nitrate; std. state, m = 1	aq	—	—	−24.33	−8.18	—	52.4	−15.5
$Ag(NH_3)_2^+$								
std. state, m = 1	aq	141.9312	—	−26.60	−4.12	—	58.6	—
$Ag(NH_3)_2NO_3$	c	203.9361	—	−85.0	—	—	—	—
std. state, m = 1	aq	—	—	−76.16	−30.73	—	93.6	—
$Ag(NH_3)_2Cl$								
std. state, m = 1	aq	177.3842	—	−66.55	−35.49	—	72.1	—
undissoc.; std. state, m = 1	aq	—	—	—	−35.01	—	—	—
$Ag(NH_3)_2Br$								
std. state, m = 1	aq	221.8402	—	−55.65	−28.97	—	78.3	—
Ag_3PO_4	c	418.5814	—	—	−210.	—	—	—
AgCN	c	133.8878	34.354	34.9	37.5	3.206	25.62	15.95
std. state, m = 1	aq	—	—	61.2	59.6	—	39.9	—
$Ag(CN)_2$								
std. state, m = 1	aq	159.9057	—	64.6	73.0	—	46.	—
AgONC								
silver fulminate	c	149.8872	—	43.	—	—	—	—
AgSCN	c	165.9518	—	21.0	24.23	—	31.3	15.
std. state, m = 1	aq	—	—	43.50	40.58	—	51.9	−4.4
SODIUM								
Na	c	22.9898	0	0	0	1.54	12.24	6.75
	g	22.9898	25.709	25.65	18.354	1.481	36.712	4.968
Na^+	g	22.9898	—	145.55	—	—	—	—
Na^{2+}	g	22.9898	—	1237.48	—	—	—	—
Na^{3+}	g	22.9898	—	2891.0	—	—	—	—
Na^{4+}	g	22.9898	—	5173.5	—	—	—	—
Na^{5+}	g	22.9898	—	8366.	—	—	—	—
Na^+	a	22.9898	—	−57.39	−62.593	—	14.1	11.1
NaO	g	38.9892	25.4	25.	19.7	2.22	54.6	8.3
NaO_2	c	54.9886	−62.96	−62.2	−52.2	4.37	27.7	17.24
Na_2O	c	61.9790	−97.85	−99.00	−89.74	2.964	17.94	16.52
Na_2O_2	c	77.9784	−120.70	−122.10	−107.00	3.75	22.70	21.33
NaOH	c	39.9972	−100.641	−101.723	−90.709	2.507	15.405	14.23
	g	39.9972	−48.6	−49.5	−50.2	2.72	54.57	11.56
in ∞ H_2O	a	39.9972	—	−112.36	—	—	—	—
NaF	c	41.9882	−136.542	−137.105	−129.902	2.031	12.30	11.20
	a	41.9882	—	−136.89	−129.23	—	10.8	−14.4
NaCl	c	58.4428	−98.168	−98.268	−91.815	2.536	17.24	12.07
	g	58.4428	−41.9	−42.22	−47.00	2.298	54.90	8.55
	a	58.4428	—	−97.34	−93.965	—	27.6	−21.5
in ∞ H_2O	a	58.4428	—	−97.34	—	—	—	—
NaClO	a	74.4422	—	−83.0	−71.4	—	24.	—
in 400 H_2O		74.4422	—	−82.8	—	—	—	—
$NaClO_2$	c	90.4416	—	−73.38	—	—	—	—
	a	90.4416	—	−73.3	−58.5	—	38.3	—
$NaClO_3$	c	106.4410	—	−87.422	−62.697	—	29.5	—
	a	106.4410	—	−82.24	−64.51	—	52.9	—
$NaClO_4$	c	122.4404	—	−91.61	−60.93	—	34.0	—
	a	122.4404	—	−88.30	−64.65	—	57.6	—

Substance			0 K	298.15 K (25°C)				
Formula and Description	State	Formula weight	ΔHf°_0 kcal/mol	ΔHf°	ΔGf° kcal/mol	$H^\circ_{298} - H^\circ_0$	S° cal/deg mol	C°_p
NaBr	c	102.8988	−84.596	−86.296	−83.409	2.770	20.75	12.28
	g	102.8988	−32.08	−34.2	−42.31	2.346	57.62	8.68
	a	102.8988	—	−86.440	−87.440	—	33.8	−22.8
NaBrO	a	118.8982	—	−79.9	−70.6	—	24.	—
$NaBrO_3$	c	150.8970	—	−79.85	−58.04	—	30.8	—
	a	150.8970	—	−73.42	−58.16	—	52.8	—
$NaBrO_4$	a	166.8964	—	−54.30	−34.40	—	61.80	—
NaI	c	149.8942	−68.593	−68.78	−68.37	2.93	23.55	12.45
	a	149.8942	—	−70.58	−74.92	—	40.7	−22.9
NaI_3	c	403.7030	—	−56.2	—	—	—	—
NaI_3	a	407.7030	—	−69.7	−74.9	—	71.3	—
$NaIO_3$	c	197.8924	—	−115.150	—	—	—	22.0
	[illegible]	[illegible]		[illegible]	[illegible]		[illegible]	
$NaIO_3 \cdot H_2O$	c	215.9078	—	−186.30	−151.56	—	38.8	—
$NaIO_3 \cdot 5H_2O$	c	287.9694	—	−466.60	—	—	—	—
$NaIO_4$	c	213.8918	—	−102.60	−77.22	—	39.0	—
	a	213.8918	—	−93.60	−76.60	—	67.	—
in 2,000 H_2O		213.8918	—	−93.808	—	—	—	—
$NaIO_3 \cdot 3H_2O$	c	267.9380	—	—	−247.8	—	—	—
NaH_4IO_6								
from $H_4IO_6^-$	au	249.9226	—	−237.8	—	—	—	—
$Na_2H_3IO_6$								
from $H_3IO_6^{2-}$	au	271.9044	—	−294.4	—	—	—	—
$NaIO_2F_2$	c	219.8898	—	−202.50	—	—	—	—
$NaICl_2$	c	220.8002	—	−96.0	—	—	—	—
	a	220.8002	—	—	−101.1	—	—	—
$NaICl_4$	c	291.7062	—	−112.0	—	—	—	—
$NaIBr_2$	c	309.7122	—	−83.0	—	—	—	—
	a	309.7122	—	—	−92.0	—	—	—
$NaBrI_2$	a	356.7076	—	−88.0	−88.9	—	61.3	—
Na_2S	c	78.0436	—	−87.2	−83.6	—	20.0	—
	a	78.0436	—	−106.9	−104.7	—	24.7	—
Na_2S_2	c	110.1076	—	−94.9	−90.5	—	25.	—
Na_2S_2	a	110.1076	—	−107.6	−106.2	—	35.0	—
Na_2SO_3	c	126.0418	−261.21	−263.1	−242.0	5.36	34.88	28.74
	a	126.0418	—	−266.70	−241.50	—	21.0	—
Na_2SO_4	c	142.0412	−328.789	−331.52	−303.59	5.551	35.75	30.64
	a	142.0412	—	−332.10	−303.16	—	33.0	−48.
$Na_2SO_4 \cdot 10H_2O$	c	322.1952	—	−1034.24	−871.75	—	141.5	—
$Na_2S_2O_3$	c	158.1058	—	−268.4	−245.7	—	37.	—
	a	158.1058	—	−270.65	−250.0	—	44.0	—
$Na_2S_2O_3 \cdot 5H_2O$	c	248.1828	—	−623.31	−533.0	—	89.0	—
$Na_2S_2O_4$	c	174.1052	—	−294.5	—	—	—	—
$NaHSO_4$	c	120.0594	—	−269.0	−237.3	—	27.0	—
from HSO_4	a	120.0594	—	−269.47	−243.28	—	45.6	−9.
Na_2SeO_4	c	188.9372	—	255.5	—	—	—	—
	a	188.9372	—	258.0	−230.7	—	41.1	—
NaN_3	c	65.0099	6.316	5.19	22.41	3.522	23.15	18.31
	a	65.0099	—	8.37	20.6	—	39.9	—
$NaNO_2$	c	68.9953	—	−85.72	−68.02	—	24.8	—
	a	68.9953	—	−82.4	−70.3	—	43.5	−12.2
$NaNO_3$	c	84.9947	−110.248	−111.82	−87.73	4.115	27.85	22.20
	a	84.9947	—	−106.95	−89.20	—	49.1	−9.6
$NaNH_2$	c	39.0125	−27.84	−29.6	−15.3	2.842	18.38	15.81
$NaNH_3$	c	40.0205	—	−16.27	−2.94	—	37.2	—
Na_3PO_4	c	163.9408	−454.79	−458.27	−427.55	6.566	41.54	36.68
	a	163.9408	—	−477.5	−431.3	—	−11.	—
$(NaPO_3)_3$	c	305.8854	−865.9	−873.	−808.	10.720	68.47	62.00
$Na_4P_2O_7$	c	265.9026	−756.2	−762.	−709.7	10.180	64.60	57.63
$Na_5P_3O_{10}$								
form I, quenched	c	367.8644	−1043.57	−1051.4	−978.5	14.070	91.25	78.16
form II	c2	367.8644	−1045.76	−1054.0	−980.0	13.67	87.37	77.72
in 5,200 H_2O		367.8644	—	−1068.1	—	—	—	—
NaH_2PO_2	c	87.9784	—	−200.5	—	—	—	—
NaH_2PO_3	c	103.9778	—	−288.0	—	—	—	—
NaH_2PO_4	c	119.9772	−363.06	−367.3	−331.3	4.75	30.47	27.93
$NaH_2PO_4 \cdot H_2O$	c	137.9926	—	−438.1	—	—	—	—
Na_2HPO_3	c	125.9596	—	−336.8	—	—	—	—
Na_2HPO_4	c	141.9590	−413.92	−417.8	−384.4	5.646	35.97	32.34
$Na_2HPO_4 \cdot 2H_2O$	c	177.9898	—	−560.7	−499.2	—	52.9	—
$Na_2HPO_4 \cdot 7H_2O$	c	268.0668	—	−913.4	−784.0	—	103.87	—
$Na_2HPO_4 \cdot 12H_2O$	c	358.1438	—	−1266.2	−1068.0	—	151.49	—
$Na_2H_2P_2O_7$	c	221.9390	−654.06	−660.8	−602.9	8.184	52.63	47.36
$NaAsO_2$	c	129.9102	—	−157.87	—	—	—	
	a	129.9102	—	−159.93	−146.25	—	23.8	—
Na_3AsO_4	c	207.8886	—	−368.	—	—	—	—
	a	207.8886	—	−384.44	−342.78	—	3.4	—
Na_3BiO_4	c	341.9470	—	−291.	—	—	—	—
Na_2CO_3	c	105.9890	−268.76	−270.24	−249.64	4.959	32.26	26.84
	a	105.9890	—	−276.62	−251.36	—	14.6	—
$Na_2CO_3 \cdot H_2O$	c	124.0044	−338.87	−342.08	−307.22	6.296	40.18	34.80
$Na_2CO_3 \cdot 7H_2O$	c	232.0968	—	−764.81	−648.8	—	102.	—
$Na_2CO_3 \cdot 10H_2O$	c	286.1430	−959.62	−975.46	−819.36	21.21	134.8	131.53
$Na_2C_2O_4$	c	133.9996	—	−315.0	—	—	—	34.
HCOONa	c	68.0078	−158.19	−159.3	−143.4	3.767	24.80	19.76
$NaHCO_3$	c	84.0072	−225.15	−227.25	−203.4	3.81	24.3	20.94
$NaOCH_3$	c	54.0244	−85.41	−87.9	−70.46	3.374	26.43	16.60
$NaC_2H_3O_2$	c	82.0350	—	−169.41	−145.14	—	29.4	19.1
	a	82.0350	—	−173.55	−150.88	—	34.8	9.6
$NaC_2H_3O_2 \cdot 3H_2O$	c	136.0812	—	−383.2	−317.6	—	58.	—
$NaOC_2H_5$	c	68.0516	—	−98.90	—	—	—	—
	a	68.0516	—	−113.	—	—	—	—
CCl_3COONa								
sodium trichloroacetate	c	185.3700	—	−178.9	—	—	—	—
NaCN								
(C,I,cubic)	c	49.0077	−22.56	−20.91	−18.27	4.480	27.63	16.82
(c,II,orthorhombic)	c2	49.0077	—	−21.69	—	—	—	—

Substance			0 K	298.15 K (25°C)				
Formula and Description	State	Formula weight	ΔHf°_0 kcal/mol	ΔHf°	ΔGf° kcal/mol	$H^\circ_{298} - H^\circ_0$	S° cal/deg mol	C°_p
$NaCN \cdot 1/2H_2O$	c	58.0154	—	−56.35	—	—	—	—
$NaCN \cdot 2H_2O$	c	85.0385	—	−162.47	—	—	—	—
NaCNO	c	65.0071	—	−96.89	−85.6	—	23.1	20.7
	a	65.0071	—	−92.3	−85.9	—	39.6	—
NaCNS	c	81.0717	—	−40.75	—	—	—	—
	a	81.0717	—	−39.12	−40.44	—	48.6	1.5
Na_2SiO_3								
sodium metasilicate	c	122.0638	—	−371.63	−349.19	—	27.21	—
	gl	122.0638	—	−368.1	—	—	—	—
$Na_2SiO_3 \cdot 5H_2O$	c	212.1408	—	−728.6	—	—	—	—
$Na_2SiO_3 \cdot 9H_2O$	c	284.2024	—	−1010.7	—	—	—	—
$Na_2Si_2O_5$	c	182.1486	—	−589.8	−555.0	—	39.21	—
sodium disilicate, stable up to 951K [formerly β]								
	c2	182.1486	—	−589.22	—	—	—	—
stable 951K to m.pt.(1147K)[formerly α]								
unstable	c3	182.1486	—	−587.28	—	—	—	—
	gl	122.0638	—	−368.1	—	—	—	—
Na_4SiO_4	c	184.0428	—	—	—	—	46.76	—
$NaBO_2$	c	65.7996	−232.38	−233.5	−220.06	2.780	17.57	15.76
$NaBO_2 \cdot 2H_2O$	c	101.8304	—	−378.	−337.0	—	37.	—
$NaBO_2 \cdot 4H_2O$	c	137.8612	—	−520.	−451.3	—	55.	—
$NaBO_3 \cdot 4H_2O$	c	153.8606	—	−505.3	—	—	—	—
$Na_2B_4O_7$	c	201.2194	−782.36	−786.6	−740.0	7.262	45.30	44.64
	am	201.2194	—	−781.8	−735.4	—	46.1	—
	a	201.2194	—	—	−747.8	—	—	—
$Na_2B_4O_7 \cdot 4H_2O$	c	273.2810	—	−1077.3	—	—	—	—
$Na_2B_4O_7 \cdot 5H_2O$	c	291.2964	—	−1147.8	—	—	—	—
$Na_2B_4O_7 \cdot 10H_2O$								
borax	c	381.3734	—	−1503.0	−1318.5	—	140.	147.
$NaBH_4$	c	37.8328	−43.100	−45.08	−29.62	3.890	24.21	20.74
	a	37.8328	—	−45.88	−35.28	—	40.5	—
$NaBF_4$	c	109.7944	−440.04	−440.9	−418.30	5.191	34.73	28.74
$NaBF_4$	a	109.7944	—	−433.8	−418.0	—	58.	—
$NaAlO_2$	c	81.9701	—	−271.30	−256.06	—	16.90	17.52
	a	81.9701	—	−277.0	−259.4	—	9.	—
$NaAl(SO_4)_2 \cdot 12H_2O$ alum	c	458.2793	—	−1434.69	—	—	—	—
$NaAlSiO_4$								
nepheline, nephelite	c	142.0549	—	−500.2	−472.8	—	29.7	
$NaAlSi_2O_6$								
jadeite	c	202.1397	—	−724.4	−681.7	—	31.9	—
dehydrated analcite	c2	202.1397	—	−713.5	−673.8	—	41.9	39.30
NaHg	c	223.5798	—	−11.3	−9.76	—	25.	—
$NaHg_2$	c	424.1698	—	−18.2	−16.2	—	42.	—
$NaHg_4$	c	825.3498	—	−21.2	−17.7	—	73.	—
Na_3Hg	c	269.5594	—	−11.0	−10.84	—	54.	—
Na_3Hg_2	c	470.1494	—	−21.9	−20.7	—	69.	—
Na_5Hg_2	c	516.1290	—	—	−21.7	—	—	—
Na_7Hg_8	c	1765.6486	—	−84.4	−76.0	—	203.	—
$Na_3Fe(CN)_6$	a	280.9238	—	−37.9	−13.5	—	106.9	—
$Na_4Fe(CN)_6$	a	303.9136	—	−120.7	−84.28	—	79.1	—
$NaMnO_4$	a	141.9254	—	−186.8	−169.5	—	59.8	—
Na_2MnO_4	c	164.9152	—	−276.3	—	—	—	—
	a	164.9152	—	−271.	−244.9	—	42.	—
Na_2CrO_4	c	161.9732	−318.2	−320.8	−295.17	6.323	42.21	33.97
	a	161.9732	—	−325.38	−299.15	—	40.2	—
$Na_2Cr_2O_7$	c	261.9674	—	−472.9	—	—	—	—
	a	216.9674	—	−471.0	−436.2	—	90.8	—
Na_2MoO_4	c	205.9172	−348.63	−350.89	−323.71	6.070	38.17	33.87
$Na_2MoO_4 \cdot 2H_2O$	c	241.9480	—	−492.1	−437.46	—	57.5	—
Na_2WO_4	c	293.8272	−367.83	−370.2	−342.86	6.05	38.6	33.41
$Na_2W_2O_7$	c	525.6754	−571.44	−574.8	−529.84	9.36	60.8	51.36
$NaVO_3$	c	121.9300	−272.31	−273.85	−254.33	4.217	27.17	23.32
Na_3VO_4	c	183.9090	−417.36	−420.14	−391.45	7.093	45.4	39.40
$Na_4V_2O_7$	c	305.8390	−693.73	−697.62	−650.46	11.75	76.1	64.47
$NaNbO_3$	c	163.8940	—	−314.5	−294.7	—	28.	—
Na_2TiO_3	c	141.8778	−377.88	−380.3	−357.6	4.917	29.08	30.03
$Na_2Ti_2O_5$	c	221.7766	—	—	—	6.910	41.56	41.68
$NaUO_3$	c	309.0170	—	−360.	—	—	—	—
Na_2UO_4								
α form	c	348.0062	−450.02	−452.5	−424.90	6.268	39.68	35.05
β form	c2	348.0062	—	−450.2	—	—	—	—
Na_3UO_4	c	370.9960	−481.15	−484.0	−454.4	7.435	47.37	41.35
STRONTIUM								
Sr	c	87.62	0	0	0	—	12.5	6.3
	g	—	—	39.3	31.3	1.481	39.32	4.968
Sr^+	g	—	—	172.11	—	—	—	—
Sr^{2+}	g	—	—	427.96	—	—	—	—
std. state, m = 1	aq	—	—	−130.45	−133.71	—	−7.8	—
Sr^{3+}	g	—	—	1435.	—	—	—	—
Sr^{4+}	g	—	—	2751.	—	—	—	—
SrO	c	103.619	—	−141.5	−134.3	—	13.0	10.76
	g	—	—	−2.	—	—	—	—
SrO_2	c	119.619	—	−151.4	—	—	—	—
$SrO_2 \cdot 8H_2O$	c	263.742	—	−722.6	—	—	—	—
Sr_2O	c	191.239	—	−154.7	—	—	—	—
SrH	g	88.628	—	+52.	45.	2.080	50.80	7.17
SrH_2	c	89.636	—	−43.1	—	—	—	—
SrOH	g	104.627	—	−41.	—	—	—	—
$Sr(OH)_2$	c	121.635	—	−229.2	—	—	—	—
	g	—	—	−135.	—	—	—	—
in 800 H_2O	aq	—	—	−240.1	—	—	—	—
SrF	g	106.618	—	−69.0	−75.1	2.219	57.31	8.27
SrF_2	c	125.617	—	−290.7	−278.4	3.125	19.63	16.73
	g	—	—	−182.7	−185.3	3.192	69.54	12.66
SrCl	g	123.073	—	−30.2	−36.5	2.338	60.2	8.73
$SrCl_2$								
α	c	158.526	—	−198.1	−186.7	3.880	27.45	18.07
	g	—	—	−116.1	−118.6	3.454	74.26	13.33
$SrCl_2 \cdot H_2O$	c	176.541	—	−271.7	−247.7	—	41.	28.7
$SrCl_2 \cdot 2H_2O$	c	194.557	—	−343.7	−306.4	—	52.	38.3

SELECTED VALUES OF CHEMICAL THERMODYNAMIC PROPERTIES (Continued)

Substance			0 K	298.15 K (25°C)				
Formula and Description	State	Formula weight	ΔHf°_0 kcal/mol	ΔHf° kcal/mol	ΔGf° kcal/mol	$H^\circ_{298} - H^\circ_0$	S° cal/deg mol	C°_p cal/deg mol
$SrCl_2 \cdot 6H_2O$	c	266.618	—	−627.1	−535.67	—	93.4	—
$Sr(ClO_4)_2$	c	286.521	—	−182.31	—	—	—	—
std. state, m = 1	aq	—	—	−192.27	−137.83	—	79.2	—
∞ H_2O	aq	—	—	−192.27	—	—	—	—
$SrBr_2$	c	247.438	—	−171.5	−166.6	4.261	32.29	18.01
	g	—	—	−98.	−106.	3.8	77.3	14.5
std. state, m = 1	aq	—	—	−188.55	−183.41	—	31.6	—
∞ H_2O	aq	—	—	−188.55	—	—	—	—
$SrBr_2$	c	247.438	—	−171.5	−166.6	4.261	32.29	18.01
	g	—	—	−98.	−106	3.8	77.3	14.5
std. state, m = 1	aq	—	—	−188.55	−183.41	—	31.6	—
$Sr(IO_3)_2$	[illegible]	[illegible]		[illegible]	[illegible]		[illegible]	
$Sr(IO_3)_2 \cdot H_2O$	c	455.441	—	−313.2	−260.4	—	66.	—
$Sr(IO_3)_2 \cdot 6H_2O$	c	545.517	—	−666.8	−543.7	—	109.	—
SrS	c	119.684	—	−112.9	−111.8	—	16.3	11.64
$SrSO_3$	c	167.682	—	−281.3	—	—	—	—
$SrSO_4$	c	183.682	—	−347.3	−320.5	—	28.	—
precipitate	c	—	—	−346.5	—	—	—	—
std. state, m = 1	aq	—	—	−347.77	−311.68	—	−3.0	—
$SrSeO_3$	c	214.578	—	−250.4	—	—	—	—
$SrSeO_4$	c	230.578	—	−273.1	—	—	—	—
$Sr(N_3)_2$	c	171.660	—	+2.1	—	—	—	—
Sr_3N_2	c	290.873	—	−93.5	—	—	—	—
$Sr(NO_2)_2$	c	179.631	—	−182.3	—	—	—	—
In 800 H_2O	aq	—	—	−180.5	—	—	—	—
$Sr(NO_3)_2$	c	211.630	—	−233.80	−186.46	6.854	46.50	35.83
std. state, m = 1	aq	—	—	−229.57	−186.93	—	62.2	—
$Sr(NO_3)_2$								
in ∞ H_2O	aq	—	—	−229.57	—	—	—	—
Sr_3P_2	c	324.808	—	−152.	—	—	—	—
$Sr_3(PO_4)_2$	c	452.803	—	−985.4	—	—	—	—
$SrHPO_4$	c	183.599	—	−435.4	−403.6	—	29.	—
$Sr(H_2PO_4)_2$	c	281.595	—	−749.2	—	—	—	—
$Sr_3(AsO_4)_2$	c	540.698	—	−792.8	−736.2	—	61.	—
hydrated precipitate	—	—	—	−803.	—	—	—	—
$SrHAsO_4$	aq	227.547	—	−345.7	—	—	—	—
$Sr(H_2AsO_4)_2$	aq	369.490	—	−563.2	—	—	—	—
SrC_2	c	111.642	—	−18.	—	—	—	—
$SrCO_3$								
strontianite	c	147.629	—	−291.6	−272.5	—	23.2	19.46
std. state, m = 1	aq	—	—	−292.29	−259.88	—	−21.4	—
SrC_2O_4	c	175.640	—	−327.6	—	—	—	—
std. state, m = 1	aq	—	—	−327.6	−294.8	—	+2.1	—
$SrCN_2$	c	127.645	—	−72.5	—	—	—	—
$Sr(CN)_2$	aq	139.656	—	−57.0	—	—	—	—
$Sr(CN)_2 \cdot 4H_2O$	c	211.717	—	−333.7	—	—	—	—
$SrSiO_3$	c	163.704	—	−390.5	−370.4	—	23.1	21.16
Sr_2SiO_4	c	267.324	—	−550.8	−523.7	—	36.6	32.09
$SrMoO_3$	c	231.558	—	−306.	—	—	—	—
$SrMoO_4$	c	247.558	—	−370.	—	—	—	—
$SrWO_4$	c	335.468	—	−391.9	−366.	—	33.	—
$SrTiO_3$	c	183.518	—	−399.71	−379.64	—	26.0	23.51
Sr_2TiO_4	c	287.138	—	−546.7	−520.7	—	38.0	34.34
$SrZrO_3$	c	226.838	—	−422.4	−402.2	—	27.5	24.71
SULFUR								
S								
rhombic	c	32.064	0	0	0	1.054	7.60	5.41
monoclinic	c	—	—	0.08	—	—	—	—
	g	—	66.1	66.636	56.951	1.591	40.084	5.658
S^+	g	—	304.95	306.97	—	—	—	—
S^{2+}	g	—	844.8	848.2	—	—	—	—
S^{3+}	g	—	1653.0	1657.9	—	—	—	—
S^{4+}	g	—	2743.6	2750.0	—	—	—	—
S^{5+}	g	—	4415.	4422.	—	—	—	—
S^{6+}	g	—	6445.	6654.	—	—	—	—
S_2	g	64.128	30.647	30.68	18.96	2.141	54.51	7.76
S_2^{2-}								
std. state, m = 1	aq	—	—	7.2	19.0	—	6.8	—
S_3	g	96.192	—	31.7	—	—	—	—
S_4	g	128.256	—	32.7	—	—	—	—
S_5	g	160.320	—	29.6	—	—	—	—
S_6	g	192.384	—	24.5	—	—	—	—
SO	g	48.0634	1.5	1.496	−4.741	2.087	53.02	7.21
SO_2	liq	64.0628	—	−76.6	—	—	—	—
	g	—	−70.336	−70.944	−71.748	2.521	59.30	9.53
undissoc.; std. state, m = 1	aq	—	—	−77.194	−71.871	—	38.7	—
SO_2								
in 10,000 H_2O	aq	—	—	−80.584	—	—	—	—
SO_3								
I, β	c	80.0622	—	−108.63	−88.19	—	12.5	—
	liq	—	—	−105.41	−88.04	—	22.85	—
	g	—	−93.21	−94.58	−88.69	2.796	61.34	12.11
SO_3^{2-}								
std. state, m = 1	aq	—	—	−151.9	−116.3	—	−7.	—
SO_4^{2-}								
std. state, m = 1	aq	96.0616	—	−217.32	−177.97	—	4.8	−70.
$S_2O_3^{2-}$	aq	112.1262	—	−155.9	—	—	—	—
$S_2O_4^{2-}$								
std. state, m = 1	aq	128.1256	—	−180.1	−143.5	—	22.	—
$S_2O_7^{2-}$	aq	176.1238	—	−334.9	—	—	—	—
$S_2O_8^{2-}$								
std. state, m = 1	aq	192.1232	—	−320.0	−265.4	—	59.3	—
H_2S	g	34.0799	−4.232	−4.93	−8.02	2.379	49.16	8.18
std. state, m = 1	aq	—	—	−9.5	−6.66	—	29.	—
HSO_3^-								
std. state, m = 1	aq	81.0702	—	−149.67	−126.15	—	33.4	—
HSO_4^-								
std. state, m = 1	aq	97.0696	—	−212.08	−180.69	—	31.5	−20.
H_2SO_3								
undissoc.; std. state, m = 1	aq	82.0781	—	−145.51	−128.56	—	55.5	—
H_2SO_4	c	98.0775	−194.069	—	—	—	—	—

Substance: Formula and Description	State	Formula weight	0 K: ΔHf_0° kcal/mol	298.15 K (25°C): ΔHf° kcal/mol	ΔGf° kcal/mol	$H_{298}^\circ - H_0^\circ$ kcal/mol	S° cal/deg mol	C_p° cal/deg mol
	liq	—	—	−194.548	−164.938	6.748	37.501	33.20
std. state, m = 1	aq	—	—	−217.32	−177.97	—	4.8	−70.
H_2SO_4								
in ∞ H_2O	aq	—	—	−217.32	—	—	—	—
$H_2SO_4 \cdot 1H_2O$	liq	116.0929	—	−269.508	−227.182	—	50.56	51.35
$H_2SO_4 \cdot 2H_2O$	liq	134.1082	—	−341.085	−286.770	—	66.06	62.34
$H_2SO_4 \cdot 3H_2O$	liq	152.1236	—	−411.186	−345.178	—	82.55	76.23
$H_2SO_4 \cdot 4H_2O$	liq	170.1389	—	−480.688	−403.001	—	99.09	91.35
$H_2SO_4 \cdot 6.5\ H_2O$	liq	215.1772	—	−653.264	−546.403	—	140.51	136.30
$H_2S_2O_4$								
std. state, m = 1	aq	130.1415	—	—	−147.4	—	—	—
undissoc.								
$H_2S_2O_6$	aq	162.1403	—	−286.4	—	—	—	—
$H_2S_2O_7$	c	178.1397	—	−304.4	—	—	—	27.
$H_2S_2O_8$								
std. state, m = 1	aq	194.1391	—	−320.0	−265.4	—	59.3	—
SF_4	g	108.0576	−183.4	−185.2	−174.8	3.482	69.77	17.45
SF_6	g	146.0544	−285.7	−289.	−264.2	4.056	69.72	23.25
std. state, m = 1	aq	—	—	−293.0	−259.3	—	39.8	—
$SOCl_2$	liq	118.9694	—	−58.7	—	—	—	29.
	g	—	−50.07	−50.8	−47.4	3.559	74.01	15.9
in C_6H_6	—	—	—	−59.7	—	—	—	—
SO_2Cl_2	liq	134.9688	—	−94.2	—	—	—	32.
	g	—	−85.50	−87.0	−76.5	3.825	74.53	18.4
in C_6H_6	—	—	—	−94.2	—	—	—	—
TANTALUM								
Ta	c	180.948	0	0	0	1.347	9.92	6.06
	g	—	186.765	186.9	176.7	1.482	44.241	4.985
Ta^+	g	—	368.72	370.33	—	—	—	—
TaO	g	196.9474	60.3	60.	53.	2.10	57.6	7.31
TaO_2	g	212.9468	−40.7	−41.	−43.	3.1	64.	12.5
Ta_2O_5								
β	c	441.8930	—	−489.0	−456.8	—	34.2	32.30
Ta_2H	c	362.9040	−6.93	−7.8	−16.5	2.84	18.9	21.7
TaF_5	c	275.9400	—	−454.97	—	—	—	—
$TaCl_3$	c	287.307	—	−132.2	—	—	—	—
$TaCl_4$	c	322.760	—	−167.7	—	—	—	—
$TaCl_5$	c	358.213	—	−205.3	—	—	—	—
$TaBr_5$	c	580.493	—	−143.0	—	—	—	—
TaS_2	c	245.076	—	−111.	—	—	—	—
TaN	c	194.9547	—	−60.1	—	—	—	9.7
Ta_2N	c	375.9027	—	−65.	—	—	—	—
TaC	c	192.9592	−34.96	−35.0	−34.6	1.56	10.11	8.79
Ta_2C	c	373.9072	—	−51.0	−50.8	—	20.7	—
$TaSi_2$	c	237.120	—	−28.	—	—	—	—
Ta_5Si_3	c	988.998	—	−76.	—	—	—	—
TaB_2	c	202.570	—	−46.	—	—	—	—
TECHNETIUM								
Tc	c	98.906	0	0	0	—	—	—
	g	—	—	162.	—	1.481	43.25	4.97
Tc^+	g	—	—	331.	—	—	—	—
Tc_2O_7	c	309.8078	—	−266.	—	—	—	—
$HTcO_4$	c	163.9115	—	−167.	—	—	—	—
std. state, m = 1	aq	—	—	−173.	—	—	—	—
TELLURIUM								
Te	c	127.60	0	0	0	1.463	11.88	6.15
	g	—	47.	47.02	37.55	1.481	43.65	4.968
	amorp	—	—	2.7	—	—	—	—
Te^+	g	—	254.76	256.26	—	—	—	—
Te^{2+}	g	—	684.	687.	—	—	—	—
Te^{3+}	g	—	1390.	1394.	—	—	—	—
Te^{4+}	g	—	2262.	2268.	—	—	—	—
Te^{5+}	g	—	3652.	3659.	—	—	—	—
TeO	g	143.599	16.	15.6	9.2	2.093	57.7	7.19
TeO_2	c	159.599	—	−77.1	−64.6	—	19.0	—
H_2Te	g	129.616	—	23.8	—	—	—	—
H_2TeO_3								
std. state, m = 1	aq	177.614	—	—	−76.2	—	—	—
H_6TeO_6	c	229.644	—	−310.4	—	—	—	—
TeF_6	g	241.590	—	−315.	—	—	—	—
$TeCl_4$	c	269.412	—	−78.0	—	—	—	33.1
$TeBr_4$	c	447.236	—	−45.5	—	—	—	—
TERBIUM								
Tb	c	158.924	0	0	0	2.250	17.50	6.91
	g	—	93.36	92.9	83.6	1.79	48.63	5.87
Tb^+	g	—	228.3	229.3	—	—	—	—
Tb^{2+}	g	—	494.	496.	—	—	—	—
Tb^{3+}								
std. state, m = 1	aq	—	—	−163.2	−155.8	—	−54.	4.
TbO	g	174.923	−19.	—	—	—	—	—
TbO_2	c	190.923	—	−232.2	—	—	—	—
Tb_2O_3	c	365.846	—	−445.8	—	—	—	27.7
$TbCl_3$	c	265.283	—	−238.3	—	—	—	—
std. state, m = 1	aq	—	—	−283.0	−249.9	—	−14.	−94.
$TbCl_3$								
in ∞ H_2O	aq	—	—	−283.0	—	—	—	—
$TbCl_3 \cdot 6H_2O$	c	373.375	—	−683.4	−583.4	—	96.4	—
$Tb(BrO_3)_3$								
in 39.1 H_2O (satd)	aq	542.646	—	−214.1	—	—	—	—
$Tb(IO_3)_3$	c	683.632	—	−326.	—	—	—	—
TbC_2	g	182.946	212.	211.7	198.7	2.47	64.	10.5
$Tb_2(CO_3)_3$	c	497.876	—	−795.7	—	—	—	—
THALLIUM								
Tl	c	204.37	0	0	0	1.632	15.45	6.29
	g	—	43.701	43.55	35.24	1.481	43.225	4.968
in Hg	liq	—	—	0.076	−0.062	—	15.80	—
Tl^+	g	—	184.553	185.883	—	—	—	—
std. state, m = 1	aq	—	—	1.28	−7.74	—	30.0	—
Tl^{2+}	g	—	655.636	658.447	—	—	—	—
Tl^{3+}	g	—	1343.5	1347.8	—	—	—	—
std. state, m = 1	aq	—	—	47.0	51.3	—	−46.	—

Substance			0 K	298.15 K (25°C)				
Formula and Description	State	Formula weight	ΔHf_0° kcal/mol	ΔHf° kcal/mol	ΔGf° kcal/mol	$H_{298}^\circ - H_0^\circ$ kcal/mol	S° cal/deg mol	C_p° cal/deg mol
Tl_2O	c	424.7394	—	−42.7	−35.2	—	30.	—
Tl_2O_3	c	456.7382	—	—	−74.5	—	—	—
Tl_2O_4	c	472.7376	—	—	−83.0	—	—	—
TlOH	c	221.3774	—	−57.1	−46.8	—	21.	—
TlOH								
std. state, m = 1	aq	—	—	−53.69	−45.33	—	27.4	—
$Tl(OH)_3$	c	255.3922	—	—	−121.2	—	—	—
TlF	c	223.3684	—	−77.6	—	—	—	—
std. state, m = 1	aq	—	—	−78.22	−74.38	—	26.7	—
TlCl	c	239.8230	−49.091	−48.79	−44.20	3.030	26.59	12.17
std. state, m = 1	aq	—	—	−38.67	−39.11	—	43.5	—
[illegible]	[illegible]	[illegible]		[illegible]		[illegible]	[illegible]	[illegible]
$TlCl_3$								
std. state, m = 1, $Tl^{3+} + 3Cl^-$	aq	—	—	−72.9	−42.8	—	−5.5	—
$TlClO_3$								
std. state, m = 1	aq	287.8212	—	−22.4	−8.5	—	68.8	
TlBr	c	284.2790	—	−41.4	−40.00	—	28.8	—
std. state, m = 1	aq	—	—	−27.77	−32.59	—	49.7	—
$TlBr_3$	aq	444.0970	—	−59.7	—	—	—	—
std. state, m = 1	aq	—	—	−40.2	−23.2	—	13.	—
$TlBrO_3$	c	332.2772	—	−32.6	−12.70	—	40.3	—
std. state, m = 1	aq	—	—	−18.7	−7.3	—	69.0	—
TlI	c	331.2744	—	−29.6	−29.97	—	30.5	—
std. state, m = 1	aq	—	—	−11.91	−20.07	—	56.6	—
$TlIO_3$	c	379.2726	—	−63.9	−45.86	—	42.2	—
std. state,m = 1	aq	—	—	−51.6	−38.3	—	58.3	—
Tl_2S	c	440.8040	—	−23.2	−22.4	—	36.	—
Tl_2SO_4	c	504.8016	—	−222.7	−198.49	—	55.1	—
std. state, m = 1	aq	—	—	−214.76	−193.45	—	64.8	—
TlN_3	c	246.3901	—	55.8	70.38	—	35.1	—
$TlNO_3$	c	266.3749	—	−58.30	−36.44	—	38.4	23.78
undissoc.; std. state, m = 1	aq	—	—	−48.93	−34.85	—	64.5	—
Tl_2CO_3	c	468.7493	—	−167.3	−146.9	—	37.1	—
TlCNS	c	262.4518	—	6.8	9.21	—	39.	—
undissoc.; std. state, m = 1	aq	—	—	16.59	13.32	—	58.2	—
THORIUM								
Th	c	232.0381	0	0	0	1.556	12.76	6.53
	g	232.0381	143.08	143.0	133.26	1.481	45.420	4.97
Th^+	g	232.0381	—	283.1	—	—	—	—
Th^{2+}	g	232.0381	—	550.6	—	—	—	—
Th^{3+}	g	232.0381	—	1012.	—	—	—	—
Th^{4+}	g	232.0381	—	1677.	—	—	—	—
	a	232.0381	—	−183.8	−168.5	—	−101.0	—
ThO	g	248.0375	−5.5	−6.0	−12.0	2.109	57.350	7.47
ThO_2	c	264.0369	−292.01	−293.12	−279.35	2.523	15.590	14.76
ThH_2	c	234.0541	−31.4	−33.4	−23.9	1.616	12.120	8.77
ThF	g	251.0365	—	—	—	2.23	61.50	8.28
ThF^{3+}	a	251.0365	—	−264.5	−246.1	—	−71.7	—
ThF_4	c	308.0317	−499.24	−499.90	−477.30	5.114	33.950	26.420
	g	308.0317	−417.1	−418.0	−409.7	4.90	81.7	22.2
ThCl	g	267.4911	—	—	—	2.35	64.3	8.71
$ThCl_2$	g	302.9441	—	—	—	3.36	75.8	13.2
$ThCl_4$	c	373.8501	—	−283.6	−261.6	—	45.5	—
$ThCl_4 \cdot 2H_2O$	c	409.8809	—	−436.0	—	—	—	—
$ThCl_4 \cdot 4H_2O$	c	445.9117	—	−587.8	—	—	—	—
$ThCl_4 \cdot 7H_2O$	c	499.9579	—	−804.4	—	—	—	—
$ThOCl_2$	c	318.9435	—	−294.5	−276.3	—	29.5	—
$ThBr_4$	c	551.6741	—	−230.7	−221.6	—	55.	—
	g	551.6741	−175.26	−182.3	−187.5	6.23	103.	25.1
$ThBr_4 \cdot 7H_2O$	c	677.7819	—	−757.2	—	—	—	—
$ThBr_4 \cdot 10H_2O$	c	731.8281	—	−975.1	—	—	—	—
$ThOBr_2$	c	407.8555	—	−283.	—	—	—	—
ThI	g	358.9425	—	—	—	2.49	68.9	8.98
ThI_2	g	485.8469	—	—	—	3.69	85.0	13.7
ThI_3	g	612.7513	—	—	—	5.12	102.7	19.6
ThI_4	c	739.6557	—	−158.9	−156.6	—	61.	—
	g	739.6557	−108.89	−110.1	−123.1	6.65	112.	25.4
ThS	c	264.1021	—	−94.5	−93.4	—	16.68	—
ThS_2	c	296.1661	—	−149.7	−148.2	—	23.0	—
Th_2S_3	c	560.2682	—	−259.0	−257.4	—	43.	—
ThN	c	246.0448	−92.93	−93.5	−86.9	2.020	13.40	10.8
	g	246.0448	—	118.	—	—	—	—
Th_3N_4	c	752.1411	—	−314.3	−289.9	—	48.	—
$Th(NO_3)_4$	c	480.0577	—	−344.5	—	—	—	—
ThP	c	263.0119	—	−83.2	−81.54	—	17.0	—
	g	263.0119	128.5	128.	116.	2.29	63.9	8.5
Th_3P_4	c	820.0095	—	−273.0	−266.0	—	53.0	—
ThC	c	244.0493	—	−29.60	—	—	—	—
$ThC_{1.94}$	c	255.3398	−35.4	−35.	−35.3	2.447	16.37	13.55
ThC_2	g	256.0605	—	173.	—	—	—	—
THULIUM								
Tm	c	168.934	0	0	0	1.767	17.69	6.46
	g	—	55.786	55.5	47.2	1.481	45.412	4.968
Tm^+	g	—	198.3	199.5	—	—	—	—
Tm^{2+}	g	—	476.	479.	—	—	—	—
Tm^{3+}	g	—	1023.	1027.	—	—	—	—
std. state, m = 1	aq	—	—	−166.8	−158.2	—	−58.	6.
TmO	g	184.933	−19.	—	—	—	—	—
Tm_2O_3	c	385.866	−449.75	−451.4	−428.9	4.99	33.4	27.9
$TmCl_3$	c	275.293	—	−235.8	—	—	—	—
	g	—	—	−166.	—	—	—	—
std. state, m = 1	aq	—	—	−286.6	−252.3	—	−18.	−92.
$TmCl_3$								
in ∞ H_2O	aq	—	—	−286.6	—	—	—	—
$TmCl_3 \cdot 6H_2O$	c	383.385	—	—	—	—	95.5	—
TmOCl	c	220.386	—	−236.0	—	—	—	—
TmI_3	c	549.647	—	−143.8	—	—	—	—
$Tm(IO_3)_3$	c	693.642	—	−328.	—	—	—	—
TmC_2	c	192.956	—	−22.	—	—	—	—

SELECTED VALUES OF CHEMICAL THERMODYNAMIC PROPERTIES (Continued)

Substance			0 K	298.15 K (25°C)				
Formula and Description	State	Formula weight	ΔHf_0° kcal/mol	ΔHf° kcal/mol	ΔGf° kcal/mol	$H_{298}^\circ - H_0^\circ$ kcal/mol	S° cal/deg mol	C_p° cal/deg mol
TIN								
Sn								
I, white	c	118.69	0	0	0	1.505	12.32	6.45
II, gray	c	—	−0.371	−0.50	0.03	1.376	10.55	6.16
	g	—	72.18	72.2	63.9	1.485	40.243	5.081
Sn^{+}	g	—	241.54	243.04	—	—	—	—
Sn^{2+}	g	—	578.97	581.95	—	—	—	—
in aq HCl, std. state, m = 1	aq	—	—	−2.1	−6.5	—	−4.	—
Sn^{3+}	g	—	1282.9	1287.4	—	—	—	—
SnO	c	134.689	—	−68.3	−61.4	—	13.5	10.59
	g	—	—	—	—	2.117	55.45	7.55
SnO_2	c	150.689	—	−138.8	−124.2	—	12.5	12.57
SnH_4	g	122.722	41.78	38.9	45.0	2.669	54.39	11.70
$Sn(OH)_2$								
precipitated	c	152.705	—	−134.1	−117.5	—	37.	—
$Sn(OH)_4$								
precipitated	c	186.719	—	−265.3	—	—	—	—
$SnCl_2$	c	189.596	—	−77.7	—	—	—	—
un-ionized, in aq HCl; std. state, m = 1	aq	—	—	−78.8	−71.6	—	41.	—
$SnCl_2 \cdot 2H_2O$	c	225.627	—	−220.2	—	—	—	—
$SnCl_4$	liq	260.502	—	−122.2	−105.2	—	61.8	39.5
in aq HCl; std. state, m = 1	aq	—	—	−152.5	−124.9	—	26.	—
$SnBr_2$	c	278.508	—	−58.2	—	—	—	—
	aq	—	—	−56.8	—	—	—	—
$SnBr_4$	c	438.326	—	−90.2	−83.7	—	63.2	—
SnI_2	c	372.499	—	−34.3	—	—	—	—
SnI_4	c	626.308	—	—	—	—	—	20.3
SnS	c	150.754	—	−24.	−23.5	—	18.4	11.77
SnS_2	c	182.818	—	—	—	—	20.9	16.76
TITANIUM								
Ti	c	47.90	0	0	0	1.149	7.32	5.98
	g	—	111.65	112.3	101.6	1.802	43.066	5.839
Ti^{+}	g	—	268.9	271.1	—	—	—	—
Ti^{2+}	g	—	582.1	585.7	—	—	—	—
Ti^{3+}	g	—	1216.0	1221.1	—	—	—	—
Ti^{4+}	g	—	2213.8	2220.4	—	—	—	—
TiO								
α	c	63.899	−123.49	−124.2	−118.3	1.48	8.31	12.
TiO_2								
anatase	c	79.899	−223.44	−224.6	−211.4	2.062	11.93	13.26
Brookite	c	—	—	−225.1	—	—	—	—
rutile	c	—	−224.6	−225.8	−212.6	2.065	12.03	13.15
	amorp	—	—	−210.	—	—	—	—
Ti_2O_3	c	143.798	−361.52	−363.5	−342.8	3.431	18.83	23.27
TiH_2	c	49.916	−26.61	−28.6	−19.2	1.18	7.1	7.2
$TiCl_2$	c	118.806	−122.64	−122.8	−111.0	3.18	20.9	16.69
$TiCl_3$	c	154.259	−172.86	−172.3	−156.2	5.00	33.4	23.22
$TiCl_4$	c	189.712	−195.75	—	—	—	—	—
	liq	—	—	−192.2	−176.2	—	60.31	34.70
$TiBr_2$	c	207.718	—	−96.	—	—	—	—
$TiBr_3$	c	287.627	−126.65	−131.1	−125.2	5.49	42.2	24.31
$TiBr_4$	c	367.536	−141.36	−147.4	−140.9	6.825	58.2	31.43
	g	—	−124.14	−131.3	−135.8	5.71	95.2	24.1
TiI_2	c	301.709	—	−63.	—	—	—	—
TiI_4	c	555.518	—	−89.8	−88.8	—	59.6	30.03
	g	—	—	−66.4	—	—	—	—
TiS	c	79.964	—	−57.	—	—	—	—
TiS_2	c	112.028	—	—	—	—	18.73	16.23
TiN	c	61.907	−79.93	−80.8	−74.0	1.311	7.23	8.86
TiP	c	78.874	—	−67.6	—	—	—	—
TiC	c	59.911	−43.80	−44.1	−43.2	1.101	5.79	8.04
TiB_2	c	69.522	−77.0	−77.4	−76.4	1.333	6.81	10.58
TUNGSTEN								
W	c	183.85	0	0	0	1.190	7.80	5.80
	g	—	202.70	203.0	192.9	1.486	41.549	5.093
W^{+}	g	—	386.8	388.6	—	—	—	—
WO	g	199.849	—	108.	—	—	—	—
WO_2	c	215.849	−139.762	−140.94	−127.61	2.087	12.08	13.41
WO_3	c	231.848	−200.100	−201.45	−182.62	2.952	18.14	17.63
WO_4^{2-}								
std. state, m = 1	aq	247.848	—	−257.1	—	—	—	—
H_2WO_4	c	249.864	—	−270.5	—	—	—	—
WS_2	c	247.978	—	−50.	—	—	—	—
WC	c	195.861	—	−9.69	—	—	—	—
W_2C	c	379.711	—	−6.3	—	—	—	—
$W(CO)_6$	c	351.913	—	−227.9	—	—	—	—
Fe_7W_6	c	1494.029	—	—	—	—	—	77.7
$FeWO_4$	c	303.695	—	−276.	−252.	—	31.5	27.39
$MnWO_4$	c	302.786	—	−311.9	—	—	—	29.7
URANIUM								
U	c	238.0290	0	0	0	1.521	12.00	6.612
	g	238.0290	127.97	128.0	117.4	1.553	47.72	5.663
U^{+}	g	238.0290	270.81	272.32	—	—	—	—
U^{2+}	g	238.0290	520.	—	—	—	—	—
U^{3+}	g	238.0290	960.	—	—	—	—	—
	g	238.0290	—	−116.9	−113.6	—	−46.	—
U^{4+}	a	238.0290	—	−141.3	−126.9	—	−98.	—
UO	g	254.0284	—	5.	—	—	—	—
UO_2	c	270.0278	−258.40	−259.3	−246.6	2.696	18.41	15.20
	g	270.0278	−110.85	−111.3	−112.7	3.15	65.6	12.28
UO_2	a	270.0278	—	—	−230.1	—	—	—
UO_2^{2+}	a	270.0278	—	−243.7	−227.9	—	−23.3	—
UO_3								
γ, orthorhombic	c	286.0272	−291.35	−292.5	−273.9	3.486	22.97	19.52
ε form, triclinic, red	c2	286.0272	—	−291.0	—	—	—	—
	c3	286.0272	−289.96	−291.0	−272.6	3.596	23.76	19.57
α, orthorhombic, prev. described as hexagonal	c4	286.0272	−290.526	−291.65	−273.02	3.509	23.02	19.44

Substance			0 K	298.15 K (25°C)				
Formula and Description	State	Formula weight	ΔHf°_0 kcal/mol	ΔHf°	ΔGf° kcal/mol	$H^\circ_{298} - H^\circ_0$	S° cal/deg mol	C°_p
β, orthorhombic, orange-red;								
δ, cubic, dark red	c5	286.0272	—	−291.5	—	—	—	—
amorphous, orange	am	286.0272	—	−288.8	—	—	—	—
	g	286.0272	−195.	—	—	—	—	—
$UO_3 \cdot H_2O$								
β, orthorhombic	c	304.0426	—	−366.6	−333.4	—	30.	—
ε form, monoclinic	c2	304.0426	—	−366.0	—	—	—	—
α, transition to β 278.3k	c3	304.0426	—	−365.2	—	—	—	—
U_3O_7								
β, tetragonal	c	826.0828	−816.38	−819.1	−775.1	9.108	59.88	51.51
α, tetragonal	c2	826.0828	—	—	—	9.009	59.19	51.09
U_3O_8								
α, orthorhombic	c	842.0822	−851.75	−854.4	−805.4	10.216	67.54	56.97
UF	g	257.0274	−5.7	−6.	−13.	2.28	60	9.04
UF_2	g	276.0258	−134.6	−135.	−138.	3.28	71.	16.0
UF_3	c	295.0242	−360.3	−360.6	−344.2	4.392	29.50	22.73
UF_4								
monoclinic	c	314.0226	−458.75	−459.1	−437.4	5.390	36.25	27.73
UF_4	g	314.0226	−382.72	−383.7	−377.5	4.76	88.	21.8
	g	333.0210	−462.8	−464.	−452.	5.6	93.	26.2
UF_6	c	352.0194	−524.80	−525.1	−494.4	7.545	54.4	39.86
UOF_2	c	292.0252	—	−358.3	−341.5	—	28.5	—
UCl_3	c	344.3880	−207.61	−207.1	−191.	5.318	38.0	24.5
UCl_4	c	379.8410	−243.97	−243.6	−222.3	6.28	47.1	29.16
UCl_6	c	450.7470	−261.8	−261.	−230.	8.90	68.3	42.0
$UOCl_2$	c	324.9344	−254.83	−255.0	−238.1	4.586	33.06	22.72
UO_2Cl_2	c	340.9338	−296.67	−297.3	−274.0	5.157	35.98	25.78
UBr_3	c	477.7560	—	−167.1	−161.0	—	46.	26.0
UBr_4	c	557.6650	—	−191.8	−183.5	—	57.0	30.6
$UOBr_2$	c	413.8464	−229.27	−232.7	−222.2	4.989	37.66	23.42
UO_2Br_2	c	429.8458	—	−271.9	−254.9	—	40.5	—
UI_3	c	618.7422	—	−110.1	−109.9	—	53.	26.8
UI_4	c	745.6466	—	−122.4	−121.1	—	63.	32.1
US_2								
β	c	302.1570	−126.1	−126.	−125.8	3.698	26.39	17.84
UO_2SO_4								
β	c	366.0894	—	−441.0	−402.4	—	37.0	34.7
$U(SO_4)_2$	c	430.1522	—	−554.	—	—	—	—
UN	c	252.0357	−69.12	−69.5	−63.5	2.173	14.92	11.37
$UO_2(NO_3)_2$	c	394.0376	—	−322.5	−264.1	—	58.	—
UP	c	269.0028	−63.8	−64.	−63.	2.58	18.7	11.9
UP_2	c	299.9766	−72.6	−73.	−71.	3.679	24.3	19.12
U_3P_4	c	837.9822	−199.2	−200.	−196.	8.87	61.8	41.8
UC	c	250.0402	−23.90	−23.5	−23.7	2.176	14.15	11.98
U_2C_3	c	512.0916	−44.43	−43.4	−44.8	4.829	32.93	25.66
UO_2CO_3	c	330.0372	—	−404.2	−373.5	—	33.	—
USi	c	266.1150	—	−19.2	—	—	—	—
USi_2	c	294.2010	—	−31.2	—	—	—	—
USi_3	c	322.2870	—	−31.6	—	—	—	—
VANADIUM								
V	c	50.942	0	0	0	1.109	6.91	5.95
	g	—	122.12	122.90	180.32	1.8898	43.544	6.217
V^+	g	—	277.550	279.811	—	—	—	—
V^{2+}	g	—	615.39	619.13	—	—	—	—
V^{3+}	g	—	1292.69	1297.91	—	—	—	—
V^{4+}	g	—	2369.8	2376.5	—	—	—	—
VO	c	66.9414	—	−103.2	−96.6	—	9.3	10.86
VO^{2+}								
std. state	aq	—	—	−116.3	−106.7	—	−32.0	—
VO_2	g	82.9408	—	−57.	—	—	—	—
V_2O_3	c	149.8822	—	−291.3	−272.3	—	23.5	24.67
V_2O_4								
α	c	165.8816	—	−341.1	−315.1	—	24.5	27.96
V_2O_5	c	181.8810	—	−370.6	−339.3	—	31.3	30.51
V_3O_5	c	232.8230	—	−462.	−434.	—	39.	—
V_4O_7	c	315.7638	—	−631.	−587.	—	52.	—
V_6O_{13}	c	513.6442	—	−1062.	—	—	—	—
HVO_4^{2-}								
std. state	aq	115.9476	—	−277.0	−233.0	—	4.	—
$H_2VO_4^-$								
std. state	aq	116.9555	—	−280.6	−244.0	—	29.	—
$HV_2O_7^{3-}$								
std. state	aq	214.8878	—	—	−428.4	—	—	—
$H_3V_2O_7^-$								
std. state	aq	216.9037	—	—	−445.5	—	—	—
VF_3	c	107.9372	—	—	—	—	23.18	21.62
VF_4	c	126.9356	—	−335.4	—	—	—	—
VF_5	liq	145.9340	—	−353.8	−328.2	—	42.0	—
VCl_2	c	121.848	—	−108.	−97.	—	23.2	17.26
VCl_3	c	157.301	—	−138.8	−122.2	—	31.3	22.27
VCl_4	liq	192.754	—	−136.1	−120.4	—	61.	—
VOCl	c	102.3944	—	−145.	−133.	—	18.	—
VBr_2	c	210.760	—	−87.3	—	—	—	—
VBr_3	c	290.669	—	−103.6	—	—	—	—
VI_2	c	304.7508	—	−60.1	—	—	—	—
VI_3	c	431.6552	—	−64.7	—	—	—	—
VI_4	g	558.5596	—	−29.3	—	—	—	—
V_2S_3	c	198.076	—	−227.	—	—	—	—
VN	c	64.9487	—	−51.9	−45.7	—	8.91	9.08
XENON								
Xeg	g	131.30	0	0	0	1.481	40.5290	4.9679
std. state, m = 1	aq	—	—	−4.2	3.2	—	15.7	—
Xe^+	g	—	279.72	281.20	—	—	—	—
Xe^{2+}	g	—	768.8	771.8	—	—	—	—
Xe^{3+}	g	—	1509.6	1514.1	—	—	—	—
YTTERBIUM								
Yb	c	173.04	0	0	0	1.604	14.31	6.39
	g	—	36.479	36.4	28.3	1.481	41.352	4.968
Yb^+	g	—	180.70	182.05	—	—	—	—
Yb^{2+}	g	—	462.	464.8	—	—	—	—
std. state, m = 1	aq	—	—	—	−126.	—	—	—

SELECTED VALUES OF CHEMICAL THERMODYNAMIC PROPERTIES (Continued)

Substance: Formula and Description	State	Formula weight	0 K: ΔHf°_0 kcal/mol	298.15 K (25°C): ΔHf° kcal/mol	ΔGf° kcal/mol	$H^\circ_{298} - H^\circ_0$ kcal/mol	S° cal/deg mol	C°_p cal/deg mol
Yb^{3+}	g	—	1043.	1047.3	—	—	—	—
std. state, m = 1	aq	—	—	−161.2	−153.9	—	−57.	6.
Yb_2O_3	c	394.078	−432.08	−433.7	−412.7	4.69	31.8	27.57
YbH	g	174.048	—	—	—	2.083	52.66	7.178
$YbCl_2$	c	243.946	—	−191.1	—	—	—	—
$YbCl_3$	c	279.399	—	−229.4	—	—	—	—
std. state, m = 1	aq	—	—	−281.1	−248.0	—	−17.	−92.
$YbCl_3$								
in ∞ H_2O	aq	—	—	−281.1	—	—	—	—
$YbCl_3 \cdot 6H_2O$	c	387.491	—	−680.2	−580.6	—	94.6	81.6
$YbOCl$	c	224.492	—	−229.9	—	—	—	—
$Yb(IO_3)_3$	c	697.748	—	−322.	—	—	—	—
$Yb(NO_3)_3$								
in 200 H_2O	aq	359.055	—	−308.997	—	—	—	—
YbC_2	c	197.062	—	−17.9	−18.5	—	19.	—
YTTRIUM								
Y	c	88.905	0	0	0	1.426	10.62	6.34
	g	—	100.49	100.7	91.1	1.639	42.87	6.18
Y^+	g	—	247.62	—	—	—	—	—
Y^{2+}	g	—	529.88	—	—	—	—	—
Y^{3+}	g	—	1003.1	—	—	—	—	—
std. state, m = 1	aq	—	—	−172.9	−165.8	—	−60.	—
Y^{4+}	g	—	2428.	—	—	—	—	—
YO	g	104.9044	−9.	−9.3	−15.5	2.115	55.88	7.53
Y_2O	g	193.8094	2.	1.0	—	2.9	—	11.7
Y_2O_2	g	209.8088	−126.	−127.4	—	3.5	—	15.8
Y_2O_3	c	225.8082	−453.40	−455.38	−434.19	3.983	23.68	24.50
YH_2	c	90.9209	−51.95	−54.0	−44.3	1.403	9.18	8.24
YH_3	c	91.9289	−61.15	−64.0	−49.9	1.613	10.02	10.36
YF	g	107.9034	−32.7	−33.	−39.	2.163	55.38	7.92
YF_2	g	126.9018	—	—	—	2.9	69.3	11.7
YF_3	c	145.900	—	−410.8	−393.1	—	24.	—
YCl	g	124.358	48.	47.8	41.5	2.286	58.33	8.56
YCl^{2+}								
std. state, m = 1	aq	—	—	−214.0	−198.7	—	−46.	—
YCl_3	c	195.264	—	−239.0	—	—	—	—
$YCl_3 \cdot 6H_2O$	c	303.356	—	−691.3	−592.1	—	92.	—
YBr_2	g	248.723	—	—	—	3.4	80.0	13.2
YI_2	g	342.7138	—	—	—	3.5	84.0	13.5
YI_3	c	469.6182	—	−147.4	—	—	—	—
$Y(IO_3)_3$	c	613.6128	—	—	−271.2	—	—	—
YS	g	120.969	42.	41.7	29.7	2.2	58.	8.2
$Y_2(SO_4)_3$	c	465.995	—	—	−866.8	—	—	138.
YC_2	c	112.9274	—	−26.	−26.	—	13.	—
$Y_2(CO_3)_3$	c	357.8379	—	—	−752.4	—	—	—
ZINC								
Zn	c	65.37	0	0	0	1.350	9.95	6.07
	g	—	31.114	31.245	22.748	1.481	38.450	4.968
Zn^+	g	—	247.740	249.352	—	—	—	—
Zn^{2+}	g	—	662.00	665.09	—	—	—	—
std. state, m = 1	aq	—	—	−36.78	−35.14	—	−26.8	11.
ZnO	c	81.369	—	−83.24	−76.08	—	10.43	9.62
ZnO_2^{2-}								
std. state, m = 1	aq	97.369	—	—	−91.85	—	—	—
$HZnO_2^-$								
std. state, m = 1	aq	98.377	—	—	−109.26	—	—	—
$Zn(OH)_2$								
γ	c	99.385	—	—	−132.38	—	—	—
β	c	—	—	−153.42	−132.31	—	19.4	—
e	c	—	—	−153.74	−132.68	—	19.5	17.3
precipitated	—	—	—	−153.5	—	—	—	—
$Zn(OH)_2$								
std. state, m = 1	aq	—	—	−146.72	−110.33	—	−31.9	−60.
$Zn(OH)_4^{2-}$								
std. state, m = 1	aq	133.399	—	—	−205.23	—	—	—
ZnF_2	c	103.367	−182.06	−182.7	−170.5	2.821	17.61	15.69
$ZnCl_2$	c	136.276	−99.255	−99.20	−88.296	3.598	26.64	17.05
std. state, m = 1	aq	—	—	−116.68	−97.88	—	0.2	−54.
in ∞ H_2O	aq	—	—	−116.68	—	—	—	—
$Zn(ClO_4)_2$								
std. state, m = 1	aq	264.271	—	−98.60	−39.26	—	60.2	—
$Zn(ClO_4)_2 \cdot 6H_2O$	c	372.363	—	−509.89	—	—	—	—
$ZnBr_2$	c	225.188	—	−78.55	−74.60	—	33.1	—
std. state, m = 1	aq	—	—	−94.88	−84.84	—	12.6	−57.
$ZnBr_2 \cdot 2H_2O$	c	261.219	—	−224.0	−191.1	—	47.5	—
ZnI_2	c	319.179	—	−49.72	−49.94	—	38.5	—
std. state, m = 1	aq	—	—	−63.16	−59.80	—	25.2	−57.
$Zn(IO_3)_2$	c	415.175	—	—	−103.68	—	—	—
std. state, m = 1	aq	—	—	−142.6	−96.3	—	29.8	—
ZnS								
wurtzite	c	97.434	—	−46.04	—	—	—	—
sphalerite	c	—	—	−49.23	−48.11	—	13.8	11.0
$ZnSO_4$	c	161.432	—	−234.9	−209.0	—	28.6	—
$ZnSO_4$								
std. state, m = 1	aq	—	—	−254.10	−213.11	—	−22.0	−59.
in ∞ H_2O	aq	—	—	−254.10	—	—	—	—
$ZnSO_4 \cdot H_2O$	c	179.447	—	−311.78	−270.58	—	33.1	—
$ZnSO_4 \cdot 6H_2O$	c	269.524	—	−663.83	−555.64	—	86.9	85.49
$ZnSO_4 \cdot 7H_2O$	c	287.539	—	−735.60	−612.59	—	92.9	91.64
$Zn(N_3)_2$	c	149.410	—	52.	—	—	—	—
undissoc.; std. state, m = 1	aq	—	—	—	129.6	—	—	—
Zn_3N_2	c	224.123	—	−5.4	—	—	—	26.
$Zn(NO_3)_2$	c	189.380	—	−115.6	—	—	—	—
ionized; std. state, m = 1	aq	—	—	−135.90	−88.36	—	43.2	−30.
$Zn(NO_3)_2$								
in ∞ H_2O	aq	—	—	−135.90	—	—	—	—
$Zn(NO_3)_2 \cdot H_2O$	c	207.395	—	−192.4	—	—	—	—
$Zn(NO_3)_2 \cdot 2H_2O$	c	225.410	—	−265.36	—	—	—	—
$Zn(NO_3)_2 \cdot 4H_2O$	c	261.441	—	−406.10	—	—	—	—
$Zn(NO_3)_2 \cdot 6H_2O$	c	279.472	—	−551.30	−423.79	—	109.2	77.2

SELECTED VALUES OF CHEMICAL THERMODYNAMIC PROPERTIES (Continued)

Substance			0 K	298.15 K (25°C)				
Formula and Description	State	Formula weight	ΔHf°_0 kcal/mol	ΔHf°	ΔGf° kcal/mol	$H^\circ_{298} - H^\circ_0$	S° cal/deg mol	C°_p
Zn_3P_2	c	258.058	—	−113.	—	—	—	—
$Zn(PO_3)_2$	c	223.314	—	−497.9	—	—	—	—
$Zn_2(P_2O_7)$	c	304.683	—	−600.0	—	—	—	—
$Z_3(PO_4)_2$	c	386.053	—	−691.3	—	—	—	—
$ZnCO_3$	c	125.379	—	−194.26	−174.85	—	19.7	19.05
$ZnCO_3 \cdot H_2O$	c	143.395	—	—	−232.0	—	—	—
ZnC_2O_4								
std. state, m = 1	aq	153.390	—	−234.0	−196.2	—	−15.9	—
$Zn(CH_3)_2$	liq	95.440	—	5.6	—	—	—	—
$Zn(C_2H_5)_2$	liq	123.494	—	2.5	—	—	—	—
$Zn(CN)_2$	c	117.406	—	22.9	—	—	—	—
$ZnSiO_3$	c	141.454	—	−301.2	—	—	—	—
Zn_2SiO_4	c	222.824	—	−391.19	−364.06	—	31.4	29.48
ZIRCONIUM								
Zr								
α, hexagonal	c	91.22	0	0	0	1.322	9.32	6.06
	g	—	145.19	145.5	135.4	1.629	43.32	6.37
Zr^+	g	—	302.9	304.7	—	—	—	—
Zr^{2+}	g	—	605.7	609.0	—	—	—	—
Zr^{3+}	g	—	1135.9	1140.6	—	—	—	—
Zr^{4+}	g	—	1927.8	1934.0	—	—	—	—
ZrO	g	107.219	—	15.	—	—	—	—
ZrO_2								
α, monoclinic	c	123.219	−261.734	−263.04	−249.24	2.091	12.04	13.43
ZrO_2								
hydrated ppt	—	—	—	−260.4	—	—	—	—
ZrO_3								
ppt	c	139.218	—	−241.	—	—	—	—
ZrH_2								
zirconium hydride	c	93.236	−38.34	−40.4	−30.8	1.284	8.37	7.40
ZrF_4								
β, monoclinic	c	167.214	−455.44	−456.8	−432.6	4.183	25.00	24.79
ZrCl	c	126.673	—	−63.	—	—	—	—
$ZrCl_2$	c	162.126	—	−120.	—	—	—	—
$ZrCl_3$	c	197.579	—	−179.	—	—	—	—
$ZrCl_4$	c	233.032	−234.60	−234.35	−212.7	5.957	43.4	28.63
$ZrOCl_2$	aq	178.125	—	−280.3	—	—	—	—
$ZrBr_4$	c	410.856	—	−181.8	—	—	—	—
$ZrOBr_2$	aq	267.037	—	−259.9	—	—	—	—
ZrI_4	c	598.838	—	−115.1	—	—	—	—
ZrS_2	c	155.348	—	−135.3	—	—	—	—
$Zr(SO_4)_2$	c	283.343	—	−529.9	—	—	—	41.
$Zr(SO_4)_2 \cdot H_2O$	c	301.359	—	−610.4	—	—	—	—
$Zr(SO_4)_2 \cdot 4H_2O$	c	355.405	—	−825.6	—	—	—	—
ZrN	c	105.227	−86.42	−87.2	−80.4	1.575	9.29	9.66
ZrC	c	103.231	−48.33	−48.5	−47.7	1.401	7.96	9.06
ZrSi	c	119.306	—	−37	—	—	—	—
$ZrSi_2$	c	147.392	—	−38.	—	—	—	—
Zr_2Si	c	210.526	—	−50.	—	—	—	—
Zr_3Si	c	301.746	—	−52.	—	—	—	—
Zr_3Si_2	c	329.832	—	−92.	—	—	—	—
$ZrSiO_4$	c	183.304	−483.32	−486.0	−458.7	3.562	20.1	23.58
ZrB_2	c	112.842	−77.69	−78.0	−77.0	1.590	8.59	11.53

THERMODYNAMIC FUNCTIONS OF COPPER, SILVER AND GOLD

1 cal = 4.1840*J* H°_0 is the enthalpy of the solid at 0°K and 1 atm pressure

T °K	C°_p J/deg-mol			$H^\circ_T - H^\circ_0$ J/mol			$(H^\circ_T - H^\circ_0)/T$ J/deg-mol		
	Cu	Ag	Au	Cu	Ag	Au	Cu	Ag	Au
1.00	0.000743	0.000818	0.00118	0.000359	0.000367	0.000478	0.000359	0.000367	0.000478
2.00	0.00177	0.00265	0.00504	0.00158	0.00197	0.00326	0.000790	0.000987	0.00163
3.00	0.00337	0.00650	0.0141	0.00409	0.00633	0.0123	0.00136	0.00211	0.00410
4.00	0.00582	0.0134	0.0306	0.00860	0.0160	0.0340	0.00215	0.00399	0.00849
5.00	0.00943	0.0243	0.0570	0.0161	0.0344	0.0768	0.00322	0.00689	0.0154
6.00	0.0145	0.0403	0.0955	0.0279	0.0663	0.152	0.00466	0.0110	0.0253
7.00	0.0213	0.0626	0.149	0.0456	0.117	0.273	0.00652	0.0167	0.0390
8.00	0.0301	0.0927	0.220	0.0712	0.194	0.456	0.00889	0.0243	0.0570
9.00	0.0414	0.132	0.313	0.107	0.306	0.720	0.0119	0.0340	0.0800
10.00	0.0555	0.183	0.431	0.155	0.462	1.090	0.0155	0.0462	0.109
11.00	0.0727	0.247	0.577	0.219	0.676	1.592	0.0199	0.0614	0.145
12.00	0.0936	0.325	0.755	0.302	0.961	2.255	0.0251	0.0801	0.188
13.00	0.119	0.421	0.963	0.407	1.332	3.112	0.0313	0.102	0.239
14.00	0.149	0.535	1.203	0.541	1.809	4.193	0.0386	0.129	0.299
15.00	0.184	0.670	1.474	0.706	2.409	5.529	0.0471	0.161	0.369
16.00	0.225	0.826	1.772	0.910	3.155	7.149	0.0569	0.197	0.447
17.00	0.273	1.002	2.096	1.158	4.067	9.081	0.0681	0.239	0.534
18.00	0.328	1.199	2.442	1.458	5.166	11.35	0.0810	0.287	0.630
19.00	0.390	1.414	2.807	1.816	6.471	13.97	0.0956	0.341	0.735
20.00	0.462	1.647	3.187	2.242	8.001	16.97	0.112	0.400	0.848
25.00	0.963	3.066	5.245	5.703	19.62	37.97	0.228	0.785	1.519
30.00	1.693	4.774	7.375	12.25	39.14	69.53	0.408	1.305	2.318
35.00	2.638	6.612	9.395	22.99	67.58	111.5	0.657	1.931	3.186
40.00	3.740	8.419	11.22	38.89	105.2	163.2	0.972	2.630	4.079
45.00	4.928	10.11	12.86	60.54	151.6	223.4	1.345	3.368	4.965
50.00	6.154	11.66	14.29	88.23	206.1	291.4	1.765	4.121	5.828
55.00	7.385	13.04	15.52	122.1	267.9	366.0	2.220	4.871	6.654
60.00	8.595	14.27	16.59	162.0	336.2	446.3	2.701	5.604	7.438
65.00	9.759	15.35	17.51	208.0	410.4	531.6	3.199	6.313	8.179
70.00	10.86	16.30	18.31	259.5	489.5	621.2	3.708	6.993	8.874
75.00	11.89	17.14	19.01	316.4	573.2	714.6	4.219	7.642	9.528
80.00	12.85	17.87	19.63	378.4	660.7	811.2	4.729	8.259	10.14
85.00	13.74	18.53	20.17	444.9	751.8	910.7	5.234	8.844	10.71
90.00	14.56	19.11	20.64	515.7	845.9	1013.	5.730	9.399	11.25
95.00	15.31	19.63	21.06	590.4	942.8	1117.	6.215	9.924	11.76
100.00	16.01	20.10	21.44	668.7	1042.	1223.	6.687	10.42	12.23
105.00	16.64	20.52	21.77	750.3	1144.	1331.	7.146	10.89	12.68
110.00	17.22	20.89	22.06	835.0	1247.	1441.	7.591	11.34	13.10
115.00	17.76	21.23	22.33	922.5	1353.	1552.	8.021	11.76	13.49
120.00	18.25	21.54	22.56	1013.	1460.	1664.	8.438	12.16	13.87
125.00	18.70	21.82	22.78	1105.	1568.	1777.	8.839	12.54	14.22
130.00	19.12	22.07	22.97	1199.	1678.	1892.	9.227	12.91	14.55
135.00	19.51	22.31	23.15	1296.	1789.	2007.	9.601	13.25	14.87
140.00	19.87	22.52	23.31	1395.	1901.	2123.	9.961	13.58	15.17
145.00	20.20	22.72	23.45	1495.	2014.	2240.	10.31	13.89	15.45
150.00	20.51	22.90	23.59	1597.	2128.	2358.	10.64	14.19	15.72
155.00	20.79	23.07	23.70	1700.	2243.	2476.	10.97	14.47	15.97
160.00	21.05	23.22	23.81	1804.	2358.	2595.	11.28	14.74	16.22
165.00	21.30	23.37	23.91	1910.	2475.	2714.	11.58	15.00	16.45
170.00	21.53	23.50	24.00	2017.	2592.	2834.	11.87	15.25	16.67
175.00	21.74	23.63	24.08	2125.	2710.	2954.	12.15	15.49	16.88
180.00	21.94	23.75	24.15	2235.	2828.	3075.	12.42	15.71	17.08
185.00	22.13	23.86	24.22	2345.	2947.	3196.	12.68	15.93	17.27
190.00	22.31	23.96	24.29	2456.	3067.	3317.	12.93	16.14	17.46
195.00	22.47	24.06	24.35	2568.	3187.	3438.	13.17	16.34	17.63
200.00	22.63	24.16	24.41	2681.	3308.	3650.	13.40	16.54	17.80
205.00	22.77	24.24	24.48	2794.	3429.	3683.	13.63	16.72	17.96
210.00	22.91	24.33	24.54	2908.	3550.	3805.	13.85	16.90	18.12
215.00	23.04	24.41	24.60	3023.	3672.	3928.	14.06	17.08	18.27
220.00	23.17	24.49	24.65	3139.	3794.	4051.	14.27	17.25	18.41
225.00	23.28	24.56	24.71	3255.	3917.	4174.	14.47	17.41	18.55
230.00	23.39	24.63	24.76	3372.	4040.	4298.	14.66	17.56	18.69
235.00	23.50	24.69	24.82	3489.	4163.	4422.	14.85	17.71	18.82
240.00	23.60	24.76	24.87	3607.	4287.	4546.	15.03	17.86	18.94
245.00	23.69	24.82	24.92	3725.	4411.	4671.	15.20	18.00	19.06
250.00	23.78	24.88	24.97	3844.	4535.	4796.	15.37	18.14	19.18
255.00	23.86	24.93	25.02	3963.	4659.	4921.	15.54	18.27	19.30
260.00	23.94	24.99	25.07	4082.	4784.	5046.	15.70	18.40	19.41
265.00	24.02	25.04	25.12	4202.	4909.	5171.	15.86	18.53	19.51
270.00	24.09	25.09	25.17	4322.	5035.	5297.	16.01	18.65	19.62
273.15	24.13	25.12	25.20	4398.	5114.	5376.	16.10	18.72	19.68
275.00	24.15	25.14	25.21	4443.	5160.	5423.	16.16	18.76	19.72
280.00	24.22	25.19	25.26	4564.	5286.	5549.	16.30	18.88	19.82
285.00	24.28	25.24	25.31	4685.	5412.	5676.	16.44	18.99	19.91
290.00	24.34	25.28	25.35	4807.	5538.	5802.	16.57	19.10	20.01
295.00	24.40	25.32	25.39	4929.	5665.	5929.	16.71	19.20	20.10
298.15	24.44	25.35	25.42	5005.	5745.	6009.	16.79	19.27	20.15
300.00	24.46	25.37	25.43	5051.	5792.	6056.	16.84	19.31	20.19

THERMODYNAMIC FUNCTIONS OF COPPER, SILVER AND GOLD (Continued)

From NSRDS-NBS George T. Furukawa, William G. Saba and Martin L. Reilly

T °K	S°_T J/deg-mol			$-(G^\circ_T - H^\circ_0)$ J/mol			$-(G^\circ_T - H^\circ_0)/T$ J/deg-mol		
	Cu	Ag	Au	Cu	Ag	Au	Cu	Ag	Au
1.00	0.000711	0.000706	0.000880	0.000351	0.000339	0.000402	0.000351	0.000339	0.000402
2.00	0.00152	0.00175	0.00266	0.00145	0.00152	0.00206	0.000727	0.000762	0.00103
3.00	0.00251	0.00347	0.00620	0.00345	0.00406	0.00631	0.00115	0.00135	0.00210
4.00	0.00379	0.00619	0.0123	0.00657	0.00879	0.0153	0.00164	0.00220	0.00383
5.00	0.00546	0.0103	0.0218	0.0112	0.0169	0.0321	0.00223	0.00338	0.00641
6.00	0.00760	0.0160	0.0354	0.0176	0.0299	0.0603	0.00294	0.00498	0.0100
7.00	0.0103	0.0238	0.0539	0.0265	0.0496	0.104	0.00379	0.00709	0.0149
8.00	0.0137	0.0341	0.0782	0.0385	0.0783	0.170	0.00481	0.00979	0.0212
9.00	0.0179	0.0472	0.109	0.0542	0.119	0.263	0.00602	0.0132	0.0292
10.00	0.0229	0.0636	0.148	0.0746	0.174	0.391	0.00746	0.0174	0.0391
11.00	0.0290	0.0839	0.196	0.100	0.247	0.562	0.00913	0.0225	0.0511
12.00	0.0362	0.109	0.253	0.133	0.343	0.786	0.0111	0.0286	0.0655
13.00	0.0447	0.138	0.322	0.173	0.466	1.073	0.0133	0.0359	0.0825
14.00	0.0545	0.174	0.402	0.223	0.622	1.434	0.0159	0.0444	0.102
15.00	0.0660	0.215	0.494	0.283	0.815	1.880	0.0189	0.0544	0.125
16.00	0.0791	0.263	0.598	0.355	1.054	2.426	0.0222	0.0659	0.152
17.00	0.0941	0.318	0.715	0.442	1.344	3.081	0.0260	0.0790	0.181
18.00	0.111	0.381	0.845	0.544	1.693	3.861	0.0302	0.0940	0.214
19.00	0.131	0.452	0.987	0.665	2.109	4.775	0.0350	0.111	0.251
20.00	0.152	0.530	1.140	0.806	2.599	5.838	0.0403	0.130	0.292
25.00	0.305	1.043	2.069	1.917	6.446	13.76	0.0767	0.258	0.550
30.00	0.541	1.750	3.214	3.995	13.35	26.89	0.133	0.445	0.896
35.00	0.871	2.623	4.505	7.487	24.22	46.14	0.214	0.692	1.318
40.00	1.294	3.625	5.881	12.86	39.79	72.08	0.322	0.995	1.802
45.00	1.802	4.715	7.299	20.57	60.61	105.0	0.457	1.347	2.334
50.00	2.385	5.862	8.729	31.01	87.04	145.1	0.620	1.741	2.902
55.00	3.029	7.040	10.15	44.52	119.3	192.3	0.809	2.169	3.496
60.00	3.724	8.228	11.55	61.38	157.5	246.5	1.023	2.624	4.109
65.00	4.458	9.414	12.91	81.82	201.6	307.7	1.259	3.101	4.734
70.00	5.222	10.59	14.24	106.0	251.6	375.6	1.514	3.594	5.366
75.00	6.007	11.74	15.53	134.1	307.4	450.0	1.788	4.099	6.001
80.00	6.806	12.87	16.78	166.1	368.9	530.8	2.076	4.612	6.635
85.00	7.612	13.97	17.98	202.1	436.1	617.7	2.378	5.130	7.267
90.00	8.421	15.05	19.15	242.2	508.6	710.6	2.691	5.652	7.895
95.00	9.229	16.10	20.28	286.4	586.5	809.1	3.014	6.174	8.517
100.00	10.03	17.12	21.37	334.5	669.6	913.3	3.345	6.696	9.133
105.00	10.83	18.11	22.42	386.7	757.7	1023.	3.683	7.216	9.740
110.00	11.62	19.07	23.44	442.8	850.6	1137.	4.025	7.733	10.34
115.00	12.39	20.01	24.43	502.8	948.3	1257.	4.372	8.246	10.93
120.00	13.16	20.92	25.38	566.7	1051.	1382.	4.723	8.755	11.51
125.00	13.91	21.80	26.31	634.4	1157.	1511.	5.075	9.260	12.09
130.00	14.66	22.66	27.20	705.8	1269.	1645.	5.429	9.759	12.65
135.00	15.39	23.50	28.07	780.9	1384.	1783.	5.785	10.25	13.21
140.00	16.10	24.32	28.92	859.7	1504.	1925.	6.140	10.74	13.75
145.00	16.80	25.11	29.74	941.9	1627.	2072.	6.496	11.22	14.29
150.00	17.49	25.88	30.54	1028.	1755.	2223.	6.851	11.70	14.82
155.00	18.17	26.64	31.31	1117.	1886.	2377.	7.206	12.17	15.34
160.00	18.84	27.37	32.07	1209.	2021.	2536.	7.559	12.63	15.85
165.00	19.49	28.09	32.80	1305.	2160.	2698.	7.910	13.09	16.35
170.00	20.13	28.79	33.52	1404.	2302.	2864.	8.260	13.54	16.85
175.00	20.75	29.47	34.21	1506.	2448.	3033.	8.608	13.99	17.33
180.00	21.37	30.14	34.89	1612.	2597.	3206.	8.954	14.43	17.81
185.00	21.97	30.79	35.55	1720.	2749.	3382.	9.298	14.86	18.28
190.00	22.57	31.43	36.20	1831.	2904.	3561.	9.639	15.29	18.74
195.00	23.15	32.05	36.83	1946.	3063.	3744.	9.978	15.71	19.20
200.00	23.72	32.66	37.45	2063.	3225.	3930.	10.31	16.12	19.65
205.00	24.28	33.26	38.05	2183.	3390.	4118.	10.65	16.54	20.09
210.00	24.83	33.85	38.64	2306.	3558.	4310.	10.98	16.94	20.52
215.00	25.37	34.42	39.22	2431.	3728.	4505.	11.31	17.34	20.95
220.00	25.90	34.98	39.79	2559.	3902.	4702.	11.63	17.74	21.37
225.00	26.42	35.53	40.34	2690.	4078.	4903.	11.96	18.12	21.79
230.00	26.94	36.07	40.89	2824.	4257.	5106.	12.28	18.51	22.20
235.00	27.44	36.60	41.42	2960.	4439.	5312.	12.59	18.89	22.60
240.00	27.94	37.12	41.94	3098.	4623.	5520.	12.91	19.26	23.00
245.00	28.42	37.63	42.46	3239.	4810.	5731.	13.22	19.63	23.39
250.00	28.90	38.14	42.96	3382.	4999.	5945.	13.53	20.00	23.78
255.00	29.37	38.63	43.46	3528.	5191.	6161.	13.83	20.36	24.16
260.00	29.84	39.11	43.94	3676.	5386.	6379.	14.14	20.71	24.53
265.00	30.30	39.59	44.42	3826.	5582.	6600.	14.44	21.07	24.91
270.00	30.75	40.06	44.89	3979.	5782.	6823.	14.74	21.41	25.27
273.15	31.02	40.35	45.18	4076.	5908.	6965.	14.92	21.63	25.50
275.00	31.19	40.52	45.35	4134.	5983.	7049.	15.03	21.76	25.63
280.00	31.62	40.97	45.81	4291.	6187.	7277.	15.32	22.10	25.99
285.00	32.05	41.42	46.25	4450.	6393.	7507.	15.61	22.43	26.34
290.00	32.48	41.86	46.69	4611.	6601.	7739.	15.90	22.76	26.69
295.00	32.89	42.29	47.13	4775.	6811.	7974.	16.19	23.09	27.03
298.15	33.15	42.56	47.40	4879.	6945.	8123.	16.36	23.29	27.24
300.00	33.30	42.72	47.56	4940.	7024.	8211.	16.47	23.41	27.37

VALUES OF CHEMICAL THERMODYNAMIC PROPERTIES OF HYDROCARBONS

The values in this table are for the ideal gas state at 298.15 K. The units for ΔHf°, ΔFf°, and $Log_{10} Kf$ are Kcal/g mol. The units for absolute entropy, S°, are cal/°K g mol.

It is frequently possible to calculate values for compounds not listed since the following increments are known for an addition of a methylene group, CH_2, to the following types of compounds:

Normal alkyl cyclohexanes
Normal alkyl benzenes
Normal alkyl cyclopentanes
Normal monoolefins (1-alkenes)
Normal acetylenes (1-alkynes)

For each of the above types of compounds the increments per CH_2 group are

ΔHf°:	−4.926 kcal/g mol
ΔFf°:	−2.048 kcal/g mol
$Log_{10} Kf$:	−1.5012 kcal/g mol
S°:	−9.183 cal/deg g mol

Relationships to SI units – The symbols cal mole^{-1} deg^{-1} and gibbs/mol are identical and refer to units of calories per degree-mole. These units can be converted to SI units of joules per degree-mole by multiplying the tabulated values by 4.184. Similarly, values in kilocalories per mole can be converted to joules per mole by multiplying with the factor 4184. For further discussions of the SI system and for conversions from other units, see *Pure and Applied Chemistry*, 21, 1, 1970.

Formula	Compound	ΔHf°	ΔFf°	$Log_{10} Kf$	S°
CH_4	Methane	−17.889	−12.140	8.8985	44.50
C_2H_2	Ethyne (acetylene)	54.194	50.000	−36.6490	47.997
C_2H_4	Ethene (ethylene)	12.496	16.282	−11.9345	52.54
C_2H_6	Ethane	−20.236	− 7.860	5.7613	54.85
C_3H_4	Propadiene (allene)	45.92	48.37	−35.4519	58.30
C_3H_4	Propyne (methyl-acetylene)	44.319	46.313	−33.9469	59.30
C_3H_6	Propene (propylene)	4.879	14.990	−10.9875	63.80
C_3H_8	Propane	−24.820	− 5.614	4.1150	64.51
C_4H_6	1,2-Butadiene	39.55	48.21	−35.3377	70.03
C_4H_6	1,3-Butadiene	26.75	36.43	−26.7004	66.62
C_4H_6	1-Butyne (ethyl acetylene)	39.70	48.52	−35.5616	69.51
C_4H_6	2-Butyne (dimethylacetylene)	35.374	44.725	−32.7823	67.71
C_4H_8	1-Butene	0.280	17.217	−12.6199	73.48
C_4H_8	cis-2-Butene	−1.362	16.046	−11.7618	71.90
C_4H_8	trans-2-Butene	−2.405	15.315	−11.2255	70.86
C_4H_8	2-Methylpropane (isobutene)	− 3.343	14.582	−10.6888	70.17
C_4H_{10}	*n*-Butane	−29.812	− 3.754	2.7516	74.10
C_4H_{10}	2-Methylpropane (isobutane)	−31.452	− 4.296	3.1489	70.42
C_5H_8	1-Pentyne	34.50	50.17	−36.7712	79.10
C_5H_8	2-Pentyne	30.80	46.41	− 34.0177	79.30
C_5H_8	3-Methyl-1-butyne	32.60	49.12	−36.0061	76.23
C_5H_8	1,2-Pentadiene	34.80	50.29	−36.861	79.7
C_5H_8	cis-1,3-Pentadiene (cis-piperylene)	18.70	34.88	−25.563	77.5
C_5H_8	trans-1,3-Pentadiene (trans-piperylene)	18.60	35.07	− 25.707	76.4
C_5H_8	1,4-Pentadiene	25.20	40.69	−29.824	79.7
C_5H_8	2,3-Pentadiene	33.10	49.22	36.074	77.6
C_5H_8	3-Methyl-1,2-butadiene	31.00	47.47	− 34.657	76.4
C_5H_8	2-Methyl-1,3-butadiene (isoprene)	18.10	34.87	25.560	75.44
C_5H_{10}	1-Pentene	− 5.000	18.787	−13.7704	83.08
C_5H_{10}	cis-2-Pentene	− 6.710	17.173	12.5874	82.76
C_5H_{10}	trans-2-Pentene	− 7.590	16.575	−12.1495	81.81
C_5H_{10}	2-Methyl-1-butene	− 8.680	15.509	−11.3680	81.73
C_5H_{10}	3-Methyl-1-butene	− 6.920	17.874	−13.1017	79.70
C_5H_{10}	2-Methyl-2-butene	−10.170	14.267	−10.4572	80.90
C_5H_{10}	Cyclopentane	−18.46	9.23	− 6.7643	70.00
C_5H_{12}	*n*-Pentane	−35.00	− 1.96	1.4366	83.27
C_6H_6	Benzene	19.820	30.989	−22.7143	64.34
C_6H_{10}	1-Hexyne	29.55	52.19	−38.258	88.27
C_6H_{12}	1-Hexene	− 9.96	20.80	−15.2491	92.25
C_6H_{12}	cis-2-Hexene	−11.56	19.18	−14.0549	92.35
C_6H_{12}	trans-2-Hexene	−12.56	18.46	−13.5291	91.40
C_6H_{12}	cis-3-Hexene	−11.56	19.66	−14.4094	90.73
C_6H_{12}	trans-3-Hexene	−12.56	18.86	−13.8262	90.04
C_6H_{12}	2-Methyl-1-pentene	−13.56	17.48	−12.8135	91.32
C_6H_{12}	3-Methyl-1-pentene	−11.02	20.28	−14.8655	90.45
C_6H_{12}	4-Methyl-1-pentene	−11.66	19.90	−14.5865	89.58
C_6H_{12}	2-Methyl-2-pentene	−14.96	16.34	−11.9780	90.45
C_6H_{12}	cis-3-Methyl-2-pentene	−14.32	16.98	−12.4471	90.45
C_6H_{12}	trans-3-Methyl-2-pentene	−14.32	16.74	−12.2697	91.26
C_6H_{12}	cis-4-Methyl-2-pentene	−13.26	18.40	−13.4903	89.23
C_6H_{12}	trans-4-Methyl-2-pentene	−14.26	17.77	−13.0216	88.02
C_6H_{12}	2-Ethyl-1-butene	−12.92	18.51	−13.5690	90.01
C_6H_{12}	2,3-Dimethyl-1-butene	−14.78	17.43	−12.7782	89.39
C_6H_{12}	3,3-Dimethyl-1-butene	−14.25	19.04	−13.9578	83.79
C_6H_{12}	2,3-Dimethyl-2-butene	−15.91	16.52	−12.1073	86.67
C_6H_{12}	Methylcyclopentane	−25.50	8.55	− 6.2649	81.24
C_6H_{12}	Cyclohexane	−29.43	7.59	− 5.5605	71.28
C_6H_{14}	*n*-Hexane	−39.96	0.05	0.037	92.45

Formula	Compound	ΔHf°	ΔFf°	Log_{10} Kf	S°
C_7H_8	Methylbenzene (toluene)	11.950	29.228	−21.4236	76.42
C_7H_{12}	1-Heptyne	24.62	54.24	−39.759	97.25
C_7H_{12}	1-Heptene	−14.85	22.84	−16.742	101.43
C_7H_{14}	Ethylcyclopentane	−30.37	10.59	− 7.7632	90.62
C_7H_{14}	1,1-Dimethylcyclopentane	−33.05	9.33	− 6.8372	85.87
C_7H_{14}	1,cis-2-Dimethylcyclopentane	−30.96	10.93	− 8.0107	87.51
C_7H_{14}	1,trans-2-Dimethylcyclopentane	−32.67	9.17	− 6.7224	87.67
C_7H_{14}	1,cis-3-Dimethylcyclopentane	−31.93	9.91	− 7.2648	87.67
C_7H_{14}	1,trans-3-Dimethylcyclopentane	−32.47	9.37	− 6.8690	87.67
C_7H_{14}	Methylcyclohexane	−36.99	6.52	− 4.7819	82.06
C_7H_{16}	*n*-Heptane	−44.89	2.09	− 1.532	101.64
C_8H_8	Ethenylbenzene (styrene)	35.22	51.10	−37.4532	82.48
C_8H_{10}	Ethylbenzene	7.120	31.208	−22.8750	86.15
C_8H_{10}	1,2-Dimethylbenzene (*o*-xylene)	4.540	29.177	−21.3860	84.31
C_8H_{10}	1,3-Dimethylbenzene (*m*-xylene)	4.120	28.405	−20.8202	85.49
C_8H_{10}	1,4-Dimethylbenzene (*p*-xylene)	4.290	28.952	−21.2214	84.23
C_8H_{14}	1-Octyne	19.70	56.29	−41.260	106.63
C_8H_{16}	1-Octene	−19.82	24.89	−18.244	110.61
C_8H_{16}	*n*-Propylcyclopentane	−35.39	12.54	− 9.195	99.80
C_8H_{16}	1,1-Dimethylcyclohexane	−43.26	8.42	− 6.174	87.24
C_8H_{16}	cis-1,2-Dimethylcyclohexane	−41.15	9.85	− 7.225	89.51
C_8H_{16}	trans-1,2-Dimethylcyclohexane	−43.02	8.24	− 6.038	88.65
C_8H_{16}	cis-1,3-Dimethylcyclohexane	−44.16	7.13	− 5.228	88.54
C_8H_{16}	trans-1,3-Dimethylcyclohexane	−42.20	8.68	− 6.363	89.92
C_8H_{16}	cis-1,4-Dimethylcyclohexane	−42.22	9.07	− 6.650	88.54
C_8H_{16}	trans-1,4-Dimethylcyclohexane	−44.12	7.58	− 5.552	87.19
C_8H_{18}	*n*-Octane	−49.82	4.14	− 3.035	110.82
C_8H_{18}	2-Methylheptane	−51.50	3.06	− 2.243	108.81
C_8H_{18}	3-Methylheptane	−50.82	3.29	− 2.412	110.32
C_8H_{18}	4-Methylheptane	−50.69	4.00	− 2.932	108.35
C_8H_{18}	3-Ethylhexane	−50.40	3.95	− 2.895	109.51
C_8H_{18}	2,2-Dimethylhexane	−53.71	2.56	− 1.876	103.06
C_8H_{18}	2,3-Dimethylhexane	−51.13	4.23	− 3.101	106.11
C_8H_{18}	2,4-Dimethylhexane	−52.44	2.80	− 2.052	106.51
C_8H_{18}	2,5-Dimethylhexane	−53.21	2.50	− 1.832	104.93
C_8H_{18}	3,3-Dimethylhexane	−52.61	3.17	− 2.324	104.70
C_8H_{18}	3,4-Dimethylhexane	−50.91	4.14	− 3.035	107.15
C_8H_{18}	2-Methyl-3-ethylpentane	−50.48	5.08	− 3.724	105.43
C_8H_{18}	3-Methyl-3-ethylpentane	−51.38	4.76	− 3.489	103.48
C_8H_{18}	2,2,3-Trimethylpentane	−52.61	4.09	− 2.998	101.62
C_8H_{18}	2,2,4-Trimethylpentane	−53.57	3.13	− 2.294	101.62
C_8H_{18}	2,3,3-Trimethylpentane	−51.73	4.52	− 3.313	103.14
C_8H_{18}	2,3,4-Trimethylpentane	−51.97	4.32	− 3.167	102.99
C_8H_{18}	2,2,3,3-Tetramethylbutane	−53.99	4.88	− 3.577	94.34
C_9H_{10}	Isopropenylbenzene (α-methylstyrene; 2-phenyl-1-propene)	27.00	49.84	−36.531	91.70
C_9H_{10}	1-Methyl-2-ethenylbenzene (*o*-Methylstyrene)	28.30	51.14	−37.484	91.70
C_9H_{10}	1-Methyl-3-ethenylbenzene (*m*-methylstyrene)	27.60	50.02	−36.665	93.1
C_9H_{10}	1-Methyl-4-ethenylbenzene (*p*-methylstyrene)	27.40	50.24	−36.825	91.7
C_9H_{12}	*n*-Propylbenzene	1.870	32.810	−24.049	95.74
C_9H_{12}	Isopropylbenzene (Cumene)	0.940	32.738	−23.996	92.87
C_9H_{12}	1,3,5-Trimethylbenzene (Mesitylene)	− 3.840	28.172	−20.6497	92.15
C_9H_{16}	1-Nonyne	14.77	58.34	−42.761	115.82
C_9H_{18}	1-Nonene	−24.74	26.94	−19.747	119.80
C_9H_{18}	*n*-Butylcyclopentane	−40.22	14.69	−10.768	108.99
C_9H_{20}	*n*-Nonane	−54.74	6.18	− 4.536	120.00
$C_{10}H_{14}$	*n*-Butylbenzene	− 3.30	34.62	−25.374	104.91
$C_{10}H_{18}$	1-Decyne	9.85	60.39	−44.262	125.00
$C_{10}H_{20}$	1-Decene	−29.67	28.99	−21.249	128.98
$C_{10}H_{22}$	*n*-Decane	−59.67	8.23	6.037	129.19
$C_{11}H_{22}$	1-Undecene	−34.60	31.03	−22.745	138.16
$C_{11}H_{24}$	*n*-Undecane	−64.60	10.28	− 7.539	138.37
$C_{12}H_{24}$	1-Dodecene	−39.52	33.08	−24.297	147.34

From Rossini, F. D., Pitzer, K. S., Arnett, R. L., Braun, R. M., and Pimentel, G. C., *Selected Values of Physical and Thermodynamic Properties of Hydrocarbons and Related Compounds,* Carnegie Press, Pittsburgh, 1953.

KEY VALUES FOR THERMODYNAMICS

The following table was prepared from data published in CODATA Bulletin 28 (April, 1978) which is a report of the CODATA Task Group on Key Values of Thermodynamics. A bibliography, references citing methods for calculations and measurements, bases for selection of data and limitations of the data are presented in Bulletin 28. One may contact CODATA by writing to CODATA Secretariat, 51 Boulevard de Montmorency, 75016 Paris, France.

The recommended values were derived from a reconsideration of all the tentative data published in CODATA Special Report 4, March 1977 ("Tentative set of key values for thermodynamics. Part VI"), which took place at a meeting of the Task Group held in Lund, Sweden, in August 1977. It was decided that values of $\Delta_f H°$ (298.15 K) for NO^-_3 (aq) and all values for Cd(g), Ge(g) and U(g) should remain tentative for the present, that P_4O_{10}(cr) and PbO(cr, yellow) should be dropped from the program, and that the remaining values given in CODATA Special Report 4 should advance to recommended status, albeit with some minor numerical adjustments.

In the following tables, the species are listed in the "Standard Order of Arrangement", as used in most modern compendia of thermodynamic data. The reference state for each element at 298.15 K is the thermodynamically stable standard state except for phosphorus, for which the "white" crystal modification has been selected, as it is the most reproducible; there is a subtlety concerning the reference state for tin, where the "white" form (thermodynamcally stable at 298.15 K) is taken as the reference state at all temperatures down to zero, even though that form is known to be metastable below 286 K.

The usual definitions of standard states have been adopted. For crystalline solids (cr) and liquids (l) the standard state is that of the pure substance (in a stated crystallographic modification, where appropriate) under a pressure of 101 325 Pa. For gases (g), the standard state is that of the ideal gas at a pressure of 101 325 Pa. For species in aqueous solution (aq), the standard state is the hypothetical ideal solution at unit activity (molality scale); the properties of ideal aqueous ionic solutions are taken equal to the sum of the properties of the individual ions. It should be noted that values of H° (298.15 K) − H° (0) for gases relate to the hypothetical ideal-gas state at zero temperature, while values of H° (298.15 K) − H° (0) for both liquids and crystalline solids relate to crystalline solids at zero temperature.

Relative atomic masses were taken from the recommendations of the IUPAC Commission on Atomic Weights, 1970. Data in the tables relate to the natural mixture of isotopic species; nuclear spin contributions have been ignored. The joule table is the primary table; values in the calorie table were derived from the corresponding values in joules by dividing by 4.184, with retention of sufficient decimal places to assure accurate reconversion.

For consistency with earlier work, the following values of the fundamental constants were employed in the calculations: gas constant, R = (8.314 33 ± 0.000 80) $J \cdot K^{-1} \cdot mol^{-1}$; Faraday constant, F = (96 487.0 ± 1.0) $J \cdot V^{-1} \cdot mol^{-1}$; constant relating wave number and energy, $N_A hc$ = (0.119 625 6 ± 0.000 002 6) $J \cdot m \cdot mol^{-1}$. These values differ very slightly from those recommended by CODATA in 1973 (CODATA Bulletin 11, December 1973). However, adoption of the 1973 values of the fundamental constants would change the values in the present tables by far less than their assigned uncertainties, so the thermodynamic data where reported may be said to be consistent with the 1973 set of fundamental constants.

CODATA RECOMMENDED KEY VALUES FOR THERMODYNAMICS, 1977
(To convert joules to calories multiply by 0.239006)

Substance	State	$\Delta_f H°$ (298.15 K) $kJ \cdot mol^{-1}$	S° (298.15 K) $J \cdot K^{-1} \cdot mol^{-1}$	H° (298.15 K) − H° (0) $kJ \cdot mol^{-1}$
O	g	249.17 ± 0.10	160.946 ± 0.020	6.728 ± 0.003
O_2	g	0	205.037 ± 0.033	8.682 ± 0.004
H	g	217.997 ± 0.006	114.604 ± 0.015	6.197 ± 0.002
H^+	aq	0	0	—
H_2	g	0	130.570 ± 0.033	8.468 ± 0.003
OH^-	aq	−230.025 ± 0.045	−10.71 ± 0.20	—
H_2O	l	−285.830 ± 0.042	69.950 ± 0.080	13.293 ± 0.020
H_2O	g	−241.814 ± 0.042	188.724 ± 0.040	9.908 ± 0.008
He	g	0	126.039 ± 0.012	6.197 ± 0.002
Ne	g	0	146.214 ± 0.016	6.197 ± 0.002
Ar	g	0	154.732 ± 0.020	6.197 ± 0.002
Kr	g	0	163.971 ± 0.020	6.197 ± 0.002
Xe	g	0	169.573 ± 0.020	6.197 ± 0.002
F	g	79.39 ± 0.30	158.640 ± 0.020	6.518 ± 0.004
F^-	aq	−335.35 ± 0.65	−13.18 ± 0.54	—
F_2	g	0	202.685 ± 0.040	8.825 ± 0.004
HF	g	−273.30 ± 0.70	173.655 ± 0.035	8.599 ± 0.004
Cl	g	121.302 ± 0.008	165.076 ± 0.020	6.272 ± 0.003
Cl^-	aq	−167.080 ± 0.088	56.73 ± 0.16	—
Cl_2	g	0	222.965 ± 0.040	9.180 ± 0.008
HCl	g	−92.31 ± 0.13	186.786 ± 0.033	8.640 ± 0.004
Br	g	111.86 ± 0.12	174.904 ± 0.020	6.197 ± 0.020
Br^-	aq	−121.50 ± 0.15	82.84 ± 0.20	—
Br_2	l	0	152.210 ± 0.040	24.52 ± 0.13
Br_2	g	30.91 ± 0.11	245.350 ± 0.054	9.724 ± 0.012
HBr	g	−36.38 ± 0.17	198.585 ± 0.033	8.648 ± 0.004
I	g	106.762 ± 0.040	180.673 ± 0.020	6.197 ± 0.002
I^-	aq	−56.90 ± 0.84	106.70 ± 0.20	—
I_2	cr	0	116.139 ± 0.080	13.196 ± 0.040
I_2	g	62.421 ± 0.080	260.567 ± 0.063	10.117 ± 0.021
HI	g	26.36 ± 0.80	206.480 ± 0.040	8.657 ± 0.006
S	cr, rhombic	0	32.054 ± 0.050	4.412 ± 0.060
S	g	276.98 ± 0.25	167.715 ± 0.035	6.657 ± 0.004
S_2	g	128.49 ± 0.30	228.055 ± 0.050	9.131 ± 0.008
SO_2	g	−296.81 ± 0	248.11 ± 0.06	10.548 ± 0.013
SO_4^{-2}	aq	−909.60 ± 0.40	18.83 ± 0.50	—
N	g	472.68 ± 0.40	153.189 ± 0.020	6.197 ± 0.002
N_2	g	0	191.502 ± 0.025	8.669 ± 0.003
NO_3^-	aq	—	146.94 ± 0.85	—
NH_3	g	−45.94 ± 0.35	192.67 ± 0.08	10.046 ± 0.008

CODATA RECOMMENDED KEY VALUES FOR THERMODYNAMICS, 1977
(To convert joules to calories multiply by 0.239006) (continued)

Substance	State	$\Delta_f H°$ (298.15 K) kJ·mol⁻¹	S° (298.15 K) J·K⁻¹·mol⁻¹	H° (298.15 K) − H° (0) kJ·mol⁻¹
NH_4^+	aq	−133.26 ± 0.25	111.17 ± 0.75	—
P	cr, white	0	41.09 ± 0.25	5.360 ± 0.015
P	g	316.5 ± 1.0	163.085 ± 0.020	6.197 ± 0.002
P_2	g	144.0 ± 2.0	218.01 ± 0.04	8.903 ± 0.0008
P_4	g	58.9 ± 0.3	279.9 ± 0.5	14.10 ± 0.24
C	cr	0	5.74 ± 0.12	1.050 ± 0.020
C	g	716.67 ± 0.44	157.988 ± 0.020	6.535 ± 0.006
CO	g	−110.53 ± 0.17	197.556 ± 0.032	8.673 ± 0.008
CO_2	g	−393.51 ± 0.13	213.677 ± 0.040	9.364 ± 0.008
Si	cr	0	18.81 ± 0.08	3.217 ± 0.008
Si	g	450 ± 8	167.870 ± 0.035	7.550 ± 0.004
SiO_2	cr, α-quartz	−910.7 ± 1.0	41.46 ± 0.20	6.916 ± 0.020
SiF_4	g	−1614.95 ± 0.85	282.65 ± 0.40	15.36 ± 0.05
Ge	Cr, cubic	0	31.09 ± 0.13	4.636 ± 0.015
GeO_2	cr, tetrag	−580.2 ± 1.2	39.71 ± 0.15	7.230 ± 0.020
GeF_4	g	−1190.15 ± 0.50	301.8 ± 1.0	17.30 ± 0.080
Sn	cr, white	0	51.18 ± 0.08	6.323 ± 0.008
Sn	g	301.2 ± 1.7	168.380 ± 0.020	6.215 ± 0.002
Sn^{+2}	aq	−8.9 ± 0.8	−15.8 ± 4.0	—
SnO	cr	−285.93 ± 0.70	57.17 ± 0.30	8.736 ± 0.022
SnO_2	cr	−580.78 ± 0.40	52.3 ± 1.2	8.76 ± 0.08
Pb	cr	0	64.80 ± 0.30	6.870 ± 0.020
Pb	g	195.20 ± 0.80	175.270 ± 0.020	6.197 ± 0.002
Pb^{+2}	aq	0.92 ± 0.25	17.7 ± 0.8	—
$PbSO_4$	cr	−919.94 ± 0.90	148.49 ± 0.40	20.050 ± 0.040
B	cr	0	5.90 ± 0.08	1.222 ± 0.008
B	g	560 ± 12	153.325 ± 0.035	6.315 ± 0.004
B_2O_3	cr	−1273.5 ± 1.4	53.97 ± 0.30	9.301 ± 0.040
BF_3	g	−1135.95 ± 0.80	254.31 ± 0.10	11.650 ± 0.020
Al	cr	0	28.35 ± 0.08	4.565 ± 0.010
Al	g	329.7 ± 4.0	164.440 ± 0.030	6.919 ± 0.004
Al_2O_3	cr, α-corundum	−1675.7 ± 1.3	50.92 ± 0.10	10.016 ± 0.020
AlF_3	cr	−1510.4 ± 1.3	66.5 ± 0.4	11.62 ± 0.04
Zn	cr	0	41.63 ± 0.13	5.657 ± 0.020
Zn	g	130.42 ± 0.20	160.875 ± 0.025	6.197 ± 0.002
Zn^{+2}	aq	−153.39 ± 0.20	−109.6 ± 0.7	—
ZnO	cr	−350.46 ± 0.27	43.64 ± 0.40	6.933 ± 0.040
Cd	cr	0	51.80 ± 0.15	6.247 ± 0.004
Cd^{+2}	aq	−75.88 ± 0.60	−72.8 ± 1.2	—
CdO	cr	−258.1 ± 0.8	54.8 ± 1.7	8.41 ± 0.08
$CdSO_4 \cdot 8/3H_2O$	cr	−1729.55 ± 0.80	229.66 ± 0.40	35.56 ± 0.04
Hg	l	0	75.90 ± 0.12	9.342 ± 0.008
Hg	g	61.38 ± 0.04	174.860 ± 0.020	6.197 ± 0.002
Hg^{+2}	aq	170.16 ± 0.20	−36.32 ± 0.80	—
Hg_2^{+2}	aq	166.82 ± 0.20	65.52 ± 0.80	—
HgO	cr, red	−90.83 ± 0.12	70.25 ± 0.30	9.117 ± 0.025
Hg_2Cl_2	cr	−265.45 ± 0.30	191.6 ± 1.5	23.25 ± 0.20
Hg_2SO_4	cr	−743.41 ± 0.50	200.71 ± 0.20	26.070 ± 0.030
Cu	cr	0	33.15 ± 0.08	5.004 ± 0.008
Cu	g	337.6 ± 1.2	166.285 ± 0.025	6.197 ± 0.002
Cu^{+2}	aq	65.69 ± 0.80	−97.1 ± 1.2	—
$CuSO_4$	cr	−771.1 ± 1.2	109.2 ± 0.4	16.86 ± 0.08
Ag	cr	0	42.55 ± 0.21	5.745 ± 0.020
Ag	g	284.9 ± 0.8	172.883 ± 0.025	6.197 ± 0.002
Ag^+	aq	105.750 ± 0.085	73.38 ± 0.40	—
AgCl	cr	−127.070 ± 0.085	96.23 ± 0.20	12.033 ± 0.040
Th	cr	0	53.39 ± 0.40	6.510 ± 0.020
Th	g	598 ± 6	190.06 ± 0.40	6.197 ± 0.002
ThO_2	cr	−1226.4 ± 3.5	65.23 ± 0.20	10.560 ± 0.020
U	cr	0	50.20 ± 0.20	6.364 ± 0.020
UO_2	cr	−1085.0 ± 1.0	77.03 ± 0.20	11.280 ± 0.020
UO_2^{+2}	aq	−1019.2 ± 2.5	−98.3 ± 4.0	—
UO_3	cr, gamma	−1223.8 ± 2.0	96.11 ± 0.40	14.585 ± 0.080
U_3O_8	cr	−3574.8 ± 2.5	282.55 ± 0.50	42.74 ± 0.10
Be	cr	0	9.50 ± 0.08	1.950 ± 0.020
Be	g	324 ± 5	136.165 ± 0.020	6.197 ± 0.002
BeO	cr	−609.4 ± 2.5	13.77 ± 0.04	2.837 ± 0.008
Mg	cr	0	32.68 ± 0.10	5.000 ± 0.030
Mg	g	147.10 ± 0.80	148.535 ± 0.020	6.197 ± 0.002
MgO	cr	−601.5 ± 0.3	26.95 ± 0.15	5.160 ± 0.020
MgF_2	cr	−1124.2 ± 1.2	57.2 ± 0.4	9.92 ± 0.06

Substance	State	$\Delta_f H°$ (298.15 K) kJ·mol⁻¹	S° (298.15 K) J·K⁻¹·mol⁻¹	H° (298.15 K) − H° (0) kJ·mol⁻¹
Ca	cr	0	41.6 ± 0.4	5.73 ± 0.04
Ca	g	177.8 ± 0.8	154.775 ± 0.020	6.197 ± 0.002
Ca^{+2}	aq	−543.10 ± 0.80	−56.4 ± 0.4	—
CaO	cr	−635.09 ± 0.90	38.1 ± 0.4	6.75 ± 0.06
Li	cr	0	29.12 ± 0.20	4.632 ± 0.040
Li^+	aq	−278.455 ± 0.090	11.30 ± 0.35	—
Na	cr	0	51.30 ± 0.20	6.460 ± 0.020
Na^+	aq	−240.300 ± 0.065	58.41 ± 0.20	—
K	cr	0	64.68 ± 0.20	7.088 ± 0.020
K^+	aq	−252.17 ± 0.10	101.04 ± 0.25	—
Rb	cr	0	76.78 ± 0.30	7.489 ± 0.020
Rb^+	aq	−251.12 ± 0.13	120.46 ± 0.40	—
Cs	cr	0	85.23 ± 0.40	7.711 ± 0.020
Cs^+	aq	−258.04 ± 0.13	132.84 ± 0.40	—

LATTICE ENERGIES

H. D. B. Jenkins

Table 1 contains calculated values of lattice energies, U_{Pot}, of crystalline salts M_aX_b. U_{Pot} is expressed in the units of kilojoules per mole, KJmol−¹. M and X can be complex or simple ions. Also cited is the lattice energy obtained from the Born-Fajans-Haber Cycle, $U^{BFHC}{}_{Pot}$ using thermochemical data published in U.S. Government publications plus certain other data which are located at the end of this table. The values quoted are of variable reliability and a full discussion of the values is to appear in a review by Jenkins and Waddington currently (1978) nearing completion.

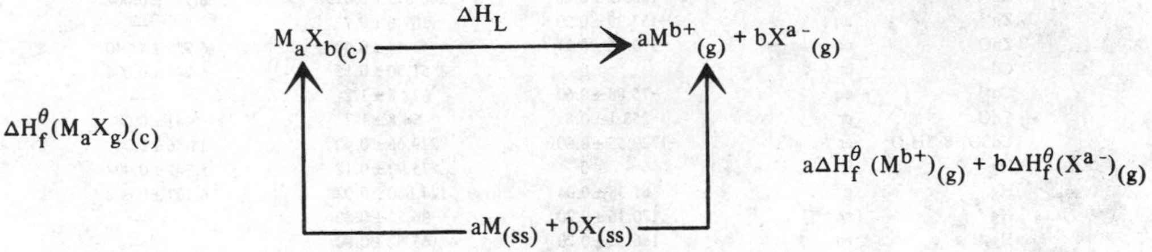

where, (ss) is the standard state of the ion or element

$$\Delta H_L = U_{Pot}(M_aX_b) + \left[a\left(\frac{n_{M^{b+}}}{2} - 2\right) + b\left(\frac{n_{X^{a-}}}{2} - 2\right)\right]RT$$

$$\Delta H_L = a\Delta H_f^\theta(M^{b+})(g) + b\Delta H_f^\theta(X^{a-})(g) - \Delta H_f^\theta(M_aX_b)(c)$$

Where n_{M^b+}, n_{X^a-} is equal to 3 for monatomic ions, 5 for linear polyatomic ions, and 6 for polyatomic nonlinear ions.

The data listed in Table 2 were employed in the calculation of the Born-Haber Cycle values in Table 1 and are not listed in Technical Notes 270 of the National Bureau of Standards.

Table 1
LATTICE ENERGIES

Salt	Calculated lattice energy ($kJmol^{-1}$)	Literature source	Thermochemical cycle lattice energy ($kJmol^{-1}$)
Acetates			
$Li(CH_3COO)$	—	—	881
$Na(CH_3COO)$	761	Morris (1959)	763
$K(CH_3COO)$	686	Morris (1959)	682
$Rb(CH_3COO)$	715	Yatsimirskii (1956)	656
$Cs(CH_3COO)$	682	Yatsimirskii (1956)	—
$NH_4(CH_3COO)$	725	Yatsimirskii (1956)	695
$Ag(CH_3COO)$	—	—	863
$Tl(CH_3COO)$	—	—	750
$Ca(CH_3COO)_2$	2431	Yatsimirskii (1956)	2294
$Sr(CH_3COO)_2$	2280	Yatsimirskii (1956)	2166
$Ba(CH_3COO)_2$	2180	Yatsimirskii (1956)	2033
$Mn(CH_3COO)_2$	2548	Yatsimirskii (1956)	2616
$Cu(CH_3COO)_2$	—	—	2835
$Zn(CH_3COO)_2$	2615	Yatsimirskii (1956)	2750
$Hg(CH_3COO)_2$	2368	Yatsimirskii (1961)	2595
$Pb(CH_3COO)_2$	2247	Yatsimirskii (1961)	2225
Acetylides			
CaC_2	2911	Vinek, Neckel, Nowotny (1967)	2904
SrC_2	2788	Vinek, Neckel, Nowotny (1967)	2784
BaC_2	2647	Vinek, Neckel, Nowotny (1967)	2654
Ammonium salts			
$NH_4(CH_3COO)$	725	Yatsimirskii (1956)	656
NH_4HF_2	705	Jenkins, Pratt (1977)	705
NH_4HCO_3	—	—	741
NH_4HSO_4	640	Yatsimirskii (1956)	(645)
NH_4BF_4	582	Ladd, Lee (1961)	—
NH_4NCO	724	Waddington (1959)	—
NH_4CN	617	Ladd (1970)	670
NH_4HCO_2	715	Morris (1959)	—
$NH_4(NH_2CH_2CO_2)$	650	Bernard, Businot, Decker (1974)	—
NH_4IO_3	—	—	808
$NH_4IO_2F_2$	678	Finch, Gates, Jenkinson (1972)	685
NH_4HS	661	Yatsimirskii (1956)	666
NH_4NO_3	661	Morris (1958)	676
NH_4ClO_4	583	Jenkins, Pratt (1978)	580
$(NH_4)_2S$	2008	Karapet'yants (1954)	(2026)
$(NH_4)_2SO_4$	1766	Jenkins, Smith (1975)	1777
$(NH_4)_2GeF_6$	1657	Jenkins, Pratt (1978)	—
$(NH_4)_2IrCl_6$	1442	Jenkins, Pratt (1978)	1440
$(NH_4)_2OsCl_6$	1433	Jenkins, Pratt (1978)	—
$(NH_4)_2PdCl_6$	1481	Jenkins, Pratt (1978)	—
$(NH_4)_2PbCl_6$	1355	Jenkins, Pratt (1978)	—
$(NH_4)_2PbCl_6$	1468	Jenkins, Pratt (1978)	—
$(NH_4)_2ReCl_6$	1402	Jenkins, Pratt (1978)	1390
$(NH_4)_2SiF_6$	1657	Jenkins, Pratt (1978)	1727
$(NH_4)_2SnCl_6$	1370	Jenkins, Pratt (1978)	1334
$(NH_4)_2SnBr_6$	1319	Jenkins, Pratt (1978)	—
$(NH_4)_2TeCl_6$	1318	Jenkins, Pratt (1978)	—
$(NH_4)_2TeBr_6$	1294	Jenkins, Pratt (1978)	—
$(NH_4)_2TiCl_6$	1413	Jenkins, Pratt (1978)	—
NH_4CNS	605	Gill, Singla, Paul, Narula (1972)	611
$(NH_4)_2SeCl_6$	1420	Jenkins, Pratt (1978)	—
$(NH_4)_2SeBr_6$	1380	Jenkins, Pratt (1978)	—
Arsenates			
$Mg_3(AsO_4)_2$	10669	Grekenschikov (1967)	10716
$Ca_3(AsO_4)_2$	9749	Grekenschikov (1967)	9653
$Sr_3(AsO_4)_2$	9330	Grekenschikov (1967)	9266
$Ba_3(AsO_4)_2$	8870	Grekenschikov (1967)	8985
$AlAsO_4$	7255	Grekenschikov (1967)	—
$GaAsO_4$	7243	Grekenschikov (1967)	—
Astatides			
LiAt	720	Ladd, Lee (1961)	—
NaAt	657	Ladd, Lee (1961)	—
KAt	615	Ladd, Lee (1961)	—
RbAt	594	Ladd, Lee (1961)	—
CsAt	586	Ladd, Lee (1961)	—
FrAt	573	Ladd, Lee (1961)	—
Azides			
LiN_3	812	Jenkins, Pratt (1977)	818
NaN_3	732	Jenkins, Pratt (1977)	731
KN_3	659	Jenkins, Pratt (1977)	658
RbN_3	637	Jenkins, Pratt (1977)	632
CsN_3	612	Jenkins, Pratt (1977)	604
AgN_3	854	Jenkins, Pratt (1977)	—
TlN_3	689	Jenkins, Pratt (1977)	686
$Ca(N_3)_2$	2186	Jenkins, Pratt (1977)	2162
$Sr(N_3)_2$	2056	Jenkins, Pratt (1977)	2066
$Ba(N_3)_2$	2021	Gora (1971)	1965
$Mn(N_3)_2$	2408	Jenkins, Pratt (1977)	2416
$Cu(N_3)_2$	2730	Jenkins, Pratt (1977)	2738
$Zn(N_3)_2$	2840	Jenkins, Pratt (1977)	2848
$Cd(N_3)_2$	2446	Jenkins, Pratt (1977)	2454
$Pb(N_3)_2$	—	Jenkins, Pratt (1977)	2173
Bihalide salts			
$LiHF_2$	893	Jenkins, Pratt (1977)	866
$NaHF_2$	788	Jenkins, Pratt (1977)	788

Table 1 (continued)
LATTICE ENERGIES

Salt	Calculated lattice energy (kJmol^{-1})	Literature source	Thermochemical cycle lattice energy (kJmol^{-1})
KHF_2	703	Jenkins, Pratt (1977)	698
$RbHF_2$	674	Jenkins, Pratt (1977)	676
$CsHF_2$	646	Jenkins, Pratt (1977)	628
NH_4HF_2	705	Jenkins, Pratt (1977)	705
$CsHCl_2$	601	Thomson, Clark, Waddington, Jenkins (1975)	—
Me_4NHCl_2	427	Thompson, Clark, Waddington, Jenkins (1975)	—
Et_4NHCl_2	346	Thompson, Clark, Waddington, Jenkins (1975)	—
Bu_4NHCl_2	290	Thompson, Clark, Waddington, Jenkins (1975)	—
Bicarbonates			
$NaHCO_3$	820	Yatsimirskii (1956)	818
$KHCO_3$	741	Yatsimirskii (1956)	736
$RbHCO_3$	707	Yatsimirskii (1956)	714
$CsHCO_3$	678	Yatsimirskii (1956)	709
NH_4HCO_3	—	—	741
$Ca(HCO_3)_2$	2402	Yatsimirskii (1956)	(2403)
$Sr(HCO_3)_2$	2255	Yatsimirskii (1956)	(2272)
$Ba(HCO_3)_2$	2155	Yatsimirskii (1956)	(2159)
Bisulphates			
NH_4HSO_4	640	Yatsimirskii (1956)	(645)
Borides			
CaB_6	5146	Samsanov, Shulishova (1956)	—
SrB_6	5104	Samsanov, Shulishova (1956)	—
BaB_6	5021	Samsanov, Shulishova (1956)	—
YB_6	7447	Samsanov, Shulishova (1956)	—
LaB_6	7406	Samsanov, Shulishova (1956)	—
CeB_6	10083	Samsanov, Shulishova (1956)	—
PrB_6	7447	Samsanov, Shulishova (1956)	—
NdB_6	7447	Samsanov, Shulishova (1956)	—
PmB_6	7406	Samsanov, Shulishova (1956)	—
SmB_6	7447	Samsanov, Shulishova (1956)	—
EuB_6	5104	Samsanov, Shulishova (1956)	—
GdB_6	7489	Samsonov, Shulishova (1956)	—
TbB_6	7489	Samsanov, Shulishova (1956)	—
DyB_6	7489	Samsanov, Shulishova (1956)	—
HoB_6	7489	Samsanov, Shulishova (1956)	—
ErB_6	7489	Samsanov, Shulishova (1956)	—
TmB_6	7489	Samsanov, Shulishova (1956)	—
YbB_6	5146	Samsanov, Shulishova (1956)	—
LuB_6	7489	Samsanov, Shulishova (1956)	—
ThB_6	10167	Samsanov, Shulishova (1956)	—
Borohydrides			
$LiBH_4$	778	Altschuller (1955)	—
$NaBH_4$	703	Altschuller (1955)	—

Salt	Calculated lattice energy (kJmol^{-1})	Literature source	Thermochemical cycle lattice energy (kJmol^{-1})
KBH_4	665	Altschuller (1955)	—
$RbBH_4$	648	Altschuller (1955)	—
$CsBH_4$	628	Altschuller (1955)	—
Borohalides			
$LiBF_4$	699	Kapustinskii, Yatsimirskii (1949)	—
$NaBF_4$	657	Kapustinskii, Yatsimirskii (1949)	619
KBF_4	611	Ladd, Lee (1961)	631
$RbBF_4$	577	Kapustinskii, Yatsimirskii (1949)	605
$CsBF_4$	556	Kapustinskii, Yatsimirskii (1949)	—
NH_4BF_4	582	Ladd, Lee (1961)	—
$Co(BF_4)_2$	2127	Bhattacharya, Das Gupta (1970)	—
$Ni(BF_4)_2$	2136	Bhattacharya, Das Gupta (1970)	—
$Zn(BF_4)_2$	2063	Bhattacharya, Das Gupta (1970)	—
$Cd(BF_4)_2$	1937	Bhattacharya, Das Gupta (1970)	—
$KBCl_4$	506	Krivtsov, Titova, Rosolovskii (1973)	—
$RbBCl_4$	489	Krivtsov, Titova, Rosolovskii (1973)	—
$CsBCl_4$	473	Krivtsov, Titova, Rosolovskii (1973)	—
Carbonates			
Li_2CO_3	2523	Yatsimirskii (1956)	2269
Na_2CO_3	2301	Yatsimirskii (1956)	2030
K_2CO_3	2084	Yatsimirskii (1956)	1858
Rb_2CO_3	2000	Yatsimirskii (1956)	1795
Cs_2CO_3	1920	Yatsimirskii (1956)	1702
$MgCO_3$	3180	Waddington (1959)	3122
$CaCO_3$	2804	Jenkins, Waddington, Pratt (1976)	2810
$SrCO_3$	2720	Waddington (1959)	2688
$BaCO_3$	2615	Waddington (1959)	2554
$MnCO_3$	3046	Waddington (1959)	3151
$FeCO_3$	3121	Waddington (1959)	3171
$CoCO_3$	3443	Yatsimirskii (1958)	3232
$NiCO_3$	—	—	3297
$CuCO_3$	3494	Yatsimirskii (1958)	3327
$ZnCO_3$	3121	Waddington (1959)	3273
$CdCO_3$	2929	Waddington (1959)	3052
$SnCO_3$	2904	Yatsimirskii (1961)	(2853)
$PbCO_3$	2728	Yatsimirskii (1961)	2750
Cyanates			
$LiNCO$	770	Waddington (1959)	—
$NaNCO$	736	Waddington (1959)	—
$KNCO$	653	Waddington (1959)	—
$RbNCO$	615	Waddington (1959)	—
$CsNCO$	586	Waddington (1959)	—
NH_4NCO	724	Waddington (1959)	—
Cyanides			
$LiCN$	—	—	849
$NaCN$	738	Jenkins, Pratt (1977)	739
KCN	674	Jenkins, Pratt (1977)	669

Table 1 (continued)
LATTICE ENERGIES

Salt	Calculated lattice energy (kJmol⁻¹)	Literature source	Thermochemical cycle lattice energy (kJmol⁻¹)
RbCN	646	Jenkins, Pratt (1977)	(640)
CsCN	612	Jenkins, Pratt (1977)	602
NH_4CN	617	Ladd (1970)	670
$Ca(CN)_2$	2268	Yatsimirskii (1956)	2191
$Sr(CN)_2$	2138	Yatsimirskii (1956)	(2076)
$Ba(CN)_2$	2046	Yatsimirskii (1956)	1960
CuCN	—	—	1035
AgCN	741	Yatsimirskii (1961)	914
$Zn(CN)_2$	2431	Ladd, Lee (1960)	2768
$Cd(CN)_2$	2284	Ladd, Lee (1960)	2542
Formates			
$Li(HCO_2)$	865	Morris (1959)	872
$Na(HCO_2)$	791	Morris (1959)	811
$K(HCO_2)$	713	Morris (1959)	729
$Rb(HCO_2)$	685	Morris (1959)	682
$Cs(HCO_2)$	651	Morris (1959)	644
NH_4HCO_2	715	Morris (1959)	—
$Mg(HCO_2)_2$	2674	Morris (1959)	—
$Ca(HCO_2)_2$	2360	Morris (1959)	2390
$Sr(HCO_2)_2$	2221	Morris (1959)	2261
$Ba(HCO_2)_2$	2092	Morris (1959)	2134
$Mn(HCO_2)_2$	2598	Morris (1959)	2701
$Co(HCO_2)_2$	—	—	2792
$Ni(HCO_2)_2$	—	—	2880
$Cu(HCO_2)_2$	2870	Morris (1959)	2913
$Zn(HCO_2)_2$	2791	Morris (1959)	2847
$Cd(HCO_2)_2$	2556	Morris (1959)	—
$Pb(HCO_2)_2$	2276	Morris (1959)	2330
Germanates			
Mg_2GeO_4	7991	Grekenschikov (1967)	—
Ca_2GeO_4	7301	Grekenschikov (1967)	7306
Sr_2GeO_4	6987	Grekenschikov (1967)	—
Ba_2GeO_4	6653	Grekenschikov (1967)	6643
Glycinates			
$Na(NH_2CH_2COO_2)$	739	Bernard, Businot, Decker (1974)	—
$K(NH_2CH_2COO_2)$	668	Bernard, Businot, Decker (1974)	—
$Rb(NH_2CH_2COO_2)$	648	Bernard, Businot, Decker (1974)	—
$NH_4(NH_2CH_2COO_2)$	650	Bernard, Businot, Decker (1974)	—
$Cu(NH_2CH_2CO_2)_2$	2694	Bernard, Businot, Decker (1974)	—
Halates			
$LiBrO_3$	883	Finch, Gardner (1965)	897
$NaBrO_3$	803	Finch, Gardner (1965)	814
$KBrO_3$	740	Finch, Gardner (1965)	745
$RbBrO_3$	720	Finch, Gardner (1965)	742
$CsBrO_3$	694	Finch, Gardner (1965)	681
$NaClO_3$	770	Morris (1958)	770
$KClO_3$	711	Morris (1958)	706

Salt	Calculated lattice energy (kJmol⁻¹)	Literature source	Thermochemical cycle lattice energy (kJmol⁻¹)
$RbClO_3$	690	Morris (1958)	687
$CsClO_3$	—	—	647
$LiIO_3$	975	Finch, Gardner (1965)	—
$NaIO_3$	883	Finch, Gardner (1965)	882
KIO_3	820	Finch, Gardner (1965)	806
$RbIO_3$	791	Finch, Gardner (1965)	—
$CsIO_3$	761	Finch, Gardner (1965)	769
NH_4IO_3	—	Finch, Gardner (1965)	808
$Mg(ClO_3)_2$	2535	Finch, Gardner (1965)	(2475)
$Ca(ClO_3)_2$	2259	Finch, Gardner (1965)	2286
$Sr(ClO_3)_2$	2138	Finch, Gardner (1965)	2155
$Ba(ClO_3)_2$	2021	Finch, Gardner (1965)	2027
Halides			
LiF	1030	Jenkins, Pratt (1977)	1036
LiCL	834	Jenkins, Pratt (1977)	853
LiBr	788	Jenkins, Pratt (1977)	807
LiI	730	Jenkins, Pratt (1977)	757
NaF	910	Jenkins, Pratt (1977)	923
NaCl	769	Jenkins, Pratt (1977)	786
NaBr	732	Jenkins, Pratt (1977)	747
NaI	682	Jenkins, Pratt (1977)	704
KF	808	Jenkins, Pratt (1977)	821
KCl	701	Jenkins, Pratt (1977)	715
KBr	671	Jenkins, Pratt (1977)	682
KI	632	Jenkins, Pratt (1977)	649
RbF	774	Jenkins, Pratt (1977)	785
RbCl	680	Jenkins, Pratt (1977)	689
RbBr	651	Jenkins, Pratt (1977)	660
RbI	617	Jenkins, Pratt (1977)	630
CsF	744	Jenkins, Pratt (1977)	740
CsCl	657	Jenkins, Pratt (1977)	659
CsBr	632	Jenkins, Pratt (1977)	631
CsI	600	Jenkins, Pratt (1977)	604
FrF	715	Ladd, Lee (1961)	—
FrCl	632	Ladd, Lee (1961)	—
FrBr	611	Ladd, Lee (1961)	—
FrI	582	Ladd, Lee (1961)	—
CuCl	921	Sharma, Madan (1964)	996
CuBr	879	Sharma, Madan (1964)	979
CuI	835	Sharma, Madan (1964)	966
AgF	953	Sharma, Madan (1964)	967
AgCl	864	Sharma, Madan (1964)	915
AgBr	830	Sharma, Madan (1964)	904
AgI	808	Sharma, Madan (1964)	889
AuCl	1013	Mamulov (1961)	1066
AuBr	1015	Mamulov (1961)	1061
AuI	1015	Mamulov (1961)	1070

Table 1 (continued)
LATTICE ENERGIES

Salt	Calculated lattice energy ($kJmol^{-1}$)	Literature source	Thermochemical cycle lattice energy ($kJmol^{-1}$)
InCl	—	—	763
InBr	—	—	767
InI	—	—	732
TlF	—	—	845
TlCl	782	Jenkins, Alcock (1975)	751
TlBr	713	Sharma, Madan (1964)	735
TlI	687	Sharma, Madan (1964)	709
BeF_2	3150	Krasnov (1961)	3505
$BeCl_2$	3004	Krasnov (1961)	3020
$BeBr_2$	2950	Brackett, Brackett (1965)	2914
BeI_2	2653	Krasnov (1961)	2800
MgF_2	2913	Brackett, Brackett (1965)	2957
$MgCl_2$	2326	Brackett, Brackett (1965)	2526
$MgBr_2$	2097	Brackett, Brackett (1965)	2440
MgI_2	1944	Brackett, Brackett (1965)	2327
CaF_2	2609	Brackett, Brackett (1965)	2630
$CaCl_2$	2223	Brackett, Brackett (1965)	2258
$CaBr_2$	2132	Brackett, Brackett (1965)	2176
CaI_2	1905	Brackett, Brackett (1965)	2074
SrF_2	2476	Brackett, Brackett (1965)	2492
$SrCl_2$	2127	Brackett, Brackett (1965)	2156
SrF_2	2008	Brackett, Brackett (1965)	2075
SrI_2	1937	Karapet'yants (1954)	1963
BaF_2	2341	Brackett, Brackett (1965)	2352
$BaCl_2$	2033	Brackett, Brackett (1965)	2056
$BaBr_2$	1950	Brackett, Brackett (1965)	1985
BaI_2	1831	Brackett, Brackett (1965)	1877
RaF_2	2284	Yatsimirskii, Krestov (1960)	—
$RaCl_2$	2004	Yatsimirskii, Krestov (1960)	—
$RaBr_2$	1929	Yatsimirskii, Krestov (1960)	—
RaI_2	1803	Yatsimirskii, Krestov (1960)	—
$ScCl_2$	2380	Nelson, Sharpe (1966)	—
$ScBr_2$	2291	Nelson, Sharpe (1966)	—
ScI_2	2201	Nelson, Sharpe (1966)	—
TiF_2	2724	Mamulov (1961)	—
$TiCl_2$	2431	Mamulov (1961)	2501
$TiBr_2$	2360	Mamulov (1961)	2419
TiI_2	2259	Mamulov (1961)	2329
VCl_2	2607	Mamulov (1961)	2579
VBr_2	—	Mamulov (1961)	2523
VI_2	—	Mamulov (1961)	2456
CrF_2	2778	Mamulov (1961)	2917
$CrCl_2$	2455	Mamulov (1961)	2586
$CrBr_2$	2377	Mamulov (1961)	2523
CrI_2	2269	Mamulov (1961)	2425
$MoCl_2$	2485	Mamulov (1961)	2733
$MoBr_2$	2448	Mamulov (1961)	2742

Salt	Calculated lattice energy ($kJmol^{-1}$)	Literature source	Thermochemical cycle lattice energy ($kJmol^{-1}$)
MoI_2	2422	Mamulov (1961)	2630
MnF_2	2644	Mamulov (1961)	—
$MnCl_2$	2368	Mamulov (1961)	2537
$MnBr_2$	2304	Mamulov (1961)	2471
MnI_2	2212	Mamulov (1961)	—
FeF_2	2769	Mamulov (1961)	—
$FeCl_2$	2525	Mamulov (1961)	2631
$FeBr_2$	2464	Mamulov (1961)	2569
FeI_2	2382	Mamulov (1961)	2480
CoF_2	2878	Morris (1957)	3018
$CoCl_2$	2709	Mamulov (1961)	2691
$CoBr_2$	2648	Yatsimirskii (1958)	2629
CoI_2	2569	Yatsimirskii (1958)	2545
NiF_2	2845	VonBaur (1961)	3066
$NiCl_2$	2753	Yatsimirskii (1958)	2772
$NiBr_2$	2699	Yatsimirskii (1958)	2709
NiI_2	2607	Yatsimirskii (1958)	2623
$PdCl_2$	2766	Yatsimirskii (1958)	2778
$PdBr_2$	—	—	2741
PdI_2	—	—	2748
CuF_2	3046	Perret (1961)	3082
$CuCl_2$	2774	Perret (1961)	2811
$CuBr_2$	2711	Perret (1961)	2763
CuI_2	2640	Perret (1961)	—
AgF_2	2919	Mamulov (1961)	2942
ZnF_2	2930	Mamulov (1961)	3032
$ZnCl_2$	2690	Mamulov (1961)	2734
$ZnBr_2$	2632	Mamulov (1961)	2678
ZnI_2	2549	Mamulov (1961)	2605
CdF_2	2740	Karapet'yants (1954)	2809
$CdCl_2$	2226	Krasnov (1961)	2552
$CdBr_2$	2468	Karapet'yants (1954)	2507
CdI_2	2406	Karapet'yants (1954)	2441
HgF_2	2757	Karapet'yants (1954)	(2798)
$HgCl_2$	2569	Yatsimirskii (1961)	2651
$HgBr_2$	2598	Karapet'yants (1954)	2628
HgI_2	2569	Karapet'yants (1954)	2610
SnF_2	2551	Mamulov (1961)	—
$SnCl_2$	2276	Mamulov (1961)	2297
$SnBr_2$	2211	Mamulov (1961)	2245
SnI_2	2123	Mamulov (1961)	2193
PbF_2	2460	Mamulov (1961)	2522
$PbCl_2$	2229	Mamulov (1961)	2269
$PbBr_2$	2169	Mamulov (1961)	2219
PbI_2	2086	Mamulov (1961)	2163
ScF_3	5096	Mamulov (1961)	5492
$ScCl_3$	4874	Kapustinskii, Yatsimirskii (1956)	4866

Table 1 (continued)
LATTICE ENERGIES

Salt	Calculated lattice energy (kJmol⁻¹)	Literature source	Thermochemical cycle lattice energy (kJmol⁻¹)	Salt	Calculated lattice energy (kJmol⁻¹)	Literature source	Thermochemical cycle lattice energy (kJmol⁻¹)
$ScBr_3$	4711	Kapustinskii, Yatsimirskii (1956)	4729	NiF_3	(6111)	Van Gool, Picken (1969)	—
ScI_3	4640	Mamulov (1961)	—	AuF_3	(5777)	Hoppe (1970)	—
YF_3	4983	Mamulov (1961)	—	$AuCl_3$	(4605)	Hoppe (1970)	—
YCl_3	4447	Mamulov (1961)	4506	$ZnCl_3$	5832	Nelson, Sharpe (1963)	—
YBr_3	4410	Wen, Sho (1975)	—	$ZnBr_3$	5732	Nelson, Sharpe (1963)	—
YI_3	4125	Mamulov (1961)	4240	ZnI_3	5636	Nelson, Sharpe (1963)	—
TiF_3	5644	Mamulov (1961)	—	AlF_3	5924	Mamulov (1961)	5215
$TiCl_3$	5134	Mamulov (1961)	5134	$AlCl_3$	5376	Mamulov (1961)	5492
$TiBr_3$	5012	Mamulov (1961)	5007	$AlBr_3$	5247	Mamulov (1961)	5361
TiI_3	4845	Mamulov (1961)	—	AlI_3	5070	Mamulov (1961)	5218
ZrF_3	—	—	(5400)	GaF_3	—	—	6205
$ZrCl_3$	—	—	4791	$GaCl_3$	5217	Mamulov (1961)	5645
$ZrBr_3$	—	—	4758	$GaBr_3$	4966	Mamulov (1961)	5552
ZrI_3	—	—	(4591)	GaI_3	4611	Mamulov (1961)	5476
VF_3	5895	Cavell, Clark (1965)	—	$InCl_3$	4736	Yatsimirskii (1961)	5187
VCl_3	5322	Mamulov (1961)	5315	$InBr_3$	4535	Yatsimirskii (1961)	5124
VBr_3	5192	Nelson, Sharpe (1966)	5214	InI_3	4234	Yatsimirskii (1961)	5005
VI_3	5058	Nelson, Sharpe (1966)	5121	TlF_3	5493	Mamulov (1961)	—
$NbCl_3$	5062	Jenkins, Waddington[a]	—	$TlCl_3$	5252	Mamulov (1961)	5258
$NbBr_3$	4980	Jenkins, Waddington[a]	—	$TlBr_3$	5171	Mamulov (1961)	—
NbI_3	4860	Jenkins, Waddington[a]	—	TlI_3	5088	Mamulov (1961)	—
CrF_3	5958	Mamulov (1961)	6033	$AsBr_3$	—	—	5497
$CrCl_3$	5473	Mamulov (1961)	5509	AsI_3	(3758)	Van Gool, Picken (1969)	4824
$CrBr_3$	5355	Mamulov (1961)	—	SbF_3	—	—	5295
CrI_3	5201	Mamulov (1961)	5274	$SbCl_3$	—	—	5032
MoF_3	(6459)	Van Gool, Picken (1964)	5230	$SbBr_3$	—	—	4954
$MoCl_3$	—	Van Gool, Picken (1964)	5230	SbI_3	—	—	4867
$MoBr_3$	—	Van Gool, Picken (1964)	5156	$BiCl_3$	—	—	4689
MoI_3	—	Van Gool, Picken (1964)	5073	$BiBr_3$	—	—	—
MnF_3	6017	Cavell, Clark (1965)	—	BiI_3	(3774)	Van Gool, Picken (1969)	—
$MnCl_3$	5544	Nelson, Sharpe (1966)	—	LaF_3	4682	Mamulov (1961)	—
$MnBr_3$	5448	Nelson, Sharpe (1966)	—	$LaCl_3$	4343	Yatsimirskii (1961)	4242
MnI_3	5330	Nelson, Sharpe (1966)	—	$LaBr_3$	4209	Wen, Sho (1975)	—
$TcCl_3$	5270	Jenkins, Waddington[a]	—	LaI_3	3916	Yatsimirskii (1961)	3986
$TcBr_3$	5215	Jenkins, Waddington[a]	—	$CeCl_3$	4297	Ladd, Lee (1961)	4284
TcI_3	5188	Jenkins, Waddington[a]	—	CeI_3	—	—	4029
FeF_3	5870	Perret (1961)	—	$PrCl_3$	4322	Ladd, Lee (1961)	4326
$FeCl_3$	5364	Perret (1961)	5359	PrI_3	—	—	4071
$FeBr_3$	5268	Perret (1961)	5333	$NdCl_3$	4343	Ladd, Lee (1961)	—
FeI_3	5117	Perret (1961)	—	$SmCl_3$	4376	Ladd, Lee (1961)	—
$RuCl_3$	—	—	5245	$EuCl_3$	4393	Ladd, Lee (1961)	—
$RuBr_3$	—	—	5223	$GdCl_3$	4406	Ladd, Lee (1961)	—
RuI_3	—	—	5222	$DyCl_3$	4481	Ladd, Lee (1961)	—
CoF_3	5991	Cavell, Clark (1965)	6118	$HoCl_3$	4501	Ladd, Lee (1961)	—
$RhCl_3$	—	—	5641	$ErCl_3$	4527	Ladd, Lee (1961)	
IrF_3	(6112)	Hoppe (1970)	—	$TmCl_3$	4548	Ladd, Lee (1961)	4550
$IrBr_3$	(4794)	Hoppe (1970)	—	TmI_3	—	—	4314

Table 1 (continued)
LATTICE ENERGIES

Salt	Calculated lattice energy (kJmol⁻¹)	Literature source	Thermochemical cycle lattice energy (kJmol⁻¹)
$YbCl_3$	—	—	4546
$AcCl_3$	4096	Ladd, Lee (1961)	—
UCl_3	4243	Ladd, Lee (1961)	—
$NpCl_3$	4268	Ladd, Lee (1961)	—
$PuCl_3$	4289	Ladd, Lee (1961)	—
$PuBr_3$	(3959)	Van Gool, Picken (1969)	—
$AmCl_3$	4293	Van Gool, Picken (1969)	—
TiF_4	10012	Mamulov (1961)	9908
$TiCl_4$	9431	Mamulov (1961)	—
$TiBr_4$	9288	Manulov (1961)	9039
TiI_4	9108	Mamulov (1961)	8893
ZrF_4	8853	Mamulov (1961)	8971
$ZrCl_4$	8096	Mamulov (1961)	8144
$ZrBr_4$	7916	Mamulov (1961)	7984
ZrI_4	7661	Mamulov (1961)	7801
MoF_4	8795	Mamulov (1961)	—
$MoCl_4$	8556	Mamulov (1961)	9573
$MoBr_4$	8510	Mamulov (1961)	9475
MoI_4	8427	Mamulov (1961)	—
$SnCl_4$	8355	Mamulov (1961)	—
$SnBr_4$	7970	Mamulov (1961)	8833
PbF_4	—	—	9461
CrF_2Cl	5795	Perret (1961)	—
CrF_2Br	5753	Perret (1961)	—
CrF_2I	5669	Perret (1961)	—
$CrCl_2Br$	5448	Perret (1961)	—
$CrCl_2I$	5381	Perret (1961)	—
$CrBr_2I$	5330	Perret (1961)	—
CuFCl	2891	Perret (1961)	—
CuFBr	2853	Perret (1961)	—
CuFI	2803	Perret (1961)	—
CuClBr	2753	Perret (1961)	—
CuClI	2694	Perret (1961)	—
CuBrI	2669	Perret (1961)	—
FeF_2Cl	5711	Perret (1961)	—
FeF_2Br	5653	Perret (1961)	—
FeF_2I	5569	Perret (1961)	—
$FeCl_2Br$	5339	Perret (1961)	—
$FeCl_2I$	5272	Perret (1961)	—
FBr_2I	5209	Perret (1961)	—
$LiIO_2F_2$	845	Finch, Gates, Jenkinson (1972)	—
$NaIO_2F_2$	766	Finch, Gates, Jenkinson (1972)	764
KIO_2F_2	699	Finch, Gates, Jenkinson (1972)	697
$RbIO_2F_2$	674	Finch, Gates, Jenkinson (1972)	671
$CsIO_2F_2$	636	Finch, Gates, Jenkinson (1972)	—
$NH_4IO_2F_2$	678	Finch, Gates, Jenkinson (1972)	—
$AgIO_2F_2$	736	Finch, Gates, Jenkinson (1972)	685

Salt	Calculated lattice energy (kJmol⁻¹)	Literature source	Thermochemical cycle lattice energy (kJmol⁻¹)
Hydrides			
LiH	858	Tsai (1964)	920
NaH	782	Tsai (1964)	808
KH	699	Tsai (1964)	714
RbH	674	Tsai (1964)	685
CsH	648	Tsai (1964)	644
TiH	—	—	1407
ZrH	—	—	1590
VH	—	—	(1344)
NbH	—	—	(1633)
PdH	—	—	(1368)
CuH	—	—	1254
TiH	996	Gibb (1962)	1407
ZrH	916	Gibb (1962)	1590
HfH	904	Gibb (1962)	—
LaH	828	Gibb (1962)	—
VH	1184	Gibb (1962)	(1344)
NbH	1163	Gibb (1962)	(1633)
TaH	1021	Gibb (1962)	—
CrH	1050	Gibb (1962)	—
NiH	929	Gibb (1962)	—
PdH	979	Gibb (1962)	1368
PtH	937	Gibb (1962)	—
CuH	828	Gibb (1962)	1254
AgH	941	Karapet'yants (1954)	—
AuH	1033	Karapet'yants (1954)	—
TlH	745	Karapet'yants (1954)	—
GeH	950	Gibb (1962)	—
PbH	778	Gibb (1962)	—
BeH_2	3205	Gibb (1962)	3295
MgH_2	2791	Gibb (1962)	2706
CaH_2	2410	Karapet'yants (1954)	2394
SrH_2	2250	Karapet'yants (1954)	2253
BaH_2	2121	Karapet'yants (1954)	2121
ScH_2	2711	Gibb (1962)	2659
YH_2	(2598)	Gibb (1962)	2670
LaH_2	2380	Gibb (1962)	2522
CeH_2	2414	Gibb (1962)	2484
PrH_2	2448	Gibb (1962)	2398
NdH_2	2464	Gibb (1962)	2367
PmH_2	2519	Gibb (1962)	—
SmH_2	2510	Gibb (1962)	2389
GdH_2	2494	Gibb (1962)	2706
AcH_2	2372	Gibb (1962)	—
ThH_2	2711	Gibb (1962)	—
PuH_2	2519	Gibb (1962)	—
AmH_2	2544	Gibb (1962)	—

Table 1 (continued)
LATTICE ENERGIES

Salt	Calculated lattice energy (kJmol⁻¹)	Literature source	Thermochemical cycle lattice energy (kJmol⁻¹)
TiH_2	2866	Gibb (1962)	2845
ZrH_2	2711	Gibb (1962)	2999
CuH_2	2941	Karapet'yants (1954)	—
ZnH_2	2870	Karapet'yants (1954)	—
HgH_2	2707	Karapet'yants (1954)	—
AlH_3	5924	Gibb (1962)	—
FeH_3	5724	Gibb (1962)	—
ScH_3	5439	Gibb (1962)	—
YH_3	5063	Gibb (1962)	—
LaH_3	4895	Gibb (1962)	4493
FeH_3	5724	Gibb (1962)	—
GaH_3	5690	Gibb (1962)	—
InH_3	5092	Gibb (1962)	—
TlH_3	5092	Gibb (1962)	—
Hydroselenides			
NaHse	703	Yatsimirskii (1956)	—
KHSe	644	Yatsimirskii (1956)	—
RbHSe	623	Yatsimirskii (1956)	—
CsHSe	598	Yatsimirskii (1956)	—
Hydrosulphides			
LiHS	759	Waddington (1959)	821
NaHS	704	Waddington (1959)	747
KHS	650	Waddington (1959)	659
RbHS	623	Waddington (1959)	637
CsHS	582	Waddington (1959)	595
NH_4HS	661	Yatsimirskii (1956)	666
$Ca(HS)_2$	2184	Yatsimirskii (1956)	(2171)
$Sr(HS)_2$	2063	Yatsimirskii (1956)	—
$Ba(HS)_2$	1979	Yatsimirskii (1956)	(1956)
Hydroxides			
LiOH	1021	Saloman (1970)	1039
NaOH	887	Saloman (1970)	900
KOH	789	Saloman (1970)	804
RbOH	766	Yatsimirskii (1956)	773
CsOH	721	Yatsimirskii (1956)	724
$Be(OH)_2$	3477	Finch, Gardner (1965)	3629
$Mg(OH)_2$	2870	Finch, Gardner (1965)	3006
$Ca(OH)_2$	2506	Finch, Gardner (1965)	2645
$Sr(OH)_2$	2330	Finch, Gardner (1965)	2483
$Ba(OH)_2$	2141	Finch, Gardner (1965)	2339
$Ti(OH)_2$	—	—	2962
$Mn(OH)_2$	2909	Wen, Sho (1975)	3008
$Fe(OH)_2$	2653	Karapet'yants (1954)	3055
$Co(OH)_2$	2786	Karapet'yants (1954)	3115
$Ni(OH)_2$	2832	Karapet'yants (1954)	3193
$Pd(OH)_2$	—	—	3175
$Cu(OH)_2$	2870	Karapet'yants (1954)	3237

Salt	Calculated lattice energy (kJmol⁻¹)	Literature source	Thermochemical cycle lattice energy (kJmol⁻¹)
CuOH	1006	Karapet'yants (1954)	—
AgOH	918	Karapet'yants (1954)	—
AuOH	1033	Karapet'yants (1954)	—
TlOH	705	Karapet'yants (1954)	—
$Zn(OH)_2$	2795	Yatsimirskii (1961)	3158
$Cd(OH)_2$	2607	Yatsimirskii (1961)	2918
$Hg(OH)_2$	2669	Karapet'yants (1954)	—
$Sn(OH)_2$	2489	Yatsimirskii (1961)	2729
$Pb(OH)_2$	2376	Yatsimirskii (1961)	2623
$Sc(OH)_3$	5063	Karapet'yants (1954)	—
$Y(OH)_3$	4707	Karapet'yants (1954)	—
$La(OH)_3$	4443	Karapet'yants (1954)	—
$Cr(OH)_3$	5556	Karapet'yants (1954)	—
$Mn(OH)_3$	6213	Wen,Sho (1975)	—
$Al(OH)_3$	5627	Karapet'yants (1954)	—
$Ga(OH)_3$	5732	Karapet'yants (1954)	—
$In(OH)_3$	5280	Karapet'yants (1954)	—
$Tl(OH)_3$	5314	Karapet'yants (1954)	—
$Ti(OH)_4$	9456	Karapet'yants (1954)	—
$Zr(OH)_4$	8619	Karapet'yants (1954)	—
$Mn(OH)_4$	10933	Brenet, Goelfier, Cabano (1963)	—
$Sn(OH)_4$	9188	Karapet'yants (1954)	—
Imides			
CaNH	3293	Altschuller (1955)	—
SrNH	3146	Altschuller (1955)	—
BaNH	2975	Altschuller (1955)	—
Metaniobates			
$NaNbO_3$	789	Lapitskii, Nebyhtsyn (1961)	—
$Ca(NbO_3)_2$	2315	Lapitskii, Nebyhtsyn (1961)	—
$Fe(NBO_3)_2$	2502	Lapitskii, Nebyhtsyn (1961)	—
Metatantalates			
$NaTaO_3$	789	Lapitskii, Nebyhtsyn (1961)	—
$Ca(TaO_3)_2$	2315	Lapitskii, Nebyhtsyn (1961)	—
$Fe(TaO_3)_2$	2502	Lapitskii, Nebyhtsyn (1961)	—
Metavanadates			
Li_3VO_4	3945	Golvkin, Fotiev (1970)	—
Na_3VO_4	3766	Golvkin, Fotiev (1970)	—
K_3VO_4	3376	Golvkin, Fotiev (1970)	—
Rb_3VO_4	3243	Golvkin, Fotiev (1970)	—
Cs_3VO_4	3137	Golvkin, Fotiev (1970)	—
Nitrates			
$LiNO_3$	848	Jenkins, Morris (1977)	848
$NaNO_3$	755	Jenkins, Morris (1977)	756
KNO_3	685	Jenkins, Morris (1977)	687
$RbNO_3$	662	Jenkins, Morris (1977)	658
$CsNO_3$	648	Jenkins, Morris (1977)	625
NH_4NO_3	661	Morris (1958)	676

Table 1 (continued)
LATTICE ENERGIES

Salt	Calculated lattice energy ($kJmol^{-1}$)	Literature source	Thermochemical cycle lattice energy ($kJmol^{-1}$)
$AgNO_3$	820	Morris (1958)	822
$TlNO_3$	690	Morris (1958)	700
$Mg(NO_3)_2$	2468	Finch, Gardner (1965)	2503
$Ca(NO_3)_2$	2209	Finch, Gardner (1965)	2228
$Sr(NO_3)_2$	2092	Finch, Gardner (1965)	2132
$Ba(NO_3)_2$	1975	Finch, Gardner (1965)	2016
$Mn(NO_3)_2$	2318	Yatsimirskii (1961)	2519
$Fe(NO_3)_2$	—	—	(2563)
$CO(NO_3)_2$	2560	Wen, Sho (1975)	2626
$Ni(NO_3)_2$	—	—	2709
$Cu(NO_3)_2$	—	—	2720
$Zn(NO_3)_2$	2376	Yatsimirskii (1961)	2628
$Cd(NO_3)_2$	2238	Yatsimirskii (1961)	2443
$Ha(NO_3)_2$	2255	Yatsimirskii (1961)	—
$Sn(NO_3)_2$	2155	Yatsimirskii (1961)	2254
$Pb(NO_3)_2$	2067	Yatsimirskii (1961)	2189
Nitrides			
ScN	7547	Baughan (1959)	7506
LaN	6876	Baughan (1959)	6793
TiN	8130	Baughan (1959)	8033
ZrN	7633	Baughan (1959)	7723
VN	8283	Baughan (1959)	8233
NbN	7939	Baughan (1959)	8022
CrN	8269	Baughan (1959)	8358
Nitrites			
$NaNO_2$	774	Morris (1958)	748
KNO_2	660	Jenkins (1977)	664
$RbNO_2$	638	Jenkins (1977)	765
$CsNO_2$	598	Jenkins (1977)	—
$Ca(NO_2)_2$	2460	Yatsimirskii (1956)	2225
$Sr(NO_2)_2$	2305	Yatsimirskii (1956)	2111
$Ba(NO_2)_2$	2205	Yatsimirskii (1956)	1987
Oxides			
Li_2O	2799	Baughan (1959)	
Na_2O	2481	Baughan (1959)	
K_2O	2238	Baughan (1959)	
Rb_2O	2163	Baughan (1959)	
Cu_2O	3273	Mamulov (1961)	
Ag_2O	3002	Mamulov (1961)	
Tl_2O	2659	Mamulov (1961)	
LiO_2	(878)	D'Orazio, Wood (1965)	(872)
NaO_2	799	Yatsimirskii (1959)	796
KO_2	741	D'Orazio, Wood (1965)	725
RbO_2	706	D'Orazio, Wood (1965)	695
CsO_2	679	D'Orazio, Wood (1965)	668
Li_2O_2	2592	Wood, D'Orazio (1965)	256
Na_2O_2	2309	Wood, D'Orazio (1965)	2305

Salt	Calculated lattice energy ($kJmol^{-1}$)	Literature source	Thermochemical cycle lattice energy ($kJmol^{-1}$)
K_2O_2	2114	Wood, D'Orazio (1965)	2078
Rb_2O_2	2025	Wood, D'Orazio (1965)	2006
Cs_2O_2	1948	Wood, D'Orazio (1965)	1861
MgO_2	3356	Wood, D'Orazio (1965)	3526
CaO_2	3144	Wood, D'Orazio (1965)	3133
SrO_2	3037	Wood, D'Orazio (1965)	2849
KO_3	697	Wood, D'Orazio (1965)	—
BeO	4293	Huggins, Sakamoto (1957)	4443
MgO	3795	Huggins, Sakomoto (1957)	3791
CaO	3414	Huggins, Sakamoto (1957)	3401
SrO	3217	Huggins, Sakamoto (1957)	3223
BaO	3029	Huggins, Sakamoto (1957)	3054
TiO	3832	Huggins, Sakamoto (1957)	3811
VO	3932	Ladd, Lee (1961)	3863
MnO	3724	Ladd, Lee (1961)	3745
FeO	3795	Ladd, Lee (1961)	3865
CoO	3837	Ladd, Lee (1961)	3910
NiO	3908	Ladd, Lee (1961)	4010
PdO	3736	Ladd, Lee (1961)	—
CuO	4135	Mamulov (1961)	4050
ZnO	4142	Ladd, Lee (1961)	3971
CdO	3806	Ladd, Lee (1961)	—
HgO	3907	Ladd, Lee (1961)	—
GeO	3919	Ladd, Lee (1961)	—
SnO	3652	Ladd, Lee (1961)	—
PbO	3520	Ladd, Lee (1961)	—
Sc_2O_3	13557	Gasharov, Sovers (1970)	13708
Y_2O_3	12705	Gasharov, Sovers (1970)	—
La_2O_3	12452	Johnson (1969)	—
Ce_2O_3	12661	Johnson (1969)	—
Pr_2O_3	12703	Johnson (1969)	—
Nd_2O_3	12736	Johnson (1969)	—
Pm_2O_3	12811	Johnson (1969)	—
Sm_2O_3	12878	Johnson (1969)	—
Eu_2O_3	12945	Johnson (1969)	—
Gd_2O_3	12996	Johnson (1969)	—
Tb_2O_3	13071	Johnson (1969)	—
Dy_2O_3	13138	Johnson (1969)	—
Ho_2O_3	13180	Johnson (1969)	—
Er_2O_3	13263	Johnson (1969)	—
Tm_2O_3	13322	Johnson (1969)	—
Yb_2O_3	13380	Johnson (1969)	—
Lu_2O_3	13665	Ladd, Lee (1961)	—
Ac_2O_3	12573	Krestov, Krestova (1969)	—
Ti_2O_3	—	—	14149
V_2O_3	15096	Mamulov (1961)	14520
Cr_2O_3	15276	Mamulov (1961)	114957

Table 1 (continued)
LATTICE ENERGIES

Salt	Calculated lattice energy ($kJmol^{-1}$)	Literature source	Thermochemical cycle lattice energy ($kJmol^{-1}$)
Mn_2O_3	15146	Mamulov (1961)	15035
Fe_2O_3	14309	Mamulov (1961)	14774
Al_2O_3	15916	Yatsimirskii (1961)	—
Ga_2O_3	15590	Yatsimirskii (1961)	15220
In_2O_3	13928	Yatsimirskii (1961)	—
Ti_2O_3	14702	Mamulov (1961)	—
Pb_2O_3	(14841)	Van Gool, Picken (1969)	—
CeO_2	9627	VanBaur (1961)	—
ThO_2	10397	Ladd, Lee (1961)	—
PaO_2	10573	Ladd, Lee (1961)	—
VO_2	10644	Ladd, Lee (1961)	—
NpO_2	10707	Ladd, Lee (1961)	—
PuO_2	10786	Ladd, Lee (1961)	—
AmO_2	10799	Ladd, Lee (1961)	—
CmO_2	10832	Ladd, Lee (1961)	—
TiO_2	12150	Ladd, Lee (1961)	—
ZrO_2	11188	Ladd, Lee (1961)	—
MoO_2	11648	Ladd, Lee (1961)	—
MnO_2	12970	Ladd, Lee (1961)	—
SiO_2	13125	Ladd, Lee (1961)	—
GeO_2	12828	Ladd, Lee (1961)	—
SnO_2	11807	Ladd, Lee (1961)	—
PbO_2	11217	—	
Perchlorates			
$LiClO_4$	709	Gill, Singla, Paul, Narula (1972)	723
$NaClO_4$	643	Jenkins, Pratt (1978)	648
$KClO_4$	599	Jenkins, Pratt (1978)	602
$RbClO_4$	564	Gill, Singla, Paul, Narula (1972)	582
$CsClO_4$	636	Morris (1958)	(542)
NH_4ClO_4	583	Jenkins, Pratt (1978)	580
$Ca(ClO_4)_2$	1958	Yatsimirskii (1956)	1971
$Sr(ClO_4)_2$	1862	Yatsimirskii (1956)	1862
$Ba(ClO_4)_2$	1795	Yatsimirskii (1958)	1769
$NaMnO_4$	661	Yatsimirskii (1956)	—
$KMnO_4$	607	Yatsimirskii (1956)	—
$RbMnO_4$	586	Yatsimirskii (1956)	—
$CsMnO_4$	565	Yatsimirskii (1956)	—
$Ca(MnO_4)_2$	1937	Yatsimirskii (1956)	—
$Sr(MnO_4)_2$	1845	Yatsimirskii (1956)	—
$Ba(MnO_4)_2$	1778	Yatsimirskii (1956)	—
Phosphates			
$Mg_3(PO_4)_2$	11632	Grekenschikov (1967)	11407
$Ca_3(PO_4)_2$	10602	Grekenschikov (1967)	10479
$Sr_3(PO_4)_2$	10125	Grekenschikov (1967)	10075
$Ba_3(PO_4)_2$	9652	Grekenschikov (1967)	9654
$MnPO_4$	7397	Grekenschikov (1967)	—
$FePO_4$	7251	Grekenschikov (1967)	7303

Salt	Calculated lattice energy ($kJmol^{-1}$)	Literature source	Thermochemical cycle lattice energy ($kJmol^{-1}$)
BPO_4	8201	Grekenschikov (1967)	—
$AlPO_4$	7427	Grekenschikov (1967)	7509
$GaPO_4$	7381	Grekenschikov (1967)	—
Phosphonium salts			
PH_4Br	616	Waddington (1965)	—
PH_4I	590	Waddington (1965)	—
Selenides			
Li_2Se	2364	Bevan, Morris (1960)	—
Na_2Se	2130	Bevan, Morris (1960)	—
K_2Se	1933	Bevan, Morris (1960)	—
Rb_2Se	1837	Bevan, Morris (1960)	—
CS_2Se	1745	Bevan, Morris (1960)	—
Ag_2Se	2686	Mamulov (1961)	—
Tl_2Se	2209	Mamulov (1961)	—
BeSe	3431	Huggins, Sakamoto (1957)	—
MgSe	3071	Huggins, Sakamoto (1957)	—
CaSe	2858	Huggins, Sakamoto (1957)	2862
SrSe	2736	Huggins, Sakamoto (1957)	—
BaSe	2611	Huggins, Sakamoto (1957)	—
MnSe	3176	Mamulov (1961)	3194
FeSe	3499	Mamulov (1961)	3396
CoSe	3554	Mamulov (1961)	3471
NiSe	3658	Mamulov (1961)	3558
CuSe	3736	Mamulov (1961)	3662
ZnSe	3502	Lambardi (1969)	3514
CdSe	3330	Mamulov (1961)	—
HgSe	3501	Mamulov (1961)	—
SnSe	3058	Mamulov (1961)	—
PbSe	3050	Mamulov (1961)	—
Selenites			
Li_2SeO_3	2171	Klushina, Selivanova (1972)	—
Na_2SeO_3	1950	Klushina, Selivanova (1972)	1931
K_2SeO_3	1774	Klushina, Selivanova (1972)	—
Rb_2SeO_3	1715	Klushina, Selivanova (1972)	1675
Cs_2SeO_3	1640	Klushina, Selivanova (1972)	—
Tl_2SeO_3	1879	Klushina, Selivanova (1972)	—
Ag_2SeO_3	2113	Klushina, Selivanova (1972)	—
$BeSeO_3$	3322	Klushina, Selivanova (1972)	—
$MgSeO_3$	3012	Klushina, Selivanova (1972)	2996
$CaSeO_3$	2732	Klushina, Selivanova (1972)	—
$SrSeO_3$	2586	Klushina, Selivanova (1972)	2586
$BaSeO_3$	2460	Klushina, Selivanova (1972)	2448
$RaSeO_3$	2456	Klushina, Selivanova (1972)	—
$MnSeO_3$	2975	Klushina, Selivanova (1972)	—
$FeSeO_3$	2895	Klushina, Selivanova (1972)	—
$CoSeO_3$	3155	Klushina, Selivanova (1972)	—
$NiSeO_3$	2945	Klushina, Selivanova (1972)	—

Table 1 (continued)
LATTICE ENERGIES

Salt	Calculated lattice energy (kJmol⁻¹)	Literature source	Thermochemical cycle lattice energy (kJmol⁻¹)
$CuSeO_3$	3209	Klushina, Selivanova (1972)	—
$ZnSeO_3$	3167	Klushina, Selivanova (1972)	—
$CdSeO_3$	2962	Klushina, Selivanova (1972)	—
$PbSeO_3$	2669	Klushina, Selivanova (1972)	—
Selenates			
Li_2SeO_4	2054	Selivanova, Karapet'yants (1963)	—
Na_2SeO_4	1879	Selivanova, Karapet'yants (1963)	—
K_2SeO_4	1732	Selivanova, Katapet'yants (1963)	—
Rb_2SeO_4	1686	Selivanova, Karapet'yants (1963)	—
Cs_2SeO_4	1615	Selivanova, Karapet'yants (1963)	—
Cu_2SeO_4	2201	Selivanova, Karapet'yants (1963)	—
Ag_2SeO_4	2033	Selivanova, Karapet'yants (1963)	—
Tl_2SeO_4	1766	Selivanova, Karapet'yants (1963)	—
Hg_2SeO_4	2163	Selivanova, Karapet'yants (1963)	—
$BeSeO_4$	3448	Selivanova, Karapet'yants (1963)	—
$MgSeO_4$	2895	Selivanova, Karapet'yants (1963)	—
$CaSeO_4$	2632	Selivanova, Karapet'yants (1963)	—
$SrSeO_4$	2489	Selivanova, Karapet'yants (1963)	—
$BaSeO_4$	2385	Selivanova, Karapet'yants (1963)	—
$RaSeO_4$	2364	Selivanova, Karapet'yants (1963)	—
$GdSeO_4$	2753	Selivanova, Karapet'yants (1963)	—
$MnSeO_4$	2837	Selivanova, Karapet'yants (1963)	—
$FeSeO_4$	3008	Selivanova, Karapet'yants (1963)	—
$CoSeO_4$	2912	Selivanova, Karapet'yants (1963)	—
$NiSeO_4$	3079	Selivanova, Karapet'yants (1963)	—
$CuSeO_4$	3104	Selivanova, Karapet'yants (1963)	—
$ZnSeO_4$	3021	Selivanova, Karapet'yants (1963)	—
$CdSeO_4$	2833	Selivanova, Karapet'yants (1963)	—
$HgSeO_4$	2845	Selivanova, Karapet'yants (1963)	—
$PbSeO_4$	2561	Selivanova, Karapet'yants (1963)	—
Sulphides			
Li_2S	2464	Morris (1958)	2472
Na_2S	2192	Morris (1958)	2203
K_2S	1979	Morris (1958)	(2052)
Rb_2S	1929	Morris (1958)	1949
Cs_2S	1892	Karapet'yants (1954)	1850
$(NH_4)_2S$	2008	Karapet'yants (1954)	(2026)
Cu_2S	2786	Karapet'yants (1954)	2865
Ag_2S	2606	Karapet'yants (1954)	2677
Au_2S	2908	Karapet'yants (1954)	—
Tl_2S	2298	Mamulov (1961)	2258
Sulphates			
Li_2SO_4	2229	Jenkins (1975)	2142
Na_2SO_4	1827	Jenkins (1975)	1938
K_2SO_4	1700	Jenkins (1975)	1796
Rb_2SO_4	1636	Jenkins (1975)	1748
Cs_2SO_4	1596	Jenkins (1975)	1658
$(NH_4)_2SO_4$	1766	Jenkins, Smith (1975)	1777
Cu_2SO_4	2276	Selivanova, Karapet'yants (1963)	2166
Ag_2SO_4	2104	Selivanova, Karapet'yants (1963)	1989
Tl_2SO_4	1828	Selivanova, Karapet'yants (1963)	1722
Hg_2SO_4	—	Selivanova, Karapet'yants (1963)	2127
$CaSO_4$	2489	Ladd, Lee (1968)	2480
$SrSO_4$	2577	Selivanova, Karapet'yants (1963)	2484
$BaSO_4$	2469	Selivanova, Karapet'yants (1963)	2374
$MnSO_4$	2920	Selivanova, Karapet'yants (1963)	2825
$FeSO_4$	2983	Selivanova, Karapet'yants (1963)	2921
$CoSO_4$	3088	Selivanova, Karapet'yants (1963)	2917
$NiSO_4$	3167	Selivanova, Karapet'yants (1963)	3044
$CuSO_4$	3167	Selivanova, Karapet'yants (1963)	3066
$ZnSO_4$	3100	Selivanova, Karapet'yants (1963)	3006
$CdSO_4$	2891		
$PbSO_4$	2635	Selivanova, Karapet'yants (1963)	2534
$Sn(SO_4)_2$	—	—	9616
Ternary salts			
CS_2CuCl_4	1393	Blake, Cotton (1964)	—
Rb_2ZnCl_4	1529	Paoletti (1965)	—
Cs_2ZnCl_4	1492	Paoletti (1965)	—
Rb_2ZnBr_4	1498	Paoletti (1965)	—
Cs_2ZnBr_4	1454	Paoletti (1965)	—
$(Me_4N)_2ZnBr_4$	1364	Paoletti (1965)	—
Cs_2ZnI_4	1386	Paoletti (1965)	—
$CsGaCl_4$	494	Jenkins (1976)	—
$NaAlCl_4$	556	Jenkins (1976)	—
$CsAlCl_4$	486	Jenkins (1976)	—
$NaFeCl_4$	492	Blake, Cotton (1963)	—
Rb_2CoCl_4	1447	Lister, Nyburg, Poyntz (1974)	—
Cs_2CoCl_4	1391	Lister, Nyburg, Poyntz (1974)	—
K_2PtCl_6	1574	Hartley (1972)	1550
Cs_2GeF_6	1573	Jenkins, Pratt (1978)	—
$(NH_4)_2GeF_6$	1657	Jenkins, Pratt (1978)	—
Cs_2GeCl_6	1375	Jenkins, Pratt (1078)	1375
K_2HfCl_6	1345	Jenkins, Pratt (1978)	1461
K_2IrCl_6	1442	Jenkins, Pratt (1978)	1440
$(NH_4)_2IrCl_6$	1442	Jenkins, Pratt (1978)	—
Na_2MoCl_6	1526	Jenkins, Pratt (1978)	—
K_2MoCl_6	1418	Jenkins, Pratt (1978)	1433
Rb_2MoCl_6	1399	Jenkins, Pratt (1978)	1399
Cs_2MoCl_6	1347	Jenkins, Pratt (1978)	1347
K_2NbCl_6	1375	Jenkins, Pratt (1978)	1398
Rb_2NbCl_6	1371	Jenkins, Pratt (1978)	1385
Cs_2NbCl_6	1381	Jenkins, Pratt (1978)	1344

Table 1 (continued)
LATTICE ENERGIES

Salt	Calculated lattice energy (kJmol^{-1})	Literature source	Thermochemical cycle lattice energy (kJmol^{-1})
K_2OsCl_6	1447	Jenkins, Pratt (1978)	1447
Cs_2OsCl_6	1409	Jenkins, Pratt (1978)	—
$(NH_4)_2OsCl_6$	1433	Jenkins, Pratt (1978)	—
K_2OsBr_6	1396	Jenkins, Pratt (1978)	—
K_2PdCl_6	1450	Jenkins, Pratt (1978)	1466
Rb_2PdCl_6	1449	Jenkins, Pratt (1978)	—
Cs_2PdCl_6	1426	Jenkins, Pratt (1978)	—
$(NH_4)_2PdCl_6$	1481	Jenkins, Pratt (1978)	—
Rb_2PbCl_6	1343	Jenkins, Pratt (1978)	1343
Cs_2PbCl_6	1344	Jenkins, Pratt (1978)	—
$(NH_4)_2PbCl_6$	1355	Jenkins, Pratt (1978)	—
K_2PtCl_6	1468	Jenkins, Pratt (1978)	
Rb_2PtCl_6	1464	Jenkins, Pratt (1978)	
Cs_2PtCl_6	1444	Jenkins, Pratt (1978)	
$(NH_4)_2PtCl_6$	1468	Jenkins, Pratt (1978)	
Tl_2PtCl_6	1546	Jenkins, Pratt (1978)	
Ag_2PtCl_6	1773	Jenkins, Pratt (1978)	
$BaPtCl_6$	2047	Jenkins, Pratt (1978)	
K_2PtBr_6	1423	Jenkins, Pratt (1978)	
Ag_2PtBr_6	1791	Jenkins, Pratt (1978)	
K_2PtI_6	1421	Lister, Nyburg, Poyntz (1974)	
K_2ReCl_6	1416	Jenkins, Pratt (1978)	1442
Rb_2ReCl_6	1414	Jenkins, Pratt (1978)	—
Cs_2ReCl_6	1398	Jenkins, Pratt (1978)	—
$(NH_4)_2ReCl_6$	1402	Jenkins, Pratt (1978)	1390
K_2ReBr_6	1375	Jenkins, Pratt (1978)	1375
K_2SiF_6	1670	Jenkins, Pratt (1978)	1628
Rb_2SiF_6	1639	Jenkins, Pratt (1978)	1621
Cs_2SiF_6	1604	Jenkins, Pratt (1978)	1498
Tl_2SiF_6	1675	Jenkins, Pratt (1978)	—
$(NH_4)_2SiF_6$	1657	Jenkins, Pratt (1978)	1727
K_2SnCl_6	1363	Jenkins, Pratt (1978)	1390
Rb_2SnCl_6	1361	Jenkins, Pratt (1978)	1363
Cs_2SnCl_6	1358	Jenkins, Pratt (1978)	—
Tl_2SnCl_6	1437	Jenkins, Pratt (1978)	—
$(NH_4)_2SnCl_6$	1370	Jenkins, Pratt (1978)	1334
Rb_2SnBr_6	1309	Jenkins, Pratt (1978)	—
Cl_2SnBr_6	1306	Jenkins, Pratt (1978)	—
$(NH_4)_2SnBr_6$	1319	Jenkins, Pratt (1978)	—
Rb_2SnI_6	1226	Jenkins, Pratt (1978)	—
Cs_2SnI_6	1243	Jenkins, Pratt (1978)	—
K_2TeCl_6	1318	Jenkins, Pratt (1978)	1320
Rb_2TeCl_6	1321	Jenkins, Pratt (1978)	—
Cs_2TeCl_6	1323	Jenkins, Pratt (1978)	—
Tl_2TeCl_6	1392	Jenkins, Pratt (1978)	—
$(NH_4)_2TeCl_6$	1318	Jenkins, Pratt (1978)	—

Salt	Calculated lattice energy (kJmol^{-1})	Literature source	Thermochemical cycle lattice energy (kJmol^{-1})
K_2RuCl_6	1451	Jenkins, Pratt (1978)	—
Rb_2CoF_6	1688	Jenkins, Pratt (1978)	—
Cs_2CoF_6	1632	Jenkins, Pratt (1978)	—
K_2NiF_6	1721	Jenkins, Pratt (1978)	—
Rb_2NiF_6	1688	Jenkins, Pratt (1978)	—
Rb_2SbCl_6	1357	Jenkins, Pratt (1978)	—
Rb_2SeCl_6	1409	Jenkins, Pratt (1978)	—
Cs_2SeCl_6	1397	Jenkins, Pratt (1978)	—
$(NH_4)_2SeCl_6$	1420	Jenkins, Pratt (1978)	—
K_2SeBr_6	1379	Jenkins, Pratt (1978)	—
$(NH_4)_2SeBr_6$	1380	Jenkins, Pratt (1978)	—
$(NH_4)_2PoCl_6$	1338	Jenkins, Pratt (1978)	—
Cs_2PoBr_6	1286	Jenkins, Pratt (1978)	—
$(NH_4)_2PoBr_6$	1292	Jenkins, Pratt (1978)	—
Cs_2CrF_6	1603	Jenkins, Pratt (1978)	—
Rb_2MnF_6	1688	Jenkins, Pratt (1978)	—
Cs_2MnF_6	1620	Jenkins, Pratt (1978)	—
K_2MnCl_6	1462	Jenkins, Pratt (1978)	—
Rb_2MnCl_6	1451	Jenkins, Pratt (1978)	—
$(NH_4)_2MnCl_6$	1464	Jenkins, Pratt (1978)	—
Cs_2TeBr_6	1306	Jenkins, Pratt (1978)	—
$(NH_4)_2TeBr_6$	1294	Jenkins, Pratt (1978)	—
Cs_2TeI_6	1246	Jenkins, Pratt (1978)	—
K_2TiCl_6	1412	Jenkins, Pratt (1978)	1443
Rb_2TiCl_6	1415	Jenkins, Pratt (1978)	1425
Cs_2TiCl_6	1402	Jenkins, Pratt (1978)	1370
Tl_2TiCl_6	1560	Jenkins, Pratt (1978)	1553
$(NH_4)_2TiCl_6$	1413	Jenkins, Pratt (1978)	—
K_2TiBr_6	1379	Jenkins, Pratt (1978)	1379
Rb_2TiBr_6	1341	Jenkins, Pratt (1978)	1365
Cs_2TiBr_6	1339	Jenkins, Pratt (1978)	1316
Na_2UBr_6	1504	Vdorenko, Suglobova, Chirkst (1974)	—
K_2UBr_6	1484	Vdorenko, Suglobova, Chirkst (1974)	—
Rb_2UBr_6	1473	Vdorenko, Suglobova, Chirkst (1974)	—
Cs_2UBr_6	1459	Vdorenko, Suglobova, Chirkst (1974)	—
K_2WCl_6	1398	Jenkins, Pratt (1978)	1423
Rb_2WCl_6	1397	Jenkins, Pratt (1978)	1434
Cs_2WCl_6	1392	Jenkins, Pratt (1978)	1366
K_2WBr_6	1408	Jenkins, Pratt (1978)	1408
Rb_2WBr_6	1361	Jenkins, Pratt (1978)	1391
Cs_2WBr_6	1362	Jenkins, Pratt (1978)	1332
K_2ZrCl_6	1339	Jenkins, Pratt (1978)	1371
Rb_2ZrCl_6	1341	Jenkins, Pratt (1978)	—
Cs_2ZrCl_6	1339	Jenkins, Pratt (1978)	1308
Tetraalkyl ammonium salts			
Me_4NCl	566	Wilson (1976)	—

Table 1 (continued)
LATTICE ENERGIES

Salt	Calculated lattice energy ($kJmol^{-1}$)	Literature source	Thermochemical cycle lattice energy ($kJmol^{-1}$)
Me_4NBr	553	Wilson (1976)	—
Me_4NI	544	Wilson (1976)	—
Tellurides			
Li_2Te	2212	Bevan, Morris (1960)	—
Na_2Te	1997	Bevan, Morris (1960)	2095
K_2Te	1830	Bevans, Morris (1960)	—
Rb_2Te	1837	Bevan, Morris (1960)	—
Cs_2Te	1745	Bevan, Morris (1960)	—
Cu_2Te	2706	Mamulov (1961)	2683
Ag_2Te	2607	Mamulov (1961)	2600
Tl_2Te	2084	Mamulov (1961)	2172
BeTe	3319	Das, Keer, Rao (1963)	—
MgTe	2878	Das, Keer, Rao (1963)	3081
CaTe	2721	Das, Keer, Rao (1963)	—
SrTe	2599	Das, Keer, Rao (1963)	—
BaTe	2473	Das, Keer, Rao (1963)	—
RaTe	2481	Yatsimirskii, Krestov (1960)	—
MnTe	3041	Mamulov (1961)	—
FeTe	3399	Mamulov (1961)	—
CoTe	3429	Mamulov (1961)	—
NiTe	3534	Mamulov (1961)	—
CuTe	3639	Mamulov (1961)	—
ZnTe	3416	Mamulov (1961)	—
CdTe	3212	Mamulov (1961)	—

Salt	Calculated lattice energy ($kJmol^{-1}$)	Literature source	Thermochemical cycle lattice energy ($kJmol^{-1}$)
PbTe	2930	Mamulov (1961)	—
Thiocyanates			
LiCNS	764	Gill, Singla, Paul, Narula (1972)	(765)
NaCNS	682	Morris (1958)	682
KCNS	623	Morris (1958)	616
RbCNS	623	Morris (1958)	619
CsCNS	623	Yatsimirskii (1956)	568
NH_4CNS	605	Gill, Singla, Paul, Narula (1972)	611
$Ca(CNS)_2$	2184	Yatsimirskii (1956)	2118
$Sr(CNS)_2$	2063	Yatsimirskii (1956)	1957
$Ba(CNS)_2$	1979	Yatsimirskii (1956)	1852
$Mn(CNS)_2$	2280	Yatsimirskii (1961)	2351
$Zn(CNS)_2$	2335	Yatsimirskii (1961)	2560
$Cd(CNS)_2$	2201	Yatsimirskii (1961)	2374
$Hg(CNS)_2$	2146	Yatsimirskii (1961)	2492
$Sn(CNS)_2$	2117	Yatsimirskii (1961)	2142
$Pb(CNS)_2$	2058	Morris (1958)	—
Vanadates			
$LiVO_3$	810	Golovkin, Fotiev (1970)	—
$NaVO_3$	761	Golovkin, Fotiev (1970)	—
KVO_3	686	Golovkin, Fotiev (1970)	—
$RbVO_3$	657	Golovkin, Fotiev (1970)	—
$CsVO_3$	628	Golovkin, Fotiev (1970)	—

Jenkins and Waddington refers to a forthcoming review article: "The Use of Lattice Energy in Inorganic Chemistry"; publication details can be obtained by writing to H. D. B. Jenkins, Department of Chemistry, University of Warwick, Coventry CV 4 7AL, Warwickshire.

Table 2
ANCILLARY THERMODYNAMIC DATA

Salt or ion	State	Source	ΔH_f^θ (kJmol⁻¹)
Li^+	g	JANAF	687.16
Na^+	g	JANAF	609.84
K^+	g	JANAF	514.19
Cs^+	g	JANAF	452.3
O_2^-	g	Jenkins, Waddington[a]	−74
CN^-	g	Jenkins, Pratt, Waddington (1977)	41
NH_4^+	g	Jenkins, Morris (1977)	630.2
C_2^{-2}	g	Jenkins, Waddington[a]	918
O_2^{-2}	g	Jenkins, Waddington[a]	553
NO_3^-	g	Jenkins, Morris (1977)	−320.1
$IO_2F_2^-$	g	Jenkins, Waddington[a]	−693
IO_3^-	g	Jenkins, Waddington[a]	−208
ClO_3^-	g	Jenkins, Waddington[a]	−200
BrO_3^-	g	Jenkins, Waddington[a]	−145
HSO_4^-	g	Jenkins, Waddington[a]	(−1012)
HCO_3^-	g	Jenkins, Waddington[a]	−738
$NH_2CH_2CO_2^-$	g	Bernard	−564
N^{-3}	g	Jenkins, Waddington[a]	(2588)
CO_3^{-2}	g	Jenkins, Waddington, Pratt	−321
PO_4^{-3}	g	Jenkins, Waddington[a]	(291)
AsO_4^{-3}	g	Jenkins, Waddington[a]	(289)
GeO_4^{-4}	g	Jenkins, Waddington[a]	(1460)
CH_3COO^-	g	Jenkins, Waddington[a]	−554
ClO_4^-	g	Jenkins, Pratt (1978)	−344
NO_2^-	g	Jenkins (1977)	−219
SeO_3^{-2}	g	Jenkins, Waddington[a]	(−249)
HCO_2^-	g	Jenkins, Waddington[a]	−463
BH_4^-	g	Waddington (1959)	−96
HS^-	g	Jenkins, Waddington[a]	−120
$TiCl_6^{-2}$	g	Jenkins, Pratt (1978)	−1330
$TiBr_6^{-2}$	g	Jenkins, Pratt (1978)	−1142
$ZrCl_6^{-2}$	g	Jenkins, Pratt (1978)	−1526
$HfCl_6^{-2}$	g	Jenkins, Pratt (1978)	−1640
$NbCl_6^{-2}$	g	Jenkins, Pratt (1978)	−1224
$TaCl_6^{-2}$	g	Jenkins, Pratt (1978)	−1275
$MoCl_6^{-2}$	g	Jenkins, Pratt (1978)	−1070
WCl_6^{-2}	g	Jenkins, Pratt (1978)	−985
WBr_6^{-2}	g	Jenkins, Pratt (1978)	−705
$ReCl_6^{-2}$	g	Jenkins, Pratt (1978)	−919
$ReBr_6^{-2}$	g	Jenkins, Pratt (1978)	−689
$OsCl_6^{-2}$	g	Jenkins, Pratt (1978)	−752
$IrCl_6^{-2}$	g	Jenkins, Pratt (1978)	−785
$PdCl_6^{-2}$	g	Jenkins, Pratt (1978)	−749
$PtCl_6^{-2}$	g	Jenkins, Pratt (1978)	−774
$PtBr_6^{-2}$	g	Jenkins, Pratt (1978)	−645
SiF_6^{-2}	g	Jenkins, Pratt (1978)	−2207
$GeCl_6^{-2}$	g	Jenkins, Pratt (1978)	−981
$SuCl_6^{-2}$	g	Jenkins, Pratt (1978)	−1156
$PbCl_6^{-2}$	g	Jenkins, Pratt (1978)	−940
$TeCl_6^{-2}$	g	Jenkins, Pratt (1978)	−902
NO_2^-	g	Jenkins, Pratt (1978)	−202
LiO_2	c	D'Orazio, Wood (1965)	−259.4
NaO_2	c	JANAF	−260.7
KO_2	c	JANAF 1974 Suppl.	−284.5
CsO_2	c	D'Orazio, Wood (1965)	−289.5
Li_2O_2	c	JANAF	−632.6
Na_2O_2	c	JANAF	−531.2
K_2O_2	c	JANAF	−495.8
Cs_2O_2	c	NBS Circular 500	−402.5
ThO_2	c	Smirnov, Ivanovskii (1957)	−1230.5
MgO_2	c	NBS Circular 500	−623.0
$NaIO_2F_2$	c	Finch, Gates, Jenkinson	−848.7
KIO_2F_2	c	Finch, Gates, Jenkinson	−876.6
$RbIO_2F_2$	c	Finch, Gates, Jenkinson	−875.3
$NH_4IO_2F_2$	c	Finch, Gates, Jenkinson	−747.7
$LiBrO_3$	c	Boyd, Vaslow (1962)	−356.4
$CsBrO_3$	c	Boyd, Vaslow (1962)	−374.5
$CsHCO_3$	c	NBS Circular 500	−995.6
$Ca(HCO_3)_2$	c	Wilcox, Bromley (1963)	(−1950)
$Sr(HCO_3)_2$	c	Wilcox, Bromley (1963)	(−1954)
$Ba(HCO_3)_2$	c	Wilcox, Bromley (1963)	(−1971)
CH_3COONH_4	c	NBS Circular 500	−618.4
CH_3COOLi	c	Rudnitskii (1961)	−748.9
CH_3COORb	c	Morris (1959)	−720.9

Salt or ion	State	Source	ΔH_f^θ (kJmol⁻¹)
$LiClO_4$	c	JANAF	−380.7
$CsClO_4$	c	Wilcox, Bromley (1963)	(−435.1)
LiH	c	JANAF	−90.62
NaH	c	JANAF	−56.4
KH	c	JANAF	−57.8
CsH	c	Smith, Bass (1963)	−49.9
LiN_3	c	Gray, Waddington (1956)	10.8
NaN_3	c	Gray, Waddington (1956)	21.25
KN_3	c	Gray, Waddington (1956)	−1.38
RbN_3	c	Gray, Waddington (1956)	−0.29
CsN_3	c	Gray, Waddington (1956)	−9.92
TlN_3	c	Gray, Waddington (1956)	233.4
AgN_3	c	Gray, Waddington (1956)	310.3
$CsHS$	c	NBS Circular 500	−263.2
$Ca(HS)_2$	c	Wilcox, Bromley (1963)	(−481)
$Ba(HS)_2$	c	Wilcox, Bromley (1956)	(−531.4)
$LiHS$	c	Juza, Laurer (1953)	−254.4
K_2TiCl_6	c	Jenkins, Pratt (1978)[b]	−1747
Rb_2TiCl_6	c	Jenkins, Pratt (1978)[b]	−1767
Cs_2TiCl_6	c	Jenkins, Pratt (1978)[b]	−1797
Tl_2TiCl_6	c	Jenkins, Pratt (1978)[b]	−1330
K_2TlBr_6	c	Jenkins, Pratt (1978)[b]	−1493
Rb_2TiBr_6	c	Jenkins, Pratt (1978)[b]	−1517
Cs_2TiBr_6	c	Jenkins, Pratt (1978)[b]	−1553
K_2ZrCl_6	c	Jenkins, Pratt (1978)[b]	−1932
Cs_2ZrCl_6	c	Jenkins, Pratt (1978)[b]	−1992
K_2HfCl_6	c	Jenkins, Pratt (1978)[b]	−1957
K_2NbCl_6	c	Jenkins, Pratt (1978)[b]	−1594
Rb_2NbCl_6	c	Jenkins, Pratt (1978)[b]	−1619
Cs_2NbCl_6	c	Jenkins, Pratt (1978)[b]	−1663
K_2ReCl_6	c	Jenkins, Pratt (1978)[b]	−1333
$(NH_4)_2ReCl_6$	c	Jenkins, Pratt (1978)[b]	−1036
K_2ReBr_6	c	Jenkins, Pratt (1978)[b]	−1036
K_2OsCl_6	c	Jenkins, Pratt (1978)[b]	−1171
K_2IrCl_6	c	Jenkins, Pratt (1978)[b]	−1197
K_2SiF_6	c	Jenkins, Pratt (1978)[b]	−2807
Rb_2SiF_6	c	Jenkins, Pratt (1978)[b]	−2838
Cs_2SiF_6	c	Jenkins, Pratt (1978)[b]	−2801
$(NH_4)_2SiF_6$	c	Jenkins, Pratt (1978)[b]	−2681
Rb_2PbCl_6	c	Jenkins, Pratt (1978)[b]	−1293
$Mn(CN)_2$	c	Karapet'yants, Karapet'yants	60.7
$RbCN$	c	Pritchard (1953)	(−108.8)
$CsCN$	c	Pritchard (1953)	(−108.7)
$LiCN$	c	Jenkins, Pratt, Waddington (1977)	−121
$Sr(CN)_2$	c	Wilcox, Bromley (1963)	(−205)
$LiNo_3$	c	NBS Circular 500	−482.3
$RbNO_3$	c	NBS Circular 500	−489.7
$CsNO_3$	c	NBS Circular 500	−494.2
$Fe(NO_3)_2$	c	Wilcox, Bromley (1963)	(−447.7)
$Sn(NO_3)_2$	c	Yatsimirskii (1961)	−456.1
$NaBF_4$	c	Altschuller (1955)	−1774
$CsIO_3$	c	Shedlovskii, Voskresenskii (1966)	−526.3
$CsClO_3$	c	Volodina, Shidlovskii, Voskresenskii (1966)	−395.8
$Mg(ClO_3)_2$	c	Wilcox, Bromley (1963)	(−523)
$Ca(ClO_3)_2$	c	Rudnitskii (1961)	−757
$Sr(ClO_3)_2$	c	Rudnitskii (1961)	−766
ScN	c	Neuman, Kroger, Kunz	−284.5
Li_2CO_3	c	JANAF	−1216.0
Cs_2CO_3	c	NBS Circular 500	−1118.8
$NiCO_3$	c	Karapet'yants (1955)	−689.1
$CuCO_3$	c	NBS Circular 500	−595.0
$SnCO_3$	c	Wilcox, Bromley (1963)	(−740.6)
HCO_2Li	c	Morris (1959)	−649.3
HCO_2Rb	c	Morris(1959)	−656.0
TiH	c	Gibb (1962)	−130.5
ZrH	c	Gibb (1962)	−173.5
VH	c	Gibb (1962)	(−32)
NbH	c	Gibb (1962)	(−100)
PdH	c	Gibb (1962)	(−37)
CuH	c	Gibb (1962)	(−21.4)
LaH_2	c	Gibb (1962)	−230.1
CeH_2	c	Gibb (1962)	−141.8

Table 2 (continued)
ANCILLARY THERMODYNAMIC DATA

Salt or ion	State	Source	ΔH_f^θ (kJmol⁻¹)	Salt or ion	State	Source	ΔH_f^θ (kJmol⁻¹)
PrH_2	c	Gibb (1962)	−200.0	Rb_2WCl_6	c	Jenkins, Pratt (1978)[b]	−1429
NdH_2	c	Gibb (1962)	−187.4	Cs_2WCl_6	c	Jenkins, Pratt (1978)[b]	−1446
SmH_2	c	Gibb (1962)	−187.4	K_2WBr_6	c	Jenkins, Pratt (1978)[b]	−1085
GdH_2	c	Gibb (1962)	−196.2	Rb_2WBr_6	c	Jenkins, Pratt (1978)[b]	−1106
CsOH	c	NBS Circular 500	−406.7	Cs_2WBr_6	c	Jenkins, Pratt (1978)[b]	−1132
$Ti(OH)_2$	c	Wilcox, Bromley (1963)	(−778.0)	K_2PdCl_6	c	Jenkins, Pratt (1978)[b]	−1187
LiOH	c	NBS Circular 500	−487.2	Ag_2PtCl_6	c	Jenkins, Pratt (1978)[b]	−527
K_2TaCl_6	c	Jenkins, Pratt (1978)[b]	−1648	$BaPtCl_6$	c	Jenkins, Pratt (1978)[b]	−1180
Rb_2TaCl_6	c	Jenkins, Pratt (1978)[b]	−1669	K_2PtBr_6	c	Jenkins, Pratt (1978)[b]	−1040
Cs_2TaCl_6	c	Jenkins, Pratt (1978)[b]	−1711	Ag_2PtBr_6	c	Jenkins, Pratt (1978)[b]	−398
Na_2MoCl_6	c	Jenkins, Pratt (1978)[b]	−1376	Cs_2GeCl_6	c	Jenkins, Pratt (1978)[b]	−1451
K_2MoCl_6	c	Jenkins, Pratt (1978)[b]	−1475	K_2SnCl_6	c	Jenkins, Pratt (1978)[b]	−1518
Rb_2MoCl_6	c	Jenkins, Pratt (1978)[b]	−1479	Rb_2SnCl_6	c	Jenkins, Pratt (1978)[b]	−1529
Cs_2MoCl_6	c	Jenkins, Pratt (1978)[b]	−1512	$(NH_4)_2SnCl_6$	c	Jenkins, Pratt (1978)[b]	−1237
K_2WCl_6	c	Jenkins, Pratt (1978)[b]	−1380	Rb_2TeCl_6	c	Jenkins, Pratt (1978)[b]	−1237

[a] Jenkins and Waddington refers to a forthcoming review article: "The Use of Lattice Energy in Inorganic Chemistry"; publication details can be obtained by writing to H. D. B. Jenkins, Department of Chemistry, University of Warwick, Coventry CV 4 7AL, Warwickshire.

[b] For sources of data used in this ancillary table, refer to *Adv. Inorg. Chem. Radiochem.*, 22, 1, 1978.

THERMODYNAMIC PROPERTIES OF ALKANE HYDROCARBONS[a]

Symbols and Units

$(G^\circ - H^\circ_0)/T$	=	Gibbs energy function, cal K⁻¹ mol⁻¹
$(H^\circ - H^\circ_0)/T$	=	enthalpy function, cal K⁻¹ mol⁻¹
S°	=	entropy, cal K⁻¹ mol⁻¹
C_p	=	heat capacity at constant pressure, cal K⁻¹ mol⁻¹
ΔHf°	=	enthalpy of formation, kcal mol⁻¹
ΔGf°	=	Gibbs energy of formation, kcal mol⁻¹
$\log_{10} Kf$	=	common logarithm of the equilibrium constant of formation
T	=	temperature, kelvins
$H^\circ - H^\circ_0$	=	enthalpy, kcal mol⁻¹

Conversion Factors

The tabulated values may be converted to SI units by use of the relations:

calories × 4.184 = joules
kilocalories × 4184 = joules

T, K	$(G^\circ - H^\circ_0)/T$, cal K⁻¹ mol⁻¹	$(H^\circ - H^\circ_0)/T$, cal K⁻¹ mol⁻¹	$H^\circ - H^\circ_0$, kcal mol⁻¹	S°, cal K⁻¹ mol⁻¹	C_p°, cal K⁻¹ mol⁻¹	ΔH_f°, kcal mol⁻¹	ΔG_f°, kcal mol⁻¹	$\log_{10} K_f$
				Methane (CH_4)				
0	0	0	0	0	0	-15.9_5	-15.9_5	∞
200	-33.3_1	7.9_4	1.58_8	41.2_5	8.0_1	-17.1_8	-13.8_8	15.1_7
273.15	-35.7_9	7.9_9	2.18_4	43.7_8	8.3_3	-17.6_8	-12.6_0	10.0_8
298.15	-36.4_9	8.0_3	2.39	44.5_2	8.5_3	-17.8_6	−12.1	8.8_8
400	-38.8_8	8.3_0	3.32	47.1_8	9.7_1	-18.6_0	-10.0_4	5.4_9
600	−42.4	9.2	5.5_4	51.6	12.5	−19.9	−5.5	1.9_9
800	−45.2	10.4	8.3	55.6	15.2	−20.8	−0.5	0.1_4
1,000	−47.7	11.6	11.6	59.3	$17._4$	−21.4	4.6	-1.0_0
1,200	$-49._9$	$12._7$	$15._3$	$62._6$	$19._2$	$-21._7$	$9._8$	-1.7_9
1,500	$-52._9$	$14._2$	$21._3$	$67._1$	21.	$-21._8$	$17._7$	-2.5_9
				Ethane (C_2H_6)				
0	0	0	0	0	0	-16.2_6	-16.2_6	∞
200	-41.6_3	8.6_7	1.73_5	50.3_0	10.1_0	-18.8_0	-11.4_6	12.5_2
273.15	-44.4_2	9.2_8	2.53_4	53.7_0	11.8_5	-19.7_0	-8.6_3	6.9_1
298.15	-45.2_4	9.5_2	2.84	54.7_6	12.5_7	-20.0_0	-7.6_0	5.5_7
400	-48.2_0	10.7_1	4.28	58.9_1	15.8_3	-21.1_8	-3.1_8	1.7_4
600	−53.1	13.5	8.0_9	66.6	22.2	−23.0	6.2	-2.2_7
800	−57.3	16.7	13.3	74.0	27.4	−23.7	16.1	-4.4_1
1,000	−61.3	19.0	19.0	80.3	$31._7$	−24.3	26.2	-5.7_3
1,200	$-65._0$	$21._4$	$25._7$	$86._4$	$35._1$	$-24._1$	$36._3$	-6.6_1
1,500	$-70._1$	$24._6$	$36._9$	$94._7$	39.	$-23._1$	$51._3$	-7.4_7
				Propane (C_3H_8)				
0	0	0	0	0	0	-19.7_4	-19.7_4	∞
200	-48.3_6	10.2_1	2.04_1	58.5_7	12.6_4	-23.4_2	-11.8_8	12.9_8
273.15	-51.7_0	11.3_3	3.09_6	63.0_3	16.2_2	-24.6_6	-7.4_7	5.9_7

THERMODYNAMIC PROPERTIES OF ALKANE HYDROCARBONS (Continued)

T, K	$(G° - H°_0)/T$, cal K^{-1} mol^{-1}	$(H° - H°_0)/T$, cal K^{-1} mol^{-1}	$H° - H°_0$, kcal mol^{-1}	S°, cal K^{-1} mol^{-1}	$C_p°$, cal K^{-1} mol^{-1}	$\Delta H_f°$, kcal mol^{-1}	$\Delta G_f°$, kcal mol^{-1}	$\log_{10} K_f$
298.15	-52.7_1	11.8_0	3.52	64.5_1	17.4_4	-25.0_7	-5.8_7	4.3_0
400	-56.4_6	13.8_8	5.55	70.3_4	22.4_6	-26.6_1	0.9_4	-0.5_1
600	-62.9	18.2	10.9_5	81.1	31.2	-28.9	15.2	-5.5_5
800	-68.7	22.4	17.9	91.1	38.0	-30.2	30.2	-8.2_5
1,000	-74.1	26.1	26.1	100.2	$43._4$	-30.8	45.4	-9.9_2
1,200	$-79._2$	$29._3$	$35._2$	$108._5$	$47._6$	$-30._7$	$60._6$	-11.0_4
1,500	$-86._2$	$33._5$	$50._2$	$119._7$	52.	$-29._7$	$83._3$	-12.1_4

n-Butane (C_4H_{10})

T, K	$(G° - H°_0)/T$, cal K^{-1} mol^{-1}	$(H° - H°_0)/T$, cal K^{-1} mol^{-1}	$H° - H°_0$, kcal mol^{-1}	S°, cal K^{-1} mol^{-1}	$C_p°$, cal K^{-1} mol^{-1}	$\Delta H_f°$, kcal mol^{-1}	$\Delta G_f°$, kcal mol^{-1}	$\log_{10} K_f$
0	0	0	0	0	0	-23.6_3	-23.6_3	∞
200	-52.5_7	13.4_1	2.68_1	65.9_8	18.3_6	-28.1_3	-12.2_3	13.3_6
273.15	-57.0_1	15.1_8	4.14_5	72.1_9	21.8_9	-29.5_7	-6.1_9	4.9_5
298.15	-58.3_7	15.8_0	4.71	74.1_7	23.3_0	-30.0_5	-4.0_2	2.9_5
400	-63.3_8	18.4_9	7.40	81.8_7	29.4_5	-31.9_0	5.1_7	-2.8_2
600	-71.9	24.0	14.4_2	95.9	40.4	-34.7	24.3	-8.8_7
800	-79.6	29.2	23.4	108.8	48.8	-36.3	44.3	-12.1_0
1,000	-86.6	33.8	33.8	120.4	$55._2$	-37.0	64.5	-14.1_0
1,200	$-93._1$	$37._8$	$45._4$	$130._9$	$60._1$	$-37._0$	$84._8$	-15.4_5
1,500	$-102._1$	$42._8$	$64._3$	$144._9$	66.	$-36._1$	$115._2$	-16.7_8

2-Methylpropane (C_4H_{10})

T, K	$(G° - H°_0)/T$, cal K^{-1} mol^{-1}	$(H° - H°_0)/T$, cal K^{-1} mol^{-1}	$H° - H°_0$, kcal mol^{-1}	S°, cal K^{-1} mol^{-1}	$C_p°$, cal K^{-1} mol^{-1}	$\Delta H_f°$, kcal mol^{-1}	$\Delta G_f°$, kcal mol^{-1}	$\log_{10} K_f$
0	0	0	0	0	0	-25.3_4	-25.3_4	∞
200	-50.7_9	11.7_2	2.34_4	62.5_1	16.8_5	-30.1_7	-13.5_8	14.8_4
273.15	-54.7_3	13.7_1	3.74_4	68.4_4	21.5_2	-31.6_7	-7.2_7	5.8_2
298.15	-55.9_7	14.4_4	4.30	70.4_1	23.1_3	-32.1_6	-5.0_0	3.6_7
400	-60.6_3	17.4_9	6.99	78.1_2	29.6_2	-34.0_0	4.5_6	-2.4_9
600	-68.9	23.4	14.0_7	92.3	40.8	-36.7	24.5	-8.9_2
800	-76.4	28.9	23.1	105.3	49.5	-38.2	45.1	-12.3_3
1,000	-83.3	33.8	33.8	117.1	$56._4$	-38.7	66.1	-14.4_4
1,200	$-89._9$	$38._0$	$45._6$	$127._9$	$61._7$	$-38._4$	$87._0$	-15.8_4
1,500	$-99._0$	$43._3$	$65._0$	$142._3$	68.	$-37._0$	$118._2$	-17.2_2

n-Pentane (C_5H_{12})

T, K	$(G° - H°_0)/T$, cal K^{-1} mol^{-1}	$(H° - H°_0)/T$, cal K^{-1} mol^{-1}	$H° - H°_0$, kcal mol^{-1}	S°, cal K^{-1} mol^{-1}	$C_p°$, cal K^{-1} mol^{-1}	$\Delta H_f°$, kcal mol^{-1}	$\Delta G_f°$, kcal mol^{-1}	$\log_{10} K_f$
0	0	0	0	0	0	-27.3_5	-27.3_5	∞
200	-57.0_3	16.4_6	3.29_1	73.4_9	22.3_6	-32.7_0	-12.4_5	13.6_0
273.15	-62.4_7	18.6_1	5.08_3	81.0_8	26.9_0	-34.4_1	-4.7_7	3.8_2
298.15	-64.1_4	19.3_8	5.78	83.5_2	28.6_9	-34.9_8	-2.0_3	1.4_9
400	-70.2_9	22.7_4	9.10	93.0_3	36.4_6	-37.1_6	9.5_8	-5.2_3
600	-80.8	29.7	17.7_9	110.5	49.9	-40.3	33.7	-12.2_7
800	-90.3	36.0	28.8	126.3	59.9	-42.2	58.7	-16.0_3
1,000	-98.9	41.6	41.6	140.5	$67._3$	-43.0	84.0	-18.3_6
1,200	$-106._9$	$46._3$	$55._6$	$153._2$	$72._8$	$-43._0$	$109._4$	-19.9_2
1,500	$-117._9$	$52._3$	$78._4$	$170._2$	79.	$-42._1$	$147._4$	-21.4_8

2-Methybutane (C_5H_{12})

T, K	$(G° - H°_0)/T$, cal K^{-1} mol^{-1}	$(H° - H°_0)/T$, cal K^{-1} mol^{-1}	$H° - H°_0$, kcal mol^{-1}	S°, cal K^{-1} mol^{-1}	$C_p°$, cal K^{-1} mol^{-1}	$\Delta H_f°$, kcal mol^{-1}	$\Delta G_f°$, kcal mol^{-1}	$\log_{10} K_f$
0	0	0	0	0	0	-28.4_4	-28.4_4	∞
200	-58.1_6	14.3_4	2.86_7	72.5_0	20.3_0	-34.2_0	-13.7_6	15.0_3
273.15	-62.9_8	16.7_4	4.57_3	79.7_2	26.3_8	-36.0_0	-6.0_0	4.8_0
298.15	-64.4_9	17.6_4	5.26	82.1_3	28.4_1	-36.5_8	-3.2_1	2.3_6
400	-70.2_0	21.4_3	8.57	91.6_3	36.5_4	-38.7_6	8.5_3	-4.6_6
600	-80.3	28.8	17.3_0	109.1	50.2	-41.9	32.9	-11.9_9
800	-89.5	35.5	28.4	125.0	60.5	-43.7	58.2	-15.8_9
1,000	-98.1	41.3	41.3	139.4	$68._4$	-44.3	83.7	-18.3_0
1,200	$-106._1$	$46._3$	$55._6$	$152._4$	$74._4$	$-44._1$	$109._3$	-19.9_1
1,500	$-117._1$	$52._7$	$79._0$	$169._8$	81.	$-42._6$	$147._5$	-21.4_9

2,2-Dimethylpropane (C_5H_{12})

T, K	$(G° - H°_0)/T$, cal K^{-1} mol^{-1}	$(H° - H°_0)/T$, cal K^{-1} mol^{-1}	$H° - H°_0$, kcal mol^{-1}	S°, cal K^{-1} mol^{-1}	$C_p°$, cal K^{-1} mol^{-1}	$\Delta H_f°$, kcal mol^{-1}	$\Delta G_f°$, kcal mol^{-1}	$\log_{10} K_f$
0	0	0	0	0	0	-32.3_8	-32.3_8	∞
200	-48.3_2	13.4_3	2.68_5	61.7_5	19.2_5	-38.3_3	-15.7_4	17.2_0
273.15	-53.0_1	17.4_6	4.77_0	70.4_7	26.6_8	-39.7_6	-7.2_2	5.7_8
298.15	-54.5_3	18.5_8	5.54	73.1_1	28.8_8	-40.2_5	-4.1_9	3.0_7
400	-60.3_3	22.7_2	9.09	83.0_5	37.2_8	-42.2_0	8.5_3	-4.6_6
600	-70.7	30.2	18.1_1	100.9	51.3	-45.0	34.7	-12.6_5
800	-80.2	36.9	29.6	117.1	62.4	-46.5	61.7	-16.8_6
1,000	-88.9	43.0	43.0	131.9	$71._2$	-46.6	88.9	-19.4_3
1,200	$-97._2$	$48._3$	$57._9$	$145._5$	$78._2$	$-45._7$	$116._0$	-21.1_3
1,500	$-108._7$	$55._1$	$82._7$	$163._8$	86.	$-42._9$	$156._2$	-22.7_6

n-Hexane (C_6H_{14})

T, K	$(G° - H°_0)/T$, cal K^{-1} mol^{-1}	$(H° - H°_0)/T$, cal K^{-1} mol^{-1}	$H° - H°_0$, kcal mol^{-1}	S°, cal K^{-1} mol^{-1}	$C_p°$, cal K^{-1} mol^{-1}	$\Delta H_f°$, kcal mol^{-1}	$\Delta G_f°$, kcal mol^{-1}	$\log_{10} K_f$
0	0	0	0	0	0	-31.0_9	-31.0_9	∞
200	-61.4_5	19.5_6	3.91_1	81.0_1	26.4_3	-37.2_6	-12.6_8	13.8_5
273.15	-67.9_2	22.1_0	6.03_5	90.0_2	31.9_2	-39.2_5	-3.3_7	2.7_0
298.15	-69.9_0	23.0_1	6.86	92.9_1	34.0_8	-39.9_1	-0.0_5	0.0_3
400	-77.2_1	27.0_1	10.80	104.2_2	43.3_9	-42.4_2	13.9_7	-7.6_3

THERMODYNAMIC PROPERTIES OF ALKANE HYDROCARBONS (Continued)

T, K	$(G° - H°_0)/T$, cal K^{-1} mol^{-1}	$(H° - H°_0)/T$, cal K^{-1} mol^{-1}	$H° - H°_0$, kcal mol^{-1}	S°, cal K^{-1} mol^{-1}	$C_p°$, cal K^{-1} mol^{-1}	$\Delta H_f°$, kcal mol^{-1}	$\Delta G_f°$, kcal mol^{-1}	$\log_{10} K_f$
600	-89.7	35.3	21.1$_5$	125.0	59.3	-46.0	43.0	-15.6$_7$
800	-100.9	42.8	34.2	143.7	70.8	-48.1	73.1	-19.9$_6$
1,000	-111.2	49.3	49.3	160.5	79.$_2$	-49.0	103.5	-22.6$_2$
1,200	-120.$_7$	54.$_8$	65.$_8$	175.$_5$	85.$_4$	-49.$_1$	134.$_0$	-24.4$_0$
1,500	-133.$_7$	61.$_7$	92.$_5$	195.$_4$	93.	-48.$_2$	179.$_7$	-26.1$_8$
2-Methylpentane (C_6H_{14})								
0	0	0	0	0	0	-32.1$_6$	-32.1$_6$	∞
200	-62.3$_5$	17.1$_8$	3.43$_6$	79.5$_3$	24.2$_2$	-38.8$_1$	-13.9$_3$	15.2$_2$
273.15	-68.1$_2$	20.0$_4$	5.47$_3$	88.1$_6$	31.5$_2$	-40.8$_9$	-4.5$_0$	3.6$_0$
298.15	-69.9$_2$	21.1$_1$	6.29	91.0$_3$	33.9$_9$	-41.5$_5$	-1.1$_3$	0.8$_3$
400	-76.7$_5$	25.6$_7$	10.27	102.4$_2$	43.8$_6$	-44.0$_3$	13.0$_8$	-7.1$_5$
600	-88.9	34.5	20.7$_3$	123.4	60.0	-47.5	42.5	-15.4$_7$
800	-99.9	42.5	34.0	142.4	71.8	-49.5	72.8	-19.8$_8$
1,000	-110.1	49.3	49.3	159.4	80.$_6$	-50.1	103.5	-22.6$_1$
1,200	-119.$_6$	55.$_1$	66.$_1$	174.$_7$	87.$_2$	-49.$_9$	134.$_2$	-24.4$_3$
1,500	-132.$_7$	62.$_3$	93.$_4$	195.$_0$	95.	-48.$_4$	180.$_0$	-26.2$_3$
3-Methylpentane (C_6H_{14})								
0	0	0	0	0	0	-31.4$_8$	-31.4$_8$	∞
200	-63.1$_3$	17.1$_0$	3.41$_9$	80.2$_3$	23.6$_4$	-38.1$_4$	-13.4$_0$	14.6$_4$
273.15	-68.8$_5$	19.8$_4$	5.41$_9$	88.6$_9$	31.0$_3$	-40.2$_6$	-4.0$_1$	3.2$_1$
298.15	-70.6$_3$	20.8$_9$	6.23	91.5$_2$	33.4$_9$	-40.9$_3$	-0.6$_5$	0.4$_8$
400	-77.3$_9$	25.3$_6$	10.14	102.7$_5$	43.3$_0$	-43.4$_6$	13.5$_1$	-7.3$_8$
600	-89.3	34.2	20.5$_0$	123.5	59.5	-47.1	42.9	-15.6$_1$
800	-100.3	42.1	33.6	142.4	71.4	-49.1	73.2	-19.9$_9$
1,000	-110.4	48.9	48.9	159.3	80.$_3$	-49.8	103.9	-22.7$_0$
1,200	-119.$_9$	54.$_7$	65.$_6$	174.$_6$	87.$_0$	-49.$_7$	134.$_6$	-24.5$_1$
1,500	-132.$_9$	61.$_9$	92.$_9$	194.$_8$	95.	-48.$_2$	180.$_5$	-26.3$_0$
2,2-Dimethylbutane (C_6H_{14})								
0	0	0	0	0	0	-34.2$_8$	-34.2$_8$	∞
200	-58.3$_5$	15.8$_2$	3.16$_3$	74.1$_7$	24.2$_5$	-41.2$_0$	-15.2$_4$	16.6$_6$
273.15	-63.7$_5$	19.0$_0$	5.19$_0$	82.7$_5$	31.3$_3$	-43.2$_8$	-5.4$_2$	4.3$_3$
298.15	-65.4$_7$	20.1$_5$	6.01	85.6$_2$	33.8$_1$	-43.9$_5$	-1.9$_1$	1.4$_0$
400	-72.0$_5$	24.9$_1$	9.96	96.9$_6$	43.7$_7$	-46.4$_4$	12.8$_5$	-7.0$_2$
600	-83.9	34.1	20.4$_6$	118.0	60.5	-49.9	43.3	-15.7$_8$
800	-94.9	42.3	33.9	137.2	73.3	-51.7	74.7	-20.4$_1$
1,000	-105.1	49.6	49.6	154.7	83.$_2$	-51.9	106.4	-23.2$_5$
1,200	-114.$_7$	55.$_8$	67.$_0$	170.$_5$	90.$_9$	-51.$_1$	137.$_9$	-25.1$_2$
1,500	-128.$_1$	63.$_8$	95.$_7$	191.$_9$	100.	-48.$_3$	184.$_9$	-26.9$_9$
2,3-Dimethylbutane (C_6H_{14})								
0	0	0	0	0	0	-32.2$_9$	-32.2$_9$	∞
200	-60.8$_6$	15.5$_6$	3.11$_1$	76.4$_2$	22.2$_6$	-39.2$_7$	-13.7$_6$	15.0$_4$
273.15	-66.1$_2$	18.5$_0$	5.05$_1$	84.6$_2$	30.6$_7$	-41.4$_4$	-4.0$_8$	3.2$_7$
298.15	-67.8$_0$	19.6$_3$	5.85	87.4$_3$	33.3$_2$	-42.1$_2$	-0.6$_2$	0.4$_6$
400	-74.2$_3$	24.4$_3$	9.77	98.6$_6$	43.4$_3$	-44.6$_5$	13.9$_6$	-7.6$_3$
600	-85.9	33.6	20.1$_7$	119.5	59.8	-48.2	44.1	-16.0$_7$
800	-96.7	41.8	33.4	138.5	72.1	-50.1	75.2	-20.5$_5$
1,000	-106.8	48.8	48.8	155.6	81.$_4$	-50.7	106.7	-23.3$_1$
1,200	-116.$_3$	54.$_8$	65.$_8$	171.$_1$	88.$_6$	-50.$_3$	138.$_1$	-25.1$_5$
1,500	-129.$_4$	62.$_4$	93.$_7$	191.$_8$	97.	-48.$_3$	185.$_1$	-26.9$_5$
***n*-Heptane (C_7H_{16})**								
0	0	0	0	0	0	-34.8$_3$	-34.8$_3$	∞
200	-65.8$_6$	22.6$_7$	4.53$_4$	88.5$_3$	30.5$_1$	-41.8$_3$	-12.9$_0$	14.1$_0$
273.15	-73.3$_5$	25.5$_8$	6.98$_8$	98.9$_4$	36.9$_6$	-44.0$_9$	-1.9$_6$	1.5$_7$
298.15	-75.6$_4$	26.6$_5$	7.94	102.2$_9$	39.4$_8$	-44.8$_4$	1.9$_4$	-1.4$_2$
400	-84.1$_1$	31.3$_0$	12.52	115.4$_1$	50.3$_5$	-47.6$_7$	18.3$_7$	-10.0$_4$
600	-98.6	40.9	24.5$_2$	139.5	68.7	-51.7	52.4	-19.0$_7$
800	-111.6	49.6	39.6	161.2	81.8	-54.1	87.4	-23.8$_8$
1,000	-123.5	57.0	57.0	180.5	91.$_7$	-55.1	123.0	-26.8$_7$
1,200	-134.$_4$	63.$_3$	76.$_0$	197.$_7$	98.$_1$	-55.$_2$	158.$_6$	-28.8$_8$
1,500	-149.$_4$	71.$_1$	106.$_6$	220.$_5$	106.	-54.$_4$	212.$_0$	-30.8$_8$
2-Methylhexane (C_7H_{16})								
0	0	0	0	0	0	-35.9$_1$	-35.9$_1$	∞
200	-66.7$_7$	20.3$_5$	4.07$_0$	87.1$_2$	28.3$_4$	-43.3$_8$	-14.1$_7$	15.4$_8$
273.15	-73.5$_8$	23.5$_8$	6.43$_9$	97.1$_6$	36.5$_1$	-45.7$_3$	-3.1$_1$	2.4$_9$
298.15	-75.7$_0$	24.7$_8$	7.39	100.4$_8$	39.3$_2$	-46.4$_8$	0.8$_4$	-0.6$_1$
400	-83.6$_9$	29.9$_4$	11.98	113.6$_3$	50.6$_6$	-49.3$_0$	17.4$_5$	-9.5$_4$
600	-97.8	40.1	24.0$_6$	137.9	69.2	-53.3	51.8	-18.8$_6$
800	-110.6	49.1	39.3	159.7	82.6	-55.5	87.2	-23.8$_1$
1,000	-122.4	56.9	56.9	179.3	92.$_4$	-56.3	123.0	-26.8$_7$

T, K	$(G° - H°_0)/T$, cal K^{-1} mol^{-1}	$(H° - H°_0)/T$, cal K^{-1} mol^{-1}	$H° - H°_0$, kcal mol^{-1}	S°, cal K^{-1} mol^{-1}	$C_p°$, cal K^{-1} mol^{-1}	$\Delta H_f°$, kcal mol^{-1}	$\Delta G_f°$, kcal mol^{-1}	$\log_{10} K_f$
1,200	$-133._4$	$63._4$	$76._1$	$196._8$	$99._8$	$-56._2$	$158._8$	-28.9_2
1,500	$-148._4$	$71._6$	$107._4$	$220._0$	108.	$-54._7$	$212._4$	-30.9_5

3-Methylhexane (C_7H_{16})

T, K	$(G° - H°_0)/T$	$(H° - H°_0)/T$	$H° - H°_0$	S°	$C_p°$	$\Delta H_f°$	$\Delta G_f°$	$\log_{10} K_f$
0	-1.38	0	0	1.38	0	-35.2_2	-35.2_2	∞
200	-68.6_8	19.9_5	3.98_9	88.6_3	27.5_8	-42.7_7	-13.8_5	15.1_4
273.15	-75.3_6	23.1_4	6.32_2	98.5_0	36.2_0	-45.1_5	-2.9_0	2.3_2
298.15	-77.4_4	24.3_6	7.26	101.8_0	39.1_0	-45.9_1	1.0_1	-0.7_4
400	-85.3_2	29.6_0	11.84	114.9_2	50.6_0	-48.7_4	17.5_0	-9.5_6
600	[illegible]	[illegible]	[illegible]	[illegible]	[illegible]	[illegible]	[illegible]	[illegible]
800	-112.0	49.0	39.2	161.0	82.6	-54.9	86.7	-23.6_8
1,000	-123.8	56.7	56.7	180.5	$92._4$	-55.8	122.2	-26.7_1
1,200	$-134._8$	$63._3$	$76._0$	$198._1$	$99._8$	$-55._6$	$157._8$	-28.7_4
1,500	$-149._8$	$71._5$	$107._2$	$221._3$	108.	$-54._1$	$211._0$	-30.7_5

3-Ethylpentane (C_7H_{16})

T, K	$(G° - H°_0)/T$	$(H° - H°_0)/T$	$H° - H°_0$	S°	$C_p°$	$\Delta H_f°$	$\Delta G_f°$	$\log_{10} K_f$
0	0	0	0	0	0	-34.8_4	-34.8_4	∞
200	-64.1_4	20.5_0	4.09_9	84.6_4	29.9_6	-42.2_8	-12.5_6	13.7_3
273.15	-71.0_4	23.9_5	6.54_1	94.9_9	37.0_6	-44.5_5	-1.3_4	1.0_7
298.15	-73.2_0	25.1_6	7.50	98.3_6	39.6_7	-45.2_9	2.6_6	-1.9_5
400	-81.2_9	30.2_5	12.10	111.5_4	50.5_6	-48.1_0	19.4_9	-10.6_5
600	-95.5	40.2	24.1_3	135.7	68.9	-52.1	54.2	-19.7_6
800	-108.3	49.1	39.3	157.4	82.2	-54.4	90.1	-24.6_1
1,000	-120.1	56.8	56.8	176.9	$92._1$	-55.3	126.3	-27.6_1
1,200	$-131._0$	$63._3$	$76._0$	$194._3$	$99._5$	$-55._2$	$162._6$	-29.6_2
1,500	$-146._1$	$71._4$	$107._2$	$217._5$	108.	$-53._8$	$217._0$	-31.6_2

2,2-Dimethylpentane (C_7H_{16})

T, K	$(G° - H°_0)/T$	$(H° - H°_0)/T$	$H° - H°_0$	S°	$C_p°$	$\Delta H_f°$	$\Delta G_f°$	$\log_{10} K_f$
0	0	0	0	0	0	-38.0_2	-38.0_2	∞
200	-62.1_9	18.2_8	3.65_5	80.4_7	27.9_3	-45.9_0	-15.3_6	16.7_8
273.15	-68.4_3	22.0_3	6.01_7	90.4_6	36.7_9	-48.2_6	-3.8_1	3.0_5
298.15	-70.4_3	23.4_0	6.98	93.8_3	39.8_4	-49.0_0	0.3_0	-0.2_2
400	-78.1_0	29.1_4	11.66	107.2_4	51.8_1	-51.7_3	17.5_8	-9.6_1
600	-92.0	40.0	24.0_2	132.0	70.9	-55.4	53.1	-19.3_5
800	-104.9	49.6	39.7	154.5	84.9	-57.2	89.6	-24.4_8
1,000	-116.8	57.8	57.8	174.6	$95._6$	-57.5	126.4	-27.6_3
1,200	$-128._0$	$64._8$	$77._8$	$192._8$	$103._9$	$-56._6$	$163._1$	-29.7_1
1,500	$-143._5$	$73._6$	$110._4$	$217._1$	114.	$-53._8$	$217._8$	-31.7_3

2,3-Dimethylpentane (C_7H_{16})

T, K	$(G° - H°_0)/T$	$(H° - H°_0)/T$	$H° - H°_0$	S°	$C_p°$	$\Delta H_f°$	$\Delta G_f°$	$\log_{10} K_f$
0	-1.38	0	0	1.38	0	-35.2_0	-35.2_0	∞
200	-68.3_2	18.1_6	3.63_1	86.4_8	25.0_9	-43.1_1	-13.7_7	15.0_5
273.15	-74.4_4	21.4_0	5.84_6	95.8_4	35.2_8	-45.6_1	-2.6_4	2.1_1
298.15	-76.3_7	22.7_1	6.77	99.0_8	38.4_4	-46.3_9	1.3_4	-0.9_8
400	-83.8_1	28.2_8	11.31	112.0_9	50.4_4	-49.2_6	18.1_1	-9.9_0
600	-97.3	39.0	23.3_8	136.3	69.3	-53.3	52.7	-19.2_1
800	-109.9	48.4	38.7	158.3	83.1	-55.4	88.4	-24.1_6
1,000	-121.5	56.4	56.4	177.9	$93._5$	-56.1	124.5	-27.2_1
1,200	$-132._4$	$63._3$	$75._9$	$195._7$	$101._3$	$-55._7$	$160._6$	-29.2_4
1,500	$-147._5$	$71._8$	$107._7$	$219._3$	110.	$-53._6$	$214._4$	-31.2_4

2,4-Dimethylpentane (C_7H_{16})

T, K	$(G° - H°_0)/T$	$(H° - H°_0)/T$	$H° - H°_0$	S°	$C_p°$	$\Delta H_f°$	$\Delta G_f°$	$\log_{10} K_f$
0	0	0	0	0	0	-37.2_1	-37.2_1	∞
200	-63.1_8	18.1_2	3.62_3	81.3_0	27.5_9	-45.1_3	-14.7_5	16.1_2
273.15	-69.3_9	22.0_2	6.01_4	91.4_1	37.6_6	-47.4_5	-3.2_7	2.6_1
298.15	-71.3_9	23.4_7	7.00	94.8_6	40.8_1	-48.1_7	0.8_2	-0.6_0
400	-79.1_1	29.4_3	11.77	108.5_4	52.6_0	-50.8_0	17.9_8	-9.8_3
600	-93.2	40.3	24.2_1	133.5	70.8	-54.4	53.2	-19.3_9
800	-106.1	49.7	39.8	155.8	84.2	-56.3	89.5	-24.4_4
1,000	-118.0	57.7	57.7	175.7	$94._2$	-56.8	126.0	-27.5_4
1,200	$-129._2$	$64._4$	$77._3$	$193._6$	$101._9$	$-56._3$	$162._5$	-29.6_0
1,500	$-144._5$	$72._8$	$109._2$	$217._3$	111.	$-54._1$	$217._0$	-31.6_1

3,3-Dimethylpentane (C_7H_{16})

T, K	$(G° - H°_0)/T$	$(H° - H°_0)/T$	$H° - H°_0$	S°	$C_p°$	$\Delta H_f°$	$\Delta G_f°$	$\log_{10} K_f$
0	0	0	0	0	0	-36.6_4	-36.6_4	∞
200	-63.7_0	18.2_4	3.64_8	81.9_4	27.1_8	-44.5_3	-14.2_8	15.6_0
273.15	-69.9_1	21.9_0	5.98_2	91.8_1	36.5_6	-46.9_1	-2.8_4	2.2_7
298.15	-71.8_9	23.2_7	6.94	95.1_6	39.6_2	-47.6_6	1.2_4	-0.9_1
400	-79.5_1	28.9_7	11.59	108.4_8	51.4_7	-50.4_2	18.3_9	-10.0_5
600	-93.3	39.8	23.8_9	133.1	70.7	-54.2	53.7	-19.5_6
800	-106.1	49.4	39.5	155.5	84.9	-56.0	90.0	-24.5_8
1,000	-118.1	57.6	57.6	175.7	$95._7$	-56.3	126.6	-27.6_6
1,200	$-129._2$	$64._7$	$77._6$	$193._9$	$104._1$	$-55._4$	$163._0$	-29.6_9
1,500	$-144._7$	$73._6$	$110._4$	$218._3$	114.	$-52._4$	$217._3$	-31.6_6

THERMODYNAMIC PROPERTIES OF ALKANE HYDROCARBONS (Continued)

T, K	$(G° - H°_0)/T$, cal K^{-1} mol^{-1}	$(H° - H°_0)/T$, cal K^{-1} mol^{-1}	$H° - H°_0$, kcal mol^{-1}	$S°$, cal K^{-1} mol^{-1}	$C_p°$, cal K^{-1} mol^{-1}	$\Delta H_f°$, kcal mol^{-1}	$\Delta G_f°$, kcal mol^{-1}	$\log_{10} K_f$
2,2,3-Trimethylbutane (C_7H_{16})								
0	0	0	0	0	0	-37.4$_2$	-37.4$_2$	∞
200	-61.3$_4$	17.3$_0$	3.45$_9$	78.6$_4$	26.6$_4$	-45.5$_0$	-14.5$_9$	15.9$_4$
273.15	-67.2$_7$	21.0$_5$	5.74$_9$	88.3$_2$	35.9$_6$	-47.9$_3$	-2.9$_0$	2.3$_2$
298.15	-69.1$_8$	22.4$_4$	6.69	91.6$_2$	39.0$_2$	-48.6$_9$	1.2$_7$	-0.9$_3$
400	-76.5$_7$	28.2$_0$	11.28	104.7$_7$	50.8$_7$	-51.5$_1$	18.7$_9$	-10.2$_7$
600	-90.1	39.1	23.4$_7$	129.2	70.2	-55.4	54.9	-19.9$_8$
800	-102.7	48.8	39.0	151.5	84.9	-57.3	91.9	-25.1$_2$
1,000	-114.5	57.2	57.2	171.7	96.$_3$	-57.5	129.3	-28.2$_6$
1,200	-125.$_6$	64.$_5$	77.$_4$	190.$_1$	105.$_2$	-56.$_4$	166.$_6$	-30.3$_4$
1,500	-141.$_0$	73.$_7$	110.$_6$	214.$_7$	115.	-53.$_0$	222.$_0$	-32.3$_4$
n-Octane (C_8H_{18})								
0	0	0	0	0	0	-38.5$_7$	-38.5$_7$	∞
200	-70.2$_6$	25.7$_9$	5.15$_8$	96.0$_5$	34.6$_0$	-46.4$_0$	-13.1$_2$	14.3$_4$
273.15	-78.7$_7$	29.0$_9$	7.94$_5$	107.8$_6$	41.9$_9$	-48.9$_4$	-0.5$_6$	0.4$_5$
298.15	-81.3$_8$	30.2$_9$	9.03	111.6$_7$	44.8$_8$	-49.7$_7$	3.9$_2$	-2.8$_7$
400	-91.0$_0$	35.5$_9$	14.24	126.5$_9$	57.3$_0$	-52.9$_3$	22.7$_7$	-12.4$_4$
600	-107.5	46.5	27.8$_8$	154.0	78.1	-57.4	61.7	-22.4$_7$
800	-122.3	56.3	45.1	178.6	92.8	-60.0	101.8	-27.8$_2$
1,000	-135.7	64.7	64.7	200.4	103.$_1$	-61.2	142.4	-31.1$_3$
1,200	-148.$_2$	71.$_8$	86.$_1$	220.$_0$	110.$_7$	-61.$_4$	183.$_2$	-33.3$_6$
1,500	-165.$_2$	80.$_5$	120.$_7$	245.$_7$	119.	-60.$_5$	244.$_3$	-35.5$_9$
2-Methylheptane (C_8H_{18})								
0	0	0	0	0	0	-39.6$_5$	-39.6$_5$	∞
200	-71.1$_5$	23.4$_7$	4.69$_3$	94.6$_2$	32.4$_2$	-47.9$_5$	-14.3$_9$	15.7$_2$
273.15	-78.9$_9$	27.0$_7$	7.39$_4$	106.0$_6$	41.5$_6$	-50.5$_7$	-1.7$_0$	1.3$_6$
298.15	-81.4$_2$	28.4$_2$	8.47	109.8$_4$	44.7$_5$	-51.4$_1$	2.8$_2$	-2.0$_7$
400	-90.5$_7$	34.2$_4$	13.70	124.8$_1$	57.6$_6$	-54.5$_5$	21.8$_6$	-11.9$_4$
600	-106.7	45.7	27.4$_4$	152.4	78.7	-59.0	61.1	-22.2$_6$
800	-121.3	55.9	44.8	177.2	93.6	-61.4	101.5	-27.7$_4$
1,000	-134.7	64.6	64.6	199.3	104.$_4$	-62.4	142.4	-31.1$_3$
1,200	-147.$_1$	72.$_0$	86.$_3$	219.$_1$	112.$_5$	-62.$_2$	183.$_4$	-33.4$_0$
1,500	-164.$_2$	81.$_0$	121.$_5$	245.$_2$	122.	-60.$_8$	244.$_6$	-35.6$_4$
3-Methylheptane (C_8H_{18})								
0	-1.38	0	0	1.38	0	-38.9$_7$	-38.9$_7$	∞
200	-73.1$_0$	23.1$_3$	4.62$_5$	96.2$_3$	31.7$_1$	-47.3$_4$	-14.0$_9$	15.4$_0$
273.15	-80.8$_2$	26.6$_9$	7.28$_9$	107.5$_1$	41.1$_8$	-50.0$_0$	-1.5$_2$	1.2$_1$
298.15	-83.2$_2$	28.0$_4$	8.36	111.2$_6$	44.4$_1$	-50.8$_4$	2.9$_7$	-2.1$_8$
400	-92.2$_6$	33.8$_8$	13.55	126.1$_4$	57.4$_0$	-54.0$_1$	21.8$_6$	-11.9$_5$
600	-108.2	45.4	27.2$_5$	153.6	78.4	-58.5	60.9	-22.1$_7$
800	-122.7	55.7	44.5	178.4	93.5	-60.9	101.0	-27.6$_1$
1,000	-136.1	64.4	64.4	200.5	104.$_3$	-61.9	141.7	-30.9$_7$
1,200	-148.$_5$	71.$_7$	86.$_1$	220.$_2$	112.$_4$	-61.$_8$	182.$_4$	-33.2$_2$
1,500	-165.$_5$	80.$_8$	121.$_2$	246.$_3$	122.	-60.$_4$	243.$_4$	-35.4$_6$
4-Methylheptane (C_8H_{18})								
0	0	0	0	0	0	-38.9$_6$	-38.9$_6$	∞
200	-71.4$_7$	22.8$_0$	4.56$_0$	94.2$_7$	31.5$_2$	-47.3$_9$	-13.7$_6$	15.0$_4$
273.15	-79.1$_0$	26.4$_5$	7.22$_5$	105.5$_5$	41.3$_7$	-50.0$_5$	-1.0$_4$	0.8$_3$
298.15	-81.4$_8$	27.8$_4$	8.30	109.3$_2$	44.7$_0$	-50.8$_9$	3.5$_0$	-2.5$_7$
400	-90.4$_8$	33.8$_5$	13.54	124.3$_3$	57.9$_2$	-54.0$_2$	22.5$_8$	-12.3$_4$
600	-106.4	45.6	27.3$_4$	152.0	79.0	-58.4	61.9	-22.5$_5$
800	-121.0	55.9	44.7	176.9	93.9	-60.8	102.4	-27.9$_8$
1,000	-134.5	64.6	64.6	199.1	104.$_6$	-61.7	143.4	-31.3$_3$
1,200	-146.$_9$	72.$_0$	86.$_4$	218.$_9$	112.$_6$	-61.$_5$	184.$_3$	-33.5$_7$
1,500	-164.$_0$	81.$_0$	121.$_6$	245.$_0$	122.	-60.$_0$	245.$_7$	-35.7$_9$
2,2-Dimethylhexane (C_8H_{18})								
0	0	0	0	0	0	-41.7$_7$	-41.7$_7$	∞
200	-66.6$_3$	21.5$_8$	4.31$_5$	88.2$_1$	32.2$_0$	-50.4$_5$	-15.6$_1$	17.0$_5$
273.15	-73.9$_5$	25.6$_7$	7.01$_0$	99.6$_2$	41.6$_8$	-53.0$_8$	-2.4$_4$	1.9$_6$
298.15	-76.2$_7$	27.1$_5$	8.10	103.4$_2$	45.0$_0$	-53.9$_1$	2.2$_4$	-1.6$_4$
400	-85.1$_1$	33.4$_2$	13.37	118.5$_3$	58.2$_8$	-57.0$_0$	21.9$_2$	-11.9$_8$
600	-101.0	45.5	27.2$_8$	146.5	79.8	-61.2	62.4	-22.7$_3$
800	-115.6	56.1	44.9	171.7	95.6	-63.4	104.0	-28.4$_0$
1,000	-129.1	65.3	65.3	194.4	107.$_4$	-63.8	145.9	-31.8$_9$
1,200	-141.$_7$	73.$_1$	87.$_7$	214.$_8$	116.$_4$	-63.$_0$	187.$_8$	-34.2$_0$
1,500	-159.$_1$	82.$_8$	124.$_2$	241.$_9$	127.	-60.$_2$	250.$_2$	-36.4$_5$
3,3-Dimethylhexane (C_8H_{18})								
0	0	0	0	0	0	-40.3$_8$	-40.3$_8$	∞
200	-68.9$_2$	20.7$_0$	4.14$_0$	89.6$_2$	30.8$_6$	-49.2$_4$	-14.6$_7$	16.0$_3$
273.15	-75.9$_7$	24.9$_3$	6.80$_7$	100.9$_0$	42.0$_0$	-51.8$_9$	-1.6$_1$	1.2$_8$
298.15	-78.2$_3$	26.5$_1$	7.90	104.7$_4$	45.6$_2$	-52.7$_1$	3.0$_5$	-2.2$_3$
400	-86.9$_4$	33.1$_8$	13.27	120.1$_2$	59.4$_4$	-55.7$_1$	22.5$_8$	-12.3$_4$
600	-102.8	45.7	27.4$_3$	143.5	81.0	-59.7	62.7	-22.8$_3$
800	-117.5	56.6	45.2	174.1	96.5	-61.6	103.8	-28.3$_6$
1,000	-131.1	65.8	65.8	196.9	108.$_1$	-61.9	145.3	-31.7$_5$

T, K	$(G° - H°_0)/T$, cal K⁻¹ mol⁻¹	$(H° - H°_0)/T$, cal K⁻¹ mol⁻¹	$H° - H°_0$, kcal mol⁻¹	$S°$, cal K⁻¹ mol⁻¹	$C_p°$, cal K⁻¹ mol⁻¹	$\Delta H_f°$, kcal mol⁻¹	$\Delta G_f°$, kcal mol⁻¹	$\log_{10} K_f$
1,200	$-143._8$	$73._6$	$88._3$	$217._4$	$117._0$	$-61._0$	$186._6$	-33.9_8
1,500	$-161._3$	$83._4$	$125._0$	$244._7$	127.	$-58._0$	$248._2$	-36.1_6
2,2,4-Trimethylpentane (Isooctane) (C_8H_{18})								
0	0	0	0	0	0	-41.2_3	-41.2_3	∞
200	-66.3_0	19.7_3	3.94_5	86.0_3	31.2_1	-50.2_8	-15.0_0	16.3_9
273.15	-73.1_0	24.1_8	6.60_6	97.2_8	41.5_8	-52.9_4	-1.6_7	1.3_3
298.15	-75.2_9	25.8_0	7.69	101.0_9	45.0_3	-53.7_7	3.0_7	-2.2_5
400	-83.7_9	32.4_4	12.98	116.2_3	58.4_6	-56.8_5	22.9_9	-12.5_6
600	[illegible]	[illegible]	26.0_4	[illegible]	[illegible]	61.0	60.9	[illegible]
800	-113.8	55.8	44.7	169.6	96.4	-63.1	105.9	-28.9_4
1,000	-127.3	65.2	65.2	192.5	$108._7$	-63.3	148.2	-32.4_0
1,200	$-139._9$	$73._3$	$88._0$	$213._2$	$118._2$	$-62._2$	$190._4$	-34.6_8
1,500	$-157._4$	$83._4$	$125._1$	$240._8$	129.	$-58._7$	$253._2$	-36.9_0
***n*-Nonane (C_9H_{20})**								
0	0	0	0	0	0	-42.3_2	-42.3_2	∞
200	-74.6_4	28.9_2	5.78_3	103.5_6	38.7_0	-50.9_8	-13.3_6	14.6_0
273.15	-84.2_0	32.5_9	8.90_2	116.7_9	47.0_3	-53.7_9	0.8_4	-0.6_7
298.15	-87.1_0	33.9_4	10.12	121.0_4	50.2_9	-54.7_1	5.9_0	-4.3_2
400	-97.8_9	39.8_8	15.95	137.7_7	64.2_5	-58.2_0	27.1_6	-14.8_4
600	-116.4	52.1	31.2_5	168.5	87.5	-63.2	71.0	-25.8_6
800	-132.9	63.1	50.5	196.0	103.7	-66.0	116.2	-31.7_4
1,000	-148.0	72.4	72.4	220.4	$115._1$	-67.3	161.9	-35.3_9
1,200	$-161._9$	$80._3$	$96._3$	$242._2$	$123._4$	$-67._5$	$207._8$	-37.8_4
1,500	$-180._1$	$89._9$	$134._8$	$270._0$	133.	$-66._7$	$276._5$	-40.2_9
2-Methyloctane (C_9H_{20})								
0	0	0	0	0	0	-43.4_0	-43.4_0	∞
200	-75.5_4	26.5_9	5.31_8	102.1_3	36.5_2	-52.5_3	-14.6_2	15.9_8
273.15	-84.4_0	30.5_8	8.35_2	114.9_8	46.6_0	-55.4_2	-0.3_0	0.2_4
298.15	-87.1_5	32.0_7	9.56	119.2_2	50.1_5	-56.3_5	4.8_0	-3.5_2
400	-97.4_5	38.5_4	15.41	135.9_9	64.6_1	-59.8_2	26.2_5	-14.3_4
600	-115.5	51.4	30.8_1	166.9	88.1	-64.7	70.4	-25.6_6
800	-131.9	62.7	50.2	194.6	104.6	-67.4	115.9	-31.6_1
1,000	-146.9	72.4	72.4	219.3	$116._4$	-68.4	161.9	-35.3_8
1,200	$-160._9$	$80._4$	$96._5$	$241._3$	$125._1$	$-68._4$	$208._0$	-37.8_7
1,500	$-180._0$	$90._4$	$135._6$	$270._4$	135.	$-66._9$	$276._9$	-40.3_5
3-Methyloctane (C_9H_{20})								
0	-1.38	0	0	1.38	0	-42.7_1	-42.7_1	∞
200	-77.4_8	26.2_4	5.24_8	103.7_2	35.7_9	-51.9_1	-14.3_2	15.6_4
273.15	-86.2_3	30.1_8	8.24_4	116.4_1	46.2_4	-54.8_3	-0.1_0	0.0_8
298.15	-88.9_3	31.6_9	9.45	120.6_2	49.8_5	-55.7_7	4.9_6	-3.6_4
400	-99.1_4	38.1_8	5.00	137.3_2	64.4_0	-59.2_6	26.2_7	-14.3_5
600	-117.1	51.0	30.6_3	168.1	87.9	-64.2	70.2	-25.5_7
800	-133.4	62.4	50.0	195.8	104.5	-66.9	115.4	-31.5_4
1,000	-148.4	72.1	72.1	220.5	$116._3$	-68.0	161.1	-35.2_2
1,200	$-162._3$	$80._2$	$96._3$	$242._5$	$125._0$	$-67._9$	$207._0$	-37.7_0
1,500	$-181._3$	$90._2$	$135._4$	$271._5$	135.	$-66._5$	$275._6$	-40.1_6
4-Methyloctane (C_9H_{20})								
0	-1.38	0	0	1.38	0	-42.7_0	-42.7_0	∞
200	-77.2_7	25.9_7	5.19_4	103.2_4	35.6_4	-51.9_5	-14.2_6	15.5_9
273.15	-85.9_4	29.9_8	8.19_0	115.9_3	46.3_5	-54.8_8	-0.0_2	0.0_1
298.15	-88.6_3	31.5_2	9.40	120.1_5	50.0_2	-55.8_1	5.0_6	-3.7_1
400	-98.8_0	38.1_2	15.25	136.9_2	64.7_1	-59.2_8	26.4_2	-14.4_3
600	-116.8	51.1	30.6_7	167.9	88.2	-64.1	70.4	-25.6_5
800	-133.1	62.5	50.1	195.6	104.7	-66.8	115.7	-31.6_0
1,000	-148.1	72.2	72.2	220.3	$116._5$	-67.8	161.5	-35.2_9
1,200	$-162._0$	$80._4$	$96._4$	$242._4$	$125._2$	$-67._8$	$207._3$	-37.7_6
1,500	$-181._0$	$90._4$	$135._5$	$271._4$	135.	$-66._3$	$276._0$	-40.2_1
3-Ethylheptane (C_9H_{20})								
0	0	0	0	0	0	-42.3_4	-42.3_4	∞
200	-74.9_0	26.5_0	5.30_0	101.4_0	38.0_7	-51.4_8	-13.4_3	14.6_7
273.15	-83.8_0	30.7_9	8.41_0	114.5_9	47.2_7	-54.3_0	0.9_3	-0.7_4
298.15	-86.5_6	32.3_2	9.64	118.8_7	50.6_4	-55.2_1	6.0_4	-4.4_3
400	-96.9_4	38.7_8	15.51	135.7_2	64.6_2	-58.6_6	27.5_2	-15.0_4
600	-115.1	51.4	30.8_6	166.5	87.7	-63.6	71.8	-26.1_4
800	-131.5	62.6	50.1	194.1	104.2	-66.3	117.3	-32.0_5
1,000	-146.5	72.2	72.2	218.7	$116._0$	-67.5	163.4	-35.7_2
1,200	$-160._4$	$80._3$	$96._3$	$240._7$	$124._8$	$-67._5$	$209._6$	-38.1_7
1,500	$-179._5$	$90._2$	$135._4$	$269._7$	135.	$-66._1$	$278._8$	-40.6_1

THERMODYNAMIC PROPERTIES OF ALKANE HYDROCARBONS (Continued)

4-Ethylheptane (C_9H_{20})

T, K	$(G° - H°_0)/T$, cal K^{-1} mol^{-1}	$(H° - H°_0)/T$, cal K^{-1} mol^{-1}	$H° - H°_0$, kcal mol^{-1}	S°, cal K^{-1} mol^{-1}	$C_p°$, cal K^{-1} mol^{-1}	$\Delta H_f°$, kcal mol^{-1}	$\Delta G_f°$, kcal mol^{-1}	$\log_{10} K_f$
0	0	0	0	0	0	-42.3$_4$	-42.3$_4$	∞
200	-74.6$_3$	26.1$_7$	5.23$_3$	100.8$_0$	37.9$_3$	-51.5$_6$	-13.3$_8$	14.6$_2$
273.15	-83.4$_3$	30.5$_7$	8.34$_9$	114.0$_0$	47.5$_2$	-54.3$_6$	1.0$_2$	-0.8$_2$
298.15	-86.1$_8$	32.1$_3$	9.58	118.3$_1$	50.9$_8$	-55.2$_7$	6.1$_5$	-4.5$_1$
400	-96.5$_3$	38.7$_5$	15.50	135.2$_8$	65.0$_9$	-58.6$_7$	27.6$_8$	-15.1$_2$
600	-114.7	51.6	30.9$_4$	166.3	88.1	-63.5	72.0	-26.2$_2$
800	-131.1	62.9	50.3	194.0	104.5	-66.2	117.6	-32.1$_3$
1,000	-146.2	72.4	72.4	218.6	116.$_2$	-67.3	163.7	-35.7$_8$
1,200	-160.$_1$	80.$_5$	96.$_6$	240.$_6$	124.$_9$	-67.$_3$	209.$_9$	-38.2$_3$
1,500	-179.$_2$	90.$_4$	135.$_6$	269.$_6$	135.	-65.$_9$	279.$_1$	-40.6$_6$

n-Decane

T, K	$(G° - H°_0)/T$, cal K^{-1} mol^{-1}	$(H° - H°_0)/T$, cal K^{-1} mol^{-1}	$H° - H°_0$, kcal mol^{-1}	S°, cal K^{-1} mol^{-1}	$C_p°$, cal K^{-1} mol^{-1}	$\Delta H_f°$, kcal mol^{-1}	$\Delta G_f°$, kcal mol^{-1}	$\log_{10} K_f$
0	0	0	0	0	0	-46.0$_6$	-46.0$_6$	∞
200	-79.0$_2$	32.0$_5$	6.40$_8$	111.0$_7$	42.8$_0$	-55.5$_6$	-13.5$_8$	14.8$_4$
273.15	-89.6$_1$	36.1$_0$	9.86$_1$	125.7$_1$	52.0$_8$	-58.6$_3$	2.2$_5$	-1.8$_0$
298.15	-92.8$_3$	37.5$_9$	11.21	130.4$_2$	55.7$_0$	-59.6$_4$	7.8$_8$	-5.7$_8$
400	-104.7$_7$	44.1$_8$	17.67	148.9$_5$	71.2$_2$	-63.4$_5$	31.5$_6$	-17.2$_4$
600	-125.3	57.7	34.6$_3$	183.0	97.0	-68.8	80.3	-29.2$_6$
800	-143.6	69.9	55.9	213.5	114.7	-71.9	130.6	-35.6$_7$
1,000	-160.3	80.2	80.2	240.5	127.$_1$	-73.3	181.4	-39.6$_4$
1,200	-175.$_7$	88.$_8$	106.$_5$	264.$_5$	136.$_1$	-73.$_6$	232.$_4$	-42.3$_2$
1,500	-196.$_7$	99.$_3$	149.$_0$	296.$_0$	146.	-72.$_8$	308.$_8$	-44.9$_9$

2-Methylnonane ($C_{10}H_{22}$)

T, K	$(G° - H°_0)/T$, cal K^{-1} mol^{-1}	$(H° - H°_0)/T$, cal K^{-1} mol^{-1}	$H° - H°_0$, kcal mol^{-1}	S°, cal K^{-1} mol^{-1}	$C_p°$, cal K^{-1} mol^{-1}	$\Delta H_f°$, kcal mol^{-1}	$\Delta G_f°$, kcal mol^{-1}	$\log_{10} K_f$
0	0	0	0	0	0	-47.1$_5$	-47.1$_5$	∞
200	-79.9$_2$	29.7$_2$	5.94$_4$	109.6$_4$	40.6$_2$	-57.1$_0$	-14.8$_5$	16.2$_2$
273.15	-89.8$_1$	34.0$_9$	9.31$_0$	123.9$_0$	51.6$_5$	-60.2$_6$	1.1$_0$	-0.8$_8$
298.15	-92.8$_7$	35.7$_2$	10.65	128.5$_9$	55.5$_6$	-61.2$_8$	6.7$_8$	-4.9$_7$
400	-104.3$_4$	42.8$_4$	17.13	147.1$_8$	71.5$_8$	-65.0$_8$	30.6$_4$	-16.7$_4$
600	-124.4	57.0	34.1$_9$	181.4	97.5	-70.4	79.8	-29.0$_5$
800	-142.6	69.5	55.6	212.1	115.6	-73.3	130.3	-35.6$_0$
1,000	-159.2	80.1	80.1	239.3	128.$_4$	-74.5	181.4	-39.6$_4$
1,200	-174.$_7$	88.$_9$	106.$_8$	263.$_6$	137.$_8$	-74.$_5$	232.$_5$	-42.3$_5$
1,500	-195.$_7$	99.$_9$	149.$_8$	295.$_6$	149.	-73.$_0$	309.$_2$	-45.0$_4$

3-Methylnonane ($C_{10}H_{22}$)

T, K	$(G° - H°_0)/T$, cal K^{-1} mol^{-1}	$(H° - H°_0)/T$, cal K^{-1} mol^{-1}	$H° - H°_0$, kcal mol^{-1}	S°, cal K^{-1} mol^{-1}	$C_p°$, cal K^{-1} mol^{-1}	$\Delta H_f°$, kcal mol^{-1}	$\Delta G_f°$, kcal mol^{-1}	$\log_{10} K_f$
0	-1.38	0	0	1.38	0	-46.4$_5$	-46.4$_5$	∞
200	-81.8$_6$	29.3$_7$	5.87$_3$	111.2$_3$	39.8$_8$	-56.4$_8$	-14.5$_4$	15.8$_9$
273.15	-91.6$_4$	33.6$_9$	9.20$_2$	125.3$_3$	51.2$_8$	-59.6$_7$	1.3$_0$	-1.0$_4$
298.15	-94.6$_6$	35.3$_3$	10.54	129.9$_9$	55.2$_5$	-60.7$_0$	6.9$_5$	-5.0$_9$
400	-106.0$_2$	42.4$_8$	16.99	148.5$_0$	71.3$_5$	-64.5$_2$	30.6$_7$	-16.7$_6$
600	-125.9	56.7	34.0$_0$	182.6	97.3	-69.9	79.5	-28.9$_7$
800	-144.1	69.2	55.4	213.3	115.4	-72.8	129.8	-35.4$_7$
1,000	-160.7	79.8	79.8	240.5	128.$_3$	-74.0	180.7	-39.4$_8$
1,200	-176.$_0$	88.$_7$	106.$_5$	264.$_7$	137.$_7$	-74.$_0$	231.$_6$	-42.1$_8$
1,500	-197.$_1$	99.$_6$	149.$_5$	296.$_7$	149.	-72.$_6$	307.$_9$	-44.8$_6$

4-Methylnonane ($C_{10}H_{22}$)

T, K	$(G° - H°_0)/T$, cal K^{-1} mol^{-1}	$(H° - H°_0)/T$, cal K^{-1} mol^{-1}	$H° - H°_0$, kcal mol^{-1}	S°, cal K^{-1} mol^{-1}	$C_p°$, cal K^{-1} mol^{-1}	$\Delta H_f°$, kcal mol^{-1}	$\Delta G_f°$, kcal mol^{-1}	$\log_{10} K_f$
0	-1.38	0	0	1.38	0	-46.4$_5$	-46.4$_5$	∞
200	-81.6$_5$	29.0$_9$	5.81$_7$	110.7$_4$	39.7$_2$	-56.5$_3$	-14.5$_0$	15.8$_4$
273.15	-91.3$_5$	33.4$_8$	9.14$_6$	124.8$_3$	51.4$_1$	-59.7$_3$	1.3$_8$	-1.1$_1$
298.15	-94.3$_5$	35.1$_6$	10.48	129.5$_1$	55.4$_5$	-60.7$_5$	7.0$_4$	-5.1$_6$
400	-105.6$_8$	42.4$_2$	16.97	148.1$_0$	71.7$_1$	-64.5$_4$	30.8$_1$	-16.8$_3$
600	-125.6	56.8	34.0$_5$	182.4	97.7	-69.8	79.7	-29.0$_4$
800	-143.7	69.4	55.5	213.1	115.7	-72.7	130.1	-35.5$_3$
1,000	-160.4	80.0	80.0	240.4	128.$_5$	-73.9	180.9	-39.5$_4$
1,200	-175.$_8$	88.$_9$	106.$_7$	264.$_7$	137.$_8$	-73.$_8$	231.$_9$	-42.2$_3$
1,500	-196.$_8$	99.$_8$	149.$_7$	296.$_6$	149.	-72.$_4$	308.$_2$	-44.9$_0$

5-Methylnonane ($C_{10}H_{22}$)

T, K	$(G° - H°_0)/T$, cal K^{-1} mol^{-1}	$(H° - H°_0)/T$, cal K^{-1} mol^{-1}	$H° - H°_0$, kcal mol^{-1}	S°, cal K^{-1} mol^{-1}	$C_p°$, cal K^{-1} mol^{-1}	$\Delta H_f°$, kcal mol^{-1}	$\Delta G_f°$, kcal mol^{-1}	$\log_{10} K_f$
0	0	0	0	0	0	-46.4$_3$	-46.4$_3$	∞
200	-80.3$_3$	29.1$_5$	5.82$_9$	109.4$_8$	39.7$_5$	-56.5$_1$	-14.2$_2$	15.5$_4$
273.15	-90.0$_4$	33.5$_2$	9.15$_5$	123.5$_6$	51.3$_1$	-59.7$_0$	1.7$_6$	-1 4$_0$
298.15	-93.0$_5$	35.1$_8$	10.49	128.2$_3$	55.3$_2$	-60.7$_3$	7.4$_4$	-5.4$_6$
400	-104.3$_8$	42.3$_9$	16.96	146.7$_7$	71.5$_2$	-64.5$_4$	31.3$_4$	-17.1$_3$
600	-124.3	56.7	34.0$_0$	181.0	97.5	-69.8	80.5	-29.3$_4$
800	-142.4	69.3	55.4	211.7	115.6	-72.8	131.2	-35.8$_3$
1,000	-159.0	79.9	79.9	238.9	128.$_4$	-73.9	182.3	-39.8$_4$
1,200	-174.$_4$	88.$_8$	106.$_6$	263.$_2$	137.$_8$	-73.$_9$	233.$_6$	-42.5$_4$
1,500	-195.$_4$	99.$_7$	149.$_6$	295.$_1$	149.	-72.$_5$	310.$_3$	-45.2$_1$

Data from Scott, D. W., U.S. Bureau of Mines Bulletin 666 (stock number 2404-01547), U.S. Government Printing Office, Washington, D.C., 1974.

[a]The original source of the data in this table contains the thermodynamic properties of all the alkane hydrocarbons through decane. Including the *meso*, *(d, l)*, *cis*, and *trans* isomers, there are 182 compounds in the original source.

HEAT OF DILUTION OF ACIDS

From National Standards Reference Data Systems NSRDS-NBS 2

Vivian B. Parker

ΔH_{diln}, the integral heat of dilution, is the change in enthalpy, per mole of solute, when a solution of concentration m_1 is diluted to a final finite concentration m_2. When the dilution is carried out by addition of an infinite amount of solvent, so the final solution is infinitely dilute, the enthalpy change is the integral heat of dilution to infinite dilution. Since Φ_L, the relative apparent molal enthalpy, is equal to and opposite in sign to this, only Φ_L is referred to here.

Φ_L, cal/mole, at 25°C

n	*m*	HF	HCl	$HClO_4$	HBr	HI	HNO_3	CH_2O_2	$C_2H_4O_2$
∞	0.00	0	0	0	0	0	0	0	0
500,000	.000111	300	5	5	5	5	5	9	40
100,000	.000555	900	10	10	9	9	11	13	50
50,000	.00111	1,300	16	14	13	12	15	20	53
20,000	.00278	1,800	25	22	22	20	23	23	55
10,000	.00555	2,130	34	30	31	29	31	25	58
7,000	.00793	2,250	40	35	37	34	36	26	59
5,000	.01110	2,360	47	40	44	41	42	26	61
4,000	.01388	2,450	54	43	49	46	46	27	62
3,000	.01850	2,550	60	47	56	52	51	28	62
2,000	.02775	2,700	74	54	68	63	59	28	63
1,500	.03700	2,812	85	58	77	71	65	29	64
1,110	.05000	2,927	97	62	89	81	73	29	65
1,000	.05551	2,969	102	62	92	84	76	29	65
900	.0617	2,989	107	63	97	88	78	30	66
800	.0694	3,015	113	64	102	92	81	31	67
700	.0793	3,037	120	65	108	96	84	32	68
600	.0925	3,057	129	65	115	102	88	32	68
555.1	.1000	3,060	133	65	119	105	89	32	69
500	.1110	3,077	140	65	124	108	92	32	70
400	.1388	3,097	156	64	135	116	97	33	72
300	.1850	3,126	176	61	150	125	103	34	76
277.5	.2000	3,129	182	59	155	128	105	35	79
200	.2775	3,142	212	50	176	140	117	36	82
150	.3700	3,148	242	36	197	154	118	39	88
111.0	.5000	3,156	280	18	225	170	119	42	97
100	.5551	3,160	295	+12	235	176	120	44	101
75	.7401	3,167	343	−14	270	194	121	49	113
55.51	1.0000	3,179	405	−48	314	223	121	54	130
50	1.1101	3,184	431	−61	331	234	121	56	147
40	1.3877	3,192	493	−91	379	260	121	60	155
37.00	1.5000	3,194	518	−103	398	269	121	62	162
30	1.8502	3,200	595	−138	455	301	124	65	183
27.75	2.0000	3,203	627	−149	477	315	126	66	192
25	2.2202	3,208	674	−162	510	336	130	67	204
22.20	2.5000	3,211	732	−173	550	365	139	68	218
20	2.7753	3,214	792	−182	590	396	149	69	233
18.50	3.0000	3,216	838	−187	624	427	159	69	245
15.86	3.500	3,221	946	−196	709	503	189	69	268
15	3.7004	3,227	988	−195	743	536	203	69	277
13.88	4.0000	3,234	1,052	−188	796	588	229	69	291
12.33	4.5000	3,246	1,171	−175	887	676	265	69	313
12	4.6255	3,249	1,190	−170	911	700	277	69	318
11.10	5.0000	3,256	1,271	−150	983	764	313	69	333
10	5.5506	3,265	1,396	−117	1,097	855	368	68	353
9.5	5.8427	3,269	1,462	−97	1,156	920	400	68	363
9.251	6.0000	3,272	1,498	−84	1,196	950	418	67	368
9.0	6.1674	3,274	1,535	−72	1,230	980	437	67	373
8.5	6.5301	3,278	1,618	−40	1,313	1,050	480	66	383
8.0	6.9383	3,282	1,710	+4	1,401	1,115	530	65	392
7.929	7.0000	3,283	1,725	11	1,416	1,130	538	65	394
7.5	7.4008	3,286	1,820	61	1,497	1,210	595	63	402
7.0	7.9295	3,290	1,942	135	1,608	1,325	661	61	411
6.938	8.0000	3,291	1,960	146	1,622	1,340	667	61	412
6.5	8.5394	3,296	2,090	229	1,738	1,450	745	58	420
6.167	9.0000	3,302	2,202	306	1,845	1,570	805	55	426
6.0	9.2510	3,305	2,265	348	1,903	1,630	840	53	429
5.551	10.0000	3,316	2,447	481	2,078	1,820	940	49	436
5.5	10.0920	3,317	2,472	499	2,102	1,850	950	49	437
5.0	11.1012	3,335	2,721	730	2,344	2,100	1,098	43	445
4.5	12.3346	3,362	3,025	1,144	2,655	2,460	1,270	37	453
4.0	13.8765	3,400	3,404	1,574	3,089	2,960	1,495	29	462
3.700	15.0000	3,428	3,680	1,893	3,415	3,350	1,645	26	469
3.5	15.8589	3,450	3,882	2,150	3,668	3,660	1,770	21	473
3.25	17.0788	3,483	4,160	2,460	4,005	4,110	1,920	17	481
3.0	18.5020	3,520	4,460	2,880	4,370	4,630	2,101	13	488
2.775	20.0000	3,557	4,750	3,300	4,760	5,190	2,270	9	496
2.5	22.2024	3,607	5,180	4,000	5,300	6,000	2,520	+4	506
2.0	27.7530	3,712	6,260	5,500	6,650	.....	3,060	−5	528
1.5	37.0040	.....	8,240	.....	8,530	.....	3,770	−13	532
1.0	55.506	.....	10,900	.....	11,670	.....	4,715	+11	518
0.5	111.012	.....	.....	.....	.....	.....	.....	77	495
0.25	222.02	.....	.....	.....	.....	.....	.....	129	.....

HEATS OF SOLUTION

From National Standards Reference Data Systems NSRDS-NBS 2

Vivian B. Parker

ΔH°_∞ 25°C for uni-univalent electrolytes in H_2O

Substance	State	ΔH°_∞
		cal/mole
HF	g	−14,700
HCl	g	−17,888
$HClO_4$	l	−21,215
$HClO_4 \cdot H_2O$	c	−7,875
HBr	g	−20,350
HI	g	−19,520
HIO_3	c	2,100
HNO_3	l	−7,954
HCOOH	l	−205
CH_3COOH	l	−360
NH_3	g	−7,290
NH_4Cl	c	3,533
NH_4ClO_4	c	8,000
NH_4Br	c	4,010
NH_4I	c	3,280
NH_4IO_3	c	7,600
NH_4NO_2	c	4,600
NH_4NO_3	c	6,140
$NH_4C_2H_3O_2$	c	−570
NH_4CN	c	4,200
NH_4CNS	c	5,400
CH_3NH_3Cl	c	1,378
$(CH_3)_3NHCl$	c	350
$N(CH_3)_4Cl$	c	975
$N(CH_3)_4Br$	c	5,800
$N(CH_3)_4I$	c	10,055
$AgClO_4$	c	1,760
$AgNO_2$	c	8,830
$AgNO_3$	c	5,400
LiOH	c	−5,632
$LiOH \cdot H_2O$	c	−1,600
LiF	c	1,130
LiCl	c	−8,850
$LiCl \cdot H_2O$	c	−4,560
$LiClO_4$	c	−6,345
$LiClO_4 \cdot 3H_2O$	c	7,795
LiBr	c	−11,670
$LiBr \cdot H_2O$	c	−5,560

Substance	State	ΔH°_∞
		cal/mole
$LiBr \cdot 2H_2O$	c	−2,250
$LiBrO_3$	c	340
LiI	c	−15,130
$LiI \cdot H_2O$	c	−7,090
$LiI \cdot 2H_2O$	c	−3,530
$LiI \cdot 3H_2O$	c	140
$LiNO_2$	c	−2,630
$LiNO_2 \cdot H_2O$	c	1,680
$LiNO_3$	c	−600
NaOH	c	−10,637
$NaOH \cdot H_2O$	c	−5,118
NaF	c	218
NaCl	c	928
$NaClO_2$	c	80
$NaClO_2 \cdot 3H_2O$	c	6,830
$NaClO_3$	c	5,191
$NaClO_4$	c	3,317
$NaClO_4 \cdot H_2O$	c	5,380
NaBr	c	−144
$NaBr \cdot 2H_2O$	c	4,454
$NaBrO_3$	c	6,430
NaI	c	−1,800
$NaI \cdot 2H_2O$	c	3,855
$NaIO_3$	c	4,850
$NaNO_2$	c	3,320
$NaNO_3$	c	4,900
$NaC_2H_3O_2$	c	−4,140
$NaC_2H_3O_2 \cdot 3H_2O$	c	4,700
NaCN	c	290
$NaCN \cdot \frac{1}{2}H_2O$	c	790
$NaCN \cdot 2H_2O$	c	4,440
NaCNO	c	4,590
NaCNS	c	1,632
KOH	c	−13,769
$KOH \cdot H_2O$	c	−3,500
$KOH \cdot 1\frac{1}{2}H_2O$	c	−2,500
KF	c	−4,238
$KF \cdot 2H_2O$	c	1,666

Substance	State	ΔH°_∞
		cal/mole
KCl	c	4,115
$KClO_3$	c	9,890
$KClO_4$	c	12,200
KBr	c	4,750
$KBrO_3$	c	9,830
KI	c	4,860
KIO_3	c	6,630
KNO_2	c	3,190
KNO_3	c	8,340
$KC_2H_3O_2$	c	−3,665
KCN	c	2,800
KCNO	c	4,840
KCNS	c	5,790
$KMnO_4$	c	10,410
RbOH	c	−14,900
$RbOH \cdot H_2O$	c	−4,310
$RbOH \cdot 2H_2O$	c	210
RbF	c	−6,240
$RbF \cdot H_2O$	c	−100
$RbF \cdot 1\frac{1}{2}H_2O$	c	320
RbCl	c	4,130
$RbClO_3$	c	11,410
$RbClO_4$	c	13,560
RbBr	c	5,230
$RbBrO_3$	c	11,700
RbI	c	6,000
$RbNO_3$	c	8,720
CsOH	c	−17,100
$CsOH \cdot H_2O$	c	−4,900
CsF	c	−8,810
$CsF \cdot H_2O$	c	−2,500
$CsF \cdot 1\frac{1}{2}H_2O$	c	−1,300
CsCl	c	4,250
$CsClO_4$	c	13,250
CsBr	c	6,210
$CsBrO_3$	c	12,060
CsI	c	7,970
$CsNO_3$	c	9,560

HEAT CAPACITY OF AQUEOUS SOLUTIONS OF VARIOUS ACIDS

From National Standards Reference Data Systems NSRDS-NBS 2

Vivian B. Parker

Φ_C is the apparent molal heat capacity of the solute, equal to $[(1000 + mM_2)C - 1000C^\circ]/m$ where C and C° are the specific heats (per unit mass) of the solution and pure solvent, respectively, m is the molality, and M_2 is the molecular weight of the solute.

Φ_C, cal/deg mole, at 25°C

n	*m*	HF	HCl	HBr	HI	HIO_3	HNO_3	CH_2O_2	$C_2H_4O_2$	$C_3H_6O_2$
∞	0.00	−25.5	−32.6	−33.9	−34.0	−29.6	−20.7	−21.0	−1.5	+26.7
500,000	.000111	−23.0	.....	.....	.....	.....	.....	−9.8	+25.8	38.0
100,000	.000555	−18.8	−32.4	−33.8	−33.9	−29.4	−20.6	−1.2	32.6	45.1
50,000	.00111	−16.6	−32.4	−33.7	−33.8	−29.3	−20.5	+3.1	34.0	48.3
20,000	.00278	−12.4	−32.3	−33.6	−33.7	−29.1	−20.4	10.2	35.7	52.2
10,000	.00555	−8.6	−32.2	−33.4	−33.5	−28.5	−20.3	12.7	36.9	54.1
7,000	.00793	−6.7	−32.1	−33.4	−33.4	−28.4	−20.2	13.7	37.4	54.8
5,000	.01110	−4.9	−32.0	−33.3	−33.3	−28.1	−20.1	14.3	37.8	55.3
4,000	.01388	−3.6	−31.9	−33.2	−33.2	−27.7	−20.1	14.7	37.9	55.6
3,000	.01850	−2.2	−31.8	−33.1	−33.1	−27.2	−20.0	15.1	38.1	56.1
2,000	.02775	−0.5	−31.6	−32.9	−32.9	−25.8	−19.8	15.8	38.5	56.7
1,500	.03700	+0.4	−31.4	−32.7	−32.7	−24.9	−19.6	16.4	38.7	57.1
1,000	.05551	1.6	−31.2	−32.5	−32.5	−23.0	−19.3	16.8	39.0	57.7
900	.0617	1.8	−31.2	−32.4	−32.4	−22.3	−19.2	17.0	39.0	57.8
800	.0694	2.0	−31.1	−32.3	−32.3	−21.5	−19.1	17.2	39.1	57.9
700	.0793	2.2	−30.9	−32.2	−32.1	−20.2	−19.0	17.3	39.2	58.0
600	.0925	2.4	−30.8	−32.1	−32.0	−18.7	−18.9	17.5	39.3	58.3
500	.1110	2.7	−30.6	−31.9	−31.7	−16.7	−18.7	17.7	39.4	58.4
400	.1388	2.8	−30.3	−31.6	−31.4	−13.9	−18.4	18.0	39.4	58.6
300	.1850	3.2	−30.0	−31.2	−30.9	−10.1	−17.9	18.3	39.4	58.8
200	.2775	3.3	−29.3	−30.6	−30.2	−4.7	−17.2	18.7	39.3	58.9
150	.3700	3.5	−28.8	−30.1	−29.6	−0.7	−16.4	18.9	39.2	58.8
100	.5551	3.8	−27.8	−29.1	−28.4	+4.8	−15.0	19.0	39.1	58.7
75	.7401	4.1	−27.0	−28.5	−27.5	9.2	−13.7	19.2	38.9	58.6
50	1.1101	4.5	−25.6	−26.8	−25.9	15.7	−11.5	19.8	38.6	58.3
40	1.3877	4.9	−24.8	−25.9	−24.8	19.0	−9.9	19.8	38.4	58.0
30	1.8502	5.3	−23.6	−24.5	−23.3	23.1	−7.3	20.0	38.0	57.1
25	2.2202	5.6	−22.7	−23.6	−22.2	25.7	−5.4	20.1	37.7	56.2
20	2.7753	5.7	−21.5	−22.2	−20.8	.....	−2.7	20.2	37.1	55.0
15	3.7004	6.0	−19.8	−20.3	.....	.....	+1.4	20.4	36.3	53.1
12	4.6255	6.2	−18.2	−18.5	.....	.....	5.0	20.4	35.4	51.3
10	5.5506	6.3	−16.8	−16.8	.....	.....	8.5	20.6	34.8	49.7
9.5	5.8427	6.4	−16.4	−16.3	.....	.....	9.2	20.6	34.7	.....
9.0	6.1674	6.4	−15.8	−15.7	.....	.....	10.3	20.6	34.5	.....
8.5	6.5301	6.5	−15.4	−15.1	.....	.....	11.4	20.7	34.4	.....
8.0	6.9383	6.6	−14.8	−14.4	.....	.....	12.5	20.7	34.2	.....
7.5	7.4008	6.7	−14.2	−13.2	.....	.....	13.7	20.8	34.0	.....
7.0	7.9295	6.9	−13.5	−12.7	.....	.....	14.9	20.8	33.8	.....
6.5	8.5394	7.0	−12.7	−11.9	.....	.....	16.1	20.8	33.5	.....
6.0	9.2510	7.1	−11.8	−10.8	.....	.....	17.1	20.9	33.3	.....
5.5	10.0920	7.2	−10.8	−9.6	.....	.....	18.3	20.9	33.0	.....
5.0	11.1012	7.3	−9.6	−8.2	.....	.....	19.3	21.0	32.8	.....
4.5	12.3346	7.4	−8.7	−6.8	.....	.....	20.4	21.1	32.5	.....
4.0	13.8765	7.5	−6.6	−5.5	.....	.....	21.3	21.2	32.2	.....
3.5	15.8589	7.6	−4.7	−4.0	.....	.....	22.1	21.3	31.8	.....
3.25	17.0788	7.6	.....	−3.2	.....	.....	22.6	21.4	31.7	.....
3.0	18.5020	7.7	.....	−2.3	.....	.....	23.0	21.5	31.5	.....
2.5	22.2024	7.8	.....	−0.4	.....	.....	23.8	21.6	31.2	.....
2.0	27.7530	7.9	.....	.....	.....	.....	24.6	21.7	30.8	.....
1.5	37.0040	.....	.....	.....	.....	.....	25.2	21.8	30.4	.....
1.0	55.506	.....	.....	.....	.....	.....	25.7	22.0	30.1	.....

THERMODYNAMIC FORMULAS

Compiled by Doctor E. A. Coomes

Legend:

p = Pressure	Cp = Molal specific heat at constant pressure
V = Volume	β = Coefficient volume expansion
T = Temperature	K = Compressibility
n = Number of mols	H = U + pV = Total heat or enthalpy
S = Entropy	A = U − TS = Helmholtz free energy
U = Internal energy (some books use E)	G = H − TS = Gibbs' free energy (some books use F).

Use of Table — Partial derivatives of the first order for the eight fundamental thermodynamic variables, namely, p, V, T, U, S, H, A, G, may be obtained in terms of $(\partial V/\partial T)_p$, $(\partial V/\partial p)_T$ and $(\partial H/\partial T)_p$ the latter three are connected to measurable quantities as follows:

$$\frac{1}{V}\left(\frac{\partial V}{\partial T}\right)_p = \beta; \quad -\frac{1}{V}\left(\frac{\partial V}{\partial p}\right)_T = K; \quad \left(\frac{\partial H}{\partial T}\right)_p = nC_p$$

Computation by the Table — The method of using the table will become apparent in working several examples. Suppose it is desired to know $(\partial H/\partial p)_s$ in terms of p, V, and T. Under caption "Constant" move horizontally to column marked "S"; across from "H" beside caption "differential" find "$-VnCp/T$." Across from "p" in column "S" find "$-nCp/T$"; $(\partial H/\partial p)_s$ is found by taking the ratio of the two:

$$(\partial H/\partial p)_S = (-VnC_p/T)/(-nC_p/T) = V$$

To find $(\partial S/\partial V)_T$ in terms of p, V, T, move to column "T" under "constant." Opposite "S" beside "differential" find "$(\partial V/\partial T)_p$"; opposite "V" find "$-(\partial V/\partial p)_T$".

Taking the ratio:

$$(\partial S/\partial V)_T = (\partial V/\partial T)_p / -(\partial V/\partial p)_T = (\partial p/\partial T)_V$$

Differential \ Constant	T	p	V	S
T	0	1	$\left(\frac{\partial V}{\partial p}\right)_T$	$-\left(\frac{\partial V}{\partial T}\right)_p$
p	−1	0	$-\left(\frac{\partial V}{\partial T}\right)_p$	$-\frac{nC_p}{T}$
V	$-\left(\frac{\partial V}{\partial p}\right)_T$	$\left(\frac{\partial V}{\partial T}\right)_p$	0	$\left(-\frac{1}{T}\right)\left[nC_p\left(\frac{\partial V}{\partial p}\right)_T + T\left(\frac{\partial V}{\partial T}\right)_p^2\right]$
S	$\left(\frac{\partial V}{\partial T}\right)_p$	$\frac{nC_p}{T}$	$\left(\frac{1}{T}\right)\left[nC_p\left(\frac{\partial V}{\partial p}\right)_T + T\left(\frac{\partial V}{\partial T}\right)_p^2\right]$	0
U	$T\left(\frac{\partial V}{\partial T}\right)_p + p\left(\frac{\partial V}{\partial p}\right)_T$	$nC_p - p\left(\frac{\partial V}{\partial T}\right)_p$	$nC_p\left(\frac{\partial V}{\partial p}\right)_T + T\left(\frac{\partial V}{\partial T}\right)_p^2$	$\left(\frac{p}{T}\right)\left[nC_p\left(\frac{\partial V}{\partial p}\right)_T + T\left(\frac{\partial V}{\partial T}\right)_p^2\right]$
H	$-V + T\left(\frac{\partial V}{\partial T}\right)_p$	nC_p	$nC_p\left(\frac{\partial V}{\partial p}\right)_T + T\left(\frac{\partial V}{\partial T}\right)_p^2 - V\left(\frac{\partial V}{\partial T}\right)$	$-\frac{VnC_p}{T}$
A	$p\left(\frac{\partial V}{\partial p}\right)_T$	$-S - p\left(\frac{\partial V}{\partial T}\right)_p$	$-S\left(\frac{\partial V}{\partial p}\right)_T$	$\left(\frac{1}{T}\right)\left[pnC_p\left(\frac{\partial V}{\partial p}\right)_T + pT\left(\frac{\partial V}{\partial T}\right)_p^2 + TS\left(\frac{\partial V}{\partial T}\right)_p\right]$
G	−V	−S	$-V\left(\frac{\partial V}{\partial T}\right)_p - S\left(\frac{\partial V}{\partial p}\right)_T$	$\left(-\frac{1}{T}\right)\left[nC_pV - TS\left(\frac{\partial V}{\partial T}\right)_p\right]$

Differential \ Constant	U	H	A	G
T	$-T\left(\frac{\partial V}{\partial T}\right)_p - p\left(\frac{\partial V}{\partial p}\right)_T$	$V - T\left(\frac{\partial V}{\partial T}\right)_p$	$-p\left(\frac{\partial V}{\partial p}\right)_T$	V
p	$-nC_p + p\left(\frac{\partial V}{\partial T}\right)_p$	$-nC_p$	$S + p\left(\frac{\partial V}{\partial T}\right)_p$	S
V	$-nC_p\left(\frac{\partial V}{\partial p}\right)_T - T\left(\frac{\partial V}{\partial T}\right)_p^2$	$-nC_p\left(\frac{\partial V}{\partial p}\right)_T - T\left(\frac{\partial V}{\partial T}\right)_p^2 + V\left(\frac{\partial V}{\partial T}\right)_p$	$S\left(\frac{\partial V}{\partial p}\right)_T$	$V\left(\frac{\partial V}{\partial T}\right)_p + S\left(\frac{\partial V}{\partial p}\right)_T$
S	$\left(-\frac{p}{T}\right)\left[nC_p\left(\frac{\partial V}{\partial p}\right)_T + T\left(\frac{\partial V}{\partial T}\right)_p^2\right]$	$\frac{VnC_p}{T}$	$\left(-\frac{1}{T}\right)\left[pnC_p\left(\frac{\partial V}{\partial p}\right)_T + pT\left(\frac{\partial V}{\partial T}\right)_p^2 + TS\left(\frac{\partial V}{\partial T}\right)_p\right]$	$\left(\frac{1}{T}\right)\left[nC_pV - TS\left(\frac{\partial V}{\partial T}\right)_p\right]$
U	0	$V\left[nC_p - p\left(\frac{\partial V}{\partial T}\right)_p\right] + p\left[nC_p\left(\frac{\partial V}{\partial p}\right)_T + T\left(\frac{\partial V}{\partial T}\right)_p^2\right]$	$-p\left[nC_p\left(\frac{\partial V}{\partial p}\right)_T + T\left(\frac{\partial V}{\partial T}\right)_p^2\right] - S\left[T\left(\frac{\partial V}{\partial T}\right)_p + p\left(\frac{\partial V}{\partial p}\right)_T\right]$	$V\left[nC_p - p\left(\frac{\partial V}{\partial T}\right)_p\right] - S\left[T\left(\frac{\partial V}{\partial T}\right)_p + p\left(\frac{\partial V}{\partial p}\right)_T\right]$
H	$-V\left[nC_p - p\left(\frac{\partial V}{\partial T}\right)_p\right] - p\left[nC_p\left(\frac{\partial V}{\partial p}\right)_T + T\left(\frac{\partial V}{\partial T}\right)_p^2\right]$	0	$-\left[S + p\left(\frac{\partial V}{\partial T}\right)_p\right] \times \left[V - T\left(\frac{\partial V}{\partial T}\right)_p - pnC_p\left(\frac{\partial V}{\partial p}\right)_T\right]$	$VnC_p + VS - TS\left(\frac{\partial V}{\partial T}\right)_p$
A	$p\left[nC_p\left(\frac{\partial V}{\partial p}\right)_T + T\left(\frac{\partial V}{\partial T}\right)_p^2\right] + S\left[T\left(\frac{\partial V}{\partial T}\right)_p + p\left(\frac{\partial V}{\partial p}\right)_T\right]$	$-\left[S + p\left(\frac{\partial V}{\partial T}\right)_p\right] \times \left[V - T\left(\frac{\partial V}{\partial T}\right)_p - pnC_p\left(\frac{\partial V}{\partial p}\right)_T\right]$	0	$-S\left[V + p\left(\frac{\partial V}{\partial p}\right)_T\right] + pV\left(\frac{\partial V}{\partial T}\right)_p$
G	$-V\left[nC_p - p\left(\frac{\partial V}{\partial T}\right)_p\right] + S\left[T\left(\frac{\partial V}{\partial T}\right)_p + p\left(\frac{\partial V}{\partial p}\right)_T\right]$	$-VnC_p - VS + TS\left(\frac{\partial V}{\partial T}\right)_p$	$S\left[V + p\left(\frac{\partial V}{\partial p}\right)_T\right] + pV\left(\frac{\partial V}{\partial T}\right)_p$	0

LIMITS OF INFLAMMABILITY

Reprinted from "Combustion Flame and Explosions of Gases", B. Lewis and G. von Elbe, authors, Academic Press (1951), publishers, by special permission.

The limits of inflammability given in the following tables were all determined at atmospheric pressure and room* temperature for upward propagation in a tube or bomb 2 inches or more in diameter. Values are on a percentage-by-volume basis.

LIMITS OF INFLAMMABILITY OF GASES AND VAPORS IN AIR

Compound	Empirical formula	Limits of inflammability Lower	Upper
Paraffin hydrocarbons			
Methane	CH_4	5.00	15.00
Ethane	C_2H_6	3.00	12.50
Propane	C_3H_8	2.12	9.35
Butane	C_4H_{10}	1.86	8.41
Isobutane	C_4H_{10}	1.80	8.44
Pentane	C_5H_{12}	1.40	7.80
Isopentane	C_5H_{12}	1.32	—
2,2-Dimethylpropane	C_5H_{12}	1.38	7.50
Hexane	C_6H_{14}	1.18	7.40
Heptane	C_7H_{16}	1.10	6.70
2,3-Dimethylpentane	C_7H_{16}	1.12	6.75
Octane	C_8H_{18}	0.95	—
Nonane	C_9H_{20}	0.83	—
Decane	$C_{10}H_{22}$	0.77	5.35
Olefins			
Ethylene	C_2H_4	2.75	28.60
Propylene	C_3H_6	2.00	11.10
Butene-1	C_4H_8	1.65	9.95
Butene-2	C_4H_8	1.75	9.70
Amylene	C_5H_{10}	1.42	8.70
Acetylenes			
Acetylene	C_2H_2	2.50	80.00
Aromatics			
Benzene	C_6H_6	1.40	7.10
Toluene	C_7H_8	1.27	6.75
o-Xylene	C_8H_{10}	1.00	6.00
Cyclic hydrocarbons			
Cyclopropane	C_3H_6	2.40	10.40
Cyclohexane	C_6H_{12}	1.26	7.75
Methylcyclohexane	C_7H_{14}	1.15	—
Terpenes			
Turpentine	$C_{10}H_{16}$	0.80	—
Alcohols			
Methyl alcohol	CH_4O	6.72	36.50
Ethyl alcohol	C_2H_6O	3.28	18.95
Allyl alcohol	C_3H_6O	2.50	18.00
η-Propyl alcohol	C_3H_8O	2.15	13.50
Isopropyl alcohol	C_3H_8O	2.02	11.80
η-Butyl alcohol	$C_4H_{10}O$	1.45	11.25
Isobutyl alcohol	$C_4H_{10}O$	1.68	—
η-Amyl alcohol	$C_5H_{12}O$	1.19	—
Isoamyl alcohol	$C_5H_{12}O$	1.20	—
Aldehydes			
Acetaldehyde	C_2H_4O	3.97	57.00
Crotonic aldehyde	C_4H_6O	2.12	15.50
Furfural	$C_5H_4O_2$	2.10	—
Paraldehyde	$C_6H_{12}O_3$	1.30	—
Ethers			
Methylethyl ether	C_3H_8O	2.00	10.00
Diethyl ether	$C_4H_{10}O$	1.85	36.50
Divinyl ether	C_4H_6O	1.70	27.00
Ketones			
Acetone	C_3H_6O	2.55	12.80
Methylethyl ketone	C_4H_8O	1.81	9.50
Methyl propyl ketone	$C_5H_{10}O$	1.55	8.15
Methylbutyl ketone	$C_6H_{12}O$	1.35	7.60
Acids			
Acetic acid	$C_2H_4O_2$	5.40	—
Hydrocyanic acid	HCN	5.60	40.00
Esters			
Methyl formate	$C_2H_4O_2$	5.05	22.70
Ethyl formate	$C_3H_6O_2$	2.75	16.40
Methyl acetate	$C_3H_6O_2$	3.15	15.60
Ethyl acetate	$C_4H_8O_2$	2.18	11.40
Propyl acetate	$C_5H_{10}O_2$	1.77	8.00
Isopropyl acetate	$C_6H_{10}O_2$	1.78	7.80
Butyl acetate	$C_6H_{12}O_2$	1.39	7.55
Amyl acetate	$C_7H_{14}O_2$	1.10	—
Hydrogen			
Hydrogen	H_2	4.00	74.20
Nitrogen compounds			
Ammonia	NH_3	15.50	27.00
Cyanogen	C_2N_2	6.60	42.60
Pyridine	C_5H_5N	1.81	12.40
Ethyl nitrate	$C_2H_5NO_3$	3.80	—
Ethyl nitrite	$C_2H_5NO_2$	3.01	50.00
Oxides			
Carbon monoxide	CO	12.50	74.20
Ethylene oxide	C_2H_4O	3.00	80.00
Propylene oxide	C_3H_6O	2.00	22.00
Dioxan	$C_4H_8O_2$	1.97	22.25
Diethyl peroxide	$C_4H_{10}O_2$	2.34	—
Sulfides			
Carbon disulfide	CS_2	1.25	50.00
Hydrogen sulfide	H_2S	4.30	45.50
Carbon oxysulfide	COS	11.90	28.50
Chlorides			
Methyl chloride	CH_3Cl	8.25	18.70
Ethyl chloride	C_2H_5Cl	4.00	14.80
Propyl chloride	C_3H_7Cl	2.60	11.10
Butyl chloride	C_4H_9Cl	1.85	10.10
Isobutyl chloride	C_4H_9Cl	2.05	8.75
Allyl chloride	C_3H_5Cl	3.28	11.15
Amyl chloride	$C_5H_{11}Cl$	1.60	8.63
Vinyl chloride	C_2H_3Cl	4.00	21.70
Ethylene dichloride	$C_2H_4Cl_2$	6.20	15.90
Propylene dichloride	$C_3H_6Cl_2$	3.40	14.50
Bromides			
Methyl bromide	CH_3Br	13.50	14.50
Ethyl bromide	C_2H_5Br	6.75	11.25
Allyl bromide	C_3H_5Br	4.36	7.25
Amines			
Methyl amine	CH_5N	4.95	20.75
Ethyl amine	C_2H_7N	3.55	13.95
Dimethyl amine	C_2H_7N	2.80	14.40
Propyl amine	C_3H_9N	2.01	10.35
Diethyl amine	$C_4H_{11}N$	1.77	10.10
Trimethyl amine	C_3H_9N	2.00	11.60
Triethyl amine	$C_6H_{15}N$	1.25	7.90

* The upper limits of some vapors were determined at somewhat higher temperatures because of their low vapor pressures.

LIMITS OF INFLAMMABILITY OF GASES AND VAPORS IN OXYGEN

Compound	Formula	Limits of inflammability Lower	Upper
Hydrogen	H_2	4.65	93.9
Deuterium	D_2	5.00	95.0
Carbon monoxide	CO	15.50	93.9
Methane	CH_4	5.40	59.2
Ethane	C_2H_6	4.10	50.5
Ethylene	C_2H_4	2.90	79.9
Propylene	C_3H_6	2.10	52.8
Cyclopropane	C_3H_6	2.45	63.1
Ammonia	NH_3	13.50	79.0
Diethyl ether	$C_4H_{10}O$	2.10	82.0
Divinyl ether	C_4H_6O	1.85	85.5

LIMITS FOR HUMAN EXPOSURE TO AIR CONTAMINANTS

From United States Federal Register
Volume 36, Number 105

Exposures by inhalation, ingestion, skin absorption, or contact to any material or substance (1) at a concentration above those specified in the "Threshold Limit Values of Airborne Contaminants for 1970" of the American Conference of Governmental Industrial Hygienists, listed in Table 1, except for the American National Standards listed in Table 2 of this section and except for values of mineral dusts listed in Table 3 of this section, and (2) concentrations above those specified in Table 1, 2, and 3 of this section, shall be avoided, or protective equipment shall be provided and used.

Table 1

Substance	ppm[a]	mg/m³ [b]
Abate		15
Acetaldehyde	200	360
Acetic acid	10	25
Acetic anhydride	5	20
Acetone	1,000	2,400
Acetonitrile	40	70
Acetylene dichloride, see 1,2-Dichloroethylene		
Acetylene tetrabromide	1	14
Acrolein	0.1	0.25
Acrylamide–Skin		0.3
Acrylonitrile–Skin	20	45
Aldrin–Skin		0.25
Allyl alcohol–Skin	2	5
Allyl chloride	1	3
**C Allyl glycidyl ether (AGE)	10	45
Allyl propyl disulfide	2	12
2-Aminoethanol, see Ethanolamine		
2-Aminopyridine	0.5	2
**Ammonia	50	35
Ammonium sulfamate (Ammate)		15
n-Amyl acetate	100	525
sec-Amyl acetate	125	650
Aniline–Skin	5	19
Anisidine (*o,p*-isomers)–Skin		0.5
Antimony and compounds (as Sb)		0.5
ANTU (alpha naphthyl thiourea)		0.3
Arsenic and compounds (as As)		0.5
Arsine	0.05	0.2
Azinphos-methyl–Skin		0.2
Barium (soluble compounds)		0.5
p-Benzoquinone, see Quinone		
Benzoyl peroxide		5
Benzyl chloride	1	5
Biphenyl, see Diphenyl		
Bisphenol A, see Diglycidyl ether		
Boron oxide		15
Boron tribromide	1	10
C Boron trifluoride	1	3
Bromine	0.1	0.7
*Bromine pentafluoride	0.1	0.7
Bromoform–Skin	0.5	5
Butadiene (1,3-butadiene)	1,000	2,200
Butanethiol, see Butyl mercaptan		
2-Butanone	200	590
2-Butoxy ethanol (Butyl Cellosolve)–Skin	50	240
Butyl acetate (*n*-butyl acetate)	150	710
sec-Butyl acetate	200	950
tert-Butyl acetate	200	950
Butyl alcohol	100	300
sec-Butyl alcohol	150	450
tert-Butyl alcohol	100	300
C Butylamine–Skin	5	15
C *tert*-Butyl chromate (as CrO_3)–Skin		0.1
n-Butyl glycidyl ether (BGE)	50	270
*Butyl mercaptan	0.5	1.5
p-tert-Butyltoluene	10	60
Calcium arsenate		1
Calcium oxide		5
**Camphor (Synthetic)	2	
Carbaryl (Sevin®)		5
Carbon black		3.5
Carbon dioxide	5,000	9,000
Carbon monoxide	50	55
Chlordane–Skin		0.5
Chlorinated camphene–Skin		0.5
Chlorinated diphenyl oxide		0.5
*Chlorine	1	3
Chlorine dioxide	0.1	0.3
C Chlorine trifluoride	0.1	0.4
C Chloroacetaldehyde	1	3

Substance	ppm[a]	mg/m³ [b]
α-Chloroacetophenone (phenacylchloride)	0.05	0.3
Chlorobenzene (monochlorobenzene)	75	350
o-Chlorobenzylidene malononitrile (OCBM)	0.05	0.4
Chlorobromomethane	200	1,050
2-Chloro-1,3-butadiene, see Chloroprene		
Chlorodiphenyl (42 percent Chlorine)–Skin		1
Chlorodiphenyl (54 percent Chlorine)–Skin		0.5
1-Chloro-2,3-epoxypropane, see Epichlorhydrin		
2-Chloroethanol, see Ethylene chlorohydrin		
Chloroethylene, see Vinyl chloride		
C Chloroform (trichloromethane)	50	240
1-Chloro-1-nitropropane	20	100
Chloropicrin	0.1	0.7
Chloroprene (2-chloro-1,3-butadiene)–Skin	25	90
Chromium, sol. chromic, chromous salts as Cr		0.5
Metal and insol. salts		1
Coal tar pitch volatiles (benzene soluble fraction) anthracene, BaP, phenanthrene, acridine, chrysene, pyrene		0.2
Cobalt, metal fume and dust		0.1
Copper fume		0.1
Dusts and Mists		1
Cotton dust (raw)		1
Crag® herbicide		15
Cresol (all isomers)–Skin	5	22
Crotonaldehyde	2	6
Cumene–Skin	50	245
Cyanide (as CN)–Skin		5
*Cyanogen	100	
Cyclohexane	300	1,050
Cyclohexanol	50	200
Cyclohexanone	50	200
Cyclohexene	300	1,015
Cyclopentadiene	75	200
2,4-D		10
DDT–Skin		1
DDVP, see Dichlorvos		
Decaborane–Skin	0.05	0.3
Demeton®–Skin		0.1
Diacetone alcohol (4-hydroxy-4-methyl-2-pentanone)	50	240
1,2-Diaminoethane, see Ethylenediamine		
Diazomethane	0.2	0.4
Diborane	0.1	0.1
Dibutyl phosphate	1	5
Dibutylphthalate		5
*C Dichloroacetylene	0.1	0.4
C *o*-Dichlorobenzene	50	300
p-Dichlorobenzene	75	450
Dichlorodifluoromethane	1,000	4,950
1,3-Dichloro-5,5-dimethyl hydantoin		0.2
1,1-Dichloroethane	100	400
1,2-Dichloroethylene	200	790
C Dichloroethyl ether–Skin	15	90
Dichloromethane, see Methylenechloride		
Dichloromonofluoromethane	1,000	4,200
C 1,1-Dichloro-1-nitroethane	10	60
1,2-Dichloropropane, see		

Table 1 (continued)

Substance	ppm[a]	mg/m^3 [b]
Propylenedichloride		
Dichlorotetrafluoroethane	1,000	7,000
Dichlorvos (DDVP)–Skin		1
Dieldrin–Skin		0.25
Diethylamine	25	75
Diethylamino ethanol–Skin	10	50
**C Diethylene triamine–Skin	10	42
Diethylether, see Ethyl ether		
Difluorodibromomethane	100	860
C Diglycidyl ether (DGE)	0.5	2.8
Dihydroxybenzene, see Hydroquinone		
Diisobutyl ketone	50	290
Diisopropylamine–Skin	5	20
Dimethoxymethane, see Methylal		
Dimethyl acetamide–Skin	10	35
Dimethylamine	10	18
Dimethylaminobenzene, see Xylidene		
Dimethylaniline (*N*-dimethylaniline)–Skin	5	25
Dimethylbenzene, see Xylene		
Dimethyl 1,2-dibromo-2,2-dichloroethyl phosphate, (Dibrom)		3
Dimethylformamide–Skin	10	30
2,6-Dimethylheptanone, see Diisobutyl ketone		
1,1-Dimethylhydrazine–Skin	0.5	1
Dimethylphthalate		5
Dimethylsulfate–Skin	1	5
Dinitrobenzene (all isomers)–Skin		1
Dinitro-*o*-cresol–Skin		0.2
Dinitrotoluene–Skin		1.5
Dioxane (Diethylene dioxide)–Skin	100	360
Diphenyl	0.2	1
Diphenylamine		10
Diphenylmethane diisocyanate (see Methylene bisphenyl isocyanate (MDI))		
Dipropylene glycol methyl ether–Skin	100	600
Di-*sec,* octyl phthalate (Di-2-ethylhexylphthalate)		5
*Endosulfan (Thiodan®)–Skin		0.1
Endrin–Skin		0.1
Epichlorhydrin–Skin	5	19
EPN–Skin		0.5
1,2-Epoxypropane, see Propyleneoxide		
2,3-Epoxy-1-propanol, see Glycidol		
Ethanethiol, see Ethylmercaptan		
Ethanolamine	3	6
2-Ethoxyethanol–Skin	200	740
2-Ethoxyethylacetate (Cellosolve acetate)–Skin	100	540
Ethyl acetate	400	1,400
Ethyl acrylate–Skin	25	100
Ethyl alcohol (ethanol)	1,000	1,900
Ethylamine	10	18
Ethyl *sec*-amyl ketone (5-methyl-3-heptanone)	25	130
Ethyl benzene	100	435
Ethyl bromide	200	890
Ethyl butyl ketone (3-Heptanone)	50	230
Ethyl chloride	1,000	2,600
Ethyl ether	400	1,200
Ethyl formate	100	300
Ethyl mercaptan	0.5	1
Ethyl silicate	100	850
Ethylene chlorohydrin–Skin	5	16
Ethylenediamine	10	25
Ethylene dibromide, see 1,2-Dibromoethane		
Ethylene dichloride, see 1,2-Dichloroethane		
C Ethylene glycol dinitrate and/or Nitroglycerin–Skin	[d]0.2	
Ethylene glycol monomethyl ether acetate, see Methyl cellosolve acetate		
Ethylene imine–Skin	0.5	1
Ethylene oxide	50	90
Ethylidine chloride, see 1,1-Dichloroethane		
N-Ethylmorpholine–Skin	20	94
Ferbam		15
Ferrovanadium dust		1
Fluoride (as F)		2.5
Fluorine	0.1	0.2
Fluorotrichloromethane	1,000	5,600
Formic acid	5	9
Furfural–Skin	5	20
Furfuryl alcohol	50	200
Glycidol (2,3-Epoxy-1-propanol)	50	150
Glycol monoethyl ether, see 2-Ethoxyethanol		
Guthion®, see Azinphosmethyl		
Hafnium		0.5
Heptachlor–Skin		0.5
Heptane (*n*-heptane)	500	2,000
Hexachloroethane–Skin	1	10
Hexachloronaphthalene–Skin		0.2
Hexane (*n*-hexane)	500	1,800
2-Hexanone	100	410
Hexone (Methyl isobutyl ketone)	100	410
sec-Hexyl acetate	50	300
Hydrazine–Skin	1	1.3
Hydrogen bromide	3	10
C Hydrogen chloride	5	7
Hydrogen cyanide–Skin	10	11
Hydrogen peroxide	1	1.4
Hydrogen selenide	0.05	0.2
Hydroquinone		2
*Indene	10	45
Indium and compounds, as In		0.1
C Iodine	0.1	1
Iron oxide fume		10
Iron salts, soluble, as Fe		1
Isoamyl acetate	100	525
Isoamyl alcohol	100	360
Isobutyl acetate	150	700
Isobutyl alcohol	100	300
Isophorone	25	140
Isopropyl acetate	250	950
Isopropyl alcohol	400	980
Isopropylamine	5	12
Isopropylether	500	2,100
Isopropyl glycidyl ether (IGE)	50	240
Ketene	0.5	0.9
Lead arsenate		0.15
Lindane–Skin		0.5
Lithium hydride		0.025
L.P.G. (liquefied petroleum gas)	1,000	1,800
Magnesium oxide fume		15
Malathion–Skin		15
Maleic anhydride	0.25	1
C Manganese and compounds, as Mn		5
Mesityl oxide	25	100
Methanethiol, see Methyl mercaptan		
Methoxychlor		15
2-Methoxyethanol, see Methyl cellosolve		
Methyl acetate	200	610
Methyl acetylene (propyne)	1,000	1,650
Methyl acetylene-propadiene mixture (MAPP)	1,000	1,800
Methyl acrylate–Skin	10	35
Methylal (dimethoxymethane)	1,000	3,100
Methyl alcohol (methanol)	200	260
Methylamine	10	12
Methyl amyl alcohol, see Methyl isobutyl carbinol		
*Methyl isoamyl ketone	100	475
Methyl (*n*-amyl) ketone (2-Heptanone)	100	465
C Methyl bromide–Skin	20	80
Methyl butyl ketone, see 2-Hexanone		
Methyl cellosolve–Skin	25	80
Methyl cellosolve acetate–Skin	25	120
Methyl chloroform	350	1,900
Methylcyclohexane	500	2,000
Methylcyclohexanol	100	470
o-Methylcyclohexanone–Skin	100	460
Methyl ethyl ketone (MEK), see 2-Butanone		
Methyl formate	100	250
Methyl iodide–Skin	5	28
Methyl isobutyl carbinol–Skin	25	100
Methyl isobutyl ketone, see		

Table 1 (continued)

Substance	ppm[a]	mg/m³ [b]
Hexone		
Methyl isocyanate–Skin	0.02	0.05
*Methyl mercaptan	0.5	1
Methyl methacrylate	100	410
Methyl propyl ketone, see 2-Pentanone		
C Methyl silicate	5	30
C α-Methyl styrene	100	480
C Methylene bisphenyl isocyanate (MDI)	0.02	0.2
Molybdenum:		
Soluble compounds		5
Insoluble compounds		15
Monomethyl aniline–Skin	2	9
C Monomethyl hydrazine–Skin	0.2	0.35
Morpholine–Skin	20	70
Naphtha (coaltar)	100	400
Naphthalene	10	50
Nickel carbonyl	0.001	0.007
Nickel, metal and soluble cmpds, as Ni		1
Nicotine–Skin		0.5
Nitric acid	2	5
Nitric oxide	25	30
p-Nitroaniline–Skin	1	6
Nitrobenzene–Skin	1	5
p-Nitrochlorobenzene–Skin		1
Nitroethane	100	310
Nitrogen dioxide	5	9
Nitrogen trifluoride	10	29
Nitroglycerin–Skin	0.2	2
Nitromethane	100	250
1-Nitropropane	25	90
2-Nitropropane	25	90
Nitrotoluene–Skin	5	30
Nitrotrichloromethane, see Chloropicrin		
Octachloronaphthalene–Skin		0.1
*Octane	400	1,900
*Oil mist, particulate		5
Osmium tetroxide		0.002
Oxalic acid		1
Oxygen difluoride	0.05	0.1
Ozone	0.1	0.2
Paraquat–Skin		0.5
Silver, metal and soluble compounds		0.01
Sodium fluoroacetate (1080)–Skin		0.05
Sodium hydroxide		2
Stibine	0.1	0.5
*Stoddard solvent	200	1,150
Strychnine		0.15
Sulfur dioxide	5	13
Sulfur hexafluoride	1,000	6,000
Sulfuric acid		1
Sulfur monochloride	1	6
Sulfur pentafluoride	0.025	0.25
Sulfuryl fluoride	5	20
Systox, see Demeton®		
2,4,5T		10
Tantalum		5
TEDP–Skin		0.2
Tellurium		0.1
Tellurium hexafluoride	0.02	0.2
TEPP–Skin		0.05
C Terphenyls	1	9
1,1,1,2-Tetrachloro-2,2-difluoroethane	500	4,170
1,1,2,2-Tetrachloro-1,2-difluoroethane	500	4,170
1,1,2,2-Tetrachloroethane–Skin	5	35
Tetrachloroethylene, see Perchloroethylene		
Tetrachloromethane, see Carbon tetrachloride		
Tetrachloronaphthalene–Skin		2
Tetraethyl lead (as Pb)–Skin		*f*0.100
Tetrahydrofuran	200	590
Tetramethyl lead (as Pb)–Skin		*f*0.150
Tetramethyl succinonitrile–Skin	0.5	3
Tetranitromethane	1	8
Tetryl (2,4,6-trinitrophenylmethylnitramine)–Skin		1.5
Thallium (soluble compounds)–Skin as Tl		0.1
Thiram		5
Tin (inorganic cmpds, except SnH_4 and SnO_2)		2
Tin (organic cmpds)		0.1
C Toluene-2,4-diisocyanate	0.02	0.14
o-Toluidine–Skin	5	22
Toxaphene, see Chlorinated camphene		
Tributyl phosphate		5
1,1,1-Trichloroethane (see Methyl chloroform		
1,1,2-Trichloroethane–Skin	10	45
Parathion–Skin		0.1
Pentaborane	0.005	0.01
Pentachloronaphthalene–Skin		0.5
Pentachlorophenol–Skin		0.5
*Pentane	500	1,500
2-Pentanone	200	700
Perchloromethyl mercaptan	0.1	0.8
Perchloryl fluoride	3	13.5
Phenol–Skin	5	19
p-Phenylene diamine–Skin		0.1
Phenyl ether (vapor)	1	7
Phenyl ether-biphenyl mixture (vapor)	1	7
Phenylethylene, see Styrene		
Phenyl glycidyl ether (PGE)	10	60
Phenylhydrazine–Skin	5	22
Phosdrin (Mevinphos®)–Skin		0.1
Phosgene (carbonyl chloride)	0.1	0.4
Phosphine	0.3	0.4
Phosphoric acid		1
Phosphorus (yellow)		0.1
Phosphorus pentachloride		1
Phosphorus pentasulfide		1
Phosphorus trichloride	0.5	3
Phthalic anhydride	2	12
Picric acid–Skin		0.1
Pival® (2-Pivalyl-1,3-indandione)		0.1
Platinum (Soluble Salts) as Pt		0.002
Propargyl alcohol–Skin	1	
n-Propyl acetate	200	840
Propyl alcohol	200	500
n-Propyl nitrate	25	110
Propylene dichloride	75	350
Propylene imine–Skin	2	5
Propylene oxide	100	240
Propyne, see Methylacetylene		
Pyrethrum		5
Pyridine	5	15
Quinone	0.1	0.4
RDX–Skin		1.5
Rhodium, Metal fume and dusts, as Rh		0.1
Soluble salts		0.001
Ronnel		10
Rotenone (commercial)		5
Selenium compounds (as Se)		0.2
Selenium hexafluoride	0.05	0.4
Trichloromethane, see Chloroform		
Trichloronaphthalene–Skin		5
1,2,3-Trichloropropane	50	300
1,1,2-Trichloro 1,2,2-trifluoroethane	1,000	7,600
Triethylamine	25	100
Trifluoromonobromomethane	1,000	6,100
*Trimethyl benzene	25	120
2,4,6-Trinitrophenol, see Picric acid		
2,4,6-Trinitrophenylmethylnitramine, see Tetryl		
Trinitrotoluene–Skin		1.5
Triorthocresyl phosphate		0.1
Triphenyl phosphate		3
Tungsten and compounds, as W:		
Soluble		1
Insoluble		5
Turpentine	100	560
Uranium (natural) sol. and insol. compounds as U		0.2
C Vanadium:		
V_2O_5 dust		0.5
V_2O_5 fume		0.1
Vinyl benzene, see Styrene		
**C Vinyl chloride	500	1,300
Vinylcyanide, see Acrylonitrile		
Vinyl toluene	100	480

LIMITS FOR HUMAN EXPOSURE TO AIR CONTAMINANTS (*Continued*)

Table 1 (continued)

Substance	ppm[a]	mg/M³ [b]
Warfarin		0.1
Xylene (xylol)	100	435
Xylidine–Skin	5	25
Yttrium		1
Zinc chloride fume		1
Zinc oxide fume		5
Zirconium compounds (as Zr)		5

*1970 Addition.

[a]Parts of vapor or gas per million parts of contaminated air by volume at 25°C and 760 mm Hg pressure.

[b]Approximate milligrams of particulate per cubic meter of air.

(No footnote "c" is used to avoid confusion with ceiling value notations.)

[d]An atmospheric concentration of not more than 0.02 ppm, or personal protection may be necessary to avoid headache.

[e]As sampled method that does not collect vapor.

[f]For control of general room air, biologic monitoring is essential for personnel control.

Table 2

	8-hour time weighted average
Benzene (Z37.4–1969)	10 ppm
Beryllium and beryllium compounds (Z37.29–1970)	0.002 mg/M³
Cadmium dust (as Cd) (Z37.5–1970)	0.2 mg/M³
Cadmium fume (as Cd) (Z37.5–1970)	0.1 mg/M³
Carbon disulfide (Z37.3–1968)	20 ppm
Carbon tetrachloride (Z37.17–1967)	10 ppm
Ethylene dibromide (Z37.31–1970)	20 ppm
Ethylene dichloride (Z37.21–1969)	50 ppm
Formaldehyde (Z37.16–1967)	3 ppm
Hydrogen fluoride (Z37.28–1969)	3 ppm
Fluoride as dust (Z37.28–1966)	2.5 mg/M³
Lead and its inorganic compounds (Z37.11–1969)	0.2 mg/M³
Methyl chloride (Z37.18–1969)	100 ppm
Methylene chloride (Z37.23–1969)	500 ppm
Organo (alkyl) mercury (Z37.30–1969)	0.01 mg/M³
Styrene (Z37.12–1969)	100 ppm
Tetrachloroethylene (Z37.22–1967)	100 ppm
Toluene (Z37.12–1967)	200 ppm

	Acceptable ceiling concentration
Hydrogen sulfide (Z37.2–1966)	20 ppm
Chromic acid and chromates (Z37.3–1971)	1 mg/10M³
Mercury (Z37.8–1971)	1 mg/10M³

Table 3

Substance	Mppcf[a]	Mg/M³
Silica:		
Crystalline:		
Quartz (respirable)	$\frac{250^f}{\%SiO_2+5}$	$\frac{10mg/M^3}{\%SiO_2+2}$ [m]
Quartz (total dust)		$\frac{30mg/M^3}{\%SiO_2+2}$
Cristobalite: Use ½ the value calculated from the count or mass formulae for quartz.		
Trioymite: Use ½ the value calculated from the formulae for quartz.		
Amorphous, including natural diatomaceous earth	20	$\frac{80mg/M^3}{\%SiO_2}$
Tremolite	5	20mg/M³

Substance	Mppcf[a]	Mg/M³
		$\%SiO_2$
Silicates (less than 1% crystalline silica):		
Asbestos – 12 fibers per milliliter greater than 5 microns in length,[i] or	2	
Mica	20	
Soapstone	20	
Talc	20	
Portland cement	50	
Graphite (natural)	15	
Coal dust (respirable fraction less than 5% SiO_2)		2.4mg/M³ or
For more than 5% SiO_2		$\frac{10mg/M^3}{\%SiO_2+2}$
Inert or Nuisance Dust:		
Respirable fraction	15	5mg/M³
Total dust	50	15mg/M³

NOTE: Conversion factors–
mppcf x 35.3 = million particles per cubic meter
= particles per cc

[a]Millions of particles per cubic foot of air, based on impinger samples counted by light-field technics.

[f]The percentage of crystalline silica in the formula is the amount determined from air-borne samples, except in those instances in which other methods have been shown to be applicable.

[i]As determined by the membrane filter method at 430 x phase contrast magnification.

[m]Both concentration and percent quartz for the application of this limit are to be determined from the fraction passing a size-selector with the following characteristics:

Aerodynamic diameter (unit density sphere)	Percent passing selector
2	90
2.5	75
3.5	50
5.0	25
10	0

The measurements under this note refer to the use of an AEC instrument. If the respirable fraction of coal dust is determined with a MRE the figure corresponding to that of 2.4 Mg/M³ in the table for coal dust is 4.5 Mg/M³.

FLAME AND BEAD TESTS

Flame Colorations

Violet

Potassium compounds. Purple red through blue glass. Easily obscured by sodium flame. Bluish green through green glass. Rubidium and Cesium compounds impart same flame as potassium compounds

Blues

Azure.—Copper chloride. Copper bromide gives azure blue followed by green. Other copper compounds give same coloration when moistened with hydrochloric acid.

Light Blue.—Lead, Arsenic, Selenium.

Greens

Emerald.—Copper compounds except the halides, and when not moistened with hydrochloric acid.

Pure Green.—Compounds of thallium and tellurium.

Yellowish.—Barium compounds. Some molybdenum compounds. Borates, especially when treated with sulphuric acid or when burned with alcohol.

Bluish.—Phosphates with sulphuric acid.

Feeble.—Antimony compounds. Ammonium compounds.

Whitish.—Zinc.

Reds

Carmine.—Lithium compounds. Violet through blue glass. Invisible through green glass. Masked by barium flame.

Scarlet.—Strontium compounds. Violet through blue glass. Yellowish through green glass. Masked by barium flame.

Yellowish.—Calcium compounds. Greenish through blue glass. Green through green glass. Masked by barium flame.

Yellow

Yellow.—All sodium compounds. Invisible with blue glass.

Borax Beads

Abbreviations employed: s., saturated; s.s., supersaturated; n.s.; not saturated; h., hot; c., cold.

Substance	Oxidizing flame	Reducing flame
Aluminum	Colorless (h.c., n.s.); opaque (s.s.)	Colorless; opaque (s.)
Antimony	Colorless; yellow or brownish (h., s.s.)	Gray and opaque
Barium	Colorless (n.s.)	
Bismuth	Colorless; yellow or brownish (h., s.s.)	Gray and opaque
Cadmium	Colorless	Gray and opaque
Calcium	Colorless (n.s.)	
Cerium	Red (h.)	Colorless (h.c.)
Chromium	Green (c.)	Green
Cobalt	Blue (h.c.)	Blue (h.c.)
Copper	Green (h.); blue (c.)	Red (c.): opaque (s.s.): colorless (h.)
Iron	Yellow or brownish red (h., n.s.)	Green (s.s.)
Lead	Colorless; yellow or brownish (h., s.s.)	Gray and opaque
Magnesium	Colorless (n.s.)	
Manganese	Violet (h.c.)	Colorless (h.c.)
Molybdenum	Colorless	Yellow or brown (h.)
Nickel	Brown; red (c.)	Gray and opaque
Silicon	Colorless (h.c.); opaque (s.s.)	Colorless; opaque (s.)
Silver	Colorless (n.s.)	Gray and opaque
Strontium	Colorless (n.s.)	
Tin	Colorless (h.c.); opaque (s.s.)	Colorless; opaque (s.)
Titanium	Colorless	Yellow (h.); violet (c.)
Tungsten	Colorless	Brown
Uranium	Yellow or brownish (h., n.s.)	Green
Vanadium	Colorless	Green

Beads of Microcosmic Salt

$NaNH_4HPO_4$

Substance	Oxidizing flame	Reducing flame
Aluminum	Colorless; opaque (s.)	Colorless; not clear (s.s)
Antimony	Colorless (n.s.)	Gray and opaque
Barium	Colorless; opaque (s.)	Colorless; not clear (s.s.)
Bismuth	Colorless (n.s.)	Gray and opaque
Cadmium	Colorless (n.s.)	Gray and opaque
Calcium	Colorless; opaque (s.)	Colorless; not clear (s.s.)
Cerium	Yellow or brownish red (h., s.)	Colorless
Chromium	Red (h., s.); green (c.)	Green (c.)
Cobalt	Blue (h.c.)	Blue (h.c.)
Copper	Blue (c.); green (h.)	Red and opaque (c.)
Iron	Yellow or brown (h., s.)	Colorless; yellow or brownish (h.)
Lead	Colorless (n.s.)	Gray and opaque
Magnesium	Colorless; opaque (s.)	Colorless; not clear (s.s.)
Manganese	Violet (h.c.)	Colorless
Molybdenum	Colorless; green (h.)	Green (h.)
Nickel	Yellow (c.); red (h., s.)	Yellow (c.); red (h.); gray and opaque
Silicon	(Swims undissolved)	(Swims undissolved)
Silver		Gray and opaque
Strontium	Colorless; opaque (s.)	Colorless; not clear (s.s.)
Tin	Colorless; opaque (s.)	Colorless
Titanium	Colorless (n.s.)	Violet (c.); yellow or brownish (h.)
Uranium	Green; yellow or brownish (h., s.)	Green (h.)
Vanadium	Yellow	Green
Zinc	Colorless (n.s.)	Gray and opaque

Sodium Carbonate Bead

Substance	Oxidizing flame	Reducing flame
Manganese	Green	Colorless

PREPARATION OF REAGENTS

The following pages present directions for the preparation of various reagents. The collection has been prepared with the active collaboration of W. D. Bonner, R. K. Carleton, L. L. Carrick, Giles B. Cooke, E. J. Cragoe, Thos. De Vries, James L. Kassner, Thos. W. Mason, F. C. Mathers, M. G. Mellon, W. C. Pierce, J. H. Reedy, Arthur A. Vernon and S. R. Wood. Many others have contributed valuable suggestions.

Volumes have been stated in milliliters (ml) and liters (l). One milliliter is equivalent to one cubic centimeter (cm^3 or cc.). Masses are indicated in grams (g).

The relation to molar solution (*M*) or normal solution (*N*) is indicated in many cases.

Distilled water should be used.

LABORATORY REAGENTS FOR GENERAL USE

DILUTE ACIDS, 3 molar. Use the amount of concentrated acid indicated and dilute to one liter.

Acetic acid, 3 *N.* Use 172 ml of 17.4 *M* acid (99-100%).

Hydrochloric acid, 3 *N.* Use 258 ml of 11.6 *M* acid (36% HCl).

Nitric acid, 3 *N.* Use 195 ml of 15.4 *M* acid (69% HNO_3).

Phosphoric acid, 9 *N.* Use 205 ml of 14.6 *M* acid (85% H_3PO_4).

Sulfuric acid, 6 *N.* Use 168 ml of 17.8 *M* acid (95% H_2SO_4).

DILUTE BASES.

Ammonium hydroxide, 3 *M*, 3 *N.* Dilute 200 ml of concentrated solution (14.8 *M*, 28% NH_3) to 1 liter.

Barium hydroxide, 0.2 *M*, 0.4 *N.* Saturated solution, 63 g per liter of $Ba(OH)_2 \cdot 8H_2O$. Use some excess, filter off $BaCO_3$ and protect from CO_2 of the air with soda lime or ascarite in a guard tube.

Calcium hydroxide, 0.02 *M*, 0.04 *N.* Saturated solution, 1.5 g per liter of $Ca(OH)_2$. Use some excess, filter off $CaCO_3$ and protect from CO_2 of the air.

Potassium hydroxide, 3 *M*, 3 *N.* Dissolve 176 g of the sticks (95%) in water and dilute to 1 liter.

Sodium hydroxide, 3 *M*, 3 *N.* Dissolve 126 g of the sticks (95%) in water and dilute to 1 liter.

GENERAL REAGENTS (See also Decinormal Solutions of Salts and Other Reagents.)

Aluminum chloride, 0.167 *M*, 0.5 *N.* Dissolve 22 g of $AlCl_3$ in 1 liter of water.

Aluminum nitrate, 0.167 *M*, 0.5 *N.* Dissolve 58 g of $Al(NO_3)_3 \cdot 7.5H_2O$ in 1 liter of water.

Aluminum sulfate, 0.083 *M*, 0.5 *N.* Dissolve 56 g of $Al_2(SO_4)_3 \cdot 18H_2O$ in 1 liter of water.

Ammonium acetate, 3 *M*, 3 *N.* Dissolve 230 g of $NH_4C_2H_3O_2$ in water and dilute to 1 liter.

Ammonium carbonate, 1.5 *M* Dissolve 144 g of the commercial salt (mixture of $(NH_4)_2CO_3 \cdot H_2O$ and $NH_4CO_2NH_2$) in 500 ml of 3 *N* NH_4OH and dilute to 1 liter.

Ammonium chloride, 3 *M*, 3 *N.* Dissolve 160 g of NH_4Cl in water. Dilute to 1 liter.

Ammonium molybdate.

1. **0.5** *M*, 1 *N.* Mix well 72 g of pure MoO_3 (or 81 g of H_2MoO_4) with 200 ml of water, and add 60 ml of conc. ammonium hydroxide. When solution is complete, filter and pour filtrate, very slowly and with rapid stirring, into a mixture of 270 ml of conc. HNO_3 and 400 ml of water. Allow to stand over night, filter and dilute to 1 liter.

2. The reagent is prepared as two solutions which are mixed as needed, thus always providing fresh reagent of proper strength and composition. Since ammonium molybdate is an expensive reagent, and since an acid solution of this reagent as usually prepared keeps for only a few days, the method proposed will avoid loss of reagent and provide more certain results for quantitative work.

Solution 1. Dissolve 100 g of ammonium molybdate (C.P. grade) in 400 ml of water and 80 ml of 15 *M* NH_4OH. Filter if necessary, though this seldom has to be done.

Solution 2. Mix 400 ml of 16 *M* nitric acid with 600 ml of water.

For use, mix the calculated amount of solution 1 with twice its volume of solution 2, adding solution 1 to solution 2 slowly with vigorous stirring. Thus, for amounts of phosphorus up to 20 mg, 10 ml of solution 1 to 20 ml of solution 2 is adequate. Increase amount as needed.

Ammonium nitrate, 1 *M*, 1 *N.* Dissolve 80 g of NH_4NO_3 in 1 liter of water.

Ammonium oxalate, 0.25 *M*, 0.5 *N.* Dissolve 35.5 g of $(NH_4)_2C_2O_4 \cdot H_2O$ in water. Dilute to 1 liter.

Ammonium sulfate, 0.25 *M*, 0.5 *N*, Dissolve 33 g of $(NH_4)_2SO_4$ in 1 liter of water.

Ammonium sulfide, colorless.

1. 3 *M.* Treat 200 ml of conc. NH_4OH with H_2S until saturated, keeping the solution cold. Add 200 ml of conc. NH_4OH and dilute of 1 liter.

2. 6 *N.* Saturate 6 *N* ammonium hydroxide (40 ml conc. ammonia solution + 60 ml H_2O) with washed H_2S gas. The ammonium hydroxide bottle must be completely full and must be kept surrounded by ice while being saturated (about 48 hours for two liters). The reagent is best preserved in brown, completely filled, glass-stoppered bottles.

Ammonium sulfide, yellow, Treat 150 ml of conc. NH_4OH with H_2S until saturated, keeping the solution cool. Add 250 ml of conc. NH_4OH and 10 g of powdered sulfur. Shake the mixture until the sulfur is dissolved and dilute to 1 liter with water. In the solution the concentration of $(NH_4)_2S_2$, $(NH_4)_2S$ and NH_4OH are 0.625, 0.4 and 1.5 normal respectively. On standing, the concentration of $(NH_4)_2S_2$ increases and that of $(NH_4)_2S$ and NH_4OH decreases.

Antimony pentachloride, 0.1 *M*, 0.5 *N.* Dissolve 30 g of $SbCl_5$ in 1 liter of water.

Antimony trichloride, 0.167 *M*, 0.5 *N.* Dissolve 38 g of $SbCl_3$ in 1 liter of water.

Aqua regia. Mix 1 part concentrated HNO_3 with 3 parts of concentrated HCl. This formula should include one volume of water if the aqua regia is to be stored for any length of time. Without water, objectionable quantities of chlorine and other gases are evolved.

Barium chloride, 0.25 *M*, 0.5 *N.* Dissolve 61 g of $BaCl_2 \cdot 2H_2O$ in water. Dilute to 1 liter.

Barium hydroxide, 0.1 *M*, about 0.2 *N.* Dissolve 32 g of $Ba(OH)_2 8H_2O$ in 1 liter of water.

Barium nitrate, 0.25 *M*, 0.5 *N.* Dissolve 65 g of $Ba(NO_3)_2$ in 1 liter of water.

Bismuth cloride, 0.167 *M*, 0.5 *N.* Dissolve 53 g of $BiCl_3$ in 1 liter of dilute HCl. Use 1 part HCl to 5 parts water.

Bismuth nitrate, 0.083 *M*, 0.25 *N.* Dissolve 40 g of $Bi(NO_3)_3 \cdot 5H_2O$ in 1 liter of dilute HNO_3. U se 1 part of HNO_3 to 5 parts of water.

Cadmium chloride, 0.25 *M*, 0.5 *N.* Dissolve 46 g of $CdCl_2$ in 1 liter of water.

Cadmium nitrate, 0.25 *M*, - 0.5 *N.* Dissolve 77 g of $Cd(NO_3)_2 \cdot 4H_2O$ in 1 liter of water.

Cadmium sulfate, 0.25 *M*, 0.5 *N.* Dissolve 70 g of $CdSO_4 \cdot 4H_2O$ in 1 liter of water.

Calcium chloride, 0.25 *M*, 0.5 *N.* Dissolve 55 g of $CaCl_2 \cdot 6H_2O$ in water. Dilute to 1 liter.

Calcium nitrate, 0.25 *M*, 0.5 *N.* Dissolve 41 g of $Ca(NO_3)_2$ in 1 liter of water.

Chloroplatinic acid.

1. 0.0512 *M*, 0.102 *N.* Dissolve 26.53 g of $H_2PtCl_6 \cdot 6H_2O$ in water. Dilute to 100 ml. Contains 0.100 g Pt per ml.

2. Make a 10% solution by dissolving 1 g of $H_2PtCl_6 \cdot 6H_2O$ in 9 ml of water. Shake thoroughly to insure complete mixing. Keep in a dropping bottle.

Chromic chloride, 0.167 *M*, 0.5 *N.* Dissolve 26 g of $CrCl_3$ in 1 liter of water.

Chromic nitrate, 0.167 *M*, 0.5 *N.* Dissolve 40 g of $Cr(NO_3)_3$ in 1 liter of water.

Chromic sulfate, 0.083 *M*, -.5 *N.* Dissolve 60 g of $Cr_2(SO_4)_3 \cdot 18H_2O$ in 1 liter of water.

Cobaltous nitrate, 0.25 *M*, 0.5 *N.* Dissolve 73 g of $Co(NO_3)_2 \cdot 6H_2O$ in 1 liter of water.

Cobaltous sulfate, 0.25 *M*, 0.5 *N.* Dissolve 70 g of $CoSO_4 \cdot 7H_2O$ in 1 liter of water.

Cupric chloride, 0.25 *M*, 0.5 *N.* Dissolve 43 g of $CuCl_2 \cdot 2H_2O$ in 1 liter of water.

Cupric nitrate, 0.25 *M*, 0.5 *N.* Dissolve 74 g of $Cu(NO_3)_2 \cdot 6H_2O$ in 1 liter of water.

Cupric sulfate, 0.5 *M*, 1 *N.* Dissolve 124.8 g of $CuSO_4 \cdot 5H_2O$ in water to which 5 ml of H_2SO_4 has been added. Dilute to 1 liter.

Ferric chloride, 0.5 *M*, 1.5 *N.* Dissolve 135.2 g of $FeCl_3 \cdot 6H_2O$ in water containing 20 ml of conc. HCl. Dilute to 1 liter.

Ferric nitrate, 0.167 *M*, 0.5 *N.* Dissolve 67 g of $Fe(NO_3)_3 \cdot 9H_2O$ in 1 liter of water.

Ferric sulfate, 0.25 *M*, 0.5 *N*. Dissolve 140.5 g of $Fe_2(SO_4)_3 \cdot 9H_2O$ in water containing 100 ml of conc. H_2SO_4. Dilute to 1 liter.

Ferrous ammonium sulfate, 0.5 *M*, 1 *N*. Dissolve 196 g of $Fe(NH_4SO_4)_2 \cdot 6H_2O$ in water containing 10 ml of conc. H_2SO_4. Dilute to 1 liter. Prepare fresh solutions for best results.

Ferrous sulfate, 0.5 *M*, 1 *N*. Dissolve 139 g of $FeSO_4 \cdot 7H_2O$ in water containing 10 ml of conc. H_2SO_4. Dilute to 1 liter. Solution does not keep well.

Lead acetate, 0.5 *M*, 1 *N*. Dissolve 190 g of $Pb(C_2H_3O_2)_2 \cdot 3H_2O$ in water. Dilute to 1 liter.

Lead nitrate, 0.25 *M*, 0.5 *N*. Dissolve 83 g of $Pb(NO_3)_2$ in water. Dilute to one liter.

Lime water, See Calcium hydroxide.

Magnesium chloride, 0.25 *M*, 0.5 *N*. Dissolve 51 g of $MgCl_2 \cdot 6H_2O$ in 1 liter of water.

Magnesium chloride reagent. Dissolve 50 g of $MgCl_2 \cdot 6H_2O$ and 100 g of NH_4Cl in 500 ml of water. Add 10 ml of conc. NH_4OH, allow to stand over night and filter if a precipitate has formed. Make acid to methyl red with dilute HCl. Dilute to 1 liter. Solution contains 0.25 *M* $MgCl_2$ and 2 *M* NH_4Cl. Solution may also be diluted with 133 ml of conc. NH_4OH and water to make 1 liter. Such a solution will contain 2 *M* NH_4OH.

Magnesium nitrate, 0.25 *M*, 0.5 *N*. Dissolve 64 g of $Mg(NO_3)_2 \cdot 6H_2O$ in 1 liter of water.

Magnesium sulfate, 0.25 *M*, 0.5 *N*. Dissolve 62 g of $MgSO_4 \cdot 7H_2O$ in 1 liter of water.

Manganous chloride, 0.25 *M*, 0.05 *N*. Dissolve 50 g of $MnCl_2 \cdot 4H_2O$ in 1 liter of water.

Manganous nitrate, 0.25 *M*, 0.5 *N*. Dissolve 72 g of $Mn(NO_3)_2 \cdot 6H_2O$ in 1 liter of water.

Manganous sulfate, 0.25 *M*, 0.5 *N*. Dissolve 69 g of $MnSO_4 \cdot 7H_2O$ in 1 liter of water.

Mercuric chloride, 0.25 *M*, 0.5 *N*. Dissolve 68 g of $HgCl_2$ in water. Dilute to 1 liter.

Mercuric nitrate, 0.25 *M*, 0.5 *N*. Dissolve 81 g of $Hg(NO_3)_2$ in 1 liter of water.

Mercuric sulfate, 0.25 *M*, 0.5 *N*. Dissolve 74 g of $HgSO_4$ in 1 liter of water.

Mercurous nitrate. Use 1 part $HgNO_3$, 20 parts water and 1 part HNO_3.

Nickel chloride, 0.25 *M*, 0.5 *N*. Dissolve 59 g of $NiCl_2 \cdot 6H_2O$ in 1 liter of water.

Nickel nitrate, 0.25 *M*, 0.5 *N*. Dissolve 73 g of $Ni(NO_3)_2 \cdot 6H_2O$ in 1 liter of water.

Nickel sulfate, 0.25 *M*, 0.5 *N*. Dissolve 66 g of $NiSO_4 \cdot 6H_2O$ in 1 liter of water.

Potassium bromide, 0.5 *M*, 0.5 *N*. Dissolve 60 g of KBr in 1 liter of water.

Potassium carbonate, 1.5 *M*, 3 *N*. Dissolve 207 g of K_2CO_3 in 1 liter of water.

Potassium chloride, 0.5 *M*, 0.5 *N*. Dissolve 37 g of KCl in 1 liter of water.

Potassium chromate, 0.25 *M*, 0.5 *N*. Dissolve 49 g of K_2CrO_4 in 1 liter of water.

Potassium cyanide, 0.5 *M*, 0.5 *N*. Dissolve 33 g of KCN in 1 liter of water.

Potassium dichromate, 0.125 *M*. Dissolve 37 g of $K_2Cr_2O_7$ in 1 liter of water.

Potassium ferricyanide, 0.167 *M*, 0.5 *N*. Dissolve 55 g of $K_3Fe(CN)_6$ in 1 liter of water.

Potassium ferrocyanide, 0.5 *M*, 2 *N*. Dissolve 211 g of $K_4Fe(CN)_6 \cdot 3H_2O$ in water. Dilute to 1 liter.

Potassium iodide, 0.5 *M*, 0.5 *N*. Dissolve 83 g of KI in 1 liter of water.

Potassium nitrate, 0.5 *M*, 0.5 *N*. Dissolve 51 g of KNO_3 in 1 liter of water.

Potassium sulfate, 0.25 *M*, 0.5 *N*. Dissolve 44 g of K_2SO_4 in 1 liter of water.

Silver nitrate, 0.5 *M*, 0.5 *N*. Dissolve 85 g of $AgNO_3$ in water. Dilute to 1 liter.

Sodium acetate, 3 *M*, 3 *N*. Dissolve 408 g of $NaC_2H_3O_2 \cdot 3H_2O$ in water. Dilute to 1 liter.

Sodium carbonate, 1.5 *M*, 3 *N*. Dissolve 159 g of Na_2CO_3, or 430 g of $Na_2CO_3 \cdot 10H_2O$ in water. Dilute to 1 liter.

Sodium chloride, 0.5 *M*, 0.5 *N*. Dissolve 29 g of NaCl in 1 liter of water.

Sodium cobaltinitrite, 0.08 *M* (reagent for potassium). Dissolve 25 g of $NaNO_2$ in 75 ml of water, add 2 ml of glacial acetic acid and then 2.5 g of $Co(NO_3)_2 \cdot 6H_2O$. Allow to stand for several days, filter and dilute to 100 ml. Reagent is somewhat unstable.

Sodium hydrogen phosphate, 0.167 *M*, 0.5 *N*. Dissolve 60 g of $Na_2HPO_4 \cdot 12H_2O$ in 1 liter of water.

Sodium nitrate, 0.5 *M*, 0.5 *N*. Dissolve 43 g of $NaNO_3$ in 1 liter of water.

SPECIAL SOLUTIONS AND REAGENTS

Sodium Sulfate, 0.25 *M*, 0.5 *N*. Dissolve 36 g of Na_2SO_4 in 1 liter of water.

Sodium sulfide, 0.5 *M*, 1 *N*. Dissolve 120 g of $Na_2S \cdot 9H_2O$ in water and dilute to 1 liter. Or, saturate 500 ml of 1 *M* NaOH (21 g of 95% NaOH sticks) with H_2S, keeping the solution cool, and dilute with 500 ml of 1 *M* NaOH.

Stannic chloride, 0.125 *M*, 0.5 *N*. Dissolve 33 g of $SnCl_4$ in 1 liter of water.

Stannous chloride, 0.5 *M*, 1 *N*. Dissolve 113 g of $SnCl_2 \cdot 2H_2O$ in 170 ml of conc. HCl, using heat if necessary. Dilute with water to 1 liter. Add a few pieces of tin foil. Prepare solution fresh at frequent intervals.

Stannous chloride (for Bettendorf test). Dissolve 113 g of $SnCl_2 \cdot 2H_2O$ in 75 ml of conc. HCl. Add a few pieces of tin foil.

Strontium chloride, 0.25 *M*, 0.5 *N*. Dissolve 67 g of $SrCl_2 \cdot 6H_2O$ in 1 liter of water.

Zinc nitrate, 0.25 *M*, 0.5 *N*. Dissolve 74 g of $Zn(NO_3)_2 \cdot 6H_2O$ in 1 liter of water.

Zinc sulfate, 0.25 *M*, 0.5 *N*. Dissolve 72 g of $ZnSO_4 \cdot 7H_2O$ in 1 liter of water.

SPECIAL SOLUTIONS AND REAGENTS

Aluminon (qualitative test for aluminum). Aluminon is a trade name for the ammonium salt of aurin tricarboxylic acid. Dissolve 1 g of the salt in 1 liter of distilled water. Shake the solution well to insure thorough mixing.

Bang's reagent (for glucose estimation). Dissolve 100 g of K_2CO_3, 66 g of KCl and 160 g of $KHCO_3$ in the order given in about 700 ml of water at 30° C. Add 4.4 g of $CuSO_4$ and dilute to 1 liter after the CO_2 is evolved. This solution should be shaken only in such a manner as not to allow entry of air. After 24 hours 300 ml are diluted to 1 liter with saturated KCl solution, shaken gently and used after 24 hours; 50 ml equivalent to 10 mg glucose.

Barfoed's reagent (test for glucose). See Cupric acetate.

Baudisch's reagent. See Cupferron.

Benedict's solution (qualitative reagent for glucose). With the aid of heat, dissolve 173 g of sodium citrate and 100 g of Na_2CO_3 in 800 ml of water. Filter, if necessary, and dilute to 850 ml. Dissolve 17.3 g of $CuSO_4 \cdot 5H_2O$ in 100 ml of water. Pour the latter solution, with constant stirring, into the arbonate-citrate solution, and make up to 1 liter.

Benzidine hydrochloride solution (for sulfite determination). Make a paste of 8 g of benzidine hydrochloride ($C_{12}H_3(NH_2)_2 \cdot 2HCl$) and 20 ml of water, add 20 ml of HCl (sp. gr. 1.12) and dilute to 1 liter with water. Each ml of this solution is equivalent to 0.00357 g of H_2SO_4.

Bertrand's reagent (glucose estimation). Consists of the following solutions:

(a) Dissolve 200 g of Rochelle salts and 150 g of NaOH in sufficient water to make 1 liter of solution.
(b) Dissolve 40 g of $CuSO_4$ in enough water to make 1 liter of solution.
(c) Dissolve 50 g of $Fe_2(SO_4)_3$ and 200 g of H_2SO_4 (sp. gr. 1.84) in sufficient water to make 1 liter of solution.
(d) Dissolve 5 g of $KMnO_4$ in sufficient water to make 1 liter of solution.

Bial's reagent (for pentose). Dissolve 1 g of orcinol ($CH_3 \cdot C_6H_3(OH)_2$) in 500 ml of 30% HCl to which 30 drops of a 10% solution of $FeCl_3$ has been added.

Boutron-Boudet soap solution.
(a) Dissolve 100 g of pure castile soap in about 2500 ml of 56% ethyl alcohol.
(b) Dissolve 0.59 g of $Ba(NO_3)_2$ in 1 liter of water.
Adjust the castile soap solution so that 2.4 ml of it will give a permanent lather with 40 ml of solution (b). When adjusted, 2.4 ml of soap solution is equivalent to 220 parts per million of hardness (as $CaCO_3$) for a 40 ml sample.
See also Soap solution.

Brucke's reagent (protein precipitation). See Potassium iodide-mercuric iodide.

Clarke's soap solution (or A.P.H.A., standard method). Estimation of hardness in water.
(a) Dissolve 100 g of pure powdered castile soap in 1 liter of 80% ethyl alcohol and allow to stand over night.
(b) Prepare a standard solution of $CaCl_2$ by dissolving 0.5 g of $CaCO_3$ in HCl (sp. gr. 1.19), neutralize with NH_4OH and make slightly alkaline to litmus, and dilute to 500 ml. One ml is equivalent to 1 mg of $CaCO_3$.
Titrate (a) against (b) and dilute (a) with 80% ethyl alcohol until 1 ml of the resulting solution is equivalent to 1 ml of (b) after making allowance for the lather factor (the amount of standard soap solution required to produce a permanent lather in 50 ml of distilled water). One ml of the adjusted solution after subtracting the lather factor is equivalent to 1 mg of $CaCO_3$.
See also Soap solution.

Cobalticyanide paper (Rinnmann's test for Zn). Dissolve 4 g of $K_3Co(CN)_6$ and 1 g of $KClO_3$ in 100 ml of water. Soak filter paper in solution and dry at 100°C. Apply drop of zinc solution and burn in an evaporating dish. A green disk is obtained if zinc is present.

Cochineal, Extract 1 g of cochineal for four days with 20 ml of alcohol and 60 ml of distilled water. Filter.

Congo red. Dissolve 0.5 g of congo red in 90 ml of distilled water and 10 ml of alcohol.

Cupferron (Baudisch's reagent for iron analysis). Dissolve 6 g of the ammonium salt of nitroso-phenyl-hydroxyl-amine (cupferron) in 100 ml of H_2O. Reagent good for one week only and must be kept in the dark.

Cupric acetate (Barfoed's reagent for reducing monosaccharides). Dissolve 66 g of cupric acetate and 10 ml of glacial acetic acid in water and dilute to 1 liter.

Cupric oxide, ammoniacal; Schweitzer's reagent (dissolves cotton, linen and silk, but not wool).
1. Dissolve 5 g of cupric sulfate in 100 ml of boiling water, and add sodium hydroxide until precipitation is complete. Wash the precipitate well, and dissolve it in a minimum quantity of ammonium hydroxide.
2. Bubble a slow stream of air through 300 ml of strong ammonium hydroxide containing 50 g of fine copper turnings. Continue for one hour.

Cupric sulfate in glycerin-potassium hydroxide (reagent for silk). Dissolve 10 g of cupric sulfate, $CuSO_4 \cdot 5H_2O$, in 100 ml of water and add 5 g of glucerin. Add KOH solution slowly until a deep blue solution is obtained.

Cupron (benzoin oxime). Dissolve 5 g in 100 ml of 95% alcohol.

Cuprous chloride, acidic (reagent for CO in gas analysis).
1. Cover the bottom of a two-liter flask with a layer of cupric oxide about one-half inch deep, suspend a bunch of copper wire so as to reach from the bottom to the top of the solution, and fill the flask with hydrochloric acid (sp. gr. 1.10). Shake occasionally. When the solution becomes nearly colorless, transfer to reagent bottles, which should also contain copper wire. The stock bottle may be refilled with dilute hydrochloric acid until either the cupric oxide or the copper wire is used up.
Copper sulfate may be substituted for copper oxide in the above procedure.
2. Dissolve 340 g of $CuCl_2 \cdot 2H_2O$ in 600 ml of conc. HCl and reduce the cupric chloride by adding 190 ml ofa saturated solution of stannous chloride or until the solution is colorless. The stannous chloride is prepared by treating 300 g of metallic tin in a 500 ml flask with conc. HCl until no more tin goes into solution.
3. (Winkler method). Add a mixture of 86 g of CuO and 17 g of finely divided metallic Cu, made by the reduction of CuO with hydrogen, to a solution of HCl, made by diluting 650 ml of conc. HCl with 325 ml of water. After the mixture has been added slowly and with frequent stirring, a spiral of copper wire is suspended in the bottle, reaching all the way to the bottom. Shake occasionally, and when the solution becomes colorless, it is ready for use.

Cuprous chloride, ammoniacal (reagent for CO in gas analysis).
1. The acid solution of cuprous chloride as prepared above is neutralized with ammonium hydroxide until an ammonia odor persists. An excess of metallic copper must be kept in the solution.
2. Pour 800 ml of acidic cuprous chloride, prepared by the Winkler method, into about 4 liters of water. Transfer the precipitate to a 250 ml graduate. After several hours, siphon off the liquid above the 50 ml mark and refill with 7.5% NH_4OH solution which may be prepared by diluting 50 ml of conc. NH_4OH with 150 ml of water. The solution is well shaken and allowed to stand for several hours. It should have a faint odor of ammonia.

Dichlorofluorescein indicator. Dissolve 1 g in 1 liter of 70% alcohol or 1 g of the sodium salt in 1 liter of water.

Dimethylglyoxime (diacetyl dioxime), 0.01 *N*. Dissolve 0.6 g of dimethylglyoxime, $(CH_3CNOH)_2$, in 500 ml of 95% ethyl alcohol. This is an especially sensitive test for nickel, a very definite crimson color being produced.

Diphenylamine (reagent for rayon). Dissolve 0.2 g in 100 ml of concentrated sulfuric acid.

Diphenylamine sulfonate (for titration of iron with $K_2Cr_2O_7$). Dissolve 0.32 g of the barium salt of diphenylamine sulfonic acid in 100 ml of water, add 0.5 g of sodium sulfate and filter off the precipitate of $BaSO_4$

Diphenylcarbazide. Dissolve 0.2 g of diphenylcarbazide in 10 ml of glacial acetic acid and dilute to 100 ml with 95% ethyl alcohol.

Esbach's reagent (estimation of protein). To a water solution of 10 g of picric acid and 20 g of citric acid, add sufficient water to make one liter of solution.

Eschka's compound. Two parts of calcined ("light") magnesia are thoroughly mixed with one part of anhydrous sodium carbonate.

Fehling's solution (reagent for reducing sugars.)
(a) Copper sulfate solution. Dissolve 34.66 g of $CuSO_4 \cdot 5H_2O$ in water and dilute to 500 ml.
(b) Alkaline tartrate solution. Dissolve 173 g of potassium sodium tartrate (Rochelle salts, $KNaC_4H_4O_6 \cdot 4H_2O$) and 50 g of NaOH in water and dilute when cold to 500 ml.
For use, mix equal volumes of the two solutions at the time of using.

Ferric-alum indicator. Dissolve 140 g of ferric-ammonium sulfate cyrstals in 400 ml of hot water. When cool, filter, and make up to a volume of 500 ml with dilute (6 *N*) nitric acid.

Folin's mixture (for uric acid). To 650 ml of water add 500 g of $(NH_4)_2SO_4$, 5 g of uranium acetate and 6 g of glacial acetic acid. Dilute to 1 liter.

Formaldehyde-sulfuric acid (Marquis' reagent for alkaloids). Add 10 ml of formaldehyde solution to 50 ml of sulfuric acid.

Froehde's reagent. See Sulfomolybdic acid.

Fuchsin (reagent for linen). Dissolve 1 g of fuchsin in 100 ml of alcohol.

Fuchsin-sulfurous acid (Schiff's reagent for aldehydes). Dissolve 0.5 g of fuchsin and 9 g of sodium bisulfite in 500 ml of water, and add 10 ml of HCl. Keep in well-stoppered bottles and protect from light.

Gunzberg's reagent (detection of HCl in gastric juice). Prepare as needed a solution containing 4 g of phloroglucinol and 2 g of vanillin in 100 ml of absolute ethyl alcohol.

Hager's reagent. See Picric acid.

Hanus solution (for iodine number). Dissolve 13.2 g of resublimed iodine in one liter of glacial acetic acid which will pass the dichromate test for reducible matter. Add sufficient bromine to double the halogen content, determined by titration (3 ml is about the proper amount). The iodine may be dissolved by the aid of heat, but the solution should be cold when the bromine is added.

Iodine, tincture of. To 50 ml of water add 70 g of I_2 and 50 g of KI. Dilute to 1 liter with alcohol.

Iodo-potassium iodide (Wagner's reagent for alkaloids). Dissolve 2 g of iodine and 6 g of KI in 100 ml of water.

Litmus (indicator). Extract litmus powder three times with boiling alcohol, each treatment consuming an hour. Reject the alcoholic extract. Treat residue with an equal weight of cold water and filter; then exhaust with five times its weight of boiling water, cool and filter. Combine the aqueous extracts.

Magnesia mixture (reagent for phosphates and arsenates). Dissolve 55 g of magnesium chloride and 105 g of ammonium chloride in water, barely acidify with hydrochloric acid, and dilute to 1 liter. The ammonium hydroxide may be omitted until just previous to use. The reagent, if completely mixed and stored for any period of time, becomes turbid.

Magnesium reagent. See S and O reagent.

Magnesium uranyl acetate. Dissolve 100 g of $UO_2(C_2H_3O_2)_2 \cdot 2H_2O$ in 60 ml of glacial acetic acid and dilute to 500 ml. Dissolve 330 g of $Mg(C_2H_3O_2)_2 \cdot 4H_2O$ in 60 ml of glacial acetic acid and dilute to 200 ml. Heat solutions to te boiling point until clear, pour the magnesium solution into the uranyl solution, cool and dilute to 1 liter. Let stand over night and filter if necessary.

Marme's reagent. See Potassium-cadmium iodide.

Marquis' reagent. See Formaldehyde-sulfuric acid.

Mayer's reagent (white precipitate with most alkaloids in slightly acid solutions). Dissolve 1.358 g of $HgCl_2$ in 60 ml of water and pour into a solution of 5 g of KI in 10 ml of H_2O. Add sufficient water to make 100 ml.

Methyl orange indicator. Dissolve 1 g of methyl orange in 1 liter of water. Filter, if necessary.

Methyl orange, modified. Dissolve 2 g of methyl orange and 2.8 g of xylene cyanole FF in 1 liter of 50% alcohol.

Methyl red indicator. Dissolve 1 g of methyl red in 600 ml of alcohol and dilute with 400 ml of water.

Methyl red, modified. Dissolve 0.50 g of methyl red and 1.25 g of xylene cyanole FF in 1 liter of 90% alcohol. Or, dissolve 1.25 g of methyl red and 0.825 g of methylene blue in 1 liter of 90% alcohol.

Millon's reagent (for albumins and phenols). Dissolve 1 part of mercury in 1 part of cold fuming nitric acid. Dilute with twice the volume of water and decant the clear solution after several hours.

Mixed indicator. Prepared by adding about 1.4 g of xylene cyanole FF to 1 g of methyl orange. The dye is seldom pure enough for these proportions to be satisfactory. Each new lot of dye should be tsted by adding additional amounts of the dye until a test portion gives the proper color change. The acid color of this indicator is like that of permanganate; the neutral color is gray; and the alkaline color is green. Described by Hickman and Linstead, J. Chem. Soc. (Lon.), 121, 2502 (1922).

Molisch's reagent. See α-Naphthol.

α-Naphthol (Molisch's reagent for wool). Dissolve 15 g of α-naphthol in 100 ml of alcohol or chloroform.

Nessler's reagent (for ammonia). Dissolve 50 g of KI the smallest possible quantity of cold water (50 ml). Add a saturated solution of mercuric chloride (about 22 g in 350 ml of water will be needed) until an excess is indicated by the formation of a precipitate. Then add 200 ml of 5 *N* NaOH and dilute to 1 liter. Let settle, and draw off the clear liquid.

Nickel oxide, ammoniacal (reagent for silk). Dissolve 5 g of nickel sulfate in 100 ml of water, and add sodium hydroxide solution until nickel hydroxide is completely precipitated. Wash the precipitate well and dissolve in 25 ml of concentrated ammonium hydroxide and 25 ml of water.

ρ-Nitrobenzene-azo-resorcinol (reagent for magnesium). Dissolve 1 g of the dye in 10 ml of *N*NaOH and dilute to 1 liter.

Nitron (detection of nitrate radical). Dissolve 10 g of nitron ($C_{20}H_{16}N_4$, 4, 5-dihydro-1, 4-diphenyl-3, 5-phenylimino-1, 2, 4-triazole) in 5 ml of glacial acetic acid and 95 ml of water. The solution may be filtered with slight suction through an alumdum crucible and kept in a dark bottle.

α-Nitroso-β-naphthol. Make a saturated solution in 50% acetic acid (1 part of glacial acetic acid with 1 part of water). Does not keep well.

Nylander's solution (carbohydrates). Dissolve 20 g of bismuth subnitrate and 40 g of Rochelle salts in 1 liter of 8% NaOH solution. Cool and filter.

Obermayer's reagent (for indoxyl in urine). Dissolve 4 g of $FeCl_3$ in one liter of HCl (sp. gr. 1.19).

Oxine. Dissolve 14 g of HC_9H_6ON in 30 ml of glacial acetic acid. Warm slightly, if necessary. Dilute to 1 liter.

Oxygen absorbent. Dissolve 300 g of ammonium chloride in one liter of water and add one liter of concentrated ammonium hydroxide solution. Shake the solution thorughly. For use as n oxygen absorbent, a bottle half full of copper turnings is filled nearly full with the NH_4Cl-NH_4OH solution and the gas passed through.

Pasteur's salt solution. To one liter of distilled water add 2.5 g of potassium phosphate, 0.25 g of calcium phosphate, 0.25 g of magnesium sulfate and 12.00 g of ammonium tartrate.

Pavy's solution (glucose reagent). To 120 ml of Fehling's solution, add 300 ml of NH_4OH (sp. gr. 0.88) and dilute to 1 liter with water.

Phenathroline ferrous ion indicator. Dissolve 1.485 g of phenanthroline monohydrate in 100 ml of 0.025 *M* ferrous sulfate solution.

Phenolphthalein. Dissolve 1 g of phenolphthalein in 50 ml of alcohol and add 50 ml of water.

Phenolsulfonic acid (determination of nitrogen as nitrate). Dissolve 25 g of phenol in 150 ml of conc. H_2SO_4, add 75 ml of fuming H_2SO_4 (15% SO_3), stir well and heat for two hours at 100° C.

Phloroglucinol solution (pentosans). Make a 3% phloroglucinol solution in alcohol. Keep in a dark bottle.

Phosphomolybdic acid (Sonnenschein's reagent for alkaloids).

1. Prepare ammonium phosphomolybdate and after washing with water, boil with nitric acid and expel NH_3; evaporate to dryness and dissolve in 2 *N* nitric acid.

2. Dissolve ammonium molybdate in HNO_3 and treat with phosphoric acid. Filter, wash the precipitate, and boil with aqua regia until the ammonium salt is decomposed. Evaporate to dryness. The residue dissolved in 10% HNO_3 constitutes Sonnenschein's reagent.

Phosphoric acid—sulfuric acid mixture. Dilute 150 ml of conc. H_2SO_4 and 100 ml of conc. H_3PO_4 (85%) with water to a volume of 1 liter.

Phosphotungstic acid (Scheibler's reagent for alkaloids).

1. Dissolve 20 g of sodium tungstate and 15 g of sodium phosphate in 100 ml of water containing a little nitric acid.

2. The reagent is a 10% solution of phosphotungstic acid in water. The phosphotungstic acid is prepared by evaporating a mixture of 10 g of sodium tungstate dissolved in 5 g of phosphoric acid (sp. gr. 1.13) and enough boiling water to effect solution. Crystals of phosphotungstic acid separate.

Picric acid (Hager's reagent for alkaloids, wool and silk). Dissolve 1 g of picric acid in 100 ml of water.

Potassium antimonate (reagent for sodium). Boil 22 g of potassium antimonate with 1 liter of water until nearly all of the salt has dissolved, cool quickly, and add 35 ml of 10% potassium hydroxide. Filter after standing over night.

Potassium-cadmium iodide (Marme's reagent for alkaloids). Add 2 g of CdI_2 to a boiling solution of 4 g of KI in 12 ml of water, and then mix with 12 ml of saturated KI solution.

Potassium hydroxide (for CO_2 absorption). Dissolve 360 gof KOH in water and dilute to 1 liter.

Potassium iodide-mercric iodide (Brucke's reagent for proteins). Dissolve 50 g of KI in 500 ml of water, and saturate with mercuric iodide (about 120 g). Dilute to 1 liter.

Potassium pyrogallate (for oxygen absorption). For mixtures of gases containing less than 28% oxygen, add 100 ml of KOH solution (50 g of KOH to 100 ml of water) to 5 g of pyrogallol. For mixtures containing more than 28% orygen the KOH solution should contain 120 g of KOH to 100 ml of water.

Pyrogallol, alkaline.

(a) Dissolve 75 g of pyrogallic acid in 75 ml of water.

(b) Dissolve 500 g of KOH in 250 ml of water. When cool, adjust until sp. gr. is 1.55.

For use, add 270 ml of solution (b) to 30 ml of solution (a).

Rosolic acid (indicator). Dissolve 1 g of rosolic acid in 10 ml of alcohol and add 100 ml of water.

S and O reagent (Suitsu and Okuma's test for Mg). Dissolve 0.5 g of the dye (*o*-*p*-dihydroxy-monazo-*p*-nitrobenzene) in 100 ml of 0.25 *N* NaOH.

Scheibler's reagent. See Phosphotungstic acid.

Schiff's reagent. See Fuchsin-sulfurous acid.

Schweitzer's reagent. See Cupric oxide, ammoniacal.

Soap solution (reagent for hardness in water). D issolve 100 g of dry castile soap in 1 liter of 80% alcohol (5 parts alcohol to 1 part water). Allow to stand several days and dilute with 70% to 80% alcohol until 6.4 ml produces a permanent lather with 20 ml of standard calcium solution. The latter solution is made by dissolving 0.2 g of $CaCO_3$ in a small amount of dilute HCl, evaporating to dryness and making up to 1 liter.

Sodium bismuthate (oxidation of manganese). Heat 20 parts of NaOH nearly to redness in an iron or nickel crucible and add slowly 10 parts of basic bismuth nitrate which has been previously dried. Add two parts of sodium peroxide, and pour the brownish-yellow fused mass on an iron plate to cool. When cold, break up in a mortar, extract with water, and collect on an asbestos filter.

Sodium hydroxide (for CO_2 absorption). Dissolve 330 g of NaOH in water and dilute to 1 liter.

Sodium nitroprusside (reagent for hydrogen sulfide and wool). Use a freshly prepared solution of 1 g of sodium nitroprusside in 10 ml of water.

SPECIAL SOLUTIONS AND REAGENTS (Continued)

Sodium oxalate, according to Sorensen (primary standard). Dissolve 30 g of the commercial salt in 1 liter of water, make slightly alkaline with sodium hydroxide, and let stand until perfectly clear. Filter and evaporate the filtrate to 100 ml. Cool and filter. Pulverize the residue and wash it several times with small volumes of water. The procedure is repeated until the mother liquor is free from sulfate and is neutral to phenolphthalein.

Sodium plumbite (reagent for wool). Dissolve 5 g of sodium hydroxide in 100 ml of water. Add 5 g of litharge and boil until dissolved.

Sodium polysulfide. Dissolve 480 g of $Na_2S \cdot 9H_2O$ in 50G ml of water, add 40 g of NaOH and 18 g of sulfur. Stir thoroughly and dilute to 1 liter with water.

Sonnenschein's reagent. See Phosphomolybdic acid.

Starch solution.

1. Make a paste with 2 g of soluble starch and 0.01 g of HgI_2 with a small amount of water. Add the mixture slowly to 1 liter of boiling water and boil for a few minutes. Keep in a glass stoppered bottle. If other than soluble starch is used, the solution will not clear on boiling; it should be allowed to stand and the clear liquid decanted.

2. A solution of starch which keeps indefinitely is made as follows: Mix 500 ml of saturated NaCl solution (filtered), 80 ml of glacial acetic acid, 20 ml of water and 3 g of starch. Bring slowly to a boil and boil for two minutes.

3. Make a paste with 1 g of soluble starch and 5 mg of HgI_2, using as little cold water as possible. Then pour about 200 ml of boiling water on the paste and stir immediately. This will give a clear solution if the paste is prepared correctly and the water actually boiling. Cool and add 4 g of KI. Starch solution decomposes on standing due to bacterial action, but this solution will keep a long time if stored under a layer of toluene.

Stoke's reagent. Dissolve 30 g of $FeSO_4$ and 20 g of tartaric acid in water and dilute to 1 liter. Just before using, add concentrated NH_4OH until the precipitate first formed is redissolved.

Sulfanilic acid (reagent for nitrites). Dissolve 0.5 g of sulfanilic acid in a mixture of 15 ml of glacial acetic acid and 135 ml of recently boiled water.

Sulfomolybdic acid (Froehde's reagent for alkaloids and glucosides). Dissolve 10 g of molybdic acid or sodium molybdate in 100 ml of conc. H_2SO_4.

Tannic acid (reagent for albumen, alkaloids and gelatin). Dissolve 10 g of tannic acid in 10 ml of alcohol and dilute with water to 100 ml.

Titration mixture. (residual chlorine in water analysis). Prepare 1 liter of dilute HCl (100 ml of HCl (sp. gr. 1.19) in sufficient water to make 1 liter). Dissolve 1 g of *o*-tolidine in 100 ml of the dilute HCl and dilute to 1 liter with dilute HCl solution.

Trinitrophenol solution. See Picric acid.

Turmeric paper. Impregnant white, unsized paper with the tincture, and dry.

Turmeric tincture (reagent for borates). Digest ground turmeric root with several quantities of water which are discarded. Dry the residue and digest it several days with six times its weight of alcohol. Filter.

Uffelmann's reagent (turns yellow in presence of a lactic acid). To a 2% solution of pure phenol in water, add a water solution of $FeCl_3$ until the phenol solution becomes a violet in color.

Wagner's reagent. See Iodo-potassium iodide.

Wagner's solution (used in phosphate rock analysis to prevent precipitation of iron and aluminum). Dissolve 25 g of citric acid and 1 g of salicylic acid in water and dilute to 1 liter. Use 50 ml of the reagent.

Wij's iodine monochloride solution (for iodine number). Dissolve 13 g of resublimed iodine in 1 liter of glacial acetic acid which will pass the dichromate test for reducible matter. Set aside 25 ml of this solution. Pass into the remainder of the solution dry chlorine gas (dried and washed by passing through H_2SO_4 (sp. gr. 1.84)) until the characteristic color of free iodine has been discharged. Now add the iodine solution which was reserved, until all free chlorine has been destroyed. A slight excess of iodine does little or no harm, but an excess of chlorine must be avoided. Preserve in well stoppered, amber colored bottles. Avoid use of solutions which have been prepared for more than 30 days.

Wij's special solution (for iodine number—Analyst 58, 523-7, 1933). To 200 ml of glacial acetic acid that will pass the dichromate test for reducible matter, add 12 g of dichloroamine T (paratoluene-sulfonedichloroamide), and 16.6 g of dry KI (in small quantities with continual shaking until all the KI has dissolved). Make up to 1 liter with the same quality of acetic acid used above and preserve in a dark colored bottle.

Zimmermann-Reinhardt reagent (determination of iron). Dissolve 70 g of $MnSO_4 \cdot 4H_2O$ in 500 ml of water, add 125 ml of conc. H_2SO_4 and 125 ml of 85% H_3PO_4, and dilute to 1 liter.

Zinc chloride solution, basic (reagent for silk). Dissolve 1000 g of zinc chloride in 850 ml of water, and add 40 g of zinc oxide. Heat until solution is complete.

Zinc uranyl acetate (reagent for sodium). Dissolve 10 g of$UO_2(C_2H_3O_2)_2 \cdot 2H_2O$ in 6 g of 30% acetic acid with heat, if necessary, and dilute to 50 ml. Dissolve 30 g of $Zn(C_2H_3O_2)_2 \cdot 2H_2O$ in 3 g of 30% acetic acid and dilute to 50 ml. Mix the two solutions, add 50 mg of NcCl, allow to stand over night and filter.

DECI-NORMAL SOLUTIONS OF SALTS AND OTHER REAGENTS

Atomic and molecular weights in the following table are based upon the 1965 atomic weight scale and the isotope C-12. The weight in grams of the compound in 1 cc of the following deci-normal solutions is found by dividing the H equivalent in the last column by 1000.

Name	Formula	Atomic or molecular weight	Hydrogen equivalent	0.1 Hydrogen equivalent in g
Acetic acid	$HC_2H_3O_2$	60.0530	$HC_2H_3O_2$	6.0053
Ammonia	NH_3	17.0306	NH_3	1.7031
Ammonium ion	NH_4^+	18.0386	NH_4	1.8039
Ammonium chloride	NH_4Cl	53.4916	NH_4Cl	5.3492
Ammonium sulfate	$(NH_4)_2SO_4$	132.1388	$\frac{1}{2}(NH_4)_2SO_4$	6.6069
Ammonium thiocyanate	NH_4CNS	76.1201	NH_4CNS	7.6120
Barium	Ba	137.34	$\frac{1}{2}Ba$	6.867
Barium carbonate	$BaCO_3$	197.3494	$\frac{1}{2}BaCO_3$	9.8675
Barium chloride hydrate	$BaCl_2 \cdot 2H_2O$	244.2767	$\frac{1}{2}BaCl_2 \cdot 2H_2O$	12.2138
Barium hydroxide	$Ba(OH)_2$	171.3547	$\frac{1}{2}Ba(OH)_2$	8.5677
Barium oxide	BaO	153.3394	$\frac{1}{2}BaO$	7.6670
Bromine	Br	79.909	Br	7.9909
Calcium	Ca	40.08	$\frac{1}{2}Ca$	2.004
Calcium carbonate	$CaCO_3$	100.0894	$\frac{1}{2}CaCO_3$	5.0045
Calcium chloride	$CaCl_2$	110.9860	$\frac{1}{2}CaCl_2$	5.5493
Calcium chloride hydrate	$CaCl_2.6H_2O$	219.0150	$\frac{1}{2}CaCl_2.6H_2O$	10.9508
Calcium hydroxide	$Ca(OH)_2$	74.0947	$\frac{1}{2}Ca(OH)_2$	3.7047
Calcium oxide	CaO	56.0794	$\frac{1}{2}CaO$	2.8040
Chlorine	Cl	35.453	Cl	3.5453
Citric acid	$C_6H_8O_7 \cdot H_2O$	210.1418	$\frac{1}{3}C_6H_8O_7 \cdot H_2O$	7.0047
Cobalt	Co	58.9332	$\frac{1}{2}Co$	2.9466
Copper	Cu	63.54	$\frac{1}{2}Cu$	3.177
Copper oxide (cupric)	CuO	79.5394	$\frac{1}{2}CuO$	3.9770
Copper sulfate hydrate	$CuSO_4 \cdot 5H_2O$	249.6783	$\frac{1}{2}CuSO_4 \cdot 5H_2O$	12.4839
Cyanogen	$(CN)_2$	26.0179	CN	2.6018
Hydrochloric acid	HCl	36.4610	HCl	3.6461
Hydrocyanic acid	HCN	27.0258	HCN	2.7026
Iodine	I	126.9044	I	12.6904
Lactic acid	$C_3H_6O_3$	90.0795	$C_3H_6O_3$	9.0080
Malic acid	$C_4H_6O_5$	134.0894	$\frac{1}{2}C_4H_6O_5$	6.7045
Magnesium	Mg	24.312	$\frac{1}{2}Mg$	1.2156
Magnesium carbonate	$MgCO_3$	84.3214	$\frac{1}{2}MgCO_3$	4.2161
Magnesium chloride	$MgCl_2$	95.2180	$\frac{1}{2}MgCl_2$	4.7609
Magnesium chloride hydrate	$MgCl_2 \cdot 6H_2O$	203.2370	$\frac{1}{2}MgCl_2 \cdot 6H_2O$	10.1623
Magnesium oxide	MgO	40.3114	$\frac{1}{2}MgO$	2.0156
Manganese	Mn	54.938	$\frac{1}{2}Mn$	2.7469
Manganese sulfate	$MnSO_4$	150.9996	$\frac{1}{2}MnSO_4$	7.5500
Mercuric chloride	$HgCl_2$	271.4960	$\frac{1}{2}HgCl_2$	13.5748
Nickel	Ni	58.71	$\frac{1}{2}Ni$	2.9356
Nitric acid	HNO_3	63.0129	HNO_3	6.3013
Nitrogen	N	14.0067	N	1.4007
Nitrogen pentoxide	N_2O_5	108.0104	$\frac{1}{2}N_2O_5$	5.4005
Oxalic acid	$H_2C_2O_4$	90.0358	$\frac{1}{2}H_2C_2O_4$	4.5018
Oxalic acid hydrate	$H_2C_2O_4 \cdot 2H_2O$	126.0665	$\frac{1}{2}H_2C_2O_4 \cdot 2H_2O$	6.3033
Oxalic acid anhydride	C_2O_3	72.0205	$\frac{1}{2}C_2O_3$	3.6010
Phosphoric acid	H_3PO_4	97.9953	$\frac{1}{3}H_3PO_4$	3.2665
Potassium	K	39.102	K	3.9102
Potassium bicarbonate	$KHCO_3$	100.1193	$KHCO_3$	10.0119
Potassium carbonate	K_2CO_3	138.2134	$\frac{1}{2}K_2CO_3$	6.9106
Potassium chloride	KCl	74.5550	KCl	7.4555
Potassium cyanide	KCN	65.1199	KCN	6.5120
Potassium hydroxide	KOH	56.1094	KOH	5.6109
Potassium oxide	K_2O	94.2034	$\frac{1}{2}K_2O$	4.7102
Potassium permanganate for Co estimation	$KMnO_4$	158.0376	$\frac{1}{6}KMnO_4$	2.6339
Potassium permanganate for Mn estimation	$KMnO_4$	158.0376	$\frac{1}{3}KMnO_4$	5.2678
Potassium tartrate	$K_2H_4C_4O_6$	226.2769	$\frac{1}{2}K_2H_4C_4O_6$	11.3139
Silver	Ag	107.87	Ag	10.787
Silver nitrate	$AgNO_3$	169.8749	$AgNO_3$	16.9875
Sodium	Na	22.9898	Na	2.2990
Sodium bicarbonate	$NaHCO_3$	84.0071	$NaHCO_3$	8.4007
Sodium carbonate	Na_2CO_3	105.9890	$\frac{1}{2}Na_2CO_3$	5.2995
Sodium chloride	NaCl	58.4428	NaCl	5.8443
Sodium hydroxide	NaOH	39.9972	NaOH	3.9997
Sodium oxide	Na_2O	61.9790	$\frac{1}{2}Na_2O$	3.0990
Sodium sulfide	Na_2S	78.0436	$\frac{1}{2}Na_2S$	3.9022
Succinic acid	$H_2C_4H_4O_4$	118.0900	$\frac{1}{2}H_2C_4H_4O_4$	5.9045
Sulfuric acid	H_2SO_4	98.0775	$\frac{1}{2}H_2SO_4$	4.9039
Sulfur trioxide	SO_3	80.0622	$\frac{1}{2}SO_3$	4.0031
Tartaric acid	$C_4H_6O_6$	150.0888	$\frac{1}{2}C_4H_6O_6$	7.5044
Zinc	Zn	65.37	$\frac{1}{2}Zn$	3.269
Zinc sulfate	$ZnSO_4 \cdot 7H_2O$	287.5390	$\frac{1}{2}ZnSO_4 .7 \cdot H_2O$	14.3769

DECI-NORMAL SOLUTIONS OF OXIDATION AND REDUCTION REAGENTS

Atomic and molecular weights in the following table are based upon the 1965 atomic weight scale and the isotope C-12. The weight in grams of the compound in 1 cc of the following deci-normal solutions is found by dividing the H equivalent in the last column by 1000.

Name	Formula	Atomic or molecular weight	Hydrogen equivalent	0.1 Hydrogen equivalent in g
Antimony	Sb	121.75	$\frac{1}{2}Sb$	6.0875
Arsenic	As	74.9216	$\frac{1}{2}As$	3.7461
Arsenic trisulfide	As_2S_3	246.0352	$\frac{1}{4}As_2S_3$	6.1509
Arsenous oxide	As_2O_3	197.8414	$\frac{1}{4}As_2O_3$	4.9460
Barium peroxide	BaO_2	169.3388	$\frac{1}{2}BaO_2$	8.4669
Barium peroxide hydrate	$BaO_2{\cdot}8H_2O$	313.4615	$\frac{1}{2}BaO_2{\cdot}8H_2O$	15.6730
Calcium	Ca	40.08	$\frac{1}{2}Ca$	2.004
Calcium carbonate	$CaCO_3$	100.0894	$\frac{1}{2}CaCO_3$	5.0045
Calcium hypochlorite	$Ca(OCl)_2$	142.9848	$\frac{1}{4}Ca(OCl)_2$	3.5746
Calcium oxide	CaO	56.0794	$\frac{1}{2}CaO$	2.8040
Chlorine	Cl	35.453	Cl	3.5453
Chromium trioxide	CrO_3	99.9942	$\frac{1}{3}CrO_3$	3.3331
Ferrous ammonium sulfate	$FeSO_4(NH_4)SO_4{\cdot}6H_2O$	392.0764	$FeSO_4(NH_4)_2SO_4{\cdot}6H_2O$	39.2076
Hydroferrocyanic acid	$H_4Fe(CN)_6$	215.9860	$H_4Fe(CN)_6$	21.5986
Hydrogen peroxide	H_2O_2	34.0147	$\frac{1}{2}H_2O_2$	1.7007
Hydrogen sulfide	H_2S	34.0799	$\frac{1}{2}H_2S$	1.7040
Iodine	I	126.9044	I	12.6904
Iron	Fe	55.847	Fe	5.5847
Iron oxide (ferrous)	FeO	71.8464	FeO	7.1846
Iron oxide (ferric)	Fe_2O_3	159.6922	$\frac{1}{2}Fe_2O_3$	7.9846
Lead peroxide	PbO_2	239.1888	$\frac{1}{2}PbO_2$	11.9594
Manganese dioxide	MnO_2	86.9368	$\frac{1}{2}MnO_2$	4.3468
Nitric acid	HNO_3	63.0129	$\frac{1}{3}HNO_3$	2.1004
Nitrogen trioxide	N_2O_3	76.0116	$\frac{1}{4}N_2O_3$	1.9002
Nitrogen pentoxide	N_2O_5	108.0104	$\frac{1}{5}N_2O_5$	1.8001
Oxalic acid	$C_2H_2O_4$	90.0358	$\frac{1}{2}C_2H_2O_4$	4.5018
Oxalic acid hydrate	$C_2H_2O_4{\cdot}2H_2O$	126.0665	$\frac{1}{2}C_2H_2O_4{\cdot}2H_2O$	6.3033
Oxygen	O	15.9994	$\frac{1}{2}O$	0.8000
Potassium dichromate	$K_2Cr_2O_7$	294.1918	$\frac{1}{6}K_2Cr_2O_7$	4.9032
Potassium chlorate	$KClO_3$	122.5532	$\frac{1}{6}KClO_3$	2.0425
Potassium chromate	K_2CrO_4	194.1076	$\frac{1}{3}K_2CrO_4$	6.4733
Potassium ferrocyanide	$K_4Fe(CN)_6$	368.3621	$K_4Fe(CN)_6$	36.8362
Potassium ferrocyanide	$K_4Fe(CN)_6{\cdot}3H_2O$	422.4081	$K_4Fe(CN)_6{\cdot}3H_2O$	42.2408
Potassium iodide	KI	166.0064	KI	16.6006
Potassium nitrate	KNO_3	101.1069	$\frac{1}{3}KNO_3$	3.3702
Potassium perchlorate	$KClO_4$	138.5526	$\frac{1}{8}KClO_4$	1.7319
Potassium permanganate	$KMnO_4$	158.0376	$\frac{1}{5}KMnO_4$	3.1608
Sodium chlorate	$NaClO_3$	106.4410	$\frac{1}{6}NaClO_3$	1.7740
Sodium nitrate	$NaNO_3$	84.9947	$\frac{1}{3}NaNO_3$	2.8332
Sodium thiosulfate	$Na_2S_2O_3{\cdot}5H_2O$	248.1825	$Na_2S_2O_3{\cdot}5H_2O$	24.8183
Stannous chloride	$SnCl_2$	189.5960	$\frac{1}{2}SnCl_2$	9.4798
Stannous oxide	SnO	134.6894	$\frac{1}{2}SnO$	6.7345
Sulfur dioxide	SO_2	64.0628	$\frac{1}{2}SO_2$	3.2031
Tin	Sn	118.69	$\frac{1}{2}Sn$	5.935

ORGANIC ANALYTICAL REAGENTS

Compiled by John H. Yoe

Determination	Reagent	Reference
Acetate	*o*-Nitrobenzaldehyde	Feigl, p. 342 (3)
Aldehydes	Dimethyl-dihydroresorcin (Dimedon)	Ind. Eng. Chem., Anal. Ed. **3**, 365 (1931)
Aluminum	Alizarin S	J. Am. Chem. Soc. **50**, 748 (1928)
	Ammonium salt of aurin tricarboxylic acid ("Aluminon")	J. Am. Chem. Soc. **49**, 2395 (1927); J. Am. Chem. Soc. **55**, 2437 (1933)
	Ammonium salt of nitrosophenyl hydroxylamine ("Cupferron")	Bull. soc. chim. Belg. **36**, 288 (1927)
	Eriochrome cyanine	Z. anal. Chem. **96**, 91 (1934)
	Hematoxylin	Ind. Eng. Chem. **16**, 233 (1924)
	8-Hydroxyquinoline	J. Am. Chem. Soc. **50**, 1900 (1928)
	Morin	Feigl, p. 182 (2); Ind. Eng. Chem., Anal. Ed. **12**, 229 (1940)
	Quinalizarine	J. Am. Pharm. Assoc. **17**, 260 (1928)
	Sodium or zinc salt of 4-sulfo-2 hydroxy-α-naphthalene-azo-β-naphthol (Pontachrome Blue Black R)	Sandell, p. 241 (6)
	Urea	Ind. Eng. Chem., Anal. Ed. **9**, 357 (1937)
Ammonia	Hematoxylin	Helv. Chim. Acta **12**, 730 (1929)
	p-Nitrobenzenediazonium chloride	Feigl, p. 235 (2)
	Phenol and sodium hypochloride	Snell, Vol. II, p. 818 (8)
	Tannin—$AgNO_3$	Snell, Vol. II, p. 819 (8)
	Thymol	J. Biol. Chem. **131**, 309 (1939)
	Zinc oxinate	Feigl, p. 674 (9)
Antimony	Hexamethylenetetramine	Z. anal. Chem. **67**, 298 (1925)
	9-Methyl-2,3,7-tryhydroxyfluorone	Helv. Chim. Acta **20**, 1427 (1937)
	Phenylthiohydantoic acid	Compt. rend. **176**, 1221 (1923)
	Pyridine	Analyst **53**, 373 (1928)
	Pyrogallol	Z. anal. Chem. **64**, 44 (1924)
	Rhodamine B	Z. anal. Chem. **70**, 400 (1927)
Arsenic	Cocaine-molybdate	Biochem. Z. **185**, 14 (1927)
	N-Ethyl-8-hydroxytetrahydroquinoline hydrochloride	Z. anal. Chem. **99**, 180 (1934)
	Quinine arsenomolybdate	Analyst **47**, 317 (1922)
	Strychnine-molybdate	Ann. Chim. applicata **23**, 517 (1933)
Barium	Sodium rhodizonate	Feigl, p. 216 (2)
	Tetrahydroxyquinone	Welcher, Vol. I, p. 225 (9)
Beryllium	1-Amino-4-hydroxyanthraquinone	Ind. Eng. Chem., Anal. **13**, 809 (1941)
	Aurin trycarboxylic	J. Am. Chem. Soc. **48**, 2125 (1926); ibid., **50**, 353 (1928)
	Curcumin	J. Am. Chem. Soc. **50**, 393 (1928)
	1,4-Dihydroxyanthraquinone (Quinizarin)	Ind. Eng. Chem. Anal. Ed., **18**, 179 (1946)
	1.4-Dihydroxyanthraquinone-2-sulfonic acid (quinizarine-2-sulfonic acid)	Sandell, p. 318 (6)
	8-Hydroxyquinoline	Bur. Standards J. Research **3**, 91 (1929)
	Morin	Ind. Eng. Chem., Anal. Ed. **12**, 674, 762 (1940)
	Naphthochrome Azurine 2B	Sandell, p. 318 (6)
	Naphthochrome Green G	Sandell, p. 318 (6)
	p-Nitrobenzeneazo-orcinol	Mikrochemie **14**, 315 (1934)
	1,2,5,8-Tetrahydroxyanthraquinone (Quinalizarin)	Siemens-Konzerns, Beryllium, p. 25 (1932)
Bismuth	Caffeine (as sulfate or nitrate)	Ind. Eng. Chem., Anal. Ed. **14**, 43 (1942)
	Cinchonine	Scott, p. 158 (7)
	Diethyldithiocarbamate	Sandell, p. 338 (6)
	Dimercaptothiodiazole	Z. anal. Chem. **98**, 184 (1934); ibid., **100**, 408 (1935)
	Dimethylglyoxime	Z. anal. Chem. **72**, 11 (1927)
	Diphenylthiocarbazone	Z. Angew. Chem. **47**, 685 (1934); Ind. Eng. Chem., Anal. Ed. **7**, 285 (1935)
	8-Hydroxyquinoline	Z. anal. Chem. **72**, 177 (1927)
	Phenyldithiobiazolonethiol	J. Indian Chem. Soc., **21**, 240, 347 (1944)
	Pyrogallol	Z. anal. Chem. **65**, 448 (1925)
	Thiourea	Z. anal. Chem. **94**, 161 (1933)
Boron	Curcumin	Chem. News **87**, 27 (1903)
	1,1-Dianthramide	Boltz, p. 346 (1)
	Diaminochrysazin	Anal. Chem. **29**, 1251 (1957)
	Diaminoanthrarufin	Ibid.
	Tribromoanthrarufin	Ibid.
	Mannitol	Scott, p. 168 (7)
	Methyl alcohol	J. Am. Chem. Soc. **50**, 1385 (1928)
	p-Nitrobenzeneazo-chromotropic acid (Chromotrope 2B)	Feigl, p. 341 (2)
	Quinalizarin	Ind. Eng. Chem., Anal. Ed. **11**, 540 (1939)
	Titan Yellow (Clayton Yellow)	Compt. rend. **138**, 1046 (1904)
	Turmeric	Ind. Eng. Chem., Anal. Ed. **4**, 180 (1932)
Bromine	Fluorescein	Snell, Vol. II, p. 725 (8)
	Fuchsin	Snell, Vol. II, p. 724 (8)
	Phenol red	Snell, Vol. II, p. 725 (8)
Cadmium	Allythiourea	Helvetica Chim. Acta. **12**, 718 (1929)
	Di-β-naphthylthiocarbazone	Ind. Eng. Chem., Anal. Ed., **16**, 333, (1944)
	Dinitrodiphenylcarbazide	Feigl, p. 96 (2)
	Diphenylcarbazide	Feigl, p. 99 (2)
	Diphenylthiocarbazone	Z. angew. Chem. **47**, 685 (1934); Ind. Eng. Chem., Anal. Ed. **11**, 364 (1939)
	Ethylenediamine	Z. anal. Chem. **77**, 340 (1929)
	Hexamethylenetetramine alliodide	C. A., **24**, 311 (1930)
	β-Naphthoquinoline	Analyst, **58**, 667 (1933)
	Nitrophenolarsinic acid	Mikrochemie **8**, 277 (1930)
	4-Nitrophthalene-diazo-aminobenzene-4-azobenzene (Cadion 2B)	C. A. **32**, 2871 (1938)
	Phenyl-trimethyl-ammonium iodide	Analyst **58**, 667 (1933)
	Pyridine	Z. Anal. Chem. **73**, 279 (1928)
Calcium	Alizarin	Biochem. J. **16**, 494 (1922); Yoe, Vol. I, p. 139 (2)
	1-amino-2-naphthol-4-sulfonic acid	J. Biol. Chem. **81**, 1 (1929)
	Ammonium oxalate	Snell, Vol. II, p. 592 (8)
	Ammonium stearate	J. Biol. Chem. **29**, 169 (1917); Yoe, Vol. II, p. 119 (3)
	2,5-Dichloro-3,6-dihydroxyquinone (Chloranilic acid)	Anal. Chem., **20**, 76 (1948)
	Dihydroxytartaric acid osazone (sodium salt)	Feigl, p. 221 (2)
	Picrolonic acid	Biochem. Z. **265**, 85 (1933)
	Potassium oleate	Sandell, p. 377 (6)
	Pyrogallol carboxylic acid	Sandell, p. 380 (6)
	Sodium sulforicinate	Biochem. Z. **137**, 157 (1923); Yoe, Vol. II, p. 125 (3)
Cerium	Benzidine	Feigl, p. 211 (2)
	Brucine	Sandell, p. 386 (6)
	Gallic acid	Snell, Vol. II, p. 608 (8)
	8-Hydroxyquinoline	Analyst, **75**, 275 (1948)
	Malachite Green	C. A. **30**, 5143 (1936), **31**, 626 (1937)
	Morphine	Sandell, p. 386 (6)
	Sulfanilic acid	Sandell, p. 386 (6)
Cesium	Dipicrylamine (Hexanitrodiphenylamine)	Mikrochemie 18, 175 (1935)
Chlorate	Aniline hydrochloride	Snell, Vol. II, p. 717 (8)
Chlorine	Benzidine hydrochloride	Ind. Eng. Chem., Anal. Ed. **4**, 2 (1932)
	Dimethyl-*p*-phenylene-diamine	Chem. Weekblad. **23**, 203 (1926)
	Oleic acid	J. Soc. Chem. Ind. **42**, 427A (1923)
	Sodium sulforicinate	Biochem. Z., **137**, 157 (1923); Yoe, Vol. II, p. 125 (3)
	Thymolphthalein	Ind. Eng. Chem. **19**, 112 (1927)
	o-Tolidine	Yoe, Vol. I, p. 157 (2)
Columbium	Benzidine	Feigl, p. 171 (2)
	1,8-Dihydroxynaphthalene-3,6-Disulfonate	Ind. Eng. Chem. **5**, 298 (1913)
	s-Diphenylcarbazide	J. Am. Chem. Soc. **50**, 2363 (1928)
	Pyrogallol dimethyl ether	C. A. **4**, 3178 (1910)
	Serichrome Blue R	Ind. Eng. Chem., Amal. Ed., **4**, 245 (1932)

ORGANIC ANALYTICAL REAGENTS (Continued)

Determination	Reagent	Reference
Cobalt	Anthranilic acid (*o*-aminobenzoic acid)	Z. anal. Chem. **93**, 241 (1933)
	Cysteine hydrochloride	J. Biol. Chem. **83**, 367 (1929)
	Dimethylglyoxime	J. Am. Chem. Soc. **43**, 482 (1921)
	3,5-Dimethylpyrazole	Ind. Eng. Chem., Anal. Ed. **2**, 38 (1930)
	Dinitrosoresorcinol	J. Am. Chem. Soc. **45**, 1439 (1923)
	Formaldoxime reagent	C. A. **27**, 927 (1933)
	o-Nitrosocresol	Sandell, p. 422 (6)
	α-Nitroso-β-naphthol	Chem. Zeit. **46**, 430 (1922)
	Nitrose-R-salt	J. Am. Chem. Soc. **43**, 746 (1921)
	β-Nitroso-α-naphthol	Ind. Eng. Chem., Anal. Ed. **12**, 405 (1940)
	o-Nitrosoresorcinol	Ind. Eng. Chem., Anal. Ed. **15**, 310 (1943)
	o-Nitrosophenol	Sandell, p. 422 (6)
	Phenylthiohydantoic acid	J. Am. Chem. Soc. **44**, 2219 (1922)
	Rubeanic acid (Dithio-oxamide)	Anal. Chim. Acta **20**, 332, 435 (1959)
	2,2′,2″-Terpyridyl	Ind. Eng. Chem., Anal. Ed., **15**, 74 (1943)
	Niobium	
Columbium	See **Nobium**	
Copper	Ammonium salt of nitrosophenyl-hydroxylamine ("Cupferron")	Ind. Eng. Chem. **3**, 629 (1911)
	m-Benzamino-semicarbazide	Snell, Vol. II, p. 129 (6)
	Benzidine	Z. anal. Chem. **67**, 31 (1925)
	α-Benzionoxime (Cupron)	Ber. **56**, 2083 (1923)
	Benzotriazole	Ind. Eng. Chem., Anal. Ed. **13**, 349 (1941)
	2-Carboxy-a′-hydroxy-5′ sulfoformazylbenzene (Zincon)	Anal. Chem. **26**, 1345 (1954)
	Diacetyl-dioxime	Analyst **54**, 333 (1929)
	Dibenzyldithiocarbamate	Sandell, p. 443 (6)
	Dihydroxyethyldithiocarbamic acid	Sandell, p. 443 (6)
	p-Dimethylaminobenzalrhodanine	J. Am. Chem. Soc. **52**, 2222 (1930) J. Am. Chem. Soc. **47**, 1268 (1925)
	s-Diphenylcarbazide	Chem. Weekblad. **21**, 20 (1924)
	Diphenylthiocarbazone (Dithizone)	J. Assoc. Official Agr. Chem. **18**, 192 (1935)
	Hydroquinone	Bull. soc. chim. **31**, 1176 (1922)
	Isatin	Rec trav chim. **42**, 199 (1923)
	Mercaptobenzothiazole	Z. anal. Chem. **102**, 24, 108 (1935)
	α-Naphthol	Bull. soc. chim. **31**, 1176 (1922)
	β-Naphthol	Am. J. Pharm. **105**, 62 (1933)
	Phenolphthalein	Compt. rend. **173**, 1082 (1921)
	Phenylthiohydantoic acid	J. Am. Chem. Soc. **44**, 225 (1922)
	Piperidinium piperidyl-dithioformate	Analyst **56**, 736 (1931)
	Potassium ethyl xanthate	Yoe, Vol. I, p. 184 (2)
	Pyridine	Z. anal. Chem. **67**, 27 (1925)
	Rubeanic acid (Dithio-oxamide)	Anal. Chim. Acta **20**, 332, 435 (1959)
	Salicylaldoxime	J. Chem. Soc. (1933) 314; Feigl, p. 40 (8)
	Salicylic acid	Yoe, Vol. I, p. 183 (2)
	Sodium diethyldithiocarbamate	Analyst **54**, 650 (1929)
	o-Toluidine	Z. anal. Chem. **67**, 31 (1925)
	Urobilin	Chem. Weekblad. **27**, 552 (1930)
Cyanide	Benzidine	Feigl, p. 276 (2)
	Phenolphthalin	Analyst **60**, 294 (1935)
	Picric acid	Helv. Chim. Acta **12**, 713 (1929); J. Am. Chem. Soc. **51**, 1171 (1929)
Fluoride	Acetylacetone	Ind. Eng. Chem., Anal. Ed. **5**, 300 (1933)
	Alizarin sodium sulfonate—$Zr(NO_3)_4$	Ind. Eng. Chem., Anal. Ed. **7**, 23 (1935)
	p-Dimethylaminoazophenylarsonic acid	Feigl, p. 271 (2)
	Triphenyltin chloride	J. Am. Chem. Soc. **54**, 4625 (1932)
	Quinalizarine—$Zr(NO_3)_4$	Ind. Eng. Chem., Anal. Ed. **6**, 61 (1934)
Gallium	Camphoric acid	Welcher, Vol. II, p. 31 (9)
	8-Hydroxyquinoline	Ind. Eng. Chem., Anal. Ed. **13**, 844 (1941)
	Morin	Mikrochemie **20**, 194 (1936)
	Quinalizarin	J. Am. Chem. Soc. **59**, 40 (1937)
Germanium	Benzidine	Mikrochemie **18**, 66 (1935)
	Quinalizarin	Mikrochemie **18**, 48 (1935)
	Tannin	Welcher, Vol. II, p. 160 (9)

Determination	Reagent	Reference
Gold	Benzidine	Bull. Chim. Farm. **52**, 461 (1912)
	o-Dianisidine	Sandell, p. 507 (6)
	Dimethylaminobenzylidene rhodanine	Feigl, p. 127 (2)
	Formaldehyde	Bull. soc. chim. **31**, 717 (1922)
	p-Fuchsine	Ind. Eng. Chem., Anal. Ed., **18**, 400 (1946)
	Malachite Green	Sandell, p. 506 (6)
	m-Phenylenediamine sulfate	Chem. Zeit. **36**, 934 (1912)
	Phenylhydrazine	Ann. chim. anal. **12**, 90 (1907)
	o-Toluidine	Analyst **44**, 94 (1919)
Hafnium	2-Hydroxy-5-methylazobenzene-4-sulfonic acid	Feigl, p. 288 (4)
Hydrogen sulfide	*p*-phenylenedimethyldiamine sulfate	Yoe, Vol. I, p. 375 (2)
Indium	Diphenylthiocarbazone (Dithizone)	Sandell, p. 516 (6)
	8-Hydroxyquinoline	Z. anorg. allgem. Chem. **209**, 129 (1932)
	Morin	Mikrochim. Acta **2**, 287 (1937)
Iodine	Starch	Snell, Vol. II, p. 740 (8)
	o-Toluidine	J. Am. Chem. Soc. **47**, 1000 (1925)
Iridium	Benzidine	Snell, Vol. II, p. 526 (8)
	Malachite Green, leuco base	C. A. **24**, 2689 (1930)
Iron	Acetylacetone	J. Am. Chem. Soc. **26**, 967 (1904)
	Alloxantin	Compt. rend. **180**, 519 (1925)
	Ammonium salt of nitrosophenyl hydroxylamine ("Cupferon")	Ind. Eng. Chem. **3**, 629 (1911)
	Bis-*p*-chlorophenylphosphoric acid	Feigl, p. 294 (4)
	Cysteine	Biochem. Z. **187**, 255 (1927)
	Dimethyl glyoxime	Z. anorg. Chemie **89**, 401 (1914)
	Dinitrosoresorcinol	J. Am. Chem. Soc. **47**, 1268 (1925)
	Diisonitrosoacetone	Chem. Listy **23**, 496 (1929); C. A. **24**, 801 (1930)
	Dioximes (various)	Anal. Chem. **19**, 1017 (1947)
	Diphenylamine	J. Am. Chem. Soc. **46**, 263 (1924)
	2,2-Bipyridyl	Snell, Vol. II, p. 316 (8)
	Disodium-1,2-dihydroxybenzene-3,5-disulfonate ("Tiron")	Ind. Eng. Chem., Anal. Ed. **16**, 111 (1944)
	Hexamethylenetetramine	Bull soc. chim. Rom. **2**, 89 (1921)
	4 Hydroxybiphenyl-3-carboxylic acid	J. Am. Chem. Soc. **70**, 648 (1948)
	8-Hydroxyquinoline	Bull. soc. chim. biol., **17**, 432 (1935)
	4-Hydroxybiphenyl-3-carboxylic acid	J. Am. Chem. Soc. **70**, 648 (1948)
	7-Iodo-8-hydroxyquinoline-5-sulfonic acid (Ferron)	J. Am. Chem. Soc. **59**, 872 (1937)
	Isonitrosoacetophenone	Ber. **60**, 527 (1927)
	Isonitrosodimethyldihydroresorcinol	Anal. Chem., **20**, 1205 (1948)
	Kojic acid	Ind. Eng. Chem., Anal. Ed., **13**, 612 (1941)
	α-Nitrose-β-naphthol	Bull. soc. chim. **35**, 641 (1924)
	Nitroso-R salt	Ind. Chem., Anal. Ed., **14**, 756 (1942); **16**, 276 (1944)
	o-Phenanthroline	Ind. Eng. Chem., Anal. Ed. **9**, 67 (1937)
	Protacatechvic acid	J. Biol. Chem., **137**, 417 (1941)
	Pyramidone	Pharm.Weekblad, **63**, 1121 (1926)
	Pyrocatechol	Helv. chim. Acta **9**, 835 (1926)
	Salicylaldoxime	Ind Eng. Chem., Anal. Ed., **12**, 448 (1940)
	Salicylic acid	J. Chem. Soc. **93**, 93 (1908)
	Salicylsulfonic acid	Snell, Vol. II, p. 321 (8)
	Sulfosalicylic acid	Biochem. Z. **181**, 391 (1927)
	Thioglycollic acid	J. Am. Chem. Soc. **49**, 1916 (1927)
Lanthanum	8-Hydroxyquinoline	Z. anal. Chem. **107**, 191 (1936)
	Sodium alizarinesulfonate	J. Am. Chem. Soc., **53**, 1217 (1931)
Lead	Ammonium thiocyanate and pyridine	Z. anal. Chem. **72**, 289 (1927)
	Aniline	Ind. Eng. Chem. **11**, 1055 (1919); Yoe, Vol. I, p. 257 (2)
	Anthranilic acid	Z. anal. Chem. **101**, 85 (1935)
	Carminic acid	Mikrochemie **7**, 301 (1929)
	s-Diphenylcarbazide	Yoe, Vol. I, p. 255 (2)
	Diphenylthiocarbazone (Dithizone)	Snell, Vol. IIA, p. 10 (8)
	Hematein	Yoe, Vol. I, p. 257 (2)
	Salicylaldoxime	Ind. Eng. Chem., Anal. Ed. **14**, 359 (1942)
	Tetramethyldiamidodiphenylmethane	Snell, Vol. II, p. 43 (8)
	Thiourea	Z. anorg. Chem., **234**, 224 (1937)

ORGANIC ANALYTICAL REAGENTS (Continued)

Determination	Reagent	Reference
Lithium	Ammonium stearate	J. Am. Chem. Soc. **52,** 2754 (1930)
	Hexamethylenetetramine	Welcher, Vol. III, p. 131 (9)
	Pyridine	Welcher, Vol. III, p. 33 (9)
Magnesium	Brilliant yellow	Sandell, p. 598 (6)
	Curcumin	Ind. Eng. Chem., Anal. Ed. **4,** 426 (1932)
	Dimethylamine	Z. anorg. Chem. **26,** 347 (1901)
	Hydroquinone	Yoe, Vol. I, p. 264 (2)
	8-Hydroxyquinoline	Z. anal. Chem. **71,** 122 (1927)
	p-Nitrobenzeneazo-α-naphthol	Feigl, p. 225 (2)
	p-Nitrobenzeneazo-resorcinol	J. Am. Chem. Soc. **51,** 1456 (1929)
	Oleic acid	Yoe, Vol. I, p. 270 (2)
	Quinalizarin	Feigl, p. 224 (2)
	Sodium 1-azo-2-hydroxy-3-(2,4-dimethylcarboxaniliodonaphthalene-1′(2-hydroxybenzene-5-sulfonate)	Anal. Chem. **28,** 202 (1956); cf. Anal. Chim. Acta **16,** 155 (1957)
	Thiazole yellow	Sandell, p. 598 (6)
	Titan yellow	Sandell, p. 591 (6)
	Tropeoline OO	Rec. trav. chim., **61,** 849 (1942)
Manganese	Benzidine	Snell, Vol. II, p. 397 (8)
	Formaldoxime	Ind. Eng. Chem., Anal. Ed. **9,** 445 (1937); **12,** 307 (1940)
	Tetramethyldiaminodiphenylmethane	Feigl, p. **174** (2); Snell, Vol. IIA, p. 311 (8)
	4,4-Tetramethyldiaminotriphenylmethane	J. Biol. Chem., **168,** 537 (1947)
Mercury	Anthranilic acid	Z. anal. Chem. **101,** 88 (1935)
	p-Dimethylaminobenzalrhodamine	J. Am. Chem. Soc. **52,** 2222 (1930)
	Di-β-naphthylthiocarbazone	Sandell, p. 630 (6)
	s-Diphenylcarbazide	Z. angew. Chem. **39,** 791 (1926)
	Diphenylthiocarbazone (Dithizone)	Z. anal. Chem. **103,** 241 (1935)
	Potassium *s*-diphenylcarbazone	Snell, Vol. II 78 (8)
	Strychnine	" II 76 (8)
Molybdenum	α-Benzoin-oxime (Cupron)	B. S. J. Research **9,** 1 (1932)
	Disodium-1,2-dihydroxy-benzene-3,5-disulfonate (Tiron)	Anal. Chim. Acta **8,** 546 (1953)
	4-Methyl-1,2-dimercaptobenzene (Dithiol)	J. Am. Pharm. Assoc., **37,** 255 (1948)
	Phenylhydrazine	Ber. **36,** 512 (1903)
	Potassium ethyl xanthate and chloroform	J. Am. Chem. Soc. **44,** 1462 (1922)
	Tannic acid	Chem. Eng. Mining Rev. **11,** 258 (1919)
Nickel	α-Benzil-dioxime	Analyst **38,** 316 (1913)
	Cyclohexanedione-dioxime	Ann. **437,** 148 (1925)
	Dicyandiamidine sulfate	Chem. Zeit. **31,** 335, 911 (1907)
	Diethyldithiocarbamate	Ind. Eng. Chem., Anal. Ed., **18,** 206 (1946)
	Dimethylglyoxime	Chem. Weekblad. **21,** 358 (1924)
	Formaldoxime	Snell, Vol. IIA, p. 265 (8)
	α-Furildioxime	J. Am. Chem. Soc. **47,** 918 (1925)
	Potassium dithiooxalate	J. Am. Chem. Soc. **54,** 1866 (1932)
	Rubianic acid (Dithio-oxamide)	Anal. Chim. Acta **20,** 332 435 (1959)
Niobium (Columbium)	Ammonium salt of nitrosophenyl hydroxylamine ("Cupferron")	Hillebrand et al., p. 120 (5)
	8 Hydroxyquinolin	Wechler, Vol. I, p. 302 (2)
Nitrate	Brucine	Yoe, Vol. I, p. 318 (2)
	Diphenylamine sulfonic acid	J. Am. Chem. Soc. **55,** 1448 (1933)
	Diphenylbenzidine	Yoe, Vol. I, p. 316 (2)
	Diphenyl-endo-anilo-hydrotriazole ("Nitron")	Fales and Kenny, Inorg. Quant. Anal., p. 347 (1938)
	Phenoldisulfonic acid	Yoe, Vol. I, p. 313 (2)
	Pyrogallol	" p. 319 (2)
	Strychnine sulfate	" p. 320 (2)
	2:4-Xylenol	J. Assoc. Off. Agri. Chem. **18,** 459 (1935)
Nitrite	Antipyrin	Yoe, Vol. I, p. 311 (2)
	Dimethylaniline	" p. 311 (2)
	Dimethyl-α-Naphthylamine	Ind. Eng. Chem., Anal. Ed. **1,** 28 (1929)
	Diphenylamine sulfate	Yoe, Vol. I, p. 654 (2)
	α-Naphthylamine and β-Naphthylamine-6,8-Disulfonic acid	J. Pharmacol. **51,** 398 (1934)
	α-Naphthylamine hydrochloride	Yoe, Vol. I, p. 309 (2)
	m-Phenylenediamine	Yoe, Vol. I, p. 310 (2)
	Sulfanilic acid and α-naphthylamine	Yoe, Vol. I, p. 308 (2)
Osmium	*s*-Diphenylthiourea	Yoe and Sarver, p. 155 (12)
	1-Naphthylamine-4,6,8-trisulfonic acid	Anal. Chim. Acta **20,** 205 (1959)
	Strychnine	Welcher, Vol. IV, p. 272 (9)
	Thiourea	Compt. rend. **167,** 235 (1918)
Oxygen	Indigo carmine	Snell, Vol. IIA, p. 726 (8)
	Pyrogallol	Dennis, Gas Analysis, p. 174 (1929)
Palladium	Dimethylglyoxime	J. Am. Chem. Soc. **57,** 2565 (1935)
	p-Fuchsine	Ind. Eng. Chem., Anal. Ed., **18,** 400 (1946)
	β-Furfuraldoxime	Ind. Eng. Chem., Anal. Ed. **14,** 491 (1942)
	6-Nitroquinoline	J. Am. Chem. Soc. **50,** 3018 (1928)
	p-Nitrosodiphenylamine	J. Am. Chem. Soc. **61,** 2058 (1939)
	p-Nitrosodimethylaniline	J. Am. Chem. Soc. **63,** 3224 (1941)
	p-Nitrosodiethylaniline	Ibid.
	Thiomalic acid	Talanta **2,** 223 (1959)
Phosphate	1,2,4-Aminonaphthosulfonic acid	Yoe, Vol. I, p. 348 (2)
	Hydroquinone	Yoe, Vol. I, pp. 346 and 353 (2)
	Quinine-molybdate	Yoe, Vol. I, p. 343 (2)
	Strychnine-molybdate	Yoe, Vol. II, p. 142 (3)
Phosphorus	Hydrazine sulfate	Yoe, Vol. I, p. 341 (2)
Platinum	*p*-Nitrosodimethylaniline	Anal. Chem. **26,** 1335, 1340 (1954)
Potassium	6-Chloro-5-nitrotoluene-3-sulfonic acid	Mikrochem. **14,** 368 (1934)
	Dipicrylamine	Z. angew. Chem. **49,** 827 (1936)
	Picric acid	J. Am. Chem. Soc. **53,** 539 (1931)
Rhodium	Thiomalic acid	Talanta **2,** 239 (1959)
Ruthenium	*s*-Diphenylthiourea	Yoe and Sarver, p. 155 (12)
	5-Hydroxyquinoline-8-carboxylic acid	Canadian J. Research B25, **49,** (1945)
	1-Naphthylamine-3,5,7-trisulfonic acid	Anal Chim Acta **20,** 211 (1959)
	Rubianic acid (Dithio-oxamide)	Mikrochemie **15,** 295 (1934)
	Thioglycolyl-β-amidonaphthalide (Thionalid)	Ind. Eng. Chem., Anal. Ed. **12.** 5611 (1940)
	Thiourea	Sandell, p. 781 (6)
Scandium	Morin	Mikrochem. Acta **2,** 9, 287 (1937)
Selenium	Codeine phosphate	Arch Pharm. **252,** 161 (1914)
	Hydrazine	Boltz, p. 321 (1)
	Hydroquinone	Am. J. Sci. **15,** 253 (1928)
	Hydroxylamine hydrochloride	J. Am. Chem. Soc. **47,** 2456 (1925)
	Pyrrol	Snell, Vol. II, p. 779 (8)
	Thiourea	Ann. chim. applicata **17,** 357 (1927); Feigl, p. 231 (8)
Silver	Chromotropic acid	Helvetica Chim. Acta **12,** 714 (1929)
	Dichlorofluorescein	J. Am. Chem. Soc. **51,** 3273 (1929)
	p-Dimethylaminobenzalrhodamine	J. Am. Chem. Soc. **52,** 2222 (1930)
	Diphenylthiocarbazone (Dithizone)	Z. anal. Chem. **101,** 1 (1935)
	Formazylcarboxylic acid	J. Anal. Chem. Russ., **2,** 131 (1934); abs. in Analyst, **73,** 352 (1948)
	Methylamine	Mikrochemie **7,** 233 (1929)
	2-Thio-5-keto-4-carbethoxy-1,3-dihydropyrimidine	Ind. Eng. Chem., Anal. Ed. **14,** 148 (1942)
Sodium	6,8-Dichlorobenzoylurea	J. Org. Chem. **3,** 414 (1938)
	Dihydroxy-tartaric acid	J. Russ. Phys. Chem. Soc. **60,** 661 (1928)
	Uranyl zinc acetate	J. Am. Chem. Soc. **51,** 1664 (1929)
Strontium	Sodium rhodizonate	Mikrochemie **2,** 187 (1924)
Sulfide	*p*-Aminodimethylaniline	Snell, Vol. IIA, p. 658 (8)
Sulfur	*p*-Phenylenedimethyldiamine-hydrochloride	Yoe, Vol. I, p. 373 (2)

ORGANIC ANALYTICAL REAGENTS

Determination	Reagent	Reference
Tantalum	Ammonium salt of nitrosophenyl hydroxylamine ("Cupferron")	Hillebrand et. al., p. 120 (5)
	Pyrogallol	Sandell, p. 695 (6)
Tellurium	Hydrazine hydrochloride	J. Am. Chem. Soc. **47,** 2456 (1925)
	Hydroquinone	Am. J. Sci. **15,** 253 (1928)
	Thiourea	Ann. chim. applicata **17,** 359 (1927)
Thallium	Diphenylthiocarbazone (Dithizone)	Analyst **60,** 394 (1935)
	Thionalid (Thioglycolyl-β-amido-naphthalide)	Z. angew. Chem. **48,** 430, 597 (1935)
Thorium	1-Amino-4-hydroxyanthraquinone	Ind. Eng. Chem., Anal. Ed. **13,** 809 (1941)
	Cupferron (Ammonium nitrosophenylhydroxylamine)	Chem. Ztg. **33,** 1298 (1908)
	8-Hydroxyquinoline	Z. anal. Chem. **100,** 98 (1935)
	Phenylarsonic acid	J. Am. Chem. Soc. **48,** 895 (1926)
Tin	Ammonium salt of nitrosophenyl hydroxylamine ("Cupferron")	Hillebrand et al., p. 120 (5)
	Cacotheline	Ind. Eng. Chem., Anal. Ed. **7,** 26 (1935)
	4-Chloro-1,2-dimercaptobenzene (1-chloro-benzene-3,4-dithiol)	J. Chem. Soc. **149,** 175 (1936)
	Hematoxylin	Sandell, p. 866 (6)
	Quinalizarin	Sandell, p. 866 (6)
	Toluene-3,4-dithiol (4-methyl-1,2-dimercaptobenzene)	Analyst **61,** 242 (1936); Ibid., **62,** 661 (1937)
Titanium	Ammonium salt of nitrosophenyl hydroxylamine ("Cupferron")	Hillebrand et al., p. 119 (5); Z. anal. Chem. **83,** 345 (1931)
	Chromotropic acid	Feigl, p. 197 (2)
	5,7-Dibromo-8-hydroxyquinoline	Z. anorg. Chem. **204,** 215 (1932)
	Dihydroxymaleic acid	Snell, Vol. II, p. 445 (8)
	Disodium 1,2-dihydroxybenzene-3,5-disulfonate ("Tiron")	Ind. Eng. Chem., Anal. Ed. **19,** 100 (1947)
	Gallic acid	Snell, Vol. II, p. 444 (8)
	p-Hydroxyphenylarsonic acid	Ind. Eng. Chem., Anal. Ed. **10,** 642 (1938)
	8-Hydroxyquinoline	Z. anal. Chem. **81,** 1 (1930)
	Resoflavine (Color Index 1015)	Anal. Chim. Acta, **1,** 244 (1947)
	Tannic acid	Analyst **55,** 605 (1930)
	Thymol	Yoe, Vol. I, p. 381 (2)
Tungsten (Wolfram)	Anti-1,5-di-(p-methoxyphenyl)-1-hydroxylamino-3-oximino-4-pentene ("Wolfron")	Ind. Eng. Chem., Anal. Ed. **16,** 45 (1944)
	Benzidine	Ber. **38,** 783 (1905)
	α-Benzoinoxime ("Cupron")	Bur. Std. J. Research **9,** 1 (1932)
	Cinchonine	Hillebrand et al., p. 689 (5)
	Hydroquinone	Z. angew. Chem. **44,** 237 (1931)
	Phenylhydrazine	Bull. soc. chim. Belg. **38,** 385 (1929)
	Rhodamine B.	Snell, Vol. II, p. 469 (8)
Tungsten (Wolfram) (Cont.)	Toluene-3,4-dithiol	Analyst, **69,** 109 (1944); J. Am. Pharm. Assoc., **37,** 255 (1948)
	Uric acid	Ann. chim. anal. **9,** 371 (1904)
	α-Benzoinoxime ("Cupron")	Bur. Std. J. Research **9,** 1 (1932)
	Cinchonine	Hillebrand et al., p. 689 (5)
	Hydroquinone	Z. angew. Chem. **44,** 237 (1931)
	Phenylhydrazine	Bull. soc. chim. Belg. **38,** 385 (1929)
	Rhodamine B.	Snell, Vol. II, p. 469 (8)
	Toluene-3,4-dithiol	Analyst, **69,** 109 (1944); J. Am. Pharm. Assoc., **37,** 255 (1948)
	Uric acid	Ann. chim. anal. **9,** 371 (1904)
Uranium	Dibenzoylmethane	Anal. Chem. **25,** 1200 (1953)
	o-Hydroxybenzoic acid	Snell, Vol. II, p. 492 (8)
	α-Quinaldinic acid	Z. anal. Chem. **95,** 400 (1933)
	Sodium diethyldithiocarbamate	Sandell, p. 921 (6)
	Sodium salicylate	Chem. Zeit. **43,** 739 (1919)
	Thioglycolic acid (Mercapto acetic acid)	Anl. Chem., **21,** 1093 (1949)!
Urea	Xanthydrol	Mikrochem. **14,** 132 (1934)
Vanadium	Ammonium benzoate	C. A. **29,** 2880 (1935)
	Aniline	C. A. **24,** 567 (1930)
	Benzidine	J. Applied Chem. (U.S.S.R.) **17,** 83 (1944); C. A. **39,** 1115 (1945)
	Diphenylamine	Yoe, Vol. I, p. 715 (2)
	Diphenylbenzidine	Ind. Eng. Chem. **20,** 764 (1928)
	8-Hydroxyquinoline	Feigl, p. 125 (2)
	Safranine	Vol. Anal., Vol. II, p. 326 (6)
	Strychnine	Yoe, Vol. I, p. 393 (2)
Wolfram	See Tungsten	
Zinc	Anthranilic acid	Z. anal. Chem. **91,** 332 (1933)
	2 Carboxy-2'hydroxy-5'-sulfoformazybenzene (Zincon)	Anal. Chem. **26,** 1345 (1954)
	Di-β-napthylthiocarbazone	Sandell, pp. 945, 962 (6)
	Diphenylamine	J. Am. Chem. Soc. **49,** 2214 (1927)
	Diphenylbenzidine	J. Am. Chem. Soc. **49,** 356 (1927)
	Diphenylthiocarbazone (Dithizone)	Ind. Eng. Chem., Anal. Ed. **9,** 127 (1937)
	8-Hydroxyquinoline	Z. anal. Chem. **71,** 171 (1927)
	5-Nitroquinaldic acid	Ind. Eng. Chem., Anal. Ed. **10,** 335 (1938)
	Pyridine	Z. anal. Chem. **73,** 356 (1928)
	α-Quinaldinic acid	Z. anal. Chem. **100,** 324 (1935)
	Resorcinol	Yoe, Vol. I, p. 396 (2)
	Urobilin	J. Ind. Hyg. **7,** 273 (1925)
Zirconium	Ammonium salt of nitrosophenyl hydroxylamine ("Cupferron")	Hillebrand et al., p. 572 (5)
	5-Chlorobromamine acid	Ind. Eng. Chem., Anal. Ed. **15,** 73 (1943)
	p-Dimethylaminoazophenylarsenic acid	Ind. Eng. Chem., Anal. Ed., **13,** 603 (1941)
	2-Hydroxy-5-methylazobenzene-4-sulfonic acid	Feigl, p. 288 (4)
	Morin	Sandell, p. 971 (6)
	Phenylarsenic acid	J. Am. Chem. Soc. **48,** 895 (1926)
	Sodium alizarin sulfonate (Alizarin S)	Chem. Weekblad **20,** 404 (1924); Feigl, p. 200 (2)

(1) Colorimetric Determination of Nonmetals, Boltz, Edition 1958.
(2) Feigl, Spot Tests in Inorganic Analysis, 5th English Ed., translated by Oesper, 1958.
(3) Feigl, Spot Tests in Organic Analysis, 5th English Ed., translated by Oesper, 1956.
(4) Feigl, Chemistry of Specific, Selective and Sensitive Reactions translated by Oesper, 1949.
(5) Hillebrand, Lundell, Bright and Hoffman, Applied Inorganic Analysis, 2nd Edition, 1953.
(6) Sandell, Colorimetric Determination of Traces of Metals, 3rd Edition, 1959.
(7) Scott's Standard Methods of Chemical Analysis, Furman, 5th Edition 1939.
(8) Snell, Snell and Snell, Colorimetric Methods of Analysis, Volume II (1949), II A (1959).
(9) Welcher, Organic Analytical Reagents, Vols. I–IV, 1947–1948.
(10) Yoe, Photometric Chemical Analysis, Vol. I, Colorimetry, 1928.
(11) Yoe, Photometric Chemical Analysis Vol. II, Nephelometry, 1929
(12) Yoe and Sarver, Organic Analytical Reagents, 1941

CALIBRATION OF VOLUMETRIC GLASSWARE FROM THE WEIGHT OF THE CONTAINED WATER OR MERCURY WHEN WEIGHED IN AIR WITH BRASS WEIGHTS

D. F. Swinehart

A borosilicate glass vessel containing g_t grams of water at a temperature of t°C has, at the same temperature, a volume $V_t = W_t \times g_t$ cubic centimeters. Similarly when filled with G_t grams of mercury at a temperature of t°C the volume at the same temperature is given by $V_t = M_t \times G_t$ cubic centimeters.

When filled with g_t grams of water at a temperature of t°C the volume of the vessel at 18°C is given by $V_{18} = W_{18} \times g_t$ cubic centimeters and the true volume at 25°C is given by $V_{25} = W_{25} \times g_t$. The volumes at 18°C and 25°C are given similarly when using mercury by using the values under M_{18} and M_{25}, respectively.

The data on water are adapted from the data of G. S. Kell, *Journal of Chemical and Engineering Data*, 12, 67–68 (1967) (Table on p. F5, 52nd Edition, this handbook) and the data on mercury are adapted from *Smithsonian Tables*, Ninth Revised Edition, Volume 120, Publication No. 4169. The coefficient of linear expansion for borosilicate glass used here is 32.5×10^{-7} deg^{-1} and the volume coefficient of expansion is 97.5×10^{-7} deg^{-1}.

t°C	W_t	W_{18}	W_{25}	M_t	M_{18}	M_{25}
0	1.001 220	1.001 396	1.001 466	0.0735 519	0.0735 648	0.0735 698
1	1.001 161	1.001 327	1.001 395	0.0735 653	0.0735 775	0.0735 825
2	1.001 120	1.001 276	1.001 345	0.0735 787	0.0735 902	0.0735 952
3	1.001 096	1.001 242	1.001 311	0.0735 920	0.0736 028	0.0736 078
4	1.001 088	1.001 225	1.001 293	0.0736 054	0.0736 154	0.0736 205
5	1.001 096	1.001 223	1.001 291	0.0736 188	0.0736 281	0.0736 332
6	1.001 120	1.001 237	1.001 306	0.0736 322	0.0736 408	0.0736 458
7	1.001 158	1.001 265	1.001 334	0.0736 456	0.0736 535	0.0736 585
8	1.001 211	1.001 309	1.001 377	0.0736 590	0.0736 662	0.0736 712
9	1.001 279	1.001 367	1.001 435	0.0736 724	0.0736 789	0.0736 839
10	1.001 360	1.001 438	1.001 506	0.0736 858	0.0736 915	0.0736 966
11	1.001 455	1.001 523	1.001 592	0.0736 992	0.0737 042	0.0737 093
12	1.001 563	1.001 622	1.001 690	0.0737 125	0.0737 168	0.0737 218
13	1.001 684	1.001 733	1.001 801	0.0737 259	0.0737 295	0.0737 345
14	1.001 816	1.001 855	1.001 923	0.0737 393	0.0737 422	0.0737 472
15	1.001 961	1.001 990	1.002 059	0.0737 526	0.0737 548	0.0737 598
16	1.002 118	1.002 138	1.002 206	0.0737 660	0.0737 674	0.0737 725
17	1.002 286	1.002 296	1.002 364	0.0737 794	0.0737 801	0.0737 852
18	1.002 466	1.002 466	1.002 534	0.0737 928	0.0737 928	0.0737 978
19	1.002 658	1.002 648	1.002 717	0.0738 062	0.0738 055	0.0738 105
20	1.002 859	1.002 839	1.002 908	0.0738 196	0.0738 182	0.0738 232
21	1.003 072	1.003 043	1.003 111	0.0738 330	0.0738 308	0.0738 359
22	1.003 294	1.003 255	1.003 323	0.0738 463	0.0738 434	0.0738 485
23	1.003 528	1.003 479	1.003 548	0.0738 597	0.0738 561	0.0738 611
24	1.003 771	1.003 712	1.003 781	0.0738 731	0.0738 688	0.0738 738
25	1.004 024	1.003 955	1.004 024	0.0738 864	0.0738 814	0.0738 864
26	1.004 287	1.004 209	1.004 277	0.0738 998	0.0738 940	0.0738 991
27	1.004 560	1.004 472	1.004 540	0.0739 132	0.0739 067	0.0739 118
28	1.004 842	1.004 744	1.004 813	0.0739 266	0.0739 194	0.0739 244
29	1.005 133	1.005 025	1.005 094	0.0739 400	0.0739 321	0.0739 371
30	1.005 434	1.005 316	1.005 385	0.0739 534	0.0739 447	0.0739 498
31	1.005 743	1.005 615	1.005 684	0.0739 669	0.0739 575	0.0739 626
32	1.006 060	1.005 923	1.005 991	0.0739 801	0.0739 700	0.0739 750
33	1.006 388	1.006 241	1.006 310	0.0739 934	0.0739 826	0.0739 876
34	1.006 723	1.006 566	1.006 635	0.0740 068	0.0739 953	0.0740 003
35	1.007 066	1.006 899	1.006 968	0.0740 202	0.0740 079	0.0740 130
36	1.007 418	1.007 242	1.007 311	0.0740 335	0.0740 205	0.0740 256
37	1.007 780	1.007 593	1.007 669	0.0740 469	0.0740 332	0.0740 382
38	1.008 149	1.007 952	1.008 021	0.0740 603	0.0740 459	0.0740 509
39	1.008 525	1.008 318	1.008 387	0.0740 737	0.0740 585	0.0740 636
40	1.008 910	1.008 694	1.008 762	0.0740 871	0.0740 712	0.0740 763
41	1.009 303	1.009 077	1.009 146	0.0741 007	0.0740 841	0.0740 891
42	1.009 703	1.009 467	1.009 536	0.0741 139	0.0740 966	0.0741 016
43	1.010 112	1.009 866	1.009 935	0.0741 273	0.0741 092	0.0741 143
44	1.010 528	1.010 272	1.010 341	0.0741 407	0.0741 219	0.0741 270
45	1.010 951	1.010 685	1.010 754	0.0741 541	0.0741 346	0.0741 396
46	1.011 382	1.011 106	1.011 175	0.0741 675	0.0741 473	0.0741 523
47	1.011 820	1.011 534	1.011 603	0.0741 810	0.0741 600	0.0741 651
48	1.012 266	1.011 970	1.012 039	0.0741 944	0.0741 727	0.0741 778
49	1.012 719	1.012 413	1.012 482	0.0742 078	0.0741 854	0.0741 904
50	1.013 180	1.012 864	1.012 933	0.0742 213	0.0741 981	0.0742 031

SOLUBILITY CHART

Abbreviations: W, soluble in water; A, insoluble in water but soluble in acids; w, sparingly soluble in water but soluble in acids; a, insoluble in water and only sparingly soluble in acids; I, insoluble in both water and acids; d, decomposes in water. * Certain salts occur in two modifications.

No.		Al	NH_4	Sb	Ba	Bi	Cd	Ca	Cr	Co	Cu	Au′	Au‴	H	Fe″	Fe‴
1	Acetates −($C_2H_3O_2$)	W $Al(-)_3$	W $NH_4(-)$	....	W $Ba(-)_2$	W $Bi(-)_3$	W $Cd(-)_2$	W $Ca(-)_2$	W $Cr(-)_3$	W $Co(-)_2$	W $Cu(-)_2$	....	....	W $C_2H_4O_2$	W $Fe(-)_2$	W $Fe_2(-)_6$
2	Arsenate −(AsO_4)	a $Al(-)$	W $(NH_4)_3(-)$	A $Sb(-)$	w $Ba_3(-)_2$	A $Bi(-)$	A $Cd_3(-)_2$	w $Ca_3(-)_2$	....	A $Co_3(-)_2$	A $Cu_3(-)_2$	....	....	W H_3AsO_4	A $Fe_3(-)_2$	A $Fe(-)$
3	Arsenite −(AsO_3)	....	W NH_4AsO_2	A $Sb(-)$	....	....	....	w $Ca_3(-)_2$	....	A $Co_3H_6(-)_4$	A $CuH(-)$	....	....	....	....	....
4	Benzoate −($C_7H_5O_2$)	....	W $NH_4(-)$	....	W $Ba(-)_2$	A $Bi(-)_3$	W $Cd(-)_2$	W $Ca(-)_2$	....	W $Co(-)_2$	w $Cu(-)_2$	....	....	W $C_7H_6O_2$	W $Fe(-)_2$	A $Fe_2(-)_6$
5	Bromide	W $AlBr_3$	W NH_4Br	d $SbBr_3$	W $BaBr_2$	d $BiBr_3$	W $CdBr_2$	W $CaBr_2$	W(I)* $CrBr_3$	W $CoBr_2$	W $CuBr_2$	w $AuBr$	W $AuBr_3$	W HBr	W $FeBr_2$	W $FeBr_3$
6	Carbonate	....	W $(NH_4)_2CO_3$	....	w $BaCO_3$	....	A $CdCO_3$	w $CaCO_3$	W $CrCO_3$	A $CoCO_3$	....	....	....	....	w $FeCO_3$	....
7	Chlorate −(ClO_3)	W $Al(-)_3$	W $NH_4(-)$	....	W $Ba(-)_2$	W $Bi(-)_3$	W $Cd(-)_2$	W $Ca(-)_2$	....	W $Co(-)_2$	W $Cu(-)_2$	....	....	W $HClO_3$	W $Fe(-)_2$	W $Fe(-)_3$
8	Chloride	W $AlCl_3$	W NH_4Cl	W $SbCl_3$	W $BaCl_2$	d $BiCl_3$	W $CdCl_2$	W $CaCl_2$	I $CrCl_3$	W $CoCl_2$	W $CuCl_2$	w $AuCl$	W $AuCl_3$	W HCl	W $FeCl_2$	W $FeCl_3$
9	Chromate −(CrO_4)	....	W $(NH_4)_2(-)$	....	A $Ba(-)$	....	A $Cd(-)$	W $Ca(-)$	....	A $Co(-)$	....	....	....	....	....	A $Fe_2(-)_3$
10	Citrate −($C_6H_5O_7$)	W $Al(-)$	W $(NH_4)_3(-)$	....	w $Ba_3(-)_2$	A $Bi(-)$	A $Cd_3(-)_2$	w $Ca_3(-)_2$	....	w $Co_3(-)_2$	....	....	....	W $C_6H_8O_7$	....	W $Fe(-)$
11	Cyanide	....	W NH_4CN	....	W $Ba(CN)_2$	w $Bi(CN)_3$	W $Cd(CN)_2$	W $Ca(CN)_2$	A $Cr(CN)_2$	A $Co(CN)_2$	A $Cu(CN)_2$	w $AuCN$	W $Au(CN)_3$	W HCN	a $Fe(CN)_2$	....
12	Ferricy'de −($Fe(CN)_6$)	....	W $(NH_4)_3(-)$	....	w $Ba_3(-)_2$	....	A $Cd_3(-)_2$	W $Ca_3(-)_2$	....	I $Co_3(-)_2$	I $Cu_3(-)_2$	....	....	W $H_3(-)$	I $Fe_3(-)_2$	....
13	Ferrocy'de −($Fe(CN)_6$)	w $Al_4(-)_3$	W $(NH_4)_4(-)$	....	W $Ba_2(-)$	....	A $Cd_2(-)$	W $Ca_2(-)$	....	I $Co_2(-)$	I $Cu_2(-)$	....	....	W $H_4(-)$	I $Fe_2(-)$	a $Fe_4(-)_3$
14	Fluoride	W AlF_3	W NH_4F	W SbF_3	w BaF_2	W BiF_3	W CdF_2	w CaF_2	W(a)* CrF_3	W CoF_2	w CuF_2	....	....	W HF	w FeF_2	w FeF_3
15	Formate −(CHO_2)	W $Al(-)_3$	W $NH_4(-)$	....	W $Ba(-)_2$	W $Bi(-)_3$	W $Cd(-)_2$	W $Ca(-)_2$	....	W $Co(-)_2$	W $Cu(-)_2$	....	....	W CH_2O_2	W $Fe(-)_2$	W $Fe(-)_3$
16	Hydroxide	A $Al(OH)_3$	W NH_4OH	....	W $Ba(OH)_2$	A $Bi(OH)_3$	A $Cd(OH)_2$	W $Ca(OH)_2$	A $Cr(OH)_3$	A $Co(OH)_2$	A $Cu(OH)_2$	W $AuOH$	A $Au(OH)_3$	....	A $Fe(OH)_2$	A $Fe(OH)_3$
17	Iodide	W AlI_3	W NH_4I	d SbI_3	W BaI_2	A BiI_3	W CdI_2	W CaI_2	W CrI_2	W CoI_2	a CuI	a AuI	a AuI_3	W HI	W FeI_2	W FeI_3
18	Nitrate	W $Al(NO_3)_3$	W NH_4NO_3	....	W $Ba(NO_3)_2$	d $Bi(NO_3)_3$	W $Cd(NO_3)_2$	W $Ca(NO_3)_2$	W $Cr(NO_3)_3$	W $Co(NO_3)_2$	W $Cu(NO_3)_2$	....	....	W HNO_3	W $Fe(NO_3)_2$	W $Fe(NO_3)_3$
19	Oxalate −(C_2O_4)	A $Al_2(-)_3$	W $(NH_4)_2(-)$	....	w $Ba(-)$	A $Bi_2(-)_3$	w $Cd(-)$	A $Ca(-)$	W $Cr(-)$	A $Co(-)$	A $Cu(-)$	....	....	W $C_2H_2O_4$	A $Fe(-)$	W $Fe_2(-)_3$
20	Oxide	a Al_2O_3	....	w Sb_2O_3	W BaO	A Bi_2O_3	A CdO	w CaO	a Cr_2O_3	A CoO	A CuO	Au_2O	A Au_2O_3	W H_2O_2	A FeO	A Fe_2O_3
21	Phosphate	A $AlPO_4$	W $NH_4H_2PO_4$	....	A $Ba_3(PO_4)_2$	A $BiPO_4$	A $Cd_3(PO_4)_2$	w $Ca_3(PO_4)_2$	w $Cr_2(PO_4)_2$	A $Co_3(PO_4)_2$	A $Cu_3(PO_4)_2$	....	....	W H_3PO_4	A $Fe_3(PO_4)_2$	w $FePO_4$
22	Silicate, −(SiO_3)	I $Al_2(-)_3$	....	....	W $Ba(-)$	....	A $Cd(-)$	w $Ca(-)$	....	A Co_2SiO_4	A $Cu(-)$	....	....	I H_2SiO_3	....	....
23	Sulfate	W $Al_2(SO_4)_3$	W $(NH_4)_2SO_4$	A $Sb_2(SO_4)_3$	a $BaSO_4$	d $Bi_2(SO_4)_3$	W $CdSO_4$	w $CaSO_4$	W(I)* $Cr_2(SO_4)_3$	W $CoSO_4$	W $CuSO_4$	....	....	W H_2SO_4	W $FeSO_4$	w $Fe_2(SO_4)_3$
24	Sulfide	d Al_2S_3	W $(NH_4)_2S$	A Sb_2S_3	d BaS	A Bi_2S_3	A CdS	w CaS	d Cr_2S_3	A CoS	A CuS	I Au_2S	I Au_2S_3	W H_2S	A FeS	d Fe_2S_3
25	Tartrate −($C_4H_4O_6$)	w $Al_2(-)_3$	W $(NH_4)_2(-)$	W $Sb_2(-)_3$	w $Ba(-)$	A $Bi_2(-)_3$	A $Cd(-)$	w $Ca(-)$	....	w $Co(-)$	w $Cu(-)$	....	....	W $C_4H_6O_6$	w $Fe(-)$	W $Fe_2(-)_3$
26	Thiocy'te	....	W NH_4CNS	....	W $Ba(CNS)_2$	....	....	W $Ca(CNS)_2$	....	W $Co(CNS)_2$	d $CuCNS$	....	....	W $CNSH$	W $Fe(CNS)_2$	W $Fe(CNS)_3$

No.		Pb	Mg	Mn	Hg′	Hg″	Ni	K	Ag	Na	Sn⁗	Sn″	Sr	Zn	Pt
1	Acetate −($C_2H_3O_2$)	W $Pb(-)_2$	W $Mg(-)_2$	W $Mn(-)_2$	w $Hg(-)$	W $Hg(-)_2$	W $Ni(-)_2$	W $K(-)$	w $Ag(-)$	W $Na(-)$	W $Sn(-)_4$	d $Sn(-)_2$	W $Sr(-)_2$	W $Zn(-)_2$	....
2	Arsenate −(AsO_4)	A $PbH(-)$	A $Mg_3(-)$	w $MnH(-)$	A $Hg_3(-)$	w $Hg_3(-)_2$	A $Ni_3(-)_2$	W $K_3(-)$	A $Ag_3(-)$	W $Na_3(-)$	....	....	w $SrH(-)$	A $Zn_3(-)_2$	....
3	Arsenite −(AsO_3)	....	W $Mg_3(-)_2$	A $Mn_3H_6(-)_4$	A $Hg_3(-)$	A $Hg_3(-)$	A $Ni_3H_6(-)_4$	W K_3AsO_3	A $Ag_3(-)$	W $Na_2H(-)$	....	A $Sn_3(-)_2$	w $Sr_3(-)_2$	....	....
4	Benzoate −($C_7H_5O_2$)	w $Pb(-)_2$	W $Mg(-)_2$	W $Mn(-)_2$	A $Hg_2(-)_2$	w $Hg(-)_2$	w $Ni(-)_2$	W $K(-)$	w $Ag(-)$	W $Na(-)$	....	....	....	W $Zn(-)_2$	....
5	Bromide	W $PbBr_2$	W $MgBr_2$	W $MnBr_2$	A $HgBr$	W $HgBr_2$	W $NiBr_2$	W KBr	a $AgBr$	W $NaBr$	W $SnBr_4$	W $SnBr_2$	W $SrBr_2$	W $ZnBr_2$	w $PtBr_4$
6	Carbonate	A $PbCO_3$	w $MgCO_3$	w $MnCO_3$	A Hg_2CO_3	....	w $NiCO_3$	W K_2CO_3	A Ag_2CO_3	W Na_2CO_3	....	....	w $SrCO_3$	w $ZnCO_3$	....
7	Chlorate −(ClO_3)	W $Pb(-)_2$	W $Mg(-)_2$	W $Mn(-)_2$	W $Hg(-)$	W $Hg(-)_2$	W $Ni(-)_2$	W $K(-)$	W $Ag(-)$	W $Na(-)$	....	W $Sn(-)_2$	W $Sr(-)_2$	W $Zn(-)_2$	....
8	Chloride	W $PbCl_2$	W $MgCl_2$	W $MnCl_2$	a $HgCl$	W $HgCl_2$	W $NiCl_2$	W KCl	a $AgCl$	W $NaCl$	W $SnCl_4$	W $SnCl_2$	W $SrCl_2$	W $ZnCl_2$	W $PtCl_4$
9	Chromate −(CrO_4)	A $Pb(-)$	W $Mg(-)$	....	w $Hg_2(-)$	w $Hg(-)$	A $Ni(-)$	W $K_2(-)$	w $Ag_2(-)$	W $Na_2(-)$	W $Sn(-)_2$	A $Sn(-)$	w $Sr(-)$	w $Zn(-)$	....
10	Citrate −($C_6H_5O_7$)	W $Pb_3(-)_2$	W $Mg_3(-)_2$	w $MnH(-)$	w $Hg_3(-)$	....	W $Ni_3(-)_2$	W $K_3(-)$	w $Ag_3(-)$	W $Na_3(-)$	....	....	A $SrH(-)$	w $Zn_3(-)_2$	....

SOLUBILITY CHART (Continued)

No.		Pb	Mg	Mn	Hg′	Hg″	Ni	K	Ag	Na	Sn''''	Sn″	Sr	Zn	Pt
11	Cyanide	w $Pb(CN)_2$	W $Mg(CN)_2$	....	A $HgCN$	W $Hg(CN)_2$	a $Ni(CN)_2$	W KCN	a $AgCN$	W $NaCN$	....	....	W $Sr(CN)_2$	A $Zn(CN)_2$	I $Pt(CN)_2$
12	Ferricy'de $-Fe(CN)_6$	w $Pb_3(-)_2$	W $Mg_3(-)_2$	....	......	A $Hg_3(-)_2$	I $Ni_3(-)_2$	W $K_3(-)$	I $Ag_3(-)$	W $Na_3(-)$	....	A $Sn_3(-)_2$	W $Sr_3(-)_2$	A $Zn_3(-)_2$	....
13	Ferrocy'de $-Fe(CN)_6$	a $Pb_2(-)$	W $Mg_2(-)$	A $Mn_2(-)$	......	I $Hg_2(-)$	I $Ni_2(-)$	W $K_4(-)$	I $Ag_4(-)$	W $Na_4(-)$	....	a $Sn_2(-)$	W $Sr_2(-)$	I $Zn_2(-)$	....
14	Fluoride	w PbF_2	w MgF_2	A MnF_2	d HgF	d HgF_2	w NiF_2	W KF	W AgF	W NaF	W SnF_4	W SnF_2	w SrF_2	w ZnF_2	W PtF_4
15	Formate $-(CHO_2)$	W $Pb(-)_2$	W $Mg(-)_2$	W $Mn(-)_2$	w $Hg(-)$	W $Hg(-)_2$	W $Ni(-)_2$	W $K(-)$	W $Ag(-)$	W $Na(-)$	....	....	W $Sr(-)_2$	W $Zn(-)_2$	....
16	Hydroxide	w $Pb(OH)_2$	A $Mg(OH)_2$	A $Mn(OH)_2$	......	A $Hg(OH)_2$	w $Ni(OH)_2$	W KOH	....	W $NaOH$	w $Sn(OH)_4$	A $Sn(OH)_2$	W $Sr(OH)_2$	A $Zn(OH)_2$	A $Pt(OH)_4$
17	Iodide	w PbI_2	W MgI_2	W MnI_2	A HgI	w HgI_2	W NiI_2	W KI	I AgI	W NaI	d SnI_4	W SnI_2	W SrI_2	W ZnI_2	I PtI_2
18	Nitrate	W $Pb(NO_3)_2$	W $Mg(NO_3)_2$	W $Mn(NO_3)_2$	W $HgNO_3$	W $Hg(NO_3)_2$	W $Ni(NO_3)_2$	W KNO_3	W $AgNO_3$	W $NaNO_3$		d $Sn(NO_3)_2$	W $Sr(NO_3)_2$	W $Zn(NO_3)_2$	W $Pt(NO_3)_4$
19	Oxalate $-(C_2O_4)$	A $Pb(-)$	w $Mg(-)$	w $Mn(-)$	a $Hg_2(-)$	A $Hg(-)$	A $Ni(-)$	W $K_2(-)$	a $Ag_2(-)$	W $Na_2(-)$	....	A $Sn(-)$	w $Sr(-)$	A $Zn(-)$	....
20	Oxide	w PbO	A MgO	A MnO	A Hg_2O	w HgO	A NiO	W K_2O	w Ag_2O	d Na_2O	A SnO_2	A SnO	W SrO	w ZnO	A PtO
21	Phosphate	A $Pb_3(PO_4)_2$	w $Mg_3(PO_4)_2$	w $Mn_3(PO_4)_2$	A Hg_3PO_4	A $Hg_3(PO_4)_2$	A $Ni_3(PO_4)_2$	W K_3PO_4	A Ag_3PO_4	W Na_3PO_4	....	A $Sn_3(PO_4)_2$	A $Sr_3(PO_4)_2$	A $Zn_3(PO_4)_2$	....
22	Silicate $--(SiO_3)$	A $Pb(-)$	A $Mg(-)$	I $Mn(-)$	......	....	....	W $K_2(-)$	....	W $Na_2(-)$	....	....	A $Sr(-)$	A $Zn(-)$	....
23	Sulfate	w $PbSO_4$	W $MgSO_4$	W $MnSO_4$	w Hg_2SO_4	d $HgSO_4$	W $NiSO_4$	W K_2SO_4	w Ag_2SO_4	W Na_2SO_4	W $Sn(SO_4)_2$	W $SnSO_4$	w $SrSO_4$	W $ZnSO_4$	W $Pt(SO_4)_2$
24	Sulfide	A PbS	d MgS	A MnS	I Hg_2S	I HgS	A NiS	W K_2S	A Ag_2S	W Na_2S	A SnS_2	A SnS	W SrS	A ZnS	I PtS
25	Tartrate $-(C_4H_4O_6)$	A $Pb(-)$	w $Mg(-)$	w $Mn(-)$	I $Hg_2(-)$	....	A $Ni(-)$	W $K_2(-)$	w $Ag_2(-)$	W $Na_2(-)$	....	W $Sn(-)$	w $Sr(-)$	w $Zn(-)$	....
26	Thiocy'te	w $Pb(CNS)_2$	W $Mg(CNS)_2$	W $Mn(CNS)_2$	A $HgCNS$	w $Hg(CNS)_2$	....	W $KCNS$	I $AgCNS$	W $NaCNS$	......	......	W $Sr(CNS)_2$	W $Zn(CNS)_2$	....

REDUCTIONS OF WEIGHINGS IN AIR TO VACUO

When the weight M in grams of a body is determined in air, a correction is necessary for the buoyancy of the air. The following table is computed for an air density of 0.0012. The corrected weight $=M+kM/1000$. values of k being found in the table.

Density of body weighed	Correction factor, k.			Density of body weighed	Correction factor, k.		
	Pt Ir weights	Brass weights	Quartz or Al weights		Pt Ir weights	Brass weights	Quartz or Al weights
.5	+2.34	+2.26	+1.95	1.6	+ .69	+ .61	+ .30
.6	+1.94	+1.86	+1.55	1.7	+ .65	+ .56	+ .25
.7	+1.66	+1.57	+1.26	1.8	+ .62	+ .52	+ .21
.75	+1.55	+1.46	+1.15	1.9	+ .58	+ .49	+ .18
.80	+1.44	+1.36	+1.05	2.0	+ .54	+ .46	+ .15
.85	+1.36	+1.27	+0.96	2.5	+ .43	+ .34	+ .03
.90	+1.28	+1.19	+ .88	3.0	+ .34	+ .26	— .05
.95	+1.21	+1.12	+ .81	4.0	+ .24	+ .16	— .15
1.00	+1.14	+1.06	+ .75	6.0	+ .14	+.06	— .25
1.1	+1.04	+0.95	+ .64	8.0	+ .09	+ .01	— .30
1.2	+0.94	+ .86	+ .55	10.0	+ .06	— .02	— .33
1.3	+ .87	+ .78	+ .47	15.0	+ .03	— .06	— .37
1.4	+ .80	+ .71	+ .40	20.0	+ .004	— .08	— .39
1.5	+ .75	+ .66	+ .35	22.0	— .001	— .09	— .40

BUFFER SOLUTIONS

OPERATIONAL DEFINITIONS OF pH

Prepared by R. A. Robinson

The operational definition of pH is:

$$pH = pH(s) + E/k$$

where E is the e.m.f. of the cell:

$$H_2|\text{Solution, pH}|\text{Saturated KCl}|\text{Solution, pH(s)}|H_2$$

the half cell on the left containing the solution whose pH is being measured and that on the right a standard buffer mixture of known pH; $k = 2.303RT/F$, where R is the gas constant, T the temperature in degrees Kelvin and F the value of the faraday.

Alternatively, the cell:

Glass electrode|Solution, pH|Saturated calomel electrode

can be used, the glass electrode being calibrated using a standard buffer mixture or, if possible, two standard buffer mixtures whose pH values lie on either side of that of the solution which is being measured. Suitable standard buffer mixtures are:

0.05 M potassium hydrogen phthalate (pH = 4.008 at 25°C)
0.025 M potassium dihydrogen phosphate
0.025 M disodium hydrogen phosphate (pH = 6.865 at 25°C)
0.01 M borax (pH = 9.180 at 25°C)

For most purposes pH can be equated to $-\log_{10} \gamma_{H^+} m_{H^+}$, i.e., to the negative logarithm of the hydrogen ion activity. There is a small difference between those two quantities if pH > 9.2 or pH < 4.0, given by:

$$-\log \gamma_{H^+} m_{H^+} = pH + 0.014(pH - 9.2) \text{ for } pH > 9.2$$
$$= pH + 0.009(4.0 - pH) \text{ for } pH < 4.0$$

It should be noted that in the table titled "Solutions giving Round Values of pH at 25°C" it is $-\log \gamma_{H^+} m_{H^+}$ and not pH which is quoted when there is a difference between them.

References:

R. G. Bates, "Electrometric pH Determinations: Theory and Practice" Wiley, New York, 1954.

R. A. Robinson and R. H. Stokes, "Electrolyte Solutions," 2nd edition, Butterworths, London; Academic Press, Inc. New York, 1959. R. C. Bates, J. Res. of N.B.S. 66 A, 179 (1962).

National Bureau of Standards

R. G. Bates and S. F. Acree, Res. **34,** 373 (1945); W. J. Hammer, C. D. Pinching and S. F. Acree, ibid. **36,** 47 (1946); G. G. Manor, N. J. DeLollis, P. W. Lindwall and S. F. Acree, ibid., **36,** 543 (1946); R. G. Bates, ibid., **39,** 411 (1947); R. G. Bates, V. E. Bower, R. G. Miller and E. R. Smith, ibid., **47,** 433 (1951); V. E. Bower, R. G. Bates and E. R. Smith, ibid., **51,** 189 (1953); V. E. Bower and R. G. Bates, ibid., **55,** 197 (1955); R. G. Bates, V. E. Bower and E. R. Smith, ibid., **56,** 305 (1956); V. E. Bower and R. G. Bates, ibid., **59,** 261; R. G. Bates and V. E. Bower, Anal. Chem., **28,** 1322 (1956).

PROPERTIES OF STANDARD AQUEOUS BUFFER SOLUTIONS AT 25°C

Solution	Buffer substance	Molality m	Weight of salt in air per liter solution	Density g/ml	Molarity M	Dilution value $\Delta pH_{\frac{1}{2}}$	ΔpH_s^a	Buffer value, equiv. per pH	Temp coeff., dpH_s/dt. Units per °C
Tetroxalate	$KH_3(C_2O_4)_2 \cdot 2H_2O$	0.05	12.61	1.0032	0.04962	+0.186	−0.0028	0.070	+0.001
Tartrate	$KHC_4H_4O_6$, sat. sol'n. at 25°C	0.0341	......	1.0036	0.034	+0.049	−0.0003	0.027	−0.0014
Phthalate	$KHC_8O_4H_4$	0.05	10.12	1.0017	0.04958	+0.052	−0.0009	0.016	+0.0012
Phosphate	KH_2PO_4 + Na_2HPO_4	0.025[b]	3.39 3.53	1.0028	0.0249[b]	+0.080	−0.0006	0.029	−0.0028
Phosphate	KH_2PO_4 + Na_2HPO_4	0.008695[c] 0.03043[d]	1.179 4.30	1.0020	0.008665[c] 0.03032[d]	+0.07[e]	−0.0005	0.016	−0.0028
Borax	$Na_2B_4O_7 \cdot 10H_2O$	0.01	3.80	0.9996	0.009971	+0.01	−0.0001	0.020	−0.0082
Calcium hydroxide	$Ca(OH)_2$, sat. sol'n. at 25°C	0.0203	......	0.9991	0.02025	−0.28	+0.0014	0.09	−0.033

[a] ΔpH_s = pHs (M Molar solution) − pH_s (m molal solution).
[b] Concentration of each phosphate salt.
[c] KH_2PO_4.
[d] Na_2HPO_4.
[e] Calculated value.

SOLUTIONS GIVING ROUND VALUES OF pH AT 25°C

Reproduced from "Electrolyte Solutions" by permission from Robinson and Stokes, authors, and Butterworth's Scientific Publications.

A*		B*		C*		D*		E*	
pH	x	pH	x	pH	x	pH	x	pH	x
1.00	67.0	2.20	49.5	4.10	1.3	5.80	3.6	7.00	46.6
1.10	52.8	2.30	45.8	4.20	3.0	5.90	4.6	7.10	45.7
1.20	42.5	2.40	42.2	4.30	4.7	6.00	5.6	7.20	44.7
1.30	33.6	2.50	38.8	4.40	6.6	6.10	6.8	7.30	43.4
1.40	26.6	2.60	35.4	4.50	8.7	6.20	8.1	7.40	42.0
1.50	20.7	2.70	32.1	4.60	11.1	6.30	9.7	7.50	40.3
1.60	16.2	2.80	28.9	4.70	13.6	6.40	11.6	7.60	38.5
1.70	13.0	2.90	25.7	4.80	16.5	6.50	13.9	7.70	36.6
1.80	10.2	3.00	22.3	4.90	19.4	6.60	16.4	7.80	34.5
1.90	8.1	3.10	18.8	5.00	22.6	6.70	19.3	7.90	32.0
2.00	6.5	3.20	15.7	5.10	25.5	6.80	22.4	8.00	29.2
2.10	5.1	3.30	12.9	5.20	28.8	6.90	25.9	8.10	26.2
2.20	3.9	3.40	10.4	5.30	31.6	7.00	29.1	8.20	22.9
		3.50	8.2	5.40	34.1	7.10	32.1	8.30	19.9
		3.60	6.3	5.50	36.6	7.20	34.7	8.40	17.2
		3.70	4.5	5.60	38.8	7.30	37.0	8.50	14.7
		3.80	2.9	5.70	40.6	7.40	39.1	8.60	12.2
		3.90	1.4	5.80	42.3	7.50	40.9	8.70	10.3
		4.00	0.1	5.90	43.7	7.60	42.4	8.80	8.5
						7.70	43.5	8.90	7.0
						7.80	44.5	9.00	5.7
						7.90	45.3		
						8.00	46.1		

F*		G*		H*		I*		J*	
pH	x	pH	x	pH	x	pH	x	pH	x
8.00	20.5	9.20	0.9	9.60	5.0	10.90	3.3	12.00	6.0
8.10	19.7	9.30	3.6	9.70	6.2	11.00	4.1	12.10	8.0
8.20	18.8	9.40	6.2	9.80	7.6	11.10	5.1	12.20	10.2
8.30	17.7	9.50	8.8	9.90	9.1	11.20	6.3	12.30	12.8
8.40	16.6	9.60	11.1	10.00	10.7	11.30	7.6	12.40	16.2
8.50	15.2	9.70	13.1	10.10	12.2	11.40	9.1	12.50	20.4
8.60	13.5	9.80	15.0	10.20	13.8	11.50	11.1	12.60	25.6
8.70	11.6	9.90	16.7	10.30	15.2	11.60	13.5	12.70	32.2
8.80	9.6	10.00	18.3	10.40	16.5	11.70	16.2	12.80	41.2
8.90	7.1	10.10	19.5	10.50	17.8	11.80	19.4	12.90	53.0
9.00	4.6	10.20	20.5	10.60	19.1	11.90	23.0	13.00	66.0
9.10	2.0	10.30	21.3	10.70	20.2	12.00	26.9		
		10.40	22.1	10.80	21.2				
		10.50	22.7	10.90	22.0				
		10.60	23.3	11.00	22.7				
		10.70	23.8						
		10.80	24.25						

*A. 25 ml of 0.2 molar KCl + x ml of 0.2 molar HCl.
*B. 50 ml of 0.1 molar potassium hydrogen phthalate + x ml of 0.1 molar HCl.
*C. 50 ml of 0.1 molar potassium hydrogen phthalate + x ml of 0.1 molar NaOH.
*D. 50 ml of 0.1 molar potassium dihydrogen phosphate + x ml 0.1 molar NaOH.
*E. 50 ml of 0.1 molar tris(hydroxymethyl) aminomethane + x ml of 0.1 M HCl.
*F. 50 ml of 0.025 molar borax + x ml of 0.1 molar HCl.
*G. 50 ml of 0.025 molar borax + x ml of 0.1 molar NaOH.
*H. 50 ml of 0.05 molar sodium bicarbonate + x ml of 0.1 molar NaOH.
*I. 50 ml of 0.05 molar disodium hydrogen phosphate + x ml of 0.1 molar NaOH.
*J. 25 ml of 0.2 molar KCl + x ml of 0.2 molar NaOH.
Final Volume of Mixtures = 100 ml

STANDARD VALUES OF pH AT TEMPERATURE 0–95°C

Temperature	Tetroxalate 0.05 molal	Tartrate 0.0341 molal (sat'd at 25°C)	Phthalate 0.05 molal	Phosphate[a]	Phosphate[b]	Borax 0.01 molal	Calcium hydroxide (sat'd at 25°C)
0	1.666	.....	4.003	6.984	7.534	9.464	13.423
5	1.668	.....	3.999	6.951	7.500	9.395	13.207
10	1.670	.....	3.998	6.923	7.472	9.332	13.003
15	1.672	.....	3.999	6.900	7.448	9.276	12.810
20	1.675	.....	4.002	6.881	7.429	9.225	12.627
25	1.679	3.557	4.008	6.865	7.413	9.180	12.454
30	1.683	3.552	4.015	6.853	7.400	9.139	12.289
35	1.688	3.549	4.024	6.844	7.389	9.102	12.133
38	1.691	3.548	4.030	6.840	7.384	9.081	12.043
40	1.694	3.547	4.035	6.838	7.380	9.068	11.984
45	1.700	3.547	4.047	6.834	7.373	9.038	11.841
50	1.707	3.549	4.060	6.833	7.367	9.011	11.705
55	1.715	3.554	4.075	6.834	.....	8.985	11.574
60	1.723	3.560	4.091	6.836	.....	8.962	11.449
70	1.743	3.580	4.126	6.845	.....	8.921	......
80	1.766	3.609	4.164	6.859	.....	8.885	......
90	1.792	3.650	4.205	6.877	.....	8.850	......
95	1.806	3.674	4.227	6.886	.....	8.833	......

[a] Solution 0.025 m KH_2PO_4 and 0.025 m Na_2HPO_4.
[b] Solution 0.008695 m KH_2PO_4 and 0.03043 m Na_2HPO_4.

APPROXIMATE pH VALUES

The following tables give approximate pH values for a number of substances such as acids, bases, foods, biological fluids, etc. All values are rounded off to the nearest tenth and are based on measurements made at 25° C. A few buffer systems with their pH values are also given.

From Modern pH and Chlorine Control, W. A. Taylor & Co., by permission

ACIDS

Hydrochloric, N	0.1	Oxalic, 0.1N	1.6	Acetic, 0.01N	3.4
Hydrochloric, 0.1N	1.1	Tartaric, 0.1N	2.2	Benzoic, 0.01N	3.1
Hydrochloric, 0.01N	2.0	Malic, 0.1N	2.2	Alum, 0.1N	3.2
Sulfuric, N	0.3	Citric, 0.1N	2.2	Carbonic (saturated)	3.8
Sulfuric, 0.1N	1.2	Formic, 0.1N	2.3	Hydrogen sulfide, 0.1N	4.1
Sulfuric, 0.01N	2.1	Lactic, 0.1N	2.4	Arsenious (saturated)	5.0
Orthophosphoric, 0.1N	1.5	Acetic, N	2.4	Hydrocyanic, 0.1N	5.1
Sulfurous, 0.1N	1.5	Acetic, 0.1N	2.9	Boric, 0.1N	5.2

BASES

Sodium hydroxide, N	14.0	Lime (saturated)	12.4	Magnesia (saturated)	10.5
Sodium hydroxide, 0.1N	13.0	Trisodium phosphate, 0.1N	12.0	Sodium sesquicarbonate, 0.1M	10.1
Sodium hydroxide, 0.01N	12.0	Sodium carbonate, 0.1N	11.6	Ferrous hydroxide (saturated)	9.5
Potassium hydroxide, N	14.0	Ammonia, N	11.6	Calcium carbonate (saturated)	9.4
Potassium hydroxide, 0.1N	13.0	Ammonia, 0.1N	11.1	Borax, 0.1N	9.2
Potassium hydroxide, 0.01N	12.0	Ammonia, 0.01N	10.6	Sodium bicarbonate, 0.1N	8.4
Sodium metasilicate, 0.1N	12.6	Potassium cyanide, 0.1N	11.0		

BIOLOGIC MATERIALS

Blood, plasma, human	7.3–7.5	Gastric contents, human	1.0–3.0	Milk, human	6.6–7.6
Spinal fluid, human	7.3–7.5	Duodenal contents, human	4.8–8.2	Bile, human	6.8–7.0
Blood, whole, dog	6.9–7.2	Feces, human	4.6–8.4		
Saliva, human	6.5–7.5	Urine, human	4.8–8.4		

FOODS

Apples	2.9–3.3	Gooseberries	2.8–3.0	Potatoes	5.6–6.0
Apricots	3.6–4.0	Grapefruit	3.0–3.3	Pumpkin	4.8–5.2
Asparagus	5.4–5.8	Grapes	3.5–4.5	Raspberries	3.2–3.6
Bananas	4.5–4.7	Hominy (lye)	6.8–8.0	Rhubarb	3.1–3.2
Beans	5.0–6.0	Jams, fruit	3.5–4.0	Salmon	6.1–6.3
Beers	4.0–5.0	Jellies, fruit	2.8–3.4	Sauerkraut	3.4–3.6
Beets	4.9–5.5	Lemons	2.2–2.4	Shrimp	6.8–7.0
Blackberries	3.2–3.6	Limes	1.8–2.0	Soft drinks	2.0–4.0
Bread, white	5.0–6.0	Maple syrup	6.5–7.0	Spinach	5.1–5.7
Butter	6.1–6.4	Milk, cows	6.3–6.6	Squash	5.0–5.4
Cabbage	5.2–5.4	Olives	3.6–3.8	Strawberries	3.0–3.5
Carrots	4.9–5.3	Oranges	3.0–4.0	Sweet potatoes	5.3–5.6
Cheese	4.8–6.4	Oysters	6.1–6.6	Tomatoes	4.0–4.4
Cherries	3.2–4.0	Peaches	3.4–3.6	Tuna	5.9–6.1
Cider	2.9–3.3	Pears	3.6–4.0	Turnips	5.2–5.6
Corn	6.0–6.5	Peas	5.8–6.4	Vinegar	2.4–3.4
Crackers	6.5–8.5	Pickles, dill	3.2–3.6	Water, drinking	6.5–8.0
Dates	6.2–6.4	Pickles, sour	3.0–3.4	Wines	2.8–3.8
Eggs, fresh white	7.6–8.0	Pimento	4.6–5.2		
Flour, wheat	5.5–6.5	Plums	2.8–3.0		

Indicator	Approximate pH range	Color-change	Preparation
Methyl Violet	0.0–1.6	yel to bl	0.01–0.05% in water
Crystal Violet	0.0–1.8	yel to bl	0.02% in water
Ethyl Violet	0.0–2.4	yel to bl	0.1 g in 50 ml of MeOH + 50 ml of water
Malachite Green	0.2–1.8	yel to bl grn	water
Methyl Green	0.2–1.8	yel to bl	0.1% in water
2-(p-dimethylaminophenylazo)pyridine	0.2–1.8 4.4–5.6	yel to bl red to yel	0.1% in EtOH
o-Cresolsulfonephthalein (Cresol Red)	0.4–1.8 7.0–8.8	yel to red yel to red	0.1 g in 26.2 ml 0.01N NaOH + 223.8 ml water
Quinaldine Red	1.0–2.2	col to red	1% in EtOH
p-(p-dimethylaminophenylazo)-benzoic acid, Na-salt (Paramethyl Red)	1.0–3.0	red to yel	EtOH
m-(p-anilnophenylazo)benzene sulfonic acid, Na-salt (Metanil Yellow)	1.2–2.4	red to yel	0.01% in water
4-Phenylazodiphenylamine	1.2–2.6	red to yel	0.01 g in 1 ml 1N HCl + 50 ml EtOH + 49 ml water
Thymolsulfonephthalein (Thymol Blue)	1.2–2.8 8.0–9.6	red to yel yel to bl	0.1 g in 21.5 ml 0.01N NaOH + 229.5 ml water
m-Cresolsulfonephthalein (Metacresol Purple)	1.2–2.8 7.4–9.0	red to yel yel to purp	0.1 g in 26.2 ml 0.01N NaOH + 223.8 ml water
p-(p-anilinophenylazo)benzenesulfonic acid, Na-salt (Orange IV)	1.4–2.8	red to yel	0.01% in water
4-o-Tolylazo-o-toluidine	1.4–2.8	or to yel	water
Erythrosine, disodium salt	2.2–3.6	or to red	0.1% in water
Benzopurpurine 48	2.2–4.2	vt to red	0.1% in water
N,N-dimethyl-p-(m-tolylazo)aniline	2.6–4.8	red to yel	0.1% in water
4,4′-Bix(2-amino-1-naphthylazo)2,2′-stilbenedisulfonic acid	3.0–4.0	purp to red	0.1 g in 5.9 ml 0.05N NaOH + 94.1 ml water
Tetrabromophenolphthaleinethyl ester, K-salt	3.0–4.2	yel to bl	0.1% in EtOH
3′,3″,5′,5″-tetrabromophenolsulfonephthalein (Bromophenol Blue)	3.0–4.6	yel to bl	0.1 g in 14.9 ml 0.01N NaOH + 235.1 ml water
2,4-Dinitrophenol	2.8–4.0	col to yel	saturated water solution
N,N-Dimethyl-p-phenylazoaniline (p-Dimethylaminoazobenzene)	2.8–4.4	red to yel	0.1 g in 90 ml in EtOH + 10 ml water
Congo Red	3.0–5.0	blue to red	0.1% in water
Methyl Orange-Xylene Cyanole solution	3.2–4.2	purp to grn	ready solution
Methyl Orange	3.2–4.4	red to yel	0.01% in water
Ethyl Orange	3.4–4.8	red to yel	0.05–0.2% in water or aqueous EtOH
4-(4-Dimethylamino-1-naphthylazo)-3-methoxybenzenesulfonic acid	3.5–4.8	vt to yel	0.1% in 60% EtOH
3′,3″,5′,5″-Tetrabromo-m-cresolsulfonephthalein (Bromocresol Green)	3.8–5.4	yel to blue	0.1 g in 14.3 ml 0.01N NaOH + 235.7 ml water
Resazurin	3.8–6.4	or to vt	water
4-Phenylazo-1-naphthylamine	4.0–5.6	red to yel	0.1% in EtOH
Ethyl Red	4.0–5.8	col to red	0.1 g in 50 ml MeOH + 50 ml water
2-(p-Dimethylaminophenylazo)-pyridine	0.2–1.8 4.4–5.6	yel to red red to yel	0.1% in EtOH
4-(p-ethoxyphenylazo)-m-phenylenediamine monohydrochloride	4.4–5.8	or to yel	0.1% in water
Lacmoid	4.4–6.2	red to bl	0.2% in EtOH
Alizarin Red S	4.6–6.0	yel to red	dilute solution in water
Methyl Red	4.8–6.0	red to yel	0.02 g in 60 ml EtOH + 40 ml water

Indicator	Approximate pH range	Color-change	Preparation
Propyl Red	4.8–6.6	red to yel	EtOH
5′,5″-Dibromo-o-cresolsulfone-phthalein (Bromocresol Purple)	5.2–6.8	yel to purp	0.1 g in 18.5 ml 0.01N NaOH + 231.5 ml water
3′,3″-Dichlorophenolsulfonephthalein (Chlorophenol Red)	5.2–6.8	yel to red	0 1 g in 23.6 ml 0.01N NaOH + 226.4 ml water
p-Nitrophenol	5.4–6.6	col to yel	0.1% in water
Alizarin	5.6–7.2 11.0–12.4	yel to red red to purp	0.1% in MeOH
2-(2,4-Dinitrophenylazo)-1-naphthol-3, 6-disulfonic acid, di-Na salt	6.0–7.0	yel to bl	0.1% in water
3′,3″-Dibromothymolsulfonephthalein (Bromothymol Blue)	6.0–7.6	yel to bl	0.1 g in 16 ml 0.01N NaOH + 234 ml water
6,8-Dinitro-2,4-(1H)quinazolinedione (m-Dinitrobenzoylene urea)	6.4–8.0	col to yel	25 g in 115 ml M NaOH + 50 ml boiling water 0.292 g of NaCl in 100 ml water
Brilliant Yellow	6.6–7.8	yel to or	1% in water
Phenolsulfonephthalein (Phenol Red)	6.6–8.0	yel to red	0.1 g in 28.2 ml 0.01N NaOH + 221.8 ml water
Neutral Red	6.8–8.0	red to amb	0.01 g in 50 ml EtOH + 50 ml water
m-Nitrophenol	6.8–8.6	col to yel	0.3% in water
o-Cresolsulfonephthalein (Cresol Red)	0.0–1.0 7.0–8.8	red to yel yel to red	0.1 g in 26.2 ml 0.01N NaOH + 223.8 ml water
Curcumin	7.4–8.6 10.2–11.8	yel to red	EtOH
m-Cresolsulfonephthalein (Metacresol Purple)	1.2–2.8 7.4–9.0	red to yel yel to purp	0.1 g in 26.2 ml 0.01N NaOH + 223.8 ml water
4,4′-Bis(4-amino-1-naphthylazo) 2,2′stilbene disulfonic acid	8.0–9.0	bl to red	0.1 g in 5.9 ml 0.05N NaOH + 94.1 ml water
Thymolsulfonephthalein (Thymol Blue)	1.2–2.8 8.0–9.6	red to yel	0.1 g in 21.5 ml 0.01N NaOH + 228.5 ml water
o-Cresolphthalein	8.2–9.8	col to red	0.04% in EtOH
p-Naphtholbenzene	8.2–10.0	or to bl	1% in dil. alkali
Phenolphthalein	8.2–10.0	col to pink	0.05 g in 50 ml EtOH + 50 ml water
Ethyl-bis(2,4-dimethylphenyl)acetate	8.4–9.6	col to bl	saturated solution in 50% acetone alcohol
Thymolphthalein	9.4–10.6	col to bl	0.04 g in 50 ml EtOH + 50 ml water
5-(p-Nitrophenylazo)salicylic acid, Na-salt (Alizarin Yellow R)	10.1–12.0	yel to red	0.01% in water
p-(2,4-Dihydroxyphenylazo)benzene-sulfonic acid, Na-salt	11.4–12.6	yel to or	0.1% in water
5,5′-Indigodisulfonic acid, di-Na-salt	11.4–13.0	bl to yel	water
2,4,6-Trinitrotoluene	11.5–13.0	col to or	0.1–0.5% in EtOH
1,3,5-Trinitrobenzene	12.0–14.0	col to or	0.1–0.5% in EtOH
Clayton Yellow	12.2–13.2	yel to amb	0.1% in water

FLUORESCENT INDICATORS

Jack DeMent

Fluorescent indicators are substances which show definite changes in fluorescence with change in pH. Some fluorescent materials are not suitable for indicators since their change in fluorescence is too gradual. Fluorescent indicators find greatest utility in the titration of opaque, highly turbid or deeply colored solutions. A long wavelength ultraviolet ("black light") lamp in a dimly lighted room provides the best environment for titrations involving fluorescent indicators, although bright daylight is sometimes sufficient to evoke a response in the bright green, yellow and orange fluorescent indicators. Titrations are carried out in non-fluorescent glassware. One should check the glassware prior to use to make certain that it does not fluoresce due to the wavelengths of light involved in the titration. The meniscus of the liquid in the burette can be followed when a few particles of an insoluble fluorescent solid are dropped onto its surface.

In this table the indicators are arranged by approximate pH range covered. In the case of some of the dyestuffs the end point may vary slightly with the source or manufacturer.

pH 0 to 2

Indicator	C.I.	From pH	To pH
Benzoflavine	—	0.3, yellow fl.	1.7, green fl.
3,6-Dioxyphthalimide	—	0, blue fl.	2.4, green fl.
Eosine YS	768	0, yellow colored	3.0, yellow fl.
Erythrosine	772	0, yellow colored	3.6, yellow fl.
Esculin	—	1.5, colorless	2, blue fl.
4-Ethoxyacridone	—	1.2, green fl.	3.2, blue fl.
3,6-Tetramethyldiaminooxanthone	—	1.2, green fl.	3.4, blue fl.

pH 2 to 4

Indicator	C.I.	From pH	To pH
Chromotropic acid	—	3.5, colorless	4.5, blue fl.
Fluorescein	766	4, colorless	4.5, green fl.
Magdala Red	—	3.0, purple colored	4.0, fl.
α-Naphthylamine	—	3.4, colorless	4.8, blue fl.
β-Naphthylamine	—	2.8, colorless	4.4, violet fl.
Phloxine	774	3.4, colorless	5.0, bright yellow fl.
Salicylic acid	—	2.5, colorless	3.5, blue fl.

pH 4 to 6

Indicator	C.I.	From pH	To pH
Acridine	788	4.9, green fl.	5.1, violet colored
Dichlorofluorescein	—	4.0, colorless	5.0, green fl.
3,6-Dioxyxanthone	—	5.4, colorless	7.6, blue-violet fl.
Erythrosine	772	4.0, colorless	4.5, yellow-green fl.
β-Methylesculetin	—	4.0, colorless	6.2, blue fl.
Neville-Winther acid	—	6.0, colorless	6.5, blue fl.
Resorufin	—	4.4, yellow fl.	6.4, weak orange fl.
Quininic acid	—	4.0, yellow colored	5.0, blue fl.
Quinine [first end point]	—	5.0, blue fl.	6.1, violet fl.

pH 6 to 8

Indicator	C.I.	From pH	To pH
Acid R Phosphine	—	(claimed for range pH 6.0–7.0)	
Brilliant Diazol Yellow	—	6.5, colorless	7.5, violet fl.
Cleves acid	—	6.5, colorless	7.5, green fl.
Coumaric acid	—	7.2, colorless	9.0, green fl.
3,6-Dioxyphthalic dinitrile	—	5.8, blue fl.	8.2, green fl.
Magnesium 8-hydroxyquinolinate	—	6.5, colorless	7.5, golden fl.
β-Methylumbelliferone	—	7.0, colorless	7.5, blue fl.
1-Naphthol-4-sulfonic acid	—	6.0, colorless	6.5, blue fl.
Orcinaurine	—	6.5, colorless	8.0, green fl.
Patent Phosphine	789	(for the range pH 6.0–7.0, green-yellow fl.)	
Thioflavine	816	(for the region pH 6.5–7.0, yellow fl.)	
Umbelliferone	—	6.5, colorless	7.6, blue fl.

FLUORESCENT INDICATORS (Continued)

pH 8 to 10

Indicator	C.I.	From pH	To pH
Acridine Orange	788	8.4, orange colored	10.4, green fl.
Ethoxyphenylnaphthostilbazonium chloride	—	9, green fl.	11, non-fl.
G Salt	—	9.0, dull blue fl.	9.5, bright blue fl.
Naphthazol derivatives	—	8.2, colorless	10.0, yellow or green fl.
α-Naphthionic acid	—	9, blue fl.	11, green fl.
2-Naphthol-3,6-disulfonic acid	—	9.5, dark blue fl.	Light blue fl. at higher pH
β-Naphthol	—	8.6, colorless	Blue fl. at higher pH
α-Naphtholsulfonic acid	—	8.0, dark blue fl.	9.0, bright violet fl.
1,4-Naphtholsulfonic acid	—	8.2, dark blue fl.	Light blue fl. at higher pH
Orcinsulfonphthalein	—	8.6, yellow colored	10.0 fl.
Quinine [second end point]	—	9.5, violet fl.	10.0, colorless
R-Salt	—	9.0, dull blue fl.	9.5, bright blue fl.
Sodium 1-naphthol-2-sulfonate	—	9.0, dark blue fl.	10.0, bright violet fl.

pH 10 to 12

Indicator	C.I.	From pH	To pH
Coumarin	—	9.8, deep green fl.	12, light green fl.
Eosine BN	771	10.5, colorless	14.0, yellow fl.
Papaverine (permanganate oxidized)	—	9.5, yellow fl.	11.0, blue fl.
Schaffers Salt	—	5.0, violet fl.	11.0, green-blue fl.
SS-Acid (sodium salt)	—	10.0, violet fl.	12.0, yellow colored

pH 12 to 14

Indicator	C.I.	From pH	To pH
Cotarnine	—	12.0, yellow fl.	13.0, white fl.
α-Naphthionic acid	—	12, blue fl.	13, green fl.
β-Naphthionic acid	—	12, blue fl.	13, violet fl.

FORMULAS FOR CALCULATING TITRATION DATA, *p*H VS. ML. OF REAGENT

A Substance Titrated V_0 ml. of solution M_0 its molarity	B Initial $[H^+]$ or $[OH^-]$	C Intermediate Points 10, 50, 99, etc., per cent neutralized, V_1 ml. of reagent of M_1 molarity added	D Equivalence Point	E Excess of Reagent V_1 volume reagent M_1 its molarity V_T total volume
(1) Strong Acid	$[H^+] = M_0$	$[H^+] = \frac{V_0M_0 - V_1M_1}{V_0 + V_1}$	$\sqrt{K_w}$	$[OH^-] = \frac{V_1M_1}{V_T}$
(2) Strong Base	$[OH^-] = M_0$	$[OH^-] = \frac{V_0M_0 - V_1M_1}{V_0 + V_1}$	$\sqrt{K_w}$	$[H^+] = \frac{V_1M_1}{V_T}$
(3) Weak Acid ($K_a = 10^{-5}$ to 10^{-8})	$[H^+] = \sqrt{M_0K_a}$	$[H^+] = \frac{[Acid]}{[Salt]} K_a$	$[OH^-] = \sqrt{\frac{K_w}{K_a} c}$	$[OH^-] = \frac{V_1M_1}{V_T}$ (Value in column D to be added)
(4) Weak Base ($K_b = 10^{-5}$ to 10^{-8})	$[OH^-] = \sqrt{M_0K_b}$	$[OH^-] = \frac{[Base]}{[Salt]} K_b$	$[H^+] = \sqrt{\frac{K_w}{K_b} c}$	$[H^+] = \frac{V_1M_1}{V_T}$ (Value in column D to be added)
(5) Salt of a Very Weak Acid (e.g. KCN)	$[OH^-] = \sqrt{\frac{K_w}{K_a} c}$	$[H^+] = \frac{[Acid]}{[Salt]} K_a$	$[H^+] = \sqrt{[Acid]K_a}$	$[H^+] = \frac{V_1M_1}{V_T}$ (Correct for value in column D)
(6) Salt of a Very Weak Base	$[H^+] = \sqrt{\frac{K_w}{K_b} c}$	$[OH^-] = \frac{[Base]}{[Salt]} K_b$	$[OH^-] = \sqrt{[Base]K_b}$	$[OH^-] = \frac{V_1M_1}{V_T}$ (Add to $[OH^-]$ found in column D)

CONVERSION FORMULAE FOR SOLUTIONS HAVING CONCENTRATIONS EXPRESSED IN VARIOUS WAYS

A = Weight per cent of solute
B = Molecular weight of solvent
E = Molecular weight of solute
F = Grams of solute per liter of solution
G = Molality
M = Molarity
N = Mole fraction
R = Density of solution grams per cc

Concentration of solute— SOUGHT	Concentration of solute—GIVEN				
	A	N	G	M	F
A	—	$\frac{100N \times E}{N \times E + (1 - N)B}$	$\frac{100G \times E}{1000 + G \times E}$	$\frac{M \times E}{10R}$	$\frac{F}{10R}$
N	$\frac{\frac{A}{E}}{\frac{A}{E} + \frac{100 - A}{B}}$	—	$\frac{B \times G}{B \times G + 1000}$	$\frac{B \times M}{M(B - E) + 1000R}$	$\frac{B \times F}{F(B - E) + 1000R \times E}$
G	$\frac{1000A}{E(100 - A)}$	$\frac{1000N}{B - N \times B}$	—	$\frac{1000M}{1000R - (M \times E)}$	$\frac{1000F}{E(1000R - F)}$
M	$\frac{10R \times A}{E}$	$\frac{1000R \times N}{N \times E + (1 - N)B}$	$\frac{1000R \times G}{1000 + E \times G}$	—	$\frac{F}{E}$
F	10AR	$\frac{1000R \times N \times E}{N \times E + (1 - N)B}$	$\frac{1000R \times G \times E}{1000 + G \times E}$	$M \times E$	—

ELECTROCHEMICAL SERIES

Petr Vanýsek

There are three tables for this Electrochemical Series. Each table lists standard reduction potentials, E° values, at 298.15 K (25°C), and at a pressure of 101.325 kPa (1 atm.). Table 1 is an alphabetical listing of the elements according to the symbols for the elements. Thus, data for Silver (Ag) precedes those for Aluminum (Al). Table 2 lists only those reduction reactions which have E° values positive to the potential of the Standard Hydrogen Electrode. In Table 2, the reactions are listed in the order of increasing positive potential and range from 0.000 V to +3.053 V. Table 3 lists only those reduction reactions which have E° values negative to the potential of the Standard Hydrogen Electrode. In Table 3, reactions are listed in the order of increasing negative potential and range from −0.017 to −4.10 V.

Table 1
ALPHABETICAL LISTING

Reaction	E°, V
$Ag^+ + e \rightleftharpoons Ag$	0.7996
$Ag^{2+} + e \rightleftharpoons Ag^+$	1.980
$Ag(ac) + e \rightleftharpoons Ag + (ac)^-$	0.643
$AgBr + e \rightleftharpoons Ag + Br^-$	0.07133
$AgBrO_3 + e \rightleftharpoons Ag + BrO_3^-$	0.546
$Ag_2C_2O_4 + 2\,e \rightleftharpoons 2Ag + C_2O_4^{2-}$	0.4647
$AgCl + e \rightleftharpoons Ag + Cl^-$	0.22233
$AgCN + e \rightleftharpoons Ag + CN^-$	−0.017
$Ag_2CO_3 + 2\,e \rightleftharpoons 2\,Ag + CO_3^{2-}$	0.47
$Ag_2CrO_4 + 2\,e \rightleftharpoons 2\,Ag + CrO_4^{2-}$	0.4470
$AgF + e \rightleftharpoons Ag + F^-$	0.779
$Ag_4[Fe(CN)_6] + 4\,e \rightleftharpoons 4\,Ag + [Fe(CN)_6]^{4-}$	0.1478
$AgI + e \rightleftharpoons Ag + I^-$	−0.15224
$AgIO_3 + e \rightleftharpoons Ag + IO_3^-$	0.354
$Ag_2MoO_4 + 2\,e \rightleftharpoons 2\,Ag + MoO_4^{2-}$	0.4573
$AgNO_2 + e \rightleftharpoons Ag + NO_2^-$	0.564
$Ag_2O + H_2O + 2\,e \rightleftharpoons 2\,Ag + 2\,OH^-$	0.342
$Ag_2O_3 + H_2O + 2\,e \rightleftharpoons 2\,AgO + 2\,OH^-$	0.739
$2\,AgO + H_2O + 2\,e \rightleftharpoons Ag_2O + 2\,OH^-$	0.607
$AgOCN + e \rightleftharpoons Ag + OCN^-$	0.41
$Ag_2S + 2\,e \rightleftharpoons 2\,Ag + S^{2-}$	−0.691
$Ag_2S + 2\,H^+ + 2\,e \rightleftharpoons 2\,Ag + H_2S$	−0.0366
$AgSCN + e \rightleftharpoons Ag + SCN^-$	0.08951
$Ag_2SeO_3 + 2\,e \rightleftharpoons 2\,Ag + SeO_3^{2-}$	0.3629
$Ag_2SO_4 + 2\,e \rightleftharpoons 2\,Ag + SO_4^{2-}$	0.654
$Ag_2WO_4 + 2\,e \rightleftharpoons 2\,Ag + WO_4^{2-}$	0.4660
$Al^{3+} + 3\,e \rightleftharpoons Al$	−1.662
$H_2AlO_3^- + H_2O + 3\,e \rightleftharpoons Al + 4\,OH^-$	−2.33
$AlF_6^{3-} + 3\,e \rightleftharpoons Al + 6\,F^-$	−2.069
$As + 3H^+ + 3\,e \rightleftharpoons AsH_3$	−0.608
$As_2O_3 + 6\,H^+ + 6\,e \rightleftharpoons 2\,As + 3\,H_2O$	0.234
$HAsO_2 + 3\,H^+ + 3\,e \rightleftharpoons As + 2\,H_2O$	0.248
$AsO_2^- + 2\,H_2O + 3\,e \rightleftharpoons As + 4\,OH^-$	−0.68
$H_3AsO_4 + 2\,H^+ + 2\,e^- \rightleftharpoons HAsO_2 + 2\,H_2O$	0.560
$AsO_4^{3-} + 2\,H_2O + 2\,e \rightleftharpoons AsO_2^- + 4\,OH^-$	−0.71
$Au^+ + e \rightleftharpoons Au$	1.692
$Au^{3+} + 2\,e \rightleftharpoons Au^+$	1.401
$Au^{3+} + 3\,e \rightleftharpoons Au$	1.498
$AuBr_2^- + e \rightleftharpoons Au + 2\,Br^-$	0.959
$AuBr_4^- + 3\,e \rightleftharpoons Au + 4\,Br^-$	0.854
$AuCl_4^- + 3\,e \rightleftharpoons Au + 4\,Cl^-$	1.002
$Au(OH)_3 + 3\,H^+ + 3\,e \rightleftharpoons Au + 3\,H_2O$	1.45
$H_2BO_3^- + 5\,H_2O + 8\,e \rightleftharpoons BH_4^- + 8\,OH^-$	−1.24
$H_2BO_3^- + H_2O + 3\,e \rightleftharpoons B + 4\,OH^-$	−1.79
$H_3BO_3 + 3\,H^+ + 3\,e \rightleftharpoons B + 3\,H_2O$	−0.8698
$Ba^{2+} + 2\,e \rightleftharpoons Ba$	−2.912
$Ba^{2+} + 2\,e \rightleftharpoons Ba(Hg)$	−1.570
$Ba(OH)_2 + 2\,e \rightleftharpoons Ba + 2\,OH^-$	−2.99
$Be^{2+} + 2\,e \rightleftharpoons Be$	−1.847
$Be_2O_3^{2-} + 3\,H_2O + 4\,e \rightleftharpoons 2\,Be + 6\,OH^-$	−2.63

Table 1 (continued)
ALPHABETICAL LISTING

Reaction	E°, V
p-benzoquinone + $2 H^+ + 2 e \rightleftharpoons$ hydroquinone	0.6992
$BiCl_4^- + 3 e \rightleftharpoons Bi + 4 Cl^-$	0.16
$Bi_2O_3 + 3 H_2O + 6 e \rightleftharpoons 2 Bi + 6 OH^-$	−0.46
$Bi_2O_4 + 4 H^+ + 2 e \rightleftharpoons 2 BiO^+ + 2 H_2O$	1.593
$BiO^+ + 2 H^+ + 3 e \rightleftharpoons Bi + H_2O$	0.320
$BiOCl + 2 H^+ + 3 e \rightleftharpoons Bi + Cl^- + H_2O$	0.1583
$Br_2(aq) + 2 e \rightleftharpoons 2 Br^-$	1.0873
$Br_2(1) + 2 e \rightleftharpoons 2 Br^-$	1.066
$HBrO + H^+ + 2 e \rightleftharpoons Br^- + H_2O$	1.331
$HBrO + H^+ + e \rightleftharpoons 1/2Br_2(aq) + H_2O$	1.574
$HBrO + H^+ + e \rightleftharpoons 1/2Br_2(\ell) + H_2O$	1.596
$BrO^- + H_2O + 2 e \rightleftharpoons Br^- + 2 OH^-$	0.761
$BrO_3^- + 6 H^+ + 5 e \rightleftharpoons 1/2Br_2 + 3 H_2O$	1.482
$BrO_3^- + 6 H^+ + 6 e \rightleftharpoons Br^- + 3 H_2O$	1.423
$BrO_3^- + 3 H_2O + 6 e \rightleftharpoons Br^- + 6 OH^-$	0.61
$Ca^+ + e \rightleftharpoons Ca$	−3.80
$Ca^{2+} + 2 e \rightleftharpoons Ca$	−2.868
Calomel electrode, 1 molal KCl	0.2800
Calomel electrode, 1 mol/ℓ KCl (NCE)	0.2801
Calomel electrode, 0.1 mol/ℓ KCl	0.3337
Calomel electrode, saturated KCl (SCE)	0.2412
Calomel electrode, saturated NaCl (SSCE)	0.2360
$Ca(OH)_2 + 2 e \rightleftharpoons Ca + 2 OH^-$	−3.02
$Cd^{2+} + 2 e \rightleftharpoons Cd$	−0.4030
$Cd^{2+} + 2 e \rightleftharpoons Cd(Hg)$	−0.3521
$Cd(OH)_2 + 2 e \rightleftharpoons Cd(Hg) + 2 OH^-$	−0.809
$CdSO_4 + 2 e \rightleftharpoons Cd + SO_4^{2-}$	−0.246
$Ce^{3+} + 3 e \rightleftharpoons Ce$	−2.483
$Ce^{3+} + 3 e \rightleftharpoons Ce(Hg)$	−1.4373
$Ce^{4+} + e \rightleftharpoons Ce^{3+}$	1.61
$CeOH^{3+} + H^+ + e \rightleftharpoons Ce^{3+} + H_2O$	1.715
$Cl_2(g) + 2 e \rightleftharpoons Cl^-$	1.35827
$HClO + H^+ + e \rightleftharpoons 1/2Cl_2 + H_2O$	1.611
$HClO + H^+ 2 e \rightleftharpoons Cl^- + H_2O$	1.482
$ClO^- + H_2O + 2 e \rightleftharpoons Cl^- + 2 OH^-$	0.81
$ClO_2 + H^+ + e \rightleftharpoons HClO_2$	1.277
$HClO_2 + 2 H^+ + 2 e \rightleftharpoons HClO + H_2O$	1.645
$HClO_2 + 3 H^+ + 3 e \rightleftharpoons 1/2Cl_2 + 2 H_2O$	1.628
$HClO_2 + 3 H^+ + 4 e \rightleftharpoons Cl^- + 2 H_2O$	1.570
$ClO_2^- + H_2O + 2 e \rightleftharpoons ClO^- + 2 OH^-$	0.66
$ClO_2^- + 2 H_2O + 4 e \rightleftharpoons Cl^- + 4 OH^-$	0.76
$ClO_2(aq) + e \rightleftharpoons ClO_2^-$	0.954
$ClO_3^- + 2 H^+ + e \rightleftharpoons ClO_2 + H_2O$	1.152
$ClO_3^- + 3 H^+ + 2 e \rightleftharpoons HClO_2 + H_2O$	1.214
$ClO_3^- + 6 H^+ + 5 e \rightleftharpoons 1/2Cl_2 + 3 H_2O$	1.47
$ClO_3^- + 6 H^+ + 6 e \rightleftharpoons Cl^- + 3 H_2O$	1.451
$ClO_3^- + H_2O + 2 e \rightleftharpoons ClO_2^- + 2 OH^-$	0.33
$ClO_3^- + 3 H_2O + 6 e \rightleftharpoons Cl^- + 6 OH^-$	0.62
$ClO_4^- + 2 H^+ + 2 e \rightleftharpoons ClO_3^- + H_2O$	1.189
$ClO_4^- + 8 H^+ + 7 e \rightleftharpoons 1/2Cl_2 + 4 H_2O$	1.39
$ClO_4^- + 8 H^+ + 8 e \rightleftharpoons Cl^- + 4 H_2O$	1.389
$ClO_4^- + H_2O + 2 e \rightleftharpoons ClO_3^- + 2 OH^-$	0.36
$(CN)_2 + 2 H^+ + 2 e \rightleftharpoons 2 HCN$	0.373
$2 HCNO + 2 H^+ + 2 e \rightleftharpoons (CN)_2 + 2 H_2O$	0.330
$(CNS)_2 + 2 e \rightleftharpoons 2 CNS^-$	0.77
$Co^{2+} + 2 e \rightleftharpoons Co$	−0.28
$Co^{3+} + e \rightleftharpoons Co^{2+}$ (2 mol/ℓ H_2SO_4)	1.83
$[Co(NH_3)_6]^{3+} + e \rightleftharpoons [Co(NH_3)_6]^{2+}$	0.108
$Co(OH)_2 + 2 e \rightleftharpoons Co + 2 OH^-$	−0.73
$Co(OH)_3 + e \rightleftharpoons Co(OH)_2 + OH^-$	0.17
$CO_2 + 2 H^+ + 2 e \rightleftharpoons HCOOH$	−0.199
$Cr^{2+} + 2 e \rightleftharpoons Cr$	−0.913
$Cr^{3+} + e \rightleftharpoons Cr^{2+}$	−0.407
$Cr^{3+} + 3 e \rightleftharpoons Cr$	−0.744
$Cr_2O_7^{2-} + 14 H^+ + 6 e \rightleftharpoons 2 Cr^{3+} + 7 H_2O$	1.232
$CrO_2^- + 2 H_2O + 3 e \rightleftharpoons Cr + 4 OH^-$	−1.2
$HCrO_4^{2-} + 7 H^+ + 3 e \rightleftharpoons Cr^{3+} + 4 H_2O$	1.350
$CrO_4^- + 4 H_2O + 3 e \rightleftharpoons Cr(OH)_3 + 5 OH^-$	−0.13
$Cr(OH)_3 + 3 e \rightleftharpoons Cr + 3 OH^-$	−1.48
$Cs^+ + e \rightleftharpoons Cs$	−2.92
$Cu^+ + e \rightleftharpoons Cu$	0.521
$Cu^{2+} + e \rightleftharpoons Cu^+$	0.153
$Cu^{2+} + 2 e \rightleftharpoons Cu$	0.3419
$Cu^{2+} + 2 e \rightleftharpoons Cu(Hg)$	0.345
$Cu^{2+} + 2 CN^- + e \rightleftharpoons [Cu(CN)_2]^-$	1.103
$CuI_2^- + e \rightleftharpoons Cu + 2 I^-$	0.00
$Cu_2O + H_2O + 2 e \rightleftharpoons 2 Cu + 2 OH^-$	−0.360
$Cu(OH)_2 + 2 e \rightleftharpoons Cu + 2 OH$	−0.222
$2 Cu(OH)_2 + 2 e \rightleftharpoons Cu_2O + 2 OH^- + H_2O$	−0.080
$D^+ + e \rightleftharpoons 1/2D_2$	−0.0034
$2 D^+ + 2 e \rightleftharpoons D_2$	−0.044
$Eu^{2+} + 2 e \rightleftharpoons Eu$	−3.395
$Eu^{3+} + 3 e \rightleftharpoons Eu$	−2.407
$Eu^{3+} + e \rightleftharpoons Eu^{2+}$	−0.36
$F_2 + 2 H^+ + 2 e \rightleftharpoons 2 HF$	3.053
$F_2 + 2 e \rightleftharpoons 2 F^-$	2.866
$F_2O + 2 H^+ + 4 e \rightleftharpoons H_2O + 2 F^-$	2.153
$Fe^{2+} + 2 e \rightleftharpoons Fe$	−0.447
$Fe^{3+} + 3 e \rightleftharpoons Fe$	−0.037
$Fe^{3+} + e \rightleftharpoons Fe^{2+}$	0.771
$[Fe(CN)_6]^{3-} + e \rightleftharpoons [Fe(CN)_6]^{4-}$	0.358
$FeO_4^{2-} + 8 H^+ + 3 e \rightleftharpoons Fe^{3+} + 4 H_2O$	2.20
$Fe(OH)_3 + e \rightleftharpoons Fe(OH)_2 + OH^-$	−0.56
$[Fe(phenanthroline)_3]^{3+} + e \rightleftharpoons [Fe(phen)_3]^{2+}$	1.147
$[Fe(phen)_3]^{3+} + e \rightleftharpoons [Fe(phen)_3]^{2+}$ (1 mol/ℓ H_2SO_4)	1.06
$[Ferricinium]^+ + e \rightleftharpoons$ ferrocene	0.400
$Ga^{3+} + 3 e \rightleftharpoons Ga$	−0.560
$H_2GaO^-_3 + H_2O + 3 e \rightleftharpoons Ga + 4 OH^-$	−1.219
$Ge^{2+} + 2 e \rightleftharpoons Ge$	0.24
$Ge^{4+} + 4 e \rightleftharpoons Ge$	0.124
$Ge^{4+} + 2 e \rightleftharpoons Ge^{2+}$	0.00
$GeO_2 + 2 H^+ + 2 e \rightleftharpoons GeO + H_2O$	−0.118
$H_2GeO_3 + 4 H^+ + 4 e \rightleftharpoons Ge + 3 H_2O$	−0.182
$2 H^+ + 2 e \rightleftharpoons H_2$	0.00000
$H_2 + 2 e \rightleftharpoons 2 H^-$	−2.23
$HO_2 + H^+ + e \rightleftharpoons H_2O_2$	1.495
$2 H_2O + 2 e \rightleftharpoons H_2 + 2 OH^-$	−0.8277
$H_2O_2 + 2 H^+ + 2 e \rightleftharpoons 2 H_2O$	1.776
$HfO^{2+} + 2 H^+ + 4 e \rightleftharpoons Hf + H_2O$	−1.724
$HfO_2 + 4 H^+ + 4 e \rightleftharpoons Hf + 2 H_2O$	−1.505
$HfO(OH)_2 + H_2O + 4 e \rightleftharpoons Hf + 4 OH^-$	−2.50
$Hg^{2+} + 2 e \rightleftharpoons Hg$	0.851
$2 Hg^{2+} + 2 e \rightleftharpoons Hg_2^{2+}$	0.920
$Hg_2^{2+} + 2 e \rightleftharpoons 2 Hg$	0.7973

Table 1 (continued)
ALPHABETICAL LISTING

Reaction	E°, V
$Hg_2(ac)_2 + 2e \rightleftharpoons 2Hg + 2(ac)^-$	0.51163
$Hg_2Br_2 + 2e \rightleftharpoons 2Hg + 2Br^-$	0.13923
$Hg_2Cl_2 + 2e \rightleftharpoons 2Hg + 2Cl^-$	0.26808
$Hg_2HPO_4 + 2e \rightleftharpoons 2Hg + HPO_4^{2-}$	0.6359
$Hg_2I_2 + 2e \rightleftharpoons 2Hg + 2I^-$	−0.0405
$Hg_2O + H_2O + 2e \rightleftharpoons 2Hg + 2OH^-$	0.123
$HgO + H_2O + 2e \rightleftharpoons Hg + 2OH^-$	0.0977
$Hg_2SO_4 + 2e \rightleftharpoons 2Hg + SO_4^{2-}$	0.6125
$I_2 + 2e \rightleftharpoons 2I^-$	0.5355
$I_3^- + 2e \rightleftharpoons 3I^-$	0.536
$H_3IO_6 + 2e \rightleftharpoons IO_3^- + 3OH^-$	0.7
$H_5IO_6 + H^+ + 2e \rightleftharpoons IO_3^- + 3H_2O$	1.601
$2HIO + 2H^+ + 2e \rightleftharpoons I_2 + 2H_2O$	1.439
$HIO + H^+ + 2e \rightleftharpoons I^- + H_2O$	0.987
$IO^- + H_2O + 2e \rightleftharpoons I^- + 2OH^-$	0.485
$2IO_3^- + 12H^+ + 10e \rightleftharpoons I_2 + 6H_2O$	1.195
$IO_3^- + 6H^+ + 6e \rightleftharpoons I^- + 3H_2O$	1.085
$IO_3^- + 2H_2O + 4e \rightleftharpoons IO^- + 4OH^-$	0.15
$IO_3^- + 3H_2O + 6e \rightleftharpoons I^- + 6OH^-$	0.26
$In^+ + e \rightleftharpoons In$	−0.14
$In^{2+} + e \rightleftharpoons In^+$	−0.40
$In^{3+} + e \rightleftharpoons In^{2+}$	−0.49
$In^{3+} + 2e \rightleftharpoons In^+$	−0.443
$In^{3+} + 3e \rightleftharpoons In$	−0.3382
$Ir^{3+} + 3e \rightleftharpoons Ir$	1.156
$[IrCl_6]^{2-} + e \rightleftharpoons [IrCl_6]^{3-}$	0.8665
$[IrCl_6]^{3-} + 3e \rightleftharpoons Ir + 6Cl^-$	0.77
$Ir_2O_3 + 3H_2O + 6e \rightleftharpoons 2Ir + 6OH^-$	0.098
$K^+ + e \rightleftharpoons K$	−2.931
$La^{3+} + 3e \rightleftharpoons La$	−2.522
$La(OH)_3 + 3e \rightleftharpoons La + 3OH^-$	−2.90
$Li^+ + e \rightleftharpoons Li$	−3.0401
$Mg^+ + e \rightleftharpoons Mg$	−2.70
$Mg^{2+} + 2e \rightleftharpoons Mg$	−2.372
$Mg(OH)_2 + 2e \rightleftharpoons Mg + 2OH^-$	−2.690
$Mn^{2+} + 2e \rightleftharpoons Mn$	−1.185
$Mn^{3+} + 3 \rightleftharpoons Mn^{2+}$	1.5415
$MnO_2 + 4H^+ + 2e \rightleftharpoons Mn^{2+} + 2H_2O$	1.224
$MnO_4^- + e \rightleftharpoons MnO_4^{2-}$	0.558
$MnO_4^- + 4H^+ + 3e \rightleftharpoons MnO_2 + 2H_2O$	1.679
$MnO_4^- + 8H^+ + 5e \rightleftharpoons Mn^{2+} + 4H_2O$	1.507
$MnO_4^- + 2H_2O + 3e \rightleftharpoons MnO_2 + 4OH^-$	0.595
$MnO_4^{2-} + 2H_2O + 2e \rightleftharpoons MnO_2 + 4OH^-$	0.60
$Mn(OH)_2 + 2e \rightleftharpoons Mn + 2OH^-$	−1.56
$Mn(OH)_3 + e \rightleftharpoons Mn(OH)_2 + OH^-$	0.15
$Mo^{3+} + 3e \rightleftharpoons Mo$	−0.200
$N_2 + 2H_2O + 6H^+ + 6e \rightleftharpoons 2NH_4OH$	0.092
$3N_2 + 2H^+ + 2e \rightleftharpoons 2NH_3$	−3.09
$N_5^+ + 3H^+ + 2e \rightleftharpoons 2NH_4^+$	1.275
$N_2O + 2H^+ + 2e \rightleftharpoons N_2 + H_2O$	1.766
$H_2N_2O_2 + 2H^+ + 2e \rightleftharpoons N_2 + 2H_2O$	2.65
$N_2O_4 + 2e \rightleftharpoons 2NO_2^-$	0.867
$N_2O_4 + 2H^+ + 2e \rightleftharpoons 2HNO_2$	1.065
$N_2O_4 + 4H^+ + 4e \rightleftharpoons 2NO + 2H_2O$	1.035
$2NH_3OH^+ + H^+ + 2e \rightleftharpoons N_2H_5^+ + 2H_2O$	1.42
$2NO + 2e \rightleftharpoons N_2O_2^{2-}$	0.10
$2NO + 2H^+ + 2e \rightleftharpoons N_2O + H_2O$	1.591
$2NO + H_2O + 2e \rightleftharpoons N_2O + 2OH^-$	0.76

Reaction	E°, V
$HNO_2 + H^+ + e \rightleftharpoons NO + H_2O$	0.983
$2HNO_2 + 4H^+ + 4e \rightleftharpoons H_2N_2O_2 + 2H_2O$	0.86
$2HNO_2 + 4H^+ + 4e \rightleftharpoons N_2O + 3H_2O$	1.297
$NO_2^- + H_2O + 3e \rightleftharpoons NO + 2OH^-$	−0.46
$2NO_2^- + 2H_2O + 4e \rightleftharpoons N_2^{2-} + 4OH^-$	−0.18
$2NO_2^- + 3H_2O + 4e \rightleftharpoons N_2O + 6OH^-$	0.15
$NO_3^- + 3H^+ + 2e \rightleftharpoons HNO_2 + H_2O$	0.934
$NO_3^- + 4H^+ + 3e \rightleftharpoons NO + 2H_2O$	0.957
$2NO_3^- + 4H^+ + 2e \rightleftharpoons N_2O_4 + 2H_2O$	0.803
$NO_3^- + H_2O + 2e \rightleftharpoons NO_2^- + 2OH^-$	0.01
$2NO_3^- + 2H_2O + 2e \rightleftharpoons N_2O_4 + 4OH^-$	−0.85
$Na^+ + e \rightleftharpoons Na$	−2.71
$Nb^{3+} + 3e \rightleftharpoons Nb$	−1.099
$Nb_2O_5 + 10H^+ + 10e \rightleftharpoons 2Nb + 5H_2O$	−0.644
$Nd^{3+} + 3e \rightleftharpoons Nd$	−2.431
$Ni^{2+} + 2e \rightleftharpoons Ni$	−0.257
$Ni(OH)_2 + 2e \rightleftharpoons Ni + 2OH^-$	−0.72
$NiO_2 + 4H^+ + 2e \rightleftharpoons Ni^{2+} + 2H_2O$	1.678
$NiO_2 + 2H_2O + 2e \rightleftharpoons Ni(OH)_2 + 2OH^-$	−0.490
$Np^{3+} + 3e \rightleftharpoons Np$	−1.856
$Np^{4+} + e \rightleftharpoons Np^{3+}$	0.147
$NpO_2 + H_2O + H^+ + e \rightleftharpoons Np(OH)_3$	−0.962
$O_2 + 2H^+ + 2e \rightleftharpoons H_2O_2$	0.695
$O_2 + 4H^+ + 4e \rightleftharpoons 2H_2O$	1.229
$O_2 + H_2O + 2e \rightleftharpoons HO_2^- + OH^-$	−0.076
$O_2 + 2H_2O + 2e \rightleftharpoons H_2O_2 + 2OH^-$	−0.146
$O_2 + 2H_2O + 4e \rightleftharpoons 4OH^-$	0.401
$O_3 + 2H^+ + 2e \rightleftharpoons O_2 + H_2O$	2.076
$O_3 + H_2O + 2e \rightleftharpoons O_2 + 2OH^-$	1.24
$O(g) + 2H^+ + 2e \rightleftharpoons H_2O$	2.421
$OH + e \rightleftharpoons OH^-$	2.02
$HO_2^- + H_2O + 2e \rightleftharpoons 3OH^-$	0.878
$OsO_4 + 8H^+ + 8e \rightleftharpoons Os + 4H_2O$	0.85
$P(red) + 3H^+ + 3e \rightleftharpoons PH_3(g)$	−0.111
$P(white) + 3H^+ + 3e \rightleftharpoons PH_3(g)$	−0.063
$P + 3H_2O + 3e \rightleftharpoons PH_3(g) + 3OH^-$	−0.87
$H_2PO_2^- + e \rightleftharpoons P + 2OH^-$	−1.82
$H_3PO_2 + H^+ + 3e \rightleftharpoons P + 2H_2O$	−0.508
$H_3PO_3 + 2H^+ + 2e \rightleftharpoons H_3PO_2 + H_2O$	−0.499
$H_3PO_3 + 3H^+ + 3e \rightleftharpoons P + 3H_2O$	−0.454
$HPO_3^{2-} + 2H_2O + 2e \rightleftharpoons H_2PO_2^- + 3OH^-$	−1.65
$HPO_3^{2-} + 2H_2O + 3e \rightleftharpoons P + 5OH^-$	−1.71
$H_3PO_4 + 2H^+ + 2e \rightleftharpoons H_3PO_3 + H_2O$	−0.276
$PO_4^{3-} + 2H_2O + 2e \rightleftharpoons HPO_3^{2-} + 3OH^-$	−1.05
$Pb^{2+} + 2e \rightleftharpoons Pb$	−0.1262
$Pb^{2+} + 2e \rightleftharpoons Pb(Hg)$	−0.1205
$PbBr_2 + 2e \rightleftharpoons Pb + 2Br^-$	−0.284
$PbCl_2 + 2e \rightleftharpoons Pb + 2Cl^-$	−0.2675
$PbF_2 + 2e \rightleftharpoons Pb + 2F^-$	−0.3444
$PbHPO_4 + 2e \rightleftharpoons Pb + HPO_4^{2-}$	−0.465
$PbI_2 + 2e \rightleftharpoons Pb + 2I^-$	−0.365
$PbO + H_2O + 2e \rightleftharpoons Pb + 2OH^-$	−0.580
$PbO_2 + 4H^+ + 2e \rightleftharpoons Pb^{2+} + 2H_2O$	1.455
$HPbO_2^- + H_2O + 2e \rightleftharpoons Pb + 3OH^-$	−0.537
$PbO_2 + H_2O + 2e \rightleftharpoons PbO + 2OH^-$	0.247
$PbO_2 + SO_4^{2-} + 4H^+ + 2e \rightleftharpoons PbSO_4 + 2H_2O$	1.6913
$PbSO_4 + 2e \rightleftharpoons Pb + SO_4^{2-}$	−0.3588

Table 1 (continued)
ALPHABETICAL LISTING

Reaction	E°, V
$PbSO_4 + 2\,e \rightleftharpoons Pb(Hg) + SO_4^{2-}$	−0.3505
$Pd^{2+} + 2\,e \rightleftharpoons Pd$	0.951
$[PdCl_4]^{2-} + 2\,e \rightleftharpoons Pd + 4\,Cl^-$	0.591
$[PdCl_6]^{2-} + 2\,e \rightleftharpoons [PdCl_4]^{2-} + 2\,Cl^-$	1.288
$Pd(OH)_2 + 2\,e \rightleftharpoons Pd + 2\,OH^-$	0.07
$Pt^{2+} + 2\,e \rightleftharpoons Pt$	1.118
$[PtCl_4]^{2-} + 2\,e \rightleftharpoons Pt + 4\,Cl^-$	0.755
$[PtCl_6]^{2-} + 2\,e \rightleftharpoons [PtCl_4]^{2-} + 2\,Cl^-$	0.68
$Pt(OH)_2 + 2\,e \rightleftharpoons Pt + 2\,OH^-$	0.14
$Pu^{3+} + 3\,e \rightleftharpoons Pu$	−2.031
$Pu^{4+} + e \rightleftharpoons Pu^{3+}$	1.006
$Pu^{5+} + e \rightleftharpoons Pu^{4+}$	1.099
$PuO_2(OH)_2 + 2\,H^+ + 2\,e \rightleftharpoons Pu(OH)_4$	1.325
$PuO_2(OH)_2 + H^+ + e \rightleftharpoons PuO_2OH + H_2O$	1.062
$Rb^+ + e \rightleftharpoons Rb$	−2.98
$Re^{3+} + 3\,e \rightleftharpoons Re$	0.300
$ReO_4^- + 4\,H^+ + 3\,e \rightleftharpoons ReO_2 + 2\,H_2O$	0.510
$ReO_2 + 4\,H^+ + 4\,e \rightleftharpoons Re + 2\,H_2O$	0.2513
$ReO_4^- + 2\,H^+ + e \rightleftharpoons ReO_3 + H_2O$	0.768
$ReO_4^- + 4\,H_2O + 7\,e \rightleftharpoons Re + 8\,OH^-$	−0.584
$ReO_4^- + 8\,H^+ + 7\,e \rightleftharpoons Re + 4\,H_2O$	0.368
$Rh^+ + e \rightleftharpoons Rh$	0.600
$Rh^{2+} + 2\,e \rightleftharpoons Rh$	0.600
$Rh^{3+} + 3\,e \rightleftharpoons Rh$	0.758
$[RhCl_6]^{3-} + 3\,e \rightleftharpoons Rh + 6\,Cl^-$	0.431
$Ru^{2+} + 2\,e \rightleftharpoons Ru$	0.455
$Ru^{3+} + e \rightleftharpoons Ru^{2+}$	0.2487
$RuO_2 + 4\,H^+ + 2\,e \rightleftharpoons Ru^{2+} + 2\,H_2O$	1.120
$RuO_4^- + e \rightleftharpoons RuO_4^{2-}$	0.59
$RuO_4 + e \rightleftharpoons RuO_4^-$	1.00
$S + 2\,e \rightleftharpoons S^{2-}$	−0.47627
$S + 2H^+ + 2\,e \rightleftharpoons H_2S(aq)$	0.142
$S + H_2O + 2\,e \rightleftharpoons HS^- + OH^-$	−0.478
$2\,S + 2\,e \rightleftharpoons S_2^{2-}$	−0.42836
$S_2O_6^{2-} + 4\,H^+ + 2\,e \rightleftharpoons 2\,H_2SO_3$	0.564
$S_2O_8^{2-} + 2\,e \rightleftharpoons 2\,SO_4^{2-}$	2.010
$S_2O_8^{2-} + 2\,H^+ + 2\,e \rightleftharpoons 2\,HSO_4^-$	2.123
$S_4O_6^{2-} + 2\,e \rightleftharpoons 2\,S_2O_3^{2-}$	0.08
$2\,H_2SO_3 + H^+ + 2\,e \rightleftharpoons HS_2O_4^- + 2\,H_2O$	−0.056
$H_2SO_3 + 4\,H^+ + 4\,e \rightleftharpoons S + 3\,H_2O$	0.449
$2\,SO_3^{2-} + 2\,H_2O + 2\,e \rightleftharpoons S_2O_4^{2-} + 4\,OH^-$	−1.12
$2\,SO_3^{2-} + 3\,H_2O + 4\,e \rightleftharpoons S_2O_3^{2-} + 6\,OH^-$	−0.571
$SO_4^{2-} + 4\,H^+ + 2\,e \rightleftharpoons H_2SO_3 + H_2O$	0.172
$2\,SO_4^{2-} + 4\,H^+ + 2\,e \rightleftharpoons S_2O_6^{2-} + H_2O$	−0.22
$SO_4^{2-} + H_2O + 2\,e \rightleftharpoons SO_3^{2-} + 2\,OH^-$	−0.93
$Sb + 3\,H^+ + 3\,e \rightleftharpoons SbH_3$	−0.510
$Sb_2O_3 + 6\,H^+ + 6\,e \rightleftharpoons 2\,Sb + 3\,H_2O$	0.152
Sb_2O_5 (senarmontite) $+ 4\,H^+ + 4\,e \rightleftharpoons Sb_2O_3 + 2\,H_2O$	0.671
Sb_2O_5 (valentinite) $+ 4\,H^+ + 4\,e \rightleftharpoons Sb_2O_3 + 2\,H_2O$	0.649
$Sb_2O_5 + 6\,H^+ + 4\,e \rightleftharpoons 2\,SbO^+ + 3\,H_2O$	0.581
$SbO^+ + 2\,H^+ + 3\,e \rightleftharpoons Sb + 2\,H_2O$	0.212
$SbO_2^- + 2\,H_2O + 3\,e \rightleftharpoons Sb + 4\,OH^-$	−0.66
$SbO_3^- + H_2O + 2\,e \rightleftharpoons SbO_2^- + 2\,OH^-$	−0.59
$Sc^{3+} + 3\,e \rightleftharpoons Sc$	−2.077
$Se + 2\,e \rightleftharpoons Se^{2-}$	−0.924
$Se + 2\,H^+ + 2\,e \rightleftharpoons H_2Se(aq)$	−0.399
$H_2SeO_3 + 4\,H^+ + 4\,e \rightleftharpoons Se + 3\,H_2O$	−0.74
$SeO_3^{2-} + 3\,H_2O + 4\,e \rightleftharpoons Se + 6\,OH^-$	−0.366
$SeO_4^{2-} + 4\,H^+ + 2\,e \rightleftharpoons H_2SeO_3 + H_2O$	1.151
$SeO_4^{2-} + H_2O + 2\,e \rightleftharpoons SeO_3^{2-} + 2\,OH^-$	0.05
$SiF_6^{2-} + 4\,e \rightleftharpoons Si + 6\,F^-$	−1.24
SiO_2 (quartz) $+ 4\,H^+ + 4\,e \rightleftharpoons Si + 2\,H_2O$	0.857
$SiO_3^{2-} + 3\,H_2O + 4\,e \rightleftharpoons Si + 6\,OH^-$	−1.697
$Sn^{2+} + 2\,e \rightleftharpoons Sn$	−0.1375
$Sn^{4+} + 2\,e \rightleftharpoons Sn^{2+}$	0.151
$HSnO_2^- + H_2O + 2\,e \rightleftharpoons Sn + 3\,OH^-$	−0.909
$Sn(OH)_6^{2-} + 2\,e \rightleftharpoons HSnO_2^- + 3\,OH^- + H_2O$	−0.93
$Sr^+ + e \rightleftharpoons Sr$	−4.10
$Sr^{2+} + 2\,e \rightleftharpoons Sr$	−2.89
$Sr^{2+} + 2\,e \rightleftharpoons Sr(Hg)$	−1.793
$Sr(OH)_2 + 2\,e \rightleftharpoons Sr + 2\,OH^-$	−2.88
$Ta_2O_5 + 10\,H^+ + 10\,e \rightleftharpoons 2\,Ta + 5\,H_2O$	−0.750
$Tc^{2+} + 2\,e \rightleftharpoons Tc$	0.400
$TcO_4^- + 4\,H^+ + 3\,e \rightleftharpoons TcO_2 + 2\,H_2O$	0.782
$Te + 2\,e \rightleftharpoons Te^{2-}$	−1.143
$Te + 2\,H^+ + 2\,e \rightleftharpoons H_2Te$	−0.793
$Te^{4+} + 4\,e \rightleftharpoons Te$	0.568
$TeO_2 + 4\,H^+ + 4\,e \rightleftharpoons Te + 2\,H_2O$	0.593
$TeO_3^{2-} + 3\,H_2O + 4\,e \rightleftharpoons Te + 6\,OH^-$	−0.57
$TeO_4^- + 8\,H^+ + 7\,e \rightleftharpoons Te + 4\,H_2O$	0.472
$H_6TeO_6 + 2\,H^+ + 2\,e \rightleftharpoons TeO_2 + 4\,H_2O$	1.02
$Th^{4+} + 4\,e \rightleftharpoons Th$	−1.899
$ThO_2 + 4\,H^+ + 4\,e \rightleftharpoons Th + 2\,H_2O$	−1.789
$Th(OH)_4 + 4\,e \rightleftharpoons Th + 4\,OH^-$	−2.48
$Ti^{2+} + 2\,e \rightleftharpoons Ti$	−1.630
$Ti^{3+} + e \rightleftharpoons Ti^{2+}$	−0.368
$TiO_2 + 4\,H^+ + 2\,e \rightleftharpoons Ti^{2+} + 2\,H_2O$	−0.502
$TiOH^{3+} + H^+ + e \rightleftharpoons Ti^{3+} + H_2O$	−0.055
$Tl^+ + e \rightleftharpoons Tl$	−0.336
$Tl^+ + e \rightleftharpoons Tl(Hg)$	−0.3338
$Tl^{3+} + 2\,e \rightleftharpoons Tl^+$	1.252
$TlBr + e \rightleftharpoons Tl + Br^-$	−0.658
$TlCl + e \rightleftharpoons Tl + Cl^-$	−0.5568
$TlI + e \rightleftharpoons Tl + I^-$	−0.752
$Tl_2O_3 + 3\,H_2O + 4\,e \rightleftharpoons 2\,Tl^+ + 6\,OH^-$	0.02
$TlOH + e \rightleftharpoons Tl + OH^-$	−0.34
$Tl(OH)_3 + 2\,e \rightleftharpoons TlOH + 2\,OH^-$	−0.05
$Tl_2SO_4 + 2\,e \rightleftharpoons Tl + SO_4^{2-}$	−0.4360
$U^{3+} + 3\,e \rightleftharpoons U$	−1.798
$U^{4+} + e \rightleftharpoons U^{3+}$	−0.607
$UO_2^+ + 4\,H^+ + e \rightleftharpoons U^{4+} + 2\,H_2O$	0.612
$UO_2^{2+} + e \rightleftharpoons UO^{+}_{2}$	0.062
$UO_2^{2+} + 4\,H^+ + 2\,e \rightleftharpoons U^{4+} + 2\,H_2O$	0.327
$UO_2^{2+} + 4\,H^+ + 6\,e \rightleftharpoons U + 2\,H_2O$	−1.444
$V^{2+} + 2\,e \rightleftharpoons V$	−1.175
$V^{3+} + e \rightleftharpoons V^{2+}$	−0.255
$VO^{2+} + 2\,H^+ + e \rightleftharpoons V^{3+} + H_2O$	0.337
$VO_2^+ + 2\,H^+ + e \rightleftharpoons VO^{2+} + H_2O$	0.991
$V(OH)_4^+ + 2\,H^+ + e \rightleftharpoons VO^{2+} + 3\,H_2O$	1.00
$V(OH)_4^+ + 4\,H^+ + 5\,e \rightleftharpoons V + 4\,H_2O$	−0.254
$W_2O_5 + 2\,H^+ + 2\,e \rightleftharpoons 2\,WO_2 + H_2O$	−0.031

Table 1 (continued)
ALPHABETICAL LISTING

Reaction	E°, V
$WO_2 + 4\,H^+ + 4\,e \rightleftharpoons W + 2\,H_2O$	−0.119
$WO_3 + 6\,H^+ + 6\,e \rightleftharpoons W + 3\,H_2O$	−0.090
$2\,WO_3 + 2\,H^+ + 2\,e \rightleftharpoons W_2O_5 + H_2O$	−0.029
$Y^{3+} + 3\,e \rightleftharpoons Y$	−2.372
$Zn^{2+} + 2\,e \rightleftharpoons Zn$	−0.7618
$Zn^{2+} + 2\,e \rightleftharpoons Zn(Hg)$	−0.7628
$ZnO_2^{2-} + 2\,H_2O + 2\,e \rightleftharpoons Zn + 4\,OH^-$	−1.215
$ZnSO_4 \cdot 7H_2O + 2\,e \rightleftharpoons Zn(Hg) + SO_4^{2-}$ (Sat'd $ZnSO_4$)	−0.7993
$ZrO_2 + 4\,H^+ + 4\,e \rightleftharpoons Zr + 2\,H_2O$	−1.553
$ZrO(OH)_2 + H_2O + 4\,e \rightleftharpoons Zr + 4\,OH^-$	−2.36

Table 2
REDUCTION REACTIONS HAVING E° VALUES MORE POSITIVE THAN THAT OF THE STANDARD HYDROGEN ELECTRODE

Reaction	E°, V
$2\,H^+ + 2\,e \rightleftharpoons H_2$	0.00000
$CuI_2^- + e \rightleftharpoons Cu + 2\,I^-$	0.00
$Ge^{4+} + 2\,e \rightleftharpoons Ge^{2+}$	0.00
$NO_3^- + H_2O + 2\,e \rightleftharpoons NO_2^- + 2\,OH^-$	0.01
$Tl_2O_3 + 3\,H_2O + 4\,e \rightleftharpoons 2\,Tl^+ + 6\,OH^-$	0.02
$SeO_4^{2-} + H_2O + 2\,e \rightleftharpoons SeO_3^{2-} + 2\,OH^-$	0.05
$UO_2^{2+} + e \rightleftharpoons UO_2^+$	0.062
$Pd(OH)_2 + 2\,e \rightleftharpoons Pd + 2\,OH^-$	0.07
$AgBr + e \rightleftharpoons Ag + Br^-$	0.07133
$S_4O_6^{2-} + 2\,e \rightleftharpoons 2\,S_2O_3^{2-}$	0.08
$AgSCN + e \rightleftharpoons Ag + SCN^-$	0.8951
$N_2 + 2\,H_2O + 6\,H^+ + 6\,e \rightleftharpoons 2\,NH_4OH$	0.092
$HgO + H_2O + 2\,e \rightleftharpoons Hg + 2\,OH^-$	0.0977
$Ir_2O_3 + 3\,H_2O + 6\,e \rightleftharpoons 2\,Ir + 6\,OH^-$	0.098
$2\,NO + 2\,e \rightleftharpoons N_2O_2^{2-}$	0.10
$[Co(NH_3)_6]^{3+} + e \rightleftharpoons [Co(NH_3)_6]^{2+}$	0.108
$Hg_2O + H_2O + 2\,e \rightleftharpoons 2\,Hg + 2\,OH^-$	0.123
$Ge^{4+} + 4\,e \rightleftharpoons Ge$	0.124
$Hg_2Br_2 + 2\,e \rightleftharpoons 2\,Hg + 2\,Br^-$	0.13923
$Pt(OH)_2 + 2\,e \rightleftharpoons Pt + 2\,OH^-$	0.14
$S + 2H^+ + 2\,e \rightleftharpoons H_2S(aq)$	0.142
$Np^{4+} + e \rightleftharpoons Np^{3+}$	0.147
$Ag_4[Fe(CN)_6] + 4\,e \rightleftharpoons 4\,Ag + [Fe(CN)_6]^{4-}$	0.1478
$IO_3^- + 2\,H_2O + 4\,e \rightleftharpoons IO^- + 4\,OH^-$	0.15
$Mn(OH)_3 + e \rightleftharpoons Mn(OH)_2 + OH^-$	0.15
$2\,NO_2^- + 3\,H_2O + 4\,e \rightleftharpoons N_2O + 6\,OH^-$	0.15
$Sn^{4+} + 2\,e \rightleftharpoons Sn^{2+}$	0.151
$Sb_2O_3 + 6\,H^+ + 6\,e \rightleftharpoons 2\,Sb + 3\,H_2O$	0.152
$Cu^{2+} + e \rightleftharpoons Cu^+$	0.153
$BiOCl + 2\,H^+ + 3\,e \rightleftharpoons Bi + Cl^- + H_2O$	0.1583
$Bi(Cl)_4^- + 3\,e \rightleftharpoons Bi + 4\,Cl^-$	0.16
$Co(OH)_3 + e \rightleftharpoons Co(OH)_2 + OH^-$	0.17
$SO_4^{2-} + 4\,H^+ + 2\,e \rightleftharpoons H_2SO_3 + H_2O$	0.172
$SbO^+ + 2\,H^+ + 3\,e \rightleftharpoons Sb + 2\,H_2O$	0.212
$AgCl + e \rightleftharpoons Ag + Cl^-$	0.22233
$As_2O_3 + 6\,H^+ + 6\,e \rightleftharpoons 2\,As + 3\,H_2O$	0.234
Calomel electrode, saturated NaCl (SSCE)	0.2360
$Ge^{2+} + 2\,e \rightleftharpoons Ge$	0.24
Calomel electrode, saturated KCl	0.2412
$PbO_2 + H_2O + 2\,e \rightleftharpoons PbO + 2\,OH^-$	0.247
$HAsO_2 + 3\,H^+ + 3_e \rightleftharpoons As + 2\,H_2O$	0.248
$Ru^{3+} + e \rightleftharpoons Ru^{2+}$	0.2487
$ReO_2 + 4\,H^+ + 4\,e \rightleftharpoons Re + 2\,H_2O$	0.2513
$IO_3^- + 3\,H_2O + 6\,e \rightleftharpoons I^- + OH^-$	0.26
$Hg_2Cl_2 + 2\,e \rightleftharpoons 2\,Hg + 2\,Cl^-$	0.26808
Calomel electrode, molal KCl	0.2800
Calomel electrode, 1 mol/l KCl (NCE)	0.2801
$Re^{3+} + 3\,e \rightleftharpoons Re$	0.300
$BiO^+ + 2\,H^+ + 3\,e \rightleftharpoons Bi + H_2O$	0.320
$UO_2^{2+} + 4\,H^+ + 2\,e \rightleftharpoons U^{4+} + 2\,H_2O$	0.327
$ClO_3^- + H_2O + 2\,e \rightleftharpoons ClO_2^- + 2\,OH^-$	0.33
$2\,HCNO + 2\,H^+ + 2\,e \rightleftharpoons (CN)_2 + 2\,H_2O$	0.330
Calomel electrode, 0.1 mol/l KCl	0.3337
$VO^{2+} + 2\,H^+ + e \rightleftharpoons V^{3+} + H_2O$	0.337
$Cu^{2+} + 2\,e \rightleftharpoons Cu$	0.3419
$Ag_2O + H_2O + 2\,e \rightleftharpoons 2\,Ag + 2\,OH^-$	0.342
$Cu^{2+} + 2\,e \rightleftharpoons Cu(Hg)$	0.345
$AgIO_3 + e \rightleftharpoons Ag + IO^-3$	0.354
$[Fe(CN)_6]^{3-} + e \rightleftharpoons [Fe(CN)_6]^{4-}$	0.358
$ClO_4^- + H_2O + 2\,e \rightleftharpoons ClO^-_3 + 2\,OH^-$	0.36
$Ag_2SeO_3 + 2\,e \rightleftharpoons 2\,Ag + SeO_3^{2-}$	0.3629
$ReO_4^- + 8\,H^+ + 7\,e \rightleftharpoons Re + 4\,H_2O$	0.368
$(CN)_2 + 2\,H^+ + 2\,e \rightleftharpoons 2\,HCN$	0.373
$[Ferricinium]^+ + e \rightleftharpoons$ ferrocene	0.400
$Tc^{2+} + 2\,e \rightleftharpoons Tc$	0.400
$O_2 + 2\,H_2O + 4\,e \rightleftharpoons 4\,OH^-$	0.401
$AgOCN + e \rightleftharpoons Ag + OCN^-$	0.41
$[RhCl_6]^{3-} + 3\,e \rightleftharpoons Rh + 6\,Cl^-$	0.431
$Ag_2CrO_4 + 2\,e \rightleftharpoons 2\,Ag + CrO_4^{2-}$	0.4470
$H_2SO_3 + 4\,H^+ + 4\,e \rightleftharpoons S + 3\,H_2O$	0.449
$Ru^{2+} + 2\,e \rightleftharpoons Ru$	0.455
$Ag_2MoO_4 + 2\,e \rightleftharpoons 2\,Ag + MoO_4^{2-}$	0.4573
$Ag_2C_2O_4 + 2\,e \rightleftharpoons 2\,Ag + C_2O_4^{2-}$	0.4647
$Ag_2WO_4 + 2\,e \rightleftharpoons 2\,Ag + WO_4^{2-}$	0.4660
$Ag_2CO_3 + 2\,e \rightleftharpoons 2\,Ag + CO_3^{2-}$	0.47
$TeO_4^- + 8\,H^+ + 7\,e \rightleftharpoons Te + 4\,H_2O$	0.472
$IO^- + H_2O + 2\,e \rightleftharpoons I^- + 2\,OH^-$	0.485
$ReO_4^- + 4\,H^+ + 3\,e \rightleftharpoons ReO_2 + 2\,H_2O$	0.510
$Hg_2(ac)_2 + 2\,e \rightleftharpoons 2\,Hg + 2\,(ac)^-$	0.51163
$Cu^+ + e \rightleftharpoons Cu$	0.521
$I_2 + 2\,e \rightleftharpoons 2\,I^-$	0.5355
$I_3^- + 2\,e \rightleftharpoons 3\,I^-$	0.536
$AgBrO_3 + e \rightleftharpoons Ag + BrO_3^-$	0.546
$MnO_4^- + e \rightleftharpoons MnO_4^{2-}$	0.558
$H_3AsO_4 + 2\,H^+ + 2\,e^- \rightleftharpoons HAsO_2 + 2\,H_2O$	0.560
$S_2O_6^{2-} + 4\,H^+ + 2\,e \rightleftharpoons 2\,H_2SO_3$	0.564
$AgNO_2 + e \rightleftharpoons Ag + NO_2^-$	0.564
$Te^{4+} + 4\,e \rightleftharpoons Te$	0.568
$Sb_2O_5 + 6\,H^+ + 4\,e \rightleftharpoons 2\,SbO^+ + 3\,H_2O$	0.581
$RuO_4^- + e \rightleftharpoons RuO_4^{2+}$	0.59
$[PdCl_4]^{2-} + 2\,e \rightleftharpoons Pd + 4\,Cl^-$	0.591

Table 2 (continued)
REDUCTION REACTIONS HAVING E° VALUES MORE POSITIVE THAN THAT OF THE STANDARD HYDROGEN ELECTRODE

Reaction	E°, V
$TeO_2 + 4 H^+ + 4 e \rightleftharpoons Te + 2 H_2O$	0.593
$MnO_4^- + 2 H_2O + 3 e \rightleftharpoons MnO_2 + 4 OH^-$	0.595
$Rh^{2+} + 2 e \rightleftharpoons Rh$	0.600
$Rh^+ + e \rightleftharpoons Rh$	0.600
$MnO_4^{2-} + 2 H_2O + 2 e \rightleftharpoons MnO_2 + 4 OH^-$	0.60
$2 AgO + H_2O + 2 e \rightleftharpoons Ag_2O + 2 OH^-$	0.607
$BrO_3^- + 3 H_2O + 6 e \rightleftharpoons Br^- + 6 OH^-$	0.61
$UO_2^+ + 4 H^+ + e \rightleftharpoons U^{4+} + 2 H_2O$	0.612
$Hg_2SO_4 + 2 e \rightleftharpoons 2 Hg + SO_4^{2-}$	0.6125
$ClO_3^- + 3 H_2O + 6 e \rightleftharpoons Cl^- + 6 OH^-$	0.62
$Hg_2HPO_4 + 2 e \rightleftharpoons 2 Hg + HPO_4^{2-}$	0.6359
$Ag(ac) + e \rightleftharpoons Ag + (ac)^-$	0.643
Sb_2O_5(valentinite) $+ 4 H^+ + 4 e \rightleftharpoons Sb_2O_3 + 2 H_2O$	0.649
$Ag_2SO_4 + 2 e \rightleftharpoons 2 Ag + SO_4^{2-}$	0.654
$ClO_2^- + H_2O + 2 e \rightleftharpoons ClO^- + 2 OH^-$	0.66
Sb_2O_5(senarmontite) $+ 4 H^+ + 4 e \rightleftharpoons Sb_2O_3 + 2 H_2O$	0.671
$[PtCl_6]^{2-} + 2 e \rightleftharpoons [PtCl_4]^{2-} + 2 Cl^-$	0.68
$O_2 + 2 H^+ + 2 e \rightleftharpoons H_2O_2$	0.695
p-benzoquinone $+ 2 H^+ + 2 e \rightleftharpoons$ hydroquinone	0.6992
$H_3IO_6 + 2 e \rightleftharpoons IO_3^- + 3 OH^-$	0.7
$Ag_2O_3 + H_2O + 2 e \rightleftharpoons 2 AgO + 2 OH^-$	0.739
$[PtCl_4]^{2-} + 2 e \rightleftharpoons Pt + 4 Cl^-$	0.755
$Rh^{3+} + 3 e \rightleftharpoons Rh$	0.758
$ClO_2^- + 2 H_2O + 4 e \rightleftharpoons Cl^- + 4 OH^-$	0.76
$2 NO + H_2O + 2 e \rightleftharpoons N_2O + 2 OH^-$	0.76
$BrO^- + H_2O + 2 e \rightleftharpoons Br^- + 2 OH^-$	0.761
$ReO_4^- + 2 H^+ + e \rightleftharpoons ReO_3 + H_2O$	0.768
$(CNS)_2 + 2 e \rightleftharpoons 2 CNS^-$	0.77
$[IrCl_6]^{3-} + 3 e \rightleftharpoons Ir + 6 Cl^-$	0.77
$Fe^{3+} + e \rightleftharpoons Fe^{2+}$	0.771
$Ag(F) + e \rightleftharpoons Ag + F^-$	0.779
$TcO_4^- + 4 H^+ + 3 e \rightleftharpoons TcO_2 + 2 H_2O$	0.782
$Hg_2^{2+} + 2 e \rightleftharpoons 2 Hg$	0.7973
$Ag^+ + e \rightleftharpoons Ag$	0.7996
$2 NO_3^- + 4 H^+ + 2 e \rightleftharpoons N_2O_4 + 2 H_2O$	0.803
$ClO^- + H_2O + 2 e \rightleftharpoons Cl^- + 2 OH^-$	0.841
$OsO_4 + 8 H^+ + 8 e \rightleftharpoons Os + 4 H_2O$	0.85
$Hg^{2+} + 2 e \rightleftharpoons Hg$	0.851
$AuBr_4^- + 3 e \rightleftharpoons Au + 4 Br^-$	0.854
SiO_2(quartz) $+ 4 H^+ + 4 e \rightleftharpoons Si + 2 H_2O$	0.857
$2 HNO_2 + 4 H^+ + 4 e \rightleftharpoons H_2N_2O_2 + H_2O$	0.86
$[IrCl_6]^{2-} + e \rightleftharpoons [IrCl_6]^{3-}$	0.8665
$N_2O_4 + 2 e \rightleftharpoons 2 NO_2^-$	0.867
$HO_2^- + H_2O + 2 e \rightleftharpoons 3 OH^-$	0.878
$2 Hg^{2+} + 2 e \rightleftharpoons Hg_2^{2+}$	0.920
$NO_3^- + 3 H^+ + 2 e \rightleftharpoons HNO_2 + H_2O$	0.934
$Pd^{2+} + 2 e \rightleftharpoons Pd$	0.951
$ClO_2(aq) + e \rightleftharpoons ClO_2^-$	0.954
$NO_3^- + 4 H^+ + 3 e \rightleftharpoons NO + 2 H_2O$	0.957
$AuBr_2^- + e \rightleftharpoons Au + 2 Br^-$	0.959
$HNO_2 + H^+ + e \rightleftharpoons NO + H_2O$	0.983
$HIO + H^+ + 2 e \rightleftharpoons I^- + H_2O$	0.987

Reaction	E°, V
$VO_2^+ + 2 H^+ + e \rightleftharpoons VO^{2+} + H_2O$	0.991
$RuO_4 + e \rightleftharpoons RuO_4^-$	1.00
$V(OH)_4^+ + 2 H^+ + e \rightleftharpoons VO^{2+} + 3 H_2O$	1.00
$AuCl_4^- + 3 e \rightleftharpoons Au + 4 Cl^-$	1.002
$Pu^{4+} + e \rightleftharpoons Pu^{3+}$	1.006
$H_6TeO_6 + 2 H^+ + 2 e \rightleftharpoons TeO_2 + 4 H_2O$	1.02
$N_2O_4 + 4 H^+ + 4 e \rightleftharpoons 2 NO + 2 H_2O$	1.035
$[Fe(phen)_3]^{3+} + e \rightleftharpoons [Fe(phen)_3]^{2+}$ (1 (mol/ℓ H_2SO_4)	1.06
$PuO_2(OH)_2 + H^+ + e \rightleftharpoons PuO_2OH + H_2O$	1.062
$N_2O_4 + 2 H^+ + 2 e \rightleftharpoons 2 HNO_2$	1.065
$Br_2(l) + 2 e \rightleftharpoons 2 Br^-$	1.066
$IO_3^- + 6 H^+ + 6 e \rightleftharpoons I^- + 3 H_2O$	1.085
$Br_2(aq) + 2 e \rightleftharpoons 2 Br^-$	1.0873
$Pu^{5+} + e \rightleftharpoons Pu^{4+}$	1.099
$Cu^{2+} + 2 CN^- + e \rightleftharpoons [Cu(CN)_2]^-$	
$Pt^{2+} + 2 e \rightleftharpoons Pt$	1.118
$RuO_2 + 4 H^+ + 2 e \rightleftharpoons Ru^{2+} + 2 H_2O$	1.120
$[Fe(phenanthroline)_3]^{3+} + e \rightleftharpoons [Fe(phen)_3]^{2+}$	1.147
$SeO_4^{2-} + 4 H^+ + 2 e \rightleftharpoons H_2SeO_3 + H_2O$	1.151
$ClO_3^- + 2 H^+ + e \rightleftharpoons ClO_2 + H_2O$	1.152
$Ir^{3+} + 3 e \rightleftharpoons Ir$	1.156
$ClO_4^- + 2 H^+ + 2 e \rightleftharpoons ClO_3^- + H_2O$	1.189
$2 IO_3^- + 12 H^+ + 10 e \rightleftharpoons I_2\ 6 H_2O$	1.195
$ClO^-_3 + 3 H^+ + 2 e \rightleftharpoons HClO_2 + H_2O$	1.214
$MnO_2 + 4 H^+ + 2 e \rightleftharpoons Mn^{2+} + 2 H_2O$	1.224
$O_2 + 4 H^+ + 4 e \rightleftharpoons 2 H_2O$	1.229
$Cr_2O_7^{2-} + 14 H^+ + 6 e \rightleftharpoons 2 Cr^{3+} + 7 H_2O$	1.232
$O_3 + H_2O + 2 e \rightleftharpoons O_2 + 2OH^-$	1.24
$Tl^{3+} + 2 e \rightleftharpoons Tl^+$	1.252
$N_2H_5^+ + 3 H^+ + 2 e \rightleftharpoons 2 NH_4^+$	1.275
$ClO_2 + H^+ + e \rightleftharpoons HClO_2$	1.277
$[PdCl_6]^{2-} + 2 e \rightleftharpoons [PdCl_4]^{2-} + 2 Cl^-$	1.288
$2 HNO_2 + 4 H^+ + 4 e \rightleftharpoons N_2O + 3 H_2O$	1.297
$PuO_2(OH)_2 + 2 H^+ + 2 e \rightleftharpoons Pu(OH)_4$	1.325
$HBrO + H^+ + 2 e \rightleftharpoons Br^- + H_2O$	1.331
$HCrO_4^- + 7 H^+ + 3 e \rightleftharpoons Cr^{3+} + 4 H_2O$	1.350
$Cl_2(g) + 2 e \rightleftharpoons Cl^-$	1.35827
$ClO_4^- + 8 H^+ + 8 e \rightleftharpoons Cl^- + 4 H_2O$	1.389
$ClO_4^- + 8 H^+ + 7 e \rightleftharpoons {}^1/_2Cl_2 + 4 H_2O$	1.39
$Au^{3+} + 2 e \rightleftharpoons Au^+$	1.401
$2 NH_3OH^+ + H^+ + 2 e \rightleftharpoons N_2H_5^+ + 2 H_2O$	1.42
$BrO_3^- + 6 H^+ + 6 e \rightleftharpoons Br^- + 3 H_2O$	1.423
$2 HIO + 2 H^+ + 2 e \rightleftharpoons I_2 + 2 H_2O$	1.439
$Au(OH)_3 + 3 H^+ + 3 e \rightleftharpoons Au^- + 3 H_2O$	1.45
$31O_3^- + 6 H^+ + 6 e \rightleftharpoons Cl^- + 3 H_2O$	1.451
$PbO_2 + 4 H^+ + 2 e \rightleftharpoons Pb^{2+} + 2 H_2O$	1.455
$ClO_3^- + 6 H^+ + 5 e \rightleftharpoons {}^1/_2Cl_2 + 3 H_2O$	1.47
$BrO_3^- + 6 H^+ + 5 e \rightleftharpoons {}^1/_2Br_2 + 3 H_2O$	1.482
$HClO + H^+ + 2 e \rightleftharpoons Cl^- + H_2O$	1.482
$HO_2 + H^+ + e \rightleftharpoons H_2O_2$	1.495
$Au^{3+} + 3 e \rightleftharpoons Au$	1.498
$MnO_4^- + 8 H^+ + 5 e \rightleftharpoons Mn^{2+} + 4 H_2O$	1.507
$Mn^{3+} + e \rightleftharpoons Mn^{2+}$	1.5415
$HClO_2 + 3 H^+ + 4 e \rightleftharpoons Cl^- + 2 H_2O$	1.570
$HBrO + H^+ + e \rightleftharpoons {}^1/_2Br_2(aq) + H_2O$	1.574

Table 2 (continued)
REDUCTION REACTIONS HAVING E° VALUES MORE POSITIVE THAN THAT OF THE STANDARD HYDROGEN ELECTRODE

Reaction	E°, V
$2\ NO + 2\ H^+ + 2\ e \rightleftharpoons N_2O + H_2O$	1.591
$Bi_2O_4 + 4\ H^+ + 2\ e \rightleftharpoons 2\ BiO^+ + 2\ H_2O$	1.593
$HBrO + H^+ + e \rightleftharpoons 1/2\ Br_2(\ell) + H_2O$	1.596
$H_5IO_6 + H^+ + 2\ e \rightleftharpoons IO_3^- + 3\ H_2O$	1.601
$Ce^{4+} + e \rightleftharpoons Ce^{3+}$	1.61
$HClO + H^+ + e \rightleftharpoons 1/2Cl_2 + H_2O$	1.611
$HClO_2 + 3\ H^+ + 3\ e \rightleftharpoons 1/2Cl_2 + 2\ H_2O$	1.628
$HClO_2 + 2\ H^+ + 2\ e \rightleftharpoons HClO + H_2O$	1.645
$NiO_2 + 4\ H^+ + 2\ e \rightleftharpoons Ni^{2+} + 2\ H_2O$	1.678
$MnO_4^- + 4\ H^+ + 3\ e \rightleftharpoons MnO_2 + 2\ H_2O$	1.679
$PbO_2 + SO_4^{2-} + 4\ H^+ + 2\ e \rightleftharpoons PbSO_4 + 2\ H_2O$	1.6913
$Au^+ + e \rightleftharpoons Au$	1.692
$CeOH^{3+} + H^+ + e \rightleftharpoons Ce^{3+} + H_2O$	1.715
$N_2O + 2\ H^+ + 2\ e \rightleftharpoons N_2 + H_2O$	1.766
$H_2O_2 + 2\ H^+ + 2\ e \rightleftharpoons 2\ H_2O$	1.776
$Co^{3+} + e \rightleftharpoons Co^{2+}(2\ mol/\ell\ H_2SO_4)$	1.83
$Ag^{2+} + e \rightleftharpoons Ag^+$	1.980
$S_2O_8^{2-} + 2\ e \rightleftharpoons 2\ SO_4^{2-}$	2.010
$OH + e \rightleftharpoons OH^-$	2.02
$O_3 + 2\ H^+ + 2\ e \rightleftharpoons O_2 + H_2O$	2.076
$S_2O_8^{2-} + 2\ H^+ + 2\ e \rightleftharpoons 2\ HSO_4^-$	2.123
$F_2O + 2\ H^+ + 4\ e \rightleftharpoons H_2O + 2\ F^-$	2.153
$FeO_4^{2-} + 8\ H^+ + 3\ e \rightleftharpoons Fe^{3+} + 4\ H_2O$	2.20
$O(g) + 2\ H^+ + 2\ e \rightleftharpoons H_2O$	2.421
$H_2N_2O_2 + 2\ H^+ + 2\ e \rightleftharpoons N_2 + 2\ H_2O$	2.65
$F_2 + 2\ e \rightleftharpoons 2\ F^-$	2.866
$F_2 + 2\ H^+ + 2\ e \rightleftharpoons 2\ HF$	3.053

Table 3
REDUCTION REACTIONS HAVING E° VALUES MORE NEGATIVE THAN THAT OF THE STANDARD HYDROGEN ELECTRODE

Reaction	E°, V
$2\ H^+ + 2\ e \rightleftharpoons H_2$	0.00000
$AgCN + e \rightleftharpoons Ag + CN^-$	−0.017
$2\ WO_3 + 2\ H^+ + 2\ e \rightleftharpoons W_2O_5 + H_2O$	−0.029
$W_2O_5 + 2\ H^+ + 2\ e \rightleftharpoons 2\ WO_2 + H_2O$	−0.031
$D^+ + e \rightleftharpoons 1/2D_2$	−0.0034
$Ag_2S + 2\ H^+ + 2\ e \rightleftharpoons 2\ Ag + H_2S$	−0.0366
$Fe^{3+} + 3\ e \rightleftharpoons Fe$	−0.037
$Hg_2I_2 + 2\ e \rightleftharpoons 2\ Hg + 2\ I^-$	−0.0405
$2\ D^+ + 2\ e \rightleftharpoons D_2$	−0.044
$Tl(OH)_3 + 2\ e \rightleftharpoons TlOH + 2\ OH^-$	−0.05
$TiOH^{3+} + H^+ + e \rightleftharpoons Ti^{3+} + H_2O$	−0.055
$2\ H_2SO_3 + H^+ + 2\ e \rightleftharpoons HS_2O_4^- + 2\ H_2O$	−0.056
$P(white) + 3\ H^+ + 3\ e \rightleftharpoons PH_3(g)$	−0.063
$O_2^- + H_2O + 2\ e \rightleftharpoons HO_2^- + OH^-$	−0.076
$2\ Cu(OH)_2 + 2\ e \rightleftharpoons Cu_2O + 2\ OH^- + H_2O$	−0.080
$WO_3 + 6\ H^+ + 6\ e \rightleftharpoons W + 3\ H_2O$	−0.090
$P(red) + 3\ H^+ + 3\ e \rightleftharpoons PH_3(g)$	−0.111
$GeO_2 + 2\ H^+ + 2\ e \rightleftharpoons GeO + H_2O$	−0.118
$WO_2 + 4\ H^+ + 4\ e \rightleftharpoons W + 2\ H_2O$	−0.119
$Pb^{2+} + 2\ e \rightleftharpoons Pb(Hg)$	−0.1205
$Pb^{2+} + 2\ e \rightleftharpoons Pb$	−0.1262
$CrO_4^{2-} + 4\ H_2O + 3\ e \rightleftharpoons Cr(OH)_3 + 5\ OH^-$	−0.13
$Sn^{2-} + 2\ e \rightleftharpoons Sn$	−0.1375
$In^+ + e \rightleftharpoons In$	−0.14
$O_2 + 2\ H_2O + 2\ e \rightleftharpoons H_2O_2 + 2\ OH^-$	−0.146
$AgI + e \rightleftharpoons Ag + I^-$	−0.15224
$2\ NO_2^- + 2\ H_2O + 4\ e \rightleftharpoons N_2O_2^{2-} + 4\ OH^-$	−0.18
$H_2GeO_3 + 4\ H^+ + 4\ e \rightleftharpoons Ge + 3\ H_2O$	−0.182
$CO_2 + 2\ H^+ + 2\ e \rightleftharpoons HCOOH$	−0.199
$Mo^{3+} + 3\ e \rightleftharpoons Mo$	−0.200
$2\ SO_2^{2-} + 4\ H^+ + 2\ e \rightleftharpoons S_2O_6^{2-} + H_2O$	−0.22
$Cu(OH)_2 + 2\ e \rightleftharpoons Cu + 2\ OH^-$	−0.222
$CdSO_4 + 2\ e \rightleftharpoons Cd + SO_4^{2-}$	−0.246
$V(OH)_4^+ + 4\ H^+ + 5\ e \rightleftharpoons V + 4\ H_2O$	−0.254
$V^{3+} + e \rightleftharpoons V^{2+}$	−0.255
$Ni^{2+} + 2\ e \rightleftharpoons Ni$	−0.257
$PbCl_2 + 2\ e \rightleftharpoons Pb + 2\ Cl^-$	−0.2675
$H_3PO_4 + 2\ H^+ + 2\ e \rightleftharpoons H_3PO_3 + H_2O$	−0.276
$Co^{2+} + 2\ e \rightleftharpoons Co$	−0.28
$PbBr_2 + 2\ e \rightleftharpoons Pb + 2\ Br^-$	−0.284
$Tl^+ + e \rightleftharpoons Tl(Hg)$	−0.3338
$Tl^+ + e \rightleftharpoons Tl$	−0.336
$In^{3+} + 3\ e \rightleftharpoons In$	−0.3382
$TlOH + e \rightleftharpoons Tl + OH^-$	−0.34
$PbF_2 + 2\ e \rightleftharpoons Pb + 2\ F^-$	−0.3444
$PbSO_4 + 2\ e \rightleftharpoons Pb(Hg) + SO_4^{2-}$	−0.3505
$Cd^{2+} + 2\ e \rightleftharpoons Cd(Hg)$	−0.3521
$PbSO_4 + 2\ e \rightleftharpoons Pb + SO_4^{2-}$	−0.3588
$Cu_2O + H_2O + 2\ e \rightleftharpoons 2\ Cu + 2\ OH^-$	−0.360
$Eu^{3+} + e \rightleftharpoons Eu^{2+}$	−0.36
$PbI_2 + 2\ e \rightleftharpoons Pb + 2\ I^-$	−0.365
$SeO_3^{2-} + 3\ H_2O + 4\ e \rightleftharpoons Se + 6\ OH^-$	−0.366
$Ti^{3+} + e \rightleftharpoons Ti^{2+}$	−0.368
$Se + 2\ H^+ + 2\ e \rightleftharpoons H_2Se(aq)$	−0.399
$In^{2+} + e \rightleftharpoons In^+$	−0.40
$Cd^{2+} + 2\ e \rightleftharpoons Cd$	−0.4030
$Cr^{3+} + e \rightleftharpoons Cr^{2+}$	−0.407
$2\ S + 2\ e \rightleftharpoons S_2^{2-}$	−0.42836
$Tl_2SO_4 + 2\ e \rightleftharpoons Tl + SO_4^{2-}$	−0.4360
$In^{3+} + 2\ e \rightleftharpoons In^+$	−0.443
$Fe^{2+} + 2\ e \rightleftharpoons Fe$	−0.447
$H_3PO_3 + 3\ H^+ + 3\ e \rightleftharpoons P + 3\ H_2O$	−0.454
$Bi_2O_3 + 3\ H_2O + 6\ e \rightleftharpoons 2\ Bi + 6\ OH^-$	−0.46
$NO_2^- + H_2O + e \rightleftharpoons NO + 2\ OH$	−0.46
$PbHPO_4 + 2\ e \rightleftharpoons Pb + HPO_4^{2-}$	−0.465
$S + 2\ e \rightleftharpoons S^{2-}$	−0.47627
$S + H_2O + 2\ e \rightleftharpoons HS^- + OH^-$	−0.478
$NiO_2 + 2\ H_2O + 2\ e \rightleftharpoons Ni\ (OH)_2 + 2\ OH^-$	−0.490
$In^{3+} + e \rightleftharpoons In^{2+}$	−0.49
$H_3PO_3 + 2\ H^+ + 2\ e \rightleftharpoons H_3PO_2 + H_2O$	−0.499
$TiO_2 + 4\ H^+ + 2\ e \rightleftharpoons Ti^{2+} + 2\ H_2O$	−0.502
$H_3PO_2 + H^+ + e \rightleftharpoons P + 2\ H_2O$	−0.508

Table 3 (continued)
REDUCTION REACTIONS HAVING E° VALUES MORE NEGATIVE THAN THAT OF THE STANDARD HYDROGEN ELECTRODE

Reaction	E°, V
$Sb + 3 H^+ + 3 e \rightleftharpoons SbH_3$	−0.510
$HPbO_2^- + H_2O + 2 e \rightleftharpoons Pb + 3 OH^-$	−0.537
$TlCl + e \rightleftharpoons Tl + Cl^-$	−0.5568
$Ga^{3+} + 3 e \rightleftharpoons Ga$	−0.560
$Fe(OH)_3 + e \rightleftharpoons Fe(OH)_2 + OH^-$	−0.56
$TeO_3^{2-} + 3 H_2O + 4 e \rightleftharpoons Te + 6 OH^-$	−0.57
$2 SO_3^{2-} + 3 H_2O + 4 e \rightleftharpoons S_2O_3^{2-} + 6 OH^-$	−0.571
$PbO + H_2O + 2 e \rightleftharpoons Pb + 2 OH^-$	−0.580
$ReO_2^- + 4 H_2O + 7 e \rightleftharpoons Re + 8 OH^-$	−0.584
$SbO_3^- + H_2O + 2 e \rightleftharpoons SbO_2^- + 2 OH^-$	−0.59
$U^{4+} + e \rightleftharpoons U^{3+}$	−0.607
$As + 3 H^+ + 3 e \rightleftharpoons AsH_3$	−0.608
$Nb_2O_5 + 10 H^+ + 10 e \rightleftharpoons 2 Nb + 5 H_2O$	−0.644
$TlBr + e \rightleftharpoons Tl + Br^-$	−0.658
$SbO_2^- + 2 H_2O + 3 e \rightleftharpoons Sb + 4 OH^-$	−0.66
$AsO_2^- + 2 H_2O + 3 e \rightleftharpoons As + 4 OH^-$	−0.68
$Ag_2S + 2 e \rightleftharpoons 2 Ag + S^{2-}$	−0.691
$AsO_4^{3-} + 2 H_2O + 2 e \rightleftharpoons AsO_2^- + 4 OH^-$	−0.71
$Ni(OH)_2 + 2 e \rightleftharpoons Ni + 2 OH^-$	−0.72
$Co(OH)_2 + 2 e \rightleftharpoons Co + 2 OH^-$	−0.73
$H_2SeO_3 + 4 H^+ + 4 e \rightleftharpoons Se + 3 H_2O$	−0.74
$Cr^{3+} + 3 e \rightleftharpoons Cr$	−0.744
$Ta_2O_5 + 10 H^+ + 10 e \rightleftharpoons 2 Ta + 5 H_2O$	−0.750
$TlI + e \rightleftharpoons Tl + I^-$	−0.752
$Zn^{2+} + 2 e \rightleftharpoons Zn$	−0.7618
$Zn^{2+} + 2 e \rightleftharpoons Zn(Hg)$	−0.7628
$Te + 2 H^+ + 2 e \rightleftharpoons H_2Te$	−0.793
$ZnSO_4 \cdot 7H_2O + 2 e \rightleftharpoons Zn(Hg) + SO_4^{2-}$ (Sat'd $ZnSO_4$)	−0.7993
$Cd(OH)_2 + 2 e \rightleftharpoons Cd(Hg) + 2 OH^-$	−0.809
$2 H_2O + 2 e \rightleftharpoons H_2 + 2 OH^-$	−0.8277
$2 NO_3^- + 2 H_2O + 2 e \rightleftharpoons N_2O_4 + 4 OH^-$	−0.85
$H_3BO_3 + 3 H^+ + 3 e \rightleftharpoons B + 3 H_2O$	−0.8698
$P + 3 H_2O + 3 e \rightleftharpoons PH_3(g) + 3 OH^-$	−0.87
$HSnO_2^- + H_2O + 2 e \rightleftharpoons Sn + 3 OH^-$	−0.909
$Cr^{2+} + 2 e \rightleftharpoons Cr$	−0.913
$Se + 2 e \rightleftharpoons Se^{2-}$	−0.924
$SO_4^{2-} + H_2O + 2 e \rightleftharpoons SO_3^{2-} + 2 OH^-$	−0.93
$Sn(OH)_6^{2-} + 2 e \rightleftharpoons HSnO_2^- + 3 OH^- + H_2O$	−0.93
$NpO_2 + H_2O + H^+ + e \rightleftharpoons Np(OH)_3$	−0.962
$PO_4^{3-} + 2 H_2O + 2 e \rightleftharpoons HPO_3^{2-} + 3 OH^-$	−1.05
$Nb^{3+} + 3 e \rightleftharpoons Nb$	−1.099
$2 SO_3^{2-} + 2 H_2O + 2 e \rightleftharpoons S_2O_4^{2-} + 4 OH^-$	−1.12
$Te + 2 e \rightleftharpoons Te^{2-}$	−1.143
$V^{2+} + 2 e \rightleftharpoons V$	−1.175
$Mn^{2+} + 2 e \rightleftharpoons Mn$	−1.185
$CrO_2^- + 2 H_2O + 3 e \rightleftharpoons Cr + 4 OH^-$	−1.2
$ZnO_2^- + 2 H_2O + 2 e \rightleftharpoons Zn + 4 OH^-$	−1.215
$H_2GaO_3^- + H_2O + 3 e \rightleftharpoons Ga + 4 OH^-$	−1.219
$H_2BO_3^- + 5 H_2O + 8 e \rightleftharpoons BH_4^- + 8 OH^-$	−1.24
$SiF_6^{2-} + 4 e \rightleftharpoons Si + 6 F^-$	−1.24
$Ce^{3+} + 3 e \rightleftharpoons Ce(Hg)$	−1.4373
$UO_2^{2+} + 4 H^+ + 6 e \rightleftharpoons U + 2 H_2O$	−1.444
$Cr(OH)_3 + 3 e \rightleftharpoons Cr + 3 OH^-$	−1.48
$HfO_2 + 4 H^+ + 4 e \rightleftharpoons Hf + 2 H_2O$	−1.505
$ZrO_2 + 4 H^+ + 4 e \rightleftharpoons Zr + 2 H_2O$	−1.553
$Mn(OH)_2 + 2 e \rightleftharpoons Mn + 2 OH^-$	−1.56
$Ba^{2+} + 2 e \rightleftharpoons Ba(Hg)$	−1.570
$Ti^{2+} + 2 e \rightleftharpoons Ti$	−1.630
$HPO_3^{2-} + 2 H_2O + 2 e \rightleftharpoons H_2PO_2^- + 3 OH^-$	−1.65
$Al^{3+} + 3 e \rightleftharpoons Al$	−1.662
$SiO_3^{2-} H_2O + 4 e \rightleftharpoons Si + 6 OH^-$	−1.697
$HPO_3^{2-} + 2 H_2O + 3 e \rightleftharpoons P + 5 OH^-$	−1.71
$HfO^{2+} + 2 H^+ + 4 e \rightleftharpoons Hf + H_2O$	−1.724
$ThO_2 + 4 H^+ + 4 e \rightleftharpoons Th + 2 H_2O$	−1.789
$H_2BO_3^- + H_2O + 3 e \rightleftharpoons B + 4 OH^-$	−1.79
$Sr^{2+} + 2 e \rightleftharpoons Sr(Hg)$	−1.793
$U^{3+} + 3 e \rightleftharpoons U$	−1.798
$H_2PO_2^- + e \rightleftharpoons P + 2 OH^-$	−1.82
$Be^{2+} + 2 e \rightleftharpoons Be$	−1.847
$Np^{3+} + 3 e \rightleftharpoons Np$	−1.856
$Th^{4+} + 4 e \rightleftharpoons Th$	−1.899
$Pu^{3+} + 3 e \rightleftharpoons Pu$	−2.031
$AlF_6^{3-} + 3 e \rightleftharpoons Al + 6 F^-$	−2.069
$Sc^{3+} + 3 e \rightleftharpoons Sc$	−2.077
$H_2 + 2 e \rightleftharpoons 2 H^-$	−2.23
$H_2AlO_3^- + H_2O + 3 e \rightleftharpoons Al + 4 OH^-$	−2.33
$ZrO(OH)_2 + H_2O + 4 e \rightleftharpoons Zr + 4 OH^-$	−2.36
$Mg^{2+} + 2 e \rightleftharpoons Mg$	−2.372
$Y^{3+} + 3 e \rightleftharpoons Y$	−2.372
$Eu^{3+} + 3 e \rightleftharpoons Eu$	−2.407
$Nd^{3+} + 3 e \rightleftharpoons Nd$	−2.431
$Th(OH)_4 + 4 e \rightleftharpoons Th + 4 OH^-$	−2.48
$Ce^{3+} + 3 e \rightleftharpoons Ce$	−2.483
$HfO(OH)_2 + H_2O + 4 e \rightleftharpoons Hf + 4 OH^-$	−2.50
$La^{3+} + 3 e \rightleftharpoons La$	−2.522
$Be_2O_3^{2-} + 3 H_2O + 4 e \rightleftharpoons 2 Be + 6 OH^-$	−2.63
$Mg(OH)_2 + 2 e \rightleftharpoons Mg + 2 OH^-$	−2.690
$Mg^+ + e \rightleftharpoons Mg$	−2.70
$Na^+ + e \rightleftharpoons Na$	−2.71
$Ca^{2+} + 2 e \rightleftharpoons Ca$	−2.868
$Sr(OH)_2 + 2 e \rightleftharpoons Sr + 2 OH^-$	−2.88
$Sr^{2+} + 2 e \rightleftharpoons Sr$	−2.89
$La(OH)_3 + 3 e \rightleftharpoons La + 3 OH^-$	−2.90
$Ba^{2+} + 2 e \rightleftharpoons Ba$	−2.912
$Cs^+ + e \rightleftharpoons Cs$	−2.92
$K^+ + e \rightleftharpoons K$	−2.931
$Rb^+ + e \rightleftharpoons Rb$	−2.98
$Ba(OH)_2 + 2 e \rightleftharpoons Ba + 2 OH^-$	−2.99
$Ca(OH)_2 + 2 e \rightleftharpoons Ca + 2 OH^-$	−3.02
$Li^+ + e \rightleftharpoons Li$	−3.0401
$3 N_2 + 2 H^+ + 2 e \rightleftharpoons 2 NH_3$	−3.09
$Eu^{2+} + 2 e \rightleftharpoons Eu$	−3.395
$Ca^+ + e \rightleftharpoons Ca$	−3.80
$Sr^+ + e \rightleftharpoons Sr$	−4.10

DISSOCIATION CONSTANTS OF ORGANIC BASES IN AQUEOUS SOLUTION

No.	Compound	Temp. °C	Step	pK_a	K_a
a1	Acetamide............	25		0.63	2.34×10^{-1}
a2	Acridine.............	20		5.58	2.63×10^{-6}
a3	α-Alanine	25		2.345	4.52×10^{-3}
a4	Alanine, glycyl-.......	25		3.153	7.03×10^{-4}
a5	Alanine, methoxy- (DL).............	25		2.037	9.18×10^{-3}
a6	Alanine, phenyl.......	25	2	9.18	6.61×10^{-10}
a7	Allothreonine.........	25	1	2.108	7.80×10^{-3}
	Allothreonine.........	25	2	9.096	8.02×10^{-10}
a8	n-Amylamine.........	25		10.63	2.34×10^{-11}
a9	Aniline..............	25		4.63	2.34×10^{-5}
a10	Aniline, n-allyl.........	25		4.17	6.76×10^{-5}
a11	Aniline, 4- (p-aminobenzoyl)...	25	1	2.932	1.17×10^{-3}
a12	Aniline, 4-benzyl......	25		2.17	6.76×10^{-3}
a13	Aniline, 2-bromo......	25		2.53	2.95×10^{-3}
a14	Aniline, 3-bromo......	25		3.58	2.63×10^{-4}
a15	Aniline, 4-bromo......	25		3.86	1.38×10^{-4}
a16	Aniline, 4-bromo- N,N-dimethyl......	25		4.232	5.86×10^{-5}
a17	Aniline, o-chloro......	25		2.65	2.24×10^{-3}
a18	Aniline, m-chloro......	25		3.46	3.47×10^{-4}
a19	Aniline, p-chloro......	25		4.15	7.08×10^{-5}
a20	Aniline, 3-chloro- N,N-dimethyl......	20		3.837	1.46×10^{-4}
a21	Aniline, 4-chloro- N,N-dimethyl......	20		4.395	4.03×10^{-5}
a22	Aniline, 3,5-dibromo...	25		2.34	4.57×10^{-3}
a23	Aniline, 2,4-dichloro...	22		2.05	8.91×10^{-3}
a24	Aniline, N,N-diethyl...	22		6.61	2.46×10^{-7}
a25	Aniline, N,N-dimethyl.	25		5.15	7.08×10^{-6}
a26	Aniline, N,N- dimethyl-3-nitro....	25		2.626	2.37×10^{-3}
a27	Aniline, N-ethyl......	24		5.12	7.59×10^{-6}
a28	Aniline, 2-fluoro......	25		3.20	6.31×10^{-4}
a29	Aniline, 3-fluoro......	25		3.50	3.16×10^{-4}
a30	Aniline, 4-fluoro......	25		4.65	2.24×10^{-5}
a31	Aniline, 2-iodo........	25		2.60	2.51×10^{-3}
a32	Aniline, N-methyl.....	25		4.848	1.41×10^{-5}
a33	Aniline, 4-methylthio..	25		4.35	4.46×10^{-5}
a34	Aniline, 3-nitro.......	25		2.466	3.42×10^{-3}
a35	Aniline, 4-nitro.......	25		1.0	1.00×10^{-1}
a36	Aniline, 2-sulfonic acid	25	2	2.459	3.47×10^{-3}
a37	Aniline, 3-sulfonic acid	25	2	3.738	1.82×10^{-4}
a38	Aniline, 4-sulfonic acid	25	2	3.227	5.92×10^{-4}
a39	o-Anisidine...........	25		4.52	3.02×10^{-5}
a40	m-Anisidine	25		4.23	5.89×10^{-5}
a41	p-Anisidine...........	25		5.34	4.57×10^{-6}
a42	Arginine	25	1	1.8217	1.51×10^{-2}
		25	2	8.9936	1.01×10^{-9}
a43	Asparagine...........	20	1	2.213	6.12×10^{-3}
		20	2	8.85	1.41×10^{-9}
a44	Asparagine, glycyl.....	25	1	2.942	1.14×10^{-3}
		18	2	8.44	3.63×10^{-9}
a45	DL-Aspartic acid......	1	1	2.122	7.55×10^{-3}
		1	2	4.006	1.00×10^{-4}
a46	Azetidine (Trimethylimidine)..	25		11.29	5.12×10^{-12}
a47	Aziridine............	25		8.01	9.77×10^{-9}
b1	Benzene, 4-aminoazo...	25		2.82	1.51×10^{-3}
b2	Benzene, 2-aminoethyl (β-Phenylamine)....	25		9.84	1.45×10^{-10}
b3	Benzene, 4-dimethylaminoazo.	25		3.226	5.94×10^{-4}
b4	Benzidine...........	30	1	4.66	2.19×10^{-5}
		30	2	3.57	2.69×10^{-4}
b5	Benzimidazole........	25		5.532	2.94×10^{-6}
b6	Benzimidazole, 2-ethyl.	25		6.18	6.61×10^{-7}
b7	Benzimidazole, 2-methyl...........	25		6.19	6.46×10^{-7}
b8	Benzimidazole, 2-phenyl...........	25	1	5.23	5.89×10^{-6}
		25	2	11.91	1.23×10^{-12}
b9	Benzoic acid, 2-amino (Anthranilic acid)...	25	1	2.108	7.80×10^{-3}
		25	2	4.946	1.13×10^{-5}
b10	Benzoic acid, 4-amino..	25	1	2.501	3.15×10^{-3}
		25	2	4.874	1.33×10^{-5}
b11	Benzylamine.........	25		9.33	4.67×10^{-10}
b12	Betaine.............	0		1.83	1.48×10^{-2}
b13	Biphenyl, 2-amino.....	22		3.82	1.51×10^{-4}
b14	Bornylamine(trans-)...	25		10.17	6.76×10^{-11}
b15	Brucine.............	25	1	8.28	5.24×10^{-9}
b16	Butane, 1-amino- 3-methyl...........	25		10.60	2.51×10^{-11}
b17	Butane, 2-amino- 2-methyl...........	19		10.85	1.41×10^{-11}
b18	Butane, 1,4-diamino (Putrescine)........	10	1	11.15	7.08×10^{-12}
		10	2	9.71	1.95×10^{-10}
b19	n-Butylamine.........	20		10.77	1.69×10^{-11}
b20	t-Butylamine.........	18		10.83	1.48×10^{-11}
b21	Butyric acid, 4-amino..	25	1	4.0312	9.31×10^{-5}
		25	2	10.5557	2.78×10^{-11}

No.	Compound	Temp. °C	Step	pK_a	K_a
b22	n-Butyric acid, glycyl-2-amino......	25	1	3.1546	7.01×10^{-4}
c1	Cacodylic acid........	25	1	1.57	2.69×10^{-2}
			2	6.27	5.37×10^{-7}
c2	β-Chlortriethyl- ammonium.........	25		8.80	1.59×10^{-9}
c3	Cinnoline............	20		2.37	4.27×10^{-3}
c4	Codeine.............	25		8.21	6.15×10^{-9}
c5	Cyclohexaneamine, n-butyl...........	25		11.23	5.89×10^{-12}
c6	Cyclohexylamine	24		10.66	2.19×10^{-11}
c7	Cystine.............	30	1	1.90	1.25×10^{-2}
		30	2	8.24	5.76×10^{-9}
d1	n-Decylamine.........	25		10.64	2.29×10^{-11}
d2	Diethylamine.........	40		10.489	3.24×10^{-11}
d3	Diisobutylamine......	21		10.91	1.23×10^{-11}
d4	Diisopropylamine.....	28.5		10.96	1.09×10^{-11}
d5	Dimethylamine.......	25		10.732	1.85×10^{-11}
d6	n-Diphenylamine......	25		0.79	1.62×10^{-1}
d7	n-Dodecaneamine (Laurylamine)......	25		10.63	2.35×10^{-11}
e1	d-Ephedrine..........	10		10.139	7.26×10^{-11}
e2	l-Ephedrine..........	10		9.958	1.10×10^{-10}
e3	Ethane, 1-amino- 3-methoxy..........	10		9.89	1.29×10^{-10}
e4	Ethane, 1,2-bismethylamino..	25	1	10.40	3.98×10^{-11}
		25	2	8.26	5.50×10^{-9}
e5	Ethanol, 2-amino......	25		9.50	3.16×10^{-10}
e6	Ethylamine..........	20		10.807	1.56×10^{-11}
e7	Ethylenediamine......	0	1	10.712	1.94×10^{-11}
		0	2	7.564	2.73×10^{-8}
g1	l-Glutamic acid.......	25	1	2.13	7.41×10^{-3}
		25	2	4.31	4.90×10^{-5}
e2	Glutamic acid, α-monoethyl........	25	1	3.846	1.42×10^{-4}
		25	2	7.838	1.45×10^{-8}
e3	l-Glutamine..........	—		9.28	5.25×10^{-10}
e4	l-Glutathione.........	25	2	3.59	2.57×10^{-4}
e5	Glycine.............	25	1	2.3503	4.46×10^{-3}
		25	2	9.7796	1.68×10^{-10}
c6	Glycinc, n-acetyl......	25		3.6608	2.11×10^{-4}
e7	Glycine, dimethyl.....	5		10.3371	4.60×10^{-11}
e8	Glycine, glycyl.......	25		3.1397	7.25×10^{-4}
e9	Glycine, glycylglycyl...	25	1	3.225	5.96×10^{-4}
		25	2	8.090	8.13×10^{-9}
e10	Glycine, leucyl........	25	1	3.25	5.62×10^{-4}
		25	2	8.28	5.25×10^{-9}
e11	Glycine, methyl (Sarcosine).........	25	1	2.21	6.16×10^{-3}
		25	2	10.12	7.58×10^{-11}
g12	Glycine, phenyl.......	25	1	1.83	1.48×10^{-2}
		25	2	4.39	4.07×10^{-5}
g13	Glycine, N,n-propyl...	25	1	2.35	4.46×10^{-3}
		25	2	10.19	6.46×10^{-11}
g14	Glycine, tetraglycyl...	20	1	3.10	7.94×10^{-4}
		20	2	8.02	9.55×10^{-9}
g15	Glycylserine..........	25	1	2.9808	1.04×10^{-3}
		25	2	8.38	4.17×10^{-9}
h1	Hexadecaneamine.....	25		10.63	2.35×10^{-11}
h2	Heptane, 1-amino.....	25		10.66	2.19×10^{-11}
h3	Heptane, 2-amino.....	19		10.88	1.58×10^{-11}
h4	Heptane, 2-methylamino......	17		10.99	1.02×10^{-11}
h5	Hexadecaneamine.....	25		10.61	2.46×10^{-11}
h6	Hexamethylene- diamine............	0	1	11.857	1.39×10^{-12}
		0	2	0.762	1.73×10^{-11}
h7	Hexanoic acid, 6-amino.	25	1	4.373	4.23×10^{-5}
		25	2	10.804	1.57×10^{-11}
h8	n-Hexylamine.........	25		10.56	2.75×10^{-11}
h9	dl-Histidine..........	25	1	1.80	1.58×10^{-2}
		25	2	6.04	9.12×10^{-7}
		25	3	9.33	4.67×10^{-10}
h10	Histidine, β-alanyl (Carnosine).........	20	1	2.73	1.86×10^{-3}
		20	2	6.87	1.35×10^{-7}
		20	3	9.73	1.48×10^{-10}
i1	Imidazol............	25		6.953	1.11×10^{-7}
i2	Imidazol, 2,4-dimethyl.	25		8.359	5.50×10^{-9}
i3	Imidazol, 1-methyl (Oxalmethyline).....	25		6.95	1.12×10^{-7}
i4	Indane, 1-amino (d-1-Hydrindamine	22.5		9.21	6.17×10^{-10}
i5	Isobutyric acid, 2-amino...........	25	1	2.357	4.30×10^{-3}
		25	2	10.205	6.23×10^{-11}
i6	Isoleucine...........	25	1	2.318	4.81×10^{-3}

No.	Compound	Temp. °C	Step	pK_a	K_a
		25	2	9.758	1.74×10^{-10}
i7	Isoquinoline (Leucoline)	20		5.42	3.80×10^{-6}
i8	Isoquinoline, 1-amino	20		7.59	2.57×10^{-8}
i9	Isoquinoline, 7-hydroxy	20	1	5.68	2.09×10^{-6}
		20	2	8.90	1.26×10^{-9}
l1	*L*-Leucine	25	1	2.328	4.70×10^{-3}
		25	2	9.744	1.80×10^{-10}
l2	Leucine, glycyl	25		3.18	6.61×10^{-4}
m1	Methionine	25	1	2.22	6.02×10^{-3}
		25	2	9.27	5.37×10^{-10}
m2	Methylamine	25		10.657	2.70×10^{-11}
m3	Morphine	25		8.21	6.16×10^{-9}
m4	Morpholine	25		8.33	4.67×10^{-9}
n1	Naphthalene, 1-amino-6-hydroxy	25		3.97	1.07×10^{-4}
n2	Naphthalene, dimethylamino	25		4.566	2.72×10^{-5}
n3	α-Naphthylamine	25		3.92	1.20×10^{-4}
n4	β-Naphthylamine	25		4.16	6.92×10^{-5}
n5	α-Naphthylamine, *n*-methyl	27		3.67	2.13×10^{-4}
n6	Neobornylamine(cis-)	25		10.01	9.77×10^{-11}
n7	Nicotine	25	1	8.02	9.55×10^{-9}
		25	2	3.12	7.59×10^{-4}
n8	*n*-Nonylamine	25		10.64	2.29×10^{-11}
n9	Norleucine	25		2.335	4.62×10^{-3}
o1	Octadecaneamine	25		10.60	2.51×10^{-11}
o2	Octylamine	25		10.65	2.24×10^{-11}
o3	Ornithine	25	1	1.705	1.97×10^{-2}
		25	2	8.690	2.04×10^{-9}
p1	Papaverine	25		6.40	3.98×10^{-7}
p2	Pentane, 3-amino	17		10.59	2.57×10^{-11}
p3	Pentane, 3-amino-3-methyl	16		11.01	9.77×10^{-12}
p4	*n*-Pentadecylamine	25		10.61	2.46×10^{-11}
p5	Pentanoic acid, 5-amino(Valeric acid)	25	1	4.270	5.37×10^{-5}
		25	2	10.766	1.71×10^{-11}
p6	Perimidine	20		6.35	4.47×10^{-7}
p7	Phenanthridine	20		5.58	2.63×10^{-6}
p8	1,10-Phenanthroline	25		4.84	1.44×10^{-5}
p9	*o*-Phenetidine (2-Ethoxyaniline)	28		4.43	3.72×10^{-5}
p10	*m*-Phenetidine (3-Ethoxyaniline)	25		4.18	6.60×10^{-5}
p11	*p*-Phenetidine (4-Ethoxyaniline)	28		5.20	6.31×10^{-6}
p12	α-Picoline	20		5.97	1.07×10^{-6}
p13	β-Picoline	20		5.68	2.09×10^{-6}
p14	γ-Picoline	20		6.02	9.55×10^{-7}
p15	Pilocarpine	30		6.87	1.35×10^{-7}
p16	Piperazine	23.5	1	9.83	1.48×10^{-10}
		23.5	2	5.56	2.76×10^{-6}
p17	Piperazine, 2,5-dimethyl(trans-)	25	1	9.66	2.19×10^{-10}
		25	2	5.20	6.31×10^{-6}
p18	Piperidine	25		11.123	7.53×10^{-12}
p19	Piperidine, 3-acetyl	25		3.18	6.61×10^{-4}
p20	Piperidine, 1-*n*-butyl	23		10.47	3.39×10^{-11}
p21	Piperidine, 1,2-dimethyl	25		10.22	6.03×10^{-11}
p22	Piperidine, 1-ethyl	23		10.45	3.55×10^{-11}
p23	Piperidine, 1-methyl	25		10.08	8.32×10^{-11}
p24	Piperidine, 2,2,6,6-tetramethyl	25		11.07	8.51×10^{-12}
p25	Piperidine, 2,2,4-trimethyl	30		11.04	9.12×10^{-12}
p26	Proline	25	1	1.952	1.11×10^{-2}
		25	2	10.640	2.29×10^{-11}
p27	Proline, hydroxy	25	1	1.818	1.52×10^{-2}
		25	2	9.662	2.18×10^{-10}
p28	Propane, 1-amino-2,2-dimethyl	25		10.15	7.08×10^{-11}
p29	Propane, 1,2-diamino	25	1	9.82	1.52×10^{-10}
		25	2	6.61	2.46×10^{-7}
p30	Propane, 1,3-diamino	10	1	10.94	1.15×10^{-11}
		10	2	9.03	9.33×10^{-10}
p31	Propane, 1,2,3-triamino	20	1	9.59	2.57×10^{-10}
		20	2	7.95	1.12×10^{-8}
p32	Propanoic acid, 3-amino (β-Alanine)	25	1	3.551	2.81×10^{-4}
		25	2	10.238	5.78×10^{-11}
p33	Propylamine	20		10.708	1.96×10^{-11}
p34	Pteridine	20		4.05	8.91×10^{-5}
p35	Pteridine, 2-amino-4,6-dihydroxy	20	2	6.59	2.57×10^{-7}
		20	3	9.31	4.90×10^{-10}
p36	Pteridine, 2-amino-4-hydroxy	20	1	2.27	5.37×10^{-3}
		20	2	7.96	1.10×10^{-8}
p37	Pteridine, 6-chloro	20		3.68	2.09×10^{-4}
p38	Pteridine, 6-hydroxy-4-methyl	20	1	4.08	8.32×10^{-5}
			2	6.41	3.89×10^{-7}
p39	Purine	20	1	2.30	5.01×10^{-3}
		20	2	8.96	1.10×10^{-9}
p40	Purine, 6-amino (Adenine)	25	1	4.12	7.59×10^{-5}
		25	2	9.83	1.48×10^{-10}
p41	Purine, 2-dimethylamino	20	1	4.00	1.00×10^{-4}
		20	2	10.24	5.75×10^{-11}
p42	Purine, 8-hydroxy	20	1	2.56	2.75×10^{-3}
		20	2	8.26	9.49×10^{-9}
p43	Pyrazine	27		0.65	2.24×10^{-1}
p44	Pyrazine, 2-methyl	27		1.45	3.54×10^{-2}
p45	Pyrazine, methylamino.	25		3.39	4.07×10^{-4}
p46	Pyridazine	20		2.24	5.76×10^{-3}
p47	Pyrimidine, 2-amino	20		3.45	3.54×10^{-4}
p48	Pyrimidine, 2-amino-4,6-dimethyl	20		4.82	1.51×10^{-5}
p49	Pyrimidine, 2-amino-5-nitro	20		0.35	4.46×10^{-1}
p50	Pyridine	25		5.25	5.62×10^{-6}
p51	Pyridine, 2-aldoxime	20	1	3.59	2.57×10^{-4}
		20	2	10.18	6.61×10^{-11}
p52	Pyridine, 2-amino	20		6.82	1.51×10^{-7}
p53	Pyridine, 4-amino	25		9.1141	7.69×10^{-10}
p54	Pyridine, 2-benzyl	25		5.13	7.41×10^{-6}
p55	Pyridine, 3-bromo	25		2.84	1.45×10^{-3}
p56	Pyridine, 3-chloro	25		2.84	1.45×10^{-3}
p57	Pyridine, 2,5-diamino	20		6.48	3.31×10^{-7}
p58	Pyridine, 2,3-dimethyl (2,3-Lutidine)	25		6.57	2.69×10^{-7}
p59	Pyridine, 2,4-dimethyl (2,4-Lutidine)	25		6.99	1.02×10^{-7}
p60	Pyridine, 3,5-dimethyl (3,5-Lutidine)	25		6.15	7.08×10^{-7}
p61	Pyridine, 2-ethyl	25		5.89	1.28×10^{-6}
p62	Pyridine, 2-formyl	20		3.80	1.59×10^{-4}
p63	Pyridine, 2-hydroxy (2-Pyridol)	20	1	0.75	9.82×10^{-1}
		20	2	11.65	2.24×10^{-12}
p64	Pyridine, 4-hydroxy	20	1	3.20	6.31×10^{-4}
		20	2	11.12	7.59×10^{-12}
p65	Pyridine, methoxy	25		6.47	3.30×10^{-7}
p66	Pyridine, 4-methylamino	20		9.65	2.24×10^{-10}
p67	Pyridine, 2,4,6-trimethyl	25		7.43	3.72×10^{-8}
p68	Pyrrolidine	25		11.27	5.37×10^{-12}
p69	Pyrrolidine, 1,2-dimethyl	26		10.20	6.31×10^{-11}
p70	Pyrrolidine, *n*-methyl	25		10.32	4.79×10^{-11}
q1	Quinazoline	20		3.43	3.72×10^{-4}
q2	Quinazoline, 5-hydroxy.	20	1	3.62	2.40×10^{-4}
		20	2	7.41	3.89×10^{-8}
q3	Quinine	25	1	8.52	3.02×10^{-9}
		25	2	4.13	7.41×10^{-5}
q4	Quinoline	20		4.90	1.25×10^{-5}
q5	Quinoline, 3-amino	20		4.91	1.23×10^{-5}
q6	Quinoline, 3-bromo	25		2.69	2.04×10^{-3}
q7	Quinoline, 8-carboxy	25		1.82	1.51×10^{-2}
q8	Quinoline, 3-hydroxy (3-Quinolinol)	20	1	4.28	5.25×10^{-5}
		20	2	8.08	8.32×10^{-9}
q9	Quinoline, 8-hydroxy (8-Quinolinol)	20	1	5.017	1.21×10^{-6}
		25	2	9.812	1.54×10^{-10}
q10	Quinoline, 8-hydroxy-5-sulfo	25	1	4.112	7.73×10^{-5}
		25	2	8.757	1.75×10^{-9}
q11	Quinoline, 6-methoxy	20		5.03	9.33×10^{-6}
q12	Quinoline, 2-methyl (Quinaldine)	20		5.83	1.48×10^{-6}
q13	Quinoline, 4-methyl (Lepidine)	20		5.67	2.14×10^{-6}
q14	Quinoline, 5-methyl	20		5.20	6.31×10^{-6}
q15	Quinoxaline (Quinazine)	20		0.56	3.63×10^{-1}
s1	Serine (2-amino-3-hydroxypropanoic acid)	25	1	2.186	5.49×10^{-3}
		25	2	9.208	6.19×10^{-10}
s2	Strychnine	25		8.26	5.49×10^{-9}
t1	Taurine (2-Aminoethane sulfonic acid)	25	2	9.0614	8.69×10^{-10}
t2	Tetradecaneamine (Myristilamine)	25		10.62	2.40×10^{-11}
t3	Thiazole	20		2.44	3.63×10^{-3}
t4	Thiazole, 2-amino	20		5.36	4.36×10^{-5}
t5	Threonine	25	1	2.088	8.16×10^{-3}
		25	2	9.10	7.94×10^{-10}
t6	*o*-Toluidine	25		4.44	3.63×10^{-5}
t7	*m*-Toluidine	25		4.73	1.86×10^{-5}
t8	*p*-Toluidine	25		5.08	8.32×10^{-6}

DISSOCIATION CONSTANTS OF ORGANIC BASES IN AQUEOUS SOLUTION (*Continued*)

No.	Compound	Temp. °C	Step	pK_a	K_a
t9	1,3,5-Triazine, 2,4,6-triamino.......	25		5.00	1.00×10^{-5}
t10	Tridecaneamine........	25		10.63	2.35×10^{-11}
t11	Triethylamine........	18		11.01	9.77×10^{-12}
t12	Trimethylamine.......	25		9.81	1.55×10^{-10}
t13	Tryptophan..........	25	1	2.43	3.72×10^{-3}
		25	2	9.44	3.63×10^{-10}
t14	Tyrosine............	25	2	9.11	7.76×10^{-10}
		25	3	10.13	7.41×10^{-11}
t15	Tyrosineamide........	25		7.33	4.68×10^{-8}
u1	Urea................	21		0.10	7.94×10^{-1}
v1	Valine..............	25	1	2.286	5.17×10^{-3}
		25	2	9.719	1.91×10^{-10}

DISSOCIATION CONSTANTS OF INORGANIC BASES IN AQUEOUS SOLUTIONS AT 298K

There is some arbitrariness about the designation of bases, rather than their conjugate acids, for tabulation. There are, nonetheless, a number of substances which are usually thought of as bases: some are listed here. Bases with pK less than zero are shown as "strong". These values describe the thermodynamic quotient of the first ionization of the base dissolved in aqueous solution at "infinite dilution". Concentration quotients may be profoundly affected by concentration, and by the nature and concentrations of other solutes. Many hydroxo-complexes are susceptible to polymerization. Consequently, the pK of $M(OH)_n$ must be used with some caution, especially if the cation has a large effective charge:radius radio. For more specific information, consult current specialist monographs.

NOTE: A = may react with excess strong base. B = Bronsted base: proton acceptor. I = basic properties may be obscured by low solubility. P = equilibria may be obscured by formation of polynuclear complexes. S = very sensitive to ionic medium. T = approximate value derived from kinetic results.

Base	pK_b	Notes
NH_3	4.75	B
N_2H_4	6.05	B
$N_2H_5^+$	14	B
NH_2OH	8.04	B
PH_3	28	B,T
LiOH	0.2	S
MOH	Strong	(M = Na, K, Rb, Cs)
$Be(OH)_2$	5.75	A,P
$BeOH^+$	8.6	A,P
$Mg(OH)_2$	I	I,P
$MgOH^+$	2.6	S,P
$M(OH)_2$	Strong	I (M = Ca, Sr, Ba)
$CaOH^+$	1.2	S
$SrOH^+$	0.7	S
$BaOH^+$	0.6	S
$Al(OH)_3$	8.3	A,P,S
$Al(OH)_2^+$	9.7	A,P,S
$Zn(OH)_2$	6.1	A,I,P,S
$Cd(OH)_2$	10.3	A,I,P,S
$Hg(OH)_2$	11.2	I,S
AgOH	2	A,I

DISSOCIATION CONSTANTS OF ORGANIC ACIDS IN AQUEOUS SOLUTIONS

Compound	T°C	Step	K	pK
Acetic	25		1.76×10^{-5}	4.75
Acetoacetic	18		2.62×10^{-4}	3.58
Acrylic	25		5.6×10^{-5}	4.25
Adipamic	25		2.35×10^{-5}	4.63
Adipic	25	1	3.71×10^{-5}	4.43
Adipic	25	2	3.87×10^{-6}	5.41
d-Alanine	25		1.35×10^{-10}	9.87
Allantoin	25		1.10×10^{-9}	8.96
Alloxanic	25		2.3×1^{-7}	6.64
α-Aminoacetic (glycine)	25		1.67×10^{-10}	9.78
o-Aminobenzoic	25		1.07×10^{-7}	6.97
m-Aminobenzoic	25		1.67×10^{-5}	4.78
p-Aminobenzoic	25		1.2×10^{-5}	4.92
o-Aminobenzosulfonic	25		3.3×10^{-3}	2.48
m-Aminobenzosulfonic	25		1.85×10^{-4}	3.73
p-Aminobenzosulfonic	25		5.81×10^{-4}	3.24
Anisic	25		3.38×10^{-5}	4.47
o-*β*-Anisylpropionic	25		1.59×10^{-5}	4.80
m-*β*-Anisylpropionic	25		2.24×10^{-5}	4.65
p-*β*-Anisylpropionic	25		2.04×10^{-5}	4.69
Ascorbic	24	1	7.94×10^{-5}	4.10
Ascorbic	16	2	1.62×10^{-12}	11.79
DL-Aspartic	25	1	1.38×10^{-4}	3.86
DL-Aspartic	25	2	1.51×10^{-10}	9.82
Barbituric	25		9.8×10^{-5}	4.01
Benzoic	25		6.46×10^{-5}	4.19
Benzosulfonic	25		2×10^{-1}	0.70
Bromoacetic	25		2.05×10^{-3}	2.69
o-Bromobenzoic	25		1.45×10^{-3}	2.84
m-Bromobenzoic	25		1.37×10^{-4}	3.86
n-Butyric	20		1.54×10^{-5}	4.81
iso-Butyric	18		1.44×10^{-5}	4.84
Cacodylic	25		6.4×10^{-7}	6.19
n-Caproic	18		1.43×10^{-5}	4.83
iso-Caproic	18		1.46×10^{-5}	4.84
Chloroacetic	25		1.40×10^{-3}	2.85
o-Chlorobenzoic	25		1.20×10^{-3}	2.92
m-Chlorobenzoic	25		1.51×10^{-4}	3.82
p-Chlorobenzoic	25		1.04×10^{-4}	3.98
α-Chlorobutyric	R.T.		1.39×10^{-3}	2.86
β-Chlorobutyric	R.T.		8.9×10^{-5}	4.05
γ-Chlorobutyric	R.T.		3.0×10^{-5}	4.52
o-Chlorocinnamic	25		5.89×10^{-5}	4.23
m-Chlorocinnamic	25		5.13×10^{-5}	4.29
p-Chlorocinnamic	25		3.89×10^{-5}	4.41
o-Chlorophenoxyacetic	25		8.91×10^{-4}	3.05
m-Chlorophenoxyacetic	25		7.94×10^{-4}	3.10
o-Chlorophenylacetic	25		1.18×10^{-5}	4.07
m-Chlorophenylacetic	25		7.25×10^{-5}	4.14
p-Chlorophenylacetic	25		6.46×10^{-5}	4.19
β-(*o*-Chlorophenyl) propionic	25		2.63×10^{-4}	4.58
β-(*m*-Chlorophenyl) propionic	25		2.57×10^{-5}	4.59
β-(*p*-Chlorophenyl) propionic	25		2.46×10^{-5}	4.61
α-Chloropropinic	25		1.47×10^{-3}	2.83
β-Chloropropionic	25		1.04×10^{-4}	3.98
cis-Cinnamic	25		1.3×10^{-4}	3.89
trans-Cinnamic	25		3.65×10^{-5}	4.44
Citric	20	1	7.10×10^{-4}	3.14
Citric	20	2	1.68×10^{-5}	4.77
Citric	20	3	6.4×10^{-7}	6.39
o-Cresol	25		6.3×10^{-11}	10.20
m-Cresol	25		9.8×10^{-11}	10.01
p-Cresol	25		6.7×10^{-11}	10.17
Crotonic (*trans-*)	25		2.03×10^{-5}	4.69
Cyanoacetic	25		3.65×10^{-3}	2.45
γ-Cyanobutyric	25		3.80×10^{-3}	2.42
o-Cyanophenoxyacetic	25		1.05×10^{-3}	2.98
m-Cyanophenoxyacetic	25		9.33×10^{-4}	3.03
p-Cyanophenoxyacetic	25		1.18×10^{-3}	2.93
Cyanopropionic	25		3.6×10^{-3}	2.44
Cyclohexane-1:1-dicarboxylic	25	1	3.55×10^{-4}	3.45
Cyclohexane-1:1-dicarboxylic	25	2	7.76×10^{-7}	6.11
Cyclopropane-1:1-dicarboxylic	25	1	1.51×10^{-2}	1.82
Cyclopropane-1:1-dicarboxylic	25	2	3.72×10^{-8}	7.43
DL-Cysteine	30	1	7.25×10^{-9}	8.14
DL-Cysteine	30	2	4.6×10^{-11}	10.34
L-Cystine	25	1	1.4×10^{-8}	7.85
LCystine	25	2	1.4×10^{-10}	9.85
Deuteroacetic (in D_2O)	25		5.5×10^{-4}	5.25
Dichloroacetic	25		3.32×0^{-2}	1.48

DISSOCIATION CONSTANTS OF ORGANIC ACIDS IN AQUEOUS SOLUTIONS (continued)

Compound	T°C	Step	K	pK
Dichloroacetylacetic	?		7.8×10^{-3}	2.11
Dichlorophenol (2,3-)	25		3.6×10^{-8}	7.44
Dihydroxybenzoic (2,2-)	25		1.14×10^{-3}	2.94
Dihydroxybenzoic (2,5-)	25		1.08×10^{-3}	2.97
Dihydroxybenzoic (3,4-)	25		3.3×10^{-5}	4.48
Dihydroxybenzoic (3,5-)	25		9.1×10^{-5}	4.04
Dihydroxymalic	25		1.12×10^{-2}	1.92
Dihydroxytartaric	25		1.2×10^{-2}	1.92
Dimethylglycine	25		1.3×10^{-10}	9.89
Dimethylmalic	25	1	6.83×10^{-4}	3.17
Dimethylmalic	25	2	8.72×10^{-7}	6.06
Dimethylmalonic	25		7.08×10^{-4}	3.15
Dinicotinic	25		1.6×10^{-3}	2.80
Dinitrophenol (2,4-)	15		1.1×10^{-4}	3.96
Dinitrophenol (3,6-)	15		7.1×10^{-6}	5.15
Diphenylacetic	25		1.15×10^{-4}	3.94
Ethylbenzoic	25		4.47×10^{-5}	4.35
Ethylphenylacetic	25		4.27×10^{-5}	4.37
Fluorobenzoic	17		1.25×10^{-3}	2.90
Formic	20		1.77×10^{-4}	3.75
Fumaric *(trans-)*	18	1	9.30×10^{-4}	3.03
Fumaric *(trans-)*	18	2	3.62×10^{-5}	4.44
Furancarboxylic	25		7.1×10^{-4}	3.15
Furoic	25		6.76×10^{-4}	3.17
Gallic	25		3.9×10^{-5}	4.41
Glutaramic	25		3.98×10^{-5}	4.60
Glutaric	25	1	4.58×10^{-5}	4.31
Glutaric	25	2	3.89×10^{-6}	5.41
Glycerol	25		7×10^{-15}	14.15
Glycine	25		1.67×10^{-10}	9.78
Glycol	25		6×10^{-15}	14.22
Glycolic	25		1.48×10^{-4}	3.83
Heptanoic	25		1.28×10^{-5}	4.89
Hexahydrobenzoic	25		1.26×10^{-5}	4.90
Hexanoic	25		1.31×10^{-5}	4.88
Hippuric	25		$1.57 \times 10\text{um}^{4}$	3.80
Histidine	25		6.7×10^{-10}	9.17
Hydroquinone	20		4.5×10^{-11}	10.35
o-Hydroxybenzoic	19	1	1.07×10^{-3}	2.97
o-Hydroxybenzoic	18	2	4×10^{-14}	13.40
m-Hydroxybenzoic	19	1	8.7×10^{-5}	4.06
m-Hydroxybenzoic	19	2	1.2×10^{-10}	9.92
p-Hydroxybenzoic	19	1	3.3×10^{-5}	4.48
p-Hydroxybenzoic	19	2	4.8×10^{-10}	9.32
β-Hydroxybutyric	25		2×10^{-5}	4.70
γ-Hydroxybutyric	25		1.9×10^{-5}	4.72
β-Hydroxypropionic	25		3.1×10^{-5}	4.51
γ-Hydroxyquinoline	20		3.1×10^{-10}	9.51
Iodoacetic	25		7.5×10^{-4}	3.12
o-Iodobenzoic	25		1.4×10^{-3}	2.85
m-Iodobenzoic	25		1.6×10^{-4}	3.80
Itaconic	25	1	1.40×10^{-4}	3.85
Itaconic	25	2	3.56×10^{-6}	5.45
Lactic	100		8.4×10^{-4}	3.08
Lutidinic	25		7.0×10^{-3}	2.15
Lysine	25		2.95×10^{-11}	10.53
Maleic	25	1	1.42×10^{-2}	1.83
Maleic	25	2	8.57×10^{-7}	6.07
Malic	25	1	3.9×10^{-4}	3.40
Malic	25	2	7.8×10^{-6}	5.11
Malonic	25	1	1.49×10^{-2}	2.83
Malonic	25	2	2.03×10^{-6}	5.69
DL-Mandelic	25		1.4×10^{-4}	3.85
Mesaconic	25	1	8.22×10^{-4}	3.09
Mesaconic	25	2	1.78×10^{-5}	4.75
Mesitylenic	25		4.8×10^{-5}	4.32
Methyl-*o*-aminobenzoic	25		4.6×10^{-5}	5.34
Methyl-*m*-aminobenzoic	25		8×10^{-6}	5.10
Methyl-*p*-aminobenzoic	25		9.2×10^{-6}	5.04
o-Methylcinnamic	25		3.16×10^{-5}	4.50
m-Methylcinnamic	25		3.63×10^{-5}	4.44
p-Methylcinnamic	25		2.76×10^{-5}	4.56
β-Methylglutaric	25		5.75×10^{-5}	4.24
n-Methylglycine	18		1.2×10^{-10}	9.92
Methylmalonic	25		1.17×10^{-4}	3.07
Methylsuccinic	25	1	7.4×10^{-5}	4.13

Compound	T°C	Step	K	pK
Methylsuccinic	25	2	2.3×10^{-4}	5.64
o-Monochlorophenol	25		3.2×10^{-9}	8.49
m-Monochlorophenol	25		1.4×10^{-9}	8.85
p-Monochlorophenol	25		6.6×10^{-10}	9.18
Naphthalenesulfonic	25		2.7×10^{-1}	0.57
α-Naphthoic	25		2×10^{-4}	3.70
β-Naphthoic	25		6.8×10^{-5}	4.17
α-Naphthol	25		4.6×10^{-10}	9.34
β-Naphthol	25		3.1×10^{-10}	9.51
Nitrobenzene	0		1.05×10^{-4}	3.98
o-Nitrobenzoic	18		6.95×10^{-3}	2.16
m-Nitrobenzoic	25		3.4×10^{-4}	3.47
p-Nitrobenzoic	25		3.93×10^{-4}	3.41
o-Nitrophenol	25		6.8×10^{-8}	7.17
m-Nitrophenol	25		5.3×10^{-9}	8.28
p-Nitrophenol	25		7×10^{-8}	7.15
o-Nitrophenylacetic	25		1.00×10^{-4}	4.00
m-Nitrophenylacetic	25		1.07×10^{-4}	3.97
p-Nitrophenylacetic	25		1.41×10^{-4}	3.85
o-*β*-Nitrophenylpropionic	25		3.16×10^{-5}	4.50
p-*β*-Nitrophenylopropionic	25		3.39×10^{-5}	4.47
Nonanic	25		1.09×10^{-5}	4.96
Octanoic	25		1.28×10^{-5}	4.89
Oxalic	25	1	5.90×10^{-2}	1.23
Oxalic	25	2	6.40×10^{-5}	4.19
Phenol	20		1.28×10^{-10}	9.89
Phenylacetic	18		5.2×10^{-5}	4.28
o-Phenylbenzoic	25		3.47×10^{-4}	3.46
γ-Phenylbutyric	25		1.74×10^{-5}	4.76
α-Phenylpropionic	25		2.27×10^{-5}	4.64
β-Phenylpropionic	25		4.25×10^{-5}	4.37
o-Phthalic	25	1	1.3×10^{-3}	2.89
o-Phthalic	25	2	3.9×10^{-6}	5.51
m-Phthalic	25	1	2.9×10^{-4}	3.54
m-Phthalic	18	2	2.5×10^{-5}	4.60
p-Phthalic	25	1	3.1×10^{-4}	3.51
p-Phthalic	16	2	1.5×10^{-5}	4.82
Picric	25		4.2×10^{-1}	0.38
Pimelic	25		3.09×10^{-5}	4.71
Propionic	25		1.34×10^{-5}	4.87
iso-Propylbenzoic	25		3.98×10^{-5}	4.40
2-Pyridinecarboxylic	25		3×10^{-6}	5.52
3-Pyridinecarboxylic	25		1.4×10^{-5}	4.85
4-Pyridinecariboxylic	25		1.1×10^{-5}	4.96
Pyrocatcchol	20		1.4×10^{-10}	9.85
Quinolinic	25		3×10^{-3}	2.52
Resorcinol	25		1.55×10^{-10}	9.81
Saccharin	18		2.1×10^{-12}	11.68
Suberic	25		2.99×10^{-5}	4.52
Succinic	25	1	6.89×10^{-5}	4.16
Succinic	25	2	2.47×10^{-6}	5.61
Sulfanilic	25		5.9×10^{-4}	3.23
α-Tataric	25	1	1.04×10^{-3}	2.98
α-Tartaric	25	2	4.55×10^{-5}	4.34
meso-Tartaric	25	1	6×10^{-4}	3.22
meso-Tartaric	25	2	1.53×10^{-5}	4.82
Theobromine	18		1.3×10^{-8}	7.89
Terephthalic	25		3.1×10^{-4}	3.51
Thioacetic	25		4.7×10^{-4}	3.33
Thiophenecarboxylic	25		3.3×10^{-4}	3.48
o-Toluic	25		1.22×10^{-4}	3.91
m-Toluic	25		5.32×10^{-5}	4.27
p-Toluic	25		4.33×10^{-5}	4.36
Trichloroacetic	25		2×10^{-1}	0.70
Trichlorophenol	25		1×10^{-6}	6.00
Trihydroxybenzoic (2,4,6-)	25		2.1×10^{-2}	1.68
Trimethylacetic	18		9.4×10^{-6}	5.03
Trinitrophenol (2,4,6-)	25		4.2×10^{-1}	0.38
Tryptophan	25		4.2×10^{-10}	9.38
Tyrosine	17		3.98×10^{-9}	8.40
Uric	12		1.3×10^{-4}	3.89
n-Valeric	18		1.51×10^{-5}	4.82
iso-Valeric	25		1.7×10^{-5}	4.77
Veronal	25		3.7×10^{-8}	7.43
Vinylacetic	25		4.57×10^{-5}	4.34
Xanthine	40		1.24×10^{-10}	9.91

DISSOCIATION CONSTANTS OF ACIDS IN WATER AT VARIOUS TEMPERATURES

Acids		0°	5°	10°	15°	20°	25°	30°	35°	40°	45°	50°
Formic	$K_A \cdot 10^4$	1.638	1.691	1.728	1.749	1.765	1.772	1.768	1.747	1.716	1.685	1.650
Acetic	$K_A \cdot 10^5$	1.657	1.700	1.729	1.745	1.753	1.754	1.750	1.728	1.703	1.670	1.633
Propionic	$K_A \cdot 10^5$	1.274	1.305	1.326	1.336	1.338	1.336	1.326	1.310	1.280	1.257	1.229
n-Butyric	$K_A \cdot 10^5$	1.563	1.574	1.576	1.569	1.542	1.515	1.484	1.439	1.395	1.347	1.302
Chloracetic	$K_A \cdot 10^3$	1.528	—	1.488	—	—	1.379	—	—	1.230	—	—
Lactic	$K_A \cdot 10^4$	1.287	—	—	—	—	1.374	—	—	—	—	1.270
Glycollic	$K_A \cdot 10^4$	1.334	—	—	—	—	1.475	—	—	—	—	1.415
Oxalic	$K_{1A} \cdot 10^5$	5.91	5.82	5.70	5.55	5.40	5.18	4.92	4.67	4.41	4.09	3.83
Malonic	$K_{2A} \cdot 10^6$	2.140	2.165	2.152	2.124	2.076	2.014	1.948	1.863	1.768	1.670	1.575
Phosphoric	$K_A \cdot 10^3$	8.968	—	—	—	—	7.516	—	—	—	—	5.495
Phosphoric	$K_{2A} \cdot 10^8$	4.85	5.24	5.57	5.89	6.12	6.34	6.46	6.53	6.58	6.59	6.55
Boric	$K_A \cdot 10^{10}$	—	3.63	4.17	4.72	5.26	5.79	6.34	6.86	7.38	—	8.32
Carbonic	$K_{1A} \cdot 10^7$	2.64	3.04	3.44	3.81	4.16	4.45	4.71	4.90	5.04	5.13	5.19
Phenol-sulfonic	$K_{2A} \cdot 10^{10}$	4.45	5.20	6.03	6.92	7.85	8.85	9.89	10.94	12.00	13.09	14.16
Glycine	$K_{1A} \cdot 10^7$	—	3.82	3.99	4.17	4.32	4.46	4.57	4.66	4.73	4.77	4.79
Citric	$K_{1A} \cdot 10^4$	6.03	6.31	6.69	6.92	7.21	7.45	7.66	7.78	7.96	7.99	8.04
	$K_{2A} \cdot 10^5$	1.45	1.54	1.60	1.65	1.70	1.73	1.76	1.77	1.78	1.76	1.75
	$K_{3A} \cdot 10^7$	4.05	4.11	4.14	4.13	4.09	4.02	3.99	3.78	3.69	3.45	3.28

Reproducibility between various workers is about $\pm (0.01—0.02) \cdot 10^5$.
All values are on the m-scale.

DISSOCIATION CONSTANTS OF INORGANIC ACIDS IN AQUEOUS SOLUTIONS

(Approximately 0.1—0.01 N)

Compound	T°C	Step	K	pK
Arsenic	18	1	5.62×10^{-3}	2.25
Arsenic	18	2	1.70×10^{-7}	6.77
Arsenic	18	3	3.95×10^{-12}	11.60
Arsenious	25		6×10^{-10}	9.23
o-Boric	20	1	7.3×10^{-10}	9.14
o-Boric	20	2	1.8×10^{-13}	12.74
o-Boric	20	3	1.6×10^{-14}	13.80
Carbonic	25	1	4.30×10^{-7}	6.37
Carbonic	25	2	5.61×10^{-11}	10.25
Chromic	25	1	1.8×10^{-1}	0.74
Chromic	25	2	3.20×10^{-7}	6.49
Germanic	25	1	2.6×10^{-9}	8.59
Germanic	25	2	1.9×10^{-13}	12.72
Hyrocyanic	25		4.93×10^{-10}	9.31
Hydrofluoric	25		3.53×10^{-4}	3.45
Hydrogen sulfide	18	1	9.1×10^{-3}	7.04
Hydrogen sulfide	18	2	1.1×10^{-12}	11.96
Hydrogen peroide	25		2.4×10^{-12}	11.62
Hypobromous	25		2.06×10^{-9}	8.69
Hypochlorous	18		2.95×10^{-5}	4.53
Hypoiodous	25		2.3×10^{-11}	10.64
Iodic	25		1.69×10^{-1}	0.77
Nitrous	12.5		4.6×10^{-4}	3.37
Periodic	25		2.3×10^{-2}	1.64
o-Phosphoric	25	1	7.52×10^{-3}	2.12
o-Phosphoric	25	2	6.23×10^{-8}	7.21
o-Phosphoric	18	3	2.2×10^{-13}	12.67
Phosphorous	18	1	1.0×10^{-2}	2.00
Phosphorous	18	2	2.6×10^{-7}	6.59
Pyrophosphoric	18	1	1.4×10^{-1}	0.85
Pyrophosphoric	18	2	3.2×10^{-2}	1.49
Pyrophosphoric	18	3	1.7×10^{-6}	5.77
Pyrophosphoric	18	4	6×10^{-9}	8.22
Selenic	25	2	1.2×10^{-2}	1.92
Selenious	25	1	3.5×10^{-2}	2.46
Selenious	25	2	5×10^{-3}	7.31
m-Silicic	R.T.	1	2×10^{-10}	9.70
m-Silicic	R.T.	2	1×10^{-12}	12.00
o-Silicic	30	1	2.2×10^{-10}	9.66
o-Silicic	30	2	2×10^{-12}	11.70
o-Silicic	30	3	1×10^{-12}	12.00
o-Silicic	30	4	1×10^{-12}	12.00
Sulfuric	25	2	1.20×10^{-2}	1.92
Sulfurous	18	1	1.54×10^{-2}	1.81
Sulfurous	18	2	1.02×10^{-7}	6.91
Telluric	18	1	2.09×10^{-8}	7.68
Telluric	18	2	6.46×10^{-12}	11.29
Tellurous	25	1	3×10^{-8}	2.48
Tellurous	25	2	2×10^{-8}	7.70
Tetraboric	25	1	$\sim 10^{-4}$	~ 4.00
Tetraboric	25	2	$\sim 10^{-9}$	~ 9.00

DISSOCIATION CONSTANTS (K_b) OF AQUEOUS AMMONIA FROM 0 TO 50°C

Temperature (°C)	pK_b	K_b
0	4.862	1.374×10^{-5}
5	4.830	1.479×10^{-5}
10	4.804	1.570×10^{-5}
15	4.782	1.652×10^{-5}
20	4.767	1.710×10^{-5}
25	4.751	1.774×10^{-5}
30	4.740	1.820×10^{-5}
35	4.733	1.849×10^{-5}
40	4.730	1.862×10^{-5}
45	4.726	1.879×10^{-5}
50	4.723	1.892×10^{-5}

Values of K_b accurate to ±0.005; determined by e.m.f. method by: Bates, R. G. and Pinching, G. D., *J. Am. Chem. Soc.*, 72, 1393, 1950.

ION PRODUCT OF WATER SUBSTANCE

William L. Marshall and E. U. Franck

Pressure (bars)	Temperature (°C) 0	25	50	75	100	150	200	250	300	350	400	450	500	600	700	800	900	1000
Sat'd Vapor	14.938	13.995	13.275	12.712	12.265	11.638	11.289	11.191	11.406	12.30	—	—	—	—	—	—	—	—
250	14.83	13.90	13.19	12.63	12.18	11.54	11.16	11.01	11.14	11.77	19.43	21.59	22.40	23.27	23.81	24.23	24.59	24.93
500	14.72	13.82	13.11	12.55	12.10	11.45	11.05	10.85	10.86	11.14	11.88	13.74	16.13	18.30	19.29	19.92	20.39	20.80
750	14.62	13.73	13.04	12.48	12.03	11.36	10.95	10.72	10.66	10.79	11.17	11.89	13.01	15.25	16.55	17.35	17.93	18.39
1,000	14.53	13.66	12.96	12.41	11.96	11.29	10.86	10.60	10.50	10.54	10.77	11.19	11.81	13.40	14.70	15.58	16.22	16.72
1,500	14.34	13.53	12.85	12.29	11.84	11.16	10.71	10.43	10.26	10.22	10.29	10.48	10.77	11.59	12.50	13.30	13.97	14.50
2,000	14.21	13.40	12.73	12.18	11.72	11.04	10.57	10.27	10.08	9.98	9.98	10.07	10.23	10.73	11.36	11.98	12.54	12.97
2,500	14.08	13.28	12.62	12.07	11.61	10.92	10.45	10.12	9.91	9.79	9.74	9.77	9.86	10.18	10.63	11.11	11.59	12.02
3,000	13.97	13.18	12.53	11.98	11.53	10.83	10.34	9.99	9.76	9.61	9.54	9.53	9.57	9.78	10.11	10.49	10.89	11.24
3,500	13.87	13.09	12.44	11.90	11.44	10.74	10.24	9.88	9.63	9.47	9.37	9.33	9.34	9.48	9.71	10.02	10.35	10.62
4,000	13.77	13.00	12.35	11.82	11.37	10.66	10.16	9.79	9.52	9.34	9.22	9.16	9.15	9.23	9.41	9.65	9.93	10.13
5,000	13.60	12.83	12.19	11.66	11.22	10.52	10.00	9.62	9.34	9.13	8.99	8.90	8.85	8.85	8.95	9.11	9.30	9.42
6,000	13.44	12.68	12.05	11.53	11.09	10.39	9.87	9.48	9.18	8.96	8.80	8.69	8.62	8.57	8.61	8.72	8.86	8.97
7,000	13.31	12.55	11.93	11.41	10.97	10.27	9.75	9.35	9.04	8.81	8.64	8.51	8.42	8.34	8.34	8.40	8.51	8.64
8,000	13.18	12.43	11.82	11.30	10.86	10.17	9.64	9.24	8.93	8.68	8.50	8.36	8.25	8.13	8.10	8.13	8.21	8.38
9,000	13.04	12.31	11.71	11.20	10.77	10.07	9.54	9.13	8.82	8.57	8.37	8.22	8.10	7.95	7.89	7.89	7.95	8.12
10,000	12.91	12.21	11.62	11.11	10.68	9.98	9.45	9.04	8.71	8.46	8.25	8.09	7.96	7.78	7.70	7.68	7.70	7.85

Note: Data in the following table were calculated from the equation, $\log_{10} K^*_w = A + B/T + C/T^2 + D/T^3 + (E + F/T\ G/T^2) \log_{10} \varrho^*_w$, where, $K^*_w = K_w/(\text{mol kg}^{-1})$, and $\varrho^*_w = \varrho_w/(\text{g cm}^{-3})$. The parameters are:

$A = -4.098$ $\quad C = +2.2362 \times 10^5 K^2$ $\quad E = +13.957$ $\quad G = +8.5641 \times 10^5 K^2$
$B = -3245.2K$ $\quad D = -3.984 \times 10^7 K^3$ $\quad F = -1262.3K$

Users of this table may wish to refer to the reference cited for the background for this international formulation of the ion product of water substance.

From *J. Phys. Chem. Ref. Data*, 10, 295, 1981. With permission.

IONIZATION CONSTANT FOR WATER (K_w)

$-\log_{10} K_w$	Temperature °C.	$-\log_{10} K_w$	Temperature °C.
14.9435	0	13.8330	30
14.7338	5	13.6801	35
14.5346	10	13.5348	40
14.3463	15	13.3960	45
14.1669	20	13.2617	50
14.0000	24	13.1369	55
13.9965	25	13.0171	60

IONIZATION CONSTANTS FOR DEUTERIUM OXIDE FROM 10 TO 50°C

From NBS Technical Note 400

The subscript *m* indicates values on the molal scale, whereas the subscript *c* indicates values on the molar scale.

t(°C)	pK_m	pK_c
10	15.526	15.439
20	15.136	15.049
25	14.955	14.869
30	14.784	14.699
40	14.468	14.385
50	14.182	14.103

STANDARD SOLUTIONS FOR CALIBRATING CONDUCTIVITY CELLS

Grams of KCl per 1 kg of solution in vacuum	$\kappa\ [10^{-3} \times S^{-1}\ m^{-1}]$ at 0°C	18°C	25°C
71.135 2	0.651 44	0.977 90	1.112 87
7.419 13	0.071 344	0.111 612	0.128 496
0.745 263	0.007 732 6	0.012 199 2	0.014 080 7

From Jones, G. and Bradshaw, B. C., *J. Am. Chem. Soc.*, 55, 1780, 1933, converted from (int. ohm)$^{-1}$ cm^{-1}.

EQUIVALENT CONDUCTANCES OF AQUEOUS SOLUTIONS OF HYDROHALOGEN ACIDS

From NSRDS-NBS 33

Walter J. Hamer and Harold J. DeWane

One may wish to refer to the above document which defines terms relating to conductance of electrolytic solutions and also discusses some general considerations of the migration of ions under applied potential gradients. In NSRDS-NBS 33 conductance equations are given and some treatment of the Debye–Hückel–Onsager–Fuoss theories is presented.

Equivalent conductances (Ω^{-1} cm^2 equiv^{-1} of aqueous solutions of HF at 0, 16, 18, 20, and 25°C

c	0°C	16°C	18°C	20°C	25°C
mol l^{-1}					
0.004	106.7	—	—	—	140.5
0.005	97.7	—	—	—	128.1
0.006	90.9	112.9	114.6	116.3	118.8
0.007	85.5	105.7	107.3	108.9	111.4
0.008	81.1	99.9	101.3	102.9	105.4
0.009	77.3	95.0	96.4	97.8	100.4
0.01	74.2	90.8	92.1	93.5	96.1
0.02	56.2	67.7	68.8	69.8	72.2
0.05	39.3	46.8	47.6	48.3	50.1
0.07	34.7	41.3	41.9	42.6	44.3
0.10	30.7	36.4	37.0	37.7	39.1
0.20	24.8	29.6	30.1	30.7	31.7
0.50	20.4	—	—	—	26.3
0.70	19.5	—	—	—	25.1
1.0	18.8	—	—	—	24.3

Equivalent conductances (Ω^{-1} cm^2 equiv^{-1}) of aqueous solutions of HCl from −20 to 65°C

c	−20°C	−10°C	0°C	5°C	10°C	15°C	20°C	25°C	30°C	35°C	40°C	45°C	50°C	55°C	65°C
mol l^{-1}															
0.0001	—	—	—	296.4	—	360.8	—	424.5	—	487.0	—	547.9	577.7	606.6	662.9
0.0005	—	—	—	295.2	—	359.2	—	422.6	—	484.7	—	545.2	575.1	603.5	660.0
0.001	—	—	—	294.3	—	358.0	—	421.2	—	483.1	—	543.2	573.1	601.3	657.8
0.005	—	—	—	291.0	—	353.5	—	415.7	—	476.7	—	535.5	564.4	592.6	647.3
0.01	—	—	—	288.6	—	350.3	—	411.9	—	472.2	—	530.3	558.7	586.5	641.2
0.05	—	—	—	280.3	—	339.9	—	398.9	—	456.7	—	512.4	—	565.6	616.9
0.1	—	—	—	275.0	—	333.3	—	391.1	—	446.8	—	501.1	—	552.8	602.8
0.5	—	—	228.7	254.8	283.0	308.1	336.4	360.7	386.8	411.9	436.9	461.1	482.4	508.0	552.3
1.0	—	—	211.7	235.2	261.6	283.9	312.2	332.2	359.0	379.4	402.9	424.8	445.3	468.1	509.3
1.5	—	—	196.2	216.9	241.5	261.5	287.5	305.8	331.1	349.4	371.6	391.5	410.8	431.7	469.9
2.0	—	—	182.0	199.9	222.7	240.7	262.9	281.4	303.3	321.6	342.4	360.6	378.2	398.0	433.6
2.5	—	131.7	168.5	184.3	205.1	221.4	239.8	258.9	277.0	295.8	315.2	332.0	347.6	366.7	399.9
3.0	—	120.8	154.6	169.5	188.5	203.4	219.3	237.6	253.3	271.5	289.3	304.8	319.0	336.9	368.0
3.5	85.5	111.3	139.6	155.6	172.2	186.5	201.6	218.3	232.9	248.6	263.9	279.4	292.1	308.6	337.2
4.0	79.3	102.7	129.2	143.4	158.1	171.5	185.6	200.0	214.2	228.4	242.2	256.6	268.2	283.6	310.1
4.5	73.7	94.9	119.5	132.0	145.4	157.4	170.6	183.1	196.6	209.5	222.5	235.2	246.7	260.2	284.7
5.0	68.5	87.8	110.3	121.3	133.5	144.4	156.6	167.4	180.2	191.9	204.1	215.4	226.5	238.4	261.0
5.5	63.6	81.1	101.7	111.4	122.5	132.3	143.6	152.9	165.0	175.6	187.1	197.1	207.7	218.3	239.1
6.0	58.9	74.9	93.7	102.2	112.3	121.2	131.5	139.7	151.0	160.6	171.3	180.2	190.3	199.7	218.9
6.5	54.4	69.1	86.2	93.8	103.0	111.0	120.4	127.7	138.2	146.8	156.9	164.8	174.3	182.7	200.4
7.0	50.2	63.7	79.3	86.0	94.4	101.7	110.2	116.9	126.4	134.3	143.3	150.8	159.7	167.2	183.5
7.5	46.3	58.6	73.0	78.9	86.5	93.3	100.9	107.0	115.7	122.9	131.6	138.1	146.2	153.1	168.1
8.0	42.7	54.0	67.1	72.4	79.4	85.6	92.4	98.2	106.1	112.6	120.6	126.7	134.0	140.4	154.1
8.5	39.4	49.8	61.7	66.5	72.9	78.7	84.7	90.3	97.3	103.2	110.7	116.4	123.0	128.8	141.5
9.0	36.4	45.9	56.8	61.2	67.1	72.5	77.8	83.1	89.4	94.8	101.7	107.1	112.9	118.3	130.1
9.5	33.6	42.3	52.3	56.4	61.8	66.8	71.5	76.6	82.3	87.3	93.6	98.7	103.9	108.8	119.7
10.0	31.2	39.1	48.2	52.0	57.0	61.6	65.8	70.7	75.9	80.5	86.3	91.1	95.7	—	—
10.5	28.9	36.1	44.5	48.0	52.7	56.9	60.7	65.3	70.1	74.3	79.6	84.1	88.4	—	—
11.0	26.8	33.4	41.1	44.4	48.8	52.6	56.1	60.2	64.9	68.7	73.6	77.7	81.7	—	—
11.5	24.9	31.0	38.0	41.1	45.3	48.5	51.9	55.3	60.1	63.6	68.0	71.7	75.6	—	—
12.0	23.1	28.7	35.3	38.0	42.0	—	48.0	—	55.6	—	62.8	—	70.0	—	—
12.5	21.4	26.7	32.7	—	39.0	—	44.4	—	51.4	—	57.9	—	64.8	—	—
13.0	20.0	24.9	—	—	—	—	—	—	—	—	—	—	—	—	—

Equivalent conductances (Ω^{-1} cm^2 equiv^{-1}) of aqueous solutions of HBr at 25°C

c	Λ	*c*	Λ	*c*	Λ	*c*	Λ	*c*	Λ	*c*	Λ
mol l^{-1}		mol l^{-1}		mol l^{-1}		mol l^{-1}		mol l^{-1}		mol l^{-1}	
0.0002	425.5	0.003	419.7	0.04	402.7	0.30	374.0	0.80	345.3	4.0	199.4
0.0003	425.0	0.004	418.5	0.05	400.4	0.35	370.8	0.85	342.6	4.5	182.4
0.0004	424.6	0.005	417.6	0.06	398.4	0.40	367.7	0.90	339.9	5.0	166.5
0.0005	424.3	0.006	416.7	0.07	396.5	0.45	364.7	0.95	337.2	5.5	151.8
0.0006	424.0	0.007	415.8	0.08	394.9	0.50	361.9	1.0	334.5	6.0	138.2
0.0007	423.7	0.008	415.1	0.09	393.4	0.55	359.0	1.5	307.6	6.5	125.7
0.0008	423.4	0.009	414.4	0.10	391.9	0.60	356.2	2.0	281.7	7.0	114.2
0.0009	423.2	0.01	413.7	0.15	386.0	0.65	353.5	2.5	257.8	7.5	103.8
0.001	422.9	0.02	408.9	0.20	381.4	0.70	350.8	3.0	236.8	8.0	94.4
0.002	421.1	0.03	405.4	0.25	377.5	0.75	348.0	3.5	217.5	8.5	85.8

EQUIVALENT CONDUCTANCES OF AQUEOUS SOLUTIONS OF HYDROHALOGEN ACIDS (Continued)

Equivalent conductances (Ω^{-1} cm^2 equiv^{-1}) of aqueous solutions of HBr from − 20 to 50°C

c	− 20°C	− 10°C	0°C	10°C	20°C	30°C	40°C	50°C
mol l^{-1}								
0.50	—	—	240.9	295.9	347.0	398.9	453.6	496.8
0.75	—	—	234.7	284.9	339.0	387.2	433.8	480.6
1.00	—	—	229.6	276.0	329.0	380.4	418.6	465.2
1.25	—	—	221.7	265.8	314.9	362.8	401.8	442.9
1.50	—	—	209.5	254.9	298.9	340.6	381.8	421.4
1.75	—	—	198.3	243.1	284.6	327.3	366.2	404.8
2.00	—	150.8	188.6	231.3	271.8	314.1	350.5	387.4
2.25	—	143.4	180.1	219.6	258.3	296.9	332.3	367.0
2.50	—	136.8	171.7	208.3	244.8	281.7	316.0	349.1
2.75	—	131.1	164.1	198.3	232.9	267.6	301.2	333.2
3.00	—	125.7	157.2	189.5	222.2	255.0	287.8	318.6
3.25	—	120.6	150.8	181.7	212.4	244.3	275.4	304.9
3.50	—	116.1	144.1	174.6	203.2	234.4	263.7	291.9
3.75	87.1	112.0	137.6	167.4	194.7	224.2	252.2	279.4
4.00	84.0	107.5	132.3	160.2	186.8	214.2	239.7	266.9
4.25	80.9	103.1	127.7	153.2	179.0	204.5	228.8	254.6
4.50	78.0	99.0	123.0	146.4	171.2	195.1	218.8	242.6
4.75	75.1	95.1	117.7	139.9	163.1	186.1	209.3	231.3
5.00	72.3	91.4	112.6	134.0	155.7	178.2	199.6	221.3
5.25	69.6	87.8	107.8	128.3	148.7	170.5	190.0	211.4
5.50	67.0	84.2	103.1	122.7	142.1	162.8	181.4	201.8
5.75	64.4	80.6	98.6	117.3	135.7	155.3	173.2	192.4
6.00	61.8	77.2	94.3	112.0	129.6	148.0	165.4	183.4
6.25	59.3	73.8	90.1	106.9	123.7	140.9	157.8	174.7
6.50	56.8	70.7	86.0	102.0	118.0	134.1	150.5	166.3
6.75	54.4	67.7	82.1	97.2	112.5	127.6	143.3	158.4
7.00	51.9	64.6	78.4	92.6	107.1	121.4	136.3	150.8

Equivalent conductances (Ω^{-1} cm^2 equiv^{-1}) of aqueous solutions of HI at 25°C

c	Λ	c	Λ	c	Λ	c	Λ	c	Λ
mol l^{-1}		mol l^{-1}		mol l^{-1}		mol l^{-1}		mol l^{-1}	
0.00045	423.2								
0.00050	423.0	0.0035	417.9	0.025	406.6	0.10	394.0	0.90	349.2
0.00055	422.9	0.0040	417.3	0.030	405.2	0.15	389.5	0.95	346.6
0.00060	422.7	0.0045	416.8	0.035	403.9	0.20	385.9	1.0	343.9
0.00065	422.6	0.0050	416.4	0.040	402.8	0.25	382.9	1.5	316.4
0.00070	422.4	0.0055	415.9	0.045	401.7	0.30	380.1	2.0	288.9
0.00075	422.3	0.0060	415.5	0.050	400.8	0.40	374.9	2.5	262.5
0.00080	422.2	0.0065	415.1	0.055	399.9	0.45	372.3	3.0	237.9
0.00085	422.0	0.0070	414.8	0.060	399.1	0.50	369.8	3.5	215.4
0.00090	421.9	0.0075	414.4	0.065	398.3	0.55	367.3	4.0	195.1
0.00095	421.8	0.0080	414.1	0.070	397.6	0.60	364.8	4.5	176.8
0.0010	421.7	0.0090	413.4	0.075	396.9	0.65	362.2	5.0	160.4
0.0015	420.7	0.0095	413.1	0.080	396.3	0.70	359.7	5.5	145.5
0.0020	419.8	0.010	412.8	0.085	395.6	0.75	357.1	6.0	131.7
0.0025	419.1	0.015	410.3	0.090	395.1	0.80	354.5	6.5	118.6
0.0030	418.5	0.020	408.3	0.095	394.5	0.85	351.9	7.0	105.7

Equivalent conductances (Ω^{-1} cm^2 equiv^{-1}) of aqueous solutions of HI from − 20to 50°C

c	− 20°C	− 10°C	0°C	10°C	20°C	30°C	40°C	50°C
mol l^{-1}								
0.4	—	—	253.9	300.7	354.5	411.9	453.6	501.8
0.6	—	—	249.1	293.6	344.6	401.1	441.5	488.4
0.8	—	—	242.4	285.5	333.9	389.1	429.3	474.8
1.0	—	—	234.8	276.9	322.7	376.3	416.5	460.8
1.2	—	—	226.6	267.9	311.4	363.1	403.4	446.3
1.4	—	—	218.2	258.8	300.2	349.8	390.0	431.4
1.6	—	—	209.9	249.7	289.1	336.6	376.5	416.3
1.8	—	—	201.8	240.7	278.4	323.7	363.0	401.2
2.0	—	—	194.0	231.9	268.1	311.1	349.6	386.1
2.2	—	147.7	186.6	223.4	258.2	299.0	336.3	371.2
2.4	—	143.5	179.5	215.1	248.8	287.3	323.3	356.5
2.6	—	138.8	172.8	207.2	239.7	276.1	310.5	342.2
2.8	—	133.9	166.5	199.5	231.0	265.4	298.1	328.2
3.0	99.9	129.0	160.5	192.0	222.7	255.1	286.1	314.7
3.2	97.4	124.1	154.7	184.8	214.6	245.2	274.5	301.7
3.4	94.2	119.5	149.1	177.8	206.7	235.7	263.4	289.2
3.6	90.6	115.0	143.7	171.0	198.9	226.4	252.6	277.3
3.8	86.9	110.9	138.3	164.3	191.2	217.4	242.3	266.0
4.0	83.6	106.9	132.8	157.6	183.5	208.4	232.5	255.3
4.2	80.9	103.0	127.3	151.0	175.6	199.6	223.1	245.3
4.4	79.3	99.2	121.6	144.3	167.5	190.7	214.1	235.9

EQUIVALENT CONDUCTIVITIES, λ, OF SOME ELECTROLYTES IN AQUEOUS SOUTION AT 25°C

Petr Vanýsek

The values of λ are given in 10^{-4} m² S mol⁻¹

Compound	Concentration, mol⁻¹ Infinite dilution	0.0005	0.001	0.005	0.01	0.02	0.05	0.1
	λ							
$AgNO_3$	133.29	131.29	130.45	127.14	124.70	121.35	115.18	109.09
$1/2BaCl_2$	139.91	135.89	134.27	127.96	123.88	119.03	111.42	105.14
$1/2CaCl_2$	135.77	131.86	130.30	124.19	120.30	115.59	108.42	102.41
$1/2Ca(OH)_2$	258	—	—	233	226	214	—	—
$1/2CuSO_4$	133.6	121.6	115.20	94.02	83.08	72.16	59.02	50.55
HCl	425.95	422.53	421.15	415.59	411.80	407.04	398.89	391.13
KBr	151.9	—	—	146.02	143.36	140.41	135.61	131.32
KCl	149.79	147.74	146.88	143.48	141.20	138.27	133.30	128.90
$KClO_4$	139.97	138.69	137.80	134.09	131.39	127.86	121.56	115.14
$1/3K_3Fe(CN)_6$	174.5	166.4	163.1	150.7	—	—	—	—
$1/4K_4Fe(CN)_6$	184		167.16	146.02	134.76	122.76	107.65	97.82
$KHCO_3$	117.94	116.04	115.28	112.18	110.03	107.17	—	—
KI	150.31	—	—	144.30	142.11	139.38	134.90	131.05
KIO_4	127.86	125.74	124.88	121.18	118.45	114.08	106.67	98.2
KNO_3	144.89	142.70	141.77	138.41	132.75	132.34	126.25	120.34
$KMnO_4$	134.8	—	133.3	—	126.5	—	—	113
KOH	271.5	—	234	230	228	—	219	213
$KReO_4$	128.20	126.03	125.12	121.31	118.49	114.49	106.40	97.40
$1/3LaCl_3$	145.9	139.6	137.0	127.5	121.8	115.3	106.2	99.1
LiCl	114.97	113.09	112.34	109.35	107.27	104.60	100.06	95.81
$LiClO_4$	105.93	104.13	103.39	100.52	98.56	96.13	92.15	88.52
$1/2MgCl_2$	129.34	125.55	124.15	118.25	114.49	109.99	103.03	97.05
NH_4Cl	149.6	—	146.7	134.4	141.21	138.25	133.22	128.69
NaCl	126.39	124.44	123.68	120.59	118.45	115.70	111.01	106.69
$NaClO_4$	117.42	115.58	114.82	111.70	109.54	106.91	102.35	98.38
NaI	126.88	125.30	124.19	121.19	119.18	116.64	112.73	108.73
$NaOOCCH_3$	91.0	89.2	88.5	85.68	83.72	81.20	76.88	72.76
$NaOOC_2H_5$	85.88	84.20	83.50	80.86	79.01	76.59	—	—
$NaOOC_3H_7$	82.66	81.00	80.27	77.54	75.72	73.35	69.29	65.24
NaOH	247.7	245.5	244.6	240.7	237.9	—	—	—
Na picrate	80.45	—	78.6	757	73.7	—	66.3	61.8
$1/2Na_2SO_4$	129.8	125.68	124.09	117.09	112.38	106.73	97.70	89.94
$1/2SrCl_2$	135.73	131.84	130.27	124.18	120.23	115.48	108.20	102.14
$1/2ZnSO_4$	132.7	121.3	114.47	95.44	84.87	74.20	61.17	52.61

EQUIVALENT IONIC CONDUCTIVITIES EXTRAPOLATED TO INFINITE DILUTION IN AQUEOUS SOLUTIONS AT 25°C

Petr Vanýsek

Ion	Λ_o (10^{-4} m² S mol⁻¹)	Ion	Λ_o (10^{-4} m² S mol⁻¹)	Ion	Λ_o (10^{-4} m² S mol⁻¹)
Inorganic cations		**Inorganic anions**		**Inorganic anions**	
Ag^+	61.9	$1/3Dy^{3+}$	65.6	NH_4^+	73.5
$1/3Al^{3+}$	61	$1/3Er^{3+}$	65.9	$N_2H_5^+$	59
$1/2Ba^{2+}$	63.6	$1/3Eu^{3+}$	67.8	Na^+	50.08
$1/2Be^{2+}$	45	$1/2Fe^{2+}$	54	$1/3Nd^{3+}$	69.4
$1/2Ca^{2+}$	59.47	$1/3Fe^{3+}$	68	$1/2Ni^{2+}$	50
$1/2Cd^{2+}$	54	$1/3Gd^{3+}$	67.3	$1/4[Ni_2(trien)_3]^{4+}$	52
$1/3Ce^{3+}$	69.8	H^+	349.65	$1/2Pb^{2+}$	71
$1/2Co^{2+}$	55	$1/2Hg_2^{2+}$	68.6	$1/3Pr^{3+}$	69.5
$1/3[Co(NH_3)_6]^{3+}$	101.9	$1/2Hg^{2+}$	63.6	$1/2Ra^{2+}$	66.8
$1/3[Co(en)_3]^{3+}$	74.7	$1/3Ho^{3+}$	66.3	Rb^+	77.8
$1/6[Co_2(trien)_3]^{6+}$	69	K^+	73.48	$1/3Sc^{3+}$	64.7
$1/3Cr^{3+}$	67	$1/3La^{3+}$	69.7	$1/3Sm^{3+}$	68.5
Cs^+	77.2	Li^+	38.66	$1/2Sr^{2+}$	59.4
$1/2Cu^{2+}$	53.6	$1/2Mg^{2+}$	53.0	Tl^+	74.7
D^+ (deuterium)	213.7 (18°)	$1/2Mn^{2+}$	53.5	$1/3Tm^{3+}$	65.4

EQUIVALENT IONIC CONDUCTIVITIES EXTRAPOLATED TO INFINITE DILUTION IN AQUEOUS SOLUTIONS AT 25°C (continued)

Ion	Λ_o (10^{-4} m^2 S mol^{-1})
Inorganic cations	
$1/2UO_2^{2+}$	32
$1/3Y^{3+}$	62
$1/3Yb^{3+}$	65.6
$1/2Zn^{2+}$	52.8
$Au(CN)_2^-$	50
$Au(CN)_4^-$	36
$B(C_6H_5)_4^-$	21
Br^-	78.1
Br_3^-	43
BrO_3^-	55.7
CN^-	78
CNO^-	64.6
$1/2CO_3^{2-}$	69.3
Cl^-	76.31
ClO_2^-	52
ClO_3^-	64.6
ClO_4^-	67.3
$1/3[Co(CN)_6]^{3-}$	98.9
$1/2CrO_4^{2-}$	85
F^-	55.4
$1/4[Fe(CN)_6]^{4-}$	110.4

Ion	Λ_o (10^{-4} m^2 S mol^{-1})
Inorganic cations	
$1/3[Fe(CN)_6]^{3-}$	100.9
$H_2AsO_4^-$	34
HCO_3^-	44.5
HF_2^-	75
$1/2HPO_4^{2-}$	33
$H_2PO_4^-$	33
$H_2PO_2^-$	46
HS^-	65
HSO_3^-	50
HSO_4^-	50
$H_2SbO_4^-$	31
I^-	76.8
IO_3^-	40.5
IO_4^-	54.5
MnO_4^-	61.3
MoO_4^-	74.5
$N(CN)_2^-$	54.5
NO_2^-	71.8
NO_3^-	71.42
$NH_2SO_3^-$	48.6
N_3^-	69

Ion	Λ_o (10^{-4} m^2 S mol^{-1})
Inorganic cations	
OCN^-	64.6
OH^-	198
PF_6^-	56.9
$1/2PO_3F^{2-}$	63.3
$1/2PO_4^{3-}$	69.0
$1/4P_2O_7^{4-}$	96
$1/3P_3O_9^{3-}$	83.6
$1/5P_3O_{10}^{5-}$	109
$1/3P_4O_{13}^{2-}$	94
ReO_4^-	54.9
SCN^-	66
$1/2SO_3^{2-}$	79.9
$1/2SO_4^{2-}$	80.0
$1/2S_2O_3^{2-}$	85.0
$1/2S_2O_4^{2-}$	66.5
$1/2S_2O_6^{2-}$	93
$1/2S_2O_8^{2-}$	86
$Sb(OH)_6^-$	31.9
$SeCN^-$	64.7
$1/2SeO_4^{2-}$	75.7
$1/2WO_4^{2-}$	69

Ion	Λ_o	Ion	Λ_o
Organic cations		**Organic cations**	
Benzyltrimethylammonium	34.6	Octadecyltriethylammonium	17.9
i-Butylammonium	38	Octadecyltrimethylammonium	19.9
Butyltrimethylammonium	33.6	Octadecyltripropylammonium	17.2
n-Decylpyridinium	29.5	Octyltrimethylammonium	26.5
Decyltrimethylammonium	24.4	Pentylammonium	37
Diethylammonium	42.0	Piperidinium	37.2
Dimethylammonium	51.8	Propylammonium	40.8
Dipropylammonium	30.1	Pyrilammonium	24.3
n-Dodecylammonium	23.8	Tetra-*n*-butylammonium	19.5
Dodecyltrimethylammonium	22.6	Tetradecyltrimethylammonium	21.5
Ethanolammonium	42.2	Tetraethylammonium	32.6
Ethylammonium	47.2	Tetramethylammonium	44.9
Ethyltrimethylammonium	40.5	Tetra-*i*-pentylammonium	17.9
Hexadecyltrimethylammonium	20.9	Tetra-*n*-pentylammonium	17.5
Hexyltrimethylammonium	29.6	Tetra-*n*-propylammonium	23.4
Histidyl	23.0	Triethylammonium	34.3
Hydroxyethyltrimethylarsonium	39.4	Triethylsulfonium	36.1
Methylammonium	58.7	Trimethylammonium	47.23
Octadecylpyridinium	20	Trimethylhexylammonium	34.6
Octadecyltributylammoniuim	16.6	Trimethylsulfonium	51.4
		Tripropylammonium	26.1
Organic anions		**Organic anions**	
Acetate	40.9	$1/2$Diethylbarbiturate^{2-}	26.3
p-Anisate	29.0	Dihydrogencitrate	30
$1/2$Azelate^{2-}	40.6	$1/2$Dimethylmalonate^{2-}	49.4
Benzoate	32.4	3,5-Dinitrobenzoate	28.3
Bromoacetate	39.2	Dodecylsulfate	24
Bromobenzoate	30	Ethylmalonate	49.3
n-Butyrate	32.6	Ethylsulfate	39.6
Chloroacetate	42.2	Fluoroacetate	44.4
m-Chlorobenzoate	31	Fluorobenzoate	33
o-Chlorobenzoate	30.2	Formate	54.6
$1/3$Citrate^{3-}	70.2	$1/2$Fumarate^{2-}	61.8
Crotonate	33.2	$1/2$Glutarate^{2-}	52.6
Cyanoacetate	43.4	Hydrogenoxalate	40.2
Cyclohexane carboxylate	28.7	Isovalerate	32.7
$1/2$ 1,1-Cyclopropane-dicarboxylate^{2-}	53.4	Iodoacetate	40.6
Decylsulfate	26	Lactate	38.8
Dichloroacetate	38.3	$1/2$Malate^{2-}	58.8

EQUIVALENT IONIC CONDUCTIVITIES EXTRAPOLATED TO INFINITE DILUTION IN AQUEOUS SOLUTIONS AT 25°C (continued)

Organic anions		Organic anions	
1/2Maleate^{2-}	61.9	Picrate	30.37
1/2Malonate^{2-}	63.5	Pivalate	31.9
Methylsulfate	48.8	Propionate	35.8
Naphtylacetate	28.4	Propylsulfate	37.1
1/2Oxalate^{2-}	74.11	Salicylate	36
Octylsulfate	29	1/2Suberate^{2-}	36
Phenylacetate	30.6	1/2Succinate^{2-}	58.8
1/2*o*-Phtalate^{2-}	52.3	1/2Tartarate^{2-}	59.6
1/2*m*-Phtalate^{2-}	54.7	Trichloroacetate	36.6

ACTIVITY COEFFICIENTS OF ACIDS, BASES AND SALTS

Petr Vanýsek

The following coefficients are valid at 25°C. The concentrations are expressed as molalities.

	0.1	**0.2**	**0.3**	**0.4**	**0.5**	**0.6**	**0.7**	**0.8**	**0.9**	**1.0**
$AgNO_3$	0.734	0.657	0.606	0.567	0.536	0.509	0.485	0.464	0.446	0.429
$AlCl_3$	0.337	0.305	0.302	0.313	0.331	0.356	0.388	0.429	0.479	0.539
$Al_2(SO_4)_3$	0.035	0.0225	0.0176	0.0153	0.0143	0.014	0.0142	0.0149	0.0159	0.0175
$BaCl_2$	0.500	0.444	0.419	0.405	0.397	0.391	0.391	0.391	0.392	0.395
$BeSO_4$	0.150	0.109	0.0885	0.0769	0.0692	0.0639	0.0600	0.0570	0.0546	0.0530
$CaCl_2$	0.518	0.472	0.455	0.448	0.448	0.453	0.460	0.470	0.484	0.500
$CdCl_2$	0.2280	0.1638	0.1329	0.1139	0.1006	0.0905	0.0827	0.0765	0.0713	0.0669
$Cd(NO_3)_2$	0.513	0.464	0.442	0.430	0.425	0.423	0.423	0.425	0.428	0.433
$CdSO_4$	0.150	0.103	0.0822	0.0699	0.0615	0.0553	0.0505	0.0468	0.0438	0.0415
$CoCl_2$	0.522	0.479	0.463	0.459	0.462	0.470	0.479	0.492	0.511	0.531
$CrCl_3$	0.331	0.298	0.294	0.300	0.314	0.335	0.362	0.397	0.436	0.481
$Cr(NO_3)_3$	0.319	0.285	0.279	0.281	0.291	0.304	0.322	0.344	0.371	0.401
$Cr_2(SO_4)_3$	0.0458	0.0300	0.0238	0.0207	0.0190	0.0182	0.0181	0.0185	0.0194	0.0208
CsBr	0.754	0.694	0.654	0.626	0.603	0.586	0.571	0.558	0.547	0.538
CsCl	0.756	0.694	0.656	0.628	0.606	0.589	0.575	0.563	0.553	0.544
CsI	0.754	0.692	0.651	0.621	0.599	0.581	0.567	0.554	0.543	0.533
$CsNO_3$	0.733	0.655	0.602	0.561	0.528	0.501	0.478	0.458	0.439	0.422
CsOH	0.795	0.761	0.744	0.739	0.739	0.742	0.748	0.754	0.762	0.771
CsOAc	0.799	0.771	0.761	0.759	0.762	0.768	0.776	0.783	0.792	0.802
Cs_2SO_4	0.456	0.382	0.338	0.311	0.291	0.274	0.262	0.251	0.242	0.235
$CuCl_2$	0.508	0.455	0.429	0.417	0.411	0.409	0.409	0.410	0.413	0.417
$Cu(NO_3)_2$	0.511	0.460	0.439	0.429	0.426	0.427	0.431	0.437	0.445	0.455
$CuSO_4$	0.150	0.104	0.0829	0.0704	0.0620	0.0559	0.0512	0.0475	0.0446	0.0423
$FeCl_2$	0.5185	0.473	0.454	0.448	0.450	0.454	0.463	0.473	0.488	0.506
HBr	0.805	0.782	0.777	0.781	0.789	0.801	0.815	0.832	0.850	0.871
HCl	0.796	0.767	0.756	0.755	0.757	0.763	0.772	0.783	0.795	0.809
$HClO_4$	0.803	0.778	0.768	0.766	0.769	0.776	0.785	0.795	0.808	0.823
HI	0.818	0.807	0.811	0.823	0.839	0.860	0.883	0.908	0.935	0.963
HNO_3	0.791	0.754	0.735	0.725	0.720	0.717	0.717	0.718	0.721	0.724
H_2SO_4	0.2655	0.2090	0.1826	—	0.1557	—	0.1417	—	—	0.1316
KBr	0.772	0.722	0.693	0.673	0.657	0.646	0.636	0.629	0.622	0.617
KCl	0.770	0.718	0.688	0.666	0.649	0.637	0.626	0.618	0.610	0.604
$KClO_3$	0.749	0.681	0.635	0.599	0.568	0.541	0.518	—	—	—
K_2CrO_4	0.456	0.382	0.340	0.313	0.292	0.276	0.263	0.253	0.243	0.235
KF	0.775	0.727	0.700	0.682	0.670	0.661	0.654	0.650	0.646	0.645
$K_3Fe(CN)_6$	0.268	0.212	0.184	0.167	0.155	0.146	0.140	0.135	0.131	0.128
$K_4Fe(CN)_6$	0.139	0.0993	0.0808	0.0693	0.0614	0.0556	0.0512	0.0479	0.0454	—
KH_2PO_4	0.731	0.653	0.602	0.561	0.529	0.501	0.477	0.456	0.438	0.421
KI	0.778	0.733	0.707	0.689	0.676	0.667	0.660	0.654	0.649	0.645
KNO_3	0.739	0.663	0.614	0.576	0.545	0.519	0.496	0.476	0.459	0.443
KOAc	0.796	0.766	0.754	0.750	0.751	0.754	0.759	0.766	0.774	0.783
KOH	0.798	0.760	0.742	0.734	0.732	0.733	0.736	0.742	0.749	0.756
KSCN	0.769	0.716	0.685	0.663	0.646	0.633	0.623	0.614	0.606	0.599
K_2SO_4	0.441	0.360	0.316	0.286	0.264	0.246	0.232	—	—	—
LiBr	0.796	0.766	0.756	0.752	0.753	0.758	0.767	0.777	0.789	0.803
LiCl	0.790	0.757	0.744	0.740	0.739	0.743	0.748	0.755	0.764	0.774
$LiClO_4$	0.812	0.794	0.792	0.798	0.808	0.820	0.834	0.852	0.869	0.887
LiI	0.815	0.802	0.804	0.813	0.824	0.838	0.852	0.870	0.888	0.910
$LiNO_3$	0.788	0.752	0.736	0.728	0.726	0.727	0.729	0.733	0.737	0.743
LiOH	0.760	0.702	0.665	0.638	0.617	0.599	0.585	0.573	0.563	0.554
LiOAc	0.784	0.742	0.721	0.709	0.700	0.691	0.689	0.688	0.688	0.689
Li_2SO_4	0.468	0.398	0.361	0.337	0.319	0.307	0.297	0.289	0.282	0.277
$MgCl_2$	0.529	0.489	0.477	0.475	0.481	0.491	0.506	0.522	0.544	0.570

ACTIVITY COEFFICIENTS OF ACIDS, BASES AND SALTS (continued)

	0.1	0.2	0.3	0.4	0.5	0.6	0.7	0.8	0.9	1.0
$MgSO_4$	0.150	0.107	0.0874	0.0756	0.0675	0.0616	0.0571	0.0536	0.0508	0.0485
$MnCl_2$	0.516	0.469	0.450	0.442	0.440	0.443	0.448	0.455	0.466	0.479
$MnSO_4$	0.150	0.105	0.0848	0.0725	0.0640	0.0578	0.0530	0.0493	0.0463	0.0439
NH_4Cl	0.770	0.718	0.687	0.665	0.649	0.636	0.625	0.617	0.609	0.603
NH_4NO_3	0.740	0.677	0.636	0.606	0.582	0.562	0.545	0.530	0.516	0.504
$(NH_4)_2SO_4$	0.439	0.356	0.311	0.280	0.257	0.240	0.226	0.214	0.205	0.196
NaBr	0.782	0.741	0.719	0.704	0.697	0.692	0.689	0.687	0.687	0.687
NaCl	0.778	0.735	0.710	0.693	0.681	0.673	0.667	0.662	0.659	0.657
$NaClO_3$	0.772	0.720	0.688	0.664	0.645	0.630	0.617	0.606	0.597	0.589
$NaClO_4$	0.775	0.729	0.701	0.683	0.668	0.656	0.648	0.641	0.635	0.629
Na_2CrO_4	0.464	0.394	0.353	0.327	0.307	0.292	0.280	0.269	0.261	0.253
NaF	0.765	0.710	0.676	0.651	0.632	0.616	0.603	0.592	0.582	0.573
NaH_2PO_4	0.744	0.675	0.629	0.593	0.563	0.539	0.517	0.499	0.483	0.468
NaI	0.787	0.751	0.735	0.727	0.723	0.723	0.724	0.727	0.731	0.736
$NaNO_3$	0.762	0.703	0.666	0.638	0.617	0.599	0.583	0.570	0.558	0.548
NaOAc	0.791	0.757	0.744	0.737	0.735	0.736	0.740	0.745	0.752	0.757
NaOH	0.766	0.727	0.708	0.697	0.690	0.685	0.681	0.679	0.678	0.678
NaSCN	0.787	0.750		0.720	0.715	0.712	0.710	0.710	0.711	0.712
Na_2SO_4	0.445	0.365	0.320	0.289	0.266	0.248	0.233	0.221	0.210	0.201
$NiCl_2$	0.522	0.479	0.463	0.460	0.464	0.471	0.482	0.496	0.515	0.563
$NiSO_4$	0.150	0.105	0.0841	0.0713	0.0627	0.0562	0.0515	0.0478	0.0448	0.0425
$Pb(NO_3)_3$	0.395	0.308	0.260	0.228	0.205	0.187	0.172	0.160	0.150	0.141
RbBr	0.763	0.706	0.673	0.650	0.632	0.617	0.605	0.595	0.586	0.578
RbCl	0.764	0.709	0.675	0.652	0.634	0.620	0.608	0.599	0.590	0.583
RbI	0.762	0.705	0.671	0.647	0.629	0.614	0.602	0.591	0.583	0.575
$RbNO_3$	0.734	0.658	0.606	0.565	0.534	0.508	0.485	0.465	0.446	0.430
RbOAc	0.796	0.767	0.756	0.753	0.755	0.759	0.766	0.773	0.782	0.792
Rb_2SO_4	0.451	0.374	0.331	0.301	0.279	0.263	0.249	0.238	0.228	0.219
$SrCl_2$	0.511	0.462	0.442	0.433	0.430	0.431	0.434	0.441	0.449	0.461
$TlClO_4$	0.730	0.652	0.599	0.559	0.527	—	—	—	—	—
$TlNO_3$	0.702	0.606	0.545	0.500	—	—	—	—	—	—
UO_2Cl_2	0.544	0.510	0.520	0.505	0.517	0.532	0.549	0.571	0.595	0.620
UO_2SO_4	0.150	0.102	0.0807	0.0689	0.0611	0.0566	0.0515	0.0483	0.0458	0.0439
$ZnCl_2$	0.515	0.462	0.432	0.411	0.394	0.380	0.369	0.357	0.348	0.339
$Zn(NO_3)_2$	0.531	0.489	0.474	0.469	0.473	0.480	0.489	0.501	0.518	0.535
$ZnSO_4$	0.150	0.140	0.0835	0.0714	0.0630	0.0569	0.0523	0.0487	0.0458	0.0435

SPECIFIC HEAT OF WATER

Heat Capacity of Air-free Water 0°—100°C at 1 Atmosphere Pressure

The heat capacity of air-free water is given in international steam table calories per gram and in absolute joules per gram. (1 absolute joule — 0.238846 I.T. Cal.).
The enthalpy or heat content is given for air-free water in I.T. Cal. per gram and in absolute joules per gram.
From Osborne, Stimson and Ginnings; B. of S. Jour. Res. 23, 238, 1939.

Temp. °C.	Thermal Capacity		Enthalpy		Temp. °C	Thermal Capacity		Enthalpy	
	Cal./g/°C	Joules/g/°C	Cal./g	Joules/g		Cal/g/°C	Joules/g/°C	Cal/g	Joules/g
0	1.00790	4.0177	0.0045	0.1036	50	.99854	4.1807	50.0070	209.2700
1	1.00652	4.2141	1.0314	4.3184	51	.99862	4.1810	51.0065	213.5538
2	1.00571	4.2107	2.0376	8.5308	52	.99871	4.1814	52.0051	217.7350
3	1.00499	4.2077	3.0429	12.7400	53	.99878	4.1817	53.0039	221.9166
4	1.00430	4.2048	4.0475	16.9462	54	.99885	4.1820	54.0027	226.0984
5	1.00368	4.2022	5.0515	21.1498	55	.99895	4.1824	55.0016	230.2806
6	1.00313	4.1999	6.0549	25.3508	56	.99905	4.1828	56.0006	234.4632
7	1.00260	4.1977	7.0578	29.5496	57	.99914	4.1832	56.9997	238.6462
8	1.00213	4.1957	8.0602	33.7463	58	.99924	4.1836	57.9989	242.8296
9	1.00170	4.1939	9.0621	37.9410	59	.99933	4.1840	58.9982	247.0134
10	1.00129	4.1922	10.0636	42.1341	60	.99943	4.1844	59.9975	251.1976
11	1.00093	4.1907	11.0647	46.3255	61	.99955	4.1849	60.9970	255.3822
12	1.00060	4.1893	12.0654	50.5155	62	.99964	4.1853	61.9966	259.5673
13	1.00029	4.1880	13.0659	54.7041	63	.99976	4.1858	62.9963	263.7529
14	1.00002	4.1869	14.0660	58.8916	64	.99988	4.1863	63.9962	267.9390
15	.99976	4.1858	15.0659	63 0779	65	1.00000	4.1868	64.9961	272.1256
16	.99955	4.1849	16.0655	67.2632	66	1.00014	4.1874	65.9962	276.3127
17	.99933	4.1840	17.0650	71.4476	67	1.00026	4.1879	66.9964	280.5003
18	.99914	4.1832	18.0642	75.6312	68	1.00041	4.1885	67.9967	284.6885
19	.99897	4.1825	19.0633	79.8141	69	1.00053	4.1890	68.9972	288.8772
20	.99883	4.1819	20.0622	83.9963	70	1.00067	4.1896	69.9977	293.0665
21	.99869	4.1813	21.0609	88.1778	71	1.00081	4.1902	70.9985	297.2564
22	.99857	4.1808	22.0596	92.3589	72	1.00096	4.1908	71.9994	301.4469
23	.99847	4.1804	23.0581	96.5395	73	1.00112	4.1915	73.0004	305.6381
24	.99838	4.1800	24.0565	100.7196	74	1.00127	4.1921	74.0016	309.8299
25	.99828	4.1796	25.0548	104.8994	75	1.00143	4.1928	75.0030	314.0224
26	.99821	4.1793	26.0530	109.0788	76	1.00160	4.1935	76.0045	318.2155
27	.99814	4.1790	27.0512	113.2580	77	1.00177	4.1942	77.0062	322.4094
28	.99809	4.1788	28.0493	117.4369	78	1.00194	4.1949	78.0080	326.6039
29	.99804	4.1786	29.0474	121.6157	79	1.00213	4.1957	79.0101	330.7992
30	.99802	4 1785	30.0455	125 7943	80	1.00229	4.1964	80.0123	334.9952
31	.99799	4.1784	31.0435	129.9727	81	1.00248	4.1972	81.0147	339.1920
32	.99797	4.1783	32.0414	134.1510	82	1.00268	4.1980	82.0172	343.3897
33	.99797	4.1783	33.0394	138.3293	83	1.00287	4.1988	83.0200	347.5881
34	.99795	4.1782	34.0374	142.5076	84	1.00308	4.1997	84.0230	351.7873
35	.99795	4.1782	35.0353	146.6858	85	1.00327	4.2005	85.0262	355.9874
36	.99797	4.1783	36.0333	150.8641	86	1.00349	4.2014	86.0295	360.1883
37	.99797	4.1783	37.0312	155.0423	87	1.00370	4.2023	87.0331	364.3902
38	.99799	4.1784	38.0292	159.2207	88	1.00392	4.2032	88.0369	368.5929
39	.99802	4.1785	39.0272	163.3991	89	1.00416	4.2042	89.0410	372.7966
40	.99804	4.1786	40.0253	167.5777	90	1.00437	4.2051	90.0452	377.0012
41	.99807	4.1787	41.0233	171.7563	91	1.00461	4.2061	91.0497	381.2068
42	.99811	4.1789	42.0214	175.9351	92	1.00485	4.2071	92.0545	385.4135
43	.99816	4.1791	43.0195	180.1141	93	1.00509	4.2081	93.0594	389.6211
44	.99819	4.1792	44.0177	184.2933	94	1.00535	4.2092	94.0647	393.8297
45	.99826	4.1795	45.0159	188.4726	95	1.00561	4.2103	95.0701	398.0395
46	.99830	4.1797	46.0142	192.6522	96	1.00588	4.2114	96.0759	402.2503
47	.99835	4.1799	47.0125	196.8320	97	1.00614	4.2125	97.0819	406.4622
48	.99842	4.1802	48.0109	201.0120	98	1.00640	4.2136	98.0882	410.6753
49	.99847	4.1804	49.0094	205.1923	99	1.00669	4.2148	99.0947	414.8895
					100	1.00697	4.2160	100.1015	419.1049

Enthalpy of Air-saturated Water
1 Atmosphere Pressure 0-100°C

Temp. °C	Enthalpy		Temp. °C	Enthalpy	
	Cal/g	Joules/g		Cal/g	Joules/g
0	0	0	50	49.9896	209.2964
5	5.0276	21.0496	55	54.9842	230.2077
10	10.0402	42.0363	60	59.9811	251.1289
15	15.0431	62.9826	65	64.9808	272.0619
20	20.0400	83.9034	70	69.9839	293.0087
25	25.0332	104.8089	75	74.9907	313.9712
30	30.0244	125.7063	80	80.0019	334.9519
35	35.0149	146.6003	85	85.0180	355.9532
40	40.0055	167.4949	90	90.0395	376.9773
45	44.9968	188.3928	95	95.0671	398.0270
			100	100.1016	419.1053

SPECIFIC HEAT OF WATER (Continued)

Specific Heat of Super-heated Steam

Specific heat of steam under constant pressure given in atmospheres and at temperatures above saturation in Cal./g/°C

Temp. °C	Pressure in atmospheres 1	2	4	6	8	10	12
110	0.481						
120	0.477	0.498					
130	0.475	0.494					
140	0.473	0.489					
150	0.472	0.486	0.519				
160	0.471	0.483	0.512	0.549			
170	0.470	0.481	0.507	0.538			
180	0.469	0.479	0.502	0.528	0.561	0.602	
190	0.469	0.478	0.498	0.522	0.549	0.583	0.625
200	0.469	0.478	0.495	0.515	0.539	0.567	0.601
210	0.470	0.477	0.493	0.510	0.531	0.555	0.584
220	0.470	0.477	0.491	0.506	0.524	0.545	0.569
230	0.471	0.477	0.489	0.504	0.519	0.537	0.557
240	0.472	0.477	0.488	0.501	0.515	0.530	0.548
250	0.473	0.477	0.488	0.499	0.512	0.525	0.540
260	0.474	0.478	0.587	0.498	0.509	0.521	0.534
270	0.474	0.478	0.487	0.497	0.507	0.518	0.529
280	0.475	0.479	0.487	0.496	0.505	0.515	0.525
290	0.476	0.480	0.487	0.495	0.504	0.513	0.523
300	0.477	0.481	0.488	0.495	0.503	0.511	0.519
310	0.478	0.482	0.408	0.495	0.502	0.510	0.518
320	0.480	0.483	0.489	0.496	0.502	0.509	0.516
330	0.482	0.484	0.490	0.496	0.502	0.508	0.515
340	0.483	0.485	0.491	0.496	0.502	0.507	0.513
350	0.484	0.486	0.492	0.497	0.502	0.507	0.512
360	0.485	0.487	0.492	0.497	0.502	0.507	0.511
370	0.486	0.488	0.493	0.498	0.503	0.507	0.511
380	0.488	0.490	0.494	0.498	0.503	0.507	0.511
390	0.489	0.491	0.495	0.499	0.503		
400	0.490	0.492	0.496	0.500	0.504		
410	0.492	0.494	0.497	0.501	0.505		
420	0.494	0.496	0.498	0.502	0.506		
430	0.495	0.497	0.500	0.504	0.507		
440	0.497	0.499	0.501	0.505	0.508		
450	0.498	0.500	0.503	0.506	0.509		
460	0.500	0.501	0.505	0.507	0.510		
470	0.502	0.503	0.506	0.508	0.512		
480	0.504	0.505	0.507	0.509	0.513		
490	0.505	0.506	0.509	0.511	0.514		
500	0.506	0.508	0.510	0.512	0.515		

Specific Heat of Water Above 100°C

Mean specific heat of water in 15°C calories between 0°C and the temperature stated.
Heat content (Enthalpy) in joules per gram between 0°C and the temperature stated.
From data by Osborne, Stimson and Fiock, B of S Jour. Res. 5, 411, 1930.

Temp. °C	Specific heat mean 0-t°C	Heat content 0-t joules/g
100	1.0008	418.75
110	1.0015	460.97
120	1.0025	503.36
130	1.0037	545.93
140	1.0050	588.71
150	1.0067	631.75
160	1.0083	675.06
170	1.0103	718.66
180	1.0127	762.72
190	1.0153	807.15
200	1.0181	852.02
210	1.0212	897.35
220	1.0247	943.24
230	1.0285	989.75
240	1.0326	1036.97
250	1.0376	1084.97
260	1.0423	1133.87
270	1.0483	1184.32

Specific Heat of Ice — Cal./g/°C

Temp. °C	Specific heat	Observer
−252 to −188	.146	Dieterici, 1903
−250	.0361	
−200	.162	Mean
−188 to −78	.285	Dieterici, 1903
−180	.199	Nernst, 1910
−160	.230	Nernst, 1910
−150	.246	
−140	.262	Nernst, 1910
−100	.329	Mean
− 78 to −18	.463	Dieterici, 1903
−31.8	.4454	Dickinson-Osborne, 1915
−23.7	.4599	Dickinson-Osborne, 1915
−24.5	.4605	Dickinson-Osborne, 1915
−20.8	.4668	Dickinson-Osborne, 1915
−14.8	.4782	Dickinson-Osborne, 1915
−14.6	.4779	Dickinson-Osborne, 1915
−11.0	.4861	Dickinson-Osborne, 1915
− 8.1	.4896	Dickinson-Osborne, 1915
− 4.3	.4989	Dickinson-Osborne, 1915
− 4.5	.4984	Dickinson-Osborne, 1915
− 60	.392	
− 38.3	.4346	Dickinson-Osborne, 1915
− 34.3	.4411	Dickinson-Osborne, 1915
− 30.6	.4488	Dickinson-Osborne, 1915
− 4.9	.4932	Dickinson-Osborne, 1915
− 2.6	.5003	Dickinson-Osborne, 1915
− 2.2	.5018	Dickinson-Osborne, 1915

Water Below 0°C

Temp. °C	Specific heat	Observer
− 6	1.0119	Martinetti, 1890
− 5	1.0155	Barnes, 1902
− 5	1.0113	Martinetti, 1890
− 4	1.0105	Martinetti, 1890
− 3	1.0102	Martinetti, 1890
− 2	1.0097	Martinetti, 1890
− 1	1.0092	Martinetti, 1890

HEAT CAPACITY OF MERCURY

The specific heat of solid mercury is given in relation to water at 15°C. The values are from Carpenter and Stoodley, Phil. Mag. 10, 249, 1930. Heat capacity is given in calories per gram and in calories per gram atom (1 cal = 4.1840 absolute joules and the atomic weight of mercury 200.61). Values for the liquid and vapor are from Douglas, Ball and Ginnings, Jour. of Res. Bureau of Standards 46, 334, 1951.

Temp. °C	Specific heat Solid	Heat capacity cal/g-atom Solid
−75.6	.0319	6.3995
−72.9	.0324	6.4998
−65.4	.0324	6.4998
−59.5	.0324	6.4998
−44.9	.0336	6.7405
−42.2	.0336	6.7405
−40.0	.0337	6.7606

Temp. °C	Heat capacity cal/g Liquid	Heat capacity cal/g-atom Liquid
−38.88	.033686	6.7578
−20	.033534	6.7272
0	.033382	6.6967
20	.033240	6.6683
25	.033206	6.6615
40	.033109	6.6419
60	.032987	6.6176
80	.032877	6.5954
100	.032776	6.5752
120	.032686	6.5571
140	.032606	6.5410
160	.032535	6.5270
180	.032476	6.5150
200	.032426	6.5050
220	.032386	6.4970
240	.032356	6.4910
260	.032336	6.4869
280	.032325	6.4847
300	.032322	6.4843
320	.032330	6.4858
340	.032346	6.4890
356.58	.032366	6.4930
360	.032371	6.4940
380	.032404	6.5005
400	.032445	6.5087
420	.032494	6.5186
440	.032550	6.5298
460	.032614	6.5426
480	.032684	6.5567
500	.032762	6.5723

Temp. °C	Heat capacity cal/g Vapor	Heat capacity cal/g-atom Vapor
0	.02476	4.968
100	.02476	4.968
200	.02477	4.969
300	.02480	4.975
400	.02489	4.993
500	.02507	5.030

HEAT CAPACITY (C_p) OF ORGANIC LIQUIDS AND VAPORS AT 25C*

The values in this table are expressed as cal/(g mole) (°K) at 1 atm pressure. Calories, g × 4.184 = joules.

Robert Shaw

	C_p(l)	C_p(g)
Alkanes		
2-Methyl butane	39.4	28.5
n-Hexane	46.6	34.2
2-Methyl pentane	46.2	34.5
3-Methyl pentane	[illegible]	[illegible]
2,3-Dimethyl butane	45.0	33.6
2,2-Dimethyl butane	45.0	33.9
n-Heptane	53.7	39.7
2-Methyl hexane	53.1	
3-Ethyl pentane	52.5	
2,2-Dimethyl pentane	52.8	
2,3-Dimethyl pentane	51.8	
2,4-Dimethyl pentane	53.5	
3,3-Dimethyl pentane	51.4	
2,2,3-Trimethyl butane	51.0	39.3
n-Octane	60.7	45.1
2,2,4-Trimethyl pentane	56.5	
2,2,3,3-Tetramethyl butane	55.9	46.0
n-Nonane	68.0	50.6
n-Decane	75.2	56.1
2-Methyl nonane	75.6	
n-Undecane	82.5	61.5
n-Dodecane	89.7	67.0
n-Tridecane	97.2	72.5
n-Tetradecane	104.8	77.9
n-Pentadecane	112.2	83.4
n-Hexadecane	119.9	88.9
Olefins		
Pentene-1	37.1	26.2
cis-Pentene-2	36.3	24.3
trans-Pentene-2	37.6	25.9
2-Methyl butene-1	37.5	26.7
3-Methyl butene-1	37.3	28.4
2-Methyl butene-2	36.4	25.1
Hexene-1	43.8	31.6
2,3-Dimethyl butene-2	41.7	30.5
Heptene-1	50.5	37.1
Octene-1	57.7	42.6
Decene-1	71.8	53.5
Undecene-1	78.9	59.0
Dodecene-1	86.2	64.4
Hexadecene-1	115.9	86.3
Butadiene-1,3	29.5	19.0
Pentadiene-1,4	34.8	25.1
2-Methyl butadiene-1,3	36.7	25.0
Butyne-2	29.4	18.6
Aromatics		
Benzene	32.4	19.5
Toluene	37.3	24.8
Ethyl benzene	44.5	30.7
p-Xylene	43.6	30.3
1,2,3-Trimethyl benzene	51.8	36.9
1,2,4-Trimethyl benzene	51.6	37.1
1,3,5-Trimethyl benzene	50.1	35.9
Oxygen-Containing		
Methanol	19.5	10.5
Ethanol	27.0	15.7
Propanol-1	34.3	20.9
Propanol-2	37.1	21.3
2-Methyl propanol-2	53.2	27.2
estimated alternative	42.2	
Di-*n*-butyl sulfide	68.0	49.7
Methyl isopropyl sulfide	41.2	28.8
Ethyl mercaptan	28.2	17.4
N-butyl mercaptan	41.3	28.4
Iso-butyl mercaptan	41.1	28.4
Amyl mercaptan	48.2	33.9
Iso-propyl mercaptan	34.7	23.1
Sec-butyl mercaptan	40.9	28.6
Cyclo-pentyl mercaptan	39.5	25.9
Tert-butyl mercaptan	41.8	29.0
Dimethyl disulfide	34.9	22.6
Diethyl ether	41.1	25.8
Acetone	30.2	18.0
Formic acid	23.6	10.8
Acetic acid	29.4	16.0
Ethyl acetate	40.0	
Benzyl alcohol	52.1	
Diphenyl ether	64.2	
Furan	27.4	
Nitrogen-Containing		
Hydrogen cyanide	16.9	8.6
Hydrazine	22.2	12.2
Unsym. dimethyl hydrazine	39.2	
Trimethyl hydrazine	44.5	
Aniline	46.2	26.1
Quinoline	47.6	
Acrylonitrile	26.5	14.9
Perfluoro piperidene	71.0	
Methyl nitrate	37.5	
Tetrahydropyrole	37.4	
Nitromethane	25.4	13.8
Ring Compounds		
Cyclopentane	30.4	19.8
Methyl cyclopentane	38.0	26.2
Ethyl cyclopentane	44.5	31.5
1,1-Dimethyl cyclopentane	44.6	31.9
cis-1,2-Dimethyl cyclopentane	45.1	32.1
trans-1,2-Dimethyl cyclopentane	45.1	32.1
trans-1,3-Dimethyl cyclopentane	45.1	32.1
Cyclopentane	29.4	18.0
Cyclohexane	36.4	25.4
Methyl cyclohexane	44.2	32.3
Ethyl cyclohexane	50.0	38.0
1,2-Dimethyl cyclohexane	50.4	37.4
cis-1,3-Dimethyl cyclohexane	50.0	37.6
trans-1,3-Dimethyl cyclohexane	50.8	37.6
cis-1,4-Dimethyl cyclohexane	50.5	37.6
trans-1,4-Dimethyl cyclohexane	50.3	37.7
Cyclohexane	35.6	25.1
Cyclooctatetraene	44	
Tetralin $C_6H_4(CH_2)_4$	52	
cis-Decalin	55.4	
trans-Decalin	54.5	
Halogen-Containing		
Chloroform	27.6	15.7
Carbon tetrachloride	31.7	19.9
1,2-Dibromoethane	32.2	18.2
1,1,1-Trichloro ethane	34.4	22.4
1,1-Dichloro ethylene	26.6	16.1
Fluoro benzene	35.1	22.7
Chloro benzene	34.9	23.6
Bromo benzene	37.1	24.2
Sulfur-Containing		
Carbon disulfide[a]	18.2	10.9
Dimethyl sulfide	28.3	17.8
Methylethyl sulfide	34.6	22.8
Methyl *n*-propyl sulfide	41.0	28.2
Methyl *n*-butyl sulfide	48.0	33.8
Diethyl sulfide	40.9	28.1
Ethyl *n*-propyl sulfide	47.4	33.4
Di-*n*-propyl sulfide	54.0	38.7
Diethyl disulfide	48.7	33.9
$(CH_2)_3S$	27.1	16.7
$(CH_2)_4S$	33.5	21.9
$(CH_2)_5S$	39.0	26.0
Phenyl mercaptan	39.1	25.2
Silicon-Containing		
Tetramethylsilane	49.8	33.5
Hexamethyldisiloxane	74.4	

*See also the table which begins on the following page.

HEAT CAPACITY (C_P) OF ORGANIC LIQUIDS AND VAPORS AT 25° C

William E. Acree, Jr.

The values in this table are expressed as J/mol-K at 1 atmosphere pressure. To convert to calories/mol-K multiply at 0.2390057. Compounds in this table are arranged in order of increasing number of carbon atoms in the molecules.

Formula	Name	C_p(1)	C_p(g)	Ref.
H_4N_2	Hydrazine	92.9	51.0	1
CCl_4	Carbon tetrachloride	132.6	83.3	1
CS_2	Carbon disulfide	76.1	45.6	1
$CHCl_3$	Chloroform	115.5	65.7	1
CHN	Hydrogen cyanide	70.7	36.0	1
CH_2O_2	Formic acid	98.7	45.2	1
CH_3NO_2	Nitromethane	106.3	57.7	1
CH_3NO_3	Methyl nitrate	156.9		1
CH_4O	Methanol	81.6	43.9	1
C_2Cl_4	Tetrachloroethene	146.5		2
$C_2H_2Cl_2$	1,1-Dichloroethylene	111.3	67.4	1
$C_2H_3Cl_3$	1,1,1-Trichloroethane	143.9	93.7	1
$C_2H_4Br_2$	1,2-Dibromoethane	134.7	76.1	1
$C_2H_4O_2$	Methylformate	119.7		3
$C_2H_4O_2$	Acetic acid	123.0	66.9	1
C_2H_6O	Ethanol	113.0	65.7	1
C_2H_6S	Ethanethiol	118.0	72.8	1
C_2H_6S	Dimethyl sulfide	118.4	74.5	1
$C_2H_6S_2$	Dimethyl disulfide	146.0	94.6	1
C_3H_3N	Acrylonitrile	110.9	62.3	1
C_3H_6O	Acetone	126.4	75.3	1
C_3H_6O	Propionaldehyde	137.2		4
$C_3H_6O_2$	Ethyl formate	144.3	—	3
$C_3H_6O_2$	Methyl acetate	143.9	—	3
C_3H_6S	$\overline{(CH_2)_3S}$ (Trimethylene sulfide)	113.4	69.9	1
C_3H_8O	Propanol-1	144.4	87.4	5,1
C_3H_8O	Propanol-2	155.2	89.1	1
C_3H_8S	Methyl ethyl sulfide	144.8	95.4	1
C_3H_8S	Propanethiol-2	145.2	96.6	1
$C_3H_{10}N_2$	Tetramethylhydrazine	186.2	—	1
C_4H_4O	Furan	114.6	—	1
C_4H_5N	Pyrrole	127.7	—	6
C_4H_6	Butadiene-1,3	123.4	79.5	1
C_4H_6	Butyne-2	123.0	77.8	1
$C_4H_6O_2$	γ-Butyrolactone	141.4	—	3
$C_4H_6O_2$	Methyl acrylate	161.5	—	3
C_4H_8O	Butanone-2	159.2	—	7
$C_4H_8O_2$	*p*-Dioxane	150.7	—	8
$C_4H_8O_2$	Ethyl acetate	167.4	—	3
C_4H_8S	$\overline{(CH_2)_4S}$ (Tetrahydrothiophene)	140.2	91.6	1
C_4H_9N	Pyrrolidine	156.5	—	1
$C_4H_{10}O$	Butanol-1	177.1	—	5
$C_4H_{10}O$	Diethyl ether	172.0	107.9	1
$C_4H_{10}S$	Butanethiol-1	172.8	118.8	1
$C_4H_{10}S$	Methyl *n*-propyl sulfide	171.5	118.0	1
$C_4H_{10}S$	Diethyl sulfide	171.1	117.6	1
$C_4H_{10}S$	Methyl isopropyl sulfide	172.4	120.5	1
$C_4H_{10}S$	2-Methylpropanethiol-2	174.9	121.3	1
$C_4H_{10}S$	2-Methylpropanethiol-1	172.0	118.8	1
$C_4H_{10}S$	1-Methylpropanethiol-1	171.1	119.7	1
$C_4H_{10}S_2$	Diethyl disulfide	203.8	141.8	1
$C_4H_{12}Si$	Tetramethylsilane	208.4	140.2	1
C_5H_8	Pentadiene-1,4	145.6	105.0	1
C_5H_8	2-Methyl butadiene-1,3	153.6	104.6	1
C_5H_{10}	Pentene-1	155.2	109.6	1
C_5H_{10}	*cis*-Pentene-2	151.9	101.7	1
C_5H_{10}	*trans*-Pentene-2	157.3	108.4	1
C_5H_{10}	2-Methyl butene-1	156.9	111.7	1
C_5H_{10}	3-Methyl butene-1	156.1	118.8	1
C_5H_{10}	2-Methyl butene-2	152.3	105.0	1
C_5H_{10}	Cyclopentane	127.2	82.8	1
$C_5H_{10}O$	Pentanone-2	185.5	—	4
$C_5H_{10}O$	Pentanone-3	190.1	—	4
$C_5H_{10}O$	3-Methyl butanone-2	180.6	—	4
$C_5H_{10}O$	Valeraldehyde	189.2	—	4
$C_5H_{10}O$	Trimethylacetaldehyde	186.2	—	4
$C_5H_{10}S$	$\overline{(CH_2)_5S}$ (Thiane)	163.2	108.8	1
C_5H_{12}	2-Methylbutane	164.8	119.2	1
$C_5H_{12}O$	Pentanol-1	209.0	—	5
$C_5H_{12}S$	Methyl *n*-butyl sulfide	200.8	141.4	1
$C_5H_{12}S$	Ethyl *n*-propyl sulfide	198.3	139.7	1
$C_5H_{12}S$	Pentanethiol-1	201.7	141.8	1
$C_6H_3Cl_3$	1,2,4-Trichlorobenzene	194.6	—	9
C_6H_5Br	Bromobenzene	155.2	101.3	1
C_6H_5Cl	Chlorobenzene	146.0	98.7	1
C_6H_5F	Fluorobenzene	146.9	95.0	1
C_6H_6	Benzene	135.6	81.6	1
C_6H_6S	Benzenethiol	163.6	105.4	1
C_6H_7N	Aniline	193.3	109.2	1
$C_6H_{10}O$	Cyclohexanone	179.5	—	4
C_6H_{12}	Hexene-1	183.3	132.2	1
C_6H_{12}	2,3-Dimethyl butene-2	174.5	127.6	1
C_6H_{12}	Cyclohexane	152.3	106.3	1
C_6H_{12}	Methylcyclopentane	159.0	109.6	1
$C_6H_{12}O$	3,3-Dimethyl butanone-2	210.0	—	4
$C_6H_{12}O$	4-Methyl pentanone-2	215.8	—	4
$C_6H_{12}O$	Hexanone-2	214.8	—	4
$C_6H_{12}O$	Hexanone-3	216.9	—	4
$C_6H_{12}O_2$	Methyl pentanoate	229.3	—	3
$C_6H_{12}O_2$	*n*-Butyl acetate	228.4	—	3
$C_6H_{12}O_2$	*t*-Butyl acetate	231.0	—	3
C_6H_{14}	*n*-Hexane	195.0	143.1	1
C_6H_{14}	2-Methyl pentane	193.3	144.3	1
C_6H_{14}	3-Methyl pentane	190.8	143.1	1
C_6H_{14}	2,3-Dimethyl butane	188.3	140.6	1
C_6H_{14}	2,2-Dimethyl butane	188.3	141.8	1
$C_6H_{14}O$	Hexanol-1	241.3	—	5
$C_6H_{14}O$	2-Methyl pentanol-1	247.6	—	10
$C_6H_{14}O$	3-Methyl pentanol-2	275.9	—	10
$C_6H_{14}O$	4-Methyl pentanol-2	273.0	—	10
$C_6H_{14}O$	3-Methyl pentanol-3	293.4	—	10
$C_6H_{14}S$	Di-*n*-propyl sulfide	225.9	161.9	1
$C_6H_{18}OSi_2$	Hexamethyl disiloxane	311.3	—	1
C_7H_6O	Benzaldehyde	177.9	—	4
C_7H_8	Toluene	156.1	103.8	1
C_7H_8O	Benzyl alcohol	218.0	—	1
C_7H_9N	Benzyl amine	207.2	—	11
$C_7H_{12}O$	Cycloheptanone	211.8	—	4
C_7H_{14}	Heptene-1	211.3	155.2	1
C_7H_{14}	Methyl cyclohexane	184.9	135.1	1
C_7H_{14}	Ethyl cyclopentane	186.2	131.8	1
C_7H_{14}	1,1-Dimethyl cyclopentane	186.6	133.5	1
C_7H_{14}	*cis*-1,2-Dimethyl cyclopentane	188.7	134.3	1
C_7H_{14}	*trans*-1,2-Dimethyl cyclopentane	188.7	134.3	1
C_7H_{14}	*trans*-1,3-Dimethyl cyclopentane	188.7	134.3	1
$C_7H_{14}O$	Heptanone-2	243.6	—	4
$C_7H_{14}O$	Heptanone-4	246.3	—	4
$C_7H_{14}O$	2,4-Dimethyl pentanone-3	233.7	—	4
$C_7H_{14}O$	2-Methyl hexanone-3	241.5	—	4
C_7H_{16}	*n*-Heptane	224.7	166.1	1
C_7H_{16}	2-Methyl hexane	222.2	—	1
C_7H_{16}	3-Ethyl pentane	219.7	—	1
C_7H_{16}	2,2-Dimethyl pentane	220.9	—	1
C_7H_{16}	2,3-Dimethyl pentane	216.7	—	1
C_7H_{16}	2,4-Dimethyl pentane	223.8	—	1
C_7H_{16}	3,3-Dimethyl pentane	215.1	—	1
C_7H_{16}	2,2,3-Trimethyl butane	213.4	164.4	1
$C_7H_{16}O$	Heptanol-1	273.7	—	5
C_8H_8	Cyclooctatetraene	184	—	1
C_8H_8O	Acetophenone	204.6	—	4
$C_8H_8O_2$	Methyl benzoate	221.3	—	3
C_8H_{10}	Ethylbenzene	186.2	128.4	1
C_8H_{10}	*p*-Xylene	182.4	126.8	1
C_8H_{10}	*o*-Xylene	187.6	—	12
C_8H_{10}	*m*-Xylene	181.8	—	12
$C_8H_{10}O$	2-Phenyl ethanol	252.6	—	11
$C_8H_{11}N$	2-Phenyl ethylamine	239.2	—	11
$C_8H_{14}O_4$	Diethyl succinate	330.5	—	3
C_8H_{16}	Ethyl cyclohexane	209.2	159.0	1
C_8H_{16}	Octene-1	241.4	178.2	1
C_8H_{16}	1,2-Dimethyl cyclohexane	210.9	156.5	1
C_8H_{16}	*cis*-1,3-Dimethyl cyclohexane	209.2	157.3	1
C_8H_{16}	*trans*-1,3-Dimethyl cyclohexane	212.5	157.3	1
C_8H_{16}	*cis*-1,4-Dimethyl cyclohexane	211.3	157.3	1
C_8H_{16}	*trans*-1,4-Dimethyl cyclohexane	210.5	157.7	1
$C_8H_{16}O$	Octanone-2	276.9	—	4
C_8H_{18}	*n* Octane	254.0	188.7	1
C_8H_{18}	2,2,4-Trimethyl pentane	236.4	—	1
C_8H_{18}	2,2,3,3-Tetramethyl butane	233.9	192.5	1
$C_8H_{18}O$	Octanol-1	305.6	—	5
$C_8H_{18}S$	Di-*n*-butyl sulfide	284.5	207.9	1
C_9H_7N	Quinoline	199.2	—	1
$C_9H_{10}O$	Propiophenone	243.8	—	4
$C_9H_{10}O_2$	Ethyl benzoate	246.0	—	3
C_9H_{12}	1,2,3-Trimethyl benzene	216.7	154.4	1
C_9H_{12}	1,2,4-Trimethyl benzene	215.9	155.2	1
C_9H_{12}	1,3,5-Trimethyl benzene	209.6	150.2	1
$C_9H_{12}O$	3-Phenyl propanol-1	280.7	—	11
$C_9H_{13}N$	3-Phenyl propylamine-1	265.6	—	11
$C_9H_{18}O$	Nonanone-5	305.9	—	4
C_9H_{20}	*n*-Nonane	284.5	211.7	1

Formula	Name	$C_p(l)$	$C_p(g)$	Ref.
$C_{10}H_{12}$	Tetralin	218	—	1
$C_{10}H_{12}O$	4-Ethyl acetophenone	260.7	—	4
$C_{10}H_{18}$	*cis*-Decalin	231.8	—	1
$C_{10}H_{18}$	*trans*-Decalin	228.0	—	1
C_{10}-H_{20}	Decene-1	300.4	223.8	1
$C_{10}H_{22}$	*n*-Decane	314.6	234.7	1
$C_{10}H_{22}$	2-Methyl nonane	316.3	—	1
$C_{11}H_{22}$	Undecene-1	330.1	246.9	1
$C_{11}H_{24}$	*n*-Undecane	345.2	257.3	1
$C_{12}H_{10}O$	Diphenyl ether	268.6	—	1
$C_{12}H_{12}O$	4-Methyl benzophenone	325.1	—	4
$C_{12}H_{14}O_4$	Diethyl phthalate	357.7	—	3
$C_{12}H_{24}$	Dodecene-1	360.7	269.4	1
$C_{12}H_{26}$	*n*-Dodecane	375.3	280.3	1
$C_{13}H_{28}$	*n*-Tridecane	406.7	303.3	1
$C_{14}H_{30}$	*n*-Tetradecane	438.5	325.9	1
$C_{15}H_{32}$	*n*-Pentadecane	469.4	348.9	1
$C_{16}H_{32}$	Hexadecene-1	484.9	361.1	1
$C_{16}H_{34}$	*n*-Hexadecane	501.7	372.0	1

REFERENCES

1. **Shaw, R.,** *J. Chem. Eng. Data*, 14, 461, 1969.
2. **Grolier, J.-P. E., Inglese, A., and Wilhelm, E.,** *J. Chem. Thermodyn.*, 14, 523, 1982.
3. **Fuchs, R.,** *J. Chem. Thermodyn.*, 11, 959, 1979.
4. **Fuchs, R.,** *Can. J. Chem.*, 58, 2305, 1980.
5. **Zegers, H. C. and Somsen, G.,** *J. Chem. Thermodyn.*, 16, 225, 1984.
6. **Scott, D. W., Berg, W. T., Hossenlopp, I. A., Hubbard, W. N., Messerly, J. F., Todd, S. S., Douslin, D. R., McCullough, J. P., and Waddington, G.,** *J. Phys. Chem.*, 71, 2263, 1967.
7. **Grolier, J.-P. E., Benson, G. C., and Picker, P.,** *J. Chem. Eng. Data*, 20, 243, 1975.
8. **Grolier, J.-P. E., Inglese, A., and Wilhelm, E.,** *J. Chem. Thermodyn.*, 16, 67, 1984.
9. **Wilhelm, E., Inglese, A., Quint, J. R., and Grolier, J.-P. E.,** *J. Chem. Thermodyn.*, 14, 303, 1982.
10. **Bravo, R., Pintos, M., Baluja, M. C., Paz Andrade, M. I., Roux-Desgranges, G., and Grolier, J.-P. E.,** *J. Chem. Thermodyn.*, 16, 73, 1984.
11. **Nichols, N. and Wadso, I.,** *J. Chem. Thermodyn.*, 7, 329, 1975.
12. **Fortier, J.-L. and Benson, G. C.,** *J. Chem. Eng. Data*, 25, 47, 1980.

HEAT CAPACITY (C_p) OF ORGANIC GASES AT 300 and 800 K

These data are a portion of those contained in "Additivity Rules for the Estimation of Thermochemical Properties", authored by S. W. Benson, F. R. Cruickshank, D. M. Golden, G. R. Haugen, H. E. O'Neal, A. S. Rodgers, R. Shaw and R. Walsh in *Chemical Reviews*, 69, 279 (1969). These data are reproduced here with permission of the copyright owner, the American Chemical Society.

	C_p	
	300 K	800 K
Alkanes		
Ethane	12.65	25.83
Propane	17.66	37.08
n-Butane	23.40	48.23
n-Hexane	34.37	70.36
n-Octane	43.35	92.50
n-Decane	(56.34)	(114.63)
n-Dodecane	(67.33)	(136.76)
2-Methylpropane	23.25	48.49
2-Methylbutane	28.54	59.71
2,2-Dimethylpropane	29.21	60.78
2,3-Dimethylbutane	33.76	70.7
2,2,3-Trimethylbutane	39.54	82.73
2,2,3,3-Tetramethylbutane	46.29	96.18
Alkenes		
Ethene	10.45	20.20
Propene	15.34	30.68
But-1-ene	20.57	41.80
cis-But-2-ene	18.96	40.87
trans-But-2-ene	21.08	41.50
2-Methylpropene	21.39	41.86
Pent-1-ene	26.31	52.95
cis-Pent-2-ene	(24.45)	(52.29)
trans-Pent-2-ene	(26.04)	(52.45)
2-Methylbut-1-ene	26.41	53.15
3-Methylbut-1-ene	(28.47)	(53.85)
2-Methylbut-2-ene	25.22	52.05
Allenic Dienes		
Allene	14.16	25.42
1,2-Butadiene	19.23	36.01
1,2-Pentadiene	(25.3)	(47.7)
2,3-Pentadiene	(24.3)	(46.6)
3-Methyl-1,2-butadiene	(25.3)	(47.2)
Nonallenic Dienes		
1,3-Butadiene	19.11	36.84
cis-1,3-Pentadiene	(22.7)	(47.0)
trans-1,3-Pentadiene	(24.9)	(47.7)
1,4-Pentadiene	(25.2)	(47.6)
2-Methyl-1,3-butadiene (isoprene)	(25.2)	(48.0)
Alkynes		
Ethyne	10.53	14.93
Propyne	14.55	25.14
But-1-yne	19.54	35.95
But-2-yne	18.70	35.14
Pent-1-yne	(25.65)	(47.1)
Pent-2-yne	(23.69)	(45.9)
Aromatics		
Benzene	19.65	45.06
Toluene	24.94	56.61
Ethylbenzene	30.88	67.15
1,2-Dimethylbenzene	32.10	66.50
1,3-Dimethylbenzene	30.66	66.41
1,4-Dimethylbenzene	30.49	66.14
n-Propylbenzene	(36.99)	(78.30)
n-Butylbenzene	(42.09)	(89.37)
1-Propylbenzene	(36.47)	(78.6)
1-Methyl-2-ethylbenzene	(37.94)	(78.1)
1-Methyl-3-ethylbenzene	(36.59)	(77.8)
1-Methyl-4-ethylbenzene	(36.42)	(77.6)
1,2,3-Trimethylbenzene	37.82	77.56
1,2,4-Trimethylbenzene	36.99	76.93
1,3,5-Trimethylbenzene	36.10	76.84
1,2,3,4-Tetramethylbenzene	45.50	89.42
1,2,3,5-Tetramethylbenzene	44.57	88.79
1,2,4,5-Tetramethylbenzene	44.77	88.41
Pentamethylbenzene	51.99	101.29
Hexamethylbenzene	59.73	113.51
Unsaturated Benzenes		
Styrene	29.35	61.40
S-Methylstyrene	(34.9)	(71.8)
cis-β-Methylstyrene	(34.9)	(71.8)
trans-β-Methylstyrene	(35.1)	(72.2)
o-Methylstyrene	(34.9)	(71.8)
m-Methylstyrene	(34.9)	(71.8)
p-Methylstyrene	(34.9)	(71.8)
Ethynylbenzene	27.63	55.79
Polyaromatics		
Biphenyl	39.05	86.92
Nonaromatic Rings: Ring Corrections		
Cyclopropane	13.44	31.44
Cyclobutane	17.37	42.42
Cyclobutene	16.03	32.26
Cyclopentane	19.98	52.44
Cyclopentene	18.08	45.78

	C_p	
	300 K	800 K
Methylcyclopentane	26.43	63.8
1-Methylcyclopentene	(24.3)	(57.0)
1,1-Dimethylcyclopentane	(32.16)	(76.18)
cis-1,2-Dimethylcyclopentane	(32.34)	74.98
trans-1,2-Dimethylcyclopentane	(32.44)	(75.84)
cis-1,3-Dimethylcyclopentane	32.72	75.68
trans-1,3-Dimethylcyclopentane	(32.72)	(75.68)
Cyclohexane	25.58	66.76
Cyclohexene	25.28	59.49
Methylcyclohexane	32.51	78.74
Ethylcyclohexane	38.23	90.1
1,1-Dimethylcyclohexane	(37.2)	(90.7)
cis-1,2-Dimethylcyclohexane	(37.7)	(90.1)
trans-1,2-Dimethylcyclohexane	(38.3)	(90.5)
cis-1,3-Dimethylcyclohexane	(37.9)	(90.5)
trans-1,3-Dimethylcyclohexane	(37.9)	(89.8)
cis-1,4-Dimethylcyclohexane	(37.9)	(89.8)
trans-1,4-Dimethylcyclohexane	(38.0)	(90.6)
Spiropentane	21.19	47.91
Alcohols and Phenols		
Methanol, CH_3OH	10.5	19.0
Ethanol, C_2H_5OH	15.7	30.3
1-Propanol, C_3H_7OH	20.9	41.0
2-Propanol $(CH_3)_2CHOH$	21.3	42.1
2-Butanol $(CH_3CH_2)(CH_3)CHOH$	27.2	52.7
2-Methyl-2-propanol, $(CH_3)_3COH$	27.2	53.3
Phenol, C_6H_5OH	24.8	50.7
o-Cresol, $CH_3C_6H_4OH$	31.2	61.6
m-Cresol, $CH_3C_6H_4OH$	29.8	61.3
p-Cresol, $CH_3C_6H_4OH$	29.8	61.1
Ethers		
Dimethyl ether, CH_3OCH_3	15.8	30.4
Peroxides and Hydroperoxides		
Hydrogen peroxide, HOOH	10.3	14.3
Cyclics		
Ethylene oxide	11.5	24.6
Propylene oxide	17.4	35.7
Trimethylene oxide	14.3	35.1
Aldehydes and Ketones		
Acetaldehyde, CH_3CHO	13.2	24.2
Propionaldehyde, C_2H_5CHO	18.8	35.2
Acetone, CH_3COCH_3	18.0	34.9
Methyl ethyl ketone, $CH_3CH_2COCH_3$	24.7	46.1
Acids, Esters, and Anhydrides		
Formic acid, HCOOH	10.8	18.3
Acetic acid, CH_3COOH	16.0	29.1
Aliphatic and Aromatic Amines		
CH_3NH_2	11.91	22.43
$(CH_3)_2NH$	16.58	33.94
$(CH_3)_3N$	22.05	45.62
$C_6H_5NH_2$	26.07	53.79
Cyanides		
C_2H_5CN	17.30	32.7
C_6H_5CN	26.2	52.1
c-C_3H_5CN	18.95	37.68
$CH_2{=}CHCN$	15.30	26.43
trans-$CH_3CH{=}CHCN$	19.71	36.97
$C_2(CN)_2$	20.58	27.26
Nitrites		
CH_3ONO	15.36	26.27
Hydrazines, Azo Compounds, and Tetrazines		
N_2H_4	12.19	21.08
CH_3NHNH_2	17.1	31.3
NH=NH	8.75	13.55
Amides		
$HCONH_2$	11.10	21.12
CH_3CONH_2	15.63	32.60
Heterocyclic N-Atom-Containing Ring Compounds in the Ideal Gas State		
Pyrrole	17.16	...
Pyridine	18.80	...
Picolines, α	24.05	...
β	23.94	...
Unique Groups		
NH_3	8.53	12.23

	C_p 300 K	C_p 800 K
HCN	8.61	11.45
CH_3CN	12.55	21.32
$(CN)_2$	13.59	17.42
$CH_2(CN)_2$	17.38	27.72
HNC	9.33	11.49
CH_3NC	12.83	21.34
HN_3	10.46	15.10
CH_2N_2	12.58	18.73
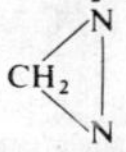	10.23	18.35
HCNO	10.57	15.19
HNO	8.29	10.77
$HONO_2$	12.80	20.33
CH_3NO_2	13.76	25.56
N_2	6.96	7.51
NO	7.13	7.83
NO_2	8.85	11.88
NO_3 sym	11.26	17.51
N_2O	9.25	12.49
N_2O_3	15.72	21.39
N_2O_4	18.52	27.11
N_2O_5	23.09	32.74
$CH_2{=}CHNO_2$	17.43	31.7
Haloalkanes		
CF_3CCl_3	28.1	39.6
CF_2ClCF_2Cl	28.9	···
CF_3CF_3	25.5	38.3
CCl_3CCl_3	32.7	41.3
CBr_3CBr_3	33.4	39.9
$CF_3CHClBr$	25.1	37.2
CF_3CHCl_2	24.4	37.3
CF_3CH_2Cl	21.4	34.8
CF_3CH_2Br	21.8	35.1
CF_3CH_2I	21.9	35.1
CF_3CH_2F	20.9	34.5
CF_3CH_3	18.9	···
$CHCl_2CH_2Cl$	21.3	33.3
CH_3CCl_3	22.4	33.3
CH_3CHF_2	16.1	···
CH_3CHCl_2	18.2	30.9
CH_3CH_2F	14.1	···
CH_3CH_2Cl	15.1	28.3
CH_3CH_2Br	15.5	28.6
$CF_2ClCF_2CF_2Cl$	37.4	···
$CF_3CF_2CH_3$	28.8	···
$CH_3CH_2CH_2Cl$	20.3	39.2
$CH_3CHClCH_3$	21.3	···
$CH_3CH_2CH_2Br$	20.7	39.4
$(CH_3)_3CCl$	27.9	···
CH_2ICH_2I	19.2	
CH_3CHICH_2I	(24.1)	
Haloalkenes, Alkynes, and Arenes		
$CFCl{=}CCl_2$	21.8	28.8
$CF_2{=}CClBr$	21.8	28.5
$CFCl{=}CFCl$	21.0	28.4
$CF_2{=}CBr_2$	22.2	···
$CF_2{=}CFCl$	20.0	28.0
$CF_2{=}CFBr$	20.5	28.1
$CF_2{=}CF_2$	19.3	27.6
$CCl_2{=}CCl_2$	22.7	29.2
$CBr_2{=}CBr_2$	24.5	29.7
$CHF{=}CCl$	18.3	26.5
$CF_2{=}CHBr$	17.6	26.2
$CF_2{=}CHF$	16.5	25.6
$CHCl{=}CCl_2$	19.3	26.9
$CHBr{=}CBr_2$	20.5	···
$CH_2{=}CFCl$	15.3	24.4
$CH_2{=}CCl_2$	16.1	24.7
$CHCl{=}CHCl$ (av)	15.8	24.6
$CHBr{=}CHBr$ *cis*	16.5	···
$CH_2{=}CHCl$	12.9	22.4
$CH_2{=}CHBr$	13.4	22.5
$CH_2{=}CHI$	13.9	22.7
$CH_3C{\equiv}CCl$	17.2	27.0
$CH_3C{\equiv}CBr$	17.6	27.1
$CH_3C{\equiv}CI$	17.8	27.2
C_6H_5F	22.7	47.6
o-$C_6H_4F_2$	25.6	49.5
m-$C_6H_4F_2$	25.5	49.7
p-$C_6H_4F_2$	25.7	49.7
p-$FC_6H_4CH_3$	27.9	58.6
$C_6H_5CF_3$	31.4	62.8
C_6H_5Cl	23.6	47.9
C_6H_5Br	24.0	48.0
C_6H_5I	24.2	48.0
Organosulfur Compounds		
CH_3SH	12.05	20.32

	C_p 300 K	C_p 800 K
CH_3CH_2SH	17.44	31.83
$CH_3(CH_2)_2SH$	22.75	43.60
$CH_3(CH_2)_3SH$	28.37	55.68
$CH_3(CH_2)_4SH$	33.91	66.78
$((CH_3)_2CH)SH$	23.04	43.26
$(CH_3)_2CHCH_2SH$	28.41	53.77
$(CH_3)(CH_3CH_2)CHSH$	28.64	54.29
$(CH_3)_3CSH$	29.04	55.53
$(CH_3)_2(CH_3CH_2)CSH$	34.46	66.28
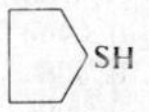	25.94	58.61
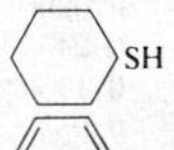	32.13	75.19
	25.22	51.59
$(CH_3)_2S$	17.77	31.59
$CH_3CH_2S_3CH$	22.82	42.93
$CH_3S(CH_2)_2CH_3$	28.17	54.45
$(CH_3)_2CHSCH_3$	28.82	54.89
$CH_3S(CH_2)_3CH_3$	33.79	66.53
$CH_3SC(CH_3)_3$	34.66	66.74
$CH_3CH_2SCH_2CH_3$	28.09	54.91
$CH_3CH_2SCH_2CH_2CH_3$	33.40	66.68
$CH_3(CH_2)_2S(CH_2)_2CH_3$	38.71	78.45
$(CH_3)_2CHSCH(CH_3)_2$	40.65	77.12
$CH_3(CH_2)_3S(CH_2)_3CH_3$	49.70	100.58
CH_3SSCH_3	22.61	37.66
$CH_3CH_2SSCH_2CH_3$	33.91	60.19
$CH_3(CH_2)_2SS(CH_2)_2CH_3$	44.50	83.70
$CH_3(SO)CH_3$	21.26	37.02
$CH_3(SO_2)CH_3$	23.9	
NH_2CSNH_2	17.72	29.06
CH_3NCS	15.70	25.87
S_8	37.25	42.62
	12.90	25.61
	16.66	36.40
	21.85	47.66
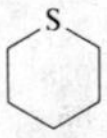	26.03	64.00
	17.52	36.01
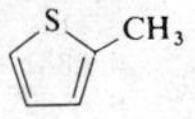	22.92	46.43
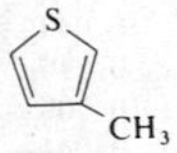	22.80	45.95
Compounds Containing Unique Groups		
SOF	9.97	12.79
$SOCl_2$	15.94	18.88
SOCl	10.88	13.03
S_2F_2	15.33	18.94
SF_2	10.44	13.14
SF	7.57	8.68
S_2Cl_2	17.44	19.46
SCl_2	12.19	13.61
SCl	8.20	8.95
$CSCl_2$	15.34	18.78
NSF_3	17.23	23.56
BO_2	10.08	13.45
H_3SiBr	12.69	20.02
H_3SiI	12.86	
H_2SiBr_2	15.68	21.73
$SiBr_4$	23.21	25.39
$ClSiBr_3$	22.78	25.31
Cl_2SiBr_2	22.39	25.23
Cl_3SiBr	21.72	25.13
$ISiBr_3$	23.39	25.42
$Si(CH_3)_4$	33.52	60.55
$(CH_3)_3SiCl$	31.36	52.15
$(CH_3)_2SiCl_2$	26.17	42.54
$(CH_3)_3SiF$	29.42	51.30
$(CH_3)_2SiF_2$	25.26	42.16
CH_3SiF_3	22.73	
$[(CH_3)_3Si]_2O$	57.22	99.95
$(CH_3)_3SiH$	28.36	50.50
$(CH_3)_2SiH_2$	22.03	39.79

SPECIFIC HEAT OF THE ELEMENTS AT 25°C

C_p = cal g^{-1} °K^{-1}

Element	Kelly: Bureau of Mines Bulletin 592 (1961)	Hultgren: Selected values of Thermodynamic properties of Metals and Alloys (1963)	N.B.S. Circular #500 Part 1 (1952)
Aluminum	0.215	0.215	0.2154
Antimony	0.049	0.0495	0.0501
Argon	0.124	......	0.124
Arsenic	0.0785	......	0.0796
Barium	0.046	0.0362	0.0458
Beryllium	0.436	0.436	0.4733
Bismuth	0.0296	0.0238	0.0292
Boron	0.245	......	0.2463
Bromine (Br_2)	0.113	0.0537	
Cadmium	0.0555	0.0552	0.0554
Calcium	0.156	0.155	0.1566
Carbon (Diamond)	0.124	......	0.120
" (Graphite)	0.170	......	0.172
Cerium	0.049	0.0459	0.0442
Cesium	0.057	0.0575	0.0558
Chlorine (Cl_2)	0.114	......	0.114
Chromium	......	0.107	0.1073
Cobalt	0.109	0.107	0.1037
Columbium	See Niobium		
Copper	0.092	0.0924	0.0920
Dysprosium	0.0414	0.0414	
Erbium	0.0401	0.0401	
Europium	0.0421	0.0326	
Fluorine (F_2)	0.197	0.197	
Gadolinium	0.055	0.056	
Gallium	0.089	0.088	0.0911
Germanium	0.077		
Gold	0.0308	0.0308	0.0305
Hafnium	0.035	0.028	0.0344
Helium	1.24	......	1.242
Holmium	0.0393	0.0394	
Hydrogen (H_2)	3.41	......	3.42
Indium	0.056	0.0556	0.0570
Iodine (I_2)	0.102	......	0.034
Iridium	0.0317	0.0312	0.0305
Iron (α)	0.106	0.1075	0.1078
Krypton	0.059	......	0.059
Lanthanum	0.047	0.0479	0.0475
Lead	0.038	0.0305	0.0308
Lithium	0.85	0.834	0.814
Lutetium	0.037	0.0285	
Magnesium	0.243	0.245	0.235
Manganese (α)	0.114	0.114	0.1147
" (β)	0.119	......	0.1120
Mercury	0.0331	0.0333	0.0331
Molybdenum	0.0599	0.0597	0.0584
Neodymium	0.049	0.0453	0.0499
Neon	0.246	......	0.246
Nickel	0.106	0.1061	0.1057
Niobium	0.064	0.0633	
Nitrogen (N_2)	0.249	......	0.249
Osmium	0.03127	0.0310	0.0310
Oxygen (O_2)	0.219	0.219	
Palladium	0.0584	0.0583	0.0590
Phosphorus, white	0.181	......	0.178
" red, triclinic	0.160		
Platinum	0.0317	0.0317	0.0325
Polonium	0.030		
Potassium	0.180	0.180	0.1787
Praseodymium	0.046	(0.0467)	0.0482
Promethium	0.0442		
Protactinium	0.029		
Radium	0.0288		
Radon	0.0224	......	0.0224
Rhenium	0.0329	0.0330	0.0327
Rhodium	0.0583	0.0580	0.0592
Rubidium	0.0861	0.0860	0.0850
Ruthenium	0.057	0.0569	
Samarium	0.043	0.0469	
Scandium	0.133	0.1173	
Selenium (Se_2)	0.0767	......	0.0535
Silicon	0.168	......	0.169
Silver	0.0566	0.0562	0.0564
Sodium	0.293	0.292	0.2952

Element	Kelly: Bureau of Mines Bulletin 592 (1961)	Hultgren: Selected values of Thermodynamic properties of Metals and Alloys (1963)	N.B.S. Circular #500 Part 1 (1952)
Strontium	0.0719	0.0719	0.0684
Sulfur, yellow	0.175	......	0.177
Tantalum	0.0334	0.0334	0.0335
Technetium	0.058		
Tellurium	0.0481	......	0.482
Terbium	0.0437	0.0435	
Thallium	0.0307	0.0307	0.0310
Thorium	0.0271	0.0281	0.0331
Thulium	0.0382		
Tin (α)	0.0510	0.0519	0.0518
Tin (β)	0.0530	0.0543	0.0530
Titanium	0.125	0.1248	0.1231
Tungsten	0.0317	0.0322	0.0324
Uranium	0.0276	0.0278	0.0276
Vanadium	0.116	0.116	0.1147
Xenon	0.0378	......	0.0379
Ytterbium	0.0346	0.0287	
Yttrium	0.068	0.0713	
Zinc	0.0928	0.0922	0.0916
Zirconium	0.0671	0.0660	

SPECIFIC HEAT AND ENTHALPY OF SOME SOLIDS AT LOW TEMPERATURES

R. J. Corruccini and J. J. Gniewek

For a more extensive listing of data one is referred to N.B.S. Monograph 21 (1960)

Joules/g × 453.6 = joules/lb × 0.239 = cal/g × 0.4299 = Btu/lb

Metals

T	Aluminum C_p	Aluminum $H - H_o$	Beryllium C_p	Beryllium $H - H_o$	Bismuth C_p	Bismuth $H - H_o$	Cadmium C_p	Cadmium $H - H_o$
°K	jg^{-1} deg^{-1} K	jg^{-1}	jg^{-1} deg^{-1} K	jg^{-1}	jg^{-1} deg^{-1} K	jg^{-1}	jg^{-1} deg^{-1} K	jg^{-1}
1	0.00010[a]	—			—	—	—	—
1	0.000051	0.000025	0.000025	0.000013	0.00000598	0.00000158	0.000008	0.000003
2	0.000108	0.000105	0.000051	0.000051	0.0000461	0.0000233	0.000033	0.000022
3	0.000176	0.000246	0.000079	0.000116	0.000170	0.000123	0.000090	0.000082
4	0.000261	0.000463	0.000109	0.000209	0.000493	0.000432	0.00021	0.00022
6	0.00050	0.00121	0.000180	0.000496	0.00214	0.00288	0.00130	0.0015
8	0.00088	0.0026	0.000271	0.000944	0.00547	0.0102	0.0043	0.0070
10	0.0014	0.0049	0.000389	0.00160	0.0104	0.0259	0.0080	0.109
15	0.0040	0.018	0.000842	0.00457	0.0238	0.111	0.025	0.102
20	0.0089	0.048	0.00161	0.0105	0.0363	0.262	0.046	0.28
25	0.0175	0.112	0.00279	0.0212	0.0477	0.472	0.066	0.56
30	0.0315	0.232	0.00450	0.0392	0.0572	0.734	0.086	0.94
35	0.0515	0.436	—	—	—	—	—	—
40	0.0775	0.755	0.00996	0.109	0.0727	1.38	0.117	1.96
50	0.142	1.85	0.0192	0.253	0.0846	2.17	0.141	3.26
60	0.214	3.64	0.0341	0.523	0.0935	3.06	0.159	4.76
70	0.287	6.15	0.0562	0.971	0.100	4.03	0.172	6.43
80	0.357	9.37	0.0906	1.69	0.105	5.05	0.182	8.20
90	0.422	13.25	0.139	2.82	0.108	6.12	0.190	10.1
100	0.481	17.76	0.199	4.51	0.111	7.21	0.196	12.0

T	Chromium C_p	Chromium $H - H_o$	Copper C_p	Copper $H - H_o$	Germanium[b] C_p	Germanium[b] $H - H_o$	Gold C_p	Gold $H - H_o$
°K	jg^{-1} deg^{-1} K	jg^{-1}	jg^{-1} deg^{-1} K	jg^{-1}	g^{-1} deg^{-1} K	jg^{-1}	jg^{-1} deg^{-1} K	jg^{-1}
1	0.0000285	0.0000142	0.000012	0.000006	0.000000528	0.000000132	0.000006	0.000002
2	0.000058	0.0000573	0.000028	0.000025	0.00000423	0.00000211	0.000025	0.000016
3	0.000089	0.000131	0.000053	0.000064	0.0000144	0.0000107	0.000070	0.000061
4	0.00014	0.000237	0.000091	0.00013	0.0000344	0.0000343	0.00016	0.00017
6	0.000206	0.000567	0.00023	0.00044	0.000125	0.000179	0.00050	0.00078
6	0.000206	0.000567	0.00023	0.00044	0.000125	0.000179	0.00050	0.00078
8	0.000312	0.00107	0.00047	0.00112	0.000335	0.000612	0.0012	0.0024
10	0.000451	0.00182	0.00086	0.0024	0.000813	0.00169	0.0022	0.0056
15	0.00102	0.00528	0.0027	0.0107	0.00445	0.0136	0.0074	0.028
20	0.00210	0.0128	0.0077	0.034	0.0125	0.0540	0.0159	0.086
25	0.00392	0.0274	0.016	0.090	0.0240	0.145	0.0263	0.191
30	0.00683	0.0532	0.027	0.195	0.0366	0.296	0.0371	0.349
40	0.0171	0.163	0.060	0.61	0.0617	0.786	0.0572	0.821
50	0.0358	0.421	0.099	1.40	0.0858	1.52	0.0726	1.47

T	Chromium		Copper		Germanium[b]		Gold	
	C_p	$H - H_o$	C_p	$H - H_o$	C_p	$H - H_o$	C_p	$H - H_o$
°K	jg^{-1} deg^{-1} K	jg^{-1}	jg^{-1} deg^{-1} K	jg^{-1}	g^{-1} deg^{-1} K	jg^{-1}	jg^{-1} deg^{-1} K	jg^{-1}
60	0.0621	0.904	0.137	2.58	0.108	2.50	0.0842	2.25
70	0.093	1.68	0.173	4.13	0.131	3.70	0.0928	3.14
80	0.127	2.77	0.205	6.02	0.153	5.12	0.0992	4.10
90	0.161	4.21	0.232	8.22	0.173	6.74	0.1043	5.12
100	0.193	5.98	0.254	10.6	0.191	8.55	0.1083	6.18

T	Indium		α-Iron[c]		γ-Iron[d]		Lead	
	C_o	$H - H_o$	C_p	$H - H_o$	C_p	$H - H_o$	C_p	$H - H_o$
°K	jg^{-1} deg^{-1} K	jg^{-1}	jg^{-1} deg^{-1} K	jg^{-1}	jg^{-1} deg^{-1} K	jg^{-1}	jg^{-1} deg^{-1} K	jg^{-1}
1	0.000029	0.000011	0.000090	0.000045	—	—	0.000026	0.000010
1	0.000019[a]	0.000006[a]	—	—	—	—	0.000012[a]	0.000003[a]
2	0.000138	0.000085	0.000183	0.000181	—	—	0.00012	0.00007
2	0.000141[a]	0.000073[a]	—	—	—	—	0.00009[a]	0.00005[a]
								0.00028
3	0.000410	0.000341	0.000279	0.000412	—	—	0.00033	0.00023[a]
3	0.000464[a]	0.000357[a]	—	—	—	—	0.00031[a]	
3.40[e]	0.000584	0.000537	—	—	—	—	—	—
3.40	0.000669[a]	0.000581[a]	—	—	—	—	—	—
4	0.00095	0.00099	0.000382	0.000742	—	—	0.0007	0.0008
4	—	—	—	—	—	—	0.0007[a]	0.0007[a]
5	—	—	—	—	—	—	0.0015	0.0018
5	—	—	—	—	—	—	0.0015[a]	0.0018[a]
6	0.00359	0.00520	0.000615	0.00173	—	—	0.0029	0.0039
6	—	—	—	—	—	—	0.0030[a]	0.0040[a]
7	—	—	—	—	—	—	0.0048	0.008
7	—	—	—	—	—	—	0.0050[a]	0.008[a]
8	0.00855	0.0170	0.00090	0.003233	—	—	0.0073	0.014
10	0.0155	0.0408	0.00124	0.00537	—	—	0.0137	0.034
15	0.036	0.170	0.00249	0.0145	—	—	0.0335	0.150
20	0.0608	0.413	0.0045	0.0316	0.007	0	0.0531	0.368
25	0.0857	0.778	0.0075	0.061	—	—	0.0681	0.672
30	0.108	1.265	0.0124	0.110	0.016	0.11	0.0796	1.042
40	0.141	2.52	0.029	0.31	0.041	0.39	0.0944	1.920
50	0.162	4.04	0.055	0.73	0.090	1.0^{2}	0.103	2.91
60	0.176	5.73	0.087	1.43	0.13^{7}	2.1^{6}	0.108	3.97
70	0.186	7.53	0.121	2.46	0.18^{0}	3.7^{5}	0.112	5.07
80	0.193	9.42	0.154	3.84	0.21^{8}	5.74^{4}	0.114	6.20
90	0.198	11.38	0.186	5.55	0.25^{5}	8.1^{1}	0.116	7.35
100	0.203	13.39	0.216	7.56	0.28^{8}	10^{8}	0.118	853

T	Molybdenum		Nickel		Palladium	
	C_p	$H - H_o$	C_p	$H - H_o$	C_p	$H - H_o$
°K	jg^{-1} deg^{-1} K	jgum[1]	jg^{-1} deg^{-1} K	jg^{-1}	jg^{-1} deg^{-1} K	jg^{-1}
1	0.0000229	0.0000105	0.000120	0.000060	0.000099	0.0000493
2	0.0000472	0.0000445	0.000242	0.000241	0.000203	0.000200
2	—	—	—	—	—	—
3	0.0000745	0.000105	0.000369	0.000546	0000318	0.000459
3	—	—	—	—	—	—
4	0.000106	0.000194	0.000503	0.00098	0.000447	0.000840
4	—	—	—	—	—	—
5	—	—	—	—	—	—
5	—	—	—	—	—	—
6	0.000191	0.000484	0.00082	0.00228	0.000891	0.00231
6	—	—	—	—	—	—
7	—	—	—	—	—	—
7	—	—	—	—	—	—
8	0.000317	0.000981	0.00119	0.00428	0.00141	0.00460
8	—	—	—	—	—	—
9	—	—	—	—	—	—
9	—	—	—	—	—	—
10	0.000498	0.00178	0.00162	0.0071	0.00210	0.00807
15	0.00131	0.00610	0.0031	0.0185	0.00471	0.0245
20	0.00287	0.0161	0.0058	0.041	0.00922	0.0586
25	0.00577	0.0374	0.0101	0.079	0.0160	0.120
30	0.00960	0.0729	0.0167	0.145	0.0258	0.223
40	0.0236	0.232	0.0381	0.413	0.0507	0.600
50	0.0410	0.554	0.0682	0.937	0.0777	1.24
60	0.0619	1.07	0.103	1.79	0.101	2.14

SPECIFIC HEAT AND ENTHALPY OF SOME SOLIDS AT LOW TEMPERATURES (continued)

	Molybdenum		Nickel		Palladium	
T	C_p	$H - H_o$	C_p	$H - H_o$	C_p	$H - H_o$
°K	jg^{-1} deg^{-1} K	jgum[1]	jg^{-1} deg^{-1} K	jg^{-1}	jg^{-1} deg^{-1} K	jg^{-1}
70	0.0838	1.80	0.139	3.00	0.122	3.26
80	0.104	2.74	0.173	4.56	0.139	4.56
90	0.123	3.88	0.204	6.45	0.154	6.03
100	0.139	5.20	0.232	8.63	0.167	7.63

	Platinum		Rhodium		Silicon[l]		Silver	
T	C_p	$H - H_o$	C_p	$H - H_o$	C_p	$H - H_o$	C_p	$H - H_o$
°K	jg^{-1} deg^{-1} K	jg^{-1}	jg^{-1} deg^{-1} K	jg^{-1}	jg^{-1} deg^{-1} K	jg^{-1}	jg^{-1} deg^{-1} K	jg^{-1}
1	0.000035	0.0000175	0.000048	0.000024	0.000000263	0.0000000658	0.0000072	0.0000032
2	0.000074	0.000071	0.000097	0.000096	0.00000210	0.00000105	0.0000239	0.0000176
3	0.000122	0.000168	0.000147	0.000218	0.00000709	0.00000532	0.0000595	0.0000574
4	0.000186	0.000320	0.000201	0.000392	0.0000168	0.0000168	0.000124	0.000146
6	0.00037	0.00085	0.00032	0.00091	0.0000596	0.0000853	0.00039	0.00062
8	0.00067	0.00188	0.00047	0.00170	0.000140	0.000279	0.00091	0.00187
10	0.00112	0.00365	0.00065	0.00281	0.000275	0.000679	0.0018	0.00452
15	0.0033	0.0135	0.00135	0.00765	0.00109	0.00374	0.0064	0.0233
20	0.0074	0.0395	0.00271	0.0174	0.00337	0.0138	0.0155	0.076
25	0.0137	0.092	0.00561	0.0373	0.00849	0.0423	0.0287	0.185
30	0.0212	0.182	0.0106	0.0071	0.0171	0.105	0.0442	0.368
40	0.038	0.048	0.266	0.256	0.0440	0.400	0.078	0.979
50	0.055	0.95	0.0489	0.633	0.0785	1.00	0.108	1.91
60	0.068	1.56	0.9724	1.238	0.115	1.97	0.133	3.12
70	0.079	2.29	0.094	2.07	0.152	3.31	0.151	4.54
80	0.088	3.12	0.114	3.11	0.188	5.01	0.166	6.13
90	0.094	4.02	0.132	4.34	0.224	7.06	0.177	7.85
100	0.100	5.01	0.147	5.74	0.259	9.47	0.187	9.67

	Sodium[m]		Tantalum		Tin (white)		Titanium	
T	C_p	$H - H_o$	C_p	$H - H_o$	C_p	$H - H_o$	C_p	$H - H_o$
°K	jg^{-1} deg^{-1} K	jg^{-1}	jg^{-1} deg^{-1} K	jg^{-1}	jg^{-1} deg^{-1}	jg^{-1}	jg^{-1} deg^{-1} K	jg^{-1}
1	0.000081	0.000035	0.000032	0.000016	0.0000170	0.0000079	0.000071	0.000035
1	—	—	0.0000063[a]	0.0000021[a]	0.0000041[a]	0.0000009[a]	—	—
2	0.000289	0.000204	0.000068	0.000065	0.000047	0.0000383	0.000146	0.000143
2	—	—	0.000054[a]	0.000026[a]	0.000048[a]	0.0000228[a]	—	—
3	0.00076	0.00070	0.000112	0.000155	0.000109	0.000113	0.000226	0.000329
3	—	—	0.000178[a]	0.000138[a]	0.000151[a]	0.000116[a]	—	—
3.72[n]	—	—	—	—	0.000198	0.000221	—	—
3.72	—	—	—	—	0.000285[a]	0.000270[a]	—	—
4	0.00160	0.00184	0.000171	0.000295	0.000245	0.000283	0.000317	0.000599
4	—	—	0.000352[a]	0.000400[a]	—	—	—	—
4.39[o]	—	—	0.000201	0.000368	—	—	—	—
4.39	—	—	0.000433[a]	0.000553[a]	—	—	—	—
5	0.00298	0.00408	—	—	0.00054	0.00065	—	—
6	0.0051	0.0081	0.000333	0.000776	0.00127	0.00151	0.00054	0.00145
8	0.0122	0.0247	0.000648	0.00173	0.0042	0.0068	0.00084	0.00281
10	0.0238	0.0602	0.00117	0.00352	0.0081	0.0190	0.00126	0.00489
12	0.0397	0.123	—	—	—	—	—	—
14	0.063	0.225	—	—	—	—	—	—
15	—	—	0.00360	0.0145	0.226	0.093	0.0033	0.0156
16	0.093	0.380	—	—	—	—	—	—
18	0.124	0.597	—	—	—	—	—	—
20	0.155	0.875	0.00823	0.0432	0.040	0.251	0.0070	0.040
25	0.259	1.90	0.0153	0.102	0.058	0.498	0.0134	0.090
30	0.364	3.45	0.0240	0.202	0.076	0.834	0.0245	0.182
40	0.544	8.03	0.0430	0.540	0.106	1.75	0.0571	0.581
50	0.695	14.2	0.0604	1.06	0.130	2.93	0.0992	1.358
60	0.793	21.7	0.0754	1.74	0.148	4.33	0.1467	2.592
70	0.86	30.0	00.189	4.27	0.162	5.88	0.189	4.27
80	0.91	38.9	0.0976	3.49	0.173	7.55	0.230	6.37
90	0.95	48.2	0.105	4.50	0.182	9.33	0.267	8.86
100	0.98	57.9	0.111	5.58	0.189	11.18	0.300	11.69

	Tungsten		Zinc	
T	C_p	$H - H_o$	C_p	$H - H_o$
°K	jg^{-1} deg^{-1} K	jg^{-1}	jg^{-1} deg^{-1} K	jg^{-1}
1	0.0000074	0.0000037	0.000011	0.000005
2	0.0000158	0.0000152	0.000028	0.000023
3	0.0000262	0.0000360	0.000058	0.000065

SPECIFIC HEAT AND ENTHALPY OF SOME SOLIDS AT LOW TEMPERATURES (continued)

	Tungsten		Zinc	
T	C_p	$H - H_0$	C_p	$H - H_0$
°K	jg^{-1} deg^{-1} K	jg^{-1}	jg^{-1} deg^{-1} K	jg^{-1}
4	0.0000393	0.0000685	0.00011	0.00014
6	0.0000783	0.000182	0.00029	0.00053
8	0.000141	0.000396	0.00096	0.0016
10	0.000234	0.000765	0.0025	0.0050
15	0.000725	0.00297	0.011	0.034
20	0.00189	0.00927	0.026	0.125
25	0.00421	0.0237	0.049	0.31
30	0.00783	0.0534	0.076	0.62
40	0.0184	0.181	0.125	1.62
50	0.0332	0.436	0.171	3.11
60	0.0483	0.843	0.208	5.01
70	0.0605	1.39	0.236	7.23
80	0.0715	2.05	0.258	9.70
90	0.0810	2.81	0.277	12.38
100	0.0888	3.66	0.293	15.24

a Superconducting.
b In germanium the electronic specific heat is markedly dependent on impurities. The values given are for pure germanium (negligible electronic specific heat).
c α-Iron is the form that is thermodynamically stable at low temperatures. It has the body-centered cubic lartice which is the basis of the ferritic steels.
d γ-Iron is stable between 910 and 1400°C. It has the face-centered cubic structure which is the basis of the austenitic steels.Since pure γ-iron is not stable at low temperatures the above values were calculated by application of the Kopp-Neumann rule to experimental data on two austenitic Fe-Mn alloys and are of uncertain accuracy.
e Superconducting transition temperature.
i Superconducting transition temperature of mercury.
j Melting temperature of mercury.
l In silicon the electronic specific heat, γT, is markedly dependent on impurities. Values of the coefficient, γa, from zero to 2.4 × 10^{-6} jg^{-1} deg^{-2} K have been reported. The values in the above table are for pure silicon (γ = 0).
m It has been shown (Barrett 1956, Hull & Rosenberg 1959) that sodium partially transforms at low temperatures from the normal body-centered cubic structure to close-packed hexagonal. The transformation is of the martensitic type and is promoted by cold-working at the low temperatures. Inasmuch as none of the calorimetric measurements on sodium were accompanied by crystallographic analysis, the tabulated data below 100°K are to some degree ambiguous.
n Superconducting transition temperature of tin.
o Superconducting transition temperature of tantalum.

CONSTANTS OF DEBYE-SOMMERFELD EQUATION

$C_v = \gamma T + \alpha T^3$; $\alpha = 12\pi^4 R/5\theta^{03}$; $0 \leq T \leq T_{max}$; T_{max} = maximum temperature to which the equation can be used with the limiting value of θ.

Substance	$10^6\gamma$	γ	$10^6\alpha$	θ^0	T^{max}
Metals	jg^{-1} deg^{-2} K	mjg-atom^{-1} deg^{-2} K	jg^{-1} deg^{-4} K	deg K	deg K
Aluminum	50.4	1.36	0.93	425	4
Beryllium	25	0.226	0.138	1160	20
Bismuth	0.32	0.067	5.66	118	2
Cadmium	5.6	0.63	2.69	186	3
Chromium	28.3	1.47	0.165	610	4
Copper	10.81	0.687	0.746	344.5	10
Germanium	a	a	0.528	370	2
Gold	3.75	0.74	2.19	165	15
Indium	15.8	1.81	13.1	109	2
α-Iron	90	5.0	0.349	464	10
Lead	15.1	3.1	10.6	96	4
Magnesium	54	1.32	1.19	406	4
α-Manganese	251	13.8	0.328	476	12
Molybdenum	23	2.18	0.238	440	4
Nickel	120	7.0	0.39	440	4
Niobium	85	7.9	0.64	320	1
Palladium	98	10.5	0.89	274	4
Platinum	34.1	6.7	0.72	240	3
Rhodium	48	4.9	0.173	478	4
Silicon	a	a	0.263	640	4
Silver	5.65	0.610	1.58	225	4
Sodium[a]	60	1.37	21.4	158	4
Tantalum	31.5	5.7	0.69	250	4
Tin (white)	14.7	1.75	2.21	195	2
Titanium	71	3.4	0.54	420	10
Tungsten	7	1.3	0.16	405	4
Zinc	9.6	0.63	1.10	300	4
Alloys					
Constantan[a]	113	6.9	0.56	384	15
Monel[a]	108	6.5	0.62	374	20
Other inorganic substances					
Diamond	—	—	0.0152	2200	50
Ice	—	—	15.2	192	10

CONSTANTS OF DEBYE-SOMMERFELD EQUATION (continued)

Substance	$10^6\gamma$	γ	$10^6\alpha$	0°	T^{max}
Pyrex	—	—	3.14	—	5
Organic substances					
Glyptal	—	—	27	—	4
Lucite	—	—	35	—	4
Polystyrene	—	—	63	—	4

[a] Superconducting.

BOILING POINT OF WATER (Hydrogen Scale)

Pressure mm	Tenths of millimeters .0	.1	.2	.3	.4	.5	.6	.7	.8	.9
700	97.714	718	722	725	729	733	737	741	745	749
701	753	757	761	765	769	773	777	781	785	789
702	792	796	800	804	808	812	816	820	824	828
703	832	836	840	844	847	851	855	859	863	867
704	871	875	879	883	887	891	895	899	902	906
705	97.910	914	918	922	926	930	934	938	942	946
706	949	953	957	961	965	969	973	977	981	985
707	989	993	996	000[a]	004[a]	008[a]	012[a]	016[a]	020[a]	024[a]
708	98.028	032	036	040	043	047	051	055	059	063
709	067	071	075	079	082	086	090	094	098	102
710	98.106	110	114	118	121	125	129	133	137	141
711	145	149	153	157	160	164	168	172	176	180
712	184	188	192	195	199	203	207	211	215	219
713	223	227	230	234	238	242	246	250	254	258
714	261	265	269	273	277	281	285	289	292	296
715	98.300	304	308	312	316	320	323	327	331	335
716	339	343	347	351	355	358	362	366	370	374
717	378	382	385	389	393	397	401	405	409	412
718	416	420	424	428	432	436	440	443	447	451
719	455	459	463	467	470	474	478	482	486	490
720	98.493	497	501	505	509	513	517	520	524	528
721	532	530	540	544	547	551	555	559	563	567
722	570	574	578	582	586	590	593	597	601	605
723	609	613	617	620	624	628	632	636	640	643
724	647	651	655	659	662	666	670	674	678	682
725	98.686	689	693	697	701	705	709	712	716	720
726	724	728	732	735	739	743	747	751	755	758
727	762	766	770	774	777	781	785	789	793	797
728	800	804	808	812	816	819	823	827	831	835
729	838	842	846	850	854	858	861	865	869	873
730	98.877	880	884	888	892	896	899	903	907	911
731	915	918	922	926	930	934	937	941	945	949
732	953	956	960	964	968	972	975	979	983	987
733	991	994	998	002[a]	006[a]	010[a]	013[a]	017	021[a]	025[a]
734	99.029	032	036	040	044	048	051	055	059	063
735	99.067	070	074	078	082	085	089	093	097	101
736	104	108	112	116	119	123	127	131	135	138
737	142	146	150	153	157	161	165	169	172	176
738	180	184	187	191	195	199	203	206	210	214
739	218	221	225	229	233	236	240	244	248	252
740	99.255	259	263	267	270	274	278	282	285	289
741	293	297	300	304	308	312	316	319	323	327
742	331	334	338	342	346	349	353	357	361	364
743	368	372	376	379	383	387	391	394	398	402
744	406	409	413	417	421	424	428	432	436	439
745	99.443	447	451	454	458	462	466	469	473	477
746	481	484	488	492	495	499	503	507	510	514
747	518	522	525	529	533	537	540	544	548	551
748	555	559	563	566	570	574	578	581	585	589
749	592	596	600	604	607	611	615	619	622	626
750	99.630	633	637	641	645	648	652	656	659	663
751	667	671	674	678	682	686	689	693	697	700
752	704	708	712	715	719	723	726	730	734	738
753	741	745	749	752	756	760	764	767	771	775
754	778	782	786	790	793	797	801	804	808	812
755	99.815	819	823	827	830	834	838	841	845	849
756	852	856	860	863	867	871	875	878	882	886
757	889	893	897	900	904	908	911	915	919	923
758	926	930	934	937	941	945	948	952	956	959
759	963	967	970	974	978	982	985	989	993	996
760	100.000	004	007	011	015	018	022	026	029	033
761	037	040	044	048	052	055	059	063	066	070
762	074	077	081	085	088	092	096	099	103	107
763	110	114	118	121	125	129	132	136	140	143
764	147	151	154	158	162	165	169	173	176	180
765	100.184	187	191	195	198	202	206	209	213	216
766	220	224	227	231	235	238	242	246	249	253
767	257	260	264	268	271	275	279	283	286	290
768	293	297	300	304	308	311	315	319	322	326
769	330	333	337	341	344	348	352	355	359	363
770	100.366	370	373	377	381	384	388	392	395	399
771	403	406	410	414	417	421	424	428	432	435
772	439	442	446	450	453	457	461	464	468	472
773	475	479	483	486	490	493	497	501	504	508
774	511	515	519	522	526	530	533	537	540	544
775	100.548	551	555	559	562	566	569	573	577	580
776	584	588	591	595	598	602	606	609	613	616
777	620	624	627	631	634	638	642	645	649	653
778	656	660	663	667	671	674	678	681	685	689
779	692	696	699	703	700	710	714	718	721	725
780	100.728	732	735	739	743	746	750	753	757	761
781	764	768	772	775	779	782	786	789	793	797
782	800	804	807	811	815	818	822	825	829	833
783	836	840	843	847	851	854	858	861	865	869
784	872	876	879	883	886	890	894	897	901	904
785	100.908	912	915	919	922	926	929	933	937	940
786	944	947	951	954	958	962	965	969	972	976
787	979	983	987	990	994	997	001[a]	005[a]	008[a]	012[a]
788	101.015	019	022	026	029	033	037	040	044	047
789	051	054	058	062	065	069	072	076	079	083
790	101.087	090	094	097	101	104	108	112	115	119
791	122	126	129	133	136	140	144	147	151	154
792	158	161	165	168	172	176	179	183	186	190
793	193	197	200	204	207	211	215	218	222	225
794	229	232	236	239	243	246	250	254	257	261
795	101.264	268	271	275	278	282	286	289	293	296
796	300	303	307	310	314	317	321	324	328	332
797	335	339	342	346	349	353	356	360	363	367
798	370	374	377	381	385	388	392	395	399	402
799	406	409	413	416	420	423	427	430	434	437
800	101.441	—	—	—	—	—	—	—	—	—

[a] For lower pressures see under Vapor Tension of Water

MELTING POINTS OF MIXTURES OF METALS

(Smithsonian Physical Tables)

Melting-points (°C)

Metals	Percentage of metal in second column (%) 0	10	20	30	40	50	60	70	80	90	100
Pb. Sn.	326	295	276	262	240	220	190	185	200	216	232
Bi.	322	290	—	—	179	145	126	168	205	—	268
Te.	322	710	790	880	917	760	600	480	410	425	446
Ag.	328	460	545	590	620	650	705	775	840	905	959
Na.	—	360	420	400	370	330	290	250	200	130	96
Cu.	326	870	920	925	945	950	955	985	1005	1020	1084
Sb.	326	250	275	330	395	440	490	525	560	600	632
Al. Sb.	650	750	840	925	945	950	970	1000	1040	1010	632
Cu.	650	630	600	560	540	580	610	755	930	1055	1084
Au.	655	675	740	800	855	915	970	1025	1055	675	1062
Ag.	650	625	615	600	590	580	575	570	650	750	954
Zn.	654	640	620	600	580	560	530	510	475	425	419
Fe.	653	860	1015	1110	1145	1145	1220	1315	1425	1500	1515
Sn.	650	645	635	625	620	605	590	570	560	540	232
Sb. Bi.	632	610	590	575	555	540	520	470	405	330	268
Ag.	630	595	570	545	520	500	503	545	680	850	959
Sn.	622	600	570	525	480	430	395	350	310	255	232
Zn.	632	555	510	540	570	565	540	525	510	470	419

MELTING POINTS OF MIXTURES OF METALS (continued)

Percentage of metal in second column (%)

Metals	0	10	20	30	40	50	60	0	80	90	100
Ni. Sn.	1455	1380	1290	1200	1235	1290	1305	1230	1060	800	232
Na. Bi.	96	425	520	590	645	690	720	730	715	570	268
Cd.	96	125	185	245	285	325	330	340	360	390	322
Cd. Ag.	322	420	520	610	700	760	805	850	895	940	954
Ti.	321	300	285	270	262	258	245	230	210	235	302
Zn.	322	280	270	295	313	327	340	355	370	390	419
Au. Cu.	1063	910	890	895	905	925	975	1000	1025	1060	1084
Ag.	1064	1062	1061	1058	1054	1049	1039	1025	1006	982	963
Pt.	1075	1125	1190	1250	1320	1380	1455	1530	1610	1685	1775
K. Na.	62	17.5	−10	−3.5	5	11	26	41	58	77	97.5
Hg.	—	—	—	—	—	90	110	135	162	265	—
Tl.	62.5	133	165	188	205	215	220	240	280	305	301
Cu. Ni.	1080	1180	1240	1290	1320	1355	1380	1410	1430	1440	1455
Ag.	1082	1035	990	945	910	870	830	788	814	875	960
Sn.	1084	1005	890	755	725	680	630	580	530	440	232
Zn.	1084	1040	995	930	900	880	820	780	700	580	419
Ag. Zn.	959	850	755	705	690	660	630	610	570	505	419
Sn.	959	870	750	630	550	495	450	420	375	300	232
Na. Hg.	96.5	90	80	70	60	45	22	55	95	215	—

[a] The data in this table are compiled from various sources — hence variations in the melting point of the metals as shown in this column. The triple point of water, 0.01°C (273.16°K) is the thermodynamic point for temperature measurements.

COMMERCIAL METALS AND ALLOYS

Miscellaneous Properties

Properties (Typical Only)

Common name and classification	Thermal conductivity J/sec cm °K	Thermal conductivity Btu/hr ft °F	Thermal conductivity kcal/sec cm °C	Specific gravity	Coeff. of linear expansion, μ in./in. °F	Electrical resistivity, microhm-cm	Modulus of elasticity, millions of psi	Approximate melting point °F	Approximate melting point °C
Ingot iron (included for comparison)	1.3	77	0.32	7.86	6.8	9	30	2800	1538
Plain carbon steel AISI–SAE 1020	1.0	56	0.23	7.86	6.7	10	30	2760	1515
Stainless steel type 304	0.3	19	0.08	8.02	9.6	72	28	2600	1427
Cast gray iron ASTM A48–48. Class 25	0.8	48	0.20	7.2	6.7	67	13	2150	1177
Malleable iron ASTM A47				7.32	6.6	30	25	2250	1232
Ductile cast iron ASTM A339, A395	0.6	34	0.14	7.2	7.5	60	25	2100	1149
Ni-resist cast iron, type 2	0.7	41	0.17	7.3	9.6	170	15.6	2250	1232
Cast 28-7 alloy (IID) ASTM A297–63T	0.04	2	0.01	7.6	9.2	41	27	2700	1482
Hastelloy C	0.2	10	0.04	3.94	6.3	139	30	2350	1288
Inconel X, annealed	0.3	17	0.07	8.25	6.7	122	31	2550	1399
Haynes Stellite alloy 25 (L605)	0.2	10	0.04	9.15	7.61	88	34	2500	1371
Aluminum alloy 3003, rolled ASTM B221	2.8	164	0.68	2.73	12.9	4	10	1200	649
Aluminum alloy 2017, annealed ASTM B221	3.0	174	0.72	2.8	12.7	4	10.5	1185	641
Aluminum alloy 380 ASTM SC84B	1.8	102	0.42	2.7	11.6	7.5	10.3	1050	566
Copper ASTM B152, B124, B133, B1, B2, B3	7.1	411	1.70	8.91	9.3	1.7	17	1980	1082
Yellow brass (high brass) ASTM B36, B134, B135	2.2	126	0.52	8.47	10.5	7	15	1710	932
Aluminum bronze ASTM B169, alloy A; ASTM B124, B150	1.3	75	0.31	7.8	9.2	12	17	1900	1038
Beryllium copper 25 ASTM B194	0.2	12	0.05	8.25	9.3	–	19	1700	927
Nickel silver 18% alloy A (wrought) ASTM B122, No. 2	0.6	34	0.14	8.8	9.0	29	18	2030	1110
Cupronickel 30%	0.5	31	0.13	8.95	8.5	35	22	2240	1227
Red brass (cast) ASTM B30, No. 4A	1.3	77	0.32	8.7	10	11	13	1825	996
Chemical lead	0.6	36	0.15	11.35	16.4	21	2	621	327
Antimonial lead (hard lead)	0.5	31	0.13	10.9	15.1	23	3	554	290
Solder 50–50	0.8	48	0.20	8.89	13.1	15	–	420	216
Magnesium alloy AZ31B	1.4	82	0.34	1.77	14.5	9	6.5	1160	627
K Monel	0.3	19	0.08	8.47	7.4	58	26	2430	1332
Nickel ASTM B160, B161, B162	1.1	63	0.26	8.89	6.6	10	30	2625	1441
Cupronickel 55–45 (Constantan)	0.4	24	0.10	8.9	8.1	49	24	2300	1260
Commercial titanium	0.3	19	0.08	5	4.9	80	16.5	3300	1816
Zinc ASTM B69	2.0	114	0.47	7.14	18	6	–	785	418
Zirconium, commercial	0.3	19	0.08	6.5	2.9	41	12	3350	1843

THERMAL PROPERTIES OF PURE METALS

From Handbook of Tables for Applied Engineering Science by R. E. Bolz and G. L. Tuve, The Chemical Rubber Co., 1970

	At Atmospheric Pressure							Liquid Metal			
		At 100°K		At 25°C (77°F)					Vapor pressure		
Metal	Latent heat of fusion, cal/g	Thermal conductivity, watts cm°C	Specific heat, cal/g°C	Specific heat, cal/g°C	Coeff. of linear expansion, ($\times 10^6$) $(°C)^{-1}$	Thermal conductivity, watts/cm°C*	Specific heat (liquid) at 2000°K cal/g°C	10^{-3} atm	10^{-6} atm	10^{-9} atm
								Boiling point temperatures, °K		
Aluminum	95	3.00	0.115	0.215	25	2.37	0.26	1,782	1,333	1,063
Antimony	38.5		0.010	0.050	9	0.105	0.062	1,007	741	612
Beryllium	324	—	0.049	0.436	12	2.18	0.78	1,793	1,347	1,085
Bismuth	12.4	—	0.026	0.030	13	0.084	0.036	1,155	851	677
Cadmium	13.2	1.03	0.047	0.055	30	0.93	0.063	655	486	388
Chromium	79	1.58	0.046	0.110	6	0.91	0.224	1,992	1,530	1,247
Cobalt	66	—	0.057	0.10	12	0.69	0.164	2,167	1,652	1,345
Copper	49	4.83	0.061	0.092	16.6	3.98	0.118	1,862	1,391	1,120
Gold	15	3.45	0.026	0.031	14.2	3.15	0.0355	2,023	1,510	1,211
Iridium	33	—	0.022	0.031	6	1.47	0.0434	3,253	2,515	2,062
Iron	65	1.32	0.052	0.108	12	0.803	0.197	2,093	1,594	1,297
Lead	5.5	0.396	0.028	0.031	29	0.346	0.033	1,230	889	698
Magnesium	88.0	1.69	0.016	0.243	25	1.59	0.32	857	638	509
Manganese	64	—	0.064	0.114	22	—	0.20	1,495	1,131	913
Mercury	2.7	—	0.029	0.033	—	0.0839	—	393	287	227
Molybdenum	69	1.79	0.033	0.060	5	1.4	0.089	3,344	2,558	2,079
Nickel	71	1.58	0.055	0.106	13	0.899	0.175	2,156	1,646	1,343
Niobium (Columbium)	68	0.552	0.045	0.064	7	0.52	0.083	3,523	2,721	2,232
Osmium	34	—	—	0.031	5	0.61	0.039	—	—	—
Platinum	24	0.79	0.024	0.032	9	0.73	0.043	2,817	2,155	1,757
Plutonium	3	—	0.019	0.032	54	0.08	0.041	2,200	1,596	1,252
Potassium	14.5	—	0.150	0.180	83	0.99	—	606	430	335
Rhodium	50	—	—	0.058	8	1.50	0.092	—	—	—
Selenium	16	—	—	0.077	37	0.005	—	—	—	—
Silicon	430	—	0.062	0.17	3	0.835	0.217	2,340	1,749	1,427
Silver	26.5	4.50	0.045	0.057	19	4.27	0.068	1,582	1,179	952
Sodium	27	—	0.234	0.293	70	1.34	—	701	504	394
Tantalum	41	0.592	0.026	0.034	6.5	0.54	0.040	3,959	3,052	2,495
Thorium	17	—	0.024	0.03	12	0.41	0.047	3,251	2,407	1,919
Tin	14.1	0.85	0.039	0.054	20	0.64	0.058	1,857	1,366	1,080
Titanium	100	0.312	0.072	0.125	8.5	0.2	0.188	2,405	1,827	1,484
Tungsten	46	2.35	0.021	0.032	4.5	1.78	0.040	4,139	3,228	2,656
Uranium	12	—	0.022	0.028	13.4	0.25	0.048	2,861	2,128	1,699
Vanadium	98	—	0.061	0.116	8	0.60	0.207	2,525	1,948	1,591
Zinc	27	1.32	0.063	0.093	35	1.15	—	752	559	449

* (watts/cm°C) × 860.421 = Cal(gm)$hr^{-1}cm^{-1}°C^{-1}$
(watts/cm°C) × 57.818 = Btu hr^{-1} ft^{-1} °F.

BOILING POINTS AND TRIPLE POINTS OF SOME LOW BOILING ELEMENTS

Element	Boiling point °K	Triple point °K	λ point °K
He	4.216		2.174
p-H_2	20.27	13.81	
n-H_2	20.379	13.95	
N_2	77.35		
O_2	90.188	54.34	
A	87.45		

MOLECULAR ELEVATION OF THE BOILING POINT*

Molecular elevation of the boiling point showing the elevation of the boiling point in degrees C due to the addition of one gram molecular weight of the dissolved substance to 1000 g of any one of the solvents below. The correction in the last column gives the number of degrees to be subtracted for each mm. of difference between the barometric reading and 760 mm.

Solvent	K_B	Barometric correction per mm.	Solvent	K_B	Barometric correction per mm.
Acetic acid	3.07	0.0008	Ethyl acetate	2.77	0.0007
Acetone	1.71	0.0004	Ethyl ether	2.02	0.0005
Aniline	3.52	0.0009	*n*-Hexane	2.75	0.0007
Benzene	2.53	0.0007	Methanol (methyl alcohol)	0.83	0.0002
Bromobenzene	6.26	0.0016	Methyl acetate	2.15	0.0005
Carbon bisulfide	2.34	0.0006	Nitrobenzene	5.24	0.0013
Carbon tetrachloride	5.03	0.0013	*n*-Octane	4.02	0.0010
Chloroform	3.63	0.0009	Phenol	3.56	0.0009
Cyclohexane	2.79	0.0007	Toluene	3.33	0.0008
Ethanol (ethyl alcohol)	1.22	0.0003	Water	0.512	0.0001

* Most values from Hoyt, C. S. and Fink, C. K., Journal of Physical Chemistry, Vol. 41, No. 3., March, 1937.

MOLECULAR DEPRESSION OF THE FREEZING POINT

Showing the depression of the freezing point due to the addition of one gram molecular weight of dissolved substance, for various solvents.

Solvent	Depression for one gram molecular weight dissolved in 100 g °C	Solvent	Depression for one gram molecular weight dissolved in 100 g °C
Acetic acid	39.0	Naphthalene	68—69
Benzene	49.0	Nitrobenzene	70.0
Benzophenone	98.0	Phenol	74.0
Diphenyl	80.0	Stearic acid	45.0
Diphenylamine	86.0	Triphenyl methane	124.5
Ethylene dibromide	118.0	Urethane	51.4
Formic acid	27.7	Water	18.5—18.7

CORRECTION OF BOILING POINTS TO STANDARD PRESSURE

By H. B. Hass and R. F. Newton

This correction may be made by using the equation:

$$\Delta t = \frac{(273.1 + t)(2.8808 - \log p)}{\phi + .15(2.8808 - \log p)} \qquad (1)$$

where Δt = degrees C to be added to the observed boiling point.

t = the observed boiling point.

$\log p$ = the logarithm of the observed pressure in millimeters of mercury.

$$\phi = \frac{\Delta H_{vap}}{2.303RT_b} = \frac{\Delta S_{vap}}{2.303R}$$

The value of ϕ may be estimated from the graph and the table. Substances not included in the table may be classified by grouping them with compounds which bear a close physical or structural resemblance to them.

Example 1. Benzene boils at 20°C. at 75 mm pressure. What is its normal boiling point? We do not find benzene in the table but we find hydrocarbons in group 2, and a group 2 compound with a boiling point of 20° has a ϕ of 4.6.

Substituting in the equation

$$\Delta t = \frac{(273.1 + 20)(2.8808 - 1.8751)}{4.60 + .15(2.8808 - 1.8751)} = 62°$$

Adding this to 20° gives 82° as a first approximation.

The graph shows that the ϕ for a compound of group 2 boiling at 82° is 4.72 instead of 4.60 which we originally used. Since ϕ is in the denominator, this increase will lower our Δt by the ratio, 4.60/4.72, or the corrected Δt is $62 \times 4.60/4.72 = 60.4$. Adding Δt to t, gives 80.4° as a second approximation.

The formula can best be used in a slightly different form when the reverse calculation is desired, *i.e.*, when one calculates the vapor pressure at a given temperature, lower than the normal boiling point.

$$2.8808 - \log p = \frac{\phi \Delta t}{273.1 + t - .15\Delta t} \qquad (2)$$

Example 2. Alcohol boils at 78.4°C. What is its vapor pressure at 20°C.? Substituting in equation 2:

$$2.8808 - \log p = \frac{6.06 \times 58.4}{293.1 - (.15 \times 58.4)} = 1.245$$
$$\log p = 2.8808 - 1.245 = 1.6358$$
$$p = 43.2 \text{ mm.}$$

Here no second approximation is necessary, since the correct value of ϕ was taken immediately, the normal boiling point having been known.

Compound	Group	Compound	Group
Acetaldehyde	3	Amines	3
Acetic acid	4	*n*-Amyl alcohol	8
Acetic anhydride	6	Anthracene	1
Acetone	3	Anthraquinone	1
Acetophenone	4	Benzaldehyde	2
Benzoic acid	5	Hydrogen cyanide	3
Benzonitrile	2	Isoamyl alcohol	7
Benzophenone	2	Isobutyl alcohol	8
Benzyl alcohol	5	Isobutyric acid	6
Butylethylene	1	Isocaproic acid	7
Butyric acid	7	Methane	1
Camphor	2	Methanol	7
Carbon monoxide	1	Methyl amine	5
Carbon oxysulfide	2	Methyl benzoate	3
Carbon suboxide	2	Methyl ether	3
Carbon sulfoselenide	2	Methyl ethyl ether	3
m.p. Chloroanilines	3	Methyl ethyl ketone	2
Chlorinated derivatives	Same group as though Cl were H	Methyl fluoride	3
		Methyl formate	4
o.m.p. Cresols	4	Methyl salicylate	2
Cyanogen	4	Methyl silicane	1
Cyanogen chloride	3	α, β Naphthols	3
Dibenzyl ketone	2	Nitrobenzene	3
Dimethyl amine	4	Nitromethane	3
Dimethyl oxalate	4	o.m.p. Nitrotoluenes	2
Dimethyl silicane	2	o.m.p. Nitrotoluidines	2
Esters	3	Phenanthrene	1
Ethanol	8	Phenol	5
Ethers	2	Phosgene	2
Ethylamine	4	Phthalic anhydride	2
Ethylene glycol	7	Propionic acid	5
Ethylene oxide	3	*n*-Propyl alcohol	8
Formic acid	3	Quinoline	2
Glycol diacetate	4	Sulfides	2
Halogen derivatives	Same group as though halogen were hydrogen.	Tetranitromethane	3
		Trichloroethylene	1
		Valeric acid	7
Heptylic acid	7	Water	6
Hydrocarbons	2		

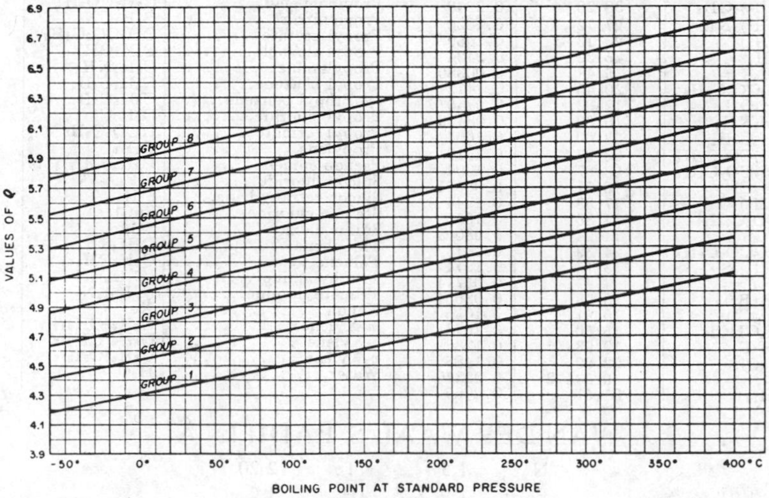

VAN DER WAALS' CONSTANTS FOR GASES

(Calculated from Amagat units in Landolt-Bornstein Physical Chemical Tables)

Van der Waals' equation is an equation of state for real gases. It may be written

$$\left(P + \frac{a}{V^2}\right)(V - b) = RT \text{ for one mole.} \qquad \text{or} \qquad \left(P + \frac{n^2a}{V^2}\right)(V - nb) = nRT \text{ for } n \text{ moles.}$$

The term a is a measure of the attractive force between the molecules. The term b is due to the finite volume of the molecules and to their general incompressibility. It is known that a and b vary to some extent with temperature.

The values for a and b in the following table are those to be used when the pressure is in atmospheres and the volume is in liters. Thus R in the above equation will be 0.08206 liter atmospheres per mole per degree. T is degrees Kelvin.

Name	Formula	a	b
		$\frac{(\text{liters})^2 \times \text{atm.}}{(\text{mole})^2}$	$\frac{\text{liters}}{\text{mole}}$
Acetic acid	CH_3CO_2H	17.59	0.1068
Acetic anhydride	$(CH_3CO)_2O$	19.90	0.1263
Acetone	$(CH_3)_2CO$	13.91	0.0994
Acetonitrile	CH_3CN	17.58	0.1168
Acetylene	C_2H_2	4.390	0.05136
Ammonia	NH_3	4.170	0.03707
Amyl formate	$HCO_2C_5H_{11}$	27.58	0.1730
Amylene	C_5H_{10}	15.90	0.1207
*Iso*amylene	C_5H_{10}	18.08	0.1405
Aniline	$C_6H_5NH_2$	26.50	0.1369
Argon	A	1.345	0.03219
Benzene	C_6H_6	18.00	0.1154
Benzonitrile	C_6H_5CN	33.39	0.1724
Bromobenzene	C_6H_5Br	28.56	0.1539
n-Butane	C_4H_{10}	14.47	0.1226
iso-Butane	C_4H_{10}	12.87	0.1142
iso-Butyl acetate	$CH_3CO_2C_4H_9$	28.50	0.1833
iso-Butyl alcohol	C_4H_9OH	17.03	0.1143
iso-Butyl benzene	$C_6H_5C_4H_9$	38.59	0.2144
iso-Butyl formate	$HCO_2C_4H_9$	22.54	0.1476
Butyronitrile	C_3H_7CN	25.72	0.1596
Capronitrile	$C_5H_{11}CN$	34.16	0.1984
Carbon dioxide	CO_2	3.592	0.04267
Carbon disulfide	CS_2	11.62	0.07685
Carbon monoxide	CO	1.485	0.03985
Carbon oxysulfide	COS	3.933	0.05817
Carbon tetrachloride	CCl_4	20.39	0.1383
Chlorine	Cl_2	6.493	0.05622
Chlorobenzene	C_6H_5Cl	25.43	0.1453
Chloroform	$CHCl_3$	15.17	0.1022
m-Cresol	$CH_3C_6H_4OH$	31.38	0.1607
Cyanogen	C_2N_2	7.667	0.06901
Cyclohexane	C_6H_{12}	22.81	0.1424
Cymene	$C_{10}H_{14}$	42.16	0.2336
Decane	$C_{10}H_{12}$	48.55	0.2905
Di-isobutyl	C_8H_{18}	34.97	0.2296
Diethylamine	$(C_2H_5)_2NH$	19.15	0.1392
Dimethylamine	$(CH_3)_2NH$	10.38	0.08570
Dimethylaniline	$C_6H_5N(CH_3)_2$	37.49	0.1970
Diphenyl	$(C_6H_5)_2$	52.79	0.2480
Diphenyl methane	$(C_6H_5)_2CH_2$	38.20	0.2240
Dipropylamine	$(C_3H_7)_2NH$	27.72	0.1820
Di-isopropyl	$(C_3H_7)_2$	23.13	0.1669
Durene	$C_{10}H_{14}$	45.32	0.2424
Ethane	C_2H_6	5.489	0.06380
Ethyl acetate	$CH_3CO_2C_2H_5$	20.45	0.1412
Ethyl alcohol	C_2H_5OH	12.02	0.08407
Ethylamine	$C_2H_5NH_2$	10.60	0.08409
Ethyl benzene	$C_2H_5C_6H_5$	28.60	0.1667
Ethyl butyrate	$C_3H_7CO_2C_2H_5$	30.07	0.1919
Ethyl isobutyrate	$C_3H_7CO_2C_2H_5$	28.87	0.1994
Ethyl chloride	C_2H_5Cl	10.91	0.08651
Ethyl ether	$(C_2H_5)_2O$	17.38	0.1344
Ethyl formate	$HCO_2C_2H_5$	14.80	0.1056
Ethyl mercaptan	C_2H_5SH	11.24	0.08098
Ethyl propionate	$C_2H_5CO_2C_2H_5$	24.39	0.1615
Ethyl sulfide	$(C_2H_5)_2S$	18.75	0.1214
Ethylene	C_2H_4	4.471	0.05714
Ethylene bromide	$(CH_2Br)_2$	13.98	0.08664
Ethylene chloride	$(CH_2Cl)_2$	16.91	0.1086
Ethylidene chloride	CH_3CHCl_2	15.50	0.1073
Fluorobenzene	C_6H_5F	19.93	0.1286
Germanium tetrachloride	$GeCl_4$	22.60	0.1485
Helium	He	0.03412	0.02370
n-Heptane	C_7H_{16}	31.51	0.2065
n-Hexane	C_6H_{14}	24.39	0.1735
Hydrogen	H_2	0.2444	0.02661
Hydrogen bromide	HBr	4.451	0.04431
Hydrogen chloride	HCl	3.667	0.04081
Hydrogen selenide	H_2Se	5.268	0.04637
Hydrogen sulfide	H_2S	4.431	0.04287
Iodobenzene	C_6H_5I	33.08	0.1656
Krypton	Kr	2.318	0.03978
Mercury	Hg	8.093	0.01696
Mesitylene	$(CH_3)_3C_6H_3$	34.32	0.1979
Methane	CH_4	2.253	0.04278
Methyl acetate	$CH_3CO_2CH_3$	15.29	0.1091
Methyl alcohol	CH_3OH	9.523	0.06702
Methylamine	CH_3NH_2	7.130	0.05992
Methyl butyrate	$C_3H_7CO_2CH_3$	23.94	0.1569
Methyl isobutyrate	$C_3H_7CO_2CH_3$	24.50	0.1637
Methyl chloride	CH_3Cl	7.471	0.06483
Methyl ether	$(CH_3)_2O$	8.073	0.07246
Methyl ethyl ether	$CH_3OC_2H_5$	11.95	0.09775
Methyl ethyl sulfide	$CH_3SC_2H_5$	19.23	0.1304
Methyl fluoride	CH_3F	4.631	0.05264
Methyl formate	HCO_2CH_3	10.84	0.08068
Methyl propionate	$C_2H_5CO_2CH_3$	19.91	0.1360
Methyl sulfide	$(CH_3)_2S$	12.87	0.09213
Methyl valerate	$C_4H_9CO_2CH_3$	28.96	0.1845
Naphthalene	$C_{10}H_8$	39.74	0.1937
Neon	Ne	0.2107	0.01709
Nitric oxide	NO	1.340	0.02789
Nitrogen	N_2	1.390	0.03913
Nitrogen dioxide	NO_2	5.284	0.04424
Nitrous oxide	N_2O	3.782	0.04415
n-Octane	C_8H_{18}	37.32	0.2368
Oxygen	O_2	1.360	0.03183
n-Pentane	C_5H_{12}	19.01	0.1460
iso-Pentane	C_5H_{12}	18.05	0.1417
Phenetole	$C_6H_5OC_2H_5$	35.16	0.1963
Phosphine	PH_3	4.631	0.05156
Phosphonium chloride	PH_4Cl	4.054	0.04545
Phosphorus	P	52.94	0.1566
Propane	C_3H_8	8.664	0.08445
Propionic acid	$C_2H_5CO_2H$	20.11	0.1187
Propionitrile	C_2H_5CN	16.44	0.1064
Propyl acetate	$CH_3CO_2C_3H_7$	24.63	0.1619
Propyl alcohol	C_3H_7OH	14.92	0.1019
Propylamine	$C_3H_7NH_2$	14.99	0.1090
Propyl benzene	$C_6H_5C_3H_7$	35.85	0.2028
iso-Propyl benzene	$C_6H_5C_3H_7$	35.64	0.2025
Propyl chloride	C_3H_7Cl	15.91	0.1141
Propyl formate	$HCO_2C_3H_7$	18.95	0.1280
Propylene	C_3H_6	8.379	0.08272
Pseudo-cumene	$C_6H_3(CH_3)_3$	36.61	0.2021
Silicon fluoride	SiF_4	4.195	0.05571
Silicon tetrahydride	SiH_4	4.320	0.05786
Stannic chloride	$SnCl_4$	26.91	0.1642
Sulfur dioxide	SO_2	6.714	0.05636
Thiophene	C_4H_4S	20.72	0.1270
Toluene	$C_6H_5CH_3$	24.06	0.1463
Triethylamine	$(C_2H_5)_3N$	27.17	0.1831
Trimethylamine	$(CH_3)_3N$	13.02	0.1084
Xenon	Xe	4.194	0.05105
m-Xylene	$C_6H_4(CH_3)_2$	30.36	0.1772
o-Xylene	$C_6H_4(CH_3)_2$	29.98	0.1755
p-Xylene	$C_6H_4(CH_3)_2$	30.93	0.1809
Water	H_2O	5.464	0.03049

VAN DER WAAL'S RADII IN Å

N	1.5	Te	2.20
P	1.9	H	1.2
As	2.0	F	1.35
Sb	2.2	Cl	1.80
O	1.40	Br	1.95
S	1.85	I	2.15
Se	2.00		

EMERGENT STEM CORRECTION FOR LIQUID-IN-GLASS THERMOMETERS

Accurate thermometers are calibrated with the entire stem immersed in the bath which determines the temperature of the thermometer bulb. However, for reasons of convenience it is common practice when using a thermometer to permit its stem to extend out of the apparatus. Under these conditions both the stem and the mercury in the exposed stem are at a temperature different from that of the bulb. This introduces an error into the observed temperature. Since the coefficient of thermal expansion of glass is less than that of mercury, the observed temperature will be less than the true temperature if the bulb is hotter than the stem and greater than the true temperature, providing the thermal gradient is reversed. For exact work the magnitude of this error can only be determined by experiment. However, for most purposes it is sufficiently accurate to apply the following equation which takes into account the difference of the thermal expansion of glass and mercury:

$$T_c = T_o + F \times L(T_o - T_m)$$

Where

T_c = corrected temperature
T_o = observed temperature
T_m = mean temperature of exposed stem. The mean temperature of the exposed stem may be determined by fastening the bulb of a second thermometer against the midpoint of the exposed liquid column.
L = the length of the exposed column in degrees above the surface of the substance whose temperature is being determined.
F = correction factor. For approximate work and when the liquid in the thermometer is mercury a value for F of 0.00016 is generally used. For more accurate work with mercury filled thermometers values as given in the following table are used. For thermometers filled with organic liquids it is customary to use 0.001 for the value of F.

Values of F for various glasses

Tm°C.	Corning 0041	Corning 8800	Corning 8810	Jena 16 III	Jena 59 III
50	0.000157	0.000166	0.000156	0.000158	0.000164
150	0.000159	0.000167	0.000157	0.000158	0.000165
250	0.000163	0.000168	0.000161	0.000161	0.000170
350	0.000168	0.000173	0.000166	—	0.000177

PRESSURE OF AQUEOUS VAPOR

Vapor Pressure of Ice

Pressure of aqueous vapor over ice in mm of Hg for temperatures from −98 to 0°C.

Temp. °C	0.0	0.2	0.4	0.6	0.8
−90	0.000070	0.000048	0.000033	0.000022	0.000015
−80	0.00040	0.00029	0.00020	0.00014	0.00010
−70	0.00194	0.00143	0.00105	0.00077	0.00056
−60	0.00808	0.00614	0.00464	0.00349	0.00261
−50	0.02955	0.0230	0.0178	0.0138	0.0106
−40	0.0966	0.0768	0.0609	0.0481	0.0378
−30	0.2859	0.2318	0.1873	0.1507	0.1209
−29	0.317	0.311	0.304	0.298	0.292
−28	0.351	0.344	0.337	0.330	0.324
−27	0.389	0.381	0.374	0.366	0.359
−26	0.430	0.422	0.414	0.405	0.397
−25	0.476	0.467	0.457	0.448	0.439
−24	0.526	0.515	0.505	0.495	0.486
−23	0.580	0.569	0.558	0.547	0.536
−22	0.640	0.627	0.615	0.603	0.592
−21	0.705	0.691	0.678	0.665	0.652
−20	0.776	0.761	0.747	0.733	0.719
−19	0.854	0.838	0.822	0.806	0.791
−18	0.939	0.921	0.904	0.887	0.870
−17	1.031	1.012	0.993	0.975	0.956
−16	1.132	1.111	1.091	1.070	1.051
−15	1.241	1.219	1.196	1.175	1.153
−14	1.361	1.336	1.312	1.288	1.264
−13	1.490	1.464	1.437	1.411	1.386
−12	1.632	1.602	1.574	1.546	1.518
−11	1.785	1.753	1.722	1.691	1.661
−10	1.950	1.916	1.883	1.849	1.817
−9	2.131	2.093	2.057	2.021	1.985
−8	2.326	2.285	2.246	2.207	2.168
−7	2.537	2.493	2.450	2.408	2.367
−6	2.765	2.718	2.672	2.626	2.581
−5	3.013	2.962	2.912	2.862	2.813
−4	3.280	3.225	3.171	3.117	3.065
−3	3.568	3.509	3.451	3.393	3.336
−2	3.880	3.816	3.753	3.691	3.630
−1	4.217	4.147	4.079	4.012	3.946
−0	4.579	4.504	4.431	4.359	4.287

VAPOR PRESSURE OF WATER BELOW 100°C

Pressure of aqueous vapor over water in mm of Hg for temperatures from −15.8 to 100°C. Values for fractional degrees between 50 and 89 were obtained by interpolation.

Temp. °C	0.0	0.2	0.4	0.6	0.8
−15	1.436	1.414	1.390	1.368	1.345
−14	1.560	1.534	1.511	1.485	1.460
−13	1.691	1.665	1.637	1.611	1.585
−12	1.834	1.804	1.776	1.748	1.720
−11	1.987	1.955	1.924	1.893	1.863
−10	2.149	2.116	2.084	2.050	2.018
−9	2.326	2.289	2.254	2.219	2.184
−8	2.514	2.475	2.437	2.399	2.362
−7	2.715	2.674	2.633	2.593	2.553
−6	2.931	2.887	2.843	2.800	2.757
−5	3.163	3.115	3.069	3.022	2.976
−4	3.410	3.359	3.309	3.259	3.211
−3	3.673	3.620	3.567	3.514	3.461
−2	3.956	3.898	3.841	3.785	3.730
−1	4.258	4.196	4.135	4.075	4.016
−0	4.579	4.513	4.448	4.385	4.320
0	4.579	4.647	4.715	4.785	4.855
1	4.926	4.998	5.070	5.144	5.219
2	5.294	5.370	5.447	5.525	5.605
3	5.685	5.766	5.848	5.931	6.015

VAPOR PRESSURE OF WATER BELOW 100°C (continued)

Temp. °C	0.0	0.2	0.4	0.6	0.8
4	6.101	6.187	6.274	6.363	6.453
5	6.543	6.635	6.728	6.822	6.917
6	7.013	7.111	7.209	7.309	7.411
7	7.513	7.617	7.722	7.828	7.936
8	8.045	8.155	8.267	8.380	8.494
9	8.609	8.727	8.845	8.965	9.086
10	9.209	9.333	9.458	9.585	9.714
11	9.844	9.976	10.109	10.244	10.380
12	10.518	10.658	10.799	10.941	11.085
13	11.231	11.379	11.528	11.680	11.833
14	11.987	12.144	12.302	12.462	12.624
15	12.788	12.953	13.121	13.290	13.461
16	13.634	13.809	13.987	14.166	14.347
17	14.530	14.715	14.903	15.092	15.284
18	15.477	15.673	15.871	16.071	16.272
19	16.477	16.685	16.894	17.105	17.319
20	17.535	17.753	17.974	18.197	18.422
21	18.650	18.880	19.113	19.349	19.587
22	19.827	20.070	20.316	20.565	20.815
23	21.068	21.324	21.583	21.845	22.110
24	22.377	22.648	22.922	23.198	23.476
25	23.756	24.039	24.326	24.617	24.912
26	25.209	25.509	25.812	26.117	26.426
27	26.739	27.055	27.374	27.696	28.021
28	28.349	28.680	29.015	29.354	29.697
29	30.043	30.392	30.745	31.102	31.461
30	31.824	32.191	32.561	32.934	33.312
31	33.695	34.082	34.471	34.864	35.261
32	35.663	36.068	36.477	36.891	37.308
33	37.729	38.155	38.584	39.018	39.457
34	39.898	40.344	40.796	41.251	41.710
35	41.175	42.644	43.117	43.595	44.078
36	44.563	45.054	45.549	46.050	46.556
37	47.067	47.582	48.102	48.627	49.157
38	49.692	50.231	50.774	51.323	51.879
39	52.442	53.009	53.580	54.156	54.737
40	55.324	55.91	56.51	57.11	57.72
41	58.34	58.96	59.58	60.22	60.86
42	61.50	62.14	62.80	63.46	64.12
43	64.80	65.48	66.16	66.86	67.56
44	68.26	68.97	69.69	70.41	71.14
45	71.88	72.62	73.36	74.12	74.88
46	75.65	76.43	77.21	78.00	78.80
47	79.60	80.41	81.23	82.05	82.87
48	83.71	84.56	85.42	86.28	87.14
49	88.02	88.90	89.79	90.69	91.59
50	92.51	93.5	94.4	95.3	96.3
51	97.20	98.2	99.1	100.1	101.1
52	102.09	103.1	104.1	105.1	106.2
53	107.20	108.2	109.3	110.4	111.4
54	112.51	113.6	114.7	115.8	116.9
55	118.04	119.1	120.3	121.5	122.6
56	123.80	125.0	126.2	127.4	128.6
57	129.82	131.0	132.3	133.5	134.7
58	136.08	137.3	138.5	139.9	141.2
59	142.60	143.9	145.2	146.6	148.0
60	149.38	150.7	152.1	153.5	155.0
61	156.43	157.8	159.3	160.8	162.3
62	163.77	165.2	166.8	168.3	169.8
63	171.38	172.9	174.5	176.1	177.7
64	179.31	180.9	182.5	184.2	185.8
65	187.54	189.2	190.9	192.6	194.3
66	196.09	197.8	199.5	201.3	203.1
67	204.96	206.8	208.6	210.5	212.3
68	214.17	216.0	218.0	219.9	221.8
69	223.73	225.7	227.7	229.7	231.7
70	233.7	235.7	237.7	239.7	241.8
71	243.9	246.0	248.2	250.3	252.4
72	254.6	256.8	259.0	261.2	263.4
73	265.7	268.0	270.2	272.6	274.8
74	277.2	279.4	281.8	284.2	286.6
75	289.1	291.5	294.0	296.4	298.8
76	301.4	303.8	306.4	308.9	311.4
77	314.1	316.6	319.2	322.0	324.6
78	327.3	330.0	332.8	335.6	338.2
79	341.0	343.8	346.6	349.4	352.2
80	355.1	358.0	361.0	363.8	366.8
81	369.7	372.6	375.6	378.8	381.8
82	384.9	388.0	391.2	394.4	397.4
83	400.6	403.8	407.0	410.2	413.6
84	416.8	420.2	423.6	426.8	430.2
85	433.6	437.0	440.4	444.0	447.5
86	450.9	454.4	458.0	461.6	465.2
87	468.7	472.4	476.0	479.8	483.4
88	487.1	491.0	494.7	498.5	502.2
89	506.1	510.0	513.9	517.8	521.8
90	525.76	529.77	533.80	537.86	541.95
91	546.05	550.18	554.35	558.53	562.75
92	566.99	571.26	575.55	579.87	584.22
93	588.60	593.00	597.43	601.89	606.38
94	610.90	615.44	620.01	624.61	629.24
95	633.90	638.59	643.30	648.05	652.82
96	657.62	662.45	667.31	672.20	677.12
97	682.07	687.04	692.05	697.10	702.17
98	707.27	712.40	717.56	722.75	727.98
99	733.24	738.53	743.85	749.20	754.58
100	760.00	765.45	770.93	776.44	782.00
101	787.57	793.18	798.82	804.50	810.21

VAPOR PRESSURE OF WATER ABOVE 100°C.

Based on values given by Keyes in the International Critical Tables.

Temp. °C	Pressure mm	Pressure Pounds per sq. in.	Temp. °F
100	760.	14.696	212.0
101	787.51	15.228	213.8
102	815.86	15.776	215.6
103	845.12	16.342	217.4
104	875.06	16.921	219.2
105	906.07	17.521	221.0
106	937.92	18.136	222.8
107	970.60	18.768	224.6
108	1004.42	19.422	226.4
109	1038.92	20.089	228.2
110	1074.56	20.779	230.0
111	1111.20	21.487	231.8
112	1148.74	22.213	233.6
113	1187.42	22.961	235.4
114	1227.25	23.731	237.2
115	1267.98	24.519	239.0
116	1309.94	25.330	240.8
117	1352.95	26.162	242.6
118	1397.18	27.017	244.4
119	1442.63	27.896	246.2
120	1489.14	28.795	248.0
121	1536.80	29.717	249.8
122	1586.04	30.669	251.6
123	1636.36	31.642	253.4
124	1687.81	32.637	255.2
125	1740.93	33.664	257.0
126	1795.12	34.712	258.8
127	1850.83	35.789	260.6
128	1907.83	36.891	262.4
129	1966.35	38.023	264.2
130	2026.10	39.180	266.0
131	2087.42	40.364	267.8
132	2150.42	41.582	269.6
133	2214.64	42.824	271.4
134	2280.76	44.103	273.2
135	2347.26	45.389	275.0
136	2416.34	46.724	276.8
137	2488.16	48.113	278.6
138	2560.67	49.515	280.4
139	2634.84	50.950	282.2
140	2710.92	52.421	284.0
141	2788.44	53.920	285.8
142	2867.48	55.448	287.6
143	2948.80	57.020	289.4
144	3031.64	58.622	291.2
145	3116.76	60.268	293.0
146	3203.40	61.944	294.8
147	3292.32	63.663	296.6
148	3382.76	65.412	298.4
149	3476.24	67.220	300.2
150	3570.48	69.042	302.0
151	3667.00	70.908	303.8
152	3766.56	72.833	305.6
153	3866.88	74.773	307.4
154	3970.24	76.772	309.2
155	4075.88	78.815	311.0
156	4183.80	80.901	312.8
157	4293.24	83.018	314.6
158	4404.96	85.178	316.4
159	4519.72	87.397	318.2
160	4636.00	89.646	320.0
161	4755.32	91.953	321.8
162	4876.92	94.304	323.6
163	5000.04	96.685	325.4
164	5126.96	99.139	327.2
165	5256.16	101.638	329.0
166	5386.88	104.165	330.8
167	5521.40	106.766	332.6
168	5658.20	109.412	334.4
169	5798.04	112.116	336.2
170	5940.92	114.879	338.0
171	6085.32	117.671	339.8
172	6233.52	120.537	341.6
173	6383.24	123.432	343.4
174	6538.28	126.430	345.2
175	6694.08	129.442	347.0
176	6852.92	132.514	348.8
177	7015.56	135.659	350.6
178	7180.48	138.848	352.4
179	7349.20	142.110	354.2
180	7520.20	145.417	356.0
181	7694.24	148.782	357.8
182	7872.08	152.221	359.6
183	8052.96	155.719	361.4
184	8236.88	159.275	363.2
185	8423.84	162.890	365.0
186	8616.12	166.609	366.8
187	8809.92	170.356	368.6
188	9007.52	174.177	370.4
189	9208.16	178.057	372.2
190	9413.36	182.025	374.0
191	9620.08	186.022	375.8
192	9831.36	190.107	377.6
193	10047.20	194.281	379.4
194	10265.32	198.499	381.2
195	10488.76	202.819	383.0
196	10715.24	207.199	384.8
197	10944.76	211.637	386.6
198	11179.60	216.178	388.4
199	11417.48	220.778	390.2
200	11659.16	225.451	392.0
201	11905.40	230.213	393.8
202	12155.44	235.048	395.6
203	12408.52	239.942	397.4
204	12666.16	244.924	399.2
205	12929.12	250.008	401.0
206	13197.40	255.196	402.8
207	13467.96	260.428	404.6

VAPOR PRESSURE OF WATER ABOVE 100°C (continued)

Temp. °C	Pressure mm	Pressure Pounds per sq. in.	Temp. °F
208	13742.32	265.733	406.4
209	14022.76	271.156	408.2
210	14305.48	276.623	410.0
211	14595.04	282.222	411.8
212	14888.40	287.895	413.6
213	15184.80	293.626	415.4
214	15488.04	299.490	417.2
215	15792.80	305.383	419.0
216	16104.40	311.408	420.8
217	16420.56	317.522	422.6
218	16742.04	323.738	424.4
219	17067.32	330.028	426.2
220	17395.64	336.377	428.0
221	17731.56	342.872	429.8
222	18072.80	349.471	431.6
223	18417.84	356.143	433.4
224	18766.68	362.888	435.2
225	19123.12	369.781	437.0
226	19482.60	376.732	438.8
227	19848.92	383.815	440.6
228	20219.80	390.987	442.4
229	20596.76	398.276	444.2
230	20978.28	405.654	446.0
231	21365.12	413.134	447.8
232	21757.28	420.717	449.6
233	22154.00	428.388	451.4
234	22558.32	436.207	453.2
235	22967.96	444.128	455.0
236	23382.92	452.152	456.8
237	23802.44	460.264	458.6
238	24229.56	468.523	460.4
239	24661.24	476.871	462.2
240	25100.52	485.365	464.0
241	25543.60	493.933	465.8
242	25994.28	502.647	467.6
243	26449.52	511.450	469.4
244	26912.36	520.400	471.2
245	27381.28	529.467	473.0
246	27855.52	538.638	474.8
247	28335.84	547.926	476.6
248	28823.76	557.360	478.4
249	29317.00	566.898	480.2
250	29817.84	576.583	482.0
251	30324.00	586.370	483.8
252	30837.76	596.305	485.6
253	31356.84	606.342	487.4
254	31885.04	616.556	489.2
255	32417.80	626.858	491.0
256	32957.40	637.292	492.8
257	33505.36	647.888	494.6
258	34059.40	658.601	496.4
259	34618.76	669.417	498.2
260	35188.00	680.425	500.0
261	35761.80	691.520	501.8
262	36343.20	702.763	503.6
263	36932.20	714.152	505.4

Temp. °C	Pressure mm	Pressure Pounds per sq. in.	Temp. °F
264	37529.56	725.703	507.2
265	38133.00	737.372	509.0
266	38742.52	749.158	510.8
267	39361.92	761.135	512.6
268	39986.64	773.215	514.4
269	40619.72	785.457	516.2
270	41261.16	797.861	518.0
271	41910.20	810.411	519.8
272	42566.08	823.094	521.6
273	43229.56	835.923	523.4
274	43902.16	848.929	525.2
275	44580.84	862.053	527.0
276	45269.40	875.367	528.8
277	45964.04	888.799	530.6
278	46669.32	902.437	532.4
279	47382.20	916.222	534.2
280	48104.20	930.183	536.0
281	48833.80	944.291	537.8
282	49570.24	958.532	539.6
283	50316.56	972.963	541.4
284	51072.76	987.586	543.2
285	51838.08	1002.385	545.0
286	52611.76	1017.345	546.8
287	53395.32	1032.497	548.6
288	54187.24	1047.810	550.4
289	54989.04	1063.314	552.2
290	55799.20	1078.980	554.0
291	56612.40	1094.705	555.8
292	57448.40	1110.871	557.6
293	58284.40	1127.036	559.4
294	59135.60	1143.496	561.2
295	59994.40	1160.102	563.0
296	60860.80	1176.856	564.8
297	61742.40	1193.903	566.6
298	62624.00	1210.950	568.4
299	63528.40	1228.439	570.2
300	64432.80	1245.927	572.0
301	65352.40	1263.709	573.8
302	66279.60	1281.638	575.6
303	67214.40	1299.714	577.4
304	68156.00	1317.937	579.2
305	69114.40	1336.454	581.0
306	70072.00	1354.971	582.8
307	71052.40	1373.929	584.6
308	72048.00	1393.181	586.4
309	73028.40	1412.139	588.2
310	74024.00	1431.390	590.0
311	75042.40	1451.083	591.8
312	76076.00	1471.070	593.6
313	77117.20	1491.203	595.4
314	78166.00	1511.484	597.2
315	79230.00	1532.058	599.0
316	80294.00	1552.632	600.8
317	81373.20	1573.501	602.6
318	82467.60	1594.663	604.4
319	83569.60	1615.972	606.2

Temp. °C	Pressure mm	Pressure Pounds per sq. in.	Temp. °F
320	84686.80	1637.575	608.0
321	85819.20	1659.472	609.8
322	86959.20	1681.516	611.6
323	88114.40	1703.854	613.4
324	89277.20	1726.339	615.2
325	90447.60	1748.971	617.0
326	91633.20	1771.897	618.8
327	92826.40	1794.969	620.6
328	94042.40	1818.483	622.4
329	95273.60	1842.291	624.2
330	96512.40	1866.245	626.0
331	97758.80	1890.346	627.8
332	99020.40	1914.742	629.6
333	100297.20	1939.431	631.4
334	101581.60	1964.267	633.2
335	102881.20	1989.398	635.0
336	104196.00	2014.822	636.8
337	105526.00	2040.540	638.6
338	106871.20	2066.552	640.4
339	108224.00	2092.710	642.2
340	109592.00	2119.163	644.0
341	110967.60	2145.763	645.8
342	112358.40	2172.657	647.6
343	113749.20	2199.550	649.4
344	115178.00	2227.179	651.2
345	116614.40	2254.954	653.0
346	118073.60	2283.171	654.8
347	119532.80	2311.387	656.6
348	121014.80	2340.044	658.4
349	122504.40	2368.848	660.2
350	124001.60	2397.799	662.0
351	125521.60	2427.191	663.8
352	127049.20	2456.730	665.6
353	128599.60	2486.710	667.4
354	130157.60	2516.837	669.2
355	131730.80	2547.258	671.0
356	133326.80	2578.119	672.8
357	134945.60	2609.422	674.6
358	136579.60	2641.018	676.4
359	138228.80	2672.908	678.2
360	139893.20	2705.093	680.0
361	141572.80	2737.571	681.8
362	143275.20	2770.490	683.6
363	144992.80	2803.703	685.4
364	146733.20	2837.357	687.2
365	148519.20	2871.892	689.0
366	150320.40	2906.722	690.8
367	152129.20	2941.698	692.6
368	153960.80	2977.116	694.4
369	155815.20	3012.974	696.2
370	157692.40	3049.273	698.0
371	159584.80	3085.866	699.8
372	161507.60	3123.047	701.6
373	163468.40	3160.963	703.4
374	165467.20	3199.613	705.2

VAPOR PRESSURE

The following table is an abridged form of the very extensive compilation by Daniel R. Stull and published in *Industrial and Engineering Chemistry,* 39, 517, 1947.

The table gives the temperatures in degrees centigrade at which the vapor of the compound listed at the left has the pressure indicated at the top of the column. Organic compounds are listed in the order of their empirical formulae and inorganic compounds in the alphabetic order of their names. Pressures greater than one atmosphere are listed in separate tables.

Abbreviations: d = decomposes; *d* = dextrorotatory; *dl* = inactive (50% *d* and 50% *l*); e = explodes; *l* = levorotatory; M.P. = melting point; p = polymerizes; s = solid.

INORGANIC COMPOUNDS
Pressures Less than 1 Atmosphere

Name	Formula	Temperature (°C)						M.P.
		1 mm	10 mm	40 mm	100 mm	400 mm	760 mm	
Aluminum	Al	1284	1487	1635	1749	1947	2056	660
Aluminum borohydride	AlB_3H_{12}	s	−42.9	−20.9	−3.9	+28.1	45.9	−64.5
Aluminum bromide	$AlBr_3$	81.3s	118.0	150.6	176.1	227.0	256.3	97.5
Aluminum chloride	$AlCl_3$	100.0s	123.8s	139.9s	152.0s	171.6s	180.2s	192.4
Aluminum fluoride	AlF_3	1238	1324	1378	1422	1496	1537	1040
Aluminum iodide	AlI_3	178.0s	225.8	265.0	294.5	354.0	385.5	—
Aluminum oxide	Al_2O_3	2148	2385	2549	2665	2874	2977	2050
Ammonia	NH_3	−109.1s	−91.9s	−79.2s	−68.4	−45.4	−33.6	−77.7
Deutero ammonia	ND_3	s	s	s	−67.4	−45.4	−33.4	−74.0
Ammonium axide	NH_4N_3	29.2s	59.2s	80.1s	95.2s	120.4s	133.8s	—
Ammonium bromide	NH_4Br	198.3s	252.0s	290.0s	320.0s	370.9s	396.0s	—
Ammonium carbamate	$NH_4CO_2NH_2$	−26.1s	−2.9s	+14.0s	26.7s	48.0s	58.3s	—
Ammonium chloride	NH_4Cl	160.4s	209.8s	245.0s	271.5s	316.5s	337.8s	520
Ammonium hydrogen sulfide	NH_4HS	−51.1	−28.7	−12.3	0.0	+21.8	33.3	—
Ammonium iodide	NH_4I	210.9s	263.5s	302.8s	331.8s	381.0s	404.9s	—
Ammonium cyanide	NH_4CN	−50.6s	−28.6s	−12.6s	−0.5s	+20.5s	31.7s	36
Antimony	Sb	886	1033	1141	1223	1364	1440	630.5
Antimony tribromide	$SbBr_2$	93.9	142.7	177.4	203.5	250.2	275.0	96.6
Antimony trichloride	$SbCl_3$	49.2s	85.2	117.8	143.3	192.2	219.0	73.4
Antimony pentachloride	$SbCl_5$	22.7	61.8	91.0	114.1	—	—	2.8
Antimony triiodide	SbI_3	163.6s	223.5	267.8	303.5	368.5	401.0	167
Antimony trioxide	Sb_2O_3	574s	666	812	957	1242	1425	656
Argon	A	−218.2s	−210.9s	−204.9s	−200.5s	190.6s	−185.6	−189.2
Arsenic (metallic)	As	372s	437s	483s	518s	579s	610s	814
Arsenic tribromide	$AsBr_3$	41.8	85.2	118.7	145.2	193.6	220.0	—
Arsenic trichloride	$AsCl_3$	−11.4	+23.5	50.0	70.9	109.7	130.4	−18
Arsenic trifluoride	AsF_3	s	s	−2.5	+13.2	41.5	56.3	−5.9
Arsenic pentafluoride	AsF_5	−117.9s	−103.1s	−92.4s	−84.3s	−64.0	−52.8	−79.8
Arsenic hydride (arsine)	AsH_3	−142.6s	−124.7s	−110.2	−98.0	−75.2	−62.1	−116.3
Arsenic trioxide	As_2O_2	212.5s	259.7s	299.2s	332.5	412.2	457.2	312.8
Barium	Ba	s	1049	1195	1301	1518	1638	850
Beryllium borohydride	BeB_2H_8	+1.0s	28.1s	46.2s	58.6s	79.7s	90.0s	123
Beryllium bromide	$BeBr_2$	289s	342s	379s	405s	451s	474s	490
Beryllium chloride	$BeCl_2$	291s	346s	384s	411	461	487	405
Beryllium iodide	BeI_2	283s	341s	382s	411s	461s	487s	488
Bismuth	Bi	1021	1136	1217	1271	1370	1420	271
Bismuth tribromide	$BiBr_3$	s	282	327	360	425	461	218
Bismuth trichloride	$BiCl_3$	s	264	311	343	405	441	230
Borine carbonyl	BH_3CO	−139.2	−121.1	−106.6	−95.3	−74.8	−64.0	−137.0
Boron tribromide	BBr_3	−41.4	−10.1	+14.0	33.5	70.0	91.7	−45
Boron trichloride	BCl_3	−91.5	−66.9	−47.8	−32.4	−3.6	+12.7	−107
Boron trifluoride	BF_3	−154.6	−141.3s	−131.0s	−123.0	−108.3	−110.7	−126.8
Dihydrodiborane	B_2H_6	−159.7	−144.3	−131.6	−120.9	−99.6	−86.5	−169
Diborane hydrobromide	B_2BrH_5	−93.3	−66.3	−45.4	−29.0	0.0	+16.3	−104.2
Triborine triamine	$B_3H_6N_3$	−63.0s	−35.3	−13.2	+4.0	34.3	50.6	−58.2
Tetrahydrotetraborane	B_4H_{10}	−90.9	−64.3	−44.3	−28.1	+0.8	16.1	−119.9
Dihydropentaborane	B_5H_9	s	−30.7	−8.0	+9.6	40.8	58.1	−47.0
Tetrahydropentaborane	B_5H_{11}	−50.2	−19.9	+2.7	20.1	51.2	67.0	—
Dihydrodecaborane	$B_{10}H_{14}$	60.0s	90.2s	117.4	142.3	d	—	99.6
Bromine	Br_2	−48.7s	−25.0s	−8.0s	+9.3	41.0	58.2	−7.3
Bromine pentafluoride	BrF_5	−69.3s	−41.9	−21.0	−4.5	+25.7	40.0	−61.4
Cadmium	Cd	394	484	553	611	711	765	320.9
Cadmium chloride	$CdCl_2$	s	656	736	797	908	967	568
Cadmium fluoride	CdF_2	1112	1286	1400	1486	1651	1751	520
Cadmium iodide	CdI_2	416	512	584	640	742	796	385
Cadmium oxide	CdO	1000s	1149s	1257s	1341s	1484s	1559s	—
Calcium	Ca	s	983	1111	1207	1388	1487	851
Carbon	C	3586s	3946s	4196s	4373s	4660s	4827s	—
Carbon tetrabromide	CBr_4	s	s	96.3	119.7	163.5	189.5	90.1
Carbon tetrachloride	CCl_4	−50.0s	−19.6	+4.3	23.0	57.8	76.7	−22.6
Carbon tetrafluoride	CF_4	−184.6s	−169.3	−158.8	−150.7	−135.5	−127.7	−183.7
Carbon dioxide	CO_2	−134.3s	−119.5s	−108.6s	−100.2s	−85.7s	−78.2s	−57.5
Carbon suboxide	C_3O_2	−94.8	−71.0	−52.0	−36.9	−8.9	+6.3	−107
Carbon disulfide	CS_2	−73.8	−44.7	−22.5	−5.1	+28.0	46.5	−110.8
Carbon subsulfide	C_3S_2	14.0	54.9	85.6	109.9	p	—	+0.4
Carbon selenosulfide	CSSe	−47.3	−16.0	+8.6	28.3	65.2	85.6	−75.2
Carbon monoxide	CO	−222.0s	−215.0s	−210.0s	−205.7s	−196.3	−191.3	−205.0

INORGANIC COMPOUNDS
Pressures Less than One Atmosphere (Continued)

Name	Formula	Temperature (°C) 1 mm	10 mm	40 mm	100 mm	400 mm	760 mm	M.P.
Carbonyl chloride	$COCl_2$	−92.9	−69.3	−50.3	−35.6	−7.6	+8.3	−104
Carbonyl selenide	$COSe$	−117.1	−95.0	−76.4	−61.7	−35.6	−21.9	—
Carbonyl sulfide	COS	−132.4	−113.3	−98.3	−85.9	−62.7	−49.9	−138.8
Chloropierin	CCl_3NO_2	−25.5	+7.8	33.8	53.8	91.8	111.9	−64
Chlorotrifluoromethane	$CClF_3$	−149.5	−134.1	−121.9	−111.7	−92.7	−81.2	—
Cyanogen	C_2N_2	−95.8s	−76.8s	−62.7s	−51.8s	−33.0	−21.0	−34.4
Cyanogen bromide	$CBrN$	−35.7s	−10.0s	+8.6s	22.6s	46.0s	61.5	58
Cyanogen chloride	$CClN$	−76.7s	−53.8s	−37.5s	−24.9s	−2.3	+13.1	−6.5
Cyanogen fluoride	CFN	−134.4s	−118.5s	−106.4s	−97.0s	−80.5s	−72.6s	—
Cyanogen iodide	CIN	25.2s	57.7s	80.3s	97.6s	126.1s	141.1s	—
Deuterocyanic acid	CDN	−68.9s	−46.7s	−30.1s	−17.5s	+10.0	26.2	−12
Dichlorodifluoromethane	CCl_2F_2	−118.5	−97.8	−81.6	−68.6	−43.9	−29.8	—
Dichlorofluoromethane	$CHCl_2F$	−91.3	−67.5	−48.8	−33.9	−6.2	+8.9	−135
Chlorodifluoromethane	$CHClF_2$	−122.8	−103.7	−88.6	−76.4	−53.6	−40.8	−160
Trichlorofluoromethane	CCl_3F	−84.3	−59.0	−39.0	−23.0	+6.8	23.7	—
Cesium	Cs	279	375	449	509	624	690	28.5
Cesium bromide	$CsBr$	748	887	993	1072	1221	1300	636
Cesium chloride	$CsCl$	744	884	989	1069	1217	1300	646
Cesium fluoride	CsF	712	844	947	1025	1170	1251	683
Cesium iodide	CsI	738	873	976	1055	1200	1280	621
Chlorine	Cl_2	−118.0s	−101.6s	−84.5	−71.7	−47.3	−33.8	−100.7
Chlorine fluoride	ClF	s	−139.0	−128.8	−120.8	−107.0	−100.5	−145
Chlorine trifluoride	ClF_3	s	−71.8	−51.3	−34.7	−4.9	+11.5	−83
Chlorine monoxide	Cl_2O	−98.5	−73.1	−54.3	−39.4	−12.5	+2.2	−116
Chlorine dioxide	ClO_2	s	−59.0	−42.8	−29.4	−4.0	+11.1	−59
Dichlorine hexoxide	Cl_2O_6	+7.5	42.0	68.0	87.7	123.8	142.0	3.5
Chlorine heptoxide	Cl_2O_7	−45.3	−13.2	+10.3	29.1	62.2	78.8	−91
Chlorosulfonic acid	HSO_3Cl	32.0	64.0	87.6	105.3	136.1	151.0d	−80
Chromium	Cr	1616	1845	2013	2139	2361	2482	1615
Chromium carbonyl	$Cr(CO)_6$	36.0	68.3	91.2	108.0	137.2	151.0	—
Chromyl chloride	CrO_2Cl_2	−18.4	+13.8	38.5	58.0	95.2	117.1	—
Cobaltous chloride	$COCl_2$	s	s	770	843	974	1050	735
Cobalt nitrosyl tricarbonyl	$Co(CO)_3NO$	s	s	+11.0	29.0	62.0	80.0	−11
Columbium pentafluoride	CbF_5	s	86.3	121.5	148.5	198.0	225.0	75.5
Copper	Cu	1628	1879	2067	2207	2465	2595	1083
Cuprous bromide	Cu_2Br_2	572	718	844	951	1189	1355	504
Cuprous chloride	Cu_2Cl_2	546	702	838	960	1249	1490	422
Cuprous iodide	Cu_2L_2	s	656	786	907	1158	1336	605
Ferric chloride	$FeCl_2$	194.0s	235.5s	256.8s	272.5s	298.0s	319.0	304
Ferrous chloride	$FeCl_2$	s	700	779	842	961	1026	—
Fluorine	F_2	−223.0	−214.1	−207.7	−202.7	−193.2	−187.9	−223
Fluorine monoxide	F_2O	−196.1	−182.3	−173.0	−165.8	−151.9	−144.6	−223.9
Gallium	Ga	1349	1541	1680	1784	1974	2071	30
Gallium trichloride	$GaCl_3$	48.0s	76.5s	107.5	132.0	176.3	200.0	77.0
Germanium hydride	GeH_4	−163.0	−145.3	−131.6	−120.4	−100.2	−88.9	−165
Germanium bromide	$GeBr_4$	s	56.8	88.1	113.2	161.6	189.0	26.1
Germanium chloride	$GeCl_4$	−45.0	−15.0	+8.0	27.5	63.8	84.0	−49.5
Trichlorogermane	$GeHCl_3$	−41.3	−13.0	+8.8	26.5	58.3d	75.0d	−71.1
Tetramethylgermanium	$Ge(CH_3)_4$	−73.2	−45.2	−23.4	−6.3	+26.0	44.0	−88
Digermane	Ge_2H_6	−88.7	−60.1	−38.2	−20.3	+13.3	31.5	−109
Trigermane	Ge_3H_8	−36.9	−0.9	+26.3	47.9	88.6	110.8	−105.6
Gold	Au	1869	2154	2363	2521	2807	2966	1063
Helium	He	−271.7	−271.3	−270.7	−270.3	−269.3	−268.6	—
Hydrogen	H_2	−263.3s	−261.3	−259.6s	−257.9	−254.5	−252.5	−259.1
Hydrogen deuteride	HD		−259.8	−258.2	−256.6	−253.0	−251.0	—
Hydrogen bromide	HBr	−138.8s	−121.8s	−108.3s	−97.7s	−78.0	−66.5	−87.0
Hydrogen chloride	HCl	−150.8s	−135.6s	−123.8s	−114.0	−95.3	−84.8	−114.3
Hydrogen cyanide	HCN	−71.0s	−47.7s	−30.9s	−17.8s	+10.2	25.9	−13.2
Hydrogen fluoride	HF	s	−65.8	−45.0	−28.2	+2.5	19.7	−83.7
Hydrogen iodide	HI	−123.3s	−102.3s	−85.6s	−72.1s	−48.3	−35.1	−50.9
Hydrogen peroxide	H_2O_2	15.3	50.4	77.0	97.9	137.4d	158.0d	−0.9
Hydrogen selenide	H_2Se	−115.3s	−97.9s	−84.7s	−74.2s	−53.6	−41.1	−64
Hydrogen sulfide	H_2S	−134.3s	−116.3s	−102.3s	−91.6s	−71.8	−60.4	−85.5
Hydrogen disulfide	H_2S_2	−43.2	−15.2	+6.0	22.0	49.6	64.0	−89.7
Hydrogen teluride	H_2Te	−96.4s	−75.4s	−59.1s	−45.7	−17.2	−2.0	−49.0
Hydroxylamine	NH_2OH	s	47.2	64.6	77.5	99.2	110.0	34.0
Iodine	I_2	38.7s	73.2s	97.5s	116.5	159.8	183.0	112.9
Iodine pentafluoride	IF_5	−15.2	+8.5	32.3	50.0	81.2	97.0	8.0
Iodine heptafluoride	IF_7	−87.0s	−63.0s	−45.3s	31.9s	−8.3s	+4.0s	5.5
Iron	Fe	1787	2039	2224	2360	2605	2735	1535
Iron pentacarbonyl	$Fe(CO)_5$	—	+4.6	30.3	50.3	86.1	105.0	−21
Krypton	Kr	−199.3s	−187.2s	−178.4s	−171.8s	−159.0s	−152.0	−156.7
Lead	Pb	973	1162	1309	1421	1630	1744	327.5
Lead bromide	$PbBr_2$	513	610	686	745	856	914	373
Lead chloride	$PbCl_2$	547	648	725	784	893	954	501
Lead fluoride	PbF_2	s	904	1003	1080	1219	1293	855
Lead iodide	PbI	479	571	644	701	807	872	402
Lead oxide	PbO	943	1085	1189	1265	1402	1472	890
Lead sulfide	PbS	852s	975s	1048s	1108s	1221	1281	1114
Lithium	Li	723	881	1003	1097	1273	1372	186
Lithium bromide	$LiBr$	748	888	994	1076	1226	1310	547

Pressures Less than One Atmosphere (Continued)

Name	Formula	Temperature (°C) 1 mm	10 mm	40 mm	100 mm	400 mm	760 mm	M.P.
Lithium chloride	$LiCl$	783	932	1045	1129	1290	1382	614
Lithium fluoride	LiF	1047	1211	1333	1425	1591	1681	870
Lithium iodide	LiI	723	841	927	993	1110	1171	446
Magnesium	Mg	621s	743	838	909	1034	1107	651
Magnesium chloride	$MgCl_2$	778	930	1050	1142	1316	1418	712
Manganese	Mn	1292	1505	1666	1792	2029	2151	1260
Manganous chloride	$MnCl_2$	s	778	879	960	1108	1190	650
Mercury	Hg	126.2	184.0	228.8	261.7	323.0	357.0	−38.9
Mercuric bromide	$HgBr_2$	136.5s	179.8s	211.5s	237.8	290.0	319.0	237
Mercuric chloride	$HgCl_2$	136.2s	180.2s	212.5s	237.0s	275.5s	304.0	277
Mercuric iodide	HgI_2	157.5s	204.5s	238.2s	261.8	324.2	354.0	259
Molybdenum	Mo	3102	3535	3859	4109	4553	5560	2622
Molybdenum hexafluoride	MoF_6	−65.5s	−40.8s	−22.1s	−8.0s	+17.2	36.0	17
Molybdenum trioxide	MoO_3	734s	814	892	955	1082	1151	795
Neon	Ne	−257.3s	−254.6s	−252.6s	−251.0s	−248.1	−246.0	−248.7
Nickel	Ni	1810	2057	2234	2364	2603	2732	1452
Nickel chloride	$NiCl_2$	671s	759s	821s	866s	945s	987s	1001
Nickel carbonyl	$Ni(CO)_4$	s	s	−23.0	−6.0	+25.8	42.5	−25
Nitrogen	N_2	−226.1s	−219.1s	−214.0s	−209.7	−200.9	−195.8	−210.0
Nitrogen trifluoride	NF_3	s	−170.7	−160.2	−152.3	−137.4	−129.0	−183.7
Nitric oxide	NO	−184.5s	−178.2s	−171.7s	−166.0s	−156.8s	−151.7	−161
Nitrous oxide	N_2O	−143.4s	−128.7s	−118.3s	−110.3s	−96.2s	−88.5	−90.9
Nitrogen tetroxide	N_2O_4	−55.6s	−36.7s	−23.9s	−14.7s	+8.0	21.0	−9.3
Nitrogen pentoxide	N_2O_5	−36.8s	−16.7s	−2.9s	+7.4s	24.4s	32.4	30
Nitrosyl chloride	$NOCl$	s	s	−60.2	−46.3	−20.3	−6.4	−64.5
Nitrosyl fluoride	NOF	−132.0	−114.3	−100.3	−88.8	−68.2	−56.0	−134
Nitroxyl fluoride	NO_2F	−143.7s	−126.2	−112.8	−102.3	−83.2	−72.0	−139
Osmium tetroxide (white)	OsO_4	−5.6s	+26.0s	50.5	71.5	109.3	130.0	40.6
Oxygen	O_2	−219.1s	−210.6	−204.1	−198.8	−188.8	−182.96	−218.4
Ozone	O_3	−180.4	−163.2	−150.7	−141.0	−122.5	−111.9	−192.1 ±a1
Phosphorus (yellow)	P	76.6	128.0	166.7	197.3	251.0	280.0	44.1
Phosphorus (violet)	P	237s	287s	323s	349s	391s	417s	590
Phosphorus (black)	P	290s	338s	371s	393s	432s	453s	—
Phosphorus tribromide	PBr_3	7.8	47.8	79.0	103.6	149.7	175.3	−40
Phosphorus trichloride	PCl_3	−51.6	−21.3	+2.3	21.0	56.9	74.2	−111.8
Phosphorus pentachloride	PCl_5	55.5s	83.2s	102.5s	117.0s	147.2s	162.0s	—
Phosphorus hydride (Phosphene)	PH_3	s	s	−129.4	−118.8	−98.3	−87.5	−132.5
Phosphonium bromide	PH_4Br	−43.7s	−21.2s	−5.0s	+7.4s	28.0s	38.3d	
Phosphonium chloride	PH_4Cl	−91.0s	−74.0s	−61.5s	−52.0s	−35.4s	−27.0s	−28.5
Phosphonium iodide	PH_4I	−25.2s	−1.1s	16.1s	29.3s	51.6s	62.3s	—
Phosphorus trioxide	P_2O_3	s	53.0	84.0	108.3	150.3	173.1	22.5
Phosphorus oxychloride	$POCl_3$	s	2.0	27.3	47.4	84.3	105.1	2
Phosphorus pentoxide (stable form)	P_2O_5	384s	442s	481s	510s	556s	591	569
Phosphorus pentoxide (Metastable form)	P_2O_5	189	236	270	294	336	358	
Phosphorus thiobromide	$PSBr_3$	50.0	83.6	108.0	126.3	157.8	175.0d	38
Phosphorus thiochloride	$PSCl_3$	−18.3	+16.1	42.7	63.8	102.3	124.0	−36.2
Platinum	Pt	2730	3146	3469	3714	4169	4407	1755
Potassium	K	341	443	524	586	708	774	62.3
Potassium bromide	KBr	795	982	1050	1137	1297	1383	730
Potassium chloride	KCl	821	968	1078	1164	1322	1407	790
Potassium fluoride	KF	885	1039	1156	1245	1411	1502	880
Potassium hydroxide	KOH	719	863	976	1064	1233	1327	380
Potassium iodide	KI	745	887	995	1080	1238	1324	723
Radon	Rn	−144.2s	−126.3s	−111.3s	−99.0s	−75.0s	−61.8	−71
Rhenium heptoxide	Re_2O_7	212.5s	248.0s	272.0s	289.0s	336.0	362.4	296
Rubidium	Rb	297	389	459	514	620	679	38.5
Rubidium bromide	$RbBr$	781	923	1031	1114	1267	1352	682
Rubidium chloride	$RbCl$	792	937	1047	1133	1294	1381	715
Rubidium fluoride	RbF	921	1016	1096	1168	1322	1408	760
Rubidium iodide	RbI	748	884	991	1072	1223	1304	642
Selenium	Se	356	442	506	554	637	680	217
Selenium dioxide	SeO_2	157.0s	202.5s	234.1s	258.0s	297.7s	317.0s	340
Selenium hexafluoride	SeF_6	−118.6s	−98.9s	−84.7	−73.9s	−55.2s	−45.8s	−34.7
Selenium oxychloride	$SeOCl_2$	34.8	71.9	98.0	118.0	151.7	168.0	8.5
Selenium tetrachloride	$SeCl_4$	74.0s	107.4s	130.1s	147.5s	176.4s	191.5d	—
Silane	SiH_4	−179.3	−163.0	−150.3	−140.5	−122.0	−111.5	−185
Silicon	Si	1724	1888	2000	2083	2220	2287	1420
Silicon dioxide	SiO_2	s	1732	1867	1969	2141	2227	1710
Silicon tetrachloride	$SiCl_4$	−63.4	−34.4	−12.1	+5.4	38.4	56.8	−68.8
Silicon tetrafluoride	SiF_4	−144.0s	−130.4s	−120.8s	−113.3s	−100.7s	−94.8s	−90
Bromosilane	SiH_3Br	s	−77.3	−57.8	−42.3	−13.3	+2.4	−93.9
Chlorosilane	SiH_3Cl	−117.8	−97.7	−81.8	−68.5	−44.5	−30.4	—
Fluorosilane	SiH_3F	−153.0	−141.2	−130.8	−122.4	−106.8	−98.0	—
Iodosilane	SiH_3I	s	−43.7	−21.8	−4.4	+27.9	45.4	−57.0
Bromodichlorofluorosilane	$SiBrCl_2F$	−86.5	−59.0	−37.0	−19.5	+15.4	35.4	−112.3
Bromotrifluorosilane	$SiBrF_3$	s	s	s	s	−55.9	−41.7	−70.5
Chlorotrifluorosilane	$SiClF_3$	−144.0s	−127.0	−112.8	−101.7	−81.0	−70.0	−142
Dibromochlorofluorosilane	$SiBr_2ClF$	−65.2	−35.6	−12.0	+6.3	43.0	59.5	−99.3
Dibromodifluorosilane	$SiBr_2F_2$	s	−66.8	−47.4	−31.9	−2.6	+13.7	−66.9

INORGANIC COMPOUNDS
Pressures Less than One Atmosphere (Continued)

Name	Formula	1 mm	10 mm	40 mm	100 mm	400 mm	760 mm	M.P.
		Temperature (°C)						
Dibromosilane	SiH_2Br_2	−60.9	−29.4	−5.2	+14.1	50.7	70.5	−70.2
Dichlorodifluorosilane	$SiCl_2F_2$	−124.7	−102.9	−85.0	−70.3	−45.0	−31.8	−139.7
Difluorosilane	SiH_2F_2	−146.7	−130.4	−117.6	−107.3	−87.6	−77.8	—
Diiodosilane	SiH_2I_2	s	18.0	52.6	79.4	125.5	149.5	−1.0
Disilane	Si_2H_6	−114.8	−91.4	−72.8	−57.5	−29.0	−14.3	−132.6
Disiloxane	$(SiH_3)_2O$	−112.5	−88.2	−70.4	−55.9	−29.3	−15.4	−144.2
Fluorotrichlorosilane	$SiCl_3F$	−92.6	−68.3	−48.8	−33.2	−4.0	+12.2	−120.8
Hexachlorodisilane	Si_2Cl_6	+4.0	38.8	65.3	85.4	120.6	139.0	−1.2
Hexachlorodisiloxane	$(SiCl_3)_2O$	−5.0	+29.4	55.2	75.4	113.6	135.6	−33.2
Hexafluorodisilane	Si_2F_6	−81.0s	−63.1s	−50.6s	−41.7s	−26.4s	−18.9s	−18.6
Octachlorotrisilane	Si_3Cl_8	46.3	89.3	121.5	146.0	189.5	211.4	—
Tetrasilane	Si_4H_{10}	−27.7	+4.3	28.4	47.4	81.7	100.0	−93.6
Tribromofluorosilane	$SiBr_3F$	−46.1	−15.1	+9.2	28.6	64.6	83.8	−82.5
Tribromosilane	$SiHBr_3$	−30.5	+3.4	30.0	51.6	90.2	111.8	−73.5
Trichlorosilane	$SiHCl_3$	−80.7	−53.4	−32.9	−16.4	+14.5	31.8	−126.6
Trifluorosilane	$SiHF_3$	−152.0s	−138.2s	−127.3	−118.7	−102.8	−95.0	−131.4
Trisilane	Si_3H_8	−68.9	−40.0	−16.9	+1.6	35.5	53.1	−117.2
Disilazane	$(SiH_3)_3N$	−68.7	−40.4	−18.5	−1.1	+31.0	48.7	−105.7
Silver	Ag	1357	1575	1743	1865	2090	2212	960.5
Silver chloride	$AgCl$	912	1074	1200	1297	1467	1564	455
Silver iodide	AgI	820	983	1111	1210	1400	1506	552
Sodium	Na	439	952	633	701	823	892	97.5
Sodium bromide	$NaBr$	806	549	1063	1148	1304	1392	755
Sodium chloride	$NaCl$	865	1017	1131	1220	1379	1465	800
Sodium cyanide	$NaCN$	817	983	1115	1214	1401	1497	564
Sodium fluoride	NaF	1077	1240	1363	1455	1617	1704	992
Sodium hydroxide	$NaOH$	739	897	1017	1111	1286	1378	318
Sodium iodide	NaI	767	903	1005	1083	1225	1304	651
Stannic bromide	$SnBr_4$	s	72.7	105.5	131.0	177.7	204.7	31.0
Stannic chloride	$SnCl_4$	−22.7	+10.0	35.2	54.7	92.1	113.0	−30.2
Stannic hydride	SnH_4	−140.0	−118.5	−102.3	−80.2	−65.2	−52.3	−149.9
Stannic iodide	SnI_4	s	175.8	218.8	254.2	315.5	348.0	144.5
Stannous chloride	$SnCl_2$	316	391	450	493	577	623	246.8
Strontium	Sr	s	898	1018	1111	1285	1384	800
Strontium oxide	SrO	2068s	2262s	2410s	—	—	—	2430
Sulfur	S	183.8	243.8	288.3	327.2	399.6	444.6	112.8
Sulfur hexafluoride	SF_6	−132.7s	−114.7s	−101.5s	−90.9s	−72.6s	−63.5s	−50.2
Sulfur dioxide	SO_2	−95.5s	−76.8s	−60.5	−46.9	−23.0	−10.0	−73.2
Sulfur monochloride	S_2Cl_2	−7.4	+27.5	54.1	75.3	115.4	138.0	−80
Sulfuryl chloride	SO_2Cl_2	s	−24.8	−1.0	+17.8	51.3	69.2	−54.1
Sulfur trioxide (α)	SO_3	−39.0s	−16.5s	−1.0s	+10.5s	32.6	44.8	16.8
Sulfur trioxide (β)	SO_3	−34.0s	−12.3s	+3.2s	14.3s	32.6	44.8	32.3
Sulfur trioxide (γ)	SO_3	−15.3s	+4.3s	17.9s	28.0s	44.0s	51.6s	62.1
Sulfuric acid	H_2SO_4	145.8	194.2	229.7	257.0	305.0	330.0d	10.5
Thionyl bromide	$SOBr_2$	−6.7	+31.0	58.8	80.6	119.2	139.5	−52.2
Thionyl chloride	$SOCl_2$	−52.9	−21.9	+2.2	21.4	56.5	75.4	−104.5
Tantalum pentafluoride	TaF_5	s	s	s	130.0	194.0	230.0	96.8
Tellurium	Te	520	650	753	838	997	1087	452
Tellurium tetrachloride	$TeCl_4$	s	233	273	304	360	392	224
Tellurium hexafluoride	TeF_6	−111.3s	−92.4s	−78.4s	−67.9s	−48.2s	−38.6s	−37.8
Thallium	Tl	825	983	1103	1196	1364	1457	303.5
Thallium bromide	$TlBr$	s	522	598	653	759	819	460
Thallium chloride	$TlCl$	s	517	589	645	748	807	430
Thallium iodide	TlI	440	531	607	663	763	823	440
Tin	Sn	1492	1703	1855	1968	2169	2270	231.9
Titanium tetrachloride	$TiCl_4$	−13.9	+21.3	48.4	71.0	112.7	136.0	−30
Tungsten	W	3990	4507	4886	5168	5666	5927	3370
Tungsten hexafluoride	WF_6	−71.4s	−49.2s	−33.0s	−20.3s	+1.2	17.3	−0.5
Uranium hexafluoride	UF_6	−38.8s	−13.8s	+4.4s	18.2s	42.7s	55.7s	69.2
Vanadyl trichloride	$VOCl_3$	−23.2	+12.2	40.0	62.5	103.5	127.2	—
Water	H_2O	−17.3s	+11.3	34.1	51.6	83.0	100.0	0.0
Xenon	Xe	−168.5s	−152.8s	−141.2s	−132.8s	−117.1s	−108.0	−111.6
Zinc	Zn	487	593	673	736	844	907	419.4
Zinc chloride	$ZnCl_2$	428	508	566	610	689	732	365
Zinc fluoride	ZnF_2	970	1086	1175	1254	1417	1497	872
Zirconium tetrabromide	$ZrBr_4$	207s	250s	281s	301s	337s	357s	450
Zirconium tetrachloride	$ZrCl_4$	190s	230s	259s	279s	312s	331s	437
Zirconium tetraiodide	ZrI_4	264s	311s	344s	369s	409s	431s	499

Pressures Greater than 1 Atmosphere

Name	Formula	1 atm.	2 atm.	5 Atm.	10 atm.	20 atm.	40 atm.	60 atm.
		Temperature, °C						
Ammonia	NH_3	−33.6	−18.7	+4.7	25.7	50.1	78.9	98.3
Argon	A	−185.6	−179.0	−166.7	−154.9	−141.3	−124.9	
Boron trichloride	BCl_3	12.7	33.2	66.0	96.7	135.4	—	—
Boron trifluoride	BF_3	−100.7	−89.4	−72.6	−57.7	−40.0	−19.0	—
Bromine	Br_2	58.2	78.8	110.3	139.8	174.0	216.0	243.5

INORGANIC COMPOUNDS

Pressures Greater than One Atmosphere (Continued)

Name	Formula	Temperature, °C						
		1 atm.	2 atm.	5 atm.	10 atm.	20 atm.	40 atm.	60 atm.
Carbontetrachloride	CCl_4	76.7	102.0	141.7	178.0	222.0	276.0	—
Carbon dioxide	CO_2	−78.2s	−69.1s	−56.7	−39.5	−18.9	+5.9	22.4
Carbon disulfide	CS_2	46.5	69.1	104.8	136.3	175.5	222.8	256.0
Carbon monoxide	CO	−191.3	−183.5	−170.7	−161.0	−149.7	—	—
Carbonyl chloride	$COCl_2$	+8.3	27.3	57.2	85.0	119.0	159.8	—
Chlorotrifluoromethane	$CClF_3$	−81.2	−66.7	−42.7	−18.5	+12.0	52.8	—
Cyanogen	C_2N_2	−21.0	−4.4	+21.4	44.6	72.6	106.5	—
Dichlorodifluoromethane	CCl_2F_2	−29.8	−12.2	+16.1	42.4	74.0	—	—
Dichlorofluoromethane	$CHCl_2F$	8.9	28.4	59.0	87.0	121.2	162.6	—
Chlorodifluoromethane	$CHClF_2$	−40.8	−24.7	+0.3	24.0	52.0	85.3	—
Trichlorofluoromethane	CCl_3F	23.7	44.1	77.3	108.2	146.7	194.0	—
Chlorine	Cl_2	−33.8	−16.9	+10.3	35.6	65.0	101.6	127.1
Helium	He	−268.6	−268.0	—	—	—	—	—
Hydrogen	H_2	−252.5	−250.2	−246.0	−241.8	—	—	—
Hydrogen bromide	HBr	−66.5	−51.5	−29.1	−8.4	+16.8	48.1	70.6
Hydrogen chloride	HCl	−84.8	−71.4	−50.5	−31.7	−8.8	+17.8	36.2
Hydrogen cyanide	HCN	25.9	45.8	75.8	102.7	135.0	169.9	—
Hydrogen iodide	HI	−35.1	−18.9	+7.3	32.0	62.2	100.7	127.5
Hydrogen sulfide	H_2S	−60.4	−45.9	−22.3	−0.4	+25.5	55.8	76.3
Hydrogen selenide	H_2Se	−41.1	−25.2	0.0	+23.4	50.8	84.6	108.7
Krypton	Kr	−152.0	−143.5	−130.0	−118.0	−101.7	−78.4	—
Neon	Ne	−246.0	−243.8	−239.9	−236.0	−230.8	—	—
Nitrogen	N_2	−195.8	−189.2	−179.1	−169.8	−157.6	—	—
Nitric oxide	NO	−151.7	−145.1	−135.7	−127.3	−116.8	−103.2	−94.8
Nitrous oxide	N_2O	−88.5	−76.8	−58.0	−40.7	−18.8	+8.0	27.4
Nitrogen tetroxide	N_2O_4	21.0	37.3	59.8	79.4	100.3	121.4	132.2
Oxygen	O_2	−183.1	−176.0	−164.5	−153.2	−140.0	−124.1	—
Silicon tetrafluoride	SiF_4	−94.8s	−84.4	−67.9	−52.6	−33.4	—	—
Chlorotrifluorosilane	$SiClF_3$	−70.0	−57.3	−37.2	−18.6	+4.1	—	—
Dichlorodifluorosilane	$SiCl_2F_2$	−31.8	−15.1	+11.6	36.6	66.2	—	—
Fluorotrichlorosilane	$SiCl_3F$	12.2	32.4	64.6	94.2	131.8	—	—
Stannic chloride	$SnCl_4$	113.0	141.3	184.3	223.0	270.0	—	—
Sulfur dioxide	SO_2	−10.0	+6.3	32.1	55.5	83.8	118.0	141.7
Sulfur trioxide	SO_3	44.8	60.0	82.5	104.0	138.0	175.0	198.0
Water	H_2O	100.0	120.1	152.4	180.5	213.1	251.1	276.5

ORGANIC COMPOUNDS

Pressures Less than 1 Atmosphere

Name	Formula	Temperature °C						M.P.
		1 mm	10 mm	40 mm	100 mm	400 mm	760 mm	
Cyanogen bromide	$CBrN$	−35.7s	−10.0s	+8.6s	22.6s	46.0s	61.5	58
Carbon tetrabromide	CBr_4	s	s	96.3	119.7	163.5	189.5	90.1
Chlorotrifluoromethane	$CClF_3$	−149.5	−134.1	−121.9	−111.7	−92.7	−81.2	—
Cyanogen chloride	$CClN$	−76.7s	−53.8s	−37.5s	−24.9s	−2.3	+13.1	−6.5
Dichlorodifluoromethane	CCl_2F_2	−118.5	−97.8	−81.6	−68.6	−43.9	−29.8	—
Carbonyl chloride	CCl_2O	−92.9	−69.3	−50.3	−35.6	−7.6	+8.3	−104
Trichlorofluoromethane	CCl_3F	−84.3	−59.0	−39.0	−23.0	+6.8	23.7	—
Trichloronitromethane	CCl_3NO_2	−25.5	+7.8	−33.8	53.8	91.8	111.9	−64
Carbontetrachloride	CCl_4	−50.0s	−19.6	+4.3	23.0	57.8	76.7	−22.6
Cyanogen fluoride	CFN	−134.4s	−118.5s	−106.4s	−97.0s	−80.5s	−72.6s	—
Carbontetrafluoride	CF_4	−184.6s	−169.3	−158.8	−150.7	−135.5	−127.7	—
Tribromomethane	$CHBr_3$	s	34.0	63.6	85.9	127.9	150.5	8.5
Chlorodifluoromethane	$CHClF_2$	−122.8	−103.7	−88.6	−76.4	−53.6	−40.8	−160
Dichlorofluoromethane	$CHCl_2F$	−91.3	−67.5	−48.8	−33.9	−6.2	+8.9	−135
Trichloromethane	$CHCl_3$	−58.0	−29.7	−7.1	+10.4	42.7	61.3	−63.5
Hydrocyanic acid	CHN	−70.8s	−48.2s	−31.3s	−18.8s	+9.8	25.8	−14
Dibromomethane	CH_2Br_2	−35.1	−2.4	+23.3	42.3	79.0	98.6	−52.8
Dichloromethane	CH_2Cl_2	−70.0	−43.3	−22.3	−6.3	24.1	40.7	−96.7
Formaldehyde	CH_2O	s	−88.0	−70.6	−57.3	−33.0	−19.5	−92
Formic acid	CH_2O_2	−20.0s	+2.1s	24.0	43.8	80.3	100.6	8.2
Dichloromethylarsine	CH_3ASCl_2	−11.1	+24.3	51.5	73.0	112.7	134.5	−59
Borine carbonyl	CH_3BO	−139.2	−121.1	−106.6	−95.3	−74.8	−64.0	—
Methyl bromide	CH_3Br	−96.3s	−72.8	−54.2	−39.4	−11.9	+3.6	−93
Methyl chloride	CH_3Cl	s	−92.4	−76.0	−63.0	−38.0	−24.0	−97.7
Trichloromethylsilane	CH_3Cl_3Si	−60.8	−30.7	−7.0	+12.1	47.0	66.4	−90
Methyl fluoride	CH_3F	−147.3	−131.6	−119.1	−109.0	−89.5	78.2	—
Methyl iodide	CH_3I	s	−45.8	−24.2	−7.0	+25.3	42.4	−64.4
Formamide	CH_3NO	70.5	109.5	137.5	157.5	193.5	210.5d	—
Nitromethane	CH_3NO_2	−29.0	+2.8	27.5	46.6	82.0	101.2	−29
Methane	CH_4	−205.9s	−195.5s	−187.7s	−181.4	−168.8	−161.5	−182.5
Dichloromethylsilane	CH_4Cl_2Si	−75.0	−47.8	−26.2	−9.0	+23.7	41.9	—
Methanol	CH_4O	−44.0	−16.2	+5.0	21.2	49.9	64.7	−97.8
Methanethiol	CH_4S	−90.7	−67.5	−49.2	−34.8	−7.9	+6.8	−121
Chloromethylsilane	CH_5ClSi	−95.0	−71.0	−51.7	−36.4	−7.8	+8.7	—
Methylamine	CH_5N	−95.8s	−73.8	−56.9	−43.7	−19.7	−6.3	−93.5
Methylsilane	CH_6Si	−138.5	−120.0	−104.8	−93.0	−70.3	−56.9	—

ORGANIC COMPOUNDS
Pressures Less than One Atmosphere (Continued)

Name	Formula	Temperature °C						M.P.
		1 mm	10 mm	40 mm	100 mm	400 mm	760 mm	
2-Methyldisilazane	CH_9NSi_2	−76.3	−50.1	−29.6	−13.1	+17.2	34.0	—
Cyanogen iodide	CIN	25.2s	57.7s	80.3s	97.6s	126.1s	141.1s	—
Tetranitromethane	CN_4O_8	s	22.7	48.4	68.9	105.9	125.7d	13
Carbon monoxide	CO	−222.0s	−215.0s	−210.0s	−205.7s	−196.3	−191.3	−205.0
Carbonyl sulfide	COS	−132.4	−113.3	−98.3	−85.9	−62.7	−49.9	−138.8
Carbonyl selenide	COSe	−117.1	−95.0	−76.4	−61.7	−35.6	−21.9	—
Carbon dioxide	CO_2	−134.3s	−119.5s	−108.6s	−100.2s	−85.7s	−78.2s	−57.5
Carbon Selenosulfide	CSSe	−47.3	−16.0	+8.6	28.3	65.2	85.6	−75.2
Carbon disulfide	CS_2	−73.8	−44.7	−22.5	−5.1	+28.0	46.5	−110.8
Trichloroacetyl bromide	C_2BrCl_3O	−7.4	+29.3	57.2	79.5	120.2	143.0	—
1-Chloro-1,2,2-trifluoroethylene	C_2ClF_3	−116.0	−95.9	−79.7	−66.7	−41.7	−27.9	−157.5
1,2-Dichloro-1,2-difluoroethylene	$C_2Cl_2F_2$	−82.0	−57.3	−38.2	−23.0	+5.0	20.9	−112
1,2-Dichloro-1,1,2,2-tetrafluoroethane	$C_2Cl_2F_4$	−95.4	−72.3	−53.7	−39.1	−12.0	+3.5	−94
1,1,2-Trichloro-1,2,2-trifluoroethane	$C_2Cl_3F_3$	−68.0s	40.3s	−18.5	−1.7	+30.2	47.6	−35
Tetrachloroethylene	C_2Cl_4	−20.6s	+13.8	40.1	61.3	100.0	120.8	−19.0
1,1,2,2-Tetrachloro-1,2-difluoroethane	$C_2Cl_4F_2$	−37.5s	−5.0s	+19.8s	38.6	73.1	92.0	26.5
Hexachloroethane	C_2Cl_6	32.7s	73.5s	102.3s	124.2s	163.8s	185.6s	186.6
Tribromoacetaldehyde	C_2HBr_3O	18.5	58.0	87.8	110.2	151.6	174.0d	—
Trichloroethylene	C_2HCl_3	−43.8	−12.4	+11.9	31.4	67.0	86.7	−73
Trichloroacetaldehyde	C_2HCl_3O	−37.8	−5.0	20.2	40.2	77.5	97.7	−57
Trichloroacetic acid	$C_2HCl_3O_2$	51.0s	88.2	116.3	137.8	175.2	195.6	57
Pentachloroethane	C_2HCl_5	+1.0	39.8	69.9	93.5	137.2	160.5	−22
Acetylene	C_2H_2	−142.9s	−128.2s	−116.7s	−2107.9s	−92.0s	−84.0s	−81.5
1,1,1,2 Tetrabomoethane	$C_2H_2Br_4$	58.0	95.7	123.2	144.0	181.0	200.0d	—
1,1,2,2-Tetrabromoethane	$C_2H_2Br_4$	65.0	110.0	144.0	170.0	217.5	243.5	—
cis-1,2-Dichloroethylene	$C_2H_2Cl_2$	−58.4	−29.9	−7.9	+9.5	41.0	59.0	−80.5
trans-1,2-Dichloroethylene	$C_3H_2Cl_2$	−65.4s	−38.0	−17.0	−0.2	+30.8	47.8	−50.0
1,1-Dichloroethane	$C_2H_2Cl_2$	−77.2	−51.2	−31.1	−15.0	+14.8	31.7	−122.5
Dichloroacetic acid	$C_2H_2Cl_2O_2$	44.0	82.6	111.8	134.0	173.7	194.4	9.7
1,1,1,2-Tetrachloroethane	$C_2H_2Cl_4$	−16.3	+19.3	46.7	68.0	108.2	130.5	−68.7
1,1,2,2-Tetrachloroethane	$C_2H_2Cl_4$	−3.8	+33.0	60.8	83.2	124.0	145.9	−36
1-Bromoethylene	C_2H_3Br	−95.4	−68.8	−48.1	−31.9	−1.1	+15.8	−138
Bromoacetic acid	$C_2H_3BrO_2$	54.7	94.1	124.0	146.3	186.7	208.0	49.5
1,1,2-Tribromoethane	$C_2H_3Br_3$	32.6	70.6	100.0	123.5	165.4	188.4	−26
1 Chloroethylene	C_2H_3Cl	−105.6	−83.7	−66.8	−53.2	−28.0	−13.8	−153.7
Chloroacetic acid	$C_2H_3ClO_2$	43.0s	81.0	109.2	130.7	169.0	189.5	61.2
1,1,1-Trichloroethane	$C_2H_3Cl_3$	−52.0	−21.9	+1.6	20.0	54.6	74.1	−30.6
1,1,2-Trichloroethane	$C_2H_3Cl_3$	−24.0	+8.3	35.2	55.7	93.0	113.9	−36.7
Trichloroacetaldehyde hydrate	$C_2H_3Cl_3O_2$	−9.8s	+19.5s	39.7s	55.0	82.1	96.2d	51.7
1-Fluoroethylene	C_2H_3F	−149.3	−132.2	−118.0	−106.2	−84.0	−72.2	−160.5
Acetonitrile	C_2H_3N	−47.0s	−16.3	+7.7	27.0	62.5	81.8	−41
Methyl thiocyanate	C_2H_3NS	−14.0	+21.6	49.0	70.4	110.8	132.9	−51
Methyl isothiocyanate	C_2H_3NS	−34.7s	+5.4s	38.2	59.3	97.8	119.0	35.5
Ethylene	C_2H_4	−168.3	−153.2	−141.3	−131.8	−113.9	−103.7	−169
1-Bromo-1-chloroethane	C_2H_4BrCl	−36.0s	−9.4s	+10.4s	28.0	63.4	82.7	16.6
1-Bromo-2-chloroethane	C_2H_4BrCl	−28.8s	+4.1	29.7	49.5	86.0	106.7	−16.6
1,2-Dibromoethane	$C_2H_4Br_2$	−27.0s	+18.6	48.0	70.4	110.1	131.5	10
1,1-Dichloroethane	$C_2H_4Cl_2$	−60.7	−32.3	−10.2	+7.2	39.8	57.4	−96.7
1,2-Dichloroethane	$C_2H_4Cl_2$	−44.5s	−13.6	+10.0	29.4	64.0	82.4	−35.3
1,1-Difluoroethane	$C_2H_4F_2$	−112.5	−91.7	−75.8	−63.2	−39.5	−26.5	−117
Acetaldehyde	C_2H_4O	−81.5	−56.8	−37.8	−22.6	+4.9	20.2	−123.5
Ethylene oxide	C_2H_4O	−89.7	−65.7	−46.9	−32.1	−4.9	+10.7	−111.3
Acetic acid	$C_2H_4O_2$	−17.2s	+17.5	43.0	63.0	99.0	118.1	16.7
Methyl formate	$C_2H_4O_2$	−74.2	−48.6	−28.7	−12.9	16.0	32.0	−99.8
Mercaptoacetic acid	$C_2H_4O_2S$	−60.0	101.5	131.8	154.0d	—	—	−16.5
Ethyl bromide	C_2H_5Br	−74.3	−47.5	−26.7	−10.0	+21.0	38.4	−117.8
Ethyl chloride	C_2H_5Cl	−89.8	−65.8	−47.0	−32.0	−3.9	+12.3	−139
2-Chloroethanol	C_2H_5ClO	−4.0	+30.3	56.0	75.0	110.0	128.8	−69
Trichloroethylsilane	$C_2H_5Cl_3Si$	−27.9	+3.6	27.9	46.3	80.3	99.5	−40
Trichloroethyoxysilane	$C_2H_5Cl_3OSi$	−32.4	0.0	+25.3	45.2	82.2	102.4	—
Ethyl fluoride	C_2H_5F	−117.0	−97.7	−81.8	−69.3	−45.5	−32.0	—
Ethyltrifluorosilane	$C_2H_5F_3Si$	−95.4	−73.7	−56.8	−43.6	−19.1	−5.4	—
Ethyl Iodide	C_2H_5I	−54.4	−24.3	−0.9	+18.0	52.3	72.4	−105
Acetamide	C_2H_5NO	65.0s	105.0	135.8	158.0	200.0	222.0	81
Acetaldoxime	C_2H_5NO	−5.8s	+25.8	48.6	66.2	98.0	115.0	47
Nitroethane	$C_2H_5NO_2$	−21.0	+12.5	38.0	57.8	94.0	114.0	−90
Di(nitrosomethyl)amine	$C_2H_5N_3O_2$	+3.2	40.0	68.2	90.3	131.3	153.0	—
Ethane	C_2H_6	−159.5	−142.9	−129.8	−119.3	−99.7	−88.6	−183.2
Dichlorodimethylsilane	$C_2H_6Cl_2Si$	−53.5	−23.8	−0.4	+17.5	51.9	70.3	−86.0
Ethanol	C_2H_6O	−31.3	−2.3	+19.0	34.9	63.5	78.4	−112
Dimethyl ether	C_2H_6O	−115.7	−93.3	−76.2	−62.7	−37.8	−23.7	−138.5
1,2-Ethanediol	$C_2H_6O_2$	53.0	92.1	120.0	141.8	178.5	197.3	−15.6
Dimethyl sulfide	C_2H_6S	−75.6	−49.2	−28.4	−12.0	+18.7	36.0	−83.2
Ethanethiol	C_2H_6S	−76.7	−50.2	−29.8	−13.0	+17.7	35.5	−121
Dimethylantimony	C_2H_6Sb	44.0	86.0	118.3	143.5	187.2	211.0	—
Ethylamine	C_2H_7N	−82.3s	−58.3	−39.8	−25.1	+2.0	16.6	−80.6
Dimethylamine	C_2H_7N	−87.7	−64.6	−46.7	−32.6	−7.1	+7.4	−96
1,2-Ethanediamine	$C_2H_8N_2$	−11.0s	+21.5	45.8	62.5	99.0	117.2	8.5
Dimethylsilane	C_2H_8Si	−115.0	−93.1	−75.7	−61.4	−35.0	−20.1	—
Dimethyldiborane	$C_2H_{10}B_2$	−106.5	−82.1	−62.4	−47.0	−18.8	−2.6	−150.2
2-Ethyldisilazane	$C_2H_{11}NSi_2$	−62.0	−32.2	−8.3	+10.4	45.9	65.9	−127

ORGANIC COMPOUNDS
Pressures Less than One Atmosphere (Continued)

Name	Formula	Temperature °C 1 mm	10 mm	40 mm	100 mm	400 mm	760 mm	M.P.
Cyanogen	C_2N_2	−95.8s	−76.8s	−62.7s	−51.8s	−33.0	−21.0	−34.4
Acrylonitrile	C_3H_3N	−51.0	−20.3	+3.8	22.8	58.3	78.5	−82
Propadiene	C_3H_4	−120.6	−101.0	−85.2	−72.5	−48.5	−35.0	−136
Propyne	C_3H_4	−111.0s	−90.5	−74.3	−61.3	−37.2	−23.3	−102.7
2,3-Dibromopropene	$C_3H_4Br_2$	−6.0	+30.0	57.8	79.5	119.5	141.2	
Methyl dichloroacetate	$C_3H_4Cl_2O_2$	3.2	38.1	64.7	85.4	122.6	143.0	—
2-Propenal	C_3H_4O	−64.5	−36.7	−15.0	+2.5	34.5	52.5	−87.7
Acrylic acid	$C_3H_4O_2$	+3.5s	39.0	66.2	86.1	122.0	141.0	14
Pyruvic acid	$C_3H_4O_3$	21.4	57.9	85.3	106.5	144.7	165.0d	13.6
1,2,3-Tribromopropane	$C_3H_5Br_3$	47.5	90.0	122.8	148.0	195.0	220.0	16.5
1-Chloropropene	C_3H_5Cl	−81.3	−54.1	−32.7	−15.1	+18.0	37.0	−99.0
3-Chloropropene	C_3H_5Cl	−70.0	−42.9	−21.2	−4.5	+27.5	44.6	−136.4
Epichlorohydrine	C_3H_5ClO	−16.5	+16.6	42.0	62.0	98.0	117.9	−25.6
Methyl chloroacetate	$C_3H_5ClO_2$	−2.9	+30.0	54.5	73.5	109.5	130.3	−31.9
1,1,1-Trichloropropane	$C_3H_5Cl_3$	−28.8	+4.2	29.9	50.0	87.5	108.2	−77.7
1,2,3-Trichloropropane	$C_3H_5Cl_3$	+9.0	46.0	74.0	96.1	137.0	158.0	−14.7
Allyltrichlorosilane	$C_3H_5Cl_3Si$	−20.7	+13.2	39.2	59.3	97.1	118.0	
Propionitrile	C_3H_5N	−35.0	−3.0	+22.0	41.4	77.7	97.1	−91.9
3-Hydroxypropionitrile	C_3H_5NO	58.7	102.0	134.1	157.7	200.0	221.0	
Ethylisothiocyanate	C_3H_5NS	−13.2s	+22.8	50.8	71.9	110.1	131.0	−5.9
Nitroglycerine	$C_3H_5N_3O_9$	127	188	235	e	—	—	11
Propylene	C_3H_6	−131.9	−112.1	−96.5	−84.1	−60.9	−47.7	−185
Cyclopropane	C_3H_6	−116.8	−97.5	−82.3	−70.0	−46.9	−33.5	−126.6
2-Bromo-2-nitrosopropane	C_3H_6BrNO	−33.5	−4.3	+17.9	35.2	66.2	83.0	
1,2-Dibromopropane	$C_3H_6Br_2$	−7.0	+29.4	57.2	78.7	118.5	141.6	−55.5
1,3-Dibromopropane	$C_3H_6Br_2$	+9.7	48.0	77.8	101.3	144.1	167.5	−34.4
2,3-Dibromo-1-propanol	$C_3G_6Br_2O$	57.0	98.2	129.8	153.0	196.0	219.0	—
1,2-Dichloropropane	$C_3H_6Cl_2$	−38.5	−6.1	+19.4	39.4	76.0	96.8	—
1,3-Dichloro-2-propanol	$C_3H_6Cl_2O$	28.0	64.7	93.0	114.8	153.5	174.3	—
Acetone	C_3H_6O	−59.4	−31.1	−9.4	+7.7	39.5	56.5	−94.6
Allyl alcohol	C_3H_6O	−20.0	+10.5	33.4	50.0	80.2	96.6	−129
Propylene oxide	C_3H_6O	−75.0	−49.0	−28.4	−12.0	+17.8	34.5	−112.1
Propionie acid	$C_3H_6O_2$	4.6	39.7	65.8	85.8	122.0	141.1	−22
Methyl acetate	$C_3H_6O_2$	−57.2	−29.3	−7.9	+9.4	40.0	57.8	−98.7
Ethyl formate	$C_3H_6O_2$	−60.5	−33.0	−11.5	+5.4	37.1	54.3	−79
Methyl glycolate	$C_3H_6O_3$	+9.6	45.3	72.3	93.7	131.7	151.5	—
Methoxyacetic acid	$C_3H_6O_3$	52.5	92.0	122.0	144.5	184.2	204.0	—
1-Bromopropane	C_3H_7Br	−53.0	−23.3	−0.3	+18.0	52.0	71.0	−109.9
2-Bromopropane	C_3H_7Br	−61.8	−32.8	−10.1	+8.0	41.5	60.0	−89.0
1-Chloropropane	C_3H_7Cl	−68.3	−41.0	−19.5	−2.5	+29.4	46.4	−122.8
2-Chloropropane	C_3H_7Cl	−78.8	−52.0	−31.0	−13.7	+18.0	36.5	−117
Trichloroisopropylsilane	$C_3H_7Cl_3Si$	−24.3	+9.9	36.5	57.8	96.8	118.5	—
1-Iodopropane	C_3H_7I	−36.0	−2.4	+23.6	43.8	81.8	102.5	−98.8
2-Iodopropane	C_3H_7I	−43.3	−11.7	+13.2	32.8	69.5	89.5	−90
Propionamide	C_3H_7NO	65.0s	105.0	134.8	156.0	194.0	213.0	79
1-Nitropropane	$C_3H_7NO_2$	−9.6	+25.3	51.8	72.3	110.6	131.6	−108
2-Nitropropane	$C_3H_7NO_2$	−18.8	+15.8	41.8	62.0	99.8	120.3	−93
Ethyl Carbamate	$C_3H_7NO_2$	s	77.8	105.6	126.2	164.0	184.0	49
Propane	C_3H_8	−128.9	−108.5	−92.4	−79.6	−55.6	−42.1	−187.1
Dichloroethoxymethylsilane	$C_3H_8Cl_2OSi$	−33.8	−1.3	+24.4	44.1	80.3	100.6	—
1-Propanol	C_3H_8O	−15.0	+14.7	36.4	52.8	82.0	97.8	−127
2-Propanol	C_3H_8O	−26.1	+2.4	23.8	39.5	67.8	82.5	−85.8
Ethyl methyl ether	C_3H_8O	−91.0	−67.8	−49.4	−34.8	−7.8	+7.5	—
1,2-Propanediol	$C_3H_8O_2$	45.5	83.2	111.2	132.0	168.1	188.2	—
1,3-Propanediol	$C_3H_8O_2$	59.4	100.6	131.0	153.4	193.8	214.2	—
2-Methoxyethanol	$C_3H_8O_2$	−13.5	+22.0	47.8	68.0	104.3	124.4	—
Glycerol	$C_3H_8O_3$	125.5	167.2	198.0	220.1	263.0	290.0	17.9
1-Propanethiol	C_3H_8S	−56.0	−26.3	−3.2	+15.3	49.2	67.4	−112
Trimethylborine	C_3H_9B	−118.0	−92.4	−74.7	−60.8	−34.7	−20.1	—
Chlorotrimethylsilane	C_3H_9ClSi	−62.8	−34.0	−11.4	+6.0	39.4	57.9	—
Trimethylgallium	C_3H_9Ga	−62.3s	−31.7s	−9.0	+8.0	39.0	55.6	−19
Propylamine	C_3H_9N	−64.4	−37.2	−16.0	+0.5	31.5	48.5	−83
Trimethylamine	C_3H_9N	−97.1	−73.8	−55.2	−40.3	−12.5	+2.9	−117.1
Trimethyl phosphate	$C_3H_9O_4P$	26.0	67.8	100.0	124.0	167.8	192.7	—
Trimethyldiborane	$C_3H_{12}B_2$	−74.0	−44.8	−22.0	−4.4	+27.8	45.5	−122.9
Carbon suboxide	C_3O_2	−94.8	−71.0	−52.0	−36.9	−8.9	+6.3	−107
Carbon subsulfide	C_3S_2	14.0	54.9	85.6	109.9	—	—	+0.4
Trichloroacetic anhydride	$C_4Cl_6O_3$	56.2	99.6	131.2	155.2	199.8	223.0	—
1,3-Butadiyne	C_4H_2	−82.5s	−61.2s	−45.9s	−34.0	−6.1	+9.7	−34.9
α,*β*-Dibromomaleic achydride	$C_4H_2Br_2O_3$	50.0	92.0	123.5	147.7	192.0	214.0	—
trans-Fumaryl chloride	$C_4H_2Cl_2O_2$	+15.0	51.8	79.5	101.0	140.0	100.0	—
Maleic anhydride	$C_4H_2O_3$	44.0s	78.7	111.8	135.8	179.5	202.0	58
2-Nitrothiophene	$C_4H_3NO_2S$	48.2	92.0	125.8	151.5	199.6	224.5	46
Butenyne	C_4H_4	−93.2	−70.0	−51.7	−37.1	−10.1	+5.3	—
Succinyl chloride	$C_4H_4Cl_2O_2$	39.0	78.0	107.5	130.0	170.0	192.5	17
Chloroaccetic anhydride	$C_4H_4Cl_2O_3$	67.2	108.0	138.2	159.8	197.0	217.0	46
Succinic anhydride	$C_4H_4O_3$	92.0s	128.2	163.0	189.0	237.0	261.0	119.6
1,4-Dioxane-2,6-dione	$C_4H_4O_4$	s	116.6	148.6	173.2	217.0	240.0	97
Thiophene	C_4H_4S	−40.7s	−10.9	+12.5	30.5	64.7	84.4	−38.3
Selenophene	$C_4\ H_4Se$	−39.0	−4.0	+24.1	47.0	89.8	114.3	—
α-Chlorocrotonic acid	$C_4H_5ClO_2$	70.0	108.0	135.6	155.9	193.2	212.0	—

ORGANIC COMPOUNDS
Pressures Less than One Atmosphere (Continued)

Name	Formula	Temperature °C						M.P.
		1 mm	10 mm	40 mm	100 mm	400 mm	760 mm	
Ethyl chloroglyoxylate	$C_4H_5ClO_3$	−5.1	+29.9	56.0	76.6	114.7	135.0	—
Ethyl trichloroacetate	$C_4H_5Cl_3O_2$	20.7	57.7	85.5	107.4	146.0	67.0	—
3-Butenenitrile	C_4H_5N	−19.6	+14.1	40.0	60.2	98.0	119.0	—
Methacrylonitrile	C_4H_5N	−44.5	−12.5	+12.8	32.8	32.8	90.3	—
cis-Crotononitrile	C_4H_5N	−29.0	+4.0	30.0	50.0	88.0	108.0	—
trans-Crotononitrile	C_4H_5N	−19.5	+15.0	41.8	62.8	101.5	122.8	—
Succinimide	$C_4H_5NO_2$	115.0s	157.0	192.0	217.4	263.5	287.5	125.5
Allylisothiocyanate	C_4H_5NS	−2.0	+38.3	67.4	89.5	129.8	156.7	−80
1,2-Butadiene	C_4H_6	−89.0	−64.2	−44.3	−28.3	+1.8	18.5	—
1,3-Butadiene	C_4H_6	−102.8	−79.7	−61.3	−46.8	−19.3	−4.5	−108.9
Cyclobutene	C_4H_6	−99.1	−75.4	−56.4	−41.2	−12.2	+2.4	—
1 Butyne	C_4H_6	−92.5	−68.7	−50.0	−34.9	−6.9	+8.7	−130
2-Butyne	C_4H_6	−73.0s	−50.5s	−33.9s	−18.8	+10.6	27.2	−32.5
Ethyl dichloroacctate	$C_4H_6Cl_2O_2$	9.6	46.3	74.0	96.1	135.9	156.5	—
2-Chloroethyl chloroacetate	$C_4H_6Cl_2O_2$	46.0	86.0	116.0	140.0	182.2	205.5	—
cis-Crotonic acid	$C_4H_6O_2$	33.5	69.0	96.0	116.3	152.2	171.9d	15.5
trans-Crotonic acid	$C_4H_6O_2$	s	80.0	107.8	128.0	165.5	185.0	72
Methyl acrylate	$C_4H_6O_2$	−43.7	−13.5	+9.2	28.0	61.8	80.2p	—
Methacrylic acid	$C_4H_6O_2$	25.5	60.0	86.4	106.6	142.5	161.0	15
Vinyl acetate	$C_4H_6O_2$	−48.0	−18.9	+5.3	23.3	55.5	72.5	—
Acetic anhydride	$C_4H_6O_3$	1.7	36.0	62.1	82.2	119.8	139.6	−73
Dimethyl oxalate	$C_4H_6O_4$	20.0	56.0	83.6	104.8	143.3	163.3	—
cis-1 Bromo-1-butene	C_4H_7Br	−44.0	−12.8	+11.5	30.8	66.8	86.2	—
trans-1-Bromo-1-butene	C_4H_7Br	−38.4	−6.4	+18.4	38.1	75.0	94.7	−100.3
2-Bromo-1-butene	C_4H_7Br	−47.3	−16.8	+7.2	26.3	61.9	81.0	−133.4
cis-2-Bromo-2-butene	C_4H_7Br	−39.0	−7.2	+17.7	37.5	74.0	93.0	−111.2
trans-2-Bromo-2-butene	C_4H_7Br	−45.0	−13.8	+10.5	29.0	66.0	85.5	−114.6
1-Bromo-2-Butanone	C_4H_7BrO	+6.2	41.8	68.2	89.2	126.3	147.0	—
2-Methylpropionyl bromide	C_4H_7BrO	13.5	50.6	79.4	101.6	141.7	163.0	—
1,1,2-Tribromobutane	$C_4H_7Br_3$	45.0	87.8	120.2	146.0	192.0	216.2	—
1,2,2-Tribromobutane	$C_4H_7Br_3$	41.0	83.2	116.0	141.8	188.0	213.8	—
2,2,3-Tribromobutane	$C_4H_7Br_3$	38.2	79.8	111.8	136.3	182.2	206.5	—
Ethyl chloroacetate	$C_4H_7ClO_2$	+1.0	37.5	65.2	86.0	123.8	144.2	−26
1,2,3-Trichlorobutane	$C_4H_7Cl_3$	+0.5	40.0	71.5	96.2	143.0	169.0	—
Butyronitrile	C_4H_7N	−20.0	+13.4	38.4	59.0	96.8	117.5	—
Diacetamide	$C_4H_7NO_2$	70.0s	108.0	138.2	160.6	202.0	223.0	78.5
1-Butene	C_4H_8	−104.8	−81.6	−63.4	−48.9	−21.7	−6.3	−130
cis-2-Butene	C_4H_8	−96.4	−73.4	−54.7	−39.8	−12.0	+3.7	−138.9
trans-2-Butene	C_4H_8	−99.4	−76.3	−57.6	−42.9	−14.8	+0.9	−105.4
2-Methylpropene	C_4H_8	−105.1	−81.9	−63.8	−49.3	−22.2	−6.9	−140.3
Cyclobutane	C_4H_8	−92.0s	−67.9s	−48.4	−32.8	−3.4	+12.9	−50
Methylcyclopropane	C_4H_8	−96.0	−72.8	−54.2	−39.3	−11.3	+4.5	—
2-Bromoethyl 2-chloroethyl ether	C_4H_8BrClO	36.5	76.3	106.6	129.8	172.3	195.8	—
1,2-Dibromobutane	$C_4H_8Br_2$	7.5	46.1	76.0	99.8	143.5	166.3	−64.5
dl-2,3-Dibromobutane	$C_4H_8Br_2$	+5.0	41.6	72.0	95.3	138.0	160.5	—
meso-2,3-Dibromobutane	$C_4H_8Br_2$	+1.5	39.3	68.2	91.7	134.2	157.3	−34.5
1,4-Dibromobutane	$C_4H_8Br_2$	32.0	72.4	104.0	128.7	173.8	197.5	−20
1,2-Dibromo-2-methylpropane	$C_4H_8Br_2$	−28.8	+10.5	42.3	68.8	119.8	149.0	−70.3
1,3-Dibromo-2-methylpropane	$C_4H_8Br_2$	14.0	53.0	83.5	107.4	150.6	174.6	—
Di(2-bromoethyl) ether	$C_4H_8Br_2O$	47.7	88.5	119.8	144.0	188.0	212.5	—
1,2-Dichlorobutane	$C_4H_8Cl_2$	−23.6	+11.5	37.7	60.2	100.8	123.5	—
2,3-Dichlorobutane	$C_4H_8Cl_2$	−25.2	+8.5	35.0	56.0	94.2	116.0	−80.4
1,1-Dichloro-2-methylpropane	$C_4H_8Cl_2$	−31.0	+2.6	28.2	48.2	85.4	−106.0d	—
1,2-Dichloro-2-methylpropane	$C_4H_8Cl_2$	−25.8	+6.7	32.0	51.7	87.8	108.0	—
1,3-Dichloro-2-methylpropane	$C_4H_8Cl_2$	−3.0	+32.0	58.6	78.8	115.4	135.0	—
di(2-chloroethyl) ether	$C_4H_8Cl_2O$	23.5	62.0	91.5	114.5	155.4	178.5	—
1,2-Epoxy-2-methylpropane	C_4H_8O	−69.0	−40.3	−17.3	+1.2	36.0	55.5	—
2-Butanone	C_4H_8O	−48.3	−17.7	+6.0	25.0	60.0	79.6	−85.9
1,4-Dioxane	$C_4H_8O_2$	−35.8s	−1.2s	+25.2	45.1	81.8	101.1	10
Butyric acid	$C_4H_8O_2$	25.5	61.5	88.0	108.0	144.5	163.5	−4.7
Isobutyric acid	$C_4H_8O_2$	14.7	51.2	77.8	98.0	134.5	154.5	−47
Ethyl acetate	$C_4H_8O_2$	−43.4	−13.5	+9.1	27.0	59.3	77.1	−82.4
Methyl propionate	$C_4H_8O_2$	−42.0	−11.8	+11.0	29.0	61.8	79.8	−87.5
Propyl formate	$C_4H_8O_2$	−43.0	−12.6	+10.8	29.5	62.6	81.3	−92.9
Isopropyl formate	$C_4H_8O_2$	−52.0	−22.7	0.2	+17.8	50.5	68.3	—
α-Hydroxyisobutyric acid	$C_4H_8O_3$	73.5s	110.5	138.0	157.7	193.8	212.0	79
Ethyl glycolate	$C_4H_8O_3$	14.3	50.5	78.1	99.8	138.0	158.2	—
1-Bromobutane	C_4H_9Br	−33.0	−0.3	+24.8	44.9	81.7	101.6	−112.4
1-Bromo-2-butanol	C_4H_9BrO	23.7	55.8	79.5	97.6	128.3	145.0	—
1-Chlorobutane	C_4H_9Cl	−49.0	−18.6	+5.0	24.0	58.8	77.8	−123.1
sec-Butyl chloride	C_4H_9Cl	−60.2	−29.2	−5.0	+14.2	50.0	68.0	−131.3
Isobutyl chloride	C_4H_9Cl	−53.8	−24.5	−1.0	+16.0	50.0	68.9	−131.2
tert-Butyl chloride	C_4H_9Cl	s	s	19.0	−1.0	+32.6	51.0	−26.5
2-(2-Chloroethoxy) ethanol	$C_4H_9ClO_2$	53.0	90.7	118.4	139.5	176.5	196.0	—
1-Iodo-2-methylpropane	C_4H_9I	−17.0	+17.0	42.8	63.5	100.3	120.4	−90.7
Ethyl methylcarbamate	$C_4H_9NO_2$	26.5	63.2	91.0	112.0	149.8	170.0	—
Propyl carbamate	$C_4H_9NO_2$	52.4	90.0	117.7	138.3	175.8	195.0	—
Di(nitrosoethyl)amine	$C_4H_9N_2O_2$	18.5	57.7	87.6	111.0	153.5	176.9	—
Butane	C_4H_{10}	−101.5	−77.8	−59.1	−44.2	−16.3	−0.5	−135
2-Methylpropane	C_4H_{10}	−109.2	−86.4	−68.4	−54.1	−27.1	−11.7	−145
Dichlorodiethylsilane	$C_4H_{10}Cl_2Si$	−9.2	+25.4	51.6	71.8	110.0	130.4	—

ORGANIC COMPOUNDS
Pressures Less than One Atmosphere (Continued)

Name	Formula	Temperature °C 1 mm	10 mm	40 mm	100 mm	400 mm	760 mm	M.P.
Diethyldifluorosilane	$C_4H_{10}F_2Si$	−56.8	−28.8	−7.3	+9.8	40.5	58.0	—
Butyl alcohol	$C_4H_{10}O$	−1.2	+30.2	53.4	70.1	100.8	117.5	−79.9
sec-Butyl alcohol	$C_4H_{10}O$	−12.2	+16.9	38.1	54.1	83.9	99.5	−114.7
Isobutyl alcohol	$C_4H_{10}O$	−9.0	+21.7	44.1	61.5	91.4	108.0	−108
tert-Butyl alcohol	$C_4H_{10}O$	−20.4s	+5.5s	24.5s	39.8	68.0	82.9	25.3
Diethyl ether	$C_4H_{10}O$	−74.3	−48.1	−27.7	−11.5	+17.9	34.6	−116.3
Methyl propyl ether	$C_4H_{10}O$	−72.2	−45.4	−24.3	−8.1	+22.5	39.1	—
1,3-Butanediol	$C_4H_{10}O_2$	22.2s	85.3	117.4	141.2	183.8	206.5	77
2,3-Butanediol	$C_4H_{10}O_2$	44.0	80.3	107.8	127.8	164.0	182.0	22.5
1,2-Dimethoxyethane	$C_4H_{10}O_2$	−48.0	−15.3	+10.7	31.8	70.8	93.0	—
2,2′-Thiodiethanol	$C_4H_{10}O_2S$	42.0	128.0	210.0d	285d	—	—	—
Diethylene glycol	$C_4H_{10}O_3$	91.8	133.8	164.3	187.5	226.5	244.8	—
1,2,3-Butanetriol	$C_4H_{10}O_3$	102.0	146.0	178.0	202.5	243.5	264.0	—
Diethyl sulfite	$C_4H_{10}O_3S$	10.0	46.4	74.2	96.3	137.0	159.0	—
Diethyl sulfate	$C_4H_{10}O_4S$	47.0	87.7	118.0	142.5	185.5	209.5d	−25.0
Diethl sulfide	$C_4H_{10}S$	−39.6	−8.0	+16.1	35.0	69.7	88.0	−99.5
Diethyl selenide	$C_4H_{10}Se$	−25.7	+7.0	31.2	51.8	88.0	108.0	—
Diethylzinc	$C_4H_{10}Zn$	−22.4	+11.7	38.0	59.1	97.3	118.0	−28
Diethylamine	$C_4H_{11}N$	s	−33.0	−11.3	+6.0	38.0	55.5	−38.9
Isobutylamine	$C_4H_{11}N$	−50.0	−21.0	+1.3	18.8	50.7	68.6	−85.0
1,3-Dichlorotetramethyldisiloxane	$C_4H_{12}Cl_2Si_2$	−7.4	+28.3	55.7	76.9	116.3	138.0	−37
Tetramethyllead	$C_4H_{12}Pb$	−29.0s	+4.4	30.3	50.8	89.0	110.0	−27.5
Tetramethylsilane	$C_4H_{12}Si$	−83.8	−58.0	−37.4	−20.9	+10.0	27.0	−102.1
Tetramethyltin	$C_4H_{12}Sn$	−51.3	−20.6	+3.5	22.8	58.5	78.0	—
Tetramethyldiborane	$C_4H_{14}B_2$	−59.6	−27.4	−3.4	+15.3	49.8	68.6	−72.5
3-Bromopyridine	C_5H_4BrN	16.8	55.2	84.1	107.8	150.0	173.4	—
2-Chloropyridine	C_5H_4ClN	13.3	41.7	81.7	104.6	147.7	170.2	—
2-Furaldehyde	$C_5H_4O_2$	18.5	54.8	82.1	103.4	141.8	151.8	−36.5
Citraconic anhydride	$C_5H_4O_2$	47.1	88.9	120.3	145.4	189.8	213.5	—
Pyridine	C_5H_5N	−18.9	+13.2	38.0	57.8	95.6	115.4	−42
Glutaryl chloride	$C_5H_6Cl_2O_2$	56.1	97.8	128.3	141.8	195.3	217.0	—
Glutaronitrile	$C_5H_6N_2$	91.3	140.0	176.4	205.5	257.3	286.2	—
Furfuryl alcohol	$C_5H_6O_2$	31.8	68.0	95.7	115.9	151.8	170.0	—
Glutaric anhydride	$C_6H_6O_3$	100.8	149.5	185.5	212.5	261.0	287.0	—
Pyrotartaric anhydride	$C_5H_6O_3$	69.7	114.2	147.8	173.8	221.0	247.4	—
2-Methylthiophene	C_5H_6S	−27.4	+6.0	32.3	53.1	91.8	112.5	−63.5
3-Methylthiophene	C_5H_6S	−24.5	+9.1	35.4	55.8	93.8	115.4	−68.9
Propyl chloroglyoxylate	$C_5H_7ClO_2$	9.7	43.5	68.8	88.0	123.0	150.0	—
Tiglonitrile	C_5H_7N	−25.5	+9.4	36.7	58.2	99.7	122.0	—
Angelonitrile	C_5H_7N	−8.0	+28.0	55.8	77.5	117.7	140.0	—
α-Ethylacrylonitrile	C_5H_7N	−29.0	+5.0	31.8	53.0	92.2	114.0	—
Ethyl cyanoacetate	$C_5H_7NO_2$	67.8	106.0	133.8	152.8	187.8	206.0	—
Isoprene	C_5H_8	−79.8	−53.3	−32.6	−16.0	+15.4	32.6	−146.7
1,3-Pentadiene	C_5H_8	−71.8	−45.0	−23.4	−6.7	+24.7	42.0	—
1,4-Pentadiene	C_5H_8	−83.5	−57.1	−37.0	−20.6	8.3	26.1	—
Tiglaldehyde	C_5H_8O	−25.0	+10.0	37.0	57.7	95.5	116.4	—
Levulinaldehyde	$C_5H_8O_2$	28.1	68.0	98.3	121.8	164.0	187.0	—
Tiglic acid	$C_5H_8O_2$	52.0s	90.2	119.0	140.5	179.2	198.5	64.5
α-Valerolactone	$C_5H_8O_2$	37.5	79.8	101.9	136.5	182.3	207.5	—
α-Ethylacrylic acid	$C_5H_8O_2$	47.0	82.0	108.1	127.5	160.7	179.2	—
Ethyl acrylate	$C_5H_8O_2$	−29.5	+2.0	26.0	44.5	80.0	99.5	−71.2
Methyl methacrylate	$C_5H_8O_2$	−30.5	+1.0	25.5	47.0	82.0	101.0	—
Levulinic acid	$C_5H_8O_3$	102.0	141.8	169.5	190.2	227.4	256.8d	33.5
Glutaric acid	$C_5H_8O_4$	155.5	196.0	226.3	247.0	283.5	303.0	97.5
Dimethyl malonate	$C_5H_8O_4$	35.0	72.0	100.0	121.9	159.8	180.7	−62
Ethyl α-chloropropionate	$C_5H_9ClO_2$	+6.6	41.0	68.2	89.3	126.2	146.5	—
Isopropyl chloroacetate	$C_5H_9ClO_2$	+3.8	40.2	68.7	90.3	128.0	148.6	—
Valeronitrile	C_5H_9N	−6.0	+30.0	57.8	78.6	118.7	140.8	—
α-Hydroxybutyronitrile	C_5H_9NO	41.0	77.8	104.8	125.0	159.8	178.8	—
1-Pentene	C_5H_{10}	−80.4	−54.5	−34.1	−17.7	+12.8	30.1	—
2-Methyl-2-butene	C_5H_{10}	−75.4	−47.9	−26.7	−9.9	+21.6	38.5	−133
2-Methyl-1-butene	C_5H_{10}	−89.1	−64.3	−44.1	−28.0	+2.5	20.2	−135
Cyclopentane	C_5H_{10}	−68.0	−40.4	−18.6	−1.3	+31.0	49.3	−93.7
1,2-Dibromopentane	$C_5H_{10}Br_2$	19.8	58.0	87.4	110.1	151.8	175.0	—
2-Chloroethyl 2-chloroisopropyl ether	$C_5H_{10}Cl_2O$	24.7	63.0	92.4	115.8	156.5	180.0	—
2-Chloroethyl 2-chloropropyl ether	$C_5H_{10}Cl_2O$	29.8	70.0	101.5	125.6	169.8	194.1	—
Di(2-chloroethoxy)methane	$C_5H_{10}Cl_2O_2$	53.0	94.0	125.5	149.6	192.0	215.0	—
Allyldiehloroethylsilane	$C_5H_{10}Cl_2Si$	−3.0	34.2	62.7	85.2	127.0	150.3	—
3-Pentanone	$C_5H_{10}O$	−12.7	+17.2	39.4	56.2	86.3	102.7	−42
2-Pentanone	$C_5H_{10}O$	−12.0	+17.9	39.8	56.8	86.8	103.3	−77.8
3-Methyl-2-butanone	$C_5H_{10}O$	−19.9	+8.3	29.6	45.5	73.8	88.9	−92
4-Hydroxy-3-methyl-2-butanone	$C_5H_{10}O_2$	44.6	81.0	108.2	129.0	165.5	185.0	—
Valeric acid	$C_5H_{10}O_2$	42.2	79.8	107.8	128.3	165.0	184.4	−34.5
Isovaleric acid	$C_5H_{10}O_2$	34.5	91.3	98.0	118.9	155.2	175.1	−37.6
Ethyl propionate	$C_5H_{10}O_2$	−28.0	+3.4	27.2	45.2	79.8	99.1	−72.6
Propyl acetate	$C_5H_{10}O_2$	−26.7	+5.0	28.8	47.8	82.0	101.8	−92.5
Isopropyl acetate	$C_5H_{10}O_2$	−38.3	−7.2	+17.0	35.7	69.8	89.0	—
Methyl butyrate	$C_5H_{10}O_2$	−26.8	+5.0	29.6	48.0	83.1	102.3	—
Methyl isobutyrate	$C_5H_{10}O_2$	−34.1	−2.9	+21.0	39.6	73.6	92.6	−84.7
Butyl formate	$C_5H_{10}O_2$	−26.4	+6.1	31.6	51.0	86.2	106.0	—
Isobutyl formate	$C_5H_{10}O_2$	−32.7	−0.8	+24.1	43.4	79.0	98.2	−95.3

ORGANIC COMPOUNDS
Pressures Less than One Atmosphere (Continued)

Name	Formula	Temperature °C 1 mm	10 mm	40 mm	100 mm	400 mm	760 mm	M.P.
sec-Butyl formate	$C_5H_{10}O_2$	−34.4	−3.1	+21.3	40.2	75.2	93.6	—
Diethyl carbonate	$C_5H_{10}O_3$	−10.1	+23.8	49.5	69.7	105.8	125.8	−43
1-Bromo-3-methylbutane	$C_5H_{11}Br$	−20.4	+13.6	39.8	60.4	99.4	120.4	—
1-Iodo-3-methylbutane	$C_5H_{11}I$	−2.5	+34.1	62.3	84.4	125.8	148.2	—
Piperidine	$C_5H_{11}N$	s	+3.9	29.2	49.0	85.7	106.0	−9
Isobutyl carbamate	$C_5H_{11}NO_2$	s	96.4	123.3	147.2	186.0	206.5	65
Isoamyl nitrate	$C_5H_{11}NO_3$	+5.2	40.3	67.6	88.6	126.5	147.5	—
Pentane	C_5H_{12}	−76.6	−50.1	−29.2	−12.6	+18.5	36.1	−129.7
2-Methylbutane	C_5H_{12}	−82.9	−57.0	−36.5	−20.2	+10.5	27.8	−159.7
2,2-Dimethylpropane	C_5H_{12}	−102.0	−76.7s	−56.1s	−39.1s	−7.1	+9.5	−16.6
Amyl alcohol	$C_5H_{12}O$	+13.6	44.9	68.0	85.8	119.8	137.8	—
Isoamyl alcohol	$C_5H_{12}O$	+10.0	40.8	63.4	80.7	113.7	130.6	−17.2
2-Pentanol	$C_5H_{12}O$	+1.5	32.2	54.1	70.7	102.3	119.7	—
tert-Amyl alcohol	$C_5H_{12}O$	−12.9s	+17.2	38.8	55.3	85.7	101.7	−11.9
Ethyl propyl ether	$C_5H_{12}O$	−64.3	−35.0	−12.0	+6.8	41.6	61.7	—
2,3,4-Pentanetriol	$C_5H_{12}O_3$	155.0	204.5	239.6	263.5	307.0	327.2	—
Ethoxytrimethylsilane	$C_5H_{14}OSi$	−50.9	−20.7	+3.7	22.1	56.3	75.7	—
Ethyltrimethylsilane	$C_5H_{14}Si$	−60.6	−31.8	−9.0	+9.2	42.8	62.0	—
Ethyltrimethyltin	$C_5H_{14}Sn$	−30.0	+3.8	30.0	50.0	87.6	108.8	—
Chloranil	$C_6Cl_4O_2$	70.7s	97.8s	116.1s	129.5s	151.3s	162.6s	290
Hexachlorobenzene	C_6Cl_6	114.4s	166.4s	206.0s	235.5	283.5	309.4	230
Pentachlorobenzene	C_6HCl_5	98.6	144.3	178.5	205.5	251.6	276.0	85.5
Pentachlorophenol	C_6HCl_5O	s	s	211.2	239.6	285.0	309.3d	188.5
3-Bromo-2,4,6-trichlorophenol	$C_6H_2BrCl_3O$	112.4	163.2	200.5	229.3	278.0	305.8	—
1,2,3,4-Tetrachlorobenzene	$C_6H_2Cl_4$	68.5	114.7	149.2	175.7	225.5	254.0	46.5
1,2,3,5-Tetrachlorobenzene	$C_6H_2Cl_4$	58.2	104.1	140.0	168.0	220.0	246.0	54.5
1,2,4,5-Tetrachlorobenzene	$C_6H_2Cl_4$	s	s	146.0	173.5	220.5	256.0	139
2,3,4,6-Tetrachlorophenol	$C_6H_2Cl_4O$	100.0	145.3	179.1	205.2	250.4	275.0	69.5
2-Bromo-4,6-dichlorophenol	$C_6H_3BrCl_2O$	84.0	130.8	165.8	193.2	242.0	268.0	68
1,2,3-Trichlorobenzene	$C_6H_3Cl_3$	40.0s	85.6	119.8	146.0	193.5	218.5	52.5
1,2,4-Trichlorobenzene	$C_6H_3Cl_3$	38.4	81.7	114.8	140.0	187.7	213.0	17
1,3,5-Trichlorobenzene	$C_6H_3Cl_3$	s	78.0	110.8	136.0	183.0	208.4	63.5
2,4,5-Trichlorophenol	$C_6H_3Cl_3O$	72.0	117.3	151.5	178.0	226.5	251.8	62
2,4,6-Trichlorophenol	$C_6H_3Cl_3O$	76.5	120.2	152.2	177.8	222.5	246.0	68.5
1,4-Dibromobenzene	$C_6H_4Br_2$	61.0s	87.7	120.8	146.5	192.5	218.6	87.5
1,4-Bromochlorobenzene	C_6H_4BrCl	32.0	72.7	103.8	128.0	172.6	196.9	—
1,2-Dichlorobenzene	$C_6H_4Cl_2$	20.0	59.1	89.4	112.9	155.8	179.0	−17.6
1,3-Dichlorobenzene	$C_6H_4Cl_2$	12.1	52.0	82.0	105.0	149.0	173.0	−24.2
1,4-Dichlorobenzene	$C_6H_4Cl_2$	s	54.8	84.8	108.4	150.2	173.9	53.0
2,4-Dichlorophenol	$C_6H_4Cl_2O$	53.0	92.8	123.4	146.0	187.5	210.0	45.0
2,6-Dichlorophenol	$C_6H_4Cl_2O$	59.5	101.0	131.6	154.6	197.7	220.0	—
2,4,6-Trielfloroaniline	$C_6H_4Cl_3N$	134.0	170.0	195.8	214.6	246.4	262.0	78
Dichlorophenylarsine	$C_6H_5AsCl_2$	61.8	116.0	151.0	178.9	228.8	256.5	—
Bromobenzene	C_6H_5Br	+2.9	40.0	68.6	90.8	132.3	156.2	30.7
Chlorobenzene	C_6H_5Cl	−13.0	+22.2	49.7	70.7	110.0	132.2	−45.2
2-Chlorophenol	C_6H_5ClO	12.1	51.2	82.0	106.0	149.8	174.5	7.0
3-Chlorophenol	C_6H_5ClO	44.2	86.1	118.0	143.0	188.7	214.0	32.5
4-Chlorophenol	C_6H_5ClO	49.8	92.2	125.0	150.0	196.0	220.0	42
Benzenesulfonylchloride	$C_6H_5ClO_2S$	65.9	112.0	147.7	174.5	224.0	251.5d	14.5
Phenyl dichlorophosphate	$C_6H_5Cl_2O_2P$	66.7	110.0	143.4	168.0	213.0	239.5	—
Trichlorophenylsilane	$C_6H_5Cl_3Si$	33.0	74.2	105.8	130.5	175.7	201.0	—
Fluorobenzene	C_6H_5F	−43.4s	−12.4	+11.5	30.4	65.7	84.7	−42.1
Trifluorophenylsilane	$C_6H_5F_3Si$	−31.0	+0.8	25.4	44.2	78.7	98.3	—
Iodobenzene	C_6H_5I	24.1	64.0	94.4	118.3	163.9	188.6	−28.5
Nitrobenzene	$C_6H_5NO_2$	44.4	84.9	115.4	139.9	185.8	210.6	+5.7
2-Nitrophenol	$C_6H_5NO_3$	49.3	90.4	122.1	146.4	191.0	214.5	45
1,5-Hexadiene-3-yne	C_6H_6	−45.1	−14.0	+10.0	29.5	64.4	84.0	—
Benzene	C_6H_6	−36.7	−11.5s	+7.6	26.1	60.6	80.1	+5.5
2-Chloroaniline	C_6H_6ClN	46.3	84.8	115.6	139.5	183.7	208.8	—
3-Chloroaniline	C_6H_6ClN	63.5	102.0	133.6	158.0	203.5	228.5	−10.4
4-Chloroaniline	C_6H_6ClN	59.3s	102.1	135.0	159.9	2.6.6	230.5	70.5
4-Chlorophenol	C_6H_6ClO	54.3	95.8	125.8	150.6	194.3	217.0	42
2-Nitroaniline	$C_6H_6N_2O_2$	104.0	150.4	186.0	213.0	260.0	284.5d	71.5
3-Nitroaniline	$C_6H_6N_2O_2$	119.3	167.8	204.2	232.1	280.2	305.7d	114
4-Nitroaniline	$C_6H_6N_2O_2$	142.4s	194.4	234.2	261.8	310.2d	336.0d	156.5
Phenol	C_6H_6O	40.1s	73.8	100.1	121.4	160.0	181.9	40.6
Pyorcatechol	$C_6H_6O_2$	s	118.3	150.6	176.0	221.5	245.5	105
Resorcinol	$C_6H_6O_2$	108.4s	152.1	185.3	209.8	253.4	276.5	110.7
Hydroquinone	$C_6H_6O_2$	132.4s	163.5s	192.0	216.5	262.5	286.2	170.3
Pyrogallol	$C_6H_6O_3$	s	167.7	204.2	232.0	281.5	309.9d	133
Benzenethiol	C_6H_6S	18.6	56.0	84.2	106.6	146.7	168.0	—
Aniline	C_6H_7N	34.8	69.4	96.7	119.9	161.9	184.4	−6.2
2-Picoline	C_6H_7N	−11.1	+24.4	51.2	71.4	108.4	128.8	−70
Ethylene-bis-(chloroacetate)	$C_6H_8Cl_2O_4$	112.0	158.0	191.0	215.0	259.5	283.5	—
1,3-Phenylenediamine	$C_6H_8N_2$	99.8	147.0	182.5	209.9	259.0	285.5	62.8
Phenylhydrazine	$C_6H_8N_2$	71.8	115.8	148.2	173.5	218.2	243.5d	19.6
α-Methylglutaric anhydride	$C_6H_8O_3$	93.8	141.8	177.5	205.0	255.5	282.5	—
α, α-Dimethylsuccinic anhydride	$C_6H_8O_3$	61.4	102.0	132.3	155.3	197.5	219.5	—
Dimethyl maleate	$C_6H_8O_4$	45.7	86.4	117.2	140.3	182.2	205.0	—
Isobutyl dichloroacetate	$C_6H_{10}Cl_2O_2$	28.6	67.5	96.7	119.8	160.0	183.0	—
Diallyldichlorosilane	$C_6H_{10}Cl_2Si$	+9.5	47.4	76.4	99.7	142.0	165.3	—

ORGANIC COMPOUNDS
Pressures Less than One Atmosphere (Continued)

Name	Formula	Temperature °C 1 mm	10 mm	40 mm	100 mm	400 mm	760 mm	M.P.
Cyclohexanone	$C_6H_{10}O$	+1.4	38.7	67.8	90.4	132.5	155.6	−45.0
Mesityl oxide	$C_6H_{10}O$	−8.7	+26.0	51.7	72.1	109.8	130.0	−59
Isocaprolactone	$C_6H_{10}O_2$	38.3	80.3	112.3	137.2	182.1	207.0	—
Propionic anhydride	$C_6H_{10}O_3$	20.6	57.7	85.6	107.2	146.0	167.0	−45
Ethyl acetoacetate	$C_6H_{10}O_3$	28.5	67.3	96.2	118.5	158.2	180.8	−45
Methyl levulinate	$C_6H_{10}O_3$	39.8	79.7	109.5	133.0	175.8	197.7	—
Adipic acid	$C_6H_{10}O_4$	159.5	205.5	240.5	265.0	312.5	337.5	152
Diethyl oxalate	$C_6H_{10}O_4$	47.4	83.8	110.6	130.8	166.2	185.7	−40.6
Glycol diacetate	$C_6H_{10}O_4$	38.3	77.1	106.1	128.0	108.3	190.5	−31
Dimethyl-l-malate	$C_6H_{10}O_5$	75.4	118.3	150.1	175.1	219.5	242.6	—
Dimethyl-d-tartrate	$C_6H_{10}O_6$	102.1	148.2	182.4	208.8	255.0	280.0	61.5
Dimethyl-dl-tartrate	$C_6H_{10}O_6$	100.4	147.5	182.4	209.5	257.4	282.0	89
Diallyl sulfide	$C_6H_{10}S$	−9.5	+26.6	54.2	75.8	116.1	138.6	−83
Ethyl α-bromoisobutyrate	$C_6H_{11}BrO_2$	10.6	48.0	77.0	99.8	141.2	163.6	—
sec-Butyl chloroacetate	$C_6H_{11}ClO_2$	17.0	54.6	83.6	105.5	147.0	167.8	—
Capronitrile	$C_6H_{11}N$	+9.2	47.5	76.9	99.8	141.0	163.7	—
1-Hexene	C_6H_{12}	−57.5	−28.1	−5.0	+13.0	46.8	66.0	−98.5
Cyclohexane	C_6H_{12}	−45.3s	−15.9s	+6.7	25.5	60.8	80.7	+6.6
Methylcyclopentane	C_6H_{12}	−53.7	−23.7	−0.6	+17.9	42.3	71.8	−142.4
Dichlorodiisopropyl ether	$C_6H_{12}Cl_2O$	29.6	68.2	97.3	119.7	159.8	182.7	—
Bis-2-chloroethyl acetal	$C_6H_{12}Cl_2O_2$	56.2	97.6	127.8	150.7	190.5	212.6	—
2-Hexanone	$C_6H_{12}O$	+7.7	38.8	62.0	79.8	111.0	127.5	−56.9
4-Methyl-2-pentanone	$C_6H_{12}O$	−1.4	+30.0	52.8	70.4	102.0	119.0	−84.7
Allyl propyl ether	$C_6H_{12}O$	−39.0	−7.9	+16.4	35.8	71.4	90.5	—
Allyl isopropyl ether	$C_6H_{12}O$	−43.7	−12.9	+10.9	29.0	61.7	79.5	—
Cyclohexanol	$C_6H_{12}O$	21.0s	56.0	83.0	103.7	141.4	161.0	23.9
Caproic acid	$C_6H_{12}O_2$	71.4	99.5	125.0	144.0	181.0	202.0	−1.5
Isocaproic acid	$C_6H_{12}O_2$	66.2	94.0	120.4	141.4	181.0	207.7	−35
4-Hydroxy-4-methyl-2-pentanone	$C_6H_{12}O_2$	22.0	58.8	86.7	108.2	147.5	167.9	−47
Methyl isovalerate	$C_6H_{12}O_2$	−19.2	+14.0	39.8	59.8	96.7	116.7	—
Ethyl butyrate	$C_6H_{12}O_2$	−18.4	+15.3	41.5	62.0	100.0	121.0	−93.3
Ethyl isobutyrate	$C_6H_{12}O_2$	−24.3	+8.4	33.8	53.5	90.0	110.1	−88.2
Propyl propionate	$C_6H_{12}O_2$	−14.2	+19.4	45.0	64.2	102.0	122.4	−76
Isobutyl acetate	$C_6H_{12}O_2$	−21.2	+12.8	39.2	59.7	97.5	118.0	−98.9
Isoamyl formate	$C_6H_{12}O_2$	−17.5	+17.1	44.0	65.4	102.7	123.3	—
Paraformaldehyde	$C_6H_{12}O_3$	−9.4s	+24.1s	49.5s	69.0s	104.3s	124.0s	155 ± 5
sec-Butyl glycolate	$C_6H_{12}O_3$	28.3	66.0	94.2	116.4	155.6	177.5	—
Hexane	C_6H_{14}	−53.9	−25.0	−2.3	+15.8	49.6	68.7	−95.3
2-Methylpentane	C_6H_{14}	−60.9	−32.1	−9.7	+8.1	41.6	60.3	−154
3-Methylpentane	C_6H_{14}	−59.0	−30.1	−7.3	+10.5	44.2	63.3	−118
2,2-Dimethylbutane	C_6H_{14}	−69.3	−41.5	−19.5	−2.0	+31.0	49.7	−99.8
2,3-Dimethylbutane	C_6H_{14}	−63.6	−34.9	−12.4	+5.4	39.0	58.0	−128.2
1-Hexanol	$C_6H_{14}O$	24.4	58.2	83.7	102.8	138.8	157.0	−51.6
2-Hexanol	$C_6H_{14}O$	14.6	45.0	67.9	87.3	121.8	139.9	—
3-Hexanol	$C_6H_{14}O$	+2.5	36.7	62.2	81.8	117.0	135.5	—
2-Methyl-1-pentanol	$C_6H_{14}O$	15.4	49.6	74.7	94.2	129.8	147.9	—
2-Methyl-2-pentanol	$C_6H_{14}O$	−4.5	+27.6	51.3	69.2	102.6	121.1	−103
2-Methyl-4-pentanol	$C_6H_{14}O$	−0.3	+33.3	58.2	78.0	113.5	131.7	—
Dipropyl ether	$C_6H_{14}O$	−43.3	−11.8	+13.2	33.0	69.5	89.5	−122
Diisopropyl ether	$C_6H_{14}O$	−57.0	−27.4	−4.5	+13.7	48.2	67.5	−60
Acetal	$C_6H_{14}O_2$	−23.0	+8.0	31.9	50.1	84.0	102.0	—
1,2-Diethoxyethane	$C_6H_{14}O_2$	−33.5	+1.6	29.7	51.8	92.1	119.5	—
Di(2-methoxyethyl) ether	$C_6H_{14}O_3$	13.0	50.0	77.5	99.5	138.5	159.8	—
Diethylene glycol, ethyl ether	$C_6H_{14}O_3$	45.3	85.8	116.7	140.3	180.3	201.9	—
Dipropyleneglycol	$C_6H_{14}O_3$	73.8	116.2	147.4	169.9	210.5	231.8	—
Triethyleneglycol	$C_6H_{14}O_4$	114.0	158.1	191.3	214.6	256.6	278.3	—
Triethylboron	$C_6H_{15}B$	—	−148.0	−131.4	−116.0	−81.0	−56.2	—
Chlorotriethylsilane	$C_6H_{15}ClSi$	−4.9	+32.0	60.2	82.3	123.6	146.3	—
Triethyl phosphate	$C_6H_{15}O_4P$	39.6	82.1	115.7	141.6	187.0	211.0	—
Triethylthallium	$C_6H_{15}Tl$	+9.3	51.7	85.4	112.1	163.5	192.9d	−63.0
Diethoxydimethylsilane	$C_6H_{16}O_2Si$	−19.1	+13.3	38.0	57.6	93.2	113.5	—
Trimethylpropylsilane	$C_6H_{16}Si$	−46.0	−13.9	+11.3	31.6	69.2	90.0	—
Trimethylpropyltin	$C_6H_{16}Sn$	−12.0	+21.8	48.5	69.8	109.6	131.7	—
1,5-Dichlorohexamethyltrisiloxane	$C_6H_{18}Cl_2O_2Si_3$	26.0	65.1	94.8	118.2	160.2	184.0	−53
Hexamethylcyclotrisiloxane	$C_6H_{18}O_3Si_2$	s	s	s	78.7	114.7	134.0	64
Hexamethyldisiloxane	$C_6H_{18}OSi_2$	−29.0	+2.8	26.7	45.6	80.0	99.2	—
3,4-Dichloro-α,α,α-trifluorotoluene	$C_7H_3Cl_2F_3$	11.0	52.2	84.0	109.2	150.5	172.8	−12.1
2-Chloro-α,α,α-trifluorotoluene	$C_7H_4ClF_3$	0.0	37.1	65.9	88.3	130.0	152.2	−6.0
2-α,α,α-tetrachlorotoluene	$C_7H_4Cl_4$	69.0	117.9	155.0	185.0	233.0	262.1	28.7
Benzoyl bromide	C_7H_5BrO	47.0	89.8	122.6	147.7	193.7	218.5	0
Benzoyl chloride	C_7H_5ClO	32.1	73.0	103.8	128.0	172.8	197.2	−0.5
α,α,α-Trichlorotoluene	$C_7H_5Cl_3$	45.8	87.6	119.8	144.3	189.2	213.5	−21.2
α,α,α-Trifluorotoluene	$C_7H_5F_3$	−32.0s	+0.4	25.7	45.3	82.0	102.2	−29.3
Benzonitrile	C_7H_5N	28.2	69.2	99.6	123.5	166.7	109.6	−12.9
Phenyl isocyanide	C_7H_5N	12.0	49.7	78.3	101.0	142.3	165.0d	—
Phenyl isocyanate	C_7H_5NO	10.6	48.5	77.7	100.6	142.7	165.6	—
2-Nitrobenzaldehyde	$C_7H_5NO_3$	85.8	133.4	168.8	196.2	246.8	273.5	40.9
3-Nitrobenzaldehyde	$C_7H_5NO_3$	96.2	142.8	177.7	204.3	252.1	278.3	58
Phenyl isothiocyanate	C_7H_5NS	47.2	89.8	122.5	147.7	194.0	218.5	−21.0
α,α-dichlorotoluene	$C_7H_6Cl_2$	35.4	78.7	112.1	138.3	187.0	214.0	−16.1
Benzaldehyde	C_7H_6O	26.2	62.0	90.1	112.5	154.1	179.0	−26

ORGANIC COMPOUNDS
Pressures Less than One Atmosphere (Continued)

Name	Formula	Temperature °C 1 mm	10 mm	40 mm	100 mm	400 mm	760 mm	M.P.
Benzoic acid	$C_7H_6O_2$	96.0s	132.1	162.6	186.2	227.0	249.2	121.7
Salicylaldehyde	$C_7H_6O_2$	33.0	73.8	105.2	129.4	173.7	196.5	−7
4-Hydroxybenzaldihyde	$C_7H_6O_2$	121.2	169.7	206.0	233.5	282.6	310.0	115.5
Salicylic acid	$C_7H_6O_3$	113.7s	146.2s	172.2	193.4	230.5	256.0	159
α-Bromotoluene	C_7H_7Br	32.2	73.4	104.8	129.8	175.2	198.5	−4
2-Bromotoluene	C_7H_7Br	24.4	62.3	91.0	112.0	157.3	181.8	−28
3-Bromotoluene	C_7H_7Br	14.8	64.0	93.9	117.8	160.0	183.7	−39.8
4-Bromotoluene	C_7H_7Br	10.3	61.1	91.8	116.4	160.2	184.5	28.5
4-Bromoanisole	C_7H_7BrO	48.8	91.9	125.0	150.1	197.5	223.0	12.5
α-Chlorotoluene	C_7H_7Cl	22.0	60.8	90.7	114.2	155.8	179.4	−39
2-Chlorotoluene	C_7H_7Cl	+5.4	43.2	72.0	94.7	137.1	159.3	—
3-Chlorotoluene	C_7H_7Cl	+4.8	43.2	73.0	96.3	139.7	162.3	—
4-Chlorotoluene	C_7H_7Cl	+5.5	43.8	73.5	96.6	139.8	162.3	+7.3
2-Fluorotoluene	C_7H_7F	−24.2	+8.9	34.7	55.3	92.8	114.0	−80
30Fluorotoluene	C_7H_7F	−22.4	+11.0	37.0	57.5	95.4	116.0	−110.8
4-Fluorotoluene	C_7H_7F	−21.8	+11.8	37.8	58.1	96.1	117.0	—
2-Iodotoluene	C_7H_7I	37.2	79.8	112.4	138.1	185.7	211.0	—
2-Nitrotoluene	$C_7H_7NO_2$	50.0	93.8	126.3	151.5	197.7	222.3	−4.1
3-Nitrotoluene	$C_7H_7NO_2$	50.2	96.0	130.7	156.9	206.8	231.9	15.1
4-Nitrotoluene	$C_7H_7NO_2$	53.7	100.5	136.0	163.0	212.5	238.3	51.9
Toluene	C_7H_8	−26.7	+6.4	31.8	51.9	89.5	110.6	−95.0
Benzyldichlorosilane	$C_7H_8Cl_2Si$	45.3	83.2	111.8	133.5	173.0	194.3	—
Dichloromethylphenylsilane	$C_7H_8Cl_2Si$	35.7	77.4	109.5	134.2	180.2	205.5	—
Dichloro-4-tolylsilane	$C_7H_8Cl_2Si$	46.2	84.2	113.2	135.5	175.2	196.3	—
Anisole	C_7H_8O	+5.4	42.2	70.7	93.0	133.8	155.5	−37.3
Benzyl alcohol	C_7H_8O	58.0	92.6	119.8	141.9	183.0	204.7	−15.3
2-Cresol	C_7H_8O	38.2	76.7	105.8	127.4	168.4	190.8	30.8
3-Cresol	C_7H_8O	52.0	87.8	116.0	138.0	179.0	202.8	10.9
4-Cresol	C_7H_8O	53.0	88.6	117.7	140.0	179.4	201.8	35.5
3,5-Dimethyl-1,2-pyrone	$C_7H_8O_2$	78.6	122.0	152.7	177.5	221.0	245.0	51.5
2-Methoxyphenol	$C_7H_8O_2$	52.4	92.0	121.6	144.0	184.1	205.0	28.3
Ethyl 2-furoate	$C_7H_8O_3$	37.6	77.1	107.5	130.4	172.5	195.0	34
Benzylamine	C_7H_9N	29.0	67.7	97.3	120.0	161.3	184.5	
N-Methylaniline	C_7H_9N	36.0	76.2	106.0	129.8	172.0	195.5	−57
2-Toluidine	C_7H_9N	44.0	81.4	110.0	133.0	176.2	199.7	−16.3
3-Toluidine	C_7H_9N	41.0	82.0	113.5	136.7	180.6	203.3	−31.5
4-Toluidine	C_7H_9N	42.0	81.8	111.5	133.7	176.9	200.4	44.5
2-Methoxyaniline	C_7H_9NO	61.0	101.7	132.0	155.2	197.3	218.5	5.2
Toluene-2,4-diamine	$C_7H_{10}N_2$	106.5	151.7	185.7	211.5	256.0	280.0	99
4-Tolylhydrazine	$C_7H_{10}N_2$	82.0	123.8	154.1	178.0	219.5	242.0d	65.5
Trimethysuccinic anhydride	$C_7H_{10}O_3$	53.5	97.4	131.0	145.4	205.5	231.0	—
Dimethyl citraconate	$C_7H_{10}O_4$	50.8	91.8	122.6	145.8	188.0	210.5	—
Dimethyl itaconate	$C_7H_{10}O_4$	69.3	106.6	133.7	153.7	189.8	208.0	38
trans-Dimethyl mesaconate	$C_7H_{10}O_4$	46.8	87.8	118.0	141.5	183.5	206.0	—
2-Cyano-2-butyl acetate	$C_7H_{11}NO_2$	42.0	82.0	111.8	133.8	173.4	195.2	—
Butyl acrylate	$C_7H_{12}O_2$	−0.5	+35.5	63.4	85.1	125.2	147.4	−64.6
Ethyl levulinate	$C_7H_{12}O_3$	47.3	87.3	117.7	141.3	183.0	206.2	—
Pimelic acid	$C_7H_{12}O_4$	163.4	212.0	247.0	272.0	318.5	342.1	103
Diethyl malonate	$C_7H_{12}O_4$	40.0	81.3	113.3	136.2	176.8	198.9	−49.8
Enanthyl chloride	$C_7H_{13}ClO$	34.2	64.6	86.4	102.7	130.7	145.0	—
Enanthonitrile	$C_7H_{13}N$	21.0	61.6	92.6	116.8	160.0	184.6	—
Ethylcyclopentane	C_7H_{14}	−32.2	−0.1	+25.0	45.0	82.3	103.4	−138.6
Methylcyclohexane	C_7H_{14}	−35.9	−3.2	+22.0	42.1	79.6	100.9	−126.4
2-Heptene	C_7H_{14}	−35.8	−3.5	+21.5	41.3	78.1	98.5	—
Enanthaldehyde	$C_7H_{14}O$	12.0	43.0	66.3	84.0	125.5	155.0	−42
2-Heptanone	$C_7H_{14}O$	19.3	55.5	81.2	100.0	133.2	150.2	—
4-Heptanone	$C_7H_{14}O$	23.0	55.0	78.1	96.0	127.3	143.7	−32.6
2,4-Dimethyl-3-pentanone	$C_7H_{14}O$	+5.2	36.7	59.6	77.0	108.0	123.7	—
Enanthic acid	$C_7H_{14}O_2$	78.0	113.2	139.5	160.0	199.6	221.5	−10
Methyl caproate	$C_7H_{14}O_2$	+5.0	42.0	70.0	91.4	129.8	150.0d	—
Ethyl isovalerate	$C_7H_{14}O_2$	−6.1	+28.7	55.2	7.9	114.0	134.3	−99.3
Propyl butyrate	$C_7H_{14}O_2$	−1.6	+34.0	61.5	82.6	121.7	142.7	−95.2
Propyl isobutyrate	$C_7H_{14}O_2$	−6.2	+28.3	54.3	73.9	112.0	133.9	—
Isopropyl isobutyrate	$C_7H_{14}O_2$	−16.3	+17.0	42.4	62.3	100.0	120.5	—
Isobutyl propionate	$C_7H_{14}O_2$	−2.3	+32.3	58.5	79.5	116.4	136.8	−71
Isoamyl acetate	$C_7H_{14}O_2$	0.0	+35.2	62.1	83.2	121.5	142.0	—
Heptane	C_7H_{16}	−34.0	−2.1	+22.3	41.8	78.0	98.4	−90.6
2-Methylhexane	C_7H_{16}	−40.4	−9.1	+14.9	34.1	69.8	90.0	−118.2
3-Methylhexane	C_7H_{16}	−39.0	−7.8	+16.4	35.6	71.6	91.9	−119.4
3-Ethylpentane	C_7H_{16}	−37.8	−6.8	+17.5	36.9	73.0	93.5	−118.6
2,2-Dimethylpentane	C_7H_{16}	−49.0	−18.7	+5.0	23.9	59.2	79.2	−123.7
2,3-Dimethylpentane	C_7H_{16}	−42.0	−10.3	+13.9	33.3	69.4	89.8	−135
2,4-Dimethylpentane	C_7H_{16}	−48.0	−17.1	+6.5	25.4	60.6	80.5	−119.5
3,3-Dimethylpentane	C_7H_{16}	−45.9	−14.4	+9.9	29.3	65.5	86.1	−135.0
2,3,3-Trimethylbutane	C_7H_{16}	s	−18.8	+5.2	24.4	60.4	80.9	−25.0
1-Heptanol	$C_7H_{16}O$	42.4	74.7	99.8	119.5	155.6	175.8	34.6
Triethyl orthoformate	$C_7H_{16}O_3$	+5.5	40.5	67.5	88.0	125.7	146.0	—
Triethoxymethylsilane	$C_7H_{18}O_3Si$	−1.5	+34.6	61.7	82.7	121.8	143.5	—
Butyltrimethylsilane	C_7H_8Si	−23.4	+9.9	35.9	56.3	93.8	115.0	—
Triethylmethylsilane	$C_7H_{18}Si$	−18.2	+16.6	44.0	65.6	105.3	127.0	—
Phthaloyl chloride	$C_8H_4Cl_2O_2$	86.3s	134.2	170.0	197.8	248.3	275.8	88.5

ORGANIC COMPOUNDS
Pressures Less than One Atmosphere (Continued)

Name	Formula	Temperature °C						M.P.
		1 mm	10 mm	40 mm	100 mm	400 mm	760 mm	
Phthalic anhydride	$C_8H_4O_3$	96.5s	134.0	172.0	202.3	256.8	284.5	130.8
α,α-Dichlorophenylacetonitrile	$C_8H_5Cl_2N$	56.0	98.1	130.0	154.5	199.5	223.5	—
Pentachloroethylbenzene	$C_8H_5Cl_5$	96.2	148.0	186.2	216.0	269.3	299.0	—
Phenylglyoxylonitrile	C_8H_5NO	44.5	85.5	116.6	141.0	185.0	208.0	33.5
2,3-Dichlorostyrene	$C_8H_6Cl_2$	61.0	104.6	137.8	163.5	210.0p	235.0p	—
2,4-Dichlorostyrene	$C_8H_6Cl_2$	53.5	97.4	129.2	153.8	200.0p	225.0p	—
2,5-Dichlorostyrene	$C_8H_6Cl_2$	55.5	98.2	131.0	155.8	202.5p	227.0p	—
2,6-Dichlorostyrene	$C_8H_6Cl_2$	47.8	90.0	122.4	147.6	193.5p	217.0p	—
3,4-Dichlorostyrene	$C_8H_6Cl_2$	57.2	100.4	133.7	158.2	205.7p	230.0p	—
3,5-Dichlorostyrene	$C_8H_6Cl_2$	53.5	97.4	129.2	153.8	200.0p	225.0p	—
3,4,5,6-Tetrachloro-1,2-xylene	$C_8H_6Cl_4$	94.4	140.3	174.2	200.5	248.3	273.5	—
1,2,3,5-Tetrachloro-4-ethylbenzene	$C_8H_6Cl_4$	77.0	126.0	162.1	191.6	243.0	270.0	—
Phenylglyoxal	$C_8H_6O_2$	s	87.8	115.5	136.2	173.5	193.5	73
Phthalide	$C_8H_6O_2$	95.5	144.0	181.0	210.0	261.8	290.0	73
Piperonal	$C_8H_6O_3$	87.0	132.0	165.7	191.7	238.5	263.0	37
3-Chlorostyrene	C_8H_7Cl	25.3	65.2	96.5	121.2	165.7p	190.0p	—
4-Chlorostyrene	C_8H_7Cl	28.0	67.5	98.0	122.0	166.0p	191.0p	−15.0
Phenylacetyl chloride	C_8H_7ClO	48.0	89.0	119.8	143.5	186.0	210.0	—
2-Tolunitrile	C_8H_7N	36.7	77.9	110.0	135.0	180.0	205.2	−13
4-Tolunitrile	C_8H_7N	42.5	85.8	109.5	145.2	193.0	217.6	29.5
Phenylacetonitrile	C_8H_7N	60.0	103.5	136.3	161.8	208.5	233.5	−23.8
2-Tolyl isocyanide	C_8H_7N	25.2	64.0	94.0	117.7	159.9	183.5	—
2-Nitrophenyl acetate	$C_8H_7NO_4$	100.0	142.0	172.8	194.1	233.5	253.0d	—
2-Methylbenzothiazole	C_8H_7NS	70.0	111.2	141.2	163.9	204.5	225.5	15.4
Benzylisothiocyanate	C_8H_7NS	79.5	121.8	153.0	177.7	220.4	243.0	—
Styrene	C_8H_8	−7.0	+ 30.8	59.8	82.0p	122.5p	145.2p	−30.6
(1,2-Dibromoethyl) benzene	$C_8H_8Br_2$	86.0	129.8	161.8	186.3	230.0	254.0	—
1,2-Dichloro-3-ethylbenzene	$C_8H_8Cl_2$	46.0	90.0	123.8	149.8	197.0	222.1	−40.8
1,2-Dichloro-4-ethylbenzene	$C_8H_8Cl_2$	47.0	92.3	127.5	153.3	201.7	226.6	−76.4
1,4-Dichloro-2-ethylbenzene	$C_8H_8Cl_2$	38.5	83.2	118.0	144.0	191.5	216.3	−61.2
Acetophenone	C_8H_8O	37.1	78.0	109.4	133.6	178.0	202.4	20.5
Phenyl acetate	$C_8H_8O_2$	38.2	78.0	108.1	131.6	173.5	195.9	—
Phenylacetic acid	$C_8H_8O_2$	97.0	141.3	173.6	198.2	243.0	265.5	76.5
Anisaldehyde	$C_8H_8O_2$	73.2	117.8	150.5	176.7	223.0	248.0	2.5
Methyl benzoate	$C_8H_8O_2$	39.0	77.3	107.8	130.8	174.7	199.5	−12.5
Methyl salicylate	$C_8H_8O_3$	54.0	95.3	126.2	150.0	197.5	223.2	−8.3
Vanillin	$C_8H_8O_3$	107.0	154.0	188.7	214.5	260.0	285.0	81.5
Dehydroacetic acid	$C_8H_8O_4$	91.7	137.3	171.0	197.5	244.5	269.0	—
2-Bromo-1,4-xylene	C_8H_9Br	37.5	78.8	110.6	135.7	181.0	206.7	+ 9.5
1-Bromo-4-ethylbenzene	C_8H_9Br	30.4	74.0	108.5	135.5	182.0	206.0	−45.0
(2-Bromoethyl) benzene	C_8H_9Br	48.0	90.5	123.2	148.2	194.0	219.0	—
1-Chloro-2-ethylbenzene	C_8H_9Cl	17.2	56.1	86.2	110.0	152.2	177.6	−80.2
1-Chloro-3-ethylbenzene	C_8H_9Cl	18.6	58.2	89.2	113.6	156.7	181.1	−53.3
1-Chloro-4-ethylbenzene	C_8H_9Cl	19.2	60.0	91.8	116.0	159.8	184.3	−62.6
1-Chloro-2-ethoxybenzene	C_8H_9ClO	45.8	86.5	117.8	141.8	185.5	208.0	—
4-Chlorophenethyl alcohol	C_8H_9ClO	84.0	129.0	162.0	188.1	234.5	259.3	—
Acetanilide	C_8H_9NO	114.0	162.0	199.6	227.2	277.0	303.8	113.5
Methyl anthranilate	$C_8H_9NO_2$	77.6	124.2	159.7	187.8	238.5	266.5	24
4-Nitro-1,3-xylene	$C_8H_9NO_2$	65.6	109.8	143.3	168.5	217.5	244.0	+ 2
Ethylbenzene	C_8H_{10}	−9.8	+ 25.9	52.8	74.1	113.8	136.2	−94.0
2-Xylene	C_8H_{10}	−3.8	+ 32.1	59.5	81.3	121.7	144.4	−25.2
3-Xylene	C_8H_{10}	−6.9	+ 28.3	55.3	76.8	116.7	139.1	−47.9
4-Xylene	C_8H_{10}	−8.1	+ 27.3	54.4	75.9	115.9	138.3	+ 13.3
Dichloroethoxyphenylsilane	$C_8H_{10}Cl_2OSi$	52.4	94.6	126.2	151.4	197.2	222.2	—
Dichloroethylpenylsilane	$C_8H_{10}Cl_2Si$	48.5	92.4	126.7	153.3	203.5	230.0	—
2-Ethylphenol	$C_8H_{10}O$	46.2	87.0	117.9	141.8	184.5	207.5	−5
3-Ethylphenol	$C_8H_{10}O$	60.0	100.2	130.0	152.0	193.3	214.0	−4
4-Ethylphenol	$C_8H_{10}O$	59.3	100.2	131.3	154.2	197.4	219.0	46.5
2,3-Xylenol	$C_8H_{10}O$	56.0s	97.6	129.2	152.2	196.0	218.0	75
2,4-Xylenol	$C_8H_{10}O$	51.8	92.3	121.5	143.0	184.2	211.5	25.5
2,5-Xylenol	$C_8H_{10}O$	51.8s	91.3	121.5	143.0	184.2	211.5	74.5
3,4-Xylenol	$C_8H_{10}O$	66.2	107.7	138.0	161.0	203.6	225.2	62.5
3,5-Xylenol	$C_8H_{10}O$	62.0s	102.4	133.3	156.0	197.8	219.5	68
Phenetole	$C_8H_{10}O$	18.1	56.4	86.6	108.4	149.8	172.0	−30.2
α-Methyl benxyl alcohol	$C_8H_{10}O$	49.0	88.0	117.8	140.3	180.7	204.0	—
Phenethylacohol	$C_8H_{10}O$	58.2	100.0	130.5	154.0	197.5	219.5	—
4,6-Dimethylresorcinol	$C_8H_{10}O_2$	49.0	90.7	122.5	147.3	193.0	215.0	—
2-Phenoxyethanol	$C_8H_{10}O_2$	78.0	121.2	152.2	176.5	221.0	245.3	11.6
Diethyl dioxosuccinate	$C_8H_{10}O_6$	70.0	112.0	143.8	167.7	210.8	233.5	—
Chlorodimethylphenylsilane	$C_8H_{11}ClSi$	29.8	70.0	101.2	124.7	168.6	193.5	—
N-Ethylaniline	$C_8H_{11}N$	38.5	80.6	113.2	137.3	180.8	204.0	−63.5
N,N-Dimethylaniline	$C_8H_{11}N$	29.5	70.0	101.6	125.8	169.2	193.1	+ 2.5
4-Ethylaniline	$C_8H_{11}N$	52.0	93.8	125.7	149.8	194.2	217.4	−4
2,4-Xylidine	$C_8H_{11}N$	52.6	93.0	123.8	146.8	188.3	211.5	—
2,6-Xylidine	$C_8H_{11}N$	44.0	87.0	120.2	146.0	193.7	217.9	—
2-Phenetidine	$C_8H_{11}NO$	67.0	108.6	139.9	163.5	207.0	228.0	—
2-Anilinoethanol	$C_8H_{11}NO$	104.0	149.6	183.7	209.5	254.5	279.6	—
Dimethyl arsanilate	$C_8H_{12}AsNO_2$	15.0	51.8	79.7	101.0	140.3	160.5	—
Diethyleneglycol-bis-chloroacetate	$C_8H_{12}Cl_2O_5$	148.3	195.8	229.0	252.0	291.8	313.0	—
Diethyl maleate	$C_8H_{12}O_4$	57.3	100.0	131.8	156.0	201.7	225.0	—
Diethyl fumarate	$C_8H_{12}O_4$	53.2	95.3	126.7	151.1	195.8	218.5	+ 0.6

ORGANIC COMPOUNDS
Pressures Less than One Atmosphere (Continued)

Name	Formula	Temperature °C						M.P.
		1 mm	10 mm	40 mm	100 mm	400 mm	760 mm	
Dimethylphenylsilane	$C_8H_{12}Si$	+5.3	42.6	71.4	94.2	136.4	159.3	—
Ethyl-*α*-ethylacetoacetate	$C_8H_{14}O_3$	40.5	80.2	110.3	133.8	175.6	198.0	—
Propyl levulinate	$C_8H_{14}O_3$	59.7	99.9	130.1	154.0	198.0	221.2	—
Isopropyl levulinate	$C_8H_{14}O_3$	48.0	88.0	118.1	141.8	185.2	208.2	—
Dipropyl oxalate	$C_8H_{14}O_4$	53.4	93.9	124.6	148.1	190.3	213.5	—
Diisopropyl oxalate	$C_8H_{14}O_4$	43.2	81.9	110.5	132.6	171.8	193.5	—
Diethyl succinate	$C_8H_{14}O_4$	54.6	96.6	127.8	151.1	193.8	216.5	−20.8
Diethyl isosuccinate	$C_8H_{14}O_4$	39.8	80.0	111.0	134.8	177.7	201.3	—
Suberic acid	$C_8H_{14}O_4$	172.8	219.5	254.6	279.0	322.8	345.5	142
Diethyl malate	$C_8H_{14}O_5$	80.7	125.6	157.8	183.9	229.5	253.4	—
Diethyl-*dl*-tartrate	$C_8H_{14}O_6$	100.0	147.2	181.7	208.0	254.3	280.0	—
Diethyl-*d*-tartrate	$C_8H_{14}O_6$	102.0	148.0	182.3	208.5	254.8	280.0	—
(2-Bromoethyl) cyclohexane	$C_8H_{15}Br$	38.7	80.5	113.0	138.0	186.2	213.0	—
Caprylonitrile	$C_8H_{15}N$	43.0	80.4	110.6	134.8	179.5	204.5	—
Ethyl *N,N*-diethyloxamate	$C_8H_{15}NO_3$	76.0	121.7	154.4	180.3	226.5	252.0	—
2-Methyl-2-heptene	C_8H_{16}	−16.1	+17.8	44.0	64.6	102.2	122.5	—
1,1-Dimethylcyclohexane	C_8H_{16}	−24.4	+10.3	37.3	57.9	97.2	119.5	−34
cis-1,2-Dimethylcyclohexane	C_8H_{16}	−15.9	+18.4	45.3	66.8	107.0	129.7	−50.0
trans-1,2-Dimethylcyclohexane	C_8H_{16}	−21.1	+13.0	39.7	61.0	100.9	123.4	−88.0
cis-1,3-Dimethylcyclohexane	C_8H_{16}	−22.7	+11.2	37.5	58.5	97.8	120.1	−76.2
trans-1,3-Dimethylcyclohexane	C_8H_{16}	−19.4	+14.9	41.4	62.5	102.1	124.4	−92.0
cis-1,4-Dimethylcyclohexane	C_8H_{16}	−20.0	+14.5	41.1	62.3	101.9	124.3	−87.4
trans-1,4-Dimethylcyclohexane	C_8H_{16}	−24.3	+10.1	36.5	57.6	97.0	119.3	−36.9
Ethylcyclohexane	C_8H_{16}	−14.5	+20.6	47.6	69.0	109.1	131.8	−111.3
Caprylaldehyde	$C_8H_{16}O$	73.4	101.2	120.0	133.9	156.5	168.5	—
Cyclohexaneethanol	$C_8H_{16}O$	50.4	90.0	119.8	142.7	183.5	205.4	—
6-Methyl-3-hepten-2-ol	$C_8H_{16}O$	41.6	76.7	102.7	122.6	156.6	175.5	—
6-Methyl-5-hepten-2-ol	$C_8H_{16}O$	41.9	77.8	104.1	123.8	156.6	174.3	—
2-Octanone	$C_8H_{16}O$	23.6	60.9	89.8	111.7	151.0	172.9	−16
2,2,4-Trimethyl-3-pentanone	$C_8H_{16}O$	14.7	46.4	69.8	87.6	118.4	135.0	—
Caprylic acid	$C_8H_{16}O_2$	92.3	124.0	150.6	172.2	213.9	237.5	16
Ethyl isocaproate	$C_8H_{16}O_2$	11.0	48.0	76.3	98.4	139.2	160.4	—
Propyl isovalerate	$C_8H_{16}O_2$	+8.0	45.1	72.8	95.0	135.0	155.9	—
Isobutyl butyrate	$C_8H_{16}O$	+4.6	42.2	71.7	94.0	135.7	156.9	—
Isobutyl isobutyrate	$C_8H_{16}O_2$	+4.1	39.9	67.2	88.0	126.3	147.5	−80.7
Amyl propionate	$C_8H_{16}O_2$	+8.5	46.3	75.5	97.6	138.4	160.2	—
Tetraethyleneglycol cholorhydrin	$C_8H_{16}ClO_4$	10.1	156.1	190.0	214.7	258.2	281.5	—
1-Iodooctane	$C_8H_{17}I$	45.8	90.0	123.3	150.0	199.3	225.5	−45.9
Ethyl-*l*-leucinate	$C_8H_{17}NO_2$	27.8	72.1	106.0	131.8	167.3	184.0	—
Octane	C_8H_{18}	−14.0	+19.2	45.1	65.7	104.0	125.0	−56.8
2-Methylheptane	C_8H_{18}	−21.0	+12.3	37.9	58.3	96.2	117.6	−109.5
3-Methylheptane	C_8H_{18}	−19.8	+13.3	38.9	59.4	97.4	118.9	−120.8
4-Methylheptane	C_8H_{18}	−20.4	+12.4	38.0	58.3	96.3	117.7	−121.1
2,2-Dimethylhexane	C_8H_{18}	−29.7	+3.1	28.2	48.2	85.6	106.8	—
2,3-Dimethylhexane	C_8H_{18}	−23.0	+9.9	35.6	56.0	94.1	115.6	—
2,4-Dimethylhexane	C_8H_{18}	−26.9	+5.2	30.5	50.6	88.2	109.4	—
2,5-Dimethylhexane	C_8H_{18}	−26.7	+5.3	30.4	50.5	97.9	109.1	−90.7
3,3-Dimethylhexane	C_8H_{18}	−25.8	+6.1	31.7	52.5	90.4	112.0	—
3,4-Dimethylhexane	C_8H_{18}	−22.1	+11.3	37.1	57.7	96.0	117.7	—
3-Ethylhexane	C_8H_{18}	−20.0	+12.8	38.5	58.9	97.0	118.5	—
2,2,3-Trimethylpentane	C_8H_{18}	−29.0	+3.9	29.5	49.9	88.2	109.8	−112.3
2,2,4-Trimethylpentane	C_8H_{18}	−36.5	−4.3	+20.7	40.7	78.0	99.2	−107.3
2,3,3-Trimethylpentane	C_8H_{18}	−25.8	+6.9	33.0	53.8	92.7	114.8	−101.5
2,3,4-Trimethylpentane	C_8H_{18}	−26.3	7.1	32.0	53.4	91.8	113.5	−109.2
2-Methyl-3-ethylpentane	C_8H_{18}	−24.0	+9.5	35.2	55.7	94.0	115.6	−114.5
3-Methyl-3-ethylpentane	C_8H_{18}	−23.9	+9.9	36.2	57.1	96.2	118.3	−90
2,2,3,3-Tetramethylbutane	C_8H_{18}	−17.4	+13.5	36.8	54.8	87.4	106.3	+100.7
Tetramethylpiperazine	$C_8H_{18}N_2$	23.7	61.7	90.0	113.8	157.8	183.5	—
1-Octanol	$C_8H_{18}O$	54.0	88.3	115.2	135.2	173.8	195.2	−15.4
2-Octanol	$C_8H_{18}O$	32.8	70.0	98.0	119.8	157.5	178.5	−38.6
1,2-Dipropoxy ethane	$C_8H_{18}O_2$	−38.8	+5.0	42.3	74.2	140.0	180.0	—
Diethylene glycol butyl ether	$C_8H_{18}O_3$	70.0	107.8	135.5	159.8	205.0	231.2	—
Tetraethylene glycol	$C_8H_{18}O_5$	153.9	197.1	228.0	250.0	228.0	307.8	—
Dibutyl sulfide	$C_8H_{18}S$	+21.7	66.4	96.0	118.6	159.0	182.0	−79.7
Dibutyl disulfide	$C_8H_{18}S_2$	+34.6	94.0	145.1	188.0	275.5	330.5	—
Diisobutylamine	$C_8H_{19}N$	−5.1	+30.6	57.8	79.2	118.0	139.5	−70
Tetraethoxysilane	$C_8H_{20}O_4Si$	16.0	52.6	81.4	103.6	146.2	168.5	—
Tetraethyllead	$C_8H_{20}Pb$	38.4	74.8	102.4	123.8	161.8	183.0	−136
Amyltrimethylsilane	$C_8H_{20}Si$	−9.2	+26.7	54.4	76.2	116.6	139.0	—
Tetraethylsilane	$C_8H_{20}Si$	−1.0	+36.3	65.3	88.0	130.2	153.0	—
Tetraethyl bistibine	$C_8H_{20}Sb_2$	97.0	151.2	193.2	225.6	286.2	320.3	—
1,3-Diethoxytetramethyldisiloxane	$C_8H_{22}O_3Si_2$	14.8	51.2	78.7	100.3	139.8	160.7	—
1,7-Dichlorooctamethyltetrasiloxane	$C_8H_{24}Cl_2O_3Si_4$	53.3	95.8	127.8	152.7	197.8	222.0	−62
Octamethyltrisiloxane	$C_8H_{24}O_2Si_2$	7.4	43.1	70.0	91.1	129.4	150.2	—
Octamethylcyclotetrasiloxane	$C_8H_{24}O_4Si_4$	21.7	59.0	87.4	110.0	149.6	171.2	17.4
Coumarin	$C_9H_6O_2$	106.0	153.4	189.0	216.5	264.7	291.0	70
Quinoline	C_9H_7N	59.7	103.8	136.7	163.2	212.3	237.7	−15
Isoquinoline	C_9H_7N	63.5	107.8	141.6	167.6	214.5	240.5	24.6
Indene	C_9H_3	16.4	58.5	90.7	114.7	157.8	181.6	−2
Cinnamylaldehyde	C_9H_8O	76.1	120.0	152.2	177.7	222.4	246.0	−7.5
trans-Cinnamic acid	$C_9H_8O_2$	127.5s	173.0	207.1	232.4	276.7	300.0	133

ORGANIC COMPOUNDS

Pressures Less than One Atmosphere (Continued)

Name	Formula	Temperature °C 1 mm	10 mm	40 mm	100 mm	400 mm	760 mm	M.P.
Skatole	C_9H_9N	95.0	139.6	171.9	197.4	242.5	266.2	95
Ethyl 3-nitrobenzoate	$C_9H_9NO_4$	108.1	155.0	192.6	220.3	270.6	298.0	47
α-Methyl styrene	C_9H_{10}	7.4	47.1	77.8	102.2	143.0	165.4p	−23.2
β-Methyl styrene	C_9H_{10}	17.5	57.0	87.7	111.7	154.7	179.0	−30.1
4-Methyl styrene	C_9H_{10}	16.0	55.1	85.0	108.6	151.2	175.0p	—
Propenylbenzene	C_9H_{10}	17.5	57.0	87.7	111.7	154.7	179.0	−30.1
2,4-Xylaldehyde	$C_9H_{10}O$	59.9s	99.0	129.7	152.2	194.1	215.5	75
Cinnamyl alcohol	$C_9H_{10}O$	72.6	117.8	151.0	177.8	224.6	250.0	33
Propiophenone	$C_9H_{10}O$	50.0	92.2	124.3	149.3	194.2	218.0	21
2-Vinylanisole	$C_9H_{10}O$	41.9	81.0	110.0	132.3	172.1	194.0p	—
3-Vinylanisole	$C_9H_{10}O$	43.4	83.0	112.5	135.3	175.8	197.5p	—
4-Vinylanisole	$C_9H_{10}O$	45.2	85.7	116.0	139.7	182.0	204.5p	—
Benzyl acetate	$C_9H_{10}O_2$	45.0	87.6	119.6	144.0	189.0	213.5	−51.5
Ethyl Benzoate	$C_9H_{10}O_2$	44.0	86.0	118.2	143.2	188.4	213.4	−34.6
Hydrocinnamic acid	$C_9H_{10}O_2$	102.2	148.7	183.3	209.0	255.0	279.8	48.5
Ethyl salicylate	$C_9H_{10}O_3$	61.2	104.2	136.7	161.5	207.0	231.5	1.3
N-Methylacetanilide	$C_9H_{11}NO$	s	118.6	152.2	179.8	227.4	253.0	102
Ethyl carbanilate	$C_9H_{11}NO_2$	107.8	143.7	168.8	187.9	220.0	237.0	52.5
1,2,3-Trimethylbenzene	C_9H_{12}	16.8	55.9	85.4	108.8	152.0	176.1	−25.5
1,2,4-Trimethylbenzene	C_9H_{12}	13.6	50.7	79.8	102.8	145.4	169.2	−44.1
1,3,5-Trimethylbenzene	C_9H_{12}	9.6	47.4	76.1	98.9	141.0	164.7	−44.8
2-Ethyltoluene	C_9H_{12}	9.4	47.6	76.4	99.0	141.4	165.1	—
3-Ethyltoluene	C_9H_{12}	7.2	44.7	73.3	95.9	137.8	161.3	−95.5
4-Ethyltoluene	C_9H_{12}	7.6	44.9	73.6	96.3	136.4	162.0	—
Cumene	C_9H_{12}	2.9	38.3	66.1	88.1	129.2	152.4	−96.0
Propylbenzene	C_9H_{12}	6.3	43.4	71.6	94.0	135.7	159.2	−99.5
2-Ethylanisole	$C_9H_{12}O$	29.7	69.0	98.8	122.3	164.2	187.1	—
3-Ethylanisole	$C_9H_{12}O$	33.7	73.9	104.8	129.2	172.8	196.5	—
4-Ethylanisole	$C_9H_{12}O$	33.5	73.9	104.7	128.4	172.3	196.5	—
3-Phenyl-1-propanol	$C_9H_{12}O$	74.7	116.0	147.4	170.3	212.8	235.0	—
2-Isopropylphenol	$C_9H_{12}O$	56.6	97.0	127.5	150.3	192.6	214.5	15.5
3-Isopropylphenol	$C_9H_{12}O$	62.0	104.1	136.2	160.2	205.0	228.0	26
4-Isopropylphenol	$C_9H_{12}O$	67.0	108.0	139.8	163.3	206.1	228.2	61
Benzyl ethyl ether	$C_9H_{12}O$	26.0	65.0	95.4	118.9	161.5	185.0	—
Chloroethoxymethylphenylsilane	$C_9H_{13}ClOSi$	44.8	94.6	117.8	142.6	187.7	212.0	—
2,4,5-Trimethylaniline	$C_9H_{13}N$	68.4	109.0	139.8	162.0	203.7	234.5	67
N,N-Dimethyl-2-toluidine	$C_9H_{13}N$	28.8	66.2	95.0	118.1	161.5	184.8	−61
N,N-Dimethyl-4-toluidine	$C_9H_{13}N$	50.1	86.7	116.3	140.3	185.4	209.5	—
4-Cumidine	$C_9H_{13}N$	60.0	102.2	134.2	158.0	203.2	227.0	—
Phorone	$C_9H_{14}O$	42.0	81.5	111.3	134.0	175.3	197.2	28
Isophorone	$C_9H_{14}O$	38.0	81.2	114.5	140.6	188.7	215.2	—
cis-Diethyl citraconate	$C_9H_{14}O_4$	59.8	103.0	135.7	160.0	206.5	230.3	—
Diethyl itaconate	$C_9H_{14}O_4$	51.3	95.2	128.2	154.3	203.1	227.9	—
Diethyl mesaconate	$C_9H_{14}O_4$	62.8	105.3	137.3	161.6	205.8	229.0	—
Trimethyl citrate	$C_9H_{14}O_7$	106.2	160.4	194.2	219.6	264.2	287.0d	78.5
Isobutyl levulinate	$C_9H_{16}O_3$	65.0	105.9	136.2	160.2	205.5	229.9	—
Azelaic acid	$C_9H_{16}O_4$	178.3	225.5	260.0	286.5	332.8	356.5d	106.5
Diethyl ethylmalonate	$C_9H_{16}O_4$	50.8	91.6	122.4	146.0	188.7	211.5	—
Diethyl glutarate	$C_9H_{16}O_4$	65.6	109.7	142.8	167.8	212.8	237.0	—
2-Nonanone	$C_9H_{18}O$	32.1	72.3	103.4	127.4	171.2	195.0	−19
Azelaldehyde	$C_9H_{19}O$	33.3	71.6	100.2	123.0	163.4	185.0	—
Pelargonic acid	$C_9H_{18}O_2$	108.2	137.4	163.7	184.4	227.5	253.5	12.5
Methyl caprylate	$C_9H_{18}O_2$	34.2	74.9	105.3	128.0	170.0	193.0d	−40
Isobutyl isovalerate	$C_9H_{18}O_2$	16.0	53.8	82.7	105.2	146.4	168.7	—
Isoamyl butyrate	$C_9H_{18}O_2$	21.2	59.9	90.0	113.1	155.3	178.6	—
Isoamyl isobutyrate	$C_9H_{18}O_2$	14.8	52.8	81.8	104.4	146.0	168.8	—
Iodononane	$C_9H_{19}I$	70.0	109.0	138.1	159.8	199.3	219.5	—
Nonane	C_9H_{20}	+1.4	38.0	66.0	88.1	128.2	150.8	−53.7
1-Nonanol	$C_9H_{20}O$	59.5	99.7	129.0	151.3	192.1	213.5	−5
Dipropyleneglycol, isopropyl ether	$C_9H_{20}O_3$	46.0	86.2	117.0	140.3	183.1	205.6	—
Tripropyleneglycol	$C_9H_{20}O_4$	96.0	140.5	173.7	199.0	244.3	267.2	—
Hexyltrimethylsilane	$C_9H_{22}Si$	+6.7	44.8	74.0	97.2	139.9	163.0	—
Triethylpropylsilane	$C_9H_{22}Si$	15.2	54.0	83.7	107.4	149.8	173.0	—
1-Bromonaphthalene	$C_{10}H_7Br$	84.2	133.6	170.2	198.8	252.0	281.1	5.5
1-Chloronaphthalene	$C_{10}H_7Cl$	80.6	118.6	153.2	180.4	230.8	259.3	−20
Dicyclopentadiene	$C_{10}H_{12}$	s	47.6	77.9	101.7	144.2	166.6d	32.9
Naphthalene	$C_{10}H_8$	52.6s	85.8	119.3	145.5	193.2	217.9	80.2
Dichloro-1-naphthylsilane	$C_{10}H_8Cl_2Si$	106.2	149.2	181.7	205.9	249.7	273.3	—
1-Naphthol	$C_{10}H_8O$	94.0s	142.0	177.8	206.0	255.8	288.0	96
2-Naphthol	$C_{10}H_8O$	s	145.5	181.7	209.8	260.6	300.8	122.5
1-Naphthylamine	$C_{10}H_9N$	104.3	153.8	191.5	220.0	272.2	306.1	50
2-Naphthylamine	$C_{10}H_9N$	108.0	157.6	195.7	224.3	277.4	246.5	111.5
2-Methylquinoline	$C_{10}H_9N$	75.3	119.0	150.8	176.2	211.7	199.5p	−1
1,3-Divinylbenzene	$C_{10}H_{10}$	32.7	73.8	105.5	130.3	175.2	261.0	−66.9
4-Phenyl-3-buten-2-one	$C_{10}H_{10}O$	81.7	127.4	161.3	187.8	235.4	288.0	41.5
α-Methylcinnamic acid	$C_{10}H_{10}O_2$	125.7	169.8	201.8	224.8	266.8	263.0	—
Methyl cinnamate	$C_{10}H_{10}O_2$	77.4	123.0	157.9	185.8	235.0	263.0	33.4
Safrole	$C_{10}H_{10}O_2$	63.8	107.6	140.1	165.1	210.0	233.0	11.2
1,2-Phenylene diacetate	$C_{10}H_{10}O_4$	98.0	145.7	179.8	206.5	253.3	278.0	—
Dimethyl phthalate	$C_{10}H_{10}O_4$	100.3	147.6	182.8	210.0	257.8	283.7	—
2,4-Dimethylstyrene	$C_{10}H_{12}$	34.2	75.8	107.7	132.3	177.5	202.0p	—

ORGANIC COMPOUNDS
Pressures Less than One Atmosphere (Continued)

Name	Formula	Temperature °C 1 mm	10 mm	40 mm	100 mm	400 mm	760 mm	M.P.
2,5-Dimethylstyrene	$C_{10}H_{12}$	29.0	69.0	100.2	124.7	168.7	193.0p	—
3-Ethylstyrene	$C_{10}H_{12}$	28.3	68.3	99.2	123.2	167.2	191.5p	—
4-Ethylstyrene	$C_{10}H_{12}$	26.0	66.3	97.3	121.5	165.0	189.0p	—
Tetralin	$C_{10}H_{12}$	38.0	79.0	110.4	135.3	181.8	207.2	−31.0
Anethole	$C_{10}H_{12}O$	62.6	106.0	139.3	164.2	210.5	235.3	22.5
4-Methylpropiophenone	$C_{10}H_{12}O$	59.6	103.8	138.0	164	212.7	228.5	
Estragole	$C_{10}H_{12}O$	52.6	93.7	124.6	148.5	192.0	215.0	—
Cuminal	$C_{10}H_{12}O$	58.0	102.0	135.2	160.0	206.7	232.0	—
4-Vinylphenetole	$C_{10}H_{12}O$	64.0	105.6	136.3	159.8	202.8	225.0p	—
Eugenol	$C_{10}H_{12}O_2$	78.4	123.0	155.8	182.2	228.3	253.5	—
Isoeugenol	$C_{10}H_{12}O_2$	86.3	132.4	167.0	194.0	242.3	267.5	−10
Chavibetol	$C_{10}H_{12}O_2$	83.6	127.0	159.8	185.5	229.8	254.0	—
Propyl benzoate	$C_{10}H_{12}O_2$	54.6	98.0	131.8	157.4	205.2	231.0	−51.6
2-Phenoxyethyl acetate	$C_{10}H_{12}O_3$	82.6	128.0	162.3	189.2	235.0	259.7	−6.7
2-Chloroethyl α-methylbenzyl ether	$C_{10}H_{13}ClO$	62.3	106.0	139.6	164.8	210.8	235.0	—
4-*tert*-Butylphenyl dichlorophosphate	$C_{10}H_{13}Cl_2O_2P$	96.0	146.0	184.3	214.3	268.2	299.0	—
1,2,3,4-Tetramethylbenzene	$C_{10}H_{14}$	42.6	81.8	111.5	135.7	180.0	204.4	−6.2
1,2,3,5-Tetramethylbenzene	$C_{10}H_{14}$	40.6	77.8	105.8	128.3	173.7	197.9	−24.0
1,2,4,5-Tetramethylbenzene	$C_{10}H_{14}$	45.0s	74.6s	104.2	128.1	172.1	195.9	79.5
4-Ethyl-1,3-xylene	$C_{10}H_{14}$	26.3	66.4	97.2	121.2	164.4	188.4	—
5-Ethyl-1,3-xylene	$C_{10}H_{14}$	22.1	62.1	92.6	116.5	159.6	183.7	—
2-Ethyl-1,4-xylene	$C_{10}H_{14}$	25.7	65.6	96.0	120.0	163.1	186.9	—
1,2-Diethylbenzene	$C_{10}H_{14}$	22.3	62.0	92.5	116.2	159.0	183.5	−31.4
1,3-Diethylbenzene	$C_{10}H_{14}$	20.7	59.9	90.4	114.4	156.9	181.1	−83.9
1,4-Diethylbenzene	$C_{10}H_{14}$	20.7	60.3	91.1	115.3	159.0	183.8	−43.2
Cymene	$C_{10}H_{14}$	17.3	57.0	87.0	110.8	153.5	177.2	−68.2
Butylbenzene	$C_{10}H_{14}$	22.7	62.0	92.4	116.2	159.2	183.1	−88.0
Isobutylbenzene	$C_{10}H_{14}$	14.1	53.7	83.3	107.0	149.6	172.8	−51.5
sec-Butylbenzene	$C_{10}H_{14}$	18.6	57.0	86.2	109.5	150.3	173.5	−75.5
tert-Butylbenzene	$C_{10}H_{14}$	13.0	51.7	80.8	103.8	145.8	168.5	−58
Carvacrol	$C_{10}H_{14}O$	70.0	113.2	145.2	109.7	213.8	237.0	+0.5
Carbone	$C_{10}H_{14}O$	57.4	100.4	133.0	157.3	203.5	227.5	—
Cuminyl alcohol	$C_{10}H_{14}O$	74.2	118.0	150.3	176.2	221.7	246.6	—
4-Ethylphenetole	$C_{10}H_{14}O$	48.5	89.5	119.8	143.5	185.7	208.0	—
Thymol	$C_{10}H_{14}O$	64.3	107.4	139.8	164.1	209.2	231.8	51.5
4-Isobutylphenol	$C_{10}H_{14}O$	72.1	113.3	147.2	171.2	214.7	237.0	—
4-*sec*-Butylphenol	$C_{10}H_{14}O$	71.4	114.8	147.8	172.4	217.6	242.1	—
2-*sec*-Butylphenol	$C_{10}H_{14}O$	57.4	100.8	133.4	157.3	203.8	228.0	—
2-*tert*-Butylphenol	$C_{10}H_{14}O$	56.6	98.1	129.2	153.3	196.3	219.5	—
4-*tert*-Butylphenol	$C_{10}H_{14}O$	70.0	114.0	146.0	170.2	214.0	238.0	99
Nicotine	$C_{10}H_{14}N_2$	61.8	107.2	142.1	169.5	219.8	247.3	—
N-Diethylaniline	$C_{10}N_{15}N$	49.7	91.9	123.6	147.3	192.4	215.5	−34.4
N-Phenyliminodiethano	$C_{10}H_{15}NO_2$	145.0	195.8	233.0	260.6	311.3	337.8	—
Camphene	$C_{10}H_{16}$	s	47.2s	75.7	97.9	138.7	160.5	50
Dipentene	$C_{10}H_{16}$	14.0	53.8	84.3	108.3	150.5	174.6	—
d-Limonene	$C_{10}H_{16}$	14.0	53.8	84.3	108.3	151.4	175.0	−96.9
Myrcene	$C_{10}H_{16}$	14.5	53.2	82.6	106.0	148.3	171.5	—
α-Phellandrene	$C_{10}H_{16}$	20.0	58.0	87.8	110.6	152.0	175.0	—
α-Pinene	$C_{10}H_{16}$	−1.0	+37.3	66.8	90.1	132.3	155.0	−55
β-Pinene	$C_{10}H_{16}$	+4.2	42.3	71.5	94.0	136.1	158.3	—
Terpenoline	$C_{10}H_{16}$	32.3	70.6	100.0	122.7	163.5	185.0	—
Diethyl arsanilate	$C_{10}H_{16}AsNO_2$	38.0	74.8	102.6	123.8	161.0	181.0	—
d-Camphor	$C_{10}H_{16}O$	41.5s	82.3s	114.0s	138.0s	182.0	209.2	178.5
l-Dihydrocarvone	$C_{10}H_{16}O$	46.6	90.0	123.7	149.7	197.0	223.0	—
α-Citral	$C_{10}H_{16}O$	61.7	103.9	135.9	160.0	205.0	228.0d	—
d-Fenchone	$C_{10}H_{16}O$	28.0	68.3	99.5	123.6	166.8	191.0	5
Pulegone	$C_{10}H_{16}O$	58.3	94.0	121.7	143.1	189.8	221.0	—
α-Thyjone	$C_{10}H_{16}O$	38.3	79.3	110.0	134.0	177.8	201	—
Ethoxydimethylphenylsilane	$C_{10}H_{16}OSi$	36.3	76.2	107.2	131.4	175.0	199.5	—
Campholenic acid	$C_{10}H_{16}O_2$	97.6	139.8	170.0	193.7	234.0	256.0	—
Diosphenol	$C_{10}H_{16}O_2$	66.7	109.0	141.2	165.6	209.5	232.0	—
Fencholic acid	$C_{10}H_{16}O_2$	101.7	142.3	171.8	194.0	237.8	264.1	19
cis-Decalin	$C_{10}H_{18}$	22.5	64.2	97.2	123.2	169.9	194.6	−43.3
trans-Decalin	$C_{10}H_{18}$	−0.8	+47.2	85.7	114.6	160.1	186.7	−30.7
d-Citronellal	$C_{10}H_{18}O$	44.0	84.8	116.1	140.1	183.8	206.5	—
Cineol	$C_{10}H_{18}O$	15.0	54.1	84.2	108.2	151.6	176.0	−1
Dihydrocarveol	$C_{10}H_{18}O$	63.9	105.0	136.1	159.8	202.8	225.0	—
dl-Fenchyl alcohol	$C_{10}H_{18}O$	45.8	82.1	110.8	132.3	173.2	202.0	35
Geraniol	$C_{10}H_{18}O$	69.2	110.0	141.8	165.3	207.8	230.0	—
d-Linalool	$C_{10}H_{18}O$	40.0	79.8	109.9	133.8	175.6	198.0	—
Nerol	$C_{10}H_{18}O$	61.7	104.0	136.1	159.8	203.5	226.0	—
α-Terpineol	$C_{10}H_{18}O$	52.8	94.3	126.0	150.1	194.3	217.5	35
Citronellic acid	$C_{10}H_{18}O_2$	99.5	141.4	171.9	195.4	236.6	257.0	—
Amyl levulinate	$C_{10}H_{18}O_3$	81.3	124.0	155.8	180.5	227.4	253.2	—
Isoamyl levulinate	$C_{10}H_{18}O_3$	75.6	118.8	151.7	177.0	222.7	247.9	—
Diethyl ethylmethylmalonate	$C_{10}H_{18}O_4$	44.7	85.7	116.7	140.8	184.1	207.5	—
Diethyl adipate	$C_{10}H_{18}O_4$	74.0	123.0	154.6	179.0	219.1	240.0	−21
Diisobutyl oxalate	$C_{10}H_{18}O_4$	63.2	105.3	137.5	161.8	205.8	229.5	—
Dipropyl succinate	$C_{10}H_{18}O_4$	77.5	122.2	154.8	180.3	226.5	250.8	—
Sebacic acid	$C_{10}H_{18}O_4$	183.0	232.0	268.2	294.5	332.8	352.3d	134.5
Dipropyl-*d*-tartrate	$C_{10}H_{18}O_6$	115.6	163.5	199.7	227.0	275.6	303.0	—

ORGANIC COMPOUNDS
Pressures Less than One Atmosphere (Continued)

Name	Formula	Temperature °C 1 mm	10 mm	40 mm	100 mm	400 mm	760 mm	M.P.
Diisopropyl-*d*-Tartrate	$C_{10}H_{18}O_6$	103.7	148.2	181.8	207.3	251.8	275.0	—
Camphylamine	$C_{10}H_{19}N$	45.3	83.7	112.5	134.6	173.8	195.0	—
Menthane	$C_{10}H_{20}$	+9.7	48.3	78.3	102.1	146.0	169.0	—
1-Decene	$C_{10}H_{20}$	14.7	53.7	83.3	106.5	149.2	172.0	—
1,2-Dibromodecane	$C_{10}H_{20}Br_2$	95.7	137.3	167.4	190.2	229.8	250.4	—
Citronellol	$C_{10}H_{20}O$	66.4	107.0	137.2	159.8	201.0	221.5	—
Capraldehyde	$C_{10}H_{20}O$	51.9	92.0	122.2	145.3	186.3	208.5	—
l-Menthol	$C_{10}H_{20}O$	56.0	96.0	126.1	149.4	190.2	212.0	42.5
Decan-2-one	$C_{10}H_{20}O$	44.2	85.8	117.1	142.4	186.7	211.0	+3.5
Capric acid	$C_{10}H_{20}O_2$	125.0	152.2	179.9	200.0	240.3	208.4	31.5
Isoamyl isovalerate	$C_{10}H_{20}O_2$	27.0	68.6	100.6	125.1	169.5	194.0	—
Decane	$C_{10}H_{22}$	16.5	55.7	85.5	108.6	150.6	174.1	−29.7
2,7-Dimethyloctane	$C_{10}H_{22}$	+6.3	42.3	71.2	93.9	136.0	159.7	−52.8
Decyl alcohol	$C_{10}H_{22}O$	69.5	111.3	142.1	105.8	208.8	231.0	+7
Diisoamyl ether	$C_{10}H_{22}O$	18.6	57.0	86.3	109.6	150.3	173.4	—
2-Butyl-2-ethylbutane-1,3-diol	$C_{10}H_{22}O_2$	94.1	136.8	167.8	191.9	233.5	255.0	—
Dihydrocitronellol	$C_{10}H_{22}O$	68.0	103.0	127.6	145.9	176.8	193.5	—
Dipropylene glycol monobutyl ether	$C_{10}H_{22}O_3$	64.7	106.0	136.3	159.8	203.8	227.0	—
Diisoamyl sulfide	$C_{10}H_{22}S$	43.0	87.6	120.0	145.3	191.0	216.0	—
Heptyltrimethylsilane	$C_{10}H_{24}Si$	22.3	62.1	92.4	116.5	159.8	184.0	—
Butyltriethylsilane	$C_{10}H_{24}Si$	27.1	67.5	98.3	123.2	167.5	192.0	—
1,5-Diethoxyhexamethyltrisiloxane	$C_{10}H_{28}O_4Si_3$	41.8	80.7	110.0	133.2	174.0	196.6	—
Decamethyltetrasiloxane	$C_{10}H_{30}O_3Si_4$	35.3	74.3	104.0	127.3	169.8	193.5	—
Decamethylcyclopentasiloxane	$C_{10}H_{30}O_5Si_5$	45.2	86.2	117.7	142.0	186.0	210.0	−38.0
1-Naphthoic acid	$C_{11}H_8O_2$	156.0	196.8	225.0	245.8	281.4	300.0	160.5
2-Naphthoic acid	$C_{11}H_8O_2$	160.8	202.8	231.5	252.7	289.5	308.5	184
Ethyl-*trans*-cinnamate	$C_{11}H_{12}O_2$	87.6	123.0	169.1	196.0	245.0	271.0	12
1-Phenyl-1,3-pentanedione	$C_{11}H_{12}O_2$	98.0	144.0	178.0	204.5	251.2	276.5	—
Ethyl benzoylacetate	$C_{11}H_{12}O_3$	107.6	150.3	181.8	205.0	244.7	265.0d	—
Myristicine	$C_{11}H_{12}O_3$	95.2	142.0	177.7	205.0	253.5	280.0	—
2,4,5-Trimethylstyrene	$C_{11}H_{14}$	48.1	91.6	124.2	149.8	196.1	221.2p	—
2,4,6-Trimethylstyrene	$C_{11}H_{14}$	37.5	79.7	111.8	136.8	182.3	207.0p	—
4-Isopropylstyrene	$C_{11}H_{14}$	34.7	76.0	108.0	132.8	178.0	202.5p	—
Isovalerophenone	$C_{11}H_{14}O$	58.3	101.4	133.8	158.0	204.2	228.0	—
Pivalophenone	$C_{11}H_{14}O$	57.8	99.0	13.4	154.0	197.7	220.0	—
2,3,5-Trimethylacetophenone	$C_{11}H_{14}O$	79.0	122.3	154.2	179.7	224.3	247.5	—
Isobutyl benzoate	$C_{11}H_{14}O_2$	64.0	108.6	141.8	166.4	212.8	237.0	—
4-Allylveratrole	$C_{11}H_{14}O_2$	85.0	127.0	158.3	183.7	226.2	248.0	—
3,5-Diethyltoluene	$C_{11}H_{16}$	34.0	75.3	107.0	131.7	176.5	200.7	—
1,2,4-Trimethyl-5-ethylbenzene	$C_{11}H_{16}$	43.7	84.6	106.0	140.3	184.5	208.1	—
1,3,5-Trimethyl-2-ethylbenzene	$C_{11}H_{16}$	38.8	80.5	113.2	137.9	183.5	208.0	—
3-Ethylcumene	$C_{11}H_{16}$	28.3	68.8	99.9	124.3	168.2	193.0	—
4-Ethylcumene	$C_{11}H_{16}$	31.5	72.0	103.3	127.2	171.8	195.8	—
sec-Amylbenzene	$C_{11}H_{16}$	29.0	69.2	100.0	124.1	168.0	193.0	—
4-*tert*-Butyl-2-cresol	$C_{11}H_{16}O$	74.3	118.0	150.8	176.2	221.8	247.0	—
2-*tert*-Butyl-4-cresol	$C_{11}H_{16}O$	70.0	112.0	143.9	167.0	210.0	232.6	—
4-*tert*-Amylphenol	$C_{11}H_{16}O$	s	125.5	160.3	189.0	239.5	266.0	93
Ethylcamphoronic anhydride	$C_{11}H_{16}O_5$	118.2	165.0	199.8	226.6	272.8	298.0	—
Bornyl formate	$C_{11}H_{18}O_2$	47.0	89.3	121.2	145.8	190.2	214.0	—
Geranyl formate	$C_{11}H_{18}O_2$	61.8	104.3	136.2	160.7	205.8	230.0	—
Neryl formate	$C_{11}H_{18}O_2$	57.3	99.7	131.5	155.6	200.0	224.5	—
Diethoxymethylphenylsilane	$C_{11}H_{18}O_2Si$	56.5	97.2	127.5	151.2	193.8	216.5	—
Diethyl-γ-oxoazelate	$C_{11}H_{18}O_5$	121.0	165.7	197.7	221.6	264.5	286.0	—
10-Hendecenoic acid	$C_{11}H_{20}O_2$	114.0	156.3	188.7	213.5	254.0	275.0	24.5
Menthyl formate	$C_{11}H_{20}O_2$	47.3	90.0	123.0	148.0	194.2	219.0	—
2-Ethylhexyl acrylate	$C_{11}H_{20}O_2$	50.0	91.8	123.7	147.9	192.2	216.0	—
Octyl acrylate	$C_{11}H_{20}O_2$	58.5	102.0	135.6	159.1	204.0	227.0	—
Hexyl levulinate	$C_{11}H_{20}O_2$	90.0	134.7	107.8	193.6	241.0	266.8	—
Hendecan-2-one	$C_{11}H_{22}O$	68.2	108.9	139.0	161.0	202.3	224.0	15
Methyl caprate	$C_{11}H_{22}O_2$	63.7	108.0	139.0	161.5	202.9	224.0d	−18
Hendecanoic acid	$C_{11}H_{22}O_2$	101.4	149.0	185.6	212.5	262.8	290.0	29.5
Hendecane	$C_{11}H_{24}$	32.7	73.9	104.4	128.1	171.9	195.8	−25.6
Nendecan-2-ol	$C_{11}H_{24}O$	71.1	112.8	143.7	167.2	209.8	232.0	—
Trimethyloctylsilane	$C_{11}H_{26}Si$	41.8	82.3	113.0	136.5	179.5	202.0	—
Amyltriethylsilane	$C_{11}H_{26}Si$	41.8	83.8	110.0	141.2	186.3	211.0	—
1-Bromobiphenyl	$C_{12}H_9Br$	98.0	150.6	190.8	221.8	277.7	310.0	90.5
2-Bromo-4-Phenylphenol	$C_{12}H_9BrO$	100.0	152.3	193.8	224.5	280.2	311.0	95
2-Chlorobiphenyl	$C_{12}H_9Cl$	89.3	134.7	109.9	197.0	243.8	267.5	34
4-Chlorobiphenyl	$C_{12}H_9Cl$	96.4	146.0	183.8	212.5	204.5	292.5	75.5
2-Chloro-3-phenylphenol	$C_{12}H_9ClO$	118.0	169.7	207.4	237.0	289.4	317.5	+6
2-Chloro-6-phenylphenol	$C_{12}H_9ClO$	119.8	170.7	208.2	237.1	289.5	317.0	—
2-Xenyl dichlorophosphate	$C_{12}H_9Cl_2PO$	138.2	187.0	223.8	251.5	301.5	328.5	—
Carbazole	$C_{12}H_9N$	s	s	s	265.0	323.0	354.8	244.8
Acenaphthene	$C_{12}H_{10}$	s	131.2	168.2	197.5	250.0	277.5	95
Biphenyl	$C_{12}H_{10}$	70.6	117.0	152.5	180.7	229.4	254.9	69.5
Diphenyl chlorophosphate	$C_{12}H_{10}ClPO_3$	121.5	182.0	227.9	265.0	337.2	278.0	—
Dichlorodiphenylsilane	$C_{12}H_{10}Cl_2Si$	109.6	158.0	195.5	223.8	275.5	304.0	—
Difluourodiphenylsilane	$C_{12}H_{10}F_2Si$	68.4	115.5	149.8	176.3	225.4	252.5	—
Azobenzene	$C_{12}H_{10}N_2$	103.5	151.5	187.9	216.0	266.1	293.0	68
1-Acetonaphthone	$C_{12}H_{10}O$	115.6	161.5	196.8	223.8	270.5	295.5	—
Diphenyl ether	$C_{12}H_{10}O$	120.2	168.5	203.8	229.8	275.8	301.0	55.5

ORGANIC COMPOUNDS
Pressures Less than One Atmosphere (Continued)

Name	Formula	Temperature °C						M.P.
		1 mm	10 mm	40 mm	100 mm	400 mm	760 mm	
2-Phenylphenol	$C_{12}H_{10}O$	66.1	114.0	150.0	178.8	230.7	258.5	27
4-Phenylphenol	$C_{12}H_{10}O$	100.0	146.2	180.3	205.9	251.8	275.0	56.5
Diphenyl sulfide	$C_{12}H_{10}O$	s	176.2	213.0	240.9	285.5	308.0	164.5
Diphenyl disulfide	$C_{12}H_{10}S$	96.1	145.0	182.8	211.8	263.9	292.5	—
Diphenyl selenide	$C_{12}H_{10}S$	131.6	180.0	214.8	241.3	285.8	310.0	61
Diphenylamine	$C_{12}H_{10}Se$	105.7	154.4	192.2	220.8	273.2	301.5	+2.5
1-Ethylnaphthalene	$C_{12}H_{11}N$	108.3	157.0	194.3	222.8	274.1	302.0	52.9
1,1-Diphenylhydrazine	$C_{12}H_{12}$	70.0	116.8	152.0	180.0	230.8	258.1d	−27
1,1-Diphenylhydrazine	$C_{12}H_{12}N_2$	126.0	176.1	213.5	242.5	294.0	322.2	44
2-Cyclohexyl-4,6-dinitrophenol	$C_{12}H_{14}N_2O_5$	132.8	175.9	206.7	229.0	269.8	291.5	—
Eugenyl acetate	$C_{12}H_{14}O_3$	101.6	148.0	183.0	209.7	257.4	282.0	29.5
Apiole	$C_{12}H_{14}O_4$	116.0	160.2	193.7	218.0	262.1	285.0	30
Diethyl phthalate	$C_{12}H_{14}O_4$	108.8	156.0	192.1	219.5	267.5	294.0	—
2,5-Diethylstyrene	$C_{12}H_{16}$	49.7	92.6	125.8	151.0	198.0	223.0p	—
Phenylcyclohexane	$C_{12}H_{16}$	67.5	111.3	144.0	169.3	214.6	240.0	+7.5
Isoamyl benzoate	$C_{12}H_{16}O_2$	72.0	121.6	158.3	186.8	235.8	262.0	—
1,2,4-Triethylbenzene	$C_{12}H_{18}$	46.0	88.5	121.7	146.8	193.7	218.0	—
1,4-Diisopropylbenzene	$C_{12}H_{18}$	40.0	81.8	114.0	138.7	184.3	209.0	—
1,3-Diisopropylbenzene	$C_{12}H_{18}$	34.7	76.0	107.9	132.3	177.6	202.0	−105
2-*tert*-Butyl-4-ethylphenol	$C_{12}H_{18}O$	76.3	121.0	154.0	179.0	223.8	247.8	—
4-*tert*-Butyl-2,5-xylenol	$C_{12}H_{18}O$	88.2	135.0	169.8	195.0	241.8	265.3	—
4-*tert*-Butyl-2,6-xylenol	$C_{12}H_{18}O$	74.0	119.0	152.2	176.0	217.8	239.8	—
6-*tert*-Butyl-2,4-xylenol	$C_{12}H_{18}O$	70.3	115.0	148.5	172.0	214.2	236.5	—
6-*tert*-Butyl-3,4-xylenol	$C_{12}H_{13}O$	83.9	127.0	159.7	184.0	226.7	249.5	—
d-Bornyl acetate	$C_{12}H_{20}O_2$	46.9	90.2	123.7	149.8	197.5	223.0	29.
Geranyl acetate	$C_{12}H_{20}O_2$	73.5	117.9	150.0	175.2	219.8	243.3d	—
Linalyl acetate	$C_{12}H_{20}O_2$	55.4	96.0	127.7	151.8	196.2	220.0d	—
Triethoxyphenylsilane	$C_{12}H_{20}O_3Si$	71.0	112.6	143.5	167.5	210.5	233.5	—
Triethyl citrate	$C_{12}H_{20}O_7$	107.0	144.0	190.4	217.8	267.5	294.0d	—
Trimethallyl phosphate	$C_{12}H_{21}PO_4$	93.7	149.8	192.0	225.7	288.5	324.0	—
Citronellyl acetate	$C_{12}H_{22}O_2$	74.7	113.0	140.5	161.0	197.8	217.0	—
Menthyl acetate	$C_{12}H_{22}O_2$	57.4	100.0	132.1	156.7	202.8	227.0	—
Dimethyl sebacate	$C_{12}H_{22}O_4$	104.0	156.2	196.0	222.6	269.6	293.5	38
Diisoamyl oxalate	$C_{12}H_{22}O_4$	85.4	131.4	165.7	192.2	240.0	165.0	—
Diisobutyl-*d*-tartrate	$C_{12}H_{22}O_6$	117.8	169.0	208.5	239.5	294.0	324.0	73.5
1-Dodecene	$C_{12}H_{24}$	47.2	87.8	118.6	142.3	183.3	208.0	−31.5
Triisobutylene	$C_{12}H_{24}$	18.0	56.5	86.7	110.0	153.0	179.0	—
Dodecan-2-one	$C_{12}H_{24}O$	77.1	120.4	152.4	177.5	222.5	246.5	—
Lauraldehyde	$C_{12}H_{24}O$	77.7	123.7	157.8	184.5	231.8	257.0	44.5
Lauric acid	$C_{12}H_{24}O_2$	121.0	166.0	201.4	227.5	273.8	299.2	48.
Dodecane	$C_{12}H_{26}$	47.8	90.0	121.7	146.2	191.0	216.2	−9.6
Dodecyl alcohol	$C_{12}H_{26}O$	91.0	134.7	167.2	192.0	235.7	259.0	24
Tripropylene glycol monoisopropyl ether	$C_{12}H_{26}O_4$	82.4	127.3	161.4	187.8	232.8	256.6	—
Triisobutylamine	$C_{12}H_{27}N$	32.3	69.8	97.8	119.7	157.8	179.0	−22
Dodecylamine	$C_{12}H_{27}N$	82.8	127.8	157.4	182.1	225.0	248.0	—
Triethylhexylsilane	$C_{12}H_{28}Si$	52.4	96.4	130.0	156.0	204.6	230.0	—
1,7-Diethoxyoctamethyltetrasiloxane	$C_{12}H_{34}O_5Si_4$	67.7	108.6	319.0	162.0	204.0	227.5	—
Dodecamethylpentasiloxane	$C_{12}H_{36}O_4Si_5$	56.6	98.0	128.8	162.8	196.5	220.5	—
Dodecamethylcyclohexasiloxane	$C_{12}H_{36}O_6Si_6$	67.3	110.0	141.8	166.3	210.6	236.0	−3.0
Acridine	$C_{13}H_9N$	129.4	184.0	224.2	256.0	314.3	346.0	110.5
Fluorene	$C_{13}H_{10}$	s	146.0	185.2	214.7	268.6	295.0	113
Benzophenone	$C_{13}H_{10}O$	108.2	157.6	195.7	224.4	276.8	305.4	48.5
Phenyl benzoate	$C_{13}H_{10}O_2$	106.8	157.8	197.6	227.8	283.5	314.0	70.5
Salol	$C_{13}H_{10}O_3$	117.8	167.0	205.0	233.8	284.8	313.0	42.5
Diphenylmethane	$C_{13}H_{12}$	76.0	122.8	157.8	186.3	237.5	264.0	26.5
Benzhydrol	$C_{13}H_{12}O$	110.0	162.0	200.0	227.5	275.6	301.0	68.5
Benzyl phenyl ether	$C_{13}H_{12}O$	95.4	144.0	180.1	209.2	259.8	287.0	—
1-Propionaphthone	$C_{13}H_{12}O$	124.0	171.0	206.9	233.5	280.2	306.0	—
Chloromethyldiphenylsilane	$C_{13}H_{13}ClSi$	105.0	152.7	189.2	216.0	266.5	295.0	—
Methyldiphenylamine	$C_{13}H_{13}N$	103.5	149.7	184.0	210.1	257.0	282.0	−7.6
2-Isopropylnap hthalene	$C_{13}H_{14}$	76.0	123.4	159.0	187.6	238.5	266.0	—
Methyldiphenylsilane	$C_{13}H_{14}Si$	88.0	132.8	166.4	193.7	241.5	266.8	—
Enanthophenone	$C_{13}H_{18}O$	100.0	145.5	178.9	204.2	248.3	271.3	—
Heptylbenzene	$C_{13}H_{20}$	64.0	110.0	144.0	170.2	217.8	244.0	—
α-Ionone	$C_{13}H_{20}O$	79.5	123.0	155.6	181.2	225.2	250.0	—
Bornyl propionate	$C_{13}H_{22}O_2$	64.6	108.0	140.4	165.7	211.2	235.0	—
2-Tridecanone	$C_{13}H_{26}O$	86.8	131.8	165.7	191.5	238.3	262.5	28.5
Methyl laurate	$C_{13}H_{26}O_2$	87.8	133.2	166.0	190.8	d	d	5
Tridecanoic acid	$C_{13}H_{26}O_2$	137.8	181.0	212.4	236.0	276.5	299.0	41
Tridecane	$C_{13}H_{28}$	59.4	104.0	137.7	162.5	209.4	234.0	−6.2
Tripropyleneglycol, monobutyl ether	$C_{13}H_{28}O_4$	101.5	147.0	179.8	204.4	247.0	269.5	—
Decyltrimethylsilane	$C_{13}H_{30}Si$	67.4	111.0	144.0	169.5	215.5	240.0	—
Triethylheptylsilane	$C_{13}H_{30}Si$	70.0	114.6	148.0	174.0	221.0	247.0	—
Anthraquinone	$C_{14}H_8O_2$	190.0s	234.2s	264.3s	285.0s	346.2	379.9	286
1,4-Dihydroxyanthraquinone	$C_{14}H_8O_4$	196.7	259.8	307.4	344.5	413.0d	450.0d	194
Anthracene	$C_{14}H_{10}$	145.0s	187.2s	217.5s	250.0	310.2	342.0	217.5
Phenanthrene	$C_{14}H_{10}$	118.2	173.0	215.8	249.0	308.0	340.2	99.5
Benzil	$C_{14}H_{10}O_2$	128.4	183.0	224.5	255.8	314.3	347.0	95
Benzoic anhydride	$C_{14}H_{10}O_3$	143.8	198.0	239.8	270.4	328.8	360.0	42
1,1-Diphenylethylene	$C_{14}H_{12}$	87.4	135.0	170.8	198.6	249.8	277.0	—
trans-Diphenylethylene	$C_{14}H_{12}$	113.2	161.0	199.0	227.4	287.3	306.5	124
Desoxybenzoin	$C_{14}H_{12}O$	123.3	173.5	212.0	241.3	293.0	321.0	60

ORGANIC COMPOUNDS
Pressures Less than One Atmosphere (Continued)

Name	Formula	Temperature °C 1 mm	10 mm	40 mm	100 mm	400 mm	760 mm	M.P.
Benzoin	$C_{14}H_{12}O$	135.6	188.0	227.6	258.0	313.5	343.0	132
Dibenzyl	$C_{14}H_{14}$	86.8	136.0	173.7	202.8	255.0	284.0	51.5
2-Isobutyronaphthone	$C_{14}H_{14}O$	133.2	181.0	215.6	242.3	288.2	313.0	—
Dibenzylamine	$C_{14}H_{15}O$	118.3	165.6	200.2	227.3	274.3	300.0	−26
Ethyldiphenylamine	$C_{14}H_{15}N$	98.3	146.0	182.0	209.8	258.8	286.0	—
1,2-Dichlorotetraethylbenzene	$C_{14}H_{20}Cl_2$	105.6	155.0	192.2	220.7	272.8	302.0	—
1,4-Dichlorotetraethylbenzene	$C_{14}H_{20}Cl_2$	91.7	143.8	183.2	212.0	265.8	296.5	—
2-(4-*tert*-Butylphenoxy) ethyl acetate	$C_{14}H_{20}O_3$	118.0	165.8	201.5	228.0	277.6	304.4	—
1,2,4,5-Tetraethylbenzene	$C_{14}H_{22}$	65.7	111.6	145.8	172.4	221.4	248.0	11.6
2,4-Di-*tert*-butylphenol	$C_{14}H_{22}O$	84.5	130.0	164.3	190.0	237.0	260.8	—
Bornyl butyrate	$C_{14}H_{24}O_2$	74.0	118.0	150.7	176.4	222.2	247.0	—
Bornyl isobutyrate	$C_{14}H_{24}O_2$	70.0	114.0	147.2	172.2	218.2	243.0	—
Geranyl butyrate	$C_{14}H_{24}O_2$	96.8	139.0	170.1	193.8	235.0	257.4	—
Geranyl isobutyrate	$C_{14}H_{24}O_2$	90.7	133.0	164.0	187.7	228.5	251.0	—
Diethyl sebacate	$C_{14}H_{26}O_4$	125.3	172.1	207.5	234.4	280.3	305.5	1.3
2-Tetradecanone	$C_{14}H_{28}O$	99.3	145.5	179.8	206.0	253.0	278.0	—
Myristaldehyde	$C_{14}H_{28}O$	99.0	148.3	186.0	214.5	267.9	297.8	23.5
Myristic acid	$C_{14}H_{28}O_2$	142.0	190.8	223.5	250.5	294.6	318.0	57.5
1-Chlorotetradecane	$C_{14}H_{29}Cl$	98.5	148.2	187.0	215.5	267.5	296.0	+0.9
Tetradecane	$C_{14}H_{30}$	76.4	120.7	152.7	178.5	226.8	252.5	5.5
Tetradecylamine	$C_{14}H_{31}N$	102.6	152.0	189.0	215.7	264.6	291.2	—
Triethyloctylsilane	$C_{14}H_{32}Si$	73.7	120.6	155.7	184.3	235.0	262.0	—
1,9-Diethoxydecamethylpentasiloxane	$C_{14}H_{40}O_6Si_5$	89.0	131.5	162.2	187.0	230.0	253.3	—
Tetradecamethylhexasiloxane	$C_{14}H_{42}O_5Si_6$	73.7	117.6	149.8	175.2	220.5	245.5	—
Tetradecamethylcycloheptasiloxane	$C_{14}H_{42}O_7Si_7$	86.3	131.5	165.3	191.8	239.2	264.0	−32
1,3-Diphenyl-2-propanone	$C_{15}H_{14}O$	125.5	177.6	216.6	246.6	301.7	330.5	34.5
1-Biphenyloxy-2,3-epoxypropane	$C_{15}H_{14}O_2$	135.3	187.2	226.3	255.0	309.8	340.0	—
1-Isovaleronaphthone	$C_{15}H_{16}O$	136.0	184.0	219.7	246.7	294.0	320.0	—
4,4-Isopropylidenebisphenol	$C_{15}H_{16}O_2$	193.0	240.8	273.0	297.0	339.0	360.5	—
Ethoxymethyldiphenylsilane	$C_{15}H_{18}OSi$	109.0	152.7	186.0	211.8	256.8	282.0	—
Helenin	$C_{15}H_{20}O_2$	157.7	192.1	215.2	232.6	260.6	275.0	76
Cadinene	$C_{15}H_{24}$	101.3	146.0	179.8	205.6	250.7	275.0	—
2,6-Di-*tert*-butyl-4-cresol	$C_{15}H_{24}O$	85.8	131.0	164.1	190.0	237.6	262.5	—
4,6-Di-*tert*-butyl-2-cresol	$C_{15}H_{24}O$	86.2	132.4	167.4	194.0	243.4	269.3	—
4,6-Di-*tert*-butyl-3-cresol	$C_{15}H_{24}O$	103.7	150.0	185.3	211.0	257.1	282.0	—
Champacol	$C_{15}H_{26}O$	100.0	148.0	184.0	211.9	261.2	288.0	91
Triethyl camphoronate	$C_{15}H_{26}O_6$	s	166.0	201.8	228.6	276.0	301.0	135
Methyl myristate	$C_{15}H_{30}O_2$	115.0	160.8	195.8	222.6	269.8	295.8	18.5
Pentadecane	$C_{15}H_{32}$	91.6	135.4	167.7	194.0	242.8	270.5	10
Tetrapropylene glycol monoisopropyl ether	$C_{15}H_{32}O_5$	116.6	163.0	197.7	223.3	268.3	292.7	—
Dodecyltrimethylsilane	$C_{15}H_{34}Si$	91.2	137.7	172.1	199.5	248.0	273.0	—
Benzyl cinnamate	$C_{16}H_{14}O_2$	173.8	221.5	255.8	281.5	326.7	350.0	39
Di(α-methylbenzyl) ether	$C_{16}H_{18}O$	96.7	144.0	179.6	206.8	254.8	281.0	—
Diethoxydiphenylsilane	$C_{16}H_{20}O_2Si$	111.5	157.6	193.2	220.0	259.7	296.0	—
Dibutyl phthalate	$C_{16}H_{22}O_4$	148.2	198.2	235.8	263.7	313.5	340.0	—
Pentaethylchlorobenzene	$C_{16}H_{25}Cl$	90.0	140.7	178.2	208.0	257.2	285.0	—
Pentaethylbenzene	$C_{16}H_{26}$	86.0	135.8	171.9	200.0	250.2	277.0	—
2,6-Di-*tert*-butyl-4-ethylphenol	$C_{16}H_{26}O$	89.1	137.0	172.1	198.0	244.0	268.6	—
4,6-Di-*tert*-butyl-3-ethylphenol	$C_{16}H_{26}O$	111.5	157.4	192.3	218.0	264.6	290.0	—
Muscone	$C_{16}H_{30}C$	118.0	170.0	210.0	241.5	297.2	328.0	—
Palmitonitrile	$C_{16}H_{31}N$	134.3	185.8	223.8	251.5	304.5	332.0	31
1-Hexadecene	$C_{16}H_{32}$	101.6	146.2	178.8	205.3	250.0	274.0	4
Tetraisobutylene	$C_{16}H_{32}$	63.8	108.5	142.2	167.5	214.6	240.0	—
2-Hexadecanone	$C_{16}H_{32}O$	109.8	167.3	203.7	230.5	279.8	307.0	—
Palmitaldehyde	$C_{16}H_{32}O$	121.6	171.8	210.0	239.5	292.3	321.0	34
Palmitic acid	$C_{16}H_{32}O_2$	153.6	205.8	244.4	271.5	326.0	353.8	64.0
Hexadecane	$C_{16}H_{34}$	105.3	149.8	181.3	208.5	258.3	287.5	18.5
Cetyl alcohol	$C_{16}H_{34}O$	122.7	177.8	219.8	251.7	312.7	344.0	49.3
Cetylamine	$C_{16}H_{36}N$	123.6	176.0	215.7	245.8	300.4	330.0	—
Decyltriethylsilane	$C_{16}H_{36}Si$	108.5	155.6	191.7	218.3	267.5	293.0	—
1,1,1-Diethoxydodecamethylhexasiloxane	$C_{16}H_{46}O_7Si_6$	103.6	147.5	180.0	205.5	250.0	273.5	—
Hexadecamethylheptasiloxane	$C_{16}H_{48}O_6Si_7$	93.2	138.5	171.8	198.0	244.7	270.0	—
Hexadecamethylcyclooctasiloxane	$C_{16}H_{48}O_8Si_8$	103.5	150.5	186.3	213.8	263.0	290.0	31.5
Benzanthrone	$C_{17}H_{10}O$	225.0	297.2	350.0	390.0	—	—	174
4-*tert*-Butylphenyl salicylate	$C_{17}H_{18}O_3$	166.2	225.0	270.7	305.8	370.6	404.0d	—
Menthyl benzoate	$C_{17}H_{24}O_2$	123.2	170.0	204.3	230.4	277.1	301.0	54.5
2-Heptadecanone	$C_{17}H_{34}O$	129.6	178.0	214.3	242.0	291.7	319.5	—
Methyl palmitate	$C_{17}H_{34}O_2$	134.3	184.3	d	—	—	—	30
Heptadecane	$C_{17}H_{36}$	115.0	160.0	195.8	223.0	274.5	303.0	22.5
Tetradecyltrimethylsilane	$C_{17}H_{38}Si$	120.0	166.2	201.5	227.8	275.0	300.0	—
Tri-2-chlorophenylthiophosphate	$C_{18}H_{12}Cl_3O_3PS$	188.2	231.2	261.7	283.8	322.0	341.3	—
Triphenyl phosphate	$C_{18}H_{15}O_4P$	193.5	249.8	290.3	322.5	379.2	413.5	49.4
Hexaethylbenzene	$C_{18}H_{30}$	s	150.3	187.7	216.0	268.5	298.3	130
2,4,6-Tri-*tert*-butylphenol	$C_{18}H_{30}O$	95.2	142.0	177.4	203.0	250.6	276.3	—
Oleic acid	$C_{18}H_{34}O_2$	176.5	223.0	257.2	286.0	334.7	360.0d	14
Elaidic acid	$C_{18}H_{34}O_2$	171.3	223.5	260.8	288.0	337.0	362.0	51.5
Stearaldehyde	$C_{18}H_{36}O$	140.0	192.1	230.8	260.0	313.8	342.5	63.5
Stearic acid	$C_{18}H_{36}O_2$	173.7	225.0	263.3	291.0	343.0	370.0d	69.3
Octadecane	$C_{18}H_{38}$	119.6	169.6	207.4	236.0	288.0	317.0	28
2-Methylheptadecane	$C_{18}H_{38}$	119.8	168.7	204.8	231.5	279.8	306.5	—
1-Octadecanol	$C_{18}H_{38}O$	150.3	202.0	240.4	269.4	320.3	349.5	58.5

ORGANIC COMPOUNDS

Pressures Less than One Atmosphere (Continued)

Name	Formula	Temperature °C 1 mm	10 mm	40 mm	100 mm	400 mm	760 mm	M.P.
Ethylcetylamine	$C_{18}H_{39}N$	133.2	186.0	226.5	256.8	313.0	342.0d	—
1,1,3-Diethoxytetradecamethylheptasiloxane	$C_{18}H_{52}O_8Si_7$	119.0	163.5	197.0	223.2	268.3	293.5	—
Octadecamethyloctasiloxane	$C_{18}H_{54}O_7Si_9$	105.8	152.3	187.5	214.5	263.5	290.0	—
Triphenylmethane	$C_{19}H_{16}$	169.7	197.0	215.5	228.4	249.8	259.2	93.4
Diphenyl-2-tolyl thiophosphate	$C_{19}H_{17}O_3PS$	159.7	201.6	230.6	252.5	290.0	310.0	—
Nonadecane	$C_{19}H_{40}$	133.2	183.5	220.0	248.0	299.8	330.0	32
Ethoxytriphenylsilane	$C_{20}H_{20}OSi$	167.0	213.5	247.0	273.5	319.5	344.0	—
Diethylhexadecylamine	$C_{20}H_{43}N$	139.8	194.0	235.0	265.5	324.6	355.0	—
1,1,5-Diethoxyhexadecamethyloctasiloxane	$C_{20}H_{58}O_9Si_8$	133.7	179.7	213.8	240.0	286.0	311.5	—
Eicosamethylnonasiloxane	$C_{20}H_{60}O_8Si_9$	144.0	189.0	220.5	244.3	286.0	307.5	—
Tritolyl phosphate	$C_{21}H_{21}O_4P$	154.6	198.0	229.7	252.2	292.7	313.0	—
Heneicosane	$C_{21}H_{44}$	152.6	205.4	243.4	272.0	323.8	350.5	40.4
Erucic acid	$C_{22}H_{42}O_2$	206.7	254.5	289.1	314.4	358.8	381.5d	33.5
Brassidic acid	$C_{22}H_{42}O_2$	209.6	256.0	290.0	316.2	359.6	382.5d	61.5
Docosane	$C_{22}H_{46}$	157.8	213.0	254.5	286.0	343.5	376.0	44.5
Docosamethyldecasiloxane	$C_{22}H_{56}O_9Si_{10}$	160.3	202.8	233.8	255.0	293.8	314.0	—
Tricosane	$C_{23}H_{48}$	170.0	223.0	261.3	289.8	339.8	366.5	47.7
Tetracosane	$C_{24}H_{50}$	183.8	237.6	276.3	305.2	358.0	386.4	51.1
Tetracosamethylhendecasiloxane	$C_{24}H_{72}O_{10}Si_{11}$	175.2	216.7	246.2	266.3	303.7	322.8	—
Pentacosane	$C_{25}H_{52}$	194.2	248.2	285.6	314.0	365.4	390.3	53.3
Hexacosane	$C_{26}H_{54}$	204.0	257.4	295.2	323.2	374.6	399.8	56.6
Dicarvacryl-2-tolyl phosphate	$C_{27}H_{33}O_4P$	180.2	221.8	251.5	272.5	309.8	330.0	—
Heptacosane	$C_{27}H_{56}$	211.7	266.8	305.7	333.5	385.0	410.6	59.5
Octacosane	$C_{28}H_{58}$	226.5	277.4	314.2	341.8	388.9	412.5	61.6
Nonacosane	$C_{29}H_{60}$	234.2	286.4	323.2	350.0	397.2	421.8	63.8
Dicarvacryl-mono-(6-chloro-2-xenyl) phosphate	$C_{32}H_{34}ClO_4P$	204.2	249.3	280.5	304.9	342.0	361.0	—

Pressures Greater than One Atmosphere

Name	Formula	Temperature °C 1 atm.	2 atm.	5 atm.	10 atm.	20 atm.	40 atm.	60 atm.
Chlorotrifluoromethane	$CClF_3$	−81.2	−66.7	−42.7	−18.5	+12.0	52.8	—
Dichlorodifluoromethane	CCl_2F_2	29.0	12.2	+16.1	42.4	74.0	—	—
Carbonyl chloride	CCl_2O	8.3	27.3	57.2	85.0	119.0	159.8	—
Trichlorofluoromethane	CCl_3F	23.7	44.1	77.3	108.2	146.7	194.0	—
Carbontetrachloride	CCl_4	76.7	102.0	141.7	178.0	222.0	276.0	—
Chlorodifluoromethane	$CHClF_2$	−40.8	−24.7	+0.3	24.0	52.0	85.3	—
Dichlorofluoromethane	$CHCl_2F$	8.9	28.4	59.0	87.0	121.2	162.6	—
Trichloromethane	$CHCl_3$	61.3	83.9	120.0	152.3	191.8	237.5	—
Hydrocyanic acid	CHN	25.8	45.5	75.5	103.5	134.2	170.2	—
Methyl bromide	CH_3Br	3.6	23.3	54.8	84.0	121.7	170.2	—
Methyl chloride	CH_3Cl	−24.0	−6.4	+22.0	47.3	77.3	113.8	137.5
Methyl fluoride	CH_3F	−78.2	−64.5	−42.0	−21.0	+2.6	26.5	43.5
Methyl iodide	CH_3I	42.4	65.5	101.8	138.0	176.5	228.5	—
Methane	CH_4	−161.5	−152.3	−138.3	−124.8	−108.5	−80.3	—
Methanol	CH_4O	64.7	84.0	112.5	138.0	167.8	203.5	224.0
Methanethiol	CH_4S	6.8	26.1	55.9	83.4	117.5	157.7	185.0
Methylamine	CH_5N	−6.3	+10.1	36.0	59.5	87.8	121.8	144.6
Carbon monoxide	CO	−191.3	−183.5	−170.7	−161.0	−149.7	—	—
Carbon dioxide	CO_2	−78.2	−69.1s	−56.7	−39.5	−18.9	+5.9	22.4
Carbon disulfide	CS_2	46.5	69.1	104.8	136.3	175.5	222.8	256.0
1-Chloro-1,2,2-trifluoroethylene	C_2ClF_3	−27.9	−11.1	+15.5	40.0	71.1	—	—
1,2-Dichloro-1,1,2,2-tetrafluoroethane	$C_2Cl_2F_4$	3.5	22.8	54.0	82.3	117.5	—	—
1,1,2-Trichloro-1,2,2-trifluoroethane	$C_2Cl_3F_3$	47.6	70.0	105.5	138.0	177.7	—	—
Acetylene	C_2H_2	−84.0s	−71.6	−50.2	−32.7	−10.0	+16.8	34.8
cis-1,2-Dichloroethylene	$C_2H_2Cl_2$	59.0	82.1	119.3	152.3	194.0	244.5	—
trans-1,2-Dichloroethylene	$C_2H_2Cl_2$	47.8	69.8	104.0	135.7	174.0	220.0	—
Ethylene	C_2H_4	−103.7	−90.8	−71.1	−52.8	−29.1	−1.5	—
1,2-Dibromoethane	$C_2H_4Br_2$	131.5	157.7	200.0	237.0	269.0	295.0	304.5
1,1-Dichloroethane	$C_2H_4Cl_2$	57.3	80.2	117.3	150.3	192.7	243.0	—
1,2-Dichloroethane	$C_2H_4Cl_2$	83.7	108.1	147.8	183.5	226.5	272.0	—
Acetic acid	$C_2H_4O_2$	118.1	143.5	180.3	214.0	252.0	297.0	—
Methyl formate	$C_2H_4O_2$	32.0	51.9	83.5	112.0	147.2	188.5	—
Ethyl bromide	C_2H_5Br	38.4	60.2	95.0	126.8	164.3	206.5	229.5
Ethyl chloride	C_2H_5Cl	12.3	32.5	64.0	92.6	127.3	167.0	—
Ethyl fluoride	C_2H_5F	−32.0	−16.7	+7.7	30.2	57.5	90.0	—
Ethane	C_2H_6	−88.6	−75.0	−52.8	−32.0	−6.4	+23.6	—
Ethanol	C_2H_6O	78.4	97.5	126.0	151.8	183.0	218.0	242.0
Dimethyl ether	C_2H_6O	−23.7	−6.4	+20.8	45.5	75.7	112.1	—
Ethanethiol	C_2H_6S	35.0	56.6	90.7	121.9	159.5	204.7	—
Dimethyl sulfide	C_2H_6S	36.0	57.8	92.3	124.5	163.8	209.0	—
Ethylamine	C_2H_7N	16.6	35.7	65.3	91.8	124.0	163.0	—
Dimethylamine	C_2H_7N	7.4	25.0	53.9	80.0	111.7	149.8	—
Cyanogen	C_2N_2	−21.0	−4.4	+21.4	44.6	72.6	106.5	—
Propadiene	C_3H_4	−35.0	−18.4	+8.0	33.2	64.5	103.5	—
Propyne	C_3H_4	−23.3	−7.1	+19.5	43.8	74.0	111.5	—
Propylene	C_3H_6	−47.7	−31.4	−4.8	+19.8	49.5	85.0	—
Acetone	C_3H_6O	56.5	78.6	113.0	144.5	181.0	214.5	—
Propionic acid	$C_3H_6O_2$	141.1	160.0	186.0	203.5	220.0	233.0	—

ORGANIC COMPOUNDS
Pressures Greater than One Atmosphere (Continued)

Name	Formula	Temperature °C						
		1 atm.	2 atm.	5 atm.	10 atm.	20 atm.	40 atm.	60 atm.
Methyl acetate	$C_3H_6O_2$	57.8	79.5	113.1	144.2	181.0	225.0	—
Ethyl formate	$C_3H_6O_2$	54.3	76.0	110.5	142.2	180.0	225.0	—
Propane	C_3H_8	−42.1	−25.6	+1.4	26.9	58.1	94.8	—
1-Propanol	C_3H_8O	97.8	117.0	149.0	177.0	210.8	250.0	—
2-Propanol	C_3H_8O	82.5	101.3	130.2	155.7	186.0	220.2	—
Ethyl methyl ether	C_3H_8O	7.5	26.5	56.4	84.0	108.0	160.0	—
Propylamine	C_3H_9N	48.5	69.8	102.8	133.4	170.0	214.5	—
1,3-Butadiene	C_4H_6	−4.5	+15.3	47.0	76.0	114.0	158.0	—
Acetic anhydride	$C_4H_6O_3$	139.6	162.0	194.0	221.5	253.0	288.5	—
Dimethyl oxalate	$C_4H_6O_4$	163.3	189.6	228.7	—	—	—	—
Butyric acid	$C_4H_8O_2$	163.5	188.3	225.0	257.0	295.0	338.0	—
Isobutyric acid	$C_4H_8O_2$	154.4	179.8	217.0	250.0	289.0	336.0	—
Ethyl acetate	$C_4H_8O_2$	77.1	100.6	136.6	169.7	209.5	—	—
Methyl propionate	$C_4H_8O_2$	79.8	103.0	139.8	172.6	212.5	—	—
Propyl formate	$C_4H_8O_2$	81.3	104.3	142.0	176.4	217.5	—	—
Butane	C_4H_{10}	−0.5	+18.8	50.0	79.5	116.0	—	—
2-Methylpropane	C_4H_{10}	−11.7	+7.5	39.0	66.8	99.5	—	—
Butyl alcohol	$C_4H_{10}O$	117.5	139.8	172.5	203.0	237.0	277.0	—
sec-Butyl alcohol	$C_4H_{10}O$	99.5	118.2	147.5	172.0	204.0	251.0	—
Isobutyl alcohol	$C_4H_{10}O$	108.0	127.3	156.2	182.0	212.5	251.0	—
tert-Butyl alcohol	$C_4H_{10}O$	82.9	102.0	130.0	154.2	184.5	222.5	—
Diethyl ether	$C_4H_{10}O$	34.6	56.0	90.0	122.0	159.0	—	—
Diethyl sulfide	$C_4H_{10}S$	88.0	112.0	153.8	190.2	234.0	—	—
Diethylamine	$C_4H_{11}N$	55.5	77.8	113.0	145.3	184.5	—	—
Tetramethylsilane	$C_4H_{12}Si$	27.0	48.0	82.0	113.0	152.0	—	—
Ethyl propionate	$C_5H_{10}O_2$	99.1	123.8	162.7	197.8	240.0	—	—
Propyl acetate	$C_5H_{10}O_2$	101.8	126.8	165.7	200.5	242.8	—	—
Isobutyl formate	$C_5H_{10}O_2$	98.2	121.8	157.8	192.4	234.0	—	—
Methyl butyrate	$C_5H_{10}O_2$	102.3	127.5	166.7	203.0	244.5	—	—
Methyl isobutyrate	$C_5H_{10}O_2$	92.6	116.7	155.2	190.2	232.0	—	—
Pentane	C_5H_{12}	36.1	58.0	92.4	124.7	164.3	—	—
2-Methylbutane	C_5H_{12}	27.8	48.8	82.8	114.5	154.0	—	—
2,2-Dimethylpropane	C_5H_{12}	+9.5	29.5	61.1	90.7	127.6	—	—
Ethyl propyl ether	$C_5H_{12}O$	61.7	85.3	123.1	156.2	197.2	—	—
Bromobenzene	C_6H_5Br	156.2	186.2	232.5	274.5	327.0	387.5	—
Chlorobenzene	C_6H_5Cl	132.2	160.2	205.0	245.3	292.8	349.8	—
Fluorobenzene	C_6H_5F	84.7	109.9	148.5	184.4	227.6	279.3	—
Iodobenzene	C_6H_5I	188.6	220.0	270.0	315.7	371.5	437.2	—
Benzene	C_6H_6	80.1	103.8	142.5	178.8	221.5	—	—
Phenol	C_6H_6O	181.9	208.0	248.2	283.8	328.7	272.3	418.7
Aniline	C_6H_7N	184.4	212.8	254.8	292.7	342.0	382.1	—
Cyclohexane	C_6H_{12}	80.7	106.0	146.4	184.0	228.4	400.0	—
Ethyl isobutyrate	$C_6H_{12}O_2$	110.1	135.5	174.2	210.0	253.0	—	—
Hexane	C_6H_{14}	68.7	93.0	131.7	166.6	209.4	—	—
2,3-Dimethylbutane	C_6H_{14}	58.0	82.0	120.3	155.7	198.7	—	—
Toluene	C_7H_8	110.6	136.5	178.0	215.8	262.5	—	—
Heptane	C_7H_{16}	98.4	124.8	165.7	202.8	247.5	319.0	—
Ethylbenzene	C_8H_{10}	136.2	163.5	207.5	246.3	294.5	—	—
Octane	C_8H_{18}	125.6	152.7	196.2	235.8	181.4	—	—
Dodecane	$C_{12}H_{26}$	216.2	249.2	300.0	345.8	—	—	—

VAPOR PRESSURE
Variation with Temperature

The following table gives the value of the constants *a* and *b* in the following equation:

$$\log_{10} p = -\frac{0.05223a}{T} + b$$

where *p* is the pressure in mm of mercury of the saturated vapor at the absolute temperature *T*. (T = t°C + 273.1). The values obtained by the use of the equation given above are valid within the temperature ranges indicated for each of the compounds.

Elements and Inorganic Compounds

Compound	Formula	Temp. range °C	*a*	*b*
Aluminum oxide	Al_2O_3	1840 to 2200 liq.	540,000	14.22
Ammonia	NH_3	−127 to −78 sol.	31,211	9.9974
Ammonium bromide	NH_4Br	250 to 400 sol.	90,208	9.9404
Ammonium chloride	NH_4Cl	100 to 400 sol.	83,486	10.0164
Ammonium cyanide	NH_4CN	7 to 17 sol.	41,484	9.978
Ammonium iodide	NH_4I	300 to 400 sol.	95,730	10.2700
Ammonium sulfhydrate	NH_4HS	6 to 40 sol.	46,025	10.7500
Antimony	Sb	1070 to 1325 liq.	189,000	9.051

VAPOR PRESSURE (Continued)

Compound	Formula	Temp. range (°C)	*a*	*b*
Argon	A	−208 to −189 sol.	7,814.5	7.5741
		−189 to −183 liq.	6,826.0	6.9605
Arsenic	As	800 to 860 liq.	47,100	6.692
		440 to 815 sol.	133,000	10.800
Arsenous oxide	As_2O_3	100 to 310 sol.	111,350	12.127
		315 to 490 liq.	52,120	6.513
Barium	Ba	930 to 1130 liq.	350,000	15.765
Bismuth	Bi	1210 to 1420 liq.	200,000	8.876
Bismuth trichloride	$BiCl_3$	91 to 213 sol.	13,125	2.681
Cadmium	Cd	150 to 320.9 sol.	109,000	8.564
		500 to 840 liq.	99,900	7.897
Cadmium iodide	CdI_2	385 to 450 liq.	122,200	9.269
Cesium	Cs	200 to 350 liq.	73,400	6.949
Cesium chloride	CsCl	986 to 1295 liq.	163,200	8.340
Calcium	Ca	960 to 1110 liq.	370,000	16.240
Carbon	C	3880 to 4430 liq.	540,000*	9.596*
Carbon dioxide	CO_2	−135 to −56.7 sol.	26,179.3	9.9082
Carbon monoxide	CO	−220 to −206 liq.	6,354	6.976
Chlorine	Cl	−154 to −103 sol.	29,293	9.950
Cobalt	Co	2375 liq.	309,000	7.571
Copper	Cu	2100 to 2310 liq.	468,000	12.344
Cuprous chloride	Cu_2Cl_2	878 to 1369 liq.	80,700	5.454
Cyanogen	$(CN)_2$	−72 to −28 sol.	32,437	9.6539
		−32 to −6 liq.	23,750	7.808
Ferrous chloride	$FeCl_2$	700 to 930 sol.	135,200	8.33
Gold	Au	2315 to 2500 liq.	385,000	9.853
Hydriodic acid	HI	−97 to −51 sol	24,160	8.259
		−50 to −34 liq.	21,580	7.630
Hydrobromic acid	HBr	−114 to −86 sol.	22,420	8.734
		−86 to −66 liq.	17,960	7.427
Hydrochloric acid	HCl	−158 to −110 sol.	19,588	8.4430
Hydrocyanic acid	HCN	−8 to +27 liq.	27,830	7.7446
Hydrofluoric acid	HF	−83 to +48 liq.	25,180	7.370
Hydrogen peroxide	H_2O_2	10 to 90 liq.	48,530	8.853
Hydrogen sulfide	H_2S	−110 to −83 sol.	20,690	7.880
Iron	Fe	2220 to 2450 liq.	309,000	7.482
Krypton	Kr	−189 to −169 sol.	10,065	7.1770
		−169 to −150 liq.	9,377.0	6.92387
Lead	Pb	525 to 1325 liq.	188,500	7.827
Lead bromide	$PbBr_2$	735 to 918 liq.	118,000	8.064
Lead chloride	$PbCl_2$	500 to 950 liq.	141,900	8.961
Lithium bromide	LiBr	1010 to 1265 liq.	152,700	8.068
Lithium chloride	LiCl	1045 to 1325 liq.	155,900	7.939
Lithium fluoride	LiF	1398 to 1666 liq.	218,400	8.753
Lithium iodide	LiI	940 to 1140 liq.	143,600	8.011
Magnesium	Mg	900 to 1070 liq.	260,000	12.993
Manganese	Mn	1510 to 1900 liq.	267,000	9.300
Mercuric bromide	$HgBr_2$	111 to 235 sol.	79,800	10.181
		238 to 331 liq.	61,250	8.284
Mercuric chloride	$HgCl_2$	60 to 130 sol.	85,030	10.888
		130 to 270 sol.	78,850	10.094
		275 to 309 liq.	61,020	8.409
Mercuric iodide	HgI_2	100 to 250 sol.	82,340	10.057
		266 to 360 liq.	62,770	8.115
Mercury	Hg	−80 to −38.87 sol.	73,000	10.383
		400 to 1300 liq.	58,700	7.752
Molybdenum	Mo	1800 to 2240 sol.	680,000	10.844
Nitrogen	N_2	−215 to −210 sol.	6,881.3	7.66558
Nitrogen dioxide	NO	−200 to −161 sol.	16,423	10.048
		−163.7 to −148 liq.	13,040	8.440
Nitrogen monoxide	N_2O	−144 to −90 sol.	23,590	9.579
		−90.1 to −88.7 liq.	16,440	7.535
Nitrogen pentoxide	N_2O_3	−30 to +30 sol.	57,180	12.647
Nitrogen tetroxide	N_2O_4	−100 to −40 sol.	55,160	13.400
		−40 to −10 sol.	45,440	11.214
		−8 to +43.2 liq.	33,430	8.814
Nitrogen trioxide	N_2O_3	−25 to 0 liq.	39,400	10.30
Phosphorus (white)	P	20 ro 44.1 sol.	63,123	9.6511
Phosphorus (violet)	P	380 to 590 sol.	108,510	11.0842
Platinum	Pt	1425 to 1765 sol.	486,000	7.786
Potassium	K	260 to 760 liq.	84,900	7.183
Potassium bromide	KBr	906 to 1063 liq.	168,100	8.2470
		1095 to 1375 liq.	163,800	7.936
Potassium chloride	KCl	906 to 1105 liq.	174,500	8.3526
		1116 to 1418 liq.	169,700	8.130
Potassium fluoride	KF	1278 to 1500 liq.	207,500	9.000
Potassium hydroxide	KOH	1170 to 1327 liq.	136,000	7.330
Potassium iodide	KI	843 to 1028 liq.	157,600	8.0957
		1063 to 1333 liq.	155,700	7.949
Rubidium	Rb	250 to 370 liq.	76,000	6.976
Rubidium chloride	RbCl	1142 to 1395 liq.	198,600	9.111
Silicon	Si	1200 to 1320 sol.	170,000	5.950
Silicon dioxide	SiO_2	1860 to 2230 liq.	506,000	13.43
Silver	Ag	1650 to 1950 liq.	250,000	8.762

* Based on boiling point of 3927°C or 4200° absolute.

VAPOR PRESSURE (Continued)

Compound	Formula	Temp. range °C	*a*	*b*
Silver chloride	$AgCl$	1255 to 1442 liq.	185,500	8.179
Sodium	Na	180 to 883 liq.	103,300	7.553
Sodium bromide	$NaBr$	1138 to 1394 liq.	161,600	7.948
Sodium chloride	$NaCl$	976 to 1155 liq.	180,300	8.3297
		1156 to 1430 liq.	185,800	8.548
Sodium cyanide	$NaCN$	800 to 1360 liq.	155,520	7.472
Sodium fluoride	NaF	1562 to 1701 liq.	218,200	8.640
Sodium hydroxide	$NaOH$	1010 to 1402 liq.	132,000	7.030
Sodium iodide	NaI	1063 to 1307 liq.	165,100	8.371
Stannic chloride	$SnCl_4$	−52 to −38 sol.	46,740	9.824
Strontium	Sr	940 to 1140 liq.	360,000	16.056
Sulfur dioxide	SO_2	−95 to −75 sol.	35,827	10.5916
Sulfur trioxide	SO_3	24 to 48 liq.	43,450	10.022
Thallium	Tl	950 to 1200 liq.	120,000	6.140
Thallium chloride	$TlCl$	665 to 807 liq.	105,200	7.974
Tin	Sn	1950 to 2270 liq.	328,000	9.643
Tungsten	W	2230 to 2770 sol.	897,000	9.920
Zinc	Zn	250 to 419.4 sol.	133,000	9.200
		600 to 985 liq.	118,000	8.108

Organic Compounds

Compound	Formula	Temp. range °C	*a*	*b*
Acetaldehyde	C_2H_4O	−24.3 to +27.5 liq.	27,707	7.8206
Acetic acid	$C_2H_4O_2$	−35 to 10 sol.	41,689	8.502
Acetic anhydride	$C_4H_6O_3$	100 to 140 liq.	45,585	8.688
Acetylene	C_2H_2	−140 to −82 sol.	21,914	8.933
Aniline	C_6H_7N	145 to 185 liq.	45,951.6	8.1278
Anthracene	$C_{14}H_{10}$	100 to 600 liq.	70,390	8.706
		100 to 160 liq.	72,000	8.91
		223 to 342 liq.	59,219	7.910
Anthraquinone	$C_{14}H_8O_2$	224 to 286 sol.	110,040	12.305
Benzene	C_6H_6	−58 to −30 sol.	42,904	9.556
		−30 to +5 sol.	44,222	9.846
		0 to 42 liq.	34,172	7.9622
		42 to 100 liq.	32,295	7.6546
Benzoic acid	$C_7H_6O_2$	60 to 110 sol.	63,820	9.033
Benzophenone	$C_{13}H_{10}O$	260 to 308 liq.	58,221	8.137
Benzoyl chloride	C_7H_5ClO	140 to 200 liq.	45,416	7.9245
Benzyl alcohol	C_7H_8O	100 to 135 liq.	59,491	9.5152
		135 to 205 liq.	53,118	8.6977
Butane	C_4H_{10}	−100 to +12 liq.	23,450	7.395
iso-Butane	C_4H_{10}	−115 to −34 liq.	21,273	7.25
n-Butyl alcohol	$C_4H_{10}O$	75 to 117.5 liq.	46,774	9.1362
Butyric acid	$C_4H_8O_2$	80 to 165 liq.	51,103	9.010
Bromobenzene	C_6H_5Br	−26 to −15 liq.	42,500	8.075
p-Bromochlorobenzene	C_6H_4BrCl	23 to 63 sol.	69,755	11.629
Camphor	$C_{10}H_{16}O$	0 to 180 sol.	53,559	8.799
Carbon tetrachloride	CCl_4	−70 to −50 sol.	34,608	8.05
		−19 to +20 liq.	33,914	8.004
Chlorobenzene	C_6H_5Cl	−35 to −15 liq.	42,250	8.500
Cyclohexane	C_6H_{12}	−5 to +5 sol.	37.394	8.594
p-Dichlorobenzene	$C_6H_4Cl_2$	30 to 50 sol.	72,218	12.480
Dichloroethane-1,1	$C_2H_4Cl_2$	0 to 30 liq.	31,706	7.909
Dichloroethane-1,2	$C_2H_4Cl_2$	0 to 30 liq.	35,598	8.126
Diphenylamine	$C_{12}H_{11}N$	278 to 284 liq.	57,350	8.088
Ethyl chloride	C_2H_5Cl	−30 to +30 liq.	26,319	7.691
Ethylene	C_2H_4	−160 to −104 liq.	14,396	7.330
Ethylene bromide	$C_2H_4Br_2$	10 to 150 liq.	38,082	7.792
Heptane	C_7H_{16}	−63 to −40 liq.	37.358	8.2585
Hexane	C_6H_{14}	−10 to +90 liq.	31,679	7.724
Iodobenzene	C_6H_5I	−30 to +18 liq.	43,000	7.500
Methane	CH_4	−194 to −184 sol.	9,896.2	7.6509
		−174 to −163 liq.	8,516.9	6.8626
Methylalcohol	CH_4O	−62 to −44 liq.	39,234	8.9547
		−10 to +80 liq.	38,324	8.8017
Methyl chloride	CH_3Cl	−47 to −10 liq.	21,988	7.481
Methyl ether	C_2H_6O	−70 to −20 liq.	23,025	7.720
Methyl fluoride	CH_3F	−102 to −76 liq.	17,053	7.445
Methyl salicylate	$C_8H_8O_3$	175 to 215 liq.	48,670	8.008
Naphthalene	$C_{10}H_8$	0 to 80 sol.	71,401	11.450
		120 to 200 liq.	47,362	7.927
o-Nitroaniline	$C_6H_6N_2O_2$	150 to 260 liq.	63,881	8.8684
m-Nitroaniline	$C_6H_6N_2O_2$	170 to 260 liq.	65,880	8.8188
p-Nitroaniline	$C_6H_6N_2O_2$	190 to 260 liq.	77,345	9.5595
Nitrobenzene	$C_6H_5NO_2$	112 to 209 liq.	48,955	8.192
Nitromethane	CH_3NO_2	47 to 100 liq.	36,914	8.033
Oxalic acid	$C_2H_2O_4$	55 to 105 sol.	90,502.6	12.2229
n-Pentane	C_5H_{12}	−20 to +50 liq.	27,691	7.558
Phenol	C_6H_6O	116 to 180 liq.	49,644	8.587
Phthalic anhydride	$C_8H_4O_3$	160 to 285 liq.	54,920	8.022
Propane	C_3H_8	−136 to −40 liq.	19,037	7.217
Propionic acid	$C_3H_6O_2$	20 to 140 liq.	46,150	8.715
n-Propyl alcohol	C_3H_8O	−45 to −10 liq.	47,274	9.5180
Propyl bromide	C_3H_7Br	0 to 30 liq.	32,430	7.821
Propyl chloride	C_3H_7Cl	0 to 50 liq.	28,894	7.593

VAPOR PRESSURE (Continued)

Compound	Formula	Temp. range (°C)	*a*	*b*
Propylene	C_3H_6	−95 to −48 liq.	19,693	7.4463
Quinoline	C_9H_7N	180 to 240 liq.	49,720	7.969
Tetrachloroethane-1,1,1,2	$C_2H_2Cl_4$	105 to 145 liq.	36,508	7.605
Tetrachloroethane-1,1,2,2	$C_2H_2Cl_4$	26 to 145 liq.	39,729	7.846
Toluene	C_7H_8	−92 to +15 liq.	39,198	8.330

VAPOR PRESSURE OF NITRIC ACID

Temperature (°C)	Vapor Pressure, mm. of Hg: 100% HNO_3	90% of HNO_3
0	14.4	5.5
10	26.6	11.
20	47.9	20.
30	81.3	37.3
40	133.	64.4
50	208.	107.
70	467.	242.
80	670.	352.
90	937.	504.
100	1282.	710.

VAPOR PRESSURE OF THE ELEMENTS

Rudolf Loebel

This table lists the temperature in degrees Celsius at which an element has a vapor pressure indicated by the headings of the columns. For pressures of one atmosphere and lower, the pressures are given in millimeters of mercury; for pressures above one atmosphere, the pressures are given in atmospheres.

Element		mm Hg					atm.				
		1	10	100	400	760	2	5	10	20	40
Aluminum	Al	1540	1780	2080	2320	2467	2610	2850	3050	3270	3530
Antimony	Sb		960	1280	1570	1750	1960	2490			
Arsenic	As	380	440	510	580	610					
Barium	Ba	860	1050	1300	1520	1610	1700	2030	2230		
Beryllium	Be	1520	1860	2300	2770	2970	3240	3730	4110	4720	5610
Bismuth	Bi		1060	1280	1450	1560	1660	1850	2000	2180	
Boron	B	2660	3030	3460	3810	4000					
Bromine	Br	−60	−30	+9	39	59	78	110			
Cadmium	Cd	393	486	610	710	765	830	930	1030	1120	1240
Calcium	Ca	800	970	1200	1390	1490	1630	1850	2020	2290	
Cesium	Cs		373	513	624	690					
Chlorine	Cl	−123	−101	−71	−46	−34	−17	+9	30	55	97
Chromium	Cr	1610	1840	2140	2360	2480	2630	2850	3010	3180	
Cobalt	Co	1910	2170	2500	2760	2870	3040	3270			
Copper	Cu		1870	2190	2440	2600	2760	3010	3500	3460	3740
Fluorine	F			−203	−193	−188	−180.7	−169.1	−159.6		
Gallium	Ga	1350	1570	1850	2060	2180	2320	2560	2730		
Germanium	Ge		2080	2440	2710	2830	2970	3200	3430		
Gold	Au	1880	2160	2520	2800	2940	3120	3490	3630	3890	
Indium	In				1960	2080	2230	2440	2600		
Iridium	Ir	2830	3170	3630	3960	4130	4310	4650			
Iron	Fe	1780	2040	2370	2620	2750	2900	3150	3360	3570	
Iodine	I	40	72	115	160	185	216	265			
Lanthanum	La				3230	3420	3620	3960	4270		
Lead	Pb	970	1160	1420	1630	1740	1880	2140	2320	2620	
Lithium	Li	750	890	1080	1240	1310	1420	1518			
Magnesium	Mg	620	740	900	1040	1110	1190	1330	1430	1560	
Manganese	Mn		1510	1810	2050	2100	2360	2580	2850		
Mercury	Hg			260	330	356.9	398	465	517	581	657
Molybdenum	Mo	3300	3770	4200	4580	4830	5050	5340	5680	5980	
Neodymium	Nd				2870	3100	3300	3680	3990		
Nickel	Ni	1800	2090	2370	2620	2730	2880	3120	3300	3310	
Palladium	Pd	1470	2290	2670	2950	3140	3270	3560	3840		
Phosphorus	P		127	199	253	283	319				
Platinum	Pt	2600	2940	3360	3650	3830	4000	4310	4570	4860	
Polonium	Po	472	587	752	890	960	1060	1200	1340		
Potassium	K			590	710	770	850	950	1110	1240	1420
Rhodium	Rh	2330	2850	3260	3590	3760	3930	4230	4440		
Rubidium	Rb		390	527	640	700					
Selenium	Se		429	547	640	685	750	850	920	1010	1120
Silver	Ag	1310	1540	1850	2060	2210	2360	2600	2850	3050	3300
Sodium	Na	440	546	700	830	890	980	1120	1230	1370	
Strontium	Sr	740	900	1100	1280	1380	1480	1670	1850	2030	
Sulfur	S		246	333	407	445	493	574	640	720	
Tellurium	Te	520	633	792	900	962	1030	1160	1250		
Thallium	Tl		1000	1210	1370	1470	1560	1750	1900	2050	2260
Tin	Sn	1610	1890	2270	2580	2750	2950	3270	3540	3890	
Titanium	Ti	2180	2480	2860	3100	3260	3400	3650	3800		
Tungsten	W	3980	4490	5160	5470	5940	6260	6670	7250	7670	
Uranium	U	2450	2800	3270	3620	3800	4040	4420			
Vanadium	V	2290	2570	2950	3220	3380	3540	3800			
Zinc	Zn		590	730	840	907	970	1090	1180	1290	

VAPOR PRESSURE OF ELEMENTS THAT ARE GASEOUS AT STANDARD CONDITIONS

The following tables contain vapor pressure data for helium, hydrogen, neon, nitrogen, and oxygen.

Vapor Pressure (atm) vs. Temperature (K) for Helium-4

Temperature, K	Pressure atm
2.177	0.04969
2.20	0.05256
2.25	0.05916
2.30	0.06629
2.35	0.07399
2.40	0.08228
2.45	0.09120
2.50	0.1008
2.55	0.1110
2.60	0.1219
2.65	0.1336
2.70	0.1460
2.75	0.1591
2.80	0.1730
2.85	0.1878
2.90	0.2033
2.95	0.2198
3.00	0.2371
3.05	0.2553
3.10	0.2744
3.15	0.2945
3.20	0.3156
3.25	0.3376
3.30	0.3607
3.35	0.3848
3.40	0.4100
3.45	0.4363
3.50	0.4637
3.55	0.4923
3.60	0.5220
3.65	0.5528
3.70	0.5849
3.75	0.6182
3.80	0.6528
3.85	0.6886
3.90	0.7257
3.95	0.7642
4.00	0.8040
4.05	0.8452
4.10	0.8878
4.15	0.9318
4.20	0.9772
4.224	1.000
4.25	1.024
4.30	1.073
4.35	1.123
4.40	1.174
4.45	1.227
4.50	1.282
4.55	1.339
4.60	1.397
4.65	1.457
4.70	1.519
4.75	1.582
4.80	1.648
4.85	1.715
4.90	1.784
4.95	1.856
5.00	1.929
5.05	2.004
5.10	2.082
5.201	2.245

Vapor Pressure (atm) vs. Temperature (K) for Equilibrium Hydrogen

Temperature, K	Pressure, atm
13.803	0.069$_5$
14	0.077$_8$
15	0.133
16	0.213
17	0.325
18	0.476
19	0.673
20	0.923
20.268	1.000
21	1.233
22	1.613
23	2.069
24	2.611
25	3.245
26	3.982
27	4.829
28	5.794
29	6.887
30	8.118
31	9.501
32	11.051
32.976	12.759

Vapor Pressure (atm) vs. Temperature (K) for Neon

Temperature, K	Pressure, atm
25	0.50366
26	0.70902
27	0.97255
28	1.3037
29	1.7124
30	2.2088
31	2.8031
32	3.5061
33	4.3286
34	5.2818
35	6.3773
36	7.6271
37	9.0439
38	10.641
39	12.432
40	14.434
41	16.661
42	19.133
43	21.867
44	24.887
44.4	26.19

Vapor Pressure (atm) vs. Temperature (K) for Nitrogen

Temperature, K	Pressure, atm
63.148	0.1237
64	0.1443
66	0.2037
68	0.2813
70	0.3807
72	0.5059
74	0.6610
76	0.8506
77.347	1.0000
78	1.0793
80	1.3520
82	1.6739
84	2.0503
86	2.4865
88	2.9882
90	3.5607
92	4.2099
94	4.9415
98	6.6748
100	7.6885
102	8.8083
104	10.041
106	11.392
108	12.870
110	14.481
112	16.233
114	18.133
116	20.190
118	22.411
120	24.806
122	27.386
124	30.174
126	33.227
126.200	33.555

Vapor Pressure (atm) vs. Temperature (K) for Oxygen

Temperature, K	Pressure, atm
54.351	0.001
56	0.002
58	0.004
60	0.007
62	0.012
64	0.018
66	0.028
68	0.042
70	0.061
72	0.087
74	0.122
76	0.167
78	0.224
80	0.297
82	0.387
84	0.497
86	0.631
88	0.791
90	0.981
90.180	1.000
92	1.205
94	1.466
96	1.768
98	2.114
100	2.509
102	2.957
104	3.462
106	4.029
108	4.661
110	5.363

Vapor Pressure (atm) vs. Temperature (K) for Oxygen (Continued)

Temperature, K	Pressure atm	Temperature, K	Pressure, atm
112	6.139	136	22.986
114	6.995	138	25.170
116	7.934	140	27.501
118	8.961		
120	10.082	142	[illegible]
		144	32.631
122	11.300	146	35.448
124	12.621	148	38.446
126	14.049	150	41.638
128	15.591		
130	17.249	152	45.041
		154	48.675
132	19.031	154.576	49.767
134	20.942		

Data from Sparks, L. L., ASRDI Oxygen Technology Survey, Vol. IV: Low Temperature Measurement, NASA-3073, 1974.

VAPOR PRESSURE OF NITRIC ACID

Temperature °C	Vapor Pressure, mm. of Hg	
	100% HNO_3	90% of HNO_3
0	14.4	5.5
10	26.6	11.
20	47.9	20.
30	81.3	37.3
40	133.	64.4
50	208.	107.
70	467.	242.
80	670.	352.
90	937.	504.
100	1282.	710.

LOW TEMPERATURE LIQUID BATHS

Liquid thermostat baths suitable for many physical measurements can be produced by using a stirred solid–liquid mixture at its melting point. Dry-ice or liquid air can be used to produce the solid. A Dewar flask is preferable as a container and, with adequate insulation or immersion in another somewhat colder bath, good temperature constancy can be maintained over several hours. Such baths are especially useful over the temperature range between dry-ice (−78C) and liquid air (−190). The following table gives the melting and normal boiling points of some readily available organic liquids suitable for this purpose. The compounds are listed in order of their increasing melting points. Temperatures are in degrees Celcius.

Compound	M.P.	B.P.
Isopentane (2-Methyl butane)	−159.9	27.85
Methyl cyclopentane	−142.4	71.8
Allyl chloride	−134.5	45
n-Pentane	−129.7	36.1
Allyl alcohol	−129	97
Ethyl alcohol	−117.3	78.5
Carbon disulfide	−110.8	46.3
Isobutyl alcohol	−108	108.1
Acetone	−95.4	56.2
Toluene	−95	110.6
Ethyl acetate	−84	77
(Dry-ice + acetone)	−78	—
p-Cymene	−67.9	177.1
Chloroform	−63.5	61.7
N-Methyl aniline	−57	196
Chlorobenzene	−45.6	132
Anisole	−37.5	155
Bromobenzene	−30.8	156
Carbon tetrachloride	−23	76.5
Benzonitrile	−13	205

FATS AND OILS

These data for fats and oils were compiled originally for the *Biology Data Book* by H. J. Harwood, and R. P. Geyer, 1964. Data are reproduced here by permission of the copyright owners of the above publication, the Federation of American Societies for Experimental Biology, Washington, D.C. pp. 380—82.

Values are typical rather than average, and frequently were derived from specific analyses for particular samples (especially the constituent fatty acids). Extreme variations may occur, depending on a number of variables such as source, treatment, and age of a fat or oil. **Specific Gravity** (column D) was calculated at the specified temperature (degrees centigrade) and referred to water at the same temperature, unless otherwise specified. **Density,** shown in parentheses (column D), was measured at the specified temperature (degrees centigrade). **Refractive Index** (column E) was measured at 50°C, unless otherwise specified.

		Constants					Constituent fatty acids (g/100 g total fatty acids)										
							Saturated						Unsaturated				
Fat or oil	Source	Melting (or solidification) Point (°C)	Specific gravity (or density)	Refractive index ($n_D^{40°C}$)	Iodine value	Saponification value	Lauric	Myristic	Palmitic	Stearic	Arachidic	Other	Palmitoleic	Oleic	Linoleic	Linolenic	Other
(A)	(B)	(C)	(D)	(E)	(F)	(G)	(H)	(I)	(J)	(K)	(L)	(M)	(N)	(O)	(P)	(Q)	(R)
Land animals																	
Butterfat	*Bos taurus*	32.2	0.911$^{40°/15°}$	1.4548	36.1	227	2.5	11.1	29.0	9.2	2.4	2.0;[a] 0.5;[b] 2.3[c]	4.6	26.7	3.6	—	3.6;[d] 0.1;[e] 0.1;[f] 0.9;[g] 1.4;[h] 1.0;[i] 1.0;[j] 0.4[k]
Depot fat	*Homo sapiens*	(15)	0.918$^{15°}$	1.4602	67.6	196.2	—	2.7	24.0	8.4	—	—	5	46.9	10.2	—	2.5[h]
Lard oil	*Sus scrofa*	(30.5)	0.919$^{15°}$	1.4615	58.6	194.6	—	1.3	28.3	11.9	—	—	2.7	47.5	6	—	0.2;[g] 2.1[h]
Neat's-foot oil	*B. taurus*	—	0.910$^{25°}$	1.464$^{25°}$	69—76	190—199	—	—	17—18	2—3	—	—	—	74—76	—	—	—
Tallow, beef	*B. taurus*	—	—	—	49.5	197	—	6.3	27.4	14.1	—	—	—	49.6	2.5	—	—
Tallow, mutton	*Ovis aries*	(42.0)	0.945$^{15°}$	1.4565	40	194	—	4.6	24.6	30.5	—	—	—	36.0	4.3	—	—
Marine animals																	
Cod-liver oil	*Gadus morhua*	—	0.925^{25}	1.481$^{25°}$	165	186	—	5.8	8.4	0.6	—	—	20.0	← 29.1 →		—	25.4;[l] 9.6[m]
Herring oil	*Clupea harengus*	—	0.900$^{60°}$	1.4610$^{60°}$	140	192	—	7.3	13.0	Trace	—	—	4.9	—	—	20.7	30.1;[l] 23.2[m]
Menhaden oil	*Brevoortia tyrannus*	—	0.903$^{60°}$	1.4645$^{60°}$	170	191	—	5.9	16.3	0.6	0.6	—	15.5	—	—	29.6	19.0;[l] 11.7[m] 0.8[n]
Sardine oil	*Sardinops caerulea*	—	0.905$^{60°}$	1.4660$^{60°}$	185	191	—	5.1	14.6	3.2	—	—	11.8	← 17.8 →		—	18.1;[l] 14.0;[m] trace;[g] 15.4[o]
Sperm oil(body)	*Physeter macrocephalus*	—	—	—	76—88	122—130	1	5	6.5	—	—	—	26.5	37	19	—	1;[m] 4;[g] 19[p]
Sperm oil(head)	*P. macrocephalus*	—	—	—	70	140—144	16	14	8	2	—	3.5[c]	15	17	6.5	—	4;[f] 14;[g] 6.5[p]
Whale oil	*Balaena mysticetus*	—	0.892^{60}	1.460$^{60°}$	120	195	0.2	9.3	15.6	2.8	—	—	14.4	35.2	—	—	13.6;[l] 5.9;[m] 2.5;[g] 0.2[q]
Plants																	
Babassu oil	*Attalea funifera*	22—26	(0.893$^{60°}$)	1.443$^{60°}$	15.5	247	44.1	15.4	8.5	2.7	0.2	0.2;[a] 4.8;[b] 6.6[c]	—	16.1	1.4	—	—
Castor oil	*Ricinus communis*	(−18.0)	0.961$^{15°}$	1.4770	85.5	180.3			← 2.4 →		—	—	—	7.4	3.1	—	87[r]
Cocoa butter	*Theobroma cacao*	34.1	0.964$^{15°}$	1.4568	36.5	193.8	—	—	24.4	35.4	—	—	—	38.1	2.1	—	—
Coconut oil	*Cocos nucifera*	25.1	0.924$^{15°}$	1.4493	10.4	268	45.4	18.0	10.5	2.3	0.4^{19}	0.8;[a] 5.4;[b] 8.4[c]	0.4	7.5	Trace	—	—
Corn oil	*Zea mays*	(−20.0)	0.922$^{15°}$	1.4734	122.6	192.0	—	1.4	10.2	3.0	—	—	1.5	49.6	34.3	—	—
Cotton seed oil	*Gossypium hirsutum*	(−1.0)	0.917$^{25°}$	1.4735	105.7	194.3	—	1.4	23.4	1.1	1.3	—	2.0	22.9	47.8	—	—
Linseed oil	*Linum usitatissimum*	(−24.0)	0.938$^{15°}$	1.4782$^{25°}$	178.7	190.3	—	—	6.3	2.5	0.5	—	—	19.0	24.1	47.4	0.2[n]
Mustard oil	*Brassica hirta*	—	0.9145$^{15°}$	1.475	102	174	—	1.3^{20}	—	—	—	—	—	27.2[t]	16.6[t]	401.8[t]	1.1;[n] 1.0;[u] 51.0[v]
Neem oil	*Melia azadirachta*	−3	0.017$^{15°}$	1.4615	71	194.5	—	2.6^{20}	14.1^{20}	24.0^{20}	0.8^{20}	—	—	58.5[t]	—	—	—
Niger-seed oil	*Guizotia abyssinica*	—	0.925$^{15°}$	1.471	128.5	190	—	3.3^{20}	8.2^{20}	4.8^{20}	0.5^{20}	—	—	30.3[t]	57.3[t]	—	—
Oiticica oil	*Licania rigida*	—	0.974$^{25°}$	—	140—180	—			← 11.3^{23} →			—	—	6.2	—	—	82.5[x]
Olive oil	*Olea europaea sativa*	(−6.0)	0.918$^{15°}$	1.4679	81.1	189.7	—	Trace	6.9	2.3	0.1	—	—	84.4	4.6	—	—
Palm oil	*Elaeis guineensis*	35.0	0.915$^{15°}$	1.4578	54.2	199.1	—	1.4	40.1	5.5	—	—	—	42.7	10.3	—	—
Palm-kernel oi!	*E. guineensis*	24.1	0.923$^{15°}$	1.4569	37.0	219.9	46.9	14.1	8.8	1.3	—	2.7;[b] 7.0[c]	—	18.5	0.7	—	—
Peanut oil	*Arachis hypogaea*	(3.0)	0.914$^{15°}$	1.4691	93.4	192.1	—	—	8.3	3.1	2.4	—	—	56.0	26.0	—	3.1;[n] 1.1[u]
Perilla oil	*Perilla frutescens*	—	(0.935$^{15°}$)	1.481$^{25°}$	195	192			← 9.6^{23} →			—	—	17.8	—	17.5	—
Poppy-seed oil	*Papaver somniferum*	(−15)	0.925$^{15°}$	1.4685	135	194	—	—	4.8^{20}	2.9^{20}	—	—	—	30.1[t]	62.2[t]	—	—
Rapeseed oil	*Brassica campestris*	(−10)	0.915$^{15°}$	1.4706	98.6	174.7	—	—	1	—	—	—	—	32	15	1	50[v]
Safflower oil	*Carthamus tinctorius*	—	(0.900$^{60°}$)	1.462$^{60°}$	145	192			← 6.8^{23} →			—	—	18.6	70.1	3.4	—
Sesame oil	*Sesamum indicum*	(−6.0)	0.919$^{25°}$	1.4646	106.6	187.9	—	—	9.1	4.3	0.8	—	—	45.4	40.4	—	—
Soybean oil	*Glycine soja*	(−16.0)	0.927$^{15°}$	1.4729	130.0	190.6	0.2	0.1	9.8	2.4	0.9	—	0.4	28.9	50.7	6.5	0.1[g]
Sunflower-seed oil	*Helianthus annuus*	(−17.0)	0.923$^{15°}$	1.4694	125.5	188.7	—	—	5.6	2.2	0.9	—	—	25.1	66.2	—	—
Tung oil	*Aleurites fordi*	(−2.5)	0.934$^{15°}$	1.5174$^{25°}$	168.2	193.1			← 4.6^{23} →			—	—	4.1	0.6	—	90.7[y]
Wheat-germ oil	*Triticum aestivum*	—	—	—	125	—			← 16.0^{23} →			—	—	28.1	52.3	3.6	—

[a] Caproic
[b] Capryli
[c] Capric
[d] Butyric
[e] Decenoic
[f] C_{12} monoethenoic
[g] C_{14} monoethenoic
[h] Gadoleic plus erucic
[i] C_{12} n-pentadecanoic.
[j] C_{17} margaric
[k] 12-Methyl tetradecanoic
[l] C_{20} polyethenoic
[m] C_{22} polyethenoic
[n] Behenic
[o] C_{14} polyethenoic
[p] Gadoleic
[q] C_{24} polyethenoic
[r] Ricinoleic
[s] Includes behenic and lignoceric
[t] Percent by weight
[u] Lignoceric
[v] Erucic
[w] Includes behenic
[x] Licanic.
[y] Eleostearic

CONCENTRATIVE PROPERTIES OF AQUEOUS SOLUTIONS: CONVERSION TABLES

A. V. WOLF, MORDEN G. BROWN AND PHOEBE G. PRENTISS

The table columns are:

$A\%$ = anhydrous solute weight per cent, g solute/100 g solution.
$H\%$ = hydrated solute weight per cent, g solute/100 g solution.
ρ or D_4^{20} = relative density at 20°C, kg/l.
D_{20}^{20} = specific gravity at 20°C.
C_s = anhydrous solute concentration, g/l.
M = molar concentration, g-mol/l.
C_w = total water concentration, g/l.
$(C_0 - C_w)$ = water displaced by anhydrous solute, g/l.
$(n - n_0) \times 10^4$ = index of refraction increment above index of refraction of pure water $\times 10^4$; refraction at 20°C.
n = index of refraction at 20°C relative to air for sodium yellow light.
Δ = freezing point depression, °C.
O = osmolality, Os/kg water.
S = osmosity, molar concentration of NaCl solution having same freezing point or osmotic pressure as given solution, g-mol/l.
η/η_0 = relative viscosity, ratio of the absolute viscosity of a solution at 20°C to the absolute viscosity of water at 20°C.
η/ρ = kinematic viscosity, ratio of absolute viscosity at 20°C (centipoise, cP) to relative density, (centistokes, cS).
ϕ = fluidity, reciprocal of absolute viscosity at 20°C, (poise^{-1}), rhe.
γ = specific conductance (electrical) at 20°C, millsiemans/cm or mS/cm.
T = condosity, molar concentration of NaCl solution having same specific conductance (electrical) at 20°C as given solution, g-mol/l.

Some related measures are:

Specific gravity, $D_{20}^{20} = D_4^{20}/.99823$
Relative density, $D_4^{20} = (C_s + C_w)/1000$
Molality, g-mol/kg water = $1000 \times M/C_w$
Osmolality, $O = \Delta/1.86$
Absolute viscosity (cP) = kinematic viscosity (cS) $\times D_4^{20}$
Specific viscosity = $\eta/\eta_0 - 1$
Specific refraction = $(n - 1)/\rho$
Specific refractive increment = $1000(n - n_0)/C_w$
Molar conductance (ohm^{-1} cm^2 g-mol^{-1}), $\Lambda_M = \lambda/M$
Relative salinity, $\Sigma = T/S$
Ratios such as S/M, T/M, r/A%, $(D_{20}^{20} - 1)/S$, etc.
G water displaced/g solute = $(C_0 - C_w)/C_s$
Ml water displaced/g solute = $(C_0 - C_w)/(.99823 \times C_s)$
G water displaced/mol solute = $(C_0 - C_w)/M$
Ml water displaced/mol solute = $(C_0 - C_w)/(.99823 \times M)$
Mols water displaced/mol solute = $(C_0 - C_w)/(18.015 \times M)$
Concentration of solute relative to water, g/kg water = $1000 \times C_s/C_w$

Relative specific refractivity = contribution of unit weight of solute relative to unit weight of water to the imaginary concentration of water with same refractive index as solution, equivalent concentration of water (ECW) = $3694.1788 \times n - 300 \times n^2 - 3394.1788$. The refractive index for this equation is relative to a vacuum whereas the tabulated values are relative to air. The above equation for the equivalent concentration of water provides better constancy of relative specific refractivity than Lorentz and Lorenz relation, ECW = $4848.3431 - 14545.029/(n^2 + 2)$.

Tables are based upon *Handbook of Chemistry and Physics,* 51st Edition (1970); G. F. Hewitt, Tables of the resistivity of aqueous chloride solutions, Chem. Eng. Div., U.K.A.E.A. Res. Group, Harwell, Gr. Brit. Oct. 1960, HL 60/5450 (S.C. 2), AERE-R 3497; A. V. Wolf, *Aqueous Solutions and Body Fluids,* Hoeber, 1966; and new data.

LIST OF TABLES

1 ACETIC ACID, CH_3COOH

MOLECULAR WEIGHT = 60.05
RELATIVE SPECIFIC REFRACTIVITY = 1.065

0.00 % by wt. data are the same for all compounds.
For Values of 0.00 wt. % solutions see Table 1, Acetic Acid.

A % by wt.	ρ D_4^{20}	D_{20}^{20}	C_s g/l	M g-mol/l	C_w g/l	$(C_o - C_w)$ g/l	$(n - n_o) \times 10^4$	n	Δ °C	O Os/kg	S g-mol/l	η/η_o	η/ρ cS	φ rhe	γ mmho/cm	T g-mol/l
0.00	0.9982	1.0000	0.0	0.000	998.2	0.0	0	1.3330	0.000	0.000	0.000	1.000	1.004	99.80	0.0	0.000
0.50	0.9989	1.0007	5.0	0.083	993.9	4.3	4	1.3334	0.159	0.086	0.045	1.010	1.013	98.81	0.3	0.003
1.00	0.9996	1.0014	10.0	0.166	989.6	8.6	7	1.3337	0.317	0.170	0.091	1.020	1.022	97.84	0.6	0.006
1.50	1.0003	1.0021	15.0	0.250	985.3	12.9	11	1.3341	0.474	0.255	0.137	1.030	1.032	96.90	0.7	0.007
2.00	1.0011	1.0028	20.0	0.333	981.0	17.2	15	1.3345	0.630	0.339	0.183	1.040	1.041	95.96	0.8	0.008
2.50	1.0018	1.0035	25.0	0.417	976.7	21.5	18	1.3348	0.786	0.423	0.229	1.050	1.051	95.02	0.9	0.009
3.00	1.0025	1.0042	30.1	0.501	972.4	25.9	22	1.3352	0.942	0.507	0.275	1.061	1.060	94.07	1.0	0.010
3.50	1.0031	1.0049	35.1	0.585	968.0	30.2	26	1.3355	1.099	0.591	0.321	1.072	1.070	93.14	1.1	0.010
4.00	1.0038	1.0056	40.2	0.669	963.7	34.5	29	1.3359	1.256	0.675	0.367	1.082	1.080	92.24	1.1	0.011
4.50	1.0045	1.0063	45.2	0.753	959.3	38.9	33	1.3363	1.415	0.761	0.414	1.092	1.090	91.36	1.2	0.011
5.00	1.0052	1.0070	50.3	0.837	955.0	43.3	36	1.3366	1.576	0.847	0.461	1.103	1.099	90.50	1.2	0.012
5.50	1.0059	1.0077	55.3	0.921	950.6	47.6	40	1.3370	1.737	0.934	0.508	1.113	1.109	89.66	1.3	0.012
6.00	1.0066	1.0084	60.4	1.006	946.2	52.0	44	1.3373	1.899	1.021	0.555	1.123	1.118	88.87	1.3	0.013
6.50	1.0073	1.0091	65.5	1.090	941.8	56.4	47	1.3377	2.062	1.108	0.602	1.132	1.126	88.14	1.4	0.013
7.00	1.0080	1.0098	70.6	1.175	937.4	60.8	51	1.3381	2.225	1.196	0.650	1.141	1.134	87.46	1.4	0.013
7.50	1.0087	1.0105	75.7	1.260	933.0	65.2	54	1.3384	2.390	1.285	0.697	1.150	1.142	86.78	1.4	0.014
8.00	1.0093	1.0111	80.7	1.345	928.6	69.6	58	1.3388	2.555	1.374	0.745	1.160	1.152	86.03	1.4	0.014
8.50	1.0100	1.0118	85.9	1.430	924.2	74.1	62	1.3391	2.722	1.463	0.792	1.171	1.162	85.20	1.5	0.014
9.00	1.0107	1.0125	91.0	1.515	919.7	78.5	65	1.3395	2.889	1.553	0.840	1.184	1.173	84.31	1.5	0.014
9.50	1.0114	1.0132	96.1	1.600	915.3	82.9	69	1.3399	3.057	1.644	0.888	1.196	1.185	83.43	1.5	0.015
10.00	1.0121	1.0138	101.2	1.685	910.8	87.4	72	1.3402	3.226	1.734	0.935	1.208	1.196	82.62	1.5	0.015
11.00	1.0134	1.0152	111.5	1.856	901.9	96.3	79	1.3409	3.567	1.918	1.031	1.229	1.215	81.19	1.6	0.015
12.00	1.0147	1.0165	121.8	2.028	893.0	105.3	86	1.3416	3.911	2.103	1.127	1.250	1.234	79.84	1.6	0.016
13.00	1.0161	1.0178	132.1	2.200	884.0	114.3	93	1.3423	4.259	2.290	1.223	1.272	1.255	78.43	1.6	0.016
14.00	1.0174	1.0192	142.4	2.372	874.9	123.3	100	1.3430	4.611	2.479	1.320	1.295	1.275	77.07	1.7	0.016
15.00	1.0187	1.0205	152.8	2.545	865.9	132.4	107	1.3437	4.967	2.670	1.415	1.317	1.295	75.79	1.7	0.016
16.00	1.0200	1.0218	163.2	2.718	856.8	141.5	114	1.3444	5.33	2.86	1.511	1.338	1.314	74.59	1.7	0.016
17.00	1.0213	1.0231	173.6	2.891	847.6	150.6	121	1.3451	5.69	3.06	1.607	1.357	1.331	73.54	1.7	0.017
18.00	1.0225	1.0243	184.1	3.065	838.5	159.8	128	1.3458	6.06	3.26	1.703	1.377	1.349	72.48	1.7	0.017
19.00	1.0238	1.0256	194.5	3.239	829.3	169.0	135	1.3465	6.43	3.46	1.798	1.402	1.372	71.18	1.7	0.017
20.00	1.0250	1.0269	205.0	3.414	820.0	178.2	142	1.3472	6.81	3.66	1.894	1.428	1.396	69.89	1.7	0.017
22.00	1.0275	1.0293	226.1	3.764	801.5	196.8	155	1.3485	7.57	4.07	2.084	1.475	1.438	67.66	1.7	0.017
24.00	1.0299	1.0318	247.2	4.116	782.8	215.5	169	1.3498	8.36	4.49	2.273	1.522	1.481	65.57	1.7	0.016
26.00	1.0323	1.0341	268.4	4.470	763.9	234.3	182	1.3512	9.17	4.93	2.462	1.569	1.523	63.61	1.6	0.015
28.00	1.0346	1.0365	289.7	4.824	744.9	253.3	195	1.3525	10.00	5.38	2.650	1.610	1.559	61.99	1.5	0.015
30.00	1.0369	1.0388	311.1	5.180	725.8	272.4	207	1.3537	10.84	5.83	2.835	1.666	1.610	59.90	1.4	0.014
32.00	1.0391	1.0410	332.5	5.537	706.6	291.6	220	1.3550	11.70	6.29	3.017	1.712	1.651	58.29	1.4	0.013
34.00	1.0413	1.0431	354.0	5.896	687.3	311.0	232	1.3562	12.55	6.75	3.192	1.758	1.692	56.77	1.3	0.012
36.00	1.0434	1.0452	375.6	6.255	667.8	330.5	245	1.3574	13.38	7.20	3.359	1.808	1.736	55.20	1.2	0.011
38.00	1.0454	1.0473	397.3	6.615	648.2	350.1	257	1.3586				1.848	1.771	54.00	1.1	0.011
40.00	1.0474	1.0492	419.0	6.977	628.4	369.8	268	1.3598				1.908	1.825	52.31	1.1	0.010
42.00	1.0493	1.0511	440.7	7.339	608.6	389.7	280	1.3610				1.956	1.868	51.02	1.0	0.010
44.00	1.0510	1.0529	462.5	7.701	588.6	409.7	291	1.3621				2.003	1.910	49.83	1.0	0.009
46.00	1.0528	1.0547	484.3	8.065	568.5	429.7	302	1.3632				2.048	1.949	48.73	0.9	0.009
48.00	1.0545	1.0564	506.2	8.429	548.4	449.9	312	1.3642				2.106	2.001	47.39	0.8	0.008
50.00	1.0562	1.0581	528.1	8.794	528.1	470.1	323	1.3653				2.154	2.043	46.33	0.8	0.007
52.00	1.0577	1.0596	550.0	9.159	507.7	490.5	333	1.3663				2.208	2.092	45.20	0.7	0.007
54.00	1.0592	1.0611	572.0	9.525	487.2	511.0	343	1.3673				2.260	2.138	44.16	0.7	0.006
56.00	1.0605	1.0624	593.9	9.890	466.6	531.6	352	1.3682				2.303	2.176	43.33	0.5	0.005
58.00	1.0618	1.0636	615.8	10.255	445.9	552.3	361	1.3691				2.355	2.222	42.38	0.3	0.003
60.00	1.0629	1.0648	637.7	10.620	425.2	573.1	370	1.3700				2.404	2.266	41.51	0.4	0.004
62.00	1.0640	1.0659	659.7	10.985	404.3	593.9	378	1.3708				2.451	2.308	40.72		
64.00	1.0650	1.0668	681.6	11.350	383.4	614.8	386	1.3716				2.497	2.349	39.97		
66.00	1.0659	1.0678	703.5	11.715	362.4	635.8	394	1.3724				2.548	2.395	39.17		
68.00	1.0668	1.0687	725.4	12.080	341.4	656.9	402	1.3732				2.589	2.432	38.55		
70.00	1.0673	1.0692	747.1	12.441	320.2	678.0	409	1.3738				2.624	2.463	38.03		
72.00	1.0676	1.0695	768.7	12.800	298.9	699.3	415	1.3745				2.657	2.494	37.56		
74.00	1.0678	1.0697	790.2	13.158	277.6	720.6	421	1.3751				2.682	2.517	37.21		
76.00	1.0680	1.0699	811.7	13.516	256.3	741.9	427	1.3757				2.709	2.542	36.84		
78.00	1.0681	1.0700	833.1	13.956	235.0	763.2	432	1.3762				2.715	2.547	36.76		
80.00	1.0680	1.0699	854.4	14.227	213.6	784.6	437	1.3767				2.715	2.547	36.76		
82.00	1.0677	1.0696	875.5	14.579	192.2	806.0	441	1.3770				2.691	2.525	37.09		
84.00	1.0673	1.0692	896.5	14.928	170.8	827.4	443	1.3773				2.653	2.491	37.62		
86.00	1.0666	1.0685	917.3	15.275	149.3	848.9	444	1.3774				2.591	2.434	38.52		
88.00	1.0658	1.0677	937.9	15.618	127.9	870.3	444	1.3774				2.506	2.356	39.82		
90.00	1.0644	1.0663	958.0	15.953	106.4	891.8	441	1.3771				2.381	2.241	41.92		
92.00	1.0629	1.0648	977.9	16.284	85.0	913.2	436	1.3766				2.236	2.108	44.63		
94.00	1.0606	1.0625	997.0	16.602	63.6	934.6	429	1.3759				2.032	1.920	49.11		
96.00	1.0578	1.0597	1015.5	16.912	42.3	955.9	418	1.3748				1.809	1.714	55.17		
98.00	1.0538	1.0557	1032.7	17.196	21.1	977.1	404	1.3734				1.532	1.457	65.14		
100.00	1.0477	1.0496	1047.7	17.446	0.0	998.2	386	1.3716				1.221	1.168	81.73		

2 ACETONE, CH_3COCH_3

MOLECULAR WEIGHT = 58.05
RELATIVE SPECIFIC REFRACTIVITY = 1.349

0.00 % by wt. data are the same for all compounds.
For Values of 0.00 wt. % solutions see Table 1, Acetic Acid.

A % by wt.	ρ D_4^{20}	D_{20}^{20}	C_s g/l	M g-mol/l	C_w g/l	$(C_o - C_w)$ g/l	$(n - n_o) \times 10^4$	n	Δ °C	O Os/kg	S g-mol/l	η/η_o	η/ρ cS	φ rhe	γ mmho/cm	T g-mol/l
0.50	0.9975	0.9993	5.0	0.086	992.5	5.7	4	1.3334	0.160	0.086	0.045	1.011	1.015	98.73		
1.00	0.9968	0.9985	10.0	0.172	986.8	11.4	7	1.3337	0.321	0.173	0.092	1.022	1.027	97.65		
1.50	0.9961	0.9978	14.9	0.257	981.1	17.1	11	1.3341	0.483	0.260	0.140	1.033	1.039	96.58		
2.00	0.9954	0.9971	19.9	0.343	975.5	22.8	15	1.3344	0.645	0.347	0.187	1.045	1.052	95.50		
2.50	0.9947	0.9964	24.9	0.428	969.8	28.4	18	1.3348	0.807	0.434	0.235	1.057	1.065	94.39		
3.00	0.9940	0.9957	29.8	0.514	964.1	34.1	22	1.3352	0.970	0.521	0.283	1.070	1.079	93.25		
3.50	0.9933	0.9950	34.8	0.599	958.5	39.7	25	1.3355	1.133	0.609	0.331	1.083	1.093	92.11		
4.00	0.9926	0.9943	39.7	0.684	952.9	45.4	29	1.3359	1.296	0.697	0.379	1.097	1.107	90.99		
4.50	0.9919	0.9937	44.6	0.769	947.3	51.0	33	1.3363	1.460	0.785	0.427	1.110	1.121	89.90		
5.00	0.9912	0.9930	49.6	0.854	941.7	56.6	36	1.3366	1.625	0.874	0.475	1.123	1.135	88.87		
5.50	0.9906	0.9923	54.5	0.939	936.1	62.1	40	1.3370	1.790	0.962	0.523	1.136	1.149	87.89		
6.00	0.9899	0.9917	59.4	1.023	930.5	67.7	44	1.3373	1.955	1.051	0.571	1.148	1.162	86.94		
6.50	0.9893	0.9910	64.3	1.108	925.0	73.3	47	1.3377	2.120	1.140	0.619	1.160	1.175	86.03		
7.00	0.9886	0.9904	69.2	1.192	919.4	78.8	51	1.3381	2.286	1.229	0.667	1.172	1.188	85.14		
7.50	0.9880	0.9897	74.1	1.276	913.9	84.4	54	1.3384	2.452	1.318	0.715	1.184	1.201	84.28		
8.00	0.9874	0.9891	79.0	1.361	908.4	89.9	58	1.3388	2.619	1.408	0.763	1.196	1.214	83.44		
8.50	0.9867	0.9885	83.9	1.445	902.9	95.4	62	1.3392	2.785	1.498	0.811	1.208	1.226	82.64		
9.00	0.9861	0.9879	88.8	1.529	897.4	100.9	65	1.3395	2.952	1.587	0.858	1.219	1.239	81.85		
9.50	0.9855	0.9873	93.6	1.613	891.9	106.3	69	1.3399	3.119	1.677	0.905	1.231	1.251	81.09		
10.00	0.9849	0.9867	98.5	1.697	886.4	111.8	73	1.3402	3.287	1.767	0.953	1.242	1.264	80.35		

3 AMMONIA, NH_3, AND AMMONIUM HYDROXIDE, NH_4OH

MOLECULAR WEIGHT, NH_3 = 17.03 FORMULA WEIGHT, NH_4OH = 35.05
RELATIVE SPECIFIC REFRACTIVITY = 1.582

0.00 % by wt. data are the same for all compounds.
For Values of 0.00 wt. % solutions see Table 1, Acetic Acid.

NH_3 % by wt.	NH_4OH % by wt.	ρ D_4^{20}	D_{20}^{20}	C_s (NH_3) g/l	M g-mol/l	C_w g/l	$(C_o - C_w)$ g/l	$(n - n_o) \times 10^4$	n	Δ °C	O Os/kg	S g-mol/l	η/η_o	η/ρ cS	φ rhe	γ mmho/cm	T g-mol/l
0.50	1.03	0.9960	0.9978	5.0	0.292	991.0	7.2	2	1.3332	0.550	0.296	0.160	1.007	1.013	99.14	0.5	0.005
1.00	2.06	0.9938	0.9956	9.9	0.584	983.9	14.3	5	1.3335	1.135	0.610	0.332	1.013	1.022	98.48	0.7	0.007
1.50	3.09	0.9917	0.9934	14.9	0.873	976.8	21.5	7	1.3337	1.725	0.927	0.504	1.020	1.031	97.82	0.9	0.008
2.00	4.12	0.9895	0.9913	19.8	1.162	969.7	28.5	9	1.3339	2.319	1.247	0.677	1.027	1.040	97.17	1.0	0.009
2.50	5.15	0.9874	0.9891	24.7	1.449	962.7	35.5	12	1.3342	2.921	1.570	0.849	1.034	1.049	96.52	1.0	0.010
3.00	6.17	0.9853	0.9870	29.6	1.736	955.7	42.5	14	1.3344	3.531	1.898	1.021	1.041	1.059	95.88	1.1	0.011
3.50	7.20	0.9832	0.9849	34.4	2.021	948.8	49.5	17	1.3347	4.150	2.231	1.193	1.048	1.068	95.24	1.1	0.011
4.00	8.23	0.9811	0.9828	39.2	2.304	941.8	56.4	19	1.3349	4.781	2.570	1.365	1.055	1.077	94.60	1.1	0.011
4.50	9.26	0.9790	0.9808	44.1	2.587	935.0	63.3	22	1.3352	5.42	2.92	1.537	1.062	1.087	93.98	1.1	0.011
5.00	10.29	0.9770	0.9787	48.8	2.868	928.1	70.1	24	1.3354	6.08	3.27	1.708	1.069	1.096	93.36	1.1	0.011
5.50	11.32	0.9750	0.9767	53.6	3.149	921.3	76.9	27	1.3357	6.75	3.63	1.879	1.076	1.106	92.75	1.1	0.011
6.00	12.35	0.9730	0.9747	58.4	3.428	914.6	83.6	30	1.3359	7.43	4.00	2.049	1.083	1.115	92.15	1.1	0.011
6.50	13.38	0.9710	0.9727	63.1	3.706	907.9	90.4	32	1.3362	8.13	4.37	2.219	1.090	1.125	91.55	1.1	0.011
7.00	14.41	0.9690	0.9707	67.8	3.983	901.2	97.1	35	1.3365	8.85	4.76	2.388	1.097	1.134	90.96	1.1	0.011
7.50	15.44	0.9671	0.9688	72.5	4.259	894.5	103.7	38	1.3367	9.58	5.15	2.557	1.104	1.144	90.38	1.1	0.011
8.00	16.47	0.9651	0.9668	77.2	4.534	887.9	110.3	40	1.3370	10.34	5.56	2.725	1.111	1.154	89.81	1.1	0.011
8.50	17.49	0.9632	0.9649	81.9	4.807	881.3	116.9	43	1.3373	11.11	5.97	2.892	1.118	1.163	89.25	1.1	0.010
9.00	18.52	0.9613	0.9630	86.5	5.080	874.8	123.5	46	1.3376	11.90	6.40	3.059	1.125	1.173	88.70	1.1	0.010
9.50	19.55	0.9594	0.9611	91.1	5.352	868.3	130.0	48	1.3378	12.71	6.84	3.225	1.132	1.182	88.16	1.0	0.010
10.00	20.58	0.9575	0.9592	95.8	5.623	861.8	136.5	51	1.3381	13.55	7.28	3.391	1.139	1.192	87.63	1.0	0.010
11.00	22.64	0.9538	0.9555	104.9	6.161	848.9	149.3	57	1.3387	15.29	8.22	3.719	1.153	1.211	86.57	1.0	0.009
12.00	24.70	0.9502	0.9519	114.0	6.695	836.2	162.1	63	1.3393	17.13	9.21	4.044	1.167	1.230	85.53	0.9	0.009
13.00	26.76	0.9466	0.9483	123.1	7.226	823.5	174.7	68	1.3398	19.07	10.25	4.365	1.180	1.250	84.55	0.8	0.008
14.00	28.81	0.9431	0.9447	132.0	7.753	811.0	187.2	74	1.3404	21.13	11.36		1.193	1.268	83.64	0.8	0.007
15.00	30.87	0.9396	0.9412	140.9	8.276	798.6	199.6	80	1.3410	23.32	12.54		1.205	1.285	82.84	0.7	0.007
16.00	32.93	0.9361	0.9378	149.8	8.795	786.4	211.9	86	1.3416	25.63	13.78		1.216	1.301	82.10	0.7	0.006
17.00	34.99	0.9327	0.9344	158.6	9.311	774.2	224.1	92	1.3422	28.09	15.10		1.226	1.317	81.43	0.6	0.006
18.00	37.05	0.9294	0.9310	167.3	9.823	762.1	236.1	98	1.3428	30.70	16.51		1.235	1.331	80.82	0.6	0.005
19.00	39.10	0.9261	0.9277	176.0	10.332	750.1	248.1	104	1.3434	33.47	18.00		1.243	1.345	80.26	0.5	0.005
20.00	41.16	0.9228	0.9245	184.6	10.838	738.3	260.0	110	1.3440	36.42	19.58		1.251	1.359	79.75	0.5	0.005
22.00	45.28	0.9164	0.9181	201.6	11.839	714.8	283.4	123	1.3453	43.36	23.31		1.265	1.383	78.88	0.4	0.004
24.00	49.40	0.9102	0.9118	218.4	12.827	691.7	306.5	135	1.3465	51.38	27.62		1.277	1.405	78.18	0.4	0.004
26.00	53.51	0.9040	0.9056	235.0	13.802	669.0	329.3	148	1.3477	60.77	32.67		1.285	1.425	77.64	0.4	0.004
28.00	57.63	0.8980	0.8996	251.4	14.764	646.5	351.7	160	1.3490	71.66	38.53						
30.00	61.74	0.8920	0.8936	267.6	15.713	624.4	373.8	172	1.3502	84.06	45.19						

4 AMMONIUM CHLORIDE, NH_4Cl

MOLECULAR WEIGHT = 53.50
RELATIVE SPECIFIC REFRACTIVITY = 1.241

0.00 % by wt. data are the same for all compounds.
For Values of 0.00 wt. % solutions see Table 1, Acetic Acid.

A% by wt.	ρ D_4^{20}	D_{20}^{20}	C_s g/l	M g-mol/l	C_w g/l	$(C_o - C_w)$ g/l	$(n - n_o)$ × 10^4	n	Δ °C	O Os/kg	S g-mol/l	η/η_o	η/ρ cS	φ rhe	γ mmho/cm	T g-mol/l
0.50	0.9998	1.0016	5.0	0.093	994.8	3.4	10	1.3340	0.322	0.173	0.092	0.997	0.999	100.13	10.5	0.110
1.00	1.0014	1.0032	10.0	0.187	991.4	6.9	19	1.3349	0.637	0.343	0.185	0.994	0.995	100.40	20.4	0.225
1.50	1.0030	1.0047	15.0	0.281	987.9	10.3	29	1.3359	0.953	0.512	0.278	0.992	0.991	100.62	30.3	0.347
2.00	1.0045	1.0063	20.1	0.376	984.4	13.8	39	1.3369	1.270	0.683	0.371	0.990	0.988	100.81	40.3	0.474
2.50	1.0061	1.0079	25.2	0.470	980.9	17.3	48	1.3378	1.590	0.855	0.465	0.988	0.984	101.00	49.9	0.601
3.00	1.0076	1.0094	30.2	0.565	977.4	20.8	58	1.3388	1.913	1.029	0.559	0.986	0.981	101.18	59.2	0.728
3.50	1.0092	1.0110	35.3	0.660	973.8	24.4	68	1.3397	2.240	1.205	0.654	0.985	0.978	101.35	68.3	0.858
4.00	1.0107	1.0125	40.4	0.755	970.3	28.0	77	1.3407	2.571	1.383	0.749	0.983	0.975	101.52	77.2	0.993
4.50	1.0122	1.0140	45.6	0.851	966.7	31.6	87	1.3417	2.907	1.563	0.845	0.982	0.972	101.68	86.3	1.13
5.00	1.0138	1.0155	50.7	0.947	963.1	35.2	96	1.3426	3.246	1.745	0.941	0.980	0.969	101.84	95.3	1.27
5.50	1.0153	1.0171	55.8	1.044	959.4	38.8	106	1.3436	3.591	1.931	1.038	0.979	0.966	101.99	104.	1.42
6.00	1.0168	1.0186	61.0	1.140	955.8	42.5	115	1.3445	3.941	2.119	1.135	0.977	0.963	102.15	113.	1.58
6.50	1.0183	1.0201	66.2	1.237	952.1	46.1	125	1.3455	4.296	2.310	1.233	0.976	0.960	102.30	122.	1.74
7.00	1.0198	1.0216	71.4	1.334	948.4	49.8	135	1.3464	4.657	2.504	1.332	0.974	0.957	102.44	131.	1.91
7.50	1.0213	1.0231	76.6	1.432	944.7	53.6	144	1.3474	5.02	2.70	1.431	0.973	0.955	102.58	139.	2.09
8.00	1.0227	1.0246	81.8	1.529	940.9	57.3	154	1.3483	5.40	2.90	1.530	0.972	0.952	102.71	147.	2.27
8.50	1.0242	1.0260	87.1	1.627	937.2	61.1	163	1.3493	5.77	3.10	1.629	0.971	0.950	102.83	156.	2.46
9.00	1.0257	1.0275	92.3	1.725	933.4	64.9	172	1.3502	6.16	3.31	1.728	0.970	0.947	102.93	164.	2.66
9.50	1.0272	1.0290	97.6	1.824	929.6	68.7	182	1.3512	6.55	3.52	1.828	0.969	0.945	103.02	172.	2.86
10.00	1.0286	1.0304	102.9	1.923	925.8	72.5	191	1.3521	6.95	3.73	1.928	0.968	0.943	103.10	180.	3.07
11.00	1.0315	1.0333	113.5	2.121	918.1	80.2	210	1.3540	7.76	4.17	2.129	0.967	0.939	103.20	200.	3.71
12.00	1.0344	1.0362	124.1	2.320	910.3	88.0	229	1.3559	8.60	4.62	2.330	0.967	0.936	103.23	220.	4.84
13.00	1.0373	1.0391	134.8	2.520	902.4	95.8	248	1.3578	9.47	5.09	2.531	0.967	0.934	103.22		
14.00	1.0401	1.0419	145.6	2.722	894.5	103.8	267	1.3596				0.967	0.932	103.17		
15.00	1.0429	1.0447	156.4	2.924	886.5	111.8	285	1.3615				0.968	0.930	103.10		
16.00	1.0457	1.0475	167.3	3.127	878.4	119.8	304	1.3634				0.969	0.928	103.01		
17.00	1.0485	1.0503	178.2	3.332	870.2	128.0	322	1.3652				0.970	0.927	102.90		
18.00	1.0512	1.0531	189.2	3.537	862.0	136.2	341	1.3671				0.971	0.926	102.75		
19.00	1.0540	1.0558	200.3	3.743	853.7	144.5	360	1.3689				0.973	0.925	102.54		
20.00	1.0567	1.0586	211.3	3.950	845.3	152.9	378	1.3708				0.976	0.925	102.25		
22.00	1.0621	1.0640	233.7	4.367	828.4	169.8	415	1.3745				0.984	0.928	101.46		
24.00	1.0674	1.0693	256.2	4.788	811.2	187.0	452	1.3782				0.994	0.933	100.40		

5 AMMONIUM SULFATE, $(NH_4)_2SO_4$

MOLECULAR WEIGHT = 132.14
RELATIVE SPECIFIC REFRACTIVITY = 0.886

0.00 % by wt. data are the same for all compounds.
For Values of 0.00 wt. % solutions see Table 1, Acetic Acid.

A% by wt.	ρ D_4^{20}	D_{20}^{20}	C_s g/l	M g-mol/l	C_w g/l	$(C_o - C_w)$ g/l	$(n - n_o)$ × 10^4	n	Δ °C	O Os/kg	S g-mol/l	η/η_o	η/ρ cS	φ rhe	γ mmho/cm	T g-mol/l
0.50	1.0012	1.0030	5.0	0.038	996.2	2.1	8	1.3338	0.173	0.093	0.049	1.006	1.007	99.22	7.4	0.077
1.00	1.0042	1.0059	10.0	0.076	994.1	4.1	16	1.3346	0.331	0.178	0.095	1.012	1.010	98.62	14.2	0.151
2.00	1.0101	1.0119	20.2	0.153	989.9	8.3	33	1.3363	0.631	0.339	0.183	1.025	1.017	97.37	25.7	0.290
3.00	1.0160	1.0178	30.5	0.231	985.6	12.7	49	1.3379	0.921	0.495	0.269	1.039	1.025	96.03	36.7	0.428
4.00	1.0220	1.0238	40.9	0.309	981.1	17.1	65	1.3395	1.208	0.649	0.353	1.055	1.034	94.63	47.2	0.565
5.00	1.0279	1.0297	51.4	0.389	976.5	21.7	82	1.3411	1.489	0.800	0.435	1.071	1.044	93.18	57.4	0.704
6.00	1.0338	1.0356	62.0	0.469	971.8	26.5	98	1.3428	1.768	0.951	0.517	1.088	1.055	91.71	67.4	0.845
7.00	1.0397	1.0416	72.8	0.551	966.9	31.3	114	1.3444	2.047	1.101	0.598	1.106	1.066	90.21	77.0	0.987
8.00	1.0456	1.0475	83.6	0.633	962.0	36.3	130	1.3460	2.327	1.251	0.679	1.125	1.079	88.67	86.3	1.13
9.00	1.0515	1.0534	94.6	0.716	956.9	41.4	146	1.3476	2.608	1.402	0.760	1.145	1.091	87.13	95.5	1.28
10.00	1.0574	1.0593	105.7	0.800	951.6	46.6	162	1.3492	2.892	1.555	0.841	1.166	1.105	85.59	105.	1.43
12.00	1.0691	1.0710	128.3	0.971	940.8	57.4	193	1.3523	3.473	1.867	1.005	1.208	1.133	82.58	122.	1.75
14.00	1.0808	1.0827	151.3	1.145	929.5	68.8	225	1.3555	4.073	2.190	1.171	1.253	1.162	79.63	139.	2.09
16.00	1.0924	1.0943	174.8	1.323	917.6	80.6	256	1.3586	4.689	2.521	1.341	1.302	1.194	76.66	156.	2.44
18.00	1.1039	1.1059	198.7	1.504	905.2	93.0	287	1.3616				1.356	1.231	73.59	171.	2.82
20.00	1.1154	1.1174	223.1	1.688	892.3	105.9	317	1.3647				1.418	1.274	70.38	185.	3.21
22.00	1.1269	1.1289	247.9	1.876	879.0	119.3	347	1.3677				1.487	1.323	67.10	198.	3.67
24.00	1.1383	1.1403	273.2	2.067	865.1	133.2	377	1.3707				1.563	1.376	63.84	210.	4.18
26.00	1.1496	1.1516	298.9	2.262	850.7	147.5	407	1.3737				1.647	1.435	60.60	220.	4.84
28.00	1.1609	1.1629	325.0	2.460	835.8	162.4	436	1.3766				1.740	1.502	57.37		
30.00	1.1721	1.1742	351.6	2.661	820.5	177.7	465	1.3795				1.843	1.576	54.15		
32.00	1.1833	1.1854	378.7	2.866	804.7	193.6	494	1.3824				1.957	1.657	50.99		
34.00	1.1945	1.1966	406.1	3.073	788.3	209.9	523	1.3853				2.082	1.747	47.93		
36.00	1.2056	1.2077	434.0	3.284	771.6	226.7	551	1.3881				2.218	1.844	44.99		
38.00	1.2166	1.2188	462.3	3.499	754.3	243.9	580	1.3909				2.366	1.948	42.19		
40.00	1.2277	1.2298	491.1	3.716	736.6	261.6	608	1.3938				2.525	2.061	39.52		

6 BARIUM CHLORIDE, $BaCl_2 \cdot 2H_2O$

MOLECULAR WEIGHT = 208.27 FORMULA WEIGHT, HYDRATE = 244.31
RELATIVE SPECIFIC REFRACTIVITY = 0.552

0.00 % by wt. data are the same for all compounds.
For Values of 0.00 wt. % solutions see Table 1, Acetic Acid.

A% by wt.	H% by wt.	ρ D_4^{20}	D_{20}^{20}	C_s g/l	M g-mol/l	C_w g/l	$(C_o - C_w)$ g/l	$(n - n_o)$ × 10^4	n	Δ °C	O Os/kg	S g-mol/l	η/η_o	η/ρ cS	φ rhe	γ mmho/cm	T g-mol/l
0.50	0.59	1.0026	1.0044	5.0	0.024	997.6	0.6	7	1.3337	0.119	0.064	0.033	1.007	1.007	99.06	4.7	0.047
1.00	1.17	1.0070	1.0088	10.1	0.048	996.9	1.3	15	1.3345	0.233	0.125	0.066	1.014	1.009	98.42	9.1	0.094
2.00	2.35	1.0159	1.0177	20.3	0.098	995.6	2.6	30	1.3360	0.461	0.248	0.133	1.024	1.010	97.46	17.4	0.189
3.00	3.52	1.0249	1.0267	30.7	0.148	994.2	4.0	45	1.3375	0.691	0.372	0.201	1.035	1.012	96.41	25.2	0.284
4.00	4.69	1.0341	1.0359	41.4	0.199	992.7	5.5	61	1.3391	0.928	0.499	0.271	1.047	1.015	95.28	32.8	0.379
5.00	5.87	1.0434	1.0452	52.2	0.250	991.2	7.0	76	1.3406	1.177	0.633	0.344	1.060	1.018	94.15	40.4	0.476
6.00	7.04	1.0528	1.0547	63.2	0.303	989.6	8.6	92	1.3422	1.435	0.772	0.420	1.073	1.021	93.05	47.9	0.575
7.00	8.21	1.0624	1.0643	74.4	0.357	988.0	10.2	108	1.3438	1.704	0.916	0.498	1.085	1.024	91.95	55.3	0.675
8.00	9.38	1.0721	1.0740	85.8	0.412	986.3	11.9	124	1.3454	1.984	1.067	0.580	1.099	1.027	90.85	62.6	0.777
9.00	10.56	1.0820	1.0839	97.4	0.468	984.6	13.6	140	1.3470	2.274	1.222	0.664	1.112	1.030	89.72	69.8	0.880
10.00	11.73	1.0921	1.0940	109.2	0.524	982.9	15.4	157	1.3487	2.575	1.385	0.751	1.127	1.034	88.55	76.7	0.983
12.00	14.08	1.1128	1.1148	133.5	0.641	979.3	19.0	191	1.3520	3.218	1.730	0.933	1.159	1.043	86.14	90.1	1.19
14.00	16.42	1.1342	1.1362	158.8	0.762	975.4	22.8	225	1.3555	3.921	2.108	1.129	1.193	1.054	83.63	103.0	1.40
16.00	18.77	1.1564	1.1584	185.0	0.888	971.4	26.9	261	1.3591	4.689	2.521	1.341	1.232	1.067	81.03	115.0	1.61
18.00	21.11	1.1793	1.1814	212.3	1.019	967.0	31.2	297	1.3627				1.274	1.083	78.32	126.0	1.82
20.00	23.46	1.2031	1.2052	240.6	1.155	962.5	35.8	334	1.3664				1.322	1.101	75.49	137.0	2.04
22.00	25.81	1.2277	1.2299	270.1	1.297	957.6	40.6	373	1.3703				1.375	1.122	72.57	147.0	2.26
24.00	28.15	1.2531	1.2553	300.7	1.444	952.4	45.9	411	1.3741				1.434	1.147	69.57	157.0	2.48
26.00	30.50	1.2793	1.2816	332.6	1.597	946.7	51.6	451	1.3781				1.500	1.175	66.53	166.0	2.70

7 CADMIUM CHLORIDE, $CdCl_2$

MOLECULAR WEIGHT = 183.32

A% by wt.	ρ D_4^{20}	D_{20}^{20}	C_s gl	M g-mol/l	C_w g/l	$(C_o - C_w)$ g/l	$(n - n_o)$ × 10^4	n	Δ °C	O Os/kg	S g-mol/l	η/η_o	η/ρ cS	φ rhe	γ mmho/cm	T g-mol/l
1.00	1.0062	1.0080	10.1	0.055	996.1	2.1	16.	1.3346	0.186	0.100	0.053	1.012	1.008	98.60	5.3	0.053
2.00	1.0148	1.0166	20.3	0.111	994.5	3.7	31.	1.3361	0.368	0.198	0.106	1.028	1.015	97.09	9.0	0.093
3.00	1.0237	1.0255	30.7	0.168	993.0	5.2	47.	1.3377	0.555	0.299	0.161	1.044	1.022	95.59	11.8	0.124
4.00	1.0325	1.0344	41.3	0.225	991.2	7.0	62.	1.3392	0.715	0.384	0.208	1.059	1.028	94.21	14.1	0.150
5.00	1.0415	1.0434	52.1	0.284	989.4	8.8	78.	1.3408	0.878	0.472	0.256	1.076	1.035	92.77	16.0	0.172
6.00	1.0508	1.0527	63.0	0.344	987.8	10.4	94.	1.3424	1.035	0.557	0.302	1.092	1.041	91.42	17.5	0.190
7.00	1.0600	1.0619	74.2	0.405	985.8	12.4	110.	1.3440	1.191	0.640	0.348	1.109	1.048	90.02	18.9	0.207
8.00	1.0692	1.0711	85.5	0.467	983.7	14.5	127.	1.3457	1.351	0.726	0.395	1.125	1.054	88.74	20.2	0.222
9.00	1.0794	1.0813	97.1	0.530	982.3	15.9	143.	1.3473	1.491	0.802	0.436	1.146	1.064	87.08	21.2	0.235
10.00	1.0897	1.0917	109.0	0.594	980.7	17.5	159.	1.3489	1.634	0.878	0.478	1.166	1.072	85.60	22.1	0.246
12.00	1.1103	1.1123	133.2	0.727	977.1	21.1	193.	1.3523	1.936	1.041	0.566	1.208	1.090	82.63	23.7	0.265
14.00	1.1308	1.1328	158.3	0.864	972.5	25.7	228.	1.3558	2.236	1.203	0.653	1.264	1.120	78.96	24.7	0.278
16.00	1.1531	1.1552	184.5	1.006	968.6	29.6	264.	1.3594	2.564	1.378	0.747	1.322	1.149	75.48	25.5	0.287
18.00	1.1749	1.1770	211.5	1.154	963.4	34.8	301.	1.3631	2.889	1.553	0.840	1.386	1.182	72.01	25.9	0.293
20.00	1.1988	1.2010	239.8	1.308	959.0	39.2	338.	1.3668	3.246	1.745	0.941	1.455	1.216	68.60	26.3	0.297
24.00	1.2476	1.2498	299.4	1.633	948.2	50.0	420.	1.3750	3.984	2.142	1.147	1.596	1.282	62.52	25.8	0.291
28.00	1.3003	1.3026	364.1	1.986	936.2	62.0	506.	1.3836	4.750	2.554	1.357	1.832	1.412	54.47	24.5	0.275
32.00	1.3558	1.3582	433.8	2.367	922.0	76.2	597.	1.3927				2.135	1.578	46.74	22.5	0.251
36.00	1.4178	1.4204	510.4	2.784	907.4	90.8	696.	1.4026				2.556	1.806	39.05	20.2	0.222
40.00	1.4843	1.4870	593.7	3.239	890.6	107.6	803.	1.4133				3.259	2.200	30.62	17.2	0.187
44.00	1.5563	1.5591	684.8	3.735	871.5	126.7	918.	1.4248				4.155	2.675	24.02	14.3	0.152
48.00	1.6345	1.6374	784.5	4.280	850.0	148.2	1044.	1,4374				5.740	3.519	17.39	11.3	0.119
52.00	1.7206	1.7237	894.7	4.881	825.9	172.3	1180.	1.4510				8.811	5.131	11.33	8.4	0.087
56.00	1.8131	1.8164	1015.4	5.539	797.7	200.5	1326.	1.4656				14.67	8.109	6.802	5.9	0.060

8 CADMIUM SULFATE, $CdSO_4$

Molecular Weight = 208.46

A% by wt.	ρ D_4^{20}	D_{20}^{20}	C_s g/l	M g-mol/l	C_w g/l	$(C_o - C_w)$ g/l	$(n - n_o)$ × 10^4	n	Δ °C	O Os/kg	S g-mol/l	η/η_o	η/ρ cS	φ rhe	γ mmo/cm	T g-mol/l
1.00	1.0078	1.0096	10.1	0.048	997.7	0.5	14.	1.3344	0.135	0.053	0.028	1.029	1.023	97.00	4.2	0.042
2.00	1.0175	1.0193	20.3	0.098	997.2	1.0	28.	1.3358	0.194	0.104	0.055	1.059	1.043	94.23	7.4	0.075
3.00	1.0271	1.0290	30.8	0.148	996.3	1.9	42.	1.3372	0.290	0.156	0.083	1.089	1.062	91.67	10.2	0.105
4.00	1.0375	1.0394	41.5	0.199	996.0	2.2	56.	1.3386	0.382	0.205	0.110	1.125	1.086	88.75	12.6	0.133
5.00	1.0478	1.0497	52.4	0.251	995.4	2.8	69.	1.3399	0.477	0.256	0.138	1.162	1.111	85.90	14.9	0.160
6.00	1.0579	1.0598	63.5	0.304	994.4	3.8	84.	1.3414	0.569	0.306	0.165	1.197	1.134	83.36	17.1	0.186
7.00	1.0684	1.0703	74.8	0.359	993.6	4.6	98	1.3428	0.657	0.354	0.191	1.238	1.161	80.62	20.1	0.211
8.00	1.0789	1.0808	86.3	0.414	992.6	5.6	114	1.3444	0.742	0.399	0.216	1.280	1.189	77.96	21.2	0.235
9.00	1.0894	1.0914	98.0	0.470	991.4	6.8	129.	1.3459	0.831	0.447	0.242	1.325	1.219	75.30	23.2	0.259
10.00	1.1004	1.1024	110.0	0.528	990.4	7.8	143.	1.3473	0.933	0.502	0.272	1.375	1.252	72.58	25.1	0.282
12.00	1.1233	1.1253	134.8	0.647	988.5	9.7	173	1.3503	1.137	0.611	0.332	1.480	1.320	67.44	28.7	0.327
14.00	1.1470	1.1491	160.6	0.770	986.4	11.8	203.	1.3533	1.354	0.728	0.396	1.603	1.400	62.27	32.0	0.369
16.00	1.1715	1.1736	187.4	0.899	984.1	14.1	234.	1.3564	1.604	0.862	0.469	1.742	1.490	57.29	35.6	0.405
18.00	1.1967	1.1989	215.4	1.033	981.3	16.9	266.	1.3596	1.854	0.997	0.542	1.909	1.598	52.29	37.5	0.439
20.00	1.2229	1.2251	244.6	1.173	978.3	19.9	299.	1.3629	2.140	1.150	0.625	2.081	1.705	47.96	39.8	0.469
24.00	1.2784	1.2807	306.8	1.472	971.6	26.6	369.	1.3699	2.864	1.540	0.833	2.515	1.971	39.69	43.2	0.513
28.00	1.3385	1.3409	374.8	1.798	963.7	34.5	445.	1.3775	3959	2.129	1.140	3.150	2.358	31.68	45.2	0.539
32.00	1.3996	1.4021	447.9	2.148	951.7	46.5	520.	1.3850				4.083	2.923	24.44	45.5	0.542
36.00	1.4655	1.4681	527.6	2.531	937.9	60.3	597.	1.3927				5.334	3.647	18.71	44.1	0.524
40.00	1.5456	1.5484	618.2	2.966	927.4	70.8	676.	1.4006				7.381	4.785	13.52	40.3	0.475

9 CALCIUM CHLORIDE, $CaCl_2 \cdot 2H_2O$

MOLECULAR WEIGHT = 110.99 FORMULA WEIGHT, HYDRATE = 147.03
RELATIVE SPECIFIC REFRACTIVITY = 0.863

0.00 % by wt. data are the same for all compounds.
For Values of 0.00 wt. % solutions see Table 1, Acetic Acid.

A% by wt.	H% by wt.	ρ D_4^{20}	D_{20}^{20}	C_s g/l	M g-mol/l	C_w g/l	$(C_o - C_w)$ g/l	$(n - n_o)$ × 10^4	n	Δ °C	O Os/kg	S g-mol/l	η/η_o	η/ρ cS	φ rhe	γ mmho/cm	T g-mol/l
0.50	0.66	1.0024	1.0041	5.0	0.045	997.3	0.9	12	1.3342	0.222	0.119	0.063	1.013	1.013	98.47	8.1	0.083
1.00	1.32	1.0065	1.0083	10.1	0.091	996.4	1.8	24	1.3354	0.440	0.237	0.127	1.026	1.021	97.27	15.7	0.169
1.50	1.99	1.0106	1.0124	15.2	0.137	995.5	2.7	36	1.3366	0.661	0.355	0.192	1.037	1.028	96.24	22.7	0.254
2.00	2.65	1.0148	1.0166	20.3	0.183	994.5	3.7	48	1.3378	0.880	0.473	0.257	1.048	1.035	95.23	29.4	0.337
2.50	3.31	1.0190	1.0208	25.5	0.230	993.5	4.7	60	1.3390	1.102	0.593	0.322	1.061	1.043	94.06	36.1	0.421
3.00	3.97	1.0232	1.0250	30.7	0.277	992.5	5.8	72	1.3402	1.330	0.715	0.389	1.076	1.053	92.79	42.6	0.504
3.50	4.64	1.0274	1.0292	36.0	0.324	991.4	6.8	84	1.3414	1.567	0.843	0.458	1.091	1.064	91.45	48.9	0.588
4.00	5.30	1.0316	1.0334	41.3	0.372	990.3	7.9	96	1.3426	1.815	0.976	0.531	1.108	1.076	90.09	55.1	0.672
4.50	5.96	1.0358	1.0377	46.6	0.420	989.2	9.0	109	1.3438	2.074	1.115	0.606	1.125	1.088	88.75	61.1	0.756
5.00	6.62	1.0401	1.0419	52.0	0.469	988.1	10.2	121	1.3451	2.345	1.261	0.684	1.141	1.099	87.47	67.0	0.840
5.50	7.29	1.0443	1.0462	57.4	0.518	986.9	11.3	133	1.3463	2.630	1.414	0.766	1.157	1.110	86.25	72.8	0.924
6.00	7.95	1.0486	1.0505	62.9	0.567	985.7	12.5	145	1.3475	2.930	1.575	0.852	1.173	1.121	85.07	78.3	1.01
6.50	8.61	1.0529	1.0548	68.4	0.617	984.5	13.8	158	1.3487	3.244	1.744	0.941	1.189	1.132	83.91	83.6	1.09
7.00	9.27	1.0572	1.0591	74.0	0.667	983.2	15.0	170	1.3500	3.573	1.921	1.033	1.206	1.143	82.77	88.7	1.17
7.50	9.94	1.0615	1.0634	79.6	0.717	981.9	16.3	182	1.3512	3.917	2.106	1.128	1.222	1.154	81.64	93.6	1.25
8.00	10.60	1.0659	1.0678	85.3	0.768	980.6	17.6	195	1.3525	4.275	2.299	1.227	1.240	1.165	80.51	98.4	1.33
8.50	11.26	1.0703	1.0722	91.0	0.820	979.3	18.9	207	1.3537	4.649	2.499	1.330	1.257	1.177	79.38	103.	1.41
9.00	11.92	1.0747	1.0766	96.7	0.871	977.9	20.3	220	1.3549	5.04	2.71	1.434	1.276	1.190	78.22	108.	1.49
9.50	12.58	1.0791	1.0810	102.5	0.924	976.6	21.7	232	1.3562	5.44	2.92	1.541	1.295	1.203	77.04	112.	1.57
10.00	13.25	1.0835	1.0854	108.3	0.976	975.1	23.1	245	1.3575	5.86	3.15	1.651	1.316	1.217	75.84	117.	1.65
11.00	14.57	1.0923	1.0943	120.2	1.083	972.2	26.1	270	1.3600	6.74	3.63	1.877	1.359	1.247	73.41	125.	1.81
12.00	15.90	1.1014	1.1033	132.2	1.191	969.2	29.0	295	1.3625	7.70	4.14	2.113	1.405	1.278	71.03	133.	1.97
13.00	17.22	1.1105	1.1125	144.4	1.301	966.2	32.1	321	1.3651	8.72	4.69	2.358	1.454	1.311	68.66	141.	2.13
14.00	18.55	1.1198	1.1218	156.8	1.412	963.0	35.2	347	1.3677	9.83	5.28	2.611	1.505	1.347	66.30	148.	2.28
15.00	19.87	1.1292	1.1312	169.4	1.526	959.8	38.4	374	1.3704	11.01	5.92	2.872	1.561	1.385	63.93	154.	2.42
16.00	21.20	1.1386	1.1407	182.2	1.641	956.5	41.8	400	1.3730	12.28	6.60	3.138	1.622	1.427	61.54	160.	2.55
17.00	22.52	1.1482	1.1502	195.2	1.759	953.0	45.2	427	1.3757	13.65	7.34	3.410	1.688	1.473	59.14	165.	2.67
18.00	23.84	1.1579	1.1599	208.4	1.878	949.5	48.8	454	1.3784	15.11	8.12	3.686	1.760	1.523	56.72	169.	2.78
19.00	25.17	1.1677	1.1697	221.9	1.999	945.8	52.4	482	1.3812	16.7	8.98	3.97	1.839	1.578	54.27	173.	2.88
20.00	26.49	1.1775	1.1796	235.5	2.122	942.0	56.2	509	1.3839	18.3	9.84	4.24	1.926	1.639	51.82	177.	2.97
22.00	29.14	1.1976	1.1997	263.5	2.374	934.1	64.1	565	1.3895	21.7	11.67		2.123	1.777	47.00	182.	3.11
24.00	31.79	1.2180	1.2201	292.3	2.634	925.7	72.6	621	1.3951	25.3	13.60		2.351	1.934	42.45	183.	3.16
26.00	34.44	1.2388	1.2410	322.1	2.902	916.7	81.5	678	1.4008	29.7	15.97		2.640	2.135	32.81	182.	3.12
28.00	37.09	1.2600	1.2622	352.8	3.179	907.2	91.0	736	1.4066	34.7	18.66		2.994	2.381	33.33	179.	3.03
30.00	39.74	1.2816	1.2838	384.5	3.464	897.1	101.1	794	1.4124	41.0	22.04		3.460	2.705	28.85	172.	2.85
32.00	42.39	1.3036	1.3059	417.1	3.758	886.4	111.8	853	1.4183	49.7	26.72		4.027	3.095	24.79	162.	2.61
34.00	45.04	1.3260	1.3283	450.8	4.062	875.1	123.1	912	1.4242				4.810	3.635	20.75	150.	2.34
36.00	47.69	1.3488	1.3512	485.6	4.375	863.2	135.0	971	1.4301				5.795	4.305	17.22	137.	2.05
38.00	50.34	1.3720	1.3745	521.4	4.697	850.7	147.6	1031	1.4361				7.306	5.336	13.66	123.	1.76
40.00	52.99	1.3957	1.3982	558.3	5.030	837.4	160.8	1090	1.4420				8.979	6.446	11.12	106.	1.45

10 CESIUM CHLORIDE, CsCl

MOLECULAR WEIGHT = 168.37
RELATIVE SPECIFIC REFRACTIVITY = 0.464

0.00 % by wt. data are the same for all compounds.
For Values of 0.00 wt. % solutions see Table 1, Acetic Acid.

A% by wt.	ρ D_4^{20}	D_{20}^{20}	C_s g/l	M g-mol/l	C_w g/l	$(C_o - C_w)$ g/l	$(n - n_o)$ × 10^4	n	Δ °C	O Os/kg	S g-mol/l	η/η_o	η/ρ cS	φ rhe	γ mmho/cm	T g-mol/l
0.50	1.0020	1.0038	5.0	0.030	997.0	1.2	4	1.3334	0.100	0.054	0.028	0.998	0.997	100.05	3.8	0.038
1.00	1.0058	1.0076	10.1	0.060	995.8	2.5	8	1.3337	0.201	0.108	0.057	0.995	0.991	100.30	7.4	0.075
1.50	1.0097	1.0114	15.1	0.090	994.5	3.7	11	1.3341	0.302	0.162	0.087	0.992	0.985	100.56	10.6	0.111
2.00	1.0135	1.0153	20.3	0.120	993.3	5.0	15	1.3345	0.403	0.217	0.116	0.990	0.979	100.81	13.8	0.147
2.50	1.0174	1.0192	25.4	0.151	992.0	6.2	19	1.3349	0.505	0.272	0.146	0.988	0.973	101.04	17.0	0.184
3.00	1.0214	1.0232	30.6	0.182	990.7	7.5	23	1.3353	0.607	0.326	0.176	0.986	0.967	101.26	20.2	0.222
3.50	1.0253	1.0271	35.9	0.213	989.4	8.8	27	1.3357	0.709	0.381	0.206	0.984	0.961	101.46	23.3	0.260
4.00	1.0293	1.0311	41.2	0.245	988.2	10.1	31	1.3361	0.812	0.436	0.236	0.982	0.956	101.65	26.5	0.300
4.50	1.0334	1.0352	46.5	0.276	986.9	11.4	35	1.3365	0.915	0.492	0.267	0.980	0.950	101.84	29.7	0.340
5.00	1.0374	1.0392	51.9	0.308	985.5	12.7	39	1.3369	1.018	0.547	0.297	0.978	0.945	102.02	32.9	0.380
5.50	1.0415	1.0433	57.3	0.340	984.2	14.0	43	1.3373	1.121	0.602	0.327	0.976	0.939	102.20	36.1	0.421
6.00	1.0456	1.0475	62.7	0.373	982.9	15.3	47	1.3377	1.223	0.658	0.358	0.975	0.934	102.37	39.3	0.462
6.50	1.0498	1.0516	68.2	0.405	981.6	16.7	52	1.3382	1.326	0.713	0.388	0.973	0.929	102.53	42.5	0.504
7.00	1.0540	1.0558	73.8	0.438	980.2	18.0	56	1.3386	1.428	0.768	0.418	0.972	0.924	102.67	45.8	0.546
7.50	1.0582	1.0601	79.4	0.471	978.8	19.4	60	1.3390	1.531	0.823	0.448	0.971	0.919	102.82	49.0	0.589
8.00	1.0625	1.0643	85.0	0.505	977.5	20.8	64	1.3394	1.635	0.879	0.478	0.969	0.914	102.95	52.3	0.633
8.50	1.0668	1.0686	90.7	0.539	976.1	22.2	69	1.3399	1.739	0.935	0.509	0.968	0.909	103.09	55.6	0.678
9.00	1.0711	1.0730	96.4	0.573	974.7	23.5	73	1.3403	1.845	0.992	0.539	0.967	0.904	103.22	59.0	0.724
9.50	1.0754	1.0773	102.2	0.607	973.3	25.0	77	1.3407	1.952	1.049	0.571	0.966	0.900	103.36	62.4	0.772
10.00	1.0798	1.0818	108.0	0.641	971.9	26.4	82	1.3412	2.060	1.108	0.602	0.964	0.895	103.51	65.8	0.822
11.00	1.0887	1.0907	119.8	0.711	969.0	29.3	91	1.3421	2.281	1.227	0.666	0.961	0.885	103.80	72.8	0.926
12.00	1.0978	1.0997	131.7	0.782	966.1	32.2	100	1.3430	2.507	1.348	0.731	0.959	0.875	104.09	80.0	1.03
13.00	1.1070	1.1089	143.9	0.855	963.1	35.2	109	1.3439	2.737	1.472	0.797	0.956	0.865	104.38	87.3	1.15
14.00	1.1163	1.1183	156.3	0.928	960.0	38.2	118	1.3448	2.972	1.598	0.864	0.953	0.856	104.67	94.8	1.27
15.00	1.1258	1.1278	168.9	1.003	956.9	41.3	128	1.3458	3.212	1.727	0.932	0.951	0.846	104.96	102.	1.40
16.00	1.1355	1.1375	181.7	1.079	953.8	44.5	138	1.3468	3.457	1.859	1.000	0.948	0.837	105.27	110.	1.54
17.00	1.1453	1.1473	194.7	1.156	950.6	47.7	147	1.3477	3.708	1.993	1.070	0.945	0.827	105.58	118.	1.68
18.00	1.1552	1.1573	207.9	1.235	947.3	51.0	157	1.3487	3.963	2.131	1.141	0.943	0.817	105.89	126.	1.83
19.00	1.1653	1.1674	221.4	1.315	943.9	54.3	167	1.3497	4.224	2.271	1.213	0.940	0.808	106.19	134.	1.99
20.00	1.1756	1.1777	235.1	1.396	940.5	57.7	178	1.3507	4.491	2.415	1.287	0.937	0.799	106.49	142.	2.17
22.00	1.1967	1.1989	263.3	1.564	933.5	64.8	198	1.3528				0.932	0.781	107.05	159.	2.53
24.00	1.2185	1.2207	292.4	1.737	926.1	72.1	220	1.3550				0.928	0.763	107.56	176.	2.95
26.00	1.2411	1.2433	322.7	1.916	918.4	79.8	242	1.3572				0.924	0.746	107.97	192.	3.48

10 CESIUM CHLORIDE, CsCl—(Continued)

A % by wt.	ρ D_4^{20}	D_{20}^{20}	C_s g/l	M g-mol/l	C_w g/l	$(C_o - C_w)$ g/l	$(n - n_o)$ × 10^4	n	Δ °C	O Os/kg	S g-mol/l	η/η_o	η/ρ cS	φ rhe	γ mmho/cm	T g-mol/l
28.00	1.2644	1.2666	354.0	2.103	910.4	87.9	264	1.3594				0.922	0.730	108.27	207.	4.15
30.00	1.2885	1.2908	386.6	2.296	902.0	96.3	288	1.3617				0.920	0.716	108.44	221.	4.95
32.00	1.3135	1.3158	420.3	2.496	893.1	105.1	311	1.3641				0.920	0.702	108.43		
34.00	1.3393	1.3417	455.4	2.705	883.9	114.3	336	1.3666				0.922	0.690	108.25		
36.00	1.3661	1.3685	491.8	2.921	874.3	124.0	361	1.3691				0.924	0.678	107.97		
38.00	1.3938	1.3963	529.6	3.146	864.2	134.1	387	1.3717				0.928	0.667	107.56		
40.00	1.4226	1.4251	569.0	3.380	853.5	144.7	414	1.3744				0.932	0.657	107.02		
42.00	1.4525	1.4550	610.0	3.623	842.4	155.8	442	1.3771				0.938	0.647	106.40		
44.00	1.4835	1.4861	652.7	3.877	830.8	167.5	470	1.3800				0.945	0.638	105.63		
46.00	1.5158	1.5185	697.3	4.141	818.5	179.7	500	1.3829				0.954	0.630	104.64		
48.00	1.5495	1.5522	743.7	4.417	805.7	192.5	530	1.3860				0.965	0.624	103.45		
50.00	1.5846	1.5874	792.3	4.706	792.3	206.0	562	1.3892				0.979	0.619	101.94		
52.00	1.6212	1.6241	843.0	5.007	778.2	220.0	595	1.3925				0.998	0.617	100.02		
54.00	1.6596	1.6625	896.2	5.323	763.4	234.8	630	1.3960				1.021	0.616	97.76		
56.00	1.6999	1.7029	951.9	5.654	747.9	250.3	666	1.3996				1.048	0.618	95.25		
58.00	1.7422	1.7453	1010.5	6.001	731.7	266.5	705	1.4035				1.078	0.620	92.54		
60.00	1.7868	1.7900	1072.1	6.367	714.7	283.5	746	1.4076				1.118	0.627	89.27		
62.00	1.8340	1.8373	1137.1	6.754	696.9	301.3	790	1.4120				1.170	0.639	85.28		
64.00	1.8842	1.8875	1205.9	7.162	678.3	319.9	837	1.4167				1.236	0.657	80.74		

11 CITRIC ACID, $(COOH)CH_2C(OH)(COOH)CH_2COOH \cdot 1H_2O$

MOLECULAR WEIGHT = 192.12 FORMULA WEIGHT, HYDRATE = 210.14
RELATIVE SPECIFIC REFRACTIVITY = 0.951

0.00 % by wt. data are the same for all compounds.
For Values of 0.00 wt. % solutions see Table 1, Acetic Acid.

A % by wt.	H % by wt.	ρ D_4^{20}	D_{20}^{20}	C_s g/l	M g-mol/l	C_w g/l	$(C_o - C_w)$ g/l	$(n - n_o)$ × 10^4	n	Δ °C	O Os/kg	S g-mol/l	η/η_o	η/ρ cS	φ rhe	γ mmho/cm	T g-mol/l
0.50	0.55	1.0002	1.0020	5.0	0.026	995.2	3.0	7	1.3336	0.052	0.028	0.014	1.011	1.013	98.74	1.2	0.012
1.00	1.09	1.0022	1.0040	10.0	0.052	992.2	6.0	13	1.3343	0.105	0.056	0.029	1.022	1.022	97.65	2.1	0.021
2.00	2.19	1.0063	1.0081	20.1	0.105	986.2	12.0	26	1.3356	0.211	0.113	0.060	1.046	1.041	95.41	3.0	0.030
3.00	3.28	1.0105	1.0123	30.3	0.158	980.2	18.1	38	1.3368	0.318	0.171	0.091	1.071	1.062	93.21	3.7	0.037
4.00	4.38	1.0147	1.0165	40.6	0.211	974.1	24.1	51	1.3381	0.428	0.230	0.123	1.096	1.082	91.05	4.2	0.042
5.00	5.47	1.0189	1.0207	50.9	0.265	968.0	30.3	64	1.3394	0.539	0.290	0.156	1.123	1.104	88.87	4.7	0.047
6.00	6.56	1.0232	1.0250	61.4	0.320	961.8	36.5	77	1.3407	0.651	0.350	0.189	1.151	1.128	86.68	5.1	0.051
7.00	7.66	1.0274	1.0292	71.9	0.374	955.5	42.8	90	1.3420	0.764	0.411	0.222	1.181	1.152	84.50	5.5	0.055
8.00	8.75	1.0316	1.0335	82.5	0.430	949.1	49.1	103	1.3433	0.879	0.473	0.256	1.212	1.177	82.32	5.7	0.058
9.00	9.84	1.0359	1.0377	93.2	0.485	942.7	55.6	116	1.3446	0.997	0.536	0.291	1.245	1.204	80.15	6.0	0.060
10.00	10.94	1.0402	1.0420	104.0	0.541	936.1	62.1	129	1.3459	1.120	0.602	0.327	1.280	1.233	77.97	6.2	0.063
12.00	13.13	1.0490	1.0509	125.9	0.655	923.2	75.1	156	1.3486	1.379	0.741	0.403	1.354	1.293	73.72	6.6	0.067
14.00	15.31	1.0580	1.0599	148.1	0.771	909.9	88.3	184	1.3514	1.656	0.890	0.484	1.433	1.358	69.62	6.9	0.070
16.00	17.50	1.0672	1.0690	170.7	0.889	896.4	101.8	211	1.3541	1.950	1.048	0.570	1.522	1.429	65.58	7.1	0.072
18.00	19.69	1.0764	1.0783	193.8	1.009	882.7	115.6	239	1.3569	2.255	1.213	0.658	1.622	1.510	61.53	7.1	0.072
20.00	21.88	1.0858	1.0877	217.2	1.130	868.6	129.6	268	1.3598	2.567	1.380	0.748	1.737	1.603	57.46	7.2	0.073
22.00	24.06	1.0953	1.0972	241.0	1.254	854.3	143.9	296	1.3626	2.884	1.550	0.839	1.868	1.709	53.43	7.2	0.073
24.00	26.25	1.1049	1.1069	265.2	1.380	839.8	158.5	325	1.3655	3.210	1.726	0.931	2.013	1.826	49.57	7.2	0.073
26.00	28.44	1.1147	1.1167	289.8	1.509	824.9	173.4	355	1.3684	3.547	1.907	1.025	2.174	1.954	45.91	7.1	0.072
28.00	30.63	1.1246	1.1266	314.9	1.639	809.7	188.5	384	1.3714	3.893	2.093	1.122	2.351	2.094	42.46	6.8	0.069
30.00	32.81	1.1346	1.1366	340.4	1.772	794.2	204.0	414	1.3744	4.250	2.285	1.220	2.544	2.247	39.23	6.5	0.066

12 COBALTOUS CHLORIDE, $CoCl_2 \cdot 6H_2O$

MOLECULAR WEIGHT = 129.85 FORMULA WEIGHT, HYDRATE = 237.95
RELATIVE SPECIFIC REFRACTIVITY = 0.733

0.00 % by wt. data are the same for all compounds.
For Values of 0.00 wt. % solutions see Table 1, Acetic Acid.

A % by wt.	H % by wt.	ρ D_4^{20}	D_{20}^{20}	C_s g/l	M g-mol/l	C_w g/l	$(C_o - C_w)$ g/l	$(n - n_o)$ × 10^4	n	Δ °C	O Os/kg	S g-mol/l	η/η_o	η/ρ cS	φ rhe	γ mmho/cm	T g-mol/l
0.50	0.92	1.0027	1.0045	5.0	0.039	997.7	0.5	11	1.3341	0.191	0.103	0.054	1.017	1.017	98.09	6.7	0.068
1.00	1.83	1.0073	1.0090	10.1	0.078	997.2	1.1	23	1.3352	0.380	0.204	0.110	1.034	1.029	96.52	12.7	0.134
2.00	3.66	1.0164	1.0182	20.3	0.157	996.0	2.2	45	1.3375	0.771	0.415	0.224	1.064	1.049	93.80	23.4	0.262
3.00	5.50	1.0256	1.0274	30.8	0.237	994.8	3.4	68	1.3398	1.187	0.638	0.347	1.096	1.071	91.05	33.6	0.389
4.00	7.33	1.0349	1.0368	41.4	0.319	993.5	4.7	91	1.3421	1.627	0.875	0.476	1.130	1.094	88.30	43.3	0.514
5.00	9.16	1.0444	1.0462	52.2	0.402	992.2	6.1	114	1.3444	2.099	1.128	0.613	1.166	1.119	85.59	52.5	0.636
6.00	10.99	1.0540	1.0559	63.2	0.487	990.8	7.5	137	1.3467	2.611	1.404	0.761	1.203	1.143	82.97	61.0	0.754
7.00	12.83	1.0637	1.0656	74.5	0.573	989.3	9.0	161	1.3491	3.164	1.701	0.918	1.241	1.169	80.43	69.0	0.868
8.00	14.66	1.0736	1.0755	85.9	0.661	987.7	10.5	185	1.3515	3.761	2.022	1.085	1.281	1.195	77.91	76.4	0.979
9.00	16.49	1.0837	1.0856	97.5	0.751	986.2	12.1	209	1.3539				1.324	1.224	75.40	83.4	1.08
10.00	18.32	1.0939	1.0958	109.4	0.842	984.5	13.7	234	1.3563				1.370	1.255	72.85	89.9	1.18
12.00	21.99	1.1148	1.1168	133.8	1.030	981.1	17.2	283	1.3613				1.474	1.324	67.73	102.	1.38
14.00	25.65	1.1364	1.1384	159.1	1.225	977.3	20.9	334	1.3664				1.592	1.403	62.71	112.	1.55
16.00	29.32	1.1587	1.1607	185.4	1.428	973.3	24.9	386	1.3716				1.725	1.492	57.85	120.	1.70
18.00	32.98	1.1816	1.1836	212.7	1.638	968.9	29.4	439	1.3769				1.876	1.591	53.19	126.	1.82
20.00	36.65	1.2050	1.2071	241.0	1.856	964.0	34.3	492	1.3822				2.046	1.701	48.78	130.	1.90

13 CREATININE, $CH_3NC(:NH)NHCOCH_2$

MOLECULAR WEIGHT = 113.12
RELATIVE SPECIFIC REFRACTIVITY = 1.301

0.00 % by wt. data are the same for all compounds.
For Values of 0.00 wt. % solutions see Table 1, Acetic Acid.

A % by wt.	D_4^{20}	D_{20}^{20}	C_s g/l	M g-mol/l	C_w g/l	$(C_o - C_w)$ g/l	$(n - n_o) \times 10^4$	n	Δ °C	O Os/kg	S g-mol/l	η/η_o	η/ρ cS	φ rhe	γ mmho/cm	T g-mol/l
0.50	0.9995	1.0012	5.0	0.044	994.5	3.8	9	1.3339	0.080	0.043	0.022	1.009	1.011	98.95		
1.00	1.0007	1.0025	10.0	0.088	990.7	7.5	19	1.3349	0.150	0.081	0.042	1.019	1.020	97.97		
1.50	1.0019	1.0037	15.0	0.133	986.9	11.3	28	1.3358	0.220	0.118	0.063	1.030	1.030	96.89		
2.00	1.0032	1.0050	20.1	0.177	983.1	15.1	38	1.3368	0.300	0.161	0.086	1.043	1.041	95.73		
2.50	1.0045	1.0063	25.1	0.222	979.4	18.9	48	1.3378	0.380	0.204	0.110	1.056	1.053	94.52		
3.00	1.0058	1.0075	30.2	0.267	975.6	22.6	57	1.3387	0.460	0.247	0.133	1.070	1.066	93.29		
3.50	1.0071	1.0088	35.2	0.312	971.8	26.4	67	1.3397				1.084	1.079	92.05		
4.00	1.0084	1.0102	40.3	0.357	968.0	30.2	77	1.3407				1.099	1.092	90.83		
4.50	1.0097	1.0115	45.4	0.402	964.3	34.0	87	1.3417				1.113	1.105	89.65		
5.00	1.0110	1.0128	50.6	0.447	960.5	37.7	97	1.3427				1.128	1.117	88.51		
5.50	1.0124	1.0142	55.7	0.492	956.7	41.5	107	1.3437				1.141	1.130	87.45		
6.00	1.0138	1.0156	60.8	0.538	952.9	45.3	117	1.3447				1.154	1.141	86.46		
6.50	1.0151	1.0169	66.0	0.583	949.2	49.1	127	1.3457				1.166	1.151	85.57		
7.00	1.0165	1.0183	71.2	0.629	945.4	52.8	137	1.3467				1.177	1.160	84.78		
7.50	1.0180	1.0198	76.3	0.675	941.6	56.6	148	1.3478				1.186	1.168	84.11		
8.00	1.0194	1.0212	81.6	0.721	937.8	60.4	158	1.3488				1.194	1.174	83.58		

14 CUPRIC SULFATE, $CuSO_4 \cdot 5H_2O$

MOLECULAR WEIGHT = 159.61 FORMULA WEIGHT, HYDRATE = 249.69
RELATIVE SPECIFIC REFRACTIVITY = 0.517

0.00 % by wt. data are the same for all compounds.
For Values of 0.00 wt. % solutions see Table 1, Acetic Acid.

A % by wt.	H % by wt.	ρ D_4^{20}	D_{20}^{20}	C_s g/l	M g-mol/l	C_w g/l	$(C_o - C_w)$ g/l	$(n - n_o) \times 10^4$	n	Δ °C	O Os/kg	S g-mol/l	η/η_o	η/ρ cS	φ rhe	γ mmho/cm	T g-mol/l
0.50	0.78	1.0033	1.0051	5.0	0.031	998.3	−0.1	9	1.3339	0.075	0.040	0.021	1.015	1.014	98.30	2.9	0.028
1.00	1.56	1.0085	1.0103	10.1	0.063	998.4	−0.2	18	1.3348	0.140	0.075	0.039	1.034	1.027	96.52	5.4	0.054
1.50	2.35	1.0137	1.0155	15.2	0.095	998.5	−0.3	28	1.3358	0.200	0.108	0.057	1.057	1.045	94.40	7.4	0.076
2.00	3.13	1.0190	1.0208	20.4	0.128	998.6	−0.4	37	1.3367	0.259	0.139	0.074	1.082	1.064	92.24	9.3	0.096
2.50	3.91	1.0243	1.0261	25.6	0.160	998.7	−0.4	47	1.3377	0.316	0.170	0.091	1.105	1.081	90.31	11.1	0.116
3.00	4.69	1.0296	1.0314	30.9	0.194	998.7	−0.5	56	1.3386	0.372	0.200	0.107	1.127	1.097	88.54	12.8	0.135
3.50	5.48	1.0349	1.0368	36.2	0.227	998.7	−0.5	66	1.3396	0.428	0.230	0.124	1.149	1.112	86.86	14.4	0.154
4.00	6.26	1.0403	1.0421	41.6	0.261	998.7	−0.5	75	1.3405	0.484	0.260	0.140	1.171	1.128	85.23	16.0	0.172
4.50	7.04	1.0457	1.0475	47.1	0.295	998.6	−0.4	85	1.3414	0.539	0.290	0.156	1.194	1.144	83.58	17.5	0.190
5.00	7.82	1.0511	1.0530	52.6	0.329	998.6	−0.3	94	1.3424	0.594	0.319	0.172	1.219	1.162	81.87	19.0	0.208
5.50	8.60	1.0565	1.0584	58.1	0.364	998.4	−0.2	104	1.3433	0.648	0.349	0.188	1.246	1.181	80.12	20.5	0.226
6.00	9.39	1.0620	1.0639	63.7	0.399	998.3	0.0	113	1.3443	0.703	0.378	0.205	1.273	1.201	78.37	21.9	0.243
6.50	10.17	1.0675	1.0694	69.4	0.435	998.1	0.1	123	1.3452	0.759	0.408	0.221	1.302	1.222	76.63	23.3	0.260
7.00	10.95	1.0730	1.0749	75.1	0.471	997.9	0.3	132	1.3462	0.816	0.439	0.238	1.333	1.244	74.90	24.6	0.277
7.50	11.73	1.0786	1.0805	80.9	0.507	997.7	0.5	142	1.3472	0.874	0.470	0.255	1.364	1.267	73.17	26.0	0.293
8.00	12.52	1.0842	1.0861	86.7	0.543	997.5	0.8	151	1.3481	0.933	0.501	0.272	1.397	1.291	71.46	27.2	0.309
8.50	13.30	1.0898	1.0918	92.6	0.580	997.2	1.0	161	1.3491	0.992	0.533	0.289	1.431	1.315	69.77	28.5	0.325
9.00	14.08	1.0955	1.0975	98.6	0.618	996.9	1.3	171	1.3501	1.052	0.566	0.307	1.466	1.341	68.09	29.7	0.341
9.50	14.86	1.1012	1.1032	104.6	0.655	996.6	1.6	181	1.3511	1.113	0.599	0.325	1.502	1.367	66.44	31.0	0.356
10.00	15.64	1.1070	1.1090	110.7	0.694	996.3	1.9	191	1.3520	1.177	0.633	0.344	1.540	1.394	64.81	32.2	0.371
11.00	17.21	1.1186	1.1206	123.0	0.771	995.6	2.7	210	1.3540	1.308	0.703	0.382	1.617	1.449	61.70	34.4	0.400
12.00	18.77	1.1304	1.1324	135.6	0.850	994.7	3.5	230	1.3560	1.447	0.778	0.423	1.698	1.505	58.77	36.6	0.427
13.00	20.34	1.1424	1.1444	148.5	0.930	993.8	4.4	251	1.3581	1.596	0.858	0.467	1.786	1.566	55.88	38.6	0.453
14.00	21.90	1.1545	1.1566	161.6	1.013	992.9	5.3	271	1.3601	1.753	0.942	0.513	1.885	1.636	52.95	40.5	0.478
15.00	23.47	1.1669	1.1690	175.0	1.097	991.9	6.4	292	1.3622				2.000	1.717	49.90	42.3	0.501
16.00	25.03	1.1796	1.1817	188.7	1.182	990.9	7.4	314	1.3644				2.132	1.811	46.81	44.0	0.523
17.00	26.59	1.1926	1.1947	202.7	1.270	989.8	8.4	336	1.3666				2.280	1.916	43.77	45.6	0.543
18.00	28.16	1.2059	1.2080	217.1	1.360	988.8	9.4	359	1.3689				2.444	2.031	40.83	47.0	0.562

15 DEXTRAN, $(C_6H_{10}O_5)_x$

MOLECULAR WEIGHT, AVERAGE = 72,000
RELATIVE SPECIFIC REFRACTIVITY = 1.045

0.00 % by wt. data are the same for all compounds.
For Values of 0.00 wt. % solutions see Table 1, Acetic Acid.

A % by wt.	ρ D_4^{20}	d D_{20}^{20}	C_s g/l	M g-mol/l	C_w g/l	$(C_o - C_w)$ g/l	$(n - n_o) \times 10^4$	n	Δ °C	O Os/kg	S g-mol/l	η/η_o	η/ρ cS	φ rhe	γ mmho/cm	T g-mol/l
0.50	1.0001	1.0019	5.0		995.1	3.1	7	1.3337				1.129	1.131	88.38		
1.00	1.0020	1.0037	10.0		992.0	6.3	15	1.3345				1.280	1.280	77.97		
1.50	1.0039	1.0056	15.1		988.8	9.4	22	1.3352				1.453	1.450	68.69		
2.00	1.0057	1.0075	20.1		985.6	12.6	29	1.3359				1.647	1.641	60.60		
2.50	1.0076	1.0094	25.2		982.5	15.8	37	1.3367				1.858	1.848	53.71		
3.00	1.0096	1.0113	30.3		979.3	19.0	44	1.3374				2.086	2.071	47.83		
3.50	1.0115	1.0133	35.4		976.1	22.2	52	1.3382				2.336	2.314	42.73		
4.00	1.0134	1.0152	40.5		972.9	25.4	59	1.3389				2.611	2.581	38.23		
4.50	1.0153	1.0171	45.7		969.6	28.6	67	1.3397				2.915	2.877	34.24		
5.00	1.0173	1.0191	50.9		966.4	31.8	74	1.3404				3.254	3.205	30.67		
5.50	1.0192	1.0210	56.1		963.2	35.1	82	1.3412				3.627	3.566	27.52		
6.00	1.0212	1.0230	61.3		959.9	38.3	89	1.3419				4.031	3.955	24.76		
6.50	1.0232	1.0250	66.5		956.7	41.6	97	1.3427				4.466	4.373	22.35		
7.00	1.0251	1.0270	71.8		953.4	44.8	105	1.3435				4.933	4.821	20.23		
7.50	1.0271	1.0290	77.0		950.1	48.1	112	1.3442				5.431	5.299	18.37		
8.00	1.0291	1.0310	82.3		946.8	51.4	120	1.3450				5.963	5.805	16.74		
8.50	1.0312	1.0330	87.6		943.5	54.7	128	1.3458				6.527	6.342	15.29		
9.00	1.0332	1.0350	93.0		940.2	58.0	135	1.3465				7.124	6.909	14.01		
9.50	1.0352	1.0371	98.3		936.9	61.4	143	1.3473				7.754	7.506	12.87		
10.00	1.0373	1.0391	103.7		933.5	64.7	151	1.3481				8.419	8.133	11.85		

16 ETHANOL, CH_3CH_2OH

MOLECULAR WEIGHT = 46.07
RELATIVE SPECIFIC REFRACTIVITY = 1.363

0.00 % by wt. data are the same for all compounds.
For Values of 0.00 wt. % solutions see Table 1, Acetic Acid.

A % by wt.	ρ D_4^{20}	D_{20}^{20}	C_s g/l	M g-mol/l	C_w g/l	$(C_o - C_w)$ g/l	$(n - n_o)$ × 10^4	n	Δ °C	O Os/kg	S g-mol/l	η/η_o	η/ρ cS	φ rhe	γ mmho/cm	T g-mol/l
0.50	0.9973	0.9991	5.0	0.108	992.3	5.9	3	1.3333	0.20	0.11	0.057	1.021	1.026	97.71		
1.00	0.9963	0.9981	10.0	0.216	986.4	11.8	6	1.3336	0.40	0.22	0.116	1.044	1.050	95.59		
1.50	0.9954	0.9972	14.9	0.324	980.5	17.8	9	1.3339	0.60	0.32	0.175	1.068	1.075	93.42		
2.00	0.9945	0.9962	19.9	0.432	974.6	23.6	12	1.3342	0.81	0.44	0.236	1.093	1.101	91.34		
2.50	0.9936	0.9953	24.8	0.539	968.7	29.5	15	1.3345	1.02	0.55	0.297	1.116	1.126	89.42		
3.00	0.9927	0.9945	29.8	0.646	962.9	35.3	18	1.3348	1.23	0.66	0.359	1.138	1.149	87.70		
3.50	0.9918	0.9936	34.7	0.754	957.1	41.1	21	1.3351	1.44	0.77	0.421	1.159	1.171	86.08		
4.00	0.9910	0.9927	39.6	0.860	951.4	46.9	24	1.3354	1.65	0.89	0.484	1.181	1.194	84.52		
4.50	0.9902	0.9919	44.6	0.967	945.6	52.6	27	1.3357	1.87	1.01	0.547	1.203	1.217	82.98		
5.00	0.9893	0.9911	49.5	1.074	939.9	58.4	31	1.3360	2.09	1.12	0.611	1.226	1.242	81.40		
5.50	0.9885	0.9903	54.4	1.180	934.2	64.1	34	1.3364	2.31	1.24	0.675	1.250	1.267	79.81		
6.00	0.9878	0.9895	59.3	1.286	928.5	69.7	37	1.3367	2.54	1.36	0.739	1.276	1.294	78.24		
6.50	0.9870	0.9887	64.2	1.393	922.8	75.4	41	1.3370	2.76	1.49	0.804	1.301	1.321	76.69		
7.00	0.9862	0.9880	69.0	1.498	917.2	81.0	44	1.3374	2.99	1.61	0.870	1.328	1.349	75.16		
7.50	0.9855	0.9872	73.9	1.604	911.6	86.7	48	1.3377	3.23	1.74	0.936	1.355	1.378	73.66		
8.00	0.9847	0.9865	78.8	1.710	906.0	92.3	51	1.3381	3.47	1.86	1.003	1.382	1.407	72.19		
8.50	0.9840	0.9857	83.6	1.816	900.4	97.9	55	1.3384	3.71	2.00	1.071	1.411	1.436	70.75		
9.00	0.9833	0.9850	88.5	1.921	894.8	103.4	58	1.3388	3.96	2.13	1.140	1.439	1.467	69.34		
9.50	0.9826	0.9843	93.3	2.026	889.2	109.0	62	1.3392	4.21	2.27	1.210	1.468	1.497	67.97		
10.00	0.9819	0.9836	98.2	2.131	883.7	114.5	65	1.3395	4.47	2.40	1.282	1.498	1.529	66.62		
11.00	0.9805	0.9822	107.9	2.341	872.6	125.6	73	1.3403	5.00	2.69	1.426	1.560	1.594	63.99		
12.00	0.9792	0.9809	117.5	2.550	861.7	136.6	80	1.3410	5.56	2.99	1.572	1.624	1.662	61.44		
13.00	0.9778	0.9796	127.1	2.759	850.7	147.5	87	1.3417	6.13	3.30	1.722	1.691	1.732	59.03		
14.00	0.9765	0.9782	136.7	2.967	839.8	158.4	95	1.3425	6.73	3.62	1.874	1.757	1.803	56.80		
15.00	0.9752	0.9769	146.3	3.175	828.9	169.3	102	1.3432	7.36	3.96	2.030	1.822	1.872	54.78		
16.00	0.9739	0.9756	155.8	3.382	818.1	180.2	110	1.3440	8.01	4.31	2.189	1.886	1.941	52.91		
17.00	0.9726	0.9743	165.3	3.589	807.3	191.0	117	1.3447	8.69	4.67	2.351	1.951	2.010	51.16		
18.00	0.9713	0.9730	174.8	3.795	796.5	201.8	125	1.3455	9.40	5.05	2.516	2.015	2.078	49.54		
19.00	0.9700	0.9717	184.3	4.000	785.7	212.5	132	1.3462	10.14	5.45	2.683	2.077	2.146	48.04		
20.00	0.9687	0.9704	193.7	4.205	774.9	223.3	140	1.3469	10.92	5.87	2.853	2.138	2.212	46.68		
22.00	0.9660	0.9677	212.5	4.613	753.5	244.8	154	1.3484	12.60	6.78	3.203	2.254	2.338	44.27		
24.00	0.9632	0.9649	231.2	5.018	732.0	266.2	168	1.3498	14.47	7.78	3.568	2.365	2.460	42.20		
26.00	0.9602	0.9619	249.7	5.419	710.6	287.7	181	1.3511	16.41	8.82	3.920	2.471	2.579	40.39		
28.00	0.9571	0.9588	268.0	5.817	689.1	309.1	194	1.3524	18.43	9.91	4.262	2.576	2.696	38.75		
30.00	0.9539	0.9556	286.2	6.211	667.7	330.5	205	1.3535	20.47	11.01	4.583	2.662	2.796	37.49		
32.00	0.9504	0.9521	304.1	6.601	646.3	352.0	216	1.3546	22.44	12.07		2.721	2.869	36.68		
34.00	0.9468	0.9485	321.9	6.988	624.9	373.3	227	1.3557	24.27	13.05		2.762	2.923	36.13		
36.00	0.9431	0.9447	339.5	7.369	603.6	394.7	236	1.3566	25.98	13.97		2.797	2.971	35.69		
38.00	0.9392	0.9408	356.9	7.747	582.3	415.9	245	1.3575	27.62	14.85		2.823	3.012	35.35		
40.00	0.9352	0.9369	374.1	8.120	561.1	437.1	253	1.3583	29.26	15.73		2.840	3.043	35.14		
42.00	0.9311	0.9328	391.1	8.488	540.0	458.2	260	1.3590	30.98	16.66		2.846	3.063	35.06		
44.00	0.9269	0.9286	407.8	8.853	519.1	479.2	268	1.3598	32.68	17.57		2.844	3.074	35.09		
46.00	0.9227	0.9243	424.4	9.213	498.2	500.0	274	1.3604	34.36	18.47		2.837	3.081	35.18		
48.00	0.9183	0.9199	440.8	9.568	477.5	520.7	280	1.3610	36.04	19.38		2.826	3.083	35.32		
50.00	0.9139	0.9155	457.0	9.919	457.0	541.3	286	1.3616	37.67	20.25		2.807	3.078	35.55		
52.00	0.9095	0.9111	472.9	10.265	436.5	561.7	291	1.3621	39.20	21.08		2.783	3.066	35.87		
54.00	0.9049	0.9065	488.7	10.607	416.3	582.0	296	1.3626	40.65	21.85		2.749	3.044	36.31		
56.00	0.9004	0.9020	504.2	10.944	396.2	602.1	300	1.3630	42.06	22.61		2.696	3.000	37.02		
58.00	0.8958	0.8974	519.5	11.277	376.2	622.0	304	1.3634	43.49	23.38		2.627	2.938	38.00		
60.00	0.8911	0.8927	534.7	11.606	356.5	641.8	308	1.3638	44.93	24.15		2.542	2.858	39.26		
62.00	0.8865	0.8880	549.6	11.930	336.9	661.4	311	1.3641	46.28	24.88		2.474	2.796	40.34		
64.00	0.8818	0.8833	564.3	12.250	317.4	680.8	315	1.3644	47.52	25.55		2.410	2.739	41.41		
66.00	0.8771	0.8786	578.9	12.565	298.2	700.0	317	1.3647	48.64	26.15		2.342	2.676	42.61		
68.00	0.8724	0.8739	593.2	12.876	279.2	719.1	320	1.3650	49.52	26.62		2.276	2.614	43.85		
70.00	0.8676	0.8692	607.3	13.183	260.3	737.9	322	1.3652				2.210	2.552	45.17		
72.00	0.8629	0.8644	621.3	13.486	241.6	756.6	324	1.3654				2.144	2.490	46.55		
74.00	0.8581	0.8596	635.0	13.784	223.1	775.1	325	1.3655				2.078	2.426	48.03		
76.00	0.8533	0.8549	648.5	14.077	204.8	793.4	327	1.3657				2.011	2.361	49.63		
78.00	0.8485	0.8500	661.8	14.366	186.7	811.6	327	1.3657				1.944	2.296	51.33		
80.00	0.8436	0.8451	674.9	14.650	168.7	829.5	328	1.3658				1.877	2.229	53.15		
82.00	0.8387	0.8401	687.7	14.927	151.0	847.3	328	1.3657				1.804	2.155	55.31		
84.00	0.8335	0.8350	700.2	15.198	133.4	864.9	326	1.3656				1.738	2.089	57.42		
86.00	0.8284	0.8299	712.4	15.464	116.0	882.3	325	1.3655				1.671	2.021	59.74		
88.00	0.8232	0.8247	724.4	15.725	98.8	899.4	323	1.3653				1.603	1.951	62.26		
90.00	0.8180	0.8194	736.2	15.979	81.8	916.4	320	1.3650				1.539	1.885	64.85		
92.00	0.8125	0.8140	747.5	16.226	65.0	933.2	316	1.3646				1.472	1.815	67.80		
94.00	0.8070	0.8084	758.6	16.466	48.4	949.8	312	1.3642				1.404	1.743	71.07		
96.00	0.8013	0.8027	769.2	16.697	32.1	966.2	306	1.3636				1.339	1.674	74.53		
98.00	0.7954	0.7968	779.5	16.920	15.9	982.3	300	1.3630				1.270	1.600	78.59		
100.00	0.7893	0.7907	789.3	17.133	0.0	998.2	284	1.3614				1.201	1.525	83.10		

17 (ETHYLENEDINITRILO)TETRAACETIC ACID DISODIUM SALT, EDTA DISODIUM*, $Na_2C_{10}H_{14}O_8N_2 \cdot 2H_2O$

MOLECULAR WEIGHT = 336.21 FORMULA WEIGHT, HYDRATE = 372.24
RELATIVE SPECIFIC REFRACTIVITY = 0.977

0.00 % by wt. data are the same for all compounds.
For Values of 0.00 wt. % solutions see Table 1, Acetic Acid.

A% by wt.	H% by wt.	ρ D_4^{20}	D_{20}^{20}	C_s g/l	M g-mol/l	C_w g/l	$(C_o - C_w)$ g/l	$(n - n_o) \times 10^4$	n	Δ °C	O Os/kg	S g-mol/l	η/η_o	η/ρ cS	φ rhe	γ mmho/cm	T g-mol/l
0.50	0.55	1.0009	1.0027	5.0	0.015	995.9	2.3	9	1.3339	0.07	0.04	0.020	1.015	1.016	98.35	1.9	0.018
1.00	1.11	1.0036	1.0053	10.0	0.030	993.5	4.7	18	1.3348	0.14	0.07	0.039	1.030	1.028	96.94	3.5	0.034
1.50	1.66	1.0062	1.0080	15.1	0.045	991.1	7.1	26	1.3356	0.21	0.11	0.059	1.044	1.040	95.55	5.0	0.049
2.00	2.21	1.0089	1.0107	20.2	0.060	988.7	9.5	35	1.3365	0.27	0.15	0.078	1.060	1.052	94.19	6.3	0.064
2.50	2.77	1.0115	1.0133	25.3	0.075	986.2	12.0	44	1.3374	0.33	0.18	0.096	1.075	1.065	92.84	7.6	0.077
3.00	3.32	1.0142	1.0160	30.4	0.090	983.8	14.5	53	1.3383	0.40	0.21	0.115	1.091	1.078	91.51	8.8	0.091
3.50	3.88	1.0169	1.0187	35.6	0.106	981.3	16.9	62	1.3392	0.46	0.25	0.133	1.107	1.091	90.18	10.1	0.105
4.00	4.43	1.0196	1.0214	40.8	0.121	978.8	19.5	70	1.3400	0.52	0.28	0.151	1.123	1.104	88.85	11.3	0.119
4.50	4.98	1.0223	1.0241	46.0	0.137	976.3	22.0	79	1.3409	0.58	0.31	0.169	1.140	1.118	87.53	12.5	0.133
5.00	5.54	1.0250	1.0268	51.2	0.152	973.7	24.5	88	1.3418	0.65	0.35	0.187	1.158	1.132	86.20	13.7	0.145
5.50	6.09	1.0277	1.0295	56.5	0.168	971.2	27.0	97	1.3427	0.71	0.38	0.206	1.176	1.147	84.86	14.6	0.156
6.00	6.64	1.0305	1.0323	61.8	0.184	968.6	29.6	106	1.3436	0.77	0.42	0.225	1.195	1.162	83.51	15.3	0.164

* Sodium(Di)Ethylenediamine Tetraacetate.

18 ETHYLENE GLYCOL, CH_2OHCH_2OH

MOLECULAR WEIGHT = 62.07
RELATIVE SPECIFIC REFRACTIVITY = 1.147

0.00 % by wt. data are the same for all compounds.
For Values of 0.00 wt. % solutions see Table 1, Acetic Acid.

A% by wt.	ρ D_4^{20}	D_{20}^{20}	C_s g/l	M g-mol/l	C_w g/l	$(C_o - C_w)$ g/l	$(n - n_o) \times 10^4$	n	Δ °C	O Os/kg	S g-mol/l	η/η_o	η/ρ cS	φ rhe	γ mmho/cm	T g-mol/l
0.50	0.9988	1.0006	5.0	0.080	993.8	4.4	5	1.3335	0.15	0.08	0.042	1.008	1.011	99.03		
1.00	0.9995	1.0012	10.0	0.161	989.5	8.8	9	1.3339	0.30	0.16	0.086	1.018	1.021	98.04		
2.00	1.0007	1.0025	20.0	0.322	980.7	17.6	19	1.3348	0.61	0.33	0.176	1.046	1.047	95.41		
3.00	1.0019	1.0037	30.1	0.484	971.9	26.4	28	1.3358	0.92	0.50	0.269	1.072	1.072	93.09		
4.00	1.0032	1.0049	40.1	0.646	963.0	35.2	37	1.3367	1.24	0.67	0.364	1.097	1.096	90.98		
5.00	1.0044	1.0062	50.2	0.809	954.2	44.0	47	1.3377	1.58	0.85	0.461	1.123	1.120	88.87		
6.00	1.0057	1.0075	60.3	0.972	945.3	52.9	56	1.3386	1.91	1.03	0.560	1.151	1.147	86.72		
7.00	1.0070	1.0087	70.5	1.136	936.5	61.8	66	1.3396	2.26	1.22	0.660	1.180	1.174	84.60		
8.00	1.0082	1.0100	80.7	1.299	927.6	70.6	75	1.3405	2.62	1.41	0.763	1.210	1.202	82.49		
9.00	1.0095	1.0113	90.9	1.464	918.7	79.6	85	1.3415	2.99	1.61	0.868	1.241	1.232	80.41		
10.00	1.0108	1.0126	101.1	1.629	909.8	88.5	95	1.3425	3.37	1.81	0.975	1.274	1.263	78.34		
12.00	1.0134	1.0152	121.6	1.959	891.8	106.4	114	1.3444	4.16	2.24	1.196	1.345	1.330	74.21		
14.00	1.0161	1.0179	142.3	2.292	873.8	124.4	134	1.3464	5.01	2.69	1.426	1.421	1.401	70.25		
16.00	1.0188	1.0206	163.0	2.626	833.8	142.5	154	1.3484	5.91	3.18	1.665	1.497	1.473	66.65		
18.00	1.0214	1.0232	183.9	2.962	837.6	160.7	174	1.3503	6.89	3.70	1.913	1.575	1.545	63.35		
20.00	1.0241	1.0259	204.8	3.300	819.3	178.9	194	1.3523	7.93	4.27	2.171	1.658	1.622	60.19		
24.00	1.0296	1.0314	247.1	3.981	782.5	215.8	234	1.3564	10.28	5.53	2.714	1.839	1.790	54.26		
28.00	1.0350	1.0369	289.8	4.669	745.2	253.0	275	1.3605	13.03	7.01	3.289	2.043	1.978	48.84		
32.00	1.0405	1.0424	333.0	5.364	707.6	290.7	316	1.3646	16.23	8.73	3.889	2.275	2.191	43.87		
36.00	1.0460	1.0478	376.6	6.067	669.4	328.8	357	1.3687	19.82	10.66	4.483	2.532	2.425	39.42		
40.00	1.0514	1.0532	420.6	6.775	630.8	367.4	398	1.3728	23.84	12.82		2.826	2.693	35.32		
44.00	1.0567	1.0586	465.0	7.491	591.8	406.5	440	1.3769	28.32	15.23		3.160	2.996	31.59		
48.00	1.0619	1.0638	509.7	8.212	552.2	446.0	481	1.3811	33.30	17.90		3.537	3.337	28.22		
52.00	1.0670	1.0689	554.8	8.939	512.2	486.1	521	1.3851	38.81	20.87		3.973	3.731	25.12		
56.00	1.0719	1.0738	600.3	9.671	471.6	526.6	562	1.3892	44.83	24.10		4.466	4.175	22.35		
60.00	1.0765	1.0784	645.9	10.406	430.6	567.6	602	1.3931	51.23	27.54		5.016	4.669	19.90		

19 FERRIC CHLORIDE, $FeCl_3 \cdot 6H_2O$

MOLECULAR WEIGHT = 162.22 FORMULA WEIGHT, HYDRATE = 270.32
RELATIVE SPECIFIC REFRACTIVITY = 0.946

0.00 % by wt. data are the same for all compounds.
For Values of 0.00 wt. % solutions see Table 1, Acetic Acid.

A% by wt.	H% by wt.	ρ D_4^{20}	D_{20}^{20}	C_s g/l	M g-mol/l	C_w g/l	$(C_o - C_w)$ g/l	$(n - n_o) \times 10^4$	n	Δ °C	O Os/kg	S g-mol/l	η/η_o	η/ρ cS	φ rhe	γ mmho/cm	T g-mol/l
0.50	0.83	1.0025	1.0043	5.0	0.031	997.5	0.7	14	1.3344	0.206	0.111	0.059	1.022	1.022	97.61		
1.00	1.67	1.0068	1.0086	10.1	0.062	996.7	1.5	28	1.3358	0.385	0.207	0.111	1.045	1.040	95.49		
2.00	3.33	1.0153	1.0171	20.3	0.125	995.0	3.3	56	1.3386	0.753	0.405	0.219	1.091	1.077	91.48		
3.00	5.00	1.0238	1.0256	30.7	0.189	993.1	5.2	83	1.3413	1.146	0.616	0.335	1.137	1.113	87.79		
4.00	6.67	1.0323	1.0341	41.3	0.255	991.0	7.3	111	1.3441	1.562	0.840	0.457	1.185	1.150	84.22		
5.00	8.33	1.0408	1.0426	52.0	0.321	988.7	9.5	138	1.3468	2.002	1.076	0.585	1.236	1.190	80.76		
6.00	10.00	1.0493	1.0512	63.0	0.388	986.4	11.9	166	1.3496	2.475	1.331	0.722	1.289	1.231	77.41		
7.00	11.66	1.0580	1.0599	74.1	0.457	984.0	14.3	194	1.3524	2.994	1.609	0.870	1.347	1.275	74.11		
8.00	13.33	1.0668	1.0687	85.3	0.526	981.5	16.7	222	1.3552	3.566	1.917	1.031	1.409	1.324	70.82		
9.00	15.00	1.0760	1.0779	96.8	0.597	979.2	19.0	251	1.3581	4.185	2.250	1.202	1.477	1.376	67.55		
10.00	16.66	1.0853	1.0872	108.5	0.669	976.8	21.5	281	1.3611	4.854	2.610	1.385	1.550	1.431	64.39		
12.00	20.00	1.1040	1.1059	132.5	0.817	971.5	26.8	340	1.3670	6.38	3.43	1.785	1.704	1.547	58.57		
14.00	23.33	1.1228	1.1248	157.2	0.969	965.6	32.6	400	1.3730	8.22	4.42	2.240	1.875	1.673	53.22		
16.00	26.66	1.1420	1.1440	182.7	1.126	959.3	39.0			10.45	5.62	2.750	2.076	1.822	48.07		
18.00	29.99	1.1615	1.1636	209.1	1.289	952.5	45.8			13.08	7.03	3.299	2.306	1.990	43.27		
20.00	33.33	1.1816	1.1837	236.3	1.457	945.3	53.0			16.14	8.68	3.873	2.565	2.175	38.91		
24.00	39.99	1.2234	1.2256	293.6	1.810	929.8	68.4			23.79	12.79		3.172	2.598	31.47		
28.00	46.66	1.2679	1.2702	355.0	2.189	912.9	85.3			33.61	18.07		4.030	3.185	24.76		
32.00	53.32	1.3153	1.3176	420.9	2.595	894.4	103.8			49.16	26.43		5.263	4.010	18.96		
36.00	59.99	1.3654	1.3678	491.5	3.030	873.9	124.4						7.116	5.222	14.03		
40.00	66.66	1.4176	1.4201	567.0	3.496	850.6	147.7						9.655	6.824	10.34		

20 FORMIC ACID, HCOOH

MOLECULAR WEIGHT = 46.03
RELATIVE SPECIFIC REFRACTIVITY = 0.916

0.00 % by wt. data are the same for all compounds.
For Values of 0.00 wt. % solutions see Table 1, Acetic Acid.

A % by wt.	ρ D^{20}_{4}	D^{20}_{20}	C_s g/l	M g-mol/l	C_w g/l	$(C_o - C_w)$ g/l	$(n - n_o)$ × 10^4	n	Δ °C	O Os/kg	S g-mol/l	η/η_o	η/ρ cS	φ rhe	γ mmho/cm	T g-mol/l
0.50	0.9994	1.0012	5.0	0.109	994.4	3.8	3	1.3333	0.210	0.113	0.060	1.004	1.007	99.36	1.4	0.013
1.00	1.0006	1.0023	10.0	0.217	990.6	7.7	6	1.3336	0.418	0.225	0.121	1.009	1.010	98.96	2.4	0.024
2.00	1.0029	1.0047	20.1	0.436	982.9	15.3	12	1.3342	0.824	0.443	0.240	1.015	1.015	98.28	3.5	0.034
3.00	1.0053	1.0071	30.2	0.655	975.2	23.1	18	1.3348	1.243	0.668	0.363	1.023	1.019	97.59	4.3	0.043
4.00	1.0077	1.0095	40.3	0.876	967.4	30.8	24	1.3354	1.672	0.899	0.489	1.030	1.024	96.90	5.0	0.050
5.00	1.0102	1.0119	50.5	1.097	959.6	38.6	29	1.3359	2.103	1.131	0.615	1.037	1.029	96.24	5.6	0.056
6.00	1.0126	1.0144	60.8	1.320	951.8	46.4	35	1.3365	2.534	1.363	0.739	1.044	1.033	95.63	6.1	0.062
7.00	1.0150	1.0168	71.1	1.544	944.0	54.3	41	1.3371	2.967	1.595	0.862	1.050	1.036	95.06	6.6	0.067
8.00	1.0175	1.0193	81.4	1.768	936.1	62.2	46	1.3376	3.402	1.829	0.985	1.056	1.040	94.51	7.0	0.071
9.00	1.0199	1.0217	91.8	1.994	928.1	70.1	52	1.3382	3.838	2.063	1.106	1.062	1.043	93.98	7.4	0.075
10.00	1.0224	1.0242	102.2	2.221	920.1	78.1	57	1.3387	4.267	2.294	1.225	1.068	1.047	93.45	7.8	0.079
12.00	1.0273	1.0291	123.3	2.678	904.0	94.2	67	1.3397	5.19	2.79	1.475	1.080	1.053	92.41	8.4	0.086
14.00	1.0322	1.0340	144.5	3.139	887.7	110.5	78	1.3408	6.11	3.28	1.715	1.092	1.060	91.40	8.8	0.091
16.00	1.0371	1.0389	165.9	3.605	871.1	127.1	88	1.3418	7.06	3.80	1.957	1.104	1.067	90.38	9.3	0.096
18.00	1.0419	1.0438	187.5	4.074	854.4	143.9	98	1.3428	8.08	4.35	2.207	1.117	1.074	89.35	9.6	0.100
20.00	1.0467	1.0486	209.3	4.548	837.4	160.9	107	1.3437	9.11	4.90	2.448	1.130	1.081	88.36	9.9	0.103
24.00	1.0562	1.0580	253.5	5.507	802.7	195.5	126	1.3456	11.10	5.97	2.891	1.154	1.094	86.52	10.3	0.108
28.00	1.0654	1.0673	298.3	6.481	767.1	231.1	145	1.3475	13.10	7.04	3.303	1.177	1.107	84.79	10.5	0.109
32.00	1.0746	1.0765	343.9	7.471	730.7	267.5	163	1.3493	15.28	8.21	3.717	1.201	1.120	83.09	10.5	0.109
36.00	1.0839	1.0858	390.2	8.477	693.7	304.6	181	1.3511	17.65	9.49	4.132	1.225	1.133	81.44	10.3	0.107
40.00	1.0935	1.0955	437.4	9.503	656.1	342.1	199	1.3529	20.18	10.85	4.539	1.251	1.146	79.78	9.9	0.103
44.00	1.1015	1.1035	484.7	10.529	616.9	381.4	217	1.3547	22.93	12.33		1.278	1.162	78.11	9.5	0.098
48.00	1.1097	1.1117	532.7	11.572	577.1	421.2	235	1.3565	26.06	14.01		1.306	1.179	76.42	8.9	0.092
52.00	1.1183	1.1203	581.5	12.634	536.8	461.4	251	1.3581	29.69	15.96		1.337	1.198	74.65	8.3	0.085
56.00	1.1273	1.1293	631.3	13.714	496.0	502.2	267	1.3597	33.81	18.18		1.371	1.218	72.81	7.6	0.078
60.00	1.1364	1.1385	681.9	14.813	454.6	543.7	282	1.3612	38.26	20.57		1.407	1.241	70.93	6.9	0.070
64.00	1.1456	1.1476	733.2	15.928	412.4	585.8	296	1.3626	43.02	23.13		1.446	1.265	69.03	6.2	0.062
68.00	1.1544	1.1565	785.0	17.055	369.4	628.8	311	1.3641				1.487	1.291	67.11	5.4	0.054
70.00	1.1586	1.1607	811.1	17.620	347.6	650.6	318	1.3648				1.508	1.305	66.16	5.1	0.051

21 D-FRUCTOSE (LEVULOSE), $C_6H_{12}O_6$

MOLECULAR WEIGHT = 180.16
RELATIVE SPECIFIC REFRACTIVITY = 1.021

0.00 % by wt. data are the same for all compounds.
For Values of 0.00 wt. % solutions see Table 1, Acetic Acid.

A % by wt.	ρ D^{20}_{4}	D^{20}_{20}	C_s g/l	M g-mol/l	C_w g/l	$(C_o - C_w)$ g/l	$(n - n_o)$ × 10^4	n	Δ °C	O Os/kg	S g-mol/l	η/η_o	η/ρ cS	φ rhe	γ mmho/cm	T g-mol/l
0.50	1.0002	1.0020	5.0	0.028	995.2	3.0	7	1.3337	0.052	0.028	0.014	1.013	1.015	98.53		
1.00	1.0021	1.0039	10.0	0.056	992.1	6.1	14	1.3344	0.104	0.056	0.029	1.026	1.026	97.28		
1.50	1.0041	1.0059	15.1	0.084	989.1	9.2	21	1.3351	0.157	0.085	0.045	1.039	1.037	96.06		
2.00	1.0061	1.0079	20.1	0.112	986.0	12.3	28	1.3358	0.211	0.113	0.060	1.052	1.048	94.87		
2.50	1.0081	1.0098	25.2	0.140	982.9	15.4	36	1.3366	0.265	0.142	0.076	1.065	1.059	93.71		
3.00	1.0101	1.0118	30.3	0.168	979.7	18.5	43	1.3373	0.319	0.172	0.092	1.078	1.069	92.59		
3.50	1.0120	1.0138	35.4	0.197	976.6	21.6	50	1.3380	0.374	0.201	0.108	1.091	1.080	91.50		
4.00	1.0140	1.0158	40.6	0.225	973.5	24.8	57	1.3387	0.430	0.231	0.124	1.104	1.091	90.40		
4.50	1.0160	1.0178	45.7	0.254	970.3	27.9	65	1.3395	0.487	0.262	0.141	1.118	1.102	89.30		
5.00	1.0181	1.0199	50.9	0.283	967.2	31.1	72	1.3402	0.544	0.292	0.158	1.132	1.114	88.16		
5.50	1.0201	1.0219	56.1	0.311	964.0	34.3	79	1.3409	0.602	0.324	0.175	1.147	1.127	87.00		
6.00	1.0221	1.0239	61.3	0.340	960.8	37.5	87	1.3417	0.660	0.355	0.192	1.163	1.140	85.83		
6.50	1.0241	1.0259	66.6	0.369	957.6	40.7	94	1.3424	0.720	0.387	0.209	1.179	1.153	84.66		
7.00	1.0262	1.0280	71.8	0.399	954.3	43.9	101	1.3431	0.780	0.419	0.227	1.196	1.167	83.48		
7.50	1.0282	1.0300	77.1	0.428	951.1	47.1	109	1.3439	0.841	0.452	0.245	1.213	1.182	82.29		
8.00	1.0303	1.0321	82.4	0.457	947.8	50.4	116	1.3446	0.902	0.485	0.263	1.230	1.197	81.11		
8.50	1.0323	1.0342	87.7	0.487	944.6	53.7	124	1.3454	0.965	0.519	0.282	1.249	1.212	79.93		
9.00	1.0344	1.0362	93.1	0.517	941.3	56.9	131	1.3461	1.028	0.553	0.300	1.267	1.228	78.75		
9.50	1.0365	1.0383	98.5	0.547	938.0	60.2	139	1.3469	1.092	0.587	0.319	1.286	1.244	77.58		
10.00	1.0385	1.0404	103.9	0.576	934.7	63.5	147	1.3476	1.158	0.622	0.338	1.306	1.260	76.42		
11.00	1.0427	1.0446	114.7	0.637	928.0	70.2	162	1.3492	1.290	0.694	0.377	1.346	1.294	74.13		
12.00	1.0469	1.0488	125.6	0.697	921.3	76.9	177	1.3507	1.427	0.767	0.417	1.388	1.329	71.89		
13.00	1.0512	1.0530	136.7	0.758	914.5	83.7	193	1.3522	1.567	0.842	0.458	1.432	1.365	69.67		
14.00	1.0554	1.0573	147.8	0.820	907.7	90.6	208	1.3538	1.710	0.920	0.500	1.480	1.405	67.45		
15.00	1.0597	1.0616	159.0	0.882	900.8	97.5	224	1.3554	1.858	0.999	0.543	1.530	1.447	65.23		
16.00	1.0640	1.0659	170.2	0.945	893.8	104.5	240	1.3569	2.009	1.080	0.587	1.584	1.491	63.02		
17.00	1.0684	1.0703	181.6	1.008	886.8	111.5	255	1.3585	2.162	1.163	0.632	1.640	1.538	60.84		
18.00	1.0728	1.0747	193.1	1.072	879.7	118.6	272	1.3601	2.319	1.247	0.677	1.700	1.588	58.69		
19.00	1.0772	1.0791	204.7	1.136	872.5	125.7	288	1.3618	2.479	1.333	0.723	1.764	1.641	56.56		
20.00	1.0816	1.0835	216.3	1.201	865.3	133.0	304	1.3634	2.640	1.419	0.769	1.833	1.698	54.45		
22.00	1.0906	1.0925	239.9	1.332	850.6	147.6	337	1.3667	3.054	1.642	0.887	1.982	1.821	50.35		
24.00	1.0996	1.1016	263.9	1.465	835.7	162.5	370	1.3700	3.431	1.845	0.993	2.150	1.959	46.41		
26.00	1.1089	1.1108	288.3	1.600	820.5	177.7	404	1.3734	3.815	2.051	1.100	2.343	2.117	42.60		
28.00	1.1182	1.1202	313.1	1.738	805.1	193.1	438	1.3768	4.196	2.256	1.205	2.557	2.291	39.03		
30.00	1.1276	1.1296	338.3	1.878	789.3	208.9	473	1.3803				2.811	2.498	35.50		
32.00	1.1372	1.1392	363.9	2.020	773.3	224.9	509	1.3839				3.106	2.737	32.13		
34.00	1.1469	1.1490	390.0	2.165	757.0	241.3	545	1.3874				3.455	3.018	28.89		
36.00	1.1568	1.1588	416.4	2.312	740.3	257.9	581	1.3911				3.891	3.370	25.65		
38.00	1.1668	1.1688	443.4	2.461	723.4	274.8	618	1.3948				4.409	3.786	22.64		
40.00	1.1769	1.1790	470.8	2.613	706.1	292.1	655	1.3985				5.036	4.288	19.82		
42.00	1.1871	1.1892	498.6	2.767	688.5	309.7	694	1.4023				5.761	4.863	17.32		
44.00	1.1975	1.1996	526.9	2.925	670.6	327.6	732	1.4062				6.631	5.548	15.05		
46.00	1.2080	1.2102	555.7	3.084	652.3	345.9	771	1.4101				7.738	6.419	12.90		
48.00	1.2187	1.2208	585.0	3.247	633.7	364.5	811	1.4141				9.042	7.434	11.04		

21 D-FRUCTOSE (LEVULOSE), $C_6H_{12}O_6$—(Continued)

A% by wt.	ρ D_4^{20}	D_{20}^{20}	C_s g/l	M g-mol/l	C_w g/l	$(C_o - C_w)$ g/l	$(n - n_o) \times 10^4$	n	Δ °C	O Os/kg	S g-mol/l	η/η_o	η/ρ cS	φ rhe	γ mmho/cm	T g-mol/l
50.00	1.2295	1.2316	614.7	3.412	614.7	383.5	851	1.4181				10.80	8.80	9.24		
52.00	1.2404	1.2426	645.0	3.580	595.4	402.9	892	1.4222				12.99	10.49	7.69		
54.00	1.2514	1.2536	675.8	3.751	575.7	422.6	934	1.4264				15.84	12.68	6.30		
56.00	1.2626	1.2649	707.1	3.925	555.6	442.7	976	1.4306				20.05	15.91	4.98		
58.00	1.2739	1.2762	738.9	4.101	535.1	463.2	1019	1.4348				25.60	20.13	3.90		
60.00	1.2854	1.2877	771.2	4.281	514.2	484.1	1062	1.4392				32.47	25.31	3.07		
62.00	1.2970	1.2992	804.1	4.463	492.8	505.4	1105	1.4435				42.97	33.20	2.32		
64.00	1.3086	1.3110	837.5	4.649	471.1	527.1	1150	1.4480				58.17	44.54	1.72		
66.00	1.3204	1.3228	871.5	4.837	448.9	549.3	1194	1.4524				82.19	62.37	1.21		
68.00	1.3323	1.3347	906.0	5.029	426.3	571.9	1240	1.4570				118.4	89.08	0.84		
70.00	1.3443	1.3467	941.0	5.223	403.3	394.9	1285	1.4615				177.8	132.5	0.56		

22 D-GLUCOSE (DEXTROSE), $C_6H_{12}O_6 \cdot 1H_2O$

MOLECULAR WEIGHT = 180.16 FORMULA WEIGHT, HYDRATE = 198.17
RELATIVE SPECIFIC REFRACTIVITY = 1.031

0.00 % by wt. data are the same for all compounds.
For Values of 0.00 wt. % solutions see Table 1, Acetic Acid.

A% by wt.	H% by wt.	ρ D_4^{20}	D_{20}^{20}	C_s g/l	M g-mol/l	C_w g/l	$(C_o - C_w)$ g/l	$(n - n_o) \times 10^4$	n	Δ °C	O Os/kg	S g-mol/l	η/η_o	η/ρ cS	φ rhe	γ mmho/cm	T g-mol/l
0.50	0.55	1.0001	1.0019	5.0	0.028	995.1	3.1	7	1.3337	0.047	0.025	0.013	1.008	1.010	99.01		
1.00	1.10	1.0020	1.0038	10.0	0.056	992.0	6.2	14	1.3344	0.107	0.057	0.030	1.019	1.019	97.94		
1.50	1.65	1.0039	1.0057	15.1	0.084	988.9	9.4	21	1.3351	0.158	0.085	0.045	1.034	1.032	96.54		
2.00	2.20	1.0058	1.0076	20.1	0.112	985.7	12.5	28	1.3358	0.214	0.115	0.061	1.050	1.046	95.05		
2.50	2.75	1.0078	1.0095	25.2	0.140	982.6	15.7	36	1.3366	0.270	0.145	0.077	1.065	1.059	93.68		
3.00	3.30	1.0097	1.0115	30.3	0.168	979.4	18.8	43	1.3373	0.323	0.174	0.093	1.081	1.072	92.36		
3.50	3.85	1.0116	1.0134	35.4	0.197	976.2	22.0	50	1.3380	0.377	0.202	0.108	1.096	1.085	91.08		
4.00	4.40	1.0136	1.0154	40.5	0.225	973.0	25.2	57	1.3387	0.433	0.233	0.125	1.111	1.098	89.82		
4.50	4.95	1.0155	1.0173	45.7	0.254	969.8	28.4	65	1.3395	0.490	0.264	0.142	1.127	1.112	88.57		
5.00	5.50	1.0175	1.0193	50.9	0.282	966.6	31.6	72	1.3402	0.549	0.295	0.159	1.143	1.126	87.31		
5.50	6.05	1.0194	1.0212	56.1	0.311	963.4	34.9	79	1.3409	0.608	0.327	0.176	1.160	1.140	86.06		
6.00	6.60	1.0214	1.0232	61.3	0.340	960.1	38.1	87	1.3417	0.668	0.359	0.194	1.177	1.154	84.82		
6.50	7.15	1.0234	1.0252	66.5	0.369	956.9	41.4	94	1.3424	0.729	0.392	0.212	1.194	1.169	83.59		
7.00	7.70	1.0254	1.0272	71.8	0.398	953.6	44.6	102	1.3432	0.790	0.425	0.230	1.212	1.184	82.37		
7.50	8.25	1.0274	1.0292	77.1	0.428	950.3	47.9	109	1.3439	0.851	0.458	0.248	1.230	1.199	81.17		
8.00	8.80	1.0294	1.0312	82.4	0.457	947.0	51.2	117	1.3447	0.913	0.491	0.266	1.248	1.215	79.97		
8.50	9.35	1.0314	1.0332	87.7	0.487	943.7	54.5	124	1.3454	0.975	0.524	0.284	1.267	1.231	78.77		
9.00	9.90	1.0334	1.0352	93.0	0.516	940.4	57.8	132	1.3462	1.038	0.558	0.303	1.286	1.247	77.58		
9.50	10.45	1.0354	1.0373	98.4	0.546	937.1	61.2	139	1.3469	1.102	0.592	0.322	1.306	1.264	76.39		
10.00	11.00	1.0375	1.0393	103.7	0.576	933.7	64.5	147	1.3477	1.167	0.627	0.341	1.327	1.282	75.21		
11.00	12.10	1.0416	1.0434	114.6	0.636	927.0	71.2	162	1.3492	1.303	0.700	0.381	1.369	1.317	72.89		
12.00	13.20	1.0457	1.0475	125.5	0.697	920.2	78.0	178	1.3508	1.443	0.776	0.422	1.413	1.354	70.63		
13.00	14.30	1.0498	1.0517	136.5	0.758	913.4	84.9	193	1.3523	1.586	0.853	0.464	1.459	1.393	68.40		
14.00	15.40	1.0540	1.0559	147.6	0.819	906.4	91.8	209	1.3539	1.731	0.931	0.506	1.509	1.434	66.15		
15.00	16.50	1.0582	1.0601	158.7	0.881	899.5	98.8	225	1.3555	1.880	1.011	0.550	1.563	1.480	63.85		
16.00	17.60	1.0624	1.0643	170.0	0.944	892.5	105.8	241	1.3571	2.033	1.093	0.594	1.622	1.530	61.53		
17.00	18.70	1.0667	1.0686	181.3	1.007	885.4	112.9	257	1.3587	2.190	1.178	0.640	1.685	1.583	59.21		
18.00	19.80	1.0710	1.0729	192.8	1.070	878.2	120.0	273	1.3603	2.353	1.265	0.687	1.753	1.640	56.93		
19.00	20.90	1.0753	1.0772	204.3	1.134	871.0	127.2	289	1.3619	2.521	1.355	0.735	1.825	1.700	54.70		
20.00	22.00	1.0797	1.0816	215.9	1.199	863.7	134.5	306	1.3635	2.696	1.449	0.785	1.900	1.763	52.53		
22.00	24.20	1.0884	1.0904	239.5	1.329	849.0	149.3	339	1.3668	3.073	1.652	0.892	2.059	1.896	48.47		
24.00	26.40	1.0973	1.0993	263.4	1.462	834.0	164.3	372	1.3702	3.480	1.871	1.007	2.238	2.044	44.59		
26.00	28.60	1.1063	1.1083	287.6	1.597	818.7	179.6	406	1.3736	3.900	2.097	1.124	2.453	2.222	40.69		
28.00	30.80	1.1154	1.1174	312.3	1.734	803.1	195.1	440	1.3770	4.337	2.332	1.244	2.702	2.427	36.94		
30.00	33.00	1.1246	1.1266	337.4	1.873	787.2	211.0	475	1.3805	4.794	2.577	1.369	2.992	2.666	33.36		
32.00	35.20	1.1340	1.1360	362.9	2.014	771.1	227.1	511	1.3840				3.317	2.931	30.09		
34.00	37.40	1.1434	1.1454	388.8	2.158	754.6	243.6	546	1.3876				3.697	3.240	26.99		
36.00	39.60	1.1529	1.1550	415.1	2.304	737.9	260.3	582	1.3912				4.185	3.637	23.85		
38.00	41.80	1.1626	1.1647	441.8	2.452	720.8	277.4	619	1.3949				4.776	4.116	20.90		
40.00	44.00	1.1724	1.1745	469.0	2.603	703.5	294.8	656	1.3986				5.482	4.685	18.21		
42.00	46.20	1.1823	1.1844	496.6	2.756	685.8	312.5	694	1.4024				6.275	5.318	15.90		
44.00	48.40	1.1924	1.1945	524.7	2.912	667.7	330.5	732	1.4062				7.221	6.068	13.82		
46.00	50.60	1.2026	1.2047	553.2	3.071	649.4	348.8	771	1.4101				8.437	7.029	11.83		
48.00	52.80	1.2130	1.2151	582.2	3.232	630.7	367.5	811	1.4141				9.863	8.148	10.12		
50.00	55.00	1.2235	1.2257	611.7	3.396	611.7	386.5	851	1.4181				11.86	9.713	8.41		
52.00	57.20	1.2342	1.2364	641.8	3.562	592.4	405.8	892	1.4222				14.46	11.74	6.90		
54.00	59.40	1.2451	1.2473	672.4	3.732	572.8	425.5	933	1.4263				17.88	14.39	5.58		
56.00	61.60	1.2562	1.2585	703.5	3.905	552.7	445.5	976	1.4306				22.84	18.22	4.37		
58.00	63.80	1.2676	1.2699	735.2	4.081	532.4	465.8	1020	1.4349				29.33	23.18	3.40		
60.00	66.00	1.2793	1.2816	767.6	4.261	511.7	486.5	1064	1.4394				37.37	29.27	2.67		

23 GLYCEROL, $CH_2OHCHOHCH_2OH$

MOLECULAR WEIGHT = 92.09
RELATIVE SPECIFIC REFRACTIVITY = 1.109

0.00 % by wt. data are the same for all compounds.
For Values of 0.00 wt. % solutions see Table 1, Acetic Acid.

A% by wt.	ρ D_4^{20}	D_{20}^{20}	C_s g/l	M g-mol/l	C_w g/l	$(C_o - C_w)$ g/l	$(n - n_o)$ × 10^4	n	Δ °C	O Os/kg	S g-mol/l	η/η_o	η/ρ cS	φ rhe	γ mmho/cm	T g-mol/l
0.50	0.9994	1.0011	5.0	0.054	994.4	3.9	6	1.3336	0.072	0.039	0.020	1.009	1.012	98.89		
1.00	1.0005	1.0023	10.0	0.109	990.5	7.7	12	1.3342	0.180	0.097	0.051	1.020	1.022	97.84		
2.00	1.0028	1.0046	20.1	0.218	982.7	15.5	23	1.3353	0.411	0.221	0.119	1.046	1.045	95.41		
3.00	1.0051	1.0069	30.2	0.327	974.9	23.3	35	1.3365	0.627	0.337	0.182	1.072	1.068	93.12		
4.00	1.0074	1.0092	40.3	0.438	967.1	31.2	46	1.3376	0.849	0.456	0.247	1.098	1.092	90.93		
5.00	1.0097	1.0115	50.5	0.548	959.2	39.0	58	1.3388	1.078	0.580	0.315	1.125	1.116	88.71		
6.00	1.0120	1.0138	60.7	0.659	951.3	46.9	70	1.3400	1.316	0.708	0.385	1.155	1.143	86.44		
7.00	1.0144	1.0162	71.0	0.771	943.4	54.9	82	1.3412	1.561	0.839	0.457	1.186	1.171	84.17		
8.00	1.0167	1.0185	81.3	0.883	935.4	62.9	94	1.3424	1.811	0.974	0.530	1.218	1.201	81.90		
9.00	1.0191	1.0209	91.7	0.996	927.4	70.9	106	1.3436	2.064	1.110	0.603	1.253	1.232	79.67		
10.00	1.0215	1.0233	102.1	1.109	919.3	78.9	118	1.3448	2.323	1.249	0.678	1.288	1.263	77.48		
12.00	1.0262	1.0281	123.1	1.337	903.1	95.1	142	1.3472	2.880	1.548	0.837	1.362	1.330	73.28		
14.00	1.0311	1.0329	144.4	1.568	886.7	111.5	167	1.3496	3.469	1.865	1.004	1.442	1.401	69.22		
16.00	1.0360	1.0378	165.8	1.800	870.2	128.0	191	1.3521	4.094	2.201	1.177	1.530	1.480	65.22		
18.00	1.0409	1.0428	187.4	2.035	853.6	144.7	217	1.3547	4.756	2.557	1.359	1.627	1.566	61.34		
20.00	1.0459	1.0478	209.2	2.272	836.8	161.5	242	1.3572	5.46	2.93	1.546	1.734	1.661	57.56		
24.00	1.0561	1.0580	253.5	2.752	802.6	195.6	294	1.3624	7.01	3.77	1.944	1.984	1.882	50.31		
28.00	1.0664	1.0683	298.6	3.243	767.8	230.4	347	1.3676	8.77	4.71	2.370	2.274	2.136	43.89		
32.00	1.0770	1.0789	344.6	3.742	732.3	265.9	400	1.3730	10.74	5.78	2.814	2.632	2.449	37.91		
36.00	1.0876	1.0896	391.5	4.252	696.1	302.2	455	1.3785	12.96	6.97	3.276	3.082	2.839	32.38		
40.00	1.0984	1.1003	439.4	4.771	659.0	339.2	511	1.3841	15.50	8.33	3.757	3.646	3.326	27.37		
44.00	1.1092	1.1112	488.1	5.300	621.2	377.1	567	1.3897				4.434	4.005	22.51		
48.00	1.1200	1.1220	537.6	5.838	582.4	415.8	624	1.3954				5.402	4.833	18.47		
52.00	1.1308	1.1328	588.0	6.385	542.8	455.4	681	1.4011				6.653	5.895	15.00		
56.00	1.1419	1.1439	639.4	6.944	502.4	495.8	739	1.4069				8.332	7.311	11.98		
60.00	1.1530	1.1551	691.8	7.513	461.2	537.0	799	1.4129				10.66	9.264	9.36		
64.00	1.1643	1.1663	745.1	8.091	419.1	579.1	859	1.4189				13.63	11.73	7.32		
68.00	1.1755	1.1775	799.3	8.680	376.1	622.1	919	1.4249				18.42	15.70	5.42		
72.00	1.1866	1.1887	854.3	9.277	332.2	666.0	980	1.4310				27.57	23.28	3.62		
76.00	1.1976	1.1997	910.2	9.883	287.4	710.8	1040	1.4370				40.49	33.88	2.46		
80.00	1.2085	1.2106	966.8	10.498	241.7	756.5	1101	1.4431				59.78	49.57	1.67		
84.00	1.2192	1.2214	1024.2	11.121	195.1	803.2	1162	1.4492				84.17	69.18	1.19		
88.00	1.2299	1.2320	1082.3	11.752	147.6	850.7	1223	1.4553				147.2	119.9	0.68		
92.00	1.2404	1.2426	1141.1	12.392	99.2	899.0	1284	1.4613				383.7	310.0	0.26		
96.00	1.2508	1.2530	1200.7	13.039	50.0	948.2	1344	1.4674				778.9	624.0	0.13		
100.00	1.2611	1.2633	1261.1	13.694	0.0	998.2	1405	1.4735				1487.0	1181.3	0.06		

24 HYDROCHLORIC ACID, HCl

MOLECULAR WEIGHT = 36.47
RELATIVE SPECIFIC REFRACTIVITY = 1.152

0.00 % by wt. data are the same for all compounds.
For Values of 0.00 wt. % solutions see Table 1, Acetic Acid.

A% by wt.	ρ D_4^{20}	D_{20}^{20}	C_s g/l	M g-mol/l	C_w g/l	$(C_o - C_w)$ g/l	$(n - n_o)$ × 10^4	n	Δ °C	O Os/kg	S g-mol/l	η/η_o	η/ρ cS	φ rhe	γ mmho/cm	T g-mol/l
0.50	1.0007	1.0025	5.0	0.137	995.7	2.5	12	1.3341	0.486	0.261	0.141	1.006	1.008	99.16	45.1	0.537
1.00	1.0031	1.0049	10.0	0.275	993.1	5.1	23	1.3353	0.989	0.532	0.289	1.013	1.012	98.50	92.9	1.23
1.50	1.0056	1.0074	15.1	0.414	990.5	7.7	35	1.3365	1.519	0.817	0.444	1.020	1.016	97.84	140.	2.10
2.00	1.0081	1.0098	20.2	0.553	987.9	10.3	46	1.3376	2.076	1.116	0.607	1.027	1.021	97.17	183.	3.14
2.50	1.0105	1.0123	25.3	0.693	985.3	13.0	58	1.3388	2.662	1.431	0.775	1.034	1.026	96.49	220.	4.85
3.00	1.0130	1.0148	30.4	0.833	982.6	15.6	69	1.3399	3.276	1.761	0.949	1.042	1.030	95.81		
3.50	1.0154	1.0172	35.5	0.975	979.9	18.3	81	1.3411	3.916	2.105	1.128	1.049	1.035	95.12		
4.00	1.0179	1.0197	40.7	1.116	977.2	21.0	92	1.3422	4.579	2.462	1.311	1.057	1.040	94.43		
4.50	1.0204	1.0222	45.9	1.259	974.4	23.8	104	1.3434	5.27	2.83	1.496	1.065	1.046	93.73		
5.00	1.0228	1.0246	51.1	1.402	971.7	26.5	115	1.3445	5.98	3.22	1.683	1.073	1.051	93.04		
5.50	1.0253	1.0271	56.4	1.546	968.9	29.3	127	1.3457	6.73	3.62	1.874	1.081	1.056	92.34		
6.00	1.0278	1.0296	61.7	1.691	966.1	32.1	138	1.3468	7.52	4.04	2.070	1.089	1.062	91.64		
6.50	1.0302	1.0321	67.0	1.836	963.3	35.0	150	1.3480	8.34	4.49	2.269	1.097	1.067	90.94		
7.00	1.0327	1.0345	72.3	1.982	960.4	37.8	162	1.3491	9.22	4.96	2.474	1.106	1.073	90.24		
7.50	1.0352	1.0370	77.6	2.129	957.5	40.7	173	1.3503	10.14	5.45	2.681	1.115	1.079	89.54		
8.00	1.0377	1.0395	83.0	2.276	954.6	43.6	185	1.3515	11.10	5.97	2.890	1.123	1.085	88.85		
8.50	1.0401	1.0420	88.4	2.424	951.7	46.5	196	1.3526	12.10	6.51	3.101	1.132	1.091	88.15		
9.00	1.0426	1.0445	93.8	2.573	948.8	49.5	208	1.3538	13.15	7.07	3.313	1.141	1.097	87.46		
9.50	1.0451	1.0469	99.3	2.722	945.8	52.4	219	1.3549	14.25	7.66	3.526	1.150	1.103	86.77		
10.00	1.0476	1.0494	104.8	2.872	942.8	55.4	231	1.3561	15.40	8.28	3.740	1.159	1.109	86.08		
11.00	1.0526	1.0544	115.8	3.175	936.8	61.4	254	1.3584	17.85	9.60	4.166	1.178	1.122	84.71		
12.00	1.0576	1.0594	126.9	3.480	930.7	67.6	277	1.3607	20.51	11.03	4.590	1.197	1.134	83.36		
13.00	1.0626	1.0645	138.1	3.788	924.4	73.8	300	1.3630				1.217	1.148	82.01		
14.00	1.0676	1.0695	149.5	4.098	918.1	80.1	323	1.3653				1.237	1.161	80.68		
15.00	1.0726	1.0745	160.9	4.412	911.8	86.5	347	1.3676				1.258	1.175	79.35		
16.00	1.0777	1.0796	172.4	4.728	905.3	93.0	370	1.3700				1.279	1.189	78.03		
17.00	1.0828	1.0847	184.1	5.047	898.7	99.5	393	1.3723				1.301	1.204	76.72		
18.00	1.0878	1.0898	195.8	5.369	892.0	106.2	416	1.3746				1.323	1.219	75.42		
19.00	1.0929	1.0949	207.7	5.694	885.3	113.0	439	1.3769				1.347	1.235	74.11		
20.00	1.0980	1.1000	219.6	6.022	878.4	119.8	462	1.3792				1.371	1.251	72.80		
22.00	1.1083	1.1102	243.8	6.686	864.5	133.8	509	1.3838				1.423	1.286	70.15		
24.00	1.1185	1.1205	268.4	7.361	850.1	148.1	555	1.3884				1.480	1.326	67.44		
26.00	1.1288	1.1308	293.5	8.047	835.3	162.9	600	1.3930				1.544	1.371	64.63		
28.00	1.1391	1.1411	318.9	8.745	820.1	178.1	646	1.3976				1.617	1.423	61.71		
30.00	1.1492	1.1513	344.8	9.454	804.5	193.8	691	1.4020				1.702	1.484	58.64		
32.00	1.1594	1.1614	371.0	10.173	788.4	209.9	736	1.4066				1.795	1.551	55.61		
34.00	1.1693	1.1714	397.6	10.901	771.8	226.5	782	1.4112				1.896	1.625	52.63		
36.00	1.1791	1.1812	424.5	11.639	754.6	243.6	828	1.4158				1.998	1.698	49.95		
38.00	1.1886	1.1907	451.7	12.385	736.9	261.3	874	1.4204				2.101	1.771	47.51		
40.00	1.1977	1.1999	479.1	13.137	718.6	279.6	920	1.4250								

25 INULIN, $(C_6H_{10}O_5)_x$*

MOLECULAR WEIGHT, AVERAGE = 5200
RELATIVE SPECIFIC REFRACTIVITY = 1.034

0.00 % by wt. data are the same for all compounds.
For Values of 0.00 wt. % solutions see Table 1, Acetic Acid.

A % by wt.	ρ D_4^{20}	D_{20}^{20}	C_s g/l	M g-mol/l	C_w g/l	$(C_o - C_w)$ g/l	$(n - n_o)$ × 10^4	n	Δ °C	O Os/kg	S g-mol/l	η/η_o	η/ρ cS	φ rhe	γ mmho/cm	T g-mol/l
0.50	1.0001	1.0019	5.0	0.001	995.1	3.1	7	1.3337	0.003	0.001	0.001	1.030	1.032	96.93		
1.00	1.0020	1.0038	10.0	0.002	992.0	6.3	14	1.3344	0.005	0.003	0.001	1.059	1.059	94.24		
1.50	1.0038	1.0056	15.1	0.003	988.8	9.5	22	1.3351	0.007	0.004	0.002	1.088	1.086	91.74		
2.00	1.0057	1.0074	20.1	0.004	985.6	12.7	29	1.3359	0.010	0.005	0.003	1.117	1.113	89.35		
2.50	1.0075	1.0093	25.2	0.005	982.3	15.9	36	1.3366	0.012	0.007	0.003	1.147	1.140	87.03		
3.00	1.0093	1.0111	30.3	0.006	979.0	19.2	43	1.3373	0.015	0.008	0.004	1.177	1.168	84.81		
3.50	1.0112	1.0129	35.4	0.007	975.8	22.5	51	1.3381	0.017	0.009	0.005	1.208	1.197	82.65		
4.00	1.0130	1.0148	40.5	0.008	972.5	25.8	58	1.3388	0.019	0.010	0.005	1.240	1.226	80.51		
4.50	1.0148	1.0166	45.7	0.009	969.2	29.1	65	1.3395	0.022	0.012	0.006	1.274	1.258	78.36		
5.00	1.0167	1.0185	50.8	0.010	965.9	32.4	73	1.3403	0.024	0.013	0.007	1.310	1.291	76.18		
5.50	1.0186	1.0204	56.0	0.011	962.6	35.7	80	1.3410	0.027	0.015	0.007	1.349	1.327	74.00		
6.00	1.0205	1.0223	61.2	0.012	959.3	38.9	88	1.3418	0.030	0.016	0.008	1.389	1.364	71.83		
6.50	1.0225	1.0243	66.5	0.013	956.0	42.2	95	1.3425	0.033	0.018	0.009	1.432	1.403	69.69		
7.00	1.0245	1.0263	71.7	0.014	952.7	45.5	103	1.3433	0.036	0.019	0.010	1.477	1.444	67.59		
7.50	1.0265	1.0283	77.0	0.015	949.5	48.7	110	1.3440	0.039	0.021	0.011	1.523	1.487	65.52		
8.00	1.0286	1.0304	82.3	0.016	946.3	52.0	118	1.3448	0.043	0.023	0.012	1.572	1.531	63.49		
8.50	1.0307	1.0325	87.6	0.017	943.1	55.2	125	1.3455	0.047	0.025	0.013	1.623	1.578	61.50		
9.00	1.0329	1.0347	93.0	0.018	939.9	58.3	133	1.3463	0.051	0.027	0.014	1.676	1.626	59.56		
9.50	1.0351	1.0370	98.3	0.019	936.8	61.4	141	1.3471	0.055	0.030	0.015	1.731	1.675	57.67		
10.00	1.0374	1.0393	103.7	0.020	933.7	64.5	148	1.3478	0.060	0.032	0.017	1.788	1.727	55.82		

* Supersaturated solutions.

26 LACTIC ACID, $CH_3CHOHCOOH$, (D-PROPANOIC ACID, 2-HYDROXY)

MOLECULAR WEIGHT = 90.08
RELATIVE SPECIFIC REFRACTIVITY = 1.058

0.00 % by wt. data are the same for all compounds.
For Values of 0.00 wt. % solutions see Table 1, Acetic Acid.

A % by wt.	ρ D_4^{20}	D_{20}^{20}	C_s g/l	M g-mol/l	C_w g/l	$(C_o - C_w)$ g/l	$(n - n_o)$ × 10^4	n	Δ °C	O Os/kg	S g-mol/l	η/η_o	η/ρ cS	φ rhe	γ mmho/cm	T g-mol/l
0.50	0.9992	1.0010	05.0	0.055	994.2	4.0	5	1.3335	0.095	0.051	0.027	1.012	1.015	98.62	1.0	0.010
1.00	1.0002	1.0020	10.0	0.111	990.2	8.0	10	1.3340	0.190	0.102	0.054	1.025	1.027	97.37	1.8	0.018
2.00	1.0023	1.0040	20.0	0.223	982.2	16.0	20	1.3350	0.378	0.203	0.109	1.054	1.053	94.69	2.6	0.026
3.00	1.0043	1.0061	30.1	0.334	974.2	24.0	30	1.3360	0.567	0.305	0.165	1.082	1.078	92.27	3.2	0.031
4.00	1.0065	1.0083	40.3	0.447	966.2	32.0	40	1.3370	0.758	0.407	0.221	1.108	1.102	90.05	3.5	0.035
5.00	1.0086	1.0104	50.4	0.560	958.2	40.0	50	1.3380	0.953	0.512	0.278	1.136	1.127	87.85	3.8	0.038
6.00	1.0108	1.0126	60.6	0.673	950.2	48.1	60	1.3390	1.155	0.621	0.337	1.165	1.154	85.64	4.1	0.041
7.00	1.0131	1.0149	70.9	0.787	942.1	56.1	70	1.3400	1.361	0.732	0.398	1.196	1.181	83.47	4.4	0.043
8.00	1.0153	1.0171	81.2	0.902	934.1	64.1	80	1.3410	1.573	0.846	0.460	1.227	1.210	81.34	4.6	0.045
9.00	1.0176	1.0194	91.6	1.017	926.0	72.2	90	1.3420	1.791	0.963	0.524	1.259	1.239	79.24	4.7	0.047
10.00	1.0199	1.0217	102.0	1.132	917.9	80.3	100	1.3430	2.015	1.084	0.589	1.293	1.269	77.19	4.9	0.049
12.00	1.0246	1.0264	123.0	1.365	901.6	96.6	120	1.3450	2.491	1.339	0.726	1.363	1.332	73.20	5.1	0.051
14.00	1.0294	1.0312	144.1	1.600	885.2	113.0	140	1.3470	2.988	1.607	0.868	1.438	1.399	69.38	5.2	0.052
16.00	1.0342	1.0360	165.5	1.837	868.7	129.5	161	1.3491	3.478	1.870	1.006	1.519	1.471	65.71	5.2	0.052
18.00	1.0390	1.0408	187.0	2.076	852.0	146.3	181	1.3511	3.962	2.130	1.141	1.604	1.546	62.22	5.2	0.052
20.00	1.0439	1.0457	208.8	2.318	835.1	163.1	202	1.3532	4.440	2.387	1.273	1.696	1.628	58.84	5.1	0.051
24.00	1.0536	1.0554	252.9	2.807	800.7	197.5	243	1.3573				1.898	1.806	52.57	4.9	0.048
28.00	1.0632	1.0651	297.7	3.305	765.5	232.7	285	1.3615				2.132	2.010	46.80	4.5	0.045
32.00	1.0728	1.0747	343.3	3.811	729.5	268.8	327	1.3657				2.409	2.251	41.43	4.1	0.041
36.00	1.0822	1.0841	389.6	4.325	692.6	305.7	370	1.3700				2.725	2.524	36.63	3.7	0.036
40.00	1.0915	1.0934	436.6	4.847	654.9	343.4	413	1.3743				3.108	2.853	32.11	3.2	0.032
44.00	1.1008	1.1028	484.4	5.377	616.5	381.8	456	1.3786				3.559	3.239	28.04	2.8	0.027
48.00	1.1105	1.1125	533.0	5.917	577.5	420.8	498	1.3828				4.098	3.699	24.35	2.3	0.022
52.00	1.1201	1.1221	582.4	6.466	537.6	460.6	541	1.3871				4.779	4.276	20.89	1.8	0.018
56.00	1.1297	1.1317	632.6	7.023	497.0	501.2	584	1.3914				5.568	4.940	17.93	1.5	0.014
60.00	1.1392	1.1412	683.5	7.588	455.7	542.6	628	1.3958				6.666	5.865	14.97	1.1	0.011
64.00	1.1486	1.1506	735.1	8.160	413.5	584.7	671	1.4001				8.008	6.988	12.46	0.8	0.008
68.00	1.1579	1.1599	787.4	8.741	370.5	627.7	715	1.4045				9.843	8.519	10.14	0.5	0.005
72.00	1.1670	1.1691	840.3	9.328	326.8	671.5	758	1.4088				12.84	11.02	7.77	0.3	0.003
76.00	1.1760	1.1781	893.8	9.922	282.2	716.0	801	1.4131				16.94	14.43	5.89	0.1	0.001
80.00	1.1848	1.1869	947.9	10.523	237.0	761.3	843	1.4173				22.12	18.70	4.51	0.1	0.001

27 LACTOSE, $C_{12}H_{22}O_{11} \cdot 1H_2O$

MOLECULAR WEIGHT = 342.30 FORMULA WEIGHT, HYDRATE = 360.31
RELATIVE SPECIFIC REFRACTIVITY = 1.036

0.00 % by wt. data are the same for all compounds.
For Values of 0.00 wt. % solutions see Table 1, Acetic Acid.

A% by wt.	H% by wt.	ρ D^{20}_4	D^{20}_{20}	C, g/l	M g-mol/l	C_w g/l	$(C_o - C_w)$ g/l	$(n - n_o)$ × 10^4	n	Δ °C	O Os/kg	S g-mol/l	η/η_o	η/ρ cS	φ rhe	γ mmho/cm	T g-mol/l
0.50	0.53	1.0002	1.0019	5.0	0.015	995.2	3.1	7	1.3337	0.027	0.015	0.007	1.011	1.013	98.71		
1.00	1.05	1.0021	1.0039	10.0	0.029	992.1	6.1	15	1.3345	0.055	0.030	0.015	1.024	1.024	97.46		
1.50	1.58	1.0041	1.0059	15.1	0.044	989.0	9.2	22	1.3352	0.083	0.045	0.023	1.039	1.037	96.01		
2.00	2.11	1.0061	1.0079	20.1	0.059	986.0	12.3	30	1.3359	0.112	0.060	0.031	1.056	1.052	94.51		
2.50	2.63	1.0081	1.0099	25.2	0.074	982.9	15.3	37	1.3367	0.140	0.075	0.040	1.072	1.065	93.12		
3.00	3.16	1.0102	1.0119	30.3	0.089	979.8	18.4	45	1.3375	0.169	0.091	0.048	1.087	1.078	91.80		
3.50	3.68	1.0122	1.0140	35.4	0.103	976.8	21.5	53	1.3382	0.198	0.107	0.056	1.103	1.091	90.52		
4.00	4.21	1.0143	1.0160	40.6	0.119	973.7	24.6	60	1.3390	0.228	0.122	0.065	1.118	1.105	89.25		
4.50	4.74	1.0163	1.0181	45.7	0.134	970.6	27.7	68	1.3398	0.258	0.139	0.074	1.135	1.119	87.96		
5.00	5.26	1.0184	1.0202	50.9	0.149	967.5	30.8	76	1.3406	0.288	0.155	0.083	1.152	1.133	86.63		
5.50	5.79	1.0204	1.0223	56.1	0.164	964.3	33.9	84	1.3413	0.319	0.172	0.092	1.170	1.149	85.27		
6.00	6.32	1.0225	1.0243	61.4	0.179	961.2	37.1	91	1.3421	0.351	0.189	0.101	1.189	1.166	83.91		
6.50	6.84	1.0246	1.0264	66.6	0.195	958.0	40.2	99	1.3429	0.385	0.207	0.111	1.209	1.183	82.54		
7.00	7.37	1.0267	1.0285	71.9	0.210	954.8	43.4	107	1.3437	0.420	0.226	0.121	1.230	1.200	81.16		
7.50	7.89	1.0287	1.0305	77.2	0.225	951.6	46.7	115	1.3445	0.456	0.245	0.132	1.251	1.218	79.79		
8.00	8.42	1.0308	1.0326	82.5	0.241	948.3	49.9	123	1.3453	0.495	0.266	0.143	1.273	1.237	78.42		
8.50	8.95	1.0329	1.0347	87.8	0.256	945.1	53.2	130	1.3460				1.295	1.257	77.05		
9.00	9.47	1.0349	1.0367	93.1	0.272	941.8	56.5	138	1.3468				1.318	1.277	75.69		
9.50	10.00	1.0370	1.0388	98.5	0.288	938.5	59.8	146	1.3476				1.342	1.297	74.34		
10.00	10.53	1.0390	1.0409	103.9	0.304	935.1	63.1	154	1.3484				1.367	1.318	73.01		
11.00	11.58	1.0432	1.0450	114.7	0.335	928.4	69.8	170	1.3500				1.418	1.362	70.37		
12.00	12.63	1.0473	1.0492	125.7	0.367	921.7	76.6	185	1.3515				1.473	1.409	67.77		
13.00	13.68	1.0515	1.0534	136.7	0.399	914.8	83.4	201	1.3531				1.530	1.458	65.24		
14.00	14.74	1.0558	1.0577	147.8	0.432	908.0	90.2	218	1.3548				1.590	1.509	62.76		
15.00	15.79	1.0602	1.0621	159.0	0.465	901.2	97.0	234	1.3564				1.654	1.563	60.35		
16.00	16.84	1.0648	1.0667	170.4	0.498	894.4	103.8	252	1.3582				1.721	1.619	58.01		
17.00	17.89	1.0696	1.0714	181.8	0.531	887.7	110.5	270	1.3600				1.791	1.678	55.73		
18.00	18.95	1.0746	1.0765	193.4	0.565	881.2	117.1	289	1.3619				1.865	1.739	53.51		

28 LANTHANUM NITRATE, $La(NO_3)_3 \cdot 6H_2O$

MOLECULAR WEIGHT = 324.93 FORMULA WEIGHT, HYDRATE = 433.02

A% by wt.	H% by wt	ρ D^{20}_4	D^{20}_{20}	C, g/l	M g-mol/ℓ	C_w g/l	$(C_0 - C_w)$ g/l	$(n - n_0)$ × 10^4	n	Δ °C	0 os/kg	S g-mol/ℓ	η/η_0	η/ρ cS	φ rhe	γ mmho/cm	T g-mol/l
1.00	1.33	1.0062	1.0080	10.1	0.031	996.1	2.1	14	1.3344	0.180	0.097	0.051	1.013	1.009	98.50	7.8	0.080
2.00	2.67	1.0142	1.0160	20.3	0.062	993.9	4.3	28	1.3358	0.358	0.192	0.103	1.025	1.013	97.33	14.2	0.151
3.00	4.00	1.0226	1.0244	30.7	0.094	991.9	6.3	42	1.3372	0.525	0.283	0.152	1.040	1.019	95.97	20.1	0.221
4.00	5.33	1.0309	1.0328	41.2	0.127	989.7	8.5	57	1.3387	0.698	0.375	0.203	1.056	1.026	94.54	25.6	0.289
5.00	6.66	1.0394	1.0413	52.0	0.160	987.4	10.8	71	1.3401	0.889	0.478	0.259	1.074	1.035	92.96	31.4	0.358
6.00	8.00	1.0483	1.0502	62.9	0.194	985.4	12.8	86	1.3416	1.079	0.580	0.315	1.091	1.043	91.46	36.0	0.420
7.00	9.33	1.0575	1.0594	74.0	0.228	983.5	14.7	101	1.3431	1.266	0.681	0.370	1.110	1.052	89.89	40.6	0.479
8.00	10.66	1.0667	1.0686	85.3	0.263	981.4	16.8	115	1.3445	1.463	0.787	0.428	1.130	1.061	88.36	45.0	0.535
9.00	11.99	1.0761	1.0780	96.8	0.298	979.3	18.9	130	1.3460	1.676	0.901	0.490	1.149	1.070	86.85	49.5	0.596
10.00	13.33	1.0853	1.0873	108.5	0.334	976.8	21.4	145	1.3475	1.874	1.007	0.548	1.170	1.080	85.32	53.4	0.648
12.00	15.99	1.1055	1.1075	132.7	0.408	972.8	25.4	176	1.3506	2.340	1.258	0.683	1.214	1.100	82.23	60.7	0.750
14.00	18.66	1.1251	1.1271	157.5	0.485	967.6	30.6	208	1.3538	2.822	1.517	0.821	1.262	1.124	79.08	67.4	0.845
16.00	21.32	1.1456	1.1477	183.3	0.564	962.3	35.9	241	1.3571	3.331	1.791	0.965	1.322	1.156	75.51	73.2	0.930
18.00	23.99	1.1667	1.1688	210.0	0.646	956.7	41.5	274	1.3604	3.869	2.080	1.115	1.390	1.194	71.79	77.7	0.999
20.00	26.65	1.1892	1.1914	237.9	0.732	951.3	46.9	308	1.3638	4.418	2.375	1.267	1.463	1.233	68.20	81.4	1.055
24.00	31.98	1.2337	1.2359	296.1	0.911	937.6	60.6	379	1.3709				1.641	1.333	60.81	86.7	1.137
28.00	37.31	1.2829	1.2852	359.2	1.105	923.7	74.5	453	1.3783				1.854	1.448	53.83	86.8	1.138
32.00	42.64	1.3319	1.3343	426.2	1.312	905.7	92.5	530	1.3860				2.156	1.622	46.29	84.0	1.095
36.00	47.98	1.3874	1.3899	499.5	1.537	887.9	110.3	613	1.3943				2.580	1.863	38.69	81.4	1.055
40.00	53.31	1.4477	1.4503	579.1	1.782	868.6	129.6	700	1.4030				3.205	2.218	31.14	71.9	0.911
44.00	58.64	1.5084	1.5111	663.7	2.043	844.7	153.5	788	1.4118				4.105	2.727	23.31	61.9	0.767

29 LEAD NITRATE, $Pb(NO_3)_2$

MOLECULAR WEIGHT = 331.23
RELATIVE SPECIFIC REFRACTIVITY = 0.479

0.00 % by wt. data are the same for all compounds.
For Values of 0.00 wt. % solutions see Table 1, Acetic Acid.

A% by wt.	ρ D_4^{20}	D_{20}^{20}	C_s g/l	M g-mol/l	C_w g/l	$(C_o - C_w)$ g/l	$(n - n_o)$ × 10^4	n	Δ °C	O Os/kg	S g-mol/l	η/η_o	η/ρ cS	φ rhe	γ mmho/cm	T g-mol/l
0.50	1.0025	1.0043	5.0	0.015	997.5	0.8	6	1.3336	0.072	0.039	0.020	1.003	1.003	99.49	2.8	0.028
1.00	1.0068	1.0086	10.1	0.030	996.7	1.5	12	1.3342	0.137	0.074	0.039	1.006	1.001	99.21	5.4	0.054
2.00	1.0155	1.0173	20.3	0.061	995.2	3.1	24	1.3354	0.258	0.139	0.074	1.011	0.998	98.71	9.6	0.099
3.00	1.0243	1.0261	30.7	0.093	993.6	4.6	36	1.3366	0.372	0.200	0.107	1.016	0.993	98.27	13.3	0.141
4.00	1.0333	1.0352	41.3	0.125	992.0	6.2	49	1.3379	0.482	0.259	0.140	1.020	0.989	97.85	16.7	0.181
5.00	1.0425	1.0444	52.1	0.157	990.4	7.8	61	1.3391	0.587	0.316	0.170	1.025	0.985	97.37	19.9	0.219
6.00	1.0519	1.0537	63.1	0.191	988.7	9.5	73	1.3403	0.690	0.371	0.201	1.031	0.982	96.78	23.0	0.256
7.00	1.0614	1.0632	74.3	0.224	987.1	11.2	86	1.3415	0.792	0.426	0.231	1.038	0.980	96.12	25.9	0.292
8.00	1.0710	1.0729	85.7	0.259	985.4	12.9	98	1.3428	0.891	0.479	0.260	1.046	0.979	95.40	28.6	0.326
9.00	1.0809	1.0828	97.3	0.294	983.6	14.6	110	1.3440	0.987	0.531	0.288	1.054	0.978	94.64	31.3	0.360
10.00	1.0909	1.0929	109.1	0.329	981.8	16.4	124	1.3454	1.081	0.581	0.316	1.063	0.976	93.89	33.9	0.393
12.00	1.1115	1.1135	133.4	0.403	978.1	20.1	151	1.3481	1.265	0.680	0.370	1.081	0.974	92.33	38.9	0.457
14.00	1.1329	1.1349	158.6	0.479	974.3	24.0	180	1.3510	1.450	0.780	0.424	1.101	0.973	90.69	43.4	0.515
16.00	1.1550	1.1571	184.8	0.558	970.2	28.0	209	1.3539	1.639	0.881	0.479	1.122	0.973	88.94	47.5	0.570
18.00	1.1779	1.1800	212.0	0.640	965.9	32.3	239	1.3569	1.825	0.981	0.534	1.146	0.975	87.10	51.4	0.621
20.00	1.2017	1.2038	240.3	0.726	961.4	36.9	271	1.3601	2.008	1.080	0.587	1.172	0.977	85.15	55.2	0.673
24.00	1.2519	1.2541	300.5	0.907	951.5	46.8	337	1.3667	2.361	1.269	0.689	1.233	0.987	80.97	62.5	0.772
28.00	1.3059	1.3082	365.7	1.104	940.3	58.0	407	1.3737	2.714	1.459	0.790	1.306	1.002	76.43	68.4	0.858
32.00	1.3640	1.3664	436.5	1.318	927.5	70.7	481	1.3811				1.398	1.027	71.39	72.5	0.920
34.00	1.3947	1.3972	474.2	1.432	920.5	77.7	520	1.3850				1.459	1.048	68.42	74.8	0.954

30 LITHIUM CHLORIDE, LiCl

MOLECULAR WEIGHT = 42.40
RELATIVE SPECIFIC REFRACTIVITY = 1.023

0.00 % by wt. data are the same for all compounds.
For Values of 0.00 wt. % solutions see Table 1, Acetic Acid.

A% by wt.	ρ D_4^{20}	D_{20}^{20}	C_s g/l	M g-mol/l	C_w g/l	$(C_o - C_w)$ g/l	$(n - n_o)$ × 10^4	n	Δ °C	O Os/kg	S g-mol/l	η/η_o	η/ρ cS	φ rhe	γ mmho/cm	T g-mol/l
0.50	1.0012	1.0029	5.0	0.118	996.2	2.1	11	1.3341	0.415	0.223	0.120	1.017	1.018	98.08	10.1	0.105
1.00	1.0041	1.0059	10.0	0.237	994.0	4.2	21	1.3351	0.836	0.450	0.244	1.035	1.033	96.43	19.0	0.208
2.00	1.0099	1.0117	20.2	0.476	989.7	8.5	43	1.3373	1.724	0.927	0.504	1.070	1.062	93.27	34.9	0.406
3.00	1.0157	1.0175	30.5	0.719	985.2	13.0	64	1.3394	2.684	1.443	0.782	1.106	1.091	90.22	49.8	0.600
4.00	1.0215	1.0233	40.9	0.964	980.6	17.6	85	1.3415	3.727	2.004	1.075	1.144	1.122	87.26	63.6	0.791
5.00	1.0272	1.0290	51.4	1.211	975.9	22.4	106	1.3436	4.859	2.612	1.386	1.183	1.154	84.36	76.4	0.978
6.00	1.0330	1.0348	62.0	1.462	971.0	27.2	127	1.3457	6.14	3.30	1.723	1.224	1.187	81.55	88.3	1.16
7.00	1.0387	1.0405	72.7	1.715	966.0	32.2	148	1.3478	7.56	4.07	2.081	1.266	1.221	78.83	99.5	1.34
8.00	1.0444	1.0463	83.6	1.971	960.9	37.3	169	1.3499	9.11	4.90	2.449	1.310	1.257	76.16	110.	1.52
9.00	1.0502	1.0521	94.5	2.229	955.7	42.6	190	1.3520	10.79	5.80	2.825	1.357	1.295	73.52	119.	1.69
10.00	1.0560	1.0578	105.6	2.490	950.4	47.9	211	1.3541	12.61	6.78	3.205	1.408	1.336	70.88	127.	1.84
12.00	1.0675	1.0694	128.1	3.021	939.4	58.8	253	1.3583	16.59	8.92	3.951	1.519	1.426	65.68	140.	2.12
14.00	1.0792	1.0811	151.1	3.563	928.1	70.1	295	1.3625	21.04	11.31	4.670	1.644	1.527	60.70	151.	2.34
16.00	1.0910	1.0929	174.6	4.117	916.4	81.8	338	1.3668				1.783	1.637	55.99	160.	2.56
18.00	1.1029	1.1048	198.5	4.682	904.4	93.9	381	1.3711				1.938	1.761	51.50	167.	2.73
20.00	1.1150	1.1170	223.0	5.260	892.0	106.2	425	1.3755				2.124	1.909	46.99	170.	2.80
22.00	1.1274	1.1294	248.0	5.850	879.3	118.9	469	1.3799				2.336	2.076	42.72	170.	2.79
24.00	1.1399	1.1419	273.6	6.452	866.3	131.9	514	1.3844				2.595	2.281	38.46	167.	2.73
26.00	1.1527	1.1548	299.7	7.069	853.0	145.2	560	1.3890				2.919	2.537	34.19	163.	2.63
28.00	1.1658	1.1678	326.4	7.699	839.4	158.9	606	1.3936				3.311	2.846	30.14	156.	2.47
30.00	1.1791	1.1812	353.7	8.343	825.4	172.9	653	1.3983				3.777	3.210	26.42	146.	2.25

31 MAGNESIUM CHLORIDE, $MgCl_2 \cdot 6H_2O$

MOLECULAR WEIGHT = 95.23 FORMULA WEIGHT, HYDRATE = 203.33
RELATIVE SPECIFIC REFRACTIVITY = 0.893

0.00 % by wt. data are the same for all compounds.
For Values of 0.00 wt. % solutions see Table 1, Acetic Acid.

A% by wt.	H% by wt.	ρ D_4^{20}	D_{20}^{20}	C_s g/l	M g-mol/l	C_w g/l	$(C_o - C_w)$ g/l	$(n - n_o)$ × 10^4	n	Δ °C	O Os/kg	S g-mol/l	η/η_o	η/ρ cS	φ rhe	γ mmho/cm	T g-mol/l
0.50	1.07	1.0022	1.0040	5.0	0.053	997.2	1.0	13	1.3343	0.255	0.137	0.073	1.022	1.022	97.66	8.6	0.092
1.00	2.14	1.0062	1.0080	10.1	0.106	996.2	2.0	26	1.3356	0.517	0.278	0.150	1.044	1.040	95.59	16.6	0.183
2.00	4.27	1.0144	1.0162	20.3	0.213	994.1	4.1	51	1.3381	1.063	0.572	0.311	1.089	1.076	91.60	31.2	0.356
3.00	6.41	1.0226	1.0244	30.7	0.322	991.9	6.3	76	1.3406	1.653	0.889	0.483	1.137	1.114	87.80	44.5	0.530
4.00	8.54	1.0309	1.0328	41.2	0.433	989.7	8.5	102	1.3432	2.297	1.235	0.671	1.186	1.153	84.13	56.4	0.690
5.00	10.68	1.0394	1.0412	52.0	0.546	987.4	10.8	127	1.3457	3.012	1.619	0.875	1.239	1.194	80.57	66.9	0.840
6.00	12.81	1.0479	1.0497	62.9	0.660	985.0	13.2	153	1.3483				1.295	1.238	77.09	77.2	0.990
7.00	14.95	1.0564	1.0583	74.0	0.777	982.5	15.7	178	1.3508				1.355	1.285	73.67	86.4	1.13
8.00	17.08	1.0651	1.0670	85.2	0.895	979.9	18.3	204	1.3534				1.420	1.336	70.30	94.5	1.26
9.00	19.22	1.0738	1.0757	96.6	1.015	977.2	21.1	231	1.3560				1.490	1.390	66.98	102.0	1.38
10.00	21.35	1.0826	1.0845	108.3	1.137	974.4	23.9	257	1.3587				1.567	1.450	63.70	108.0	1.49
12.00	25.62	1.1005	1.1025	132.1	1.387	968.5	29.8	311	1.3641				1.742	1.586	57.30	119.0	1.69
14.00	29.89	1.1189	1.1209	156.6	1.645	962.3	36.0	365	1.3695				1.952	1.748	51.13	127.0	1.84
16.00	34.16	1.1372	1.1392	181.9	1.911	955.2	43.0	419	1.3749				2.203	1.941	45.30	132.0	1.94
18.00	38.43	1.1553	1.1574	208.0	2.184	947.4	50.8	474	1.3804				2.502	2.170	39.89	134.0	1.99
20.00	42.70	1.1742	1.1763	234.8	2.466	939.4	58.9	529	1.3859				2.861	2.441	34.89	134.0	1.99
22.00	46.97	1.1938	1.1959	262.6	2.758	931.2	67.1	585	1.3915				3.316	2.783	30.10	131.0	1.93
24.00	51.24	1.2140	1.2162	291.4	3.060	922.6	75.6	642	1.3972				3.909	3.226	25.53	126.0	1.81
26.00	55.51	1.2346	1.2368	321.0	3.371	913.6	84.6	700	1.4030				4.685	3.802	21.30	118.0	1.67
28.00	59.78	1.2555	1.2577	351.5	3.691	903.9	94.3	759	1.4089				5.698	4.547	17.52	109.0	1.50
30.00	64.05	1.2763	1.2786	382.9	4.021	893.4	104.8	818	1.4148				7.003	5.498	14.25	97.9	1.32

32 MAGNESIUM SULFATE, $MgSO_4 \cdot 7H_2O$

MOLECULAR WEIGHT = 120.37 FORMULA WEIGHT, HYDRATE = 246.48
RELATIVE SPECIFIC REFRACTIVITY = 0.572

0.00 % by wt. data are the same for all compounds.
For Values of 0.00 wt. % solutions see Table 1, Acetic Acid.

A% by wt.	H% by wt.	ρ D_4^{20}	D_{20}^{20}	C_s g/l	M g-mol/l	C_w g/l	$(C_o - C_w)$ g/l	$(n - n_o) \times 10^4$	n	Δ °C	O Os/kg	S g-mol/l	η/η_o	η/ρ cS	φ rhe	γ mmho/cm	T g-mol/l
0.50	1.02	1.0033	1.0051	5.0	0.042	998.3	0.0	10	1.3340	0.103	0.055	0.029	1.025	1.024	97.35	4.1	0.041
1.00	2.05	1.0084	1.0102	10.1	0.084	998.3	−0.1	20	1.3350	0.192	0.103	0.055	1.052	1.045	94.88	7.6	0.078
2.00	4.10	1.0186	1.0204	20.4	0.169	998.2	0.0	41	1.3371	0.361	0.194	0.104	1.110	1.092	89.91	13.3	0.141
3.00	6.15	1.0289	1.0307	30.9	0.256	998.0	0.2	61	1.3391	0.522	0.281	0.151	1.175	1.144	84.94	18.4	0.201
4.00	8.19	1.0392	1.0411	41.6	0.345	997.6	0.6	81	1.3411	0.692	0.372	0.201	1.247	1.202	80.05	23.1	0.258
5.00	10.24	1.0497	1.0515	52.5	0.436	997.2	1.1	101	1.3431	0.868	0.467	0.253	1.325	1.265	75.32	27.4	0.311
6.00	12.29	1.0602	1.0621	63.6	0.528	996.6	1.6	121	1.3451	1.048	0.564	0.306	1.408	1.330	70.90	31.1	0.358
7.00	14.34	1.0708	1.0727	75.0	0.623	995.9	2.3	142	1.3471	1.235	0.664	0.361	1.495	1.399	66.77	34.4	0.399
8.00	16.39	1.0816	1.0835	86.5	0.719	995.1	3.2	162	1.3492	1.431	0.770	0.419	1.590	1.473	62.75	37.3	0.436
9.00	18.44	1.0924	1.0944	98.3	0.817	994.1	4.1	182	1.3512	1.635	0.879	0.478	1.699	1.558	58.75	40.0	0.471
10.00	20.48	1.1034	1.1053	110.3	0.917	993.0	5.2	202	1.3532	1.849	0.994	0.541	1.825	1.657	54.69	42.7	0.506
12.00	24.57	1.1257	1.1276	135.1	1.122	990.6	7.7	242	1.3572	2.313	1.243	0.675	2.100	1.869	47.53	48.4	0.581
14.00	28.67	1.1484	1.1504	160.8	1.335	987.6	10.6	283	1.3613	2.861	1.538	0.832	2.407	2.100	41.46	53.3	0.647
16.00	32.76	1.1717	1.1737	187.5	1.557	984.2	14.0	324	1.3654	3.673	1.974	1.061	2.803	2.397	35.61	55.2	0.673
18.00	36.86	1.1955	1.1976	215.2	1.787	980.3	17.9	364	1.3694				3.353	2.810	29.77	53.7	0.652
20.00	40.95	1.2198	1.2220	244.0	2.026	975.9	22.4	405	1.3735				4.139	3.400	24.11	51.1	0.617
22.00	45.05	1.2447	1.2469	273.8	2.275	970.9	27.4	446	1.3776				5.189	4.177	19.23	48.8	0.586
24.00	49.14	1.2701	1.2724	304.8	2.532	965.3	32.9	487	1.3817				6.485	5.116	15.39	45.9	0.547
26.00	53.24	1.2961	1.2984	337.0	2.799	959.1	39.1	528	1.3858				8.050	6.223	12.40	42.3	0.500

33 MALTOSE, $C_{12}H_{22}O_{11} \cdot 1H_2O$

MOLECULAR WEIGHT = 342.29 FORMULA WEIGHT, HYDRATE = 360.31
RELATIVE SPECIFIC REFRACTIVITY = 1.038

0.00 % by wt. data are the same for all compounds.
For Values of 0.00 wt. % solutions see Table 1, Acetic Acid.

A% by wt.	H% by wt.	ρ D_4^{20}	D_{20}^{20}	C_s g/l	M g-mol/l	C_w g/l	$(C_o - C_w)$ g/l	$(n - n_o) \times 10^4$	n	Δ °C	O Os/kg	S g-mol/l	η/η_o	η/ρ cS	φ rhe	γ mmho/cm	T g-mol/l
0.50	0.53	1.0003	1.0020	5.0	0.015	995.3	3.0	7	1.3337	0.027	0.015	0.007	1.014	1.016	98.45		
1.00	1.05	1.0023	1.0041	10.0	0.029	992.3	6.0	15	1.3345	0.055	0.030	0.015	1.028	1.028	97.08		
1.50	1.58	1.0043	1.0061	15.1	0.044	989.2	9.0	22	1.3352	0.083	0.045	0.023	1.043	1.040	95.71		
2.00	2.11	1.0063	1.0081	20.1	0.059	986.2	12.0	29	1.3359	0.112	0.060	0.031	1.058	1.053	94.33		
2.50	2.63	1.0083	1.0101	25.2	0.074	983.1	15.1	37	1.3367	0.140	0.075	0.040	1.074	1.067	92.95		
3.00	3.16	1.0104	1.0121	30.3	0.089	980.0	18.2	44	1.3374	0.169	0.091	0.048	1.090	1.081	91.56		
3.50	3.68	1.0124	1.0142	35.4	0.104	976.9	21.3	52	1.3382	0.199	0.107	0.057	1.107	1.095	90.17		
4.00	4.21	1.0144	1.0162	40.6	0.119	973.8	24.4	59	1.3389	0.229	0.123	0.065	1.124	1.110	88.78		
4.50	4.74	1.0164	1.0182	45.7	0.134	970.7	27.6	67	1.3397	0.259	0.139	0.074	1.142	1.126	87.41		
5.00	5.26	1.0184	1.0202	50.9	0.149	967.5	30.7	74	1.3404	0.290	0.156	0.083	1.160	1.141	86.03		
5.50	5.79	1.0204	1.0222	56.1	0.164	964.3	33.9	82	1.3412	0.321	0.172	0.092	1.179	1.157	84.68		
6.00	6.32	1.0224	1.0243	61.3	0.179	961.1	37.1	90	1.3420	0.352	0.189	0.101	1.198	1.174	83.34		
6.50	6.84	1.0245	1.0263	66.6	0.195	957.9	40.4	97	1.3427	0.384	0.206	0.111	1.217	1.190	82.01		
7.00	7.37	1.0265	1.0283	71.9	0.210	954.6	43.6	105	1.3435	0.416	0.224	0.120	1.237	1.207	80.69		
7.50	7.89	1.0285	1.0303	77.1	0.225	951.3	46.9	113	1.3443	0.449	0.241	0.130	1.257	1.225	79.38		
8.00	8.42	1.0305	1.0323	82.4	0.241	948.0	50.2	120	1.3450	0.482	0.259	0.139	1.278	1.243	78.07		
8.50	8.95	1.0325	1.0343	87.8	0.256	944.7	53.5	128	1.3458	0.516	0.277	0.149	1.300	1.262	76.77		
9.00	9.47	1.0345	1.0363	93.1	0.272	941.4	56.8	136	1.3466	0.550	0.296	0.159	1.322	1.281	75.48		
9.50	10.00	1.0365	1.0383	98.5	0.288	938.0	60.2	144	1.3474	0.584	0.314	0.169	1.345	1.300	74.19		
10.00	10.53	1.0385	1.0403	103.8	0.303	934.6	63.6	152	1.3482	0.619	0.333	0.180	1.369	1.321	72.90		
11.00	11.58	1.0425	1.0443	114.7	0.335	927.8	70.4	167	1.3497	0.691	0.371	0.201	1.419	1.364	70.34		
12.00	12.63	1.0465	1.0483	125.6	0.367	920.9	77.3	183	1.3513	0.765	0.411	0.223	1.471	1.409	67.83		
13.00	13.68	1.0505	1.0523	136.6	0.399	913.9	84.3	199	1.3529	0.841	0.452	0.245	1.527	1.456	65.37		
14.00	14.74	1.0545	1.0563	147.6	0.431	906.8	91.4	216	1.3546	0.919	0.494	0.268	1.585	1.506	62.97		
15.00	15.79	1.0585	1.0603	158.8	0.464	899.7	98.6	232	1.3562	1.000	0.538	0.292	1.646	1.558	60.63		
16.00	16.84	1.0629	1.0648	170.1	0.497	892.8	105.4	248	1.3578	1.080	0.581	0.315	1.712	1.614	58.31		
17.00	17.89	1.0672	1.0691	181.4	0.530	885.8	112.4	265	1.3595	1.165	0.626	0.340	1.780	1.672	56.05		
18.00	18.95	1.0716	1.0735	192.9	0.563	878.7	119.5	282	1.3612	1.250	0.672	0.365	1.855	1.735	53.79		
19.00	20.00	1.0759	1.0778	204.4	0.597	871.5	126.7	298	1.3628	1.338	0.719	0.391	1.936	1.803	51.55		
20.00	21.05	1.0801	1.0820	216.0	0.631	864.1	134.1	314	1.3644	1.430	0.769	0.418	2.017	1.871	49.48		
22.00	23.16	1.0894	1.0913	239.7	0.700	849.7	148.5	348	1.3678	1.635	0.879	0.478	2.212	2.034	45.13		
24.00	25.26	1.0984	1.1004	263.6	0.770	834.8	163.4	384	1.3714	1.848	0.993	0.540	2.458	2.242	40.60		
26.00	27.37	1.1080	1.1100	288.1	0.841	819.9	178.3	419	1.3749	2.080	1.118	0.608	2.748	2.485	36.32		
28.00	29.47	1.1171	1.1191	312.8	0.913	804.3	193.9	455	1.3785	2.335	1.255	0.681	3.060	2.745	32.62		
30.00	31.58	1.1269	1.1289	338.1	0.987	788.8	209.4	491	1.3821	2.615	1.406	0.762	3.420	3.041	29.18		
32.00	33.68	1.1367	1.1387	363.7	1.062	773.0	225.2	528	1.3858	2.925	1.572	0.850	3.910	3.447	25.52		
34.00	35.79	1.1463	1.1483	389.7	1.138	756.6	241.6	566	1.3896	3.252	1.748	0.943	4.438	3.879	22.49		
36.00	37.89	1.1561	1.1582	416.2	1.215	739.9	258.3	605	1.3935	3.600	1.935	1.040	5.040	4.368	19.80		
38.00	40.00	1.1663	1.1684	443.2	1.294	723.1	275.1	644	1.3974	3.990	2.145	1.148	5.820	5.000	17.15		
40.00	42.10	1.1769	1.1790	470.8	1.375	706.1	292.1	683	1.4013	4.408	2.370	1.264	6.912	5.885	14.44		
42.00	44.21	1.1878	1.1899	498.9	1.457	688.9	309.3	721	1.4051	4.878	2.622	1.390	8.175	6.896	12.21		
44.00	46.31	1.1979	1.2000	527.1	1.539	670.8	327.4	764	1.4094	5.352	2.877	1.518	9.630	8.055	10.36		
46.00	48.42	1.2084	1.2105	555.9	1.623	652.5	345.7	806	1.4136				11.45	9.492	8.72		
48.00	50.52	1.2194	1.2216	585.3	1.709	634.1	364.1	847	1.4177				14.12	11.60	7.07		
50.00	52.63	1.2304	1.2326	615.2	1.796	615.2	383.0	887	1.4217				17.75	14.46	5.62		
52.00	54.74	1.2416	1.2438	645.6	1.885	596.0	402.2	930	1.4260				21.99	17.74	4.54		
54.00	56.84	1.2528	1.2550	676.5	1.975	576.3	421.9	978	1.4308				28.70	22.96	3.48		
56.00	58.94	1.2638	1.2660	707.7	2.066	556.1	442.1	1020	1.4350				38.15	30.25	2.62		
58.00	61.05	1.2746	1.2769	739.3	2.159	535.3	462.9	1064	1.4394				49.20	38.68	2.03		
60.00	63.16	1.2855	1.2878	771.3	2.252	514.2	484.0	1110	1.4440								

34 MANGANOUS SULFATE, $MnSO_4 \cdot 1H_2O$

MOLECULAR WEIGHT = 151.00 FORMULA WEIGHT, HYDRATE = 169.01
RELATIVE SPECIFIC REFRACTIVITY = 0.541

0.00 % by wt. data are the same for all compounds.
For Values of 0.00 wt. % solutions see Table 1, Acetic Acid.

A % by wt.	H % by wt.	ρ D_4^{20}	D_{20}^{20}	C_s g/l	M g-mol/l	C_w g/l	$(C_o - C_w)$ g/l	$(n - n_o) \times 10^4$	n	Δ °C	O Os/kg	S g-mol/l	η/η_o	η/ρ cS	φ rhe	γ mmho/cm	T g-mol/l
1.00	1.12	1.0080	1.0098	10.1	0.067	997.9	0.3	18	1.3348	0.160	0.086	0.045	1.044	1.038	95.59	6.2	0.063
2.00	2.24	1.0178	1.0196	20.4	0.135	997.5	0.8	36	1.3366	0.306	0.164	0.088	1.088	1.071	91.73	10.6	0.111
3.00	3.36	1.0277	1.0296	30.8	0.204	996.9	1.3	54	1.3384	0.437	0.235	0.126	1.135	1.107	87.93	14.6	0.157
4.00	4.48	1.0378	1.0396	41.5	0.275	996.3	2.0	72	1.3402	0.569	0.306	0.165	1.185	1.145	84.19	18.3	0.199
5.00	5.60	1.0480	1.0498	52.4	0.347	995.6	2.7	90	1.3420	0.701	0.377	0.204	1.240	1.186	80.48	21.6	0.240
6.00	6.72	1.0583	1.0602	63.5	0.421	994.8	3.4	108	1.3438	0.838	0.451	0.244	1.298	1.229	76.87	24.7	0.278
7.00	7.83	1.0688	1.0707	74.8	0.495	994.0	4.3	127	1.3457	0.978	0.526	0.285	1.360	1.275	73.36	27.4	0.312
8.00	8.95	1.0794	1.0813	86.4	0.572	993.1	5.2	145	1.3475	1.122	0.603	0.328	1.428	1.325	69.90	29.9	0.343
9.00	10.07	1.0902	1.0921	98.1	0.650	992.1	6.1	164	1.3494	1.275	0.685	0.373	1.502	1.380	66.46	32.3	0.372
10.00	11.19	1.1012	1.1031	110.1	0.729	991.1	7.2	183	1.3513	1.439	0.773	0.421	1.584	1.441	63.01	34.5	0.401
11.00	12.31	1.1123	1.1143	122.4	0.810	990.0	8.3	202	1.3532	1.612	0.867	0.472	1.675	1.509	59.57	36.7	0.428
12.00	13.43	1.1236	1.1256	134.8	0.893	988.8	9.5	221	1.3551	1.797	0.966	0.526	1.775	1.583	56.23	38.7	0.454
13.00	14.55	1.1351	1.1371	147.6	0.977	987.5	10.7	240	1.3570	1.995	1.073	0.583	1.883	1.663	52.99	40.5	0.478
14.00	15.67	1.1467	1.1487	160.5	1.063	986.2	12.1	259	1.3589	2.207	1.187	0.645	2.001	1.749	49.87	42.2	0.499
15.00	16.79	1.1585	1.1606	173.8	1.151	984.7	13.5	279	1.3609	2.431	1.307	0.709	2.129	1.841	46.88	43.7	0.518
16.00	17.91	1.1705	1.1726	187.3	1.240	983.2	15.0	299	1.3629	2.669	1.435	0.777	2.267	1.940	44.03	44.9	0.534
17.00	19.03	1.1827	1.1847	201.1	1.331	981.6	16.6	318	1.3648	2.922	1.571	0.850	2.415	2.046	41.32	45.9	0.548
18.00	20.15	1.1950	1.1971	215.1	1.425	979.9	18.3	338	1.3668	3.194	1.717	0.926	2.575	2.160	38.75	46.7	0.559
19.00	21.27	1.2075	1.2097	229.4	1.519	978.1	20.1	358	1.3688	3.486	1.874	1.008	2.747	2.280	36.32	47.3	0.566
20.00	22.39	1.2203	1.2224	244.1	1.616	976.2	22.0	378	1.3708	3.801	2.043	1.096	2.932	2.408	34.04	47.6	0.570

35 D-MANNITOL, $CH_2OH(CHOH)_4CH_2OH$

MOLECULAR WEIGHT = 182.17
RELATIVE SPECIFIC REFRACTIVITY = 1.063

0.00 % by wt. data are the same for all compounds.
For Values of 0.00 wt. % solutions see Table 1, Acetic Acid.

A % by wt.	ρ D_4^{20}	D_{20}^{20}	C_s g/l	M g-mol/l	C_w g/l	$(C_o - C_w)$ g/l	$(n - n_o) \times 10^4$	n	Δ °C	O Os/kg	S g-mol/l	η/η_o	η/ρ cS	φ rhe	γ mmho/cm	T g-mol/l
0.50	1.0000	1.0018	5.0	0.027	995.0	3.2	7	1.3337	0.051	0.028	0.014	1.015	1.017	98.30		
1.00	1.0017	1.0035	10.0	0.055	991.7	6.5	15	1.3345	0.103	0.055	0.029	1.030	1.030	96.89		
1.50	1.0035	1.0053	15.1	0.083	988.5	9.8	22	1.3352	0.156	0.084	0.044	1.044	1.042	95.59		
2.00	1.0053	1.0071	20.1	0.110	985.2	13.1	29	1.3359	0.209	0.112	0.059	1.058	1.055	94.33		
2.50	1.0070	1.0088	25.2	0.138	981.9	16.4	37	1.3367	0.262	0.141	0.075	1.072	1.067	93.07		
3.00	1.0088	1.0106	30.3	0.166	978.5	19.7	44	1.3374	0.317	0.170	0.091	1.087	1.079	91.84		
3.50	1.0106	1.0124	35.4	0.194	975.2	23.0	52	1.3381	0.371	0.200	0.107	1.101	1.092	90.61		
4.00	1.0124	1.0141	40.5	0.222	971.9	26.4	59	1.3389	0.427	0.230	0.123	1.117	1.105	89.38		
4.50	1.0141	1.0159	45.6	0.251	968.5	29.7	66	1.3396	0.483	0.260	0.140	1.132	1.119	88.13		
5.00	1.0159	1.0177	50.8	0.279	965.1	33.1	73	1.3403	0.540	0.290	0.156	1.149	1.133	86.86		
5.50	1.0177	1.0195	56.0	0.307	961.7	36.5	81	1.3411	0.598	0.321	0.173	1.166	1.148	85.56		
6.00	1.0195	1.0213	61.2	0.336	958.3	39.9	88	1.3418	0.656	0.353	0.191	1.185	1.164	84.24		
6.50	1.0212	1.0230	66.4	0.364	954.9	43.4	95	1.3425	0.715	0.384	0.208	1.204	1.181	82.92		
7.00	1.0230	1.0248	71.6	0.393	951.4	46.8	103	1.3433	0.774	0.416	0.225	1.223	1.198	81.59		
7.50	1.0248	1.0266	76.9	0.422	947.9	50.3	110	1.3440	0.835	0.449	0.243	1.244	1.216	80.26		
8.00	1.0266	1.0284	82.1	0.451	944.5	53.8	117	1.3447	0.896	0.482	0.261	1.264	1.234	78.93		
8.50	1.0284	1.0302	87.4	0.480	941.0	57.2	125	1.3455	0.958	0.515	0.279	1.286	1.253	77.62		
9.00	1.0302	1.0320	92.7	0.509	937.5	60.7	132	1.3462	1.020	0.549	0.298	1.308	1.272	76.31		
9.50	1.0320	1.0338	98.0	0.538	934.0	64.3	139	1.3469	1.084	0.583	0.316	1.330	1.291	75.03		
10.00	1.0338	1.0357	103.4	0.568	930.5	67.8	147	1.3477	1.148	0.617	0.335	1.353	1.311	73.76		
11.00	1.0375	1.0393	114.1	0.626	923.4	74.8	161	1.3491	1.279	0.688	0.374	1.400	1.352	71.28		
12.00	1.0412	1.0431	124.9	0.686	916.3	81.9	176	1.3506	1.413	0.760	0.413	1.450	1.395	68.85		
13.00	1.0450	1.0469	135.9	0.746	909.2	89.0	191	1.3521	1.551	0.834	0.454	1.501	1.440	66.47		
14.00	1.0489	1.0508	146.9	0.806	902.1	96.1	207	1.3536	1.692	0.910	0.495	1.555	1.486	64.16		
15.00	1.0529	1.0548	157.9	0.867	895.0	103.2	222	1.3552	1.837	0.988	0.537	1.612	1.534	61.91		

36 METHANOL, CH_3OH

MOLECULAR WEIGHT = 32.03
RELATIVE SPECIFIC REFRACTIVITY = 1.247

0.00 % by wt. data are the same for all compounds.
For Values of 0.00 wt. % solutions see Table 1, Acetic Acid.

A% by wt.	ρ D_4^{20}	D_{20}^{20}	C_s g/l	M g-mol/l	C_w g/l	$(C_o - C_w)$ g/l	$(n - n_o)$ × 10^4	n	Δ °C	O Os/kg	S g-mol/l	η/η_o	η/ρ cS	φ rhe	γ mmho/cm	T g-mol/l
0.50	0.9973	0.9991	5.0	0.156	992.4	5.9	1	1.3331	0.278	0.149	0.080	1.020	1.025	97.84		
1.00	0.9964	0.9982	10.0	0.311	986.5	11.7	2	1.3332	0.560	0.301	0.162	1.038	1.044	96.15		
1.50	0.9956	0.9973	14.9	0.466	980.6	17.6	3	1.3333	0.847	0.455	0.247	1.053	1.060	94.74		
2.00	0.9947	0.9965	19.9	0.621	974.8	23.4	4	1.3334	1.140	0.613	0.333	1.068	1.076	93.45		
2.50	0.9938	0.9956	24.8	0.776	969.0	29.2	5	1.3335	1.442	0.776	0.422	1.083	1.092	92.14		
3.00	0.9930	0.9947	29.8	0.930	963.2	35.0	6	1.3336	1.750	0.941	0.512	1.098	1.108	90.85		
3.50	0.9921	0.9939	34.7	1.084	957.4	40.8	8	1.3337	2.058	1.106	0.601	1.114	1.125	89.59		
4.00	0.9913	0.9930	39.7	1.238	951.6	46.6	9	1.3339	2.370	1.274	0.692	1.129	1.142	88.36		
4.50	0.9904	0.9922	44.6	1.392	945.9	52.4	10	1.3340	2.691	1.447	0.784	1.145	1.159	87.15		
5.00	0.9896	0.9914	49.5	1.545	940.1	58.1	11	1.3341	3.020	1.624	0.877	1.161	1.176	85.96		
5.50	0.9888	0.9905	54.4	1.698	934.4	63.8	12	1.3342	3.362	1.807	0.974	1.177	1.193	84.79		
6.00	0.9880	0.9897	59.3	1.851	928.7	69.5	14	1.3343	3.710	1.995	1.071	1.194	1.210	83.62		
6.50	0.9872	0.9889	64.2	2.003	923.0	75.2	15	1.3345	4.058	2.182	1.167	1.210	1.228	82.46		
7.00	0.9864	0.9881	69.0	2.156	917.3	80.9	16	1.3346	4.410	2.371	1.265	1.227	1.247	81.33		
7.50	0.9855	0.9873	73.9	2.308	911.6	86.6	17	1.3347	4.769	2.564	1.362	1.244	1.265	80.22		
8.00	0.9848	0.9865	78.8	2.460	906.0	92.3	19	1.3348	5.13	2.76	1.459	1.261	1.283	79.15		
8.50	0.9840	0.9857	83.6	2.611	900.3	97.9	20	1.3350	5.49	2.95	1.554	1.278	1.301	78.11		
9.00	0.9832	0.9849	88.5	2.763	894.7	103.5	21	1.3351	5.85	3.15	1.649	1.294	1.319	77.12		
9.50	0.9824	0.9841	93.3	2.914	889.1	109.2	23	1.3352	6.22	3.35	1.745	1.310	1.336	76.17		
10.00	0.9816	0.9833	98.2	3.065	883.5	114.8	24	1.3354	6.60	3.55	1.841	1.326	1.354	75.26		
11.00	0.9801	0.9818	107.8	3.366	872.3	126.0	27	1.3356	7.36	3.96	2.032	1.357	1.387	73.57		
12.00	0.9785	0.9803	117.4	3.666	861.1	137.1	29	1.3359	8.14	4.38	2.221	1.386	1.419	72.00		
13.00	0.9770	0.9787	127.0	3.965	850.0	148.2	32	1.3362	8.93	4.80	2.407	1.415	1.451	70.53		
14.00	0.9755	0.9772	136.6	4.264	838.9	159.3	35	1.3365	9.72	5.23	2.588	1.443	1.482	69.15		
15.00	0.9740	0.9757	146.1	4.561	827.9	170.3	38	1.3367	10.53	5.66	2.767	1.471	1.513	67.85		
16.00	0.9725	0.9742	155.6	4.858	816.9	181.3	40	1.3370	11.36	6.11	2.946	1.498	1.544	66.61		
17.00	0.9710	0.9727	165.1	5.154	805.9	192.3	43	1.3373	12.23	6.57	3.126	1.525	1.574	65.45		
18.00	0.9695	0.9712	174.5	5.449	795.0	203.2	46	1.3376	13.13	7.06	3.308	1.551	1.603	64.35		
19.00	0.9680	0.9698	183.9	5.742	784.1	214.1	49	1.3379	14.06	7.56	3.490	1.576	1.632	63.31		
20.00	0.9666	0.9683	193.3	6.035	773.2	225.0	51	1.3381	15.02	8.07	3.669	1.601	1.660	62.34		
22.00	0.9636	0.9653	212.0	6.618	751.6	246.6	57	1.3387	16.98	9.13	4.019	1.649	1.715	60.52		
24.00	0.9606	0.9623	230.5	7.198	730.1	268.2	62	1.3392	19.04	10.23	4.359	1.694	1.767	58.91		
26.00	0.9576	0.9593	249.0	7.773	708.6	289.6	67	1.3397	21.23	11.42		1.732	1.813	57.61		
28.00	0.9545	0.9562	267.3	8.344	687.2	311.0	72	1.3402	23.59	12.68		1.765	1.853	56.55		
30.00	0.9514	0.9531	285.4	8.911	666.0	332.3	77	1.3407	25.91	13.93		1.791	1.886	55.72		
32.00	0.9482	0.9499	303.4	9.473	644.8	353.5	81	1.3411	28.15	15.13		1.810	1.913	55.13		
34.00	0.9450	0.9466	321.3	10.031	623.7	374.6	85	1.3415	30.48	16.39		1.823	1.933	54.74		
36.00	0.9416	0.9433	339.0	10.583	602.6	395.6	89	1.3419	32.97	17.73		1.831	1.949	54.49		
38.00	0.9382	0.9399	356.5	11.131	581.7	416.6	92	1.3422	35.60	19.14		1.835	1.959	54.40		
40.00	0.9347	0.9363	373.9	11.672	560.8	437.4	95	1.3425	38.6	20.8		1.833	1.965	54.45		
42.00	0.9311	0.9327	391.0	12.209	540.0	458.2	97	1.3427	41.5	22.3		1.827	1.966	54.62		
44.00	0.9273	0.9290	408.0	12.739	519.3	478.9	99	1.3429	44.5	23.9		1.817	1.963	54.93		
46.00	0.9235	0.9251	424.8	13.263	498.7	499.5	100	1.3430	47.8	25.7		1.801	1.955	55.40		
48.00	0.9196	0.9212	441.4	13.781	478.2	520.1	101	1.3431	51.2	27.5		1.781	1.940	56.05		
50.00	0.9156	0.9172	457.8	14.292	457.8	540.5	101	1.3431	54.5	29.3		1.757	1.923	56.80		
52.00	0.9114	0.9130	473.9	14.797	437.5	560.7	101	1.3431	58.1	31.2		1.732	1.905	57.61		
54.00	0.9072	0.9088	489.9	15.295	417.3	580.9	100	1.3430	62.0	33.3		1.705	1.883	58.52		
56.00	0.9030	0.9046	505.7	15.787	397.3	600.9	99	1.3429	66.0	35.5		1.673	1.857	59.64		
58.00	0.8987	0.9003	521.2	16.274	377.5	620.8	98	1.3428	70.0	37.6		1.638	1.826	60.94		
60.00	0.8944	0.8960	536.6	16.754	357.8	640.5	96	1.3426	74.5	40.0		1.597	1.789	62.49		
62.00	0.8901	0.8917	551.9	17.230	338.3	660.0	95	1.3425	79.3	42.6		1.550	1.744	64.40		
64.00	0.8856	0.8872	566.8	17.695	318.8	679.4	92	1.3422	84.4	45.4		1.500	1.697	66.54		
66.00	0.8810	0.8825	581.4	18.153	299.5	698.7	89	1.3419	89.6	48.2		1.453	1.653	68.66		
68.00	0.8763	0.8778	595.9	18.603	280.4	717.8	85	1.3415	96.3	51.8		1.410	1.612	70.78		
70.00	0.8715	0.8730	610.0	19.046	261.4	736.8	81	1.3411				1.365	1.569	73.11		
72.00	0.8667	0.8682	624.0	19.482	242.7	755.6	77	1.3407				1.315	1.521	75.87		
74.00	0.8618	0.8633	637.7	19.910	224.1	774.2	72	1.3402				1.264	1.470	78.93		
76.00	0.8568	0.8583	651.2	20.331	205.6	792.6	67	1.3397				1.216	1.422	82.05		
78.00	0.8518	0.8533	664.4	20.744	187.4	810.8	61	1.3391				1.170	1.377	85.28		
80.00	0.8468	0.8483	677.4	21.149	169.4	828.9	55	1.3385				1.126	1.332	88.63		
82.00	0.8416	0.8431	690.2	21.547	151.5	846.7	49	1.3379				1.072	1.276	93.11		
84.00	0.8365	0.8380	702.6	21.937	133.8	864.4	42	1.3372				1.021	1.223	97.75		
86.00	0.8312	0.8327	714.9	22.318	116.4	881.9	35	1.3365				0.968	1.167	103.1		
88.00	0.8259	0.8274	726.8	22.691	99.1	899.1	27	1.3357				0.914	1.109	109.2		
90.00	0.8204	0.8219	738.4	23.053	82.0	916.2	19	1.3348				0.859	1.049	116.2		
92.00	0.8148	0.8163	749.6	23.404	65.2	933.0	9	1.3339				0.804	0.989	124.1		
94.00	0.8089	0.8103	760.3	23.738	48.5	949.7	−2	1.3328				0.748	0.927	133.4		
96.00	0.8034	0.8048	771.3	24.081	32.1	966.1	−14	1.3316				0.694	0.866	143.8		
98.00	0.7976	0.7990	781.6	24.402	16.0	982.2	−26	1.3304				0.638	0.801	156.4		
100.0	0.7917	0.7931	791.7	24.717	0.0	998.2	−40	1.3290				0.585	0.740	170.6		

37 NICKEL SULFATE, $NiSO_4 \cdot 6H_2O$

MOLECULAR WEIGHT = 154.75 FORMULA WEIGHT, HYDRATE = 262.85
RELATIVE SPECIFIC REFRACTIVITY = 0.524

0.00 % by wt. data are the same for all compounds.
For Values of 0.00 wt. % solutions see Table 1, Acetic Acid.

A % by wt.	H % by wt.	D_4^{20}	D_{20}^{20}	C_s g/l	M g-mol/l	C_w g/l	$(C_o - C_w)$ g/l	$(n - n_o) \times 10^4$	n	Δ °C	O Os/kg	S g-mol/l	η/η_o	η/ρ cS	φ rhe	γ mmho/cm	T g-mol/l
0.50	0.85	1.0035	1.0053	5.0	0.032	998.5	− 0.3	10	1.3340	0.081	0.044	0.023	1.022	1.020	97.66	3.1	0.031
1.00	1.70	1.0089	1.0107	10.1	0.065	998.8	− 0.6	20	1.3349	0.150	0.081	0.043	1.044	1.037	95.59	5.8	0.058
1.50	2.55	1.0142	1.0160	15.2	0.098	999.0	− 0.8	29	1.3359	0.217	0.117	0.062	1.066	1.053	93.60	8.0	0.081
2.00	3.40	1.0196	1.0214	20.4	0.132	999.2	− 1.0	39	1.3369	0.282	0.152	0.081	1.089	1.070	91.64	9.9	0.103
2.50	4.25	1.0250	1.0268	25.6	0.166	999.4	− 1.2	49	1.3379	0.346	0.186	0.099	1.113	1.088	89.67	11.9	0.125
3.00	5.10	1.0304	1.0323	30.9	0.200	999.5	− 1.3	60	1.3389	0.410	0.220	0.118	1.138	1.107	87.70	13.7	0.146
3.50	5.94	1.0359	1.0377	36.3	0.234	999.6	− 1.4	70	1.3400	0.475	0.255	0.137	1.164	1.126	85.75	15.5	0.167
4.00	6.79	1.0413	1.0431	41.7	0.269	999.6	− 1.4	80	1.3410	0.540	0.290	0.156	1.191	1.146	83.80	17.3	0.187
4.50	7.64	1.0467	1.0486	47.1	0.304	999.6	− 1.4	90	1.3420	0.605	0.325	0.176	1.219	1.167	81.87	18.9	0.207
5.00	8.49	1.0521	1.0540	52.6	0.340	999.5	− 1.3	100	1.3430	0.671	0.361	0.195	1.248	1.189	79.95	20.5	0.226
5.50	9.34	1.0575	1.0594	58.2	0.376	999.4	− 1.1	110	1.3440	0.736	0.396	0.214	1.279	1.211	78.06	22.0	0.245
6.00	10.19	1.0629	1.0648	63.8	0.412	999.1	− 0.9	120	1.3450	0.803	0.432	0.234	1.310	1.235	76.18	23.5	0.263

38 NITRIC ACID, HNO_3

MOLECULAR WEIGHT = 63.02
RELATIVE SPECIFIC REFRACTIVITY = 0.818

0.00 % by wt. data are the same for all compounds.
For Values of 0.00 wt. % solutions see Table 1, Acetic Acid.

A % by wt.	ρ D_4^{20}	d D_{20}^{20}	C_s g/l	M g-mol/l	C_w g/l	$(C_o - C_w)$ g/l	$(n - n_o) \times 10^4$	n	Δ °C	O Os/kg	S g-mol/l	η/η_o	η/ρ cS	φ rhe	γ mmho/cm	T g-mol/l
0.50	1.0009	1.0027	5.0	0.079	995.9	2.3	6	1.3336	0.281	0.151	0.080	1.002	1.003	99.64	28.4	0.323
1.00	1.0037	1.0054	10.0	0.159	993.6	4.6	13	1.3343	0.558	0.300	0.162	1.003	1.001	99.50	56.1	0.686
1.50	1.0064	1.0082	15.1	0.240	991.3	6.9	19	1.3349	0.837	0.450	0.244	1.004	1.000	99.40	84.7	1.10
2.00	1.0091	1.0109	20.2	0.320	988.9	9.3	26	1.3356	1.120	0.602	0.327	1.005	0.998	99.30	108.	1.50
2.50	1.0119	1.0137	25.3	0.401	986.6	11.7	32	1.3362	1.408	0.757	0.412	1.006	0.997	99.17	138.	1.97
3.00	1.0146	1.0164	30.4	0.483	984.2	14.0	39	1.3368	1.704	0.916	0.498	1.008	0.995	99.01	160.	2.57
3.50	1.0174	1.0192	35.6	0.565	981.8	16.5	45	1.3375	2.006	1.078	0.586	1.010	0.995	98.83	184.	3.18
4.00	1.0202	1.0220	40.8	0.648	979.4	18.9	51	1.3381	2.315	1.245	0.676	1.012	0.994	98.64	213.	4.31
4.50	1.0230	1.0248	46.0	0.730	976.9	21.3	58	1.3388	2.632	1.415	0.767	1.014	0.993	98.43		
5.00	1.0257	1.0276	51.3	0.814	974.5	23.8	64	1.3394	2.958	1.590	0.859	1.016	0.992	98.23		
5.50	1.0286	1.0304	56.6	0.898	972.0	26.3	71	1.3401	3.290	1.769	0.953	1.018	0.992	98.02		
6.00	1.0314	1.0332	61.9	0.982	969.5	28.7	78	1.3407	3.629	1.951	1.048	1.020	0.991	97.81		
6.50	1.0342	1.0360	67.2	1.067	967.0	31.3	84	1.3414	3.974	2.137	1.144	1.023	0.991	97.59		
7.00	1.0370	1.0389	72.6	1.152	964.4	33.8	91	1.3421	4.327	2.326	1.241	1.025	0.990	97.36		
7.50	1.0399	1.0417	78.0	1.238	961.9	36.3	97	1.3427	4.687	2.520	1.340	1.028	0.990	97.12		
8.00	1.0427	1.0446	83.4	1.324	959.3	38.9	104	1.3434	5.05	2.72	1.439	1.030	0.990	96.88		
8.50	1.0456	1.0475	88.9	1.410	956.7	41.5	110	1.3440	5.43	2.92	1.538	1.033	0.990	96.62		
9.00	1.0485	1.0504	94.4	1.497	954.1	44.1	117	1.3447	5.81	3.12	1.639	1.036	0.990	96.35		
9.50	1.0514	1.0533	99.9	1.585	951.5	46.7	124	1.3454	6.20	3.33	1.740	1.039	0.990	96.07		
10.00	1.0543	1.0562	105.4	1.673	948.9	49.4	130	1.3460	6.60	3.55	1.841	1.042	0.990	95.78		
11.00	1.0602	1.0620	116.6	1.850	943.5	54.7	144	1.3474	7.42	3.99	2.045	1.049	0.991	95.15		
12.00	1.0660	1.0679	127.9	2.030	938.1	60.1	157	1.3487	8.27	4.45	2.251	1.056	0.993	94.48		
13.00	1.0720	1.0739	139.4	2.211	932.6	65.6	170	1.3500	9.15	4.92	2.459	1.064	0.995	93.76		
14.00	1.0780	1.0799	150.9	2.395	927.1	71.2	184	1.3514	10.08	5.42	2.667	1.073	0.997	93.00		
15.00	1.0840	1.0859	162.6	2.580	921.4	76.8	198	1.3527	11.04	5.93	2.877	1.082	1.001	92.20		
16.00	1.0901	1.0921	174.4	2.768	915.7	82.5	211	1.3541	12.04	6.47	3.087	1.092	1.004	91.35		
17.00	1.0963	1.0982	186.4	2.957	909.9	88.3	225	1.3555	13.08	7.03	3.298	1.103	1.008	90.47		
18.00	1.1025	1.1044	198.4	3.149	904.0	94.2	239	1.3569	14.16	7.61	3.509	1.114	1.013	89.55		
19.00	1.1087	1.1107	210.7	3.343	898.0	100.2	253	1.3582	15.30	8.22	3.720	1.126	1.018	88.60		
20.00	1.1150	1.1170	223.0	3.538	892.0	106.2	266	1.3596				1.139	1.024	87.62		
22.00	1.1277	1.1297	248.1	3.937	879.6	118.6	294	1.3624				1.167	1.037	85.55		
24.00	1.1406	1.1426	273.7	4.344	866.8	131.4	322	1.3652				1.197	1.052	83.36		
26.00	1.1536	1.1557	299.9	4.759	853.7	144.6	350	1.3680				1.231	1.069	81.06		
28.00	1.1668	1.1688	326.7	5.184	840.1	158.1	378	1.3708				1.268	1.089	78.70		
30.00	1.1801	1.1822	354.0	5.618	826.0	172.2	406	1.3736				1.308	1.111	76.30		
32.00	1.1934	1.1955	381.9	6.060	811.5	186.7	433	1.3763				1.351	1.134	73.87		
34.00	1.2068	1.2090	410.3	6.511	796.5	201.7	460	1.3790				1.397	1.160	71.42		
36.00	1.2202	1.2224	439.3	6.970	780.9	217.3	487	1.3817				1.447	1.188	68.96		
38.00	1.2335	1.2357	468.7	7.438	764.8	233.5	513	1.3842				1.501	1.219	66.50		
40.00	1.2466	1.2489	498.7	7.913	748.0	250.2	537	1.3867				1.558	1.252	64.06		

39 OXALIC ACID, HO_2CCO_2H

MOLECULAR WEIGHT = 90.04

A% by wt.	ρ D_4^{20}	D_{20}^{20}	C_s g/l	M g-mol/l	C_w g/l	$(C_o - C_w)$ g/l	$(n - n_o)$ × 10^4	n	Δ °C	O Os/kg	S g-mol/l	η/η_o	η/ρ cS	φ rhe	γ mmho/cm	T g-mol/l
0.50	1.0006	1.0024	5.0	0.556	995.6	2.6	6.	1.3336	0.159	0.086	0.045	1.011	1.012	98.76	14.0	0.149
1.00	1.0030	1.0048	10.0	0.111	993.0	5.2	12.	1.3342	0.304	0.163	0.087	1.021	1.020	97.75	21.8	0.242
1.50	1.0054	1.0072	15.1	0.167	990.3	7.9	17.	1.3347	0.440	0.237	0.127	1.031	1.028	96.75	29.3	0.335
2.00	1.0079	1.0097	20.2	0.224	987.7	10.5	23.	1.3353	0.572	0.308	0.166	1.042	1.036	95.77	35.3	0.411
2.50	1.0103	1.0121	25.3	0.281	985.0	13.2	29.	1.3359	0.705	0.379	0.205	1.053	1.044	94.81	41.2	0.486
3.00	1.0126	1.0144	30.4	0.337	982.2	16.0	34.	1.3364	0.838	0.450	0.244	1.063	1.052	93.87	46.9	0.561
3.50	1.0150	1.0168	35.5	0.395	979.5	18.7	40.	1.3370	0.967	0.520	0.282	1.074	1.060	92.95	51.7	0.626
4.00	1.0174	1.0192	40.7	0.452	976.7	21.5	45.	1.3375	1.093	0.587	0.319	1.084	1.068	92.03	57.1	0.700
4.50	1.0197	1.0215	45.9	0.510	973.8	24.4	51.	1.3381				1.095	1.076	91.14	61.2	0.757
5.00	1.0220	1.0238	51.1	0.568	970.9	27.3	56.	1.3386				1.106	1.084	90.27	65.6	0.819
5.50	1.0244	1.0262	56.3	0.626	968.1	30.1	62.	1.3392				1.116	1.092	89.39	69.7	0.879
6.00	1.0265	1.0284	61.6	0.684	964.9	33.3	67.	1.3397				1.127	1.100	88.56	73.7	0.937
6.50	1.0288	1.0307	66.9	0.743	961.9	36.3	72.	1.3402				1.138	1.108	87.73	77.2	0.990
7.00	1.0310	1.0329	72.2	0.802	958.8	39.4	77.	1.3407				1.148	1.116	86.91	80.4	1.038
7.50	1.0332	1.0351	77.5	0.861	955.7	42.5	83.	1.3413				1.159	1.124	86.11	83.4	1.085
8.00	1.0355	1.0374	82.8	0.920	952.7	45.5	88.	1.3418				1.170	1.132	85.31	86.1	1.128

40 PHOSPHORIC ACID, H_3PO_4

MOLECULAR WEIGHT = 98.00
RELATIVE SPECIFIC REFRACTIVITY = 0.727

0.00 % by wt. data are the same for all compounds.
For Values of 0.00 wt. % solutions see Table 1. Acetic Acid.

A% by wt.	ρ D_4^{20}	D_{20}^{20}	C_s g/l	M g-mol/l	C_w g/l	$(C_o - C_w)$ g/l	$(n - n_o)$ × 10^4	n	Δ °C	O Os/kg	S g-mol/l	η/η_o	η/ρ cS	φ rhe	γ mmho/cm	T g-mol/l
0.50	1.0010	1.0028	5.0	0.051	996.0	2.2	5	1.3335	0.124	0.067	0.035	1.008	1.009	99.06	5.5	0.057
1.00	1.0038	1.0056	10.0	0.102	993.7	4.5	10	1.3340	0.243	0.131	0.069	1.018	1.016	98.04	10.1	0.105
1.50	1.0065	1.0083	15.1	0.154	991.4	6.8	15	1.3345	0.354	0.190	0.102	1.032	1.028	96.68	13.4	0.142
2.00	1.0092	1.0110	20.2	0.206	989.0	9.2	19	1.3349	0.464	0.249	0.134	1.048	1.041	95.23	16.2	0.175
2.50	1.0119	1.0137	25.3	0.258	986.6	11.6	24	1.3354	0.578	0.311	0.168	1.063	1.052	93.91	19.0	0.208
3.00	1.0146	1.0164	30.4	0.311	984.2	14.1	28	1.3358	0.694	0.373	0.202	1.077	1.064	92.64	21.6	0.239
3.50	1.0173	1.0191	35.6	0.363	981.7	16.5	33	1.3363	0.811	0.436	0.236	1.092	1.075	91.41	24.1	0.270
4.00	1.0200	1.0218	40.8	0.416	979.2	19.0	37	1.3367	0.929	0.499	0.271	1.106	1.087	90.22	26.5	0.300
4.50	1.0227	1.0245	46.0	0.470	976.7	21.6	42	1.3372	1.045	0.562	0.305	1.121	1.098	89.03	29.0	0.331
5.00	1.0254	1.0272	51.3	0.523	974.1	24.1	46	1.3376	1.160	0.624	0.339	1.136	1.110	87.85	31.5	0.363
5.50	1.0281	1.0299	56.5	0.577	971.6	26.7	51	1.3381	1.271	0.683	0.372	1.151	1.122	86.69	34.2	0.397
6.00	1.0309	1.0327	61.9	0.631	969.0	29.2	55	1.3385	1.381	0.742	0.404	1.167	1.134	85.54	36.9	0.431
6.50	1.0336	1.0354	67.2	0.686	966.4	31.9	60	1.3390	1.496	0.804	0.438	1.182	1.146	84.42	39.6	0.466
7.00	1.0363	1.0381	72.5	0.740	963.8	34.5	64	1.3394	1.619	0.871	0.474	1.198	1.158	83.32	42.4	0.502
7.50	1.0391	1.0409	77.9	0.795	961.1	37.1	69	1.3399	1.748	0.940	0.511	1.214	1.170	82.23	45.2	0.539
8.00	1.0418	1.0437	83.3	0.850	958.5	39.7	73	1.3403	1.882	1.012	0.550	1.230	1.183	81.14	48.0	0.576
8.50	1.0446	1.0465	88.8	0.906	955.8	42.4	78	1.3408	2.020	1.086	0.590	1.247	1.196	80.05	50.8	0.614
9.00	1.0474	1.0493	94.3	0.962	953.2	45.1	83	1.3413	2.160	1.161	0.631	1.264	1.209	78.96	53.7	0.653
9.50	1.0503	1.0521	99.8	1.018	950.5	47.7	87	1.3417	2.305	1.239	0.673	1.282	1.223	77.87	56.5	0.692
10.00	1.0531	1.0550	105.3	1.075	947.8	50.4	92	1.3422	2.450	1.317	0.715	1.300	1.237	76.77	59.4	0.731
11.00	1.0589	1.0607	116.5	1.189	942.4	55.9	101	1.3431	2.721	1.463	0.792	1.338	1.267	74.56	65.1	0.812
12.00	1.0647	1.0665	127.8	1.304	936.9	61.3	111	1.3441	3.010	1.618	0.874	1.379	1.298	72.38	70.9	0.896
13.00	1.0705	1.0724	139.2	1.420	931.4	66.9	120	1.3450	3.382	1.818	0.979	1.421	1.330	70.22	76.7	0.983
14.00	1.0765	1.0784	150.7	1.538	925.8	72.5	130	1.3460	3.760	2.022	1.085	1.466	1.364	68.08	82.5	1.07
15.00	1.0825	1.0844	162.4	1.657	920.1	78.1	140	1.3470	4.075	2.191	1.172	1.513	1.400	65.98	88.4	1.17
16.00	1.0885	1.0905	174.2	1.777	914.4	83.9	150	1.3480	4.450	2.392	1.276	1.562	1.438	63.90	94.3	1.26
17.00	1.0947	1.0966	186.1	1.899	908.6	89.7	160	1.3489	4.820	2.591	1.376	1.613	1.477	61.86	100.	1.36
18.00	1.1009	1.1028	198.2	2.022	902.7	95.5	170	1.3500	5.25	2.82	1.491	1.668	1.518	59.85	106.	1.46
19.00	1.1071	1.1091	210.4	2.146	896.8	101.5	180	1.3510	5.72	3.07	1.614	1.724	1.561	57.88	112.	1.56
20.00	1.1135	1.1154	222.7	2.272	890.8	107.5	190	1.3520	6.23	3.35	1.747	1.784	1.605	55.94	118.	1.67
22.00	1.1263	1.1283	247.8	2.528	878.5	119.7	211	1.3540	7.38	3.97	2.037	1.910	1.699	52.26	129.	1.89
24.00	1.1395	1.1415	273.5	2.790	866.0	132.2	231	1.3561	8.69	4.67	2.350	2.045	1.798	48.80	141.	2.12
26.00	1.1528	1.1549	299.7	3.059	853.1	145.1	253	1.3582	10.12	5.44	2.677	2.194	1.907	45.50	152.	2.37
28.00	1.1665	1.1685	326.6	3.333	839.9	158.4	274	1.3604	11.64	6.26	3.004	2.360	2.027	42.29	163.	2.63
30.00	1.1804	1.1825	354.1	3.613	826.3	172.0	295	1.3625	13.23	7.11	3.328	2.548	2.163	39.17	173.	2.88
32.00	1.1945	1.1966	382.2	3.900	812.3	186.0	317	1.3647	14.94	8.03	3.656	2.760	2.315	36.16	183.	3.13
34.00	1.2089	1.2111	411.0	4.194	797.9	200.3	339	1.3669	16.81	9.04	3.989	2.995	2.482	33.33	191.	3.38
36.00	1.2236	1.2257	440.5	4.495	783.1	215.1	361	1.3691	18.85	10.13	4.329	3.253	2.664	30.68	199.	3.63
38.00	1.2385	1.2407	470.6	4.802	767.8	230.4	383	1.3713	21.09	11.34	4.677	3.537	2.862	28.21	205.	3.88
40.00	1.2536	1.2558	501.4	5.117	752.2	246.1	405	1.3735	23.58	12.68		3.848	3.076	25.94	209.	4.13

41 POTASSIUM BICARBONATE, $KHCO_3$

MOLECULAR WEIGHT = 100.12
RELATIVE SPECIFIC REFRACTIVITY = 0.667

0.00 % by wt. data are the same for all compounds.
For Values of 0.00 wt. % solutions see Table 1, Acetic Acid.

A% by wt.	ρ D_4^{20}	D_{20}^{20}	C_s g/l	M g-mol/l	C_w g/l	$(C_o - C_w)$ g/l	$(n - n_o) \times 10^4$	n	Δ °C	O Os/kg	S g-mol/l	η/η_o	η/ρ cS	φ rhe	γ mmho/cm	T g-mol/l
0.50	1.0014	1.0031	5.0	0.050	996.4	1.9	5	1.3335	0.175	0.094	0.050	1.007	1.007	99.14	4.6	0.046
1.00	1.0046	1.0064	10.0	0.100	994.6	3.7	11	1.3341	0.344	0.185	0.099	1.013	1.010	98.52	8.9	0.092
2.00	1.0114	1.0132	20.2	0.202	991.1	7.1	23	1.3353	0.668	0.359	0.194	1.025	1.016	97.37	17.0	0.184
3.00	1.0181	1.0199	30.5	0.305	987.6	10.6	35	1.3365	0.983	0.528	0.287	1.038	1.021	96.17	24.6	0.275
4.00	1.0247	1.0265	41.0	0.409	983.7	14.5	46	1.3376	1.291	0.694	0.378	1.051	1.028	94.95	31.7	0.365
5.00	1.0310	1.0329	51.6	0.515	979.5	18.8	56	1.3386	1.600	0.860	0.468	1.065	1.035	93.71	38.8	0.455
6.00	1.0379	1.0397	62.3	0.622	975.6	22.6	68	1.3397	1.911	1.028	0.559	1.079	1.042	92.46	45.8	0.547
7.00	1.0446	1.0465	73.1	0.730	971.5	26.7	79	1.3409	2.222	1.195	0.649	1.094	1.049	91.21	52.7	0.639
8.00	1.0514	1.0532	84.1	0.840	967.3	31.0	90	1.3419	2.533	1.362	0.738	1.110	1.057	89.95	59.4	0.732
9.00	1.0581	1.0600	95.2	0.951	962.9	35.3	100	1.3430	2.844	1.529	0.827	1.126	1.066	88.65	65.9	0.825
10.00	1.0650	1.0668	106.5	1.064	958.5	39.8	111	1.3441	3.157	1.697	0.916	1.143	1.075	87.31	72.4	0.918
12.00	1.0788	1.0807	129.5	1.293	949.3	48.9	132	1.3462	3.785	2.035	1.092	1.181	1.097	84.54	84.5	1.10
14.00	1.0929	1.0948	153.0	1.528	939.9	58.4	154	1.3484	4.414	2.373	1.266	1.222	1.120	81.68	95.9	1.29
16.00	1.1073	1.1093	177.2	1.770	930.1	68.1	176	1.3506				1.267	1.147	78.77	107.0	1.47
18.00	1.1221	1.1241	202.0	2.017	920.1	78.1	198	1.3528				1.316	1.175	75.83	118.0	1.67
20.00	1.1372	1.1392	227.4	2.272	909.7	88.5	220	1.3550				1.370	1.207	72.85	128.0	1.86
22.00	1.1527	1.1547	253.6	2.533	899.1	99.2	242	1.3572				1.429	1.242	69.83	137.0	2.05
24.00	1.1685	1.1706	280.4	2.801	888.1	110.2	265	1.3595				1.494	1.281	66.80	145.0	2.23

42 POTASSIUM BIPHTHALATE, $KHC_8H_4O_4$

MOLECULAR WEIGHT = 204.23
RELATIVE SPECIFIC REFRACTIVITY = 1.097

0.00 % by wt. data are the same for all compounds.
For Values of 0.00 wt. % solutions see Table 1, Acetic Acid.

A% by wt.	ρ D_4^{20}	D_{20}^{20}	C_s g/l	M g-mol/l	C_w g/l	$(C_o - C_w)$ g/l	$(n - n_o) \times 10^4$	n	Δ °C	O Os/kg	S g-mol/l	η/η_o	η/ρ cS	φ rhe	γ mmho/cm	T g-mol/l
0.50	1.0004	1.0021	5.0	0.024	995.4	2.9	9	1.3339	0.085	0.046	0.024	1.012	1.013	98.67	2.0	0.020
1.00	1.0025	1.0043	10.0	0.049	992.5	5.7	19	1.3349	0.169	0.091	0.048	1.023	1.022	97.56	4.0	0.040
1.50	1.0047	1.0065	15.1	0.074	989.6	8.6	28	1.3358	0.254	0.136	0.073	1.034	1.032	96.48	5.9	0.059
2.00	1.0069	1.0087	20.1	0.099	986.8	11.5	38	1.3368	0.337	0.181	0.097	1.046	1.041	95.41	7.7	0.079
2.50	1.0092	1.0109	25.2	0.124	983.9	14.3	47	1.3377	0.421	0.226	0.121	1.058	1.050	94.35	9.4	0.098
3.00	1.0114	1.0132	30.3	0.149	981.1	17.2	57	1.3387	0.503	0.271	0.146	1.070	1.060	93.29	11.0	0.115
3.50	1.0137	1.0155	35.5	0.174	978.2	20.0	66	1.3396	0.585	0.315	0.170	1.082	1.070	92.24	12.5	0.132
4.00	1.0160	1.0178	40.6	0.199	975.3	22.9	76	1.3406	0.667	0.359	0.194	1.094	1.079	91.19	14.0	0.149
4.50	1.0183	1.0201	45.8	0.224	972.5	25.8	86	1.3416	0.749	0.403	0.218	1.107	1.089	90.15	15.3	0.165
5.00	1.0206	1.0224	51.0	0.250	969.6	28.7	95	1.3425	0.831	0.447	0.242	1.120	1.100	89.11	16.7	0.181
5.50	1.0229	1.0247	56.3	0.275	966.7	31.6	105	1.3435	0.913	0.491	0.266	1.133	1.110	88.07	18.1	0.197
6.00	1.0252	1.0270	61.5	0.301	963.7	34.5	115	1.3444	0.995	0.535	0.290	1.147	1.121	87.04	19.4	0.213
6.50	1.0275	1.0294	66.8	0.327	960.8	37.5	124	1.3454	1.076	0.579	0.314	1.160	1.132	86.01	20.7	0.229
7.00	1.0298	1.0317	72.1	0.353	957.8	40.5	133	1.3463	1.158	0.622	0.338	1.174	1.143	84.99	21.9	0.244
7.50	1.0321	1.0339	77.4	0.379	954.7	43.5	143	1.3473	1.239	0.666	0.362	1.189	1.154	83.97	23.1	0.258
8.00	1.0344	1.0362	82.7	0.405	951.6	46.6	152	1.3482	1.320	0.710	0.386	1.203	1.165	82.96	24.2	0.272

43 POTASSIUM BROMIDE, KBr

MOLECULAR WEIGHT = 119.01
RELATIVE SPECIFIC REFRACTIVITY = 0.627

0.00 % by wt. data are the same for all compounds.
For Values of 0.00 wt. % solutions see Table 1, Acetic Acid.

A% by wt.	ρ D_4^{20}	D_{20}^{20}	C_s g/l	M g-mol/l	C_w g/l	$(C_o - C_w)$ g/l	$(n - n_o)$ × 10^4	n	Δ °C	O Os/kg	S g-mol/l	η/η_o	η/ρ cS	φ rhe	γ mmho/cm	T g-mol/l
0.50	1.0018	1.0036	5.0	0.042	996.8	1.4	6	1.3336	0.148	0.080	0.042	0.998	0.998	100.00	5.2	0.053
1.00	1.0054	1.0072	10.1	0.084	995.4	2.9	12	1.3342	0.294	0.158	0.084	0.996	0.993	100.20	10.2	0.106
1.50	1.0090	1.0108	15.1	0.127	993.9	4.3	18	1.3348	0.439	0.236	0.127	0.994	0.987	100.40	14.9	0.160
2.00	1.0127	1.0145	20.3	0.170	992.4	5.8	24	1.3354	0.585	0.315	0.170	0.992	0.982	100.61	19.5	0.214
2.50	1.0164	1.0182	25.4	0.214	990.9	7.3	30	1.3360	0.731	0.393	0.213	0.990	0.976	100.82	24.2	0.270
3.00	1.0200	1.0218	30.6	0.257	989.4	8.8	36	1.3366	0.878	0.472	0.256	0.988	0.970	101.04	28.8	0.328
3.50	1.0238	1.0256	35.8	0.301	987.9	10.3	42	1.3372	1.026	0.552	0.300	0.985	0.965	101.27	33.5	0.387
4.00	1.0275	1.0293	41.1	0.345	986.4	11.9	49	1.3379	1.175	0.632	0.343	0.983	0.959	101.50	38.3	0.447
4.50	1.0312	1.0331	46.4	0.390	984.8	13.4	55	1.3385	1.325	0.712	0.387	0.981	0.953	101.73	43.0	0.509
5.00	1.0350	1.0368	51.8	0.435	983.3	15.0	61	1.3391	1.476	0.794	0.432	0.979	0.948	101.94	47.7	0.572
5.50	1.0388	1.0406	57.1	0.480	981.7	16.6	67	1.3397	1.629	0.876	0.477	0.977	0.942	102.15	52.5	0.636
6.00	1.0426	1.0445	62.6	0.526	980.1	18.2	74	1.3403	1.784	0.959	0.522	0.975	0.937	102.34	57.2	0.701
6.50	1.0465	1.0483	68.0	0.572	978.4	19.8	80	1.3410	1.939	1.043	0.567	0.973	0.932	102.54	62.0	0.767
7.00	1.0503	1.0522	73.5	0.618	976.8	21.4	86	1.3416	2.097	1.127	0.613	0.972	0.927	102.72	66.7	0.834
7.50	1.0542	1.0561	79.1	0.664	975.1	23.1	92	1.3422	2.255	1.213	0.658	0.970	0.922	102.90	71.5	0.903
8.00	1.0581	1.0600	84.6	0.711	973.5	24.8	99	1.3429	2.416	1.299	0.705	0.968	0.917	103.08	76.3	0.974
8.50	1.0620	1.0639	90.3	0.759	971.8	26.5	105	1.3435	2.578	1.386	0.751	0.967	0.912	103.25	81.1	1.05
9.00	1.0660	1.0679	95.9	0.806	970.1	28.2	112	1.3441	2.741	1.474	0.798	0.965	0.907	103.41	85.9	1.12
9.50	1.0700	1.0719	101.6	0.854	968.3	29.9	118	1.3448	2.907	1.563	0.845	0.964	0.902	103.58	90.7	1.20
10.00	1.0740	1.0759	107.4	0.902	966.6	31.6	124	1.3454	3.074	1.653	0.893	0.962	0.898	103.74	95.6	1.28
11.00	1.0821	1.0840	119.0	1.000	963.1	35.2	137	1.3467	3.415	1.836	0.989	0.959	0.888	104.06	105.	1.45
12.00	1.0903	1.0922	130.8	1.099	959.5	38.8	151	1.3481	3.764	2.024	1.086	0.956	0.879	104.36	115.	1.63
13.00	1.0986	1.1005	142.8	1.200	955.8	42.5	164	1.3494	4.122	2.216	1.185	0.954	0.870	104.64	125.	1.82
14.00	1.1070	1.1090	155.0	1.302	952.0	46.2	177	1.3507	4.487	2.412	1.286	0.951	0.861	104.90	134.	2.02
15.00	1.1155	1.1175	167.3	1.406	948.2	50.0	191	1.3521	4.862	2.614	1.387	0.949	0.853	105.13	144.	2.23
16.00	1.1242	1.1262	179.9	1.511	944.3	53.9	205	1.3535	5.25	2.82	1.490	0.947	0.844	105.35	154.	2.46
17.00	1.1330	1.1350	192.6	1.618	940.4	57.9	218	1.3548	5.64	3.03	1.593	0.946	0.836	105.54	164.	2.69
18.00	1.1419	1.1439	205.5	1.727	936.4	61.9	232	1.3562	6.04	3.25	1.698	0.944	0.829	105.70	174.	2.95
19.00	1.1509	1.1530	218.7	1.837	932.2	66.0	247	1.3577	6.46	3.47	1.804	0.943	0.821	105.84	184.	3.22
20.00	1.1601	1.1621	232.0	1.950	928.1	70.2	261	1.3591	6.88	3.70	1.912	0.942	0.814	105.95	194.	3.50
22.00	1.1788	1.1809	259.3	2.179	919.4	78.8	290	1.3620	7.76	4.17	2.130	0.941	0.800	106.07	216.	4.55
24.00	1.1980	1.2002	287.5	2.416	910.5	87.7	320	1.3650	8.70	4.68	2.352	0.941	0.787	106.07		
26.00	1.2179	1.2200	316.6	2.661	901.2	97.0	350	1.3680	9.68	5.20	2.579	0.942	0.775	105.92		
28.00	1.2383	1.2405	346.7	2.913	891.5	106.7	382	1.3711	10.72	5.76	2.809	0.945	0.765	105.59		
30.00	1.2593	1.2615	377.8	3.174	881.5	116.7	413	1.3743	11.82	6.35	3.042	0.950	0.756	105.05		
32.00	1.2810	1.2832	409.9	3.444	871.0	127.2	446	1.3776	12.98	6.98	3.279	0.957	0.748	104.30		
34.00	1.3033	1.3056	443.1	3.723	860.2	138.1	479	1.3809				0.966	0.743	103.33		
36.00	1.3263	1.3287	477.5	4.012	848.8	149.4	513	1.3843				0.977	0.738	102.13		
38.00	1.3501	1.3525	513.0	4.311	837.0	161.2	548	1.3878				0.991	0.736	100.69		
40.00	1.3746	1.3770	549.8	4.620	824.8	173.5	584	1.3914				1.008	0.735	99.01		

44 POTASSIUM CARBONATE, $K_2CO_3 \cdot 1\frac{1}{2}H_2O$

MOLECULAR WEIGHT = 138.20 FORMULA WEIGHT, HYDRATE = 165.23
RELATIVE SPECIFIC REFRACTIVITY = 0.608

0.00 % by wt. data are the same for all compounds.
For Values of 0.00 wt. % solutions see Table 1, Acetic Acid.

A % by wt.	H % by wt.	ρ D_4^{20}	D_{20}^{20}	C_s g/l	M g-mol/l	C_w g/l	$(C_o - C_w)$ g/l	$(n - n_o)$ × 10^4	n	Δ °C	O Os/kg	S g-mol/l	η/η_o	η/ρ cS	φ rhe	γ mmho/cm	T g-mol/l
0.50	0.60	1.0027	1.0045	5.0	0.036	997.7	0.5	9	1.3339	0.176	0.095	0.050	1.011	1.011	98.68	7.0	0.072
1.00	1.20	1.0072	1.0090	10.1	0.073	997.2	1.1	17	1.3347	0.339	0.182	0.097	1.023	1.018	97.57	13.6	0.144
2.00	2.39	1.0163	1.0181	20.3	0.147	995.9	2.3	35	1.3365	0.661	0.355	0.192	1.046	1.031	95.41	25.4	0.287
3.00	3.59	1.0254	1.0272	30.8	0.223	994.6	3.6	52	1.3382	0.986	0.530	0.288	1.069	1.045	93.36	36.7	0.428
4.00	4.78	1.0345	1.0363	41.4	0.299	993.1	5.1	69	1.3399	1.322	0.711	0.387	1.092	1.058	91.37	47.4	0.568
5.00	5.98	1.0437	1.0455	52.2	0.378	991.5	6.7	86	1.3416	1.668	0.897	0.488	1.117	1.072	89.35	58.0	0.712
6.00	7.17	1.0529	1.0548	63.2	0.457	989.7	8.5	103	1.3433	2.025	1.089	0.592	1.144	1.088	87.27	68.5	0.862
7.00	8.37	1.0622	1.0640	74.4	0.538	987.8	10.4	120	1.3450	2.395	1.287	0.699	1.172	1.105	85.17	78.8	1.02
8.00	9.56	1.0715	1.0734	85.7	0.620	985.8	12.5	137	1.3467	2.774	1.492	0.807	1.202	1.124	83.06	88.9	1.17
9.00	10.76	1.0809	1.0828	97.3	0.704	983.6	14.6	154	1.3484	3.167	1.702	0.919	1.233	1.143	80.94	98.8	1.34
10.00	11.96	1.0904	1.0923	109.0	0.789	981.3	16.9	171	1.3501	3.574	1.921	1.033	1.266	1.163	78.83	109.	1.50
12.00	14.35	1.1095	1.1115	133.1	0.963	976.4	21.8	205	1.3535	4.445	2.390	1.274	1.336	1.206	74.73	127.	1.83
14.00	16.74	1.1291	1.1311	158.1	1.144	971.0	27.2	239	1.3569	5.39	2.90	1.529	1.411	1.252	70.75	144.	2.17
16.00	19.13	1.1490	1.1510	183.8	1.330	965.2	33.1	273	1.3603	6.42	3.45	1.794	1.494	1.303	66.79	160.	2.53
18.00	21.52	1.1692	1.1713	210.5	1.523	958.8	39.5	307	1.3637	7.55	4.06	2.077	1.591	1.363	62.74	175.	2.90
20.00	23.91	1.1898	1.1919	238.0	1.722	951.9	46.4	341	1.3671	8.82	4.74	2.381	1.704	1.435	58.57	188.	3.31
24.00	28.69	1.2320	1.2342	295.7	2.140	936.3	61.9	409	1.3739	11.96	6.43	3.071	1.974	1.605	50.57	211.	4.24
28.00	33.48	1.2755	1.2778	357.1	2.584	918.4	79.9	477	1.3807	16.01	8.61	3.849	2.326	1.828	42.90		
32.00	38.26	1.3204	1.3227	422.5	3.057	897.8	100.4	544	1.3874	21.46	11.54		2.828	2.146	35.29		
36.00	43.04	1.3665	1.3690	492.0	3.560	874.6	123.6	610	1.3940	28.58	15.37		3.496	2.563	28.55		
40.00	47.82	1.4142	1.4167	565.7	4.093	848.5	149.7	676	1.4006	37.55	20.19		4.351	3.083	22.94		
44.00	52.61	1.4633	1.4659	643.9	4.659	819.5	178.8	741	1.4071				5.709	3.909	17.48		
48.00	57.39	1.5142	1.5169	726.8	5.259	787.4	210.8	806	1.4136				7.749	5.128	12.88		
50.00	59.78	1.5404	1.5431	770.2	5.573	770.2	228.0	838	1.4168				9.350	6.082	10.67		

45 POTASSIUM CHLORIDE, KCl

MOLECULAR WEIGHT = 74.55
RELATIVE SPECIFIC REFRACTIVITY = 0.758

0.00 % by wt. data are the same for all compounds.
For Values of 0.00 wt. % solutions see Table 1. Acetic Acid.

A % by wt.	D_4^{20}	D_{20}^{20}	C_s g/l	M g-mol/l	C_w g/l	$(C_o - C_w)$ g/l	$(n - n_o)$ × 10^4	n	Δ °C	O Os/kg	S g-mol/l	η/η_o	η/ρ cS	φ rhe	γ mmho/cm	T g-mol/l
0.50	1.0014	1.0032	5.0	0.067	996.4	1.8	7	1.3337	0.234	0.126	0.067	0.998	0.999	99.97	8.2	0.084
1.00	1.0046	1.0064	10.0	0.135	994.6	3.7	14	1.3343	0.463	0.249	0.134	0.997	0.995	100.08	15.7	0.169
1.50	1.0078	1.0096	15.1	0.203	992.7	5.5	20	1.3350	0.691	0.372	0.201	0.997	0.991	100.11	22.7	0.253
2.00	1.0110	1.0128	20.2	0.271	990.8	7.4	27	1.3357	0.920	0.495	0.268	0.997	0.988	100.12	29.5	0.338
2.50	1.0142	1.0160	25.4	0.340	988.9	9.4	34	1.3364	1.149	0.618	0.336	0.996	0.984	100.16	36.5	0.426
3.00	1.0174	1.0192	30.5	0.409	986.9	11.3	41	1.3371	1.380	0.742	0.404	0.996	0.981	100.20	43.6	0.517
3.50	1.0207	1.0225	35.7	0.479	984.9	13.3	47	1.3377	1.613	0.867	0.472	0.995	0.977	100.26	50.6	0.610
4.00	1.0239	1.0257	41.0	0.549	982.9	15.3	54	1.3384	1.847	0.993	0.540	0.995	0.974	100.32	57.6	0.707
4.50	1.0271	1.0289	46.2	0.620	980.9	17.3	61	1.3391	2.083	1.120	0.609	0.994	0.970	100.38	64.8	0.807
5.00	1.0304	1.0322	51.5	0.691	978.8	19.4	68	1.3398	2.322	1.248	0.678	0.994	0.966	100.44	71.9	0.911
5.50	1.0336	1.0354	56.8	0.763	976.8	21.5	74	1.3404	2.561	1.377	0.747	0.993	0.963	100.51	79.1	1.01
6.00	1.0369	1.0387	62.2	0.835	974.7	23.6	81	1.3411	2.803	1.507	0.816	0.992	0.959	100.60	86.2	1.12
6.50	1.0402	1.0420	67.6	0.907	972.5	25.7	88	1.3418	3.047	1.638	0.885	0.991	0.955	100.69	93.3	1.24
7.00	1.0434	1.0453	73.0	0.980	970.4	27.8	95	1.3425	3.294	1.771	0.955	0.990	0.951	100.79	100.	1.36
7.50	1.0467	1.0486	78.5	1.053	968.2	30.0	102	1.3431	3.544	1.905	1.025	0.989	0.947	100.89	108.	1.48
8.00	1.0500	1.0519	84.0	1.127	966.0	32.2	108	1.3438	3.797	2.041	1.095	0.988	0.943	100.99	115.	1.61
8.50	1.0533	1.0552	89.5	1.201	963.8	34.4	115	1.3445	4.050	2.177	1.165	0.987	0.939	101.07	122.	1.74
9.00	1.0566	1.0585	95.1	1.276	961.6	36.7	122	1.3452	4.300	2.312	1.234	0.987	0.936	101.14	129.	1.88
9.50	1.0600	1.0619	100.7	1.351	959.3	39.0	129	1.3459	4.550	2.446	1.303	0.986	0.932	101.18	136.	2.02
10.00	1.0633	1.0652	106.3	1.426	957.0	41.2	136	1.3466	4.805	2.583	1.372	0.986	0.929	101.20	143.	2.17
11.00	1.0700	1.0719	117.7	1.579	952.3	45.9	149	1.3479	5.33	2.87	1.513	0.987	0.924	101.15	157.	2.49
12.00	1.0768	1.0787	129.2	1.733	947.6	50.7	163	1.3493	5.88	3.16	1.657	0.988	0.919	101.01	172.	2.84
13.00	1.0836	1.0855	140.9	1.890	942.7	55.5	177	1.3507	6.45	3.47	1.804	0.990	0.915	100.82	185.	3.21
14.00	1.0905	1.0924	152.7	2.048	937.8	60.4	191	1.3521				0.992	0.912	100.58	197.	3.61
15.00	1.0974	1.0993	164.6	2.208	932.8	65.5	205	1.3535				0.995	0.908	100.34	208.	4.05
16.00	1.1043	1.1063	176.7	2.370	927.7	70.6	219	1.3549				0.997	0.905	100.11	220.	4.85
17.00	1.1114	1.1133	188.9	2.534	922.4	75.8	233	1.3563				0.999	0.901	99.87		
18.00	1.1185	1.1204	201.3	2.701	917.1	81.1	247	1.3577				1.002	0.898	99.60		
19.00	1.1256	1.1276	213.9	2.869	911.7	86.5	262	1.3592				1.005	0.895	99.26		
20.00	1.1328	1.1348	226.6	3.039	906.2	92.0	276	1.3606				1.010	0.893	98.81		
22.00	1.1474	1.1494	252.4	3.386	895.0	103.2	305	1.3635				1.022	0.893	97.64		
24.00	1.1623	1.1643	278.9	3.742	883.3	114.9	335	1.3665				1.038	0.895	96.15		

46 POTASSIUM CHROMATE, K_2CrO_4

MOLECULAR WEIGHT = 194.20
RELATIVE SPECIFIC REFRACTIVITY = 0.820

0.00 % by wt. data are the same for all compounds.
For Values of 0.00 wt. % solutions see Table 1. Acetic Acid.

A % by wt.	ρ D_4^{20}	D_{20}^{20}	C_s g/l	M g-mol/l	C_w g/l	$(C_o - C_w)$ g/l	$(n - n_o)$ × 10^4	n	Δ °C	O Os/kg	S g-mol/l	η/η_o	η/ρ cS	φ rhe	γ mmho/cm	T g-mol/l
0.50	1.0023	1.0040	5.0	0.026	997.3	1.0	11	1.3341	0.118	0.063	0.033	1.004	1.004	99.35	5.6	0.056
1.00	1.0063	1.0081	10.1	0.052	996.2	2.0	22	1.3352	0.235	0.126	0.067	1.009	1.005	98.91	10.8	0.113
2.00	1.0144	1.0162	20.3	0.104	994.1	4.2	44	1.3374	0.464	0.249	0.134	1.018	1.006	98.04	20.5	0.226
3.00	1.0224	1.0242	30.7	0.158	991.7	6.5	66	1.3396	0.682	0.367	0.198	1.027	1.006	97.21	29.8	0.340
4.00	1.0305	1.0323	41.2	0.212	989.2	9.0	87	1.3417	0.892	0.480	0.260	1.035	1.007	96.41	38.8	0.455
5.00	1.0386	1.0404	51.9	0.267	986.6	11.6	109	1.3439	1.103	0.593	0.322	1.044	1.007	95.59	47.8	0.573
6.00	1.0467	1.0486	62.8	0.323	983.9	14.3	131	1.3461	1.314	0.707	0.384	1.053	1.008	94.73	56.7	0.695
7.00	1.0550	1.0568	73.8	0.380	981.1	17.1	153	1.3483	1.524	0.819	0.446	1.063	1.010	93.85	65.6	0.819
8.00	1.0633	1.0652	85.1	0.438	978.2	20.0	175	1.3505	1.734	0.932	0.507	1.074	1.012	92.95	74.3	0.947
9.00	1.0718	1.0736	96.5	0.497	975.3	22.9	197	1.3527	1.946	1.047	0.569	1.085	1.014	92.02	83.0	1.08
10.00	1.0803	1.0823	108.0	0.556	972.3	25.9	220	1.3550	2.164	1.163	0.632	1.096	1.017	91.06	91.6	1.22
12.00	1.0980	1.0999	131.8	0.678	966.2	32.0	267	1.3597	2.613	1.405	0.761	1.120	1.022	89.07	109.	1.50
14.00	1.1163	1.1183	156.3	0.805	960.0	38.2	315	1.3645	3.076	1.654	0.893	1.147	1.030	87.00	125.	1.81
16.00	1.1344	1.1364	181.5	0.935	952.9	45.3	363	1.3693	3.555	1.911	1.028	1.176	1.039	84.84	141.	2.14
18.00	1.1527	1.1547	207.5	1.068	945.2	53.0	410	1.3740	4.050	2.177	1.165	1.209	1.051	82.56	157.	2.49
20.00	1.1720	1.1740	234.4	1.207	937.6	60.7	459	1.3789	4.561	2.452	1.306	1.245	1.064	80.16	172.	2.86
22.00	1.1920	1.1941	262.2	1.350	929.7	68.5	509	1.3839				1.285	1.081	77.64	189.	3.32
24.00	1.2125	1.2146	291.0	1.498	921.5	76.7	561	1.3891				1.330	1.099	75.04	201.	3.78
26.00	1.2335	1.2356	320.7	1.651	912.8	85.5	614	1.3944				1.379	1.120	72.37	214.	4.40
28.00	1.2547	1.2570	351.3	1.809	903.4	94.8	668	1.3998				1.433	1.144	69.64		
30.00	1.2763	1.2785	382.9	1.972	893.4	104.8	724	1.4054				1.492	1.171	66.89		

47 POTASSIUM DICHROMATE, $K_2Cr_2O_7$

MOLECULAR WEIGHT = 294.21
RELATIVE SPECIFIC REFRACTIVITY = 0.835

0.00 % by wt. data are the same for all compounds.
For Values of 0.00 wt. % solutions see Table 1. Acetic Acid.

A % by wt.	ρ D_4^{20}	D_{20}^{20}	C_s g/l	M g-mol/l	C_w g/l	$(C_o - C_w)$ g/l	$(n - n_o)$ × 10^4	n	Δ °C	O Os/kg	S g-mol/l	η/η_o	η/ρ cS	φ rhe	γ mmho/cm	T g-mol/l
0.50	1.0017	1.0035	5.0	0.017	996.7	1.5	9	1.3339	0.101	0.054	0.028	1.000	1.001	99.75	3.7	0.037
1.00	1.0052	1.0070	10.1	0.034	995.1	3.1	18	1.3348	0.192	0.103	0.055	1.001	0.998	99.70	7.3	0.074
1.50	1.0087	1.0105	15.1	0.051	993.6	4.7	27	1.3357	0.278	0.149	0.080	1.001	0.995	99.65	10.6	0.111
2.00	1.0122	1.0140	20.2	0.069	992.0	6.3	36	1.3366	0.359	0.193	0.103	1.002	0.992	99.60	13.8	0.147
2.50	1.0157	1.0175	25.4	0.086	990.3	7.9	46	1.3375	0.435	0.234	0.126	1.003	0.989	99.55	17.0	0.184
3.00	1.0193	1.0211	30.6	0.104	988.7	9.5	55	1.3385				1.003	0.986	99.50	20.1	0.221
3.50	1.0228	1.0246	35.8	0.122	987.0	11.2	64	1.3394				1.003	0.983	99.45	23.1	0.258
4.00	1.0264	1.0282	41.1	0.140	985.3	12.9	73	1.3403				1.004	0.980	99.40	26.2	0.296
4.50	1.0300	1.0318	46.3	0.158	983.6	14.6	83	1.3413				1.004	0.977	99.35	29.2	0.334
5.00	1.0336	1.0354	51.7	0.176	981.9	16.3	92	1.3422				1.005	0.974	99.30	32.2	0.372

47 POTASSIUM DICHROMATE, $K_2Cr_2O_7$—(Continued)

A % by wt.	ρ D_4^{20}	D_{20}^{20}	C_s g/l	M g-mol/l	C_w g/l	(C_o – C_w) g/l	(n – n_o) × 10^4	n	Δ °C	O Os/kg	S g-mol/l	η/η_o	η/ρ cS	φ rhe	γ mmho/cm	T g-mol/l
5.50	1.0372	1.0390	57.0	0.194	980.1	18.1	102	1.3431				1.006	0.971	99.25	35.3	0.410
6.00	1.0408	1.0426	62.4	0.212	978.3	19.9	111	1.3441				1.006	0.969	99.21	38.3	0.449
6.50	1.0444	1.0463	67.9	0.231	976.5	21.7	121	1.3450				1.007	0.966	99.16	41.4	0.488
7.00	1.0481	1.0499	73.4	0.249	974.7	23.5	130	1.3460				1.007	0.963	99.11	44.4	0.528
7.50	1.0517	1.0536	78.9	0.268	972.8	25.4	140	1.3470				1.008	0.960	99.06	47.4	0.567
8.00	1.0554	1.0573	84.4	0.287	971.0	27.3	149	1.3479				1.008	0.957	99.01	50.3	0.607
8.50	1.0591	1.0610	90.0	0.306	969.1	29.2	159	1.3489				1.009	0.954	98.96	53.3	0.647
9.00	1.0628	1.0647	95.7	0.325	967.2	31.1	169	1.3499				1.009	0.951	98.91	56.3	0.688
9.50	1.0665	1.0684	101.3	0.344	965.2	33.0	179	1.3509				1.010	0.948	98.86	59.2	0.729
10.00	1.0703	1.0722	107.0	0.364	963.3	35.0	189	1.3519				1.010	0.946	98.81	62.1	0.770

48 POTASSIUM FERRICYANIDE, $K_3Fe(CN)_6$

MOLECULAR WEIGHT = 329.26
RELATIVE SPECIFIC REFRACTIVITY = 0.929

0.00 % by wt. data are the same for all compounds.
For Values of 0.00 wt. % solutions see Table 1, Acetic Acid.

A % by wt.	ρ D_4^{20}	D_{20}^{20}	C_s g/l	M g-mol/l	C_w g/l	(C_o – C_w) g/l	(n – n_o) × 10^4	n	Δ °C	O Os/kg	S g-mol/l	η/η_o	η/ρ cS	φ rhe	γ mmho/cm	T g-mol/l
0.50	1.0008	1.0026	5.0	0.015	995.8	2.4	8	1.3338	0.093	0.050	0.026	1.003	1.004	99.49	5.0	0.050
1.00	1.0034	1.0052	10.0	0.030	993.4	4.8	17	1.3347	0.180	0.097	0.051	1.006	1.005	99.21	9.7	0.100
1.50	1.0061	1.0079	15.1	0.046	991.0	7.3	25	1.3355	0.258	0.139	0.074	1.009	1.005	98.95	14.0	0.150
2.00	1.0087	1.0105	20.2	0.061	988.5	9.7	34	1.3364	0.334	0.180	0.096	1.011	1.004	98.71	18.2	0.199
2.50	1.0114	1.0132	25.3	0.077	986.1	12.1	42	1.3372	0.412	0.222	0.119	1.013	1.004	98.51	22.4	0.249
3.00	1.0140	1.0158	30.4	0.092	983.6	14.6	50	1.3380	0.491	0.264	0.142	1.015	1.003	98.33	26.5	0.298
3.50	1.0167	1.0185	35.6	0.108	981.1	17.1	59	1.3389	0.570	0.306	0.165	1.017	1.002	98.17	30.5	0.348
4.00	1.0194	1.0212	40.8	0.124	978.6	19.6	67	1.3397	0.649	0.349	0.189	1.018	1.001	98.02	34.5	0.399
4.50	1.0221	1.0239	46.0	0.140	976.1	22.1	75	1.3405	0.728	0.392	0.212	1.020	1.000	97.85	38.4	0.450
5.00	1.0249	1.0267	51.2	0.156	973.6	24.6	84	1.3414	0.807	0.434	0.235	1.022	0.999	97.65	42.4	0.502
5.50	1.0276	1.0294	56.5	0.172	971.1	27.2	92	1.3422	0.885	0.476	0.258	1.024	0.999	97.43	46.4	0.555
6.00	1.0304	1.0322	61.8	0.188	968.5	29.7	100	1.3430	0.962	0.517	0.281	1.027	0.999	97.19	50.5	0.609
6.50	1.0331	1.0350	67.2	0.204	966.0	32.3	109	1.3438	1.039	0.559	0.303	1.030	0.999	96.94	54.5	0.663
7.00	1.0359	1.0377	72.5	0.220	963.4	34.8	117	1.3447	1.115	0.600	0.326	1.032	0.999	96.67	58.5	0.719
7.50	1.0387	1.0406	77.9	0.237	960.8	37.4	125	1.3455	1.192	0.641	0.348	1.035	0.999	96.39	62.5	0.775
8.00	1.0415	1.0434	83.3	0.253	958.2	40.0	134	1.3464	1.268	0.682	0.371	1.039	0.999	96.10	66.5	0.832
8.50	1.0444	1.0462	88.8	0.270	955.6	42.6	142	1.3472	1.345	0.723	0.393	1.042	0.999	95.80	70.4	0.890
9.00	1.0472	1.0491	94.2	0.286	953.0	45.3	150	1.3480	1.422	0.764	0.416	1.045	1.000	95.49	74.4	0.949
9.50	1.0501	1.0519	99.8	0.303	950.3	47.9	159	1.3489	1.500	0.807	0.439	1.049	1.000	95.18	78.4	1.01
10.00	1.0530	1.0548	105.3	0.320	947.7	50.6	167	1.3497	1.580	0.849	0.462	1.052	1.001	94.87	82.4	1.07
11.00	1.0588	1.0607	116.5	0.354	942.3	55.9	184	1.3514	1.740	0.935	0.509	1.059	1.002	94.22	90.4	1.20
12.00	1.0647	1.0665	127.8	0.388	936.9	61.3	201	1.3531	1.901	1.022	0.556	1.067	1.004	93.55	98.4	1.33
13.00	1.0706	1.0725	139.2	0.423	931.4	66.8	219	1.3549	2.066	1.110	0.604	1.075	1.006	92.85	106.0	1.46
14.00	1.0766	1.0785	150.7	0.458	925.9	72.4	236	1.3566	2.235	1.202	0.653	1.084	1.008	92.11	114.0	1.60
15.00	1.0826	1.0845	162.4	0.493	920.2	78.0	254	1.3583	2.413	1.298	0.704	1.093	1.012	91.31	122.0	1.74
16.00	1.0887	1.0907	174.2	0.529	914.5	83.7	271	1.3601	2.600	1.398	0.758	1.103	1.015	90.46	129.0	1.88
17.00	1.0949	1.0968	186.1	0.565	908.8	89.5	289	1.3619	2.795	1.503	0.813	1.114	1.020	89.58	136.0	2.03
18.00	1.1011	1.1031	198.2	0.602	902.9	95.3	307	1.3637	2.997	1.611	0.871	1.126	1.024	88.65	143.0	2.17
19.00	1.1074	1.1093	210.4	0.639	897.0	101.3	325	1.3655	3.208	1.725	0.930	1.138	1.030	87.67	150.0	2.33
20.00	1.1137	1.1156	222.7	0.676	890.9	107.3	344	1.3674	3.427	1.843	0.992	1.152	1.036	86.63	157.0	2.50
22.00	1.1263	1.1283	247.8	0.753	878.6	119.7	381	1.3711				1.182	1.052	84.41	171.0	2.88
24.00	1.1391	1.1411	273.4	0.830	865.7	132.5	419	1.3749				1.217	1.070	82.02	185.0	3.31
26.00	1.1519	1.1540	299.5	0.910	852.4	145.8	457	1.3787				1.256	1.092	79.49	198.0	3.79
28.00	1.1647	1.1668	326.1	0.990	838.6	159.6	495	1.3825				1.299	1.117	76.83	210.0	4.34
30.00	1.1773	1.1794	353.2	1.073	824.1	174.1	533	1.3863				1.347	1.146	74.09	222.0	4.96

49 POTASSIUM FERROCYANIDE, $K_4Fe(CN)_6 \cdot 3H_2O$

MOLECULAR WEIGHT = 368.36 FORMULA WEIGHT, HYDRATE = 422.41
RELATIVE SPECIFIC REFRACTIVITY = 0.921

0.00 % by wt. data are the same for all compounds.
For Values of 0.00 wt. % solutions see Table 1, Acetic Acid.

A % by wt.	H % by wt.	ρ D_4^{20}	D_{20}^{20}	C_s g/l	M g-mol/l	C_w g/l	(C_o – C_w) g/l	(n – n_o) × 10^4	n	Δ °C	O Os/kg	S g-mol/l	η/η_o	η/ρ cS	φ rhe	γ mmho/cm	T g-mol/l
0.50	0.57	1.0016	1.0034	5.0	0.014	996.6	1.6	10	1.3340	0.083	0.044	0.023	1.007	1.008	99.07	4.8	0.048
1.00	1.15	1.0049	1.0067	10.0	0.027	994.9	3.3	21	1.3351	0.162	0.087	0.046	1.014	1.011	98.42	9.3	0.096
1.50	1.72	1.0083	1.0101	15.1	0.041	993.1	5.1	31	1.3361	0.239	0.129	0.068	1.020	1.013	97.87	13.4	0.142
2.00	2.29	1.0116	1.0134	20.2	0.055	991.4	6.9	41	1.3371	0.314	0.169	0.090	1.025	1.015	97.37	17.3	0.188
2.50	2.87	1.0149	1.0167	25.4	0.069	989.5	8.7	51	1.3381	0.385	0.207	0.111	1.031	1.017	96.84	21.3	0.235
3.00	3.44	1.0182	1.0200	30.5	0.083	987.7	10.6	62	1.3391	0.453	0.243	0.131	1.036	1.020	96.33	25.2	0.282
3.50	4.01	1.0215	1.0233	35.8	0.097	985.8	12.5	72	1.3402	0.518	0.278	0.150	1.042	1.022	95.81	29.0	0.331
4.00	4.59	1.0248	1.0266	41.0	0.111	983.8	14.4	82	1.3412	0.582	0.313	0.169	1.047	1.024	95.30	32.9	0.379
4.50	5.16	1.0281	1.0300	46.3	0.126	981.9	16.4	92	1.3422	0.645	0.347	0.187	1.053	1.026	94.77	36.7	0.428
5.00	5.73	1.0315	1.0333	51.6	0.140	979.9	18.4	102	1.3432	0.709	0.381	0.206	1.059	1.029	94.24	40.6	0.478
5.50	6.31	1.0348	1.0366	56.9	0.155	977.9	20.4	112	1.3442	0.773	0.415	0.225	1.065	1.031	93.70	44.4	0.528
6.00	6.88	1.0381	1.0400	62.3	0.169	975.8	22.4	122	1.3452	0.836	0.450	0.244	1.071	1.034	93.15	48.2	0.578
6.50	7.45	1.0415	1.0433	67.7	0.184	973.8	24.5	132	1.3462	0.899	0.483	0.262	1.078	1.037	92.60	51.9	0.629
7.00	8.03	1.0449	1.0467	73.1	0.199	971.7	26.5	143	1.3473	0.961	0.517	0.281	1.084	1.040	92.04	55.7	0.680
7.50	8.60	1.0483	1.0501	78.6	0.213	969.6	28.6	153	1.3483	1.023	0.550	0.299	1.091	1.043	91.47	59.4	0.731
8.00	9.17	1.0517	1.0535	84.1	0.228	967.5	30.7	163	1.3493	1.084	0.583	0.317	1.098	1.046	90.90	63.1	0.784
8.50	9.75	1.0553	1.0572	89.7	0.244	965.6	32.6	175	1.3505	1.144	0.615	0.334	1.105	1.049	90.32	66.8	0.837
9.00	10.32	1.0588	1.0607	95.3	0.259	963.5	34.7	185	1.3515	1.204	0.647	0.352	1.112	1.052	89.74	70.6	0.891
9.50	10.89	1.0623	1.0641	100.9	0.274	961.3	36.9	196	1.3525	1.262	0.679	0.369	1.119	1.056	89.15	74.3	0.947
10.00	11.47	1.0657	1.0676	106.6	0.289	959.2	39.1	206	1.3536	1.320	0.710	0.386	1.127	1.060	88.55	78.1	1.00
11.00	12.61	1.0727	1.0746	118.0	0.320	954.7	43.5	227	1.3557	1.433	0.771	0.419	1.143	1.067	87.35	85.8	1.12

49 POTASSIUM FERROCYANIDE, $K_4Fe(CN)_6 \cdot 3H_2O$—(Continued)

A% by wt.	H% by wt.	ρ D_4^{20}	D_{20}^{20}	C_s g/l	M g-mol/l	C_w g/l	$(C_o - C_w)$ g/l	$(n - n_o) \times 10^4$	n	Δ °C	O Os/kg	S g-mol/l	η/η_o	η/ρ cS	φ rhe	γ mmho/cm	T g-mol/l
12.00	13.76	1.0798	1.0817	129.6	0.352	950.2	48.0	248	1.3578				1.159	1.075	86.13	93.7	1.25
13.00	14.91	1.0869	1.0889	141.3	0.384	945.6	52.6	270	1.3600				1.176	1.084	84.89	101.	1.38
14.00	16.05	1.0941	1.0961	153.2	0.416	941.0	57.3	291	1.3621				1.193	1.093	83.63	109.	1.51
15.00	17.20	1.1014	1.1034	165.2	0.449	936.2	62.0	313	1.3643				1.212	1.103	82.34	117.	1.64
16.00	18.35	1.1088	1.1107	177.4	0.482	931.4	66.9	335	1.3665				1.232	1.113	81.03	124.	1.78
17.00	19.49	1.1162	1.1182	189.8	0.515	926.4	71.8	357	1.3687				1.252	1.124	79.70	131.	1.91
18.00	20.64	1.1237	1.1257	202.3	0.549	921.4	76.8	379	1.3709				1.274	1.136	78.35	137.	2.05
19.00	21.79	1.1312	1.1332	214.9	0.583	916.3	81.9	401	1.3731				1.296	1.148	76.98	143.	2.18
20.00	22.93	1.1389	1.1409	227.8	0.618	911.1	87.1	424	1.3754				1.320	1.161	75.61	149.	2.32

50 POTASSIUM HYDROXIDE, KOH

MOLECULAR WEIGHT = 56.11
RELATIVE SPECIFIC REFRACTIVITY = 0.680

0.00 % by wt. data are the same for all compounds.
For Values of 0.00 wt. % solutions see Table 1, Acetic Acid.

A% by wt.	ρ D_4^{20}	D_{20}^{20}	C_s g/l	M g-mol/l	C_w g/l	$(C_o - C_w)$ g/l	$(n - n_o) \times 10^4$	n	Δ °C	O Os/kg	S g-mol/l	η/η_o	η/ρ cS	φ rhe	γ mmho/cm	T g-mol/l
0.50	1.0025	1.0043	5.0	0.089	997.5	0.7	10	1.3340	0.299	0.161	0.086	1.008	1.008	98.98	20.0	0.220
1.00	1.0068	1.0086	10.1	0.179	996.7	1.5	20	1.3350	0.609	0.327	0.177	1.017	1.012	98.13	38.5	0.452
1.50	1.0111	1.0129	15.2	0.270	995.9	2.3	29	1.3359	0.924	0.497	0.269	1.026	1.017	97.25	56.9	0.697
2.00	1.0155	1.0172	20.3	0.362	995.1	3.1	39	1.3369	1.242	0.668	0.363	1.036	1.022	96.36	75.0	0.957
2.50	1.0198	1.0216	25.5	0.454	994.3	3.9	49	1.3379	1.562	0.840	0.457	1.046	1.027	95.45	92.8	1.23
3.00	1.0242	1.0260	30.7	0.548	993.5	4.8	58	1.3388	1.886	1.014	0.551	1.056	1.033	94.52	110.	1.53
3.50	1.0286	1.0304	36.0	0.642	992.6	5.6	68	1.3398	2.217	1.192	0.648	1.066	1.039	93.58	128.	1.85
4.00	1.0330	1.0348	41.3	0.736	991.7	6.5	78	1.3408	2.565	1.379	0.748	1.077	1.045	92.63	144.	2.20
4.50	1.0374	1.0393	46.7	0.832	990.8	7.5	87	1.3417	2.940	1.581	0.854	1.089	1.051	91.68	161.	2.58
5.00	1.0419	1.0437	52.1	0.928	989.8	8.4	97	1.3427	3.356	1.804	0.972	1.100	1.058	90.71	178.	3.00
5.50	1.0464	1.0482	57.6	1.026	988.8	9.4	106	1.3436	3.747	2.014	1.081	1.112	1.065	89.75	193.	3.47
6.00	1.0509	1.0527	63.1	1.124	987.8	10.4	116	1.3445	4.144	2.228	1.191	1.124	1.072	88.78	206.	3.97
6.50	1.0554	1.0572	68.6	1.223	986.8	11.5	125	1.3455	4.521	2.431	1.295	1.137	1.079	87.81	216.	4.52
7.00	1.0599	1.0618	74.2	1.322	985.7	12.5	134	1.3464	4.921	2.649	1.403	1.149	1.086	86.84		
7.50	1.0644	1.0663	79.8	1.423	984.6	13.6	144	1.3474				1.162	1.094	85.87		
8.00	1.0690	1.0709	85.5	1.524	983.5	14.8	153	1.3483				1.175	1.102	84.91		
8.50	1.0736	1.0755	91.3	1.626	982.3	15.9	162	1.3492				1.189	1.110	83.95		
9.00	1.0781	1.0801	97.0	1.729	981.1	17.1	172	1.3502				1.203	1.118	82.98		
9.50	1.0827	1.0847	102.9	1.833	979.9	18.4	181	1.3511				1.217	1.126	82.03		
10.00	1.0873	1.0893	108.7	1.938	978.6	19.6	190	1.3520				1.231	1.134	81.07		
11.00	1.0966	1.0985	120.6	2.150	976.0	22.3	209	1.3539				1.261	1.152	79.17		
12.00	1.1059	1.1079	132.7	2.365	973.2	25.0	228	1.3558				1.291	1.170	77.28		
13.00	1.1153	1.1172	145.0	2.584	970.3	28.0	246	1.3576				1.324	1.189	75.39		
14.00	1.1246	1.1266	157.5	2.806	967.2	31.0	265	1.3595				1.358	1.210	73.50		
15.00	1.1341	1.1361	170.1	3.032	964.0	34.3	284	1.3614				1.394	1.232	71.59		
16.00	1.1435	1.1456	183.0	3.261	960.6	37.7	302	1.3632				1.433	1.255	69.66		
17.00	1.1531	1.1551	196.0	3.493	957.0	41.2	321	1.3651				1.474	1.281	67.69		
18.00	1.1626	1.1647	209.3	3.730	953.3	44.9	340	1.3670				1.518	1.308	65.75		
19.00	1.1722	1.1743	222.7	3.969	949.5	48.8	358	1.3688				1.565	1.338	63.77		
20.00	1.1818	1.1839	236.4	4.212	945.4	52.8	377	1.3707				1.616	1.370	61.76		
22.00	1.2014	1.2035	264.3	4.710	937.1	61.2	414	1.3744				1.729	1.442	57.72		
24.00	1.2210	1.2231	293.0	5.223	927.9	70.3	451	1.3781				1.857	1.524	53.74		
26.00	1.2408	1.2430	322.6	5.750	918.2	80.0	488	1.3818				2.002	1.617	49.85		
28.00	1.2609	1.2632	353.1	6.292	907.9	90.4	524	1.3854				2.166	1.721	46.09		
30.00	1.2813	1.2836	384.4	6.851	896.9	101.3	559	1.3889				2.352	1.839	42.44		
32.00	1.3020	1.3043	416.6	7.425	885.4	112.9	593	1.3923				2.565	1.974	38.90		
34.00	1.3230	1.3254	449.8	8.017	873.2	125.0	628	1.3957				2.814	2.131	35.47		
36.00	1.3444	1.3468	484.0	8.626	860.4	137.8	664	1.3993				3.105	2.315	32.14		
38.00	1.3661	1.3685	519.1	9.252	847.0	151.3	700	1.4030				3.453	2.533	28.90		
40.00	1.3881	1.3906	555.2	9.896	832.9	165.4	738	1.4068				3.871	2.795	25.78		
42.00	1.4104	1.4129	592.4	10.558	818.1	180.2	776	1.4106				4.380	3.112	22.78		
44.00	1.4331	1.4356	630.6	11.238	802.5	195.7	813	1.4143				5.003	3.498	19.95		
46.00	1.4560	1.4586	669.8	11.936	786.2	212.0	849	1.4179				5.769	3.970	17.30		
48.00	1.4791	1.4817	710.0	12.653	769.1	229.1	884	1.4214				6.713	4.547	14.87		
50.00	1.5024	1.5050	751.2	13.388	751.2	247.0	917	1.4247				7.876	5.253	12.67		

51 POTASSIUM IODIDE, KI

MOLECULAR WEIGHT = 166.03
RELATIVE SPECIFIC REFRACTIVITY = 0.647

0.00 % by wt. data are the same for all compounds.
For Values of 0.00 wt. % solutions see Table 1, Acetic Acid.

A% by wt.	ρ D_4^{20}	D_{20}^{20}	C_s g/l	M g-mol/l	C_w g/l	$(C_o - C_w)$ g/l	$(n - n_o) \times 10^4$	n	Δ °C	O Os/kg	S g-mol/l	η/η_o	η/ρ cS	φ rhe	γ mmho/cm	T g-mol/l
0.50	1.0019	1.0037	5.0	0.030	996.9	1.3	7	1.3337	0.111	0.060	0.031	0.9972	0.9973	100.1	3.8	0.038
1.00	1.0056	1.0074	10.1	0.061	995.5	2.7	13	1.3343	0.220	0.118	0.063	0.9945	0.9910	100.3	7.5	0.076
2.00	1.0131	1.0149	20.3	0.122	992.8	5.4	27	1.3357	0.431	0.232	0.124	0.9892	0.9784	100.9	14.2	0.152
3.00	1.0206	1.0224	30.6	0.184	990.0	8.2	40	1.3370	0.642	0.345	0.187	0.9840	0.9661	101.4	21.1	0.233
4.00	1.0282	1.0301	41.1	0.248	987.1	11.1	54	1.3384	0.858	0.461	0.250	0.9788	0.9538	102.0	28.1	0.320
5.00	1.0360	1.0378	51.8	0.312	984.2	14.1	67	1.3397	1.079	0.580	0.315	0.9736	0.9417	102.5	35.2	0.409
6.00	1.0438	1.0457	62.6	0.377	981.2	17.1	81	1.3411	1.304	0.701	0.381	0.9670	0.9283	103.2	42.3	0.501
7.00	1.0517	1.0536	73.6	0.443	978.1	20.1	95	1.3425	1.534	0.825	0.449	0.9608	0.9153	103.9	49.6	0.597
8.00	1.0598	1.0617	84.8	0.511	975.0	23.2	110	1.3440	1.769	0.951	0.517	0.9549	0.9028	104.5	56.9	0.697
9.00	1.0679	1.0698	96.1	0.579	971.8	26.4	124	1.3454	2.010	1.081	0.588	0.9494	0.8908	105.1	64.3	0.801
10.00	1.0762	1.0781	107.6	0.648	968.6	29.6	139	1.3469	2.257	1.213	0.659	0.9442	0.8791	105.7	71.8	0.909
12.00	1.0931	1.0950	131.2	0.790	961.9	36.3	168	1.3498	2.767	1.488	0.805	0.9348	0.8569	106.8	86.8	1.14

51 POTASSIUM IODIDE, KI—(Continued)

A% by wt.	ρ D_4^{20}	D_{20}^{20}	C_s g/l	M g-mol/l	C_w g/l	$(C_o - C_w)$ g/l	$(n - n_o) \times 10^4$	n	Δ °C	O Os/kg	S g-mol/l	η/η_o	η/ρ cS	φ rhe	γ mmho/cm	T g-mol/l
14.00	1.1105	1.1125	155.5	0.936	955.0	43.2	199	1.3529	3.304	1.776	0.957	0.9265	0.8360	107.7	102.	1.39
16.00	1.1284	1.1304	180.5	1.087	947.9	50.4	231	1.3560	3.869	2.080	1.115	0.9193	0.8163	108.6	118.	1.66
18.00	1.1469	1.1489	206.4	1.243	940.4	57.8	263	1.3593	4.464	2.400	1.279	0.9131	0.7977	109.3	133.	1.96
20.00	1.1659	1.1680	233.2	1.405	932.8	65.5	296	1.3626	5.09	2.74	1.449	0.9077	0.7801	109.9	149.	2.30
22.00	1.1856	1.1877	260.8	1.571	924.8	73.4	331	1.3661	5.76	3.09	1.624	0.9031	0.7632	110.5	164.	2.66
24.00	1.2060	1.2081	289.4	1.743	916.5	81.7	366	1.3696	6.46	3.47	1.805	0.8991	0.7470	111.0	180.	3.05
26.00	1.2270	1.2291	319.0	1.921	908.0	90.3	403	1.3733	7.21	3.87	1.993	0.8956	0.7314	111.4	195.	3.53
28.00	1.2487	1.2509	349.6	2.106	899.1	99.2	441	1.3771	8.01	4.30	2.188	0.8928	0.7164	111.8	210.	4.16
30.00	1.2712	1.2734	381.4	2.297	889.8	108.4	480	1.3810	8.86	4.76	2.390	0.8904	0.7019	112.1	224.	5.20
32.00	1.2944	1.2967	414.2	2.495	880.2	118.0	521	1.3851	9.76	5.25	2.597	0.8888	0.6880	112.3		
34.00	1.3185	1.3208	448.3	2.700	870.2	128.0	563	1.3893	10.72	5.76	2.808	0.8880	0.6748	112.4		
36.00	1.3434	1.3458	483.6	2.913	859.8	138.5	606	1.3936	11.73	6.31	3.024	0.8884	0.6626	112.3		
38.00	1.3692	1.3716	520.3	3.134	848.9	149.3	651	1.3981	12.81	6.89	3.246	0.8905	0.6517	112.1		
40.00	1.3959	1.3984	558.4	3.363	837.6	160.7	697	1.4027	13.97	7.51	3.472	0.8952	0.6426	111.5		

52 POTASSIUM NITRATE, KNO_3

MOLECULAR WEIGHT = 101.10
RELATIVE SPECIFIC REFRACTIVITY = 0.651

0.00 % by wt. data are the same for all compounds.
For Values of 0.00 wt. % solutions see Table 1, Acetic Acid.

A% by wt.	ρ D_4^{20}	D_{20}^{20}	C_s g/l	M g-mol/l	C_w g/l	$(C_o - C_w)$ g/l	$(n - n_o) \times 10^4$	n	Δ °C	O Os/kg	S g-mol/l	η/η_o	η/ρ cS	φ rhe	γ mmho/cm	T g-mol/l
0.50	1.0014	1.0031	5.0	0.050	996.4	1.9	5	1.3335	0.171	0.092	0.049	0.997	0.998	100.10	5.5	0.056
1.00	1.0045	1.0063	10.0	0.099	994.5	3.8	10	1.3339	0.333	0.179	0.096	0.994	0.992	100.40	10.7	0.112
2.00	1.0108	1.0126	20.2	0.200	990.6	7.7	19	1.3349	0.642	0.345	0.187	0.988	0.979	101.01	20.1	0.222
3.00	1.0171	1.0189	30.5	0.302	986.6	11.7	28	1.3358	0.938	0.504	0.274	0.984	0.969	101.46	29.3	0.334
4.00	1.0234	1.0252	40.9	0.405	982.5	15.7	38	1.3368	1.223	0.657	0.357	0.981	0.960	101.77	38.3	0.448
5.00	1.0298	1.0317	51.5	0.509	978.3	19.9	47	1.3377	1.498	0.805	0.438	0.978	0.952	102.05	47.0	0.563
6.00	1.0363	1.0381	62.2	0.615	974.1	24.1	57	1.3386	1.763	0.948	0.516	0.975	0.943	102.33	55.5	0.678
7.00	1.0428	1.0447	73.0	0.722	969.8	28.4	66	1.3396	2.022	1.087	0.591	0.973	0.935	102.59	63.7	0.793
8.00	1.0494	1.0512	84.0	0.830	965.4	32.8	76	1.3405	2.273	1.222	0.664	0.971	0.927	102.81	71.7	0.908
9.00	1.0560	1.0579	95.0	0.940	961.0	37.3	85	1.3415	2.517	1.353	0.734	0.969	0.920	102.98	79.5	1.03
10.00	1.0627	1.0646	106.3	1.051	956.4	41.8	95	1.3425	2.754	1.481	0.802	0.968	0.913	103.10	87.3	1.15
12.00	1.0762	1.0781	129.1	1.277	947.0	51.2	114	1.3444				0.968	0.901	103.13	103.	1.40
14.00	1.0899	1.0918	152.6	1.509	937.3	60.9	133	1.3463				0.970	0.892	102.90	117.	1.66
16.00	1.1039	1.1058	176.6	1.747	927.3	71.0	152	1.3482				0.974	0.884	102.44	131.	1.94
18.00	1.1181	1.1201	201.3	1.991	916.9	81.4	172	1.3502				0.980	0.879	101.79	145.	2.22
20.00	1.1326	1.1346	226.5	2.241	906.1	92.2	191	1.3521				0.988	0.874	101.01	157.	2.49
22.00	1.1473	1.1493	252.4	2.497	894.9	103.3	211	1.3541				0.997	0.871	100.10	168.	2.75
24.00	1.1623	1.1644	279.0	2.759	883.3	114.9	231	1.3561				1.008	0.869	99.01	178.	3.00

53 POTASSIUM OXALATE, $K_2C_2O_4 \cdot 1H_2O$

MOLECULAR WEIGHT = 166.21 FORMULA WEIGHT, HYDRATE = 184.23
RELATIVE SPECIFIC REFRACTIVITY = 0.676

0.00 % by wt. data are the same for all compounds.
For Values of 0.00 wt. % solutions see Table 1, Acetic Acid.

A% by wt.	H% by wt.	ρ D_4^{20}	D_{20}^{20}	C_s g/l	M g-mol/l	C_w g/l	$(C_o - C_w)$ g/l	$(n - n_o) \times 10^4$	n	Δ °C	O Os/kg	S g-mol/l	η/η_o	η/ρ cS	φ rhe	γ mmho/cm	T g-mol/l
0.50	0.55	1.0018	1.0035	5.0	0.030	996.8	1.5	8	1.3337	0.137	0.074	0.039	1.005	1.005	99.33	5.8	0.059
1.00	1.11	1.0053	1.0071	10.1	0.060	995.3	3.0	15	1.3345	0.273	0.147	0.078	1.010	1.007	98.81	11.4	0.119
1.50	1.66	1.0089	1.0107	15.1	0.091	993.8	4.5	22	1.3352	0.407	0.219	0.118	1.016	1.009	98.25	16.6	0.180
2.00	2.22	1.0125	1.0143	20.3	0.122	992.3	6.0	29	1.3359	0.540	0.290	0.156	1.022	1.011	97.65	21.7	0.241
2.50	2.77	1.0161	1.0179	25.4	0.153	990.7	7.5	36	1.3366	0.669	0.360	0.194	1.028	1.014	97.04	26.8	0.302
3.00	3.33	1.0198	1.0216	30.6	0.184	989.2	9.0	43	1.3373	0.797	0.429	0.232	1.035	1.017	96.41	31.7	0.364
3.50	3.88	1.0235	1.0253	35.8	0.216	987.6	10.6	50	1.3380	0.925	0.497	0.270	1.042	1.020	95.77	36.6	0.426
4.00	4.43	1.0272	1.0290	41.1	0.247	986.1	12.2	56	1.3386	1.055	0.567	0.308	1.049	1.024	95.11	41.4	0.489
4.50	4.99	1.0309	1.0327	46.4	0.279	984.5	13.8	63	1.3393	1.186	0.638	0.347	1.057	1.027	94.45	46.2	0.552
5.00	5.54	1.0346	1.0364	51.7	0.311	982.9	15.4	70	1.3400	1.318	0.709	0.385	1.064	1.030	93.80	50.9	0.615
5.50	6.10	1.0383	1.0402	57.1	0.344	981.2	17.0	77	1.3407	1.451	0.780	0.425	1.071	1.034	93.14	55.6	0.679
6.00	6.65	1.0421	1.0439	62.5	0.376	979.6	18.7	84	1.3414	1.585	0.852	0.464	1.079	1.038	92.48	60.3	0.743
6.50	7.20	1.0459	1.0477	68.0	0.409	977.9	20.3	90	1.3420	1.720	0.925	0.503	1.087	1.041	91.82	64.8	0.807
7.00	7.76	1.0497	1.0516	73.5	0.442	976.2	22.0	97	1.3427	1.856	0.998	0.543	1.095	1.045	91.15	69.3	0.872
7.50	8.31	1.0535	1.0554	79.0	0.475	974.5	23.7	104	1.3434	1.994	1.072	0.583	1.103	1.049	90.48	73.8	0.937
8.00	8.87	1.0574	1.0592	84.6	0.509	972.8	25.5	111	1.3441	2.132	1.146	0.623	1.111	1.053	89.81	78.1	1.00
8.50	9.42	1.0612	1.0631	90.2	0.543	971.0	27.2	118	1.3448	2.273	1.222	0.663	1.120	1.057	89.13	82.4	1.07
9.00	9.98	1.0651	1.0670	95.9	0.577	969.3	29.0	126	1.3456	2.414	1.298	0.704	1.128	1.061	88.45	86.6	1.13
9.50	10.53	1.0691	1.0709	101.6	0.611	967.5	30.7	133	1.3463	2.557	1.375	0.745	1.137	1.066	87.77	90.8	1.20
10.00	11.08	1.0730	1.0749	107.3	0.646	965.7	32.5	140	1.3470	2.701	1.452	0.786	1.146	1.070	87.09	94.8	1.27
11.00	12.19	1.0810	1.0829	118.9	0.715	962.1	36.2	154	1.3484	2.993	1.609	0.870	1.166	1.081	85.58	104.	1.42
12.00	13.30	1.0891	1.0910	130.7	0.786	958.4	39.8	168	1.3498	3.291	1.769	0.954	1.188	1.093	84.01	112.	1.57
13.00	14.41	1.0973	1.0993	142.7	0.858	954.7	43.6	182	1.3512	3.594	1.932	1.039	1.211	1.106	82.40	121.	1.72
14.00	15.52	1.1057	1.1077	154.8	0.931	950.9	47.3	196	1.3526	3.903	2.098	1.125	1.234	1.118	80.89	129.	1.87

54 POTASSIUM PERMANGANATE, $KMnO_4$

MOLECULAR WEIGHT = 158.04
RELATIVE SPECIFIC REFRACTIVITY = 0.811

0.00 % by wt. data are the same for all compounds.
For Values of 0.00 wt. % solutions see Table 1, Acetic Acid.

A% by wt.	ρ D_4^{20}	D_{20}^{20}	C_s g/l	M g-mol/l	C_w g/l	$(C_o - C_w)$ g/l	$(n - n_o)$ × 10^4	n	Δ °C	O Os/kg	S g-mol/l	η/η_o	η/ρ cS	φ rhe	γ mmho/cm	T g-mol/l
0.50	1.0017	1.0034	5.0	0.032	996.7	1.6			0.110	0.059	0.031	0.999	0.999	99.93	3.5	0.035
1.00	1.0051	1.0068	10.1	0.064	995.0	3.2			0.217	0.117	0.062	0.998	0.995	100.04	6.9	0.070
1.50	1.0085	1.0102	15.1	0.096	993.3	4.9			0.323	0.174	0.093	0.997	0.990	100.14	10.0	0.104
2.00	1.0118	1.0136	20.2	0.128	991.6	6.6			0.426	0.229	0.123	0.996	0.986	100.24	13.0	0.138
2.50	1.0152	1.0170	25.4	0.161	989.9	8.4						0.994	0.981	100.36	16.0	0.173
3.00	1.0186	1.0204	30.6	0.193	988.1	10.2						0.993	0.977	100.49	19.0	0.209
3.50	1.0220	1.0238	35.8	0.226	986.3	12.0						0.992	0.972	100.63	22.0	0.244
4.00	1.0254	1.0272	41.0	0.260	984.4	13.8						0.990	0.968	100.78	24.9	0.280
4.50	1.0288	1.0306	46.3	0.293	982.5	15.7						0.989	0.963	100.94	27.7	0.315
5.00	1.0322	1.0340	51.6	0.327	980.6	17.7						0.987	0.958	101.12	30.5	0.350
5.50	1.0356	1.0374	57.0	0.360	978.6	19.6						0.985	0.953	101.30	33.3	0.385
6.00	1.0390	1.0408	62.3	0.394	976.6	21.6						0.983	0.948	101.51	36.0	0.420

55 POTASSIUM PHOSPHATE, DIHYDROGEN (MONOBASIC), KH_2PO_4

MOLECULAR WEIGHT = 136.13
RELATIVE SPECIFIC REFRACTIVITY = 0.632

0.00 % by wt. data are the same for all compounds.
For Values of 0.00 wt. % solutions see Table 1, Acetic Acid.

A% by wt.	ρ D_4^{20}	D_{20}^{20}	C_s g/l	M g-mol/l	C_w g/l	$(C_o - C_w)$ g/l	$(n - n_o)$ × 10^4	n	Δ °C	O Os/kg	S g-mol/l	η/η_o	η/ρ cS	φ rhe	γ mmho/cm	T g-mol/l
0.50	1.0018	1.0035	5.0	0.037	996.8	1.5	6	1.3336	0.130	0.070	0.037	1.008	1.008	98.98	3.0	0.030
1.00	1.0053	1.0071	10.1	0.074	995.3	3.0	12	1.3342	0.252	0.136	0.072	1.017	1.014	98.13	5.9	0.059
1.50	1.0089	1.0107	15.1	0.111	993.8	4.5	18	1.3348	0.371	0.200	0.107	1.026	1.019	97.25	8.5	0.087
2.00	1.0125	1.0143	20.2	0.149	992.2	6.0	24	1.3354	0.488	0.262	0.141	1.036	1.025	96.33	11.0	0.115
2.50	1.0161	1.0179	25.4	0.187	990.7	7.5	29	1.3359	0.603	0.324	0.175	1.046	1.032	95.37	13.5	0.143
3.00	1.0197	1.0215	30.6	0.225	989.1	9.1	35	1.3365	0.720	0.387	0.209	1.058	1.039	94.37	15.9	0.171
3.50	1.0233	1.0251	35.8	0.263	987.5	10.7	41	1.3371	0.838	0.451	0.244	1.069	1.047	93.34	18.3	0.199
4.00	1.0269	1.0288	41.1	0.302	985.9	12.4	47	1.3377	0.957	0.515	0.279	1.081	1.055	92.30	20.6	0.227
4.50	1.0306	1.0324	46.4	0.341	984.2	14.0	52	1.3382	1.075	0.578	0.314	1.094	1.063	91.26	22.8	0.255
5.00	1.0342	1.0360	51.7	0.380	982.5	15.8	58	1.3388	1.190	0.640	0.348	1.106	1.072	90.24	25.0	0.282
5.50	1.0378	1.0396	57.1	0.419	980.7	17.5	64	1.3394	1.303	0.700	0.381	1.119	1.080	89.23	27.2	0.309
6.00	1.0414	1.0432	62.5	0.459	978.9	19.3	70	1.3400	1.414	0.760	0.414	1.131	1.088	88.22	29.4	0.335
6.50	1.0450	1.0469	67.9	0.499	977.1	21.2	75	1.3405	1.523	0.819	0.446	1.144	1.097	87.21	31.5	0.361
7.00	1.0486	1.0505	73.4	0.539	975.2	23.0	81	1.3411	1.631	0.877	0.477	1.158	1.106	86.21	33.5	0.387
7.50	1.0522	1.0541	78.9	0.580	973.3	24.9	87	1.3417	1.736	0.933	0.508	1.171	1.115	85.21	35.5	0.412
8.00	1.0558	1.0577	84.5	0.620	971.4	26.9	92	1.3422	1.839	0.989	0.538	1.185	1.125	84.22	37.5	0.437
8.50	1.0594	1.0613	90.1	0.662	969.4	28.9	98	1.3428	1.941	1.043	0.567	1.199	1.134	83.23	39.3	0.461
9.00	1.0630	1.0649	95.7	0.703	967.4	30.9	104	1.3434	2.039	1.096	0.596	1.213	1.144	82.25	41.2	0.485
9.50	1.0667	1.0686	101.3	0.744	965.3	32.9	109	1.3439	2.136	1.148	0.624	1.228	1.154	81.27	42.9	0.508
10.00	1.0703	1.0722	107.0	0.786	963.3	35.0	115	1.3445	2.230	1.199	0.651	1.243	1.164	80.29	44.6	0.531

56 POTASSIUM PHOSPHATE, MONOHYDROGEN (DIBASIC), $K_2HPO_4 \cdot 3H_2O$

MOLECULAR WEIGHT = 174.18   FORMULA WEIGHT, HYDRATE = 228.23
RELATIVE SPECIFIC REFRACTIVITY = 0.589

0.00 % by wt. data are the same for all compounds.
For Values of 0.00 wt. % solutions see Table 1, Acetic Acid.

A% by wt.	H% by wt.	ρ D_4^{20}	D_{20}^{20}	C_s g/l	M g.mol/l	C_w g/l	$(C_o - C_w)$ g/l	$(n - n_o)$ × 10^4	n	Δ °C	O Os/kg	S g-mol/l	η/η_o	η/ρ cS	φ rhe	γ mmho/cm	T g-mol/l
0.50	0.66	1.0025	1.0043	5.0	0.029	997.5	0.7	8.	1.3338	0.126	0.068	0.036	1.011	1.010	98.76	5.2	0.052
1.00	1.31	1.0068	1.0086	10.1	0.058	996.7	1.5	15.	1.3345	0.250	0.134	0.071	1.021	1.016	97.76	9.9	0.102
1.50	1.97	1.0110	1.0128	15.2	0.087	995.8	2.4	23.	1.3353	0.371	0.199	0.107	1.032	1.023	96.69	14.2	0.151
2.00	2.62	1.0153	1.0171	20.3	0.117	994.9	3.3	31.	1.3361	0.490	0.263	0.142	1.044	1.030	95.62	18.3	0.200
2.50	3.28	1.0195	1.0213	25.5	0.146	994.0	4.2	38.	1.3368	0.613	0.329	0.178	1.055	1.037	94.59	23.0	0.248
3.00	3.93	1.0238	1.0256	30.7	0.176	993.1	5.2	46.	1.3376	0.732	0.394	0.213	1.067	1.044	93.56	26.8	0.294
3.50	4.59	1.0281	1.0299	36.0	0.207	992.1	6.1	54.	1.3384	0.855	0.457	0.249	1.079	1.052	92.46	29.7	0.340
4.00	5.24	1.0324	1.0342	41.3	0.237	991.1	7.1	62.	1.3392	0.974	0.523	0.284	1.092	1.060	92.18	33.3	0.385
4.50	5.90	1.0368	1.0386	46.7	0.268	990.1	8.1	69.	1.3399	1.096	0.589	0.320	1.105	1.068	90.31	36.8	0.430
5.00	6.55	1.0412	1.0430	52.1	0.299	989.1	9.1	77.	1.3407	1.218	0.655	0.356	1.118	1.076	89.26	40.3	0.475
5.50	7.21	1.0456	1.0474	57.5	0.330	988.1	10.2	85.	1.3415	1.337	0.719	0.391	1.131	1.084	88.23	43.8	0.520
6.00	7.86	1.0500	1.0519	63.0	0.362	987.0	11.2	92.	1.3422	1.460	0.785	0.427	1.145	1.093	87.13	47.2	0.565
6.50	8.52	1.0545	1.0564	68.5	0.394	986.0	12.3	100.	1.3430	1.580	0.850	0.462	1.160	1.102	86.05	50.6	0.610
7.00	9.17	1.0590	1.0609	74.1	0.426	984.9	13.3	108.	1.3438	1.703	0.915	0.498	1.175	1.112	84.92	53.8	0.654
7.50	9.83	1.0635	1.0654	79.8	0.458	983.8	14.5	115.	1.3445	1.823	0.980	0.533	1.191	1.122	83.80	57.0	0.698
8.00	10.48	1.0680	1.0699	85.4	0.491	982.6	15.7	123.	1.3453	1.947	1.046	0.569	1.207	1.132	82.71	60.0	0.740

57 POTASSIUM SULFATE, K_2SO_4

MOLECULAR WEIGHT = 174.26
RELATIVE SPECIFIC REFRACTIVITY = 0.565

0.00 % by wt. data are the same for all compounds.
For Values of 0.00 wt. % solutions see Table 1, Acetic Acid.

A% by wt.	ρ D_4^{20}	D_{20}^{20}	C_s g/l	M g-mol/l	C_w g/l	$(C_o - C_w)$ g/l	$(n - n_o)$ × 10^4	n	Δ °C	O Os/kg	S g-mol/l	η/η_o	η/ρ cS	φ rhe	γ mmho/cm	T g-mol/l
0.50	1.0022	1.0040	5.0	0.029	997.2	1.0	6	1.3336	0.136	0.073	0.039	1.004	1.004	99.37	5.8	0.059
1.00	1.0062	1.0080	10.1	0.058	996.2	2.1	13	1.3343	0.262	0.141	0.075	1.009	1.005	98.91	11.2	0.117
1.50	1.0102	1.0120	15.2	0.087	995.1	3.1	19	1.3349	0.382	0.205	0.110	1.014	1.006	98.42	16.2	0.175
2.00	1.0143	1.0161	20.3	0.116	994.0	4.2	25	1.3355	0.499	0.268	0.144	1.019	1.007	97.92	21.0	0.232
2.50	1.0183	1.0201	25.5	0.146	992.9	5.4	32	1.3362	0.614	0.330	0.178	1.025	1.008	97.39	25.6	0.289
3.00	1.0224	1.0242	30.7	0.176	991.7	6.5	38	1.3368	0.726	0.390	0.211	1.031	1.010	96.84	30.2	0.347

57 POTASSIUM SULFATE, K_2SO_4—(Continued)

A % by wt.	D_4^{20}	D_{20}^{20}	C_s g/l	M g-mol/l	C_w g/l	$(C_o - C_w)$ g/l	$(n - n_o) \times 10^4$	n	Δ °C	O Os/kg	S g-mol/l	η/η_o	η/ρ cS	φ rhe	γ mmho/cm	T g-mol/l
3.50	1.0265	1.0283	35.9	0.206	990.6	7.7	44	1.3374	0.839	0.451	0.244	1.037	1.012	96.27	34.7	0.404
4.00	1.0306	1.0324	41.2	0.237	989.4	8.9	50	1.3380	0.950	0.511	0.277	1.043	1.014	95.69	39.2	0.461
4.50	1.0347	1.0365	46.6	0.267	988.1	10.1	57	1.3386	1.061	0.571	0.310	1.049	1.016	95.10	43.7	0.518
5.00	1.0388	1.0406	51.9	0.298	986.8	11.4	63	1.3393	1.172	0.630	0.342	1.056	1.019	94.50	48.0	0.576
5.50	1.0429	1.0447	57.4	0.329	985.5	12.7	69	1.3399				1.063	1.021	93.89	52.3	0.634
6.00	1.0470	1.0489	62.8	0.361	984.2	14.0	75	1.3405				1.070	1.024	93.27	56.6	0.692
6.50	1.0512	1.0530	68.3	0.392	982.8	15.4	81	1.3411				1.077	1.027	92.64	60.8	0.751
7.00	1.0553	1.0572	73.9	0.424	981.4	16.8	87	1.3417				1.085	1.030	92.02	64.9	0.810
7.50	1.0595	1.0614	79.5	0.456	980.0	18.2	93	1.3422				1.092	1.033	91.39	69.0	0.869
8.00	1.0637	1.0655	85.1	0.488	978.6	19.7	98	1.3428				1.100	1.036	90.77	73.1	0.928
8.50	1.0679	1.0697	90.8	0.521	977.1	21.1	104	1.3434				1.107	1.039	90.14	77.0	0.988
9.00	1.0721	1.0740	96.5	0.554	975.6	22.6	110	1.3440				1.115	1.042	89.53	80.9	1.04
9.50	1.0763	1.0782	102.3	0.587	974.1	24.2	116	1.3446				1.122	1.045	88.92	84.8	1.10
10.00	1.0806	1.0825	108.1	0.620	972.5	25.7	122	1.3452				1.130	1.048	88.32	88.6	1.16

58 POTASSIUM THIOCYANATE, KSCN

MOLECULAR WEIGHT = 97.18
RELATIVE SPECIFIC REFRACTIVITY = 1.025

0.00 % by wt. data are the same for all compounds.
For Values of 0.00 wt. % solutions see Table 1, Acetic Acid.

A % by wt.	ρ D_4^{20}	D_{20}^{20}	C_s g/l	M g-mol/l	C_w g/l	$(C_o - C_w)$ g/l	$(n - n_o) \times 10^4$	n	Δ °C	O Os/kg	S g-mol/l	η/η_o	η/ρ cS	φ rhe	γ mmho/cm	T g-mol/l
0.50	1.0007	1.0025	5.0	0.051	995.7	2.5	9	1.3339	0.177	0.095	0.050	0.9974	0.9987	100.1	5.8	0.058
1.00	1.0032	1.0049	10.0	0.103	993.1	5.1	19	1.3349	0.355	0.191	0.102	0.9947	0.9936	100.3	11.4	0.119
2.00	1.0081	1.0099	20.2	0.207	987.9	10.3	37	1.3367	0.709	0.381	0.206	0.9887	0.9827	100.9	21.9	0.243
3.00	1.0130	1.0148	30.4	0.313	982.6	15.6	55	1.3385	1.061	0.570	0.310	0.9833	0.9726	101.5	32.1	0.370
4.00	1.0180	1.0198	40.7	0.419	977.3	21.0	74	1.3404	1.414	0.760	0.413	0.9785	0.9632	102.0	42.1	0.499
5.00	1.0229	1.0248	51.1	0.526	971.8	26.4	93	1.3423	1.771	0.952	0.518	0.9741	0.9542	102.5	52.2	0.632
6.00	1.0279	1.0298	61.7	0.635	966.3	32.0	111	1.3441	2.136	1.148	0.624	0.9700	0.9456	102.9	62.4	0.773
7.00	1.0330	1.0348	72.3	0.744	960.7	37.6	130	1.3460	2.506	1.347	0.731	0.9664	0.9374	103.3	72.7	0.922
8.00	1.0380	1.0398	83.0	0.854	955.0	43.3	149	1.3479	2.882	1.549	0.838	0.9632	0.9298	103.6	82.9	1.08
9.00	1.0431	1.0449	93.9	0.966	949.2	49.0	168	1.3498	3.264	1.755	0.946	0.9604	0.9225	103.9	93.	1.24
10.00	1.0482	1.0500	104.8	1.079	943.3	54.9	187	1.3517	3.653	1.964	1.055	0.9579	0.9157	104.2	103.	1.40
12.00	1.0585	1.0603	127.0	1.307	931.4	66.8	226	1.3556	4.451	2.393	1.276	0.9541	0.9032	104.6	121.	1.72
14.00	1.0689	1.0708	149.6	1.540	919.2	79.0	265	1.3595	5.28	2.84	1.498	0.9521	0.8925	104.8	138.	2.06
16.00	1.0795	1.0814	172.7	1.777	906.8	91.5	304	1.3634				0.9523	0.8840	104.8	156.	2.47
18.00	1.0902	1.0922	196.2	2.019	894.0	104.3	345	1.3674				0.9550	0.8777	104.5	176.	2.95
20.00	1.1011	1.1031	220.2	2.266	880.9	117.3	385	1.3715				0.9597	0.8733	104.0	197.	3.60
24.00	1.1236	1.1255	269.7	2.775	853.9	144.3	469	1.3798				0.9755	0.8700	102.3		
28.00	1.1467	1.1488	321.1	3.304	825.6	172.6	554	1.3884				1.001	0.8747	99.69		
32.00	1.1706	1.1727	374.6	3.855	796.0	202.2	643	1.3972				1.037	0.8874	96.27		
36.00	1.1949	1.1970	430.2	4.427	764.7	233.5	732	1.4062				1.081	0.9069	92.28		
40.00	1.2199	1.2220	488.0	5.021	731.9	266.3	824	1.4154				1.150	0.9446	86.78		
44.00	1.2455	1.2477	548.0	5.639	697.5	300.7	918	1.4248				1.253	1.008	79.65		
48.00	1.2719	1.2741	610.5	6.282	661.4	336.9	1014	1.4343				1.389	1.094	71.86		
52.00	1.2990	1.3013	675.5	6.951	623.5	374.7	1112	1.4442				1.552	1.197	64.31		
56.00	1.3269	1.3292	743.1	7.646	583.8	414.4	1212	1.4542				1.735	1.310	57.52		
60.00	1.3556	1.3580	813.4	8.370	542.2	456.0	1315	1.4645				2.013	1.488	49.58		
64.00	1.3853	1.3877	886.6	9.123	498.7	499.5	1420	1.4750				2.434	1.761	41.00		

59 PROCAINE HYDROCHLORIDE, $C_6H_4[COOCH_2CH_2N(C_2H_5)_2][NH_2] \cdot HCl$-1,4

MOLECULAR WEIGHT = 272.78
RELATIVE SPECIFIC REFRACTIVITY = 1.460

0.00 % by wt. data are the same for all compounds.
For Values of 0.00 wt. % solutions see Table 1, Acetic Acid.

A % by wt.	ρ D_4^{20}	D_{20}^{20}	C_s g/l	M g-mol/l	C_w g/l	$(C_o - C_w)$ g/l	$(n - n_o) \times 10^4$	n	Δ °C	O Os/kg	S g-mol/l	η/η_o	η/ρ cS	φ rhe	γ mmho/cm	T g-mol/l
0.50	0.9991	1.0008	5.0	0.018	994.1	4.2	11	1.3341	0.062	0.033	0.017	1.014	1.017	98.42	1.4	0.014
1.00	0.9999	1.0017	10.0	0.037	989.9	8.3	22	1.3352	0.126	0.068	0.035	1.028	1.030	97.08	2.7	0.027
2.00	1.0016	1.0034	20.0	0.073	981.6	16.7	44	1.3374	0.237	0.127	0.068	1.056	1.056	94.51	5.2	0.052
3.00	1.0033	1.0051	30.1	0.110	973.2	25.0	66	1.3396	0.343	0.184	0.099	1.083	1.082	92.16	7.4	0.076
4.00	1.0050	1.0068	40.2	0.147	964.8	33.4	89	1.3419	0.433	0.233	0.125	1.110	1.106	89.95	9.4	0.098
5.00	1.0067	1.0085	50.3	0.185	956.4	41.8	111	1.3441	0.519	0.279	0.150	1.138	1.133	87.70	11.4	0.119
6.00	1.0085	1.0103	60.5	0.222	948.0	50.2	133	1.3463	0.593	0.319	0.172	1.169	1.161	85.38	13.1	0.139
7.00	1.0103	1.0120	70.7	0.259	939.5	58.7	156	1.3486	0.675	0.363	0.196	1.202	1.192	83.06	14.7	0.157
8.00	1.0120	1.0138	81.0	0.297	931.1	67.2	179	1.3508	0.756	0.406	0.220	1.236	1.224	80.73	16.2	0.174
9.00	1.0138	1.0156	91.2	0.334	922.6	75.6	201	1.3531	0.824	0.443	0.240	1.273	1.258	78.40	17.5	0.191
10.00	1.0156	1.0174	101.6	0.372	914.1	84.2	224	1.3554	0.889	0.478	0.259	1.312	1.294	76.07	18.8	0.206
12.00	1.0193	1.0211	122.3	0.448	897.0	101.3	270	1.3600	1.026	0.552	0.300	1.397	1.373	71.45	21.1	0.233
14.00	1.0230	1.0248	143.2	0.525	879.8	118.5	316	1.3646	1.164	0.626	0.340	1.491	1.460	66.94	23.0	0.256
16.00	1.0267	1.0285	164.3	0.602	862.4	135.8	362	1.3692	1.297	0.698	0.379	1.595	1.557	62.57	24.6	0.277
18.00	1.0305	1.0323	185.5	0.680	845.0	153.2	409	1.3739	1.425	0.766	0.417	1.708	1.661	58.44	26.0	0.294
20.00	1.0343	1.0361	206.9	0.758	827.4	170.8	456	1.3786	1.545	0.831	0.452	1.838	1.781	54.30	27.2	0.309
24.00	1.0420	1.0439	250.1	0.917	791.9	206.3	552	1.3881				2.143	2.061	46.57	29.0	0.331
28.00	1.0499	1.0517	294.0	1.078	755.9	242.3	649	1.3978				2.542	2.426	39.26	29.8	0.342
32.00	1.0578	1.0596	338.5	1.241	719.3	278.9	747	1.4077				3.089	2.927	32.30	30.2	0.346
36.00	1.0658	1.0676	383.7	1.407	682.1	316.2	848	1.4178				3.746	3.522	26.64	29.5	0.338
40.00	1.0738	1.0757	429.5	1.575	644.3	354.0	951	1.4280				4.757	4.439	20.98	28.3	0.323
44.00	1.0819	1.0838	476.0	1.745	605.9	392.4	1055	1.4385				6.028	5.583	16.56	26.8	0.304
48.00	1.0900	1.0920	523.2	1.918	566.8	431.4	1162	1.4491				7.888	7.251	12.65	24.7	0.278
52.00	1.0982	1.1002	571.1	2.094	527.2	471.1	1270	1.4600				11.22	10.24	8.89	22.3	0.248
56.00	1.1066	1.1085	619.7	2.272	486.9	511.3	1381	1.4711				15.97	14.46	6.25	19.5	0.214
60.00	1.1151	1.1171	669.1	2.453	446.0	552.2	1493	1.4823				22.12	19.88	4.51	16.1	0.174

60 1-PROPANOL, $CH_3CH_2CH_2OH$

MOLECULAR WEIGHT = 60.09

A% by wt.	ρ D_4^{20}	D_{20}^{20}	C_s g/l	M g-mol/l	C_w g/l	$(C_0 - C_w)$ g/l	$(n - n_0)$ × 10^4	n	Δ °C	O Os/kg	S g-mol/l	η/η_0	η/ρ cS	φ rhe	γ mmho/cm	T g-mol/l
1.00	0.9963	0.9981	10.0	0.166	986.3	11.9	9.	1.3339	0.307	0.165	0.088	1.049	1.055	95.14	0.0	0.000
2.00	0.9946	0.9964	19.9	0.331	974.7	23.5	18.	1.3348	0.613	0.329	0.178	1.098	1.106	90.91		
3.00	0.9928	0.9946	29.8	0.496	963.0	35.2	27.	1.3357	0.933	0.502	0.272	1.150	1.161	86.76		
4.00	0.9911	0.9929	39.6	0.660	951.5	46.7	36.	1.3366	1.239	0.666	0.362	1.206	1.219	82.77		
5.00	0.9896	0.9914	49.5	0.823	940.1	58.1	46.	1.3376	1.573	0.846	0.460	1.264	1.280	78.94		
6.00	0.9882	0.9900	59.3	0.987	928.9	69.3	55.	1.3385	1.906	1.024	0.557	1.322	1.340	75.52		
7.00	0.9868	0.9886	69.1	1.150	[illegible]	[illegible]	[illegible]	[illegible]	[illegible]	[illegible]	[illegible]	[illegible]	[illegible]	[illegible]		
8.00	0.9855	0.9873	78.8	1.312	906.7	91.5	74.	1.3404	2.609	1.403	0.760	1.446	1.470	69.03		
9.00	0.9842	0.9860	88.6	1.474	895.6	102.6	84.	1.3414	2.988	1.606	0.868	1.511	1.538	66.06		
10.00	0.9829	0.9847	98.3	1.636	884.6	113.6	93.	1.3423	3.356	1.805	0.972	1.574	1.605	63.39		
12.00	0.9804	0.9822	117.6	1.958	862.8	135.4	112.	1.3442	3.477	2.014	1.181	1.707	1.745	58.45		
14.00	0.9779	0.9797	136.9	2.278	841.0	157.2	130.	1.3460	4.914	2.642	1.401	1.845	1.890	54.10		
16.00	0.9749	0.9767	156.0	2.596	818.9	179.3	147.	1.3477	5.776	3.108	1.629	1.982	2.037	50.35		
18.00	0.9719	0.9737	174.9	2.911	797.0	201.2	164.	1.3494	6.667	3.588	1.859	2.102	2.167	47.48		
20.00	0.9686	0.9703	193.7	3.224	774.9	223.3	180.	1.3510	7.555	4.062	2.079	2.214	2.290	45.09		
24.00	0.9612	0.9629	230.7	3.839	730.5	267.7	209.	1.3539	9.115	4.901	2.450	2.427	2.530	41.12		
28.00	0.9533	0.9550	266.9	4.442	686.4	311.8	236.	1.3566	10.17	5.47	2.688	2.607	2.740	38.28		
32.00	0.9452	0.9469	302.5	5.034	642.7	355.5	262.	1.3592	10.66	5.73	2.797	2.759	2.925	36.17		
36.00	0.9370	0.9387	337.3	5.614	599.7	398.5	284.	1.3614				2.894	3.095	34.48		
40.00	0.9288	0.9305	371.5	6.183	557.3	440.9	305.	1.3635				3.004	3.241	33.22		
44.00	0.9206	0.9223	405.1	6.741	515.5	482.7	328.	1.3658				3.096	3.370	32.23		
48.00	0.9127	0.9143	438.1	7.291	483.6	514.6	348.	1.3678				3.162	3.472	31.56		
52.00	0.9043	0.9059	470.2	7.826	434.1	564.1	367.	1.3697				3.195	3.540	31.24		
56.00	0.8959	0.8975	501.7	8.349	394.2	604.0	385.	1.3715				3.197	3.576	31.21		
60.00	0.8875	0.8891	532.5	8.862	355.0	643.2	404.	1.3734				3.180	3.590	31.39		
64.00	0.8790	0.8806	562.6	9.362	316.4	681.8	422.	1.3752				3.135	3.574	31.83		
68.00	0.8706	0.8722	592.0	9.852	278.6	719.6	437.	1.3767				3.075	3.539	32.46		
72.00	0.8623	0.8639	620.9	10.332	241.4	756.8	453.	1.3783				2.997	3.482	33.30		
76.00	0.8549	0.8564	649.7	10.813	205.2	793.0	467.	1.3797				2.909	3.410	34.30		
80.00	0.8470	0.8485	677.6	11.276	169.4	828.8	482.	1.3812				2.816	3.332	35.43		
84.00	0.8390	0.8405	704.8	11.728	134.2	864.0	495.	1.3825				2.720	3.249	36.69		
88.00	0.8306	0.8321	730.9	12.164	99.7	898.5	505.	1.3835				2.590	3.125	38.53		
92.00	0.8218	0.8233	756.1	12.582	65.7	932.5	513.	1.3843				2.467	3.008	40.45		
96.00	0.8130	0.8145	780.5	12.989	32.5	965.7	518.	1.3848				2.345	2.890	42.56		
100.00	0.8034	0.8048	803.4	13.370	0.0	998.2	522.	1.3852				2.223	2.773	44.89		

61 2-PROPANOL, $CH_3CHOHCH_3$

MOLECULAR WEIGHT = 60.09

A% by wt.	ρ D_4^{20}	D_{20}^{20}	C_s g/l	M g-mol/l	C_w g/l	$(C_0 - C_w)$ g/l	$(n - n_0)$ × 10^4	n	Δ °C	O Os/kg	S g-mol/l	η/η_0	η/ρ cS	φ rhe	γ mmho/cm	T g-mol/l
1.00	0.9960	0.9978	10.0	0.166	986.0	12.2	8.	1.3338	0.304	0.163	0.087	1.054	1.060	94.72	0.0	0.000
2.00	0.9939	0.9957	19.9	0.331	974.0	24.2	16.	1.3346	0.603	0.324	0.175	1.110	1.119	89.91		
3.00	0.9920	0.9938	29.8	0.495	962.2	36.0	25.	1.3355	0.926	0.498	0.270	1.164	1.176	85.72		
4.00	0.9902	0.9920	39.6	0.659	950.6	47.6	34.	1.3364	1.259	0.677	0.368	1.223	1.238	81.57		
5.00	0.9884	0.9902	49.4	0.822	939.0	59.2	43.	1.3373	1.610	0.866	0.471	1.284	1.302	77.71		
6.00	0.9871	0.9889	59.2	0.986	927.9	70.3	52.	1.3382	1.960	1.054	0.573	1.349	1.369	74.00		
7.00	0.9855	0.9873	69.0	1.148	916.5	81.7	62.	1.3392	2.320	1.247	0.677	1.414	1.438	70.56		
8.00	0.9843	0.9861	78.7	1.310	905.6	92.6	71.	1.3401	2.679	1.440	0.780	1.482	1.509	67.33		
9.00	0.9831	0.9849	88.5	1.472	894.6	103.6	80.	1.3410	3.062	1.646	0.889	1.550	1.580	64.38		
10.00	0.9816	0.9834	98.2	1.634	883.4	114.8	90.	1.3420	3.481	1.871	1.007	1.626	1.660	61.37		
12.00	0.9793	0.9811	117.5	1.956	861.8	136.4	109.	1.3439	4.433	2.383	1.271	1.790	1.831	55.77		
14.00	0.9772	0.9790	136.8	2.277	840.4	157.8	129.	1.3459	5.29	2.85	1.503	1.966	2.016	50.76		
16.00	0.9751	0.9769	156.0	2.596	819.1	179.1	148.	1.3478	6.36	3.42	1.78	2.156	2.215	46.30		
18.00	0.9725	0.9743	175.1	2.913	797.4	200.8	166.	1.3496	7.40	3.98	2.04	2.347	2.418	42.53		
20.00	0.9696	0.9713	193.9	3.227	775.7	222.5	184.	1.3514	8.52	4.58	2.31	2.545	2.630	39.21		
24.00	0.9630	0.9647	231.1	3.846	731.9	266.3	217.	1.3547	10.91	5.87	2.85	2.851	2.966	35.01		
28.00	0.9555	0.9572	267.5	4.452	688.0	310.2	245.	1.3575				3.115	3.267	32.03		
32.00	0.9478	0.9495	303.3	5.047	644.5	353.7	270.	1.3600				3.330	3.520	29.97		
36.00	0.9393	0.9410	338.2	5.627	601.1	397.1	292.	1.3622				3.499	3.733	28.52		
40.00	0.9302	0.9319	372.1	6.192	558.1	440.1	312.	1.3642				3.630	3.910	27.49		
44.00	0.9207	0.9224	405.1	6.742	515.6	482.6	330.	1.3660				3.732	4.061	26.75		
48.00	0.9113	0.9129	437.4	7.279	473.9	524.3	346.	1.3676				3.772	4.147	26.46		
52.00	0.9019	0.9035	469.0	7.805	432.9	565.3	362.	1.3692				3.789	4.210	26.34		
56.00	0.8920	0.8936	499.5	8.313	392.5	605.7	375.	1.3705				3.781	4.247	26.40		
60.00	0.8824	0.8840	529.4	8.811	353.0	645.2	387.	1.3717				3.719	4.223	26.84		
64.00	0.8729	0.8745	558.7	9.297	314.2	684.0	400.	1.3730				3.636	4.174	27.45		
68.00	0.8632	0.8648	587.0	9.769	276.2	722.0	411.	1.3741				3.528	4.095	28.29		
72.00	0.8535	0.8550	614.5	10.226	239.0	759.2	421.	1.3751				3.391	3.981	29.43		
76.00	0.8438	0.8453	641.3	10.672	202.5	795.7	428.	1.3758				3.235	3.841	30.85		
80.00	0.8341	0.8356	667.3	11.105	166.8	831.4	435.	1.3765				3.072	3.690	32.49		
84.00	0.8243	0.8258	692.4	11.523	131.9	866.3	440.	1.3770				2.886	3.508	34.58		
88.00	0.8145	0.8160	716.8	11.929	97.7	900.5	445.	1.3775				2.684	3.302	37.18		
92.00	0.8046	0.8061	740.3	12.319	64.3	933.9	447.	1.3777				2.530	3.151	39.44		
96.00	0.7949	0.7963	763.1	12.699	31.8	966.4	446.	1.3776				2.424	3.055	41.18		
100.00	0.7848	0.7862	784.8	13.060	0.0	998.2	442.	1.3742				2.428	3.100	41.10		

62 PROPYLENE GLYCOL, $CH_2OHCHOHCH_3$

MOLECULAR WEIGHT = 76.09
RELATIVE SPECIFIC REFRACTIVITY = 1.236

0.00% by wt. data are the same for all compounds.
For Values of 0.00 wt % solutions see Table 1. Acetic Acid

A% by wt	ρ D_4^{20}	D_{20}^{20}	C_s g/l	M g-mol/l	C_w g/l	$(C_o - C_w)$ g/l	$(n - n_o)$ × 10^4	n	Δ °C	θ Os/kg	S g-mol/l	η/η_o	η/ρ cS	φ rhe	? mmho/cm	T g-mol/l
0.50	0.9985	1.0003	5.0	0.066	993.5	4.7	5	1.3335	0.131	0.071	0.037	1.013	1.017	98.52		
1.00	0.9988	1.0006	10.0	0.131	988.8	9.4	10	1.3340	0.259	0.139	0.074	1.028	1.031	97.08		
2.00	0.9994	1.0012	20.0	0.263	979.4	18.8	20	1.3350	0.501	0.269	0.145	1.064	1.067	93.80		
3.00	1.0001	1.0019	30.0	0.394	970.1	28.2	31	1.3361	0.758	0.407	0.221	1.101	1.104	90.61		
4.00	1.0008	1.0025	40.0	0.526	960.7	37.5	41	1.3371	1.027	0.552	0.300	1.140	1.142	87.52		
5.00	1.0015	1.0032	50.1	0.658	951.4	46.8	52	1.3382	1.303	0.701	0.381	1.181	1.182	84.50		
6.00	1.0022	1.0040	60.1	0.790	942.1	56.2	63	1.3393	1.583	0.851	0.463	1.223	1.223	81.59		
7.00	1.0030	1.0047	70.2	0.923	932.8	65.5	74	1.3403	1.870	1.005	0.547	1.267	1.266	78.77		
8.00	1.0037	1.0055	80.3	1.055	923.4	74.8	83	1.3414	2.164	1.163	0.632	1.313	1.310	76.03		
9.00	1.0045	1.0063	90.4	1.188	914.1	84.1	95	1.3425	2.465	1.325	0.719	1.361	1.357	73.34		
10.00	1.0054	1.0071	100.5	1.321	904.8	93.4	107	1.3436	2.773	1.491	0.807	1.412	1.407	70.68		
12.00	1.0070	1.0088	120.8	1.588	886.2	112.0	129	1.3459	3.534	1.900	1.022	1.524	1.516	65.50		
14.00	1.0088	1.0106	141.2	1.856	867.6	130.7	152	1.3482	4.393	2.362	1.260	1.646	1.635	60.65		
16.00	1.0106	1.0124	161.7	2.125	848.9	149.4	175	1.3504	5.44	2.92	1.540	1.776	1.761	56.19		
18.00	1.0124	1.0142	182.2	2.395	830.2	168.1	198	1.3528				1.915	1.895	52.12		
20.00	1.0142	1.0160	202.8	2.666	811.4	186.9	221	1.3551				2.066	2.041	48.31		
24.00	1.0178	1.0196	244.3	3.210	773.6	224.7	267	1.3597				2.410	2.372	41.42		
28.00	1.0214	1.0232	286.0	3.759	735.4	262.8	314	1.3644				2.811	2.758	35.50		
32.00	1.0248	1.0266	327.9	4.310	696.8	301.4	360	1.3690				3.280	3.207	30.43		
36.00	1.0279	1.0297	370.1	4.863	657.9	340.4	406	1.3736				3.811	3.715	26.19		
40.00	1.0308	1.0326	412.3	5.419	618.5	379.8	451	1.3780				4.432	4.308	22.52		
44.00	1.0333	1.0352	454.7	5.975	578.7	419.6	494	1.3824				5.144	4.988	19.40		
48.00	1.0356	1.0374	497.1	6.533	538.5	459.7	537	1.3867				5.963	5.769	16.74		
52.00	1.0377	1.0395	539.6	7.091	498.1	500.2	580	1.3910				6.939	6.700	14.38		
56.00	1.0396	1.0415	582.2	7.651	457.4	540.8	622	1.3952				8.067	7.775	12.37		
60.00	1.0418	1.0436	625.1	8.215	416.7	581.5	665	1.3995				9.348	8.991	10.68		

63 SEA WATER*

RELATIVE SPECIFIC REFRACTIVITY = 0.789

0.00% by wt data are the same for all compounds.
For values of 0.00 wt × solutions see Table 1. Acetic Acid

A% by wt	Salinity ‰	Chlorinity ‰	ρ D_4^{20}	D_{20}^{20}	C_s g/l	C_w g/l	$(C_o - C_w)$ g/l	$(n - n_o)$ × 10^4	n	Δ °C	O Os/kg	S g-mol/l	η/η_o	η/ρ cS	φ rhe	γ mmho/cm	T g-mol/l
0.50	4.94	2.72	1.0019	1.0037	5.0	996.9	1.3	9	1.3339	0.269	0.144	0.077	1.008	1.008	99.02	7.8	0.082
1.00	9.92	5.48	1.0057	1.0075	10.1	995.5	2.7	18	1.3348	0.537	0.289	0.156	1.016	1.012	98.23	15.2	0.163
1.50	14.91	8.24	1.0094	1.0112	15.1	994.1	4.1	27	1.3357	0.806	0.433	0.235	1.024	1.017	97.45	22.0	0.244
2.00	19.89	11.00	1.0132	1.0150	20.3	992.7	5.6	36	1.3366	1.075	0.578	0.314	1.033	1.022	96.61	28.4	0.324
2.50	24.87	13.76	1.0170	1.0188	25.4	991.2	7.0	45	1.3375	1.346	0.724	0.394	1.044	1.029	95.64	34.5	0.401
3.00	29.86	16.53	1.0207	1.0225	30.6	989.7	8.5	54	1.3384	1.616	0.869	0.473	1.055	1.036	94.58	40.4	0.476
3.50	34.84	19.29	1.0245	1.0263	35.8	988.1	10.1	63	1.3393	1.877	1.009	0.549	1.067	1.044	93.55	46.3	0.552
4.00	39.82	22.05	1.0283	1.0301	41.1	986.6	11.7	72	1.3402	2.129	1.145	0.622	1.077	1.050	92.67	52.4	0.635
4.50	44.81	24.81	1.0320	1.0339	46.4	985.0	13.3	81	1.3411	2.374	1.277	0.693	1.087	1.056	91.83	58.6	0.721
5.00	49.79	27.57	1.0358	1.0376	51.8	983.3	14.9	90	1.3420	2.612	1.404	0.761	1.100	1.065	90.75	64.2	0.799
5.50	54.78	30.33	1.0396	1.0414	57.1	981.6	16.6	99	1.3429	2.842	1.528	0.827	1.117	1.077	89.38	68.7	0.865
6.00	59.76	33.09	1.0434	1.0452	62.5	979.9	18.3	108	1.3438	3.064	1.647	0.890	1.137	1.093	87.80	72.6	0.922
6.50	64.74	35.85	1.0472	1.0490	68.0	978.1	20.1	118	1.3448								
7.00	69.73	38.61	1.0509	1.0528	73.5	976.3	21.9	127	1.3457								
7.50	74.71	41.37	1.0547	1.0566	79.0	974.6	23.6	136	1.3466								
8.00	79.69	44.13	1.0585	1.0604	84.6	972.9	25.3	146	1.3476								
8.50	84.68	46.90	1.0623	1.0642	90.2	971.2	27.1	155	1.3485								
9.00	89.66	49.66	1.0662	1.0680	95.9	969.3	28.9	164	1.3494								
9.50	94.64	52.42	1.0700	1.0719	101.6	967.5	30.8	173	1.3503								
10.00	99.63	55.18	1.0738	1.0757	107.3	965.6	32.6	182	1.3512								
11.00	109.6	60.70	1.0814	1.0833	118.9	961.7	36.5	200	1.3530								
12.00	119.6	66.22	1.0891	1.0910	130.6	957.8	40.5	219	1.3548								
13.00	129.5	71.74	1.0968	1.0987	142.5	953.7	44.5	237	1.3567								
14.00	139.5	77.27	1.1045	1.1065	154.6	949.5	48.7	255	1.3585								
15.00	149.5	82.79	1.1122	1.1142	166.8	945.2	53.0	273	1.3602								

* Values corresponding to A% from 0.00 6.00 are based upon measurements of artificial sea water as formulated by Lyman and Fleming. *J. Marine Res.*, 3, 134, 194. Values corresponding to A% above 6.00 are based upon natural sea water from the Gulf of Mexico at Aransas Pass. Texas, concentrated by evaporation, according Behrns, ibid. 23: 165, 1965.

64 SERUM OR PLASMA, HUMAN

0.00% by wt. data are the same for all compounds
For Values of 0.00 wt % solutions see Table 1. Acetic Acid.

A% by wt	ρ D_4^{20}	D_{20}^{20}	C_S g/l	C_{Pr} g/100 ml	C_{NPr} g/100 ml	C_w g/l	$(C_o - C_w)$ g/l	$(n - n_o)$ × 10^4	n
1.00	1.0029	1.0047	10.0	0.0	1.00	993	5.0	16	1.3346
1.50	1.0051	1.0069	15.0	0.0	1.50	990	8.0	24	1.3354
2.00	1.0069	1.0087	20.0	0.3	1.69	987	11.0	33	1.3363
2.50	1.0082	1.0100	25.0	0.8	1.68	983	15.0	42	1.3372
3.00	1.0096	1.0114	30.0	1.4	1.68	979	19.0	52	1.3382
3.50	1.0110	1.0128	35.0	1.9	1.67	976	22.0	61	1.3391
4.00	1.0123	1.0141	40.0	2.4	1.66	972	26.0	71	1.3401
4.50	1.0137	1.0155	46.0	2.9	1.66	968	30.0	80	1.3410
5.00	1.0150	1.0168	51.0	3.4	1.65	964	34.0	90	1.3420
5.50	1.0164	1.0182	56.0	3.9	1.64	960	38.0	99	1.3429

64 SERUM OR PLASMA, HUMAN—(Continued)

A % by wt.	ρ D_4^{20}	D_{20}^{20}	C_s g/l	C_{Pr} g/100 ml	C_{NPr} g/100 ml	C_w g/l	$(C_o - C_w)$ g/l	$(n - n_o)$ × 10^4	n
6.00	1.0177	1.0195	61.	4.5	1.64	957.	41.	108	1.3438
6.50	1.0190	1.0208	66.	5.0	1.63	953.	45.	118	1.3448
7.00	1.0203	1.0221	71.	5.5	1.62	949.	49.	127	1.3457
7.50	1.0216	1.0234	77.	6.0	1.62	945.	53.	137	1.3467
8.00	1.0229	1.0247	82.	6.6	1.61	941.	57.	146	1.3476
8.50	1.0242	1.0260	87.	7.1	1.60	937.	61.	156	1.3486
9.00	1.0255	1.0273	92.	7.6	1.60	933.	65.	165	1.3495
9.50	1.0267	1.0286	98.	8.2	1.59	929.	69.	175	1.3505
10.00	1.0279	1.0298	103.	8.7	1.58	925.	73.	184	1.3514
11.00	1.0305	1.0324	113.	9.8	1.57	917.	81.	203	1.3533
12.00	1.0329	1.0348	124.	10.8	1.55	909.	89.	222	1.3552
13.00	1.0354	1.0373	133.	11.9	1.54	901.	97.	241	1.3571
14.00	1.0378	1.0397	145.	13.0	1.53	893.	105.	260	1.3590
15.00	1.0401	1.0420	156.	14.1	1.51	884.	114.	279	1.3609

Notes: C_{Pr} = g protein/100 ml; C_{NPr} = g nonprotein solute/100 ml.

65 SERUM OR PLASMA, RABBIT OR GUINEA PIG

0.00 % by wt. data are the same for all compounds.
For Values of 0.00 wt. % solutions see Table 1, Acetic Acid.

A % by wt.	ρ D_4^{20}	D_{20}^{20}	C_s g/l	C_{Pr} g/100 ml	C_w g/l	$(C_o - C_w)$ g/l	$(n - n_o)$ × 10^4	n
4.0	1.0125	1.0143	41.	3.3	972.	28.	71	1.3401
4.5	1.0139	1.0156	46.	3.8	968.	30.	80	1.3410
5.0	1.0152	1.0170	51.	4.2	964.	34.	89	1.3419
5.5	1.0165	1.0183	56.	4.6	961.	38.	98	1.3428
6.0	1.0179	1.0197	61.	5.0	957.	41.	107	1.3437
6.5	1.0192	1.0210	66.	5.4	953.	45.	116	1.3446
7.0	1.0206	1.0224	71.	5.8	949.	49.	125	1.3455
7.5	1.0219	1.0238	77.	6.2	945.	53.	134	1.3464
8.0	1.0233	1.0251	82.	6.6	941.	57.	143	1.3473
8.5	1.0247	1.0265	87.	7.0	938.	61.	152	1.3482
9.0	1.0260	1.0278	92.	7.4	934.	65.	161	1.3491
9.5	1.0274	1.0292	98.	7.8	930.	68.	170	1.3500
10.0	1.0287	1.0306	103.	8.2	926.	72.	179	1.3509
11.0	1.0315	1.0333	113.	9.0	918.	80.	197	1.3527
12.0	1.0342	1.0360	124.	9.8	910.	88.	215	1.3545

Note: C_{Pr} = g protein/100 ml.

66 SILVER NITRATE, $AgNO_3$

MOLECULAR WEIGHT = 169.89
RELATIVE SPECIFIC REFRACTIVITY = 0.475

0.00 % by wt. data are the same for all compounds.
For Values of 0.00 wt. % solutions see Table 1, Acetic Acid.

A % by wt.	ρ D_4^{20}	D_{20}^{20}	C_s g/l	M g-mol/l	C_w g/l	$(C_o - C_w)$ g/l	$(n - n_o)$ × 10^4	n	Δ °C	O Os/kg	S g-mol/l	η/η_o	η/ρ cS	φ rhe	γ mmho/cm	T g-mol/l
0.50	1.0027	1.0045	5.0	0.030	997.7	0.5	6	1.3336	0.103	0.056	0.029	1.001	1.001	99.66	3.1	0.031
1.00	1.0070	1.0088	10.1	0.059	996.9	1.3	12	1.3342	0.202	0.109	0.058	1.003	0.998	99.50	6.1	0.062
2.00	1.0154	1.0172	20.3	0.120	995.1	3.1	22	1.3352	0.395	0.213	0.114	1.007	0.994	99.11	12.0	0.126
3.00	1.0239	1.0257	30.7	0.181	993.2	5.1	33	1.3363	0.586	0.315	0.170	1.011	0.989	98.74	17.2	0.187
4.00	1.0327	1.0345	41.3	0.243	991.4	6.8	44	1.3374	0.775	0.416	0.226	1.014	0.984	98.39	22.0	0.245
5.00	1.0417	1.0436	52.1	0.307	989.6	8.6	55	1.3385	0.961	0.517	0.281	1.018	0.979	98.04	26.7	0.303
6.00	1.0506	1.0524	63.0	0.371	987.5	10.7	66	1.3396	1.146	0.616	0.335	1.022	0.974	97.69	31.4	0.362
7.00	1.0597	1.0616	74.2	0.437	985.5	12.7	78	1.3407	1.329	0.715	0.389	1.025	0.969	97.34	36.1	0.421
8.00	1.0690	1.0709	85.5	0.503	983.5	14.7	89	1.3419	1.511	0.812	0.442	1.029	0.964	97.00	40.7	0.480
9.00	1.0785	1.0804	97.1	0.571	981.4	16.8	101	1.3431	1.690	0.909	0.494	1.033	0.959	96.63	45.3	0.540
10.00	1.0882	1.0901	108.8	0.641	979.3	18.9	113	1.3443	1.868	1.004	0.546	1.037	0.955	96.24	49.8	0.600
12.00	1.1079	1.1099	133.0	0.783	975.0	23.2	138	1.3467	2.213	1.190	0.646	1.047	0.947	95.35	58.8	0.723
14.00	1.1284	1.1304	158.0	0.930	970.4	27.8	163	1.3493	2.546	1.369	0.742	1.058	0.939	94.35	67.6	0.849
16.00	1.1496	1.1516	183.9	1.083	965.7	32.6	189	1.3519	2.862	1.539	0.832	1.070	0.933	93.25	76.3	0.976
18.00	1.1715	1.1736	210.9	1.241	960.6	37.6	216	1.3546				1.084	0.927	92.07	84.7	1.10
20.00	1.1942	1.1963	238.8	1.406	955.4	42.9	244	1.3574				1.099	0.922	90.81	92.8	1.23
22.00	1.2177	1.2198	267.9	1.577	949.8	48.5	273	1.3602				1.115	0.918	89.49	101.0	1.36
24.00	1.2420	1.2442	298.1	1.755	943.9	54.3	302	1.3632				1.133	0.914	88.09	108.0	1.49
26.00	1.2672	1.2694	329.5	1.939	937.7	60.5	332	1.3662				1.152	0.911	86.60	115.0	1.61
28.00	1.2933	1.2956	362.1	2.132	931.2	67.1	364	1.3694				1.174	0.910	85.01	122.0	1.74
30.00	1.3204	1.3228	396.1	2.332	924.3	73.9	397	1.3726				1.198	0.909	83.31	129.0	1.88
32.00	1.3487	1.3510	431.6	2.540	917.1	81.1	430	1.3760				1.225	0.910	81.49	136.0	2.02
34.00	1.3780	1.3805	468.5	2.758	909.5	88.7	465	1.3795				1.254	0.912	79.58	142.0	2.16
36.00	1.4087	1.4112	507.1	2.985	901.6	96.7	502	1.3832				1.287	0.915	77.56	149.0	2.31
38.00	1.4407	1.4433	547.5	3.223	893.2	105.0	541	1.3871				1.323	0.920	75.44	156.0	2.46
40.00	1.4743	1.4769	589.7	3.471	884.6	113.7	581	1.3911				1.363	0.926	73.22	162.0	2.61

67 SODIUM ACETATE, CH_3COONa

MOLECULAR WEIGHT = 82.04
RELATIVE SPECIFIC REFRACTIVITY = 0.884

0.00 % by wt. data are the same for all compounds.
For Values of 0.00 wt. % solutions see Table 1, Acetic Acid.

A % by wt.	ρ D_4^{20}	D_{20}^{20}	C_s g/l	M g-mol/l	C_w g/l	$(C_o - C_w)$ g/l	$(n - n_o)$ × 10^4	n	Δ °C	O Os/kg	S g-mol/l	η/η_o	η/ρ cS	φ rhe	γ mmho/cm	T g-mol/l
0.50	1.0008	1.0026	5.0	0.061	995.8	2.4	7	1.3337	0.217	0.116	0.062	1.019	1.020	97.96	3.9	0.039
1.00	1.0034	1.0052	10.0	0.122	993.4	4.9	14	1.3344	0.433	0.233	0.125	1.038	1.037	96.15	7.6	0.078
2.00	1.0085	1.0103	20.2	0.246	988.3	9.9	28	1.3358	0.877	0.471	0.256	1.078	1.071	92.58	14.4	0.154
3.00	1.0135	1.0153	30.4	0.371	983.1	15.2	42	1.3372	1.339	0.720	0.391	1.122	1.109	88.98	20.4	0.225
4.00	1.0184	1.0202	40.7	0.497	977.7	20.6	56	1.3386	1.820	0.978	0.532	1.169	1.150	85.38	25.7	0.291
5.00	1.0234	1.0252	51.2	0.624	972.2	26.0	70	1.3400	2.323	1.249	0.678	1.220	1.195	81.80	30.9	0.355
6.00	1.0283	1.0302	61.7	0.752	966.6	31.6	85	1.3414	2.848	1.531	0.828	1.275	1.242	78.27	35.9	0.419
7.00	1.0334	1.0352	72.3	0.882	961.1	37.2	99	1.3428	3.398	1.827	0.984	1.334	1.294	74.79	40.8	0.482
8.00	1.0386	1.0404	83.1	1.013	955.5	42.7	112	1.3442	3.975	2.137	1.144	1.398	1.348	71.41	45.4	0.542
9.00	1.0440	1.0458	94.0	1.145	950.0	48.2	126	1.3456	4.566	2.455	1.307	1.465	1.406	68.13	49.6	0.598
10.00	1.0495	1.0514	105.0	1.279	944.6	53.6	140	1.3470				1.536	1.466	64.97	53.4	0.648
12.00	1.0607	1.0625	127.3	1.551	933.4	64.9	168	1.3498				1.685	1.592	59.22	59.0	0.725
14.00	1.0718	1.0737	150.1	1.829	921.8	76.5	196	1.3526				1.851	1.731	53.90	62.7	0.777
16.00	1.0830	1.0849	173.3	2.112	909.7	88.5	224	1.3554				2.050	1.897	48.68	65.5	0.818
18.00	1.0940	1.0960	196.9	2.400	897.1	101.1	253	1.3583				2.279	2.087	43.79	67.8	0.852
20.00	1.1050	1.1070	221.0	2.694	884.0	114.2	281	1.3611				2.562	2.323	38.95	69.3	0.873
22.00	1.1159	1.1179	245.5	2.993	870.4	127.8	309	1.3639				2.942	2.642	33.92	69.9	0.882
24.00	1.1268	1.1288	270.4	3.296	856.4	141.9	336	1.3666				3.393	3.017	29.41	69.7	0.878
26.00	1.1377	1.1397	295.8	3.606	841.9	156.3	363	1.3693				3.869	3.408	25.79	68.8	0.865
28.00	1.1488	1.1508	321.7	3.921	827.1	171.1	390	1.3720				4.379	3.819	22.79	67.0	0.840
30.00	1.1602	1.1623	348.1	4.243	812.2	186.1	418	1.3748				4.930	4.258	20.24	64.3	0.800

68 SODIUM BICARBONATE, $NaHCO_3$

MOLECULAR WEIGHT = 84.01
RELATIVE SPECIFIC REFRACTIVITY = 0.687

0.00 % by wt. data are the same for all compounds.
For Values of 0.00 wt. % solutions see Table 1, Acetic Acid.

A % by wt.	ρ D_4^{20}	D_{20}^{20}	C_s g/l	M g-mol/l	C_w g/l	$(C_o - C_w)$ g/l	$(n - n_o)$ × 10^4	n	Δ °C	O Os/kg	S g-mol/l	η/η_o	η/ρ cS	φ rhe	γ mmho/cm	T g-mol/l
0.50	1.0018	1.0036	5.0	0.060	996.8	1.4	7	1.3337	0.199	0.107	0.057	1.013	1.013	98.52	4.2	0.041
1.00	1.0054	1.0072	10.1	0.120	995.3	2.9	14	1.3344	0.396	0.213	0.114	1.026	1.023	97.23	8.2	0.083
1.50	1.0089	1.0107	15.1	0.180	993.8	4.4	21	1.3351	0.591	0.318	0.172	1.040	1.033	95.94	11.7	0.123
2.00	1.0125	1.0143	20.2	0.241	992.2	6.0	27	1.3357	0.784	0.422	0.228	1.055	1.044	94.64	15.0	0.161
2.50	1.0160	1.0178	25.4	0.302	990.6	7.6	34	1.3364	0.975	0.524	0.284	1.069	1.055	93.33	18.0	0.197
3.00	1.0196	1.0214	30.6	0.364	989.0	9.3	41	1.3370	1.164	0.626	0.340	1.084	1.066	92.03	20.8	0.230
3.50	1.0231	1.0249	35.8	0.426	987.3	11.0	47	1.3377	1.350	0.726	0.395	1.100	1.077	90.73	23.5	0.263
4.00	1.0266	1.0284	41.1	0.489	985.5	12.7	54	1.3383	1.535	0.825	0.449	1.116	1.089	89.44	26.1	0.295
4.50	1.0301	1.0320	46.4	0.552	983.8	14.5	60	1.3390	1.718	0.924	0.502	1.132	1.101	88.16	28.7	0.327
5.00	1.0337	1.0355	51.7	0.615	982.0	16.2	67	1.3396	1.899	1.021	0.555	1.149	1.114	86.88	31.4	0.361
5.50	1.0372	1.0391	57.0	0.679	980.2	18.1	73	1.3403	2.078	1.117	0.607	1.166	1.126	85.61	34.2	0.397
6.00	1.0408	1.0426	62.4	0.743	978.4	19.9	79	1.3409	2.256	1.213	0.659	1.183	1.139	84.36	37.3	0.437

69 SODIUM BROMIDE, NaBr

MOLECULAR WEIGHT = 102.91
RELATIVE SPECIFIC REFRACTIVITY = 0.626

0.00 % by wt. data are the same for all compounds.
For Values of 0.00 wt. % solutions see Table 1, Acetic Acid.

A % by wt.	ρ D_4^{20}	D_{20}^{20}	C_s g/l	M g-mol/l	C_w g/l	$(C_o - C_w)$ g/l	$(n - n_o)$ × 10^4	n	Δ °C	O Os/kg	S g-mol/l	η/η_o	η/ρ cS	φ rhe	γ mmho/cm	T g-mol/l
0.50	1.0021	1.0039	5.0	0.049	997.1	1.1	7	1.3337	0.173	0.093	0.049	1.002	1.002	99.55	5.0	0.050
1.00	1.0060	1.0078	10.1	0.098	995.9	2.3	14	1.3344	0.344	0.185	0.099	1.005	1.001	99.30	9.7	0.100
1.50	1.0099	1.0117	15.1	0.147	994.8	3.5	21	1.3351	0.515	0.277	0.149	1.007	1.000	99.06	14.1	0.150
2.00	1.0139	1.0157	20.3	0.197	993.6	4.7	28	1.3358	0.687	0.369	0.200	1.010	0.998	98.81	18.4	0.201
2.50	1.0178	1.0196	25.4	0.247	992.4	5.9	35	1.3365	0.860	0.463	0.251	1.013	0.997	98.56	22.7	0.253
3.00	1.0218	1.0236	30.7	0.298	991.1	7.1	42	1.3372	1.036	0.557	0.303	1.015	0.995	98.31	26.9	0.305
3.50	1.0258	1.0276	35.9	0.349	989.9	8.3	49	1.3379	1.214	0.653	0.355	1.018	0.994	98.06	31.2	0.359
4.00	1.0298	1.0317	41.2	0.400	988.6	9.6	57	1.3386	1.394	0.750	0.408	1.020	0.993	97.80	35.5	0.413
4.50	1.0339	1.0357	46.5	0.452	987.4	10.9	64	1.3394	1.577	0.848	0.461	1.023	0.992	97.54	39.8	0.468
5.00	1.0380	1.0398	51.9	0.504	986.1	12.2	71	1.3401	1.761	0.947	0.515	1.026	0.990	97.27	44.0	0.523
5.50	1.0421	1.0439	57.3	0.557	984.8	13.5	78	1.3408	1.949	1.048	0.570	1.029	0.989	97.00	48.2	0.578
6.00	1.0462	1.0481	62.8	0.610	983.5	14.8	86	1.3415	2.139	1.150	0.625	1.032	0.988	96.73	52.4	0.634
6.50	1.0504	1.0522	68.3	0.663	982.1	16.1	93	1.3423	2.332	1.254	0.681	1.035	0.987	96.46	56.5	0.691
7.00	1.0546	1.0564	73.8	0.717	980.8	17.5	100	1.3430	2.528	1.359	0.737	1.038	0.986	96.18	60.6	0.748
7.50	1.0588	1.0607	79.4	0.772	979.4	18.8	108	1.3438	2.728	1.466	0.794	1.041	0.985	95.90	64.7	0.806
8.00	1.0630	1.0649	85.0	0.826	978.0	20.2	115	1.3445	2.929	1.575	0.851	1.044	0.984	95.61	68.7	0.864
8.50	1.0673	1.0692	90.7	0.882	976.6	21.6	123	1.3453	3.133	1.685	0.909	1.047	0.983	95.31	72.7	0.923
9.00	1.0716	1.0735	96.4	0.937	975.2	23.1	130	1.3460	3.342	1.797	0.968	1.051	0.982	95.00	76.7	0.983
9.50	1.0760	1.0779	102.2	0.993	973.7	24.5	138	1.3468	3.554	1.911	1.027	1.054	0.982	94.67	80.6	1.04
10.00	1.0803	1.0822	108.0	1.050	972.3	25.9	145	1.3475	3.768	2.026	1.087	1.058	0.981	94.33	84.6	1.10
11.00	1.0892	1.0911	119.8	1.164	969.3	28.9	161	1.3491	4.206	2.261	1.208	1.066	0.981	93.61	92.4	1.22
12.00	1.0981	1.1000	131.8	1.280	966.3	31.9	176	1.3506	4.665	2.508	1.334	1.075	0.981	92.85	100.	1.35
13.00	1.1072	1.1091	143.9	1.399	963.2	35.0	192	1.3522	5.150	2.770	1.463	1.084	0.981	92.05	108.	1.48
14.00	1.1164	1.1184	156.3	1.519	960.1	38.1	208	1.3538	5.650	3.040	1.597	1.094	0.982	91.21	115.	1.61
15.00	1.1257	1.1277	168.9	1.641	956.9	41.4	224	1.3554	6.180	3.320	1.734	1.105	0.984	90.32	122.	1.75
16.00	1.1352	1.1372	181.6	1.765	953.6	44.7	241	1.3570	6.740	3.620	1.875	1.117	0.986	89.38	129.	1.89
17.00	1.1448	1.1469	194.6	1.891	950.2	48.0	257	1.3587	7.320	3.930	2.021	1.129	0.988	88.40	136.	2.03
18.00	1.1546	1.1566	207.8	2.019	946.8	51.5	274	1.3604				1.142	0.991	87.37	143.	2.18
19.00	1.1645	1.1665	221.3	2.150	943.2	55.0	291	1.3621				1.157	0.995	86.29	150.	2.33
20.00	1.1745	1.1766	234.9	2.283	939.6	58.6	308	1.3638				1.172	1.000	85.15	157.	2.50
22.00	1.1951	1.1972	262.9	2.555	932.2	66.1	343	1.3673				1.205	1.010	82.81	172.	2.84

69 SODIUM BROMIDE, NaBr—(Continued)

A% by wt.	ρ D_4^{20}	D_{20}^{20}	C_s g/l	M g-mol/l	C_w g/l	$(C_o - C_w)$ g/l	$(n - n_o)$ × 10^4	n	Δ °C	O Os/kg	S g-mol/l	η/η_o	η/ρ cS	φ rhe	γ mmho/cm	T g-mol/l
24.00	1.2163	1.2184	291.9	2.837	924.4	73.9	378	1.3708				1.242	1.023	80.37	185.0	3.21
26.00	1.2382	1.2404	321.9	3.128	916.2	82.0	415	1.3745				1.284	1.039	77.75	197.0	3.60
28.00	1.2608	1.2630	353.0	3.430	907.8	90.5	453	1.3783				1.333	1.059	74.88	207.0	4.03
30.00	1.2842	1.2864	385.2	3.744	889.9	99.3	492	1.3822				1.392	1.086	71.70	216.0	4.50
32.00	1.3083	1.3106	418.7	4.068	889.7	108.6	532	1.3862				1.462	1.120	68.26		
34.00	1.3333	1.3357	453.3	4.405	880.0	118.2	573	1.3903				1.543	1.160	64.68		
36.00	1.3592	1.3616	489.3	4.755	869.9	128.3	616	1.3946				1.636	1.206	61.01		
38.00	1.3860	1.3885	526.7	5.118	859.3	138.9	660	1.3990				1.742	1.259	57.30		
40.00	1.4138	1.4163	565.5	5.495	848.3	150.0	705	1.4035				1.862	1.320	53.60		

70 SODIUM CARBONATE, $Na_2CO_3 \cdot 10H_2O$

MOLECULAR WEIGHT = 106.00 FORMULA WEIGHT, HYDRATE = 286.16
RELATIVE SPECIFIC REFRACTIVITY = 0.619

0.00 % by wt. data are the same for all compounds.
For Values of 0.00 wt. % solutions see Table 1, Acetic Acid.

A% by wt.	H% by wt.	ρ D_4^{20}	D_{20}^{20}	C_s g/l	M g-mol/l	C_w g/l	$(C_o - C_w)$ g/l	$(n - n_o)$ × 10^4	n	Δ °C	O Os/kg	S g-mol/l	η/η_o	η/ρ cS	φ rhe	γ mmho/cm	T g-mol/l
0.50	1.35	1.0034	1.0052	5.0	0.047	998.4	−0.2	11	1.3341	0.222	0.119	0.063	1.023	1.021	97.58	7.0	0.072
1.00	2.70	1.0086	1.0104	10.1	0.095	998.5	−0.3	23	1.3352	0.425	0.229	0.123	1.047	1.040	95.32	13.1	0.139
1.50	4.05	1.0138	1.0156	15.2	0.143	998.6	−0.4	34	1.3364	0.598	0.322	0.174	1.073	1.060	93.03	18.4	0.201
2.00	5.40	1.0190	1.0208	20.4	0.192	998.6	−0.4	45	1.3375	0.751	0.404	0.219	1.100	1.082	90.73	23.3	0.260
2.50	6.75	1.0242	1.0260	25.6	0.242	998.6	−0.4	56	1.3386	0.915	0.492	0.267	1.128	1.104	88.46	27.8	0.316
3.00	8.10	1.0294	1.0312	30.9	0.291	998.5	−0.3	67	1.3397	1.082	0.582	0.316	1.157	1.127	86.23	32.0	0.369
3.50	9.45	1.0346	1.0364	36.2	0.342	998.4	−0.2	78	1.3408	1.250	0.672	0.365	1.188	1.150	84.01	36.0	0.420
4.00	10.80	1.0398	1.0416	41.6	0.392	998.2	0.0	89	1.3419	1.421	0.764	0.416	1.220	1.176	81.81	39.8	0.468
4.50	12.15	1.0450	1.0468	47.0	0.444	998.0	0.3	100	1.3430	1.593	0.857	0.466	1.253	1.202	79.62	43.5	0.515
5.00	13.50	1.0502	1.0521	52.5	0.495	997.7	0.5	110	1.3440	1.768	0.951	0.517	1.289	1.230	77.42	47.0	0.562
5.50	14.85	1.0554	1.0573	58.0	0.548	997.3	0.9	121	1.3451	1.945	1.046	0.569	1.326	1.259	75.26	50.4	0.608
6.00	16.20	1.0606	1.0625	63.6	0.600	997.0	1.3	132	1.3462	2.125	1.142	0.621	1.364	1.289	73.16	53.6	0.652
6.50	17.55	1.0658	1.0677	69.3	0.654	996.5	1.7	142	1.3472				1.404	1.320	71.09	56.7	0.694
7.00	18.90	1.0711	1.0730	75.0	0.707	996.1	2.1	153	1.3483				1.445	1.352	69.05	59.7	0.735
7.50	20.25	1.0763	1.0782	80.7	0.762	995.6	2.6	164	1.3494				1.489	1.386	67.04	62.4	0.774
8.00	21.60	1.0816	1.0835	86.5	0.816	995.1	3.1	174	1.3504				1.535	1.422	65.03	65.1	0.812
8.50	22.95	1.0869	1.0888	92.4	0.872	994.5	3.7	185	1.3515				1.583	1.460	63.03	67.6	0.848
9.00	24.30	1.0922	1.0942	98.3	0.927	993.9	4.3	196	1.3525				1.635	1.500	61.03	70.0	0.883
9.50	25.65	1.0975	1.0995	104.3	0.984	993.3	5.0	206	1.3536				1.691	1.543	59.03	72.2	0.916
10.00	27.00	1.1029	1.1048	110.3	1.040	992.6	5.6	217	1.3547				1.750	1.590	57.03	74.4	0.948
11.00	29.70	1.1136	1.1156	122.5	1.156	991.1	7.1	238	1.3568				1.880	1.691	53.09	78.3	1.00
12.00	32.40	1.1244	1.1264	134.9	1.273	989.5	8.8	259	1.3589				2.024	1.803	49.32	81.7	1.05
13.00	35.10	1.1353	1.1373	147.6	1.392	987.7	10.5	280	1.3610				2.182	1.926	45.74	84.6	1.10
14.00	37.79	1.1463	1.1483	160.5	1.514	985.8	12.4	301	1.3631				2.356	2.059	42.36	86.9	1.14
15.00	40.49	1.1574	1.1595	173.6	1.638	983.8	14.4	322	1.3652				2.546	2.204	39.20	88.6	1.16

71 SODIUM CHLORIDE, NaCl

MOLECULAR WEIGHT = 58.44
RELATIVE SPECIFIC REFRACTIVITY = 0.794

0.00 % by wt. data are the same for all compounds.
For Values of 0.00 wt. % solutions see Table 1, Acetic Acid.

A% by wt.	ρ D_4^{20}	D_{20}^{20}	C_s g/l	M g-mol/l	C_w g/l	$(C_o - C_w)$ g/l	$(n - n_o)$ × 10^4	n	Δ °C	O Os/kg	S g-mol/l	η/η_o	η/ρ cS	φ rhe	γ mmho/cm	T g-mol/l
0.10	0.9989	1.0007	1.0	0.017	997.9	0.3	2	1.3332	0.062	0.033	0.017	1.002	1.005	99.61	1.7	0.017
0.20	0.9997	1.0014	2.0	0.034	997.7	0.6	4	1.3333	0.121	0.065	0.034	1.004	1.006	99.43	3.3	0.034
0.30	1.0004	1.0021	3.0	0.051	997.4	0.9	5	1.3335	0.181	0.097	0.051	1.006	1.007	99.24	5.0	0.051
0.40	1.0011	1.0028	4.0	0.069	997.1	1.2	7	1.3337	0.240	0.129	0.069	1.007	1.008	99.06	6.6	0.069
0.50	1.0018	1.0036	5.0	0.086	996.8	1.5	9	1.3339	0.299	0.161	0.086	1.009	1.009	98.89	8.2	0.086
0.60	1.0025	1.0043	6.0	0.103	996.5	1.8	11	1.3340	0.358	0.192	0.103	1.011	1.011	98.71	9.8	0.103
0.70	1.0032	1.0050	7.0	0.120	996.2	2.1	12	1.3342	0.417	0.224	0.120	1.013	1.012	98.54	11.4	0.120
0.80	1.0039	1.0057	8.0	0.137	995.9	2.4	14	1.3344	0.475	0.256	0.137	1.015	1.013	98.37	12.9	0.137
0.90	1.0046	1.0064	9.0	0.155	995.6	2.7	16	1.3346	0.534	0.287	0.155	1.016	1.014	98.20	14.4	0.155
1.00	1.0053	1.0071	10.1	0.172	995.3	3.0	18	1.3347	0.593	0.319	0.172	1.018	1.015	98.04	16.0	0.172
1.10	1.0060	1.0078	11.1	0.189	995.0	3.3	19	1.3349	0.652	0.351	0.189	1.020	1.016	97.87	17.5	0.189
1.20	1.0068	1.0085	12.1	0.207	994.7	3.6	21	1.3351	0.711	0.382	0.207	1.021	1.017	97.72	18.9	0.207
1.30	1.0075	1.0093	13.1	0.224	994.4	3.9	23	1.3353	0.770	0.414	0.224	1.023	1.017	97.56	20.4	0.224
1.40	1.0082	1.0100	14.1	0.241	994.1	4.2	25	1.3354	0.829	0.446	0.241	1.025	1.018	97.41	21.8	0.241
1.50	1.0089	1.0107	15.1	0.259	993.8	4.5	26	1.3356	0.888	0.478	0.259	1.026	1.019	97.26	23.2	0.259
1.60	1.0096	1.0114	16.2	0.276	993.5	4.8	28	1.3358	0.948	0.509	0.276	1.028	1.020	97.11	24.6	0.276
1.70	1.0103	1.0121	17.2	0.294	993.1	5.1	30	1.3360	1.007	0.541	0.294	1.029	1.021	96.96	26.0	0.294
1.80	1.0110	1.0128	18.2	0.311	992.8	5.4	32	1.3362	1.067	0.573	0.311	1.031	1.022	96.82	27.4	0.311
1.90	1.0117	1.0135	19.2	0.329	992.5	5.7	33	1.3363	1.126	0.605	0.329	1.032	1.022	96.67	28.8	0.329
2.00	1.0125	1.0143	20.2	0.346	992.2	6.0	35	1.3365	1.186	0.638	0.346	1.034	1.023	96.52	30.2	0.346
2.10	1.0132	1.0150	21.3	0.364	991.9	6.3	37	1.3367	1.246	0.670	0.364	1.036	1.024	96.37	31.6	0.364
2.20	1.0139	1.0157	22.3	0.382	991.6	6.6	39	1.3369	1.306	0.702	0.382	1.037	1.025	96.22	33.0	0.382
2.30	1.0146	1.0164	23.3	0.399	991.3	7.0	40	1.3370	1.366	0.734	0.399	1.039	1.026	96.07	34.3	0.399
2.40	1.0153	1.0171	24.4	0.418	991.0	7.3	42	1.3372	1.426	0.767	0.418	1.040	1.027	95.93	35.7	0.418
2.50	1.0160	1.0178	25.4	0.435	990.6	7.6	44	1.3374	1.486	0.799	0.435	1.042	1.028	95.78	37.1	0.435
2.60	1.0168	1.0185	26.4	0.452	990.3	7.9	46	1.3376	1.547	0.832	0.452	1.044	1.028	95.64	38.5	0.452
2.70	1.0175	1.0193	27.5	0.470	990.0	8.2	47	1.3377	1.607	0.864	0.470	1.045	1.029	95.49	39.9	0.470
2.80	1.0182	1.0200	28.5	0.488	989.7	8.6	49	1.3379	1.668	0.897	0.488	1.047	1.030	95.35	41.2	0.488
2.90	1.0189	1.0207	29.5	0.505	989.4	8.9	51	1.3381	1.729	0.930	0.505	1.048	1.031	95.21	42.6	0.505
3.00	1.0196	1.0214	30.6	0.523	989.0	9.2	53	1.3383	1.790	0.962	0.523	1.050	1.032	95.06	44.0	0.523
3.10	1.0203	1.0221	31.6	0.541	988.7	9.5	54	1.3384	1.851	0.995	0.541	1.051	1.033	94.92	45.3	0.541
3.20	1.0211	1.0229	32.7	0.559	988.4	9.9	56	1.3386	1.913	1.028	0.559	1.053	1.033	94.78	46.7	0.559

71 SODIUM CHLORIDE, NaCl—(Continued)

A% by wt.	ρ D_4^{20}	D_{20}^{20}	C_s g/l	M g-mol/l	C_w g/l	$(C_o - C_w)$ g/l	$(n - n_o)$ × 10^4	n	Δ °C	O Os/kg	S g-mol/l	η/η_o	η/ρ cS	φ rhe	γ mmho/cm	T g-mol/l
3.30	1.0218	1.0236	33.7	0.577	988.0	10.2	58	1.3388	1.974	1.061	0.577	1.055	1.034	94.64	48.0	0.577
3.40	1.0225	1.0243	34.8	0.595	987.7	10.5	60	1.3390	2.036	1.094	0.595	1.056	1.035	94.50	49.4	0.595
3.50	1.0232	1.0250	35.8	0.613	987.4	10.8	61	1.3391	2.098	1.128	0.613	1.058	1.036	94.35	50.7	0.613
3.60	1.0239	1.0257	36.9	0.631	987.1	11.2	63	1.3393	2.160	1.161	0.631	1.059	1.037	94.21	52.0	0.631
3.70	1.0246	1.0265	37.9	0.649	986.7	11.5	65	1.3395	2.222	1.194	0.649	1.061	1.037	94.07	53.3	0.649
3.80	1.0254	1.0272	39.0	0.667	986.4	11.8	67	1.3397	2.284	1.228	0.667	1.063	1.038	93.93	54.7	0.667
3.90	1.0261	1.0279	40.0	0.685	986.1	12.2	68	1.3398	2.347	1.262	0.685	1.064	1.039	93.78	56.0	0.685
4.00	1.0268	1.0286	41.1	0.703	985.7	12.5	70	1.3400	2.409	1.295	0.703	1.066	1.040	93.64	57.3	0.703
4.10	1.0275	1.0293	42.1	0.721	985.4	12.8	72	1.3402	2.472	1.329	0.721	1.067	1.041	93.49	58.6	0.721
4.20	1.0282	1.0301	43.2	0.739	985.1	13.2	74	1.3404	2.535	1.363	0.739	1.069	1.042	93.35	59.9	0.739
4.30	1.0290	1.0308	44.2	0.757	984.7	13.5	75	1.3405	2.598	1.397	0.757	1.071	1.043	93.20	61.2	0.757
4.40	1.0297	1.0315	45.3	0.775	984.4	13.9	77	1.3407	2.662	1.431	0.775	1.072	1.044	93.06	62.5	0.775
4.50	1.0304	1.0322	46.4	0.794	984.0	14.2	79	1.3409	2.725	1.465	0.794	1.074	1.045	92.91	63.8	0.794
4.60	1.0311	1.0330	47.4	0.812	983.7	14.5	81	1.3411	2.789	1.500	0.812	1.076	1.046	92.76	65.1	0.812
4.70	1.0318	1.0337	48.5	0.830	983.4	14.9	82	1.3412	2.853	1.534	0.830	1.078	1.046	92.61	66.3	0.830
4.80	1.0326	1.0344	49.6	0.848	983.0	15.2	84	1.3414	2.917	1.569	0.848	1.079	1.047	92.46	67.6	0.848
4.90	1.0333	1.0351	50.6	0.866	982.7	15.6	86	1.3416	2.982	1.603	0.866	1.081	1.048	92.31	68.9	0.866
5.00	1.0340	1.0358	51.7	0.885	982.3	15.9	88	1.3418	3.046	1.638	0.885	1.083	1.049	92.15	70.1	0.885
5.20	1.0355	1.0373	53.8	0.921	981.6	16.6	91	1.3421	3.176	1.708	0.921	1.087	1.052	91.84	72.6	0.921
5.40	1.0369	1.0387	56.0	0.958	980.9	17.3	95	1.3425	3.307	1.778	0.958	1.090	1.054	91.53	75.1	0.958
5.60	1.0384	1.0402	58.1	0.995	980.2	18.0	98	1.3428	3.438	1.848	0.995	1.094	1.056	91.22	77.5	0.995
5.80	1.0398	1.0417	60.3	1.032	979.5	18.7	102	1.3432	3.570	1.919	1.032	1.098	1.058	90.90	80.0	1.032
6.00	1.0413	1.0431	62.5	1.069	978.8	19.4	105	1.3435	3.703	1.991	1.069	1.102	1.060	90.59	82.4	1.069
6.20	1.0427	1.0446	64.6	1.106	978.1	20.2	109	1.3439	3.837	2.063	1.106	1.106	1.062	90.27	84.7	1.106
6.40	1.0442	1.0460	66.8	1.144	977.4	20.9	112	1.3442	3.972	2.135	1.144	1.109	1.065	89.95	87.1	1.144
6.60	1.0456	1.0475	69.0	1.181	976.6	21.6	116	1.3446	4.107	2.208	1.181	1.113	1.067	89.63	89.4	1.181
6.80	1.0471	1.0490	71.2	1.218	975.9	22.3	119	1.3449	4.244	2.282	1.218	1.117	1.069	89.31	91.8	1.218
7.00	1.0486	1.0504	73.4	1.256	975.2	23.1	123	1.3453	4.378	2.354	1.256	1.122	1.072	88.99	94.1	1.256
7.20	1.0500	1.0519	75.6	1.294	974.4	23.8	126	1.3456	4.516	2.428	1.294	1.126	1.074	88.66	96.4	1.294
7.40	1.0515	1.0533	77.8	1.331	973.7	24.6	130	1.3460	4.655	2.503	1.331	1.130	1.077	88.33	98.7	1.331
7.60	1.0530	1.0548	80.0	1.369	972.9	25.3	133	1.3463	4.795	2.578	1.369	1.134	1.079	88.00	101.	1.369
7.80	1.0544	1.0563	82.2	1.407	972.2	26.1	137	1.3467	4.937	2.654	1.407	1.138	1.082	87.67	103.	1.407
8.00	1.0559	1.0578	84.5	1.445	971.4	26.8	140	1.3470	5.079	2.731	1.445	1.143	1.084	87.33	105.	1.445
8.20	1.0574	1.0592	86.7	1.484	970.7	27.6	144	1.3474	5.222	2,808	1.484	1.147	1.087	86.99	107.	1.484
8.40	1.0588	1.0607	88.9	1.522	969.9	28.3	147	1.3477	5.367	2.885	1.522	1.152	1.090	86.65	109.	1.522
8.60	1.0603	1.0622	91.2	1.560	969.1	29.1	151	1.3481	5.512	2.964	1.560	1.156	1.093	86.30	112.	1.560
8.80	1.0618	1.0637	93.4	1.599	968.3	29.9	154	1.3484	5.659	3.043	1.599	1.161	1.096	85.95	114.	1.599
9.00	1.0633	1.0651	95.7	1.637	967.6	30.7	158	1.3488	5.807	3.122	1.637	1.166	1.099	85.60	116.	1.637
9.20	1.0647	1.0666	98.0	1.676	966.8	31.4	161	1.3491	5.956	3.202	1.676	1.171	1.102	85.25	118.	1.676
9.40	1.0662	1.0681	100.2	1.715	966.0	32.2	165	1.3495	6.106	3.283	1.715	1.176	1.105	84.89	120.	1.715
9.60	1.0677	1.0696	102.5	1.754	965.2	33.0	168	1.3498	6.258	3.364	1.754	1.181	1.108	84.53	122.	1.754
9.80	1.0692	1.0711	104.8	1.793	964.4	33.8	172	1.3502	6.410	3.446	1.793	1.186	1.111	84.16	124.	1.793
10.00	1.0707	1.0726	107.1	1.832	963.6	34.6	175	1.3505	6.564	3.529	1.832	1.191	1.115	83.80	126.	1.832
10.50	1.0744	1.0763	112.8	1.930	961.6	36.6	184	1.3514	6.954	3.739	1.930	1.204	1.123	82.86	131.	1.930
11.00	1.0781	1.0801	118.6	2.029	959.5	38.7	193	1.3523	7.353	3.953	2.029	1.218	1.132	81.90	136.	2.029
11.50	1.0819	1.0838	124.4	2.129	957.5	40.8	202	1.3532	7.760	4.172	2.129	1.233	1.142	80.94	140.	2.129
12.00	1.0857	1.0876	130.3	2.229	955.4	42.9	211	1.3541	8.176	4.396	2.229	1.248	1.152	79.95	145.	2.229
12.50	1.0894	1.0914	136.2	2.330	953.3	45.0	220	1.3549	8.602	4.625	2.330	1.264	1.162	78.96	149.	2.330
13.00	1.0932	1.0952	142.1	2.432	951.1	47.1	228	1.3558	9.038	4.859	2.432	1.280	1.173	77.97	154.	2.432
13.50	1.0970	1.0990	148.1	2.534	948.9	49.3	237	1.3567	9.484	5.099	2.534	1.297	1.184	76.97	158.	2.534
14.00	1.1008	1.1028	154.1	2.637	946.7	51.5	246	1.3576	9.940	5.344	2.637	1.314	1.196	75.97	163.	2.637
14.50	1.1047	1.1066	160.2	2.741	944.5	53.7	255	1.3585	10.408	5.596	2.741	1.331	1.207	74.97	167.	2.741
15.00	1.1085	1.1105	166.3	2.845	942.2	56.0	264	1.3594	10.888	5.854	2.845	1.349	1.219	73.98	171.	2.845
16.00	1.1162	1.1182	178.6	3.056	937.6	60.6	282	1.3612	11.885	6.390	3.056	1.385	1.243	72.07	179.	3.056
17.00	1.1240	1.1260	191.1	3.270	932.9	65.3	300	1.3630	12.935	6.954	3.270	1.421	1.267	70.21	186.	3.270
18.00	1.1319	1.1339	203.7	3.486	928.1	70.1	318	1.3648	14.044	7.550	3.486	1.460	1.293	68.34	193.	3.486
19.00	1.1398	1.1418	216.6	3.706	923.2	75.0	336	1.3666	15.216	8.181	3.706	1.504	1.322	66.36	199.	3.705
20.00	1.1478	1.1498	229.6	3.928	918.2	80.0	354	1.3684	16.458	8.849	3.928	1.554	1.357	64.22	204.	3.928
21.00	1.1558	1.1579	242.7	4.153	913.1	85.1	372	1.3702	17.776	9.557	4.153	1.611	1.396	61.95	209.	4.153
22.00	1.1640	1.1660	256.1	4.382	907.9	90.4	391	1.3721	19.176	10.310	4.382	1.673	1.441	59.64	213.	4.382
23.00	1.1721	1.1742	269.6	4.613	902.6	95.7	409	1.3739	20.667	11.111	4.613	1.742	1.489	57.29	217.	4.613
24.00	1.1804	1.1825	283.3	4.848	897.1	101.1	428	1.3757			4.848	1.817	1.542	54.93	220.	4.848
25.00	1.1887	1.1909	297.2	5.085	891.6	106.7	446	1.3776			5.085	1.898	1.600	52.58	222.	5.085
26.00	1.1972	1.1993	311.3	5.326	885.9	112.3	465	1.3795			5.326	1.986	1.662	50.26	225.	5.326

72 SODIUM CITRATE, $Na_3C_6H_5O_7 \cdot 2H_2O$

MOLECULAR WEIGHT = 258.068 FORMULA WEIGHT, HYDRATE = 294.10

A% by wt.	H% by wt.	ρ D_4^{20}	D_{20}^{20}	C_S g/l	M g-mol/l	C_W g/l	$(C_0 - C_W)$ g/l	$(n - n_0)$ × 10^4	n	Δ °C	O Os/kg	S g-mol/l	η/η_0	η/ρ cS	φ rhe	γ mmho/cm	T g-mol/l
1.00	1.14	1.0049	1.0067	10.0	0.039	994.9	3.3	18.	1.3348	0.197	0.106	0.056	1.041	1.038	95.87	7.4	0.075
2.00	2.28	1.0120	1.0138	20.2	0.078	991.8	6.4	36.	1.3366	0.389	0.209	0.112	1.079	1.068	92.52	12.8	0.136
3.00	3.42	1.0186	1.0204	30.6	0.118	988.0	10.2	53.	1.3383	0.586	0.315	0.170	1.120	1.102	89.09	17.9	0.195
4.00	4.56	1.0260	1.0278	41.0	0.159	985.0	13.2	71.	1.3401	0.787	0.423	0.229	1.164	1.137	85.72	22.1	0.246
5.00	5.70	1.0331	1.0350	51.7	0.200	981.4	16.8	89.	1.3419	0.974	0.523	0.284	1.208	1.172	82.59	26.2	0.296
6.00	6.84	1.0405	1.0424	62.4	0.242	978.1	20.1	107.	1.3437	1.171	0.630	0.342	1.260	1.213	79.23	29.9	0.343
7.00	7.98	1.0482	1.0501	73.4	0.284	974.8	23.4	125.	1.3455	1.361	0.732	0.398	1.311	1.253	76.14	33.2	0.384
8.00	9.12	1.0557	1.0576	84.5	0.327	971.2	27.0	143.	1.3473	1.566	0.842	0.458	1.368	1.298	73.00	36.5	0.420
9.00	10.26	1.0632	1.0651	95.7	0.371	967.5	30.7	161.	1.3491	1.768	0.951	0.517	1.424	1.342	70.09	39.4	0.463
10.00	11.40	1.0708	1.0727	107.1	0.415	963.7	34.5	179.	1.3509	1.957	1.052	0.572	1.496	1.400	66.71	42.1	0.497
12.00	13.68	1.0861	1.0881	130.3	0.505	955.8	42.4	216.	1.3546	2.376	1.277	0.693	1.646	1.519	60.61	48.9	0.588
14.00	15.95	1.1019	1.1039	154.3	0.598	947.6	50.6	253.	1.3583	2.819	1.515	0.820	1.828	1.662	54.60	50.4	0.608
16.00	18.23	1.1173	1.1193	178.8	0.693	938.5	59.7	288.	1.3618	3.274	1.760	0.949	2.041	1.830	48.91	53.5	0.650
18.00	20.51	1.1327	1.1347	203.9	0.790	928.8	69.4	326.	1.3656	3.815	2.051	1.100	2.285	2.021	43.68	55.7	0.680
20.00	22.79	1.1492	1.1513	229.8	0.891	919.4	78.8	363.	1.3693	4.393	2.362	1.260	2.591	2.259	38.52	57.1	0.700
24.00	27.35	1.1813	1.1834	283.5	1.099	897.8	100.4	437.	1.3767				3.402	2.886	29.33	57.7	0.708
28.00	31.91	1.2151	1.2173	340.2	1.318	874.9	123.3	515.	1.3845				4.577	3.774	21.81	55.9	0.683
32.00	36.47	1.2487	1.2510	399.6	1.548	849.1	149.1	593.	1.3923				6.528	5.238	15.29	51.0	0.616
36.00	41.03	1.2843	1.2866	462.3	1.792	822.0	176.2	671.	1.4001				9.768	7.621	10.22	44.5	0.529

73 SODIUM DIATRIZOATE, $(CH_3CONH)_2C_6I_3COONa$, HYPAQUE®

MOLECULAR WEIGHT = 635.92
RELATIVE SPECIFIC REFRACTIVITY = 0.823

0.00 % by wt. data are the same for all compounds.
For Values of 0.00 wt. % solutions see Table 1, Acetic Acid.

A% by wt.	ρ D_4^{20}	D_{20}^{20}	C_s g/l	M g-mol/l	C_w g/l	$(C_o - C_w)$ g/l	$(n - n_o)$ × 10^4	n	Δ °C	O Os/kg	S g-mol/l	η/η_o	η/ρ cS	φ rhe	γ mmho/cm	T g-mol/l
1.00	1.0041	1.0059	10.0	0.016	994.1	4.1	16	1.3346	0.058	0.031	0.016	1.021	1.019	97.75	0.8	0.008
2.00	1.0102	1.0120	20.2	0.032	990.0	8.2	31	1.3361	0.110	0.059	0.031	1.041	1.033	95.87	1.6	0.016
3.00	1.0164	1.0182	30.5	0.048	985.9	12.3	48	1.3377	0.164	0.088	0.047	1.061	1.046	94.05	2.4	0.024
4.00	1.0227	1.0245	40.9	0.064	981.8	16.4	64	1.3394	0.220	0.118	0.063	1.082	1.060	92.26	3.1	0.031
5.00	1.0292	1.0310	51.5	0.081	977.7	20.5	80	1.3410	0.276	0.148	0.079	1.104	1.075	90.40	3.8	0.038
6.00	1.0357	1.0376	62.1	0.098	973.6	24.6	97	1.3427	0.332	0.178	0.095	1.128	1.092	88.44	4.5	0.045
7.00	1.0424	1.0443	73.0	0.115	969.5	28.8	114	1.3444	0.389	0.209	0.112	1.155	1.110	86.42	5.2	0.052
8.00	1.0493	1.0511	83.9	0.132	965.3	32.9	131	1.3461	0.446	0.240	0.129	1.183	1.130	84.38	5.9	0.059
9.00	1.0562	1.0581	95.1	0.150	961.1	37.1	148	1.3478	0.504	0.271	0.146	1.212	1.150	82.32	6.5	0.066
10.00	1.0632	1.0651	106.3	0.167	956.9	41.3	165	1.3495	0.562	0.302	0.163	1.243	1.171	80.29	7.2	0.073
12.00	1.0776	1.0796	129.3	0.203	948.3	49.9	201	1.3530	0.683	0.367	0.199	1.305	1.213	76.50	8.4	0.086
14.00	1.0925	1.0944	152.9	0.240	939.5	58.7	237	1.3567	0.808	0.434	0.235	1.375	1.261	72.60	9.6	0.099
16.00	1.1077	1.1097	177.2	0.279	930.5	67.7	274	1.3603	0.936	0.503	0.273	1.470	1.329	67.91	10.6	0.110
18.00	1.1234	1.1254	202.2	0.318	921.2	77.0	313	1.3643	1.067	0.574	0.312	1.592	1.420	62.70	11.5	0.120
20.00	1.1395	1.1415	227.9	0.358	911.6	86.6	353	1.3683	1.205	0.648	0.352	1.724	1.516	57.89	12.2	0.129
22.00	1.1560	1.1581	254.3	0.400	901.7	96.5	394	1.3724	1.351	0.726	0.395	1.847	1.601	54.03	13.0	0.138
24.00	1.1730	1.1751	281.5	0.443	891.5	106.7	436	1.3766	1.506	0.810	0.441	1.965	1.678	50.79	13.7	0.146
26.00	1.1906	1.1927	309.5	0.487	881.0	117.2	480	1.3810	1.671	0.898	0.489	2.094	1.762	47.66	14.4	0.153
28.00	1.2086	1.2108	338.4	0.532	870.2	128.0	525	1.3854	1.846	0.993	0.540	2.254	1.869	44.27	15.0	0.160
30.00	1.2273	1.2295	368.2	0.579	859.1	139.1	571	1.3901	2.033	1.093	0.594	2.469	2.016	40.42	15.5	0.167
32.00	1.2467	1.2489	399.0	0.627	847.8	150.5	619	1.3949	2.232	1.200	0.652	2.742	2.204	36.39	16.2	0.174
34.00	1.2670	1.2692	430.8	0.677	836.2	162.0	669	1.3999	2.444	1.314	0.713	3.066	2.425	32.55	16.8	0.182
36.00	1.2882	1.2904	463.7	0.729	824.4	173.8	722	1.4052	2.672	1.437	0.778	3.449	2.683	28.93	17.2	0.187
38.00	1.3105	1.3128	498.0	0.783	812.5	185.8	778	1.4108	2.917	1.568	0.848	3.902	2.984	25.58	17.2	0.187
40.00	1.3286	1.3310	531.5	0.836	797.2	201.0	834	1.4164				4.400	3.318	22.68	16.6	0.180

®Registered trademark, sodium 3,5-diacetamido-2,4,6 tri-iodo benzoate.

74 SODIUM DICHROMATE, $Na_2Cr_2O_7 \cdot 2H_2O$

MOLECULAR WEIGHT = 261.97 FORMULA WEIGHT, HYDRATE = 298.00
RELATIVE SPECIFIC REFRACTIVITY = 0.859

0.00 % by wt. data are the same for all compounds.
For Values of 0.00 wt. % solutions see Table 1, Acetic Acid.

A% by wt.	H% by wt.	ρ D_4^{20}	D_{20}^{20}	C_s g/l	M g-mol/l	C_w g/l	$(C_o - C_w)$ g/l	$(n - n_o)$ × 10^4	C_s n	Δ °C	O Os/kg	S g-mol/l	η/η_o	η/ρ cS	φ rhe	γ mmho/cm	T g-mol/l
0.50	0.57	1.0019	1.0037	5.0	0.019	996.9	1.4	10	1.3340	0.103	0.056	0.029	1.005	1.005	99.33	3.5	0.035
1.00	1.14	1.0056	1.0073	10.1	0.038	995.5	2.7	21	1.3351	0.204	0.110	0.058	1.009	1.005	98.91	6.9	0.070
2.00	2.28	1.0130	1.0148	20.3	0.077	992.7	5.5	42	1.3371	0.396	0.213	0.114	1.016	1.005	98.23	13.1	0.139
3.00	3.41	1.0204	1.0222	30.6	0.117	989.8	8.4	63	1.3392	0.581	0.312	0.169	1.024	1.005	97.47	19.1	0.209
4.00	4.55	1.0280	1.0298	41.1	0.157	986.9	11.4	84	1.3414	0.762	0.410	0.222	1.033	1.007	96.64	24.9	0.280
5.00	5.69	1.0356	1.0374	51.8	0.198	983.8	14.4	105	1.3435	0.946	0.509	0.276	1.042	1.008	95.78	30.5	0.350
6.00	6.83	1.0433	1.0452	62.6	0.239	980.7	17.5	127	1.3457	1.135	0.610	0.331	1.052	1.010	94.91	36.0	0.419
7.00	7.96	1.0511	1.0530	73.6	0.281	977.5	20.7	149	1.3479	1.325	0.712	0.387	1.061	1.012	94.02	41.2	0.486
8.00	9.10	1.0590	1.0609	84.7	0.323	974.3	23.9	171	1.3501	1.519	0.817	0.444	1.072	1.014	93.10	46.3	0.553
9.00	10.24	1.0670	1.0689	96.0	0.367	971.0	27.2	193	1.3523	1.720	0.924	0.503	1.083	1.017	92.15	51.5	0.622
10.00	11.38	1.0751	1.0770	107.5	0.410	967.6	30.6	216	1.3546	1.929	1.037	0.564	1.095	1.021	91.14	56.7	0.694
12.00	13.65	1.0916	1.0936	131.0	0.500	960.6	37.6	262	1.3592	2.388	1.284	0.697	1.121	1.029	89.01	67.8	0.851
14.00	15.93	1.1086	1.1105	155.2	0.592	953.4	44.8	310	1.3640	2.876	1.546	0.836	1.151	1.040	86.72	78.9	1.02
16.00	18.20	1.1260	1.1280	180.2	0.688	945.9	52.4	358	1.3688	3.356	1.804	0.972	1.184	1.054	84.28	89.2	1.18
18.00	20.48	1.1439	1.1460	205.9	0.786	938.0	60.2	408	1.3738	3.830	2.059	1.104	1.221	1.070	81.71	98.6	1.33
20.00	22.75	1.1624	1.1644	232.5	0.887	929.9	68.3	460	1.3790	4.301	2.312	1.234	1.264	1.090	78.96	107.	1.48
24.00	27.30	1.2009	1.2031	288.2	1.100	912.7	85.5	567	1.3897				1.364	1.138	73.18	121.	1.73
28.00	31.85	1.2418	1.2440	347.7	1.327	894.1	104.1	681	1.4011				1.494	1.206	66.79	131.	1.93
32.00	36.40	1.2851	1.2874	411.2	1.570	873.9	124.4	801	1.4131				1.677	1.308	59.51	140.	2.10
36.00	40.95	1.3302	1.3326	478.9	1.828	851.4	146.9	925	1.4255				1.914	1.442	52.13	145.	2.22
40.00	45.50	1.3786	1.3810	551.4	2.105	827.1	171.1	1058	1.4388				2.246	1.632	44.43	148.	2.29
44.00	50.05	1.4304	1.4330	629.4	2.403	801.0	197.2	1200	1.4530				2.646	1.853	37.72	147.	2.26
48.00	54.60	1.4860	1.4886	713.3	2.723	772.7	225.5	1353	1.4683				3.225	2.174	30.95	141.	2.12
52.00	59.15	1.5454	1.5482	803.6	3.068	741.8	256.4						4.297	2.786	23.23	131.	1.94
56.00	63.70	1.6088	1.6117	900.9	3.439	707.9	290.3						5.933	3.695	16.82	117.	1.67
60.00	68.25	1.6763	1.6793	1005.8	3.839	670.5	327.7						8.221	4.914	12.14	93.9	1.25

75 SODIUM FERROCYANIDE, $Na_4Fe(CN)_6 \cdot 10H_2O$

MOLECULAR WEIGHT = 303.91 FORMULA WEIGHT, HYDRATE = 484.07
RELATIVE SPECIFIC REFRACTIVITY = 0.991

0.00 % by wt. data are the same for all compounds.
For Values of 0.00 wt. % solutions see Table 1, Acetic Acid.

A% by wt.	H% by wt.	ρ D_4^{20}	D_{20}^{20}	C_s g/l	M g-mol/l	C_w g/l	$(C_0 - C_w)$ g/l	$(n - n_o)$ × 10^4	n	Δ °C	O Os/kg	S g-mol/l	η/η_o	η/ρ cS	φ rhe	γ mmho/cm	T g-mol/l
0.50	0.80	1.0018	1.0036	5.0	0.016	996.8	1.4	13	1.3342	0.091	0.049	0.026	1.016	1.016	98.23	5.0	0.050
1.00	1.59	1.0055	1.0073	10.1	0.033	995.4	2.8	25	1.3355	0.192	0.103	0.055	1.032	1.028	96.73	9.8	0.102
1.50	2.39	1.0092	1.0110	15.1	0.050	994.0	4.2	38	1.3368	0.302	0.162	0.087	1.047	1.040	95.28	14.3	0.153
2.00	3.19	1.0129	1.0147	20.3	0.067	992.6	5.6	51	1.3381	0.394	0.212	0.114	1.063	1.052	93.87	18.7	0.204
2.50	3.98	1.0166	1.0184	25.4	0.084	991.2	7.0	64	1.3394	0.486	0.261	0.141	1.079	1.063	92.50	22.8	0.254
3.00	4.78	1.0204	1.0222	30.6	0.101	989.8	8.5	77	1.3407	0.576	0.310	0.167	1.095	1.075	91.16	26.7	0.303
3.50	5.57	1.0239	1.0258	35.8	0.118	988.1	10.1	89	1.3419	0.666	0.358	0.194	1.111	1.087	89.83	30.5	0.350
4.00	6.37	1.0276	1.0294	41.1	0.135	986.5	11.7	102	1.3432	0.755	0.406	0.220	1.127	1.099	88.52	34.1	0.396
4.50	7.17	1.0313	1.0331	46.4	0.153	984.9	13.3	115	1.3445	0.842	0.453	0.245	1.144	1.112	87.22	37.7	0.441
5.00	7.96	1.0350	1.0368	51.8	0.170	983.3	15.0	128	1.3457	0.929	0.499	0.271	1.161	1.124	85.93	41.1	0.485
5.50	8.76	1.0387	1.0406	57.1	0.188	981.6	16.6	140	1.3470	1.014	0.545	0.296	1.179	1.138	84.63	44.5	0.529
6.00	9.56	1.0425	1.0443	62.5	0.206	979.9	18.3	153	1.3483	1.098	0.590	0.321	1.198	1.151	83.33	47.7	0.572
6.50	10.35	1.0462	1.0481	68.0	0.224	978.2	20.0	166	1.3496	1.181	0.635	0.345	1.217	1.165	82.02	51.0	0.616
7.00	11.15	1.0500	1.0519	73.5	0.242	976.5	21.7	179	1.3509	1.262	0.679	0.369	1.237	1.180	80.70	54.2	0.659
7.50	11.95	1.0539	1.0557	79.0	0.260	974.8	23.4	192	1.3522	1.342	0.722	0.393	1.257	1.195	79.38	57.3	0.702
8.00	12.74	1.0577	1.0596	84.6	0.278	973.1	25.1	205	1.3535	1.421	0.764	0.416	1.279	1.211	78.04	60.5	0.746
8.50	13.54	1.0616	1.0634	90.2	0.297	971.3	26.9	218	1.3548	1.498	0.806	0.438	1.301	1.228	76.70	63.6	0.790
9.00	14.34	1.0655	1.0673	95.9	0.316	969.6	28.7	232	1.3562	1.574	0.846	0.460	1.325	1.246	75.34	66.7	0.835
9.50	15.13	1.0694	1.0713	101.6	0.334	967.8	30.4	245	1.3575	1.648	0.886	0.482	1.349	1.264	73.98	69.7	0.879
10.00	15.93	1.0733	1.0752	107.3	0.353	966.0	32.2	258	1.3588	1.720	0.925	0.503	1.375	1.283	72.61	72.8	0.925
11.00	17.52	1.0813	1.0832	118.9	0.391	962.4	35.9	285	1.3615				1.429	1.324	69.85	78.8	1.02
12.00	19.11	1.0894	1.0913	130.7	0.430	958.6	39.6	312	1.3642				1.488	1.368	67.09	84.7	1.10
13.00	20.71	1.0976	1.0995	142.7	0.469	954.9	43.4	340	1.3670				1.551	1.416	64.34	90.0	1.19
14.00	22.30	1.1058	1.1078	154.8	0.509	951.0	47.2	367	1.3697				1.619	1.467	61.63	94.7	1.27
15.00	23.89	1.1142	1.1162	167.1	0.550	947.1	51.1	395	1.3725				1.692	1.522	58.98	98.1	1.32

76 SODIUM HYDROXIDE, NaOH

MOLECULAR WEIGHT = 40.01
RELATIVE SPECIFIC REFRACTIVITY = 0.692

0.00 % by wt. data are the same for all compounds.
For Values of 0.00 wt. % solutions see Table 1, Acetic Acid.

A % by wt.	ρ D_4^{20}	D_{20}^{20}	C_s g/l	M g-mol/l	C_w g/l	$(C_o - C_w)$ g/l	$(n - n_o)$ × 10^4	n	Δ °C	O Os/kg	S g-mol/l	η/η_o	η/ρ cS	φ rhe	γ mmho/cm	T g-mol/l
0.50	1.0039	1.0057	5.0	0.125	998.9	− 0.6	14	1.3344	0.429	0.231	0.124	1.025	1.023	97.34	24.8	0.283
1.00	1.0095	1.0113	10.1	0.252	999.4	− 1.2	29	1.3358	0.860	0.462	0.251	1.052	1.044	94.87	48.6	0.584
1.50	1.0151	1.0169	15.2	0.381	999.9	− 1.7	43	1.3373	1.294	0.695	0.378	1.080	1.066	92.39	71.3	0.900
2.00	1.0207	1.0225	20.4	0.510	1000.3	− 2.0	56	1.3386	1.735	0.933	0.507	1.110	1.090	89.91	93.1	1.24
2.50	1.0262	1.0281	25.7	0.641	1000.6	− 2.4	70	1.3400	2.184	1.174	0.638	1.141	1.114	87.44	114.	1.61
3.00	1.0318	1.0336	31.0	0.774	1000.8	− 2.6	84	1.3414	2.642	1.420	0.770	1.174	1.140	84.98	134.	2.02
3.50	1.0373	1.0391	36.3	0.907	1001.0	− 2.8	97	1.3427	3.109	1.671	0.902	1.209	1.168	82.54	153.	2.45
4.00	1.0428	1.0446	41.7	1.043	1001.1	− 2.9	111	1.3441	3.587	1.929	1.037	1.246	1.197	80.11	171.	2.92
4.50	1.0483	1.0502	47.2	1.179	1001.1	− 2.9	124	1.3454	4.074	2.100	1.172	1.285	1.228	77.69	189.	3.42
5.00	1.0538	1.0557	52.7	1.317	1001.1	− 2.9	138	1.3467	4.569	2.457	1.308	1.326	1.261	75.26	206.	3.96
5.50	1.0593	1.0612	58.3	1.456	1001.0	− 2.8	151	1.3481	5.08	2.73	1.445	1.369	1.295	72.90	222.	4.54
6.00	1.0648	1.0667	63.9	1.597	1000.9	− 2.7	164	1.3494	5.60	3.01	1.582	1.413	1.330	70.62		
6.50	1.0703	1.0722	69.6	1.739	1000.7	− 2.5	177	1.3507	6.13	3.30	1.722	1.459	1.366	68.40		
7.00	1.0758	1.0777	75.3	1.882	1000.5	− 2.3	190	1.3520	6.69	3.60	1.864	1.507	1.404	66.21		
7.50	1.0813	1.0833	81.1	2.027	1000.2	− 2.0	203	1.3533	7.27	3.91	2.008	1.558	1.444	64.05		
8.00	1.0869	1.0888	86.9	2.173	999.9	− 1.7	216	1.3546	7.87	4.23	2.155	1.613	1.487	61.89		
8.50	1.0924	1.0943	92.9	2.321	999.5	− 1.3	229	1.3559	8.48	4.56	2.303	1.671	1.533	59.72		
9.00	1.0979	1.0998	98.8	2.470	999.1	− 0.9	242	1.3572	9.12	4.91	2.452	1.734	1.583	57.54		
9.50	1.1034	1.1054	104.8	2.620	998.6	− 0.3	255	1.3585	9.78	5.26	2.602	1.803	1.637	55.35		
10.00	1.1089	1.1109	110.9	2.772	998.0	0.2	268	1.3597	10.47	5.63	2.753	1.878	1.697	53.14		
11.00	1.1199	1.1219	123.2	3.079	996.7	1.5	293	1.3623	11.89	6.39	3.058	2.035	1.821	49.04		
12.00	1.1309	1.1329	135.7	3.392	995.2	3.0	318	1.3648	13.42	7.21	3.365	2.197	1.947	45.42		
13.00	1.1419	1.1440	148.5	3.710	993.5	4.8	343	1.3673	15.04	8.08	3.673	2.371	2.080	42.09		
14.00	1.1530	1.1550	161.4	4.034	991.6	6.7	367	1.3697	16.76	9.01	3.981	2.563	2.228	38.93		
15.00	1.1640	1.1661	174.6	4.364	989.4	8.8	392	1.3722				2.783	2.395	35.87		
16.00	1.1751	1.1771	188.0	4.699	987.0	11.2	416	1.3746				3.037	2.590	32.86		
17.00	1.1861	1.1882	201.6	5.040	984.5	13.8	440	1.3770				3.337	2.819	29.91		
18.00	1.1971	1.1993	215.5	5.386	981.7	16.6	463	1.3793				3.691	3.089	27.04		
19.00	1.2082	1.2103	229.6	5.737	978.6	19.6	487	1.3817				4.111	3.410	24.27		
20.00	1.2192	1.2214	243.8	6.094	975.4	22.9	510	1.3840				4.610	3.789	21.65		
22.00	1.2412	1.2434	273.1	6.825	968.1	30.1	555	1.3885				5.753	4.644	17.35		
24.00	1.2631	1.2653	303.1	7.576	959.9	38.3	599	1.3929				7.086	5.621	14.08		
26.00	1.2848	1.2871	334.0	8.349	950.8	47.5	641	1.3971				8.727	6.806	11.44		
28.00	1.3064	1.3087	365.8	9.142	940.6	57.6	682	1.4012				10.81	8.29	9.23		
30.00	1.3277	1.3301	398.3	9.956	929.4	68.8	721	1.4051				13.49	10.18	7.40		
32.00	1.3488	1.3512	431.6	10.788	917.2	81.0	758	1.4088				16.81	12.49	5.94		
34.00	1.3697	1.3721	465.7	11.639	904.0	94.3	793	1.4123				20.71	15.15	4.82		
36.00	1.3901	1.3926	500.5	12.508	889.7	108.5	826	1.4156				25.24	18.19	3.95		
38.00	1.4102	1.4127	535.9	13.394	874.3	123.9	856	1.4186				30.40	21.60	3.28		
40.00	1.4299	1.4324	571.9	14.295	857.9	140.3	885	1.4215				36.24	25.40	2.75		

77 SODIUM MOLYBDATE, $Na_2MoO_4 \cdot 2H_2O$

MOLECULAR WEIGHT = 205.94 FORMULA WEIGHT, HYDRATE = 241.98
RELATIVE SPECIFIC REFRACTIVITY = 0.642

0.00 % by wt. data are the same for all compounds.
For Values of 0.00 wt. % solutions see Table 1, Acetic Acid.

A % by wt.	H % by wt.	ρ D_4^{20}	D_{20}^{20}	C_s g/l	M g-mol/l	C_w g/l	$(C_o - C_w)$ g/l	$(n - n_o)$ × 10^4	n	Δ °C	O Os/kg	S g-mol/l	η/η_o	η/ρ cS	φ rhe	γ mmho/cm	T g-mol/l
0.50	0.59	1.0024	1.0042	5.0	0.024	997.4	0.8	8	1.3338	0.096	0.052	0.027	1.013	1.012	98.56	4.1	0.041
1.00	1.18	1.0067	1.0084	10.1	0.049	996.6	1.6	17	1.3347	0.203	0.109	0.058	1.025	1.020	97.37	7.8	0.080
1.50	1.76	1.0110	1.0128	15.2	0.074	995.8	2.4	25	1.3355	0.322	0.173	0.092	1.037	1.028	96.24	11.2	0.118
2.00	2.35	1.0153	1.0171	20.3	0.099	995.0	3.2	34	1.3364	0.440	0.237	0.127	1.049	1.035	95.14	14.4	0.154
2.50	2.94	1.0197	1.0215	25.5	0.124	994.2	4.0	43	1.3373	0.558	0.300	0.162	1.061	1.043	94.05	17.5	0.190
3.00	3.53	1.0242	1.0260	30.7	0.149	993.4	4.8	52	1.3382	0.670	0.360	0.195	1.073	1.050	92.98	20.5	0.226
3.50	4.11	1.0286	1.0305	36.0	0.175	992.6	5.6	61	1.3391	0.782	0.420	0.228	1.086	1.058	91.91	23.3	0.261
4.00	4.70	1.0332	1.0350	41.3	0.201	991.8	6.4	70	1.3400	0.900	0.484	0.263	1.098	1.065	90.85	26.1	0.296
4.50	5.29	1.0377	1.0395	46.7	0.227	991.0	7.2	79	1.3409	1.015	0.546	0.296	1.112	1.073	89.79	28.9	0.330
5.00	5.88	1.0423	1.0441	52.1	0.253	990.1	8.1	88	1.3418	1.126	0.606	0.329	1.125	1.082	88.71	31.6	0.364
5.50	6.46	1.0468	1.0487	57.6	0.280	989.3	9.0	97	1.3427	1.238	0.666	0.362	1.139	1.090	87.62	34.3	0.398
6.00	7.05	1.0514	1.0533	63.1	0.306	988.4	9.9	106	1.3436	1.350	0.726	0.395	1.153	1.099	86.53	36.9	0.431
6.50	7.64	1.0561	1.0579	68.6	0.333	987.4	10.8	115	1.3445				1.168	1.108	85.44	39.5	0.465
7.00	8.23	1.0607	1.0626	74.2	0.361	986.4	11.8	124	1.3454				1.183	1.118	84.35	42.1	0.497
7.50	8.81	1.0653	1.0672	79.9	0.388	985.4	12.8	133	1.3463				1.199	1.127	83.25	44.5	0.530
8.00	9.40	1.0700	1.0719	85.6	0.416	984.4	13.8	142	1.3472				1.215	1.138	82.16	46.9	0.562
8.50	9.99	1.0747	1.0766	91.3	0.444	983.3	14.9	151	1.3481				1.231	1.148	81.06	49.3	0.593
9.00	10.58	1.0794	1.0813	97.1	0.472	982.2	16.0	160	1.3490				1.248	1.159	79.97	51.6	0.624

78 SODIUM NITRATE, $NaNO_3$

MOLECULAR WEIGHT = 85.01
RELATIVE SPECIFIC REFRACTIVITY = 0.657

0.00 % by wt. data are the same for all compounds.
For Values of 0.00 wt. % solutions see Table 1, Acetic Acid.

A % by wt.	ρ D_4^{20}	D_{20}^{20}	C_s g/l	M g-mol/l	C_w g/l	$(C_o - C_w)$ g/l	$(n - n_o)$ × 10^4	n	Δ °C	O Os/kg	S g-mol/l	η/η_o	η/ρ cS	φ rhe	γ mmho/cm	T g-mol/l
0.50	1.0016	1.0034	5.0	0.059	996.6	1.6	6	1.3336	0.204	0.110	0.058	1.002	1.003	99.55	5.4	0.055
1.00	1.0050	1.0067	10.0	0.118	994.9	3.3	11	1.3341	0.403	0.216	0.116	1.005	1.002	99.30	10.6	0.111
2.00	1.0117	1.0135	20.2	0.238	991.5	6.7	23	1.3353	0.793	0.426	0.231	1.010	1.000	98.81	20.4	0.225
3.00	1.0185	1.0203	30.6	0.359	988.0	10.3	34	1.3364	1.177	0.633	0.344	1.016	0.999	98.24	29.5	0.336
4.00	1.0254	1.0272	41.0	0.482	984.4	13.9	46	1.3375	1.561	0.839	0.456	1.023	0.999	97.60	38.0	0.445
5.00	1.0322	1.0341	51.6	0.607	980.6	17.6	57	1.3387	1.942	1.044	0.568	1.030	1.000	96.89	46.2	0.552
6.00	1.0392	1.0410	62.4	0.733	976.8	21.4	68	1.3398	2.323	1.249	0.678	1.038	1.001	96.14	54.2	0.659
7.00	1.0462	1.0480	73.2	0.861	972.9	25.3	79	1.3409	2.703	1.453	0.787	1.047	1.003	95.32	61.7	0.764

78 SODIUM NITRATE, $NaNO_3$—(Continued)

A% by wt.	ρ D_4^{20}	D_{20}^{20}	C_s g/l	M g-mol/l	C_w g/l	$(C_o - C_w)$ g/l	$(n - n_o) \times 10^4$	n	Δ °C	O Os/kg	S g-mol/l	η/η_o	η/ρ cS	φ rhe	γ mmho/cm	T g-mol/l
8.00	1.0532	1.0550	84.3	0.991	968.9	29.3	91	1.3421	3.083	1.657	0.895	1.057	1.005	94.45	68.9	0.868
9.00	1.0603	1.0622	95.4	1.123	964.8	33.4	102	1.3432	3.462	1.862	1.002	1.067	1.009	93.51	75.8	0.971
10.00	1.0674	1.0693	106.7	1.256	960.7	37.6	113	1.3443	3.841	2.065	1.107	1.079	1.013	92.49	82.6	1.07
12.00	1.0819	1.0838	129.8	1.527	952.1	46.1	136	1.3466	4.599	2.472	1.316	1.105	1.024	90.28	95.3	1.27
14.00	1.0967	1.0986	153.5	1.806	943.2	55.1	159	1.3489	5.37	2.89	1.523	1.136	1.038	87.85	106.	1.46
18.00	1.1272	1.1292	202.9	2.387	924.3	73.9	206	1.3536	6.98	3.75	1.937	1.213	1.078	82.30	125.	1.82
20.00	1.1429	1.1449	228.6	2.689	914.3	83.9	230	1.3559	7.81	4.20	2.142	1.260	1.105	79.21	134.	1.99
24.00	1.1752	1.1772	282.0	3.318	893.1	105.1	277	1.3607	9.52	5.12	2.542	1.374	1.172	72.64	149.	2.30
28.00	1.2085	1.2106	338.4	3.981	870.1	128.1	325	1.3654	11.28	6.07	2.929	1.519	1.259	65.71	160.	2.56
30.00	1.2256	1.2278	367.7	4.325	858.0	140.3	348	1.3678				1.606	1.313	62.14	165.	2.68
34.00	1.2610	1.2632	428.7	5.043	832.2	166.0	397	1.3726				1.814	1.442	55.01	173.	2.87
40.00	1.3175	1.3198	527.0	6.199	790.5	207.7	472	1.3802				2.222	1.690	44.91	178.	3.00

79 SODIUM PHOSPHATE (TRIBASIC), $Na_3PO_4 \cdot 12H_2O$

MOLECULAR WEIGHT = 163.96   FORMULA WEIGHT, HYDRATE = 380.14
RELATIVE SPECIFIC REFRACTIVITY = 0.555

0.00 % by wt. data are the same for all compounds.
For Values of 0.00 wt. % solutions see Table 1, Acetic Acid.

A% by wt.	H% by wt.	ρ D_4^{20}	D_{20}^{20}	C_s g/l	M g-mol/l	C_w g/l	$(C_o - C_w)$ g/l	$(n - n_o) \times 10^4$	n	Δ °C	O Os/kg	S g-mol/l	η/η_o	η/ρ cS	φ rhe	γ mmho/cm	T g-mol/l
0.50	1.16	1.0042	1.0059	5.0	0.031	999.1	−0.9	13	1.3343	0.192	0.103	0.055	1.031	1.029	96.80	7.3	0.075
1.00	2.32	1.0100	1.0118	10.1	0.062	999.9	−1.7	26	1.3356	0.368	0.198	0.106	1.062	1.054	93.97	14.1	0.150
1.50	3.48	1.0158	1.0176	15.2	0.093	1000.6	−2.3	39	1.3369	0.527	0.283	0.153	1.092	1.078	91.36	18.8	0.206
2.00	4.64	1.0216	1.0234	20.4	0.125	1001.2	−2.9	51	1.3381	0.668	0.359	0.194	1.124	1.102	88.79	22.7	0.253
2.50	5.80	1.0275	1.0293	25.7	0.157	1001.8	−3.6	64	1.3394	0.790	0.425	0.230	1.159	1.130	86.13	26.7	0.302
3.00	6.96	1.0335	1.0353	31.0	0.189	1002.5	−4.2	76	1.3406				1.196	1.160	83.45	30.4	0.349
3.50	8.11	1.0395	1.0413	36.4	0.222	1003.1	−4.9	89	1.3419				1.236	1.191	80.76	34.0	0.394
4.00	9.27	1.0456	1.0474	41.8	0.255	1003.8	−5.5	102	1.3432				1.278	1.225	78.06	37.3	0.436
4.50	10.43	1.0517	1.0536	47.3	0.289	1004.4	−6.2	114	1.3444				1.324	1.261	75.39	40.5	0.477
5.00	11.59	1.0579	1.0598	52.9	0.323	1005.0	−6.8	127	1.3457				1.372	1.299	72.74	43.5	0.516
5.50	12.75	1.0642	1.0661	58.5	0.357	1005.7	−7.4	140	1.3470				1.423	1.340	70.13	46.4	0.554
6.00	13.91	1.0705	1.0724	64.2	0.392	1006.3	−8.0	152	1.3482				1.477	1.383	67.56	49.0	0.589
6.50	15.07	1.0768	1.0787	70.0	0.427	1006.8	−8.6	165	1.3495				1.535	1.428	65.03	51.5	0.622
7.00	16.23	1.0832	1.0851	75.8	0.462	1007.4	−9.1	177	1.3507				1.595	1.475	62.57	53.7	0.653
7.50	17.39	1.0896	1.0915	81.7	0.498	1007.9	−9.7	189	1.3519				1.659	1.525	60.16	55.8	0.681
8.00	18.55	1.0961	1.0980	87.7	0.535	1008.4	−10.1	202	1.3532				1.726	1.578	57.82	57.6	0.706

80 SODIUM PHOSPHATE, DIHYDROGEN (MONOBASIC), $NaH_2PO_4 \cdot 1H_2O$

MOLECULAR WEIGHT = 119.97   FORMULA WEIGHT, HYDRATE = 137.99
RELATIVE SPECIFIC REFRACTIVITY = 0.637

0.00 % by wt. data are the same for all compounds.
For Values of 0.00 wt. % solutions see Table 1, Acetic Acid.

A% by wt.	H% by wt.	ρ D_4^{20}	D_{20}^{20}	C_s g/l	M g-mol/l	C_w g/l	$(C_o - C_w)$ g/l	$(n - n_o) \times 10^4$	n	Δ °C	O Os/kg	S g-mol/l	η/η_o	η/ρ cS	φ rhe	γ mmho/cm	T g-mol/l
0.50	0.58	1.0019	1.0037	5.0	0.042	996.9	1.3	7	1.3336	0.140	0.075	0.040	1.016	1.017	98.18	2.2	0.021
1.00	1.15	1.0056	1.0074	10.1	0.084	995.6	2.7	13	1.3343	0.279	0.150	0.080	1.033	1.029	96.61	4.4	0.044
1.50	1.73	1.0094	1.0111	15.1	0.126	994.2	4.0	20	1.3349	0.418	0.225	0.120	1.049	1.042	95.10	6.8	0.068
2.00	2.30	1.0131	1.0149	20.3	0.169	992.8	5.4	26	1.3356	0.556	0.299	0.161	1.066	1.054	93.62	9.1	0.094
2.50	2.88	1.0168	1.0186	25.4	0.212	991.4	6.8	33	1.3362	0.695	0.374	0.202	1.083	1.067	92.13	11.3	0.119
3.00	3.45	1.0206	1.0224	30.6	0.255	990.0	8.3	39	1.3369	0.835	0.449	0.243	1.101	1.081	90.65	13.4	0.143
3.50	4.03	1.0244	1.0262	35.9	0.299	988.5	9.7	45	1.3375	0.976	0.525	0.285	1.119	1.095	89.17	15.4	0.166
4.00	4.60	1.0281	1.0300	41.1	0.343	987.0	11.2	52	1.3382	1.116	0.600	0.326	1.138	1.109	87.69	17.4	0.189
4.50	5.18	1.0319	1.0338	46.4	0.387	985.5	12.7	58	1.3388	1.254	0.674	0.367	1.158	1.124	86.21	19.2	0.211
5.00	5.75	1.0358	1.0376	51.8	0.432	984.0	14.3	65	1.3395	1.388	0.746	0.406	1.178	1.140	84.72	21.0	0.232
5.50	6.33	1.0396	1.0414	57.2	0.477	982.4	15.8	71	1.3401	1.519	0.817	0.444	1.199	1.156	83.23	22.6	0.252
6.00	6.90	1.0434	1.0453	62.6	0.522	980.8	17.4	78	1.3408	1.645	0.884	0.481	1.221	1.172	81.74	24.1	0.270
6.50	7.48	1.0473	1.0491	68.1	0.567	979.2	19.0	84	1.3414	1.767	0.950	0.517	1.243	1.190	80.26	25.5	0.287
7.00	8.05	1.0511	1.0530	73.6	0.613	977.6	20.7	91	1.3421	1.887	1.014	0.552	1.267	1.208	78.78	26.7	0.303
7.50	8.63	1.0550	1.0569	79.1	0.660	975.9	22.3	97	1.3427	2.004	1.077	0.586	1.291	1.226	77.31	27.9	0.317
8.00	9.20	1.0589	1.0608	84.7	0.706	974.2	24.0	104	1.3434	2.119	1.139	0.619	1.316	1.245	75.85	29.0	0.330
8.50	9.78	1.0628	1.0647	90.3	0.753	972.5	25.7	110	1.3440	2.234	1.201	0.652	1.342	1.265	74.39	30.0	0.344
9.00	10.35	1.0668	1.0686	96.0	0.800	970.8	27.5	117	1.3447	2.350	1.263	0.686	1.368	1.285	72.93	31.0	0.357
9.50	10.93	1.0707	1.0726	101.7	0.848	969.0	29.2	123	1.3453	2.466	1.326	0.719	1.396	1.307	71.48	32.1	0.370
10.00	11.50	1.0747	1.0766	107.5	0.896	967.2	31.0	130	1.3460	2.584	1.389	0.753	1.425	1.329	70.04	33.2	0.384
11.00	12.65	1.0826	1.0846	119.1	0.993	963.6	34.7	143	1.3473	2.821	1.517	0.821	1.485	1.374	67.20	35.4	0.412
12.00	13.80	1.0907	1.0926	130.9	1.091	959.8	38.4	156	1.3486	3.056	1.643	0.887	1.549	1.423	64.44	37.5	0.439
13.00	14.95	1.0988	1.1007	142.8	1.191	955.9	42.3	169	1.3499	3.291	1.769	0.954	1.617	1.474	61.73	39.6	0.465
14.00	16.10	1.1070	1.1089	155.0	1.292	952.0	46.2	182	1.3512	3.529	1.897	1.021	1.691	1.530	59.04	41.5	0.490
15.00	17.25	1.1152	1.1172	167.3	1.394	947.9	50.3	195	1.3525	3.775	2.029	1.089	1.771	1.591	56.35	43.3	0.513
16.00	18.40	1.1236	1.1255	179.8	1.498	943.8	54.4	209	1.3538	4.028	2.166	1.159	1.857	1.656	53.74	44.8	0.534
17.00	19.55	1.1320	1.1340	192.4	1.604	939.5	58.7	222	1.3552	4.287	2.305	1.230	1.948	1.724	51.24	46.2	0.552
18.00	20.70	1.1404	1.1425	205.3	1.711	935.2	63.1	235	1.3565	4.551	2.447	1.303	2.046	1.798	48.77	47.4	0.568
19.00	21.85	1.1490	1.1510	218.3	1.820	930.7	67.5	248	1.3578	4.820	2.592	1.376	2.155	1.880	46.30	48.5	0.583
20.00	23.00	1.1576	1.1597	231.5	1.930	926.1	72.1	262	1.3592	5.10	2.74	1.450	2.278	1.972	43.81	49.6	0.597
22.00	25.30	1.1752	1.1773	258.5	2.155	916.6	81.6	289	1.3618				2.545	2.170	39.21	51.4	0.622
24.00	27.60	1.1931	1.1952	286.3	2.387	906.7	91.5	316	1.3646				2.844	2.389	35.09	52.7	0.639
26.00	29.91	1.2113	1.2134	314.9	2.625	896.3	101.9	343	1.3673				3.208	2.654	31.11	53.5	0.650
28.00	32.21	1.2299	1.2320	344.4	2.870	885.5	112.7	370	1.3700				3.675	2.994	27.16	53.9	0.655
30.00	34.51	1.2488	1.2510	374.6	3.123	874.2	124.1	398	1.3728				4.291	3.443	23.26	54.0	0.656
32.00	36.81	1.2682	1.2704	405.8	3.383	862.4	135.9	426	1.3756				5.069	4.005	19.69	53.7	0.653
34.00	39.11	1.2879	1.2902	437.9	3.650	850.0	148.2	454	1.3784				5.996	4.665	16.64	52.9	0.642
36.00	41.41	1.3080	1.3103	470.9	3.925	837.1	161.1	482	1.3812				7.084	5.427	14.09	51.5	0.622
38.00	43.71	1.3285	1.3308	504.8	4.208	823.6	174.6	511	1.3840				8.346	6.295	11.96	49.3	0.593
40.00	46.01	1.3493	1.3517	539.7	4.499	809.6	188.6	539	1.3869				9.794	7.273	10.19	46.1	0.551

81 SODIUM PHOSPHATE, MONOHYDROGEN (DIBASIC), $Na_2HPO_4 \cdot 7H_2O$

MOLECULAR WEIGHT = 141.97 FORMULA WEIGHT, HYDRATE = 268.09
RELATIVE SPECIFIC REFRACTIVITY = 0.576

0.00 % by wt. data are the same for all compounds.
For Values of 0.00 wt. % solutions see Table 1, Acetic Acid.

A% by wt.	H% by wt.	ρ D_4^{20}	D_{20}^{20}	C_s g/l	M g-mol/l	C_w g/l	$(C_o - C_w)$ g/l	$(n - n_o)$ × 10^4	n	Δ °C	O Os/kg	S g-mol/l	η/η_o	η/ρ cS	φ rhe	γ mmho/cm	T g-mol/l
0.50	0.94	1.0032	1.0050	5.0	0.035	998.2	0.0	10	1.3340	0.167	0.090	0.047	1.019	1.018	97.91	4.6	0.046
1.00	1.89	1.0082	1.0100	10.1	0.071	998.1	0.1	19	1.3349	0.319	0.172	0.092	1.040	1.034	95.96	8.7	0.090
1.50	2.83	1.0131	1.0149	15.2	0.107	997.9	0.3	28	1.3358	0.462	0.248	0.133	1.062	1.051	93.94	12.3	0.130
2.00	3.78	1.0180	1.0198	20.4	0.143	997.7	0.6	38	1.3368				1.086	1.069	91.90	15.6	0.168
2.50	4.72	1.0229	1.0247	25.6	0.180	997.4	0.9	47	1.3377				1.111	1.088	89.86	18.7	0.205
3.00	5.67	1.0279	1.0297	30.8	0.217	997.0	1.2	56	1.3386				1.136	1.108	87.82	21.6	0.240
3.50	6.61	1.0328	1.0346	36.1	0.255	996.7	1.6	66	1.3396				1.163	1.129	85.80	24.3	0.273
4.00	7.55	1.0378	1.0396	41.5	0.292	996.3	1.9	75	1.3405				1.191	1.150	83.78	26.8	0.304
4.50	8.50	1.0428	1.0446	46.9	0.331	995.9	2.4	84	1.3414				1.221	1.173	81.77	29.2	0.334
5.00	9.44	1.0478	1.0496	52.4	0.369	995.4	2.8	94	1.3424				1.251	1.196	79.78	31.4	0.362
5.50	10.39	1.0528	1.0546	57.9	0.408	994.9	3.4	103	1.3433				1.283	1.221	77.80	33.5	0.388

82 SODIUM SULFATE, $Na_2SO_4 \cdot 10H_2O$

MOLECULAR WEIGHT = 142.06 FORMULA WEIGHT, HYDRATE = 322.22
RELATIVE SPECIFIC REFRACTIVITY = 0.548

0.00 % by wt. data are the same for all compounds.
For Values of 0.00 wt. % solutions see Table 1, Acetic Acid.

A% by wt.	H% by wt.	ρ D_4^{20}	D_{20}^{20}	C_s g/l	M g-mol/l	C_w g/l	$(C_o - C_w)$ g/l	$(n - n_o)$ × 10^4	n	Δ °C	O Os/kg	S g-mol/l	η/η_o	η/ρ cS	φ rhe	γ mmho/cm	T g-mol/l
0.50	1.13	1.0027	1.0044	5.0	0.035	997.7	0.6	8	1.3338	0.165	0.089	0.047	1.011	1.010	98.71	5.9	0.060
1.00	2.27	1.0071	1.0089	10.1	0.071	997.1	1.2	15	1.3345	0.320	0.172	0.092	1.024	1.019	97.46	11.2	0.117
1.50	3.40	1.0116	1.0134	15.2	0.107	996.4	1.8	23	1.3353	0.466	0.251	0.135	1.039	1.030	96.02	15.7	0.169
2.00	4.54	1.0161	1.0179	20.3	0.143	995.8	2.4	30	1.3360	0.606	0.326	0.176	1.056	1.041	94.51	19.8	0.218
2.50	5.67	1.0206	1.0225	25.5	0.180	995.1	3.1	38	1.3368	0.742	0.399	0.216	1.072	1.053	93.06	23.9	0.268
3.00	6.80	1.0252	1.0270	30.8	0.216	994.4	3.8	46	1.3376	0.873	0.469	0.254	1.089	1.065	91.63	27.9	0.317
3.50	7.94	1.0298	1.0316	36.0	0.254	993.7	4.5	53	1.3383	1.001	0.538	0.292	1.106	1.077	90.21	31.8	0.365
4.00	9.07	1.0343	1.0362	41.4	0.291	993.0	5.3	61	1.3391	1.125	0.605	0.329	1.124	1.089	88.80	35.5	0.413
4.50	10.21	1.0389	1.0408	46.8	0.329	992.2	6.0	68	1.3398	1.245	0.669	0.364	1.142	1.101	87.38	39.2	0.460
5.00	11.34	1.0436	1.0454	52.2	0.367	991.4	6.8	76	1.3406	1.359	0.731	0.397	1.161	1.115	85.96	42.7	0.506
5.50	12.48	1.0481	1.0499	57.6	0.406	990.4	7.8	83	1.3413	1.465	0.788	0.429	1.180	1.129	84.54	46.1	0.551
6.00	13.61	1.0526	1.0545	63.2	0.445	989.5	8.8	90	1.3420	1.560	0.839	0.456	1.200	1.143	83.14	49.4	0.594
6.50	14.74	1.0572	1.0591	68.7	0.484	988.5	9.7	97	1.3427				1.221	1.157	81.75	52.5	0.636
7.00	15.88	1.0619	1.0638	74.3	0.523	987.5	10.7	105	1.3435				1.242	1.172	80.36	55.5	0.676
7.50	17.01	1.0666	1.0684	80.0	0.563	986.6	11.7	112	1.3442				1.264	1.187	78.98	58.3	0.716
8.00	18.15	1.0713	1.0732	85.7	0.603	985.6	12.7	120	1.3449				1.286	1.203	77.59	61.1	0.755
8.50	19.28	1.0760	1.0779	91.5	0.644	984.6	13.7	127	1.3457				1.310	1.220	76.20	63.7	0.793
9.00	20.41	1.0808	1.0827	97.3	0.685	983.5	14.7	134	1.3464				1.334	1.237	74.80	66.3	0.830
9.50	21.55	1.0856	1.0875	103.1	0.726	982.5	15.7	142	1.3472				1.360	1.255	73.38	68.8	0.866
10.00	22.68	1.0905	1.0924	109.0	0.768	981.4	16.8	149	1.3479				1.387	1.274	71.95	71.3	0.902
11.00	24.95	1.1002	1.1022	121.0	0.852	979.2	19.0	164	1.3494				1.444	1.315	69.11	75.9	0.970
12.00	27.22	1.1101	1.1121	133.2	0.938	976.9	21.3	179	1.3509				1.505	1.359	66.30	80.1	1.03
13.00	29.49	1.1201	1.1221	145.6	1.025	974.5	23.8	194	1.3524				1.571	1.406	63.52	83.9	1.09
14.00	31.75	1.1301	1.1321	158.2	1.114	971.9	26.3	209	1.3539				1.643	1.457	60.74	87.5	1.15
15.00	34.02	1.1402	1.1422	171.0	1.204	969.2	29.1	223	1.3553				1.722	1.513	57.96	91.1	1.21
16.00	36.29	1.1503	1.1523	184.0	1.296	966.2	32.0	237	1.3567				1.808	1.575	55.20	94.9	1.27
17.00	38.56	1.1604	1.1625	197.3	1.389	963.1	35.1	251	1.3581				1.901	1.641	52.50	98.5	1.33
18.00	40.83	1.1705	1.1726	210.7	1.483	959.8	38.4	265	1.3595				2.001	1.713	49.88	102.0	1.39
19.00	43.10	1.1806	1.1827	224.3	1.579	956.3	41.9	278	1.3608				2.108	1.789	47.34	105.0	1.44
20.00	45.36	1.1907	1.1928	238.1	1.676	952.6	45.7	290	1.3620				2.223	1.871	44.89	109.0	1.50
22.00	49.90	1.2106	1.2127	266.3	1.875	944.2	54.0	313	1.3643				2.476	2.050	40.30	114.0	1.61

83 SODIUM TARTRATE, $NaOOC(CHOH)_2COONa \cdot 2H_2O$

MOLECULAR WEIGHT = 194.06 FORMULA WEIGHT, HYDRATE = 230.10
RELATIVE SPECIFIC REFRACTIVITY = 0.781

0.00 % by wt. data are the same for all compounds.
For Values of 0.00 wt. % solutions see Table 1, Acetic Acid.

A% by wt.	H% by wt.	ρ D_4^{20}	D_{20}^{20}	C_s g/l	M g-mol/l	C_w g/l	$(C_o - C_w)$ g/l	$(n - n_o)$ × 10^4	n	Δ °C	O Os/kg	S g-mol/l	η/η_o	η/ρ cS	φ rhe	γ mmho/cm	T g-mol/l
0.50	0.59	1.0017	1.0035	5.0	0.026	996.7	1.5	8	1.3338	0.111	0.060	0.031	1.015	1.016	98.28	3.8	0.037
1.00	1.19	1.0052	1.0070	10.1	0.052	995.2	3.1	17	1.3347	0.224	0.121	0.064	1.031	1.028	96.80	7.1	0.072
2.00	2.37	1.0123	1.0141	20.2	0.104	992.0	6.2	33	1.3363	0.457	0.246	0.132	1.062	1.051	93.97	12.6	0.133
3.00	3.56	1.0194	1.0212	30.6	0.158	988.8	9.4	50	1.3380	0.678	0.365	0.197	1.094	1.075	91.21	17.5	0.191
4.00	4.74	1.0266	1.0284	41.1	0.212	985.5	12.7	67	1.3397	0.891	0.479	0.260	1.128	1.101	88.49	22.1	0.245
5.00	5.93	1.0338	1.0356	51.7	0.266	982.1	16.1	84	1.3414	1.103	0.593	0.322	1.164	1.128	85.74	26.2	0.297
6.00	7.11	1.0410	1.0428	62.5	0.322	978.5	19.7	101	1.3431	1.317	0.708	0.385	1.203	1.157	82.99	30.1	0.346
7.00	8.30	1.0482	1.0501	73.4	0.378	974.9	23.4	118	1.3448	1.529	0.822	0.447	1.243	1.188	80.28	33.7	0.390
8.00	9.49	1.0555	1.0574	84.4	0.435	971.1	27.2	135	1.3465	1.742	0.937	0.509	1.287	1.222	77.56	37.0	0.432
9.00	10.67	1.0628	1.0647	95.7	0.493	967.2	31.1	152	1.3482	1.955	1.051	0.572	1.334	1.258	74.81	40.0	0.471
10.00	11.86	1.0702	1.0721	107.0	0.551	963.2	35.1	169	1.3499	2.171	1.167	0.634	1.386	1.298	72.01	43.0	0.509
12.00	14.23	1.0851	1.0870	130.2	0.671	954.9	43.4	203	1.3533	2.607	1.402	0.760	1.499	1.384	66.58	48.1	0.577
14.00	16.60	1.1002	1.1022	154.0	0.794	946.2	52.0	237	1.3567	3.048	1.639	0.885	1.625	1.480	61.43	52.2	0.632
15.00	17.79	1.1079	1.1098	166.2	0.856	941.7	56.5	254	1.3584	3.269	1.758	0.948	1.695	1.533	58.88	53.9	0.655
18.00	21.34	1.1313	1.1333	203.6	1.049	927.6	70.6	307	1.3637	3.938	2.117	1.134	1.951	1.728	51.14	57.9	0.710
20.00	23.71	1.1471	1.1492	229.4	1.182	917.7	80.5	342	1.3672	4.385	2.358	1.258	2.174	1.899	45.91	59.9	0.738
22.00	26.09	1.1633	1.1653	255.9	1.319	907.4	90.9	378	1.3708				2.444	2.105	40.84	61.4	0.759
24.00	28.46	1.1797	1.1818	283.1	1.459	896.6	101.7	414	1.3744				2.757	2.342	36.20	62.0	0.768
26.00	30.83	1.1963	1.1984	311.0	1.603	885.3	113.0	450	1.3780				3.117	2.611	32.01	61.6	0.762
28.00	33.20	1.2132	1.2153	339.7	1.750	873.5	124.7	487	1.3817				3.528	2.914	28.29	60.1	0.741

84 SODIUM THIOCYANATE, NaCNS

MOLECULAR WEIGHT = 81.07

A% by wt	ρ D_4^{20}	D_{20}^{20}	C_s g/l	M g-mol/l	C_w g/l	$(C_o - C_w)$ g/l	$(n - n_o)$ × 10^4	n	Δ C	O Os/kg	S g-mol/l	η/η_o	η/ρ cS	φ rhe	γ mmho/cm	T g-mol/l
1.00	1.0030	1.0048	10.0	0.124	993.0	5.2	22	1.3352	0.426	0.229	0.123	1.006	1.005	99.21	10.0	0.113
2.00	1.0078	1.0096	20.2	0.249	987.6	10.6	44	1.3374	0.848	0.456	0.247	1.012	1.006	98.64	20.5	0.226
3.00	1.0127	1.0145	30.4	0.375	982.3	15.9	66	1.3396	1.280	0.688	0.374	1.019	1.008	97.96	29.8	0.341
4.00	1.0176	1.0194	40.7	0.502	976.9	21.3	88	1.3418	1.734	0.932	0.507	1.025	1.009	97.40	39.0	0.458
5.00	1.0226	1.0244	51.1	0.631	971.5	26.7	110	1.3440	2.202	1.184	0.643	1.032	1.011	96.73	48.0	0.576
6.00	1.0275	1.0294	61.7	0.760	965.8	32.4	132	1.3462	2.679	1.440	0.780	1.039	1.013	96.07	56.7	0.694
7.00	1.0327	10.346	72.3	0.892	960.4	37.8	155	1.3485	3.168	1.703	0.919	1.047	1.016	95.31	65.1	0.812
8.00	1.0381	1.0400	83.1	1.024	955.0	43.2	178	1.3508	3.671	1.974	1.060	1.057	1.020	94.44	73.3	0.932
9.00	1.0435	1.0454	93.9	1.158	949.6	48.6	201	1.3531	4.184	2.250	1.202	1.067	1.025	93.49	81.2	1.052
10.00	1.0491	1.0510	104.9	1.294	944.2	54.0	224	1.3554	4.706	2.530	1.345	1.079	1.031	92.45	89.0	1.174
12.00	1.0603	1.0622	127.2	1.569	933.1	65.1	270	1.3600				1.106	1.045	90.25	104	1.418
14.00	1.0715	1.0734	150.0	1.850	921.5	76.7	318	1.3648				1.138	1.064	87.72	118	1.664
16.00	1.0828	1.0848	173.3	2.137	909.5	88.7	366	1.3696				1.175	1.087	84.96	130	1.912
18.00	1.0942	1.0962	197.0	2.430	897.2	101.0	414	1.3744				1.217	1.114	82.04	142	2.162
20.00	1.1056	1.1076	221.1	2.728	884.5	113.7	464	1.3794				1.268	1.149	78.72	154	2.412
24.00	1.1284	1.1304	270.8	3.340	857.6	140.6	564	1.3894				1.396	1.240	71.47	170	2.786
28.00	1.1515	1.1536	322.4	3.977	829.1	169.1	666	1.3996				1.546	1.345	64.57	178	3.016
32.00	1.1763	1.1784	376.4	4.643	799.9	198.3	774	1.4104				1.764	1.503	56.56	184	3.194
36.00	1.2018	1.2040	432.7	5.337	769.1	229.1	882	1.4212				2.101	1.752	47.49	185	3.227
40.00	1.2282	1.2304	491.3	6.060	736.9	261.3	990	1.4320				2.483	2.026	40.19	177	2.986
44.00	1.2549	1.2572	552.2	6.811	702.7	295.5	1105	1.4435				3.151	2.516	31.67	165	2.678
48.00	1.2821	1.2844	615.4	7.591	666.7	331.5	1223	1.4553				3.936	3.076	25.36	148	2.290
52.00	1.3100	1.3124	681.2	8.403	628.8	369.4	1342	1.4672				5.279	4.038	18.90	130	1.900
56.0	1.3388	1.3412	749.7	9.248	589.1	409.1	1461	1.4791				7.554	5.654	13.21	108	1.493

85 SODIUM THIOSULFATE, $Na_2S_2O_3 \cdot 5H_2O$

MOLECULAR WEIGHT = 158.13 FORMULA WEIGHT, HYDRATE = 248.21
RELATIVE SPECIFIC REFRACTIVITY = 0.773

0.00 % by wt. data are the same for all compounds.
For Values of 0.00 wt. % solutions see Table 1, Acetic Acid.

A% by wt.	H% by wt.	ρ D_4^{20}	D_{20}^{20}	C_s g/l	M g-mol/l	C_w g/l	$(C_o - C_w)$ g/l	$(n - n_o)$ × 10^4	n	Δ °C	O Os/kg	S g-mol/l	η/η_o	η/ρ cS	φ rhe	γ mmho/cm	T g-mol/l
0.50	0.78	1.0024	1.0041	5.0	0.032	997.4	0.9	10	1.3340	0.139	0.075	0.039	1.010	1.010	98.76	5.7	0.057
1.00	1.57	1.0065	1.0083	10.1	0.064	996.4	1.8	21	1.3351	0.279	0.150	0.080	1.021	1.016	97.75	10.7	0.112
2.00	3.14	1.0148	1.0166	20.3	0.128	994.5	3.7	42	1.3371	0.566	0.304	0.164	1.042	1.029	95.78	19.5	0.214
3.00	4.71	1.0231	1.0249	30.7	0.194	992.4	5.8	62	1.3392	0.835	0.449	0.243	1.064	1.042	93.77	27.7	0.315
4.00	6.28	1.0315	1.0333	41.3	0.261	990.2	8.0	83	1.3413	1.090	0.586	0.318	1.088	1.057	91.74	35.6	0.415
5.00	7.85	1.0399	1.0417	52.0	0.329	987.9	10.3	104	1.3434	1.341	0.721	0.392	1.113	1.072	89.67	43.3	0.513
6.00	9.42	1.0483	1.0502	62.9	0.398	985.4	12.8	124	1.3454	1.588	0.854	0.464	1.139	1.089	87.59	50.6	0.610
7.00	10.99	1.0568	1.0587	74.0	0.468	982.8	15.4	145	1.3475	1.826	0.982	0.534	1.167	1.107	85.50	57.5	0.705
8.00	12.56	1.0654	1.0673	85.2	0.539	980.1	18.1	166	1.3496	2.062	1.108	0.602	1.197	1.126	83.38	64.2	0.799
9.00	14.13	1.0740	1.0759	96.7	0.611	977.3	20.9	187	1.3517	2.299	1.236	0.671	1.229	1.147	81.21	70.5	0.891
10.00	15.70	1.0827	1.0846	108.3	0.685	974.4	23.8	208	1.3538	2.546	1.369	0.742	1.264	1.170	78.96	76.7	0.982
12.00	18.84	1.1003	1.1023	132.0	0.835	968.3	29.9	251	1.3581	3.064	1.647	0.890	1.342	1.222	74.35	88.2	1.16
14.00	21.98	1.1182	1.1202	156.6	0.990	961.7	36.5	294	1.3624	3.604	1.938	1.041	1.432	1.283	69.71	98.8	1.33
16.00	25.11	1.1365	1.1385	181.8	1.150	954.6	43.6	337	1.3667	4.168	2.241	1.197	1.534	1.353	65.04	108.	1.49
18.00	28.25	1.1551	1.1571	207.9	1.315	947.1	51.1	381	1.3711	4.758	2.558	1.359	1.654	1.435	60.34	117.	1.64
20.00	31.39	1.1740	1.1760	234.8	1.485	939.2	59.1	426	1.3756	5.37	2.89	1.524	1.794	1.531	55.63	123.	1.77
22.00	34.53	1.1932	1.1953	262.5	1.660	930.7	67.5	471	1.3801				1.954	1.641	51.08	129.	1.87
24.00	37.67	1.2128	1.2150	291.1	1.841	921.8	76.5	517	1.3847				2.137	1.765	46.70	132.	1.94
26.00	40.81	1.2328	1.2350	320.5	2.027	912.3	85.9	563	1.3893				2.351	1.911	42.45	135.	2.00
28.00	43.95	1.2532	1.2554	350.9	2.219	902.3	95.9	610	1.3940				2.591	2.072	38.52	136.	2.02
30.00	47.09	1.2739	1.2762	382.2	2.417	891.7	106.5	657	1.3987				2.897	2.279	34.45	136.	2.03
32.00	50.23	1.2950	1.2973	414.4	2.621	880.6	117.6	705	1.4035				3.298	2.552	30.26	135.	1.99
34.00	53.37	1.3164	1.3188	447.6	2.831	868.8	129.4	754	1.4084				3.784	2.881	26.37	132.	1.93
36.00	56.51	1.3382	1.3406	481.8	3.047	856.5	141.8	802	1.4132				4.350	3.257	22.95	128.	1.86
38.00	59.65	1.3603	1.3627	516.9	3.269	843.4	154.8	851	1.4181				5.001	3.684	19.96	124.	1.77
40.00	62.79	1.3827	1.3852	553.1	3.498	829.6	168.6	899	1.4229				5.747	4.165	17.37	118.	1.66

86 SODIUM TUNGSTATE, $Na_2WO_4 \cdot 2H_2O$

MOLECULAR WEIGHT = 293.91 FORMULA WEIGHT, HYDRATE = 329.95
RELATIVE SPECIFIC REFRACTIVITY = 0.414

0.00 % by wt. data are the same for all compounds.
For Values of 0.00 wt. % solutions see Table 1, Acetic Acid.

A % by wt.	H % by wt.	ρ D_4^{20}	D_{20}^{20}	C_s g/l	M g-mol/l	C_w g/l	$(C_o - C_w)$ g/l	$(n - n_o)$ × 10^4	n	Δ °C	O Os/kg	S g-mol/l	η/η_o	η/ρ cS	φ rhe	γ mmho/cm	T g-mol/l
0.50	0.56	1.0028	1.0046	5.0	0.017	997.8	0.4	5	1.3335	0.081	0.044	0.023	1.007	1.006	99.13	2.8	0.028
1.00	1.12	1.0074	1.0091	10.1	0.034	997.3	0.9	11	1.3341	0.161	0.086	0.046	1.014	1.009	98.42	5.5	0.055
1.50	1.68	1.0119	1.0137	15.2	0.052	996.8	1.5	16	1.3346	0.239	0.129	0.068	1.022	1.012	97.67	7.9	0.081
2.00	2.25	1.0166	1.0184	20.3	0.069	996.2	2.0	21	1.3351	0.317	0.171	0.091	1.030	1.015	96.89	10.3	0.107
2.50	2.81	1.0212	1.0230	25.5	0.087	995.7	2.5	27	1.3357	0.396	0.213	0.114	1.039	1.019	96.08	12.5	0.132
3.00	3.37	1.0259	1.0278	30.8	0.105	995.2	3.1	33	1.3362	0.474	0.255	0.137	1.048	1.024	95.23	14.7	0.157
3.50	3.93	1.0307	1.0326	36.1	0.123	994.7	3.6	38	1.3368	0.553	0.297	0.160	1.058	1.028	94.35	16.8	0.182
4.00	4.49	1.0356	1.0374	41.4	0.141	994.1	4.1	44	1.3374	0.633	0.340	0.184	1.068	1.033	93.46	18.8	0.206
4.50	5.05	1.0405	1.0423	46.8	0.159	993.6	4.6	50	1.3380	0.714	0.384	0.208	1.078	1.038	92.58	20.8	0.230
5.00	5.61	1.0453	1.0472	52.3	0.178	993.1	5.2	56	1.3386	0.796	0.428	0.232	1.088	1.043	91.73	22.8	0.254
5.50	6.17	1.0502	1.0520	57.8	0.197	992.4	5.8	62	1.3392	0.879	0.472	0.256	1.098	1.048	90.89	24.7	0.278
6.00	6.74	1.0550	1.0569	63.3	0.215	991.7	6.5	67	1.3397	0.962	0.517	0.281	1.108	1.052	90.06	26.7	0.302
6.50	7.30	1.0599	1.0618	68.9	0.234	991.0	7.2	73	1.3403	1.046	0.562	0.305	1.118	1.057	89.24	28.6	0.326
7.00	7.86	1.0648	1.0667	74.5	0.254	990.3	8.0	79	1.3409	1.130	0.607	0.330	1.129	1.062	88.43	30.5	0.350
7.50	8.42	1.0698	1.0717	80.2	0.273	989.5	8.7	85	1.3414	1.212	0.652	0.354	1.139	1.067	87.62	32.4	0.374
8.00	8.98	1.0748	1.0767	86.0	0.293	988.8	9.4	91	1.3420	1.293	0.695	0.378	1.150	1.072	86.81	34.3	0.398
8.50	9.54	1.0799	1.0818	91.8	0.312	988.1	10.1	97	1.3427	1.371	0.737	0.401	1.160	1.077	86.02	36.1	0.422
9.00	10.10	1.0852	1.0871	97.7	0.332	987.5	10.7	103	1.3433	1.445	0.777	0.423	1.171	1.081	85.23	38.0	0.445

87 STRONTIUM CHLORIDE, $SrCl_2 \cdot 6H_2O$

MOLECULAR WEIGHT = 158.54 FORMULA WEIGHT, HYDRATE = 266.64
RELATIVE SPECIFIC REFRACTIVITY = 0.635

0.00 % by wt. data are the same for all compounds.
For Values of 0.00 wt. % solutions see Table 1, Acetic Acid.

A % by wt.	H % by wt.	ρ D_4^{20}	D_{20}^{20}	C_s g/l	M g-mol/l	C_w g/l	$(C_o - C_w)$ g/l	$(n - n_o)$ × 10^4	n	Δ °C	O Os/kg	S g-mol/l	η/η_o	η/ρ cS	φ rhe	γ mmho/cm	T g-mol/l
0.50	0.84	1.0027	1.0044	5.0	0.032	997.6	0.6	9	1.3339	0.158	0.085	0.045	1.010	1.009	98.85	5.9	0.060
1.00	1.68	1.0071	1.0089	10.1	0.064	997.0	1.2	18	1.3348	0.310	0.167	0.089	1.019	1.014	97.94	11.4	0.120
2.00	3.36	1.0161	1.0179	20.3	0.128	995.8	2.5	36	1.3366	0.619	0.333	0.180	1.037	1.023	96.24	22.0	0.244
3.00	5.05	1.0252	1.0270	30.8	0.194	994.4	3.8	54	1.3384	0.934	0.502	0.272	1.055	1.032	94.56	31.4	0.362
4.00	6.73	1.0344	1.0362	41.4	0.261	993.0	5.2	73	1.3402	1.261	0.678	0.369	1.074	1.041	92.89	40.4	0.475
5.00	8.41	1.0437	1.0456	52.2	0.329	991.5	6.7	91	1.3421	1.607	0.861	0.470	1.094	1.050	91.23	49.1	0.591
6.00	10.09	1.0532	1.0551	63.2	0.399	990.0	8.2	110	1.3440	1.978	1.064	0.578	1.114	1.060	89.59	58.0	0.712
7.00	11.77	1.0628	1.0647	74.4	0.469	988.4	9.8	129	1.3459	2.377	1.278	0.693	1.134	1.069	87.98	66.8	0.836
8.00	13.45	1.0726	1.0745	85.8	0.541	986.8	11.4	148	1.3478	2.800	1.505	0.815	1.155	1.079	86.37	75.3	0.962
9.00	15.14	1.0825	1.0844	97.4	0.614	985.0	13.2	168	1.3498	3.252	1.748	0.943	1.178	1.090	84.73	83.6	1.08
10.00	16.82	1.0925	1.0944	109.3	0.689	983.3	15.0	188	1.3518	3.736	2.009	1.078	1.202	1.102	83.03	91.5	1.21
12.00	20.18	1.1131	1.1150	133.6	0.842	979.5	18.7	228	1.3558	4.811	2.587	1.374	1.255	1.130	79.53	107.	1.46
14.00	23.55	1.1342	1.1362	158.8	1.002	975.4	22.9	269	1.3599	6.03	3.24	1.696	1.314	1.160	75.98	120.	1.71
16.00	26.91	1.1558	1.1579	184.9	1.166	970.9	27.3	311	1.3641	7.41	3.99	2.044	1.380	1.196	72.32	133.	1.96
18.00	30.27	1.1780	1.1801	212.0	1.337	966.0	32.3	354	1.3684	8.98	4.83	2.418	1.457	1.239	68.52	144.	2.19
20.00	33.64	1.2008	1.2029	240.2	1.515	960.6	37.6	398	1.3728	10.74	5.78	2.814	1.546	1.290	64.55	153.	2.40
22.00	37.00	1.2241	1.2263	269.3	1.699	954.8	43.4	442	1.3772	12.74	6.85	3.230	1.647	1.348	60.58	160.	2.56
24.00	40.36	1.2481	1.2503	299.5	1.889	948.6	49.7	487	1.3817	14.99	8.06	3.664	1.761	1.414	56.67	165.	2.68
26.00	43.73	1.2728	1.2751	330.9	2.087	941.9	56.3	534	1.3864				1.893	1.491	52.71	170.	2.79
28.00	47.09	1.2983	1.3006	363.5	2.293	934.8	63.4	582	1.3911				2.052	1.584	48.64	174.	2.90
30.00	50.46	1.3248	1.3271	397.4	2.507	927.3	70.9	631	1.3961				2.241	1.695	44.53	178.0	3.02
32.00	53.82	1.3523	1.3547	432.7	2.729	919.6	78.7	683	1.4013				2.522	1.869	39.57	178.0	3.02
34.00	57.18	1.3811	1.3835	469.6	2.962	911.5	86.7	737	1.4067				2.840	2.060	35.14	175.0	2.92
36.00	60.55	1.4114	1.4139	508.1	3.205	903.3	94.9	794	1.4124				3.200	2.272	31.19	166.0	2.70

88 SUCROSE, $C_{12}H_{22}O_{11}$

MOLECULAR WEIGHT = 342.30
RELATIVE SPECIFIC REFRACTIVITY = 1.031

0.00 % by wt. data are the same for all compounds.
For Values of 0.00 wt. % solutions see Table 1, Acetic Acid.

A% by wt.	ρ D_4^{20}	D_{20}^{20}	C_s g/l	M g-mol/l	C_w g/l	$(C_o - C_w)$ g/l	$(n - n_o)$ × 10^4	n	Δ °C	O Os/kg	S g-mol/l	η/η_o	η/ρ cS	φ rhe	γ mmho/cm	T g-mol/l
0.50	1.0002	1.0019	5.0	0.015	995.2	3.1	7	1.3337	0.027	0.015	0.007	1.013	1.015	98.53		
1.00	1.0021	1.0039	10.0	0.029	992.1	6.2	14	1.3344	0.055	0.030	0.015	1.026	1.026	97.27		
1.50	1.0040	1.0058	15.1	0.044	989.0	9.3	22	1.3351	0.083	0.045	0.023	1.039	1.037	96.03		
2.00	1.0060	1.0078	20.1	0.059	985.9	12.4	29	1.3359	0.112	0.060	0.031	1.053	1.049	94.78		
2.50	1.0079	1.0097	25.2	0.074	982.7	15.5	36	1.3366	0.140	0.076	0.040	1.067	1.061	93.51		
3.00	1.0099	1.0117	30.3	0.089	979.6	18.6	43	1.3373	0.170	0.091	0.048	1.082	1.074	92.24		
3.50	1.0119	1.0137	35.4	0.103	976.5	21.8	51	1.3381	0.199	0.107	0.057	1.097	1.086	90.99		
4.00	1.0139	1.0156	40.6	0.118	973.3	24.9	58	1.3388	0.229	0.123	0.066	1.112	1.099	89.75		
4.50	1.0158	1.0176	45.7	0.134	970.1	28.1	65	1.3395	0.260	0.140	0.074	1.128	1.112	88.49		
5.00	1.0178	1.0196	50.9	0.149	966.9	31.3	73	1.3403	0.291	0.156	0.083	1.144	1.126	87.24		
5.50	1.0198	1.0216	56.1	0.164	963.7	34.5	80	1.3410	0.322	0.173	0.093	1.160	1.140	86.02		
6.00	1.0218	1.0236	61.3	0.179	960.5	37.7	88	1.3418	0.354	0.190	0.102	1.177	1.154	84.79		
6.50	1.0238	1.0257	66.5	0.194	957.3	40.9	95	1.3425	0.386	0.208	0.111	1.195	1.169	83.54		
7.00	1.0259	1.0277	71.8	0.210	954.1	44.2	103	1.3433	0.419	0.225	0.121	1.213	1.185	82.28		
7.50	1.0279	1.0297	77.1	0.225	950.8	47.4	110	1.3440	0.452	0.243	0.131	1.232	1.201	81.02		
8.00	1.0299	1.0317	82.4	0.241	947.5	50.7	118	1.3448	0.485	0.261	0.140	1.251	1.217	79.78		
8.50	1.0320	1.0338	87.7	0.256	944.2	54.0	125	1.3455	0.520	0.279	0.150	1.271	1.234	78.54		
9.00	1.0340	1.0358	93.1	0.272	940.9	57.3	133	1.3463	0.554	0.298	0.161	1.291	1.251	77.30		
9.50	1.0361	1.0379	98.4	0.288	937.6	60.6	141	1.3471	0.589	0.317	0.171	1.312	1.269	76.09		
10.00	1.0381	1.0400	103.8	0.303	934.3	63.9	148	1.3478	0.625	0.336	0.181	1.333	1.287	74.87		
11.00	1.0423	1.0441	114.7	0.335	927.6	70.6	164	1.3494	0.698	0.375	0.203	1.378	1.325	72.42		
12.00	1.0465	1.0483	125.6	0.367	920.9	77.4	180	1.3509	0.773	0.415	0.225	1.426	1.365	69.99		
13.00	1.0507	1.0525	136.6	0.399	914.1	84.2	195	1.3525	0.850	0.457	0.248	1.477	1.409	67.57		
14.00	1.0549	1.0568	147.7	0.431	907.2	91.0	211	1.3541	0.930	0.500	0.271	1.531	1.454	65.19		
15.00	1.0592	1.0610	158.9	0.464	900.3	97.9	227	1.3557	1.012	0.544	0.295	1.589	1.503	62.81		
16.00	1.0635	1.0653	170.2	0.497	893.3	104.9	243	1.3573	1.097	0.590	0.320	1.650	1.555	60.49		
17.00	1.0678	1.0697	181.5	0.530	886.3	112.0	259	1.3589	1.185	0.637	0.346	1.716	1.610	58.16		
18.00	1.0722	1.0741	193.0	0.564	879.2	119.1	276	1.3606	1.275	0.686	0.373	1.786	1.669	55.88		
19.00	1.0766	1.0785	204.5	0.598	872.0	126.2	292	1.3622	1.369	0.736	0.400	1.861	1.732	53.63		
20.00	1.0810	1.0829	216.2	0.632	864.8	133.5	309	1.3639	1.465	0.788	0.429	1.941	1.799	51.42		
22.00	1.0899	1.0918	239.8	0.701	850.1	148.1	342	1.3672	1.668	0.897	0.488	2.120	1.949	47.08		
24.00	1.0990	1.1009	263.8	0.771	835.2	163.0	376	1.3706	1.886	1.014	0.551	2.326	2.121	42.91		
26.00	1.1082	1.1102	288.1	0.842	820.1	178.2	411	1.3741	2.120	1.140	0.619	2.568	2.322	38.86		
28.00	1.1175	1.1195	312.9	0.914	804.6	193.6	446	1.3776	2.371	1.275	0.692	2.849	2.554	35.03		
30.00	1.1270	1.1290	338.1	0.988	788.9	209.3	482	1.3812	2.644	1.421	0.770	3.181	2.828	31.37		
32.00	1.1366	1.1386	363.7	1.063	772.9	225.3	518	1.3848	2.942	1.582	0.855	3.754	3.309	26.59		
34.00	1.1464	1.1484	389.8	1.139	756.6	241.6	555	1.3885	3.268	1.757	0.947	4.044	3.535	24.68		
36.00	1.1562	1.1583	416.2	1.216	740.0	258.2	592	1.3922	3.625	1.949	1.047	4.612	3.997	21.64		
38.00	1.1663	1.1683	443.2	1.295	723.1	275.1	630	1.3960	4.018	2.160	1.156	5.304	4.557	18.82		
40.00	1.1765	1.1785	470.6	1.375	705.9	292.4	669	1.3999	4.452	2.394	1.276	6.150	5.238	16.23		
42.00	1.1868	1.1889	498.4	1.456	688.3	309.9	708	1.4038	4.932	2.652	1.406	7.220	6.096	13.82		
44.00	1.1972	1.1994	526.8	1.539	670.5	327.8	748	1.4078				8.579	7.180	11.63		
46.00	1.2079	1.2100	555.6	1.623	652.2	346.0	788	1.4118				10.28	8.53	9.71		
48.00	1.2186	1.2208	584.9	1.709	633.7	364.6	829	1.4159				12.49	10.27	7.99		
50.00	1.2295	1.2317	614.8	1.796	614.8	383.5	871	1.4201				15.40	12.55	6.48		
52.00	1.2406	1.2428	645.1	1.885	595.5	402.7	913	1.4243				19.30	15.59	5.17		
54.00	1.2518	1.2540	676.0	1.975	575.8	422.4	956	1.4286				24.63	19.71	4.05		
56.00	1.2632	1.2654	707.4	2.067	555.8	442.4	1000	1.4330				32.06	25.43	3.11		
58.00	1.2747	1.2770	739.3	2.160	535.4	462.8	1044	1.4374				42.69	33.56	2.34		
60.00	1.2864	1.2887	771.9	2.255	514.6	483.7	1089	1.4419				58.37	45.46	1.71		
62.00	1.2983	1.3006	804.9	2.352	493.3	504.9	1135	1.4465				82.26	63.49	1.21		
64.00	1.3103	1.3126	838.6	2.450	471.7	526.5	1181	1.4511				119.9	91.69	0.83		
66.00	1.3224	1.3248	872.8	2.550	449.6	548.6	1228	1.4558				181.7	137.6	0.55		
68.00	1.3348	1.3371	907.6	2.652	427.1	571.1	1276	1.4606				287.9	216.1	0.35		
70.00	1.3472	1.3496	943.1	2.755	404.2	594.1	1324	1.4654				480.6	357.4	0.21		
72.00	1.3599	1.3623	979.1	2.860	380.8	617.5	1373	1.4703				853.2	628.6	0.12		
74.00	1.3726	1.3751	1015.7	2.967	356.9	641.4	1423	1.4753				1628.	1188.	0.06		
76.00	1.3855	1.3880	1053.0	3.076	332.5	665.7	1473	1.4803								
78.00	1.3986	1.4011	1090.9	3.187	307.7	690.5	1524	1.4854								
80.00	1.4117	1.4142	1129.4	3.299	282.3	715.9	1576	1.4906								
82.00	1.4250	1.4275	1168.5	3.414	256.5	741.7	1628	1.4958								
84.00	1.4383	1.4409	1208.2	3.530	230.1	768.1	1681	1.5010								

89 SULFURIC ACID, H_2SO_4

MOLECULAR WEIGHT = 98.08
RELATIVE SPECIFIC REFRACTIVITY = 0.685

0.00 % by wt. data are the same for all compounds.
For Values of 0.00 wt. % solutions see Table 1, Acetic Acid.

A% by wt.	ρ D_4^{20}	D_{20}^{20}	C_s g/l	M g-mol/l	C_w g/l	$(C_o - C_w)$ g/l	$(n - n_o) \times 10^4$	n	Δ °C	O Os/kg	S g-mol/l	η/η_o	η/ρ cS	φ rhe	γ mmho/cm	T g-mol/l
0.50	1.0016	1.0034	5.0	0.051	996.6	1.7	6	1.3336	0.210	0.113	0.060	1.008	1.009	98.96	24.3	0.277
1.00	1.0049	1.0067	10.0	0.102	994.9	3.3	13	1.3342	0.423	0.227	0.122	1.017	1.014	98.13	47.8	0.573
1.50	1.0083	1.0101	15.1	0.154	993.2	5.1	19	1.3349	0.662	0.356	0.192	1.025	1.019	97.34	70.3	0.886
2.00	1.0116	1.0134	20.2	0.206	991.4	6.8	25	1.3355	0.796	0.428	0.232	1.034	1.024	96.52	92.	1.22
2.50	1.0150	1.0168	25.4	0.259	989.6	8.6	31	1.3361	1.004	0.540	0.292	1.044	1.031	95.66	113.	1.50
3.00	1.0183	1.0201	30.6	0.311	987.8	10.4	37	1.3367	1.172	0.630	0.343	1.057	1.040	94.45	134.	1.98
3.50	1.0217	1.0235	35.8	0.365	985.9	12.3	43	1.3373	1.354	0.728	0.396	1.070	1.049	93.29	155.	2.42
4.00	1.0250	1.0269	41.0	0.418	984.0	14.2	49	1.3379	1.599	0.860	0.468	1.083	1.059	92.11	175.	2.93
4.50	1.0284	1.0302	46.3	0.472	982.1	16.1	55	1.3385	1.855	0.998	0.543	1.097	1.069	90.97	194.	3.57
5.00	1.0318	1.0336	51.6	0.526	980.2	18.0	61	1.3391	2.047	1.101	0.598	1.110	1.078	89.91	211.	4.25
5.50	1.0352	1.0370	56.9	0.580	978.2	20.0	67	1.3397	2.259	1.214	0.659	1.122	1.086	88.93		
6.00	1.0385	1.0404	62.3	0.635	976.2	22.0	73	1.3403	2.495	1.341	0.727	1.134	1.094	88.00		
6.50	1.0419	1.0438	67.7	0.691	974.2	24.0	79	1.3409	2.730	1.468	0.795	1.146	1.102	87.10		
7.00	1.0453	1.0472	73.2	0.746	972.2	26.1	85	1.3415	2.952	1.587	0.858	1.157	1.109	86.24		
7.50	1.0488	1.0506	78.7	0.802	970.1	28.1	91	1.3421	3.197	1.719	0.927	1.169	1.117	85.39		
8.00	1.0522	1.0541	84.2	0.858	968.0	30.2	97	1.3427	3.493	1.878	1.010	1.180	1.124	84.56		
8.50	1.0556	1.0575	89.7	0.915	965.9	32.3	103	1.3433	3.801	2.043	1.096	1.192	1.131	83.74		
9.00	1.0591	1.0610	95.3	0.972	963.8	34.4	109	1.3439	4.083	2.195	1.174	1.204	1.139	82.92		
9.50	1.0626	1.0645	100.9	1.029	961.6	36.6	115	1.3445	4.360	2.344	1.250	1.216	1.146	82.10		
10.00	1.0661	1.0680	106.6	1.087	959.5	38.8	121	1.3451	4.644	2.497	1.328	1.228	1.154	81.27		
11.00	1.0731	1.0750	118.0	1.204	955.1	43.2	133	1.3463	5.25	2.82	1.490	1.253	1.170	79.63		
12.00	1.0802	1.0821	129.6	1.322	950.6	47.6	145	1.3475	5.93	3.19	1.669	1.279	1.187	78.02		
13.00	1.0874	1.0893	141.4	1.441	946.0	52.2	158	1.3488	6.67	3.59	1.859	1.306	1.203	76.43		
14.00	1.0947	1.0966	153.3	1.563	941.4	56.8	170	1.3500	7.49	4.03	2.063	1.334	1.221	74.82		
15.00	1.1020	1.1039	165.3	1.685	936.7	61.5	183	1.3513	8.35	4.49	2.270	1.364	1.240	73.17		
16.00	1.1094	1.1114	177.5	1.810	931.9	66.3	195	1.3525	9.26	4.98	2.483	1.396	1.261	71.47		
17.00	1.1169	1.1189	189.9	1.936	927.0	71.2	208	1.3538	10.23	5.50	2.702	1.431	1.284	69.74		
18.00	1.1245	1.1265	202.4	2.064	922.1	76.2	221	1.3551	11.29	6.07	2.932	1.467	1.308	68.01		
19.00	1.1321	1.1341	215.1	2.193	917.0	81.2	233	1.3563	12.43	6.68	3.169	1.505	1.332	66.32		
20.00	1.1398	1.1418	228.0	2.324	911.9	86.4	246	1.3576	13.64	7.33	3.409	1.543	1.356	64.68		
22.00	1.1554	1.1575	254.2	2.592	901.2	97.0	272	1.3602	16.48	8.86	3.932	1.621	1.405	61.58		
24.00	1.1714	1.1735	281.1	2.866	890.3	108.0	298	1.3628	19.85	10.67	4.488	1.703	1.457	58.60		
26.00	1.1872	1.1893	308.7	3.147	878.5	119.7	323	1.3653	24.29	13.06		1.793	1.513	55.67		
28.00	1.2031	1.2052	336.9	3.435	866.2	132.0	347	1.3677	29.65	15.94		1.890	1.574	52.81		
30.00	1.2191	1.2213	365.7	3.729	853.4	144.9	371	1.3701	36.21	19.47		1.997	1.641	49.98		
32.00	1.2353	1.2375	395.3	4.030	840.0	158.2	395	1.3725	44.76	24.07		2.118	1.718	47.12		
34.00	1.2518	1.2540	425.6	4.339	826.2	172.1	419	1.3749	55.28	29.72		2.250	1.801	44.36		
36.00	1.2685	1.2707	456.7	4.656	811.8	186.4	443	1.3773				2.387	1.885	41.82		
38.00	1.2855	1.2878	488.5	4.981	797.0	201.2	467	1.3797				2.528	1.970	39.48		
40.00	1.3028	1.3051	521.1	5.313	781.7	216.5	491	1.3821				2.685	2.065	37.17		
42.00	1.3205	1.3229	554.6	5.655	765.9	232.3	516	1.3846				2.866	2.174	34.83		
44.00	1.3386	1.3410	589.0	6.005	749.6	248.6	540	1.3870				3.067	2.296	32.53		
46.00	1.3570	1.3594	624.2	6.365	732.8	265.4	565	1.3895				3.292	2.431	30.32		
48.00	1.3759	1.3783	660.4	6.734	715.5	282.8	590	1.3920				3.539	2.577	28.20		
50.00	1.3952	1.3977	697.6	7.113	697.6	300.6	616	1.3945				3.818	2.742	26.14		
52.00	1.4149	1.4174	735.8	7.502	679.2	319.1	641	1.3971				4.134	2.927	24.14		
54.00	1.4351	1.4377	775.0	7.901	660.2	338.1	667	1.3997				4.490	3.135	22.23		
56.00	1.4558	1.4584	815.3	8.312	640.6	357.7	694	1.4024				4.896	3.370	20.38		
58.00	1.4770	1.4796	856.7	8.734	620.3	377.9	720	1.4050				5.343	3.625	18.68		
60.00	1.4987	1.5013	899.2	9.168	599.5	398.8	747	1.4077				5.905	3.948	16.90		
62.00	1.5200	1.5227	942.4	9.608	577.6	420.6										
64.00	1.5421	1.5448	986.9	10.062	555.2	443.0										
66.00	1.5646	1.5674	1032.6	10.528	532.0	466.2										
68.00	1.5874	1.5902	1079.4	11.005	508.0	490.2										
70.00	1.6105	1.6134	1127.4	11.495	483.1	515.1										
72.00	1.6338	1.6367	1176.3	11.993	457.5	540.7										
74.00	1.6574	1.6603	1226.5	12.505	430.9	567.3										
76.00	1.6810	1.6840	1277.6	13.026	403.4	594.8										
78.00	1.7043	1.7073	1329.4	13.554	374.9	623.3										
80.00	1.7272	1.7303	1381.8	14.088	345.4	652.8										
82.00	1.7491	1.7522	1434.3	14.624	314.8	683.4										
84.00	1.7693	1.7724	1486.2	15.153	283.1	715.1										
86.00	1.7872	1.7904	1537.0	15.671	250.2	748.0										
88.00	1.8022	1.8054	1585.9	16.169	216.3	781.9										
90.00	1.8144	1.8176	1633.0	16.650	181.4	816.8										
92.00	1.8240	1.8272	1678.1	17.110	145.9	852.3										
94.00	1.8312	1.8344	1721.3	17.550	109.9	888.3										
96.00	1.8355	1.8388	1762.1	17.966	73.4	924.8										
98.00	1.8361	1.8394	1799.4	18.346	36.7	961.5										
100.00	1.8305	1.8337	1830.5	18.663	0.0	998.2										

90 TARTARIC ACID, $HO_2C(CHOH)_2CO_2H$

MOLECULAR WEIGHT = 150.09

A% by wt.	ρ D^{20}_4	D^{20}_{20}	C_s g/l	M g-mol/l	C_w g/l	$(C_o - C_w)$ g/l	$(n - n_o)$ × 10^4	n	Δ °C	O Os/kg	S g-mol/l	η/η_o	η/ρ cS	φ rhe	γ mmho/cm	T g-mol/l
1.00	1.0029	1.0047	10.0	0.067	992.9	5.3	12	1.3342	0.132	0.070	0.037	1.021	1.020	97.76	2.5	0.025
2.00	1.0072	1.0090	20.1	0.134	987.1	11.1	25	1.3355	0.262	0.141	0.075	1.044	1.039	95.56	3.9	0.039
3.00	1.0121	1.0139	30.4	0.202	981.7	16.5	37	1.3367	0.402	0.216	0.116	1.069	1.058	93.39	4.9	0.049
4.00	1.0163	1.0181	40.7	0.271	975.6	22.6	49	1.3379	0.538	0.290	0.156	1.092	1.077	91.36	5.5	0.056
5.00	1.0208	1.0226	51.0	0.340	969.8	28.4	62	1.3392	0.681	0.366	0.198	1.119	1.098	89.22	6.2	0.063
6.00	1.0255	1.0273	61.5	0.410	964.0	34.2	73	1.3403	0.824	0.443	0.240	1.147	1.121	86.99	6.9	0.070
7.00	1.0310	1.0329	72.2	0.481	958.8	39.4	88	1.3418	0.974	0.523	0.284	1.176	1.143	84.86	7.3	0.075
8.00	1.0348	1.0367	82.8	0.552	952.0	46.2	99	1.3429	1.130	0.607	0.330	1.207	1.169	82.67	7.8	0.080
9.00	1.0393	1.0412	93.5	0.623	945.8	52.4	111	1.3441	1.283	0.690	0.375	1.236	1.192	80.72	8.2	0.084
10.00	1.0443	1.0462	104.4	0.696	939.9	58.3	124	1.3454	1.443	0.776	0.422	1.272	1.220	78.49	8.5	0.088
12.00	1.0540	1.0559	126.5	0.843	927.5	70.7	151	1.3481	1.768	0.951	0.517	1.343	1.277	74.30	9.2	0.095
14.00	1.0639	1.0658	148.9	0.992	915.0	83.2	178	1.3508	2.130	1.145	0.622	1.413	1.331	70.62	9.7	0.100
16.00	1.0738	1.0757	171.8	1.145	902.0	96.2	205	1.3535	2.504	1.346	0.730	1.498	1.398	66.61	10.1	0.104
18.00	1.0839	1.0859	195.1	1.300	888.8	109.4	231	1.3561	2.917	1.568	0.848	1.591	1.471	62.72	10.3	0.107
20.00	1.0942	1.0962	218.8	1.458	875.4	122.8	260	1.3590	3.360	1.807	0.973	1.701	1.558	58.66	10.5	0.109
24.00	1.1152	1.1172	267.6	1.783	847.6	150.6	316	1.3646	4.309	2.316	1.236	1.957	1.758	51.01	10.6	0.110
28.00	1.1366	1.1387	318.3	2.120	818.3	179.9	375	1.3705	5.46	2.93	1.546	2.296	2.024	43.47	10.4	0.108
32.00	1.1588	1.1609	370.8	2.471	788.0	210.2	436	1.3766	6.73	3.62	1.874	2.695	2.330	37.04	10.1	0.104
36.00	1.1813	1.1834	425.3	2.833	756.0	242.2	499	1.3829	8.30	4.47	2.259	3.342	2.750	30.78	9.3	0.096
40.00	1.2049	1.2071	482.0	3.211	722.9	275.3	561	1.3891	10.20	5.48	2.695	4.002	3.328	24.94	8.2	0.085
44.00	1.2290	1.2312	540.8	3.603	688.2	310.0	627	1.3957				5.088	4.148	19.62	7.3	0.074
48.00	1.2535	1.2558	601.7	4.009	651.8	346.4	692	1.4022				6.605	5.280	15.11	6.2	0.063
52.00	1.2787	1.2810	664.9	4.430	613.8	384.3	762	1.4092				8.961	7.022	11.14	5.0	0.050
56.00	1.3058	1.3082	731.3	4.872	574.5	423.7	834	1.4164				12.60	9.670	7.92	3.7	0.037
58.00	1.3196	1.3220	765.4	5.099	554.2	444.0	869	1.4199				15.64	11.87	6.38	3.0	0.030

91 TETRACAINE HYDROCHLORIDE, $C_{15}H_{24}N_2O_2 \cdot HCl$

MOLECULAR WEIGHT = 300.84

A% by wt.	ρ D^{20}_4	D^{20}_{20}	C_s g/l	M g-mol/l	C_w g/l	$(C_o - C_w)$ g/l	$(n - n_o)$ × 10^4	n	Δ °C	O Os/kg	S g-mol/l	η/η_o	η/ρ cS	φ rhe	γ mmho/cm	T g-mol/l
0.50	0.9988	1.0006	5.0	0.017	993.8	4.4	11	1.3341	0.058	0.031	0.016	1.013	1.016	98.54	1.2	0.012
1.00	0.9995	1.0013	10.0	0.033	989.5	8.7	23	1.3353	0.110	0.059	0.031	1.029	1.032	96.95	2.4	0.024
1.50	1.0001	1.0019	15.0	0.050	985.1	13.1	34	1.3364	0.159	0.086	0.045	1.045	1.047	95.50	3.6	0.036
2.00	1.0008	1.0026	20.0	0.067	980.8	17.4	46	1.3376	0.204	0.109	0.058	1.063	1.064	93.91	4.7	0.047
2.50	1.0014	1.0032	25.0	0.083	976.4	21.8	57	1.3387				1.081	1.082	92.29	5.6	0.057
3.00	1.0021	1.0039	30.1	0.100	972.0	26.2	69	1.3399				1.100	1.100	90.72	6.6	0.067
3.50	1.0027	1.0045	35.1	0.117	967.6	30.6	80	1.3410				1.121	1.120	89.05	7.4	0.076
4.00	1.0034	1.0052	40.1	0.133	963.3	34.9	92	1.3422				1.142	1.140	87.42	8.2	0.084
4.50	1.0040	1.0058	45.2	0.150	958.8	39.4	103	1.3433				1.163	1.161	85.79	8.9	0.092
5.00	1.0046	1.0064	50.2	0.167	954.4	43.8	115	1.3445				1.187	1.184	84.07	9.6	0.099
5.50	1.0053	1.0071	55.3	0.184	950.0	48.2	126	1.3456				1.211	1.207	82.41	10.2	0.106
6.00	1.0059	1.0077	60.4	0.201	945.5	52.7	138	1.3468				1.237	1.232	80.69	10.8	0.113
6.50	1.0066	1.0084	65.4	0.217	941.2	57.0	149	1.3479				1.263	1.257	79.03	11.4	0.120
7.00	1.0072	1.0090	70.5	0.234	936.7	61.5	160	1.3490				1.291	1.284	77.33	11.9	0.126
7.50	1.0079	1.0097	75.6	0.251	932.3	65.9	172	1.3502				1.319	1.311	75.68	12.5	0.132
8.00	1.0085	1.0103	80.7	0.268	927.8	70.4	183	1.3513				1.348	1.339	74.05	13.0	0.138
8.50	1.0092	1.0110	85.8	0.285	923.4	74.8	195	1.3525				1.379	1.369	72.38	13.6	0.144
9.00	1.0098	1.0116	90.9	0.302	918.9	79.3	206	1.3536				1.410	1.399	70.79	14.0	0.149
9.50	1.0105	1.0123	96.0	0.319	914.5	83.7	218	1.3548				1.441	1.429	69.25	14.4	0.154
10.00	1.0111	1.0129	101.1	0.336	910.0	88.2	299	1.3559				1.477	1.464	67.56	14.7	0.158
11.00	1.0124	1.0142	111.4	0.370	901.0	97.2	252	1.3582				1.548	1.532	64.48	15.8	0.170
12.00	1.0137	1.0155	121.6	0.404	892.1	106.1	275	1.3605				1.623	1.604	61.50	16.8	0.182

92 TRICHLOROACETIC ACID, CCl_3COOH

MOLECULAR WEIGHT = 163.38
RELATIVE SPECIFIC REFRACTIVITY = 0.865

0.00 % by wt. data are the same for all compounds.
For Values of 0.00 wt. % solutions see Table 1, Acetic Acid.

A % by wt.	ρ D_4^{20}	D_{20}^{20}	C_s g/l	M g-mol/l	C_w g/l	$(C_o - C_w)$ g/l	$(n - n_o) \times 10^4$	n	Δ °C	O Os/kg	S g-mol/l	$(n - n_o)$ η/η_o	η/ρ cS	φ rhe	γ mmho/cm	T g-mol/l
0.50	1.0008	1.0026	5.0	0.031	995.8	2.4	7	1.3337	0.105	0.057	0.030	1.009	1.010	98.91	10.3	0.107
1.00	1.0034	1.0051	10.0	0.061	993.3	4.9	13	1.3343	0.211	0.113	0.060	1.019	1.018	97.94	19.6	0.215
2.00	1.0083	1.0101	20.2	0.123	988.2	10.1	26	1.3356	0.423	0.227	0.122	1.042	1.035	95.78	37.2	0.435
3.00	1.0133	1.0151	30.4	0.186	982.9	15.4	39	1.3369	0.638	0.343	0.185	1.067	1.055	93.53	54.0	0.656
4.00	1.0182	1.0200	40.7	0.249	977.4	20.8	52	1.3381	0.857	0.461	0.250	1.094	1.077	91.24	69.8	0.880
5.00	1.0230	1.0248	51.2	0.313	971.9	26.4	64	1.3394	1.079	0.580	0.315	1.121	1.098	89.03	84.7	1.10
6.00	1.0279	1.0297	61.7	0.377	966.2	32.0	76	1.3406	1.303	0.700	0.381	1.148	1.119	86.95	99.	1.33
7.00	1.0328	1.0346	72.3	0.443	960.3	37.7	89	1.3418	1.529	0.822	0.447	1.175	1.140	84.95	113.	1.57
8.00	1.0378	1.0396	83.0	0.508	954.8	43.5	101	1.3431	1.758	0.945	0.514	1.202	1.161	83.01	125.	1.80
9.00	1.0428	1.0446	93.9	0.574	948.9	49.3	114	1.3444	1.992	1.071	0.582	1.231	1.182	81.10	137.	2.04
10.00	1.0479	1.0497	104.8	0.641	943.1	55.1	126	1.3456	2.233	1.201	0.652	1.260	1.205	79.21	148.	2.28
12.00	1.0583	1.0602	127.0	0.777	931.3	66.9	153	1.3483	2.734	1.470	0.796	1.323	1.252	75.44	169.	2.76
14.00	1.0692	1.0710	149.7	0.916	919.5	78.8	180	1.3510	3.261	1.753	0.945	1.390	1.302	71.81	186.	3.24
16.00	1.0806	1.0825	172.9	1.058	907.7	90.6	209	1.3539	3.815	2.051	1.100	1.459	1.353	68.41	200.	3.73
18.00	1.0921	1.0941	196.6	1.203	895.5	102.7	239	1.3568				1.530	1.404	65.21	211.	4.22
20.00	1.1035	1.1055	220.7	1.351	882.8	115.4	267	1.3597				1.605	1.457	62.18	221.	4.90
24.00	1.1260	1.1280	270.2	1.654	855.8	142.5	322	1.3652				1.764	1.570	56.58		
28.00	1.1485	1.1505	321.6	1.968	826.9	171.3	376	1.3705				1.931	1.685	51.68		
32.00	1.1713	1.1734	374.8	2.294	796.5	201.8	429	1.3759				2.114	1.808	47.22		
36.00	1.1947	1.1968	430.1	2.633	764.6	233.6	483	1.3813				2.315	1.942	43.11		
40.00	1.2188	1.2210	487.5	2.984	731.3	266.9	538	1.3868				2.540	2.088	39.29		
44.00	1.2435	1.2457	547.1	3.349	696.3	301.9	593	1.3923				2.791	2.249	35.75		
48.00	1.2682	1.2704	608.7	3.726	659.5	338.8	647	1.3977				3.070	2.425	32.51		
50.00	1.2803	1.2826	640.2	3.918	640.2	358.1	673	1.4003				3.219	2.519	31.00		

93 TRIS(HYDROXYMETHYL)AMINOMETHANE,* THAM,® $H_2NC(CH_2OH)_3$

MOLECULAR WEIGHT = 121.14
RELATIVE SPECIFIC REFRACTIVITY = 1.169

0.00 % by wt. data are the same for all compounds.
For Values of 0.00 wt. % solutions see Table 1, Acetic Acid.

A % by wt.	ρ D_4^{20}	D_{20}^{20}	C_s g/l	M g-mol/l	C_w g/l	$(C_o - C_w)$ g/l	$(n - n_o) \times 10^4$	n	Δ °C	O Os/kg	S g-mol/l	η/η_o	η/ρ cS	φ rhe	γ mmho/cm	T g-mol/l
0.50	0.9994	1.0012	5.0	0.041	994.4	3.8	7	1.3337	0.080	0.043	0.022	1.012	1.015	98.59		
1.00	1.0006	1.0024	10.0	0.083	990.6	7.6	15	1.3344	0.159	0.085	0.045	1.025	1.026	97.37		
2.00	1.0030	1.0048	20.1	0.166	982.9	15.3	29	1.3359	0.314	0.169	0.090	1.052	1.051	94.87		
3.00	1.0054	1.0072	30.2	0.249	975.2	23.0	44	1.3374	0.473	0.254	0.137	1.081	1.078	92.29		
4.00	1.0078	1.0096	40.3	0.333	967.5	30.7	58	1.3388	0.636	0.342	0.185	1.113	1.106	89.67		
5.00	1.0103	1.0121	50.5	0.417	959.8	38.4	73	1.3403	0.804	0.432	0.234	1.146	1.137	87.09		
6.00	1.0128	1.0146	60.8	0.502	952.0	46.2	88	1.3418	0.974	0.524	0.284	1.180	1.168	84.57		
7.00	1.0153	1.0171	71.1	0.587	944.3	54.0	103	1.3433	1.149	0.618	0.336	1.216	1.200	82.10		
8.00	1.0179	1.0197	81.4	0.672	936.4	61.8	118	1.3448	1.328	0.714	0.388	1.253	1.233	79.66		
9.00	1.0204	1.0222	91.8	0.758	928.6	69.6	133	1.3463	1.510	0.812	0.442	1.292	1.269	77.24		
10.00	1.0230	1.0248	102.3	0.844	920.7	77.5	148	1.3478	1.696	0.912	0.496	1.334	1.307	74.81		
12.00	1.0282	1.0301	123.4	1.019	904.8	93.4	178	1.3508	2.076	1.116	0.606	1.424	1.388	70.09		
14.00	1.0335	1.0354	144.7	1.194	888.8	109.4	209	1.3539	2.473	1.329	0.721	1.524	1.478	65.48		
16.00	1.0389	1.0407	166.2	1.372	872.7	125.6	240	1.3570	2.899	1.559	0.843	1.639	1.581	60.88		
18.00	1.0443	1.0462	188.0	1.552	856.4	141.9	271	1.3601	3.357	1.805	0.972	1.768	1.696	56.45		
20.00	1.0498	1.0517	210.0	1.733	839.9	158.4	303	1.3633	3.847	2.068	1.109	1.916	1.829	52.09		
22.00	1.0554	1.0572	232.2	1.917	823.2	175.0	335	1.3665				2.079	1.974	48.00		
24.00	1.0610	1.0628	254.6	2.102	806.3	191.9	367	1.3697				2.256	2.131	44.23		
26.00	1.0666	1.0685	277.3	2.289	789.3	208.9	400	1.3730				2.459	2.310	40.59		
28.00	1.0723	1.0742	300.2	2.479	772.1	226.2	433	1.3763				2.700	2.523	36.97		
30.00	1.0781	1.0800	323.4	2.670	754.6	243.6	467	1.3797				2.992	2.781	33.36		
32.00	1.0839	1.0858	346.8	2.863	737.0	261.2	501	1.3831				3.337	3.085	29.91		
34.00	1.0897	1.0916	370.5	3.058	719.2	279.0	535	1.3865				3.729	3.429	26.76		
36.00	1.0956	1.0976	394.4	3.256	701.2	297.0	570	1.3900				4.169	3.812	23.94		
38.00	1.1016	1.1035	418.6	3.456	683.0	315.2	605	1.3935				4.658	4.237	21.43		
40.00	1.1076	1.1096	443.1	3.657	664.6	333.6	640	1.3970				5.198	4.702	19.20		

* 1,2-Propanediol, 2-amino-2(hydroxymethyl). ® Registered trademark.

94 UREA, NH_2CONH_2

MOLECULAR WEIGHT = 60.06
RELATIVE SPECIFIC REFRACTIVITY = 1.147

0.00 % by wt. data are the same for all compounds.
For Values of 0.00 wt. % solutions see Table 1, Acetic Acid.

A% by wt.	ρ D_4^{20}	D_{20}^{20}	C_s g/l	M g-mol/l	C_w g/l	$(C_o - C_w)$ g/l	$(n - n_o)$ × 10^4	n	Δ °C	O Os/kg	S g-mol/l	η/η_o	η/ρ cS	φ rhe	γ mmho/cm	T g-mol/l
0.50	0.9995	1.0012	5.0	0.083	994.5	3.8	7	1.3337	0.155	0.083	0.044	1.005	1.007	99.33		
1.00	1.0007	1.0025	10.0	0.167	990.7	7.5	14	1.3344	0.310	0.167	0.089	1.008	1.009	99.01		
1.50	1.0020	1.0038	15.0	0.250	987.0	11.3	21	1.3351	0.465	0.250	0.134	1.009	1.009	98.89		
2.00	1.0033	1.0050	20.1	0.334	983.2	15.0	28	1.3358	0.620	0.333	0.180	1.010	1.009	98.81		
2.50	1.0045	1.0063	25.1	0.418	979.4	18.8	35	1.3365	0.775	0.417	0.226	1.012	1.010	98.60		
3.00	1.0058	1.0076	30.2	0.502	975.7	22.6	42	1.3372	0.928	0.499	0.271	1.015	1.011	98.30		
3.50	1.0071	1.0089	35.3	0.587	971.9	26.3	50	1.3379	1.082	0.582	0.316	1.019	1.014	97.95		
4.00	1.0085	1.0102	40.3	0.672	968.1	30.1	57	1.3387	1.237	0.665	0.362	1.023	1.016	97.56		
4.50	1.0098	1.0116	45.4	0.757	964.3	33.9	64	1.3394	1.393	0.749	0.407	1.027	1.019	97.17		
5.00	1.0111	1.0129	50.6	0.842	960.6	37.7	71	1.3401	1.552	0.835	0.454	1.031	1.022	96.80		
5.50	1.0125	1.0142	55.7	0.927	956.8	41.5	79	1.3409	1.715	0.922	0.501	1.035	1.024	96.45		
6.00	1.0138	1.0156	60.8	1.013	953.0	45.3	86	1.3416	1.880	1.011	0.550	1.039	1.027	96.09		
6.50	1.0151	1.0169	66.0	1.099	949.2	49.1	93	1.3423	2.048	1.101	0.598	1.043	1.029	95.72		
7.00	1.0165	1.0183	71.2	1.185	945.4	52.9	101	1.3431	2.218	1.192	0.648	1.047	1.032	95.35		
7.50	1.0179	1.0197	76.3	1.271	941.5	56.7	108	1.3438	2.389	1.285	0.697	1.051	1.034	94.98		
8.00	1.0192	1.0210	81.5	1.358	937.7	60.5	116	1.3446	2.562	1.378	0.747	1.055	1.037	94.60		
8.50	1.0206	1.0224	86.8	1.444	933.9	64.4	123	1.3453	2.737	1.471	0.797	1.059	1.040	94.23		
9.00	1.0220	1.0238	92.0	1.531	930.0	68.2	131	1.3461	2.911	1.565	0.846	1.063	1.043	93.85		
9.50	1.0234	1.0252	97.2	1.619	926.2	72.1	138	1.3468	3.086	1.659	0.896	1.068	1.045	93.47		
10.00	1.0248	1.0266	102.5	1.706	922.3	75.9	146	1.3476	3.260	1.753	0.945	1.072	1.048	93.10		
11.00	1.0276	1.0294	113.0	1.882	914.5	83.7	161	1.3491	3.606	1.939	1.042	1.081	1.054	92.36		
12.00	1.0304	1.0322	123.6	2.059	906.7	91.5	176	1.3506	3.952	2.125	1.138	1.089	1.059	91.64		
13.00	1.0332	1.0350	134.3	2.236	898.9	99.4	192	1.3521	4.301	2.312	1.234	1.098	1.065	90.91		
14.00	1.0360	1.0378	145.0	2.415	891.0	107.3	207	1.3537	4.656	2.503	1.332	1.107	1.071	90.15		
15.00	1.0388	1.0407	155.8	2.594	883.0	115.2	222	1.3552	5.02	2.70	1.430	1.117	1.077	89.35		
16.00	1.0417	1.0435	166.7	2.775	875.0	123.2	238	1.3568	5.40	2.90	1.531	1.128	1.085	88.50		
17.00	1.0445	1.0463	177.6	2.956	866.9	131.3	253	1.3583	5.79	3.11	1.633	1.139	1.093	87.62		
18.00	1.0473	1.0492	188.5	3.139	858.8	139.4	269	1.3599	6.19	3.33	1.736	1.151	1.101	86.72		
19.00	1.0502	1.0520	199.5	3.322	850.6	147.6	284	1.3614	6.59	3.54	1.839	1.163	1.110	85.80		
20.00	1.0530	1.0549	210.6	3.506	842.4	155.8	300	1.3629	7.00	3.76	1.941	1.176	1.119	84.86		
22.00	1.0586	1.0605	232.9	3.878	825.7	172.5	331	1.3661	7.81	4.20	2.141	1.203	1.139	82.93		
24.00	1.0643	1.0662	255.4	4.253	808.9	189.4	362	1.3692	8.64	4.65	2.339	1.233	1.161	80.95		
26.00	1.0699	1.0718	278.2	4.632	791.7	206.5	393	1.3723	9.52	5.12	2.543	1.263	1.183	79.00		
28.00	1.0756	1.0775	301.2	5.014	774.4	223.8	424	1.3754	10.45	5.62	2.751	1.295	1.206	77.09		
30.00	1.0812	1.0831	324.4	5.401	756.8	241.4	455	1.3785	11.40	6.13	2.955	1.329	1.232	75.09		
32.00	1.0869	1.0888	347.8	5.791	739.1	259.2	487	1.3817	12.34	6.63	3.150	1.368	1.261	72.97		
34.00	1.0926	1.0945	371.5	6.185	721.1	277.1	518	1.3848	13.27	7.14	3.337	1.410	1.293	70.77		
36.00	1.0984	1.1004	395.4	6.584	703.0	295.2	551	1.3881	14.20	7.63	3.516	1.456	1.329	68.52		
38.00	1.1044	1.1064	419.7	6.988	684.7	313.5	583	1.3913	15.11	8.12	3.686	1.506	1.367	66.26		
40.00	1.1106	1.1126	444.2	7.397	666.4	331.9	617	1.3947	15.99	8.60	3.845	1.562	1.409	63.89		
42.00	1.1171	1.1191	469.2	7.812	647.9	350.3	652	1.3982	16.83	9.05	3.993	1.626	1.458	61.40		
44.00	1.1239	1.1259	494.5	8.234	629.4	368.8	688	1.4018	17.62	9.47	4.127	1.697	1.512	58.83		
46.00	1.1313	1.1333	520.4	8.665	610.9	387.3	726	1.4056				1.776	1.573	56.21		

95 URINE, CAT

RELATIVE SPECIFIC REFRACTIVITY = 1.08

0.00 % by wt. data are the same for all compounds.
For Values of 0.00 wt. % solutions see Table 1, Acetic Acid.

A % by wt.	ρ D_4^{20}	D_{20}^{20}	C_s g/l	C_w g/l	$(C_o - C_w)$ g/l	$(n - n_o)$ × 10^4	n
0.50	1.000	1.002	5	995	3	8	1.3338
1.00	1.002	1.004	10	992	6	16	1.3346
1.50	1.004	1.006	15	989	9	24	1.3354
2.00	1.006	1.008	20	986	12	33	1.3363
2.50	1.008	1.010	25	983	15	41	1.3371
3.00	1.010	1.012	30	979	19	49	1.3379
3.50	1.011	1.013	35	976	22	57	1.3387
4.00	1.013	1.015	41	973	25	65	1.3395
4.50	1.015	1.017	46	970	28	73	1.3403
5.00	1.017	1.019	51	966	32	81	1.3411
5.50	1.019	1.021	56	963	35	89	1.3419
6.00	1.021	1.023	61	960	38	97	1.3427
6.50	1.023	1.025	67	956	42	105	1.3435
7.00	1.025	1.027	72	953	45	113	1.3443
7.50	1.026	1.028	77	950	48	121	1.3451
8.00	1.028	1.030	82	946	52	129	1.3459
8.50	1.030	1.032	88	943	55	136	1.3466
9.00	1.032	1.034	93	939	59	144	1.3474
9.50	1.034	1.036	98	936	62	152	1.3482
10.00	1.036	1.038	104	932	66	160	1.3490
11.00	1.039	1.041	114	925	73	175	1.3505
12.00	1.043	1.045	125	918	80	191	1.3521
13.00	1.047	1.049	136	911	87	206	1.3536
14.00	1.050	1.052	147	903	95	222	1.3552
15.00	1.054	1.056	158	896	102	237	1.3567
16.00	1.058	1.060	169	888	110	252	1.3582
17.00	1.061	1.063	180	881	117	267	1.3597
18.00	1.065	1.067	192	873	125	282	1.3612
19.00	1.068	1.070	203	865	133	296	1.3626
20.00	1.072	1.074	214	857	141	311	1.3641

96 URINE, GUINEA PIG

RELATIVE SPECIFIC REFRACTIVITIES: CLEAR* = 0.984; TURBID = 0.955

0.00 % by wt. data are the same for all compounds.
For Values of 0.00 wt. % solutions see Table 1, Acetic Acid.

CLEAR								TURBID					
A % by wt.	ρ D_4^{20}	D_{20}^{20}	C_s g/l	C_w g/l	$(C_o - C_w)$ g/l	$(n - n_o)$ × 10^4	n	A % by wt.	ρ D_4^{20}	D_{20}^{20}	C_s g/l	C_w g/l	$(C_o - C_w)$ g/l
0.5	1.001	1.002	5	996	2	8	1.3338	0.5	1.001	1.002	5	996	2
1.0	1.003	1.005	10	993	5	16	1.3346	1.0	1.003	1.005	10	993	5
1.5	1.005	1.007	15	990	8	23	1.3353	1.5	1.005	1.007	15	990	8
2.0	1.008	1.009	20	988	10	31	1.3361	2.1	1.008	1.010	21	987	11
2.5	1.010	1.012	25	985	13	39	1.3369	2.6	1.010	1.012	26	984	14
3.0	1.012	1.014	30	982	16	47	1.3377	3.1	1.013	1.015	31	982	16
3.5	1.015	1.016	36	979	19	55	1.3385	3.6	1.015	1.017	37	978	20
4.0	1.017	1.019	41	976	22	63	1.3393	4.1	1.018	1.020	42	976	22
4.5	1.019	1.021	46	973	25	70	1.3400	4.6	1.020	1.022	47	973	25
5.0	1.022	1.024	51	971	27	78	1.3408	5.2	1.023	1.025	53	970	28
5.5	1.024	1.026	56	968	30	86	1.3416	5.7	1.026	1.027	58	968	30
6.0	1.026	1.028	62	964	34	94	1.3424	6.2	1.029	1.030	64	965	33
6.5	1.029	1.031	67	962	36	102	1.3432	6.7	1.031	1.032	69	962	36
7.0	1.031	1.033	72	959	39	110	1.3440	7.2	1.034	1.035	74	960	38
7.5	1.033	1.035	77	956	42	117	1.3447	7.8	1.036	1.038	81	955	43
8.0	1.036	1.038	83	953	45	125	1.3455	8.3	1.038	1.040	86	952	46
8.5	1.038	1.040	88	950	48	133	1.3463	8.8	1.040	1.042	92	948	50
9.0	1.040	1.042	94	946	52	141	1.3471	9.3	1.043	1.045	97	946	52
9.5	1.043	1.045	99	944	54	149	1.3479	9.8	1.046	1.047	103	943	55
10.0	1.045	1.047	104	941	57	157	1.3487	10.3	1.049	1.050	108	941	57
11.0	1.050	1.052	116	934	64	172	1.3502	11.4	1.054	1.055	120	934	64
12.0	1.054	1.056	126	928	70	188	1.3518	12.4	1.058	1.060	131	927	71
13.0	1.059	1.061	138	921	77	204	1.3534	13.4	1.064	1.066	143	921	77
14.0	1.064	1.066	149	915	83	220	1.3550	14.5	1.069	1.071	155	914	84
15.0	1.069	1.070	160	909	89	235	1.3565	15.5	1.074	1.076	166	908	90
16.0	1.073	1.075	172	901	97	251	1.3581	16.5	1.078	1.080	178	900	98
17.0	1.078	1.080	183	895	103	266	1.3596	17.6	1.084	1.086	191	893	105
18.0	1.083	1.085	195	888	110	282	1.3612	18.6	1.089	1.091	203	886	112
19.0	1.087	1.089	207	880	118	298	1.3628	19.6	1.094	1.096	214	880	118
20.0	1.092	1.094	218	874	124	314	1.3644	20.7	1.099	1.101	227	872	126

* Cleared by centrifuging 15 minutes at 25,000 G.

97 URINE, HUMAN

RELATIVE SPECIFIC REFRACTIVITY = 1.02

0.00 % by wt. data are the same for all compounds.
For Values of 0.00 wt. % solutions see Table 1, Acetic Acid.

A% by wt.	ρ D_4^{20}	D_{20}^{20}	C_s g/l	C_w g/l	$(C_o - C_w)$ g/l	$(n - n_o) \times 10^4$	n
0.50	1.000	1.002	5	996	2	8	1.3338
1.00	1.003	1.005	10	993	5	16	1.3346
1.50	1.005	1.007	15	990	8	24	1.3354
2.00	1.007	1.009	20	987	11	32	1.3362
2.50	1.009	1.011	24	984	14	40	1.3370
3.00	1.012	1.014	31	981	17	47	1.3377
3.50	1.014	1.016	36	978	20	55	1.3385
4.00	1.016	1.018	41	975	23	63	1.3393
4.50	1.018	1.020	46	972	26	71	1.3401
5.00	1.020	1.022	51	969	29	78	1.3408
5.50	1.022	1.024	56	966	32	86	1.3416
6.00	1.024	1.026	61	962	36	94	1.3424
6.50	1.026	1.028	67	959	39	101	1.3431
7.00	1.027	1.029	72	956	42	109	1.3439
7.50	1.029	1.031	77	952	46	116	1.3446
8.00	1.031	1.033	83	949	49	123	1.3453
8.50	1.033	1.035	88	945	53	131	1.3461
9.00	1.034	1.036	93	941	57	138	1.3468
9.50	1.036	1.038	98	937	61	145	1.3475
10.00	1.037	1.039	104	934	64	152	1.3482

98 URINE, RABBIT

RELATIVE SPECIFIC REFRACTIVITIES: CLEAR* = 0.997; TURBID = 0.958

0.00 % by wt. data are the same for all compounds.
For Values of 0.00 wt. % solutions see Table 1, Acetic Acid.

CLEAR								TURBID					
A% by wt.	ρ D_4^{20}	D_{20}^{20}	C_s g/l	C_w g/l	$(C_o - C_w)$ g/l	$(n - n_o) \times 10^4$	n	A% by wt.	ρ D_4^{20}	D_{20}^{20}	C_s g/l	C_w g/l	$(C_o - C_w)$ g/l
0.5	1.000	1.002	5	995	3	7	1.3337	0.6	1.002	1.003	6	996	2
1.0	1.003	1.004	10	993	5	15	1.3345	1.1	1.003	1.005	11	992	6
1.5	1.005	1.007	15	990	8	22	1.3352	1.7	1.006	1.008	17	989	9
2.0	1.007	1.009	20	987	11	30	1.3360	2.2	1.008	1.010	22	986	12
2.5	1.009	1.011	25	984	14	38	1.3368	2.7	1.011	1.012	27	984	14
3.0	1.011	1.013	30	981	17	45	1.3375	3.2	1.013	1.014	32	981	17
3.5	1.014	1.015	35	979	19	53	1.3383	3.8	1.016	1.017	39	977	21
4.0	1.016	1.018	41	975	23	60	1.3390	4.3	1.018	1.019	44	974	24
4.5	1.018	1.020	46	972	26	68	1.3398	4.9	1.020	1.021	50	970	28
5.0	1.020	1.022	51	969	29	76	1.3406	5.4	1.022	1.024	55	967	31
5.5	1.022	1.024	56	966	32	83	1.3413	5.9	1.024	1.026	60	964	34
6.0	1.025	1.026	62	963	35	91	1.3421	6.5	1.027	1.029	67	960	38
6.5	1.027	1.029	67	960	38	98	1.3428	7.0	1.029	1.031	72	957	41
7.0	1.029	1.031	72	957	41	106	1.3436	7.5	1.032	1.034	77	955	43
7.5	1.031	1.033	77	954	44	113	1.3443	8.0	1.034	1.036	83	951	47
8.0	1.033	1.035	83	950	48	121	1.3451	8.6	1.037	1.039	89	948	50
8.5	1.036	1.037	88	948	50	128	1.3458	9.1	1.039	1.041	95	944	54
9.0	1.038	1.040	93	945	53	136	1.3466	9.7	1.042	1.044	101	941	57
9.5	1.040	1.042	99	941	57	143	1.3473	10.2	1.044	1.046	106	938	60
10.0	1.042	1.044	104	938	60	151	1.3481	10.8	1.047	1.049	113	934	64
11.0	1.046	1.048	115	931	67	166	1.3496	11.8	1.051	1.053	124	927	71
12.0	1.051	1.053	126	925	73	181	1.3511	12.9	1.056	1.058	136	920	78
13.0	1.055	1.057	137	918	80	196	1.3526	14.0	1.061	1.063	149	912	86
14.0	1.060	1.062	148	912	86	211	1.3541	15.1	1.066	1.068	161	905	93
15.0	1.064	1.066	160	904	94	226	1.3556	16.2	1.071	1.073	174	897	101
16.0	1.068	1.070	171	897	101	242	1.3572	17.2	1.076	1.077	185	891	107
17.0	1.073	1.075	182	891	107	257	1.3587	18.3	1.080	1.082	198	882	116

* Cleared by centrifuging 15 minutes at 25,000 G.

99 ZINC SULFATE, $ZnSO_4 \cdot 7H_2O$

MOLECULAR WEIGHT = 161.44 FORMULA WEIGHT, HYDRATE = 287.56
RELATIVE SPECIFIC REFRACTIVITY = 0.493

0.00 % by wt. data are the same for all compounds.
For Values of 0.00 wt. % solutions see Table 1, Acetic Acid.

A% by wt.	H% by wt.	ρ D_4^{20}	D_{20}^{20}	C, g/l	M g-mol/l	C_w g/l	$(C_o - C_w)$ g/l	$(n - n_o)$ × 10^4	n	Δ °C	O Os/kg	S g-mol/l	η/η_o	η/ρ cS	φ rhe	γ mmho/cm	T g-mol/l
0.50	0.89	1.0034	1.0051	5.0	0.031	998.3	−0.1	9	1.3339	0.079	0.043	0.022	1.019	1.017	97.97	2.8	0.028
1.00	1.78	1.0085	1.0103	10.1	0.062	998.4	−0.2	18	1.3348	0.150	0.081	0.042	1.038	1.031	96.15	5.4	0.054
1.50	2.67	1.0137	1.0155	15.2	0.094	998.5	−0.3	27	1.3357	0.217	0.117	0.062	1.058	1.046	94.32	7.8	0.080
2.00	3.56	1.0190	1.0208	20.4	0.126	998.6	−0.4	36	1.3366	0.282	0.152	0.081	1.079	1.061	92.49	10.0	0.104
2.50	4.45	1.0243	1.0261	25.6	0.159	998.7	−0.4	45	1.3375	0.346	0.186	0.099	1.101	1.077	90.65	12.1	0.127
3.00	5.34	1.0296	1.0314	30.9	0.191	998.7	−0.5	54	1.3384	0.408	0.219	0.118	1.124	1.094	88.80	13.9	0.148
3.50	6.23	1.0349	1.0368	36.2	0.224	998.7	−0.5	64	1.3393	0.470	0.253	0.136	1.148	1.111	86.91	15.7	0.169
4.00	7.12	1.0403	1.0421	41.6	0.258	998.7	−0.5	73	1.3403	0.531	0.286	0.154	1.173	1.130	85.10	17.3	0.188
4.50	8.02	1.0457	1.0475	47.1	0.291	998.6	−0.4	82	1.3412	0.591	0.318	0.172	1.198	1.148	83.27	18.9	0.207
5.00	8.91	1.0511	1.0530	52.6	0.326	998.5	−0.3	91	1.3421	0.651	0.350	0.189	1.225	1.168	81.47	20.5	0.226
5.50	9.80	1.0565	1.0584	58.1	0.360	998.4	−0.2	100	1.3430	0.711	0.382	0.207	1.252	1.188	79.70	22.0	0.245
6.00	10.69	1.0620	1.0639	63.7	0.395	998.3	0.0	109	1.3439	0.772	0.415	0.225	1.280	1.208	77.98	23.5	0.263
6.50	11.58	1.0675	1.0694	69.4	0.430	998.1	0.1	118	1.3448	0.833	0.448	0.243	1.308	1.228	76.28	24.9	0.280
7.00	12.47	1.0730	1.0749	75.1	0.465	997.9	0.3	127	1.3457	0.893	0.480	0.260	1.338	1.249	74.60	26.3	0.297
7.50	13.36	1.0786	1.0805	80.9	0.501	997.7	0.5	136	1.3466	0.953	0.512	0.278	1.368	1.271	72.95	27.6	0.314
8.00	14.25	1.0842	1.0861	86.7	0.537	997.5	0.8	146	1.3475	1.013	0.544	0.296	1.400	1.294	71.30	28.9	0.330
8.50	15.14	1.0899	1.0918	92.6	0.574	997.2	1.0	155	1.3485	1.073	0.577	0.313	1.433	1.317	69.65	30.1	0.345
9.00	16.03	1.0956	1.0975	98.6	0.611	997.0	1.3	164	1.3494	1.136	0.611	0.332	1.467	1.342	68.01	31.3	0.360
9.50	16.92	1.1013	1.1032	104.6	0.648	996.7	1.6	173	1.3503	1.200	0.645	0.351	1.504	1.368	66.37	32.5	0.375
10.00	17.81	1.1071	1.1091	110.7	0.686	996.4	1.8	183	1.3513	1.267	0.681	0.370	1.542	1.396	64.72	33.7	0.390
11.00	19.59	1.1188	1.1208	123.1	0.762	995.8	2.5	202	1.3532	1.407	0.756	0.411	1.624	1.454	61.45	35.9	0.418
12.00	21.37	1.1308	1.1328	135.7	0.841	995.1	3.1	221	1.3551	1.553	0.835	0.454	1.713	1.518	58.25	38.0	0.445
13.00	23.16	1.1429	1.1450	148.6	0.920	994.4	3.9	241	1.3570	1.711	0.920	0.500	1.810	1.587	55.14	39.9	0.470
14.00	24.94	1.1553	1.1573	161.7	1.002	993.6	4.7	260	1.3590	1.889	1.015	0.552	1.914	1.660	52.14	41.7	0.492
15.00	26.72	1.1679	1.1699	175.2	1.085	992.7	5.6	280	1.3610	2.087	1.122	0.610	2.027	1.739	49.24	43.3	0.513
16.00	28.50	1.1806	1.1827	188.9	1.170	991.7	6.5	300	1.3630	2.306	1.240	0.673	2.148	1.823	46.46	44.6	0.531

OSMOTIC PARAMETERS AND ELECTRICAL CONDUCTIVITIES OF AQUEOUS SOLUTIONS

For any aqueous solutions within the range, the two right hand columns provide values for either condosity (T) or specific conductance (γ) where the other value is given; the three left hand columns provide values for freezing point depression (Δ), osmolality (O) or osmosity (S) where any one of these values is given. Values in all four columns are interconvertible *only* for solutions of sodium chloride, in which case, by definition, osmosity, molarity (M) and condosity are always equal. Other concentrative properties of specific solutions may be found in conversion tables on preceding pages.

Table columns are:

- Δ = freezing point depression, °C
- O = osmolality (=Δ/1.86, Os/kg
- S = osmosity, molar concentration of NaCl solution having same freezing point or osmotic pressure as given solution, g-mol/l
- M = molarity of NaCl solution, g-mol/l
- T = condosity, molar concentration of NaCl solution having same specific conductance at 20°C as given solution, g-mol/l
- γ = specific conductance (electrical) at 20°C, mmho/cm.

Δ °C	O Os/Kg	S,M,T, g-mol/l	γ mmho/cm
0.000	0.000	0.000	0.0
0.037	0.020	0.010	1.0
0.072	0.039	0.020	2.0
0.107	0.057	0.030	3.0
0.142	0.076	0.040	4.0
0.176	0.095	0.050	5.0
0.211	0.113	0.060	5.9
0.245	0.132	0.070	6.9
0.279	0.150	0.080	7.8
0.314	0.169	0.090	8.7
0.348	0.187	0.100	9.7
0.382	0.205	0.110	10.6
0.416	0.224	0.120	11.4
0.450	0.242	0.130	12.3
0.484	0.260	0.140	13.2
0.518	0.279	0.150	14.1
0.552	0.297	0.160	14.9
0.586	0.315	0.170	15.8
0.620	0.333	0.180	16.6
0.654	0.352	0.190	17.5
0.688	0.370	0.200	18.3
0.722	0.388	0.210	19.2
0.756	0.407	0.220	20.0
0.790	0.425	0.230	20.8
0.824	0.443	0.240	21.6
0.858	0.461	0.250	22.4
0.892	0.480	0.260	23.3
0.926	0.498	0.270	24.1
0.960	0.516	0.280	24.9
0.994	0.534	0.290	25.7
1.028	0.553	0.300	26.5
1.062	0.571	0.310	27.3
1.096	0.589	0.320	28.1
1.130	0.607	0.330	28.9
1.164	0.626	0.340	29.7
1.198	0.644	0.350	30.5
1.232	0.662	0.360	31.3
1.266	0.681	0.370	32.1
1.300	0.699	0.380	32.9
1.334	0.717	0.390	33.7
1.368	0.736	0.400	34.4
1.402	0.754	0.410	35.2
1.436	0.772	0.420	36.0
1.470	0.791	0.430	36.8
1.505	0.809	0.440	37.6
1.539	0.827	0.450	38.4
1.573	0.846	0.460	39.2
1.607	0.864	0.470	39.9
1.641	0.882	0.480	40.7
1.676	0.901	0.490	41.5
1.710	0.919	0.500	42.3
1.744	0.938	0.510	43.0
1.778	0.956	0.520	43.8
1.813	0.975	0.530	44.6
1.847	0.993	0.540	45.3
1.881	1.011	0.550	46.1
1.916	1.030	0.560	46.8
1.950	1.048	0.570	47.6
1.985	1.067	0.580	48.3
2.019	1.086	0.590	49.1
2.054	1.104	0.600	49.8
2.088	1.123	0.610	50.6
2.123	1.141	0.620	51.3
2.157	1.160	0.630	52.0
2.192	1.178	0.640	52.8
2.226	1.197	0.650	53.5
2.261	1.216	0.660	54.2
2.296	1.234	0.670	55.0
2.330	1.253	0.680	55.7
2.365	1.271	0.690	56.4
2.400	1.290	0.700	57.1
2.434	1.309	0.710	57.9
2.469	1.328	0.720	58.6
2.504	1.346	0.730	59.3
2.539	1.365	0.740	60.0
2.574	1.384	0.750	60.7
2.609	1.403	0.760	61.4
2.644	1.421	0.770	62.1

Δ °C	O Os/Kg	S,M,T, g-mol/l	γ mmho/cm
2.679	1.440	0.780	62.9
2.714	1.459	0.790	63.6
2.749	1.478	0.800	64.3
2.784	1.497	0.810	65.0
2.819	1.515	0.820	65.7
2.854	1.534	0.830	66.3
2.889	1.553	0.840	67.0
2.924	1.572	0.850	67.7
2.959	1.591	0.860	68.4
2.995	1.610	0.870	69.1
3.030	1.629	0.880	69.8
3.065	1.648	0.890	70.5
3.101	1.667	0.900	71.2
3.136	1.686	0.910	71.8
3.171	1.705	0.920	72.5
3.207	1.724	0.930	73.2
3.242	1.743	0.940	73.9
3.278	1.762	0.950	74.5
3.313	1.781	0.960	75.2
3.349	1.801	0.970	75.9
3.385	1.820	0.980	76.5
3.420	1.839	0.990	77.2
3.456	1.858	1.000	77.8
3.492	1.877	1.010	78.5
3.527	1.896	1.020	79.2
3.563	1.916	1.030	79.8
3.599	1.935	1.040	80.5
3.635	1.954	1.050	81.1
3.671	1.974	1.060	81.8
3.707	1.993	1.070	82.4
3.743	2.012	1.080	83.1
3.779	2.032	1.090	83.7
3.815	2.051	1.100	84.3
3.851	2.070	1.110	85.0
3.887	2.090	1.120	85.6
3.923	2.109	1.130	86.2
3.959	2.129	1.140	86.9
3.995	2.148	1.150	87.5
4.032	2.168	1.160	88.1
4.068	2.187	1.170	88.8
4.104	2.207	1.180	89.4
4.141	2.226	1.190	90.0
4.177	2.246	1.200	90.6
4.214	2.265	1.210	91.3
4.250	2.285	1.220	91.9
4.287	2.305	1.230	92.5
4.323	2.324	1.240	93.1
4.360	2.344	1.250	93.7
4.393	2.362	1.260	94.3
4.429	2.381	1.270	94.9
4.466	2.401	1.280	95.6
4.503	2.421	1.290	96.2
4.539	2.441	1.300	96.8
4.576	2.460	1.310	97.4
4.613	2.480	1.320	98.0
4.650	2.500	1.330	98.6
4.687	2.520	1.340	99.2
4.724	2.540	1.350	99.8
4.761	2.560	1.360	100
4.798	2.580	1.370	101
4.835	2.600	1.380	102
4.872	2.620	1.390	102
4.910	2.640	1.400	103
4.947	2.660	1.410	103
4.984	2.680	1.420	104
5.02	2.70	1.430	104
5.06	2.72	1.440	105
5.10	2.74	1.450	106
5.13	2.76	1.460	106
5.17	2.78	1.470	107
5.21	2.80	1.480	107
5.25	2.82	1.490	108
5.28	2.84	1.500	109
5.32	2.86	1.510	109
5.36	2.88	1.520	110
5.40	2.90	1.530	110
5.44	2.92	1.540	111
5.47	2.94	1.550	111
5.51	2.96	1.560	112
5.55	2.98	1.570	112
5.59	3.00	1.580	113
5.63	3.02	1.590	114
5.66	3.05	1.600	114
5.70	3.07	1.610	115
5.74	3.09	1.620	115
5.78	3.11	1.630	116
5.82	3.13	1.640	116

Δ °C	O Os/Kg	S,M,T, g-mol/l	γ mmho/cm
5.86	3.15	1.650	117
5.89	3.17	1.660	117
5.93	3.19	1.670	118
5.97	3.21	1.680	119
6.01	3.23	1.690	119
6.05	3.25	1.700	120
6.09	3.27	1.710	120
6.13	3.29	1.720	121
6.16	3.31	1.730	121
6.20	3.34	1.740	122
6.24	3.36	1.750	122
6.28	3.38	1.760	123
6.32	3.40	1.770	123
6.36	3.42	1.780	124
6.40	3.44	1.790	125
6.44	3.46	1.800	125
6.48	3.48	1.810	126
6.52	3.50	1.820	126
6.56	3.52	1.830	127
6.60	3.55	1.840	127
6.64	3.57	1.850	128
6.67	3.59	1.860	128
6.71	3.61	1.870	129
6.75	3.63	1.880	129
6.79	3.65	1.890	129
6.83	3.67	1.900	130
6.87	3.70	1.910	130
6.91	3.72	1.920	131
6.95	3.74	1.930	131
6.99	3.76	1.940	132
7.03	3.78	1.950	132
7.07	3.80	1.960	133
7.11	3.82	1.970	133
7.15	3.85	1.980	134
7.19	3.87	1.990	134
7.23	3.89	2.000	135
7.27	3.91	2.010	135
7.32	3.93	2.020	136
7.36	3.95	2.030	136
7.40	3.98	2.040	137
7.44	4.00	2.050	137
7.48	4.02	2.060	137
7.52	4.04	2.070	138
7.56	4.06	2.080	138
7.60	4.09	2.090	139
7.64	4.11	2.100	139
7.68	4.13	2.110	140
7.72	4.15	2.120	140
7.76	4.17	2.130	141
7.81	4.20	2.140	141
7.85	4.22	2.150	142
7.89	4.24	2.160	142
7.93	4.26	2.170	142
7.97	4.29	2.180	143
8.01	4.31	2.190	143
8.05	4.33	2.200	144
8.10	4.35	2.210	144
8.14	4.38	2.220	145
8.18	4.40	2.230	145
8.22	4.42	2.240	146
8.26	4.44	2.250	146
8.31	4.47	2.260	147
8.35	4.49	2.270	147
8.39	4.51	2.280	148
8.43	4.53	2.290	148
8.47	4.56	2.300	148
8.52	4.58	2.310	149
8.56	4.60	2.320	149
8.60	4.62	2.330	150
8.64	4.65	2.340	150
8.69	4.67	2.350	151
8.73	4.69	2.360	151
8.77	4.72	2.370	152
8.82	4.74	2.380	152
8.86	4.76	2.390	153
8.90	4.79	2.400	153
8.94	4.81	2.410	153
8.99	4.83	2.420	154
9.03	4.85	2.430	154
9.07	4.88	2.440	155
9.12	4.90	2.450	155
9.16	4.92	2.460	156
9.20	4.95	2.470	156
9.25	4.97	2.480	156
9.29	5.00	2.490	157

Δ °C	O Os/Kg	S,M,T, g-mol/l	γ mmho/cm
9.33	5.02	2.500	157
9.38	5.04	2.510	158
9.42	5.07	2.520	158
9.47	5.09	2.530	159
9.51	5.11	2.540	159
9.55	5.14	2.550	160
9.60	5.16	2.560	160
9.64	5.18	2.570	160
9.69	5.21	2.580	161
9.73	5.23	2.590	161
9.78	5.26	2.600	162
9.82	5.28	2.610	162
9.86	5.30	2.620	163
9.91	5.33	2.630	163
9.95	5.35	2.640	163
10.00	5.38	2.650	164
10.04	5.40	2.660	164
10.09	5.42	2.670	165
10.13	5.45	2.680	165
10.18	5.47	2.690	165
10.22	5.50	2.700	166
10.27	5.52	2.710	166
10.31	5.55	2.720	167
10.36	5.57	2.730	167
10.41	5.59	2.740	168
10.45	5.62	2.750	168
10.50	5.64	2.760	168
10.54	5.67	2.770	169
10.59	5.69	2.780	169
10.63	5.72	2.790	170
10.68	5.74	2.800	170
10.73	5.77	2.810	170
10.77	5.79	2.820	171
10.82	5.82	2.830	171
10.86	5.84	2.840	172
10.91	5.87	2.850	172
10.96	5.89	2.860	172
11.00	5.92	2.870	173
11.05	5.94	2.880	173
11.10	5.97	2.890	173
11.14	5.99	2.900	174
11.19	6.02	2.910	174
11.24	6.04	2.920	175
11.28	6.07	2.930	175
11.33	6.09	2.940	175
11.38	6.12	2.950	176
11.43	6.14	2.960	176
11.47	6.17	2.970	176
11.52	6.19	2.980	177
11.57	6.22	2.990	177
11.62	6.25	3.000	178
11.71	6.30	3.020	178
11.81	6.35	3.040	179
11.90	6.40	3.060	180
12.00	6.45	3.080	181
12.10	6.50	3.100	181
12.20	6.56	3.120	182
12.29	6.61	3.140	183
12.39	6.66	3.160	183
12.49	6.71	3.180	184
12.59	6.77	3.200	185
12.69	6.82	3.220	185
12.79	6.87	3.240	186
12.89	6.93	3.260	187
12.99	6.98	3.280	187
13.09	7.04	3.300	188
13.19	7.09	3.320	189
13.29	7.15	3.340	189
13.39	7.20	3.360	190
13.49	7.26	3.380	190
13.60	7.31	3.400	191
13.70	7.37	3.420	192
13.80	7.42	3.440	192
13.91	7.48	3.460	193
14.01	7.53	3.480	193
14.12	7.59	3.500	194
14.22	7.65	3.520	195
14.33	7.70	3.540	195
14.43	7.76	3.560	196
14.54	7.82	3.580	196
14.65	7.87	3.600	197
14.75	7.93	3.620	197
14.86	7.99	3.640	198
14.97	8.05	3.660	198
15.08	8.11	3.680	199

OSMOTIC PARAMETERS AND ELECTRICAL CONDUCTIVITIES OF AQUEOUS SOLUTIONS

(Continued)

Δ °C	O Os/Kg	S.M.T. g-mol/l	γ mmho/cm
15.19	8.16	3.700	199
15.30	8.22	3.720	200
15.41	8.28	3.740	200
15.52	8.34	3.760	201
15.63	8.40	3.780	201
15.74	8.46	3.800	202
15.85	8.52	3.820	202
15.96	8.58	3.840	203
16.07	8.64	3.860	203
16.19	8.70	3.880	204
16.30	8.76	3.900	204
16.41	8.82	3.920	205
16.53	8.89	3.940	205
16.64	8.95	3.960	206
16.76	9.01	3.980	206
16.87	9.07	4.000	206
16.99	9.13	4.020	207
17.11	9.20	4.040	207
17.22	9.26	4.060	208
17.34	9.32	4.080	208
17.46	9.39	4.100	209

Δ °C	O Os/Kg	S.M.T. g-mol/l	γ mmho/cm
17.58	9.45	4.120	209
17.70	9.51	4.140	209
17.82	9.58	4.160	210
17.94	9.64	4.180	210
18.06	9.71	4.200	210
18.18	9.77	4.220	211
18.30	9.84	4.240	211
18.42	9.90	4.260	212
18.55	9.97	4.280	212
18.67	10.04	4.300	212
18.79	10.10	4.320	213
18.92	10.17	4.340	213
19.04	10.24	4.360	213
19.17	10.30	4.380	214
19.29	10.37	4.400	214
19.42	10.44	4.420	214
19.55	10.51	4.440	215
19.67	10.58	4.460	215
19.80	10.65	4.480	215
19.93	10.72	4.500	216
20.06	10.78	4.520	216

Δ °C	O Os/Kg	S.M.T. g-mol/l	γ mmho/cm
20.19	10.85	4.540	216
20.32	10.92	4.560	216
20.45	10.99	4.580	217
20.58	11.07	4.600	217
20.71	11.14	4.620	217
20.85	11.21	4.640	218
20.98	11.28	4.660	218
21.11	11.35	4.680	218
		4.700	218
		4.750	219
		4.800	220
		4.850	220
		4.900	221
		4.950	221
		5.000	222
		5.050	222
		5.100	223
		5.150	223
		5.200	224
		5.250	224
		5.300	225

DENSITY OF AQUEOUS INVERT SUGAR SOLUTIONS

(From NBS Monograph 64, June 7, 1963) To convert density to specific gravity multiply listed values by 0.998234.

t=20 C

Invert sugar is the equimolar mixture of fructose and glucose obtained by hydrolyzing sucrose.

Wt % (Vacuum)	Specific Gravity	G solute per liter of solution	lb solute per cu ft of solution	lb solute per gallon (U.S.) of solution	Wt % (Vacuum)	Specific Gravity	G solute per liter of solution	lb solute per cu ft of solution	lb solute per gallon (U.S.) of solution
1	1.00211	10.021	0.62561	0.08363	44	1.19539	525.972	32.83643	4.38924
2	1.00601	20.120	1.25609	0.16790	45	1.20057	540.257	33.72824	4.50844
3	1.00994	30.298	1.89150	0.25284	46	1.20577	554.654	34.62705	4.62859
4	1.01389	40.556	2.53191	0.33844	47	1.21101	569.175	35.53360	4.74977
5	1.01786	50.893	3.17725	0.42470	48	1.21629	583.819	36.44782	4.87197
6	1.02186	61.312	3.82771	0.51165	49	1.22159	598.579	37.36929	4.99514
7	1.02589	71.812	4.48322	0.59927	50	1.22693	613.465	38.29862	5.11937
8	1.02995	82.396	5.14398	0.68759	51	1.23230	628.473	39.23557	5.24461
9	1.03403	93.063	5.80992	0.77661	52	1.23771	643.609	40.18051	5.37092
10	1.03814	103.814	6.48111	0.86633	53	1.24315	658.870	41.13325	5.49827
11	1.04227	114.649	7.15754	0.95675	54	1.24861	674.249	42.09337	5.62661
12	1.04644	125.573	7.83952	1.04791	55	1.25412	689.766	43.06209	5.75610
13	1.05063	136.582	8.52681	1.13978	56	1.25965	705.404	44.03837	5.88660
14	1.05484	147.678	9.21954	1.23237	57	1.26522	721.175	45.02296	6.01821
15	1.05909	158.864	9.91788	1.32572	58	1.27082	737.076	46.01565	6.15090
16	1.06337	170.139	10.62178	1.41981	59	1.27646	753.111	47.01672	6.28471
17	1.06767	181.504	11.33129	1.51465	60	1.28112	768.672	47.98819	6.41457
18	1.07200	192.960	12.04649	1.61025	61	1.28782	785.570	49.04310	6.55558
19	1.07637	204.510	12.76756	1.70664	62	1.29355	802.001	50.06892	6.69270
20	1.08076	216.152	13.49437	1.80379	63	1.29932	818.572	51.10345	6.83098
21	1.08518	227.888	14.22705	1.90173	64	1.30511	835.270	52.14591	6.97033
22	1.08963	239.719	14.96566	2.00046	65	1.31094	852.111	53.19729	7.11087
23	1.09411	251.645	15.71020	2.09998	66	1.31680	869.088	54.25716	7.25254
24	1.09862	263.669	16.46086	2.20032	67	1.32269	886.202	55.32559	7.39536
25	1.10316	275.790	17.21757	2.30147	68	1.32862	903.462	56.40313	7.53939
26	1.10773	288.010	17.98046	2.40344	69	1.33458	920.860	57.48929	7.68458
27	1.11233	300.329	18.74954	2.50625	70	1.34057	938.399	58.58425	7.83094
28	1.11697	312.752	19.52511	2.60992	71	1.34659	956.079	59.68801	7.79848
29	1.12163	325.273	20.30679	2.71440	72	1.35264	973.901	60.80064	8.12720
30	1.12633	337.899	21.09503	2.81977	73	1.35872	991.866	61.92219	8.27712
31	1.13105	350.626	21.88958	2.92597	74	1.36484	1009.982	63.05318	8.42830
32	1.13581	363.459	22.69075	3.03307	75	1.37099	1028.243	64.19321	8.58069
33	1.14060	376.398	23.49853	3.14104	76	1.37717	1046.649	65.34230	8.73429
34	1.14542	389.443	24.31293	3.24990	77	1.38337	1065.195	66.50012	8.88905
35	1.15027	402.595	25.13401	3.35966	78	1.38962	1083.904	67.66813	9.04518
36	1.15516	415.858	25.96201	3.47034	79	1.39589	1102.753	68.84487	9.20247
37	1.16007	429.226	26.79658	3.58199	80	1.40219	1121.752	70.03098	9.36102
38	1.16502	442.708	27.63826	3.69440	81	1.40852	1140.901	71.22645	9.52082
39	1.17000	456.300	28.48681	3.80728	82	1.41488	1160.202	72.43141	9.68189
40	1.17502	470.008	29.34260	3.92222	83	1.42128	1179.662	73.64630	9.84428
41	1.18006	483.825	30.20519	4.03752	84	1.42770	1199.268	74.87030	10.00790
42	1.18514	497.759	31.07509	4.15380	85	1.43415	1219.028	76.10392	10.17279
43	1.19025	511.808	31.95217	4.27104					

HEAT OF COMBUSTION
For Organic Compounds

The heat of combustion is given in kilogram calores per gram molecular weight of the substance when combustion takes place at atmospheric pressure and at either 20°C or 25°C. If the data are for 20°C there is no asterisk for the numerical value of the heat of combustion. If the numerical value is for 25°C there is an asterisk marking the numerical value of the heat of combustion. The final products of combustion are gaseous carbon dioxide, liquid water and nitrogen gas for C, H, N compounds. For method of computing heats of formation see statement following this table.

Name	Formula	Physical state	Heat of combustion, kg. calories
Acetaldehyde	CH_3CHO	liquid	278.77*
Acetamide	CH_3CONH_2	solid	282.6
Acetanilide	$CH_3CONHC_6H_5$	solid	1010.4
Acetic acid	CH_3CO_2H	liquid	209.02*
Acetic anhydride	$(CH_3CO)_2O$	liquid	431.70*
Acetone	$(CH_3)_2CO$	liquid	427.92*
Acetonitrile	CH_3CN	liquid	302.4
Acetophenone	$C_6H_5COCH_3$	liquid	991.60*
Acetylacetone	$CH_3COCH_2COCH_3$	liquid	615.9
Acetylene	$(CH)_2$	gas	310.61*
Acrolein	CH_2CHO	liquid	389.6
Acrylic acid	CH_2CO_2H	liquid	327.0*
Adipic acid	$(CH_2)_4(CH_2H)_2$	solid	668.29*
Alanine	$CH_3CH(NH_2)CO_2H$	solid	387.7
Aldol, see β-hydroxybutyr-aldehyde			
Alizarin, see Dihydroxyanthraquinone			
Allyl alcohol	CH_2CH_2OH	liquid	442.4
Allylene	CH_3C	gas	465.1
p-Aminoazobenzene	$H_2NC_6H_4N_2C_6H_5$	solid	1574.0
p-Aminophenol	$HOC_6H_4NH_2$	solid	760.0
Amygdalin	$C_{20}H_{27}O_{11}N$	solid	2348.4
Amyl acetate	$C_4H_9CO_2C_2H_5$	liquid	1042.5
Amyl aocohol	$(CH_3)_2CHCH_2CH_2OH$	liquid	793.7
Amylene	C_5H_{10}	liquid	803.4
Anethole	$C_{10}H_{12}O$	solid	1324.4
Aniline	$C_6H_5NH_2$	liquid	811.7
p-Anisidine	$CH_3OC_6H_4NH_2$	solid	924.0
Anisole	$C_6H_50CH_3$	liquid	905.1
Anthracene	$C_{14}H_{10}$	solid	1712.0*
Anthraquinone	$C_{14}H_8O_2$	solid	1544.5
Arabinose	$C_5H_{10}O_5$	solid	559.9
Arabitol	$C_5H_{12}O_5$	solid	661.2
Arachidic acid	$C_{20}H_{40}O_2$	solid	3025.9
Azelaic acid	$(CH_2)_7(CO_2H)_2$	solid	1141.7
Azobenzene	$(C_6H_5N)_2$	solid	1545.9
Azoxybenzene	$(C_6H_5N)_2O$	solid	1534.5
Behenic acid	$C_{22}H_{44}O_2$	solid	3338.4
Benzalacetone	$C_6H_5CHCOCH_3$	solid	1257.4
Benzaldehyde	C_6H_5CHO	liquid	843.2*
Benzamide	$C_6H_5CONH_2$	solid	847.6
Benzanilide	$C_6H_5CONHC_6H_5$	solid	1575.5
Benzene	C_6H_6	liquid	780.96*
Benzenediazonium nitrate	$C_6H_5N_2NO_3$	solid	782.6
Benzidine	$(C_6H_4NH_2)_2$	solid	1560.9
Benzil	$(C_6H_5CO)_2$	solid	1624.6
Benzoic acid*	$C_6H_5CO_2H$	solid	771.24*
Benzoic anhydride	$(C_6H_5CO)_2O$	solid	1555.1
Benzoin	$C_6H_5.CHOH.COC_6H_5$	solid	1671.4
Benzonitrile	C_6H_5CN	liquid	865.5
Benzophenone	$(C_6H_5)_2CO$	solid	1556.5
Benzoyl chloride	C_6H_5COCl	liquid	782.8
Benzoyl peroxide	$(C_6H_5CO)_2O_2$	solid	1551.7
Benzyl alcohol	$C_6H_5CH_2OH$	liquid	894.3
Benzylamine	$C_6H_5CH_2NH_2$	liquid	969.4
Benzyl carbylamine	$C_6H_5CH_2NC$	liquid	1046.5
Benzyl chloride	$C_6H_5CH_2Cl$	liquid	886.4
Benzyl cyanide	$C_6H_5CH_2CN$	liquid	1023.5
Borneol	$C_{10}H_{18}O$	liquid	1469.6
Brucine	$C_{23}H_{26}O_4N_2$	gas	687.68*
n-Butyl alcohol	C_4H_9OH	liquid	639.53*
tert-Butyl alcohol, see Trimethyl carbinol			
n-Butylamine	$C_4H_9NH_2$	liquid	710.6
sec-Butylamine	$(CH_3)(C_2H_5)NH_2$	liquid	713.0
tert-Butylamine	$(CH_3)_3CNH_2$	liquid	716.0
tert-Butylbenzene	$C_6H_5C(CH_3)_3$	liquid	1400.4
n-Butyramide	$C_3H_7CONH_2$	solid	596.0
n-Butyric acid	$C_3H_7CO_2H$	liquid	521.87*
n-Butyronitrile	C_3H_7CN	liquid	613.3
Caffeine	$C_8H_{10}O_2N_4$	solid	1014.2
Camphene	$C_{10}H_{16}$	solid	1468.8
Camphor	$C_{10}H_{16}O$	solid	1411.0
Cane sugar, see Sucrose			
Capric acid	$C_9H_{18}O_2$	solid	1453.07*
Caproic acid	$C_5H_{11}CO_2H$	liquid	834.49*
Carbon disulfide	CS_2	liquid	246.6
Carbon subnitride	$(C.CN)_2$	solid	514.8

* 25°C

HEAT OF COMBUSTION
For Organic Compounds (Continued)

Name	Formula	Physical state	Heat of combustion. kg. calories
Carbon tetrachloride	CCl_4	liquid	37.3
Carbonyl sulfide	COS	gas	130.5
Carvacrol	$C_{10}H_{14}O$	liquid	1354.5
Cetyl alcohol	$C_{16}H_{34}O$	solid	2504.4
Cetyl palmitate	$C_{32}H_{64}O$	solid	4872.8
Chloracetic acid	$ClCH_2CO_2H$	solid	171.0
o-Chlorobenzoic acid	$ClC_6H_4CO_2H$	solid	734.5
Chloroform	$CHCl_3$	liquid	89.2
Chrysene	$C_{18}H_{12}$	solid	2139.1
Cinnamic acid *(trans)*	$C_6H_5CH{:}CHCO_2H$	solid	1040.2
Cinnamic aldehyde	$C_6H_5CH{:}CHCHO$	liquid	1112.3
Cinnamic anhydride	$C_{18}H_{14}O_3$	solid	2091.3
d-Citrene	$C_{10}H_{16}$	liquid	1483.0
Citric acid (anhydr)	$C_6H_8O_7$	solid	468.6*
Codeine	$C_{18}H_{21}O_3N.H_2O$	solid	2327.6
Coniine	$C_8H_{17}N$	liquid	1275.5
Creatine (anhydr)	$C_4H_9O_2N_3$	solid	559.8
Creatinine	$C_4H_7ON_3$	solid	563.4
o-Cresol	$CH_3C_6H_4OH$	liquid	882.6
o-Cresol	$CH_3C_6H_4OH$	solid	882.72*
m-Cresol	$CH_3C_6H_4OH$	liquid	880.5
p-Cresol	$CH_3C_6H_4OH$	liquid	882.5
p-Cresol	$CH_3C_6H_4OH$	solid	883.99*
m-Cresolmethyl ether	$CH_3C_6H_4OCH_3$	liquid	1057.0
Crotonaldehyde	C_3H_5CHO	liquid	542.1
Cyanoacetic acid	$NCCH_2CO_2H$	solid	298.8
Cyanogen	$(CN)_2$	gas	258.3
Cyclobutane	C_4H_8	liquid	650.22*
Cycloheptane	$(CH_2)_7$	liquid	1087.3
Cycloheptanol	$CH_2(CH_2)_5CHOH$	liquid	1050.2
Cycloheptene	C_7H_{12}	liquid	1099.09*
Cyclohexane	$(CH_2)_6$	liquid	936.87*
Cyclohe xanol	$CH_2(CH_2)_4CHOH$	liquid	890.7
Cyclohexene, see Tetrahydrobenzene			
Cyclopentane	$(CH_2)_5$	liquid	786.55*
Cyclopropane, see Trimethylene			
Cymene	$C_6H_4(CH_3)(CH_3COCO_3)$-(1,4)	liquid	1402.8
Decahydronaphthalene *(cis)*	$C_{10}H_{18}$	liquid	1502.5
Decahydronaphthalene *(trons)*	$C_{10}H_{18}$	liquid	1499.5
Decane	$C_{10}H_{22}$	liquid	1610.2
Dextrose, see Glucose			
Diallyl	$(CH_2CHCH_2)_2$	vapor	903.4
Diamyl ether	$(C_5H_{11})_2O$	liquid	1609.3
Diamylene	$C_{10}H_{20}$	liquid	1582.2
Dibenzyl	$(C_6H_5CH_2)_2$	solid	1810.6
Dibenzyl amine	$(C_6H_5CH_2)_2NH$	solid	1853.0
o-Dichlorobenzene	$C_6H_4Cl_2$	liquid	671.8
Diethylacetic acid	$(C_2H_5)_2CHCO_2H$	liquid	830.8
Diethyl amine	$(C_2H_5)_2NH$	liquid	716.9
Diethylaniline	$C_6H_5(C_2H_5)_2$	liquid	1451.6
Diethyl carbonate	$CO(OC_2H_5)_2$	liquid	647.9
Diethyl ether	$(C_2H_5)_2O$	liquid	657.52*
Diethyl ketone	$(C_2H_5)_2CO$	liquid	735.6
Diethyl malonate	$CH_2(CO_2C_2H_5)_2$	liquid	860.4
Diethyl oxalate	$(CO_2C_2H_5)_2$	liquid	716.0
Diethyl succintae	$(CH_2CO_2C_2H_5)$	liquid	1007.3
Dihydrobenzene	C_6H_8	liquid	847.8
δ_1-Dihydronaphthalene	$C_{10}H_{10}$	liquid	1296.3
δ_1-Dihydronaphthalene	$C_{10}H_{10}$	solid	1298.3
Dihydroxyanthraquinone	$C_{14}H_6O_2(OH)_2$-(1,2)	solid	1448.9
Diisoamyl	$[(CH_3)_2CHCH_2CH_2]_2$	liquid	1615.8
Diisobutylene	$[(CH_3)_2CHCH_2]_2$	liquid	1252.4
Diisopropyl	$[(CH_3)_2CH]_2$	vapor	993.9
Diisopropyl ketone	$[(CH_3)_2CH]_2CO$	liquid	1045.5
Dimethyl amine	$(CH_3)_2NH$	liquid	416.7
Dimethylaniline	$C_6H_5N(CH_3)_2$	liquid	1142.7
Dimethyl carbonate	$CO(OCH_3)_2$	liquid	340.8
Dimethyl ether	$(CH_3)_2O$	gas	
Dimethylethyl carbinol	$C_2H_5(CH_3)_2CHOH$	liquid	784.6
Dimethyl fumarate	$(CHCO_2CH_3)_2$	solid	663.3*
2,5-Dimethylhexane	$(CH_3)_2CH.C_2H_4CH(CH_3)_2$	liquid	1303.3
3,4-Dimethylhexane	$[(C_2H_5)(CH_3)CH]_2$	liquid	1303.7
Dimethyl maleate	$(CHCO_2CH_3)_2$	liquid	669.4*
Dimethyl oxalate	$(CO_2CH_3)_2$	liquid	400.2*
2,2-Dimethylpentane	$(CH_3)_2C.C_3H_7$	liquid	1148.9
2,3-Dimethylpentane	$(CH_3)_7CHCH(CH_3)C_2H_5$	liquid	1148.9
2,4-Dimethylpentane	$(CH_3)_2CHCH_2CH(CH_3)_2$	liquid	1148.9
3,3-Dimethylpentane	$(CH_3)_7C(C_2H_5)_2$	liquid	1147.9
Dimethyl phthalate	$C_6H_4(CO_7CH_3)_2$	liquid	1119.7
Dimethyl succinate	$(CH_2CO_7CH_3)_2$	solid	706.3*
m-Dinitrobenzene	$C_6H_4(NO_2)_2$	solid	696.8

HEAT OF COMBUSTION
For Organic Compounds (Continued)

Name	Formula	Physical state	Heat of combustion. kg. calories
Dinitrophenol	$C_6H_3(OH)(NO_2)_2$—(1, 2,4)	solid	648.0
Dinitrotoluene	$C_6H_3)(NO_2)_2$—(1,2,4)	solid	852.8
Diphenyl	$(C_6H_5)_2$	solid	1493.6
Diphenyl amine	$(C_6H_5)_2NH$	solid	1536.2
Diphenyl carbinol	$(C_6H_5)_2CHOH$	solid	1615.4
Diphenylmethane	$(C_6H_5)_2CH_2$	solid	1655.0
Diphenylnitrosamine	$(C_6H_5)_2N.NO$	solid	1532.6
Dipropargyl	$(CH: C.CH_2)_2$	vapor	882.9
Dipropyl ketone	$(C_3H_7)_2CO$	liquid	1050.5
Dulcitol	$C_6H_{14}O_6$	solid	729.1
Durene	$C_6H_2(CH_3)_4$—(1,2,4,5)	solid	1393.6
Eicosane	$C_{20}H_{42}$	solid	3183.1
Erythritol	$C_4H_{10}O_4$	solid	504.1
Ethane	C_2H_6	gas	372.81*
Ethine, see Acetylene			
Ethyl acetate	$CH_3CO_7C_2H_5$	liquid	536.9
Ethyl acetoacetate	$CH_3COCH_2CO_2C_2H_5$	liquid	690.8
Ethyl alcohol	C_2H_6OH	liquid	326.68*
Ethyl amine	$C_2H_5NH_2$	liquid	409.5*
Ethylaniline	$C_6H_5NHC_2H_5$	liquid	1121.5
Ethylbenzene	$C_2H_5C_6H_5$	liquid	1091.2
Ethyl benzoate	$C_6H_5CO_2C_2H_5$	liquid	1098.7
Ethyl bromide	C_2H_5Br	vapor	340.5
Ethyl n-butyrate	$C_3H_7CO_2C_2H_5$	liquid	851.2
Ethyl carbylamine	C_2H_5NC	liquid	477.1
Ethyl chloride	C_2H_5Cl	vapor	316.7
Ethylcycloheptane	$C_2H_5C_7H_{13}$	liquid	1406.8
Ethyl formate	$HCO_2C_2H_5$	liquid	391.7
3-Ethylhexane	$(C_2H_5)_2CH.C_3H_7$	liquid	1302.3
Ethyl iodide	C_2H_5I	liquid	356.0
Ethyl isobutyrate	$(CH_3)_2CHCH_2CO_2C_2H_5$	liquid	845.7
Ethyl isocyanate	C_2H_5NCO	liquid	424.5
Ethyl nitrate	$C_2H_5ONO_2$	vapor	322.4
Ethyl nitrite	C_2H_5ONO	vapor	332.6
3-Ethylpentane	$(C_2H_5)_3CH$	liquid	1149.9
Ethyl propionate	$C_2H_5CO_2C_2H_5$	liquid	690.8
Ethyl salicylate	$HOC_6H_4CO_2C_2H_5$	liquid	1051.2
Ethyl valerate	$C_4H_9CO_2C_2H_5$	liquid	1017.5
Ethylene	CH_2CH_2	gas	337.23*
Ethylene chloride	$(CH_7Cl)_2$	vapor	271.0
Ethylene diamine	$(CH_2NH_2)_3$	liquid	452.6
Ethylene glycol	$(CH_2OH)_2$	liquid	281.9
Ethylene iodide	$(CH_2I)_2$	solid	324.8
Ethylene oxide	CH_2CH_2O	liquid	302.1
Ethylidene chloride	CH_3CHCl_2	liquid	267.1
Eugenol	$C_{10}H_{12}O_2$	liquid	1286.6
Fenchane	$C_{10}H_{18}$	liquid	1502.6
Fluorene	$(C_6H_4)_2$: CH_2	solid	1584.9
Fluorobenzene	C_6H_5F	liquid	747.2
Formaldehyde	CH_2O	gas	136.42*
Formamide	$HCONH_2$	solid	134.9
Formic acid	HCO_2H	liquid	60.86*
β-D-Fructose	$C_6H_{12}O_6$	solid	672.0*
Fumaric acid *(trans)*	$(CHCO_2H)_2$	solid	318.99*
Furfural	C_4H_3OCHO	liquid	559.5
a-D-Galactose	$C_6H_{12}O_6$	solid	670.1*
Gallic acid	$C_6H_2(OH)_3CO_2H$—(1,3,5,6)	solid	633.7
a-D-Glucose	$C_6H_{12}O_6$	solid	669.94*
Glutaric acid	$(CH_2)_3(CO_2H)_2$	solid	514.08*
Glycerol	$(CH_2OH)_2CHOH$	liquid	397.0
Glyceryl tributyrate	$C_{15}H_{26}O_6$	liquid	1941.1
Glycine	$H_2NCH_2CO_2H$	solid	232.67*
Glycogen	$(C_6H_{10}O_5)$x per kg	solid	4186.8
Glycollic acid	CH_7OHCO_2H	solid	166.1*
Glycylglycine	$C_4H_8O_3N_2$	solid	470.7
n-Heptaldehyde	$CH_3(CH_2)_5CHO$	liquid	1062.2*
n-Heptane	C_7H_{16}	liquid	1149.9
Heptine-l	CH: $C(CH_2)_4CH_3$	liquid	1091.2
n-Heptyl alcohol	$CH_3(CH_2)_5CH_2OH$	liquid	1104.9
Heptyl amine	$C_7H_{15}NH_2$	liquid	1178.9
Heptylic acid	$C_7H_{14}O_2$	liquid	986.1
n-Hexane	C_6H_{14}	liquid	995.01*
Hexachlorbenzene	C_6Cl_6	solid	509.0
Hexachlorethane	C_2Cl_6	solid	110.0
Hexadecane	$C_{16}II_{34}$	solid	2559.1
Hexahydronaphthalene	$C_{10}H_{14}$	liquid	1419.3
Hexamethylbenzene	$C_6(CH_3)_6$	solid	1711.9
Hexamethylenetetramine	$(CH_2)_6N_4$	solid	1006.7
Hexamethylethane	$(CH_3)_3Cl_2$	solid	1301.8
Hexyl amine	$C_6H_{13}NH_2$	liquid	1022.2
Hexylene	C_6H_{12}	liquid	952.6

HEAT OF COMBUSTION
For Organic Compounds (Continued)

Name	Formula	Physical state	Heat of combustion. kg. calories
Hippuric acid	$C_6H_5CONHCH_2CO_2H$	solid	1012.4
Hydantoic acid	$C_3H_6O_3N_2$	solid	308.6
Hydrazobenzene	$(C_6H_5NH)_2$	solid	1597.3
Hydroquinol	$C_6H_4(OH)_2$	solid	681.78*
Hydroquinoldimethyl ether	$(CH_3O)_2C_6H_4$	solid	1014.7
p-Hydroxyazobenzene	$HOC_6H_4N_2C_6H_5$	solid	1502.0
o-Hydroxybenzaldehyde	$C_6H_4(OH)CHO$	liquid	796.4*
m-Hydroxybenzaldehyde	$C_6H_4(OH)CHO$	solid	788.7
p-Hydroxybenzaldehyde	$C_6H_4(OH)CHO$	solid	792.7
m-Hydroxybenzoic acid	$HOC_6H_4CO_2H$	solid	726.1
p-Hydroxybenzoic acid	$HOC_6H_4CO_2H$	solid	725.4
β-Hydroxybutyraldehyde	$CH_3CHOHCH_2CHO$	liquid	546.6
Indigo	$C_{16}H_{10}O_2N_2$	solid	1815.0
Indole	C_8H_7N	solid	1022.2
Inositol	$C_6H_{12}O_6$	solid	662.1
Iodoform	CHI_3	solid	161.9
Isoamyl amine	$(CH_3)_2CHC_2H_4NH_2$	liquid	866.8
Isobutane	$(CH_3)_2CH$	gas	683.4
Isobutyl alcohol	$(CH_3)_2CHCH_2OH$	liquid	638.2
Isobutyl amine	$C_4H_9NH_2$	liquid	713.6
Isobutylene	$(CH_3)_2C:CH_2$	gas	647.2
Isobutyraldehyde	$(CH_3)_2CHCHO$	vapor	596.8
Isobutyramide	$(CH_3)_2CHCONH_2$	solid	595.9
Isobutyric acid	$(CH_3)_2CHCO_2H$	liquid	517.4
Isoeugenol	$C_{10}H_{12}O_2$	liquid	1277.6
Isopentane	C_5H_{12}	gas	843.5(?)
Isopentane	C_5H_{12}	liquid	838.3(?)
Isophthalic acid	$C_6H_4(CO_2H)_2$	solid	768.3
Isopropyl alcohol	$(CH_3)_2CHOH$	liquid	474.8
Isopropylbenzene	$(CH_3)_2CHC_6H_5$	liquid	1247.3
Isopropyltoluene	$C_6H_4(CH_3)(CH_3CHC_3)$—(1,3)	liquid	1409.5
Isopropyltoluene, see Cymene			
Isosafrole	$C_{10}H_{10}O_2$	liquid	1233.9
Lactic acid, DL	$CH_3CHOHCO_2H$	liquid	326.8*
Lactose (anhydr.)	$C_{12}H_{22}O_{11}$	solid	1350.0*
Lauric acid	$C_{12}H_{24}O_2$	solid	1763.25*
Leucine	$C_6H_{13}O_2N$	solid	855.6
d-Limonene	$C_{10}H_{16}$	liquid	1471.2
Maleic acid *(cis)*	$(CHCO_2H)_2$	solid	323.89*
Maleic anhydride	$(CHCO)_2O$	solid	332.10*
l-Malic acid	$(CHOHCH_2):(CO_2H)_2$	solid	317.37*
Malonic acid	$CH_2(CO_2H)_2$	solid	205.82*
Maltose	$C_{12}H_{22}O_{11}$	solid	1349.3*
Mandelic acid	$C_6H_5CHOHCO_2H$	solid	890.3
d-Mannitol	$C_6H_{14}O_6$	solid	727.6
Menthene	$C_{10}H_{18}$	liquid	1523.2
Menthol	$C_{10}H_{20}O$	solid	1508.8
Mesitylene	$(CH_3)_3C_6H_3$—(1,3,5)	liquid	1243.6
Mesityl oxide	$(CH_3)_2C:CHCOCH_3$	liquid	846.7
Mesotartaric acid	$(CHOH)_2(CO_2H)_2$	solid	276.0
Methane	CH_4	gas	212.79*
Methyl acetate	$CH_3CO_2CH_3$	liquid	381.2
Methyl alcohol	CH_3OH	liquid	173.64*
Methyl amine	CH_3NH_2	liquid	253.5*
Methylaniline	$C_6H_5NHCH_3$	liquid	973.5
Methyl benzoate	$C_6H_5CO_2CH_3$	liquid	945.9*
Methyl bromide	CH_3Br	vapor	184.0
Methyl butyl ketone	$CH_3COC_4H_9$	liquid	895.2
Methyl *tert*-butyl ketone, see Pinacoline			
Methyl butyrate	$C_3H_7COOCH_3$	liquid	692.8
Methyl carbylamine	CH_3NC	liquid	320.1
Methyl chloride	CH_3Cl	gas	164.2
Methyl cinnamate	$C_{10}H_{10}O_2$	solid	1213.0
Methylcyclobutane	$CH_3CHCH_2CH_2CH_2$	liquid	784.2
Methylcycloheptane	$CH_3C_7H_{13}$	liquid	1244.5
Methylcyclohexane	$CH_3C_6H_{11}$	liquid	1091.8
Methylcyclopentane	$CH_3CH.C_3H_6CH_2$	liquid	937.9
Methyldiethyl carbinol	$CH_3(C_2H_5)_2CHOH$	liquid	927.0
Methylene chloride	CH_2Cl_2	vapor	106.8
Methylene iodide	CH_2I_2	liquid	178.4
Methylethyl ether	$CH_3OC_2H_5$	vapor	503.69*
Methylethyl ketone	$CH_3COC_2H_5$	liquid	584.17*
Methyl formate	HCO_2CH_3	liquid	234.1*
2-Methylheptane	$(CH_3)_2CH.C_5H_{11}$	liquid	1306.1
2-Methylhexane	$(CH_3)_2CHC_4H_9$	liquid	1148.9
3-Methylhexane	$(C_2H_5)(CH_3)CHC_3H_7$	liquid	1148.9
Methylhexyl ketone	$CH_3COC_6H_{13}$	liquid	1205.1
Methyl iodide	CH_3I	liquid	194.7
Methyl isobutyrate	$(CH_3)_2CHCO_2CH_3$	liquid	694.2
Methyl isocyanate	CH_3NCO	liquid	269.4
Methylisopropyl ketone	$CH_3COCH(CH_3)_2$	liquid	733.9
Methyl lactate	$CH_3CHOHCO_2CH_3$	liquid	497.2

HEAT OF COMBUSTION
For Organic Compounds (Continued)

Name	Formula	Physical state	Heat of combustion. kg. calories
Methyl propionate	$C_2H_5CO_2CH_3$	vapor	552.3
Methylpropyl ketone	$CH_3COC_3H_7$	liquid	740.78*
Methyl salicylate	$HOC_6H_4CO_2CH_3$	liquid	898.6*
Milk sugar, see Lactose			
Morphine	$C_{17}H_{19}O_3N.H_2O$	solid	2146.3
Mucic acid	$C_6H_{10}O_8$	solid	483.0*
Myristic acid	$C_{14}H_{28}O_2$	solid	2073.91*
Naphthalene	$C_{10}H_8$	solid	1231.8*
α-Naphthoic acid	$C_{10}H_7CO_2H$	solid	1231.8
β-Naphthoic acid	$C_{10}H_7CO_2H$	solid	1227.6
α-Naphthol	$C_{10}H_7OH$	solid	1185.4
β-Naphthol	$C_{10}H_7OH$	solid	1187.2
α-Naphthonitrile	$C_{10}H_7CN$	solid	1326.2
β-Naphthonitrile	$C_{10}H_7CN$	solid	1321.0
α-Naphthoquinone	$C_{10}H_6O_2$	solid	1100.8
β-Naphthoquinone	$C_{10}H_6O_2$	solid	1106.4
α-Naphthyl amine	$C_{10}H_7NH_2$	solid	1263.5
β-Naphthyl amine	$C_{10}H_7NH_2$	solid	1261.0
Narceine	$C_{23}H_{27}O_8N.2H_2O$	solid	2802.9
Narcotine	$C_{22}H_{23}O_7N$	solid	2644.5
Nicotine	$C_{10}H_{14}N_2$	liquid	1427.7
o-Nitraniline	$C_6H_4(NH_2)(NO_2)$	solid	765.8
m-Nitraniline	$C_6H_4(NH_2)(NO_2)$	solid	765.2
p-Nitraniline	$C_6H_4(NH_2)(NO_2)$	solid	761.0
m-Nitrobenzaldehyde	$O_2NC_6H_4CHO$	solid	800.4
Nitrobenzene	$C_6H_5NO_2$	liquid	739.2
m-Nitrobenzoic acid	$O_2NC_6H_4CO_2H$	solid	729.1
Nitroethane	$C_2H_5NO_2$	liquid	322.2
Nitroglycerine, see Trinitroglycerol			
Nitromethane	CH_3NO_2	liquid	169.4
o-Nitrophenol	$HOC_6H_6NO_2$	solid	689.1
m-Nitrophenol	$HOC_5H_4NO_2$	solid	684.4
p-Nitrophenol	$HOC_6H_4NO_2$	solid	688.8
Nitropropane	$C_3H_7NO_2$	liquid	477.9
p-Nitrotoluene	$CH_3C_6H_4NO_2$	liquid	897.0
p-Nitrotoluene	$CH_3C_6H_6NO_2$	solid	888.6
Octahydronaphthalene	$C_{10}H_{16}$	liquid	1461.7
n-Octane	C_8H_{18}	liquid	1302.7
Octyl alcohol	$C_8H_{18}O$	liquid	1262.0
Oleic acid	$C_{18}H_{34}O_2$	liquid	2657.4*
Oxalic acid, *a*	$(CO_2H)_2$	solid	58.7*
Oxamide	$(CONH_2)_2$	solid	203.2
Palmitic acid	$C_{16}H_{32}O_2$	solid	2384.76*
Papaverine	$C_{20}H_{21}O_4N$	solid	2478.1
Pentamethylbenzene	$C_6H(CH_3)_5$	solid	1554.0
n-Pentane	C_5H_{12}	gas	845.16*
n-Pentane	C_8H_{12}	liquid	838.78*
Phenacetin	$C_{10}H_{13}O_2N$	solid	1285.2
Phenanthraquinone	$C_{14}H_8O_2$	solid	1544.0
Phenanthrene	$C_{14}H_{10}$	solid	1685.6*
Phenetole	$C_6H_5OC_2H_5$	liquid	1060.3
Phenol	C_6H_5OH	solid	729.80*
Phenylacetic acid	$C_6H_5CH_7CO_2H$	solid	930.4*
Phenylacetylene	C_6H_5C	liquid	1024.2
Phenylalanine	$C_9H_{11}O_2N$	solid	1111.3
p-Phenylenediamine	$C_6H_4(NH_2)_2$	solid	843.4
Phenylethylene, see Styrene			
Phenylglycine	$C_2H_5NHCH_2CO_2H$	solid	955.1
Phenylhydrazine	$C_6H_5N_2H_3$	solid	875.4
Phenylhydroxylamine	C_6H_5NHOH	liquid	803.7
Phenyl iodide	C_6H_5I	liquid	770.7
Phloroglucinol	$C_6H_3(OH)_3$	solid	635.7
o-Phthalic acid	$C_6H_4(CO_2H)_2$	solid	770.44*
Phthalic anhydride	$C_6H_4(CO_2O$	solid	783.4
Phthalimide	$C_8H_5O_2N$	solid	849.5
Picric acid	$C_6H_2(OH)(NO_2)_3$—(1,2,4,6)	solid	611.8
Pinacoline	$CH_3COC(CH_3)_3$	solid	891.8
Piperidine	$C_5H_{11}N$	liquid	826.6
Piperonal	$C_8H_6O_3$	solid	870.7
Propane	C_3H_8	gas	530.57*
Propine, see Allylene			
Propionaodehyde	C_2H_5CHO	liquid	434.1*
Propionamide	$C_2H_5CONH_2$	solid	439.9
Propionic acid	$C_2H_5CO_2H$	liquid	365.03*
Propionic anhydride	$(C_2H_5CO)_2O$	liquid	746.6
Propionitrile	C_2H_5CN	liquid	456.4
n-Propyl alcohol	C_3H_7OH	liquid	482.75*
Propyl amine	$C_3H_7NH_2$	liquid	565.3*
n-Propylbenzene	$C_3H_7C_6H_5$	liquid	1246.4
Propyl bromide	C_3H_7Br	vapor	497.3
Propyl carbylamine	C_3H_7NC	liquid	639.6

HEAT OF COMBUSTION
For Organic Compounds (Continued)

Name	Formula	Physical state	Heat of combustion. kg. calories
Propyl chloride	C_3H_7Cl	vapor	478.3
Propylene	CH_3CH_2	gas	490.2
Propylene glycol	$CH_3CHOHCH_2OH$	liquid	431.0
n-Propyl iodide	C_3H_7I	liquid	514.3
n-Propyltoluene	$C_6H_4(CH_3)(C_3H_7)$—(1,3)	liquid	1405.4
Pseudocumene	$C_6H_3(CH_3)_3$—(1,2,4)	liquid	1241.7
Pyridine	C_5H_5N	liquid	665.0*
Pyrocatechol	$C_6H_4(OH)_2$	solid	683.0*
Pyrogallol	$C_6H_3(OH)_3$	solid	638.7
Pyrrole	C_4H_5N	liquid	567.7
Quercitol	$C_6H_{12}O_5$	solid	704.2
Quinoline	C_9H_7N	liquid	1123.5
Quinone	$O:C_6H_4:O$	solid	656.6
Raffinose	$C_{18}H_{32}O_{16}$	solid	2025.5
Retene	$C_{18}H_{18}$	solid	2306.8
Resorcinol	$C_6H_4(OH)_2$	solid	681.30*
Resorcinoldimethyl ether	$(CH_3O)_2C_6H_4$	liquid	1022.6
Rhamnose	$C_6H_{12}O_6$	solid	718.3
Safrole	$C_{10}H_{10}O_2$	liquid	1244.1
Salicylaldehyde, see o-Hydroxybenzaldehyde			
Salicylic acid*	$HOC_6H_4CO_2H$—(1,2)	solid	722.4**
Sarcosine	$CH_3NHCH_2CO_2H$	solid	401.1
Sebacic acid	$(CH_2)_8(CO_2H)_2$	solid	1297.3
Skatole	C_9H_9N	liquid	1170.5
d-Sorbose	$C_6H_{12}O_6$	solid	668.3
Starch	$(C_6H_{10}O_5)x$ per kg	solid	4178.8
Stearic acid	$C_{18}H_{36}O_2$	solid	2696.12*
Strychnine	$C_{21}H_{22}O_2N_2$	solid	2685.7
Styrene	$C_6H_5CH_2$	liquid	1047.1
Suberic acid	$(CH_2)_6(CO_2H)_2$	solid	985.2
Succinic acid	$(CH_2CO_2H)_2$	solid	356.36*
Succinic acid nitrile	$(CH_2CN)_2$	liquid	545.7
Succinic anhydride	$(CH_2CO)_2O$	solid	369.0*
Succinimide	$C_4H_5O_2N$	solid	437.9
Sucrose	$C_{12}H_{22}O_{11}$	solid	1348.2*
Sylvestrene	$C_{10}H_{15}$	liquid	1464.7
l-Tartaric acid	$(CHOH)_2(CO_2H)_2$	solid	271.7*
d,l-Tartaric acid (anhydr)	$(CHOH)_2(CO_2H)_2$	solid	272.6*
Terephthalic acid	$C_6H_4(CO_2H)_2$	solid	770.4
Terpin hydrate	$C_{10}H_{22}O_3$	solid	1451.0
Terpineol	$C_{10}H_{18}O$	solid	1469.5
Tetrahydrobenzene	C_6H_{10}	liquid	891.9
Tetrahydronaphthalene	$C_{10}H_{12}$	liquid	1352.4
Tetramethylmethane	$(CH_3)_4C$	gas	842.6
Tetraphenylmethane	$(C_6H_5)_4C$	solid	3102.4
Tetryl	$C_6H_5N_5O_8$	solid	842.3
Thebaine	$C_{19}H_{21}O_3N$	solid	2441.3
Thiophene	C_4H_4S	liquid	670.5
Thujane	$C_{10}H_{18}$	liquid	1506.4
Thymol	$C_{10}H_{14}O$	liquid	1353.4
Thymol	$C_{10}H_{14}O$	solid	1349.7
Thymoquinone	$C_{10}H_{12}O_2$	solid	1271.3
Toluene	$CH_2C_6H_5$	liquid	934.2
o-Toluic acid	$CH_3C_6H_4CO_2H$	solid	928.9
m-Toluic acid	$CH_3C_6H_4CO_2H$	solid	928.6
p-Toluic acid	$CH_3C_6H_4CO_2H$	solid	926.9
o-Toluidine	$CH_3C_6H_4NH_2$	liquid	964.3
m-Toluidine	$CH_3C_6H_4NH_2$	liquid	965.3
p-Toluidine	$Ch_3C_6H_4NH_2$	solid	958.4
o-Tolunitrile	$CH_3C_6H_4CN$	liquid	1030.3
Toluquinone	$C_7H_6O_3$	solid	803.2
Triaminotriphenyl carbinol	$(C_6H_4NH_2)_2COH$	solid	2483.5
Tribenzyl amine	$(C_6H_5CH_2)_3N$	solid	2762.1
Trichloracetic acid	$Cl_2C.CO_2H$	solid	92.8
Triethyl amine	$(C_2H_5)_2N$	liquid	1036.8
Triethyl carbinol	$(C_2H_5)_3CHOH$	liquid	1080.0
Triisoamyl amine	$[(CH_3)_2CHCH_2CH_2]_3N$	liquid	2459.3
Triisobutyl amine	$[(CH_3)_2CHCH_2]_3N$	liquid	1973.6
Trimethyl amine	$(CH_3)_3N$	liquid	578.6
2,2,3-Trimethylbutane	$(CH_3)_2C.CH(CH_3)_2$	liquid	1147.9
Trimethyl carbinol	$(CH_3)_2COH$	liquid	629.3
Trimethylene	$CH_2CH_2CH_2$	gas	499.89*
Trimethylethylene	$(CH_3)_2CCH_3$	liquid	796.0
Trimethylethylene	$(CH_3)_2CCH_3$	vapor	803.6
2,2,4-Trimethylpentane	$(CH_3)_3C.CH_2CH(CH_3)_2$	liquid	1303.9
Trinitrobenzene	$C_6H_3(NO_2)_3$—(1,3,5)	solid	663.7
Trinitroglycerol	$C_3H_5(NO_3)_3$	liquid	368.4
Trinitrotoluene	$C_5H_2(CH_3)(NO_2)_3$—(1,2,4,6)	solid	820.7
Triphenyl amine	$(C_6H_5)_3N$	solid	2267.8
Triphenylbenzene	$C_6H_3(C_6H_5)_3$—(1,3,5)	solid	2936.7
Triphenyl carbinol	$(C_6H_5)_3CHOH$	sed	2340.8

** Recommended as a secondary thermochemical standard.

HEAT OF COMBUSTION
For Organic Compounds (Continued)

Name	Formula	Physical state	Heat of combustion. kg. calories
Triphenylmethane	$(C_6H_5)_2CH$	solid	2388.7
Triphenyl methyl	$(C_6H_5)_3C$	solid	2378.5
Tyrosine	$C_9H_{11}O_3N$	solid	1070.2
Undecyclic acid	$C_{11}H_{22}O_2$	solid	1615.9
Urea	$(NH_2)_2CO$	solid	150.97*
Urethane	$NH_2CO_2C_2H_5$	solid	397.2
Uric acid	$C_5H_4O_3N_4$	solid	460.2
n-Valeric acid	$C_4H_9CO_2H$	liquid	678.12*
Vanillin	$C_6H_3(OH)(OCH_3)CHO$ (1,2,4)	solid	914.1
o-Xylene	$(CH_3)_2C_6H_4$	liquid	1091.7
m-Xylene	$(CH_3)_2C_6H_4$	liquid	1088.4
p-Xylene	$(CH_3)_2C_6H_4$	liquid	1089.1
Xylose	$C_5H_{10}O_5$	solid	559.0*

HEAT OF FORMATION

The thermal change involved in the formation of 1 mol of substance from its elements is the heat of formation, ΔH, of the substance. If all of the substances involved in the reaction are each in their standard states and each substance is at unit activity, the thermal change is the standard heat of formation, ΔH°. By definition, all elements in their standard state have a heat of formation of zero. By further definition, the sign of ΔH is negative if heat is evolved and is positive if heat is absorbed. Thus, for the thermochemical reaction:

$$7C\,(s) + 3H_2(g) + O_2(g) = C_6H_5COOH\,(s) \quad \Delta H^\circ = -771.24 \text{ kcal}$$

(benzoic acid)

Heats of formation may be calculated from heats of combustion. The heat of formation of compound "A" is equal to the sum of the heats of formation of the products of combustion of compound "A" minus the heat of combustion of compound "A". Some heats of formation of some products of combustion of organic compounds are:

Substance	ΔH, heat of formation (kcal/g mole)
Free Elements	0
CO	− 26.416
CO_2	− 93.963
½ H_2O (1), from 1 H	− 34.158
H_2O (1)	− 68.317
HF (Dilute aqueous solution)	− 76.531
HCl (Dilute aqueous solution)	− 39.850
HBr (Dilute aqueous solution)	− 28.958
HI (Dilute aqueous solution)	− 13.106
HNO_3 (Dilute aqueous solution)	− 48.484
H_2SO_4 (Dilute aqueous solution)	−213.552
SO_2 (g)	− 70.336

Two examples of calculations are:

Example I

Calculate the heat of formation of methane (CH_4) where

Heat of combustion of CH_4 = −210.8 kcal/g mol
Heat of formation of CO_2 = −93.963 kcal/g mol
Heat of formation of H_2O = −68.317 kcal/g mol

and where the combustion reaction occurs according to the following equation:

$$CH_4(gas) + 2O_2(gas) = CO_2(gas) + 2H_2O\,(liquid)$$

Heat of formation of CH_4 equals:

	Heat of formation of CO_2	=	−93.963 kcal
+	Two times heat of formation of H_2O	=	2(−68.317)kcal
−	Heat of combustion of CH_4	=	−(−210.8) kcal
			−19.8 kcal/g mol

Example II

Calculate the heat of formation of ethylene (C_2H_4) where

Heat of combustion of C_2H_4 = −337.23 kcal/g mol

and where the combustion reaction is as follows:

$$C_2H_4(gas) + 3O_2(gas) = 2CO_2(gas) + 2H_2O\,(liquid)$$

	Two times heat of formation of CO_2	=	2(−93.963)kcal
+	Two times heat of formation of H_2O	=	2(−68.317)kcal
−	Heat of combustion of C_2H_4	=	−(−337.23) kcal
			+12.67 kcal/g mol

RECOMMENDED DAILY DIETARY ALLOWANCES

Data in this table are from the 1980 report of the Food and Nutrition Board, National Academy of Sciences-National Research Council. They are reproduced by permission of the National Academy Press. Customarily, these values are revised each 5 years by a panel of scientists and nutritional experts. Some reasons for there not being a revised set of values published since 1980 are presented in *Science*, 23–, 420, 1985. The allowances are intended to provide for individual variations among most normal persons as they live in the U.S. under usual environmental stresses. Diets should be based on a variety of common foods in order to provide other nutrients for which human requirements have been less well defined.

		Weight		Height			Fat-soluble vitamins			Water-soluble vitamins							Minerals					
	Age (years)	kg	lb.	cm	in.	Protein (g)	Vitamin A activity (RE)[a]	Vitamin D (µg)[b]	Vitamin (mg α T.E.)	Ascorbic acid (mg)	Folacin (µg)[c]	Niacin (mg N.E.)[d]	Riboflavin (mg)	Thiamine (mg)	Vitamin B6 (mg)	Vitamin B12 (µg)[e]	Calcium (mg)	Phosphorus (mg)	Iodine (µg)	Iron (mg)	Magnesium (mg)	Zinc (mg)
Infants	0.0—0.5	6	13	60	24	kg × 2.2	420	10	3	35	30	6	0.4	0.3	0.3	0.5	360	240	40	10	50	3
	0.5—1.0	9	20	71	28	kg × 2.0	400	10	4	35	45	8	0.6	0.5	0.6	1.5	540	360	50	15	70	5
Children	1—3	13	29	90	35	23	400	10	5	45	100	9	0.8	0.7	0.9	2.0	800	800	70	15	150	10
	4—6	20	44	112	44	30	500	10	6	45	200	11	1.0	0.9	1.3	2.5	800	800	90	10	200	10
	7—10	28	62	132	52	34	700	10	7	45	300	16	1.4	1.2	1.6	3.0	800	800	120	10	250	10
Males	11—14	45	99	157	62	45	1000	10	8	50	400	18	1.6	1.4	1.8	3.0	1200	1200	150	18	350	15
	15—18	66	145	176	69	56	1000	10	10	60	400	18	1.7	1.4	2.0	3.0	1200	1200	150	18	400	15
	19—22	70	154	177	70	56	1000	7.5	10	60	400	19	1.7	1.4	2.2	3.0	800	800	150	10	350	15
	23—50	70	154	178	70	56	1000	5	10	60	400	18	1.6	1.4	2.2	3.0	800	800	150	10	350	15
	51+	70	154	178	70	56	1000	5	10	60	400	16	1.4	1.2	2.2	3.0	800	800	150	10	350	15
Females	11—14	46	101	157	62	46	800	10	8	50	400	15	1.3	1.1	1.8	3.0	1200	1200	150	18	300	15
	15—18	55	120	163	64	46	800	10	8	60	400	14	1.3	1.1	2.0	3.0	1200	1200	150	18	300	15
	19—22	55	120	163	64	44	800	7.5	8	60	400	14	1.3	1.1	2.0	3.0	800	800	150	18	300	15
	23—50	55	120	163	64	44	800	5	8	60	400	13	1.2	1.0	2.0	3.0	800	800	150	18	300	15
	51+	55	120	163	64	44	800	5	8	60	400	13	1.2	1.0	2.0	3.0	800	800	150	10	300	15
Pregnant						+30	+200	+5	+2	+20	+400	+2	+0.3	+0.4	+0.6	+1.0	+400	+400	+25	f	+150	+5
Lactating						+20	+400	+5	+3	+40	+100		+0.5	+0.5	+0.5	+1.0	+400	+400	+50	f	+150	+10

[a] Retinol equivalents. 1 Retinol equivalent = 1 µg retinol or 6 µg β-carotene.

[b] As cholecalciferol. 10 µg cholecalciferol = 400 IU vitamin D.

[c] The folacin allowances refer to dietary sources as determined by *Lactobacillus casei* assay after treatment with enzymes ("conjugases") to make polyglutamyl forms to the vitamin available to the test organism.

[d] 1 N.E. (niacin equivalent) is equal to 1 mg of niacin or 60 mg of dietary tryptophan.

[e] The RDA for vitamin B_{12} in infants is based on average concentration of vitamin in human milk. The allowances after weaning are based on energy intake (as recommended by the American Academy of Pediatrics) and consideration of other factors such as intestinal absorption.

[f] The increased requirement during pregnancy cannot be met by the iron content of habitual American diets nor by the existing iron stores of many women; therefore the use of 30—60 mg of supplemental iron is recommended. Iron needs during lactation are not substantially different from those of nonpregnant women, but continued supplementation of the mother 3—4 months after parturition is advisable in order to replenish stores depleted by pregnancy.

NUTRITIVE VALUE OF THE EDIBLE PART OF FOODS

From U.S. Department of Agriculture, Human Nutrition and Information Service, Home and Garden Bulletin No. 72 (1985). Susan E. Gebhardt and Ruth Matthews

										Nutrients in indicated quantity												
							Fatty acids										Vitamin A value					
Item No.	Foods, approximate measures, units, and weight (weight of edible portion only)	Grams	Water (%)	Food energy (cal)	Protein (g)	Fat (g)	Saturated (g)	Mono-unsaturated (g)	Poly-unsaturated (g)	Cholesterol (mg)	Carbo-hydrate (g)	Calcium (mg)	Phos-phorus (mg)	Iron (mg)	Potassium (mg)	Sodium (mg)	(IU)[1]	(RE)[2]	Thiamin (mg)	Riboflavin (mg)	Niacin	Ascorbic acid (mg)
	Beverages																					
	Alcoholic																					
	Beer																					
1	Regular 12 fl oz	360	92	150	1	0	0.0	0.0	0.0	0	13	14	50	0.1	115	18	0	0	0.02	0.09	1.8	0
2	Light 12 fl oz	355	95	95	1	0	0.0	0.0	0.0	0	5	14	43	0.1	64	11	0	0	0.03	0.11	1.4	0
	Gin, rum, vodka, whiskey																					
3	80-proof 1½ fl oz	42	67	95	0	0	0.0	0.0	0.0	0	Tr	Tr	Tr	Tr	1	Tr	0	0	Tr	Tr	Tr	0
4	86-proof 1½ fl oz	42	64	105	0	0	0.0	0.0	0.0	0	Tr	Tr	Tr	Tr	1	Tr	0	0	Tr	Tr	Tr	0
5	90-proof 1½ fl oz	42	62	110	0	0	0.0	0.0	0.0	0	Tr	Tr	Tr	Tr	1	Tr	0	0	Tr	Tr	Tr	0
	Wines																					
6	Dessert 3½ fl oz	103	77	140	Tr	0	0.0	0.0	0.0	0	8	8	9	0.2	95	9	[3]	[3]	0.01	0.02	0.2	0
	Table																					
7	Red 3½ fl oz	102	88	75	Tr	0	0.0	0.0	0.0	0	3	8	18	0.4	113	5	[3]	[3]	0.00	0.03	0.1	0
8	White 3½ fl oz	102	87	80	Tr	0	0.0	0.0	0.0	0	3	9	14	0.3	83	5	[3]	[3]	0.00	0.01	0.1	0
	Carbonated																					
9	Club soda 12 fl oz	355	100	0	0	0	0.0	0.0	0.0	0	0	18	0	Tr	0	78	0	0	0.00	0.00	0.0	0
	Cola type																					
10	Regular 12 fl oz	369	89	160	0	0	0.0	0.0	0.0	0	41	11	52	0.2	7	18	0	0	0.00	0.00	0.0	0
11	Diet, artificially sweetened 12 fl oz	355	100	Tr	0	0	0.0	0.0	0.0	0	Tr	14	39	0.2	7	[3]32	0	0	0.00	0.00	0.0	0
12	Ginger ale 12 fl oz	366	91	125	0	0	0.0	0.0	0.0	0	32	11	0	0.1	4	29	0	0	0.00	0.00	0.0	0
13	Grape 12 fl oz	372	88	180	0	0	0.0	0.0	0.0	0	46	15	0	0.4	4	48	0	0	0.00	0.00	0.0	0
14	Lemon-lime 12 fl oz	372	89	155	0	0	0.0	0.0	0.0	0	39	7	0	0.4	4	33	0	0	0.00	0.00	0.0	0
15	Orange 12 fl oz	372	88	180	0	0	0.0	0.0	0.0	0	46	15	4	0.3	7	52	0	0	0.00	0.00	0.0	0
16	Pepper type 12 fl oz	369	89	160	0	0	0.0	0.0	0.0	0	41	11	41	0.1	4	37	0	0	0.00	0.00	0.0	0
17	Root beer 12 fl oz	370	89	165	0	0	0.0	0.0	0.0	0	42	15	0	0.2	4	48	0	0	0.00	0.00	0.0	0

Item No.	Foods, approximate measures, units, and weight (weight of edible portion only)	Grams	Water (%)	Food energy (cal)	Protein (g)	Fat (g)	Fatty acids: Saturated (g)	Fatty acids: Mono-unsaturated (g)	Fatty acids: Poly-unsaturated (g)	Cholesterol (mg)	Carbo-hydrate (g)	Calcium (mg)	Phos-phorus (mg)	Iron (mg)	Potassium (mg)	Sodium (mg)	Vitamin A value (IU)[1]	Vitamin A value (RE)[2]	Thiamin (mg)	Riboflavin (mg)	Niacin	Ascorbic acid (mg)
	Cocoa and chocolate-flavored beverages; see Dairy Products (items 95—98)																					
	Coffee																					
18	Brewed 6 fl oz	180	100	Tr	Tr	Tr	Tr	Tr	Tr	0	Tr	4	2	Tr	124	2	0	0	0.00	0.02	0.4	0
19	Instant, prepared (2 tsp powder plus 6 fl oz water) 6 fl oz	182	99	Tr	Tr	Tr	Tr	Tr	Tr	0	1	2	6	0.1	71	Tr	0	0	0.00	0.03	0.6	0
	Fruit drinks, noncarbonated																					
	Canned																					
20	Fruit punch drink 6 fl oz	190	88	85	Tr	0	0.0	0.0	0.0	0	22	15	2	0.4	48	15	20	2	0.03	0.04	Tr	61[6]
21	Grape drink 6 fl oz	187	86	100	Tr	0	0.0	0.0	0.0	0	26	2	2	0.3	9	11	Tr	Tr	0.01	0.01	Tr	64[6]
22	Pineapple-grapefruit juice drink 6 fl oz	187	87	90	Tr	Tr	Tr	Tr	Tr	0	23	13	7	0.9	97	24	60	6	0.06	0.04	0.5	110[6]
	Frozen																					
	Lemonade concentrate																					
23	Undiluted 6-fl-oz can	219	49	425	Tr	Tr	Tr	Tr	Tr	0	112	9	13	0.4	153	4	40	4	0.04	0.07	0.7	66
24	Diluted with $4^1/_3$ parts water by volume 6 fl oz	185	89	80	Tr	Tr	Tr	Tr	Tr	0	21	2	2	0.1	30	1	10	1	0.01	0.02	0.2	13
	Limeade concentrate																					
25	Undiluted 6-fl-oz can	218	50	410	Tr	Tr	Tr	Tr	Tr	0	108	11	13	0.2	129	Tr	Tr	Tr	0.02	0.02	0.2	26
26	Diluted with $4^1/_3$ parts water by volume 6 fl oz	185	89	75	Tr	Tr	Tr	Tr	Tr	0	20	2	2	Tr	24	Tr	Tr	Tr	Tr	Tr	Tr	4
	Fruit juices; see type under Fruits and Fruit Juices																					
	Milk beverages; see Dairy Products (items 92—105)																					
	Tea																					
27	Brewed 8 fl oz	240	100	Tr	Tr	Tr	Tr	Tr	Tr	0	Tr	0	2	Tr	36	1	0	0	0.00	0.03	Tr	0
	Instant, powder, prepared																					
28	Unsweetened (1 tsp powder plus 8 fl oz water) 8 fl oz	241	100	Tr	Tr	Tr	Tr	Tr	Tr	0	1	1	4	Tr	61	1	0	0	0.00	0.02	0.1	0
29	Sweetened (3 tsp powder plus 8 fl oz water) 8 fl oz	262	91	85	Tr	Tr	Tr	Tr	Tr	0	22	1	3	Tr	49	Tr	0	0	0.00	0.04	0.1	0

Item No.	Foods, approximate measures, units, and weight (weight of edible portion only)	Grams	Water (%)	Food energy (cal)	Protein (g)	Fat (g)	Fatty acids: Saturated (g)	Fatty acids: Mono-unsaturated (g)	Fatty acids: Poly-unsaturated (g)	Cholesterol (mg)	Carbohydrate (g)	Calcium (mg)	Phosphorus (mg)	Iron (mg)	Potassium (mg)	Sodium (mg)	Vitamin A value (IU)[1]	Vitamin A value (RE)[2]	Thiamin (mg)	Riboflavin (mg)	Niacin	Ascorbic acid (mg)
	Dairy Products																					
	Butter; see Fats and Oils (items 128—130)																					
	Cheese																					
	Natural																					
30	Blue																					
	1 oz	28	42	100	6	8	5.3	2.2	0.2	21	1	150	110	0.1	73	396	200	65	0.01	0.11	0.3	0
31	Camembert (3 wedges per 4-oz container)																					
	1 wedge	38	52	115	8	9	5.8	2.7	0.3	27	Tr	147	132	0.1	71	320	350	96	0.01	0.19	0.2	0
	Cheddar																					
32	Cut pieces																					
	1 oz	28	37	115	7	9	6.0	2.7	0.3	30	Tr	204	145	0.2	28	176	300	86	0.01	0.11	Tr	0
33	1 cubic inch	17	37	70	4	6	3.6	1.6	0.2	18	Tr	123	87	0.1	17	105	180	52	Tr	0.06	Tr	0
34	Shredded																					
	1 cup	113	37	455	28	37	23.8	10.6	1.1	119	1	815	579	0.8	111	701	1200	342	0.03	0.42	0.1	0
	Cottage (curd not pressed down)																					
	Creamed (cottage cheese, 4% fat)																					
35	Large curd																					
	1 cup	225	79	235	28	10	6.4	2.9	0.3	34	6	135	297	0.3	190	911	370	108	0.05	0.37	0.3	Tr
36	Small curd																					
	1 cup	210	79	215	26	9	6.0	2.7	0.3	31	6	126	277	0.3	177	850	340	101	0.04	0.34	0.3	Tr
37	With fruit																					
	1 cup	226	72	280	22	8	4.9	2.2	0.2	25	30	108	236	0.2	151	915	280	81	0.04	0.29	0.2	Tr
38	Lowfat (2%)																					
	1 cup	226	79	205	31	4	2.8	1.2	0.1	19	8	155	340	0.4	217	918	160	45	0.05	0.42	0.3	Tr
39	Uncreamed (cottage cheese dry curd, less than 1/2% fat)																					
	1 cup	145	80	125	25	1	0.4	0.2	Tr	10	3	46	151	0.3	47	19	40	12	0.04	0.21	0.2	0
40	Cream																					
	1 oz	28	54	100	2	10	6.2	2.8	0.4	31	1	23	30	0.3	34	84	400	124	Tr	0.06	Tr	0
41	Feta																					
	1 oz	28	55	75	4	6	4.2	1.3	0.2	25	1	140	96	0.2	18	316	130	36	0.04	0.24	0.3	0
	Mozzarella, made with																					
42	Whole milk																					
	1 oz	28	54	80	6	6	3.7	1.9	0.2	22	1	147	105	0.1	19	106	220	68	Tr	0.07	Tr	0
43	Part skim milk (low moisture)																					
	1 oz	28	49	80	8	5	3.1	1.4	0.1	15	1	207	149	0.1	27	150	180	54	0.01	0.10	Tr	0
44	Muenster																					
	1 oz	28	42	105	7	9	5.4	2.5	0.2	27	Tr	203	133	0.1	38	178	320	90	Tr	0.09	Tr	0
	Parmesan, grated																					
45	Cup, not pressed down																					
	1 cup	100	18	455	42	30	19.1	8.7	0.7	79	4	1376	807	1.0	107	1861	700	173	0.05	0.39	0.3	0
46	Tablespoon																					
	1 tbsp	5	18	25	2	2	1.0	0.4	Tr	4	Tr	69	40	Tr	5	93	40	9	Tr	0.02	Tr	0

Item No.	Foods, approximate measures, units, and weight (weight of edible portion only)	Grams	Water (%)	Food energy (cal)	Protein (g)	Fat (g)	Fatty acids: Saturated (g)	Fatty acids: Mono-unsaturated (g)	Fatty acids: Poly-unsaturated (g)	Cholesterol (mg)	Carbo-hydrate (g)	Calcium (mg)	Phos-phorus (mg)	Iron (mg)	Potassium (mg)	Sodium (mg)	Vitamin A value (IU)[1]	Vitamin A value (RE)[2]	Thiamin (mg)	Riboflavin (mg)	Niacin	Ascorbic acid (mg)
										Nutrients in indicated quantity												
47	Ounce 1 oz	28	18	130	12	9	5.4	2.5	0.2	22	1	390	229	0.3	30	528	200	49	0.01	0.11	0.1	0
48	Provolone 1 oz	28	41	100	7	8	4.8	2.1	0.2	20	1	214	141	0.1	39	248	230	75	0.01	0.09	Tr	0
	Ricotta, made with																					
49	Whole milk 1 cup	246	72	430	28	32	20.4	8.9	0.9	124	7	509	389	0.9	257	207	1210	330	0.03	0.48	0.3	0
50	Part skim milk 1 cup	246	74	340	28	19	12.1	5.7	0.6	76	13	669	449	1.1	307	307	1060	278	0.05	0.46	0.2	0
51	Swiss 1 oz	28	37	105	8	8	5.0	2.1	0.3	26	1	272	171	Tr	31	74	240	72	0.01	0.10	Tr	0
	Pasteurized process cheese																					
52	American 1 oz	28	39	105	6	9	5.6	2.5	0.3	27	Tr	174	211	0.1	46	406	340	82	0.01	0.10	Tr	0
53	Swiss 1 oz	28	42	95	7	7	4.5	2.0	0.2	24	1	219	216	0.2	61	388	230	65	Tr	0.08	Tr	0
54	Pasteurized process cheese food, American 1 oz	28	43	95	6	7	4.4	2.0	0.2	18	2	163	130	0.2	79	337	260	62	0.01	0.13	Tr	0
55	Pasteurized process cheese spread, American 1 oz	28	48	80	5	6	3.8	1.8	0.2	16	2	159	202	0.1	69	381	220	54	0.01	0.12	Tr	0
	Cream, sweet																					
56	Half-and-half (cream and milk) 1 cup	242	81	315	7	28	17.3	8.0	1.0	89	10	254	230	0.2	314	98	1050	259	0.08	0.36	0.2	2
57	1 tbsp	15	81	20	Tr	2	1.1	0.5	0.1	6	1	16	14	Tr	19	6	70	16	0.01	0.02	Tr	Tr
58	Light, coffee, or table 1 cup	240	74	470	6	46	28.8	13.4	1.7	159	9	231	192	0.1	292	95	1730	437	0.08	0.36	0.1	2
59	1 tbsp	15	74	30	Tr	3	1.8	0.8	0.1	10	1	14	12	Tr	18	6	110	27	Tr	0.02	Tr	Tr
	Whipping, unwhipped (volume about double when whipped)																					
60	Light 1 cup	239	64	700	5	74	46.2	21.7	2.1	265	7	166	146	0.1	231	82	2690	705	0.06	0.30	0.1	1
61	1 tbsp	15	64	45	Tr	5	2.9	1.4	0.1	17	Tr	10	9	Tr	15	5	170	44	Tr	0.02	Tr	Tr
62	Heavy 1 cup	238	58	820	5	88	54.8	25.4	3.3	326	7	154	149	0.1	179	89	3500	1002	0.05	0.26	0.1	1
63	1 tbsp	15	58	50	Tr	6	3.5	1.6	0.2	21	Tr	10	9	Tr	11	6	220	63	Tr	0.02	Tr	Tr
64	Whipped topping (pressurized) 1 cup	60	61	155	2	13	8.3	3.9	0.5	46	7	61	54	Tr	88	78	550	124	0.02	0.04	Tr	0
65	1 tbsp	3	61	10	Tr	1	0.4	0.2	Tr	2	Tr	3	3	Tr	4	4	30	6	Tr	Tr	Tr	0
66	Cream, sour 1 cup	230	71	495	7	48	30.0	13.9	1.8	102	10	268	195	0.1	331	123	1820	448	0.08	0.34	0.2	2
67	1 tbsp	12	71	25	Tr	3	1.6	0.7	0.1	5	1	14	10	Tr	17	6	90	23	Tr	0.02	Tr	Tr

Item No.	Foods, approximate measures, units, and weight (weight of edible portion only)	Grams	Water (%)	Food energy (cal)	Protein (g)	Fat (g)	Fatty acids: Saturated (g)	Fatty acids: Mono-unsaturated (g)	Fatty acids: Poly-unsaturated (g)	Cholesterol (mg)	Carbo-hydrate (g)	Calcium (mg)	Phos-phorus (mg)	Iron (mg)	Potassium (mg)	Sodium (mg)	Vitamin A value (IU)[1]	Vitamin A value (RE)[2]	Thiamin (mg)	Riboflavin (mg)	Niacin	Ascorbic acid (mg)
	Nutrients in indicated quantity																					
	Cream products, imitation (made with vegetable fat)																					
	Sweet																					
	Creamers																					
	Liquid (frozen)																					
68	1 tbsp	15	77	20	Tr	1	1.4	Tr	Tr	0	2	1	10	Tr	29	12	10[7]	1[5]	0.00	0.00	0.0	0
	Powdered																					
69	1 tsp	2	2	10	Tr	1	0.7	Tr	Tr	0	1	Tr	8	Tr	16	4	Tr	Tr	0.00	Tr	0.0	0
	Whipped topping																					
	Frozen																					
70	1 cup	75	50	240	1	19	16.3	1.2	0.4	0	17	5	6	0.1	14	19	650[7]	65[7]	0.00	0.00	0.0	0
71	1 tbsp	4	50	15	Tr	1	0.9	0.1	Tr	0	1	Tr	Tr	Tr	1	1	30[7]	3[7]	0.00	0.00	0.0	0
	Powdered, made with whole milk																					
72	1 cup	80	67	150	3	10	8.5	0.7	0.2	8	13	72	69	Tr	121	53	290[7]	39[7]	0.02	0.09	Tr	1
73	1 tbsp	4	67	10	Tr	Tr	0.4	Tr	Tr	Tr	1	4	3	Tr	6	3	10[7]	2[7]	Tr	Tr	Tr	Tr
	Pressurized																					
74	1 cup	70	60	185	1	16	13.2	1.3	0.2	0	11	4	13	Tr	13	43	330[7]	33[7]	0.00	0.00	0.0	0
75	1 tbsp	4	60	10	Tr	1	0.8	0.1	Tr	0	1	Tr	1	Tr	1	2	20[7]	2[7]	0.00	0.00	0.0	0
	Sour dressing (filled cream type product, nonbutterfat)																					
76	1 cup	235	75	415	8	39	31.2	4.6	1.1	13	11	266	205	0.1	380	113	20	5	0.09	0.38	0.2	2
77	1 tbsp	12	75	20	Tr	2	1.6	0.2	0.1	1	1	14	10	Tr	19	6	Tr	Tr	Tr	0.02	Tr	Tr
	Ice cream; see Milk desserts, frozen (items 106—111)																					
	Ice milk; see Milk desserts, frozen (items 112—114)																					
	Milk																					
	Fluid																					
	Whole (3.3% fat)																					
78	1 cup	244	88	150	8	8	5.1	2.4	0.3	33	11	291	228	0.1	370	120	310	76	0.09	0.40	0.2	2
	Lowfat (2%)																					
	No milk solids added																					
79	1 cup	244	89	120	8	5	2.9	1.4	0.2	18	12	297	232	0.1	377	122	500	139	0.10	0.40	0.2	2
	Milk solids added, label claim less than 10 g of protein per cup																					
80	1 cup	245	89	125	9	5	2.9	1.4	0.2	18	12	313	245	0.1	397	128	500	140	0.10	0.42	0.2	2
	Lowfat (1%)																					
	No milk solids added																					
81	1 cup	244	90	100	8	3	1.6	0.7	0.1	10	12	300	235	0.1	381	123	500	144	0.10	0.41	0.2	2
	Milk solids added, label claim less than 10 g of protein per cup																					
82	1 cup	245	90	105	9	2	1.5	0.7	0.1	10	12	313	245	0.1	397	128	500	145	0.10	0.42	0.2	2

Item No.	Foods, approximate measures, units, and weight (weight of edible portion only)	Grams	Water (%)	Food energy (cal)	Protein (g)	Fat (g)	Fatty acids: Saturated (g)	Fatty acids: Monounsaturated (g)	Fatty acids: Polyunsaturated (g)	Cholesterol (mg)	Carbohydrate (g)	Calcium (mg)	Phosphorus (mg)	Iron (mg)	Potassium (mg)	Sodium (mg)	Vitamin A value (IU)[1]	Vitamin A value (RE)[2]	Thiamin (mg)	Riboflavin (mg)	Niacin	Ascorbic acid (mg)
	Nonfat (skim)																					
83	No milk solids added 1 cup	245	91	85	8	Tr	0.3	0.1	Tr	4	12	302	247	0.1	406	126	500	149	0.09	0.34	0.2	2
84	Milk solids added, label claim less than 10 g of protein per cup 1 cup	245	90	90	9	1	0.4	0.2	Tr	5	12	316	255	0.1	418	130	500	149	0.10	0.43	0.2	2
85	Buttermilk 1 cup	245	90	100	8	2	1.3	0.6	0.1	9	12	285	219	0.1	371	257	80	20	0.08	0.38	0.1	2
	Canned																					
86	Condensed, sweetened 1 cup	306	27	980	24	27	16.8	7.4	1.0	104	166	868	775	0.6	1136	389	1000	248	0.28	1.27	0.6	8
	Evaporated																					
87	Whole milk 1 cup	252	74	340	17	19	11.6	5.9	0.6	74	25	657	510	0.5	764	267	610	136	0.12	0.80	0.5	5
88	Skim milk 1 cup	255	79	200	19	1	0.3	0.2	Tr	9	29	738	497	0.7	845	293	1000	298	0.11	0.79	0.4	3
	Dried																					
	Buttermilk 1 cup	120	3	465	41	7	4.3	2.0	0.3	83	59	1421	1119	0.4	1910	621	260	65	0.47	1.89	1.1	7
	Nonfat, instantized																					
90	Envelope, 3.2 oz, net wt.[8] 1 envelope	91	4	325	32	1	0.4	0.2	Tr	17	47	1120	896	0.3	1552	499	2160[9]	646[9]	0.38	1.59	0.8	5
91	Cup 1 cup	68	4	245	24	Tr	0.3	0.1	Tr	12	35	837	670	0.2	1160	373	1610[9]	483[9]	0.28	1.19	0.6	4
	Milk beverages																					
	Chocolate milk (commercial)																					
92	Regular 1 cup	250	82	210	8	8	5.3	2.5	0.3	31	26	280	251	0.6	417	149	300	73	0.09	0.41	0.3	2
93	Lowfat (2%) 1 cup	250	84	180	8	5	3.1	1.5	0.2	17	26	284	254	0.6	422	151	500	143	0.09	0.41	0.3	2
94	Lowfat (1%) 1 cup	250	85	160	8	3	1.5	0.8	0.1	7	26	287	256	0.6	425	152	500	148	0.10	0.42	0.3	2
	Cocoa and chocolate-flavored beverages																					
95	Powder containing nonfat dry milk 1 oz	28	1	100	3	1	0.6	0.3	Tr	1	22	90	88	0.3	223	139	Tr	Tr	0.03	0.17	0.2	Tr
96	Prepared (6 oz water plus 1 oz powder) 1 serving	206	86	100	3	1	0.6	0.3	Tr	1	22	90	88	0.3	223	139	Tr	Tr	0.03	0.17	0.2	Tr
97	Powder without nonfat dry milk 3/4 oz	21	1	75	1	1	0.3	0.2	Tr	0	19	7	26	0.7	135	56	Tr	Tr	Tr	0.03	0.1	Tr

Item No.	Foods, approximate measures, units, and weight (weight of edible portion only)	Grams	Water (%)	Food energy (cal)	Protein (g)	Fat (g)	Fatty acids: Saturated (g)	Fatty acids: Mono-unsaturated (g)	Fatty acids: Poly-unsaturated (g)	Cholesterol (mg)	Carbo-hydrate (g)	Calcium (mg)	Phos-phorus (mg)	Iron (mg)	Potassium (mg)	Sodium (mg)	Vitamin A value (IU)[1]	Vitamin A value (RE)[2]	Thiamin (mg)	Riboflavin (mg)	Niacin	Ascorbic acid (mg)
98	Prepared (8 oz whole milk plus 3/4 oz powder) 1 serving	265	81	225	9	9	5.4	2.5	0.3	33	30	298	254	0.9	508	176	310	76	0.10	0.43	0.3	3
99	Eggnog (commercial) 1 cup	254	74	340	10	19	11.3	5.7	0.9	149	34	330	278	0.5	420	138	890	203	0.09	0.48	0.3	4
	Malted milk																					
	Chocolate																					
100	Powder 3/4 oz	21	2	85	1	1	0.5	0.3	0.1	1	18	13	37	0.4	130	49	20	5	0.04	0.04	0.4	0
101	Prepared (8 oz whole milk plus 3/4 oz powder) 1 serving	265	81	235	9	9	5.5	2.7	0.4	34	29	304	265	0.5	500	168	330	80	0.14	0.43	0.7	2
	Natural																					
102	Powder 3/4 oz	21	3	85	3	2	0.9	0.5	0.3	4	15	56	79	0.2	159	96	70	17	0.11	0.14	1.1	0
103	Prepared (8 oz whole milk plus 3/4 oz powder) 1 serving	265	81	235	11	10	6.0	2.9	0.6	37	27	347	307	0.3	529	215	380	93	0.20	0.54	1.3	2
	Shakes, thick																					
104	Chocolate 10-oz container	283	72	335	9	8	4.8	2.2	0.3	30	60	374	357	0.9	634	314	240	59	0.13	0.63	0.4	0
105	Vanilla 10-oz container	283	74	315	11	9	5.3	2.5	0.3	33	50	413	326	0.3	517	270	320	79	0.08	0.55	0.4	0
	Milk desserts, frozen																					
	Ice cream, vanilla																					
	Regular (about 11% fat)																					
106	Hardened 1/2 gal	1064	61	2155	38	115	71.3	33.1	4.3	476	254	1406	1075	1.0	2052	929	4340	1064	0.42	2.63	1.1	6
107	1 cup	133	61	270	5	14	8.9	4.1	0.5	59	32	176	134	0.1	257	116	540	133	0.05	0.33	0.1	1
108	3 fl oz	50	61	100	2	5	3.4	1.6	0.2	22	12	66	51	Tr	96	44	200	50	0.02	0.12	0.1	Tr
109	Soft serve (frozen custard) 1 cup	173	60	375	7	23	13.5	6.7	1.0	153	38	236	199	0.4	338	153	790	199	0.08	0.45	0.2	1
110	Rich (about 16% fat), hardened 1/2 gal	1188	59	2805	33	190	118.3	54.9	7.1	703	256	1213	927	0.8	1771	868	7200	1758	0.36	2.27	0.9	5
111	1 cup	148	59	350	4	24	14.7	6.8	0.9	88	32	151	115	0.1	221	108	900	219	0.04	0.28	0.1	1
	Ice milk, vanilla																					
112	Hardened (about 4% fat) 1/2 gal	1048	69	1470	41	45	28.1	13.0	1.7	146	232	1409	1035	1.5	2117	836	1710	419	0.61	2.78	0.9	6
113	1 cup	131	69	185	5	6	3.5	1.6	0.2	18	29	176	129	0.2	265	105	210	52	0.08	0.35	0.1	1
114	Soft serve (about 3% fat) 1 cup	175	70	225	8	5	2.9	1.3	0.2	13	38	274	202	0.3	412	163	175	44	0.12	0.54	0.2	1

Item No.	Foods, approximate measures, units, and weight (weight of edible portion only)	Grams	Water (%)	Food energy (cal)	Protein (g)	Fat (g)	Fatty acids: Saturated (g)	Fatty acids: Mono-unsaturated (g)	Fatty acids: Poly-unsaturated (g)	Cholesterol (mg)	Carbo-hydrate (g)	Calcium (mg)	Phos-phorus (mg)	Iron (mg)	Potassium (mg)	Sodium (mg)	Vitamin A value (IU)[1]	Vitamin A value (RE)[2]	Thiamin (mg)	Riboflavin (mg)	Niacin	Ascorbic acid (mg)
115	Sherbet (about 2% fat)																					
	1/2 gal	1542	66	2160	17	31	19.0	8.8	1.1	113	469	827	594	2.5	1585	706	1480	308	0.26	0.71	1.0	31
116	1 cup	193	66	270	2	4	2.4	1.1	0.1	14	59	103	74	0.3	198	88	190	39	0.03	0.09	0.1	4
	Yogurt																					
	With added milk solids																					
	Made with lowfat milk																					
117	Fruit-flavored[10]																					
	8-oz container	227	74	230	10	2	1.6	0.7	0.1	10	43	345	271	0.2	442	133	100	25	0.08	0.40	0.2	1
118	Plain																					
	8-oz container	227	85	145	12	4	2.3	1.0	0.1	14	16	415	326	0.2	531	159	150	36	0.10	0.49	0.3	2
119	Made with nonfat milk																					
	8-oz container	227	85	125	13	Tr	0.3	0.1	Tr	4	17	452	355	0.2	579	174	20	5	0.11	0.53	0.3	2
	Without added milk solids																					
120	Made with whole milk																					
	8-oz container	227	88	140	8	7	4.8	2.0	0.2	29	11	274	215	0.1	351	105	280	68	0.07	0.32	0.2	1
	Eggs																					
	Eggs, large (24 oz/dozen)																					
	Raw																					
121	Whole, without shell																					
	1 egg	50	75	80	6	6	1.7	2.2	0.7	274	1	28	90	1.0	65	69	260	78	0.04	0.15	Tr	0
122	White																					
	1 white	33	88	15	3	Tr	0.0	0.0	0.0	0	Tr	4	4	Tr	45	50	0	0	Tr	0.09	Tr	0
123	Yolk																					
	1 yolk	17	49	65	3	6	1.7	2.2	0.7	272	Tr	25	86	0.9	15	8	310	94	0.04	0.07	Tr	0
	Cooked																					
124	Fried in butter																					
	1 egg	46	68	95	6	7	2.7	2.7	0.8	278	1	29	91	1.1	66	162	320	94	0.04	0.14	Tr	0
125	Hard-cooked, shell removed																					
	1 egg	50	75	80	6	6	1.7	2.2	0.7	274	1	28	90	1.0	65	69	260	78	0.04	0.14	Tr	0
126	Poached																					
	1 egg	50	74	80	6	6	1.7	2.2	0.7	273	1	28	90	1.0	65	146	260	78	0.03	0.13	Tr	0
127	Scrambled (milk added) in butter. Also omelet																					
	1 egg	64	73	110	7	8	3.2	2.9	0.8	282	2	54	109	1.0	97	176	350	102	0.04	0.18	Tr	Tr
	Fats and Oils																					
	Butter (4 sticks/lb)																					
128	Stick																					
	1/2 cup	113	16	810	1	92	57.1	26.4	3.4	247	Tr	27	26	0.2	29	933[11]	3460[12]	852[12]	0.01	0.04	Tr	0
129	Tablespoon (1/8 stick)																					
	1 tbsp	14	16	100	Tr	11	7.1	3.3	0.4	31	Tr	3	3	Tr	4	116[11]	430[12]	106[12]	Tr	Tr	Tr	0
130	Pat (1 in. square, 1/3 in. high; 90/lb)																					
	1 pat	5	16	35	Tr	4	2.5	1.2	0.2	11	Tr	1	1	Tr	1	41[11]	150[12]	38[12]	Tr	Tr	Tr	0

Item No.	Foods, approximate measures, units, and weight (weight of edible portion only)	Grams	Water (%)	Food energy (cal)	Protein (g)	Fat (g)	Fatty acids: Saturated (g)	Fatty acids: Monounsaturated (g)	Fatty acids: Polyunsaturated (g)	Cholesterol (mg)	Carbohydrate (g)	Calcium (mg)	Phosphorus (mg)	Iron (mg)	Potassium (mg)	Sodium (mg)	Vitamin A value (IU)[1]	Vitamin A value (RE)[2]	Thiamin (mg)	Riboflavin (mg)	Niacin	Ascorbic acid (mg)
131	Fats, cooking (vegetable shortenings)																					
	1 cup	205	0	1810	0	205	51.3	91.2	53.5	0	0	0	0	0.0	0	0	0	0	0.00	0.00	0.0	0
132	1 tbsp	13	0	115	0	13	3.3	5.8	3.4	0	0	0	0	0.0	0	0	0	0	0.00	0.00	0.0	0
133	Lard																					
	1 cup	205	0	1850	0	205	80.4	92.5	23.0	195	0	0	0	0.0	0	0	0	0	0.00	0.00	0.0	0
134	1 tbsp	13	0	115	0	13	5.1	5.9	1.5	12	0	0	0	0.0	0	0	0	0	0.00	0.00	0.0	0
	Margarine																					
135	Imitation (about 40% fat), soft																					
	8-oz container	227	58	785	1	88	17.5	35.6	31.3	0	1	40	31	0.0	57	2178[13]	7510[14]	2254[14]	0.01	0.05	Tr	Tr
136	1 tbsp	14	58	50	Tr	5	1.1	2.2	1.9	0	Tr	2	2	0.0	4	134[13]	460[14]	139[14]	Tr	Tr	Tr	Tr
	Regular (about 80% fat)																					
	Hard (4 sticks/lb)																					
137	Stick																					
	1/2 cup	113	16	810	1	91	17.9	40.5	28.7	0	1	34	26	0.1	48	1066[13]	3740[14]	1122[14]	0.01	0.04	Tr	Tr
138	Tablespoon (1/8 stick)																					
	1 tbsp	14	16	100	Tr	11	2.2	5.0	3.6	0	Tr	4	3	Tr	6	132[13]	460[14]	139[14]	Tr	0.01	Tr	Tr
139	Pat (1 in. square, 1/3 in. high; 90/lb)																					
	1 pat	5	16	35	Tr	4	0.8	1.8	1.3	0	Tr	1	1	Tr	2	47[13]	170[14]	50[14]	Tr	Tr	Tr	Tr
140	Soft																					
	8-oz container	227	16	1625	2	183	31.3	64.7	78.5	0	1	60	46	0.0	86	2449[13]	7510[14]	2254[14]	0.02	0.07	Tr	Tr
141	1 tbsp	14	16	100	Tr	11	1.9	4.0	4.8	0	Tr	4	3	0.0	5	151[13]	460[14]	139[14]	Tr	Tr	Tr	Tr
	Spread (about 60% fat)																					
	Hard (4 sticks/lb)																					
142	Stick																					
	s1/2 cup	113	37	610	1	69	15.9	29.4	20.5	0	0	24	18	0.0	34	1123[13]	3740[14]	1122[14]	0.01	0.03	Tr	Tr
143	Tablespoon (1/8 stick)																					
	1 tbsp	14	37	75	Tr	9	2.0	3.6	2.5	0	0	3	2	0.0	4	139[13]	460[14]	139[14]	Tr	Tr	Tr	Tr
144	Pat (1 in. square, 1/3 in. high; 90/lb)																					
	1 pat	5	37	25	Tr	3	0.7	1.3	0.9	0	0	1	1	0.0	1	50[13]	170[14]	50[14]	Tr	Tr	Tr	Tr
145	Soft																					
	8-oz container	227	37	1225	1	138	29.1	71.5	31.3	0	0	47	37	0.0	68	2256[13]	7510[14]	2254[14]	0.02	0.06	Tr	Tr
146	1 tbsp	14	37	75	Tr	9	1.8	4.4	1.9	0	0	3	2	0.0	4	139[13]	460[14]	139[14]	Tr	Tr	Tr	Tr
	Oils, salad or cooking																					
147	Corn																					
	1 cup	218	0	1925	0	218	27.7	52.8	128.0	0	0	0	0	0.0	0	0	0	0	0.00	0.00	0.0	0
148	1 tbsp	14	0	125	0	14	1.8	3.4	8.2	0	0	0	0	0.0	0	0	0	0	0.00	0.00	0.0	0
149	Olive																					
	1 cup	216	0	1910	0	216	29.2	159.2	18.1	0	0	0	0	0.0	0	0	0	0	0.00	0.00	0.0	0
150	1 tbsp	14	0	125	0	14	1.9	10.3	1.2	0	0	0	0	0.0	0	0	0	0	0.00	0.00	0.0	0
151	Peanut																					
	1 cup	216	0	1910	0	216	36.5	99.8	69.1	0	0	0	0	0.0	0	0	0	0	0.00	0.00	0.0	0
152	1 tbsp	14	0	125	0	14	2.4	6.5	4.5	0	0	0	0	0.0	0	0	0	0	0.00	0.00	0.0	0
153	Safflower																					
	1 cup	218	0	1925	0	218	19.8	26.4	162.4	0	0	0	0	0.0	0	0	0	0	0.00	0.00	0.0	0
154	1 tbsp	14	0	125	0	14	1.3	1.7	10.4	0	0	0	0	0.0	0	0	0	0	0.00	0.00	0.0	0

							Fatty acids										Vitamin A value					
Item No.	Foods, approximate measures, units, and weight (weight of edible portion only)	Grams	Water (%)	Food energy (cal)	Protein (g)	Fat (g)	Saturated (g)	Mono-unsaturated (g)	Poly-unsaturated (g)	Cholesterol (mg)	Carbohydrate (g)	Calcium (mg)	Phosphorus (mg)	Iron (mg)	Potassium (mg)	Sodium (mg)	(IU)[1]	(RE)[2]	Thiamin (mg)	Riboflavin (mg)	Niacin	Ascorbic acid (mg)
155	Soybean oil, hydrogenated (partially hardened)																					
	1 cup	218	0	1925	0	218	32.5	93.7	82.0	0	0	0	0	0.0	0	0	0	0	0.00	0.00	0.0	0
156	1 tbsp	14	0	125	0	14	2.1	6.0	5.3	0	0	0	0	0.0	0	0	0	0	0.00	0.00	0.0	0
157	Soybean-cottonseed oil blend, hydrogenated																					
	1 cup	218	0	1925	0	218	39.2	64.3	104.9	0	0	0	0	0.0	0	0	0	0	0.00	0.00	0.0	0
158	1 tbsp	14	0	125	0	14	2.5	4.1	6.7	0	0	0	0	0.0	0	0	0	0	0.00	0.00	0.0	0
159	Sunflower																					
	1 cup	218	0	1925	0	218	22.5	42.5	143.2	0	0	0	0	0.0	0	0	0	0	0.00	0.00	0.0	0
160	1 tbsp	14	0	125	0	14	1.4	2.7	9.2	0	0	0	0	0.0	0	0	0	0	0.00	0.00	0.0	0
	Salad dressings																					
	Commercial																					
161	Blue cheese																					
	1 tbsp	15	32	75	1	8	1.5	1.8	4.2	3	1	12	11	Tr	6	164	30	10	Tr	0.02	Tr	Tr
	French																					
162	Regular																					
	1 tbsp	16	35	85	Tr	9	1.4	4.0	3.5	0	1	2	1	Tr	2	188	Tr	Tr	Tr	Tr	Tr	Tr
163	Low calorie																					
	1 tbsp	16	75	25	Tr	2	0.2	0.3	1.0	0	2	6	5	Tr	3	306	Tr	Tr	Tr	Tr	Tr	Tr
	Italian																					
164	Regular																					
	1 tbsp	15	34	80	Tr	9	1.3	3.7	3.2	0	1	1	1	Tr	5	162	30	3	Tr	Tr	Tr	Tr
165	Low calorie																					
	1 tbsp	15	86	5	Tr	Tr	Tr	Tr	Tr	0	2	1	1	Tr	4	136	Tr	Tr	Tr	Tr	Tr	Tr
	Mayonnaise																					
166	Regular																					
	1 tbsp	14	15	100	Tr	11	1.7	3.2	5.8	8	Tr	3	4	0.1	5	80	40	12	0.00	0.00	Tr	0
167	Imitation																					
	1 tbsp	15	63	35	Tr	3	0.5	0.7	1.6	4	2	Tr	Tr	0.0	2	75	0	0	0.00	0.00	0.0	0
168	Mayonnaise type																					
	1 tbsp	15	40	60	Tr	5	0.7	1.4	2.7	4	4	2	4	Tr	1	107	30	13	Tr	Tr	Tr	0
169	Tartar sauce																					
	1 tbsp	14	34	75	Tr	8	1.2	2.6	3.9	4	1	3	4	0.1	11	182	30	9	Tr	Tr	0.0	Tr
	Thousand island																					
170	Regular																					
	1 tbsp	16	46	60	Tr	6	1.0	1.3	3.2	4	2	2	3	0.1	18	112	50	15	Tr	Tr	Tr	0
171	Low calorie																					
	1 tbsp	15	69	25	Tr	2	0.2	0.4	0.9	2	2	2	3	0.1	17	150	50	14	Tr	Tr	Tr	0
	Prepared from home recipe																					
172	Cooked type[15]																					
	1 tbsp	16	69	25	1	2	0.5	0.6	0.3	9	2	13	14	0.1	19	117	70	20	0.01	0.02	Tr	Tr
173	Vinegar and oil																					
	1 tbsp	16	47	70	0	8	1.5	2.4	3.9	0	Tr	0	0	0.0	1	Tr	0	0	0.00	0.00	0.0	0

Item No.	Foods, approximate measures, units, and weight (weight of edible portion only)	Grams	Water (%)	Food energy (cal)	Protein (g)	Fat (g)	Fatty acids: Saturated (g)	Fatty acids: Mono-unsaturated (g)	Fatty acids: Poly-unsaturated (g)	Cholesterol (mg)	Carbohydrate (g)	Calcium (mg)	Phosphorus (mg)	Iron (mg)	Potassium (mg)	Sodium (mg)	Vitamin A value (IU)[1]	Vitamin A value (RE)[2]	Thiamin (mg)	Riboflavin (mg)	Niacin	Ascorbic acid (mg)
	Fish and Shellfish																					
	Clams																					
174	Raw, meat only 3 oz	85	82	65	11	1	0.3	0.3	0.3	43	2	59	138	2.6	154	102	90	26	0.09	0.15	1.1	9
175	Canned, drained solids 3 oz	85	77	85	13	2	0.5	0.5	0.4	54	2	47	116	3.5	119	102	90	26	0.01	0.09	0.9	3
176	Crabmeat, canned 1 cup	135	77	135	23	3	0.5	0.8	1.4	135	1	61	246	1.1	149	1350	50	14	0.11	0.11	2.6	0
177	Fish sticks, frozen, reheated (stick, 4 by 1 by 1/2 in.) 1 fish stick	28	52	70	6	3	0.8	1.4	0.8	26	4	11	58	0.3	94	53	20	5	0.03	0.05	0.6	0
	Flounder or Sole, baked, with lemon juice																					
178	With butter 3 oz	85	73	120	16	6	3.2	1.5	0.5	68	Tr	13	187	0.3	272	145	210	54	0.05	0.08	1.6	1
179	With margarine 3 oz	85	73	120	16	6	1.2	2.3	1.9	55	Tr	14	187	0.3	273	151	230	69	0.05	0.08	1.6	1
180	Without added fat 3 oz	85	78	80	17	1	0.3	0.2	0.4	59	Tr	13	197	0.3	286	101	30	10	0.05	0.08	1.7	1
181	Haddock, breaded, fried[16] 3 oz	85	61	175	17	9	2.4	3.9	2.4	75	7	34	183	1.0	270	123	70	20	0.06	0.10	2.9	0
182	Halibut, broiled, with butter and lemon juice 3 oz	85	67	140	20	6	3.3	1.6	0.7	62	Tr	14	206	0.7	441	103	610	174	0.06	0.07	7.7	1
183	Herring, pickled 3 oz	85	59	190	17	13	4.3	4.6	3.1	85	0	29	128	0.9	85	850	110	33	0.04	0.18	2.8	0
184	Ocean perch, breaded, fried[16] 1 fillet	85	59	185	16	11	2.6	4.6	2.8	66	7	31	191	1.2	241	138	70	20	0.10	0.11	2.0	0
	Oysters																					
185	Raw, meat only (13—19 medium Selects) 1 cup	240	85	160	20	4	1.4	0.5	1.4	120	8	226	343	15.6	290	175	740	223	0.34	0.43	6.0	24
186	Breaded, fried[16] 1 oyster	45	65	90	5	5	1.4	2.1	1.4	35	5	49	73	3.0	64	70	150	44	0.07	0.10	1.3	4
	Salmon																					
187	Canned (pink), solids and liquid 3 oz	85	71	120	17	5	0.9	1.5	2.1	34	0	167[17]	243	0.7	307	443	60	18	0.03	0.15	6.8	0
188	Baked (red) 3 oz	85	67	140	21	5	1.2	2.4	1.4	60	0	26	269	0.5	305	55	290	87	0.18	0.14	5.5	0
189	Smoked 3 oz	85	59	150	18	8	2.6	3.9	0.7	51	0	12	208	0.8	327	1700	260	77	0.17	0.17	6.8	0
190	Sardines, Atlantic, canned in oil, drained solids 3 oz	85	62	175	20	9	2.1	3.7	2.9	85	0	371[17]	424	2.6	349	425	190	56	0.03	0.17	4.6	0
191	Scallops, breaded, frozen, reheated 6 scallops	90	59	195	15	10	2.5	4.1	2.5	70	10	39	203	2.0	369	298	70	21	0.11	0.11	1.6	0

Item No.	Foods, approximate measures, units, and weight (weight of edible portion only)	Grams	Water (%)	Food energy (cal)	Protein (g)	Fat (g)	Fatty acids: Saturated (g)	Fatty acids: Monounsaturated (g)	Fatty acids: Polyunsaturated (g)	Cholesterol (mg)	Carbohydrate (g)	Calcium (mg)	Phosphorus (mg)	Iron (mg)	Potassium (mg)	Sodium (mg)	Vitamin A value (IU)[1]	Vitamin A value (RE)[2]	Thiamin (mg)	Riboflavin (mg)	Niacin	Ascorbic acid (mg)
	Shrimp																					
192	Canned, drained solids																					
	3 oz	85	70	100	21	1	0.2	0.2	0.4	128	1	98	224	1.4	1	1955	50	15	0.01	0.03	1.5	0
193	French fried (7 medium)[18]																					
	3 oz	85	55	200	16	10	2.5	4.1	2.6	168	11	61	154	2.0	189	384	90	26	0.06	0.09	2.8	0
194	Trout, broiled, with butter and lemon juice																					
	3 oz	85	63	175	21	9	4.1	2.9	1.6	71	Tr	26	259	1.0	297	122	230	60	0.07	0.07	2.3	1
	Tuna, canned, drained solids																					
195	Oil pack, chunk light																					
	3 oz	85	61	165	24	7	1.4	1.9	3.1	55	0	7	199	1.6	298	303	70	20	0.04	0.09	10.1	0
196	Water pack, solid white																					
	3 oz	85	63	135	30	1	0.3	0.2	0.3	48	0	17	202	0.6	255	468	110	32	0.03	0.10	13.4	0
197	Tuna salad[19]																					
	1 cup	205	63	375	33	19	3.3	4.9	9.2	80	19	31	281	2.5	531	877	230	53	0.06	0.14	13.3	6

Fruits and Fruit Juices

Item No.	Foods, approximate measures, units, and weight (weight of edible portion only)	Grams	Water (%)	Food energy (cal)	Protein (g)	Fat (g)	Fatty acids: Saturated (g)	Fatty acids: Monounsaturated (g)	Fatty acids: Polyunsaturated (g)	Cholesterol (mg)	Carbohydrate (g)	Calcium (mg)	Phosphorus (mg)	Iron (mg)	Potassium (mg)	Sodium (mg)	Vitamin A value (IU)[1]	Vitamin A value (RE)[2]	Thiamin (mg)	Riboflavin (mg)	Niacin	Ascorbic acid (mg)
	Apples																					
	Raw																					
	Unpeeled, without cores																					
198	2 3/4-in. diam. (about 3/lb with cores)																					
	1 apple	138	84	80	Tr	Tr	0.1	Tr	0.1	0	21	10	10	0.2	159	Tr	70	7	0.02	0.02	0.1	8
199	3 1/4-in. diam. (about 2/lb with cores)																					
	1 apple	212	84	125	Tr	1	0.1	Tr	0.2	0	32	15	15	0.4	244	Tr	110	11	0.04	0.03	0.2	12
200	Peeled, sliced																					
	1 cup	110	84	65	Tr	Tr	0.1	Tr	0.1	0	16	4	8	0.1	124	Tr	50	5	0.02	0.01	0.1	4
201	Dried, sulfured																					
	10 rings	64	32	155	1	Tr	Tr	Tr	0.1	0	42	9	24	0.9	288	56[20]	0	0	0.00	0.10	0.6	2
202	Apple juice, bottled or canned[21]																					
	1 cup	248	88	115	Tr	Tr	Tr	Tr	0.1	0	29	17	17	0.9	295	7	Tr	Tr	0.05	0.04	0.2	2[22]
	Applesauce, canned																					
203	Sweetened																					
	1 cup	255	80	195	Tr	Tr	0.1	Tr	0.1	0	51	10	18	0.9	156	8	30	3	0.03	0.07	0.5	4[22]
204	Unsweetened																					
	1 cup	244	88	105	Tr	Tr	Tr	Tr	Tr	0	28	7	17	0.3	183	5	70	7	0.03	0.06	0.5	3[22]
	Apricots																					
205	Raw, without pits (about 12/lb with pits)																					
	3 apricots	106	86	50	1	Tr	Tr	0.2	0.1	0	12	15	20	0.6	314	1	2770	277	0.03	0.04	0.6	11
	Canned (fruit and liquid)																					
	Heavy syrup pack																					
206	1 cup	258	78	215	1	Tr	Tr	0.1	Tr	0	55	23	31	0.8	361	10	3170	317	0.05	0.06	1.0	8
207	3 halves	85	78	70	Tr	Tr	Tr	Tr	Tr	0	18	8	10	0.3	119	3	1050	105	0.02	0.02	0.3	3

Item No.	Foods, approximate measures, units, and weight (weight of edible portion only)	Grams	Water (%)	Food energy (cal)	Protein (g)	Fat (g)	Fatty acids: Saturated (g)	Fatty acids: Mono-unsaturated (g)	Fatty acids: Poly-unsaturated (g)	Cholesterol (mg)	Carbo-hydrate (g)	Calcium (mg)	Phos-phorus (mg)	Iron (mg)	Potassium (mg)	Sodium (mg)	Vitamin A value (IU)[1]	Vitamin A value (RE)[2]	Thiamin (mg)	Riboflavin (mg)	Niacin	Ascorbic acid (mg)
208	Juice pack																					
	1 cup	248	87	120	2	Tr	Tr	Tr	Tr	0	31	30	50	0.7	409	10	4190	419	0.04	0.05	0.9	12
209	3 halves	84	87	40	1	Tr	Tr	Tr	Tr	0	10	10	17	0.3	139	3	1420	142	0.02	0.02	0.3	4
	Dried																					
210	Uncooked (28 large or 37 medium halves/cup)																					
	1 cup	130	31	310	5	1	Tr	0.3	0.1	0	80	59	152	6.1	1791	13	9410	941	0.01	0.20	3.9	3
211	Cooked, unsweetened, fruit and liquid																					
	1 cup	250	76	210	3	Tr	Tr	0.2	0.1	0	55	40	103	4.2	1222	8	5910	591	0.02	0.08	2.4	4
212	Apricot nectar, canned																					
	1 cup	251	85	140	1	Tr	Tr	0.1	Tr	0	36	18	23	1.0	286	8	3300	330	0.02	0.04	0.7	2[22]
	Avocados, raw, whole, without skin and seed																					
213	California (about 2/lb with skin and seed)																					
	1 avocado	173	73	305	4	30	4.5	19.4	3.5	0	12	19	73	2.0	1097	21	1060	106	0.19	0.21	3.3	14
214	Florida (about 1/lb with skin and seed)																					
	1 avocado	304	80	340	5	27	5.3	14.8	4.5	0	27	33	119	1.6	1484	15	1860	186	0.33	0.37	5.8	24
	Bananas, raw, without peel																					
215	Whole (about $2^1/_2$/lb with peel)																					
	1 banana	114	74	105	1	1	0.2	Tr	0.1	0	27	7	23	0.4	451	1	90	9	0.05	0.11	0.6	10
216	Sliced																					
	1 cup	150	74	140	2	1	0.3	0.1	0.1	0	35	9	30	0.5	594	2	120	12	0.07	0.15	0.8	14
217	Blackberries, raw																					
	1 cup	144	86	75	1	1	0.2	0.1	0.1	0	18	46	30	0.8	282	Tr	240	24	0.04	0.06	0.6	30
	Blueberries																					
218	Raw																					
	1 cup	145	85	80	1	1	Tr	0.1	0.3	0	20	9	15	0.2	129	9	150	15	0.07	0.07	0.5	19
219	Frozen, sweetened																					
	10-oz container	284	77	230	1	Tr	Tr	0.1	0.2	0	62	17	20	1.1	170	3	120	12	0.06	0.15	0.7	3
220	1 cup	230	77	185	1	Tr	Tr	Tr	0.1	0	50	14	16	0.9	138	2	100	10	0.05	0.12	0.6	2
	Cantaloupe; see Melons (item 251)																					
	Cherries																					
221	Sour, red, pitted, canned, water pack																					
	1 cup	244	90	90	2	Tr	0.1	0.1	0.1	0	22	27	24	3.3	239	17	1840	184	0.04	0.10	0.4	5
222	Sweet, raw, without pits and stems																					
	10 cherries	68	81	50	1	1	0.1	0.2	0.2	0	11	10	13	0.3	152	Tr	150	15	0.03	0.04	0.3	5
223	Cranberry juice cocktail, bottled, sweetened																					
	1 cup	253	85	145	Tr	Tr	Tr	Tr	0.1	0	38	8	3	0.4	61	10	10	1	0.01	0.04	0.1	108[23]
224	Cranberry sauce, sweetened, canned, strained																					
	1 cup	277	61	420	1	Tr	Tr	0.1	0.2	0	108	11	17	0.6	72	80	60	6	0.04	0.06	0.3	6

Item No.	Foods, approximate measures, units, and weight (weight of edible portion only)	Grams	Water (%)	Food energy (cal)	Protein (g)	Fat (g)	Fatty acids: Saturated (g)	Fatty acids: Mono-unsaturated (g)	Fatty acids: Poly-unsaturated (g)	Cholesterol (mg)	Carbohydrate (g)	Calcium (mg)	Phosphorus (mg)	Iron (mg)	Potassium (mg)	Sodium (mg)	Vitamin A value (IU)[1]	Vitamin A value (RE)[2]	Thiamin (mg)	Riboflavin (mg)	Niacin	Ascorbic acid (mg)
	Dates																					
225	Whole, without pits 10 dates	83	23	230	2	Tr	0.1	0.1	Tr	0	61	27	33	1.0	541	2	40	4	0.07	0.08	1.8	0
226	Chopped 1 cup	178	23	490	4	1	0.3	0.2	Tr	0	131	57	71	2.0	1161	5	90	9	0.16	0.18	3.9	0
227	Figs, dried 10 figs	187	28	475	6	2	0.4	0.5	1.0	0	122	269	127	4.2	1331	21	250	25	0.13	0.16	1.3	1
	Fruit cocktail, canned, fruit and liquid																					
228	Heavy syrup pack 1 cup	255	80	185	1	Tr	Tr	Tr	0.1	0	48	15	28	0.7	224	15	520	52	0.05	0.05	1.0	5
229	Juice pack 1 cup	248	87	115	1	Tr	Tr	Tr	Tr	0	29	20	35	0.5	236	10	760	76	0.03	0.04	1.0	7
	Grapefruit																					
230	Raw, without peel, membrane and seeds ($3^3/_4$-in. diam., 1 lb 1 oz, whole, with refuse) $^1/_2$ grapefruit	120	91	40	1	Tr	Tr	Tr	Tr	0	10	14	10	0.1	167	Tr	10[24]	1[24]	0.04	0.02	0.3	41
231	Canned, sections with syrup 1 cup	254	84	150	1	Tr	Tr	Tr	0.1	0	39	36	25	1.0	328	5	Tr	Tr	0.10	0.05	0.6	54
	Grapefruit juice																					
232	Raw 1 cup	247	90	95	1	Tr	Tr	Tr	0.1	0	23	22	37	0.5	400	2	20	2	0.10	0.05	0.5	94
	Canned																					
233	Unsweetened 1 cup	247	90	95	1	Tr	Tr	Tr	0.1	0	22	17	27	0.5	378	2	20	2	0.10	0.05	0.6	72
234	Sweetened 1 cup	250	87	115	1	Tr	Tr	Tr	0.1	0	28	20	28	0.9	405	5	20	2	0.10	0.06	0.8	67
	Frozen concentrate, unsweetened																					
235	Undiluted 6-fl-oz can	207	62	300	4	1	0.1	0.1	0.2	0	72	56	101	1.0	1002	6	60	6	0.30	0.16	1.6	248
236	Diluted with 3 parts water by volume 1 cup	247	89	100	1	Tr	Tr	Tr	0.1	0	24	20	35	0.3	336	2	20	2	0.10	0.05	0.5	83
	Grapes, European type (adherent skin), raw																					
237	Thompson Seedless 10 grapes	50	81	35	Tr	Tr	0.1	Tr	0.1	0	9	6	7	0.1	93	1	40	4	0.05	0.03	0.2	5
238	Tokay and Emperor, seeded types 10 grapes	57	81	40	Tr	Tr	0.1	Tr	0.1	0	10	6	7	0.1	105	1	40	4	0.05	0.03	0.2	6
	Grape juice																					
239	Canned or bottled 1 cup	253	84	155	1	Tr	0.1	Tr	0.1	0	38	23	28	0.6	334	8	20	2	0.07	0.09	0.7	Tr[22]

Item No.	Foods, approximate measures, units, and weight (weight of edible portion only)	Grams	Water (%)	Food energy (cal)	Protein (g)	Fat (g)	Fatty acids: Saturated (g)	Fatty acids: Mono-unsaturated (g)	Fatty acids: Poly-unsaturated (g)	Cholesterol (mg)	Carbo-hydrate (g)	Calcium (mg)	Phos-phorus (mg)	Iron (mg)	Potassium (mg)	Sodium (mg)	Vitamin A value (IU)[1]	Vitamin A value (RE)[2]	Thiamin (mg)	Riboflavin (mg)	Niacin	Ascorbic acid (mg)
	Frozen concentrate, sweetened																					
240	Undiluted 6-fl-oz can	216	54	385	1	1	0.2	Tr	0.2	0	96	28	32	0.8	160	15	60	6	0.11	0.20	0.9	179[23]
241	Diluted with 3 parts water by volume 1 cup	250	87	125	Tr	Tr	0.1	Tr	0.1	0	32	10	10	0.3	53	5	20	2	0.04	0.07	0.3	60[23]
242	Kiwifruit, raw, without skin (about 5/lb with skin) 1 kiwifruit	76	83	45	1	Tr	Tr	0.1	0.1	0	11	20	30	0.3	252	4	130	13	0.02	0.04	0.4	74
243	Lemons, raw, without peel and seeds (about 4/lb with peel and seeds) 1 lemon	58	89	15	1	Tr	Tr	Tr	0.1	0	5	15	9	0.3	80	1	20	2	0.02	0.01	0.1	31
	Lemon juice																					
244	Raw 1 cup	244	91	60	1	Tr	Tr	Tr	Tr	0	21	17	15	0.1	303	2	50	5	0.07	0.02	0.2	112
245	Canned or bottled, unsweetened 1 cup	244	92	50	1	1	0.1	Tr	0.2	0	16	27	22	0.3	249	51[25]	40	4	0.10	0.02	0.5	61
246	1 tbsp	15	92	5	Tr	Tr	Tr	Tr	Tr	0	1	2	1	Tr	15	3[25]	Tr	Tr	0.01	Tr	Tr	4
247	Frozen, single-strength, unsweetened 6-fl-oz can	244	92	55	1	1	0.1	Tr	0.2	0	16	20	20	0.3	217	2	30	3	0.14	0.03	0.3	77
	Lime juice																					
248	Raw 1 cup	246	90	65	1	Tr	Tr	Tr	0.1	0	22	22	17	0.1	268	2	20	2	0.05	0.02	0.2	72
249	Canned, unsweetened 1 cup	246	93	50	1	1	0.1	0.1	0.2	0	16	30	25	0.6	185	39[25]	40	4	0.08	0.01	0.4	16
250	Mangos, raw, without skin and seed (about 1 1/2/lb with skin and seed) 1 mango	207	82	135	1	1	0.1	0.2	0.1	0	35	21	23	0.3	323	4	8060	806	0.12	0.12	1.2	57
	Melons, raw, without rind and cavity contents																					
251	Cantaloupe, orange-fleshed (5-in. diam., 2 1/3 lb, whole, with rind and cavity contents) 1/2 melon	267	90	95	2	1	0.1	0.1	0.3	0	22	29	45	0.6	825	24	8610	861	0.10	0.06	1.5	113
252	Honeydew (6 1/2-in. diam., 5 1/4 lb, whole, with rind and cavity contents) 1/10 melon	129	90	45	1	Tr	Tr	Tr	0.1	0	12	8	13	0.1	350	13	50	5	0.10	0.02	0.8	32
253	Nectarines, raw, without pits (about 3/lb with pits) 1 nectarine	136	86	65	1	1	0.1	0.2	0.3	0	16	7	22	0.2	288	Tr	1000	100	0.02	0.06	1.3	7

Item No.	Foods, approximate measures, units, and weight (weight of edible portion only)	Grams	Water (%)	Food energy (cal)	Protein (g)	Fat (g)	Fatty acids: Saturated (g)	Fatty acids: Monounsaturated (g)	Fatty acids: Polyunsaturated (g)	Cholesterol (mg)	Carbohydrate (g)	Calcium (mg)	Phosphorus (mg)	Iron (mg)	Potassium (mg)	Sodium (mg)	Vitamin A value (IU)[1]	Vitamin A value (RE)[2]	Thiamin (mg)	Riboflavin (mg)	Niacin	Ascorbic acid (mg)
	Oranges, raw																					
254	Whole, without peel and seeds ($2^{6}/_{8}$-in. diam., about $2^{1}/_{2}$/lb, with peel and seeds) 1 orange	131	87	60	1	Tr	Tr	Tr	Tr	0	15	52	18	0.1	237	Tr	270	27	0.11	0.05	0.4	70
255	Sections without membranes 1 cup	180	87	85	2	Tr	Tr	Tr	Tr	0	21	72	25	0.2	326	Tr	370	37	0.16	0.07	0.5	96
	Orange juice																					
256	Raw, all varieties 1 cup	248	88	110	2	Tr	0.1	0.1	0.1	0	26	27	42	0.5	496	2	500	50	0.22	0.07	1.0	124
257	Canned, unsweetened 1 cup	249	89	105	1	Tr	Tr	0.1	0.1	0	25	20	35	1.1	436	5	440	44	0.15	0.07	0.8	86
258	Chilled 1 cup	249	88	110	2	1	0.1	0.1	0.2	0	25	25	27	0.4	473	2	190	19	0.28	0.05	0.7	82
	Frozen concentrate																					
259	Undiluted 6-fl-oz can	213	58	340	5	Tr	0.1	0.1	0.1	0	81	68	121	0.7	1436	6	590	59	0.60	0.14	1.5	294
260	Diluted with 3 parts water by volume 1 cup	249	88	110	2	Tr	Tr	Tr	Tr	0	27	22	40	0.2	473	2	190	19	0.20	0.04	0.5	97
261	Orange and grapefruit juice, canned 1 cup	247	89	105	1	Tr	Tr	Tr	Tr	0	25	20	35	1.1	390	7	290	29	0.14	0.07	0.8	72
262	Papayas, raw, $^{1}/_{2}$-in. cubes 1 cup	140	86	65	1	Tr	0.1	0.1	Tr	0	17	35	12	0.3	247	9	400	40	0.04	0.04	0.5	92
	Peaches																					
	Raw																					
263	Whole, $2^{1}/_{2}$-in. diam., peeled, pitted (about 4/lb with peels and pits) 1 peach	87	88	35	1	Tr	Tr	Tr	Tr	0	10	4	10	0.1	171	Tr	470	47	0.01	0.04	0.9	6
264	Sliced 1 cup	170	88	75	1	Tr	Tr	0.1	0.1	0	19	9	20	0.2	335	Tr	910	91	0.03	0.07	1.7	11
	Canned, fruit and liquid																					
	Heavy syrup pack																					
265	1 cup	256	79	190	1	Tr	Tr	0.1	0.1	0	51	8	28	0.7	236	15	850	85	0.03	0.06	1.6	7
266	1 half	81	79	60	Tr	Tr	Tr	Tr	Tr	0	16	2	9	0.2	75	5	270	27	0.01	0.02	0.5	2
	Juice pack																					
267	1 cup	248	87	110	2	Tr	Tr	Tr	Tr	0	29	15	42	0.7	317	10	940	94	0.02	0.04	1.4	9
268	1 half	77	87	35	Tr	Tr	Tr	Tr	Tr	0	9	5	13	0.2	99	3	290	29	0.01	0.01	0.4	3
	Dried																					
269	Uncooked 1 cup	160	32	380	6	1	0.1	0.4	0.6	0	98	45	190	6.5	1594	11	3460	346	Tr	0.34	7.0	8
270	Cooked, unsweetened, fruit and liquid 1 cup	258	78	200	3	1	0.1	0.2	0.3	0	51	23	98	3.4	826	5	510	51	0.01	0.05	3.9	10

Item No.	Foods, approximate measures, units, and weight (weight of edible portion only)	Grams	Water (%)	Food energy (cal)	Protein (g)	Fat (g)	Fatty acids: Saturated (g)	Fatty acids: Mono-unsaturated (g)	Fatty acids: Poly-unsaturated (g)	Cholesterol (mg)	Carbohydrate (g)	Calcium (mg)	Phosphorus (mg)	Iron (mg)	Potassium (mg)	Sodium (mg)	Vitamin A value (IU)[1]	Vitamin A value (RE)[2]	Thiamin (mg)	Riboflavin (mg)	Niacin (mg)	Ascorbic acid (mg)
	Frozen, sliced, sweetened																					
271	10-oz container	284	75	265	2	Tr	Tr	0.1	0.2	0	68	9	31	1.1	369	17	810	81	0.04	0.10	1.9	268[23]
272	1 cup	250	75	235	2	Tr	Tr	0.1	0.2	0	60	8	28	0.9	325	15	710	71	0.03	0.09	1.6	236[23]
	Pears																					
	Raw, with skin, cored																					
	Bartlett, $2^1/_2$-in. diam. (about $2^1/_2$/lb with cores and stems)																					
273	1 pear	166	84	100	1	1	Tr	0.1	0.2	0	25	18	18	0.4	208	Tr	30	3	0.03	0.07	0.2	7
	Bosc, $2^1/_2$-in. diam. (about 3/lb with cores and stems)																					
274	1 pear	141	84	85	1	1	Tr	0.1	0.1	0	21	16	16	0.4	176	Tr	30	3	0.03	0.06	0.1	6
	D'Anjou, 3-in. diam. (about 2/lb with cores and stems)																					
275	1 pear	200	84	120	1	1	Tr	0.2	0.2	0	30	22	22	0.5	250	Tr	40	4	0.04	0.08	0.2	8
	Canned, fruit and liquid																					
	Heavy syrup pack																					
276	1 cup	255	80	190	1	Tr	Tr	0.1	0.1	0	49	13	18	0.6	166	13	10	1	0.03	0.06	0.6	3
277	1 half	79	80	60	Tr	Tr	Tr	Tr	Tr	0	15	4	6	0.2	51	4	Tr	Tr	0.01	0.02	0.2	1
	Juice pack																					
278	1 cup	248	86	125	1	Tr	Tr	Tr	Tr	0	32	22	30	0.7	238	10	10	1	0.03	0.03	0.5	4
279	1 half	77	86	40	Tr	Tr	Tr	Tr	Tr	0	10	7	9	0.2	74	3	Tr	Tr	0.01	0.01	0.2	1
	Pineapple																					
	Raw, diced																					
280	1 cup	155	87	75	1	1	Tr	0.1	0.2	0	19	11	11	0.6	175	2	40	4	0.14	0.06	0.7	24
	Canned, fruit and liquid																					
	Heavy syrup pack																					
	Crushed, chunks, tidbits																					
281	1 cup	255	79	200	1	Tr	Tr	Tr	0.1	0	52	36	18	1.0	265	3	40	4	0.23	0.06	0.7	19
	Slices																					
282	1 slice	58	79	45	Tr	Tr	Tr	Tr	Tr	0	12	8	4	0.2	60	1	10	1	0.05	0.01	0.2	4
	Juice pack																					
	Chunks or tidbits																					
283	1 cup	250	84	150	1	Tr	Tr	Tr	0.1	0	39	35	15	0.7	305	3	100	10	0.24	0.05	0.7	24
	Slices																					
284	1 slice	58	84	35	Tr	Tr	Tr	Tr	Tr	0	9	8	3	0.2	71	1	20	2	0.06	0.01	0.2	6
	Pineapple juice, unsweetened, canned																					
285	1 cup	250	86	140	1	Tr	Tr	Tr	0.1	0	34	43	20	0.7	335	3	10	1	0.14	0.06	0.6	27
	Plantains, without peel																					
	Raw																					
286	1 plantain	179	65	220	2	1	0.3	0.1	0.1	0	57	5	61	1.1	893	7	2020	202	0.09	0.10	1.2	33
	Cooked, boiled, sliced																					
287	1 cup	154	67	180	1	Tr	0.1	Tr	0.1	0	48	3	43	0.9	716	8	1400	140	0.07	0.08	1.2	17

		Nutrients in indicated quantity																				
							Fatty acids										Vitamin A value					
Item No.	Foods, approximate measures, units, and weight (weight of edible portion only)	Grams	Water (%)	Food energy (cal)	Protein (g)	Fat (g)	Saturated (g)	Mono-unsatu-rated (g)	Poly-unsatu-rated (g)	Cholesterol (mg)	Carbo-hydrate (g)	Calcium (mg)	Phos-phorus (mg)	Iron (mg)	Potassium (mg)	Sodium (mg)	(IU)[1]	(RE)[2]	Thiamin (mg)	Riboflavin (mg)	Niacin	Ascorbic acid (mg)
	Plums, without pits																					
	Raw																					
288	$2^{1}/_{8}$-in. diam. (about $6^{1}/_{2}$/lb with pits)																					
	1 plum	66	85	35	1	Tr	Tr	0.3	0.1	0	9	3	7	0.1	114	Tr	210	21	0.03	0.06	0.3	6
289	$1^{1}/_{2}$-in. diam. (about 15/lb with pits)																					
	1 plum	28	85	15	Tr	Tr	Tr	0.1	Tr	0	4	1	3	Tr	48	Tr	90	9	0.01	0.03	0.1	3
	Canned, purple, fruit and liquid																					
290	Heavy syrup pack																					
	1 cup	258	76	230	1	Tr	Tr	0.2	0.1	0	60	23	34	2.2	235	49	670	67	0.04	0.10	0.8	1
291	3 plums	133	76	120	Tr	Tr	Tr	0.1	Tr	0	31	12	17	1.1	121	25	340	34	0.02	0.05	0.4	1
292	Juice pack																					
	1 cup	252	84	145	1	Tr	Tr	Tr	Tr	0	38	25	38	0.9	388	3	2540	254	0.06	0.15	1.2	7
293	3 plums	95	84	55	Tr	Tr	Tr	Tr	Tr	0	14	10	14	0.3	146	1	960	96	0.02	0.06	0.4	3
	Prunes, dried																					
294	Uncooked																					
	4 extra large or 5 large prunes	49	32	115	1	Tr	Tr	0.2	0.1	0	31	25	39	1.2	365	2	970	97	0.04	0.08	1.0	2
295	Cooked, unsweetened, fruit and liquid																					
	1 cup	212	70	225	2	Tr	Tr	0.3	0.1	0	60	49	74	2.4	708	4	650	65	0.05	0.21	1.5	6
296	Prune juice, canned or bottled																					
	1 cup	256	81	180	2	Tr	Tr	0.1	Tr	0	45	31	64	3.0	707	10	10	1	0.04	0.18	2.0	10
	Raisins, seedless																					
297	Cup, not pressed down																					
	1 cup	145	15	435	5	1	0.2	Tr	0.2	0	115	71	141	3.0	1089	17	10	1	0.23	0.13	1.2	5
298	Packet, $^{1}/_{2}$ oz ($1^{1}/_{2}$ tbsp)																					
	1 packet	14	15	40	Tr	Tr	Tr	Tr	Tr	0	11	7	14	0.3	105	2	Tr	Tr	0.02	0.01	0.1	Tr
	Raspberries																					
299	Raw																					
	1 cup	123	87	60	1	1	Tr	0.1	0.4	0	14	27	15	0.7	187	Tr	160	16	0.04	0.11	1.1	31
300	Frozen, sweetened																					
	10-oz container	284	73	295	2	Tr	Tr	Tr	0.3	0	74	43	48	1.8	324	3	170	17	0.05	0.13	0.7	47
301	1 cup	250	73	255	2	Tr	Tr	Tr	0.2	0	65	38	43	1.6	285	3	150	15	0.05	0.11	0.6	41
302	Rhubarb, cooked, added sugar																					
	1 cup	240	68	280	1	Tr	Tr	Tr	0.1	0	75	348	19	0.5	230	2	170	17	0.04	0.06	0.5	8
	Strawberries																					
303	Raw, capped, whole																					
	1 cup	149	92	45	1	1	Tr	0.1	0.3	0	10	21	28	0.6	247	1	40	4	0.03	0.10	0.3	84
304	Frozen, sweetened, sliced																					
	10-oz container	284	73	275	2	Tr	Tr	0.1	0.2	0	74	31	37	1.7	278	9	70	7	0.05	0.14	1.1	118
305	1 cup	255	73	245	1	Tr	Tr	Tr	0.2	0	66	28	33	1.5	250	8	60	6	0.04	0.13	1.0	106

Item No.	Foods, approximate measures, units, and weight (weight of edible portion only)	Grams	Water (%)	Food energy (cal)	Protein (g)	Fat (g)	Fatty acids: Saturated (g)	Fatty acids: Monounsaturated (g)	Fatty acids: Polyunsaturated (g)	Cholesterol (mg)	Carbohydrate (g)	Calcium (mg)	Phosphorus (mg)	Iron (mg)	Potassium (mg)	Sodium (mg)	Vitamin A value (IU)[1]	Vitamin A value (RE)[2]	Thiamin (mg)	Riboflavin (mg)	Niacin	Ascorbic acid (mg)
	Tangerines																					
306	Raw, without peel and seeds (2 3/8-in. diam., about 4/lb, with peel and seeds) 1 tangerine	84	88	35	1	Tr	Tr	Tr	Tr	0	9	12	8	0.1	132	1	770	77	0.09	0.02	0.1	26
307	Canned, light syrup, fruit and liquid 1 cup	252	83	155	1	Tr	Tr	Tr	0.1	0	41	18	25	0.9	197	15	2120	212	0.13	0.11	1.1	50
308	Tangerine juice, canned, sweetened 1 cup	249	87	125	1	Tr	Tr	Tr	0.1	0	30	45	35	0.5	443	2	1050	105	0.15	0.05	0.2	55
	Watermelon, raw, without rind and seeds																					
309	Piece (4 by 8 in. wedge with rind and seeds; 1/16 of 32 2/3-lb melon, 10 by 16 in.) 1 piece	482	92	155	3	2	0.3	0.2	1.0	0	35	39	43	0.8	559	10	1760	176	0.39	0.10	1.0	46
310	Diced 1 cup	160	92	50	1	1	0.1	0.1	0.3	0	11	13	14	0.3	186	3	590	59	0.13	0.03	0.3	15

Grain Products

Item No.	Foods, approximate measures, units, and weight (weight of edible portion only)	Grams	Water (%)	Food energy (cal)	Protein (g)	Fat (g)	Fatty acids: Saturated (g)	Fatty acids: Monounsaturated (g)	Fatty acids: Polyunsaturated (g)	Cholesterol (mg)	Carbohydrate (g)	Calcium (mg)	Phosphorus (mg)	Iron (mg)	Potassium (mg)	Sodium (mg)	Vitamin A value (IU)[1]	Vitamin A value (RE)[2]	Thiamin (mg)	Riboflavin (mg)	Niacin	Ascorbic acid (mg)
311	Bagels, plain or water, enriched, 3 1/2-in. diam.[26] 1 bagel	68	29	200	7	2	0.3	0.5	0.7	0	38	29	46	1.8	50	245	0	0	0.26	0.20	2.4	0
312	Barley, pearled, light, uncooked 1 cup	200	11	700	16	2	0.3	0.2	0.9	0	158	32	378	4.2	320	6	0	0	0.24	0.10	6.2	0
	Biscuits, baking powder, 2-in. diam. (enriched flour, vegetable shortening)																					
313	From home recipe 1 biscuit	28	28	100	2	5	1.2	2.0	1.3	Tr	13	47	36	0.7	32	195	10	3	0.08	0.08	0.8	Tr
314	From mix 1 biscuit	28	29	95	2	3	0.8	1.4	0.9	Tr	14	58	128	0.7	56	262	20	4	0.12	0.11	0.8	Tr
315	From refrigerated dough 1 biscuit	20	30	65	1	2	0.6	0.9	0.6	1	10	4	79	0.5	18	249	0	0	0.08	0.05	0.7	0
	Breadcrumbs, enriched																					
316	Dry, grated 1 cup	100	7	390	13	5	1.5	1.6	1.0	5	73	122	141	4.1	152	736	0	0	0.35	0.35	4.8	0
	Soft; see White bread (item 351)																					
	Breads																					
317	Boston brown bread, canned, slice, 3 1/4 in. by 1/2 in.[27] 1 slice	45	45	95	2	1	0.3	0.1	0.1	3	21	41	72	0.9	131	113	0[28]	0[28]	0.06	0.04	0.7	0

Item No.	Foods, approximate measures, units, and weight (weight of edible portion only)	Grams	Water (%)	Food energy (cal)	Protein (g)	Fat (g)	Fatty acids: Saturated (g)	Fatty acids: Mono-unsaturated (g)	Fatty acids: Poly-unsaturated (g)	Cholesterol (mg)	Carbo-hydrate (g)	Calcium (mg)	Phos-phorus (mg)	Iron (mg)	Potassium (mg)	Sodium (mg)	Vitamin A value (IU)[1]	Vitamin A value (RE)[2]	Thiamin (mg)	Riboflavin (mg)	Niacin	Ascorbic acid (mg)
	Cracked-wheat bread (3/4 enriched wheat flour, 1/4 cracked wheat flour)[27]																					
318	Loaf, 1 lb																					
	1 loaf	454	35	1190	42	16	3.1	4.3	5.7	0	227	295	581	12.1	608	1966	Tr	T	1.73	1.73	15.3	Tr
319	Slice (18/loaf)																					
	1 slice	25	35	65	2	1	0.2	0.2	0.3	0	12	16	32	0.7	34	106	Tr	T	0.10	0.09	0.8	Tr
320	Toasted																					
	1 slice	21	26	65	2	1	0.2	0.2	0.3	0	12	16	32	0.7	34	106	Tr	T	0.07	0.09	0.8	Tr
	French or vienna bread, enriched[27]																					
321	Loaf, 1 lb																					
	1 loaf	454	34	1270	43	18	3.8	5.7	5.9	0	230	499	386	14.0	409	2633	Tr	T	2.09	1.59	18.2	Tr
	Slice																					
322	French, 5 by 2 1/2 by 1 in.																					
	1 slice	35	34	100	3	1	0.3	0.4	0.5	0	18	39	30	1.1	32	203	Tr	T	0.16	0.12	1.4	Tr
323	Vienna, 4 3/4 by 4 by 1/2 in.																					
	1 slice	25	34	70	2	1	0.2	0.3	0.3	0	13	28	21	0.8	23	145	Tr	T	0.12	0.09	1.0	Tr
	Italian bread, enriched																					
324	Loaf, 1 lb																					
	1 loaf	454	32	1255	41	4	0.6	0.3	1.6	0	256	77	350	12.7	336	2656	0	[illegible]	1.80	1.10	15.0	0
325	Slice, 4 1/2 by 3 1/4 by 3/4 in.																					
	1 slice	30	32	85	3	Tr	Tr	Tr	0.1	0	17	5	23	0.8	22	176	0	[illegible]	0.12	0.07	1.0	0
	Mixed grain bread, enriched[27]																					
326	Loaf, 1 lb																					
	1 loaf	454	37	1165	45	17	3.2	4.1	6.5	0	212	472	962	14.8	990	1870	Tr	T	1.77	1.73	18.9	Tr
327	Slice (18/loaf)																					
	1 slice	25	37	65	2	1	0.2	0.2	0.4	0	12	27	55	0.8	56	106	Tr	T	0.10	0.10	1.1	Tr
328	Toasted																					
	1 slice	23	27	65	2	1	0.2	0.2	0.4	0	12	27	55	0.8	56	106	Tr	T	0.08	0.10	1.1	Tr
	Oatmeal bread, enriched[27]																					
329	Loaf, 1 lb																					
	1 loaf	454	37	1145	38	20	3.7	7.1	8.2	0	212	267	563	12.0	707	2231	0	[illegible]	2.09	1.20	15.4	0
330	Slice (18/loaf)																					
	1 slice	25	37	65	2	1	0.2	0.4	0.5	0	12	15	31	0.7	39	124	0	[illegible]	0.12	0.07	0.9	0
331	Toasted																					
	1 slice	23	30	65	2	1	0.2	0.4	0.5	0	12	15	31	0.7	39	124	0	[illegible]	0.09	0.07	0.9	0
332	Pita bread, enriched, white, 6 1/2-in. diam.																					
	1 pita	60	31	165	6	1	0.1	0.1	0.4	0	33	49	60	1.4	71	339	0	[illegible]	0.27	0.12	2.2	0
	Pumpernickel (2/3 rye flour, 1/3 enriched wheat flour)[27]																					
333	Loaf, 1 lb																					
	1 loaf	454	37	1160	42	16	2.6	3.6	6.4	0	218	322	990	12.4	1966	2461	0	[illegible]	1.54	2.36	15.0	0

Item No.	Foods, approximate measures, units, and weight (weight of edible portion only)	Grams	Water (%)	Food energy (cal)	Protein (g)	Fat (g)	Fatty acids: Saturated (g)	Fatty acids: Mono-unsaturated (g)	Fatty acids: Poly-unsaturated (g)	Cholesterol (mg)	Carbohydrate (g)	Calcium (mg)	Phosphorus (mg)	Iron (mg)	Potassium (mg)	Sodium (mg)	Vitamin A value (IU)[1]	Vitamin A value (RE)[2]	Thiamin (mg)	Riboflavin (mg)	Niacin	Ascorbic acid (mg)
334	Slice, 5 by 4 by 3/8 in. 1 slice	32	37	80	3	1	0.2	0.3	0.5	0	16	23	71	0.9	141	177	0	0	0.11	0.17	1.1	0
335	Toasted 1 slice	29	28	80	3	1	0.2	0.3	0.5	0	16	23	71	0.9	141	177	0	0	0.09	0.17	1.1	0
	Raisin bread, enriched[27]																					
336	Loaf, 1 lb 1 loaf	454	33	1260	37	18	4.1	6.5	6.7	0	239	463	395	14.1	1058	1657	Tr	Tr	1.50	2.81	18.6	Tr
337	Slice (18/loaf) 1 slice	25	33	65	2	1	0.2	0.3	0.4	0	13	25	22	0.8	59	92	Tr	Tr	0.08	0.15	1.0	Tr
338	Toasted 1 slice	21	24	65	2	1	0.2	0.3	0.4	0	13	25	22	0.8	59	92	Tr	Tr	0.06	0.15	1.0	Tr
	Rye bread, light (2/3 enriched wheat flour, 1/3 rye flour)[27]																					
339	Loaf, 1 lb 1 loaf	454	37	1190	38	17	3.3	5.2	5.5	0	218	363	658	12.3	926	3164	0	0	1.86	1.45	15.0	0
340	Slice, 4 3/4 by 3 3/4 by 7/16 in. 1 slice	25	37	65	2	1	0.2	0.3	0.3	0	12	20	36	0.7	51	175	0	0	0.10	0.08	0.8	0
341	Toasted 1 slice	22	28	65	2	1	0.2	0.3	0.3	0	12	20	36	0.7	51	175	0	0	0.08	0.08	0.8	0
	Wheat bread, enriched[27]																					
342	Loaf, 1 lb 1 loaf	454	37	1160	43	19	3.9	7.3	4.5	0	213	572	835	15.8	627	2447	Tr	Tr	2.09	1.45	20.5	Tr
343	Slice (18/loaf) 1 slice	25	37	65	2	1	0.2	0.4	0.3	0	12	32	47	0.9	35	138	Tr	Tr	0.12	0.08	1.2	Tr
344	Toasted 1 slice	23	28	65	3	1	0.2	0.4	0.3	0	12	32	47	0.9	35	138	Tr	Tr	0.10	0.08	1.2	Tr
	White bread, enriched[27]																					
345	Loaf, 1 lb 1 loaf	454	37	1210	38	18	5.6	6.5	4.2	0	222	572	490	12.9	508	2334	Tr	Tr	2.13	1.41	17.0	Tr
346	Slice (18/loaf) 1 slice	25	37	65	2	1	0.3	0.4	0.2	0	12	32	27	0.7	28	129	Tr	Tr	0.12	0.08	0.9	Tr
347	Toasted 1 slice	22	28	65	2	1	0.3	0.4	0.2	0	12	32	27	0.7	28	129	Tr	Tr	0.09	0.08	0.9	Tr
348	Slice (22/loaf) 1 slice	20	37	55	2	1	0.2	0.3	0.2	0	10	25	21	0.6	22	101	Tr	Tr	0.09	0.06	0.7	Tr
349	Toasted 1 slice	17	28	55	2	1	0.2	0.3	0.2	0	10	25	21	0.6	22	101	Tr	Tr	0.07	0.06	0.7	Tr
350	Cubes 1 cup	30	37	80	2	1	0.4	0.4	0.3	0	15	38	32	0.9	34	154	Tr	Tr	0.14	0.09	1.1	Tr
351	Crumbs, soft 1 cup	45	37	120	4	2	0.6	0.6	0.4	0	22	57	49	1.3	50	231	Tr	Tr	0.21	0.14	1.7	Tr
	Whole-wheat bread[27]																					
352	Loaf, 1 lb 1 loaf	454	38	1110	44	20	5.8	6.8	5.2	0	206	327	1180	15.5	799	2887	Tr	Tr	1.59	0.95	17.4	Tr
353	Slice (16/loaf) 1 slice	28	38	70	3	1	0.4	0.4	0.3	0	13	20	74	1.0	50	180	Tr	Tr	0.10	0.06	1.1	Tr
354	Toasted 1 slice	25	29	70	3	1	0.4	0.4	0.3	0	13	20	74	1.0	50	180	Tr	Tr	0.08	0.06	1.1	Tr

Item No.	Foods, approximate measures, units, and weight (weight of edible portion only)	Grams	Water (%)	Food energy (cal)	Protein (g)	Fat (g)	Fatty acids: Saturated (g)	Fatty acids: Mono-unsaturated (g)	Fatty acids: Poly-unsaturated (g)	Cholesterol (mg)	Carbo-hydrate (g)	Calcium (mg)	Phos-phorus (mg)	Iron (mg)	Potassium (mg)	Sodium (mg)	Vitamin A value (IU)[1]	Vitamin A value (RE)[2]	Thiamin (mg)	Riboflavin (mg)	Niacin	Ascorbic acid (mg)
	Nutrients in indicated quantity																					
	Bread stuffing (from enriched bread), prepared from mix																					
355	Dry type 1 cup	140	33	500	9	31	6.1	13.3	9.6	0	50	92	136	2.2	126	1254	910	273	0.17	0.20	2.5	0
356	Moist type 1 cup	203	61	420	9	26	5.3	11.3	8.0	67	40	81	134	2.0	118	1023	850	255	0.10	0.18	1.6	0
	Breakfast cereals																					
	Hot type, cooked																					
	Corn (hominy) grits																					
357	Regular and quick, enriched 1 cup	242	85	145	3	Tr	Tr	0.1	0.2	0	31	0	29	1.5[29]	53	0[30]	0[31]	0 [illegible]	0.24[29]	0.15[29]	2.0[29]	0
358	Instant, plain 1 pkt	137	85	80	2	Tr	Tr	Tr	0.1	0	18	7	16	1.0[29]	29	343	0	[illegible]	0.18[29]	0.08[29]	1.3[29]	0
	Cream of Wheat®																					
359	Regular, quick, instant 1 cup	244	86	140	4	Tr	0.1	Tr	0.2	0	29	54[32]	43[33]	10.9[32]	46	5[33,34]	0	[illegible]	0.24[32]	0.07[32]	1.5[32]	0
360	Mix'n Eat, plain 1 pkt	142	82	100	3	Tr	Tr	Tr	0.1	0	21	20[32]	20[32]	8.1[32]	38	241	1250[32]	376[2]	0.43[32]	0.28[32]	5.0[32]	0
361	Malt-O-Meal® 1 cup	240	88	120	4	Tr	Tr	Tr	0.1	0	26	5	24[32]	9.6[32]	31	2[35]	0	[illegible]	0.48[32]	0.24[32]	5.8[32]	0
	Oatmeal or rolled oats																					
362	Regular, quick, instant, nonfortified 1 cup	234	85	145	6	2	0.4	0.8	1.0	0	25	19	178	1.6	131	2[36]	40	4	0.26	0.05	0.3	0
	Instant, fortified																					
363	Plain 1 pkt	177	86	105	4	2	0.3	0.6	0.7	0	18	163[29]	133	6.3[29]	99	285[29]	1510[29]	453[9]	0.53[29]	0.28[29]	5.5[29]	0
364	Flavored 1 pkt	164	76	160	5	2	0.3	0.7	0.8	0	31	168[29]	148	6.7[29]	137	254[29]	1530[29]	460[9]	0.53[29]	0.38[29]	5.9[29]	Tr
	Ready to eat																					
365	All-Bran® (about $^1/_3$ cup) 1 oz	28	3	70	4	1	0.1	0.1	0.3	0	21	23	264	4.5[32]	350	320	1250[32]	375[2]	0.37[32]	0.43[32]	5.0[32]	15[32]
366	Cap'n Crunch® (about $^3/_4$ cup) 1 oz	28	3	120	1	3	1.7	0.3	0.4	0	23	5	36	7.5[29]	37	213	40	4	0.50[29]	0.55[29]	6.6[29]	0
367	Cheerios® (about $1^1/_4$ cup) 1 oz	28	5	110	4	2	0.3	0.6	0.7	0	20	48	134	4.5[32]	101	307	1250[32]	375[2]	0.37[32]	0.43[32]	5.0[32]	15[32]
	Corn Flakes (about $1^1/_4$ cup)																					
368	Kellogg's® 1 oz	28	3	110	2	Tr	Tr	Tr	Tr	0	24	1	18	1.8[32]	26	351	1250[32]	375[2]	0.37[32]	0.43[32]	5.0[32]	15[32]
369	Toasties® 1 oz	28	3	110	2	Tr	Tr	Tr	Tr	0	24	1	12	0.7[29]	33	297	1250[32]	375[2]	0.37[32]	0.43[32]	5.0[32]	0

Item No.	Foods, approximate measures, units, and weight (weight of edible portion only)	Grams	Water (%)	Food energy (cal)	Protein (g)	Fat (g)	Fatty acids: Saturated (g)	Fatty acids: Mono-unsaturated (g)	Fatty acids: Poly-unsaturated (g)	Cholesterol (mg)	Carbohydrate (g)	Calcium (mg)	Phosphorus (mg)	Iron (mg)	Potassium (mg)	Sodium (mg)	Vitamin A value (IU)[1]	Vitamin A value (RE)[2]	Thiamin (mg)	Riboflavin (mg)	Niacin	Ascorbic acid (mg)
	40% Bran Flakes																					
370	Kellogg's® (about 3/4 cup) 1 oz	28	3	90	4	1	0.1	0.1	0.3	0	22	14	139	8.1[32]	180	264	1250[32]	375[32]	0.37[32]	0.43[32]	5.0[32]	0
371	Post® (about 2/3 cup) 1 oz	28	3	90	3	Tr	0.1	0.1	0.2	0	22	12	179	4.5[32]	151	260	1250[32]	375[32]	0.37[32]	0.43[32]	5.0[32]	0
372	Froot Loops® (about 1 cup) 1 oz	28	3	110	2	1	0.2	0.1	0.1	0	25	3	24	4.5[32]	26	145	1250[32]	375[32]	0.37[32]	0.43[32]	5.0[32]	15[32]
373	Golden Grahams® (about 3/4 cup) 1 oz	28	2	110	2	1	0.7	0.1	0.2	Tr	24	17	41	4.5[32]	63	346	1250[32]	375[32]	0.37[32]	0.43[32]	5.0[32]	15[32]
374	Grape-Nuts® (about 1/4 cup) 1 oz	28	3	100	3	Tr	Tr	Tr	0.1	0	23	11	71	1.2	95	197	1250[32]	375[32]	0.37[32]	0.43[32]	5.0[32]	0
375	Honey Nut Cheerios® (about 3/4 cup) 1 oz	28	3	105	3	1	0.1	0.3	0.3	0	23	20	105	4.5[32]	99	257	1250[32]	375[32]	0.37[32]	0.43[32]	5.0[32]	15[32]
376	Lucky Charms® (about 1 cup) 1 oz	28	3	110	3	1	0.2	0.4	0.4	0	23	32	79	4.5[32]	59	201	1250[32]	375[32]	0.37[32]	0.43[32]	5.0[32]	15[32]
377	Nature Valley® Granola (about 1/3 cup) 1 oz	28	4	125	3	5	3.3	0.7	0.7	0	19	18	89	0.9	98	58	20	2	0.10	0.05	0.2	0
378	100% Natural Cereal (about 1/4 cup) 1 oz	28	2	135	3	6	4.1	1.2	0.5	Tr	18	49	104	0.8	140	12	20	2	0.09	0.15	0.6	0
379	Product 19® (about 3/4 cup) 1 oz	28	3	110	3	Tr	Tr	Tr	0.1	0	24	3	40	18.0[32]	44	325	5000[32]	1501[32]	1.50[32]	1.70[32]	20.0[32]	60[32]
	Raisin Bran																					
380	Kellogg's® (about 3/4 cup) 1 oz	28	8	90	3	1	0.1	0.1	0.3	0	21	10	105	3.5[32]	147	207	960[32]	288[32]	0.28[32]	0.34[32]	3.9[32]	0
381	Post® (about 1/2 cup) 1 oz	28	9	85	3	1	0.1	0.1	0.3	0	21	13	119	4.5[32]	175	185	1250[32]	375[32]	0.37[32]	0.43[32]	5.0[32]	0
382	Rice Krispies® (about 1 cup) 1 oz	28	2	110	2	Tr	Tr	Tr	0.1	0	25	4	34	1.8[32]	29	340	1250[32]	375[32]	0.37[32]	0.43[32]	5.0[32]	15[32]
383	Shredded Wheat (about 2/3 cup) 1 oz	28	5	100	3	1	0.1	0.1	0.3	0	23	11	100	1.2	102	3	0	0	0.07	0.08	1.5	0
384	Special K® (about 1 1/3 cup) 1 oz	28	2	110	6	Tr	Tr	Tr	Tr	Tr	21	8	55	4.5[32]	49	265	1250[32]	375[32]	0.37[32]	0.43[32]	5.0[32]	15[32]
385	Super Sugar Crisp® (about 7/8 cup) 1 oz	28	2	105	2	Tr	Tr	Tr	0.1	0	26	6	52	1.8[32]	105	25	1250[32]	375[32]	0.37[32]	0.43[32]	5.0[32]	0

Item No.	Foods, approximate measures, units, and weight (weight of edible portion only)	Grams	Water (%)	Food energy (cal)	Protein (g)	Fat (g)	Fatty acids: Saturated (g)	Fatty acids: Mono-unsaturated (g)	Fatty acids: Poly-unsaturated (g)	Cholesterol (mg)	Carbohydrate (g)	Calcium (mg)	Phosphorus (mg)	Iron (mg)	Potassium (mg)	Sodium (mg)	Vitamin A value (IU)[1]	Vitamin A value (RE)	Thiamin (mg)	Riboflavin (mg)	Niacin	Ascorbic acid (mg)
386	Sugar Frosted Flakes, Kellogg's® (about 3/4 cup) 1 oz	28	3	110	1	Tr	Tr	Tr	Tr	0	26	1	21	1.8[32]	18	230	1250[32]	375[32]	0.37[32]	0.43[32]	5.0[32]	15[32]
387	Sugar Smacks® (about 3/4 cup) 1 oz	28	3	105	2	1	0.1	0.1	0.2	0	25	3	31	1.8[32]	42	75	1250[32]	375[32]	0.37[32]	0.43[32]	5.0[32]	15[32]
388	Total® (about 1 cup) 1 oz	28	4	100	3	1	0.1	0.1	0.3	0	22	48	118	18.0[32]	106	352	5000[32]	1501[32]	1.50[32]	1.70[32]	20.0[32]	60[32]
389	Trix® (about 1 cup) 1 oz	28	3	110	2	Tr	0.2	0.1	0.1	0	25	6	19	4.5[32]	27	181	1250[32]	375[32]	0.37[32]	0.43[32]	5.0[32]	15[32]
390	Wheaties® (about 1 cup) 1 oz	28	5	100	3	Tr	0.1	Tr	0.2	0	23	43	98	4.5[32]	106	354	1250[32]	375[32]	0.37[32]	0.43[32]	5.0[32]	15[32]
391	Buckwheat flour, light, sifted 1 cup	98	12	340	6	1	0.2	0.4	0.4	0	78	11	86	1.0	314	2	0	0	0.08	0.04	0.4	0
392	Bulgur, uncooked 1 cup	170	10	600	19	3	1.2	0.3	1.2	0	129	49	575	9.5	389	7	0	0	0.48	0.24	7.7	0
	Cakes prepared from cake mixes with enriched flour[37]																					
	Angelfood																					
393	Whole cake, 9 3/4-in. diam. tube cake 1 cake	635	38	1510	38	2	0.4	0.2	1.0	0	342	527	1086	2.7	845	3226	0	0	0.32	1.27	1.6	0
394	Piece, 1/12 of cake 1 piece	53	38	125	3	Tr	Tr	Tr	0.1	0	29	44	91	0.2	71	269	0	0	0.03	0.11	0.1	0
	Coffeecake, crumb																					
395	Whole cake, 7 3/4 by 5 5/8 by 1 1/4 in. 1 cake	430	30	1385	27	41	11.8	16.7	9.6	279	225	262	748	7.3	469	1853	690	194	0.82	0.90	7.7	1
396	Piece, 1/6 of cake 1 piece	72	30	230	5	7	2.0	2.8	1.6	47	38	44	125	1.2	78	310	120	32	0.14	0.15	1.3	Tr
	Devil's food with chocolate frosting																					
397	Whole, 2-layer cake, 8- or 9-in. diam. 1 cake	1107	24	3755	49	136	51.4	19.7	598	645	653	1162	22.1	1439	2900	1660	498	11	1.66	10.0	1	
398	Piece, 1/16 of cake 1 piece	69	24	235	3	8	3.5	3.2	1.2	37	40	41	72	1.4	90	181	100	31	0.07	0.10	0.6	Tr
399	Cupcake, 2 1/2-in. diam. 1 cupcake	35	24	120	2	4	1.8	1.6	0.6	19	20	21	37	0.7	46	92	50	16	0.04	0.05	0.3	Tr
	Gingerbread																					
400	Whole cake, 8 in. square 1 cake	570	37	1575	18	39	9.6	16.4	10.5	6	291	513	570	10.8	1562	1733	0	0	0.86	1.03	7.4	1
401	Piece, 1/9 of cake 1 piece	63	37	175	2	4	1.1	1.8	1.2	1	32	57	63	1.2	173	192	0	0	0.09	0.11	0.8	Tr

Item No.	Foods, approximate measures, units, and weight (weight of edible portion only)	Grams	Water (%)	Food energy (cal)	Protein (g)	Fat (g)	Fatty acids: Saturated (g)	Fatty acids: Mono-unsaturated (g)	Fatty acids: Poly-unsaturated (g)	Cholesterol (mg)	Carbohydrate (g)	Calcium (mg)	Phosphorus (mg)	Iron (mg)	Potassium (mg)	Sodium (mg)	Vitamin A value (IU)[1]	Vitamin A value (RE)[2]	Thiamin (mg)	Riboflavin (mg)	Niacin	Ascorbic acid (mg)
	Yellow with chocolate frosting																					
402	Whole, 2-layer cake, 8- or 9-in. diam. 1 cake	1108	26	3735	45	125	47.8	48.8	21.8	576	638	1008	2017	15.5	1208	2515	1550	465	1.22	1.66	11.1	1
403	Piece, $^1/_{16}$ of cake 1 piece	69	26	235	3	8	3.0	3.0	1.4	36	40	63	126	1.0	75	157	100	29	0.08	0.10	0.7	Tr
	Cakes prepared from home recipes using enriched flour																					
	Carrot, with cream cheese frosting[38]																					
404	Whole cake, 10-in. diam. tube cake 1 cake	1536	23	6175	63	328	66.0	135.2	107.5	1183	775	707	998	21.0	1720	4470	2240	246	1.83	1.97	14.7	23
405	Piece, $^1/_{16}$ of cake 1 piece	96	23	385	4	21	4.1	8.4	6.7	74	48	44	62	1.3	108	279	140	15	0.11	0.12	0.9	1
	Fruitcake, dark[38]																					
406	Whole cake, $7^1/_2$-in. diam., $2^1/_4$-in. high tube cake 1 cake	1361	18	5185	74	228	47.6	113.0	51.7	640	783	1293	1592	37.6	6138	2123	1720	422	2.41	2.55	17.0	504
407	Piece, $^1/_{32}$ of cake, $^2/_3$-in. arc 1 piece	43	18	165	2	7	1.5	3.6	1.6	20	25	41	50	1.2	194	67	50	13	0.08	0.08	0.5	16
	Plain sheet cake[39]																					
	Without frosting																					
408	Whole cake, 9-in. square 1 cake	777	25	2830	35	108	29.5	45.1	25.6	552	434	497	793	11.7	614	2331	1320	373	1.24	1.40	10.1	2
409	Piece, $^1/_9$ of cake 1 piece	86	25	315	4	12	3.3	5.0	2.8	61	48	55	88	1.3	68	258	150	41	0.14	0.15	1.1	Tr
	With uncooked white frosting																					
410	Whole cake, 9-in. square 1 cake	1096	21	4020	37	129	41.6	50.4	26.3	636	694	548	822	11.0	669	2488	2190	647	1.21	1.42	9.9	2
411	Piece, $^1/_9$ of cake 1 piece	121	21	445	4	14	4.6	5.6	2.9	70	77	61	91	1.2	74	275	240	71	0.13	0.16	1.1	Tr
	Pound[40]																					
412	Loaf, $8^1/_2$ by $3^1/_2$ by $3^1/_4$ in. 1 loaf	514	22	2025	33	94	21.1	40.9	26.7	555	265	339	473	9.3	483	1645	3470	1033	0.93	1.08	7.8	1
413	Slice, $^1/_{17}$ of loaf 1 slice	30	22	120	2	5	1.2	2.4	1.6	32	15	20	28	0.5	28	96	200	60	0.05	0.06	0.5	Tr
	Cakes, commercial, made with enriched flour																					
	Pound																					
414	Loaf, $8^1/_2$ by $3^1/_2$ by 3 in. 1 loaf	500	24	1935	26	94	52.0	30.0	4.0	1100	257	146	517	8.0	443	1857	2820	715	0.96	1.12	8.1	0

Item No.	Foods, approximate measures, units, and weight (weight of edible portion only)	Grams	Water (%)	Food energy (cal)	Protein (g)	Fat (g)	Fatty acids: Saturated (g)	Fatty acids: Mono-unsaturated (g)	Fatty acids: Poly-unsaturated (g)	Cholesterol (mg)	Carbohydrate (g)	Calcium (mg)	Phosphorus (mg)	Iron (mg)	Potassium (mg)	Sodium (mg)	Vitamin A value (IU)[1]	Vitamin A value (RE)[2]	Thiamin (mg)	Riboflavin (mg)	Niacin	Ascorbic acid (mg)
415	Slice, $^1/_{17}$ of loaf 1 slice	29	24	110	2	5	3.0	1.7	0.2	64	15	8	30	0.5	26	108	160	41	0.06	0.06	0.5	0
	Snack cakes																					
416	Devil's food with creme filling (2 small cakes per pkg) 1 small cake	28	20	105	1	4	1.7	1.5	0.6	15	17	21	26	1.0	34	105	20	4	0.06	0.09	0.7	0
417	Sponge with creme filling (2 small cakes per pkg) 1 small cake	42	19	155	1	5	2.3	2.1	0.5	7	27	14	44	0.6	37	155	30	9	0.07	0.06	0.6	0
	White with white frosting																					
418	Whole, 2-layer cake, 8- or 9-in. diam. 1 cake	1140	24	4170	43	148	33.1	61.6	42.2	46	670	536	1585	15.5	832	2827	640	194	3.19	2.05	27.6	0
419	Piece, $^1/_{16}$ of cake 1 piece	71	24	260	3	9	2.1	3.8	2.6	3	42	33	99	1.0	52	176	40	12	0.20	0.13	1.7	0
	Yellow with chocolate frosting																					
420	Whole, 2-layer cake, 8- or 9-in. diam. 1 cake	1108	23	3895	40	175	92.0	58.7	10.0	609	620	366	1884	19.9	1972	3080	1850	488	0.78	2.22	10.0	0
421	Piece, $^1/_{16}$ of cake 1 piece	69	23	245	2	11	5.7	3.7	0.6	38	39	23	117	1.2	123	192	120	30	0.05	0.14	0.6	0
	Cheesecake																					
422	Whole cake, 9-in. diam. 1 cake	1110	46	3350	60	213	119.9	65.5	14.4	2053	317	622	977	5.3	1088	2464	2820	833	0.33	1.44	5.1	56
423	Piece, $^1/_{12}$ of cake 1 piece	92	46	280	5	18	9.9	5.4	1.2	170	26	52	81	0.4	90	204	230	69	0.03	0.12	0.4	5
	Cookies made with enriched flour																					
	Brownies with nuts																					
424	Commercial, with frosting, $1^1/_2$ by $1^3/_4$ by $^7/_8$ in. 1 brownie	25	13	100	1	4	1.6	2.0	0.6	14	16	13	26	0.6	50	59	70	18	0.08	0.07	0.3	Tr
425	From home recipe, $1^3/_4$ by $1^3/_4$ by $^7/_8$ in.[38] 1 brownie	20	10	95	1	6	1.4	2.8	1.2	18	11	9	26	0.4	35	51	20	6	0.05	0.05	0.3	Tr
	Chocolate chip																					
426	Commercial, $2^1/_4$-in. diam., $^3/_8$ in. thick 4 cookies	42	4	180	2	9	2.9	3.1	2.6	5	28	13	41	0.8	68	140	50	15	0.10	0.23	1.0	Tr
427	From home recipe, $2^1/_3$-in. diam.[27] 4 cookies	40	3	185	2	11	3.9	4.3	2.0	18	26	13	34	1.0	82	82	20	5	0.06	0.06	0.6	0
428	From refrigerated dough, $2^1/_4$-in. diam., $^3/_8$ in. thick 4 cookies	48	5	225	2	11	4.0	4.4	2.0	22	32	13	34	1.0	62	173	30	8	0.06	0.10	0.9	0

Item No.	Foods, approximate measures, units, and weight (weight of edible portion only)	Grams	Water (%)	Food energy (cal)	Protein (g)	Fat (g)	Fatty acids: Saturated (g)	Fatty acids: Mono-unsaturated (g)	Fatty acids: Poly-unsaturated (g)	Cholesterol (mg)	Carbohydrate (g)	Calcium (mg)	Phosphorus (mg)	Iron (mg)	Potassium (mg)	Sodium (mg)	Vitamin A value (IU)[1]	Vitamin A value (RE)[2]	Thiamin (mg)	Riboflavin (mg)	Niacin	Ascorbic acid (mg)
							Nutrients in indicated quantity															
429	Fig bars, square, $1^5/_8$ by $1^5/_8$ by $^3/_8$ in. or rectangular, $1^1/_2$ by $1^3/_4$ by $^1/_2$ in.																					
	4 cookies	56	12	210	2	4	1.0	1.5	1.0	27	42	40	34	1.4	162	180	60	6	0.08	0.07	0.7	Tr
330	Oatmeal with raisins, $2^5/_8$-in. diam., $^1/_4$ in. thick																					
	4 cookies	52	4	245	3	10	2.5	4.5	2.8	2	36	18	58	1.1	90	148	40	12	0.09	0.08	1.0	0
431	Peanut butter cookie, from home recipe, $2^5/_8$-in. diam.[27]																					
	4 cookies	48	3	245	4	14	4.0	5.8	2.8	0	28	21	60	1.1	110	142	20	5	0.07	0.07	1.9	0
432	Sandwich type (chocolate or vanilla), $1^3/_4$-in. diam., $^3/_8$ in. thick																					
	4 cookies	40	2	195	2	8	2.0	3.6	2.2	0	29	12	40	1.4	66	189	0	0	0.09	0.07	0.8	0
	Shortbread																					
433	Commercial																					
	4 small cookies	32	6	155	2	8	2.9	3.0	1.1	27	20	13	39	0.8	38	123	30	8	0.10	0.09	0.9	0
434	From home recipe[40]																					
	2 large cookies	28	3	145	2	8	1.3	2.7	3.4	0	17	6	31	0.6	18	125	300	89	0.08	0.06	0.7	Tr
435	Sugar cookie, from refrigerated dough, $2^1/_2$-in. diam., $^1/_4$ in. thick																					
	4 cookies	48	4	235	2	12	2.3	5.0	3.6	29	31	50	91	0.9	33	261	40	11	0.09	0.06	1.1	0
436	Vanilla wafers, $1^3/_4$-in. diam., $^1/_4$ in. thick																					
	10 cookies	40	4	185	2	7	1.8	3.0	1.8	25	29	16	36	0.8	50	150	50	14	0.07	0.10	1.0	0
437	Corn chips																					
	1-oz package	28	1	155	2	9	1.4	2.4	3.7	0	16	35	52	0.5	52	233	110	11	0.04	0.05	0.4	1
	Cornmeal																					
438	Whole-ground, unbolted, dry form																					
	1 cup	122	12	435	11	5	0.5	1.1	2.5	0	90	24	312	2.2	346	1	620	62	0.46	0.13	2.4	0
439	Bolted (nearly whole-grain), dry form																					
	1 cup	122	12	440	11	4	0.5	0.9	2.2	0	91	21	272	2.2	303	1	590	59	0.37	0.10	2.3	0
	Degermed, enriched																					
440	Dry form																					
	1 cup	138	12	500	11	2	0.2	0.4	0.9	0	108	8	137	5.9	166	1	610	61	0.61	0.36	4.8	0
441	Cooked																					
	1 cup	240	88	120	3	Tr	Tr	0.1	0.2	0	26	2	34	1.4	38	0	140	14	0.14	0.10	1.2	0
	Crackers[41]																					
	Cheese																					
442	Plain, 1 in. square																					
	10 crackers	10	4	50	1	3	0.9	1.2	0.3	6	6	11	17	0.3	17	112	20	5	0.05	0.04	0.4	0
443	Sandwich type (peanut butter)																					
	1 sandwich)	8	3	40	1	2	0.4	0.8	0.3	1	5	7	25	0.3	17	90	Tr	Tr	0.04	0.03	0.6	0

Item No.	Foods, approximate measures, units, and weight (weight of edible portion only)	Grams	Water (%)	Food energy (cal)	Protein (g)	Fat (g)	Fatty acids: Saturated (g)	Fatty acids: Mono-unsaturated (g)	Fatty acids: Poly-unsaturated (g)	Cholesterol (mg)	Carbohydrate (g)	Calcium (mg)	Phosphorus (mg)	Iron (mg)	Potassium (mg)	Sodium (mg)	Vitamin A value (IU)[1]	Vitamin A value (RE)[2]	Thiamin (mg)	Riboflavin (mg)	Niacin (mg)	Ascorbic acid (mg)
444	Graham, plain, $2^1/_2$ in. square 2 crackers	14	5	60	1	1	0.4	0.6	0.4	0	11	6	20	0.4	36	86	0	[illegible]	0.02	0.03	0.6	0
445	Melba toast, plain 1 piece	5	4	20	1	Tr	0.1	0.1	0.1	0	4	6	10	0.1	11	44	0	[illegible]	0.01	0.01	0.1	0
446	Rye wafers, whole-grain, $1^7/_8$ by $3^1/_2$ in. 2 wafers	14	5	55	1	1	0.3	0.4	0.3	0	10	7	44	0.5	65	115	0	[illegible]	0.06	0.03	0.5	0
447	Saltines[42] 4 crackers	12	4	50	1	1	0.5	0.4	0.2	4	9	3	12	0.5	17	165	0	[illegible]	0.06	0.05	0.6	0
448	Snack-type, standard 1 round cracker	3	3	15	Tr	1	0.2	0.4	0.1	0	2	3	6	0.1	4	30	Tr	[illegible]	0.01	0.01	0.1	0
449	Wheat, thin 4 crackers	8	3	35	1	1	0.5	0.5	0.4	0	5	3	15	0.3	17	69	Tr	[illegible]	0.04	0.03	0.4	0
450	Whole-wheat wafers 2 crackers	8	4	35	1	2	0.5	0.6	0.4	0	5	3	22	0.2	31	59	0	[illegible]	0.02	0.03	0.4	0
451	Croissants, made with enriched flour, $4^1/_2$ by 4 by $1^3/_4$ in. 1 croissant	57	22	235	5	12	3.5	6.7	1.4	13	27	20	64	2.1	68	452	50	[illegible]	0.17	0.13	1.3	0
	Danish pastry, made with enriched flour Plain without fruit or nuts																					
452	Packaged ring, 12 oz 1 ring	340	27	1305	21	71	21.8	28.6	15.6	292	152	360	347	6.5	316	1302	360	[illegible]	0.95	1.02	8.5	Tr
453	Round recipe, about $4^1/_4$-in. diam., 1 in. high 1 pastry	57	27	220	4	12	3.6	4.8	2.6	49	26	60	58	1.1	53	218	60	[illegible]	0.16	0.17	1.4	Tr
454	Ounce 1 oz	28	27	110	2	6	1.8	2.4	1.3	24	13	30	29	0.5	26	109	30	[illegible]	0.08	0.09	0.7	Tr
455	Fruit, round piece 1 pastry	65	30	235	4	13	3.9	5.2	2.9	56	28	17	80	1.3	57	233	40	[illegible]	0.16	0.14	1.4	Tr
	Doughnuts, made with enriched flour																					
456	Cake type, plain, $3^1/_4$-in. diam., 1 in. high 1 doughnut	50	21	210	3	12	2.8	5.0	3.0	20	24	22	111	1.0	58	192	20	[illegible]	0.12	0.12	1.1	Tr
457	Yeast-leavened, glazed, $3^3/_4$-in. diam., $1^1/_4$ in. high 1 doughnut	60	27	235	4	13	5.2	5.5	0.9	21	26	17	55	1.4	64	222	Tr	[illegible]	0.28	0.12	1.8	0
458	English muffins, plain, enriched 1 muffin	57	42	140	5	1	0.3	0.2	0.3	0	27	96	67	1.7	331	378	0	[illegible]	0.26	0.19	2.2	0
459	Toasted 1 muffin	50	29	140	5	1	0.3	0.2	0.3	0	27	96	67	1.7	331	378	0	[illegible]	0.23	0.19	2.2	0
460	French toast, from home recipe 1 slice	65	53	155	6	7	1.6	2.0	1.6	112	17	72	85	1.3	86	257	110	[illegible]	0.12	0.16	1.0	Tr

Item No.	Foods, approximate measures, units, and weight (weight of edible portion only)	Grams	Water (%)	Food energy (cal)	Protein (g)	Fat (g)	Fatty acids: Saturated (g)	Fatty acids: Mono-unsaturated (g)	Fatty acids: Poly-unsaturated (g)	Cholesterol (mg)	Carbo-hydrate (g)	Calcium (mg)	Phos-phorus (mg)	Iron (mg)	Potassium (mg)	Sodium (mg)	Vitamin A value (IU)[1]	Vitamin A value (RE)[2]	Thiamin (mg)	Riboflavin (mg)	Niacin	Ascorbic acid (mg)
	Macaroni, enriched, cooked (cut lengths, elbows, shells)																					
461	Firm stage (hot) 1 cup	130	64	190	7	1	0.1	0.1	0.3	0	39	14	85	2.1	103	1	0	0	0.23	0.13	1.8	0
	Tender stage																					
462	Cold 1 cup	105	72	115	4	Tr	0.1	0.1	0.2	0	24	8	53	1.3	64	1	0	0	0.15	0.08	1.2	0
463	Hot 1 cup	140	72	155	5	1	0.1	0.1	0.2	0	32	11	70	1.7	85	1	0	0	0.20	0.11	1.5	0
	Muffins made with enriched flour, 2 1/2-in. diam., 1 1/2 in. high																					
	From home recipe																					
464	Blueberry[27] 1 muffin	45	37	135	3	5	1.5	2.1	1.2	19	20	54	46	0.9	47	198	40	9	0.10	0.11	0.9	1
465	Bran[38] 1 muffin	45	35	125	3	6	1.4	1.6	2.3	24	19	60	125	1.4	99	189	230	30	0.11	0.13	1.3	3
466	Corn (enriched, degermed cornmeal and flour)[27] 1 muffin	45	33	145	3	5	1.5	2.2	1.4	23	21	66	59	0.9	57	169	80	15	0.11	0.11	0.9	Tr
	From commercial mix (egg and water added)																					
467	Blueberry 1 muffin	45	33	140	3	5	1.4	2.0	1.2	45	22	15	90	0.9	54	225	50	11	0.10	0.17	1.1	Tr
468	Bran 1 muffin	45	28	140	3	4	1.3	1.6	1.0	28	24	27	182	1.7	50	385	100	14	0.08	0.12	1.9	0
469	Corn 1 muffin	45	30	145	3	6	1.7	2.3	1.4	42	22	30	128	1.3	31	291	90	16	0.09	0.09	0.8	Tr
470	Noodles (egg noodles), enriched, cooked 1 cup	160	70	200	7	2	0.5	0.6	0.6	50	37	16	94	2.6	70	3	110	34	0.22	0.13	1.9	0
471	Noodles, chow mein, canned 1 cup	45	11	220	6	11	2.1	7.3	0.4	5	26	14	41	0.4	33	450	0	0	0.05	0.03	0.6	0
	Pancakes, 4-in. diam.																					
472	Buckwheat, from mix (with buckwheat and enriched flours), egg and milk added 1 pancake	27	58	55	2	2	0.9	0.9	0.5	20	6	59	91	0.4	66	125	60	17	0.04	0.05	0.2	Tr
	Plain																					
473	From home recipe using enriched flour 1 pancake	27	50	60	2	2	0.5	0.8	0.5	16	9	27	38	0.5	33	115	30	10	0.06	0.07	0.5	Tr
474	From mix (with enriched flour), egg, milk, and oil added 1 pancake	27	54	60	2	2	0.5	0.9	0.5	16	8	36	71	0.7	43	160	30	7	0.09	0.12	0.8	Tr

Item No.	Foods, approximate measures, units, and weight (weight of edible portion only)	Grams	Water (%)	Food energy (cal)	Protein (g)	Fat (g)	Fatty acids: Saturated (g)	Fatty acids: Mono-unsaturated (g)	Fatty acids: Poly-unsaturated (g)	Cholesterol (mg)	Carbo-hydrate (g)	Calcium (mg)	Phos-phorus (mg)	Iron (mg)	Potassium (mg)	Sodium (mg)	Vitamin A value (IU)[1]	Vitamin A value (RE)[2]	Thiamin (mg)	Riboflavin (mg)	Niacin	Ascorbic acid (mg)
	Piecrust, made with enriched flour and vegetable shortening, baked																					
475	From home recipe, 9-in. diam. 1 pie shell	180	15	900	11	60	14.8	25.9	15.7	0	79	25	90	4.5	90	1100	0	0	0.54	0.40	5.0	0
476	From mix, 9-in. diam. Piecrust for 2-crust pie	320	19	1485	20	93	22.7	41.0	25.0	0	141	131	272	9.3	179	2602	0	0	1.06	0.80	9.9	0
	Pies, piecrust made with enriched flour, vegetable shortening, 9-in. diam.																					
	Apple																					
477	Whole 1 pie	945	48	2420	21	105	27.4	44.4	26.5	0	360	76	208	9.5	756	2844	280	28	1.04	0.76	9.5	9
478	Piece, 1/6 of pie 1 piece	158	48	405	3	18	4.6	7.4	4.4	0	60	13	35	1.6	126	476	50	5	0.17	0.13	1.6	2
	Blueberry																					
479	Whole 1 pie	945	51	2285	23	102	25.5	44.4	27.4	0	330	104	217	12.3	945	2533	850	85	1.04	0.85	10.4	38
480	Piece, 1/6 of pie 1 piece	158	51	380	4	17	4.3	7.4	4.6	0	55	17	36	2.1	158	423	140	14	0.17	0.14	1.7	6
	Cherry																					
481	Whole 1 pie	945	47	2465	25	107	28.4	46.3	27.4	0	363	132	236	9.5	992	2873	4160	416	1.13	0.85	9.5	0
482	Piece, 1/6 of pie 1 piece	158	47	410	4	18	4.7	7.7	4.6	0	61	22	40	1.6	166	480	700	70	0.19	0.14	1.6	0
	Creme																					
483	Whole 1 pie	910	43	2710	20	139	90.1	23.7	6.4	46	351	273	919	6.8	796	2207	1250	391	0.36	0.89	6.4	0
484	Piece, 1/6 of pie 1 piece	152	43	455	3	23	15.0	4.0	1.1	8	59	46	154	1.1	133	369	210	65	0.06	0.15	1.1	0
	Custard																					
485	Whole 1 pie	910	58	1985	56	101	33.7	40.0	19.1	1010	213	874	1028	9.1	1247	2612	2090	573	0.82	1.91	5.5	0
486	Piece, 1/6 of pie 1 piece	152	58	330	9	17	5.6	6.7	3.2	169	36	146	172	1.5	208	436	350	96	0.14	0.32	0.9	0
	Lemon meringue																					
487	Whole 1 pie	840	47	2140	31	86	26.0	34.4	17.6	857	317	118	412	8.4	420	2369	1430	395	0.59	0.84	5.0	25
488	Piece, 1/6 of pie 1 piece	140	47	355	5	14	4.3	5.7	2.9	143	53	20	69	1.4	70	395	240	66	0.10	0.14	0.8	4
	Peach																					
489	Whole 1 pie	945	48	2410	24	101	24.6	43.5	26.5	0	361	95	274	11.3	1408	2533	6900	690	1.04	0.95	14.2	28
490	Piece, 1/6 of pie 1 piece	158	48	405	4	17	4.1	7.3	4.4	0	60	16	46	1.9	235	423	1150	115	0.17	0.16	2.4	5
	Pecan																					
491	Whole 1 pie	825	20	3450	42	189	28.1	101.5	47.0	569	423	388	850	27.2	1015	1823	1320	322	1.82	0.99	6.6	0

Item No.	Foods, approximate measures, units, and weight (weight of edible portion only)	Grams	Water (%)	Food energy (cal)	Protein (g)	Fat (g)	Fatty acids: Saturated (g)	Fatty acids: Mono-unsaturated (g)	Fatty acids: Poly-unsaturated (g)	Cholesterol (mg)	Carbohydrate (g)	Calcium (mg)	Phosphorus (mg)	Iron (mg)	Potassium (mg)	Sodium (mg)	Vitamin A value (IU)[1]	Vitamin A value (RE)[2]	Thiamin (mg)	Riboflavin (mg)	Niacin	Ascorbic acid (mg)
492	Piece, $^1/_6$ of pie																					
	1 piece	138	20	575	7	32	4.7	17.0	7.9	95	71	65	142	4.6	170	305	220	54	0.30	0.17	1.1	0
	Pumpkin																					
493	Whole																					
	1 pie	910	59	1920	36	102	38.2	40.0	18.2	655	223	464	628	8.2	1456	1947	22480	2493	0.82	1.27	7.3	0
494	Piece, $^1/_6$ of pie																					
	1 piece	152	59	320	6	17	6.4	6.7	3.0	109	37	78	105	1.4	243	325	3750	416	0.14	0.21	1.2	0
	Pies, fried																					
495	Apple																					
	1 pie	85	43	255	2	14	5.8	6.6	0.6	14	31	12	34	0.9	42	326	30	3	0.09	0.06	1.0	1
496	Cherry																					
	1 pie	85	42	250	2	14	5.8	6.7	0.6	13	32	11	41	0.7	61	371	190	19	0.06	0.06	0.6	1
	Popcorn, popped																					
497	Air-popped, unsalted																					
	1 cup	8	4	30	1	Tr	Tr	0.1	0.2	0	6	1	22	0.2	20	Tr	10	1	0.03	0.01	0.2	0
498	Popped in vegetable oil, salted																					
	1 cup	11	3	55	1	3	0.5	1.4	1.2	0	6	3	31	0.3	19	86	20	2	0.01	0.02	0.1	0
499	Sugar syrup coated																					
	1 cup	35	4	135	2	1	0.1	0.3	0.6	0	30	2	47	0.5	90	Tr	30	3	0.13	0.02	0.4	0
	Pretzels, made with enriched flour																					
500	Stick, $2^1/_4$ in. long																					
	10 pretzels	3	3	10	Tr	Tr	Tr	Tr	Tr	0	2	1	3	0.1	3	48	0	0	0.01	0.01	0.1	0
501	Twisted, dutch, $2^3/_4$ by $2^5/_8$ in.																					
	1 pretzel	16	3	65	2	1	0.1	0.2	0.2	0	13	4	15	0.3	16	258	0	0	0.05	0.04	0.7	0
502	Twisted, thin, $3^1/_4$ by $2^1/_4$ by $^1/_4$ in.																					
	10 pretzels	60	3	240	6	2	0.4	0.8	0.6	0	48	16	55	1.2	61	966	0	0	0.19	0.15	2.6	0
	Rice																					
503	Brown, cooked, served hot																					
	1 cup	195	70	230	5	1	0.3	0.3	0.4	0	50	23	142	1.0	137	0	0	0	0.18	0.04	2.7	0
	White, enriched																					
	Commercial varieties, all types																					
504	Raw																					
	1 cup	185	12	670	12	1	0.2	0.2	0.3	0	149	44	174	5.4	170	9	0	0	0.81	0.06	6.5	0
505	Cooked, served hot																					
	1 cup	205	73	225	4	Tr	0.1	0.1	0.1	0	50	21	57	1.8	57	0	0	0	0.23	0.02	2.1	0
506	Instant, ready-to-serve, hot																					
	1 cup	165	73	180	4	0	0.1	0.1	0.1	0	40	5	31	1.3	0	0	0	0	0.21	0.02	1.7	0
	Parboiled																					
507	Raw																					
	1 cup	185	10	685	14	1	0.1	0.1	0.2	0	150	111	370	5.4	278	17	0	0	0.81	0.07	6.5	0
508	Cooked, served hot																					
	1 cup	175	73	185	4	Tr	Tr	Tr	0.1	0	41	33	100	1.4	75	0	0	0	0.19	0.02	2.1	0

Item No.	Foods, approximate measures, units, and weight (weight of edible portion only)	Grams	Water (%)	Food energy (cal)	Protein (g)	Fat (g)	Fatty acids: Saturated (g)	Fatty acids: Mono-unsaturated (g)	Fatty acids: Poly-unsaturated (g)	Cholesterol (mg)	Carbo-hydrate (g)	Calcium (mg)	Phos-phorus (mg)	Iron (mg)	Potassium (mg)	Sodium (mg)	Vitamin A value (IU)[1]	Vitamin A value (RE)[2]	Thiamin (mg)	Riboflavin (mg)	Niacin	Ascorbic acid (mg)
	Rolls, enriched																					
	Commercial																					
509	Dinner, $2^1/_2$-in. diam., 2 in. high																					
	1 roll	28	32	85	2	2	0.5	0.8	0.6	Tr	14	33	44	0.8	36	155	Tr	Tr	0.14	0.09	1.1	Tr
510	Frankfurter and hamburger (8 per $11^1/_2$-oz pkg.)																					
	1 roll	40	34	115	3	2	0.5	0.8	0.6	Tr	20	54	44	1.2	56	241	Tr	Tr	0.20	0.13	1.6	Tr
511	Hard, $3^3/_4$-in. diam., 2 in. high																					
	1 roll	50	25	155	5	2	0.4	0.5	0.6	Tr	30	24	46	1.4	49	313	0	0	0.20	0.12	1.7	0
512	Hoagie or submarine, $11^1/_2$ by 3 by $2^1/_2$ in.																					
	1 roll	135	31	400	11	8	1.8	3.0	2.2	Tr	72	100	115	3.8	128	683	0	0	0.54	0.33	4.5	0
	From home recipe																					
513	Dinner, $2^1/_2$-in. diam., 2 in. high																					
	1 roll	35	26	120	3	3	0.8	1.2	0.9	12	20	16	36	1.1	41	98	30	8	0.12	0.12	1.2	0
	Spaghetti, enriched, cooked																					
514	Firm stage, "al dente," served hot																					
	1 cup	130	64	190	7	1	0.1	0.1	0.3	0	39	14	85	2.0	103	1	0	0	0.23	0.13	1.8	0
515	Tender stage, served hot																					
	1 cup	140	73	155	5	1	0.1	0.1	0.2	0	32	11	70	1.7	85	1	0	0	0.20	0.11	1.5	0
516	Toaster pastries																					
	1 pastry	54	13	210	2	6	1.7	3.6	0.4	0	38	104	104	2.2	91	248	520	52	0.17	0.18	2.3	4
517	Tortillas, corn																					
	1 tortilla	30	45	65	2	1	0.1	0.3	0.6	0	13	42	55	0.6	43	1	80	8	0.05	0.03	0.4	0
	Waffles, made with enriched flour, 7-in. diam.																					
518	From home recipe																					
	1 waffle	75	37	245	7	13	4.0	4.9	2.6	02	26	154	135	1.5	129	445	140	39	0.18	0.24	1.5	Tr
519	From mix, egg and milk added																					
	1 waffle	75	42	205	7	8	2.7	2.9	1.5	59	27	179	257	1.2	146	515	170	49	0.14	0.23	0.9	Tr
	Wheat flours																					
	All-purpose or family flour, enriched																					
520	Sifted, spooned																					
	1 cup	115	12	420	12	1	0.2	0.1	0.5	0	88	18	100	5.1	109	2	0	0	0.73	0.46	6.1	0
521	Unsifted, spooned																					
	1 cup	125	12	455	13	1	0.2	0.1	0.5	0	95	20	109	5.5	119	3	0	0	0.80	0.50	6.6	0
522	Cake or pastry flour, enriched, sifted, spooned																					
	1 cup	96	12	350	7	1	0.1	0.1	0.3	0	76	16	70	4.2	91	2	0	0	0.58	0.38	5.1	0
523	Self-rising, enriched, unsifted, spooned																					
	1 cup	125	12	440	12	1	0.2	0.1	0.5	0	93	331	583	5.5	113	1349	0	0	0.80	0.50	6.6	0

Item No.	Foods, approximate measures, units, and weight (weight of edible portion only)	Grams	Water (%)	Food energy (cal)	Protein (g)	Fat (g)	Fatty acids: Saturated (g)	Fatty acids: Mono-unsaturated (g)	Fatty acids: Poly-unsaturated (g)	Cholesterol (mg)	Carbo-hydrate (g)	Calcium (mg)	Phos-phorus (mg)	Iron (mg)	Potassium (mg)	Sodium (mg)	Vitamin A value (IU)[1]	Vitamin A value (RE)[2]	Thiamin (mg)	Riboflavin (mg)	Niacin	Ascorbic acid (mg)
524	Whole-wheat, from hard wheats, stirred, 1 cup	120	12	400	16	2	0.3	0.3	1.1	0	85	49	446	5.2	444	4	0	0	0.66	0.14	5.2	0
	Legumes, Nuts, and Seeds																					
	Almonds, shelled																					
525	Slivered, packed, 1 cup	135	4	795	27	70	6.7	45.8	14.8	0	28	359	702	4.9	988	15	0	0	0.28	1.05	4.5	1
526	Whole, 1 oz	28	4	165	6	15	1.4	9.6	3.1	0	6	75	147	1.0	208	3	0	0	0.06	0.22	1.0	Tr
	Beans, dry																					
	Cooked, drained																					
527	Black, 1 cup	171	66	225	15	1	0.1	0.1	0.5	0	41	47	239	2.9	608	1	Tr	Tr	0.43	0.05	0.9	0
528	Great Northern, 1 cup	180	69	210	14	1	0.1	0.1	0.6	0	38	90	266	4.9	749	13	0	0	0.25	0.13	1.3	0
529	Lima, 1 cup	190	64	260	16	1	0.2	0.1	0.5	0	49	55	293	5.9	1163	4	0	0	0.25	0.11	1.3	0
530	Pea (navy), 1 cup	190	69	225	15	1	0.1	0.1	0.7	0	40	95	281	5.1	790	13	0	0	0.27	0.13	1.3	0
531	Pinto, 1 cup	180	65	265	15	1	0.1	0.1	0.5	0	49	86	296	5.4	882	3	Tr	Tr	0.33	0.16	0.7	0
	Canned, solids and liquid																					
	White with																					
532	Frankfurters (sliced), 1 cup	255	71	365	19	18	7.4	8.8	0.7	30	32	94	303	4.8	668	1374	330	33	0.18	0.15	3.3	Tr
533	Pork and tomato sauce, 1 cup	255	71	310	16	7	2.4	2.7	0.7	10	48	138	235	4.6	536	1181	330	33	0.20	0.08	1.5	5
534	Pork and sweet sauce, 1 cup	255	66	385	16	12	4.3	4.9	1.2	10	54	161	291	5.9	536	969	330	33	0.15	0.10	1.3	5
535	Red kidney, 1 cup	255	76	230	15	1	0.1	0.1	0.6	0	42	74	278	4.6	673	968	10	1	0.13	0.10	1.5	0
536	Black-eyed peas, dry, cooked (with residual cooking liquid), 1 cup	250	80	190	13	1	0.2	Tr	0.3	0	35	43	238	3.3	573	20	30	3	0.40	0.10	1.0	0
537	Brazil nuts, shelled, 1 oz	28	3	185	4	19	4.6	6.5	6.8	0	4	50	170	1.0	170	1	Tr	Tr	0.28	0.03	0.5	Tr
538	Carob flour, 1 cup	140	3	255	6	Tr	Tr	0.1	0.1	0	126	390	102	5.7	1275	24	Tr	Tr	0.07	0.07	2.2	Tr
	Cashew nuts, salted																					
	Dry roasted																					
539	1 cup	137	2	785	21	63	12.5	37.4	10.7	0	45	62	671	8.2	774	877[43]	0	0	0.27	0.27	1.9	0
540	1 oz	28	2	165	4	13	2.6	7.7	2.2	0	9	13	139	1.7	160	181[43]	0	0	0.06	0.06	0.4	0
	Roasted in oil																					
541	1 cup	130	4	750	21	63	12.4	36.9	10.6	0	37	53	554	5.3	689	814[44]	0	0	0.55	0.23	2.3	0
542	1 oz	28	4	165	5	14	2.7	8.1	2.3	0	8	12	121	1.2	150	177[44]	0	0	0.12	0.05	0.5	0

Item No.	Foods, approximate measures, units, and weight (weight of edible portion only)	Grams	Water (%)	Food energy (cal)	Protein (g)	Fat (g)	Fatty acids: Saturated (g)	Fatty acids: Mono-unsaturated (g)	Fatty acids: Poly-unsaturated (g)	Cholesterol (mg)	Carbo-hydrate (g)	Calcium (mg)	Phos-phorus (mg)	Iron (mg)	Potassium (mg)	Sodium (mg)	Vitamin A value (IU)[1]	Vitamin A value (RE)[2]	Thiamin (mg)	Riboflavin (mg)	Niacin	Ascorbic acid (mg)
	Nutrients in indicated quantity																					
543	Chestnuts, European (Italian), roasted, shelled																					
	1 cup	143	40	350	5	3	0.6	1.1	1.2	0	76	41	153	1.3	847	3	30	3	0.35	0.25	1.9	37
544	Chickpeas, cooked, drained																					
	1 cup	163	60	270	15	4	0.4	0.9	1.9	0	45	80	273	4.9	475	11	Tr	Tr	0.18	0.09	0.9	0
	Coconut																					
	Raw																					
545	Piece, about 2 by 2 by 1/2 in.																					
	1 piece	45	47	160	1	15	13.4	0.6	0.2	0	7	6	51	1.1	160	9	0	0	0.03	0.01	0.2	1
546	Shredded or grated																					
	1 cup	80	47	285	3	27	23.8	1.1	0.3	0	12	11	90	1.9	285	16	0	0	0.05	0.02	0.4	3
547	Dried, sweetened, shredded																					
	1 cup	93	13	470	3	33	29.3	1.4	0.4	0	44	14	99	1.8	313	244	0	0	0.03	0.02	0.4	1
548	Filberts (hazelnuts), chopped																					
	1 cup	115	5	725	15	72	5.3	56.5	6.9	0	18	216	359	3.8	512	3	80	8	0.58	0.13	1.3	1
549	1 oz	28	5	180	4	18	1.3	13.9	1.7	0	4	53	88	0.9	126	1	20	2	0.14	0.03	0.3	Tr
550	Lentils, dry, cooked																					
	1 cup	200	72	215	16	1	0.1	0.2	0.5	0	38	50	238	4.2	498	26	40	4	0.14	0.12	1.2	0
551	Macadamia nuts, roasted in oil, salted																					
	1 cup	134	2	960	10	103	15.4	80.9	1.8	0	17	60	268	2.4	441	348[45]	10	1	0.29	0.15	2.7	0
552	1 oz	28	2	205	2	22	3.2	17.1	0.4	0	4	13	57	0.5	93	74[45]	Tr	Tr	0.06	0.03	0.6	0
	Mixed nuts, with peanuts, salted																					
553	Dry roasted																					
	1 oz	28	2	170	5	15	2.0	8.9	3.1	0	7	20	123	1.0	169	190[46]	Tr	Tr	0.06	0.06	1.3	0
554	Roasted in oil																					
	1 oz	28	2	175	5	16	2.5	9.0	3.8	0	6	31	131	0.9	165	185[46]	10	[illegible]	0.14	0.06	1.4	Tr
555	Peanuts, roasted in oil, salted																					
	1 cup	145	2	840	39	71	9.9	35.5	22.6	0	27	125	734	2.8	1019	626[47]	0	0	0.42	0.15	21.5	0
556	1 oz	28	2	165	8	14	1.9	6.9	4.4	0	5	24	143	0.5	199	122[47]	0	0	0.08	0.03	4.2	0
557	Peanut butter																					
	1 tbsp	16	1	95	5	8	1.4	4.0	2.5	0	3	5	60	0.3	110	75	0	0	0.02	0.02	2.2	0
558	Peas, split, dry, cooked																					
	1 cup	200	70	230	16	1	0.1	0.1	0.3	0	42	22	178	3.4	592	26	80	[illegible]	0.30	0.18	1.8	0
559	Pecans, halves																					
	1 cup	108	5	720	8	73	5.9	45.5	18.1	0	20	39	314	2.3	423	1	140	[illegible]	0.92	0.14	1.0	2
560	1 oz	28	5	190	2	19	1.5	12.0	4.7	0	5	10	83	0.6	111	Tr	40	[illegible]	0.24	0.04	0.3	1
561	Pine nuts (pinyons), shelled																					
	1 oz	28	6	160	3	17	2.7	6.5	7.3	0	5	2	10	0.9	178	20	10	[illegible]	0.35	0.06	1.2	1
562	Pistachio nuts, dried, shelled																					
	1 oz	28	4	165	6	14	1.7	9.3	2.1	0	7	38	143	1.9	310	2	70	[illegible]	0.23	0.05	0.3	Tr
563	Pumpkin and squash kernels, dry, hulled																					
	1 oz	28	7	155	7	13	2.5	4.0	5.9	0	5	12	333	4.2	229	5	110	[illegible]	0.06	0.09	0.5	Tr

							Fatty acids											Vitamin A value					
Item No.	Foods, approximate measures, units, and weight (weight of edible portion only)	Grams	Water (%)	Food energy (cal)	Protein (g)	Fat (g)	Saturated (g)	Mono-unsaturated (g)	Poly-unsaturated (g)	Cholesterol (mg)	Carbo-hydrate (g)	Calcium (mg)	Phos-phorus (mg)	Iron (mg)	Potassium (mg)	Sodium (mg)	(IU)[1]	(RE)[2]	Thiamin (mg)	Riboflavin (mg)	Niacin (mg)	Ascorbic acid (mg)
564	Refried beans, canned																					
	1 cup	290	72	295	18	3	0.4	0.6	1.4	0	51	141	245	5.1	1141	1228	0	0	0.14	0.16	1.4	17
565	Sesame seeds, dry, hulled																					
	1 tbsp	8	5	45	2	4	0.6	1.7	1.9	0	1	11	62	0.6	33	3	10	1	0.06	0.01	0.4	0
566	Soybeans, dry, cooked, drained																					
	1 cup	180	71	235	20	10	1.3	1.9	5.3	0	19	131	322	4.9	972	4	50	5	0.38	0.16	1.1	0
	Soy products																					
567	Miso																					
	1 cup	276	53	470	29	13	1.8	2.6	7.3	0	65	188	853	4.7	922	8142	110	11	0.17	0.28	0.8	0
568	Tofu, piece 2 1/2 by 2 3/4 by 1 in.																					
	1 piece	120	85	85	9	5	0.7	1.0	2.9	0	3	108	151	2.3	50	8	0	0	0.07	0.04	0.1	0
569	Sunflower seeds, dry, hulled																					
	1 oz	28	5	160	6	14	1.5	2.7	9.3	0	5	33	200	1.9	195	1	10	1	0.65	0.07	1.3	Tr
570	Tahini																					
	1 tbsp	15	3	90	3	8	1.1	3.0	3.5	0	3	21	119	0.7	69	5	10	1	0.24	0.02	0.8	1
	Walnuts																					
571	Black, chopped																					
	1 cup	125	4	760	30	71	4.5	15.9	46.9	0	15	73	580	3.8	655	1	370	37	0.27	0.14	0.9	Tr
572	1 oz	28	4	170	7	16	1.0	3.6	10.6	0	3	16	132	0.9	149	Tr	80	8	0.06	0.03	0.2	Tr
573	English or Persian, pieces or chips																					
	1 cup	120	4	770	17	74	6.7	17.0	47.0	0	22	113	380	2.9	602	12	150	15	0.46	0.18	1.3	4
574	1 oz	28	4	180	4	18	1.6	4.0	11.1	0	5	27	90	0.7	142	3	40	4	0.11	0.04	0.3	1

Meat and Meat Products

	Beef, cooked[48]																					
	Cuts braised, simmered, or pot roasted																					
	Relatively fat such as chuck blade																					
575	Lean and fat, piece, 2 1/2 by 2 1/2 by 3/4 in.																					
	3 oz	85	43	325	22	26	10.8	11.7	0.9	87	0	11	163	2.5	163	53	Tr	Tr	0.06	0.19	2.0	0
576	Lean only from item 575																					
	2.2 oz	62	53	170	19	9	3.9	4.2	0.3	66	0	8	146	2.3	163	44	Tr	Tr	0.05	0.17	1.7	0
	Relatively lean, such as bottom round																					
577	Lean and fat, piece, 4 1/8 by 2 1/4 by 1/2 in.																					
	3 oz	85	54	220	25	13	4.8	5.7	0.5	81	0	5	217	2.8	248	43	Tr	Tr	0.06	0.21	3.3	0
578	Lean only from item 577																					
	2.8 oz	78	57	175	25	8	2.7	3.4	0.3	75	0	4	212	2.7	240	40	Tr	Tr	0.06	0.20	3.0	0

Item No.	Foods, approximate measures, units, and weight (weight of edible portion only)	Grams	Water (%)	Food energy (cal)	Protein (g)	Fat (g)	Fatty acids: Saturated (g)	Fatty acids: Monounsaturated (g)	Fatty acids: Polyunsaturated (g)	Cholesterol (mg)	Carbohydrate (g)	Calcium (mg)	Phosphorus (mg)	Iron (mg)	Potassium (mg)	Sodium (mg)	Vitamin A value (IU)[1]	Vitamin A value (RE)[2]	Thiamin (mg)	Riboflavin (mg)	Niacin	Ascorbic acid (mg)
	Ground beef, broiled, patty, 3 by $^5/_8$ in.																					
579	Lean, 3 oz	85	56	230	21	16	6.2	6.9	0.6	74	0	9	134	1.8	256	65	Tr	Tr	0.04	0.18	4.4	0
580	Regular, 3 oz	85	54	245	20	18	6.9	7.7	0.7	76	0	9	144	2.1	248	70	Tr	Tr	0.03	0.16	4.9	0
581	Heart, lean, braised, 3 oz	85	65	150	24	5	1.2	0.8	1.6	164	0	5	213	6.4	198	54	Tr	Tr	0.12	1.31	3.4	5
582	Liver, fried, slice, $6^1/_2$ by $2^3/_8$ by $^3/_8$ in.[49], 3 oz	85	56	185	23	7	2.5	3.6	1.3	410	7	9	392	5.3	309	90	30690[50]	9120[50]	0.18	3.52	12.3	23
	Roast, oven cooked, no liquid added																					
	Relatively fat, such as rib																					
583	Lean and fat, 2 pieces, $4^1/_8$ by $2^1/_4$ by $^1/_4$ in., 3 oz	85	46	315	19	26	10.8	11.4	0.9	72	0	8	145	2.0	246	54	Tr	Tr	0.06	0.16	3.1	0
584	Lean only from item 583, 2.2 oz	61	57	150	17	9	3.6	3.7	0.3	49	0	5	127	1.7	218	45	Tr	Tr	0.05	0.13	2.7	0
	Relatively lean, such as eye of round																					
585	Lean and fat, 2 pieces, $2^1/_2$ by $2^1/_2$ by $^3/_8$ in., 3 oz	85	57	205	23	12	4.9	5.4	0.5	62	0	5	177	1.6	308	50	Tr	Tr	0.07	0.14	3.0	0
586	Lean only from item 585, 2.6 oz	75	63	135	22	5	1.9	2.1	0.2	52	0	3	170	1.5	297	46	Tr	Tr	0.07	0.13	2.8	0
	Steak																					
	Sirloin, broiled																					
587	Lean and fat, piece, $2^1/_2$ by $2^1/_2$ by $^3/_4$ in., 3 oz	85	53	240	23	15	6.4	6.9	0.6	77	0	9	186	2.6	306	53	Tr	Tr	0.10	0.23	3.3	0
588	Lean only from item 587, 2.5 oz	72	59	150	22	6	2.6	2.8	0.3	64	0	8	176	2.4	290	48	Tr	Tr	0.09	0.22	3.1	0
589	Beef, canned, corned, 3 oz	85	59	185	22	10	4.2	4.9	0.4	80	0	17	90	3.7	51	802	Tr	Tr	0.02	0.20	2.9	0
590	Beef, dried, chipped, 2.5 oz	72	48	145	24	4	1.8	2.0	0.2	46	0	14	287	2.3	142	3053	Tr	Tr	0.05	0.23	2.7	0
	Lamb, cooked																					
	Chops, (3/lb with bone)																					
	Arm, braised																					
591	Lean and fat, 2.2 oz	63	44	220	20	15	6.9	6.0	0.9	77	0	16	132	1.5	195	46	Tr	Tr	0.04	0.16	4.4	0

Item No.	Foods, approximate measures, units, and weight (weight of edible portion only)	Grams	Water (%)	Food energy (cal)	Protein (g)	Fat (g)	Fatty acids: Saturated (g)	Fatty acids: Monounsaturated (g)	Fatty acids: Polyunsaturated (g)	Cholesterol (mg)	Carbohydrate (g)	Calcium (mg)	Phosphorus (mg)	Iron (mg)	Potassium (mg)	Sodium (mg)	Vitamin A value (IU)[1]	Vitamin A value (RE)[2]	Thiamin (mg)	Riboflavin (mg)	Niacin	Ascorbic acid (mg)
592	Lean only from item 591 1.7 oz	48	49	135	17	7	2.9	2.6	0.4	59	0	12	111	1.3	162	36	Tr	Tr	0.03	0.13	3.0	0
	Loin, broiled																					
593	Lean and fat 2.8 oz	80	54	235	22	16	7.3	6.4	1.0	78	0	16	162	1.4	272	62	Tr	Tr	0.09	0.21	5.5	0
594	Lean only from item 593 2.3 oz	64	61	140	19	6	2.6	2.4	0.4	60	0	12	145	1.3	241	54	Tr	Tr	0.08	0.18	4.4	0
	Leg, roasted																					
595	Lean and fat, 2 pieces, 4 1/8 by 2 1/4 by 1/4 in. 3 oz	85	59	205	22	13	5.6	4.9	0.8	78	0	8	162	1.7	273	57	Tr	Tr	0.09	0.24	5.5	0
596	Lean only from item 595 2.6 oz	73	64	140	20	6	2.4	2.2	0.4	65	0	6	150	1.5	247	50	Tr	Tr	0.08	0.20	4.6	0
	Rib, roasted																					
597	Lean and fat, 3 pieces, 2 1/2 by 2 1/2 by 1/4 in. 3 oz	85	47	315	18	26	12.1	10.6	1.5	77	0	19	139	1.4	224	60	Tr	Tr	0.08	0.18	5.5	0
598	Lean only from item 597 2 oz	57	60	130	15	7	3.2	3.0	0.5	50	0	12	111	1.0	179	46	Tr	Tr	0.05	0.13	3.5	0
	Pork, cured, cooked																					
	Bacon																					
599	Regular 3 medium slices	19	13	110	6	9	3.3	4.5	1.1	16	Tr	2	64	0.3	92	303	0	0	0.13	0.05	1.4	6
600	Canadian-style 2 slices	46	62	85	11	4	1.3	1.9	0.4	27	1	5	136	0.4	179	711	0	0	0.38	0.09	3.2	10
	Ham, light cure, roasted																					
601	Lean and fat, 2 pieces, 4 1/8 by 2 1/4 by 1/4 in. 3 oz	85	58	205	18	14	5.1	6.7	1.5	53	0	6	182	0.7	243	1009	0	0	0.51	0.19	3.8	0
602	Lean only from item 601 2.4 oz	68	66	104	17	4	1.3	1.7	0.4	37	0	5	154	0.6	215	902	0	0	0.46	0.17	3.4	0
603	Ham, canned, roasted, 2 pieces, 4 1/8 by 2 1/4 by 1/4 in. 3 oz	85	67	140	18	7	2.4	3.5	0.8	35	Tr	6	188	0.9	298	908	0	0	0.82	0.21	4.3	19[51]
	Luncheon meat																					
604	Canned, spiced or unspiced, slice, 3 by 2 by 1/2 in. 2 slices	42	52	140	5	13	4.5	6.0	1.5	26	1	3	34	0.3	90	541	0	0	0.15	0.08	1.3	Tr
605	Chopped ham (8 slices/6-oz pkg) 2 slices	42	64	95	7	7	2.4	3.4	0.9	21	0	3	65	0.3	134	576	0	0	0.27	0.09	1.6	8[51]

Item No.	Foods, approximate measures, units, and weight (weight of edible portion only)	Grams	Water (%)	Food energy (cal)	Protein (g)	Fat (g)	Fatty acids: Saturated (g)	Fatty acids: Mono-unsaturated (g)	Fatty acids: Poly-unsaturated (g)	Cholesterol (mg)	Carbo-hydrate (g)	Calcium (mg)	Phos-phorus (mg)	Iron (mg)	Potassium (mg)	Sodium (mg)	Vitamin A value (IU)[1]	Vitamin A value (RE)[2]	Thiamin (mg)	Riboflavin (mg)	Niacin	Ascorbic acid (mg)
	Cooked ham (8 slices/8-oz pkg)																					
606	Regular 2 slices	57	65	105	10	6	1.9	2.8	0.7	32	2	4	141	0.6	189	751	0	0	0.49	0.14	3.0	16[51]
607	Extra lean 2 slices	57	71	75	11	3	0.9	1.3	0.3	27	1	4	124	0.4	200	815	0	0	0.53	0.13	2.8	15[51]
	Pork, fresh, cooked																					
	Chop, loin (cut 3/lb with bone)																					
	Broiled																					
608	Lean and fat 3.1 oz	87	50	275	24	19	7.0	8.8	2.2	84	0	3	184	0.7	312	61	10	3	0.87	0.24	4.3	Tr
609	Lean only from item 608 2.5 oz	72	57	165	23	8	2.6	3.4	0.9	71	0	4	176	0.7	302	56	10	1	0.83	0.22	4.0	Tr
	Pan fried																					
610	Lean and fat 3.1 oz	89	45	335	21	27	9.8	12.5	3.1	92	0	4	190	0.7	323	64	10	3	0.91	0.24	4.6	Tr
611	Lean only from item 610 2.4 oz	67	54	180	19	11	3.7	4.8	1.3	72	0	3	178	0.7	305	57	10	1	0.84	0.22	4.0	Tr
	Ham (leg), roasted																					
612	Lean and fat, piece, $2^1/_2$ by $2^1/_2$ by $^3/_4$ in. 3 oz	85	53	250	21	18	6.4	8.1	2.0	79	0	5	210	0.9	280	50	10	2	0.54	0.27	3.9	Tr
613	Lean only from item 612 2.5 oz	72	60	160	20	8	2.7	3.6	1.0	68	0	5	202	0.8	269	46	10	1	0.50	0.25	3.6	Tr
	Rib, roasted																					
614	Lean and fat, piece $2^1/_2$ by $^3/_4$ in. 3 oz	85	51	270	21	20	7.2	9.2	2.3	69	0	9	190	0.8	313	37	10	3	0.50	0.24	4.2	Tr
615	Lean only from item 614 2.5 oz	71	57	175	20	10	3.4	4.4	1.2	56	0	8	182	0.7	300	33	10	2	0.45	0.22	3.8	Tr
	Shoulder cut, braised																					
616	Lean and fat, 3 pieces, $2^1/_2$ by $2^1/_2$ by $^1/_4$ in. 3 oz	85	47	295	23	22	7.9	10.0	2.4	93	0	6	162	1.4	286	75	10	3	0.46	0.26	4.4	Tr
617	Lean only from item 616 2.4 oz	67	54	165	22	8	2.8	3.7	1.0	76	0	5	151	1.3	271	68	10	1	0.40	0.24	4.0	Tr
	Sausages (See also Luncheon meats, items 604—607)																					
618	Bologna, slice (8/8-oz pkg) 2 slices	57	54	180	7	16	6.1	7.6	1.4	31	2	7	52	0.9	103	581	0	0	0.10	0.08	1.5	12[51]

Item No.	Foods, approximate measures, units, and weight (weight of edible portion only)	Grams	Water (%)	Food energy (cal)	Protein (g)	Fat (g)	Fatty acids: Saturated (g)	Fatty acids: Mono-unsaturated (g)	Fatty acids: Poly-unsaturated (g)	Cholesterol (mg)	Carbo-hydrate (g)	Calcium (mg)	Phos-phorus (mg)	Iron (mg)	Potassium (mg)	Sodium (mg)	Vitamin A value (IU)[1]	Vitamin A value (RE)[2]	Thiamin (mg)	Riboflavin (mg)	Niacin	Ascorbic acid (mg)
	Braunschweiger, slice (6/6-oz pkg) 2 slices	57	48	205	8	18	6.2	8.5	2.1	89	2	5	96	5.3	113	652	8010	2405	0.14	0.87	4.8	6[51]
620	Brown and serve (10—11/8-oz pkg), browned 1 link	13	45	50	2	5	1.7	2.2	0.5	9	Tr	1	14	0.1	25	105	0	0	0.05	0.02	0.4	0
621	Frankfurter (10/1-lb pkg), cooked (reheated) 1 frankfurter	45	54	145	5	13	4.8	6.2	1.2	23	1	5	39	0.5	75	504	0	0	0.09	0.05	1.2	12[51]
622	Pork link (16/1-lb pkg), cooked[52] 1 link	13	45	50	3	4	1.4	1.8	0.5	11	Tr	4	24	0.2	47	168	0	0	0.10	0.03	0.6	Tr
	Salami																					
623	Cooked type, slice (8/8-oz pkg) 2 slices	57	60	145	8	11	4.6	5.2	1.2	37	1	7	66	1.5	113	607	0	0	0.14	0.21	2.0	7[51]
624	Dry type, slice (12/4-oz pkg) 2 slices	20	35	85	5	7	2.4	3.4	0.6	16	1	2	28	0.3	76	372	0	0	0.12	0.06	1.0	5[51]
625	Sandwich spread (pork, beef) 1 tbsp	15	60	35	1	3	0.9	1.1	0.4	6	2	2	9	0.1	17	152	10	1	0.03	0.02	0.3	0
626	Vienna sausage (7/4-oz can) 1 sausage	16	60	45	2	4	1.5	2.0	0.3	8	Tr	2	8	0.1	16	152	0	0	0.01	0.02	0.3	0
	Veal, medium fat, cooked, bone removed																					
627	Cutlet, $4^1/_8$ by $2^1/_4$ by $^1/_2$ in., braised or broiled 3 oz	85	60	185	23	9	4.1	4.1	0.6	109	0	9	196	0.8	258	56	Tr	Tr	0.06	0.21	4.6	0
628	Rib, 2 pieces, $4^1/_8$ by $2^1/_4$ by $^1/_4$ in., roasted 3 oz	85	55	230	23	14	6.0	6.0	1.0	109	0	10	211	0.7	259	57	Tr	Tr	0.11	0.26	6.6	0

Mixed Dishes and Fast Foods

Item No.	Foods, approximate measures, units, and weight (weight of edible portion only)	Grams	Water (%)	Food energy (cal)	Protein (g)	Fat (g)	Fatty acids: Saturated (g)	Fatty acids: Mono-unsaturated (g)	Fatty acids: Poly-unsaturated (g)	Cholesterol (mg)	Carbo-hydrate (g)	Calcium (mg)	Phos-phorus (mg)	Iron (mg)	Potassium (mg)	Sodium (mg)	Vitamin A value (IU)[1]	Vitamin A value (RE)[2]	Thiamin (mg)	Riboflavin (mg)	Niacin	Ascorbic acid (mg)
	Mixed dishes																					
629	Beef and vegetable stew, from home recipe 1 cup	245	82	220	16	11	4.4	4.5	0.5	71	15	29	184	2.9	613	292	5690	568	0.15	0.17	4.7	17
630	Beef potpie, from home recipe, baked, piece, $^1/_3$ of 9-in. diam. pie[53] 1 piece	210	55	515	21	30	7.9	12.9	7.4	42	39	29	149	3.8	334	596	4220	517	0.29	0.29	4.8	6
631	Chicken a la king, cooked, from home recipe 1 cup	245	68	470	27	34	12.9	13.4	6.2	221	12	127	358	2.5	404	760	1130	272	0.10	0.42	5.4	12

Item No.	Foods, approximate measures, units, and weight (weight of edible portion only)	Grams	Water (%)	Food energy (cal)	Protein (g)	Fat (g)	Fatty acids: Saturated (g)	Fatty acids: Mono-unsaturated (g)	Fatty acids: Poly-unsaturated (g)	Cholesterol (mg)	Carbohydrate (g)	Calcium (mg)	Phosphorus (mg)	Iron (mg)	Potassium (mg)	Sodium (mg)	Vitamin A value (IU)[1]	Vitamin A value (RE)[2]	Thiamin (mg)	Riboflavin (mg)	Niacin (mg)	Ascorbic acid (mg)
632	Chicken and noodles, cooked, from home recipe																					
	1 cup	240	71	365	22	18	5.1	7.1	3.9	103	26	26	247	2.2	149	600	430	130	0.05	0.17	4.3	Tr
	Chicken chow mein																					
633	Canned																					
	1 cup	250	89	95	7	Tr	0.1	0.1	0.8	8	18	45	85	1.3	418	725	150	28	0.05	0.10	1.0	13
634	From home recipe																					
	1 cup	250	78	255	31	10	4.1	4.9	3.5	75	10	58	293	2.5	473	718	280	50	0.08	0.23	4.3	10
635	Chicken potpie, from home recipe, baked, piece, 1/3 of 9-in. diam. pie[53]																					
	1 piece	232	57	545	23	31	10.3	15.5	6.6	56	42	70	232	3.0	343	594	7220	735	0.32	0.32	4.9	5
636	Chile con carne with beans, canned																					
	1 cup	255	72	340	19	16	5.8	7.2	1.0	28	31	82	321	4.3	594	1354	150	15	0.08	0.18	3.3	8
637	Chop suey with beef and pork, from home recipe																					
	1 cup	250	75	300	26	17	4.3	7.4	4.2	68	13	60	248	4.8	425	1053	600	60	0.28	0.38	5.0	33
	Macaroni (enriched) and cheese																					
638	Canned[54]																					
	1 cup	240	80	230	9	10	4.7	2.9	1.3	24	26	199	182	1.0	139	730	260	72	0.12	0.24	1.0	Tr
639	From home recipe[40]																					
	1 cup	200	58	430	17	22	9.8	7.4	3.6	44	40	362	322	1.8	240	1086	860	232	0.20	0.40	1.8	1
640	Quiche Lorraine, 1/8 of 8-in. diam. quiche[53]																					
	1 slice	176	47	600	13	48	23.2	17.8	4.1	285	29	211	276	1.0	283	653	1640	454	0.11	0.32	Tr	Tr
	Spaghetti (enriched) in tomato sauce with cheese																					
641	Canned																					
	1 cup	250	80	190	6	2	0.4	0.4	0.5	3	39	40	88	2.8	303	955	930	120	0.35	0.28	4.5	10
642	From home recipe																					
	1 cup	250	77	260	9	9	3.0	3.6	1.2	8	37	80	135	2.3	408	955	1080	140	0.25	0.18	2.3	13
	Spaghetti (enriched) with meatballs and tomato sauce																					
643	Canned																					
	1 cup	250	78	260	12	10	2.4	3.9	3.1	23	29	53	113	3.3	245	1220	1000	100	0.15	0.18	2.3	5
644	From home recipe																					
	1 cup	248	70	330	19	12	3.9	4.4	2.2	89	39	124	236	3.7	665	1009	1590	159	0.25	0.30	4.0	22
	Fast food entrees																					
	Cheeseburger																					
645	Regular																					
	1 sandwich	112	46	300	15	15	7.3	5.6	1.0	44	28	135	174	2.3	219	672	340	65	0.26	0.24	3.7	1
646	4 oz patty																					
	1 sandwich	194	46	525	30	31	15.1	12.2	1.4	104	40	236	320	4.5	407	1224	670	128	0.33	0.48	7.4	3

Item No.	Foods, approximate measures, units, and weight (weight of edible portion only)	Grams	Water (%)	Food energy (cal)	Protein (g)	Fat (g)	Fatty acids: Saturated (g)	Fatty acids: Monounsaturated (g)	Fatty acids: Polyunsaturated (g)	Cholesterol (mg)	Carbohydrate (g)	Calcium (mg)	Phosphorus (mg)	Iron (mg)	Potassium (mg)	Sodium (mg)	Vitamin A value (IU)[1]	Vitamin A value (RE)[2]	Thiamin (mg)	Riboflavin (mg)	Niacin	Ascorbic acid (mg)
	Chicken, fried; see Poultry and Poultry Products (items 656—659)																					
647	Enchilada, 1 enchilada	230	72	235	20	16	7.7	6.7	0.6	19	24	322	662	11.0	2180	4451	2720	352	0.18	0.26	Tr	Tr
648	English muffin, egg, cheese, and bacon, 1 sandwich	138	49	360	18	18	8.0	8.0	0.7	213	31	197	290	3.1	201	832	650	160	0.46	0.50	3.7	1
	Fish sandwich																					
649	Regular, with cheese, 1 sandwich	140	43	420	16	23	6.3	6.9	7.7	56	39	132	223	1.8	274	667	160	25	0.32	0.26	3.3	2
650	Large, without cheese, 1 sandwich	170	48	470	18	27	6.3	8.7	9.5	91	41	61	246	2.2	375	621	110	15	0.35	0.23	3.5	1
	Hamburger																					
651	Regular, 1 sandwich	98	46	245	12	11	4.4	5.3	0.5	32	28	56	107	2.2	202	463	80	14	0.23	0.24	3.8	1
652	4 oz patty, 1 sandwich	174	50	445	25	21	7.1	11.7	0.6	71	38	75	225	4.8	404	763	160	28	0.38	0.38	7.8	1
653	Pizza, cheese, $^{1}/_{8}$ of 15-in. diam. pizza[53], 1 slice	120	46	290	15	9	4.1	2.6	1.3	56	39	220	216	1.6	230	699	750	106	0.34	0.29	4.2	2
654	Roast beef sandwich, 1 sandwich	150	52	345	22	13	3.5	6.9	1.8	55	34	60	222	4.0	338	757	240	32	0.40	0.33	6.0	2
655	Taco, 1 taco	81	55	195	9	11	4.1	5.5	0.8	21	15	109	134	1.2	263	456	420	57	0.09	0.07	1.4	1

Poultry and Poultry Products

Item No.	Foods, approximate measures, units, and weight (weight of edible portion only)	Grams	Water (%)	Food energy (cal)	Protein (g)	Fat (g)	Fatty acids: Saturated (g)	Fatty acids: Monounsaturated (g)	Fatty acids: Polyunsaturated (g)	Cholesterol (mg)	Carbohydrate (g)	Calcium (mg)	Phosphorus (mg)	Iron (mg)	Potassium (mg)	Sodium (mg)	Vitamin A value (IU)[1]	Vitamin A value (RE)[2]	Thiamin (mg)	Riboflavin (mg)	Niacin	Ascorbic acid (mg)
	Chicken																					
	Fried, flesh, with skin[55]																					
	Batter dipped																					
656	Breast, $^{1}/_{2}$ breast (5.6 oz with bones), 4.9 oz	140	52	365	35	18	4.9	7.6	4.3	119	13	28	259	1.8	281	385	90	28	0.16	0.20	14.7	0
657	Drumstick (3.4 oz with bones), 2.5 oz	72	53	195	16	11	3.0	4.6	2.7	62	6	12	106	1.0	134	194	60	19	0.08	0.15	3.7	0
	Flour coated																					
658	Breast, $^{1}/_{2}$ breast (4.2 oz with bones), 3.5 oz	98	57	220	31	9	2.4	3.4	1.9	87	2	16	228	1.2	254	74	50	15	0.08	0.13	13.5	0
659	Drumstick (2.6 oz with bones), 1.7 oz	49	57	120	13	7	1.8	2.7	1.6	44	1	6	86	0.7	112	44	40	12	0.04	0.11	3.0	0
	Roasted, flesh only																					
660	Breast, $^{1}/_{2}$ breast (4.2 oz with bones and skin), 3.0 oz	86	65	140	27	3	0.9	1.1	0.7	73	0	13	196	0.9	220	64	20	5	0.06	0.10	11.8	0

Item No.	Foods, approximate measures, units, and weight (weight of edible portion only)	Grams	Water (%)	Food energy (cal)	Protein (g)	Fat (g)	Fatty acids: Saturated (g)	Fatty acids: Monounsaturated (g)	Fatty acids: Polyunsaturated (g)	Cholesterol (mg)	Carbohydrate (g)	Calcium (mg)	Phosphorus (mg)	Iron (mg)	Potassium (mg)	Sodium (mg)	Vitamin A value (IU)[1]	Vitamin A value (RE)[2]	Thiamin (mg)	Riboflavin (mg)	Niacin (mg)	Ascorbic acid (mg)
661	Drumstick (2.9 oz with bones and skin) 1.6 oz	44	67	75	12	2	0.7	0.8	0.6	41	0	5	81	0.6	108	42	30	8	0.03	0.10	2.7	0
662	Stewed, flesh only, light and dark meat, chopped or diced 1 cup	140	67	250	38	9	2.6	3.3	2.2	116	0	20	210	1.6	252	98	70	21	0.07	0.23	8.6	0
663	Chicken liver, cooked 1 liver	20	68	30	5	1	0.4	0.3	0.2	126	Tr	3	62	1.7	28	10	3270	983	0.03	0.35	0.9	3
664	Duck, roasted, flesh only 1/2 duck	221	64	445	52	25	9.2	8.2	3.2	197	0	27	449	6.0	557	144	170	51	0.57	1.04	11.3	0
	Turkey, roasted, flesh only																					
665	Dark meat, piece, 2 1/2 by 1 5/8 by 1/4 in. 4 pieces	85	63	160	24	6	2.1	1.4	1.8	72	0	27	173	2.0	246	67	0	0	0.05	0.21	3.1	0
666	Light meat, piece, 4 by 2 by 1/4 in. 2 pieces	85	66	135	25	3	0.9	0.5	0.7	59	0	16	186	1.1	259	54	0	0	0.05	0.11	5.8	0
	Light and dark meat																					
667	Chopped or diced 1 cup	140	65	240	41	7	2.3	1.4	2.0	106	0	35	298	2.5	417	98	0	0	0.09	0.25	7.6	0
668	Pieces (1 slice white meat, 4 by 2 by 1/4 in. and 2 slices dark meat, 2 1/2 by 1 5/8 by 1/4 in.) 3 pieces	85	65	145	25	4	1.4	0.9	1.2	65	0	21	181	1.5	253	60	0	0	0.05	0.15	4.6	0
	Poultry food products																					
	Chicken																					
669	Canned, boneless 5 oz	142	69	235	31	11	3.1	4.5	2.5	88	0	20	158	2.2	196	714	170	48	0.02	0.18	9.0	3
670	Frankfurter (10/1-lb pkg) 1 frankfurter	45	58	115	6	9	2.5	3.8	1.8	45	3	43	48	0.9	38	616	60	17	0.03	0.05	1.4	0
671	Roll, light (6 slices/6-oz pkg) 2 slices	57	69	90	11	4	1.1	1.7	0.9	28	1	24	89	0.6	129	331	50	14	0.04	0.07	3.0	0
	Turkey																					
672	Gravy and turkey, frozen 5-oz package	142	85	95	8	4	1.2	1.4	0.7	26	7	20	115	1.3	87	787	60	18	0.03	0.18	2.6	0
673	Ham, cured turkey thigh meat (8 slices/8-oz pkg) 2 slices	57	71	75	11	3	1.0	0.7	0.9	32	Tr	6	108	1.6	184	565	0	0	0.03	0.14	2.0	0
674	Loaf, breast meat (8 slices/6-oz pkg) 2 slices	42	72	45	10	1	0.2	0.2	0.1	17	0	3	97	0.2	118	608	0	0	0.02	0.05	3.5	0[56]

Item No.	Foods, approximate measures, units, and weight (weight of edible portion only)	Grams	Water (%)	Food energy (cal)	Protein (g)	Fat (g)	Fatty acids: Saturated (g)	Fatty acids: Monounsaturated (g)	Fatty acids: Polyunsaturated (g)	Cholesterol (mg)	Carbohydrate (g)	Calcium (mg)	Phosphorus (mg)	Iron (mg)	Potassium (mg)	Sodium (mg)	Vitamin A value (IU)[1]	Vitamin A value (RE)[2]	Thiamin (mg)	Riboflavin (mg)	Niacin	Ascorbic acid (mg)
675	Patties, breaded, battered, fried (2.25 oz) 1 patty	64	50	180	9	12	3.0	4.8	3.0	40	10	9	173	1.4	176	512	20	7	0.06	0.12	1.5	0
676	Roast, boneless, frozen, seasoned, light and dark meat, cooked 3 oz	85	68	130	18	5	1.6	1.0	1.4	45	3	4	207	1.4	253	578	0	0	0.04	0.14	5.3	0

Soups, Sauces, and Gravies

Item No.	Foods, approximate measures, units, and weight (weight of edible portion only)	Grams	Water (%)	Food energy (cal)	Protein (g)	Fat (g)	Fatty acids: Saturated (g)	Fatty acids: Monounsaturated (g)	Fatty acids: Polyunsaturated (g)	Cholesterol (mg)	Carbohydrate (g)	Calcium (mg)	Phosphorus (mg)	Iron (mg)	Potassium (mg)	Sodium (mg)	Vitamin A value (IU)[1]	Vitamin A value (RE)[2]	Thiamin (mg)	Riboflavin (mg)	Niacin	Ascorbic acid (mg)
	Soups																					
	Canned, condensed																					
	Prepared with equal volume of milk																					
677	Clam chowder, New England 1 cup	248	85	165	9	7	3.0	2.3	1.1	22	17	186	156	1.5	300	992	160	40	0.07	0.24	1.0	3
678	Cream of chicken 1 cup	248	85	190	7	11	4.6	4.5	1.6	27	15	181	151	0.7	273	1047	710	94	0.07	0.26	0.9	1
679	Cream of mushroom 1 cup	248	85	205	6	14	5.1	3.0	4.6	20	15	179	156	0.6	270	1076	150	37	0.08	0.28	0.9	2
680	Tomato 1 cup	248	85	160	6	6	2.9	1.6	1.1	17	22	159	149	1.8	449	932	850	109	0.13	0.25	1.5	68
	Canned, condensed																					
	Prepared with equal volume of water																					
681	Bean with bacon 1 cup	253	84	170	8	6	1.5	2.2	1.8	3	23	81	132	2.0	402	951	890	89	0.09	0.03	0.6	2
682	Beef broth, bouillon, consomme 1 cup	240	98	15	3	1	0.3	0.2	Tr	Tr	Tr	14	31	0.4	130	782	0	0	Tr	0.05	1.9	0
683	Beef noodle 1 cup	244	92	85	5	3	1.1	1.2	0.5	5	9	15	46	1.1	100	952	630	63	0.07	0.06	1.1	Tr
684	Chicken noodle 1 cup	241	92	75	4	2	0.7	1.1	0.6	7	9	17	36	0.8	55	1106	710	71	0.05	0.06	1.4	Tr
685	Chicken rice 1 cup	241	94	60	4	2	0.5	0.9	0.4	7	7	17	22	0.7	101	815	660	66	0.02	0.02	1.1	Tr
686	Clam chowder, Manhattan 1 cup	244	90	80	4	2	0.4	0.4	1.3	2	12	34	59	1.9	261	1808	920	92	0.06	0.05	1.3	3
687	Cream of chicken 1 cup	244	91	115	3	7	2.1	3.3	1.5	10	9	34	37	0.6	88	986	560	56	0.03	0.06	0.8	Tr
688	Cream of mushroom 1 cup	244	90	130	2	9	2.4	1.7	4.2	2	9	46	49	0.5	100	1032	0	0	0.05	0.09	0.7	1
689	Minestrone 1 cup	241	91	80	4	3	0.6	0.7	1.1	2	11	34	55	0.9	313	911	2340	234	0.05	0.04	0.9	1

Item No.	Foods, approximate measures, units, and weight (weight of edible portion only)	Grams	Water (%)	Food energy (cal)	Protein (g)	Fat (g)	Fatty acids: Saturated (g)	Fatty acids: Monounsaturated (g)	Fatty acids: Polyunsaturated (g)	Cholesterol (mg)	Carbohydrate (g)	Calcium (mg)	Phosphorus (mg)	Iron (mg)	Potassium (mg)	Sodium (mg)	Vitamin A value (IU)[1]	Vitamin A value (RE)[2]	Thiamin (mg)	Riboflavin (mg)	Niacin	Ascorbic acid (mg)
690	Pea, green, 1 cup	250	83	165	9	3	1.4	1.0	0.4	0	27	28	125	2.0	190	988	200	20	0.11	0.07	1.2	2
691	Tomato, 1 cup	244	90	85	2	2	0.4	0.4	1.0	0	17	12	34	1.8	264	871	690	69	0.09	0.05	1.4	66
692	Vegetable beef, 1 cup	244	92	80	6	2	0.9	0.8	0.1	5	10	17	41	1.1	173	956	1890	189	0.04	0.05	1.0	2
693	Vegetarian, 1 cup	241	92	70	2	2	0.3	0.8	0.7	0	12	22	34	1.1	210	822	3010	301	0.05	0.05	0.9	1
	Dehydrated																					
	Unprepared																					
694	Bouillon, 1 pkt	6	3	15	1	1	0.3	0.2	Tr	1	1	4	19	0.1	27	1019	Tr	Tr	Tr	0.01	0.3	0
695	Onion, 1 pkt	7	4	20	1	Tr	0.1	0.2	Tr	Tr	4	10	23	0.1	47	627	Tr	Tr	0.02	0.04	0.4	Tr
	Prepared with water																					
696	Chicken noodle, 1 pkt (6-fl-oz)	188	94	40	2	1	0.2	0.4	0.3	2	6	24	24	0.4	23	957	50	5	0.05	0.04	0.7	Tr
697	Onion, 1 pkt (6-fl-oz)	184	96	20	1	Tr	0.1	0.2	0.1	0	4	9	22	0.1	48	635	Tr	Tr	0.02	0.04	0.4	Tr
698	Tomato vegetable, 1 pkt (6-fl-oz)	189	94	40	1	1	0.3	0.2	0.1	0	8	6	23	0.5	78	856	140	14	0.04	0.03	0.6	5
	Sauces																					
	From dry mix																					
699	Cheese, prepared with milk, 1 cup	279	77	305	16	17	9.3	5.3	1.6	53	23	569	438	0.3	552	1565	390	117	0.15	0.56	0.3	2
700	Hollandaise, prepared with water, 1 cup	259	84	240	5	20	11.6	5.9	0.9	52	14	124	127	0.9	124	1564	730	220	0.05	0.18	0.1	Tr
701	White sauce, prepared with milk, 1 cup	264	81	240	10	13	6.4	4.7	1.7	34	21	425	256	0.3	444	797	310	92	0.08	0.45	0.5	3
	From home recipe																					
702	White sauce, medium[57], 1 cup	250	73	395	10	30	9.1	11.9	7.2	32	24	292	238	0.9	381	888	1190	340	0.15	0.43	0.8	2
	Ready to serve																					
703	Barbecue, 1 tbsp	16	81	10	Tr	Tr	Tr	0.1	0.1	0	2	3	3	0.1	28	130	140	14	Tr	Tr	0.1	1
704	Soy, 1 tbsp	18	68	10	2	0	0.0	0.0	0.0	0	2	3	38	0.5	64	1029	0	0	0.01	0.02	0.6	0
	Gravies																					
	Canned																					
705	Beef, 1 cup	233	87	125	9	5	2.7	2.3	0.2	7	11	14	70	1.6	189	117	0	0	0.07	0.08	1.5	0
706	Chicken, 1 cup	238	85	190	5	14	3.4	6.1	3.6	5	13	48	69	1.1	259	1373	880	264	0.04	0.10	1.1	0
707	Mushroom, 1 cup	238	89	120	3	6	1.0	2.8	2.4	0	13	17	36	1.6	252	1357	0	0	0.08	0.15	1.6	0

Item No.	Foods, approximate measures, units, and weight (weight of edible portion only)	Grams	Water (%)	Food energy (cal)	Protein (g)	Fat (g)	Fatty acids: Saturated (g)	Fatty acids: Mono-unsaturated (g)	Fatty acids: Poly-unsaturated (g)	Cholesterol (mg)	Carbo-hydrate (g)	Calcium (mg)	Phos-phorus (mg)	Iron (mg)	Potassium (mg)	Sodium (mg)	Vitamin A value (IU)[1]	Vitamin A value (RE)[2]	Thiamin (mg)	Riboflavin (mg)	Niacin	Ascorbic acid (mg)
	From dry mix																					
708	Brown																					
	1 cup	261	91	80	3	2	0.9	0.8	0.1	2	14	66	47	0.2	61	1147	0	0	0.04	0.09	0.9	0
709	Chicken																					
	1 cup	260	91	85	3	2	0.5	0.9	0.4	3	14	39	47	0.3	62	1134	0	0	0.05	0.15	0.8	3

Sugars and Sweets

Item No.	Foods, approximate measures, units, and weight (weight of edible portion only)	Grams	Water (%)	Food energy (cal)	Protein (g)	Fat (g)	Fatty acids: Saturated (g)	Fatty acids: Mono-unsaturated (g)	Fatty acids: Poly-unsaturated (g)	Cholesterol (mg)	Carbo-hydrate (g)	Calcium (mg)	Phos-phorus (mg)	Iron (mg)	Potassium (mg)	Sodium (mg)	Vitamin A value (IU)[1]	Vitamin A value (RE)[2]	Thiamin (mg)	Riboflavin (mg)	Niacin	Ascorbic acid (mg)
	Candy																					
710	Caramels, plain or chocolate																					
	1 oz	28	8	115	1	3	2.2	0.3	0.1	1	22	42	35	0.4	54	64	Tr	Tr	0.01	0.05	0.1	Tr
	Chocolate																					
711	Milk, plain																					
	1 oz	28	1	145	2	9	5.4	3.0	0.3	6	16	50	61	0.4	96	23	30	10	0.02	0.10	0.1	Tr
712	Milk, with almonds																					
	1 oz	28	2	150	3	10	4.8	4.1	0.7	5	15	65	77	0.5	125	23	30	8	0.02	0.12	0.2	Tr
713	Milk, with peanuts																					
	1 oz	28	1	155	4	11	4.2	3.5	1.5	5	13	49	83	0.4	138	19	30	8	0.07	0.07	1.4	Tr
714	Milk, with rice cereal																					
	1 oz	28	2	140	2	7	4.4	2.5	0.2	6	18	48	57	0.2	100	46	30	8	0.01	0.08	0.1	Tr
715	Semisweet, small pieces (60/oz)																					
	1 cup or 6 oz	170	1	860	7	61	36.2	19.9	1.9	0	97	51	178	5.8	593	24	30	3	0.10	0.14	0.9	Tr
716	Sweet (dark)																					
	1 oz	28	1	150	1	10	5.9	3.3	0.3	0	16	7	41	0.6	86	5	10	1	0.01	0.04	0.1	Tr
717	Fondant, uncoated (mints, candy corn, other)																					
	1 oz	28	3	105	Tr	0	0.0	0.0	0.0	0	27	2	Tr	0.1	1	57	0	0	Tr	Tr	Tr	0
718	Fudge, chocolate, plain																					
	1 oz	28	8	115	1	3	2.1	1.0	0.1	1	21	22	24	0.3	42	54	Tr	Tr	0.01	0.03	0.1	Tr
719	Gum drops																					
	1 oz	28	12	100	Tr	Tr	Tr	Tr	0.1	0	25	2	Tr	0.1	1	10	0	0	0.00	Tr	Tr	0
720	Hard																					
	1 oz	28	1	110	0	0	0.0	0.0	0.0	0	28	Tr	2	0.1	1	7	0	0	0.10	0.00	0.0	0
721	Jelly beans																					
	1 oz	28	6	105	Tr	Tr	Tr	Tr	0.1	0	26	1	1	0.3	11	7	0	0	0.00	Tr	Tr	0
722	Marshmallows																					
	1 oz	28	17	90	1	0	0.0	0.0	0.0	0	23	1	2	0.5	2	25	0	0	0.00	Tr	Tr	0
723	Custard, baked																					
	1 cup	265	77	305	14	15	6.8	5.4	0.7	278	29	297	310	1.1	387	209	530	146	0.11	0.50	0.3	1
724	Gelatin dessert prepared with gelatin dessert powder and water																					
	1/2 cup	120	84	70	2	0	0.0	0.0	0.0	0	17	2	23	Tr	Tr	55	0	0	0.00	0.00	0.0	0
725	Honey, strained or extracted																					
	1 cup	339	17	1030	1	0	0.0	0.0	0.0	0	279	17	20	1.7	173	17	0	0	0.02	0.14	1.0	3

Item No.	Foods, approximate measures, units, and weight (weight of edible portion only)	Grams	Water (%)	Food energy (cal)	Protein (g)	Fat (g)	Fatty acids: Saturated (g)	Fatty acids: Monounsaturated (g)	Fatty acids: Polyunsaturated (g)	Cholesterol (mg)	Carbohydrate (g)	Calcium (mg)	Phosphorus (mg)	Iron (mg)	Potassium (mg)	Sodium (mg)	Vitamin A value (IU)[1]	Vitamin A value (RE)[2]	Thiamin (mg)	Riboflavin (mg)	Niacin	Ascorbic acid (mg)
726	1 tbsp	21	17	65	Tr	0	0.0	0.0	0.0	0	17	1	1	0.1	11	1	0	0	Tr	0.01	0.1	Tr
727	Jams and preserves																					
	1 tbsp	20	29	55	Tr	Tr	0.0	Tr	Tr	0	14	4	2	0.2	18	2	Tr	Tr	Tr	0.01	Tr	Tr
728	1 packet	14	29	40	Tr	Tr	0.0	Tr	Tr	0	10	3	1	0.1	12	2	Tr	Tr	Tr	Tr	Tr	Tr
729	Jellies																					
	1 tbsp	18	28	50	Tr	Tr	Tr	Tr	Tr	0	13	2	Tr	0.1	16	5	Tr	Tr	Tr	0.01	Tr	1
730	1 packet	14	28	40	Tr	Tr	Tr	Tr	Tr	0	10	1	Tr	Tr	13	4	Tr	Tr	Tr	Tr	Tr	1
731	Popsicle, 3-fl-oz size																					
	1 popsicle	95	80	70	0	0	0.0	0.0	0.0	0	18	0	0	Tr	4	11	0	0	0.00	0.00	0.0	0
	Puddings																					
	Canned																					
732	Chocolate																					
	5-oz can	142	68	205	3	11	9.5	0.5	0.1	1	30	74	117	1.2	254	285	100	31	0.04	0.17	0.6	Tr
733	Tapioca																					
	5-oz can	142	74	160	3	5	4.8	Tr	Tr	Tr	28	119	113	0.3	212	252	Tr	Tr	0.03	0.14	0.4	Tr
734	Vanilla																					
	5-oz can	142	69	220	2	10	9.5	0.2	0.1	1	33	79	94	0.2	155	305	Tr	Tr	0.03	0.12	0.6	Tr
	Dry mix, prepared with whole milk																					
	Chocolate																					
735	Instant																					
	1/2 cup	130	71	155	4	4	2.3	1.1	0.2	14	27	130	329	0.3	176	440	130	33	0.04	0.18	0.1	1
736	Regular (cooked)																					
	1/2 cup	130	73	150	4	4	2.4	1.1	0.1	15	25	146	120	0.2	190	167	140	34	0.05	0.20	0.1	1
737	Rice																					
	1/2 cup	132	73	155	4	4	2.3	1.1	0.1	15	27	133	110	0.5	165	140	140	33	0.10	0.18	0.6	1
738	Tapioca																					
	1/2 cup	130	75	145	4	4	2.3	1.1	0.1	15	25	131	103	0.1	167	152	140	34	0.04	0.18	0.1	1
	Vanilla																					
739	Instant																					
	1/2 cup	130	73	150	4	4	2.2	1.1	0.2	15	27	129	273	0.1	164	375	140	33	0.04	0.17	0.1	1
740	Regular (cooked)																					
	1/2 cup	130	74	145	4	4	2.3	1.0	0.1	15	25	132	102	0.1	166	178	140	34	0.04	0.18	0.1	1
	Sugars																					
741	Brown, pressed down																					
	1 cup	220	2	820	0	0	0.0	0.0	0.0	0	199	3	Tr	0.1	7	5	0	0	0.00	0.00	0.0	0
	White																					
742	Granulated																					
	1 cup	200	1	770	0	0	0.0	0.0	0.0	0	12	Tr	Tr	Tr	Tr	Tr	0	0	0.00	0.00	0.0	0
743	1 tbsp	12	1	45	0	0	0.0	0.0	0.0	0	6	Tr	Tr	Tr	Tr	Tr	0	0	0.00	0.00	0.0	0
744	1 packet	6	1	25	0	0	0.0	0.0	0.0	0	100	1	Tr	Tr	4	2	0	0	0.00	0.00	0.0	0
745	Powdered, sifted, spooned into cup																					
	1 cup	100	1	385	0	0	0.0	0.0	0.0	0	22	6	49	0.8	85	36	Tr	Tr	Tr	0.02	0.1	0
	Syrups																					
	Chocolate-flavored syrup or topping																					
746	Thin type																					
	2 tbsp	38	37	85	1	Tr	0.2	0.1	0.1	0	21	38	60	0.5	82	42	40	13	0.02	0.08	0.1	0

Item No.	Foods, approximate measures, units, and weight (weight of edible portion only)	Grams	Water (%)	Food energy (cal)	Protein (g)	Fat (g)	Fatty acids: Saturated (g)	Fatty acids: Mono-unsaturated (g)	Fatty acids: Poly-unsaturated (g)	Cholesterol (mg)	Carbohydrate (g)	Calcium (mg)	Phosphorus (mg)	Iron (mg)	Potassium (mg)	Sodium (mg)	Vitamin A value (IU)[1]	Vitamin A value (RE)[2]	Thiamin (mg)	Riboflavin (mg)	Niacin	Ascorbic acid (mg)
747	Fudge type 2 tbsp	38	25	125	2	5	3.1	1.7	0.2	0	22	274	34	10.1	1171	38	0	0	0.04	0.08	0.8	0
748	Molasses, cane, blackstrap 2 tbsp	40	24	85	0	0	0.0	0.0	0.0	0	32	1	4	Tr	7	19	0	0	0.00	0.00	0.0	0
749	Table syrup (corn and maple) 2 tbsp	42	25	122	0	0	0.0	0.0	0.0	0	212	187	56	4.8	757	97	0	0	0.02	0.07	0.2	0

Vegetables and Vegetable Products

Item No.	Foods, approximate measures, units, and weight (weight of edible portion only)	Grams	Water (%)	Food energy (cal)	Protein (g)	Fat (g)	Fatty acids: Saturated (g)	Fatty acids: Mono-unsaturated (g)	Fatty acids: Poly-unsaturated (g)	Cholesterol (mg)	Carbohydrate (g)	Calcium (mg)	Phosphorus (mg)	Iron (mg)	Potassium (mg)	Sodium (mg)	Vitamin A value (IU)[1]	Vitamin A value (RE)[2]	Thiamin (mg)	Riboflavin (mg)	Niacin	Ascorbic acid (mg)
750	Alfalfa seeds, sprouted, raw 1 cup	33	91	10	1	Tr	Tr	Tr	0.1	0	1	11	23	0.3	26	2	50	5	0.03	0.04	0.2	3
751	Artichokes, globe or French, cooked, drained 1 artichoke	120	87	55	3	Tr	Tr	Tr	0.1	0	12	47	72	1.6	316	79	170	17	0.07	0.06	0.7	9
	Asparagus, green Cooked, drained From raw																					
752	Cuts and tips 1 cup	180	92	45	5	1	0.1	Tr	0.2	0	8	43	110	1.2	558	7	1490	149	0.18	0.22	1.9	49
753	Spears, 1/2-in. diam. at base 4 spears	60	92	15	2	Tr	Tr	Tr	0.1	0	3	14	37	0.4	186	2	500	50	0.06	0.07	0.6	16
	From frozen																					
754	Cuts and tips 1 cup	180	91	50	5	1	0.2	Tr	0.3	0	9	41	99	1.2	392	7	1470	147	0.12	0.19	1.9	44
755	Spears, 1/2-in. diam. at base 4 spears	60	91	15	2	Tr	0.1	Tr	0.1	0	3	14	33	0.4	131	2	490	49	0.04	0.06	0.6	15
756	Canned, spears, 1/2-in. diam. at base 4 spears	80	95	10	1	Tr	Tr	Tr	0.1	0	2	11	30	0.5	122	278[58]	380	38	0.04	0.07	0.7	13
757	Bamboo shoots, canned, drained 1 cup	131	94	25	2	1	0.1	Tr	0.2	0	4	10	33	0.4	105	9	10	1	0.03	0.03	0.2	1
	Beans Lima, immature seeds, frozen, cooked, drained																					
758	Thick-seeded types (Ford-hooks) 1 cup	170	74	170	10	1	0.1	Tr	0.3	0	32	37	107	2.3	694	90	320	32	0.13	0.10	1.8	22
759	Thin-seeded types (baby limas) 1 cup	180	72	190	12	1	0.1	Tr	0.3	0	35	50	202	3.5	740	52	300	30	0.13	0.10	1.4	10
	Snap Cooked, drained																					
760	From raw (cut and French style) 1 cup	125	89	45	2	Tr	0.1	Tr	0.2	0	10	58	49	1.6	374	4	830[59]	83[59]	0.09	0.12	0.8	12

Item No.	Foods, approximate measures, units, and weight (weight of edible portion only)	Grams	Water (%)	Food energy (cal)	Protein (g)	Fat (g)	Fatty acids: Saturated (g)	Fatty acids: Mono-unsaturated (g)	Fatty acids: Poly-unsaturated (g)	Cholesterol (mg)	Carbo-hydrate (g)	Calcium (mg)	Phos-phorus (mg)	Iron (mg)	Potassium (mg)	Sodium (mg)	Vitamin A value (IU)[1]	Vitamin A value (RE)[2]	Thiamin (mg)	Riboflavin (mg)	Niacin	Ascorbic acid (mg)
	Nutrients in indicated quantity																					
761	From frozen (cut) 1 cup	135	92	35	2	Tr	Tr	Tr	0.1	0	8	61	32	1.1	151	18	710[60]	71[illegible]	0.06	0.10	0.6	11
762	Canned, drained solids (cut) 1 cup	135	93	25	2	Tr	Tr	Tr	0.1	0	6	35	26	1.2	147	339[61]	470[62]	47[illegible]	0.02	0.08	0.3	6
	Beans, mature; see Beans, dry (items 527—535) and Black-eyed peas, dry (item 536)																					
	Bean sprouts (mung)																					
763	Raw 1 cup	104	90	30	3	Tr	Tr	Tr	0.1	0	6	14	56	0.9	155	6	20	[illegible]	0.09	0.13	0.8	14
764	Cooked, drained 1 cup	124	93	25	3	Tr	Tr	Tr	Tr	0	5	15	35	0.8	125	12	20	[illegible]	0.06	0.13	1.0	14
	Beets																					
	Cooked, drained																					
765	Diced or sliced 1 cup	170	91	55	2	Tr	Tr	Tr	Tr	0	11	19	53	1.1	530	83	20	[illegible]	0.05	0.02	0.5	9
766	Whole beets, 2-in. diam. 2 beets	100	91	30	1	Tr	Tr	Tr	Tr	0	7	11	31	0.6	312	49	10		0.03	0.01	0.3	6
767	Canned, drained solids, diced or sliced 1 cup	170	91	55	2	Tr	Tr	Tr	0.1	0	12	26	29	3.1	252	466[63]	20	[illegible]	0.02	0.07	0.3	7
768	Beet greens, leaves and stems, cooked, drained 1 cup	144	89	40	4	Tr	Tr	0.1	0.1	0	8	164	59	2.7	1309	347	7340	73[illegible]	0.17	0.42	0.7	36
	Black-eyed peas, immature seeds, cooked and drained																					
769	From raw 1 cup	165	72	180	13	1	0.3	0.1	0.6	0	30	46	196	2.4	693	7	1050	10[illegible]	0.11	0.18	1.8	3
770	From frozen 1 cup	170	66	225	14	1	0.3	0.1	0.5	0	40	39	207	3.6	638	9	130	1[illegible]	0.44	0.11	1.2	4
	Broccoli																					
771	Raw 1 spear	151	91	40	4	1	0.1	Tr	0.3	0	8	72	100	1.3	491	41	2330	23[illegible]	0.10	0.18	1.0	141
	Cooked, drained																					
	From raw																					
772	Spear, medium 1 spear	180	90	50	5	1	0.1	Tr	0.2	0	10	205	86	2.1	293	20	2540	25[illegible]	0.15	0.37	1.4	113
773	Spears, cut into $^1/_2$-in. pieces 1 cup	155	90	45	5	Tr	0.1	Tr	0.2	0	9	177	74	1.8	253	17	2180	21[illegible]	0.13	0.32	1.2	97
	From frozen																					
774	Piece, $4^1/_2$ to 5 in. long 1 piece	30	91	10	1	Tr	Tr	Tr	Tr	0	2	15	17	0.2	54	7	570	5[illegible]	0.02	0.02	0.1	12
775	Chopped 1 cup	185	91	50	6	Tr	Tr	Tr	0.1	0	10	94	102	1.1	333	44	3500	35[illegible]	0.10	0.15	0.8	74

Item No.	Foods, approximate measures, units, and weight (weight of edible portion only)	Grams	Water (%)	Food energy (cal)	Protein (g)	Fat (g)	Fatty acids: Saturated (g)	Fatty acids: Mono-unsaturated (g)	Fatty acids: Poly-unsaturated (g)	Cholesterol (mg)	Carbo-hydrate (g)	Calcium (mg)	Phos-phorus (mg)	Iron (mg)	Potassium (mg)	Sodium (mg)	Vitamin A value (IU)[1]	Vitamin A value (RE)[2]	Thiamin (mg)	Riboflavin (mg)	Niacin	Ascorbic acid (mg)
	Brussels sprouts, cooked, drained																					
776	From raw, 7—8 sprouts, $1^1/_4$ to $1^1/_2$-in. diam.																					
	1 cup	155	87	60	4	1	0.2	0.1	0.4	0	13	56	87	1.9	491	33	1110	111	0.17	0.12	0.9	96
777	From frozen																					
	1 cup	155	87	65	6	1	0.1	Tr	0.3	0	13	37	84	1.1	504	36	910	91	0.16	0.18	0.8	71
	Cabbage, common varieties																					
778	Raw, coarsely shredded or sliced																					
	1 cup	70	93	15	1	Tr	Tr	Tr	0.1	0	4	33	16	0.4	172	13	90	9	0.04	0.02	0.2	33
779	Cooked, drained																					
	1 cup	150	94	30	1	Tr	Tr	Tr	0.2	0	7	50	38	0.6	308	29	130	13	0.09	0.08	0.3	36
	Cabbage, Chinese																					
780	Pak-choi, cooked, drained																					
	1 cup	170	96	20	3	Tr	Tr	Tr	0.1	0	3	158	49	1.8	631	58	4370	437	0.05	0.11	0.7	44
781	Pe-tsai, raw, 1-in. pieces																					
	1 cup	76	94	10	1	Tr	Tr	Tr	0.1	0	2	59	22	0.2	181	7	910	91	0.03	0.04	0.3	21
782	Cabbage, red, raw, coarsely shredded or sliced																					
	1 cup	70	92	20	1	Tr	Tr	Tr	0.1	0	4	36	29	0.3	144	8	30	3	0.04	0.02	0.2	40
783	Cabbage, savoy, raw, coarsely shredded or sliced																					
	1 cup	70	91	20	1	Tr	Tr	Tr	Tr	0	4	25	29	0.3	161	20	700	70	0.05	0.02	0.2	22
	Carrots																					
	Raw, without crowns and tips, scraped																					
784	Whole, $7^1/_2$ by $1^1/_8$ in., or strips, $2^1/_2$ to 3 in. long																					
	1 carrot or 18 strips	72	88	30	1	Tr	Tr	Tr	0.1	0	7	19	32	0.4	233	25	20250	2025	0.07	0.04	0.7	7
785	Grated																					
	1 cup	110	88	45	1	Tr	Tr	Tr	0.1	0	11	30	48	0.6	355	39	30940	3094	0.11	0.06	1.0	10
	Cooked, sliced, drained																					
786	From raw																					
	1 cup	156	87	70	2	Tr	0.1	Tr	0.1	0	16	48	47	1.0	354	103	38300	3830	0.05	0.09	0.8	4
787	From frozen																					
	1 cup	146	90	55	2	Tr	Tr	Tr	0.1	0	12	41	38	0.7	231	86	25850	2585	0.04	0.05	0.6	4
788	Canned, sliced, drained solids																					
	1 cup	146	93	35	1	Tr	0.1	Tr	0.1	0	8	37	35	0.9	261	352[64]	20110	2011	0.03	0.04	0.8	4
	Cauliflower																					
789	Raw (flowerets)																					
	1 cup	100	92	25	2	Tr	Tr	Tr	0.1	0	5	29	46	0.6	355	15	20	2	0.08	0.06	0.6	72
	Cooked, drained																					
790	From raw (flowerets)																					
	1 cup	125	93	30	2	Tr	Tr	Tr	0.1	0	6	34	44	0.5	404	8	20	2	0.08	0.07	0.7	69
791	From frozen (flowerets)																					
	1 cup	180	94	35	3	Tr	0.1	Tr	0.2	0	7	31	43	0.7	250	32	40	4	0.07	0.10	0.6	56

Item No.	Foods, approximate measures, units, and weight (weight of edible portion only)	Grams	Water (%)	Food energy (cal)	Protein (g)	Fat (g)	Fatty acids: Saturated (g)	Fatty acids: Mono-unsaturated (g)	Fatty acids: Poly-unsaturated (g)	Cholesterol (mg)	Carbohydrate (g)	Calcium (mg)	Phosphorus (mg)	Iron (mg)	Potassium (mg)	Sodium (mg)	Vitamin A value (IU)[1]	Vitamin A value (RE)[2]	Thiamin (mg)	Riboflavin (mg)	Niacin	Ascorbic acid (mg)
	Celery, pascal type, raw																					
792	Stalk, large outer, 8 by $1^1/_2$ in. (at root end) 1 stalk	40	95	5	Tr	Tr	Tr	Tr	Tr	0	1	14	10	0.2	114	35	50	5	0.01	0.01	0.1	3
793	Pieces, diced 1 cup	120	95	20	1	Tr	Tr	Tr	0.1	0	4	43	31	0.6	341	106	150	15	0.04	0.04	0.4	8
	Collards, cooked, drained																					
794	From raw (leaves without stems) 1 cup	190	96	25	2	Tr	0.1	Tr	0.2	0	5	148	19	0.8	177	36	4220	422	0.03	0.08	0.4	19
795	From frozen (chopped) 1 cup	170	88	60	5	1	0.1	0.1	0.4	0	12	357	46	1.9	427	85	10170	1017	0.08	0.20	1.1	45
	Corn, sweet																					
	Cooked, drained																					
796	From raw, ear 5 by $1^3/_4$ in. 1 ear	77	70	85	3	1	0.2	0.3	0.5	0	19	2	79	0.5	192	13	170[65]	17[65]	0.17	0.06	1.2	5
	From frozen																					
797	Ear, trimmed to about $3^1/_2$ in. long 1 ear	63	73	60	2	Tr	0.1	0.1	0.2	0	14	2	47	0.4	158	3	130[65]	13[65]	0.11	0.04	1.0	3
798	Kernels 1 cup	165	76	135	5	Tr	Tr	Tr	0.1	0	34	3	78	0.5	229	8	410[65]	41[65]	0.11	0.12	2.1	4
	Canned																					
799	Cream style 1 cup	256	79	185	4	1	0.2	0.3	0.5	0	46	8	131	1.0	343	730[66]	250[65]	25[65]	0.06	0.14	2.5	12
800	Whole kernel, vacuum pack 1 cup	210	77	165	5	1	0.2	0.3	0.5	0	41	11	134	0.9	391	571[67]	510[65]	51[65]	0.09	0.15	2.5	17
	Cowpeas; see Black-eyed peas, immature (items 769, 770), mature (item 536)																					
801	Cucumber, with peel, slices, $^1/_8$ in. thick (large, $2^1/_8$-in. diam.; small, $1^3/_4$-in. diam.) 6 large or 8 small slices	28	96	5	Tr	Tr	Tr	Tr	Tr	0	1	4	5	0.1	42	1	10	1	0.01	0.01	0.1	1
802	Dandelion greens, cooked, drained 1 cup	105	90	35	2	1	0.1	Tr	0.3	0	7	147	44	1.9	244	46	12290	1229	0.14	0.18	0.5	19
803	Eggplant, cooked, steamed 1 cup	96	92	25	1	Tr	Tr	Tr	0.1	0	6	6	21	0.3	238	3	60	6	0.07	0.02	0.6	1
804	Endive, curly (including escarole), raw, small pieces 1 cup	50	94	10	1	Tr	Tr	Tr	Tr	0	2	26	14	0.4	157	11	1030	103	0.04	0.04	0.2	3
805	Jerusalem-artichoke, raw, sliced 1 cup	150	78	115	3	Tr	0.0	Tr	Tr	0	26	21	117	5.1	644	6	30	3	0.30	0.09	2.0	6

Item No.	Foods, approximate measures, units, and weight (weight of edible portion only)	Grams	Water (%)	Food energy (cal)	Protein (g)	Fat (g)	Fatty acids: Saturated (g)	Fatty acids: Monounsaturated (g)	Fatty acids: Polyunsaturated (g)	Cholesterol (mg)	Carbohydrate (g)	Calcium (mg)	Phosphorus (mg)	Iron (mg)	Potassium (mg)	Sodium (mg)	Vitamin A value (IU)[1]	Vitamin A value (RE)[2]	Thiamin (mg)	Riboflavin (mg)	Niacin	Ascorbic acid (mg)
	Kale, cooked, drained																					
806	From raw, chopped 1 cup	130	91	40	2	1	0.1	Tr	0.3	0	7	94	36	1.2	296	30	9620	962	0.07	0.09	0.7	53
807	From frozen, chopped 1 cup	130	91	40	4	1	0.1	Tr	0.3	0	7	179	36	1.2	417	20	8260	826	0.06	0.15	0.9	33
808	Kohlrabi, thickened bulb-like stems, cooked, drained, diced 1 cup	165	90	50	3	Tr	Tr	Tr	0.1	0	11	41	74	0.7	561	35	60	6	0.07	0.03	0.6	89
	Lettuce, raw																					
	Butterhead, as Boston types																					
809	Head, 5-in. diam. 1 head	163	96	20	2	Tr	Tr	Tr	0.2	0	4	52	38	0.5	419	8	1580	158	0.10	0.10	0.5	13
810	Leaves 1 outer or 2 inner leaves	15	96	Tr	Tr	Tr	Tr	Tr	Tr	0	Tr	5	3	Tr	39	1	150	15	0.01	0.01	Tr	1
	Crisphead, as iceberg																					
811	Head, 6-in. diam. 1 head	539	96	70	5	1	0.1	Tr	0.5	0	11	102	108	2.7	852	49	1780	178	0.25	0.16	1.0	21
812	Wedge, 1/4 of head 1 wedge	135	96	20	1	Tr	Tr	Tr	0.1	0	3	26	27	0.7	213	12	450	45	0.06	0.04	0.3	5
813	Pieces, chopped or shredded 1 cup	55	96	5	1	Tr	Tr	Tr	0.1	0	1	10	11	0.3	87	5	180	18	0.03	0.02	0.1	2
814	Looseleaf (bunching varieties including romaine or cos), chopped or shredded pieces 1 cup	56	94	10	1	Tr	Tr	Tr	0.1	0	2	38	14	0.8	148	5	1060	106	0.03	0.04	0.2	10
	Mushrooms																					
815	Raw, sliced or chopped 1 cup	70	92	20	1	Tr	Tr	Tr	0.1	0	3	4	73	0.9	259	3	0	0	0.07	0.31	2.9	2
816	Cooked, drained 1 cup	156	91	40	3	1	0.1	Tr	0.3	0	8	9	136	2.7	555	3	0	0	0.11	0.47	7.0	6
817	Canned, drained solids 1 cup	156	91	35	3	Tr	0.1	Tr	0.2	0	8	17	103	1.2	201	663	0	0	0.13	0.03	2.5	0
818	Mustard greens, without stems and midribs, cooked, drained 1 cup	140	94	20	3	Tr	Tr	0.2	0.1	0	3	104	57	1.0	283	22	4240	424	0.06	0.09	0.6	35
819	Okra pods, 3 by 5/8 in., cooked 8 pods	85	90	25	2	Tr	Tr	Tr	Tr	0	6	54	48	0.4	274	4	490	49	0.11	0.05	0.7	14
	Onions																					
	Raw																					
820	Chopped 1 cup	160	91	55	2	Tr	0.1	0.1	0.2	0	12	40	46	0.6	248	3	0	0	0.10	0.02	0.2	13

Item No.	Foods, approximate measures, units, and weight (weight of edible portion only)	Grams	Water (%)	Food energy (cal)	Protein (g)	Fat (g)	Fatty acids: Saturated (g)	Fatty acids: Monounsaturated (g)	Fatty acids: Polyunsaturated (g)	Cholesterol (mg)	Carbohydrate (g)	Calcium (mg)	Phosphorus (mg)	Iron (mg)	Potassium (mg)	Sodium (mg)	Vitamin A value (IU)[1]	Vitamin A value (RE)[2]	Thiamin (mg)	Riboflavin (mg)	Niacin	Ascorbic acid (mg)
821	Sliced 1 cup	115	91	40	1	Tr	0.1	Tr	0.1	0	8	29	33	0.4	178	2	0	0	0.07	0.01	0.1	10
822	Cooked (whole or sliced), drained 1 cup	210	92	60	2	Tr	0.1	Tr	0.1	0	13	57	48	0.4	319	17	0	0	0.09	0.02	0.2	12
823	Onions, spring, raw, bulb (3/8-in. diam.) and white portion of top 6 onions	30	92	10	1	Tr	Tr	Tr	Tr	0	2	18	10	0.6	77	1	1500	150	0.02	0.04	0.1	14
824	Onion rings, breaded, par-fried, frozen, prepared 2 rings	20	29	80	1	5	1.7	2.2	1.0	0	8	6	16	0.3	26	75	50	5	0.06	0.03	0.7	Tr
	Parsley																					
825	Raw 10 sprigs	10	88	5	Tr	Tr	Tr	Tr	Tr	0	1	13	4	0.6	54	4	520	52	0.01	0.01	0.1	9
826	Freeze-dried 1 tbsp	0.4	2	Tr	Tr	Tr	Tr	Tr	Tr	0	Tr	1	2	0.2	25	2	250	25	Tr	0.01	Tr	1
827	Parsnips, cooked (diced or 2 in. lengths), drained 1 cup	156	78	125	2	Tr	0.1	0.2	0.1	0	30	58	108	0.9	573	16	0	0	0.13	0.08	1.1	20
828	Peas, edible pod, cooked, drained 1 cup	160	89	65	5	Tr	0.1	Tr	0.2	0	11	67	88	3.2	384	6	210	21	0.20	0.12	0.9	77
	Peas, green																					
829	Canned, drained solids 1 cup	170	82	115	8	1	0.1	0.1	0.3	0	21	34	114	1.6	294	372[68]	1310	131	0.21	0.13	1.2	16
830	Frozen, cooked, drained 1 cup	160	80	125	8	Tr	0.1	Tr	0.2	0	23	38	144	2.5	269	139	1070	107	0.45	0.16	2.4	16
	Peppers																					
831	Hot chili, raw 1 pepper	45	88	20	1	Tr	Tr	Tr	Tr	0	4	8	21	0.5	153	3	4840[69]	484[69]	0.04	0.04	0.4	109
	Sweet (about 5/lb, whole), stem and seeds removed																					
832	Raw 1 pepper	74	93	20	1	Tr	Tr	Tr	0.2	0	4	4	16	0.9	144	2	390[70]	39[70]	0.06	0.04	0.4	95[71]
833	Cooked, drained 1 pepper	73	95	15	Tr	Tr	Tr	Tr	0.1	0	3	3	11	0.6	94	1	280[72]	28[72]	0.04	0.03	0.3	81[73]
	Potatoes, cooked Baked (about 2/lb, raw)																					
834	With skin 1 potato	202	71	220	5	Tr	0.1	Tr	0.1	0	51	20	115	2.7	844	16	0	0	0.22	0.07	3.3	26
835	Flesh only 1 potato	156	75	145	3	Tr	Tr	Tr	0.1	0	34	8	78	0.5	610	8	0	0	0.16	0.03	2.2	20
	Boiled (about 3/lb, raw)																					
836	Peeled after boiling 1 potato	136	77	120	3	Tr	Tr	Tr	0.1	0	27	7	60	0.4	515	5	0	0	0.14	0.03	2.0	18
837	Peeled before boiling 1 potato	135	77	115	2	Tr	Tr	Tr	0.1	0	27	11	54	0.4	443	7	0	0	0.13	0.03	1.8	10

Item No.	Foods, approximate measures, units, and weight (weight of edible portion only)	Grams	Water (%)	Food energy (cal)	Protein (g)	Fat (g)	Fatty acids: Saturated (g)	Fatty acids: Monounsaturated (g)	Fatty acids: Polyunsaturated (g)	Cholesterol (mg)	Carbohydrate (g)	Calcium (mg)	Phosphorus (mg)	Iron (mg)	Potassium (mg)	Sodium (mg)	Vitamin A value (IU)[1]	Vitamin A value (RE)[2]	Thiamin (mg)	Riboflavin (mg)	Niacin	Ascorbic acid (mg)
	French fried, strip, 2 to $3^1/_2$ in. long, frozen																					
838	Oven heated 10 strips	50	53	110	2	4	2.1	1.8	0.3	0	17	5	43	0.7	229	16	0	0	0.06	0.02	1.2	5
839	Fried in vegetable oil 10 strips	50	38	160	2	8	2.5	1.6	3.8	0	20	10	47	0.4	366	108	0	0	0.09	0.01	1.6	5
	Potato products, prepared																					
	Au gratin																					
840	From dry mix 1 cup	245	79	230	6	10	6.3	2.9	0.3	12	31	203	233	0.8	537	1076	520	76	0.05	0.20	2.3	8
841	From home recipe 1 cup	245	74	325	12	19	11.6	5.3	0.7	56	28	292	277	1.6	970	1061	650	93	0.16	0.28	2.4	24
842	Hashed brown, from frozen 1 cup	156	56	340	5	18	7.0	8.0	2.1	0	44	23	112	2.4	680	53	0	0	0.17	0.03	3.8	10
	Mashed																					
	From home recipe																					
843	Milk added 1 cup	210	78	160	4	1	0.7	0.3	0.1	4	37	55	101	0.6	628	636	40	12	0.18	0.08	2.3	14
844	Milk and margarine added 1 cup	210	76	225	4	9	2.2	3.7	2.5	4	35	55	97	0.5	607	620	360	42	0.18	0.08	2.3	13
845	From dehydrated flakes (without milk), water, milk, butter, and salt added 1 cup	210	76	235	4	12	7.2	3.3	0.5	29	32	103	118	0.5	489	697	380	44	0.23	0.11	1.4	20
846	Potato salad, made with mayonnaise 1 cup	250	76	360	7	21	3.6	6.2	9.3	170	28	48	130	1.6	635	1323	520	83	0.19	0.15	2.2	25
	Scalloped																					
847	From dry mix 1 cup	245	79	230	5	11	6.5	3.0	0.5	27	31	88	137	0.9	497	835	360	51	0.05	0.14	2.5	8
848	From home recipe 1 cup	245	81	210	7	9	5.5	2.5	0.4	29	26	140	154	1.4	926	821	330	47	0.17	0.23	2.6	26
849	Potato chips 10 chips	20	3	105	1	7	1.8	1.2	3.6	0	10	5	31	0.2	260	94	0	0	0.03	Tr	0.8	8
	Pumpkin																					
850	Cooked from raw, mashed 1 cup	245	94	50	2	Tr	0.1	Tr	Tr	0	12	37	74	1.4	564	2	2650	265	0.08	0.19	1.0	12
851	Canned 1 cup	245	90	85	3	1	0.4	0.1	Tr	0	20	64	86	3.4	505	12	54040	5404	0.06	0.13	0.9	10
852	Radishes, raw, stem ends, rootlets cut off 4 radishes	18	95	5	Tr	Tr	Tr	Tr	Tr	0	1	4	3	0.1	42	4	Tr	Tr	Tr	0.01	0.1	4
853	Sauerkraut, canned, solids and liquid 1 cup	236	93	45	2	Tr	0.1	Tr	0.1	0	10	71	47	3.5	401	1560	40	4	0.05	0.05	0.3	35

Item No.	Foods, approximate measures, units, and weight (weight of edible portion only)	Grams	Water (%)	Food energy (cal)	Protein (g)	Fat (g)	Fatty acids: Saturated (g)	Fatty acids: Mono-unsaturated (g)	Fatty acids: Poly-unsaturated (g)	Cholesterol (mg)	Carbohydrate (g)	Calcium (mg)	Phosphorus (mg)	Iron (mg)	Potassium (mg)	Sodium (mg)	Vitamin A value (IU)[1]	Vitamin A value (RE)[2]	Thiamin (mg)	Riboflavin (mg)	Niacin	Ascorbic acid (mg)
	Seaweed																					
854	Kelp, raw 1 oz	28	82	10	Tr	Tr	0.1	Tr	Tr	0	3	48	12	0.8	25	66	30	3	0.01	0.04	0.1	1
855	Spirulina, dried 1 oz	28	5	80	16	2	0.8	0.2	0.6	0	7	34	33	8.1	386	297	160	16	0.67	1.04	3.6	3
	Southern peas; see Black-eyed peas, immature (items 769, 770), mature (item 536)																					
	Spinach																					
856	Raw, chopped 1 cup	55	92	10	2	Tr	Tr	Tr	0.1	0	2	54	27	1.5	307	43	3690	369	0.04	0.10	0.4	15
	Cooked, drained																					
857	From raw 1 cup	180	91	40	5	Tr	0.1	Tr	0.2	0	7	245	101	6.4	839	126	14740	1474	0.17	0.42	0.9	18
858	From frozen (leaf) 1 cup	190	90	55	6	Tr	0.1	Tr	0.2	0	10	277	91	2.9	566	163	14790	1479	0.11	0.32	0.8	23
859	Canned, drained solids 1 cup	214	92	50	6	1	0.2	Tr	0.4	0	7	272	94	4.9	740	683[74]	18780	1878	0.03	0.30	0.8	31
860	Spinach souffle 1 cup	136	74	220	11	18	7.1	6.8	3.1	184	3	230	231	1.3	201	763	3460	675	0.09	0.30	0.5	3
	Squash, cooked																					
861	Summer (all varieties), sliced, drained 1 cup	180	94	35	2	1	0.1	Tr	0.2	0	8	49	70	0.6	346	2	520	52	0.08	0.07	0.9	10
862	Winter (all varieties), baked, cubes 1 cup	205	89	80	2	1	0.3	0.1	0.5	0	18	29	41	0.7	896	2	7290	729	0.17	0.05	1.4	20
	Sunchoke; see Jerusalem-artichoke (item 805)																					
	Sweetpotatoes																					
	Cooked (raw, 5 by 2 in., about $2^1/_2$/lb)																					
863	Baked in skin, peeled 1 potato	114	73	115	2	Tr	Tr	Tr	0.1	0	28	32	63	0.5	397	11	24880	2488	0.08	0.14	0.7	28
864	Boiled, without skin 1 potato	151	73	160	2	Tr	0.1	Tr	0.2	0	37	32	41	0.8	278	20	25750	2575	0.08	0.21	1.0	26
865	Candied, $2^1/_2$ by 2-in. piece 1 piece	105	67	145	1	3	1.4	0.7	0.2	8	29	27	27	1.2	198	74	4400	440	0.02	0.04	0.4	7
	Canned																					
866	Solid pack (mashed) 1 cup	255	74	260	5	1	0.1	Tr	0.2	0	59	77	133	3.4	536	191	38570	3857	0.07	0.23	2.4	13
867	Vacuum pack, piece $2^3/_4$ by 1 in. 1 piece	40	76	35	1	Tr	Tr	Tr	Tr	0	8	9	20	0.4	125	21	3190	319	0.01	0.02	0.3	11
	Tomatoes																					
868	Raw, $2^3/_5$-in. diam. (3/12-oz pkg.) 1 tomato	123	94	25	1	Tr	Tr	Tr	0.1	0	5	9	28	0.6	255	10	1390	139	0.07	0.06	0.7	22

Item No.	Foods, approximate measures, units, and weight (weight of edible portion only)	Grams	Water (%)	Food energy (cal)	Protein (g)	Fat (g)	Fatty acids: Saturated (g)	Fatty acids: Mono-unsaturated (g)	Fatty acids: Poly-unsaturated (g)	Cholesterol (mg)	Carbo-hydrate (g)	Calcium (mg)	Phos-phorus (mg)	Iron (mg)	Potassium (mg)	Sodium (mg)	Vitamin A value (IU)[1]	Vitamin A value (RE)[2]	Thiamin (mg)	Riboflavin (mg)	Niacin	Ascorbic acid (mg)
869	Canned, solids and liquid																					
	1 cup	240	94	50	2	1	0.1	0.1	0.2	0	10	62	46	1.5	530	391[75]	1450	145	0.11	0.07	1.8	36
870	Tomato juice, canned																					
	1 cup	244	94	40	2	Tr	Tr	Tr	0.1	0	10	22	46	1.4	537	881[76]	1360	136	0.11	0.08	1.6	45
	Tomato products, canned																					
871	Paste																					
	1 cup	262	74	220	10	2	0.3	0.4	0.9	0	49	92	207	7.8	2442	170[77]	6470	647	0.41	0.50	8.4	111
872	Puree																					
	1 cup	250	87	105	4	Tr	Tr	Tr	0.1	0	25	38	100	2.3	1050	50[78]	3400	340	0.18	0.14	4.3	88
873	Sauce																					
	1 cup	245	89	75	3	Tr	0.1	0.1	0.2	0	18	34	78	1.9	909	1482[79]	2400	240	0.16	0.14	2.8	32
874	Turnips, cooked, diced																					
	1 cup	156	94	30	1	Tr	Tr	Tr	0.1	0	8	34	30	0.3	211	78	0	0	0.04	0.04	0.5	18
	Turnip greens, cooked, drained																					
875	From raw (leaves and stems)																					
	1 cup	144	93	30	2	Tr	0.1	Tr	0.1	0	6	197	42	1.2	292	42	7920	792	0.06	0.10	0.6	39
876	From frozen (chopped)																					
	1 cup	164	90	50	5	1	0.2	Tr	0.3	0	8	249	56	3.2	367	25	13080	1308	0.09	0.12	0.8	36
877	Vegetable juice cocktail, canned																					
	1 cup	242	94	45	2	Tr	Tr	Tr	0.1	0	11	27	41	1.0	467	883	2830	283	0.10	0.07	1.8	67
	Vegetables, mixed																					
878	Canned, drained solids																					
	1 cup	163	87	75	4	Tr	0.1	Tr	0.2	0	15	44	68	1.7	474	243	18990	1899	0.08	0.08	0.9	8
879	Frozen, cooked, drained																					
	1 cup	182	83	105	5	Tr	0.1	Tr	0.1	0	24	46	93	1.5	308	64	7780	778	0.13	0.22	1.5	6
880	Waterchestnuts, canned																					
	1 cup	140	86	70	1	Tr	Tr	Tr	Tr	0	17	6	27	1.2	165	11	10	1	0.02	0.03	0.5	2

Miscellaneous Items

Item No.	Foods, approximate measures, units, and weight (weight of edible portion only)	Grams	Water (%)	Food energy (cal)	Protein (g)	Fat (g)	Fatty acids: Saturated (g)	Fatty acids: Mono-unsaturated (g)	Fatty acids: Poly-unsaturated (g)	Cholesterol (mg)	Carbo-hydrate (g)	Calcium (mg)	Phos-phorus (mg)	Iron (mg)	Potassium (mg)	Sodium (mg)	Vitamin A value (IU)[1]	Vitamin A value (RE)[2]	Thiamin (mg)	Riboflavin (mg)	Niacin	Ascorbic acid (mg)
	Baking powders for home use																					
	Sodium aluminum sulfate																					
881	With monocalcium phosphate monohydrate																					
	1 tsp	3	2	5	Tr	0	0.0	0.0	0.0	0	1	58	87	0.0	5	329	0	0	0.00	0.00	0.0	0
882	With monocalcium phosphate monohydrate, calcium sulfate																					
	1 tsp	2.9	1	5	Tr	0	0.0	0.0	0.0	0	1	183	45	0.0	4	290	0	0	0.00	0.00	0.0	0
883	Straight phosphate																					
	1 tsp	3.8	2	5	Tr	0	0.0	0.0	0.0	0	1	239	359	0.0	6	312	0	0	0.00	0.00	0.0	0

							Nutrients in indicated quantity															
							Fatty acids										Vitamin A value					
Item No.	Foods, approximate measures, units, and weight (weight of edible portion only)	Grams	Water (%)	Food energy (cal)	Protein (g)	Fat (g)	Saturated (g)	Mono-unsatu-rated (g)	Poly-unsatu-rated (g)	Cholesterol (mg)	Carbo-hydrate (g)	Calcium (mg)	Phos-phorus (mg)	Iron (mg)	Potassium (mg)	Sodium (mg)	(IU)[1]	(RE)[2]	Thiamin (mg)	Riboflavin (mg)	Niacin	Ascorbic acid (mg)
884	Low sodium 1 tsp	4.3	1	5	Tr	0	0.0	0.0	0.0	0	1	207	314	0.0	891	Tr	0	0	0.00	0.00	0.0	0
885	Catsup 1 cup	273	69	290	5	1	0.2	0.2	0.4	0	69	60	137	2.2	991	2845	3820	382	0.25	0.19	4.4	41
886	1 tbsp	15	69	15	Tr	Tr	Tr	Tr	Tr	0	4	3	8	0.1	54	156	210	21	0.01	0.01	0.2	2
887	Celery seed 1 tsp	2	6	10	Tr	1	Tr	0.3	0.1	0	1	35	11	0.9	28	3	Tr	Tr	0.01	0.01	0.1	Tr
888	Chili powder 1 tsp	2.6	8	10	Tr	Tr	0.1	0.1	0.2	0	1	7	8	0.4	50	26	910	91	0.01	0.02	0.2	2
	Chocolate																					
889	Bitter or baking 1 oz	28	2	145	3	15	9.0	4.9	0.5	0	8	22	109	1.9	235	1	10	1	0.01	0.07	0.4	0
	Semisweet; see Candy (item 715)																					
890	Cinnamon 1 tsp	2.3	10	5	Tr	Tr	Tr	Tr	Tr	0	2	28	1	0.9	12	1	10	1	Tr	Tr	Tr	1
891	Curry powder 1 tsp	2	10	5	Tr	Tr	[3]	[3]	[3]	0	1	10	7	0.6	31	1	20	2	0.01	0.01	0.1	Tr
892	Garlic powder 1 tsp	2.8	6	10	Tr	Tr	Tr	Tr	Tr	0	2	2	12	0.1	31	1	0	0	0.01	Tr	Tr	Tr
893	Gelatin, dry 1 envelope	7	13	25	6	Tr	Tr	Tr	Tr	0	0	1	0	0.0	2	6	0	0	0.00	0.00	0.0	0
894	Mustard, prepared, yellow 1 tsp or individual packet	5	80	5	Tr	Tr	Tr	0.2	Tr	0	Tr	4	4	0.1	7	63	0	0	Tr	0.01	Tr	Tr
	Olives, canned																					
895	Green 4 medium or 3 extra large	13	78	15	Tr	2	0.2	1.2	0.1	0	Tr	8	2	0.2	7	312	40	4	Tr	Tr	Tr	0
896	Ripe, Mission, pitted 3 small or 2 large	9	73	15	Tr	2	0.3	1.3	0.2	0	Tr	10	2	0.2	2	68	10	1	Tr	Tr	Tr	0
897	Onion powder 1 tsp	2.1	5	5	Tr	Tr	Tr	Tr	Tr	0	2	8	7	0.1	20	1	Tr	Tr	0.01	Tr	Tr	Tr
898	Oregano 1 tsp	1.5	7	5	Tr	Tr	Tr	Tr	0.1	0	1	24	3	0.7	25	Tr	100	10	0.01	Tr	0.1	1
899	Paprika 1 tsp	2.1	10	5	Tr	Tr	Tr	Tr	0.2	0	1	4	7	0.5	49	1	1270	127	0.01	0.04	0.3	1
900	Pepper, black 1 tsp	2.1	11	5	Tr	Tr	Tr	Tr	Tr	0	1	9	4	0.6	26	1	Tr	Tr	Tr	0.01	Tr	0
	Pickles, cucumber																					
901	Dill, medium, whole, 3 3/4 in. long, 1 1/4-in. diam. 1 pickle	65	93	5	Tr	Tr	Tr	Tr	0.1	0	1	17	14	0.7	130	928	70	7	Tr	0.01	Tr	4
902	Fresh-pack, slices 1 1/2-in. diam., 1/4 in. thick 2 slices	15	79	10	Tr	Tr	Tr	Tr	Tr	0	3	5	4	0.3	30	101	20	2	Tr	Tr	Tr	1
903	Sweet, gherkin, small, whole, about 2 1/2 in. long, 3/4-in. diam. 1 pickle	15	61	20	Tr	Tr	Tr	Tr	Tr	0	5	2	2	0.2	30	107	10	1	Tr	Tr	Tr	1

Item No.	Foods, approximate measures, units, and weight (weight of edible portion only)	Grams	Water (%)	Food energy (cal)	Protein (g)	Fat (g)	Fatty acids: Saturated (g)	Fatty acids: Mono-unsaturated (g)	Fatty acids: Poly-unsaturated (g)	Cholesterol (mg)	Carbohydrate (g)	Calcium (mg)	Phosphorus (mg)	Iron (mg)	Potassium (mg)	Sodium (mg)	Vitamin A value (IU)[1]	Vitamin A value (RE)[2]	Thiamin (mg)	Riboflavin (mg)	Niacin (mg)	Ascorbic acid (mg)
	Popcorn; see Grain Products (items 497—499)																					
904	Relish, finely chopped, sweet																					
	1 tbsp	15	63	20	Tr	Tr	Tr	Tr	Tr	0	5	3	2	0.1	30	107	20	2	Tr	Tr	0.0	1
905	Salt																					
	1 tsp	5.5	0	0	0	0	0.0	0.0	0.0	0	0	14	3	Tr	Tr	2132	0	0	0.00	0.00	0.0	0
906	Vinegar, cider																					
	1 tbsp	15	94	Tr	Tr	0	0.0	0.0	0.0	0	1	1	1	0.1	15	Tr	0	0	0.00	0.00	0.0	0
	Yeast																					
907	Baker's, dry, active																					
	1 pkg	7	5	20	3	Tr	Tr	0.1	Tr	0	3	3	90	1.1	140	4	Tr	Tr	0.16	0.38	2.6	Tr
908	Brewer's, dry																					
	1 tbsp	8	5	25	3	Tr	Tr	Tr	0.0	0	3	17[80]	140	1.4	152	10	Tr	Tr	1.25	0.34	3.0	Tr

NOTE: Tr indicates the nutrient is present in trace amounts.

1. International units.
2. Retinol equivalents.
3. Value not determined.
4. Mineral content varies with source of water.
5. Blend of aspartame and saccharin; if only saccharin is used, sodium is 75 mg; if only aspartame is used, sodium is 23 mg.
6. With added ascorbic acid.
7. Vitamin A value is largely from beta-carotene used for coloring.
8. Yields 1 qt of fluid milk when reconstituted according to package directions.
9. With added Vitamin A.
10. Carbohydrate content varies widely because of the amount of sugar added and the amount and solids content of added flavoring. Consult the label if more precise values for carbohydrates and calories are needed.
11. For salted butter; unsalted butter contains 12 mg sodium per stick, 2 mg per tablespoon, or 1 mg per pat.
12. Values for Vitamin A are year-round averages.
13. For salted margarine.
14. Based on average Vitamin A content for fortified margarine. Federal specifications for fortified margarine require a minimum of 15,000 IU per pound.
15. Fatty acid values apply to product made with regular margarine.
16. Dipped in egg, milk, and breadcrumbs; fried in vegetable shortening.
17. If bones are discarded, value for calcium will be greatly reduced.
18. Dipped in egg, breadcrumbs, and flour; fried in vegetable shortening.
19. Made with drained tuna, celery, onion, pickle relish, and mayonnaise type dressing.
20. Sodium bisulfite used to preserve color; unsulfited product would contain less sodium.
21. Also applies to pressurized apple cider.
22. Without ascorbic acid. For value with added ascorbic acid, refer to label.
23. With added ascorbic acid.
24. For white grapefruit; pink grapefruit have about 310 IU or 31 RE.
25. Sodium benzoate and sodium bisulfite added as preservatives.
26. Egg bagels have 44 mg cholesterol and 22 IU or 7 RE vitamin A per bagel.
27. Made with vegetable shortening.
28. Made with white cornmeal. If made with yellow cornmeal, value is 32 IU or 3 RE.
29. Nutrient added.
30. Cooked without salt. If salt is added according to label recommendations, sodium content is 540 mg.
31. For white corn grits. Cooked yellow grits contain 145 IU or 14 RE.
32. Value based on label declaration for added nutrients.

[33] For regular and instant cereal. For quick cereal, phosphorus is 102 mg and sodium is 142 mg.
[34] Cooked without salt. If salt is added according to label recommendations, sodium content is 390 mg.
[35] Cooked without salt. If salt is added according to label recommendations, sodium content is 324 mg.
[36] Cooked without salt. If salt is added according to label recommendations, sodium content is 374 mg.
[37] Excepting angelfood cake, cakes were made from mixes containing vegetable shortening and frostings were made with margarine.
[38] Made with vegetable oil.
[39] Cake made with vegetable shortening; frosting with margarine.
[40] Made with margarine.
[41] Crackers made with enriched flour except for rye wafers and whole-wheat wafers.
[42] Made with lard.
[43] Cashews without salt contain 21 mg sodium per cup or 4 mg per oz.
[44] Cashews without salt contain 22 mg sodium per cup or 5 mg per oz.
[45] Macadamia nuts without salt contain 9 mg sodium per cup or 2 mg per oz.
[46] Mixed nuts without salt contain 3 mg sodium per oz.
[47] Peanuts without salt contain 22 mg sodium per cup or 4 mg per oz.
[48] Outer layer of fat was removed to within approximately $^1/_2$ in. of the lean. Deposits of fat within the cut were not removed.
[49] Fried in vegetable shortening.
[50] Value varies widely.
[51] Contains added sodium ascorbate. If sodium ascorbate is not added, ascorbic acid content is negligible.
[52] One patty (8 per pound) of bulk sausage is equivalent to 2 links.
[53] Crust made with vegetable shortening and enriched flour.
[54] Made with corn oil.
[55] Fried in vegetable shortening.
[56] If sodium ascorbate is added, product contains 11 mg ascorbic acid.
[57] Made with enriched flour, margarine, and whole milk.
[58] For regular pack; special dietary pack contains 3 mg sodium.
[59] For green varieties; yellow varieties contain 101 IU or 10 RE.
[60] For green varieties; yellow varieties contain 151 IU or 15 RE.
[61] For regular pack; special dietary pack contains 3 mg sodium.
[62] For green varieties; yellow varieties contain 142 IU or 14 RE.
[63] For regular pack; special dietary pack contains 78 mg sodium.
[64] For regular pack; special dietary pack contains 61 mg sodium.
[65] For yellow varieties; white varieties contain only a trace of vitamin A.
[66] For regular pack; special dietary pack contains 8 mg sodium.
[67] For regular pack; special dietary pack contains 6 mg sodium.
[68] For regular pack; special dietary pack contains 3 mg sodium.
[69] For red peppers; green peppers contain 350 IU or 35 RE.
[70] For green peppers; red peppers contain 4220 IU or 422 RE.
[71] For green peppers; red peppers contain 141 mg ascorbic acid.
[72] For green peppers; red peppers contain 2740 IU or 274 RE.
[73] For green peppers; red peppers contain 121 mg ascorbic acid.
[74] With added salt; if none is added, sodium content is 58 mg.
[75] For regular pack; special dietary pack contains 31 mg sodium.
[76] With added salt; if none is added, sodium content is 24 mg.
[77] With no added salt; if salt is added, sodium content is 2070 mg.
[78] With no added salt; if salt is added, sodium content is 998 mg.
[79] With salt added.
[80] Value may vary from 6 to 60 mg.

LOWERING OF VAPOR PRESSURE BY SALTS IN AQUEOUS SOLUTIONS

The table gives the reduction of the vapor pressure in millimeters due to the presence of the number of grammolecules of salt per liter of water given at the head of the columns, at the temperature 100° C, at which temperature the vapor pressure of pure water is 760 millimeters.

(From Smithsonian Tables.)

Substance	0.5	1.0	2.0	3.0	4.0	5.0	6.0	8.0	10.0
$Al_2(SO_4)_3$	12.8	36.5							
$AlCl_3$	22.5	61.0	179.0	318.0					
BaS_2O_6	6.6	15.4	34.4						
$Ba(OH)_2$	12.3	22.5	30.0						
$Ba(NO_3)_2$	13.5	27.0							
$Ba(ClO_3)_2$	15.8	33.3	70.5	108.2					
$BaCl_2$	16.4	36.7	77.6						
$BaBr_2$	16.8	38.8	91.4	150.0	204.7				
CaS_2O_8	9.9	23.0	56.0	106.0					
$Ca(NO_3)_2$	16.4	34.8	74.6	139.3	161.7	205.4			
$CaCl_2$	17.0	39.8	95.3	166.6	241.5	319.5			
$CaBr_2$	17.7	44.2	105.8	191.0	283.3	368.5			
$CdSO_4$	4.1	8.9	18.1						
CdI_2	7.6	14.8	33.5	52.7					
$CdBr_2$	8.6	17.8	36.7	55.7	80.0				
$CdCl_2$	9.6	18.8	36.7	57.0	77.3	99.0			
$Cd(NO_3)_2$	15.9	36.1	78.0	122.2					
$Cd(ClO_3)_2$	17.5								
$CoSO_4$	5.5	10.7	22.9	45.5					
$CoCl_2$	15.0	34.8	83.0	136.0	186.4				
$Co(NO_3)_2$	17.3	39.2	89.0	152.0	218.7	282.0	332.0		
$FeSO_4$	5.8	10.7	24.0	42.4					
H_3BO_3	6.0	12.3	25.1	38.0	51.0				
H_3PO_4	6.6	14.0	28.6	45.2	62.0	81.5	103.0	146.9	189.5
H_3AsO_4	7.3	15.0	30.2	46.4	64.9				
H_2SO_4	12.9	26.5	62.8	104.0	148.0	198.4	247.0	343.2	
KH_2PO_4	10.2	19.5	33.3	47.8	60.5	73.1	85.2		
KNO_3	10.3	21.1	40.1	57.6	74.5	88.2	102.1	126.3	148.0
$KClO_3$	10.6	21.6	42.8	62.1	80.0				
$KBrO_3$	10.9	22.4	45.0						
$KHSO_4$	10.9	21.9	43.3	65.3	85.5	107.8	129.9	170.0	
KNO_2	11.1	22.8	44.8	67.0	90.0	110.5	130.7	167.0	198.8
$KClO_4$	11.5	22.3							
KCl	12.2	24.4	48.8	74.1	100.9	128.5	152.2		
$KHCO_3$	11.6	23.6	59.0	77.6	104.2	132.0	160.0	210.0	255.0
KI	12.5	25.3	52.2	82.6	112.2	141.5	171.8	225.5	278.5
$K_2C_2O_4$	13.9	28.3	59.8	94.2	131.0				
K_2WO_4	13.9	33.0	75.0	123.8	175.4	226.4			
K_2CO_3	14.4	31.0	68.3	105.5	152.0	209.0	258.5	350.0	
KOH	15.0	29.5	64.0	99.2	140.0	181.8	223.0	309.5	387.8
K_2CrO_4	16.2	29.5	60.0						
$LiNO_3$	12.2	25.9	55.7	88.9	122.2	155.1	188.0	253.4	309.2
$LiCl$	12.1	25.5	57.1	95.0	132.5	175.5	219.5	311.5	393.5
$LiBr$	12.2	26.2	60.0	97.0	140.0	186.3	241.5	341.5	438.0
Li_2SO_4	13.3	28.1	56.8	89.0					
$LiHSO_4$	12.8	27.0	57.0	93.0	130.0	168.0			
LiI	13.6	28.6	64.7	105.2	154.5	206.0	264.0	357.0	445.0
Li_2SiFl_6	15.4	34.0	70.0	106.0					

Substance	0.5	1.0	2.0	3.0	4.0	5.0	6.0	8.0	10.0
$LiOH$	15.9	37.4	78.1						
Li_2CrO_4	16.4	32.6	74.0	120.0	171.0				
$MgSO_4$	6.5	12.0	24.5	47.5					
$MgCl_2$	16.8	39.0	100.5	183.3	277.0	377.0			
$Mg(NO_3)_2$	17.6	42.0	101.0	174.8					
$MgBr_2$	17.9	44.0	115.8	205.3	298.5				
$MgH_2(SO_4)_2$	18.3	46.0	116.0						
$MnSO_4$	6.0	10.5	21.0						
$MnCl_2$	15.0	34.0	76.0	122.3	167.0	209.0			
NaH_2PO_4	10.5	20.0	36.5	51.7	66.8	82.0	96.5	126.7	157.1
$NaHSO_4$	10.9	22.1	47.3	75.0	100.2	126.1	148.5	189.7	231.4
$NaNO_3$	10.6	22.5	46.2	68.1	90.3	111.5	131.7	167.8	198.8
$NaClO_3$	10.5	23.0	48.4	73.5	98.5	123.3	147.5	196.5	223.5
$(NaPO_3)_6$	11.6								
$NaOH$	11.8	22.8	48.2	77.3	107.5	139.1	172.5	243.3	314.0
$NaNO_2$	11.6	24.4	50.0	75.0	98.2	122.5	146.5	189.0	226.2
Na_2HPO_4	12.1	23.5	43.0	60.0	78.7	99.8	122.1		
$NaHCO_3$	12.9	24.1	48.2	77.6	102.2	127.8	152.0	198.0	239.4
Na_2SO_4	12.6	25.0	48.9	74.2					
$NaCl$	12.3	25.2	52.1	80.0	111.0	143.0	176.5		
$NaBrO_3$	12.1	25.0	54.1	81.3	108.8	130.0			
$NaBr$	12.6	25.9	57.0	89.2	124.2	159.5	197.5	268.0	
NaI	12.1	25.6	60.2	99.5	136.7	177.5	221.0	301.5	370.0
$Na_4P_2O_7$	13.2	22.0							
Na_2CO_3	14.3	27.3	53.5	80.2	111.0				
$Na_2C_2O_4$	14.5	30.0	65.8	105.8	146.0				
Na_2WO_4	14.8	33.6	71.6	115.7	162.6				
Na_3PO_4	16.5	30.0	52.5						
$(NaPO_3)_3$	17.1	36.5							
NH_4NO_3	12.8	22.0	42.1	62.7	82.9	103.8	121.0	152.2	180.0
$(NH_4)_2SiFl_6$	11.5	25.0	44.5						
NH_4Cl	12.0	23.7	45.1	69.3	94.2	118.5	138.2	179.0	213.8
NH_4HSO_4	11.5	22.0	46.8	71.0	94.5	118.	139.0	181.2	218.0
$(NH_4)_2SO_4$	11.0	24.0	46.5	69.5	93.0	117.0	141.8		
NH_4Br	11.9	23.9	48.8	74.1	99.4	121.5	145.5	190.2	228.5
NH_4I	12.9	25.1	49.8	78.5	104.5	132.3	156.0	200.0	243.5
$NiSO_4$	5.0	10.2	21.5						
$NiCl_2$	16.1	37.0	86.7	147.0	212.8				
$Ni(NO_3)_2$	16.1	37.3	91.3	156.2	235.0				
$Pb(NO_3)_2$	12.3	23.5	45.0	63.0					
$Sr(SO_3)_2$	7.2	20.3	47.0						
$Sr(NO_3)_2$	15.8	31.0	64.0	97.4	131.4				
$SrCl_2$	16.8	38.8	91.4	156.8	223.3	281.5			
$SrBr_2$	17.8	42.0	101.1	179.0	267.0				
$ZnSO_4$	4.9	10.4	21.5	42.1	66.2				
$ZnCl_2$	9.2	18.7	46.2	75.0	107.0	153.0	195.0		
$Zn(NO_3)_2$	16.6	39.0	93.5	157.5	223.8				

THERMAL CONDUCTIVITY OF GASES

The values in this table are given as cal/(sec)(cm^2)(°C/cm) $\times 10^{-6}$. To convert these values to Btu/(hr)(ft^2)(°F/ft) $\times 10^{-6}$ multiply by 241.909.

Gas °F °C	−400 −240	−300 −184.4	−200 −128.9	−100 −73.3	−40 −40	−20 −28.9	0 −17.8	20 −6.7	40 4.4	60 15.6	80 26.7	100 37.8	120 48.9	200 93.3
Acetylene	....	....	....	28.10	34.71	37.19	39.67	42.15	45.04	47.94	50.83	53.72	56.62	69.43
Air	....	....	....	....	50.09	52.15	54.22	56.24	58.31	60.34	62.20	64.22	66.04	....
Ammonia	....	....	....	....	43.39	45.87	48.35	50.83	53.31	55.79	58.68	61.58	64.47	....
Argon	....	....	....	....	34.30	35.95	37.19	38.85	40.09	41.33	42.57	44.22	45.46	....
Bromine	....	....	....	....	....	....	9.09	....	....	....	....	11.57	....	....
n-Butane	....	....	....	....	....	....	....	30.99	33.06	35.54	38.02	40.91	43.39	54.14
i-Butane	....	....	....	....	....	....	....	32.65	33.89	36.37	38.85	41.74	44.22	55.79
Carbon dioxide	....	....	....	....	27.90	29.75	31.70	33.68	35.62	37.61	39.67	41.74	43.81	....
Carbon disulfide	....	....	....	....	....	....	14.05	15.29	16.53	17.77	19.01	19.84	....	....
Carbon monoxide	....	....	....	....	47.94	50.00	51.95	53.85	55.87	57.86	59.92	61.99	63.89	....
Chlorine	....	....	....	....	15.29	16.53	17.36	18.18	19.01	20.25	21.08	21.90	23.14	....
Deuterium	....	....	....	....	274.82	285.15	295.07	305.81	309.95	322.34	334.74	343.01	355.40	....
Ethane	....	....	....	23.97	32.65	35.54	38.43	41.33	44.63	47.94	51.24	54.55	58.27	74.39
Ethanol	....	....	....	....	....	....	....	29.34	30.99	32.65	34.71	36.78	....	....
Ethylamine	....	....	....	....	....	....	....	31.41	33.47	35.54	37.61	39.67	42.15	....
Ethylene	....	....	....	26.86	33.06	35.54	38.02	40.50	43.39	46.29	49.18	52.07	54.96	68.19
Fluorine	....	18.18	30.58	43.39	50.83	52.90	55.38	57.86	59.92	61.99	64.06	66.12	68.19	76.04
Helium	84.31	163.24	221.51	274.8	304.99	314.49	324.00	333.50	343.42	352.10	360.36	368.63	376.07	....
Hydrogen	59.92	142.57	227.29	308.7	357.47	371.93	388.46	405.00	417.39	433.92	446.32	458.72	471.11	....
Hydrogen bromide	....	....	....	....	15.29	16.11	16.49	17.77	18.60	19.84	20.66	21.49	....	....
Hydrogen chloride	....	....	....	....	25.62	26.86	28.51	29.75	30.99	32.23	33.89	35.12	....	....
Hydrogen cyanide	....	....	....	....	....	....	23.97	25.62	26.86	28.10	29.75	30.99	32.65	....
Hydrogen sulfide	....	....	....	....	....	....	28.10	29.75	31.41	33.47	....	36.78	....	....
Krypton	....	....	....	....	....	....	19.84	....	....	....	....	23.56	....	....
Methane	....	22.32	36.86	52.07	61.37	64.55	67.86	71.08	74.39	78.11	81.83	85.54	89.26	106.62
Neon	....	....	....	....	97.94	100.84	104.14	107.03	109.93	112.82	115.71	118.19	121.09	....
Nitric oxide	....	....	30.91	42.40	49.01	51.24	53.39	55.54	57.65	59.76	61.99	64.06	66.12	74.39
Nitrogen	....	20.25	33.06	44.22	50.42	52.48	54.55	56.20	58.27	60.34	62.40	64.06	65.71	....
Nitrous oxide	....	....	....	....	28.93	30.91	32.90	35.04	37.15	39.30	41.45	43.81	46.08	....
Oxygen	....	18.84	31.66	43.72	50.54	52.81	54.96	57.24	59.43	61.58	63.64	65.91	68.19	76.87
n-Propane	....	....	....	....	27.69	29.75	32.23	34.71	37.19	39.67	42.47	45.46	48.35	60.75
R-11(CCl_3F)	....	....	....	....	....	12.81	13.64	14.88	15.70	16.53	17.77	18.60	....	....
R-12(CCl_2F_2)	....	....	....	....	....	17.36	18.60	19.42	20.66	21.49	22.73	23.56	....	....
R-21($CHCl_2F$)	....	....	....	....	....	....	21.90	22.32	22.73	23.14	23.56	23.97	....	....
R-22($CHClF_2$)	....	....	....	....	....	....	24.80	25.62	26.45	27.28	28.10	28.93	....	....
Water	....	....	....	....	....	....	34.71	36.78	38.85	40.50	42.57	44.63	46.70	54.96

THERMAL CONDUCTIVITY OF GASEOUS HELIUM, NITROGEN AND WATER

From NSRDS-NBS 8

R. W. Powell, C. Y. Ho, and P. E. Liley

The thermal conductivity, k, is given in the units Milliwatt cm^{-1} °K^{-1}. To convert to Cal(*gm*) hr^{-1} cm^{-1} °K^{-1} multiply the values listed in the table by 0.860421

T (K)	k: He	k: N_2
0.08	0.00044	
0.09	0.00053	
0.10	0.00064	
0.15	0.00130	
0.20	0.00231	
0.25	0.0039	
0.30	0.0062	
0.35	0.0089	
0.40	0.0120	
0.45	0.0154	
0.5	0.0187	
0.6	0.0231	
0.7	0.0252	
0.8	0.0262	
0.9	0.0266	
1.0	0.0269	
1.25	0.0281	
1.5	0.0306	
2.0	0.0393	
2.5	0.0502	
3.0	0.0607	
3.5	0.0732	
4.0	0.0803	
4.5	0.0879	
5.0	0.0962	
6	0.1113	
7	0.1247	
8	0.1393	
9	0.1523	
10	0.1640	
12	0.1866	
14	0.2067	
16	0.2259	
18	0.2435	
20	0.2582	
25	0.2962	
30	0.3330	
35	0.3669	
40	0.4000	
45	0.4314	
50	0.4623	(0.0485)*
60	0.521	(0.0578)*
70	0.578	(0.0670)*
80	0.631	0.0762
90	0.679	0.0852
100	0.730	0.0941
110	0.776	0.1030
120	0.819	0.1119
130	0.863	0.1208
140	0.907	0.1296

T (K)	k: He	k: N_2	k: H_2O
150	0.950	0.1385	
160	0.992	0.1474	
170	1.033	0.1562	
180	1.072	0.1651	
190	1.112	0.1739	
200	1.151	0.1826	
210	1.190	0.1908	
220	1.228	0.1989	
230	1.266	0.2067	
240	1.304	0.2145	
250	1.338	0.2222	(0.140)*
260	1.372	0.2298	(0.148)*
270	1.405	0.2374	(0.156)*
280	1.437	0.2449	0.164
290	1.468	0.2524	0.172
300	1.499	0.2598	0.181
310	1.530	0.2671	0.189
320	1.560	0.2741	0.197
330	1.590	0.2808	0.205
340	1.619	0.2874	0.214
350	1.649	0.2939	0.222
360	1.678	0.3002	0.231
370	1.708	0.3065	0.239
380	1.737	0.3127	0.248
390	1.766	0.3189	0.256
400	1.795	0.3252	0.264
410	1.824	0.3314	0.273
420	1.853	0.3376	0.282
430	1.882	0.3438	0.291
440	1.914	0.3501	0.300
450	1.947	0.3564	0.307
460	1.980	0.3626	0.317
470	2.013	0.3688	0.327
480	2.046	0.3749	0.337
490	2.080	0.3808	0.347
500	2.114	0.3864	0.357
510	2.15	0.392	0.368
520	2.18	0.398	0.378
530	2.22	0.403	0.389
540	2.25	0.408	0.400
550	2.29	0.414	0.411
560	2.33	0.420	0.422
570	2.36	0.425	0.432
580	2.40	0.431	0.443
590	2.43	0.436	0.454
600	2.47	0.441	0.464
610	2.51	0.446	0.475
620	2.54	0.452	0.486
630	2.58	0.457	0.497
640	2.61	0.462	0.508

T (K)	k: He	k: N_2	k: H_2O
650	2.64	0.467	0.518
660	2.67	0.472	0.529
670	2.69	0.478	0.540
680	2.72	0.483	0.551
690	2.75	0.488	0.562
700	2.78	0.493	0.572
710	2.81	0.498	0.58
720	2.84	0.503	0.59
730	2.87	0.508	0.60
740	2.90	0.513	0.62
750	2.92	0.517	0.63
760	2.95	0.522	0.64
770	2.98	0.526	0.65
780	3.01	0.531	0.66
790	3.04	0.536	0.67
800	3.07	0.541	0.68
810	3.09	0.546	0.69
820	3.12	0.551	0.70
830	3.15	0.555	0.71
840	3.18	0.559	0.72
850	3.21	0.564	0.73
860	3.23	0.569	0.74
870	3.26	0.574	0.75
880	3.29	0.578	0.76
890	3.32	0.583	0.77
900	3.35	0.587	0.78
910	3.37	0.592	
920	3.40	0.596	
930	3.43	0.600	
940	3.46	0.605	
950	3.49	0.609	
960	3.52	0.613	
970	3.54	0.618	
980	3.57	0.622	
990	3.60	0.626	
1000	3.63	0.631	
1050	3.76	0.651	
1100	3.89	0.672	
1150	4.03	0.693	
1200	4.16	0.713	
1250	4.29	0.733	
1300	4.43	0.754	
1350	4.55	0.775	
1400	4.69	0.797	
1450	4.82	0.819	
1500	4.94	0.842	
1550	5.07	0.867	
1600	5.21	0.893	
1650	5.33	0.921	
1700	5.45	0.950	

T (K)	k: He	k: N_2
1750	5.57	0.981
1800	5.70	1.013
1850	5.83	1.046
1900	5.96	1.080
1950	6.08	1.113
2000	6.20	1.146
2100	6.44	1.207
2200	6.69	1.263
2300	6.93	1.314
2400	7.16	1.361
2500	7.39	1.406
2600	7.62	1.449
2700	7.85	1.494
2800	8.07	1.542
2900	8.29	1.590
3000	8.51	1.640
3100	8.72	1.691
3200	8.95	1.743
3300	9.16	1.795
3400	9.37	1.853
3500	9.58	1.915
3600	9.79	
3700	10.00	
3800	10.22	
3900	10.43	
4000	10.64	
4100	10.85	
4200	11.06	
4300	11.27	
4400	11.48	
4500	11.69	
4600	11.90	
4700	12.11	
4800	12.31	
4900	12.51	
5000	12.71	

THERMAL CONDUCTIVITY OF DIELECTRIC CRYSTALS

Name	Remarks	Conductivity nw/cm deg K 83°K	273°K
Marble	Small crystals, 99.9% $CaCO_3$	42	33
Do	99.99% $CaCo_3$	54	38
Do	Large crystals	50	33
Calcite	Main crystal axis perpendicular to rod axis	180	46
Do	Main crystal axis parallel to rod axis	293	54
Sylvite	Natural crystal	159	75
KCl	Pressed at 8000 atm	314	88
KCl	From a melt	402	92
NaCl	Do	343	92
NaCl	Pressed at 8000 atm	251	71
Rock salt	Do	180	63
Sylvite	Do	343	84
KCl	Pressed at 1250 atm	243	75
KCl	Pressed at 2500 atm	368	92
KCl	Pressed at 8900 atm	402	96
KBr	Pressed at 8000 atm	92	38
NaBr	Do	50	25
KI	Do	121	29
KF	Do	234	71
NaF	Do	519	105
RbI	Do	59	33
RbCl	Do	29	21
90% KBr, 10% KCl	Do	50	29
75% KBr, 25% KCl	Do	29	21
50% KBr, 50% KCl	Do	25	25
25% KBr, 75% KCl	Pressed at 8000 atm	46	33
10% KBr, 90% KCl	Do	80	50
50% KCl, 50% NaCl	Do	188	71
KNO_3	Do	17	21
Mercuric chloride	Do	17	13
NH_4Cl	Do	109	25
NH_4Br	Do	67	25
$Ba(NO_3)_2$	Do	33	13
Copper sulfate		29	21
Magnesium sulfate		25	25
$K_4Fe(CN_6)$		17	17
Chrom alum		13	21
Potassium alum		13	21
Potassium bichromate	Main crystal axis perpendicular to rod axis	17	21
Do	Main crystal axis parallel to rod axis	17	17
Topaz	Mineral		234
Zincblend	Do	63	264
Beryll	Do	88	84
Tourmaline	Do	38	46

THERMAL CONDUCTIVITY OF ORGANIC COMPOUNDS[a]

Substance	k	t, °C	t, °F
Acetaldehyde	0.0004089	21	69.8
Acetic acid	0.0004109	20	68
Acetic anhydride	0.0005286	21	69.8
Acetone	0.0004750	−80	−112
	0.0004543	16	61
	0.0004031	75	167
Allyl alcohol	0.0004295	30	86
Amyl acetate(n)	0.0003085	20	68
(iso)	0.000310	20	68
Amyl alcohol(n)	0.0003874	30—100	86—212
(iso)	0.0003531	30	86
Amyl bromide (n)	0.0002350	18.64.4	
Aniline	0.0004237	16.5	61.5
Benzene	0.0003780	22.5	72.5
	0.0003275	50	122
	0.0003630	60	140
	0.0002870	140	284
Bromobenzene	0.0002664	20	68
Butyl acetate (n)	0.000327	20	68
Butyl alcohol (n)	0.0003663	20	68
Carbon tetrachloride	0.0002470	20	68
	0.0002333	50	122
Chlorobenzene	0.0003457	30—100	86—212
Chlorotoluene (p)	0.000310	20	68
Chloroform	0.0002891	16	61
	0.000246	20	68
Cresol(m)	0.0003581	20	68
(p)	0.000345	20.1	68.2
Cumene	0.000298	20	68
Cymene (p)	0.0003217	30	86
Decane	0.0003349	30	86
Diethyl ether	0.0003283	30	86
Dichloroethane, 1—2	0.000302	20	68
Di-isopropyl ether	0.000262	20	68
Ethyl acetate	0.0003560	16	60.8
Ethyl alcohol	0.0003995	20	68
Ethyl benzene	0.0003160	20	68
Ethyl bromide	0.0002862	30	86
Ethyl ether	0.0003283	30	86
Ethyl iodide	0.0002651	30	86
Ethylene glycol	0.0006236	20	68
	0.0006323	15	122
	0.0006443	80	176
Freon-12 (CCl_2F_2)	0.0002310	0—75	32—167
Freon-21 ($CHCl_2F$)	0.0003180	0—75	32—167
Freon-22 ($CHClF_2$)	0.0002309	40	104
Freon-113 (CCl_2FCCCl_2F)	0.0002379	0—80	32—176
Freon-114 ($C_2H_2F_4$)	0.0002127	0—75	32—167
Glycerol	0.000703	20	68
Heptane (n)	0.0003354	30	86
Heptyl alcohol	0.0003882	70—100	86—212
Hexane (n)	0.0003287	30—100	86—212
Hexyl alcohol (n)	0.0003857	30—100	86—212
Iodobenzene	0.0002874	30—100	86—212
Mesitylene	0.0003246	20	68
Methyl alcohol	0.0004832	20	68
Methyl aniline	0.0004419	21.5	70.5
Methyl chloride	0.0004597	−15(−)+30	5—86
Methyl cyclohexane	0.0003052	30	86
Methylene chloride	0.0002908	0	32
Nitrobenzene	0.0003907	30—100	86—212
Nitromethane	0.0005142	30	86
Nonane (n)	0.0003374	30—100	86—212
Nonyl alcohol (n)	0.0004014	30—100	86-212
Octane (n)	0.0003469	30	86
Octyl alcohol (n)	0.0003973	30—100	86—212
Oleic acid	0.0005514	26.5	79.7
Palmitic acid	0.0004097	72.5	162.5
Pentachloroethane	0.0002994	20	68
Pentane (n)	0.0003221	30	86
Phenetole	0.0003577	−20	−4
Phenyl hydrazine	0.0004121	25	69.8
Propyl acetate (sio)	0.000321	20	68
Propyl alcohol (iso)	0.0003362	20	68
Propylene chloride	0.0002994	20—50	68—122
Propylene glycol (1—2)	0.0004799	20—80	68—176
Stearic acid	0.0003824	72.5	162.5
Tetrachloroethane (sym)	0.000272	20	68
Tetrachloroethylene	0.0003866	20	68
Toluene	0.0003804	−80	−112
	0.0003221	20	68
	0.0002808	80	176
Trichloroethylene	0.0003246	−60	−76
	0.0002775	20	68
Triethylamine	0.0003498	−80	−112
	0.0002891	20	68
	0.0002664	44.4	112
Xylene (o)	0.0003411	−20(−)+80	(−4)−176
Xylene (m)	0.0003767	25	77

[a]The values in this table are given as cal/(sec)(cm²)(°C/cm). To convert these values to Btu/(hr)(ft₂)(°F/ft) multiply by 242.08 .

THERMAL CONDUCTIVITY OF INORGANIC COMPOUNDS

Ammonia	0.0001198	−15(−)+30	5—86
Argon	0.0002895	−183	−297
	0.0001677	−133	−207
	0.000553	−105	−157.5
	0.0000409	−75	−102.5
Carbon dioxide	0.0002040	−50	−58
	0.0002412	−40	−40
	0.0002664	−30	−22
	0.0002746	−20	−4
	0.0002495	0	32
	0.0001677	30	86
Nitrogen	0.0003400	−196	−321.5
	0.0002028	−158	−253
	0.0000640	−105	−155
Oxygen	0.0000500	(−207)—(−191)	(−340)—(−312)
	0.0000504	(−178)—(−182)	(−288)—(−295)
Water	0.001326	−3	27
	0.001372	+7	45
	0.001456	27	81
	0.001522	47	117
	0.001575	67	153
	0.001625	97	207
	0.001635	107	225
	0.001628	157	315
	0.001580	197	387
	0.001463	217	111
	0.001288	297	567
	0.001004	347	657

THERMAL CONDUCTIVITY OF MISCELLANEOUS SUBSTANCES

Chlorinated diphenyl 1242	0.0002936	30—100	86—212
Chlorinated diphenyl 1248	0.0002808	30—100	86—212
Kerosene	0.0003572	30	86
Light heat transfer oil	0.0003159	30—100	86—212
Petroleum ether	0.0003118	30	86
Red oil	0.0003366	30	86
Transformer oil	0.0004242	70—100	86—212

THERMAL CONDUCTIVITY OF MATERIALS
(Bureau of Standards Letter Circular No. 227)

D = Density in pound per cubic foot.
K = Thermal conductivity in B.T.U. per hour, square foot, and temperature gradient of 1 degree Fahrenheit per inch thickness. The lower the conductivity, the greater the insulating values.

Soft Flexible Materials in Sheet Form

		D	K
Dry zero	Kapok between burlap or paper	1.0	0.24
		2.0	0.25
Cabots quilt	Eel grass between kraft paper	3.4	0.25
		4.6	0.26
Hair felt	Felted cattle hair	11.0	0.26
		13.0	0.26
Balsam wool	Chemically treated wood fiber	2.2	0.27
Hairinsul	75% hair 25% jute	6.3	0.27
	50% hair 50% jute	6.1	0.26
Linofelt	Flax fibers between paper	4.9	0.28
Thermofelt	Jute and asbestos fibers, felted	10.0	0.37
	Hair and asbestos fibers, felted	7.8	0.28

Loose Materials

		D	K
Rock wool	Fibrous material made from rock also made in sheet form, felted and confined with wire netting	6.0	0.26
		10.0	0.27
		14.0	0.28
		18.0	0.29
Glass wool	Pyrex glass, curled	4.0	0.29
		10.0	0.29
Sil-O-Cel	Powdered diatomaceous earth	10.6	0.31
Regranulated cork	Fine particles	9.4	0.30
	about 3/16 inch particles	8.1	0.31
Thermofill	Gypsum in powdered form	26	0.52
		34	0.60
Sawdust	Various	12.0	0.41
	redwood	10.9	0.42
Savings	Various, from planer	8.8	0.41
Charcoal	From maple, beech and birch, coarse	13.2	0.36
	6 mesh	15.2	0.37
	20 mesh	19.2	0.39

Semiflexible Materials in Sheet Form

		D	K
Flaxlinum	Flax fiber	13.0	0.31
Fibrofelt	Flax and rye fiber	13.6	0.32

Semiflexible Materials in Sheet Form

		D	K
Flaxlinum	Flax fiber	13.0	0.31
Fibrofelt	Flax and rye fiber	13.6	0.32

Semirigid Materials in Board Form

		D	K
Corkboard	No added binder; very low density	5.4	0.25
Corkboard	No added binder; low density	7.0	0.27
Corkboard	No added binder; medium density	10.6	0.30
Corkboard	No added binder; High density	14.0	0.34
Eureka	Corkboard with asphaltic binder	14.5	0.32
Rock Cork	Rock wood block with binder Also called "Tucork"	14.5	0.326
Lith	Board containing rock wool, flax and straw pulp	14.3	0.40

Stiff Fibrous Materials in Sheet Form

		D	K
Insulite	Wood pulp	16.2	0.34
		16.9	0.34
Celotex	Sugar cane fiber	13.2	0.34
		14.8	0.34
*Masonite		K =	0.33
*Inso-board			0.33
*Malzewood			0.33 to 0.39
*Cornstalk Pith Board			0.24 to 0.30
*Maftex			0.34

Cellular Gypsum

	D	K
Insulex or Pyrocell	8	0.35
	12	0.44
	18	0.59
	24	0.77
	30	1.00

Woods (Across Grain)

	D	K
Balsa	7.3	0.33
	8.8	0.38
	20	0.58
Cypress	29	0.67
White pine	32	0.78
Mahogany	34	0.90
Virginia pine	34	0.98
Oak	38	1.02
Maple	44	1.10

Miscellaneous Bulding Materials
(Data taken from various sources)

	K		K
Cinder concrete	2 to 3	Limestone	4 to 9
Building gypsum	About 3	Concrete	6 to 9
Plaster	2 to 5	Sandstone	8 to 16
Building brick	3 to 6	Marble	14 to 20
Glass	5 to 6	Granite	13 to 28

* From various commercial laboratories and the work of O. R. Sweeney at Iowa State College.

THERMAL CONDUCTIVITY DATA ON CERAMIC MATERIALS

Description[a]	Class[b]	Water Abs. %	Bulk Density g/cc	Thermal[c] Conductivity 100°F	200°F	300°F
Single Crystals						
Silicon carbide	5	—	—	52.0	50.0	49.0
Periclase	5	—	—	26.7	22.5	19.5
Sapphire, c-axis	5	—	—	20.2	16.0	14.0
Sapphire, a-axis	5	—	—	18.7	15.0	12.9
Topaz, a-axis	5	—	—	10.8	9.4	7.9
Kyanite, c-axis	5	—	—	10.00	8.6	7.4
Kyanite, b-axis	5	—	—	9.6	8.3	7.1
Spinel, $MgO \cdot Al_2O_2$	5	—	—	6.80	6.20	5.50
Quartz, c-axis	4	—	—	6.40	5.40	5.20
Quartz, A-axis	4	—	—	3.40	3.00	2.60
Rutile, c-axis	5	—	—	5.60	4.80	4.40
Rutile, a-axis	5	—	—	3.20	3.20	3.20
Fluorite	5	—	—	5.30	4.37	3.45
Beryl, aquamarine, c-axis	4	—	—	3.18	3.15	3.12
Beryl, aquamarine, a-axis	4	—	—	2.52	2.52	2.52
Zircon, a-axis	4	—	—	2.45	2.45	2.45
Zircon, c-axis	4	—	—	2.34	2.34	2.35
Polycrystalline Single Oxide Ceramics						
Pure BeO, hot pressed	2	0.03	2.97	125.0	104.0	92.0
MgO (spec. pure)	1	0.83	3.21	21.2	18.4	16.0
$SnO_2$98%	1	0.03	6.62	17.5	15.0	12.7
ZnO (yellow)	1	0.00	5.28	16.8	14.6	12.5
ZnO gray)	1	0.03	5.20	13.6	11.8	10.2
CuO (100%)	1	0.04	6.76	10.2	9.00	7.80
ThO_2, hot pressed	2	—	9.58	8.00	7.02	6.50
CeO_2	1	0.00	6.20	6.63	6.29	5.20
Mn_2O_4	1	0.02	4.21	4.18	3.80	3.41
PbO (100%)	1	0.38	7.98	1.6	1.25	0.98

[a] Composition: 90%, MgO, 10%, Al_2O_2 designates weight percent. $Li_2O{:}4B_2O_3$ designates mole composition, does not indicate compound formation.

[b] Classification: 1 = research body; 2 = industrial research body; 3 = commercial body; 4 = natural mineral; 5 = synthetic mineral.

[c] Thermal conductivity: Units in Btu/(hr) (sq ft) (°F/ft); to convert to cal/(sec) (sq cm)(°C/cm) multiply by 0.00413. (I) = determination made with high vacuum apparatus, inconel thermodes. No letters following value, determination made with high vacuum thermal conductivity apparatus, co-per thermodes.

By permission from Engineering Research Bulletin No. 40 Rutgers University, 1958.

THERMAL CONDUCTIVITIES OF GLASSES BETWEEN −150 AND +100°C

E. H. Ratcliffe

Type of glass	Approximate silica contents (wt. %)	Approximate contents other oxides normally present in quantity (wt. %)		Estimated approximate thermal conductivity at various temperatures: Temperature (°C)	Thermal conductivity $\left(\frac{\text{cal cm}}{\text{cm}^2\text{ s deg C}}\right) \times 10^4$
(a) Vitreous silica	100	....		−150	20.0
				−100	25.0
				− 50	28.8
				0	31.5
				50	33.7
				100	35.4
(b) 'Vycor' glass	96	B_2O_3	3	−100	24
				0	30
				100	34
(c) General information					
'Crown' glasses	50–75	Various		−100	12–20.5
				30	19–26
				100	21–29
'Flint' glasses	20–55	Various		−100	9–15
				30	13–21
				100	15–23
(d) Pyrex type chemically-resistant borosilicate glasses	80–81	B_2O_3	12–13	−100	21
		Na_2O	4	0	26
		Al	2	100	30
(e) Borosilicate crown glasses	60–65	B_2O_3	15–20	−100	16–17.5
				0	21–22.5
				100	24–25.5
	65–70	B_2O_3	10–15	−100	17.5–19
				0	22.5–24
				100	25.5–27
	70–75	B_2O_3	5–10	−100	19–20.5
				0	24.5–26
				100	27.5–29
(f) (i) Zinc crown glasses	55–65	ZnO	5–15	−100	21–22
		Remainder		0	26–27
		B_2O_3, Al_2O_3		100	28–30
		ZnO	5–15	−100	14–17
		Remainder		0	17 21
		Na_2O, K_2O		100	20–23
		ZnO	15–25	−100	21–22
		Remainder		0	26–27
		B_2O_3, Al_2O_3		100	27–29
		ZnO	15–25	−100	16–19
		Remainder		0	20–23
		Na_2O, K_2O		100	22–25

THERMAL CONDUCTIVITIES OF GLASSES BETWEEN −150 AND +100°C (Continued)

Type of glass	Approximate silica contents (wt. %)	Approximate contents other oxides normally present in quantity (wt. %)		Estimated approximate thermal conductivity at various temperatures: Temperature (°C)	Thermal conductivity $\left(\frac{\text{cal cm}}{\text{cm}^2\text{ s deg C}}\right) \times 10^4$
(f) (ii) Zinc crown glasses	65–75	ZnO	5–15	−100	21–22
		Remainder		0	27–28
		B_2O_3, Al_2O_3		100	29–31
		ZnO	5–15	−100	17–20
		Remainder		0	21–25
		Na_2O, K_2O		100	24–27
		ZnO	15–25	−100	21–23
		Remainder		0	27–28
		B_2O_3, Al_2O_3		100	29–30
		ZnO	15–25	−100	16–20
		Remainder		0	20–24
		Na_2O, K_2O		100	25–29
(g) Barium crown glasses	31	B_2O_3	12	−100	13
		Al_2O_3	8	0	17
		BaO	48	100	19
	41	B_2O_3	6	−100	14
		Al_2O_3	2	0	18
		ZnO	8	100	20
		BaO	43		
	47	B_2O_3	4	−100	15
		Na_2O	1	0	18
		K_2O	7	100	21
		ZnO	8		
		BaO	32		
	65	B_2O_3	2	−100	17
		Na_2O	5	0	21
		K_2O	15	100	24
		ZnO	2		
		BaO	10		
(h) Borate glasses					
Borate flint glass	9	B_2O_3	36	−100	13
		Na_2O	1	0	16
		K_2O	2	100	19
		PbO	36		
		Al_2O_3	10		
		ZnO	6		
Borate flint glass	.....	B_2O_3	56	−100	12
		Al_2O_3	12	0	16
		PbO	32	100	20
Borate flint glass	.....	B_2O_3	43	−100	9
		Al_2O_3	5	0	13
		PbO	52	100	17
Borate glass	4	B_2O_3	55	−100	15
		Al_2O_3	14	0	19
		PbO	11	100	21
		K_2O	4		
		ZnO	12		
Borate crown glass	.....	B_2O_3	64	−100	12
		Na_2O	8	0	16
		K_2O	3	100	20
		BaO	4		
		PbO	3		
		Al_2O_3	18		
Light borate crown glass	.....	B_2O_3	69	−100	13
		Na_2O	8	0	17
		BaO	5	100	21
		Al_2O_3	18		
Zinc borate glass	.....	B_2O_3	40	−100	16
		ZnO	60	0	18
				100	20
(i) Phosphate crown glasses					
Potash phosphate glass	.....	P_2O_5	70	0	18
		B_2O_3	3	100	20
		K_2O	12		
	.....	Al_2O_3	10		
		MgO	4		
Baryta phosphate glass	.....	P_2O_5	60	45	18
		B_2O_3	3		
		Al_2O_3	8		
		BaO	28		

THERMAL CONDUCTIVITIES OF GLASSES BETWEEN −150 AND +100°C (Continued)

Type of glass	Approximate silica contents (wt. %)	Approximate contents other oxides normally present in quantity (wt. %)		Estimated approximate thermal conductivity at various temperatures: Temperature (°C)	Thermal conductivity $\left(\frac{\text{cal cm}}{\text{cm}^2\text{ s deg C}}\right) \times 10^4$
(j) Soda-lime glasses	75	Na_2O	17	−100	18
		CaO	8	0	23
				100	26
	75	Na_2O	12	−100	21
		CaO	13	0	26
				100	28
	72	Na_2O	15	−100	19
		CaO	11	0	24
		Al_2O_3	2	100	27
	65	Na_2O	25	−100	16
		CaO	10	0	20
				100	23
	65	Na_2O	15	−100	20
		CaO	20	0	24
				100	26
	60	Na_2O	20	−100	18
		CaO	20	0	22
				100	24
(k) Other crown glasses					
Crown glass	75	Na_2O	9	−100	19
		K_2O	11	0	24
		CaO	5	100	26
High dispersion crown glass	68	Na_2O	16	−100	16
		ZnO	3	0	20
		PbO	13	100	24
(l) Miscellaneous flint glasses					
(i) Silicate flint glasses					
Light flint glasses	65	PbO	25	−100	16–17
		Others	10	0	21–22
				100	24–25
	55	PbO	35	−100	14–16
		Others	10	0	18–20
				100	21–22
Ordinary flint glass	45	PbO	45	−100	12–14
		Others	10	0	16–18
				100	19–20
Heavy flint glass	35	PbO	60	−100	11–12
		Others	5	0	14–15
				100	17–18
Very heavy flint glasses	25	PbO	73	−100	10–11
		Others	2	0	13–14
				100	15–16
	20	PbO	80	−100	10
				0	12
				100	14
(ii) Borosilicate flint glass	33	B_2O_3	31	−100	15
		PbO	25	0	20
		Al_2O_3	7	100	23
		K_2O	3		
		Na_2O	1		
(iii) Barium flint glass	50	BaO	24	−100	14
		PbO	6	0	17
		K_2O	8	100	20
		Na_2O	3		
		ZnO	8		
		Sb_2O_3	1		
(m) Other glasses					
(i) Potassium glass	59	K_2O	33	50	21–22
		CaO	8		
(ii) Iron glasses	63	Fe_2O_3	10	−100	19
		Na_2O	17	0	23
		MgO	4	100	25
		CaO	3		
		Al_2O_3	2		
	67	Fe_2O_3	15	0	21–22
		Na_2O_3	18	100	24–25
	62	Fe_2O_3	20	0	20.5–21.5
		Na_2O	18	100	23–24
(ii) Rock glasses					
Obsidian				0	32
				100	35
Artificial diabase				0	27
				100	30

THERMAL CONDUCTIVITY OF CERTAIN METALS

From NSRDS-NBS 8

R. W. Powell, C. Y. Ho, and P. E. Liley

The thermal conductivity, k, is given in the units Watt cm^{-1} $°K^{-1}$.

To convert to Cal(*gm*) hr^{-1} cm^{-1} $°C^{-1}$ multiply the values listed in the tables by 860.421

To convert to Btu hr^{-1} ft^{-1} $°F^{-1}$ multiply the values listed in the tables by 57.818.

ρ_0 is the residual electrical resistivity and the value of ρ at 4.2°K is used approximately as ρ_0.

T,K	Aluminum 99.996+% ρ_0 = 0.00315 μohm cm	Copper 99.999+% ρ_0 = 0.000851 μohm cm	Gold 99.999+% ρ_0 = 0.0055 μohm cm	Iron 99.998+% ρ_0 = 0.0327 μohm cm	Manganin	Platinum 99.999% ρ_0 = 0.0106 μohm cm	Silver 99.999+% ρ_0 = 0.00062 μohm cm	Tungsten 99.99+% ρ_0 = 0.0017 μohm cm
0	0	0	0	0	0	0	0	0
1	7.8	28.7	4.4	0.75	0.0007	2.31	39.4	14.4
2	15.5	57.3	8.9	1.49	0.0018	4.60	78.3	28.7
3	23.2	85.5	13.1	2.24	0.0031	6.79	115	42.6
4	30.8	113	17.1	2.97	0.0046	8.8	147	55.6
5	38.1	138	20.7	3.71	0.0062	10.5	172	67.1
6	45.1	159	23.7	4.42	0.0078	11.8	187	76.2
7	51.5	177	26.0	5.13	0.0095	12.6	193	82.4
8	57.3	189	27.5	5.80	0.0111	12.9	190	85.3
9	62.2	195	28.2	6.45	0.0128	12.8	181	85.1
10	66.1	196	28.2	7.05	0.0145	12.3	168	82.4
11	69.0	193	27.7	7.62	0.0162	11.7	154	77.9
12	70.8	185	26.7	8.13	0.0180	10.9	139	72.4
13	71.5	176	25.5	8.58	0.0197	10.1	124	66.4
14	71.3	166	24.1	8.97	0.0215	9.3	109	60.4
15	70.2	156	22.6	9.30	0.0232	8.4	96	54.8
16	68.4	145	20.9	9.56	0.0250	7.6	85	49.3
18	63.5	124	17.7	9.88	0.0285	6.1	66	40.0
20	56.5	105	15.0	9.97	0.0322	4.9	51	32.6
25	40.0	68	10.2	9.36	0.0410	3.15	29.5	20.4
30	28.5	43	7.6	8.14	0.0497	2.28	19.3	13.1
35	21.0	29	6.1	6.81	0.0583	1.80	13.7	8.9
40	16.0	20.5	5.2	5.55	0.067	1.51	10.5	6.5
45	12.5	15.3	4.6	4.50	0.075	1.32	8.4	5.07
50	10.0	12.2	4.2	3.72	0.082	1.18	7.0	4.17
60	6.7	8.5	3.8	2.65	0.097	1.01	5.5	3.18
70	5.0	6.7	3.58	2.04	0.110	0.90	4.97	2.76
80	4.0	5.7	3.52	1.68	0.120	0.84	4.71	2.56
90	3.4	5.14	3.48	1.46	0.127	0.81	4.60	2.44
100	3.0	4.83	3.45	1.32	0.133	0.79	4.50	2.35
150	2.47	4.28	3.35	1.04	0.156	0.762	4.32	2.10
200	2.37	4.13	3.27	0.94	0.172	0.748	4.30	1.97
250	2.35	4.04	3.20	0.865	0.193	0.737	4.28	1.86
273	2.36	4.01	3.18	0.835	0.206	0.734	4.28	1.82
300	2.37	3.98	3.15	0.803	0.222	0.730	4.27	1.78
350	2.40	3.94	3.13	0.744	0.250	0.726	4.24	1.70
400	2.40	3.92	3.12	0.694	(0.279)	0.722	4.20	1.62
500	2.37	3.88	3.09	0.613	(0.338)	0.719	4.13	1.49
600	2.32	3.83	3.04	0.547	(0.397)	0.720	4.05	1.39
700	2.26	3.77	2.98	0.487		0.723	3.97	1.33
800	2.20	3.71	2.92	0.433		0.729	3.89	1.28
900	2.13	3.64	2.85	0.380		0.737	3.82	1.24
1000	[0.93]**	3.57	(2.78)	0.326		0.748	(3.74)	1.21
1100	[0.96]	3.50	(2.71)	0.297		0.760	(3.66)	1.18
1200	[0.99]	3.42	(2.62)	0.282		0.775	(3.58)	1.15
1300	[1.02]	(3.34)†	(2.51)	0.299		0.791		1.13
1400				0.309		0.807		1.11
1500				0.318		0.824		1.09
1600				(0.327)		0.842		1.07
						0.860		1.05
						0.877		1.03
						(0.895)		1.02
						(0.913)		1.00
								0.98
								0.96
								0.94
								0.925
								0.915
								0.905
								0.900
								(0.895)

* In the table the third significant figure is given only for the purpose of comparison and for smoothness and is not indicative of the degree of accuracy.

** Values in square brackets are for liquid state.

† Values in parentheses are extrapolated.

‡ Estimated.

THERMAL CONDUCTIVITY OF CERTAIN LIQUIDS

From NSRDS-NBS 8
R. W. Powell, C. Y. Ho, and P. E. Liley

The thermal conductivity, k, is given in the units Milliwatt cm^{-1} °K^{-1}. To convert to Cal(*gm*) hr^{-1} cm^{-1} °K^{-1} multiply the values listed in the table by 0.860421

T (K)	Helium	Nitrogen	Argon	Carbon tetra-chloride	Diphenyl	m-Terphenyl	Toluene	Water
2.4	0.192							
2.6	0.193							
2.8	0.197							
3.0	0.204							
3.2	0.214							
3.4	0.227							
3.6	0.241							
3.8	0.260							
4.0	0.282							
4.2	0.307							
4.4	(0.335)‡							
4.6	(0.366)‡							
4.8	(0.400)‡							
5.0	(0.437)‡							
5.2	(0.477)‡							
60		1.692†						
65		1.598						
70		1.504						
75		1.411						
80		1.320‡	1.315†					
85		1.229‡	1.258					
90		1.140‡	1.200‡					
95		1.051‡	1.141‡					
100		0.965‡	1.082‡					
105		0.879‡	1.023‡					
110		0.794‡	0.963‡					
115		0.710‡	0.903‡					
120		0.627‡	0.842‡					
125		0.544‡	0.780‡					
130			0.717‡					
135			0.654‡					
140			0.591‡					
145			0.527‡					
150			0.463‡				(1.719)†	
160							(1.694)†	
170							(1.669)†	
180							1.644	
190							1.619	
200							1.594	
210							1.569	
220							1.543	
230				(1.169)†			1.518	
240				(1.150)†			1.492	
250				1.131			1.467	5.22†
260				1.112			1.442	5.39†
270				1.093			1.416	5.55†
280				1.074			1.391	5.74
290				1.055			1.365	5.92
300				1.036			1.340	6.09
310				1.017			1.315	6.23
320				0.997			1.289	6.37
330				0.978	(1.402)†		1.264	6.48
340				0.959	(1.387)†		1.238	6.59
350				0.940	1.373	(1.361)†	1.213	6.68
360				(0.921)	1.359	(1.356)†	1.188	6.75
370				(0.902)	1.345	1.351	1.162	6.80
380				(0.882)	1.331	1.346	1.137	6.84‡
390				(0.863)	1.316	1.341	(1.112)‡	6.86‡
400				(0.844)	1.302	1.335	(1.086)‡	6.86‡
410				(0.825)	1.288	1.329	(1.061)‡	6.86‡
420				(0.806)	1.274	1.323	(1.036)‡	6.84‡
430				(0.787)	1.259	1.317	(1.013)‡	6.81‡
440				(0.768)	1.245	1.310	(0.985)‡	6.78‡
450				(0.749)	1.231	1.304	(0.959)‡	6.73‡
460					1.217	1.297	(0.933)‡	6.67‡
470					1.202	1.290	(0.908)‡	6.61‡
480					1.188	1.283	(0.885)‡	6.53‡
490					1.174	1.276	(0.862)‡	6.45‡
500					1.160	1.268	(0.839)‡	6.35‡
510					1.146	1.261		6.24‡
520					1.131	1.254		6.12‡
530					1.117‡	1.246		5.99‡
540					1.103‡	1.238		5.86‡
550					1.089‡	1.230		5.71‡
560					1.074‡	1.222		5.55‡
570					1.060‡	1.213		5.39‡
580					1.046‡	1.205		5.20‡
590					1.032‡	1.197		5.01‡
600					1.018‡	1.188		4.81‡
610						1.180		4.60‡
620						1.172		4.40‡
630						1.163		(4.20)‡
640						1.155‡		(4.01)‡
650						1.146‡		

† Extrapolated for the supercooled liquid. [Approximate n.m.p. in K: N_2, 63; A, 84; CCl_4, 250; $C_{12}H_{10}$, 342; m-$C_{18}H_{14}$, 361; p-$C_{18}H_{14}$, 486; C_7H_{10}, 178; H_2O, 273.1].

‡ Under saturation vapor pressure [Approximate n.b.p. in K: He, 4.3; N_2, 77; A, 88; CCl_4, 350; $C_{12}H_{10}$, 528; m-$C_{18}H_{14}$, 637; p-$C_{18}H_{14}$, 658; C_7H_{10}, 384; H_2O, 373].

THERMAL CONDUCTIVITY OF THE ELEMENTS

Data contained in the following table were extracted from the extensive compilation prepared by C. Y. Ho, R. W. Powell, and P. E. Liley under the National Standard Reference Data System (NSRDS) of the National Bureau of Standards project at the Thermophysical Properties Research Center (TPRC) at Purdue University and published in the Journal of Physical and Chemical Reference Data, *1*, 279-421 (1972). The data in the table below are used with the permission of the authors and the copyright owners, the American Institute of Physics and the American Chemical Society. Users are referred to their more extensive compilation for conductivities at temperatures other than those listed in the table below, and also to obtain an understanding of the basis of selection of recommended and provisional values. Temperatures are in kelvins (K) and conductivities, k, in watt per centimeter kelvin (W cm^{-1} K^{-1}), except as noted. If the numerical value of k has a superscript m, the units of k are milliwatt per centimeter kelvin (mW cm^{-1} K^{-1}). To convert the listed conductivities to units other than those in the tables, one should make use of the conversion factors listed in the table "Conversion Factors for Units of Thermal Conductivity", which follows this table. Conductivity values listed with an asterisk*, are provisional values.

Element	State or Condition	Conductivity at 273.2K	298.2K	373.2K
Aluminum	Solid	2.36	2.37	2.40
Antimony	Polycrystalline	0.255	0.244	0.219
Argon	Gas at 1 atm.	0.1619^m (270K)	0.1772^m (300K)	0.2103^m (370K)
Arsenic	Solid, Gray, Polycrystalline	0.539*	0.502*	0.427*
Barium	Solid	0.185*	0.184* (295K)	
Beryllium	Polycrystalline	2.18*	2.01	1.68
Bismuth	Solid			
	// to triagonal axis	0.0554	0.0530	0.0481
	⊥ to triagonal axis	0.0953	0.0919	0.0844
	Polycrystalline	0.0822	0.0792	0.0722
Boron	Solid	0.318	0.274	0.188
Bromine	Saturated liquid	1.30^m* (270K)	1.22^m* (300K)	1.06^m* (370K)
	Saturated vapor		0.048^m* (300K)	
	Gas	0.042^m* (270K)	0.048^m* (300K)	0.057^m* (350K)
Cadmium	Solid			
	// to c-axis	0.835	0.830	0.816
	⊥ to c-axis	1.04	1.04	1.02
	Polycrystalline	0.975	0.969	0.953
Calcium	Solid	2.06*	2.01*	1.92*
Carbon	Solid, Amorphous	0.0150	0.0159	0.0182
	Solid, Type I (Diamond)	9.94	9.90	7.03*
	Solid, Type IIa (Diamond)	26.2	23.2	17.0*
	Solid, Type IIb (Diamond)	15.2	13.6	10.2*
	Solid, Acheson graphite			
	// to axis of extrusion	1.69	1.65	1.50
	⊥ to axis of extrusion	1.21	1.19	1.11
	Solid, AGOT graphite			
	// to axis of extrusion	2.28	2.21	1.95
	⊥ to axis of extrusion	1.41	1.38	1.22
	Solid, ATJ graphite			
	// to molding pressure	0.984	0.982	0.933
	⊥ to molding pressure	1.31	1.29	1.21
	Solid, AWG graphite			
	// to molding pressure	0.807	0.796	0.733
	⊥ to molding pressure	1.32	1.28	1.16
	Solid, Pyrolytic graphite			
	// to layer planes	21.3	19.6	15.1
	⊥ to layer planes	0.0636	0.0573	0.0442
	Solid, 875S graphite			
	// to axis of extrusion	1.97*	1.92*	1.75*
	⊥ to axis of extrusion	1.49*	1.46*	1.34*
	Solid, 890S graphite			
	// to axis of extrusion	1.87*	1.83*	1.66*
	⊥ to axis of extrusion	1.51*	1.48*	1.36*
Cerium	Solid, Polycrystalline	0.108*	0.113	0.128*
Cesium	Solid	0.361*	0.359 (301.9K)	
	Liquid		0.197 (301.9K)	0.201
Chlorine	Saturated liquid	1.49^m* (270K)	1.34^m* (300K)	0.95^m* (370K)
	Saturated vapor	0.082^m* (270K)	0.097^m* (300K)	0.155^m* (370K)

Element	State or Condition	Conductivity at 273.2K	298.2K	373.2K
	Gas, 1 atm.	0.078^m (270K)	0.089^m (300K)	0.114^m (370K)
Chromium	Solid, Polycrystalline	0.965	0.939	0.921
Cobalt	Solid, Polycrystalline	1.05	1.00	0.890
Copper	Solid	4.03	4.01	3.95
Dysprosium	Solid			
	// to c-axis	0.114*	0.117*	
	⊥ to c-axis	0.101*	0.103*	
	Polycrystalline	0.105*	0.107*	0.108*
Erbium	Solid			
	// to c-axis	0.187*	0.184*	
	⊥ to c-axis	0.127*	0.126*	
	Polycrystalline	0.147*	0.145*	0.140*
Europium	Solid	0.140*	0.139*	
Fluorine	Gas, 1 atm.	0.251^m (270K)	0.279^m (300K)	0.344^m (370K)
Gadolinium	Solid			
	// to c-axis	0.104*	0.108*	
	⊥ to c-axis	0.103*	0.103*	
	Polycrystalline	0.103*	0.105*	
Gallium	Solid			
	// to a-axis	0.410	0.408	
	// to b-axis	0.884	0.883	
	// to c-axis	0.160	0.159	
	Liquid		0.281 (302.93K)	0.328
Germanium	Solid	0.667	0.602	0.465
Gold	Solid	3.19	3.18	3.13
Hafnium	Solid, Polycrystalline	0.233*	0.230	0.224
Helium	Solid, ^{3}He	0.033 (0.9K)	0.020 (1K)	0.0021 (2K)
	Solid, ^{4}He	0.650 (0.9K)	0.245 (1K)	0.0018 (2K)
	Liquid, saturated; He-I	0.191^m (2.5K)	0.232^m (3.5K)	0.434^m (5K)
	Gas, 1 atm.	1.411^m	1.520^m	1.766^m
Holmium	Solid			
	// to c-axis	0.215*	0.222*	
	⊥ to c-axis	0.136*	0.138*	
	Polycrystalline	0.159*	0.162*	0.170*
Hydrogen	Solid, Normal Hydrogen	2.30 (4K)	0.0158 (10K)	0.0090 (15K)
	Liquid, saturated; Normal Hydrogen	1.022^m (15K)	1.269^m (25K)	0.60^m* (33K)
	Gas, 1 atm. Normal Hydrogen	1.665^m	1.815^m	2.106^m
	Liquid, saturated; para-Hydrogen	0.824^m (14K)	0.998^m (25K)	0.58^m (32K)
	Vapor, saturated; para-Hydrogen	0.081^m* (10K)	0.242^m* (25K)	0.58^m* (32K)
	Gas, 1 atm.; para-Hydrogen	1.768^m*	1.880^m*	2.126^m*
	Deuterium:			
	Liquid, saturated	1.26^m (20K)	1.37^m (30K)	0.83^m* (38K)
	Vapor, saturated	0.084^m* (20K)	0.26^m* (30K)	0.83^m* (38K)
	Gas, 1 atm.	1.294^m* (270K)	1.406^m* (300K)	1.66^m* (370K)
	Tritium:			
	Liquid, saturated	1.25 (21K)	1.34 (30K)	0.68 (44K)
Indium	Solid, Polycrystalline	0.837	0.818	0.762
Iodine	Solid	4.81^m*	4.49^m* (300K)	3.75^m* (386.8K)
	Liquid, saturated			1.16^m* (386.8K)
Iridium	Solid	1.48	1.47	1.45
Iron	Solid	0.865	0.804	0.720
	Armco Iron	0.747	0.728	0.676
Krypton	Solid	0.4^m* (1K)	17^m (10K)	2.5^m (116K)
	Liquid, saturated			0.931^m (116K)
	Vapor, saturated	0.0406^m* (120K)	0.0554^m* (150K)	0.21^m* (210K)
	Gas	0.0860^m (270K)	0.0949^m (300K)	0.1145^m (370K)

THERMAL CONDUCTIVITY OF THE ELEMENTS (*Continued*)

Element	State or Condition	Conductivity at 273.2K	298.2K	373.2K
Lanthanum	Solid, Polycrystalline	0.131	0.134	0.145
Lead	Solid	0.356	0.353	0.344
Lithium	Solid	0.859	0.848	0.818
Lutetium	Solid			
	∥ to c-axis	0.236*	0.232*	
	⊥ to c-axis	0.140*	0.138*	
	Polycrystalline	0.167*	0.164*	
Magnesium	Solid, Polycrystalline	1.57	1.56	1.54
Manganese	Solid	0.0768*	0.0781*	
Mercury	Liquid	0.0782	0.0830	0.0947
Molybdenum	Solid	1.39*	1.38*	1.35*
Neodymium	Solid, Polycrystalline	0.165*	0.165*	0.167*
Neon	Gas	0.461^{m}*	0.493^{m}*	0.563^{m}*
		(270K)	(300K)	(370K)
Neptunium	Solid		0.063*	
			(300K)	
Nickel	Solid	0.941	0.909	0.827
Niobium	Solid	0.533	0.537	0.548
Nitrogen	Solid	56^{m}	17^{m}	3.2^{m}
		(4K)	(10K)	(25K)
	Liquid, saturated	1.60^{m}	0.966^{m}	0.37^{m}*
		(65K)	(100K)	(126K)
	Vapor, saturated	0.061^{m}*	0.111^{m}*	0.37^{m}*
		(65K)	(100K)	(126K)
	Gas, 1 atm.	0.2374^{m}	0.2598^{m}	0.3065^{m}
		(270K)	(300K)	(370K)
Cesium	Solid			
	∥ to c-axis	2.93	14.3	15.4
		(2K)	(10K)	(30K)
	⊥ to c-axis	1.76	8.65	11.1
		(2K)	(10K)	(30K)
	Polycrystalline	2.09	10.2	12.4
		(2K)	(10K)	(30K)
	Polycrystalline	0.880*	0.876*	0.870*
Oxygen	Liquid, saturated	1.501^{m}	1.023^{m}	0.41^{m}
		(90K)	(125K)	(155K)
	Vapor, saturated	0.081^{m}*	0.135^{m}*	0.41^{m}*
		(90K)	(125K)	(155K)
	Gas, 1 atm.	0.2424	0.2674	0.3204
		(270K)	(300K)	(370K)
Palladium	Solid	0.716*	0.718	0.730
Phosphorus	Solid			
	Black (Polycrystalline)	0.132	0.121	
	White	0.00250*	0.00236*	
	Liquid, White			0.00181
Platinum	Solid	0.717	0.716	0.717
Plutonium	Solid, polycrystalline	0.0616*	0.0670*	0.0790*
				(350K)
Potassium	Solid	1.036*	1.025	
	Liquid			0.532
Praeseodymium	Solid, polycrystalline	0.120	0.125	0.134
Promethium	Solid, polycrystalline		0.179*	0.184*
Radium	Solid		0.186	
			(293.2K)	
Radon	Gas, 1 atm.	0.0327^{m}*	0.0364^{m}*	0.0445^{m}*
		(270K)	(300K)	(370K)
Rhenium	Solid, polycrystalline	0.486	0.480	0.466
Rhodium	Solid	1.51	1.50	1.47
Rubidium	Solid	0.583*	0.582	0.581
				(312.04K)
	Liquid			0.333
				(312.04K)
Ruthenium	Solid, polycrystalline	1.17	1.17	1.15
Samarium	Solid, polycrystalline	0.133*	0.133*	0.133*
Scandium	Solid, polycrystalline	0.157	0.158	
Selenium	Solid			
	∥ to c-axis	0.0481	0.0452	0.0483
	⊥ to c-axis	0.0137	0.0131	0.0139
	Amorphous	0.00428	0.00519	0.00818
				(323.2K)
Silicon	Solid	1.68	1.49	1.08
Silver	Solid	4.29	4.29	4.26
Sodium	Solid	1.42	1.42	1.32
				(371K)
Strontium	Solid, polycrystalline	0.364*	0.354*	0.325*

THERMAL CONDUCTIVITY OF THE ELEMENTS (*Continued*)

Element	State or Condition	Conductivity at 273.2K	298.2K	373.2K
Sulfur	Solid, polycrystalline	0.00287	0.00270	0.00154
	Solid, amorphous	0.00200	0.00205	0.00216* (350K)
	Liquid			0.00129 (392.2K)
Tantalum	Solid	0.574	0.575	0.577
Technetium	Solid, polycrystalline	0.509*	0.506	0.501
Tellurium	Solid			
	// to c-axis	0.0360	0.0338	0.0292
	⊥ to c-axis	0.0208	0.197	0.173
Terbium	Solid			
	// to c-axis	0.138*	0.147*	
	⊥ to c-axis	0.0900*	0.0956*	
	Polycrystalline	0.104*	0.111*	
Thallium	Solid, polycrystalline	0.469	0.461	0.443
Thorium	Solid	0.540*	0.540*	0.543*
Thulium	Solid			
	// to c-axis	0.242*	0.242*	
	⊥ to c-axis	0.140*	0.141*	
	Polycrystalline	0.168*	0.169*	
Tin	Solid			
	// to c-axis	0.527	0.516	0.489
	⊥ to c-axis	0.759	0.743	0.704
	Polycrystalline	0.682	0.668	0.632
Titanium	Solid, polycrystalline	0.224	0.219	0.207
Tungsten	Solid	1.77	1.73	1.63
Uranium	Solid, polycrystalline	0.270	0.275	0.291
Vanadium	Solid	0.307*	0.307	0.310
Xenon	Liquid, saturated	0.31^{m} (270K)	0.16^{m}* (290K)	
	Vapor, saturated	0.084^{m}*	0.16^{m}*	
	Gas, 1 atm.	0.0514^{m} (270K)	0.0569^{m} (300K)	0.0695^{m} (370K)
Ytterbium	Solid	0.354*	0.349*	0.343*
Yttrium	Solid, polycrystalline	0.170*	0.172*	0.177*
Zinc	Solid, polycrystalline	1.17	1.16	1.12
Zirconium	Solid, polycrystalline	0.232*	0.227 (300K)	0.218

THERMAL CONDUCTIVITY OF ROCKS

Rock	Temperature, °C	Conductivity, Kcal m^{-1} hr^{-1} deg^{-1}	Heat Capacity, cal g^{-1} deg^{-1}
Granite	0	3.02	0.192
	50	2.81	–
	100	2.59	–
	200	2.34	0.228
	300	2.12	–
	400	–	0.258
Marble	118	1.44	0.21
	196	1.29	0.24
	245	1.19	–
	360	0.95	0.271
Dolomitic limestone	130	1.41	–
	181	1.37	–
	268	1.29	–
	377	1.15	–
Shale	0	1.65	0.17
	100	1.51	–
	120	1.33	–
	188	1.41	0.24
	304	1.26	0.245
Sandstone (quartzitic)	0	4.9	–
	100	3.82	0.26
	200	3.24	–

CONVERSION FACTORS FOR UNITS OF THERMAL CONDUCTIVITY

MULTIPLY by appropriate factor to OBTAIN→	$Btu_{IT}\ h^{-1}\ ft^{-1}\ F^{-1}$	Btu_{IT} in. $h^{-1}\ ft^{-2}\ F^{-1}$	$Btu_{th}\ h^{-1}\ ft^{-1}\ F^{-1}$	Btu_{th} in. $h^{-1}\ ft^{-2}\ F^{-1}$	$cal_{IT}\ s^{-1}\ cm^{-1}\ C^{-1}$	$cal_{th}\ s^{-1}\ cm^{-1}\ C^{-1}$
$Btu_{IT}\ h^{-1}\ ft^{-1}\ F^{-1}$	1	12	1.00067	12.0080	4.13379×10^{-3}	4.13656×10^{-3}
Btu_{IT} in. $h^{-1}\ ft^{-2}\ F^{-1}$	8.33333×10^{-2}	1	8.33891×10^{-2}	1.00067	3.44482×10^{-4}	3.44713×10^{-4}
$Btu_{th}\ h^{-1}\ ft^{-1}\ F^{-1}$	0.999331	11.9920	1	12	4.13102×10^{-3}	4.13379×10^{-3}
Btu_{th} in. $h^{-1}\ ft^{-2}\ F^{-1}$	8.32776×10^{-2}	0.999331	8.33333×10^{-2}	1	3.44252×10^{-4}	3.44482×10^{-4}
$cal_{IT}\ s^{-1}\ cm^{-1}\ C^{-1}$	2.41909×10^{2}	2.90291×10^{3}	2.42071×10^{2}	2.90485×10^{3}	1	1.00067
$cal_{th}\ s^{-1}\ cm^{-1}\ C^{-1}$	2.41747×10^{2}	2.90096×10^{3}	2.41909×10^{2}	2.90291×10^{3}	0.999331	1
$kcal_{th}\ h^{-1}\ m^{-1}\ C^{-1}$	0.671520	8.05824	0.671969	8.06363	2.77592×10^{-3}	2.77778×10^{-3}
$J\ s^{-1}\ cm^{-1}\ K^{-1}$	57.7789	6.93347×10^{2}	57.8176	6.93811×10^{2}	0.238846	0.239006
$W\ cm^{-1}\ K^{-1}$	57.7789	6.93347×10^{2}	57.8176	6.93811×10^{2}	0.238846	0.239006
$W\ m^{-1}\ K^{-1}$	0.577789	6.93347	0.578176	6.93811	2.38846×10^{-3}	2.39006×10^{-3}
$mW\ cm^{-1}\ K^{-1}$	5.77789×10^{-2}	0.693347	5.78176×10^{-2}	0.693811	2.38846×10^{-4}	2.39006×10^{-4}

MULTIPLY by appropriate factor to OBTAIN→	$kcal_{th}\ h^{-1}\ m^{-1}\ C^{-1}$	$J\ s^{-1}\ cm^{-1}\ K^{-1}$	$W\ cm^{-1}\ K^{-1}$	$W\ m^{-1}\ K^{-1}$	$mW\ cm^{-1}\ K^{-1}$
$Btu_{IT}\ h^{-1}\ ft^{-1}\ F^{-1}$	1.48916	1.73073×10^{-2}	1.73073×10^{-2}	1.73073	17.3073
Btu_{IT} in. $h^{-1}\ ft^{-2}\ F^{-1}$	0.124097	1.44228×10^{-3}	1.44228×10^{-3}	0.144228	1.44228
$Btu_{th}\ h^{-1}\ ft^{-1}\ F^{-1}$	1.48816	1.72958×10^{-2}	1.72958×10^{-2}	1.72958	17.2958
Btu_{th} in. $h^{-1}\ ft^{-2}\ F^{-1}$	0.124014	1.44131×10^{-3}	1.44131×10^{-3}	0.144131	1.44131
$cal_{IT}\ s^{-1}\ cm^{-1}\ C^{-1}$	3.60241×10^{2}	4.1868	4.1868	4.1868×10^{2}	4.1868×10^{3}
$cal_{th}\ s^{-1}\ cm^{-1}\ C^{-1}$	3.6×10^{2}	4.184	4.184	4.184×10^{2}	4.184×10^{3}
$kcal_{th}\ h^{-1}\ m^{-1}\ C^{-1}$	1	1.16222×10^{-2}	1.16222×10^{-2}	1.16222	11.6222
$J\ s^{-1}\ cm^{-1}\ K^{-1}$	86.0421	1	1	1×10^{2}	1×10^{3}
$W\ cm^{-1}\ K^{-1}$	86.0421	1	1	1×10^{2}	1×10^{3}
$W\ m^{-1}\ K^{-1}$	0.860421	1×10^{-2}	1×10^{-2}	1	10
$mW\ cm^{-1}\ K^{-1}$	8.60421×10^{-2}	1×10^{-3}	1×10^{-3}	0.1	1

Properties of Saturated Steam and Saturated Water

Temp. F	Press. psia	Volume, ft³/lbm Water v_f	Evap. v_{fg}	Steam v_g	Enthalpy, Btu/lbm Water h_f	Evap. h_{fg}	Steam h_g	Entropy, Btu/lbm ×F Water s_f	Evap. s_{fg}	Steam s_g	Temp. F
705.47	3208.2	0.05078	0.00000	0.05078	906.0	0.0	906.0	1.0612	0.0000	1.0612	705.47
705.0	3198.3	0.04427	0.01304	0.05730	873.0	61.4	934.4	1.0329	0.0527	1.0856	705.0
704.5	3187.8	0.04233	0.01822	0.06055	861.9	85.3	947.2	1.0234	0.0732	1.0967	704.5
704.0	3177.2	0.04108	0.02192	0.06300	854.2	102.0	956.2	1.0169	0.0876	1.1046	704.0
703.5	3166.8	0.04015	0.02489	0.06504	848.2	115.2	963.5	1.0118	0.0991	1.1109	703.5
703.0	3156.3	0.03940	0.02744	0.06684	843.2	126.4	969.6	1.0076	0.1087	1.1163	703.0
702.5	3145.9	0.03878	0.02969	0.06847	838.9	136.1	974.9	1.0039	0.1171	1.1210	702.5
702.0	3135.5	0.03824	0.03173	0.06997	835.0	144.7	979.7	1.0006	0.1246	1.1252	702.0
701.5	3125.2	0.03777	0.03361	0.07138	831.5	152.6	984.0	0.9977	0.1314	1.1291	701.5
701.0	3114.9	0.03735	0.03536	0.07271	828.2	159.8	988.0	0.9949	0.1377	1.1326	701.0
700.5	3104.6	0.03697	0.03701	0.07397	825.2	166.5	991.7	0.9924	0.1435	1.1359	700.5
700.0	3094.3	0.03662	0.03857	0.07519	822.4	172.7	995.2	0.9901	0.1490	1.1390	700.0
699.0	3073.9	0.03600	0.04149	0.07749	817.3	184.2	1001.5	0.9858	0.1590	1.1447	699.0
698.0	3053.6	0.03546	0.04420	0.07966	812.6	194.6	1007.2	0.9818	0.1681	1.1499	698.0
697.0	3033.5	0.03498	0.04674	0.08172	808.4	204.0	1012.4	0.9783	0.1764	1.1547	697.0
696.0	3013.4	0.03455	0.04916	0.08371	804.4	212.8	1017.2	0.9749	0.1841	1.1591	696.0
695.0	2993.5	0.03415	0.05147	0.08563	800.6	221.0	1021.7	0.9718	0.1914	1.1632	695.0
694.0	2973.7	0.03379	0.05370	0.08749	797.1	228.8	1025.9	0.9689	0.1983	1.1671	694.0
693.0	2954.0	0.03345	0.05587	0.08931	793.8	236.1	1029.9	0.9660	0.2048	1.1708	693.0
692.0	2934.5	0.03313	0.05797	0.09110	790.5	243.1	1033.6	0.9634	0.2110	1.1744	692.0
690.0	2895.7	0.03256	0.06203	0.09459	784.5	256.1	1040.6	0.9583	0.2227	1.1810	690.0
688.0	2857.4	0.03204	0.06595	0.09799	778.8	268.2	1047.0	0.9535	0.2337	1.1872	688.0
686.0	2819.5	0.03157	0.06976	0.10133	773.4	279.5	1052.9	0.9490	0.2439	1.1930	686.0
684.0	2782.1	0.03114	0.07349	0.10463	768.2	290.2	1058.4	0.9447	0.2537	1.1984	684.0
682.0	2745.1	0.03074	0.07716	0.10790	763.3	300.4	1063.6	0.9406	0.2631	1.2036	682.0
680.0	2708.6	0.03037	0.08080	0.11117	758.5	310.1	1068.5	0.9365	0.2720	1.2086	680.0
678.0	2672.5	0.03002	0.08440	0.11442	753.8	319.4	1073.2	0.9326	0.2807	1.2133	678.0
676.0	2636.8	0.02970	0.08799	0.11769	749.2	328.5	1077.6	0.9287	0.2892	1.2179	676.0
674.0	2601.5	0.02939	0.09156	0.12096	744.7	337.2	1081.9	0.9249	0.2974	1.2223	674.0
672.0	2566.6	0.02911	0.09514	0.12424	740.2	345.7	1085.9	0.9212	0.3054	1.2266	672.0
670.0	2532.2	0.02884	0.09871	0.12755	735.8	354.0	1089.8	0.9174	0.3133	1.2307	670.0
668.0	2498.1	0.02858	0.10229	0.13087	731.5	362.1	1093.5	0.9137	0.3210	1.2347	668.0
666.0	2464.4	0.02834	0.10588	0.13421	727.1	370.0	1097.1	0.9100	0.3286	1.2387	666.0
664.0	2431.1	0.02811	0.10947	0.13757	722.9	377.7	1100.6	0.9064	0.3361	1.2425	664.0
662.0	2398.2	0.02789	0.11306	0.14095	718.8	385.1	1103.9	0.9028	0.3434	1.2462	662.0
660.0	2365.7	0.02768	0.11663	0.14431	714.9	392.1	1107.0	0.8995	0.3502	1.2498	660.0
658.0	2333.5	0.02748	0.12023	0.14771	711.1	399.0	1110.1	0.8963	0.3570	1.2533	658.0
656.0	2301.7	0.02728	0.12387	0.15115	707.4	405.7	1113.1	0.8931	0.3637	1.2567	656.0
654.0	2270.3	0.02709	0.12754	0.15463	703.7	412.2	1115.9	0.8899	0.3702	1.2601	654.0
652.0	2239.2	0.02691	0.13124	0.15816	700.0	418.7	1118.7	0.8868	0.3767	1.2634	652.0
650.0	2208.4	0.02674	0.13499	0.16173	696.4	425.0	1121.4	0.8837	0.3830	1.2667	650.0
648.0	2178.1	0.02657	0.13876	0.16534	692.9	431.1	1124.0	0.8806	0.3893	1.2699	648.0
646.0	2148.0	0.02641	0.14258	0.16899	689.4	437.2	1126.6	0.8776	0.3954	1.2730	646.0
644.0	2118.3	0.02625	0.14644	0.17269	685.9	443.1	1129.0	0.8746	0.4015	1.2761	644.0
642.0	2088.9	0.02610	0.15033	0.17643	682.5	448.9	1131.4	0.8716	0.4075	1.2791	642.0
640.0	2059.9	0.02595	0.15427	0.18021	679.1	454.6	1133.7	0.8686	0.4134	1.2821	640.0
638.0	2031.2	0.02580	0.15824	0.18405	675.8	460.2	1136.0	0.8657	0.4193	1.2850	638.0
636.0	2002.8	0.02566	0.16226	0.18792	672.4	465.7	1138.1	0.8628	0.4251	1.2879	636.0
634.0	1974.7	0.02553	0.16633	0.19185	669.1	471.1	1140.2	0.8599	0.4307	1.2907	634.0
632.0	1947.0	0.02539	0.17044	0.19583	665.9	476.4	1142.2	0.8571	0.4364	1.2934	632.0
630.0	1919.5	0.02526	0.17459	0.19986	662.7	481.6	1144.2	0.8542	0.4419	1.2962	630.0
628.0	1892.4	0.02514	0.17880	0.20394	659.5	486.7	1146.1	0.8514	0.4474	1.2988	628.0
626.0	1865.6	0.02501	0.18306	0.20807	656.3	491.7	1148.0	0.8486	0.4529	1.3015	626.0
624.0	1839.0	0.02489	0.18737	0.21226	653.1	496.6	1149.8	0.8458	0.4583	1.3041	624.0
622.0	1812.8	0.02477	0.19173	0.21650	650.0	501.5	1151.5	0.8430	0.4636	1.3066	622.0
620.0	1786.9	0.02466	0.19615	0.22081	646.9	506.3	1153.2	0.8403	0.4689	1.3092	620.0
618.0	1761.2	0.02455	0.20063	0.22517	643.8	511.0	1154.8	0.8375	0.4742	1.3117	618.0
616.0	1735.9	0.02444	0.20516	0.22960	640.8	515.6	1156.4	0.8348	0.4794	1.3141	616.0
614.0	1710.8	0.02433	0.20976	0.23409	637.8	520.2	1158.0	0.8321	0.4845	1.3166	614.0
612.0	1686.1	0.02422	0.21442	0.23865	634.8	524.7	1159.5	0.8294	0.4896	1.3190	612.0
610.0	1661.6	0.02412	0.21915	0.24327	631.8	529.2	1160.9	0.8267	0.4947	1.3214	610.0
608.0	1637.3	0.02402	0.22394	0.24796	628.8	533.6	1162.4	0.8240	0.4997	1.3238	608.0
606.0	1613.4	0.02392	0.22881	0.25273	625.9	537.9	1163.8	0.8214	0.5048	1.3261	606.0
604.0	1589.7	0.02382	0.23374	0.25757	622.9	542.2	1165.1	0.8187	0.5097	1.3284	604.0
602.0	1566.3	0.02373	0.23875	0.26248	620.0	546.4	1166.4	0.8161	0.5147	1.3307	602.0
600.0	1543.2	0.02364	0.24384	0.26747	617.1	550.6	1167.7	0.8134	0.5196	1.3330	600.0
598.0	1520.4	0.02354	0.24900	0.27255	614.3	554.7	1169.0	0.8108	0.5245	1.3353	598.0
596.0	1497.8	0.02345	0.25425	0.27770	611.4	558.8	1170.2	0.8082	0.5293	1.3375	596.0
594.0	1475.4	0.02337	0.25958	0.28294	608.6	562.8	1171.4	0.8056	0.5342	1.3398	594.0
592.0	1453.3	0.02328	0.26499	0.28827	605.7	566.8	1172.6	0.8030	0.5390	1.3420	592.0
590.0	1431.5	0.02319	0.27049	0.29368	602.9	570.8	1173.7	0.8004	0.5437	1.3442	590.0
588.0	1410.0	0.02311	0.27608	0.29919	600.1	574.7	1174.8	0.7978	0.5485	1.3464	588.0
586.0	1388.6	0.02303	0.28176	0.30478	597.3	578.5	1175.9	0.7953	0.5532	1.3485	586.0
584.0	1367.6	0.02295	0.28753	0.31048	594.6	582.4	1176.9	0.7927	0.5580	1.3507	584.0
582.0	1346.7	0.02287	0.29340	0.31627	591.8	586.1	1178.0	0.7902	0.5627	1.3528	582.0
580.0	1326.2	0.02279	0.29937	0.32216	589.1	589.9	1179.0	0.7876	0.5673	1.3550	580.0

Quantities for saturated liquid v_f h_f s_f

Quantities for saturated vapor v_g h_g s_g

Increment for evaporation v_{fg} h_{fg} s_{fg}

STEAM TABLES (Continued)

Properties of Saturated Steam and Saturated Water

Temp. F	Press. psia	Volume, ft³/lbm Water v_f	Evap. v_{fg}	Steam v_g	Enthalpy, Btu/lbm Water h_f	Evap. h_{fg}	Steam h_g	Entropy, Btu/lbm ×F Water s_f	Evap. s_{fg}	Steam s_g	Temp. F
580.0	1326.17	0.02279	0.29937	0.32216	589.1	589.9	1179.0	0.7876	0.5673	1.3550	580.0
578.0	1305.84	0.02271	0.30544	0.32816	586.4	593.6	1179.9	0.7851	0.5720	1.3571	578.0
576.0	1285.74	0.02264	0.31162	0.33426	583.7	597.2	1180.9	0.7825	0.5766	1.3592	576.0
574.0	1265.89	0.02256	0.31790	0.34046	581.0	600.9	1181.8	0.7800	0.5813	1.3613	574.0
572.0	1246.26	0.02249	0.32429	0.34678	578.3	604.5	1182.7	0.7775	0.5859	1.3634	572.0
570.0	1226.88	0.02242	0.33079	0.35321	575.6	608.0	1183.6	0.7750	0.5905	1.3654	570.0
568.0	1207.72	0.02235	0.33741	0.35975	572.9	611.5	1184.5	0.7725	0.5950	1.3675	568.0
566.0	1188.80	0.02228	0.34414	0.36642	570.3	615.0	1185.3	0.7699	0.5996	1.3696	566.0
564.0	1170.10	0.02221	0.35099	0.37320	567.6	618.5	1186.1	0.7674	0.6041	1.3716	564.0
562.0	1151.63	0.02214	0.35797	0.38011	565.0	621.9	1186.9	0.7650	0.6087	1.3736	562.0
560.0	1133.38	0.02207	0.36507	0.38714	562.4	625.3	1187.7	0.7625	0.6132	1.3757	560.0
558.0	1115.36	0.02201	0.37230	0.39431	559.8	628.6	1188.4	0.7600	0.6177	1.3777	558.0
556.0	1097.55	0.02194	0.37966	0.40160	557.2	632.0	1189.2	0.7575	0.6222	1.3797	556.0
554.0	1079.96	0.02188	0.38715	0.40903	554.6	635.3	1189.9	0.7550	0.6267	1.3817	554.0
552.0	1062.59	0.02182	0.39479	0.41660	552.0	638.5	1190.6	0.7525	0.6311	1.3837	552.0
550.0	1045.43	0.02176	0.40256	0.42432	549.5	641.8	1191.2	0.7501	0.6356	1.3856	550.0
548.0	1028.49	0.02169	0.41048	0.43217	546.9	645.0	1191.9	0.7476	0.6400	1.3876	548.0
546.0	1011.75	0.02163	0.41855	0.44018	544.4	648.1	1192.5	0.7451	0.6445	1.3896	546.0
544.0	995.22	0.02157	0.42677	0.44834	541.8	651.3	1193.1	0.7427	0.6489	1.3915	544.0
542.0	978.90	0.02151	0.43514	0.45665	539.3	654.4	1193.7	0.7402	0.6533	1.3935	542.0
540.0	962.79	0.02146	0.44367	0.46513	536.8	657.5	1194.3	0.7378	0.6577	1.3954	540.0
538.0	946.88	0.02140	0.45237	0.47377	534.2	660.6	1194.8	0.7353	0.6621	1.3974	538.0
536.0	931.17	0.02134	0.46123	0.48257	531.7	663.6	1195.4	0.7329	0.6665	1.3993	536.0
534.0	915.66	0.02129	0.47026	0.49155	529.2	666.6	1195.9	0.7304	0.6708	1.4012	534.0
532.0	900.34	0.02123	0.47947	0.50070	526.8	669.6	1196.4	0.7280	0.6752	1.4032	532.0
530.0	885.23	0.02118	0.48886	0.51004	524.3	672.6	1196.9	0.7255	0.6796	1.4051	530.0
528.0	870.31	0.02112	0.49843	0.51955	521.8	675.5	1197.3	0.7231	0.6839	1.4070	528.0
526.0	855.58	0.02107	0.50819	0.52926	519.3	678.4	1197.8	0.7206	0.6883	1.4089	526.0
524.0	841.04	0.02102	0.51814	0.53916	516.9	681.3	1198.2	0.7182	0.6926	1.4108	524.0
522.0	826.69	0.02097	0.52829	0.54926	514.4	684.2	1198.6	0.7158	0.6969	1.4127	522.0
520.0	812.53	0.02091	0.53864	0.55956	512.0	687.0	1199.0	0.7133	0.7013	1.4146	520.0
518.0	798.55	0.02086	0.54920	0.57006	509.6	689.9	1199.4	0.7109	0.7056	1.4165	518.0
516.0	784.76	0.02081	0.55997	0.58079	507.1	692.7	1199.8	0.7085	0.7099	1.4183	516.0
514.0	771.15	0.02076	0.57096	0.59173	504.7	695.4	1200.2	0.7060	0.7142	1.4202	514.0
512.0	757.72	0.02072	0.58218	0.60280	502.3	608.2	1200.5	0.7000	0.7185	1.4221	512.0
510.0	744.47	0.02067	0.59362	0.61429	499.9	700.9	1200.8	0.7012	0.7228	1.4240	510.0
508.0	731.40	0.02062	0.60530	0.62592	497.5	703.7	1201.1	0.6987	0.7271	1.4258	508.0
506.0	718.50	0.02057	0.61722	0.63779	495.1	706.3	1201.4	0.6963	0.7314	1.4277	506.0
504.0	705.78	0.02053	0.62938	0.64991	492.7	709.0	1201.7	0.6939	0.7357	1.4296	504.0
502.0	693.23	0.02048	0.64180	0.66228	490.3	711.7	1202.0	0.6915	0.7400	1.4314	502.0
500.0	680.86	0.02043	0.65448	0.67492	487.9	714.3	1202.2	0.6890	0.7443	1.4333	500.0
498.0	668.65	0.02039	0.66743	0.68782	485.6	716.9	1202.5	0.6866	0.7486	1.4352	498.0
496.0	656.61	0.02034	0.68065	0.70100	483.2	719.5	1202.7	0.6842	0.7528	1.4370	496.0
494.0	644.73	0.02030	0.69415	0.71445	480.8	722.1	1202.9	0.6818	0.7571	1.4389	494.0
492.0	633.03	0.02026	0.70794	0.72820	478.5	724.6	1203.1	0.6793	0.7614	1.4407	492.0
490.0	621.48	0.02021	0.72203	0.74224	476.1	727.2	1203.3	0.6769	0.7657	1.4426	490.0
488.0	610.10	0.02017	0.73641	0.75658	473.8	729.7	1203.5	0.6745	0.7700	1.4444	488.0
486.0	598.87	0.02013	0.75111	0.77124	471.5	732.2	1203.7	0.6721	0.7742	1.4463	486.0
484.0	587.81	0.02009	0.76613	0.78622	469.1	734.7	1203.8	0.6696	0.7785	1.4481	484.0
482.0	576.90	0.02004	0.78148	0.80152	466.8	737.2	1204.0	0.6672	0.7828	1.4500	482.0
480.0	566.15	0.02000	0.79716	0.81717	464.5	739.6	1204.1	0.6648	0.7871	1.4518	480.0
478.0	555.55	0.01996	0.81319	0.83315	462.2	742.1	1204.2	0.6624	0.7913	1.4537	478.0
476.0	545.11	0.01992	0.82958	0.84950	459.9	744.5	1204.3	0.6599	0.7956	1.4555	476.0
474.0	534.81	0.01988	0.84632	0.86621	457.5	746.9	1204.4	0.6575	0.7999	1.4574	474.0
472.0	524.67	0.01984	0.86345	0.88329	455.2	749.3	1204.5	0.6551	0.8042	1.4592	472.0
470.0	514.67	0.01980	0.88095	0.90076	452.9	751.6	1204.6	0.6527	0.8084	1.4611	470.0
468.0	504.83	0.01976	0.89885	0.91862	450.7	754.0	1204.6	0.6502	0.8127	1.4629	468.0
466.0	495.12	0.01973	0.91716	0.93689	448.4	756.3	1204.7	0.6478	0.8170	1.4648	466.0
464.0	485.56	0.01969	0.93588	0.95557	446.1	758.6	1204.7	0.6454	0.8213	1.4667	464.0
462.0	476.14	0.01965	0.95504	0.97469	443.8	761.0	1204.8	0.6429	0.8256	1.4685	462.0
460.0	466.87	0.01961	0.97463	0.99424	441.5	763.2	1204.8	0.6405	0.8299	1.4704	460.0
458.0	457.73	0.01958	0.99467	1.01425	439.3	765.5	1204.8	0.6381	0.8342	1.4722	458.0
456.0	448.73	0.01954	1.01518	1.03472	437.0	767.8	1204.8	0.6356	0.8385	1.4741	456.0
454.0	439.87	0.01950	1.03616	1.05567	434.7	770.0	1204.8	0.6332	0.8428	1.4759	454.0
452.0	431.14	0.01947	1.05764	1.07711	432.5	772.3	1204.8	0.6308	0.8471	1.4778	452.0
450.0	422.55	0.01943	1.07962	1.09905	430.2	774.5	1204.7	0.6283	0.8514	1.4797	450.0
448.0	414.09	0.01940	1.10212	1.12152	428.0	776.7	1204.7	0.6259	0.8557	1.4815	448.0
446.0	405.76	0.01936	1.12515	1.14452	425.7	778.9	1204.6	0.6234	0.8600	1.4834	446.0
444.0	397.56	0.01933	1.14874	1.16806	423.5	781.1	1204.6	0.6210	0.8643	1.4853	444.0
442.0	389.49	0.01929	1.17288	1.19217	421.3	783.2	1204.5	0.6185	0.8686	1.4872	442.0
440.0	381.54	0.01926	1.19761	1.21687	419.0	785.4	1204.4	0.6161	0.8729	1.4890	440.0
438.0	373.72	0.01923	1.22293	1.24216	416.8	787.5	1204.3	0.6136	0.8773	1.4909	438.0
436.0	366.03	0.01919	1.24887	1.26806	414.6	789.7	1204.2	0.6112	0.8816	1.4928	436.0
434.0	358.46	0.01916	1.27544	1.29460	412.4	791.8	1204.1	0.6087	0.8859	1.4947	434.0
432.0	351.00	0.01913	1.30266	1.32179	410.1	793.9	1204.0	0.6063	0.8903	1.4966	432.0
430.0	343.67	0.01909	1.33055	1.34965	407.9	796.0	1203.9	0.6038	0.8946	1.4985	430.0

STEAM TABLES (Continued)

Properties of Saturated Steam and Saturated Water

Temp. F	Press. psia	Volume, ft³/lbm Water v_f	Evap. v_{fg}	Steam v_g	Enthalpy, Btu/lbm Water h_f	Evap. h_{fg}	Steam h_g	Entropy, Btu/lbm ×F Water s_f	Evap. s_{fg}	Steam s_g	Temp. F
430.0	343.674	0.01909	1.3306	1.3496	407.9	796.0	1203.9	0.6038	0.8946	1.4985	430.0
428.0	336.463	0.01906	1.3591	1.3782	405.7	798.0	1203.7	0.6014	0.8990	1.5004	428.0
426.0	329.369	0.01903	1.3884	1.4075	403.5	800.1	1203.6	0.5989	0.9034	1.5023	426.0
424.0	322.391	0.01900	1.4184	1.4374	401.3	802.2	1203.5	0.5964	0.9077	1.5042	424.0
422.0	315.529	0.01897	1.4492	1.4682	399.1	804.2	1203.3	0.5940	0.9121	1.5061	422.0
420.0	308.780	0.01894	1.4808	1.4997	396.9	806.2	1203.1	0.5915	0.9165	1.5080	420.0
418.0	302.143	0.01890	1.5131	1.5320	394.7	808.2	1202.9	0.5890	0.9209	1.5099	418.0
416.0	295.617	0.01887	1.5463	1.5651	392.5	810.2	1202.8	0.5866	0.9253	1.5118	416.0
414.0	289.201	0.01884	1.5803	1.5991	390.3	812.2	1202.6	0.5841	0.9297	1.5137	414.0
412.0	282.894	0.01881	1.6152	1.6340	388.1	814.2	1202.4	0.5816	0.9341	1.5157	412.0
410.0	276.694	0.01878	1.6510	1.6697	386.0	816.2	1202.1	0.5791	0.9385	1.5176	410.0
408.0	270.600	0.01875	1.6877	1.7064	383.8	818.2	1201.9	0.5766	0.9429	1.5195	408.0
406.0	264.611	0.01872	1.7253	1.7441	381.6	820.1	1201.7	0.5742	0.9473	1.5215	406.0
404.0	258.725	0.01870	1.7640	1.7827	379.4	822.0	1201.5	0.5717	0.9518	1.5234	404.0
402.0	252.942	0.01867	1.8037	1.8223	377.3	824.0	1201.2	0.5692	0.9562	1.5254	402.0
400.0	247.259	0.01864	1.8444	1.8630	375.1	825.9	1201.0	0.5667	0.9607	1.5274	400.0
398.0	241.677	0.01861	1.8862	1.9048	372.9	827.8	1200.7	0.5642	0.9651	1.5293	398.0
396.0	236.193	0.01858	1.9291	1.9477	370.8	829.7	1200.4	0.5617	0.9696	1.5313	396.0
394.0	230.807	0.01855	1.9731	1.9917	368.6	831.6	1200.2	0.5592	0.9741	1.5333	394.0
392.0	225.516	0.01853	2.0184	2.0369	366.5	833.4	1199.9	0.5567	0.9786	1.5352	392.0
390.0	220.321	0.01850	2.0649	2.0833	364.3	835.3	1199.6	0.5542	0.9831	1.5372	390.0
388.0	215.220	0.01847	2.1126	2.1311	362.2	837.2	1199.3	0.5516	0.9876	1.5392	388.0
386.0	210.211	0.01844	2.1616	2.1801	360.0	839.0	1199.0	0.5491	0.9921	1.5412	386.0
384.0	205.294	0.01842	2.2120	2.2304	357.9	840.8	1198.7	0.5466	0.9966	1.5432	384.0
382.0	200.467	0.01839	2.2638	2.2821	355.7	842.7	1198.4	0.5441	1.0012	1.5452	382.0
380.0	195.729	0.01836	2.3170	2.3353	353.6	844.5	1198.0	0.5416	1.0057	1.5473	380.0
378.0	191.080	0.01834	2.3716	2.3900	351.4	846.3	1197.7	0.5390	1.0103	1.5493	378.0
376.0	186.517	0.01831	2.4279	2.4462	349.3	848.1	1197.4	0.5365	1.0148	1.5513	376.0
374.0	182.040	0.01829	2.4857	2.5039	347.2	849.8	1197.0	0.5340	1.0194	1.5534	374.0
372.0	177.648	0.01826	2.5451	2.5633	345.0	851.6	1196.7	0.5314	1.0240	1.5554	372.0
370.0	173.339	0.01823	2.6062	2.6244	342.9	853.4	1196.3	0.5289	1.0286	1.5575	370.0
368.0	169.113	0.01821	2.6691	2.6873	340.8	855.1	1195.9	0.5263	1.0332	1.5595	368.0
366.0	164.968	0.01818	2.7337	2.7519	338.7	856.9	1195.6	0.5238	1.0378	1.5616	366.0
364.0	160.903	0.01816	2.8002	2.8184	336.5	858.6	1195.2	0.5212	1.0424	1.5637	364.0
362.0	156.917	0.01813	2.8687	2.8868	334.4	860.4	1194.8	0.5187	1.0471	1.5658	362.0
360.0	153.010	0.01811	2.9392	2.9573	332.3	862.1	1194.4	0.5161	1.0517	1.5678	360.0
358.0	149.179	0.01809	3.0117	3.0298	330.2	863.8	1194.0	0.5135	1.0564	1.5699	358.0
356.0	145.424	0.01806	3.0863	3.1044	328.1	865.5	1193.6	0.5110	1.0611	1.5721	356.0
354.0	141.744	0.01804	3.1632	3.1812	326.0	867.2	1193.2	0.5084	1.0658	1.5742	354.0
352.0	138.138	0.01801	3.2423	3.2603	323.9	868.9	1192.7	0.5058	1.0705	1.5763	352.0
350.0	134.604	0.01799	3.3238	3.3418	321.8	870.6	1192.3	0.5032	1.0752	1.5784	350.0
348.0	131.142	0.01797	3.4078	3.4258	319.7	872.2	1191.9	0.5006	1.0799	1.5806	348.0
346.0	127.751	0.01794	3.4943	3.5122	317.6	873.9	1191.4	0.4980	1.0847	1.5827	346.0
344.0	124.430	0.01792	3.5834	3.6013	315.5	875.5	1191.0	0.4954	1.0894	1.5849	344.0
342.0	121.177	0.01790	3.6752	3.6931	313.4	877.2	1190.5	0.4928	1.0942	1.5871	342.0
340.0	117.992	0.01787	3.7699	3.7878	311.3	878.8	1190.1	0.4902	1.0990	1.5892	340.0
338.0	114.873	0.01785	3.8675	3.8853	309.2	880.5	1189.6	0.4876	1.1038	1.5914	338.0
336.0	111.820	0.01783	3.9681	3.9859	307.1	882.1	1189.1	0.4850	1.1086	1.5936	336.0
334.0	108.832	0.01781	4.0718	4.0896	305.0	883.7	1188.7	0.4824	1.1134	1.5958	334.0
332.0	105.907	0.01779	4.1788	4.1966	302.9	885.3	1188.2	0.4798	1.1183	1.5981	332.0
330.0	103.045	0.01776	4.2892	4.3069	300.8	886.9	1187.7	0.4772	1.1231	1.6003	330.0
328.0	100.245	0.01774	4.4030	4.4208	298.7	888.5	1187.2	0.4745	1.1280	1.6025	328.0
326.0	97.506	0.01772	4.5205	4.5382	296.6	890.1	1186.7	0.4719	1.1329	1.6048	326.0
324.0	94.826	0.01770	4.6418	4.6595	294.6	891.6	1186.2	0.4692	1.1378	1.6071	324.0
322.0	92.205	0.01768	4.7669	4.7846	292.5	893.2	1185.7	0.4666	1.1427	1.6093	322.0
320.0	89.643	0.01766	4.8961	4.9138	290.4	894.8	1185.2	0.4640	1.1477	1.6116	320.0
318.0	87.137	0.01764	5.0295	5.0471	288.3	896.3	1184.7	0.4613	1.1526	1.6139	318.0
316.0	84.688	0.01761	5.1673	5.1849	286.3	897.9	1184.1	0.4586	1.1576	1.6162	316.0
314.0	82.293	0.01759	5.3096	5.3272	284.2	899.4	1183.6	0.4560	1.1626	1.6185	314.0
312.0	79.953	0.01757	5.4566	5.4742	282.1	901.0	1183.1	0.4533	1.1676	1.6209	312.0
310.0	77.667	0.01755	5.6085	5.6260	280.0	902.5	1182.5	0.4506	1.1726	1.6232	310.0
308.0	75.433	0.01753	5.7655	5.7830	278.0	904.0	1182.0	0.4479	1.1776	1.6256	308.0
306.0	73.251	0.01751	5.9277	5.9452	275.9	905.5	1181.4	0.4453	1.1827	1.6279	306.0
304.0	71.119	0.01749	6.0955	6.1130	273.8	907.0	1180.9	0.4426	1.1877	1.6303	304.0
302.0	69.038	0.01747	6.2689	6.2864	271.8	908.5	1180.3	0.4399	1.1928	1.6327	302.0
300.0	67.005	0.01745	6.4483	6.4658	269.7	910.0	1179.7	0.4372	1.1979	1.6351	300.0
298.0	65.021	0.01743	6.6339	6.6513	267.7	911.5	1179.2	0.4345	1.2031	1.6375	298.0
296.0	63.084	0.01741	6.8259	6.8433	265.6	913.0	1178.6	0.4317	1.2082	1.6400	296.0
294.0	61.194	0.01739	7.0245	7.0419	263.5	914.5	1178.0	0.4290	1.2134	1.6424	294.0
292.0	59.350	0.01738	7.2301	7.2475	261.5	915.9	1177.4	0.4263	1.2186	1.6449	292.0
290.0	57.550	0.01736	7.4430	7.4603	259.4	917.4	1176.8	0.4236	1.2238	1.6473	290.0
288.0	55.795	0.01734	7.6634	7.6807	257.4	918.8	1176.2	0.4208	1.2290	1.6498	288.0
286.0	54.083	0.01732	7.8916	7.9089	255.3	920.3	1175.6	0.4181	1.2342	1.6523	286.0
284.0	52.414	0.01730	8.1280	8.1453	253.3	921.7	1175.0	0.4154	1.2395	1.6548	284.0
282.0	50.786	0.01728	8.3729	8.3902	251.2	923.2	1174.4	0.4126	1.2448	1.6574	282.0
280.0	49.200	0.01726	8.6267	8.6439	249.2	924.6	1173.8	0.4098	1.2501	1.6599	280.0

STEAM TABLES (Continued)

Properties of Saturated Steam and Saturated Water

Temp. F	Press. psia	Volume, ft³/lbm Water v_f	Evap. v_{fg}	Steam v_g	Enthalpy, Btu/lbm Water h_f	Evap. h_{fg}	Steam h_g	Entropy, Btu/lbm ×F Water s_f	Evap. s_{fg}	Steam s_g	Temp. F
280.0	49.200	0.017264	8.627	8.644	249.17	924.6	1173.8	0.4098	1.2501	1.6599	280.0
278.0	47.653	0.017246	8.890	8.907	247.13	926.0	1173.2	0.4071	1.2554	1.6625	278.0
276.0	46.147	0.017228	9.162	9.180	245.08	927.5	1172.5	0.4043	1.2607	1.6650	276.0
274.0	44.678	0.017210	9.445	9.462	243.03	928.9	1171.9	0.4015	1.2661	1.6676	274.0
272.0	43.249	0.017193	9.738	9.755	240.99	930.3	1171.3	0.3987	1.2715	1.6702	272.0
270.0	41.856	0.017175	10.040	10.060	[illegible]	[illegible]	[illegible]	[illegible]	[illegible]	[illegible]	270.0
268.0	40.500	0.017157	10.358	10.375	236.91	933.1	1170.0	0.3932	1.2823	1.6755	268.0
266.0	39.179	0.017140	10.685	10.703	234.87	934.5	1169.3	0.3904	1.2878	1.6781	266.0
264.0	37.894	0.017123	11.025	11.042	232.83	935.9	1168.7	0.3876	1.2933	1.6808	264.0
262.0	36.644	0.017106	11.378	11.395	230.79	937.3	1168.0	0.3847	1.2988	1.6835	262.0
260.0	35.427	0.017089	11.745	11.762	228.76	938.6	1167.4	0.3819	1.3043	1.6862	260.0
258.0	34.243	0.017072	12.125	12.142	226.72	940.0	1166.7	0.3791	1.3098	1.6889	258.0
256.0	33.091	0.017055	12.520	12.538	224.69	941.4	1166.1	0.3763	1.3154	1.6917	256.0
254.0	31.972	0.017039	12.931	12.948	222.65	942.7	1165.4	0.3734	1.3210	1.6944	254.0
252.0	30.883	0.017022	13.358	13.375	220.62	944.1	1164.7	0.3706	1.3266	1.6972	252.0
250.0	29.825	0.017006	13.802	13.819	218.59	945.4	1164.0	0.3677	1.3323	1.7000	250.0
248.0	28.796	0.016990	14.264	14.281	216.56	946.8	1163.4	0.3649	1.3379	1.7028	248.0
246.0	27.797	0.016974	14.744	14.761	214.53	948.1	1162.7	0.3620	1.3436	1.7056	246.0
244.0	26.826	0.016958	15.243	15.260	212.50	949.5	1162.0	0.3591	1.3494	1.7085	244.0
242.0	25.883	0.016942	15.763	15.780	210.48	950.8	1161.3	0.3562	1.3551	1.7113	242.0
240.0	24.968	0.016926	16.304	16.321	208.45	952.1	1160.6	0.3533	1.3609	1.7142	240.0
238.0	24.079	0.016910	16.867	16.884	206.42	953.5	1159.9	0.3505	1.3667	1.7171	238.0
236.0	23.216	0.016895	17.454	17.471	204.40	954.8	1159.2	0.3476	1.3725	1.7201	236.0
234.0	22.379	0.016880	18.065	18.082	202.38	956.1	1158.5	0.3446	1.3784	1.7230	234.0
232.0	21.567	0.016864	18.701	18.718	200.35	957.4	1157.8	0.3417	1.3842	1.7260	232.0
230.0	20.779	0.016849	19.364	19.381	198.33	958.7	1157.1	0.3388	1.3902	1.7290	230.0
229.0	20.394	0.016842	19.707	19.723	197.32	959.4	1156.7	0.3373	1.3931	1.7305	229.0
228.0	20.015	0.016834	20.056	20.073	196.31	960.0	1156.3	0.3359	1.3961	1.7320	228.0
227.0	19.642	0.016827	20.413	20.429	195.30	960.7	1156.0	0.3344	1.3991	1.7335	227.0
226.0	19.274	0.016819	20.777	20.794	194.29	961.3	1155.6	0.3329	1.4021	1.7350	226.0
225.0	18.912	0.016812	21.149	21.166	193.28	962.0	1155.3	0.3315	1.4051	1.7365	225.0
224.0	18.556	0.016805	21.529	21.545	192.27	962.6	1154.9	0.3300	1.4081	1.7380	224.0
223.0	18.206	0.016797	21.917	21.933	191.26	963.3	1154.5	0.3285	1.4111	1.7396	223.0
222.0	17.860	0.016790	22.313	22.330	190.25	963.9	1154.2	0.3270	1.4141	1.7411	222.0
221.0	17.521	0.016783	22.718	22.735	189.24	964.6	1153.8	0.3255	1.4171	1.7427	221.0
220.0	17.186	0.016775	23.131	23.148	188.23	965.2	1153.4	0.3241	1.4201	1.7442	220.0
219.0	16.857	0.016768	23.554	23.571	187.22	965.8	1153.1	0.3226	1.4232	1.7458	219.0
218.0	16.533	0.016761	23.986	24.002	186.21	966.5	1152.7	0.3211	1.4262	1.7473	218.0
217.0	16.214	0.016754	24.427	24.444	185.21	967.1	1152.3	0.3196	1.4293	1.7489	217.0
216.0	15.901	0.016747	24.878	24.894	184.20	967.8	1152.0	0.3181	1.4323	1.7505	216.0
215.0	15.592	0.016740	25.338	25.355	183.19	968.4	1151.6	0.3166	1.4354	1.7520	215.0
214.0	15.289	0.016733	25.809	25.826	182.18	969.0	1151.2	0.3151	1.4385	1.7536	214.0
213.0	14.990	0.016726	26.290	26.307	181.17	969.7	1150.8	0.3136	1.4416	1.7552	213.0
212.0	14.696	0.016719	26.782	26.799	180.17	970.3	1150.5	0.3121	1.4447	1.7568	212.0
211.0	14.407	0.016712	27.285	27.302	179.16	970.9	1150.1	0.3106	1.4478	1.7584	211.0
210.0	14.123	0.016705	27.799	27.816	178.15	971.6	1149.7	0.3091	1.4509	1.7600	210.0
209.0	13.843	0.016698	28.324	28.341	177.14	972.2	1149.4	0.3076	1.4540	1.7616	209.0
208.0	13.568	0.016691	28.862	28.878	176.14	972.8	1149.0	0.3061	1.4571	1.7632	208.0
207.0	13.297	0.016684	29.411	29.428	175.13	973.5	1148.6	0.3046	1.4602	1.7649	207.0
206.0	13.031	0.016677	29.973	29.989	174.12	974.1	1148.2	0.3031	1.4634	1.7665	206.0
205.0	12.770	0.016670	30.547	30.564	173.12	974.7	1147.9	0.3016	1.4665	1.7681	205.0
204.0	12.512	0.016664	31.135	31.151	172.11	975.4	1147.5	0.3001	1.4697	1.7698	204.0
203.0	12.259	0.016657	31.736	31.752	171.10	976.0	1147.1	0.2986	1.4728	1.7714	203.0
202.0	12.011	0.016650	32.350	32.367	170.10	976.6	1146.7	0.2971	1.4760	1.7731	202.0
201.0	11.766	0.016643	32.979	32.996	169.09	977.2	1146.3	0.2955	1.4792	1.7747	201.0
200.0	11.526	0.016637	33.622	33.639	168.09	977.9	1146.0	0.2940	1.4824	1.7764	200.0
199.0	11.290	0.016630	34.280	34.297	167.08	978.5	1145.6	0.2925	1.4856	1.7781	199.0
198.0	11.058	0.016624	34.954	34.970	166.08	979.1	1145.2	0.2910	1.4888	1.7798	198.0
197.0	10.830	0.016617	35.643	35.659	165.07	979.7	1144.8	0.2894	1.4920	1.7814	197.0
196.0	10.605	0.016611	36.348	36.364	164.06	980.4	1144.4	0.2879	1.4952	1.7831	196.0
195.0	10.385	0.016604	37.069	37.086	163.06	981.0	1144.0	0.2864	1.4985	1.7848	195.0
194.0	10.168	0.016598	37.808	37.824	162.05	981.6	1143.7	0.2848	1.5017	1.7865	194.0
193.0	9.956	0.016591	38.564	38.580	161.05	982.2	1143.3	0.2833	1.5050	1.7882	193.0
192.0	9.747	0.016585	39.337	39.354	160.05	982.8	1142.9	0.2818	1.5082	1.7900	192.0
191.0	9.541	0.016578	40.130	40.146	159.04	983.5	1142.5	0.2802	1.5115	1.7917	191.0
190.0	9.340	0.016572	40.941	40.957	158.04	984.1	1142.1	0.2787	1.5148	1.7934	190.0
189.0	9.141	0.016566	41.771	41.787	157.03	984.7	1141.7	0.2771	1.5180	1.7952	189.0
188.0	8.947	0.016559	42.621	42.638	156.03	985.3	1141.3	0.2756	1.5213	1.7969	188.0
187.0	8.756	0.016553	43.492	43.508	155.02	985.9	1140.9	0.2740	1.5246	1.7987	187.0
186.0	8.568	0.016547	44.383	44.400	154.02	986.5	1140.5	0.2725	1.5279	1.8004	186.0
185.0	8.384	0.016541	45.297	45.313	153.02	987.1	1140.2	0.2709	1.5313	1.8022	185.0
184.0	8.203	0.016534	46.232	46.249	152.01	987.8	1139.8	0.2694	1.5346	1.8040	184.0
183.0	8.025	0.016528	47.190	47.207	151.01	988.4	1139.4	0.2678	1.5379	1.8057	183.0
182.0	7.850	0.016522	48.172	48.189	150.01	989.0	1139.0	0.2662	1.5413	1.8075	182.0
181.0	7.679	0.016516	49.178	49.194	149.00	989.6	1138.6	0.2647	1.5446	1.8093	181.0
180.0	7.511	0.016510	50.208	50.225	148.00	990.2	1138.2	0.2631	1.5480	1.8111	180.0

STEAM TABLES (Continued)

Properties of Saturated Steam and Saturated Water

Temp. F	Press. psia	Volume, ft³/lbm Water v_f	Evap. v_{fg}	Steam v_g	Enthalpy, Btu/lbm Water h_f	Evap. h_{fg}	Steam h_g	Entropy, Btu/lbm×F Water s_f	Evap. s_{fg}	Steam s_g	Temp. F
180.0	7.5110	0.016510	50.21	50.22	148.00	990.2	1138.2	0.2631	1.5480	1.8111	180.0
179.0	7.3460	0.016504	51.26	51.28	147.00	990.8	1137.8	0.2615	1.5514	1.8129	179.0
178.0	7.1840	0.016498	52.35	52.36	145.99	991.4	1137.4	0.2600	1.5548	1.8147	178.0
177.0	7.0250	0.016492	53.46	53.47	144.99	992.0	1137.0	0.2584	1.5582	1.8166	177.0
176.0	6.8690	0.016486	54.59	54.61	143.99	992.6	1136.6	0.2568	1.5616	1.8184	176.0
175.0	6.7159	0.016480	55.76	55.77	142.99	993.2	1136.2	0.2552	1.5650	1.8202	175.0
174.0	6.5656	0.016474	56.95	56.97	141.98	993.8	1135.8	0.2537	1.5684	1.8221	174.0
173.0	6.4182	0.016468	58.18	58.19	140.98	994.4	1135.4	0.2521	1.5718	1.8239	173.0
172.0	6.2736	0.016463	59.43	59.45	139.98	995.0	1135.0	0.2505	1.5753	1.8258	172.0
171.0	6.1318	0.016457	60.72	60.74	138.98	995.6	1134.6	0.2489	1.5787	1.8276	171.0
170.0	5.9926	0.016451	62.04	62.06	137.97	996.2	1134.2	0.2473	1.5822	1.8295	170.0
169.0	5.8562	0.016445	63.39	63.41	136.97	996.8	1133.8	0.2457	1.5857	1.8314	169.0
168.0	5.7223	0.016440	64.78	64.80	135.97	997.4	1133.4	0.2441	1.5892	1.8333	168.0
167.0	5.5911	0.016434	66.21	66.22	134.97	998.0	1133.0	0.2425	1.5926	1.8352	167.0
166.0	5.4623	0.016428	67.67	67.68	133.97	998.6	1132.6	0.2409	1.5961	1.8371	166.0
165.0	5.3361	0.016423	69.17	69.18	132.96	999.2	1132.2	0.2393	1.5997	1.8390	165.0
164.0	5.2124	0.016417	70.70	70.72	131.96	999.8	1131.8	0.2377	1.6032	1.8409	164.0
163.0	5.0911	0.016412	72.28	72.30	130.96	1000.4	1131.4	0.2361	1.6067	1.8428	163.0
162.0	4.9722	0.016406	73.90	73.92	129.96	1001.0	1131.0	0.2345	1.6103	1.8448	162.0
161.0	4.8556	0.016401	75.56	75.58	128.96	1001.6	1130.6	0.2329	1.6138	1.8467	161.0
160.0	4.7414	0.016395	77.27	77.29	127.96	1002.2	1130.2	0.2313	1.6174	1.8487	160.0
159.0	4.6294	0.016390	79.02	79.04	126.96	1002.8	1129.8	0.2297	1.6210	1.8506	159.0
158.0	4.5197	0.016384	80.82	80.83	125.96	1003.4	1129.4	0.2281	1.6245	1.8526	158.0
157.0	4.4122	0.016379	82.66	82.68	124.95	1004.0	1129.0	0.2264	1.6281	1.8546	157.0
156.0	4.3068	0.016374	84.56	84.57	123.95	1004.6	1128.6	0.2248	1.6318	1.8566	156.0
155.0	4.2036	0.016369	86.50	86.52	122.95	1005.2	1128.2	0.2232	1.6354	1.8586	155.0
154.0	4.1025	0.016363	88.50	88.52	121.95	1005.8	1127.7	0.2216	1.6390	1.8606	154.0
153.0	4.0035	0.016358	90.55	90.57	120.95	1006.4	1127.3	0.2199	1.6426	1.8626	153.0
152.0	3.9065	0.016353	92.66	92.68	119.95	1007.0	1126.9	0.2183	1.6463	1.8646	152.0
151.0	3.8114	0.016348	94.83	94.84	118.95	1007.6	1126.5	0.2167	1.6500	1.8666	151.0
150.0	3.7184	0.016343	97.05	97.07	117.95	1008.2	1126.1	0.2150	1.6536	1.8686	150.0
149.0	3.6273	0.016337	99.33	99.35	116.95	1008.7	1125.7	0.2134	1.6573	1.8707	149.0
148.0	3.5381	0.016332	101.68	101.70	115.95	1009.3	1125.3	0.2117	1.6610	1.8727	148.0
147.0	3.4508	0.016327	104.10	104.11	114.95	1009.9	1124.9	0.2101	1.6647	1.8748	147.0
146.0	3.3653	0.016322	106.58	106.59	113.95	1010.5	1124.5	0.2084	1.6684	1.8769	146.0
145.0	3.2816	0.016317	109.12	109.14	112.95	1011.1	1124.0	0.2068	1.6722	1.8789	145.0
144.0	3.1997	0.016312	111.74	111.76	111.95	1011.7	1123.6	0.2051	1.6759	1.8810	144.0
143.0	3.1195	0.016308	114.44	114.45	110.95	1012.3	1123.2	0.2035	1.6797	1.8831	143.0
142.0	3.0411	0.016303	117.21	117.22	109.95	1012.9	1122.8	0.2018	1.6834	1.8852	142.0
141.0	2.9643	0.016298	120.05	120.07	108.95	1013.4	1122.4	0.2001	1.6872	1.8873	141.0
140.0	2.8892	0.016293	122.98	123.00	107.95	1014.0	1122.0	0.1985	1.6910	1.8895	140.0
139.0	2.8157	0.016288	125.99	126.01	106.95	1014.6	1121.6	0.1968	1.6948	1.8916	139.0
138.0	2.7438	0.016284	129.09	129.11	105.95	1015.2	1121.1	0.1951	1.6986	1.8937	138.0
137.0	2.6735	0.016279	132.28	132.29	104.95	1015.8	1120.7	0.1935	1.7024	1.8959	137.0
136.0	2.6047	0.016274	135.55	135.57	103.95	1016.4	1120.3	0.1918	1.7063	1.8980	136.0
135.0	2.5375	0.016270	138.93	138.94	102.95	1016.9	1119.9	0.1901	1.7101	1.9002	135.0
134.0	2.4717	0.016265	142.40	142.41	101.95	1017.5	1119.5	0.1884	1.7140	1.9024	134.0
133.0	2.4074	0.016260	145.97	145.98	100.95	1018.1	1119.1	0.1867	1.7178	1.9046	133.0
132.0	2.3445	0.016256	149.64	149.66	99.95	1018.7	1118.6	0.1851	1.7217	1.9068	132.0
131.0	2.2830	0.016251	153.42	153.44	98.95	1019.3	1118.2	0.1834	1.7256	1.9090	131.0
130.0	2.2230	0.016247	157.32	157.33	97.96	1019.8	1117.8	0.1817	1.7295	1.9112	130.0
129.0	2.1642	0.016243	161.32	161.34	96.96	1020.4	1117.4	0.1800	1.7335	1.9134	129.0
128.0	2.1068	0.016238	165.45	165.47	95.96	1021.0	1117.0	0.1783	1.7374	1.9157	128.0
127.0	2.0507	0.016234	169.70	169.72	94.96	1021.6	1116.5	0.1766	1.7413	1.9179	127.0
126.0	1.9959	0.016229	174.08	174.09	93.96	1022.2	1116.1	0.1749	1.7453	1.9202	126.0
125.0	1.9424	0.016225	178.58	178.60	92.96	1022.7	1115.7	0.1732	1.7493	1.9224	125.0
124.0	1.8901	0.016221	183.23	183.24	91.96	1023.3	1115.3	0.1715	1.7533	1.9247	124.0
123.0	1.8390	0.016217	188.01	188.03	90.96	1023.9	1114.9	0.1697	1.7573	1.9270	123.0
122.0	1.7891	0.016213	192.94	192.95	89.96	1024.5	1114.4	0.1680	1.7613	1.9293	122.0
121.0	1.7403	0.016208	198.01	198.03	88.96	1025.0	1114.0	0.1663	1.7653	1.9316	121.0
120.0	1.6927	0.016204	203.25	203.26	87.97	1025.6	1113.6	0.1646	1.7693	1.9339	120.0
119.0	1.6463	0.016200	208.64	208.66	86.97	1026.2	1113.2	0.1629	1.7734	1.9362	119.0
118.0	1.6009	0.016196	214.20	214.21	85.97	1026.8	1112.7	0.1611	1.7774	1.9386	118.0
117.0	1.5566	0.016192	219.93	219.94	84.97	1027.3	1112.3	0.1594	1.7815	1.9409	117.0
116.0	1.5133	0.016188	225.84	225.85	83.97	1027.9	1111.9	0.1577	1.7856	1.9433	116.0
115.0	1.4711	0.016184	231.93	231.94	82.97	1028.5	1111.5	0.1559	1.7897	1.9457	115.0
114.0	1.4299	0.016180	238.21	238.22	81.97	1029.1	1111.0	0.1542	1.7938	1.9480	114.0
113.0	1.3898	0.016177	244.69	244.70	80.98	1029.6	1110.6	0.1525	1.7980	1.9504	113.0
112.0	1.3505	0.016173	251.37	251.38	79.98	1030.2	1110.2	0.1507	1.8021	1.9528	112.0
111.0	1.3123	0.016169	258.26	258.28	78.98	1030.8	1109.8	0.1490	1.8063	1.9552	111.0
110.0	1.2750	0.016165	265.37	265.39	77.98	1031.4	1109.3	0.1472	1.8105	1.9577	110.0
109.0	1.2385	0.016162	272.71	272.72	76.98	1031.9	1108.9	0.1455	1.8146	1.9601	109.0
108.0	1.2030	0.016158	280.28	280.30	75.98	1032.5	1108.5	0.1437	1.8188	1.9626	108.0
107.0	1.1684	0.016154	288.09	288.11	74.99	1033.1	1108.1	0.1419	1.8231	1.9650	107.0
106.0	1.1347	0.016151	296.16	296.18	73.99	1033.6	1107.6	0.1402	1.8273	1.9675	106.0
105.0	1.1017	0.016147	304.49	304.50	72.99	1034.2	1107.2	0.1384	1.8315	1.9700	105.0

STEAM TABLES (Continued)

Properties of Saturated Steam and Saturated Water

Temp. F	Press. psia	Volume, ft³/lbm Water v_f	Volume, ft³/lbm Evap. v_{fg}	Volume, ft³/lbm Steam v_g	Enthalpy, Btu/lbm Water h_f	Enthalpy, Btu/lbm Evap. h_{fg}	Enthalpy, Btu/lbm Steam h_g	Entropy, Btu/lbm×F Water s_f	Entropy, Btu/lbm×F Evap. s_{fg}	Entropy, Btu/lbm×F Steam s_g	Temp. F
105.0	1.10174	0.016147	304.5	304.5	72.990	1034.2	1107.2	0.1384	1.8315	1.9700	105.0
104.0	1.06965	0.016144	313.1	313.1	71.992	1034.8	1106.8	0.1366	1.8358	1.9725	104.0
103.0	1.03838	0.016140	322.0	322.0	70.993	1035.4	1106.3	0.1349	1.8401	1.9750	103.0
102.0	1.00789	0.016137	331.1	331.1	69.995	1035.9	1105.9	0.1331	1.8444	1.9775	102.0
101.0	0.97818	0.016133	340.6	340.6	68.997	1036.5	1105.5	0.1313	1.8487	1.9800	101.0
100.0	0.94924	0.016130	350.4	350.4	67.999	1037.1	1105.1	0.1295	1.8530	1.9825	100.0
99.0	0.92103	0.016127	360.5	360.5	67.001	1037.6	1104.6	0.1278	1.8573	1.9851	99.0
98.0	0.89356	0.016123	370.9	370.9	66.003	1038.2	1104.2	0.1260	1.8617	1.9876	98.0
97.0	0.86679	0.016120	381.7	381.7	65.005	1038.8	1103.8	0.1242	1.8660	1.9902	97.0
96.0	0.84072	0.016117	392.8	392.9	64.006	1039.3	1103.3	0.1224	1.8704	1.9928	96.0
95.0	0.81534	0.016114	404.4	404.4	63.008	1039.9	1102.9	0.1206	1.8748	1.9954	95.0
94.0	0.79062	0.016111	416.3	416.3	62.010	1040.5	1102.5	0.1188	1.8792	1.9980	94.0
93.0	0.76655	0.016108	428.6	428.6	61.012	1041.0	1102.1	0.1170	1.8837	2.0006	93.0
92.0	0.74313	0.016105	441.3	441.3	60.014	1041.6	1101.6	0.1152	1.8881	2.0033	92.0
91.0	0.72032	0.016102	454.5	454.5	59.016	1042.2	1101.2	0.1134	1.8926	2.0059	91.0
90.0	0.69813	0.016099	468.1	468.1	58.018	1042.7	1100.8	0.1115	1.8970	2.0086	90.0
89.0	0.67653	0.016096	482.2	482.2	57.020	1043.3	1100.3	0.1097	1.9015	2.0112	89.0
88.0	0.65551	0.016093	496.8	496.8	56.022	1043.9	1099.9	0.1079	1.9060	2.0139	88.0
87.0	0.63507	0.016090	511.9	511.9	55.024	1044.4	1099.5	0.1061	1.9105	2.0166	87.0
86.0	0.61518	0.016087	527.5	527.5	54.026	1045.0	1099.0	0.1043	1.9151	2.0193	86.0
85.0	0.59583	0.016085	543.6	543.6	53.027	1045.6	1098.6	0.1024	1.9196	2.0221	85.0
84.0	0.57702	0.016082	560.3	560.3	52.029	1046.1	1098.2	0.1006	1.9242	2.0248	84.0
83.0	0.55872	0.016079	577.6	577.6	51.031	1046.7	1097.7	0.0988	1.9288	2.0275	83.0
82.0	0.54093	0.016077	595.5	595.6	50.033	1047.3	1097.3	0.0969	1.9334	2.0303	82.0
81.0	0.52364	0.016074	614.1	614.1	49.035	1047.8	1096.9	0.0951	1.9380	2.0331	81.0
80.0	0.050683	0.016072	633.3	633.3	48.037	1048.4	1096.4	0.0932	1.9426	2.0359	80.0
79.0	0.49049	0.016070	653.2	653.2	47.038	1049.0	1096.0	0.0914	1.9473	2.0387	79.0
78.0	0.47461	0.016067	673.8	673.9	46.040	1049.5	1095.6	0.0895	1.9520	2.0415	78.0
77.0	0.45919	0.016065	695.2	695.2	45.042	1050.1	1095.1	0.0877	1.9567	2.0443	77.0
76.0	0.44420	0.016063	717.4	717.4	44.043	1050.7	1094.7	0.0858	1.9614	2.0472	76.0
75.0	0.42964	0.016060	740.3	740.3	43.045	1051.2	1094.3	0.0839	1.9661	2.0500	75.0
74.0	0.41550	0.016058	764.1	764.1	42.046	1051.8	1093.8	0.0821	1.9708	2.0529	74.0
73.0	0.40177	0.016056	788.8	788.8	41.048	1052.4	1093.4	0.0802	1.9756	2.0558	73.0
72.0	0.38844	0.016054	814.3	814.3	40.049	1052.9	1093.0	0.0783	1.9804	2.0587	72.0
71.0	0.37510	0.016052	840.8	840.9	39.050	1053.5	1092.5	0.0764	1.9852	2.0616	71.0
70.0	0.36292	0.016050	868.3	868.4	38.052	1054.0	1092.1	0.0745	1.9900	2.0645	70.0
69.0	0.35073	0.016048	896.9	896.9	37.053	1054.6	1091.7	0.0727	1.9948	2.0675	69.0
68.0	0.33889	0.016046	926.5	926.5	36.054	1055.2	1091.2	0.0708	1.9996	2.0704	68.0
67.0	0.32740	0.016044	957.2	957.2	35.055	1055.7	1090.8	0.0689	2.0045	2.0734	67.0
66.0	0.31626	0.016043	989.0	989.1	34.056	1056.3	1090.4	0.0670	2.0094	2.0764	66.0
65.0	0.30545	0.016041	1022.1	1022.1	33.057	1056.9	1089.9	0.0651	2.0143	2.0794	65.0
64.0	0.29497	0.016039	1056.5	1056.5	32.058	1057.4	1089.5	0.0632	2.0192	2.0824	64.0
63.0	0.28480	0.016038	1092.1	1092.1	31.058	1058.0	1089.0	0.0613	2.0242	2.0854	63.0
62.0	0.27494	0.016036	1129.2	1129.2	30.059	1058.5	1088.6	0.0593	2.0291	2.0885	62.0
61.0	0.26538	0.016035	1167.6	1167.6	29.059	1059.1	1088.2	0.0574	2.0341	2.0915	61.0
60.0	0.25611	0.016033	1207.6	1207.6	28.060	1059.7	1087.7	0.0555	2.0391	2.0946	60.0
59.0	0.24713	0.016032	1249.1	1249.1	27.060	1060.2	1087.3	0.0536	2.0441	2.0977	59.0
58.0	0.23843	0.016031	1292.2	1292.2	26.060	1060.8	1086.9	0.0516	2.0491	2.1008	58.0
57.0	0.23000	0.016029	1337.0	1337.0	25.060	1061.4	1086.4	0.0497	2.0542	2.1039	57.0
56.0	0.22183	0.016028	1383.6	1383.6	24.059	1061.9	1086.0	0.0478	2.0593	2.1070	56.0
55.0	0.21392	0.016027	1432.0	1432.0	23.059	1062.5	1085.6	0.0458	2.0644	2.1102	55.0
54.0	0.20625	0.016026	1482.4	1482.4	22.058	1063.1	1085.1	0.0439	2.0695	2.1134	54.0
53.0	0.19883	0.016025	1534.7	1534.8	21.058	1063.6	1084.7	0.0419	2.0746	2.1165	53.0
52.0	0.19165	0.016024	1589.2	1589.2	20.057	1064.2	1084.2	0.0400	2.0798	2.1197	52.0
51.0	0.18469	0.016023	1645.9	1645.9	19.056	1064.7	1083.8	0.0380	2.0849	2.1230	51.0
50.0	0.17796	0.016023	1704.8	1704.8	18.054	1065.3	1083.4	0.0361	2.0901	2.1262	50.0
49.0	0.17144	0.016022	1766.2	1766.2	17.053	1065.9	1082.9	0.0341	2.0953	2.1294	49.0
48.0	0.16514	0.016021	1830.0	1830.0	16.051	1066.4	1082.5	0.0321	2.1006	2.1327	48.0
47.0	0.15904	0.016021	1896.5	1896.5	15.049	1067.0	1082.1	0.0301	2.1058	2.1360	47.0
46.0	0.15314	0.016020	1965.7	1965.7	14.047	1067.6	1081.6	0.0282	2.1111	2.1393	46.0
45.0	0.14744	0.016020	2037.7	2037.8	13.044	1068.1	1081.2	0.0262	2.1164	2.1426	45.0
44.0	0.14192	0.016019	2112.8	2112.8	12.041	1068.7	1080.7	0.0242	2.1217	2.1459	44.0
43.0	0.13659	0.016019	2191.0	2191.0	11.038	1069.3	1080.3	0.0222	2.1271	2.1493	43.0
42.0	0.13143	0.016019	2272.4	2272.4	10.035	1069.8	1079.9	0.0202	2.1325	2.1527	42.0
41.0	0.12645	0.016019	2357.3	2357.3	9.031	1070.4	1079.4	0.0182	2.1378	2.1560	41.0
40.0	0.12163	0.016019	2445.8	2445.8	8.027	1071.0	1079.0	0.0162	2.1432	2.1594	40.0
39.0	0.11698	0.016019	2538.0	2538.0	7.023	1071.5	1078.5	0.0142	2.1487	2.1629	39.0
38.0	0.11249	0.016019	2634.1	2634.2	6.018	1072.1	1078.1	0.0122	2.1541	2.1663	38.0
37.0	0.10815	0.016019	2734.4	2734.4	5.013	1072.7	1077.7	0.0101	2.1596	2.1697	37.0
36.0	0.10395	0.016020	2839.0	2839.0	4.008	1073.2	1077.2	0.0081	2.1651	2.1732	36.0
35.0	0.09991	0.016020	2948.1	2948.1	3.002	1073.8	1076.8	0.0061	2.1706	2.1767	35.0
34.0	0.09600	0.016021	3061.9	3061.9	1.996	1074.4	1076.4	0.0041	2.1762	2.1802	34.0
33.0	0.09223	0.016021	3180.7	3180.7	0.989	1074.9	1075.9	0.0020	2.1817	2.1837	33.0
32.018	0.08865	0.016022	3302.4	3302.4	0.0003	1075.5	1075.5	0.0000	2.1872	2.1872	32.018
*32.0	0.08859	0.016022	3304.7	3304.7	−0.0179	1075.5	1075.5	−0.0000	2.1873	2.1873	32.0

*The states here shown are metastable

STEAM TABLES (Continued)

Specific Heat at constant pressure of Steam and of Water

Temp. F	c_p, Btu/lbm × F															Temp. F
Press., psia	1	1.5	2	3	4	6	8	10	15	20	30	40	60	80	100	Press., psia
Sat. Water	0.998	0.998	0.999	1.000	1.000	1.002	1.003	1.004	1.007	1.010	1.014	1.019	1.026	1.033	1.039	Sat. Water
Sat. Steam	0.450	0.452	0.454	0.458	0.461	0.466	0.471	0.475	0.485	0.493	0.508	0.521	0.543	0.564	0.582	Sat. Steam
1500	0.559	0.559	0.559	0.559	0.559	0.559	0.559	0.559	0.559	0.559	0.560	0.560	0.560	0.561	0.561	1500
1480	0.557	0.557	0.557	0.557	0.557	0.557	0.557	0.558	0.558	0.558	0.558	0.558	0.559	0.559	0.559	1480
1460	0.556	0.556	0.556	0.556	0.556	0.556	0.556	0.556	0.556	0.556	0.556	0.557	0.557	0.557	0.558	1460
1440	0.554	0.554	0.554	0.554	0.554	0.554	0.554	0.554	0.554	0.554	0.555	0.555	0.555	0.556	0.556	1440
1420	0.552	0.552	0.552	0.552	0.552	0.552	0.553	0.553	0.553	0.553	0.553	0.553	0.554	0.554	0.555	1420
1400	0.551	0.551	0.551	0.551	0.551	0.551	0.551	0.551	0.551	0.551	0.551	0.552	0.552	0.553	0.553	1400
1380	0.549	0.549	0.549	0.549	0.549	0.549	0.549	0.549	0.549	0.549	0.550	0.550	0.550	0.551	0.551	1380
1360	0.547	0.547	0.547	0.547	0.547	0.547	0.547	0.547	0.548	0.548	0.548	0.548	0.549	0.549	0.550	1360
1340	0.546	0.546	0.546	0.546	0.546	0.546	0.546	0.546	0.546	0.546	0.546	0.547	0.547	0.548	0.548	1340
1320	0.544	0.544	0.544	0.544	0.544	0.544	0.544	0.544	0.544	0.544	0.545	0.545	0.545	0.546	0.546	1320
1300	0.542	0.542	0.542	0.542	0.542	0.542	0.542	0.542	0.542	0.543	0.543	0.543	0.544	0.544	0.545	1300
1280	0.540	0.540	0.540	0.540	0.540	0.540	0.540	0.541	0.541	0.541	0.541	0.541	0.542	0.543	0.543	1280
1260	0.538	0.539	0.539	0.539	0.539	0.539	0.539	0.539	0.539	0.539	0.539	0.540	0.540	0.541	0.541	1260
1240	0.537	0.537	0.537	0.537	0.537	0.537	0.537	0.537	0.537	0.537	0.538	0.538	0.539	0.539	0.540	1240
1220	0.535	0.535	0.535	0.535	0.535	0.535	0.535	0.535	0.535	0.536	0.536	0.536	0.537	0.537	0.538	1220
1200	0.533	0.533	0.533	0.533	0.533	0.533	0.533	0.533	0.534	0.534	0.534	0.534	0.535	0.536	0.536	1200
1180	0.531	0.531	0.531	0.531	0.531	0.531	0.532	0.532	0.532	0.532	0.532	0.533	0.533	0.534	0.535	1180
1160	0.529	0.529	0.530	0.530	0.530	0.530	0.530	0.530	0.530	0.530	0.530	0.531	0.532	0.532	0.533	1160
1140	0.528	0.528	0.528	0.528	0.528	0.528	0.528	0.528	0.528	0.528	0.529	0.529	0.530	0.531	0.531	1140
1120	0.526	0.526	0.526	0.526	0.526	0.526	0.526	0.526	0.526	0.527	0.527	0.527	0.528	0.529	0.530	1120
1100	0.524	0.524	0.524	0.524	0.524	0.524	0.524	0.524	0.525	0.525	0.525	0.526	0.526	0.527	0.528	1100
1080	0.522	0.522	0.522	0.522	0.522	0.522	0.522	0.523	0.523	0.523	0.523	0.524	0.525	0.525	0.526	1080
1060	0.520	0.520	0.520	0.520	0.520	0.521	0.521	0.521	0.521	0.521	0.522	0.522	0.523	0.524	0.524	1060
1040	0.518	0.519	0.519	0.519	0.519	0.519	0.519	0.519	0.519	0.519	0.520	0.520	0.521	0.522	0.523	1040
1020	0.517	0.517	0.517	0.517	0.517	0.517	0.517	0.517	0.517	0.518	0.518	0.518	0.519	0.520	0.521	1020
1000	0.515	0.515	0.515	0.515	0.515	0.515	0.515	0.515	0.515	0.516	0.516	0.517	0.518	0.519	0.519	1000
980	0.513	0.513	0.513	0.513	0.513	0.513	0.513	0.513	0.514	0.514	0.514	0.515	0.516	0.517	0.518	980
960	0.511	0.511	0.511	0.511	0.511	0.511	0.512	0.512	0.512	0.512	0.513	0.513	0.514	0.515	0.516	960
940	0.509	0.509	0.509	0.509	0.509	0.510	0.510	0.510	0.510	0.510	0.511	0.511	0.512	0.514	0.515	940
920	0.507	0.508	0.508	0.508	0.508	0.508	0.508	0.508	0.508	0.509	0.509	0.510	0.511	0.512	0.513	920
900	0.506	0.506	0.506	0.506	0.506	0.506	0.506	0.506	0.506	0.507	0.507	0.508	0.509	0.510	0.512	900
880	0.504	0.504	0.504	0.504	0.504	0.504	0.504	0.504	0.505	0.505	0.506	0.506	0.508	0.509	0.510	880
860	0.502	0.502	0.502	0.502	0.502	0.502	0.503	0.503	0.503	0.503	0.504	0.505	0.506	0.507	0.509	860
840	0.500	0.500	0.500	0.500	0.500	0.501	0.501	0.501	0.501	0.502	0.502	0.503	0.504	0.506	0.507	840
820	0.498	0.498	0.499	0.499	0.499	0.499	0.499	0.499	0.499	0.500	0.501	0.501	0.503	0.504	0.506	820
800	0.497	0.497	0.497	0.497	0.497	0.497	0.497	0.497	0.498	0.498	0.499	0.500	0.501	0.503	0.505	800
780	0.495	0.495	0.495	0.495	0.495	0.495	0.495	0.496	0.496	0.496	0.497	0.498	0.500	0.502	0.503	780
760	0.493	0.493	0.493	0.493	0.493	0.494	0.494	0.494	0.494	0.495	0.496	0.497	0.499	0.500	0.502	760
740	0.491	0.491	0.491	0.492	0.492	0.492	0.492	0.492	0.493	0.493	0.494	0.495	0.497	0.499	0.501	740
720	0.490	0.490	0.490	0.490	0.490	0.490	0.490	0.491	0.491	0.492	0.493	0.494	0.496	0.498	0.500	720
700	0.488	0.488	0.488	0.488	0.488	0.488	0.489	0.489	0.490	0.490	0.491	0.492	0.495	0.497	0.500	700
680	0.486	0.486	0.486	0.486	0.487	0.487	0.487	0.487	0.488	0.489	0.490	0.491	0.494	0.496	0.499	680
660	0.484	0.485	0.485	0.485	0.485	0.485	0.485	0.486	0.486	0.487	0.489	0.490	0.493	0.496	0.499	660
640	0.483	0.483	0.483	0.483	0.483	0.484	0.484	0.484	0.485	0.486	0.487	0.489	0.492	0.495	0.499	640
620	0.481	0.481	0.481	0.481	0.482	0.482	0.482	0.483	0.483	0.484	0.486	0.488	0.491	0.495	0.499	620
600	0.479	0.480	0.480	0.480	0.480	0.480	0.481	0.481	0.482	0.483	0.485	0.487	0.491	0.495	0.499	600
580	0.478	0.478	0.478	0.478	0.478	0.479	0.479	0.480	0.481	0.482	0.484	0.486	0.491	0.495	0.500	580
560	0.476	0.476	0.476	0.477	0.477	0.477	0.478	0.478	0.479	0.481	0.483	0.485	0.490	0.496	0.501	560
540	0.475	0.475	0.475	0.475	0.475	0.476	0.476	0.477	0.478	0.480	0.482	0.485	0.491	0.497	0.503	540
520	0.473	0.473	0.473	0.474	0.474	0.475	0.475	0.476	0.477	0.479	0.482	0.485	0.491	0.498	0.505	520
500	0.472	0.472	0.472	0.472	0.473	0.473	0.474	0.475	0.476	0.478	0.481	0.485	0.492	0.500	0.508	500
480	0.470	0.470	0.470	0.471	0.471	0.472	0.473	0.473	0.475	0.477	0.481	0.485	0.493	0.502	0.511	480
460	0.469	0.469	0.469	0.469	0.470	0.471	0.472	0.472	0.475	0.477	0.481	0.486	0.495	0.505	0.516	460
440	0.467	0.467	0.468	0.468	0.469	0.470	0.470	0.471	0.474	0.476	0.481	0.487	0.498	0.509	0.522	440
420	0.466	0.466	0.466	0.467	0.467	0.468	0.470	0.471	0.473	0.476	0.482	0.488	0.501	0.514	0.528	420
400	0.464	0.465	0.465	0.466	0.466	0.467	0.469	0.470	0.473	0.476	0.483	0.490	0.504	0.520	0.536	400
380	0.463	0.463	0.464	0.464	0.465	0.466	0.468	0.469	0.473	0.477	0.484	0.492	0.509	0.527	0.546	380
360	0.462	0.462	0.462	0.463	0.464	0.466	0.467	0.469	0.473	0.477	0.486	0.495	0.515	0.536	0.558	360
340	0.460	0.461	0.461	0.462	0.463	0.465	0.467	0.469	0.473	0.478	0.488	0.499	0.521	0.546	0.572	340
320	0.459	0.460	0.460	0.461	0.462	0.464	0.467	0.469	0.474	0.480	0.491	0.504	0.530	0.558	1.036	320
300	0.458	0.459	0.459	0.460	0.462	0.464	0.466	0.469	0.475	0.482	0.495	0.509	0.539	1.029	1.029	300
280	0.457	0.458	0.458	0.460	0.461	0.464	0.467	0.469	0.477	0.484	0.500	0.516	1.022	1.022	1.022	280
260	0.456	0.457	0.457	0.459	0.461	0.464	0.467	0.470	0.478	0.487	0.505	1.017	1.017	1.017	1.016	260
240	0.455	0.456	0.457	0.458	0.460	0.464	0.468	0.471	0.481	0.491	1.012	1.012	1.012	1.012	1.012	240
220	0.454	0.455	0.456	0.458	0.460	0.464	0.468	0.473	0.484	1.008	1.008	1.008	1.008	1.008	1.008	220
200	0.453	0.454	0.455	0.458	0.460	0.465	0.470	0.475	1.005	1.005	1.005	1.005	1.005	1.005	1.005	200
180	0.452	0.454	0.455	0.458	0.460	0.466	1.003	1.003	1.003	1.003	1.003	1.003	1.003	1.002	1.002	180
160	0.451	0.453	0.455	0.458	0.461	1.001	1.001	1.001	1.001	1.001	1.001	1.001	1.001	1.001	1.001	160
140	0.451	0.453	0.454	1.000	1.000	1.000	1.000	1.000	0.999	0.999	0.999	0.999	0.999	0.999	0.999	140
120	0.450	0.452	0.999	0.999	0.999	0.999	0.999	0.999	0.999	0.999	0.998	0.998	0.998	0.998	0.998	120
100	0.998	0.998	0.998	0.998	0.998	0.998	0.998	0.998	0.998	0.998	0.998	0.998	0.998	0.998	0.998	100
80	0.998	0.998	0.998	0.998	0.998	0.998	0.998	0.998	0.998	0.998	0.998	0.998	0.998	0.998	0.998	80
60	1.000	1.000	1.000	1.000	1.000	1.000	1.000	1.000	1.000	1.000	1.000	1.000	0.999	0.999	0.999	60
40	1.004	1.004	1.004	1.004	1.004	1.004	1.004	1.004	1.004	1.004	1.004	1.004	1.004	1.004	1.003	40
32	1.007	1.007	1.007	1.007	1.007	1.007	1.007	1.007	1.007	1.007	1.007	1.007	1.007	1.007	1.006	32

STEAM TABLES (Continued)

Specific Heat at constant pressure of Steam and of Water

Temp. F	c_p, Btu/lbm × F															Temp. F
Press., psia	150	200	300	400	600	800	1000	1500	2000	3000	4000	6000	8000	10000	15000	Press., psia
Sat. Water	1.054	1.067	1.093	1.118	1.168	1.224	1.286	1.492	1.841	7.646	—	—	—	—	--	Sat. Water
Sat. Steam	0.624	0.661	0.729	0.792	0.915	1.046	1.191	1.667	2.557	13.66	—	—	—	—	—	Sat. Steam
1500	0.562	0.563	0.565	0.567	0.571	0.576	0.580	0.590	0.601	0.623	0.645	0.691	0.737	0.780	0.868	1500
1480	0.561	0.562	0.564	0.566	0.570	0.575	0.579	0.590	0.601	0.623	0.647	0.694	0.742	0.786	0.878	1480
1460	0.560	0.560	0.562	0.565	0.569	0.573	0.578	0.589	0.601	0.624	0.648	0.698	0.747	0.793	0.888	1460
1440	0.557	0.559	0.561	0.563	0.568	0.572	0.577	0.589	0.600	0.625	0.650	0.701	0.753	0.800	0.900	1440
1420	0.556	0.557	0.559	0.562	0.566	0.571	0.576	0.588	0.600	0.625	0.651	0.705	0.759	0.808	0.909	1420
1400	0.554	0.555	0.558	0.560	0.565	0.570	0.575	0.587	0.600	0.626	0.653	0.709	0.765	0.817	0.926	1400
1380	0.553	0.554	0.556	0.559	0.564	0.569	0.574	0.587	0.600	0.627	0.655	0.714	0.773	0.827	0.939	1380
1360	0.551	0.552	0.555	0.558	0.563	0.568	0.573	0.586	0.600	0.628	0.657	0.719	0.781	0.838	0.953	1360
1340	0.549	0.551	0.553	0.556	0.561	0.567	0.572	0.586	0.600	0.629	0.660	0.725	0.790	0.850	0.968	1340
1320	0.548	0.549	0.552	0.555	0.560	0.566	0.571	0.585	0.600	0.630	0.663	0.731	0.800	0.864	0.983	1320
1300	0.546	0.548	0.550	0.553	0.559	0.565	0.570	0.585	0.600	0.632	0.666	0.738	0.811	0.879	0.998	1300
1280	0.545	0.546	0.549	0.552	0.558	0.564	0.570	0.585	0.600	0.634	0.669	0.746	0.824	0.897	1.014	1280
1260	0.543	0.544	0.547	0.550	0.556	0.563	0.569	0.585	0.601	0.636	0.673	0.755	0.838	0.918	1.033	1260
1240	0.541	0.543	0.546	0.549	0.555	0.562	0.568	0.584	0.601	0.638	0.678	0.765	0.855	0.942	1.053	1240
1220	0.540	0.541	0.544	0.548	0.554	0.561	0.567	0.584	0.602	0.641	0.683	0.777	0.875	0.969	1.072	1220
1200	0.538	0.540	0.543	0.546	0.553	0.560	0.567	0.584	0.603	0.644	0.689	0.790	0.897	1.000	1.095	1200
1180	0.536	0.538	0.541	0.545	0.552	0.559	0.566	0.584	0.604	0.647	0.696	0.805	0.922	1.033	1.117	1180
1160	0.535	0.536	0.540	0.544	0.551	0.558	0.565	0.585	0.606	0.652	0.704	0.823	0.952	1.070	1.143	1160
1140	0.533	0.535	0.539	0.542	0.550	0.557	0.565	0.585	0.607	0.656	0.713	0.843	0.986	1.107	1.167	1140
1120	0.531	0.533	0.537	0.541	0.549	0.557	0.565	0.586	0.609	0.662	0.723	0.866	1.025	1.149	1.190	1120
1100	0.530	0.532	0.536	0.540	0.548	0.556	0.564	0.587	0.612	0.668	0.735	0.893	1.070	1.193	1.220	1100
1080	0.528	0.530	0.534	0.538	0.547	0.555	0.564	0.588	0.615	0.676	0.749	0.924	1.120	1.242	1.240	1080
1060	0.527	0.529	0.533	0.537	0.546	0.555	0.564	0.590	0.618	0.685	0.765	0.960	1.176	1.295	1.260	1060
1040	0.525	0.527	0.532	0.536	0.545	0.555	0.565	0.592	0.622	0.695	0.783	1.002	1.238	1.351	1.282	1040
1020	0.523	0.526	0.530	0.535	0.545	0.555	0.565	0.594	0.627	0.707	0.804	1.051	1.306	1.399	1.298	1020
1000	0.522	0.524	0.529	0.534	0.544	0.555	0.566	0.597	0.633	0.721	0.829	1.110	1.382	1.471	1.306	1000
980	0.520	0.523	0.528	0.533	0.544	0.555	0.567	0.601	0.640	0.737	0.858	1.180	1.475	1.531	1.312	980
960	0.519	0.521	0.527	0.532	0.543	0.556	0.568	0.605	0.648	0.756	0.893	1.267	1.598	1.595	1.310	960
940	0.517	0.520	0.526	0.531	0.543	0.556	0.570	0.610	0.658	0.778	0.934	1.376	1.708	1.639	1.299	940
920	0.516	0.519	0.525	0.531	0.544	0.558	0.573	0.617	0.669	0.803	0.984	1.520	1.819	1.667	1.281	920
900	0.515	0.518	0.524	0.530	0.544	0.550	0.576	0.621	0.683	0.834	1.040	1.710	1.932	1.000	1.259	900
880	0.513	0.516	0.523	0.530	0.545	0.561	0.580	0.633	0.699	0.872	1.130	1.993	2.000	1.633	1.232	880
860	0.512	0.515	0.523	0.530	0.546	0.564	0.584	0.644	0.718	0.918	1.240	2.316	2.019	1.593	1.212	860
840	0.511	0.514	0.522	0.530	0.548	0.568	0.590	0.657	0.740	0.977	1.395	2.653	1.978	1.547	1.192	840
820	0.510	0.514	0.522	0.531	0.550	0.572	0.597	0.672	0.767	1.054	1.620	2.886	1.888	1.503	1.175	820
800	0.509	0.513	0.522	0.532	0.553	0.577	0.605	0.690	0.800	1.160	1.967	2.872	1.768	1.459	1.157	800
780	0.508	0.513	0.522	0.533	0.557	0.584	0.615	0.712	0.840	1.312	2.550	2.547	1.670	1.416	1.142	780
760	0.507	0.512	0.523	0.535	0.561	0.592	0.628	0.738	0.892	1.542	4.462	2.156	1.576	1.370	1.126	760
740	0.507	0.512	0.524	0.537	0.567	0.602	0.642	0.770	0.960	1.913	8.119	1.886	1.493	1.332	1.114	740
720	0.506	0.512	0.525	0.540	0.574	0.613	0.660	0.811	1.052	2.584	3.458	1.696	1.421	1.290	1.100	720
700	0.506	0.513	0.528	0.544	0.582	0.627	0.681	0.861	1.181	6.145°	2.237	1.557	1.358	1.250	1.089	700
680	0.506	0.514	0.530	0.549	0.592	0.644	0.707	0.927	1.365	2.469	1.789	1.450	1.303	1.217	1.079	680
660	0.507	0.515	0.534	0.555	0.604	0.665	0.738	1.015	1.639	1.851	1.587	1.369	1.256	1.187	1.071	660
640	0.507	0.517	0.538	0.562	0.619	0.690	0.777	1.135	2.219	1.601	1.454	1.303	1.216	1.157	1.063	640
620	0.509	0.519	0.543	0.571	0.637	0.720	0.826	1.308	1.614	1.455	1.362	1.252	1.184	1.136	1.056	620
600	0.510	0.522	0.550	0.582	0.659	0.757	0.888	1.586	1.453	1.358	1.295	1.211	1.157	1.118	1.052	600
580	0.513	0.526	0.558	0.595	0.685	0.804	0.969	1.393	1.351	1.289	1.243	1.178	1.134	1.102	1.046	580
560	0.516	0.531	0.568	0.611	0.717	0.862	1.079	1.309	1.281	1.237	1.202	1.151	1.115	1.087	1.039	560
540	0.519	0.538	0.580	0.630	0.756	0.937	1.272	1.249	1.229	1.196	1.169	1.128	1.098	1.074	1.031	540
520	0.524	0.545	0.594	0.653	0.804	1.035	1.221	1.204	1.189	1.164	1.142	1.109	1.083	1.062	1.024	520
500	0.530	0.554	0.611	0.680	0.865	1.187	1.181	1.169	1.157	1.137	1.120	1.092	1.069	1.051	1.017	500
480	0.537	0.565	0.632	0.714	1.159	1.154	1.150	1.140	1.131	1.115	1.101	1.077	1.057	1.041	1.010	480
460	0.545	0.578	0.657	0.755	1.132	1.128	1.125	1.117	1.110	1.096	1.084	1.064	1.047	1.033	1.004	460
440	0.556	0.594	0.687	1.113	1.110	1.107	1.104	1.098	1.092	1.080	1.070	1.052	1.038	1.025	0.999	440
420	0.568	0.614	0.724	1.094	1.091	1.089	1.087	1.081	1.076	1.067	1.058	1.042	1.029	1.018	0.994	420
400	0.583	0.636	1.079	1.078	1.076	1.074	1.072	1.067	1.063	1.055	1.047	1.034	1.022	1.011	0.990	400
380	0.601	1.066	1.065	1.065	1.063	1.061	1.059	1.056	1.052	1.044	1.038	1.026	1.015	1.006	0.986	380
360	0.622	1.054	1.054	1.053	1.052	1.050	1.049	1.045	1.042	1.036	1.030	1.019	1.009	1.001	0.982	360
340	1.045	1.044	1.044	1.043	1.042	1.040	1.039	1.036	1.033	1.028	1.022	1.013	1.004	0.996	0.979	340
320	1.036	1.036	1.035	1.034	1.033	1.032	1.031	1.028	1.026	1.021	1.016	1.007	0.999	0.992	0.976	320
300	1.028	1.028	1.028	1.027	1.026	1.025	1.024	1.022	1.019	1.015	1.010	1.002	0.995	0.988	0.973	300
280	1.022	1.022	1.021	1.021	1.020	1.019	1.018	1.016	1.014	1.009	1.005	0.998	0.991	0.985	0.971	280
260	1.016	1.016	1.016	1.015	1.014	1.013	1.013	1.011	1.009	1.005	1.001	0.994	0.988	0.982	0.968	260
240	1.012	1.011	1.011	1.011	1.010	1.009	1.008	1.006	1.004	1.001	0.997	0.991	0.985	0.979	0.966	240
220	1.008	1.008	1.007	1.007	1.006	1.005	1.005	1.003	1.001	0.998	0.994	0.988	0.982	0.977	0.964	220
200	1.005	1.004	1.004	1.004	1.003	1.002	1.002	1.000	0.998	0.995	0.992	0.986	0.980	0.975	0.963	200
180	1.002	1.002	1.002	1.001	1.001	1.000	0.999	0.998	0.996	0.993	0.989	0.983	0.978	0.973	0.961	180
160	1.000	1.000	1.000	0.999	0.999	0.998	0.997	0.996	0.994	0.991	0.987	0.981	0.976	0.971	0.959	160
140	0.999	0.999	0.998	0.998	0.997	0.997	0.996	0.994	0.992	0.989	0.986	0.980	0.974	0.969	0.958	140
120	0.998	0.998	0.997	0.997	0.996	0.996	0.995	0.993	0.991	0.988	0.984	0.978	0.972	0.967	0.957	120
100	0.997	0.997	0.997	0.996	0.996	0.995	0.994	0.992	0.990	0.986	0.983	0.976	0.970	0.965	0.955	100
80	0.998	0.997	0.997	0.996	0.995	0.994	0.994	0.991	0.989	0.985	0.981	0.974	0.968	0.962	0.951	80
60	0.999	0.999	0.998	0.997	0.996	0.995	0.994	0.991	0.989	0.984	0.979	0.970	0.963	0.956	0.942	60
40	1.003	1.003	1.002	1.001	1.000	0.998	0.997	0.993	0.989	0.983	0.976	0.965	0.954	0.945	0.920	40
32	1.006	1.006	1.005	1.004	1.002	1.000	0.999	0.994	0.990	0.983	0.975	0.962	0.949	0.937	0.904	32

°Critical point.

Thermal Conductivity of Steam and Water

Temp. F	k, (Btu/hr × ft. × F) × 10³												
Press., psia	1	2	5	10	20	50	100	200	500	1000	2000	5000	7500
Sat. Water	364.0	373.1	383.8	390.4	395.2	397.4	394.7	386.2	361.7	327.6	271.8	—	—
Sat. Steam	11.6	12.2	13.0	13.8	14.8	16.6	18.4	21.1	27.2	36.5	61.3	—	—
1500	63.7	63.7	63.7	63.7	63.7	63.8	64.0	64.3	65.4	67.1	70.7	82.0	92.2
1450	61.4	61.4	61.5	61.5	61.5	61.6	61.8	62.1	63.2	64.9	68.5	80.1	90.6
1400	59.2	59.2	59.2	59.2	59.3	59.4	59.6	59.9	60.9	62.7	66.3	78.2	89.2
1350	57.0	57.0	57.0	57.0	57.1	57.2	57.3	57.7	58.7	60.5	64.2	76.3	87.9
1300	54.8	54.8	54.8	54.8	54.8	54.9	55.1	55.5	56.5	58.3	62.0	74.6	86.9
1250	52.6	52.6	52.6	52.6	52.6	52.7	52.9	53.2	54.3	56.1	59.9	73.0	86.3
1200	50.4	50.4	50.4	50.4	50.4	50.5	50.7	51.0	52.1	53.9	57.8	71.6	86.2
1150	48.2	48.2	48.2	48.2	48.2	48.3	48.5	48.9	49.9	51.8	55.7	70.5	87.0
1100	46.0	46.0	46.0	46.0	46.1	46.2	46.3	46.7	47.8	49.6	53.7	69.8	89.0
1050	43.9	43.9	43.9	43.9	43.9	44.0	44.2	44.6	45.6	47.5	51.8	69.7	93.4
1000	41.7	41.7	41.8	41.8	41.8	41.9	42.1	42.4	43.5	45.5	50.0	70.7	102.9
950	39.6	39.6	39.7	39.7	39.7	39.8	40.0	40.3	41.4	43.5	48.3	73.5	115.5
900	37.6	37.6	37.6	37.6	37.6	37.7	37.9	38.3	39.4	41.5	46.8	80.2	138.7
850	35.5	35.6	35.6	35.6	35.6	35.7	35.9	36.3	37.4	39.7	45.6	96.7	178.8
800	33.6	33.6	33.6	33.6	33.6	33.7	33.9	34.3	35.5	37.9	44.9	129.6	223.2
750	31.6	31.6	31.6	31.6	31.7	31.8	32.0	32.3	33.6	36.3	45.2	202.5	258.3
700	29.7	29.7	29.7	29.7	29.8	29.9	30.1	30.4	31.8	35.0	47.5°	262.8	295.1
650	27.8	27.8	27.9	27.9	27.9	28.0	28.2	28.6	30.1	34.1	55.7	304.3	326.7
600	26.0	26.0	26.1	26.1	26.1	26.2	26.4	26.9	28.7	34.1	301.9	333.7	349.3
550	24.3	24.3	24.3	24.3	24.4	24.5	24.7	25.2	27.5	36.1	333.7	356.1	368.0
500	22.6	22.6	22.6	22.6	22.7	22.8	23.0	23.6	26.9	350.8	357.4	373.8	383.6
450	21.0	21.0	21.0	21.0	21.0	21.2	21.4	22.3	368.1	370.6	375.3	387.9	396.5
400	19.4	19.4	19.4	19.4	19.5	19.6	20.0	21.3	383.0	384.9	388.5	398.6	406.4
350	17.9	17.9	17.9	17.9	18.0	18.2	18.8	392.0	392.9	394.4	397.4	406.1	413.2
300	16.5	16.5	16.5	16.5	16.6	16.9	396.9	397.2	398.0	399.3	402.0	409.9	416.4
250	15.1	15.1	15.1	15.2	15.3	396.9	397.0	397.3	398.1	399.4	402.1	409.7	415.8
200	13.8	13.8	13.9	14.0	391.6	391.6	391.8	392.1	393.0	394.4	397.2	404.9	410.6
150	12.7	12.7	380.5	380.5	380.6	380.7	380.8	381.1	382.1	383.7	386.7	394.7	400.3
100	363.3	363.3	363.3	363.3	363.3	363.4	363.6	363.9	365.0	366.6	369.8	378.3	384.1
50	339.1	339.1	339.1	339.1	339.2	339.3	339.4	339.8	340.8	342.5	345.7	354.6	361.0
32	328.6	328.6	328.6	328.6	328.6	328.7	328.9	329.2	330.3	331.9	335.1	344.1	350.8

° Critical point.

THERMODYNAMIC PROPERTIES

Ammonia (NH_3)

Temp. (°F)	Abs. press. sat. vap		Heat content abv. −40°F BTU/lb		Ht. of vaporiz. BTU/lb	Heat content abv. −40°C g-cal/g		Ht. of vaporiz. g-cal/g	Spec. vol. sat. vap.		Density sat. vap.		Dens. liq. lb/ft³	Entropy from −40°F BTU/lb/°F		Temp. (°C)
	lb/in.²	kg/cm²	Liq.	Vap.		Liq.	Vap.		ft³/lb	m³/kg	lb/ft³	kg/m³		Liq.	Vap.	
−60	5.55	0.390	−21.2	589.6	610.8	−11.8	327.6	339.3	44.73	2.792	0.02235	0.3580	43.91	−0.0517	1.4769	−51.11
−58	5.93	0.417	−19.1	590.4	609.5	−10.6	328.0	338.6	42.05	2.625	0.02378	.3809		−0.0464	1.4713	−50.00
−56	6.33	0.445	−17.0	591.2	608.2	−9.44	328.4	337.9	39.56	2.470	0.02528	.4049		−0.0412	1.4658	−48.89
−54	6.75	0.475	−14.8	592.1	606.9	−8.22	328.9	337.2	37.24	2.325	0.02685	.4301		−0.0360	1.4604	−47.78
−52	7.20	0.506	−12.7	592.9	605.6	−7.06	329.4	336.4	35.09	2.191	0.02850	.4565		−0.0307	1.4551	−46.67
−50	7.67	0.539	−10.6	593.7	604.3	−5.89	329.8	335.7	33.08	2.065	0.03023	0.4842	43.49	−0.0256	1.4497	−45.56
−48	8.16	0.574	−8.5	594.4	602.9	−4.7	330.2	334.9	31.20	1.948	0.03205	.5134		−0.0204	1.4445	−44.44
−46	8.68	0.610	−6.4	595.2	601.6	−3.6	330.7	334.2	29.45	1.839	0.03395	.5438		−0.0153	1.4393	−43.33
−44	9.23	0.649	−4.3	596.0	600.3	−2.4	331.1	333.5	27.82	1.737	0.03595	.5758		−0.0102	1.4342	−42.22
−42	9.81	0.690	−2.1	596.8	598.9	−1.2	331.6	332.7	26.29	1.641	0.03804	.6093		−0.0051	1.4292	−41.11
−40	10.41	0.7319	0.0	597.6	597.6	0.0	332.0	332.0	24.86	1.552	0.04022	0.6442	43.08	0.0000	1.4242	−40.00
−38	11.04	0.7762	2.1	598.3	596.2	1.2	332.4	331.2	23.53	1.469	0.04251	.6809		0.0051	1.4193	−38.89
−36	11.71	0.8233	4.3	599.1	594.8	2.4	332.8	330.4	22.27	1.390	0.04489	.7190		0.0101	1.4144	−37.78
−34	12.41	0.8725	6.4	599.9	593.5	3.6	333.3	329.7	21.10	1.317	0.04739	.7591		0.0151	1.4096	−36.67
−32	13.14	0.9238	8.5	600.6	592.1	4.7	333.7	328.9	20.00	1.249	0.04999	.8007		0.0201	1.4048	−35.56
−30	13.90	0.9773	10.7	601.4	590.7	5.94	334.1	328.2	18.97	1.184	0.05271	0.8443	42.65	0.0250	1.4001	−34.44
−28	14.71	1.034	12.8	602.1	589.3	7.11	334.5	327.4	18.00	1.124	0.05555	.8898		0.0300	1.3955	−33.33
−26	15.55	1.093	14.9	602.8	587.9	8.28	334.9	326.6	17.09	1.067	0.05850	.9371		0.0350	1.3909	−32.22
−24	16.42	1.154	17.1	603.6	586.5	9.50	335.3	325.8	16.24	1.014	0.06158	.9864		0.0399	1.3863	−31.11
−22	17.34	1.219	19.2	604.3	585.1	10.7	335.7	325.1	15.43	0.9633	0.06479	1.038		0.0448	1.3818	−30.00
−20	18.30	1.287	21.4	605.0	583.6	11.9	336.1	324.2	14.68	0.9164	0.06813	1.091	42.22	0.0497	1.3774	−28.89
−18	19.30	1.357	23.5	605.7	582.2	13.1	336.5	323.4	13.97	0.8721	0.07161	1.147		0.0545	1.3729	−27.78
−16	20.34	1.430	25.6	606.4	580.8	14.2	336.9	322.7	13.29	0.8297	0.07522	1.205		0.0594	1.3686	−26.67
−14	21.43	1.507	27.8	607.1	579.3	15.4	337.3	321.8	12.66	0.7903	0.07898	1.265		0.0642	1.3643	−25.56
−12	22.56	1.586	30.0	607.8	577.8	16.7	337.7	321.0	12.06	0.7529	0.08289	1.328		0.0690	1.3600	−24.44
−10	23.74	1.669	32.1	608.5	576.4	17.8	338.1	320.2	11.50	0.7179	0.08695	1.393	41.78	0.0738	1.3558	−23.33
−8	24.97	1.756	34.3	609.2	574.9	19.1	338.4	319.4	10.97	0.6848	0.09117	1.460		0.0786	1.3516	−22.22
−6	26.26	1.846	36.4	609.8	573.4	20.2	338.8	318.6	10.47	0.6536	0.09555	1.531		0.0833	1.3474	−21.11
−4	27.59	1.940	38.6	610.5	571.9	21.4	339.2	317.7	9.991	0.6237	0.1001	1.603		0.0880	1.3433	−20.00
−2	28.98	2.037	40.7	611.1	570.4	22.6	339.5	316.9	9.541	0.5956	0.1048	1.679		0.0928	1.3393	−18.89
0	30.42	2.139	42.9	611.8	568.9	23.8	339.9	316.1	9.116	0.5691	0.1097	1.757	41.34	0.0975	1.3352	−17.78
2	31.92	2.244	45.1	612.4	567.3	25.1	340.2	315.2	8.714	0.5440	0.1148	1.839		0.1022	1.3312	−16.67
4	33.47	2.353	47.2	613.0	565.8	26.2	340.6	314.3	8.333	0.5202	0.1200	1.922		0.1069	1.3273	−15.56
6	35.00	2.467	49.4	613.6	564.2	27.4	340.9	313.4	7.971	0.4976	0.1254	2.009		0.1115	1.3234	−14.44

THERMODYNAMIC PROPERTIES (continued)

Temp. (°F)	Abs. press. sat. vap lb/in.²	Abs. press. sat. vap kg/cm²	Heat content abv. −40°F BTU/lb Liq.	Heat content abv. −40°F BTU/lb Vap.	Ht. of vaporiz. BTU/lb	Heat content abv. −40°C g-cal/g Liq.	Heat content abv. −40°C g-cal/g Vap.	Ht. of vaporiz. g-cal/g	Spec. vol. sat. vap. ft³/lb	Spec. vol. sat. vap. m³/kg	Density sat. vap. lb/ft³	Density sat. vap. kg/m³	Dens. liq. lb/ft³	Entropy from −40°F BTU/lb/°F Liq.	Entropy from −40°F BTU/lb/°F Vap.	Temp. (°C)
8	36.77	2.585	51.6	614.3	562.7	28.7	341.3	312.6	7.629	0.4763	0.1311	2.100		0.1162	1.3195	−13.33
10	38.51	2.708	53.8	614.9	561.1	29.9	341.6	311.7	7.304	0.4560	0.1369	2.193	40.89	0.1208	1.3157	−12.22
12	40.31	2.834	56.0	615.5	559.5	31.1	341.9	310.8	6.996	0.4367	0.1429	2.289		0.1254	1.3118	−11.11
14	42.18	2.966	58.2	616.1	557.9	32.3	342.3	309.9	6.703	0.4185	0.1492	2.390		0.1300	1.3081	−10.00
16	44.12	3.102	60.3	616.6	556.3	33.5	342.6	309.1	6.425	0.4011	0.1556	2.492		0.1346	1.3043	−8.89
18	[illegible]	[illegible]	[illegible]	[illegible]	554.7	34.7	342.9	308.2	6.161	0.3846	0.1623	2.600		0.1392	1.3006	−7.78
20	48.21	3.390	64.7	617.8	553.1	35.9	343.2	307.3	5.910	0.3690	0.1692	2.710	40.43	0.1437	1.2969	−6.67
22	50.36	3.541	66.9	618.3	551.4	37.2	343.5	306.3	5.671	0.3540	0.1763	2.824		0.1483	1.2933	−5.56
24	52.59	3.697	69.1	618.9	549.8	38.4	343.8	305.4	5.443	0.3398	0.1837	2.943		0.1528	1.2897	−4.44
26	54.90	3.860	71.3	619.4	548.1	39.6	344.1	304.5	5.227	0.3263	0.1913	3.064		0.1573	1.2861	−3.33
28	57.28	4.027	73.5	619.9	546.4	40.8	344.4	303.6	5.021	0.3135	0.1992	3.191		0.1618	1.2825	−2.22
30	59.74	4.200	75.7	620.5	544.8	42.1	344.7	302.7	4.825	0.3012	0.2073	3.321	39.96	0.1663	1.2790	−1.11
32	62.29	4.379	77.9	621.0	543.1	43.3	345.0	301.7	4.637	0.2895	0.2156	3.453		0.1708	1.2755	0.00
34	64.91	4.564	80.1	621.5	541.4	44.5	345.3	300.8	4.459	0.2784	0.2243	3.593		0.1753	1.2721	1.11
36	67.63	4.755	82.3	622.0	539.7	45.7	345.6	299.8	4.289	0.2678	0.2332	3.735		0.1797	1.2686	2.22
38	70.43	4.952	84.6	622.5	537.9	47.0	345.8	298.8	4.126	0.2576	0.2423	3.881		0.1841	1.2652	3.33
40	73.32	5.155	86.8	623.0	536.2	48.2	346.1	297.9	3.971	0.2479	0.2518	4.003	39.49	0.1885	1.2618	4.44
42	76.31	5.365	89.0	623.4	534.4	49.4	346.3	296.9	3.823	0.2387	0.2616	4.190		0.1930	1.2585	5.56
44	79.38	5.581	91.2	623.9	532.7	50.7	346.6	295.9	3.682	0.2299	0.2716	4.350		0.1974	1.2552	6.67
46	82.55	5.804	93.5	624.4	530.9	51.9	346.9	294.9	3.547	0.2214	0.2819	4.515		0.2018	1.2519	7.78
48	85.82	6.034	95.7	624.8	529.1	53.2	347.1	293.9	3.418	0.2134	0.2926	4.687		0.2062	1.2486	8.89
50	89.19	6.271	97.9	625.2	527.3	54.4	347.3	292.9	3.294	0.2056	0.3036	4.863	39.00	0.2105	1.2453	10.00
52	92.66	6.515	100.2	625.7	525.5	55.67	347.6	291.9	3.176	0.1983	0.3149	5.044		0.2149	1.2421	11.11
54	96.23	6.766	102.4	626.1	523.7	56.89	347.8	290.9	3.063	0.1912	0.3265	5.230		0.2192	1.2389	12.22
56	99.91	7.024	104.7	626.5	521.8	58.17	348.1	289.9	2.954	0.1844	0.3385	5.422		0.2236	1.2357	13.33
58	103.7	7.291	106.9	626.9	520.0	59.39	348.3	288.9	2.851	0.1780	0.3508	5.619		0.2279	1.2325	14.44
60	107.6	7.565	109.2	627.3	518.1	60.67	348.5	287.8	2.751	0.1717	0.3635	5.823	38.50	0.2322	1.2294	15.56
62	111.6	7.846	111.5	627.7	516.2	61.94	348.7	286.8	2.656	0.1658	0.3765	6.031		0.2365	1.2262	16.67
64	115.7	8.135	113.7	628.0	514.3	63.17	348.9	285.7	2.565	0.1601	0.3899	6.245		0.2408	1.2231	17.78
66	120.0	8.437	116.0	628.4	512.4	64.44	349.1	284.7	2.477	0.1546	0.4037	6.466		0.2451	1.2201	18.89
68	124.3	8.739	118.3	628.8	510.5	65.72	349.3	283.6	2.393	0.1494	0.4179	6.694		0.2404	1.2170	20.00
70	128.8	9.056	120.5	629.1	508.6	66.94	349.5	282.6	2.312	0.1443	0.4325	6.928	38.00	0.2537	1.2140	21.11
72	133.4	9.379	122.8	629.4	506.6	68.22	349.7	281.4	2.235	0.1395	0.4474	7.166		0.2579	1.2110	22.22
74	138.1	9.709	125.1	629.8	504.7	69.50	349.9	280.4	2.161	0.1349	0.4628	7.413		0.2622	1.2080	23.33
76	143.0	10.05	127.4	630.1	502.7	70.78	350.1	279.3	2.089	0.1304	0.4786	7.666		0.2664	1.2050	24.44
78	147.9	10.40	129.7	630.4	500.7	72.06	350.2	278.2	2.021	0.1262	0.4949	7.927		0.2706	1.2020	25.56
80	153.0	10.76	132.0	630.7	498.7	73.33	350.4	277.1	1.955	0.1220	0.5115	8.193	37.48	0.2749	1.1991	26.67
82	158.3	11.13	134.3	631.0	496.7	74.61	350.6	275.9	1.892	0.1181	0.5287	8.469		0.2791	1.1962	27.78
84	163.7	11.51	136.6	631.3	494.7	75.89	350.7	274.8	1.831	0.1143	0.5462	8.749		0.2833	1.1933	28.89
86	169.2	11.90	138.9	631.5	492.6	77.17	350.8	273.7	1.772	0.1106	0.5643	9.039		0.2875	1.1904	30.00
88	174.8	12.29	141.2	631.8	490.6	78.44	351.0	272.6	1.716	0.1071	0.5828	9.335		0.2917	1.1875	31.11
90	180.6	12.70	143.5	632.0	488.5	79.72	351.1	271.4	1.661	0.1037	0.6019	9.641	36.95	0.2958	1.1846	32.22
92	186.6	13.12	145.8	632.2	486.4	81.00	351.2	270.2	1.609	0.1004	0.6214	9.954		0.3000	1.1818	33.33
94	192.7	13.55	148.2	632.5	484.3	82.33	351.4	269.1	1.559	0.09733	0.6415	10.28		0.3041	1.1789	34.44
96	198.9	13.98	150.5	632.6	482.1	83.61	351.4	267.8	1.510	0.09427	0.6620	10.60		0.3083	1.1761	35.56
98	205.3	14.43	152.9	632.9	480.0	84.94	351.6	266.7	1.464	0.09140	0.6832	10.94		0.3125	1.1733	36.67
100	211.9	14.90	155.2	633.0	477.8	86.22	351.7	265.4	1.419	0.08859	0.7048	11.29	36.40	0.3166	1.1705	37.78
102	218.6	15.37	157.6	633.2	475.6	87.56	351.8	264.2	1.375	0.08584	0.7270	11.65		0.3207	1.1677	38.89
104	225.4	15.85	159.9	633.4	473.5	88.83	351.9	263.1	1.334	0.08328	0.7498	12.01		0.3248	1.1649	40.00
106	232.5	16.35	162.3	633.5	471.2	90.17	351.9	261.8	1.293	0.08072	0.7732	12.39		0.3289	1.1621	41.11
108	239.7	16.85	164.6	633.6	469.0	91.44	352.0	260.6	1.254	0.07829	0.7972	12.77		0.3330	1.1593	42.22
110	247.0	17.37	167.0	633.7	466.7	92.78	352.1	259.3	1.217	0.07598	0.8219	13.17	35.84	0.3372	1.1566	43.33
112	254.5	17.89	169.4	633.8	464.4	94.11	352.1	258.0	1.180	0.07367	0.8471	13.57		0.3413	1.1538	44.44
114	262.2	18.43	171.8	633.9	462.1	95.44	352.2	256.7	1.145	0.07148	0.8730	13.98		0.3453	1.1510	45.56
116	270.1	18.99	174.2	634.0	459.8	96.78	352.2	255.4	1.112	0.06942	0.8996	14.41		0.3495	1.1483	46.67
118	278.2	19.56	176.6	634.0	457.4	98.11	352.2	254.1	1.079	0.06736	0.9269	14.85		0.3535	1.1455	47.78
120	286.4	20.14	179.0	634.0	455.0	99.45	352.2	252.8	1.047	0.06536	0.9549	15.30	35.26	0.3576	1.1427	48.89
122	294.8	20.73	181.4	634.0	452.6	100.8	352.2	251.4	1.017	0.06349	0.9837	15.76		0.3618	1.1400	50.00
124	303.4	21.33	183.9	634.0	450.1	102.2	352.2	250.1	0.987	0.0616	1.0132	16.229		0.3659	1.1372	51.11

Carbon Dioxide (CO_2)

Temp. (°F)	Abs. press sat. vap lb/in.²	Abs. press sat. vap kg/cm²	Heat content abv. 32°F BTU/lb Liq.	Heat content abv. 32°F BTU/lb Vap	Heat of vaporiz. BTU/lb	Heat content abv. 0°C g-cal/g Liq.	Heat content abv. 0°C g-cal/g Vap	Heat of vaporiz. g-cal/g	Spec. vol. sat. vap. ft³/lb	Spec. vol. sat. vap. m³/kg	Density sat. vap. lb/ft³	Density sat. vap. kg/m³	Dens. liq. lb/ft³	Entropy from 32°F BTU/lb/°F Liq.	Entropy from 32°F BTU/lb/°F Vap.	Temp. (°C)
−20	220.6	15.51	−23.96	102.0	126.0	−13.31	56.67	70.00	0.4166	0.02601	2.401	38.46	64.34	−0.0514	0.2353	−28.89
−18	228.4	16.06	−23.13	102.1	125.2	−12.85	56.72	69.56	0.4018	0.02508	2.489	39.87	64.15	−0.0495	0.2342	−27.78
−16	236.4	16.62	−22.30	102.2	124.5	−12.39	56.78	69.17	0.3876	0.02420	2.580	41.33	63.94	−0.0476	0.2331	−26.67
−14	244.6	17.20	−21.46	102.2	123.7	−11.92	56.78	68.72	0.3739	0.02334	2.674	42.83	63.73	−0.0458	0.2319	−25.56

Temp. (°F)	Abs. press sat. vap lb/in.²	Abs. press sat. vap kg/cm²	Heat content abv. 32°F BTU/lb Liq.	Heat content abv. 32°F BTU/lb Vap	Heat of vaporiz. BTU/lb	Heat content abv. 0°C g-cal/g Liq.	Heat content abv. 0°C g-cal/g Vap	Heat of vaporiz. g-cal/g	Spec. vol. sat. vap. ft³/lb	Spec. vol. sat. vap. m³/kg	Density sat. vap. lb/ft³	Density sat. vap. kg/m³	Dens. liq. lb/ft³	Entropy from 32°F BTU/lb/°F Liq.	Entropy from 32°F BTU/lb/°F Vap.	Temp. (°C)
−12	253.0	17.79	−20.61	102.3	122.9	−11.45	56.83	68.28	0.3608	0.02252	2.772	44.40	63.49	−0.0439	0.2307	−24.44
−10	261.7	18.40	−19.76	102.3	122.0	−10.98	56.83	67.78	0.3482	0.02174	2.872	46.00	63.25	−0.0420	0.2296	−23.33
−8	270.6	19.03	−18.90	102.3	121.2	−10.50	56.83	67.33	0.3360	0.02098	2.976	47.67	63.01	−0.0401	0.2284	−22.22
−6	279.7	19.66	−18.04	102.3	120.3	−10.02	56.83	66.83	0.3243	0.02025	3.083	49.38	62.76	−0.0382	0.2273	−21.11
−4	289.1	20.33	−17.17	102.3	119.5	−9.539	56.83	66.39	0.3131	0.01955	3.194	51.16	62.50	−0.0362	0.2261	−20.00
−2	298.7	21.00	−16.29	102.3	118.6	−9.050	56.83	65.89	0.3022	0.01887	3.309	53.00	62.23	−0.0343	0.2249	−18.89
0	308.6	21.70	−15.41	102.2	117.7	−8.561	56.78	65.39	0.2918	0.01822	3.427	54.89	61.95	−0.0324	0.2237	−17.78
2	318.7	22.41	−14.51	102.2	116.7	−8.061	56.78	64.83	0.2817	0.01759	3.550	56.86	61.65	−0.0304	0.2225	−16.67
4	329.1	23.14	−13.61	102.1	115.8	−7.561	56.72	64.33	0.2720	0.01698	3.676	58.88	61.36	−0.0285	0.2213	−15.56
6	339.8	23.89	−12.71	102.1	114.8	−7.061	56.72	63.78	0.2627	0.01640	3.807	60.98	61.07	−0.0266	0.2201	−14.44
8	350.7	24.66	−11.79	102.0	113.8	−6.550	56.67	63.22	0.2537	0.01584	3.942	63.14	60.77	−0.0246	0.2189	−13.33
10	361.8	25.44	−10.87	101.9	112.8	−6.039	56.61	62.67	0.2450	0.01529	4.082	65.39	60.48	−0.0226	0.2176	−12.22
12	373.3	26.25	−9.934	101.8	111.7	−5.519	56.56	62.06	0.2366	0.01477	4.227	67.71	60.18	−0.0206	0.2164	−11.11
14	385.0	27.07	−8.992	101.7	110.7	−4.996	56.50	61.50	0.2285	0.01426	4.377	70.11	59.88	−0.0186	0.2151	−10.00
16	397.1	27.92	−8.038	101.5	109.6	−4.466	56.39	60.89	0.2207	0.01378	4.532	72.59	59.58	−0.0166	0.2139	−8.89
18	409.4	28.78	−7.076	101.4	108.5	−3.931	56.33	60.28	0.2131	0.01330	4.692	75.16	59.27	−0.0146	0.2126	−7.78
20	422.0	29.67	−6.102	101.2	107.3	−3.390	56.22	59.61	0.2058	0.01285	4.859	77.83	58.95	−0.0126	0.2113	−6.67
22	434.9	30.58	−5.117	101.0	106.1	−2.843	56.11	58.94	0.1987	0.01240	5.031	80.59	58.64	−0.0105	0.2100	−5.56
24	448.1	31.50	−4.121	100.8	104.9	−2.289	56.00	58.28	0.1919	0.01198	5.211	83.47	58.31	−0.0085	0.2087	−4.44
25	454.8	31.98	−3.618	100.7	104.3	−2.010	55.94	57.94	0.1886	0.01177	5.303	84.94	58.14	−0.0074	0.2080	−3.89
27	468.5	32.94	−2.601	100.5	103.1	−1.445	55.83	57.28	0.1821	0.01137	5.492	87.97	57.81	−0.0053	0.2066	−2.78
29	482.5	33.92	−1.570	100.2	101.8	−0.8722	55.67	56.56	0.1758	0.01097	5.688	91.11	57.47	−0.0032	0.2053	−1.67
31	496.8	34.93	−0.525	99.98	100.5	−0.292	55.54	55.83	0.1697	0.01059	5.892	94.38	57.12	−0.0011	0.2039	−0.56
33	511.4	35.95	0.531	99.69	99.16	0.295	55.38	55.09	0.1639	0.01023	6.103	97.76	56.77	0.0011	0.2025	0.56
35	526.4	37.01	1.604	99.38	97.77	0.8911	55.21	54.32	0.1581	0.009870	6.323	101.3	56.41	0.0033	0.2010	1.67
37	541.7	38.09	2.697	99.05	96.35	1.498	55.03	53.53	0.1526	0.009527	6.553	105.0	56.03	0.0055	0.1996	2.78
39	557.4	39.19	3.806	98.69	94.88	2.114	54.83	52.71	0.1472	0.009189	6.792	108.8	55.65	0.0077	0.1981	3.89
41	573.4	40.31	4.932	98.31	93.37	2.740	54.62	51.87	0.1420	0.008865	7.040	112.8	55.25	0.0099	0.1965	5.00
43	589.8	41.47	6.080	97.90	91.82	3.378	54.39	51.01	0.1370	0.008553	7.300	116.9	54.84	0.0122	0.1950	6.11
45	606.5	42.64	7.251	97.46	90.21	4.028	54.14	50.12	0.1321	0.008247	7.571	121.3	54.41	0.0146	0.1934	7.22
47	623.6	43.84	8.443	96.99	88.55	4.691	53.88	49.19	0.1273	0.007947	7.854	125.8	53.97	0.0169	0.1918	8.33
49	641.1	45.07	9.664	95.50	86.83	5.369	53.61	48.24	0.1227	0.007660	8.151	130.6	53.51	0.0193	0.1901	9.44
51	659.0	46.33	10.91	95.97	85.06	6.061	53.32	47.26	0.1182	0.007379	8.461	135.5	53.04	0.0218	0.1884	10.56
53	677.3	47.62	12.19	95.40	83.21	6.772	53.00	46.23	0.1138	0.007104	8.787	140.8	52.55	0.0243	0.1867	11.67
55	695.9	48.93	13.49	94.78	81.29	7.494	52.66	45.16	0.1095	0.006836	9.132	146.3	52.05	0.0268	0.1849	12.78
57	714.9	50.26	14.84	94.13	79.30	8.244	52.29	44.06	0.1053	0.006574	9.497	152.1	51.53	0.0294	0.1830	13.89
59	734.3	51.63	16.22	93.44	77.22	9.011	51.91	42.90	0.1012	0.006318	9.880	158.3	50.99	0.0321	0.1811	15.00
61	754.2	53.03	17.65	92.69	75.04	9.806	51.49	41.69	0.0972	0.00607	10.29	164.8	50.42	0.0348	0.1790	16.11
63	774.5	54.45	19.13	91.88	72.75	10.63	51.04	40.42	0.0933	0.00582	10.72	171.7	49.80	0.0377	0.1770	17.22
65	795.1	55.90	20.66	91.01	70.35	11.48	50.56	39.08	0.0894	0.00558	11.18	179.1	49.14	0.0406	0.1748	18.33
67	816.2	57.38	22.25	90.07	67.81	12.36	50.04	37.67	0.0856	0.00534	11.67	186.9	48.44	0.0436	0.1725	19.44
69	837.8	58.90	23.92	89.04	65.12	13.29	49.47	36.18	0.0819	0.00511	12.21	195.6	47.69	0.0468	0.1701	20.56
71	859.8	60.45	25.67	87.92	62.25	14.26	48.84	34.58	0.0782	0.00488	12.82	205.4	46.87	0.0501	0.1676	21.67
73	882.2	62.02	27.52	86.69	59.17	15.29	48.16	32.87	0.0745	0.00465	13.43	215.1	45.99	0.0535	0.1647	22.78
75	905.1	63.63	29.50	85.33	55.83	16.39	47.41	31.02	0.0708	0.00442	14.13	226.3	45.05	0.0573	0.1618	23.89
77	928.4	65.27	31.62	83.80	52.17	17.57	46.56	28.98	0.0671	0.00419	14.90	238.7	44.06	0.0613	0.1585	25.00
79	952.2	66.95	33.95	82.06	48.11	18.86	45.59	26.73	0.0633	0.00395	15.81	253.2	43.04	0.0656	0.1550	26.11
81	976.5	68.65	36.54	80.03	43.49	20.30	44.46	24.16	0.0592	0.00370	16.90	270.7	41.95	0.0704	0.1509	27.22
83	1001.0	70.377	39.53	77.60	38.07	21.96	43.11	21.15	0.0548	0.00342	18.25	292.3	40.62	0.0759	0.1461	28.33
85	1027.0	72.205	43.18	74.47	31.29	23.99	41.37	17.38	0.0500	0.00312	20.00	320.4	38.76	0.0826	0.1401	29.44
86	1039.0	73.049	45.45	72.46	27.00	25.25	40.26	15.00	0.0474	0.00296	21.09	337.8	37.41	0.0868	0.1363	30.00
87	1052.0	73.963	48.32	69.84	21.52	26.84	38.80	11.96	0.0446	0.00278	22.42	359.1	35.34	0.0921	0.1314	30.56
88	1065.0	74.877	52.78	65.62	12.84	29.32	36.46	7.133	0.0401	0.00250	24.95	399.6	32.79	0.1002	0.1237	31.11

Sulfur Dioxide (SO_2)

Temp. (°F)	Abs. press. sat. vap. lb/in.²	Abs. press. sat. vap. kg/cm²	Heat content abv. −40°F BTU/lb Liq.	Heat content abv. −40°F BTU/lb Vap.	Ht. of vaporiz BTU/lb	Heat content abv. −40°C g-cal/g Liq.	Heat content abv. −40°C g-cal/g Vap.	Ht. of vaporiz. g-cal/g	Spec. vol. sat. vap. ft³/lb	Spec. vol. sat. vap. m³/kg	Density sat. vap. lb/ft³	Density sat. vap. kg/m³	Dens. liq. lb/ft³	Entropy from −40°F BTU/lb/°F Liq.	Entropy from −40°F BTU/lb/°F Vap.	Temp. (°C)
−40	3.136	0.2205	0.00	178.61	178.61	0.00	99.228	99.228	22.42	1.400	0.04460	0.7144	95.79	0.00000	0.42562	−40.00
−30	4.331	0.3045	2.93	179.90	176.97	1.63	99.945	98.317	16.56	1.034	0.06039	0.9673	94.94	0.00674	0.41864	−34.44
−20	5.883	0.4136	5.98	181.07	175.09	3.32	100.59	97.272	12.42	0.7754	0.08119	1.301	94.10	0.01366	0.41192	−28.89
−10	7.863	0.5528	9.16	182.13	172.97	5.09	101.18	96.095	9.44	0.5893	0.1025	1.642	93.27	0.02075	0.40544	−23.33
0	10.35	0.7277	12.44	183.07	170.63	6.911	101.71	94.795	7.280	0.4545	0.1374	2.201	92.42	0.02795	0.39917	−17.78
2	10.91	0.7670	13.12	183.25	170.13	7.289	101.81	94.517	6.923	0.4322	0.1444	2.313	92.25	0.02941	0.39794	−16.67
4	11.50	0.8085	13.78	183.41	169.63	7.656	101.89	94.239	6.584	0.4110	0.1501	2.404	92.08	0.03084	0.39670	−15.56
5	11.81	0.8303	14.11	183.49	169.38	7.839	101.94	94.100	6.421	0.4009	0.1558	2.496	92.00	0.03155	0.39609	−15.00
6	12.12	0.8521	14.45	183.57	169.12	8.028	101.98	93.956	6.266	0.3912	0.1596	2.556	91.91	0.03228	0.39547	−14.44
8	12.75	0.8964	15.13	183.73	168.60	8.406	102.07	93.667	5.967	0.3725	0.1676	2.685	91.74	0.03373	0.39426	−13.33

Temp. (°F)	Abs. press. sat. vap. lb/in.²	Abs. press. sat. vap. kg/cm²	Heat content abv. −40°F BTU/lb Liq.	Heat content abv. −40°F BTU/lb Vap.	Ht. of vaporiz BTU/lb	Heat content abv. −40°C g-cal/g Liq.	Heat content abv. −40°C g-cal/g Vap.	Ht. of vaporiz. g-cal/g	Spec. vol. sat. vap. ft³/lb	Spec. vol. sat. vap. m³/kg	Density sat. vap. lb/ft³	Density sat. vap. kg/m³	Dens. liq. lb/ft³	Entropy from −40°F BTU/lb/°F Liq.	Entropy from −40°F BTU/lb/°F Vap.	Temp. (°C)
10	13.42	0.9435	15.80	183.87	168.07	8.778	102.15	93.372	5.682	0.3547	0.1760	2.819	91.58	0.03519	0.39306	−12.22
12	14.12	0.9927	16.48	184.01	167.53	9.156	102.23	93.072	5.417	0.3382	0.1846	2.957	91.41	0.03664	0.39185	−11.11
14	14.84	1.043	17.15	184.14	166.97	9.528	102.30	92.761	5.164	0.3224	0.1936	3.101	91.24	0.03808	0.39065	−10.00
16	15.59	1.096	17.84	184.28	166.44	9.911	102.38	92.467	4.926	0.3075	0.2030	3.252	91.07	0.03953	0.38946	−8.89
18	16.37	1.1509	18.52	184.40	165.88	10.29	102.44	92.156	4.701	0.2935	0.2127	3.407	90.89	0.04098	0.38827	−7.78
20	17.18	1.208	19.20	184.52	165.32	10.67	102.51	91.845	4.487	0.2801	0.2228	3.569	90.71	0.04241	0.38707	−6.67
22	18.03	1.268	19.90	184.64	164.74	11.06	102.58	91.522	4.287	0.2676	0.2332	3.735	90.53	0.04385	0.38589	−5.56
24	18.89	1.328	20.58	184.74	164.16	11.43	102.63	91.200	4.096	0.2557	0.2441	3.910	90.33	0.04528	0.38471	−4.44
26	19.80	1.392	21.26	184.84	163.58	11.81	102.69	90.878	3.915	0.2444	0.2559	4.099	90.15	0.04671	0.38354	−3.33
28	20.73	1.457	21.96	184.94	162.98	12.20	102.74	90.545	3.744	0.2337	0.2671	4.278	89.96	0.04814	0.38237	−2.22
30	21.70	1.526	22.64	185.02	162.38	12.58	102.79	90.211	3.581	0.2236	0.2800	4.485	89.76	0.04956	0.38119	−1.11
32	22.71	1.597	23.33	185.10	161.77	12.96	102.83	89.872	3.437	0.2146	0.2909	4.660	89.58	0.05099	0.38003	0.00
34	23.75	1.670	24.03	185.18	161.15	13.35	102.88	89.528	3.283	0.2050	0.3046	4.879	89.39	0.05242	0.37887	1.11
36	24.82	1.745	24.72	185.25	160.53	13.73	102.92	89.183	3.144	0.1963	0.3181	5.095	89.18	0.05384	0.37772	2.22
38	25.95	1.824	25.42	185.31	159.89	14.12	102.95	88.828	3.013	0.1881	0.3319	5.316	89.00	0.05527	0.37657	3.33
40	27.10	1.905	26.12	185.37	159.25	14.51	102.98	88.472	2.887	0.1802	0.3464	5.549	88.81	0.05668	0.37541	4.44
42	28.29	1.989	26.81	185.42	158.61	14.89	103.01	88.117	2.769	0.1729	0.3611	5.784	88.62	0.05809	0.37425	5.56
44	29.52	2.075	27.51	185.46	157.95	15.28	103.03	87.750	2.656	0.1658	0.3765	6.031	88.43	0.05949	0.37311	6.67
46	30.79	2.165	28.21	185.50	157.29	15.67	103.06	87.383	2.548	0.1591	0.3925	6.287	88.24	0.06090	0.37197	7.78
48	32.10	2.257	28.92	185.54	156.62	16.07	103.08	87.011	2.446	0.1527	0.4088	6.548	88.05	0.06231	0.37083	8.89
50	33.45	2.352	29.61	185.56	155.95	16.45	103.09	86.639	2.348	0.1466	0.4259	6.822	87.87	0.06370	0.36969	10.00
52	34.86	2.451	30.31	185.58	155.27	16.84	103.10	86.261	2.256	0.1408	0.4433	7.101	87.67	0.06509	0.36857	11.11
54	36.31	2.553	31.00	185.59	154.59	17.22	103.11	85.883	2.167	0.1353	0.4615	7.392	87.51	0.06646	0.36743	12.22
56	37.80	2.658	31.72	185.61	153.89	17.62	103.12	85.495	2.083	0.1300	0.4801	7.690	87.31	0.06785	0.36629	13.33
58	39.33	2.765	32.42	185.61	153.19	18.01	103.12	85.106	2.003	0.1250	0.4992	7.996	87.13	0.06923	0.36517	14.44
60	40.93	2.878	33.10	185.59	152.49	18.39	103.11	84.717	1.926	0.1202	0.5194	8.320	86.95	0.07066	0.36405	15.56
62	42.58	2.994	33.79	185.57	151.78	18.77	103.09	84.322	1.853	0.1157	0.5396	8.643	86.77	0.07196	0.36293	16.67
64	44.27	3.112	34.49	185.55	151.06	19.16	103.08	83.922	1.783	0.1113	0.5609	8.984	86.59	0.07333	0.36181	17.78
66	46.00	3.234	35.19	185.53	150.34	19.55	103.07	83.522	1.716	0.1071	0.5827	9.334	86.41	0.07469	0.36070	18.89
68	47.78	3.359	35.88	185.50	149.62	19.93	103.06	83.122	1.652	0.1031	0.6054	9.697	86.22	0.07602	0.35958	20.00
70	49.62	3.489	36.58	185.46	148.88	20.32	103.03	82.711	1.590	0.09926	0.6290	10.08	86.02	0.07736	0.35846	21.11
72	51.54	3.624	37.28	185.42	148.14	20.71	103.01	82.300	1.532	0.09564	0.6527	10.45	85.82	0.07871	0.35736	22.22
74	53.48	3.760	37.97	185.37	147.40	21.09	102.98	81.889	1.476	0.09214	0.6777	10.86	85.62	0.08003	0.35624	23.33
76	55.48	3.901	38.67	185.31	146.64	21.48	102.95	81.467	1.422	0.08877	0.7030	11.26	85.42	0.08135	0.35512	24.44
78	57.56	4.047	39.36	185.24	145.88	21.87	102.91	81.045	1.371	0.08559	0.7295	11.69	85.23	0.08268	0.35401	25.56
80	59.68	4.196	40.05	185.17	145.12	22.25	102.87	80.622	1.321	0.08247	0.7570	12.13	85.03	0.8399	0.35291	26.67
82	61.88	4.351	40.73	185.09	144.36	22.63	102.83	80.200	1.274	0.07953	0.7850	12.57	84.84	0.08525	0.35177	27.78
84	64.14	4.509	41.43	185.01	143.58	23.02	102.78	79.767	1.229	0.07672	0.8140	13.04	84.64	0.08653	0.35065	28.89
86	66.45	4.672	42.12	184.92	142.80	23.40	102.73	79.333	1.185	0.07398	0.8440	13.52	84.44	0.08783	0.34954	30.00
88	68.84	4.840	42.80	184.82	142.02	23.78	102.68	78.900	1.144	0.07142	0.8740	14.00	84.25	0.08910	0.34843	31.11
90	71.25	5.009	43.50	184.72	141.22	24.17	102.62	78.456	1.104	0.06892	0.9058	14.51	84.05	0.09038	0.34731	32.22
92	73.70	5.182	44.19	184.61	140.42	24.55	102.56	78.011	1.065	0.06649	0.9390	15.04	83.86	0.09165	0.34620	33.33
94	76.30	5.364	44.86	184.49	139.62	24.92	102.49	77.567	1.028	0.06418	0.9730	15.59	83.67	0.09389	0.34508	34.44
96	79.03	5.556	45.54	184.37	138.83	25.30	102.43	77.128	0.9931	0.06200	1.007	16.13	83.47	0.09411	0.34397	35.56
98	81.77	5.749	46.22	184.25	138.03	25.68	102.36	76.683	0.9591	0.05987	1.043	16.71	83.27	0.09532	0.34285	36.67
100	84.52	5.942	46.90	184.10	137.20	26.06	102.28	76.222	0.9262	0.05782	1.080	17.30	83.07	0.09657	0.34173	37.78

Butane ($CH_3(CH_2)_2CH_3$)

Temp. (°F)	Abs. press. sat. vap. lb/in.²	Abs. press. sat. vap. kg/cm²	Heat content abv. 32°F BTU/lb Liq.	Heat content abv. 32°F BTU/lb Vap.	Ht. of vaporiz. BTU/lb	Heat content abv. 0°C g-cal/g Liq.	Heat content abv. 0°C g-cal/g Vap.	Ht. of vaporiz. g-cal/g	Spec. vol. sat. vap. ft³/lb	Spec. vol. sat. vap. m³/kg	Density of sat. vap. lb/ft³	Density of sat. vap. kg/m³	Density of liq. lb/ft³	Density of liq. kg/m³	Temp. (°C)
0	7.3	0.51	−17.2	153.3	170.5	−9.56	85.17	94.72	11.1	0.693	0.0901	1.44	38.59	618.1	−17.78
10	9.2	0.65	−11.7	156.8	168.5	−6.50	87.11	93.61	8.95	0.559	0.112	1.79	38.24	612.5	−12.22
20	11.6	0.816	−6.7	160.3	167.0	−3.7	89.06	92.78	7.23	0.451	0.138	2.21	37.89	606.9	−6.67
30	14.4	1.01	−1.2	164.3	165.5	−0.67	91.28	91.94	5.90	0.368	0.169	2.71	37.54	601.3	−1.11
40	17.7	1.24	4.3	167.8	163.5	2.4	93.22	90.83	4.88	0.305	0.205	3.28	37.19	595.7	4.44
50	21.6	1.52	9.8	171.3	161.5	5.4	95.17	89.72	4.07	0.254	0.246	3.94	36.82	589.8	10.00
60	26.3	1.85	15.8	175.3	159.5	8.78	97.39	88.61	3.40	0.212	0.294	4.71	36.45	583.9	15.56
70	31.6	2.22	21.3	178.8	157.5	11.8	99.33	87.50	2.88	0.180	0.347	5.56	36.06	577.6	21.11
80	37.6	2.64	27.3	182.3	155.0	15.2	101.3	86.11	2.46	0.154	0.407	6.52	35.65	571.0	26.67
90	44.5	3.13	33.8	185.8	152.0	18.8	103.2	84.44	2.10	0.131	0.476	7.62	35.24	564.5	32.22
100	52.2	3.67	39.8	189.3	149.5	22.1	105.2	83.06	1.81	0.113	0.552	8.84	34.84	558.1	37.78
110	60.8	4.27	46.3	193.3	147.0	25.7	107.4	81.67	1.58	0.0986	0.633	10.1	34.41	551.2	43.33
120	70.8	4.98	52.8	196.3	143.5	29.3	109.1	79.72	1.38	0.0862	0.725	11.6	33.96	544.0	48.89
130	81.4	5.72	59.3	199.8	140.5	32.9	111.0	78.06	1.21	0.0755	0.826	13.2	33.49	536.4	54.44
140	92.6	6.51	66.3	203.8	137.5	36.8	113.2	76.39	1.07	0.0668	0.934	15.0	32.98	528.3	60.00

THERMODYNAMIC PROPERTIES (continued)

Isobutane $(CH_3)_3CH$

Temp.	Abs. press. sat. vap.		Heat content abv. 32°F BTU/lb		Ht. of vaporiz.	Heat content abv. 0°C g-cal/g		Ht. of vaporiz.	Spec. vol. sat. vap.		Density of sat. vap.		Density of liq.		Temp.
(°F)	lb/in.²	kg/cm²	Liq.	Vap.	BTU/lb	Liq.	Vap.	g-cal/g	ft³/lb	m³/kg	lb/ft³	kg/m³	lb/ft³	kg/m³	(°C)
−20	7.50	0.527	−25.5	140.0	165.5	−14.2	77.78	91.94	10.5	0.655	0.0952	1.52	38.35	614.3	−28.89
−10	9.28	0.652	−21.0	142.0	163.0	−11.7	78.89	90.56	8.91	0.556	0.112	1.79	37.95	607.9	−23.33
0	11.6	0.816	−16.5	144.0	160.5	−9.17	80.00	89.17	7.17	0.448	0.139	2.23	37.60	602.3	−17.78
10	14.6	1.03	−11.5	147.0	158.5	−6.39	81.67	88.06	5.75	0.359	0.174	2.79	37.20	595.9	−12.22
20	18.2	1.28	−6.5	149.5	156.0	−3.6	83.06	86.67	4.68	0.292	0.214	3.43	36.80	589.5	−6.67
30	22.3	1.57	−1.0	152.5	153.5	−0.56	84.72	85.28	3.86	0.241	0.259	4.15	36.40	583.1	−1.11
40	26.9	1.89	4.5	155.5	151.0	2.5	86.39	83.89	3.22	0.201	0.311	4.98	36.00	576.6	4.44
50	32.5	2.28	10.5	159.0	148.5	5.83	88.33	82.50	2.71	0.169	0.369	5.91	35.60	570.2	10.00
60	38.7	2.72	16.5	162.5	146.0	9.17	90.28	81.11	2.28	0.142	0.439	7.03	35.20	563.8	15.56
70	45.8	3.22	23.0	166.5	143.5	12.8	92.50	79.72	1.94	0.121	0.515	8.25	34.80	557.4	21.11
80	53.9	3.79	30.0	170.5	140.5	16.7	94.72	78.06	1.66	0.104	0.602	9.64	34.35	550.2	26.67
90	63.3	4.45	37.0	174.5	137.5	20.6	96.94	76.39	1.42	0.0886	0.704	11.3	33.90	543.0	32.22
100	73.7	5.18	44.5	179.0	134.5	24.7	99.44	74.72	1.23	0.0768	0.813	13.0	33.35	535.8	37.78
110	85.1	5.98	52.5	183.5	131.0	29.2	101.9	72.78	1.07	0.0668	0.935	15.0	33.00	528.6	43.33
120	98.0	6.89	60.5	188.0	127.5	33.6	104.4	70.83	0.926	0.0578	1.08	17.3	32.50	520.6	48.89
130	112.0	7.87	69.5	193.5	124.0	38.6	107.5	68.89	0.811	0.0506	1.23	19.7	32.00	512.6	54.44
140	126.8	8.915	78.5	199.0	120.5	43.6	110.6	66.94	0.716	0.0447	1.32	21.1	31.80	509.4	60.00

Propane (C_3H_8)

Temp.	Abs. press. sat. vap.		Heat content abv. 32°F BTU/lb		Ht. of vaporiz.	Heat content abv. 0°C g-cal/g		Ht. of vaporiz.	Spec. vol. sat. vap.		Density of sat. vap.		Density of liq.		Temp.
(°F)	lb/in.²	kg/cm²	Liq.	Vap.	BTU/lb	Liq.	Vap.	g-cal/g	ft³/lb	m³/kg	lb/ft³	kg/m³	lb/ft³	kg/m³	(°C)
−70	7.37	0.518	−55.2	134.3	189.5	−30.7	74.61	105.3	12.9	0.805	0.0775	1.24	37.40	599.1	−56.67
−60	9.72	0.683	−50.2	136.8	187.0	−27.9	76.00	103.9	9.93	0.620	0.111	1.78	37.00	592.7	−51.11
−50	12.6	0.886	−44.7	139.8	184.5	−24.8	77.67	102.5	7.74	0.483	0.129	2.07	36.60	586.3	−45.56
−40	16.2	1.14	−39.7	141.8	181.5	−22.1	78.78	100.8	6.13	0.383	0.163	2.61	36.19	579.7	−40.00
−30	20.3	1.43	−34.2	144.8	179.0	−19.0	80.44	99.44	4.93	0.308	0.203	3.25	35.78	573.1	−34.44
−20	25.4	1.79	−29.2	146.8	176.0	−16.2	81.56	97.78	4.00	0.250	0.250	4.00	35.37	566.6	−28.89
−10	31.4	2.21	−23.7	149.8	173.5	−13.2	83.22	96.39	3.26	0.204	0.307	4.92	34.96	560.0	−23.33
0	38.2	2.69	−18.2	152.3	170.5	−10.1	84.61	94.72	2.71	0.169	0.369	5.91	34.54	553.3	−17.78
10	46.0	3.23	−12.7	155.3	168.0	−7.06	86.28	93.33	2.27	0.142	0.441	7.06	34.12	546.5	−12.22
20	55.5	3.90	−7.2	157.8	165.0	−4.0	87.67	91.67	1.90	0.119	0.526	8.43	33.67	539.3	−6.67
30	66.3	4.66	−1.2	160.8	162.0	−0.67	89.33	90.00	1.60	0.0999	0.625	10.0	33.20	531.8	−1.11
40	78.0	5.48	4.8	163.8	159.0	2.7	91.00	88.33	1.37	0.0855	0.730	11.7	32.73	524.3	4.44
50	91.8	6.45	10.8	166.8	156.0	6.00	92.67	86.67	1.18	0.0737	0.847	13.6	32.24	516.4	10.00
60	107.1	7.530	16.8	169.8	153.0	9.33	94.33	85.00	1.01	0.0631	0.990	15.9	31.75	508.6	15.56
70	124.0	8.718	22.8	172.3	149.5	12.7	95.72	83.06	0.883	0.0551	1.13	18.1	31.24	500.4	21.11
80	142.8	10.04	29.3	175.3	146.0	16.3	97.39	81.11	0.770	0.0481	1.30	20.8	30.70	491.8	26.67
90	164.0	11.53	35.8	178.3	142.5	19.9	99.06	79.17	0.673	0.0420	1.49	23.9	30.15	482.9	32.22
100	187.0	13.15	42.3	180.8	138.5	23.5	100.4	76.94	0.591	0.0369	1.69	27.1	29.58	473.8	37.78
110	213.0	14.98	48.8	182.8	134.0	27.1	101.6	74.44	0.519	0.0324	1.96	31.4	28.85	462.1	43.33
120	240.0	16.87	55.3	184.3	129.0	30.7	102.4	71.67	0.459	0.0287	2.18	34.9	28.30	453.3	48.89

Difluorodichloromethane $(CCl_2F_2$("F-12"))

Temp.	Abs. press. sat. vap.		Heat content abv. −40°F BTU/lb		Ht. of vaporiz.	Heat content abv. −40°C g-cal/g		Ht. of vaporiz.	Spec. vol sat. vap.		Density of vap.		Dens. liq. lb/	Entropy from −40°F BTU/lb/°F		Temp.
(°F)	lb/in.²	kg/cm²	Liq.	Vap.	BTU/lb	Liq.	Vap.	g-cal/g	ft³/lb	m³/kg	lb/ft³	kg/m³	ft³	Liq.	Vap.	(°C)
−40	9.32	0.655	0	73.50	73.50	0	40.83	40.83	3.911	0.2442	0.2557	4.096	94.58	0	0.17517	−40.00
−30	12.02	0.845	2.03	74.70	72.67	1.13	41.50	40.37	3.088	0.1928	0.3238	5.187	93.59	0.00471	0.17387	−34.44
−20	15.28	1.074	4.07	75.87	71.80	2.26	42.15	39.89	2.474	0.1544	0.4042	6.474	92.58	0.00940	0.17275	−28.89
−10	19.20	1.350	6.14	77.05	70.91	3.41	42.81	39.39	2.003	0.1250	0.4993	7.998	91.57	0.01403	0.17175	−23.33
0	23.87	1.678	8.25	78.21	69.96	4.58	43.45	38.87	1.637	0.1022	0.6109	9.785	90.52	0.01869	0.17091	−17.78
10	29.35	2.064	10.39	79.36	68.97	5.771	44.09	38.32	1.351	0.08434	0.7402	11.86	89.45	0.02328	0.17015	−12.22
20	35.75	2.513	12.55	80.49	67.94	6.972	44.72	37.74	1.121	0.06998	0.8921	14.29	88.37	0.02783	0.16949	−6.67
30	43.16	3.034	14.76	81.61	66.85	8.200	45.34	37.14	0.939	0.0586	1.065	17.06	87.24	0.03233	0.16887	−1.11
40	51.68	3.633	17.00	82.71	65.71	9.444	45.95	36.51	0.792	0.0494	1.263	20.23	86.10	0.03680	0.16833	4.44
50	61.39	4.316	19.27	83.78	64.51	10.71	46.54	35.84	0.673	0.0420	1.485	23.79	84.94	0.04126	0.16785	10.00
60	72.41	5.091	21.57	84.82	63.25	11.98	47.12	35.14	0.575	0.0359	1.740	27.87	83.78	0.04568	0.16741	15.56
70	84.82	5.963	23.90	85.82	61.92	13.28	47.68	34.40	0.493	0.0308	2.028	32.48	82.60	0.05009	0.16701	21.11
80	98.76	6.944	26.28	86.80	60.52	14.60	48.22	33.62	0.425	0.0265	2.353	37.69	81.39	0.05446	0.16662	26.67
90	114.3	8.036	28.70	87.74	59.04	15.94	48.74	32.80	0.368	0.0230	2.721	43.58	80.11	0.05882	0.16624	32.22
100	131.6	9.252	31.16	88.62	57.46	17.31	49.23	31.92	0.319	0.0199	3.135	50.22	78.80	0.06316	0.16584	37.78
110	150.7	10.60	33.65	89.43	55.78	18.69	49.68	30.99	0.277	0.0173	3.610	57.82	77.46	0.06749	0.16542	43.33
120	171.8	12.08	36.16	90.15	53.99	20.09	50.08	29.99	0.240	0.0150	4.167	66.75	76.02	0.07180	0.16495	48.89

THERMODYNAMIC PROPERTIES (continued)

Carbon Disulfide (CS_2)

Temp.	Abs. press. sat. vap.		Heat content abv. 32°F BTU/lb		Ht. of vaporiz.	Heat content abv. 0°C g-cal/g		Ht. of vaporiz. g-cal/	Spec. vol. sat. vap.		Density sat. vap.		Temp.
(°F)	lb/in.²	kg/cm²	Liq.	Vap.	BTU/lb	Liq.	Vap.	g	ft³/lb	m³/kg	lb/ft³	kg/m³	(°C)
0	1.10	0.0773	−8.60	156.90	165.5	−4.78	87.167	91.94	53.76	3.356	0.0186	0.2979	−17.78
10	1.46	0.103	−5.60	158.90	164.5	−3.11	88.278	91.39	43.47	2.714	0.0230	0.3684	−12.22
20	1.89	0.133	−3.00	160.20	163.2	−1.67	89.000	90.67	34.84	2.175	0.0287	0.4597	−6.67
30	2.36	0.166	−0.50	161.70	162.2	−0.28	89.833	90.11	29.49	1.841	0.0339	0.5430	−1.11
40	3.03	0.213	2.05	163.25	161.2	1.14	90.695	89.56	23.52	1.468	0.0425	0.6808	4.44
50	3.90	0.274	4.24	164.24	160.0	2.36	91.245	88.89	20.60	1.286	0.0482	0.7721	10.00
60	4.95	0.348	7.20	166.40	159.2	4.00	92.445	88.44	18.00	1.124	0.0555	0.8890	15.56
70	5.85	0.411	9.80	167.90	158.1	5.44	92.278	87.83	13.20	0.824	0.0758	1.214	21.11
80	7.30	0.513	11.70	168.60	156.9	6.500	93.667	87.17	10.40	0.649	0.0961	1.539	26.67
90	9.15	0.643	13.80	169.40	155.6	7.667	94.111	86.44	8.30	0.518	0.1204	1.920	32.22
100	11.08	0.7790	16.15	170.55	154.4	8.972	94.750	85.78	7.03	0.439	0.1369	2.193	37.78
110	13.50	0.9491	18.30	171.50	153.2	10.17	95.278	85.11	5.80	0.362	0.1724	2.762	43.33
120	16.10	1.132	20.01	172.01	152.0	11.12	95.561	84.44	5.10	0.318	0.1960	3.140	48.89

Carbon Tetrachloride (CCl_4)

Temp.	Abs. press. sat. vap.		Heat content abv. 32°F BTU/lb		Ht. of vaporiz.	Heat content abv. 0°C g-cal/g		Ht. of vaporiz. g-cal/	Spec. vol. sat. vap.		Density sat. vap.		Temp.
(°F)	lb/in.²	kg/cm²	Liq.	Vap.	BTU/lb	Liq.	Vap.	g	ft³/lb	m³/kg	lb/ft³	kg/m³	(°C)
20	0.40	0.028	−2.00	92.45	94.45	−1.11	51.36	52.47	69.5	4.34	0.01438	0.2303	−6.67
30	0.60	0.042	−0.25	93.45	93.70	−0.14	51.92	52.06	53.0	3.31	0.01886	0.3021	−1.11
40	0.84	0.059	1.60	94.80	93.20	0.889	52.67	51.78	40.0	2.50	0.02500	0.4005	4.44
60	1.42	0.100	5.95	98.15	92.20	3.31	54.53	51.22	24.0	1.50	0.04166	0.6673	15.56
70	1.85	0.130	8.20	99.53	91.40	4.56	55.29	50.78	19.5	1.22	0.05128	0.8214	21.11
80	2.40	0.169	9.80	99.87	90.07	5.44	55.48	50.04	16.0	0.999	0.06345	1.016	26.67
90	3.12	0.219	11.60	101.62	90.02	6.444	56.46	50.01	13.0	0.812	0.07692	1.232	32.22
100	4.00	0.281	13.40	102.80	89.40	7.444	57.11	49.67	10.0	0.624	0.1000	1.602	37.78
110	4.89	0.344	15.80	104.50	88.70	8.778	58.06	49.28	8.5	0.53	0.1176	1.884	43.33
120	5.95	0.418	18.06	105.90	87.90	10.03	58.83	48.83	7.5	0.47	0.1333	2.135	48.89

Ethyl Ether ($(C_2H_5)_2O$)

Temp.	Abs. press. sat. vap.		Heat content abv. 32°F BTU/lb		Ht. of vaporiz.	Heat content abv. 0°C g-cal/g		Ht. of vaporiz. g-cal/	Spec. vol. sat. vap.		Density sat. vap.		Temp.
(°F)	lb/in.²	kg/cm²	Liq.	Vap.	BTU/lb	Liq.	Vap.	g	ft³/lb	m³/kg	lb/ft³	kg/m³	(°C)
0	1.3	0.091	−18.00	153.00	171.0	−10.00	85.000	95.00	38.0	2.37	0.0263	0.4213	−17.78
10	1.8	0.13	−12.0	158.43	170.4	−6.67	88.017	94.67	32.5	2.03	0.0332	0.5318	−12.22
20	2.5	0.18	−6.50	163.50	170.0	−3.61	90.833	94.44	27.0	1.69	0.0372	0.5959	−6.67
30	3.4	0.24	−1.50	167.90	169.4	−0.833	93.278	94.11	21.4	1.34	0.0468	0.7496	−1.11
40	4.4	0.31	4.00	172.40	168.4	2.22	95.778	93.56	17.0	1.06	0.0588	0.9419	4.44
50	5.5	0.39	9.57	177.17	167.6	5.32	98.428	93.11	13.2	0.824	0.0757	1.213	10.00
70	8.8	0.62	20.04	185.44	165.4	11.13	103.02	91.89	7.8	0.49	0.1280	2.050	21.11
80	10.9	0.766	26.40	190.60	164.2	14.67	105.89	91.22	6.2	0.39	0.1620	2.595	26.67
90	13.4	0.942	31.50	194.50	163.0	17.50	108.60	90.56	5.1	0.32	0.1960	3.140	32.22
100	16.0	1.12	36.50	197.50	161.5	20.28	109.72	89.72	4.5	0.28	0.2220	3.556	37.78

THERMODYNAMIC PROPERTIES (continued)

Methyl Chloride (CH_3Cl)

Temp. (°F)	Abs. press. sat. vap. lb/in.²	Abs. press. sat. vap. kg/cm²	Heat content abv. 32°F BTU/lb Liq.	Heat content abv. 32°F BTU/lb Vap.	Ht. of vaporiz. BTU/lb	Heat content abv. 0°C g-cal/g Liq.	Heat content abv. 0°C g-cal/g Vap.	Ht. of vaporiz. g-cal/g	Spec. vol. sat. vap. ft³/lb	Spec. vol. sat. vap. m³/kg	Spec. vol. liq. ft³/lb	Spec. vol. liq. m³/lb	Density sat. vap. lb/ft³	Density sat. vap. kg/m³	Density of liq. lb/ft³	Density of liq. kg/m³	Temp. (°C)
−20	11.75	0.8261	−19.0	167.36	186.36	−10.6	92.978	103.53	8.09	0.505	0.015827	0.0009880	0.124	1.98	63.185	1012.1	−28.89
−10	15.0	1.055	−15.38	168.83	184.21	−8.544	93.795	102.34	6.46	0.403	0.015985	0.0009979	0.155	2.48	62.560	1002.1	−23.33
−5	16.79	1.180	−13.58	169.54	183.12	−7.544	94.189	101.73	5.80	0.362	0.016013	0.0009997	0.172	2.76	62.450	1000.3	−20.56
0	18.8	1.32	−11.75	170.23	181.98	−6.528	94.572	101.10	5.18	0.323	0.016146	0.001008	0.193	3.09	61.936	992.09	−17.78
5	21.0	1.48	−9.93	170.96	180.84	−5.517	94.978	100.47	4.68	0.292	0.016228	0.001013	0.214	3.42	61.623	987.08	−15.00
10	23.3	1.64	−8.06	171.58	179.65	−4.478	95.322	99.806	4.18	0.261	0.016310	0.001018	0.239	3.83	61.311	982.08	−12.22
15	25.9	1.82	−6.74	172.24	178.47	−3.744	95.689	99.150	3.88	0.242	0.016388	0.001023	0.258	4.13	61.022	977.45	−9.44
20	28.8	2.02	−4.32	172.95	177.27	−2.400	96.083	98.483	3.41	0.213	0.016474	0.001028	0.293	4.70	60.702	972.32	−6.67
25	31.8	2.24	−2.48	173.63	176.10	−1.378	96.461	97.833	3.09	0.193	0.016552	0.001033	0.324	5.18	60.415	967.73	−3.89
30	35.2	2.47	−0.62	174.28	174.90	−0.344	96.822	97.167	2.81	0.175	0.016645	0.001039	0.356	5.70	60.077	962.31	−1.11
35	38.7	2.72	1.75	174.92	173.77	0.972	97.178	96.539	2.50	0.156	0.016746	0.001045	0.400	6.41	59.715	956.51	1.67
40	42.6	3.00	3.15	175.57	172.42	1.75	97.539	95.789	2.31	0.144	0.016809	0.001049	0.433	6.93	59.492	952.94	4.44
45	46.9	3.30	5.04	176.20	171.16	2.80	97.889	95.089	2.10	0.131	0.016929	0.001057	0.476	7.63	59.069	946.17	7.22
50	51.5	3.62	6.88	176.78	169.90	3.82	98.211	94.389	1.93	0.120	0.017023	0.001063	0.518	8.30	58.745	940.98	10.00
55	56.4	3.97	8.80	177.45	168.65	4.89	98.583	93.695	1.75	0.109	0.017118	0.001069	0.571	9.16	58.419	935.76	12.78
60	61.6	4.33	10.70	178.05	167.35	5.944	98.917	92.972	1.61	0.101	0.017219	0.001075	0.621	9.95	58.077	930.28	15.56
65	67.3	4.73	12.62	178.64	166.02	7.011	99.245	92.233	1.47	0.0918	0.017318	0.001081	0.680	10.9	57.742	924.91	18.33
70	73.3	5.15	14.52	179.17	164.65	8.067	99.539	91.472	1.34	0.0837	0.017421	0.001088	0.746	12.0	57.403	919.48	21.11
75	79.2	5.57	16.46	179.78	163.28	9.144	99.878	90.711	1.24	0.0774	0.017526	0.001094	0.806	12.9	57.058	913.96	23.89
80	85.3	6.00	18.36	180.24	161.88	10.20	100.13	89.933	1.14	0.0712	0.017632	0.001101	0.877	14.1	56.714	908.44	26.67
85	94.1	6.62	20.12	180.74	160.48	11.18	100.41	89.156	1.05	0.0655	0.017740	0.001108	0.952	15.3	56.369	902.92	29.44
90	102.1	7.178	22.13	181.22	159.09	12.29	100.68	88.383	0.98	0.061	0.017850	0.001114	1.02	16.3	56.022	897.36	32.22
95	110.3	7.755	24.07	181.76	157.70	13.37	100.98	87.611	0.91	0.057	0.017961	0.001121	1.10	17.6	55.675	891.80	35.00
100	118.8	8.352	26.06	182.36	156.30	14.48	101.31	86.833	0.85	0.053	0.018074	0.001128	1.18	18.8	55.327	886.23	37.78

Mercury (Hg)

Temp. (°F)	Abs. press. sat. vap. lb/in.²	Abs. press. sat. vap. kg/cm²	Heat content abv. 32°F BTU/lb Liq.	Heat content abv. 32°F BTU/lb Vap.	Ht. of vaporiz. BTU/lb	Heat conent abv. 0°C g-cal/g Liq.	Heat conent abv. 0°C g-cal/g Vap.	Ht. of vaporix. g-cal/g	Spec. vol. sat. vap. ft³/lb	Spec. vol. sat. vap. m³/kg	Density of sat. vap. lb/ft³	Density of sat. vap. kg/m³	Entropy above 32°F BTU/lb/°F Liq.	Entropy above 32°F BTU/lb/°F Vap.	Entropy above 32°F BTU/lb/°F Evap.	Temp (°C)
402	0.4	0.03	13.81	141.96	128.15	7.672	78.867	71.195	114.50	7.1480	0.008733	0.1399	0.0209	0.1696	0.1487	205.56
444	0.8	0.06	15.36	142.60	127.24	8.533	79.222	70.689	59.72	3.728	0.016745	0.26822	0.0227	0.1635	0.1408	228.89
458	1.0	0.07	15.89	142.81	126.92	8.828	79.339	70.511	48.45	3.025	0.02064	0.3306	0.0233	0.1616	0.1383	236.67
485	1.5	0.11	16.90	143.23	126.33	9.389	79.572	70.183	33.14	2.069	0.03017	0.4833	0.0244	0.1581	0.1337	251.67
505	2.0	0.14	17.65	143.54	125.89	9.806	79.745	69.939	25.32	1.581	0.03948	0.6324	0.0251	0.1556	0.1305	262.78
558	4.0	0.28	19.62	144.34	124.72	10.90	80.189	69.289	13.26	0.8278	0.07540	1.208	0.0271	0.1497	0.1226	292.22
591	6.0	0.42	20.87	144.86	123.99	11.59	80.478	68.883	9.096	0.5678	0.10993	1.7609	0.0283	0.1462	0.1179	310.56
617	8.0	0.56	21.81	145.24	123.43	12.12	80.689	68.572	6.9630	0.43469	0.14361	2.3003	0.0292	0.1439	0.1147	325.00
637	10.0	0.703	22.58	145.56	122.98	12.54	80.867	68.322	5.6610	0.35341	0.17664	2.8294	0.0299	0.1420	0.1121	336.11
676	15.0	1.05	24.04	146.16	122.12	13.36	81.200	67.844	3.8923	0.24299	0.25691	4.1152	0.0312	0.1387	0.1075	357.78
706	20.0	1.41	25.15	146.61	121.46	13.97	81.450	67.478	2.983	0.1862	0.3352	5.369	0.0322	0.1364	0.1042	374.44
730	25.0	1.76	26.05	146.98	120.93	14.47	81.656	67.183	2.429	0.1516	0.4117	6.595	0.0330	0.1346	0.1016	387.78
751	30.0	2.11	26.81	147.29	120.48	14.89	81.828	66.933	2.053	0.1282	0.4871	7.802	0.0336	0.1331	0.0995	399.44
769	35.0	2.46	27.49	147.57	120.08	15.27	81.983	66.711	1.7815	0.11122	0.5613	8.991	0.0342	0.1319	0.0977	409.44
785	40.0	2.81	28.08	147.81	119.73	15.60	82.117	66.517	1.5762	0.098399	0.6344	10.16	0.0346	0.1308	0.0962	418.33
799	45.0	3.16	28.62	148.04	119.42	15.90	82.245	66.344	1.4147	0.088317	0.7069	11.32	0.0351	0.1300	0.0949	426.11
812	50	3.5	29.11	148.24	119.13	16.17	82.356	66.183	1.284	0.08016	0.7788	12.47	0.0355	0.1291	0.0936	433.33
836	60	4.2	29.99	148.60	118.61	16.66	82.556	65.894	1.086	0.06780	0.9204	14.74	0.0361	0.1276	0.0915	446.67
857	70	4.9	30.75	148.90	118.15	17.08	82.722	65.639	0.9436	0.05891	1.0597	16.974	0.0367	0.1265	0.0898	458.33
875	80	5.6	31.44	149.19	117.75	17.47	82.883	65.417	0.8349	0.05212	1.1977	19.185	0.0372	0.1254	0.0882	468.33
892	90	6.3	32.06	149.44	117.38	17.81	83.022	65.211	0.7497	0.04680	1.3338	21.365	0.0377	0.1247	0.0870	477.78
907	100	7.03	32.63	149.68	117.05	18.13	83.156	65.028	0.6811	0.04252	1.4682	23.518	0.0381	0.1237	0.0856	486.11
921	110	7.73	33.16	149.90	116.74	18.42	83.278	64.856	0.6242	0.03897	1.6020	25.661	0.0385	0.1230	0.0845	493.89
934	120	8.44	33.66	150.10	116.44	18.70	83.389	64.689	0.5767	0.03600	1.7340	27.775	0.0389	0.1224	0.0835	501.11
947	130	9.14	34.12	150.29	116.17	18.96	82.495	64.539	0.5360	0.03346	1.8656	29.883	0.0392	0.1218	0.0826	508.33
958	140	9.84	34.55	150.47	115.92	19.19	82.595	64.400	0.5012	0.03129	1.9952	31.959	0.0395	0.1213	0.0818	514.44
969	150	10.5	34.96	150.63	115.67	19.42	83.683	64.261	0.4706	0.02938	2.125	34.04	0.0398	0.1207	0.0809	520.56
1000	180	12.7	36.09	151.10	115.01	20.05	83.945	63.894	0.3990	0.02491	2.506	40.14	0.0406	0.1194	0.0788	537.78

PHYSICAL PROPERTIES OF FLUOROCARBON REFRIGERANTS

Property No.	Refrigerant name		11	12	13	13B1	14	21
1	Formula		CCl_3F	CCl_2F_2	$CClF_3$	$CBrF_3$	CF_4	$CHCl_2F$
2	Molecular weight		137.37	120.91	104.46	148.92	88.01	102.92
3	Normal boiling point; °C		23.82	−29.79	−81.4	−57.75	−127.96	8.92
4	Normal freezing point; °C		−111	−158	−181	−168	−184	−135
5	Critical temperature; °C		198	112	28.9	67	−45.67	178.5
6	Critical pressure; atm		43.5	40.6	38.2	39.1	36.96	51
7	Critical volume; cm/mol		247	217	181	200	141	197
8	Critical density; g/cm		0.554	0.558	0.578	0.745	0.626	0.522
9	Density of liquid at 25°C; g/cm		1.467	1.311	$1.298^{-30°C}$	1.538	$1.317^{-80°C}$	1.366
10	Density of saturated vapor at B.P.; g/liter		5.86	6.33	7.01	8.71	7.62	4.57
11	Specific heat of liquid at 25°C; cal/g		0.208	0.232	$0.247^{-30°C}$	0.208	$0.294^{-80°C}$	0.256
12	Specific heat of vapor at 25°C and 1 atm; cal/g		$0.142^{38°C}$	0.145	0.158	0.112	0.169	0.140
13	Heat of vaporization at B.P.; cal/g		43.10	39.47	35.47	28.38	32.49	57.86
14	Thermal conductivity at 25°C; Btu/(hr)(ft)(°F)	liquid	0.050	0.041	0.020	0.025	$0.040^{-100°F}$	0.063
		vapor; 1 atm	0.00484	0.00557	—	—	—	0.0569
15	Viscosity at 25°C; centipoise	liquid	0.42	0.26	0.016	0.15	0.020	0.34
		vapor; 1 atm	0.011	0.013	—	0.016	—	0.011
16	Surface tension at 25°C; dyne cm		18	9	$14^{-73.3°C}$	4	$14^{-73.3°C}$	18
17	Refractive index of liquid at 25°C		1.374	1.287	$1.199^{-73.3°C}$	1.238	$1.151^{-73.3°C}$	1.354
18	Dielectric constant	liquid	$2.28^{29°C}$	$2.13^{29°C}$	—	—	—	$5.34^{28°C}$
		vapor at 0.5 atm	$1.10019^{29°C}$	$1.0016^{29°C}$	$1.0013^{29°C}$	—	$1.0006^{24.5°C}$	$1.0035^{30°C}$
19	Solubility in water at 25°C and 1 atm; wt %		0.011	0.028	0.009	0.03	0.0015	0.95
20	Solubility of water in compound at 25°C; wt %		0.011	0.009	—	$0.0095^{21.1°C}$	—	0.13
21	Toxicity; Group number (See separate table for definition of group number.)		5a	6	probably 6	6	probably 6	<4 >5

Property No.	22	23	112	113	114	114B2	115	116	500	502
1	$CHClF_2$	CHF_3	$C_2Cl_4F_2$	$C_2Cl_3F_3$	$C_2Cl_2F_4$	$C_2Br_2F_4$	C_2ClF_5	C_2F_6	*	**
2	80.47	70.01	203.83	187.38	170.91	259.83	154.47	138.01	105.5	120.7
3	−40.75	−82.03	92.8	47.57	3.77	47.26	−38.7	−78.2	−33.5	−45.6
4	−160	−155.2	26	−35	−94	−110.5	−106	−100.6	−158.9	—
5	96	25.9	278	214.1	145.7	214.5	80	19.7	105.4	179.89
6	49.12	47.7	34	33.7	32.2	34	30.8	29.4	43.7	590.3
7	165	133	370	325	293	329	259	225	—	290
8	0.525	0.525	0.55	0.576	0.582	0.790	0.596	0.612	0.497	0.56
9	1.194	0.670	$1.634^{30°C}$	1.565	1.456	2.163	1.291	$1.587^{-73.3°C}$	$1.138^{30°C}$	1.242
10	4.72	4.66	7.02	7.38	7.83	—	8.37	9.01	—	6.05
11	0.300	1.553	—	0.218	0.243	0.166	0.285	$0.232^{-73.3°C}$	$0.161^{30°C}$	—
12	0.157	0.176	—	$0.161^{60°C}$	0.170	—	0.164	0.182^{0mm}	—	—
13	55.81	57.23	37 (est)	35.07	32.51	25 (est)	30.11	27.97	—	42.48
14	0.052	0.008	0.040	0.038	0.034	0.027	0.026	$0.045^{-100°F}$	—	0.038
	0.00678	—	—	$0.0045^{0.5\ atm}$	0.00646	—	0.00803	0.0098 (est)	—	—
15	0.23	0.016 (est)	1.21	0.68	0.38	0.72	0.26	—	$0.292^{-15°C}$	0.25
	0.013	—	—	$0.01^{0.1\ atm}$	0.011	—	0.013	—	—	—
16	8	$15^{-73.3°C}$	$23^{30°C}$	19	12	18	5	$16^{-73.3°C}$	—	8
17	1.256	$1.251^{-73.3°C}$	1.413	1.354	1.288	1.367	1.214	$1.206^{-73.3°C}$	—	1.235
18	$6.11^{24°C}$	—	$2.52^{25°C}$	$2.41^{25°C}$	$2.26^{25°C}$	$2.34^{25°C}$	—	—	—	—
	$1.0035^{25.4°C}$	—	—	—	$1.0021^{26.5°C}$	—	$1.0018^{27.4°C}$	—	—	—
19	0.30	0.10	0.12 (sat. pr)	0.017 (sat. pr)	0.013	—	0.006	—	—	—
20	0.13	—	—	0.011	0.009	—	—	—	—	0.560
21	5a	probably 6	<4 >5	<4 >5	6	5a	6	probably 6	5a	5a

* Azeotrope of CCl_2F_2 (73.8 wt %) and $C_2H_4F_2$ (26.2 wt %) ** Azeotrope of $CHClF_2$ (48.8 wt %) and C_2ClF_5 (51.2 wt %).

UNDERWRITERS' LABORATORIES' CLASSIFICATION OF COMPARATIVE LIFEHAZARD OF GASES AND VAPORS

(Group number definition)

Group	Definition	Examples
1	Gases or vapors which in concentrations of the order of 1/2 to 1% for durations of exposure of the order of 5 min are lethal or produce serious injury.	Sulfur dioxide
2	Gases or vapors which in concentrations of the order of 1/2 to 1 % for duration of exposure of the order of 1/2 hr are lethal or produce serious injury	Ammonia, methyl bromide
3	Gases or vapors which in concentrations of the order of 2 to 2 1/2 % for durations of exposure of the order of 1 hr are lethal or produce serious injury,	Bromochloromethane carbon tetrachloride, chloroform, methyl formate
4	Gases or vapors which in concentrations of the order of 2 to 2 1/2 % for durations of exposure of the order of 2 hr are lethal or produce serious injury.	Dichloroethylene methyl chloride, ethylbromide
Between 4 and 5	Appear to classify as somewhat less toxic than Group 4.	Methylene chloride, ethyl chloride.
		Refrigerant 112[a]
	Much less toxic than group 4 but somewhat more toxic than Group 5.	Refrigerant 113
		Refrigerant 21
5a	Gases or vapors much less toxic than Group 4 but more toxic than Group 6.	Refrigerant 11
		Refrigerant 22
		Refrigerant 114B2
		Refrigerant 502
		Carbon dioxide
5b	Gases or vapors which available data indicate would classify as either Group 5a or Group 6.	Ethane, propane, butane
6	Gases or vapors which in concentrations up to at least about 20% by volume for duration of exposure of the order of 2 hr do not appear to produce injury.	Refrigerant 13B1
		Refrigerant 12
		Refrigerant 114
		Refrigerant 115
		Refrigerant 13[a]
		Refrigerant 14[a]
		Refrigerant 23[a]
		Refrigerant 116[a]
		Refrigerant C318[a]

[a] Not tested by U.L. but estimated to belong in group indicated.

THERMAL CONDUCTIVITY OF LIQUID FLUOROCARBONS

To convert from W/(m)(K) to Btu/(hr)(ft)(°F) divided by 1.7296.
To convert from W/(m)(K) to cal/(sec)(cm)(°C) divide by 418.4

Fluorocarbon	Formula	Temperature, K	Conductivity, W/(m)(K)
12	CCl_2F_2	277.2	94.14
		298.1	97.49
		303.8	103.76
		317.9	110.04
		329.9	117.15
		346.8	126.36
22	$CHClF_3$	289.5	107.11
		302.8	114.64
		327.8	126.76
		346.8	140.16
114	$C_2Cl_2F_4$	303.6	113.39
		316.8	122.59
		328.7	130.96
		343.0	140.58
13B1	$CBrF_3$	277.3	91.21
		282.3	93.30
		303.5	103.34
		318.7	111.71
		331.9	118.41
		342.8	123.01
		346.9	128.03
C-318	C_4F_8	280.1	112.97
		287.3	117.57
		298.0	130.96
		310.7	141.42
		323.2	148.11
		332.2	156.48
		342.9	158.99
		348.4	165.69
		350.8	166.94

MISCELLANEOUS PROPERTIES OF COMMON REFRIGERANTS

Refrigerant	Formula	Flash Point °F.	Ignition Temp. °F.	Explosive Limits % by Volume: Lower	Explosive Limits % by Volume: Upper	Vapor Density (Air = 1)	Boiling Point °F	Threshold Limit Value* Parts per Million in Air	Water Soluble	Odor
Ammonia	NH_3	—	1204	16	25	0.59	−28	100	yes	yes
Bromotrifluoromethane (Kulene-131)	CF_3Br		nonflammable			5.25	−73.6	—	no	yes
Butane	C_4H_{10}	−76	806	1.8	8.4	2.04	33	—	no	no
Carbon dioxide	CO_2		nonflammable			1.53	−108	5000	yes	no
Carbon tetrachloride	CCL_4		nonflammable			5.32	170	25	no	yes
Dichlorodifluoromethane (Freon-12)	CCl_2F_2		nonflammable			4.17	−21.6	—	no	no
Dichlorodifluoromethane, 73.8% Ethylidene fluoride, 26.2% (Carrene-7)	CCl_2F_2 CH_3CHF_2		nonflammable			3.24	−28.0	—	no	yes
Dichloromonofluoromethane (Freon-21)	$CHCl_2F$		practically nonflammable			3.55	48	—	no	yes
Dichlorotetrafluoroethane (Freon-114)	$C_2Cl_2F_4$		practically nonflammable			5.89	38	—	no	no
Ethane	C_2H_6	<20	950	3.0	12.5	1.04	−128	—	no	no
Ethylene	C_2H_4	<20	842	3.1	32	0.972	−155	—	yes	yes
Isobutane	$(CH_3)_3CH$	<20	1010	1.8	8.4	2.01	14	—	no	no
Methyl chloride	CH_3Cl	632	1170	10.7	11.4	1.78	−11	100	yes	yes
Monochlorodifluoromethane (Freon-22)	$CHClF_2$		practically nonflammable			2.9	−41	—	yes	no
Monochlorotrifluoromethane (Freon-13)	$CClF_3$		nonflammable			3.6	−112	—	—	no
Propane	C_3H_8	<20	871	2.2	9.5	1.56	−45	—	no	no
Propylene	C_3H_6	<20	927	2.4	10.3	1.49	−53	—	yes	yes
Sulfur dioxide	SO_2		nonflammable			2.2	14	10	yes	yes
Tetrafluoromethane (Freon-14)	CF_4		nonflammable			3.0	−198	—	no	no
Trichloroethylene	C_2HCl_3		nonflammable at normal temperature			4.53	189	200	no	yes
Trichloromonofluoromethane (Freon-11)(Carrene-2)	CCL_3F		nonflammable			4.7	75.3	—	no	no
Trichlorotrifluoroethane (Freon-113)	$C_2Cl_3F_3$		practically nonflammable			6.4	118	—	no	no

* Maximum average atmospheric concentration of contaminants to which workers may be exposed for an eight-hour work day without injury to health. (American Conference of Governmental Industrial Hygienists: "Threshold Limit Values for 1954.") Where blanks appear in this column no published information was available on threshold limit values.

HYGROMETRIC AND BAROMETRIC TABLES

CONVERSION TABLE FOR BAROMETRIC READINGS

U.S. inches to cm.

Inches	.00	.01	.02	.03	.04	.05	.06	.07	.08	.09
27.0	68.580	.606	.631	.656	.682	.707	.733	.758	.783	.809
27.1	.834	.860	.885	.910	.936	.961	.987	*.012	*.037	*.063
27.2	69.088	.114	.139	.164	.190	.215	.241	.266	.291	.317
27.3	.342	.368	.393	.418	.444	.469	.495	.520	.545	.571
27.4	.596	.622	.647	.672	.698	.723	.749	.774	.799	.825
27.5	.850	.876	.901	.926	.952	.977	*.002	*.028	*.053	*.079
27.6	70.104	.130	.155	.180	.206	.231	.257	.282	.307	.333
27.7	.358	.384	.409	.434	.460	.485	.511	.536	.561	.587
27.8	.612	.638	.663	.688	.714	.739	.765	.790	.815	.841
27.9	.866	.892	.917	.942	.968	.993	*.018	*.044	*.069	*.095
28.0	71.120	.146	.171	.196	.222	.247	.273	.298	.323	.349
28.1	.374	.400	.425	.450	.476	.501	.527	.552	.577	.603
28.2	.628	.654	.679	.704	.730	.755	.781	.806	.831	.857
28.3	.882	.908	.933	.958	.984	*.009	*.035	*.060	*.085	*.111
28.4	72.136	.162	.187	.212	.238	.263	.289	.314	.339	.365
28.5	.390	.416	.441	.466	.492	.517	.543	.568	.593	.619
28.6	.644	.670	.695	.720	.746	.771	.797	.822	.847	.873
28.7	.898	.924	.949	.974	*.000	*.025	*.051	*.076	*.101	*.127
28.8	73.152	.178	.203	.228	.254	.279	.305	.330	.355	.381
28.9	.406	.432	.457	.482	.508	.533	.559	.584	.609	.635

Inches	.00	.01	.02	.03	.04	.05	.06	.07	.08	.09
29.0	.660	.686	.711	.736	.762	.787	.813	.838	.863	.889
29.1	.914	.940	.965	.990	*.016	*.041	*.067	*.092	*.117	*.143
29.2	74.168	.194	.219	.244	.270	.295	.321	.346	.371	.397
29.3	.422	.448	.473	.498	.524	.549	.575	.600	.625	.651
29.4	.676	.702	.727	.752	.778	.803	.829	.854	.879	.905
29.5	.930	.956	.981	*.006	*.032	*.057	*.083	*.108	*.133	*.159
29.6	75.184	.210	.235	.260	.286	.311	.337	.362	.387	.413
29.7	.438	.464	.489	.514	.540	.565	.591	.616	.641	.667
29.8	.692	.718	.743	.768	.794	.819	.845	.870	.895	.921
29.9	.946	.972	.997	*.022	*.048	*.073	*.099	*.124	*.149	*.175
30.0	76.200	.226	.251	.277	.302	.327	.353	.378	.404	.429
30.1	.454	.480	.505	.531	.556	.581	.607	.632	.658	.683
30.2	.708	.734	.759	.785	.810	.835	.861	.886	.912	.937
30.3	.962	.988	*.013	*.039	*.064	*.089	*.115	*.140	*.166	*.191
30.4	77.216	.242	.267	.293	.318	.343	.369	.394	.420	.445
30.5	.470	.496	.521	.547	.572	.597	.623	.648	.674	.699
30.6	.724	.750	.775	.801	.826	.851	.877	.902	.928	.953
30.7	.978	*.004	*.029	*.055	*.080	*.105	*.131	*.156	*.182	*.207
30.8	78.232	.258	.283	.309	.334	.359	.385	.410	.436	.461
30.9	.486	.512	.537	.563	.588	.613	.639	.664	.690	.715

U.S. Inches to Millibars

Based on the relation 1 inch of mercury at 32°F represents a pressure of 33.8639 millibars.

Note: Figures in last nine columns to be preceded by 7, 8, 9 or 10 as indicated in column 2.

Inches	.00	.01	.02	.03	.04	.05	.06	.07	.08	.09
23.0	7 78.87	79.21	79.55	79.89	80.22	80.56	80.90	81.24	81.58	81.92
23.1	7 82.26	82.59	82.93	83.27	83.61	83.95	84.29	84.63	84.97	85.30
23.2	7 85.64	85.98	86.32	86.66	87.00	87.34	87.67	88.01	88.35	88.69
23.3	7 89.03	89.37	89.71	90.04	90.38	90.72	91.06	91.40	91.74	92.08
23.4	7 92.42	92.75	93.09	93.43	93.77	94.11	94.45	94.79	95.12	95.46
23.5	7 95.80	96.14	96.48	96.82	97.16	97.49	97.83	98.17	98.51	98.85
23.6	7 99.19	99.53	99.87	*00.20	*00.54	*00.88	*01.22	*01.56	*01.90	*02.24

Inches	.00	.01	.02	.03	.04	.05	.06	.07	.08	.09
23.7	8 02.57	02.91	03.25	03.59	03.93	04.27	04.61	04.94	05.28	05.62
23.8	8 05.96	06.30	06.64	06.98	07.32	07.65	07.99	08.33	08.67	09.01
23.9	8 09.35	09.69	10.02	10.36	10.70	11.04	11.38	11.72	12.06	12.39
24.0	8 12.73	13.07	13.41	13.75	14.09	14.43	14.77	15.10	15.44	15.78
24.1	8 16.12	16.46	16.80	17.14	17.47	17.81	18.15	18.49	18.83	19.17
24.2	8 19.51	19.85	20.18	20.52	20.86	21.20	21.54	21.88	22.22	22.55
24.3	8 22.89	23.23	23.57	23.91	24.25	24.59	24.92	25.26	25.60	25.94

U.S. Inches to Millibars (Continued)

Inches	.00	.01	.02	.03	.04	.05	.06	.07	.08	.09
24.4	8 26.28	26.62	26.96	27.30	27.63	27.97	28.31	28.65	28.99	29.33
24.5	8 29.67	30.00	30.34	30.68	31.02	31.36	31.70	32.04	32.37	32.71
24.6	8 33.05	33.39	33.73	34.07	34.41	34.75	35.08	35.42	35.76	36.10
24.7	8 36.44	36.78	37.12	37.45	37.79	38.13	38.47	38.81	39.15	39.49
24.8	8 39.82	40.16	40.50	40.84	41.18	41.52	41.86	42.20	42.53	42.87
24.9	8 43.21	43.55	43.89	44.23	44.57	44.90	45.24	45.58	45.92	46.26
25.0	8 46.60	46.94	47.27	47.61	47.95	48.29	48.63	48.97	49.31	49.65
25.1	8 49.98	50.32	50.66	51.00	51.34	51.68	52.02	52.35	52.69	53.03
25.2	8 53.37	53.71	54.05	54.39	54.72	55.06	55.40	55.74	56.08	56.42
25.3	8 56.76	57.10	57.43	57.77	58.11	58.45	58.79	59.13	59.47	59.80
25.4	8 60.14	60.48	60.82	61.16	61.50	61.84	62.17	[illegible]	[illegible]	[illegible]
25.5	8 63.53	63.87	64.21	64.55	64.88	65.22	65.56	65.90	66.24	66.58
25.6	8 66.92	67.25	67.59	67.93	68.27	68.61	68.95	69.29	69.62	69.96
25.7	8 70.30	70.64	70.98	71.32	71.66	72.00	72.33	72.67	73.01	73.35
25.8	8 73.69	74.03	74.37	74.70	75.04	75.38	75.72	76.06	76.40	76.74
25.9	8 77.08	77.41	77.75	78.09	78.43	78.77	79.11	79.45	79.78	80.12
26.0	8 80.46	80.80	81.14	81.48	81.82	82.15	82.49	82.83	83.17	83.51
26.1	8 83.85	84.19	84.53	84.86	85.20	85.54	85.88	86.22	86.56	86.90
26.2	8 87.23	87.57	87.91	88.25	88.59	88.93	89.27	89.60	89.94	90.28
26.3	8 90.62	90.96	91.30	91.64	91.98	92.31	92.65	92.99	93.33	93.67
26.4	8 94.01	94.35	94.68	95.02	95.36	95.70	96.04	96.38	96.72	97.05
26.5	8 97.39	97.73	98.07	98.41	98.75	99.09	99.43	99.76	*00.10	*00.44
26.6	9 00.78	01.12	01.46	01.80	02.13	02.47	02.81	03.15	03.49	03.83
26.7	9 04.17	04.50	04.84	05.18	05.52	05.86	06.20	06.54	06.88	07.21
26.8	9 07.55	07.89	08.23	08.57	08.91	09.25	09.58	09.92	10.26	10.60
26.9	9 10.94	11.28	11.62	11.95	12.29	12.63	12.97	13.31	13.65	13.99
27.0	9 14.33	14.66	15.00	15.34	15.68	16.02	16.36	16.70	17.03	17.37
27.1	9 17.71	18.05	18.39	18.73	19.07	19.40	19.74	20.08	20.42	20.76
27.2	9 21.10	21.44	21.78	22.11	22.45	22.79	23.13	23.47	23.81	24.15
27.3	9 24.48	24.82	25.16	25.50	25.84	26.18	26.52	26.85	27.19	27.53
27.4	9 27.87	28.21	28.55	28.89	29.23	29.56	29.90	30.24	30.58	30.92
27.5	9 31.26	31.60	31.93	32.27	32.61	32.95	33.29	33.63	33.97	34.31
27.6	9 34.64	34.98	35.32	35.66	36.00	36.34	36.68	37.01	37.35	37.69
27.7	9 38.03	38.37	38.71	39.05	39.38	39.72	40.06	40.40	40.74	41.08
27.8	9 41.42	41.76	42.09	42.43	42.77	43.11	43.45	43.79	44.13	44.46
27.9	9 44.80	45.14	45.48	45.82	46.16	46.50	46.83	47.17	47.51	47.85
28.0	9 48.19	48.53	48.87	49.21	49.54	49.88	50.22	50.56	50.90	51.24
28.1	9 51.58	51.91	52.25	52.59	52.93	53.27	53.61	53.95	54.28	54.62

Inches	.00	.01	.02	.03	.04	.05	.06	.07	.08	.09
28.2	9 54.96	55.30	55.64	55.98	56.32	56.66	56.99	57.33	57.67	58.01
28.3	9 58.35	58.69	59.03	59.36	59.70	60.04	60.38	60.72	61.06	61.40
28.4	9 61.73	62.07	62.41	62.75	63.09	63.43	63.77	64.11	64.44	64.78
28.5	9 65.12	65.46	65.80	66.14	66.48	66.81	67.15	67.49	67.83	68.17
28.6	9 68.51	68.85	69.18	69.52	69.86	70.20	70.54	70.88	71.22	71.56
28.7	9 71.89	72.23	72.57	72.91	73.25	73.59	73.93	74.26	74.60	74.94
28.8	9 75.28	75.62	75.96	76.30	76.63	76.97	77.31	77.65	77.99	78.33
28.9	9 78.67	79.01	79.34	79.68	80.02	80.36	80.70	81.04	81.38	81.71
29.0	9 82.05	82.39	82.73	83.07	83.41	83.75	84.08	84.42	84.76	85.10
29.1	9 85.44	85.78	86.12	86.46	86.79	87.13	87.47	87.81	88.15	88.49
29.2	9 88.83	89.16	89.50	89.84	90.18	90.52	90.86	91.20	91.53	91.87
29.3	9 92.21	92.55	92.89	[illegible]	[illegible]	[illegible]	[illegible]	[illegible]	[illegible]	[illegible]
29.4	9 95.60	95.94	96.28	96.61	96.95	97.29	97.63	97.97	98.31	98.65
29.5	9 98.99	99.32	99.66	00.00	*00.34	*00.68	*01.02	*01.36	*01.69	*02.03
29.6	10 02.37	02.71	03.05	03.39	03.73	04.06	04.40	04.74	05.08	05.42
29.7	10 05.76	06.10	06.44	06.77	07.11	07.45	07.79	08.13	08.47	08.81
29.8	10 09.14	09.48	09.82	10.16	10.50	10.84	11.18	11.51	11.85	12.19
29.9	10 12.53	12.87	13.21	13.55	13.89	14.22	14.56	14.90	15.24	15.58
30.0	10 15.92	16.26	16.59	16.93	17.27	17.61	17.95	18.29	18.63	18.96
30.1	10 19.30	19.64	19.98	20.32	20.66	21.00	21.34	21.67	22.01	22.35
30.2	10 22.69	23.03	23.37	23.71	24.04	24.38	24.72	25.06	25.40	25.74
30.3	10 26.08	26.41	26.75	27.09	27.43	27.77	28.11	28.45	28.79	29.12
30.4	10 29.46	29.80	30.14	30.48	30.82	31.16	31.49	31.83	32.17	32.51
30.5	10 32.85	33.19	33.53	33.86	34.20	34.54	34.88	35.22	35.56	35.90
30.6	10 36.24	36.57	36.91	37.25	37.59	37.93	38.27	38.61	38.94	39.28
30.7	10 39.62	39.96	40.30	40.64	40.98	41.31	41.65	41.99	42.33	42.67
30.8	10 43.01	43.35	43.69	44.02	44.36	44.70	45.04	45.38	45.72	46.06
30.9	10 46.39	46.73	47.07	47.41	47.75	48.09	48.43	48.76	49.10	49.44
31.0	10 49.78	50.12	50.46	50.80	51.14	51.47	51.81	52.15	52.49	52.83
31.1	10 53.17	53.51	53.84	54.18	54.52	54.86	55.20	55.54	55.88	56.22
31.2	10 56.55	56.89	57.23	57.57	57.91	58.25	58.59	58.92	59.26	59.60
31.3	10 59.94	60.28	60.62	60.96	61.29	61.63	61.97	62.31	62.65	62.99
31.4	10 63.33	63.67	64.00	64.34	64.68	65.02	65.36	65.70	66.04	66.37
31.5	10 66.71	67.05	67.39	67.73	68.07	68.41	68.74	69.08	69.42	69.76
31.6	10 70.10	70.44	70.78	71.12	71.45	71.79	72.13	72.47	72.81	73.15
31.7	10 73.49	73.82	74.16	74.50	74.84	75.18	75.52	75.86	76.19	76.53
31.8	10 76.87	77.21	77.55	77.89	78.23	78.57	78.90	79.24	79.58	79.92
31.9	10 80.26	80.60	80.94	81.27	81.61	81.95	82.29	82.63	82.97	83.31

Centimeters to Millibars

Based on the relation 1 centimeter of mercury at 0°C represents a pressure of 13.3322 millibars.

Note: Figures in last nine columns to be preceded by 9.

Centimeters	.00	.01	.02	.03	.04	.05	.06	.07	.08	.09
68.0	9 06.59	06.72	06.86	06.99	07.12	07.26	07.39	07.52	07.66	07.79
68.1	9 07.92	08.06	08.19	08.32	08.46	08.59	08.72	08.86	08.99	09.12
68.2	9 09.26	09.39	09.52	09.66	09.79	09.92	10.06	10.19	10.32	10.46
68.3	9 10.59	10.72	10.86	10.99	11.12	11.26	11.39	11.52	11.66	11.79
68.4	9 11.92	12.06	12.19	12.32	12.46	12.59	12.72	12.86	12.99	13.12
68.5	9 13.26	13.39	13.52	13.66	13.79	13.92	14.06	14.19	14.32	14.46
68.6	9 14.59	14.72	14.86	14.99	15.12	15.26	15.39	15.52	15.66	15.79
68.7	9 15.92	16.06	16.19	16.32	16.46	16.59	16.72	16.86	16.99	17.12
68.8	9 17.26	17.39	17.52	17.66	17.79	17.92	18.06	18.19	18.32	18.46
68.9	9 18.59	18.72	18.86	18.99	19.12	19.26	19.39	19.52	19.66	19.79
69.0	9 19.92	20.06	20.19	20.32	20.46	20.59	20.72	20.86	20.99	21.12
69.1	9 21.26	21.39	21.52	21.65	21.79	21.92	22.05	22.19	22.32	22.45
69.2	9 22.59	22.72	22.85	22.99	23.12	23.25	23.39	23.52	23.65	23.79
69.3	9 23.92	24.05	24.19	24.32	24.45	24.59	24.72	24.85	24.99	25.12
69.4	9 25.25	25.39	25.52	25.65	25.79	25.92	26.05	26.19	26.32	26.45
69.5	9 26.59	26.72	26.85	26.99	27.12	27.25	27.39	27.52	27.65	27.79
69.6	9 27.92	28.05	28.19	28.32	28.45	28.59	28.72	28.85	28.99	29.12
69.7	9 29.25	29.39	29.52	29.65	29.79	29.92	30.05	30.19	30.32	30.45
69.8	9 30.59	30.72	30.85	30.99	31.12	31.25	31.39	31.52	31.65	31.79
69.9	9 31.92	32.05	32.19	32.32	32.45	32.59	32.72	32.85	32.99	33.12
70.0	9 33.25	33.39	33.52	33.65	33.79	33.92	34.05	34.19	34.32	34.45
70.1	9 34.59	34.72	34.85	34.99	35.12	35.25	35.39	35.52	35.65	35.79
70.2	9 35.92	36.05	36.19	36.32	36.45	36.59	36.72	36.85	36.99	37.12
70.3	9 37.25	37.39	37.52	37.65	37.79	37.92	38.05	38.19	38.32	38.45
70.4	9 38.59	38.72	38.85	38.99	39.12	39.25	39.39	39.52	39.65	39.79
70.5	9 39.92	40.05	40.19	40.32	40.45	40.59	40.72	40.85	40.99	41.12
70.6	9 41.25	41.39	41.52	41.65	41.79	41.92	42.05	42.19	42.32	42.45
70.7	9 42.59	42.72	42.85	42.99	43.12	43.25	43.39	43.52	43.65	43.79
70.8	9 43.92	44.05	44.19	44.32	44.45	44.59	44.72	44.85	44.99	45.12
70.9	9 45.25	45.39	45.52	45.65	45.79	45.92	46.05	46.19	46.32	46.45
71.0	9 46.59	46.72	46.85	46.99	47.12	47.25	47.39	47.52	47.65	47.79
71.1	9 47.92	48.05	48.19	48.32	48.45	48.59	48.72	48.85	48.99	49.12
71.2	9 49.25	49.39	49.52	49.65	49.79	49.92	50.05	50.19	50.32	50.45
71.3	9 50.59	50.72	50.85	50.99	51.12	51.25	51.39	51.52	51.65	51.79
71.4	9 51.92	52.05	52.19	52.32	52.45	52.59	52.72	52.85	52.99	53.12
71.5	9 53.25	53.39	53.52	53.65	53.79	53.92	54.05	54.19	54.32	54.45
71.6	9 54.59	54.72	54.85	54.99	55.12	55.25	55.39	55.52	55.65	55.79
71.7	9 55.92	56.05	56.19	56.32	56.45	56.59	56.72	56.85	56.99	57.12
71.8	9 57.25	57.39	57.52	57.65	57.79	57.92	58.05	58.19	58.32	58.45
71.9	9 58.59	58.72	58.85	58.99	59.12	59.25	59.39	59.52	59.65	59.79
72.0	9 59.92	60.05	60.19	60.32	60.45	60.59	60.72	60.85	60.98	61.12
72.1	9 61.25	61.38	61.52	61.65	61.78	61.92	62.05	62.18	62.32	62.45
72.2	9 62.58	62.72	62.85	62.98	63.12	63.25	63.38	63.52	63.65	63.78
72.3	9 63.92	64.05	64.18	64.32	64.45	64.58	64.72	64.85	64.98	65.12
72.4	9 65.25	65.38	65.52	65.65	65.78	65.92	66.05	66.18	66.32	66.45
72.5	9 66.58	66.72	66.85	66.98	67.12	67.25	67.38	67.52	67.65	67.78
72.6	9 67.92	68.05	68.18	68.32	68.45	68.58	68.72	68.85	68.98	69.12
72.7	9 69.25	69.38	69.52	69.65	69.78	69.92	70.05	70.18	70.32	70.45
72.8	9 70.58	70.72	70.85	70.98	71.12	71.25	71.38	71.52	71.65	71.78
72.9	9 71.92	72.05	72.18	72.32	72.45	72.58	72.72	72.85	72.98	73.12

Centimeters	.00	.01	.02	.03	.04	.05	.06	.07	.08	.09
73.0	9 73.25	73.38	73.52	73.65	73.78	73.92	74.05	74.18	74.32	74.45
73.1	9 74.58	74.72	74.85	74.98	75.12	75.25	75.38	75.52	75.65	75.78
73.2	9 75.92	76.05	76.18	76.32	76.45	76.58	76.72	76.85	76.98	77.12
73.3	9 77.25	77.38	77.52	77.65	77.78	77.92	78.05	78.18	78.32	78.45
73.4	9 78.58	78.72	78.85	78.98	79.12	79.25	79.38	79.52	79.65	79.78
73.5	9 79.92	80.05	80.18	80.32	80.45	80.58	80.72	80.85	80.98	81.12
73.6	9 81.25	81.38	81.52	81.65	81.78	81.92	82.05	82.18	82.32	82.45
73.7	9 82.58	82.72	82.85	82.98	83.12	83.25	83.38	83.52	83.65	83.78
73.8	9 83.92	84.05	84.18	84.32	84.45	84.58	84.72	84.85	84.98	85.12
73.9	9 85.25	85.38	85.52	85.65	85.78	85.92	86.05	86.18	86.32	86.45
74.0	9 86.58	86.72	86.85	86.98	87.12	87.25	87.38	87.52	87.65	87.78
74.1	9 87.92	88.05	88.18	88.32	88.45	88.58	88.72	88.85	88.98	89.12
74.2	9 89.25	89.38	89.52	89.65	89.78	89.92	90.05	90.18	90.32	90.45
74.3	9 90.58	90.72	90.85	90.98	91.12	91.25	91.38	91.52	91.65	91.78
74.4	9 91.92	92.05	92.18	92.32	92.45	92.58	92.72	92.85	92.98	93.12
74.5	9 93.25	93.38	93.52	93.65	93.78	93.92	94.05	94.18	94.32	94.45
74.6	9 94.58	94.72	94.85	94.98	95.12	95.25	95.38	95.52	95.65	95.78
74.7	9 95.92	96.05	96.18	96.32	96.45	96.58	96.72	96.85	96.98	97.12
74.8	9 97.25	97.38	97.52	97.65	97.78	97.92	98.05	98.18	98.32	98.45
74.9	9 98.58	98.72	98.85	98.98	99.12	99.25	99.38	99.52	99.65	99.78
75.0	9 99.92	*00.05	*00.18	*00.31	*00.45	*00.58	*00.71	*00.85	*00.98	*01.11
75.1	10 01.25	01.38	01.51	01.65	01.78	01.91	02.05	02.18	02.31	02.45
75.2	10 02.58	02.71	02.85	02.98	03.11	03.25	03.38	03.51	03.65	03.78
75.3	10 03.91	04.05	04.18	04.31	04.45	04.58	04.71	04.85	04.98	05.11
75.4	10 05.25	05.38	05.51	05.65	05.78	05.91	06.05	06.18	06.31	06.45
75.5	10 06.58	06.71	06.85	06.98	07.11	07.25	07.38	07.51	07.65	07.78
75.6	10 07.91	08.05	08.18	08.31	08.45	08.58	08.71	08.85	08.98	09.11
75.7	10 09.25	09.38	09.51	09.65	09.78	09.91	10.05	10.18	10.31	10.45
75.8	10 10.58	10.71	10.85	10.98	11.11	11.25	11.38	11.51	11.65	11.78
75.9	10 11.91	12.05	12.18	12.31	12.45	12.58	12.71	12.85	12.98	13.11
76.0	10 13.25	13.38	13.51	13.65	13.78	13.91	14.05	14.18	14.31	14.45
76.1	10 14.58	14.71	14.85	14.98	15.11	15.25	15.38	15.51	15.65	15.78
76.2	10 15.91	16.05	16.18	16.31	16.45	16.58	16.71	16.85	16.98	17.11
76.3	10 17.25	17.38	17.51	17.65	17.78	17.91	18.05	18.18	18.31	18.45
76.4	10 18.58	18.71	18.85	18.98	19.11	19.25	19.38	19.51	19.65	19.78
76.5	10 19.91	20.05	20.18	20.31	20.45	20.58	20.71	20.85	20.98	21.11
76.6	10 21.25	21.38	21.51	21.65	21.78	21.91	22.05	22.18	22.31	22.45
76.7	10 22.58	22.71	22.85	22.98	23.11	23.25	23.38	23.51	23.65	23.78
76.8	10 23.91	24.05	24.18	24.31	24.45	24.58	24.71	24.85	24.98	25.11
76.9	10 25.25	25.38	25.51	25.65	25.78	25.91	26.05	26.18	26.31	26.45
77.0	10 26.58	26.71	26.85	26.98	27.11	27.25	27.38	27.51	27.65	27.78
77.1	10 27.91	28.05	28.18	28.31	28.45	28.58	28.71	28.85	28.98	29.11
77.2	10 29.25	29.38	29.51	29.65	29.78	29.91	30.05	30.18	30.31	30.45
77.3	10 30.58	30.71	30.85	30.98	31.11	31.25	31.38	31.51	31.65	31.78
77.4	10 31.91	32.05	32.18	32.31	32.45	32.58	32.71	32.85	32.98	33.11
77.5	10 33.25	33.38	33.51	33.65	33.78	33.91	34.05	34.18	34.31	34.45
77.6	10 34.58	34.71	34.85	34.98	35.11	35.25	35.38	35.51	35.65	35.78
77.7	10 35.91	36.05	36.18	36.31	36.45	36.58	36.71	36.85	36.98	37.11
77.8	10 37.25	37.38	37.51	37.65	37.78	37.91	38.05	38.18	38.31	38.45
77.9	10 38.58	38.71	38.85	38.98	39.11	39.24	39.38	39.51	39.64	39.78

TEMPERATURE CORRECTION FOR BAROMETER READINGS

Brass Scale — Metric Units

To reduce readings of a mercurial barometer with a brass scale to 0°C subtract the appropriate quantity as found in the table. These values are based on the coefficient of expansion of mercury (181792 + 0.175t + 0.035116t²) × 10⁻⁹, and of brass 0.0000184 per °C. Corrections are in millimeters.

Temp. °C	Observed height in millimeters 620	630	640	650	660	670	680	690	700	710	720	730	740	750	760	770	780	790
0	0.00	0.00	0.00	0.00	0.00	0.00	0.00	0.00	0.00	0.00	0.00	0.00	0.00	0.00	0.00	0.00	0.00	0.00
1	.10	.10	.10	.11	.11	.11	.11	.11	.11	.12	.12	.12	.12	.12	.12	.13	.13	.13
2	.20	.21	.21	.21	.22	.22	.22	.23	.23	.23	.24	.24	.24	.25	.25	.25	.25	.26
3	.30	.31	.31	.32	.32	.33	.33	.34	.34	.35	.35	.36	.36	.37	.37	.38	.38	.39
4	.40	.41	.42	.42	.43	.44	.44	.45	.46	.46	.47	.48	.48	.49	.50	.50	.51	.52
5	0.51	0.51	0.52	0.53	0.54	0.55	0.56	0.56	0.57	0.58	0.59	0.60	0.60	0.61	0.62	0.63	0.64	0.64
6	.61	.62	.63	.64	.65	.66	.67	.68	.69	.70	.71	.71	.72	.73	.74	.75	.76	.77
7	.71	.72	.73	.74	.75	.77	.78	.79	.80	.81	.82	.83	.85	.86	.87	.88	.89	.90
8	.81	.82	.84	.85	.86	.87	.89	.90	.91	.93	.94	.95	.97	.98	.99	1.01	1.02	1.03
9	.91	.92	.94	.95	.97	.98	1.00	1.01	1.03	1.04	1.06	1.07	1.09	1.10	1.12	1.13	1.15	1.16
10	1.01	1.03	1.04	1.06	1.08	1.09	1.11	1.13	1.14	1.16	1.17	1.19	1.21	1.22	1.24	1.26	1.27	1.29
11	1.11	1.13	1.15	1.17	1.18	1.20	1.22	1.24	1.26	1.27	1.29	1.31	1.33	1.35	1.36	1.38	1.40	1.42
12	1.21	1.23	1.25	1.27	1.29	1.31	1.33	1.35	1.37	1.39	1.41	1.43	1.45	1.47	1.49	1.51	1.53	1.55
13	1.31	1.34	1.36	1.38	1.40	1.42	1.44	1.46	1.48	1.50	1.53	1.55	1.57	1.59	1.61	1.63	1.65	1.67
14	1.41	1.44	1.46	1.48	1.51	1.53	1.55	1.57	1.60	1.62	1.64	1.67	1.69	1.71	1.73	1.76	1.78	1.80
15	1.52	1.54	1.56	1.59	1.61	1.64	1.66	1.69	1.71	1.74	1.76	1.78	1.81	1.83	1.86	1.88	1.91	1.93
16	1.62	1.64	1.67	1.69	1.72	1.75	1.77	1.80	1.82	1.85	1.88	1.90	1.93	1.96	1.98	2.01	2.03	2.06
17	1.72	1.74	1.77	1.80	1.83	1.86	1.88	1.91	1.94	1.97	1.99	2.02	2.05	2.08	2.10	2.13	2.16	2.19
18	1.82	1.85	1.88	1.91	1.93	1.96	1.99	2.02	2.05	2.08	2.11	2.14	2.17	2.20	2.23	2.26	2.29	2.32
19	1.92	1.95	1.98	2.01	2.04	2.07	2.10	2.13	2.17	2.20	2.23	2.26	2.29	2.32	2.35	2.38	2.41	2.44
20	2.02	2.05	2.08	2.12	2.15	2.18	2.21	2.25	2.28	2.31	2.34	2.38	2.41	2.44	2.47	2.51	2.54	2.57
21	2.12	2.15	2.19	2.22	2.26	2.29	2.32	2.36	2.39	2.43	2.46	2.50	2.53	2.56	2.60	2.63	2.67	2.70
22	2.22	2.26	2.29	2.33	2.36	2.40	2.43	2.47	2.51	2.54	2.58	2.61	2.65	2.69	2.72	2.76	2.79	2.83
23	2.32	2.36	2.40	2.43	2.47	2.51	2.54	2.58	2.62	2.66	2.69	2.73	2.77	2.81	2.84	2.88	2.92	2.96
24	2.42	2.46	2.50	2.54	2.58	2.62	2.66	2.69	2.73	2.77	2.81	2.85	2.89	2.93	2.97	3.01	3.05	3.08
25	2.52	2.56	2.60	2.64	2.68	2.72	2.77	2.81	2.85	2.89	2.93	2.97	3.01	3.05	3.09	3.13	3.17	3.21
26	2.62	2.66	2.71	2.75	2.79	2.83	2.88	2.92	2.96	3.00	3.04	3.09	3.13	3.17	3.21	3.26	3.30	3.34
27	2.72	2.77	2.81	2.85	2.90	2.94	2.99	3.03	3.07	3.12	3.16	3.20	3.25	3.29	3.34	3.38	3.42	3.47
28	2.82	2.87	2.91	2.96	3.00	3.05	3.10	3.14	3.19	3.23	3.28	3.32	3.37	3.41	3.46	3.51	3.55	3.60
29	2.92	2.97	3.02	3.06	3.11	3.16	3.21	3.25	3.30	3.35	3.39	3.44	3.49	3.54	3.58	3.63	3.68	3.72
30	3.02	3.07	3.12	3.17	3.22	3.27	3.32	3.36	3.41	3.46	3.51	3.56	3.61	3.66	3.71	3.75	3.80	3.85
31	3.12	3.17	3.22	3.27	3.32	3.37	3.43	3.48	3.53	3.58	3.63	3.68	3.73	3.78	3.83	3.88	3.93	3.98
32	3.22	3.28	3.33	3.38	3.43	3.48	3.54	3.59	3.64	3.69	3.74	3.79	3.85	3.90	3.95	4.00	4.05	4.11
33	3.32	3.38	3.43	3.48	3.54	3.59	3.64	3.70	3.75	3.81	3.86	3.91	3.97	4.02	4.07	4.13	4.18	4.23
34	3.42	3.48	3.53	3.59	3.64	3.70	3.75	3.81	3.87	3.92	3.98	4.03	4.09	4.14	4.20	4.25	4.31	4.36
35	3.52	3.58	3.64	3.69	3.75	3.81	3.86	3.92	3.98	4.03	4.09	4.15	4.21	4.26	4.32	4.38	4.43	4.49

Brass Scale — English Units

Standard Temperature of scale 62°F; of mercury, 32°F. Zero correction at 28.5°F; subtract corrections above, add below. Owing to the difference in the standard temperature of English and metric scales, readings taken in inches to be reduced to centimeters should first be corrected for temperature.

Temp. °F	Observed height in inches 23.0 in.	23.5 in.	24.0 in.	24.5 in.	25.0 in.	25.5 in.	26.0 in.	26.5 in.	27.0 in.	27.5 in.	28.0 in.	28.5 in.	29.0 in.	29.5 in.	30.0 in.	30.5 in.	31.0 in.	31.5 in.
0	+.060	+.061	+.063	+.064	+.065	+.067	+.068	+.069	+.070	.072	.073	.075	.076	.077	.078	.080	.081	.082
2	.056	.057	.058	.060	.061	.062	.063	.065	.065	.067	.068	.069	.070	.072	.073	.074	.075	.077
4	.052	.053	.054	.055	.056	.057	.058	.060	.061	.062	.063	.064	.065	.066	.067	.069	.070	.071
6	.047	.048	.049	.051	.052	.053	.054	.055	.056	.057	.058	.059	.060	.061	.062	.063	.064	.065
8	.043	.044	.045	.046	.047	.048	.049	.050	.051	.052	.053	.054	.054	.056	.056	.057	.058	.059
10	.039	.040	.041	.042	.042	.043	.044	.045	.046	.047	.047	.048	.049	.050	.051	.052	.053	.054
12	.035	.036	.036	.037	.038	.039	.039	.040	.041	.042	.042	.043	.044	.045	.045	.046	.047	.048
14	.031	.031	.032	.033	.033	.034	.035	.035	.036	.037	.037	.038	.039	.039	.040	.041	.041	.042
16	.026	.027	.028	.028	.029	.029	.030	.031	.031	.032	.032	.033	.033	.034	.034	.035	.036	.036
18	.022	.023	.023	.024	.024	.025	.025	.026	.026	.027	.027	.028	.028	.029	.029	.030	.030	.031
20	.018	.018	.019	.019	.020	.020	.020	.021	.021	.022	.022	.022	.023	.023	.024	.024	.024	.025
22	.014	.014	.014	.015	.015	.015	.016	.016	.016	.017	.017	.017	.017	.018	.018	.018	.019	.019
24	.010	.010	.010	.010	.011	.011	.011	.011	.011	.012	.012	.012	.012	.012	.013	.013	.013	.013
26	.005	.006	.006	.006	.006	.006	.006	.006	.006	.007	.007	.007	.007	.007	.007	.007	.007	.003
28	+.001	+.001	+.001	+.001	+.001	+.001	+.001	+.002	+.002	+.002	+.002	+.002	+.002	+.002	+.002	+.002	+.002	+.002
30	−.003	−.003	−.003	−.003	−.003	−.003	−.003	−.003	−.003	−.003	−.003	−.004	−.004	−.004	−.004	−.004	−.004	−.004
32	.007	.007	.007	.008	.008	.008	.008	.008	.008	.008	.009	.009	.009	.009	.009	.009	.009	.010
34	.011	.011	.012	.012	.012	.012	.013	.013	.013	.013	.014	.014	.014	.014	.015	.015	.015	.015
36	.015	.016	.016	.016	.017	.017	.017	.018	.018	.018	.019	.019	.019	.020	.020	.020	.021	.021
38	.020	.020	.020	.021	.021	.022	.022	.023	.023	.023	.024	.024	.025	.025	.026	.026	.026	.027
40	.024	.024	.025	.025	.026	.026	.027	.027	.028	.028	.029	.030	.030	.031	.031	.032	.032	.033
42	.028	.029	.029	.030	.030	.031	.032	.032	.033	.033	.034	.035	.035	.036	.036	.037	.038	.038
44	.032	.033	.033	.034	.035	.036	.036	.037	.038	.038	.039	.040	.040	.041	.042	.043	.043	.044
46	.036	.037	.038	.039	.039	.040	.041	.042	.043	.043	.044	.045	.046	.047	.047	.048	.049	.050
48	.040	.041	.042	.043	.044	.045	.046	.047	.047	.048	.049	.050	.051	.052	.053	.054	.054	.055
50	.045	.046	.046	.048	.048	.050	.050	.052	.052	.053	.054	.055	.056	.057	.058	.059	.060	.061
52	.049	.050	.051	.052	.053	.054	.055	.056	.057	.058	.059	.061	.061	.063	.064	.065	.066	.067
54	.053	.054	.055	.057	.057	.059	.060	.061	.062	.063	.064	.066	.067	.068	.069	.070	.071	.073
56	.057	.058	.060	.061	.062	.063	.064	.066	.067	.068	.069	.071	.072	.073	.074	.076	.077	.078
58	.061	.063	.064	.065	.066	.068	.069	.071	.072	.073	.074	.076	.077	.079	.080	.081	.082	.084
60	.065	.067	.068	.070	.071	.073	.074	.076	.077	.078	.080	.081	.082	.084	.085	.087	.088	.090
62	.069	.071	.073	.074	.076	.077	.079	.080	.082	.083	.085	.086	.088	.089	.091	.092	.094	.095
64	.074	.075	.077	.079	.080	.082	.083	.085	.086	.088	.090	.092	.093	.095	.096	.098	.099	.101
66	.078	.079	.081	.083	.085	.087	.088	.090	.091	.093	.095	.097	.098	.100	.101	.103	.105	.107
68	.082	.084	.085	.088	.089	.091	.093	.095	.096	.098	.100	.102	.103	.105	.107	.109	.110	.113
70	.086	.088	.090	.092	.094	.096	.097	.100	.101	.103	.105	.107	.109	.111	.112	.115	.116	.118

TEMPERATURE CORRECTION FOR BAROMETER READINGS (continued)

Observed height in inches

Temp. °F	23.0 in.	23.5 in.	24.0 in.	24.5 in.	25.0 in.	25.5 in.	26.0 in.	26.5 in.	27.0 in.	27.5 in.	28.0 in.	28.5 in.	29.0 in.	29.5 in.	30.0 in.	30.5 in.	31.0 in.	31.5 in.
72	.090	.092	.094	.096	.098	.100	.102	.104	.106	.108	.110	.112	.114	.116	.118	.120	.122	.124
74	.094	.096	.098	.101	.103	.105	.107	.109	.111	.113	.115	.117	.119	.121	.123	.126	.127	.130
76	.098	.101	.103	.105	.107	.110	.111	.114	.116	.118	.120	.122	.124	.127	.128	.131	.133	.135
78	.103	.105	.107	.110	.112	.114	.116	.119	.120	.123	.125	.128	.129	.132	.134	.137	.138	.141
80	.107	.109	.111	.114	.116	.119	.121	.123	.125	.128	.130	.133	.135	.137	.139	.142	.144	.147
82	.111	.113	.116	.119	.121	.123	.125	.128	.130	.133	.135	.138	.140	.143	.145	.148	.149	.152
84	.115	.118	.120	.123	.125	.128	.130	.133	.135	.138	.140	.143	.145	.148	.150	.153	.155	.158
86	.119	.122	.124	.127	.130	.133	.135	.138	.140	.143	.145	.148	.150	.153	.155	.159	.161	.164
88	.123	.126	.129	.132	.134	.137	.139	.143	.145	.148	.150	.153	.155	.159	.161	.164	.166	.169
90	.127	.130	.133	.136	.138	.142	.144	.147	.150	.153	.155	.158	.161	.164	.166	.170	.172	.175
92	.132	.134	.137	.141	.143	.146	.149	.152	.154	.158	.160	.163	.166	.169	.172	.175	.177	.181
94	.136	.139	.142	.145	.147	.151	.153	.157	.159	.163	.165	.169	.171	.175	.177	.180	.183	.186
96	.140	.143	.146	.150	.152	.155	.158	.161	.164	.168	.170	.174	.176	.180	.182	.186	.188	.192
98	.144	.147	.150	.154	.156	.160	.163	.166	.169	.172	.175	.179	.181	.185	.188	.191	.194	.197
100	.148	.151	.154	.158	.161	.164	.167	.171	.174	.177	.180	.184	.187	.190	.193	.197	.200	.203

TEMPERATURE CORRECTION, GLASS SCALE

Metric

To reduce readings of a mercurial barometer with a glass scale to 0°C. subtract the appropriate quantity as found in table.

Temp °C.	Observed height in centimeters. 70 cm.	71 cm.	72 cm.	73 cm.	74 cm.	75 cm.	76 cm.	77 cm.	78 cm.	Temp. °C.	Observed height in centimeters. 70 cm.	71 cm.	72 cm.	73 cm.	74 cm.	75 cm.	76 cm.	77 cm.	78 cm.
0	0.000	0.000	0.000	0.000	0.000	0.000	0.000	0.000	0.000	15	0.181	0.184	0.186	0.189	0.191	0.193	0.196	0.198	0.201
1	.012	.012	.013	.013	.013	.013	.013	.013	.014	16	.194	.196	.199	.201	.204	.207	.209	.212	.214
2	.025	.025	.025	.026	.026	.026	.026	.027	.027	17	.205	.208	.210	.213	.216	.219	.221	.224	.227
3	.036	.036	.037	.037	.038	.038	.039	.039	.040	18	.217	.220	.223	.226	.229	.232	.235	.238	.241
4	.048	.049	.049	.050	.051	.051	.052	.053	.053	19	.230	.233	.236	.239	.242	.245	.248	.251	.254
5	.060	.061	.062	.063	.064	.064	.065	.066	.067	20	.242	.245	.248	.252	.255	.258	.261	.264	.268
6	.073	.074	.074	.076	.077	.077	.078	.079	.080	21	.254	.258	.261	.264	.268	.271	.275	.278	.281
7	.085	.086	.087	.088	.089	.091	.092	.093	.094	22	.266	.269	.273	.276	.280	.283	.287	.290	.294
8	.096	.098	.099	.100	.101	.103	.104	.105	.107	23	.278	.282	.285	.289	.293	.296	.300	.304	.308
9	.109	.110	.111	.113	.114	.116	.117	.119	.120	24	.290	.294	.298	.302	.306	.310	.313	.317	.321
10	.121	.122	.124	.126	.127	.129	.130	.132	.134	25	.303	.307	.311	.315	.319	.323	.327	.331	.335
11	.133	.135	.137	.138	.140	.142	.144	.146	.147	26	.315	.319	.323	.327	.332	.336	.340	.344	.348
12	.144	.146	.148	.150	.152	.154	.156	.158	.160	27	.326	.331	.335	.339	.344	.348	.352	.357	.361
13	.157	.159	.161	.163	.165	.167	.169	.171	.174	28	.339	.343	.348	.352	.357	.361	.366	.370	.375
14	.169	.171	.174	.176	.178	.180	.183	.185	.187	29	.351	.356	.360	.365	.370	.374	.379	.384	.388
										30	.363	.368	.373	.378	.383	.387	.392	.397	.402

WEIGHT IN GRAMS OF A CUBIC METER OF SATURATED AQUEOUS VAPOR

(From Smithsonian Tables)

Mass in grams per cubic meter.

Temp. °C	0.0	1.0	2.0	3.0	4.0	5.0	6.0	7.0	8.0	9.0
−20	1.074	.988	.909	.836	.768	.705	.646	.592	.542	.496
−10	2.358	2.186	2.026	1.876	1.736	1.605	1.483	1.369	1.264	1.165
− 0	4.847	4.523	4.217	3.930	3.660	3.407	3.169	2.946	2.737	2.541
+ 0	4.847	5.192	5.559	5.947	6.360	6.797	7.260	7.750	8.270	8.819
+10	9.399	10.01	10.66	11.35	12.07	12.83	13.63	14.48	15.37	16.31
+20	17.30	18.34	19.43	20.58	21.78	23.05	24.38	25.78	27.24	28.78
+30	30.38	32.07	33.83	35.68	37.61	39.63	41.75	43.96	46.26	28.67

EFFICIENCY OF DRYING AGENTS

Compiled by John H. Yoe

A. Drying agents depending upon chemical action (absorption) for their efficiency:*

Substance	Residual water, mg per liter of dry air**	Reference
P_2O_5	<1 mg in 40,000 1,	Morley, Am. J. Sci., 34, 199 (1887); J.A.C.S., 26, 1171 (1904).
$Mg(ClO_4)_2$ anhyd.	"Unweighable" in 2101,	Willard and Smith, J.A.C.S., 44, 2255 (1922).
BaO	0.00065	Bower, Bur. Std. J. Res., 12, 241 (1934).
KOH fused	0.002	Baxter and Starkweather, J.A.C.S., 38, 2038 (1916).
CaO	0.003	Bower, loc. cit.
H_2SO_4	0.003	Baxter and Starkweather, loc. cit.
$CaSO_4$ anhyd.	0.005	Bower loc. cit.
Al_2O_3	0.005	Ibid.
KOH sticks	0.014	Ibid.
NaOH fused	0.16	Baxter and Starkweather, loc. cit.
$CaBr_2$	0.18	Baxter and Warren, J.A.C.S., 33, 340 (1911).
$CaCl_2$ fused	0.34	Baxter and Starkweather, loc. cit.
NaOH sticks	0.80	Bower loc. cit.
$Ba(ClO_4)_2$	0.82	Ibid.
$ZnCl_2$	0.85	Baxter and Warren, loc. cit.
$ZnBr_2$	1.16	Ibid.
$CaCl_2$ granular	1.5	Bower, loc. cit.
$CuSO_4$ anhyd.	2.8	Ibid.

B. Drying agents depending upon physical action (adsorption) for their efficiency:* Alumina (low temperature fired), asbestos, charcoal, clay and porcelain (low temperature fired), glass wool, kieselguhr, silica gel, refrigeration.

* It should be noted that the efficiency of some drying agents (e.g. Al_2O_3 and anhydrous $CaCl_2$, and probably also BaO, anhydrous $Mg(ClO_4)_2$, $Mg(ClO_4)_2 \cdot 3H_2O$, anhydrous $Ba(ClO_4)_2$, and $CaSO_4$) depends upon both adsorption and absorption.

** 30°C. for Bower's values; others 25°C. or room temp.

REDUCTION OF BAROMETER TO SEA LEVEL

The correction to be added to reduce barometric readings to "sea level" values depends principally on three factors: The temperature of the air column (assumed) from the station to sea level, the altitude of the station, and the value of the reading itself. Two tables are provided. Table I is entered with the altitude and assumed temperature and is a fator "2000 m" taken out. Table II is entered with the above factor and the approximate barometer reading and the final correction taken out.

The correction is to be added. If B_0 is the corrected or sea level value; B the barometer reading at the station; C the correction,—

$$C = B_0 - B = B(10^m - 1)$$

The actual barometer reading at the station should be corrected for temperature of the mercury column by the usual methods before entering the tables or applying the sea level correction.

A complete explanation of the theory of the corrections and a more extended set of tables will be found in the Smithsonian Meterological Tables.

Latitude Factor

The influence of the latitude on the value of the correction is usually negligible, being overshadowed by uncertainties in the assumed temperature of the air column. For cases where this correction is desirable the table below is provided. The value of the temperature-altitude factor "2000 m" obtained in Table I is corrected for latitude by subtracting for latitudes 0-45° and adding for latitudes from 45-90° the values found. With this corrected value of "2000 m" Table II is entered for the value of the correction.

LATITUDE FACTOR

To be used in connection with Tables I and II, either English or metric units, to obtain latitude corrections to temperature-altitude factor. For latitudes 0-45° subtract the correction. For latitudes 45-90° add the correction.

Temp.-Alt. from Table I	Latitude			
	0°	15°	30°	45°
100	0.3	0.2	0.1	0.0
200	0.5	0.5	0.3	0.0
300	0.8	0.7	0.4	0.0
	90°	75°	60°	45°

METRIC UNITS—TABLE I

Values of the temperature-altitude factor (2000 m.) for entering table II.

Altitude in meters	Assumed temperature of air column °C									
	−16°	−8°	0°	+4°	+8°	+12°	+16°	+20°	+24°	+28°
10	1.2	1.1	1.1	1.1	1.0	1.0	1.0	1.0	1.0	1.0
50	5.8	5.6	5.4	5.3	5.2	5.2	5.1	5.0	4.9	4.9
100	11.5	11.2	10.8	10.7	10.5	10.3	10.2	10.0	9.9	9.7
150	17.3	16.7	16.2	16.0	15.7	15. 5	15.3	15.0	14.8	14.6
200	23.0	22.3	21.6	21.3	21.0	20.7	20.3	20.0	19.7	19.5
250	28.8	27.9	27.0	26.6	26.2	25.8	25.4	25.0	24.7	24.3
300	34.5	33.5	32.5	32.0	31.5	31.0	30.5	30.1	29.6	29.2
350	40.3	39.0	37.9	37.3	36.7	36.2	35.6	35.1	34.6	34.0
400	46.0	44.6	43.3	42.6	42.0	41.3	40.7	40.1	39.5	38.9
450	51.8	50.2	48.7	47.9	47.2	46.5	45.8	45.1	44.4	43.8
500	57.5	55.8	54.1	53.3	52.4	51.6	50.9	50.1	49.4	48.6
550	63.3	61.4	59.5	58.6	57.7	56.8	55.9	55.1	54.3	53.5
600	69.0	66.9	64.9	63.9	62.9	62.0	61.0	60.1	59.2	58.3
650	74.8	72.5	70.3	69.2	68.2	67.1	66.1	65.1	64.2	63.2
700	80.6	78.1	75.7	74.6	73.4	72.3	71.2	70.1	69.1	68.1
750	86.3	83.7	81.1	79.9	78.7	77.5	76.3	75.1	74.0	72.9
800	92.1	89.2	86.5	85.2	83.9	82.6	81.4	80.1	79.0	77.8
850	97.8	94.8	92.0	90.5	89.2	87.8	86.4	85.2	83.9	82.7
900	103.6	100.4	97.4	95.9	94.4	93.0	91.5	90.2	88.8	87.5
950	109.3	106.0	102.8	101.2	99.6	98.1	96.6	95.2	93.8	92.4
1000	115.1	111.5	108.2	106.5	104.9	103.3	101.7	100.2	98.7	97.3
1050	120.8	117.1	113.6	111.8	110.1	108.4	106.8	105.2	103.6	102.1
1100	126.6	122.7	119.0	117.2	115.4	113.6	111.9	110.2	108.6	107.0
1150	132.3	128.3	124.4	122.5	120.6	118.8	117.0	115.2	113.5	111.8
1200	138.1	133.8	129.8	127.8	125.9	123.9	122.0	120.2	118.4	116.7
1250	143.8	139.4	135.2	133.1	131.1	129.1	127.1	125.2	123.4	121.6
1300	149.6	145.0	140.6	138.5	136.3	134.3	132.2	130.2	128.3	126.4
1350	155.3	150.6	146.0	143.8	141.6	139.4	137.3	135.2	133.2	131.3
1400	161.1	156.2	151.4	149.1	146.8	144.6	142.4	140.2	138.2	136.2
1450	166.8	161.7	156.8	154.5	152.1	149.7	147.5	145.3	143.1	141.0

Altitude in meters	Assumed temperature of air column °C									
	−16°	−8°	0°	+4°	+8°	+12°	+16°	+20°	+24°	+28°
1500	172.6	167.3	162.3	159.8	157.3	154.9	152.5	150.3	148.0	145.9
1550	178.3	172.9	167.7	165.1	162.6	160.1	157.6	155.3	153.0	150.7
1600	184.1	178.5	173.1	170.4	167.8	165.2	162.7	160.3	157.9	155.6
1650	189.8	184.0	178.5	175.7	173.0	170.4	167.8	165.3	162.8	160.5
1700	195.6	189.6	183.9	181.1	178.3	175.6	172.9	170.3	167.8	165.3
1750	201.4	195.2	189.3	186.4	183.5	180.7	178.0	175.3	172.7	170.2
1800	207.1	200.8	194.7	191.7	188.8	185.9	183.1	180.3	177.6	175.0
1850	212.9	206.3	200.1	197.0	194.0	191.0	188.1	185.3	182.6	179.9
1900	218.6	211.9	205.5	202.4	199.3	196.2	193.2	190.3	187.5	184.8
1950	224.4	217.5	210.9	207.7	204.5	201.4	198.3	195.3	192.4	189.6
2000	230.1	223.0	216.3	213.0	209.7	206.5	203.4	200.3	197.4	194.5
2050	235.9	228.6	221.7	218.3	215.0	211.7	208.5	205.3	202.3	199.3
2100	241.6	234.2	227.1	223.7	220.2	216.8	213.5	210.4	207.2	204.2
2150	247.4	239.8	232.5	229.0	225.5	222.0	218.6	215.4	212.2	209.1
2200	253.1	245.4	237.9	234.3	230.7	227.2	223.7	220.4	217.1	213.9
2250	258.9	250.9	243.4	239.6	235.9	232.3	228.8	225.4	222.0	218.8
2300	264.6	256.5	248.8	245.0	241.2	237.5	233.9	230.4	227.0	223.6
2350	270.4	262.1	254.2	250.3	246.4	242.7	239.0	235.4	231.9	228.5
2400	276.1	267.7	259.6	255.6	251.7	247.8	244.0	240.4	236.8	233.4
2450	281.9	273.2	265.0	260.9	256.9	253.0	249.1	245.4	241.8	238.2
2500	287.6	278.8	270.4	266.2	262.2	258.1	254.2	250.4	246.7	243.1
2550	293.4	284.4	275.8	271.6	267.4	263.3	259.3	255.4	251.6	247.9
2600	299.1	290.0	281.2	276.9	272.6	268.5	264.4	260.4	256.6	252.8
2650	304.9	295.5	286.6	282.2	277.9	273.6	269.5	265.4	261.5	257.7
2700	310.6	301.1	292.0	287.5	283.1	278.8	274.5	270.4	266.4	262.5
2750	316.4	306.7	297.4	292.9	288.4	283.9	279.6	275.4	271.4	267.4
2800	322.1	312.3	302.8	298.2	293.6	289.1	284.7	280.4	276.3	272.2
2850	327.9	317.8	308.2	303.5	298.8	294.3	289.8	285.4	281.2	277.1
2900	333.6	323.4	313.6	308.8	304.1	299.4	294.9	290.4	286.2	282.0
2950	339.4	329.0	319.0	314.2	309.3	304.6	299.9	295.5	291.1	286.8
3000	345.1	334.5	324.4	319.5	314.6	309.7	305.0	300.5	296.0	291.7

METRIC UNITS — TABLE II

Values of Correction to be Added

Temp.-alt.-factor	Barometer reading 780 mm	760 mm	740 mm	720 mm	700 mm	
1	0.9	0.9	0.9	0.8	0.8	
5	4.5	4.4	4.3	4.2	4.0	
10	9.0	8.8	8.6	8.3	8.1	
15	13.6	13.2	12.9	12.5	12.2	
20	18.2	17.7	17.2	16.8	16.3	
25	22.8	22.2	21.6	21.0	20.4	
30	27.4	26.7	26.0	25.3	24.6	
35	—	31.2	30.4	29.6	28.8	
	760 mm	740 mm	720 mm	700 mm	680 mm	660 mm
40	35.8	34.9	33.9	33.0	32.0	31.1
45	40.4	39.3	38.3	37.2	36.2	35.1
50	45.0	43.8	42.7	41.5	40.3	39.1
55	49.7	48.4	47.1	45.8	44.5	43.1
60	—	52.9	51.5	50.1	48.6	47.2
65	—	57.5	55.9	54.4	52.8	51.3
70	—	62.1	60.4	58.7	57.1	55.4
75	—	66.7	64.9	63.1	61.3	59.5
	720 mm	700 mm	680 mm	660 mm	640 mm	
80	69.5	67.5	65.6	63.7	61.7	
85	74.0	72.0	69.9	67.9	65.8	
90	78.6	76.4	74.2	72.1	69.9	
95	83.2	80.9	78.6	76.3	74.0	
100	87.9	85.4	83.0	80.5	78.1	
105	—	89.9	87.4	84.8	82.2	
110	—	94.5	91.8	89.1	86.4	
115	—	99.1	96.3	93.4	90.6	
120	—	103.7	100.7	97.8	94.8	
125	—	108.3	105.3	102.2	99.1	
	680 mm	660 mm	640 mm	620 mm	600 mm	
125	105.3	102.2	99.1	96.0	92.9	
130	109.8	106.6	103.3	100.1	96.9	
135	114.3	111.0	107.6	104.3	100.9	
140	118.9	115.4	111.9	108.4	104.9	
145	123.5	119.9	116.3	112.6	109.0	
150	128.2	124.4	120.6	116.9	113.1	
155	—	128.9	125.0	121.1	117.2	
160	—	133.5	129.4	125.4	121.4	
165	—	138.1	133.9	129.7	125.5	
170	—	142.7	138.4	134.0	129.7	

Temp.-alt.-factor	Barometer reading 640 mm	620 mm	600 mm	580 mm	560 mm
170	138.4	134.0	129.7	125.4	121.1
175	142.9	138.4	133.9	129.5	125.0
180	147.4	142.8	138.2	133.6	129.0
185	151.9	147.2	142.4	137.7	132.9
190	156.5	151.6	146.7	141.8	136.0
195	161.1	156.1	151.0	146.0	141.0
200	165.7	160.5	155.4	150.2	145.0
205	170.4	165.0	159.7	154.4	149.1
210	—	169.6	164.1	158.6	153.2
215	—	174.1	168.5	162.9	157.3
	620 mm	600 mm	580 mm	560 mm	540 mm
215	174.1	168.5	162.9	157.3	151.7
220	178.7	172.9	167.2	161.4	155.7
225	183.3	177.4	171.5	165.6	159.7
230	188.0	181.9	175.8	169.8	163.7
235	192.6	186.4	180.2	174.0	167.8
240	—	191.0	184.6	178.2	171.9
245	—	195.5	189.0	182.5	176.0
250	—	200.1	193.4	186.8	180.1
255	—	204.7	197.9	191.1	184.3
260	—	209.4	202.4	195.4	188.4
	580 mm	560 mm	540 mm	520 mm	
260	202.4	195.4	188.4	181.5	
265	206.9	199.8	192.6	185.5	
270	211.5	204.2	196.9	189.6	
275	216.0	208.6	201.1	193.7	
280	220.6	213.0	205.4	197.8	
285	225.2	217.5	209.7	201.9	
290	229.9	222.0	214.0	206.1	
295	—	226.5	218.4	210.3	
300	—	231.0	222.8	214.5	
	560 mm	540 mm	520 mm	500 mm	480 mm
305	235.6	227.2	218.8	210.3	201.9
310	240.2	231.6	223.0	214.4	205.9
315	244.8	236.0	227.3	218.6	209.8
320	249.4	240.5	231.6	222.7	213.8
325	254.1	245.0	236.0	226.9	217.8
330	—	249.6	240.3	231.1	221.8
335	—	254.1	244.7	235.3	225.9
340	—	258.7	249.1	239.6	230.0
345	—	263.3	253.6	243.8	234.1

ENGLISH UNITS — TABLE I

Values of the temperature-altitude factor (2000 m.) for entering table II

Altitude feet	Assumed temperature of air column °F −20	0	+10	+20	+30	+40	+50	+60	+70	+80
200	7.4	7.1	6.9	6.8	6.6	6.5	6.3	6.2	6.1	6.0
400	14.8	14.1	13.8	13.5	13.2	13.0	12.7	12.4	12.2	11.9
600	22.2	21.2	20.7	20.3	19.9	19.5	19.0	18.6	18.2	17.9
800	29.6	28.3	27.7	27.1	26.5	25.9	25.4	24.8	24.3	23.8
1000	37.0	35.3	34.6	33.8	33.1	32.4	31.7	31.1	30.4	29.8
1200	44.3	42.4	41.5	40.6	39.7	38.9	38.1	37.3	36.5	35.8
1400	51.7	49.5	48.4	47.4	46.4	45.4	44.4	43.5	42.6	41.7
1600	59.1	56.5	55.3	54.1	53.0	51.9	50.8	49.7	48.7	47.7
1800	66.5	63.6	62.2	60.9	59.6	58.4	57.1	55.9	54.7	53.6
2000	73.9	70.6	69.1	67.7	66.2	64.8	63.4	62.1	60.8	59.6
2200	81.3	77.7	76.0	74.4	72.9	71.3	69.8	68.3	66.9	65.5
2400	88.7	84.8	82.9	81.2	79.5	77.8	76.1	74.5	73.0	71.5
2600	96.1	91.8	89.9	87.9	86.1	84.3	82.5	80.7	79.1	77.5
2800	103.5	98.9	96.8	94.7	92.7	90.8	88.8	87.0	85.1	83.4
3000	110.9	106.0	103.7	101.5	99.3	97.2	95.2	93.2	91.2	89.4
3200	118.2	113.0	110.6	108.2	106.0	103.7	101.5	99.4	97.3	95.3
3400	125.6	120.1	117.5	115.0	112.6	110.2	107.9	105.6	103.4	101.3
3600	133.0	127.2	124.4	121.8	119.2	116.7	114.2	111.8	109.5	107.2
3800	140.4	134.2	131.3	128.5	125.8	123.2	120.5	118.0	115.5	113.2
4000	147.8	141.3	138.2	135.3	132.4	129.6	126.9	124.2	121.6	119.2
4200	155.2	148.3	145.1	142.1	139.1	136.1	133.2	130.4	127.7	125.1
4400	162.6	155.4	152.0	148.8	145.7	142.6	139.6	136.6	133.8	131.1
4600	170.0	162.5	159.0	155.6	152.3	149.1	145.9	142.8	139.9	137.0

Altitude feet	Assumed temperature of air column °F −20	0	+10	+20	+30	+40	+50	+0	+70	+80
4800	177.3	169.5	165.9	162.3	158.9	155.6	152.2	149.0	145.9	143.0
5000	184.7	176.6	172.8	169.1	165.6	162.0	158.6	155.2	152.0	148.9
5200	192.1	183.7	179.7	175.9	172.2	168.5	164.9	161.5	158.1	154.9
5400	199.5	190.7	186.6	182.6	178.8	175.0	171.3	167.7	164.2	160.8
5600	206.9	197.8	193.5	189.4	185.4	181.5	177.6	173.9	170.3	166.8
5800	214.3	204.8	200.4	196.2	192.0	188.0	184.0	180.1	176.3	172.8
6000	221.7	211.9	207.3	202.9	198.7	194.4	190.3	186.3	182.4	178.7
6200	229.1	219.0	214.2	209.7	205.3	200.9	196.6	192.5	188.5	184.7
6400	236.4	226.0	221.1	216.4	211.9	207.4	203.0	198.7	194.6	190.6
6600	243.8	233.1	228.0	223.2	218.5	213.0	209.3	204.9	200.7	196.6
6800	251.2	240.1	235.0	230.0	225.1	220.4	215.7	211.1	206.7	202.5
7000	258.6	247.2	241.9	236.7	231.8	226.8	222.0	217.3	212.8	208.5
7200	266.0	254.3	248.8	243.5	238.4	233.3	228.4	223.5	218.9	214.4
7400	273.4	261.3	255.7	250.2	245.0	239.8	234.7	229.7	225.0	220.4
7600	280.8	268.4	262.6	257.0	251.6	246.3	241.0	235.9	231.1	226.4
7800	288.1	275.4	269.5	263.8	258.2	252.8	247.4	242.2	237.1	232.3
8000	295.5	282.5	276.4	270.5	264.8	259.2	253.7	248.4	243.2	238.3
8200	302.9	289.6	283.3	277.3	271.5	265.7	260.1	254.6	249.3	244.2
8400	310.3	296.6	290.2	284.0	278.1	272.2	266.4	260.8	255.4	250.2
8600	317.7	303.7	297.1	290.8	284.7	278.7	272.7	267.0	261.4	256.1
8800	325.1	310.7	304.0	297.6	291.3	285.2	279.1	273.2	267.5	262.1
9000	332.5	317.8	310.9	304.3	297.9	291.6	285.4	279.4	273.6	268.0

ENGLISH UNITS — TABLE II

Value of Correction to be Added.

Temp alt. factor	Barometer reading 31	30	29	28	27	Temp. alt. factor	Barometer reading 26	25	24	23	22
	in.	in.	in.	in.	in.		in.	in.	in.	in.	in.
1	0.04	0.03	0.03	—	—	165	5.44	5.23	5.02	—	—
5	0.18	0.17	0.17	—	—	170	5.62	5.40	5.19	—	—
10	0.36	0.35	0.34	0.32	—	175	—	5.58	5.36	—	—
15	0.54	0.52	0.51	0.49	—	180	—	5.76	5.53	5.30	—
20	0.72	0.70	0.68	0.65	—	185	—	5.93	5.70	5.46	—
25	—	0.88	0.85	0.82	—	190	—	6.11	5.87	5.62	—
30	—	1.05	1.02	0.98	—	195	—	6.29	6.04	5.79	—
35	—	1.23	1.19	1.15	—	200	—	—	6.21	5.96	—
40	—	1.41	1.37	1.32	1.27	205	—	—	6.39	6.12	—
45	—	1.60	1.54	1.49	1.44	210	—	—	6.56	6.29	—
50	—	—	1.72	1.66	1.60	215	—	—	6.74	6.46	—
55	—	—	1.90	1.83	1.76	220	—	—	6.92	6.63	6.34
60	—	—	2.07	2.00	1.93	225	—	—	7.10	6.80	6.51
65	—	—	2.25	2.18	2.10	230	—	—	7.28	6.97	6.67
70	—	—	2.43	2.35	2.27	235	—	—	7.46	7.15	6.84
75	—	—	—	2.53	2.43	240	—	—	—	7.32	7.00
80	—	—	—	2.70	2.60	245	—	—	—	7.49	7.17

Temp alt. factor	Barometer reading 28	27	26	25	24	Temp. alt. factor	Barometer reading 23	22	21	20
	in.	in.	in.	in.	in.		in.	in.	in.	in.
75	2.53	2.43	2.34	—	—	250	7.67	7.34	—	—
80	2.70	2.60	2.51	—	—	255	7.85	7.51	—	—
85	2.88	2.78	2.67	—	—	260	8.03	7.68	7.33	—
90	3.06	2.95	2.84	—	—	265	8.21	7.85	7.49	—
95	3.24	3.12	3.01	—	—	270	8.39	8.02	7.66	—
100	3.42	3.29	3.17	—	—	275	8.57	8.19	7.82	—
105	3.60	3.47	3.34	3.21	—	280	—	8.37	7.99	—
110	—	3.85	3.51	3.38	—	285	—	8.54	8.16	—
115	—	3.82	3.68	3.54	—	290	—	8.72	8.32	—
120	—	4.00	3.85	3.70	—	295	—	8.90	8.49	8.09
125	—	4.18	4.02	3.87	—	300	—	9.08	8.66	8.25
130	—	4.36	4.20	4.04	—	305	—	9.26	8.83	8.41
135	—	4.54	4.37	4.20	—	310	—	9.44	9.01	8.58
140	—	—	4.55	4.37	4.20	315	—	9.62	9.18	8.74
145	—	—	4.72	4.54	4.36	320	—	9.80	9.35	8.91
150	—	—	4.90	4.71	4.52	325	—	—	9.53	9.08
155	—	—	5.08	4.88	4.69	330	—	—	9.71	9.24
160	—	—	5.26	5.06	4.85	—	—	—	—	

REDUCTION OF BAROMETER TO GRAVITY AT SEA LEVEL

Metric Units
Correction to be subtracted given in millimeters
(From Smithsonian Physical Tables)

Height above sea level in meters	Observed Height of Barometer in Millimeters 500	550	600	650	700	750	800
100	—	—	—	—	.02	.02	.02
200	—	—	—	—	.04	.05	.05
300	—	—	—	—	.07	.07	.07
400	—	—	—	—	.09	.10	.10
500	—	—	—	—	.11	.12	.13
600	—	—	—	.12	.13	.14	—
700	—	—	—	.14	.15	.16	—
800	—	—	—	.16	.18	.19	—
900	—	—	—	.18	.20	.22	—
1000	—	.18	.19	.20	.22	.24	—
1100	—	.19	.21	.22	.24	—	—
1200	—	.21	.23	.24	.26	—	—
1300	—	.22	.24	.26	.29	—	—
1400	—	.24	.26	.28	.31	—	—
1500	.24	.26	.28	.30	.33	—	—
1600	.25	.28	.30	.32	—	—	—
1700	.27	.30	.32	.34	—	—	—
1800	.28	.31	.34	.36	—	—	—
1900	.30	.33	.36	.39	—	—	—
2000	.31	.34	.38	.41	—	—	—
2100	.33	.36	.40	—	—	—	—
2200	.35	.38	.41	—	—	—	—
2300	.36	.40	.43	—	—	—	—
2400	.38	.42	.45	—	—	—	—
2500	.39	.43	.47	—	—	—	—

English Units

Height above sea level in feet	Observed Height in Inches 18	20	22	24	26	28	30
1000	—	—	—	—	.003	.003	.003
2000	—	—	—	.004	.005	.005	.006
3000	—	—	.007	.007	.008	.008	—
4000	—	—	.009	.009	.010	—	—
4500	—	—	0.10	0.10	0.11	—	—
5000	—	0.10	0.11	0.11	0.12	—	—
5500	—	0.11	0.12	0.13	—	—	—
6000	—	0.11	0.13	0.14	—	—	—
6500	0.11	0.12	0.14	0.15	—	—	—
7000	0.12	0.13	0.15	0.16	—	—	—
7500	0.13	0.14	0.16	0.17	—	—	—
8000	0.14	0.15	0.17	—	—	—	—
8500	0.15	0.16	0.18	—	—	—	—
9000	0.16	0.17	0.19	—	—	—	—
9500	0.16	0.18	0.20	—	—	—	—

REDUCTION OF BAROMETER TO LATITUDE 45°

Metric Scale
For latitudes below 45°, subtract the correction; for latitudes greater than 45° it is to be added. Corrections in cm.
(From Smithsonian Meterological Tables.)

Latitude		Observed Height of Barometer in Centimeters 68	70	72	74	76	78
25°	65°	0.116	0.120	0.123	0.127	0.130	0.133
26	64	.111	.115	.118	.121	.125	.128
27	63	.106	.110	.113	.116	.119	.122
28	62	.101	.104	.107	.110	.113	.116
29	61	.096	.099	.102	.104	.107	.110
30	60	.091	.094	.096	.098	.101	.104
31	59	.085	.087	.090	.092	.095	.097
32	58	.079	.082	.084	.086	.089	.091
33	57	.074	.076	.078	.080	.082	.084
34	56	.068	.070	.072	.074	.076	.078
35	55	.062	.064	.066	.067	.069	.071
36	54	.056	.058	.059	.061	.063	.064
37	53	.050	.051	.053	.054	.056	.057
38	52	.044	.045	.046	.048	.049	.050
39	51	.038	.039	.040	.041	.042	.043
40	50	.031	.032	.033	.034	.035	.036
41	49	.025	.026	.027	.027	.028	.029
42	48	.019	.019	.020	.021	.021	.022
43	47	.013	.013	.013	14	.014	.014
44	46	.006	.007	.007	.007	.007	.007

English Scale (Corrections in inches)

Latitude		Observed Height in Inches 25	26	27	28	29	30
25°	65°	0.043	0.044	0.046	0.048	0.050	0.051
26	64	.041	.043	.044	.046	.048	.049
27	63	.039	.041	.042	.044	.045	.047
28	62	.037	.039	.040	.042	.043	.045
29	61	.035	.037	.038	.039	.041	.042
30	60	.033	.035	.036	.037	.039	.040
31	59	.031	.032	.034	.035	.036	.037
32	58	.029	.030	.032	.033	.034	.035
33	57	.027	.028	.029	.030	.031	.032
34	56	.025	.026	.027	.028	.029	.030
35	55	.023	.024	.025	.025	.026	.027
36	54	.021	.021	.022	.023	.024	.025
37	53	.018	.019	.020	.021	.021	.022
38	52	.016	.017	.017	.018	.019	.019
39	51	.014	.014	.015	.015	.016	.017
40	50	.012	.012	.012	.013	.013	.014
41	49	.009	.010	.010	.010	.011	.011
42	48	.007	.007	.008	.008	.008	.008
43	47	.005	.005	.005	.005	.005	.006
44	46	.002	.002	.003	.003	.003	.003

RELATIVE HUMIDITY — DEW-POINT

The table gives the relative humidity of the air for temperature t and dewpoint d.
(From Smithsonian Meterological Tables.)

Depression of dew-point t-d, °C	Dew-Point (d) −10	0	+10	+20	+30
0.0	100%	100%	100%	100%	100%
0.2	98	99	99	99	99
0.4	97	97	97	98	98
0.6	95	96	96	96	97
0.8	94	94	95	95	96
1.0	92	93	94	94	94
1.2	91	92	92	93	93
1.4	90	90	91	92	92
1.6	88	89	90	91	91
1.8	87	88	89	90	90
2.0	86	87	88	88	89
2.2	84	85	86	87	88
2.4	83	84	85	86	87
2.6	82	83	84	85	86
2.8	80	82	83	84	85
3.0	79	81	82	83	84
3.2	78	80	81	82	83
3.4	77	79	80	81	82
3.6	76	77	79	80	82
3.8	75	76	78	79	81
4.0	73	75	77	78	80
4.2	72	74	76	77	79
4.4	71	73	75	77	78
4.6	70	72	74	76	77
4.8	69	71	73	75	76
5.0	68	70	72	74	75
5.2	67	69	71	73	75

Depression of dew-point t-d, °C	Dew-Point (d) −10	0	+10	+20	+30
5.4	66	68	70	72	74
5.6	65	67	69	71	73
5.8	64	66	69	70	72
6.0	63	66	68	70	71
6.2	62	65	67	69	71
6.4	61	64	66	68	70
6.6	60	63	65	67	69
6.8	60	62	64	66	68
7.0	59	61	63	66	68
7.2	58	60	63	65	67
7.4	57	60	62	64	66
7.6	56	59	61	63	65
7.8	55	58	60	63	65
8.0	54	57	60	62	64
8.2	54	56	59	61	63
8.4	53	56	58	60	63
8.6	52	55	57	60	62
8.8	51	54	57	59	61
9.0	51	53	56	58	61
9.2	50	53	55	58	60
9.4	49	52	55	57	59
9.6	48	51	54	56	59
9.8	48	51	53	56	58
10.0	47	50	53	55	57
10.5	45	48	51	54	—
11.0	44	47	49	52	—
11.5	42	45	48	51	—

Depression of dew-point t-d, °C	Dew-Point (d) −10	0	+10	+20	+30
12.0	41	44	47	49	—
12.5	39	42	45	48	—
13.0	38	41	44	46	—
13.5	37	40	43	45	—
14.0	35	38	41	44	—
14.5	34	37	40	43	—
15.0	33	36	39	42	—
15.5	32	35	38	40	—
16.0	31	34	37	39	—
16.5	30	33	36	38	—
17.0	29	32	35	37	—
17.5	28	31	34	36	—
18.0	27	30	33	35	—
18.5	26	29	32	34	—
19.0	25	28	31	33	—
19.5	24	27	30	33	—
20.0	24	26	29	32	—
21.0	22	25	27	—	—
22.0	21	23	26	—	—
23.0	19	22	24	—	—
24.0	18	21	23	—	—
25.0	17	19	22	—	—
26.0	16	18	21	—	—
27.0	15	17	20	—	—
28.0	14	16	19	—	—
29.0	13	15	18	—	—
30.0	12	14	17	—	—

RELATIVE HUMIDITY FROM WET AND DRY BULB THERMOMETER (CENT. SCALE)

This table gives the approximate relative humidity directly from the reading of the air temperature (dry bulb) (t′°C) and the wet bulb (t′°C). It is computed for a barometric pressure of 74.27 cm Hg. Errors resulting from the use of this table for air temperatures above −10°C and between 77.5 and 71 cm Hg will usually be within the errors of observation.

Condensed from Bulletin of the U.S. Weather Bureau No. 1071

t-t′t	0.2	0.4	0.6	0.8	1.0	1.2	1.4	1.6	1.8	2.0	2.2	2.4	2.6	2.8	3.0	3.2	3.4	3.6	3.8	4.0	4.5	5.0	5.5	6.0	6.5	7.0	7.5	8.0		9.0	9.5	10.0	10.5	11.0
−10	93	87	80	74	67	61	54	48	41	35	28	22	16	9																				
−9	94	88	81	75	69	63	57	51	45	39	33	27	21	15	9																			
−8	94	88	83	77	71	65	60	54	48	43	37	32	26	20	15	10																		
−7	95	89	84	78	73	67	62	57	52	46	41	36	31	25	20	15	10	5																
−6	95	90	85	79	74	69	64	59	54	49	45	40	35	30	25	20	15	11	6															
−5	95	90	86	81	76	71	66	62	57	52	48	43	39	34	29	25	20	16	11	7														
−4	95	91	86	82	77	73	68	64	59	55	51	46	42	38	33	29	25	21	17	12														
−3	96	91	87	82	78	74	70	66	62	57	53	49	45	41	37	33	29	25	21	17	8													
−2	96	92	88	84	79	75	71	68	64	60	56	52	48	44	40	37	33	29	25	22	12													
−1	96	92	88	84	81	77	73	69	66	62	58	54	51	47	43	40	36	33	29	26	17	8												
0	96	93	89	85	81	78	74	71	67	64	60	57	53	50	46	43	40	36	33	29	21	13	5											
1	97	93	90	86	83	80	76	73	70	66	63	59	56	53	49	46	43	40	36	33	25	17	10											
2	97	93	90	87	84	81	78	74	71	68	65	62	59	55	52	49	46	43	40	37	29	22	14	7										
3	97	94	91	88	84	82	78	76	72	70	67	64	61	58	55	52	49	46	43	40	33	26	19	12	5									
4	97	94	91	88	85	82	79	77	74	71	68	65	62	60	57	54	51	48	46	43	36	29	22	16	9									
5	97	94	91	88	86	83	80	77	75	72	69	67	64	61	58	56	53	51	48	45	39	33	26	20	13	7								
6	97	94	92	89	86	84	81	78	76	73	70	68	65	63	60	58	55	53	50	48	41	35	29	24	17	11	5							
7	97	95	92	89	87	84	82	79	77	74	72	69	67	64	62	59	57	54	52	50	44	38	32	26	21	15	10							
8	97	95	92	90	87	85	82	80	77	75	73	70	68	65	63	61	58	56	54	51	46	40	35	29	24	19	14	8						
9	98	95	93	90	88	85	83	81	78	76	74	71	69	67	64	62	60	58	55	53	48	42	37	32	27	22	17	12	7					
10	98	95	93	90	88	86	83	81	79	77	74	72	70	68	66	63	61	59	57	55	50	44	39	34	29	24	20	15	10	6				
11	98	95	93	91	89	86	84	82	80	78	75	73	71	69	67	65	62	60	58	56	51	46	41	36	32	27	22	18	13	9	5			
12	98	96	93	91	89	87	85	82	80	78	76	74	72	70	68	66	64	62	60	58	53	48	43	39	34	29	25	21	16	12	8			
13	98	96	93	91	89	87	85	83	81	79	77	75	73	71	69	67	65	63	61	59	54	50	45	41	36	32	28	23	19	15	11	7		
14	98	96	94	92	90	88	86	84	82	79	78	76	74	72	70	68	66	64	62	60	56	51	47	42	38	34	30	26	22	18	14	10	6	
15	98	96	94	92	90	88	86	84	82	80	78	76	74	73	71	69	67	65	63	61	57	53	48	44	40	36	32	27	24	20	16	13	9	6

	0.5	1.0	1.5	2.0	2.5	3.0	3.5	4.0	4.5	5.0	5.5	6.0	6.5	7.0	7.5	8.0	8.5	9.0	9.5	10.0	10.5	11.0	11.5	12.0	12.5	13.0	13.5	14.0	14.5	15.0	16.0	17.0	18.0	19.0	20.0
16	95	90	85	81	76	71	67	63	58	54	50	46	42	38	34	30	26	23	19	15	12	8	5												
17	95	90	86	81	76	72	68	64	60	55	51	47	43	40	36	32	28	25	21	18	14	11	8												
18	95	91	86	82	77	73	69	65	61	57	53	49	45	41	38	34	30	27	23	20	17	14	10	7											
19	95	91	87	82	78	74	70	65	62	58	54	50	46	43	39	36	32	29	26	22	19	16	13	10	7										
20	96	91	87	83	78	74	70	66	63	59	55	51	48	44	41	37	34	31	28	24	21	18	15	12	9	6									
21	96	91	87	83	79	75	71	67	64	60	56	53	49	46	42	39	36	32	29	26	23	20	17	14	12	9	6								
22	96	92	87	83	80	76	72	68	64	61	57	54	50	47	44	40	37	34	31	28	25	22	19	17	14	11	8	6							
23	96	92	88	84	80	76	72	69	65	62	58	55	52	48	45	42	39	36	33	30	27	24	21	19	16	13	11	8	6						
24	96	92	88	84	80	77	73	69	66	62	59	56	53	49	46	43	40	37	34	31	29	26	23	20	18	15	13	10	8	5					
25	96	92	88	84	81	77	74	70	67	63	60	57	54	50	47	44	41	39	36	33	30	28	25	22	20	17	15	12	10	8					
26	96	92	88	85	81	78	74	71	67	64	61	58	54	51	49	46	43	40	37	34	32	29	26	24	21	19	17	14	12	10	5				
27	96	92	89	85	82	78	75	71	68	65	62	58	56	52	50	47	44	41	38	36	33	31	28	26	23	21	18	16	14	12	7				
28	96	93	89	85	82	78	75	72	69	65	62	59	56	53	51	48	45	42	40	37	34	32	29	27	25	22	20	18	16	13	9	5			
29	96	93	89	86	82	79	76	72	69	66	63	60	57	54	52	49	46	43	41	38	36	33	31	28	26	24	22	19	17	15	11	7			
30	96	93	89	86	83	79	76	73	70	67	64	61	58	55	52	50	47	44	42	39	37	35	32	30	28	25	23	21	19	17	13	9	5		
31	96	93	90	86	83	80	77	73	70	67	64	61	59	56	53	51	48	45	43	40	38	36	33	31	29	27	25	22	20	18	14	11	7		
32	96	93	90	86	83	80	77	74	71	68	65	62	60	57	54	51	49	46	44	41	39	37	35	32	30	28	26	24	22	20	16	12	9	5	
33	97	93	90	87	83	80	77	74	71	68	66	63	60	57	55	52	50	47	45	42	40	38	36	33	31	29	27	25	23	21	17	14	10	7	
34	97	93	90	87	84	81	78	75	72	69	66	63	61	58	56	53	51	48	46	43	41	39	37	35	32	30	28	26	24	22	20	16	12	9	5
35	97	94	90	87	84	81	78	75	72	69	67	64	61	59	56	54	51	49	47	44	42	40	38	36	33	31	29	27	25	23	21	17	14	10	7
36	97	94	90	87	84	81	78	75	73	70	67	64	62	59	57	54	52	50	48	45	43	41	39	37	35	32	30	28	26	24	23	19	15	12	8
37	97	94	91	87	84	82	79	76	73	70	68	65	63	60	58	55	53	51	48	46	44	42	40	38	36	34	32	30	28	26	24	20	17	13	10
38	97	94	91	88	84	82	79	76	74	71	68	66	63	61	58	56	54	51	49	47	45	43	41	39	37	35	33	31	29	27	25	21	18	15	11
39	97	94	91	88	85	82	79	77	74	71	69	66	64	61	59	57	54	52	50	48	46	44	42	40	38	36	34	32	30	28	26	23	19	16	13
40	97	94	91	88	85	82	80	77	74	72	69	67	64	62	59	57	54	53	51	48	46	43	42	39	38	36	34	32	30	28	25	22	18	15	12

REDUCTION OF PSYCHROMETRIC OBSERVATION

For the reduction of observations with the wet and dry bulb thermometer. Assuming the relative velocity of the air to the thermometer bulbs is at least three meters per second; if t is the temperature of the air as indicated by the dry bulb, t_w, the temperature of the wet bulb, B, the barometric pressure, and E_w, the vapor tension of water corresponding to t_2, then the actual vapor tension is

$$E = E_w - 0.00066B(t - t_w)\,[1 + 0.00115(t - t_w)] \text{ millimeters}$$

The value of the term

$$0.00066B(t - t_w)\,[1 + 0.00115\,(t - t_w)]$$

is given in the following table.

(From Miller's Laboratory Physics, Ginn & Co., Publishers, by permission.)

Barometric Pressure B in Millimeters

$t - t_w$	700	710	720	730	740	750	760	770	$t - t_2$	700	710	720	730	740	750	760	770
°C	mm	mm	mm	mm	mm	mm	mm	mm	°C	mm	mm	mm	mm	mm	mm	mm	mm
1	0.463	0.469	0.476	0.482	0.489	0.496	0.502	0.509	11	5.146	5.220	5.293	5.367	5.440	5.514	5.587	5.661
2	0.926	0.939	0.953	0.966	0.979	0.992	1.006	1.019	12	5.621	5.701	5.781	5.861	5.942	6.022	6.102	6.183
3	1.391	1.411	1.431	1.450	1.470	1.490	1.510	1.530	13	6.096	6.183	6.270	6.357	6.444	6.531	6.618	6.705
4	1.857	1.883	1.910	1.936	1.963	1.989	2.016	2.042	14	6.572	6.666	6.760	6.854	6.948	7.042	7.135	7.229
5	2.323	2.356	2.390	2.423	2.456	2.489	2.522	2.556	15	7.050	7.150	7.251	7.352	7.452	7.553	7.654	7.754
6	2.791	2.831	2.871	2.911	2.951	2.990	3.030	3.070	16	7.528	7.636	7.743	7.851	7.958	8.066	8.173	8.281
7	3.260	3.307	3.353	3.400	3.446	3.493	3.539	3.586	17	8.008	8.122	8.236	8.351	8.465	8.580	8.694	8.808
8	3.730	3.783	3.837	3.890	3.943	3.996	4.050	4.103	18	8.488	8.609	8.731	8.852	8.973	9.094	9.216	9.337
9	4.201	4.261	4.321	4.381	4.441	4.501	4.561	4.621	19	8.970	9.098	9.226	9.354	9.482	9.610	9.739	9.867
10	4.673	4.740	4.807	4.873	4.940	5.007	5.074	5.140	20	9.453	9.588	9.723	9.858	9.993	10.128	10.263	10.398

CONSTANT HUMIDITY

The following table shows the % humidity and the aqueous tension at the given temperature within a closed space when an excess of the substance indicated is in contact with a saturated aqueous solution of the given solid phase.

Solid phase	t°C.	% humidity	Aq. tension mm Hg	Solid phase	t°C.	% humidity	Aq. tension mm Hg
$H_3PO_4.\frac{1}{2}H_2O$	24	9	1.99	$NaClO_3$	20	75	13.0
$KC_2H_3O_2$	168	13	738	$(NH_4)_2SO_4$	108	75	754
$LiCl.H_2O$	20	15	2.60	$NaC_2H_3O_2.3H_2O$	20	76	13.2
$KC_2H_3O_2$	20	20	3.47	$H_2C_2O_4.2H_2O$	20	76	13.2
KF	100	22.9	174	$Na_2S_2O_3.5H_2O$	20	78	13.5
NaBr	100	22.9	174	NH_4Cl	20	79.5	13.8
NaCl,KNO_3 and $NaNO_3$	16.39	30.49	4.23	NH_4Cl	25	79.3	18.6
$CaCl_2.6H_2O$	24.5	31	7.08	NH_4Cl	30	77.5	24.4
$CaCl_2.6H_2O$	20	32.3	5.61	$(NH_4)_2SO_4$	20	81	14.1
$CaCl_2.6H_2O$	18.5	35	5.54	$(NH_4)_2SO_4$	25	81.1	19.1
CrO_3	20	35	6.08	$(NH_4)_2SO_4$	30	81.1	25.6
$CaCl_2.6H_2O$	10	38	3.47	KBr	20	84	14.6
$CaCl_2 6H_2O$	5	39.8	2.59	Tl_2SO_4	104.7	84.8	768
$Zn(NO_3)_2.6H_2O$	20	42	7.29	$KHSO_4$	20	86	14.9
$K_2CO_3.2H_2O$	24.5	43	9.82	$Na_2CO_3.10H_2O$	24.5	87	20.9
$K_2CO_3.2H_2O$	18.5	44	6.96	$BaCl_2.2H_2O$	24.5	88	20.1
KNO_2	20	45	7.81	K_2CrO_4	20	88	15.3
KCNS	20	47	8.16	$Pb(NO_3)_2$	103.5	88.4	760
NaI	100	50.4	383	$ZnSO_4.7H_2O$	20	90	15.6
$Ca(NO_3)_2.4H_2O$	24.5	51	11.6	$Na_2CO_3.10H_2O$	18.5	92	14.6
$NaHSO_4.H_2O$	20	52	9.03	$NaBrO_3$	20	92	16.0
$Na_2Cr_2O_7.2H_2O$	20	52	9.03	K_2HPO_4	20	92	16.0
$Mg(NO_3)_2.6H_2O$	24.5	52	11.9	$NH_4H_2PO_4$	30	92.9	29.3
$NaClO_3$	100	54	410	$NH_4H_2PO_4$	25	93	21.9
$Ca(NO_3)_2.4H_2O$	18.5	56	8.86	$Na_2SO_4.10H_2O$	20	93	16.1
$Mg(NO_3)_2.6H_2O$	18.5	56	8.86	$NH_4H_2PO_4$	20	93.1	16.2
KI	100	56.2	427	$ZnSO_4.7H_2O$	5	94.7	6.10
$NaBr.2H_2O$	20	58	10.1	$Na_2SO_3.7H_2O$	20	95	16.5
$Mg(C_2H_3O_2)_2.4H_2O$	20	65	11.3	$Na_2HPO_4.12H_2O$	20	95	16.5
$NaNO_2$	20	66	11.5	NaF	100	96.6	734
NH_4Cl and KNO_3	30	68.6	21.6	$Pb(NO_3)_2$	20	98	17.0
KBr	100	69.2	526	$CuSO_4.5H_2O$	20	98	17.0
NH_4Cl and KNO_3	25	71.2	16.7	$TlNO_3$	100.3	98.7	759
NH_4Cl and KNO_3	20	72.6	12.6	TlCl	100.1	99.7	761

CONSTANT HUMIDITY WITH SULFURIC ACID SOLUTIONS

The relative humidity and pressure of aqueous vapor of air in equilibrium conditions above aqueous solutions of sulfuric acid are given below.

Density of acid solution	Relative humidity	Vapor pressure at 20°C	Density of acid solution	Relative humidity	Vapor pressure at 20°C
1.00	100.0	17.4	1.30	58.3	10.1
1.05	97.5	17.0	1.35	47.2	8.3
1.10	93.9	16.3	1.40	37.1	6.5
1.15	88.8	15.4	1.50	18.8	3.3
1.20	80.5	14.0	1.60	8.5	1.5
1.25	70.4	12.2	1.70	3.2	0.6

For concentration of sulfuric acid solution refer to tables relating density to percent composition.

VELOCITY OF SOUND

Compiled by Gordon E. Becker, Bell Telephone Laboratories

The data for the Velocity of Sound is Various Materials were compiled from a variety of sources. For more extensive tables one is referred to the following books:
AIP Handbook, Smithsonian Tables.
Mason: Physical Acoustics and the Properties of Solids (1958).
Chalmers and Quarrell: Physical Examination of Metals (1960).
Mason: Piezoelectric Crystals and their Application to Ultrasonics (1950).
Bergmann: Der Ultraschall (Hirzel, 1954).

Definition of Terms: V_l = Velocity of plane longitudinal wave in bulk material
V_s = Velocity of plane transverse (shear) wave
V_{ext} = Velocity of longitudinal wave (extensional wave) in thin rods.

SOLIDS

Substance	Density g/cc	V_l m/sec	V_s m/sec	V_{ext} m/sec
Metals				
Aluminum, rolled	2.7	6420	3040	5000
Berylium	1.87	12890	8880	12870
Brass (70 Cu, 30 Zn)	8.6	4700	2110	3480
Copper, annealed	8.93	4760	2325	3810
Copper, rolled	8.93	5010	2270	3750
Duralumin 17S	2.79	6320	3130	5150
Gold, hard-drawn	19.7	3240	1200	2030
Iron, electrolytic	7.9	5950	3240	5120
Iron, Armco	7.85	5960	3240	5200
Lead, annealed	11.4	2160	700	1190
Lead, rolled	11.4	1960	690	1210
Magnesium, drawn, annealed	1.74	5770	3050	4940
Molybdenum	10.1	6250	3350	5400
Monel metal	8.90	5350	2720	4400
Nickel (unmagnetized)	8.85	5480	2990	4800
Nickel	8.9	6040	3000	4900
Platinum	21.4	3260	1730	2800
Silver	10.4	3650	1610	2680
Steel, mild	7.85	5960	3235	5200
Steel, 347 Stainless	7.9	5790	3100	5000
Steel (1%C)	7.84	5940	3220	5180
Steel (1%C, hardened)	7.84	5854	3150	5070
Tin, rolled	7.3	3320	1670	2730
Titanium	4.5	6070	3125	5080
Tungsten, annealed	19.3	5220	2890	4620
Tungsten, drawn	19.3	5410	2640	4320
Tungsten Carbide	13.8	6655	3980	6220
Zinc, rolled	7.1	4210	2440	3850
Various				
Fused silica	2.2	5968	3764	5760
Glass, pyrex	2.32	5640	3280	5170
Glass, heavy silicate flint	3.88	3980	2380	3720
Glass, light borate crown	2.24	5100	2840	4540
Lucite	1.18	2680	1100	1840
Nylon 6-6	1.11	2620	1070	1800
Polyethylene	0.90	1950	540	920
Polystyrene	1.06	2350	1120	2240
Rubber, butyl	1.07	1830		
Rubber, gum	0.95	1550		
Rubber neoprene	1.33	1600		
Brick	1.8			3650
Clay rock	2.2			3480
Cork	0.25			500
Marble	2.6			3810
Paraffin	0.9			1300
Tallow				390
Woods				
Ash, along the fiber				4670
Ash, across the rings				1390
Ash, along the rings				1260
Beech, along the fiber				3340
Elm, along the fiber				4120
Maple, along the fiber				4110
Oak, along the fiber				3850

LIQUIDS

Substance	Formula	Density g/cc	Velocity at 25°C m/sec	−δv/δt m/sec °C
Acetone	C_3H_6O	0.79	1174	4.5
Benzene	C_6H_6	0.870	1295	4.65
Carbon disulphide	CS_2	1.26	1149	—
Carbon tetrachloride	CCl_4	1.595	926	2.7
Castor oil	$C_{11}H_{10}O_{10}$	0.969	1477	3.6
Chloroform	$CHCl_3$	1.49	987	3.4
Ethanol	C_2H_6O	0.79	1207	4.0
Ethanol amide	C_2H_7NO	1.018	1724	3.4
Ethyl ether	$C_4H_{10}O$	0.713	985	4.87
Ethylene glycol	$C_2H_6O_2$	1.113	1658	2.1
Glycerol	$C_3H_8O_3$	1.26	1904	2.2
Kerosene	—	0.81	1324	3.6
Mercury	Hg	13.5	1450	—
Methanol	CH_4O	0.791	1103	3.2
Nitrobenzene	$C_6H_5NO_2$	1.20	1463	3.6
Turpentine	—	0.88	1255	—
Water (distilled)	H_2O	0.998	1496.7+ 2	−2.4
Water (sea)	—	1.025	1531	−2.4
Xylene hexafluoride	$C_8H_4F_6$	1.37	879	—

GASES AND VAPORS

Substance	Formula	Density g/l	Velocity m/sec	δv/δt m/sec °C
Gases (0°C)				
Air, dry		1.293	331.45	0.59
Ammonia	NH_3	0.771	415	
Argon	Ar	1.783	319	0.56
Carbon monoxide	CO	1.25	338	0.6
Carbon dioxide	CO_2	1.977	259	0.4
Chlorine	CL_2	3.214	206	
Deuterium	D_2		890	1.6
Ethane (10°C)	C_2H_6	1.356	308	
Ethylene	C_2H_4	1.260	317	
Helium	He	0.178	965	0.8
Hydrogen	H_2	0.0899	1284	2.2
Hydrogen bromide	HBr	3.50	200	
Hydrogen chloride	HCl	1.639	296	
Hydrogen iodide	HI	5.66	157	
Hydrogen sulfide	H_2S	1.539	289	
Illuminating (Coal gas)			453	
Methane	CH_4	0.7168	430	
Neon	Ne	0.900	435	0.8
Nitric oxide (10°C)	NO	1.34	324	
Nitrogen	N_2	1.251	334	0.6
Nitrous oxide	N_2O	1.977	263	0.5
Oxygen	O_2	1.429	316	0.56
Sulfur dioxide	SO_2	2.927	213	0.47
Vapors (97.1°C)				
Acetone	C_3H_6O		239	0.32
Benzene	C_6H_6		202	0.3
Carbon tetrachloride	CCl_4		145	
Chloroform	$CHCl_3$		171	0.24
Ethanol	C_2H_6O		269	0.4
Ethyl ether	$C_4H_{10}O$		206	0.3
Methanol	CH_4O		335	0.46
Water vapor (134°C)	H_2O		494	

SOUND VELOCITY IN WATER ABOVE 212°F

By permission from the Acoustical Society of America, Volume 31 (1959) and J. C. McDade, D. R. Pardue, A. L. Gedrich and F. Vrataric.

Temperature°F	Velocity m/sec	Velocity ft/sec
186.8	1552	5092
200	1548	5079
210	1544	5066
220	1538	5046
230	1532	5026
240	1524	5000
250	1516	4974
260	1507	4944
270	1497	4911
280	1487	4879
290	1476	4843
300	1465	4806
310	1453	4767
320	1440	4724
330	1426	4678
340	1412	4633
350	1398	4587
360	1383	4537

Temperature °F	Velocity m/sec	Velocity ft/sec
370	1368	4488
380	1353	4439
390	1337	4386
400	1320	4331
410	1302	4272
420	1283	4209
430	1264	4147
440	1244	4081
450	1220	4010
460	1200	3940
470	1180	3880
480	1160	3800
490	1140	3730
500	1110	3650
510	1090	3570
520	1070	3500
530	1040	3410
540	1010	3320
550	980	3230

MUSICAL SCALES

EQUAL TEMPERED CHROMATIC SCALE

$A_4 = 440$

American Standard pitch. Adopted by the American Standards Association in 1936

Note	Frequency	Note	Frequency	Note	Frequency	Note	Frequency
C_0	16.35	C_2	65.41	C_4	261.63	C_6	1046.50
$C\#_0$	17.32	$C\#_2$	69.30	$C\#_4$	277.18	$C\#_6$	1108.73
D_0	18.35	D_2	73.42	D_4	293.66	D_6	1174.66
$D\#_0$	19.45	$D\#_2$	77.78	$D\#_4$	311.13	$D\#_6$	1244.51
E_0	20.60	E_2	82.41	E_4	329.63	E_6	1318.51
F_0	21.83	F_2	87.31	F_4	349.23	F_6	1396.91
$F\#_0$	23.12	$F\#_2$	92.50	$F\#_4$	369.99	$F\#_6$	1479.98
G_0	24.50	G_2	98.00	G_4	392.00	G_6	1567.98
$G\#_0$	25.96	$G\#_2$	103.83	$G\#_4$	415.30	$G\#_6$	1661.22
A_0	27.50	A_2	110.00	A_4	440.00	A_6	1760.00
$A\#_0$	29.14	$A\#_2$	116.54	$A\#_4$	466.16	$A\#_6$	1864.66
B_0	30.87	B_2	123.47	B_4	493.88	B_6	1975.53
C_1	32.70	C_3	130.81	C_5	523.25	C_7	2093.00
$C\#_1$	34.65	$C\#_3$	138.59	$C\#_5$	554.37	$C\#_7$	2217.46
D_1	36.71	D_3	146.83	D_5	587.33	D_7	2349.32
$D\#_1$	38.89	$D\#_3$	155.56	$D\#_5$	622.25	$D\#_7$	2489.02
E_1	41.20	E_3	164.81	E_5	659.26	E_7	2637.02
F_1	43.65	F_3	174.61	F_5	698.46	F_7	2793.83
F#1	46.25	$F\#_3$	185.00	$F\#_5$	739.99	$F\#_7$	2959.96
G_1	49.00	G_3	196.00	G_5	783.99	G_7	3135.96
$G\#_1$	51.91	$G\#_3$	207.65	$G\#_5$	830.61	$G\#_7$	3322.44
A_1	55.00	A_3	220.00	A_5	880.00	A_7	3520.00
$A\#_1$	58.27	$A\#_3$	233.08	$A\#_5$	932.33	$A\#_7$	3729.31
B_1	61.74	B_3	246.94	B_5	987.77	B_7	3951.07
						C_8	4186.01

EQUAL TEMPERED CHROMATIC SCALE

$A_4 = 435$

International Pitch, adopted 1891

Note	Frequency	Note	Frequency	Note	Frequency	Note	Frequency
C_0	16.17	C_2	64.66	C_4	258.65	C_6	1034.61
$C\#_0$	17.13	$C\#_2$	68.51	$C\#_4$	274.03	$C\#_6$	1096.13
D_0	18.15	D_2	72.58	D_4	290.33	D_6	1161.31
$D\#_0$	19.22	$D\#_2$	76.90	$D\#_4$	307.59	$D\#_6$	1230.37
E_0	20.37	E_2	81.47	E_4	325.88	E_6	1303.53
F_0	21.58	F_2	86.31	F_4	345.26	F_6	1381.04
$F\#_0$	22.86	$F\#_2$	91.45	$F\#_4$	365.79	$F\#_6$	1463.16
G_0	24.22	G_2	96.89	G_4	387.54	G_6	1550.16
$G\#_0$	25.66	$G\#_2$	102.65	$G\#_4$	410.59	$G\#_6$	1642.34
A_0	27.19	A_2	108.75	A_4	435.00	A_6	1740.00
$A\#_0$	28.80	$A\#_2$	115.22	$A\#_4$	460.87	$A\#_6$	1843.47
B_0	30.52	B_2	122.07	B_4	488.27	B_6	1953.08
C_1	32.33	C_3	129.33	C_5	517.31	C_7	2069.22
$C\#_1$	34.25	$C\#_3$	137.02	$C\#_5$	548.07	$C\#_7$	2192.26
D_1	36.29	D_3	145.16	D_5	580.66	D_7	2322.62
$D\#_1$	38.45	$D\#_3$	153.80	$D\#_5$	615.18	$D\#_7$	2460.73
E_1	40.74	E_3	162.94	E_5	651.76	E_7	2607.05
F_1	43.16	F_3	172.63	F_5	690.52	F_7	2762.08
$F\#_1$	45.72	$F\#_3$	182.89	$F\#_5$	731.58	$F\#_7$	2926.32
G_1	48.44	G_3	193.77	G_5	775.08	G_7	3100.33
$G\#_1$	51.32	$G\#_3$	205.29	$G\#_5$	821.17	$G\#_7$	3284.68
A_1	54.38	A_3	217.50	A_5	870.00	A_7	3480.00
$A\#_1$	57.61	$A\#_3$	230.43	$A\#_5$	921.73	$A\#_7$	3686.93
B_1	61.03	B_3	244.14	B_5	976.54	B_7	3906.17
						C_8	4138.44

SCIENTIFIC OR JUST SCALE

$C_4 = 256$

Note	Frequency	Note	Frequency	Note	Frequency	Note	Frequency
C_0	16	C_2	64	C_4	256	C_6	1024
D_0	18	D_2	72	D_4	288	D_6	1152
E_0	20	E_2	80	E_4	320	E_6	1280
F_0	21.33	F_2	85.33	F_4	341.33	F_6	1365.33
G_0	24	G_2	96	G_4	384	G_6	1536
A_0	26.67	A_2	106.67	A_4	426.67	A_6	1706.67
B_0	30	B_2	120	B_4	480	B_6	1920
C_1	32	C_3	128	C_5	512	C_7	2048
D_1	36	D_3	144	D_5	576	D_7	2304
E_1	40	E_3	160	E_5	640	E_7	2560
F_1	42.67	F_3	170.67	F_5	682.67	F_7	2730.67
G_1	48	G_3	192	G_5	768	G_7	3072
A_1	53.33	A_3	213.33	A_5	853.33	A_7	3413.33
B_1	60	B_3	240	B_5	960	B_7	3840
						C_8	4096

ABSORPTION AND VELOCITY OF SOUND IN STILL AIR

The following data refer only to the temperature 20C (68F). They were abstracted from an extensive compilation prepared by L. B. Evans and H. E. Bass. The entire report, Tables of Absorption and Velocity of Sound in Still Air at 68F (20C), AD-738 576 is available from National Technical Information Service, U. S. Department of Commerce, 5285 Port Royal Road, Springfield, Va. 22151.

Frequency (Hz)	Absorption (dB/1000 ft)	Absorption (dB/Km)	Absorption (dB/sec)	Velocity (1000 ft/sec)
Relative Humidity = 0%				
20.	0.154	0.51	0.174	1.126892
40.	0.327	1.07	0.368	1.127013
50.	0.384	1.26	0.433	1.127050
63.	0.436	1.43	0.491	1.127085
100.	0.509	1.67	0.573	1.127131
200.	0.560	1.84	0.631	1.127161
400.	0.596	1.96	0.672	1.127169
630.	0.645	2.11	0.727	1.127171
800.	0.692	2.27	0.780	1.127172
1250.	0.861	2.82	0.970	1.127172
2000.	1.262	4.14	1.423	1.127173
4000.	2.696	8.84	3.039	1.127178
6300.	4.541	14.89	5.118	1.127182
10000.	8.013	26.28	9.032	1.127184
12500.	10.918	35.81	12.306	1.127186
16000.	15.901	52.15	17.923	1.127187
20000.	22.978	75.37	25.901	1.127187
40000.	81.405	267.01	91.759	1.127188
63000.	196.544	644.66	221.542	1.127188
80000.	314.677	1032.14	354.700	1.127188
Relative Humidity = 5%				
20.	0.031	0.10	0.034	1.126973
40.	0.074	0.24	0.083	1.126996
50.	0.092	0.30	0.104	1.127004
63.	0.114	0.37	0.129	1.127009
100.	0.179	0.59	0.202	1.127019
200.	0.449	1.47	0.506	1.127028
400.	1.451	4.76	1.635	1.127043
630.	3.211	10.53	3.619	1.127067
800.	4.774	15.66	5.380	1.127088
1250.	9.164	30.06	10.329	1.127147
2000.	15.175	49.77	17.106	1.127228
4000.	22.685	74.41	25.573	1.127321
6300.	26.245	86.08	29.587	1.127352
10000.	30.781	100.96	34.701	1.127365
12500.	34.306	112.52	38.676	1.127369
16000.	40.263	132.06	45.391	1.127372
20000.	48.653	159.58	54.850	1.127373
40000.	115.903	380.16	130.666	1.127378
63000.	242.070	793.99	272.905	1.127381
80000.	367.063	1203.97	413.821	1.127383
Relative Humidity = 10%				
20.	0.021	0.07	0.024	1.127167
40.	0.064	0.21	0.072	1.127183
50.	0.084	0.28	0.095	1.127191
63.	0.108	0.35	0.122	1.127199
100.	0.161	0.53	0.181	1.127213
200.	0.289	0.95	0.326	1.127225
400.	0.706	2.32	0.796	1.127230
630.	1.501	4.92	1.692	1.127234
800.	2.297	7.54	2.590	1.127238
1250.	5.155	16.91	5.811	1.127254
2000.	11.658	38.24	13.141	1.127287
4000.	31.023	101.76	34.975	1.127386
6300.	47.085	154.44	53.087	1.127462
10000.	61.578	201.98	69.431	1.127522
12500.	68.146	223.52	76.837	1.127540
16000.	76.231	250.04	85.955	1.127555
20000.	85.605	280.78	96.525	1.127563
40000.	151.938	498.36	171.321	1.127576
63000.	277.191	909.19	312.555	1.127581
80000.	403.662	1324.01	455.162	1.127583

ABSORPTION AND VELOCITY OF SOUND IN STILL AIR (*Continued*)

Frequency (Hz)	Absorption (dB/1000 ft)	Absorption (dB/Km)	Absorption (dB/sec)	Velocity (1000 ft/sec)
Relative Humidity = 20%				
20.	0.013	0.04	0.014	1.127568
40.	0.045	0.15	0.051	1.127577
50.	0.066	0.22	0.074	1.127582
63.	0.093	0.30	0.105	1.127587
100.	0.164	0.54	0.185	1.127603
200.	0.285	0.93	0.321	1.127624
400.	0.476	1.56	0.537	1.127633
630.	0.789	2.59	0.890	1.127636
800.	1.103	3.62	1.244	1.127637
1250.	2.277	7.47	2.568	1.127640
2000.	5.310	17.42	5.987	1.127645
4000.	18.991	62.29	21.416	1.127670
6300.	41.151	134.98	46.406	1.127710
10000.	79.657	261.28	89.836	1.127778
12500.	103.004	337.85	116.170	1.127817
16000.	130.465	427.93	147.146	1.127859
20000.	155.809	511.05	175.736	1.127893
40000.	251.952	826.40	284.192	1.127960
63000.	382.062	1253.16	430.957	1.127977
80000.	508.369	1667.45	573.431	1.127982
Relative Humidity = 30%				
20.	0.009	0.03	0.010	1.127976
40.	0.034	0.11	0.038	1.127980
50.	0.051	0.17	0.057	1.127984
63.	0.075	0.25	0.085	1.127987
100.	0.151	0.50	0.170	1.127999
200.	0.309	1.01	0.349	1.128023
400.	0.484	1.59	0.546	1.128037
630.	0.682	2.24	0.770	1.128041
800.	0.868	2.85	0.979	1.128044
1250.	1.552	5.09	1.751	1.128045
2000.	3.333	10.93	3.760	1.128047
4000.	11.856	38.89	13.374	1.128056
6300.	27.626	90.61	31.164	1.128072
10000.	62.493	204.98	70.498	1.128105
12500.	89.659	294.08	101.148	1.128131
16000.	128.814	422.51	145.324	1.128166
20000.	171.847	563.66	193.879	1.128204
40000.	338.710	1110.97	382.172	1.128316
63000.	499.838	1639.47	563.998	1.128361
80000.	635.085	2083.08	716.614	1.128375
Relative Humidity = 40%				
20.	0.007	0.02	0.008	1.128386
40.	0.027	0.09	0.030	1.128388
50.	0.041	0.13	0.046	1.128390
63.	0.062	0.20	0.070	1.128392
100.	0.134	0.44	0.151	1.128402
200.	0.318	1.04	0.359	1.128424
400.	0.524	1.72	0.592	1.128443
630.	0.692	2.27	0.781	1.128449
800.	0.829	2.72	0.935	1.128451
1250.	1.309	4.29	1.477	1.128453
2000.	2.544	8.34	2.870	1.128456
4000.	8.523	27.96	9.618	1.128460
6300.	19.995	65.58	22.564	1.128467
10000.	47.390	155.44	53.478	1.128484
12500.	70.857	232.41	79.962	1.128499
16000.	108.308	355.25	122.228	1.128521
20000.	154.838	507.87	174.742	1.128549
40000.	380.371	1247.62	429.310	1.128663
63000.	596.091	1955.18	672.827	1.128733
80000.	753.514	2471.53	850.536	1.128758
Relative Humidity = 50%				
20.	0.006	0.02	0.006	1.128795
40.	0.022	0.07	0.025	1.128797
50.	0.034	0.11	0.038	1.128798
63.	0.052	0.17	0.058	1.128800
100.	0.117	0.38	0.132	1.128807
200.	0.313	1.03	0.353	1.128826
400.	0.563	1.85	0.636	1.128849

Frequency (Hz)	Absorption (dB/1000 ft)	Absorption (dB/Km)	Absorption (dB/sec)	Velocity (1000 ft/sec)
630.	0.734	2.41	0.828	1.128858
800.	0.851	2.79	0.961	1.128860
1250.	1.231	4.04	1.390	1.128862
2000.	2.176	7.14	2.457	1.128865
4000.	6.752	22.15	7.622	1.128867
6300.	15.643	51.31	17.659	1.128871
10000.	37.521	123.07	42.357	1.128881
12500.	57.009	186.99	64.357	1.128890
16000.	89.594	293.87	101.143	1.128903
20000.	132.719	435.32	149.830	1.128922
40000.	382.722	1255.33	432.101	1.129020
63000.	654.606	2147.11	739.115	1.129099
80000.	844.024	2768.40	953.016	1.129133
Relative Humidity = 60%				
20.	0.005	0.02	0.005	1.129207
40.	0.018	0.06	0.021	1.129208
50.	0.029	0.09	0.032	1.129209
63.	0.044	0.15	0.050	1.129210
100.	0.103	0.34	0.117	1.129215
200.	0.301	0.99	0.339	1.129232
400.	0.593	1.94	0.669	1.129254
630.	0.782	2.57	0.884	1.129266
800.	0.896	2.94	1.012	1.129269
1250.	1.223	4.01	1.381	1.129273
2000.	1.997	6.55	2.256	1.129275
4000.	5.711	18.73	6.449	1.129277
6300.	12.962	42.51	14.638	1.129279
10000.	31.050	101.84	35.064	1.129286
12500.	47.462	155.67	53.598	1.129292
16000.	75.544	247.78	85.312	1.129300
20000.	113.958	373.78	128.694	1.129313
40000.	364.443	1195.37	411.598	1.129389
63000.	677.023	2220.64	764.675	1.129467
80000.	899.912	2951.71	1016.458	1.129508
Relative Humidity = 70%				
20.	0.004	0.01	0.005	1.129618
40.	0.016	0.05	0.018	1.129619
50.	0.025	0.08	0.028	1.129619
63.	0.039	0.13	0.044	1.129620
100.	0.092	0.30	0.104	1.129623
200.	0.284	0.93	0.321	1.129638
400.	0.611	2.01	0.691	1.129662
630.	0.829	2.72	0.937	1.129673
800.	0.947	3.11	1.070	1.129678
1250.	1.250	4.10	1.412	1.129683
2000.	1.915	6.28	2.163	1.129685
4000.	5.056	16.58	5.712	1.129687
6300.	11.197	36.73	12.649	1.129689
10000.	26.624	87.33	30.077	1.129693
12500.	40.758	133.69	46.044	1.129697
16000.	65.246	214.01	73.708	1.129704
20000.	99.365	325.92	112.254	1.129712
40000.	339.153	1112.42	383.165	1.129770
63000.	673.667	2209.63	761.137	1.129842
80000.	924.481	3032.30	1044.556	1.129884
Relative Humidity = 80%				
20.	0.004	0.01	0.004	1.130030
40.	0.014	0.05	0.016	1.130030
50.	0.022	0.07	0.025	1.130031
63.	0.034	0.11	0.039	1.130032
100.	0.082	0.27	0.093	1.130034
200.	0.267	0.88	0.302	1.130047
400.	0.620	2.03	0.701	1.130069
630.	0.870	2.85	0.983	1.130082
800.	0.998	3.27	1.128	1.130088
1250.	1.293	4.24	1.461	1.130094
2000.	1.887	6.19	2.132	1.130096
4000.	4.626	15.17	5.228	1.130098
6300.	9.976	32.72	11.274	1.130100
10000.	23.466	76.97	26.519	1.130102
12500.	35.896	117.74	40.566	1.130106
16000.	57.588	188.89	65.081	1.130110
20000.	88.144	289.11	99.612	1.130116

ABSORPTION AND VELOCITY OF SOUND IN STILL AIR (*Continued*)

Frequency (Hz)	Absorption (dB/1000 ft)	Absorption (dB/Km)	Absorption (dB/sec)	Velocity (1000 ft/sec)
40000.	313.525	1028.36	354.334	1.130161
63000.	655.282	2149.33	740.615	1.130223
80000.	925.566	3035.86	1046.134	1.130264
Relative Humidity = 90%				
20.	0.003	0.01	0.004	1.130443
40.	0.013	0.04	0.014	1.130443
50.	0.019	0.06	0.022	1.130444
63.	0.031	0.10	0.035	1.130444
100.	0.074	0.24	0.084	1.130446
200.	0.250	0.82	0.283	1.130457
400.	0.621	2.04	0.702	1.130478
630.	0.903	2.96	1.021	1.130493
800.	1.045	3.43	1.182	1.130498
1250.	1.344	4.41	1.520	1.130505
2000.	1.892	6.21	2.139	1.130508
4000.	4.337	14.22	4.903	1.130511
6300.	9.099	29.84	10.286	1.130512
10000.	21.133	69.32	23.891	1.130514
12500.	32.258	105.81	36.468	1.130516
16000.	51.762	169.78	58.518	1.130520
20000.	79.424	260.51	89.791	1.130526
40000.	290.117	951.58	327.995	1.130561
63000.	629.701	2065.42	711.948	1.130612
80000.	911.286	2989.02	1030.346	1.130651
Relative Humidity = 100%				
20.	0.003	0.01	0.003	1.130856
40.	0.011	0.04	0.013	1.130856
50.	0.018	0.06	0.020	1.130857
63.	0.028	0.09	0.031	1.130857
100.	0.068	0.22	0.076	1.130858
200.	0.234	0.77	0.265	1.130868
400.	0.616	2.02	0.696	1.130888
630.	0.928	3.05	1.050	1.130902
800.	1.087	3.57	1.230	1.130909
1250.	1.399	4.59	1.582	1.130917
2000.	1.917	6.29	2.168	1.130921
4000.	4.140	13.58	4.682	1.130924
6300.	8.451	27.72	9.557	1.130925
10000.	19.357	63.49	21.891	1.130926
12500.	29.461	96.63	33.318	1.130929
16000.	47.227	154.90	53.410	1.130931
20000.	72.538	237.93	82.036	1.130935
40000.	269.598	884.28	304.905	1.130963
63000.	601.714	1973.62	680.543	1.131007
80000.	888.113	2913.01	1004.492	1.131042

VELOCITY OF SOUND IN DRY AIR

Data in this table apply only to dry air. These data have been calculated with air being treated as a perfect gas.

Temp °C	0 m sec^{-1}	1 m sec^{-1}	2 m sec^{-1}	3 m sec^{-1}	4 m sec^{-1}	5 m sec^{-1}	6 m sec^{-1}	7 m sec^{-1}	8 m sec^{-1}	9 m sec^{-1}
60	366.05	366.60	367.14	367.69	368.24	368.78	369.33	369.87	370.42	370.96
50	360.51	361.07	361.62	362.18	362.74	363.29	363.84	364.39	364.95	365.50
40	354.89	355.46	356.02	356.58	357.15	357.71	358.27	358.83	359.39	359.95
30	349.18	349.75	350.33	350.90	351.47	352.04	352.62	353.19	353.75	354.32
20	343.37	343.95	344.54	345.12	345.70	346.29	346.87	347.44	348.02	348.60
10	337.46	338.06	338.65	339.25	339.84	340.43	341.02	341.61	342.20	342.78
0	331.45	332.06	332.66	333.27	333.87	334.47	335.07	335.67	336.27	336.87
−10	325.33	324.71	324.09	323.47	322.84	322.22	321.60	320.97	320.34	319.72
−20	319.09	318.45	317.82	317.19	316.55	315.92	315.28	314.64	314.00	313.36
−30	312.72	312.08	311.43	310.78	310.14	309.49	308.84	308.19	307.53	306.88
−40	306.22	305.56	304.91	304.25	303.58	302.92	302.26	301.59	300.92	300.25
−50	299.58	298.91	298.24	297.56	296.89	296.21	295.53	294.85	294.16	293.48
−60	292.79	292.11	291.42	290.73	290.03	289.34	288.64	287.95	287.25	286.55
−70	285.84	285.14	284.43	283.73	283.02	282.30	281.59	280.88	280.16	279.44
−80	278.72	278.00	277.27	276.55	275.82	275.09	274.36	273.62	272.89	272.15
−90	271.41	270.67	269.92	269.18	268.43	267.68	266.93	266.17	265.42	264.66

SPARK-GAP VOLTAGES

Based on results of the American Institute of Electric Engineers Air at 760 mm, 25°C.

Peak voltage, kilovolts	Diameter of spherical electrodes, cm: 2.5	5	10	25	Needle points
	Length of spark gap cm				
5	0.13	0.15	0.15	0.16	0.42
10	0.27	0.29	0.30	0.32	0.85
15	0.42	0.44	0.46	0.48	1.30
20	0.58	0.60	0.62	0.64	1.75
25	0.76	0.77	0.78	0.81	2.20
30	0.95	0.94	0.95	0.98	2.69
35	1.17	1.12	1.12	1.15	2.69
40	1.41	1.30	1.29	1.32	3.81
45	1.68	1.50	1.47	1.49	4.49
50	2.00	1.71	1.65	1.66	5.20
60	2.82	2.17	2.02	2.01	6.81
70	4.05	2.68	2.42	2.37	8.81
80	—	3.26	2.84	2.74	11.1
90	—	3.94	3.28	3.11	13.3
100	—	4.77	3. 75	3.49	15.5
110	—	5.79	4.25	3.88	17.7
120	—	7.07	4.78	4.28	19.8
130	—	—	5.35	4.69	22.0
140	—	—	5.97	5.10	24.1
150	—	—	6.64	5.52	26.1
160	—	—	7.37	5.95	28.1
170	—	—	8.16	6.39	30.1
180	—	—	9.03	6.84	32.0
190	—	—	10.0	7.30	33.9
200	—	—	11.1	7.76	35.7
210	—	—	12.3	8.24	37.6
220	—	—	13.7	8.73	39.5
230	—	—	15.3	9.24	41.4
240	—	—	—	9.76	43.4
250	—	—	—	10.3	45.2
300	—	—	—	13.3	54.7

CORRECTIONS FOR TEMPERATURE AND PRESSURE

Values found in the above table may be corrected for temperature and pressure by multiplying the values given by the appropriate correction factor found below:

Temp. °C	Pressure mm: 720	740	760	780
0	1.04	1.06	1.09	1.12
10	1.00	1.02	1.05	1.08
20	0.96	0.99	1.02	1.04
30	0.93	0.96	0.98	1.01

DIELECTRIC CONSTANTS

Dielectric Constants of Pure Liquids

The values listed in the following table were obtained from National Bureau of Standards Circular 514

The dielectric constants are intended to be the limiting values at low frequencies, the so-called static values. Temperature is the only variable considered explicitly. Usually pressure is atmospheric or insignificantly different with respect to its effect on the dielectric constant. Where data are listed above the normal boiling point, the pressure corresponds to the vapor pressure of the liquid unless otherwise noted in the footnote.

Symbols

- ε = dielectric constant (ε vacuum −1)
- t = temperature, Centigrade (°C)
- T = temperature, Absolute (°K)
- a = $-d\varepsilon/dt$
- α = $-d\log_{10}\varepsilon \cdot dt$
- t_1, t_2 = the limits of temperature between which a or α is considered applicable
- mp = melting point
- bp = boiling point
- f = frequency of alternating current in cycles per second

Standard Liquids

Formula	Liquid	ε20°C	ε25°C	a(or α)*
C_6H_{12}	Cyclohexane	2.023	2.015	0.0016
CCl_4	Carbon tetrachloride	2.238	2.228	0.0020
C_6H_6	Benzene	2.284	2.274	0.0020
C_6H_5Cl	Chlorobenzene	5.708	5.621	0.00133α
$C_2H_4Cl_2$	1,2-Dichloroethane	10.65	10.36	0.00240(α)
CH_4O	Methanol	33.62	32.63	0.00260(α)
$C_6H_5NO_2$	Nitrobenzene	35.74	34.82	0.00225(α)
H_2O	Water	80.37	78.54	0.00200(α)
H_2	Hydrogen	1.228 at 20.4°K		0.0034
O_2	Oxygen	1.507 at 80.0°K		0.0024

* The values of a or α given in this table were derived from data in the vicinity of room temperature and are not necessarily identical with the values listed in the following sections of this table. They may be used to calculate values of dielectric constant between 15° and 30°C without introducing significant error.

Dielectric Constants of Pure Liquids

Inorganic Liquids

	Substance	ε	°C	a(or α) × 10²	Range t_1, t_2
A	Argon	1.53_8	−191	0.34	−191.−184
$AlBr_3$	Aluminum bromide	3.38	100	0.33	100,240
AsH_3	Arsine	2.50	−100	0.43	−116,−72
BBr_3	Boron bromide	2.58	0	0.28	−70,80
Br_2	Bromine	3.09	20	0.7	0,50
CO_2	Carbon dioxide	1.60[c]	20	—	—
Cl_2	Chlorine	2.10_1	−50	0.31	−65,−33
		1.91	14	0.32	−22,14
		1.7_2	77		
		1.5_4	142		
D_2	Deuterium	1.277	20°K	0.4	18.8,21.2° K
D_2O	Deuterium oxide	78.25	25	[d]	0.4,98
F_2	Fluorine	1.54	−202	0.19	−216,−190
$GeCl_4$	Germanium tetrachloride	2.43[a]	25	0.240	0.55
HBr	Hydrogen bromide	7.00	−85	0.26(α)	−85,−70
HCl	Hydrogen chloride	6.35	−15	0.288(α)	−85,−15
		12.	−113	—	—
		4.6	28	—	—
HF	Hydrogen fluoride	17_5	−73	—	—
		13_4	−42		
		11_1	−27		
		84	0		
HI	Hydrogen iodide	3.39	−50	0.8	−51,−37
H_2	Hydrogen	1.228	20.4°K	0.34	14,21°K
H_2O	Water	78.54	25	[e]	0,100
		34.5_9	200	[f]	100,370
H_2O_2	Hydrogen peroxide	84.2	0	[f]	−30,20
H_2S	Hydrogen sulfide	9.26	−85.5	—	—
		9.05	−78.5	—	—
He	Helium	1.055_4	2.06°K	—	—
		1.055_9	2.30[f]		
		1.055_3	2.63		
		1.053_9	3.09		
		1.051_8	3.58		
		1.048	4.19		
I_2	Iodine	6.8	400	—	ε_r = 13.3—0.0 0.016 (±0.002) T
		$11._7$	140		
		$13._0$	168		
NH_3	Ammonia	25.	−77.7	—	—
		22.4	−33.4	—	—
		18.9	5	—	—
		17.8	15		
		16.9	25		
		16.3	35		
NOBr	Nitrosyl bromide	$13._4$	15	—	—
NOCl	Nitrosyl chloride	$18._2$	12	—	—
N_2	Nitrogen	1.454	−203	0.29	−210,−195
N_2H_4	Hydrazine	$52._9$	20	0.21(α)	0,25
N_2O	Dinitrogen oxide	1.97	−90	—	—
		1.61	0	0.6	−6,14
N_2O_4	Dinitrogen tetroxide	2.5[b]	15	—	—
O_2	Oxygen	1.507	−193	0.24	−218,−183
P	Phosphorus	4.10	34	—	—
		4.06	46		
		3.86	85		
PCl_3	Phosphorus trichloride	3.43	25	0.84	17,60
PCl_4	Phosphorus pentachloride	2.8_4	160	—	—
$POCl_3$	Phosphoryl chloride	$13._3$	22	—	—
$PSCl_3$	Thophosphoryl chloride	5.8	22	—	—
$PbCl_4$	Lead tetrachloride	2.78	20	—	—
S	Sulfur	3.52	118	[g]	—
		3.48	231		
$SOBr_2$	Thionyl bromide	9.06	20	3.0	at 20
$SOCl_2$	Thionyl chloride	9.25	20	3.9	at 20
SO_2	Sulfur dioxide	17.6	−20	0.287(α)	−65,−15
		15.0_8	0	—	—
		$14._1$	20	7.7	14,140
		2.1_0	154[h]		
SO_2	Sulfur trioxide	3.11	18	—	—
S_2Cl_2	Sulfur monochloride	4.79	15	0.146(α)	−41,15
SO_2Cl_2	Sulfuryl chloride	$10._0$	22	—	—
$SbCl_5$	Antimony pentachloride	3.22	20	0.46	2,47
Se	Selenium	5.40	250	0.25	237,301
$SiCl_4$	Silicon tetrachloride	2.4[a]	16	—	—
$SnCl_4$	Tin tetrachloride	2.87	20	0.30	−30,20
$TiCl_4$	Titanium tetrachloride	2.80	20	0.20	−20,20

Organic Liquids

	Substance	ε	°C	a(or α) × 10²	Range t_1, t_2
CCl_4	Carbon tetrachloride	2.238	20	0.200	−10,60
CN_2O_8	Tetranitromethane	2.52_1	25	—	—
CO_2	Carbon dioxide	1.60_4[c]	0	—	—
CS_2	Carbon disulfide	2.641	20	0.268	−90,130
		3.001	−110		
		2.19	180		
$CHBr_2$	Bromoform	4.39	20	0.105(α)	10,70
$CHCl_3$	Chloroform	4.806	20	0.160(α)	0,50
		6.76	−60	—	—
		6.12	−40		
		5.61	−20		
		3.7_1	100	—	—
		3.3_2	140		
		2.9_2	180		
CHN	Hydrocyanic acid	$158._1$	0	[i]	−13,18
		$114._9$	20	0.63(α)	18,26
CH_2Br_2	Dibromomethane	7.77	10	—	—
		6.68	40		
CH_2Cl_2	Dichloromethane	9.08	20	[j]	−80,25
CH_2I_2	Diiodomethane	5.32	25	—	—
CH_1O_2	Formic acid	$58._5$[a]	16	—	—
CH_2Br	Bromomethane	9.82	0	[k]	−80,0
CH_2Cl	Chloromethane	12.6	−20	[l]	−70,−20
CII_3I	Iodomethane	7.00	20	[m]	−70,40
CH_3NO	Formamide	109.	20	72.	18,25
CH_4	Methane	1.70	−173	0.2	−181,−159
CH_4O	Methanol	32.63	25	0.264(α)	5,55
		64.	−113	—	—
		54.	−80		
		40.	−20		
CH_5N	Methylamine	11.4	−10	0.26(α)	−30,−10
		9.4	25	—	—
C_2					
C_2HCl_3	Trichloroethylene	3.4_2	ca 16	—	—
C_2HCl_3O	Chloral	4.9_4	20	0.17(α)	15,45
		7.6	−40		
		4.2	62		
$C_2HCl_3O_2$	Trichloroacetic acid	4.6	60	—	—
$C_2H_2Br_2$	*cis*-1,2Dibromoethylene	7.7_2	0	—	—
		7.0_5	25		
	trans-1,2-Dibromoethylene	2.9_7	0	—	—
		2.8_8	25		
$C_2H_2Br_4$	1,1,2,3-Tetrabromoethane	8.6	3	—	—
		7.0	22		
$C_2H_2Cl_2$	1,1-Dichloroethylene	4.6_7	16	—	—
	cis-1,2-Dichloroethylene	9.20	25	—	—
	trans-1,2-Dichloroethylene	2.14	25	—	—
$C_2H_2Cl_2O_2$	Dichloroacetic acid	8.2	22	—	—
		7.8	61		
C_2H_3ClO	Acetyl chloride	$16._9$	2	—	—
		$15._8$	22		
$C_2H_3ClO_2$	Chloroacetic acid	12.3	60	2.	60,80
C_2H_4O	Ethylene oxide	$13._9$	−1	—	—
	Acetaldehyde	$21._8$[a]	10	—	—
$C_2H_4O_2$	Acetic acid	6.15	20	—	—
		6.29	40		
		6.62	70		
	Methyl formate	8.5	20	5.	0,20
C_2H_4ClO	2-Chloroethanol (ethylene chlorohydrin)	$25._8$	25	—	—
		$13._2$	132		
C_3H_5NO	Acetamide	50.[a]	83	—	—
$C_2H_5NO_2$	Nitroethane	28.0_6	30	11.4	30,35
C_2H_6O	Ethanol	24.30	25	—	—
		24.3[5]	25	0.270(α)	−5,70
		41.0[5]	−60	0.297(α)	−110,−20
	Methyl ether	5.02	25	2.38	25,100
		2.97	110		
		2.64	120		
		2.37	125		
		2.26	126.1		
		1.90	127.6		
$(C_2H_6OSi)_n$ $n = 4$	Octamethylcyclotetrasiloxane	2.39	20	—	—

Dielectric Constants of Pure Liquids

Organic Liquids

	Substance	ε	°C	a(or α) × 10^2	Range t_1, t_2
$n = 5$	Decamethylcyclopentasiloxane	2.50	20	—	—
$n = 6$	Dodecamethylcyclohexasiloxane	2.59	20	—	—
$n = 7$	Tetradecamethylcycloheptasiloxane	2.68	20	—	—
$n = 8$	Hexadecamethylcyclooctasiloxane	2.74	20	—	—
$C_2H_6O_2$	Glycol	$37._7$	25	0.224(α)	20,100
C_2H_7N	Dimethylamine	6.32	0	—	—
	C_3	5.26	25		
C_3H_6	Propene	1.87_5	20	—	—
		1.79^5	45		
		1.69	65		
		1.53_0	85		
		1.44_1	90		
		1.33_1	91.9[h]		
C_3H_6O	2-Propen-1-ol(Allylalcohol)	$21._6$	15	—	—
	Acetone	20.7_0	25	0.205(α)	−60,40
		17.7	56	—	—
	Propionaldehyde	$18._5$[a]	17	—	—
$C_3H_6O_2$	Propionic acid	3.30	10	—	—
		3.44	40		
	Ethyl formate	7.1_6	25	—	—
	Methyl acetate	6.68	25	2.2	25,40
$C_3H_6O_3$	*dl*-Lactic acid	22.	17	—	—
$C_3H_6NO_2$	Ethyl carbamate (Urethan)	14.2	50	5.2	50,70
C_3H_8	Propane	1.61	0	0.20	−90,15
C_3H_8O	1-Propanol	20.1	25	0.293(α)	20,90
		38.	−80	—	—
		29.	−34		
	2-Propanol	18.3	25	0.310(α)	20,70
$C_3H_8O_2$	1,2-Propanediol	$32._0$	20	0.27(α)	at 20
	1,3-Propanediol	$35._0$	20	0.23(α)	at 20
$C_3H_8O_3$	Glycerol	42.5	25	0.208(α)	0,100
C_3H_9N	Isopropylamine	5.5[h]	20	—	—
	Trimethylamine	2.44	25	0.52	0,25
	C_4				
$C_4H_2O_3$	Maleic anhydride	50[a]	60	—	—
C_4H_4O	Furan	2.95	25	—	—
C_4H_4S	Thiophene	2.76	16	—	—
C_4H_5N	Pyrrole	7.48	18	—	—
C_4H_5NS	Allyl isothiocyanate	$17._2$[b]	18	—	—
C_4H_6O	Vinyl ether	3.94	20	—	—
$C_4H_6O_3$	Acetic anhydride	$22.^4$	1	—	—
		$20._7$	19		
C_4H_8O	2-Butanone	18.5_1	20	0.207(α)	−60,60
	Butyraldehyde	13.4	26	—	—
		10.8	77		
$C_4H_8O_2$	Butyric acid	2.97	20	−0.23	10,70
	Isobutyric acid	2.71	10	—	—
		2.73	40		
	Propyl formate	7.7_2[a]	19	—	—
	Ethylacetate	6.02	25	1.5	at 25
		5.3_0	77	—	—
	Methyl propionate	5.5[n]	19	—	—
	1,4-Dioxane	2.209	25	0.170	20,50
C_4H_9NO	Morpho line	7.33	25	—	—
$C_4H_{10}O$	1-Butanol	17.8	20	0.300(α)	−40,20
		17.1	25	0.335(α)	25,70
		8.2	118	—	—
	2-Methyl-1-propanol	17.7	25	0.377(α)	20,90
		34.	−80	—	—
		26.	−34		
	2-Butanol	15.8	25	—	—
	2-Methyl-2-propanol	10.9	30	—	—
		8.49	50		
		6.89	70		
	Ethyl ether	4.335	20	2.0	at 20
		4.34^5	20	0.217(α)	−40,30
		10.4	−116	—	
		3.97	40	0.170(α)	40,140
		2.1_2	180		
		1.8^9	190		
		1.5_3	193.3[h]		
$C_4H_{10}Zn$	Diethyl zinc	2.5_5	20	—	—
	C_5				
$C_5H_4O_2$	Furfural	$46.^9$	1	—	—
		$41._9$	20		
		$34._9$	50		
C_5H_5N	Pyridine	12.3	25	—	—
		9.4	116	—	—
C_5H_8	1,3-Pentadiene[r]	2.32	25	—	—
	2-Methyl-1,3-butadiene (Isoprene)	2.10	25	0.24	−75,25
$C_5H_8O_2$	2,4-Pentanedione (Acetylacetone)	$25._7$[a]	20	—	—
C_5H_{10}	1-Pentene	2.100	20	—	—
	2-Methyl-1-butene	2.197	20	—	—
	Cyclopentane	1.965	20	—	—
	Ethylcyclopropane	1.933	20	—	—
$C_5H_{10}O$	2-Pentanone	15.4_5	20	0.195(α)	−40.80
		22.0	−60		
	3-Pentanone	17.0_0	20	0.225(α)	0,80
		19.4	−20		
		19.8	−40		
$C_5H_{10}O_2$	Valeric acid	2.6_6	20	—	—
	Isovaleric acid	2.6_4	20	—	—
	Methyl butyrate	5.6[n]	20	—	—
$C_5H_{11}N$	Piperidine	5.8[b]	22	—	—
C_5H_{12}	*n*-Pentane	1.844	20	0.160	−50,30
		2.011	−90		
		1.984	−70		
	2-Methylbutane	1.843	20	—	—
$C_5H_{12}O$	1-Pentanol	13.9	25	0.23(α)	15,35
	3-Methyl-1-butanol	14.7	25	—	—
		5.8^2	132		
	2-Methyl-2-butanol	5.82	25	—	—
	C_6				
C_6H_4Cl	*o*-Dichlorobenzene	9.93	25	0.194(α)	0.50
	m-Dichlorobenzene	5.04	25	0.120(α)	0,50
	p-Dichlorobenzene	2.41	50	0.18	50,80
C_6H_5Br	Bromobenzene	5.40	25	0.115(α)	0,70
C_6H_5Cl	Chlorobenzene	5.708	20	—	—
		5.621	25		
		5.71	20	0.130(α)	0,80
		7.28	−50		
		6.30	−20		
		4.21	130		
C_6H_5Cl	*o*-Chlorophenol	6.31	25	2.7	25,58
	p-Chlorophenol	9.47	55	3.7	55,65
C_6H_5I	Iodobenzene	4.63	20	—	—
$C_6H_5NO_2$	Nitrobenzene	34.82	25	0.225(α)	10,80
		20.8	130	0.164(α)	130,211
		24.9	90		
		22.7	110		
$C_6H_5NO_3$	*o*-Nitrophenol	$17._3$	50	6.4	50,60
C_6H_6	Benzene	2.284	20	0.200	10,60
		2.073	129	—	—
		1.966	182		
C_6H_6BrN	*m*-Bromoaniline	$13._0$[n]	19	—	—
C_6H_6ClN	*m*-Chloroaniline	$13.^4$[n]	19	—	—
$C_6H_6N_2O_2$	*o*-Nitroaniline	$34._5$	90	3.	90,110
	p-Nitroaniline	$56._3$	160	6.	160,180
C_6H_6O	Phenol	9.78	60	0.32(α)	40,70
C_6H_7N	Aniline	6.89	20	0.148(α)	0,50
		5.93	70	—	—
		4.54	184.6	—	—
	2-Methylpyridine (α-Picoline)	9.8[b]	20	—	—
$C_6H_8N_2$	Phenylhydrazine	7.2	23	—	—
C_6H_{10}	Cyclohexene	2.220	25	—	—
		2.6_0	−105	—	—
$C_6H_{10}O$	Cyclohexanone	18.3	20	—	—
		$19._9$	−40	—	—
$C_6H_{10}O$	Ethyl acetoacetate	$15._7$[a]	22	—	—
C_6H_{12}	Cyclohexane	2.023	20	0.160	10,60
	Methylcyclopentane	1.985	20	—	—
	Ethylcyclobutane	1.965	20	—	—
$C_6H_{12}O$	Cyclohexanol	15.0	25	0.437(α)	20,66
		7.2_4	100		
		4.8_8	150		
$C_6H_{12}O_2$	Butyl acetate	5.01	20	1.4	20,40
		6.8_4	−73		
	Ethyl butyrate	5.10	18	1.0	at 20
$C_6H_{12}O_3$	Paraldehyde	13.9	25	—	—
		6.29	128		
C_6H_{14}	*n*-Hexane	1.890	20	0.155	−10,50
		2.044	−90		
		1.990	−50		

Dielectric Constants of Pure Liquids

Organic Liquids

	Substance	ε	°C	a (or α) × 10^2	Range t_1, t_2
$C_6H_{14}O$	1-Hexanol	13.3	25	0.35(α)	15,35
		8.5_6	75		
	Propyl ether	3.3_9	26	—	—
	Isopropyl ether	3.88	25	1.8	0,25
$C_6H_{15}Al$	Triethyl aluminum	2.9	20	—	—
$C_6H_{15}N$	Dipropylamine	2.9[b]	21	—	—
	Triethylamine	2.42	25	—	—
$C_6H_{15}N$	Dipropylamine	2.9[b]	21	—	—
	Triethylamine	2.42	25	—	—
$C_6H_{18}OSi_2$	$(CH_3)_3Si[OSi(CH_3)_2]nCH_3$				
n = 1	Hexamethyldisiloxane	2.17	20	—	—
n = 2	Octamethyltrisiloxane	2.30	20	—	—
n = 3	Decamethyltetrasiloxane	2.39	20	—	—
n = 4	Dodecamethylpentasiloxane	2.46	20	—	—
n = 5	Tetradecamethylhexasiloxane	2.50	20	—	—
n = 66[w]		2.72	20	—	—
	C_7				
C_7H_5ClO	Benzoyl chloride	29.	0	—	—
		23.	20		
C_7H_5NO	Phenyl isocyanate	8.8[b]	20	—	—
C_7H_5NS	Phenyl isothiocyanate	$10._4$[e]	20	—	—
C_7H_6O	Benzaldehyde	$19._7$	0	—	—
		$17._8$	20		
$C_7H_6O_2$	Salicylaldehyde	$17._1$	30	7.	30,40
C_7H_7Br	*o*-Bromotoluene	4.23	58	—	—
	m-Bromotoluene	5.36	58	—	—
	p-Bromotoluene	5.49	58	—	—
C_7H_7Cl	*o*-Chlorotoluene	4.45	20	—	—
		4.16	58	—	—
	m-Chlorotoluene	5.55	20	—	—
		5.04	58	—	—
	p-Chlorotoluene	6.08	20	—	—
		5.55	58	—	—
	α-Chlorotoluene	7.0	13	—	—
$C_7H_7NO_2$	*o*-Nitrotoluene	27.4	20	15.	at 20
		$21._6$	58	—	—
		11.8	222	—	—
	m-Nitrotoluene	$23._8$	20	—	—
		$21._9$	58	—	—
	p-Nitrotoluene	$22._2$	58	—	—
C_7H_8	Toluene	2.438	0	0.0455(α)	−90,0
		2.379	25	0.243	0,90
		2.15_7	127		
		2.04_2	181		
C_7H_8O	Benzyl alcohol	13.1	20	—	—
		9.47	70	—	—
		6.6	132	—	—
	o-Cresol	11.5	25	11	25,30
	m-Cresol	11.8	25	0.41(α)	15,50
	p-Cresol	9.9_1	58	—	—
C_7H_9N	*o*-Toluidine	6.34	18	—	—
		5.71	58	—	—
		4.00	200	—	—
	m-Toluidine	5.95	18	—	—
		5.45	58	—	—
	p-Toluidine	4.98	54	—	—
	N-Methylaniline	5.97	22	—	—
	1-Heptene	2.05	20	—	—
	C_8				
C_8H_8	Styrene (Phenylethylene)	2.43	25	—	—
		2.32	75		
	Acetophenone	17.39	25	4.	at 25
		8.64	202	—	—
$C_8H_8O_2$	Phenyl acetate	5.23	20	0.7	at 20
	Methyl benzoate	6.59	20	0.14(α)	20,50
$C_8H_8O_3$	Methyl salicylate	9.41	30	3.1	30,40
C_8H_{10}	Ethylbenzene	2.412	20	—	—
	o-Xylene	2.568	20	0.266	−20,130
	m-Xylene	2.374	20	0.195	−40,180
	p-Xylene	2.270	20	0.160	20,130
C_8H_{18}	*n*-Octane	1.948	20	0.130	−50,50
		1.879	70		
		1.817	110		
	2,2,3-Trimethylpentane	1.96	20	—	—
	2,2,4-Trimethylpentane	1.940	20	0.142	−100,100
$C_8H_{18}O$	1-Octanol	10.3_4	20	0.410(α)	20,60
$C_8H_{20}O_4Si$	Tetraethyl silicate	4.1[b]	ca 20	—	—
	C_9				
C_9H_7N	Quinoline	9.00	25	—	—
		5.05	238		
	Isoquinoline	10.7	25	—	—
C_9H_8O	Cinnamaldehyde	16.9	24	—	—
$C_9H_{10}O_2$	Benzyl acetate	5.1[n]	21	—	—
	Ethyl benzoate	6.02	20	2.1	20,40
$C_9H_{10}O_3$	Ethyl salicylate	7.99	30	2.	30,40
C_9H_{12}	Isopropylbenzene (Cumene)	2.38_0	20	—	—
	1,3,5-Trimethylbenzene (Mesitylene)	2.27_9	20	—	—
C_9H_{20}	*n*-Nonane	1.972	20	0.135	−10,90
		2.059	−50		
		1.847	110		
		1.787	150		
	C_{10}				
$C_{10}H_8$	Naphthalene	2.54	85	—	—
$C_{10}H_{10}O_4$	Dimethyl phthalate	8.5	24	—	—
$C_{10}H_{16}$	*d*-Camphene	2.33	ca 40	—	—
	d-Pinene	2.64	25	—	—
	l-Pinene	2.76	20	—	—
	Terpinene	2.7[b]	21	—	—
	d-Limonene	2.3_6	20	—	—
	d-Limonene (Dipentene)	2.3[o]	20	—	—
$C_{10}H_{22}$	*n*-Decane	1.991	20	0.130	10,110
		2.050	−30		
		1.844	130		
		1.783	170		
$C_{10}H_{22}$	1-Decanol	8.1	20	—	—
$C_{11}H_{24}$	*n*-Undecane	2.005	20	0.125	10,130
		2.039	−10		
		1.838	150		
		1.781	190		
	C_{12}				
$C_{12}H_{10}$	Diphenyl	2.53	75	0.18	75,155
$C_{12}H_{10}O$	Phenyl ether	3.65	30	0.7	30,50
$C_{12}H_{11}N$	Diphenylamine	3.3	52	—	—
$C_{12}H_{26}$	*n*-Dodecane	2.015	20	0.120	10,150
		2.047	−10		
		1.776	210		
$C_{13}H_{10}$	Benzophenone	11.4	50	—	—
	C_{14}				
$C_{14}H_{15}N$	Dibenzylamine	3.6[b]	20	—	—
	C_{16}				
$C_{16}H_{32}O_2$	Palmitic acid	2.30	71	—	—
	C_{18}				
$C_{18}H_{32}O_2$	Linoleic acid	2.61	0	—	—
		2.71	20		
		2.70	70		
		2.60	120		
$C_{18}H_{34}O_2$	Oleic acid	2.46	20	—	—
		2.45	60		
		2.41	100		
$C_{18}H_{36}O_2$	Stearic acid	2.29	70	—	—
		2.26	100	—	—
	Ethyl palmitate	3.20	20	0.4	20,40
		2.71	104	—	—
		2.46	182		
	C_{19}				
$C_{19}H_{16}$	Triphenylmethane	2.45	100	0.14	94,175
$C_{19}H_{38}O_4$	Monopalmitin	5.34	67	—	—
		5.09	80		
	C_{20}				
$C_{20}H_{38}O_2$	Ethyl Oleate	3.17	28	0.48	28,122
$C_{20}H_{40}O_2$	Ethyl Stearate	2.98	40	0.6	32,50
		2.69	100		
		2.48	167		
	C_{21}				
$C_{21}H_{21}O_4P$	Tricresyl phosphate	6.9	40	—	—
	C_{22}				
$C_{22}H_{42}O_3$	Butyl oleate	4.0	25	—	—
$C_{22}H_{44}O_2$	Butyl stearate	3.11_1	30	0.53	30,35

Dielectric Constants of Pure Liquids

[a] $f = 4 \times 10^8$ cycles/sec.
[b] $f = 3.6 \times 10^8$ cycles/sec.
[c] At pressure of 50 atmospheres.
[d] $\varepsilon = 78.25\,[1 - 4.617\,(10^{-3})\,t - 25) + 1.22\,(10^{-5})\,(t-25)^2 - 1.7\,(10^{-8})\,(t-25)^3]$; average deviation ± 0.04%.
[e] $\varepsilon = 78.54\,[1 - 4.579\,(10^{-3})\,(t-25) + 1.19\,(10^{-5})\,(t-25)^2 - 2.8\,(10^{-8})\,(t-25)^3]$; average deviation ± 0.03%.
[f] Liquid transition and discontinuity in variation of dielectric constant with temperature at 2.295°K.
[g] Graphical data in the range 118—350°C show a miniumum near 160° and a broad maximum near 200°.
[h] Critical temperature.
[i] $\text{Log}_{10}\,\varepsilon = 2.199 - 0.0079\,t + 0.00005\,t^2$.
[j] $\varepsilon = (3320/T) - 2.24$.
[k] $\varepsilon = (3320/T) - 2.34$.
[l] $\varepsilon = 12.6 - 0.061\,(t+20) + 0.0005\,(t+20)^2$.
[m] $\varepsilon = (2160/T) - 0.39$.
[n] $f = 5 \times 10^8$ cycles/sec.
[p] Critical temperature = 126.9°C.
[s] $\varepsilon = 84.2 - 0.62\,t + 0.0032\,t^2$.
[t] Mixture of *cis-trans* isomers.
[w] Silicone oil of average molecular weight corresponding to this formula.
[x] Value chosen to conform with the remainder of the tabulated data for this substance.
[y] $\varepsilon = 5321/T + 233.76 - 0.9297\,T + 0.001417\,T^2 - 0.0000008292\,T^3$.

Dielectric Constants of Solids
Compiled by Earle C. Gregg, Jr.

Solids[a] (17 to 22°C)

Material	Frequency	Dielectric constant
Acetamide	4×10^8	4.0
Acetanilide	—	2.9
Acetic acid (2°C)	4×10^8	4.1
Aluminum oleate	4×10^8	2.40
Ammonium bromide	10^8	7.1
Ammonium chloride	10^8	7.0
Antimony trichloride	10^8	5.34
Apatite ⊥ optic axis	3×10^8	9.50
Apatite ∥ optic axis	3×10^8	7.41
Asphalt	$<3 \times 10^6$	2.68
Barium chloride (anhyd.)	6×10^7	11.4
Barium chloride ($2H_2O$)	6×10^7	9.4
Barium nitrate	6×10^7	5.9
Barium sulfate (15°C)	10^8	11.4
Beryl ⊥ optic axis	10^4	7.02
Beryl ∥ optic axis	10^4	6.08
Calcite ⊥ optic axis	10^4	8.5
Calcite ∥ optic axis	10^4	8.0
Calcium carbonate	10^6	6.14
Calcium fluoride	10^4	7.36
Calcium sulfate ($2H_2O$)	10^4	5.66
Cassiterite ⊥ optic axis	10^{12}	23.4
Cassiterite ∥ optic axis	10^{12}	24
d-Cocaine	5×10^8	3.10
Cupric oleate	4×10^8	2.80
Cupric oxide (15°C)	10^8	18.1
Cupric sulfate (anhyd.)	6×10^7	10.3
Cupric sulfate ($5H_2O$)	6×10^7	7.8
Diamond	10^8	5.5
Diphenylmethane	4×10^8	2.7
Dolomite ⊥ optic axis	10^8	8.0
Dolomite ∥	10^8	6.8
Ferrous oxide (15°C)	10^8	14.2
Iodine	10^8	4
Lead acetate	10^6	2.6
Lead carbonate (15°C)	10^8	18.6
Lead chloride	10^6	4.2
Lead monoxide (15°C)	10^8	25.9
Lead nitrate	6×10^7	37.7
Lead oleate	4×10^8	3.27
Lead sulfate	10^6	14.3
Lead sulfide (15°)	16^6	17.9
Malachite (mean)	10^{12}	7.2
Mercuric chloride	10^6	3.2
Mercurous chloride	10^6	9.4
Naphthalene	4×10^8	2.52
Phenanthrene	4×10^8	2.80
Phenol (10°C)	4×10^8	4.3
Phosphorus, red	10^8	4.1
Phosphorus, yellow	10^8	3.6
Potassium aluminum sulfate	10^6	3.8
Potassium carbonate (15°C)	10^8	5.6
Potassium chlorate	6×10^7	5.1
Potassium chloride	10^4	5.03
Potassium chromate	6×10^7	7.3
Potassium iodide	6×10^7	5.6
Potassium nitrate	6×10^7	5.0
Potassium sulfate	6×10^7	5.9
Quartz ⊥ optic axis	3×10^7	4.34
Quartz ∥ optic axis	3×10^7	4.27
Resorcinol	4×10^8	3.2
Ruby ⊥ optic axis	10^4	13.27
Ruby ∥ optic axis	10^4	11.28
Rutile ⊥ optic axis	10^8	86
Rutile ∥ optic axis	10^8	170
Selenium	10^8	6.6
Silver bromide	10^6	12.2
Silver chloride	10^6	11.2
Silver cyanide	10^6	5.6
Smithsonite ⊥ optic axis	10^{12}	9.3
Smithsonite ∥ optic axis	10^{10}	9.4
Sodium carbonate (anhyd.)	6×10^7	8.4
Sodium carbonate ($10H_2O$)	6×10^7	5.3
Sodium chloride	10^4	6.12
Sodium nitrate	—	5.2
Sodium oleate	4×10^8	2.75
Sodium perchlorate	6×10^7	5.4
Sucrose (mean)	3×10^8	3.32
Sulfur (mean)	—	4.0
Thallium chloride	10^6	46.9
p-Toluidine	4×10^8	3.0
Tourmaline ⊥ optic axis	10^4	7.10
Tourmaline ∥ optic axis	10^4	6.3
Urea	4×10^8	3.5
Zircon ⊥, ∥	10^8	12

[a] For plastics and other insulating materials, refer to table on Properties of Dielectrics.

DIELECTRIC CONSTANTS (Continued)

Table of Dielectric Constants of Reference Gases at 20°C and 1 Atmosphere

The listed values ($\varepsilon - 1$) refer to the gas at a temperature of 20°C and pressure of 1 atmosphere. The values can be adjusted to slightly different conditions without introducing more than 0.1% error by use of the following equation:

$$\frac{(\varepsilon - 1)_{t,p}}{(\varepsilon - 1)_{20°,1\text{ atm}}} = \frac{p}{760[1 + 0.003411(t - 20)]}$$

where p = pressure in mm Hg
t = degrees C
t should be between 10 and 30°C and p between 700 and 800 mm Hg. From National Bureau of Standards Circular 537

Substance	$(\varepsilon - 1)\cdot 10^6$	Ref.
Helium	Radio frequency	
	67.8	Watson
	63.7	Hector
	64.5	Jelatis
	Microwave	
	65.6	Birnbaum
	65.2	Essen
	Optical	
	64.6	Koch
	64.5	Cuthbertson
Hydrogen	Radio frequency	
	254.0	Watson
	Microwave	
	253.4	Essen
	Optical	
	254.1	Cuthbertson
	253.6	Koch
	253.7	Kirn
	254.3	Tausz
Oxygen	Radio frequency	
	$494._3$	Watson
	496.2	Jelatis
	Microwave[a]	
	494.9	Birnbaum
	495.0	Essen
	494.9	Essen
	Optical	
	494.5	Cuthbertson
	493.5	Lowery
	494.7	Tausz
	494.4	Ladenberg
Argon	Radio frequency	
	513.0	Watson
	516.4	Jelatis
	Microwave	
	517.7	Essen
	Optical	
	516.8	Cuthbertson
	517.8	Quarder
	517.0	Tausz
	516.7	Damkohler
Air (dry, CO_2 Free)	Radio frequency	
	$537._0$	Watson
	Microwave[a]	
	536.6	Birnbaum
	536.6	Essen
	536.6	Essen
	Optical	
	536.9	Koch
	535.8	Meggers
	536.0	Traub
	536.7	Quarder
	536.4	Lowery
	536.5	Tausz
	536.1	Koster
	536.3	Perard
	535.8	Barrell
Nitrogen	Radio frequency	
	$547._2$	Watson
	Microwave	
	547.3	Birnbaum
	548.0	Essen
	548.0	Essen
	Optical	
	548.9	Cuthbertson
	548.7	Koch
	547.2	Tausz
Carbon dioxide	Radio frequency	
	$921._5$	Watson
	Microwave	
	922.4	Birnbaum
	920.6	Essen

[a] These values were derived from measurements of the refractive index after making allowance for the magnetic permeability of oxygen.

DIELECTRIC CONSTANTS OF GASES AT 760 MM PRESSURE

Compiled by Earl C. Gregg, Jr.

Material	Temperature °C	Frequency cycles/sec	Dielectric Constant
Acetaldehyde	100	$<3 \times 10^6$	1.0213
	0	2×10^6	
Acetone			1.0159
Acetyl chloride	0	$<3 \times 10^6$	1.0217
Acetylene	0	$<3 \times 10^6$	1.00134
Air			1.000590
Ammonia	0	$<3 \times 10^6$	1.0072
β-Amylene			1.0028
Argon	23	10^{10}	1.000545
Benzene	400	3×10^8	1.0028
Bromine	0	$<3 \times 10^6$	1.0128
Butylene			1.00319
Carbon dioxide	100	$<3 \times 10^6$	1.000985
Carbon disulfide	23	10^{10}	1.0029
Carbon monoxide	23	10^{10}	1.00070
Carbon tetrachloride	100	$<3 \times 10^6$	1.0030
Chloroform	100	$<3 \times 10^6$	1.0042
Dichlorodifluoromethane	100	$<3 \times 10^6$	1.00029
	0	$<3 \times 10^6$	
	0	$<3 \times 10^6$	
Dichlorofluoromethane	100	$<3 \times 10^6$	1.00049
Dimethylamine	0	2×10^6	1.00040
Ethane	0	$<3 \times 10^6$	1.00150
Ethylalcohol	100	$<3 \times 10^6$	1.0061
Ethylamine			1.00053
Ethyl bromide	20	$<3 \times 10^6$	1.0139
Ethyl chloride	20	$<3 \times 10^6$	1.0132
Ethyl ether	23	10^{10}	1.0049
Ethyl formate	126	$<3 \times 10^6$	1.0083
Ethyl iodide	20	$<3 \times 10^6$	1.0140
			1.0089
Ethylene	110	$<3 \times 10^6$	1.00144
Helium	140	$<3 \times 10^6$	1.0000684
n-Heptane	20	$<3 \times 10^6$	1.0035
Hydrogen	100	$<3 \times 10^6$	1.000264
Hydrogen bromide	20	$<3 \times 10^6$	1.00313
Hydrogen chloride	0	$<3 \times 10^6$	1.0046
	0	$<3 \times 10^6$	
Hydrogen iodide	0	$<10^6$	1.00234
	100	$<3 \times 10^6$	
Hydrogen sulfide	0	$<10^6$	1.00030
	100	$<3 \times 10^6$	
Mercury	180	$<3 \times 10^6$	1.00074
Methane	0	$<3 \times 10^6$	1.000944
Methyl alcohol	0	$<10^6$	1.0057
Methyl bromide	0	$<3 \times 10^6$	1.00095
Methyl chloride	0	$<3 \times 10^6$	1.00094
Methyl iodide	110	$<3 \times 10^6$	1.0063
Methylamine	120	$<3 \times 10^6$	1.0038
Methylene chloride	23	10^{10}	1.0065
Neon			1.000127
Nitrogen			1.000580
Nitromethane	23	10^{10}	1.0247
Nitrous oxide (N_2O)	23	2.5×10^{10}	1.00113
	0	$<3 \times 10^6$	
Oxygen	100	$<3 \times 10^6$	1.000523
n-Pentane	23	10^{10}	1.0025
n-Propyl chloride	20	$<3 \times 10^6$	1.0143
iso-Propyl chloride	20	$<3 \times 10^6$	1.0152
Sulfur dioxide	100	$<3 \times 10^6$	1.00075
Toluene	100	$<3 \times 10^6$	1.0043
Vinyl bromide	20	$<3 \times 10^6$	1.0081
	100	$<3 \times 10^6$	
Water (steam)	0	$<3 \times 10^6$	1.0126
	0	$<3 \times 10^6$	1.00785

PROPERTIES OF DIELECTRICS

In most cases properties have been determined by A.S.T.M. (American Society for Testing Materials) test methods at room temperature under standard conditions. Values will in general change considerably with temperature.

DIELECTRIC CONSTANTS OF SOME PLASTICS AND RUBBERS

Name	°C	Frequency (hertz) 1×10^3	1×10^6	1×10^8
Plastics				
Phenol-formaldehyde	25—27	5.15—8.61	4.45—5.05	4.1—4.5
	57	6.35	4.90	4.5
	88	8.5	5.2	4.7
Phenol-aniline-formaldehyde	25	4.50	4.31	4.11
	79	4.75	4.51	4.35
Melamine-formaldehyde	24—28	6.0—6.90	5.82—6.20	5.5—5.55
	57	6.95	5.40	4.90
	88	11.8	6.0	5.5
Urea-formaldehyde	24	6.7	6.0	5.2
	80	7.8	6.8	—
Polyamide resins				
Nylon 66	25	3.75	3.33	3.16
Nylon 610	25	3.50	3.14	3.0
	84	11.2	4.4	3.4
Cellulose acetate	26	3.50—4.48	3.28—3.90	3.05—3.40
Cellulose nitrate	27	8.4	6.6	5.2
	78	7.5	6.2	5.2
Methyl cellulose	22	6.8	5.7	4.3
Ethyl cellulose	25	3.09	3.01	2.90
Silicone resins	25	3.79—3.91	3.79—3.82	3.82
Polyethylene	−12	2.37	2.35	2.33
	23	2.26	2.26	2.26
Polyisobutylene	25	2.23	2.23	2.23
Vinylite QYNA	20	3.10	2.88	2.85
	76	3.83	3.0	2.8
	110	8.6		
Vinylite 5544	25	7.20	4.13	3.05
Vinylite 5901	25	5.5	3.4	3.0
Vinylite VU	24	5.65	3.30	2.80
	79	8.15	5.5	3.4
Vinylite VYHW	20	3.12	2.91	2.83
Vinylite VYNW	20	3.15	2.90	2.8
Polyvinyl chloride	25	4.55 (1×10^4)	3.3	—

Name	°C	Frequency (hertz) 1×10^3	1×10^6	1×10^8
Polyvinylidene and vinyl	23	4.65	3.18	2.82
chloride	84	4.94	4.40	3.2
Polychlorotrifluoroethylene	25	2.76	2.48	2.36
Polytetrafluoroethylene	22	2.1	2.1	2.1
(Teflon)	100	2.04	2.04	—
Polyvinylalcohol acetate	25	7.8	5.2	—
	85	100	10	—
Polyvinylacetals	26—27	3.02—3.12	2.86—2.92	2.67
	88	3.5	3.1	2.85
Polyacrylates				
Lucite	−12	2.9	2.63	2.50
	23	2.84	2.63	2.58
	81	3.45	2.72	2.59
Plexiglas	27	3.12	2.76	—
Polystyrene	25	2.54—2.56	2.54—2.56	2.55
	80	2.54	2.54	2.54
Styrene copolymers	25	2.55—2.95	2.55—2.80	2.55—2.77
Polyesters	25	3.22—4.3	3.12—4.0	2.94—2.98
Alkyd resins				
Alkyd isocyanate foam	25	1.223	1.218	1.20
Plaskon, clay filled	25	5.26	4.92	4.77
Plaskon, glass filled	25	5.04	4.73	4.50
Epoxy resins	25	3.63—3.67	3.52—3.62	3.32—3.35
Rubbers				
Hevea, vulcanized	27	2.94	2.74	2.42
Hevea compound	27	36	9	6.8
Gutta percha	25	2.60	2.53	2.47
Balata	25	2.50	2.50	2.42
Buna S	20	2.66	2.56	2.52
Butyl rubber oompound	25	2.42	2.40	2.39
Neoprene	24	6.60	6.26	4.5
Silicon rubber	25	3.12—3.30	3.10—3.20	3.06—3.18

DIELECTRIC CONSTANTS OF CERAMICS

Material	Dielectric constant 10^6 cycles	Dielectric strength volts mil	Volume resistivity Ohms-cm (23°C)	Loss factor[a]
Alumina	4.5—8.4	40—160	10^{11}—10^{14}	0.0002—0.01
Corderite	4.5—5.4	40—250	10^{12}—10^{14}	0.004—0.012
Forsterite	6.2	240	10^{14}	0.0004
Porcelain (dry process)	6.0—8.0	40—240	10^{12}—10^{14}	0.0003—0.02
Porcelain (wet process)	6.0—7.0	90—400	10^{12}—10^{14}	0.006—0.01
Porcelain, zircon	7.1—10.5	250—400	10^{13}—10^{15}	0.0002—0.008
Steatite	5.5—7.5	200—400	10^{13}—10^{15}	0.0002—0.004
Titanates (Ba, Sr, Ca, Mg, and Pb)	15—12.000	50—300	10^{8}—10^{15}	0.0001—0.02
Titanium dioxide	14—110	100—210	10^{13}—10^{18}	0.0002—0.005

DIELECTRIC CONSTANTS OF WAXES

Material	Dielectric constant 10^6 Cycles	Dielectric strength volts mil	Volume resistivity Ohms-cm (23°C)	Loss factor[a]
Acrawax C	2.4	—	—	0.005
Beeswax, white	2.75—3.0	—	5×10^{14}	0.025
Beeswax, yellow	2.9	—	8×10^{14}	0.029
Candelilla	2.25—2.50			
Carnauba	2.75—3.0			
Cerese, brown G	2.27	—	—	0.0025
Ceresine	2.25—2.50	—	$>5 \times 10^{18}$	0.0011
Halowax 1001	~4.10	—	2×10^{13}	0.014
Halowax 1013	~4.75	—	—	0.036
Halowax 1014	~4.40	—	—	0.035
Halowax 11-314	2.94	—	—	0.00094
Microcrystalline waxes	2.2—2.5			
Opalwax	3.1	—	—	0.34

[a] Power factor × dielectric constant equals loss factor.

DIELECTRIC CONSTANTS OF WAXES (continued)

Material	Dielectric constant 10^6 Cycles	Dielectric strength volts mil	Volume resistivity Ohms-cm (23°C)	Loss factor[a]
Ozokerite wax	2.3	100—150	5×10^{14}	0.0018
Paraffin	2.0—2.5	250	10^{15}—10^{19}	0.003 (900 cps)
Parawax	2.25	—	10^{16}	0.00045
135 A.M.P. wax	2.25	—	—	0.00023

DIELECTRIC CONSTANTS OF GLASSES

Type	Dielectric constant at 100 mc (20°C)	Volume resistivity (350°C megohm-cm)	Loss factor[a]
Corning 0010	6.32	10	0.015
Corning 0080	6.75	0.13	0.058
Corning 0120	6.65	100	0.012
Pyrex 1710	6.00	2,500	0.025
Pyrex 3320	4.71	—	0.019
Pyrex 7040	4.65	80	0.013
Pyrex 7050	4.77	16	0.017
Pyrex 7052	5.07	25	0.019
Pyrex 7060	4.70	13	0.018
Pyrex 7070	4.00	1,300	0.0048
Vycor 7230	3.83	—	0.0061
Pyrex 7720	4.50	16	0.014
Pyrex 7740	5.00	4	0.040
Pyrex 7750	4.28	50	0.011
Pyrex 7760	4.50	50	0.0081
Vycor 7900	3.9	130	0.0023
Vycor 7910	3.8	1,600	0.00091
Vycor 7911	3.8	4,000	0.00072
Corning 8870	9.5	5,000	0.0085
G. E. Clear (silica glass)	3.81	4,000—30,000	0.00038
Quartz (fused)	3.75 4.1 (1 mc)	—	0.0002 (1 mc)

[a] Power factor × dielectric constant equals loss factor.

STATIC DIELECTRIC CONSTANT OF WATER SUBSTANCE[a]

The temperatures are in degrees kelvin and the pressures in megapascals. Some conversion factors for pressure which may be useful are:

Megapascals × 9.8692 = atmospheres (/60 mm Hg)
Megapascals × 14.504 = pounds per square inch
Megapascals × 10^{-6} = newtons per square meter (pascals)
Megapascals × 10.1972 = kilograms per square centimeter
Megapascals × 10 = bars
Megapascals × 7.501 × 10^3 = mm Hg at 0°C
Megapascals × 4.014 × 10^3 = inches of H_2O at 4°C

P/T	273.15	298.15	323.15	348.15	373.15	398.15	423.15	448.15	473.15	498.15	523.15	548.15	573.15	623.15	673.15	723.15	773.15	823.15
10	88.28	78.85	70.27	62.59	55.76	49.70	44.30	39.47	35.11	31.13	27.43	23.90	20.39	1.23	1.17	1.14	1.11	1.10
20	88.75	79.24	70.63	62.94	56.11	50.05	44.66	39.85	35.52	31.58	27.95	24.54	21.24	14.07	1.64	1.42	1.32	1.26
30	89.20	79.63	70.98	63.28	56.44	50.39	45.01	40.22	35.91	32.01	28.43	25.11	21.95	15.66	5.91	2.07	1.68	1.51
40	89.64	80.00	71.32	63.61	56.77	50.72	45.34	40.56	36.28	32.40	28.87	25.61	22.56	16.72	10.46	3.84	2.34	1.90
50	90.07	80.36	71.66	63.93	57.08	51.03	45.67	40.89	36.63	32.78	29.28	26.08	23.10	17.55	12.16	6.57	3.45	2.48
60	90.49	80.72	71.98	64.24	57.39	51.34	45.98	41.21	36.96	33.13	29.67	26.50	23.58	18.24	13.28	8.53	4.90	3.26
70	90.90	81.07	72.30	64.54	57.69	51.64	46.28	41.52	37.28	33.47	30.03	26.90	24.02	18.84	14.16	9.87	6.31	4.20
80	91.29	81.42	72.62	64.84	57.98	51.93	46.57	41.82	37.59	33.79	30.37	27.27	24.43	19.37	14.88	10.88	7.50	5.16
90	91.67	81.75	72.92	65.13	58.27	52.21	46.86	42.11	37.89	34.10	30.70	27.62	24.81	19.85	15.50	11.70	8.47	6.06
100	92.04	82.08	73.22	65.42	58.55	52.49	47.14	42.39	38.17	34.40	31.01	27.95	25.17	20.29	16.05	12.39	9.29	6.88
125	92.89	82.84	73.93	66.09	59.19	53.12	47.78	43.05	38.86	35.13	31.78	28.76	26.03	21.26	17.21	13.77	10.88	8.53
150	93.71	83.57	74.62	66.74	59.82	53.75	48.40	43.68	39.50	35.78	32.46	29.47	26.77	22.09	18.16	14.85	12.07	9.80
175	94.48	84.28	75.27	67.36	60.42	54.34	48.98	44.27	40.10	36.39	33.09	30.12	27.45	22.83	18.98	15.74	13.04	10.81
200	95.20	84.94	75.89	67.95	61.00	54.90	49.54	44.83	40.66	36.97	33.67	30.72	28.07	23.49	19.69	16.51	13.86	11.65
225	95.87	85.58	76.50	68.53	61.55	55.44	50.08	45.36	41.20	37.51	34.22	31.28	28.64	24.09	20.33	17.19	14.56	12.38
250	96.51	86.20	77.08	69.08	62.08	55.96	50.59	45.87	41.70	38.02	34.74	31.81	29.17	24.65	20.91	17.80	15.19	13.01
300	97.69	87.34	78.17	70.14	63.10	56.94	51.55	46.82	42.65	38.97	35.69	32.77	30.15	25.65	21.94	18.85	16.25	14.07
350	98.75	88.40	79.19	71.12	64.05	57.86	52.45	47.70	43.52	39.83	36.56	33.64	31.02	26.53	22.83	19.74	17.14	14.93
400	99.72	89.39	80.13	72.03	64.94	58.74	53.30	48.53	44.33	40.64	37.36	34.43	31.81	27.32	23.62	20.52	17.89	15.66
450	100.60	90.30	81.02	72.89	65.78	59.56	54.10	49.31	45.10	41.38	38.09	35.16	32.54	28.04	24.32	21.20	18.55	16.28
500	101.42	91.16	81.84	73.69	66.57	60.33	54.85	50.05	45.82	42.09	38.78	35.84	33.21	28.70	24.96	21.82	19.14	16.83

[a] Prepared by International Association for the Properties of Steam.

PROPERTIES OF DIELECTRICS (Continued)

DIELECTRIC CONSTANT OF DEUTERIUM OXIDE

t	ϵ	$-\frac{d\epsilon}{dt}$	$-\frac{1}{\epsilon}\frac{d\epsilon}{dt}$
°C			
4	85.877	0.3974	4.627×10^{-3}
5	85.480	.3956	4.628
10	83.526	.3862	4.624
15	81.618	.3771	4.620
20	79.755	.3681	4.615
25	77.936	.3594	4.611
30	76.161	.3509	4.607
35	74.427	.3425	4.602
40	72.735	.3344	4.597
45	71.083	.3265	4.593
50	69.470	.3187	4.587
55	67.896	.3112	4.583
60	66.358	.3038	4.578
65	64.857	.2967	4.575
70	63.391	.2898	4.571
75	61.959	.2830	4.567
80	60.561	.2765	4.565
85	59.194	.2701	4.563
90	57.859	.2640	4.563
95	56.554	.2581	4.564
100	55.278	.2523	4.564

DIELECTRIC CONSTANTS (Continued)

Dielectric Constant of Liquid Parahydrogen vs. Temperature (°K) and Pressure (atm)

R. J. Corruccini

P atm \ T, °K	20	21	22	23	24	25	26	27	28	29	30	31	32
1	1.2297												
2	1.2302	1.2260	1.2216										
3	1.2306	1.2265	1.2221	1.2174	1.2122								
4	1.2311	1.2270	1.2227	1.2180	1.2129	1.2073	1.2010						
5	1.2315	1.2275	1.2233	1.2186	1.2136	1.2081	1.2020	1.1950					
6	1.2320	1.2280	1.2238	1.2192	1.2143	1.2089	1.2029	1.1962	1.1883				
7	1.2324	1.2285	1.2243	1.2198	1.2150	1.2097	1.2039	1.1973	1.1897	1.1805			
8	1.2329	1.2290	1.2249	1.2204	1.2157	1.2105	1.2048	1.1984	1.1911	1.1824			
9	1.2333	1.2295	1.2254	1.2210	1.2163	1.2112	1.2056	1.1994	1.1924	1.1842	1.1734		
10	1.2337	1.2300	1.2259	1.2216	1.2169	1.2119	1.2065	1.2004	1.1936	1.1857	1.1758	1.1621	
15	1.2358	1.2322	1.2284	1.2243	1.2200	1.2153	1.2103	1.2049	1.1990	1.1924	1.1847	1.1758	1.1645
20	1.2378	1.2343	1.2307	1.2268	1.2227	1.2184	1.2137	1.2088	1.2034	1.1976	1.1913	1.1839	1.1757
25	1.2396	1.2363	1.2328	1.2291	1.2253	1.2211	1.2168	1.2122	1.2073	1.2021	1.1964	1.1903	1.1832
30	1.2414	1.2382	1.2349	1.2313	1.2276	1.2237	1.2196	1.2153	1.2107	1.2059	1.2008	1.1952	1.1891
35	1.2431	1.2400	1.2368	1.2334	1.2298	1.2261	1.2222	1.2181	1.2138	1.2093	1.2046	1.1995	1.1942
40	1.2448	1.2418	1.2386	1.2354	1.2319	1.2284	1.2246	1.2208	1.2167	1.2124	1.2080	1.2033	1.1984
45	1.2464	1.2434	1.2404	1.2372	1.2339	1.2305	1.2269	1.2232	1.2193	1.2153	1.2111	1.2067	1.2021
50	1.2479	1.2450	1.2421	1.2390	1.2358	1.2325	1.2291	1.2255	1.2218	1.2179	1.2139	1.2098	1.2055
60	1.2508	1.2481	1.2453	1.2424	1.2394	1.2363	1.2331	1.2297	1.2263	1.2227	1.2191	1.2153	1.2114
70	1.2535	1.2510	1.2483	1.2455	1.2427	1.2397	1.2367	1.2336	1.2303	1.2270	1.2236	1.2201	1.2165
80	1.2561	1.2536	1.2511	1.2484	1.2457	1.2429	1.2400	1.2371	1.2340	1.2309	1.2277	1.2244	1.2211
90	1.2585	1.2561	1.2537	1.2512	1.2486	1.2459	1.2431	1.2403	1.2374	1.2345	1.2315	1.2284	1.2252
100	1.2608	1.2586	1.2562	1.2538	1.2513	1.2487	1.2461	1.2434	1.2406	1.2378	1.2349	1.2320	1.2290
120	1.2652	1.2631	1.2609	1.2586	1.2563	1.2539	1.2514	1.2489	1.2464	1.2438	1.2412	1.2385	1.2357
140	1.2693	1.2672	1.2651	1.2630	1.2608	1.2586	1.2563	1.2540	1.2516	1.2492	1.2467	1.2442	1.2417
160	1.2730	1.2711	1.2691	1.2671	1.2650	1.2629	1.2607	1.2585	1.2563	1.2540	1.2517	1.2494	1.2470
180	1.2766	1.2747	1.2728	1.2709	1.2689	1.2669	1.2649	1.2628	1.2606	1.2585	1.2563	1.2541	1.2518
200	1.2799	1.2781	1.2763	1.2745	1.2726	1.2707	1.2687	1.2667	1.2647	1.2626	1.2605	1.2584	1.2563
220	1.2831	1.2814	1.2796	1.2779	1.2760	1.2742	1.2723	1.2704	1.2685	1.2665	1.2645	1.2625	1.2605
240		1.2845	1.2828	1.2811	1.2793	1.2775	1.2757	1.2739	1.2720	1.2701	1.2682	1.2663	1.2643
260		1.2874	1.2858	1.2841	1.2824	1.2807	1.2790	1.2772	1.2754	1.2736	1.2717	1.2699	1.2680
280			1.2886	1.2870	1.2853	1.2837	1.2821	1.2803	1.2786	1.2768	1.2751	1.2733	1.2714
300			1.2914	1.2898	1.2882	1.2866	1.2850	1.2833	1.2817	1.2800	1.2782	1.2765	1.2747
320				1.2925	1.2910	1.2894	1.2878	1.2862	1.2846	1.2829	1.2813	1.2796	1.2779
340				1.2951	1.2936	1.2921	1.2905	1.2890	1.2874	1.2858	1.2842	1.2825	1.2809

Note: Values below the stepped line represent an extrapolation of p with density.

Selected Values of Electric Dipole Moments for Molecules in the Gas Phase

Ralph D. Nelson, Jr., David R. Lide, Jr., and Arthur A. Maryott

The following table was abstracted from the publication, "Selected Values of Electric Dipole Moments for Molecules in the Gas Phase" compiled by Nelson, Lide and Maryott and published as part of the National Reference Data Series—National Bureau of Standards (NSRDS—NBS 10). The publication is available from the Superintendent of Documents, U.S. Government Printing Office, Washington, D.C., 20402. Those desiring a complete listing of all compounds in the NSRDS—NBS 10, discussion of the bibliographic procedure and the principal methods of dipole moment measurement should obtain the publication.

Values of the dipole moment, μ, are expressed in the cgs system of units, since this is the system universally used by workers in the field. The numerical values are in debye units, D, ($1\ D = 10^{-18}$ electrostatic units of charge$\times$centimeters). The conversion factor to the Système International is $1\ D = 3.33564 \times 10^{-30}$ coulomb-meter.

Code symbol	Estimated accuracy of value
A	$\pm$ 1% or, for $\mu < 1.0$ D, $\pm$ 0.01 D
B	$\pm$ 2% or, for $\mu < 1.0$ D, $\pm$ 0.02 D
C	$\pm$ 5% or, for $\mu < 1.0$ D, $\pm$ 0.05 D
S	$\mu \equiv 0$ on grounds of molecular symmetry
i	The significance of these values may involve some ambiguity because of the possibility of different conformations or spatial isomers.

Compounds not containing carbon

Formula	Compound name	Selected moment (debyes)	
$AgCl$	Silver chloride	5.73	C
$AsCl_3$	Arsenic trichloride	1.59	C
AsF_3	Arsenic trifluoride	2.59	B
AsH_3	Arsine	0.20	C
BCl_3	Boron trichloride	0	S
BF_3	Boron trifluoride	0	S
B_2H_6	Diborane	0	S
$B_3H_6N_3$	Triborotriazine (Borazine)	0	S
B_5H_9	Pentaborane	2.13	B
BaO	Barium oxide	7.95	A
BrH	Hydrogen bromide	0.82	B
BrH_3Si	Bromosilane	1.33	B
BrK	Potassium bromide	10.41	B
$BrLi$	Lithium bromide	7.27	A
Br_2Hg	Mercury dibromide	0	S
Br_4Sn	Tin tetrabromide	0	S
$ClCs$	Cesium chloride	10.42	A
ClF	Chlorine fluoride	0.88	C
$ClFO_3$	Perchloryl fluoride	0.023	A
$ClGeH_3$	Chlorogermane	2.13	A
ClH	Hydrogen chloride	1.08	B
ClH_3Si	Chlorosilane	1.31	A
ClK	Potassium chloride	10.27	A
$ClLi$	Lithium chloride	7.13	A
$ClNa$	Sodium chloride	9.00	A
$ClNO_2$	Nitryl chloride	0.53	A
$ClTl$	Thallium chloride	4.44	B
Cl_2F_3P	Dichlorotrifluorophosphorus	0.68	C
Cl_2H_2Si	Dichlorosilane	1.17	B
Cl_2Hg	Mercury dichloride	0	S
Cl_2OS	Thionyl chloride	1.45	B
Cl_2O_2S	Sulfuryl chloride	1.81	B
Cl_3F_2P	Trichlorodifluorophosphorus	0	S
Cl_3HSi	Trichlorosilane	0.86	B
Cl_3P	Phosphorus trichloride	0.78	C
Cl_4FP	Tetrachlorofluorophosphorus	0.21	B
Cl_4Ge	Germanium tetrachloride	0	S
Cl_4Si	Silicon tetrachloride	0	S
Cl_4Sn	Tin tetrachloride	0	S
Cl_4Ti	Titanium tetrachloride	0	S
CsF	Cesium fluoride	7.88	A
FH	Hydrogen fluoride	1.82	A
FH_3Si	Fluorosilane	1.27	B
FH_5Si_2	Fluorodisilane	1.26	A
FK	Potassium fluoride	8.60	A
FLi	Lithium fluoride	6.33	A
FNO	Nitrosyl fluoride	1.81	B
FNa	Sodium fluoride	8.16	A
FRb	Rubidium fluoride	8.55	A
FTl	Thallium fluoride	4.23	A
F_2HN	Difluoramine	1.92	A
F_2H_2Si	Difluorosilane	1.55	A
F_2N_2	*cis*-Difluorodiazine	0.16	A
F_2O	Oxygen difluoride	0.297	A
F_2OS	Thionyl fluoride	1.63	A
F_2O_2	Dioxygen difluoride	1.44	C
F_2O_2S	Sulfuryl fluoride	1.12	B
F_2S_2	Sulfur monofluoride ($S = SF_2$ isomer)	1.03	C
F_2S_2	Sulfur monofluoride (FSSF isomer)	1.45	B
F_2Si	Silicon difluoride	1.23	B
F_3HSi	Trifluorosilane	1.27	B
F_3N	Nitrogen trifluoride	0.235	A
F_3NS	Nitridotrifluorosulfur	1.91	B
F_3OP	Phosphoryl fluoride	1.76	B
F_3P	Phosphorus trifluoride	1.03	A
F_3PS	Thiophosphoryl fluoride	0.64	B
F_4N_2	Tetrafluorohydrazine, *gauche* conformation	0.26	B
F_4S	Sulfur tetrafluoride	0.632	A
F_4Si	Silicon tetrafluoride	0	S
F_5P	Phosphorus pentafluoride	0	S
F_5I	Iodine pentafluoride	2.18	C
F_6S	Sulfur hexafluoride	0	S
F_6Se	Selenium hexafluoride	0	S
F_6Te	Tellurium hexafluoride	0	S
F_6U	Uranium hexafluoride	0	S
HI	Hydrogen iodide	0.44	B
HLi	Lithium hydride	5.88	A
HN	Imidyl radical		
HNO_3	Nitric acid	2.17	A
HO	Hydroxyl radical	1.66	A
H_2O	Water	1.85	A
H_2O_2	Hydrogen peroxide	2.2	D
H_2S	Hydrogen sulfide	0.97	A
H_3N	Ammonia	1.47	A
H_3P	Phosphine	0.58	A
H_3Sb	Stibine	0.12	C
H_4N_2	Hydrazine	1.75	C
H_4Si	Silane	0	S
H_6OSi_2	Disilyl ether (disiloxane)	0.24	B
H_6Si_2	Disilane	0	S
HgI_2	Mercury diiodide	0	S
ILi	Lithium iodide	7.43	A
I_4Sn	Tin tetraiodide	0	S
NO	Nitrogen monoxide (nitric oxide)	0.153	A
NO_2	Nitrogen dioxide	0.316	A
N_2O	Dinitrogen oxide (nitrous oxide)	0.167	A
OS	Sulfur monoxide	1.55	A
OS_2	Disulfur monoxide	1.47	B
OSr	Strontium oxide	8.90	A
O_2S	Sulfur dioxide	1.63	A
O_3	Ozone	0.53	B
O_3S	Sulfur trioxide	0	S
O_4Os	Osmium tetroxide	0	S

Compounds containing carbon

Formula	Compound name	Selected moment (debyes)	
$CBrF_3$	Bromotrifluoromethane	0.65	C
CBr_2F_2	Dibromodifluoromethane	0.66	C
$CClF_3$	Chlorotrifluoromethane	0.50	A
$CClN$	Cyanogen chloride	2.82	B
CCl_2F_2	Dichlorodifluoromethane	0.51	C
CCl_2O	Carbonyl chloride (phosgene)	1.17	A
CCl_2S	Thiocarbonyl chloride	0.29	C
CCl_3F	Trichlorofluoromethane	0.45	C
CCl_3NO_2	Trichloronitromethane	1.89	C
CCl_4	Carbon tetrachloride	0	S
CFN	Cyanogen fluoride	2.17	C
CF_2	Carbon difluoride	0.46	B
CF_2O	Carbonyl fluoride	0.95	A
CF_3I	Iodotrifluoromethane	0.92	C
CF_3NO_2	Trifluoronitromethane	1.44	C
CF_4	Carbon tetrafluoride	0	S
CN_4O_8	Tetranitromethane	0	S
CO	Carbon monoxide	0.112	A
COS	Carbonyl sulfide	0.712	A
$COSe$	Carbonyl selenide	0.73	B
CO_2	Carbon dioxide	0	S
CS	Carbon monosulfide	1.98	A
$CSTe$	Thiocarbonyl telluride	0.17	A
CS_2	Carbon disulfide	0	S
$CHBr_3$	Tribromomethane	0.99	B
$CHClF_2$	Chlorodifluoromethane	1.42	B

Selected Values of Electric Dipole Moments for Molecules in the Gas Phase (Continued)

Compounds containing carbon—Continued

Formula	Compound name	Selected moment (debyes)	
$CHCl_2F$	Dichlorofluoromethane	1.29	B
$CHCl_3$	Trichloromethane (chloroform)	1.01	B
$CHFO$	Formyl fluoride	2.02	A
CHF_3	Trifluoromethane	1.65	A
CHN	Hydrogen cyanide	2.98	A
CHP	Methylidyne phosphide (methinophosphide)	0.390	A
CH_2Br_2	Dibromomethane	1.43	B
CH_2ClF	Chlorofluoromethane	1.82	B
CH_2ClNO_2	Chloronitromethane	2.01	B
CH_2Cl_2	Dichloromethane	1.60	B
CH_2F_2	Difluoromethane	1.97	A
CH_2N_2	Cyanogen amide (cyanamide)	4.27	C
CH_2N_2	Diazomethane	1.50	A
CH_2N_2	Diazirine	1.59	C
CH_2O	Methanal (formaldehyde)	2.33	A
CH_2O_2	Methanoic acid (formic acid)	1.41	A
CH_3BF_2	Methyl difluoroborane	1.66	B
CH_3BO	Carbonyl borane	1.80	B
CH_3Br	Bromomethane	1.81	A
CH_3Cl	Chloromethane	1.87	A
CH_3F	Fluoromethane	1.85	A
CH_3I	Iodomethane	1.62	B
CH_3NO	Hydroxyliminomethane (formaldoxime)	0.44	A
CH_3NO	Formyl amide (formamide)	3.73	B
CH_3NOS	Methyl sulfinylamine	1.70	B
CH_3NO_2	Nitromethane	3.46	A
CH_3NO_3	Methyl nitrate	3.12	B
CH_3N_3	Methyl azide	2.17	B
CH_4	Methane	0	S
CH_4F_2Si	Methyl difluorosilane	2.11	A
CH_4O	Methanol	1.70	A
CH_4S	Methanethiol (methyl mercaptan)	1.52	C
CH_5FSi	Methyl monofluorosilane	1.71	A
CH_5N	Methyl amine	1.31	B
CH_5P	Methyl phosphine	1.10	A
CH_6Ge	Methyl germane	0.643	A
CH_6OSi	Methoxysilane	1.17	B
CH_6Si	Methyl silane	0.735	A
CH_6Sn	Methyl stannane	0.68	C
C_2ClF_5	Chloropentafluorethane	0.52	C
C_2F_6	Hexafluorethane	0	S
C_2N_2	Dicyanogen (cyanogen)	0	S
C_2N_2S	Dicyano sulfide	3.02	A
C_2HCl	Chloroacetylene	0.44	A
C_2HCl_5	Pentachloroethane	0.92	C
C_2HF	Fluoroacetylene	0.73	C
C_2HF_3	Trifluoroethylene	1.40	C
C_2HF_5	Pentafluoroethane	1.54	C
C_2H_2	Acetylene	0	S
$C_2H_2Cl_2$	1, 1-Dichloroethylene	1.34	A
$C_2H_2Cl_2$	*cis*-1, 2-Dichloroethylene	1.90	B
$C_2H_2Cl_2O$	Chloroacetyl chloride	2.23	Ci
$C_2H_2Cl_4$	1,1,2,2-Tetrachloroethane	1.32	Ci
C_2H_2FN	Fluorocyanomethane	3.43	C
$C_2H_2F_2$	1,1-Difluoroethylene	1.38	A
$C_2H_2F_2$	*cis*-1,2-Difluoroethylene	2.42	A
$C_2H_2N_2O$	1,2,5-Oxadiazole	3.38	A
$C_2H_2N_2O$	1,3,4-Oxadiazole	3.04	B
$C_2H_2N_2S$	1,2,5-Thiadiazole	1.56	A
$C_2H_2N_2S$	1.3.4-Thiadiazole	3.29	B
C_2H_2O	Methylene carbonyl (ketene)	1.42	B
C_2H_3Br	Bromoethylene	1.42	B
C_2H_3Cl	Chloroethylene	1.45	B
$C_2H_3ClF_2$	1-Chloro-1,1-difluoroethane	2.14	B
C_2H_3ClO	Acetyl chloride	2.72	C
$C_2H_3Cl_3$	1,1,1-Trichloroethane	1.78	B
C_2H_3F	Fluoroethylene	1.43	A
C_2H_3FO	Acetyl fluoride	2.96	A
$C_2H_3F_3$	1,1,1-Trifluoroethane	2.32	B
$C_2H_3F_3$	1,1,2-Trifluoroethane	1.58	B
C_2H_3N	Cyanomethane (acetonitrile)	3.92	A
C_2H_3N	Isocyanomethane	3.85	B
C_2H_4	Ethylene	0	S
C_2H_4ClF	1-Chloro-2-fluoroethane, *gauche* conformation	2.72	C
$C_2H_4ClNO_2$	1-Chloro-1-nitroethane	3.27	B
$C_2H_4Cl_2$	1,1-Dichloroethane	2.06	B
$C_2H_4F_2$	1,1-Difluoroethane	2.27	B
C_2H_4Ge	Germyl acetylene	0.136	A
C_2H_4O	Oxirane (ethylene oxide)	1.89	A
C_2H_4O	Ethanal (acetaldehyde)	2.69	B
$C_2H_4O_2$	Ethanoic acid (acetic acid)	1.74	C
$C_2H_4O_2$	Methyl methanoate (methly formate)	1.77	B
C_2H_4S	Thiirane (ethylene sulfide)	1.85	A
C_2H_4Si	Silyl acetylene	0.316	A
C_2H_5Br	Bromoethane	2.03	A
C_2H_5BrO	Bromomethoxymethane	2.05	Ci
C_2H_5Cl	Chloroethane	2.05	A
C_2H_5ClO	2-Chloroethanol	1.78	Ci
C_2H_5F	Fluoroethane	1.94	B
C_2H_5I	Iodoethane	1.91	B
C_2H_5N	Iminoethane (ethyleneimine)	1.90	A
C_2H_5N	Methyliminomethane ($CH_3N = CH_2$)	1.53	B

Compounds containing carbon—Continued

Formula	Compound name	Selected moment (debyes)	
C_2H_5NO	Acetyl amine (acetamide)	3.76	Bi
C_2H_5NO	Methylaminomethanal (*N*-methylformamide)	3.83	Bi
$C_2H_5NO_2$	Nitritoethane (ethyl nitrite)	2.40	Ci
$C_2H_5NO_2$	Nitroethane	3.65	B
C_2H_6	Ethane	0	S
C_2H_6BF	Dimethyl fluoroborane	1.32	C
C_2H_6O	Ethanol	1.69	Bi
C_2H_6O	Dimethyl ether	1.30	A
C_2H_6OS	Dimethylsulfoxide	3.96	A
$C_2H_6O_2$	1,2-Ethanediol (ethylene glycol)	2.28	Ci
$C_2H_6O_2S$	Dimethyl sulfoxylate (dimethyl sulfone)	4.49	B
C_2H_6S	Ethanethiol	1.58	Bi
C_2H_6S	Dimethyl sulfide	1.50	A
C_2H_6Si	Silyl ethylene	0.66	A
$C_2H_7B_5$	2,4-Dicarbaheptaborane	1.32	B
C_2H_7N	Aminoethane (ethyl amine)	1.22	Ci
C_2H_7N	Dimethyl amine	1.03	B
C_2H_7P	Ethyl phosphine	1.17	Bi
C_2H_7P	Dimethyl phosphine	1.23	A
$C_2H_8N_2$	1,2-Diaminoethane	1.99	Ci
C_2H_8Si	Dimethyl silane	0.75	A
C_2H_8Si	Ethyl silane	0.81	B
C_3O_2	Dicarbonyl carbon (carbon suboxide)	0	S
C_3HF_3	3,3,3-Trifluoropropyne	2.36	B
C_3HN	Cyanoacetylene	3.72	A
$C_3H_2N_2$	Dicyanomethane	3.73	A
C_3H_2O	Propynal	2.47	B
$C_3H_2O_3$	Vinylene carbonate	4.55	A
C_3H_3Br	3-Bromopropyne	1.54	C
C_3H_3Cl	3-Chloropropyne	1.68	C
$C_3H_3F_3$	3,3,3-Trifluoropropene	2.45	B
C_3H_3N	Cyanoethylene	3.87	B
C_3H_3NO	Acetyl cyanide	3.45	B
C_3H_3NS	Thiazole	1.62	B
C_3H_4	Cyclopropene	0.45	A
C_3H_4	Propyne	0.781	A
C_3H_4	Propadiene (allene)	0	S
$C_3H_4Cl_2$	1,1-Dichlorocyclopropane	1.58	B
C_3H_4O	Ethylidene carbonyl (methyl ketene)	1.79	B
C_3H_4O	Propenal, *trans* conformation (acrolein)	3.12	B
$C_3H_4O_2$	2-Oxoöxetane (β-propiolactone)	4.18	A
$C_3H_4O_2$	Vinyl formate	1.49	A
C_3H_5Cl	2-Chloropropene	1.66	B
C_3H_5Cl	*cis*-1-Chloropropene	1.67	C
C_3H_5Cl	*trans*-1-Chloropropene	1.97	C
C_3H_5Cl	3-Chloropropene	1.94	Ci
C_3H_5F	*cis*-1-Fluoropropene	1.46	B
C_3H_5F	2-Fluoropropene	1.61	B
C_3H_5F	3-Fluoropropene, *cis* conformation	1.76	A
C_3H_5F	3-Fluoropropene, *gauche* conformation.	1.94	A
C_3H_5N	Cyanoethane (propionitrile)	4.02	A
C_3H_6	Cyclopropane	0	S
C_3H_6	Propene	0.366	A
$C_3H_6ClNO_2$	1-Chloro-1-nitropropane	3.48	Bi
$C_3H_6Cl_2$	1,2-Dichloropropane		i
$C_3H_6Cl_2$	1,3-Dichloropropane	2.08	Bi
$C_3H_6Cl_2$	2,2-Dichloropropane	2.27	C
C_3H_6O	Oxetane (trimethylene oxide)	1.94	A
C_3H_6O	Methyl oxirane (propylene oxide)	2.01	A
C_3H_6O	Propanone (acetone)	2.88	A
C_3H_6O	2-Propen-1-ol (allyl alcohol)	1.60	C
C_3H_6O	Propanal, *cis* conformation (propionaldehyde)	2.52	B
$C_3H_6O_2$	Propanoic acid	1.75	Ci
$C_3H_6O_2$	Methyl acetate	1.72	Ci
$C_3H_6O_2$	Ethyl formate	1.93	Ci
$C_3H_6O_3$	1,3,5-Trioxane	2.08	A
C_3H_6S	Thietane (trimethylene sulfide)	1.85	C
C_3H_6S	Methyl thiirane (propylene sulfide)	1.95	A
C_3H_7Br	1-Bromopropane	2.18	Ci
C_3H_7Br	2-Bromopropane	2.21	C
C_3H_7Cl	1-Chloropropane	2.05	Bi
C_3H_7Cl	2-Chloropropane	2.17	C
C_3H_7F	1-Fluoropropane, *gauche* conformation	1.90	C
C_3H_7F	1-Fluoropropane, *trans* conformation	2.05	B
C_3H_7I	1-Iodopropane	2.04	Ci
C_3H_7NO	N,N-Dimethylformamide	3.82	Bi
C_3H_7NO	Acetyl methylamine (*N*-Methylacetamide)	3.73	Bi
$C_3H_7NO_2$	1-Nitropropane	3.66	Bi
$C_3H_7NO_2$	2-Nitropropane	3.73	B
C_3H_8	Propane	0.084	A
C_3H_8O	1-Propanol	1.68	Bi
C_3H_8O	2-Propanol	1.66	Bi
C_3H_8O	Methoxyethane (methyl ethyl ether)	1.23	Ci
C_3H_9As	Trimethyl arsine	0.86	B
C_3H_9N	Trimethyl amine	0.612	A
C_3H_9N	1-Aminopropane (*n*-propylamine)	1.17	Ci
C_3H_9P	Trimethyl phosphine	1.19	A

Selected Values of Electric Dipole Moments for Molecules in the Gas Phase (Continued)

Compounds containing carbon—Continued

Formula	Compound name	Selected moment (debyes)	
$C_3H_{10}Si$	Trimethyl silane	0.525	A
C_4F_8	Perfluorocyclobutane	0	S
$C_4H_2N_2$	*trans*-1,2-Dicyanoethylene	0	S
$C_4H_4Cl_2$	1,4-Dichloro-2-butyne	2.10	Bi
$C_4H_4F_2$	1,1-Difluoro-1,3-butadiene (*trans* conformation)	1.29	A
C_4H_4O	Furan	0.66	A
$C_4H_4O_2$	Diketene	3.53	B
C_4H_4S	Thiophene	0.55	C
C_4H_5Cl	4-Chloro-1,2-butadiene	2.02	Ci
C_4H_5Cl	1-Chloro-2-butyne	2.19	C
C_4H_5F	2-Fluoro-1,3-butadiene (*trans* conformation)	1.42	A
C_4H_5N	Pyrrole	1.84	C
C_4H_5N	*cis*-1-Cyanopropene	4.08	B
C_4H_5N	*trans*-1-Cyanopropene	4.50	B
C_4H_5N	2-Cyanopropene (methacrylonitrile)	3.69	C
C_4H_6	Cyclobutene	0.132	A
C_4H_6	1-Butyne	0.80	C
C_4H_6	1,2-Butadiene	0.403	A
C_4H_6	1,3-Butadiene	0	S
C_4H_6O	Cyclobutanone	2.99	B
C_4H_6O	*trans*-2-Butenal (crotonaldehyde)	3.67	Bi
C_4H_6O	2-Methylpropenal (methacrolein)	2.68	Ci
C_4H_6O	3-Butene-2-one	3.16	B
C_4H_7Cl	1-Chloro-2-methylpropene	1.95	Bi
$C_4H_7Cl_3$	1,1,2-Trichloro-2-methylpropane	1.86	Ci
C_4H_7F	Fluorocyclobutane	1.94	A
C_4H_7N	1-Cyanopropane	4.07	Bi
C_4H_8	1-Butene	0.34	Ci
C_4H_8	*trans*-2-Butene	0	S
C_4H_8	2-Methylpropene	0.50	A
$C_4H_8Cl_2$	1,4-Dichlorobutane	2.22	Ci
C_4H_8O	Tetrahydrofuran	1.63	C
C_4H_8O	*cis*-2,3-Dimethyloxirane	2.03	A
C_4H_8O	Butanal	2.72	Bi
$C_4H_8O_2$	1,4-Dioxane	0	S
$C_4H_8O_2$	Ethyl acetate	1.78	Ci
C_4H_9Br	1-Bromobutane	2.08	Ci
C_4H_9Br	2-Bromobutane	2.23	Ci
C_4H_9Cl	1-Chlorobutane	2.05	Bi
C_4H_9Cl	2-Chlorobutane	2.04	Ci
C_4H_9Cl	1-Chloro-2-methylpropane	2.00	Ci
C_4H_9Cl	2-Chloro-2-methylpropane	2.13	B
C_4H_9F	2-Fluoro-2-methylpropane	1.96	A
C_4H_9I	1-Iodobutane	2.12	Ci
C_4H_9NO	Propanoyl methylamine (N-methylpropionamide)	3.61	Bi
C_4H_9NO	Acetyl dimethylamine (N,N-dimethylacetamide)	3.81	Bi
$C_4H_9NO_2$	2-Nitrito-2-methylpropane (t-butyl nitrite)	2.74	Ci
$C_4H_9NO_2$	1-Nitrobutane	3.59	Bi
$C_4H_9NO_2$	2-Nitro-2-methylpropane	3.71	B
C_4H_{10}	Butane	$\leqslant$0.05	Ci
C_4H_{10}	2-Methylpropane	0.132	A
$C_4H_{10}O$	1-Butanol	1.66	Bi
$C_4H_{10}O$	2-Methylpropan-1-ol (isobutanol)	1.64	C
$C_4H_{10}O$	Diethyl ether	1.15	Bi
$C_4H_{10}S$	Diethyl sulfide	1.54	Ci
$C_4H_{11}N$	Diethyl amine	0.92	Ci
C_5H_5N	Pyridine	2.19	B
C_5H_5N	1-Cyano-1,3-butadiene	3.90	Ci
C_5H_6	1,3-Cyclopentadiene	0.419	A
C_5H_8	Cyclopentene	0.20	B
C_5H_8	1-Pentyne	0.81	Ci
C_5H_8	2-Methyl-1,3-butadiene (*trans* conformation)	0.25	A
$C_5H_8O_2$	Acetylacetone		Ci
C_5H_9N	1-Cyanobutane	4.12	Bi
C_5H_9N	2-Cyano-2-methylpropane	3.95	A
$C_5H_{10}O_3$	Diethyl carbonate	1.10	Ci
$C_5H_{11}Br$	1-Bromopentane	2.20	Ci
$C_5H_{11}Cl$	1-Chloropentane	2.16	Ci
C_5H_{12}	2-Methylbutane	0.13	C
C_5H_{12}	2,2-Dimethylpropane	0	S
$C_6H_2Cl_2O_2$	2,5-Dichloro-1,4-cyclohexadienedione	0	S
$C_6H_4ClNO_2$	*o*-Chloronitrobenzene	4.64	B
$C_6H_4ClNO_2$	*m*-Chloronitrobenzene	3.73	B
$C_6H_4ClNO_2$	*p*-Chloronitrobenzene	2.83	B
$C_6H_4Cl_2$	*o*-Dichlorobenzene	2.50	B
$C_6H_4Cl_2$	*m*-Dichlorobenzene	1.72	C
$C_6H_4Cl_2$	*p*-Dichlorobenzene	0	S
$C_6H_4FNO_2$	*p*-Fluoronitrobenzene	2.87	B
$C_6H_4F_2$	*m*-Difluorobenzene	1.58	B
$C_6H_4N_2O_4$	*p*-Dinitrobenzene	0	S
$C_6H_4O_2$	1,4-Cyclohexadienedione (*p*-benzoquinone)	0	S
C_6H_5Br	Bromobenzene	1.70	B
C_6H_5Cl	Chlorobenzene	1.69	B
C_6H_5ClO	*p*-Chlorophenol	2.11	C
C_6H_5F	Fluorobenzene	1.60	C
C_6H_5I	Iodobenzene	1.70	C
$C_6H_5NO_2$	Nitrobenzene	4.22	B
C_6H_6	Benzene	0	S

Compounds containing carbon—Continued

Formula	Compound name	Selected moment (debyes)	
C_6H_6O	Phenol	1.45	C
C_6H_7N	Aminobenzene (aniline)	1.53	C
C_6H_8	1,3-Cyclohexadiene	0.44	B
C_6H_{10}	1-Hexyne	0.83	Ci
C_6H_{10}	3, 3-Dimethyl-1-butyne	0.66	A
$C_6H_{10}Cl_2$	*cis*-le,2a-Dichlorocyclohexane	3.11	C
$C_6H_{12}N_2$	Diisopropylidene hydrazine (dimethyl ketazine)	1.53	Bi
$C_6H_{12}O_2$	Pentyl formate (n-amyl formate)	1.90	Ci
$C_6H_{12}O_3$	2,4,6-Trimethyl-1,3,5-trioxane (paraldehyde)	1.43	C
$C_6H_{14}O$	Dipropyl ether	1.21	Ci
$C_6H_{14}O_2$	1,1-Diethoxyethane		i
$C_6H_{15}N$	Triethyl amine	0.66	Ci
$C_7H_4ClF_3$	*o*-Chloro(trifluoromethyl)benzene	3.46	B
$C_7H_4ClF_3$	*p*-Chloro(trifluoromethyl)benzene	1.58	C
$C_7H_5F_3$	(Trifluoromethyl)benzene	2.86	B
C_7H_5N	Cyanobenzene (benzonitrile)	4.18	B
C_7H_7Cl	*o*-Chlorotoluene	1.56	C
C_7H_7Cl	*p*-Chlorotoluene	2.21	B
C_7H_7F	*o*-Fluorotoluene	1.37	C
C_7H_7F	*m*-Fluorotoluene	1.86	C
C_7H_7F	*p*-Fluorotoluene	2.00	C
$C_7H_7NO_3$	*o*-Nitro(methoxy)benzene	4.83	Bi
$C_7H_7NO_3$	*m*-Nitro(methoxy)benzene	4.55	Bi
$C_7H_7NO_3$	*p*-Nitro(methoxy)benzene	5.26	B
C_7H_8	1,3,5-Cycloheptatriene	0.25	C
C_7H_8	Toluene	0.36	C
C_7H_8O	Phenylmethanol (benzyl alcohol)	1.71	C
C_7H_8O	Methoxybenzene (anisole)	1.38	C
C_7H_9NO	*o*-Amino(methoxy)benzene	1.61	Ci
$C_7H_{14}O_2$	Pentyl acetate (*n*-amyl acetate)	1.75	Ci
$C_7H_{15}Br$	1-Bromoheptane	2.16	Ci
$C_8H_4N_2$	*p*-Dicyanobenzene	0	S
C_8H_8O	Acetylbenzene (acetophenone)	3.02	B
$C_8H_8O_2$	2,5-Dimethyl-1,-4-cyclohexadienedione	0	S
C_8H_{10}	Ethylbenzene	0.59	C
C_8H_{10}	*o*-Xylene	0.62	C
C_8H_{10}	*p*-Xylene	0	S
$C_8H_{12}O_2$	Tetramethylcyclobutane-1,3-dione	0	S
$C_8H_{18}O$	Dibutyl ether	1.17	Ci
C_9H_7N	Quinoline	2.29	C
C_9H_7N	Isoquinoline	2.73	C
$C_9H_{10}O_2$	Ethyl phenylformate (ethyl benzoate)	2.00	Ci
$C_{10}H_8$	Azulene	0.80	B
$C_{10}H_{14}BeO_4$	Bis(2,4-pentanedionato) beryllium	0	S
$C_{12}H_9BrO$	*p*-Bromophenoxybenzene	1.98	C
$C_{12}H_9NO_3$	*p*-Nitrophenoxybenzene	4.54	B
$C_{12}H_{10}$	Phenylbenzene (diphenyl)	0	S
$C_{13}H_{11}BrO$	*p*-Bromophenoxy-*p*-toluene	2.45	C
$C_{14}H_{14}O$	Bis(*p*-tolyl) ether	1.54	C
$C_{15}H_{21}AlO_6$	Tris(2,4-pentanedionato) aluminum	0	S
$C_{15}H_{21}CrO_6$	Tris(2,4-pentanedionato) chromium (III)	0	S
$C_{15}H_{21}FeO_6$	Tris(2,4-pentanedionato) iron (III)	0	S
$C_{20}H_{28}O_8Th$	Tetrakis(2,4-pentanedionato) thorium	0	S

DIPOLE MOMENTS

The method of measurement of the dipole moments is indicated in the following **two tables** by the symbols:

- B benzene solution
- C carbon tetrachloride solution
- D 1,4-dioxane solution
- H n-heptane solution
- St measurement of Stark effect in microwave spectrum of gas

Dipole Moments for Some Inorganic Compounds

Compound	Dipole Moment × 10^{-18} e. s. u.	Method
$AlBr_3$	5.14	B
AlI_3	2.48	B
CsCl	10.42	St
CsF	7.875	St
HF	1.92 ± 0.02	..
HCl	1.084 ± 0.003–0.007	..
NBr	0.78	..
HDSe	0.62	St
HI	0.38	..
DCl	1.084 ± 0.003–0.007	..
HNO_3	2.16	St
$HgBr_2$	0.95	B
$HgCl_2$	1.23	B
H_2O	1.87	..
H_2O_2	2.13 ± 0.05	..
H_2S	1.10	..
SO_2	1.60	..
SO_3	0.00	..
SO_2F_2	1.110	St
NH_3	1.3	..
N_2H_4	1.84	..
NO	0.16	..
NO_2	0.29	..
N_2O_4	0.37	..
NOCl	1.83	..
NOBr	1.87	..
PCl_3	0.90–1.16	..
PCl_5	0.0	..
CO	0.10	..
CO_2	0.0	..
SiD_2F_2	1.53	St
SiH_2F_2	1.54	St
$SnCl_4$	0.95	B
SnI_4	0	B
$TiCl_4$	0	C

Dipole Moments for Some Organo-metallic Compounds

Compound	Dipole Moment × 10^{-18} e. s. u.	Method
Beryllium diethyl	1.0	H
Cadmium diethyl	0.3	H
Mercury diethyl	0.0	H
Magnesium diethyl	4.8	D
Zinc diethyl	0.0	H
Beryllium diphenyl	1.6	B
Cadmium diphenyl	0.6	B
Mercury diphenyl	0.2	B
Magnesium diphenyl	4.9	D
Zinc diphenyl	0.8	B
Chromium (0), diphenyl	0	B
Chromium, ditolyl	0	B
Cobalt, mononitrosyl tricarbonyl	0.72	B
Cyclopentadienyl, chromium dicarbonyl mono nitrosyl	3.23	B
Cyclopentadienyl, manganese tricarbonyl	3.30	B
Cyclopentadienyl, cobalt ducarbonyl	2.87	B
Cyclopentadienyl, vanadium tetracarbonyl	3.17	B
Penta cyclopentadienyl, dicobalt	0	B
Dicyclopentadienyl, iron (II)	0	B
Dicyclopentadienyl, lead (II)	1.63	B
Dicyclopentadienyl, tin (II)	1.02	B
Ethyl lithium	0.87	B
Iron, dinitrosyl dicarbonyl	0.95	B
Iron, tetracarbonyl-diiodide	3.68	B
Iron, tetracarbonyl mono-(methyl isonitrile)	5.07	B
Iron, pentacarbonyl	0.63	B
Iron, bis(p-chlorophenyl cyclopentadienyl	3.12	B
Ruthenium (II), di-indenyl	0	B

Dipole Moments

Dipole Moments of Amino Acid Esters

Substance	$\mu \cdot 10^{18}$ e. s. u.
Glycine ethyl ester	2.11
α-Alanine ethyl ester	2.09
α-Aminobutyric acid ethyl ester	2.13
α-Aminovaleric acid ethyl ester	2.13
Valine ethyl ester	2.11
α-Aminocaproic acid ethyl ester	2.13
β-Alanine ethyl ester	2.14
β-Aminobutyric acid ethyl ester	2.11

Accurate to $\pm 0.01 \cdot 10^{-18}$ e. s. u.
J. Wyman, Chem. Rev., 1936, **19**, 213.

Dipole Moments of Amides

Urea	4.56
Thiourea	4.89
Symm.-dimethylurea	4.8
Tetraethylurea	3.3
Propylurea	4.1
Acetamide	3.6
Sulfamide	3.9
Benzamide	3.6
Valeramide	3.7
Caproamide	3.9

For comprehensive list of dipole moments see Trans. Faraday Soc., 1934, **30**, General Discussion.

Dipole Moments of Some Hormones and Related Compounds in Dioxan

Cholestane-3(β) : 7(α)-diol	2.31
Cholestane-3(β) : 7(β)-diol	2.55
Cholestane	2.98
Δ^5-Cholestane-3(β)ol-7 one	3.79
Androsterone	3.70
β-Androsterone	2.95
Δ^5-Androstene-3(β) : 17(α)-diol	2.89
Δ^5-Androstene-3(β) : 17(β)-diol	2.69
Δ^5-Androstene-3(β)ol-17 one	2.46
Testosterone	4.32
cis-Testosterone	5.17
Δ^4-Androstene-3 : 17 dione	3.32
Isophorone	3.96

Ethylenic >C═C< in a six membered ring and conjugated with >C═O increases the dipole moment approximately by 1 Debye. Non-conjugated >C═C< in sterols decreases the dipole moment by approximately 0.49. Biological activity is not correlated with dipole moment. W. D. Kumler and G. M. Fohlen, J. Am. Chem. Soc., 1945, **67**, 437.

FINE-STRUCTURE SEPARATIONS IN ATOMIC NEGATIVE IONS

From the *Journal of Physical and Chemical Reference Data,* Volume 14, No. 3, 731, 1985. Reproduced by permission of the copyright owners, the American Chemical Society and the American Institute of Physics, and the authors, H. Hotop and W. C. Lineberger. Users of data in this table are referred to the original publication for references for the data.

Z	Negative ion	Fine-structure interval[a]	Separation (cm^{-1})	Method[b]
2	He^- (1s2s2p $^4P^o$)	$5/2 \rightarrow 3/2$	0.027508(27)	rf
		$5/2 \rightarrow 1/2$	0.2888(18)	rf
5	$B^-(^3P)$	$0 \rightarrow 1$	4(1)	RIE
		$0 \rightarrow 2$	9(1)	RIE
6	$C^-(^2D)$	$3/2 \rightarrow 5/2$	3(1)	LIE
8	$O^-(^2P)$	$3/2 \rightarrow 1/2$	177.08(5)	LPT
13	$Al^-(^3P)$	$0 \rightarrow 1$	26(3)	RIE
		$0 \rightarrow 2$	76(7)	RIE
14	$Si^-(^2D)$	$3/2 \rightarrow 5/2$	7(2)	LIE
15	$P^-(^3P)$	$2 \rightarrow 1$	181(2)	LPT
		$2 \rightarrow 0$	263(2)	LPT
16	$S^-(^2P)$	$3/2 \rightarrow 1/2$	483.54(1)	LPT
22	$Ti^-(^4F)$	$3/2 \rightarrow 5/2$	72(7)	LIE
		$5/2 \rightarrow 7/2$	99(10)	LIE
		$7/2 \rightarrow 9/2$	124(12)	LIE
		$3/2 \rightarrow 9/2$	295(15)	LIE
23	$V^-(^5D)$	$0 \rightarrow 1$	35(4)	RIE
		$1 \rightarrow 2$	70(7)	RIE
		$2 \rightarrow 3$	100(10)	RIE
		$3 \rightarrow 4$	125(13)	RIE
		$0 \rightarrow 4$	330(17)	RIE
26	$Fe^-(^4F)$	$9/2 \rightarrow 7/2$	540(50)	RIE
		$7/2 \rightarrow 5/2$	390(40)	RIE
		$5/2 \rightarrow 3/2$	270(30)	RIE
		$9/2 \rightarrow 3/2$	1200(60)	RIE
27	$Co^-(^3F)$	$4 \rightarrow 3$	910(50)	LPES
		$3 \rightarrow 2$	650(50)	LPES
		$4 \rightarrow 2$	1560(50)	LPES
28	$Ni^-(^2D)$	$5/2 \rightarrow 3/2$	1470(100)	LPES
31	$Ga^-(^3P)$	$0 \rightarrow 1$	220(20)	RIE; QIE
		$0 \rightarrow 2$	580(50)	RIE; QIE
32	$Ge^-(^2D)$	$3/2 \rightarrow 5/2$	160(30)	LIE
33	$As^-(^3P)$	$2 \rightarrow 1$	1100(200)	LIE; QIE
		$2 \rightarrow 0$	1500(200)	LIE; QIE
		$2 \rightarrow 0$	≈1370	PT
34	$Se^-(^2P)$	$3/2 \rightarrow 1/2$	2279(2)	LPT
40	$Zr^-(^4F)$	$3/2 \rightarrow 5/2$	250(50)	RIE
		$5/2 \rightarrow 7/2$	330(70)	RIE
		$7/2 \rightarrow 9/2$	370(70)	RIE
		$3/2 \rightarrow 9/2$	950(100)	RIE
41	$Nb^-(^5D)$	$0 \rightarrow 1$	110(20)	RIE
		$1 \rightarrow 2$	200(40)	RIE
		$2 \rightarrow 3$	250(40)	RIE
		$3 \rightarrow 4$	310(60)	RIE
		$0 \rightarrow 4$	860(90)	RIE
45	$Rh^-(^3F)$	$4 \rightarrow 3$	2370(65)	LPES
		$3 \rightarrow 2$	1000(65)	LPES
		$4 \rightarrow 2$	3370(65)	LPES
46	$Pd^-(^2D)$	$5/2 \rightarrow 3/2$[c]	3450(350)	RIE
49	$In^-(^3P)$	$0 \rightarrow 1$	680(70)	RIE; QIE
		$0 \rightarrow 2$	1550(150)	RIE; QIE
50	$Sn^-(^2D)$	$3/2 \rightarrow 5/2$	800(200)	LIE
51	$Sb^-(^3P)$	$2 \rightarrow 1$	2700(500)	LIE; QIE
		$2 \rightarrow 0$	3000(500)	LIE; QIE
		$2 \rightarrow (1,0)$	≈2740	PT
52	$Te^-(^2P)$	$3/2 \rightarrow 1/2$	5008(5)	LPT
73	$Ta^-(^5D)$	$0 \rightarrow 1$	1070(110)	LPES
		$1 \rightarrow 2$	1170(120)	LPES
		$2 \rightarrow 3$[d]	980(200)	RIE
77	$Ir^-(^3F)$	$4 \rightarrow 3$	7600(1500)	RIE
		$3 \rightarrow 2$	4400(900)	RIE
		$4 \rightarrow 2$	12000(1200)	RIE
78	$Pt^-(^2D)$	$5/2 \rightarrow 3/2$	10000(1000)	RIE; QIE

[a] Total angular momentum of lower (left) and upper fine structure levels are listed.

[b] Abbreviations used: rf — radio frequency resonance technique; RIE — isoelectronic extrapolation of ratios of fine structure separations; LIE — isoelectronic extrapolation from logarithmic plot; QIE — quadratic isoelectronic extrapolation; LPT — tunable laser photodetachment threshold; LPES — laser photodetachment electron spectrometry; and PT — photodetachment threshold using conventional light sources.

[c] $J = 3/2$ not bound.

[d] $J = 3$ not bound.

ELECTRON AFFINITIES

Thomas M. Miller

Electron affinity is defined as the energy difference between the lowest (ground) state of the neutral and the lowest state of the corresponding negative ion. The accuracy of electron affinity measurements has been greatly improved since the advent of laser photodetachment experiments with negative ions. Electron affinities can be determined with optical precision, though a detailed understanding of atomic and molecular states and splittings is required to specify the photodetachment threshold corresponding to the electron affinity.

Atomic and molecular electron affinities are discussed in two excellent articles reviewing photodetachment studies which appear in *Gas Phase Ion Chemistry*, Vol. 3, M. T. Bowers, Ed., Academic Press, Orlando, 1984, Chapter 21 by P. S. Drzaic, J. Marks, and J. I. Brauman, "Electron Photodetachment from Gas Phase Negative Ions," p. 167; and Chapter 22 by R. D. Mead, A. E. Stevens, and W. C. Lineberger, "Photodetachment in Negative Ion Beams," p. 213. Persons interested in photodetachment details should consult these articles. For simplicity in the tables below, any electron affinity which was discussed in the articles by Drzaic et al. and Mead et al. is referenced to those sources, where original references are given. A great many additional electron affinities have been provided here by G. B. Ellison, W. C. Lineberger, H. Hotop, and D. G. Leopold.

Electron affinities for the rare earths have not been measured, but theoretical estimates have been made by S. G. Bratsch and J. J. Lagowski, *Chem. Phys. Lett.*, 107, 136, 1984. Abbreviations used in the tables: calc = calculated value; PT = photodetachment threshold using a lamp as a light source; LPT = laser photodetachment threshold; LPES = laser photoelectron spectroscopy; DA = dissociative attachment; LOGS = laser optogalvanic spectroscopy; e-scat = electron scattering; plasma = absorption or emission in a plasma; CT = charge transfer; and CD = collisional detachment.

Table 1
ATOMIC ELECTRON AFFINITIES

Atomic number	Atom	Electron affinity in eV	Uncertainty in eV	Method	Ref.
1	H	0.754209	0.000003	calc	1
2	He	Not stable[a]	—	calc	1
3	Li	0.6180	0.0005	LPT	1
4	Be	Not stable	—	calc	1
5	B	0.277	0.010	LPES	1
6	C	1.2629	0.0003	LPT	1
7	N	Not stable	—	DA	1
8	O	1.461125	0.000001	LPT	4
9	F	3.399	0.003	plasma	1
10	Ne	Not stable	—	calc	1

Table 1 (continued)
ATOMIC ELECTRON AFFINITIES

Atomic number	Atom	Electron affinity in eV	Uncertainty in eV	Method	Ref.	Atomic number	Atom	Electron affinity in eV	Uncertainty in eV	Method	Ref.
11	Na	0.547930	0.000025	LPT	1	42	Mo	0.746	0.010	LPES	1
12	Mg	Not stable	—	e-scat	1	43	Tc	0.55	0.15	calc	1
13	Al	0.441	0.010	LPES	1	44	Ru	1.05	0.10	calc	1
14	Si	1.385	0.005	LPES	1	45	Rh	1.137	0.008	LPES	1
15	P	0.7465	0.0003	LPT	1	46	Pd	0.557	0.0008	LPES	1
16	S	2.077120	0.000001	LPT	1	47	Ag	1.302	0.009	LPES	1
17	Cl	3.617	0.001	plasma	1	48	Cd	Not stable		calc	1
18	Ar	Not stable	—	calc	1	49	In	0.30	0.15	PT	1
19	K	0.50147	0.0001	LPT	1	50	Sn	1.112	0.004	LPES	28
20	Ca	Not stable	—	calc	1	51	Sb	1.07	0.05	PT	1
21	Sc	0.188	0.020	LPES	1	52	Te	1.9708	0.0003	LPT	1
22	Ti	0.079	0.014	LPES	1	53	I	3.0591	0.001	LOGS	1
23	V	0.525	0.012	LPES	1	54	Xe	Not stable	—	calc	1
24	Cr	0.666	0.012	LPES	1	55	Cs	0.471630	0.000025	LPT	1
25	Mn	Not stable	—	calc	1	56	Ba	Not stable	—	calc	1
26	Fe	0.151	0.003	LPES	27	57	La	0.5	0.25	calc	1
27	Co	0.662	0.003	LPES	27	72	Hf	≈0	—	calc	1
28	Ni	1.156	0.010	LPES	1	73	Ta	0.322	0.012	LPES	1
29	Cu	1.228	0.010	LPES	1	74	W	0.815	0.003	LPES	5
30	Zn	Not stable	—	e-scat	1	75	Re	0.15	0.10	calc	1
31	Ga	0.3	0.15	PT	1	76	Os	1.10	0.15	calc	1
32	Ge	1.233	0.003	LPES	28	77	Ir	1.565	0.008	LPES	1
33	As	0.81	0.03	PT	1	78	Pt	2.128	0.002	LPT	1
34	Se	2.0206	0.0003	LPT	1	79	Au	2.30863	0.00003	LPT	1
35	Br	3.365	0.004	plasma	1	80	Hg	Not stable	—	e-scat	1
36	Kr	Not stable	—	calc	1	81	Tl	0.2	0.2	PT	1
37	Rb	0.48592	0.00002	LPT	1	82	Pb	0.364	0.008	LPES	1
38	Sr	Not stable	—	calc	1	83	Bi	0.946	0.010	LPES	1
39	Y	0.307	0.012	LPES	1	84	Po	1.9	0.3	calc	1
40	Zr	0.426	0.014	LPES	1	85	At	2.8	0.2	calc	1
41	Nb	0.893	0.025	LPES	1	86	Rn	Not stable	—	calc	1

[a] Not stable refers to the negative ion.

Table 2
ELECTRON AFFINITIES FOR DIATOMIC MOLECULES

Molecule	Electron affinity in eV	Uncertainty in eV	Method	Ref.	Molecule	Electron affinity in eV	Uncertainty in eV	Method	Ref.
As_2	0	—	PT	2	LiCl	0.594	0.009	LPES	30
AsH	1.0	0.1	PT	2	MgCl	1.589	0.011	LPES	31
BO	3.12	0.09	PT	6	MgH	1.05	0.06	PT	2
BeH	0.7	0.1	PT	2	MgI	1.899	0.018	LPES	31
Br_2	2.55	0.10	CT	2	MnD	0.866	0.010	LPES	9
C_2	3.39	0.15	LPT	2	MnH	0.869	0.010	LPES	9
CH	1.238	0.008	LPES	2	NH	0.370	0.004	LPT	32
CN	3.821	0.004	LOGS	7	NO	0.024	0.010	LPES	2
CS	0.205	0.021	LPES	2	NS	1.194	0.011	LPES	2
CaH	0.93	0.05	PT	2	NaBr	0.788	0.008	LPES	30
Cl_2	2.38	0.10	CT	2	NaCl	0.727	0.008	LPES	30
ClO	2.17	0.10	LPT	8	NaF	0.521	0.008	LPES	30
Co_2	1.110	0.008	LPES	27	NaI	0.867	0.011	LPES	30
CoD	0.680	0.010	LPES	29	NiD	0.477	0.007	LPES	29
CoH	0.671	0.010	LPES	29	NiH	0.481	0.007	LPES	29
CrD	0.568	0.010	LPES	29	O_2	0.440	0.008	LPES	2
CrH	0.563	0.010	LPES	29	OD	1.825548	0.000037	LPT	3
CrO	1.222	0.010	LPES	5	OH	1.827670	0.000021	LPT	3
CsCl	0.451	0.015	LPES	30	P_2	≤0.65	—	PT	2
F_2	3.08	0.10	CT	2	PH	1.028	0.010	LPES	2
Fe_2	0.902	0.008	LPES	27	PO	1.092	0.010	LPES	2
FeD	0.932	0.015	LPES	9	RbCl	0.544	0.009	LPES	30
FeH	0.934	0.011	LPES	9	Re_2	1.571	0.008	LPES	33
FeO	1.499	0.010	LPES	5	S_2	1.663	0.040	LPES	2
I_2	2.55	0.05	CT	2	SD	2.315	0.002	LPES	10
IBr	2.55	0.10	CT	2	SH	2.317	0.002	LPES	10
KBr	0.642	0.012	LPES	30	SO	1.126	0.013	LPES	2
KCl	0.582	0.007	LPES	30	SeH	2.21	0.03	PT	2
KI	0.727	0.012	LPES	30	SiH	1.277	0.009	LPES	2
					ZnH	≤0.95	—	PT	2

Table 3
ELECTRON AFFINITIES FOR TRIATOMIC MOLECULES

Molecule	Electron affinity in eV	Uncertainty in eV	Method	Ref.
AsH_2	1.27	0.03	PT	2
C_3	1.981	0.019	LPES	11
CD_2	0.64	0.01	LPES	12
CH_2	0.67	0.01	LPES	12
C_2H	2.94	0.10	PT	2
C_2O	1.848	0.027	LPES	13
COS	0.46	0.20	CD	2
CS_2	1.0	0.2	CD	2
CoD_2	1.465	0.013	LPES	34
CoH_2	1.450	0.014	LPES	34
DCO	0.301	0.005	LPES	35
DNO	0.330	0.015	LPES	14
DO_2	1.089	0.017	LPES	15
FeCO	1.26	0.02	LPES	2
GeH_2	1.097	0.015	LPES	28
HCO	0.313	0.005	LPES	35
HNO	0.338	0.015	LPES	14
HO_2	1.133	0.024	LPES	15
MnD_2	0.465	0.014	LPES	34
MnH_2	0.444	0.016	LPES	34
N_3	2.70	0.12	PT	2
NH_2	0.744	0.022	PT	2
N_2O	≤1.465	—	CT	2
NO_2	2.275	0.025	LPT	2
NiCO	0.804	0.012	LPES	2
NiD_2	1.926	0.007	LPES	34
NiH_2	1.934	0.008	LPES	34
O_3	2.1028	0.0025	LPT	2
PH_2	1.27	0.010	LPES	2
S_3	2.093	0.025	LPES	16
SO_2	1.095	0.008	LPES	16
S_2O	1.878	0.008	LPES	16
SeD_2	1.038	0.013	LPES	34
SeH_2	1.049	0.014	LPES	34
SiH_2	1.124	0.020	LPES	2

Table 4
ELECTRON AFFINITIES FOR LARGER POLYATOMIC MOLECULES

Molecule	Electron affinity in eV	Uncertainty in eV	Method	Ref.
$Fe(CO)_2$	1.22	0.02	LPES	2
$Fe(CO)_3$	1.8	0.2	LPES	2
$Fe(CO)_4$	2.4	0.3	LPES	2
GeH_3	≤1.74	0.04	PT	2
HNO_3	0.57	0.15	CD	2
$Ni(CO)_2$	0.643	0.014	LPES	2
$Ni(CO)_3$	1.077	0.013	LPES	2
$OH(H_2O)$	<2.95	0.15	PT	2
PBr_3	1.59	0.15	CD	2
PBr_2Cl	1.63	0.20	CD	2
PCl_2Br	1.52	0.20	CD	2
PCl_3	0.82	0.10	CD	2
$POCl_2$	3.83	0.25	CD	2
$POCl_3$	1.41	0.20	CD	2
SF_4	2.35	0.10	CT	2
SF_6	0.46	0.20	CD	2
SO_3	≥1.70	0.15	CD	2
SeF_6	2.9	0.2	CD	2
SiF_3	≤2.95	0.10	PT	17
SiH_3	≤1.44	0.03	PT	2
TeF_6	3.34	0.17	CD	2
UF_5	4.4	0.4	CD	2
UF_6	≥5.1	—	CD	2
WF_5	1.25	0.3	CD	18
WF_6	3.36	0.10	CT	19
	3.72	0.20	CD	18
CH_3	0.08	0.03	LPES	2
CO_3	2.69	0.14	LPES	2
CF_3Br	0.91	0.2	CD	2
CF_3I	1.57	0.2	CD	2
CH_3I	0.2	0.1	CT	2
$CO_3(H_2O)$	2.1	0.2	PT	2
CH_3NO_2	0.44	0.2	CD	2
CD_3O	1.552	0.022	LPES	2
CH_3O	1.570	0.022	LPES	2
CD_3S	1.856	0.006	LPT	2
CH_3S	1.861	0.004	LPT	2
C_2DO	2.350	0.020	LPES	13
C_2HO	2.350	0.020	LPES	13
C_2D_2 (vinylidene-d_2)	0.49	0.02	LPES	20
C_2H_2 (vinylidene)	0.47	0.02	LPES	20
C_2H_2FO (acetyl fluoride enolate)	2.22	0.09	PT	2
C_2D_2N (cyanomethyl-d_2 radical)	1.532	0.014	LPES	21
C_2H_2N (cyanomethyl radical)	1.534	0.014	LPES	21
	1.507	0.018	PT	2
C_2H_2N (isocyanomethyl radical)	1.067	0.024	LPES	231
C_2D_3O (acetaldehyde-d_3 enolate)	1.81899	0.00012	LPT	22
C_2H_3O (acetaldehyde enolate)	1.82478	0.00012	LPT	22
C_2H_5N (ethyl nitrene)	0.56	0.01	PT	2
C_2D_5O (ethoxide-d_5)	1.702	0.033	LPES	23
C_2H_5O (ethoxide)	1.726	0.033	LPES	23
C_2H_5S (ethyl sulfide)	1.953	0.006	LPT	2
C_3	1.981	0.020	LPES	11
C_3H	1.858	0.023	LPES	11
C_3H_2	1.794	0.021	LPES	11
$C_3H_2F_3O$ (1,1,1-trifluoroacetone enolate)	2.58	0.12	PT	2
C_3H_3 (propargyl radical, $CH_2C{\equiv}CH$)	0.893	0.025	LPES	24
C_3H_2D ($CH_2C{\equiv}CD$)	0.88	0.15	LPES	24
C_3D_2H ($CD_2C{\equiv}CH$)	0.907	0.023	LPES	24
C_3H_3N (CH_3CH-CN)	1.247	0.012	LPES	21
C_3D_5 (allyl-d_5)	0.381	0.025	LPES	25
C_3H_5 (allyl)	0.362	0.019	LPES	25
C_3H_4D ($CH_2{=}CD$-CH_2)	0.373	0.019	LPES	25
C_3H_5O (acetone enolate)	1.757	0.033	LPES	23
C_3H_5O (propionaldehyde enolate)	1.611	0.023	LPES	23
$C_3H_5O_2$ (methyl acetate enolate)	1.80	0.06	PT	2

Table 4 (continued)
ELECTRON AFFINITIES FOR LARGER POLYATOMIC MOLECULES

Molecule	Electron affinity in eV	Uncertainty in eV	Method	Ref.
C_3H_7O (*n*-propyl oxide)	1.789	0.033	LPES	23
C_3H_7O (isopropyl oxide)	1.839	0.029	LPES	23
C_3H_7S (*n*-propyl sulfide)	2.00	0.02	PT	2
C_3H_7S (*i*-propyl sulfide)	2.02	0.02	PT	2
C_3O	1.34	0.15	LPES	11
C_3O_2	0.85	0.15	LPES	11
$C_4F_4O_3$ (tetrafluorosuccinic anhydride)	0.5	0.2	CD	2
$C_4H_2O_3$ (maleic anhydride)	1.4	0.2	CD	2
C_4H_4N (pyrrolate)	2.39	0.13	PT	2
C_4H_5O (cyclobutane enolate)	1.84	0.06	PT	2
C_4H_7O (butyraldehyde enolate)	1.67	0.05	PT	2
C_4H_6DO (2-butanone-3-d_1 enolate)	1.67	0.05	PT	2
$C_4H_5D_2O$ (2-butanone-3,3-d_2 enolate)	1.75	0.06	PT	2
C_4H_9O (*t*-butoxyl)	1.912	0.054	LPES	23
C_4H_9S (*n*-butyl sulfide)	2.03	0.02	PT	2
C_4H_9S (*t*-butyl sulfide)	2.07	0.02	PT	2
C_4O	2.05	0.15	LPES	11
C_4O_2	2.0	0.2	LPES	11
$C_5F_6O_3$ (hexafluoroglutaric anhydride)	1.5	0.2	CD	2
C_5D_5 (cyclopentadienyl-d_5)	1.790	0.008	LPES	11
C_5H_5 (cyclopentadienyl)	1.804	0.007	LPES	11
C_5H_7 (pentadienyl)	0.91	0.03	PT	2
C_5H_7O (cyclopentanone enolate)	1.62	0.06	PT	2
C_5H_9O (3-pentanone enolate)	1.69	0.05	PT	2
$C_5H_{11}O$ (neopentoxyl)	1.93	0.05	LPT	2
$C_5H_{11}S$ (*n*-pentyl sulfide)	2.09	0.02	PT	2
C_5O_2	1.2	0.2	LPES	11
$C_6Br_4O_2$ (*p*-bromanil)	2.44	0.20	CT	2
$C_6Cl_4O_2$ (*p*-chloranil)	2.76	0.20	CT	2
$C_6F_4O_2$ (*p*-fluoranil)	2.92	0.20	CT	2
C_6F_6 (hexafluorobenzene)	>1.8	0.3	CT	2
C_6F_{10} (perfluorocyclohexene)	>1.4	0.3	CT	2
C_6D_4 (*o*-benzyne-d_4)	0.551	0.010	LPES	36
C_6H_4 (*o*-benzyne)	0.560	0.010	LPES	36
C_6H_4ClO (*o*-chloroperoxide)	≤2.58	0.08	PT	2
$C_6H_4O_2$ (*p*-benzoquinone)	1.89	0.3	CD	2
C_6D_5 (phenyl-d_5)	1.092	0.020	LPES	26
C_6H_5 (phenyl)	1.102	0.020	LPES	26
C_6H_5N (phenyl nitrene)	1.46	0.02	PT	2
$C_6H_5NO_2$ (nitrobenzene)	>0.7	0.2	CT	2
C_6H_5O (phenoxide)	≤2.36	0.06	PT	2
C_6H_5S (thiophenoxide)	≤2.47	0.06	PT	2
C_6H_5NH (anilide)	1.70	0.03	PT	2
C_6H_7 (methylcyclopentadienyl)	<1.67	0.04	PT	2
C_6H_9O (cyclohexanone enolate)	1.55	0.05	PT	2
$C_6H_{11}O$ (pinacolone enolate)	1.88	0.06	PT	2
$C_6H_{11}O$ (3,3-dimethylbutanal enolate)	1.82	0.06	PT	2
C_6N_4 (TCNE)	2.3	0.3	PT	2
C_7F_8 (octafluorotoluene)	>1.7	0.3	CT	2
C_7H_6FO (*m*-fluoroacetophenone enolate)	2.218	0.010	LPT	2
C_7H_6FO (*p*-fluoroacetophenone enolate)	2.176	0.010	LPT	2
C_7H_7O (*o*-methyl phenoxide)	≤2.36	0.06	PT	2
C_7H_9 (heptatrienyl)	1.27	0.03	PT	2
C_7H_9O (2-norbornanone enolate)	1.61	0.05	PT	2
$C_7H_{11}O$ (cycloheptanone enolate)	1.48	0.04	PT	2
$C_7H_{11}O$ (2,5-dimethylcyclopentanone enolate)	1.49	0.04	PT	2
$C_7H_{13}O$ (4-heptanone enolate)	1.72	0.06	PT	2
$C_7H_{13}O$ (di-isopropyl ketone enolate)	1.46	0.04	PT	2
C_8H_7O (acetophenone enolate)	2.057	0.010	LPT	2
C_8H_7O (phenylacetaldehyde enolate)	2.10	0.08	PT	2
$C_8H_{13}O$ (cyclooctanone enolate)	1.63	0.06	PT	2
C_9H_9O (*m*-methylacetophenone enolate)	2.030	0.010	LPT	2
C_9H_9SiN (trimethylsilylnitrene)	1.43	0.10	PT	2
$C_9H_{15}O$ (cyclononanone enolate)	1.69	0.06	PT	2
$C_{10}H_{17}O$ (cyclodecanone enolate)	1.83	0.06	PT	2
$C_{12}H_4N_4$ (TCNQ)	2.8	0.3	CD	2
$C_{12}H_9$ (perinaphthenyl)	1.07	0.10	PT	2
$C_{12}H_{15}O$ (*t*-butylacetophenone enolalte)	2.032	0.010	LPT	2
$C_{12}H_{21}O$ (cyclododecanone enolate)	1.90	0.07	PT	2

REFERENCES

1. **Mead, R. D., Stevens, A. E., and Lineberger, W. C.**, in *Gas Phase Ion Chemistry*, Vol. 3, Bowers, M. T., Ed., Academic Press, Orlando, 1984, p. 213. The present tabulation contains many adjustments to the atomic electron affinities which have been made by H. Hotop and W. C. Lineberger, *J. Phys. Chem. Ref. Data*, 14, 731, 1985. Most of the adjustments simply reflect the improvement in the electron affinities for O and OH which are often used to calibrate energy scales.
2. **Drzaic, P. S., Marks, J., and Brauman, J. I.**, in *Gas Phase Ion Chemistry*, Vol. 3, Bowers, M. T., Ed., Academic Press, Orlando, 1984, p. 167.
3. The correct electron affinities of OH and OD are (14741.03 ± 0.17) cm^{-1} and (14723.92 ± 0.30) cm^{-1}, respectively. See Reference 1. For the present tabulation the value e/hc = 8065.4786 ± 0.0208 cm^{-1}/eV [Taylor, B. N., *Ref. Mod. Phys.*, 56, S31, 1984] has been used to convert electron affinities from cm^{-1} into eV. Only in the cases of O, S, and C_2H_3O is the 2.6 ppm uncertainty in e/hc significant. However, the uncertainties in the electron affinities given in the tables do not include the uncertainty in e/hc.
4. **Neumark, D. M., Lykke, K. R., Andersen, T., and Lineberger, W. C.**, *Phys. Rev. A*, 32, 1890, 1985.
5. **Leopold, D. G., Murray, K. M., Miller, T. M., and Lineberger, W. C.**, to be published 1985.
6. **Sinnott, G. and Beaty, E. C.**, IVth Int. Conf. Phys. Electronic At. Collisions, Amsterdam, 1971. *Book of Abstracts*, North-Holland, Amsterdam, 1971, p. 176.
7. **Klein, R., McGinnis, R. P., and Leone, S. R.**, *Chem. Phys. Lett.*, 100, 475, 1983.
8. **Lee, L. C., Smith, G. P., Moseley, J. T., Cosby, P. C., and Guest, J. A.**, *J. Chem. Phys.*, 70, 3237, 1979.
9. **Stevens, A. E., Fiegerle, C. S., and Lineberger, W. C.**, *J. Chem. Phys.*, 78, 5420, 1983.
10. **Breyer, F., Frey, P., and Hotop, H.**, *Z. Phys.*, A 300, 7, 1981.
11. **Oakes, J. M. and Ellison, G. B.**, to be published.
12. **Leopold, D. G., Murray, K. M., and Lineberger, W. C.**, *J. Chem. Phys.*, 81, 1048, 1984.
13. **Oakes, J. M., Jones, M. E., Bierbaum, V. M., and Ellison, G. B.**, *J. Phys. Chem.*, 87, 4810, 1983.
14. **Ellis, H. B., Jr. and Ellison, G. B.**, *J. Chem. Phys.*, 78, 6541, 1983.
15. **Oakes, J. M., Harding, L. B., and Ellison, G. B.**, *J. Chem. Phys.*, submitted, 1985.
16. **Nimlos, M. E. and Ellison, G. B.**, to be published.
17. **Richardson, L. M., Stephenson, L. M., and Brauman, J. I.**, *Chem. Phys. Lett.*, 30, 17, 1975.
18. **Dispert, H. and Lacmann, K.**, *Chem. Phys. Lett.*, 45, 311, 1977.
19. **Viggiano, A. A. and Paulson, J. F.**, 37th Gaseous Electronics Conference, Boulder, CO, 1984. *Bull. Am. Phys. Soc.*, to be published, 1985.
20. **Burnett, S. M., Stevens, A. E., Fiegerle, C. S., and Lineberger, W. C.**, *Chem. Phys. Lett.*, 100, 124, 1983.
21. **Moran, S., Ellis, H. B., Jr., and Ellison, G. B.**, *J. Am. Chem. Soc.*, submitted, 1985.
22. **Mead, R. D., Lykke, K. R., Lineberger, W. C., Marks, J., and Brauman, J. I.**, *J. Chem. Phys.*, 81, 4883, 1984. See Lykke, K. R., Mead, R. D., and Lineberger, W. C., *Phys. Rev. Lett.*, 52, 2221, 1984. The electron affinities are (14717.7 ± 1.0) cm^{-1} for acetaldehyde enolate and (14671.0 ± 1.0) cm^{-1} for acetaldehyde-d_3 enolate.
23. **Ellison, G. B., Engelking, P. C., and Lineberger, W. C.**, *J. Phys. Chem.*, 86, 4873, 1982.
24. **Oakes, J. M. and Ellison, G. B.**, *J. Am. Chem. Soc.*, 105, 2969, 1983.
25. **Ellison, G. B. and Oakes, J. M.**, *J. Am. Chem. Soc.*, 106, 7734, 1984.
26. **Miller, A. E. S. and Lineberger, W. C.**, to be published.
27. **Leopold, D. G. and Lineberger, W. C.**, *J. Am. Chem. Soc.*, submitted, 1985.
28. **Miller, T. M., Miller, A. E. S., and Lineberger, W. C.**, *Phys. Rev. A*, submitted, 1985.
29. **Miller, A. E. S., Fiegerle, C. S., and Lineberger, W. C.**, *J. Chem. Phys.*, submitted, 1985.
30. **Miller, T. M., Leopold, D. G., Murray, K. K., and Lineberger, W. C.**, *J. Chem. Phys.*, submitted, 1986.
31. **Miller, T. M. and Lineberger, W. C.**, to be published.
32. **Neumark, D. M., Lykke, K. R., Andersen, T., and Lineberger, W. C.**, *J. Chem. Phys.*, 83, 4365, 1985.
33. **Leopold, D. G., Miller, T. M., and Lineberger, W. C.**, *J. Am. Chem. Soc.*, submitted, 1985.
34. **Miller, A. E. S., Fiegerle, C. S., and Lineberger, W. C.**, *J. Chem. Phys.*, submitted, 1985.
35. **Murray, K. K., Miller, T. M., Leopold, D. G., and Lineberger, W. C.**, *J. Chem. Phys.*, submitted, 1985.
36. **Leopold, D. G., Miller, T. M., Miller, A. E. S., and Lineberger, W. C.**, *J. Am. Chem. Soc.*, submitted, 1985.

ATOMIC AND MOLECULAR POLARIZABILITIES

Thomas M. Miller

The *polarizability* of an atom or molecule describes the response of the electron cloud to an external electric field. The atomic or molecular energy shift ΔW due to an external electric field E is proportional to the electric field squared, for external fields which are weak compared to the internal electric fields between the nucleus and electron cloud. The *electric dipole polarizability* α is the constant of proportionality defined by $\Delta W = -{}^1/_2\, \alpha\, E^2$. The induced electric dipole moment is αE. *Hyperpolarizabilities*, coefficients of higher powers of E, are less often required. Technically, the polarizability is a tensor quantity but for spherically symmetric charge distributions reduces to a single number. In any case, an *average polarizability* is usually adequate in calculations. Frequency-dependent or *dynamic polarizabilities* are needed for electric fields which vary in time, except for frequencies which are much lower than electron orbital frequencies, where *static polarizabilities* suffice.

Polarizabilities for atoms and molecules in excited states are found to be larger than for ground states and may be positive or negative. Molecular polarizabilities are very slightly temperature dependent since the size of the molecule depends on its vibrational and rotational quantum numbers. Only in the case of hydrogen has this effect been studied enough to warrant consideration in Table 3.

Tabulated here are static average electric dipole polarizabilities for ground state atoms and molecules.

Polarizabilities are normally expressed in cgs units of cm^3. Ground state polarizabilities are in the range of 10^{-24} cm^3 = 1 $Å^3$ and hence are often given in $Å^3$ units. Theorists tend to use atomic units of a_o^3 where a_o is the Bohr radius. The conversion is $\alpha(cm^3) = 0.148184 \times \alpha(a_o^3)$. Polarizabilities are rarely encountered in SI units, $C{\cdot}m^2/V = J/(V/m)^2$. The conversion from cgs units to SI units is $\alpha(C{\cdot}m^2/V) = 4\pi\epsilon_o \times 10^{-6} \times \alpha(cm^3)$, where ϵ_o is the permittivity of free space in SI units and the factor 10^{-6} simply converts cm^3 into m^3. Thus, $\alpha(C{\cdot}m^2/V) = 1.11265 \times 10^{-16} \times \alpha(cm^3)$. Persons measuring excited state polarizabilities by optical methods tend to use units of $MHz/(V/cm)^2$, where the energy shift, ΔW, is expressed in frequency units with a factor of h understood. The polarizability is $-2\,\Delta W/E^2$. The conversion into cgs units is $\alpha(cm^3) = 5.95531 \times 10^{-16} \times \alpha[MHz/(V/cm)^2]$.

The polarizability appears in many formulas for low energy processes involving the valence electrons of atoms or molecules. These formulas are given below in cgs units: the polarizability α is in cm^3; masses m or μ are in grams; energies are in ergs; and electric charges are in esu, where $e = 4.8032 \times 10^{-10}$ esu. The symbol α(ν) denotes a frequency (ν) dependent polarizability, where α(ν) reduces to the static polarizability α for ν = 0. For further information and references, see T. M. Miller and B. Bederson, *Advances in Atomic and Molecular Physics*, Vol. 13, pp. 1—55, 1977. Details on polarizability-related interactions, especially in regard to hyperpolarizabilities and nonlinear optical phenomena, are given by M. P. Bogaard and B. J. Orr, in *Physical Chemistry*, Series Two, Vol. 2, Molecular Structure and Properties, A. D. Buckingham, Ed., Butterworths, London, 1975, pp. 149—194. A tabulation of tensor and hyperpolarizabilities is included.

Table 1
FORMULA INVOLVING POLARIZABILITY

Description	Formula	Remarks
Lorentz-Lorenz relation	$\alpha(\nu) = \frac{3}{4\pi n}\left[\frac{\eta^2(\nu) - 1}{\eta^2(\nu) + 2}\right]$	For a gas of atoms or nonpolar molecules; the gas number density is n, and the index of refraction is $\eta(\nu)$
Refraction by polar molecules	$\alpha(\nu) + \frac{d^2}{3kT} = \frac{3}{4\pi n}\left[\frac{\eta^2(\nu)-1}{\eta^2(\nu)+2}\right]$	The dipole moment is d, in esu·cm ($= 10^{-18}$ D)
Dielectric constant (dimensionless)	$K(\nu) = 1 + 4\pi n\,\alpha(\nu)$	From the Lorentz-Lorenz relation for the usual case of $K(\nu) \approx 1$
Index of refraction (dimensionless)	$\eta(\nu) = 1 + 2\pi n\,\alpha(\nu)$	From $\eta^2(\nu) = K(\nu)$
Diamagnetic susceptibility	$\chi_m \simeq e^2(a_o N_e \alpha)^{1/2}/4m_e c^2$	From the approximation that the static polarizability is given by the variational formula $\alpha = (4/9a_o)\sum_i (N_i r_i^2)^2$; N_e is the number of electrons, m_e is the electron mass, and a_o is the Bohr radius, 5.292×10^{-9} cm; a crude approximation is $\chi_m = (E_i/4m_e c^2)\alpha$ where E_i is the ionization energy
Long-range electron- or ion-molecule interaction energy	$V(r) = -e^2\alpha/2r^4$	The target molecule polarizability is α
Ion mobility in a gas	$K = 13.87/(\alpha\mu)^{1/2}$ cm^2/(volt·sec)	This one formula is not in cgs units. It gives mobility in usual units of cm^2/(volt·sec) when the gas molecule α is in Å^3 or 10^{-24} cm^3 units and the reduced mass μ of the ion-molecule pair is in amu; classical limit; pure polarization potential
Langevin capture cross section	$\sigma(v_o) = \frac{2\pi e}{v_o}\left(\frac{\alpha}{\mu}\right)^{1/2}$	The relative velocity of approach for an ion-molecular pair is v_o; the target molecular polarizability is α and the reduced mass of the ion-molecular pair is μ
Langevin reaction rate coefficient	$k = 2\pi e\left(\frac{\alpha}{\mu}\right)^{1/2}$	"Gas kinetic" rate coefficient for an ion-molecule reaction
Rate coefficient for polar molecules	$k_d = 2\pi e\left[\left(\frac{\alpha}{\mu}\right)^{1/2} + cd\left(\frac{2}{\mu\pi kT}\right)^{1/2}\right]$	The dipole moment of the neutral is d is esu·cm; the number c is a "locking factor" that depends on α and d, and is between 0 and 1
Modified effective range cross section for electron-neutral scattering	$\sigma(k) = 4\pi A^2 + \frac{8\pi^2\mu}{3\hbar^2}e^2\alpha A k + \ldots$	Here, k is the wavenumber of the electron, equal to its momentum divided by $\hbar$, and $\hbar$ is Planck's constant divided by 2π; A is called the "scattering length;" The reduced mass is μ
van der Waals constant between two systems A,B	$C_6 = \frac{3}{2}\left(\frac{\alpha^A\alpha^B E^A E^B}{E^A + E^B}\right)$	For the interaction potential term $V_6(r) = -C_6 r^{-6}$; the dipole polarizability is α and E is an average dipole transition energy
Dipole-quadrupole constant between two systems A,B	$C_8 = \frac{15}{4}\left(\frac{\alpha^A\alpha_q^B E^A E_q^B}{E^A + E_q^B}\right) + \frac{15}{4}\left(\frac{\alpha_q^A\alpha^B E_q^A E^B}{E_q^A + E^B}\right)$	For the interaction potential term $V_8(r) = -C_8 r^{-8}$; the dipolar polarizability is α and the quadrupole polarizability (the next higher order polarizability beyond dipole) is α_q; E and E_q are average dipole and quadrupole transition energies, respectively
Relation between $\alpha(\nu)$ and oscillator strengths	$\alpha(\nu) = \frac{e^2\hbar^2}{m_e}\sum_k \frac{f_k}{(E_k)^2 - (h\nu)^2}$	Here, f_k is the oscillator strength from the ground state to an excited state k; E_k is the excitation energy of state k; this formula is especially useful for static polarizabilities ($\nu = 0$)
Dynamic polarizability	$\alpha(\nu) = \frac{\alpha E_r^2}{E_r^2 - (h\nu)^2}$	Approximate variation of the frequency-dependent polarizability $\alpha(\nu)$ from zero frequency up to the first dipole-allowed electronic transition, of energy E_r; the static dipole polarizability is α, i.e., for $\nu = 0$; ignoring infrared contributions
Rayleigh scattering cross section	$\sigma(\nu) = \left(\frac{8\pi}{9c^4}\right)(2\pi\nu)^4 \times [3\alpha^2(\nu) + \frac{2}{3}\gamma^2(\nu)]$	The photon frequency is ν; the average polarizability is $\alpha(\nu)$ and the polarizability anisotropy (the difference between polarizabilities parallel and perpendicular to the applied field) is $\gamma(\nu)$

Table 1 (continued)
FORMULA INVOLVING POLARIZABILITY

Description	Formula	Remarks
Casimir-Polder effect	$V(r) = -g\left(\frac{3\alpha^2\bar{E}}{4r^6}\right)$	$V(r)$ is the interaction energy for two nonpolar molecules at very large distances $r \gg cE_i/h$, where E is an average over the valence excitation energies of the molecules; g is a numerical factor which depends on $\bar{E}$ but is typically between 1 and 0.25
Verdet constant	$V(\nu) = \frac{\nu n}{2mc^2}\left(\frac{d\alpha(\nu)}{d\nu}\right)$	Defined from $\theta = V(\nu)B$, where θ is the angle of rotation of linearly polarized light through a medium of number density n, per unit length, for a longitudinal magnetic field strength B (Faraday effect)

Table 2
STATIC AVERAGE ELECTRIC DIPOLE POLARIZABILITIES FOR GROUND STATE ATOMS

Atomic Number	Atom	Polarizability (units of 10^{-24} cm^3)	Estimated accuracy (%)	Method	Ref.
1	H	0.666793	"exact"	Calc	MB77
2	He	0.204956	"exact"	Calc	MB77
		0.2050	0.1	Index/diel	NB65/OC67
3	Li	24.3	2	Beam	MB77
4	Be	5.60	2	Calc	MB77
5	B	3.03	2	Calc	MB77
6	C	1.76	2	Calc	MB77
7	N	1.10	2	Calc/index	MB77
8	O	0.802	2	Calc/index	MB77
9	F	0.557	2	Calc	MB77
10	Ne	0.3956	0.1	Diel	OC67
11	Na	23.6	2	Beam	MB77
12	Mg	10.6	2	Calc	MB77
13	Al	8.34	2	Calc	MB77
14	Si	5.38	2	Calc	MB77
15	P	3.63	2	Calc	MB77
16	S	2.90	2	Calc	MB77
17	Cl	2.18	2	Calc	MB77
18	Ar	1.6411	0.05	Index/diel	NB65/OC67
19	K	43.4	2	Beam	MB77
20	Ca	22.8	2	Calc	MB77
		25.0	8	Beam	MB77
21	Sc	17.8	25	Calc	D84
22	Ti	14.6	25	Calc	D84
23	V	12.4	25	Calc	D84
24	Cr	11.6	25	Calc	D84
25	Mn	9.4	25	Calc	D84
26	Fe	8.4	25	Calc	D84
27	Co	7.5	25	Calc	D84
28	Ni	6.8	25	Calc	D84
29	Cu	6.1	25	Calc	D84
		7.31	25	Calc	G84
30	Zn	7.1	25	Calc	MB77
		5.6	25	Calc	D84
31	Ga	8.12	2	Calc	MB77
32	Ge	6.07	2	Calc	MB77
33	As	4.31	2	Calc	MB77
34	Se	3.77	2	Calc	MB77
35	Br	3.05	2	Calc	MB77
36	Kr	2.4844	0.05	Diel	OC67
37	Rb	47.3	2	Beam	MB77
38	Sr	27.6	8	Beam	MB77
39	Y	22.7	25	Calc	D84
40	Zr	17.9	25	Calc	D84
41	Nb	15.7	25	Calc	D84
42	Mo	12.8	25	Calc	D84
43	Tc	11.4	25	Calc	D84
44	Ru	9.6	25	Calc	D84
45	Rh	8.6	25	Calc	D84
46	Pd	4.8	25	Calc	D84
47	Ag	7.2	25	Calc	D84
		8.56	25	Calc	G84
48	Cd	7.2	25	Calc	D84

Table 2 (continued)
STATIC AVERAGE ELECTRIC DIPOLE POLARIZABILITIES FOR GROUND STATE ATOMS

Atomic Number	Atom	Polarizability (units of 10^{-24} cm^3)	Estimated accuracy (%)	Method	Ref.
49	In	10.2	12	Beam	GMBSJ84
		9.1	25	Calc	D84
50	Sn	7.7	25	Calc	D84
51	Sb	6.6	25	Calc	D84
52	Te	5.5	25	Calc	D84
53	I	5.35	25	Index	A56
		4.7	25	Calc	D84
54	Xe	4.044	0.5	Diel	MB77
55	Cs	59.6	2	Beam	MB77
56	Ba	39.7	8	Beam	MB77
57	La	31.1	25	Calc	D84
58	Ce	29.6	25	Calc	D84
59	Pr	28.2	25	Calc	D84
60	Nd	31.4	25	Calc	D84
61	Pm	30.1	25	Calc	D84
62	Sm	28.8	25	Calc	D84
63	Eu	27.7	25	Calc	D84
64	Gd	23.5	25	Calc	D84
65	Tb	25.5	25	Calc	D84
66	Dy	24.5	25	Calc	D84
67	Ho	23.6	25	Calc	D84
68	Er	22.7	25	Calc	D84
69	Tm	21.8	25	Calc	D84
70	Yb	21.0	25	Calc	D84
71	Lu	21.9	25	Calc	D84
72	Hf	16.2	25	Calc	D84
73	Ta	13.1	25	Calc	D84
74	W	11.1	25	Calc	D84
75	Re	9.7	25	Calc	D84
76	Os	8.5	25	Calc	D84
77	Ir	7.6	25	Calc	D84
78	Pt	6.5	25	Calc	D84
79	Au	5.8	25	Calc	D84
		6.48	25	Calc	G84
80	Hg	5.7	25	Calc	D84
		5.1	15	Diel	MB77
81	Tl	7.6	15	Beam	NYU84
		7.5	25	Calc	D84
82	Pb	6.8	25	Calc	D84
83	Bi	7.4	25	Calc	D84
84	Po	6.8	25	Calc	D84
85	At	6.0	25	Calc	D84
86	Rn	5.3	25	Calc	D84
87	Fr	48.7	25	Calc	D84
88	Ra	38.3	25	Calc	D84
89	Ac	32.1	25	Calc	D84
90	Th	32.1	25	Calc	D84
91	Pa	25.4	25	Calc	D84
92	U	27.4	25	Calc	D84
93	Np	24.8	25	Calc	D84
94	Pu	24.5	25	Calc	D84
95	Am	23.3	25	Calc	D84
96	Cm	23.0	25	Calc	D84
97	Bk	22.7	25	Calc	D84
98	Cf	20.5	25	Calc	D84
99	Es	19.7	25	Calc	D84
100	Fm	23.8	25	Calc	D84
101	Md	18.2	25	Calc	D84
102	No	17.5	25	Calc	D84

Note: Calc = calculated value; Beam = atomic beam deflection technique; Index = determination based on the measured index of refraction; Diel = determination based on the measured dielectric constant.

REFERENCES

A56. **Atoji, M.,** *J. Chem. Phys.*, 25, 174, 1956. Semiempirical method based on molecular polarizabilities and atomic radii.

D84. **Doolen, G. D.,** Los Alamos National Laboratory, unpublished. A relativistic linear response method was used. The method is that described by A. Zangwill and P. Soven, *Phys. Rev. A*, 21, 1561, 1980. Adjustments of less than 10% have been made to these results to bring them into agreement with accurate experimental values where available, for the purpose of presenting "recommended" polarizabilities in Table 2. (T. M. Miller.)

G84. **Gollisch, H.,** *J. Phys. B*, 17, 1463, 1984. Other results and useful references are contained in this paper.

GMBSJ84. **Guella, T. P., Miller, T. M., Bederson, B., Stockdale, J. A. D., and Jaduszliwer, B.,** *Phys. Rev. A*, 29, 2977, 1984.

MB77. **Miller, T. M. and Bederson, B.**, *Adv. At. Mol. Phys.*, 13, 1, 1977. For simplicity, any value in Table 2 which has not changed since this 1977 review is referenced as MB77. Persons interested in original references and further details should consult MB77.

NB65. **Newell, A. C. and Baird, R. D.**, *J. Appl. Phys.*, 36, 3751, 1965.

NYU84. Preliminary value from the New York University group. See GMBSJ84.

OC67. **Orcutt, R. H. and Cole, R. H.**, *J. Chem. Phys.*, 46, 697, 1967. See also the later references from this group, given following Table 2.

Table 3

Average Electric Dipole Polarizabilities for Ground State Diatomic Molecules (in Units of 10^{-24} cm^3)

Molecule	Polarizability	Ref.	Molecule	Polarizability	Ref.
BH	3.32*	1	HF	2.46	3
Br_2	7.02	2	HI	5.44	3
CO	1.95	3		5.35	2
Cl_2	4.61	3	HgCl	7.4*	9
Cs_2	91.	4	ICl	12.3	2
D_2			K_2	61.	4
(v=0, J=0)	0.7921*	5	Li_2	34.	4
(293°K)	0.7954	6	LiCl	3.46*	10
DCl	2.84	2	LiF	10.8*	11
F_2	1.38*	7	LiH	3.84*	12
H_2				3.68*	13
(v=0, J=0)	0.8023*	5		3.88*	14
(293°K)	0.8045*	5	N_2	1.7403	6, 8
(293°K)	0.8042	6	NO	1.70	2
(313°K)	0.8059	8	Na_2	30.	4
HBr	3.61	3	NaLi	40.	4
HCl	2.63	3	O_2	1.5812	6
	2.77	2	Rb_2	68.	4
HD					
(v=0, J=0)	0.7976*	5			

Average Electric Dipole Polarizabilities for Ground State Triatomic Molecules (in Units of 10^{-24} cm^3)

Molecule	Polarizability	Ref.	Molecule	Polarizability	Ref.
BeH_2	4.34*	14	$HgBr_2$	14.5	2
CO_2	2.911	8	$HgCl_2$	11.6	2
CS_2	8.74	3	HgI_2	19.1	2
	8.86	2	N_2O	3.03	8
D_2O	1.26	2	NO_2	3.02	2†
H_2O	1.45	2	O_3	3.21	2
H_2S	3.78	3	OCS	5.71	2
	3.95	2		5.2	15
HCN	2.59	3	SO_2	3.72	3
	2.46	2		4.28	2

Average Electric Dipole Polarizabilities for Ground State Inorganic Polyatomic Molecules (Larger Than Triatomic) (in Units of 10^{-24} cm^3)

Molecule	Polarizability	Ref.	Molecule	Polarizability	Ref.
$AsCl_3$	14.9	2	$(NaF)_2$	21.	16
AsN_3	5.75	2	$(NaI)_2$	27.	16
BCl_3	9.47	2	OsO_4	8.17	2
BF_3	3.31	2	PCl_3	12.8	2
$(BN_3)_2$	5.73	2	PF_5	6.10	2
$(BH_2N)_3$	8.	2†	PH_3	4.84	2
ClF_3	6.32	2	$(RbBr)_2$	48.	16
$(CsBr)_2$	54.	16	$(RbCl)_2$	43.	16
$(CsCl)_2$	42.	16	$(RbF)_2$	41.	16
$(CsF)_2$	28.	16	$(RbI)_2$	46.	16
$(CsI)_2$	52.	16	SF_6	6.54	8
$GeCl_4$	15.1	2	$(SF_5)_2$	13.2	2
GeH_3Cl	6.7	2†	SO_3	4.84	2
$(HgCl)_2$	14.7	9	SO_2Cl_2	10.5	2
$(KBr)_2$	42.	16	SeF_6	7.33	2
$(KCl)_2$	32.	16	SiF_4	5.45	2
$(KF)_2$	25.	16	SiH_4	5.44	2
$(KI)_2$	36	16	$(SiH_3)_2$	11.1	2
(LiBr)	19.	16	$SiHCl_3$	10.7	2
$(LiCl)_2$	13.	16	SiH_2Cl_2	8.92	2
$(LiF)_2$	7.	16	SiH_3Cl	7.02	2
$(LiI)_2$	23.	16	$SnBr_4$	22.0	2
ND_3	1.70	2	$SnCl_4$	18.0	2
NF_3	3.62	2		13.8	15
NH_3	2.26	3	SnI_4	32.3	2
	2.10	2	TeF_6	9.00	2
$(NO_2)_2$	6.69	2	$TiCl_4$	16.4	2
$(NaBr)_2$	27.	16	UF_6	12.5	2
$(NaCl)_2$	23.	16			

Average Electric Dipole Polarizabilities for Ground State Hydrocarbon Molecules (in Units of 10^{-24} cm^3)

Molecule	Polarizability	Ref.
CH_4		
Methane	2.593	8
C_2H_2		
Acetylene	3.33	3
	3.93	2
C_2H_4		
Ethylene	4.252	8
C_2H_6		
Ethane	4.47	3
	4.43	2
C_3H_4		
Propyne	6.18	2
C_3H_6		
Propene	6.26	2
Cyclopropane	5.66	2
C_3H_8		
Propane	6.29	3
	6.37	2
C_4H_6		
1-butyne	7.41	2†
1,3-butadiene	8.64	2
C_4H_8		
1-butene	7.97	2
	8.52	2
trans-2-butene	8.49	2
trans-2,3-epoxy butane	8.22*	17
2-methylpropene	8.29	2
C_4H_{10}		
n-butane	8.20	2
C_5H_6		
1,3-cyclopentadiene	8.64	2
C_5H_8		
1-pentyne	9.12	2
trans-1,3-pentadiene	10.0	2
isoprene	9.99	2
C_5H_{10}		
Cyclopentane	9.15	18
C_5H_{12}		
Pentane	9.99	2
Neopentane	10.20	18
C_6H_6		
Benzene	10.32	3
	10.74	2
C_6H_{10}		
1-hexyne	10.9	2†
2-ethyl-1,3-butadiene	11.8	2†
3-methyl-1,3-pentadiene	11.8	2†
2-methyl-1,3-pentadiene	12.1	2†
2,3-dimethyl-1,3-butadiene	11.8	2†
Cyclohexene	10.7	2†
C_6H_{12}		
Cyclohexane	11.0	18
	10.87	15
C_6H_{14}		
n-hexane	11.9	2
C_7H_8		
Toluene	12.3	2
	12.26	15
C_7H_{12}		
1-heptyne	12.8	2†
C_7H_{14}		
Methylcyclohexane	13.1	2
C_7H_{16}		
n-heptane	13.7	2
C_8H_8		
Styrene	15.	2
C_8H_{10}		
Ethylbenzene	14.2	2
o-xylene	14.9	2
p-xylene	14.1	15
m-xylene	14.9	2
	14.2	15
	14.18	15
C_8H_{16}		
Ethylcyclohexane	15.9	2
C_8H_{18}		
n-octane	15.9	2
C_9H_{12}		
Isopropylbenzene	16.0	2†
C_9H_{18}		
Isopropylcyclohexane	17.2	2
$C_{10}H_8$		
Napthalene	16.5	17
$C_{10}H_{14}$		
t-butylbenzene	17.8	2†
$C_{10}H_{20}$		
t-butylcyclohexane	19.8	2
$C_{14}H_{10}$		
Anthracene	25.4	17
Phenanthracene	38.8*	17

Average Electric Dipole Polarizabilities for Ground State Organic Halides (in Units of 10^{-24} cm^3)

Molecule	Polarizability	Ref.
CBr_2F_2 Dibromodifluoromethane	9.	2†
$CClF_3$ Chlorotrifluoromethane	5.59	8
CCl_2F_2 Dichlorodifluoromethane	7.81	2
CCl_2O Phosgene	7.29	2
CCl_2S Thiophosgene	10.2	2
CCl_3F Trichlorofluoromethane	9.47	2
CCl_3NO_2 Trichloronitromethane	10.8	2†
CCl_4 Carbon tetrachloride	11.2	2
	10.5	3
CF_4 Carbon tetrafluoride	3.838	8
CF_2O Carbonylfluoride	1.88*	17
$CHBr_3$ Bromoform	11.8	17
$CHBrF_2$ Bromodifluoromethane	5.7	2†
$CHClF_2$ Chlorodifluoromethane	5.91	2
$CHCl_2F$ Dichlorofluoromethane	6.82	2
$CHCl_3$ Chloroform	9.5	8
CHF_3 Fluoroform	3.57	8
$CHFO$ Fluoroformaldehyde	1.76*	17
CHI_3 Iodoform	18.0	17
CH_2Br_2 Dibromomethane	9.32	2
CH_2ClNO_2 Chloronitromethane	6.9	2†

Molecule	Polarizability	Ref.
CH_2Cl_2		
Dichloromethane	6.48	3
	7.93	2
CH_3Br		
Bromomethane	6.03	2
	5.55	15
CH_3Cl		
Chloromethane	4.72	8
CH_3F		
Fluoromethane	2.97	8
CH_3I		
Iodomethane	7.97	2
C_2ClF_5		
Chloropentafluoroethane	6.3	2†
$C_2C\ell_2F_4$		
1,2-dichlorotetra-fluorethane	8.5	2†
C_2Cl_3N		
Trichloroacetonitrile	6.10	18
C_2F_6		
Hexafluoroethane	6.82	2
C_2HBr		
Bromoacetylene	7.39	2
C_2HCl		
Chloroacetylene	6.07	2
C_2HCl_5		
Pentachloroethane	14.0	2
$C_2H_2Cl_2$		
cis-dichloroethylene	7.89	2
$C_2H_2Cl_2F_2$		
1,1-dichloro-2,2-di-fluoroethane	8.4	2†
$C_2H_2Cl_2O$		
Chloroacetyl chloride	8.92	2
$C_2H_2Cl_3F$		
1,2,2-trichloro-1-fluoroethane	10.2	2†
$C_2H_2Cl_4$		
1,1,2,2-tetrachloro-ethane	12.1	2†
C_2H_2ClN		
Chloroacetonitrile	6.10	18
C_2H_3Br		
Bromoethylene	7.59	2
C_2H_3Cl	6.41	
Chloroethylene		2
$C_2H_3ClF_2$		
1-chloro-1,1-difluoro-ethane	8.05	2
C_2H_3ClO		
Acetyl chloride	6.62	2
$C_2H_3ClO_2$		
ethyl chloroformate	7.1	2†
$C_2H_3Cl_3$		
1,1,1-trichloroethane	10.7	2
$C_2H_3F_3$		
1,1,1-trifluoroethane	4.4	2†
C_2H_3I		
Iodoethylene	9.3	2†
C_2H_4BrCl		
1-boromo-2-chloroethane	9.5	2†
$C_2H_4Br_2$		
1,2-dibromoethane	10.7	2†
C_2H_4ClF		
1-chloro-2-fluoroethane	6.5	2†
$C_2H_4ClNO_2$		
1-chloro-1-nitroethane	10.9	2
$C_2H_4Cl_2$		
1,1-dichloroethane	8.64	2
1,2-dichloroethane	8.	2†
C_2H_5Br		
Bromoethane	8.05	2
C_2H_5Cl	8.29	2
Chloroethane	6.4	15
C_2H_5ClO		
2-chloroethanol	7.1	2†
Chloromethoxymethane	7.1	2†
C_2H_5F		
Fluoroethane	4.96	2
C_2H_5I		
Iodoethane	10.0	2
$C_3H_4Cl_2$		
Dichloropropene	10.1	2†
C_3H_5Cl		
Chloropropene	8.3	2
C_3H_5ClO		
Chloroacetone	8.4	2†
$C_3H_5ClO_2$		
Ethyl chloroformate	9.0	2†
$C_3H_6ClNO_2$		
1-chloro-1-nitropropane	10.4	2†
$C_3H_6Cl_2$		
Dichloropropane	10.9	2†
C_3H_7Br		
1-bromopropane	9.4	2†
2-bromopropane	9.6	2†
C_3H_7Cl		
Chloropropane	10.0	2
C_3H_7I		
1-iodopropane	11.5	2†
C_4H_5Cl		
4-chloro-1,2-butadiene	10.0	2†
C_4H_7Cl		
1-chloro-2-methylpropene	10.8	2
$C_4H_8Cl_2$		
1,4-dichlorobutane	12.0	2†
C_4H_9Br		
Bromobutane	13.9	2
C_4H_9Cl		
1-chlorobutane	11.3	2
1-chloro-2-methylpropane	11.1	2
2-chloro-2-methylpropane	12.5	2†
2-chlorobutane	12.4	2
C_4H_9I		
1-iodobutane	13.3	2†
$C_5H_{11}Br$		
1-bromopentane	13.1	2†
$C_5H_{11}Cl$		
1-chloropentane	12.0	2†
$C_6H_2Cl_2O_2$		
2,5-dichloro-1,4-benzoquinone	18.4	2
C_6H_4BrF		
p-bromofluorobenzene	13.4	2†
$C_6H_4ClNO_2$		
Chloronitrobenzene	14.6	2†
$C_6H_4Cl_2$	14.0	2
Dichlorobenzene	14.3	15
C_6H_4FI		
p-fluoroiodobenzene	15.5	2†
$C_6H_4FNO_2$		
p-fluoronitrobenzene	12.8	2†
$C_6H_4F_2$		
m-difluorobenzene	10.3	2†
C_6H_5Br		
Bromobenzene	14.7	2
C_6H_5Cl	14.1	2
Chlorobenzene	12.3	15
C_6H_5ClO		
Chlorophenol	13.0	2†
C_6H_5F		2
Fluorobenzene	10.3	
C_6H_5I		
Iodobenzene	15.5	2†
C_7H_7F		
Fluorotoluene	12.3	2†
$C_7H_{15}Br$		
1-bromoheptane	16.8	2†
$C_{12}H_8Br_2O$		
4,4′-dibromodiphenyl ether	27.8	2†
$C_{12}H_9BrO$		
4-bromodiphenyl ether	24.2	2†
$C_{13}H_{11}BrO$		
p-bromophenyl-p-tolyl ether	26.6	2†

Average Electric Dipole Polarizabilities for Other Ground State Organic Molecules (in Units of 10^{-24} cm^3)

Molecule	Polarizability	Ref.
CN_4O_8		
Tetranitromethane	15.3	2
CH_2O		
Formaldehyde	2.8	2†
	2.45	18
CH_2O_2		
Formic acid	3.4	2†
CH_3NO		
Foramide	4.2	2†
	4.08	18
CH_3NO_2		
Nitromethane	7.37	2
CH_4O		
Methanol	3.29	2
	3.23	15
	3.32	18
CH_5N		
ethyl amine	4.7	2
	4.01	19
C_2N_2		
Cyanogen	7.99	2
C_2H_2O		
Ketene	4.4	2†
C_2H_3N		
Acetonitrile	4.40	2†
	4.48	18
C_2H_4O		
Acetaldehyde	4.6	2†
	4.59	18
$C_2H_4O_2$		
Acetic acid	5.1	2†
$C_2H_4O_4$		
Formic acid dimer	12.7	2
C_2H_5NO		
Acetamide	5.67	18
-methyl foramide	5.91	18
$C_2H_5NO_2$		
Nitroethane	9.63	2
Ethyl nitrite	7.0	15
C_2H_5O		
Ethylene oxide	4.43	18
C_2H_6O	5.41	2
Ethanol	5.11	18
	5.84	2
Methyl ether	5.16	15
$C_2H_6O_2$		
Ethylene glycol	5.7	2†
$C_2H_6O_2S$		
Dimethyl sulfone	7.3	2†
C_2H_6S		
Ethanethiol	7.41	2
C_2H_7N		
Ethyl amine	7.10	2
Dimethyl amine	6.37	2
$C_2H_8N_2$		
Ethylene diamine	7.2	2†
$C_3H_2N_2$		
alononitrile	5.79	18
C_3H_3N		
Acrylonitrile	8.05	2
C_3H_4O		
Propenal	6.38	2†
C_3H_5N	6.70	2
Propionitrile	6.24	18
C_3H_6O		
Acetone	6.33	15
	6.4	2†
	6.39	18
Allyl alcohol	7.65	2
Propionaldehyde	6.50	2
$C_3H_6O_2$		
Propionic acid	6.9	2†
Ethyl formate	8.01	2
Methyl acetate	6.94	2
$C_3H_6O_3$		
Dimethyl carbonate	7.7	2†
C_3H_7NO		
-methyl acetamide	7.82	18
.N-dimethyl formamide	7.81	18
$C_3H_7NO_2$		
Nitropropane	8.5	2†
C_3H_8O		
2-propanol	7.61	2
	6.97	18
1-propanol	6.74	2
Methoxyethane	7.93	2†
$C_3H_8O_2$		
Dimethoxymethane	7.7	2†
C_3H_9N		
n-propylamine	9.20	2
trimethylamine	8.15	2
$C_4H_2N_2$		
Fumaronitrile	11.8	2
$C_4H_4N_2$		
Succinonitrile	8.1	2†
Pyrimidene	8.53*	17
Pyridazine	9.27*	17
$C_4H_4O_2$		
Diketene	8.0	2†
C_4H_4S		
Thiophene	9.67	2
C_4H_5N		
Methacrylonitrile	8.0	2†
Trans-crotononitrile	8.2	2†
C_4H_6O		
Crotonaldehyde	8.5	2†
Methacrylaldehyde	8.3	2†
$C_4H_6O_2$		
Biacetyl	8.2	2†
$C_4H_6O_3$		
Acetic anhydride	8.9	2†
C_4H_6S		
Divinyl sulfide	10.9	2†
C_4H_7N		
Butyronitrile	8.4	2†
Isobutyronitrile	8.05	18
C_4H_8O		
Butyraldehyde	8.2	2†
Methyl ethyl ketone	8.13	15
$C_4H_8O_2$		
Ethyl acetate	9.7	2
1,4-dioxane	10.0	2
$C_4H_8O_2$		
p-dioxane	8.60	18
2-methyl-1,3-dioxolane	9.44	15
$C_4H_9NO_2$		
1-nitrobutane	10.4	2†
2-methyl-2-nitropropane	10.3	2†
$C_4H_{10}O$		
Ethyl ether	10.2	2
	8.73	15
1-butanol	8.88	2
2-methylpropanol	8.92	2
$C_4H_{10}S$		
Ethyl sulfide	10.8	2
$C_4H_{11}N$		
n-butylamine	13.5	2
Diethylamine	10.2	2
C_5H_5N		
Pyridine	9.5	15
4-cyano-1,3-butadiene	10.5	2†
$C_5H_8O_2$		
Acetyl acetone	10.5	2†
C_5H_9N		
Valeronitrile	10.4	2
22-DMPN	9.59	18
$C_5H_{10}O$		
Diethyl ketone	9.93	15
Methyl propyl ketone	9.93	15
$C_5H_{10}O_3$		
diethyl carbonate	11.3	2
$C_5H_{12}O_2$		
Diethoxyethane	11.3	2†

Molecule	Polarizability	Ref.
$C_5H_{12}O_4$		
Tetramethyl orthocarbonate	13.	2†
$C_6H_4N_2O_4$		
p-dinitrobenzene	18.4	2
$C_6H_4O_2$		
p-benzoquinone	14.5	2
$C_6H_5NO_2$		
nitrobenzene	14.7	2
	12.92	15
C_6H_6O		
Phenol	11.1	2†
	9.94*	17
C_6H_7N		
Aniline	12.1	2†
$C_6H_8N_2$		
Phenylenediamine	13.8	2†
$C_6H_{10}O_3$		
Ethyl acetoacetate	12.9	2†
$C_6H_{12}N_2$		
Dimethylketazine	15.6	2
$C_6H_{12}O$		
Cyclohexanol	11.56	18
$C_6H_{12}O_2$		
Amyl formate	14.2	2
$C_6H_{12}O_3$		
Paraldehyde	17.9	2
$C_6H_{14}O$	12.8	2
Propyl ether	12.5	15
$C_6H_{14}O_2$		
1,1-diethoxyethane	13.2	2†
$C_6H_{15}N$		
Triethylamine	13.1	2
$C_7H_4N_2O_2$		
p-cyanonitrobenzene	19.	2
C_7H_5N		
Benzonitrile	12.5	2†
$C_7H_7NO_3$		
Nitroanisole	15.7	2†
C_7H_7O		
Anisole	13.1	2†
C_7H_9NO		
o-anisidine	14.2	2†
$C_7H_{14}O$		
Cyclohexyl methyl ether	13.4	2†
di-isopropyl ketone	13.5	15
$C_7H_{14}O_2$		
Amyl acetate	14.9	2
$C_8H_4N_2$		
p-dicyanobenzene	19.2	2
C_8H_8O		
Acetophenone	15.0	2
$C_8H_8O_2$		
2,5-dimethyl-1,4-benzoquinone	18.8	2
$C_8H_{10}O$		
Phenetole	14.9	2
$C_8H_{11}N$		
N-dimethylaniline	16.2	2†
$C_8H_{12}O_2$		
Ethyl sorbate	17.2	2†
Tetramethylcyclobutane-1,3-dione	18.6	2
$C_8H_{14}O_4$		
Diethyl succinate	16.8	2†
$C_8H_{18}O$		
n-butyl ether	17.2	2
$C_9H_{10}O_2$		
Ethyl benzoate	16.9	2†
$C_{10}H_{14}BeO_4$		
Beryllium acetylacetonate	34.1	2
$C_{12}H_9NO_3$		
4-nitrodiphenyl ether	24.7	2†
$C_{14}H_{14}O$		
di-p-tolyl ether	24.9	2†
$C_{15}H_{21}AlO_6$		
Aluminum acetylacetonate	51.9	2
$C_{15}H_{21}CrO_6$	53.7	2
Chromium acetylacetonate		
$C_{15}H_{21}FeO_6$		
Ferric acetylacetonate	58.1	2
$C_{20}H_{18}O_8Th$		
Thorium acetylacetonate	79.	2

Note: All polarizabilities in Table 3 are experimental values except those values marked by an asterisk (*), which indicates a calculated result. The experimental polarizabilities are mostly determined by measurements of a dielectric constant or refractive index which are quite accurate (0.5% or better). However, one should treat many of the results with several percent of caution because of the age of the data and because some of the results refer to optical frequencies rather than static. Comments given with the references are intended to allow one to judge the degree of caution required. Interested persons should consult these references. In many cases, the reference given is to a theoretical paper in which the experimental results are quoted. These papers, noted in the References, contain valuable information on polarizability calculations and experimental data which often includes the tensor components of the polarizability.

REFERENCES

1. **McCullough, E. A., Jr.,** *J. Chem. Phys.*, 63, 5050, 1975. This calculation is for α_{zz}, not $\bar{\alpha}$.
2. **Maryott, A. A. and Buckley, F.,** U. S. National Bureau of Standards Circular No. 537, 1953. A tabulation of dipole moments, dielectric constants, and molar refractions measured between 1910 and 1952, and used here to determine polarizabilities if no more recent result exists. The polarizability is $3/(4\pi N_o)$ times the molar polarization or molar refraction, where N_o is Avogadro's number. The value $3/(4\pi N_o) = 0.3964308 \times 10^{-24}$ cm^3 was used for this conversation. A dagger (†) following the reference number in the present table indicates that the polarizability was derived from the molar refraction and hence may not include some low-frequency contributions to the static polarizability; these "static" polarizabilities are therefore low by 1 to 30%.
3. **Hirschfelder, J. O., Curtis, C. F., and Bird, R. B.,** *Molecular Theory of Gases and Liquids*, John Wiley & Sons, New York, 1954, 950. Fundamental information on molecular polarizabilities.
4. **Miller, T. M. and Bederson, B.,** *Adv. At. Mol. Phys.*, 13, 1, 1977. Review emphasizing atomic polarizabilities and measurement techniques. The data quoted in Table 3 are accurate to 8 to 12%.
5. **Kolos, W. and Wolniewicz, L.,** *J. Chem. Phys.*, 46, 1426, 1967. Highly accurate molecular hydrogen calculations.
6. **Newell, A. C. and Baird, R. C.,** J. Appl. Phys., 36, 3751, 1965. Highly accurate refractive index measurements at 47.7 GHz (essentially static).
7. **Jao, T. C., Beebe, N. H. F., Person, W. B., and Sabin, J. R.,** *Chem. Phys. Lett.*, 26, 47, 1974. Tensor polarizabilities, derivatives, and other results are reported.
8. **Orcutt, R. H. and Cole, R. H.,** *J. Chem. Phys.*, 46, 697, 1967; Sutter, H. and Cole, R. H., *J. Chem. Phys.*, 52, 132, 1970; Bose, T. K. and Cole, R. H., *J. Chem. Phys.*, 52, 140, 1970 and 54, 3829, 1971; Nelson, R. D. and Cole, R. H., *J. Chem. Phys.*, 54, 1971; Bose, T. K., Sochanski, J. S., and Cole, R. H., *J. Chem. Phys.*, 57, 3592, 1972; Kirouac, S. and Bose, T. K., *J. Chem. Phys.*, 59, 3043, 1973 and 64, 1580, 1976. Highly accurate dielectric constant measurements. These modern data give the most accurate polarizabilities available.
9. **Huestis, D. L.,** Technical Report #MP 78—25, SRI International (project PYU 6158), Menlo Park, CA 94025. Molar refractions for mercury-chlorine compounds are analyzed.
10. **Bounds, D. G., Clarke, J. H. R., and Hinchliffe, A.,** *Chem. Phys. Lett.*, 45, 367, 1977. Theoretical tensor polarizability for LiCl.
11. **Kolker, H. J. and Karplus, M.,** *J. Chem. Phys.*, 39, 2011, 1963. Theoretical.
12. **Gutschick, V. P. and McKoy, V.,** *J. Chem. Phys.*, 58, 2397, 1973. Theoretical tensor polarizabilities.
13. **Gready, J. E., Bacskay, G. B., and Hush, N. S.,** *Chem. Phys.*, 22, 141, 1977 and 23, 9, 1977. Theoretical.
14. **Amos, A. T. and Yoffe, J. A.,** *J. Chem. Phys.*, 63, 4723, 1975. Theoretical.

15. **Stuart, H. A.,** *Landolt-Börnstein Zahlenwerte und Funktionen,* Vol. 1, Part 3, Eucken, A. and Hellwege, K. H., Eds., Springer-Verlag, Berlin, 1951, p. 511. Tabulation of molecular polarizabilities. Two misprints in the chemical symbols have been corrected.
16. **Kremens, R., Bederson, B., Jaduszliwer, B., Stockdale, J., and Tino, A.,** *J. Chem. Phys.,* 81, 1676, 1984. Guella, T. P., Miller, T. M., Bederson, B., and Jaduszliwer, B., *Bull. Am. Phys. Soc.,* 29, 797, 1984; also, to be published. Average polarizability measurements and semi-empirical calculations. The data quoted in Table 3 are accurate to about 14%.
17. **Marchese, F. T. and Jaffé,** *Theoret. Chim. Acta (Berlin),* 45, 241, 1977. Theoretical and experimental tensor polarizabilities are tabulated in this paper.
18. **Applequist, J., Carl, J. R., and Fung, K.-K.,** *J. Am. Chem. Soc.,* 94, 2952, 1972. Excellent reference on the calculation of molecular polarizabilities, including extensive tables of tensor polarizabilities, both theoretical and experimental, at 589.3 nm wavelength.
19. **Bridge, N. J. and Buckingham, A. D.,** *Proc. R. Soc. (London),* A295, 334, 1966. Measured tensor polarizabilities at 633 nm wavelength.

IONIZATION POTENTIALS[a]

Z Element	I	II	III	IV	V	VI	VII	VIII	IX	X	XI	XII	XIII	XIV	XV	XVI	XVII	XVIII	XIX	XX	XXI
1 H	13.598																				
2 He	24.587	54.416																			
3 Li	5.392	75.638	122.451																		
4 Be	9.322	18.211	153.893	217.713																	
5 B	8.298	25.154	37.930	259.368	340.217																
6 C	11.260	24.383	47.887	64.492	392.077	489.981															
7 N	14.534	29.601	47.448	77.472	97.888	552.057	667.029														
8 O	13.618	35.116	54.934	54.934	77.412	113.896	138.116	739.315	871.387												
9 F	17.422	34.970	62.707	87.138	114.240	157.161	185.182	953.886	1103.089												
10 Ne	21.564	40.962	63.45	97.11	126.21	157.93	207.27	239.09	1195.797	1362.164											
11 Na	5.139	47.286	71.64	98.91	138.39	172.15	208.47	264.18	299.87	1465.091	1648.659										
12 Mg	7.646	15.035	80.143	109.24	141.26	186.50	224.94	265.90	327.95	367.53	1761.802	1962.613									
13 Al	5.986	18.828	28.447	119.99	153.71	190.47	241.43	284.59	330.21	398.57	442.07	2085.983	2304.080								
14 Si	8.151	16.345	33.492	45.141	166.77	205.05	246.52	303.17	351.10	401.43	476.06	523.50	2437.676	2673.108							
15 P	10.486	19.725	30.18	51.37	65.023	230.43	263.22	309.41	371.73	424.50	479.57	560.41	611.85	2816.943	3069.762						
16 S	10.360	23.33	34.83	47.30	72.68	88.049	280.93	328.23	379.10	447.09	504.78	564.65	651.63	707.14	3223.836	3494.099					
17 Cl	12.967	23.81	39.61	53.46	67.8	98.03	114.193	348.28	400.05	455.62	529.26	591.97	656.69	749.74	809.39	3658.425	3946.193				
18 Ar	15.759	27.629	40.74	59.81	75.02	91.007	124.319	143.456	422.44	478.68	538.95	618.24	686.09	755.73	854.75	918	4120.778	4426.114			
19 K	4.341	31.625	45.72	60.91	82.66	100.0	117.56	154.86	175.814	503.44	564.13	629.09	714.02	787.13	861.77	968	1034	4610.955	4933.931		
20 Ca	6.113	11.871	50.908	67.10	84.41	108.78	127.7	147.24	188.54	211.270	591.25	656.39	726.03	816.61	895.12	974	1087	1157	5129.045	5469.738	
21 Sc	6.54	12.80	24.76	73.47	91.66	111.1	138.0	158.7	180.02	225.32	249.832	685.89	755.47	829.79	926.00						
22 Ti	6.82	13.58	27.491	43.266	99.22	119.36	140.8	168.5	193.2	215.91	265.23	291.497	787.33	861.33	940.36						
23 V	6.74	14.65	29.310	46.707	65.23	128.12	150.17	173.7	205.8	230.5	255.04	308.25	336.267	895.58	974.02						
24 Cr	6.766	16.50	30.96	49.1	69.3	90.56	161.1	184.7	209.3	244.4	270.8	298.0	355	384.30	1010.64						
25 Mn	7.435	15.640	33.667	51.2	72.4	95	119.27	196.46	221.8	248.3	286.0	314.4	343.6	404	435.3	1136.2					
26 Fe	7.870	16.18	30.651	54.8	75.0	99	125	151.06	235.04	262.1	290.4	330.8	361.0	392.2	457	489.5	1266.1				
27 Co	7.86	17.06	33.50	51.3	79.5	102	129	157	186.13	276	305	336	379	411	444	512	546.8	1403.0			
28 Ni	7.635	18.168	35.17	54.9	75.5	108	133	162	193	224.5	321.2	352	384	430	464	499	571	607.2	1547		
29 Cu	7.726	20.292	36.83	55.2	79.9	103	139	166	199	232	266	368.8	401	435	484	520	557	633	671	1698	
30 Zn	9.394	17.964	39.722	59.4	82.6	108	134	174	203	238	274	310.8	419.7	454	490	542	579	619	698	738	1856
31 Ga	5.999	20.51	30.71	64																	
32 Ge	7.899	15.934	34.22	45.71	93.5																
33 As	9.81	18.633	28.351	50.13	62.63	127.6															
34 Se	9.752	21.19	30.820	42.944	68.3	81.70	155.4														
35 Br	11.814	21.8	36	47.3	59.7	88.6	103.0	192.8													
36 Kr	13.999	24.359	36.95	52.5	64.7	78.5	111.0	126	230.39												
37 Rb	4.177	27.28	40	52.6	71.0	84.4	99.2	136	150	277.1											
38 Sr	5.695	11.030	43.6	57	71.6	90.8	106	122.3	162	177	324.1										
39 Y	6.38	12.24	20.52	61.8	77.0	93.0	116	129	146.52	191	206	374.0									
40 Zr	6.84	13.13	22.99	34.34	81.5																
41 Nb	6.88	14.32	25.04	38.3	50.55	102.6	125														
42 Mo	7.099	16.15	27.16	46.4	61.2	68	126.8	153													
43 Te	7.28	15.26	29.54																		
44 Ru	7.37	16.76	28.47																		
45 Rh	7.46	18.08	31.06																		
46 Pd	8.34	19.43	32.93																		
47 Ag	7.576	21.49	34.83																		

Z Element	I	II	III	IV	V	VI	VII	VIII	IX	X	XI	XII	XIII	XIV	XV	XVI	XVII	XVIII	XIX	XX	XXI
48 Cd	8.993	16.908	37.48																		
49 In	5.786	18.869	28.03	54																	
50 Sn	7.344	14.632	30.502	40.734	72.28																
51 Sb	8.641	16.53	25.3	44.2	56	108															
52 Te	9.009	18.6	27.96	37.41	58.75	70.7	137														
53 I	10.451	19.131	33																		
54 Xe	12.130	21.21	32.1																		
55 Ca	3.894	25.1																			
56 Ba	5.212	10.004																			
57 La	5.577	11.06	19.175																		
58 Ce	5.47	10.85	20.20	36.72																	
59 Pr	5.42	10.55	21.62	38.95	57.45																
60 Nd	5.49	10.72																			
61 Pm	5.55	10.90																			
62 Sm	5.63	11.07																			
63 Eu	5.67	11.25																			
64 Gd	6.14	12.1																			
65 Tb	5.85	11.52																			
66 Dy	5.93	11.67																			
67 Ho	6.02	11.80																			
68 Er	6.10	11.93																			
69 Tm	6.18	12.05	23.71																		
70 Yb	6.254	12.17	25.2																		
71 Lu	5.426	13.9																			
72 Hf	7.0	14.9	23.3	33.3																	
73 Ta	7.89																				
74 W	7.98																				
75 Re	7.88																				
76 Os	8.7																				
77 Ir	9.1																				
78 Pt	9.0	18.563																			
79 Au	9.225	20.5																			
80 Hg	10.437	18.756	34.2																		
81 Tl	6.108	20.428	29.83																		
82 Pb	7.416	15.032	31.937	42.32	68.8																
83 Bi	7.289	16.69	25.56	45.3	56.0	88.3															
84 Po	8.42																				
85 At																					
86 Ru	10.748																				
87 Fr																					
88 Ra	5.279	10.147																			
89 Ac	6.9	12.1																			
90 Th		11.5	20.0	28.8																	
91 Pa																					
92 U																					
93 Np																					
94 Pu	5.8																				
95 Am	6.0																				

[a] Numerical values in this table are expressed in electron volts. The conversion factor used in converting the spectral data to electron volts was 1 eV = 8065.73 cm^{-1}.

From Moore, C. E., *Analyses of Optical Spectra*, NSRDS-NBS 34, Office of Standard Reference Data, National Bureau of Standards, Washington, D.C.

NUCLEAR SPINS, MOMENTS, AND MAGNETIC RESONANCE FREQUENCIES

Kenneth Lee and Weston A. Anderson

1967

This table contains the published values for the nuclear spins, magnetic moments, and quadrupole moments, and the calculated values for the nuclear magnetic resonance (NMR) frequency and for the relative sensitivities. Only those isotopes with both published spin and magnetic moment values are tabulated. The magnetic and quadrupole moment values were selected from results published during the period from January, 1955 to June, 1967. Earlier references were obtained from H. E. Walchli, A Table of Nuclear Moment Data, U.S. Atomic Energy Commission Report ORNL—1469, Supplement I (1953) and Supplement II (1955), and D. Streminger, J. M. Hollander, and G. T. Seaborg, Table of Isotopes, Rev. Mod. Phys. **30**, 585 (1958). A table containing the known (1963) spin and electromagnetic moment values of nuclear ground and excited states has been compiled by I. Lindgren, Perturbed Angular Correlations; E. Karlsson, E. Mathias, and K. Siegbahn, editors; North-Holland Publishing Co. (1964). The magnetic moments given in this latter table are corrected for the diamagnetic effect. A more complete list of spin and moment results for nuclei in excited states are included in Lindgren's table.

In general, the results chosen for this table were selected with an inclination to NMR measurements and to the precision of the measurement. Only six significant figures are used in this table. Therefore, the number of figures may be less than those published. The experimental methods employed in determining the moments are indicated by the following symbols:

Ab = atomic beam magnetic resonance (hyperfine structure, double or triple resonance or other method)
E = electron spin resonance or electron-nuclear double resonance
M = microwave absorption in gases
Mb = molecular (or diamagnetic) beam magnetic resonance
Mc = miscellaneous
Mo = Mössbauer effect
N = nuclear magnetic resonance
No = nuclear orientation
O = optical spectroscopy (hyperfine structure, band structure, double resonance, or optical pumping)
Qr = quadrupole resonance

Other symbols used in the table are:

A = atomic weight (mass number)
El = element
I = nuclear spin in units of $h/2\pi$
μ = magnetic moment in units of the nuclear magneton $eh/4\pi Mc$
m = metastable excited state
Q = quadrupole moment in units of barns (10^{-24} cm²)
Z = atomic number
* = radioactive isotope
• = magnetic moment observed by NMR
( ) = assumed or estimated values

Assuming a nuclear magneton value of 5.0505×10^{-24} erg/gauss, the NMR frequency was calculated for a total field of 10^4 gauss. The sensitivities, relative to the proton, are calculated from the following expressions:

$$\text{Sensitivity at constant field} = 7.652 \times 10^{-3}\mu^3 (I+1)/I^2$$
$$\text{Sensitivity at constant frequency} = 0.2387\, \mu(I+1).$$

These expressions assume an equal number of nuclei, a constant temperature, and $T_1 = T_2$ (the longitudinal relaxation time equals the transverse relaxation time). These sensitivities represent the ideal induced voltage in the receiver coil at saturation and with a constant noise source. The calculated values are therefore determined under complete optimum conditions and should be regarded as such.

Isotope						Relative Sensitivity for Equal Number of Nuclei		Magnetic Moment μ			Electric Quadrupole Moment Q		
Z	El	A	Spin I	NMR Frequency in MHz for a 10 kilogauss field	Natural Abundance %	At constant field	At constant frequency	In multiples of the nuclear magneton (eh/4πMc)	Reference	Method	In multiples of barns (10^{-24}cm²)	Reference	Method
0	n	1*	1/2	29.167	—	0.322	0.685	−1.91315	1	Ab	—		
1	•H	1	1/2	42.5759	99.985	1.00	1.00	2.79268	2	N	—		
1	•H	2	1	6.53566	1.5x10^{-2}	9.65x10^{-3}	0.409	0.857387	3	N	2.73x10^{-3}	4	Ab
1	•H	3*	1/2	45.4129	—	1.21	1.07	2.97877	5	N	—		
2	•He	3	1/2	32.433	1.3x10^{-4}	0.442	0.762	−2.1274	6 7	N	—		
3	•Li	6	1	6.2653	7.42	8.50x10^{-3}	0.392	0.82192	8	N	6.9x10^{-4}	9	N
3	•Li	7	3/2	16.546	92.58	0.293	1.94	3.2560	8	N	−3x10^{-2}	10	O
3	Li	8*	2	6.300	—	2.59x10^{-2}	1.184	1.653	11	Ab			
4	•Be	9	3/2	5.9834	100	1.39x10^{-2}	0.703	−1.1774	12 13	N	5.2x10^{-2}	14	Ab
5	•B	10	3	4.5754	19.58	1.99x10^{-2}	1.72	1.8007	15	N	7.4x10^{-2}	16	Ab
5	•B	11	3/2	13.660	80.42	0.165	1.60	2.6880	17	N	3.55x10^{-2}	16	Ab
6	•C	13	1/2	10.7054	1.108	1.59x10^{-2}	0.251	0.702199	18	N	—		
7	N	13*	1/2	4.91	—	1.53x10^{-3}	0.115	(−)0.322	19	Ab	—		
7	•N	14	1	3.0756	99.63	1.01x10^{-3}	0.193	0.40347	20 21	N	1.6x10^{-2}		Mc
7	•N	15	1/2	4.3142	0.37	1.04x10^{-3}	0.101	−0.28298	20 21	N	—		
8	O	15*	1/2	11.0	—	1.70x10^{-2}	0.257	0.719	23	Ab	—		
8	•O	17	5/2	5.772	3.7x10^{-2}	2.91x10^{-2}	1.58	−1.8930	24	N	−2.6x10^{-2}	25	M
9	•F	17*	5/2	14.40	—	0.451	3.94	4.720	26	N			
9	•F	19	1/2	40.0541	100	0.833	0.941	2.62727	17 21	N	—		
9	•F	20*	2	7.977	—	5.26x10^{-2}	1.50	2.093	27	N			
10	Ne	19*	(1/2)	28.75	—	0.308	0.675	−1.886	28	Ab			
10	Ne	21	3/2	3.3611	0.257	2.50x10^{-3}	0.395	−0.66140	29	Ab			
11	Na	21*	3/2	12.126	—	0.116	1.42	2.3861	30	Ab			
11	Na	22*	3	4.436	—	1.81x10^{-2}	1.67	1.746	31	Ab			
11	•Na	23	3/2	11.262	100	9.25x10^{-2}	1.32	2.2161	17	N	0.14-0.15	32 33	O
11	Na	24*	4	3.221	—	1.15x10^{-2}	2.02	1.690	34	Ab			
12	•Mg	25	5/2	2.6054	10.13	2.67x10^{-3}	0.714	−0.85449	13	N			
13	•Al	27	5/2	11.094	100	0.206	3.04	3.6385	35	N	0.149	36	Ab
14	•Si	29	1/2	8.4578	4.70	7.84x10^{-3}	0.199	−0.55477	37	N	—		
15	•P	31	1/2	17.235	100	6.63x10^{-2}	0.405	1.1305	38	N	—		
15	P	32*	1	1.923	—	2.46x10^{-4}	0.120	−0.2523	39	E			
16	•S	33	3/2	3.2654	0.76	2.26x10^{-3}	0.383	0.64257	37	N	−6.4x10^{-2}	40	M
16	S	35*	3/2	5.08	—	8.50x10^{-3}	0.597	1.00	41	M	4.5x10^{-2}	40	M
17	•Cl	35	3/2	4.1717	75.53	4.70x10^{-3}	0.490	0.82091	8	N	−7.89x10^{-2}	42	Ab
17	•Cl	36*	2	4.8931	—	1.21x10^{-2}	0.919	1.2838	43	N	−1.72x10^{-2}	44	M
17	•Cl	37	3/2	3.472	24.47	2.71x10^{-3}	0.408	0.6833	45	N	−6.21x10^{-2}	42	Ab
18	Ar	37*	3/2	5.08	—	8.50x10^{-3}	0.597	1.0	240	O			
19	K	38*	3	3.491	—	8.82x10^{-3}	1.31	1.374	46	Ab			
19	•K	39	3/2	1.9868	93.10	5.08x10^{-4}	0.233	0.39097	47 48	N	0.11	49 50	O
19	K	40*	4	2.470	1.18x10^{-2}	5.21x10^{-3}	1.55	−1.296	51	Mb			
19	•K	41	3/2	1.0905	6.88	8.40x10^{-5}	0.128	0.21459	48	N			
19	K	42*	2	4.345	—	8.50x10^{-3}	0.816	−1.140	52	Ab			
19	K	43*	3/2	0.828	—	3.68x10^{-5}	9.73x10^{-2}	0.163	53	Ab			
20	•Ca	41*	7/2	3.4681	—	1.14x10^{-2}	1.71	−1.5924	54	N			
20	•Ca	43	7/2	2.8646	0.145	6.40x10^{-3}	1.41	−1.3153	55	N			
21	Sc	43*	7/2	10.04	—	0.275	4.95	4.61	56	Ab	−0.26	56	Ab
21	Sc	44*	2	9.76	—	9.63x10^{-2}	1.83	2.56	57	Ab	0.14	57	Ab
21	Sc	44$^{*}_{m}$	6	5.03	—	9.24x10^{-2}	6.62	3.96	57	Ab	0.37	57	Ab
21	•Sc	45	7/2	10.343	100	0.301	5.10	4.7492	58	N	−0.22	59	Ab
21	Sc	46*	4	5.77	—	6.65x10^{-2}	3.62	3.03	60	Ab	0.12	60	Ab
21	Sc	47*	7/2	11.6	—	0.426	5.73	5.33	56	Ab	−0.22	56	Ab
22	Ti	45*	7/2	0.207	—	2.40x10^{-6}	0.102	0.095	61	Ab	1.5x10^{-2}	61	Ab
22	•Ti	47	5/2	2.4000	7.28	2.09x10^{-3}	0.658	−0.78710	62	N			
22	•Ti	49	7/2	2.4005	5.51	3.76x10^{-3}	1.18	−1.1022	62	N			
23	V	49*	7/2	9.71	—	0.249	4.79	4.46	63	E			
23	•V	50	6	4.2450	0.24	5.55x10^{-2}	5.58	3.3413	8	N			
23	•V	51	7/2	11.19	99.76	0.382	5.52	5.139	64	N	−4x10^{-2}	65	O
24	•Cr	53	3/2	2.4065	9.55	9.03x10^{-4}	0.283	−0.47354	66	N			
25	Mn	52*	6	3.907	—	4.33x10^{-2}	5.14	3.075	67	Ab			
25	Mn	52$^{*}_{m}$	2	0.030	—	2.9x10^{-9}	5.73x10^{-3}	0.008	68	Ab			
25	Mn	53*	7/2	11.0	—	0.362	5.42	5.05	69	E			

NUCLEAR SPINS, MOMENTS, AND MAGNETIC RESONANCE FREQUENCIES (Continued)

Isotope			Spin I	NMR Frequency in MHz for a 10 kilogauss field	Natural Abundance %	Relative Sensitivity for Equal Number of Nuclei		Magnetic Moment μ			Electric Quadrupole Moment Q		
Z	El	A				At constant field	At constant frequency	In multiples of the nuclear magneton (eh/4πMc)	Reference	Method	In multiples of barns (10^{-24}cm²)	Reference	Method
25	Mn	54*	(2) (3)	8.4 6.6	—	6.11×10^{-2} 5.98×10^{-2}	1.58 2.48	(2.2) (2.6)	70	No			
25	•Mn	55	5/2	10.501	100	0.175	2.88	3.444	71	E	0.55	72	M
25	Mn	56*	3	8.233	—	0.116	3.09	3.240	73	Ab			
26	•Fe	57	1/2	1.3758	2.19	3.37×10^{-5}	3.23×10^{-2}	0.09024	74	E	—		
27	Co	55*	7/2	10.0	—	0.274	4.94	4.6	75	No			
27	Co	56*	4	7.34	—	0.136	4.60	3.85	76	E			
27	Co	57*	7/2	10.1	—	0.283	4.99	4.65	76	E			
27	Co	58*	2	15.4	—	0.381	2.90	4.05	77	E			
27	•Co	59	7/2	10.054	100	0.277	4.96	4.6163	241 242	N	0.40	78	Ab
27	Co	60*	5	5.793	—	0.101	5.44	3.800	79	E			
28	•Ni	61	3/2	3.8047	1.19	3.57×10^{-3}	0.447	−0.74868	80	N			
29	Cu	61*	3/2	10.8	—	8.22×10^{-2}	1.27	2.13	81	Ab			
29	•Cu	63	3/2	11.285	69.09	9.31×10^{-2}	1.33	2.2206	8	N	−0.16	82	E
29	Cu	64*	1	3.1	—	9.79×10^{-4}	0.191	0.40	83	Ab			
29	•Cu	65	3/2	12.089	30.91	0.114	1.42	2.3789	8	N	−0.15	82	E
29	Cu	66*	1	1.65	—	1.54×10^{-4}	0.103	−0.216	81	Ab			
30	Zn	65*	5/2	2.345	—	1.95×10^{-3}	0.643	0.7692	83	O	-2.4×10^{-2}	83	O
30	•Zn	67	5/2	2.663	4.11	2.85×10^{-3}	0.730	0.8733	37	N	0.15	83	O
31	Ga	68*	1	0.0892	—	2.45×10^{-8}	5.59×10^{-3}	0.0117	84	Ab	3.1×10^{-2}	84	Ab
31	Ga	69	3/2	10.22	60.4	6.91×10^{-2}	1.20	2.011	8	N	0.178	85	Ab
31	Ga	71	3/2	12.984	39.6	0.142	1.52	2.5549	8	N	0.112	85	Ab
31	Ga	72*	3	0.33591	—	7.80×10^{-6}	0.126	−0.13220	86	Ab	0.72	86	Ab
32	Ge	71*	1/2	8.4	—	7.64×10^{-3}	0.197	0.55	87	Ab	—		
32	•Ge	73	9/2	1.4852	7.76	1.40×10^{-3}	1.15	−0.87679	62	N	−0.2	88	M
33	•As	75	3/2	7.2919	100	2.51×10^{-2}	0.856	1.4349	15	N	0.3	89	M
33	As	76*	2	3.45	—	4.27×10^{-3}	0.649	−0.906	90	No			
34	•Se	77	1/2	8.118	7.58	6.93×10^{-3}	0.191	0.5325	8	N	—		
34	Se	79	7/2	2.22	—	2.98×10^{-3}	1.10	−1.02	91	M	0.9	40	M
35	Br	76*	1	4.18	—	2.52×10^{-3}	0.262	(−)0.548	92	Ab	0.27	92	Ab
35	•Br	79	3/2	10.667	50.54	7.86×10^{-2}	1.25	2.0990	8	N	0.33	93	Ab
35	Br	80*	1	3.92	—	2.08×10^{-3}	0.245	0.514	94	Ab	0.20	94	Ab
35	Br	80m*	5	2.008	—	4.20×10^{-3}	1.89	1.317	94	Ab	0.76	94	Ab
35	•Br	81	3/2	11.498	49.46	9.85×10^{-2}	1.35	2.2626	8	N	0.28	93	Ab
35	Br	82*	5	2.479	—	7.90×10^{-3}	2.33	(+)1.626	95	Ab	(+)0.76	95	Ab
36	•Kr	83	9/2	1.638	11.55	1.88×10^{-3}	1.27	−0.9671	96	N	0.15	97	O
36	Kr	85*	9/2	1.6956	—	2.08×10^{-3}	1.31	−1.001	98	O	0.25	98	O
37	Rb	81*	3/2	10.4	—	7.32×10^{-2}	1.22	2.05	99	Ab			
37	Rb	82m*	5	2.29	—	6.20×10^{-3}	2.15	1.50	99	Ab			
37	Rb	83*	5/2	4.33	—	1.23×10^{-2}	1.19	1.42	99	Ab			
37	Rb	84*	2	5.03	—	1.32×10^{-2}	0.945	−1.32	99	Ab			
37	•Rb	85	5/2	4.1108	72.15	1.05×10^{-2}	1.13	1.3482	100	N	0.27	101	Ab
37	Rb	86*	2	6.44	—	2.77×10^{-2}	1.21	−1.69	102	Ab			
37	•Rb	87	3/2	13.931	27.85	0.175	1.64	2.7414	103	N	0.13	101	Ab
38	•Sr	87	9/2	1.8452	7.02	2.69×10^{-3}	1.43	−1.0893	66	N	0.2	104	E
39	•Y	89	1/2	2.0859	100	1.18×10^{-4}	4.90×10^{-2}	−0.13682	47	N	—		
39	Y	90*	2	6.17	—	2.44×10^{-2}	1.16	−1.62	105	Ab	−0.16	105	Ab
39	Y	91*	1/2	2.49	—	1.99×10^{-4}	5.84×10^{-2}	0.163	60	Ab	—		
40	•Zr	91	5/2	3.97249	11.23	9.48×10^{-3}	1.09	−1.30284	106	N			
41	•Nb	93	9/2	10.407	100	0.482	8.07	6.1435	107	N	−0.2	108	O
42	•Mo	95	5/2	2.774	15.72	3.23×10^{-3}	0.760	0.9097	45	N	0.12	109	N
42	•Mo	97	5/2	2.832	9.46	3.43×10^{-3}	0.776	−0.9289	45	N	1.1	109	N
43	•Tc	99*	9/2	9.5830	—	0.376	7.43	5.6572	110	N	0.3	111	O
44	Ru	99	5/2	1.9607	12.72	1.95×10^{-4}	0.169	−0.6430	112	Mo			
44	Ru	101	5/2	2.1975	17.07	1.41×10^{-3}	0.576	−0.7207	113	O			
45	•Rh	103	1/2	1.3401	100	3.11×10^{-5}	3.15×10^{-2}	−0.08790	43 114	N	—		
46	•Pd	105	5/2	1.95	22.23	1.12×10^{-3}	0.534	−0.639	115	N			
47	Ag	104*	5	6.1	—	0.118	5.73	4.0	116	Ab			
47	Ag	104m*	2	14.0	—	0.291	2.65	3.7	116	Ab			
47	Ag	105*	1/2	1.54	—	4.73×10^{-5}	3.62×10^{-2}	0.101	117	Ab	—		
47	•Ag	107	1/2	1.7229	51.82	6.62×10^{-5}	4.05×10^{-2}	−0.113	47 118	N	—		
47	Ag	108*	1	32.0	—	1.13	2.01	4.2	119	Ab			
47	•Ag	109	1/2	1.9807	48.18	1.01×10^{-4}	4.65×10^{-2}	−0.129	47 118	N	—		
47	Ag	110m*	6	4.557	—	6.87×10^{-2}	5.99	3.587	120	Ab			
47	Ag	111*	1/2	2.21	—	1.40×10^{-4}	5.19×10^{-2}	−0.145	121	Ab	—		
47	Ag	112*	2	0.2077	—	9.00×10^{-7}	3.90×10^{-2}	0.054	122	Ab			
47	Ag	113*	1/2	2.41	—	1.81×10^{-4}	5.66×10^{-2}	0.158	122	Ab	—		
48	Cd	107*	5/2	1.879	—	1.00×10^{-3}	0.515	−0.616	123	O	0.8	123	O
48	Cd	109*	5/2	2.529	—	2.44×10^{-3}	0.693	−0.829	124	O	0.8	124	O
48	•Cd	111	1/2	9.028	12.75	9.54×10^{-3}	0.212	−0.592	125 126	N	—		
48	•Cd	113	1/2	9.445	12.26	1.09×10^{-2}	0.222	−0.619	125 126	N	—		
48	Cd	113m*	11/2	1.51	—	2.13×10^{-3}	1.69	−1.09	239	O	−0.79	239	O
48	Cd	115*	1/2	9.862	—	1.24×10^{-2}	0.232	−0.646	127	O	—		
48	Cd	115m*	11/2	1.447	—	1.87×10^{-3}	1.62	−1.044	127	O	−0.61	127	O
49	•In	113	9/2	9.3099	4.28	0.345	7.22	5.496	45	N	1.14	97	Ab
49	In	113m*	1/2	3.209	—	4.28×10^{-4}	7.54×10^{-2}	−0.210	128	Ab	—		
49	In	114m*	5	7.2	—	0.191	6.73	4.7	129	Ab			
49	•In	115*	9/2	9.3301	95.72	0.347	7.23	5.507	130	N	1.16	97	Ab
49	In	115m*	1/2	3.715	—	6.64×10^{-4}	8.73×10^{-2}	−0.243	131	Ab	—		
49	In	116*	5	6.42	—	0.137	6.03	4.21	132	Ab			
49	In	116m*	5	6.7	—	0.156	6.30	4.4	129	Ab			
50	•Sn	115	1/2	13.922	0.35	3.50×10^{-2}	0.327	−0.913	125	N	—		
50	•Sn	117	1/2	15.168	7.61	4.52×10^{-2}	0.356	−0.994	125	N	—		
50	•Sn	119	1/2	15.869	8.58	5.18×10^{-2}	0.373	−1.040	125	N	—		
51	•Sb	121	5/2	10.189	57.25	0.160	2.79	3.341	133	N	−0.5	134	O
51	Sb	122*	2	7.24	—	3.94×10^{-2}	1.36	−1.90	135 136	No			
51	•Sb	123	7/2	5.5176	42.75	4.57×10^{-2}	2.72	2.533	45	N	−0.7	134	O
52	Te	119*	1/2	4.12	—	9.04×10^{-4}	9.67×10^{-2}	0.27	137	Ab	—		
52	•Te	123	1/2	11.16	0.87	1.8×10^{-2}	0.262	−0.731	37	N	—		
52	•Te	125	1/2	13.45	6.99	3.15×10^{-2}	0.316	−0.882	37	N	—		
53	I	125*	5/2	9.0	—	0.116	2.51	3	138	M	−0.66	138	M
53	•I	127	5/2	8.5183	100	9.34×10^{-2}	2.33	2.793	139	N	−0.69	134	O
53	•I	129*	7/2	5.6694	—	4.96×10^{-2}	2.80	2.603	139	N	−0.48	140	Qr
53	I	131*	7/2	5.963	—	5.77×10^{-2}	2.94	2.738	141	Ab	−0.41	141	Ab
54	•Xe	129	1/2	11.777	26.44	2.12×10^{-2}	0.277	−0.772	142	N	—		
54	•Xe	131	3/2	3.4911	21.18	2.76×10^{-3}	0.410	0.686	142	N	−0.12	143	O
55	Cs	127*	1/2	21.8	—	0.134	0.512	1.43	144	Ab	—		
55	Cs	129*	1/2	22.4	—	0.146	0.526	1.47	144	Ab	—		
55	Cs	130*	1	10.7	—	4.20×10^{-2}	0.668	1.4	144	Ab			
55	Cs	131*	5/2	10.72	—	0.186	2.94	3.517	145	Ab			
55	Cs	132*	2	8.46	—	6.28×10^{-2}	1.59	2.22	144	Ab			
55	•Cs	133	7/2	5.58469	100	4.74×10^{-2}	2.75	2.564	8	N	-3×10^{-3}	146 147	Ab
55	Cs	134*	4	5.666	—	6.28×10^{-2}	3.55	2.973	148	Ab	0.43	149	O
55	Cs	134m*	8	1.0447	—	1.42×10^{-3}	2.36	1.096	150	Ab			
55	Cs	135*	7/2	5.9096	—	5.62×10^{-2}	2.91	2.713	148	Ab			
55	Cs	137*	7/2	6.1459	—	6.32×10^{-2}	3.03	2.821	148	Ab			
56	•Ba	135	3/2	4.2296	6.59	4.90×10^{-3}	0.497	0.832	151 152	N	0.25	153	O
56	•Ba	137	3/2	4.7315	11.32	6.86×10^{-3}	0.556	0.931	151 152	N	0.2	153	O
57	•La	138	5*	5.6171	0.089	9.19×10^{-2}	5.28	3.684	154	N	2.7	154	N
57	•La	139	7/2	6.0144	99.911	5.92×10^{-2}	2.97	2.761	107	N	0.21	155	Ab
58	Ce	137*	3/2	4.6	—	6.20×10^{-3}	0.537	0.9	156	No			
58	Ce	137m*	11/2	0.96	—	5.40×10^{-4}	1.07	0.69	157	No			
58	Ce	139*	3/2	5.1	—	8.50×10^{-3}	0.597	1.0	156	No			
58	Ce	141*	7/2	2.1	—	2.57×10^{-3}	1.04	0.97	158	E			

NUCLEAR SPINS, MOMENTS, AND MAGNETIC RESONANCE FREQUENCIES (Continued)

Isotope Z	El	A	Spin I	NMR Frequency in MHz for a 10 kilogauss field	Natural Abundance %	Relative Sensitivity for Equal Number of Nuclei: At constant field	At constant frequency	Magnetic Moment μ: In multiples of the nuclear magneton (eh/4πMc)	Reference	Method	Electric Quadrupole Moment Q: In multiples of barns ($10^{-24}cm^2$)	Reference	Method
58	Ce	143*	7/2	2.2	—	2.81×10^{-3}	1.07	1.0	156	No			
59	•Pr	141	5/2	12.5	100	0.293	3.42	4.09	159 160	O	-5.9×10^{-2}	158	Ab
59	Pr	142*	2	1.1	—	1.55×10^{-4}	0.215	0.30	161	Ab	4×10^{-2}	161	Ab
60	Nd	143	7/2	2.315	12.17	3.38×10^{-3}	1.14	−1.063	162	Ab	−0.48	162	Ab
60	Nd	145	7/2	1.42	8.30	7.86×10^{-4}	0.703	−0.654	162	Ab	−0.25	162	Ab
60	Nd	147*	5/2	1.77	—	8.32×10^{-4}	0.484	0.579	158	E			
61	Pm	143*	(5/2)	11.6	—	0.235	3.17	(3.8)	163	No			
			(7/2)	8.5		0.167	4.19	(3.9)					
61	Pm	144*	(5)	2.6	—	9.02×10^{-3}	2.43	(1.7)	164	No			
			(6)	2.3		8.68×10^{-3}	3.01	(1.8)					
61	Pm	147*	7/2	5.62	—	4.83×10^{-2}	2.77	2.58	165	O	0.7	165	O
61	Pm	148*	1	16	—	0.142	1.00	2.1	166	Ab	0.2	166	Ab
61	Pm	148^*_m	6	2.3	—	8.68×10^{-3}	3.01	1.8	163	No			
61	Pm	149*	7/2	7.2	—	0.101	3.54	3.3	163	No			
61	Pm	151*	5/2	5.5	—	2.50×10^{-2}	1.50	1.8	167	Ab	1.9	167	Ab
62	Sm	147	7/2	1.76	14.97	1.48×10^{-3}	0.867	−0.807	168	Ab	−0.208	158	Ab
62	Sm	149	7/2	1.40	13.83	7.47×10^{-4}	0.691	−0.643	158	Ab	6.0×10^{-2}	158	Ab
63	•Eu	151	5/2	10.559	47.82	0.178	2.89	3.4630	169	Ab	1.16	170	O
63	Eu	152*	3	4.858	—	2.38×10^{-2}	1.83	1.912	171	Ab			
63	•Eu	153	5/2	4.6627	52.18	1.53×10^{-2}	1.28	1.5292	169 172	Ab	2.9	170	O
63	Eu	154*	3	5.084	—	2.72×10^{-2}	1.91	2.001	158	E			
64	•Gd	155	3/2	1.6	14.73	2.79×10^{-4}	0.191	−0.32	173 174	O	1.6	173	O
64	•Gd	157	3/2	2.0	15.68	5.44×10^{-4}	0.239	−0.40	173 174	O	2	173	O
65	Tb	156*	3	3.8	—	1.15×10^{-2}	1.43	1.5	175	No	1.4	175	No
65	Tb	159	3/2	9.66	100	5.83×10^{-2}	1.13	1.90	158	E	1.3	176	O
65	Tb	160*	3	4.1	—	1.39×10^{-2}	1.53	1.6	177	No	1.9	177	No
66	Dy	155*	(3/2)	1.1	—	7.87×10^{-5}	0.125	0.21	178	No			
66	Dy	157*	(3/2)	1.6	—	2.79×10^{-4}	0.191	0.32	178	No			
66	Dy	161	5/2	1.4	18.88	4.17×10^{-4}	0.384	−0.46	158	E	1.4	158	E
66	Dy	163	5/2	2.0	24.97	1.12×10^{-3}	0.535	0.64	158	E	1.6	158	E
67	Ho	165	7/2	8.73	100	0.181	4.31	4.01	158	Ab	2.82	158	Ab
68	Er	165*	5/2	2.0	—	1.18×10^{-3}	0.543	0.65	166	Ab	2.2	166	Ab
68	Er	167	7/2	1.23	22.94	5.07×10^{-4}	0.607	−0.565	162	Ab	2.83	162	Ab
68	Er	169*	1/2	7.8	—	6.09×10^{-3}	0.183	0.51	179	Ab	—		
68	Er	171*	5/2	2.1	—	1.47×10^{-3}	0.585	0.70	180	Ab			
69	Tm	166*	2	0.19	—	7.17×10^{-7}	3.58×10^{-2}	0.05	181	Ab	4.6	181	Ab
69	•Tm	169	1/2	3.52	100	5.66×10^{-4}	8.27×10^{-2}	−0.231	182 160	Ab	—		
69	Tm	170*	1	2.0	—	2.69×10^{-4}	0.124	0.26	183	Ab	0.61	183	Ab
69	Tm	171*	1/2	3.46	—	5.37×10^{-4}	8.13×10^{-2}	0.227	180	Ab	—		
70	•Yb	171	1/2	7.4990	14.31	5.46×10^{-3}	0.176	0.49188	184 185	O	—		
70	Yb	173	5/2	2.0659	16.13	1.33×10^{-3}	0.566	−0.67755	184 185	O	2.8	186	O
70	Yb	175*	(7/2)	0.33	—	9.40×10^{-6}	0.161	−0.15	187	No			
71	•Lu	175	7/2	4.86	97.41	3.12×10^{-2}	2.40	2.23	188	N	5.68	189	Ab
71	Lu	176*	7	3.4	2.59	3.72×10^{-2}	5.92	3.1	190	Ab	8.0	190	Ab
71	Lu	177*	7/2	4.84	—	3.08×10^{-2}	2.38	2.22	191	Ab	5.51	191	Ab
72	Hf	177	7/2	1.3	18.50	6.38×10^{-4}	0.655	0.61	192 193	O	3	192 193	O
72	Hf	179	9/2	0.80	13.75	2.16×10^{-4}	0.617	−0.47	193	O	3	193	O
73	•Ta	181	7/2	5.096	99.988	3.60×10^{-2}	2.51	2.340	194	N	3	195	O
74	•W	183	1/2	1.7716	14.40	7.20×10^{-5}	4.16×10^{-2}	0.116205	196	N	—		
75	•Re	185	5/2	9.5855	37.07	0.133	2.63	3.1437	13	N	2.8	97 241	O
75	Re	186*	1	13.17	—	7.90×10^{-2}	0.825	1.728	197	Ab			
75	•Re	187*	5/2	9.6837	62.93	0.137	2.65	3.1759	13	N	2.6	97 241	O

Isotope Z	El	A	Spin I	NMR Frequency in MHz for a 10 kilogauss field	Natural Abundance %	Relative Sensitivity for Equal Number of Nuclei: At constant field	At constant frequency	Magnetic Moment μ: In multiples of the nuclear magneton (eh/4πMc)	Reference	Method	Electric Quadrupole Moment Q: In multiples of barns ($10^{-24}cm^2$)	Reference	Method
75	Re	188*	1	13.55	—	8.59×10^{-2}	0.848	1.777	197	Ab			
76	•Os	187	1/2	0.98059	1.64	1.22×10^{-5}	2.30×10^{-2}	0.06432	198	N	—		
76	•Os	189	3/2	3.3034	16.1	2.34×10^{-3}	0.388	0.65004	199	N	0.8	195	O
77	•Ir	191	3/2	0.7318	37.3	2.53×10^{-5}	8.59×10^{-2}	0.1440	200 201	N	1.5	202	O
77	•Ir	193	3/2	0.7968	62.7	3.27×10^{-5}	9.36×10^{-2}	0.1568	200 201	N	1.5	202	O
78	•Pt	195	1/2	9.153	33.8	9.94×10^{-3}	0.215	0.6004	45	N	—		
79	Au	190*	1	0.496	—	4.20×10^{-6}	3.10×10^{-2}	0.065	203	Ab			
79	Au	194*	1	0.56	—	5.95×10^{-6}	3.49×10^{-2}	0.073	204	Ab			
79	Au	195*	3/2	0.742	—	2.65×10^{-5}	8.71×10^{-2}	0.146	204	Ab			
79	Au	196*	2	2.3	—	1.24×10^{-3}	0.430	0.6	204	Ab			
79	•Au	197	3/2	0.729188	100	2.51×10^{-5}	8.56×10^{-2}	0.143489	205 201	Ab	0.59	206	Ab
79	Au	198*	2	2.227	—	1.14×10^{-3}	0.418	0.5842	207	Ab			
79	Au	199*	3/2	1.358	—	1.62×10^{-4}	0.160	0.2673	207	Ab			
80	Hg	193*	3/2	3.1	—	1.93×10^{-3}	0.364	−0.61	208	O			
80	Hg	193^*_m	13/2	1.23	—	1.57×10^{-3}	1.88	−1.05	209	O	1.37	209	O
80	Hg	195*	1/2	8.1	—	6.84×10^{-3}	0.190	0.53	210 211	O	—		
80	Hg	195^*_m	13/2	1.22	—	1.53×10^{-3}	1.86	−1.04	209	O	1.41	209	O
80	Hg	197*	1/2	7.9	—	6.46×10^{-3}	0.186	0.52	212	O	—		
80	•Hg	199	1/2	7.59012	16.84	5.67×10^{-3}	0.178	0.497859	213 45	No			
80	•Hg	201	3/2	2.8099	13.22	1.44×10^{-3}	0.330	−0.55293	214 215	O	0.50	216	O
80	Hg	203*	5/2	2.5	—	2.45×10^{-3}	0.693	0.83	217	O	0.5	217	O
81	Tl	197*	1/2	23.6	—	0.171	0.555	1.55	218	O	—		
81	Tl	199*	1/2	23.9	—	0.178	0.562	1.57	219	O	—		
81	Tl	200*	2	0.57	—	1.94×10^{-5}	0.107	(0.15)	219	O			
81	Tl	201*	1/2	24.1	—	0.181	0.566	1.58	219	O	—		
81	Tl	202*	2	0.57	—	1.94×10^{-5}	0.107	(0.15)	219	O			
81	•Tl	203	1/2	24.332	29.50	0.187	0.571	1.5960	220	N	—		
81	Tl	204*	2	0.34	—	4.05×10^{-6}	6.37×10^{-2}	0.089	221	Ab			
81	•Tl	205	1/2	24.570	70.50	0.192	0.577	1.6116	8	N	—		
82	•Pb	207	1/2	8.90771	22.6	9.16×10^{-3}	0.209	0.584284	222	N	—		
83	Bi	203*	9/2	7.78	—	0.201	6.03	4.59	223	Ab	−0.64	223	Ab
83	Bi	204*	6	5.40	—	0.114	7.10	4.25	223	Ab	−0.41	223	Ab
83	Bi	205*	9/2	9.3	—	0.346	7.22	(5.5)	223	Ab			
83	Bi	206*	6	5.79	—	0.141	7.62	4.56	223	Ab	−0.19	223	Ab
83	•Bi	209*	9/2	6.84178	100	0.137	5.30	4.03896	224	N	−0.4	97	O
83	Bi	210*	1	0.337	—	1.32×10^{-6}	2.11×10^{-2}	0.0442	225	Ab	0.13	225	Ab
84	Po	205*	5/2	0.79	—	7.55×10^{-5}	0.217	0.26	226	Ab	0.17	226	Ab
84	Po	207*	5/2	0.82	—	8.43×10^{-5}	0.226	0.27	226	Ab	0.28	226	Ab
89	Ac	227*	3/2	5.6	—	1.13×10^{-2}	0.656	1.1	227	O	−1.7	227	O
90	Th	229*	5/2	1.2	—	2.74×10^{-4}	0.334	0.4	228	O	4.6	228	O
91	Pa	231*	3/2	9.96	—	6.40×10^{-2}	1.17	1.96	229	E			
91	Pa	233*	3/2	17	—	0.334	2.03	3.4	230	Ab	−3.0	230	Ab
92	U	233*	5/2	1.6	—	6.75×10^{-4}	0.451	0.54	231	E	3.5	231	E
92	U	235*	7/2	0.76	0.72	1.21×10^{-4}	0.376	0.35	231	E	4.1	231	E
93	Np	237*	5/2	18	—	0.926	5.01	(6)	232	E			
94	Pu	239*	1/2	3.05	—	3.67×10^{-4}	7.16×10^{-2}	0.200	233	Ab	—		
94	Pu	241*	5/2	2.09	—	1.38×10^{-3}	0.573	−0.686	234	O			
95	Am	241*	5/2	4.82	—	1.69×10^{-2}	1.32	1.58	235	Ab	4.9	236	O
95	Am	242*	1	2.90	—	8.46×10^{-4}	0.182	0.381	235	Ab	−2.8	237	Ab
95	Am	243*	5/2	4.79	—	1.66×10^{-2}	1.31	1.57	238	O	4.9	236	O
Free electron with g = 2.00232			1/2	2.80246×10^{4}	—	2.84×10^{-8}	657	−1836.09	—		—		

REFERENCES

(1) V. W. Cohen, et al., Phys. Rev. **104**, 283 (1956).
(2) H. Sommer, et al., Phys. Rev. **82**, 697 (1951).
(3) T. F. Wimett, Phys. Rev. **91**, 499A (1953).
(4) H. Kopfermann, et al., Z. Physik **144**, 9 (1956).
(5) F. Bloch, et al., Phys. Rev. **71**, 551 (1947).
(6) H. L. Anderson, Phys. Rev. **76**, 1460 (1949).
(7) M. P. Klein, et al., Phys. Rev. **106**, 837 (1957).
(8) H. E. Walchli, Thesis, M.S., U. of Tenn. (1954). AEC Report ORNL-1775.
(9) N. A. Schuster, et al., Phys. Rev. **81**, 157 (1951).
(10) K. C. Brog, et al., Phys. Rev. **153**, 91 (1967).
(11) D. Connor, Phys. Rev. Letters **3**, 429 (1959).
(12) L. C. Brown, et al., J. Chem. Phys. **24**, 751 (1956).
(13) F. Alder, et al., Phys. Rev. **82**, 105 (1951).
(14) A. G. Blachman, et al., Bull Am. Phys. Soc. **11**, 343 (1966).
(15) Y. Ting, et al., Phys. Rev. **89**, 595 (1953).
(16) G. Wessel, Phys. Rev. **92**, 1581 (1953).
(17) G. Lindström, Arkiv Fysik **4**, 1 (1951).
(18) V. Royden, Phys. Rev. **96**, 543 (1954).
(19) A. M. Bernstein, et al., Phys. Rev. **136**, B27 (1964).
(20) L. W. Anderson, et al., Phys. Rev. **116**, 87 (1959).
(21) M. R. Baker, et al., Phys. Rev. **133**, A1533 (1964).
(22) A. Bassompiere, Compt. Rend. **240**, 285 (1955).
(23) E. D. Commins, et al., Phys. Rev. **131**, 700 (1963).
(24) F. Alder, et al., Phys. Rev. **81**, 1067 (1951).
(25) M. J. Stevenson, et al., Phys. Rev. **107**, 635 (1957).
(26) K. Sugimoto, et al., J. Phys. Soc. Japan **21**, 213 (1966).
(27) T. Tsang, et al., Phys. Rev. **132**, 1141 (1963).
(28) E. D. Commins, et al., Phys. Rev. Letters **10**, 347 (1963).
(29) J. T. LaTourette, et al., Phys. Rev. **107**, 1202 (1957).
(30) O. Ames, et al., Phys. Rev. **137**, B1157 (1965).
(31) L. Davis, et al., Phys. Rev. **76**, 1068 (1949).
(32) H. Ackermann, Z. Physik. **194**, 253 (1966).
(33) M. Baumann, et al., Z. Physik **194**, 270 (1966).
(34) Y. W. Chan, et al., Phys. Rev. **150**, 933 (1966).
(35) L. C. Brown, et al., J. Chem. Phys. **24**, 751 (1956).
(36) H. Lew, et al., Phys. Rev. **90**, 1 (1953).
(37) H. E. Weaver, Phys. Rev. **89**, 923 (1953).
(38) T. Kanda, et al., Phys. Rev. **85**, 938 (1952).
(39) G. Feher, et al., Phys. Rev. **107**, 1462 (1957).
(40) G. R. Bird, et al., Phys. Rev. **94**, 1203 (1954).
(41) B. F. Burke, et al., Phys. Rev. **93**, 193 (1954).
(42) V. Jaccarino, et al., Phys. Rev. **83**, 471 (1951).
(43) P. B. Sogo, et al., Phys. Rev. **98**, 1316 (1955).
(44) C. H. Townes, et al., Phys. Rev. **76**, 691 (1949).
(45) W. G. Proctor, et al., Phys. Rev. **81**, 20 (1951).
(46) E. A. Phillips, et al., Phys. Rev. **138**, B773 (1965).
(47) E. Brun, et al., Phys. Rev. **93**, 172 (1954).
(48) O. Lutz, et al., Phys. Letters **24A**, 122 (1967).
(49) G. W. Series, Phys. Rev. **105**, 1128 (1957).
(50) G. J. Ritter, et al., Proc. Roy. Soc. (London) **238**, 473 (1957).
(51) J. T. Eisinger, et al., Phys. Rev. **86**, 73 (1952).
(52) J. M. Kahn, et al., Phys. Rev. **134**, A45 (1964).
(53) F. R. Petersen, et al., Phys. Rev. **116**, 734 (1959).
(54) E. Brun, et al., Phys. Rev. Letters **9**, 166 (1962).
(55) C. D. Jeffries, Phys. Rev. **90**, 1130 (1953).
(56) R. G. Cornwall, et al., Phys. Rev. **141**, 1106 (1966).
(57) D. L. Harris, et al., Phys. Rev. **132**, 310 (1963).
(58) D. M. Hunten, Can. J. Phys. **29**, 463 (1951).
(59) G. Fricke, et al., Z. Physik, **156**, 416 (1959).
(60) F. R. Petersen, et al., Phys. Rev. **128**, 1740 (1962).
(61) R. G. Cornwall, et al., Phys. Rev. **148**, 1157 (1966).
(62) C. D. Jeffries, Phys. Rev. **92**, 1262 (1953).
(63) M. M. Weiss, et al., Bull. Am. Phys. Soc. **2**, 31 (1957).
(64) H. E. Walchli, et al., Phys. Rev. **87**, 541 (1952).
(65) K. Murakawa, et al., J. Phys. Soc. Japan **21**, 1466 (1966).
(66) C. D. Jeffries, et al., Phys. Rev. **91**, 1286 (1953).
(67) K. E. Adelroth, et al., Arkiv Fysik **31**, 549 (1966).
(68) E. A. Phillips, et al., Phys. Rev. **140**, B555 (1965).
(69) W. Dobrowolski, et al., Phys. Rev. **104**, 1378 (1956).
(70) R. W. Bauer, et al., Phys. Rev. **120**, 946 (1960).
(71) W. B. Mims, et al., Phys. Letters **24A**, 481 (1967).
(72) A. Javan, et al., Phys. Rev. **96**, 649 (1954).
(73) W. J. Childs, et al., Phys. Rev. **122**, 891 (1961).
(74) P. R. Locher, et al., Phys. Rev. **139**, A991 (1965).
(75) R. W. Bauer, et al., Nucl. Phys. **16**, 264 (1960).
(76) J. M. Baker, et al., Proc. Phys. Soc. (London), **A69**, 354 (1956).
(77) W. Dobrov, et al., Phys. Rev. **108**, 60 (1957).
(78) D. V. Ehrenstein, et al., Z. Physik **159**, 230 (1960).
(79) W. Dobrowolski, et al., Phys. Rev. **101**, 1001 (1956).
(80) L. E. Drain, Phys. Letters **11**, 114 (1964).
(81) B. M. Dodsworth, et al., Phys. Rev. **142**, 638 (1966).
(82) B. Bleaney, et al., Proc. Roy. Soc. (London) **228A**, 166 (1955).
(83) A Lemonick, et al., Phys. Rev. **95**, 1356 (1954).
(84) V. J. Ehlers, et al., Phys. Rev. **127**, 529 (1962).
(85) R. T. Daly, et al., Phys. Rev. **96**, 539 (1954).
(86) W. J. Childs, et al., Phys. Rev. **120**, 2138 (1960).
(87) W. J. Childs, et al., Phys. Rev. **141**, 15 (1966).
(88) J. M. Mays, et al., Phys. Rev. **81**, 940 (1951).
(89) B. P. Dailey, et al., Phys. Rev. **74**, 1245 (1948).
(90) F. M. Pipkin, et al., Phys. Rev. **106**, 1102 (1957).
(91) W. A. Hardy, et al., Phys. Rev. **92**, 1532 (1953).
(92) E. Lipworth, et al., Phys. Rev. **119**, 1053 (1960).
(93) J. G. King, et al., Phys. Rev. **94**, 1610 (1954).
(94) M. B. White, et al., Phys. Rev. **136**, B584 (1964).
(95) H. L. Garvin, et al., Phys. Rev. **116**, 393 (1959).
(96) E. Brun, et al., Helv. Phys. Acta **27**, 173A (1954).
(97) J. E. Mack, Rev. Mod. Phys. **22**, 64 (1950).
(98) E. Rasmussen, et al., Z. Physik **141**, 160 (1955).
(99) J. C. Hubbs, et al., Phys. Rev. **107**, 723 (1957).
(100) H. Walchli, et al., Phys. Rev. **85**, 922 (1952).
(101) B. Senitzky, et al., Phys. Rev. **103**, 315 (1956).
(102) E. H. Bellamy, et al., Phil. Mag. **44**, 33 (1953).
(103) E. Yasaitis, et al., Phys. Rev. **82**, 750 (1951).
(104) J. W. Culvahouse, et al., Phys. Rev. **140**, A1181 (1965).
(105) F. R. Petersen, et al., Phys. Rev. **125**, 284 (1962).
(106) E. Brun, et al., Phys. Rev. **105**, 1929 (1957).
(107) R. E. Sheriff, et al., Phys. Rev. **82**, 651 (1951).
(108) K. Murakawa, Phys. Rev. **98**, 1285 (1955).
(109) A. Narath, et al., Phys. Rev. **143**, 328 (1966).
(110) H. Walchli, et al., Phys. Rev. **85**, 479 (1952).
(111) K. G. Kessler, et al., Phys. Rev. **92**, 303 (1953).
(112) E. Matthias, et al., Phys. Rev. **139**, B532 (1965).
(113) K. Murakawa, J. Phys. Soc. Japan **10**, 919 (1955).
(114) J. A. Seitchik, et al., Phys. Rev. **138**, A148 (1965).
(115) J. A. Seitchik, et al., Phys. Rev. **136**, A1119 (1964).
(116) O. Ames, et al., Phys. Rev. **123**, 1793 (1960).
(117) W. B. Ewbank, et al., Phys. Rev. **129**, 1617 (1963).
(118) P. B. Sogo, et al., Phys. Rev. **93**, 174 (1954).
(119) G. K. Rochester, et al., Phys. Letters 8, **266** (1964).
(120) S. G. Schmelling, et al., Phys. Rev. **154**, 1142 (1967).
(121) G. K. Woodgate, et al., Proc. Phys. Soc. (London) **69**, 581 (1956).
(122) Y. W. Chan, et al., Phys. Rev. **133**, B1138 (1964).
(123) F. W. Byron, et al., Phys. Rev. **132**, 1181 (1963).
(124) P. Thaddeus, et al., Phys. Rev. **132**, 1186 (1963).
(125) W. G. Proctor, Phys. Rev. **79**, 35 (1950).
(126) M. Leduc, et al., Compt. Rend. **B262**, 736 (1966).
(127) M. N. McDermott, et al., Phys. Rev. **134**, B25 (1964).
(128) W. J. Childs, et al., Phys. Rev. **118**, 1578 (1960).
(129) L. S. Goodman, et al., Phys. Rev. **108**, 1524 (1957).
(130) M. Rice, et al., Phys. Rev. **106**, 953 (1957).
(131) J. A. Cameron, et al., Can. J. Phys. **40**, 931 (1962).
(132) P. B. Nutter, Phil. Mag. **1**, 587 (1956).
(133) V. W. Cohen, et al., Phys. Rev. **79**, 191 (1950).
(134) K. Murakawa, Phys. Rev. **100**, 1369 (1955).
(135) F. M. Pipkin, Bull. Am. Phys. Soc. **3**, 8 (1958).
(136) P. C. B. Fernando, et al., Phil. Mag. **5**, 1309 (1960).
(137) K. E. Adelroth, et al., Arkiv Fysik **30**, 111 (1965).
(138) P. C. Fletcher, et al., Phys. Rev. **110**, 536 (1958).
(139) H. Walchli, et al., Phys. Rev. **82**, 97 (1951).
(140) R. Livingston, et al., Phys. Rev. **90**, 609 (1953).
(141) E. Lipworth, et al., Phys. Rev. **119**, 2022 (1960).
(142) D. Brinkmann, Helv. Phys. Acta **36**, 413 (1963).
(143) A. Bohr, et al., Arkiv Fysik **4**, 455 (1952).
(144) W. A. Nierenberg, et al., Phys. Rev. **112**, 186 (1958).
(145) R. D. Worley, et al., Phys. Rev. **140**, B1483 (1965).
(146) P. Buck, et al., Phys. Rev. **104**, 553 (1956).
(147) K. H. Althoff, et al., Naturwiss. **41**, 368 (1954).
(148) H. H. Stroke, et al., Phys. Rev. **105**, 590 (1957).
(149) G. Heinzelmann, et al., Phys. Letters **21**, 162 (1966).
(150) V. W. Cohen, et al., Phys. Rev. **127**, 517 (1962).
(151) H. E. Walchli, et al., Phys. Rev. **102**, 1334 (1956).
(152) L. Olschewski, et al., Z. Physik **196**, 77 (1966).
(153) N. I. Kaliteevskii, et al., Soviet Physics—JETP **12**, 661 (1961).
(154) P. B. Sogo, et al., Phys. Rev. **99**, 613 (1955).
(155) K. Murakawa, et al., J. Phys. Soc. Japan **16**, 2533 (1961).
(156) J. N. Haag, et al., Phys. Rev. **129**, 1601 (1963).
(157) J. Blok, et al., Phys. Rev. **143**, 78 (1966).
(158) B. Bleaney, Quantum Electronics Conf., Paris, 1963; P. Grivet and N. Bloembergen, editors; Columbia U. Press (1964).
(159) J. Reader, et al., Phys. Rev. **137**, B784 (1965).
(160) E. D. Jones, Phys. Rev. Letters **19**, 432 (1967).
(161) A. Y. Cabezas, et al., Phys. Rev. **126**, 1004 (1962).
(162) K. F. Smith, et al., Proc. Phys. Soc. (London) **86**, 1249 (1965).
(163) R. W. Grant, et al., Phys. Rev. **130**, 1100 (1963).
(164) D. A. Shirley, et al., Phys. Rev. **121**, 558 (1961).
(165) J. Reader, Phys. Rev. **141**, 1123 (1966).
(166) D. Ali, et al., Phys. Rev. **138**, B1356 (1965).
(167) B. Budick, et al., Phys. Rev. **132**, 723 (1963).
(168) G. K. Woodgate, Proc. Roy. Soc. (London) **A293**, 117 (1966).
(169) L. Evans, et al., Proc. Roy. Soc. (London) **A289**, 114 (1965).
(170) W. Müller, et al., Z. Physik **183**, 303 (1965).
(171) S. S. Alpert, Phys. Rev. **129**, 1344 (1963).
(172) E. L. Boyd, Phys. Rev. **145**, 174 (1966).
(173) N. I. Kaliteevskii, et al., Soviet Phys.—JETP **37**, 629 (1960).
(174) E. L. Boyd, et al., Phys. Rev. Letters **12**, 20 (1964).
(175) C. A. Lovejoy, et al., Nucl. Phys. **30**, 452 (1962).
(176) C. Arnoult, et al., J. Opt. Soc. Am. **56**, 177 (1966).
(177) C. E. Johnson, et al., Phys. Rev. **120**, 2108 (1960).
(178) Q. O. Navarro, et al., Phys. Rev. **123**, 186 (1961).
(179) W. M. Doyle, et al., Phys. Rev. **131**, 1586 (1963).
(180) B. Budick, et al., Phys. Rev. **135**, B1281 (1964).
(181) J. C. Walker, Phys. Rev. **127**, 1739 (1962).
(182) D. Giglberger, et al., Z. Physik **199**, 244 (1967).
(183) A. Y. Cabezas, et al., Phys. Rev. **120**, 920 (1960).
(184) L. Olschewski, et al., Z. Physik **200**, 224 (1967).
(185) A. C. Gossard, et al., Phys. Rev. **133**, A881 (1964).
(186) J. S. Ross, et al., Phys. Rev. **128**, 1159 (1962).
(187) M. A. Grace, et al., Phil. Mag. **2**, 1079 (1957).
(188) A. H. Reddoch, et al., Phys. Rev. **126**, 1493 (1962).
(189) G. J. Ritter, Phys. Rev. **126**, 240 (1962).
(190) I. J. Spalding, et al., Proc. Phys. Soc. (London) **79**, 787 (1962).
(191) F. R. Petersen, et al., Phys. Rev. **126**, 252 (1962).
(192) D. R. Speck, Bull. Am. Phys. Soc. **1**, 282 (1956).
(193) D. R. Speck, et al., Phys. Rev. **101**, 1831 (1956).
(194) L. H. Bennett, et al., Phys. Rev. **120**, 1812 (1960).
(195) K. Murakawa, et al., Phys. Rev. **105**, 671 (1957).
(196) M. P. Klein, et al., Bull. Am. Phys. Soc. **6**, 104 (1961).
(197) L. Armstrong, et al., Phys. Rev. **138**, B310 (1965).
(198) J. Kaufmann, et al., Phys. Letters **24A**, 115 (1967).
(199) H. R. Loeliger, et al., Phys. Rev. **95**, 291 (1954).

REFERENCES (Continued)

(200) A. Narath, (to be published in Phys. Rev.).
(201) A. Narath, et al., Bull. Am. Phys. Soc. **12**, 314 (1967).
(202) W. von Siemens, Ann. Physik **13**, 136 (1953).
(203) Y. W. Chan, et al., Phys. Rev. **144**, 1020 (1966).
(204) Y. W. Chan, et al., Phys. Rev. **137**, B1129 (1965).
(205) H. Dahmen, et al., Z. Physik **200**, 456 (1967).
(206) W. J. Childs, et al., Phys. Rev. **141**, 176 (1966).
(207) P. A. Vanden Bout, et al., Phys. Rev. **158**, 1078 (1967).
(208) H. Kleiman, et al., Phys. Letters **13**, 212 (1964).
(209) W. J. Tomlinson, et al., Nucl. Phys. **60**, 614 (1964).
(210) H. Kleiman, et al., J. Opt. Soc. Am. **53**, 822 (1963).
(211) W. J. Tomlinson, et al., J. Opt. Soc. Am. **53**, 828 (1963).
(212) F. Bitter, et al., Phys. Rev. **96**, 1531 (1954).
(213) B. Cagnac, et al., Compt. Rend. **249**, 77 (1959).
(214) B. Cagnac, Ann. Phys. **6**, 467 (1961).
(215) J. C. Lehmann, et al., Compt. Rend. **257**, 3152 (1963).
(216) J. Blaise, et al., J. Phys. Radium **18**, 193 (1957).
(217) O. Redi, et al., Phys. Letters **8**, 257 (1964).
(218) S. P. Davis, et al., J. Opt. Soc. Am. **56**, 1604 (1966).
(219) R. J. Hull, et al., Phys. Rev. **122**, 1574 (1961).
(220) H. L. Poss, Phys. Rev. **75**, 600 (1949).
(221) G. O. Brink, et al., Phys. Rev. **107**, 189 (1957).
(222) E. B. Baker, J. Chem. Phys. **26**, 960 (1957).
(223) I. Lindgren, et al., Arkiv Fysik **15**, 445 (1959).
(224) Y. Ting, et al., Phys. Rev. **89**, 595 (1953).
(225) S. S. Alpert, et al., Phys. Rev. **125**, 256 (1962).
(226) C. M. Olsmats, et al., Arkiv Fysik **19**, 469 (1961).
(227) M. Fred, et al., Phys. Rev. **98**, 1514 (1955).
(228) V. N. Egorov, Opt. Spectry. (USSR) **16**, 301 (1964).
(229) J. D. Axe, et al., Phys. Rev. **121**, 1630 (1961).
(230) R. Marrus, et al., Nucl. Phys. **23**, 90 (1961).
(231) P. B. Dorain, et al., Phys. Rev. **105**, 1307 (1957).
(232) B. Bleaney, et al., Phil. Mag. **45**, 992 (1954).
(233) J. Faust, et al., Phys. Letters **16**, 71 (1965).
(234) R. J. Champeau, et al., Compt. Rend. **257**, 1238 (1963).
(235) L. Armstrong, et al., Phys. Rev. **144**, 994 (1966).
(236) T. E. Manning, et al., Phys. Rev. **102**, 1108 (1956).
(237) R. Marrus, et al., Phys. Rev. **124**, 1904 (1961).
(238) M. Fred, et al., J. Opt. Soc. Am. **47**, 1076 (1957).
(239) B. Perry, et al., Bull. Am. Phys. Soc. **8**, 345 (1963).
(240) M. M. Robertson, et al., Phys. Rev. **140**, B820 (1965).
(241) R. E. Walstedt, et al., Phys. Rev. **162**, 301 (1967).
(242) R. Freeman, et al., Proc. Roy Soc. (London) **242A**, 455 (1957).

IONIZATION POTENTIALS OF MOLECULES

From data published up to July, 1966
Condensed by J. L. Franklin and Pat Haug from a compilation entitled
"Ionization Potentials, Appearance Potentials and Heats of Formation of Positive Ions"
by
J. L. Franklin, J. G. Dillard, H. M. Rosenstock, J. T. Herron, K. Draxl and F. H. Field
Published by National Standard Reference Data System

The following symbols are employed for the principal important methods:

CTS = Charge Transfer Spectra, EI = Electron Impact, PE = Photoelectron Spectroscopy, PI = Photoionization, S = Optical Spectroscopy.

Molecule	Ionization Potential ev	Method	ΔH_f of Ion Kcal/mole	Reference
H_2	15.427 ±0.002	S	356	*1*
D_2	15.46 ±0.01	PI	356	*2*
BH	9.77 ±0.05	S	333	*3*
BH_3	11.4 ±0.2	EI	279	*
B_2H_6	12.0	EI	286	*4, 5*
B_5H_9	10.5	EI	262	*
B_6H_{10}	9.3 ±0.1	EI	237	*6*
C_2	12.0 ±0.6	EI	475	*7*
C_3	12.6	EI	480	*7*
CH	11.13 ±0.22	S	399	*8*
CH_2	10.396 ±0.003	EI, S, PI	333	*9, 10, 11*
CH_3	9.83	S, PI	259	*12, 13*
CD_3	9.832 ±0.002	S	259	*12*
CH_4	12.6	PI	274	*
CD_4	12.888	PI	280	*11, 14*
C_2H_2	11.4	PI, PE	317	*
C_2D_2	11.416 ±0.006	PI	317	*15*
C_2H_3	9.4	EI	269	*
C_2H_4	10.5	S, PE	253	*16, 17*
C_2H_5	8.4	PI	219	*13*
C_2H_6	11.5	PI, PE	245	*14, 17*
$HC\equiv C-CH_2$	8.25	PI, EI		*18*
$CH_2=C=CH_2$	10.16 ±0.02	EI	280	*19*
$CH_3C\equiv CH$	10.36 ±0.01	PI	283	*20*
cyclo-C_3H_4	9.95	EI	296	*21*
C_3H_5 (allyl)	8.15	EI	216	*22, 8*
cyclo-C_3H_5	8.05	EI	239	*23*
C_3H_6	9.73	S, PI	229	*
cyclo-C_3H_6	10.09 ±0.02	PI	245	*24*
n-C_3H_7	8.1	PI	209	*13*
iso-C_3H_7	7.5	PI	190	*13*
C_3H_8	11.1	PI, PE	231	*
C_4H_2	10.2 ±0.1	EI		*25*
C_4H_4	9.87	EI	294	*25, 26*
$CH_3CH=C=CH_2$	9.57 ±0.02	EI	259	*19*
$CH_2=CHCH=CH_2$	9.07	PI, PE	236	*
$C_2H_5C\equiv CH$	10.18 ±0.01	PI	274	*20*
$CH_3C\equiv CCH_3$	9.9 ±0.1	EI	263	*25*
cyclo-C_4H_7	7.88 ±0.05	EI	213	*23*
$CH_3CH=CHCH_2$	7.71 ±0.05	EI	203	*27, 8*
$CH_2=C(CH_3)CH_2$	8.03 ±0.05	EI	203	*27, 8*
1-C_4H_8	9.6	PI	221	*
cis-2-C_4H_8	9.13	PI	209	*24, 28*
trans-2-C_4H_8	9.13	PI	208	*24, 28*
iso-C_4H_8	9.23 ±0.02	PI	209	*24, 28*
cyclo-C_4H_8	10.58	EI	250	*23*
n-C_4H_9	8.64 ±0.05	EI	218	*29*
sec-C_4H_9	7.93 ±0.05	EI	192	*29*
iso-C_4H_9	8.35 ±0.05	EI	205	*29*
tert-C_4H_9	7.42 ±0.07	EI	176	*29*
n-C_4H_{10}	10.63 ±0.03	PI	215	*30*
iso-C_4H_{10}	10.57	PI	212	*24*
cyclo-C_5H_6	8.97	EI	239	*31, 32*
$C_2H_5CH=C=CH_2$	9.42	EI	252	*21*
$CH_3CH=CHCH=CH_2$	8.68	EI	219	*21*
$CH_3CH=C=CHCH_3$	9.26	EI	247	*21*
$CH_2=CHCH_2CH=CH_2$	9.58	EI	246	*21*
$CH_2=CHC(CH_3)=CH_2$	8.845 ±0.005	PI	235	*24*
cyclo-C_5H_8	9.01 ±0.01	PI	216	*24*
cyclo-C_5H_9	7.79 ±0.02	EI	194	*23*
1-C_5H_{10}	9.50 ±0.02	PI	214	*24*
cis-2-C_5H_{10}	9.11	EI	203	*21*
trans-2-C_5H_{10}	9.06	EI	201	*21*
$(CH_3)_2CHCH=CH_2$	9.51 ±0.03	PI	212	*24*
$C_2H_5C(CH_3)=CH_2$	9.12 ±0.02	PI	202	*24*
$(CH_3)_2C=CHCH_3$	8.67 ±0.02	PI	189	*24*
cyclo-$C_3H_4(CH_3)_2$	9.77 ±0.02	EI	225	*33*
cyclo-C_5H_{10}	10.53 ±0.05	PI	224	*24*
tert-C_5H_{11}	7.12 ±0.1	EI	164	*34, 35*
neo-C_5H_{11}	8.33 ±0.1	EI	196	*34*
n-C_5H_{12}	10.35	PI	204	*24*
iso-C_5H_{12}	10.32	PI	201	*24*
neo-C_5H_{12}	10.35	PI	199	*24*
C_6H_4 (benzyne)	9.6	EI		*
cyclo-C_6H_5	9.2	PI, PE	284	*36, 37*
cyclo-C_6H_6	9.24	S, PI	233	*
$CH\equiv CCH=CHCH=CH_2$	9.50	EI	307	*38*
$C_2H_5C\equiv CC\equiv CH$	9.25	EI	307	*38*
$CH_3C\equiv CCH_2C\equiv CH$	9.75	EI	319	*38*
$CH_3C\equiv CC\equiv CCH_3$	9.20	EI	301	*38*
$CH\equiv CCH_2CH_2C\equiv CH$	10.35	EI	338	*38*
C_6H_8 (1-methylcyclopentadiene)	8.43 ±0.05	EI	218	*31*
C_6H_8 (2-methylcyclopentadiene)	8.46 ±0.05	EI	219	*31*
cyclo-C_6H_{10}	8.72	PE	199	*17*
cyclo-C_6H_{11}	7.7	EI	185	*23*
1-C_6H_{12}	9.45 ±0.02	PI	208	*35, 24*
$(CH_3)_2C=C(CH_3)_2$	8.30	PI	175	*28*
cyclo-C_6H_{12}	9.8	PI, PE	197	*17, 24*
n-C_6H_{14}	10.18	PI	195	*24*
iso-C_6H_{14}	10.12	PI	192	*24*
$(C_2H_5)_2CHCH_3$	10.08	PI	191	*24*
$C_2H_5C(CH_3)_3$	10.06	PI	188	*24*
$(CH_3)_2CHCH(CH_3)_2$	10.02	PI	189	*24*
cyclo-$C_6H_5CH_2$	7.76 ±0.08	EI	216	*39*
cyclo-C_7H_7	6.240 ±0.01	S	209	*40*
cyclo-$C_6H_5CH_3$	8.82 ±0.01	PI	215	*
cyclo-C_7H_8	8.5	EI	240	*32, 41, 42*
bicyclo-(2.2.1)C_7H_8	8.67	EI	267	*42*
bicyclo-(3.2.0)C_7H_8	9.37	EI	246	*41*
C_7H_{10} (1,2-dimethylcyclopentadiene)	8.1 ±0.1	EI	204	*31*
C_7H_{10} (5,5-dimethylcyclopentadiene)	8.22 ±0.05	EI	206	*31*
C_7H_{10} (1,3-cycloheptadiene)	8.55	EI	219	*41*
C_7H_{10} (norbornene)	8.95 ±0.15	EI	237	*43*
C_7H_{12} (4-methylcyclohexane)	8.91 ±0.01	PI	198	*24*
cyclo-$C_6H_{11}CH_3$	9.85 ±0.03	PI	190	*24*
n-C_7H_{16}	9.90 ±0.05	PI	183	*37*
cyclo-$C_6H_5C\equiv CH$	8.815 ±0.005	PI	279	*24*
C_8H_8 (styrene)	8.47 ±0.02	PI	232	*24*
C_8H_8 (cyclotatetraene)	8.0	PI, PE	255	*17, 24*
C_8H_8 (cubane)	8.74 ±0.15	EI	350	*44*
m-$C_6H_4CH_3CH_2$	7.65 ±0.03	EI	206	*39*
p-$C_6H_4CH_3CH_2$	7.46 ±0.03	EI	202	*39*
cyclo-$C_6H_5C_2H_5$	8.76 ±0.01	PI	209	*24*
o-$C_6H_4(CH_3)_2$	8.56	PI	202	*
m-$C_6H_4(CH_3)_2$	8.58	PI, PE	202	*
p-$C_6H_4(CH_3)_2$	8.44	PI	199	*
C_8H_{10} (7-methylcycloheptatriene)	8.39 ±0.1	EI	231	*45*
C_8H_{10} (1-methylspiroheptadiene)	8.02 ±0.1	EI	229	*45*
C_8H_{10} (2-methylspiroheptadiene)	8.07 ±0.1	EI	230	*45*
C_8H_{10} (6-methylspiroheptadiene)	8.40 ±0.1	EI	239	*45*
C_8H_{12} (1,2,3-trimethylcyclopentadiene)	7.96 ±0.05	EI	194	*31*
C_8H_{12} (1,5,5-trimethylcyclopentadiene)	8.00 ±0.1	EI	193	*31*

* Average of several values.

IONIZATION POTENTIALS OF MOLECULES (Continued)

Molecule	Ionization Potential ev	Method	ΔH_f of Ion Kcal/mole	Reference
C_8H_{12} (4-vinylcyclohexene)	8.93 ±0.02	PI	224	*24*
cis-1,2-cyclo-$C_6H_{10}(CH_3)_2$	10.08 ±0.02	EI	191	*46*
trans-1,2-cyclo-$C_6H_{10}(CH_3)_2$	10.08 ±0.03	EI	189	*46*
C_8H_{18} (2,2,4-trimethylpentane)	9.86	PI	174	*24*
C_8H_{18} (2,2,3,3-tetramethylbutane)	9.79	EI	184	*8*
C_9H_8 (indene)	8.81	EI	246	*47*
$C_6H_5C(CH_3)=CH_2$	8.35 ±0.01	PI	220	*24*
cyclo-C_6H_5(n-C_3H_7)	8.72 ±0.01	PI	203	*24*
cyclo-C_6H_5(iso-C_3H_7)	8.69 ±0.01	PI	201	*24*
1,2,3 cyclo-$C_6H_3(CH_3)_3$	8.48	PI	193	*28*
1,2,4 cyclo-$C_6H_3(CH_3)_3$	8.27	PI	187	*28*
1,3,5 cyclo-$C_6H_3(CH_3)_3$	8.4	PI	190	*
$C_{10}H_8$ (naphthalene)	8.12	PI	220	*24, 48*
$C_{10}H_8$ (azulene)	7.42	S	243	*49, 50*
$C_6H_4C_3H_7CH_2$	7.42	EI	188	*39*
cyclo-C_6H_5 (n-C_4H_9)	8.69 ±0.01	PI	197	*24*
$C_{10}H_{14}$ (sec-butylbenzene)	8.68 ±0.01	PI	196	*24*
$C_{10}H_{14}$ (tert-butylbenzene)	8.68 ±0.01	PI	193	*24*
$C_{10}H_{14}$ (1,2,3,5-tetramethylbenzene)	8.47 ±0.05	EI	185	*24*
$C_{10}H_{14}$ (1,2,4,5-tetramethylbenzene)	8.03	PI, PE	174	*
$C_{10}H_{18}$ (cis-decaline)	9.61 ±0.02	EI	181	*51*
$C_{10}H_{18}$ (trans-decaline)	9.61 ±0.02	EI	178	*51*
$C_{11}H_9$ (1-naphthyl methyl)	7.35	EI	208	*52*
$C_{11}H_9$ (2-naphthyl methyl)	7.56 ±0.05	EI	217	*52*
$C_{11}H_{10}$ (methylnaphthalene)	7.96 ±0.01	PI	209	*24*
$C_{11}H_{10}$ (2-methylnaphthalene)	7.955 ±0.01	PI	209	*24*
$C_6H(CH_3)_5$	7.92 ±0.02	PI	155	*53*
$C_{11}H_{18}$ (hexamethylcyclopentadiene)	7.74 ±0.05	EI	165	*31*
$C_{12}H_{10}$ (biphenyl)	8.27 ±0.01	PI	230	*24*
cyclo-$C_6(CH_3)_6$	7.85 ±0.02	PI	152	*28*
$C_{13}H_{10}$ (fluorene)	8.63	EI	243	*47*
$C_{14}H_{10}$ (diphenylacetylene)	8.85 ±0.05	EI	303	*54*
$C_{14}H_{10}$ (anthracene)	7.55	EI	228	*55*
$C_{14}H_{10}$ (phenanthrene)	8.1	EI	233	*54, 55*
$C_{18}H_{12}$ (1,2-benzanthracene)	8.01	EI	251	*56*
$C_{18}H_{30}$ (1-phenyldodecane)	9.05 ±0.1	EI	165	*57*
$C_{18}H_{30}$ (3-phenyldodecane)	8.95 ±0.1	EI	163	*57*
$C_{19}H_{32}$ (7-phenyltridecane)	8.91 ±0.1	EI	157	*57*
$C_{26}H_{46}$ (1-phenylicosane)	9.34 ±0.1	EI	132	*57*
$C_{26}H_{46}$ (2-phenylicosane)	9.22 ±0.1	EI	129	*57*
$C_{26}H_{46}$ (3-phenylicosane)	8.95 ±0.1	EI	123	*57*
$C_{26}H_{46}$ (4-phenylicosane)	9.01 ±0.1	EI	125	*57*
$C_{26}H_{46}$ (5-phenylicosane)	9.04 ±0.1	EI	125	*57*
$C_{26}H_{46}$ (7-phenylicosane)	8.97 ±0.1	EI	124	*57*
$C_{26}H_{46}$ (9-phenylicosane)	9.06 ±0.1	EI	126	*57*
$(CH_3)_3B$	8.8 ±0.2	EI	173	*58*
$(C_2H_5)_3B$	9.0 ±0.2	EI	170	*58*
N_2	15.576	S	359	*59*
NH	13.10 ±0.05	EI	382	*60*
NH_2	11.4 ±0.1	EI	304	*61*
NH_3	10.2	S, PI, PE	223	*
N_2H_2	9.85 ±0.1	EI		*62*
N_2H_4	8.74 ±0.06	PI	224	*63*
CN	14.3	EI	430	*64, 65*
HCN	13.8	EI	351	*26, 66*
C_2N_2	13.6	EI	387	*8, 64*
CH_3N	8.97	PI	201	*67*
C_2H_2N	10.9	EI	298	*68*
CH_3CN	12.2	PI	302	*14, 24*
cyclo-C_2H_5N	9.94 ±0.15	EI	255	*69*
$C_2H_5NH_2$	8.86 ±0.02	PI	193	*24*
$(CH_3)_2NH$	8.24 ±0.02	PI	186	*24*
$CH_2=CHCN$	10.91 ±0.01	PI	296	*24*
C_2H_5CN	11.84 ±0.02	PI	289	*24*
C_3H_7N	9.1 ±0.15	EI	225	*70*
$(CH_2)_3NH$	9.1 ±0.15	EI	225	*70*
n-$C_3H_7NH_2$	8.78 ±0.02	PI	185	*24*
iso-$C_3H_7NH_2$	8.72 ±0.03	PI	183	*24*
$(CH_3)_3N$	7.82 ±0.02	PI	175	*24*
$CH_2=CHCH_2CN$	10.39 ±0.01	PI	281	*24*
C_4H_5N (pyrrole)	8.20 ±0.01	PI	215	*24*
$(CH_3)_2CCN$	9.15 ±0.1	EI	239	*68*
n-C_3H_7CN	11.67 ±0.05	PI	280	*24*
C_4H_9N(pyrrolidine)	8.41	PE	192	*17*
n-$C_4H_9NH_2$	8.71 ±0.03	PI	179	*24*
sec-$C_4H_9NH_2$	8.70	PI	177	*24*
iso-$C_4H_9NH_2$	8.70	PI	177	*24*
tert-$C_4H_9NH_2$	8.64	PI	173	*24*
$(C_2H_5)_2NH$	8.01 ±0.01	PI	163	*24*
C_5H_5N (pyridine)	9.3	S, PI	247	*24, 71*
C_6H_7N (aniline)	7.7	PI	202	*
C_6H_7N (2-picoline)	9.02 ±0.03	PI	232	*24*
C_6H_7N (3-picoline)	9.04 ±0.03	PI	236	*24*
C_6H_7N (4-picoline)	9.04 ±0.03	PI	233	*24*
$C_6H_{13}N$ (cyclohexylamine)	8.86	PE	181	*17*
(n-C_3H_7)$_2$NH	7.84 ±0.02	PI	153	*24*
(iso-C_3H_7)$_2$NH	7.73 ±0.03	PI	148	*24*
$(C_2H_5)_3N$	7.50 ±0.02	PI	147	*24*
cyclo-C_6H_5CN	9.705 ±0.01	PI	277	*24*
C_7H_9N (n-methylaniline)	7.32	PI	192	*72, 73*
C_7H_9N (m-toluidine)	7.50 ±0.02	PI	189	*48*
C_7H_9N (2,3-lutidine)	8.85 ±0.02	PI	218	*24*
C_7H_9N (2,4-lutidine)	8.85 ±0.03	PI	218	*24*
C_7H_9N (2,6-lutidine)	8.85 ±0.02	PI	218	*24*
$C_6H_5CH_2CN$	9.40 ±0.5	EI	259	*74*
m-$C_6H_4CH_3CN$	9.66 ±0.05	EI	271	*74*
p-$C_6H_4CH_3CN$	9.76	EI	273	*75*
cyclo-$C_6H_5NHC_2H_5$	7.56	CTS	193	*76*
cyclo-$C_6H_5N(CH_3)_2$	7.12	PI	185	*72, 73*
(n-C_4H_9)$_2$NH	7.69 ±0.03	PI	140	*24*
$C_9H_{13}N$ (N-n-propylaniline)	7.54	CTS	188	*76*
$C_9H_{13}N$ (N-ethyl-N-methylaniline)	7.37	CTS	185	*76*
$C_9H_{13}N$ (N,N-dimethyl-o-toluidine)	7.37	CTS	184	*76*
$C_9H_{13}N$ (N,N-dimethyl-m-toluidine)	7.35	CTS	181	*76*
$C_9H_{13}N$ (N,N-dimethyl-p-toluidine)	7.33	CTS	181	*76*
(n-C_3H_7)$_3$N	7.23	PI	207	*24*
$C_{10}H_{15}N$ (N-n-butylaniline)	7.53	CTS	183	*76*
$C_{10}H_{15}N$ (N,N-diethylaniline)	6.99	CTS	172	*76*
$C_{10}H_{15}N$ (N,N-dimethyl-p-ethylaniline)	7.38	CTS	177	*76*
$C_{10}H_{15}N$ (N,N-2,4-tetramethylaniline)	7.17	CTS	171	*76*
$C_{10}H_{15}N$ (N,N,2,6-tetramethylaniline)	7.22	CTS	173	*76*
$C_{10}H_{15}N$ (N,N,3,5-tetramethylaniline)	7.25	CTS	172	*76*
$C_{11}H_{17}N$ (N,N-diethyl-p-toluidene)	6.93	CTS	164	*76*
$C_{11}H_{17}N$ (N,N-dimethyl-p-isopropylaniline)	7.41	CTS	174	*76*
$(C_6H_5)_2NH$	7.25 ±0.03	PI	223	*77*
$C_{12}H_{19}N$ (N,N-di-n-propylaniline)	6.96	CTS	163	*76*
$C_{12}H_{19}N$ (N,N-dimethyl-p-tert-butylaniline)	7.43	CTS	165	*76*
$C_{14}H_{23}N$ (N,N-di-n-butylaniline)	6.95	CTS	153	*76*

* Average of several values.

Molecule	Ionization Potential ev	Method	ΔH_f of Ion Kcal/mole	Reference
$(C_6H_5)_3N$	6.86 ±0.03	PI	243	*77*
CH_2N_2 (diazirine)	10.18 ±0.05	EI	314	*78*
CH_2N_2 (diazomethane)	8.999 ±0.001	S	257	*79*
CH_6N_2 (methyl-hydrazine)	8.00 ±0.06	PI	207	*63*
$CH_3N{=}NCH_3$	8.65 ±0.1	EI	243	*80*
$C_2H_8N_2$ (1,1-dimethyl-hydrazine)	7.67 ±0.05	PI	197	*81*
$C_2H_8N_2$ (1,2-dimethyl-hydrazine)	7.75 ±0.1	EI	200	*81*
$(CH_3)_3N_2H$	7.93 ±0.1	EI	202	*82*
$o\text{-}C_4H_4N_2$ (o-diazine)	9.9	EI	275	*83*
$m\text{-}C_4H_4N_2$ (m-diazine)	9.9	EI	277	*83*
$p\text{-}C_4H_4N_2$ (p-diazine)	9.8	EI	274	*83*
$1,1(C_2H_5)_2N_2H_2$	7.59 ±0.05	PI	184	*63*
$(CH_3)_4N_2$	7.76 ±0.05	EI	196	*82*
$C_5H_4NNH_2$	8.97 ±0.05	EI	244	*84*
$C_5H_{14}N_2$ (1-methyl-1-n-butylhydrazine)	7.62 ±0.05	PI	180	*63*
$(CH_3)_2NC_6H_4N(CH_3)_2$ [p-bis(dimethylamino)-benzene]	6.9	CTS	180	*85*
CH_3N_3	9.5 ±0.1	EI	276	*86*
O_2	12.063 ±0.001	PI	278	*
O_3	12.3 ±0.1	PE	318	*87*
OH	13.17 ±0.1	EI	313	*88*
H_2O	12.6	PI	233	*
D_2O	12.6	PI	232	*37*
HO_2	11.53 ±0.02	EI	271	*89*
H_2O_2	11.0	EI	233	*90, 91*
Li_2O	6.8	EI	120	*92, 93*
CO	14.013 ±0.004	S	297	*94*
CO_2	13.769 ±0.03	S	223	*96*
NO	9.25	PI, S	235	*
N_2O	12.894	S	317	*95*
NO_2	9.79	PI	233	*97, 98*
CHO	9.8	EI	221	*
CH_2O	10.88	PI	223	*24, 48*
CH_3OH	10.84	PI, PE	202	*17, 24*
CH_3CO	10.3	PI	132	*
CH_3CHO	10.2	PI	196	*
C_2H_4O (ethylene oxide)	10.6	PI, S	231	*
C_2H_5OH	10.49	PI	185	*24, 30*
CH_3OCH_3	9.98	S, PI	186	*24, 99*
$CH_2{=}CHCHO$	10.10 ±0.01	PI	210	*24*
C_2H_5CHO	9.98	PI	181	*24*
CH_3COCH_3	9.69	PI	171	*
$CH_2{=}CHCH_2OH$	9.67 ±0.05	PI	191	*24*
$CH_2{=}CHOCH_3$	8.93 ±0.02	PI	178	*24*
C_3H_6O (propyleneoxide)	10.22 ±0.02	PI	214	*24*
C_3H_6O (trimethylene-oxide)	9.667 ±0.005	S	199	*100*
$n\text{-}C_3H_7OH$	10.1	PI	172	*24, 101*
$iso\text{-}C_3H_7OH$	10.15	PI	169	*24*
C_4H_4O (furan)	8.89	S, PI	197	*24, 102*
$CH_3CH{=}CHCHO$	9.73 ±0.01	PI	194	*24*
$n\text{-}C_3H_7CHO$	9.86 ±0.02	PI	174	*24*
$iso\text{-}C_3H_7CHO$	9.74 ±0.03	PI	169	*24*
$C_2H_5COCH_3$	9.5	PI	161	*
$cyclo\text{-}C_4H_8O$	9.42	S	174	*100*
$n\text{-}C_4H_9OH$	10.04	PI	165	*24*
$C_2H_5OC_2H_5$	9.6	PI	161	*24*
C_5H_8O (cyclo-pentanone)	9.26 ±0.01	PI	163	*24*
C_5H_8O (dihydropyran)	8.34 ±0.01	PI	164	*24*
$n\text{-}C_4H_9CHO$	9.82 ±0.05	PI	168	*24*
$iso\text{-}C_4H_9CHO$	9.71 ±0.05	PI	164	*24*
$n\text{-}C_3H_7COCH_3$	9.37 ±0.02	PI	154	*103*
$iso\text{-}C_3H_7COCH_3$	9.30 ±0.02	PI	151	*24, 103*
$(C_2H_5)_2CO$	9.32 ±0.01	PI	153	*24*
$cyclo\text{-}C_5H_{10}O$	9.25 ±0.01	S	161	*100*
C_6H_5O	8.84	EI	226	*36*
C_6H_6O (phenol)	8.51	PI	173	*24, 48*
$(CH_3)_2C{=}CHCOCH_3$	9.08 ±0.03	PI	168	*24*
$C_6H_{10}O$ (cyclo-hexanone)	9.14 ±0.01	PI	152	*24*
$n\text{-}C_4H_9COCH_3$	9.35	PI	149	*24, 103*
$iso\text{-}C_4H_9COCH_3$	9.30	PI	147	*24, 103*
$tert\text{-}C_4H_9COCH_3$	9.17 ±0.03	PI	140	*24*
$(n\text{-}C_3H_7)_2O$	9.27 ±0.05	PI	147	*24*
$(iso\text{-}C_3H_7)_2O$	9.20 ±0.05	PI	142	*24*
C_6H_5CHO	9.52	PI	209	*24*
C_7H_6O (tropone)	9.68 ±0.02	EI	240	*104*
$cyclo\text{-}C_6H_5CH_2OH$	9.14 ±0.05	EI	186	*74*
$cyclo\text{-}C_6H_5OCH_3$	8.21 ±0.02	PI	173	*24*
C_7H_8O (m-cresol)	8.52 ±0.05	EI	165	*74*
$n\text{-}C_5H_{11}COCH_3$	9.33 ±0.03	PI	143	*24*
C_8H_8O (acetophenone)	9.27 ±0.03	PI	191	*24*
C_8H_8O (p-methyl-benzaldehyde)	9.33 ±0.05	EI	194	*105*
$C_8H_{10}O$ (benzyl methyl ether)	[illegible]	PI	184	[illegible]
$C_8H_{10}O$ (phenyl ethyl ether)	8.13 ±0.02	PI	167	*24*
$C_8H_{10}O$ (m-methylanisole)	8.31 ±0.05	EI	169	*74*
$C_9H_{10}O$ (phenyl ethyl ketone)	9.27 ±0.05	EI	189	*74*
$C_9H_{10}O$ (m-methyl-acetophenone)	9.15 ±0.05	EI	182	*74*
$C_{12}H_{10}O$ (phenyl ether)	8.82 ±0.05	EI	220	*106*
$C_{13}H_{10}O$ (benzophenone)	9.4	EI	229	*106, 105*
$C_{14}H_{12}O$ (p-methyl-benzophenone)	9.13 ±0.05	EI	214	*105*
HCOOH	11.05 ±0.01	PI	164	*24*
CH_3COOH	10.69±0.03	PI	135	*24*
$HCOOCH_3$	10.815 ±0.005	PE	166	*165*
C_2H_5COOH	10.24 ±0.03	PI	127	*24*
$HCOOC_2H_5$	10.61 ±0.01	PI	156	*24*
CH_3COOCH_3	10.27 ±0.02	PI	138	*24*
$(CH_3O)_2CH_2$	10.00 ±0.05	PI	145	*24*
$CH_3COOCH{=}CH_2$	9.19 ±0.05	PI	137	*24*
$CH_3COCOCH_3$	9.24 ±0.03	PI	135	*24*
$n\text{-}C_3H_7COOH$	10.16 ±0.05	PI	121	*24*
$iso\text{-}C_3H_7COOH$	10.02 ±0.05	PI	115	*24*
$HCOOCH_2CH_2CH_3$	10.54 ±0.01	PI	149	*24*
$CH_3COOC_2H_5$	10.11 ±0.02	PI	126	*24*
$C_2H_5COOCH_3$	10.15 ±0.03	PI	127	*24*
$C_4H_8O_2$ (p-dioxane)	9.13 ±0.03	PI	126	*24*
$(CH_3O)_2CHCH_3$	9.65 ±0.03	PI	129	*24*
$C_4H_4O_2$ (2-furfur-aldehyde)	9.21 ±0.01	PI	187	*24*
$CH_3COCH_2COCH_3$	8.87 ±0.03	PI	122	*24*
$HCOO(CH_2)_3CH_3$	10.50 ±0.02	PI	144	*24*
$HCOOCH_2CH(CH_3)_2$	10.46 ±0.02	PI	139	*24*
$CH_3COOCH_2CH_2CH_3$	10.04 ±0.03	PI	121	*24*
$CH_3COOCH(CH_3)_2$	9.99 ±0.03	PI	116	*24*
$C_2H_5COOC_2H_5$	10.00 ±0.02	PI	119	*24*
$n\text{-}C_3H_7COOCH_3$	10.07 ±0.03	PI	125	*24*
$iso\text{-}C_3H_7COOCH_3$	9.98 ±0.02	PI	121	*24*
$(C_2H_5O)_2CH_2$	9.70 ±0.05	PI	134	*24*
$p\text{-}C_6H_4O_2$	9.67 ±0.02	PI	198	*48*
$CH_3COOC_4H_9$	9.56 ±0.03	PI	104	*48*
$CH_3COO(CH_2)_3CH_3$	10.01	PI	114	*24*
$CH_3COOCH_2CH(CH_3)_2$	9.97	PI	111	*24*
$CH_3COOCH(CH_3)C_2H_5$	9.91 ±0.03	PI	110	*24*
C_6H_5COOH	9.73 ±0.09	EI	152	*105*
$p\text{-}HOC_6H_4CHO$	9.32 ±0.02	EI	157	*105*
$C_8H_8O_2$ (α-hydroxy-acetophenone)	9.33 ±0.05	EI	159	*74*
$C_6H_5COOCH_3$	9.35 ±0.06	EI	144	*105*
$C_8H_8O_2$ (p-methoxy-benzaldehyde)	8.60 ±0.03	EI	150	*105*
$C_8H_8O_2$ (m-hydroxy-acetophenone)	8.67 ±0.05	EI	134	*74*
$C_8H_8O_2$ (p-hydroxy-acetophenone)	8.70 ±0.03	EI	135	*105*
$C_9H_{10}O_2$ (α-methoxy-acetophenone)	8.60 ±0.05	EI	142	*74*
$C_9H_{10}O_2$ (m-methoxy-acetophenone)	8.53 ±0.05	EI	140	*74*
$C_9H_{10}O_2$ (p-methoxy-acetophenone)	8.62 ±0.05	EI	142	*105*
$C_9H_{10}O_2$ (methyl p-toluate)	8.94 ±0.04	EI	130	*105*
$C_{13}H_{10}O_2$ (p-hydroxy-benzophenone)	8.59 ±0.05	EI	165	*105*
$C_{13}H_{10}O_2$ (phenyl benzoate)	8.98 ±0.05	EI	177	*106*
$C_{14}H_{10}O_2$ (benzil)	8.78 ±0.05	EI	181	*106*

* Average of several values.

IONIZATION POTENTIALS OF MOLECULES (Continued)

Molecule	Ionization Potential ev	Method	ΔH_f of Ion Kcal/mole	Reference
$CH_3OCH_2COOCH_3$	9.56 ±0.05	EI	88	*74*
$C_9H_{10}O_3$ (methyl p-methoxybenzoate)	8.43 ±0.04	EI	90	*105*
$(C_6H_5)_2CO$	9.01 ±0.05	EI	122	*106*
CH_3CONH_2	9.77 ±0.02	PI	171	*24*
$HCON(CH_3)_2$	9.12 ±0.02	PI	160	*24*
$CH_3CONHCH_3$	8.90 ±0.02	PI	150	*24*
$CH_3CON(CH_3)_2$	8.81 ±0.03	PI	145	*24*
C_5H_4NOH	9.70 ±0.05	EI	209	*84*
$HCON(C_2H_5)_2$	8.89 ±0.02	PI	145	*24*
C_6H_5NO (2-pyridine-carboxaldehyde)	9.75 ±0.05	EI	227	*84*
C_6H_5NO (4-pyridine-carboxaldehyde)	10.12 ±0.05	EI	235	*84*
$CH_3CON(C_2H_5)_2$	8.60 ±0.02	PI	130	*24*
C_6H_5NCO	8.77 ±0.02	PI	222	*24*
C_7H_7NO (benzamide)	9.4 ±0.2	EI	197	*107, 108*
C_7H_7NO (p-amino-benzaldehyde)	8.25 ±0.02	EI	182	*105*
C_7H_9NO (p-methoxy-analine)	7.82	EI	169	*75*
C_8H_9NO (acetanilide)	8.39 ±0.10	EI	171	*108*
C_8H_9NO (m-amino-acetophenone)	8.09 ±0.05	EI	171	*74*
C_8H_9NO (p-amino-acetophenone)	8.17 ±0.02	EI	172	*105*
C_9H_7NO (α-cyano-acetophenone)	9.56 ±0.05	EI	235	*74*
CH_3NO_2	11.1	PI	238	*14, 24*
$C_2H_5NO_2$	10.88 ±0.05	PI	226	*24*
n-$C_3H_7NO_2$	10.81 ±0.03	PI	221	*24*
iso-$C_3H_7NO_2$	10.71 ±0.05	PI	217	*24*
$C_6H_5NO_2$	9.92	PI	246	*24*
$C_7H_6NO_2$ (m-nitro-benzyl radical)	8.56 ±0.1	EI	227	*39*
$C_7H_7NO_2$ (m-nitro-toluene)	9.65 ±0.05	EI	233	*74*
$C_7H_7NO_2$ (p-nitro-toluene)	9.82	EI	237	*75*
$C_8H_9NO_2$	8.08 ±0.01	EI	122	*90*
$C_6H_6N_2O_2$ (o-nitro-aniline)	8.66	EI	215	*75*
$C_6H_6N_2O_2$ (m-nitro-aniline)	8.7	EI	216	*75*
$C_6H_6N_2O_2$ (p-nitro-aniline)	8.85	EI	219	*75*
$C_2H_5NO_3$	11.22	PI	222	*24*
n-$C_3H_7ONO_2$	11.07 ±0.02	PI	213	*24*
$C_6H_5NO_3$ (p-nitro-phenol)	9.52	EI	187	*75*
$C_7H_5NO_3$ (p-nitro-benzaldehyde)	10.27 ±0.01	EI	217	*105*
$C_8H_7NO_3$ (m-nitro-acetophenone)	9.89 ±0.05	EI	201	*74*
$C_8H_7NO_3$ (p-nitro-acetophenone)	10.07 ±0.02	EI	205	*105*
$C_8H_7NO_4$ (methyl-p-nitrobenzoate)	10.20 ±0.03	EI	160	*105*
F_2	15.7	S	362	*109*
HF	15.77 ±0.02	EI	299	*110*
BF	11.3	EI	233	*111, 112*
BF_3	15.5	EI	87	*
CF_2	11.8	EI	237	*113, 114*
C_2F_4	10.12	PI	78	*28*
C_6F_6	9.97	PI		*28*
NF_2	11.9	EI	284	*115, 116*
trans-N_2F_2	13.1 ±0.1	EI	322	*115*
NF_3	13.2 ±0.2	EI	275	*115, 123*
N_2F_4	12.04 ±0.10	EI	276	*124*
OF_2	13.6	EI	309	*117, 118*
XeF_2	11.5 ±0.2	S	239	*119*
CH_2F	9.35	EI	207	*29*
CH_3F	12.85 ±0.01	EI	229	*120*
C_2H_3F	10.37	PI	211	*28, 121*
cyclo-C_6H_5F	9.2	S, PI	186	*
C_7H_7F	8.9	PI	172	*24*
CHF_2	9.45	EI	143	*29, 122*
$C_2H_2F_2$	10.30	PI	159	*28, 121*
o-$C_6H_4F_2$	9.31	PI	147	*121*
p-$C_6H_4F_2$	9.15	PI	140	*121*
C_2HF_3	10.14	PI	122	*28*
$CH_2{=}CHCF_3$	10.9	PI	93	*28*

Molecule	Ionization Potential ev	Method	ΔH_f of Ion Kcal/mole	Reference
$C_7H_5F_3$	9.68	S, PI	84	*24, 125*
$C_7H_{11}F_3$ (trifluoro-methylcyclohexane)	10.46 ±0.02	PI	37	*24*
C_6H_5OF (o-fluoro-phenol)	8.66 ±0.01	PI	132	*24*
Na_2	4.90 ±0.01	PI	147	*126*
AlF	9.8	EI	166	*127, 128*
$(CH_3)_4Si$	9.9	EI	171	*129, 130*
PH_3	9.98	PI	231	*131*
PF_3	9.71	PI	4	*131*
CH_3PH_2	9.72 ±0.15	EI	217	*132*
$C_2H_5PH_2$	9.47 ±0.5	EI	206	*132*
$(CH_3)_3P$	8.60 ±0.2	EI	175	*132*
$(C_6H_5)_3P$	7.36 ±0.05	PI	242	*77*
S_6	9.7	EI	248	*133*
S_7	9.2 ±0.3	EI		*133*
HS	10.50 ±0.1	EI	276	*134*
H_2S	10.4	S	235	*135*
CS_2	10.080	S	261	*95*
SO_2	12.34 ±0.02	PI	214	*30*
CH_3S	8.06 ±0.1	EI	218	*134*
CH_3SH	9.440 ±0.005	PI	212	*24*
C_2H_4S (ethylene sulfide)	8.87 ±0.15	EI	224	*69*
C_2H_5SH	9.285 ±0.005	PI	203	*24*
CH_3SCH_3	8.685 ±0.005	PI	191	*24*
C_3H_6S (propylene sulfide)	8.6 ±0.2	EI	218	*136*
$(CH_2)_3S$	8.9 ±0.15	EI	220	*70*
n-C_3H_7SH	9.195 ±0.005	PI	197	*24*
$C_2H_5SCH_3$	8.55 ±0.01	PI	183	*24*
C_4H_4S (thiophene)	8.860 ±0.005	PI	229	*24*
$CH_3SCH_2{-}CH{=}CH_2$	8.70 ±0.2	EI	211	*137*
$(CH_2)_4S$	8.57 ±0.75	EI	190	*70*
n-C_4H_9SH	9.14 ±0.02	PI	191	*24*
$C_2H_5SC_2H_5$	8.430 ±0.005	PI	175	*24*
n-$C_3H_7SCH_3$	8.8 ±0.15	EI	183	*138*
iso-$C_3H_7SCH_3$	8.7 ±0.2	EI	179	*137*
C_6H_5S	8.63 ±0.1	EI	250	*134*
C_6H_5SH	8.32 ±0.01	PI	217	*138*
C_6H_8S (2-ethyl-thiophene)	8.8 ±0.2	EI	215	*139*
$(n{-}C_3H_7)_2S$	8.5	PI, EI	170	*24, 143*
$C_6H_5S_2CH_3$	8.9	EI	229	*137*
$C_7H_{10}S$ (2-propyl-thiophene)	8.6 ±0.2	EI	205	*139*
$C_8H_{12}S$	8.8	EI	221	*137*
$C_8H_{12}S$ (2-butyl-thiophene)	8.5 ±0.2	EI	198	*139*
CH_3SSCH_3	8.46 ±0.03	PI	189	*24*
$C_2H_5SSC_2H_5$	8.27 ±0.03	PI	173	*24*
CH_3SSSCH_3	8.80 ±0.15	EI	203	*140*
COS	11.17 ±0.01	PI	224	*138*
SO_2F_2	13.3 ±0.1	EI	102	*141*
CH_3NCS	9.25 ±0.03	PI	245	*24*
CH_3SCN	10.065 ±0.01	PI	270	*24*
C_2H_5NCS	9.14 ±0.03	PI	237	*24*
C_2H_5SCN	9.89 ±0.01	PI	261	*24*
C_7H_5NS (phenyl-isothiocyanate)	8.520 ±0.005	PI	260	*24*
$C_6H_5CH_2SCN$	9.06 ±0.05	EI	274	*74*
NH_2CSNH_2	8.50 ±0.05	EI	194	*142*
$NH_2CSNHCH_3$	8.29 ±0.05	EI	188	*142*
$NH_2CSNHCH{=}CH_2$	8.29 ±0.05	EI	213	*142*
$NH_2CSN(CH_3)_2$	8.34 ±0.05	EI	186	*142*
$CH_3NHCSNHCH_3$	8.17 ±0.05	EI	184	*142*
$CH_3NHCSN(CH_3)_2$	7.93 ±0.05	EI	176	*142*
$C_5H_{12}N_2S$	7.98 ±0.05	EI	170	*142*
$(CH_3)_2NCSN(CH_3)_2$	7.95 ±0.05	EI	173	*142*
CH_3COSH	10.00 ±0.02	PI	179	*24*
Cl_2	11.48 ±0.01	PI	265	*24*
HCl	12.74	PI	272	*
LiCl	10.1	EI	186	*144*
CCl_3	8.78 ±0.05	EI	214	*
CCl_4	11.47 ±0.01	PI	240	*24*
C_2Cl_4	9.32 ±0.01	PI	212	*24, 28*
PCl_3	9.91	PI	160	*131*
CH_2Cl	9.32	EI	244	*29*
CH_3Cl	11.3	S, PI	239	*14, 145*
C_2H_3Cl	9.996	S, PI	239	*
C_2H_5Cl	10.97	PI	226	*24*
$CH_3C{\equiv}CCl$	9.9 ±0.1	EI	267	*25*
n-C_3H_7Cl	10.82 ±0.03	PI	219	*24*

* Average of several values.

Molecule	Ionization Potential ev	Method	ΔH_f of Ion Kcal/mole	Reference
iso-C_3H_7Cl	10.78 ±0.02	PI	211	*24*
n-C_4H_9Cl	10.67 ±0.03	PI	210	*24*
sec-C_4H_9Cl	10.65 ±0.03	PI	210	*24*
iso-C_4H_9Cl	10.66 ±0.03	PI	208	*24*
tert-C_4H_9Cl	10.61 ±0.03	PI	202	*24*
C_6H_5Cl	9.07	PI	222	*24*
$C_6H_5CH_2Cl$	9.19 ±0.05	EI	219	*74*
o-$C_6H_4ClCH_3$	8.83 ±0.02	PI	208	*24*
m-$C_6H_4ClCH_3$	8.83 ±0.02	PI	207	*24*
p-$C_6H_4ClCH_3$	8.69 ±0.02	PI	204	*24*
C_7H_9Cl (endo-5-chloro-2-norbornene)	9.1 ±0.15	EI	233	*43*
C_7H_9Cl (exo-5-chloro-2-norbornene)	9.15 ±0.15	EI	234	*43*
C_7H_9Cl (3-chloro-nortricyclene)	9.51 ±0.15	EI	234	*43*
$CHCl_2$	9.30	EI	245	*29*
CH_2Cl_2	11.35 ±0.02	PI	240	*24*
cis-$C_2H_2Cl_2$	9.65	PI	223	*
trans-$C_2H_2Cl_2$	9.64	PI	224	*
CH_2ClCH_2Cl	11.12 ±0.05	PI	225	*24*
$CH_2=CClCH_2Cl$	9.82 ±0.03	PI	218	*24*
1,2-$C_3H_6Cl_2$	10.87 ±0.05	PI	215	*24*
1,3-$C_3H_6Cl_2$	10.85 ±0.05	PI	215	*24*
o-$C_6H_4Cl_2$	9.06	PI	217	*24, 28*
m-$C_6H_4Cl_2$	9.12 ±0.01	PI	217	*24*
p-$C_6H_4Cl_2$	8.95	PI	212	*28*
$CHCl_3$	11.42 ±0.03	PI	239	*24*
C_2HCl_3	9.45	PI	216	*
$CHCl_2CHCl_2$	11.10 ±0.05	EI	220	*146*
CNCl	12.49 ±0.04	EI	321	*147*
CF_3Cl	12.91 ±0.03	PI	132	*24*
C_2F_3Cl	10.4 ±0.2	EI	107	*148*
C_6F_5Cl	10.4 ±0.1	EI		*149*
CF_2Cl_2	12.31 ±0.05	PI	170	*24*
$CF_3CCl=CClCF_3$	10.36 ±0.1	PI		*24*
$CFCl_3$	11.77 ±0.02	PI	205	*24*
CF_3CCl_3	11.78 ±0.03	PI		*24*
$CFCl_2CF_2Cl$	11.99 ±0.02	PI	95	*24*
ClO_3F	13.6 ±0.2	EI	308	*150*
C_5H_4NCl (2-chloro-pyridene)	9.91 ±0.05	EI	255	*84*
C_5H_4NCl (4-chloro-pyridene)	10.15 ±0.05	EI	260	*84*
CH_3COCl	11.02 ±0.05	PI	196	*24*
CH_3COCH_2Cl	9.99	EI	173	*105, 151*
o-$C_6H_4(OH)Cl$	9.28	EI	181	*75*
p-$C_6H_4(OH)Cl$	9.07	EI	175	*75*
C_6H_5COCl	9.70 ±0.01	EI	195	*105*
p-C_6H_4ClCHO	9.61 ±0.01	EI	201	*105*
C_8H_7OCl (α-chloro-acetophenone)	9.5	EI	195	*74, 105*
C_8H_7OCl (p-chloro-acetophenone)	9.47 ±0.05	EI	190	*105*
C_8H_7OCl (p-methyl-benzylchloride)	9.37 ±0.01	EI	187	*105*
$C_6H_4ClCOC_6H_5$	9.68 ±0.01	EI	227	*105*
$CH_2ClCOOCH_3$	10.53 ±0.05	EI	138	*74*
p-$CH_3OC_6H_4COCl$	8.87 ±0.05	EI	149	*105*
p-ClC_6H_4COCl	9.58 ±0.03	EI	192	*105*
cis-C_2H_2FCl	9.86	PI	191	*24, 121*
trans-C_2H_2FCl	9.87	PI	191	*24, 121*
o-C_6H_4FCl	9.155 ±0.01	PI	180	*24*
m-C_6H_4FCl	9.21 ±0.01	PI	180	*24*
p-C_6H_4FCl	9.43 ±0.02	EI	185	*152*
CHF_2Cl	12.45 ±0.05	PI	174	*24*
cis-C_2HF_2Cl	9.86 ±0.02	PI	147	*121*
trans-C_2HF_2Cl	9.83 ±0.02	PI	147	*121*
CH_3CF_2Cl	11.98 ±0.01	PI		*24*
n-$C_3F_7CH_2Cl$	11.84 ±0.02	PI		*24*
$CHFCl_2$	12.39 ±0.20	EI	217	*153*
$(CH_3)_3SiCl$	10.58 ±0.04	EI	160	*130*
CH_3SiCl_3	11.36 ±0.03	EI	136	*153*
$CH_2=CHSiCl_3$	10.79 ±0.02	PI	148	*24*
$C_2H_5SiCl_3$	10.74 ±0.04	EI	117	*153*
iso-$C_3H_7SiCl_3$	10.28 ±0.1	EI	100	*153*
C_4H_3ClS	8.68 ±0.01	PI	217	*24*
$C_6H_4NO_2COCl$	10.66 ±0.01	EI	219	*105*
CaF	5.8	EI	75	*
$C_5H_5V(CO)_4$	8.2 ±0.3	EI	9	*154*
$Cr(CO)_6$	8.03 ±0.03	PI	−55	*155*
$C_5H_5Mn(CO)_3$	8.3 ±0.4	EI	83	*154*
$Fe(CO)_5$	7.95 ±0.03	PI	8	*155*
$C_5H_5Co(CO)_2$	8.3 ±0.2	EI	136	*154*
$Ni(CO)_4$	8.28 ±0.03	PI	47	*155*
$(CH_3)_4Ge$	9.2 ±0.2	EI	177	*156*
As_4	9.07 ±0.07	EI	244	*157*
AsH_3	10.03	PI	247	*131*
$AsCl_3$	11.7 ±0.1	EI	208	*158*
$(CH_3)_3As$	8.3 ±0.1	EI	202	*158*
$(C_6H_5)_3As$	7.34 ±0.07	PI	250	*77*
Br_2	10.54 ±0.03	PI	250	*24, 150*
HBr	11.62 ±0.03	PI	259	*24*
$MgBr_2$	10.65 ±0.15	EI	172	*160*
BrCl	11.1 ±0.2	EI	259	*161*
CH_3Br	10.53	S, PI	234	*
C_2H_3Br	9.80	PI	243	*
C_2H_5Br	10.29	S	222	*24, 162*
$CH_3C{\equiv}CBr$	10.1 ±0.1	EI	283	*25*
$CH_3CH=CHBr$	9.30 ±0.05	PI	224	*24*
n-C_3H_7Br	10.18 ±0.01	PI	216	*24*
iso-C_3H_7Br	10.075 ±0.01	PI	208	*24*
n-C_4H_9Br	10.125 ±0.01	PI	208	*24*
sec-C_4H_9Br	9.98 ±0.01	PI	206	*24*
iso-C_4H_9Br	10.09 ±0.02	PI	208	*24*
tert-C_4H_9Br	9.89 ±0.03	PI	201	*24*
n-$C_5H_{11}Br$	10.10 ±0.02	PI	205	*24*
C_6H_5Br	8.98 ±0.02	PI	231	*24*
o-$C_6H_4BrCH_3$	8.78 ±0.01	PI	218	*24*
m-$C_6H_4BrCH_3$	8.81 ±0.02	PI	218	*24*
p-$C_6H_4BrCH_3$	8.67 ±0.02	PI	215	*24*
CH_2Br_2	10.49 ±0.02	PI	241	*24*
cis-$C_2H_2Br_2$	9.45	PI	241	*
trans-$C_2H_2Br_2$	9.46	PI	240	*
CH_3CHBr_2	10.19 ±0.03	PI	240	*24*
1,3-$C_3H_6Br_2$	10.07 ±0.02	PI	221	*24*
$CHBr_3$	10.51 ±0.02	PI	246	*24*
C_2HBr_3	9.27	PI	240	*24, 28*
CNBr	11.95 ±0.08	EI	320	*147*
CF_3Br	11.89	EI	121	*
C_5H_4NBr (2-bromo-pyridine)	9.65 ±0.05	EI	261	*84*
C_5H_4NBr (4-bromo-pyridine)	9.94 ±0.05	EI	267	*84*
CH_3COBr	10.55 ±0.05	PI	197	*24*
C_6H_4BrOH	9.04	EI	187	*75*
$CH_2BrCOOCH_3$	10.37 ±0.05	EI	146	*74*
C_6H_4FBr	8.99 ±0.03	PI	187	*24*
CF_2BrCH_2Br	10.83 ±0.01	PI	160	*24*
C_4H_3BrS	8.63 ±0.01	PI	228	*24*
CH_2ClBr	10.77 ±0.01	PI	236	*24*
CH_2BrCH_2Cl	10.63 ±0.03	PI	227	*24*
$CHCl_2Br$	10.88 ±0.05	EI	237	*146*
$(CH_3)_3SiBr$	10.24 ±0.02	EI	171	*130*
$Mo(CO)_6$	8.12 ±0.03	PI	−31	*155*
RuO_4	12.33 ±0.23	EI	240	*163*
$(CH_3)_4Sn$	8.25 ±0.15	EI	186	*129*
SbH_3	9.58	PI	256	*131*
$(C_6H_5)_3Sb$	7.3 ±0.1	PI	255	*77*
I_2	9.28 ±0.02	PI	229	*24*
HI	10.39	PI	246	*24, 159*
ICl	10.31 ±0.02	EI	242	*164*
IBr	9.98 ±0.03	EI	240	*164*
CH_3I	9.54	S, PI	223	*24, 145*
C_2H_5I	9.33	S, PI	213	*24, 162*
n-C_3H_7I	9.26 ±0.01	PI	208	*24*
iso-C_3H_7I	9.17 ±0.02	PI	201	*24*
n-C_4H_9I	9.21 ±0.01	PI	202	*24*
sec-C_4H_9I	9.09 ±0.02	PI	198	*24*
iso-C_4H_9I	9.18 ±0.02	PI	200	*24*
tert-C_4H_9I	9.02 ±0.03	PI	193	*24*
n-$C_5H_{11}I$	9.19 ±0.01	PI	197	*24*
C_6H_5I	8.73 ±0.03	PI	238	*24*
o-$C_6H_4ICH_3$	8.62 ±0.01	PI	228	*24*
m-$C_6H_4ICH_3$	8.61 ±0.03	PI	226	*24*
p-$C_6H_4ICH_3$	8.50 ±0.01	PI	224	*24*
$W(CO)_6$	8.18 ±0.03	PI	−20	*155*
OsO_4	12.97 ±0.12	EI	219	*163*
$(CH_3)_2Hg$	9.0	EI	233	*143, 156*
$(C_2H_5)_2Hg$	8.5 ±0.1	EI	221	*143*
(iso-$C_3H_7)_2Hg$	7.6 ±0.1	EI	188	*143*
CH_3HgCl	11.5 ±0.2	EI	253	*143*
$(CH_3)_4Pb$	8.0 ±0.4	EI	217	*129*
$(C_6H_5)_3Bi$	7.3 ±0.1	PI	288	*77*

* Average of several values.

REFERENCES

1. Beutler, H. and Junger, H. O. *Zeit. f. Physik* **100**, 80 (1936).
2. Dibeler, V. H., Reese, R. M. and Krauss, M. *Adv. Mass Spectry.* **3**, 471 (1966).
3. Bauer, S. H., Herzberg, G. and Johns, J. W. C. *J. Mol. Spectr.* **13**, 256 (1964).
4. Margrave, J. L. *J. Phys. Chem.* **61**, 38 (1957).
5. Koski, W. S., Kaufman, J. J., Pachucki, C. F. and Shipko, F. J. *J. Am. Chem. Soc.* **80**, 3202 (1958).
6. Fehlner, T. P. and Koski, W. S. *J. Am. Chem. Soc.* **86**, 581 (1964).
7. Drowart, J., DeMaria, G. and Inghram, M. G. *J. Chem. Phys.* **29**, 1015 (1958).
8. Field, F. H. and Franklin, J. L. *Electron Impact Phenomena and the Properties of Gaseous Ions*, Academic Press, Inc., New York, N.Y. (1957).
9. Herzberg, G. *Can. J. Phys.* **39**, 1511 (1961).
10. Waldron, J. D., *Metropolitan Vickers Gazette* **27**, 66 (1956).
11. Dibeler, V. H., Krauss, M., Reese, R. M. and Harllee, F. N. *J. Chem. Phys.* **42**, 3791 (1965).
12. Herzberg, G. and Shoosmith, J., *Can J. Phys.* **34**, 523 (1956).
13. Elder, F. A., Giese, C., Steiner, B. and Inghram, M. *J. Chem. Phys.* **36**, 3292 (1962).
14. Nicholson, A. J. C., *J. Chem. Phys.* **43**, 1171 (1965).
15. Dibeler, V. H. and Reese, R. M. *J. Res. Natl. Bur. Std.* **A68**, 409 (1964).
16. Zelikoff, M. and Watanabe, K., *J. Opt. Soc. Am.* **43**, 756 (1953).
17. Al-Joboury, M. I. and Turner, D. W., *J. Chem. Soc. London* **4434** (1964).
18. Farmer, J. B. and Lossing, F. P., *Can. J. Chem.* **33**, 861 (1955).
19. Collin, J. and Lossing, F. P., *J. Am. Chem. Soc.* **79**, 5848 (1957).
20. Watanabe, K. and Namioka, T., *J. Chem. Phys.* **24**, 915 (1956).
21. Collin, J. and Lossing, F. P., *J. Am. Chem. Soc.* **81**, 2064 (1959).
22. Dorman, F. H. *J. Chem. Phys.* **43**, 3507 (1965).
23. Pottie, R. F., Harrison, A. G. and Lossing, F. P. *J. Am. Chem. Soc.* **83**, 3204 (1961).
24. Watanabe, K., Nakayama, T. and Mottl, J. *J. Quant. Spectrosc. Radiat. Transfer* **2**, 369 (1962).
25. Coats, F. H. and Anderson, R. C. *J. Am. Chem. Soc.* **79**, 1340 (1957).
26. Varsel, C. J., Morrell, F. A., Resnik, F. E. and Powell, W. A. *Anal. Chem.* **32**, 182 (1960).
27. McDowell, C. A., Lossing, F. P., Henderson, I. H. S. and Farmer, J. B. *Can. J. Chem.* **34**, 345 (1956).
28. Bralsford, R., Harris, P. V. and Price, W. C. *Proc. Roy. Soc.* **A258**, 459 (1960).
29. Lossing, F. P., Kebarle, P. and DeSousa, J. B. *Adv. Mass Spectrometry* **431** (1959).
30. Watanabe, K. *J. Chem. Phys.* **26**, 542 (1957).
31. Meyer, F. and Harrison, A. G. *Can. J. Chem.* **42**, 2256 (1964).
32. Harrison, A. G., Honnen, L. R., Dauben, H. J. and Lossing, F. P. *J. Am. Chem. Soc.* **82**, 5593 (1960).
33. Natalis, P. and Laune, *J. Bull. Soc. Chim. Belg.* **73**, 944 (1964).
34. Taubert, R. and Lossing, F. P. *J. Am. Chem. Soc.* **84**, 1523 (1962).
35. Steiner, B., Giese, C. F. and Inghram, M. G. *J. Chem. Phys.* **34**, 189 (1961).
36. Fisher, I. P., Palmer, T. F. and Lossing, F. P. *J. Am. Chem. Soc.* **86**, 2741 (1964).
37. Brehm, B. *Z. Naturforschg.* **21a**, 196 (1966).
38. Momigny, J., Brakier, L. and D Or, L. *Bull. Classe Sci. Acad. Roy. Belg.* **48**, 1002 (1962).
39. Harrison, A. G., Kebarle, P. and Lossing, F. P. *J. Am. Chem. Soc.* **83**, 777 (1961).
40. Thrush, B. A. and Zwolenik, J. J. *Disc. Faraday Soc.* **35**, 196 (1963).
41. Lifshitz, C. and Bauer, S. H. *J. Phys. Chem.* **67**, 1629 (1963).
42. Meyerson, S., McCollum, J. D. and Rylander, P. N. *J. Am. Chem. Soc.* **83**, 1401 (1961).
43. Steele, W. C., Jennings, B. H., Botyos, G. L. and Dudek, G. O. *J. Org. Chem.* **30**, 2886 (1965).
44. Kybett, B. D., Carroll, S., Natalis, P., Bonnell, D. W., Margrave, J. L. and Franklin, J. L. *J. Am. Chem. Soc.* **88**, 626 (1966).
45. Meyer, F., Haynes, P., McLean, S. and Harrison, A. G. *Can. J. Chem.* **43**, 211 (1965).
46. Natalis, P. *Bull. Soc. Chim. Belg.* **73**, 961 (1964).
47. Pottie, R. F. and Lossing, F. P. *J. Am. Chem. Soc.* **85**, 269 (1963).
48. Vilesov, F. I. and Terenin, A. N. *Dokl. Phys. Chem., Proc. Acad. Sci.* **USSR 115**, 539 (1957).
49. Clark, L. B. *J. Chem. Phys.* **43**, 2566 (1965).
50. Kitagawa, T., Harada, Y., Inokuchi, H. and Kodera, K. *J. Mol. Spectr.* **19**, 1 (1966).
51. Natalis, P. *Bull. Soc. Roy. Sci. Liege* **31**, 803 (1962).
52. Harrison, A. G. and Lossing, F. P. *J. Am. Chem. Soc.* **82**, 1052 (1960).
53. Vilesov, F. I. *J. Phys. Chem.* **USSR 35**, 986 (1961).
54. Natalis, P. and Franklin, J. L. *J. Phys. Chem.* **69**, 2935 (1965).
55. Wacks, M. E. and Dibeler, V. H. *J. Chem. Phys.* **31**, 1557 (1959).
56. Wacks, M. E. *J. Chem. Phys.* **41**, 1661 (1964).
57. King, A. B. *J. Chem. Phys.* **42**, 3526 (1965).
58. Law, R. W. and Margrave, J. L. *J. Chem. Phys.* **25**, 1086 (1956).
59. Lofthus, A. *The Molecular Spectrum of Nitrogen* Department of Physics, University of Oslo, Blindern, **Norway, Spectroscopic Report No. 2**, 1 (1960).
60. Reed, R. I. and Snedden, W. *J. Chem. Soc.* **4132** (1959).
61. Foner, S. N. and Hudson, R. L. *J. Chem. Phys.* **29**, 442 (1958).
62. Foner, S. N. and Hudson, R. L. *J. Chem. Phys.* **28**, 719 (1958).
63. Akopyan, M. E. and Vilesov, F. I. *Kinetics and Catalysis*, **4**, 32 (1963).
64. Dibeler, V. H., Reese, R. M. and Franklin, J. L. *J. Am. Chem. Soc.* **83**, 1813 (1961).
65. Berkowitz, J. *J. Chem. Phys.* **36**, 2533 (1962).
66. Morrison, J. D. and Nicholson, A. J. C. *J. Chem. Phys.* **20**, 1021 (1952).
67. Watanabe, K. and Mottl, J. R. *J. Chem. Phys.* **26**, 1773 (1957).
68. Pottie, R. F. and Lossing, F. P. *J. Am. Chem. Soc.* **83**, 4737 (1961)
69. Gallegos, E. and Kiser, R. W. *J. Phys. Chem.* **65**, 1177 (1961).
70. Gallegos, E. J. and Kiser, R. W. *J. Phys. Chem.* **66**, 136 (1962).
71. Amr El-Sayed, M. F., Kasha, M. and Tanaka, Y. *J. Chem. Phys.* **34**, 334 (1961).
72. Akopyan, M. E. and Vilesov, F. I. *Dokl. Phys. Chem., Proc. Acad. Sci.* **USSR 158**, 965 (1964).
73. Kurbatov, B. L., Vilesov, F. I. and Terenin, A. N. *Soviet Physics-Doklady* **6**, 883 (1962).
74. Pignataro, S., Foffani, A., Innorta, G. and Distefano, G. *Z. Phys. Chem. Neue Folge* **49**, 291 (1966).
75. Crable, G. F. and Kearns, G. L. *J. Phys. Chem.* **66**, 436 (1962).
76. Farrell, P. G. and Newton, J. *J. Phys. Chem.* **69**, 3506 (1965).
77. Vilesov, F. I. and Zaitsev, V. M. *Dokl. Phys. Chem., Proc. Acad. Sci.* **USSR 154**, 117 (1964).
78. Paulett, G. S. and Ettinger, R. *J. Chem. Phys.* **39**, 825 (1963).
79. Merer, A. J., *Can. J. Phys.* **42**, 1242 (1964).
80. Gowenlock, B. G., Majer, J. R. and Snelling, D. R. *Trans. Faraday Soc.* **58**, 670 (1962).
81. Akopyan, M. E., Vilesov, F. I. and Terenin, A. N. *Izv. Akad. Nauk USSR, Ser. Fiz.* **27**, 1083 (1963).
82. Dibeler, V. H., Franklin, J. L. and Reese, R. M. *J. Am. Chem. Soc.* **81**, 68 (1959).
83. Momigny, J., Urbain, J. and Wankenne, H. *Bull. Soc. Roy. Sci. Liege* **34**, 337 (1965).
84. Basila, M. R. and Clancy, D. J. *J. Phys. Chem.* **67**, 1551 (1963).
85. Finch, A. C. M. *J. Chem. Soc.* **2272** (1964).
86. Franklin, J. L., Dibeler, V. H., Reese, R. M. and Krauss, M. *J. Am. Chem. Soc.* **80**, 298 (1958).
87. Radwan, T. N. and Turner, D. W. *J. Chem. Soc.* **Sect. A85** (1966).
88. Foner, S. N. and Hudson, R. L. *Advances in Chem.* **Ser. 34** (1962).
89. Foner, S. N. and Hudson, R. L. *J. Chem. Phys.* **23**, 1364 (1955).
90. Lindeman, L. P. and Guffy, J. C. *J. Chem. Phys.* **29**, 247 (1958).
91. Foner, S. N. and Hudson, R. L. *J. Chem. Phys.* **36**, 2676 (1962).
92. Berkowitz, J., Chupka, W. A., Blue, G. D. and Margrave, J. L. *J. Phys. Chem.*, **63**, 644 (1959).
93. White, D., Seshadri, K. S., Dever, D. F., Mann, D. E. and Linevski, M. J. *J. Chem. Phys.* **39**, 2463 (1963).
94. Krupenie, P. H. *The Band Spectrum of Carbon Monoxide* Institute for Basic Standards, National Bureau of Standards, Washington, D.C., **NSRDS-NBS 5**, 1 (1966).
95. Tanaka, Y., Jursa, A. S. and LeBlanc, F. J. *J. Chem. Phys.* **32**, 1205 (1960).
96. Tanaka, Y., Jursa, A. S. and LeBlanc, F. J. *J, Chem. Phys.* **32**, 1199 (1960).
97. Nakayama, T., Kitamura, M. Y. and Watanabe, K. *J. Chem. Phys.* **30**, 1180 (1959).
98. Frost, D. C., Mak, D. and McDowell, C. A. *Can. J. Chem.* **40**, 1064 (1962).
99. Hernandez, G. J. *J. Chem. Phys.* **38**, 1644 (1963).
100. Hernandez, G. J. *J. Chem. Phys.* **38**, 2233 (1963).
101. Chupka, W. A. *J. Chem. Phys.* **30**, 191 (1959).
102. Watanabe, K. and Nakayama, T. *J. Chem. Phys.* **29**, 48 (1958).
103. Murad, E. and Inghram, M. G. *J. Chem. Phys*, **40**, 3263 (1964).
104. Higasi, K., Nozoe, T. and Omura, I. *Bull. Chem. Soc. Japan* **30**, 408 (1957).
105. Foffani, A., Pignataro, S., Cantone, B. and Grasso, F. *Z. Phys. Chem. Neue Folge* **42**, 221 (1964).
106. Natalis, P. and Franklin, J. L. *J. Phys. Chem.* **69**, 2943 (1965).
107. Cotter, J. L. *J. Chem. Soc.* **5742** (1965).
108. Cotter, J. L. *J. Chem. Soc.* **5477** (1964).
109. Iczkowski, R. P. and Margrave, J. L. *J. Chem. Phys.* **30**, 403 (1959).
110. Frost, D. C. and McDowell, C. A. *Can. J. Chem.* **36**, 39 (1958).
111. Hildenbrand, D. L. and Muran, E. *J. Chem. Phys.* **43**, 1400 (1965).
112. Marriott, J. and Craggs, J. D. *J. Electronics and Control* **3**, 194 (1957).
113. Fisher, I. P., Homer, J. B. and Lossing, F. P. *J. Am. Chem. Soc.* **87**, 957 (1965).
114. Pottie, R. F. *J. Chem. Phys.* **42**, 2607 (1965).
115. Herron, J. T. and Dibeler, V. H. *J. Res. Natl. Bur. Std.* **65**, 405 (1961).
116. Loughran, E. D. and Mader, C. *J. Chem. Phys.* **32**, 1578 (1960).
117. Frost, D. C. and McDowell, C. A. *The Determination of Ionization and Dissociation Potentials of Molecules by Radiation with Electrons* Department of Chemistry, University of British Columbia, Vancouver 8, B. C., Canada, **AFCRL-TR-60-423 1** (1960).
118. Dibeler, V. H., Reese, R. M. and Franklin, J. L. *J. Chem. Phys.* **27**, 1296 (1957).
119. Wilson, E. G. Jortner, J. and Rice, S. A. *J. Am. Chem. Soc.* **85**, 813 (1963).
120. Frost, D. C. and McDowell, C. A. *Proc. Roy. Soc.* **A241**, 194 (1957).
121. Momigny, J. *Nature* **199**, 1179 (1963).
122. Martin, R. H., Lampe, F. W. and Taft, R. W. *J. Am. Chem. Soc.* **88**, 1353 (1966).
123. Reese, R. M. and Dibeler, V. H. *J. Chem. Phys.* **24**, 1175 (1956).
124. Herron, J. T. and Dibeler, V. H. *J. Chem. Phys.* **33**, 1595 (1960).
125. Hammond, V. J., Price, W. C., Teegan, J. P. and Walsh, A. D. *Disc. Faraday Soc.* **9**, 53 (1950).
126. Hudson, R. D. *J. Chem. Phys.* **43**, 1790 (1965).
127. Margrave, J. L. *J. Chem. Phys.* **41**, 2250 (1964).

REFERENCES (Continued)

128. Ehlert, T. C. and Margrave, J. L. *J. Am. Chem. Soc.* **86**, 3901 (1964).
129. Hobrock, B. G. and Kiser, R. W. *J. Phys. Chem.* **65**, 2186 (1961).
130. Hess, G. G., Lampe, F. W. and Sommer, L. H. *J. Am. Chem. Soc.* **87**, 5327 (1965).
131. Price, W. C. and Passmore, T. R. *Disc. Faraday Soc.* **35**, 232 (1963).
132. Wada, Y. and Kiser, R. W. *J. Phys. Chem.* **68**, 2290 (1964).
133. Berkowitz, J. and Chupka, W. A. *J. Chem. Phys.* **40**, 287 (1964).
134. Palmer, T. F. and Lossing, F. P. *J. Am. Chem. Soc.* **84**, 4661 (1962).
135. Price, W. C. *Bull. Am. Phys. Soc.* **10**, 9 (1935).
136. Hobrock, B. G. and Kiser, R. W. *J. Phys. Chem.* **66**, 1551 (1962).
137. Hobrock, B. G. and Kiser, R. W. *J. Phys. Chem.* **67**, 648 (1963).
138. Hobrock, B. G. and Kiser, R. W. *J. Phys. Chem.* **66**, 1648 (1962).
139. Khvostenko, V. I. *Russ. J. Phys. Chem.* **36**, 197 (1962).
140. Hobrock, B. G. and Kiser, R. W. *J. Phys. Chem.* **67**, 1283 (1963).
141. Reese, R. M., Dibeler, V. H. and Franklin, J. L. *J. Chem. Phys.* **29**, 880 (1958).
142. Baldwin, M., Maccoll, A., Kirkien-Konasiewicz, A. and Saville, B. *Chem. Ind.* **286** (1966).
143. Gowenlock, B. G., Kay, J. and Majer, J. R. *Trans. Faraday Soc.* **59**, 2463 (1963).
144. Berkowitz, J., Tasman, H. A. and Chupka, W. A. *J. Chem. Phys.* **36**, 2170 (1962).
145. Price, W. C. *J. Chem. Phys.* **4**, 539 (1936).
146. Harrison, A. G. and Shannon, T. W. *Can. J. Chem.* **40**, 1730 (1962).
147. Herron, J. T. and Dibeler, V. H. *J. Am. Chem. Soc.* **82**, 1555 (1960).
148. Margrave, J. L. *J. Chem. Phys.* **31**, 1432 (1959).
149. Majer, J. R. and Patrick, C. R. *Trans. Faraday Soc.* **58**, 17 (1962).
150. Dibeler, V. H., Reese, R. M. and Mann, D. E. *J. Chem. Phys.* **27**, 176 (1957).
151. Foffani, A., Pignataro, S., Cantone, B. and Grasso, F. *Nuovo Cimento* **29**, 918 (1963).
152. Momigny, J. and Wirtz-Cordier, A. M. *Ann. Soc. Sci. Bruxelles* **76**, 164 (1962).
153. Hobrock, D. L. and Kiser, R. W. *J. Phys. Chem.* **68**, 575 (1964).
154. Winters, R. E. and Kiser, R. W. *J. Organometal. Chem.* **4**, 190 (1965).
155. Vilesov, F. I. and Kurbatov, B. L. *Dokl. Phys. Chem., Proc. Acad. Sci. USSR* **140**, 792 (1961).
156. Hobrock, B. G. and Kiser, R. W. *J. Phys. Chem.* **66**, 155 (1962).
157. Westmore, J. B., Mann, K. H. and Tickner, A. W. *J. Phys. Chem.* **68**, 606 (1964).
158. Cullen, W. R. and Frost, D. C. *Can. J. Chem.* **40**, 390 (1962).
159. Morrison, J. D., Hurzeler, H., Inghram, M. G. and Stanton, H. E. *J. Chem. Phys.* **33**, 821 (1960).
160. Berkowitz, J. and Marquart, J. R. *J. Chem. Phys.* **37**, 1853 (1962).
161. Irsa, A. P. and Friedman, L. *J. Inorg. Nucl. Chem.* **6**, 77 (1958).
162. Price, W. C. *J. Chem. Phys.* **4**, 547 (1936).
163. Dillard, J. G. and Kiser, R. W. *J. Phys. Chem.* **69**, 3893 (1965).
164. Frost, D. C. and McDowell, C. A. *Can. J. Chem.* **38**, 407 (1960).
165. Thomas, R. K., *Proc. R. Soc. London Ser. A*, 331, 249, 1972.

ELECTRON WORK FUNCTIONS OF THE ELEMENTS

Compiled by Herbert B. Michaelson, 1977

The measured values cited for polycrystalline and single-crystal specimens are selected as being the best available data at this time. The selection is based on (1) The validity of the experimental technique (e.g., vacua of 10^{-9} or 10^{-10} Torr, clean surfaces, and identification of crystal-face distribution and other surface conditions), and (2) Best agreement with preferred values and theoretical values of the true work function (given variously by Fomenko,[1] Rivière,[2] Trasatti,[3] and Lang and Kohn[4]). Experimental data that are not well substantiated according to these criteria are listed in *italics*. Crystallographic directions for single-crystal data are indicated by parentheses.

Abbreviations apply to the experimental method: T, thermionic; P, photoelectric; CPD, contact potential difference; F, field emission. Important distinctions among such measurements are discussed in the Rivière[2] paper, pp. 180 to 198.

Element	Experimental value, ϕ (eV)	Experimental method	Ref.
Ag	*4.26*	P	5
	4.64 (100)	P	5
	4.52 (110)	P	5
	4.74 (111)	P	6
Al	4.28	P	7
	4.41 (100)	P	8
	4.06 (110)	P	7
	4.24 (111)	P	8
As	*3.75*	P	9
Au	5.1	P	10
	5.47 (100)	P	11
	5.37 (110)		11
	5.31 (111)		11
B	*4.45*	T	12
Ba	*2.7*	T	13
Be	4.98	P	14
Bi	*4.22*	P	15
C	*5.0*	CPD	16
Ca	2.87	P	17
Cd	*4.22*	CPD	18
Ce	2.9	P	10
Co	5.0	P	10
Cr	4.5	P	10
Cs	2.14	P	19
Cu	4.65	P	10
	4.59 (100)	P	20
	4.48 (110)	P	20
	4.94 (111)	P	20
	4.53 (112)	P	20
Eu	2.5	P	10
Fe	4.5	P	10
	4.67 (100)	P	21
	4.81α (111)	P	22
	4.70α	P	23
	4.62β	P	23
	4.68γ	P	23
Ga	4.2	CPD	24
Ge	5.0	CPD	25
	4.80 (111)	P	26
Gd	3.1	P	10
Hf	3.9	P	10
Hg	4.49	P	27
In	4.12	P	28
Ir	*5.27*	T	29
	5.42 (110)	F	30
	5.76 (111)	F	30
	5.67 (100)	F	31
	5.00 (210)	F	31
K	2.30	P	32
La	3.5	P	10
Li	*2.9*	F	33
Lu	*3.3*	CPD	34
Mg	*3.66*	P	35
Mn	4.1	P	10
Mo	4.6	P	10
	4.53 (100)	P	36
	4.95 (110)	P	36
	4.55 (111)	P	36
	4.36 (112)	P	36
	4.50 (114)	P	36
	4.55 (332)	P	36
Na	2.75	P	37
Nb	4.3	P	10
	4.02 (001)	T	38
	4.87 (110)	T	38
	4.36 (111)	T	38
	4.63 (112)	T	38
	4.29 (113)	T	38
	3.95 (116)	T	38
	4.18 (310)	T	38
Nd	3.2	P	10
Ni	5.15	P	10
	5.22 (100)	P	39
	5.04 (110)	P	39
	5.35 (111)	P	39
Os	*4.83*	T	29
Pb	4.25	P	40
Pd	5.12	P	31
	5.6 (111)	P	41
Pt	5.65	P	10
	5.7 (111)	P	41
Rb	*2.16*	P	27

ELECTRON WORK FUNCTIONS OF THE ELEMENTS (continued)

Element	Experimental value, ϕ (eV)	Experimental method	Ref.	Element	Experimental value, ϕ (eV)	Experimental method	Ref.
Re	*4.96*	T	29	Te	4.95	P	44
	5.75 (1011)	F	33	Th	3.4	T	51
Rh	4.98	P	31	Ti	4.33	P	10
Ru	4.71	P	31	Tl	*3.84*	CPD	52
Sb	4.55 (amorph.)	–	42	U	3.63	P & CPD	53
	4.7 (100)	–	43		3.73 (100)	P & CPD	54
Sc	3.5	P	10		3.90 (110)	P & CPD	54
Se	5.9	P	44		3.67 (113)	P & CPD	54
Si	4.85n	CPD	40	V	4.3	P	10
	4.91 p (100)	CPD	45	W	4.55	CPD	55
	4.60p (111)	P	46		4.63 (100)	F	30
Sm	2.7	P	10		5.25 (110)	F	30
Sn	*4.42*	CPD	47		4.47 (111)	F	30
Sr	*2.59*	T	48		4.18 (113)	CPD	56
Ta	4.25	T	29		4.30 (116)	T	57
	4.15 (100)	T	49	Y	3.1	P	10
	4.80 (110)	T	49	Zn	*4.33*	P	15
	4.00 (111)	T	49		4.9 (0001)	CPD	58
Tb	3.0	P	50	Zr	4.05	P	10

REFERENCES

1. **Fomenko, V. S.,** *Emission Properties of Materials,* 3rd ed., Naukova Dumka, Kiev, 1970 (in Russian).
2. **Rivière, J. C.,** *Solid State Surface Science,* Green, M., Ed., Vol. 1, Marcel Dekker, 1969, chap. 4.
3. **Trasatti, S.,** *Chim. Ind.* (Milan), 53(6), 559, 1971.
4. **Lang, N. D. and Kohn, W.,** *Phys. Rev. B,* 3(4), 1215, 1971.
5. **Dweydari, A. W. and Mee, C. H. B.,** *Phys. Status Solidi A,* 27, 223, 1975.
6. **Dweydari, A. W. and Mee, C. H. B.,** *Phys. Status Solidi A,* 17, 247, 1973.
7. **Eastment, R. M. and Mee, C. H. B.,** *J. Phys. F,* 3, 1738, 1973.
8. **Grepstad, J. K., Gartland, P. O., and Slagsvold, B. J.,** *Surf. Sci.,* 57, 348, 1976.
9. **Raisin, C. and Pinchaux, R.,** *Solid State Commun.,* 16, 941, 1975.
10. **Eastman, D. E.,** *Phys. Rev. Sect. B,* 2, 1, 1970.
11. **Potter, H. C. and Blakeley, J. M.,** *J. Vac. Sci. Technol.,* 12, 635, 1975 and Potter, H. C., Ph.D. thesis Cornell University, Materials Science Center Rep. No. 1353, 1970.
12. **Adirovich, E. I. and Gol'dshtein, L. M.,** *Fiz. Tverdogo Tela* (Leningrad), 9, 1258, 1967.
13. **Bondarenko, B. V. and Makhov, V. I.,** *Sov. Phys. Solid State,* 12(7), 1522, 1971.
14. **Gustafsson, Broden, and Nilsson,** *J. Phys. F,* 4, 2351, 1974.
15. **Suhrmann, R. and Wedler, G.,** *Z. Angew. Phys.,* 14, 70, 1962.
16. **Robrieux, B., Faure, R., and Dussaulcy, J. P.,** *C. R. Acad. Sci. Ser. B,* 278(14), 659, 1974.
17. **Gaudart, L. and Riviora, R.,** *Appl. Opt.,* 10, 2336, 1971.
18. **Anderson, P. A.,** *Phys. Rev.,* 98, 1739, 1955.
19. **Boutry, G. A. and Dormont, H.,** *Philips Tech. Rev.,* 30, 225, 1969.
20. **Gartland, P. O.,** *Phys. Norv.,* 6(3,4), 201, 1972.
21. **Ueda, K. and Shimizu, R.,** *Jp. J. Appl. Phys.,* 11(6), 916, 1972.
22. **Kobayashi, H. and Kato, S.,** *Surf. Sci.,* 18(2), 341, 1969.
23. **Cardwell, A.,** *Phys. Rev.,* 92, 554, 1953.
24. **Osipova, E. V., Shurmovskaya, N. A., and Burshtein, R. Kh.,** *Elektrokhimiya,* 5(10), 1139, 1969 (in Russian).
25. **Boiko, B. A., Gorodetskii, D. A., and Yas'ko, A. A.,** *Sov. Phys. Solid State,* 15(11), 2101, 1974.
26. **Gobeli, G. W. and Allen, F. G.,** *Surf. Sci.,* 2, 402, 1964.
27. **Lazarev, V. B. and Malov, Y. I.,** *Fiz. Met. Metalloved.,* 24(3), 565, 1967.
28. **Peisner, J., Roboz, P., and Barna, P. B.,** *Phys. Stat. A,* 4, K187, 1971.
29. **Wilson, R. G.,** *J. Appl. Phys.,* 37, 3170, 1966.
30. **Strayer, R. W., Mackie, W., and Swanson, L. W.,** *Surf. Sci.,* 34, 225, 1973.
31. **Nieuwenhuys, Bouwman, and Sachtler,** *Thin Solid Films,* 21, 51, 1974.
32. **Van Oirschot, Th. G. J., van den Brink, M., and Sachtler, W. H. M.,** *Surf. Sci.,* 29, 189, 1972.
33. **Ovchinnikov, A. P. and Tsarev, B. M.,** *Sov. Phys. Solid State,* 9(12), 2766, 1968.
34. **Bondarenko, B. V. and Makhov, V. I.,** *Sov. Phys. Solid State,* 12, 2986, 1971.
35. **Garron, R.,** *C. R. Acad. Sci.,* 258, 1458, 1964.
36. **Berge, Gartland, and Slagsvold,** *Surf. Sci.,* 43, 275, 1974.
37. **Whitefield, R. J. and Brady, J. J.,** *Phys. Rev. Lett.,* 26(7), 380, 1971.
38. **Leblanc, R. P., Vanbrugghe, B. C., and Girouard, F. E.,** *Can. J. Phys.,* 52, 1589, 1974.
39. **Baker, B. G., Johnson, E. B., and Maire, G. I. C.,** *Surf. Sci.,* 24, 572, 1971.
40. **Thanailakis, A.,** *Inst. Phys. Conf. Ser.,* p. 59, 1974.
41. **Demuth, J. E.,** *Chem. Phys. Lett.,* 45, 12, 1977.
42. **Gorodetskii, D. A. and Yas'ko, A. A.,** *Sov. Phys. Solid State,* 13(11), 2928, 1972.
43. **Gorodetskii, D. A. and Yas'ko, A. A.,** *Sov. Phys. Solid State,* 13(5), 1085, 1971.
44. **Williams, R. H. and Polanco, J. I.,** *J. Phys. C,* 7, 2745, 1974.
45. **Allen, F. G.,** *J. Phys. Chem. Solids,* 8, 119, 1959.
46. **Allen, F. G. and Gobeli, G. W.,** *J. Appl. Phys.,* 35, 597, 1964.
47. **Simmons, J. G.,** *Phys. Rev. Lett.,* 10, 10, 1963.
48. **Alleau, T.,** *Surface Phenomena in Thermionic Emitters, Round Table Conf.,* Inst. Tech. Phys. Julich Nucl. Res. Establ., Julich, Germany, 1969, p. 54 (in English).
49. **Protopopov, Mikheeva, Shreinberg, and Shuppe,** *Fiz. Tverdogo Tela,* 8(4), 1140, 1966.
50. **Nemchenok, R. L., Strakovskaya, S. E., and Titenskii, A. I.,** *Fiz. Tverdogo Tela* 11(9), 2692, 1969.
51. **Estrup, P. J., Anderson, J. R., and Danforth, W. E.,** *Surf. Sci.,* 4, 286, 1966.
52. **Klein, O. and Lange, E.,** *Z. Elektrochem.,* 44, 542, 1938.
53. **Hopkins, B. J. and Sargood, A. J.,** *Nuovo Cimento,* 5, 459, 1967.
54. **Lea, C. and Mee, C. H. B.,** *J. Appl. Phys.,* 39, 5890, 1968.
55. **Hopkins, B. J. and Rivière, J. C.,** *Proc. Phys. Soc.* (London), 81, 590, 1963.
56. **Love, H. M. and Dyer, G. L.,** *Can. J. Phys.,* 40, 1837, 1962.
57. **Sultanov, V. M.,** *Radio Eng. Electron.,* 9, 252, 1964 (English translation).
58. **Baker, J. M. and Blakeley, J. M.,** *Surf. Sci.,* 32, 45, 1972.

PROPERTIES OF METALS AS CONDUCTORS

Metal.	Resistivity microhm-centimeters 20° C	Temp. coefficient 20° C.	Specific gravity.	Tensile strength, lbs./in.	Melting point ° C.
*Advance. See *constantan*					
Aluminum	2.824	0.0039	2.70	30,000	659
Antimony	41.7	.0036	6.6	....	630
Arsenic	33.3	.0042	5.73	....	....
Bismuth	120	.004	9.8	....	271
Brass	7	.002	8.6	70,000	900
Cadmium	7.6	.0038	8.6	....	321
*Calido. See *nichrome*					
Climax	87	.0007	8.1	150,000	1250
Cobalt	9.8	.0033	8.71	....	1480
Constantan	49	.00001	8.9	120,000	1190
Copper: annealed	1.7241	.00393	8.89	30,000	1083
[illegible]	1.771	[illegible]	[illegible]	[illegible]	
Eureka. See *constantan*					
Excello	92	.00016	8.9	95,000	1500
Gas Carbon	5000	−.0005	....	....	3500
German silver, 18% Ni	33	.0004	8.4	150,000	1100
Gold	2.44	.0034	19.3	20,000	1063
Ideal. See *constantan*					
Iron, 99.98% pure	10	.005	7.8	....	1530
Lead	22	.0039	11.4	3,000	327
Magnesium	4.6	.004	1.74	33,000	651
Manganin	44	.00001	8.4	150,000	910
Mercury	95.783	.00089	13.546	0	−38.9
Molybdenum, drawn	5.7	.004	9.0	....	2500
Monel metal	42	.0020	8.9	160,000	1300
*Nichrome	100	.0004	8.2	150,000	1500
Nickel	7.8	.006	8.9	120,000	1452
Palladium	11	.0033	12.2	39,000	1550
Phosphor bronze	7.8	.0018	8.9	25,000	750
Platinum	10	.003	21.4	50,000	1755
Silver	1.59	.0038	10.5	42,000	960
Steel, E. B. B.	10.4	.005	7.7	53,000	1510
Steel, B. B.	11.9	.004	7.7	58,000	1510
Steel, Siemens-Martin	18	.003	7.7	100,000	1510
Steel, manganese	70	.001	7.5	230,000	1260
Tantalum	15.5	.0031	16.6	....	2850
*Therlo	47	.00001	8.2	....	....
Tin	11.5	.0042	7.3	4,000	232
Tungsten, drawn	5.6	.0045	19	500,000	3400
Zinc	5.8	.0037	7.1	10,000	419

* Trade mark.

Superconductivity*

B.W. Roberts

General Electric Research Laboratory, Schenectady, New York

The following tables on superconductivity include superconductive properties of chemical elements, thin films, a selected list of compounds and alloys, and high-magnetic-field superconductors.

The historically first observed and most distinctive property of a superconductive body is the near total loss of resistance at a critical temperature (T_c) that is characteristic of each material. Figure 1(a) below illustrates schematically two types of possible transitions. The sharp vertical discontinuity in resistance is indicative of that found for a single crystal of a very pure element or one of a few well annealed alloy compositions. The broad transition, illustrated by broken lines, suggests the transition shape seen for materials that are not homogeneous and contain unusual strain distributions. Careful testing of the resistivity limits for superconductors shows that it is less than 4×10^{-23} ohm-cm, while the lowest resistivity observed in metals is of the order of 10^{-13} ohm-cm. If one compares the resistivity of a superconductive body to that of copper at room temperature, the superconductive body is at least 10^{17} times less resistive.

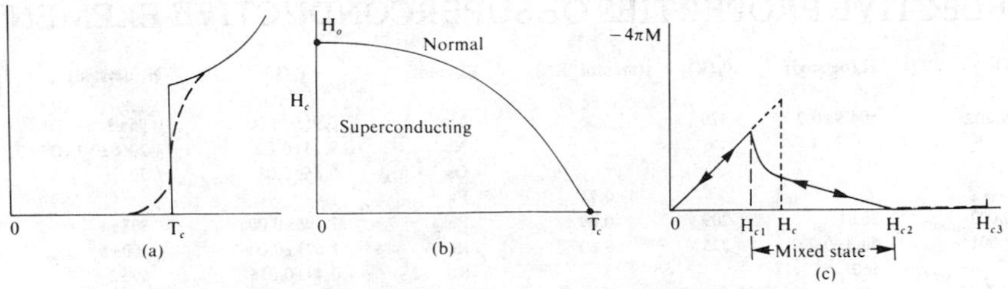

Figure 1. PHYSICAL PROPERTIES OF SUPERCONDUCTORS

(a) Resistivity versus temperature for a pure and perfect lattice (solid line). Impure and/or imperfect lattice (broken line).
(b) Magnetic-field temperature dependence for Type-I or "soft" superconductors.
(c) Schematic magnetization curve for "hard" or Type-II superconductors.

The temperature interval ΔT_c, over which the transition between the normal and superconductive states takes place, may be of the order of as little as 2×10^{-5} °K *or* several °K in width, depending on the material state. The narrow transition width was attained in 99.9999 percent pure gallium single crystals.

*Prepared for Office of Standard Reference Data, National Bureau of Standards, by Standard Reference Data Center on Superconductive Materials, Schenectady, N.Y.

A Type-I superconductor below T_c, as exemplified by a pure metal, exhibits perfect diamagnetism and excludes a magnetic field up to some critical field H_c, whereupon it reverts to the normal state as shown in the H-T diagram of Figure 1(b).

The difference in entropy near absolute zero between the superconductive and normal states relates directly to the electronic specific heat, γ: $(S_s - S_n)_{T \to 0} = -\gamma T$.

The magnetization of a typical high-field superconductor is shown in Figure 1(c). The discovery of the large current-carrying capability of Nb_3Sn and other similar alloys has led to an extensive study of the physical properties of these alloys. In brief, a high-field superconductor, or Type-II superconductor, passes from the perfect diamagnetic state at low magnetic fields to a mixed state and finally to a sheathed state before attaining the normal resistive state of the metal. The magnetic field values separating the four stages are given as H_{c1}, H_{c2}, and H_{c3}. The superconductive state below H_{c1} is perfectly diamagnetic, identical to the state of most pure metals of the "soft" or Type-I superconductor. Between H_{c1} and H_{c2} a "mixed superconductive state" is found in which fluxons (a minimal unit of magnetic flux) create lines of normal superconductor in a superconductive matrix. The volume of the normal state is proportional to $-4\pi M$ in the "mixed state" region. Thus at H_{c2} the fluxon density has become so great as to drive the interior volume of the superconductive body completely normal. Between H_{c2} and H_{c3} the superconductor has a sheath of current-carrying superconductive material at the body surface, and above H_{c3} the normal state exists. With several types of careful measurement, it is possible to determine H_{c1}, H_{c2}, and H_{c3}. Table 2-35 contains some of the available data on high-field superconductive materials.

High-field superconductive phenomena are also related to specimen dimension and configuration. For example, the Type-I superconductor, Hg, has entirely different magnetization behavior in high magnetic fields when contained in the very fine sets of filamentary tunnels found in an unprocessed Vycor glass. The great majority of superconductive materials are Type II. The elements in very pure form and a very few precisely stoichiometric and well annealed compounds are Type-I with the possible exceptions of vanadium and niobium.

Metallurgical Aspects. The sensitivity of superconductive properties to the material state is most pronounced and has been used in a reverse sense to study and specify the detailed state of alloys. The mechanical state, the homogeneity, and the presence of impurity atoms and other electron-scattering centers are all capable of controlling the critical temperature and the current-carrying capabilities in high-magnetic fields. Well annealed specimens tend to show sharper transitions than those that are strained or inhomogeneous. This sensitivity to mechanical state underlines a general problem in the tabulation of properties for superconductive materials. The occasional divergent values of the critical temperature and of the critical fields quoted for a Type-II superconductor may lie in the variation in sample preparation. Critical temperatures of materials studied early in the history of superconductivity must be evaluated in light of the probable metallurgical state of the material, as well as the availability of less pure starting elements. It has been noted that recent work has given extended consideration to the metallurgical aspects of sample preparation.

REFERENCES

References to the data presented in this section, to additional entries of superconductive materials, and to those materials specifically tested and found non-superconductive to some low temperature may be found in the following publications:

"Superconductive Materials and Some of Their Properties", *Progress in Cryogenics*, B.W. Roberts, Vol. IV, Heywood and Co., 1964, pp. 160–231.

"Superconductive Materials and Some of Their Properties", B.W. Roberts, National Bureau of Standards Technical Notes 408 and 482, U.S. Government Printing Office, 1966 and 1969.

SELECTIVE PROPERTIES OF SUPERCONDUCTIVE ELEMENTS

Element	T_c(K)	H_o(oersted)	θ_D(K)	γ(mJmol^{-1}K^{-1})	Element	T_c(K)	H_o(oersted)	θ_D(K)	γ(mJmol^{-1}K^{-1})
Al	1.175±0.002	104.9±0.3	420	1.35	Mo	0.915±0.005	96±3	460	1.83
Am* (α,?)	0.6				Nb	9.25±0.02	2060±50,HF	276	7.80
Am* (β,?)	1.0				Os	0.66±0.03	70	500	2.35
Be	0.026			0.21	Pa	1.4			
Cd	0.517±0.002	28±1	209	0.69	Pb	7.196±0.006	803±1	96	3.1
Ga	1.083±0.001	58.3±0.2	325	0.60	Re	1.697±0.006	200±5	4.5	2.35
Ga (β)	5.9,6.2	560			Ru	0.49±0.015	69±2	580	2.8
Ga (γ)	7	950,hf*			Sn	3.722±0.001	305±2	195	1.78
GA (Δ)	7.85	815,HF			Ta	4.47±0.04	829±6	258	6.15
Hf	0.128	12.7		2.21	Tc	7.8±0.1	1410,HF	411	6.28
Hg (α)	4.154±0.001	411±2	87,71.9	1.81	Th	1.38±0.02	160±3	165	4.32
Hg (β)	3.949	339	93	1.37	Ti	0.40±0.04	56	415	3.3
In	3.408±0.001	281.5±2	109	1.672	Tl	2.38±0.02	178±2	78.5	1.47
Ir	0.1125±0.001	16±0.05	425	3.19	V	5.40±0.05	1408	383	9.82
La (α)	4.88±0.02	800±10	151	9.8	W	0.0154±0.0005	1.15±0.03	383	0.90
La (β)	6.00±0.1	1096,1600	139	11.3	Zn	0.85±0.01	54±0.3	310	0.66
Lu	0.1±0.03	350±50			Zr	0.61±0.15	47	290	2.77
					Zr (ω)	0.65, 0.95			

* HF denotes high field superconductive properties.

From Roberts, B. W., Properties of Selected Superconductive Materials, 1978 Supplement, NBS Technical Note 983, U.S. Government Printing Office, Washington, D.C.

RANGE OF CRITICAL TEMPERATURES OBSERVED FOR SUPERCONDUCTIVE ELEMENTS IN THIN FILMS CONDENSED USUALLY AT LOW TEMPERATURES

Element	T_c Range (K)	H_c (oersted)	Element	T_c Range (K)	H_c (oersted)
			Mo	3.3—8.0	
Al	1.15—~5.7	HF*	Nb	2.0—10.1	
Be	5—9.75	HF	Pb	1.8—7.5	
Bi	6.17—2.6		Re	1.7—~7	
Cd			Sn	3.5—~6	
(Disordered)	0.79—0.91		Ta	<1.7—4.51	HF
			Tc	4.6—7.7	
(Ordered)	0.53—0.59		Ti	1.3 Max	
Ga	?.5—8.5	HF	Tl	2.33—2.96	
Hg	3.87—4.5		V	1.8—6.02	
In	2.2—5.6	HF	W	<1.0—4.1	
La	3.55—6.74		Zn	0.77—1.70, ~1.9	

* HF denotes high magnetic field superconductive properties.

ELEMENTS EXHIBITING SUPERCONDUCTIVITY UNDER OR AFTER APPLICATION OF HIGH PRESSURE

Element	T_c Range (K)	Pressure (kbar)*
Al	1.98—0.075	0—62
As	0.31—0.5	220—140
	0.2—0.25	~140—100
Ba II	~1—1.8	~55—85
III	1.8—5	~85—144
IV	4.5—5.4	144—190
Bi II	3.9	25—27
III	6.55—7.25	28—38
IV	7.0,8.7—6.0	43,43—62
V	6.7,8.3	48—80
VI	8.55	90,92—101
VII(?)	8.2	30
Ce (α)	0.020—0.045	20—35
Ce (α′)	1.9—1.3	45—125
Cs V	~1.5	>125
Ga II	6.38	≥35
II′	7.5	≥35 then P removed
Ge	5.35	115
La	~5.5—12.9	0—210
Lu	0.022—1.0	45—190
P	5.8	170
Pb II	3.55	160
Re II	2.3 Max.	"Plastic" compression
Sb (prepared 120 kbar, held below 77K)	2.6—2.7	
Sb III	3.55—3.40	85—~150
Se II	6.75,6.95	~130
Si	6.7—7.1	120—130
Sn II	5.2—4.85	125—160
III	5.30	113
Te II	2.4—5.1	38—55
III	4.1—4.2	~53—62
IV	4.72—4	63—80
( )	3.3—2.8	100—260
Tl (cubic form)	1.45	35
(Hexagonal form)	1.95	35
U	2.4—0.4	10—85
Y	2.3—1.7—2.5	110—125—160
Zr (omega form, metastable)	1—1.7	60—~130

* 1 kbar = 10^8 newton/meter^2 = 0.987 katm

From Roberts, B. W., Properties of Selected Superconductive Materials, 1978 Supplement, NBS Technical Note 983, U.S. Government Printing Office, Washington, D.C.

SELECTED SUPERCONDUCTIVE COMPOUNDS AND ALLOYS

All compositions are denoted on an atomic basis, i.e., AB, AB_2, or AB_3 for compounds, unless noted. Solid solutions or odd compositions may be denoted as A_zB_{1-z} or A_zB. A series of three or more alloys is indicated as A_xB_{1-x} or by actual indication of the atomic fraction range, such as $A_{0-0.6}B_{1-0.4}$. The critical temperature of such a series of alloys is denoted by a range of values or possibly the maximum value.

The selection of the critical temperature from a transition in the effective permeability, or the change in resistance, or possibly the incremental changes in frequency observed by certain techniques is not often obvious from the literature. Most authors choose the mid-point of such curves as the probable critical temperature of the idealized material, while others will choose the highest temperature at which a deviation from the normal state property is observed. In view of the previous discussion concerning the variability of the superconductive properties as a function of purity and other metallurgical aspects, it is recommended that appropriate literature be checked to determine the most probable critical temperature or critical field of a given alloy.

A very limited amount of data on critical fields, H_o, is available for these compounds and alloys; these values are given at the end of the table.

SYMBOLS: n = number of normal carriers per cubic centimeter for semiconductor superconductors.

Substance	T_c, °K	*Crystal structure type*††
$Ag_xAl_yZn_{1-x-y}$	0.5–0.845	
$Ag_7BF_4O_8$	0.15	Cubic
$AgBi_2$	3.0–2.78	
$Ag_7F_{0.25}N_{0.75}O_{10.25}$	0.85–0.90	
Ag_7FO_8	0.3	Cubic
Ag_2F	0.066	
$Ag_{0.8-0.3}Ga_{0.2-0.7}$	6.5–8	
Ag_4Ge	0.85	Hex., c.p.
$Ag_{0.438}Hg_{0.562}$	0.64	$D8_2$
$AgIn_2$	~2.4	C16
$Ag_{0.1}In_{0.9}Te$ (n = 1.40×10^{22})	1.20–1.89	B1
$Ag_{0.2}In_{0.8}Te$ (n = 1.07×10^{22})	0.77–1.00	B1
AgLa (9.5 kbar)	1.2	B2
Ag_7NO_{11}	1.04	Cubic
Ag_xPb_{1-x}	7.2 max.	
Ag_xSn_{1-x} (film)	2.0–3.8	
Ag_xSn_{1-x}	1.5–3.7	
$AgTe_3$	2.6	Cubic
$AgTh_2$	2.26	C16
$Ag_{0.03}Tl_{0.97}$	2.67	
$Ag_{0.94}Tl_{0.06}$	2.32	
Ag_xZn_{1-x}	0.5–0.845	
Al (film)	1.3–2.31	
Al (1 to 21 katm)	1.170–0.687	A1
$AlAu_4$	0.4–0.7	Like A13
Al_2CMo_3	10.0	A13
Al_2CMo_3	9.8–10.2	A13 + trace 2nd phase
Al_2CaSi	5.8	
$Al_{0.131}Cr_{0.088}V_{0.781}$	1.46	Cubic
$AlGe_2$	1.75	
$Al_{0.5}Ge_{0.5}Nb$	12.6	A15
$Al_{\sim 0.8}Ge_{\sim 0.2}Nb_3$	20.7	A15
$AlLa_3$	5.57	DO_{19}
Al_2La	3.23	C15
Al_3Mg_2	0.84	Cubic, f.c.
$AlMo_3$	0.58	A15
$AlMo_6Pd$	2.1	
AlN	1.55	B4
Al_2NNb_3	1.3	A13
$AlNb_3$	18.0	A15
Al_xNb_{1-x}	<4.2–13.5	$D8_b$
Al_xNb_{1-x}	12–17.5	A15
$Al_{0.27}Nb_{0.73-0.48}V_{0-0.25}$	14.5–17.5	A15
$AlNb_xV_{1-x}$	<4.2–13.5	
AlOs	0.39	B2
Al_3Os	5.90	
AlPb (films)	1.2–7	
Al_2Pt	0.48–0.55	C1
Al_5Re_{24}	3.35	A12

Substance	T_c, °K	*Crystal structure type*††
Al_3Th	0.75	DO_{19}
$Al_xTi_yV_{1-x-y}$	2.05–3.62	Cubic
$Al_{0.108}V_{0.892}$	1.82	Cubic
Al_xZn_{1-x}	0.5–0.845	
$AlZr_3$	0.73	$L1_2$
AsBiPb	9.0	
AsBiPbSb	9.0	
$As_{0.33}InTe_{0.67}$ (n = 1.24×10^{22})	0.85–1.15	B1
$As_{0.5}InTe_{0.5}$ (n = 0.97×10^{22})	0.44–0.62	B1
$As_{0.50}Ni_{0.06}Pd_{0.44}$	1.39	C2
AsPb	8.4	
$AsPd_2$ (low-temperature phase)	0.60	Hexagonal
$AsPd_2$ (high-temp. phase)	1.70	C22
$AsPd_5$	0.46	Complex
AsRh	0.58	B31
$AsRh_{1.4-1.6}$	<0.03–0.56	Hexagonal
AsSn	4.10	
AsSn (n = 2.14×10^{22})	3.41–3.65	B1
$As_{\sim 2}Sn_{\sim 3}$	3.5–3.6, 1.21–1.17	
As_3Sn_4 (n = 0.56×10^{22})	1.16–1.19	Rhombohedral
Au_5Ba	0.4–0.7	$D2_d$
AuBe	2.64	B20
Au_2Bi	1.80	C15
Au_5Ca	0.34–0.38	$C15_b$
AuGa	1.2	B31
$Au_{0.40-0.92}Ge_{0.60-0.08}$	<0.32–1.63	Complex
AuIn	0.4–0.6	Complex
AuLu	<0.35	B2
$AuNb_3$	11.5	A15
$AuNb_3$	1.2	A2
$Au_{0-0.3}Nb_{1-0.7}$	1.1–11.0	
$Au_{0.02-0.98}Nb_3Rh_{0.98-0.02}$	2.53–10.9	A15
$AuNb_{3(1-x)}V_{3x}$	1.5–11.0	A15
$AuPb_2$	3.15	
$AuPb_2$ (film)	4.3	
$AuPb_3$	4.40	
$AuPb_3$ (film)	4.25	
Au_2Pb	1.18, 6–7	C15
$AuSb_2$	0.58	C2
AuSn	1.25	$B8_1$
Au_xSn_{1-x} (film)	2.0–3.8	
Au_5Sn	0.7–1.1	A3
Au_3Te_5	1.62	Cubic
$AuTh_2$	3.08	C16
AuTl	1.92	
AuV_3	0.74	A15

††See key at end of table.

SELECTED SUPERCONDUCTIVE COMPOUNDS AND ALLOYS (Continued)

Substance	T_c, °*K*	*Crystal structure type††*
Au_xZn_{1-x}	0.50–0.845	
$AuZn_3$	1.21	Cubic
Au_xZr_y	1.7–2.8	A3
$AuZr_3$	0.92	A15
$BCMo_2$	5.4	Orthorhombic
$B_{0.03}C_{0.51}Mo_{0.47}$	12.5	
$BCMo_2$	5.3–7.0	Orthorhombic
[illegible]	[illegible]	Cubic
B_6La	5.7	
$B_{12}Lu$	0.48	
BMo	0.5 (extrapolated)	
BMo_2	4.74	Cl6
BNb	8.25	B_f
BRe_2	2.80, 4.6	
$B_{0.3}Ru_{0.7}$	2.58	$D10_2$
$B_{12}Sc$	0.39	
BTa	4.0	B_f
B_6Th	0.74	
BW_2	3.1	Cl6
B_6Y	6.5–7.1	
$B_{12}Y$	4.7	
BZr	3.4	Cubic
$B_{12}Zr$	5.82	
$BaBi_3$	5.69	Tetragonal
$Ba_xO_3Sr_{1-x}Ti$ ($n = 4.2–11 \times 10^{19}$)	<0.1–0.55	
$Ba_{0.13}O_3W$	1.9	Tetragonal
$Ba_{0.14}O_3W$	<1.25–2.2	Hexagonal
$BaRh_2$	6.0	Cl5
$Be_{22}Mo$	2.51	Cubic, like $Be_{22}Re$
$Be_8Nb_5Zr_2$	5.2	
$Be_{0.98-0.92}Re_{0.02-0.08}$ (quenched)	9.5–9.75	Cubic
$Be_{0.957}Re_{0.043}$	9.62	Cubic, like $Be_{22}Re$
BeTc	5.21	Cubic
$Be_{22}W$	4.12	Cubic, like $Be_{22}Re$
$Be_{13}W$	4.1	Tetragonal
Bi_3Ca	2.0	
$Bi_{0.5}Cd_{0.13}Pb_{0.25}Sn_{0.12}$ (weight fractions)	8.2	
BiCo	0.42–0.49	
Bi_2Cs	4.75	Cl5
Bi_xCu_{1-x} (electrodeposited)	2.2	
BiCu	1.33–1.40	
$Bi_{0.019}In_{0.981}$	3.86	
$Bi_{0.05}In_{0.95}$	4.65	α-phase
$Bi_{0.10}In_{0.90}$	5.05	α-phase
$Bi_{0.15-0.30}In_{0.85-0.70}$	5.3–5.4	α- and β-phases
$Bi_{0.34-0.48}In_{0.66-0.52}$	4.0–4.1	
Bi_3In_5	4.1	
$BiIn_2$	5.65	β-phase
Bi_2Ir	1.7–2.3	
Bi_2Ir (quenched)	3.0–3.96	
BiK	3.6	
Bi_2K	3.58	Cl5
BiLi	2.47	Ll_0, α-phase
$Bi_{4-9}Mg$	0.7–~1.0	
Bi_3Mo	3–3.7	
BiNa	2.25	Ll_0
$BiNb_3$ (high pressure and temperature)	3.05	A15
BiNi	4.25	$B8_1$
Bi_3Ni	4.06	Orthorhombic
$Bi_{1-0}Pb_{0-1}$	7.26–9.14	
$Bi_{1-0}Pb_{0-1}$ (film)	7.25–8.67	
$Bi_{0.05-0.40}Pb_{0.95-0.60}$	7.35–8.4	Hexagonal, c.p., to ε-phase
BiPbSb	8.9	
$Bi_{0.5}Pb_{0.31}Sn_{0.19}$ (weight fractions)	8.5	

Substance	T_c, °*K*	*Crystal structure type††*
$Bi_{0.5}Pb_{0.25}Sn_{0.25}$	8.5	
$BiPd_2$	4.0	
$Bi_{0.4}Pd_{0.6}$	3.7 4	Hexagonal, ordered
BiPd	3.7	Orthorhombic
Bi_2Pd	1.70	Monoclinic, α-phase
Bi_2Pd	4.23	Tetragonal, β-phase
BiPdSe	1.0	C2
BiPdTe	1.2	C2
BiPt	1.21	$B8_1$
BiPtSe	1.45	C2
BiPtTe	1.15	C2
Bi_2Pt	0.155	Hexagonal
Bi_2Rb	4.25	Cl5
$BiRe_2$	1.9–2.2	
BiRh	2.06	$B8_1$
Bi_3Rh	3.2	Orthorhombic, like NiB_3
Bi_4Rh	2.7	Hexagonal
Bi_3Sn	3.6–3.8	
BiSn	3.8	
Bi_xSn_y	3.85–4.18	
Bi_3Sr	5.62	Ll_2
Bi_3Te	0.75–1.0	
Bi_5Tl_3	6.4	
$Bi_{0.26}Tl_{0.74}$	4.4	Cubic, disordered
$Bi_{0.26}Tl_{0.74}$	4.15	Ll_2, ordered?
Bi_2Y_3	2.25	
Bi_3Zn	0.8–0.9	
$Bi_{0.3}Zr_{0.7}$	1.51	
$BiZr_3$	2.4–2.8	
CCs_x	0.020–0.135	Hexagonal
C_8K (gold)	0.55	
$CGaMo_2$	3.7–4.1	Hexagonal, H-phase
$CHf_{0.5}Mo_{0.5}$	3.4	Bl
$CHf_{0.3}Mo_{0.7}$	5.5	Bl
$CHf_{0.25}Mo_{0.75}$	6.6	Bl
$CHf_{0.7}Nb_{0.3}$	6.1	Bl
$CHf_{0.6}Nb_{0.4}$	4.5	Bl
$CHf_{0.5}Nb_{0.5}$	4.8	Bl
$CHf_{0.4}Nb_{0.6}$	5.6	Bl
$CHf_{0.25}Nb_{0.75}$	7.0	Bl
$CHf_{0.2}Nb_{0.8}$	7.8	Bl
$CHf_{0.9-0.1}Ta_{0.1-0.9}$	5.0–9.0	Bl
Ck (excess K)	0.55	Hexagonal
C_8K	0.39	Hexagonal
$C_{0.40-0.44}Mo_{0.60-0.56}$	9–13	
CMo	6.5, 9.26	
CMo_2	12.2	Orthorhombic
$C_{0.44}Mo_{0.56}$	1.3	Bl
$C_{0.5}Mo_xNb_{1-x}$	10.8–12.5	Bl
$C_{0.6}Mo_{4.8}Si_3$	7.6	$D8_8$
$CMo_{0.2}Ta_{0.8}$	7.5	Bl
$CMo_{0.5}Ta_{0.5}$	7.7	Bl
$CMo_{0.75}Ta_{0.25}$	8.5	Bl
$CMo_{0.8}Ta_{0.2}$	8.7	Bl
$CMo_{0.85}Ta_{0.15}$	8.9	Bl
CMo_xTi_{1-x}	10.2 max.	Bl
$CMo_{0.83}Ti_{0.17}$	10.2	Bl
CMo_xV_{1-x}	2.9–9.3	Bl
CMo_xZr_{1-x}	3.8–9.5	Bl
$C_{0.1-0.9}N_{0.9-0.1}Nb$	8.5–17.9	
$C_{0-0.38}N_{1-0.62}Ta$	10.0–11.3	
CNb (whiskers)	7.5–10.5	
$C_{0.984}Nb$	9.8	Bl
CNb (extrapolated)	~14	
$C_{0.7-1.0}Nb_{0.3-0}$	6–11	Bl
CNb_2	9.1	
CNb_xTa_{1-x}	8.2–13.9	
CNb_xTi_{1-x}	<4.2–8.8	Bl
$CNb_{0.6-0.9}W_{0.4-0.1}$	12.5–11.6	Bl
$CNb_{0.1-0.9}Zr_{0.9-0.1}$	4.2–8.4	Bl

Substance	T_c, °K	*Crystal structure type*††
CRb_x (gold)	0.023–0.151	Hexagonal
$CRe_{0.01-0.08}W$	1.3–5.0	
$CRe_{0.06}W$	5.0	
CTa	~11 (extrapolated)	
$C_{0.987}Ta$	9.7	
$C_{0.848-0.987}Ta$	2.04–9.7	
CTa (film)	5.09	B1
CTa_2	3.26	L'_3
$CTa_{0.4}Ti_{0.6}$	4.8	B1
$CTa_{1-0.4}W_{0-0.6}$	8.5–10.5	B1
$CTa_{0.2-0.9}Zr_{0.8-0.1}$	4.6–8.3	B1
CTc (excess C)	3.85	Cubic
$CTi_{0.5-0.7}W_{0.5-0.3}$	6.7–2.1	B1
CW	1.0	
CW_2	2.74	L'_3
CW_2	5.2	Cubic, f.c.
$CaIr_2$	6.15	C15
$Ca_xO_3Sr_{1-x}Ti$ (n = $3.7-11.0 \times 10^{19}$)	<0.1–0.55	
$Ca_{0.1}O_3W$	1.4–3.4	Hexagonal
CaPb	7.0	
$CaRh_2$	6.40	C15
$Cd_{0.3-0.5}Hg_{0.7-0.5}$	1.70–1.92	
CdHg	1.77, 2.15	Tetragonal
$Cd_{0.0075-0.05}In_{1-x}$	3.24–3.36	Tetragonal
$Cd_{0.97}Pb_{0.03}$	4.2	
CdSn	3.65	
$Cd_{0.17}Tl_{0.83}$	2.3	
$Cd_{0.18}Tl_{0.82}$	2.54	
$CeCo_2$	0.84	C15
$CeCo_{1.67}Ni_{0.33}$	0.46	C15
$CeCo_{1.67}Rh_{0.33}$	0.47	C15
$Ce_xGd_{1-x}Ru_2$	3.2–5.2	C15
$CeIr_3$	3.34	
$CeIr_5$	1.82	
$Ce_{0.005}La_{0.995}$	4.6	
Ce_xLa_{1-x}	1.3–6.3	
$Ce_xPr_{1-x}Ru_2$	1.4–5.3	C15
Ce_xPt_{1-x}	0.7–1.55	
$CeRu_2$	6.0	C15
$Co_xFe_{1-x}Si_2$	1.4 max.	C1
$CoHf_2$	0.56	$E9_3$
$CoLa_3$	4.28	
$CoLu_3$	~0.35	
$Co_{0-0.01}Mo_{0.8}Re_{0.2}$	2–10	
$Co_{0.02-0.10}Nb_3Rh_{0.98-0.90}$	2.28–1.90	A15
$Co_xNi_{1-x}Si_2$	1.4 max.	C1
$Co_{0.5}Rh_{0.5}Si_2$	2.5	
$Co_xRh_{1-x}Si_2$	3.65 max.	
$Co_{\sim 0.3}Sc_{\sim 0.7}$	~0.35	
$CoSi_2$	1.40, 1.22	C1
Co_3Th_7	1.83	$D10_2$
Co_xTi_{1-x}	2.8 max.	Co in α-Ti
Co_xTi_{1-x}	3.8 max.	Co in β-Ti
$CoTi_2$	3.44	$E9_3$
CoTi	0.71	A2
CoU	1.7	B2, distorted
CoU_6	2.29	$D2_c$
$Co_{0.28}Y_{0.72}$	0.34	
CoY_3	<0.34	
$CoZr_2$	6.3	C16
$Co_{0.1}Zr_{0.9}$	3.9	A3
$Cr_{0.6}Ir_{0.4}$	0.4	Hexagonal, c.p.
$Cr_{0.65}Ir_{0.35}$	0.59	Hexagonal, c.p.
$Cr_{0.7}Ir_{0.3}$	0.76	Hexagonal, c.p.
$Cr_{0.72}Ir_{0.28}$	0.83	
Cr_3Ir	0.45	A15
$Cr_{0-0.1}Nb_{1-0.9}$	4.6–9.2	A2
$Cr_{0.80}Os_{0.20}$	2.5	Cubic
Cr_xRe_{1-x}	1.2–5.2	
$Cr_{0.40}Re_{0.60}$	2.15	$D8_b$
$Cr_{0.8-0.6}Rh_{0.2-0.4}$	0.5–1.10	A3
Cr_3Ru (annealed)	3.3	A15
Cr_2Ru	2.02	$D8_b$
$Cr_{0.1-0.5}Ru_{0.9-0.5}$	0.34–1.65	A3
Cr_xTi_{1-x}	3.6 max.	Cr in α-Ti
Cr_xTi_{1-x}	4.2 max.	Cr in β-Ti
$Cr_{0.1}Ti_{0.3}V_{0.6}$	5.6	
$Cr_{0.0175}U_{0.9825}$	0.75	β-phase
$Cs_{0.32}O_3W$	1.12	Hexagonal
$Cu_{0.15}In_{0.85}$ (film)	3.75	
$Cu_{0.04-0.08}In_{1-x}$	4.4	
CuLa	5.85	
Cu_xPb_{1-x}	5.7–7.7	
CuS	1.62	B18
CuS_2	1.48–1.53	C18
CuSSe	1.5–2.0	C18
$CuSe_2$	2.3–2.43	C18
CuSeTe	1.6–2.0	C18
Cu_xSn_{1-x}	3.2–3.7	
Cu_xSn_{1-x} (film) (made at 10°K)	3.6–7	
Cu_xSn_{1-x} (film) (made at 300°K)	2.8–3.7	
$CuTe_2$	<1.25–1.3	C18
$CuTh_2$	3.49	C16
$Cu_{0-0.027}V$	3.9–5.3	A2
Cu_xZn_{1-x}	0.5–0.845	
Er_xLa_{1-x}	1.4–6.3	
$Fe_{0-0.04}Mo_{0.8}Re_{0.2}$	1–10	
$Fe_{0.05}Ni_{0.05}Zr_{0.90}$	~3.9	
Fe_3Th_7	1.86	D10
Fe_xTi_{1-x}	3.2 max.	Fe in α-Ti
Fe_xTi_{1-x}	3.7 max.	Fe in β-Ti
$Fe_xTi_{0.6}V_{1-x}$	6.8 max.	
FeU_6	3.86	$D2_c$
$Fe_{0.1}Zr_{0.9}$	1.0	A3
$Ga_{0.5}Ge_{0.5}Nb_3$	7.3	A15
$GaLa_3$	5.84	
Ga_2Mo	9.5	
$GaMo_3$	0.76	A15
Ga_4Mo	9.8	
GaN (black)	5.85	B4
$GaNb_3$	14.5	A15
$Ga_xNb_3Sn_{1-x}$	14–18.37	A15
$Ga_{0.7}Pt_{0.3}$	2.9	C1
GaPt	1.74	B20
GaSb (120 kbar, 77°K, annealed)	4.24	A5
GaSb (unannealed)	~5.9	
$Ga_{0.1}Sn_{1-0}$ (quenched)	3.47–4.18	
$Ga_{0-1}Sn_{1-0}$ (annealed)	2.6–3.85	
Ga_5V_2	3.55	Tetragonal, Mn_2Hg_5 type
GaV_3	16.8	A15
$GaV_{2.1-3.5}$	6.3–14.45	A15
$GaV_{4.5}$	9.15	
Ga_3Zr	1.38	
Gd_xLa_{1-x}	<1.0–5.5	
$Gd_xOs_2Y_{1-x}$	1.4–4.7	
$Gd_xRu_2Th_{1-x}$	3.6 max.	C15
GeIr	4.7	B31
Ge_2La	1.49, 2.2	Orthorhombic, distorted $ThSi_2$-type
$GeMo_3$	1.43	A15
$GeNb_2$	1.9	
$GeNb_3$ (quenched)	6–17	A15
$Ge_{0.29}Nb_{0.71}$	6	A15
$Ge_xNb_3Sn_{1-x}$	17.6–18.0	A15
$Ge_{0.5}Nb_3Sn_{0.5}$	11.3	
GePt	0.40	B31
Ge_3Rh_5	2.12	Orthorhombic, related to $InNi_2$
Ge_2Sc	1.3	
Ge_3Te_4 (n = 1.06×10^{22})	1.55–1.80	Rhombohedral
Ge_xTe_{1-x} (n = $8.5 - 64 \times 10^{20}$)	0.07–0.41	B1
GeV_3	6.01	A15

SELECTED SUPERCONDUCTIVE COMPOUNDS AND ALLOYS (Continued)

Substance	T_c, °K	*Crystal structure type*††
Ge_2Y	3.80	C_c
$Ge_{1.62}Y$	2.4	
$H_{0.33}Nb_{0.67}$	7.28	Cubic, b.c.
$H_{0.1}Nb_{0.9}$	7.38	Cubic, b.c.
$H_{0.05}Nb_{0.95}$	7.83	Cubic, b.c.
$H_{0.12}Ta_{0.88}$	2.81	Cubic, b.c.
$H_{0.08}Ta_{0.92}$	3.26	Cubic, b.c.
$H_{0.04}Ta_{0.96}$	3.62	Cubic, b.c.
$HfN_{0.989}$	6.6	Bl
$Hf_{0-0.5}Nb_{1-0.5}$	8.3–9.5	A2
$Hf_{0.75}Nb_{0.25}$	>4.2	
$HfOs_2$	2.69	C14
$HfRe_2$	4.80	C14
$Hf_{0.14}Re_{0.86}$	5.86	A12
$Hf_{0.99-0.96}Rh_{0.01-0.04}$	0.85–1.51	
$Hf_{0-0.55}Ta_{1-0.45}$	4.4–6.5	A2
HfV_2	8.9–9.6	C15
Hg_xIn_{1-x}	3.14–4.55	
HgIn	3.81	
Hg_2K	1.20	Orthorhombic
Hg_3K	3.18	
Hg_4K	3.27	
Hg_8K	3.42	
Hg_3Li	1.7	Hexagonal
Hg_2Na	1.62	Hexagonal
Hg_4Na	3.05	
Hg_xPb_{1-x}	4.14–7.26	
HgSn	4.2	
Hg_xTl_{1-x}	2.30–4.109	
Hg_5Tl_2	3.86	
Ho_xLa_{1-x}	1.3–6.3	
$InLa_3$	9.83, 10.4	Ll_2
$InLa_3$ (0–35, kbar)	9.75–10.55	
$In_{1-0.86}Mg_{0-0.14}$	3.395–3.363	
$InNb_3$ (high pressure and temp.)	4–8, 9.2	A15
$In_{0-0.3}Nb_3Sn_{1-0.7}$	18.0–18.19	A15
$In_{0.5}Nb_3Zr_{0.5}$	6.4	
$In_{0.11}O_3W$	<1.25–2.8	Hexagonal
$In_{0.95-0.85}Pb_{0.05-0.15}$	3.6–5.05	
$In_{0.98-0.91}Pb_{0.02-0.09}$	3.45–4.2	
InPb	6.65	
InPd	0.7	B2
InSb (quenched from 170 kbar into liquid N_2)	4.8	Like A5
InSb	2.1	
$(InSb)_{0.95-0.10}Sn_{0.05-0.90}$ (various heat treatments)	3.8–5.1	
$(InSb)_{0-0.07}Sn_{1-0.93}$	3.67–3.74	
In_3Sn	~5.5	
In_xSn_{1-x}	3.4–7.3	
$In_{0.82-1}Te$ ($n = 0.83\text{–}1.71 \times 10^{22}$)	1.02–3.45	Bl
$In_{1.000}Te_{1.002}$	3.5–3.7	Bl
In_3Te_4 ($n = 0.47 \times 10^{22}$)	1.15–1.25	Rhombohedral
In_xTl_{1-x}	2.7–3.374	
$In_{0.8}Tl_{0.2}$	3.223	
$In_{0.62}Tl_{0.38}$	2.760	
$In_{0.78-0.69}Tl_{0.22-0.31}$	3.18–3.32	Tetragonal
$In_{0.69-0.62}Tl_{0.31-0.38}$	2.98–3.3	Cubic, f.c.
Ir_2La	0.48	C15
Ir_3La	2.32	$Dl0_2$
Ir_3La_7	2.24	$Dl0_2$
Ir_5La	2.13	
Ir_2Lu	2.47	C15
Ir_3Lu	2.89	C15
IrMo	<1.0	A3
$IrMo_3$	8.8	A15
$IrMo_3$	6.8	$D8_b$
$IrNb_3$	1.9	A15
$Ir_{0.4}Nb_{0.6}$	9.8	$D8_b$
$Ir_{0.37}Nb_{0.63}$	2.32	$D8_b$
IrNb	7.9	$D8_b$
$Ir_{0.02}Nb_3Rh_{0.98}$	2.43	A15
$Ir_{0.05}Nb_3Rh_{0.95}$	2.38	A15

Substance	T_c, °K	*Crystal structure type*††
$Ir_{0.287}O_{0.14}Ti_{0.573}$	5.5	$E9_3$
$Ir_{0.265}O_{0.035}Ti_{0.65}$	2.30	$E9_3$
Ir_xOs_{1-x}	0.3–0.98 (max.)–0.6	
IrOsY	2.6	C15
$Ir_{1.5}Os_{0.5}$	2.4	C14
Ir_2Sc	2.07	C15
$Ir_{2.5}Sc$	2.46	C15
$IrSn_2$	0.65–0.78	Cl
Ir_2Sr	5.70	C15
$Ir_{0.5}Te_{0.5}$	~3	
$IrTe_3$	1.18	C2
IrTh	<0.37	B_f
Ir_2Th	6.50	C15
Ir_3Th	4.71	
Ir_3Th_7	1.52	$Dl0_2$
Ir_5Th	3.93	$D2_d$
$IrTi_3$	5.40	A15
IrV_2	1.39	A15
IrW_3	3.82	
$Ir_{0.28}W_{0.72}$	4.49	
Ir_2Y	2.18, 1.38	C15
$Ir_{0.69}Y_{0.31}$	1.98, 1.44	C15
$Ir_{0.70}Y_{0.30}$	2.16	C15
Ir_2Y	1.09	C15
Ir_2Y_3	1.61	
Ir_xY_{1-x}	0.3–3.7	
Ir_2Zr	4.10	C15
$Ir_{0.1}Zr_{0.9}$	5.5	A3
$K_{0.27-0.31}O_3W$	0.50	Hexagonal
$K_{0.40-0.57}O_3W$	1.5	Tetragonal
$La_{0.55}Lu_{0.45}$	2.2	Hexagonal, La type
$La_{0.8}Lu_{0.2}$	3.4	Hexagonal, La Type
$LaMg_2$	1.05	C15
LaN	1.35	
$LaOs_2$	6.5	C15
$LaPt_2$	0.46	C15
$La_{0.28}Pt_{0.72}$	0.54	C15
$LaRh_3$	2.60	
$LaRh_5$	1.62	
La_7Rh_3	2.58	$Dl0_2$
$LaRu_2$	1.63	C15
La_3S_4	6.5	$D7_3$
La_3Se_4	8.6	$D7_3$
$LaSi_2$	2.3	C_c
La_xY_{1-x}	1.7–5.4	
LaZn	1.04	B2
LiPb	7.2	
$LuOs_2$	3.49	C14
$Lu_{0.275}Rh_{0.725}$	1.27	C15
$LuRh_5$	0.49	
$LuRu_2$	0.86	C14
$Mg_{\sim 0.47}Tl_{\sim 0.53}$	2.75	B2
Mg_2Nb	5.6	
Mn_xTi_{1-x}	2.3 max.	Mn in α-Ti
Mn_xTi_{1-x}	1.1–3.0	Mn in β-Ti
MnU_6	2.32	$D2_c$
MoN	12	Hexagonal
Mo_2N	5.0	Cubic, f.c.
Mo_xNb_{1-x}	0.016–9.2	
Mo_3Os	7.2	A15
$Mo_{0.62}Os_{0.38}$	5.65	$D8_b$
Mo_3P	5.31	DO_e
$Mo_{0.5}Pd_{0.5}$	3.52	A3
Mo_3Re	10.0	
Mo_xRe_{1-x}	1.2–12.2	
$MoRe_3$	9.25, 9.89	A12
$Mo_{0.42}Re_{0.58}$	6.35	$D8_b$
$Mo_{0.52}Re_{0.48}$	11.1	
$Mo_{0.57}Re_{0.43}$	14.0	
$Mo_{\sim 0.60}Re_{0.395}$	10.6	
MoRh	1.97	A3
Mo_xRh_{1-x}	1.5–8.2	Cubic, b.c.
MoRu	9.5–10.5	A3

Substance	T_c, °K	*Crystal structure type*††
$Mo_{0.61}Ru_{0.39}$	7.18	$D8_b$
$Mo_{0.2}Ru_{0.8}$	1.66	A3
Mo_3Sb_4	2.1	
Mo_3Si	1.30	A15
$MoSi_{0.7}$	1.34	
Mo_xSiV_{3-x}	4.54–16.0	A15
Mo_xTc_{1-x}	10.8–15.8	
$Mo_{0.16}Ti_{0.84}$	4.18, 4.25	
$Mo_{0.913}Ti_{0.087}$	2.95	
$Mo_{0.04}Ti_{0.96}$	2.0	Cubic
$Mo_{0.025}Ti_{0.975}$	1.8	
Mo_xU_{1-x}	0.7–2.1	
Mo_xV_{1-x}	0–~5.3	
Mo_2Zr	4.27–4.75	C15
NNb (whiskers)	10–14.5	
NNb (diffusion wires)	16.10	
NNb (film)	6–9	B1
$N_{0.988}Nb$	14.9	B1
$N_{0.824-0.988}Nb$	14.4–15.3	B1
$N_{0.70-0.795}Nb$	11.3–12.9	Cubic and tetragonal
NNb_xO_y	13.5–17.0	B1
NNb_xO_y	6.0–11	
$N_{100-42\ w/o}Nb_{0-58\ w/o}Ti$†	15–16.8	
$N_{100-75\ w/o}Nb_{0-25\ w/o}Zr$†	12.5–16.35	
NNb_xZr_{1-x}	9.8–13.8	B1
$N_{0.93}Nb_{0.85}Zr_{0.15}$	13.8	B1
$N_xO_yTi_z$	2.9–5.6	Cubic
$N_xO_yV_z$	5.8–8.2	Cubic
$N_{0.34}Re$	4–5	Cubic, f.c.
NTa	12–14 (extrapolated)	B1
NTa (film)	4.84	B1
$N_{0.6-0.987}Ti$	<1.17–5.8	B1
$N_{0.82-0.99}V$	2.9–7.9	B1
NZr	9.8	B1
$N_{0.906-0.984}Zr$	3.0–9.5	B1
$Na_{0.28-0.35}O_3W$	0.56	Tetragonal
$Na_{0.28}Pb_{0.72}$	7.2	
NbO	1.25	
$NbOs_2$	2.52	A12
Nb_3Os	1.05	A15
$Nb_{0.6}Os_{0.4}$	1.89, 1.78	$D8_b$
$Nb_3Os_{0.02-0.10}Rh_{0.98-0.90}$	2.42–2.30	A15
$Nb_{0.6}Pd_{0.4}$	1.60	$D8_f$ plus cubic
$Nb_3Pd_{0.02-0.10}Rh_{0.98-0.90}$	2.49–2.55	A15
$Nb_{0.62}Pt_{0.38}$	4.21	$D8_b$
Nb_3Pt	10.9	A15
Nb_5Pt_3	3.73	$D8_b$
$Nb_3Pt_{0.02-0.98}Rh_{0.98-0.02}$	2.52–9.6	A15
$Nb_{0.38-0.18}Re_{0.62-0.82}$	2.43–9.70	A12
Nb_3Rh	2.64	A15
$Nb_{0.60}Rh_{0.40}$	4.21	$D8_b$ plus other
$Nb_3Rh_{0.98-0.90}Ru_{0.02-0.10}$	2.42–2.44	A15
Nb_xRu_{1-x}	1.2–4.8	
NbS_2	6.1–6.3	Hexagonal, $NbSe_2$ type
NbS_2	5.0–5.5	Hexagonal, three-layer type
$Nb_3Sb_{0-0.7}Sn_{1-0.3}$	6.8–18	A15
$NbSe_2$	5.15–5.62	Hexagonal, NbS_2 type
$Nb_{1-1.05}Se_2$	2.2–7.0	Hexagonal, NbS_2 type
Nb_3Si	1.5	$L1_2$
Nb_3SiSnV_3	4.0	
Nb_3Sn	18.05	A15
$Nb_{0.8}Sn_{0.2}$	18.18, 18.5	A15
Nb_xSn_{1-x} (film)	2.6–18.5	
$NbSn_2$	2.60	Orthorhombic
Nb_3Sn_2	16.6	Tetragonal
$NbSnTa_2$	10.8	A15
Nb_2SnTa	16.4	A15
$Nb_{2.5}SnTa_{0.5}$	17.6	A15
$Nb_{2.75}SnTa_{0.25}$	17.8	A15
$Nb_{3x}SnTa_{3(1-x)}$	6.0–18.0	
NbSnTaV	6.2	A15
$Nb_2SnTa_{0.5}V_{0.5}$	12.2	A15
$NbSnV_2$	5.5	A15
Nb_2SnV	9.8	A15
$Nb_{2.5}SnV_{0.5}$	14.2	A15
Nb_xTa_{1-x}	4.4–9.2	A2
$NbTc_3$	10.5	A12
Nb_xTi_{1-x}	0.6–9.8	
$Nb_{0.6}Ti_{0.4}$	9.8	
Nb_xU_{1-x}	1.95 max.	
$Nb_{0.88}V_{0.12}$	5.7	A2
$Nb_{0.75}Zr_{0.25}$	10.8	
$Nb_{0.66}Zr_{0.33}$	10.8	
$Ni_{0.3}Th_{0.7}$	1.98	$D10_2$
$NiZr_2$	1.52	
$Ni_{0.1}Zr_{0.9}$	1.5	A3
$O_3Rb_{0.27-0.29}W$	1.98	Hexagonal
O_3SrTi ($n = 1.7–12.0 \times 10^{19}$)	0.12–0.37	
O_3SrTi ($n = 10^{18}–10^{21}$)	0.05–0.47	
O_3SrTi ($n = \sim 10^{20}$)	0.47	
OTi	0.58	
$O_3Sr_{0.08}W$	2–4	Hexagonal
$O_3Tl_{0.30}W$	2.0–2.14	Hexagonal
OV_3Zr_3	7.5	$E9_3$
OW_3 (film)	3.35, 1.1	A15
OsReY	2.0	C14
Os_2Sc	4.6	C14
OsTa	1.95	A12
Os_3Th_7	1.51	$D10_2$
Os_xW_{1-x}	0.9–4.1	
OsW_3	~3	
Os_2Y	4.7	C14
Os_2Zr	3.0	C14
Os_xZr_{1-x}	1.50–5.6	
PPb	7.8	
$PPd_{3.0-3.2}$	<0.35–0.7	DO_{11}
P_3Pd_7 (high temperature)	1.0	Rhombohedral
P_3Pd_7 (low temp.)	0.70	Complex
PRh	1.22	
PRh_2	1.3	C1
PW_3	2.26	DO_e
Pb_2Pd	2.95	C16
Pb_4Pt	2.80	Related to C16
Pb_2Rh	2.66	C16
PbSb	6.6	
PbTe (plus 0.1 *w/o* Pb)†	5.19	
PbTe (plus 0.1 *w/o* Tl)†	5.24–5.27	
$PbTl_{0.27}$	6.43	
$PbTl_{0.17}$	6.73	
$PbTl_{0.12}$	6.88	
$PbTl_{0.075}$	6.98	
$PbTl_{0.04}$	7.06	
$Pb_{1-0.26}Tl_{0-0.74}$	7.20–3.68	
$PbTl_2$	3.75–4.1	
Pb_3Zr_5	4.60	$D8_8$
$PbZr_3$	0.76	A15
$Pd_{0.9}Pt_{0.1}Te_2$	1.65	C6
$Pd_{0.05}Ru_{0.05}Zr_{0.9}$	~9	
$Pd_{2.2}S$ (quenched)	1.63	Cubic
$PdSb_2$	1.25	C2
PdSb	1.50	$B8_1$
PdSbSe	1.0	C2
PdSbTe	1.2	C2
Pd_4Se	0.42	Tetragonal
$Pd_{6-7}Se$	0.66	Like Pd_4Te
$Pd_{2.8}Se$	2.3	
Pd_xSe_{1-x}	2.5 max.	

†*w/o* denotes weight percent.

Substance	T_c, °K	*Crystal structure type*††
PdSi	0.93	B31
PdSn	0.41	B31
$PdSn_2$	3.34	
Pd_2Sn	0.41	C37
Pd_3Sn_2	0.47–0.64	$B8_2$
PdTe	2.3, 3.85	$B8_1$
$PdTe_{1.02-1.08}$	2.56–1.88	$B8_1$
$PdTe_2$	1.69	C6
$PdTe_{2.1}$	1.89	C6
$PdTe_{2.3}$	1.85	C6
$Pd_{1.1}Te$	4.07	$B8_1$
$PdTh_2$	0.85	C16
$Pd_{0.1}Zr_{0.9}$	7.5	A3
PtSb	2.1	$B8_1$
PtSi	0.88	B31
PtSn	0.37	$B8_1$
PtTe	0.59	Orthorhombic
PtTh	0.44	B_f
Pt_3Th_7	0.98	$D10_2$
Pt_5Th	3.13	
$PtTi_3$	0.58	A15
$Pt_{0.02}U_{0.98}$	0.87	β-phase
$PtV_{2.5}$	1.36	A15
PtV_3	2.87–3.20	A15
$PtV_{3.5}$	1.26	A15
$Pt_{0.5}W_{0.5}$	1.45	A1
Pt_xW_{1-x}	0.4–2.7	
Pt_2Y_3	0.90	
Pt_2Y	1.57, 1.70	C15
Pt_3Y_7	0.82	$D10_2$
PtZr	3.0	A3
$Re_{0.64}Ta_{0.36}$	1.46	A12
$Re_{24}Ti_5$	6.60	A12
Re_xTi_{1-x}	6.6 max.	
$Re_{0.76}V_{0.24}$	4.52	$D8_b$
$Re_{0.92}V_{0.08}$	6.8	A3
$Re_{0.6}W_{0.4}$	6.0	
$Re_{0.5}W_{0.5}$	5.12	$D8_b$
Re_2Y	1.83	C14
Re_2Zr	5.9	C14
Re_6Zr	7.40	A12
$Rh_{17}S_{15}$	5.8	Cubic
$Rh_{\sim 0.24}Sc_{\sim 0.76}$	0.88, 0.92	
Rh_xSe_{1-x}	6.0 max.	
Rh_2Sr	6.2	C15
$Rh_{0.4}Ta_{0.6}$	2.35	$D8_b$
$RhTe_2$	1.51	C2
$Rh_{0.67}Te_{0.33}$	0.49	
Rh_xTe_{1-x}	1.51 max.	
RhTh	0.36	B_f
Rh_3Th_7	2.15	$D10_2$
Rh_5Th	1.07	
Rh_xTi_{1-x}	2.25–3.95	
$Rh_{0.02}U_{0.98}$	0.96	
RhV_3	0.38	A15
RhW	~3.4	A3
RhY_3	0.65	
Rh_2Y_3	1.48	
Rh_3Y	1.07	C15
Rh_5Y	0.56	
$RhZr_2$	10.8	C16
$Rh_{0.005}Zr$ (annealed)	5.8	

Substance	T_c, °K	*Crystal structure type*††
$Rh_{0-0.45}Zr_{1-0.55}$	2.1–10.8	
$Rh_{0.1}Zr_{0.9}$	9.0	Hexagonal, c.p.
Ru_2Sc	1.67	C14
Ru_2Th	3.56	C15
RuTi	1.07	B2
$Ru_{0.05}Ti_{0.95}$	2.5	
$Ru_{0.1}Ti_{0.9}$	3.5	
$Ru_xTi_{0.6}V_y$	6.6 max.	
$Ru_{0.45}V_{0.55}$	4.0	B2
RuW	7.5	A3
Ru_2Y	1.52	C14
Ru_2Zr	1.84	C14
$Ru_{0.1}Zr_{0.9}$	5.7	A3
SbSn	1.30–1.42, 1.42–2.37	Bl or distorted Bl
$SbTi_3$	5.8	A15
Sb_2Tl_7	5.2	
$Sb_{0.01-0.03}V_{0.99-0.97}$	3.76–2.63	A2
SbV_3	0.80	A15
Si_2Th	3.2	C_c, α-phase
Si_2Th	2.4	C32, β-phase
SiV_3	17.1	A15
$Si_{0.9}V_3Al_{0.1}$	14.05	A15
$Si_{0.9}V_3B_{0.1}$	15.8	A15
$Si_{0.9}V_3C_{0.1}$	16.4	A15
$SiV_{2.7}Cr_{0.3}$	11.3	A15
$Si_{0.9}V_3Ge_{0.1}$	14.0	A15
$SiV_{2.7}Mo_{0.3}$	11.7	A15
$SiV_{2.7}Nb_{0.3}$	12.8	A15
$SiV_{2.7}Ru_{0.3}$	2.9	A15
$SiV_{2.7}Ti_{0.3}$	10.9	A15
$SiV_{2.7}Zr_{0.3}$	13.2	A15
Si_2W_3	2.8, 2.84	
$Sn_{0.174-0.104}Ta_{0.826-0.896}$	6.5–<4.2	A15
$SnTa_3$	8.35	A15, highly ordered
$SnTa_3$	6.2	A15, partially ordered
$SnTaV_2$	2.8	A15
$SnTa_2V$	3.7	A15
Sn_xTe_{1-x} ($n = 10.5–20 \times 10^{20}$)	0.07–0.22	Bl
Sn_xTl_{1-x}	2.37–5.2	
SnV_3	3.8	A15
$Sn_{0.02-0.057}V_{0.98-0.943}$	2.87–~1.6	A2
$Ta_{0.025}Ti_{0.975}$	1.3	Hexagonal
$Ta_{0.05}Ti_{0.95}$	2.9	Hexagonal
$Ta_{0.05-0.75}V_{0.095-0.25}$	4.30–2.65	A2
$Ta_{0.8-1}W_{0.2-0}$	1.2–4.4	A2
$Tc_{0.1-0.4}W_{0.9-0.6}$	1.25–7.18	Cubic
$Tc_{0.50}W_{0.50}$	7.52	α plus σ
$Tc_{0.60}W_{0.40}$	7.88	σ plus α
Tc_6Zr	9.7	A12
$Th_{0-0.55}Y_{1-0.45}$	1.2–1.8	
$Ti_{0.70}V_{0.30}$	6.14	Cubic
Ti_xV_{1-x}	0.2–7.5	
$Ti_{0.5}Zr_{0.5}$ (annealed)	1.23	
$Ti_{0.5}Zr_{0.5}$ (quenched)	2.0	
V_2Zr	8.80	C15
$V_{0.26}Zr_{0.74}$	≈5.9	
W_2Zr	2.16	C15

CRITICAL FIELD DATA

Substance	H_o, *oersteds*
Ag_2F	2.5
Ag_7NO_{11}	57
Al_2CMo_3	1,700
$BaBi_3$	740
Bi_2Pt	10
Bi_3Sr	530
Bi_5Tl_3	>400
CdSn	>266
$CoSi_2$	105
$Cr_{0.1}Ti_{0.3}V_{0.6}$	1,360
$In_{1-0.86}Mg_{0-0.14}$	272.4–259.2

Substance	H_o, *oersteds*
InSb	1,100
In_xTl_{1-x}	252–284
$In_{0.8}Tl_{0.2}$	252
$Mg_{\sim 0.47}Tl_{\sim 0.53}$	220
$Mo_{0.16}Ti_{0.84}$	<985
$NbSn_2$	620
$PbTl_{0.27}$	756
$PbTl_{0.17}$	796
$PbTl_{0.12}$	849
$PbTl_{0.075}$	880
$PbTl_{0.04}$	864

KEY TO CRYSTAL STRUCTURE TYPES

"Struck-turbericht" type*	Example	Class	"Struck-turbericht" type*	Example	Class
A1	Cu	Cubic, f.c.	$C15_b$	$AuBe_5$	Cubic
A2	W	Cubic, b.c.	C16	$CuAl_2$	Tetragonal, b.c.
A3	Mg	Hexagonal, close packed	C18	FeS_2	Orthorhombic
A4	Diamond	Cubic, f.c.	C22	Fe_2P	Trigonal
A5	White Sn	Tetragonal, b.c.	C23	$PbCl_2$	Orthorhombic
A6	In	Tetragonal, b.c. (f.c. cell usually used)	C32	AlB_2	Hexagonal
			C36	$MgNi_2$	Hexagonal
A7	As	Rhombohedral	C37	Co_2Si	Orthorhombic
A8	Se	Trigonal	C49	$ZrSi_2$	Orthorhombic
A10	Hg	Rhombohedral	C54	$TiSi_2$	Orthorhombic
A12	α-Mn	Cubic, b.c.	C_c	Si_2Th	Tetragonal, b.c.
A13	β-Mn	Cubic	DO_3	BiF_3	Cubic, f.c.
A15	"β-W" (W_3O)	Cubic	DO_{11}	Fe_3C	Orthorhombic
B1	NaCl	Cubic, f.c.	DO_{18}	Na_3As	Hexagonal
B2	CsCl	Cubic	DO_{19}	Ni_3Sn	Hexagonal
B3	ZnS	Cubic	DO_{20}	$NiAl_3$	Orthorhombic
B4	ZnS	Hexagonal	DO_{22}	$TiAl_3$	Tetragonal
$B8_1$	NiAs	Hexagonal	DO_e	Ni_3P	Tetragonal, b.c.
$B8_2$	Ni_2In	Hexagonal	$D1_3$	Al_4Ba	Tetragonal, b.c.
B10	PbO	Tetragonal	$D1_c$	$PtSn_4$	Orthorhombic
B11	γ-CuTi	Tetragonal	$D2_1$	CaB_6	Cubic
B17	PtS	Tetragonal	$D2_c$	MnU_6	Tetragonal, b.c.
B18	CuS	Hexagonal	$D2_d$	$CaZn_5$	Hexagonal
B20	FeSi	Cubic	$D5_2$	La_2O_3	Trigonal
B27	FeB	Orthorhombic	$D5_8$	Sb_2S_3	Orthorhombic
B31	MnP	Orthorhombic	$D7_3$	Th_3P_4	Cubic, b.c.
B32	NaTl	Cubic, f.c.	$D7_b$	Ta_3B_4	Orthorhombic
B34	PdS	Tetragonal	$D8_1$	Fe_3Zn_{10}	Cubic, b.c.
B_f	δ-CrB	Orthorhombic	$D8_2$	Cu_5Zn_8	Cubic, b.c.
B_g	MoB	Tetragonal, b.c.	$D8_3$	Cu_9Al_4	Cubic
B_h	WC	Hexagonal	$D8_8$	Mn_5Si_3	Hexagonal
B_i	γ-MoC	Hexagonal	$D8_b$	CrFe	Tetragonal
C1	CaF_2	Cubic, f.c.	$D8_i$	Mo_2B_5	Rhombohedral
$C1_b$	MgAgAs	Cubic, f.c.	$D10_2$	Fe_3Th_7	Hexagonal
C2	FeS_2	Cubic	$E2_1$	$CaTiO_3$	Cubic
C6	CdI_2	Trigonal	$E9_3$	Fe_3W_3C	Cubic, f.c.
C11b	$MoSi_2$	Tetragonal, b.c.	$L1_0$	CuAu	Tetragonal
C12	$CaSi_2$	Rhombohedral	$L1_2$	Cu_3Au	Cubic
C14	$MgZn_2$	Hexagonal	$L'2_b$	ThH_2	Tetragonal, b.c.
C15	Cu_2Mg	Cubic, f.c.	L'_3	Fe_2N	Hexagonal

*See "Handbook of Lattice Spacing and Structures of Metals", W.B. Pearson, Vol. I, Pergamon Press, 1958, p. 79. and Vol. II, Pergamon Press, 1967.

HIGH CRITICAL MAGNETIC-FIELD SUPERCONDUCTIVE COMPOUNDS AND ALLOYS

With Critical Temperatures, H_{c1}, H_{c2}, H_{c3}, and the Temperature of Field Observations, T_{obs}

Substance	T_c, °K	H_{c1}, kg	H_{c2}, kg	H_{c3}, kg	T_{obs}, °K†
Al_2CMo_3	9.8–10.2	0.091	156		1.2
$AlNb_3$		0.375			
$Ba_xO_3Sr_{1-x}Ti$	<0.1–0.55	0.0039 max.			
$Bi_{0.5}Cd_{0.1}Pb_{0.27}Sn_{0.13}$			>24		3.06
Bi_xPb_{1-x}	7.35–8.4	0.122 max.	~30 max.		4.2
$Bi_{0.56}Pb_{0.44}$	8.8		15		4.2
$Bi_{7.5\ w/o}Pb_{92.5\ w/o}$‡			2.32		
$Bi_{0.099}Pb_{0.901}$		0.29	2.8		
$Bi_{0.02}Pb_{0.98}$		0.46	0.73		
$Bi_{0.53}Pb_{0.32}Sn_{0.16}$			>25		3.06
$Bi_{1-0.93}Sn_{0-0.07}$			0–0.032		3.7
Bi_5Tl_3	6.4		>5.56		3.35
C_8K (excess K)	0.55		0.160 (H⊥c)		0.32
			0.730 (H∥c)		0.32
C_8K	0.39		0.025 (H⊥c)		0.32
			0.250 (H∥c)		0.32
$C_{0.44}Mo_{0.56}$	12.5–13.5	0.087	98.5		1.2
CNb	8–10	0.12	16.9		4.2
$CNb_{0.4}Ta_{0.6}$	10–13.6	0.19	14.1		1.2
CTa	9–11.4	0.22	4.6		1.2
$Ca_xO_3Sr_{1-x}Ti$	<0.1–0.55	0.002–0.004			
$Cd_{0.1}Hg_{0.9}$ (by weight)		0.23	0.34		2.04
$Cd_{0.05}Hg_{0.95}$		0.28	0.31		2.16
$Cr_{0.10}Ti_{0.30}V_{0.60}$	5.6	0.071	84.4		0
GaN	5.85	0.725			4.2
Ga_xNb_{1-x}			>28		4.2
GaSb (annealed)	4.24		2.64		3.5
$GaV_{1.95}$	5.3		73***		
$GaV_{2.1-3.5}$	6.3–14.45		230–300**		0
GaV_3		0.4	350***		0
			500**		
$GaV_{4.5}$	9 15		121*		0
Hf_xNb_y			>52–>102		1.2
Hf_xTa_y			>28–>86		1.2
$Hg_{0.05}Pb_{0.95}$		0.235	2.3		
$Hg_{0.101}Pb_{0.899}$		0.23	4.3		4.2
$Hg_{0.15}Pb_{0.85}$	~6.75		>13		2.93
$In_{0.98}Pb_{0.02}$	3.45	0.1		0.12	2.76
$In_{0.96}Pb_{0.04}$	3.68	0.1	0.12	0.25	2.94
$In_{0.94}Pb_{0.06}$	3.90	0.095	0.18	0.35	3.12
$In_{0.913}Pb_{0.087}$	4.2	~0.17	0.55	2.65	

HIGH CRITICAL MAGNETIC-FIELD SUPERCONDUCTIVE COMPOUNDS AND ALLOYS (Continued)

Substance	T_c, °K	H_{c1}, kg	H_{c2}, kg	H_{c3}, kg	T_{obs}, °K†
$In_{0.316}Pb_{0.684}$		0.155	3.7		4.2
$In_{0.17}Pb_{0.83}$			2.8	5.5	4.2
$In_{1.000}Te_{1.002}$	3.5–3.7		1.2*		0
$In_{0.95}Tl_{0.05}$		0.263	0.263		3.3
$In_{0.90}Tl_{0.10}$		0.257	0.257		3.25
$In_{0.83}Tl_{0.17}$		0.242	0.39		3.21
$In_{0.75}Tl_{0.25}$		0.216	0.50		3.16
LaN	1.35	0.45			0.76
La_3S_4	6.5	≈0.15	>25		1.3
La_3Se_4	8.6	≈0.2	>25		1.25
$Mo_{0.52}Re_{0.48}$	11.1		14–21	22–33	4.2
			18–28	37–43	1.3
Mo[illegible]Re[illegible]	10.6		14–20	20–37	4.2
			19–26	26–37	1.3
$Mo_{\sim 0.5}Tc_{\sim 0.5}$			~75*		0
$Mo_{0.16}Ti_{0.84}$	4.18	0.028	98.7*		0
			36–38		3.0
$Mo_{0.913}Ti_{0.087}$	2.95	0.060	~15		4.2
$Mo_{0.1-0.3}U_{0.9-0.7}$	1.85–2.06		>25		
$Mo_{0.17}Zr_{0.83}$			~30		
$N_{(12.8\ w/o)}Nb$	15.2		>9.5		13.2
NNb (wires)	16.1		153*		0
			132		4.2
			95		8
			53		12
NNb_xO_{1-x}	13.5–17.0		~38		
NNb_xZr_{1-x}	9.8–13.8		4–>130		4.2
$N_{0.93}Nb_{0.85}Zr_{0.15}$	13.8		>130		4.2
$Na_{0.086}Pb_{0.914}$		0.19	6.0		
$Na_{0.016}Pb_{0.984}$		0.28	2.05		
Nb	9.15		2.020		1.4
			1.710		4.2
Nb		0.4–1.1	3–5.5		4.2
Nb (unstrained)		1.1–1.8	3.40	6–9.1	4.2
Nb (strained)		1.25–1.92	3.44	6.0–8.7	4.2
Nb (cold-drawn wire)		2.48	4.10	≈10	4.2
Nb (film)			>25		4.2
NbSc			>30		
Nb_3Sn		0.170	221		4.2
			70		14.15
			54		15
			34		16
			17		17
$Nb_{0.1}Ta_{0.9}$		0.084	0.154		4.195
$Nb_{0.2}Ta_{0.8}$			10		4.2
$Nb_{0.65-0.73}Ta_{0.02-0.10}Zr_{0.25}$			>70–>90		4.2
Nb_xTi_{1-x}			148 max.		1.2
			120 max.		4.2
$Nb_{0.222}U_{0.778}$		1.98	23		1.2
Nb_xZr_{1-x}			127 max.		1.2
			94 max.		4.2
O_3SrTi	0.43	.0049*	.504*		0
O_3SrTi	0.33	.00195*	.420*		0
$PbSb_{1\ w/o}$ (quenched)			>1.5		4.2
$PbSb_{1\ w/o}$ (annealed)			>0.7		4.2
$PbSb_{2.8\ w/o}$ (quenched)			>2.3		4.2
$PbSb_{2.8\ w/o}$ (annealed)			>0.7		4.2
$Pb_{0.871}Sn_{0.129}$		0.45	1.1		
$Pb_{0.965}Sn_{0.035}$		0.53	0.56		
$Pb_{1-0.26}Tl_{0-0.74}$	7.20–3.68		2–6.9*		0
$PbTl_{0.17}$	6.73		4.5*		0
$Re_{0.26}W_{0.74}$			>30		
$Sb_{0.93}Sn_{0.07}$			0.12		3.7
SiV_3	17.0	0.55	156***		
Sn_xTe_{1-x}		0.00043–0.00236	0.005–0.0775		0.012–0.079
Ta (99.95%)		0.425	1.850		1.3
		0.325	1.425		2.27
		0.275	1.175		2.66
		0.090	0.375		3.72
$Ta_{0.5}Nb_{0.5}$			3.55		4.2
$Ta_{0.65-0}Ti_{0.35-1}$	4.4–7.8		>14–138		1.2
$Ta_{0.5}Ti_{0.5}$			138		1.2
Te	~3.3	0.25*			0
Tc_xW_{1-x}	5.75–7.88		8–44		4.2
Ti				2.7	4.2
$Ti_{0.75}V_{0.25}$	5.3	0.029*	199*		0
$Ti_{0.775}V_{0.225}$	4.7	0.024*	172*		0
$Ti_{0.615}V_{0.385}$	7.07	0.050	~34		4.2
$Ti_{0.516}V_{0.484}$	7.20	0.062	~28		4.2
$Ti_{0.415}V_{0.585}$	7.49	0.078	~25		4.2
$Ti_{0.12}V_{0.88}$			17.3	28.1	4.2
$Ti_{0.09}V_{0.91}$			14.3	16.4	4.2
$Ti_{0.06}V_{0.94}$			8.2	12.7	4.2
$Ti_{0.03}V_{0.97}$			3.8	6.8	4.2
Ti_xV_{1-x}			108 max.		1.2
V	5.31	~0.8	~3.4		1.79
		~0.75	~3.15		2
		~0.45	~2.2		3
		~0.30	~1.2		4
$V_{0.26}Zr_{0.74}$	≈5.9	0 238			1.05
		0.227			1.78
		0.185			3.04
		0.165			3.5
W (film)	1.7–4.1		>34		1

†Temperature of critical field measurement.

‡*w/o* denotes weight percent.

*Extrapolated.

**Linear extrapolation.

***Parabolic extrapolation.

TABLES OF PROPERTIES OF SEMICONDUCTORS

Compiled by Dr. Brian Randall Pamplin

The term "semiconductor" is applied to a material in which electric current is carried by electrons or holes and whose electrical conductivity when extremely pure rises exponentially with temperature and may be increased from this low "intrinsic" value by many orders of magnitude by "doping" with electrically active impurities.

Semiconductors are characterised by an energy gap in the allowed energies of electrons in the material which separates the normally filled energy levels of the *valence band* (where "missing" electrons behave like positively charged current carriers "holes") and the *conduction band* (where electrons behave rather like a gas of free negatively charged carriers with an effective mass dependent on the material and the direction of the electrons' motion). This energy gap depends on the nature of the material and varies with direction in anisotropic crystals. It is slightly dependent on temperature and pressure, and this dependence is usually almost linear at normal temperatures and pressures.

The data is presented in three tables. Table I "General Properties of Semiconductors" lists the main crystallographic and semiconducting properties of a large number of semiconducting materials in three main categories; "Tetrahedral Semiconductors" in which every atom is tetrahedrally co-ordinated to four nearest neighbour atoms (or atomic sites) as for example in the diamond structure; "Octahedral Semiconductors" in which every atom is octahedrally co-ordinated to six nearest neighbour atoms—as for example in the halite structure; and "Other Semiconductors".

Table II gives more detailed information about some better known semiconductors, while Table III gives some information about the electronic energy band structure parameters of the best known materials.

TABLE I

GENERAL PROPERTIES OF SEMICONDUCTORS

(listed by Crystal Structure)

Substance	Lattice Parameters (A° Room temperature)	Density (gm/cc)	Melting Point (°K)	Minimum Room Temperature Energy Gap (eV)	Thermal Conductivity (mW cm⁻¹ K⁻¹)	Heat of Formation k cal/mole	Mobility (Room Temperature) (cm²/V.s) Electrons	Holes	Remarks
PART A ADAMANTINE SEMICONDUCTORS									
§A1 Diamond Structure Elements (Strukturbericht symbol A4, Space Group Fd3m-O_h^7)									
C	3.5597	3.51	4300	5.4	2000	161	1800	1400	
Si	5.43072	2.3283	1685	1.107	1240	77.5	1900	500	
Ge	5.65754	5.3234	1231	0.67	640	69.5	3800	1820	
α-Sn	6.4912	5.765	503	0.08		64	2500	2400	
§A2 Sphalerite (Zinc Blende) Structure Compounds (Strukturbericht symbol B3 Space Group F$\bar{4}$3m-T_d^2)									
I VII Compounds									
CuF	4.255								
CuCl	5.4057	3.53	695						
CuBr	5.6905	4.72	770	2.94		115			
CuI	6.0427	5.63	878			105			
AgBr						102	4000		
AgI	6.473	5.67				93	30		
II VI Compounds									
BeS	4.865	2.36							
BeSe	5.139	4.315							
BeTe	5.626	5.090							
BePo	8.38	7.3							
Zn0	4.63								see § A3
ZnS	5.4093	4.079	2100	3.54	140	114	180	5(400°C)	see also § A3
ZnSe	5.6676	5.42	1790	2.58	140	101	540	28	
ZnTe	6.101	5.72	1568	2.26	140	90	340	100	
ZnPo									
CdS	5.5818		1750						see § A3
CdSe	6.05		1512						see § A3
CdTe	6.477	5.86	1365	1.44	55	81	1200	50	
CdPo									
HgS	5.8517	7.73	~2020						usually cinnabar
HgSe	6.084	8.25	1070	-0.10	10	59	20000		
HgTe	6.460	8.17	943	-0.15	20	58	25000	350	
III V Compounds									
BN	3.615	3.49	3000	~4	200	195			
BP(L.T.)	4.538	2.9		~6			500	70	
BAs	4.777								
AlP	5.451	2.85	1770	2.5					
AlAs	5.6622	3.81	1870	2.16		150	1200	420	
AlSb	6.1355	4.218	1330	1.60	600	140	200–400	550	
GaP	5.4505	4.13	1750	2.24	1100	152	300	100	
GaAs	5.65315	5.316	1510	1.35	370	128	8800	400	
GaSb	6.0954	5.619	980	0.67	270	118	4000	1400	
InP	5.86875	4.787	1330	1.27	800	134	4600	150	
InAs	6.05838	5.66	1215	0.36	290	114	33000	460	
InSb	6.47877	5.775	798	0.165	160	107	78000	750	
Other Sphalerite Structure Compounds									
MnS	5.60								see also § A3
MnSe	5.82								see also § A3
β-SiC	4.348	3.21	3070	2.3			4000		
Ga_2Te_3	5.899	5.75	1063	~1.0	~14	65			
In_2Te_3(H.T.)	6.150	5.8	940	~1.0	~8	47.4	~10		
$MgGeP_2$	5.652								
$ZnSnP_2$	5.65			2.1					
$ZnSnAs_2$(H.T.)	5.851	5.53	1050	~0.7	70				

TABLE 1

GENERAL PROPERTIES OF SEMICONDUCTORS (Continued)

Substance	Lattice Parameters (A° Room temperature)		Density (gm/cc)	Melting Point (°K)	Minimum Room Temperature Energy Gap (eV)	Thermal Conductivity (mW cm^{-1} K^{-1})	Heat of Formation k cal/mole	Mobility (Room Temperature) (cm^2/V.s) Electrons	Holes	Remarks
§ A3 Wurtzite (Zincite) Structure Compounds (Strukturbericht symbol B4, Space Group P 6_3mc - C^4_{6v})										
I VII Compounds										
CuCl	3.91	6.42		T_c 680°K						
CuBr	4.06	6.66		T_c 658°K						
CuI	4.31	7.09								
AgI	4.580	7.494			2.63					
II VI Compounds										
BeO	2.698	4.380		2800						
MgTe	4.54	7.39	3.85	~2800						
ZnO	3.24950	5.2069	5.66	2250	3.2	6	154	180		
ZnS	3.8140	6.2576	4.1	2100	3.67		110			
ZnSe	3.996	6.626		1793						
ZnTe	4.27	6.99		1568						
CdS	4.1348	6.7490	4.82	1748	2.42		96	400		
CdSe	4.299	7.010	5.66	1512	1.74		90	650		
CdTe	4.57	7.47			1.50					
III V Compounds										
BP(H.T.)	3.562	5.900								
AlN	3.111	4.978	3.26	~2500	6.02		197			
GaN	3.190	5.189	6.10	1500	3.34		157			
InN	3.533	5.693	6.88	1200	2.0		133			
Other Wurtzite Structure Compounds										
MnS	3.985	6.45								
MnSe	4.12	6.72								
SiC	3.076	5.048								
MnTe	4.078	6.701			~1.0					
Al_2S_3	3.579	5.829	2.55		4.1		426			
Al_2Se_3	3.890	6.30	3.91		3.1		367			
§ A$\overline{4}$ Chalcopyrite Structure Compounds (Strukturbericht symbol $E1_1$, Space Group I $\overline{4}$ 2d $-D^{12}_{2d}$)										
I III VI_2 Compounds										
$CuAlS_2$	5.323	10.44	3.47		2.5					
$CuAlSe_2$	5.617	10.92	4.70	1270	1.1					
$CuAlTe_2$	5.976	11.80	5.50	1160	0.88					
$CuGaS_2$	5.360	10.49	4.35							
$CuGaSe_2$	5.618	11.01	5.56	1310	0.96, 1.63					
$CuGaTe_2$	6.013	11.93	5.99	1150	0.82, 1.0					
$CuInS_2$	5.528	11.08	4.75		1.2					
$CuInSe_2$	5.785	11.57	5.77	1250	0.86, 0.92	37				
$CuInTe_2$	6.179	12.365	6.10	970	0.95	49				
$CuTlS_2$	5.580	11.17	6.32							
$CuTlSe_2$(L.T.)	5.844	11.65	7.11	680	1.07					
$CuFeS_2$	5.25	10.32		1150	0.53					
$CuFeSe_2$				850	0.16					
$CuLaS_2$	5.65	10.86								
$AgAlS_2$	5.707	10.28	3.94							
$AgAlSe_2$	5.968	10.77	5.07	1220	0.7					
$AgAlTe_2$	6.309	11.85	6.18	1000	0.56					
$AgGaS_2$	5.755	10.28	4.72							
$AgGaSe_2$	5.985	10.90	5.84	1120	1.66					
$AgGaTe_2$	6.301	11.96	6.05	990	1.1	10				
$AgInS_2$(L.T.)	5.828	11.19	5.00		1.9					
$AgInSe_2$	6.102	11.69	5.81	1053	1.18	30				
$AgInTe_2$	6.42	12.59	6.12	965	0.96, 0.52					
$AgFeS_2$	5.66	10.30	4.53							
II IV V_2 Compounds										
$ZnSiP_2$	5.400	10.441	3.39	1640	2.3			1000		
$ZnGeP_2$	5.465	10.771	4.17	1295	2.2					
$CdSiP_2$	5.678	10.431	4.00	~1470	2.2			1000		
$CdGeP_2$	5.741	10.775	4.48	~1060	1.8					
$CdSnP_2$	5.900	11.518			1.5					
$ZnSiAs_2$	5.61	10.88	4.70	~1350	1.7				50	
$ZnGeAs_2$	5.672	11.153	5.32	~1150	0.85	110				
$ZnSnAs_2$	5.8515	11.704	5.53	~ 910	0.65	150			300	disorders at 910°K
$CdSiAs_2$	5.884	10.882								
$CdGeAs_2$	5.9427	11.2172	5.60	~ 903	0.53	40		70	25	disorders at 903°K
$CdSnAs_2$	6.0944	11.9182	5.72	880	0.26	70		22000	250	
§ A5 "Defect Chalcopyrite" Structure Compounds (Strukturbericht symbol E3, Space Group I $\overline{4}$ $- S^2_4$)										
$ZnAl_2Se_4$	5.503	10.90	4.37							
$ZnAl_2Te_4$(?)	5.904	12.05	4.95							
$ZnGa_2S_4$(?)	5.274	10.44	3.80							
$ZnGa_2Se_4$(?)	5.496	10.99	5.21							
$ZnGa_2Te_4$(?)	5.937	11.87	5.67		1.35					
$ZnIn_2Se_4$	5.711	11.42	5.44	1250	1.82					
$ZnIn_2Te_4$	6.122	12.24	5.83	1075	1.2					
$CdAl_2S_4$	5.564	10.32	3.06							
$CdAl_2Se_4$	5.747	10.68	4.54							
$CdAl_2Te_4$(?)	6.011	12.21	5.10							
$CdGa_2S_4$	5.577	10.08	4.03							
$CdGa_2Se_4$	5.743	10.73	5.32							
$CdGa_2Te_4$	6.093	11.81	5.77							
$CdIn_2Te_4$	6.205	12.41	5.9	1060	(1.26 or 0.9)					

TABLE I.

GENERAL PROPERTIES OF SEMICONDUCTORS (Continued)

Substance	Lattice Parameters (A° Room temperature)	Density (gm/cc)	Melting Point (°K)	Minimum Room Temperature Energy Gap (eV)	Thermal Conductivity (mW cm⁻¹ K⁻¹)	Heat of Formation k cal/mole	Mobility (Room Temperature) Electrons (cm²/V.s)	Holes (cm²/V.s)	Remarks
$HgAl_2S_4$	5.488 10.26	4.11							
$HgAl_2Se_4$	5.708 10.74	5.05							
$HgAl_2Te_4$(?)	6.004 12.11	5.81							
$HgGa_2S_4$	5.507 10.23	5.00							
$HgGa_2Se_4$	5.715 10.78	6.18							
$HgIn_2Se_4$	5.764 11.80	6.3	1100	0.6					
$HgIn_2Te_4$(?)	6.186 12.37	6.3	980	0.86		200			
§ A6 Other Adamantine Compounds									
αSiC	3.0817 15.1183	3.21	3070	2.86			400		6H structure
$Hg_5Ga_2Te_8$	6.235								B3 with super lattice
$Hg_5In_2Te_8$	6.328			0.7			2000		B3 with super lattice
Cd In_2Se_4 a=c=5.823				1.55					
PART B OCTAHEDRAL SEMICONDUCTORS									
Halite Structure Semiconductors (Strukturbericht symbol B1, Space Group Fm3m − O_h^5)									
SnSe	6.020		1133						
SnTe	6.313		1080(max)	0.5	91				
PbS	5.9362	7.61	1390	0.37	23	104	600	600	
PbSe	6.1243	8.15	1340	0.26	17	94	1000	900	
PbTe	6.454	8.16	1180	0.25	23	94	1600	600	
Selected Other Binary Chalcides									
BiSe	5.99	7.98	880	0.4					
BiTe	6.47								
EuSe	6.191		2300	1.8	2.4				
GdSe	5.771		2400						
NiD	4.1684	6.6	2260	2.0 or 3.7			4	–	
CdO	4.6953		1700	2.5	7	127	100		
SrS	6.0199	3.643	3000	4.1					
Selected Multineny Compounds									
$AgSbSe_2$	5.786	6.60	910	0.58	10.5				
$AgSbTe_2$ (or $Ag_{19}Sb_{29}Te_{52}$)	6.078	7.12	830	0.7, 0.27	8.6, 0.3				
$AgBiS_2$(H.T.)	5.648								
$AgBiSe_2$(H.T.)	5.82								
$AgBiTe_2$(H.T.)	6.155								
Cu_2CdSnS_4	5.586	10.83		1.16			<2		

TABLE II

SEMICONDUCTING PROPERTIES OF SELECTED MATERIALS

Substance	Minimum Energy Gap (eV) R.T.	0°K	$\frac{dE_g}{dT}$ x 10⁴ eV/°C	$\frac{dE_g}{dP}$ x 10⁶ eV.cm²/kg	Density of States Electron Effective Mass m_{d_n} (m_o)	Electron Mobility and Temperature Dependence μ_n cm²/V.s	−x	Density of States Hole Effective Mass m_{d_p} (m_o)	Hole Mobility and Temperature Dependence μ_p cm²/V.s	−x
Si	1.107	1.153	−2.3	−2.0	1.1	1,900	2.6	0.56	500	2.3
Ge	0.67	0.744	−3.7	+7.3	0.55	3,800	1.66	0.3	1,820	2.33
αSn	0.08	0.094	−0.5		0.02	2,500	1.65	0.3	2,400	2.0
Te	0.33				0.68	1,100		0.19	560	
III–V Compounds										
AlAs	2.2	2.3				1,200			420	
AlSb	1.6	1.7	−3.5	−1.6	0.09	200	1.5	0.4	500	1.8
GaP	2.24	2.40	−5.4	−1.7	0.35	300	1.5	0.5	150	1.5
GaAs	1.35	1.53	−5.0	+9.4	0.068	9,000	1.0	0.5	500	2.1
GaSb	0.67	0.78	−3.5	+12	0.050	5,000	2.0	0.23	1,400	0.9
InP	1.27	1.41	−4.6	+4.6	0.067	5,000	2.0		200	2.4
InAs	0.36	0.43	−2.8	+8	0.022	33,000	1.2	0.41	460	2.3
InSb	0.165	0.23	−2.8	+15	0.014	78,000	1.6	0.4	750	2.1
II–VI Compounds										
ZnO	3.2		−9.5	+0.6	0.38	180	1.5			
ZnS	3.54		−5.3	+5.7		180			5(400°C)	
ZnSe	2.58	2.80	−7.2	+6		540			28	
ZnTe	2.26			+6		340			100	
CdO	2.5 ± .1		−6		0.1	120				
CdS	2.42		−5	+3.3	0.165	400		0.8		
CdSe	1.74	1.85	−4.6		0.13	650	1.0	0.6		
CdTe	1.44	1.56	−4.1	+8	0.14	1,200		0.35	50	
HgSe	0.30				0.030	20,000	2.0			
HgTe	0.15		−1		0.017	25,000		0.5	350	
Halite Structure Compounds										
PbS	0.37	0.28	+4		0.16	800		0.1	1,000	2.2
PbSe	0.26	0.16	+4		0.3	1,500		0.34	1,500	2.2
PbTe	0.25	0.19	+4	−7	0.21	1,600		0.14	750	2.2

TABLE II

SEMICONDUCTING PROPERTIES OF SELECTED MATERIALS (Continued)

Substance	Minimum Energy Gap (eV) R.T.	0°K	$\frac{dE_g}{dT}$ x 10^4 eV/°C	$\frac{dE_g}{dP}$ x 10^6 eV.cm²/kg	Density of States Electron Effective Mass m_{d_n} (m_o)	Electron Mobility and Temperature Dependence μ_n cm²/V.s	−x	Density of States Hole Effective Mass m_{d_p} (m_o)	Hole Mobility and Temperature Dependence μ_p cm²/V.s	−x
Others										
ZnSb	0.50	0.56			0.15	10				1.5
CdSb	0.45	0.57	−5.4		0.15	300			2,000	1.5
Bi_2S_3	1.3					200			1,100	
Bi_2Se_3	0.27					600			675	
Bi_2Te_3	0.13		−0.95		0.58	1,200	1.68	1.07	510	1.95
Mg_2Si		0.77	−6.4		0.46	400	2.5		70	
Mg_2Ge		0.74	−9			280	2		110	
Mg_2Sn	0.21	0.33	−3.5		0.37	320			260	
Mg_3Sb_2		0.32				20			82	
Zn_3As_2	0.93					10	1.1		10	
Cd_3As_2	0.55				0.046	100,000	0.88			
GaSe	2.05		3.8						20	
GaTe	1.66	1.80	−3.6			14	−5			
InSe	1.8					900				
TlSe	0.57		−3.9		0.3	30		0.6	20	1.5
$CdSnAs_2$	0.23				0.05	25,000	1.7			
Ga_2Te_3	1.1	1.55	−4.8							
α-In_2Te_3	1.1	1.2			0.7				50	1.1
β-In_2Te_3	1.0								5	
$Hg_5In_2Te_8$	0.5								11,000	
SnO_2									78	

TABLE III

PART A. DATA ON VALENCE BANDS OF SEMICONDUCTORS (Room Temperature data)

Substance	Band Curvature Effective Mass: Heavy Holes	Light Holes	"Split-off" Band Holes	Energy Separation of "Split-off" Band (eV)	Measured (Light) Hole Mobility cm²/V.s
	(Expressed as fraction of free electron mass)				
Semiconductors with Valence Band Maximum at Centre of Brillouin Zone ("Γ")					
Si	0.52	0.16	0.25	0.044	500
Ge	0.34	0.043	0.08	0.3	1,820
Sn	0.3				2,400
AlAs					
AlSb	0.4			0.7	550
GaP				0.13	100
GaAs	0.8	0.12	0.20	0.34	400
GaSb	0.23	0.06		0.7	1,400
InP				0.21	150
InAs	0.41	0.025	0.083	0.43	460
InSb	0.4	0.015		0.85	750
CdTe	0.35				50
HgTe	0.5				350

Semiconductors with Multiple Valence Band Maxima

Substance	Number of Equivalent Valleys Direction	Curvature Effective Masses Longitudinal m_L	Transverse m_T	Anisotropy $K = \frac{m_L}{m_T}$	Measured (Light) Hole Mobility cm²/V.s
PbSe	4 "L" [111]	0.095	0.047	2.0	1,500
PbTe	4 "L" [111]	0.27	0.02	10	750
Bi_2Te_3	6	0.207	∼0.045	4.5	515

PART B. DATA ON CONDUCTION BANDS OF SEMICONDUCTORS (Room Temperature Data)

Single Valley Semiconductors

Substance	Energy Gap (eV)	Effective Mass (m_o)	Mobility (cm²/V.s)	Comments
GaAs	1.35	0.067	8,500	3(or 6?) equivalent [100] valleys 0.36 eV above this maximum with a mobility of ∼50
InP	1.27	0.067	5,000	3(or 6?) equivalent [100] valleys 0.4 eV above this minimum.
InAs	0.36	0.022	33,000	equivalent valleys ∼1.0 eV above this minimum.
InSb	0.165	0.014	78,000	
CdTe	1.44	0.11	1,000	4(or 8?) equivalent [111] valleys 0.51 eV above this minimum.

Multivalley Semiconductors

Substance	Energy Gap	Number of Equivalent Valleys and Direction	Band Curvature Effective Mass Longitudinal m_L	Transverse m_T	Anisotropy $K = \frac{m_L}{m_T}$	Comments
Si	1.107	6 in [100] "Δ"	0.90	0.192	4.7	
Ge	0.67	4 in [111] at "L"	1.588	0.0815	19.5	
GaSb	0.67	as Ge (?)	∼1.0	∼0.2	∼5	
PbSe	0.26	4 in [111] at "L"	0.085	0.05	1.7	
PbTe	0.25	4 in [111] at "L"	0.21	0.029	5.5	
Bi_2Te_3	0.13	6			∼0.05	

STANDARD CALIBRATION TABLES FOR THERMOCOUPLES

The following tables which represent the Temperature-E.M.F. functions of various thermocouples should be used with appropriate correction curves if precise results are desired. These curves must be determined for each individual couple by plotting ΔE, the difference between the observed and the standard E.M.F., against the standard E.M.F. at three or more fixed temperature points. The value ΔE as shown by such a correction curve is then subtracted algebraically from the observed E.M.F. to give the true E.M.F. reading.

In the following tables the fixed or "cold junction" is at 0°C.; when the cold junction is not maintained at 0°C. the readings of the E.M.F. must be corrected as follows: $Et = E_{(t-tc)} + Etc$ where $E_{(t-tc)}$ is the observed reading, Etc is the E.M.F. for the temperature corresponding to the cold junction temperature as read from the standard table and Et is the E.M.F. produced by the jot junction corrected to the value which would be obtained with the cold junction at 0°C. The temperature corresponding to Et is then obtained by reference to the standard table.

Since the E.M.F.-temperature function is not linear the cold junction should be maintained at a temperature very close to that at whch the thermocouple was calibrated. Otherwise considerable error will result despite the above correction.

PLATINUM VERSUS PLATINUM-10-PERCENT RHODIUM THERMOCOUPLES

(Electromotive Force in Absolute Millivolts. Temperatures in Degrees C (Int. 1948). Reference Junctions at 0°C.)

°C	0	10	20	30	40	50	60	70	80	90
	Millivolts									
0	0	0.06	0.11	0.17	0.24	0.30	0.36	0.43	0.50	0.57
100	0.64	0.72	0.79	0.87	0.95	1.03	1.11	1.19	1.27	1.35
200	1.44	1.52	1.61	1.69	1.78	1.87	1.96	2.05	2.14	2.23
300	2.32	2.41	2.50	2.59	2.69	2.78	2.87	2.97	3.06	3.16
400	3.25	3.35	3.44	3.54	3.64	3.73	3.83	3.93	4.02	4.12
500	4.22	4.32	4.42	4.52	4.62	4.72	4.82	4.92	5.02	5.12
600	5.22	5.33	5.43	5.53	5.64	5.74	5.84	5.95	6.05	6.16
700	6.26	6.37	6.47	6.58	6.68	6.79	6.90	7.01	7.11	7.22
800	7.33	7.44	7.55	7.66	7.77	7.88	7.99	8.10	8.21	8.32
900	8.43	8.55	8.66	8.77	8.88	9.00	9.11	9.23	9.34	9.46
1000	9.57	9.6	9.80	9.92	10.04	10.15	10.27	10.39	10.51	10.62
1100	10.74	10.86	10.98	11.10	11.22	11.34	11.46	11.58	11.70	11.82
1200	11.94	12.06	12.18	12.30	12.42	12.54	12.66	12.78	12.90	13.02
1300	13.14	13.26	13.38	13.50	13.62	13.74	13.86	13.98	14.10	14.22
1400	14.34	14.46	14.58	14.70	14.82	14.94	15.05	15.17	15.29	15.41
1500	15.53	15.65	15.77	15.89	16.01	16.12	16.24	16.36	16.48	16.60
1600	16.72	16.83	16.95	17.07	17.19	17.31	17.42	17.54	17.66	17.77
1700	17.89	18.01	18.12	18.24	18.36	18.47	18.59	—	—	—

(Electromotive Force in Absolute Millivolts. Temperatures in Degrees F.* Reference Junctions at 32°F.)

°F	0	10	20	30	40	50	60	70	80	90
	Millivolts									
0	—	—	—	—	0.02	0.06	0.09	0.12	0.15	0.19
100	0.22	0.26	0.29	0.33	0.36	0.40	0.44	0.48	0.52	0.56
200	0.60	0.64	0.68	0.72	0.76	0.80	0.84	0.89	0.93	0.97
300	1.02	1.06	1.11	1.15	1.20	1.24	1.29	1.33	1.38	1.43
400	1.47	1.52	1.57	1.62	1.66	1.71	1.76	1.81	1.86	1.91
500	1.96	2.01	2.06	2.11	2.16	2.21	2.26	2.31	2.36	2.41
600	2.46	2.51	2.56	2.61	2.66	2.72	2.77	2.82	2.87	2.92
700	2.98	3.03	3.08	3.14	3.19	3.24	3.29	3.35	3.40	3.45
800	3.51	3.56	3.61	3.67	3.72	3.78	3.83	3.88	3.94	3.99
900	4.05	4.10	4.16	4.21	4.26	4.32	4.37	4.43	4.49	4.54
1000	4.60	4.65	4.71	4.76	4.82	4.87	4.93	4.99	5.04	5.10
1100	5.16	5.21	5.27	5.33	5.38	5.44	5.50	5.56	5.61	5.67
1200	5.73	5.78	5.84	5.90	5.96	6.02	6.07	6.13	6.19	6.25
1300	6.31	6.37	6.42	6.48	6.54	6.60	6.66	6.72	6.78	6.84
1400	6.90	6.96	7.02	7.08	7.14	7.20	7.26	7.32	7.38	7.44
1500	7.50	7.56	7.62	7.68	7.74	7.80	7.86	7.93	7.99	8.05
1600	8.11	8.17	8.23	8.30	8.36	8.42	8.48	8.55	8.61	8.67
1700	8.73	8.80	8.86	8.92	8.98	9.05	9.11	9.17	9.24	9.30
1800	9.37	9.43	9.49	9.56	9.62	9.69	9.75	9.82	9.88	9.94
1900	10.01	10.07	10.14	1.20	10.27	10.33	10.40	10.47	10.53	10.60
2000	10.66	10.73	10.79	10.86	10.93	10.99	11.06	11.12	11.19	11.26
2100	11.32	11.39	11.46	11.52	11.59	11.66	11.72	11.79	11.86	11.92
2200	11.99	12.06	12.12	12.19	12.26	12.32	12.39	12.46	12.52	12.59
2300	12.66	12.72	12.79	12.86	12.92	12.99	13.06	13.12	13.19	13.26
2400	13.33	13.39	13.46	13.53	13.59	13.66	13.73	13.79	13.86	13.92
2500	13.99	14.06	14.12	14.19	14.26	14.32	14.39	14.46	14.52	14.59
2600	14.66	14.72	14.79	14.86	14.92	14.99	15.05	15.12	15.19	15.25
2700	15.32	15.39	15.45	15.52	15.58	15.65	15.72	15.78	15.85	15.91
2800	15.98	16.05	16.11	16.18	16.24	16.31	16.37	16.44	16.51	16.57
2900	16.64	16.70	16.77	16.83	16.90	16.97	17.03	17.10	17.16	17.23
3000	17.29	17.36	17.42	17.49	17.55	17.62	17.68	17.75	17.81	17.88
3100	17.94	18.01	18.07	18.14	18.20	18.27	18.33	18.40	18.46	18.53
3200	18.59	18.66	—	—	—	—	—	—	—	—

* Based on the International Temperature Scale of 1948.

PLATINUM VERSUS PLATINUM-13-PERCENT RHODIUM THERMOCOUPLES

(Electromotive Force in Absolute Millivolts. Temperatures in Degrees C (Int. 1948) Reference Junctions at 0°C.)

°C	0	10	20	30	40	50	60	70	80	90
	Millivolts									
0	0.00	0.06	0.11	0.17	0.23	0.30	0.36	0.43	0.50	0.57
100	0.65	0.72	0.80	0.88	0.96	1.04	1.12	1.21	1.29	1.38
200	1.47	1.55	1.64	1.73	1.83	1.92	2.01	2.11	2.20	2.30
300	2.40	2.49	2.59	2.69	2.79	2.89	2.99	3.09	3.19	3.30
400	3.40	3.50	3.61	3.71	3.82	3.92	4.03	4.13	4.24	4.35
500	4.46	4.56	4.67	4.78	4.89	5.00	5.12	5.23	5.34	5.45
600	5.56	5.68	5.79	5.91	6.02	6.14	6.25	6.37	6.49	6.60
700	6.72	6.84	6.96	7.08	7.20	7.32	7.44	7.56	7.68	7.80
800	7.92	8.05	8.17	8.29	8.42	8.54	8.67	8.80	8.92	9.05
900	9.18	9.30	9.43	9.56	9.69	9.82	9.95	10.08	10.21	10.34
1000	10.47	10.60	10.74	10.87	11.00	11.14	11.27	11.41	11.54	11.68
1100	11.82	11.94	12.09	12.23	12.37	12.50	12.64	12.78	12.92	13.06
1200	13.19	13.33	13.47	13.61	13.75	13.89	14.03	14.17	14.30	14.44
1300	14.58	14.72	14.86	15.00	15.14	15.28	15.42	15.55	15.69	15.83
1400	15.97	16.11	16.25	16.39	16.52	16.66	16.80	16.94	17.08	17.22
1500	17.36	17.49	17.63	17.77	17.91	18.04	18.18	18.32	18.45	18.59
1600	18.73	18.86	19.00	19.14	19.27	19.41	19.55	19.68	19.82	19.95
1700	20.09	—	—	—	—	—	—	—	—	—

(Electromotive Force in Absolute Millivolts. Temperatures in Degrees F* Reference Junctions at 32°F.)

°F	0	10	20	30	40	50	60	70	80	90
	Millivolts									
0	—	—	—	—	0.02	0.06	0.09	0.12	0.15	0.19
100	0.22	0.26	0.29	0.33	0.36	0.40	0.44	0.48	0.52	0.56
200	0.60	0.64	0.68	0.72	0.76	0.81	0.85	0.89	0.94	0.98
300	1.03	1.08	1.12	1.17	1.21	1.26	1.31	1.36	1.41	1.46
400	1.50	1.55	1.60	1.65	1.70	1.75	1.81	1.86	1.91	1.96
500	2.01	2.07	2.12	2.17	2.22	2.28	2.33	2.38	2.44	2.49
600	2.55	2.60	2.66	2.71	2.77	2.82	2.88	2.94	2.99	3.05
700	3.10	3.16	3.22	3.27	3.33	3.39	3.45	3.50	3.56	3.62
800	3.68	3.74	3.79	3.85	3.91	3.97	4.03	4.09	4.14	4.21
900	4.26	4.32	4.38	4.44	4.50	4.56	4.62	4.69	4.75	4.81
1000	4.87	4.93	4.99	5.05	5.12	5.18	5.24	5.30	5.36	5.43
1100	5.49	5.55	5.61	5.68	5.74	5.81	5.87	5.93	6.00	6.06
1200	6.13	6.19	6.25	6.32	6.38	6.45	6.51	6.58	6.64	6.71
1300	6.77	6.84	6.90	6.97	7.04	7.10	7.17	7.24	7.30	7.37
1400	7.44	7.50	7.57	7.64	7.71	7.77	7.84	7.91	7.98	8.05
1500	8.12	8.18	8.25	8.32	8.39	8.46	8.53	8.60	8.67	8.74
1600	8.81	8.88	8.95	9.02	9.09	9.16	9.23	9.30	9.37	9.45
1700	9.52	9.59	9.66	9.73	9.80	9.87	9.95	10.02	10.09	10.16
1800	10.24	10.31	10.38	10.46	10.53	10.60	10.68	10.75	10.82	10.90
1900	10.97	11.05	11.12	11.20	11.27	11.35	11.42	11.50	11.58	11.65
2000	11.73	11.80	11.88	11.95	12.03	12.11	12.18	12.26	12.34	12.41
2100	12.49	12.56	12.64	12.72	12.80	12.87	12.95	13.03	13.10	13.18
2200	13.26	13.33	13.41	13.49	13.56	13.64	13.72	13.80	13.87	13.95
2300	14.03	14.10	14.18	14.26	14.34	14.41	14.49	14.57	14.64	14.72
2400	14.80	14.88	15.95	15.03	15.11	15.18	15.26	15.34	15.42	15.49
2500	15.57	15.65	15.72	15.80	15.88	15.95	16.03	16.11	16.19	16.26
2600	16.34	16.42	16.49	16.57	16.65	16.73	16.80	16.88	16.96	17.03
2700	17.11	17.19	17.26	17.34	17.42	17.49	17.57	17.65	17.72	17.80
2800	17.88	17.95	18.03	18.10	18.18	18.26	18.33	18.41	18.48	18.56
2900	18.64	18.71	18.79	18.86	18.94	19.02	19.09	19.17	19.24	19.32
3000	19.39	19.47	19.55	19.62	19.70	19.77	19.85	19.92	20.00	20.08
3100	20.15	—	—	—	—	—	—	—	—	—

CHROMEL-ALUMEL THERMOCOUPLES

(Electromotive Force in Absolute Millivolts. Temperatures in Degrees C (Int. 1948) Reference Junctions at 0°C)

°C	0	1	2	3	4	5	6	7	8	9
	Millivolts									
−190	−5.60	−5.62	−5.63	−5.65	−5.67	−5.68	−5.70	−5.71	−5.73	−5.74
−180	−5.43	−5.45	−5.46	−5.48	−5.50	−5.52	−5.53	−5.55	−5.57	−5.58
−170	−5.24	−5.26	−5.28	−5.30	−5.32	−5.34	−5.35	−5.37	−5.39	−5.41
−160	−5.03	−5.05	−5.08	−5.10	−5.12	−5.14	−5.16	−5.18	−5.20	−5.22
−150	−4.81	−4.84	−4.86	−4.88	−4.90	−4.92	−4.95	−4.97	−4.99	−5.01
−140	−4.58	−4.60	−4.62	−4.65	−4.67	−4.70	−4.72	−4.74	−4.77	−4.79
−130	−4.32	−4.35	−4.37	−4.40	−4.42	−4.45	−4.48	−4.50	−4.52	−4.55
−120	−4.06	−4.08	−4.11	−4.14	−4.16	−4.19	−4.22	−4.24	−4.27	−4.30
−110	−3.78	−3.81	−3.84	−3.86	−3.89	−3.92	−3.95	−3.98	−4.00	−4.03
−100	−3.49	−3.52	−3.55	−3.58	−3.61	−3.64	−3.66	−3.69	−3.72	−3.75
−90	−3.19	−3.22	−3.25	−3.28	−3.31	−3.34	−3.37	−3.40	−3.43	−3.46
−80	−2.87	−2.90	−2.93	−2.96	−3.00	−3.03	−3.06	−3.09	−3.12	−3.16
−70	−2.54	−2.57	−2.61	−2.64	−2.67	−2.71	−2.74	−2.77	−2.80	−2.84
−60	−2.29	−2.24	−2.27	−2.30	−2.34	−2.37	−2.41	−2.33	−2.47	−2.51
−50	−1.86	−1.89	−1.93	−1.96	−2.00	−2.03	−2.07	−2.10	−2.13	−2.17
−40	−1.50	−1.54	−1.57	−1.61	−1.64	−1.68	−1.72	−1.75	−1.79	−1.82
−30	−1.14	−1.17	−1.21	−1.25	−1.28	−1.32	−1.36	−1.39	−1.43	−1.37
−20	−0.77	−0.80	−0.84	−0.88	−0.92	−0.95	−0.99	−1.03	−1.06	−1.10
−10	−0.39	−0.42	−0.46	−0.50	−0.54	−0.58	−0.62	−0.66	−0.69	−0.73
(−)0	−0.00	−0.04	−0.08	−0.12	−0.16	−0.19	−0.23	−0.27	−0.31	−0.35
(+)0	0.00	0.04	0.08	0.12	0.16	0.20	0.24	0.28	0.32	0.36
10	0.40	0.44	0.48	0.52	0.56	0.60	0.64	0.68	0.72	0.76
20	0.80	0.84	0.88	0.92	0.96	1.00	1.04	1.08	1.12	1.16
30	1.20	1.24	1.28	1.32	1.36	1.40	1.44	1.49	1.53	1.57
40	1.61	1.65	1.69	1.73	1.77	1.81	1.85	1.90	1.94	1.98
50	2.02	2.06	2.10	2.14	2.18	2.23	2.27	2.31	2.35	2.39
60	2.43	2.47	2.51	2.56	2.60	2.64	2.68	2.72	2.76	2.80
70	2.85	2.89	2.93	2.97	3.01	3.05	3.10	3.14	3.18	3.22
80	3.26	3.30	3.35	3.39	3.43	3.47	3.51	3.56	3.60	3.63
90	3.68	3.72	3.76	3.81	3.85	3.89	3.93	3.97	4.01	4.06
100	4.10	4.14	4.18	4.22	4.26	4.31	4.35	4.39	4.43	4.47
110	4.51	4.55	4.60	4.64	4.68	4.72	4.76	4.80	4.84	4.88
120	4.92	4.96	5.01	5.05	5.09	5.13	5.17	5.21	5.25	5.29
130	5.33	5.37	5.41	5.45	5.49	5.53	5.57	5.61	5.65	5.69
140	5.73	5.77	5.81	5.85	5.89	5.93	5.97	6.01	6.05	6.09
150	6.13	6.17	6.21	6.25	6.29	6.33	6.37	6.41	6.45	6.49
160	6.53	6.57	6.61	6.65	6.69	6.73	6.77	6.81	6.85	6.89
170	6.93	6.97	7.01	7.05	7.09	7.13	7.17	7.21	7.25	7.29
180	7.33	7.37	7.41	7.45	7.49	7.53	7.57	7.61	7.65	7.69
190	7.73	7.77	7.81	7.85	7.89	7.93	7.97	8.01	8.05	8.09
200	8.13	8.17	8.21	8.25	8.29	8.33	8.37	8.41	−8.46	8.50
210	8.54	8.58	8.62	8.66	8.70	8.74	8.78	8.82	8.86	8.90
220	8.94	8.98	9.02	9.06	9.10	9.14	9.18	9.22	9.26	9.30
230	9.34	9.38	9.42	9.46	9.50	9.54	9.59	9.63	9.67	9.71
240	9.75	9.79	9.83	9.87	9.91	9.95	9.99	10.03	10.07	10.11
250	10.16	10.20	10.24	10.28	10.32	10.36	10.40	10.44	10.48	10.52
260	10.57	10.61	10.65	10.69	10.73	10.77	10.81	10.85	10.89	10.93
270	10.98	11.02	11.06	11.10	11.14	11.18	11.22	11.26	11.30	11.34
280	11.39	11.43	11.47	11.51	11.55	11.59	11.63	11.67	11.72	11.76
290	11.80	11.84	11.88	11.92	11.96	12.01	12.05	12.09	12.13	12.17
300	12.21	12.25	12.29	12.34	12.38	12.42	12.46	12.50	12.54	12.58
310	12.63	12.67	12.71	12.75	12.79	12.83	12.88	12.92	12.96	13.00
320	13.04	13.08	13.12	13.17	13.21	13.25	13.29	13.33	13.37	13.42
330	13.46	13.50	13.54	13.58	13.62	13.67	13.71	13.75	13.79	13.83
340	13.88	13.92	13.96	14.00	14.04	14.09	14.13	14.17	14.21	14.25
350	14.29	14.34	14.38	14.42	14.46	14.50	14.55	14.59	14.63	14.67
360	14.71	14.76	14.80	14.84	14.88	14.92	14.97	15.01	15.05	15.09
370	15.13	15.18	15.22	15.26	15.30	15.34	15.39	15.43	15.47	15.51
380	15.55	15.60	15.64	15.68	15.72	15.76	15.81	15.85	15.89	15.93
390	15.98	16.02	16.06	16.10	16.14	16.19	16.23	16.27	16.31	16.36
400	16.40	16.44	16.48	16.52	16.57	16.61	16.65	16.69	16.74	16.78
410	16.82	16.86	16.91	16.95	16.99	17.03	17.07	17.12	17.16	17.20
420	17.24	17.29	17.33	17.37	17.41	17.46	17.50	17.54	17.58	17.62
430	17.67	17.71	17.75	17.79	17.84	17.88	17.92	17.96	18.01	18.05
440	18.09	18.13	18.17	18.22	18.26	18.30	18.34	18.39	18.43	18.47
450	18.51	18.56	18.60	18.64	18.68	18.73	18.77	18.81	18.85	18.90
460	18.94	18.98	19.02	19.07	19.11	19.15	19.19	19.24	19.28	19.32
470	19.36	19.41	19.45	19.49	19.54	19.58	19.62	19.66	19.71	19.75
480	19.79	19.84	19.88	19.92	19.96	20.01	20.03	20.09	20.13	20.18
490	20.22	20.26	20.31	20.35	20.39	20.43	20.48	20.53	20.56	20.60
500	20.65	20.69	20.73	20.77	20.82	20.86	20.90	20.94	20.99	21.03
510	21.07	21.11	21.16	21.20	21.24	21.28	21.32	21.37	21.41	21.45
520	21.50	21.54	21.58	21.63	21.67	21.71	21.75	21.80	21.84	21.88
530	21.92	21.97	22.01	22.05	22.09	22.14	22.18	22.22	22.26	22.31
540	22.35	22.39	22.43	22.48	22.52	22.56	22.61	22.65	22.69	22.73
550	22.78	22.82	22.86	22.90	22.95	22.99	23.03	23.07	23.12	23.16
560	23.20	23.25	23.29	23.33	23.38	23.42	23.46	23.50	23.54	23.59
570	23.63	23.67	23.72	23.76	23.80	23.84	23.89	23.93	23.97	24.01
580	24.06	24.10	24.14	24.18	24.23	24.27	24.31	24.36	24.40	24.44
590	24.49	24.53	24.57	24.61	25.65	24.70	24.74	24.78	24.83	24.87
600	24.91	24.95	25.00	25.04	25.08	25.12	25.17	25.21	25.25	25.29
610	25.34	25.38	25.42	25.47	25.51	25.55	25.59	25.64	25.68	25.72
620	25.76	25.81	25.85	25.89	25.93	25.98	26.02	26.06	26.10	26.15
630	26.19	26.23	26.27	26.32	26.36	26.40	26.44	26.48	26.53	26.57
640	26.61	26.65	26.70	26.74	26.78	26.82	26.86	26.91	26.95	26.99
650	27.03	27.07	27.12	27.16	27.20	27.24	27.28	27.33	27.37	27.41
660	27.45	27.49	27.54	27.58	27.62	27.66	27.71	27.75	27.79	27.83
670	27.87	27.92	27.96	28.00	28.04	28.08	28.13	28.17	28.21	28.25
680	28.29	28.34	28.38	28.42	28.46	28.50	28.55	28.59	28.63	28.67
690	28.72	28.76	28.80	28.84	28.88	28.93	28.97	29.01	29.05	29.10
700	29.14	29.18	29.22	29.26	29.30	29.35	29.39	29.43	29.47	29.52
710	29.56	29.60	29.64	29.68	29.72	29.77	29.81	29.85	29.89	29.93
720	29.97	30.02	30.06	30.10	30.14	30.18	30.23	30.27	30.31	30.35
730	30.39	30.44	30.48	30.52	30.56	30.60	30.65	30.69	30.73	30.77
740	30.81	30.85	30.90	30.94	30.98	31.02	31.06	31.10	31.15	31.19
750	31.23	31.27	31.31	31.35	31.40	31.44	31.48	31.52	31.56	31.60
760	31.65	31.69	31.73	31.77	31.81	31.85	31.90	31.94	31.98	32.02
770	32.06	32.10	32.15	32.19	32.23	32.27	32.31	32.35	32.39	32.43
780	32.48	32.52	32.56	32.60	32.64	32.68	32.72	32.76	32.81	32.85
790	32.89	32.93	32.97	33.01	33.05	33.09	33.13	33.18	33.22	33.26
800	33.30	33.34	33.38	33.42	33.46	33.50	33.54	33.59	33.63	33.67
810	33.71	33.75	33.79	33.83	33.87	33.91	33.95	33.99	34.04	34.08
820	34.12	34.16	34.20	34.24	34.28	34.32	34.36	34.40	34.44	34.48
830	34.53	34.57	34.61	34.65	34.69	34.73	34.77	34.81	34.85	34.89
840	34.93	34.97	35.02	35.06	35.10	35.14	35.18	35.22	35.26	35.30
850	35.34	35.38	35.42	35.46	35.50	35.54	35.58	35.63	35.67	35.71
860	35.75	35.79	35.83	35.87	35.91	35.95	35.99	36.03	36.07	36.11
870	36.15	36.19	36.23	36.27	36.31	36.35	36.39	36.43	36.47	36.51
880	36.55	36.59	36.63	36.67	36.72	36.76	36.80	36.84	36.88	36.92
890	36.96	37.00	37.04	37.08	37.12	37.16	37.20	37.24	37.28	37.32
900	37.36	37.40	37.44	37.48	37.52	37.56	37.60	37.64	37.68	37.72
910	37.76	37.80	37.84	37.88	37.92	37.96	38.00	38.04	38.08	38.12
920	38.16	38.20	38.24	38.28	38.32	38.36	38.40	38.44	38.48	38.52
930	38.56	38.60	38.64	38.68	38.72	38.76	38.80	38.84	38.88	38.92
940	38.95	38.99	39.03	39.07	39.11	39.15	39.19	39.23	39.27	39.31
950	39.35	39.39	39.43	39.47	39.51	39.55	39.59	39.63	39.67	39.71
960	39.75	39.79	39.83	39.86	39.90	39.94	39.98	40.02	40.06	40.10
970	40.14	40.18	40.22	40.26	40.30	40.34	40.38	40.41	40.45	40.49
980	40.53	40.57	40.61	40.65	40.69	40.73	40.77	40.81	40.85	40.89
990	30.92	40.96	41.00	41.04	41.08	41.12	41.16	41.20	41.24	41.28
1000	41.31	41.35	41.39	41.43	41.47	41.51	41.55	41.59	41.63	41.67
1010	41.70	41.74	41.78	41.82	41.86	41.90	41.94	41.98	42.02	42.05
1020	42.09	42.13	42.17	42.21	42.25	42.29	42.33	42.36	42.40	42.44
1030	42.48	42.52	42.56	42.60	42.63	42.67	42.71	42.75	42.79	42.83
1040	42.87	42.90	42.94	42.98	43.02	43.06	43.10	43.14	43.17	43.21
1050	43.25	43.29	43.33	43.37	43.41	43.44	43.48	43.52	43.56	43.60
1060	43.63	43.67	43.71	43.75	43.79	43.83	43.87	43.90	43.94	43.98
1070	44.02	44.06	44.10	44.13	44.17	44.21	44.25	44.29	44.33	44.36
1080	44.40	44.44	44.48	44.52	44.55	44.59	44.63	44.67	44.71	44.74
1090	44.78	44.82	44.86	44.90	44.93	44.97	45.01	45.05	45.09	45.12
1100	45.16	45.20	45.24	45.27	45.31	45.35	45.39	45.43	45.46	45.50
1110	45.54	45.58	45.62	45.65	45.69	45.73	45.77	45.80	45.84	45.88
1120	45.92	45.96	45.99	46.03	46.07	46.11	46.14	46.18	46.22	46.26
1130	46.29	46.33	46.37	46.41	46.44	46.48	46.52	46.56	46.59	46.63
1140	46.67	46.70	46.74	46.78	46.82	46.85	46.89	46.93	46.97	47.00
1150	47.04	47.08	47.12	47.15	47.19	47.23	47.26	47.30	47.34	47.38
1160	47.41	47.45	47.49	47.52	47.56	47.60	47.63	47.67	47.71	47.75
1170	47.78	47.82	47.86	47.89	47.93	47.97	48.00	48.04	48.08	48.12
1180	48.15	48.19	48.23	48.26	48.30	48.34	48.37	48.41	48.45	48.48
1190	48.52	48.56	48.59	48.63	48.67	48.70	48.74	48.78	48.81	48.85
1200	48.89	48.92	48.96	49.00	49.03	49.07	49.11	49.14	49.18	49.22
1210	49.25	49.29	49.32	49.36	49.40	49.43	49.47	49.51	49.54	49.58
1220	49.62	49.65	49.69	49.72	49.76	49.80	49.83	49.87	49.90	49.94
1230	49.98	50.01	50.05	50.08	50.12	50.16	50.19	50.23	50.26	50.30
1240	50.34	50.37	50.41	50.44	50.48	50.52	50.55	50.59	50.62	50.66
1250	50.69	50.73	50.77	50.80	50.84	50.87	50.91	50.94	50.98	51.02
1260	51.05	51.09	51.12	51.16	51.19	51.23	51.27	51.30	51.34	51.37
1270	51.41	51.44	51.48	51.51	51.55	51.58	51.62	51.66	51.69	51.73
1280	51.76	51.80	51.83	51.87	51.90	51.94	51.97	52.01	52.04	52.08
1290	52.11	52.15	52.18	52.22	52.25	52.29	52.32	52.36	52.39	52.43
1300	52.46	52.50	52.53	52.57	52.60	52.64	52.67	52.71	52.74	52.78
1310	52.81	52.85	52.88	52.92	52.95	52.99	53.02	53.06	53.09	53.13
1320	53.16	53.20	53.23	53.27	53.30	53.34	53.37	53.41	53.44	53.47
1330	53.51	53.54	53.58	53.61	53.65	53.68	53.72	53.75	53.79	53.82
1340	53.85	53.89	53.92	53.96	53.99	54.03	54.06	54.10	54.13	54.16
1350	54.20	54.23	54.27	54.30	54.34	54.37	54.40	54.44	54.47	54.51
1360	54.54	54.57	54.61	54.64	54.68	54.71	54.74	54.78	54.81	54.85
1370	54.88	54.91	—	—	—	—	—	—	—	—

* Based on the International Temperature Scale of 1948.

CHROMEL-ALUMEL THERMOCOUPLES (Continued)

(Electromotive Force in Absolute Millivolts. Temperatures in Degrees F* Reference Junctions at 32°F.)

°F	0	1	2	3	4	5	6	7	8	9
	Millivolts									
−300	−5.51	−5.52	−5.53	−5.54	−5.54	−5.55	−5.56	−5.57	−5.58	−5.59
−290	−5.41	−5.42	−5.43	−5.44	−5.45	−5.46	−5.47	−5.48	−5.49	−5.50
−280	−5.30	−5.31	−5.32	−5.34	−5.35	−5.36	−5.37	−5.38	−5.39	−5.40
−270	−5.20	−5.21	−5.22	−5.23	−5.24	−5.25	−5.26	−5.27	−5.28	−5.29
−260	−5.08	−5.09	−5.10	−5.12	−5.13	−5.14	−5.15	−5.16	−5.17	−5.18
−250	−4.96	−4.97	−4.99	−5.00	−5.01	−5.02	−5.03	−5.04	−5.06	−5.07
−240	−4.84	−4.85	−4.86	−4.88	−4.89	−4.90	−4.91	−4.92	−4.94	−4.95
−230	−4.71	−4.72	−4.74	−4.75	−4.76	−4.77	−4.79	−4.80	−4.81	−4.82
−220	−4.58	−4.59	−4.60	−4.62	−4.63	−4.64	−4.66	−4.67	−4.68	−4.70
−210	−4.44	−4.45	−4.46	−4.48	−4.49	−4.51	−4.52	−4.53	−4.55	−4.56
−200	−4.29	−4.31	−4.32	−4.34	−4.35	−4.36	−4.38	−4.39	−4.41	−4.42
−190	−4.15	−4.16	−4.18	−4.19	−4.21	−4.22	−4.24	−4.25	−4.26	−4.28
−180	−4.00	−4.01	−4.03	−4.04	−4.06	−4.07	−4.09	−4.10	−4.12	−4.13
−170	−3.84	−3.86	−3.88	−3.89	−3.91	−3.92	−3.94	−3.95	−3.97	−3.98
−160	−3.69	−3.70	−3.72	−3.73	−3.75	−3.76	−3.78	−3.80	−3.81	−3.83
−150	−3.52	−3.54	−3.56	−3.57	−3.59	−3.60	−3.62	−3.64	−3.65	−3.67
−140	−3.36	−3.38	−3.39	−3.41	−3.42	−3.44	−3.46	−3.47	−3.49	−3.51
−130	−3.19	−3.20	−3.22	−3.24	−3.25	−3.27	−3.29	−3.31	−3.32	−3.34
−120	−3.01	−3.03	−3.05	−3.06	−3.08	−3.10	−3.12	−3.13	−3.15	−3.17
−110	−2.83	−2.85	−2.87	−2.89	−2.90	−2.92	−2.94	−2.96	−2.98	−2.99
−100	−2.65	−2.67	−2.69	−2.71	−2.72	−2.74	−2.76	−2.78	−2.80	−2.82
−90	−2.47	−2.49	−2.50	−2.52	−2.54	−2.56	−2.58	−2.60	−2.62	−2.63
−80	−2.28	−2.30	−2.32	−2.34	−2.36	−2.37	−2.39	−2.41	−2.43	−2.45
−70	−2.09	−2.11	−2.13	−2.15	−2.17	−2.18	−2.20	−2.22	−2.24	−2.26
−60	−1.90	−1.92	−1.94	−1.96	−1.97	−1.99	−2.01	−2.03	−2.05	−2.07
−50	−1.70	−1.72	−1.74	−1.76	−1.78	−1.80	−1.82	−1.84	−1.86	−1.88
−40	−1.50	−1.52	−1.54	−1.56	−1.58	−1.60	−1.62	−1.64	−1.66	−1.68
−30	−1.30	−1.32	−1.34	−1.36	−1.38	−1.40	−1.42	−1.44	−1.46	−1.48
−20	−1.10	−1.12	−1.14	−1.16	−1.18	−1.20	−1.22	−1.24	−1.26	−1.28
−10	−0.89	−0.91	−0.93	−0.95	−0.97	−0.99	−1.01	−1.03	−1.06	−1.08
(−)0	−0.68	−0.70	−0.72	−0.75	−0.77	−0.79	−0.81	−0.83	−0.85	−0.87
(+)0	−0.68	−0.66	−0.64	−0.62	−0.60	−0.58	−0.56	−0.54	−0.52	−0.49
10	−0.47	−0.45	−0.43	−0.41	−0.39	−0.37	−0.34	−0.32	−0.30	−0.28
20	−0.26	−0.24	−0.22	−0.19	−0.17	−0.15	−0.13	−0.11	−0.09	−0.07
30	−0.04	−0.02	0.00	0.02	0.04	0.07	0.09	0.11	0.13	0.15
40	0.18	0.20	0.22	0.24	0.26	0.29	0.31	0.33	0.35	0.37
50	0.40	0.42	0.44	0.46	0.48	0.51	0.53	0.55	0.57	0.60
60	0.62	0.64	0.65	0.68	0.71	0.73	0.75	0.77	0.80	0.82
70	0.84	0.86	0.88	0.91	0.93	0.95	0.97	1.00	1.02	1.04
80	1.06	1.09	1.11	1.13	1.15	1.18	1.20	1.22	1.24	1.27
90	1.29	1.31	1.33	1.36	1.38	1.40	1.43	1.45	1.47	1.49
100	1.52	1.54	1.56	1.58	1.61	1.63	1.65	1.68	1.70	1.72
110	1.74	1.77	1.79	1.81	1.84	1.86	1.88	1.90	1.93	1.95
120	1.97	2.00	2.02	2.04	2.06	2.09	2.11	2.13	2.16	2.18
130	2.20	2.23	2.25	2.27	2.29	2.32	2.34	2.36	2.39	2.41
140	2.43	2.46	2.48	2.50	2.52	2.55	2.57	2.59	2.62	2.64
150	2.66	2.69	2.71	2.73	2.75	2.78	2.80	2.82	2.85	2.87
160	2.89	2.92	2.94	2.96	2.98	3.01	3.03	3.05	3.08	3.10
170	3.12	3.15	3.17	3.19	3.22	3.24	3.26	3.29	3.31	3.33
180	3.36	3.38	3.40	3.43	3.45	3.47	3.49	2.52	3.54	3.56
190	3.59	3.61	3.63	3.66	3.68	3.70	3.73	3.75	3.77	3.80
200	3.82	3.84	3.87	3.89	3.91	3.94	3.96	3.98	4.01	4.03
210	4.05	4.08	4.10	4.12	4.15	4.17	4.19	4.21	4.24	4.26
220	4.28	4.31	4.33	4.35	4.38	4.40	4.42	4.44	4.47	4.49
230	4.51	4.54	4.56	4.58	4.61	4.63	4.65	4.67	4.70	4.72
240	4.74	4.77	4.79	4.81	4.83	4.86	4.88	4.90	4.92	4.95
250	4.97	4.99	5.02	5.04	5.06	5.08	5.11	5.13	5.15	5.17
260	5.20	5.22	5.24	5.26	5.29	5.31	5.33	5.35	5.38	5.40
270	5.42	5.44	5.47	5.49	5.51	5.53	5.56	5.58	5.60	5.62
280	5.65	5.67	5.69	5.71	5.73	5.76	5.78	5.80	5.82	5.85
290	5.87	5.89	5.91	5.93	5.96	5.98	6.00	6.02	6.05	6.07
300	6.09	6.11	6.13	6.16	6.18	6.20	6.22	6.25	6.27	6.29
310	6.31	6.33	6.36	6.38	6.40	6.42	6.45	6.47	6.49	6.51
320	6.53	6.56	6.58	6.60	6.62	6.65	6.67	6.69	6.71	6.73
330	6.76	6.78	6.80	6.82	6.84	6.87	6.89	6.91	6.93	6.96
340	6.98	7.00	7.02	7.04	7.07	7.09	7.11	7.13	7.15	7.18
350	7.20	7.22	7.24	7.26	7.29	7.31	7.33	7.35	7.38	7.40
360	7.42	7.44	7.46	7.49	7.51	7.53	7.55	7.58	7.60	7.62
370	7.64	7.66	7.69	7.71	7.73	7.75	7.78	7.80	7.82	7.84
380	7.87	7.89	7.91	7.93	7.95	7.98	8.00	8.02	8.04	8.07
390	8.09	8.11	8.13	8.16	8.18	8.20	8.22	8.24	8.27	8.29
400	8.31	8.33	8.36	8.38	8.40	8.42	8.45	8.47	8.49	8.51
410	8.54	8.56	8.58	8.60	8.62	8.65	8.67	8.69	8.71	8.74
420	8.76	8.78	8.80	8.82	8.85	8.87	8.89	8.91	8.94	8.96
430	8.98	9.00	9.03	9.05	9.07	9.09	9.12	9.14	9.16	9.18
440	9.21	9.23	9.25	9.27	9.30	9.32	9.34	9.36	9.39	9.41
450	9.43	9.45	9.48	9.50	9.52	9.54	9.57	9.59	9.61	9.63
460	9.66	9.68	9.70	9.73	9.75	9.77	9.79	9.82	9.84	9.86
470	9.88	9.91	9.93	9.95	9.97	10.00	10.02	10.04	10.06	10.09
480	10.11	10.13	10.16	10.18	10.20	10.22	10.25	10.27	10.29	10.31
490	10.34	10.36	10.38	10.40	10.43	10.45	10.47	10.50	10.52	10.54
500	10.57	10.59	10.61	10.63	10.66	10.68	10.70	10.72	10.75	10.77

* Based on on the International Temperature Scale of 1948.

°F	0	1	2	3	4	5	6	7	8	9
	Millivolts									
510	10.79	10.82	10.84	10.86	10.88	10.91	10.93	10.95	10.98	11.00
520	11.02	11.04	11.07	11.09	11.11	11.13	11.16	11.18	11.20	11.23
530	11.25	11.27	11.29	11.32	11.34	11.36	11.39	11.41	11.43	11.45
540	11.48	11.50	11.52	11.55	11.57	11.59	11.61	11.64	11.66	11.68
550	11.71	11.73	11.75	11.78	11.80	11.82	11.84	11.87	11.89	11.91
560	11.94	11.98	11.98	12.01	12.03	12.05	12.07	12.10	12.12	12.14
570	12.17	12.19	12.21	12.24	12.26	12.28	12.30	12.33	12.35	12.37
580	12.40	12.42	12.44	12.47	12.49	12.51	12.53	12.56	12.58	12.60
590	12.63	12.65	12.67	12.70	12.72	12.74	12.76	12.79	12.81	12.83
600	12.86	12.88	12.90	12.93	12.95	12.97	13.00	13.02	13.04	13.06
610	13.09	13.11	13.13	13.16	13.18	13.20	13.23	13.25	13.27	13.30
620	13.32	13.34	13.36	13.39	13.41	13.44	13.46	13.48	13.50	13.53
630	13.55	13.57	13.60	13.62	13.63	13.67	13.69	13.71	13.74	13.76
640	13.78	13.81	13.83	13.85	13.88	13.90	13.92	13.95	13.97	13.99
650	14.02	14.04	14.06	14.09	14.11	14.13	14.15	14.18	14.20	14.22
660	14.25	14.27	14.29	14.32	14.34	14.36	14.39	14.41	14.43	14.46
670	14.48	14.50	14.53	14.55	14.57	14.60	14.62	14.64	14.67	14.69
680	14.71	14.74	14.76	14.78	14.81	14.83	14.85	14.88	14.90	14.92
690	14.95	14.97	14.99	15.02	15.04	15.06	15.09	15.11	15.13	15.16
700	15.18	15.20	15.23	15.25	15.27	15.30	15.32	15.34	15.37	15.39
710	15.41	15.44	15.46	15.48	15.51	15.53	15.55	15.58	15.60	15.62
720	15.65	15.67	15.69	15.72	15.74	15.76	15.79	15.81	15.83	15.86
730	15.88	15.90	15.93	15.95	15.98	16.00	16.02	16.05	16.07	16.09
740	16.12	16.14	16.16	16.19	16.21	16.23	16.26	16.28	16.30	16.33
750	16.35	16.37	16.40	16.42	16.45	16.47	16.49	16.52	16.54	16.56
760	16.59	16.61	16.63	16.66	16.68	16.70	16.73	16.75	16.77	16.80
770	16.82	16.84	16.87	16.89	16.92	16.94	16.96	16.99	17.01	17.03
780	17.06	17.08	17.10	17.13	17.15	17.17	17.20	17.22	17.24	17.27
790	17.29	17.31	17.34	17.36	17.39	17.41	17.43	17.46	17.48	17.50
800	17.53	17.55	17.57	17.60	17.62	17.64	17.67	17.69	17.71	17.74
810	17.76	17.78	17.81	17.83	17.86	17.88	17.90	17.93	17.95	17.97
820	18.00	18.02	18.04	18.07	18.09	18.11	18.14	18.16	18.18	18.21
830	18.23	18.25	18.28	18.30	18.33	18.35	18.37	18.40	18.42	18.44
840	18.47	18.49	18.51	18.54	18.56	18.58	18.61	18.63	18.65	18.68
850	18.70	18.73	18.75	18.77	18.80	18.82	18.84	18.87	18.89	18.91
860	18.94	18.96	18.90	19.01	19.03	19.06	19.08	19.10	19.13	19.15
870	19.18	19.20	19.22	19.25	19.27	19.29	19.32	19.34	19.36	19.39
880	19.41	19.44	19.46	19.48	19.51	19.53	19.55	19.58	19.60	19.63
890	19.65	19.67	19.70	19.72	19.75	19.77	19.79	19.82	19.84	19.86
900	19.89	19.91	19.94	19.96	19.98	20.01	20.03	20.05	20.08	20.10
910	20.13	20.15	20.17	20.20	20.22	20.24	20.27	20.29	20.32	20.24
920	20.36	20.39	20.41	20.43	20.46	20.48	20.50	20.53	20.55	20.58
930	20.60	20.62	20.65	20.67	20.69	20.72	20.74	20.76	20.79	20.81
940	20.84	20.86	20.88	20.91	20.93	20.95	20.98	21.00	21.03	21.05
950	21.07	21.10	21.12	21.14	21.17	21.19	21.21	21.24	21.26	21.28
960	21.31	21.33	21.36	21.38	21.40	21.43	21.45	21.47	21.50	21.52
970	21.54	21.57	21.59	21.62	21.64	21.66	21.69	21.71	21.73	21.76
980	21.78	21.81	21.83	21.85	21.88	21.90	21.92	21.95	21.97	21.99
990	22.02	22.04	22.07	22.09	22.11	22.14	22.16	22.18	22.21	22.23
1000	22.26	22.28	22.30	22.33	22.35	22.37	22.40	22.42	22.44	22.47
1010	22.49	22.52	22.54	22.56	22.59	22.61	22.63	22.66	22.68	22.71
1020	22.73	22.75	22.78	22.80	22.82	22.85	22.87	22.90	22.92	22.94
1030	22.97	22.99	23.01	23.04	23.06	23.08	23.11	23.13	23.16	23.18
1040	23.20	23.23	23.25	23.27	23.30	23.32	23.35	23.37	23.39	23.42
1050	23.44	23.46	23.49	23.51	23.54	23.56	23.58	23.61	23.63	23.65
1060	23.68	23.70	23.72	23.75	23.77	23.80	23.82	23.84	23.87	23.89
1070	23.91	23.94	23.96	23.99	24.01	24.03	24.06	24.08	24.10	24.13
1080	24.15	24.18	24.20	24.22	24.25	24.27	24.29	24.32	24.34	24.36
1090	24.39	24.41	24.44	24.46	24.49	24.51	24.53	24.55	24.58	24.60
1100	24.63	24.65	24.67	24.70	24.72	24.74	24.77	24.79	24.82	24.84
1110	24.86	24.89	24.91	24.93	24.96	24.98	25.01	25.03	25.05	25.08
1120	25.10	25.12	25.15	25.17	25.20	25.22	25.24	25.27	25.29	25.31
1130	25.34	25.36	25.38	25.41	25.43	25.46	25.48	25.50	25.53	25.55
1140	25.57	25.60	25.62	25.65	25.67	25.69	25.72	25.74	25.76	25.79
1150	25.81	25.83	25.86	25.88	25.91	25.93	25.95	25.98	26.00	26.02
1160	26.05	26.07	26.09	26.12	26.14	26.16	26.19	26.21	26.24	26.26
1170	26.28	26.31	26.33	26.35	26.38	26.40	26.42	26.45	26.47	26.49
1180	26.52	26.54	26.56	26.59	26.61	26.63	26.66	26.68	26.70	26.73
1190	26.75	26.77	26.80	26.82	26.85	26.87	26.89	26.91	26.94	26.96
1200	26.98	27.01	27.03	27.06	27.08	27.10	27.12	27.15	27.17	27.20
1210	27.22	27.24	27.27	27.29	27.31	27.34	27.36	27.38	27.40	27.43
1220	27.45	27.48	27.50	27.52	27.55	27.57	27.59	27.62	27.64	27.66
1230	27.69	27.71	27.73	27.76	27.78	27.80	27.83	27.85	27.87	27.90
1240	27.92	27.94	27.97	27.99	28.01	28.04	28.06	28.08	28.11	28.13
1250	28.15	28.18	28.20	28.22	28.25	28.27	28.29	28.32	28.34	28.37
1260	28.39	28.41	28.44	28.46	28.48	28.50	28.53	28.55	28.58	28.60
1270	28.62	28.65	28.67	28.69	28.72	28.74	28.76	28.79	28.81	28.83
1280	28.86	28.88	28.90	28.93	28.95	28.97	29.00	29.02	29.04	29.07
1290	29.09	29.11	29.14	29.16	29.18	29.21	29.23	29.25	29.28	29.30
1300	29.32	29.35	29.37	29.39	29.42	29.44	29.46	29.49	29.51	29.53
1310	29.56	29.58	29.60	29.63	29.65	29.67	29.70	29.72	29.74	29.77
1320	29.79	29.81	29.84	29.86	29.88	29.91	29.93	29.95	29.97	30.00

CHROMEL-ALUMEL THERMOCOUPLES (Continued)

Millivolts

°F	0	1	2	3	4	5	6	7	8	9
1330	30.02	30.05	30.07	30.09	30.11	30.14	30.16	30.18	30.21	30.23
1340	30.25	30.28	30.30	30.32	30.35	30.37	30.39	30.42	30.44	30.46
1350	30.49	30.51	30.53	30.56	30.58	30.60	30.63	30.65	30.67	30.70
1360	30.72	30.74	30.77	30.79	30.81	30.83	30.86	30.88	30.90	30.93
1370	30.95	30.97	31.00	31.02	31.04	31.07	31.09	31.11	31.14	31.16
1380	31.18	31.21	31.23	31.25	31.28	31.30	31.32	31.34	31.37	31.39
1390	31.42	31.44	31.46	31.48	31.51	31.53	31.55	31.58	31.60	31.62
1400	31.65	31.67	31.69	31.72	31.74	31.76	31.78	31.81	31.83	31.85
1410	31.88	31.90	31.92	31.95	31.97	31.99	32.02	32.04	32.06	32.08
1420	32.11	32.13	32.15	32.18	32.20	32.22	32.25	32.27	32.29	32.31
1430	32.34	32.36	32.38	32.41	32.43	32.45	32.48	32.50	32.52	32.54
1440	32.57	32.59	32.61	32.64	32.66	32.68	32.70	32.73	32.75	32.77
1450	32.80	32.82	32.84	32.86	32.89	32.91	32.93	32.96	32.98	33.00
1460	33.02	33.05	33.07	33.09	33.12	33.14	33.16	33.18	33.21	33.23
1470	33.25	33.28	33.30	33.32	33.34	33.37	33.39	33.41	33.43	33.46
1480	33.48	33.50	33.53	33.55	33.57	33.59	33.62	33.64	33.66	33.69
1490	33.71	33.73	33.75	33.78	33.80	33.82	33.84	33.87	33.89	33.91
1500	33.93	33.96	33.98	34.00	34.03	34.05	34.07	34.09	34.12	34.14
1510	34.16	34.18	34.21	34.23	34.25	34.28	34.30	34.32	34.34	34.37
1520	34.39	34.41	34.43	34.46	34.48	34.50	34.53	34.55	34.57	34.59
1530	34.62	34.64	34.66	34.68	34.71	34.73	34.75	34.77	34.80	34.82
1540	34.84	34.87	34.89	34.91	34.93	34.96	34.98	35.00	35.02	35.05
1550	35.07	35.09	35.11	35.14	35.16	35.18	35.21	35.23	35.25	35.27
1560	35.29	35.32	35.34	35.36	35.39	35.41	35.43	35.45	35.48	35.50
1570	35.52	35.54	35.57	35.59	35.61	35.63	35.66	35.68	35.70	35.72
1580	35.75	35.77	35.79	35.81	35.84	35.86	35.88	35.90	35.93	35.95
1590	35.97	35.99	36.02	36.04	36.06	36.08	36.11	36.13	36.15	36.17
1600	36.19	36.19	36.22	36.24	36.29	36.31	36.33	36.35	36.37	36.40
1610	36.42	36.44	36.46	36.49	36.51	36.53	36.55	36.58	36.60	36.62
1620	36.64	36.67	36.69	36.71	36.73	36.76	36.78	36.80	36.82	36.84
1630	36.87	36.89	36.91	36.93	36.96	36.98	37.00	37.02	37.05	37.07
1640	37.09	37.11	37.14	37.16	37.18	37.20	37.23	37.25	37.27	37.29
1650	37.31	37.34	37.36	37.38	37.40	37.43	37.45	37.47	37.49	37.52
1660	37.54	37.56	37.58	37.60	37.63	37.65	37.67	37.69	37.72	37.74
1670	37.76	37.78	37.81	37.83	37.85	37.87	37.89	37.92	37.94	37.96
1680	37.98	38.01	38.03	38.05	38.07	38.09	38.12	38.14	38.16	38.18
1690	38.20	38.23	38.25	38.27	38.29	38.32	38.34	38.36	38.38	38.40
1700	38.43	38.45	38.47	38.49	38.51	38.54	38.56	38.58	38.60	38.62
1710	38.65	38.67	38.69	38.71	38.73	38.76	38.78	38.80	38.82	38.84
1720	38.87	38.89	38.91	38.93	38.95	38.98	39.00	39.02	39.04	39.06
1730	39.09	39.11	39.13	39.15	39.17	39.20	39.22	39.24	39.26	39.28
1740	39.31	39.33	39.35	39.37	39.39	39.42	39.44	39.46	39.48	39.50
1750	39.53	39.55	39.57	39.59	39.61	39.64	39.66	39.68	39.70	39.72
1760	39.75	39.77	39.79	39.81	39.83	39.86	39.88	39.69	39.92	39.94
1770	39.96	39.99	40.01	40.03	40.05	40.07	40.10	40.12	40.14	40.16
1780	40.18	40.20	40.23	40.25	40.27	40.29	40.31	40.34	40.36	40.38
1790	40.40	40.42	40.44	40.47	40.49	40.51	40.53	40.55	40.58	40.60
1800	40.62	40.64	40.66	40.68	40.71	40.73	40.75	40.77	40.79	40.82
1810	40.84	40.86	40.88	40.90	40.92	40.95	40.97	40.99	41.01	41.03
1820	41.05	41.08	41.10	41.12	41.14	41.16	41.18	41.21	41.23	41.25
1830	41.27	41.29	41.31	41.34	41.36	41.38	41.40	41.42	41.45	41.47
1840	41.49	41.51	41.53	41.55	41.57	41.60	41.62	41.64	41.66	41.68
1850	41.70	41.73	41.75	41.77	41.79	41.81	41.83	41.85	41.88	41.90
1860	41.92	41.94	41.96	41.99	42.01	42.03	42.05	42.07	42.09	42.11
1870	42.14	42.16	42.18	42.20	42.22	42.24	42.26	42.29	42.31	42.33
1880	42.35	42.37	42.39	42.42	42.44	42.46	42.48	42.50	42.52	42.55
1890	42.57	42.59	42.61	42.63	42.65	42.67	42.69	42.72	42.74	42.76
1900	42.78	42.80	42.82	42.84	42.87	42.89	42.91	42.93	42.95	42.97
1910	42.99	43.01	43.04	43.06	43.08	43.10	43.12	43.14	43.17	43.19

Millivolts

°F	0	1	2	3	4	5	6	7	8	9
1920	43.21	43.23	43.25	43.27	43.29	43.21	43.34	43.36	43.38	43.40
1930	43.42	43.44	43.47	43.49	43.51	43.53	43.55	43.57	43.59	43.61
1940	43.63	43.66	43.68	43.70	43.72	43.75	43.76	43.78	43.81	43.83
1950	43.85	43.87	43.89	43.91	43.93	43.95	43.98	44.00	44.02	44.04
1960	44.06	44.08	44.10	44.13	44.15	44.17	44.19	44.21	44.23	44.25
1970	44.27	44.30	44.32	44.34	44.36	44.38	44.40	44.42	44.44	44.47
1980	44.49	44.51	44.53	44.55	44.57	44.59	44.61	44.63	44.66	44.68
1990	44.70	44.72	44.74	44.76	44.78	44.80	44.82	44.85	44.87	44.89
2000	44.91	44.93	44.95	44.97	44.99	45.01	45.03	45.06	45.08	45.10
2010	45.12	45.14	45.16	45.18	45.20	45.22	45.24	45.27	45.29	45.31
2020	45.33	45.35	45.87	45.39	45.41	45.43	45.45	45.48	45.50	45.52
2030	45.54	45.56	45.58	45.60	45.62	45.64	45.66	45.69	45.71	45.73
2040	45.75	45.77	45.79	45.81	45.83	45.85	45.87	45.90	45.92	45.94
2050	45.96	45.98	46.00	46.02	46.04	46.06	46.08	46.11	46.13	46.15
2060	46.17	46.19	46.21	46.23	46.25	46.27	46.29	46.31	46.33	46.36
2070	46.38	46.40	46.42	46.44	46.46	46.48	46.50	46.52	46.54	46.56
2080	46.58	46.60	46.63	46.65	46.67	46.69	46.71	46.73	46.75	46.77
2090	46.79	46.81	46.83	46.85	46.87	46.90	46.92	46.94	46.96	46.98
2100	47.00	47.02	47.04	47.06	47.08	47.10	47.12	47.14	47.17	47.19
2110	47.21	47.23	47.25	47.27	47.29	47.31	47.33	47.35	47.37	47.39
2120	47.41	47.43	47.45	47.47	47.49	47.52	47.54	47.56	47.58	47.60
2130	47.62	47.64	47.66	47.68	47.70	47.72	47.74	47.76	47.78	47.80
2140	47.82	47.84	47.86	47.89	47.91	47.93	47.95	47.97	47.99	48.01
2150	48.03	48.05	48.07	48.09	48.11	48.13	48.15	48.17	48.19	48.21
2160	48.23	48.25	48.27	48.29	48.32	48.34	48.36	48.38	48.40	48.42
2170	48.44	48.46	48.48	48.50	48.52	48.54	48.56	48.58	48.60	48.62
2180	48.64	48.66	48.68	48.70	48.72	48.74	48.76	48.79	48.81	48.83
2190	48.85	48.87	48.89	48.91	48.93	48.95	48.97	48.99	49.01	49.03
2200	49.05	49.07	49.09	49.11	49.13	49.15	49.17	49.19	49.21	49.23
2210	49.25	49.27	49.29	49.31	49.33	49.35	49.37	49.39	49.41	49.43
2220	49.45	49.47	49.49	49.51	49.53	49.55	49.57	49.59	49.61	49.63
2230	49.65	49.67	49.69	49.71	49.73	49.76	49.78	49.80	49.82	49.84
2240	49.86	49.88	49.90	49.92	49.94	49.96	49.98	50.00	50.02	50.04
2250	50.06	50.08	50.10	50.12	50.14	50.16	50.18	50.20	50.22	50.24
2260	50.26	50.28	50.30	50.32	50.34	50.36	50.38	50.40	50.42	50.44
2270	50.46	50.48	50.50	50.52	50.54	50.56	50.57	50.59	50.61	50.63
2280	50.65	50.67	50.69	50.71	50.73	50.75	50.77	50.79	50.81	50.83
2290	50.85	50.87	50.89	50.91	50.93	50.95	50.97	50.99	51.01	51.03
2300	51.05	51.07	51.09	51.11	51.13	51.15	51.17	51.19	51.21	51.23
2310	51.25	51.27	51.29	51.31	51.33	51.35	51.37	51.39	51.41	51.43
2320	51.45	51.47	51.48	51.50	51.52	51.54	51.56	51.58	51.60	51.62
2330	51.64	51.66	51.68	51.70	51.72	51.74	51.76	51.78	51.80	51.82
2340	51.84	51.86	51.88	51.90	51.92	51.94	51.96	51.98	52.00	52.01
2350	52.03	52.05	52.07	52.09	52.11	52.13	52.15	52.17	52.19	52.21
2360	52.23	52.25	52.27	52.29	52.31	52.33	52.35	52.37	52.39	52.41
2370	52.42	52.44	52.46	52.48	52.50	52.52	52.54	52.56	52.58	52.60
2380	52.62	52.64	52.66	52.68	52.70	52.72	52.74	52.76	52.77	52.79
2390	52.81	52.83	52.85	52.87	52.89	52.91	52.93	52.95	52.97	52.99
2400	53.01	53.03	53.05	53.07	53.08	53.10	53.12	53.14	53.16	53.18
2410	53.20	53.22	53.24	53.26	53.28	53.30	53.32	53.34	53.35	53.37
2420	53.39	53.41	53.43	53.45	53.47	53.49	53.51	53.53	53.55	53.57
2430	53.59	53.60	53.62	53.64	53.66	53.68	53.70	53.72	53.74	53.76
2440	53.78	53.80	53.82	53.83	53.85	53.87	53.89	53.91	53.93	53.95
2450	53.97	53.99	54.01	54.03	54.04	54.06	54.08	54.10	54.12	54.14
2460	54.16	54.18	54.20	54.22	54.24	54.25	54.27	54.29	54.31	54.33
2470	54.35	54.37	54.39	54.41	54.43	54.44	54.46	54.48	54.50	54.52
2480	54.54	54.56	54.58	54.60	54.62	54.63	54.65	54.67	54.69	54.71
2490	54.73	54.75	54.77	54.79	54.81	54.82	54.84	54.86	54.88	54.90
2500	54.92	—	—	—	—	—	—	—	—	—

IRON-CONSTANTAN THERMOCOUPLES (MODIFIED 1913)

(Electromotive Force in Absolute Millivolts. Temperatures in Degrees C (Int. 1948). Reference Junctions at 0°C.)

Millivolts

°C	0	1	2	3	4	5	6	7	8	9
−190	−7.66	−7.69	−7.71	−7.73	−7.76	−7.78				
−180	−7.40	−7.43	−7.46	−7.49	−7.51	−7.54	−7.56	−7.59	−7.61	−7.64
−170	−7.12	−7.15	−7.18	−7.21	−7.24	−7.27	−7.30	−7.32	−7.35	−7.38
−160	−6.82	−6.85	−6.88	−6.91	−6.94	−6.97	−7.00	−7.03	−7.06	−7.09
−150	−6.50	−6.53	−5.56	−6.60	−6.63	−6.66	−6.69	−6.72	−6.76	−6.79
−140	−6.16	−6.19	−6.22	−6.26	−6.29	−6.33	−6.36	−6.40	−6.43	−6.46
−130	−5.80	−5.84	−5.87	−5.91	−5.94	−5.98	−6.01	−6.05	−6.08	−6.12
−120	−5.42	−5.46	−5.50	−5.54	−5.58	−5.61	−5.65	−5.69	−5.72	−5.76
−110	−5.03	−5.07	−5.11	−5.15	−5.19	−5.23	−5.27	−5.31	−5.35	−5.38
−100	−4.63	−4.67	−4.71	−4.75	−4.79	−4.83	−4.87	−4.91	−4.95	−4.99
−90	−4.21	−4.25	−4.30	−4.34	−4.38	−4.42	−4.46	−4.50	−4.55	−4.59
−80	−3.78	−3.82	−3.87	−3.91	−3.96	−4.00	−4.04	−4.08	−4.13	−4.17
−70	−3.34	−3.38	−3.43	−3.47	−3.52	−3.56	−3.60	−3.65	−3.69	−3.74
−60	−2.89	−2.94	−2.98	−3.03	−3.07	3.12	−3.16	−3.21	−3.25	−3.30
−50	−2.43	−2.48	−2.52	−2.57	−2.62	−2.66	−2.71	−2.75	−2.80	−2.84

Millivolts

°C	0	1	2	3	4	5	6	7	8	9
−40	−1.96	−2.01	−2.06	−2.10	−2.15	−2.20	−2.24	−2.29	−2.34	−2.38
−30	−1.48	−1.53	−1.58	−1.63	−1.67	−1.72	−1.77	−1.82	−1.87	−1.91
−20	−1.00	−1.04	−1.09	−1.14	−1.19	−1.24	−1.29	−1.34	−1.39	−1.43
−10	−0.50	−0.55	−0.60	−0.65	−0.70	−0.75	−0.80	−0.85	−0.90	−0.95
(−)0	0.00	−0.05	−0.10	−0.15	−0.20	−0.25	−0.30	−0.35	−0.40	−0.45
(+)0	0.00	0.05	0.10	0.15	0.20	0.25	0.30	0.35	0.40	0.45
10	0.50	0.56	0.61	0.66	0.71	0.76	0.81	0.86	0.91	0.97
20	1.02	1.07	1.12	1.17	1.22	1.28	1.33	1.38	1.43	1.48
30	1.54	1.59	1.64	1.69	1.74	1.80	1.85	1.90	1.95	2.00
40	2.06	2.11	2.16	2.22	2.27	2.32	2.37	2.42	2.48	2.53
50	2.58	2.64	2.69	2.74	2.80	2.85	2.90	2.96	3.01	3.06
60	3.11	3.17	3.22	3.27	3.33	3.38	3.43	3.49	3.54	3.60
70	3.65	3.70	3.76	3.81	3.86	3.92	3.97	4.02	4.08	4.13
80	4.19	4.24	4.29	4.35	4.40	4.46	4.51	4.56	4.62	4.67
90	4.73	4.78	4.83	4.89	4.94	5.00	5.05	5.10	5.16	5.21

IRON-CONSTANTAN THERMOCOUPLES (MODIFIED 1913) (Continued)

Millivolts

°C	0	1	2	3	4	5	6	7	8	9
100	5.27	5.32	5.38	5.43	5.48	5.54	5.59	5.65	5.70	5.76
110	5.81	5.86	5.92	5.97	6.03	6.08	6.14	6.19	6.25	6.30
120	6.36	6.41	6.47	6.52	6.58	6.63	6.68	6.74	6.79	6.85
130	6.90	6.96	7.01	7.07	7.12	7.18	7.23	7.29	7.34	7.40
140	7.45	7.51	7.56	7.62	7.67	7.73	7.78	7.84	7.89	7.95
150	8.00	8.06	8.12	8.17	8.23	8.28	8.34	8.39	8.45	8.50
160	8.56	8.61	8.67	8.72	8.78	8.84	8.89	8.95	9.00	9.06
170	9.11	9.17	9.22	9.28	9.33	9.39	9.44	9.50	9.56	9.61
180	9.67	9.72	9.78	9.83	9.89	9.95	10.00	10.06	10.11	10.17
190	10.22	10.28	10.34	10.39	10.45	10.50	10.56	10.61	10.67	10.72
200	10.78	10.84	10.89	10.95	11.00	11.06	11.12	11.17	11.23	11.28
210	11.34	11.39	11.45	11.50	11.56	11.62	11.67	11.73	11.78	11.84
220	11.89	11.95	12.00	12.06	12.12	12.17	12.23	12.28	12.34	12.39
230	12.45	12.50	12.56	12.62	12.67	12.73	12.78	12.84	12.89	12.95
240	13.01	13.06	13.12	13.17	13.23	13.28	13.34	13.40	13.45	13.51
250	13.56	13.62	13.67	13.73	13.78	13.84	13.89	13.95	14.00	14.06
260	14.12	14.17	14.23	14.28	14.34	14.39	14.45	14.50	14.56	14.61
270	14.67	14.72	14.78	14.83	14.89	14.94	15.00	15.06	15.11	15.17
280	15.22	15.28	15.33	15.39	15.44	15.50	15.55	15.61	15.66	15.72
290	15.77	15.83	15.88	15.94	16.00	16.05	16.11	16.16	16.22	16.27
300	16.33	16.38	16.44	16.49	16.55	16.60	16.66	16.71	16.77	16.82
310	16.88	16.93	16.99	17.04	17.10	17.15	17.21	17.26	17.32	17.37
320	17.43	17.48	17.54	17.60	17.65	17.71	17.76	17.82	17.87	17.93
330	17.98	18.04	18.09	18.15	18.20	18.26	18.32	18.37	18.43	18.48
340	18.54	18.59	18.65	18.70	18.76	18.81	18.87	18.92	18.98	19.03
350	19.09	19.14	19.20	19.26	19.31	19.37	19.42	19.48	19.53	19.59
360	19.64	19.70	19.75	19.81	19.86	19.92	19.97	20.03	20.08	20.14
370	20.20	20.25	20.31	20.36	20.42	20.47	20.53	20.58	20.64	20.69
380	20.75	20.80	20.86	20.91	20.97	21.02	21.08	21.13	21.19	21.24
390	21.30	21.35	21.41	21.46	21.52	21.57	21.63	21.68	21.74	21.79
400	21.85	21.90	21.96	22.02	22.07	22.13	22.18	22.24	22.29	22.35
410	22.40	22.46	22.51	22.57	22.62	22.68	22.73	22.79	22.84	22.90
420	22.95	23.01	23.06	23.12	23.17	23.23	23.28	23.34	23.39	23.45
430	23.50	23.56	23.61	23.67	23.72	23.78	23.83	23.89	23.94	24.00

Millivolts

°C	0	1	2	3	4	5	6	7	8	9
440	24.06	24.11	24.17	24.22	24.28	24.33	24.39	24.44	24.50	24.55
450	24.61	24.66	24.72	24.77	24.83	24.88	24.94	25.00	25.05	25.11
460	25.16	25.22	25.27	25.33	25.38	25.44	25.49	25.55	25.60	25.66
470	25.72	25.77	25.83	25.88	25.94	25.99	26.05	26.10	26.16	26.22
480	26.27	26.33	26.38	26.44	26.49	26.55	26.61	26.66	26.72	26.77
490	26.83	26.89	26.94	27.00	27.05	27.11	27.17	27.22	27.28	27.33
500	27.39	27.45	27.50	27.56	27.61	27.67	27.73	27.78	27.84	27.90
510	27.95	28.01	28.07	28.12	28.18	28.23	28.29	28.35	28.40	28.46
520	28.52	28.57	28.63	28.69	28.74	28.80	28.86	28.91	28.97	29.02
530	29.08	29.14	29.20	29.25	29.31	29.37	29.42	29.48	29.54	29.59
540	29.65	29.71	29.76	29.82	29.88	29.94	29.99	30.05	30.11	30.16
550	30.22	30.28	30.34	30.39	30.45	30.51	30.57	30.62	30.68	30.74
560	30.80	30.85	30.91	30.97	31.02	31.08	31.14	31.20	31.26	31.31
570	31.37	31.43	31.49	31.54	31.60	31.66	31.72	31.78	31.83	31.89
580	31.95	32.01	32.06	32.12	32.18	32.24	32.30	32.36	32.41	32.47
590	32.53	32.59	32.65	32.71	32.76	32.82	32.88	32.94	33.00	33.06
600	33.11	33.17	33.23	33.29	33.35	33.41	33.46	33.52	33.58	33.64
610	33.70	33.76	33.82	33.88	33.94	33.99	34.05	34.11	34.17	34.23
620	34.29	34.35	34.41	34.47	34.53	34.58	34.64	34.70	34.76	34.82
630	34.88	34.94	35.00	35.06	35.12	35.18	35.24	35.30	35.36	35.42
640	35.48	35.54	35.60	36.66	35.72	35.78	35.84	35.90	35.96	36.02
650	36.08	36.14	36.20	36.26	36.32	36.38	36.44	36.50	36.56	36.62
660	36.69	36.75	36.81	36.87	36.93	36.99	37.05	37.11	37.18	37.24
670	37.30	37.36	37.42	37.48	37.54	37.60	37.66	37.73	37.79	37.85
680	37.91	37.97	38.04	38.10	38.16	38.22	38.28	38.34	38.41	38.47
690	38.53	38.59	38.66	38.72	38.78	38.84	38.90	38.97	39.03	39.09
700	39.15	39.22	39.28	39.34	39.40	39.47	39.53	39.59	39.65	39.72
710	39.78	39.84	39.91	39.97	40.03	40.10	40.16	40.22	40.28	40.35
720	40.41	40.48	40.54	40.60	40.66	40.73	40.79	40.86	40.92	40.98
730	41.05	41.11	41.17	41.24	41.30	41.36	41.43	41.49	41.56	41.62
740	41.68	41.75	41.81	41.87	41.94	42.00	42.07	42.13	42.19	42.26
750	42.32	42.38	42.45	42.51	42.58	42.64	42.70	42.77	42.83	42.90
760	42.96	—	—	—	—	—	—	—	—	—

(Electromotive Force in Absolute Millivolts. Temperatures in Degrees F* Reference Junctions at 32°F.)

Millivolts

°F	0	1	2	3	4	5	6	7	8	9
−310	−7.66	−7.68	−7.69	−7.70	−7.71	−7.73	−7.74	−7.75	−7.76	−7.78
−300	−7.52	−7.54	−7.55	−7.57	−7.58	−7.59	−7.61	−7.62	−7.64	−7.65
−290	−7.38	−7.39	−7.40	−7.42	−7.44	−7.45	−7.46	−7.48	−7.49	−7.51
−280	−7.22	−7.24	−7.25	−7.27	−7.28	−7.30	−7.31	−7.33	−7.34	−7.36
−270	−7.06	−7.07	−7.09	−7.11	−7.12	−7.14	−7.15	−7.17	−7.19	−7.20
−260	−6.89	−6.90	−6.92	−6.94	−6.96	−6.97	−6.99	−7.01	−7.02	−7.04
−250	−6.71	−6.73	−6.75	−6.77	−6.78	−6.80	−6.82	−6.84	−6.85	−6.87
−240	−6.53	−6.55	−6.57	−6.59	−6.61	−6.62	−6.64	−6.66	−6.68	−6.70
−230	−6.35	−6.37	−6.38	−6.40	−6.42	−6.44	−6.46	−6.48	−6.50	−6.52
−220	−6.16	−6.18	−6.19	−6.21	−6.23	−6.25	−6.27	−6.29	−6.31	−6.33
−210	−5.96	−5.98	−6.00	−6.02	−6.04	−6.06	−6.08	−6.10	−6.12	−6.14
−200	−5.76	−5.78	−5.80	−5.82	−5.84	−5.86	−5.88	−5.90	−5.92	−5.94
−190	−5.55	−5.57	−5.59	−5.61	−5.63	−5.65	−5.67	−5.70	−5.72	−5.74
−180	−5.34	−5.36	−5.38	−5.40	−5.42	−5.44	−5.46	−5.49	−5.51	−5.53
−170	−5.12	−5.14	−5.16	−5.19	−5.21	−5.23	−5.25	−5.27	−5.30	−5.32
−160	−4.90	−4.92	−4.94	−4.97	−4.99	−5.01	−5.03	−5.06	−5.08	−5.10
−150	−4.68	−4.70	−4.72	−4.74	−4.76	−4.79	−4.81	−4.83	−4.86	−4.88
−140	−4.44	−4.47	−4.49	−4.51	−4.54	−4.56	−4.58	−4.61	−4.63	−4.65
−130	−4.21	−4.23	−4.26	−4.28	−4.30	−4.33	−4.35	−4.38	−4.40	−4.42
−120	−3.97	−4.00	−4.02	−4.04	−4.07	−4.09	−4.12	−4.14	−4.16	−4.19
−110	−3.73	3.76	−3.78	−3.81	−3.83	−3.85	−3.88	−3.90	−3.93	−3.95
−100	−3.49	−3.51	−3.54	−3.56	−3.59	−3.61	−3.64	−3.66	−3.68	−3.71
−90	−3.24	−3.27	−3.29	−3.32	−3.34	−3.36	−3.39	−3.41	−3.44	−3.46
−80	−2.99	−3.02	−3.04	−3.07	−3.09	−3.12	−3.14	−3.17	−3.19	−3.22
−70	−2.74	−2.76	−2.79	−2.81	−2.84	−2.86	−2.89	−2.92	−2.94	−2.97
−60	−2.48	−2.51	−2.53	−2.56	−2.58	−2.61	−2.64	−2.66	−2.69	−2.71
−50	−2.22	−2.25	−2.27	−2.30	−2.33	−2.35	−2.38	−2.40	−2.43	−2.46
−40	−1.96	−1.99	−2.01	−2.04	−2.06	−2.09	−2.12	−2.14	−2.17	−2.20
−30	−1.70	−1.72	−1.75	−1.78	−1.80	−1.83	−1.86	−1.88	−1.91	−1.94
−20	−1.43	−1.46	−1.48	−1.51	−1.54	−1.56	−1.59	−1.62	−1.64	−1.67
−10	−1.16	−1.19	−1.21	−1.24	−1.27	−1.29	−1.32	−1.35	−1.38	−1.40
(−)0	−0.89	−0.91	−0.94	−0.97	−1.00	−1.02	−1.05	−1.08	−1.10	−1.13
+(0)	−0.89	−0.86	−0.83	−0.80	−0.78	−0.75	−0.72	−0.70	−0.67	−0.64
10	−0.61	−0.58	−0.56	−0.53	−0.50	−0.48	−0.45	−0.42	−0.39	−0.36
20	−0.34	−0.31	−0.28	−0.25	−0.22	−0.20	−0.17	−0.14	−0.11	−0.09
30	−0.06	−0.03	0.00	0.03	0.05	0.08	0.11	0.14	0.17	0.19
40	0.22	0.25	0.28	0.31	0.34	0.36	0.39	0.42	0.45	0.48
50	0.50	0.53	0.56	0.59	0.62	0.65	0.67	0.70	0.73	0.76
60	0.79	0.82	0.84	0.87	0.90	0.93	0.96	0.99	1.02	1.04
70	1.07	1.10	1.13	1.16	1.19	1.22	1.25	1.28	1.30	1.33
80	1.36	1.39	1.42	1.45	1.48	1.51	1.54	1.56	1.59	1.62

Millivolts

°F	0	1	2	3	4	5	6	7	8	9
90	1.65	1.68	1.71	1.74	1.77	1.80	1.83	1.85	1.88	1.91
100	1.94	1.97	2.00	2.03	2.06	2.09	2.12	2.14	2.17	2.20
110	2.23	2.26	2.29	2.32	2.35	2.38	2.41	2.44	2.47	2.50
120	2.52	2.55	2.58	2.61	2.64	2.67	2.70	2.73	2.76	2.79
130	2.82	2.85	2.88	2.91	2.94	2.97	3.00	3.03	3.06	3.08
140	3.11	3.14	3.17	3.20	3.23	3.26	3.29	3.32	3.35	3.38
150	3.41	3.44	3.47	3.50	3.53	3.56	3.59	3.62	3.65	3.68
160	3.71	3.74	3.77	3.80	3.83	3.86	3.89	3.92	3.95	3.98
170	4.01	4.04	4.07	4.10	4.13	4.16	4.19	4.22	4.25	4.28
180	4.31	4.34	4.37	4.40	4.43	4.46	4.49	4.52	4.55	4.58
190	4.61	4.64	4.67	4.70	4.73	4.76	4.79	4.82	4.85	4.88
200	4.91	4.94	4.97	5.00	5.03	5.06	5.09	5.12	5.15	5.18
210	5.21	5.24	5.27	5.30	5.33	5.36	5.39	5.42	5.45	5.48
220	5.51	5.54	5.57	5.60	5.63	5.66	5.69	5.72	5.75	5.78
230	5.81	5.84	5.87	5.90	5.93	5.96	5.99	6.02	6.05	6.08
240	6.11	6.14	6.17	6.20	6.24	6.27	6.30	6.33	6.36	6.39
250	6.42	6.45	6.48	6.51	6.54	6.57	6.60	6.63	6.66	6.69
260	6.72	6.75	6.78	6.81	6.84	6.87	6.90	6.93	6.96	7.00
270	7.03	7.06	7.09	7.12	7.15	7.18	7.21	7.24	7.27	7.30
280	7.33	7.36	7.39	7.42	7.45	7.48	7.51	7.54	7.58	7.61
290	7.64	7.67	7.70	7.73	7.76	7.79	7.82	7.85	7.88	7.91
300	7.94	7.97	8.00	8.04	8.07	8.10	8.13	8.16	8.19	8.22
310	8.25	8.28	8.31	8.34	8.37	8.40	8.44	8.47	8.50	8.53
320	8.56	8.59	8.62	8.65	8.68	8.71	8.74	8.77	8.80	8.84
330	8.87	8.90	8.93	8.96	8.99	9.02	9.05	9.08	9.11	9.14
340	9.17	9.20	9.24	9.27	9.30	9.33	9.36	9.39	9.42	9.45
350	9.48	9.51	9.54	9.58	9.61	9.64	9.67	9.70	9.73	9.76
360	9.79	9.82	9.85	9.88	9.92	9.95	9.98	10.01	10.04	10.07
370	10.10	10.13	10.16	10.19	10.22	10.25	10.28	10.32	10.35	10.38
380	10.41	10.44	10.47	10.50	10.53	10.56	10.60	10.63	10.66	10.69
390	10.72	10.75	10.78	10.81	10.84	10.87	10.90	10.94	10.97	11.00
400	11.03	11.06	11.00	11.12	11.15	11.18	11.21	11.24	11.28	11.31
410	11.34	11.37	11.40	11.43	11.46	11.49	11.52	11.55	11.58	11.62
420	11.65	11.68	11.71	11.74	11.77	11.80	11.83	11.86	11.89	11.92
430	11.96	11.99	12.02	12.05	12.08	12.11	12.14	12.17	12.20	12.23
440	12.26	12.30	12.33	12.36	12.39	12.42	12.45	12.48	12.51	12.54
450	12.57	12.60	12.64	12.67	12.70	12.73	12.76	12.79	12.82	12.85
460	12.88	12.91	12.94	12.98	13.01	13.04	13.07	13.10	13.13	13.16
470	13.19	13.22	13.25	13.28	13.31	13.34	13.38	13.41	13.44	13.47
480	13.50	13.53	13.56	13.59	13.62	13.65	13.68	13.72	13.75	13.78
490	13.81	13.84	13.87	13.90	13.93	13.96	13.99	14.02	14.05	14.08

* Based on the International Temperature Scale of 1948.

IRON-CONSTANTAN THERMOCOUPLES (MODIFIED 1913) (Continued)

°F	0	1	2	3	4	5	6	7	8	9
	Millivolts									
500	14.12	14.15	14.18	14.21	14.24	14.27	14.30	14.33	14.36	14.39
510	14.42	14.45	14.48	14.52	14.55	14.58	14.61	14.64	14.67	14.70
520	14.73	14.76	14.79	14.82	14.85	14.88	14.91	14.94	14.98	15.01
530	15.04	15.07	15.10	15.13	15.16	15.19	15.22	15.25	15.28	15.31
540	15.34	15.37	15.40	15.44	15.47	15.50	15.53	15.56	15.59	15.62
550	15.65	15.68	15.71	15.74	15.77	15.80	15.84	15.87	15.90	15.93
560	15.96	15.99	16.02	16.05	16.08	16.11	16.14	16.17	16.20	16.23
570	16.26	16.30	16.33	16.36	16.39	16.42	16.45	16.48	16.51	16.54
580	16.57	16.60	16.63	16.66	16.69	16.72	16.75	16.78	16.82	16.85
590	16.88	16.91	16.94	16.97	17.00	17.03	17.06	17.09	17.12	17.15
600	17.18	17.21	17.24	17.28	17.31	17.34	17.37	17.40	17.43	17.46
610	17.49	17.52	17.55	17.58	17.61	17.64	17.68	17.71	17.74	17.77
620	17.80	17.83	17.86	17.89	17.92	17.95	17.98	18.01	18.04	18.08
630	18.11	18.14	18.17	18.20	18.23	18.26	18.29	18.32	18.35	18.38
640	18.41	18.44	18.47	18.50	18.54	18.57	18.60	18.63	18.66	18.69
650	18.72	18.75	18.78	18.81	18.84	18.87	18.90	18.94	18.97	19.00
660	19.03	19.06	19.09	19.12	19.15	19.18	19.21	19.24	19.27	19.30
670	19.34	19.37	19.40	19.43	19.46	19.49	19.52	19.55	19.58	19.61
680	19.64	19.67	19.70	19.74	19.77	19.80	19.83	19.86	19.89	19.92
690	19.95	19.98	20.01	20.04	20.07	20.10	20.13	20.16	20.20	20.23
700	20.26	20.29	20.32	20.35	20.38	20.41	20.44	20.47	20.50	20.53
710	20.56	20.59	20.62	20.66	20.69	20.72	20.75	20.78	20.81	20.84
720	20.87	20.90	20.93	20.96	20.99	21.02	21.05	21.08	21.11	21.14
730	21.18	21.21	21.24	21.27	21.30	21.33	21.36	21.39	21.42	21.45
740	21.48	21.51	21.54	21.57	21.60	21.64	21.67	21.70	21.73	21.76
750	21.79	21.82	21.85	21.88	21.91	21.94	21.97	22.00	22.03	22.06
760	22.10	22.13	22.16	22.19	22.22	22.25	22.28	22.31	22.34	22.37
770	22.40	22.43	22.46	22.49	22.52	22.55	22.58	22.62	22.65	22.68
780	22.71	22.74	22.77	22.80	22.83	22.86	22.89	22.92	22.95	22.98
790	23.01	23.04	23.08	23.11	23.14	23.17	23.20	23.23	23.26	23.29
800	23.32	23.35	23.38	23.41	23.44	23.47	23.50	23.53	23.56	23.60
810	23.63	23.66	23.69	23.72	23.75	23.78	23.81	23.84	23.87	23.90
820	23.93	23.96	23.99	24.02	24.06	24.09	24.12	24.15	24.18	24.21
830	24.24	24.27	24.30	24.33	24.36	24.39	24.42	24.45	24.48	24.52
840	24.55	24.58	24.61	24.64	24.67	24.70	24.73	24.76	24.79	24.82
850	24.85	24.88	24.91	24.94	24.98	25.01	25.04	25.07	25.10	25.13
860	25.16	25.19	25.22	25.25	25.28	25.32	25.35	25.38	25.41	25.44
870	25.47	25.50	25.53	25.56	25.59	25.62	25.65	25.68	25.72	25.75
880	25.78	25.81	25.84	25.87	25.90	25.93	25.96	25.99	26.02	26.06
890	26.09	26.12	26.15	26.18	26.21	26.24	26.27	26.30	26.33	26.36
900	26.40	26.43	26.46	26.49	26.52	26.55	26.58	26.61	26.64	26.67
910	26.70	26.74	26.77	26.80	26.83	26.86	26.89	26.92	26.95	26.98
920	27.02	27.05	27.08	27.11	27.14	27.17	27.20	27.23	27.26	27.30
930	27.33	27.36	27.39	27.42	27.45	27.48	27.51	27.54	27.58	27.61
940	27.64	27.67	27.70	27.73	27.76	27.80	27.83	27.86	27.89	27.92
950	27.95	27.98	28.02	28.05	28.08	28.11	28.14	28.17	28.20	28.23

°F	0	1	2	3	4	5	6	7	8	9
	Millivolts									
960	28.26	28.30	28.33	28.36	28.39	28.42	28.45	28.48	28.52	28.55
970	28.58	28.61	28.64	28.67	28.70	28.74	28.77	28.80	28.83	28.86
980	28.89	28.92	28.96	28.99	29.02	29.05	29.08	29.11	29.14	29.18
990	29.21	29.24	29.27	29.30	29.33	29.37	29.40	29.43	29.46	29.49
1000	29.52	29.56	29.59	29.62	29.65	29.68	29.71	29.75	29.78	29.81
1010	29.84	29.87	29.90	29.94	29.97	30.00	30.03	30.06	30.10	30.13
1020	30.16	30.19	30.22	30.25	30.28	30.32	30.35	30.38	30.41	30.44
1030	30.48	30.51	30.54	30.57	30.60	30.64	30.67	30.70	30.73	30.76
1040	30.80	30.83	30.86	30.89	30.92	30.96	30.99	31.02	31.05	31.08
1050	31.12	31.15	31.18	31.21	31.24	31.28	31.31	31.34	31.37	31.40
1060	31.44	31.47	31.50	31.53	31.56	31.60	31.63	31.66	31.69	31.72
1070	31.76	31.79	31.82	31.85	31.88	31.92	31.95	31.98	32.01	32.05
1080	32.08	32.11	32.14	32.18	32.21	32.24	32.27	32.30	32.34	32.37
1090	32.40	32.43	32.47	32.50	32.53	32.56	32.60	32.63	32.66	32.69
1100	32.72	32.76	32.79	32.82	32.86	32.89	32.92	32.95	32.98	33.02
1110	33.05	33.08	33.11	33.15	33.18	33.21	33.24	33.28	33.31	33.34
1120	33.37	33.41	33.44	33.47	33.50	33.54	33.57	33.60	33.64	33.67
1130	33.70	33.73	33.76	33.80	33.83	33.86	33.90	33.93	33.96	33.99
1140	34.03	34.06	34.09	34.12	34.16	34.19	34.22	34.26	34.29	34.32
1150	34.36	34.39	34.42	34.45	34.49	34.52	34.55	34.58	34.62	34.65
1160	34.68	34.72	34.75	34.78	34.82	34.85	34.88	34.92	34.95	34.98
1170	35.01	35.05	35.08	35.11	35.15	35.18	35.21	35.25	35.28	35.31
1180	35.35	35.38	35.41	35.45	35.48	35.51	35.54	35.58	35.61	35.64
1190	35.68	35.71	35.74	35.78	35.81	35.84	35.88	35.91	35.94	35.98
1200	36.01	36.05	36.08	36.11	36.15	36.18	36.21	36.25	36.28	36.31
1210	36.35	36.38	36.42	36.45	36.48	36.52	36.55	36.58	36.62	36.65
1220	36.69	36.72	36.75	36.79	36.82	36.86	36.89	36.92	36.96	36.99
1230	37.02	37.06	37.09	37.13	37.16	37.20	37.23	37.26	37.30	37.33
1240	37.36	37.40	37.43	37.47	37.50	37.54	37.57	37.60	37.64	37.67
1250	37.71	37.74	37.78	37.81	37.84	37.88	37.91	37.95	37.98	38.02
1260	38.05	38.08	38.12	38.15	38.19	38.22	38.26	38.29	38.32	38.36
1270	38.39	38.43	38.46	38.50	38.53	38.57	38.60	38.64	38.67	38.70
1280	38.74	38.77	38.81	38.84	38.88	38.91	38.95	38.98	39.02	39.05
1290	39.08	39.12	39.15	39.19	39.22	39.26	39.29	39.33	39.36	39.40
1300	39.43	39.47	39.50	39.54	39.57	39.61	39.64	39.68	39.71	39.75
1310	39.78	39.82	39.85	39.89	39.92	39.96	39.99	40.03	40.06	40.10
1320	40.13	40.17	40.20	40.24	40.27	40.31	40.34	40.38	40.41	40.45
1330	40.48	40.52	40.55	40.59	40.62	40.66	40.69	40.73	40.76	40.80
1340	40.83	40.87	40.90	40.94	40.98	41.01	41.05	41.08	41.12	41.15
1350	41.19	41.22	41.26	41.29	41.33	41.36	41.40	41.43	41.47	41.50
1360	41.54	41.58	41.61	41.65	41.68	41.72	41.75	41.79	41.82	41.86
1370	41.90	41.93	41.97	42.00	42.04	42.07	42.11	42.14	42.18	42.22
1380	42.25	42.29	42.32	42.36	42.39	42.43	42.46	42.50	42.53	42.57
1390	42.61	42.64	42.68	42.71	42.75	42.78	42.82	42.85	42.89	42.92
1400	42.96	—	—	—	—	—	—	—	—	—

TEMPERATURE-E. M. F. VALUES FOR COPPER-CONSTANTAN

E. M. F. values are in millivolts; reference junctions at 0°C.; temperatures are in degrees C.
Roeser and Wensel, National Bureau of Standards

°C	0	10	20	30	40	50	60	70	80	90
	Millivolts									
−200	−5.54									
−100	−3.35	−3.62	−3.89	−4.14	−4.38	−4.60	−4.82	−5.02	−5.20	−5.38
0	0	−0.38	−0.75	−1.11	−1.47	−1.81	−2.14	−2.46	−2.77	−3.06
0	0	0.39	0.79	1.19	1.61	2.03	2.47	2.91	3.36	3.81
100	4.28	4.75	5.23	5.71	6.20	6.70	7.21	7.72	8.23	8.76
200	9.29	9.82	10.36	10.91	11.46	12.01	12.57	13.14	13.71	14.28
300	14.86	15.44	16.03	16.62	17.22	17.82	18.42	19.03	19.64	20.25
400	20.87									

E. M. F. values are in millivolts; reference junctions at 32°F.; temperatures are in degrees F.
Roeser and Wensel, National Bureau of Standards

°F	0	10	20	30	40	50	60	70	80	90
	Millivolts									
−300	−5.28									
−200	−4.11	−4.25	−4.38	−4.50	−4.63	−4.75	−4.86	−4.97	−5.08	−5.18
−100	−2.56	−2.73	−2.90	−3.06	−3.22	−3.38	−3.53	−3.68	−3.83	−3.97
0	−0.67	−0.87	−1.07	−1.27	−1.47	−1.66	−1.84	−2.03	−2.21	−2.39
0	−0.67	−0.46	−0.26	−0.04	+0.17	0.39	0.61	0.83	1.06	1.29
100	1.52	1.75	1.99	2.23	2.47	2.71	2.96	3.21	3.46	3.71
200	3.97	4.22	4.48	4.75	5.01	5.28	5.55	5.82	6.09	6.37
300	6.64	6.92	7.21	7.49	7.77	8.06	8.35	8.64	8.93	9.23
400	9.52	9.82	10.12	10.42	10.72	11.03	11.33	11.64	11.95	12.26
500	12.57	12.89	13.20	13.52	13.83	14.15	14.47	14.79	15.12	15.44
600	15.77	16.10	16.42	16.75	17.08	17.42	17.75	18.08	18.42	18.75
700	19.09	19.43	19.77	20.11	20.45	20.80				

REFERENCE TABLE FOR Pt TO Pt—10 PER CENT Rh THERMOCOUPLE

Emfs are expressed in microvolts and temperatures in °C. Cold junctions at 0°C. Roeser and Wensel, National Bureau of Standards

E(μv)	0	1,000	2,000	3,000	4,000	5,000	6,000	7,000	8,000	9,000	10,000	11,000	12,000	13,000	14,000	15,000	16,000	17,000
0	0	146.9	265.0	373.7	477.7	578.1	675.3	769.5	861.0	950.2	1,037.2	1,122.3	1,206.4	1,290.0	1,373.8	1,458.0	1,542.6	1,627.8
	17.7	12.5	11.2	10.5	10.2	9.9	9.5	9.3	9.0	8.8	8.6	8.5	8.3	8.3	8.4	8.4	8.5	8.6
100	17.7	159.4	276.2	384.2	487.9	588.0	684.8	778.8	870.0	959.0	1,045.8	1,130.8	1,214.7	1,298.3	1,382.2	1,466.4	1,551.1	1,636.4
	16.7	12.3	11.1	10.5	10.2	9.8	9.5	9.2	9.0	8.8	8.6	8.4	8.4	8.4	8.4	8.4	8.5	8.5
200	34.4	171.7	287.3	394.7	498.1	597.8	694.3	788.0	879.0	967.8	1,054.4	1,139.2	1,223.1	1,306.7	1,390.6	1,474.8	1,559.6	1,644.9
	15.8	12.1	11.0	10.5	10.1	9.8	9.5	9.2	9.0	8.7	8.5	8.4	8.3	8.4	8.4	8.5	8.5	8.6
300	50.2	183.8	298.3	405.2	508.2	607.6	703.8	797.2	888.0	976.5	1,062.9	1,147.6	1,231.4	1,315.1	1,399.0	1,483.3	1,568.1	1,653.5
	15.2	12.0	11.0	10.5	10.1	9.8	9.5	9.2	9.0	8.8	8.6	8.4	8.4	8.4	8.4	8.5	8.5	8.6
400	65.4	195.8	309.3	415.7	518.3	617.4	713.3	806.4	897.0	985.3	1,071.5	1,156.0	1,239.8	1,323.5	1,407.4	1,491.8	1,576.6	1,662.1
	14.6	11.8	10.9	10.4	10.1	9.7	9.4	9.2	8.9	8.7	8.5	8.4	8.4	8.3	8.4	8.4	8.5	8.6
500	80.0	207.6	320.2	426.1	528.4	627.1	722.7	815.6	905.9	994.0	1,080.0	1,164.4	1,248.2	1,331.8	1,415.8	1,500.2	1,585.1	1,670.7
	14.1	11.7	10.8	10.4	10.0	9.7	9.4	9.1	8.9	8.7	8.5	8.4	8.3	8.4	8.4	8.5	8.6	8.6
600	94.1	219.3	331.0	436.5	538.4	636.8	732.1	824.7	914.8	1,002.7	1,088.5	1,172.8	1,256.5	1,340.2	1,424.2	1,508.7	1,593.7	1,679.3
	13.7	11.6	10.7	10.3	10.0	9.7	9.4	9.1	8.9	8.6	8.5	8.4	8.5	8.4	8.5	8.5	8.5	8.6
700	107.8	230.9	341.7	446.8	548.4	646.5	741.5	833.8	923.7	1,011.3	1,097.0	1,181.2	1,264.9	1,348.6	1,432.7	1,517.2	1,602.2	1,687.9
	13.3	11.5	10.7	10.3	9.9	9.6	9.4	9.1	8.9	8.7	8.4	8.4	8.3	8.4	8.4	8.4	8.5	8.6
800	121.1	242.4	352.4	457.1	558.3	656.1	750.9	842.9	932.6	1,020.0	1,105.4	1,189.6	1,273.2	1,357.0	1,441.1	1,525.6	1,610.7	1,696.5
	13.0	11.3	10.7	10.3	9.9	9.6	9.3	9.1	8.8	8.6	8.5	8.4	8.4	8.4	8.4	8.5	8.6	8.6
900	134.1	253.7	363.1	467.4	568.2	665.7	760.2	852.0	941.4	1,028.6	1,113.9	1,198.0	1,281.6	1,365.4	1,449.5	1,534.1	1,619.3	1,705.1
	12.8	11.3	10.6	10.3	9.9	9.6	9.3	9.0	8.8	8.6	8.4	8.4	8.4	8.4	8.5	8.5	8.5	8.6
1,000	146.9	265.0	373.7	477.7	578.1	675.3	769.5	861.0	950.2	1,037.2	1,122.3	1,206.4	1,290.0	1,373.8	1,458.0	1,542.6	1,627.8	1,713.7

Emfs are expressed in microvolts and temperatures in °C. Cold junctions at 0°C. Roeser and Wensel, National Bureau of Standards

E(μv)	0	1,000	2,000	3,000	4,000	5,000	6,000	7,000	8,000	9,000	10,000	11,000	12,000	13,000	14,000	15,000	16,000	17,000
0	32.0	296.4	509.0	704.7	891.9	1,072.6	1,247.5	1,417.1	1,581.8	1,742.4	1,899.0	2,052.1	2,203.5	2,354.0	2,504.8	2,656.4	2,808.7	2,962.0
	31.9	22.5	20.1	19.0	18.4	17.7	17.1	16.7	16.2	15.8	15.5	15.2	15.0	15.0	15.2	15.1	15.3	15.4
100	63.9	318.9	529.1	723.7	910.3	1,090.3	1,264.6	1,433.8	1,598.0	1,758.2	1,914.5	2,067.3	2,218.5	2,369.0	2,520.0	2,671.5	2,824.0	2,977.4
	30.0	22.1	20.0	18.9	18.3	17.7	17.1	16.6	16.2	15.8	15.4	15.2	15.0	15.1	15.1	15.2	15.3	15.4
200	93.9	341.0	549.1	742.6	928.6	1,108.0	1,281.7	1,450.4	1.614.2	1,774.0	1,929.9	2,082.5	2,233.5	2,384.1	2,535.1	2,686.7	2,839.3	2,992.8
	28.5	21.8	19.9	18.8	18.2	17.7	17.1	16.6	16.2	15.7	15.4	15.2	15.0	15.1	15.1	15.2	15.3	15.5
300	122.4	362.8	569.0	761.4	946.8	1,125.7	1,298.8	1,467.0	1,630.4	1,789.7	1,945.3	2,097.7	2,248.5	2,399.2	2,550.2	2,701.9	2,854.6	3,008.3
	27.3	21.6	19.8	18.8	18.1	17.6	17.1	16.5	16.1	15.8	15.4	15.1	15.1	15.1	15.1	15.3	15.3	15.5
400	149.7	384.4	588.8	780.2	964.9	1,143.3	1,315.9	1,483.5	1,646.5	1,805.5	1,960.7	2,112.8	2,263.6	2,414.3	2,565.3	2,717.2	2,869.9	3,023.8
	26.3	21.3	19.6	18.8	18.1	17.5	17.0	16.5	16.1	15.7	15.3	15.1	15.1	15.0	15.1	15.2	15.3	15.5
500	176.0	405.7	608.4	799.0	983.0	1,160.8	1,332.9	1,500.0	1,662.6	1,821.2	1,976.0	2,127.9	2,278.7	2,429.3	2,580.4	2,732.4	2,885.2	3,039.3
	25.4	21.0	19.4	18.7	18.1	17.5	16.9	16.5	16.0	15.6	15.3	15.1	15.0	15.1	15.2	15.3	15.4	15.4
600	201.4	426.7	627.8	817.7	1,001.1	1,178.3	1,349.8	1,516.5	1,678.6	1,836.8	1,991.3	2,143.0	2,293.7	2,444.4	2,595.6	2,747.7	2,900.6	3,054.7
	24.6	20.9	19.3	18.6	18.0	17.4	16.9	16.4	16.0	15.6	15.2	15.2	15.1	15.1	15.2	15.3	15.4	15.5
700	226.0	447.6	647.1	836.3	1,019.1	1,195.7	1,366.7	1,532.9	1,694.6	1,852.4	2,006.5	2,158.2	2,308.8	2,459.5	2,610.8	2,763.0	2,916.0	3,070.2
	24.0	20.7	19.3	18.5	17.9	17.3	16.9	16.3	16.0	15.6	15.2	15.1	15.0	15.1	15.2	15.2	15.3	15.5
800	250.0	468.3	666.4	854.8	1,037.0	1,213.0	1,383.6	1,549.2	1,710.6	1,868.0	2,021.7	2,173.3	2,323.8	2,474.6	2,626.0	2,778.2	2,931.3	3,085.7
	23.4	20.4	19.2	18.6	17.8	17.3	16.8	16.3	15.9	15.5	15.2	15.1	15.1	15.1	15.2	15.2	15.4	15.5
900	273.4	488.7	685.6	873.4	1,054.8	1,230.3	1,400.4	1,565.5	1,726.5	1,883.5	2,036.9	2,188.4	2,338.9	2,489.7	2,641.2	2,793.4	2,946.7	3,101.2
	23.0	20.3	19.1	18.5	17.8	17.2	16.7	16.3	15.9	15.5	15.2	15.1	15.1	15.1	15.2	15.3	15.3	15.5
1,000	296.4	509.0	704.7	891.9	1,072.6	1,247.5	1,417.1	1,581.8	1,742.4	1,899.0	2,052.1	2,203.5	2,354.0	2,504.8	2,656.4	2,808.7	2,962.0	3,116.7

REFERENCE TABLE FOR Pt TO Pt—13 PER CENT Rh THERMOCOUPLE

Emfs are expressed in microvolts and temperatures in °C. Cold junctions at 0°C. Roeser and Wensel, National Bureau of Standards

E(μv)	0	1,000	2,000	3,000	4,000	5,000	6,000	7,000	8,000	9,000	10,000	11,000	12,000	13,000	14,000	15,000	16,000	17,000	18,000	19,000
0	0	145.3	258.8	361.0	457.4	549.8	638.3	723.5	806.0	886.1	964.1	1,040.0	1,113.9	1,186.9	1,259.3	1,331.8	1,404.3	1,476.9	1,550.0	1,623.6
	17.9	12.2	10.6	9.9	9.5	9.0	8.7	8.4	8.1	7.9	7.6	7.4	7.3	7.2	7.3	7.2	7.3	7.3	7.4	7.4
100	17.9	157.5	269.4	370.9	466.9	558.8	647.0	731.9	814.1	894.0	971.7	1,047.4	1,121.2	1.194.1	1,266.6	1,339.0	1,411.6	1,484.2	1,557.4	1,631.0
	16.7	12.0	10.5	9.8	9.4	9.0	8.6	8.4	8.1	7.8	7.7	7.5	7.4	7.3	7.2	7.2	7.3	7.3	7.4	7.4
200	34.6	169.5	279.9	380.7	476.3	567.8	655.6	740.3	822.2	901.8	979.4	1,054.9	1,128.6	1,201.4	1,273.8	1,346.2	1,418.9	1,491.5	1,564.8	1,638.4
	15.8	11.7	10.4	9.8	9.4	9.0	8.6	8.3	8.1	7.9	7.6	7.4	7.3	7.3	7.3	7.3	7.2	7.3	7.3	7.4
300	50.4	181.2	290.3	390.5	485.7	576.8	644.2	748.6	830.3	909.7	987.0	1,062.3	1,135.9	1,208.7	1,281.1	1.353.5	1,426.1	1.498.8	1,572.1	1.645.8
	15.1	11.5	10.3	9.7	9.3	8.9	8.6	8.3	8.0	7.8	7.7	7.4	7.3	7.2	7.2	7.3	7.2	7.3	7.4	7.4
400	65.5	192.7	300.6	400.2	495.0	585.7	672.8	756.9	838.3	917.5	994.7	1,069.7	1.143.2	1,215.9	1,288.3	1,360.8	1,433.3	1,506.1	1,579.5	1,653.2
	14.5	11.4	10.2	9.7	9.3	8.9	8.5	8.2	8.0	7.8	7.6	7.4	7.3	7.3	7.2	7.2	7.2	7.3	7.3	7.4
500	80.0	204.1	310.8	409.9	504.3	594.6	681.3	765.1	846.3	925.3	1,002.3	1,077.1	1,150.5	1,223.2	1,295.5	1,368.0	1,440.5	1,513.4	1,586.8	1,660.6
	13.9	11.2	10.2	9.6	9.2	8.8	8.5	8.2	8.0	7.8	7.6	7.3	7.3	7.2	7.2	7.3	7.3	7.3	7.4	7.4
600	93.9	215.3	321.0	419.5	513.5	603.4	689.8	773.3	854.3	933.1	1,009.9	1,084.4	1.157.8	1,230.4	1,302.7	1,375.3	1447.8	1,520.7	1,594.2	1,668.0
	13.4	11.1	10.1	9.5	9.1	8.8	8.5	8.2	8.0	7.8	7.6	7.4	7.3	7.2	7.2	7.3	7.2	7.3	7.3	7.4
700	107.3	226.4	331.1	429.0	522.6	612.2	698.3	781.5	862.3	940.9	1,017.5	1,091.8	1,165.1	1,237.6	1,309.9	1,382.6	1,455.0	1,528.0	1,601.5	1,675.4
	13.0	10.9	10.0	9.5	9.1	8.7	8.4	8.2	8.0	7.8	7.5	7.4	7.2	7.2	7.3	7.2	7.3	7.3	7.4	7.4
800	120.3	237.3	341.1	438.5	531.7	620.9	706.7	789.7	870.3	948.7	1,025.0	1,099.2	1,172.3	1,244.8	1,317.2	1.389.8	1,462.3	1,525.3	1,608.9	1,682.8
	12.6	10.8	10.0	9.5	9.1	8.7	8.4	8.1	7.9	7.7	7.5	7.3	7.3	7.3	7.3	7.3	7.3	7.4	7.3	7.4
900	132.9	248.1	351.1	448.0	540.8	629.6	715.1	797.8	878.2	956.4	1,032.5	1,106.5	1,179.6	1,252.1	1,324.5	1,397.1	1,469.6	1,542.7	1,616,2	1,690.2
	12.4	10.7	9.9	9.4	9.0	8.7	8.4	8.2	7.9	7.7	7.5	7.4	7.3	7.2	7.3	7.2	7.3	7.3	7.4	7.4
1,000	145.3	258.8	361.0	457.4	549.8	638.3	723.5	806.0	886.1	964.1	1,040.0	1,113.9	1,186.9	1,259.3	1,331.8	1,404.3	1,476.9	1,550.0	1,623.6	1,697.6

Emfs are expressed in microvolts and temperatures in °F. Cold junctions at 32°F. Roeser and Wensel, National Bureau of Standards

E(μv)	0	1,000	2,000	3,000	4,000	5,000	6,000	7,000	8,000	9,000	10,000	11,000	12,000	13,000	14,000	15,000	16,000	17,000	18,000	19,000
0	32.0	293.5	497.8	681.8	855.3	1,021.6	1,180.9	1334.3	1,482.8	1,627.0	1,767.4	1,904.0	2,037.0	2,168.4	2,298.7	2,429.2	2,559.7	2,690.4	2,822.0	2,954.5
	32.2	22.0	19.1	17.8	17.0	16.2	15.6	15.1	14.6	14.2	13.8	13.4	13.2	13.0	13.1	13.0	13.1	13.2	13.3	13.3
100	64.2	315.5	516.9	699.6	872.3	1,037.8	1,196.5	1,349.4	1,497.4	1,641.2	1,781.2	1,917.4	2,050.2	2,181.4	2,311.8	2,442.2	2,572.8	2,703.6	2,835.3	2,967.8
	30.1	21.6	18.9	17.7	16.2	16.2	15.6	15.1	14.6	14.1	13.7	13.4	13.2	13.1	13.0	13.0	13.1	13.1	13.3	13.3
200	94.3	337.1	535.8	717.3	889.3	1,054.0	1,212.1	1,364.5	1,512.0	1,655.3	1,794.9	1,930.8	2,063.4	2,194.5	2,324.8	2,455.2	2,585.9	2,716.7	2,848.6	2,981.1
	28.4	21.1	18.7	17.6	16.9	16.2	15.5	15.0	14.5	14.1	13.7	13.3	13.2	13.1	13.1	13.1	13.0	13.1	13.2	13.3
300	122.7	358.2	554.5	734.9	906.2	1,070.2	1,227.6	1,379.5	1,526.5	1,669.4	1,808.6	1,944.1	2,076.6	2,207.6	2,337.9	2,468.3	2,598.9	2,729.8	2,861.8	2,994.4
	27.2	20.7	18.6	17.5	16.8	16.1	15.4	14.9	14.4	14.1	13.8	13.3	13.2	13.0	13.0	13.1	13.0	13.2	13.3	13.4
400	149.9	378.9	573.1	752.4	923.0	1,086.3	1,243.0	1,394.4	1,540.9	1,683.5	1,822.4	1,957.4	2.089.8	2,220.6	2,350.9	2.481.4	2,611.9	2,743.0	2,875.1	3,007.8
	26.1	20.4	18.4	17.4	16.7	16.0	15.3	14.8	14.4	14.0	13.7	13.3	13.1	13.1	13.0	13.0	13.0	13.1	13.2	13.3
500	176.0	399.3	591.5	769.8	939.7	1,102.3	1,258.3	1,409.2	1,555.3	1,697.5	1,836.1	1,970.7	2,102.9	2.233.7	2,363.9	2,494.4	2,624.9	2,756.1	2,888.3	3,021.1
	25.0	20.2	18.3	17.3	16.6	15.9	15.3	14.8	14.4	14.1	13.7	13.3	13.1	13.0	13.0	13.1	13.1	13.2	13.3	13.3
600	201.0	419.5	609.8	787.1	956.3	1,118.2	1,273.6	1,424.0	1,569.7	1,711.6	1,849.8	1,984.0	2,116.0	2,246.7	2,376.9	2,507.5	2,638.0	2,769.3	2,901.6	3,034.4
	24.1	20.0	18.2	17.2	16.4	15.8	15.3	14.8	14.4	14.0	13.6	13.3	13.1	13.0	13.0	13.1	13.0	13.1	13.2	13.3
700	225.1	439.5	628.0	804.3	972.7	1,134.0	1,288.9	1,438.8	1,584.1	1,725.6	1,863.4	1,997.3	2,129.1	2,259.7	2,389.9	2,520.6	2,651.0	2,782.4	2,914.8	3,047.7
	23.4	19.7	18.0	17.1	16.4	15.7	15.2	14.7	14.4	14.0	13.6	13.3	13.1	13.0	13.1	13.0	13.1	13.2	13.2	13.3
800	248.5	459.2	646.0	821.4	989.1	1,149.7	1,304.1	1,453.5	1,598.5	1,739.6	1,877.0	2,010.6	2,142.2	2,272.7	2,403.0	2,533.6	2,664,1	2,795.6	2,928.0	3,061.0
	22.7	19.4	18.0	17.0	16.3	15.6	15.1	14.6	14.3	13.9	13.5	13.2	13.1	13.0	13.1	13.1	13.1	13.2	13.2	13.4
900	271.2	478.6	664.0	838.4	1,005.4	1,165.3	1,319.2	1,468.1	1,612.8	1,753.5	1,890.5	2,023.8	2,155.3	2,285.7	2,416.1	2,546.7	2,677.2	2,808.8	2,941.2	3,074.4
	22.3	19.2	17.8	16.9	16.2	15.6	15.1	14.7	14.2	13.9	13.5	13.2	13.1	13.0	13.1	13.0	13.2	13.2	13.3	13.3
1,000	293.5	497.8	681.8	855.3	1,021.6	1,180.9	1,334.3	1,482.8	1,627.0	1,767.4	1,904.0	2,037.0	2,168.4	2,298.7	2,429.2	2,559.7	2,690.4	2,822.0	2,954.5	3,087.7

VALUES FOR THE LANGEVIN FUNCTION $\mathcal{L}(u)$

Compiled by Allen L. King

Because of random thermal rotations a dipole or dipole-like element ordinarily has an average dipole moment of zero. If it is placed in an orienting field F however it tends to align itself with the field so that the average component of the dipole moment parallel to the field equals p. Classically, if the system is in thermal equilibrium,

$$\overline{p} = p_o(\coth u - 1/u) = p_o\mathcal{L}(u)$$

Here p_o is the moment of the dipole itself and $\mathcal{L}(u)$ is the Langevin function of the argument u which equals Fp_o/kT.

$$\text{For } \mu << 1 \; \mathcal{L}(\mu) = \mu/3$$

The Langevin function applies to permanent electric and magnetic dipoles.

(u)	0	1	2	3	4	5	6	7	8	9	10	Diff.
0.0	0000	0033	0066	0100	0133	0166	0200	0233	0266	0300	0333	33
0.1	0333	0366	0400	0433	0466	0499	0532	0565	0598	0631	0665	33
0.2	0665	0698	0731	0764	0797	0830	0862	0896	0929	0961	0994	33
0.3	0994	1027	1059	1092	1124	1157	1190	1222	1254	1287	1319	32
0.4	1319	1352	1384	1416	1448	1480	1512	1544	1576	1608	1640	32
0.5	1640	1671	1703	1734	1765	1797	1829	1860	1892	1923	1953	31
0.6	1953	1985	2016	2046	2077	2108	2138	2169	2200	2230	2260	31
0.7	2260	2290	2321	2351	2381	2411	2441	2471	2500	2530	2559	30
0.8	2559	2589	2618	2647	2676	2705	2734	2763	2792	2821	2850	29
0.9	2850	2878	2906	2934	2963	2991	3019	3047	3075	3103	3130	28
1.0	3130	3158	3185	3212	3240	3267	3294	3321	3348	3375	3401	27
1.1	3401	3428	3454	3480	3507	3533	3559	3585	3611	3637	3662	26
1.2	3662	3688	3713	3738	3763	3788	3813	3838	3863	3888	3913	25
1.3	3913	3937	3961	3985	4009	4033	4057	4081	4105	4129	4152	24
1.4	4152	4176	4199	4222	4245	4268	4291	4313	4336	4359	4381	23
1.5	4381	4403	4426	4448	4469	4491	4514	4536	4557	4579	4600	22
1.6	4600	4621	4642	4663	4684	4705	4726	4747	4768	4788	4808	21
1.7	4808	4828	4849	4869	4889	4909	4928	4948	4967	4987	5006	20
1.8	5006	5025	5044	5064	5083	5102	5121	5139	5158	5176	5195	19
1.9	5195	5213	5231	5249	5267	5285	5303	5321	5338	5356	5373	18
2.0	5373	5391	5408	5425	5442	5459	5476	5493	5509	5526	5542	17
2.1	5542	5559	5575	5591	5608	5624	5640	5656	5672	5688	5704	16
2.2	5704	5719	5734	5750	5765	5780	5795	5810	5825	5840	5855	15
2.3	5855	5870	5885	5899	5913	5928	5943	5957	5971	5985	6000	14
2.4	6000	6014	6027	6041	6055	6068	6082	6095	6109	6122	6136	13.5
2.5	6136	6149	6162	6175	6188	6201	6214	6227	6239	6252	6265	13
2.6	6265	6278	6290	6302	6314	6326	6339	6351	6363	6375	6387	12
2.7	6387	6399	6411	6422	6434	6446	6457	6469	6480	6492	6503	11.5
2.8	6503	6514	6525	6536	6548	6559	6570	6581	6591	6602	6613	11
2.9	6613	6624	6634	6644	6655	6665	6676	6686	6696	6707	6717	10
3.0	6717	6727	6737	6747	6757	6767	6777	6787	6796	6806	6815	10
3.1	6815	6825	6834	6844	6853	6862	6871	6880	6890	6899	6908	9
3.2	6908	6917	6926	6935	6944	6953	6962	6971	6979	6988	6997	9
3.3	6997	7006	7014	7023	7031	7040	7048	7056	7064	7073	7081	8.5
3.4	7081	7089	7097	7105	7113	7121	7129	7137	7145	7153	7161	8
3.5	7161	7169	7176	7184	7192	7200	7207	7215	7223	7230	7237	8
3.6	7237	7245	7252	7259	7267	7274	7281	7288	7295	7302	7309	7
3.7	7309	7317	7324	7330	7337	7344	7351	7358	7364	7371	7378	7
3.8	7378	7385	7392	7398	7405	7412	7418	7425	7431	7437	7444	6.5
3.9	7444	7450	7457	7463	7470	7476	7482	7488	7494	7501	7507	6

Note: Higher values of the Langevin function are nearly equal to (u − 1)/u.

MAGNETIC PROPERTIES AND COMPOSITION OF SOME PERMANENT MAGNETIC ALLOYS

Name	Composition,* Weight percent					Remanence, B_r (Gauss)	Coercive force H_c (Oersteds)	Maximum energy product, $(BH)_{max}$ (Gauss-Oersteds × 10^{-6})
	Al	Ni	Co	Cu	Other			
U.S.A.								
Alnico I	12	20–22	5	....	......	7,100	440	1.4
Alnico II	10	17	12.5	6	......	7,200	540	1.6
Alnico III	12	24–26	....	3	......	6,900	470	1.35
Alnico IV	12	27–28	5	....	......	5,500	700	1.3
Alnico V†	8	14	24	3	......	12,500	600	5.0
Alnico V DG†	8	14	24	3	......	13,100	640	6.0
Alnico VI†	8	15	24	3	1.25 Ti	10,500	750	3.75
Alnico VII†	8.5	18	24	3	5 Ti	7,200	1,050	2.75
Alnico XII	6	18	35	....	8 Ti	5,800	950	1.6
Carbon steel	....	....	....	....	1 Mn 0.9 C	10,000	50	0.2
Chromium steel	....	....	....	....	3.5 Cr 0.9 C 0.3 Mn	9,700	65	0.3
Cobalt steel	....	....	17	....	2.5 Cr 8 W 0.75 C	9,500	150	0.65

MAGNETIC PROPERTIES AND COMPOSITION OF SOME PERMANENT MAGNETIC ALLOYS (Continued)

Name	Composition,* Weight percent					Remanence, B_r (Gauss)	Coercive force H_c (Oersteds)	Maximum energy product, $(BH)_{max}$ (Gauss-Oersteds $\times 10^{-6}$)
	Al	Ni	Co	Cu	Other			
Cunico	....	21	29	50	.....	3,400	660	0.80
Cunife	....	20	....	60	......	5,400	550	1.5
Ferroxdur 1	$BaFe_{12}O_{19}$					2,200	1,800	1.0
Ferroxdur 2	$BaF_{12}O_{19}$ (oriented)					3,840	2,000	3.5
Platinum-Cobalt	....	....	23	....	77 Pt	6,000	4,300	7.5
Remalloy	....	....	12	....	17 Mo	10,500	250	1.1
Silmanol	4.4	....	....	....	86.6 Ag 8.8 Mn	550	6,000	0.075
Tungsten steel	....	....	....	....	5 W 0.3 Mn 0.7 C	10,300	70	0.32
Vicalloy I	....	....	52	....	10 V	8,800	300	1.0
Vicalloy II (wire)	....	....	52	....	14 V	10,000	510	3.5
Germany								
Alni 90	12	21	....	....	......	8,000	350	1.2
Alni 120	13	27	....	....	......	6,000	570	1.2
Alnico 130	12	23	5	....	......	6,300	620	1.4
Alnico 160	11	24	12	4	......	6,200	700	1.6
Alnico 190	12	21	15	4	......	7,000	700	1.8
Alnico 250	8	19	23	4	6 Ti	6,500	1,000	2.2
Alnico 400†	9	15	23	4	......	12,000	650	4.8
Alnico 580† (semicolumnar)	9	15	23	4	......	13,000	700	6.0
Oerstit 800	9	18	19	4	4 Ti	6,600	750	1.95
Great Britain								
Alcomax I	7.5	11	25	3	1.5 Ti	12,000	475	3.5
Alcomax II	8	11.5	24	4.5	......	12,400	575	4.7
Alcomax IISC (semicolumnar)	8	11	22	4.5	......	12,800	600	5.15
Alcomax III	8	13.5	24	3	0.8 Nb	12,500	670	5.10
Alcomax IIISC (semicolumnar)	8	13.5	24	3	0.8 Nb	13,000	700	5.80
Alcomax IV	8	13.5	24	3	2.5 Nb	11,200	750	4.30
Alcomax IVSC (semicolumnar)	8	13.5	24	3	2.5 Nb	11,700	780	5.10
Alni, high B_r	13	24	....	3.5	......	6,200	480	1.25
Alni, normal	....	....	....	....	......	5,600	580	1.25
Alni, high H_c	12	32	....	....	0–0.5 Ti	5,000	680	1.25
Alnico, high B_r	10	17	12	6	......	8,000	500	1.70
Alnico, normal	....	....	....	....	......	7,250	560	1.70
Alnico, high H_c	10	20	13.5	6	0.25 Ti	6,600	620	1.70
Columax (columnar)	similar to Alcomax III or IV					13,000–14,000	700–800	7.0–8.5
Hycomax	9	21	20	1.6	......	9,500	830	3.3

* Remainder of unlisted composition is either iron or iron plus trace impurities
† Cast anisotropic. Unmarked ones are cast isotropic.

HIGH PERMEABILITY MAGNETIC ALLOYS

(See separate table for magnetic properties)

Name	Composition,* Weight percent	Sp. gr., gm. per cc	Tensile strength		Remark	Use
			Kg/mm²**	Form		
Silicon iron AISI M 15	Si 4	7.68–7.64	44.3	....	Annealed 4 hrs 802–1093°C	Low core losses
Silicon iron AISI M 8	Si 3	7.68–7.64	44.2	Grain oriented	Annealed 4 hrs 802–1204°C	
45 Permalloy	Ni 45; Mn 0.3	8.17	....	....	....	Audio transformer, coils, relays
Monimax	Ni 47; Mo 3	8.27	....	....	....	High frequency coils
4-79 Permalloy	Ni 79; Mo 4; Mn 0.3	8.74	55.4	....	H_2 annealed 1121°C	Audio coils, transformers, magnetic shields
Sinimax	Ni 43; Si 3	7.70	....	....	....	High frequency coils
Nu-metal	Ni 75; Cr 2; Cu 5	8.58	44.8	....	H_2 annealed 1221°C	Audio coils, magnetic shields, transformers
Supermalloy	Ni 79; Mo 5; Mn 0.3	8.77	....	....	....	Pulse transformers, magnetic amplifiers, coils
2-V Permendur	Co 40; V 2	8.15	46.3	....	....	D-c electromagnets, pole tips

* Iron is additional alloying metal.
** kg/mm² × 1422.33 = lbs/in.²

CAST PERMANENT MAGNETIC ALLOYS

(See separate table for magnetic properties)

Alloy name, country of manufacture*	Composition,** weight percent	Sp. gr., gm per cc	Thermal expansion: Cm × 10^{-6} / cm × °C	Thermal expansion: Between °C	Tensile strength: *** Kg/mm²	Tensile strength: Form	Remark†	Use
Alnico I (USA)*	Al 12; Ni 20–22; Co 5	6.9	12.6	20–300	2.9	Cast	i.	Permanent magnets
Alnico II (USA)*	Al 10; Ni 17; Cu 6; Co 12.5	7.1	12.4	20–300	2.1 45.7	Cast Sintered	i.	Temperature controls, magnetic toys and novelties
Alnico III (USA)*	Al 12; Ni 24–26; Cu 3	6.9	12	20–300	8.5	Cast	i.	Tractor magnetos
Alnico IV (USA)*	Al 12; Ni 27–28; Co 5	7.0	13.1	20–300	6.3 42.1	Cast Sintered	i.	Application requiring high coercive force
Alnico V (USA)*	Al 8; Ni 14; Co 24; Cu 3	7.3	11.3	......	3.8 35	Cast Sintered	a.	Application requiring high energy
Alnico V DG (USA)*	Al 8; Ni 14; Co 24; Cu 3	7.3	11.3	......	....	....	a., c.	
Alnico VI (USA)*	Al 8; Ni 15; Co 24; Cu 3; Ti 1.25	7.3	11.4	......	16.1	Cast	a.	Application requiring high energy
Alnico VII (USA)*	Al 8.5; Ni 18; Cu 3; Co 24; Ti 5	7.17	11.4	......	....	....	a.	
Alnico XII (USA)*	Al 6; Ni 18; Co 35; Ti 8	7.2	11	20–300	....	....	.....	Permanent magnets
Comol (USA)*	Co 12; Mo 17	8.16	9.3	20–300	88.6	....	.....	Permanent magnets
Cunife (USA)*	Cu 60; Ni 20	8.52	....	.....	70.3	....	.....	Permanent magnets
Cunico (USA)*	Cu 50; Ni 21	8.31	....	.....	70.3	....	.....	Permanent magnets
Barium ferrite Feroxdur (USA)*	Ba $Fe_{12}O_{19}$	4.7	10	......	70.3	....	....	Ceramics
Alcomax I (GB)*	Al 7.5; Ni 11; Co 25; Cu 3; Ti 1.5	....	....	......	....	....	a.	Permanent magnets
Alcomax II (GB)*	Al 8; Ni 11.5; Co 24; Cu 4.5	....	....	......	....	....	a.	Permanent magnets
Alcomax II SC(GB)*	Al 8; Ni 11; Co 22; Cu 4.5	7.3	....	......	....	....	a., sc.	
Alcomax III (GB)*	Al 8; Ni 13.5; Co 24; Nb 0.8	7.3	....	......	....	....	a.	Magnets for motors, loudspeakers
Alcomax IV (GB)*	Al 8; Ni 13.5; Cu 3; Co 24; Nb 2.5	....	....	......	....	....	.....	Magnets for cycle-dynamos
Columax (GB)*	Similar to Alcomax III or IV	....	....	......	....	....	a., sc.	Permanent magnets, heat treatable
Hycomax (GB)*	Al 9; Ni 21; Co 20; Cu 1.6	....	....	......	....	....	a.	Permanent magnets
Alnico (high H_c) (GB)*	Al 10; Ni 20; Co 13.5; Cu 6; Ti 0.25	7.3	....	......	....	....	i.	
Alnico (high B_r) (GB)*	Al 10; Ni 17; Co 12; Cu 6	7.3	....	......	....	....	i.	
Alni (high H_c) (GB)*	Al 12; Ni 32; Ti 0–0.5	6.9	....	......	....	....	i.	
Alni (high B_r) (GB)*	Al 13; Ni 24; Cu 3.5	....	....	......	....	....	i.	
Alnico 580 (Ger)*	Al 9; Ni 15; Co 23; Cu 4	....	....	......	....	....	i.	
Alnico 400 (Ger)*	Al 9; Ni 15; Co 23; Cu 4	....	....	......	....	....	a.	
Oerstit 800 (Ger)*	Al 9; Ni 18; Co 19; Cu 4; Ti 4	....	....	......	....	....	i.	Permanent magnets
Alnico 250 (Ger)*	Al 8; Ni 19; Co 23; Cu 4; Ti 6	....	....	......	....	....	i.	
Alnico 190 (Ger)*	Al 12; Ni 21; Cu 4; Co 15	....	....	......	....	....	i.	
Alnico 160 (Austria)	Al 11; Ni 24; Co 12; Cu 4	....	....	......	....	....	i.	Permanent magnets, sintered
Alnico 130 (Ger)*	Al 12; Ni 23; Co 5	....	....	......	....	....	i.	
Alni 120 (Ger)*	Al 13; Ni 27	....	....	......	....	....	i.	
Alni 90 (Ger)*	Al 12; Ni 21	....	....	......	....	....	i.	

* USA—United States; GB—Great Britain; Ger—Germany.

** Iron is the additional alloying metal for each of the magnets listed.

*** kg/mm × 1422.33 = lbs/in.²

† i. = isotropic; a. = anisotropic; c. = columnar; sc. = semicolumnar.

Properties of Antiferromagnetic Compounds

Compound	Crystal Symmetry	θ_N(°K)	θ_P(°K)	$(P_A)_{eff}$ μ_B	$\frac{P_A}{\mu_B}$
$CoCl_2$	Rhombohedral	25	−38.1	5.18	3.1 ± 0.6
CoF_2	Tetragonal	38	50	5.15	3.0
CoO	Tetragonal	291	330	5.1	3.8
Cr	Cubic	475			
Cr_2O_3	Rhombohedral	307	485	3.73	3.0
CrSb	Hexagonal	723	550	4.92	2.7
$CuBr_2$	Monoclinic	189	246	1.9	
$CuCl_2 \cdot 2H_2O$	Orthorhombic	4.3	4–5	1.9	
$CuCl_2$	Monoclinic	~70	109	2.08	
$FeCl_2$	Hexagonal	24	−48	5.38	4.4 ± 0.7
FeF_2	Tetragonal	79–90	117	5.56	4.64
FeO	Rhombohedral	198	507	7.06	3.32
α-Fe_2O_3	Rhombohedral	953	2940	6.4	5.0
α-Mn	Cubic	95			
$MnBr_2 \cdot 4H_2O$	Monoclinic	2.1	$\begin{Bmatrix}2.5\\1.3\end{Bmatrix}$	5.93	
$MnCl_2 \cdot 4H_2O$	Monoclinic	1.66	1.8	5.94	
MnF_2	Tetragonal	72–75	113.2	5.71	5
MnO	Rhombohedral	122	610	5.95	5.0
β-MnS	Cubic	160	982	5.82	5.0
MnSe	Cubic	~173	361	5.67	
MnTe	Hexagonal	310–323	690	6.07	5.0
$NiCl_2$	Hexagonal	50	−68	3.32	
NiF_2	Tetragonal	78.5–83	115.6	3.5	2.0
NiO	Rhombohedral	533–650	~2000	4.6	2.0
$TiCl_3$		100			
V_2O_3		170			

1. θ_N = Néel temperature, determined from susceptibility maxima or from the disappearance of magnetic scattering.
2. θ_P = a constant in the Curie-Weiss law written in the form $\chi_A = C_A/(T + \theta_P)$, which is valid for antiferromagnetic material for $T > \theta_N$.
3. $(P_A)_{eff}$ = effective moment per atom, derived from the atomic Curie constant $C_A = (P_A)^2_{eff}(N^2/3R)$ and expressed in units of the Bohr magneton, $\mu_B = 0.9273 \times 10^{-20}$ erg gauss^{-1}.
4. P_A = magnetic moment per atom, obtained from neutron diffraction measurements in the ordered state.

SATURATION CONSTANTS AND CURIE POINTS OF FERROMAGNETIC ELEMENTS

Element	σ_s (20°C)	M_s (20°C)	σ_s (0°K)	n_B	Curie Point (°C)
Fe	218.0	1,714	221.9	2.219	770
Co	161	1,422	162.5	1.715	1,131
Ni	54.39	484.1	57.50	0.604	358
Gd	0	0	253.5	7.12	16

σ_s = saturation magnetic moment/gram; M_s = saturation magnetic moment/cm³, in cgs units. n_B = magnetic moment per atom in Bohr magnetons.

From American Institute of Physics Handbook, McGraw-Hill Company (1963) by permission.

MAGNETIC PROPERTIES OF TRANSFORMER STEELS

Ordinary Transformer Steel

B (Gauss)	H (Oersted)	Permeability = B/H
2,000	0.60	3,340
4,000	0.87	4,600
6,000	1.10	5,450
8,000	1.48	5,400
10,000	2.28	4,380
12.000	3.85	3,120
14,000	10.9	1,280
16,000	43.0	372
18,000	149	121

HIGH SILICON TRANSFORMER STEEL

B	H	Permeability
2,000	0.50	4,000
4,000	0.70	5,720
6,000	0.90	6,670
8,000	1.28	6,250
10,000	1.99	5,020
12,000	3.60	3,340
14,000	9.80	1,430
16,000	47.4	338
18,000	165	109

SATURATION CONSTANTS FOR MAGNETIC SUBSTANCES

Substance	Field intensity (For saturation)	Induced magnetization (For saturation)	Substance	Field intensity (For saturation)	Induced magnetization (For saturation)
Cobalt	9000	1300	Nickel, hard	8000	400
Iron, wrought	2000	1700	annealed	7000	515
cast	4000	1200	Vicker's steel	15000	1600
Manganese steel	7000	200			

INITIAL PERMEABILITY OF HIGH PURITY IRON FOR VARIOUS TEMPERATURES

L. Alberts and B. J. Shepstone

Temperature °C	Permeability (gauss/oersted)
0	920
200	1040
400	1440
600	2550
700	3900
770	12580

MAGNETIC MATERIALS

High-permeability Materials

Material	Form	Approximate percent composition: Fe	Ni	Co	Mo	Other	Typical heat treatment °C	Permeability at $B = 20$ gausses	Maximum permeability	Saturation flux density B gausses	Hysteresis ‡ loss, W_h ergs/cm³	Coercive ‡ force H_c oersteds	Resistivity microhm cm	Density, g/cm³
Cold rolled steel	Sheet	98.5	—	—	—	—	950 Anneal	180	2,000	21,000	—	1.8	10	7.88
Iron	Sheet	99.91	—	—	—	—	950 Anneal	200	5,000	21,500	5,000	1.0	10	7.88
Purified iron	Sheet	99.95	—	—	—	—	1480 H_2 + 880	5,000	180,000	21,500	300	.05	10	7.88
4% Silicon-iron	Sheet	96	—	—	—	4 Si	800 Anneal	500	7,000	19,700	3,500	.5	60	7.65
Grain oriented*	Sheet	97	—	—	—	3 Si	800 Anneal	1,500	30,000	20,000	—	.15	47	7.67
45 Permalloy	Sheet	54.7	45	—	—	.3 Mn	1050 Anneal	2,500	25,000	16,000	1,200	.3	45	8.17
45 Permalloy †	Sheet	54.7	45	—	—	.3 Mn	1200 H_2 Anneal	4,000	50,000	16,000	—	.07	45	8.17
Hipernik	Sheet	50	50	—	—	—	1200 H_2 Anneal	4,500	70,000	16,000	220	.05	50	8.25
Monimax	Sheet	—	—	—	—	—	1125 H_2 Anneal	2,000	35,000	15,000	—	.1	80	8.27
Sinimax	Sheet	—	—	—	—	—	1125 H_2 Anneal	3,000	35,000	11,000	—	—	90	—
78 Permalloy	Sheet	21.2	78.5	—	—	.3 Mn	1050 + 600 Q§	8,000	100,000	10,700	200	.05	16	8.60
4-79 Permalloy	Sheet	16.7	79	—	4	.3 Mn	1100 + Q	20,000	100,000	8,700	200	.05	55	8.72
Mu metal	Sheet	18	75	—	—	2 Cr, 5 Cu	1175 H_2	20,000	100,000	6,500	—	.05	62	8.58
Supermalloy	Sheet	15.7	79	—	5	.3 Mn	1300 H_2 + Q	100,000	800,000	8,000		.002	60	8.77
Permendur	Sheet	49.7	—	50	—	.3 Mn	800 Anneal	800	5,000	24,500	12,000	2.0	7	8.3
2V Permendur	Sheet	49	—	49	—	2 V	800 Anneal	800	4,500	24,000	6,000	2.0	26	8.2
Hiperco	Sheet	64	—	34	—	Cr	850 Anneal	650	10,000	24,200	—	1.0	25	8.0
2-81 Permalloy	Insulated powder	17	81	—	2	—	650 Anneal	125	130	8,000	—	<1.0	10^6	7.8
Carbonyl iron	Insulated powder	99.9	—	—	—	—	—	55	132	—	—	—	—	7.86
Ferroxcube III	Sintered powder	$MnFe_2O_4 + ZnFe_2O_4$					—	1,000	1,500	2,500	—	.1	10^8	5.0

*Properties in direction of rolling.

† Similar properties for Nicaloi, 4750 alloy, Carpenter 49, Armco 48.

‡ At saturation.

§ Q, quench or controlled cooling.

Permanent Magnet Alloys

Material	Percent composition (remainder Fe)	Heat treatment* (temperature, °C)	Magnetizing force $H_{max.}$ oersteds	Coercive force H_c oersteds	Residual induction B_r gausses	Energy product $BH_{max.}$ × 10^{-6}	Method of fabrication†	Mechanical properties‡	Weight lb/in.³
Carbon steel	1 Mn, 0.9 C	Q 800	300	50	10,000	.20	HR, M, P	H, S	.280
Tungsten steel	5 W, 0.3 Mn, 0.7 C	Q 850	300	70	10,300	.32	HR, M, P	H, S	.292
Chromium steel	3.5 Cr, 0.9 C, 0.3 Mn	Q 830	300	65	9,700	.30	HR, M, P	H, S	.280
17% Cobalt steel	17 Co, 0.75 C, 2.5 Cr, 8 W	—	1,000	150	9,500	.65	HR, M, P	H, S	—
36% Cobalt steel	36 Co, 0.7 C, 4 Cr, 5 W	Q 950	1,000	240	9,500	.97	HR, M, P	H, S	.296
Remalloy or Comol	17 Mo, 12 Co	Q 1200, B 700	1,000	250	10,500	1.1	HR, M, P	H	.295
Alnico I	12 Al, 20 Ni, 5 Co	A 1200, B 700	2,000	440	7,200	1.4	C, G	H, B	.249
Alnico II	10 Al, 17 Ni, 2.5 Co, 6 Cu	A 1200, B 600	2,000	550	7,200	1.6	C, G	H, B	.256
Alnico II (sintered)	10 Al, 17 Ni, 2.5 Co, 6 Cu	A 1300	2,000	520	6,900	1.4	Sn, G	H	.249
Alnico IV	12 Al, 28 Ni, 5 Co	Q 1200, B 650	3,000	700	5,500	1.3	Sn, C, G	H	.253
Alnico V	8 Al, 14 Ni, 24 Co, 3 Cu	AF 1300, B 600	2,000	550	12,500	4.5	C, G	H, B	.264
Alnico VI	8 Al, 15 Ni, 24 Co, 3 Cu, 1 Ti	—	3,000	750	10,000	3.5	C, G	H, B	.268
Alnico XII	6 Al, 18 Ni, 35 Co, 8 Ti	—	3,000	950	5,800	1.5	C, G	H, B	.26
Vicalloy I	52 Co, 10 V	B 600	1,000	300	8,800	1.0	C, CR, M, P	D	.295
Vicalloy II (wire)	52 Co, 14 V	CW + B 600	2,000	510	10,000	3.5	C, CR, M, P	D	.292
Cunife (wire)	60 Cu, 20 Ni	CW + B 600	2,400	550	5,400	1.5	C, CR, M, P	D, M	.311
Cunico	50 Cu, 21 Ni, 29 Co	—	3,200	660	3,400	.80	C, CR, M, P	D, M	.300
Vectolite	30 Fe_2O_3, 44 Fe_3O_4, 26 C_2O_3	—	3,000	1,000	1,600	.60	Sn, G	W	.113
Silmanal	86.8 Ag, 8.8 Mn, 4.4 Al	—	20,000	6,000[a]	550	.075	C, CR, M, P	D, M	.325
Platinum-cobalt	77 Pt, 23 Co	Q 1200, B 650	15,000	3,600	5,900	6.5	C, CR, M	D	—
Hyflux	Fine powder	—	2,000	390	6,600	.97	—	—	.176

[a] Value given is intrinsic H_c.
* Q—Quenched in oil or water. A—Air cooled. B—Baked. F—Cooled in magnetic field. CW—Cold worked.
† HR—Hot rolled or forged. CR—Cold rolled or drawn. M—Machined. G—Must be ground. P—Punched. C—Cast. Sn—Sintered.
‡ H—Hard. B—Brittle. S—Strong. D—Ductile. M—Malleable. W—Weak.

MAGNETIC SUSCEPTIBILITY OF THE ELEMENTS AND INORGANIC COMPOUNDS

The following table lists the magnetic susceptibilities of one gram formula weight of a number of paramagnetic and diamagnetic inorganic compounds as well as the magnetic susceptibilities of the elements.

In each instance the magnetic moment is expressed in egs units.

A more extensive listing of the magnetic susceptibilities of inorganic compounds as well as those for organic compounds may be found in Constantes Selectionnes Diamagnetisme et Paramagnetisme Relaxation Paramagnetique, Volume 7. This table is abridged from the above publication by permission of the publishers.

Substance	Formula	Temp. °K.	Susceptibility 10^{-6} cgs
Aluminum (s)	Al	ord.	+16.5
(*l*)	Al		+12.0
Fluoride	AlF_3	302	−13.4
Oxide	Al_2O_3	ord.	−37.0
Sulfate	$Al_2(SO_4)_3$	ord.	−93.0
	$Al_2(SO_4)_3 \cdot 18H_2O$	ord.	−323.0
Ammonia (g)	NH_3	ord.	−18.0
(aq)	NH_3	ord.	−17.0
Ammonium			
Acetate	$NH_4C_2H_3O_2$	ord.	−41.1
Bromide	NH_4Br	ord.	−47.0
Carbonate	$(NH_4)_2CO_3$	ord.	−42.50
Chlorate	NH_4ClO_3	ord.	−42.1
Chloride	NH_4Cl	ord.	−36.7
Fluoride	NH_4F	ord.	−23.0
Hydroxide (aq)	NH_4OH	ord.	−31.5
Iodate	NH_4IO_3	ord.	−62.3
Iodide	NH_4I	ord.	−66.0
Nitrate	NH_4NO_3	ord.	−33.6
Sulfate	$(NH_4)_2SO_4$	ord.	−67.0
Thiocyanate	NH_4SCN	ord.	−48.1
Americium (s)	Am	300	+1000.0
Antimony (s)	Sb	293	−99.0
(*l*)	Sb		−2.5
Bromide	$SbBr_3$	ord.	−115.0
Chloride, tri	$SbCl_3$	ord.	−86.7
Chloride, penta	$SbCl_5$	ord.	−120.0
Fluoride	SbF_3	ord.	−46.0
Iodide	SbI_3	ord.	−147.0
Oxide	Sb_2O_3	ord.	−69.4
Sulfide	Sb_2S_3	ord.	−86.0
Argon (g)	A	ord.	−19.6
Arsenic (α)	As	293	−5.5
(β)	As	293	−23.7
(γ)	As	293	−23.0
Bromide	$AsBr_3$	ord.	−106.0
Chloride	$AsCl_3$	ord.	−79.9
Iodide	AsI_3	ord.	−142.0
Sulfide	As_2S_3	ord.	−70.0
Arsenious Acid	H_3AsO_3	ord.	−51.2
Barium	Ba	ord.	+20.6
Acetate	$Ba(C_2H_3O_2)_2 \cdot H_2O$	ord.	−100.1
Bromate	$Ba(BrO_3)_2$	ord.	−105.8
Bromide	$BaBr_2$	ord.	−92.0
	$BaBr_2 \cdot 2H_2O$	ord.	−119.0
Carbonate	$Ba(CO_3)$	ord.	−58.9
Chlorate	$Ba(ClO_3)_2$	ord.	−87.5
Chloride	$BaCl_2$	ord.	−72.6
	$BaCl_2 \cdot 2H_2O$	ord.	−100.0
Fluoride	BaF_2	ord.	−51.0
Hydroxide	$Ba(OH)_2$	ord.	−53.2
	$Ba(OH)_2 \cdot 8H_2O$	ord.	−157.0
Iodate	$Ba(IO_3)_2$	ord.	−122.5
Iodide	BaI_2	ord.	−124.0
	$BaI_2 \cdot 2H_2O$	ord.	−163.0
Nitrate	$Ba(NO_3)_2$	ord.	−66.5
Oxide	BaO	ord.	−29.1
	BaO_2	ord.	−40.6
Sulfate	$BaSO_4$	ord.	−71.3
Beryllium (s)	Be	ord.	−9.0
Chloride	$BeCl_2$	ord.	−26.5
Hydroxide	$Be(OH)_2$	ord.	−23.1
Nitrate (aq)	$Be(NO_3)_2$	298	−41.0
Oxide	BeO	ord.	−11.9
Sulfate	$BeSO_4$	ord.	−37.0
Bismuth (s)	Bi	ord.	−280.1
(*l*)	Bi		−10.5
Bromide	$BiBr_3$	ord.	−147.0
Chloride	$BiCl_3$	ord.	−26.5
Chromate	$Bi_2(CrO_4)_3$	ord.	+154.0
Fluoride	BiF_3	303	−61.0
Hydroxide	$Bi(OH)_3$	ord.	−65.8
Iodide	BiI_3	ord.	−200.5
Nitrate	$Bi(NO_3)_3$	ord.	−91.0
	$Bi(NO_3)_3 \cdot 5H_2O$	ord.	−159.0
Oxide	BiO	ord.	−110.0
	Bi_2O_3	ord.	−83.0
Phosphate	$BiPO_4$	ord.	−77.0
Sulfate	$Bi_2(SO_4)_3$	ord.	−199.0
Sulfide	Bi_2S_3	ord.	−123.0
Boric Acid	H_3BO_3	ord.	−34.1
Boron (s)	B	ord.	−6.7
Chloride	BCl_3	ord.	−59.9
Oxide	B_2O_3	ord.	−39.0
Bromine (*l*)	Br_2		−56.4
(g)	Br_2		−73.5
Fluoride	BrF_3	ord.	−33.9
	BrF_5	ord.	−45.1
Cadmium (s)	Cd	ord.	−19.8
(*l*)	Cd		−18.0
Acetate	$Cd(C_2H_3O_2)_2$	ord.	−83.7
Bromide	$CdBr_2$	ord.	−87.3
	$CdBr_2 \cdot 4H_2O$	ord.	−140.0
Carbonate	$CdCO_3$	ord.	−46.7

MAGNETIC SUSCEPTIBILITY OF THE ELEMENTS AND INORGANIC COMPOUNDS (Continued)

Substance	Formula	Temp. °K.	Susceptibility 10^{-6} cgs
Chloride	$CdCl_2$	ord.	−68.7
	$CdCl_2 \cdot 2H_2O$	ord.	−99.0
Chromate	$CdCrO_4$	ord.	−16.8
Cyanide	$Cd(CN)_2$	ord.	−54.0
Fluoride	CdF_2	ord.	−40.6
Hydroxide	$Cd(OH)_2$	ord.	−41.0
Iodate	$Cd(IO_3)_2$	ord.	−108.4
Iodide	CdI_2	ord.	−117.2
Nitrate	$Cd(NO_3)_2$	ord.	−55.1
	$Cd(NO_3)_2 \cdot 4H_2O$	ord.	−140.0
Oxide	CdO	ord.	−30.0
Phosphate	$Cd_3(PO_4)_2$	ord.	−159.0
Sulfate	$CdSO_4$	ord.	−59.2
Sulfide	CdS	ord.	−50.0
Calcium (s)	Ca		+40.0
Acetate	$Ca(C_2H_3O_2)_2$	ord.	−70.5
Bromate	$Ca(BrO_3)_2$	ord.	−84.0
Bromide	$CaBr_2$	ord.	−73.8
	$CaBr_2 \cdot 3H_2O$	ord.	−115.0
Carbonate	$CaCO_3$	ord.	−38.2
Chloride	$CaCl_2$	ord.	−54.7
Fluoride	CaF_2	ord.	−28.0
Hydroxide	$Ca(OH)_2$	ord.	−22.0
Iodate	$Ca(IO_3)_2$	ord.	−101.4
Iodide	CaI_2	ord.	−109.0
Nitrate (aq)	$Ca(NO_3)_2$	ord.	−45.9
Oxide	CaO	ord.	−15.0
	CaO_2	ord.	−23.8
Sulfate	$CaSO_4$	ord.	−49.7
	$CaSO_4 \cdot 2H_2O$	ord.	−74.0
Carbon (dia)	C	ord.	−5.9
(graph)	C	ord.	−6.0
Dioxide	CO_2	ord.	−21.0
Monoxide	CO	ord.	−9.8
Cerium (α)	Ce	80.5	+5160.0
(β)	Ce	293	+2450.0
(β)	Ce	80.5	+6230.0
(γ)	Ce	287.9	+2420.0
(γ)	Ce	125.6	+4640.0
(γ)	Ce	80.5	+5200.0
Chloride	$CeCl_3$	287	+2490.0
Fluoride	CeF_3	293	+2190.0
Nitrate	$Ce(NO_3)_3 \cdot 5H_2O$	292	+2310.0
Oxide	CeO_2	293	+26.0
Sulfate	$CeSO_4$	ord	+37.0
	$Ce_2(SO_4)_3 \cdot 5H_2O$	293	+4540.0
	$Ce(SO_4)_2 \cdot 4H_2O$	293	−97.0
Sulfide	CeS	ord.	+2110.0
	Ce_2S_3	292	+5080.0
Cesium (s)	Cs	ord.	+29.0
(*l*)	Cs		+26.5
Bromate	$CsBrO_3$	ord.	−75.1
Bromide	$CsBr$	ord.	−67.2
Carbonate	Cs_2CO_3	ord.	−103.6
Chlorate	$CsClO_3$	ord.	−65.0
Chloride	$CsCl$	ord.	−86.7
Fluoride	CsF	ord.	−44.5
Iodate	$CsIO_3$	ord.	−83.1
Iodide	CsI	ord.	−82.6
Oxide	CsO_2	293	+1534.0
	CsO_2	90	+4504.0
Sulfate	Cs_2SO_4	ord.	−116.0
Sulfide	Cs_2S	ord.	−104.0
Chlorine (*l*)	Cl_2	ord.	−40.5
Fluoride, tri	ClF_3	ord.	−26.5
Chromium	Cr	273	+180.0
	Cr	1713	+224.0
Acetate	$Cr(C_2H_3O_2)_3$	293	+5104.0
Chloride	$CrCl_2$	293	+7230.0
	$CrCl_3$	293	+6890.0
Fluoride	CrF_3	293	+4370.0
Oxide	Cr_2O_3	300	+1960.0
	CrO_3	ord.	+40.0
Sulfate	$Cr_2(SO_4)_3$	293	+11,800.0
	$Cr_2(SO_4)_3 \cdot 8H_2O$	290	+12,700.0
	$Cr_2(SO_4)_3 \cdot 10H_2O$	290	+12,600.0
	$Cr_2(SO_4)_3 \cdot 14H_2O$	290	+12,160.0
	$CrSO_4 \cdot 6H_2O$	293	+9690.0
Sulfide	CrS	ord.	+2390.0
Cobalt	Co		ferro
Acetate	$Co(C_2H_3O_2)_2$	298	+11,000.0
Bromide	$CoBr_2$	293	+13,000.0
Chloride	$CoCl_2$	293	+12,660.0
	$CoCl_2 \cdot 6H_2O$	293	+9710.0
Cyanide	$Co(CN)_2$	303	+3825.0
Fluoride	CoF_2	293	+9490.0
	CoF_3	293	+1900.0
Iodide	CoI_2	293	+10,760.0
Oxide	CoO	260	+4900.0
	Co_2O_3	ord.	+4560.0
	Co_3O_4	ord.	+7380.0
Phosphate	$Co_3(PO_4)_2$	291	+28,110.0
Sulfate	$CoSO_4$	293	+10,000.0
	$Co_2(SO_4)_3$	297	+1000.0
Sulfide	CoS	688	+251.0
	CoS	293	+225.0
Thiocyanate	$Co(SCN)_2$	303	+11,090.0
Copper (s)	Cu	296	−5.46
(*l*)	Cu		−6.16
Bromide	$CuBr$	ord.	−49.0
	$CuBr_2$	341.6	+653.3
	$CuBr_2$	292.7	+685.5
	$CuBr_2$	189	+736.9
	$CuBr_2$	90	+658.7
Chloride	$CuCl$	ord.	−40.0
	$CuCl_2$	373.3	+1030.0
	$CuCl_2$	289	+1080.0
	$CuCl_2$	170	+1815.0
	$CuCl_2$	69.25	+2370.0
	$CuCl_2 \cdot 2H_2O$	293	+1420.0
Cyanide	$CuCN$	ord.	−24.0
Fluoride	CuF_2	293	+1050.0
	CuF_2	90	+1420.0
	$CuF_2 \cdot 2H_2O$	293	+1600.0
Hydroxide	$Cu(OH)_2$	292	+1170.0
Iodide	CuI	ord.	−63.0
Nitrate	$Cu(NO_3)_2 \cdot 3H_2O$	293	+1570.0
	$Cu(NO_3)_2 \cdot 6H_2O$	293	+1625.0
Oxide	Cu_2O	293	−20.0
	CuO	780	+259.6
	CuO	561	+267.3
	CuO	397	+256.9
	CuO	289.6	+238.9
	CuO	120	+156.2
Phosphide	Cu_3P	ord.	−33.0
	CuP_2	ord.	−35.0
Sulfate	$CuSO_4$	293	+1330.0
	$CuSO_4 \cdot H_2O$	293	+1520.0
	$CuSO_4 \cdot 3H_2O$	ord.	+1480.0
	$CuSO_4 \cdot 5H_2O$	293	+1460.0
Sulfide	CuS	293	−2.0
Thiocyanate	$CuSCN$	ord.	−48.0
Dysprosium	Dy	293.2	103,500.0
Oxide	Dy_2O_3	287.2	+89,600.0
Sulfate	$Dy_2(SO_4)_3$	293	+91,400.0
	$Dy_2(SO_4)_3 \cdot 8H_2O$	291.2	+92,760.0
Sulfide	Dy_2S_3	292	+95,200.0
Erbium	Er	291	+44,300.0
Oxide	Er_2O_3	286	+73,920.0
Sulfate	$Er_2(SO_4)_3 \cdot 8H_2O$	293	+74,600.0
Sulfide	Er_2S_3	292	+77,200.0
Europium	Eu	293	+34,000.0
Bromide	$EuBr_2$	292	+26,800.0
Chloride	$EuCl_2$	292	+26,500.0
Fluoride	EuF_2	292	+23,750.0
Iodide	EuI_2	292	+26,000.0
Oxide	Eu_2O_3	298	+10,100.0
Sulfate	$EuSO_4$	293	+25,730.0
	$Eu_2(SO_4)_3$	293	+10,400.0
	$Eu_2(SO_4)_3 \cdot 8H_2O$	293	+9,540.0
Sulfide	EuS	293	+23,800.0
	EuS	195	+35,400.0
Gadolinium	Gd	300.6	+755,000.0
Chloride	$GdCl_3$	293	+27,930.0
Oxide	Gd_2O_3	293	+53,200.0
Sulfate	$Gd_2(SO_4)_3$	285.5	+54,200.0
	$Gd_2(SO_4)_3 \cdot 8H_2O$	293	+53,280.0
Sulfide	Gd_2S_3	292	+55,500.0
Gallium (s)	Ga	80	−24.4
(s)	Ga	290	−21.6
(*l*)	Ga	313	+2.5
Chloride	$GaCl_3$	ord.	−63.0
Iodide	GaI_3	ord.	−149.0
Iodide	GaI_3	ord.	−149.0
Oxide	Ga_2O	ord.	−34.0
Sulfide	Ga_2S	ord.	−36.0
	GaS	ord.	−23.0
	Ga_2S_3	ord.	−80.0
Germanium	Ge	293	−76.84
Chloride	$GeCl_4$	ord.	−72.0
Fluoride	GeF_4	ord.	−50.0
Iodide	GeI_4	ord.	−174.0
Oxide	GeO	ord.	−28.8
	GeO_2	ord.	−34.3
Sulfide	GeS	ord.	−40.9
	GeS_2	ord.	−53.3
Gold (s)	Au	296	−28.0
(*l*)	Au		−34.0
Bromide	$AuBr$	ord.	−61.0
Chloride	$AuCl$	ord.	−67.0
	$AuCl_3$	ord.	−112.0
Fluoride	AuF_3	ord.	+74.0
Iodide	AuI	ord.	−91.0
Phosphide	AuP_3	ord.	−107.0
Hafnium (s)	Hf	298	+75.0
(s)	Hf	1673	+104.0
Oxide	HfO_2	ord.	−23.0
Helium (g)	He	ord.	−1.88
Holmium	Ho		
Oxide	Ho_2O_3	293	+88,100.0
Sulfate	$Ho_2(SO_4)_3$	293	+91,700.0
	$Ho_2(SO_4)_3 \cdot 8H_2O$	293	+91,600.0
Hydrogen (g)	H_2		−3.98

MAGNETIC SUSCEPTIBILITY OF THE ELEMENTS AND INORGANIC COMPOUNDS (Continued)

Substance	Formula	Temp. °K.	Susceptibility 10^{-6} cgs
Bromide (*l*)	HBr	273	
(aq)	HBr	ord.	
Chloride (*l*)	HCl	273	−22.6
(aq)	HCl	300	−22.0
Fluoride (*l*)	HF	287	−8.6
(aq)	HF	ord.	−9.3
Iodide (*l*)	HI	281	−47.7
(*l*)	HI	233	−48.3
(s)	HI	195	−47.3
(aq)	HI	ord.	−30.1
Oxide—See Water			
Peroxide	H_2O_2	ord.	−17.7
Sulfide	H_2S	ord.	−25.5
Indium	In	ord.	−64.0
Bromide	$InBr_3$	ord.	−107.0
Chloride	InCl	ord.	−30.0
	$InCl_2$	ord.	−56.0
	$InCl_3$	ord.	−86.0
Fluoride	InF_2	ord.	−61.0
Oxide	In_2O	ord.	−47.0
	In_2O_3	ord.	−56.0
Sulfide	In_2S	ord.	−50.0
	InS	ord.	−28.0
	In_2S_3	ord.	−98.0
Iodic Acid	HIO_3	ord.	−48.0
metaper	HIO_4	ord.	−56.5
orthoparaper	H_5IO_6	ord.	−71.4
Iodine (s)	I_2	ord.	−88.7
(atomic)	I	1303	+869.0
(atomic)	I	1400	+1120.0
Chloride	ICl	ord.	−54.6
	ICl_3	ord.	−90.2
Fluoride	IF_5	ord.	−58.1
Oxide	I_2O_5	ord.	−79.4
Iridium	Ir	298	+25.6
	Ir	698	+32.1
Chloride	$IrCl_3$	ord.	−14.4
Oxide	IrO_2	298	+224.0
Iron	Fe		ferro
Bromide	$FeBr_2$	ord.	+13,600.0
Carbonate	$FeCO_3$	293	+11,300.0
Chloride	$FeCl_2$	293	+14,750.0
	$FeCl_2 \cdot 4H_2O$	293	+12,900.0
	$FeCl_3$	293	+13,450.0
	$FeCl_3$	398	+9980.0
	$FeCl_3 \cdot 6H_2O$	290	+15,250.0
Fluoride	FeF_2	293	+9500.0
	FeF_3	305	+13,760.0
	$FeF_3 \cdot 3H_2O$	293	+7870.0
Iodide	FeI_2	ord.	+13,600.0
Nitrate	$Fe(NO_3)_3 \cdot 9H_2O$	293	+15,200.0
Oxide	FeO	293	+7200.0
	Fe_2O_3	1033	+3586.0
Phosphate	$FePO_4$	ord.	+11,500.0
Sulfate	$FeSO_4$	293	+10,200.0
	$FeSO_4 \cdot H_2O$	290	+10,500.0
	$FeSO_4 \cdot 7H_2O$	293	+11,200.0
Sulfide	FeS	293	+1074.0
Krypton	Kr		−28.8
Lanthanum	La	ord.	<118.0
Oxide	La_2O_3	ord.	−78.0
Sulfate	$La_2(SO_4)_3 \cdot 9H_2O$	293	−262.0
Sulfide	La_2S_3	292	−37.0
	La_2S_4	293	−100.0
Lead (s)	Pb	289	−23.0
(*l*)	Pb	330	−15.5
Acetate	$Pb(C_2H_3O_2)_2$	ord.	−89.1
Bromide	$PbBr_2$	ord.	−90.6
Carbonate	$PbCO_3$	ord.	−61.2
Chloride	$PbCl_2$	ord.	−73.8
Chromate	$PbCrO_4$	ord.	−18.0
Fluoride	PbF_2	ord.	−58.1
Iodate	$Pb(IO_3)_2$	ord.	−131.0
Iodide	PbI_2	ord.	−126.5
Nitrate	$Pb(NO_3)_2$	ord.	−74.0
Oxide	PbO	ord.	−42.0
Phosphate	$Pb_3(PO_4)_2$	ord.	−182.0
Sulfate	$PbSO_4$	ord.	−69.7
Sulfide	PbS	ord.	−84.0
Thiocyanate	$Lb(CNS)_2$	ord.	−82.0
Lithium	Li	ord.	+14.2
Acetate	$LiC_2H_3O_2$	ord.	−34.0
Bromate	$LiBrO_3$	ord.	−39.0
Bromide	LiBr	ord.	−34.7
Carbonate	Li_2CO_3	ord.	−27.0
Chlorate (aq)	$LiClO_3$	ord.	−28.8
Chloride	LiCl	ord.	−24.3
Fluoride	LiF	ord.	−10.1
Hydride	LiH	ord.	−4.6
Hydroxide	LiOH	ord.	−12.3
Iodate	$LiIO_3$	ord.	−47.0
Iodide	LiI	ord.	−50.0
Nitrate	$LiNO_3 \cdot 3H_2O$	ord.	−62.0
Sulfate	Li_2SO_4	ord.	−40.0
Lutetium	Lu	ord.	>0.0
Magnesium	Mg	ord.	+13.1

Substance	Formula	Temp. °K.	Susceptibility 10^{-6} cgs
Acetate	$Mg(C_2H_3O_2)_2 \cdot 4H_2O$	ord.	−116.0
Bromide	$MgBr_2$	ord.	−72.0
Carbonate	$MgCO_3$	ord.	−32.4
	$MgCO_3 \cdot 3H_2O$	ord.	−72.7
Chloride	$MgCl_2$	ord.	−47.4
Fluoride	MgF_2	ord.	−22.7
Hydroxide	$Mg(OH)_2$	288	−22.1
Iodide	MgI_2	ord.	−111.0
Oxide	MgO	ord.	−10.2
Phosphate	$Mg_3(PO_4)_2 \cdot 4H_2O$	ord.	−167.0
Sulfate	$MgSO_4$	294	−50.0
	$MgSO_4 \cdot H_2O$	ord.	−61.0
	$MgSO_4 \cdot 5H_2O$	ord.	−109.0
	$MgSO_4 \cdot 7H_2O$	ord.	−135.7
Manganese (α)	Mn	293	+529.0
(β)	Mn	293	+483.0
Acetate	$Mn(C_2H_3O_2)_2$	293	+13,650.0
Bromide	$MnBr_2$	294	+13,900.0
Carbonate	$MnCO_3$	293	+11,400.0
Chloride	$MnCl_2$	293	+14,350.0
	$MnCl_2 \cdot 4H_2O$	293	+14,600.0
Fluoride	MnF_2	290	+10,700.0
	MnF_3	293	+10,500.0
Hydroxide	$Mn(OH)_2$	293	+13,500.0
Iodide	MnI_2	293	+14,400.0
Oxide	MnO	293	+4850.0
	Mn_2O_3	293	+14,100.0
	MnO_2	293	+2280.0
	Mn_3O_4	298	+12,400.0
Sulfate	$MnSO_4$	293	+13,660.0
	$MnSO_4 \cdot H_2O$	293	+14,200.0
	$MnSO_4 \cdot 4H_2O$	293	+14,600.0
	$MnSO_4 \cdot 5H_2O$	293	+14,700.0
Sulfide (α)	MnS	293	+5630.0
(β)	MnS	293	+3850.0
Mercury (s)	Hg		−24.1
(*l*)	Hg	293	−33.44
(g)	Hg		−78.3
Acetate	$HgC_2H_3O_2$	ord.	−70.5
	$Hg(C_2H_3O_2)_2$	ord.	−100.0
Bromate	$HgBrO_3$	ord.	−57.7
Bromide	HgBr	ord.	−57.2
	$HgBr_2$	ord.	−94.2
Chloride	HgCl	ord.	−52.0
	$HgCl_2$	ord.	−82.0
Chromate	Hg_2CrO_4	ord.	−63.0
	$HgCrO_4$	ord.	−12.5
Cyanide	$Hg(CN)_2$	ord.	−67.0
Fluoride	HgF	ord.	−53.0
	HgF_2	302	−62.0
Hydroxide	$Hg_2(OH)_2$	ord.	−100.0
Iodate	$HgIO_3$	ord.	−92.0
Iodide	HgI	ord.	−83.0
	HgI_2	ord.	−128.6
Nitrate	$HgNO_3$	ord.	−55.9
	$Hg(NO_3)_2$	ord.	−74.0
Oxide	Hg_2O	ord.	−76.3
	HgO	ord.	−44.0
Sulfate	Hg_2SO_4	ord.	−123.0
	$HgSO_4$	ord.	−78.1
Sulfide	HgS	ord.	−55.4
Thiocyanate	$Hg(SCN)_2$	ord.	−96.5
Molybdenum	Mo	298	+89.0
	Mo	63.8	+108.0
	Mo	20.4	+149.2
Bromide	$MoBr_3$	293	+525.0
	$MoBr_4$	293	+520.0
	Mo_3Br_6	290.5	−46.0
Chloride	$MoCl_3$	290	+43.0
	$MoCl_4$	291	+1750.0
	$MoCl_5$	289	+990.0
Fluoride	MoF_6	ord.	−26.0
Oxide	Mo_2O_3	ord.	−42.0
	MoO_2	289	+41.0
	MoO_3	292.5	+3.0
	Mo_3O_8	ord.	+42.0
Sulfide	MoS_3	289	−63.0
Neodymium	Nd	287.7	5628.0
Fluoride	NdF_3	293	+4980.0
Nitrate	$Nd(NO_3)_3$	293	+5020.0
Oxide	Nd_2O_3	292.0	+10,200.0
Sulfate	$Nd_2(SO_4)_3$	293	+9990.0
Sulfide	Nd_2S_3	292	+5550.0
Neon	Ne	ord.	−6.74
Nickel	Ni		ferro
Acetate	$Ni(C_2H_3O_2)_2$	293	+4690.0
Bromide	$NiBr_2$	293	+5600.0
Chloride	$NiCl_2$	293	+6145.0
	$NiCl_2 \cdot 6H_2O$	293	4240.0
Fluoride	NiF_2	293	+2410.0
Hydroxide	$Ni(OH)_2$	ord.	+4500.0
Iodide	NiI_2	293	+3875.0
Nitrate	$Ni(NO_3)_2 \cdot 6H_2O$	293.5	+4300.0
Oxide	NiO	293	+660.0
Sulfate	$NiSO_4$	293	+4005.0
Sulfide	NiS	293	+190.0

MAGNETIC SUSCEPTIBILITY OF THE ELEMENTS AND INORGANIC COMPOUNDS (Continued)

Substance	Formula	Temp. °K.	Susceptibility 10^{-6} cgs
	Ni_3S_2	ord.	+1030.0
Niobium	Nb	298	+195.0
Oxide	Nb_2O_5	ord.	−10.0
Nitric Acid	HNO_3	ord.	−19.9
Nitrogen	N_2	ord.	−12.0
Oxide	N_2O	285	−18.9
(g)	NO	293	+1461.0
(g)	NO	203.8	+1895.0
(g)	NO	146.9	+2324.0
(*l*)	NO	117.64	+114.2
(s)	NO	90	+19.8
	N_2O_3	291	−16.0
	NO_2	408	+150.0
	N_2O_4	303.6	−22.1
	N_2O_4	295.1	−23.0
	N_2O_4	257	−25.4
(aq)	N_2O_5	289	−35.6
Osmium	Os	298	+9.9
Chloride	$OsCl_2$	ord.	+41.3
Oxygen (g)	O_2	293	+3449.0
(*l*)	O_2	90.1	+7699.0
(*l*)	O_2	70.8	+8685.0
(s, γ)	O_2	54.3	+10,200.0
(s, β)	O_2		
(s, α)	O_2	23.7	+1760.0
Ozone (*l*)	O_3		+6.7
Palladium	Pd	288	+567.4
Chloride	$PdCl_2$	291.3	−38.0
Fluoride	PdF_3	293	+1760.0
Hydride	PdH	ord.	+1077.0
	Pd_4H	ord.	+2353.0
Phosphoric Acid (aq)	H_3PO_4	ord.	−43.8
Phosphorous Acid (aq)	H_3PO_3	ord.	−42.5
Phosphorous (red)	P		−20.8
(black)	P		−26.6
Chloride	PCl_3	ord.	−63.4
Platinum	Pt	290.3	+201.9
Chloride	$PtCl_2$	298	−54.0
	$PtCl_3$	ord.	−66.7
	$PtCl_4$	ord.	−93.0
Fluoride	PtF_4	293	+455.0
Oxide	Pt_2O_3	ord.	−37.70
Plutonium	Pu	293	+610.0
Fluoride	PuF_4	301	+1760.0
	PuF_6	295	+173.0
Oxide	PuO_2	300	+730.0
Potassium	K	ord.	+20.8
Acetate (aq)	$KC_2H_3O_2$	28	−45.0
Bromate	$KBrO_3$	ord.	−52.6
Bromide	KBr	ord.	−49.1
Carbonate	K_2CO_3	ord.	−59.0
Chlorate	$KClO_3$	ord.	−42.8
Chloride	KCl	ord.	−39.0
Chromate	K_2CrO_4	ord.	−3.9
	$K_2Cr_2O_7$	293	+29.4
Cyanide	KCN	ord.	−37.0
Ferricyanide	$K_3Fe(CN)_6$	297	+2290.0
Ferrocyanide	$K_4Fe(CN)_6$	ord.	−130.0
	$K_4Fe(CN)_6 \cdot 3H_2O$	ord.	−172.3
Fluoride	KF	ord.	−23.6
Hydroxide (aq)	KOH	ord.	−22.0
Iodate	KIO_3	ord.	−63.1
Iodide	KI	ord.	−63.8
Nitrate	KNO_3	ord.	−33.7
Nitrite	KNO_2	ord.	−23.3
Oxide	KO_2	293	+3230.0
	KO_3	ord.	+1185.0
Permanganate	K_4MnO_4	ord.	+20.0
Sulfate	K_2SO_4	ord.	−67.0
	$KHSO_4$	ord.	−49.8
Sulfide	K_2S	ord.	−60.0
	K_2S_2	ord.	−71.0
	K_2S_3	ord.	−80.0
	K_2S_4	ord.	−89.0
	K_2S_5	ord.	−98.0
Sulfite	K_2SO_3	ord.	−64.0
Thiocyanate	KSCN	ord.	−48.0
Praseodymium	Pr	293	+5010.0
Chloride	$PrCl_3$	307.1	+44.5
Oxide	PrO_2	293	+1930.0
	Pr_2O_3	827	+4000.0
Oxide	Pr_2O_3	294.5	+8994.0
Sulfate	$Pr_2(SO_4)_3$	291	+9660.0
	$Pr_2(SO_4)_3 \cdot 8H_2O$	289	+9880.0
Sulfide	Pr_2S_3	292	+10,770.0
Rhenium	Re	293	+67.6
Chloride	$ReCl_5$	293	+1225.0
Oxide	ReO_2	ord.	+44.0
	$ReO_2 \cdot 2H_2O$	295	+74.0
	ReO_3	ord.	+16.0
	Re_2O_7	ord.	−16.0
Sulfide	ReS_2	ord.	+38.0
Rhodium	Rh	298	+111.0
	Rh	723	+123.0
Chloride	$RhCl_3$	298	−7.5
Fluoride	RhF_4	293	+500.0

Substance	Formula	Temp. °K.	Susceptibility 10^{-6} cgs
Oxide	Rh_2O_3	298	+104.0
Sulfate	$Rh_2(SO_4)_3 \cdot 6H_2O$	298	+104.0
	$Rh_2(SO_4)_3 \cdot 14H_2O$	298	+149.0
Rubidium	Rb	303	+17.0
Bromide	RbBr	ord.	−56.4
Carbonate	Rb_2CO_3	ord.	−75.4
Chloride	RbCl	ord.	−46.0
Fluoride	RbF	ord.	−31.9
Iodide	RbI	ord.	−72.2
Nitrate	$RbNO_3$	ord.	−41.0
Oxide	RbO_2	293	+1527.0
Sulfate	Rb_2SO_4	ord.	−88.4
Sulfide	Rb_2S	ord.	−80.0
	Rb_2S_2	ord.	−90.0
Ruthenium	Ru	298	+43.2
	Ru	723	+50.2
Chloride	$RuCl_3$	290.9	+1998.0
Oxide	RuO_2	298	+162.0
Samarium	Sm	291	+1860.0
	Sm	195	+2230.0
Bromide	$SmBr_2$	293	+5337.0
	$SmBr_3$	293	+972.0
Oxide	Sm_2O_3	292	+1988.0
	Sm_2O_3	170	+1960.0
	Sm_2O_3	85	+2282.0
Sulfate	$Sm_2(SO_4)_3 \cdot 8H_2O$	293	+1710.0
Sulfide	Sm_2S_3	292	+3300.0
Scandium	Sc	292	+315.0
Selenic Acid	H_2SeO_4	ord.	−51.2
Selenious Acid	H_2SeO_3	ord.	−45.4
Selenium (s)	Se	ord.	−25.0
(*l*)	Se	900	−24.0
Bromide	Se_2Br_2	ord.	−113.0
Chloride	Se_2Cl_2	ord.	−94.8
Fluoride	SeF_6	ord.	−51.0
Oxide	SeO_2	ord.	−27.2
Silicon	Si	ord.	−3.9
Bromide	$SiBr_4$	ord.	−128.6
Carbide	SiC	ord.	−12.8
Chloride	$SiCl_4$	ord.	−88.3
Hydroxide	$Si(OH)_4$	ord.	−42.6
Oxide	SiO_2	ord.	−29.6
Silver (s)	Ag	296	−19.5
(*l*)	Ag		−24.0
Acetate	$AgC_2H_3O_2$	ord.	−60.4
Bromide	AgBr	283	−59.7
Carbonate	Ag_2CO_3	ord.	−80.90
Chloride	AgCl	ord.	−49.0
Chromate	Ag_2CrO_4	ord.	−40.0
Cyanide	AgCN	ord.	−43.2
Fluoride	AgF	ord.	−36.5
Iodide	AgI	ord.	−80.0
Nitrate	$AgNO_3$	ord.	−45.7
Nitrite	$AgNO_2$	ord.	−42.0
Oxide	Ag_2O	ord.	−134.0
	AgO	287	−19.6
Permanganate	$AgMnO_4$	300	−63.0
Phosphate	Ag_3PO_4	ord.	−120.0
Sulfate	Ag_2SO_4	ord.	−92.90
Thiocyanate	AgSCN	ord.	−61.8
Sodium	Na	ord.	+16.0
Acetate	$NaC_2H_3O_2$	ord.	−37.6
Borate, tetra	$Na_2B_4O_7$	ord.	−85.0
Bromate	$NaBrO_3$	ord.	−44.2
Bromide	NaBr	ord.	−41.0
Carbonate	Na_2CO_3	ord.	−41.0
Chlorate	$NaClO_3$	ord.	−34.7
Chloride	NaCl	ord.	−30.3
Chromate, di	$Na_2Cr_2O_7$	ord.	+55.0
Fluoride	NaF	ord.	−16.4
Hydroxide (aq)	NaOH	300	−16.0
Iodate	$NaIO_3$	ord.	−53.0
Iodide	NaI	ord.	−57.0
Nitrate	$NaNO_3$	ord.	−25.6
Nitrite	$NaNO_2$	ord.	−14.5
Oxide	Na_2O	ord.	−19.8
	Na_2O_2	ord.	−28.10
Phosphate, meta	$NaPO_3$	ord.	−42.5
	Na_2HPO_4	ord.	−56.6
Sulfate	Na_2SO_4	ord.	−52.0
	$Na_2SO_4 \cdot 10H_2O$	ord.	−184.0
Sulfide	Na_2S	ord.	−39.0
	Na_2S_2	ord.	−53.0
	Na_2S_3	ord.	−68.0
	Na_2S_4	ord.	−84.0
	Na_2S_5	ord.	−99.0
Strontium	Sr	ord.	+92.0
Acetate	$Sr(C_2H_3O_2)_2$	ord.	−79.0
Bromate	$Sr(BrO_3)_2$	ord.	−93.5
Bromide	$SrBr_2$	ord.	−86.6
	$SrBr_2 \cdot 6H_2O$	ord.	−160.0
Carbonate	$SrCO_3$	ord.	−47.0
Chlorate	$Sr(ClO_3)_2$	ord.	−73.0
Chloride	$SrCl_2$	ord.	−63.0
	$SrCl_2 \cdot 6H_2O$	ord.	−145.0
Chromate	$SrCrO_4$	ord.	−5.1

MAGNETIC SUSCEPTIBILITY OF THE ELEMENTS AND INORGANIC COMPOUNDS (Continued)

Substance	Formula	Temp. °K.	Susceptibility 10^{-6} cgs
Fluoride	SrF_2	ord.	−37.2
Hydroxide	$Sr(OH)_2$	ord.	−40.0
	$Sr(OH)_2 \cdot 8H_2O$	ord.	−136.0
Iodate	$Sr(IO_3)_2$	ord.	−108.0
Iodide	SrI_2	ord.	−112.0
Nitrate	$Sr(NO_3)_2$	ord.	−57.2
	$Sr(NO_3)_2 \cdot 4H_2O$	ord.	−106.0
Oxide	SrO	ord.	−35.0
	SrO_2	ord.	−32.3
Sulfate	$SrSO_4$	ord.	−57.9
Sulfur (α)	S	ord.	−15.5
(β)	S	ord.	−14.9
(*l*)	S		−15.4
(g)	S	828	+700.0
(g)	S	1023	+464.0
Chloride	S_2Cl_2	ord.	−62.2
	SCl_2	ord.	−49.4
	SCl_3	ord.	−49.4
Fluoride	SF_6	ord.	−44.0
Iodide	SI	ord.	−52.7
Oxide (*l*)	SO_2	ord.	−18.2
Sulfuric Acid	H_2SO_4	ord.	−39.8
Tantalum	Ta	293	+154.0
	Ta	2143	+124.0
Chloride	$TaCl_5$	304	+140.0
Fluoride	TaF_5	293	+795.0
Oxide	Ta_2O_5	ord.	−32.0
Technetium	Tc	402	250.0
	Tc	298	270.0
	Tc	78	290.0
Oxide	$TcO_2 \cdot 2H_2O$	300	+244.0
	Tc_2O_7	298	−40.0
Tellurium (s)	Te	ord.	−39.5
(*l*)	Te		−6.4
Bromide	$TeBr_2$	ord.	−106.0
Chloride	$TeCl_2$	ord.	−94.0
Fluoride	TeF_6	ord.	−66.0
Terbium	Tb	273	+146,000.0
Oxide	Tb_2O_3	288.1	+78,340.0
Sulfate	$Tb_2(SO_4)_3$	293	+78,200.0
	$Tb_2(SO_4)_3 \cdot 8H_2O$	293	+76,500.0
Thallium (α)	Tl	ord.	−50.9
(β)	Tl	>508	−32.3
(*l*)	Tl	573	−26.8
Acetate	$TlC_2H_3O_2$	ord.	−69.0
Bromate	$TlBrO_3$	ord.	−75.9
Bromide	TlBr	ord.	−63.9
Carbonate	Tl_2CO_3	ord.	−101.6
Chlorate	$TlClO_3$	ord.	−65.5
Chloride	TlCl	ord.	−57.8
Chromate	Tl_2CrO_4	ord.	−39.3
Cyanide	TlCN	ord.	−49.0
Fluoride	TlF	ord.	−44.4
Iodate	$TlIO_3$	ord.	−86.8
Iodide	TlI	ord.	−82.2
Nitrate	$TlNO_3$	ord.	−56.5
Nitrite	$TlNO_2$	ord.	−50.8
Oxide	Tl_2O_3	ord.	+76.0
Phosphate	Tl_3PO_4	ord.	−145.2
Sulfate	Tl_2SO_4	ord.	−112.6
Sulfide	Tl_2S	ord.	−88.8
Thiocyanate	TlCNS	ord.	−66.7
Thorium	Th	293	+132.0
	Th	90	+153.0
Chloride	$ThCl_4 \cdot 8H_2O$	305.2	−180.0
Nitrate	$Th(NO_3)_4$	ord.	−108.0
Oxide	ThO_2	ord.	−16.0
Thulium	Tm	291	+25,500.0
Oxide	Tm_2O_3	296.5	+51,444.0
Tin (white)	Sn	ord.	+3.1
(gray)	Sn	280	−37.0
(gray)	Sn	100	−31.7
(*l*)	Sn		−4.5
Bromide	$SnBr_4$	ord.	−149.0
Chloride	$SnCl_2$	ord.	−69.0
	$SnCl_2 \cdot 2H_2O$	ord.	−91.4
(*l*)	$SnCl_4$	ord.	−115.0
Hydroxide	$Sn(OH)_4$	ord.	−60.0
Oxide	SnO	ord.	−19.0
	SnO_2	ord.	−41.0
Titanium	Ti	293	+153.0
	Ti	90	+150.0
Bromide	$TiBr_2$	288	+640.0
	$TiBr_3$	441	+520.0
	$TiBr_3$	291	+660.0
	$TiBr_3$	195	+680.0
	$TiBr_3$	90	+220.0
Carbide	TiC	ord.	+8.0
Chloride	$TiCl_2$	288	+570.0
	$TiCl_2$	685	+705.0
	$TiCl_3$	373	+1030.0
	$TiCl_3$	292	+1110.0
	$TiCl_3$	212	+690.0
	$TiCl_3$	90	+220.0
	$TiCl_4$	ord.	−54.0
Fluoride	TiF_3	293	+1300.0

Substance	Formula	Temp. °K.	Susceptibility 10^{-6} cgs
Iodide	TiI_2	288	+1790.0
	TiI_3	434	+221.0
	TiI_3	292	+160.0
	TiI_3	195	+159.0
	TiI_3	90	+167.0
Oxide	Ti_2O_3	382	+152.0
	Ti_2O_3	298	+125.6
	TiO_3	248	+132.4
	TiO_2	ord.	+5.9
Sulfide	TiS	ord.	+100.0
Tungsten	W	298	+59.0
Bromide	WBr_5	293	+250.0
Bromide	WC	o-d.	+10.0
Carbide	WCl_2	293	−25.0
Chloride	WCl_5	293	+387.0
	WCl_6	ord.	−71.0
	WF_6	ord.	−40.0
Fluoride	WO_2	ord.	+57.0
Oxide	WO_3	ord.	−15.8
Sulfide	WS_2	303	+5850.0
Tungstic Acid, ortho	H_2WO_4	ord.	−28.0
Uranium (α)	U	78	+395.0
(α)	U	298	+409.0
(α)	U	623	+440.0
(β)	U		
(γ)	U	1393	+514.0
Bromide	UBr_3	294	+4740.0
	UBr_4	293	+3530.0
Chloride	UCl_3	300	+3460.0
	UCl_4	294	+3680.0
Fluoride	UF_4	300	+3530.0
	UF_6	ord.	+43.0
Hydrides	UH_3	462	+2821.0
	UH_3	391	+3568.0
	UH_3	295	+6244.0
	UH_3	255	+9306.0
Iodide	UI_3	293	+4460.0
Oxide	UO	293	+1600.0
	UO_2	293	+2360.0
	UO_3	ord.	+128.0
Sulfate	$U(SO_4)_2$	ord.	+31.0
Sulfide (α)	US_2	290.5	+3137.0
(β)	US_2	290.5	+3170.0
	U_2S_3	ord.	+5206.0
	U_2S_5	ord.	+11,220.0
Vanadium	V	298	+255.0
Bromide	VBr_2	293	+3230.0
	VBr_2	195	+3760.0
	VBr_2	90	+4470.0
	VBr_3	293	+2890.0
	VBr_3	195	+4110.0
	VBr_3	90	+8540.0
Chloride	VCl_2	293	+2410.0
	VCl_3	293	+3030.0
	VCl_4	293	+1130.0
	VCl_4	195	+1700.0
	VCl_4	90	+4360.0
Fluoride	VF_3	293	+2730.0
Oxide	VO_2	290	+270.0
	V_2O_3	293	+1976.0
	V_2O_5	ord.	+128.0
Sulfide	VS	ord.	+600.0
	V_2S_3	293	+1560.0
Water (g)	H_2O	>373	−13.1
(*l*)	H_2O	373	−13.09
(*l*)	H_2O	293	−12.97
(*l*)	H_2O	273	−12.93
(s)	H_2O	273	−12.65
(s)	H_2O	223	−12.31
(*l*)	DHO	302	−12.97
(*l*)	D_2O	293	−12.76
(*l*)	D_2O	276.8	−12.66
(s)	D_2O	276.8	−12.54
(s)	D_2O	213	−12.41
Xenon	Xe	ord.	−43.9
Ytterbium	Yb	292	+249.0
	Yb	90	+639.0
Sulfide	Yb_2S_3	292	+18,300.0
Yttrium	Y	292	+2.15
	Y	90	+2.43
Oxide	Y_2O_3	293	+44.4
Sulfide	Y_2S_3	ord.	+100.0
Zinc (s)	Zn	ord.	−11.4
(*l*)	Zn		−7.8
Acetate	$Zn(C_2H_3O_2)_2 \cdot 2H_2O$	ord.	−101.0
Carbonate	$ZnCO_3$	ord.	−34.0
Chloride	$ZnCl_2$	296	−65.0
Cyanide	$Zn(CN)_2$	ord.	−46.0
Fluoride	ZnF_2	299.6	−38.2
Hydroxide	$Zn(OH)_2$	ord.	−67.0
Iodide	ZnI_2	ord.	−98.0
Nitrate (aq)	$Zn(NO_3)_2$	ord.	−63.0
Oxide	ZnO	ord.	−46.0
Phosphate	$Zn_3(PO_4)_2$	ord.	−141.0
Sulfate	$ZnSO_4$	ord.	−45.0
	$ZnSO_4 \cdot H_2O$	ord.	−63.0

MAGNETIC SUSCEPTIBILITY OF THE ELEMENTS AND INORGANIC COMPOUNDS (Continued)

Substance	Formula	Temp. °K.	Susceptibility 10^{-6} cgs	Substance	Formula	Temp. °K.	Susceptibility 10^{-6} cgs
	$ZnSO_4 \cdot 7H_2O$	ord.	−143.0	Carbide	ZrC	ord.	−26.0
Sulfide	ZnS	ord.	−25.0	Nitrate	$Zr(NO_3)_4 \cdot 5H_2O$	ord.	−77.0
Zirconium	Zr	293	−122.0	Oxide	ZrO_2	ord.	−13.8
	Zr	90	+119.0				

DIAMAGNETIC SUSCEPTIBILITIES OF ORGANIC COMPOUNDS

Compiled by George W. Smith

This table and its supplement contain values for the molar susceptibility of χ_M, specific susceptibility χ, and volumetric susceptibility K. The cgs Gaussian system of units is employed. In the Gaussian units the relation between magnetic induction B and magnetic field strength H is

$$B = H + 4\pi I \qquad (1)$$

where I is the magnetization or magnetic moment per unit volume. Actually, the quantities involved in equation (1) are all vectors, but one may assume that all three are collinear, a reasonable assumption for organic diamagnetic substances in the liquid state. For crystals I may vary with crystal orientation. Equation (I) may be written

$$B = H + 4\pi KH = (I + 4\pi K)H \qquad (2)$$

Here, K is the magnetic susceptibility, often called the volumetric susceptibility and is a unitless quantity.

Other susceptibilities of use to chemists and physicists are the specific or mass susceptibility which is defined

$$\chi = K/\rho \qquad (3)$$

where ϱ is the density of the sample in grams per cc., and the molar susceptibility which is defined as

$$\chi_M = M_\chi = MK/\rho \qquad (4)$$

where M is the molecular weight of the substance in grams.

Temperatures, when listed, are enclosed in parentheses and are listed in degrees C.

Literature references for values contained in this table may be found in General Motors Research Laboratories Bulletins GMR-317 and GMR-396.

Compound	$-\chi_M \times 10^6$	$-\chi \times 10^6$	$-K \times 10^6$
Acenaphthanthracene	184	.73	
Acenaphthene	109.3	(.709)	(.726) (99°)
Acetal	81.39	.688(32°)	(.568) (32°)
Acetaldehyde	22.70	$.515_3$	(.403) (18°)
Acetamide	34.1	.577	(.618) (20°)
Acetic acid	31.54	.525 (32°)	(.551) (32°)
Acetic anhydride	(52.8)	.517	(.562) (15°)
Acetomainofluorene	(141)	.63	
Acetone	33.7_8	$.581_4$	(.460) (20°)
Acetonitrile	28.0	(.682)	(.534) (20°)
Acetonylacetone	62.51	$.547_6$	(.531) (20°)
Acetophenone	72.05	$.599_8$	(.615) (20°)
Acetophenone oxime	79.9_0	$.592_0$	
Acetophenone oxime-O-methyl ether	92.3_1	$.618_8$	
Acetoxime	44.4_2	$.607_6$	(.480) (20°)
Acetoxime-O-benzyl ether	104.8_9	$.642_7$	
Acetoxime-O-methyl-ether	54.8_7	$.629_8$	
Acetylacetone	54.88	$.548_1$	(.535) (20°)
Acetyl chloride	38.9	.496	(.548) (20°)
Acetylene	12.5	(.480)	
Acetylphenylacetylene	86.9	.508	
Acetylthiophene	71.7	(.568)	
Acridine	(123.3)	.688	(.757) (20°)
Adonitol	91.30	.600	
Alanine	50.5	(.567)	
Allyl acetate	(56.7)	.566	(.525) (20°)
Allyl alcohol	36.70	.632	.540 (20°)
1-Allylpyrrole	73.80	.685 (20°)	
Aminoazobenzene	(118.3)	.600	
Aminoazotoluene	(142.2)	.631	
o-Aminoazotoluene	(138)	.61 ± .02	
α-Aminobutyric acid	62.1	(.602)	
Aminomethyldietyldiazine	114.8	(.696)	
4-Aminostilbene	122.5	(.628)	
2-Aminothiazole	56.0	.564	
n-Amyl acetate	89.06	684_5	.5979 (20.7°)
iso-Amyl acetate	89.40	.687	.599 (20°)
n-Amyl alcohol	(67.5)	.766	(.624) (20°)
γ-Amyl alcohol	(71.0)	.8060	(.655) (25°)
Inactive Amyl alcohol	69.06	783_4 (25°)	
iso-Amyl alcohol	68.96	$.782_3$ (25°)	(.64) (15°)
sec-Amyl alcohol	69.1	.785	(.635) (20°)
tert-Amyl alcohol	(70.9)	.804	(.654) (15°)
n-Amylamine	69.4	(.796)	(.606) (20°)
iso-Amulamine	71.6	(.821)	(.616) (20°)
n-Amylbenzene	112.55	(.759)	(.652) (20°)
iso-Amyl bromide	(88.7)	.587	(.706) (20°)
iso-Amyl-n-butyrate	113.52	$.717_4$ (25°)	(.616) (20°)
iso-Amyl chloride	(79.0)	.741	(.662) (20°)
iso-Amyl cyanide	73.4	(.755)	(.609) (20°)
iso-Amylene	53.7	.766	
Amylene bromide	(114.5)	.498	
Amylene chloride	(95.2)	.675	
iso-Amyl ether	(129)	.813	(.635) (15°)
iso-Amyl formate	78.38	$.674_8$	(.591) (25°)
Amylidene chloride	(93.4)	.662	
Amyl iodide	(118.7)	.5996 (18°)	(.910) (20°)
n-Amyl methyl ketone	80.50	$.705_3$	(.580) (15°)
Amyl nitrate	(76.4)	.574	
iso-Amyl propionate	101.73	705_4 (25°)	(.609) (25°)
n-Amyl valerate	124.55	$.723_9$	(.638) (0°)
Anethole	(96.0)	.648	(.644) (15°)
Aniline	62.95	(.676)	(.691) (20°)
Anisidine	(80.5)	.654	(.72) (20°)
Anisole	72.79	(.673)	(.672) (15°)
Anthanthrene	204.2	.739	
Anthanthrone	178.1	.581	
Anthracene	(130)	.731	(.914) (27°)
Anthracenedinitrile	154.6	(.678)	
Anthracenonitrile	142.1	(.700)	
Anthraquinone	(119.6)	.575	(.825) (20°)
Anthrazine	245.7	.646	
Arabinose	85.70	.571	(.905) (20°)
Arbutoside	158.0	(.589)	
Asarone	131.4	(.631)	(.735) (18°)
Asparagine	69.5	(.526)	(.812) (15°)
Aspartic acid	64.2±.4	(.482)	(.800) (12°)
Aurin	(161.4)	.556	
p-Azoanisole	(147.7)	.610	
Azobenzene	(106.8)	.586	.611 (70.5°)
p-Azophenetole	(171.7)	.635	
m-Azotoluene	(127.8)	.608	.643 (58°)
Azulene	98.5	.768	
Barbituric acid (Anh.)	53.8	(.420)	
Barbituric acid (.$2H_2O$)	78.6	(4.79)	
Benzalazine	(123.7)	.594	
Benzaldehyde	60.78	(.573)	(.602) (15°)
Benzaldoxime	(69.8)	.576	(.639) (20°)
Benzamide	(72.3)	.597	(.801) (4°)
Benzanthrone	142.9	.620	
Benzene	54.84	.702 (32°)	.611
Benzidine	110.9	.603	(.754) (20°)
Benzil	(118.6)	.564	.616 (100°)
Benzoic acid	70.28	(.575)	(.728) (15°)
Benzoic anhydride	(124.9)	.552	(.662) (15°)
Benzonitile	65.19	(.632)	(.638) (15°)
Benzophenone	109.60	$.601_3$	(.66) (50°)
3,4-Benzopyrene	135.7	.538	
Benzopyrene	194.0		
Benzoyl acetone	(95.0)	.586	(.639) (60°)
Benzoyl chloride	(75.8)	.539 (20°)	(.657) (15°)
Benzyl acetate	93.18	(.620)	(.655) (16°)
Benzyl alcohol	71.83	(.664)	(.697) (15°)
Benzylamine	75.26	(.702)	(.690) (19°)
Benzyl chloride	81.98	(.647)	(.713) (18°)
Benzyl formate	81.43	(.598)	(.646) (20°)
Benzylideneaniline	(100.4)	.554	
Benzylidene chloride	(97.9)	.608	(.763) (14°)
Benzylidenemethylamine	(73.1)	.613	
Benzylmethylketone	83.44	$.621_9$	(.624) (20°)
Bibenzyl	(126.8)	.696	.671 (54.5°)
3,4-Bis-(p-hydroxyphenyl)-2,4-hexadiene	(157)	.59	
m,m′-Bitolyl	(127.4)	.6993 (27.4°)	(.699) (16°)
m,m′-Bitolyl sulfide	(140.0)	.6530 (27.4°)	
Borneol	126.0	(.817)	(.826) (20°)
Bromobenzene	78.92	.5030 (20°)	(.753) (20°)

Compound	$-\chi_M \times 10^6$	$-\chi \times 10^6$	$-K \times 10^6$
Bromobenzenediazocyanide	86.88	(.414)	
Bromochloromethane	55.0 + .6	(.425)	(.846) (19°)
Bromodichloromethane	66.3 + .3	(.405)	(.812) (15°)
Bromoform	82.60	.327	.948 (20°)
Bromonaphthalene	(123.8)	.598	
α-Bromonaphthalene	115.90	.560	.840 (20°)
m-Bromotoluene	(93.4)	.546	(.770) (20°)
Bromotrichloromethane	73.1 ± .7	(.369)	(.758) (0°)
Butane	57.4	(.998)	
iso-Butane	51.7	(.890)	
1,4-Butanediol	61.5	(.682)	(.696) (20°)
2-Butene (cis)	42.6	(.759)	
2-Butene (trans)	43.3	(.772)	
1-Butene-3,4-diacetate	95.5	(.555)	
2-Butene-1,4-diacetate (cis)	95.2	(.553)	
2-Butene-1,4-diacetate (trans)	95.1	(.552)	
2-Butene-1,4-diol (cis)	54.3	(.616)	
2-Butene-1,4-diol (trans)	53.5	(.607)	
n-Butyl acetate	77.47	$.666_9$ (25°)	(.583) (25°)
iso-Butyl acetate	78.52	$.676_9$	(.584) (25°)
n-Butyl alcohol	56.536 (20°)	(.7627)	(.6176) (20°)
iso-Butyl alcohol	57.704 (20°)	(.7785)	(.624) (20°)
sec-Butyl alcohol	57.683 (20°)	(.7782)	(.629) (20°)
tert-Butyl alcohol	57.42	$.774_7$ (25°)	(.611) (20°)
n-Butylamine	58.9	(.805)	(.596) (20°)
iso-Butylamine	59.8	(.818)	(.599) (20°)
9-Butylanthracene	176.0	(.751)	
n-Butylbenzene	100.79	(.751)	(.646) (20°)
iso-Butylbenzene	101.81	(.759)	(.648) (20°)
tert-Butylbenzene	102.5	(.764)	(.662) (20°)
n-Butyl benzoate	116.69	$.654_8$	(.656) (25°)
Butyl bromide	77.14	.563 (20°)	(.730) (20°)
iso-Butyl bromide	79.88	.583 (20°)	(.737) (20°)
1-n-Butyl chloride	67.10	.725	.642 (20°)
2-n-Butyl chloride	67.40	.728	.635 (20°)
n-Butyl cyanide	(62.8)	.7558 (27.4°)	(.606) (20°)
tert-Butyl cyclohexane	115.09	.8205	.6670 (20°)
1,4-Butyl diacetate	103.4	(.594)	
n-Butyl ethyl ketone	80.73	$.707_2$	(.579) (20°)
n-Butyl formate	65.83	$.644_6$	(.571) (25°)
iso-Butyl formate	66.79	$.654_9$	(.574) (25°)
iso-Butylideneazine	(95.8)	.683	
Butyl iodide	(93.6)	.5086 (18°)	(.822) (20°)
iso-Butyl methyl ketone	70.05	$.699_5$	(.561) (20°)
tert-Butyl methyl ketone	69.86	$.697_9$	(.558) (16°)
n-Butyl perfluor-n-butyrate	126.7	(.469)	
p-tert-Butylphenol	108.0	(.719)	(.653) (114°)
Butyl sulfide	(113.7)	.7774 (27.4°)	(.652) (16°)
Butyl thiocyanate	(79.38)	.6891 (27.4°)	(.659) (25°)
2-Butyne-1,4-diacetate	95.9	(.564)	
2-Butyne-1,4-dibenzoate	169.0	(.561)	
2-Butyne-1,4-diol	50.3	(.584)	
n-Butyraldehyde	46.08	$.639_4$	(.522) (20°)
iso-Butyraldehyde	46.38	$.643_6$	(.511) (20°)
iso-Butyraldoxine	56.1_2	$.644_3$	(.576) (20°)
n-Butyric acid	55.10	.625	.598 (20°)
iso-Butyric acid	56.06	$.636_3$ (25°)	(.601) (25°)
Butyronitrile	49.4	(.715)	(.569) (15°)
Butyrylphenylacetylene	(106.4)	.618	
Cacodyl	(99.9)	.476	(.689) (15°)
Cacodylic acid	(79.9)	.579	
Camphor	(103)	.68	(.67) (25°)
Camphoric acid	129.0	(.644)	(.791) (20°)
Camphoric anhydride	(113)	.620	(.740) (20°)
n-Caproic acid	78.55	$.676_2$	(.624) (25°)
Caproylphenylacetylene	(130.4)	.651	
n-Caprylic acid	101.60	.7053	(.642) (20°)
Carbanilide	134.05	(.6316)	(.783) (20°)
Carbazole	117.4	(.702)	
Carbon disulfide	42.2	.554	(.699) (22°)
Carbon tetrabromide	93.73	.2826 (20°)	(.966)
Carbon tetrachloride	66.60	.433	.691 (20°)
Carbon tetraiodide	(136)	.261	(1.13) (20°)
Carvacrol	(109.1)	.726	(.709) (20°)
Carvone	(92.2)	.614	(.590) (20°)
Cetyl alcohol	(183.5)	.757 (17.5°)	(.619) (50°)
Cetyl mercaptan	390.4	(1.510)	
Chloral	(67.7)	.459	(.694) (20°)
Chloranil	(112.6)	.458	
Chloracetic acid	48.1	(.509)	(.804) (20°)
Chloroacetone	(50.9)	.550	(.633) (20°)
Chloroacetylchloride	53.7	(.475)	(.710) (0°)
p-Chloranisole	89.1	(.625)	
Chlorobenzene	69.97	(.6216)	(.688) (20°)
Chlorobenzene diazocyanide	65.02	(.393)	
Chlorodibromomethane	75.1 + .4	(.361)	(.883) (15°)
Chlorodifluoromethane	38.6	.446	
1-Chloro-2,3-dihydroxypropane	(77.9)	.604	
Chlorodiphenylmethane	131.9	(.651)	
Chloroethylene	35.9	.574	(.528) (liq., 15°)
Chloroform	59.30	.497	.740 (20°)
Chlorofumaric acid	67.02	(.445)	
p-Chloroiodi benzene	99.42	(.417)	(.756) (57°)
Chloromaleic acid	67.36	(.448)	
Chloromethylstilbene	144.8	(.633)	
α-Chloromaphthalene	107.60	.661	.789 (20°)
m-Chloronitrobenzene	(74.8)	.475	.638 (48°)
o-Chlorophenol	77.4	(.602)	(.747) (18°)
p-Chlorophenol	77.6	(.604)	(.789) (20°)
1-(o-Chlorophenylazo)-2-naphthol	161.0	(.570)	
1-(p-Chlorphenylazo)-2-naphthol	161.4	(.571)	
Chlorotrifluoroethylene	49.1	.422	
Chlorotrifluoromethane	45.3 + 1.5	(.434)	
Cholesterol	(284.2)	.735	(.764) (20°)
Chrysene	166.67	.731	
Chrysoidine	(126.3)	.595	
Cinnamic acid	78.36	.529	(.660) (4°)
Cinnamic acid (α-trans)	78.2	(.528)	
Cinnamic acid (β-trans)	79.0	(.533)	
Cinnamic acid (cis-MP 68°)	77.6	(.524)	
Cinnamic acid (cis-MP 58°)	77.9	(.526)	
Cinnamic acid (cis-MP 42°)	83.2	(.562)	
Cinnamic aldehyde	(74.8)	.566	(.629) (15°)
Cinnamyl alcohol	(87.2)	.650	(.679) (20°)
Cinnamylidencanileine	123.2	(.595)	
Citral	(98.9)	.650	(.577) (20°)
Coronene	(243.3)	.810	
Coumarin	82.5	(.565)	(.528) (20°)
o-Cresol	72.90	.675	(.706) (20°)
m-Cresol	72.02	.667 (26°)	(.690) (20°)
p-Cresol	72.1	.667 (26°)	(.690) (20°)
o-Creaylmethyl ether	81.94	.671 (40°)	(.661) (15°)
m-Creaylmethyl ether	77.91	.638 (40°)	(.623) (15°)
p-Cresylmethyl ether	79.13	.648 (40°)	(.629) (19°)
Cumene	89.53	$.744_9$	(.642)
Cyamelide	(56.1)	.435	(.490) (15°)
Cyameluric acid	101.1 (10°)	(.457)	
9-Cyanoanthracene	142.1	(.699)	
Cyanogen	(21.6)	.415	(.359) (liq., 17°)
Cyanuric acid	61.5	.476	(.842) (0°)
Cyclobutanecarboxylic acid	58.16	.5816 (30°)	(.613) (30°)
1,3-Cyclohexadiene	48.6	(.607)	(.510) (20°)
1,4-Cyclohexadiene	48.7	(.608)	(.515) (20°)
Cyclohexane	68.13	.8100 (27.5°)	(.627) (20°)
Cyclohexanecarboxylic acid	83.24	.6499 (30°)	(.668) (30°)
Cyclohexanol	73.40	.732	.694 (20°)
Cyclohexanone	62.0_4	$.632_3$	(.599) (20°)
Cyclohexanone oxime	71.5_2	$.632_1$	
Cyclohexanoneoxime-O-methyl ether	82.9_6	$.652_3$	
Cyclohexene	57.5	(.700)	(.567) (20°)
Cyclohexenol	64.1	(.653)	
Cyclooctane	91.4	(.815)	(.684) (20°)
Cyclooctene	84.6	(.769)	(.654) (20°)
Cyclooctatetraene	(53.9)	.518	
Cyclopentane	59.18	.8439	.6290 (20°)
Cyclopentanecarboxylic acid	73.48	.6446 (30°)	(.677) (30°)
Cyclopentanone	51.63	.6141 (30°)	(.582) (30°)
Cyclopropane	39.9	(.948)	(.683) (−79°)
Cyclopropanecarboxylic acid	45.33	.5271 (30°)	(.569) (30°)
p-Cymene	102.8	.766 (20°)	(.656) (20°)
Decalin	106,70	.7718	.6814 (20°)
cis-Decalin	(107.0)	.774	(.686) (35°)
trans-Decalm	(107.7)	.779	(.670) (35°)
n-Decane	119.74	(.8416)	(.6143) (20°)
1-Deuterio pyrrole	48.75	.716 (20°)	
Deuteroindene	80.88	.690	(.692) (13°)
Diacetal	(153.8)	.668	
Di-iso-amylamine	(133.1)	.846	(.649) (21°)
Diazoacetic ester	57	(.50)	(.54) (24°)
Dibenzocoronene	289.4	.778	
1,2,5,6-Dibenzofluorene	184	.69 ± .03	
3,4,5,6-Dibenzophenanthrene	(203)	.73 ± .03	
Dibenzphenanthrone	200.5	.716	
Dibenzypyrene	213.6	.706	
Dibenzpyrenequinone	183.1	.551	
iso-Dibenzpyrenequinone	194.6	.586	
Dibenzyl ketone	131.70	$.626_2$	
p-Dibromobenzene	(101.4)	.430	.786 (100°)
2,3-Dibromo-2-butene-1,4-diol	94.2	(.383)	
Dibromodichloromethane	81.1 ± .4	(.334)	(.808) (25°)
1,2-Dibromodiiodoethylene	(140.1)	.320	
1,2-Dibromoethylene	(71.7)	.386	(.877) (17.5°)
1,2-Dibromo-2-fluoroethane	(78.0)	.379	(.855) (17°)
Dibromo-4-nitrophenol	(167.5)	.564	
1,2-Dibromotetrachloroethane	(126.0)	.387	(1.049)
Di-n-butylamine	103.7	(.802)	(.767) (20°)
Di-iso-butylamine	105.7	(.817)	(.609) (20°)
Di-sec-butylamine	105.9	(.819)	(.641) (0°)
Di-iso-butyl ketone	104.30	$.733_5$	(.591) (20°)
di-tert-butyl ketone	104.06	$.732_4$	
2,6-Di-tert-butyl-4-methyl phenol	165.3	(.750)	
2,4-Di-tert-butyl phenol	155.6	(.754)	
Dibutyl phthalate	175.1	(.629)	(.657) (21°)
Di-iso-butyralacetylene	(125.6)	.738	
Dicetyl sulfide	401.7	(.832)	
Dichloroacetic acid	58.2	(.451)	(.705) (20°)
Dichloroacetyl chloride	69.0	(.468)	
o-Dichlorobenzene	84.26	.5734	(.748) (20°)

DIAMAGNETIC SUSCEPTIBILITIES OF ORGANIC COMPOUNDS (Continued)

Compound	$-\chi_M \times 10^6$	$-\chi \times 10^6$	$-K \times 10^6$
m-Dichlorobenzene	83.19	.5661	(.729) (20°)
p-Dichlorobenzene	82.93	.5644	(.823) (20.5°)
1,4-Dichlor-2-butyne	74.2	(.603)	
1,2-Dichloro-1,2-dibromoethane	(108.6)	.423	
1,1-Dichloro-difluoroethylene	60.0	.451	
Dichlorodifluoromethane	52.2	.432	(.642) (−30°)
1,1-Dichloroethylene	49.2	.508	(.635) (15°)
cis-1,2-Dichloroethylene	51.0	.526	(.679) (15°)
trans-1,2-Dichloroethylene	48.9	.504	(.638) (15°)
1,3-Dichloro-2-hydroxypropane	(80.1)	.621	
Dicyandiamide	44.55	(.530)	(.742) (14°)
Dicyclohexanol acetylene	(151.6)	.682	
Dicyclohexyl	129.31	.7776	.6889 (20°)
1,1-Dicyclohexylnonane	231.98	.7930	.7001 (20°)
Diethanolacetylene	(75.3)	.660	
Diethyl acetaldehyde	70.71	$.705_9$	(.576) (20°)
Diethylallylacetophenone	146.2	(.676)	(.663) (16°)
Diethyl allylmalonate	118.8	(.593)	(.602) (14°)
Diethylamine	56.8	(.777)	(.552) (18°)
Diethylcyclohexylamine	(124.5)	.802	(.699) (0°)
Diethyl ethylmalonate	115.2	(.612)	(.614) (20°)
Diethyl ketone	58.1_4	$.675_1$	(.551) (19°)
Diethyl ketoxime	68.3_1	.6754	
Diethyl malonate	(92.6)	.5782	(.611) (20°)
Diethyl-3-(1-methyl butane) ethyl-malonate	175	(.677)	
Diethyl oxalate	81.71	.5595	(.603) (15°)
Diethyl phthalate	127.5	(.574)	(.645) (25°)
Diethyl sebacate	(177.0)	.685	(.661) (20°)
Diethylstilbestrol	172.0	(.547)	
Diethylstilbestrol dipropionate	265.2	.720	
Diethyl succinate	105.07	.6035	(.628) (20°)
Diethyl sulfate	(86.8)	.563	(.667) (15°)
Diethyl sulfide	(67.9)	.753	(.630) (20°)
Diethyl tartrate	(113.4)	.550	(.662) (20°)
Difluoroacetamide	(41.2)	.433	
1-Difluoro-2-dibromoethane	(85.5)	.382	(.883) (20°)
1,1-Difluoro-2,2-dichloroethyl amyl ether	129.84	(.587)	(.694) (20°)
1,1-Difluoro-2,2-dichloroethyl butyl ether	119.48	(.577)	(.703) (20°)
1,1-Difluoro-2,2-dichloroethyl ethyl ether	96.13	(.537)	(.723) (20°)
1,1-Difluoro-2,2-dichloroethyl methyl ether	80.68	(.489)	(.696) (20°)
1,1-Difluoro-2,2-dichloroethyl propyl ether	107.19	(.555)	(.701) (20°)
Difluoroethanol	(41.3)	.503	
Di-n-heptylamine	171.5	(.805)	
Di-n-hexylamine	148.9	(.803)	
Dihydronaphthalene	(85.1)	.654	(.652) (12°)
o-Dimethoxybenzene	87.39	.6329	(.686) (25°)
m-Dimethoxybenzene	87.21	.6316	(.682) (0°)
p-Dimethoxybenzene	86.65	.6275	(.661) (55°)
o-(2,5-Dimethoxybenzoyl)-benzoic acid	161.0	(.562)	
Dimethoxymethane	(47.3)	.621	(.532)
Dimethylacetophenone	96.8	(.653)	(.645) (16°)
Dimethylallylacetophenone	122.4	(.650)	(.635) (16°)
Dimethylaniline	89.66	(.740)	
2,2-Dimethylbutane	76.24	.8848	.5744 (20°)
2,3-Dimethylbutane	76.22	.8845	.5853 (20°)
2,3-Dimethyl-2-butene	65.9	(.783)	(.557)
Dimethylcyclohexanone	(84.8)	.672	
1,2 and 1,3 Dimethylcyclopentanes	81.31	.8281	.6224 (20°)
2,5-Dimethyl-2,5-dibromo-3-hexine	(135.6)	.506	
Dimethyl diethylketo tetrahydrofurfurane	(116.2)	.753	
2,5-Dimethyl-1-ethylpyrrole	94.61	.768 (20°)	
2,5-Dimethyl-3-ethylpyrrole	93.87	.762 (20°)	
2,5-Dimethylfuran	66.37	.687 (20°)	(.620) (18°)
Dimethyl furazan	57.27	.584	
2,5-Dimethyl-4-heptene	100.6	(.797)	
2,4-Dimethyl-2,4-hexadiene	(78.7)	.714	
2,3-Dimethylhexane	98.77	.8648	.6164 (20°)
2,5-Dimethylhexane	98.15	.8593	.5969 (20°)
3,4-Dimethylhexane	99.06	.8673	.6420 (20°)
2,6-Dimethyl-4 hexanol	116.9	.812	
2,5-Dimethyl-3-hexine-2,5-diol	(103.0)	.724	
Dimethyl isoxazole	59.7	(.615)	
Dimethylketo tetrahydrofurfurane	(68.5)	.600	
Dimethyl malonate	69.69	.5277	(.609) (20°)
1,6-Dimethylnaphthalene	113.3	(.725)	
2,4-Dimethylnonane	134.68	(.862)	(.636) (20°)
3,4-Dimethylnonane	134.70	(.862)	(.647) (20°)
4,5-Dimethylnonane	134.52	(.861)	(.647) (20°)
2,6-Dimethyl-2,6,8-nonatriene	(108.8)	.724	
Dimethyl-2,4-nonatriene	(148.8)	.990	
2,6-Dimethyloctane	122.54	(.861)	(.627) (20°)
3,4-Dimethyloxadiazole	57.17	.583	
Dimethyl oxalate	(55.7)	.472	(.542) (54°)
Dimethyl oxamide	(63.2)	.544	
2,2-Dimethylpentane	86.97	.8680	.5849 (20°)
2,3-Dimethylpentane	87.51	.8733	.6070 (20°)
2,4-Dimethylpentane	87.48	.8732	.5876 (20°)
2,2-Dimethylpropane	63.1	(.875)	(.536) (0°)
2,5-Dimethyl-3-propyl-pyrrole	106.07	.773 (20°)	
2,4-Dimethylpyrrole	69.64	.732 (20°)	(.679) 14°)
2,5-Dimethylpyrrole	71.92	.756 (20°)	(.707) (20°)
α,ω-Dimethyl styrene	(90.7)	.686	
Dimethyl succinate	81.50	.5581	(.625) (18°)
Dimethyl sulfate	(62.2)	.493	(.657) (20°)
Dimethyl sulfide	(44.9)	.723	(.612) (21°)
Dimethyltrichloromethylcarbinol	(105)	.59	
N,N-Dimethyl urea	55.1	(.625)	(.784)
N,N′-D imethyl urea	56.3	(.639)	(.730)
o-Dinitrobenzene	65.98	.3921	(.614) (17°)
m-Dinitrobenzene	70.53	.4197	(.659) (0°)
p-Dinitrobenzene	68.30	.4064	(.660) (30°)
2,4-Dinitrophenol	(73.1)	.397	(.668) (24°)
Dinitroresorciniol	(62.4)	.312	
1,4-Dioxane	52.16	.592 (32°)	(.606) (32°)
Diphenyl	103.25	.6695	(.664) (73°)
1,1-Diphenylallyl-3-chloride	146.1	(.639)	
1,3-Diphenylallyl-3-chloride	140.7	(.615)	
Diphenylamine	(109.7)	.648	.686 (55.5°)
Diphenyl-bis-diazocyanide	85.03	(.327)	
Diphenylbutadiene	129.6	(.629)	
Diphenylchloroarsine	(145.5)	.550	(.871) (40°)
Diphenyldecapentaene	180.5	(.635)	
Diphenyldiacetylene	(134.6)	.640	
Diphenyldiazomethane	115	(.592)	
Diphenyldihydrotetrazine	129.9	(.545)	
1,1-Diphenylethylene	(118.0)	.655	(.680) (14°)
1,6-Diphenylhexane	171.81	.7208	.6877 (20°)
Diphenylhexatriene	146.9	(.632)	
Diphenylmethane	(115.7)	.688	.684 (35.5°)
Diphenylmethanol	119.1	.647	
1,1-Diphenylnonane	206.32	.7357	.6935 (20°)
Diphenyloctatetraene	164.3	(.636)	
Diphenylphenoxyarsine	(225.2)	.567	
N,N-Diphenyl urea	126.3	(.595)	(.759)
N,N′-Diphenyl urea	127.5	(.600)	(.743) (20°)
Di-n-propyl ketone	80.45	$.705_0$	(.576) (20°)
Di-iso-propyl ketone	81.14	$.711_0$	(.573) (20°)
Dipropyl oxalate	105.27	.6046	(.628) (o°)
Di-iso-propyl oxalate	106.02	.6089	
Dodecyl alcohol	147.70	.7849 (20.7°)	(.652) (24°)
Dulcitol	112.40	.617	(.905) (15°)
Elaidic acid	204.8	(.725)	(.619) (79°)
Erythritol	73.80	.604	(.876) (20°)
Ethane	37.3_7	(.910)	(.511) (−100°)
4-Ethoxy-3-methoxybenzyl acetate	138.5	.619	
4-Ethoxy-3-methoxybenzyl benzoate	177.3	.620	
1-Ethoxynaphthalene	119.9	(.696)	(.738) (20°)
2-Ethoxynaphthalene	119.2	(.692)	(.734) (25°)
Ethyl acetate	54.10	.614	.554 (20°)
Ethyl acetoacetate	71.67	$.550_8$	(.565) (20°)
Ethylacetophenone	95.5	(.644)	(.639) (16°)
Ethyl alcohol	33.60	.728	.575 (20°)
Ethylally acetophenone	122.5	(.651)	(.634) (16°)
Ethyl amylpropiolate	(112.7)	.670	
Ethylaniline	89.30	(.737)	(.709)
9-Ethyl anthracene	153.0	(.741)	(.771) (99°)
Ethylbenzene	77.20	.7272	.6341 (20°)
Ethyl benzoate	93.32	$.621_1$	(.648) (25°)
Ethyl benzoylacetate	(115.3)	.600	(.673) (20°)
Ethyl benzylidenecyanoacetate	(116.3)	.578	
Ethyl benzylmalonate	(154.5)	.6172	(.663) (20°)
Ethyl bromide	54.70	.502	.719 (20°)
Ethyl bromoacetate	(82.8)	.496	(.747) (20°)
Ethyl-1-isobutylacetoacetate	(121.4)	.652	
Ethyl butylmalonate	139.3	$.644_2$	(.629) (20°)
Ethyl-n-butyrate	(77.7)	.6693	(.585) (25°)
Ethyl-iso-butyrate	78.32	$.674;_3$	(.583) (25°)
Ethyl chloroacetate	(72.3)	.590	(.684) (20°)
Ethyl cinnamate	(107.5)	.610	(.640) (20°)
Ethyl-iso-cyanate	(45.6)	.642	(.582) (16°)
Ethyl cyanoacetate	(67.3)	.595	(.632) (20°)
Ethylcyclohexane	91.09	.8118	.6324 (20°)
Ethyldiallylacetophenone	147.4	(.646)	(.636) (16°)
Ethyl dibromocinnamate	174.5	.519	
Ethyl dicloracetate	85.2	(.543)	(.698) (20°)
Ethyl diethylacetoacetate	(117.9)	.6328	(.615) (20°)
Ethyl diethylmalonate	(140.4)	.6492	(.641) (20°)
Ethyl dithiolacetate	(71.0)	.5904 (27.4°)	
Ethylene	12.0	(.428)	(.242) (−102°)
Ethylene	15.30	.546 (32°)	(.309) (−102°)
Ethylene bromide	78.80	.419	.915 (20°)
Ethylene chloride	59.62	.602 (32°)	(.757) (20°)
Ethylenediamine	46.26	.771 (32°)	(.686) (20°)
Ethylene iodide	104.7	.371 (32°)	(.791) (10°)

DIAMAGNETIC SUSCEPTIBILITIES OF ORGANIC COMPOUNDS (Continued)

Compound	$-\chi_M \times 10^6$	$-\chi \times 10^6$	$-K \times 10^6$
Ethylene oxide	30.7	(.697)	(.618) (7°)
Ethyl ether	55.10	.743	.534 (20°)
Ethyl ethylacetoacetate	93.9	.5937	(.582) (20°)
Ethyl ethylbutylmalonate	(163.3)	.6683	(.650) (20°)
Ethyl ethylpropylmalonate	(152.4)	.6619	(.648) (20°)
Ethyl formate	43.00	.580	.531 (20°)
Ethyl hexylpropiolate	129.9	.713	
Ethyl hydroxylamine	(4370)	.704	
Ethylidene chloride	(57.4)	.580	(.681) (20°)
Ethyl iodide	(69.7)	.4470 (17.5)	(.884) (20°)
Ethyl iodoacetate	(97.6)	.456	(.829) (13°)
Ethyl lactate	(72.6)	.615	(.633) (25°)
Ethyl methylacetoacetate	(81.9)	.5684	(.569) (20°)
Ethyl methyl ketoxime	57.3_2	$.658_0$	(.530) (20°)
Ethyl-1-methyl-2-oxocyclohexane-carboxylate	112.1	(.608)	
Ethyl methylphenylmalonate	(153.2)	.6121	(.658) (20°)
Ethyl nitrophenylpropiolate	114.6	.523	
Ethyl oxamate	62.0	(.529)	(.427) (19°)
Ethyl perfluor-n-butyrate	103.5	(.427)	
Ethyl phenylacetate	104.27	(.635)	(.656) (20°)
Ethyl phenylmalonate	(142.2)	.6017	(.659) (20°)
Ethyl phenylpropiolate	(104.2)	.598	(.636) (13°)
Ethyl phosphate	(98.2)	.539	(.576) (25°)
Ethyl propionate	(66.5)	.6514	(.584) (15°)
Ethyl propylacetoacetate	(105.7)	.6135	(.593) (20°)
Ethyl-n-propyl ketone	69.03	$.689_1$	(.560) (22°)
Ethyl succinimide	(72.0)	.566	
Ethyl sulfide	(67.0)	.631	
Ethyl sulfite	(75.4)	.546	(.604) (0°)
Ethylsulfone ethyl ether	(81.8)	.592	
Ethyl thiocyanate	(55.7)	.6392 (27.4°)	(.637) (25°)
Ethyl isothiocyanate	(59.0)	.6772 (27.4°)	(.680) (15°)
Ethyl thiolacetate	(62.7)	.6019 (27.4°)	(.586) (25°)
Ethyl thionacetate	(63.5)	.6098 (27.4°)	
Ethyl tribromoacetate	(119.5)	.368	(.821) (20°)
Ethyl trichloracetate	(99.6)	.520	(.719) (20°)
N-Ethyl urea	55.5	(.630)	(.764) (18°)
Ethyl iso-valerate	(91.1)	.700	(.607) (20°)
Eucalyptol	(116.3)	.754	(.699) (20°)
Eugenol and iso-eugenol	(102.1)	.622	(.663) (20°)
Flavanthrone	241.0	.590	
Fluorene	110.5	.655	
Fluorenone	99.4	.552	(.623) (100°)
Fluorobenzene	(58.4)	.608	(.623) (20°)
Fluorobromacetic acid	(59.5)	.379	
Fluorodichloromethane	48.8	.474	(.676) (0°)
p-Fluorophenetole	(88.0)	.628	
Fluoro trichloroethylene	72.5	.485	(.742) (25°)
Fluorotrichloromethane	58.7	.427	(.638) (17°)
Formaldehyde	(18.6)	.62	(.51) (−20°)
Formamide	(21.9)	.486	(.551) (20°)
Formic acid	19.90	.432	.527 (20°)
β-Formylpropionic acid	55.3	(.542)	
Fructose	102.60	.570	
Fulvene (Bensene χ_M measured to be 49)	42.9	(.549)	(.452) (20°)
Fulvene χ 54.8/49	48.0	(.614)	(.505) (20°)
Fumaric acid	49.11	(.423)	(.692) (20°)
Furan	43.09	.633 (20°)	(.598) (15°)
Furfural	47.1	(.490)	(.568) (20°)
Galactose	103.00	.572	
Gallic acid	90.0	(.529)	(.896) (4°)
Geraniol formate	(119.9)	.658	(.610) (20°)
Glucose	102.60	.570	
D-Glucose	101.5	(.563)	(.869) (25°)
Glutamic acid	78.5	(.533)	
Glycerol	57.06	.619	.779 (20°)
Glycine	40.3	(.537)	(.846) (50°)
Glycol	38.80	.624	.698 (20°)
Gluaiscol	(79.2)	.638	(.720) (21°)
Helianthrone	189.9	.497	
1,2-Heptadiene	73.5	(.764)	
2,3-Heptadiene	72.1	(.749)	
Heptaldehyde	81.02	$.709_6$	(.603) (20°)
n-Heptane	85.24	.8507	.5817 (20°)
4-Heptanol	91.5	.789	(.647) (20°)
n-Heptanoic acid	88.60	.680	.626 (20°)
n-Heptyl amine	93.1	(.808)	(.628) (20°)
n-Heptyl benzene	134.41	.7625	.6528 (20°)
Heptyl cyclohexane	147.40	.8084	.6559 (20°)
n-Heptylic acid	89.74	.6900	(.630) (25°)
1-Heptyne	77.0	(.801)	(.584) (25°)
2-Heptyne	79.5	(.826)	(.615) (25°)
Hexabromoethane	(148.0)	.294	(1.124) (20°)
Hexachlorobenzene	(147.5)	.518	(1.059) (24°)
Hexachloroethane	(112.7)	.476	(.995) (20°)
Hexachlorohexatrione	(145.0)	.433	
n-Hexadecane	187.63	.8286	.6421 (20°)
1,5-Hexadiene	(55.1)	.671	(.462) (20°)
2,3-Hexadiene	60.9	(.741)	
n-Hexaldehyde	69.40	$.693_0$	
2,2,4,7,9,9-Hexamethyldecane	191.52	.8458	.6596 (20°)
Hexamethyl dialoxane	118.9	.7324	
Hexamethylene glycol	84.30	.713	
n-Hexane	(74.6)	.8654 (27.4°)	(.565)
Hexene	65.7	(.781)	
Hexestrol	(165)	.61 ± .02	
n-Hexyl alcohol	79.20	.774	.637 (20°)
n-Hexyl benzene	124.23 (20°)	(.767)	(.658) (20°)
n-Hexyl methyl ketone	91.4_2	$.713_1$	(.583)
n-Hexyl methyl ketoxime	102.5_8	$.716_2$	(.634) (20°)
Hexylpropiolanide	(103.7)	.677	
Hydrinidene	(78.5)	.664	(.639) (16°)
Hydroquinone	64.63	.587	(.797) (20°)
Hydroxyazobenzene	(99.7)	.503	
p-Hydroxybenzaldehyde	66.8	(.547)	(.618) (130°)
4-Hydroxy-2-butanone	48.5	.55	(.573) (14°)
Indene (natural)	84.79	(.730)	(.723) (25°)
Indene (synthetic)	80.89	(.696)	(.690) (25°)
Indole	85.0	(.726)	
Iodobenzene	92.00	.451	.826 (20°)
Iodoform (in sol'n)	117.1	.2974 (20°)	(1.192) (17°)
1-Iodo-2-phenylacetylene	(110.1)	.483	
o-Iodotoluene	(112.3)	.5145 (30°)	(.874) (20°)
m-Iodotoluene	(112.3)	.5152 (30°)	(.875) (20°)
p-Iodotoluene	101.31	(.465)	(.780) (40°)
Leucine	84.9	(.647)	
iso-Leucine	84.9	(.647)	
Maleic acid	49.71	(.428)	(.681) (20°)
Maleic anhydride	(35.8)	.365	(.341) (20°)
Malonic acid	(46.3)	.4453	(.726) (15°)
Mannitol	111.20	.610	(.908) (20°)
Mannose	102.90	.571	(.879)
Mesitylene	92.32	.7682 (20°)	(.665) (20°)
Methane	12.2_7	(.765)	
Methione	91.0	(.610)	
p-Methoxyazobenzene	(118.9)	.560	
o-Methoxybenzaldehyde	76.0	(.558)	(.632) (20°)
p-Methoxybenzaldehyde	78.0	(.572)	(.642) (20°)
o-Methoxybenzyl alcohol	87.9	.637	(.664) (25°)
I-Methoxynaphthalene	107.0	(.676)	(.741) (14°)
2-Methoxynaphthalene	107.6	(.680)	
1-(o-Methoxyphenylazo)-2-naphthol	163.6	(.588)	
Methoxysaligenin acetate	110.3	.613	
Methyl acetate	42.60	.575	.537 (20°)
Methyl acetoacetate	59.60	$.513_2$	(.553) (20°)
Methylacetylacetone	(65.0)	.569	
Methyl alcohol	21.40	.668	.530 (20°)
Methylallylaketone	111.9	(.1.330)	
Methylamine	(27.0)	.870	(.608) (−11°)
N-Methylaniline	82.74	(.773)	(.762) (20°)
9-Methylanthracene	146.5	.762	(.812) (99°)
Methyl benzoate	81.59	$.599_3$	(.651)(25°)
Methyl-o-benzoylbenzoate	139.4	(.580)	(.690) (19°)
Methylbenzylaniline	(132.2)	.670	
Methyl bromide	47.8	.451	(1.044) (0°)
2-Methylbutane	64.40	.8925	.5531 (20°)
2-Methyl-2-butene	54.14	(.772)	(.516) (13°)
Methyl butyl ketone	(69.1)	.690	(.563) (15°)
Methyl iso-butyl ketone	(69.3)	.692	(.554) (20°)
Methyl tert-butyl ketone	(70.4)	.703	(.562) (16°)
p-Methyl-o-tert-butylphenol	120.3	(.732)	
Methyl butyrate	(66.4)	.6498	(.588) (16°)
o-Methylcarbanilde	154.0	(.681)	
Methyl chloride	(32.0)	.633	
Methyl chloroacetate	58.1	(.535)	(.661) (20°)
Methylcholanthrene	(182)	.68 ± .04	
3-Methylcholanthrene	194.0	(.723)	
Methylcyclohexane	78.91	.8038	.6181 (20°)
2-Methylcyclohexanone	(74.0)	.660	(.610) (18°)
3-Methylcyclohexanone	(74.8)	.667	(.610) (20°)
4-Methylcyclohexanone	(63.5)	.566	(.516) (24°)
Methylcyclopentane	70.17	.8338	.6245 (20°)
4-Methyl-2,6-di-tert-butylphenol	167.6	(.761)	
Methyl dichloracetate	73.1	(.511)	
Methyldiphenoxyphosphine oxide	(152.9)	.616	
Methyldiphenyltriazine	(155.1)	.627	
Methylene bromide	65.10	.375	.935 (20°)
Methylene chloride	(46.6)	.549	(.733) (20°)
Methylene iodide	93.10	.348	1.156 (20°)
Methylene succinic acid	57.57	(.443)	(.723)
Methyl ether	26.3	.571	
Methylethylallyacetophenone	133.3	(.659)	(.643) (16°)
Methyl ethyl ketone	45.5_8	$.632_2$	(.509) (20°)
Methyl formate	(32.0)	.5327	(.519) (20°)
Methylfurmaic acid	56.98	(.438)	(.642)
3-Methylheptane	97.99	.8580	.6056 (20°)
2-Methyl-4-heptene	88.0	(.784)	
5-Methyl-1,2-hexadiene	73.6	(.765)	(.553) (19°)
2-Methylhexane	86.24	.8607	.5841 (20°)
Methyl hexyl ketone	(93.3)	.728	(.596) (20°)
Methyl-m-hydroxybenzoate	88.4	(.581)	
Methyl-p-hydroxybenzoate	88.7	(.583)	
Methyl iodide	(57.2)	.403	(.918) (20°)
Methylmaleic acid	57.84	(.446)	(.721)
9-Methyl-10-methoxyanthracene	158.1	(.711)	
Methyl-o-methoxybenzoate	95.6	(.575)	(.665) (19°)
Methyl-p-methoxybenzoate	98.6	(.593)	
Methyl-α-methoxy-isobutyrate	(81.9)	.620	
1-Methylnaphthalene	102.8	(.723)	(.741) (14°)
2-Methylnaphthalene	102.6	(.722)	(.743) (20°)
4-Methylnonane	121.39	(.853)	(.625) (20°)

DIAMAGNETIC SUSCEPTIBILITIES OF ORGANIC COMPOUNDS (Continued)

Compound	$-\chi_M \times 10^6$	$-\chi \times 10^6$	$-K \times 10^6$
5-Methyl-5-nonene	111.6	(.796)	
4-Methyloctane	109.63	(.855)	(.618) (20°)
Methylol urea	48.3	(.493)	
2-Methylpentane	75.26	.8734	.5707 (20°)
3-Methylpentane	75.52	.8764	.5823 (20°)
4-Methyl-2-pentanol	80.4	.788	(.641) (20°)
Methyl perfluor-n-butyrate	92.5	(.406)	
Methyl phenylacetate	92.73	(.618)	(.645) (16°)
Methyl phenylpropriolate	95.6	.597	
2-Methylpropene	44.4	(.791)	
Methyl propionate	(55.0)	.6240	(.571) (20°)
Methyl-n-propyl ketone	57.41	.666$_4$	(.541) (15°)
Methyl-iso-propyl ketone	58.45	.679$_0$	(.545) (20°)
1-Methylpyrrole	58.56	.722 (20°)	(.664) (10°)
2-Methylpyrrole	60.10	.741 (20°)	(.700)
Methyl salicylate	86.30	.567	.668 (20°)
Methyl silicone	(172.7)	.730	
α-Methyl styrene	(80.1)	.678	(.620) (20°)
2-Methylthiazole	59.56	.601 (20°)	
2-Methylthiophene	66.35	.676 (20°)	(.689) (20°)
Methyl trichloroacetate	84.2	(.475)	(.707) (19°)
N-Methyl urea	44.6	(.602)	(.725)
Morpholine	55.0	(.631)	(.631)
Myleran	169.7	.69	(.834) (4°)
Myristic acid	176.0	(.7741)	(.661) (60°)
Naphthalaldehydic acid	117.6	(.588)	
Naphthalene	(91.9)	.717	(.821) (20°)
Naphthalene picrate	185.9	(.523)	
2-Naphthalenesulfonylamine	127.6	(.616)	
2-Naphthalenesulfonyl chloride	121.91	(.538)	
meso-Naphthodianthrene	214.6	.612	
meso-Naphthodianthrone	221.8	.583	
1-Naphthol	98.2	.681	(.834) (4°)
	97.0	.673	(.819) (4°)
2-Naphthol	98.25	(.682)	(.829) (4°)
α-Naphthonitrile	103.3	(.674)	(.753) (5°)
β-Naphthonitrile	101.0	(.659)	(.721) (60°)
α-Naphthoquinone	73.5	(.465)	(.661)
β-Naphthoquinone	67.9	(.429)	
N-1-Naphthylacetamide	117.8	(.636)	
N-2-Naphthylacetamide	117.8	(.636)	
1-Naphthylamine	98.8	.690	.757 (54°)
2-Naphthylamine	98.00	(.684)	(.726) (98°)
1-Naphthylamine hydrochloride	(127.6)	.710	
Nicotine	113.328	(.699)	(.705) (20°)
o-Nitroaniline	66.47	(.481)	(.694) (15°)
m-Nitroaniline	70.09	(.507)	(.725) (20°)
p-Nitroaniline	66.43	(.481)	(.691) (14°)
o-Nitrobenzaldehyde	68.23	.4517	
m-Nitrobenzaldehyde	68.55	.4538	
p-Nitrobenzaldehyde	66.57	.4407	(.507) (0°)
Nitrobenzene	61.80	.502	.604 (20°)
Nitrobenzene diazo cyanide	59.22	(.336)	
o-Nitrobenzoic acid	76.11	.4556	(.718) (4°)
m-Nitrobenzoic acid	80.22	.4802	(.717) (4°)
p-Nitrobenzoic acid	78.81	.4718	(.731) (32°)
o-Nitrobemobenzene	87.3	(.432)	(.700) (80°)
m-Nitrobromobenzene	89.5	(.443)	(.755) (20°)
p-Nitrobromobenzene	89.6	(.444)	(.859) (22°)
m-Nitro carbanilide	148.1	(.576)	
Nitroethane	(35.4)	.472	(.497) (20°)
Nitromethane	21.1	.3457	(.391) (25°)
1-Nitronaphthalene	98.47	(.569)	(.696) (62°)
o-Nitrophenol	73.3	.527 (24°)	(.873) (20°)
m-Nitrophenol	70.8	.509 (25°)	(.756)
p-Nitrophenol	69.5	.500 (22°)	(.740)
1-(m-Nitrophenylazo)-2-naphthol	142.0	(.484)	
1-(p-Nirophenylazo)-2-naphthol	141.7	(.483)	
Nitrophenylfluoroform	(84.1)	.440	
2-Nitropropane	45.73	.5135	(.509) (20°)
Nitrosobenzene	59.1	(.552)	
N-Nitrosodiethylamine	59.3	(.580)	(.546) (20°)
p-Nitrosodiethylaniline	92.6	(.520)	(.644) (15°)
p-Nitrosodimethylaniline	73.3	(.488)	
N-Nitrosodiphenylamine	110.7	(.558)	
1-Nitroso-2-naphthol	83.9	(.485)	
2-Nitroso-1-naphthol	82.7	(.478)	
4-Nitroso-1-naphthol	91.8	(.530)	
m-Nitrosonitrobenzene	66.0	(.433)	
p-Nitrosonitrobenzene	65.8	(.433)	
p-Nitrosophenol	50.7	(.412)	
Nitrosopiperidine	(63.4)	.555	(.590) (20°)
p-Nitrosotoluene	70.4	(.581)	
o-Nitrotoluene	72.28	.5272	(.613) (20°)
m-Nitrotoluene	72.71	.5304	(.614) (20°)
p-Nitrotoluene (in sol'n)	72.06	.5257	(.676) (20°)
n-Nonane	108.13	.8431	.6057 (20°)
1,2-Octadiene	83.6	(.759)	
n-Octane	96.63	.8460	.5949 (20°)
Octanoxime	(102.7)	.717	
Oxtyl alcohol	102.65	.7766 (20°)	(.640) (20°)
Oxtyl chloride	(114.9)	.773	(.676) (20°)
Oxtycyclohexane	158.09	.8051	.6578 (20°)
Oxtylene	(89.5)	.798	(.576) (17°)
Oxtylene bromide	(150.4)	.553	
n-Octyl mercaptan	(115.1)	.7866 (27.4°)	
Oenanthylidene chloride	(116.5)	.689	
Oleic acid	208.5	(.738)	(.661) (18°)
Opianic acid	111.5	(.530)	
Ovalene	353.8	.888	
Oxalic acid (anh.)	33.8	(.375)	
Oxalic acid	60.05	(.4763)	(.787)
Oxamide	(39.0)	.443	(.738)
Palmitic acid	198.6	(.775)	(.661) (62°)
Paraldehyde	(86.2)	.652	(.648) (20°)
Pentabromophenol	(194.0)	.397	
Pentacene	(205.4)	.738	
Pentachloroethane	(99.1)	.490	(.819) (25°)
Pentachlorohexadione	(129.5)	.452	
2,3-Pentadiene	49.1	(.721)	(.501) (20°)
n-Pentane	63.05	.8739	.5472 (20°)
2,4-Pentanediol	70.4	.677	
Perfluoroacetic acid	43.3	(.380)	
Perfluoro-n-butyric acid	81.0	(.378)	
Perfluorobutyric anhydride	149.4	(.387)	
Perfluorocyclooctane oxide	157.6	(.379)	
Perfluoropropionic acid	61.0	(.372)	
Perhydroanthracene	146.01	.7592	.7178 (20°)
Perylene	166.8	.662	
Phenanthrene	(127.9)	.718	(.763) (100°)
Phenanthrenequinone	104.5	.502	(.698)
Phenanthrenonitrile	139.0	(.685)	
o-Phenetidine	(101.7)	.741 (25°)	
p-Phenetidine	(96.8)	.706 (25°)	(.749) (15°)
Phenetole	(84.5)	.692	(.689) (20°)
Phenol	60.21	(.640)	(.675) (45°)
Phenothiazine	114.8	(.576)	
Phenylacetaldehyde	72.01	.599$_4$	(.614) (20°)
Phenyl acetate	82.04	(.603)	(.647) (25°)
Phenylacetic acid	82.72	(.608)	(.657) (80°)
Phenylacetylene	72.01	(.705)	(.655) (20°)
1-Phenylazo-2-naphthol	137.6	(.554)	
2-Phenylbensour	130.5	(.672)	
1	(140.3)	.599	
Phenylbutadiene	(85.7)	.658	
4-Phenyl-1-butene	93.49	(.7077)	(.6239) (20°)
Phenylbutyl acetate	134.5	.653	
Phenyl n-butyrate	105.46	(.643)	
Phenyl iso-cyanate	(72.7)	.610	(.699) (20°)
o-Phenylenediamine	71.98	.6662	
m-Phenylenediamine	70.53	.6529	(.723) (58°)
p-Phenylenedianine	70.28	.6503	
Phenyl ether	(108.1)	.635	(.681) (20°)
Phenylethyl sulfide	(94.4)	.6826 (27.4°)	
Phenylfluoroform	(77.3)	.529	
Phenylhydrazine	67.82	(.627)	(.688) (23°)
Phenylhydroxylamine	(68.2)	.625	
Phenyl mercaptan	(70.8)	.6425 (27.4°)	(.693) (20°)
1-Phenyl-2-Methylbutane	113.53	(.766)	(.660) (20°)
Phenylmethyl sulfide	(83.2)	.6695 (27.4°)	
Phenylpropiolamide	(83.3)	.574	
Phenyl propionate	93.79	(.625)	(.654) (25°)
Phenylsulfone	(129.0)	.591	(.740) (20°)
Phenyl thiocyanate	(81.5)	.6027 (27.4°)	(.677) (24°)
Phenyl isothiocyanate	(86.0)	.6365 (27.4°)	(l719) (24°)
1-Phenyl-4,6,6-trimethyl-heptane	173.90	(l796)	(.682) (20°)
N-Phenyl urea	82.1	(.603)	(.785)
Phloroglucinol	(73.4)	.582	
Phthalamide	(91.3)	.556	
Phthalic acid	83.61	.5035	(.802) (20°)
iso-Phthalic acid	84.64	.5097	
tere-Phthalic acid	83.51	.5029	(.759)
Phthalic anhydride	67.31	(.454)	(.694) (4°)
Phthalimide	(78.4)	.533	
Picric acid	84.38	(.368)	(.649)
Piperazine	56.8	(.659)	
Piperidine	64.2	(.754)	(.650) (20°)
Propane	40.5	(.919)	(.538) (−45°)
Propene	31.5	(.749)	(.456) (−47°)
Propionaldehyde	34.32	.591$_0$	(.477) (20°)
Propionic acid	43.50	.586	.582 (20°)
Propionitrile	38.5	(.699)	(.547) (21°)
Propionylphenylacetylene	(95.1)	.601	
Propiophenone	83.73	.624$_0$	(.631) (20°)
n-Propyl acetate	65.91	.645$_3$ (25°)	(.569) (25°)
iso-Propyl acetate	67.04	.656$_4$	(.566) (25°)
n-Propyl alcohol	45.176 (20°)	(.7518)	(.6047) (20°)
iso-Propyl alcohol	45.794 (20°)	(.7621)	(.5985) (20°)
9-Propylanthracene	164.0	(.744)	
n-Propylbenzene	89.24	(.742)	(.640) (20°)
n-Propylbenzoate	105.00	(.640)	(.646) (25°)
n-Propl bromide	(65.6)	.533	(.721) (20°)
iso-Propyl bromide	(65.1)	.529	(.693) (20°)
Propyl butyrate	(89.4)	.6867	(.604) (15°)
prim-Propyl chloride	56.10	.715	.633 (20°)
iso-Propylcyclohexane	102.65	.8131	.6528 (20°)
Propylenediamine	(58.1)	.784	(.688) (15°)
Propylenediamine	(58.1)	.784	(.688) (15°)
Propylene oxide	42.5	(.732)	(.629) (0°)
Propyl formate	(55.0)	.6248	(.563) (20°)
Propyl hexylpropiolate	(136.8)	.697	
Propyl iodide	(84.3)	.4958 (30°)	(.864) (20°)
Propyl propionate (extrap.)	(77.95)	.6711	(.593) (20°)
Propyl sulfide	(92.1)	.7787 (27.4°)	(.634) (17°)

DIAMAGNETIC SUSCEPTIBILITIES OF ORGANIC COMPOUNDS (Continued)

Compound	$-\chi_M \times 10^6$	$-\chi \times 10^6$	$-K \times 10^6$
N-Propyl urea	67.4	(.660)	
Pseudocumene	(101.6)	.845 (20°)	(.740) (20°)
Pyramidone	149.0	(.645)	
Pyranthrene	266.9	.709	
Pyranthrone	250.3	.616	
Pyrazine	37.6	(.469)	(.484) (61°)
Pyrene	147.9	.731	(.933) (0°)
Pyridine	49.21	(.622)	(.611) (20°)
Pyrocatechol	68.76	.6248	(.857) (15°)
Pyrrole	47.6	(.709)	(.688) (20°)
Pyrrolidine	54.8	(.771)	(.657) (23°)
Quinoline	86.0	(.666)	(.729) (20°)
Quinone	38.4	(.355)	(.468) (20°)
Quinonoxime	(50.4)	.409	
Resorcinol	67.26	.6112	(.785) (15°)
Rhamnose	99.20	.605	(.890) (20°)
Safrol and iso-Safro	(97.5)	.601	(.66) (20°)
Salicylaldehyde	64.4	(.527)	(.615) (20°)
Salycylic acid	72.23	.523	(.755) (20°)
Saligenin	76.9	.620	(.720) (25°)
Salol	(123.2)	.575	.678 (45°)
Salvarsan dihydrochloride	(246.1)	.518	
Selenophene	66.82	.510	
iso-Selenophene	110-111	.84-.85	
trans-Selenophene	70-77	.53-.59	
Sorbitol	107.80	.592	
Stearic acid	220.8	(.776)	(.657) (69°)
Stilbene	(120.0)	.666	(.646) (125°)
Stilbestrol	(130)	.62, .63	
Styrene	(68.2)	.655	(.594) (20°)
Succinic acid	(57.9)	.4902	(.767) (15°)
Succinic anhydride	(47.5)	.475	(.524)
Succinimide	(47.3)	.477	(.674) (16°)
Sulfamide	44.4	(.462)	(.832)
p-Sulfanilamide	80.15	(.465)	
Terpineol	111.9	(.725)	(.678) (room temp.)
Tetrabenzylmonosilane	266.2	(.678)	
1,1,2,2-Tetrabromoethane	(123.4)	.357	(1.058) (20°)
Tetrabromoethylene	(114.8)	.334	
Tetracene	(168.0)	.736	
1,1,2,2-Tetrachloroethane	(89.8)	.535	(.856) (20°)
Tetrachloroethylene	81.6	.492	(.802) (15°)
Tetrahydroquinoline	(89.0)	.668	(.715) (4°)
Tetraiodoethylene	(164.3)	.309	(.922) (20°)
Tetraiodopyrrole	(188.9)	.331	
Tetramethylketotetrahydrofurfurane	(104.7)	.736	
Tetranitromethane	43.02	.2195	(.360) (20°)
Tetraphenylbutadiene	228.0	(.636)	
Tetraphenyldecapentaene	280.8	(.643)	
Tetraphenylhexatriene	246.4	(.641)	
Tetraphenyloctatetraene	264.1	(.643)	
Tetraphenyltubene	344.0	(.646)	
Tetra-p-tolymonosilane	276.4	(.704)	
Tetrolic acetal	(97.8)	.688	
Tetronic acid	(52.5)	.525	
Thiacoumerin	93.6	.577	
Thiazole	50.55	.595 (20°)	(.714) (17°)
Thiobarbituric acid	72.9	(.506)	
Thiophene	57.38	.682 (20°)	(.726) (20°)
Tolane	(118.9)	.667	(.644) (100°)
Toluene	66.11	.7176	.6179 (20°)
o-Toluidine	76.0	.710 (24°)	(.709) (20°)
m-Toluidine	74.6	.697 (25°)	(.689) (20°)
p-Toluidine	72.1	.673 (25°)	(.704) (20°)
α-Tolunitrile	76.87	(.656)	(.666) (18°)
1-(o-Tolylazo)-2-naphthol	148.7	(.567)	
1-(p-Tolylazo)-2-naphthol	157.6	(.601)	

Compound	$-\chi_M \times 10^6$	$-\chi \times 10^6$	$-K \times 10^6$
Triallylacetophenone	152.5	(.634)	
Tri-iso-amylamine	192	.845	(.647) (25°)
Trianilinophosphine oxide	(201.7)	.624	
1,2,3-Tribromopropane	(117.9)	.420	(1.023) (23°)
Tri-iso-butylamine	(156.8)	.846	(.646) (25°)
Trichloroacetic acid (in sol'n)	73.0	(.44)	(.723) (46°)
Trichlorobenzene	(106.5)	.587	
Trichlor-tert-butyl alcohol (in sol'n)	98.01	.552	
Trichloroethylene	65.8	.501	(.734) (20°)
Trichloronitromethane	(75.3)	.458	(.756) (20°)
Triethylamine	81.4	(.804)	(.586) (20°)
Triethyl citrate	(161.9)	.586	(.666) (20°)
Triethyl phosphate	(125.3)	.688	(.735) (20°)
Triethylphosphine	(90.0)	.762	(.610) (15°)
Triethylphosphine oxide	(91.6)	.683	
Triethyl phosphite	(104.8)	.631	(.611) (20°)
Triethyl triazinetricarbonate	(164.1)	.552	
Trifluorocresol	(83.8)	.517	
Tri-n-heptylmaine	251.3	(.806)	
Tri-n-hexylamine	221.7	(.823)	
Trimethylacetophenone	108.2	(.667)	(.648) (16°)
2,2,3-Trimethylbutane	88.36	.8818	.6086 (20°)
2,2,3-Trimethylpentane	99.86	.8743	.6261 (20°)
2,2,4-Trimethylpentane	98.34	.8610	.5958 (20°)
2,3,5-Trimethylpyrrole	82.31	.754 (20°)	
1,3,5-Trinitrobenzene	74.55	(.350)	(.591) (20°)
Triperfluorobutylamine	253.0	(.377)	
Triphenoxyarsine	(195.2)	.551	
Triphenylarsine	(177.0)	.578	
Triphenylarsine dihydroxide	(270.5)	.795	
Triphenylarsine oxide	(199.1)	.618	
Triphenylbismuthine	(196.8)	.447	(.708) (20°)
Triphenylbismuthine dinitrate	(254.5)	.451	
Triphenylcarbinol	(175.7)	.675	(.802) (20°)
Triphenylmethane	(165.6)	.678	.686 (100°)
Triphenylphosphine	(166.8)	.636	(.759)
Triphenyl phosphite	(183.7)	.592	(.701) (18°)
Triphenylstilbine	(182.2)	.516	
Triphenylstibine dihydroxide	(238.5)	.616	
N,N′,N-Triphenyl urea	176.5	(.613)	
Triquinoyl	(133.0)	.426	
Tropolone	61	.50	
Tryptophan	132.0	(.646)	
Tyrosine	105.3	(.581)	
Undecane	131.84	(.8435)	(.6247) (20°)
Urea	33.4	(.556)	(.742) (20°)
Urethan	(57)	.64	(.63) (21°)
iso-Valeraldehyde	(57.5)	.668	(.536) (17°)
n-Valeric acid	66.85	.6548	(.617) (20°)
iso-Valeric acid	(67.7)	.663	(.631) (15°)
Valerylphenylacetylene	(119.0)	.639	
Valine	74.3	(.634)	
Violanthrene	273.5	.641	
Violanthrone	204.8	.449	
iso-Violanthrone	215.9	.473	
Water	(13.00)	.7218 (20°)	(.7205) (20°)
Water (value usually used as standard)	(12.97)	.720 (20°)	(.719) (20°)
Xanthone	(108.1)	.551	
o-Xylene	77.78	.7327	.6440 (20°)
m-Xylene	76.56	.7212	.6235 (20°)
p-Xylene	76.78	.7232	.6226 (20°)
Xylose	84.80	.565	(.862) (20°)

Supplementary Table

Formula	Compound	$-\chi_M \times 10^6$	$-\chi \times 10^6$	$-K \times 10^6$
$H_6N_3B_3$	Borazine	49.6	.616	
CO	Carbon monoxide	9.8	.350	
COS	Carbon oxysulfide	32.4	.539	
$COCl_2$	Phosgene	48	.485	(.675) (19°)
CO_2	Carbon dioxide	20	.45	
$CNCl$	Cyanogen chloride	32.4	.527	(.642) (4°)
CCl_2S	Thiophosgene	50.6	.440	(.664) (15°)
CH_2N_2	Cyaniamide	24.8	.590	(.639)
CH_3SiCl_3	Methyl trichlorsilane	87.45	.584$_8$	
CH_4N_2S	Thiourea	42.4	.557	(.782) (20°)
C_2HF_3	Trifluoroethylene	32.2	.393	
$C_2HF_3Cl_2$	1,2-Dichloro-1,1,2-trifluoroethane	66.2	.433	
$C_2HF_3Br_2$	1,2-Dibromo-1,1,2-trifluoroethane	90.9	.376	
$C_2H_3O_2N_3$	Urasole	46.2	.456	
C_2H_4OS	Thioacetic acid	38.4	.505	(.542) (10°)
$C_2H_4O_2S$	Mercaptoacetic acid	50.0	.543	(.720) (20°)
C_2H_4NS	Thioacetamide	42.45	.565	
$C_2H_5SiCl_3$	Ethyl trichlorosilane	98.84	.604$_6$	
$C_2H_6O_2S_2$	Methyl thiosulfite	62.3	.494	
$C_2H_6O_3S$	Methyl sulfite	54.2	.492	
$C_2H_6N_2S$	N-Methyl thiourea	53.6	.595	
C_2H_6S	Ethyl mercaptan	47.0	.756	(.635)

Formula	Compound	$-\chi_M \times 10^6$	$-\chi \times 10^6$	$-K \times 10^6$
$C_2H_6SiCl_2$	Dimethyl dichlorasilane	82.45	$.639_2$	
$C_2H_8N_2 \cdot HCl$	1,2-Diaminoethane hydrochloride	76.2	.789	
C_3H_3ON	Pyruvic nitrile	33.2	.481	
$C_3H_5OF_3$	1-Methoxy-1,1,2-trifluoroethane	55.9	.490	
C_3H_5OCl	Propionyl chloride	51	.55	(.59)
$C_3H_5O_2Br$	Methylbromoacetate	71.1	.465	
C_3H_5Cl	1-Chloro-2-propene	47.8	.625	
C_3H_5Br	1-Bromo-2-propene	58.6	.484	
C_3H_5I	1-Iodo-2-propene	72.8	.433	
$C_3H_6N_6$	Melamine	61.8	.490	(.771) (250°)
C_3H_7N	1-Amino-2-propene	40.1	.702	
$C_3H_7SiCl_3$	n-Propyl trichlorosilane	110.2	$.620_0$	
$C_3H_8O_2N_2$	N-Hydroxymethyl-N-methylurea	68.0	.653	
$C_3H_8N_2S$	N,N-Dimethyl thiourea	64.2	.616	
$C_3H_8N_2S$	N,N′-Dimethyl thiourea	64.7	.621	
$C_3H_8N_2S$	N-Ethyl thiourea	62.8	.603	
C_3H_8S	Propyl mercaptan	58.5	.768	(.642) (25°)
C_3H_9SiCl	Trimethyl chlorosilane	77.36	$.712_2$	
$C_3H_{12}N_3B_3$	N-Trimethyl borazine	78.6	.641	
$C_4H_5ON_3$	Cytosine	55.8	.502	
$C_4H_6O_2$	Vinylacetate	46.4	.539	(.502)
$C_4H_6O_2NCl_3$	Nitro tri(chloromethyl) methane	107.8	.522	
$C_4H_6O_6$	Tartaric acid	67.5	.402	
C_4H_7OCl	Butyryl chloride	62.1	.582	(.598) (20°)
C_4H_7OCl	iso-Butryl chloride	63.9	.599	
$C_4H_8N_2S$	N-Allyl thiourea	69.0	.595	(.725) (20°)
$C_4H_9ON_3$	Acetone semicarbazone	66.29	$.575_8$	(.716) (20°)
$C_4H_{10}O_2S_2$	Ethyl thiosulfite	86.2	.559	
$C_4H_{10}S_2$	Ethyl disulfide	83.6	.684	(.679) (20°)
$C_4H_{10}SiCl_2$	Diethyl dichlorosilane	105.80	$.673_5$	
$C_5H_4O_3N_4$	Uric acid	66.2	.394	(.746)
C_5H_5	Cyclopentadiene	44.5	.673	
$C_5H_6O_2N_2$	Thymine	57.1	.453	
$C_5H_8O_2$	Methyl methacrylate	57.3	.572	(.535) (20°)
$C_5H_8N_2$	3,5-Dimethyl pyrazole	56.2	.585	
$C_5H_8Cl_4$	Tetra (chloromethyl) methane	129	.652	
$C_5H_9S_3Na$	Sodium butyl thiocarbonate	104.5	.555	
$C_5H_9Cl_3$	1,1,1-Tri(chloromethyl) ethane	114	.650	
$C_5H_{10}O$	Cyclopentanol	64.0	.743	(.705) (20°)
$C_5H_{10}O_2$	Ethyl carbonate	75.4	.738	
$C_5H_{10}O_6$	Di-Ribose	84.6	.564	
$C_5H_{10}NS_2Na$	Sodium diethyldithiocarbonate	99.2	.579	
$C_5H_{10}Cl_2$	2,2-Di-(Chloromethyl) propane	96.9	.687	
$C_5H_{11}ON$	Methyl n-propyl ketoxime	68.82	$.679_6$	(.618) (20°)
$C_5H_{11}ON_3$	Ethyl methyl ketone semicarbazone	77.93	$.603_4$	(.710) (20°)
$C_5H_{12}ON_2$	N,N′-Diethyl urea	74.1	.638	
$C_5H_{12}ON_2$	N,N,N′,N′-Tetramethyl urea	75.7	.652	(.634) (15°)
$C_5H_{14}O_2Si$	Methyl diethoxysilane	92.99	$.692_4$	
$C_6H_3O_4N_2Cl$	1-Chlor-2,4-dinitrobenzene	84.4	.417	(.708) (22°)
$C_6H_3Cl_2I$	3,4-Dichloro-1-iodobenzene	118	.432	
$C_6H_5SiCl_3$	Phenyl trichlorosilane	120.4	$.569_4$	
C_6H_6NBr	o-Bromoaniline	87.32	.508	
C_6H_6NBr	m-Bromoaniline	84.89	.494	(.780) (20.4°)
C_6H_6NBr	p-Bromoaniline	84.06	.489	(.880)
$C_6H_7O_3NS \cdot H_2O$	Sulfanlic acid	90.1	.471	
C_6H_7N	2-Methyl pyridine	60.3	.648	(.616) (15°)
C_6H_7N	3-Methyl pyridine	59.8	.642	(.617) (15°)
C_6H_7N	4-Methyl pyridine	59.8	.642	(.614) (15°)
$C_6H_7N \cdot HI$	Aniline hydroiodide	113.6	.514	
C_6H_{10}	2,3-Dimethyl-1,3-butadiene	57.2	.696	(.539) (0°)
C_6HuOCl	2-Ethyl butyryl chloride	87.2	.648	
$C_6HuSiCl_3$	Cyclohexytrichlorosilane	138.1	$.634_3$	
$C_6H_{12}N_2S_3$	Tetramethyl thiuran monosulfide	118.4	.568	(.795)
$C_6H_{12}N_2S_4$	Tetramethl thiuran disulfide	140.6	.585	(.755) (20°)
$C_6H_{13}ON$	Methyl-n-butyl ketoxime	79.97	$.694_2$	(.623) (20°)
$C_6H_{13}ON_3$	Methyl-n-propyl ketone semicarbazone	89.47	$.624_9$	(.669) (20°)
$C_6H_{13}ON_3$	Diethyl ketone semicarbazone	90.68	$.633_3$	(.730) (20°)
C_6H_{14}	3-Ethyl pentane	86.21	$.861_0$	(.601) (20°)
$C_6H_{14}O$	iso-Propyl ether	79.4	.777	(.564) (20°)
$C_6H_{14}S_2$	Propyl disulfide	106.2	.707	
$C_6H_{16}O_2Si$	Dimethyl diethoxysilane	104.6	$.705_2$	
$C_6H_{18}O_3Si_3$	Hexamethylcyclotrisolaxane	140.7	$.632_4$	
$C_6H_{18}N_3B_3$	Hexamethylborazine	119	.723	
$C_7H_5O_2Cl$	o-Chlorobenzoic acid	83.56	.534	(.824) (20°)
$C_7H_5O_4N$	Quinoleic acid	72.3	.433	
$C_7H_5NS_2$	2-Mercaptobenzothiazole	99.4	.495	(.843) (20°)
$C_7H_5F_2Cl$	Chlorodifluoromethylbenzene	87.2	.536	
$C_7H_7O_2N$	o-Aminobenzoic acid	77.18	.563	
C_7H_7Cl	o-Chlorotoluene	81.98	.648	(.701) (20°)
C_7H_7Cl	m-Chlorotoluene	80.07	.633	(.679) (20°)
C_7H_7Cl	p-Chlorotoluene	80.07	.633	(.677) (20°)
$C_7H_8N_2S$	N-phenylthiourea	87.4	.574	(.75)
C_7H_9ON	o-Anisidine	80.44	.654	(.714) (20°)
C_7H_9ON	m-Anisidine	79.95	.650	(.712) (20°)
C_7H_9ON	p-Anisidine	80.56	.655	(.702) (55°)
C_7H_9N	2,4-Dimethyl pyridine	71.50	.667	(.633) (0°)

Formula	Compound	$-\chi_M \times 10^6$	$-\chi \times 10^6$	$-K \times 10^6$
C_7H_9N	2,6-Dimethyl pyridine	71.72	.669	(.630) (0°)
$C_7H_{15}ON$	Methyl amyl ketoxime	91.24	$.706_2$	(.630) (20°)
$C_7H_{15}ON_3$	Methyl-n-butyl ketone semicarbazone	100.40	$.638_6$	(.643) (20°)
$C_7H_{16}O$	1-Heptanol	91.7	.790	(.649) (20°)
$C_7H_{18}O_3Si$	Methyl triethoxysilane	120.6	$.676_2$	
C_8H_7OCl	Phenylacetyl chloride	88.5	.572	(.668) (20°)
$C_8H_8O_2$	o-Toluic acid	80.83	.594	(.626) (112°)
$C_8H_{10}ON_2$	N-Methyl, N′Phenyl urea	93.35	.622	
$C_8H_{10}O_4$	Butyne diacetate	95.9	.563	
$C_8H_{11}ON$	m-Phenetidine	90.28	.659	
$C_8H_{11}N$	2,4,6-Trimethyl pyridine	83.22	.687	(.630) (20°)
$C_8H_{16}O_2$	Hexylacetate	100.9	.700	(.623) (0°)
$C_8H_{17}ON$	Methyl amyl ketone semicarbazone	110.05	$.633_5$	(.687) (20°)
C_8H_{18}	3-Methyl-3-ethyl pentane	99.9	.875	(.623)
C_8H_{18}	4-Methyl heptane	97.30	$.851_7$	(.614)
C_8H_{18}	3-Ethyl hexane	97.76	$.855_8$	(.614) (20°)
C_8H_{18}	2,3,4-Trimethyl pentane	99.75	$.872_3$	
$C_8H_{18}S_2$	Butyl disulfide	129.6	.727	
$C_8H_{18}S_3$	Butyl trisulfide	144.5	.687	
$C_8H_{18}S_4$	Butyl tetrasulfide	158.5	.653	
$C_8H_{20}O_4Si$	Tetraethoxysilane	137.1	$.657_9$	
$C_8H_{24}O_4Si_4$	Octamethyl cyclotetrasiloxane	187.4	$.632_2$	
C_9H_5N	Phenyl propionitrile	78.2	.615	
$C_9H_6O_2$	Phenyl propiolic acid	81.0	.554	
$C_9H_6O_3$	7-Hydroxycoumarin	88.22	$.544_9$	
C_9H_7N	iso-Quinoline	83.9	.650	(.714) (20°)
$C_9H_{10}O_4N_4$	Acetone-2,4-dinitrophenylhydrazone	110.62	$.464_4$	(.653) (20°)
$C_9H_{11}ON$	Benzylidenaemino-1-hydroxyethane	91.0	.610	
$C_9H_{11}N$	Benzylideneaminoethane	85.5	.642	
$C_9H_{12}O_3S$	Ethyl-p-toluene sulfonate	115	.574	
$C_9H_{12}O_6N_2$	Unidine	106.9	.438	
$C_9H_{13}O_5N_3$	Cytidine	123.7	.509	
$C_9H_{14}O_3$	1-Acetyl-1 ethoxycarbonyl cyclobutane	103.4	.608	
$C_9H_{14}O_4$	1,1-Di(ethoxycarbonyl) cyclopropane	110.4	.593	
$C_9H_{14}Si$	Trimethylphenylsilane	109.1	.726	
$C_9H_{19}ON_3$	Methyl-n-hexyl ketone semicarbazone	123.60	$.662_2$	(.716) (20°)
$C_9H_{20}O$	Di-isobutyl carbinol	116.9	.810	(.667) (0°)
$C_9H_{20}ON_2$	N,N,N′N′-Tetracthylurea	122.4	.710	
$C_9H_{20}N_2S$	N,N,N′,N′-Tetracthylthiourea	132.6	.704	
$C_9H_{24}N_3B_3$	B-Triethyl-N-Trimethyl borzine	146	.706	
$C_{10}H_6O_7N_2$	5,7-Dinitro-7-hydroxy-4-methylcoumarin	111.9	$.420_6$	
$C_{10}H_7O_3Br$	7-Hydroxy-4-methyl-3-bromocoumarin	127.3	$.498_7$	
$C_{10}H_7O_5N$	5-Nitro-6-hydroxy-4-methylcoumarin	105.6	$.477_9$	
$C_{10}H_7O_5N$	6-Nitro-7-hydroxy-4-met-ylcoumarin	105.5	$.477_5$	
$C_{10}H_7O_6N$	8-Nitro-7-hydroxy-4-methylcoumarin	106.0	$.479_8$	
$C_{10}H_8O_3$	6-Hydroxy-4-methylcoumarin	98.69	$.560_7$	
$C_{10}H_8O_3$	7-Hydroxy-4-methylcoumarin	99.96	$.563_7$	
$C_{10}H_8O_4$	Benzylidene malonic acid	97.5	.507	
$C_{10}H_8O_4$	5,7-Dihydroxy-4-methylcoumarin	106.9	$.556_7$	
$C_{10}H_8O_4$	7,8-Dihydroxy-4-methylcoumarin	105.1	$.550_4$	
$C_{10}H_8S$	1-Mercaptonaphthalene	109.5	.684	
$C_{10}H_8S_2$	1,8-Dimercaptonaphthalene	118.0	.614	
$C_{10}H_{10}O_4$	Methyl terephthalate	101.6	.523	
$C_{10}H_{10}O_8$	Bis cyclopentadienyl osmium	193(20°)	.602	
$C_{10}H_{12}$	1,2,3,4-T etrahydronaphthalene	93.3	.706	(.685)
$C_{10}H_{12}O_4N_4$	Ethyl methyl ketone-2,4-dinitrophenylhydrazone	124.11	$.492_1$	(.631) (20°)
$C_{10}H_{13}O_4N_5$	Adenosine	137.5	.514	
$C_{10}H_{13}O_5N_5$	Guanosine	149.1	.526	
$C_{10}H_{14}$	Durene	101.2	.754	(.632) (81°)
$C_{10}H_{14}$	see-Butyl benzene	101.31	$.754_9$	(.651) (20°)
$C_{10}H_{16}O_4$	1,1-Di(ethoxycarbonyl) cyclobutane	118	.589	
$C_{10}H_{19}N$	Camphylamine	122.0	.796	
$C_{11}H_9BrS$	1-Bromo-4-Methylthionaphthalene	147.0	.581	
$C_{11}H_{10}O_2$	4,6-Dimethylcoumarin	106.2	$.610_3$	
$C_{11}H_{10}O_2$	4,7-Dimethylcoumarin	107.6	$.619_0$	
$C_{11}H_{10}O_3$	5-Hydroxy-4,7-dimethytcoumarin	113.5	$.597_3$	
$C_{11}H_{10}O_3$	6-Methoxy-4-methylcoumarin	109.5	$.576_3$	
$C_{11}H_{10}O_3$	7-Methoxy-4-methylcoumarin	110.5	$.581_4$	
$C_{11}H_{10}S$	Methyl-α-naphthyl sulfide	120.2	.690	
$C_{11}H_{14}O_2$	Benzyl butyrate	116.3	.653	(.663) (17.5°)
$C_{11}H_{14}O_4N_4$	Diethyl ketone-2,4-dinitrophenylhydrazone	135.16	$.507_6$	(.670) (20°)
$C_{11}H_{14}O_4N_4$	Methyl-n-propyl ketone-2,4-dinitrophenylhydrazone	135.70	$.509_7$	(.623) (20°)
$C_{11}H_{16}ON_2$	N,N-Diethyl-N′-phenyl urea	126.4	$.657_3$	
$C_{12}H_6O_4N_2Cl_2$	4,4′-Diechloro-2,2′-dinitro-1,1′-biphenyl	150	.479	
$C_{12}H_8$	Acenaphthylene	111.6	.733	(.659) (16°)
$C_{12}H_8O_4N_2$	2,4′-Dinitro-1,1′-biphenyl	116.5	.477	(.703) (20°)
$C_{12}H_8Cl_2$	4,4′-Dichloro-1,1′-biphenyl	133.1	.497	(.859) (20°)
$C_{12}H_8Br_2$	4,4′-Dibromo-1,1′-biphenyl	151.3	.485	(.920) (20°)
$C_{12}H_9O_2N$	2-Nitro-1,1′-biphenyl	109	.547	(.788) (20°)
$C_{12}H_0O_2N$	5-Nitroacenaphthene	116.0	.582	
$C_{12}H_{10}O_4$	Cinnamylidene malonic acid	105.4	.483	
$C_{12}H_{10}O_4$	Quinhydrone	105	.481	(.674) (20°)
$C_{12}H_{10}S$	Phenyl sulfide	119.2	.640	(.716) (20°)
$C_{12}H_{10}S_2$	Phenyl disulfide	122.5	.561	
$C_{12}H_{10}CbSi$	Diphenyl dichlorosilane	153.5	$.606_3$	

Formula	Compound	$-\chi_M \times 10^6$	$-\chi \times 10^6$	$-K \times 10^6$
$C_{12}H_{12}O_2Si$	Diphenyl silanediol	131.6	$.608_2$	
$C_{12}H_{16}O_2$	Amyl benzoate	128.5	.668	
$C_{12}H_{16}O_4N_4$	Methyl-n-butyl ketone-2,4-dinitrophenylhydrazone	147.65	$.526_8$	(.643) (20°)
$C_{12}H_{18}$	Hexamethyl benzene	122.5	.755	
$C_{12}H_{18}$	Disopropyl benzene	124.77	$.767_1$	
$C_{12}H_{22}Ou$	Sucrose (saccharose)	189.1	.552	(.877) (15°)
$C_{12}H_{25}Cl_3Si$	Dodecyl trichlorosilane	209.2	$.688_9$	
$C_{12}H_{30}N_3B_3$	Hexaethylborazine	189	.759	
$C_{13}H_8OS$	Thioxanthone	130	.612	
$C_{13}H_{10}O_2$	Phenyl benzoate	117.3	.592	
$C_{13}H_{10}S$	Thiobenzophenone	118.1	.596	
$C_{13}H_{11}NS$	N-Thiobenzoyl aniline	123.0	.577	
$C_{13}H_{12}N_2S$	N,N′-Diphenyl thiourea	136.5	.598	
$C_{13}H_{12}N_2S$	N,N-Diphenyl thiourea	134.9	.591	
$C_{13}H_{18}O_4N_4$	Methyl amyl ketone-2,4-dinitrophenylhydrazone	156.84	$.532_9$	(.651) (20°)
$C_{14}H_8N_2S_4$	Di-2-benzothiazolyl disulfide	189.0	.568	
$C_{14}H_{10}O$	Anthrone	118	.608	
$C_{14}H_{12}O_2$	Benzyl benzoate	132.2	.622	(.693) (18°)
$C_{14}H_{12}O_2$	Diphenyl acetic acid	124.5	.587	
$C_{14}H_{14}O_2S_2$	p-Toyl thiosulfonate	157	.564	
$C_{14}H_{14}N_2$	p-Azotoluene	135.1	.643	
$C_{14}H_{16}Si$	Dimethyl diphenyl silane	146.6	.690	
$C_{14}H_{20}O_4N_4$	Methyl n-hexyl keton-2,4-dinitrophenylhydrazone	171.73	$.557_0$	(.655) (20°)
$C_{15}H_{10}O_2$	Flavone	120	.540	
$C_{15}H_{12}$	9-Ethylidene fluorene	124.8	.649	
$C_{15}H_{12}O$	Chalcone	125.7	.604	(.646) (62°)
$C_{15}H_{14}O_2$	Benzyl phenylacetate	143.7	.635	
$C_{15}H_{16}ON_2$	N,N′-Dimethyl-N,N′-Diphenyl urea	148.9	.620	
$C_{16}H_{10}$	Fluoranthene	138.0	.682	
$C_{16}H_{12}$	1-Benzylidene indene	130.5	.716	
$C_{16}H_{16}$	1,2-ibenzylamine ethane	164	.787	
$C_{16}H_{33}SiCl_3$	Hexadecyl trichlorosilane	252.4	$.701_3$	
$C_{16}H_{48}O_6Si_7$	Hexadecamethyl heptasiloxane	351.1	$.658_8$	
$C_{17}H_{16}$	9-Cutydiene fluorene	147.5	.670	
$C_{18}H_8S_4$	Tetrathioetracene	202	.573	
$C_{18}H_{10}Cl_2$	5,11-Dichloro tetracene	202	.680	
$C_{18}H_{10}Br_2$	,11-Dibromo tetracene	216	.559	
$C_{18}H_{12}$	Triphenylene	156.6	.686	
$C_{18}H_{14}$	o-Diphenyl benzene	150.4	.653	
$C_{18}H_{16}$	p-Diphenyl benzene (Terphenyl)	152	.660	
$C_{18}H_{15}SiCl$	Triphenyl chlorosilane	186.7	$.633_3$	(.814) (0°)
$C_{18}H_{16}OSi$	Triphenyl silanol	176.6	$.634_5$	
$C_{18}H_{16}Si$	Triphenylsilane	174	.668	
$C_{18}H_{29}$	5-Cyclohexyl acenaphthene	165.2	.699	
$C_{18}H_{22}$	Dimesitylene	171.8	.721	
$C_{18}H_{24}O_2$	Estradiol	186.6	.717	
$C_{18}H_{37}SiCl_3$	Octadecyl trichlorosilane	273.7	$.705_7$	
$C_{19}H_{14}O$	4-Benzoyl-1,1′-Biphenyl	158	.612	
$C_{19}H_{18}Si$	Methyl triphenylsilane	186	.678	
$C_{20}H_{14}O_2$	1,4-Dibenzoyl benzene	153	.534	
$C_{21}H_{24}N_3B_3$	B-Trimethyl-N-triphenylborazine	234	.667	
$C_{24}H_{18}$	4,4′-Diphenyl-1,1′-Biphenyl	201.3	.657	
$C_{24}H_{2S}i$	Tetraphenyl silane	224	.666	
$C_{24}H_{30}N_3B_3$	B-Triethyl-N-triphenylborazine	264	.693	
$C_{24}H_{40}O_4$	Desoxychlolic acid	272.0	.715	
$C_{24}H_{40}O_5$	Cholic acid	282.3	.691	
$C_{25}H_{20}N_2S$	N,N,N′,N′-Tetraphenylthiourea	229.7	.604	
$C_{26}H_{20}$	Tetraphenyl ethylene	217.4	.654	
$C_{28}H_{16}O_2$	Bianthrone (Dianthraquinone)	220	.572	
$C_{28}H_{18}O_2$	Dianthrone (Diant-ronol)	229	.593	
$C_{28}H_{20}$	1,1,4,4-Tetraphenyl-1,2,3-butatriene	227	.659	
$C_{28}H_{44}O$	Calciferol	273.3	.689	
$C_{28}H_{44}O$	Ergosterol	279.6	.704	
$C_{36}H_{30}O_3Si_3$	Hexaphenyl cyclotrisiloxane	364.1	$.612_9$	
$C_{48}H_{40}O_4Si_4$	Octaphenyl cyclotetrasiloxane	485.3	$.613_2$	
$C_{60}H_{36}$	Hexabenzocoronene	346.0	.457	

DIAMAGNETIC SUSCEPTIBILITY DATA OF ORGANOSILICON COMPOUNDS

R. R. Gupta

The molecular susceptibility is represented by χ_M and expressed in c.g.s. units. For some of the compounds volume or specific susceptibility is available in the literature and in such cases χ_M has been calculated by the relation[1]

$$\chi_M = M \times \chi = M \kappa/\zeta$$

where M is the molecular weight of the substance in grams, χ is the specific or mass susceptibility, κ is the volume susceptibility and ζ is the density of the substance in grams per cubic centimeter.

The compounds are arranged in the increasing order of C, H, Si, O, Cl, Br, I, S, and N.

Compound	Molecular formula	Structural formula	^{x}Mx−10^{-6}	Ref.
Methyl trichlorosilane	CH_3SiCl_3	CH_3–Si–Cl_3	85.40, 87.45	2, 8
Methyl tribromosilane	CH_3SiBr_3	CH_3–Si–Br_3	126.00, 128.00, 155.50	10, 11, 4

DIAMAGNETIC SUSCEPTIBILITY DATA OF ORGANO-SILICON COMPOUNDS (Continued)

Compound	Molecular formula	Structural formula	$-\chi_M \cdot 10^{-6}$	Ref.
Methyl dichlorosilane	CH_4SiCl_2	$CH_3-SiH-Cl_2$	66.90	4
Ethyl trichlorosilane	$C_2H_5SiCl_3$	$C_2H_5-Si-Cl_3$	98.00, 98.84, 96.40	2, 9, 7
Ethyl tribromosilane	$C_2H_5SiBr_3$	$C_2H_5-Si-Br_3$	147.50	4
Ethyl triiodosilane	$C_2H_5SiI_3$	$C_2H_5-Si-I_3$	171.50	4
Ethyl dichlorosilane	$C_2H_6SiCl_2$	$C_2H_5-SiH-Cl_2$	78.90	4
Dimethyl dichlorosilane	$C_2H_6SiCl_2$	$(CH_3)_2-Si-Cl_2$	81.30, 82.45	4, 8
Dimethyl diiodosilane	$C_2H_6SiI_2$	$(CH_3)_2-Si-I_2$	131.80	4
Dimethyl silanediol	$C_2H_8SiO_2$	$(CH_3)_2-Si-(OH)_2$	58.40	17, 18
n-Propyl trichlorosilane	$C_3H_7SiCl_3$	$CH_3-(CH_2)_2-Si-Cl_3$	108.00, 110.20	4, 8
Trimethyl chlorosilane	C_3H_9SiCl	$(CH_3)_3-SiCl$	79.00, 73.76	4, 8
Trimethyl bromosilane	C_3H_9SiBr	$(CH_3)_3-Si-Br$	91.30	4
Trimethyl iodosilane	C_3H_9SiI	$(CH_3)_3-Si-I$	104.10	4
1,4,6,9-Tetraoxa-5-silaspiro[4,4]nonane	$C_4H_8SiO_4$	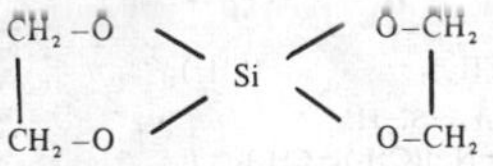	80.94	5
Diethyl dichlorosilane	$C_4H_{10}SiCl_2$	$(C_2H_5)_2-SiCl_2$	105.80	8
Tetramethylsilane	$C_4H_{12}Si$	$Si-(CH_3)_4$	74.80	1
Trimethyl methoxysilane	$C_4H_{12}SiO$	$(CH_3)_3-Si-OCH_3$	78.80	1
Dimethyl dimethoxysilane	$C_4H_{12}SiO_2$	$(CH_3)_2-Si-(OCH_3)_2$	81.70, 81.60, 81.95	1, 17, 5
Diethyl silanediol	$C_4H_{12}SiO_2$	$(C_2H_5)_2-Si-(OH)_2$	81.00	17, 18
Methyl trimethoxysilane	$C_4H_{12}SiO_3$	$CH_3-Si-(OCH_3)_3$	85.6	17
Tetramethoxysilane	$C_4H_{12}SiO_4$	$Si-(OCH_3)_4$	83.68	5
Trimethylacetoxysilane	$C_5H_{12}SiO_2$	$(CH_3)_3-Si-OCOCH_3$	85.00, 86.09	16, 5
Trimethylsilylchloromethylamidoxime	$C_5H_{13}SiOClN_2$	$NH_2-C-(CH_2Cl)=N-O-Si-(CH_3)_3$	109.60	15
Trimethylethylsilane	$C_5H_{14}Si$	$C_2H_5-Si-(CH_3)_3$	86.00	2
Trimethylethoxysilane	$C_5H_{14}SiO$	$(CH_3)_3-Si-OC_2H_5$	89.50	1, 17
Phenyl trichlorosilane	$C_6H_5SiCl_3$	$C_6H_5-Si-Cl_3$	120.4	8
Dimethyl diacetoxysilane	$C_6H_{12}SiO_4$	$(CH_3)_2-Si-(OCOCH_3)_2$	98.3, 101.18	16, 5
1,4,6,9-Tetraoxa-2,7-dimethyl-5-silaspiro[4,4]nonane	$C_6H_{12}SiO_4$	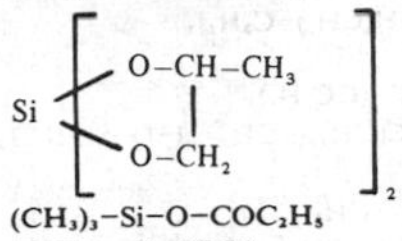	104.34	5
Trimethylpropionoxysilane	$C_6H_{14}Si-O_2$	$(CH_3)_3-Si-O-COC_2H_5$	99.09	5
Dimethyldiethylsilane	$C_6H_{16}Si$	$(C_2H_5)_2-Si-(CH_3)_2$	97.90	1
Triethylsilane	$C_6H_{16}Si$	$(C_2H_5)_3-Si-H$	95.00	3
Trimethylepropoxysilane	$C_6H_{16}SiO$	$(CH_3)_3-Si-O-(CH_2)_2-CH_3$	101.1	1
Trimethylsilylethylamidoxime	$C_6H_{16}SiON_2$	$(NH_2)-C(C_2H_5)=N-O-Si(CH_3)_3$	101.00	15
Dimethyldiethoxysilane	$C_6H_{16}SiO_2$	$(CH_3)_2-Si-(OC_2H_5)_2$	103.60, 104.70, 104.1, 104.6	1, 18, 5, 8
Trimethylthio-n-propylsilane	$C_6H_{16}SiS$	$(CH_3)_3-Si-S-(CH_2)_2-CH_3$	114.09	12
Trimethylthio-isopropylsilane	$C_6H_{16}SiS$	$(CH_3)_3-Si-S-(CH_3)_2$	115.11	12
Hexamethyldisiloxane	$C_6H_{18}Si_2O$	$(CH_3)_3-Si-O-Si-(CH_3)_3$	118.9	8
Hexamethylcyclotrisiloxane	$C_6H_{18}Si_3O_3$	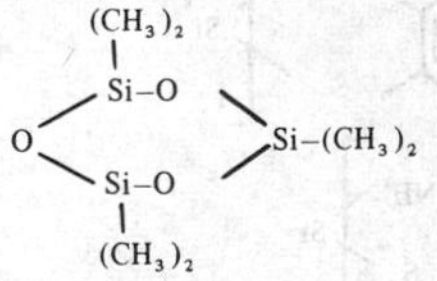	140.5	8
Methyltriacetoxysilane	$C_7H_{12}SiO_6$	$(CH_3)-Si-(OCOCH_3)_3$	112.50	5
Triethylmethylsilane	$C_7H_{18}Si$	$(C_2H_5)_3-Si-CH_3$	109.30	2
Trimethylbutylsilane	$C_7H_{18}Si$	$(CH_3)_3-Si-(CH_2)_3-CH_3$	109.00	2
Trimethylbutoxysilane	$C_7H_{18}SiO$	$(CH_3)_3-Si-O-(CH_2)_3-CH_3$	114.2	5
Trimethylsilyl-*n*-propylamidoxime	$C_7H_{18}SiON_2$	$NH_2-C-(CH_2-CH_2-CH_3)=N-O-Si-(CH_3)_3$	123.8	15
Methyltriethoxysilane	$C_7H_{18}SiO_3$	$CH_3-Si-(OC_2H_5)_3$	120.3	17
Trimethylthio-*n*-butylsilane	$C_7H_{18}SiS$	$(CH_3)_3-Si-S-(CH_2)_3CH_3$	125.73	12
Trimethylthio-*tert*-butylsilane	$C_7H_{18}SiS$	$(CH_3)_3-Si-S-C-(CH_3)_3$	126.55	12
Trimethyl*N*-dimethylsilane	$C_7H_{19}SiN$	$(CH_3)_3-Si-N-(CH_3)_2$	91.3	1
2,2-Dimethyl-2-sila-2,3-dihydrobenzothiazole	$C_8H_{11}SiN$	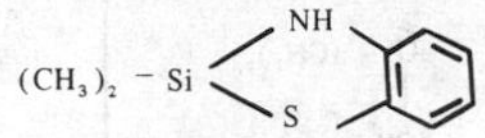	144.44	12
Tetraacetoxysilane	$C_8H_{12}SiO$	$Si-(OCOCH_3)_4$	129.26	5
Dimethyldiproprionoxysilane	$C_8H_{18}SiO$	$(CH_3)_2-Si-(OCOC_2H_5)_2$	122.0, 123.97	16, 5
1,4,6,9-Tetraoxo-2,3,7,8-tetramethyl-5-silaspiro[4,4]nonane	$C_8H_{16}SiO_4$	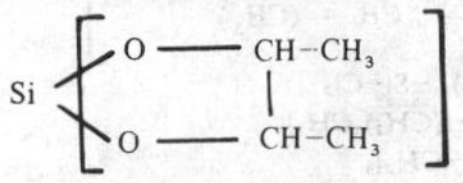	127.12	5
1,5,7,11-Tetraoxa-2,8-dimethyl-6-silaspiro[5,5]undecane	$C_8H_{16}SiO_4$	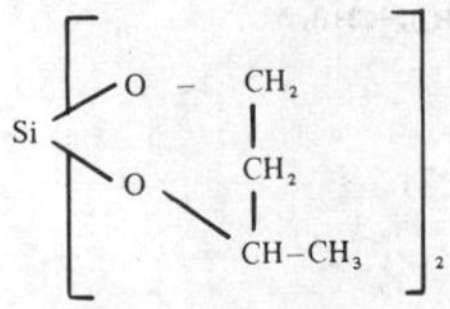	126.34	5

DIAMAGNETIC SUSCEPTIBILITY DATA OF ORGANO-SILICON COMPOUNDS (Continued)

Compound	Molecular formula	Structural formula	${}^{x}Mx-10^{-6}$	Ref.
Dimethyldi-*n*-propylsilane	C_8H_2OSi	$(CH_3)_2-Si-[(CH_2)_2-CH_3]_2$	121.70	1
Tetraethylsilane	C_8H_20Si	$Si-(C_2H_5)_4$	120.30, 120.40, 117.00	1, 4, 5
Dimethyl-di-*n*-propoxysilane	$C_8H_2OSiO_2$	$(CH_3)_2-Si-[O(CH_2)_2-CH_3]_2$	125.70, 126.90	1, 5
Tetraethoxysilane	$C_8H_2OSiO_4$	$Si-(OC_2H_5)_4$	83.68, 134.50	5, 17
Bis-(trimethylsilyl)-methylamidoxime	$C_8H_{22}SiO_2N_2$	$(CH_3)_3-Si-NH-C(CH_3)=N-O-Si-(CH_3)_3$	156.60	15
Octamethyltrisiloxane	$C_8H_{24}Si_3O_2$	$(CH_3)_3-Si-O-Si-O-Si(CH_3)_3(CH_3)_2$	165.00	16
Octamethyltetracyclosiloxane	$C_8H_{24}Si_4O_4$	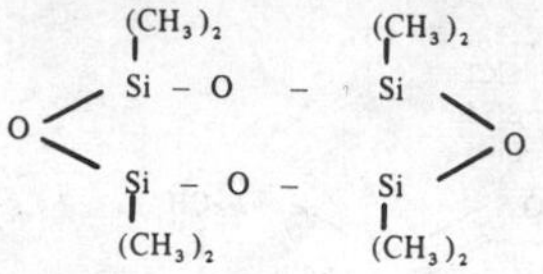	188.00	5
Tri-*n*-propylsilane	$C_9H_{22}Si$	$CH_3-(CH_2)_{23}-Si-H$	130.00	3
Trimethyl *N*-(*n*-propyl)silane	$C_9H_{23}SiN$	$(CH_3)_3-Si-N-[(CH_2)_2-CH_3]_2$	113.4	1
Bis(trimethyl)-ethylamidoxime	$C_9H_{24}Si_2ON_2$	$(CH_3)_3-Si-NH-C(C_2H_5)=N-O-Si(CH_3)_3$	168.00	15
Bis(trimethyl)-phenylamidoxime	$C_{10}H_{16}SiON_2$	$NH_2-C-(C_6H_5)=N-O-Si(CH_3)_3$	141.7	15
2,2-Diacetoxy-4,4,6,6,tetramethyl-1,3-dioxa-2-sila-cyclohexane	$C_{10}H_{20}SiO_4$	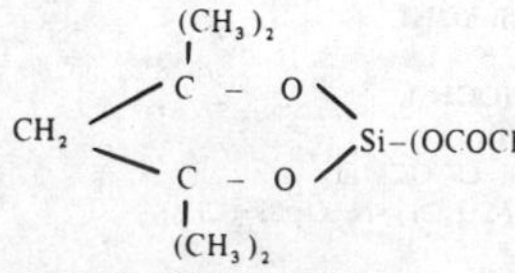	150.65	5
Dimethyl di-*n*-butylsilane	$C_{10}H_{24}Si$	$(CH_3)_2-Si-[(CH_2)_3-CH_3]_2$	145.80	1
Methyl tri-*n*-butylsilane	$C_{10}H_{24}Si$	$CH_3-Si-[(CH_2)_2-CH_3]_3$	143.90	1
Dimethyl di-*n*-butoxysilane	$C_{10}H_{24}SiO_2$	$(CH_3)_2-Si-[O(CH_2)_3CH_3]_2$	147.40, 149.40	1, 5
Dimethyl di-iso-butoxysilane	$C_{10}H_{24}SiO_2$	$(CH_3)_2-Si-[O-CH_2-CH-(CH_3)_2]_2$	150.73	5
Dimethyl di-(1-methylpropoxy)silane	$C_{10}H_{24}SiO_2$	$(CH_3)_2-Si-[O-CH(CH_3)-C_2H_5]_2$	150.83	5
Di-*n*-propyldiethoxysilane	$C_{10}H_{24}SiO_2$	$[CH_3-(CH_2)_2]_2-Si-(OC_2H_5)_2$	150.80	17, 18
Bis(trimethylsilyl) *n*-propyl amidoxime	$C_{10}H_{26}SiON_2$	$(CH_3)_3-Si-NH-C(CH_2)_2-CH_3=N-O-Si-(CH_3)_3$	179.00	15
Trimethyl-*n*-octylsilane	$C_{11}H_{26}Si$	$(CH_3)_3-Si-(CH_2)_7-CH_3$	156.00	2
Trimethyl-*n*-octoxysilane	$C_{11}H_{26}SiO$	$(CH_3)_3-Si-O(CH_2)_7-CH_3$	149.00	1
Trimethyl *N*-(di-*N*-butyl)silane	$C_{11}H_{27}SiN$	$(CH_3)_3-Si-N-[(CH_2)_3-CH_3]_2$	158.5	1
2,2′Spirobis(6-chloro-2,3-dihydrobenzothiazole)silane	$C_{12}H_8SiCl_2S_2N_2$	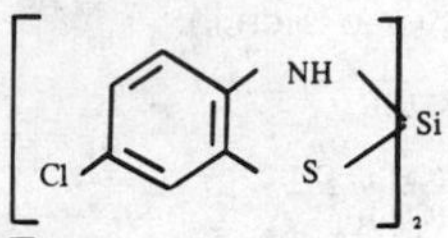	183.29	12
2,2′Spirobis(6-bromo-2,3-dihydrobenzothiazole)silane	$C_{12}H_8SiBr_2S_2N_2$	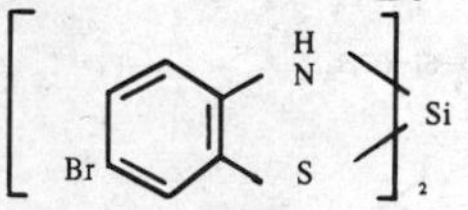	199.37	12
2,2′Spirobis(2,3-dihydrobenzothiazole)silane	$C_{12}H_{10}SiS_2N_2$	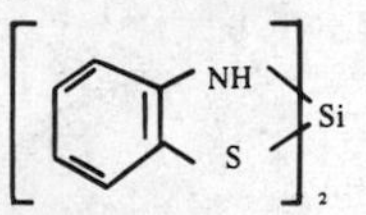	153.71	12
Diphenylsilanediol	$C_{12}H_{12}SiO_2$	$(C_6H_5)_2-Si-(OH)_2$	131.60	8
Tetrapropionoxysilane	$C_{12}H_{20}SiO_8$	$Si-(OCOC_2H_5)_4$	129.26	5
1,4,6,9-Tetraoxo-2,3,7,8-octamethyl-1-5-silaspiro[4,4]nonane	$C_{12}H_{24}SiO_4$	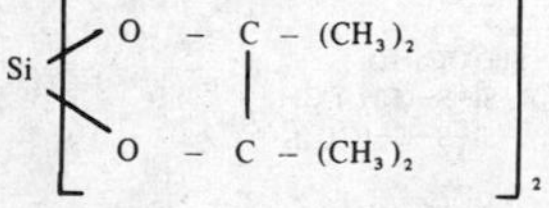	127.45	5
1,5,7,11-Tetraoxa-2,4,8,10-hexamethyl-1-6-silaspiro[5,5]undecane	$C_{12}H_{24}SiO_4$	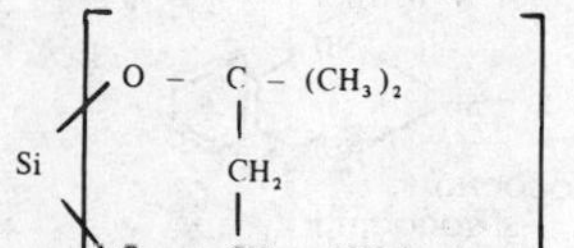	172.48	5
Dodeconyltrichlorosilane	$C_{12}H_{25}SiCl_3$	$CH_3-(CH_2)_{11}-Si-Cl_3$	210.40, 209.20	2, 8
Dimethyl di-*n*-pentylsilane	$C_{12}H_{28}Si$	$(CH_3)_2-Si-[(CH_2)_4CH_3]_2$	167.60	1
Tetra-*n*-propylsilane	$C_{12}H_{28}Si$	$Si-[(CH_2)_2-CH_3]_4$	165.80	1
Tri-*n*-butylsilane	$C_{12}H_{28}Si$	$H-Si-[(CH_2)_3-CH_3]_3$	172.00	3

DIAMAGNETIC SUSCEPTIBILITY DATA OF ORGANO-SILCON COMPOUNDS (Continued)

Compound	Molecular formula	Structural formula	$\chi_M \times 10^{-6}$	Ref.
Dimethyl di-*n*-pentoxysilane	$C_{12}H_{28}SiO_2$	$(CH_3)_2–Si[O–(CH_2)_4–CH_3]_2$	172.37	5
Hexaethoxydisiloxane	$C_{12}H_{30}Si_2O_7$	$(H_5C_2O)_3–Si–O–Si–(OC_2H_5)_3$	217.00	13
Bis-(trimethylsilyl)-phenylamidoxime	$C_{13}H_{24}Si_2ON_2$	$(CH_3)_3–Si–NH–C(C_6H_5)=N–O–Si–(CH_3)_3$	199.50	15
Methyltri-*n*-butylsilane	$C_{13}H_{30}Si$	$CH_3–Si–[(CH_2)_3–CH_3]_3$	178.50	1
2,2′-Spirobis(6-methyl-2,3-dihydrobenzothiazole)silane	$C_{14}H_{14}S_2N_2$	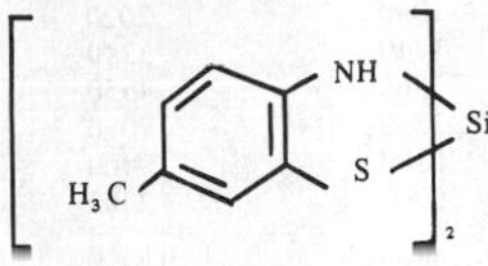	170.06	12
Diphenyldimethoxysilane	$C_{14}H_{16}SiO_2$	$(C_6H_5)_2–Si–(OCH_3)_2$	156.80	19
Trimethyldodeconylsilane	$C_{15}H_{34}Si$	$(CH_3)_3–Si–(CH_2)_{11}–CH_3$	192.00	2
Trimethyl-*n*-dodeconoxysilane	$C_{15}H_{34}SiO$	$(CH_3)_3–Si–O–(CH_2)_{11}–CH_3$	205.6	1
Diphenyl diethoxysilane	$C_{16}H_{20}SiO_2$	$(C_6H_5)_2–Si–(OC_2H_5)_2$	179.25	19
Hexadodeconyltrichlorosilane	$C_{16}H_{33}SiCl_3$	$C_{16}H_{33}–Si–Cl_3$	259.00, 252.40	2, 8
Tetra-*n*-butylsilane	$C_{16}H_{36}Si$	$Si–[(CH_2)_3–CH_3]_4$	202.50	3
Hexa-dodecamethylheptasiloxane	$C_{16}H_{48}O_6$	$(CH_3)_3–Si–(O–Si)_5–O–Si–(CH_3)_3$ \ $(CH_3)_2$	351.3	8
Triphenylhydroxysilane	$C_{18}H_{16}SiO$	$(C_6H_5)_3–Si–OH$	176.0	8
Dephenyl di-*n*-propoxysilane	$C_{18}H_{24}SiO_2$	$(C_6H_5)_2–Si–[O(CH_2)_2–CH_3]_2$	201.58	19
Diphenyl di-isopropoxysilane	$C_{18}H_{24}SiO_2$	$(C_6H_5)_2–Si–[OCH–(CH_3)]_2$	202.46	19
n-Octadeconyltrichlorosilane	$C_{18}H_7SiCl_3$	$CH_3–(CH_2)_{17}–Si–Cl_3$	278.00, 273.7	2, 8
Diphenyl di-*n*-butoxysilane	$C_{20}H_{28}SiO_2$	$(C_6H_5)_2–Si–[O–(CH_2)_3–CH_3]_2$	223.62	19
Diphenyl di-iso-butoxy silane	$C_{20}H_{28}SiO_2$	$(C_6H_5)_2–Si–[O–CH_2–CH(CH_3)]_2$	224.46	19
Trimethyl*n*-octadeconyl silane	$C_{21}H_{46}Si$	$CH_3–(CH_2)_{17}–Si–(CH_3)_3$	278.00	2
Tetraphenylsilane	$C_{24}H_{20}Si$	$Si–(C_6H_5)_4$	212.20	7
Tetrabenzylsilane	$C_{28}H_{28}Si$	$Si–(CH_2–C_6H_5)_4$	266.20	6
Tetra-*p*-tolylsilane	$C_{28}H_{28}Si$	$Si–(C_6H_4–CH_3)_4$	276.40	6
Other silicon compounds				
Carborundum (silicon carbide)	CSi	C–Si	12.80	20
Silane	SiH_4	$Si–H_4$	20.4	13
Silicon dioxide	SiO_2	O=Si=O	29.6	11
Tetrachlorosilane	$SiCl_4$	$Si–Cl_4$	87.40, 84.60, 88.30	4, 11, 10
Tetrabromosilane	$SiBr_4$	$Si–Br_4$	123.30	4
Tetraiodosilane	SiI_4	$Si–I_4$	186.20	4
Trichlorosilane	$HSiCl_3$	$H–Si–Cl_3$	70.50, 71.30, 70.60	2, 7, 4
Silicic acid	H_2SiO_3	$O=S–(OH)_2$	33.30	11
Ortho silicic acid	H_4SiO_4	$Si–(OH)_4$	42.60	21
Disilane	Si_2H_6	$H_3–Si–Si–H_3$	37.3	4
Hexachlorodisilane	Si_2Cl_6	$Cl_3–Si–Si–Cl_3$	138.00	7

REFERENCES

1. **Abel, E. W., Bush, R. P., Jenkins, C. R., and Zobel, T.** *Trans. Faraday Soc.*, 60, 1214, 1964.
2. **Abel, E. W. and Bush, R. P.** *Trans. Faraday Soc.*, 59, 630, 1963.
3. **Dorfman, Y. G. and Lependina, O. L.**, *Zh. Strukt. Khim.*, 5, 632, 1964.
4. **Lister, M. W. and Marson, R.**, *Can. J. Chem.*, 42(9), 2101, 1964.
5. **Mital, R. L. and Gupta, R. R.**, *J. Am. Chem. Soc.*, 91, 4664, 1969.
6. **Asai, K.**, *Sci. Rep. Res. Inst. Tohoku Univ.*, 2, 205, 1950.
7. **Pascal, P.**, *Compt. Rend.*, 218, 57, 1944.
8. **Mathur, R. M.**, *Trans. Faraday Soc.*, 54, 1577, 1958.
9. **Mathur, R. M.**, *J. Vikram Univ.*, 2, 55, 1958.
10. **Nevgi, M. B.**, *J. Univ. Bombay*, 7(3), 82, 1958.
11. **Fox, G., Gorter, C. J., and Smits, L. J.**, *Tables Constantes Selectioness Diamagnetisme et Paramagnetisme*, Masson et cie, Paris, 1957, 122.
12. **Mital, R. L., Goyal, R. D., and Gupta, R. R.**, *Inorg. Chem.*, 11, 1924, 1972.
13. **Barter, C., Meisenheimer, R. G., and Stevenson, D. P.**, *J. Phys. Chem.*, 64, 1312, 1960.
14. **Foex, G., Gorter, C. J., and Smits, L. J.**, *Compt. Rend.*, 175, 125, 1922.
15. **Goel, A. B., and Gupta, V. D.**, *Indian J. Chem.*, 13(2), 181, 1975.
16. **Pacault, A.**, *Compt. Rend.*, 232, 1352, 1951.
17. **Mital, R. L.**, *Bull. Chem. Soc. Jpn.*, 37(10), 1440, 1964.
18. **Mital, R. L.**, *J. Phy. Chem.*, 68, 1613, 1964.
19. **Goyal, R. D., Gupta, R. R., and Mital, R. L.**, *J. Phys. Chem.*, 76, 1579, 1972.
20. **Sigamony, A.**, *Proc. Indian Acad. Sci.*, 19A, 377, 1944.
21. **Pascal, P.**, *Compt. Rend.*, 175, 814, 1922.

MASS ATTENUATION COEFFICIENTS

Wavelength	Formvar $(C_5H_7O_2)_x$	Collodion $(C_{12}H_{11}O_{22}N_6)_x$	Polypropylene $(CH_2)_x$	Cellulose acetate $(C_{10}H_{21}O_{15})_x$	Mylar $(C_{10}H_8O_4)_x$	Teflon $(CF_2)_x$	Energy (eV)
2.0	14	20	8	19	14	28	6199.0
4.0	113	156	69	150	116	220	3099.5
6.0	372	510	234	489	384	700	2066.3
8.0	850	1140	550	1100	870	1540	1549.8
10.0	1580	2110	1040	2020	1630	2800	1239.8
12.0	2600	3450	1740	3310	2680	4520	1033.2
14.0	3920	5200	2660	4950	4040	6700	885.6
16.0	5600	7300	3830	7000	5800	9400	774.9
18.0	7500	9800	5200	9400	7800	12600	688.8
20.0	9900	12800	6900	12300	10200	2780	619.9
22.0	12500	16200	8800	15500	12900	3540	563.5
24.0	8200	6400	11000	4850	8500	4430	516.6
26.0	10100	7900	13500	5900	10400	5400	476.8
28.0	12000	9400	16200	7100	12400	6500	442.8
30.0	14300	11100	19200	8400	14700	7800	413.3
32.0	16700	8200	22400	9900	17200	9100	387.4
34.0	19300	9500	25900	11400	19900	10600	364.6
36.0	22100	10800	29600	13100	22800	12100	344.4
38.0	25000	12300	33600	14900	25800	13800	326.3
40.0	28200	13900	37800	16800	29100	15600	309.9
42.0	31500	15500	42200	18700	32500	17500	295.2
44.0	3250	4540	1940	4370	3350	6900	281.8
46.0	3640	5100	2180	4900	3760	7800	269.5
48.0	4050	5600	2430	5400	4170	8600	258.3
50.0	4450	6200	2690	6000	4590	9500	248.0
52.0	4910	6800	2960	6600	5100	10500	238.4
54.0	5400	7500	3240	7200	5600	11500	229.6
56.0	5900	8200	3540	7900	6100	12500	221.4
58.0	6400	8900	3860	8600	6600	13600	213.8
60.0	7000	9700	4190	9300	7200	14800	206.6
62.0	7500	10500	4540	10100	7800	16000	200.0
64.0	8100	11300	4880	10900	8400	17300	193.7
66.0	8700	12100	5200	11700	9000	18600	187.8
68.0	9400	13100	5700	12600	9700	19900	182.3
70.0	10000	14000	6000	13500	10300	21300	177.1
72.0	10700	14900	6400	14400	11100	22700	172.2
74.0	11400	15900	6800	15400	11800	24200	167.5
76.0	12200	17000	7300	16300	12500	25700	163.1
78.0	12900	18000	7700	17300	13300	27200	158.9
80.0	13600	19100	8100	18400	14100	28800	155.0
82.0	14500	20200	8600	19500	14900	30500	151.2
84.0	15300	21400	9100	20600	15800	32100	147.6
86.0	16100	22600	9600	21700	16600	33800	144.2
88.0	17000	23700	10100	22900	17500	35500	140.9
90.0	17800	25000	10600	24100	18400	37300	137.8
92.0	18800	26300	11100	25300	19400	39100	134.8
94.0	19700	27600	11700	26500	20300	40900	131.9
96.0	20600	28900	12200	27800	21300	43000	129.1
98.0	21600	30300	12800	29200	22300	44600	126.5
100.0	22600	31700	13400	30500	23300	46300	124.0
105.0	25200	35300	15000	34000	26000	51000	118.1
110.0	27900	39100	16600	37600	28800	56000	112.7
115.0	30700	43000	18200	41300	31600	61000	107.8
120.0	33600	47000	20000	45100	34600	67000	103.3
125.0	36800	52000	21900	49600	38000	72000	99.2
130.0	39800	56000	23900	53000	41100	78000	95.4
135.0	43200	60000	25900	58000	44600	83000	91.8
140.0	46700	65000	28100	63000	48200	89000	88.6
145.0	50000	70000	30300	67000	52000	95000	85.5
150.0	54000	75000	32600	72000	55000	101000	82.7
155.0	57000	80000	35000	76000	59000	106000	80.0
160.0	61000	85000	37600	82000	63000	112000	77.5
165.0	65000	90000	40200	87000	67000	118000	75.1
170.0	69000	96000	42800	92000	72000	124000	72.9
175.0	73000	101000	45400	97000	75000	130000	70.8
180.0	77000	107000	47900	102000	80000	136000	68.9
185.0	82000	113000	51000	108000	84000	141000	67.0
190.0	86000	118000	54000	112000	88000	147000	65.3
195.0	90000	124000	57000	118000	93000	153000	63.6
200.0	95000	130000	61000	124000	98000	159000	62.0

MASS ATTENUATION COEFFICIENTS (Continued)

Wavelength	Polystyrene $(CH)_x$	Nylon $(C_{12}H_{22}O_3N_2)_x$	Vyns $(C_{22}H_{33}O_2Cl_9)_x$	Saran $(C_2H_2Cl_2)_x$	Aluminum oxide Al_2O_3	Quartz $(SiO_2)_x$	Energy (eV)
2.0	9	12	114	164	68	77	6199.0
4.0	74	96	750	1070	480	530	3099.5
6.0	252	318	350	375	1440	1600	2066.3
8.0	590	730	780	820	860	980	1549.8
10.0	1120	1370	1440	1520	1580	1810	1239.8
12.0	1870	2270	2370	2490	2570	2950	1033.2
14.0	2870	3440	3590	3760	3840	4400	885.6
16.0	4120	4910	5100	5400	5500	6200	774.9
18.0	5600	6700	6900	7200	7300	8400	688.8
20.0	7500	8800	9000	9300	9600	11000	619.9
22.0	9500	11100	11400	11800	12100	13800	563.5
24.0	11900	10600	12800	14400	4290	4920	516.6
26.0	14600	13000	15500	17300	5200	6000	476.8
28.0	17400	15500	18400	20400	6300	7200	442.8
30.0	20700	18400	21400	23700	7500	8500	413.3
32.0	24200	17300	24700	27100	8700	9900	387.4
34.0	28000	20000	28100	30800	10100	11300	364.6
36.0	31900	22800	31600	34400	11500	12900	344.4
38.0	36200	25800	35400	38200	13100	14500	326.3
40.0	40700	29100	39100	42100	14700	16200	309.9
42.0	45500	32500	43100	46000	16300	18000	295.2
44.0	2090	2730	25700	37000	18100	19800	281.8
46.0	2350	3060	23600	33600	19900	21700	269.5
48.0	2620	3400	25300	36100	21800	23600	258.3
50.0	2900	3750	27300	38800	23700	25500	248.0
52.0	3190	4130	28900	41100	25700	27500	238.4
54.0	3500	4540	30600	43400	27700	29400	229.6
56.0	3820	4950	32200	45700	29800	31800	221.4
58.0	4160	5400	33900	47900	31800	33700	213.8
60.0	4510	5800	35100	49500	33900	35600	206.6
62.0	4890	6300			35900	37500	200.0
64.0	5300	6800			38000	39900	193.7
66.0	5600	7300			40600	41900	187.8
68.0	6100	7900			42800	44000	182.3
70.0	6500	8400			44900	46000	177.1
72.0	6900	9000			47100	48000	172.2
74.0	7400	9600			49300	50000	167.5
76.0	7800	10200			51000	52000	163.1
78.0	8300	10800			54000	54000	158.9
80.0	8800	11400			56000	56000	155.0
82.0	9300	12100			58000	52000	151.2
84.0	9800	12800			61000	53000	147.6
86.0	10300	13500			62000	56000	144.2
88.0	10900	14200			65000	57000	140.9
90.0	11400	15000			67000	59000	137.8
92.0	12000	15700			69000	61000	134.8
94.0	12600	16500			71000	63000	131.9
96.0	13200	17300			73000	65000	129.1
98.0	13800	18100			75000	67000	126.5
100.0	14500	19000			77000	69000	124.0
105.0	16100	21100			83000	73000	118.1
110.0	17900	23400			79000	78000	112.7
115.0	19600	25800			84000	82000	107.8
120.0	21600	28300			88000	86000	103.3
125.0	23600	31000			93000		99.2
130.0	25700	33600			97000		95.4
135.0	28000	36500			101000		91.8
140.0	30300	39400			106000		88.6
145.0	32700	42400			110000		85.5
150.0	35100	45600			114000		82.7
155.0	37700	48900			118000		80.0
160.0	40500	52000			122000		77.5
165.0	43300	56000			125000		75.1
170.0	46100	59000					72.9
175.0	48900	63000					70.8
180.0	52000	66000					68.9
185.0	55000	70000					67.0
190.0	58000	74000					65.3
195.0	62000	78000					63.6
200.0	65000	82000					62.0

MASS ATTENUATION COEFFICIENTS (Continued)

Wavelength	Stearate $CH_3(CH_2)_{16}COO^-$	Animal proteins C = 52.5%, H = 7%, S = 1.5%, O = 22.5%, N = 16.5%	Air O = 21%, N = 78%, Ar = 1%	P 10 CH_4 = 10%, Ar = 90%	Methane CH_4	Q Gas C_4H_{10} = 1.3%, He = 98.7%	Energy (eV)
2.0	10	16	21	230	7	1	6199.0
4.0	84	126	148	162	60	12	3099.5
6.0	281	361	481	467	205	41	2066.3
8.0	650	820	1090	1020	479	97	1549.8
10.0	1220	1530	2020	1850	910	185	1239.8
12.0	2030	2530	3310	3010	1520	313	1033.2
14.0	3080	3830	4980	4500	2330	484	885.6
16.0	4410	5500	7100	6400	3350	700	774.9
18.0	6000	7400	9500	8400	4570	970	688.8
20.0	7900	9700	12400	10900	6100	1300	619.9
22.0	10100	12300	15700	13500	7700	1670	563.5
24.0	10000	10300	14100	16400	9700	2110	516.6
26.0	12200	12600	17100	19600	11800	2620	476.8
28.0	14600	15100	20400	22800	14100	3160	442.8
30.0	17300	17800	24000	26300	16800	3780	413.3
32.0	20300	15500	2290	29700	19600	4460	387.4
34.0	23400	17900	2650	33300	22700	5200	364.6
36.0	26800	20400	3040	36900	25900	6000	344.4
38.0	30300	23100	3460	40500	29300	6900	326.3
40.0	34100	26000	3810	37600	33000	7800	309.9
42.0	38200	29100	4270	40900	36900	8800	295.2
44.0	2390	3750	4780	42600	1700	2960	281.8
46.0	2680	4180	5300	45600	1910	3380	269.5
48.0	2980	4610	5900	48900	2130	3840	258.3
50.0	3290	5000	6400	52000	2350	4340	248.0
52.0	3620	5500	6300		2590	4910	238.4
54.0	3970	5900	7000		2840	5500	229.6
56.0	4340	6400	7600		3100	6100	221.4
58.0	4730	6900	8200		3380	6700	213.8
60.0	5100	7500	8900		3660	7400	206.6
62.0	5600	8100	9700		3970	8200	200.0
64.0	6000	8700	10400		4270	8900	193.7
66.0	6400	9300	11200		4570	9700	187.8
68.0	6900	10000	12100		4940	10600	182.3
70.0	7400	10600	12900		5200	11500	177.1
72.0	7900	11300	13800		5600	12500	172.2
74.0	8400	12000	14700		6000	13500	167.5
76.0	8900	11600	15600		6400	14600	163.1
78.0	9500	12300	16600		6700	15600	158.9
80.0	10000	13000	17600		7100	16800	155.0
82.0	10600	13800	18600		7600	18000	151.2
84.0	11200	14600	19700		7900	19300	147.6
86.0	11800	15400	20800		8400	20500	144.2
88.0	12400	16200	21900		8800	21900	140.9
90.0	13100	17000	23100		9300	23300	137.8
92.0	13700	17900	24200		9700	24700	134.8
94.0	14400	18800	25400		10300	26200	131.9
96.0	15100	19700	26700		10700	27800	129.1
98.0	15800	20700	28000		11200	29400	126.5
100.0	16600	21600	29200		11800	31000	124.0
105.0	18400	24100	32700		13100	35400	118.1
110.0	20400	26700	36200		14500	40100	112.7
115.0	22500	29300	39900		15900	44800	107.8
120.0	24600	32200	43800		17500	50000	103.3
125.0	27000	35300	48000		19200	56000	99.2
130.0	29300	38200	52000		20900	62000	95.4
135.0	31800	41500	57000		22700	68000	91.8
140.0	34400	44800	61000		24600	75000	88.6
145.0	37000	48200	65000		26500	82000	85.5
150.0	39800	52000	71000		28500	89000	82.7
155.0	42600	55000	76000		30600	96000	80.0
160.0	45700	59000	80000		32900	104000	77.5
165.0	48600	63000	86000		35100	112000	75.1
170.0	52000	67000	91000		37400	121000	72.9
175.0	55000	71000	96000		39700	130000	70.8
180.0	58000	75000	102000		41900	138000	68.9
185.0	62000	79000	108000		44900	147000	67.0
190.0	65000	83000	114000		47200	157000	65.3
195.0	68000	88000	119000		50000	167000	63.6
200.0	72000	92000	125000		53000	177000	62.0

INTRODUCTION TO X-RAY CROSS SECTIONS

Alex F. Burr

These tables are part of an extensive report published by W. H. McMaster, et al. as UCRL 50174 and available from the National Technical Information Service, Springfield, Va. 22151. Section I of UCRL 50174 describes the data base and the treatment given it, Section II contains the total cross sections between 1 and 1000 keV for all the elements, Section III contains results used in producing Section II, and Section IV contains total cross sections for selected energies and is reproduced in part here. To obtain these values existing experimental x-ray total cross section data and theoretical cross section calculations were surveyed. The coherent (Rayleigh) scattering cross sections and the incoherent (Compton) scattering cross sections were computed. The photo-electric cross sections were obtained by least squares fitting of experimental data, theory, and interpolation of experiment and theory. The following table contains cross sections interpolated from the basic compilation at those wavelengths of most use to x-ray crystallographers. The wavelengths chosen were selected to correspond to those given in the International Tables for X-Ray Crystallography. The energy-to-wavelength conversion is given below.

Table I. Energy-to-wavelength conversion

Target radiation	Å	keV	Target radiation	Å	keV
Ag $K\bar{\alpha}$	0.5608	22.105	Ni $K\bar{\alpha}$	1.6591	7.472
$K\beta_1$	0.4970	24.942	$K\beta_1$	1.5001	8.265
Pd $K\bar{\alpha}$	0.5869	21.125	Co $K\bar{\alpha}$	1.7902	6.925
$K\beta_1$	0.5205	23.819	$K\beta_1$	1.6208	7.649
Rh $K\bar{\alpha}$	0.6147	20.169	Fe $K\bar{\alpha}$	1.9373	6.400
$K\beta_1$	0.5456	22.724	$K\beta_1$	1.7565	7.058
Mo $K\bar{\alpha}$	0.7107	17.444	Mn $K\bar{\alpha}$	2.1031	5.895
$K\beta_1$	0.6323	19.608	$K\beta_1$	1.9102	6.490
Zn $K\bar{\alpha}$	1.4364	8.631	Cr $K\bar{\alpha}$	2.2909	5.412
$K\beta_1$	1.2952	9.572	$K\beta_1$	2.0848	5.947
Cu $K\bar{\alpha}$	1.5418	8.041	Ti $K\bar{\alpha}$	2.7496	4.509
$K\beta_1$	1.3922	8.905	$K\beta_1$	2.5138	4.932

Table II. Total Cross Section in cm^2/g

Z KEV	1 H	2 He	3 Li	4 Be	5 B	6 C	7 N	8 O	9 F	10 Ne
4.51	.432	.661	2.10	5.63	12.9	25.6	43.0	64.8	91.8	125
4.93	.421	.550	1.62	4.25	9.69	19.4	32.6	49.4	70.3	95.9
5.41	.412	.465	1.24	3.18	7.23	14.5	24.4	37.2	53.1	72.7
5.90	.405	.405	.986	2.45	5.53	11.1	18.7	28.6	41.1	56.4
5.95	.405	.400	.964	2.39	5.39	10.8	18.2	27.9	40.0	54.9
6.40	.400	.362	.798	1.92	4.28	8.55	14.5	22.2	32.0	44.0
6.49	.400	.355	.770	1.84	4.10	8.18	13.9	21.3	30.6	42.2
6.93	.397	.329	.659	1.52	3.36	6.68	11.3	17.4	25.1	34.7
7.06	.396	.322	.631	1.44	3.17	6.30	10.7	16.5	23.7	32.8
7.47	.394	.303	.555	1.23	2.67	5.28	8.96	13.8	19.9	27.6
7.65	.393	.297	.528	1.15	2.49	4.92	8.33	12.9	18.6	25.7
8.04	.391	.284	.477	1.01	2.14	4.22	7.14	11.0	16.0	22.1
8.27	.390	.277	.452	.936	1.98	3.88	6.56	10.1	14.7	20.4
8.63	.389	.268	.417	.837	1.74	3.40	5.74	8.87	12.8	7.9
8.91	.388	.262	.394	.774	1.59	3.09	5.22	8.06	11.7	16.2
9.57	.386	.250	.349	.651	1.30	2.49	4.19	6.47	9.36	13.1
17.44	.373	.202	.197	.245	.345	.535	.790	1.15	1.58	2.21
19.61	.370	.197	.187	.222	.293	.429	.605	.855	1.15	1.60
20.17	.369	.196	.185	.217	.283	.408	.570	.799	1.07	1.48
21.13	.368	.195	.182	.210	.268	.379	.519	.717	.952	1.31
22.11	.366	.193	.179	.205	.256	.354	.476	.648	.851	1.16
22.72	.366	.192	.177	.201	.249	.340	.452	.610	.795	1.08
23.82	.364	.191	.175	.196	.239	.319	.416	.553	.711	.963
24.94	.363	.189	.173	.192	.229	.301	.385	.504	.640	.861

Z KEV	11 Na	12 Mg	13 Al	14 Si	15 P	16 S	17 Cl	18 Ar	19 K	20 Ca
4.51	168	220	264	336	386	464	518	577	679	805
4.93	129	171	206	263	304	364	41[illegible]	456	542	638
5.41	98.5	131	158	203	236	282	32[illegible]	356	427	500
5.90	76.6	102	124	160	186	223	256	282	342	398
5.95	74.7	99.6	121	156	182	217	25[illegible]	276	334	389
6.40	59.9	80.2	97.5	126	148	177	205	225	275	319
6.49	57.5	77	93.7	121	142	170	197	217	264	307
6.93	47.3	63.5	77.5	100	118	141	165	181	222	257
7.06	44.7	60.1	73.4	95.1	112	134	156	172	211	244
7.47	37.7	50.8	62.2	80.7	95.2	114	134	147	181	209
7.65	35.2	47.4	58.1	75.4	89.1	107	125	137	170	196
8.04	30.3	40.9	50.2	65.3	77.3	92.5	109	120	148	171
8.27	27.9	37.6	46.3	60.2	71.3	85.5	101	111	138	159
8.63	24.4	33.0	40.7	53.0	62.9	75.4	89.4	97.7	122	141
8.91	22.2	30.1	37.1	48.4	57.4	68.9	81.3	89.3	112	129
9.57	17.9	24.2	30.0	39.1	46.6	55.9	66.7	72.7	91.4	106
17.44	2.94	3.98	5.04	6.53	7.87	9.63	11.5	12.6	16.2	19
19.61	2.10	2.83	3.59	4.62	5.57	6.84	8.26	8.95	11.5	13.6
20.17	1.94	2.60	3.30	4.26	5.13	6.30	7.51	8.24	10.6	12.5
21.13	1.70	2.28	2.89	3.71	4.47	5.50	6.[illegible]3	7.18	9.24	10.9
22.11	1.50	2.00	2.54	3.25	3.91	4.82	5.[illegible]0	6.28	8.08	9.57
22.72	1.39	1.86	2.35	3.00	3.61	4.45	5.[illegible]5	5.79	7.45	8.84
23.82	1.23	1.63	2.06	2.62	3.15	3.88	4.[illegible]6	5.04	6.49	7.71
24.94	1.09	1.44	1.81	2.30	2.76	3.40	4.[illegible]8	4.41	5.67	6.75

X-RAY CROSS SECTIONS (Continued)

Table II. Total Cross Section in cm^2/g (Continued)

Z KEV	21 Sc	22 Ti	23 V	24 Cr	25 Mn	26 Fe	27 Co	28 Ni	29 Cu	30 Zn	31 Ga	32 Ge	33 As	34 Se	35 Br	36 Kr	37 Rb	38 Sr	39 Y	40 Zr
4.51	819	111	125	143	160	188	206	240	257	280	309	329	368	403	435	464	508	552	599	648
4.93	658	86.8	97.3	111	125	147	161	188	201	220	242	258	288	317	342	366	400	436	473	511
5.41	521	571	75.1	85.7	96.1	113	125	146	155	172	187	200	224	246	266	285	312	339	369	399
5.90	420	459	513	67.4	75.6	88.9	98.4	115	123	137	148	158	178	195	211	226	248	270	294	317
5.95	411	449	501	65.8	73.8	86.8	96.1	113	120	134	144	155	173	190	206	221	242	263	287	310
6.40	339	370	411	462	59.9	70.4	78.3	91.8	97.4	110	117	126	142	156	169	181	198	215	235	254
6.49	327	357	396	445	57.6	67.7	75.3	88.3	93.6	106	113	122	136	150	162	174	191	207	227	244
6.93	276	301	333	375	405	56.3	62.9	73.8	78.1	88.7	94.2	102	114	125	136	146	160	174	190	205
7.06	262	286	316	357	385	53.3	59.6	70.0	74.1	84.3	89.3	96.8	108	119	129	138	152	165	181	195
7.47	226	246	271	307	331	367	50.9	59.8	63.2	72.4	76.2	82.9	92.5	101	110	119	130	141	155	167
7.65	212	231	255	288	311	346	47.7	56.1	59.2	68.0	71.4	77.8	86.8	95.1	104	111	122	132	145	157
8.04	186	202	223	252	273	304	339	48.8	51.5	59.5	62.1	67.9	75.7	82.9	90.3	97	106	115	127	137
8.27	173	188	206	234	253	283	315	45.2	47.7	55.3	57.5	63.0	70.1	77.8	83.7	89.9	98.5	107	118	127
8.63	153	167	183	208	225	253	281	306	42.3	49.2	50.9	56.0	62.2	68.0	74.2	79.8	87.4	94.7	105	113
8.91	141	153	168	191	207	234	259	283	38.7	45.3	46.7	51.4	57.0	62.3	68.1	73.2	80.2	86.8	96.2	103
9.57	116	126	138	157	170	194	214	236	245	37.4	38.1	42.3	46.7	51.0	55.8	60.0	65.7	71.0	79.0	84.8
17.44	21.0	23.3	25.2	29.3	31.9	37.7	41.0	47.2	49.3	55.5	56.9	60.5	66.0	68.8	74.7	79.1	83.0	88.0	97.6	16.1
19.61	15.0	16.7	18.1	21.0	22.9	27.2	29.5	34.2	35.8	40.3	41.7	44.3	48.6	51.2	55.6	58.6	62.1	65.6	72.6	75.2
20.17	13.8	15.4	16.7	19.4	21.1	25.1	27.2	31.6	33.1	37.2	38.6	41.0	45.1	47.6	51.7	54.5	57.8	61.0	67.5	70.1
21.13	12.1	13.4	14.6	17.0	18.5	22.0	23.8	27.7	29.0	32.7	34.0	36.1	39.7	42.1	45.6	48.1	51.1	54.0	59.7	62.1
22.11	10.5	11.8	12.8	14.9	16.2	19.3	20.9	24.3	25.5	28.7	29.9	31.8	35.0	37.2	40.4	42.5	45.3	47.9	53.0	55.2
22.72	9.72	10.9	11.8	13.7	15.0	17.8	19.3	22.5	23.6	26.6	27.7	29.4	32.4	34.6	37.5	39.5	42.1	44.5	49.2	51.4
23.82	8.47	9.48	10.3	12.0	13.1	15.6	16.9	19.7	20.7	23.3	24.3	25.8	28.5	30.5	33.1	34.8	37.2	39.3	43.5	45.5
24.94	7.40	8.30	9.02	10.5	11.5	13.7	14.8	17.3	18.2	20.4	21.4	22.7	25.1	26.9	29.2	30.7	32.9	34.8	38.5	40.4

Z KEV	41 Nb	42 Mo	43 Te	44 Ru	45 Rh	46 Pd	47 Ag	48 Cd	49 In	50 Sn	51 Sb	52 Te	53 I	54 Xe	55 Cs	56 Ba	57 La	58 Ce	59 Pr	60 Nd
4.51	697	738	786	832	892	928	987	1064	1151	1128	997	753	293	300	324	334	355	383	414	433
4.93	552	585	621	660	708	739	785	842	906	899	926	843	921	683	259	266	282	304	330	344
5.41	432	457	486	518	555	581	617	659	706	709	733	769	835	755	803	587	223	240	261	271
5.90	344	365	387	414	444	466	495	526	561	569	592	617	666	701	742	660	677	521	210	218
5.95	336	357	378	404	434	455	484	514	548	557	579	603	651	685	725	645	662	509	205	213
6.40	276	293	310	333	357	375	398	422	449	460	479	497	535	565	597	615	636	592	450	464
6.49	266	282	299	320	344	361	384	407	433	443	462	479	516	545	575	593	613	571	610	447
6.93	223	237	251	269	289	304	324	342	363	374	391	404	434	459	484	499	519	559	596	532
7.06	212	225	238	256	275	289	308	325	345	356	373	385	413	437	460	475	494	532	567	506
7.47	182	193	204	220	236	249	265	279	295	307	322	331	355	376	395	408	427	459	488	506
7.65	170	181	192	207	222	234	249	262	277	289	303	312	333	353	372	384	402	432	459	476
8.04	149	158	168	181	194	205	218	229	242	253	267	273	292	310	325	336	354	379	402	418
8.27	138	147	156	168	180	190	203	213	225	236	248	254	271	288	302	312	329	352	374	389
8.63	122	130	138	149	160	169	180	189	200	210	221	226	241	256	269	278	294	314	333	347
8.91	112	120	127	137	147	156	166	174	183	193	204	208	222	236	248	256	271	290	307	320
9.57	92.0	98.2	104	113	121	128	137	143	151	159	168	172	183	195	204	211	224	240	253	265
17.44	17.0	18.4	19.8	21.3	23.1	24.4	26.4	27.7	29.1	31.2	33.0	33.9	36.3	38.3	40.4	42.4	45.3	48.6	50.8	53.3
19.61	81.2	13.3	14.3	15.4	16.7	17.6	19.1	20.1	21.2	22.6	23.9	24.7	26.5	27.9	29.5	31.0	33.1	35.5	37.1	38.9
20.17	75.6	79.3	13.2	14.2	15.4	16.3	17.7	18.6	19.6	20.9	22.1	22.8	24.6	25.8	27.3	28.7	30.7	33.0	34.4	36.0
21.13	67.1	70.3	73.0	12.5	13.5	14.3	15.5	16.4	17.3	18.4	19.4	20.1	21.7	22.7	24.1	25.4	27.0	29.1	30.3	31.7
22.11	59.6	62.5	65.0	11.0	11.9	12.6	13.7	14.5	15.3	16.2	17.1	17.7	19.2	20.0	21.3	22.5	23.9	25.7	26.8	28.0
22.72	55.5	58.2	60.6	63.8	11.0	11.6	12.7	13.4	14.1	15.0	15.8	16.4	17.8	18.6	19.8	20.9	22.1	23.9	24.9	26.0
23.82	49.1	51.5	53.7	56.6	58.3	10.2	11.1	11.8	12.4	13.2	13.8	14.4	15.7	16.3	17.4	18.4	19.5	21.0	21.9	22.9
24.94	43.6	45.7	47.7	50.3	52.0	57.0	97.8	10.4	11.0	11.6	12.1	12.7	13.9	14.4	15.4	16.3	17.2	18.6	19.3	20.2

X-RAY CROSS SECTIONS (Continued)

Table II. Total Cross Section in cm²/g (Continued)

Z KEV	61 Pm	62 Sm	63 Eu	64 Gd	65 Tb	66 Dy	67 Ho	68 Er	69 Tm	70 Yb	71 Lu	72 Hf	73 Ta	74 W	75 Re	76 Os	77 Ir	78 Pt	79 Au	80 Hg
4.51	455	473	503	510	546	568	589	615	644	664	696	720	736	753	796	824	871	934	906	958
4.93	361	375	398	405	432	449	465	485	508	524	549	568	581	598	631	653	688	734	720	760
5.41	285	295	313	319	339	352	363	380	397	410	430	445	455	470	496	512	540	572	568	598
5.90	229	237	251	256	271	281	290	303	317	327	343	355	363	378	397	410	431	455	457	480
5.95	224	232	245	250	265	275	283	296	310	319	335	347	355	369	388	401	422	444	447	469
6.40	186	192	203	207	219	227	234	244	255	263	276	286	293	306	321	332	348	365	371	388
6.49	476	185	195	200	211	219	225	235	246	254	266	276	282	295	310	320	336	351	358	375
6.93	401	412	165	170	179	185	190	198	207	214	225	233	238	250	262	270	283	295	303	317
7.06	536	392	420	161	170	176	181	189	197	204	214	222	227	238	249	257	270	281	289	302
7.47	535	475	361	369	147	152	156	163	170	175	184	191	195	206	215	222	233	241	250	261
7.65	503	446	477	347	367	143	146	153	160	165	173	180	184	194	203	209	219	226	236	246
8.04	441	454	418	427	322	337	128	134	140	145	152	158	162	171	178	184	192	198	208	216
8.27	410	422	450	397	420	313	333	125	131	135	142	147	150	159	166	171	179	184	194	202
8.63	366	376	401	410	375	392	296	318	117	120	126	131	134	142	149	153	160	164	174	180
8.91	336	347	369	377	400	360	272	292	289	111	117	121	124	132	137	141	148	151	161	167
9.57	278	287	305	312	330	344	364	336	239	232	247	236	103	110	114	118	123	125	134	139
17.44	55.5	58.0	61.2	62.8	66.8	68.9	72.1	75.6	79.0	80.2	84.2	86.3	89.5	95.8	98.7	100	103	109	111	115
19.61	40.5	42.4	44.7	46.0	48.9	50.4	52.8	55.1	57.9	59.2	62.0	64.2	66.1	70.6	72.5	74.1	77.2	80.2	82.3	85.3
20.17	37.6	39.3	41.5	42.6	45.3	46.7	48.9	51.0	53.8	55.0	57.6	59.7	61.4	65.5	67.3	68.9	71.9	74.6	76.5	79.4
21.13	33.1	34.7	36.6	37.6	40.0	41.2	43.2	45.0	47.5	48.7	51.0	52.9	54.4	58.0	59.5	61.1	63.8	66.0	67.8	70.4
22.11	29.3	30.7	32.4	33.3	35.4	36.5	38.3	39.8	42.1	43.2	45.3	46.9	48.3	51.4	52.8	54.2	56.8	58.6	60.2	62.6
22.72	27.1	28.5	30.1	30.9	32.9	33.9	35.6	37.0	39.1	40.2	42.1	43.7	44.9	47.8	49.1	50.5	52.9	54.5	56.0	58.3
23.82	23.9	25.1	26.5	27.3	29.0	29.9	31.4	32.6	34.5	35.5	37.2	38.6	39.7	42.2	43.3	44.6	46.8	48.2	49.5	51.7
24.94	21.1	22.2	23.4	24.1	25.6	26.4	27.8	28.8	30.6	31.5	33.0	34.2	35.2	37.4	38.4	39.6	41.5	42.7	43.9	45.9

Z KEV	81 Tl	82 Pb	83 Bi	86 Rn	90 Th	92 U	95 Pu
4.51	991	1035	1066	1174	1098	1084	960
4.93	785	820	847	930	993	862	1023
5.41	617	645	667	731	844	774	803
5.90	494	517	536	586	678	672	731
5.95	483	505	524	573	663	657	772
6.40	400	418	435	474	549	545	638
6.49	386	403	419	458	530	526	615
6.93	326	341	355	387	449	446	520
7.06	311	325	338	369	428	425	495
7.47	268	280	293	318	370	363	427
7.65	253	264	276	300	348	347	402
8.04	222	232	243	264	307	306	353
8.27	207	216	227	246	286	285	329
8.63	185	194	203	220	256	256	294
8.91	171	179	188	203	237	236	271
9.57	142	149	156	169	197	197	225
17.44	119	123	126	117	99.5	96.7	48.8
19.61	88.3	90.6	93.5	101	73.3	72.6	79.0
20.17	82.0	84.1	87.0	93.8	95.0	67.8	73.7
21.13	72.7	74.5	77.2	83.3	97.2	84.2	65.7
22.11	64.6	66.1	68.7	74.1	86.3	86.9	58.8
22.72	60.1	61.5	64.0	69.1	80.3	81.1	76.4
23.82	53.2	54.4	56.7	61.2	71.0	72.1	78.3
24.94	47.2	48.2	50.4	54.4	62.9	64.2	69.8

X-RAY WAVELENGTHS

J. A. Bearden

These tables were originally published as the final report to the U.S. Atomic Energy Commission as Report NYO-10586 in partial fulfillment of Contract AT(30-1)-2543. The tables were later reproduced in *Review of Modern Physics*. The data may also be obtained from the Superintendent of Documents, U.S. Government Printing Office, Washington, D. C. 20402 in the publication NSRDS-NBS 14. Persons seeking discussion of the experimental work, conventions, secondary standards, etc. will find these in *Review of Modern Physics*, Vol. 39, No. 1, 78-124, January 1967.

THE W $K\alpha_1$ WAVELENGTH STANDARD

A wavelength standard should possess characteristics which permit its ready redetermination in other laboratories by different techniques. Considering all of the factors involved in the selection of a wavelength standard, the W $K\alpha_1$ line is superior to any other x-ray or γ-ray wavelength. Its advantages as the x-ray wavelength standard are discussed in *Review of Modern Physics* Vol. 39, page 82 (1967).

$$\lambda \mathrm{W}\ K\alpha_1 = 0.2090100\ \text{Å} \pm 5\ \text{ppm}.$$

This numerical value of the wavelength of the W $K\alpha_1$ line is used to define the *x-ray wavelength standard* by the relation

$$\lambda(\mathrm{W}\ K\alpha_1) = 0.2090100\ \text{Å}^*.$$

This is a new unit of length which may differ from the angstrom by ±5 ppm (probable error), *but as a wavelength standard it has no error*. In order to clearly indicate that this unit is not exactly an angstrom, it has been designated Å*.

Wavelengths tabulated normally refer to the pure element in its solid form. However, there are many instances in which such data are not available. For example, rare gases are of necessity almost always used in the gaseous form, while the rare-earth elements were customarily used in the form of salts.

In high precision work there is some ambiguity as to exactly what feature of a line profile should be taken to be the "true wavelength." In double-crystal work the line peak is usually employed. In crystallography the centroid is widely used; in photographic work with visual observation of the plates, there is involved some subjective criterion of the observer which it is difficult to define precisely. In this survey the peak of the line profile has been adopted as the standard criterion.

In the study of the X-ray literature, the wavelengths of a number of lines were noted which appeared inconsistent with the remaining data. A Moseley-type diagram was constructed, and if the value was clearly outside estimated probable error, it was assumed that an experimental or typographical error had occurred, and the interpolated value was listed in the table. Such cases are marked with a dagger (†) as a superscript to the wavelength. For elements of atomic number 85 through 89 and 91, there are no measured lines of the *K* series and very few of other series except for 88 radium and 91 protactinium. Likewise there are very few measurements for 43 technetium and 54 xenon. In these cases, interpolated values are listed for the more prominent lines and marked with a dagger (†).

X-RAY WAVELENGTHS

X-ray wavelengths in Å* units and in keV. The probable error (p.e.) is the error in the last digit of wavelength. Designation indicates both conventional Siegbahn notation (if applicable) and transition, e.g., $\beta_1\ L_{II}M_{IV}$ denotes a transition between the L_{II} and M_{IV} levels, which is the $L\beta_1$ line in Siegbahn notation.

Designation	Å*	p.e.	keV	Å*	p.e.	keV
	3 Lithium			**4 Beryllium**		
$\alpha\ KL$	228.	1	0.0543	114.	1	0.1085
	5 Boron			**6 Carbon**		
$\alpha\ KL$	67.6	3	0.1833	44.7	3	0.277
	7 Nitrogen			**8 Oxygen**		
$\alpha\ KL$	31.6	4	0.3924	23.62	3	0.5249
	9 Fluorine			**10 Neon**		
$\alpha_{1,2}\ KL_{II,III}$	18.32	2	0.6768	14.610	3	0.8486
$\beta\ KM$				14.452	5	0.8579
	11 Sodium			**12 Magnesium**		
$\alpha_{1,2}\ KL_{II,III}$	11.9101	9	1.0410	9.8900	2	1.25360
$\beta\ KM$	11.575	2	1.0711	9.521	2	1.3022
$L_{II,III}M$	407.1	5	0.03045	251.5	5	0.0493
$L_IL_{II,III}$	376	1	0.0330	317	1	0.0392
	13 Aluminum			**14 Silicon**		
$\alpha_2\ KL_{II}$	8.34173	9	1.48627	7.12791	9	1.73938
$\alpha_1\ KL_{III}$	8.33934	9	1.48670	7.12542	9	1.73998
$\beta\ KM$	7.960	2	1.5574	6.753	1	1.8359
$L_{II,III}$	171.4	5	0.0724	135.5	4	0.0915
$L_IL_{II,III}$	290.	1	0.0428			
	15 Phosphorus			**16 Sulfur**		
$\alpha_2\ KL_{II}$	6.160†	1	2.0127	5.37496	8	2.30664
$\alpha_1\ KL_{III}$	6.157†	1	2.0137	5.37216	7	2.30784
$\beta\ KM$	5.796	2	2.1390			
$\beta_1\ KM$				5.0316	2	2.4640
$\beta_2\ KM$				5.0233	3	2.4681
$L_{II,III}M$	103.8	4	0.1194			
$l,\eta\ L_{II,III}M_I$				83.4	3	0.1487
	17 Chlorine			**18 Argon**		
$\alpha_2\ KL_{II}$	4.7307	1	2.62078	4.19474	5	2.95563
$\alpha_1\ KL_{III}$	4.7278	1	2.62239	4.19180	5	2.95770
$\beta\ KM$	4.4034	3	2.8156			
$\beta_{1,3}\ KM_{II,III}$				3.8860	2	3.1905
$\eta\ L_{II}M_I$	67.33	9	0.1841	55.9†	1	0.2217
$l\ L_{III}M_I$	67.90	9	0.1826	56.3†	1	0.2201
	19 Potassium			**20 Calcium**		
$\alpha_2\ KL_{II}$	3.7445	2	3.3111	3.36166	3	3.68809
$\alpha_1\ KL_{III}$	3.7414	2	3.3138	3.35839	3	3.69168
$\beta_{1,3}\ KM_{II,III}$	3.4539	2	3.5896	3.0897	2	4.0127
$\beta_5\ KM_{IV,V}$	3.4413	4	3.6027	3.0746	3	4.0325

Designation	Å*	p.e.	keV	Å*	p.e.	keV
	19 Potassium (*Cont.*)			**20 Calcium** (*Cont.*)		
$\eta\ L_{II}M_I$	47.24	2	0.2625	40.46	2	0.3064
β_1				35.94	2	0.3449
$l\ L_{III}M_I$	47.74	1	0.25971	40.96	2	0.3027
$\alpha_{1,2}\ L_{III}M_{IV,V}$				36.33	2	0.3413
$M_{II,III}N_I$	692	9	0.0179	525.	9	0.0236
	21 Scandium			**22 Titanium**		
$\alpha_2\ KL_{II}$	3.0342	1	4.0861	2.75216	2	4.50486
$\alpha_1\ KL_{III}$	3.0309†	1	4.0906	2.74851	2	4.51084
$\beta_{1,3}\ KM_{II,III}$	2.7796	2	4.4605	2.51391	2	4.93181
$\beta_5\ KM_{IV,V}$	2.7634	3	4.4865	2.4985	2	4.9623
$\eta\ L_{II}M_I$	35.13	2	0.3529	30.89	3	0.4013
$\beta_1\ L_{II}M_{IV}$	31.02	2	0.3996	27.05	2	0.4584
$l\ L_{III}M_I$	35.59	3	0.3483	31.36	2	0.3953
$\alpha_{1,2}\ L_{III}M_{IV,V}$	31.35	3	0.3954	27.42	2	0.4522
	23 Vanadium			**24 Chromium**		
$\alpha_2\ KL_{II}$	2.50738	2	4.94464	2.293606	3	5.40551
$\alpha_1\ KL_{III}$	2.50356	2	4.95220	2.28970	2	5.41472
$\beta_{1,3}\ KM_{II,III}$	2.28440	2	5.42729	2.08487	2	5.94671
$\beta_5\ KM_{IV,V}$	2.26951	6	5.4629	2.07087	6	5.9869
$\beta_{3,4}\ L_IM_{II,III}$	21.19†	9	0.585	18.96	2	0.654
$\eta\ L_{II}M_I$	27.34	3	0.4535	24.30	3	0.5102
$\beta_1\ L_{II}M_{IV}$	23.88	4	0.5192	21.27	1	0.5828
$l\ L_{III}M_I$	27.77	1	0.4465	24.78	1	0.5003
$\alpha_{1,2}\ L_{III}M_{IV,V}$	24.25	3	0.5113	21.64	3	0.5728
$M_{II,III}M_{IV,V}$	337.	9	0.037	309.	9	0.040
	25 Manganese			**26 Iron**		
$\alpha_2\ KL_{II}$	2.10578	2	5.88765	1.939980	9	6.39084
$\alpha_1\ KL_{III}$	2.101820	9	5.89875	1.936042	9	6.40384
$\beta_{1,3}\ KM_{II,III}$	1.91021	2	6.49045	1.75661	2	7.05798
$\beta_5\ KM_{IV,V}$	1.8971	1	6.5352	1.7442	1	7.1081
$\beta_{3,4}\ L_IM_{II,III}$	17.19	2	0.721	15.65	2	0.792
$\eta\ L_{II}M_I$	21.85	2	0.5675	19.75	4	0.628
$\beta_1\ L_{II}M_{IV}$	19.11	2	0.6488	17.26	1	0.7185
$l\ L_{III}M_I$	22.29	1	0.5563	20.15	1	0.6152
$\alpha_{1,2}\ L_{III}M_{IV,V}$	19.45	1	0.6374	17.59	2	0.7050
$M_{II,III}M_{IV,V}$	273.	6	0.045	243.	5	0.051
	27 Cobalt			**28 Nickel**		
$\alpha_2\ KL_{II}$	1.792850	9	6.91530	1.661747	8	7.46089
$\alpha_1\ KL_{III}$	1.788965	9	6.93032	1.657910	8	7.47815
$\beta_{1,3}\ KM_{II,III}$	1.62079	2	7.64943	1.500135	8	8.26466
$\beta_5\ KM_{IV,V}$	1.60891	3	7.7059	1.48862	4	8.3286
$\beta_{3,4}\ L_IM_{II,III}$	14.31	3	0.870	13.18	1	0.941
$\eta\ L_{II}M_I$	17.87	3	0.694	16.27	3	0.762
$\beta_1\ L_{II}M_{IV}$	15.666	8	0.7914	14.271	6	0.8688
$l\ L_{III}M_I$	18.292	8	0.6778	16.693	9	0.7427
$\alpha_{1,2}\ L_{III}M_{IV,V}$	15.972	6	0.7762	14.561	3	0.8515
$M_{II,III}M_{IV,V}$	214.	6	0.058	190.	2	0.0651

TABLE A (*Continued*)

Designation	Å*	p.e.	keV	Å*	p.e.	keV
	29 Copper			**30 Zinc**		
$\alpha_2\ KL_{II}$	1.544390	2	8.02783	1.439000	8	8.61578
$\alpha_1\ KL_{III}$	1.540562	2	8.04778	1.435155	7	8.63886
$\beta_3\ KM_{II}$	1.3926	1	8.9029			
$\beta_{1,3}\ KM_{II,III}$	1.392218	9	8.90529	1.29525	2	9.5720
$\beta_2\ KN_{II,III}$				1.28372	2	9.6580
$\beta_5\ KM_{IV,V}$	1.38109	3	8.9770	1.2848	1	9.6501
$\beta_{3,4}\ L_IM_{II,III}$	12.122	8	1.0228	11.200	7	1.1070
$\eta\ L_{II}M_I$	14.90	2	0.832	13.68	2	0.906
$\beta_1\ L_{II}M_{IV}$	13.053	3	0.9498	11.983	3	1.0347
$l\ L_{III}M_I$	15.286	9	0.8111	14.02	2	0.884
$\alpha_{1,2}\ L_{III}M_{IV,V}$	13.336	3	0.9297	12.254	3	1.0117
$M_{II,III}M_{IV,V}$	173.	3	0.072	157.	3	0.079

Designation	Å*	p.e.	keV	Å*	p.e.	keV
	31 Gallium			**32 Germanium**		
$\alpha_2\ KL_{II}$	1.34399	1	9.22482	1.258011	9	9.85532
$\alpha_1\ KL_{III}$	1.340083	9	9.25174	1.254054	9	9.88642
$\beta_3\ KM_{II}$	1.20835	5	10.2603	1.12936	9	10.9780
$\beta_1\ KM_{III}$	1.20789	2	10.2642	1.12894	2	10.9821
$\beta_2\ KN_{II,III}$	1.19600	2	10.3663	1.11686	2	11.1008
$\beta_5\ KM_{IV,V}$	1.1981	2	10.348	1.1195	1	11.0745
$\beta_4\ L_IM_{II}$				9.640	2	1.2861
$\beta_3\ L_IM_{III}$				9.581	2	1.2941
$\beta_{3,4}\ L_IM_{II,III}$	10.359†	8	1.197			
$\eta\ L_{II}M_I$	12.597	2	0.9842	11.609	2	1.0680
$\beta_1\ L_{II}M_{IV}$	11.023	2	1.1248	10.175	1	1.2185
$l\ L_{III}M_I$	12.953	2	0.9572	11.965	4	1.0362
$\alpha_{1,2}\ L_{III}M_{IV,V}$	11.292	1	1.09792	10.4361	8	1.18800

Designation	Å*	p.e.	keV	Å*	p.e.	keV
	33 Arsenic			**34 Selenium**		
$\alpha_2\ KL_{II}$	1.17987	1	10.50799	1.10882	2	11.1814
$\alpha_1\ KL_{III}$	1.17588	1	10.54372	1.10477	2	11.2224
$\beta_3\ KM_{II}$	1.05783	5	11.7203	0.99268	5	12.4896
$\beta_1\ KM_{III}$	1.05730	2	11.7262	0.99218	3	12.4959
$\beta_2\ KN_{II,III}$	1.04500	3	11.8642	0.97992	5	12.6522
$\beta_5\ KM_{IV,V}$	1.0488	1	11.822	0.9843	1	12.595
$\beta_{3,4}\ L_IM_{II,III}$	8.929	1	1.3884	8.321†	9	1.490
$\eta\ L_{II}M_I$	10.734	1	1.1550	9.962	1	1.2446
$\beta_1\ L_{II}M_{IV}$	9.4141	8	1.3170	8.7358	5	1.41923
$l\ L_{III}M_I$	11.072	1	1.1198	10.294	1	1.2044
$\alpha_{1,2}\ L_{III}M_{IV,V}$	9.6709	8	1.2820	8.9900	5	1.37910
M_VN_{III}				230.	2	0.0538

Designation	Å*	p.e.	keV	Å*	p.e.	keV
	35 Bromine			**36 Krypton**		
$\alpha_2\ KL_{II}$	1.04382	2	11.8776	0.9841	1	12.598
$\alpha_1\ KL_{III}$	1.03974	2	11.9242	0.9801	1	12.649
$\beta_3\ KM_{II}$	0.93327	5	13.2845	0.8790	1	14.104
$\beta_1\ KM_{III}$	0.93279	2	13.2914	0.8785	1	14.112
$\beta_2\ KN_{II,III}$	0.92046	2	13.4695	0.8661	1	14.315
$\beta_5\ KM_{IV,V}$	0.9255	1	13.396	0.8708	2	14.238
$\beta_4\ KN_{IV,V}$				0.8653	2	14.328
$\beta_4\ L_IM_{II}$				7.304	5	1.697
$\beta_3\ L_IM_{III}$				7.264	5	1.707

Designation	Å*	p.e.	keV	Å*	p.e.	keV
	35 Bromine (*Cont.*)			**36 Krypton (*Cont.*)**		
$\beta_{3,4}\ L_IM_{II,III}$	7.767†	9	1.596			
$\eta\ L_{II}M_I$	9.255	1	1.3396			
$\beta_1\ L_{II}M_{IV}$	8.1251	5	1.52590	7.576†	3	1.6366
γ_5				7.279	5	1.703
$l\ L_{III}M_I$	9.585	1	1.2935			
$\alpha_{1,2}\ L_{III}M_{IV,V}$	8.3746	5	1.48043	7.817†	3	1.5860
β_6				7.510	4	1.6510
$L_{II}N_{III}$				7.250	5	1.710
M_IM_{II}	184.6	3	0.0672			
M_IM_{III}	164.7	3	0.0753			
$M_{II}M_{IV}$	109.4	3	0.1133			
$M_{II}N_I$	76.9	2	0.1613			
$M_{III}M_{IV,V}$	113.8	3	0.1089			
	79.8	3	0.1554			
$\zeta_2\ M_{IV}N_{II}$	191.1	2	0.06488			
$M_{IV}N_{III}$	189.5	3	0.0654			
$\zeta_1\ M_VN_{III}$	192.6	2	0.06437			

Designation	Å*	p.e.	keV	Å*	p.e.	keV
	37 Rubidium			**38 Strontium**		
$\alpha_2\ KL_{II}$	0.92969	1	13.3358	0.87943	1	14.0979
$\alpha_1\ KL_{III}$	0.925553	9	13.3953	0.87526	1	14.1650
$\beta_3\ KM_{II}$	0.82921	3	14.9517	0.78345	3	15.8249
$\beta_1\ KM_{III}$	0.82868	2	14.9613	0.78292	2	15.8357
$\beta_2\ KN_{II,III}$	0.81645	3	15.1854	0.77081	3	16.0846
$\beta_5\ KM_{IV,V}$	0.8219	1	15.085	0.7764	1	15.969
$\beta_4\ KN_{IV,V}$	0.8154	2	15.205	0.76989	5	16.104
$\beta_4\ L_IM_{II}$	6.8207	3	1.81771	6.4026	3	1.93643
$\beta_3\ L_IM_{III}$	6.7876	3	1.82659	6.3672	3	1.94719
$\gamma_{2,3}\ L_IN_{II,III}$	6.0458	3	2.0507	5.6445	3	2.1965
$\eta\ L_{II}M_I$	8.0415	4	1.54177	7.5171	3	1.64933
$\beta_1\ L_{II}M_{IV}$	7.0759	3	1.75217	6.6239	3	1.87172
$\gamma_5\ L_{II}N_{IV}$	6.7553	3	1.83532	6.2961	3	1.96916
$l\ L_{III}M_I$	8.3636	4	1.48238	7.8362	3	1.58215
$\alpha_2\ L_{III}M_{IV}$	7.3251	3	1.69256	6.8697	3	1.80474
$\alpha_1\ L_{III}M_V$	7.3183	2	1.69413	6.8628	2	1.80656
$\beta_6\ L_{III}N_I$	6.9842	3	1.77517	6.5191	3	1.90181
M_IM_{III}	144.4	3	0.0859			
$M_{II}M_{IV}$	91.5	2	0.1355	85.7	2	0.1447
$M_{II}N_I$	57.0	2	0.2174	51.3	1	0.2416
$M_{III}M_{IV,V}$	96.7	2	0.1282	91.4	2	0.1357
$M_{III}N_I$	59.5	2	0.2083	53.6	1	0.2313
$\zeta_2\ M_{IV}N_{II}$	127.8	2	0.0970			
$M_{IV}N_{III}$	126.8	2	0.0978			
$\zeta_2\ M_{IV}N_{II,III}$				108.0	2	0.1148
$\zeta_1\ M_VN_{III}$	128.7	2	0.0964	108.7	1	0.1140

Designation	Å*	p.e.	keV	Å*	p.e.	keV
	39 Yttrium			**40 Zirconium**		
$\alpha_2\ KL_{II}$	0.83305	1	14.8829	0.79015	1	15.6909
$\alpha_1\ KL_{III}$	0.82884	1	14.9584	0.78593	1	15.7751
$\beta_3\ KM_{II}$	0.74126	3	16.7258	0.70228	4	17.654
$\beta_1\ KM_{III}$	0.74072	2	16.7378	0.70173	3	17.6678
$\beta_2\ KN_{II,III}$	0.72864	4	17.0154	0.68993	4	17.970
$\beta_5\ KM_{IV,V}$	0.7345	1	16.879	0.6959	1	17.815

Designation	Å*	p.e.	keV	Å*	p.e.	keV
	39 Yttrium (*Cont.*)			**40 Zirconium** (*Cont.*)		
$\beta_4\ KN_{IV,V}$	0.72776	5	17.036	0.68901	5	17.994
$\beta_4\ L_IM_{II}$	6.0186	3	2.0600	5.6681	3	2.1873
$\beta_3\ L_IM_{III}$	5.9832	3	2.0722	5.6330	3	2.2010
$\gamma_{2,3}\ L_IN_{II,III}$	5.2830	3	2.3468	4.9536	3	2.5029
$\eta\ L_IM_I$	7.0406	3	1.76095	6.6069	3	1.87654
$\beta_1\ L_{II}M_{IV}$	6.2120	3	1.99584	5.8360	3	2.1244
$\gamma_5\ L_{II}N_I$	5.8754	3	2.1102	5.4977	3	2.2551
$\gamma_1\ L_{II}N_{IV}$				5.3843	3	2.3027
$l\ L_{III}M_I$	7.3563	3	1.68536	6.9185	3	1.79201
$\alpha_2\ L_{III}M_{IV}$	6.4558	3	1.92047	6.0778	3	2.0399
$\alpha_1\ L_{III}M_V$	6.4488	2	1.92256	6.0705	2	2.04236
$\beta_6\ L_{III}N_I$	6.0942	3	2.0344	5.7101	3	2.1712
$\beta_{2,15}$				5.5863	3	2.2194
$M_{II}M_{IV}$	81.5	2	0.1522	76.7	2	0.1617
$M_{II}N_I$	46.48	9	0.267			
$M_{III}M_V$				80.9	3	0.1533
$M_{III}N_I$	48.5	2	0.256			
$M_{III}M_{IV,V}$	86.5	2	0.1434			
$\zeta\ M_{IV,V}N_{II,III}$	93.4	2	0.1328	82.1	2	0.1511
$M_{IV,V}O_{II,III}$				70.0	4	0.177

Designation	Å*	p.e.	keV	Å*	p.e.	keV
	41 Niobium			**42 Molybdenum**		
$\alpha_2\ KL_{II}$	0.75044	1	16.5210	0.713590	6	17.3743
$\alpha_1\ KL_{III}$	0.74620	1	16.6151	0.709300	1	17.47934
$\beta_3\ KM_{II}$	0.66634	3	18.6063	0.632872	9	19.5903
$\beta_1\ KM_{III}$	0.66576	2	18.6225	0.632288	9	19.6083
β_2^{II}				0.62107	5	19.963
$\beta_2\ KN_{II,III}$	0.65416	4	18.953	0.62099	2	19.9652
$\beta_4\ KN_{IV,V}$	0.65318	5	18.981			
$\beta_5^{II}\ KM_{IV}$				0.62708	5	19.771
$\beta_5^{I}\ KM_V$				0.62692	5	19.776
$\beta_4\ KN_{IV,V}$				0.62001	9	19.996
$\beta_4\ L_IM_{II}$	5.3455	3	2.3194	5.0488	3	2.4557
$\beta_3\ L_IM_{III}$	5.3102	3	2.3348	5.0133	3	2.4730
$\gamma_{2,3}\ L_IN_{II,III}$	4.6542	2	2.6638	4.3800	2	2.8306
$\eta\ L_{II}M_I$	6.2109	3	1.99620	5.8475	3	2.1202
$\beta_1\ L_{II}M_{IV}$	5.4923	3	2.2574	5.17708	8	2.39481
$\gamma_5\ L_{II}N_I$	5.1517	3	2.4066	4.8369	2	2.5632
$\gamma_1\ L_{II}N_{IV}$	5.0361	3	2.4618	4.7258	2	2.6235
$l\ L_{III}M_I$	6.5176	3	1.90225	6.1508	3	2.01568
$\alpha_2\ L_{III}M_{IV}$	5.7319	3	2.1630	5.41437	8	2.28985
$\alpha_1\ L_{III}M_V$	5.7243	2	2.16589	5.40655	8	2.29316
$\beta_6\ L_{III}N_I$	5.3613	3	2.3125	5.0488	5	2.4557
$\beta_{2,15}\ L_{III}N_{IV,V}$	5.2379	3	2.3670	4.9232	2	2.5183
$M_{II}M_{IV}$	72.1	3	0.1718	68.9	2	0.1798
$M_{II}N_I$	38.4	3	0.323	35.3	3	0.351
$M_{II}N_{IV}$	33.1	2	0.375			
$M_{III}M_V$	78.4	2	0.1582	74.9	1	0.1656
$M_{III}N_I$	40.7	2	0.305	37.5	2	0.331
$\gamma\ M_{III}N_{IV,V}$	34.9	2	0.356			
$\zeta\ M_{IV,V}N_{II,III}$	72.19	9	0.1717	64.38	7	0.1926
$M_{IV,V}O_{II,III}$	61.9	2	0.2002	54.8	2	0.2262

Designation	Å*	p.e.	keV	Å*	p.e.	keV
	43 Technetium			**44 Ruthenium**		
$\alpha_2\ KL_{II}$	0.67932†	3	18.2508	0.647408	5	19.1504
$\alpha_1\ KL_{III}$	0.67502†	3	18.3671	0.643083	4	19.2792
$\beta_3\ KM_{II}$	0.60188†	4	20.599	0.573067	4	21.6346
$\beta_1\ KM_{III}$	0.60130†	4	20.619	0.572482	4	21.6568
$\beta_2\ KN_{II,III}$	0.59024†	5	21.005	0.56166	3	22.074
$\beta_5^{II}\ KM_{IV}$				0.5680	2	21.829
$\beta_5^{I}\ KM_V$				0.56785	9	21.834
β_4				0.56089	9	22.104
$\beta_4\ L_{II}M_{IV}$				4.5230	2	2.7411
$\beta_3\ L_IM_{III}$				4.4866	3	2.7634
$\gamma_{2,3}\ L_IN_{II,III}$				3.8977	2	3.1809
$\eta\ L_{II}M_I$				5.2050	2	2.38197
$\beta_1\ L_{II}M_{IV}$	4.8873†	8	2.5368	4.62058	3	2.68323
$\gamma_5\ L_{II}N_I$				4.2873	2	2.8918
$\gamma_1\ L_{II}N_{IV}$				4.1822	2	2.9645
$l\ L_{III}M_I$				5.5035	3	2.2528
$\alpha_2\ L_{III}M_{IV}$				4.85381	7	2.55431
$\alpha_1\ L_{III}M_V$	5.1148†	3	2.4240	4.84575	5	2.55855
$\beta_6\ L_{III}N_I$				4.4866	3	2.7634
$\beta_{2,15}\ L_{III}N_{IV,V}$				4.3718	2	2.8360
$M_{II}M_{IV}$				62.2	1	0.1992
$M_{II}N_I$				32.3	2	0.384
$M_{II}N_{IV}$				25.50	9	0.486
$M_{III}M_V$				68.3	1	0.1814
$\gamma\ M_{III}N_{IV,V}$				26.9	1	0.462
$\zeta\ M_{IV,V}N_{II,III}$				52.34	7	0.2369
$M_{IV,V}O_{II,III}$				44.8	1	0.2768

Designation	Å*	p.e.	keV	Å*	p.e.	keV
	45 Rhodium			**46 Palladium**		
$\alpha_2\ KL_{II}$	0.617630	4	20.0737	0.589821	3	21.0201
$\alpha_1\ KL_{III}$	0.613279	4	20.2161	0.585448	3	21.1771
$\beta_3\ KM_{II}$	0.546200	4	22.6989	0.521123	4	23.7911
$\beta_1\ KM_{III}$	0.545605	4	22.7236	0.520520	4	23.8187
$\beta_2^{II}\ KN_{II}$	0.53513	5	23.168			
$\beta_2\ KN_{II,III}$	0.53503	2	23.1728	0.510228	4	24.2991
$\beta_5^{II}\ KM_{IV}$	0.54118	9	22.909			
$\beta_5^{I}\ KM_V$	0.54101	9	22.917			
$\beta_4\ KN_{IV,V}$	0.53401	9	23.217	0.5093	2	24.346
$\beta_5\ KM_{IV,V}$				0.51670	9	23.995
$\beta_4\ L_IM_{II}$	4.2888	2	2.8908	4.0711	2	3.0454
$\beta_3\ L_IM_{III}$	4.2522	2	2.9157	4.0346	2	3.0730
$\gamma_{2,3}\ L_IN_{II,III}$	3.6855	2	3.3640	3.4892	2	3.5533
$\eta\ L_{II}M_I$	4.9217	2	2.5191	4.6605	2	2.6603
$\beta_1\ L_{II}M_{IV}$	4.37414	4	2.83441	4.14622	5	2.99022
$\gamma_5\ L_{II}N_I$	4.0451	2	3.0650	3.8222	2	3.2437
$\gamma_1\ L_{II}N_{IV}$	3.9437	2	3.1438	3.7246	2	3.3287
$l\ L_{III}M_I$	5.2169	3	2.3765	4.9525	3	2.5034
$\alpha_2\ L_{III}M_{IV}$	4.60545	9	2.69205	4.37588	7	2.83329
$\alpha_1\ L_{III}M_V$	4.59743	9	2.69674	4.36767	5	2.83861
$\beta_6\ L_{III}N_I$	4.2417	2	2.9229	4.0162	2	3.0870
$\beta_{2,15}\ L_{III}N_{IV,V}$	4.1310	2	3.0013	3.90887	4	3.17179
$\beta_{10}\ L_IM_{IV}$				3.7988	2	3.2637

Designation	Å*	p.e.	keV	Å*	p.e.	keV
	45 Rhodium (*Cont.*)			**46 Palladium** (*Cont.*)		
$\beta_9\ L_{I}M_{V}$				3.7920	2	3.2696
$M_{I}N_{II,III}$				20.1	2	0.616
$M_{II}M_{IV}$	59.3	1	0.2090	56.5	1	0.2194
$M_{II}N_{I}$	28.1	2	0.442	26.2	2	0.474
$M_{II}N_{IV}$				22.1	1	0.560
$M_{II}M_{V}$	65.5	1	0.1892	62.9	1	0.1970
$M_{II}^{\,I}N_{I}$	29.8	1	0.417	27.9	1	0.445
$\gamma\ M_{III}N_{IV,V}$	25.01	9	0.496	23.3†	1	0.531
$\zeta\ M_{IV,V}N_{II,III}$	47.67	9	0.2601	43.6	1	0.2844
$M_{IV,V}O_{II,III}$	40.9	2	0.303	37.4	2	0.332
	47 Silver			**48 Cadmium**		
$\alpha_2\ KL_{II}$	0.563798	4	21.9903	0.539422	3	22.9841
$\alpha_1\ KL_{III}$	0.5594075	6	22.16292	0.535010	3	23.1736
$\beta_3\ KM_{II}$	0.497685	4	24.9115	0.475730	5	26.0612
$\beta_1\ KM_{III}$	0.497069	4	24.9424	0.475105	6	26.0955
$\beta_2\ KN_{II,III}$	0.487032	4	25.4564	0.465328	7	26.6438
$\beta_5\ KM_{IV,V}$	0.49306	2	25.145			
$\beta_4\ KN_{IV,V}$	0.48598	3	25.512			
$\beta_4\ L_{I}M_{II}$	3.87023	5	3.20346	3.68203	9	3.36719
$\beta_3\ L_{I}M_{III}$	3.83313	9	3.23446	3.64495	9	3.40145
$\gamma_2\ L_{I}N_{II}$	3.31216	9	3.7432	3.1377	2	3.9513
$\gamma_3\ L_{I}N_{III}$	3.30635	9	3.7498			
$\eta\ L_{II}M_{I}$	4.4183	2	2.8061	4.19315	9	2.95675
$\beta_1\ L_{II}M_{IV}$	3.93473	3	3.15094	3.73823	4	3.31657
$\gamma_5\ L_{II}N_{I}$	3.61638	9	3.42832	3.42551	9	3.61935
$\gamma_1\ L_{III}N_{IV}$	3.52260	4	3.51959	3.33564	6	3.71686
$l\ L_{III}M_{I}$	4.7076	2	2.6337	4.48014	9	2.76735
$\alpha_2\ L_{III}M_{IV}$	4.16294	5	2.97821	3.96496	6	3.12691
$\alpha_1\ L_{III}M_{V}$	4.15443	3	2.98431	3.95635	4	3.13373
$\beta_6\ L_{III}N_{I}$	3.80774	9	3.25603	3.61467	9	3.42994
$\beta_{2,15}\ L_{III}N_{IV,V}$	3.70335	3	3.34781	3.51408	4	3.52812
$\beta_{10}\ L_{I}M_{IV}$	3.61158	9	3.43287	3.4367	2	3.6075
$\beta_9\ L_{I}M_{V}$	3.60497	9	3.43917	3.43015	9	3.61445
$M_{I}N_{II,III}$	18.8	2	0.658			
$M_{II}M_{IV}$	54.0	1	0.2295	52.0	2	0.2384
$M_{II}N_{I}$				22.9	2	0.540
$M_{II}N_{IV}$	20.66	7	0.600	19.40	7	0.639
$M_{III}M_{V}$	60.5	1	0.2048	58.7	2	0.2111
$M_{III}N_{I}$	26.0	1	0.478	24.5	1	0.507
$\gamma\ M_{III}N_{IV,V}$	21.82	7	0.568	20.47	7	0.606
$M_{IV}O_{II,III}$				30.4	1	0.408
$\zeta\ M_{IV,V}N_{II,III}$	39.77	7	0.3117	36.8	1	0.3371
$M_{V}N_{I}$	24.4	2	0.509			
$M_{V}O_{III}$				30.8	1	0.403
$M_{IV,V}O_{II,III}$	33.5	3	0.370			
	49 Indium			**50 Tin**		
$\alpha_2\ KL_{II}$	0.516544	3	24.0020	0.495053	3	25.0440
$\alpha_1\ KL_{III}$	0.512113	3	24.2097	0.490599	3	25.2713
$\beta_3\ KM_{II}$	0.455181	4	27.2377	0.435877	5	28.4440

Designation	Å*	p.e.	keV	Å*	p.e.	keV
	49 Indium (*Cont.*)			**50 Tin** (*Cont.*)		
$\beta_1\ KM_{III}$	0.454545	4	27.2759	0.435236	5	28.4860
$\beta_2\ KN_{II,III}$	0.44500	1	27.8608	0.425915	8	29.1093
$KO_{II,III}$	0.44374	3	27.940	0.42467	3	29.195
$\beta_5^{II}\ KM_{IV}$	0.45098	2	27.491	0.43184	3	28.710
$\beta_5^{I}\ KM_{V}$	0.45086	2	27.499	0.43175	3	28.716
$\beta_4\ KN_{IV,V}$	0.44393	4	27.928	0.42495	3	29.175
$\beta_4\ L_{I}M_{II}$	3.50697	9	3.5353	3.34335	9	3.7083
$\beta_3\ L_{I}M_{III}$	3.46984	9	3.5731	3.30585	3	3.7500
$\gamma_{2,3}\ L_{I}N_{II,III}$	2.9800	2	4.1605	2.8327	2	4.3768
$\gamma_4\ L_{I}O_{II,III}$	2.9264	2	4.2367	2.7775	2	4.4638
$\eta\ L_{II}M_{I}$	3.98327	9	3.11254	3.78876	9	3.27234
$\beta_1\ L_{II}M_{IV}$	3.55531	4	3.48721	3.38487	3	3.66280
$\gamma_5\ L_{II}N_{I}$	3.24907	9	3.8159	3.08475	9	4.0192
$\gamma_1\ L_{II}N_{IV}$	3.16213	4	3.92081	3.00115	3	4.13112
$l\ L_{III}M_{I}$	4.26873	9	2.90440	4.07165	9	3.04499
$\alpha_2\ L_{III}M_{IV}$	3.78073	6	3.27929	3.60891	4	3.43542
$\alpha_1\ L_{III}M_{V}$	3.77192	4	3.28694	3.59994	3	3.44398
$\beta_6\ L_{III}N_{I}$	3.43606	9	3.60823	3.26901	9	3.7926
$\beta_{2,15}\ L_{III}N_{IV,V}$	3.33838	3	3.71381	3.17505	3	3.90486
$\beta_7\ L_{III}O_{I}$	3.324	4	3.730	3.1564	3	3.9279
$\beta_{10}\ L_{I}M_{IV}$	3.27404	9	3.7868	3.12170	9	3.9716
$\beta_9\ L_{I}M_{V}$	3.26763	9	3.7942	3.11513	9	3.9800
$M_{II}M_{IV}$				47.3	1	0.2621
$M_{II}N_{I}$				20.0	1	0.619
$M_{II}N_{IV}$				16.93	5	0.733
$M_{III}M_{V}$				54.2	1	0.2287
$M_{III}N_{I}$				21.5	1	0.575
$\gamma\ M_{III}N_{IV,V}$				17.94	5	0.691
$M_{IV}O_{II,III}$				25.3	1	0.491
$\zeta\ M_{IV,V}N_{II,III}$				31.24	9	0.397
$M_{V}O_{III}$				25.7	1	0.483
	51 Antimony			**52 Tellurium**		
$\alpha_2\ KL_{II}$	0.474827	3	26.1108	0.455784	3	27.2017
$\alpha_1\ KL_{III}$	0.470354	3	26.3591	0.451295	3	27.4723
$\beta_3\ KM_{II}$	0.417737	4	29.6792	0.400659	4	30.9443
$\beta_1\ KM_{III}$	0.417085	3	29.7256	0.399995	5	30.9957
$\beta_2\ KN_{II,III}$	0.407973	5	30.3895	0.391102	6	31.7004
$KO_{II,III}$	0.40666	1	30.4875	0.38974	1	31.8114
$\beta_5^{II}\ KM_{IV}$	0.41388	1	29.9560			
$\beta_5^{I}\ KM_{V}$	0.41378	1	29.9632			
$\beta_4\ KN_{IV,V}$	0.40702	1	30.4604			
$\beta_4\ L_{I}M_{II}$	3.19014	9	3.8864	3.04661	9	4.0695
$\beta_3\ L_{I}M_{III}$	3.15258	9	3.9327	3.00893	9	4.1204
$\gamma_{2,3}\ L_{I}N_{II,III}$	2.6953	2	4.5999	2.5674	2	4.8290
$\gamma_4\ L_{I}O_{II,III}$	2.6398	2	4.6967	2.5113	2	4.9369
$\eta\ L_{II}M_{I}$	3.60765	9	3.43661	3.43832	9	3.60586
$\beta_1\ L_{II}M_{IV}$	3.22567	4	3.84357	3.07677	6	4.02958
$\gamma_5\ L_{II}N_{I}$	2.93187	9	4.2287	2.79007	9	4.4437
$\gamma_1\ L_{II}N_{IV}$	2.85159	3	4.34779	2.71241	6	4.5709
$l\ L_{III}M_{I}$	3.88826	9	3.18860	3.71696	9	3.33555
$\alpha_2\ L_{III}M_{IV}$	3.44840	6	3.59532	3.29846	9	3.7588

Designation	Å*	p.e.	keV	Å*	p.e.	keV
	51 Antimony (*Cont.*)			**52 Tellurium** (*Cont.*)		
$\alpha_1\ L_{III}M_V$	3.43941	4	3.60472	3.28920	6	3.76933
$\beta_6\ L_{III}N_I$	3.11513	9	3.9800	2.97088	9	4.1732
$\beta_{2,15}\ L_{III}N_{IV,V}$	3.02335	3	4.10078	2.88217	8	4.3017
$\beta_7\ L_{III}O_I$	3.0052	3	4.1255	2.8634	3	4.3298
$\beta_{10}\ L_IM_{IV}$	2.97917	9	4.1616	2.84679	9	4.3551
$\beta_9\ L_IM_V$	2.97261	9	4.1708	2.83897	9	4.3671
$M_{II}M_{IV}$	45.2	1	0.2743			
$M_{II}N_I$	18.8	1	0.658	17.6	1	0.703
$M_{II}N_{IV}$	15.98	5	0.776			
$M_{III}M_V$	52.2	1	0.2375	50.3	1	0.2465
$M_{III}N_I$	20.2	1	0.612	19.1	1	0.648
$\gamma\ M_{III}N_{IV,V}$	16.92	4	0.733	15.93	4	0.778
$M_{IV}O_{II,III}$				21.34	5	0.581
$\zeta\ M_{IV,V}N_{II,III}$	28.88	8	0.429	26.72	9	0.464
M_VO_{III}				21.78	5	0.569
	53 Iodine			**54 Xenon**		
$\alpha_2\ KL_{II}$	0.437829	7	28.3172	0.42087†	2	29.458
$\alpha_1\ KL_{III}$	0.433318	5	28.6120	0.41634†	2	29.779
$\beta_3\ KM_{II}$	0.384564	4	32.2394	0.36941†	2	33.562
$\beta_1\ KM_{III}$	0.383905	4	32.2947	0.36872†	2	33.624
$\beta_2\ KN_{II,III}$	0.37523†	2	33.042	0.36026†	3	34.415
$\beta_4\ L_IM_{II}$	2.91207	9	4.2575			
$\beta_3\ L_IM_{III}$	2.87429	9	4.3134			
$\gamma_{2,3}\ L_IN_{II,III}$	2.4475	2	5.0657			
$\gamma_4\ L_IO_{II,III}$	2.3913	2	5.1848			
$\eta\ L_{II}M_I$	3.27979	9	3.7801			
$\beta_1\ L_{II}M_{IV}$	2.93744	6	4.22072			
$\gamma_5\ L_{II}N_I$	2.65710	9	4.6660			
$\gamma_1\ L_{II}N_{IV}$	2.58244	8	4.8009			
$l\ L_{III}M_I$	3.55754	9	3.48502			
$\alpha_2\ L_{III}M_{IV}$	3.15791	6	3.92604			
$\alpha_1\ L_{III}M_V$	3.14860	6	3.93765	3.0166†	2	4.1099
$\beta_6\ L_{III}N_I$	2.83672	9	4.3706			
$\beta_{2,15}\ L_{III}N_{IV,V}$	2.75053	8	4.5075			
$\beta_7\ L_{III}O_I$	2.7288	3	4.5435			
$\beta_{10}\ L_IM_{IV}$	2.72104	9	4.5564			
$\beta_9\ L_IM_V$	2.71352	9	4.5690			
	55 Cesium			**56 Barium**		
$\alpha_2\ KL_{II}$	0.404835	4	30.6251	0.389668	5	31.8171
$\alpha_1\ KL_{III}$	0.400290	4	30.9728	0.385111	4	32.1936
$\beta_3\ KM_{II}$	0.355050	4	34.9194	0.341507	4	36.3040
$\beta_1\ KM_{III}$	0.354364	7	34.9869	0.340811	3	36.3782
$\beta_2\ KN_{II,III}$	0.34611	2	35.822	0.33277	1	37.257
$KO_{II,III}$				0.33127	2	37.426
$\beta_5^{II}\ KM_{IV}$				0.33835	2	36.643
$\beta_5^{I}\ KM_V$				0.33814	2	36.666
$\beta_4\ KN_{IV,V}$				0.33229	2	37.311
$\beta_4\ L_IM_{II}$	2.6666	2	4.6494	2.5553	2	4.8519
$\beta_3\ L_IM_{III}$	2.6285	2	4.7167	2.5164	2	4.9269
$\gamma_2\ L_IN_{II}$	2.2371	2	5.5420	2.1387	2	5.7969
$\gamma_3\ L_IN_{III}$	2.2328	2	5.5527	2.1342	2	5.8092

Designation	Å*	p.e.	keV	Å*	p.e.	keV
	55 Cesium (*Cont.*)			**56 Barium** (*Cont.*)		
$\gamma_4\ L_IO_{II,III}$	2.1741	2	5.7026	2.0756	3	5.9733
$\eta\ L_{II}M_I$	2.9932	2	4.1421	2.8627	3	4.3309
$\beta_1\ L_{II}M_{IV}$	2.6837	2	4.6198	2.56821	5	4.82753
$\gamma_5\ L_{II}N_I$	2.4174	2	5.1287	2.3085	3	5.3707
$\gamma_1\ L_{II}N_{IV}$	2.3480	2	5.2804	2.2415	2	5.5311
$l\ L_{III}M_I$	3.2670	2	3.7950	3.1355	2	3.9541
$\alpha_2\ L_{III}M_{IV}$	2.9020	2	4.2722	2.78553	5	4.45090
$\alpha_1\ L_{III}M_V$	2.8924	2	4.2865	2.77595	5	4.46626
$\beta_6\ L_{III}N_I$	2.5932	2	4.7811	2.4826	2	4.9939
$\beta_{2,15}\ L_{III}N_{IV,V}$	2.5118	2	4.9359	2.40435	6	5.1565
$\beta_7\ L_{III}O_I$	2.4849	2	4.9893	2.3806	2	5.2079
$\beta_{10}\ L_IM_{IV}$	2.4920	2	4.9752	2.3869	2	5.1941
$\beta_9\ L_IM_V$	2.4783	2	5.0026	2.3764	2	5.2171
$\gamma\ M_{III}N_{IV,V}$				12.75	3	0.973
$M_{IV}O_{II}$				15.91	5	0.779
$M_{IV}O_{III}$				15.72	9	0.789
$\zeta\ M_VN_{III}$				20.64	4	0.601
M_VO_{III}				16.20	5	0.765
$N_{IV}O_{II}$	188.6	1	0.06574	163.3	2	0.07590
$N_{IV}O_{III}$	183.8	1	0.06746	159.0	2	0.07796
N_VO_{III}	190.3	1	0.06515	164.6	2	0.07530
	57 Lanthanum			**58 Cerium**		
$\alpha_2\ KL_{II}$	0.375313	2	33.0341	0.361683	2	34.2789
$\alpha_1\ KL_{III}$	0.370737	2	33.4418	0.357092	2	34.7197
$\beta_3\ KM_{II}$	0.328686	4	37.7202	0.316520	4	39.1701
$\beta_1\ KM_{III}$	0.327983	3	37.8010	0.315816	2	39.2573
$\beta_2\ KN_{II,III}$	0.320117	7	38.7299	0.30816	1	40.233
$KO_{II,III}$	0.31864	2	38.909	0.30668	2	40.427
$\beta_5^{II}\ KM_{IV}$	0.32563	2	38.074	0.31357	2	39.539
$\beta_5^{I}\ KM_V$	0.32546	2	38.094	0.31342	2	39.558
$\beta_4\ KN_{IV,V}$	0.31931	2	38.828	0.30737	2	40.337
$\beta_4\ L_IM_{II}$	2.4493	3	5.0620	2.3497	4	5.2765
$\beta_3\ L_IM_{III}$	2.4105	3	5.1434	2.3109	3	5.3651
$\gamma_2\ L_IN_{II}$	2.0460	4	6.060	1.9602	3	6.3250
$\gamma_3\ L_IN_{III}$	2.0410	4	6.074	1.9553	3	6.3409
$\gamma_4\ L_IO_{II,III}$	1.9830	4	6.252	1.8991	4	6.528
$\eta\ L_{II}M_I$	2.740	3	4.525	2.6203	4	4.7315
$\beta_1\ L_{II}M_{IV}$	2.45891	5	5.0421	2.3561	3	5.2622
$\gamma_5\ L_{II}N_I$	2.2056	4	5.621	2.1103	3	5.8751
$\gamma_1\ L_{II}N_{IV}$	2.1418	3	5.7885	2.0487	4	6.052
$\gamma_8\ L_{II}O_I$				2.0237	4	6.126
$l\ L_{III}M_I$	3.006	3	4.124	2.8917	4	4.2875
$\alpha_2\ L_{III}M_{IV}$	2.67533	5	4.63423	2.5706	3	4.8230
$\alpha_1\ L_{III}M_V$	2.66570	5	4.65097	2.5615	2	4.8402
$\beta_6\ L_{III}N_I$	2.3790	4	5.2114	2.2818	3	5.4334
$\beta_{2,15}\ L_{III}N_{IV,V}$	2.3030	3	5.3835	2.2087	2	5.6134
$\beta_7\ L_{III}O_I$	2.275	3	5.450	2.1701	2	5.7132
$\beta_{10}\ L_IM_{IV}$	2.290	3	5.415	2.1958	5	5.646
$\beta_9\ L_IM_V$	2.282	3	5.434	2.1885	3	5.6650
$\gamma\ M_{III}N_{IV,V}$	12.08	4	1.027	11.53	1	1.0749
$\beta\ M_{IV}N_{VI}$	14.51	5	0.854	13.75	4	0.902
$\zeta\ M_VN_{III}$	19.44	5	0.638	18.35	4	0.676
$\alpha\ M_VN_{VI,VII}$	14.88	5	0.833	14.04	2	0.883

Designation	Å*	p.e.	keV	Å*	p.e.	keV
	57 Lanthanum (*Cont.*)			**58 Cerium** (*Cont.*)		
$M_VO_{II,III}$				14.39	5	0.862
$N_{IV,V}O_{II,III}$	152.6	6	0.0812	144.4	6	0.0859
	59 Praseodymium			**60 Neodymium**		
$\alpha_2\ KL_{II}$	0.348749	2	35.5502	0.336472	2	36.8474
$\alpha_1\ KL_{III}$	0.344140	2	36.0263	0.331846	2	37.3610
$\beta_3\ KM_{II}$	0.304975	5	40.6529	0.294027	3	42.1665
$\beta_1\ KM_{III}$	0.304261	4	40.7482	0.293299	2	42.2713
$\beta_2\ KN_{II,III}$	0.29679	2	41.773	0.2861†	1	43.33
$\beta_4\ L_IM_{II}$	2.2550	4	5.4981	2.1669	3	5.7216
$\beta_3\ L_IM_{III}$	2.2172	3	5.5918	2.1268	2	5.8294
$\gamma_2\ L_IN_{II}$	1.8791	4	6.598	1.8013	4	6.883
$\gamma_3\ L_IN_{III}$	1.8740	4	6.616	1.7964	4	6.902
$\gamma_4\ L_IO_{II,III}$	1.8193	4	6.815	1.7445	4	7.107
$\eta\ L_{II}M_I$	2.512	3	4.935	2.4094	4	5.1457
$\beta_1\ L_{II}M_{IV}$	2.2588	3	5.4889	2.1669	2	5.7216
$\gamma_5\ L_{II}N_I$	2.0205	4	6.136	1.9355	4	6.406
$\gamma_1\ L_{II}N_{IV}$	1.9611	3	6.3221	1.8779	2	6.6021
$\gamma_8\ L_{II}O_I$	1.9362	4	6.403	1.8552	5	6.683
$l\ L_{III}M_I$	2.7841	4	4.4532	2.6760	4	4.6330
$\alpha_2\ L_{III}M_{IV}$	2.4729	3	5.0135	2.3807	3	5.2077
$\alpha_1\ L_{III}M_V$	2.4630	2	5.0337	2.3704	2	5.2304
$\beta_6\ L_{III}N_I$	2.1906	4	5.660	2.1039	3	5.8930
$\beta_{2,15}\ L_{III}N_{IV,V}$	2.1194	4	5.850	2.0360	3	6.0894
$\beta_7\ L_{III}O_I$	2.0919	4	5.927	2.0092	3	6.1708
$\beta_{10}\ L_IM_{IV}$	2.1071	4	5.884	2.0237	3	6.1265
$\beta_9\ L_IM_V$	2.1004	4	5.903	2.0165	3	6.1484
$\gamma\ M_{III}N_{IV,V}$	10.998	9	1.1273	10.505	9	1.180
$\beta\ M_{IV}N_{VI}$	13.06	2	0.950	12.44	2	0.997
$\zeta\ M_VN_{III}$	17.38	4	0.714	16.46	4	0.753
$\alpha\ M_VN_{VI,VII}$	13.343	5	0.9292	12.68	2	0.978
$N_{IV,V}N_{VI,VII}$	113.	1	0.1095	107.	1	0.116
$N_{IV,V}O_{II,III}$	136.5	4	0.0908	128.9	7	0.0962
	61 Promethium			**62 Samarium**		
$\alpha_2\ KL_{II}$	0.324803	4	38.1712	0.313698	2	39.5224
$\alpha_1\ KL_{III}$	0.320160	4	38.7247	0.309040	2	40.1181
$\beta_3\ KM_{II}$	0.28363†	4	43.713	0.27376	2	45.289
$\beta_1\ KM_{III}$	0.28290†	3	43.826	0.27301	2	45.413
$\beta_2\ KN_{II,III}$	0.2759†	1	44.94	0.2662	1	46.58
$KO_{II,III}$				0.26491	3	46.801
$\beta_5\ KM_{IV,V}$				0.27111	3	45.731
$\beta_4\ L_IM_{II}$				2.00095	6	6.1963
$\beta_3\ L_IM_{III}$	2.0421	4	6.071	1.96241	3	6.3180
$\gamma_2\ L_IN_{II}$				1.66044	6	7.4668
$\gamma_3\ L_IN_{III}$				1.65601	3	7.4867
$\gamma_4\ L_IO_{II,III}$				1.60728	3	7.7137
$\eta\ L_{II}M_I$				2.21824	3	5.5892
$\beta_1\ L_{II}M_{IV}$	2.0797	4	5.961	1.99806	3	6.2051
$\gamma_5\ L_{II}N_I$				1.77934	3	6.9678
$\gamma_1\ L_{II}N_{IV}$	1.7989	9	6.892	1.72724	3	7.1780
$\gamma_6\ L_{II}O_{IV}$				1.6966	9	7.3076
$l\ L_{III}M_I$				2.4823	4	4.9945

Designation	Å*	p.e.	keV	Å*	p.e.	keV
	61 Promethium (*Cont.*)			**62 Samarium** (*Cont.*)		
$\alpha_2\ L_{III}M_{IV}$	2.2926	4	5.4078	2.21062	3	5.6084
$\alpha_1\ L_{III}M_V$	2.2822	3	5.4325	2.1998	2	5.6361
$\beta_6\ L_{III}N_I$				1.94643	3	6.3697
$\beta_{2,15}\ L_{III}N_{IV,V}$	1.9559	6	6.339	1.88221	3	6.5870
$\beta_7\ L_{III}O_I$				1.85626	3	6.6791
$\beta_5\ L_{III}O_{IV,V}$				1.84700	9	6.7126
$\beta_{10}\ L_IM_{IV}$				1.86990	3	6.6304
$\beta_9\ L_IM_V$				1.86166	3	6.6597
$\gamma\ M_{III}N_{IV,V}$				9.600	9	1.291
$\beta\ M_{IV}N_{VI}$				11.27	1	1.0998
$\zeta\ M_VN_{III}$				14.91	4	0.831
$\alpha\ M_VN_{VI,VII}$				11.47	3	1.081
$N_{IV,V}N_{VI,VII}$				98.	1	0.126
$N_{IV,V}O_{II,III}$				117.4	4	0.1056
	63 Europium			**64 Gadolinium**		
$\alpha_2\ KL_{II}$	0.303118	2	40.9019	0.293038	2	42.3089
$\alpha_1\ KL_{III}$	0.298446	2	41.5422	0.288353	2	42.9962
$\beta_3\ KM_{II}$	0.264332	5	46.9036	0.25534	2	48.555
$\beta_1\ KM_{III}$	0.263577	5	47.0379	0.25460	2	48.697
$\beta_2\ KN_{II,III}$	0.256923	8	48.256	0.24816	3	49.959
$KO_{II,III}$	0.255645	7	48.497	0.24687	3	50.221
$\beta_5\ KM_{IV,V}$				0.25275	3	49.052
$\beta_4\ L_IM_{II}$	1.9255	2	6.4389	1.8540	2	6.6871
$\beta_3\ L_IM_{III}$	1.8867	2	6.5713	1.8150	2	6.8311
$\gamma_2\ L_IN_{II}$	1.5961	2	7.7677	1.5331	2	8.087
$\gamma_3\ L_IN_{III}$	1.5903	2	7.7961	1.5297	2	8.105
$\gamma_4\ L_IO_{II,III}$	1.5439	1	8.0304	1.4839	2	8.355
$\eta\ L_{II}M_I$	2.1315	2	5.8166	2.0494	1	6.0495
$\beta_1\ L_{II}M_{IV}$	1.9203	2	6.4564	1.8468	2	6.7132
$\gamma_5\ L_{II}N_I$	1.7085	2	7.2566	1.6412	2	7.5543
$\gamma_1\ L_{II}N_{IV}$	1.6574	2	7.4803	1.5924	2	7.7858
$\gamma_8\ L_{II}O_I$	1.6346	2	7.5849	1.5707	2	7.894
$\gamma_6\ L_{II}O_{IV}$	1.6282	2	7.6147	1.5644	2	7.925
$l\ L_{III}M_I$	2.3948	2	5.1772	2.3122	2	5.3621
$\alpha_2\ L_{III}M_{IV}$	2.1315	2	5.8166	2.0578	2	6.0250
$\alpha_1\ L_{III}M_V$	2.1209	2	5.8457	2.0468	2	6.0572
$\beta_6\ L_{III}N_I$	1.8737	2	6.6170	1.8054	2	6.8671
$\beta_{2,15}\ L_{III}N_{IV,V}$	1.8118	2	6.8432	1.7455	2	7.1028
$\beta_7\ L_{III}O_I$	1.7851	2	6.9453	1.7203	2	7.2071
$\beta_5\ L_{III}O_{IV,V}$	1.7772	2	6.9763	1.7130	2	7.2374
$\beta_{10}\ L_IM_{IV}$	1.7993	3	6.890	1.7315	3	7.160
$\beta_9\ L_IM_V$	1.7916	3	6.920	1.7240	3	7.192
$L_IO_{IV,V}$				1.4807	3	8.373
$\gamma\ M_{III}N_{IV,V}$	9.211	9	1.346	8.844	9	1.402
$\beta\ M_{IV}N_{VI}$	10.750	7	1.1533	10.254	6	1.2091
$\zeta\ M_VN_{III}$	14.22	2	0.872	13.57	2	0.914
$\alpha\ M_VN_{VI,VII}$	10.96	3	1.131	10.46	3	1.185
$N_{IV,V}O_{II,III}$	112.0	6	0.1107			
	65 Terbium			**66 Dysprosium**		
$\alpha_2\ KL_{II}$	0.283423	2	43.7441	0.274247	2	45.2078
$\alpha_1\ KL_{III}$	0.278724	2	44.4816	0.269533	2	45.9984
$\beta\ KM_{II}$	0.24683	2	50.229	0.23862	2	51.957

Designation	Å*	p.e.	keV	Å*	p.e.	keV
	65 Terbium (*Cont.*)			**66 Dysprosium** (*Cont.*)		
$\beta_1\ KM_{III}$	0.24608	2	50.382	0.23788	2	52.119
$\beta_2\ KN_{II,III}$	0.2397†	2	51.72	0.2317†	2	53.51
$KO_{II,III}$	0.23858	3	51.965	0.23056	3	53.774
$\beta_5\ KM_{IV,V}$				0.23618	3	52.494
$\beta_4\ L_IM_{II}$	1.7864	2	6.9403	1.72103	7	7.2039
$\beta_3\ L_IM_{III}$	1.7472	2	7.0959	1.6822	2	7.3702
$\gamma_2\ L_IN_{II}$	1.4764	2	8.398	1.42278	7	8.7140
$\gamma_3\ L_IN_{III}$	1.4718	2	8.423	1.41640	7	8.7532
$\gamma_4\ L_IO_{II,III}$	1.4276	2	8.685	1.37459	7	9.0195
$\eta\ L_{II}M_I$	1.9730	2	6.2839	1.89743	7	6.5342
$\beta_1\ L_{II}M_{IV}$	1.7768	3	6.978	1.71062	7	7.2477
$\gamma_5\ L_{II}N_I$	1.5787	2	7.8535	1.51824	7	8.1661
$\gamma_1\ L_{II}N_{IV}$	1.5303	2	8.102	1.47266	7	8.4188
$\gamma_8\ L_{II}O_I$	1.5097	2	8.212			
$\gamma_6\ L_{II}O_{IV}$	1.5035	2	8.246	1.44579	7	8.5753
$l\ L_{III}M_I$	2.2352	2	5.5467	2.15877	7	5.7431
$\alpha_2\ L_{III}M_{IV}$	1.9875	2	6.2380	1.91991	3	6.4577
$\alpha_1\ L_{II}M_V$	1.9765	2	6.2728	1.90881	3	6.4952
$\beta_6\ L_{III}N_I$	1.7422	2	7.1163	1.68213	7	7.3705
$\beta_{2,15}\ L_{III}N_{IV,V}$	1.6830	2	7.3667	1.62369	7	7.6357
$\beta_7\ L_{III}O_I$	1.6585	2	7.4753	1.60447	7	7.7272
$\beta_5\ L_{III}O_{IV,V}$	1.6510	2	7.5094	1.58837	7	7.8055
$\beta_{10}\ L_IM_{IV}$	1.6673	3	7.436	1.60743	9	7.7130
$\beta_9\ L_IM_V$				1.59973	9	7.7501
$L_IO_{IV,V}$	1.4228	3	8.714			
$\gamma\ M_{III}N_{IV,V}$	8.486	9	1.461	8.144	9	1.522
$\beta\ M_{IV}N_{VI}$	9.792	6	1.2661	9.357	6	1.3250
$\zeta\ M_VN_{III}$	12.98	2	0.955	12.43	2	0.998
$\alpha\ M_VN_{VI,VII}$	10.00	2	1.240	9.59	2	1.293
$N_{IV,V}N_{VI,VII}$	86.	1	0.144	83.	1	0.149
$N_{IV,V}O_{II,III}$	102.2	4	0.1213	97.2	8	0.128
	67 Holmium			**68 Erbium**		
$\alpha_2\ KL_{II}$	0.265486	2	46.6997	0.257110	2	48.2211
$\alpha_1\ KL_{III}$	0.260756	2	47.5467	0.252365	2	49.1277
$\beta_3\ KM_{II}$	0.23083	2	53.711	0.22341	2	55.494
$\beta_1\ KM_{III}$	0.23012	2	53.877	0.22266	2	55.681
$\beta_2\ KN_{II,III}$	0.2241†	2	55.32	0.2167†	2	57.21
$KO_{II,III}$	0.22305	3	55.584	0.21581	3	57.450
$\beta_5\ KM_{IV,V}$	0.22855	3	54.246	0.22124	3	56.040
$\beta_4\ L_IM_{II}$	1.6595	2	7.4708	1.6007	1	7.7453
$\beta_3\ L_IM_{III}$	1.6203	2	7.6519	1.5616	1	7.9392
$\gamma_2\ L_IN_{II}$	1.3698	2	9.051	1.3210	2	9.385
$\gamma_3\ L_IN_{III}$	1.3643	2	9.087	1.3146	1	9.4309
$\gamma_4\ L_IO_{II,III}$	1.3225	2	9.374	1.2752	2	9.722
$\eta\ L_{II}M_I$	1.8264	2	6.7883	1.7566	1	7.0579
$\beta_1\ L_{II}M_{IV}$	1.6475	2	7.5253	1.5873	1	7.8109
$\gamma_5\ L_{II}N_I$	1.4618	2	8.481	1.4067	3	8.814
$\gamma_1\ L_{II}N_{IV}$	1.4174	2	8.747	1.3641	2	9.089
$\gamma_8\ L_{II}O_I$	1.3983	2	8.867			
$\gamma_6\ L_{II}O_{IV}$	1.3923	2	8.905	1.3397	3	9.255
$l\ L_{III}M_I$	2.0860	2	5.9434	2.015	1	6.152
$\alpha_2\ L_{III}M_{IV}$	1.8561	2	6.6795	1.7955	2	6.9050
$\alpha_1\ L_{III}M_V$	1.8450	2	6.7198	1.78425	9	6.9487
$\beta_6\ L_{III}N_I$	1.6237	2	7.6359	1.5675	2	7.909
$\beta_{2,15}\ L_{III}N_{IV,V}$	1.5671	2	7.911	1.51399	9	8.1890
$\beta_7\ L_{III}O_I$				1.4941	3	8.298
$\beta_5\ L_{III}O_{IV,V}$	1.5378	2	8.062	1.4848	3	8.350

Designation	Å*	p.e.	keV	Å*	p.e.	keV
	67 Holmium (*Cont.*)			**68 Erbium** (*Cont.*)		
$\beta_{10}\ L_IM_{IV}$	1.5486	3	8.006	1.4941	3	8.298
$L_IO_{IV,V}$	1.3208	3	9.387			
$\beta_9\ L_IM_V$				1.4855	5	8.346
$M_{II}N_{IV}$				7.60	1	1.632
$\gamma\ M_{III}N_{IV,V}$	7.865	9	1.576			
$\gamma\ M_{III}\ _V$				7.540	8	1.043
$\beta\ M_{IV}N_{VI}$	8.965	4	1.3830	8.592	3	1.4430
$\zeta\ M_VN_{III}$	11.86	1	1.0450	11.37	1	1.0901
$\alpha\ M_VN_{VI,VII}$	9.20	2	1.348	8.82	1	1.406
$N_{IV}N_{VI}$				72.7	9	0.171
$N_VN_{VI,VII}$				76.3	7	0.163
	69 Thulium			**70 Ytterbium**		
$\alpha_2\ KL_{II}$	0.249095	2	49.7726	0.241424	2	51.3540
$\alpha_1\ KL_{III}$	0.244338	2	50.7416	0.236655	2	52.3889
$\beta_3\ KM_{II}$	0.21636	2	57.304	0.2096†	1	59.14
$\beta_1\ KM_{III}$	0.21556	2	57.517	0.20884	8	59.37
$\beta_2\ KN_{II,III}$	0.2098†	2	59.09	0.2033†	2	60.98
$KO_{II,III}$	0.20891	2	59.346	0.20226	2	61.298
$\beta_5\ KM_{IV,V}$	0.21404	2	57.923	0.20739	2	59.782
$\beta_4\ L_IM_{II}$	1.5448	2	8.026	1.49138	3	8.3132
$\beta_3\ L_IM_{III}$	1.5063	2	8.231	1.45233	5	8.5367
$\gamma_2\ L_IN_{II}$	1.2742	2	9.730	1.22879	7	10.0897
$\gamma_3\ L_IN_{III}$	1.2678	2	9.779	1.22232	5	10.1431
$\gamma_4\ L_IO_{II,III}$	1.2294	2	10.084	1.1853	1	10.4603
$\eta\ L_{II}M_I$	1.6963	2	7.3088	1.63560	5	7.5802
$\beta_1\ L_{II}M_{IV}$	1.5304	2	8.101	1.47565	5	8.4018
$\gamma_5\ L_{II}N_I$	1.3558	2	9.144	1.3063	1	9.4910
$\gamma_1\ L_{II}N_{IV}$	1.3153	2	9.426	1.26769	5	9.8701
$\gamma_8\ L_{II}O_I$				1.24923	5	9.9246
$\gamma_6\ L_{II}O_{IV}$	1.2905	2	9.607	1.24271	3	9.9766
$l\ L_{III}M_I$	1.9550	2	6.3419	1.89415	5	6.5455
$\alpha_2\ L_{III}M_{IV}$	1.7381	2	7.1331	1.68285	5	7.3673
$\alpha_1\ L_{III}M_V$	1.7268†	2	7.1799	1.67189	4	7.4156
$\beta_6\ L_{III}N_I$	1.5162	2	8.177	1.4661	1	8.4563
$\beta_{2,15}\ L_{III}N_{IV,V}$	1.4640	2	8.468	1.41550	5	8.7588
$\beta_7\ L_{III}O_I$				1.3948	1	8.8889
$\beta_5\ L_{III}O_{IV,V}$	1.4349	2	8.641	1.38696	7	8.9390
$\beta_{10}\ L_IM_{IV}$	1.4410	3	8.604	1.3915	1	8.9100
$\beta_9\ L_IM_V$	1.4336	3	8.648	1.3838	1	8.9597
L_IO_I				1.1886	1	10.4312
$L_IO_{IV,V}$	1.2263	3	10.110	1.1827	1	10.4833
$L_{II}M_{II}$				1.58844	9	7.8052
$L_{II}O_{II,III}$				1.2453	1	9.9561
$t\ L_{III}M_{II}$				1.83091	9	6.7715
$L_{III}O_{II,III}$				1.3898	1	8.9209
$M_{III}N_I$				8.470	9	1.464
$\gamma\ M_{III}N_V$				7.024	8	1.765
$\beta\ M_{IV}N_{VI}$	8.249	7	1.503	7.909	2	1.5675
$\zeta\ M_VN_{III}$				10.48	1	1.183
$\alpha\ M_VN_{VI,VII}$	8.48	1	1.462	8.149	5	1.5214
$N_{IV}N_{VI}$				65.1	7	0.190
$N_VN_{VI,VII}$				69.3	5	0.179
	71 Lutetium			**72 Hafnium**		
$\alpha_2\ KL_{II}$	0.234081	2	52.9650	0.227024	3	54.6114
$\alpha_1\ KL_{III}$	0.229298	2	54.0698	0.222227	3	55.7902
$\beta_3\ KM_{II}$	0.20309†	4	61.05	0.19686†	4	62.98

Designation	Å*	p.e.	keV	Å*	p.e.	keV
	71 Lutetium (*Cont.*)			**72 Hafnium** (*Cont.*)		
$\beta_1\ KM_{III}$	0.20231†	3	61.283	0.19607†	3	63.234
$\beta_2\ KN_{II,III}$	0.1969†	2	62.97	0.1908†	2	64.98
$KO_{II,III}$	0.19589	2	63.293			
$\beta_5\ KM_{IV,V}$	0.20084	2	61.732			
$\beta_4\ L_IM_{II}$	1.44056	5	8.6064	1.39220	5	8.9054
$\beta_3\ L_IM_{III}$	1.40140	5	8.8469	1.35300	5	9.1634
$\gamma_2\ L_IN_{II}$	1.1853	2	10.460	1.14442	5	10.8335
$\gamma_3\ L_IN_{III}$	1.17953	4	10.5110	1.13841	5	10.8907
$\gamma'_4\ L_IO_{II}$				1.10376	5	11.2326
$\gamma_4\ L_IO_{II,III}$	1.1435	1	10.8425	1.10303	5	11.2401
$\eta\ L_{II}M_I$	1.5779	1	7.8575	1.52325	5	8.1393
$\beta_1\ L_{II}M_{IV}$	1.42359	3	8.7090	1.37410	5	9.0227
$\gamma_5\ L_{II}N_I$	1.2596	1	9.8428	1.21537	5	10.2011
$\gamma_1\ L_{II}N_{IV}$	1.22228	4	10.1434	1.17900	5	10.5158
$\gamma_8\ L_{II}O_I$	1.2047	1	10.2915	1.16138	5	10.6754
$\gamma_6\ L_{II}O_{IV}$	1.1987	1	10.3431	1.15519	5	10.7325
$l\ L_{III}M_I$	1.8360	1	6.7528	1.78145	5	6.9596
$\alpha_2\ L_{III}M_{IV}$	1.63029	5	7.6049	1.58046	5	7.8446
$\alpha_1\ L_{III}M_{IV}$	1.61951	3	7.6555	1.56958	5	7.8990
$\beta_6\ L_{III}N_I$	1.4189	1	8.7376	1.37410	5	9.0227
$\beta_{15}\ L_{III}N_{IV}$	1.3715	1	9.0395	1.32783	5	9.3371
$\beta_2\ L_{III}N_V$	1.37012	3	9.0489	1.32639	5	9.3473
$\beta_7\ L_{III}O_I$	1.34949	5	9.1873	1.30564	5	9.4958
$\beta_5\ L_{III}O_{IV,V}$	1.34183	7	9.2397	1.29761	5	9.5546
L_IM_I				1.43025	9	8.6685
$\beta_{10}\ L_IM_{IV}$	1.3430	2	9.232	1.29819	9	9.5503
$\beta_9 L_IM_V$	1.3358	1	9.2816	1.29025	9	9.6090
L_IN_{IV}	1.16227	9	10.6672	1.12250	9	11.0451
$\gamma_{11}\ L_IN_V$	1.16107	9	10.6782	1.12146	9	11.0553
L_IO_I				1.10664	9	11.2034
L_IO_{IV}				1.10086	9	11.2622
$L_{II}M_{II}$	1.53333	9	8.0858	1.48064	9	8.3735
$\beta_{17}\ L_{II}M_{III}$				1.43643	9	8.6312
$L_{II}N_V$				1.17788	9	10.5258
$v\ L_{II}N_{VI}$				1.15830	9	10.7037
$L_{II}O_{II,III}$	1.2014	1	10.3198			
$t\ L_{III}M_{II}$	1.7760	1	6.9810	1.72305	9	7.1954
$s\ L_{III}M_{III}$				1.66346	9	7.4532
$L_{III}N_{II}$				1.35887	9	9.1239
$L_{III}N_{III}$				1.35053	9	9.1802
$u\ L_{III}N_{VI,VII}$				1.30165	9	9.5249
$L_{III}O_{II,III}$	1.34524	9	9.2163			
$M_{III}N_I$				7.887	9	1.572
$\gamma\ M_{III}N_V$	6.768	6	1.832	6.544	4	1.895
ζ_2				9.686	7	1.2800
$\beta\ M_{IV}N_{VI}$	7.601	2	1.6312	7.303	1	1.6976
ζ_1				9.686	7	1.2800
$\alpha\ M_VN_{VI,VII}$	7.840	2	1.5813	7.539	1	1.6446
$N_{IV}N_{VI}$	63.0	5	0.197			
$N_VN_{VI,VII}$	65.7	2	0.1886			
	73 Tantalum			**74 Tungsten**		
$\alpha_2\ KL_{II}$	0.220305	8	56.277	0.213828	2	57.9817
$\alpha_1\ KL_{III}$	0.215497	4	57.532	0.2090100	Std	59.31824
$\beta_3\ KM_{II}$	0.190890	2	64.9488	0.185181	2	66.9514
$\beta_1\ KM_{III}$	0.190089	4	65.223	0.184374	2	67.2443
$\beta_2^{II}\ KN_{II}$	0.185188	9	66.949	0.17960	1	69.031
$\beta_2^{I}\ KN_{III}$	0.185011	8	67.013	0.179421	7	69.101

Designation	Å*	p.e.	keV	Å*	p.e.	keV
	73 Tantalum (*Cont.*)			**74 Tungsten** (*Cont.*)		
$KO_{II,III}$	0.184031	7	67.370	0.178444	5	69.479
KL_I				0.21592	4	57.42
$\beta_5^{II}\ KM_{IV}$	0.188920	6	65.626	0.183264	5	67.652
$\beta_5^{I}\ KM_V$	0.188757	6	65.683	0.183092	7	67.715
$\beta_4\ KN_{IV,V}$	0.18451	1	67.194	0.17892	2	69.294
$\beta_4\ L_IM_{II}$	1.34581	3	9.2124	1.30162	5	9.5252
$\beta_3\ L_IM_{III}$	1.30678	3	9.4875	1.26269	5	9.8188
$\gamma_2\ L_IN_{II}$	1.1053	1	11.217	1.06806	3	11.6080
$\gamma_3\ L_IN_{III}$	1.09936	4	11.2776	1.06200	6	11.6743
$\gamma'_4\ L_IO_{II}$	1.06544	3	11.6366	1.02863	3	12.0530
$\gamma_4\ L_IO_{III}$	1.06467	3	11.6451	1.02775	3	12.0634
$\eta\ L_{II}M_I$	1.47106	5	8.4280	1.42110	3	8.7243
$\beta_1\ L_{II}M_{IV}$	1.32698	3	9.3431	1.281809	9	9.67235
$\gamma_5\ L_{II}N_I$	1.1729	1	10.5702	1.13235	3	10.9490
$\gamma_1\ L_{II}N_{IV}$	1.13794	3	10.8952	1.09855	3	11.2859
$\gamma_8\ L_{II}O_I$	1.1205	1	11.0646	1.08113	4	11.4677
$\gamma_6\ L_{II}O_{IV}$	1.11388	3	11.1306	1.07448	5	11.5387
$l\ L_{III}M_I$	1.72841	5	7.1731	1.6782	1	7.3878
$\alpha_2\ L_{III}M_{IV}$	1.53293	2	8.0879	1.48743	2	8.3352
$\alpha_1\ L_{III}M_V$	1.52197	2	8.1461	1.47639	2	8.3976
$\beta_6\ L_{III}N_I$	1.33094	8	9.3153	1.28989	7	9.6117
$\beta_{15}\ L_{III}N_{IV}$	1.28619	5	9.6394	1.24631	3	9.9478
$\beta_2\ L_{III}N_V$	1.28454	2	9.6518	1.24460	3	9.9615
$\beta_7\ L_{III}O_I$	1.26385	5	9.8098	1.22400	4	10.1292
$\beta_5\ L_{III}O_{IV,V}$	1.2555	1	9.8750	1.21545	3	10.2004
L_IM_I				1.3365	3	9.277
$\beta_{10}\ L_IM_{IV}$	1.2537	2	9.889	1.21218	3	10.2279
$\beta_9\ L_IM_V$	1.2466	2	9.946	1.20479	7	10.2907
L_IN_I	1.11521	9	11.1173			
L_IN_{IV}	1.08377	7	11.4398	1.0468	2	11.844
$\gamma_{11}\ L_IN_V$	1.08205	7	11.4580	1.0458	1	11.856
$L_IN_{VI,VII}$	1.06357	9	11.6570			
L_IO_I	1.06771	9	11.6118	1.0317	3	12.017
$L_IO_{IV,V}$	1.06192	9	11.6752	1.0250	2	12.095
$L_{II}M_{II}$	1.43048	9	8.6671			
$\beta_{17}\ L_{II}M_{III}$	1.3864	1	8.9428	1.3387	2	9.261
$L_{II}M_V$	1.31897	9	9.3998	1.2728	2	9.741
$L_{II}N_{II}$	1.1600	2	10.688	1.1218	3	11.052
$L_{II}N_{III}$	1.1553	1	10.7316	1.1149	2	11.120
$L_{II}N_V$	1.13687	9	10.9055			
$v\ L_{II}N_{VI}$	1.1158	1	11.1113	1.0771	1	11.510
$L_{II}O_{II}$	1.11789	9	11.0907			
$L_{II}O_{III}$	1.11693	9	11.1001	1.0792	2	11.488
$t\ L_{III}M_{II}$	1.67265	9	7.4123	1.6244	3	7.632
$s\ L_{III}M_{III}$	1.61264	9	7.6881	1.5642	3	7.926
$L_{III}N_{II}$	1.3167	1	9.4158	1.2765	2	9.712
$L_{III}N_{III}$	1.3086	1	9.4742	1.2672	2	9.784
$u\ L_{III}N_{VI,VII}$	1.25778	4	9.8572	1.21868	5	10.1733
$L_{III}O_{II,III}$	1.2601	3	9.839	1.2211	2	10.153
M_IN_{III}	5.40	2	2.295	5.172	9	2.397
$M_IO_{II,III}$				4.44	2	2.79
$M_{II}N_I$				6.28	2	1.973
$M_{II}N_{IV}$	5.570	4	2.226	5.357	4	2.314
$M_{III}N_I$	7.612	9	1.629	7.360	8	1.684
$M_{III}N_{IV}$	6.353	5	1.951	6.134	4	2.021
$\gamma\ M_{III}N_V$	6.312	4	1.964	6.092	3	2.035
$M_{III}O_I$	5.83	2	2.126	5.628	8	2.203
$M_{III}O_{IV,V}$	5.67	3	2.19			

Designation	Å*	p.e.	keV	Å*	p.e.	keV
	73 Tantalum (*Cont.*)			**74 Tungsten** (*Cont.*)		
$\zeta_2\ M_{IV}N_{II}$	9.330	5	1.3288	8.993	5	1.3787
$M_{IV}N_{III}$	8.90	2	1.393	8.573	8	1.446
$\beta\ M_{IV}N_{VI}$	7.023	1	1.7655	6.757	1	1.8349
$M_{IV}O_{II}$	7.09	2	1.740	6.806	9	1.822
$\zeta_1\ M_VN_{III}$	9.316	4	1.3308	8.962	4	1.3835
$\alpha\ M_VN_{VI,VII}$	7.252	1	1.7096			
$\alpha_2\ M_VN_{VI}$				6.992	2	1.7731
$\alpha_1\ M_VN_{VII}$				6.983	1	1.7754
M_VO_{III}	7.30	2	1.700	7.005	9	1.770
$N_{II}N_{IV}$				54.0	2	0.2295
$N_{IV}N_{VI}$	58.2	1	0.2130	55.8	1	0.2221
$N_VN_{VI,VII}$	61.1	2	0.2028			
N_VN_{VI}				59.5	3	0.208
N_VN_{VII}				58.4	1	0.2122
	75 Rhenium			**76 Osmium**		
$\alpha_2\ KL_{II}$	0.207611	1	59.7179	0.201639	2	61.4867
$\alpha_1\ KL_{III}$	0.202781	2	61.1403	0.196794	2	63.0005
$\beta_3\ KM_{II}$	0.179697	3	68.994	0.174431	3	71.077
$\beta_1\ KM_{III}$	0.178880	3	69.310	0.173611	3	71.413
$\beta_2^{II}\ KN_{II}$	0.17425	1	71.151	0.16910	1	73.318
$\beta_2^{I}\ KN_{III}$	0.174054	6	71.232	0.168906	6	73.402
$KO_{II,III}$	0.17308	1	71.633	0.16798	1	73.808
$\beta_5^{II}\ KM_{IV}$	0.17783	1	69.719	0.17262	1	71.824
$\beta_5^{I}\ KM_V$	0.17766	1	69.786	0.17245	1	71.895
$\beta_4\ KN_{IV,V}$	0.17362	2	71.410	0.16842	2	73.615
$\beta_4\ L_IM_{II}$	1.25917	5	9.8463	1.21844	5	10.1754
$\beta_3\ L_IM_{III}$	1.22031	5	10.1598	1.17955	7	10.5108
$\gamma_2\ L_IN_{II}$	1.03233	5	12.0098	0.99805	5	12.4224
$\gamma_3\ L_IN_{III}$	1.02613	7	12.0824	0.99186	5	12.4998
$\gamma'_4\ L_IO_{II}$	0.99334	5	12.4813	0.96033	8	12.910
$\gamma_4\ L_IO_{III}$	0.99249	5	12.4920	0.95938	8	12.923
$\eta\ L_{II}M_I$	1.37342	5	9.0272	1.32785	7	9.3370
$\beta_1\ L_{II}M_{IV}$	1.23858	2	10.0100	1.19727	7	10.3553
$\gamma_5\ L_{II}N_I$	1.09388	5	11.3341	1.05693	5	11.7303
$\gamma_1\ L_{II}N_{IV}$	1.06099	5	11.6854	1.02503	5	12.0953
$\gamma_8\ L_{II}O_I$	1.04398	5	11.8758	1.00788	5	12.3012
$\gamma_6\ L_{II}O_{IV}$	1.03699	9	11.956	1.00107	5	12.3848
$l\ L_{III}M_I$	1.63056	5	7.6036	1.58498	7	7.8222
$\alpha_2\ L_{III}M_{IV}$	1.44396	5	8.5862	1.40234	5	8.8410
$\alpha_1\ L_{III}M_V$	1.43290	4	8.6525	1.39121	5	8.9117
$\beta_6\ L_{III}N_I$	1.25100	5	9.9105	1.21349	5	10.2169
$\beta_{15}\ L_{III}N_{IV}$	1.20819	5	10.2617	1.17167	5	10.5816
$\beta_2\ L_{III}N_V$	1.20660	4	10.2752	1.16979	8	10.5985
$\beta_7\ L_{III}O_I$	1.18610	5	10.4529	1.14933	8	10.7872
$\beta_5\ L_{III}O_{IV,V}$	1.17721	5	10.5318	1.1405	1	10.8711
$\beta_{10}\ L_IM_{IV}$	1.17218	5	10.5770	1.13353	5	10.9376
$\beta_9\ L_IM_V$	1.16487	4	10.6433	1.12637	6	11.0071
L_IN_I	1.0420	1	11.899			
L_IN_{IV}	1.0119	1	12.252	0.9772	3	12.687
$\gamma_{11}\ L_IN_V$	1.0108	1	12.266	0.9765	3	12.696
L_IO_I	0.9965	1	12.442	0.96318	7	12.8721
$L_IO_{IV,V}$	0.9900	1	12.524	0.95603	5	12.9683
$L_{II}M_{II}$	1.3366	1	9.2761	1.2934	2	9.586
$\beta_{17}\ L_{II}M_{III}$	1.2927	1	9.5910	1.2480	2	9.934

Designation	Å*	p.e.	keV	Å*	p.e.	keV
	75 Rhenium (*Cont.*)			**76 Osmium** (*Cont.*)		
$L_{II}M_V$	1.2305	1	10.0753	1.18977	7	10.4205
$L_{II}N_{II}$	1.0839	1	11.438			
$L_{II}N_{III}$	1.0767	1	11.515	1.03973	5	11.9243
$v\ L_{II}N_{VI}$	1.0404	1	11.917	1.0050	2	12.337
$L_{II}O_{III}$	1.0397	1	11.925	1.0047	2	12.340
$t\ L_{III}M_{II}$	1.5789	1	7.8525	1.5347	2	8.079
$s\ L_{III}M_{III}$	1.5178	1	8.1682	1.4735	2	8.414
$L_{III}N_I$				1.20086	7	10.3244
$L_{III}N_{III}$	1.2283	1	10.0933			
$u\ L_{III}N_{VI,VII}$	1.1815	1	10.4931	1.14537	7	10.8245
M_IN_{III}				4.79	2	2.59
$M_{II}N_I$				5.81	2	2.133
$M_{II}N_{IV}$				4.955	4	2.502
$M_{III}N_I$				6.89	2	1.798
$M_{III}N_{IV}$	5.931	5	2.090	5.724	5	2.166
$\gamma\ M_{III}N_V$	5.885	2	2.1067	5.682	4	2.182
$\zeta_2\ M_{IV}N_{II}$	8.664	5	1.4310	8.359	5	1.4831
$M_{IV}N_{III}$	8.239	8	1.505			
$\beta\ M_{IV}N_{VI}$	6.504	1	1.9061	6.267	1	1.9783
$\zeta_1\ M_VN_{III}$	8.629	4	1.4368	8.310	4	1.4919
$\alpha\ M_VN_{VI,VII}$	6.729	1	1.8425	6.490	1	1.9102
$N_{IV}N_{VI}$				51.9	1	0.2388
$N_VN_{VI,VII}$				54.7	2	0.2266
	77 Iridium			**78 Platinum**		
$\alpha_2\ KL_{II}$	0.195904	2	63.2867	0.190381	4	65.122
$\alpha_1\ KL_{III}$	0.191047	2	64.8956	0.185511	4	66.832
$\beta_3\ KM_{II}$	0.169367	2	73.2027	0.164501	3	75.368
$\beta_1\ KM_{III}$	0.168542	2	73.5608	0.163675	3	75.748
$\beta_2^{II}\ KN_{II}$	0.16415	1	75.529	0.15939	1	77.785
$\beta_2^{I}\ KN_{III}$	0.163956	7	75.619	0.15920	1	77.878
$KO_{II,III}$	0.163019	5	76.053	0.15826	1	78.341
$\beta_5^{II}\ KM_{IV}$	0.16759	2	73.980	0.16271	2	76.199
$\beta_5^{I}\ KM_V$	0.167373	9	74.075	0.16255	3	76.27
$\beta_4\ KN_{IV,V}$	0.16352	2	75.821	0.15881	2	78.069
$\beta_4\ L_IM_{II}$	1.17958	3	10.5106	1.14223	5	10.8543
$\beta_3\ L_IM_{III}$	1.14085	3	10.8674	1.10394	5	11.2308
$\gamma_2\ L_IN_{II}$	0.96545	3	12.8418	0.93427	5	13.2704
$\gamma_3\ L_IN_{III}$	0.95931	5	12.9240	0.92791	5	13.3613
$\gamma'_4\ L_IO_{II}$	0.92831	3	13.3555	0.89747	4	13.8145
$\gamma_4\ L_IO_{III}$	0.92744	3	13.3681	0.89659	4	13.8281
$\eta\ L_{II}M_I$	1.28448	3	9.6522	1.2429	2	9.975
$\beta_1\ L_{II}M_{IV}$	1.15781	3	10.7083	1.11990	2	11.0707
$\gamma_5\ L_{II}N_I$	1.02175	5	12.1342	0.9877	2	12.552
$\gamma_1\ L_{II}N_{IV}$	0.99085	3	12.5126	0.95797	3	12.9420
$\gamma_8\ L_{II}O_I$	0.97409	3	12.7279	0.9411	1	13.173
$\gamma_6\ L_{II}O_{IV}$	0.96708	4	12.8201	0.9342	2	13.271
$l\ L_{III}M_I$	1.54094	3	8.0458	1.4995	2	8.268
$\alpha_2\ L_{III}M_{IV}$	1.36250	5	9.0995	1.32432	2	9.3618
$\alpha_1\ L_{III}M_V$	1.35128	3	9.1751	1.31304	3	9.4423
$\beta_6\ L_{III}N_I$	1.17796	3	10.5251	1.14355	5	10.8418
$\beta_{15}\ L_{III}N_{IV}$	1.13707	3	10.9036			
$\beta_2\ L_{III}N_V$	1.13532	3	10.9203	1.10200	3	11.2505
$\beta_7\ L_{III}O_I$	1.11489	3	11.1205	1.08168	3	11.4619

Designation	Å*	p.e.	keV	Å*	p.e.	keV
	77 Iridium (*Cont.*)			**78 Platinum** (*Cont.*)		
$\beta_5\ L_{III}O_{IV,V}$	1.10585	3	11.2114	1.0724	2	11.561
L_IM_I	1.2102	2	10.245	1.16962	9	10.6001
$\beta_{10}\ L_IM_{IV}$	1.09702	4	11.3016	1.06183	7	11.6762
$\beta_9\ L_IM_V$	1.08975	5	11.3770	1.05446	5	11.7577
L_IN_I	0.9766	2	12.695	0.9455	2	13.113
L_IN_{IV}	0.9459	2	13.108			
$\gamma_{11}\ L_IN_V$	0.9446	2	13.126	0.9143	2	13.560
$L_IO_{IV,V}$	0.9243	3	13.413			
L_IO_I				0.8995	2	13.784
L_IO_{IV}				0.8943	1	13.864
L_IO_V				0.8934	1	13.878
$L_{II}M_{II}$	1.2502	3	9.917	1.213	1	10.225
$\beta_{17}\ L_{II}M_{III}$	1.2069	2	10.273	1.1667	1	10.6265
$L_{II}M_V$	1.1489	2	10.791	1.1129	2	11.140
$L_{II}N_{II}$	1.0120	2	12.251	0.9792	2	12.661
$L_{II}N_{III}$	1.0054	3	12.332	0.97173	4	12.7588
$v\ L_{II}N_{VI}$	0.97161	6	12.7603	0.93931	5	13.1992
$L_{II}O_{III}$	0.96979	5	12.7843			
$t\ L_{III}M_{II}$	1.4930	3	8.304	1.4530	2	8.533
$s\ L_{III}M_{III}$	1.4318	2	8.659	1.3895	2	8.923
$L_{III}N_{II}$	1.16545	5	10.6380	1.1310	2	10.962
$L_{III}N_{III}$	1.1560	3	10.725	1.1226	2	11.044
$u\ L_{III}N_{VI,VII}$	1.11145	4	11.1549	1.07896	5	11.4908
$L_{III}O_{II,III}$	1.10923	6	11.1772	1.0761	3	11.521
M_IN_{III}	4.631†	9	2.677	4.460	9	2.780
$M_{II}N_{IV}$	4.780	4	2.594	4.601	4	2.695
$M_{III}N_I$	6.669	9	1.859	6.455	9	1.921
$M_{III}N_{IV}$	5.540	5	2.238	5.357	5	2.314
$\gamma\ M_{III}N_V$	5.500	4	2.254	5.319	4	2.331
$M_{III}O_I$				4.876	9	2.543
$M_{III}O_{IV,V}$	4.869	9	2.546	4.694	8	2.641
$\zeta_2\ M_{IV}N_{II}$	8.065	5	1.5373	7.790	5	1.592
$M_{IV}N_{III}$	7.645	8	1.622	7.371	8	1.682
$\beta\ M_{IV}N_{VI}$	6.038	1	2.0535	5.828	1	2.1273
$\zeta_1\ M_VN_{III}$	8.021	4	1.5458	7.738	4	1.6022
$\alpha_2\ M_VN_{VI}$	6.275	3	1.9758	6.058	3	2.047
$\alpha_1\ M_VN_{VII}$	6.262	1	1.9799	6.047	1	2.0505
M_VO_{III}				5.987	9	2.071
$N_{IV}N_{VI}$	50.2	1	0.2470	48.1	2	0.258
$N_VN_{VI,VII}$	52.8	1	0.2348	50.9	1	0.2436
	79 Gold			**80 Mercury**		
$\alpha_2\ KL_{II}$	0.185075	2	66.9895	0.179958	3	68.895
$\alpha_1\ KL_{III}$	0.180195	2	68.8037	0.175068	3	70.819
$\beta_3\ KM_{II}$	0.159810	2	77.580	0.155321	3	79.822
$\beta_1\ KM_{III}$	0.158982	3	77.984	0.154487	3	80.253
$\beta_2^{II}\ KN_{II}$	0.15483	2	80.08	0.15040	2	82.43
$\beta_2^{I}\ KN_{III}$	0.154618	9	80.185	0.15020	2	82.54
$KO_{II,III}$	0.153694	7	80.667	0.14931	2	83.04
KL_I	0.18672	4	66.40			
$\beta_5^{II}\ KM_{IV}$	0.158062	7	78.438			
$\beta_5^{I}\ KM_V$	0.157880	5	78.529			
$\beta_5\ KM_{IV,V}$				0.15353	2	80.75
$\beta_4\ KN_{IV,V}$	0.154224	5	80.391	0.14978	2	82.78
$\beta_4\ L_IM_{II}$	1.10651	3	11.2047	1.07222	7	11.5630
$\beta_3\ L_IM_{III}$	1.06785	9	11.6103	1.03358	7	11.9953

Designation	Å*	p.e.	keV	Å*	p.e.	keV
	79 Gold (*Cont.*)			**80 Mercury** (*Cont.*)		
$\gamma_2\ L_IN_{II}$	0.90434	3	13.7095	0.87544	7	14.162
$\gamma_3\ L_IN_{III}$	0.89783	5	13.8090	0.86915	7	14.265
$\gamma'_4\ L_IO_{II}$	0.86816	4	14.2809	0.84013	7	14.757
$\gamma_4\ L_IO_{III}$	0.86703	4	14.2996	0.83894	7	14.778
$\eta\ L_{II}M_I$	1.20273	3	10.3083	1.1640	1	10.6512
$\beta_1\ L_{II}M_{IV}$	1.08353	3	11.4423	1.04868	5	11.8226
$\gamma_5\ L_{II}N_I$	0.95559	3	12.9743	0.92453	7	13.410
$\gamma_1\ L_{II}N_{IV}$	0.92650	3	13.3817	0.89646	5	13.8301
$\gamma_8\ L_{II}O_I$	0.90989	5	13.6260	0.87995	7	14.090
$\gamma_6\ L_{II}O_{IV}$	0.90297	3	13.7304	0.87319	7	14.199
$l\ L_{III}M_I$	1.45964	9	8.4939	1.4216	1	8.7210
$\alpha_2\ L_{III}M_{IV}$	1.28772	3	9.6280	1.25264	7	9.8976
$\alpha_1\ L_{III}M_V$	1.27640	3	9.7133	1.24120	5	9.9888
$\beta_6\ L_{III}N_I$	1.11092	3	11.1602	1.07975	7	11.4824
$\beta_{15}\ L_{III}N_{IV}$	1.07188	5	11.5667	1.04151	7	11.9040
$\beta_2\ L_{III}N_V$	1.07022	3	11.5847	1.03975	7	11.9241
$\beta_7\ L_{III}O_I$	1.04974	8	11.8106	1.01937	7	12.1625
$\beta_5\ L_{III}O_{IV,V}$	1.04044	3	11.9163	1.00987	7	12.2769
L_IM_I	1.13525	5	10.9210	1.0999	2	11.272
$\beta_{10}\ L_IM_{IV}$	1.02789	7	12.0617	0.9962	2	12.446
$\beta_9\ L_IM_V$	1.02063	7	12.1474	0.9871	2	12.560
L_IN_I	0.9131	1	13.578	0.8827	2	14.045
L_IN_{IV}	0.88563	7	13.999			
$\gamma_{11}\ L_IN_V$	0.88433	7	14.020	0.85657	7	14.474
L_IO_I	0.87074	5	14.2385	0.8452	2	14.670
$L_IO_{IV,V}$	0.86400	5	14.3497	0.8350	2	14.847
$L_{II}M_{II}$	1.1708	1	10.5892	1.1387	5	10.888
$\beta_{17}\ L_{II}M_{III}$	1.12798	5	10.9915	1.0916	5	11.358
$L_{II}M_V$	1.0756	2	11.526			
$L_{II}N_{III}$	0.9402	2	13.186	0.90894	7	13.640
$v\ L_{II}N_{VI}$	0.90837	5	13.6487	0.87885	7	14.107
$L_{II}O_{II}$	0.90746	7	13.662	0.8784	1	14.114
$L_{II}O_{III}$	0.90638	7	13.679	0.8758	1	14.156
$t\ L_{III}M_{II}$	1.41366	7	8.7702	1.3746	2	9.019
$s\ L_{III}M_{III}$	1.35131	7	9.1749	1.3112	2	9.455
$L_{III}N_{II}$	1.09968	7	11.2743	1.0649	2	11.642
$L_{III}N_{III}$	1.09026	7	11.3717	1.0585	1	11.713
$u\ L_{III}N_{VI,VII}$	1.04752	5	11.8357			
$u'\ L_{III}N_{VI}$				1.01769	7	12.1826
$u\ L_{III}N_{VII}$				1.01674	7	12.1940
$L_{III}O_{II,III}$	1.0450	2	11.865			
$L_{III}O_{II}$				1.01558	7	12.2079
$L_{III}O_{III}$				1.01404	7	12.2264
$L_{III}P_{II,III}$	1.03876	7	11.9355			
M_IN_{III}	4.300	9	2.883			
$M_{II}N_{IV}$	4.432	4	2.797			
$M_{III}N_I$	6.259	9	1.981	6.09	2	2.036
$M_{III}N_{IV}$	5.186	5	2.391			
$\gamma\ M_{III}N_V$	5.145	4	2.410	4.984†	2	2.4875
$M_{III}O_I$	4.703	9	2.636			
$M_{III}O_{IV,V}$	4.522	6	2.742			
$\zeta_2\ M_{IV}N_{II}$	7.523	5	1.648			
$M_{IV}N_{III}$	7.101	8	1.746	6.87	2	1.805
$\beta\ M_{IV}N_{VI}$	5.624	1	2.2046	5.4318†	9	2.2825
$\zeta_1\ M_VN_{III}$	7.466	4	1.6605			
$\alpha_2\ M_VN_{VI}$	5.854	3	2.118			

Designation	Å*	p.e.	keV	Å*	p.e.	keV
	79 Gold (*Cont.*)			**80 Mercury** (*Cont.*)		
$\alpha_1\ M_VN_{VII}$	5.840	1	2.1229	5.6476†	9	2.1953
M_VO_{III}	5.767	9	2.150			
$N_{IV}N_{VI}$	46.8	2	0.265	45.2†	3	0.274
$N_VN_{VI,VII}$	49.4	1	0.2510	47.9†	3	0.259
	81 Thallium			**82 Lead**		
$\alpha_2\ KL_{II}$	0.175036	2	70.8319	0.170294	2	72.8042
$\alpha_1\ KL_{III}$	0.170136	2	72.8715	0.165376	2	74.9694
$\beta_3\ KM_{II}$	0.150980	6	82.118	0.146810	4	84.450
$\beta_1\ KM_{III}$	0.150142	5	82.576	0.145970	6	84.936
$\beta_2^{II}\ KN_{II}$	0.14614	1	84.836	0.14212	2	87.23
$\beta_2^{I}\ KN_{III}$	0.14595	1	84.946	0.14191	1	87.364
$KO_{II,III}$	0.14509	1	85.451	0.141012	8	87.922
KP				0.1408	1	88.06
$\beta_5\ KM_{IV,V}$	0.14917	1	83.114			
$\beta_5^{II}\ KM_{IV}$				0.14512	2	85.43
$\beta_5^{I}\ KM_V$				0.14495	3	85.53
$\beta_4\ KN_{IV,V}$	0.14553	2	85.19	0.14155	3	87.59
$\beta_4\ L_IM_{II}$	1.03918	3	11.9306	1.0075	1	12.306
$\beta_3\ L_IM_{III}$	1.00062	3	12.3904	0.96911	7	12.7933
$\gamma_2\ L_IN_{II}$	0.84773	5	14.6251	0.8210	2	15.101
$\gamma_3\ L_IN_{III}$	0.84130	4	14.7368	0.8147	1	15.218
$\gamma'_4\ L_IO_{II}$	0.81308	5	15.2482	0.78706	7	15.752
$\gamma_4\ L_IO_{III}$	0.81184	3	15.2716	0.7858	1	15.777
$\eta\ L_{II}M_I$	1.12769	3	10.9943	1.09241	7	11.3493
$\beta_1\ L_{II}M_{IV}$	1.01513	4	12.2133	0.98291	3	12.6137
$\gamma_5\ L_{II}N_I$	0.89500	4	13.8526	0.86655	5	14.3075
$\gamma_1\ L_{II}N_{IV}$	0.86752	3	14.2915	0.83973	3	14.7644
$\gamma_8\ L_{II}O_I$	0.8513	2	14.564	0.82365	5	15.0527
$\gamma_6\ L_{II}O_{IV}$	0.8442	2	14.685	0.81683	5	15.1783
$L_{II}P_I$				0.81583	5	15.1969
$l\ L_{III}M_I$	1.38477	3	8.9532	1.34990	7	9.1845
$\alpha_2\ L_{III}M_{IV}$	1.21875	3	10.1728	1.18648	5	10.4495
$\alpha_1\ L_{III}M_V$	1.20739	4	10.2685	1.17501	2	10.5515
$\beta_6\ L_{III}N_I$	1.04963	5	11.8118	1.0210	1	12.143
$\beta_{15}\ L_{III}N_{IV}$	1.01201	3	12.2510	0.98389	7	12.6011
$\beta_2\ L_{III}N_V$	1.01031	3	12.2715	0.98221	7	12.6226
$\beta_7\ L_{III}O_I$	0.99017	5	12.5212	0.9620	1	12.888
$\beta_5\ L_{III}O_{IV,V}$	0.98058	3	12.6436	0.9526	1	13.015
L_IM_I	1.0644	2	11.648	1.0323	2	12.010
$\beta_{10}\ L_IM_{IV}$	0.96389	7	12.8626	0.9339	2	13.275
$\beta_9\ L_IM_V$	0.95675	7	12.9585	0.9268	1	13.377
L_IN_I	0.8549	1	14.503	0.82859	7	14.963
L_IN_{IV}	0.83001	7	14.937	0.80364	7	15.427
$\gamma_{11}\ L_IN_V$	0.82879	5	14.9593	0.80233	9	15.453
$L_IN_{VI,VII}$				0.7884	1	15.725
L_IO_I	0.8158	1	15.198	0.7897	1	15.699
$L_IO_{IV,V}$	0.80861	5	15.3327	0.78257	7	15.843
$L_{II}M_{II}$	1.0997	1	11.274	1.0644	2	11.648
$\beta_{17}\ L_{II}M_{III}$	1.05609	7	11.7397	1.0223	1	12.127
$L_{II}M_V$	1.00722	5	12.3093	0.9747	1	12.720
$L_{II}N_{II}$	0.882	2	14.057	0.8585	3	14.442

Designation	Å*	p.e.	keV	Å*	p.e.	keV
	81 Thallium (*Cont.*)			**82 Lead** (*Cont.*)		
$L_{II}N_{III}$	0.87996	5	14.0893	0.85192	7	14.553
$L_{II}N_V$				0.8382	2	14.791
$v\ L_{II}N_{VI}$	0.85048	5	14.5777	0.82327	7	15.060
$L_{II}O_{II}$	0.8490	1	14.604			
$L_{II}O_{III}$				0.8200	1	15.120
$t\ L_{III}M_{II}$	1.34154	5	9.2417	1.30767	7	9.4811
$s\ L_{III}M_{III}$	1.27807	5	9.7007	1.24385	7	9.9675
$L_{III}N_{II}$				1.01040	7	12.2705
$L_{III}N_{III}$	1.0286	1	12.053	1.0005	1	12.392
$u\ L_{III}N_{VI,VII}$	0.9888	1	12.538	0.96133	7	12.8968
$L_{III}O_{II}$	0.98738	5	12.5566	0.9586	1	12.934
$L_{III}O_{III}$	0.98538	5	12.5820	0.9578	1	12.945
$L_{III}P_{II,III}$	0.97926	5	12.6607	0.95118	7	13.0344
M_IN_{III}	4.013	9	3.089	3.872	9	3.202
$M_{II}N_I$				4.655	8	2.664
$M_{II}N_{IV}$	4.116	4	3.013	3.968	5	3.124
$M_{III}N_I$	5.884	8	2.107	5.704	8	2.174
$M_{III}N_{IV}$	4.865	5	2.548	4.715	3	2.630
$\gamma\ M_{III}N_V$	4.823	4	2.571	4.674	1	2.6527
$M_{III}O_I$				4.244	9	2.921
$M_{II}O_{IV,V}$	4.216	6	2.941	4.069	6	3.047
$\zeta_2\ M_{IV}N_{II}$	7.032	5	1.763	6.802	5	1.823
$M_{IV}N_{III}$				6.384	7	1.942
$\beta\ M_{IV}N_{VI}$	5.249	1	2.3621	5.076	1	2.4427
$M_{IV}O_{II}$	5.196	9	2.386	5.004	9	2.477
$\zeta_1\ M_VN_{III}$	6.974	4	1.778	6.740	3	1.8395
$\alpha_2\ M_VN_{VI}$	5.472	2	2.2656	5.299	2	2.3397
$\alpha_1\ M_VN_{VII}$	5.460	1	2.2706	5.286	1	2.3455
M_VO_{III}				5.168	9	2.399
$N_{IV}N_{VI}$				42.3	2	0.293
$N_VN_{VI,VII}$	46.5	2	0.267	45.0	1	0.2756
$N_{VI}O_{IV}$	115.3	2	0.1075	102.4	1	0.1211
$N_{VI}O_V$	113.0	1	0.10968	100.2	2	0.1237
$N_{VII}O_V$	117.7	1	0.10530	104.3	1	0.1189
	83 Bismuth			**84 Polonium**		
$\alpha_2\ KL_{II}$	0.165717	2	74.8148	0.16130†	1	76.862
$\alpha_1\ KL_{III}$	0.160789	2	77.1079	0.15636†	1	79.290
$\beta_3\ KM_{II}$	0.142779	7	86.834	0.13892†	2	89.25
$\beta_1\ KM_{III}$	0.141948	3	87.343	0.13807†	2	89.80
$\beta_2^{II}\ KN_{II}$	0.13817	1	89.733	0.13438†	2	92.26
$\beta_2^{I}\ KN_{III}$	0.13797	1	89.864	0.13418†	2	92.40
$KO_{II,III}$	0.13709	1	90.435			
$\beta_5\ KM_{IV,V}$	0.14111	1	87.860			
$\beta_4\ KN_{IV,V}$	0.13759	2	90.11			
$\beta_4\ L_IM_{II}$	0.97690	4	12.6912	0.9475	3	13.086
$\beta_3\ L_IM_{III}$	0.93855	3	13.2098	0.9091	3	13.638
$\gamma_2\ L_IN_{II}$	0.79565	3	15.5824	0.772	1	16.07
$\gamma_3\ L_IN_{III}$	0.78917	5	15.7102			
$\gamma'_4\ L_IO_{II}$	0.76198	3	16.2709			
$\gamma_4\ L_IO_{III}$	0.76087	3	16.2947			
$\gamma_{13}\ L_IP_{II,III}$	0.75690	3	16.3802			
$\eta\ L_{II}M_I$	1.05856	3	11.7122			
$\beta_1\ L_{II}M_{IV}$	0.951978	9	13.0235	0.9220	2	13.447
$\gamma_5\ L_{II}N_I$	0.83923	5	14.7732			

X-Ray Wavelengths

TABLE A (*Continued*)

Designation	Å*	p.e.	keV	Å*	p.e.	keV
	83 Bismuth (*Cont.*)			84 Polonium (*Cont.*)		
$\gamma_1\ L_{II}N_{IV}$	0.81311	2	15.2477	0.78748	9	15.744
$\gamma_8\ L_{II}O_{I}$	0.7973	1	15.551			
$\gamma_6\ L_{II}O_{IV}$	0.79043	3	15.6853	0.7645	2	16.218
$l\ L_{III}M_{I}$	1.31610	7	9.4204	1.2829	5	9.664
$\alpha_2\ L_{III}M_{IV}$	1.15536	1	10.73091	1.12548†	5	11.0158
$\alpha_1\ L_{III}M_{V}$	1.14386	2	10.8388	1.11386	4	11.1308
$\beta_6\ L_{III}N_{I}$	0.99331	3	12.4816	0.9672	2	12.819
$\beta_{15}\ L_{III}N_{IV}$	0.95702	5	12.9549	0.9312	2	13.314
$\beta_2\ L_{III}N_{V}$	0.95518	4	12.9799	0.92937	5	13.3404
$\beta_7\ L_{III}O_{I}$	0.93505	5	13.2593			
$\beta_5\ L_{III}O_{IV,V}$	0.92556	3	13.3953	0.8996	2	13.782
$L_{I}M_{I}$	1.0005	9	12.39			
$\beta_{10}\ L_{I}M_{IV}$	0.90495	4	13.7002			
$\beta_9\ L_{I}M_{V}$	0.89791	3	13.8077			
$L_{I}N_{I}$	0.8022	1	15.456			
$L_{I}N_{IV}$	0.7795	5	15.904			
$\gamma_{11}\ L_{I}N_{V}$	0.77728	5	15.951			
$L_{I}N_{VI,VII}$	0.7641	5	16.23			
$L_{I}O_{IV,V}$	0.75791	5	16.358			
$L_{II}M_{II}$	1.0346	9	11.98			
$\beta_{17}\ L_{II}M_{III}$	0.98913	5	12.5344			
$L_{II}M_{V}$	0.94419	5	13.1310			
$L_{II}N_{II}$	0.8344	9	14.86			
$L_{II}N_{III}$	0.8248	1	15.031			
$v\ L_{II}N_{VI}$	0.79721	9	15.552			
$L_{II}O_{III}$	0.79384	5	15.6178			
$t\ L_{III}M_{II}$	1.2748	1	9.7252			
$s\ L_{III}M_{III}$	1.2105	1	10.2421			
$L_{III}N_{II}$	0.98280	5	12.6151			
$L_{III}N_{III}$	0.97321	5	12.7394			
$u\ L_{III}N_{VI,VII}$	0.93505	5	13.2593			
$L_{III}O_{II}$	0.9323	2	13.298			
$L_{III}O_{III}$	0.9302	2	13.328			
$L_{III}P_{II,III}$	0.92413	4	13.4159			
$M_{I}N_{II}$	3.892	9	3.185			
$M_{I}N_{III}$	3.740	9	3.315			
$M_{II}N_{IV}$	3.834	4	3.234			
$M_{III}N_{I}$	5.537	8	2.239			
$M_{III}N_{IV}$	4.571	5	2.712			
$\gamma\ M_{III}N_{V}$	4.532	2	2.735			
$M_{III}O_{I}$	4.105	9	3.021			
$M_{III}O_{IV,V}$	3.932	6	3.153			
$\zeta_2\ M_{IV}N_{II}$	6.585	5	1.883			
$M_{IV}N_{III}$	6.162	8	2.012			
$\beta\ M_{IV}N_{VI}$	4.909	1	2.5255			
$M_{IV}O_{II}$	4.823	3	2.571			
$M_{IV}P_{II,III}$	4.59	2	2.70			
$\zeta_1\ M_{V}N_{III}$	6.521	4	1.901			
$\alpha_2\ M_{V}N_{VI}$	5.130	2	2.4170			
$\alpha_1\ M_{V}N_{VII}$	5.118	1	2.4226			
$N_{I}P_{II,III}$	13.30	6	0.932			
$N_{VI}O_{IV}$	91.6	1	0.1354			
$N_{VII}O_{V}$	93.2	1	0.1330			

Designation	Å*	p.e.	keV	Å*	p.e.	keV
	85 Astatine			**86 Radon**		
$\alpha_2\ KL_{II}$	0.15705†	2	78.95	0.15294†	3	81.07
$\alpha_1\ KL_{III}$	0.15210†	2	81.52	0.14798†	3	83.78
$\beta_3\ KM_{II}$	0.13517†	4	91.72	0.13155†	5	94.24
$\beta_1\ KM_{III}$	0.13432†	4	92.30	0.13069†	5	94.87
$\beta_2^{II}\ KN_{II}$	0.13072†	4	94.84	0.12719†	5	97.47
$\beta_2^{I}\ KN_{III}$	0.13052†	4	94.99	0.12698†	5	97.64
$\beta_3\ L_{I}M_{III}$	0.88135†	9	14.067	0.85436†	9	14.512
$\beta_1\ L_{II}M_{IV}$	0.89349†	9	13.876	0.86605†	9	14.316
$\gamma_1\ L_{II}N_{IV}$	0.76289†	9	16.251	0.73928†	9	16.770
$\alpha_2\ L_{III}M_{IV}$	1.09671†	5	11.3048	1.06899†	5	11.5979
$\alpha_1\ L_{III}M_{V}$	1.08500†	5	11.4268	1.05723†	5	11.7270
	87 Francium			**88 Radium**		
$\alpha_2\ KL_{II}$	0.14896†	3	83.23	0.14512†	2	85.43
$\alpha_1\ KL_{III}$	0.14399†	3	86.10	0.14014†	2	88.47
$\beta_3\ KM_{II}$	0.12807†	5	96.81	0.12469†	3	99.43
$\beta_1\ KM_{III}$	0.12719†	5	97.47	0.12382†	3	100.13
$\beta_2^{II}\ KN_{II}$	0.12379†	5	100.16	0.12050†	3	102.89
$\beta_2^{I}\ KN_{III}$	0.12358†	5	100.33	0.12029†	3	103.07
$\beta_4\ L_{I}M_{II}$				0.84071	5	14.7472
$\beta_3\ L_{I}M_{III}$	0.82789†	9	14.976	0.80273	5	15.4449
$\gamma_2\ L_{I}N_{II}$				0.68199	5	18.179
$\gamma_3\ L_{I}N_{III}$				0.67538	5	18.357
$\gamma'_4\ L_{I}O_{II}$				0.65131	5	19.036
$\gamma_4\ L_{I}O_{III}$				0.64965	5	19.084
$\gamma_{13}\ L_{I}P_{II,III}$				0.64513	5	19.218
$\eta\ L_{II}M_{I}$				0.90742	5	13.6630
$\beta_1\ L_{II}M_{IV}$	0.83940†	9	14.770	0.81375	5	15.2358
$\gamma_5\ L_{II}N_{I}$				0.71774	5	17.274
$\gamma_1\ L_{II}N_{IV}$	0.71652†	9	17.303	0.69463	5	17.849
γ_8				0.6801	1	18.230
$\gamma_6\ L_{II}O_{IV}$				0.67328	5	18.414
$L_{II}P_{I}$				0.6724	1	18.439
$l\ L_{III}M_{I}$				1.16719	5	10.6222
$\alpha_2\ L_{III}M_{IV}$	1.04230	5	11.8950	1.01656	5	12.1962
$\alpha_1\ L_{III}M_{V}$	1.03049	5	12.0313	1.00473	5	12.3397
$\beta_6\ L_{III}N_{I}$				0.87088	5	14.2362
$\beta_{15}\ L_{III}N_{IV}$				0.83722	5	14.8086
$\beta_2\ L_{III}N_{V}$	0.858	2	14.45	0.83537	5	14.8414
$\beta_7\ L_{III}O_{I}$				0.8162	1	15.190
$\beta_5\ L_{III}O_{IV,V}$				0.80627	5	15.3771
$L_{III}P_{I}$				0.8050	1	15.402
$\beta_{10}\ L_{I}M_{IV}$				0.77546	5	15.988
$\beta_9\ L_{I}M_{V}$				0.76857	5	16.131
$L_{I}N_{I}$				0.6874	1	18.036
$L_{I}N_{IV}$				0.6666	1	18.600
$\gamma_{11}\ L_{I}N_{V}$				0.6654	1	18.633
$L_{I}O_{IV,V}$				0.6468	1	19.167
$\beta_{17}\ L_{II}M_{III}$				0.8438	1	14.692
$L_{II}N_{III}$				0.7043	1	17.604
$L_{II}N_{V}$				0.6932	1	17.884
$L_{II}O_{II}$				0.6780	1	18.286

Designation	Å*	p.e.	keV	Å*	p.e.	keV
	87 Francium (*Cont.*)			**88 Radium** (*Cont.*)		
$L_{II}O_{III}$				0.6764	1	18.330
$L_{II}P_{II,III}$				0.6714	1	18.466
$L_{III}N_{II}$				0.8618	1	14.387
$L_{III}N_{III}$				0.8512	1	14.566
u $L_{III}N_{VI,VII}$				0.8186	1	15.146
$L_{III}P_{II,III}$				0.8038	1	15.425
	89 Actinium			**90 Thorium**		
α_2 KL_{II}	0.14141†	2	87.67	0.137829	2	89.953
α_1 KL_{III}	0.136417†	8	90.884	0.132813	2	93.350
β_3 KM_{II}	0.12143†	2	102.10	0.118268	3	104.831
β_1 KM_{III}	0.12055†	2	102.85	0.117396	9	105.609
β_2^{II} KN_{II}	0.11732†	2	105.67	0.11426	1	108.511
β_2^{I} KN_{III}	0.11711†	2	105.86	0.114040	9	108.717
$KO_{II,III}$				0.11322	1	109.500
β_5 $KM_{IV,V}$				0.116667	9	106.269
β_4 $KN_{IV,V}$				0.11366	2	109.08
β_4 L_IM_{II}				0.79257	4	15.6429
β_3 L_IM_{III}	0.77822†	9	15.931	0.75479	3	16.4258
γ_2 $L_{II}N_{II}$				0.64221	4	19.305
γ_3 L_IN_{III}				0.63559	4	19.507
γ'_4 L_IO_{II}				0.61251	4	20.242
γ_4 L_IO_{III}				0.61098	4	20.292
γ_{13} $L_IP_{II,III}$				0.60705	8	20.424
η $L_{II}M_I$				0.85446	4	14.5099
β_1 $L_{II}M_{IV}$	0.78903†	9	15.713	0.765210	9	16.2022
γ_5 $L_{II}N_I$				0.67491	4	18.370
γ_1 $L_{II}N_{IV}$	0.67351†	9	18.408	0.65313	3	18.9825
γ_8 $L_{II}O_I$				0.63898	5	19.403
γ_6 $L_{II}O_{IV}$				0.63258	4	19.599
$L_{II}P_I$				0.6316	1	19.629
$L_{II}P_{IV}$				0.62991	9	19.682
l $L_{III}M_I$				1.11508	4	11.1186
α_2 $L_{III}M_{IV}$	0.99178†	5	12.5008	0.96788	2	12.8096
α_1 $L_{III}M_V$	0.97993†	5	12.6520	0.95600	3	12.9687
β_6 $L_{III}N_I$				0.82790	8	14.975
β_{15} $L_{III}N_{IV}$				0.79539	5	15.5875
β_2 $L_{III}N_V$				0.79354	3	15.6237
β_7 $L_{III}O_I$				0.77437	4	16.0105
β_5 $L_{III}O_{IV,V}$				0.76468	5	16.213
$L_{III}P_I$				0.76338	5	16.241
$L_{III}P_{IV,V}$				0.76087	9	16.295
β_{10} L_IM_{IV}				0.7301	1	16.981
β_9 L_IM_V				0.7234	1	17.139
L_IN_I				0.64755	5	19.146
L_IN_{IV}				0.6276	1	19.755
γ_{11} L_IN_V				0.62636	9	19.794
$L_IN_{VI,VII}$				0.6160	1	20.128
L_IO_I				0.6146	1	20.174
$L_IO_{IV,V}$				0.6083	1	20.383
$L_{II}M_{II}$				0.8338	1	14.869
β_{17} $L_{II}M_{III}$				0.79257	4	15.6429
$L_{II}M_V$				0.7579	1	16.359
$L_{II}N_{III}$				0.6620	1	18.729
$L_{II}N_V$				0.6521	1	19.014

Designation	Å*	p.e.	keV	Å*	p.e.	keV
	89 Actinium (*Cont.*)			**90 Thorium** (*Cont.*)		
v $L_{II}N_{VI}$				0.64064	9	19.353
$L_{II}O_{II}$				0.6369	1	19.466
$L_{II}O_{III}$				0.6356	1	19.506
$L_{II}P_{II,III}$				0.6312	1	19.642
t $L_{III}M_I$				1.08009	9	11.4788
s $L_{III}M_{II}$				1.0112	1	12.261
$L_{III}N_{II}$				0.8190	2	15.138
$L_{III}N_{III}$				0.8082	1	15.341
u $L_{III}N_{VI,VII}$				0.77661	5	15.964
$L_{III}O_{II}$				0.7713	1	16.074
$L_{III}O_{III}$				0.7690	1	16.123
$L_{III}P_{II,III}$				0.7625	2	16.260
M_IN_{III}				2.934	8	4.23
M_IO_{III}				2.442	9	5.08
$M_{II}N_I$				3.537	9	3.505
$M_{II}N_{IV}$				3.011	2	4.117
$M_{II}O_{IV}$				2.618	5	4.735
$M_{III}N_I$				4.568	5	2.714
$M_{III}N_{IV}$				3.718	3	3.335
γ $M_{III}N_V$				3.679	2	3.370
$M_{III}O_I$				3.283	9	3.78
$M_{III}O_{IV,V}$				3.131	3	3.959
ζ_2 M_VN_{II}				5.340	5	2.322
$M_{IV}N_{III}$				4.911	5	2.524
β $M_{IV}N_{VI}$				3.941	1	3.1458
$M_{IV}O_{II}$				3.808	4	3.256
ζ_1 M_VN_{III}				5.245	5	2.364
α_2 M_VN_{VI}				4.151	2	2.987
α_1 M_VN_{VII}				4.1381	9	2.9961
M_VP_{III}				3.760	9	3.298
N P_{II}				9.44	7	1.313
N P_{III}				9.40	7	1.1319
$N_{II}O_{IV}$				11.56	5	1.072
N_IP_I				11.07	7	1.120
$N_{III}O_V$				13.8	1	0.897
$N_{IV}N_{VI}$				33.57	9	0.3693
$N_VN_{VI,VII}$				36.32	9	0.3414
$N_{VI}O_{IV}$				49.5	1	0.2505
$N_{VI}O_V$				48.2	1	0.2572
$N_{VII}O_V$				50.0	1	0.2479
$O_{III}P_{IV,V}$				68.2	3	0.1817
$O_{IV,V}Q_{II,III}$				181.	5	0.068
	91 Protactinium			**92 Uranium**		
α_2 KL_{II}	0.134343†	9	92.287	0.130968	4	94.665
α_1 KL_{III}	0.129325†	3	95.868	0.125947	3	98.439
β_3 KM_{II}	0.11523†	2	107.60	0.112296	4	110.406
β_1 KM_{III}	0.114345†	8	108.427	0.111394	5	111.300
β_2^{II} KN_{II}	0.11129†	2	111.40	0.10837	1	114.40
β_2^{I} KN_{III}	0.11107†	2	111.62	0.10818	1	114.60
$KO_{II,III}$				0.10744	1	115.39
β_5 $KM_{IV,V}$				0.11069	1	112.01
β_4 $KN_{IV,V}$				0.10780	2	115.01
β_4 L_IM_I	0.7699	1	16.104	0.747985	9	16.5753
β_3 L_IM_{III}	0.73230	5	16.930	0.71029	2	17.4550

Designation	Å*	p.e.	keV	Å*	p.e.	keV
	91 Protactinium (*Cont.*)			**92 Uranium** (*Cont.*)		
$\gamma_2\ L_{I}N_{II}$	0.6239	1	19.872	0.605237	9	20.4847
$\gamma_3\ L_{I}N_{III}$	0.6169	1	20.098	0.598574	9	20.7127
$\gamma'_4\ L_{I}O_{II}$				0.576700	9	21.4984
$\gamma_4\ L_{I}O_{II,III}$	0.5937	1	20.882	0.57499	9	21.562
γ_{13}				0.5706	1	21.729
$\eta\ L_{II}M_{I}$	0.8295	1	14.946	0.80509	2	15.3997
$\beta_1\ L_{II}M_{IV}$	0.74232	5	16.702	0.719984	8	17.2200
$\gamma_5\ L_{II}N_{I}$	0.6550	1	18.930	0.63557	2	19.5072
$\gamma_1\ L_{II}N_{IV}$	0.63358†	9	19.568	0.614770	9	20.1671
$\gamma_8\ L_{II}O_{I}$				0.60125	5	20.621
$\gamma_6\ L_{II}O_{IV}$	0.6133	1	20.216	0.594845	9	20.8426
$L_{II}P_{IV}$				0.59203	5	20.942
$l\ L_{III}M_{I}$	1.0908	1	11.366	1.06712	2	11.6183
$\alpha_2\ L_{III}M_{IV}$	0.94482†	5	13.1222	0.922558	9	13.4388
$\alpha_1\ L_{III}M_{V}$	0.93284	5	13.2907	0.910639	9	13.6147
$\beta_6\ L_{III}N_{I}$	0.8079	1	15.347	0.78838	2	15.7260
$\beta_{15}\ L_{III}N_{IV}$				0.756642	9	16.3857
$\beta_2\ L_{III}N_{V}$	0.7737	1	16.024	0.754681	9	16.4283
$\beta_7\ L_{III}O_{I}$	0.7546	2	16.431	0.73602	6	16.845
$\beta_5\ L_{III}O_{IV,V}$	0.7452	2	16.636	0.726305	9	17.0701
$L_{III}P_{I}$				0.72521	5	17.096
$L_{III}P_{IV,V}$				0.72240	5	17.162
$\beta_{10}\ L_{I}M_{V}$	0.7088	2	17.492	0.68760	5	18.031
$\beta_9\ L_{I}M_{V}$	0.7018	1	17.667	0.681014	8	18.2054
$L_{I}N_{IV}$				0.59096	5	20.979
$\gamma_{11}\ L_{I}N_{V}$				0.58986	5	21.019
$L_{I}O_{IV,V}$				0.5725	1	21.657
$\beta_{17}\ L_{II}M_{III}$				0.74503	5	16.641
$L_{II}N_{III}$				0.6228	1	19.907
$v\ L_{II}N_{VI}$				0.6031	1	20.556
$L_{III}O_{III}$				0.59728	5	20.758
$L_{II}P_{II,III}$				0.5930	2	20.906
$t\ L_{III}M_{II}$				1.0347	1	11.982
$s\ L_{III}M_{III}$				0.9636	1	12.866
$L_{III}N_{II}$				0.78017	9	15.892
$L_{III}N_{III}$				0.7691	1	16.120
$u\ L_{III}N_{VI,VII}$				0.738603	9	16.7859
$L_{III}O_{II}$				0.7333	1	16.907
$L_{III}O_{III}$				0.7309	1	16.962
$L_{III}P_{II,III}$				0.72426	5	17.118
$M_{I}N_{II}$				2.92	2	4.25
$M_{I}N_{III}$				2.753	8	4.50
$M_{I}O_{III}$				2.304	7	5.38
$M_{I}P_{III}$				2.253	6	5.50
$M_{II}N_{I}$	3.441	5	3.603	3.329	4	3.724
$M_{II}N_{IV}$	2.910	2	4.260	2.817	2	4.401
$M_{II}O_{IV}$	2.527	4	4.906	2.443	4	5.075
$M_{III}N_{I}$	4.450	4	2.786	4.330	2	2.863
$M_{III}N_{IV}$	3.614	2	3.430	3.521	2	3.521
$\gamma\ M_{III}N_{V}$	3.577	1	3.4657	3.479	1	3.563
$M_{III}O_{I}$	3.245	9	3.82	3.115	7	3.980
$M_{III}O_{IV,V}$	3.038	2	4.081	2.948	2	4.205
$\zeta_2\ M_{IV}N_{II}$	5.193	2	2.3876	5.050	2	2.4548
$M_{IV}N_{III}$				4.625	5	2.681
$\beta\ M_{IV}N_{VI}$	3.827	1	3.2397	3.716	1	3.3367

Designation	Å*	p.e.	keV	Å*	p.e.	keV
	91 Protactinium (*Cont.*)			**92 Uranium** (*Cont.*)		
$M_{IV}O_{II}$	3.691	2	3.359	3.576	1	3.4666
$\zeta_1\ M_{V}N_{III}$	5.092	2	2.4350	4.946	2	2.507
$\alpha_2\ M_{V}N_{VI}$	4.035	3	3.072	3.924	1	3.1595
$\alpha_1\ M_{V}N_{VII}$	4.022	1	3.0823	3.910	1	3.1708
$N_{I}O_{III}$				10.09	7	1.229
$N_{I}P_{II}$				8.81	7	1.41
$N_{I}P_{III}$				8.76	7	1.42
$N_{II}P_{I}$				10.40	7	1.192
$N_{III}O_{V}$				12.90	9	0.961
$N_{IV}N_{VI}$				31.8	1	0.390
$N_{V}N_{VI,VII}$				34.8	1	0.357
$N_{IV}O_{IV}$				43.3	2	0.286
$N_{VI}O_{V}$				42.1	2	0.295
$N_{I}P_{IV,V}$				8.60	7	1.44
	93 Neptunium			**94 Plutonium**		
$\beta_4\ L_{I}M_{II}$	0.72671	2	17.0607	0.70620	2	17.5560
$\beta_3\ L_{I}M_{III}$	0.68920†	9	17.989	0.66871	2	18.5405
$\gamma_2\ L_{I}N_{II}$	0.5873	5	21.11	0.57068	2	21.7251
$\gamma_3\ L_{I}N_{III}$	0.5810	5	21.34	0.564001	9	21.9824
$\gamma'_4\ L_{I}O_{II}$				0.5432	1	22.823
$\gamma_4\ L_{I}O_{II,III}$	0.5585	5	22.20	0.5416	1	22.891
$\eta\ L_{II}M_{I}$	0.7809	2	15.876	0.7591	1	16.333
$\beta_1\ L_{II}M_{IV}$	0.698478	9	17.7502	0.67772	2	18.2937
$\gamma_5\ L_{II}N_{I}$	0.616	1	20.12	0.5988	1	20.704
$\gamma_1\ L_{II}N_{IV}$	0.596498	9	20.7848	0.578882	9	21.4173
γ_8				0.5658	1	21.914
$\gamma_6\ L_{II}O_{IV}$	0.57699	5	21.488	0.55973	2	22.1502
$l\ L_{III}M_{I}$	1.0428	6	11.890	1.0226	1	12.124
$\alpha_2\ L_{III}M_{IV}$	0.901045	9	13.7597	0.88028	2	14.0842
$\alpha_1\ L_{III}M_{V}$	0.889128	9	13.9441	0.86830	2	14.2786
$\beta_6\ L_{III}N_{I}$	0.769	1	16.13	0.75148	2	16.4983
$\beta_{15}\ L_{III}N_{IV}$				0.7205	1	17.208
$\beta_2\ L_{III}N_{V}$	0.736230	9	16.8400	0.71851	2	17.2553
$\beta_7\ L_{III}O_{I}$				0.7003	1	17.705
$\beta_5\ L_{III}O_{IV,V}$	0.70814	2	17.5081	0.69068	2	17.9506
$\beta_{10}\ L_{I}M_{IV}$				0.6482	1	19.126
$\beta_9\ L_{I}M_{V}$				0.6416	1	19.323
$u\ L_{III}N_{VI,VII}$				0.7031	1	17.635
	95 Americium					
$\beta_4\ L_{I}M_{II}$	0.68639	2	18.0627			
$\beta_3\ L_{I}M_{III}$	0.64891	2	19.1059			
$\gamma_2\ L_{I}N_{II}$	0.5544	2	22.361			
$\beta_1\ L_{II}M_{IV}$	0.657655	9	18.8520			
$\gamma_1\ L_{II}N_{IV}$	0.561886	9	22.0652			
$\gamma_6\ L_{II}O_{IV}$	0.54311	2	22.8282			
$l\ L_{III}M_{I}$	1.0012	6	12.384			
$\alpha_2\ L_{III}M_{IV}$	0.860266	9	14.4119			
$\alpha_1\ L_{III}M_{V}$	0.848187	9	14.6172			
$\beta_6\ L_{III}N_{I}$	0.73418	2	16.8870			
$\beta_{15}\ L_{III}N_{IV}$	0.70341	2	17.6258			
$\beta_2\ L_{III}N_{V}$	0.701390	9	17.6765			
$\beta_5\ L_{III}O_{IV,V}$	0.67383	2	18.3996			

TABLE B. Wavelengths in numerical order of the emission lines and absorption edges.

Wavelength Å*	p.e.	Element	Designation		keV
0.10723	1	92 U	K	Abs. Edge	115.62
0.10744	1	92 U		$KO_{II,III}$	115.39
0.10780	2	92 U	$K\beta_4$	$KN_{IV,V}$	115.01
0.10818	1	92 U	$K\beta_2^{I}$	KN_{III}	114.60
0.10837	1	92 U	$K\beta_2^{II}$	KN_{II}	114.40
0.11069	1	92 U	$K\beta_5$	$KM_{IV,V}$	112.01
0.11107	2	91 Pa	$K\beta_2^{I}$	KN_{III}	111.62
0.11129	2	91 Pa	$K\beta_2^{II}$	KN_{II}	111.40
0.111394	5	92 U	$K\beta_1$	KM_{III}	111.300
0.112296	4	92 U	$K\beta_3$	KM_{II}	110.406
0.11307	1	90 Th	K	Abs. Edge	109.646
0.11322	1	90 Th		$KO_{II,III}$	109.500
0.11366	2	90 Th	$K\beta_4$	$KN_{IV,V}$	109.08
0.114040	9	90 Th	$K\beta_2^{I}$	KN_{III}	108.717
0.11426	1	90 Th	$K\beta_2^{II}$	KN_{II}	108.511
0.114345	8	91 Pa	$K\beta_1$	KM_{III}	108.427
0.11523	2	91 Pa	$K\beta_3$	KM_{II}	107.60
0.116667	9	90 Th	$K\beta_5$	$KM_{IV,V}$	106.269
0.11711	2	89 Ac	$K\beta_2^{I}$	KN_{III}	105.86
0.11732	2	89 Ac	$K\beta_2^{II}$	KN_{II}	105.67
0.117396	9	90 Th	$K\beta_1$	KM_{III}	105.609
0.118268	3	90 Th	$K\beta_3$	KM_{II}	104.831
0.12029	3	88 Ra	$K\beta_2^{I}$	KN_{III}	103.07
0.12050	3	88 Ra	$K\beta_2^{II}$	KN_{II}	102.89
0.12055	2	89 Ac	$K\beta_1$	KM_{III}	102.85
0.12143	2	89 Ac	$K\beta_3$	KM_{II}	102.10
0.12358	5	87 Fr	$K\beta_2^{I}$	KN_{III}	100.33
0.12379	5	87 Fr	$K\beta_2^{II}$	KN_{II}	100.16
0.12382	3	88 Ra	$K\beta_1$	KM_{III}	100.13
0.12469	3	88 Ra	$K\beta_3$	KM_{II}	99.43
0.125947	3	92 U	$K\alpha_1$	KL_{III}	98.439
0.12698	5	86 Rn	$K\beta_2^{I}$	KN_{III}	97.64
0.12719	5	86 Rn	$K\beta_2^{II}$	KN_{II}	97.47
0.12719	5	87 Fr	$K\beta_1$	KM_{III}	97.47
0.12807	5	87 Fr	$K\beta_3$	KM_{II}	96.81
0.129325	3	91 Pa	$K\alpha_1$	KL_{III}	95.868
0.13052	4	85 At	$K\beta_2^{I}$	KN_{III}	94.99
0.13069	5	86 Rn	$K\beta_1$	KM_{III}	94.87
0.13072	4	85 At	$K\beta_2^{II}$	KN_{II}	94.84
0.130968	4	92 U	$K\alpha_2$	KL_{II}	94.665
0.13155	5	86 Rn	$K\beta_3$	KM_{II}	94.24
0.132813	2	90 Th	$K\alpha_1$	KL_{III}	93.350
0.13418	2	84 Po	$K\beta_2^{I}$	KN_{III}	92.40
0.13432	4	85 At	$K\beta_1$	KM_{III}	92.30
0.134343	9	91 Pa	$K\alpha_2$	KL_{II}	92.287
0.13438	2	84 Po	$K\beta_2^{II}$	KN_{II}	92.26
0.13517	4	85 At	$K\beta_3$	KM_{II}	91.72
0.136417	8	89 Ac	$K\alpha_1$	KL_{III}	90.884
0.13694	1	83 Bi	K	Abs. Edge	90.534
0.13709	1	83 Bi		$KO_{II,III}$	90.435
0.13759	2	83 Bi	$K\beta_4$	$KN_{IV,V}$	90.11
0.137829	2	90 Th	$K\alpha_2$	KL_{II}	89.953
0.13797	1	83 Bi	$K\beta_2^{I}$	KN_{III}	89.864
0.13807	2	84 Po	$K\beta_1$	KM_{III}	89.80
0.13817	1	83 Bi	$K\beta_2^{II}$	KN_{II}	89.733
0.13892	2	84 Po	$K\beta_3$	KM_{II}	89.25
0.14014	2	88 Ra	$K\alpha_1$	KL_{III}	88.47
0.1408	1	82 Pb		KP	88.06
0.140880	5	82 Pb	K	Abs. Edge	88.005
0.141012	8	82 Pb		$KO_{II,III}$	87.922
0.14111	1	83 Bi	$K\beta_5$	$KM_{IV,V}$	87.860
0.14141	2	89 Ac	$K\alpha_2$	KL_{II}	87.67
0.14155	3	82 Pb	$K\beta_4$	$KN_{IV,V}$	87.59
0.14191	1	82 Pb	$K\beta_2^{I}$	KN_{III}	87.361
0.141948	3	83 Bi	$K\beta_1$	KM_{III}	87.343
0.14212	2	82 Pb	$K\beta_2^{II}$	KN_{II}	87.23
0.142779	7	83 Bi	$K\beta_3$	KM_{II}	86.834
0.14399	3	87 Fr	$K\alpha_1$	KL_{III}	86.10
0.14495	1	81 Tl	K	Abs. Edge	85.533
0.14495	3	82 Pb	$K\beta_5^{I}$	KM_{V}	85.53
0.14509	1	81 Tl		$KO_{II,III}$	85.451
0.14512	2	82 Pb	$K\beta_5^{II}$	KM_{IV}	85.43
0.14512	2	88 Ra	$K\alpha_2$	KL_{II}	85.43
0.14553	2	81 Tl	$K\beta_4$	$KN_{IV,V}$	85.19
0.14595	1	81 Tl	$K\beta_2^{I}$	KN_{III}	84.946
0.145970	6	82 Pb	$K\beta_1$	KM_{III}	84.936
0.14614	1	81 Tl	$K\beta_2^{II}$	KN_{II}	84.836
0.146810	4	82 Pb	$K\beta_3$	KM_{II}	84.450
0.14798	3	86 Rn	$K\alpha_1$	KL_{III}	83.78
0.14896	3	87 Fr	$K\alpha_2$	KL_{II}	83.23
0.14917	1	81 Tl	$K\beta_5$	$KM_{IV,V}$	83.114
0.14918	1	80 Hg	K	Abs. Edge	83.109
0.14931	2	80 Hg		$KO_{II,III}$	83.04
0.14978	2	80 Hg	$K\beta_4$	$KN_{IV,V}$	82.78
0.150142	5	81 Tl	$K\beta_1$	KM_{III}	82.576
0.15020	2	80 Hg	$K\beta_2^{I}$	KN_{III}	82.54
0.15040	2	80 Hg	$K\beta_2^{II}$	KN_{II}	82.43
0.150980	6	81 Tl	$K\beta_3$	KM_{II}	82.118
0.15210	2	85 At	$K\alpha_1$	KL_{III}	81.52
0.15294	3	86 Rn	$K\alpha_2$	KL_{II}	81.07
0.15353	2	80 Hg	$K\beta_5$	$KM_{IV,V}$	80.75
0.153593	5	79 Au	K	Abs. Edge	80.720
0.153694	7	79 Au		$KO_{II,III}$	80.667
0.154224	5	79 Au	$K\beta_4$	$KN_{IV,V}$	80.391
0.154487	3	80 Hg	$K\beta_1$	KM_{III}	80.253
0.154618	9	79 Au	$K\beta_2^{I}$	KN_{III}	80.185
0.15483	2	79 Au	$K\beta_2^{II}$	KN_{II}	80.08
0.155321	3	80 Hg	$K\beta_3$	KM_{II}	79.822
0.15636	1	84 Po	$K\alpha_1$	KL_{III}	79.290
0.15705	2	85 At	$K\alpha_2$	KL_{II}	78.95
0.157880	5	79 Au	$K\beta_5^{I}$	KM_{V}	78.529
0.158062	7	79 Au	$K\beta_5^{II}$	KM_{IV}	78.438
0.15818	1	78 Pt	K	Abs. Edge	78.381
0.15826	1	78 Pt		$KO_{II,III}$	78.341
0.15881	2	78 Pt	$K\beta_4$	$KN_{IV,V}$	78.069
0.158982	3	79 Au	$K\beta_1$	KM_{III}	77.984
0.15920	1	78 Pt	$K\beta_2^{I}$	KN_{III}	77.878
0.15939	1	78 Pt	$K\beta_2^{II}$	KN_{II}	77.785
0.159810	2	79 Au	$K\beta_3$	KM_{II}	77.580
0.160789	2	83 Bi	$K\alpha_1$	KL_{III}	77.1079
0.16130	1	84 Po	$K\alpha_2$	KL_{II}	76.862
0.16255	3	78 Pt	$K\beta_5^{I}$	KM_{V}	76.27
0.16271	2	78 Pt	$K\beta_5^{II}$	KM_{IV}	76.199
0.16292	1	77 Ir	K	Abs. Edge	76.101

Wavelength Å*	p.e.	Element	Designation		keV
0.163019	5	77 Ir		$KO_{II,III}$	76.053
0.16352	2	77 Ir	$K\beta_4$	$KN_{IV,V}$	75.821
0.163675	3	78 Pt	$K\beta_1$	KM_{III}	75.748
0.163956	7	77 Ir	$K\beta_2^{I}$	KN_{III}	75.619
0.16415	1	77 Ir	$K\beta_2^{II}$	KN_{II}	75.529
0.164501	3	78 Pt	$K\beta_3$	KM_{II}	75.368
0.165376	2	82 Pb	$K\alpha_1$	KL_{III}	74.9694
0.165717	2	83 Bi	$K\alpha_2$	KL_{II}	74.8148
0.167373	9	77 Ir	$K\beta_5^{I}$	KM_{V}	74.075
0.16759	2	77 Ir	$K\beta_5^{II}$	KM_{IV}	73.980
0.16787	1	76 Os	K	Abs. Edge	73.856
0.16798	1	76 Os		$KO_{II,III}$	73.808
0.16842	2	76 Os	$K\beta_4$	$KN_{IV,V}$	73.615
0.168542	2	77 Ir	$K\beta_1$	KM_{III}	73.5608
0.168906	6	76 Os	$K\beta_2^{I}$	KN_{III}	73.402
0.16910	1	76 Os	$K\beta_2^{II}$	KN_{II}	73.318
0.169367	2	77 Ir	$K\beta_3$	KM_{II}	73.2027
0.170136	2	81 Tl	$K\alpha_1$	KL_{III}	72.8715
0.170294	2	82 Pb	$K\alpha_2$	KL_{II}	72.8042
0.17245	1	76 Os	$K\beta_5^{I}$	KM_{V}	71.895
0.17262	1	76 Os	$K\beta_5^{II}$	KM_{IV}	71.824
0.17302	1	75 Re	K	Abs. Edge	71.658
0.17308	1	75 Re		$KO_{II,III}$	71.633
0.173611	3	76 Os	$K\beta_1$	KM_{III}	71.413
0.17362	2	75 Re	$K\beta_4$	$KN_{IV,V}$	71.410
0.174054	6	75 Re	$K\beta_2^{I}$	KN_{III}	71.232
0.17425	1	75 Re	$K\beta_2^{II}$	KN_{II}	71.151
0.174431	3	76 Os	$K\beta_3$	KM_{II}	71.077
0.175036	2	81 Tl	$K\alpha_2$	KL_{II}	70.8319
0.175068	3	80 Hg	$K\alpha_1$	KL_{III}	70.819
0.17766	1	75 Re	$K\beta_5^{I}$	KM_{V}	69.786
0.17783	1	75 Re	$K\beta_5^{II}$	KM_{IV}	69.719
0.17837	1	74 W	K	Abs. Edge	69.508
0.178444	5	74 W		$KO_{II,III}$	69.479
0.178880	3	75 Re	$K\beta_1$	KM_{III}	69.310
0.17892	2	74 W	$K\beta_4$	$KN_{IV,V}$	69.294
0.179421	7	74 W	$K\beta_2^{I}$	KN_{III}	69.101
0.17960	1	74 W	$K\beta_2^{II}$	KN_{II}	69.031
0.179697	3	75 Re	$K\beta_3$	KM_{II}	68.994
0.179958	3	80 Hg	$K\alpha_2$	KL_{II}	68.895
0.180195	2	79 Au	$K\alpha_1$	KL_{III}	68.8037
0.183092	7	74 W	$K\beta_5^{I}$	KM_{V}	67.715
0.183264	5	74 W	$K\beta_5^{II}$	KM_{IV}	67.652
0.18394	1	73 Ta	K	Abs. Edge	67.403
0.184031	7	73 Ta		$KO_{II,III}$	67.370
0.184374	2	74 W	$K\beta_1$	KM_{III}	67.2443
0.18451	1	73 Ta	$K\beta_4$	$KN_{IV,V}$	67.194
0.185011	8	73 Ta	$K\beta_2^{I}$	KN_{III}	67.013
0.185075	2	79 Au	$K\alpha_2$	KL_{II}	66.9895
0.185181	2	74 W	$K\beta_3$	KM_{II}	66.9514
0.185188	9	73 Ta	$K\beta_2^{II}$	KN_{II}	66.949
0.185511	4	78 Pt	$K\alpha_1$	KL_{III}	66.832
0.18672	4	79 Au		KL_{I}	66.40
0.188757	6	73 Ta	$K\beta_5^{I}$	KM_{V}	65.683
0.188920	6	73 Ta	$K\beta_5^{II}$	KM_{IV}	65.626
0.18982	5	72 Hf	K	Abs. Edge	65.31
0.190089	4	73 Ta	$K\beta_1$	KM_{III}	65.223
0.190381	4	78 Pt	$K\alpha_2$	KL_{II}	65.122
0.1908	2	72 Hf	$K\beta_2$	$KN_{II,III}$	64.98
0.190890	2	73 Ta	$K\beta_3$	KM_{II}	64.9488
0.191047	2	77 Ir	$K\alpha_1$	KL_{III}	64.8956
0.19585	5	71 Lu	K	Abs. Edge	63.31
0.19589	2	71 Lu		$KO_{II,III}$	63.293
0.195904	2	77 Ir	$K\alpha_2$	KL_{II}	63.2867
0.19607	3	72 Hf	$K\beta_1$	KM_{III}	63.234
0.196794	2	76 Os	$K\alpha_1$	KL_{III}	63.0005
0.19686	4	72 Hf	$K\beta_3$	KM_{II}	62.98
0.1969	2	71 Lu	$K\beta_2$	$KN_{II,III}$	62.97
0.20084	2	71 Lu	$K\beta_5$	$KM_{IV,V}$	61.732
0.201639	2	76 Os	$K\alpha_2$	KL_{II}	61.4867
0.20224	5	70 Yb	K	Abs. Edge	61.30
0.20226	2	70 Yb		$KO_{II,III}$	61.298
0.20231	3	71 Lu	$K\beta_1$	KM_{III}	61.283
0.202781	2	75 Re	$K\alpha_1$	KL_{III}	61.1403
0.20309	4	71 Lu	$K\beta_3$	KM_{II}	61.05
0.2033	2	70 Yb	$K\beta_2$	$KN_{II,III}$	60.89
0.20739	2	70 Yb	$K\beta_5$	$KM_{IV,V}$	59.782
0.207611	1	75 Re	$K\alpha_2$	KL_{II}	59.7179
0.20880	5	69 Tm	K	Abs. Edge	59.38
0.20884	8	70 Yb	$K\beta_1$	KM_{III}	59.37
0.20891	2	69 Tm		$KO_{II,III}$	59.346
0.2090100	Std.	74 W	$K\alpha_1$	KL_{III}	59.31824
0.2096	1	70 Yb	$K\beta_3$	KM_{II}	59.14
0.2098	2	69 Tm	$K\beta_2$	$KN_{II,III}$	59.09
0.213828	2	74 W	$K\alpha_2$	KL_{II}	57.9817
0.21404	2	69 Tm	$K\beta_5$	$KM_{IV,V}$	57.923
0.215497	4	73 Ta	$K\alpha_1$	KL_{III}	57.532
0.21556	2	69 Tm	$K\beta_1$	KM_{III}	57.517
0.21567	1	68 Er	K	Abs. Edge	57.487
0.21581	3	68 Er		$KO_{II,III}$	57.450
0.21592	4	74 W		KL_{I}	57.42
0.21636	2	69 Tm	$K\beta_3$	KM_{II}	57.304
0.2167	2	68 Er	$K\beta_2$	$KN_{II,III}$	57.21
0.220305	8	73 Ta	$K\alpha_2$	KL_{II}	56.277
0.22124	3	68 Er	$K\beta_5$	$KM_{IV,V}$	56.040
0.222227	3	72 Hf	$K\alpha_1$	KL_{III}	55.7902
0.22266	2	68 Er	$K\beta_1$	KM_{III}	55.681
0.22291	1	67 Ho	K	Abs. Edge	55.619
0.22305	3	67 Ho		$KO_{II,III}$	55.584
0.22341	2	68 Er	$K\beta_3$	KM_{II}	55.494
0.2241	2	67 Ho	$K\beta_2$	$KN_{II,III}$	55.32
0.227024	3	72 Hf	$K\alpha_2$	KL_{II}	54.6114
0.22855	3	67 Ho	$K\beta_5$	$KM_{IV,V}$	54.246
0.229298	2	71 Lu	$K\alpha_1$	KL_{III}	54.0698
0.23012	2	67 Ho	$K\beta_1$	KM_{III}	53.877
0.23048	1	66 Dy	K	Abs. Edge	53.793
0.23056	3	66 Dy		$KO_{II,III}$	53.774
0.23083	2	67 Ho	$K\beta_3$	KM_{II}	53.711
0.2317	2	66 Dy	$K\beta_2$	$KN_{II,III}$	53.47
0.234081	2	71 Lu	$K\alpha_2$	KL_{II}	52.9650
0.23618	3	66 Dy	$K\beta_5$	$KM_{IV,V}$	52.494
0.236655	2	70 Yb	$K\alpha_1$	KL_{III}	52.3889
0.23788	2	66 Dy	$K\beta_1$	KM_{III}	52.119
0.23841	1	65 Tb	K	Abs. Edge	52.002

Wavelength Å*	p.e.	Element		Designation	keV
0.23858	3	65 Tb		$KO_{II,III}$	51.965
0.23862	2	66 Dy	$K\beta_3$	KM_{II}	51.957
0.2397	2	65 Tb	$K\beta_2$	$KN_{II,III}$	51.68
0.241424	2	70 Yb	$K\alpha_2$	KL_{II}	51.3540
0.244338	2	69 Tm	$K\alpha_1$	KL_{III}	50.7416
0.24608	2	65 Tb	$K\beta_1$	KM_{III}	50.382
0.24681	1	64 Gd	K	Abs. Edge	50.233
0.24683	2	65 Tb	$K\beta_3$	KM_{II}	50.229
0.24687	3	64 Gd		$KO_{II,III}$	50.221
0.24816	3	64 Gd	$K\beta_2$	$KN_{II,III}$	49.959
0.249095	2	69 Tm	$K\alpha_2$	KL_{II}	49.7726
0.252365	2	68 Er	$K\alpha_1$	KL_{III}	49.1277
0.25275	3	64 Gd	$K\beta_5$	$KM_{IV,V}$	49.052
0.25460	2	64 Gd	$K\beta_1$	KM_{III}	48.697
0.25534	2	64 Gd	$K\beta_3$	KM_{II}	48.555
0.25553	1	63 Eu	K	Abs. Edge	48.519
0.255645	7	63 Eu		$KO_{II,III}$	48.497
0.256923	8	63 Eu	$K\beta_2^{I}$	$KN_{II,III}$	48.256
0.257110	2	68 Er	$K\alpha_2$	KL_{II}	48.2211
0.260756	2	67 Ho	$K\alpha_1$	KL_{III}	47.5467
0.263577	5	63 Eu	$K\beta_1$	KM_{III}	47.0379
0.264332	5	63 Eu	$K\beta_3$	KM_{II}	46.9036
0.26464	5	62 Sm	K	Abs. Edge	46.849
0.26491	3	62 Sm		$KO_{II,III}$	46.801
0.265486	2	67 Ho	$K\alpha_2$	KL_{II}	46.6997
0.2662	1	62 Sm	$K\beta_2$	$KN_{II,III}$	46.57
0.269533	2	66 Dy	$K\alpha_1$	KL_{III}	45.9984
0.27111	3	62 Sm	$K\beta_5$	$KM_{IV,V}$	45.731
0.27301	2	62 Sm	$K\beta_1$	KM_{III}	45.413
0.27376	2	62 Sm	$K\beta_3$	KM_{II}	45.289
0.274247	2	66 Dy	$K\alpha_2$	KL_{II}	45.2078
0.27431	5	61 Pm	K	Abs. Edge	45.198
0.2759	1	61 Pm	$K\beta_2$	$KN_{II,III}$	44.93
0.278724	2	65 Tb	$K\alpha_1$	KL_{III}	44.4816
0.28290	3	61 Pm	$K\beta_1$	KM_{III}	43.826
0.283423	2	65 Tb	$K\alpha_2$	KL_{II}	43.7441
0.28363	4	61 Pm	$K\beta_3$	KM_{II}	43.713
0.28453	5	60 Nd	K	Abs. Edge	43.574
0.2861	1	60 Nd	$K\beta_2$	$KN_{II,III}$	43.32
0.288353	2	64 Gd	$K\alpha_1$	KL_{III}	42.9962
0.293038	2	64 Gd	$K\alpha_2$	KL_{II}	42.3089
0.293299	2	60 Nd	$K\beta_1$	KM_{III}	42.2713
0.294027	3	60 Nd	$K\beta_3$	KM_{II}	42.1665
0.29518	5	59 Pr	K	Abs. Edge	42.002
0.29679	2	59 Pr	$K\beta_2$	$KN_{II,III}$	41.773
0.298446	2	63 Eu	$K\alpha_1$	KL_{III}	41.5422
0.303118	2	63 Eu	$K\alpha_2$	KL_{II}	40.9019
0.304261	4	59 Pr	$K\beta_1$	KM_{III}	40.7482
0.304975	5	59 Pr	$K\beta_3$	KM_{II}	40.6529
0.30648	5	58 Ce	K	Abs. Edge	40.453
0.30668	2	58 Ce		$KO_{II,III}$	40.427
0.30737	2	58 Ce	$K\beta_4^{I}$	$KN_{IV,V}$	40.337
0.30816	1	58 Ce	$K\beta_2$	$KN_{II,III}$	40.233
0.309040	2	62 Sm	$K\alpha_1$	KL_{III}	40.1181
0.31342	2	58 Ce	$K\beta_5^{I}$	KM_{V}	39.558
0.31357	2	58 Ce	$K\beta_5^{II}$	KM_{IV}	39.539
0.313698	2	62 Sm	$K\alpha_2$	KL_{II}	39.5224
0.315816	2	58 Ce	$K\beta_1$	KM_{III}	39.2573
0.316520	4	58 Ce	$K\beta_3$	KM_{II}	39.1701
0.31844	5	57 La	K	Abs. Edge	38.934
0.31864	2	57 La		$KO_{II,III}$	38.909
0.31931	2	57 La	$K\beta_4^{I}$	$KN_{IV,V}$	38.828
0.320117	7	57 La	$K\beta_2$	$KN_{II,III}$	38.7299
0.320160	4	61 Pm	$K\alpha_1$	KL_{III}	38.7247
0.324803	4	61 Pm	$K\alpha_2$	KL_{II}	38.1712
0.32546	2	57 La	$K\beta_5^{I}$	KM_{V}	38.094
0.32563	2	57 La	$K\beta_5^{II}$	KM_{IV}	38.074
0.327983	3	57 La	$K\beta_1$	KM_{III}	37.8010
0.328686	4	57 La	$K\beta_3$	KM_{II}	37.7202
0.33104	1	56 Ba	K	Abs. Edge	37.452
0.33127	2	56 Ba		$KO_{II,III}$	37.426
0.331846	2	60 Nd	$K\alpha_1$	KL_{III}	37.3610
0.33229	2	56 Ba	$K\beta_4^{II}$	KN_{IV}	37.311
0.33277	1	56 Ba	$K\beta_2$	$KN_{II,III}$	37.257
0.336472	2	60 Nd	$K\alpha_2$	KL_{II}	36.8474
0.33814	2	56 Ba	$K\beta_5^{I}$	KM_{V}	36.666
0.33835	2	56 Ba	$K\beta_5^{II}$	KM_{IV}	36.643
0.340811	3	56 Ba	$K\beta_1$	KM_{III}	36.3782
0.341507	4	56 Ba	$K\beta_3$	KM_{II}	36.3040
0.344140	2	59 Pr	$K\alpha_1$	KL_{III}	36.0263
0.34451	1	55 Cs	K	Abs. Edge	35.987
0.34611	2	55 Cs	$K\beta_2$	$KN_{II,III}$	35.822
0.348749	2	59 Pr	$K\alpha_2$	KL_{II}	35.5502
0.354364	7	55 Cs	$K\beta_1$	KM_{III}	34.9869
0.355050	4	55 Cs	$K\beta_3$	KM_{II}	34.9194
0.357092	2	58 Ce	$K\alpha_1$	KL_{III}	34.7197
0.3584	5	54 Xe	K	Abs. Edge	34.59
0.36026	3	54 Xe	$K\beta_2$	$KN_{II,III}$	34.415
0.361683	2	58 Ce	$K\alpha_2$	KL_{II}	34.2789
0.36872	2	54 Xe	$K\beta_1$	KM_{III}	33.624
0.36941	2	54 Xe	$K\beta_3$	KM_{II}	33.562
0.370737	2	57 La	$K\alpha_1$	KL_{III}	33.4418
0.37381	1	53 I	K	Abs. Edge	33.1665
0.37523	2	53 I	$K\beta_2$	$KN_{II,III}$	33.042
0.375313	2	57 La	$K\alpha_2$	KL_{II}	33.0341
0.383905	4	53 I	$K\beta_1$	KM_{III}	32.2947
0.384564	4	53 I	$K\beta_3$	KM_{II}	32.2394
0.385111	4	56 Ba	$K\alpha_1$	KL_{III}	32.1936
0.389668	5	56 Ba	$K\alpha_2$	KL_{II}	31.8171
0.38974	1	52 Te		$KO_{II,III}$	31.8114
0.38974	1	52 Te	K	Abs. Edge	31.8114
0.391102	6	52 Te	$K\beta_2$	$KN_{II,III}$	31.7004
0.399995	5	52 Te	$K\beta_1$	KM_{III}	30.9957
0.400290	4	55 Cs	$K\alpha_1$	KL_{III}	30.9728
0.400659	4	52 Te	$K\beta_3$	KM_{II}	30.9443
0.404835	4	55 Cs	$K\alpha_2$	KL_{II}	30.6251
0.40666	1	51 Sb		$KO_{II,III}$	30.4875
0.40668	1	51 Sb	K	Abs. Edge	30.4860
0.40702	1	51 Sb	$K\beta_4^{I}$	$KN_{IV,V}$	30.4604
0.407973	5	51 Sb	$K\beta_2$	$KN_{II,III}$	30.3895
0.41378	1	51 Sb	$K\beta_5^{I}$	KM_{V}	29.9632
0.41388	1	51 Sb	$K\beta_5^{II}$	KM_{IV}	29.9560
0.41634	2	54 Xe	$K\alpha_1$	KL_{III}	29.779
0.417085	3	51 Sb	$K\beta_1$	KM_{III}	29.7256

Wavelength Å*	p.e.	Element	Designation		keV
0.417737	4	51 Sb	$K\beta_3$	KM_{II}	29.6792
0.42087	2	54 Xe	$K\alpha_2$	KL_{II}	29.458
0.42467	3	50 Sn		$KO_{II,III}$	29.195
0.42467	1	50 Sn	K	Abs. Edge	29.1947
0.42495	3	50 Sn	$K\beta_4^{I}$	$KN_{IV,V}$	29.175
0.425915	8	50 Sn	$K\beta_2$	$KN_{II,III}$	29.1093
0.43175	3	50 Sn	$K\beta_5^{I}$	KM_{V}	28.716
0.43184	3	50 Sn	$K\beta_5^{II}$	KM_{IV}	28.710
0.433318	5	53 I	$K\alpha_1$	KL_{III}	28.6120
0.435236	5	50 Sn	$K\beta_1$	KM_{III}	28.4860
0.435877	5	50 Sn	$K\beta_3$	KM_{II}	28.4440
0.437829	7	53 I	$K\alpha_2$	KL_{II}	28.3172
0.44371	1	49 In	K	Abs. Edge	27.9420
0.44374	3	49 In		$KO_{II,III}$	27.940
0.44393	4	49 In	$K\beta_4^{I}$	$KN_{IV,V}$	27.928
0.44500	1	49 In	$K\beta_2$	$KN_{II,III}$	27.8608
0.45086	2	49 In	$K\beta_5^{I}$	KM_{V}	27.499
0.45098	2	49 In	$K\beta_5^{II}$	KM_{IV}	27.491
0.451295	3	52 Te	$K\alpha_1$	KL_{III}	27.4723
0.454545	4	49 In	$K\beta_1$	KM_{III}	27.2759
0.455181	4	49 In	$K\beta_3$	KM_{II}	27.2377
0.455784	3	52 Te	$K\alpha_2$	KL_{II}	27.2017
0.46407	1	48 Cd	K	Abs. Edge	26.7159
0.465328	7	48 Cd	$K\beta_2$	$KN_{II,III}$	26.6438
0.470354	3	51 Sb	$K\alpha_1$	KL_{III}	26.3591
0.474827	3	51 Sb	$K\alpha_2$	KL_{II}	26.1108
0.475105	6	48 Cd	$K\beta_1$	KM_{III}	26.0955
0.475730	5	48 Cd	$K\beta_3$	KM_{II}	26.0612
0.48589	1	47 Ag	K	Abs. Edge	25.5165
0.4859	9	47 Ag	$K\beta_4$	$KN_{IV,V}$	25.512
0.487032	4	47 Ag	$K\beta_2$	$KN_{II,III}$	25.4564
0.490599	3	50 Sn	$K\alpha_1$	KL_{III}	25.2713
0.49306	2	47 Ag	$K\beta_5$	$KM_{IV,V}$	25.145
0.495053	3	50 Sn	$K\alpha_2$	KL_{II}	25.0440
0.497069	4	47 Ag	$K\beta_1$	KM_{III}	24.9424
0.497685	4	47 Ag	$K\beta_3$	KM_{II}	24.9115
0.5092	1	46 Pd	K	Abs. Edge	24.348
0.5093	2	46 Pd	$K\beta_4$	$KN_{IV,V}$	24.346
0.510228	4	46 Pd	$K\beta_2$	$KN_{II,III}$	24.2991
0.512113	3	49 In	$K\alpha_1$	KL_{III}	24.2097
0.516544	3	49 In	$K\alpha_2$	KL_{II}	24.0020
0.51670	9	46 Pd	$K\beta_5$	$KM_{IV,V}$	23.995
0.520520	4	46 Pd	$K\beta_1$	KM_{III}	23.8187
0.521123	4	46 Pd	$K\beta_3$	KM_{II}	23.7911
0.53395	1	45 Rh	K	Abs. Edge	23.2198
0.53401	9	45 Rh	$K\beta_4^{I}$	$KN_{IV,V}$	23.217
0.535010	3	48 Cd	$K\alpha_1$	KL_{III}	23.1736
0.53503	2	45 Rh	$K\beta_2$	$KN_{II,III}$	23.1728
0.53513	5	45 Rh	$K\beta_2^{II}$	KN_{II}	23.168
0.5365	1	94 Pu	L_I	Abs. Edge	23.109
0.539422	3	48 Cd	$K\alpha_2$	KL_{II}	22.9841
0.54101	9	45 Rh	$K\beta_5^{I}$	KM_{V}	22.917
0.54118	9	45 Rh	$K\beta_5^{II}$	KM_{IV}	22.909
0.5416	1	94 Pu	$L\gamma_4$	L_IO_{III}	22.891
0.54311	2	95 Am	$L\gamma_6$	$L_{II}O_{IV}$	22.8282
0.5432	1	94 Pu	$L\gamma_4'$	L_IO_{II}	22.823
0.545605	4	45 Rh	$K\beta_1$	KM_{III}	22.7236
0.546200	4	45 Rh	$K\beta_3$	KM_{II}	22.6989
0.5544	2	95 Am	$L\gamma_2$	L_IN_{II}	22.361
0.5572	1	94 Pu	L_{II}	Abs. Edge	22.253
0.5585	5	93 Np	$L\gamma_4$	$L_IO_{II,III}$	22.20
0.5594075	6	47 Ag	$K\alpha_1$	KL_{III}	22.16292
0.55973	2	94 Pu	$L\gamma_6$	$L_{II}O_{IV}$	22.1502
0.56051	1	44 Ru	K	Abs. Edge	22.1193
0.56089	9	44 Ru	$K\beta_4$	$KN_{IV,V}$	22.104
0.56166	3	44 Ru	$K\beta_2$	$KN_{II,III}$	22.074
0.561886	9	95 Am	$L\gamma_1$	$L_{II}N_{IV}$	22.0652
0.563798	4	47 Ag	$K\alpha_2$	KL_{II}	21.9903
0.564001	9	94 Pu	$L\gamma_3$	L_IN_{III}	21.9824
0.5658	1	94 Pu	$L\gamma_8$	$L_{II}O_I$	21.914
0.56785	9	44 Ru	$K\beta_5^{I}$	KM_{V}	21.834
0.5680	2	44 Ru	$K\beta_5^{II}$	KM_{IV}	21.829
0.5695	1	92 U	L_I	Abs. Edge	21.771
0.5706	1	92 U	$L\gamma_{13}$	$L_IP_{II,III}$	21.729
0.57068	2	94 Pu	$L\gamma_2$	L_IN_{II}	21.1251
0.572482	4	44 Ru	$K\beta_1$	KM_{III}	21.6568
0.5725	1	92 U		$L_IO_{IV,V}$	21.657
0.573067	4	44 Ru	$K\beta_3$	KM_{II}	21.6346
0.57499	9	92 U	$L\gamma_4$	L_IO_{III}	21.562
0.576700	9	92 U	$L\gamma_4'$	L_IO_{II}	21.4984
0.57699	5	93 Np	$L\gamma_6$	$L_{II}O_{IV}$	21.488
0.578882	9	94 Pu	$L\gamma_1$	$L_{II}N_{IV}$	21.4173
0.5810	5	93 Np	$L\gamma_3$	L_IN_{III}	21.34
0.585448	3	46 Pd	$K\alpha_1$	KL_{III}	21.1771
0.5873	5	93 Np	$L\gamma_2$	L_IN_{II}	21.11
0.58906	1	43 Te	K	Abs. Edge	21.0473
0.589821	3	46 Pd	$K\alpha_2$	KL_{II}	21.0201
0.58986	5	92 U	$L\gamma_{11}$	L_IN_V	21.019
0.59024	5	43 Tc	$K\beta_2$	$KN_{II,III}$	21.005
0.59096	5	92 U		L_IN_{IV}	20.979
0.5919	1	92 U	L_{II}	Abs. Edge	20.945
0.59203	5	92 U		$L_{II}P_{IV}$	20.942
0.5930	2	92 U		$L_{II}P_{II,III}$	20.906
0.5937	1	91 Pa	$L\gamma_4$	$L_IO_{II,III}$	20.882
0.594845	9	92 U	$L\gamma_6$	$L_{II}O_{IV}$	20.8426
0.596498	9	93 Np	$L\gamma_1$	$L_{II}N_{IV}$	20.7848
0.59728	5	92 U		$L_{II}O_{III}$	20.758
0.598574	9	92 U	$L\gamma_3$	L_IN_{III}	20.7127
0.5988	1	94 Pu	$L\gamma_5$	$L_{II}N_I$	20.704
0.60125	5	92 U	$L\gamma_8$	$L_{II}O_I$	20.621
0.60130	4	43 Tc	$K\beta_1$	KM_{III}	20.619
0.60188	4	43 Tc	$K\beta_3$	KM_{II}	20.599
0.6031	1	92 U	Lv	$L_{II}N_{VI}$	20.556
0.605237	9	92 U	$L\gamma_2$	L_IN_{II}	20.4847
0.6059	1	90 Th	L_I	Abs. Edge	20.464
0.60705	8	90 Th	$L\gamma_{13}$	$L_IP_{II,III}$	20.424
0.6083	1	90 Th		$L_IO_{IV,V}$	20.383
0.61098	4	90 Th	$L\gamma_4$	L_IO_{III}	20.292
0.61251	4	90 Th	$L\gamma_4'$	L_IO_{II}	20.242
0.6133	1	91 Pa	$L\gamma_6$	$L_{II}O_{IV}$	20.216
0.613279	4	45 Rh	$K\alpha_1$	KL_{III}	20.2161
0.6146	1	90 Th		L_IO_I	20.174
0.614770	9	92 U	$L\gamma_1$	$L_{II}N_{IV}$	20.1671
0.6160	1	90 Th		$L_IN_{VI,VII}$	20.128

Wavelength Å*	p.e.	Element	Designation		keV
0.616	1	93 Np	$L\gamma_5$	$L_{II}N_I$	20.12
0.6169	1	91 Pa	$L\gamma_3$	L_IN_{III}	20.098
0.617630	4	45 Rh	$K\alpha_2$	KL_{II}	20.0737
0.61978	1	42 Mo	K	Abs. Edge	20.0039
0.62001	9	42 Mo	$K\beta_4^I$	$KN_{IV,V}$	19.996
0.62099	2	42 Mo	$K\beta_2$	$KN_{II,III}$	19.9652
0.62107	5	42 Mo	$K\beta_2^{II}$	KN_{II}	19.963
0.6228	1	92 U		$L_{II}N_{III}$	19.907
0.6239	1	91 Pa	$L\gamma_2$	L_IN_{II}	19.872
0.62636	9	90 Th	$L\gamma_{11}$	L_IN_V	19.794
0.62692	5	42 No	$K\beta_5^I$	KM_V	19.776
0.62708	5	42 Mo	$K\beta_5^{II}$	KM_{IV}	19.771
0.6276	1	90 Th		L_IN_{IV}	19.755
0.6299	1	90 Th	L_{II}	Abs. Edge	19.683
0.62991	9	90 Th		$L_{II}P_{IV}$	19.682
0.6312	1	90 Th		$L_{II}P_{II,III}$	19.642
0.6316	1	90 Th		$L_{II}P_I$	19.629
0.632288	9	42 Mo	$K\beta_1$	KM_{III}	19.6083
0.63258	4	90 Th	$L\gamma_6$	$L_{II}O_{IV}$	19.599
0.632872	2	42 Mo	$K\beta_3$	KM_{II}	19.5903
0.63358	9	91 Pa	$L\gamma_1$	$L_{II}N_{IV}$	19.568
0.63557	2	92 U	$L\gamma_5$	$L_{II}N_I$	19.5072
0.63559	4	90 Th	$L\gamma_3$	L_IN_{III}	19.507
0.6356	1	90 Th		$L_{II}O_{III}$	19.506
0.6369	1	90 Th		$L_{II}O_{II}$	19.466
0.63898	5	90 Th	$L\gamma_8$	$L_{II}O_I$	19.403
0.64064	9	90 Th	Lv	$L_{II}N_{VI}$	19.353
0.6416	1	94 Pu	$L\beta_9$	L_IM_V	19.323
0.64221	4	90 Th	$L\gamma_2$	L_IN_{II}	19.305
0.643083	4	44 Ru	$K\alpha_1$	KL_{III}	19.2792
0.6445	1	88 Ra	L_I	Abs. Edge	19.236
0.64513	5	88 Ra	$L\gamma_{13}$	$L_IP_{II,III}$	19.218
0.6468	1	88 Ra		$L_IO_{IV,V}$	19.167
0.647408	5	44 Ru	$K\alpha_2$	KL_{II}	19.1504
0.64755	5	90 Th		L_IN_I	19.146
0.6482	1	94 Pu	$L\beta_{10}$	L_IM_{IV}	19.126
0.64891	2	95 Am	$L\beta_3$	L_IM_{III}	19.1059
0.64965	5	88 Ra	$L\gamma_4$	L_IO_{III}	19.084
0.65131	5	88 Ra	$L\gamma_4'$	L_IO_{II}	19.036
0.6521	1	90 Th		$L_{II}N_V$	19.014
0.65298	1	41 Nb	K	Abs. Edge	18.9869
0.65313	3	90 Th	$L\gamma_1$	$L_{II}N_{IV}$	18.9825
0.65318	5	41 Nb	$K\beta_4$	$KN_{IV,V}$	18.981
0.65416	4	41 Nb	$K\beta_2$	$KN_{II,III}$	18.953
0.6550	1	91 Pa	$L\gamma_5$	$L_{II}N_I$	18.930
0.657655	9	95 Am	$L\beta_1$	$L_{II}M_{IV}$	18.8520
0.6620	1	90 Th		$L_{II}N_{III}$	18.729
0.6654	1	88 Ra	$L\gamma_{11}$	L_IN_V	18.633
0.66576	2	41 Nb	$K\beta_1$	KM_{III}	18.6225
0.66634	3	41 Nb	$K\beta_3$	KM_{II}	18.6063
0.6666	1	88 Ra		L_IN_{IV}	18.600
0.66871	2	94 Pu	$L\beta_3$	L_IM_{III}	18.5405
0.6707	1	88 Ra	L_{II}	Abs. Edge	18.486
0.6714	1	88 Ra		$L_{II}P_{II,III}$	18.466
0.6724	1	88 Ra		$L_{II}P_I$	18.439
0.67328	5	88 Ra	$L\gamma_6$	$L_{II}O_{IV}$	18.414
0.67351	9	89 Ac	$L\gamma_1$	$L_{II}N_{IV}$	18.408
0.67383	2	95 Am	$L\beta_5$	$L_{III}O_{IV,V}$	18.3996
0.67491	4	90 Th	$L\gamma_5$	$L_{II}N_I$	18.370
0.67502	3	43 Tc	$K\alpha_1$	KL_{III}	18.3671
0.67538	5	88 Ra	$L\gamma_3$	L_IN_{III}	18.357
0.6764	1	88 Ra		$L_{II}O_{III}$	18.330
0.67772	2	94 Pu	$L\beta_1$	$L_{II}M_{IV}$	18.2937
0.6780	1	88 Ra		$L_{II}O_{II}$	18.286
0.67932	3	43 Tc	$K\alpha_2$	KL_{II}	18.2508
0.6801	1	88 Ra	$L\gamma_8$	$L_{II}O_I$	18.230
0.681014	8	92 U	$L\beta_9$	L_IM_V	18.2054
0.68199	5	88 Ra	$L\gamma_2$	L_IN_{II}	18.179
0.68639	2	95 Am	$L\beta_4$	L_IM_{II}	18.0627
0.6867	1	94 Pu	L_{III}	Abs. Edge	18.054
0.6874	1	88 Ra		L_IN_I	18.036
0.68760	5	92 U	$L\beta_{10}$	L_IM_{IV}	18.031
0.68883	1	40 Zr	K	Abs. Edge	17.9989
0.68901	5	40 Zr	$K\beta_4$	$KN_{IV,V}$	17.994
0.68920	9	93 Np	$L\beta_3$	L_IM_{III}	17.989
0.68993	4	40 Zr	$K\beta_2$	$KN_{II,III}$	17.970
0.69068	2	94 Pu	$L\beta_5$	$L_{III}O_{IV,V}$	17.9506
0.6932	1	88 Ra		$L_{II}N_V$	17.884
0.69463	5	88 Ra	$L\gamma_1$	$L_{II}N_{IV}$	17.849
0.6959	1	40 Zr	$K\beta_5$	$KM_{IV,V}$	17.815
0.698478	9	93 Np	$L\beta_1$	$L_{II}M_{IV}$	17.7502
0.7003	1	94 Pu	$L\beta_7$	$L_{III}O_I$	17.705
0.701390	9	95 Am	$L\beta_2$	$L_{III}N_V$	17.6765
0.70173	3	40 Zr	$K\beta_1$	KM_{III}	17.6678
0.7018	1	91 Pa	$L\beta_9$	L_IM_V	17.667
0.70228	4	40 Zr	$K\beta_3$	KM_{II}	17.654
0.7031	1	94 Pu	Lu	$L_{III}N_{VI,VII}$	17.635
0.70341	2	95 Am	$L\beta_{15}$	$L_{III}N_{IV}$	17.6258
0.7043	1	88 Ra		$L_{II}N_{III}$	17.604
0.70620	2	94 Pu	$L\beta_4$	L_IM_{II}	17.5560
0.70814	2	93 Np	$L\beta_5$	$L_{III}O_{IV,V}$	17.5081
0.7088	2	91 Pa	$L\beta_{10}$	L_IM_{IV}	17.492
0.709300	1	42 Mo	$K\alpha_1$	KL_{III}	17.47934
0.71029	2	92 U	$L\beta_3$	L_IM_{III}	17.4550
0.713590	6	42 Mo	$K\alpha_2$	KL_{II}	17.3743
0.71652	9	87 Fr	$L\gamma_1$	$L_{II}N_{IV}$	17.303
0.71774	5	88 Ra	$L\gamma_5$	$L_{II}N_I$	17.274
0.71851	2	94 Pu	$L\beta_2$	$L_{III}N_V$	17.2553
0.719984	8	92 U	$L\beta_1$	$L_{II}M_{IV}$	17.2200
0.7205	1	94 Pu	$L\beta_{15}$	$L_{III}N_{IV}$	17.208
0.7223	1	92 U	L_{III}	Abs. Edge	17.165
0.72240	5	92 U		$L_{III}P_{IV,V}$	17.162
0.7234	1	90 Th	$L\beta_9$	L_IM_V	17.139
0.72426	5	92 U		$L_{III}P_{II,III}$	17.118
0.72521	5	92 U		$L_{III}P_I$	17.096
0.726305	9	92 U	$L\beta_5$	$L_{III}O_{IV,V}$	17.0701
0.72671	2	93 Np	$L\beta_4$	L_IM_{II}	17.0607
0.72766	5	39 Y	K	Abs. Edge	17.038
0.72776	5	39 Y	$K\beta_4$	$KN_{IV,V}$	17.036
0.72864	4	39 Y	$K\beta_2$	$KN_{II,III}$	17.0154
0.7301	1	90 Th	$L\beta_{10}$	L_IM_{IV}	16.981
0.7309	1	92 U		$L_{III}O_{III}$	16.962
0.73230	5	91 Pa	$L\beta_3$	L_IM_{III}	16.930
0.7333	1	92 U		$L_{III}O_{II}$	16.907

Wavelength Å*	p.e.	Element		Designation	keV
0.73418	2	95 Am	$L\beta_6$	$L_{III}N_I$	16.8870
0.7345	1	39 Y	$K\beta_5$	$KM_{IV,V}$	16.879
0.73602	6	92 U	$L\beta_7$	$L_{III}O_I$	16.845
0.736230	9	93 Np	$L\beta_2$	$L_{III}N_V$	16.8400
0.738603	9	92 U	Lu	$L_{III}N_{VI,VII}$	16.7859
0.73928	9	86 Rn	$L\gamma_1$	$L_{II}N_{IV}$	16.770
0.74072	2	39 Y	$K\beta_1$	KM_{III}	16.7378
0.74126	3	39 Y	$K\beta_3$	KM_{II}	16.7258
0.74232	5	91 Pa	$L\beta_1$	$L_{II}M_{IV}$	16.702
0.74503	5	92 U	$L\beta_{17}$	$L_{II}M_{III}$	16.641
0.7452	2	91 Pa	$L\beta_5$	$L_{III}O_{IV,V}$	16.636
0.74620	1	41 Nb	$K\alpha_1$	KL_{III}	16.6151
0.747985	9	92 U	$L\beta_4$	L_IM_{II}	16.5753
0.75044	1	41 Nb	$K\alpha_2$	KL_{II}	16.5210
0.75148	2	94 Pu	$L\beta_6$	$L_{III}N_I$	16.4983
0.7546	2	91 Pa	$L\beta_7$	$L_{III}O_I$	16.431
0.754681	9	92 U	$L\beta_2$	$L_{III}N_V$	16.4283
0.75479	3	90 Th	$L\beta_3$	L_IM_{III}	16.4258
0.756642	9	92 U	$L\beta_{15}$	$L_{III}N_{IV}$	16.3857
0.75690	3	83 Bi	$L\gamma_{13}$	$L_IP_{II,III}$	16.3802
0.7571	1	83 Bi	L_I	Abs. Edge	16.376
0.7579	1	90 Th		$L_{II}M_V$	16.359
0.75791	5	83 Bi		$L_IO_{IV,V}$	16.358
0.7591	1	94 Pu	$L\eta$	$L_{II}M_I$	16.333
0.7607	1	90 Th	L_{III}	Abs. Edge	16.299
0.76087	9	90 Th		$L_{III}P_{IV,V}$	16.295
0.76087	3	83 Bi	$L\gamma_4$	L_IO_{III}	16.2947
0.76198	3	83 Bi	$L\gamma_4'$	L_IO_{II}	16.2709
0.7625	2	90 Th		$L_{III}P_{II,III}$	16.260
0.76289	9	85 At	$L\gamma_1$	$L_{II}N_{IV}$	16.251
0.76338	5	90 Th		$L_{III}P_I$	16.241
0.7641	5	83 Bi		$L_IN_{VI,VII}$	16.23
0.7645	2	84 Po	$L\gamma_6$	$L_{II}O_{IV}$	16.218
0.76468	5	90 Th	$L\beta_5$	$L_{III}O_{IV,V}$	16.213
0.765210	9	90 Th	$L\beta_1$	$L_{II}M_{IV}$	16.2022
0.76857	5	88 Ra	$L\beta_9$	L_IM_V	16.131
0.769	1	93 Np	$L\beta_6$	$L_{III}N_I$	16.13
0.7690	1	90 Th		$L_{III}O_{III}$	16.123
0.7691	1	92 U		$L_{III}N_{III}$	16.120
0.76973	5	38 Sr	K	Abs. Edge	16.107
0.7699	1	91 Pa	$L\beta_4$	L_IM_{II}	16.104
0.76989	5	38 Sr	$K\beta_4$	$KN_{IV,V}$	16.104
0.77081	3	38 Sr	$K\beta_2$	$KN_{II,III}$	16.0846
0.7713	1	90 Th		$L_{III}O_{II}$	16.074
0.772	1	84 Po	$L\gamma_2$	L_IN_{II}	16.07
0.7737	1	91 Pa	$L\beta_2$	$L_{III}N_V$	16.024
0.77437	4	90 Th	$L\beta_7$	$L_{III}O_I$	16.0105
0.77546	5	88 Ra	$L\beta_{10}$	L_IM_{IV}	15.988
0.7764	1	38 Sr	$K\beta_5$	$KM_{IV,V}$	15.969
0.77661	5	90 Th	Lu	$L_{III}N_{VI,VII}$	15.964
0.77728	5	83 Bi	$L\gamma_{11}$	L_IN_V	15.951
0.77822	9	89 Ac	$L\beta_3$	L_IM_{III}	15.931
0.77954	5	83 Bi		L_IN_{IV}	15.904
0.78017	9	92 U		$L_{III}N_{II}$	15.892
0.7809	2	93 Np	$L\eta$	$L_{II}M_I$	15.876
0.78196	5	82 Pb	L_I	Abs. Edge	15.855
0.78257	7	82 Pb		$L_IO_{IV,V}$	15.843
0.78292	2	38 Sr	$K\beta_1$	KM_{III}	15.8357
0.78345	3	38 Sr	$K\beta_3$	KM_{II}	15.8249
0.7858	1	82 Pb	$L\gamma_4$	L_IO_{III}	15.777
0.78593	1	40 Zr	$K\alpha_1$	KL_{III}	15.7751
0.78706	7	82 Pb	$L\gamma_4'$	L_IO_{II}	15.752
0.78748	9	84 Po	$L\gamma_1$	$L_{II}N_{IV}$	15.744
0.78838	2	92 U	$L\beta_6$	$L_{III}N_I$	15.7260
0.7884	1	82 Pb		$L_IN_{VI,VII}$	15.725
0.7887	1	83 Bi	L_{II}	Abs. Edge	15.719
0.78903	9	89 Ac	$L\beta_1$	$L_{II}M_{IV}$	15.713
0.78917	5	83 Bi	$L\gamma_3$	L_IN_{III}	15.7102
0.7897	1	82 Pb		L_IO_I	15.699
0.79015	1	40 Zr	$K\alpha_2$	KL_{II}	15.6909
0.79043	3	83 Bi	$L\gamma_6$	$L_{II}O_{IV}$	15.6853
0.79257	4	90 Th	$L\beta_4$	L_IM_{II}	15.6429
0.79257	4	90 Th	$L\beta_{17}$	$L_{II}M_{III}$	15.6429
0.79354	3	90 Th	$L\beta_2$	$L_{III}M_V$	15.6237
0.79384	5	83 Bi		$L_{II}O_{III}$	15.6178
0.79539	5	90 Th	$L\beta_{15}$	$L_{III}N_{IV}$	15.5875
0.79565	3	83 Bi	$L\gamma_2$	L_IN_{II}	15.5824
0.79721	9	83 Bi	Lv	$L_{II}N_{VI}$	15.552
0.7973	1	83 Bi	$L\gamma_8$	$L_{II}O_I$	15.551
0.8022	1	83 Bi		L_IN_I	15.456
0.80233	9	82 Pb	$L\gamma_{11}$	L_IN_V	15.453
0.80273	5	88 Ra	$L\beta_3$	L_IM_{III}	15.4449
0.8028	1	88 Ra	L_{III}	Abs. Edge	15.444
0.80364	7	82 Pb		L_IN_{IV}	15.427
0.8038	1	88 Ra		$L_{III}P_{II,III}$	15.425
0.8050	1	88 Ra		$L_{III}P_I$	15.402
0.80509	2	92 U	$L\eta$	$L_{II}M_I$	15.3997
0.80627	5	88 Ra	$L\beta_5$	$L_{III}O_{IV,V}$	15.3771
0.8079	1	91 Pa	$L\beta_6$	$L_{III}N_I$	15.347
0.8081	1	81 Tl	L_I	Abs. Edge	15.343
0.8082	1	90 Th		$L_{III}N_{III}$	15.341
0.80861	5	81 Tl		$L_IO_{IV,V}$	15.3327
0.81163	9	90 Th		L_IM_I	15.276
0.81184	5	81 Tl	$L\gamma_4$	L_IO_{III}	15.2716
0.81308	5	81 Tl	$L\gamma_4'$	L_IO_{II}	15.2482
0.81311	2	83 Bi	$L\gamma_1$	$L_{II}N_{IV}$	15.2477
0.81375	5	88 Ra	$L\beta_1$	$L_{II}M_{IV}$	15.2358
0.8147	1	82 Pb	$L\gamma_3$	L_IN_{III}	15.218
0.81538	5	82 Pb	L_{II}	Abs. Edge	15.2053
0.8154	2	37 Rb	$K\beta_4$	$KN_{IV,V}$	15.205
0.81554	5	37 Rb	K	Abs. Edge	15.2023
0.8158	1	81 Tl		L_IO_I	15.198
0.81583	5	82 Pb		$L_{II}P_I$	15.1969
0.8162	1	88 Ra	$L\beta_7$	$L_{III}O_I$	15.190
0.81645	3	37 Rb	$K\beta_2$	$KN_{II,III}$	15.1854
0.81683	5	82 Pb	$L\gamma_6$	$L_{II}O_{IV}$	15.1783
0.8186	1	88 Ra	Lu	$L_{III}N_{VI,VII}$	15.146
0.8190	2	90 Th		$L_{III}N_{II}$	15.138
0.8200	1	82 Pb		$L_{II}O_{III}$	15.120
0.8210	2	82 Pb	$L\gamma_2$	L_IN_{II}	15.101
0.8219	1	37 Rb	$K\beta_5$	$KM_{IV,V}$	15.085
0.82327	7	82 Pb	Lv	$L_{II}N_{VI}$	15.060
0.82365	5	82 Pb	$L\gamma_8$	$L_{II}O_I$	15.0527
0.8248	1	83 Bi		$L_{II}N_{III}$	15.031

Wavelength Å*	p.e.	Element		Designation	keV
0.82789	9	87 Fr	$L\beta_3$	L_IM_{III}	14.976
0.82790	8	90 Th	$L\beta_6$	$L_{III}N_I$	14.975
0.82859	7	82 Pb		L_IN_I	14.963
0.82868	2	37 Rb	$K\beta_1$	KM_{III}	14.9613
0.82879	5	81 Tl	$L\gamma_{11}$	L_IN_V	14.9593
0.82884	1	39 Y	$K\alpha_1$	KL_{III}	14.9584
0.82921	3	37 Rb	$K\beta_3$	KM_{II}	14.9517
0.8295	1	91 Pa	$L\eta$	$L_{II}M_I$	14.946
0.83001	7	81 Tl		L_IN_{IV}	14.937
0.83305	1	39 Y	$K\alpha_2$	KL_{II}	14.8829
0.8338	1	90 Th		$L_{II}M_{II}$	14.869
0.8344	9	83 Bi		$L_{II}N_{II}$	14.86
0.8350	2	80 Hg		$L_IO_{IV,V}$	14.847
0.8353	1	80 Hg	L_I	Abs. Edge	14.842
0.83537	5	88 Ra	$L\beta_2$	$L_{III}N_V$	14.8414
0.83722	5	88 Ra	$L\beta_{15}$	$L_{III}N_{IV}$	14.8086
0.8382	2	82 Pb		$L_{II}N_V$	14.791
0.83894	7	80 Hg	$L\gamma_4$	L_IO_{III}	14.778
0.83923	5	83 Bi	$L\gamma_5$	$L_{II}N_I$	14.7732
0.83940	9	87 Fr	$L\beta_1$	$L_{II}M_{IV}$	14.770
0.83973	3	82 Pb	$L\gamma_1$	$L_{II}N_{IV}$	14.7644
0.84013	7	80 Hg	$L\gamma_4'$	L_IO_{II}	14.757
0.84071	5	88 Ra	$L\beta_4$	L_IM_{II}	14.7472
0.84130	4	81 Tl	$L\gamma_3$	L_IN_{III}	14.7368
0.8434	1	81 Tl	L_{II}	Abs. Edge	14.699
0.8438	1	88 Ra	$L\beta_{17}$	$L_{II}M_{III}$	14.692
0.8442	2	81 Tl	$L\gamma_6$	$L_{II}O_{IV}$	14.685
0.8452	2	80 Hg		L_IO_I	14.670
0.84773	5	81 Tl	$L\gamma_2$	L_IN_{II}	14.6251
0.848187	9	95 Am	$L\alpha_1$	$L_{III}M_V$	14.6172
0.8490	1	81 Tl		$L_{II}O_{II}$	14.604
0.85048	5	81 Tl	Lv	$L_{II}N_{VI}$	14.5777
0.8512	1	88 Ra		$L_{III}N_{III}$	14.566
0.8513	2	81 Tl	$L\gamma_8$	$L_{II}O_I$	14.564
0.85192	7	82 Pb		L_IN_{III}	14.553
0.85436	9	86 Rn	$L\beta_3$	L_IM_{III}	14.512
0.85446	4	90 Th	$L\eta$	$L_{II}M_I$	14.5099
0.8549	1	81 Tl		L_IN_I	14.503
0.85657	7	80 Hg	$L\gamma_{11}$	L_IN_V	14.474
0.858	2	87 Fr	$L\beta_2$	$L_{III}N_V$	14.45
0.8585	3	82 Pb		$L_{II}N_{II}$	14.442
0.860266	9	95 Am	$L\alpha_2$	$L_{III}M_{IV}$	14.4119
0.8618	1	88 Ra		$L_{III}N_{II}$	14.387
0.86376	5	79 Au	L_I	Abs. Edge	14.3537
0.86400	5	79 Au		$L_IO_{IV,V}$	14.3497
0.8653	2	36 Kr	$K\beta_4$	$KN_{IV,V}$	14.328
0.86552	1	36 Kr	K	Abs. Edge	14.3244
0.86605	9	86 Rn	$L\beta_1$	$L_{II}M_{IV}$	14.316
0.8661	1	36 Kr	$K\beta_2$	$KN_{II,III}$	14.315
0.86655	5	82 Pb	$L\gamma_5$	$L_{II}N_I$	14.3075
0.86703	4	79 Au	$L\gamma_4$	L_IO_{III}	14.2996
0.86752	3	81 Tl	$L\gamma_1$	$L_{II}N_{IV}$	14.2915
0.86816	4	79 Au	$L\gamma_4'$	L_IO_{II}	14.2809
0.86830	2	94 Pu	$L\alpha_1$	$L_{III}M_V$	14.2786
0.86915	7	80 Hg	$L\gamma_3$	L_IN_{III}	14.265
0.87074	5	79 Au		L_IO_I	14.2385
0.8708	2	36 Kr	$K\beta_5$	$KM_{IV,V}$	14.238
0.87088	5	88 Ra	$L\beta_6$	$L_{III}N_I$	14.2362
0.8722	1	80 Hg	L_{II}	Abs. Edge	14.215
0.87319	7	80 Hg	$L\gamma_6$	$L_{II}O_{IV}$	14.199
0.87526	1	38 Sr	$K\alpha_1$	KL_{III}	14.1650
0.87544	7	80 Hg	$L\gamma_2$	L_IN_{II}	14.162
0.8758	1	80 Hg		$L_{II}O_{III}$	14.156
0.8784	1	80 Hg		$L_{II}O_{II}$	14.114
0.8785	1	36 Kr	$K\beta_1$	KM_{III}	14.112
0.87885	7	80 Hg	Lv	$L_{II}N_{VI}$	14.107
0.8790	1	36 Kr	$K\beta_3$	KM_{II}	14.104
0.87943	1	38 Sr	$K\alpha_2$	KL_{II}	14.0979
0.87995	7	80 Hg	$L\gamma_8$	$L_{II}O_I$	14.090
0.87996	5	81 Tl		$L_{II}N_{III}$	14.0893
0.88028	2	94 Pu	$L\alpha_2$	$L_{III}M_{IV}$	14.0842
0.88135	9	85 At	$L\beta_3$	L_IM_{III}	14.067
0.8827	2	80 Hg		L_IN_I	14.045
0.88433	7	79 Au	$L\gamma_{11}$	L_IN_V	14.020
0.88563	7	79 Au		L_IN_{IV}	13.999
0.8882	2	81 Tl		$L_{II}M_{II}$	13.959
0.889128	9	93 Np	$L\alpha_1$	$L_{III}M_V$	13.9441
0.8931	1	78 Pt	L_I	Abs. Edge	13.883
0.8934	1	78 Pt		L_IO_V	13.878
0.89349	9	85 At	$L\beta_1$	$L_{II}M_{IV}$	13.876
0.8943	1	78 Pt		L_IO_{IV}	13.864
0.89500	4	81 Tl	$L\gamma_5$	$L_{II}N_I$	13.8526
0.89646	5	80 Hg	$L\gamma_1$	$L_{II}N_{IV}$	13.8301
0.89659	4	78 Pt	$L\gamma_4$	L_IO_{III}	13.8281
0.89747	4	78 Pt	$L\gamma_4'$	L_IO_{II}	13.8145
0.89783	5	79 Au	$L\gamma_3$	L_IN_{III}	13.8090
0.89791	3	83 Bi	$L\beta_9$	L_IM_V	13.8077
0.8995	2	78 Pt		L_IO_I	13.784
0.8996	2	84 Po	$L\beta_5$	$L_{III}O_{IV,V}$	13.782
0.901045	9	93 Np	$L\alpha_2$	$L_{III}M_{IV}$	13.7597
0.90259	5	79 Au	L_{II}	Abs. Edge	13.7361
0.90297	3	79 Au	$L\gamma_6$	$L_{II}O_{IV}$	13.7304
0.90434	3	79 Au	$L\gamma_2$	L_IN_{II}	13.7095
0.90495	4	83 Bi	$L\beta_{10}$	L_IM_{IV}	13.7002
0.90638	7	79 Au		$L_{II}O_{III}$	13.679
0.90742	5	88 Ra	$L\eta$	$L_{II}M_I$	13.6630
0.90746	7	79 Au		$L_{II}O_{II}$	13.662
0.90837	5	79 Au	Lv	$L_{II}N_{VI}$	13.6487
0.90894	7	80 Hg		$L_{II}N_{III}$	13.640
0.9091	3	84 Po	$L\beta_3$	L_IM_{III}	13.638
0.90989	5	79 Au	$L\gamma_8$	$L_{II}O_I$	13.6260
0.910639	9	92 U	$L\alpha_1$	$L_{III}M_V$	13.6147
0.9131	1	79 Au		L_IN_I	13.578
0.9143	2	78 Pt	$L\gamma_{11}$	L_IN_V	13.560
0.9204	1	35 Br	K	Abs. Edge	13.470
0.92046	2	35 Br	$K\beta_2$	$KN_{II,III}$	13.4695
0.9220	2	84 Po	$L\beta_1$	$L_{II}M_{IV}$	13.447
0.922558	9	92 U	$L\alpha_2$	$L_{III}M_{IV}$	13.4388
0.9234	1	83 Bi	L_{III}	Abs. Edge	13.426
0.9236	1	77 Ir	L_I	Abs. Edge	13.423
0.92413	4	83 Bi		$L_{III}P_{II,III}$	13.4159
0.9243	3	77 Ir		$L_IO_{IV,V}$	13.413
0.92453	7	80 Hg	$L\gamma_5$	$L_{II}N_I$	13.410
0.9255	1	35 Br	$K\beta_5$	$KM_{IV,V}$	13.396

Wavelength Å*	p.e.	Element	Designation		keV
0.925553	9	37 Rb	$K\alpha_1$	KL_{III}	13.3953
0.92556	3	83 Bi	$L\beta_5$	$L_{III}O_{IV,V}$	13.3953
0.92650	3	79 Au	$L\gamma_1$	$L_{II}N_{IV}$	13.3817
0.9268	1	82 Pb	$L\beta_9$	L_IM_V	13.377
0.92744	3	77 Ir	$L\gamma_4$	L_IO_{III}	13.3681
0.92791	5	78 Pt	$L\gamma_3$	L_IN_{III}	13.3613
0.92831	3	77 Ir	$L\gamma_4'$	L_IO_{II}	13.3555
0.92937	5	84 Po	$L\beta_2$	$L_{III}N_V$	13.3404
0.92969	1	37 Rb	$K\alpha_2$	KL_{II}	13.3358
0.9302	2	83 Bi		$L_{III}O_{III}$	13.328
0.9312	2	84 Po	$L\beta_{15}$	$L_{III}N_{IV}$	13.314
0.9323	2	83 Bi		$L_{III}O_{II}$	13.298
0.93279	2	35 Br	$K\beta_1$	KM_{III}	13.2914
0.93284	5	91 Pa	$L\alpha_1$	$L_{III}M_V$	13.2907
0.93327	5	35 Br	$K\beta_3$	KM_{II}	13.2845
0.9339	2	82 Pb	$L\beta_{10}$	L_IM_{IV}	13.275
0.93414	5	78 Pt	L_{II}	Abs. Edge	13.2723
0.9342	2	78 Pt	$L\gamma_6$	$L_{II}O_{IV}$	13.271
0.93427	5	78 Pt	$L\gamma_2$	L_IN_{II}	13.2704
0.93505	5	83 Bi	$L\beta_7$	$L_{III}O_I$	13.2593
0.93505	5	83 Bi	Lu	$L_{III}N_{VI,VII}$	13.2593
0.93855	3	83 Bi	$L\beta_3$	L_IM_{III}	13.2098
0.93931	5	78 Pt	Lv	$L_{II}N_{VI}$	13.1992
0.9402	2	79 Au		$L_{II}N_{III}$	13.186
0.9411	1	78 Pt	$L\gamma_8$	$L_{II}O_I$	13.173
0.94419	5	83 Bi		$L_{II}M_V$	13.1310
0.9446	2	77 Ir	$L\gamma_{11}$	L_IN_V	13.126
0.94482	5	91 Pa	$L\alpha_2$	$L_{III}M_{IV}$	13.1222
0.9455	2	78 Pt		L_IN_I	13.113
0.9459	2	77 Ir		L_IN_{IV}	13.108
0.9475	3	84 Po	$L\beta_4$	L_IM_{II}	13.086
0.95073	5	82 Pb	L_{III}	Abs. Edge	13.0406
0.95118	7	82 Pb		$L_{III}P_{II,III}$	13.0344
0.951978	9	83 Bi	$L\beta_1$	$L_{II}M_{IV}$	13.0235
0.9526	1	82 Pb	$L\beta_5$	$L_{III}O_{IV,V}$	13.015
0.95518	4	83 Bi	$L\beta_2$	$L_{III}N_V$	12.9799
0.95559	3	79 Au	$L\gamma_5$	$L_{II}N_I$	12.9743
0.9558	1	76 Os	L_I	Abs. Edge	12.972
0.95600	3	90 Th	$L\alpha_1$	$L_{III}M_V$	12.9687
0.95603	5	76 Os		$L_IO_{IV,V}$	12.9683
0.95675	7	81 Tl	$L\beta_9$	L_IM_V	12.9585
0.95702	5	83 Bi	$L\beta_{15}$	$L_{III}N_{IV}$	12.9549
0.9578	1	82 Pb		$L_{III}O_{III}$	12.945
0.95797	3	78 Pt	$L\gamma_1$	$L_{II}N_{IV}$	12.9420
0.9586	1	82 Pb		$L_{III}O_{II}$	12.934
0.95931	5	77 Ir	$L\gamma_3$	L_IN_{III}	12.9240
0.95938	8	76 Os	$L\gamma_4$	L_IO_{III}	12.923
0.96033	8	76 Os	$L\gamma_4'$	L_IO_{II}	12.910
0.96133	7	82 Pb	Lu	$L_{III}N_{VI,VII}$	12.8968
0.9620	1	82 Pb	$L\beta_7$	$L_{III}O_I$	12.888
0.96318	7	76 Os		L_IO_I	12.8721
0.9636	1	92 U	Ls	$L_{III}M_{III}$	12.866
0.96389	7	81 Tl	$L\beta_{10}$	L_IM_{IV}	12.8626
0.96545	3	77 Ir	$L\gamma_2$	L_IN_{II}	12.8418
0.96708	4	77 Ir	$L\gamma_6$	$L_{II}O_{IV}$	12.8201
0.9671	1	77 Ir	L_{II}	Abs. Edge	12.820
0.9672	2	84 Po	$L\beta_6$	$L_{III}N_I$	12.819
0.96788	2	90 Th	$L\alpha_2$	$L_{III}M_{IV}$	12.8096
0.96911	7	82 Pb	$L\beta_3$	L_IM_{III}	12.7933
0.96979	5	77 Ir		$L_{II}O_{III}$	12.7843
0.97161	6	77 Ir	Lv	$L_{II}N_{VI}$	12.7603
0.97173	4	78 Pt		$L_{II}N_{III}$	12.7588
0.97321	5	83 Bi		$L_{III}N_{III}$	12.7394
0.97409	3	77 Ir	$L\gamma_8$	$L_{II}O_I$	12.7279
0.9747	1	82 Pb		$L_{II}M_V$	12.720
0.9765	3	76 Os	$L\gamma_{11}$	L_IN_V	12.696
0.9766	2	77 Ir		L_IN_I	12.695
0.97690	4	83 Bi	$L\beta_4$	L_IM_{II}	12.6912
0.9772	3	76 Os		L_IN_{IV}	12.687
0.9792	2	78 Pt		$L_{II}N_{II}$	12.661
0.97926	5	81 Tl		$L_{III}P_{II,III}$	12.6607
0.9793	1	81 Tl	L_{III}	Abs. Edge	12.660
0.97974	1	34 Se	K	Abs. Edge	12.6545
0.97992	5	34 Se	$K\beta_2$	$KN_{II,III}$	12.6522
0.97993	5	89 Ac	$L\alpha_1$	$L_{III}M_V$	12.6520
0.9801	1	36 Kr	$K\alpha_1$	KL_{III}	12.649
0.98058	3	81 Tl	$L\beta_5$	$L_{III}O_{IV,V}$	12.6436
0.98221	7	82 Pb	$L\beta_2$	$L_{III}N_V$	12.6226
0.98280	5	83 Bi		$L_{III}N_{II}$	12.6151
0.98291	3	82 Pb	$L\beta_1$	$L_{II}M_{IV}$	12.6137
0.98389	7	82 Pb	$L\beta_{15}$	$L_{III}N_{IV}$	12.6011
0.9841	1	36 Kr	$K\alpha_2$	KL_{II}	12.598
0.9843	1	34 Se	$K\beta_5$	$KM_{IV,V}$	12.595
0.98538	5	81 Tl		$L_{III}O_{III}$	12.5820
0.9871	2	80 Hg	$L\beta_9$	L_IM_V	12.560
0.98738	5	81 Tl		$L_{III}O_{II}$	12.5566
0.9877	2	78 Pt	$L\gamma_5$	$L_{II}N_I$	12.552
0.9888	1	81 Tl	Lu	$L_{III}N_{VI,VII}$	12.538
0.98913	5	83 Bi	$L\beta_{17}$	$L_{II}M_{III}$	12.5344
0.9894	1	75 Re	L_I	Abs. Edge	12.530
0.9900	1	75 Re		$L_IO_{IV,V}$	12.524
0.99017	5	81 Tl	$L\beta_7$	$L_{III}O_I$	12.5212
0.99085	3	77 Ir	$L\gamma_1$	$L_{II}N_{IV}$	12.5126
0.99178	5	89 Ac	$L\alpha_2$	$L_{III}M_{IV}$	12.5008
0.99186	5	76 Os	$L\gamma_3$	L_IN_{III}	12.4998
0.99218	3	34 Se	$K\beta_1$	KM_{III}	12.4959
0.99249	5	75 Re	$L\gamma_4$	L_IO_{III}	12.4920
0.99268	5	34 Se	$K\beta_3$	KM_{II}	12.4896
0.99331	3	83 Bi	$L\beta_6$	$L_{III}N_I$	12.4816
0.99334	5	75 Re	$L\gamma_4'$	L_IO_{II}	12.4813
0.9962	2	80 Hg	$L\beta_{10}$	L_IM_{IV}	12.446
0.9965	1	75 Re		L_IO_I	12.442
0.99805	5	76 Os	$L\gamma_2$	L_IN_{II}	12.4224
1.0005	1	82 Pb		$L_{III}N_{III}$	12.392
1.0005	9	83 Bi		L_IM_I	12.39
1.00062	3	81 Tl	$L\beta_3$	L_IM_{III}	12.3904
1.00107	5	76 Os	$L\gamma_6$	$L_{II}O_{IV}$	12.3848
1.0012	6	95 Am	Ll	$L_{III}M_I$	12.384
1.0014	1	76 Os	L_{II}	Abs. Edge	12.381
1.0047	2	76 Os		$L_{II}O_{III}$	12.340
1.00473	5	88 Ra	$L\alpha_1$	$L_{III}M_V$	12.3397
1.0050	2	76 Os	Lv	$L_{II}N_{VI}$	12.337
1.0054	3	77 Ir		$L_{II}N_{III}$	12.332
1.00722	5	81 Tl		$L_{II}M_V$	12.3093

Wavelength Å*	p.e.	Element		Designation	keV
1.0075	1	82 Pb	$L\beta_4$	$L_{I}M_{II}$	12.306
1.00788	5	76 Os	$L\gamma_8$	$L_{II}O_{I}$	12.3012
1.0091	1	80 Hg	L_{III}	Abs. Edge	12.286
1.00987	7	80 Hg	$L\beta_5$	$L_{III}O_{IV,V}$	12.2769
1.01031	3	81 Tl	$L\beta_2$	$L_{III}N_{V}$	12.2715
1.01040	7	82 Pb		$L_{III}N_{II}$	12.2705
1.0108	1	75 Re	$L\gamma_{11}$	$L_{I}N_{V}$	12.266
1.0112	1	90 Th	Ls	$L_{III}M_{III}$	12.261
1.0119	1	75 Re		$L_{I}N_{IV}$	12.252
1.0120	2	77 Ir		$L_{II}N_{II}$	12.251
1.01201	3	81 Tl	$L\beta_{15}$	$L_{III}N_{IV}$	12.2510
1.01404	7	80 Hg		$L_{III}O_{III}$	12.2264
1.01513	4	81 Tl	$L\beta_1$	$L_{II}M_{IV}$	12.2133
1.01558	7	80 Hg		$L_{III}O_{II}$	12.2079
1.01656	5	88 Ra	$L\alpha_2$	$L_{III}M_{IV}$	12.1962
1.01674	7	80 Hg	Lu	$L_{III}N_{VII}$	12.1940
1.01769	7	80 Hg	Lu'	$L_{III}N_{VI}$	12.1826
1.01937	7	80 Hg	$L\beta_7$	$L_{III}O_{I}$	12.1625
1.02063	7	79 Au	$L\beta_9$	$L_{I}M_{V}$	12.1474
1.0210	1	82 Pb	$L\beta_6$	$L_{III}N_{I}$	12.143
1.02175	5	77 Ir	$L\gamma_5$	$L_{II}N_{I}$	12.1342
1.0223	1	82 Pb	$L\beta_{17}$	$L_{II}M_{III}$	12.127
1.0226	1	94 Pu	Ll	$L_{III}M_{I}$	12.124
1.02467	5	74 W	L_{I}	Abs. Edge	12.0996
1.0250	2	74 W		$L_{I}O_{IV,V}$	12.095
1.02503	5	76 Os	$L\gamma_1$	$L_{II}N_{IV}$	12.0953
1.02613	7	75 Re	$L\gamma_3$	$L_{I}N_{III}$	12.0824
1.02775	3	74 W	$L\gamma_4$	$L_{I}O_{III}$	12.0634
1.02789	7	79 Au	$L\beta_{10}$	$L_{I}M_{IV}$	12.0617
1.0286	1	81 Tl		$L_{III}N_{III}$	12.053
1.02863	3	74 W	$L\gamma_4'$	$L_{I}O_{II}$	12.0530
1.03049	5	87 Fr	$L\alpha_1$	$L_{III}M_{V}$	12.0313
1.0317	3	74 W		$L_{I}O_{I}$	12.017
1.03233	5	75 Re	$L\gamma_2$	$L_{I}N_{II}$	12.0098
1.0323	2	82 Pb		$L_{I}M_{I}$	12.010
1.03358	7	80 Hg	$L\beta_3$	$L_{I}M_{III}$	11.9953
1.0346	9	83 Bi		$L_{I}M_{II}$	11.98
1.0347	1	92 U	Lt	$L_{III}M_{II}$	11.982
1.03699	9	75 Re	$L\gamma_6$	$L_{II}O_{IV}$	11.956
1.0371	1	75 Re	L_{II}	Abs. Edge	11.954
1.03876	7	79 Au		$L_{III}P_{II,III}$	11.9355
1.03918	3	81 Tl	$L\beta_4$	$L_{I}M_{II}$	11.9306
1.0397	1	75 Re		$L_{II}O_{III}$	11.925
1.03973	5	76 Os		$L_{II}N_{III}$	11.9243
1.03974	2	35 Br	$K\alpha_1$	KL_{III}	11.9242
1.03975	7	80 Hg	$L\beta_2$	$L_{III}N_{V}$	11.9241
1.04000	5	79 Au	L_{III}	Abs. Edge	11.9212
1.0404	1	75 Re	Lv	$L_{II}N_{VI}$	11.917
1.04044	3	79 Au	$L\beta_5$	$L_{III}O_{IV,V}$	11.9163
1.04151	7	80 Hg	$L\beta_{15}$	$L_{III}N_{IV}$	11.9040
1.0420	1	75 Re		$L_{I}N_{I}$	11.899
1.04230	5	87 Fr	$L\alpha_2$	$L_{III}M_{IV}$	11.8950
1.0428	6	93 Np	Ll	$L_{III}M_{I}$	11.890
1.04382	2	35 Br	$K\alpha_2$	KL_{II}	11.8776
1.04398	5	75 Re	$L\gamma_8$	$L_{II}O_{I}$	11.8758
1.0450	2	79 Au		$L_{III}O_{II,III}$	11.865
1.0450	1	33 As	K	Abs. Edge	11.865

Wavelength Å*	p.e.	Element		Designation	keV
1.04500	3	33 As	$K\beta_2$	$KN_{II,III}$	11.8642
1.0458	1	74 W	$L\gamma_{11}$	$L_{I}N_{V}$	11.856
1.0468	2	74 W		$L_{I}N_{IV}$	11.844
1.04752	5	79 Au	Lu	$L_{III}N_{VI,VII}$	11.8357
1.04868	5	80 Hg	$L\beta_1$	$L_{II}M_{IV}$	11.8226
1.0488	1	33 As	$K\beta_5$	$KM_{IV,V}$	11.822
1.04963	5	81 Tl	$L\beta_6$	$L_{III}N_{I}$	11.8118
1.04974	8	79 Au	$L\beta_7$	$L_{III}O_{I}$	11.8106
1.05446	5	78 Pt	$L\beta_9$	$L_{I}M_{V}$	11.7577
1.05609	7	81 Tl	$L\beta_{17}$	$L_{II}M_{III}$	11.7397
1.05693	5	76 Os	$L\gamma_5$	$L_{II}N_{I}$	11.7303
1.05723	5	86 Rn	$L\alpha_1$	$L_{III}M_{V}$	11.7270
1.05730	2	33 As	$K\beta_1$	KM_{III}	11.7262
1.05783	5	33 As	$K\beta_3$	KM_{II}	11.7203
1.0585	1	80 Hg		$L_{III}N_{III}$	11.713
1.05856	3	83 Bi	$L\eta$	$L_{II}M_{I}$	11.7122
1.06099	5	75 Re	$L\gamma_1$	$L_{II}N_{IV}$	11.6854
1.0613	1	73 Ta	L_{I}	Abs. Edge	11.682
1.06183	7	78 Pt	$L\beta_{10}$	$L_{I}M_{IV}$	11.6762
1.06192	9	73 Ta		$L_{I}O_{IV,V}$	11.6752
1.06200	6	74 W	$L\gamma_3$	$L_{I}N_{III}$	11.6743
1.06357	9	73 Ta		$L_{I}N_{VI,VII}$	11.6570
1.0644	2	82 Pb		$L_{II}M_{II}$	11.648
1.0644	2	81 Tl		$L_{I}M_{I}$	11.648
1.06467	3	73 Ta	$L\gamma_4$	$L_{I}O_{III}$	11.6451
1.0649	2	80 Hg		$L_{III}N_{II}$	11.642
1.06544	3	73 Ta	$L\gamma_4'$	$L_{I}O_{II}$	11.6366
1.06712	2	92 U	Ll	$L_{III}M_{I}$	11.6183
1.06771	9	73 Ta		$L_{I}O_{I}$	11.6118
1.06785	9	79 Au	$L\beta_3$	$L_{I}M_{III}$	11.6103
1.06806	3	74 W	$L\gamma_2$	$L_{I}N_{II}$	11.6080
1.06899	5	86 Rn	$L\alpha_2$	$L_{III}M_{IV}$	11.5979
1.07022	3	79 Au	$L\beta_2$	$L_{III}N_{V}$	11.5847
1.07188	5	79 Au	$L\beta_{15}$	$L_{III}N_{IV}$	11.5667
1.07222	7	80 Hg	$L\beta_4$	$L_{I}M_{II}$	11.5630
1.0723	1	78 Pt	L_{III}	Abs. Edge	11.562
1.0724	2	78 Pt	$L\beta_5$	$L_{III}O_{IV,V}$	11.561
1.07448	5	74 W	$L\gamma_6$	$L_{II}O_{IV}$	11.5387
1.0745	1	74 W	L_{II}	Abs. Edge	11.538
1.0756	2	79 Au		$L_{II}M_{V}$	11.526
1.0761	3	78 Pt		$L_{III}O_{II,III}$	11.521
1.0767	1	75 Re		$L_{I}N_{III}$	11.515
1.0771	1	74 W	Lv	$L_{II}N_{VI}$	11.510
1.07896	5	78 Pt	Lu	$L_{III}N_{VI,VII}$	11.4908
1.0792	2	74 W		$L_{II}O_{III}$	11.488
1.07975	7	80 Hg	$L\beta_6$	$L_{III}N_{I}$	11.4824
1.08009	9	90 Th	Lt	$L_{III}M_{II}$	11.4788
1.08113	4	74 W	$L\gamma_8$	$L_{II}O_{I}$	11.4677
1.08168	3	78 Pt	$L\beta_7$	$L_{III}O_{I}$	11.4619
1.08205	7	73 Ta	$L\gamma_{11}$	$L_{I}N_{V}$	11.4580
1.08353	3	79 Au	$L\beta_1$	$L_{II}M_{IV}$	11.4423
1.08377	7	73 Ta		$L_{I}N_{IV}$	11.4398
1.0839	1	75 Re		$L_{II}N_{II}$	11.438
1.08500	5	85 At	$L\alpha_1$	$L_{III}M_{V}$	11.4268
1.08975	5	77 Ir	$L\beta_9$	$L_{I}M_{V}$	11.3770
1.09026	7	79 Au		$L_{III}N_{III}$	11.3717
1.0908	1	91 Pa	Ll	$L_{III}M_{I}$	11.366

Wavelength Å*	p.e.	Element		Designation	keV
1.0916	5	80 Hg	$L\beta_{17}$	$L_{II}M_{III}$	11.358
1.09241	7	82 Pb	$L\eta$	$L_{II}M_{I}$	11.3493
1.09388	5	75 Re	$L\gamma_{5}$	$L_{II}N_{I}$	11.3341
1.09671	5	85 At	$L\alpha_{2}$	$L_{III}M_{IV}$	11.3048
1.09702	4	77 Ir	$L\beta_{10}$	$L_{I}M_{IV}$	11.3016
1.09855	3	74 W	$L\gamma_{1}$	$L_{II}N_{IV}$	11.2859
1.09936	4	73 Ta	$L\gamma_{3}$	$L_{I}N_{III}$	11.2776
1.0997	1	81 Tl		$L_{II}M_{II}$	11.274
1.0997	1	72 Hf	L_{I}	Abs. Edge	11.274
1.09968	7	79 Au		$L_{III}N_{II}$	11.2743
1.0999	2	80 Hg		$L_{I}M_{I}$	11.272
1.10086	9	72 Hf		$L_{I}O_{IV}$	11.2622
1.10200	3	78 Pt	$L\beta_{2}$	$L_{III}N_{V}$	11.2505
1.10303	5	72 Hf	$L\gamma_{4}$	$L_{I}O_{III}$	11.2401
1.10376	5	72 Hf	$L\gamma_{4}'$	$L_{I}O_{II}$	11.2326
1.10394	5	78 Pt	$L\beta_{3}$	$L_{I}M_{III}$	11.2308
1.10477	2	34 Se	$K\alpha_{1}$	KL_{III}	11.2224
1.1053	1	73 Ta	$L\gamma_{2}$	$L_{I}N_{II}$	11.217
1.1058	1	77 Ir	L_{III}	Abs. Edge	11.212
1.10585	3	77 Ir	$L\beta_{5}$	$L_{III}O_{IV,V}$	11.2114
1.10651	3	79 Au	$L\beta_{4}$	$L_{I}M_{II}$	11.2047
1.10664	9	72 Hf		$L_{I}O_{I}$	11.2034
1.10882	2	34 Se	$K\alpha_{2}$	KL_{II}	11.1814
1.10923	6	77 Ir		$L_{III}O_{II,III}$	11.1772
1.11092	3	79 Au	$L\beta_{6}$	$L_{III}N_{I}$	11.1602
1.11145	4	77 Ir	Lu	$L_{III}N_{VI,VII}$	11.1549
1.1129	2	78 Pt		$L_{II}M_{V}$	11.140
1.1137	1	73 Ta	L_{II}	Abs. Edge	11.132
1.11386	4	84 Po	$L\alpha_{1}$	$L_{III}M_{V}$	11.1308
1.11388	3	73 Ta	$L\gamma_{6}$	$L_{II}O_{IV}$	11.1306
1.11489	3	77 Ir	$L\beta_{7}$	$L_{III}O_{I}$	11.1205
1.1149	2	74 W		$L_{II}N_{III}$	11.120
1.11508	4	90 Th	Ll	$L_{III}M_{I}$	11.1186
1.11521	9	73 Ta		$L_{I}N_{I}$	11.1173
1.1158	1	73 Ta	Lv	$L_{II}N_{VI}$	11.1113
1.11658	5	32 Ge	K	Abs. Edge	11.1036
1.11686	2	32 Ge	$K\beta_{2}$	$KN_{II,III}$	11.1008
1.11693	9	73 Ta		$L_{II}O_{III}$	11.1001
1.11789	9	73 Ta		$L_{II}O_{II}$	11.0907
1.1195	1	32 Ge	$K\beta_{5}$	$KM_{IV,V}$	11.0745
1.11990	2	78 Pt	$L\beta_{1}$	$L_{II}M_{IV}$	11.0707
1.1205	1	73 Ta	$L\gamma_{8}$	$L_{II}O_{I}$	11.0646
1.12146	9	72 Hf	$L\gamma_{11}$	$L_{I}N_{V}$	11.0553
1.1218	3	74 W		$L_{II}N_{II}$	11.052
1.12250	9	72 Hf		$L_{I}N_{IV}$	11.0451
1.1226	2	78 Pt		$L_{III}N_{III}$	11.044
1.12548	5	84 Po	$L\alpha_{2}$	$L_{III}M_{IV}$	11.0158
1.12637	6	76 Os	$L\beta_{9}$	$L_{I}M_{V}$	11.0071
1.12769	3	81 Tl	$L\eta$	$L_{II}M_{I}$	10.9943
1.12798	5	79 Au	$L\beta_{17}$	$L_{II}M_{III}$	10.9915
1.12894	2	32 Ge	$K\beta_{1}$	KM_{III}	10.9821
1.12936	9	32 Ge	$K\beta_{3}$	KM_{II}	10.9780
1.1310	2	78 Pt		$L_{III}N_{II}$	10.962
1.13235	3	74 W	$L\gamma_{5}$	$L_{II}N_{I}$	10.9490
1.13353	5	76 Os	$L\beta_{10}$	$L_{I}M_{IV}$	10.9376
1.13525	5	79 Au		$L_{I}M_{I}$	10.9210
1.13532	3	77 Ir	$L\beta_{2}$	$L_{III}N_{V}$	10.9203
1.13687	9	73 Ta		$L_{II}N_{V}$	10.9055
1.13707	3	77 Ir	$L\beta_{15}$	$L_{III}N_{IV}$	10.9036
1.13794	3	73 Ta	$L\gamma_{1}$	$L_{II}N_{IV}$	10.8952
1.13841	5	72 Hf	$L\gamma_{3}$	$L_{I}N_{III}$	10.8907
1.1387	5	80 Hg		$L_{II}M_{II}$	10.888
1.1402	1	71 Lu	L_{I}	Abs. Edge	10.8740
1.1405	1	76 Os	$L\beta_{5}$	$L_{III}O_{IV,V}$	10.8711
1.1408	1	76 Os	L_{III}	Abs. Edge	10.8683
1.14085	3	77 Ir	$L\beta_{3}$	$L_{I}M_{III}$	10.8674
1.14223	5	78 Pt	$L\beta_{4}$	$L_{I}M_{II}$	10.8543
1.1435	1	71 Lu	$L\gamma_{4}$	$L_{I}O_{II,III}$	10.8425
1.14355	5	78 Pt	$L\beta_{6}$	$L_{III}N_{I}$	10.8418
1.14386	2	83 Bi	$L\alpha_{1}$	$L_{III}M_{V}$	10.8388
1.14442	5	72 Hf	$L\gamma_{2}$	$L_{I}N_{II}$	10.8335
1.14537	7	76 Os	Lu	$L_{III}N_{VI,VII}$	10.8245
1.1489	2	77 Ir		$L_{II}M_{V}$	10.791
1.14933	8	76 Os	$L\beta_{7}$	$L_{III}O_{I}$	10.7872
1.1548	1	72 Hf	L_{II}	Abs. Edge	10.7362
1.15519	5	72 Hf	$L\gamma_{6}$	$L_{II}O_{IV}$	10.7325
1.1553	1	73 Ta		$L_{II}N_{III}$	10.7316
1.15536	1	83 Bi	$L\alpha_{2}$	$L_{III}M_{IV}$	10.73091
1.1560	3	77 Ir		$L_{III}N_{III}$	10.725
1.15781	3	77 Ir	$L\beta_{1}$	$L_{II}M_{IV}$	10.7083
1.15830	9	72 Hf	Lv	$L_{II}N_{VI}$	10.7037
1.1600	2	73 Ta		$L_{II}N_{II}$	10.688
1.16107	9	71 Lu	$L\gamma_{11}$	$L_{I}N_{V}$	10.6782
1.16138	5	72 Hf	$L\gamma_{8}$	$L_{II}O_{I}$	10.6754
1.16227	9	71 Lu		$L_{I}N_{IV}$	10.6672
1.1640	1	80 Hg	$L\eta$	$L_{II}M_{I}$	10.6512
1.16487	4	75 Re	$L\beta_{9}$	$L_{I}M_{V}$	10.6433
1.16545	5	77 Ir		$L_{III}N_{II}$	10.6380
1.1667	1	78 Pt	$L\beta_{17}$	$L_{II}M_{III}$	10.6265
1.16719	5	88 Ra	Ll	$L_{III}M_{I}$	10.6222
1.16962	9	78 Pt		$L_{I}M_{I}$	10.6001
1.16979	8	76 Os	$L\beta_{2}$	$L_{III}N_{V}$	10.5985
1.1708	1	79 Au		$L_{II}M_{II}$	10.5892
1.17167	5	76 Os	$L\beta_{15}$	$L_{III}N_{IV}$	10.5816
1.17218	5	75 Re	$L\beta_{10}$	$L_{I}M_{IV}$	10.5770
1.1729	1	73 Ta	$L\gamma_{5}$	$L_{II}N_{I}$	10.5702
1.17501	2	82 Pb	$L\alpha_{1}$	$L_{III}M_{V}$	10.5515
1.17588	1	33 As	$K\alpha_{1}$	KL_{III}	10.54372
1.17721	5	75 Re	$L\beta_{5}$	$L_{III}O_{IV,V}$	10.5318
1.1773	1	75 Re	L_{III}	Abs. Edge	10.5306
1.17788	9	72 Hf		$L_{II}N_{V}$	10.5258
1.17796	3	77 Ir	$L\beta_{6}$	$L_{III}N_{I}$	10.5251
1.17900	5	72 Hf	$L\gamma_{1}$	$L_{II}N_{IV}$	10.5158
1.17953	4	71 Lu	$L\gamma_{3}$	$L_{I}N_{III}$	10.5110
1.17955	7	76 Os	$L\beta_{3}$	$L_{I}M_{III}$	10.5108
1.17958	3	77 Ir	$L\beta_{4}$	$L_{I}M_{II}$	10.5106
1.17987	1	33 As	$K\alpha_{2}$	KL_{II}	10.50799
1.1815	1	75 Re	Lu	$L_{III}N_{VI,VII}$	10.4931
1.1818	1	70 Yb	L_{I}	Abs. Edge	10.4904
1.1827	1	70 Yb		$L_{I}O_{IV,V}$	10.4833
1.1853	1	70 Yb	$L\gamma_{4}$	$L_{I}O_{II,III}$	10.4603
1.1853	2	71 Lu	$L\gamma_{2}$	$L_{I}N_{II}$	10.460
1.18610	5	75 Re	$L\beta_{7}$	$L_{III}O_{I}$	10.4529
1.18648	5	82 Pb	$L\alpha_{2}$	$L_{III}M_{IV}$	10.4495

Wavelength Å*	p.e.	Element		Designation	keV
1.4941	3	68 Er	$L\beta_7$	$L_{III}O_I$	8.298
1.4941	3	68 Er	$L\beta_{10}$	L_IM_{IV}	8.298
1.4995	2	78 Pt	Ll	$L_{III}M_I$	8.268
1.500135	8	28 Ni	$K\beta_{1,3}$	$KM_{II,III}$	8.26466
1.5023	1	65 Tb	L_{II}	Abs. Edge	8.2527
1.5035	2	65 Tb	$L\gamma_6$	$L_{II}O_{IV}$	8.246
1.5063	2	69 Tm	$L\beta_3$	L_IM_{III}	8.231
1.5097	2	65 Tb	$L\gamma_8$	$L_{II}O_I$	8.212
1.51399	9	68 Er	$L\beta_{2,15}$	$L_{III}N_{IV,V}$	8.1890
1.5162	2	69 Tm	$L\beta_6$	$L_{III}N_I$	8.177
1.5178	1	75 Re	Ls	$L_{III}M_{III}$	8.1682
1.51824	7	66 Dy	$L\gamma_5$	$L_{II}N_I$	8.1661
1.52197	2	73 Ta	$L\alpha_1$	$L_{III}M_V$	8.1461
1.52325	5	72 Hf	$L\eta$	$L_{II}M_I$	8.1393
1.5297	2	64 Gd	$L\gamma_3$	L_IN_{III}	8.105
1.5303	2	65 Tb	$L\gamma_1$	$L_{II}N_{IV}$	8.102
1.5304	2	69 Tm	$L\beta_1$	$L_{II}M_{IV}$	8.101
1.53293	2	73 Ta	$L\alpha_2$	$L_{III}M_{IV}$	8.0879
1.5331	2	64 Gd	$L\gamma_2$	L_IN_{II}	8.087
1.53333	9	71 Lu		$L_{II}M_{II}$	8.0858
1.5347	2	76 Os	Lt	$L_{III}M_{II}$	8.079
1.5368	1	67 Ho	L_{III}	Abs. Edge	8.0676
1.5378	2	67 Ho	$L\beta_5$	$L_{III}O_{IV,V}$	8.062
1.5381	1	63 Eu	L_I	Abs. Edge	8.0607
1.540562	2	29 Cu	$K\alpha_1$	KL_{III}	8.04778
1.54094	3	77 Ir	Ll	$L_{III}M_I$	8.0458
1.5439	1	63 Eu	$L\gamma_4$	$L_IO_{II,III}$	8.0304
1.544390	9	29 Cu	$K\alpha_2$	KL_{II}	8.02783
1.5448	2	69 Tm	$L\beta_4$	L_IM_{II}	8.026
1.5486	3	67 Ho	$L\beta_{10}$	L_IM_{IV}	8.006
1.5616	1	68 Er	$L\beta_3$	L_IM_{III}	7.9392
1.5632	1	64 Gd	L_{II}	Abs. Edge	7.9310
1.5642	3	74 W	Ls	$L_{III}M_{III}$	7.926
1.5644	2	64 Gd	$L\gamma_6$	$L_{II}O_{IV}$	7.925
1.5671	2	67 Ho	$L\beta_{2,15}$	$L_{III}N_{IV,V}$	7.911
1.5675	2	68 Er	$L\beta_6$	$L_{III}N_I$	7.909
1.56958	5	72 Hf	$L\alpha_1$	$L_{III}M_V$	7.8990
1.5707	2	64 Gd	$L\gamma_8$	$L_{II}O_I$	7.894
1.5779	1	71 Lu	$L\eta$	$L_{II}M_I$	7.8575
1.5787	2	65 Tb	$L\gamma_5$	$L_{II}N_I$	7.8535
1.5789	1	75 Re	Lt	$L_{III}M_{II}$	7.8525
1.58046	5	72 Hf	$L\alpha_2$	$L_{III}M_{IV}$	7.8446
1.58498	7	76 Os	Ll	$L_{III}M_I$	7.8222
1.5873	1	68 Er	$L\beta_1$	$L_{II}M_{IV}$	7.8109
1.58837	7	66 Dy	$L\beta_5$	$L_{III}O_{IV,V}$	7.8055
1.58844	9	70 Yb		$L_{II}M_{II}$	7.8052
1.5903	2	63 Eu	$L\gamma_3$	L_IN_{III}	7.7961
1.5916	1	66 Dy	L_{III}	Abs. Edge	7.7897
1.5924	2	64 Gd	$L\gamma_1$	$L_{II}N_{IV}$	7.7858
1.5961	2	63 Eu	$L\gamma_2$	L_IN_{II}	7.7677
1.59973	9	66 Dy	$L\beta_9$	L_IM_V	7.7501
1.6002	1	62 Sm	L_I	Abs. Edge	7.7478
1.6007	1	68 Er	$L\beta_4$	L_IM_{II}	7.7453
1.60447	7	66 Dy	$L\beta_7$	$L_{III}O_I$	7.7272
1.60728	3	62 Sm	$L\gamma_4$	$L_IO_{II,III}$	7.714
1.60743	9	66 Dy	$L\beta_{10}$	L_IM_{IV}	7.7130
1.60815	1	27 Co	K	Abs. Edge	7.70954
1.60891	3	27 Co	$K\beta_5$	$KM_{IV,V}$	7.7059
1.61264	9	73 Ta	Ls	$L_{III}M_{III}$	7.6881
1.61951	3	71 Lu	$L\alpha_1$	$L_{III}M_V$	7.6555
1.6203	2	67 Ho	$L\beta_3$	L_IM_{III}	7.6519
1.62079	2	27 Co	$K\beta_{1,3}$	$KM_{II,III}$	7.64943
1.6237	2	67 Ho	$L\beta_6$	$L_{III}N_I$	7.6359
1.62369	7	66 Dy	$L\beta_{2,15}$	$L_{III}N_{IV,V}$	7.6357
1.6244	3	74 W	Lt	$L_{III}M_{II}$	7.6324
1.6271	1	63 Eu	L_{II}	Abs. Edge	7.6199
1.6282	2	63 Eu	$L\gamma_6$	$L_{II}O_{IV}$	7.6147
1.63029	5	71 Lu	$L\alpha_2$	$L_{III}M_{IV}$	7.6049
1.63056	5	75 Re	Ll	$L_{III}M_I$	7.6036
1.6346	2	63 Eu	$L\gamma_8$	$L_{II}O_I$	7.5849
1.63560	5	70 Yb	$L\eta$	$L_{II}M_I$	7.5802
1.6412	2	64 Gd	$L\gamma_5$	$L_{II}N_I$	7.5543
1.6475	2	67 Ho	$L\beta_1$	$L_{II}M_{IV}$	7.5253
1.6497	1	65 Tb	L_{III}	Abs. Edge	7.5153
1.6510	2	65 Tb	$L\beta_5$	$L_{III}O_{IV,V}$	7.5094
1.65601	3	62 Sm	$L\gamma_3$	L_IN_{III}	7.487
1.6574	2	63 Eu	$L\gamma_1$	$L_{II}N_{IV}$	7.4803
1.657910	8	28 Ni	$K\alpha_1$	KL_{III}	7.47815
1.6585	2	65 Tb	$L\beta_7$	$L_{III}O_I$	7.4753
1.6595	2	67 Ho	$L\beta_4$	L_IM_{II}	7.4708
1.66044	6	62 Sm	$L\gamma_2$	L_IN_{II}	7.467
1.661747	8	28 Ni	$K\alpha_2$	KL_{II}	7.46089
1.66346	9	72 Hf	Ls	$L_{III}M_{III}$	7.4532
1.6673	3	65 Tb	$L\beta_{10}$	L_IM_{IV}	7.436
1.6674	5	61 Pm	L_I	Abs. Edge	7.436
1.67189	4	70 Yb	$L\alpha_1$	$L_{III}M_V$	7.4156
1.67265	9	73 Ta	Lt	$L_{III}M_{II}$	7.4123
1.6782	1	74 W	Ll	$L_{III}M_I$	7.3878
1.68213	7	66 Dy	$L\beta_6$	$L_{III}N_I$	7.3705
1.6822	2	66 Dy	$L\beta_3$	L_IM_{III}	7.3702
1.68285	5	70 Yb	$L\alpha_2$	$L_{III}M_{IV}$	7.3673
1.6830	2	65 Tb	$L\beta_{2,15}$	$L_{III}N_{IV,V}$	7.3667
1.6953	1	62 Sm	L_{II}	Abs. Edge	7.3132
1.6963	2	69 Tm	$L\eta$	$L_{II}M_I$	7.3088
1.6966	9	62 Sm	$L\gamma_6$	$L_{II}O_{IV}$	7.308
1.7085	2	63 Eu	$L\gamma_5$	$L_{II}N_I$	7.2566
1.71062	7	66 Dy	$L\beta_1$	$L_{II}M_{IV}$	7.2477
1.7117	1	64 Gd	L_{III}	Abs. Edge	7.2430
1.7130	2	64 Gd	$L\beta_5$	$L_{III}O_{IV,V}$	7.2374
1.7203	2	64 Gd	$L\beta_7$	$L_{III}O_I$	7.2071
1.72103	7	66 Dy	$L\beta_4$	L_IM_{II}	7.2039
1.72305	9	72 Hf	Lt	$L_{III}M_{II}$	7.1954
1.7240	3	64 Gd	$L\beta_9$	L_IM_V	7.192
1.72724	3	62 Sm	$L\gamma_1$	$L_{II}N_{IV}$	7.178
1.7268	2	69 Tm	$L\alpha_1$	$L_{III}M_V$	7.1799
1.72841	5	73 Ta	Ll	$L_{III}M_I$	7.1731
1.7315	3	64 Gd	$L\beta_{10}$	L_IM_{IV}	7.160
1.7381	2	69 Tm	$L\alpha_2$	$L_{III}M_{IV}$	7.1331
1.7390	1	60 Nd	L_I	Abs. Edge	7.1294
1.7422	2	65 Tb	$L\beta_6$	$L_{III}N_I$	7.1163
1.74346	1	26 Fe	K	Abs. Edge	7.11120
1.7442	1	26 Fe	$K\beta_5$	$KM_{IV,V}$	7.1081
1.7445	4	60 Nd	$L\gamma_4$	$L_IO_{II,III}$	7.107
1.7455	2	64 Gd	$L\beta_{2,15}$	$L_{III}N_{IV,V}$	7.1028

Wavelength Å*	p.e.	Element		Designation	keV
1.1886	1	70 Yb		$L_{I}O_{I}$	10.4312
1.18977	7	76 Os		$L_{II}M_{V}$	10.4205
1.1958	1	31 Ga	K	Abs. Edge	10.3682
1.19600	2	31 Ga	$K\beta_{2}$	$KN_{II,III}$	10.3663
1.19727	7	76 Os	$L\beta_{1}$	$L_{II}M_{IV}$	10.3553
1.1981	2	31 Ga	$K\beta_{5}$	$KM_{IV,V}$	10.348
1.1985	1	71 Lu	L_{II}	Abs. Edge	10.3448
1.1987	1	71 Lu	$L\gamma_{6}$	$L_{II}O_{IV}$	10.3431
1.20086	7	76 Os		$L_{III}N_{II}$	10.3244
1.2014	1	71 Lu		$L_{II}O_{II,III}$	10.3198
1.20273	3	79 Au	$L\eta$	$L_{II}M_{I}$	10.3083
1.2047	1	71 Lu	$L\gamma_{8}$	$L_{II}O_{I}$	10.2915
1.20479	7	74 W	$L\beta_{9}$	$L_{I}M_{V}$	10.2907
1.20660	4	75 Re	$L\beta_{2}$	$L_{III}N_{V}$	10.2752
1.2069	2	77 Ir	$L\beta_{17}$	$L_{II}M_{III}$	10.273
1.20739	4	81 Tl	$L\alpha_{1}$	$L_{III}M_{V}$	10.2685
1.20789	2	31 Ga	$K\beta_{1}$	KM_{III}	10.2642
1.20819	5	75 Re	$L\beta_{15}$	$L_{III}N_{IV}$	10.2617
1.20835	5	31 Ga	$K\beta_{3}$	KM_{II}	10.2603
1.2102	2	77 Ir		$L_{I}M_{I}$	10.245
1.2105	1	83 Bi	Ls	$L_{III}M_{III}$	10.2421
1.21218	3	74 W	$L\beta_{10}$	$L_{I}M_{IV}$	10.2279
1.213	1	78 Pt		$L_{II}M_{II}$	10.225
1.21349	5	76 Os	$L\beta_{6}$	$L_{III}N_{I}$	10.2169
1.21537	5	72 Hf	$L\gamma_{5}$	$L_{II}N_{I}$	10.2011
1.21545	3	74 W	$L\beta_{5}$	$L_{III}O_{IV,V}$	10.2004
1.2155	1	74 W	L_{III}	Abs. Edge	10.1999
1.21844	5	76 Os	$L\beta_{4}$	$L_{I}M_{II}$	10.1754
1.21868	5	74 W	Lu	$L_{III}N_{VI,VII}$	10.1733
1.21875	3	81 Tl	$L\alpha_{2}$	$L_{III}M_{IV}$	10.1728
1.22031	5	75 Re	$L\beta_{3}$	$L_{I}M_{III}$	10.1598
1.2211	2	74 W		$L_{III}O_{II,III}$	10.153
1.22228	4	71 Lu	$L\gamma_{1}$	$L_{II}N_{IV}$	10.1434
1.22232	5	70 Yb	$L\gamma_{3}$	$L_{I}N_{III}$	10.1431
1.22400	4	74 W	$L\beta_{7}$	$L_{III}O_{I}$	10.1292
1.2250	1	69 Tm	L_{I}	Abs. Edge	10.1206
1.2263	3	69 Tm		$L_{I}O_{IV,V}$	10.110
1.2283	1	75 Re		$L_{III}N_{III}$	10.0933
1.22879	7	70 Yb	$L\gamma_{2}$	$L_{I}N_{II}$	10.0897
1.2294	2	69 Tm	$L\gamma_{4}$	$L_{I}O_{II,III}$	10.084
1.2305	1	75 Re		$L_{II}M_{V}$	10.0753
1.23858	2	75 Re	$L\beta_{1}$	$L_{II}M_{IV}$	10.0100
1.24120	5	80 Hg	$L\alpha_{1}$	$L_{III}M_{V}$	9.9888
1.24271	3	70 Yb	$L\gamma_{6}$	$L_{II}O_{IV}$	9.9766
1.2428	1	70 Yb	L_{II}	Abs. Edge	9.9761
1.2429	2	78 Pt	$L\eta$	$L_{II}M_{I}$	9.975
1.24385	7	82 Pb	Ls	$L_{III}M_{III}$	9.9675
1.24460	3	74 W	$L\beta_{2}$	$L_{III}N_{V}$	9.9615
1.2453	1	70 Yb		$L_{II}O_{II,III}$	9.9561
1.24631	3	74 W	$L\beta_{15}$	$L_{III}N_{IV}$	9.9478
1.2466	2	73 Ta	$L\beta_{9}$	$L_{I}M_{V}$	9.946
1.2480	2	76 Os	$L\beta_{17}$	$L_{II}M_{III}$	9.934
1.24923	5	70 Yb	$L\gamma_{8}$	$L_{II}O_{I}$	9.9246
1.2502	3	77 Ir		$L_{II}M_{II}$	9.917
1.25100	5	75 Re	$L\beta_{6}$	$L_{III}N_{I}$	9.9105
1.25264	7	80 Hg	$L\alpha_{2}$	$L_{III}M_{IV}$	9.8976
1.2537	2	73 Ta	$L\beta_{10}$	$L_{I}M_{IV}$	9.889
1.254054	9	32 Ge	$K\alpha_{1}$	KL_{III}	9.88642
1.2553	1	73 Ta	L_{III}	Abs. Edge	9.8766
1.2555	1	73 Ta	$L\beta_{5}$	$L_{III}O_{IV,V}$	9.8750
1.25778	4	73 Ta	Lu	$L_{III}N_{VI,VII}$	9.8572
1.258011	9	32 Ge	$K\alpha_{2}$	KL_{II}	9.85532
1.25917	5	75 Re	$L\beta_{4}$	$L_{I}M_{II}$	9.8463
1.2596	1	71 Lu	$L\gamma_{5}$	$L_{II}N_{I}$	9.8428
1.2601	3	73 Ta		$L_{III}O_{II,III}$	9.839
1.26269	5	74 W	$L\beta_{3}$	$L_{I}M_{III}$	9.8188
1.26385	5	73 Ta	$L\beta_{7}$	$L_{III}O_{I}$	9.8098
1.2672	2	74 W		$L_{III}N_{III}$	9.784
1.26769	5	70 Yb	$L\gamma_{1}$	$L_{II}N_{IV}$	9.7801
1.2678	2	69 Tm	$L\gamma_{3}$	$L_{I}N_{III}$	9.779
1.2706	1	68 Er	L_{I}	Abs. Edge	9.7574
1.2728	2	74 W		$L_{II}M_{V}$	9.741
1.2742	2	69 Tm	$L\gamma_{2}$	$L_{I}N_{II}$	9.730
1.2748	1	83 Bi	Lt	$L_{III}M_{II}$	9.7252
1.2752	2	68 Er	$L\gamma_{4}$	$L_{I}O_{II,III}$	9.722
1.27640	3	79 Au	$L\alpha_{1}$	$L_{III}M_{V}$	9.7133
1.2765	2	74 W		$L_{III}N_{II}$	9.712
1.27807	5	81 Tl	Ls	$L_{III}M_{III}$	9.7007
1.281809	9	74 W	$L\beta_{1}$	$L_{II}M_{IV}$	9.67235
1.2829	5	84 Po	Ll	$L_{III}M_{I}$	9.664
1.2834	1	30 Zn	K	Abs. Edge	9.6607
1.28372	2	30 Zn	$K\beta_{2}$	$KN_{II,III}$	9.6580
1.28448	3	77 Ir	$L\eta$	$L_{II}M_{I}$	9.6522
1.28454	2	73 Ta	$L\beta_{2}$	$L_{III}N_{V}$	9.6518
1.2848	1	30 Zn	$K\beta_{5}$	$KM_{IV,V}$	9.6501
1.28619	5	73 Ta	$L\beta_{15}$	$L_{III}N_{IV}$	9.6394
1.28772	3	79 Au	$L\alpha_{2}$	$L_{III}M_{IV}$	9.6280
1.2892	1	69 Tm	L_{II}	Abs. Edge	9.6171
1.28989	7	74 W	$L\beta_{6}$	$L_{III}N_{I}$	9.6117
1.29025	9	72 Hf	$L\beta_{9}$	$L_{I}M_{V}$	9.6090
1.2905	2	69 Tm	$L\gamma_{6}$	$L_{II}O_{IV}$	9.607
1.2927	1	75 Re	$L\beta_{17}$	$L_{II}M_{III}$	9.5910
1.2934	2	76 Os		$L_{II}M_{II}$	9.586
1.29525	2	30 Zn	$K\beta_{1,3}$	$KM_{II,III}$	9.5720
1.2972	1	72 Hf	L_{III}	Abs. Edge	9.5577
1.29761	5	72 Hf	$L\beta_{5}$	$L_{III}O_{IV,V}$	9.5546
1.29819	9	72 Hf	$L\beta_{10}$	$L_{I}M_{IV}$	9.5503
1.30162	5	74 W	$L\beta_{4}$	$L_{I}M_{II}$	9.5252
1.30165	9	72 Hf	Lu	$L_{III}N_{VI,VII}$	9.5249
1.30564	5	72 Hf	$L\beta_{7}$	$L_{III}O_{I}$	9.4958
1.3063	1	70 Yb	$L\gamma_{5}$	$L_{II}N_{I}$	9.4910
1.30678	3	73 Ta	$L\beta_{3}$	$L_{I}M_{III}$	9.4875
1.30767	7	82 Pb	Lt	$L_{III}M_{II}$	9.4811
1.3086	1	73 Ta		$L_{III}N_{III}$	9.4742
1.3112	2	80 Hg	Ls	$L_{III}M_{III}$	9.455
1.31304	3	78 Pt	$L\alpha_{1}$	$L_{III}M_{V}$	9.4423
1.3146	1	68 Er	$L\gamma_{3}$	$L_{I}N_{III}$	9.4309
1.3153	2	69 Tm	$L\gamma_{1}$	$L_{II}N_{IV}$	9.426
1.31610	7	83 Bi	Ll	$L_{III}M_{I}$	9.4204
1.3167	1	73 Ta		$L_{III}N_{II}$	9.4158
1.31897	9	73 Ta		$L_{II}M_{V}$	9.3998
1.3190	1	67 Ho	L_{I}	Abs. Edge	9.3994
1.3208	3	67 Ho		$L_{I}O_{IV,V}$	9.387
1.3210	2	68 Er	$L\gamma_{2}$	$L_{I}N_{II}$	9.385

Wavelength Å*	p.e.	Element	Designation		keV
1.3225	2	67 Ho	$L\gamma_4$	$L_IO_{II,III}$	9.374
1.32432	2	78 Pt	$L\alpha_2$	$L_{III}M_{IV}$	9.3618
1.32639	5	72 Hf	$L\beta_2$	$L_{III}N_V$	9.3473
1.32698	3	73 Ta	$L\beta_1$	$L_{II}M_{IV}$	9.3431
1.32783	5	72 Hf	$L\beta_{15}$	$L_{III}N_{IV}$	9.3371
1.32785	7	76 Os	$L\eta$	$L_{II}M_I$	9.3370
1.33094	8	73 Ta	$L\beta_6$	$L_{III}N_I$	9.3153
1.3358	1	71 Lu	$L\beta_9$	L_IM_V	9.2816
1.3365	3	74 W		L_IM_I	9.277
1.3366	1	75 Re		$L_{II}M_{II}$	9.2761
1.3386	1	68 Er	L_{II}	Abs. Edge	9.2622
1.3387	2	74 W	$L\beta_{17}$	$L_{II}M_{III}$	9.261
1.3397	3	68 Er	$L\gamma_6$	$L_{II}O_{IV}$	9.255
1.340083	9	31 Ga	$K\alpha_1$	KL_{III}	9.25174
1.3405	1	71 Lu	L_{III}	Abs. Edge	9.2490
1.34154	5	81 Tl	Lt	$L_{III}M_{II}$	9.2417
1.34183	7	71 Lu	$L\beta_5$	$L_{III}O_{IV,V}$	9.2397
1.3430	2	71 Lu	$L\beta_{10}$	L_IM_{IV}	9.232
1.34399	1	31 Ga	$K\alpha_2$	KL_{II}	9.22482
1.34524	9	71 Lu		$L_{III}O_{II,III}$	9.2163
1.34581	3	73 Ta	$L\beta_4$	L_IM_{II}	9.2124
1.34949	5	71 Lu	$L\beta_7$	$L_{III}O_I$	9.1873
1.34990	7	82 Pb	Ll	$L_{III}M_I$	9.1845
1.35053	9	72 Hf		$L_{III}N_{III}$	9.1802
1.35128	3	77 Ir	$L\alpha_1$	$L_{III}M_V$	9.1751
1.35131	7	79 Au	Ls	$L_{III}M_{III}$	9.1749
1.35300	5	72 Hf	$L\beta_3$	L_IM_{III}	9.1634
1.3558	2	69 Tm	$L\gamma_5$	$L_{II}N_I$	9.144
1.35887	9	72 Hf		$L_{III}N_{II}$	9.1239
1.36250	5	77 Ir	$L\alpha_2$	$L_{III}M_{IV}$	9.0995
1.3641	2	68 Er	$L\gamma_1$	$L_{II}N_{IV}$	9.089
1.3643	2	67 Ho	$L\gamma_3$	L_IN_{III}	9.087
1.3692	1	66 Dy	L_I	Abs. Edge	9.0548
1.3698	2	67 Ho	$L\gamma_2$	L_IN_{II}	9.051
1.37012	3	71 Lu	$L\beta_2$	$L_{III}N_V$	9.0489
1.3715	1	71 Lu	$L\beta_{15}$	$L_{III}N_{IV}$	9.0395
1.37342	5	75 Re	$L\eta$	$L_{II}M_I$	9.0272
1.37410	5	72 Hf	$L\beta_1$	$L_{II}M_{IV}$	9.0227
1.37410	5	72 Hf	$L\beta_6$	$L_{III}N_I$	9.0227
1.37459	7	66 Dy	$L\gamma_4$	$L_IO_{II,III}$	9.0195
1.3746	2	80 Hg	Lt	$L_{III}M_{II}$	9.019
1.38059	5	29 Cu	K	Abs. Edge	8.9803
1.38109	3	29 Cu	$K\beta_2$	$KM_{IV,V}$	8.9770
1.3838	1	70 Yb	$L\beta_9$	L_IM_V	8.9597
1.38477	3	81 Tl	Ll	$L_{III}M_I$	8.9532
1.3862	1	70 Yb	L_{III}	Abs. Edge	8.9441
1.3864	1	73 Ta	$L\beta_{17}$	$L_{II}M_{III}$	8.9428
1.38696	7	70 Yb	$L\beta_5$	$L_{III}O_{IV,V}$	8.9390
1.3895	2	78 Pt	Ls	$L_{III}M_{III}$	8.923
1.3898	1	70 Yb		$L_{III}O_{II,III}$	8.9209
1.3905	1	67 Ho	L_{II}	Abs. Edge	8.9164
1.39121	5	76 Os	$L\alpha_1$	$L_{III}M_V$	8.9117
1.3915	1	70 Yb	$L\beta_{10}$	L_IM_{IV}	8.9100
1.39220	5	72 Hf	$L\beta_4$	L_IM_{II}	8.9054
1.392218	9	29 Cu	$K\beta_{1,3}$	$KM_{II,III}$	8.90529
1.3923	2	67 Ho	$L\gamma_6$	$L_{II}O_{IV}$	8.905
1.3926	1	29 Cu	$K\beta_3$	KM_{II}	8.9029
1.3948	1	70 Yb	$L\beta_7$	$L_{III}O_I$	8.8889
1.3983	2	67 Ho	$L\gamma_8$	$L_{II}O_I$	8.867
1.40140	5	71 Lu	$L\beta_3$	L_IM_{III}	8.8469
1.40234	5	76 Os	$L\alpha_2$	$L_{III}M_{IV}$	8.8410
1.4067	3	68 Er	$L\gamma_5$	$L_{II}N_I$	8.814
1.41366	7	79 Au	Lt	$L_{III}M_{II}$	8.7702
1.41550	5	70 Yb	$L\beta_{2,15}$	$L_{III}N_{IV,V}$	8.7588
1.41640	7	66 Dy	$L\gamma_3$	L_IN_{III}	8.7532
1.4174	2	67 Ho	$L\gamma_1$	$L_{II}N_{IV}$	8.747
1.4189	1	71 Lu	$L\beta_6$	$L_{III}N_I$	8.7376
1.42110	3	74 W	$L\eta$	$L_{II}M_I$	8.7243
1.4216	1	80 Hg	Ll	$L_{III}M_I$	8.7210
1.4223	1	65 Tb	L_I	Abs. Edge	8.7167
1.42278	7	66 Dy	$L\gamma_2$	L_IN_{II}	8.7140
1.4228	3	65 Tb		$L_IO_{IV,V}$	8.714
1.42359	3	71 Lu	$L\beta_1$	$L_{II}M_{IV}$	8.7090
1.4276	2	65 Tb	$L\gamma_4$	$L_IO_{II,III}$	8.685
1.43025	9	72 Hf		L_IM_I	8.6685
1.43048	9	73 Ta		$L_{II}M_{II}$	8.6671
1.4318	2	77 Ir	Ls	$L_{III}M_{III}$	8.659
1.43290	4	75 Re	$L\alpha_1$	$L_{III}M_V$	8.6525
1.4334	1	69 Tm	L_{III}	Abs. Edge	8.6496
1.4336	3	69 Tm	$L\beta_9$	L_IM_V	8.648
1.4349	2	69 Tm	$L\beta_5$	$L_{III}O_{IV,V}$	8.641
1.435155	7	30 Zn	$K\alpha_1$	KL_{III}	8.63886
1.43643	9	72 Hf	$L\beta_{17}$	$L_{II}M_{III}$	8.6312
1.439000	8	30 Zn	$K\alpha_2$	KL_{II}	8.61578
1.44056	5	71 Lu	$L\beta_4$	L_IM_{II}	8.6064
1.4410	3	69 Tm	$L\beta_{10}$	L_IM_{IV}	8.604
1.44396	5	75 Re	$L\alpha_2$	$L_{III}M_{IV}$	8.5862
1.4445	1	66 Dy	L_{II}	Abs. Edge	8.5830
1.44579	7	66 Dy	$L\gamma_6$	$L_{II}O_{IV}$	8.5753
1.45233	5	70 Yb	$L\beta_3$	L_IM_{III}	8.5367
1.4530	2	78 Pt	Lt	$L_{III}M_{II}$	8.533
1.45964	9	79 Au	Ll	$L_{III}M_I$	8.4939
1.4618	2	67 Ho	$L\gamma_5$	$L_{II}N_I$	8.481
1.4640	2	69 Tm	$L\beta_{2,15}$	$L_{III}N_{IV,V}$	8.468
1.4661	1	70 Yb	$L\beta_6$	$L_{III}N_I$	8.4563
1.47106	5	73 Ta	$L\eta$	$L_{II}M_I$	8.4280
1.4718	2	65 Tb	$L\gamma_3$	L_IN_{III}	8.423
1.47266	7	66 Dy	$L\gamma_1$	$L_{II}N_{IV}$	8.4188
1.4735	2	76 Os	Ls	$L_{III}M_{III}$	8.414
1.47565	5	70 Yb	$L\beta_1$	$L_{II}M_{IV}$	8.4018
1.4764	2	65 Tb	$L\gamma_2$	L_IN_{II}	8.398
1.47639	2	74 W	$L\alpha_1$	$L_{III}M_V$	8.3976
1.4784	1	64 Gd	L_I	Abs. Edge	8.3864
1.48064	9	72 Hf		$L_{II}M_{II}$	8.3735
1.4807	3	64 Gd		$L_IO_{IV,V}$	8.373
1.4835	1	68 Er	L_{III}	Abs. Edge	8.3575
1.4839	2	64 Gd	$L\gamma_4$	$L_IO_{II,III}$	8.355
1.4848	3	68 Er	$L\beta_5$	$L_{III}O_{IV,V}$	8.350
1.4855	5	68 Er	$L\beta_9$	L_IM_V	8.346
1.48743	2	74 W	$L\alpha_2$	$L_{III}M_{IV}$	8.3352
1.48807	1	28 Ni	K	Abs. Edge	8.33165
1.48862	4	28 Ni	$K\beta_5$	$KM_{IV,V}$	8.3286
1.49138	3	70 Yb	$L\beta_4$	L_IM_{II}	8.3132
1.4930	3	77 Ir	Lt	$L_{III}M_{II}$	8.304

Wavelength Å*	p.e.	Element	Designation		keV
1.7472	2	65 Tb	$L\beta_3$	L_IM_{III}	7.0959
1.75661	2	26 Fe	$K\beta_{1,3}$	$KM_{II,III}$	7.05798
1.7566	1	68 Er	$L\eta$	$L_{II}M_I$	7.0579
1.7676	5	61 Pm	L_{II}	Abs. Edge	7.014
1.7760	1	71 Lu	Lt	$L_{III}M_{II}$	6.9810
1.7761	1	63 Eu	L_{III}	Abs. Edge	6.9806
1.7768	3	65 Tb	$L\beta_1$	$L_{II}M_{IV}$	6.978
1.7772	2	63 Eu	$L\beta_5$	$L_{III}O_{IV,V}$	6.9763
1.77934	3	62 Sm	$L\gamma_5$	$L_{II}N_I$	6.968
1.78145	5	72 Hf	Ll	$L_{III}M_I$	6.9596
1.78425	9	68 Er	$L\alpha_1$	$L_{III}M_V$	6.9487
1.7851	2	63 Eu	$L\beta_7$	$L_{III}O_I$	6.9453
1.7864	2	65 Tb	$L\beta_4$	L_IM_{II}	6.9403
1.788965	9	27 Co	$K\alpha_1$	KL_{III}	6.93032
1.7916	3	63 Eu	$L\beta_9$	L_IM_V	6.920
1.792850	9	27 Co	$K\alpha_2$	KL_{II}	6.91530
1.7955	2	68 Er	$L\alpha_2$	$L_{III}M_{IV}$	6.9050
1.7964	4	60 Nd	$L\gamma_3$	L_IN_{III}	6.902
1.7989	9	61 Pm	$L\gamma_1$	$L_{II}N_{IV}$	6.892
1.7993	3	63 Eu	$L\beta_{10}$	L_IM_{IV}	6.890
1.8013	4	60 Nd	$L\gamma_2$	L_IN_{II}	6.883
1.8054	2	64 Gd	$L\beta_6$	$L_{III}N_I$	6.8671
1.8118	2	63 Eu	$L\beta_{2,15}$	$L_{III}N_{IV,V}$	6.8432
1.8141	5	59 Pr	L_I	Abs. Edge	6.834
1.8150	2	64 Gd	$L\beta_3$	L_IM_{III}	6.8311
1.8193	4	59 Pr	$L\gamma_4$	$L_IO_{II,III}$	6.815
1.8264	2	67 Ho	$L\eta$	$L_{II}M_I$	6.7883
1.83091	9	70 Yb	Lt	$L_{III}M_{II}$	6.7715
1.8360	1	71 Lu	Ll	$L_{III}M_I$	6.7528
1.8440	1	60 Nd	L_{II}	Abs. Edge	6.7234
1.8450	2	67 Ho	$L\alpha_1$	$L_{III}M_V$	6.7198
1.8457	1	62 Sm	L_{III}	Abs. Edge	6.7172
1.8468	2	64 Gd	$L\beta_1$	$L_{II}M_{IV}$	6.7132
1.84700	9	62 Sm	$L\beta_5$	$L_{III}O_{IV,V}$	6.7126
1.8540	2	64 Gd	$L\beta_4$	L_IM_{II}	6.6871
1.8552	5	60 Nd	$L\gamma_8$	$L_{II}O_I$	6.683
1.8561	2	67 Ho	$L\alpha_2$	$L_{III}M_{IV}$	6.6795
1.85626	3	62 Sm	$L\beta_7$	$L_{III}O_I$	6.679
1.86166	3	62 Sm	$L\beta_9$	L_IM_V	6.660
1.86990	3	62 Sm	$L\beta_{10}$	L_IM_{IV}	6.634
1.8737	2	63 Eu	$L\beta_6$	$L_{III}N_I$	6.6170
1.8740	4	59 Pr	$L\gamma_3$	L_IN_{III}	6.616
1.8779	2	60 Nd	$L\gamma_1$	$L_{II}N_{IV}$	6.6021
1.8791	4	59 Pr	$L\gamma_2$	L_IN_{II}	6.598
1.8821	3	62 Sm	$L\beta_{2,15}$	$L_{III}N_{IV,V}$	6.586
1.8867	2	63 Eu	$L\beta_3$	L_IM_{III}	6.5713
1.8934	5	58 Ce	L_I	Abs. Edge	6.548
1.89415	5	70 Yb	Ll	$L_{III}M_I$	6.5455
1.89643	5	25 Mn	K	Abs. Edge	6.5376
1.8971	1	25 Mn	$K\beta_5$	$KM_{IV,V}$	6.5352
1.89743	7	66 Dy	$L\eta$	$L_{II}M_I$	6.5342
1.8991	4	58 Ce	$L\gamma_4$	$L_IO_{II,III}$	6.528
1.90881	3	66 Dy	$L\alpha_1$	$L_{III}M_V$	6.4952
1.91021	2	25 Mn	$K\beta_{1,3}$	$KM_{II,III}$	6.49045
1.9191	1	61 Pm	L_{III}	Abs. Edge	6.4605
1.91991	3	66 Dy	$L\alpha_2$	$L_{III}M_{IV}$	6.4577
1.9203	2	63 Eu	$L\beta_1$	$L_{II}M_{IV}$	6.4564
1.9255	2	63 Eu	$L\beta_4$	L_IM_{II}	6.4389
1.9255	5	59 Pr	L_{II}	Abs. Edge	6.439
1.9355	4	60 Nd	$L\gamma_5$	$L_{II}N_I$	6.406
1.936042	9	26 Fe	$K\alpha_1$	KL_{III}	6.40384
1.9362	4	59 Pr	$L\gamma_8$	$L_{II}O_I$	6.403
1.939980	9	26 Fe	$K\alpha_2$	KL_{II}	6.39084
1.94643	3	62 Sm	$L\beta_6$	$L_{III}N_I$	6.3693
1.9550	2	69 Tm	Ll	$L_{III}M_I$	6.3419
1.9553	3	58 Ce	$L\gamma_3$	L_IN_{III}	6.3409
1.9559	6	61 Pm	$L\beta_{2,15}$	$L_{III}N_{IV,V}$	6.339
1.9602	3	58 Ce	$L\gamma_2$	L_IN_{II}	6.3250
1.9611	3	59 Pr	$L\gamma_1$	$L_{II}N_{IV}$	6.3221
1.96241	3	62 Sm	$L\beta_3$	L_IM_{III}	6.318
1.9730	2	65 Tb	$L\eta$	$L_{II}M_I$	6.2839
1.9765	2	65 Tb	$L\alpha_1$	$L_{III}M_V$	6.2728
1.9780	5	57 La	L_I	Abs. Edge	6.268
1.9830	4	57 La	$L\gamma_4$	$L_IO_{II,III}$	6.252
1.9875	2	65 Tb	$L\alpha_2$	$L_{III}M_{IV}$	6.2380
1.9967	1	60 Nd	L_{III}	Abs. Edge	6.2092
1.99806	3	62 Sm	$L\beta_1$	$L_{II}M_{IV}$	6.2051
2.00095	6	62 Sm	$L\beta_4$	L_IM_{II}	6.196
2.0092	3	60 Nd	$L\beta_7$	$L_{III}O_I$	6.1708
2.0124	5	58 Ce	L_{II}	Abs. Edge	6.161
2.015	1	68 Er	Ll	$L_{III}M_I$	6.152
2.0165	3	60 Nd	$L\beta_9$	L_IM_V	6.1484
2.0205	4	59 Pr	$L\gamma_5$	$L_{II}N_I$	6.136
2.0237	4	58 Ce	$L\gamma_8$	$L_{II}O_I$	6.126
2.0237	3	60 Nd	$L\beta_{10}$	L_IM_{IV}	6.1265
2.0360	3	60 Nd	$L\beta_{2,15}$	$L_{III}N_{IV,V}$	6.0894
2.0410	4	57 La	$L\gamma_3$	L_IN_{III}	6.074
2.0421	4	61 Pm	$L\beta_3$	L_IM_{III}	6.071
2.0460	4	57 La	$L\gamma_2$	L_IN_{II}	6.060
2.0468	2	64 Gd	$L\alpha_1$	$L_{III}M_V$	6.0572
2.0487	4	58 Ce	$L\gamma_1$	$L_{II}N_{IV}$	6.052
2.0494	1	64 Gd	$L\eta$	$L_{II}M_I$	6.0495
2.0578	2	64 Gd	$L\alpha_2$	$L_{III}M_{IV}$	6.0250
2.0678	5	56 Ba	L_I	Abs. Edge	5.996
2.07020	5	24 Cr	K	Abs. Edge	5.9888
2.07087	6	24 Cr	$K\beta_5$	$KM_{IV,V}$	5.9869
2.0756	3	56 Ba	$L\gamma_4$	$L_IO_{II,III}$	5.9733
2.0791	5	59 Pr	L_{III}	Abs. Edge	5.963
2.0797	4	61 Pm	$L\beta_1$	$L_{II}M_{IV}$	5.961
2.08487	2	24 Cr	$K\beta_{1,3}$	$KM_{II,III}$	5.94671
2.0860	2	67 Ho	Ll	$L_{III}M_I$	5.9434
2.0919	4	59 Pr	$L\beta_7$	$L_{III}O_I$	5.927
2.1004	4	59 Pr	$L\beta_9$	L_IM_V	5.903
2.101820	9	25 Mn	$K\alpha_1$	KL_{III}	5.89875
2.1039	3	60 Nd	$L\beta_6$	$L_{III}N_I$	5.8930
2.1053	5	57 La	L_{II}	Abs. Edge	5.889
2.10578	2	25 Mn	$K\alpha_2$	KL_{II}	5.88765
2.1071	4	59 Pr	$L\beta_{10}$	L_IM_{IV}	5.884
2.1103	3	58 Ce	$L\gamma_5$	$L_{II}N_I$	5.8751
2.1194	4	59 Pr	$L\beta_{2,15}$	$L_{III}N_{IV,V}$	5.850
2.1209	2	63 Eu	$L\alpha_1$	$L_{III}M_V$	5.8457
2.1268	2	60 Nd	$L\beta_3$	L_IM_{III}	5.8294
2.1315	2	63 Eu	$L\eta$	$L_{II}M_I$	5.8166
2.1315	2	63 Eu	$L\alpha_2$	$L_{III}M_{IV}$	5.8166

TABLE B (*Continued*)

Wavelength Å*	p.e.	Element	Designation		keV
2.1342	2	56 Ba	$L\gamma_3$	$L_{I}N_{III}$	5.8092
2.1387	2	56 Ba	$L\gamma_2$	$L_{I}N_{II}$	5.7969
2.1418	3	57 La	$L\gamma_1$	$L_{II}N_{IV}$	5.7885
2.15877	7	66 Dy	Ll	$L_{III}M_{I}$	5.7431
2.166	1	58 Ce	L_{III}	Abs. Edge	5.723
2.1669	3	60 Nd	$L\beta_4$	$L_{I}M_{II}$	5.7216
2.1669	2	60 Nd	$L\beta_1$	$L_{II}M_{IV}$	5.7216
2.1673	5	55 Cs	L_{I}	Abs. Edge	5.721
2.1701	2	58 Ce	$L\beta_7$	$L_{III}O_{I}$	5.7132
2.1741	2	55 Cs	$L\gamma_4$	$L_{I}O_{II,III}$	5.7026
2.1885	3	58 Ce	$L\beta_9$	$L_{I}M_{V}$	5.6650
2.1906	4	59 Pr	$L\beta_6$	$L_{III}N_{I}$	5.660
2.1958	5	58 Ce	$L\beta_{10}$	$L_{I}M_{IV}$	5.646
2.1998	2	62 Sm	$L\alpha_1$	$L_{III}M_{V}$	5.6361
2.2048	1	56 Ba	L_{II}	Abs. Edge	5.6233
2.2056	4	57 La	$L\gamma_5$	$L_{II}N_{I}$	5.621
2.2087	2	58 Ce	$L\beta_{2,15}$	$L_{III}N_{IV,V}$	5.6134
2.21062	3	62 Sm	$L\alpha_2$	$L_{III}M_{IV}$	5.6090
2.2172	3	59 Pr	$L\beta_3$	$L_{I}M_{III}$	5.5918
2.21824	3	62 Sm	$L\eta$	$L_{II}M_{I}$	5.589
2.2328	2	55 Cs	$L\gamma_3$	$L_{I}N_{III}$	5.5527
2.2352	2	65 Tb	Ll	$L_{III}M_{I}$	5.5467
2.2371	2	55 Cs	$L\gamma_2$	$L_{I}N_{II}$	5.5420
2.2415	2	56 Ba	$L\gamma_1$	$L_{II}N_{IV}$	5.5311
2.253	6	92 U		$M_{I}P_{III}$	5.50
2.2550	4	59 Pr	$L\beta_4$	$L_{I}M_{II}$	5.4981
2.2588	3	59 Pr	$L\beta_1$	$L_{II}M_{IV}$	5.4889
2.261	1	57 La	L_{III}	Abs. Edge	5.484
2.2691	1	23 V	K	Abs. Edge	5.4639
2.26951	6	23 V	$K\beta_5$	$KM_{IV,V}$	5.4629
2.2737	1	54 Xe	L_{I}	Abs. Edge	5.4528
2.275	3	57 La	$L\beta_7$	$L_{III}O_{I}$	5.450
2.282	3	57 La	$L\beta_9$	$L_{I}M_{V}$	5.434
2.2818	3	58 Ce	$L\beta_6$	$L_{III}N_{I}$	5.4334
2.2822	3	61 Pm	$L\alpha_1$	$L_{III}M_{V}$	5.4325
2.28440	2	23 V	$K\beta_{1,3}$	$KM_{II,III}$	5.42729
2.28970	2	24 Cr	$K\alpha_1$	KL_{III}	5.41472
2.290	3	57 La	$L\beta_{10}$	$L_{I}M_{IV}$	5.415
2.2926	4	61 Pm	$L\alpha_2$	$L_{III}M_{IV}$	5.4078
2.293606	3	24 Cr	$K\alpha_2$	KL_{II}	5.405509
2.3030	3	57 La	$L\beta_{2,15}$	$L_{III}N_{IV,V}$	5.3835
2.304	7	92 U		$M_{I}O_{III}$	5.38
2.3085	3	56 Ba	$L\gamma_5$	$L_{II}N_{I}$	5.3707
2.3109	3	58 Ce	$L\beta_3$	$L_{I}M_{III}$	5.3651
2.3122	2	64 Gd	Ll	$L_{III}M_{I}$	5.3621
2.3139	1	55 Cs	L_{II}	Abs. Edge	5.3581
2.3480	2	55 Cs	$L\gamma_1$	$L_{II}N_{IV}$	5.2804
2.3497	4	58 Ce	$L\beta_4$	$L_{I}M_{II}$	5.2765
2.3561	3	58 Ce	$L\beta_1$	$L_{II}M_{IV}$	5.2622
2.3629	1	56 Ba	L_{III}	Abs. Edge	5.2470
2.3704	2	60 Nd	$L\alpha_1$	$L_{III}M_{V}$	5.2304
2.3764	2	56 Ba	$L\beta_9$	$L_{I}M_{V}$	5.2171
2.3790	4	57 La	$L\beta_6$	$L_{III}N_{I}$	5.2114
2.3806	2	56 Ba	$L\beta_7$	$L_{III}O_{I}$	5.2079
2.3807	3	60 Nd	$L\alpha_2$	$L_{III}M_{IV}$	5.2077
2.3869	2	56 Ba	$L\beta_{10}$	$L_{I}M_{IV}$	5.1941
2.3880	5	53 I	L_{I}	Abs. Edge	5.192
2.3913	2	53 I	$L\gamma_4$	$L_{I}O_{II,III}$	5.1848
2.3948	2	63 Eu	Ll	$L_{III}M_{I}$	5.1772
2.40435	6	56 Ba	$L\beta_{2,15}$	$L_{III}N_{IV,V}$	5.1565
2.4094	4	60 Nd	$L\eta$	$L_{II}M_{I}$	5.1457
2.4105	3	57 La	$L\beta_3$	$L_{I}M_{III}$	5.1434
2.4174	2	55 Cs	$L\gamma_5$	$L_{II}N_{I}$	5.1287
2.4292	1	54 Xe	L_{II}	Abs. Edge	5.1037
2.442	9	90 Th		$M_{I}O_{III}$	5.08
2.443	4	92 U		$M_{II}O_{IV}$	5.075
2.4475	2	53 I	$L\gamma_{2,3}$	$L_{I}N_{II,III}$	5.0657
2.4493	3	57 La	$L\beta_4$	$L_{I}M_{II}$	5.0620
2.45891	5	57 La	$L\beta_1$	$L_{II}M_{IV}$	5.0421
2.4630	2	59 Pr	$L\alpha_1$	$L_{III}M_{V}$	5.0337
2.4729	3	59 Pr	$L\alpha_2$	$L_{III}M_{IV}$	5.0135
2.4740	1	55 Cs	L_{III}	Abs. Edge	5.0113
2.4783	2	55 Cs	$L\beta_9$	$L_{I}M_{V}$	5.0026
2.4823	4	62 Sm	Ll	$L_{III}M_{I}$	4.9945
2.4826	2	56 Ba	$L\beta_6$	$L_{III}N_{I}$	4.9939
2.4849	2	55 Cs	$L\beta_7$	$L_{III}O_{I}$	4.9893
2.4920	2	55 Cs	$L\beta_{10}$	$L_{I}M_{IV}$	4.9752
2.49734	5	22 Ti	K	Abs. Edge	4.96452
2.4985	2	22 Ti	$K\beta_5$	$KM_{IV,V}$	4.9623
2.50356	2	23 V	$K\alpha_1$	KL_{III}	4.95220
2.50738	2	23 V	$K\alpha_2$	KL_{II}	4.94464
2.5099	1	52 Te	L_{I}	Abs. Edge	4.9397
2.5113	2	52 Te	$L\gamma_4$	$L_{I}O_{II,III}$	4.9369
2.5118	2	55 Cs	$L\beta_{2,15}$	$L_{III}N_{IV,V}$	4.9359
2.512	3	59 Pr	$L\eta$	$L_{II}M_{I}$	4.935
2.51391	2	22 Ti	$K\beta_{1,3}$	$KM_{II,III}$	4.93181
2.5164	2	56 Ba	$L\beta_3$	$L_{I}M_{III}$	4.9269
2.527	4	91 Pa		$M_{II}O_{IV}$	4.906
2.5542	5	53 I	L_{II}	Abs. Edge	4.8540
2.5553	2	56 Ba	$L\beta_4$	$L_{I}M_{II}$	4.8519
2.5615	2	58 Ce	$L\alpha_1$	$L_{III}M_{V}$	4.8402
2.5674	2	52 Te	$L\gamma_{2,3}$	$L_{I}N_{II,III}$	4.8290
2.56821	5	56 Ba	$L\beta_1$	$L_{II}M_{IV}$	4.82753
2.5706	3	58 Ce	$L\alpha_2$	$L_{III}M_{IV}$	4.8230
2.58244	8	53 I	$L\gamma_1$	$L_{II}N_{IV}$	4.8009
2.5926	1	54 Xe	L_{III}	Abs. Edge	4.7822
2.5932	2	55 Cs	$L\beta_6$	$L_{III}N_{I}$	4.7811
2.618	5	90 Th		$M_{II}O_{IV}$	4.735
2.6203	4	58 Ce	$L\eta$	$L_{II}M_{I}$	4.7315
2.6285	2	55 Cs	$L\beta_3$	$L_{I}M_{III}$	4.7167
2.6388	1	51 Sb	L_{I}	Abs. Edge	4.6984
2.6398	2	51 Sb	$L\gamma_4$	$L_{I}O_{II,III}$	4.6967
2.65710	9	53 I	$L\gamma_5$	$L_{II}N_{I}$	4.6660
2.66570	5	57 La	$L\alpha_1$	$L_{III}M_{V}$	4.65097
2.6666	2	55 Cs	$L\beta_4$	$L_{I}M_{II}$	4.6494
2.67533	5	57 La	$L\alpha_2$	$L_{III}M_{IV}$	4.63423
2.6760	4	60 Nd	Ll	$L_{III}M_{I}$	4.6330
2.6837	2	55 Cs	$L\beta_1$	$L_{II}M_{IV}$	4.6198
2.6879	1	52 Te	L_{II}	Abs. Edge	4.6126
2.6953	2	51 Sb	$L\gamma_{2,3}$	$L_{I}N_{II,III}$	4.5999
2.71241	6	52 Te	$L\gamma_1$	$L_{II}N_{IV}$	4.5709
2.71352	9	53 I	$L\beta_9$	$L_{I}M_{V}$	4.5690
2.7196	5	53 I	L_{III}	Abs. Edge	4.5587
2.72104	9	53 I	$L\beta_{10}$	$L_{I}M_{IV}$	4.5564

Wavelength Å*	p.e.	Element	Designation		keV
2.7288	3	53 I	$L\beta_7$	$L_{III}O_I$	4.5435
2.740	3	57 La	$L\eta$	$L_{II}M_I$	4.525
2.74851	2	22 Ti	$K\alpha_1$	KL_{III}	4.51084
2.75053	8	53 I	$L\beta_{2,15}$	$L_{III}N_{IV,V}$	4.5075
2.75216	2	22 Ti	$K\alpha_2$	KL_{II}	4.50486
2.753	8	92 U		M_IN_{III}	4.50
2.762	1	21 Sc	K	Abs. Edge	4.489
2.7634	3	21 Sc	$K\beta_5$	$KM_{IV,V}$	4.4865
2.77595	5	56 Ba	$L\alpha_1$	$L_{III}M_V$	4.46626
2.7769	1	50 Sn	L_I	Abs. Edge	4.4648
2.7775	2	50 Sn	$L\gamma_4$	$L_IO_{II,III}$	4.4638
2.7796	2	21 Sc	$K\beta_{1,3}$	$KM_{II,III}$	4.4605
2.7841	4	59 Pr	Ll	$L_{III}M_I$	4.4532
2.78553	5	56 Ba	$L\alpha_2$	$L_{III}M_{IV}$	4.45090
2.79007	9	52 Te	$L\gamma_5$	$L_{II}N_I$	4.4437
2.817	2	92 U		$M_{II}N_{IV}$	4.401
2.8294	5	51 Sb	L_{II}	Abs. Edge	4.3819
2.8327	2	50 Sn	$L\gamma_{2,3}$	$L_IN_{II,III}$	4.3768
2.83672	9	53 I	$L\beta_6$	$L_{III}N_I$	4.3706
2.83897	9	52 Te	$L\beta_9$	L_IM_V	4.3671
2.84679	9	52 Te	$L\beta_{10}$	L_IM_{IV}	4.3551
2.85159	3	51 Sb	$L\gamma_1$	$L_{II}N_{IV}$	4.34779
2.8555	1	52 Te	L_{III}	Abs. Edge	4.3418
2.8627	3	56 Ba	$L\eta$	$L_{II}M_I$	4.3309
2.8634	3	52 Te	$L\beta_7$	$L_{III}O_I$	4.3298
2.87429	9	53 I	$L\beta_3$	L_IM_{III}	4.3134
2.88217	8	52 Te	$L\beta_{2,15}$	$L_{III}N_{IV,V}$	4.3017
2.884	5	92 U	M_{III}	Abs. Edge	4.299
2.8917	4	58 Ce	Ll	$L_{III}M_I$	4.2875
2.8924	2	55 Cs	$L\alpha_1$	$L_{III}M_V$	4.2865
2.9020	2	55 Cs	$L\alpha_2$	$L_{III}M_{IV}$	4.2722
2.910	2	91 Pa		$M_{II}N_{IV}$	4.260
2.91207	9	53 I	$L\beta_4$	L_IM_{II}	4.2575
2.92	2	92 U		M_IN_{II}	4.25
2.9260	1	49 In	L_I	Abs. Edge	4.2373
2.9264	2	49 In	$L\gamma_4$	$L_IO_{II,III}$	4.2367
2.93187	9	51 Sb	$L\gamma_5$	$L_{II}N_I$	4.2287
2.934	8	90 Th		M_IN_{III}	4.23
2.93744	6	53 I	$L\beta_1$	$L_{II}M_{IV}$	4.22072
2.948	2	92 U		$M_{III}O_{IV,V}$	4.205
2.97088	9	52 Te	$L\beta_6$	$L_{III}N_I$	4.1732
2.97261	9	51 Sb	$L\beta_9$	L_IM_V	4.1708
2.97917	9	51 Sb	$L\beta_{10}$	L_IM_{IV}	4.1616
2.9800	2	49 In	$L\gamma_{2,3}$	$L_IN_{II,III}$	4.1605
2.9823	1	50 Sn	L_{II}	Abs. Edge	4.1573
2.9932	2	55 Cs	$L\eta$	$L_{II}M_I$	4.1421
3.0003	1	51 Sb	L_{III}	Abs. Edge	4.1323
3.00115	3	50 Sn	$L\gamma_1$	$L_{II}N_{IV}$	4.13112
3.0052	3	51 Sb	$L\beta_7$	$L_{III}O_I$	4.1255
3.006	3	57 La	Ll	$L_{III}M_I$	4.124
3.00893	9	52 Te	$L\beta_3$	L_IM_{III}	4.1204
3.011	2	90 Th		$M_{II}N_{IV}$	4.117
3.0166	2	54 Xe	$L\alpha_1$	$L_{III}M_V$	4.1099
3.02335	3	51 Sb	$L\beta_{2,15}$	$L_{III}N_{IV,V}$	4.10078
3.0309	1	21 Sc	$K\alpha_1$	KL_{III}	4.0906
3.0342	1	21 Sc	$K\alpha_2$	KL_{II}	4.0861
3.038	2	91 Pa		$M_{III}O_{IV,V}$	4.081
3.04661	9	52 Te	$L\beta_4$	L_IM_{II}	4.0695
3.068	5	90 Th	M_{III}	Abs. Edge	4.041
3.0703	1	20 Ca	K	Abs. Edge	4.0381
3.0746	3	20 Ca	$K\beta_5$	$KM_{IV,V}$	4.0325
3.07677	6	52 Te	$L\beta_1$	$L_{II}M_{IV}$	4.02958
3.08475	9	50 Sn	$L\gamma_5$	$L_{II}N_I$	4.0192
3.0849	1	48 Cd	L_I	Abs. Edge	4.0190
3.0897	2	20 Ca	$K\beta_{1,3}$	$KM_{II,III}$	4.0127
3.094	5	83 Bi	M_I	Abs. Edge	4.007
3.11513	9	50 Sn	$L\beta_9$	L_IM_V	3.9800
3.11513	9	51 Sb	$L\beta_6$	$L_{III}N_I$	3.9800
3.115	7	92 U		$M_{III}O_I$	3.980
3.12170	9	50 Sn	$L\beta_{10}$	L_IM_{IV}	3.9716
3.131	3	90 Th		$M_{III}O_{IV,V}$	3.959
3.1355	2	56 Ba	Ll	$L_{III}M_I$	3.9541
3.1377	2	48 Cd	$L\gamma_2$	L_IN_{II}	3.9513
3.1473	1	49 In	L_{II}	Abs. Edge	3.9393
3.14860	6	53 I	$L\alpha_1$	$L_{III}M_V$	3.93765
3.15258	9	51 Sb	$L\beta_3$	L_IM_{III}	3.9327
3.1557	1	50 Sn	L_{III}	Abs. Edge	3.9288
3.1564	3	50 Sn	$L\beta_7$	$L_{III}O_I$	3.9279
3.15791	6	53 I	$L\alpha_2$	$L_{III}M_{IV}$	3.92604
3.16213	4	49 In	$L\gamma_1$	$L_{II}N_{IV}$	3.92081
3.17505	3	50 Sn	$L\beta_{2,15}$	$L_{III}N_{IV,V}$	3.90486
3.19014	9	51 Sb	$L\beta_4$	L_IM_{II}	3.8364
3.217	5	82 Pb	M_I	Abs. Edge	3.854
3.22567	4	51 Sb	$L\beta_1$	$L_{II}M_{IV}$	3.84357
3.245	9	91 Pa		$M_{III}O_I$	3.82
3.24907	9	49 In	$L\gamma_5$	$L_{II}N_I$	3.8159
3.2564	1	47 Ag	L_I	Abs. Edge	3.8072
3.2670	2	55 Cs	Ll	$L_{III}M_I$	3.7950
3.26763	9	49 In	$L\beta_9$	L_IM_V	3.7942
3.26901	9	50 Sn	$L\beta_6$	$L_{III}N_I$	3.7926
3.27404	9	49 In	$L\beta_{10}$	L_IM_{IV}	3.7868
3.27979	9	53 I	$L\eta$	$L_{II}M_I$	3.7801
3.283	9	90 Th		$M_{III}O_I$	3.78
3.28920	6	52 Te	$L\alpha_1$	$L_{III}M_V$	3.76933
3.29846	9	52 Te	$L\alpha_2$	$L_{III}M_{IV}$	3.7588
3.30585	3	50 Sn	$L\beta_3$	L_IM_{III}	3.7500
3.30635	9	47 Ag	$L\gamma_3$	L_IN_{III}	3.7498
3.31216	9	47 Ag	$L\gamma_2$	L_IN_{II}	3.7432
3.3237	1	49 In	L_{III}	Abs. Edge	3.7302
3.324	4	49 In	$L\beta_7$	$L_{III}O_I$	3.730
3.3257	1	48 Cd	L_{II}	Abs. Edge	3.7280
3.329	4	92 U		$M_{II}N_I$	3.724
3.333	5	92 U	M_{IV}	Abs. Edge	3.720
3.33564	6	48 Cd	$L\gamma_1$	$L_{II}N_{IV}$	3.71686
3.33838	3	49 In	$L\beta_{2,15}$	$L_{III}N_{IV,V}$	3.71381
3.34335	9	50 Sn	$L\beta_4$	L_IM_{II}	3.7083
3.346	5	81 Tl	M_I	Abs. Edge	3.705
3.35839	3	20 Ca	$K\alpha_1$	KL_{III}	3.69168
3.359	5	83 Bi	M_{II}	Abs. Edge	3.691
3.36166	3	20 Ca	$K\alpha_2$	KL_{II}	3.68809
3.38487	3	50 Sn	$L\beta_1$	$L_{II}M_{IV}$	3.66280
3.42551	9	48 Cd	$L\gamma_5$	$L_{II}N_I$	3.61935
3.43015	9	48 Cd	$L\beta_9$	L_IM_V	3.61445
3.43606	9	49 In	$L\beta_6$	$L_{III}N_I$	3.60823

Wavelength Å*	p.e.	Element	Designation		keV
3.4365	1	19 K	K	Abs. Edge	3.6078
3.4367	2	48 Cd	$L\beta_{10}$	$L_{I}M_{IV}$	3.6075
3.437	1	46 Pd	L_{I}	Abs. Edge	3.607
3.43832	9	52 Te	$L\eta$	$L_{II}M_{I}$	3.60586
3.43941	4	51 Sb	$L\alpha_1$	$L_{III}M_{V}$	3.60472
3.441	5	91 Pa		$M_{II}N_{I}$	3.603
3.4413	4	19 K	$K\beta_5$	$KM_{IV,V}$	3.6027
3.44840	6	51 Sb	$L\alpha_2$	$L_{III}M_{IV}$	3.59532
3.4539	2	19 K	$K\beta_{1,3}$	$KM_{II,III}$	3.5896
3.46984	9	49 In	$L\beta_3$	$L_{I}M_{III}$	3.57311
3.478	5	80 Hg	M_{I}	Abs. Edge	3.565
3.479	1	92 U	$M\gamma$	$M_{III}N_{V}$	3.563
3.4892	2	46 Pd	$L\gamma_{2,3}$	$L_{I}N_{II,III}$	3.5533
3.492	5	82 Pb	M_{II}	Abs. Edge	3.550
3.497	5	92 U	M_{V}	Abs. Edge	3.545
3.5047	1	48 Cd	L_{III}	Abs. Edge	3.5376
3.50697	9	49 In	$L\beta_4$	$L_{I}M_{II}$	3.53528
3.51408	4	48 Cd	$L\beta_{2,15}$	$L_{III}N_{IV,V}$	3.52812
3.5164	1	47 Ag	L_{II}	Abs. Edge	3.5258
3.521	2	92 U		$M_{III}N_{IV}$	3.521
3.52260	4	47 Ag	$L\gamma_1$	$L_{II}N_{IV}$	3.51959
3.537	9	90 Th		$M_{II}N_{I}$	3.505
3.55531	4	49 In	$L\beta_1$	$L_{II}M_{IV}$	3.48721
3.557	5	90 Th	M_{IV}	Abs. Edge	3.485
3.55754	9	53 I	Ll	$L_{III}M_{I}$	3.48502
3.576	1	92 U		$M_{IV}O_{II}$	3.4666
3.577	1	91 Pa	$M\gamma$	$M_{III}N_{V}$	3.4657
3.59994	3	50 Sn	$L\alpha_1$	$L_{III}M_{V}$	3.44398
3.60497	9	47 Ag	$L\beta_9$	$L_{I}M_{V}$	3.43917
3.60765	9	51 Sb	$L\eta$	$L_{II}M_{I}$	3.43661
3.60891	4	50 Sn	$L\alpha_2$	$L_{III}M_{IV}$	3.43542
3.61158	9	47 Ag	$L\beta_{10}$	$L_{I}M_{IV}$	3.43287
3.614	2	91 Pa		$M_{III}N_{IV}$	3.430
3.61467	9	48 Cd	$L\beta_6$	$L_{III}N_{I}$	3.42994
3.61638	9	47 Ag	$L\gamma_5$	$L_{II}N_{I}$	3.42832
3.616	5	79 Au	M_{I}	Abs. Edge	3.428
3.629	5	45 Rh	L_{I}	Abs. Edge	3.417
3.634	5	81 Tl	M_{II}	Abs. Edge	3.412
3.64495	9	48 Cd	$L\beta_3$	$L_{I}M_{III}$	3.40145
3.679	2	90 Th	$M\gamma$	$M_{III}N_{V}$	3.370
3.68203	9	48 Cd	$L\beta_4$	$L_{I}M_{II}$	3.36719
3.6855	2	45 Rh	$L\gamma_{2,3}$	$L_{I}N_{II,III}$	3.3640
3.691	2	91 Pa		$M_{IV}O_{II}$	3.359
3.6999	1	47 Ag	L_{III}	Abs. Edge	3.35096
3.70335	3	47 Ag	$L\beta_{2,15}$	$L_{III}N_{IV,V}$	3.34781
3.716	1	92 U	$M\beta$	$M_{IV}N_{VI}$	3.3367
3.71696	9	52 Te	Ll	$L_{III}M_{I}$	3.33555
3.718	3	90 Th		$M_{III}N_{IV}$	3.335
3.7228	1	46 Pd	L_{II}	Abs. Edge	3.33031
3.7246	2	46 Pd	$L\gamma_1$	$L_{II}N_{IV}$	3.3287
3.729	5	90 Th	M_{V}	Abs. Edge	3.325
3.73823	4	48 Cd	$L\beta_1$	$L_{II}M_{IV}$	3.31657
3.740	9	83 Bi		$M_{I}N_{III}$	3.315
3.7414	2	19 K	$K\alpha_1$	KL_{III}	3.3138
3.7445	2	19 K	$K\alpha_2$	KL_{II}	3.3111
3.760	9	90 Th		$M_{V}P_{III}$	3.298
3.762	5	78 Pt	M_{I}	Abs. Edge	3.296

Wavelength Å*	p.e.	Element	Designation		keV
3.77192	4	49 In	$L\alpha_1$	$L_{III}M_{V}$	3.28694
3.78073	6	49 In	$L\alpha_2$	$L_{III}M_{IV}$	3.27929
3.783	5	80 Hg	M_{II}	Abs. Edge	3.277
3.78876	9	50 Sn	$L\eta$	$L_{II}M_{I}$	3.27234
3.7920	2	46 Pd	$L\beta_9$	$L_{I}M_{V}$	3.2696
3.7988	2	46 Pd	$L\beta_{10}$	$L_{I}M_{IV}$	3.2637
3.80774	9	47 Ag	$L\beta_6$	$L_{III}N_{I}$	3.25603
3.808	4	90 Th		$M_{IV}O_{II}$	3.256
3.8222	2	46 Pd	$L\gamma_5$	$L_{II}N_{I}$	3.2437
3.827	1	91 Pa	$M\beta$	$M_{IV}N_{VI}$	3.2397
3.83313	9	47 Ag	$L\beta_3$	$L_{I}M_{III}$	3.23446
3.834	4	83 Bi		$M_{II}N_{IV}$	3.234
3.835	5	44 Ru	L_{I}	Abs. Edge	3.233
3.87023	5	47 Ag	$L\beta_4$	$L_{I}M_{II}$	3.20346
3.87090	5	18 A	K	Abs. Edge	3.20290
3.872	9	82 Pb		$M_{I}N_{III}$	3.202
3.8860	2	18 A	$K\beta_{1,3}$	$KM_{II,III}$	3.1905
3.88826	9	51 Sb	Ll	$L_{III}M_{I}$	3.18860
3.892	9	83 Bi		$M_{I}N_{II}$	3.185
3.8977	2	44 Ru	$L\gamma_{2,3}$	$L_{I}N_{II,III}$	3.1809
3.904	5	83 Bi	M_{III}	Abs. Edge	3.176
3.9074	1	46 Pd	L_{III}	Abs. Edge	3.17298
3.90887	4	46 Pd	$L\beta_{2,15}$	$L_{III}N_{IV,V}$	3.17179
3.910	1	92 U	$M\alpha_1$	$M_{V}N_{VII}$	3.1708
3.915	5	77 Ir	M_{I}	Abs. Edge	3.167
3.924	1	92 U	$M\alpha_2$	$M_{V}N_{VI}$	3.1595
3.932	6	83 Bi		$M_{III}O_{IV,V}$	3.153
3.93473	3	47 Ag	$L\beta_1$	$L_{II}M_{IV}$	3.15094
3.936	5	79 Au	M_{II}	Abs. Edge	3.150
3.941	1	90 Th	$M\beta$	$M_{IV}N_{VI}$	3.1458
3.9425	5	45 Rh	L_{II}	Abs. Edge	3.1448
3.9437	2	45 Rh	$L\gamma_1$	$L_{II}N_{IV}$	3.1438
3.95635	4	48 Cd	$L\alpha_1$	$L_{III}M_{V}$	3.13373
3.96496	6	48 Cd	$L\alpha_2$	$L_{III}M_{IV}$	3.12691
3.968	5	82 Pb		$M_{II}N_{IV}$	3.124
3.98327	9	49 In	$L\eta$	$L_{II}M_{I}$	3.11254
4.013	9	81 Tl		$M_{I}N_{III}$	3.089
4.0162	2	46 Pd	$L\beta_6$	$L_{III}N_{I}$	3.0870
4.022	1	91 Pa	$M\alpha_1$	$M_{V}N_{VII}$	3.0823
4.0346	2	46 Pd	$L\beta_3$	$L_{I}M_{III}$	3.0730
4.035	3	91 Pa	$M\alpha_2$	$M_{V}N_{VI}$	3.072
4.0451	2	45 Rh	$L\gamma_5$	$L_{II}N_{I}$	3.0650
4.047	1	82 Pb	M_{III}	Abs. Edge	3.0632
4.058	5	43 Te	L_{I}	Abs. Edge	3.055
4.069	6	82 Pb		$M_{III}O_{IV,V}$	3.047
4.0711	2	46 Pd	$L\beta_4$	$L_{I}M_{II}$	3.0454
4.071	5	76 Os	M_{I}	Abs. Edge	3.045
4.07165	9	50 Sn	Ll	$L_{III}M_{I}$	3.04499
4.093	5	78 Pt	M_{II}	Abs. Edge	3.029
4.105	9	83 Bi		$M_{III}O_{I}$	3.021
4.116	4	81 Tl		$M_{II}N_{IV}$	3.013
4.1299	5	45 Rh	L_{III}	Abs. Edge	3.0021
4.1310	2	45 Rh	$L\beta_{2,15}$	$L_{III}N_{IV,V}$	3.0013
4.1381	9	90 Th	$M\alpha_1$	$M_{V}N_{VII}$	2.9961
4.14622	5	46 Pd	$L\beta_1$	$L_{II}M_{IV}$	2.99022
4.151	2	90 Th	$M\alpha_2$	$M_{V}N_{VI}$	2.987
4.15443	3	47 Ag	$L\alpha_1$	$L_{III}M_{V}$	2.98431

Wavelength Å*	p.e.	Element	Designation		keV
4.16294	5	47 Ag	$L\alpha_2$	$L_{III}M_{IV}$	2.97821
4.180	1	44 Ru	L_{II}	Abs. Edge	2.9663
4.1822	2	44 Ru	$L\gamma_1$	$L_{II}N_{IV}$	2.9645
4.19180	5	18 A	$K\alpha_1$	KL_{III}	2.95770
4.19315	9	48 Cd	$L\eta$	$L_{II}M_I$	2.95675
4.19474	5	18 A	$K\alpha_2$	KL_{II}	2.95563
4.198	1	81 Tl	M_{III}	Abs. Edge	2.9535
4.216	6	81 Tl		$M_{III}O_{IV,V}$	2.941
4.236	5	75 Re	M_I	Abs. Edge	2.927
4.2417	2	45 Rh	$L\beta_6$	$L_{III}N_I$	2.9229
4.244	9	82 Pb		$M_{III}O_I$	2.921
4.2522	2	45 Rh	$L\beta_3$	L_IM_{III}	2.9157
4.260	5	77 Ir	M_{II}	Abs. Edge	2.910
4.26873	9	49 In	Ll	$L_{III}M_I$	2.90440
4.2873	2	44 Ru	$L\gamma_5$	$L_{II}N_I$	2.8918
4.2888	2	45 Rh	$L\beta_4$	L_IM_{II}	2.8908
4.300	9	79 Au		M_IN_{III}	2.883
4.304	5	42 Mo	L_I	Abs. Edge	2.881
4.330	2	92 U		$M_{III}N_I$	2.863
4.355	1	80 Hg	M_{III}	Abs. Edge	2.8469
4.36767	5	46 Pd	$L\alpha_1$	$L_{III}M_V$	2.83861
4.369	1	44 Ru	L_{III}	Abs. Edge	2.8377
4.3718	2	44 Ru	$L\beta_{2,15}$	$L_{III}N_{IV,V}$	2.8360
4.37414	4	45 Rh	$L\beta_1$	$L_{II}M_{IV}$	2.83441
4.37588	7	46 Pd	$L\alpha_2$	$L_{III}M_{IV}$	2.83329
4.3800	2	42 Mo	$L\gamma_{2,3}$	$L_IN_{II,III}$	2.8306
4.3971	1	17 Cl	K	Abs. Edge	2.81960
4.4034	3	17 Cl	$K\beta$	KM	2.8156
4.407	5	74 W	M_I	Abs. Edge	2.813
4.4183	2	47 Ag	$L\eta$	$L_{II}M_I$	2.8061
4.432	4	79 Au		$M_{II}N_{IV}$	2.797
4.433	5	76 Os	M_{II}	Abs. Edge	2.797
4.436	1	43 Te	L_{II}	Abs. Edge	2.7948
4.44	2	74 W		$M_IO_{II,III}$	2.79
4.450	4	91 Pa		$M_{III}N_I$	2.786
4.460	9	78 Pt		M_IN_{III}	2.780
4.48014	9	48 Cd	Ll	$L_{III}M_I$	2.76735
4.4866	3	44 Ru	$L\beta_3$	L_IM_{III}	2.7634
4.4866	3	44 Ru	$L\beta_6$	$L_{III}N_I$	2.7634
4.518	1	79 Au	M_{III}	Abs. Edge	2.7439
4.522	6	79 Au		$M_{III}O_{IV,V}$	2.742
4.5230	2	44 Ru	$L\beta_4$	L_IM_{II}	2.7411
4.532	2	83 Bi	$M\gamma$	$M_{III}N_V$	2.735
4.568	5	90 Th		$M_{III}N_I$	2.714
4.571	5	83 Bi		$M_{III}N_{IV}$	2.712
4.572	5	83 Bi	M_{IV}	Abs. Edge	2.711
4.575	5	41 Nb	L_I	Abs. Edge	2.710
4.585	5	73 Ta	M_I	Abs. Edge	2.704
4.59	2	83 Bi		$M_{IV}P_{II,III}$	2.70
4.59743	9	45 Rh	$L\alpha_1$	$L_{III}M_V$	2.69674
4.601	4	78 Pt		$M_{II}N_{IV}$	2.695
4.60545	9	45 Rh	$L\alpha_2$	$L_{III}M_{IV}$	2.69205
4.620	5	75 Re	M_{II}	Abs. Edge	2.684
4.62058	3	44 Ru	$L\beta_1$	$L_{II}M_{IV}$	2.68323
4.625	5	92 U		$M_{IV}N_{III}$	2.681
4.630	1	43 Tc	L_{III}	Abs. Edge	2.6780
4.631	9	77 Ir		M_IN_{III}	2.677
4.6542	2	41 Nb	$L\gamma_{2,3}$	$L_IN_{II,III}$	2.6638
4.655	8	82 Pb		$M_{II}N_I$	2.664
4.6605	2	46 Pd	$L\eta$	$L_{II}M_I$	2.6603
4.674	1	82 Pb	$M\gamma$	$M_{III}N_V$	2.6527
4.686	1	78 Pt	M_{III}	Abs. Edge	2.6459
4.694	8	78 Pt		$M_{III}O_{IV,V}$	2.641
4.703	9	79 Au		$M_{III}O_I$	2.636
4.7076	2	47 Ag	Ll	$L_{III}M_I$	2.6337
4.715	3	82 Pb		$M_{III}N_{IV}$	2.630
4.719	1	42 Mo	L_{II}	Abs. Edge	2.6274
4.7258	2	42 Mo	$L\gamma_1$	$L_{II}N_{IV}$	2.6235
4.7278	1	17 Cl	$K\alpha_1$	KL_{III}	2.62239
4.7307	1	17 Cl	$K\alpha_2$	KL_{II}	2.62078
4.757	5	82 Pb	M_{IV}	Abs. Edge	2.606
4.764	5	83 Bi	M_V	Abs. Edge	2.603
4.780	4	77 Ir		$M_{II}N_{IV}$	2.594
4.79	2	76 Os		M_IN_{III}	2.59
4.815	5	74 W	M_{II}	Abs. Edge	2.575
4.823	3	83 Bi		$M_{IV}O_{II}$	2.571
4.823	4	81 Tl	$M\gamma$	$M_{III}N_V$	2.571
4.8369	2	42 Mo	$L\gamma_5$	$L_{II}N_I$	2.5632
4.84575	5	44 Ru	$L\alpha_1$	$L_{III}M_V$	2.55855
4.85381	7	44 Ru	$L\alpha_2$	$L_{III}M_{IV}$	2.55431
4.861	1	77 Ir	M_{III}	Abs. Edge	2.5505
4.865	5	81 Tl		$M_{III}N_{IV}$	2.548
4.869	9	77 Ir		$M_{III}O_{IV,V}$	2.546
4.876	9	78 Pt		$M_{III}O_I$	2.543
4.879	5	40 Zr	L_I	Abs. Edge	2.541
4.8873	8	43 Tc	$L\beta_1$	$L_{II}M_{IV}$	2.5368
4.909	1	83 Bi	$M\beta$	$M_{IV}N_{VI}$	2.5255
4.911	5	90 Th		$M_{IV}N_{III}$	2.524
4.913	1	42 Mo	L_{III}	Abs. Edge	2.5234
4.9217	2	45 Rh	$L\eta$	$L_{II}M_I$	2.5191
4.9232	2	42 Mo	$L\beta_{2,15}$	$L_{III}N_{IV,V}$	2.5183
4.946	2	92 U	$M\zeta_1$	M_VN_{III}	2.507
4.952	5	81 Tl	M_{IV}	Abs. Edge	2.504
4.9525	3	46 Pd	Ll	$L_{III}M_I$	2.5034
4.9536	3	40 Zr	$L\gamma_{2,3}$	$L_IN_{II,III}$	2.5029
4.955	4	76 Os		$M_{II}N_{IV}$	2.502
4.955	5	82 Pb	M_V	Abs. Edge	2.502
4.984	2	80 Hg	$M\gamma$	$M_{III}N_V$	2.4875
5.004	9	82 Pb		$M_{IV}O_{II}$	2.477
5.0133	3	42 Mo	$L\beta_3$	L_IM_{III}	2.4730
5.0185	1	16 S	K	Abs. Edge	2.47048
5.020	5	73 Ta	M_{II}	Abs. Edge	2.470
5.0233	3	16 S	$K\beta_x$	KM	2.4681
5.031	1	41 Nb	L_{II}	Abs. Edge	2.4641
5.0316	2	16 S	$K\beta_1$	KM	2.46404
5.0361	3	41 Nb	$L\gamma_1$	$L_{II}N_{IV}$	2.4618
5.043	5	76 Os	M_{III}	Abs. Edge	2.458
5.0488	3	42 Mo	$L\beta_4$	L_IM_{II}	2.4557
5.0488	5	42 Mo	$L\beta_6$	$L_{III}N_I$	2.4557
5.050	2	92 U	$M\zeta_2$	$M_{IV}N_{II}$	2.4548
5.076	1	82 Pb	$M\beta$	$M_{IV}N_{VI}$	2.4427
5.092	2	91 Pa	$M\zeta_1$	M_VN_{III}	2.4350
5.1148	3	43 Tc	$L\alpha_1$	$L_{III}M_V$	2.4240
5.118	1	83 Bi	$M\alpha_1$	M_VN_{VII}	2.4226

Wavelength Å*	p.e.	Element	Designation		keV
5.130	2	83 Bi	$M\alpha_2$	M_VN_{VI}	2.4170
5.145	4	79 Au	$M\gamma$	$M_{III}N_V$	2.410
5.1517	3	41 Nb	$L\gamma_5$	$L_{II}N_I$	2.4066
5.153	5	81 Tl	M_V	Abs. Edge	2.406
5.157	5	80 Hg	M_{IV}	Abs. Edge	2.404
5.168	9	82 Pb		M_VO_{III}	2.399
5.172	9	74 W		M_IN_{III}	2.397
5.17708	8	42 Mo	$L\beta_1$	$L_{II}M_{IV}$	2.39481
5.186	5	79 Au		$M_{III}N_{IV}$	2.391
5.193	2	91 Pa	$M\zeta_2$	$M_{IV}N_{II}$	2.3876
5.196	9	81 Tl		$M_{IV}O_{II}$	2.386
5.2050	2	44 Ru	$L\eta$	$L_{II}M_I$	2.38197
5.217	5	39 Y	L_I	Abs. Edge	2.377
5.2169	3	45 Rh	Ll	$L_{III}M_I$	2.3765
5.230	1	41 Nb	L_{III}	Abs. Edge	2.3706
5.234	5	75 Re	M_{III}	Abs. Edge	2.369
5.2379	3	41 Nb	$L\beta_{2,15}$	$L_{III}N_{IV,V}$	2.3670
5.245	5	90 Th	$M\zeta_1$	M_VN_{III}	2.364
5.249	1	81 Tl	$M\beta$	$M_{IV}N_{VI}$	2.3621
5.2830	3	39 Y	$L\gamma_{2,3}$	$L_IN_{II,III}$	2.3468
5.286	1	82 Pb	$M\alpha_1$	M_VN_{VII}	2.3455
5.299	2	82 Pb	$M\alpha_2$	M_VN_{VI}	2.3397
5.3102	3	41 Nb	$L\beta_3$	L_IM_{III}	2.3348
5.319	4	78 Pt	$M\gamma$	$M_{III}N_V$	2.331
5.340	5	90 Th	$M\zeta_2$	$M_{IV}N_{II}$	2.322
5.3455	3	41 Nb	$L\beta_4$	L_IM_{II}	2.3194
5.357	4	74 W		$M_{II}N_{IV}$	2.314
5.357	5	78 Pt		$M_{III}N_{IV}$	2.314
5.36	1	80 Hg	M_V	Abs. Edge	2.313
5.3613	3	41 Nb	$L\beta_6$	$L_{III}N_I$	2.3125
5.37216	7	16 S	$K\alpha_1$	KL_{III}	2.30784
5.374	5	79 Au	M_{IV}	Abs. Edge	2.307
5.37496	8	16 S	$K\alpha_2$	KL_{II}	2.30664
5.378	1	40 Zr	L_{II}	Abs. Edge	2.3053
5.3843	3	40 Zr	$L\gamma_1$	$L_{II}N_{IV}$	2.3027
5.40	2	73 Ta		M_IN_{III}	2.295
5.40655	8	42 Mo	$L\alpha_1$	$L_{III}M_V$	2.29316
5.41437	8	42 Mo	$L\alpha_2$	$L_{III}M_{IV}$	2.28985
5.4318	9	80 Hg	$M\beta$	$M_{IV}N_{VI}$	2.2825
5.435	1	74 W	M_{III}	Abs. Edge	2.2811
5.460	1	81 Tl	$M\alpha_1$	M_VN_{VII}	2.2706
5.472	2	81 Tl	$M\alpha_2$	M_VN_{VI}	2.2656
5.4923	3	41 Nb	$L\beta_1$	$L_{II}M_{IV}$	2.2574
5.4977	3	40 Zr	$L\gamma_5$	$L_{II}N_I$	2.2551
5.500	4	77 Ir	$M\gamma$	$M_{III}N_V$	2.254
5.5035	3	44 Ru	Ll	$L_{III}M_I$	2.2528
5.537	8	83 Bi		$M_{III}N_I$	2.239
5.540	5	77 Ir		$M_{III}N_{IV}$	2.238
5.570	4	73 Ta		$M_{II}N_{IV}$	2.226
5.579	1	40 Zr	L_{III}	Abs. Edge	2.2225
5.584	5	79 Au	M_V	Abs. Edge	2.220
5.5863	3	40 Zr	$L\beta_{2,15}$	$L_{III}N_{IV,V}$	2.2194
5.59	1	78 Pt	M_{IV}	Abs. Edge	2.217
5.592	5	38 Sr	L_I	Abs. Edge	2.217
5.624	1	79 Au	$M\beta$	$M_{IV}N_{VI}$	2.2046
5.628	8	74 W		$M_{III}O_I$	2.203
5.6330	3	40 Zr	$L\beta_3$	L_IM_{III}	2.2010
5.6445	3	38 Sr	$L\gamma_{2,3}$	$L_IN_{II,III}$	2.1965
5.6476	9	80 Hg	$M\alpha_1$	M_VN_{VII}	2.1953
5.650	5	73 Ta	M_{III}	Abs. Edge	2.194
5.6681	3	40 Zr	$L\beta_4$	L_IM_{II}	2.1873
5.67	3	73 Ta		$M_{III}O_{IV,V}$	2.19
5.682	4	76 Os	$M\gamma$	$M_{III}N_V$	2.182
5.704	8	82 Pb		$M_{III}N_I$	2.174
5.7101	3	40 Zr	$L\beta_6$	$L_{III}N_I$	2.1712
5.724	5	76 Os		$M_{III}N_{IV}$	2.166
5.7243	2	41 Nb	$L\alpha_1$	$L_{III}M_V$	2.16589
5.7319	3	41 Nb	$L\alpha_2$	$L_{III}M_{IV}$	2.1630
5.756	1	39 Y	L_{II}	Abs. Edge	2.1540
5.767	9	79 Au		M_VO_{III}	2.150
5.784	1	15 P	K	Abs. Edge	2.1435
5.796	2	15 P	$K\beta$	KM	2.1391
5.81	2	76 Os		$M_{II}N_I$	2.133
5.81	1	78 Pt	M_V	Abs. Edge	2.133
5.828	1	78 Pt	$M\beta$	$M_{IV}N_{VI}$	2.1273
5.83	2	73 Ta		$M_{III}O_I$	2.126
5.83	1	77 Ir	M_{IV}	Abs. Edge	2.126
5.8360	3	40 Zr	$L\beta_1$	$L_{II}M_{IV}$	2.1244
5.840	1	79 Au	$M\alpha_1$	M_VN_{VII}	2.1229
5.8475	3	42 Mo	$L\eta$	$L_{II}M_I$	2.1202
5.854	3	79 Au	$M\alpha_2$	M_VN_{VI}	2.118
5.8754	3	39 Y	$L\gamma_5$	$L_{II}N_I$	2.1102
5.884	8	81 Tl		$M_{III}N_I$	2.107
5.885	2	75 Re	$M\gamma$	$M_{III}N_V$	2.1067
5.931	5	75 Re		$M_{III}N_{IV}$	2.090
5.962	1	39 Y	L_{III}	Abs. Edge	2.0794
5.9832	3	39 Y	$L\beta_3$	L_IM_{III}	2.0722
5.987	9	78 Pt		M_VO_{III}	2.071
6.008	5	37 Rb	L_I	Abs. Edge	2.063
6.0186	3	39 Y	$L\beta_4$	L_IM_{II}	2.0600
6.038	1	77 Ir	$M\beta$	$M_{IV}N_{VI}$	2.0535
6.0458	3	37 Rb	$L\gamma_{2,3}$	$L_IN_{II,III}$	2.0507
6.047	1	78 Pt	$M\alpha_1$	M_VN_{VII}	2.0505
6.05	1	77 Ir	M_V	Abs. Edge	2.048
6.058	3	78 Pt	$M\alpha_2$	M_VN_{VI}	2.047
6.0705	2	40 Zr	$L\alpha_1$	$L_{III}M_V$	2.04236
6.073	5	76 Os	M_{IV}	Abs. Edge	2.042
6.0778	3	40 Zr	$L\alpha_2$	$L_{III}M_{IV}$	2.0399
6.09	2	80 Hg		$M_{III}N_I$	2.036
6.092	3	74 W	$M\gamma$	$M_{III}N_V$	2.035
6.0942	3	39 Y	$L\beta_6$	$L_{III}N_I$	2.0344
6.134	4	74 W		$M_{III}N_{IV}$	2.021
6.1508	3	42 Mo	Ll	$L_{III}M_I$	2.01568
6.157	1	15 P	$K\alpha_1$	KL_{III}	2.0137
6.160	1	15 P	$K\alpha_2$	KL_{II}	2.0127
6.162	8	83 Bi		$M_{IV}N_{III}$	2.012
6.173	1	38 Sr	L_{II}	Abs. Edge	2.0085
6.2109	3	41 Nb	$L\eta$	$L_{II}M_I$	1.99620
6.2120	3	39 Y	$L\beta_1$	$L_{II}M_{IV}$	1.99584
6.259	9	79 Au		$M_{III}N_I$	1.981
6.262	1	77 Ir	$M\alpha_1$	M_VN_{VII}	1.9799
6.267	1	76 Os	$M\beta$	$M_{IV}N_{VI}$	1.9783
6.275	3	77 Ir	$M\alpha_2$	M_VN_{VI}	1.9758
6.28	2	74 W		$M_{II}N_I$	1.973

Wavelength Å*	p.e.	Element		Designation	keV
6.2961	3	38 Sr	$L\gamma_5$	$L_{II}N_{I}$	1.96916
6.30	1	76 Os	M_{V}	Abs. Edge	1.967
6.312	4	73 Ta	$M\gamma$	$M_{III}N_{V}$	1.964
6.33	1	75 Re	M_{IV}	Abs. Edge	1.958
6.353	5	73 Ta		$M_{III}N_{IV}$	1.951
6.3672	3	38 Sr	$L\beta_3$	$L_{I}M_{III}$	1.94719
6.384	7	82 Pb		$M_{IV}N_{III}$	1.942
6.387	1	38 Sr	L_{III}	Abs. Edge	1.9411
6.4026	3	38 Sr	$L\beta_4$	$L_{I}M_{II}$	1.93643
6.4488	2	39 Y	$L\alpha_1$	$L_{III}M_{V}$	1.92256
6.455	9	78 Pt		$M_{III}N_{I}$	1.921
6.4558	3	39 Y	$L\alpha_2$	$L_{III}M_{IV}$	1.92047
6.47	1	36 Kr	L_{I}	Abs. Edge	1.915
6.490	1	76 Os	$M\alpha$	$M_{V}N_{VI,VII}$	1.9102
6.504	1	75 Re	$M\beta$	$M_{IV}N_{VI}$	1.9061
6.5176	3	41 Nb	Ll	$L_{III}M_{I}$	1.90225
6.5191	3	38 Sr	$L\beta_6$	$L_{III}N_{I}$	1.90181
6.521	4	83 Bi	$M\zeta_1$	$M_{V}N_{III}$	1.901
6.544	4	72 Hf	$M\gamma$	$M_{III}N_{V}$	1.895
6.560	5	75 Re	M_{V}	Abs. Edge	1.890
6.585	5	83 Bi	$M\zeta_2$	$M_{IV}N_{II}$	1.883
6.59	1	74 W	M_{IV}	Abs. Edge	1.880
6.6069	3	40 Zr	$L\eta$	$L_{II}M_{I}$	1.87654
6.6239	3	38 Sr	$L\beta_1$	$L_{II}M_{IV}$	1.87172
6.644	1	37 Rb	L_{II}	Abs. Edge	1.8661
6.669	9	77 Ir		$M_{III}N_{I}$	1.859
6.729	1	75 Re	$M\alpha$	$M_{V}N_{VI,VII}$	1.8425
6.738	1	14 Si	K	Abs. Edge	1.8400
6.740	3	82 Pb	$M\zeta_1$	$M_{V}N_{III}$	1.8395
6.7530	1	14 Si	$K\beta$	KM	1.83594
6.755	3	37 Rb	$L\gamma_5$	$L_{II}N_{IV}$	1.83532
6.757	1	74 W	$M\beta$	$M_{IV}N_{VI}$	1.8349
6.768	6	71 Lu	$M\gamma$	$M_{III}N_{V}$	1.832
6.7876	3	37 Rb	$L\beta_3$	$L_{I}M_{III}$	1.82659
6.802	5	82 Pb	$M\zeta_2$	$M_{IV}N_{II}$	1.823
6.806	9	74 W		$M_{IV}O_{II}$	1.822
6.8207	3	37 Rb	$L\beta_4$	$L_{I}M_{II}$	1.81771
6.83	1	74 W	M_{V}	Abs. Edge	1.814
6.862	1	37 Rb	L_{III}	Abs. Edge	1.8067
6.8628	2	38 Sr	$L\alpha_1$	$L_{III}M_{V}$	1.80656
6.8697	3	38 Sr	$L\alpha_2$	$L_{III}M_{IV}$	1.80474
6.87	1	73 Ta	M_{IV}	Abs. Edge	1.804
6.87	2	80 Hg	δ	$M_{IV}N_{III}$	1.805
6.89	2	76 Os		$M_{III}N_{I}$	1.798
6.9185	3	40 Zr	Ll	$L_{III}M_{I}$	1.79201
6.959	5	35 Br	L_{I}	Abs. Edge	1.781
6.974	4	81 Tl	$M\zeta_1$	$M_{V}N_{III}$	1.778
6.983	1	74 W	$M\alpha_1$	$M_{V}N_{VII}$	1.7754
6.9842	3	37 Rb	$L\beta_6$	$L_{III}N_{I}$	1.77517
6.992	2	74 W	$M\alpha_2$	$M_{V}N_{VI}$	1.7731
7.005	9	74 W		$M_{V}O_{III}$	1.770
7.023	1	73 Ta	$M\beta$	$M_{IV}N_{VI}$	1.7655
7.024	8	70 Yb	$M\gamma$	$M_{III}N_{V}$	1.765
7.032	5	81 Tl	$M\zeta_2$	$M_{IV}N_{II}$	1.763
7.0406	3	39 Y	$L\eta$	$L_{II}M_{I}$	1.76095
7.0759	3	37 Rb	$L\beta_1$	$L_{II}M_{IV}$	1.75217
7.09	2	73 Ta		$M_{IV}O_{II,III}$	1.748
7.101	8	79 Au		$M_{IV}N_{III}$	1.746
7.11	1	73 Ta	M_{V}	Abs. Edge	1.743
7.12542	9	14 Si	$K\alpha_1$	KL_{III}	1.73998
7.12791	9	14 Si	$K\alpha_2$	KL_{II}	1.73938
7.168	1	36 Kr	L_{II}	Abs. Edge	1.7297
7.250	5	36 Kr		$L_{II}N_{III}$	1.710
7.252	1	73 Ta	$M\alpha$	$M_{V}N_{VI,VII}$	1.7096
7.264	5	36 Kr	$L\beta_3$	$L_{I}M_{III}$	1.707
7.279	5	36 Kr	$L\gamma_5$	$L_{II}N_{I}$	1.703
7.30	2	73 Ta		$M_{V}O_{III}$	1.700
7.303	1	72 Hf	$M\beta$	$M_{IV}N_{VI}$	1.6976
7.304	5	36 Kr	$L\beta_4$	$L_{I}M_{II}$	1.697
7.3183	2	37 Rb	$L\alpha_1$	$L_{III}M_{V}$	1.69413
7.3251	3	37 Rb	$L\alpha_2$	$L_{III}M_{IV}$	1.69256
7.3563	3	39 Y	Ll	$L_{III}M_{I}$	1.68536
7.360	8	74 W		$M_{III}N_{I}$	1.684
7.371	8	78 Pt		$M_{IV}N_{III}$	1.682
7.392	1	36 Kr	L_{III}	Abs. Edge	1.6772
7.466	4	79 Au	$M\zeta_1$	$M_{V}N_{III}$	1.6605
7.503	1	34 Se	L_{I}	Abs. Edge	1.6525
7.510	4	36 Kr	$L\beta_6$	$L_{III}N_{I}$	1.6510
7.5171	3	38 Sr	$L\eta$	$L_{II}M_{I}$	1.64933
7.523	5	79 Au	$M\zeta_2$	$M_{IV}N_{II}$	1.648
7.539	1	72 Hf	$M\alpha$	$M_{V}N_{VI,VII}$	1.6446
7.546	8	68 Er	$M\gamma$	$M_{III}N_{V}$	1.643
7.576	3	36 Kr	$L\beta_1$	$L_{II}M_{IV}$	1.6366
7.60	1	68 Er		$M_{III}N_{IV}$	1.632
7.601	2	71 Lu	$M\beta$	$M_{IV}N_{VI}$	1.6312
7.612	9	73 Ta		$M_{III}N_{I}$	1.629
7.645	8	77 Ir		$M_{IV}N_{III}$	1.622
7.738	4	78 Pt	$M\zeta_1$	$M_{V}N_{III}$	1.6022
7.753	5	35 Br	L_{II}	Abs. Edge	1.599
7.767	9	35 Br	$L\beta_{3,4}$	$L_{I}M_{II,III}$	1.596
7.790	5	78 Pt	$M\zeta_2$	$M_{IV}N_{II}$	1.592
7.817	3	36 Kr	$L\alpha_{1,2}$	$L_{III}M_{IV,V}$	1.5860
7.8362	3	38 Sr	Ll	$L_{III}M_{I}$	1.58215
7.840	2	71 Lu	$M\alpha$	$M_{V}N_{VI,VII}$	1.5813
7.865	9	67 Ho	$M\gamma$	$M_{III}N_{IV,V}$	1.576
7.887	9	72 Hf		$M_{III}N_{I}$	1.572
7.909	2	70 Yb	$M\beta$	$M_{IV}N_{VI}$	1.5675
7.94813	5	13 Al	K	Abs. Edge	1.55988
7.960	2	13 Al	$K\beta$	KM	1.55745
7.984	5	35 Br	L_{III}	Abs. Edge	1.5530
8.021	4	77 Ir	$M\zeta_1$	$M_{V}N_{III}$	1.5458
8.0415	4	37 Rb	$L\eta$	$L_{II}M_{I}$	1.54177
8.065	5	77 Ir	$M\zeta_2$	$M_{IV}N_{II}$	1.5373
8.107	1	33 As	L_{I}	Abs. Edge	1.5293
8.1251	5	35 Br	$L\beta_1$	$L_{II}M_{IV}$	1.52590
8.144	9	66 Dy	$M\gamma$	$M_{III}N_{IV,V}$	1.522
8.149	5	70 Yb	$M\alpha$	$M_{V}N_{VI,VII}$	1.5214
8.239	8	75 Re		$M_{IV}N_{III}$	1.505
8.249	7	69 Tm	$M\beta$	$M_{IV}N_{VI}$	1.503
8.310	4	76 Os	$M\zeta_1$	$M_{V}N_{III}$	1.4919
8.321	9	34 Se	$L\beta_{3,4}$	$L_{I}M_{II,III}$	1.490
8.33934	9	13 Al	$K\alpha_1$	KL_{III}	1.48670
8.34173	9	13 Al	$K\alpha_2$	KL_{II}	1.48627
8.359	5	76 Os	$M\zeta_2$	$M_{IV}N_{II}$	1.4831

Wavelength Å*	p.e.	Element		Designation	keV
8.3636	4	37 Rb	Ll	$L_{III}M_{I}$	1.48238
8.3746	5	35 Br	$L\alpha_{1,2}$	$L_{III}M_{IV,V}$	1.48043
8.407	1	34 Se	L_{II}	Abs. Edge	1.4747
8.470	9	70 Yb		$M_{III}N_{I}$	1.464
8.48	1	69 Tm	$M\alpha$	$M_{V}N_{VI,VII}$	1.462
8.486	9	65 Tb	$M\gamma$	$M_{III}N_{IV,V}$	1.461
8.487	5	69 Tm	M_{V}	Abs. Edge	1.4609
8.573	8	74 W		$M_{IV}N_{III}$	1.446
8.592	3	68 Er	$M\beta$	$M_{IV}N_{VI}$	1.4430
8.60	7	92 U		$N_{I}P_{IV,V}$	1.44
8.601	5	68 Er	M_{IV}	Abs. Edge	1.4415
8.629	4	75 Re	$M\zeta_{1}$	$M_{V}N_{III}$	1.4368
8.646	1	34 Se	L_{III}	Abs. Edge	1.4340
8.664	5	75 Re	$M\zeta_{2}$	$M_{IV}N_{II}$	1.4310
8.7358	5	34 Se	$L\beta_{1}$	$L_{II}M_{IV}$	1.41923
8.76	7	92 U		$N_{I}P_{III}$	1.42
8.773	1	32 Ge	L_{I}	Abs. Edge	1.4132
8.81	7	92 U		$N_{I}P_{II}$	1.41
8.82	1	68 Er	$M\alpha$	$M_{V}N_{VI,VII}$	1.406
8.844	9	64 Gd	$M\gamma$	$M_{III}N_{IV,V}$	1.402
8.847	5	68 Er	M_{V}	Abs. Edge	1.4013
8.90	2	73 Ta		$M_{IV}N_{III}$	1.393
8.929	1	33 As	$L\beta_{3,4}$	$L_{I}M_{II,III}$	1.3884
8.962	4	74 W	$M\zeta_{1}$	$M_{V}N_{III}$	1.3835
8.965	4	67 Ho	$M\beta$	$M_{IV}N_{VI}$	1.3830
8.9900	5	34 Se	$L\alpha_{1,2}$	$L_{III}M_{IV,V}$	1.37910
8.993	5	74 W	$M\zeta_{2}$	$M_{IV}N_{II}$	1.3787
9.125	1	33 As	L_{II}	Abs. Edge	1.3587
9.20	2	67 Ho	$M\alpha$	$M_{V}N_{VI,VII}$	1.348
9.211	9	63 Eu	$M\gamma$	$M_{III}N_{IV,V}$	1.346
9.255	1	35 Br	$L\eta$	$L_{II}M_{I}$	1.3396
9.316	4	73 Ta	$M\zeta_{1}$	$M_{V}N_{III}$	1.3308
9.330	5	73 Ta	$M\zeta_{2}$	$M_{IV}N_{II}$	1.3288
9.357	6	66 Dy	$M\beta$	$M_{IV}N_{VI}$	1.3250
9.367	1	33 As	L_{III}	Abs. Edge	1.3235
9.40	7	90 Th		$N_{I}P_{III}$	1.319
9.4141	8	33 As	$L\beta_{1}$	$L_{II}M_{IV}$	1.3170
9.44	7	90 Th		$N_{I}P_{II}$	1.313
9.5122	1	12 Mg	K	Abs. Edge	1.30339
9.517	5	31 Ga	L_{I}	Abs. Edge	1.3028
9.521	2	12 Mg	$K\beta$	KM	1.3022
9.581	2	32 Ge	$L\beta_{3}$	$L_{I}M_{III}$	1.2941
9.585	1	35 Br	Ll	$L_{III}M_{I}$	1.2935
9.59	2	66 Dy	$M\alpha$	$M_{V}N_{VI,VII}$	1.293
9.600	9	62 Sm	$M\gamma$	$M_{III}N_{IV,V}$	1.291
9.640	2	32 Ge	$L\beta_{4}$	$L_{I}M_{II}$	1.2861
9.6709	8	33 As	$L\alpha_{1,2}$	$L_{III}M_{IV,V}$	1.2820
9.686	7	72 Hf	$M\zeta_{2}$	$M_{IV}N_{II}$	1.2800
9.686	7	72 Hf	$M\zeta_{1}$	$M_{V}N_{III}$	1.2800
9.792	6	65 Tb	$M\beta$	$M_{IV}N_{VI}$	1.2661
9.8900	2	12 Mg	$K\alpha_{1,2}$	$KL_{II,III}$	1.25360
9.924	1	32 Ge	L_{II}	Abs. Edge	1.2494
9.962	1	34 Se	$L\eta$	$L_{II}M_{I}$	1.2446
10.00	2	65 Tb	$M\alpha$	$M_{V}N_{VI,VII}$	1.240
10.09	7	92 U		$N_{I}O_{III}$	1.229
10.175	1	32 Ge	$L\beta_{1}$	$L_{II}M_{IV}$	1.2185
10.187	1	32 Ge	L_{III}	Abs. Edge	1.2170
10.254	6	64 Gd	$M\beta$	$M_{IV}N_{VI}$	1.2091
10.294	1	34 Se	Ll	$L_{III}M_{I}$	1.2044
10.359	9	31 Ga	$L\beta_{3,4}$	$L_{I}M_{II,III}$	1.197
10.40	7	92 U		$N_{II}P_{I}$	1.192
10.4361	8	32 Ge	$L\alpha_{1,2}$	$L_{III}M_{IV,V}$	1.18800
10.46	3	64 Gd	$M\alpha$	$M_{V}N_{VI,VII}$	1.185
10.48	1	70 Yb	$M\zeta$	$M_{V}N_{III}$	1.183
10.505	9	60 Nd	$M\gamma$	$M_{III}N_{IV,V}$	1.180
10.711	5	63 Eu	M_{IV}	Abs. Edge	1.1575
10.734	1	33 As	$L\eta$	$L_{II}M_{I}$	1.1550
10.750	7	63 Eu	$M\beta$	$M_{IV}N_{VI}$	1.1533
10.828	5	31 Ga	L_{II}	Abs. Edge	1.1450
10.96	3	63 Eu	$M\alpha$	$M_{V}N_{VI,VII}$	1.131
10.998	9	59 Pr	$M\gamma$	$M_{III}N_{IV,V}$	1.1273
11.013	5	63 Eu	M_{V}	Abs. Edge	1.1258
11.023	2	31 Ga	$L\beta_{1}$	$L_{II}M_{IV}$	1.1248
11.072	1	33 As	Ll	$L_{III}M_{I}$	1.1198
11.07	7	90 Th		$N_{II}P_{I}$	1.120
11.100	1	31 Ga	L_{III}	Abs. Edge	1.1169
11.200	7	30 Zn	$L\beta_{3,4}$	$L_{I}M_{II,III}$	1.1070
11.27	1	62 Sm	$M\beta$	$M_{IV}N_{VI}$	1.0998
11.288	5	62 Sm	M_{IV}	Abs. Edge	1.0983
11.292	1	31 Ga	$L\alpha_{1,2}$	$L_{III}M_{IV,V}$	1.09792
11.37	1	68 Er	$M\zeta$	$M_{V}N_{III}$	1.0901
11.47	3	62 Sm	$M\alpha$	$M_{V}N_{VI,VII}$	1.081
11.53	1	58 Ce	$M\gamma$	$M_{III}N_{IV,V}$	1.0749
11.552	5	62 Sm	M_{V}	Abs. Edge	1.0732
11.56	5	90 Th		$N_{II}O_{IV}$	1.072
11.569	1	11 Na	K	Abs. Edge	1.07167
11.575	2	11 Na	$K\beta$	KM	1.0711
11.609	2	32 Ge	$L\eta$	$L_{II}M_{I}$	1.0680
11.862	1	30 Zn	L_{II}	Abs. Edge	1.04523
11.86	1	67 Ho	$M\zeta$	$M_{V}N_{III}$	1.0450
11.9101	9	11 Na	$K\alpha_{1,2}$	$KL_{II,III}$	1.04098
11.965	2	32 Ge	Ll	$L_{III}M_{I}$	1.0362
11.983	3	30 Zn	$L\beta_{1}$	$L_{II}M_{IV}$	1.0347
12.08	4	57 La	$M\gamma$	$M_{III}N_{IV,V}$	1.027
12.122	3	29 Cu	$L\beta_{3,4}$	$L_{I}M_{II,III}$	1.0228
12.131	1	30 Zn	L_{III}	Abs. Edge	1.02201
12.254	3	30 Zn	$L\alpha_{1,2}$	$L_{III}M_{IV,V}$	1.0117
12.43	2	66 Dy	$M\zeta$	$M_{V}N_{III}$	0.998
12.44	2	60 Nd	$M\beta$	$M_{IV}N_{VI}$	0.997
12.459	5	60 Nd	M_{IV}	Abs. Edge	0.9951
12.597	2	31 Ga	$L\eta$	$L_{II}M_{I}$	0.9842
12.68	2	60 Nd	$M\alpha$	$M_{V}N_{VI,VII}$	0.978
12.737	5	60 Nd	M_{V}	Abs. Edge	0.9734
12.75	3	56 Ba	$M\gamma$	$M_{III}N_{IV,V}$	0.973
12.90	9	92 U		$N_{III}O_{V}$	0.961
12.953	2	31 Ga	Ll	$L_{III}M_{I}$	0.9572
12.98	2	65 Tb	$M\zeta$	$M_{V}N_{III}$	0.955
13.014	1	29 Cu	L_{II}	Abs. Edge	0.95268
13.053	3	29 Cu	$L\beta_{1}$	$L_{II}M_{IV}$	0.9498
13.06	2	59 Pr	$M\beta$	$M_{IV}N_{VI}$	0.950
13.06	1	30 Zn	L_{I}	Abs. Edge	0.9495
13.122	5	59 Pr	M_{IV}	Abs. Edge	0.9448
13.18	2	28 Ni	$L\beta_{3,4}$	$L_{I}M_{II,III}$	0.941
13.288	1	29 Cu	L_{III}	Abs. Edge	0.93306

Wavelength Å*	p.e.	Element		Designation	keV
13.30	6	83 Bi		$N_IP_{II,III}$	0.932
13.336	3	29 Cu	$L\alpha_{1,2}$	$L_{III}M_{IV,V}$	0.9297
13.343	5	59 Pr	$M\alpha$	$M_VN_{VI,VII}$	0.9292
13.394	5	59 Pr	M_V	Abs. Edge	0.9257
13.57	2	64 Gd	$M\zeta$	M_VN_{III}	0.914
13.68	2	30 Zn	$L\eta$	$L_{II}M_I$	0.906
13.75	4	58 Ce	$M\beta$	$M_{IV}N_{VI}$	0.902
13.8	1	90 Th		$N_{III}O_V$	0.897
14.02	2	30 Zn	Ll	$L_{III}M_I$	0.884
14.04	2	58 Ce	$M\alpha$	$M_VN_{VI,VII}$	0.883
14.22	2	63 Eu	$M\zeta$	M_VN_{III}	0.872
14.242	5	28 Ni	L_{II}	Abs. Edge	0.8706
14.271	6	28 Ni	$L\beta_1$	$L_{II}M_{IV}$	0.8688
14.3018	1	10 Ne	K	Abs. Edge	0.866889
14.31	3	27 Co	$L\beta_{3,4}$	$L_IM_{II,III}$	0.870
14.39	5	58 Ce		$M_VO_{II,III}$	0.862
14.452	5	10 Ne	$K\beta$	KM	0.8579
14.51	5	57 La	$M\beta$	$M_{IV}N_{VI}$	0.854
14.525	5	28 Ni	L_{III}	Abs. Edge	0.8536
14.561	3	28 Ni	$L\alpha_{1,2}$	$L_{III}M_{IV,V}$	0.8515
14.610	3	10 Ne	$K\alpha_{1,2}$	$KL_{II,III}$	0.8486
14.88	5	57 La	$M\alpha$	$M_VN_{VI,VII}$	0.833
14.90	2	29 Cu	$L\eta$	$L_{II}M_I$	0.832
14.91	4	62 Sm	$M\zeta$	M_VN_{III}	0.831
15.286	9	29 Cu	Ll	$L_{III}M_I$	0.8111
15.56	1	56 Ba	M_{IV}	Abs. Edge	0.7967
15.618	5	27 Co	L_{II}	Abs. Edge	0.7938
15.65	4	26 Fe	$L\beta_{3,4}$	$L_IM_{II,III}$	0.792
15.666	8	27 Co	$L\beta_1$	$L_{II}M_{IV}$	0.7914
15.72	9	56 Ba		$M_{IV}O_{III}$	0.789
15.89	1	56 Ba	M_V	Abs. Edge	0.7801
15.91	5	56 Ba		$M_{IV}O_{II}$	0.779
15.915	5	27 Co	L_{III}	Abs. Edge	0.7790
15.93	4	52 Te	$M\gamma$	$M_{III}N_{IV,V}$	0.778
15.972	6	27 Co	$L\alpha_{1,2}$	$L_{III}M_{IV,V}$	0.7762
15.98	5	51 Sb		$M_{II}N_{IV}$	0.776
16.20	5	56 Ba		M_VO_{III}	0.765
16.27	3	28 Ni	$L\eta$	$L_{II}M_I$	0.762
16.46	4	60 Nd	$M\zeta$	M_VN_{III}	0.753
16.693	9	28 Ni	Ll	$L_{III}M_I$	0.7427
16.7	1	24 Cr	L_I	Abs. Edge	0.741
16.92	4	51 Sb	$M\gamma$	$M_{III}N_{IV,V}$	0.733
16.93	5	50 Sn		$M_{II}N_{IV}$	0.733
17.19	4	25 Mn	$L\beta_{3,4}$	$L_IM_{II,III}$	0.721
17.202	5	26 Fe	L_{II}	Abs. Edge	0.7208
17.26	1	26 Fe	$L\beta_1$	$L_{II}M_{IV}$	0.7185
17.38	4	59 Pr	$M\zeta$	M_VN_{III}	0.714
17.525	5	26 Fe	L_{III}	Abs. Edge	0.7074
17.59	2	26 Fe	$L\alpha_{1,2}$	$L_{III}M_{IV,V}$	0.7050
17.6	1	52 Te		$M_{II}N_I$	0.703
17.87	3	27 Co	$L\eta$	$L_{II}M_I$	0.694
17.94	5	50 Sn	$M\gamma$	$M_{III}N_{IV,V}$	0.691
17.9	1	24 Cr	L_{II}	Abs. Edge	0.691
18.292	8	27 Co	Ll	$L_{III}M_I$	0.6778
18.32	2	9 F	$K\alpha$	KL	0.6768
18.35	4	58 Ce	$M\zeta$	M_VN_{III}	0.676
18.8	1	51 Sb		$M_{II}N_I$	0.658
18.8	2	47 Ag		$M_IN_{II,III}$	0.658
18.96	4	24 Cr	$L\beta_{3,4}$	$L_IM_{II,III}$	0.654
19.11	2	25 Mn	$L\beta_1$	$L_{II}M_{IV}$	0.6488
19.1	1	52 Te		$M_{III}N_I$	0.648
19.40	7	48 Cd		$M_{II}N_{IV}$	0.639
19.44	5	57 La	$M\zeta$	M_VN_{III}	0.638
19.45	1	25 Mn	$L\alpha_{1,2}$	$L_{III}M_{IV,V}$	0.6374
19.66	5	53 I	$M_{IV,V}$	Abs. Edge	0.631
19.75	4	26 Fe	$L\eta$	$L_{II}M_I$	0.628
20.0	1	50 Sn		$M_{II}N_I$	0.619
20.1	2	46 Pd		$M_IN_{II,III}$	0.616
20.15	1	26 Fe	Ll	$L_{III}M_I$	0.6152
20.2	1	51 Sb		$M_{III}N_I$	0.612
20.47	7	48 Cd	$M\gamma$	$M_{III}N_{IV,V}$	0.606
20.64	4	56 Ba	$M\zeta$	M_VN_{III}	0.601
20.66	7	47 Ag		$M_{II}N_{IV}$	0.600
20.7	1	24 Cr	L_{III}	Abs. Edge	0.598
21.19	5	23 Va	$L\beta_{3,4}$	$L_IM_{II,III}$	0.585
21.27	1	24 Cr	$L\beta_1$	$L_{II}M_{IV}$	0.5828
21.34	5	52 Te		$M_{IV}O_{II,III}$	0.581
21.5	1	50 Sn		$M_{III}N_I$	0.575
21.64	3	24 Cr	$L\alpha_{1,2}$	$L_{III}M_{IV,V}$	0.5728
21.78	5	52 Te		M_VO_{III}	0.569
21.82	7	47 Ag	$M\gamma$	$M_{III}N_{IV,V}$	0.568
21.85	2	25 Mn	$L\eta$	$L_{II}M_I$	0.5675
22.1	1	46 Pd		$M_{II}N_{IV}$	0.560
22.29	1	25 Mn	Ll	$L_{III}M_I$	0.5563
22.9	2	48 Cd		$M_{II}N_I$	0.540
23.32	1	8 O	K	Abs. Edge	0.5317
23.3	1	46 Pd	$M\gamma$	$M_{III}N_{IV,V}$	0.531
23.62	3	8 O	$K\alpha$	KL	0.5249
23.88	4	23 Va	$L\beta_1$	$L_{II}M_{IV}$	0.5192
24.25	3	23 Va	$L\alpha_{1,2}$	$L_{III}M_{IV,V}$	0.5113
24.28	5	50 Sn	$M_{IV,V}$	Abs. Edge	0.511
24.30	3	24 Cr	$L\eta$	$L_{II}M_I$	0.5102
24.4	2	47 Ag		M_VN_I	0.509
24.5	1	48 Cd		$M_{III}N_I$	0.507
24.78	1	24 Cr	Ll	$L_{III}M_I$	0.5003
25.01	9	45 Rh	$M\gamma$	$M_{III}N_{IV,V}$	0.496
25.3	1	50 Sn		$M_{IV}O_{II,III}$	0.491
25.50	9	44 Ru		$M_{II}N_{IV}$	0.486
25.7	1	50 Sn		M_VO_{III}	0.483
26.0	1	47 Ag		$M_{III}N_I$	0.478
26.2	2	46 Pd		$M_{II}N_I$	0.474
26.72	9	52 Te	$M\zeta$	$M_{IV,V}N_{II,III}$	0.464
26.9	1	44 Ru	$M\gamma$	$M_{III}N_{IV,V}$	0.462
27.05	2	22 Ti	$L\beta_1$	$L_{II}M_{IV}$	0.4584
27.29	1	22 Ti	$L_{II,III}$	Abs. Edge	0.4544
27.34	3	23 Va	$L\eta$	$L_{II}M_I$	0.4535
27.42	2	22 Ti	$L\alpha_{1,2}$	$L_{III}M_{IV,V}$	0.4522
27.77	1	23 Va	Ll	$L_{III}M_I$	0.4465
27.9	1	46 Pd		$M_{III}N_I$	0.445
28.1	2	45 Rh		$M_{II}N_I$	0.442
28.13	5	48 Cd	$M_{IV,V}$	Abs. Edge	0.4408
28.88	8	51 Sb	$M\zeta$	$M_{IV,V}N_{II,III}$	0.429
29.8	1	45 Rh		$M_{III}N_I$	0.417
30.4	1	48 Cd		$M_{IV}O_{II,III}$	0.408

Wavelength Å*	p.e.	Element		Designation	keV
30.8	1	48 Cd		$M_{V}O_{III}$	0.403
30.82	5	47 Ag	M_{IV}	Abs. Edge	0.4022
30.89	3	22 Ti	$L\eta$	$L_{II}M_{I}$	0.4013
30.99	1	7 N	K	Abs. Edge	0.4000
31.02	2	21 Sc	$L\beta_1$	$L_{II}M_{IV}$	0.3996
31.14	5	47 Ag	M_{V}	Abs. Edge	0.3981
31.24	9	50 Sn	$M\zeta$	$M_{IV,V}N_{II,III}$	0.397
31.35	3	21 Sc	$L\alpha_{1,2}$	$L_{III}M_{IV,V}$	0.3954
31.36	2	22 Ti	Ll	$L_{III}M_{I}$	0.3953
31.60	4	7 N	$K\alpha$	KL	0.3924
31.8	1	92 U		$N_{IV}N_{VI}$	0.390
32.3	2	44 Ru		$M_{II}N_{I}$	0.384
33.1	2	41 Nb		$M_{II}N_{IV}$	0.375
33.5	3	47 Ag		$M_{IV,V}O_{II,III}$	0.370
33.57	9	90 Th		$N_{IV}N_{VI}$	0.3693
34.8	1	92 U		$N_{V}N_{VI,VII}$	0.357
34.9	2	41 Nb	$M\gamma$	$M_{III}N_{IV,V}$	0.356
35.13	2	21 Sc	$L\eta$	$L_{II}M_{I}$	0.3529
35.13	1	20 Ca	L_{II}	Abs. Edge	0.3529
35.3	3	42 Mo		$M_{II}N_{I}$	0.351
35.49	1	20 Ca	L_{III}	Abs. Edge	0.34931
35.59	3	21 Sc	Ll	$L_{III}M_{I}$	0.3483
35.63	1	20 Ca	$L_{II,III}$	Abs. Edge	0.34793
35.94	2	20 Ca	$L\beta_1$	$L_{II}M_{IV}$	0.3449
36.32	9	90 Th		$N_{V}N_{VI,VII}$	0.3414
36.33	2	20 Ca	$L\alpha_{1,2}$	$L_{III}M_{IV,V}$	0.3413
36.8	1	48 Cd	$M\zeta$	$M_{IV,V}N_{II,III}$	0.3371
37.4	2	46 Pd		$M_{IV,V}O_{II,III}$	0.332
37.5	2	42 Mo		$M_{III}N_{I}$	0.331
38.4	3	41 Nb		$M_{II}N_{I}$	0.323
39.77	7	47 Ag	$M\zeta$	$M_{IV,V}N_{II,III}$	0.3117
40.46	2	20 Ca	$L\eta$	$L_{II}M_{I}$	0.3064
40.7	2	41 Nb		$M_{III}N_{I}$	0.305
40.9	2	45 Rh		$M_{IV,V}O_{II,III}$	0.303
40.96	2	20 Ca	Ll	$L_{III}M_{I}$	0.3027
42.1	2	92 U		$N_{VI}O_{V}$	0.295
42.1	1	19 K	$L_{II,III}$	Abs. Edge	0.2946
42.3	2	82 Pb		$N_{IV}N_{VI}$	0.293
43.3	2	92 U		$N_{VI}O_{IV}$	0.286
43.6	1	46 Pd	$M\zeta$	$M_{IV,V}N_{II,III}$	0.2844
43.68	1	6 C	K	Abs. Edge	0.28384
44.7	3	6 C	$K\alpha$	KL	0.277
44.8	1	44 Ru		$M_{IV,V}O_{II,III}$	0.2768
45.0	1	82 Pb		$N_{V}N_{VI,VII}$	0.2756
45.2	3	80 Hg		$N_{IV}N_{VI}$	0.274
45.2	1	51 Sb		$M_{II}M_{IV}$	0.2743
46.48	9	39 Y		$M_{II}N_{I}$	0.267
46.5	2	81 Tl		$N_{V}N_{VI,VII}$	0.267
46.8	2	79 Au		$N_{IV}N_{VI}$	0.265
47.24	2	19 K	Ll	$L_{II}M_{I}$	0.2625
47.3	1	50 Sn		$M_{II}M_{IV}$	0.2621
47.67	9	45 Rh	$M\zeta$	$M_{IV,V}N_{II,III}$	0.2601
47.74	1	19 K	Ll	$L_{III}M_{I}$	0.25971
47.9	3	80 Hg		$N_{V}N_{VI,VII}$	0.259
48.1	2	78 Pt		$N_{IV}N_{VI}$	0.258
48.2	1	90 Th		$N_{VI}O_{V}$	0.2572
48.5	2	39 Y		$M_{III}N_{I}$	0.256
49.4	1	79 Au		$N_{V}N_{VI,VII}$	0.2510
49.5	1	90 Th		$N_{VI}O_{IV}$	0.2505
50.0	1	90 Th		$N_{VII}O_{V}$	0.2479
50.2	1	77 Ir		$N_{IV}N_{VI}$	0.2470
50.3	1	52 Te		$M_{III}M_{V}$	0.2465
50.9	1	78 Pt		$N_{V}N_{VI,VII}$	0.2436
51.3	1	38 Sr		$M_{II}N_{I}$	0.2416
51.9	1	76 Os		$N_{IV}N_{VI}$	0.2388
52.0	2	48 Cd		$M_{II}M_{IV}$	0.2384
52.2	1	51 Sb		$M_{III}M_{V}$	0.2375
52.34	7	44 Ru	$M\zeta$	$M_{IV,V}N_{II,III}$	0.2369
52.8	1	77 Ir		$N_{V}N_{VI,VII}$	0.2348
53.6	1	38 Sr		$M_{III}N_{I}$	0.2313
54.0	2	74 W		$N_{II}N_{IV}$	0.2295
54.0	1	47 Ag		$M_{II}M_{IV}$	0.2295
54.2	1	50 Sn		$M_{III}M_{V}$	0.2287
54.7	2	76 Os		$N_{V}N_{VI,VII}$	0.2266
54.8	2	42 Mo		$M_{IV,V}O_{II,III}$	0.2262
55.8	1	74 W		$N_{IV}N_{VI}$	0.2221
55.9	1	18 A	$L\eta$	$L_{II}M_{I}$	0.2217
56.3	1	18 A	Ll	$L_{III}M_{I}$	0.2201
56.5	1	46 Pd		$M_{II}M_{IV}$	0.2194
57.0	2	37 Rb		$M_{II}N_{I}$	0.2174
58.2	1	73 Ta		$N_{IV}N_{VI}$	0.2130
58.4	1	74 W		$N_{V}N_{VII}$	0.2122
58.7	2	48 Cd		$M_{III}M_{V}$	0.2111
59.3	1	45 Rh		$M_{II}M_{IV}$	0.2090
59.5	3	74 W		$N_{V}N_{VI}$	0.208
59.5	2	37 Rb		$M_{III}N_{I}$	0.2083
60.5	1	47 Ag		$M_{III}M_{V}$	0.2048
61.1	2	73 Ta		$N_{V}N_{VI,VII}$	0.2028
61.9	2	41 Nb		$M_{IV,V}O_{II,III}$	0.2002
62.2	1	44 Ru		$M_{II}M_{IV}$	0.1992
62.9	1	46 Pd		$M_{III}M_{V}$	0.1970
63.0	5	71 Lu		$N_{IV}N_{VI}$	0.197
64.38	7	42 Mo	$M\zeta$	$M_{IV,V}N_{II,III}$	0.1926
65.1	7	70 Yb		$N_{IV}N_{VI}$	0.190
65.5	1	45 Rh		$M_{III}M_{V}$	0.1892
65.7	2	71 Lu		$N_{V}N_{VI,VII}$	0.1886
67.33	9	17 Cl	$L\eta$	$L_{II}M_{I}$	0.1841
67.6	3	5 B	$K\alpha$	KL	0.1833
67.90	9	17 Cl	Ll	$L_{III}M_{I}$	0.1826
68.2	3	90 Th		$O_{III}P_{IV,V}$	0.1817
68.3	1	44 Ru		$M_{III}M_{V}$	0.1814
68.9	2	42 Mo		$M_{II}M_{IV}$	0.1798
69.3	5	70 Yb		$N_{V}N_{VI,VII}$	0.179
70.0	4	40 Zr		$M_{IV,V}O_{II,III}$	0.177
72.1	3	41 Nb		$M_{II}M_{IV}$	0.1718
72.19	9	41 Nb	$M\zeta$	$M_{IV,V}N_{II,III}$	0.1717
72.7	9	68 Er		$N_{IV}N_{VI}$	0.171
74.9	1	42 Mo		$M_{III}M_{V}$	0.1656
76.3	7	68 Er		$N_{V}N_{VI,VII}$	0.163
76.7	2	40 Zr		$M_{II}M_{IV}$	0.1617
76.9	2	35 Br		$M_{II}N_{I}$	0.1613
78.4	2	41 Nb		$M_{III}M_{V}$	0.1582
79.8	3	35 Br		$M_{III}N_{I}$	0.1554
80.9	3	40 Zr		$M_{III}M_{V}$	0.1533

Wavelength Å*	p.e.	Element	Designation		keV
81.5	2	39 Y		$M_{II}M_{IV}$	0.1522
82.1	2	40 Zr	$M\zeta$	$M_{IV,V}N_{II,III}$	0.1511
83.	1	66 Dy		$N_{IV,V}N_{VI,VII}$	0.149
83.4	3	16 S	Ll, η	$L_{II,III}M_{I}$	0.1487
85.7	2	38 Sr		$M_{II}M_{IV}$	0.1447
86.	1	65 Tb		$N_{IV,V}N_{VI,VII}$	0.144
86.5	2	39 Y		$M_{III}M_{IV,V}$	0.1434
91.4	2	38 Sr		$M_{III}M_{IV,V}$	0.1357
91.5	2	37 Rb		$M_{II}M_{IV}$	0.1355
91.6	1	83 Bi		$N_{VI}O_{IV}$	0.1354
93.2	1	83 Bi		$N_{VII}O_{V}$	0.1330
93.4	2	39 Y	$M\zeta$	$M_{IV,V}N_{II,III}$	0.1328
94.	1	15 P	$L_{II,III}$	Abs. Edge	0.132
96.7	2	37 Rb		$M_{III}M_{IV,V}$	0.1282
97.2	8	66 Dy		$N_{IV,V}O_{II,III}$	0.128
98.	1	62 Sm		$N_{IV,V}N_{VI,VII}$	0.126
100.2	2	82 Pb		$N_{VI}O_{V}$	0.1237
102.2	4	65 Tb		$N_{IV,V}O_{II,III}$	0.1213
102.4	1	82 Pb		$N_{VI}O_{IV}$	0.1211
103.8	4	15 P		$L_{II,III}M$	0.1194
104.3	1	82 Pb		$N_{VII}O_{V}$	0.1189
107.	1	60 Nd		$N_{IV,V}N_{VI,VII}$	0.116
108.0	2	38 Sr	$M\zeta_2$	$M_{IV}N_{II,III}$	0.1148
108.7	1	38 Sr	$M\zeta_1$	$M_{V}N_{III}$	0.1140
109.4	3	35 Br		$M_{II}M_{IV}$	0.1133
110.6	5	29 Cu	M_{I}	Abs. Edge	0.1121
111.	1	4 Be	K	Abs. Edge	0.111
112.0	6	63 Eu		$N_{IV,V}O_{II,III}$	0.1107
113.0	1	81 Tl		$N_{VI}O_{V}$	0.10968
113.	1	59 Pr		$N_{IV,V}N_{VI,VII}$	0.1095
113.8	3	35 Br		$M_{III}M_{IV,V}$	0.1089
114.	1	4 Be	$K\alpha$	KL	0.1085
115.3	2	81 Tl		$N_{VI}O_{IV}$	0.1075
117.4	4	62 Sm		$N_{IV,V}O_{II,III}$	0.1056
117.7	1	81 Tl		$N_{VII}O_{V}$	0.10530
123.	1	14 Si	$L_{II,III}$	Abs. Edge	0.1006
126.8	2	37 Rb		$M_{IV}N_{III}$	0.0978
127.8	2	37 Rb	$M\zeta_2$	$M_{IV}N_{II}$	0.0970
128.7	2	37 Rb	$M\zeta_1$	$M_{V}N_{III}$	0.0964
128.9	7	60 Nd		$N_{IV,V}O_{II,III}$	0.0962
135.5	4	14 Si		$L_{II,III}M$	0.0915
136.5	4	59 Pr		$N_{IV,V}O_{II,III}$	0.0908
137.0	5	30 Zn	M_{II}	Abs. Edge	0.0905
142.5	1	13 Al	L_{I}	Abs. Edge	0.08701
143.9	5	30 Zn	M_{III}	Abs. Edge	0.0862
144.4	6	58 Ce		$N_{IV,V}O_{II,III}$	0.0859
144.4	3	37 Rb		$M_{I}M_{III}$	0.0859
152.6	6	57 La		$N_{IV,V}O_{II,III}$	0.0812
157.	3	30 Zn		$M_{II,III}M_{IV,V}$	0.079
159.0	2	56 Ba		$N_{IV}O_{III}$	0.07796
159.5	5	29 Cu	M_{II}	Abs. Edge	0.0777
163.3	2	56 Ba		$N_{IV}O_{II}$	0.07590
164.6	2	56 Ba		$N_{V}O_{III}$	0.07530
164.7	3	35 Br		$M_{I}M_{III}$	0.0753
166.0	5	29 Cu	M_{III}	Abs. Edge	0.0747
170.4	1	13 Al	$L_{II,III}$	Abs. Edge	0.07278
171.4	5	13 Al		$L_{II,III}M$	0.0724
173.	3	29 Cu		$M_{II,III}M_{IV,V}$	0.072
181.	5	90 Th		$O_{IV,V}Q_{II,III}$	0.068
183.8	1	55 Cs		$N_{IV}O_{III}$	0.06746
184.6	3	35 Br		$M_{I}M_{II}$	0.0672
188.4	1	28 Ni	M_{III}	Abs. Edge	0.06581
188.6	1	55 Cs		$N_{IV}O_{II}$	0.06574
189.5	3	35 Br		$M_{IV}N_{III}$	0.0654
190.3	1	55 Cs		$N_{V}O_{III}$	0.06515
190.	2	28 Ni		$M_{II,III}M_{IV,V}$	0.0651
191.1	2	35 Br	$M\zeta_2$	$M_{IV}N_{II}$	0.06488
192.6	2	35 Br	$M\zeta_1$	$M_{V}N_{III}$	0.06437
197.3	1	12 Mg	L_{I}	Abs. Edge	0.06284
202.	5	27 Co	$M_{II,III}$	Abs. Edge	0.061
203.	1	16 S		$L_{I}L_{II,III}$	0.061
214.	6	27 Co		$M_{II,III}M_{IV,V}$	0.058
224.	1	53 I	$N_{IV,V}$	Abs. Edge	0.0552
226.5	1	3 Li	K	Abs. Edge	0.05475
227.8	1	34 Se	M_{V}	Abs. Edge	0.05443
228.	1	3 Li	$K\alpha$	KL	0.0543
230.	2	34 Se		$M_{V}N_{III}$	0.0538
230.	1	26 Fe	$M_{II,III}$	Abs. Edge	0.0538
243.	5	26 Fe		$M_{II,III}M_{IV,V}$	0.051
249.3	1	12 Mg	L_{II}	Abs. Edge	0.04973
250.7	1	12 Mg	L_{III}	Abs. Edge	0.04945
251.5	5	12 Mg		$L_{II,III}M$	0.04929
273.	6	25 Mn		$M_{II,III}M_{IV,V}$	0.045
290.	1	13 Al		$L_{I}L_{II,III}$	0.0428
309.	9	24 Cr		$M_{II,III}M_{IV,V}$	0.040
317.	1	12 Mg		$L_{I}L_{II,III}$	0.0392
337.	9	23 V		$M_{II,III}M_{IV,V}$	0.0368
376.	1	11 Na		$L_{I}L_{II,III}$	0.03299
399.	5	35 Br	N_{I}	Abs. Edge	0.0311
405.	5	11 Na	$L_{II,III}$	Abs. Edge	0.0306
407.1	5	11 Na		$L_{II,III}M$	0.03045
417.	5	17 Cl	M_{I}	Abs. Edge	0.0297
444.	5	53 I	O_{I}	Abs. Edge	0.0279
525.	9	20 Ca		$M_{II,III}N_{I}$	0.0236
692.	9	19 K		$M_{II,III}N_{I}$	0.0179

X-RAY ATOMIC ENERGY LEVELS*

J. A. Bearden and A. F. Burr

These tables were originally published as the final report to the U.S. Atomic Energy Commission as Report NYO-2543-1 in partial fulfillment of Contract AT(30-1)-2543. The tables were later reproduced in *Review of Modern Physics*. The data may also be obtained from the Superintendent of Documents, U.S. Government Printing Office, Washington, D. C. 20402 in the publication NSRDS-NBS 14. Persons seeking discussion of the details of calculations, sources of energy level information and the problem of properly interpreting the experimental measurements should refer to the original publication or to *Review of Modern Physics*, Vol. 39, 125–142, January 1967.

All of the x-ray emission wavelengths have recently been reevaluated and placed on a consistent Å* scale. For most elements these data give a highly overdetermined set of equations for energy level differences, which have been solved by least-squares adjustment for each case. This procedure makes "best" use of all x-ray wavelength data, and also permits calculation of the probable error for each energy difference. Photoelectron measurements of absolute energy levels are more precise than x-ray absorption edge data. These have been used to establish the absolute scale for eighty-one elements and, in many cases, to provide additional energy level difference data. The x-ray absorption wavelengths were used for eight elements and ionization measurements for two; the remaining five were interpolated by a Moseley diagram involving the output values of energy levels from adjacent elements. Probable errors are listed on an absolute energy basis. In the original source of the present data, a table of energy levels in Rydberg units is given. Difference tables in volts, Rydbergs, and milli-Å* wavelength units, with the respective probable errors, are also included there.

Recommended values of the atomic energy levels, and probable errors in eV. Where available, photoelectron direct measurements are listed in brackets [ ] immediately under the recommended values. The measured values of the x-ray absorption energies are shown in parentheses ( ). Interpolated values are enclosed in angle brackets ⟨ ⟩.

Level	1 H	2 He	3 Li	4 Be	5 B	6 C	7 N	8 O
K	13.59811[a]	24.58678[b]	54.75±0.02 (54.75)	111.0±1.0 (111.0)	188.0±0.4 [188.0][c]	283.8±0.4 [283.8][c] (283.8)	401.6±0.4 [401.6][c]	532.0±0.4 [532.0][c]
L_{I}								23.7±0.4 [23.7][d]
$L_{II,III}$					4.7±0.9	6.4±1.9	9.2±0.6	7.1±0.8
	9 F	**10 Ne**	**11 Na**	**12 Mg**	**13 Al**	**14 Si**	**15 P**	**16 S**
K	685.4±0.4 [685.4][c]	866.9±0.3 (866.9)	1072.1±0.4 [1072.1][c] (1072.)	1305.0±0.4 [1305.0][c] (1303.)	1559.6±0.4 [1559.6][c] (1559.8)	1838.9±0.4 [1838.9][c]	2145.5±0.4 [2145.5][d]	2472.0±0.4 [2472.0][c] (2470.)
L_{I}	⟨31.⟩	⟨45.⟩	63.3±0.4 [63.3][d]	89.4±0.4 [89.4][d] (63.)	117.7±0.4 [117.7][d] (87.)	148.7±0.4 [148.7][d]	189.3±0.4 [189.3][d]	229.2±0.4 [229.2][d]
$L_{II,III}$	8.6±0.8	18.3±0.4	31.1±0.4 (31.)	51.4±0.5 (50.)	73.1±0.5 (72.8)	99.2±0.5 (100.6)	132.2±0.5 (132.)	164.8±0.7
	17 Cl	**18 Ar**	**19 K**	**20 Ca**	**21 Sc**	**22 Ti**	**23 V**	**24 Cr**
K	2822.4±0.3 [2822.4][c] (2020.)	3202.9±0.3 (3202.9)	3607.4±0.4 [3607.4][c] (3607.8)	4038.1±0.4 [4038.1][c] (4038.1)	4492.8±0.4 [4492.8][c]	4966.4±0.4 [4966.4][d] (4964.5)	5465.1±0.3 [5465.1][c] (5464.)	5989.2±0.3 [5989.2][c] (5989.)
L_{I}	270.2±0.4 [270.2][d]	320. ⟨320.⟩[d]	377.1±0.4 [377.1][d]	437.8±0.4 [437.8][d]	500.4±0.4 [500.4][d]	563.7±0.4 [563.7][d]	628.2±0.4 [628.2][d]	694.6±0.4 [694.6][d]
L_{II}	201.6±0.3	247.3±0.3	296.3±0.4	350.0±0.4	406.7±0.4	461.5±0.4	520.5±0.3	583.7±0.3
L_{III}	200.0±0.3	245.2±0.3	293.6±0.4	346.4±0.4	402.2±0.4	455.5±0.4	512.9±0.3	574.5±0.3
M_{I}	17.5±0.4	25.3±0.4	33.9±0.4	43.7±0.4	53.8±0.4	60.3±0.4	66.5±0.4	74.1±0.4
$M_{II,III}$	6.8±0.4	12.4±0.3	17.8±0.4	25.4±0.4	32.3±0.5	34.6±0.4	37.8±0.3	42.5±0.3
$M_{IV,V}$					6.6±0.5	3.7	2.2±0.3	2.3±0.4
	25 Mn	**26 Fe**	**27 Co**	**28 Ni**	**29 Cu**	**30 Zn**	**31 Ga**	**32 Ge**
K	6539.0±0.4 [6539.0][c] (6538.)	7112.0±0.9 [7111.3][c,f] (7111.2)	7708.9±0.3 [7708.9][c] (7709.5)	8332.8±0.4 [8332.8][c] (8331.6)	8978.9±0.4 [8978.9][c,g] (8980.3)	9658.6±0.6 [9658.6][g] (9660.7)	10367.1±0.5 [10367.1][g] (10368.2)	11103.1±0.7 [11103.8][g] (11103.6)
L_{I}	769.0±0.4 [769.0][d]	846.1±0.4 [846.1][d]	925.6±0.4 [925.6][d]	1008.1±0.4 [1008.1][d]	1096.1±0.4 [1096.0][d]	1193.6±0.9	1297.7±1.1	1414.3±0.7 [1413.6][g]
L_{II}	651.4±0.4	721.1±0.9 (720.8)	793.6±0.3 (793.8)	871.9±0.4 (870.6)	951.0±0.4 [950.0][h] (953.)	1042.8±0.6 (1045.)	1142.3±0.5	1247.8±0.7 (1249.)
L_{III}	640.3±0.4	708.1±0.9 (707.4)	778.6±0.3 (779.0)	854.7±0.4 (853.6)	931.1±0.4 [931.4][h] (933.)	1019.7±0.6 (1022.)	1115.4±0.5 (1117.)	1216.7±0.7 (1217.0)

* Wavelengths corresponding to these energy levels may be calculated from $\lambda\ (\mathrm{nm}) = \frac{1.239852}{E(\mathrm{keV})}$.

	25 Mn	26 Fe	27 Co	28 Ni	29 Cu	30 Zn	31 Ga	32 Ge
M_{I}	83.9±0.5	92.9±0.9	100.7±0.4	111.8±0.6	119.8±0.6	135.9±1.1	158.1±0.5	180.0±0.8
M_{II}, M_{III}	48.6±0.4 (54.)	54.0±0.9 (54.)	59.5±0.3 (61.)	68.1±0.4 (66.)	73.6±0.4 (75.)	86.6±0.6 (86.)		
M_{II}							106.8±0.7	127.9±0.9
M_{III}							102.9±0.5	120.8±0.7
$M_{IV,V}$	3.3±0.5	3.6±0.9	2.9±0.3	3.6±0.4	1.6±0.4	8.1±0.6	17.4±0.5	28.7±0.7

	33 As	34 Se	35 Br	36 Kr	37 Rb	38 Sr	39 Y	40 Zr
K	11866.7±0.7 [11866.7][l] (11865.)	12657.8±0.7 [12657.8][g] (12654.5)	13473.7±0.4 (13470.)	14325.6±0.8 (14324.4)	15199.7±0.3 (15202.)	16104.6±0.3 (16107.)	17038.4±0.3 (17038.)	17997.6±0.4 (17999.)
L_{I}	1526.5±0.8 (1529.)	1653.9±3.5 (1652.5)	1782.0±0.4 [1782.0][j]	1921.0±0.6 [1921.2][k]	2065.1±0.3 [2065.4][j]	2216.3±0.3 [2216.2][l]	2372.5±0.3 [2372.7][l]	2531.6±0.3 [2531.6][l]
L_{II}	1358.6±0.7 (1358.7)	1476.2±0.7 (1474.7)	1596.0±0.4 [1596.2][j]	1727.2±0.5 [1727.2][k] (1730.)	1863.9±0.3 [1863.4][j]	2006.8±0.3 [2006.6][l] (2008.5)	2155.5±0.3 [2155.0][l] (2154.0)	2306.7±0.3 [2306.5][l] (2305.3)
L_{III}	1323.1±0.7 (1323.5)	1435.8±0.7 (1434.0)	1549.9±0.4 [1549.7][j]	1674.9±0.5 [1674.8][k] (1677.)	1804.4±0.3 [1804.6][j]	1939.6±0.3 [1939.9][l] (1941.)	2080.0±0.3 [2080.2][l] (2079.4)	2222.3±0.3 [2222.5][l] (2222.5)
M_{I}	203.5±0.7	231.5±0.7	256.5±0.4		322.1±0.3	357.5±0.3	393.6±0.3	430.3±0.3
M_{II}	146.4±1.2	168.2±1.3	189.3±0.4	222.7±1.1	247.4±0.3	279.8±0.3	312.4±0.4	344.2±0.4
M_{III}	140.5±0.8	161.9±1.0	181.5±0.4	213.8±1.1	238.5±0.3	269.1±0.3	300.3±0.4	330.5±0.4
M_{IV}, M_{V}	41.2±0.7	56.7±0.8		88.9±0.8				
M_{IV}			70.1±0.4		111.8±0.3	135.0±0.3	159.6±0.3	182.4±0.3
M_{V}			69.0±0.4		110.3±0.3	133.1±0.3	157.4±0.3	180.0±0.3
N_{I}			27.3±0.5	24.0±0.8	29.3±0.3	37.7±0.3	45.4±0.3	51.3±0.3
N_{II}, N_{III}	2.5±1.0	5.6±1.3		10.6±1.9		19.9±0.3	25.6±0.4	28.7±0.4
N_{II}			5.2±0.4		14.8±0.4			
N_{III}			4.6±0.4		14.0±0.3			

	41 Nb	42 Mo	43 Tc	44 Ru	45 Rh	46 Pd	47 Ag	48 Cd
K	18985.6±0.4 (18987.)	19999.5±0.3 (20004.)	21044.0±0.7	22117.2±0.3 (22119.)	23219.9±0.3 (23219.8)	24350.3±0.3 (24348.)	25514.0±0.3 (25516.)	26711.2±0.3 (26716.)
L_{I}	2697.7±0.3 [2697.7][l]	2865.5±0.3 [2866.0][l]	3042.5±0.4 [3042.5][l]	3224.0±0.3 [3224.3][l]	3411.9±0.3 [3412.0][l] (3417.)	3604.3±0.3 [3604.6][l] (3607.)	3805.8±0.3 [3806.2][m] (3807.)	4018.0±0.3 [4018.1][m] (4019.)
L_{II}	2464.7±0.3 [2464.7][l]	2625.1±0.3 [2624.5][l] (2627.)	2793.2±0.4 [2973.2][l]	2966.9±0.3 [2966.8][l] (2966.3)	3146.1±0.3 [3146.3][l] (3145.)	3330.3±0.3 [3330.3][l] (3330.3)	3523.7±0.3 [3523.6][g,m] (3526.)	3727.0±0.3 [3727.1][m] (3728.)
L_{III}	2370.5±0.3 [2370.6][l]	2520.2±0.3 [2520.2][l] (2523.2)	2676.9±0.4 [2676.9][l]	2837.9±0.3 [2837.7][l] (2837.7)	3003.8±0.3 [3003.5][g,l] (3002.)	3173.3±0.3 [3173.0][g,l] (3173.0)	3351.1±0.3 [3350.8][g] (3351.0)	3537.5±0.3 [3537.3][g] (3537.6)
M_{I}	468.4±0.3	504.6±0.3		585.0±0.3	627.1±0.3	669.9±0.3	717.5±0.3	770.2±0.3
M_{II}	378.4±0.4	409.7±0.4	444.9±1.5	482.8±0.3	521.0±0.3	559.1±0.3	602.4±0.3	650.7±0.3
M_{III}	363.0±0.4	392.3±0.3	425.0±1.5	460.6±0.3	496.2±0.3	531.5±0.3	571.4±0.3	616.5±0.3
M_{IV}	207.4±0.3	230.3±0.3	256.4±0.5	283.6±0.3	311.7±0.3	340.0±0.3	372.8±0.3	410.5±0.3
M_{V}	204.6±0.3	227.0±0.3	252.9±0.4	279.4±0.3	307.0±0.3	334.7±0.3	366.7±0.3	403.7±0.3
N_{I}	58.1±0.3	61.8±0.3		74.9±0.3	81.0±0.3	86.4±0.3	95.2±0.3	107.6±0.3
N_{II}, N_{III}	33.9±0.4	34.8±0.4	38.9±1.9	43.1±0.4	47.9±0.4	51.1±0.4		66.9±0.4
N_{II}							62.6±0.3	
N_{III}							55.9±0.3	
$N_{IV,V}$	3.2±0.3	1.8±0.3		2.0±0.3	2.5±0.4	1.5±0.3	3.3±0.3	9.3±0.3

	49 In	50 Sn	51 Sb	52 Te	53 I	54 Xe	55 Cs	56 Ba
K	27939.9±0.3	29200.1±0.4 (29195.)	30491.2±0.3 (30486.)	31813.8±0.3 (31811.)	33169.4±0.4 (33167.)	34561.4±1.1 (34590.)	35984.6±0.4 (35987.)	37440.6±0.4 (37452.)
L_{I}	4237.5±0.3 [4237.7][m] (4237.3)	4464.7±0.3 [4464.5][g] (4464.8)	4698.3±0.3 [4698.3][m] (4698.4)	4939.2±0.3 [4939.3][m] (4939.7)	5188.1±0.3 [5188.1][j]	5452.8±0.4 (5452.8)	5714.3±0.4 [5712.7][j] (5721.)	5988.8±0.4 [5986.8][j] (5996.)
L_{II}	3938.0±0.3 [3937.8][m] (3939.3)	4156.1±0.3 [4156.2][g] (4157.)	4380.4±0.3 [4380.6][m] (4382.)	4612.0±0.3 [4612.0][m] (4612.6)	4852.1±0.3 [4852.0][j]	5103.7±0.4 (5103.7)	5359.4±0.3 [5359.5][j] (5358.)	5623.6±0.3 [5623.6][j] (5623.3)
L_{III}	3730.1±0.3 [3730.0][g] (3730.2)	3928.8±0.3 [3928.8][g] (3928.8)	4132.2±0.3 [4132.2][g] (4132.3)	4341.4±0.3 [4341.2][g] (4341.8)	4557.1±0.3 [4557.1][j]	4782.2±0.4 (4782.2)	5011.9±0.3 [5012.0][j] (5011.3)	5247.0±0.3 [5247.3][j] (5247.0)
M_{I}	825.6±0.3	883.8±0.3	943.7±0.3	1006.0±0.3	1072.1±0.3		1217.1±0.4	1292.8±0.4
M_{II}	702.2±0.3	756.4±0.4	811.9±0.3	869.7±0.3	930.5±0.3	999.0±2.1	1065.0±0.5	1136.7±0.5

	49 In	50 Sn	51 Sb	52 Te	53 I	54 Xe	55 Cs	56 Ba
M_{III}	664.3±0.3	714.4±0.3	765.6±0.3	818.7±0.3	874.6±0.3	937.0±2.1	997.6±0.5	1062.2±0.5
M_{IV}	450.8±0.3	493.3±0.3	536.9±0.3	582.5±0.3	631.3±0.3		739.5±0.4	796.1±0.3
M_{V}	443.1±0.3	484.8±0.3	527.5±0.3	572.1±0.3	619.4±0.3	672.3±0.5	725.5±0.5	780.7±0.3
N_{I}	121.9±0.3	136.5±0.4	152.0±0.3	168.3±0.3	186.4±0.3		230.8±0.4	253.0±0.5
N_{II}	77.4±0.4	88.6±0.4	98.4±0.5	110.2±0.5	122.7±0.5	146.7±3.1	172.3±0.6	191.8±0.7
N_{III}							161.6±0.6	179.7±0.6
N_{IV}	16.2±0.3	23.9±0.3	31.4±0.3	39.8±0.3	49.6±0.3		78.8±0.5	92.5±0.5
N_{V}							76.5±0.5	89.9±0.5
O_{I}	0.1±4.5	0.9±0.5	6.7±0.5	11.6±0.6	13.6±0.6		22.7±0.5	39.1±0.6
O_{II}	0.8±0.4	1.1±0.5	2.1±0.4	2.3±0.5	3.3±0.5		13.1±0.5	16.6±0.5
O_{III}							11.4±0.5	14.6±0.5

	57 La	58 Ce	59 Pr	60 Nd	61 Pm	62 Sm	63 Eu	64 Gd
K	38924.6±0.4 (38934.)	40443.0±0.4 (40453.)	41990.6±0.5 (42002.)	43568.9±0.4 (43574.)	45184.0±0.7 (45198.)	46834.2±0.5 (46849.)	48519.0±0.4 (48519.)	50239.1±0.5 (50233.)
L_{I}	6266.3±0.5 [6266.3][a]	6548.8±0.5 [6548.5][a]	6834.8±0.5 [6834.9][a]	7126.0±0.4 [7125.8][a] (7129.)	7427.9±0.8 [7427.9][o]	7736.8±0.5 [7736.2][a] (7748.)	8052.0±0.4 [8051.7][a] (8061.)	8375.6±0.5 [8375.4][a] (8386.)
L_{II}	5890.6±0.4 [5890.7][a]	6164.2±0.4 [6164.3][a]	6440.4±0.5 [6440.2][a]	6721.5±0.4 [6721.8][a] (6723.)	7012.8±0.6 [7012.8][o]	7311.8±0.4 [7312.0][a] (7313.)	7617.1±0.4 [7617.6][a] (7620.)	7930.3±0.4 [7930.5][a] (7931.)
L_{III}	5482.7±0.4 [5482.6][a]	5723.4±0.4 [5723.6][a]	5964.3±0.4 [5964.3][a]	6207.9±0.4 [6208.0][a] (6209.)	6459.3±0.6 [6459.4][o]	6716.2±0.5 [6716.8][a] (6717.)	6976.9±0.4 [6976.7][a] (6981.)	7242.8±0.4 [7242.8][a] (7243.)
M_{I}	1361.3±0.3	1434.6±0.6	1511.0±0.8	1575.3±0.7		1722.8±0.8	1800.0±0.5	1880.8±0.5
M_{II}	1204.4±0.6	1272.8±0.6	1337.4±0.7	1402.8±0.6	1471.4±6.2	1540.7±1.2	1613.9±0.7	1688.3±0.7
M_{III}	1123.4±0.5	1185.4±0.5	1242.2±0.6	1297.4±0.5	1356.9±1.4	1419.8±1.1	1480.6±0.6	1544.0±0.8
M_{IV}	848.5±0.4	901.3±0.6	951.1±0.6	999.9±0.6	1051.5±0.9	1106.0±0.8	1160.6±0.6	1217.2±0.6
M_{V}	831.7±0.4	883.3±0.5	931.0±0.6	977.7±0.6	1026.9±1.0	1080.2±0.6	1130.9±0.6	1185.2±0.6
N_{I}	270.4±0.8	289.6±0.7	304.5±0.9	315.2±0.8		345.7±0.9	360.2±0.7	375.8±0.7
N_{II}	205.8±1.2	223.3±1.1	236.3±1.5	243.3±1.6	242.±16.	265.6±1.9	283.9±1.0	288.5±1.2
N_{III}	191.4±0.9	207.2±0.9	217.6±1.1	224.6±1.3		247.4±1.5	256.6±0.8	270.9±0.9
$N_{IV,V}$	98.9±0.8	110.0±0.6	113.2±0.7	117.5±0.7	120.4±2.0	129.0±1.2	133.2±0.6	140.5±0.8
$N_{VI,VII}$		0.1±1.2	2.0±0.6	1.5±0.9		5.5±1.1	0.0±3.2	0.1±3.5
O_{I}	32.3±7.2	37.8±1.3	37.4±1.0	37.5±0.9		37.4±1.5	31.8±0.7	36.1±0.8
$O_{II,III}$	14.4±1.2	19.8±1.2	22.3±0.7	21.1±0.8		21.3±1.5	22.0±0.6	20.3±1.2

	65 Tb	66 Dy	67 Ho	68 Er	69 Tm	70 Yb	71 Lu	72 Hf
K	51995.7±0.5 (52002.)	53788.5±0.5 (53793.)	55617.7±0.5 (55619.)	57485.5±0.5 (57487.)	59389.6±0.5	61332.3±0.5 (61300.)	63313.8±0.5 (63310.)	65350.8±0.6 (65310.)
L_{I}	8708.0±0.5 [8707.6][a] (8717.)	9045.8±0.5 [9046.5][a]	9394.2±0.4 [9394.3][a] (9399.)	9751.3±0.4 [9751.5][a] (9757.)	10115.7±0.4 [10115.6][a] (10121.)	10486.4±0.4 [10487.3][a] (10490.)	10870.4±0.4 [10870.1][a] (10874.)	11270.7±0.4 [11271.6][o] (11274.)
L_{II}	8251.6±0.4 [8251.8][a] (8253.)	8580.6±0.4 [8580.4][a] (8583.)	8917.8±0.4 [8918.2][a] (8916.)	9264.3±0.4 [9264.3][a] (9262.)	9616.9±0.4 [9617.1][a] (9617.1)	9978.2±0.4 [9977.9][a] (9976.)	10348.6±0.4 [10349.0][a] (10345.)	10739.4±0.4 [10738.9][o] (10736.)
L_{III}	7514.0±0.4 [7514.2][a] (7515.)	7790.1±0.4 [7789.6][a] (7789.7)	8071.1±0.4 [8070.6][a] (8068.)	8357.9±0.4 [8357.6][a] (8357.5)	8648.0±0.4 [8647.8][a] (8649.6)	8943.6±0.4 [8942.6][a] (8944.1)	9244.1±0.4 [9243.8][a]	9560.7±0.4 [9560.4][o] (9558.)
M_{I}	1967.5±0.6	2046.8±0.4	2128.3±0.6	2206.5±0.6	2306.8±0.7	2398.1±0.4	2491.2±0.5	2600.9±0.4
M_{II}	1767.7±0.9	1841.8±0.5	1922.8±1.0	2005.8±0.6	2089.8±1.1	2173.0±0.4	2263.5±0.4	2365.4±0.4
M_{III}	1611.3±0.8	1675.6±0.9	1741.2±0.9	1811.8±0.6	1884.5±1.1	1949.8±0.5	2023.6±0.5	2107.6±0.4
M_{IV}	1275.0±0.6	1332.5±0.4	1391.5±0.7	1453.3±0.5	1514.6±0.7	1576.3±0.4	1639.4±0.4	1716.4±0.4
M_{V}	1241.2±0.7	1294.9±0.4	1351.4±0.8	1409.3±0.5	1467.7±0.9	1527.8±0.4	1588.5±0.4	1661.7±0.4
N_{I}	397.9±0.8	416.3±0.5	435.7±0.8	449.1±1.0	471.7±0.9	487.2±0.6	506.2±0.6	538.1±0.4
N_{II}	310.2±1.2	331.8±0.6	343.5±1.4	366.2±1.5	385.9±1.6	396.7±0.7	410.1±1.8	437.0±0.5
N_{III}	385.0±1.0	292.9±0.6	306.6±0.9	320.0±0.7	336.6±1.6	343.5±0.5	359.3±0.5	380.4±0.5
N_{IV}	147.0±0.8	154.2±0.5	161.0±1.0	176.7±1.2	179.6±1.2	198.1±0.5	204.8±0.5	223.8±0.4
N_{V}				167.6±1.5		184.9±1.3	195.0±0.4	213.7±0.5

	65 Tb	66 Dy	67 Ho	68 Er	69 Tm	70 Yb	71 Lu	72 Hf
$N_{VI,VII}$	2.6±1.5	4.2±1.6	3.7±3.0	4.3±1.4	5.3±1.9	6.3±1.0	6.9±0.5	17.1±0.5
O_{I}	39.0±0.8	62.9±0.5	51.2±1.3	59.8±1.7	53.2±3.0	54.1±0.5	56.8±0.5	64.9±0.4
O_{II}, O_{III}	25.4±0.8	26.3±0.6	20.3±1.5	29.4±1.6	32.3±1.6	23.4±0.6	28.0±0.6	
O_{II}								38.1±0.6
O_{III}								30.6±0.6

	73 Ta	74 W	75 Re	76 Os	77 Ir	78 Pt	79 Au	80 Hg
K	67416.4±0.6 (67403.)	69525.0±0.3 (69508.)	71676.4±0.4 (71658.)	73870.8±0.5	76111.0±0.5	78394.8±0.7 (78381.)	80724.9±0.5 (80720.)	83102.3±0.8
L_{I}	11681.5±0.3 [11680.2]^{p} (11682.)	12099.8±0.3 [12098.2]^{p} (12099.6)	12526.7±0.4 (12530.)	12968.0±0.4 (12972.)	13418.5±0.3 (13423.)	13879.9±0.4 (13883.)	14352.8±0.4 (14353.7)	14839.3±1.0 (14842.)
L_{II}	11136.1±0.3 [11136.1]^{p} (11132.)	11544.0±0.3 [11541.4]^{p} (11538.)	11958.7±0.3 [11956.9]^{p} (11954.)	12385.0±0.4 (12381.)	12824.1±0.3 [12824.0]^{e,p} (12820.)	13272.6±0.3 [13272.6]^{e,p} (13272.3)	13733.6±0.3 [13733.5]^{e,p} (13736.)	14208.7±0.7 (14215.)
L_{III}	9881.1±0.3 [9880.3]^{p} (9877.7)	10206.8±0.3 [10204.2]^{p} (10200.)	10535.3±0.3 [10534.2]^{p} (10531.)	10870.9±0.3 [10870.7]^{p} (10868.)	11215.2±0.3 [11215.1]^{e,p} (11212.)	11563.7±0.3 [11563.7]^{e,p} (11562.)	11918.7±0.3 [11918.2]^{e,p} (11921.)	12283.9±0.4 [12284.0]^{e,p} (12286.)
M_{I}	2708.0±0.4	2819.6±0.4	2931.7±0.4	3048.5±0.4	3173.7±1.7	3296.0±0.9	3424.9±0.3 [3424.8]^{p}	3561.6±1.1
M_{II}	2468.7±0.3 [2468.6]^{p}	2574.9±0.3 [2575.0]^{p}	2681.6±0.4	2792.2±0.3 [2791.9]^{p}	2908.7±0.3 [2909.1]^{p}	3026.5±0.4 [3026.5]^{p} (3029.)	3147.8±0.4 [3149.5]^{p}	3278.5±1.3
M_{III}	2194.0±0.3 [2194.1]^{p}	2281.0±0.3 [2281.0]^{p}	2367.3±0.3 [2367.3]^{p}	2457.2±0.4 [2457.4]^{p}	2550.7±0.3 [2550.5]^{p} (2550.5)	2645.4±0.4 [2645.5]^{p} (2645.9)	2743.0±0.3 [2743.1]^{p} (2744.0)	2847.1±0.4 [2847.1]^{p}
M_{IV}	1793.2±0.3 [1793.1]^{p}	1871.6±0.3 [1871.4]^{p}	1948.9±0.3 [1948.9]^{p}	2030.8±0.3 [2031.0]^{p}	2116.1±0.3 [2116.1]^{p}	2201.9±0.3 [2201.9]^{p}	2291.1±0.3 [2291.2]^{p} (2307.)	2384.9±0.3 [2384.9]^{p}
M_{V}	1735.1±0.3 [1735.2]^{p}	1809.2±0.3 [1809.3]^{p}	1882.9±0.3 [1882.9]^{p}	1960.1±0.3 [1960.2]^{p}	2040.4±0.3 [2040.5]^{p}	2121.6±0.3 [2121.6]^{p}	2205.7±0.3 [2206.1]^{p} (2220.)	2294.9±0.3 [2294.9]^{p}
N_{I}	565.5±0.5	595.0±0.4	625.0±0.4	654.3±0.5	690.1±0.4	722.0±0.6	758.8±0.4	800.3±1.0
N_{II}	464.8±0.5	491.6±0.4	517.9±0.5	546.5±0.5	577.1±0.4	609.2±0.6	643.7±0.5	676.9±2.4
N_{III}	404.5±0.4	425.3±0.5	444.4±0.5	468.2±0.6	494.3±0.6	519.0±0.6	545.4±0.5	571.0±1.4
N_{IV}	241.3±0.4	258.8±0.4	273.7±0.5	289.4±0.5	311.4±0.4	330.8±0.5	352.0±0.4	378.3±1.0
N_{V}	229.3±0.3	245.4±0.4	260.2±0.4	272.8±0.6	294.9±0.4	313.3±0.4	333.9±0.4	359.8±1.2
N_{VI}, N_{VII}	25.0±0.4		40.6±0.4	46.3±0.6				
N_{VI}		36.5±0.4			63.4±0.4	74.3±0.4	86.4±0.4	102.2±0.5
N_{VII}		33.6±0.4			60.5±0.4	71.1±0.5	82.8±0.5	98.5±0.5
O_{I}	71.1±0.5	77.1±0.4	82.8±0.5	83.7±0.6	95.2±0.4	101.7±0.4	107.8±0.7	120.3±1.3
O_{II}	44.9±0.4	46.8±0.5	45.6±0.7	58.0±1.1	63.0±0.6	65.3±0.7	71.7±0.7	80.5±1.3
O_{III}	36.4±0.4	35.6±0.5	34.6±0.6	45.4±1.0	50.5±0.6	51.7±0.7	53.7±0.7	57.6±1.3
$O_{IV,V}$	5.7±0.4	6.1±0.4	3.5±0.5		3.8±0.4	2.2±1.3	2.5±0.5	6.4±1.4

	81 Tl	82 Pb	83 Bi	84 Po	85 At	86 Rn	87 Fr	88 Ra
K	85530.4±0.6	88004.5±0.7 (88005.)	90525.9±0.7 (90534.)	93105.0±3.8	95729.9±7.7	98404.±12.	101137.±13.	103921.9±7.2
L_{I}	15346.7±0.4 (15343.)	15860.8±0.5 (15855.)	16387.5±0.4 (16376.)	16939.3±9.8	17493.±29.	18049.±38.	18639.±40.	19236.7±1.5 (19236.0)
L_{II}	14697.9±0.3 [14697.3]^{p} (14699.)	15200.0±0.4 (15205.)	15711.1±0.3 [15708.4]^{p} (15719.)	16244.3±2.4	16784.7±2.5	17337.1±3.4	17906.5±3.5	18484.3±1.5 (18486.0)
L_{III}	12657.5±0.3 [12656.3]^{e,p} (12660.)	13035.2±0.3 [13034.9]^{e,p} (13041.)	13418.6±0.3 [13418.3]^{e,p} (13426.)	13813.8±1.0 ⟨13813.8⟩	14213.5±2.0 ⟨14213.5⟩	14619.4±3.0 ⟨14619.4⟩	15031.2±3.0 ⟨15031.2⟩	15444.4±1.5 (15444.0)
M_{I}	3704.1±0.4	3850.7±0.5	3999.1±0.3 [3999.1]^{p}	4149.4±3.9	⟨4317.⟩	⟨4482.⟩	⟨4652.⟩	4822.0±1.5
M_{II}	3415.7±0.3 [3415.7]^{p}	3554.2±0.3 [3554.2]^{p}	3696.3±0.3 [3696.4]^{p}	3854.1±9.8	4008.±28.	4159.±38.	4327.±40.	4489.5±1.8
M_{III}	2956.6±0.3 [2956.5]^{p}	3066.4±0.4 [3066.3]^{p}	3176.9±0.3 [3176.8]^{p}	3301.9±9.9	3426.±29.	3538.±38.	3663.±40.	3791.8±1.7
M_{IV}	2485.1±0.3 [2485.2]^{p}	2585.6±0.3 [2585.5]^{p} (2606.)	2687.6±0.3 [2687.4]^{p}	2798.0±1.2	2908.7±2.1	3021.5±3.1	3136.2±3.1	3248.4±1.6

	81 Tl	82 Pb	83 Bi	84 Po	85 At	86 Rn	87 Fr	88 Ra
M_{V}	2389.3±0.3 [2389.4][p]	2484.0±0.3 [2484.2][p] (2502.)	2579.6±0.3 [2579.5][p]	2683.0±1.1	2786.7±2.1	2892.4±3.1	2999.9±3.1	3104.9±1.6
N_{I}	845.5±0.5	893.6±0.7	938.2±0.3 [938.7][p]	995.3±2.9	⟨1042.⟩	⟨1097.⟩	⟨1153.⟩	1208.4±1.6
N_{II}	721.3±0.8	763.9±0.8	805.3±0.3 [805.3][p]	851.±12.	886.±30.	929.±40.	980±42.	1057.6±1.8
N_{III}	609.0±0.5	644.5±0.6	678.9±0.3 [678.0][p]	705.±14.	740.±30.	768.±40.	810±43.	879.1±1.8
N_{IV}	406.6±0.4	435.2±0.5	463.6±0.3 [463.6][p]	500.2±2.4	533.2±3.2	566.6±4.0	603.3±4.1	635.9±1.6
N_{V}	386.2±0.5	412.9±0.6	440.0±0.3 [440.1][p]	473.4±1.3			577.±34.	602.7±1.7
N_{VI}	122.8±0.4	142.9±0.4	161.9±0.5 }					298.9±2.4
N_{VII}	118.5±0.4	138.1±0.4	157.4±0.6 }					
O_{I}	136.3±0.7	147.3±0.8	159.3±0.7					254.4±2.1
O_{II}	99.6±0.6	104.8±1.0	116.8±0.7					200.4±2.0
O_{III}	75.4±0.6	86.0±1.0	92.8±0.6					152.8±2.0
O_{IV}	15.3±0.4	21.8±0.4	26.5±0.5 }	31.4±3.2				67.2±1.7
O_{V}	13.1±0.4	19.2±0.4	24.4±0.6 }					
P_{I}		3.1±1.0						43.5±2.2
$P_{II,III}$		0.7±1.0	2.7±0.7					18.8±1.8

	89 Ac	90 Th	91 Pa	92 U	93 Np	94 Pu	95 Am	96 Cm
K	106755.3±5.3	109650.9±0.9	112601.4±2.4	115606.1±1.6	118678.±33.	121818.±44.	125027.±55.	128220
L_{I}	19840.±18.	20472.1±0.5 (20464.)	21104.6±1.8 (21128.)	21757.4±0.3 (21771.)	22426.8±0.9	23097.2±1.6 (23109.)	23772.9±2.0 ⟨23772.9⟩	24460
L_{II}	19083.2±2.8	19693.2±0.4 (19683.)	20313.7±1.5 (20319.)	20947.6±0.3 (20945.)	21600.5±0.4	22266.2±0.7 (22253.)	22944.0±1.0	23779
L_{III}	15871.0±2.0 ⟨15871.0⟩	16300.3±0.3 [16299.6][q] (16299.)	16733.1±1.4 (16733.)	17166.3±0.3 [17168.5][r] (17165.)	17610.0±0.4 (17606.2)	18056.8±0.6 (18053.1)	18504.1±0.9 (18504.1)	18930
M_{I}	⟨5002.⟩	5182.3±0.3 [5182.3][q]	5366.9±1.6	5548.0±0.4	5723.2±3.6	5932.9±1.4	6120.5±7.5	6288
M_{II}	4656.±18.	4830.4±0.4 [4830.6][q]	5000.9±2.3	5182.2±0.4 [5180.9][r]	5366.2±0.7 [5366.4][s]	5541.2±1.7	5710.2±2.1	5895
M_{III}	3909.±18.	4046.1±0.4 [4046.1][q] (4041.)	4173.8±1.8	4303.4±0.3 [4303.6][r] (4299.)	4434.7±0.5 [4434.6][s]	4556.6±1.5	4667.0±2.1	4797
M_{IV}	3370.2±2.1	3490.8±0.3 [3490.7][q] (3485.)	3611.2±1.4 (3608.)	3727.6±0.3 [3728.1][r] (3720.)	3850.3±0.4 [3849.8][s]	3972.6±0.6 [3972.7][t]	4092.1±1.0	4227
M_{V}	3219.0±2.1	3332.0±0.3 [3332.1][q] (3325.)	3441.8±1.4 (3436.)	3551.7±0.3 [3551.7][r] (3545.)	3665.8±0.4 [3664.2][s]	3778.1±0.6 [3778.0][t]	3886.9±1.0	3971
N_{I}	⟨1269.⟩	1329.5±0.4 [1329.8][q]	1387.1±1.9	1440.8±0.4 [1441.3][r]	1500.7±0.8 [1500.7][s]	1558.6±0.8	1617.1±1.1	1643
N_{II}	1080.±19.	1168.2±0.4 [1168.3][q]	1224.3±1.6	1272.6±0.3 [1272.5][r]	1327.7±0.8 [1327.7][s]	1372.1±1.8	1411.8±8.3	1440
N_{III}	890.±19.	967.3±0.4 [967.6][q]	1006.7±1.7	1044.9±0.3 [1044.9][r]	1086.8±0.7 [1086.8][s]	1114.8±1.6	⟨1135.7⟩	1154
N_{IV}	674.9±3.7	714.1±0.4 [714.4][q]	743.4±2.1	780.4±0.3 [779.7][r]	815.9±0.5 [817.1][s]	848.9±0.6 [848.9][t]	878.7±1.0	
N_{V}		676.4±0.4 [676.4][q]	708.2±1.8	737.7±0.3 [737.6][r]	770.3±0.4 [773.2][s]	801.4±0.6 [801.4][t]	827.6±1.0	
N_{VI}		344.4±0.3 [344.2][q]	371.2±1.6	391.3±0.6	415.0±0.8 [415.0][s]	445.8±1.7		
N_{VII}		335.2±0.4 [335.0][q]	359.5±1.6	380.9±0.9	404.4±0.5 [404.4][s]	432.4±2.1		
O_{I}		290.2±0.8	309.6±4.3	323.7±1.1		351.9±2.4		385
O_{II}		229.4±1.1 }	222.9±3.9	{ 259.3±0.5	283.4±0.8 [283.4][s]	274.1±4.7		
O_{III}		181.8±0.4 [181.8][q] }		{ 195.1±1.3	206.1±0.7 [206.1][s]	206.5±4.7		

	89 Ac	90 Th	91 Pa	92 U	93 Np	94 Pu	95 Am	96 Cm
O_{IV}		94.3±0.4 [94.4][q]	94.1±2.8	105.0±0.5	109.3±0.7 [108.8][s]	116.0±1.2	115.8±1.3	
O_{V}		87.9±0.3 [88.1][q]		96.3±1.4	101.3±0.5 [101.4][s]	105.4±1.0	103.3±1.1	
P_{I}		59.5±1.1		70.7±1.2				
P_{II}		49.0±2.5		42.3±9.0				
P_{III}		43.0±2.5		32.3±9.0				

	97 Bk	98 Cf	99 Es	100 Fm	101 Md	102 No	103 Lw
K	[131590±40][u]	135960	139490	143090	146780	150540	154380
L_{I}	[25275±17][u]	26110	26900	27700	28530	29380	30240
L_{II}	[24385±17][u]	25250	26020	26810	27610	28440	29280
L_{III}	[19452±20][u]	19930	20410	20900	21390	21880	22360
M_{I}	[6556±21][u]	6754	6977	7205	7441	7675	7900
M_{II}	[6147±31][u]	6359	6574	6793	7019	7245	7460
M_{III}	[4977±31][u]	5109	5252	5397	5546	5688	5710
M_{IV}	4366	4497	4630	4766	4903	5037	5150
M_{V}	4132	4253	4374	4498	4622	4741	4860
N_{I}	[1755±22][u]	1799	1868	1937	2010	2078	2140
N_{II}	1554	1616	1680	1747	1814	1876	1930
N_{III}	1235	1279	1321	1366	1410	1448	1480
O_{I}	[398±22][u]	419	435	454	472	484	490

[a] J. E. Mack, 1949, as given in C. E. Moore, *Atomic Energy Levels* (U. S. National Bureau of Standards, Washington, D. C., 1949), Vol. 1, p. 1.
[b] G. Herzberg, 1957, as given in C. E. Moore, *Atomic Energy Levels* (U. S. National Bureau of Standards, Washington, D. C., 1958), Vol. 3, p. 238.
[c] See Ref. 18.
[d] A. Fahlman, D. Hamrin, R. Nordberg, C. Nordling, and K. Siegbahn, Phys. Rev. Letters **14**, 127 (1965). See also Ref. 26.
[e] See Ref. 15.
[f] See Ref. 11.
[g] C. Nordling, Arkiv Fysik **15**, 397 (1959).
[h] E. Sokolowski, C. Nordling, and K. Siegbahn, Arkiv Fysik **12**, 301 (1957).
[i] C. Nordling and S. Hagström, Arkiv Fysik **16**, 515 (1960).
[j] I. Andersson and S. Hagström, Arkiv Fysik **27**, 161 (1964).
[k] M. O. Krause, Phys. Rev. **140**, A1845 (1965).
[l] A. Fahlman, O. Hörnfeldt, and C. Nordling, Arkiv Fysik **23**, 75 (1962).
[m] P. Bergvall, O. Hörnfeldt, and C. Nordling, Arkiv Fysik **17**, 113 (1960).
[n] P. Bergvall and S. Hagström, Arkiv Fysik **17**, 61 (1960).
[o] S. Hagström, Z. Physik **178**, 82 (1964).
[p] A. Fahlman and S. Hagström, Arkiv Fysik **27**, 69 (1964).
[q] C. Nordling and S. Hagström, Z. Physik **178**, 418 (1964).
[r] C. Nordling and S. Hagström, Arkiv Fysik **15**, 431 (1959).
[s] S. Hagström, Bull. Am. Phys. Soc. **11**, 389 (1966).
[t] A. Fahlman, K. Hamrin, R. Nordberg, C. Nordling, K. Siegbahn, and L. W. Holm, Phys. Letters **19**, 643 (1966).
[u] J. M. Hollander, M. D. Holtz, T. Novakov, and R. L. Graham ,Arkiv Fysik **28**, 375 (1965).

LATTICE SPACING OF COMMON ANALYZING CRYSTALS

Crystal	Reflection plane	*d* Spacing, Å	Crystal	Reflection plane	*d* Spacing, Å
ADP[a]	101	5.31	Lead stearate		51
ADP[a]	110	5.325	LiF	200	2.014
ADP[a]	200	3.75	LiF	220	1.424
Beryl	$10\bar{1}0$	7.98	Mica	002	9.96
Calcite	100	3.036	NaCl	200	2.820
EDDT[b]	020	4.404	Oxalic acid	001	5.85
Germanium	111	3.265	PET[d]	002	4.371
Graphite	001	6.69	Quartz	$10\bar{1}0$	4.255
Gypsum	010	7.600	Quartz	$10\bar{1}1$	3.343
KAP[c]	001	13.32	Quartz	$11\bar{2}0$	2.456
KBr	200	3.29	Silicon	111	3.13
KCl	200	3.14	Topaz	303	1.356

While several of the above spacings have been measured to more than four significant figures, no more than four figures are given here because complications introduced by the index of refraction, anomalous dispersion, temperature coefficient of expansion, and crystal impurities must be considered before the additional figures are useful.

[a] Ammonium dihydrogen phosphate.
[b] Ethylenediamine d-tartrate.
[c] Potassium acid phthalate.
[d] Pentaerythritol.

RADIATIVE TRANSITION PROBABILITIES FOR *K* X-RAY LINES

$K\beta'_1 = KM_2 + KM_3 + KM_{4,5}$ $\quad K\beta'_2 = KN_{2,3} + KO_{2,3}$

$K\alpha = K\alpha_1 + K\alpha_2$ $\quad K\beta = K\beta'_1 + K\beta'_2$

Element	$K\alpha_2/K\alpha_1$	$K\alpha_3/K\alpha_1$	$K\beta_1/K\alpha_1$	$K\beta'_2/K\alpha_1$	$K\beta_4/K\alpha_1$	$K\beta_5/K\alpha_1$	$K\beta_3/K\beta_1$	$K\beta/K\alpha$
$_{12}$Mg								0.013
$_{14}$Si								0.027
$_{16}$S								0.059
$_{18}$Ar								0.105
$_{20}$Ca	0.502							0.128
$_{22}$Ti	0.503							0.134
$_{24}$Cr	0.504							0.135
$_{26}$Fe	0.506							0.135
$_{28}$Ni	0.508							0.135
$_{30}$Zn	0.510							0.138
$_{32}$Ge	0.513							0.147
$_{34}$Se	0.515							0.157
$_{36}$Kr	0.517			0.019				0.172
$_{38}$Sr	0.520			0.030				0.180
$_{40}$Br	0.523			0.037				0.190
$_{42}$Mo	0.525			0.041				0.197
$_{44}$Ru	0.527			0.045				0.204
$_{46}$Pd	0.529			0.048				0.210
$_{48}$Cd	0.532			0.053			0.519	0.213
$_{50}$Sn	0.534			0.055			0.519	0.220
$_{52}$Te	0.537			0.058			0.519	0.225
$_{54}$Xe	0.539			0.064			0.518	0.232
$_{56}$Ba	0.543			0.070			0.518	0.237
$_{58}$Ce	0.546			0.076			0.518	0.242
$_{60}$Nd	0.549	0.11×10^{-3}		0.083			0.518	0.247
$_{62}$Sm	0.552	0.14×10^{-3}		0.086			0.517	0.250
$_{64}$Gd	0.556	0.17×10^{-3}	0.192	0.089	0.85×10^{-3}	3.02×10^{-3}	0.517	0.255
$_{66}$Dy	0.560	0.21×10^{-3}	0.198	0.089	0.92×10^{-3}	3.43×10^{-3}	0.517	0.257
$_{68}$Er	0.564	0.26×10^{-3}	0.202	0.088	0.96×10^{-3}	3.85×10^{-3}	0.518	0.260
$_{70}$Yb	0.567	0.30×10^{-3}	0.207	0.087	1.04×10^{-3}	4.23×10^{-3}	0.518	0.264
$_{72}$Hf	0.572	0.36×10^{-3}	0.212	0.085	1.16×10^{-3}	4.62×10^{-3}	0.518	0.267
$_{74}$W	0.576	0.43×10^{-3}	0.216	0.086	1.28×10^{-3}	5.04×10^{-3}	0.518	0.269
$_{76}$Os	0.580	0.51×10^{-3}	0.222	0.087	1.43×10^{-3}	5.44×10^{-3}	0.519	0.273
$_{78}$Pt	0.583	0.63×10^{-3}	0.226	0.091	1.61×10^{-3}	5.84×10^{-3}	0.520	0.275
$_{80}$Hg	0.588	0.76×10^{-3}	0.228	0.096	1.80×10^{-3}	6.24×10^{-3}	0.520	0.278
$_{82}$Pb	0.593	0.91×10^{-3}	0.228	0.102	2.02×10^{-3}	6.64×10^{-3}	0.521	0.280
$_{84}$Po	0.597	1.12×10^{-3}	0.228	0.108	2.26×10^{-3}	7.05×10^{-3}	0.522	0.283
$_{86}$Rn	0.602	1.32×10^{-3}	0.228	0.113	2.52×10^{-3}	7.48×10^{-3}	0.523	0.286
$_{88}$Ra	0.608	1.58×10^{-3}	0.230	0.117	2.80×10^{-3}	7.80×10^{-3}	0.524	0.287
$_{90}$Th	0.613	1.85×10^{-3}	0.232	0.120	3.13×10^{-3}	8.25×10^{-3}	0.525	0.288
$_{92}$U	0.619	2.15×10^{-3}	0.234	0.123	3.47×10^{-3}	8.65×10^{-3}	0.527	0.289
$_{94}$Pu	0.625		0.234	0.125			0.528	0.291
$_{96}$Cm	0.632		0.234	0.128			0.529	0.293
$_{98}$Cf	0.642		0.238	0.132			0.531	0.295
$_{100}$Fm	0.648		0.240	0.135			0.533	0.297

From Salem, S. I., Panossian, S. L., and Krause, R. A., *At. Data Nucl. Data Tables,* 14, 91, 1974. Reproduced by permission of the copyright owner, Academic Press.

RADIATIVE TRANSITION PROBABILITIES FOR *L* X-RAY LINES

The following three tables present data for the radiative transition probabilities for the L_1, L_2, and L_3 X-ray lines. The data are normalized respectively to $L\beta_3 = 100$, $L\beta_1 = 100$ and $L\alpha_1 = 100$.

L_1 X-RAY LINES NORMALIZED TO $L\beta_3 = 100$

Element	$L\beta_3$	$L\beta_4$	$L\gamma_2$	$L\gamma_3$	Element	$L\beta_3$	$L\beta_4$	$L\gamma_2$	$L\gamma_3$
$_{32}$Ge	100			17.3	$_{66}$Dy	100	61.8	19.5	28.0
$_{34}$Se	100			18.0	$_{68}$Er	100	63.5	19.8	29.0
$_{36}$Kr	100			18.2	$_{70}$Yb	100	65.5	20.7	29.8
$_{38}$Sr	100			18.8	$_{72}$Hf	100	67.8	21.2	30.7
$_{40}$Zr	100			19.0	$_{74}$W	100	70.5	21.8	31.8
$_{42}$Mo	100	70.6		19.6	$_{76}$Os	100	73.2	23.0	32.8
$_{44}$Ru	100	67.8		20.2	$_{78}$Pt	100	76.5	24.5	33.8
$_{46}$Pd	100	65.5		20.6	$_{80}$Hg	100	80.3	26.3	35.0
$_{48}$Cd	100	63.5		21.3	$_{82}$Pb	100	84.2	28.6	36.0
$_{50}$Sn	100	62.1		22.0	$_{84}$Po	100	88.5	31.3	37.2
$_{52}$Te	100	60.7		22.6	$_{86}$Rn	100	93.4	34.2	38.2
$_{54}$Xe	100	59.8		23.3	$_{88}$Ra	100	98.9	37.5	39.6
$_{56}$Ba	100	59.5		24.0	$_{90}$Th	100	104.5	41.2	41.0
$_{58}$Ce	100	59.2		24.6	$_{92}$U	100	110.2	45.0	42.6
$_{60}$Nd	100	59.4		25.4	$_{94}$Pu	100	116.2	49.5	44.0
$_{62}$Sm	100	60.0		26.3	$_{96}$Cm	100	123.0	55.7	45.7
$_{64}$Gd	100	60.8	19.2	27.0					

L_2 X-RAY LINES NORMALIZED TO $L\beta_1$ = 100

Element	L_{β_1}	L_η	L_{γ_1}	L_{γ_6}	Element	L_{β_1}	L_η	L_{γ_1}	L_{γ_6}
28^{Ni}	100	7.60			64^{Gd}	100	2.35	17.00	
30^{Zn}	100	6.80			66^{Dy}	100	2.25	17.40	
32^{Ge}	100	6.28			68^{Er}	100	2.16	17.80	
34^{Se}	100	5.80			70^{Yb}	100	2.10	18.17	
36^{Kr}	100	5.35			72^{Hf}	100	2.08	18.43	
38^{Sr}	100	4.93			74^{W}	100	2.10	18.80	0.72
40^{Zr}	100	4.60	3.30		76^{Os}	100	2.12	19.34	1.65
42^{Mo}	100	4.30	5.50		78^{Pt}	100	2.18	19.73	2.40
44^{Ru}	100	4.00	7.33		80^{Hg}	100	2.25	20.35	3.10
46^{Pd}	100	3.75	10.67		82^{Pb}	100	2.30	20.93	3.65
48^{Cd}	100	3.55	10.60		84^{Po}	100	2.40	21.54	4.15
50^{Sn}	100	3.35	11.80		86^{Rn}	100	2.46	22.20	4.55
52^{Te}	100	3.20	12.70		88^{Ra}	100	2.50	22.87	4.87
54^{Xe}	100	3.00	14.00		90^{Th}	100	2.60	23.43	5.02
56^{Ba}	100	2.85	14.50		92^{U}	100	2.65	24.10	5.12
58^{Ce}	100	2.70	15.30		94^{Pu}	100	2.70	24.40	5.16
60^{Nd}	100	2.60	16.00		96^{Cm}	100	2.75	25.07	5.20
62^{Sn}	100	2.45	16.50						

L_3 X-RAY LINES NORMALIZED TO $L\alpha_1$ = 100

Element	L_{α_1}	$L_{\beta_{2,15}}$	L_{α_2}	L_{β_5}	L_{β_6}	L_ℓ
26^{Fe}	100					12.22
28^{Ni}	100					8.95
30^{Zn}	100					7.34
32^{Ge}	100					6.45
34^{Se}	100					7.76
36^{Kr}	100					5.28
38^{Sr}	100					4.92
40^{Zr}	100	0.70	11.10			4.67
42^{Mo}	100	5.17	11.10			4.45
44^{Ru}	100	9.30	11.12			4.28
46^{Pd}	100	11.80	11.12			4.11
48^{Cd}	100	14.33	11.12			4.07
50^{Sn}	100	16.00	11.13			4.00
52^{Te}	100	18.00	11.13			4.00
54^{Xe}	100	19.40	11.13			4.00
56^{Ba}	100	20.67	11.13			4.02
58^{Ce}	100	21.00	11.14			4.09
60^{Nd}	100	21.33	11.14		0.875	4.13
62^{Sm}	100	21.07	11.14		0.925	4.16
64^{Gd}	100	20.83	11.14		0.99	4.20
66^{Dy}	100	20.50	11.14		1.05	4.26
68^{Er}	100	20.04	11.15		1.12	4.33
70^{Yb}	100	19.40	11.15		1.17	4.47
72^{Hf}	100	21.33	11.15	0.30	1.21	4.59
74^{W}	100	22.74	11.16	0.50	1.25	4.76
76^{Os}	100	23.40	11.16	1.32	1.37	4.95
78^{Pt}	100	24.00	11.16	1.98	1.43	5.14
80^{Hg}	100	24.50	11.17	2.62	1.50	5.37
82^{Pb}	100	24.83	11.17	3.21	1.56	5.58
84^{Po}	100	25.13	11.17	3.73	1.62	5.80
86^{Rn}	100	25.60	11.18	4.25	1.68	6.00
88^{Ra}	100	25.92	11.18	4.73	1.76	6.26
90^{Th}	100	26.17	11.18	5.18	1.82	6.54
92^{U}	100	26.40	11.18	5.58	1.89	6.79
94^{Pu}	100	26.67	11.18	5.92	1.95	7.02
96^{Cm}	100	26.93	11.18	6.26	2.01	7.34

From Salem, S. I., Panossian, S. L., and Krause, R. A., *At. Data Nucl. Data Tables,* 14, 91, 1974.

NATURAL WIDTH, IN eV OF THE INDICATED K X-RAY LINES

Element	$K\alpha_1$	$K\alpha_2$	$K\beta_1$	$K\beta_3$
20Ca	1.00	0.98		
22Ti	1.45	2.13		
24Cr	2.05	2.64		
26Fe	2.45	3.20		
28Ni	3.00	3.70		
30Zn	3.40	3.96		
32Ge	3.75	4.18		
34Se	4.10	4.43		
36Kr	4.23	4.62		
38Sr	5.17	4.97		
40Zn	5.70	5.25		
42Mo	6.82	6.80		
44Ru	7.41	7.96		
46Pd	8.80	9.20		
48Cd	9.80	10.40		
50Sn	11.20	12.40	11.80	11.00
52Te	12.80	14.20	13.30	12.30
54Xe	14.20	15.10	15.30	13.43
56Ba	16.10	16.80	18.15	16.00
58Ce	18.60	19.50	20.60	17.95
60Nd	21.50	21.50	23.25	21.33
62Sm	26.00	24.70	25.65	24.65
64Gd	29.50	28.00	29.37	28.00
66Dy	33.90	32.20	32.73	32.00
68Er	37.40	35.50	36.20	35.70
70Yb	42.00	40.60	41.43	41.15
72Hf	45.30	44.30	46.00	46.10
74W	47.75	48.00	51.83	51.50
76Os	53.00	49.40	55.90	55.95
78Pt	60.30	54.30	59.98	62.13
80Hg	64.75	68.20	65.75	68.95
82Pb	68.30	79.00	72.20	74.90
84Po	73.20	86.30	78.60	82.85
86Rn	80.00	89.50	85.50	91.20
88Ra	87.00	91.20	94.20	98.95
90Th	94.70	97.00	99.70	105.00
92U	103.00	106.00	115.00	120.00

From Salem, S. I. and Lee, P. L., *At. Data Nucl. Data Tables,* 18, 233, 1976.

NATURAL WIDTH, IN eV OF THE INDICATED L X-RAY LINES

Element	$L\alpha_1$	$L\alpha_2$	$L\beta_1$	$L\beta_2$	$L\beta_3$	$L\beta_4$	$L\gamma_1$
40Zn	1.68	1.52	1.87	5.13	5.50	5.60	3.34
42Mo	1.86	1.80	2.03	5.30	5.90	5.78	3.76
44Ru	2.03	1.98	2.18	5.45	6.35	5.96	4.15
46Pd	2.21	2.16	2.36	5.63	6.80	6.18	4.50
48Cd	2.43	2.40	2.54	5.82	7.23	6.28	4.83
50Sn	2.62	2.62	2.75	6.10	7.70	6.60	5.23
52Tc	2.88	2.88	2.96	6.25	8.22	6.82	5.60
54Xe	3.15	3.15	3.20	6.43	8.70	7.15	5.95
56Ba	3.39	3.45	3.45	6.70	9.20	7.42	6.35
58Ce	3.70	3.78	3.73	6.86	9.70	7.82	6.75
60Nd	3.93	4.08	4.00	7.18	10.30	8.15	7.16
62Sm	4.13	4.50	4.33	7.42	10.80	8.60	7.50
64Gd	4.46	4.90	4.63	7.70	11.20	9.08	7.83
66Dy	4.81	5.35	5.03	7.90	11.50	9.60	8.30
68Er	5.17	5.73	5.45	8.28	11.85	10.03	8.75
70Yb	5.40	6.22	5.90	8.58	12.20	11.00	9.20
72Hf	5.83	6.70	6.36	8.92	12.40	12.80	9.63
74W	6.50	7.20	6.90	9.06	13.10	14.60	10.20
76Os	7.04	7.70	7.42	9.60	14.60	16.50	10.65
78Pt	7.60	8.28	8.00	9.95	16.10	18.00	11.20
80Hg	8.10	8.80	8.70	10.40	17.40	19.70	11.80
82Pb	8.82	9.35	9.35	10.75	18.65	21.30	12.30
84Po	9.50	9.95	10.10	11.25	19.90	22.70	13.05
86Rn	10.03	10.50	10.65	11.65	21.00	24.00	13.55
88Ra	11.00	11.20	11.60	12.20	22.00	25.20	14.30
90Th	11.90	11.80	12.40	12.80	22.85	26.35	15.00
92U	12.40	12.40	13.50	13.30	23.70	27.50	15.70
94Pu	13.20	13.00	14.10	13.90	24.10	28.30	16.40
96Cm	14.80	13.60	15.70	14.60	25.00	29.40	17.10

DIFFRACTION DATA FOR CUBIC ISOMORPHS

From Volume 14, pages 689, 690, and 691 of the Analytical Edition of *Industrial and Engineering Chemistry*, with permission.

X Units	Substance
A 4	
3.56	C (diamond)
5.42	Si
5.62	Ge
6.46	α-Sn
A 1	
3.517	Ni
3.554	α-Co
3.60	Taenite (57.7% Fe, 40.8% Ni, 0.5% P)
3.608	Cu
3.63	γ-Fe (1370°K.)
3.797	Rh
3.831	Ir
3.880	Pd
3.88—4.04	Pd-H
3.912	Pt
4.041	Al
4.070	Au
4.077	Ag
4.30	Co-N
4.40—4.46	Ti-H
4.52	Ne (4°K.)
4.66	Zr-H
4.84	β-Tl
4.939	Pb
5.08	Th
5.14	α-Ce
5.296	β-La
5.43	A (4°K.)
5.56	Ca
5.59	Kr (20°K.)
5.70	Kr (92°K.)
6.05	Sr
6.20	X (88°K.)
A 2	
2.861	α-Fe
2.875	α-Cr
2.90	β-Fe (1070°K.)
2.93	δ-Fe (1700°K.)
3.03	V
3.03—3.41	V-C
3.140	Mo
3.157	W
3.295	Cb
3.30	Ta
3.32	β-Ti (1200°K.)
3.46	Li (~80°K.)
3.50	Li
3.61	β-Zr (1120°K.)
4.24	Na (~80°K.)
4.29	Na
5.02	Ba
5.20	K (120°K.)
5.33	K
5.62	Rb (~80°K.)
6.05	Cs (~80°K.)
B 1	
4.018	LiF
4.065	LiD
4.08	VO
4.09	LiH
4.12	(Li_2TiO_3)
4.12—4.20	$(Li_2TiO_3$-MgO)
4.13	VN
4.14	CrN
4.14	VC
4.142	$(63Li_2Fe_2O_4 \cdot 37Li_2TiO_3)$
4.173	NiO
4.207	MgO
4.282	MgO (1570°K.)
4.225	TiN
4.235	TiO
4.24	80 TiN-20 TiC
4.27	CoO
4.28	V-N
4.283	FeO (160°K.)
4.290	FeO (299°K.)
4.30	VC (ε-phase)
4.315	TiC
4.40	CbC
4.41	CbN
4.426	MnO (117°K.)
4.436	MnO (299°K.)
4.44	ScN
4.446	TaC
4.458	HfC
4.615	NaF
4.62	ZrN
4.69	CdO
4.69	ZrC
4.80	CaO
4.82	(Na_2CeO_3)
4.84	(Na_2PrO_3)
4.88	NaH
4.92	AgF
5.006	CaNH
5.13	SrO
5.14	LiCl
5.14	NdN
5.19	MgS
5.192	MnS (130°K.)
5.210	MnS (299°K.)
5.33	KF
5.45	MgSe
5.45	MnSe
5.45	SrNH
5.49	LiBr
5.52	BaO
5.545	AgCl
5.55—5.76	AgCl-AgBr
5.627	NaCl
5.63	RbF
5.68	CaS
5.69	SnAs
5.70	KH
5.755	AgBr
5.76—5.92	AgBr-AgI
5.83	NdP
5.83	NaCN
5.84	BaNH
5.87	SrS
5.91	CaSe
5.94	PbS
5.95	NaBr
5.957	EuS
5.96	NdAs
6.00	PrAs
6.00	LiI
6.01	CsF
6.04	RbH
6.05	β-NaSH (>360°K.)
6.06	CeAs
6.13	LaAs
6.14	PbSe
6.23	SrSe
6.278	KCl
6.285	SnTe
6.31	NdSb
6.345	CaTe
6.35	PrSb
6.36	BaS
6.38	CsH
6.40	CeSb
6.44	PbTe
6.462	NaI
6.45	PrBi
6.48	LaSb
6.49	CeBi
6.53	KCN
6.53	NH_4Cl (>457°K.)
6.56	RbCl
6.57	LaBi
6.58	KBr
6.59	BaSe
6.60	β-KSH (>440°K.)
6.65	SrTe
6.82	RbCN
6.86	RbBr
6.90	NH_4Br (>411°K.)
6.93	β-RbSH (470°K.)
6.99	BaTe
7.052	KI
7.10	β-CsCl (>730°K.)
7.24	NH_4I (>255°K.)
7.325	RbI
$H0_5$	
6.96 ± 0.04	$AgClO_4$ (453 ± 20°K.)

DIFFRACTION DATA FOR CUBIC ISOMORPHS (Continued)

X Units	Substance
7.16 ± 0.10	$NaClO_4$ (618 ± 35°K.)
7.49 ± 0.02	$KClO_4$ (598 ± 15°K.)
7.65 ± 0.05	$TlClO_4$(553°K.)
7.65 ± 0.02	NH_4ClO_4(528 ± 15°K.)
7.68 ± 0.03	$RbClO_4$(583 ± 10°K.)
7.97 ± 0.01	$CsClO_4$ (513 ± 10°K.)
B 3	
4.255	CuF
4.36	CSi IV
4.855	BeS
5.10	BeSe
5.304	$(Cu, Fe, Mo, Sn)_4(S, As, Te)_2$, cousite
5.41	CuCl
5.425	β-ZnS
5.43	AlP
5.44	GaP
5.58	BeTe
5.60	MnS (red)
5.63	AlAs
5.635	GaAs
5.655	ZnSe
5.68	CuBr
5.82	β-CdS
5.84	HgS
5.86	InP
6.04	CdSe
6.04	InAs
6.05	CuI
6.07	HgSe
6.08	ZnTe
6.103	α-Cu_2HgI_4
6.12	AlSb
6.12	GaSb
6.13	SnSb
6.383	α-Ag_2HgI_4
6.40	HgTe
6.43	CdTe
6.45	InSb
6.48	AgI
B 32	
6.195	LiGa
6.209	LiZn
6.36	LiAl
6.687	LiCd
6.786	LiIn
7.297	NaIn
7.373	$(CeMg_3)$
7.373	$(PrMg_3)$
7.473	NaTl
B 20	
4.437	NiSi
4.438	FeSi
4.438	CoSi
4.548	MnSi
4.620	CrSi
B 2	
2.603	NiBe
2.606	CoBe
2.69	CuBe
2.813	PdBe
2.82	AlNi
2.945	CuZn
2.989	CuPd
3.146	AuZn
3.156	AgZn
3.168	AgLi
3.259	AuMg
3.275	AgMg
3.287	HgLi
3.325	AgCd
3.34	AuCd (670°K.)
3.424	LiTl
3.442	HgMg
3.628	MgTl
3.67	PrZn
3.70	CeZn
3.73	AlNd
3.74	α-RbCl (83°K.)
3.75	LaZn
3.82	TlCn
3.82	PrCd
3.84	TlSb
3.835	TlCl
3.847	CaTl
3.86	NH_4Cl (<457°K.)
3.86	CeCd
3.88	MgPr
3.90	LaCd
3.97	TlBr
3.98	TlBi
4.024	SrTl
4.05	NH_4Br (<411°K.)
4.112	CsCl
4.20	TlI
4.20	CsCl (<720°K.)
4.25	CsCN
4.287	CsBr
4.29	CsSH
4.37	NH_4I (290°K.)
4.56	CsI
$D 2_1$	
4.07	YB_6
4.07	ErB_6
4.10	NdB_6
4.12	GdB_6
4.12	PrB_6
4.13	CeB_6
4.13	YbB_6
4.14	CaB_6
4.15	LaB_6
4.15	ThB_6
4.19	SrB_6
4.33	BaB_6
C 1	
4.33	Be_2C
4.619	Li_2O
5.06	$(3ZrO_2 \cdot MgO)$
5.07	ZrO_2
5.08	$(95ZrO_2 \cdot 5CeO_2)$
5.13	$(95HiO_2 \cdot 5CeO_2)$
5.38	PrO_2
5.40	CeO_2
5.40	CdF_2
5.406	CuF_2
5.45	CaF_2
5.47	UO_2
5.526	$(66CaF_2 \cdot 33YF_3)$
5.53	$(91CaF_2 \cdot 9ThF_4)$
5.54	HgF_2
5.55	Na_2O
5.58	ThO_2
5.59	Cu_2S
5.704	Li_2S
5.749	Cu_2Se
5.782	SrF_2
5.796	EuF_2
5.838	$(66SrF_2 \cdot 33LaF_3$
5.91	$PtAl_2$
5.91	$PtGa_2$
5.935	β-PbF_2(520°K.)
5.99	Al_2Au
6.005	Li_2Se
6.06	$AuGa_2$
6.19	BaF_2
6.34	Mg_2Si
6.35	$PtIn_2$
6.368	RaF_2
6.379	Mg_2Ge
6.436	K_2O
6.50	Li_2Te
6.50	$AuIn_2$
6.526	Na_2S
6.763	Mg_2Sn
6.809	Na_2Se
6.81	Mg_2Pb
6.98	$SrCl_2$
7.314	Na_2Te
7.38	K_2S
7.65	RbS_2
7.676	K_2Se
8.152	K_2Te
C 15	
5.94	Be_2Cu
6.287	Be_2Ag
6.435	Be_2Ti
6.96	MgNiZn
7.03	Cu_2Mg
7.61	W_2Zr
7.79	Au_2Na

X Units	Substance
7.91	Au_2Pb
7.94	Au_2Bi
8.02	Al_2Ca
8.04	Al_2Ce
8.16	Al_2La
9.50	Bi_2K
C 2	
5.41	FeS_2
5.42	(Fe, Ni)S_2 (6.5% Ni)
5.57	RbS_2
5.57	RuS_2
5.57	Bravoite (53.8% NiS_2, 39.1% FeS_2, 7.1% CoS_2)
5.62	OsS_2
5.64	CoS_2
5.65	(Cu, Ni, Co, Fe)(S, Se)$_2$
5.68	PtP_2
5.74	NiS_2
5.85	$CoSe_2$
5.92	$RuSe_2$
5.93	$OsSe_2$
5.94	$PtAs_2$
5.97	$PdAs_2$
6.02	$NiSe_2$
6.096	MnS_2
6.36	$RuTe_2$
6.37	$OsTe_2$
6.43	$PtSb_2$
6.44	$PdSb_2$
6.64	$AuSb_2$
6.94	$MnTe_2$
C 3	
4.25	Cu_2O
4.73	Ag_2O
F 1	
5.55	CoAsS
5.68	NiAsS
5.90	NiSbS
	(Ni, Fe)AsS, plessite
	Ni(As, Sb)S, corynite
	Ni(Sb, Bi)S, kallilite
	(Co, Ni)SbS, willyamite
D 5_3	
8.13	Be_3N_2
9.37	(Mn, Fe)$_2O_3$
9.42	Mn_2O_3
9.74	Zn_3N_2
9.79	Sc_2O_3
9.94	Mg_3N_2
10.12	In_2O_3
10.15	Be_3P_2
10.37	Lu_2O_3
10.39	Yb_2O_3
10.52	Tm_2O_3
10.54	Er_2O_3
10.57	Tl_2O_3
10.58	Ho_2O_3
10.60	Y_2O_3
10.63	Dy_2O_3
10.70	Tb_2O_3
10.79	Gd_2O_3
10.79	Cd_3N_2
10.84	Eu_2O_3
10.85	Sm_2O_3
11.05	Nd_2O_3
11.40	α-Ca_3N_2
12.02	Mg_3P_2
12.33	Mg_3As_2
D 6_1	
11.05	As_4O_6
11.14	Sb_4O_6
D 1_1	
10.32	$ZrCl_4$
11.25	$TiBr_4$
(11.34)	(CBr_4) (>320°K.)
(11.62)	(CI_4)
11.89	GeI_4
11.99	SiI_4
12.00	TiI_4
12.23	SnI_4
E 2_1	
3.67	$YAlO_3$
3.75	$CdTiO_3$
3.78	$LaAlO_3$
3.80	$CaTiO_3$

X Units	Substance
3.83	$NaWO_3$
3.85	(Na, Ce, Ca)(Ti, Cb)O_3 Loparite
3.88	$NaTaO_3$
3.89	$LaGaO_3$
3.89	$NaCbO_3$
3.91	$SrTiO_3$
3.92	$CaSnO_3$
3.97	$BaTiO_3$
3.98	$KTaO_3$
3.99	$CaZrO_3$
4.00	$KMgF_3$
4.005	$KNiF_3$
4.01	$KCbO_3$
4.03	$SrSnO_3$
4.05	$KZnF_3$
4.07	$KCoF_3$
4.07	$SrHfO_3$
4.09	$SrZrO_3$
4.18	$BaZrO_3$
4.35	$BaPrO_3$
4.38	$BaCeO_3$
4.46	KIO_3
4.48	$BaThO_3$
4.5	NH_4IO_3
4.52	$RbIO_3$
4.66	$CsIO_3$
5.12	$MgZrO_3$
5.20	$CsCdCl_3$
5.33	$CsCdBr_3$
5.44	$CsHgCl_3$
5.77	$CsHgBr_3$
G 0_3	
6.57	$NaClO_3$
6.71	$NaBrO_3$
G 2_1	
7.60	$Ca(NO_3)_2$
7.81	$Sr(NO_3)_2$
7.84	$Pb(NO_3)_2$
8.11	$Ba(NO_3)_2$
H 1_1	
8.045	$NiAl_2O_4$
8.07	$CuAl_2O_4$
8.07	$CoCo_2O_4$
8.07	$MgAl_2O_4$
8.08	$CoAl_2O_4$
8.08	$ZnAl_2O_4$
8.10	$FeAl_2O_4$
8.11	(Ni, Co)(Co, Ni)$_2O_4$
8.11	(Zn, Co)Co_2O_4
8.11	$MgCo_2O_4$
8.27	$MnAl_2O_4$
8.27	(Mn, Co)(Co, Mn)$_2O_4$
H 1_1	
8.28	$MgGa_2O_4$
8.30	$NiCr_2O_4$
8.30	$MgCr_2O_4$
8.31	$ZnCr_2O_4$
8.32	$CoCr_2O_4$
8.32	$ZnGa_2O_4$
8.35	$NiFe_2O_4$
8.35	$Cu_2Cr_2O_4$
8.35	$FeCr_2O_4$
8.36	$MgFe_2O_4$
8.38	$CoFe_2O_4$
8.38	$NiMn_2O_4$
8.40	$ZnFe_2O_4$
8.40	$FeFe_2O_4$
8.42	(Mn, Mg)Fe_2O_4
8.42	$TiCo_2O_4$
8.43	$MnCr_2O_4$
8.43	$TiMg_2O_4$
8.43	$TiZn_2O_4$
8.44	$CuFe_2O_4$
8.47	FeV_2O_4
8.49	$MnCr_2O_4$
8.50	$TiFe_2O_4$
8.54	$MnFe_2O_4$
8.58	$CdCr_2O_4$
8.58	$SnMg_2O_4$
8.61	$SnCo_2O_4$
8.63	$SnZn_2O_4$
8.67	$CdFe_2O_4$
8.67	$TiMn_2O_4$
8.81	$MgIn_2O_4$
9.26	Ag_2MoO_4

X Units		Substance
9.4		$CoCo_2S_4$
9.45		$(Co, Ni)_3S_4$
9.46		$CuCo_2O_4$
9.5		NiN_2S_4
9.92		$ZnCr_2S_4$
10.05		$MnCr_2S_4$
10.19		$CdCr_2S_4$
12.54		$K_2Zn(Cn)_4$
12.76		$K_2Hg(CN)_4$
12.84		$K_2Cd(CN)_4$
	$H\ 5_8$	
10.08		$2Na_2SO_4 \cdot NaCl \cdot NaF$
	$H\ 4_{13}$	
12.11		$KCr(SO_4)_2 \cdot 12H_2O$
12.12		$KAl(SO_4)_2 \cdot 12H_2O$
12.15		$NH_4Al(SO_4)_2 \cdot 12H_2O$
12.20		$RbAl(SO_4)_2 \cdot 12H_2O$
12.21		$TlAl(SO_4)_2 \cdot 12H_2O$
12.31		$CsAl(SO_4)_2 \cdot 12H_2O$
12.44		$NH_3 \cdot CH_3Al(SO_4)_2 \cdot 12H_2O$ (β-alum)
	Langbeinite	
9.93		$K_2Mg(SO_4)_3$
10.2		$K_2(Ca_2Mg)SO_3)_3$
	$H\ 2_1$	
6.00		Ag_3PO_4
6.120		Ag_3AsO_4(90°K.)
6.130		Ag_3AsO_4 (380°K.)
	$H\ 2_4$	
5.37		Cu_3VS_4
	$J\ 1_1$	
8.17		K_2SiF_6
8.35		$(NH_4)_2SiF_6$
8.38		$Rb_2CrF_5H_2O$
8.41		$Tl_2CrF_5H_2O$
8.42		$(NH_4)_2VF_5H_2O$
8.42		$Rb_2VF_5H_2O$
8.45		$Tl_2VF_5H_2O$
8.45		Rb_2SiF_6
8.58		Tl_2SiF_6
8.87		Cs_2SiF_6
8.99		Cs_2GeF_6
9.73		K_2PtCl_6
9.73		K_2OsCl_6
9.76		Tl_2PtCl_6
9.84		$(NH_4)_2PtCl_6$
9.86		K_2ReCl_6
9.88		Rb_2PtCl_6
9.92		Rb_2TiCl_6
9.94		$(NH_4)_2SeCl_6$
9.97		K_2SnCl_6
9.97		Tl_2SnCl_6
9.98		Rb_2SeCl_6
10.02		Rb_2PdBr_6
10.04		$(NH_4)SnCl_6$
10.08		$Ni(NH_3)_6Cl_2$
10.10		Rb_2SnCl_6
10.10		$Co(NH_3)_6Cl_2$
10.11		Tl_2TeCl_6
10.14		$(NH_4)_2PBCl_6$
10.14		K_2TeCl_6
10.15		$Fe(NH_2)_6Cl_2$
10.16		$Mg(NH_3)_6Cl_2$
10.17		Cs_2PtCl_6
	$J\ 1_1$	
10.18		Rb_2ZrCl_6
10.18		$(NH_4)_2TeCl_6$
10.20		$Mn(NH_3)_6Cl_2$
10.20		Rb_2PbCl_6
10.22		Cs_2TiCl_6
10.23		Rb_2TeCl_6
10.25		$Zn(NH_2)_6(ClI_4)_3$
10.26		Cs_2SeCl_6
10.30		K_2OSBr_6
10.35		Cs_2SnCl_6
10.36		K_2SeBr_6
10.36		K_2PtBr_6
10.39		$Co(NH_3)_6Br_2$
10.4		$Ni(NH_2)_6Br_2$
10.41		Cs_2ZrCl_6
10.42		Cs_2PbCl_6
10.45		Cs_2TeCl_6
10.45		$Co(NH_2)H_2OSO_4Br$
10.46		$(NH_4)SeBr_6$
10.46		$Zn(NH_2)_6Br_2$
10.47		$Fe(NH_3)_6Br_2$

X Units		Substance
10.47		$Mg(NH_3)_6Br_2$
10.48		K_2SnBr_6
10.51		$Co(NH_3)_6SO_4Br$
10.52		$Mn(NH_3)_6Br_2$
10.54		$Sr_2Ni(NO_2)_6$
10.55		$Pb_2Ni(NO_2)_6$
10.57		$(NH_4)SnBr_6$
10.58		Rb_2SnBr_6
10.62		$Co(NH_3)_5H_2OSO_4I$
10.63		$Co(NH_3)_6SeO_4Br$
10.67		$Ba_2Ni(NO_2)_6$
10.71		$Ca(NH_3)_6Br_2$
10.71		$Co(NH_3)_6SO_4I$
10.77		Cs_2SnBr_6
10.79		$Co(NH_3)_6SeO_4I$
10.9		$Ni(NH_3)_6I_2$
10.91		$Co(NH_3)_6I_2$
10.96		$Zn(NH_3)_6I_2$
10.97		$Fe(NH_3)_6I_2$
10.98		$Mg(NH_2)_6I_2$
11.04		$Mn(NH_3)_6I_2$
11.04		$Cd(NH_3I_2$
11.24		$Ca(NH_3)_6I_2$
11.27		$Ni(NH_3)_6(BF_4)_2$
11.3		$Co(NH_3)_6(BF_4)_2$
(11.3)		$Zn(NH_3)_6(ClO_4)_2$
11.34		$Mg(NH_3)_6(BF_4)_2$
11.34		$Fe(NH_3)_6(BF_4)_2$
11.37		$Mn(NH_3)_6(BF_4)_2$
11.38		$Cd(NH_3)_6(BF_4)_2$
11.41		$Ni(NH_3)_6(ClO_4)_2$
11.43		$Co(NH_3)_6(ClO_4)_2$
11.46		$Ni(NH_3)_6(SO_3F)_2$
11.49		$Co(NH_2)_6(SO_3F)_2$
11.52		$Fe(NH_3)_6(ClO_4)_2$
11.53		$Mg(NH_3)_6(ClO_4)_2$
11.54		$Cd(NH_3)_6Br_2$
11.54		$Fe(NH_3)_6(SO_2F)_2$
11.58		$Mn(NH_3)_6(ClO_4)_2$
11.59		$Cd(NH_3)_6(ClO_4)_2$
11.59		$Mn(NH_2)_6(SO_2F)_2$
11.62		$Cd(NH_2)_6(SO_3F)_2$
11.91		$Ni(NH_3)_6(PF_6)_2$
11.94		$Co(NH_2)_6(PF_6)_2$
12.03		$Ni(NH_2 \cdot CH_3)_6I_2$
12.05		$Co(NH_2 \cdot CH_3)_6I_2$
12.19		$\{NH(CH_3)_3\}_2SnCl_6$
12.41		$\{S(CH_3)_2\}_2SnCl_6$
12.65		$\{N(CH_3)_4\}_2PtCl_6$
12.80		$\{S(CH_3)_2C_2H_5\}_2SnCl_6$
12.87		$\{N(CH_3)_4\}_2SnCl_6$
13.17		$\{N(CH_3)_2CH_5\}_2SnCl_6$
13.51		$\{N(CH_3)(C_2H_5)_3\}_2SnCl_6$
13.93		$\{P(CH_3)(C_2H_5)_3\}_2SnCl_6$
	$J\ 2_1$ and Related Structures	
8.88		Li_3FeF_6
8.90		$(NH_4)_3AlF_6$
9.01		$(NH_4)_3CrF_6$
9.04		$(NH_4)_3VF_6$
9.10		$(NH_4)_3FeF_6$
9.10		$(NH_4)MoO_3F_2$
9.26		Na_2FeF_6
9.93		K_2FeF_6
9.96		$CuLi_2Fe(CN)_6$
10.0		$CuR_2Fe(CN)_6R$ = Na, K, Rb, NH_4, Tl
10.15		$K_2CdFe(NO_2)_6$
10.17		$K_2CaCo(NO_2)_6$
10.19		$K_2CaFe(NO_2)_6$
10.2		$Fe^{III}RFe^{II}(CN)_6$ R = Na, K, Rb, NH_4
10.22		$K_2HgFe(NO_2)_6$
10.23		$K_2SrCo(NO_2)_6$
10.25		$(NH_4)_2CaFe(NO_2)_6$
10.25		$NaTl_2Co(NO_2)_6$
10.28		$K_2CdNi(NO_2)_6$
10.28		$(NH_4)_2CdFe(NO_2)_6$
10.29		$K_2HgNi(NO_2)_6$
10.30		$K_2SrFe(NO_2)_6$
10.30		$Tl_2CaFe(NO_2)_6$
10.31		$K_2PbFe(NO_2)_6$
10.32		$K_2CaNi(NO_2)_6$
10.34		$(NH_4)_2SrFe(NO_2)_6$
10.37		$(NH_4)_2PbFe(NO_2)_6$
10.37		$Tl_2CdNi(NO_2)_6$
10.39		$Tl_2PbFe(NO_2)_6$

DIFFRACTION DATA FOR CUBIC ISOMORPHS(Continued)

X Units	Substance
10.39	$NaRb_2Co(NO_2)_6$
10.40	$Tl_2SrFe(NO_2)_6$
10.4	$K_2PbCo(NO_2)_6$
10.41	$(NH_4)_2CdNi)NO_2)_6$
10.42	$Tl_2HgNi(NO_2)_6$
10.43	$K_2BaFe(NO_2)_6$
10.45	$K_2BaCo(NO_2)_6$
10.45	$K_2Co(NO_2)_6$
10.46	$(NH_4)_2HgNi(NO_2)_6$
10.47	$Rb_2HgNi(NO_2)_6$
10.49	$K_2SrNi(NO_2)_6$
10.49	$K_4Ni(NO_2)_6$
10.50	$(NH_4)_2BaFe(NO_2)_6$
10.54	$K_2LiBi(NO_2)_6$
10.55	$K_2PbNi(NO_2)_6$
10.55	$Tl_2BaFe(NO_2)_6$
10.58	$Rb_2CdNi, Cd(NO_2)_6$
10.58	$K_3Ir(NO_2)_6$
10.59	$Rb_2LiBi(NO_2)_6$
10.6	$K_2PbCu(NO_2)_6$
10.63	$K_2Rh(NO_2)_6$
10.63	$(NH_4)_2LiBi(NO_2)_6$
10.64	$Tl_2LiBi(NO_2)_6$
10.67	$K_2BaNi(NO_2)_6$
10.70	$NaCs_2Co(NO_2)_6$
10.70	$Ba_1\{Rh(NO_2)_6\}_2$
10.72	$Tl_2Co(NO_2)_6$
10.73	$Rb_3Co(NO_2)_6$
10.73	$(NH_4)_2Ir(NO_2)_6$
10.73	$Tl_2Ir(NO_2)_6$
10.77	$Rb_2Ir(NO_2)_6$
10.8	$(NH_4)_2Co(NO_2)_6$
10.81	$Cs_2Cd\{Ni, Cd(NO_2)_6\}$
10.82	$Co(NH_3)_5H_2OI_3$
10.83	$Rb_3Rh(NO_2)_6$
10.88	$K_2NaBi(NO_2)_6$
10.89	$Co(NH_3)_6I_3$
10.91	$Tl_3Rh(NO_2)_6$
10.91	$(NH_4)_3Rh (NO_2)_6$
10.94	$Cs_2LiBi(NO_2)_6$
10.95	$K_2AgBi(NO_2)_6$
10.98	$Rb_2NaBi(NO_2)_6$
10.99	$(NH_4)_2NaBi(NO_2)_6$
11.01	$Tl_2NaBi(NO_2)_6$
11.05	$Rb_2AgBi(NO_2)_6$
11.06	$Tl_2AgBi(NO_2)_6$
11.10	$(NH_4)_2AgBi(NO_2)_6$
11.15	$Cs_2NaBi(NO_2)_6$
11.15	$Cs_3Co(NO_2)_6$
11.17	$Cs_3Ir(NO_2)_6$
11.19	$Cs_3Bi(NO_2)_6$
11.19	$Cs_2AgBi(NO_2)_6$
11.21	$Co(NH_3)_6(BF_4)_3$
11.30	$Cs_3Rh(NO_2)_6$
11.32	$\{Co(NH_3)_6 \cdot H_2O\}(ClO_4)_3$
11.39	$Co(NH_3)_6(ClO_4)_3$
11.67	$Co(NH_3)_6(PF_6)$
$K\,6_1$	
7.46	SiP_2O_7
7.80	TiP_2O_7
7.98	SnP_2O_7
8.18	HfP_2O_7
8.20	ZrP_2O_7
8.61	UP_2O_7
$S\,1_4$	
11.51	$Al_2(Mg, Fe)_3(SiO_4)_3$, pyrope
11.51	$Al_2Fe_2(SiO_4)_3$, almandite
11.60	$Al_2Mn_3(SiO_4)_3$, spessartite
11.87	$Al_2Ca_2(SiO_4)_3$, grossularite
11.89	$(Al, Fe)_2Ca_3(SiO_4)_3$, hessonite
11.95	$Cr_2Ca_3(SiO_4)_3$, uvarovite
12.03	$Fe_2Ca_3(SiO_4)_3$, andradite
12.10	$(Na, Li)_3AlF_6$, cryolithionite
12.35—12.46	$(Mg, Mn)_2(Ca, Na)_3AsO_4)_3$, berzelite
$S\,6_1$	
13.68	$NaAlSi_2O_6H_2O$
$S\,0_8$	
13.82	$Al_{13}Si_5O_{20}(OH, F)_{18}Cl$, zunyite
$S\,6_2$	
8.87	$Na_4(AlSiO_4)_3Cl$, sodalite
Tetrahedrite	
10.19	$(Cu, Fe)_{12}As_4S_{13}$, binnite
10.2—10.6	$(Cu, Ag)_{10} (Zn, Fe)_2(Sb, As)_4S_{13}$

PHOTOMETRIC QUANTITIES, UNITS AND STANDARDS

Photometric quantities and units are also given in the section Quantities and Units under the sub-division Light.

Candela-cd (formerly candle): The International System (SI) unit of luminous intensity. One candela is defined as the luminous intensity of one sixtieth of one square centimeter of projected area of a black body radiator operating at the temperature of freezing platinum (2042K).

Footcandle-fc: The unit of illumination when a foot is the unit of length. Footcandle is the illumination on a surface one square foot in area on which there is a uniformly distributed flux of one lumen, or the illumination on a surface all points of which are at a distance of one foot from a directionally uniform point source of one candela. (1 footcandle = 10.76 lux.)

Lumen-lm: The unit of luminous flux. It is equal to the flux through a unit solid angle (steradian), from a uniform point source of one candela, or to the flux on a unit surface all points of which are at unit distance from a uniform source of one candela.

Lux-lx: The International System (SI) unit of illumination in which the meter is the unit of length. (See Footcandle above.) 1 Lux = 0.0929 footcandles.

Luminous Efficacy of a Source of Light: The quotient of the total luminous flux emitted by the total lamp power input. It is expressed in lumens per watt.

Luminance-L: The luminous intensity (photometric brightness) of any surface in a given direction per unit of projected area of the surface as viewed from that direction. In International System (SI) units the luminance of light sources is commonly expressed in candelas per square centimeter.

EFFICACIES AND OTHER CHARACTERISTICS OF ILLUMINANTS

Gordon D. Rowe

A. Incandescent Lamps (120 volt; Inside Frost)

Nominal lamp watts	Bulb	Base	Approximate rated life (hr)	Approximate initial lumens	Approximate initial efficacy	Chromaticity (°K)
25	A-19	Medium	2500	235	9.4	
40	A-19	Medium	1000	480	12.0	
60	A-19	Medium	1000	870	14.5	
75	A-19	Medium	750	1,190	15.9	
100	A-19	Medium	750	1,750	17.5	
150	A-21	Medium	750	2,850	19.0	
200	A-23	Medium	750	4,010	20.0	
500	PS-35	Mogul	1000	10,850	21.7	
750	PS-52	Mogul	1000	17,040	22.7	
1000	PS-52	Mogul	1000	23,740	23.7	
1500	Clear PS-52	Mogul	1000	34,400	22.9	
5000	Clear G-64	Mogul bipost	150	145,000	29.0	
10000	Clear G-96	Mogul bipost	75	335,000	33.6	

Note: The chromaticity of incandescent lamps is generally in the order of 2900—3000K; the color rendering index, 99+. These factors will vary with changes in the voltage.

B. Fluorescent Lamp (F40T12)

	Nominal lamp watts	Approximate rated life (hr)[a]	Approximate initial lumens	Approximate initial efficacy (lm/lamp W)	Chromaticity (°K)	Chromaticity coordinates X	Chromaticity coordinates Y	Color-rendering index
Cool white	40	20,000+	3150	78.75	4150	0.377	0.385	62
Deluxe cool white	40	20,000+	2250	56.25	4175	0.372	0.371	89
Warm white	40	20,000+	3200	80.00	3000	0.443	0.411	52
Deluxe warm white	40	20,000+	2200	55.00	3025	0.435	0.404	77
Daylight	40	20,000+	2600	65.00	6250	0.316	0.341	75
White	40	20,000+	3150	78.75	3450	0.410	0.400	60
Sign white	40	20,000+	2400	60.00	5200	0.340	0.354	82
Natural	40	20,000+	2100	52.50	3700	0.389	0.368	90
Chroma 50	40	20,000+	2210	55.25	5000	0.346	0.362	90
Chroma 75	40	20,000+	2000	50.00	7500	0.303	0.319	92
Regal white®	40	20,000+	2850	71.25	3000	0.434	0.403	81
Lite white	40	15,000[b]	3450	86.25	4200	0.377	0.393	49
SP30	40	15,000	3325	83.10	3000	0.438	0.403	70
SP35	40	15,000[b]	3325	83.10	3500	0.413	0.401	73
SP41	40	15,000	3265	81.60	4100	0.376	0.390	70
Green	40	20,000+	4350	108.75	6975	0.244	0.628	—
Cool green	40	20,000+	2850	71.25	6450	0.307	0.376	68
Blue	40	20,000+	1200	30.00	—	0.191	0.202	—
Gold	40	20,000+	2400	60.00	2500	0.516	0.476	38
Pink	40	20,000+	1100	27.50	—	0.538	0.343	—
Red	40	20,000+	200	5.00	—	0.691	0.303	—
Plant light	40	20,000+	850	21.25	6750	0.323	0.240	−2
Plant light (wide spectrum)	40	20,000+	1950	48.75	3050	0.412	0.360	90
Cool white (Watt-Miser® Plus)	32	15,000+	2750	86.00	4150	0.377	0.385	62
Warm white (Watt-Miser® Plus)	32	15,000+	2800	87.50	3000	0.443	0.411	52
SP30 (Watt-Miser® Plus)	32	15,000+	2900	90.60	3000	0.438	0.403	70
SP35 (Watt-Miser® Plus)	32	15,000+	2900	90.60	3500	0.413	0.401	73
SP41 (Watt-Miser® Plus)	32	15,000+	2850	89.00	4100	0.376	0.390	70
Lite white (Watt-Miser® Plus)	32	15,000+	2925	91.40	4200	0.377	0.393	49
Cool white (Watt-Miser®)	34	20,000	2750	80.90	4150	0.377	0.385	62
Deluxe cool white (Watt-Miser®)	32	20,000	1925	56.60	4175	0.372	0.371	89
Warm white (Watt-Miser®)	32	20,000	2800	82.35	3000	0.443	0.411	52
Deluxe warm white (Watt-Miser®)	32	20,000	1925	56.60	3025	0.435	0.404	77
Daylight (Watt-Miser®)	32	20,000	2300	67.65	6250	0.316	0.341	75
White (Watt-Miser®)	32	20,000	2800	82.35	3450	0.410	0.400	60

	Nominal lamp watts	Approximate rated life (hr)[a]	Approximate initial lumens	Approximate initial efficacy (lm/lamp W)	Chromaticity (°K)	Chromaticity coordinates X	Chromaticity coordinates Y	Color-rendering index
Regal White® (Watt-Miser®)	32	20,000	2450	72.00	3000	0.434	0.403	81
SP30 (Watt-Miser®)	32	20,000	2900	85.30	3000	0.438	0.403	70
SP35 (Watt-Miser®)	32	20,000	2900	85.30	3500	0.413	0.401	73
SP41 (Watt-Miser®)	32	20,000	2850	83.80	4100	0.376	0.390	70
Lite white (Watt-Miser® II)	32	20,000	2925	86.00	4200	0.377	0.393	49
Cool white	75	12,000[c]	6300	84.00				
Deluxe cool white	75	12,000[c]	4500	60.00				
Warm white	75	12,000[c]	6500	86.65				
Deluxe warm white	75	12,000[c]	4350	58.00				
Daylight	75	12,000[c]	5450	72.65				
White	75	12,000[c]	6400	85.30				
Cool white (Watt-Miser®)	60	12,000[c]	5600	93.30				
Deluxe cool white (Watt-Miser®)	60	12,000[c]	4000	66.65				
Warm white (Watt-Miser®)	60	12,000[c]	5800	96.65				
Daylight (Watt-Miser®)	60	12,000[c]	4840	80.65				
White (Watt-Miser®)	60	12,000[c]	5600	93.30				
Lite white (Watt-Miser® II)	60	12,000[c]	6000	100.00				
High-output (F96T12, 800ma)								
Cool white	110	12,000[c]	9200	83.60				
Deluxe cool white	110	12,000	6600	60.00				
Warm white	110	12,000[c]	9200	83.60				
Deluxe warm white	110	12,000[c]	6550	59.50				
Daylight	110	12,000[c]	7800	71.00				
White	110	12,000[c]	9200	83.60				
Cool white (Watt-Miser®)	95	12,000[c]	8300	87.35				
Deluxe cool white (Watt-Miser®)	95	12,000[c]	6100	64.20				
Warm white (Watt-Miser®)	95	12,000[c]	8500	89.50				
White (Watt-Miser®)	95	12,000[c]	8300	87.35				
Lite white (Watt-Miser® II)	95	12,000[c]	8800	92.60				
Cool white	215	12,000[c]	16,000	74.40				
Deluxe cool white	215	12,000[c]	11,000	51.15				
Warm white	215	12,000[c]	15,000	69.75				
Daylight	215	12,000[c]	13,300	61.85				
Cool white (Watt-Miser®)	185	12,000[c]	14,000	75.65				
Lite white (Watt-Miser® II)	185	12,000[c]	14,900	80.50				

Note: Chromaticity, Chromaticity Index, and Color Rendering Index (CRI) are the same as for the equivalent colors of the 40-W lamps listed in the F40T12 section of this table.

C. High-Intensity Discharge (HID)

	Bulb	ANSI Code	Nominal lamp watts	Approximate rated life (hr.)	Approximate initial lumens[d]	Approximate initial efficacy (lm/lamp W)[d]
Mercury-vapor						
Clear	E-37	H33CD-R400	400	24,000+	21,000	52.50
Deluxe white	E-37	H33GL-R400/DX	400	24,000+	22,500	56.25
Warm deluxe white	E-37	H33GL-R400WDX	400	24,000+	19,500	48.75
Clear	BT-56	H36GV-R1000	1000	24,000+	57,000	57.00
Deluxe white	BT-56	H36GW-R1000/DX	1000	24,000+	63,000	63.00
Warm deluxe white	BT-56	H36GW-R1000/WDX	1000	24,000+	58,000	58.00
Metal halide (Multi-Vapor®)						
Clear; vertical, base up	E-23½	M57PE-R175/XBU	175	10,000	16,600	94.85
Diffuse; vertical, base up	E-23½	M57PF-R175/XBU	175	10,000	15,750	90.00
Clear; vertical, base down	E-23½	M57PE-R175/XBD	175	10,000	16,600	94.85
Diffuse; vertical, base down	E-23½	M57PF-R175/XBD	175	10,000	15,750	90.00
Clear; "any position"	E-28	M57PE-R175/U	175	10,000	14,000	80.00
Phosphor-coated; "any position"	E-28	M57PF-R175/U	175	10,000	14,000	80.00
Clear; "any position"	E-28	M58PG-R250/U	250	10,000	20,500	82.00
Phosphor-coated; "any position"	E-28	M58PH-R250/U	250	10,000	20,500	82.00
Clear; "any position"	E-37	M59PJ-R400/U	400	20,000	36,000	90.00
Phosphor-coated; "any position"	E-37	M59PK-R400/U	400	20,000	36,000	90.00
Clear; "any position"	BT-56	M47PA-R1000/U	1000	12,000	110,000	111.00
Phosphor-coated; "any position"	BT-56	M47PB-R1000/U	1000	12,000	105,000	105.00
High-pressure sodium (Lucalox®)						
Clear; medium base	E-17	S76HA-35	35	16,000	2,250	64.80
Diffuse; medium base	E-17	S76HB-35	35	16,000	2,150	61.40
Clear; mogul base	E-23½	S68MS-50	50	24,000+	4,000	80.00
Diffuse; mogul base	E-23½	S68MY-50	50	24,000+	3,800	76.00
Clear; medium base	E-17	S68XX-50	50	24,000+	4,000	80.00
Diffuse; medium base	E-17	S68YY-50	50	24,000+	3,800	76.00
Clear; mogul base	E-23½	S62-ME70	70	24,000+	5,800	82.85
Diffuse; mogul base	E-23½	S62-MF70	70	24,000+	5,400	77.00
Clear; medium base	E-17	S62-LG70	70	24,000+	5,800	82.85
Diffuse; medium base	E-17	S62LH-70	70	24,000+	5,400	77.00
Clear; mogul base	E-23½	S54SB-100	100	24,000+	9,500	95.00
Diffuse; mogul base	E-23½	S54MC-100	100	24,000+	8,800	88.00
Clear; medium base	E-17	S54SG-100	100	24,000+	9,500	95.00
Diffuse; medium base	E-17	S54SH-100	100	24,000+	8,800	88.00
Clear; mogul base	E-23½	S55SC-150	150	24,000+	16,000	106.65
Diffuse; mogul base	E-23½	S55MD-150	150	24,000+	15,000	100.00
Clear; mogul base	E-28	S56SD-150	150	24,000+	15,000	100.00
Clear; mogul base	E-18	S66MN-200	200	24,000+	22,000	110.00
Clear; mogul base	E-18	S50VA-250	250	24,000+	27,500	110.00
Clear; mogul base	E-18	S50VA-250/S	250	24,000+	30,000	120.00
Diffuse; mogul base	E-28	S50VC-250	250	24,000+	26,000	104.00

	Bulb	ANSI Code	Nominal lamp watts	Approximate rated life (hr.)	Approximate initial lumens[d]	Approximate initial efficacy (lm/lamp W)[d]
Clear; mogul base (deluxe color)	E-18	S50VA-250/DX	250	10,000	22,500	90.00
Clear; mogul base	E-18	S67MR-310	310	24,000 +	37,000	119.35
Clear; mogul base	E-18	S51WA-400	400	24,000 +	50,000	125.00
Diffuse; mogul base	E-37	S51WB-400	400	24,000 +	47,500	118.75
Clear; mogul base	E-25	S52XB-1000	1000	24,000 +	140,000	140.00

Note: ® = Registered trademark of the General Electric Company.

[a] Life rating at 3 or more hours per start on rapid-start circuits, 15,000 hr.
[b] Life with rapid-start circuits; on preheat circuits, 12,000 hr.
[c] Rated average life at 3 hr/start; at 12 hr/start, 18,000 hr.
[d] Initial lumen rating as measured when lamp is operated in vertical position.

APPROXIMATE LUMINANCE OF VARIOUS LIGHT SOURCES

Compiled by C.J. Allen and G.D. Rowe

Luminance of source is given in candelas per square centimeter

Source		Approx. avg. luminance* ($Cd\ cm^{-2}$)
Natural sources		
Clear sky	Average luminance	0.8
Sun (as observed from earth's surface)	At meridian	160000
Sun (as observed from earth's surface)	Near horizon	600
Moon (as observed from earth's surface)	Bright spot	0.25
Combustion sources		
Candle flame (sperm)	Bright spot	1.0
Welsbach mantle	Bright spot	6.2
Acetylene flame	Mees burner	10.5
Photoflash lamps		16000-40000 peak
Incandescent electric lamps		
Carbon filament	3 Lumens per watt	52
Tungsten filament	Vacuum lamp– 10 lumens per watt	200
Tungsten filament	Gas-filled lamp– 20 lumens per watt	1200
Tungsten filament	750-watt projection lamp–26 lumens per watt	7500
Fluorescent lamps (cool white)		
Rapid start (40wT12)	430 mA	0.82
High-output (96T12)	800 mA	1.13
Grooved bulb (96T17)	1500 mA	1.50
Fluorescent lamps (other than cool white)		
Daylight (40wT12)	430 mA	0.62
Blue (40wT12)	430 mA	0.30
Green (40wT12)	430 mA	1.17
Red (40wT12)	430 mA	0.05
High-intensity discharge (HID)		
Mercury-vapor (E-37)		
Clear	400 watt	970
Color-improved	400 watt	11.0
Deluxe-white	400 watt	12.1
Mercury-vapor (BT-56)		
Clear	1000 watt	980
Color-improved	1000 watt	15.0
Deluxe-white	1000 watt	17.2
Metal halide (E-37)		
Clear	400 watt	810
Color-improved	1000 watt	930
Deluxe white	1500 watt	1620
High pressure sodium (Lucalox®)	250 watt	520
	400 watt	780
	1000 watt	810

®Registered trademark of the General Electric Company
*Luminance values perpendicular to lamp axis

FLAME STANDARDS

Value of Various Former Standards in International Candles

Standard Pentane lamp, burning pentane	10.0 candles
Standard Hefner lamp, burning amyl acetate	0.9 candles
Standard Carcel lamp, burning colza oil	9.6 candles

The *Carcel unit* is the horizontal intensity of the carcel lamp, burning 42 grams of colza oil per hour. For a consumption between 38 and 46 grams per hour the intensity may be considered proportional to the consumption.

The *Hefner unit* is the horizontal intensity of the Hefner lamp burning amyl acetate, with a flame 4 cm. high. If the flame is l mm. high, the intensity $I = 1 + 0.027(l - 40)$.

WAVE LENGTHS OF VARIOUS RADIATIONS

	Ångstroms
Gamma rays	0.005–1.40
X-rays	0.1–100
Ultra violet, below	4000
Limit of sun's U.V. at earth's surface	2920
Visible spectrum	4000–7000
Violet, representative, 4100, limits	4000–4240
Blue, representative, 4700, limits	4240–4912
Green, representative, 5200, limits	4912–5750
Maximum visibility	5560
Yellow, representative, 5800, limits	5750–5850
Orange, representative, 6000, limits	5850–6470
Red, representative, 6500, limits	6470–7000
Infra red, greater than	7000
Hertzian waves, beyond	2.20×10^6

BRIGHTNESS OF TUNGSTEN

Characteristics of Straight Tungsten Wire in a Vacuum
(Forsythe and Worthing, 1924).

Temperature °K			Brightness candles/cm²	$\frac{B\,dt}{dB\,T}$
Absolute	Brightness	Color		
1000	966	1006	0.00012	22.0
1200	1149	1210	0.006	20.0
1400	1330	1414	0.11	17.2
1600	1509	1619	0.92	15.2
1800	1684	1825	5.05	13.7
2000	1857	2033	20.0	12.3
2200	2026	2242	61.3	11.2
2400	2192	2452	157.0	10.3
2600	2356	2663	347.0	9.6
2800	2516	2878	694.0	8.9
3000	2673	3094	1257.0	8.3
3200	2827	3311	2110.0	7.8
3400	2978	3533	3370.0	7.6
3655*	3165	3817	5740.0	7.3

* Melting-point of tungsten.

WAVE LENGTHS OF THE FRAUNHOFER LINES

Sun's Spectrum

At 15°C and 76 cm pressure. Wave length in Ångström units
(Fabry and Buisson system).

Line	Due to	Wave length	Line	Due to	Wave length
U	Fe	2947.9	*h*	H	4101.750
t	Fe	2994.4	*g*	Ca	4226.742
T	Fe	3021.067	*G*	Fe	4307.914
s	Fe	3047.623		Ca	4307.749
S_1, S_2	Fe	3100.683	*G′*	H	4340.477
	Fe	3100.326	*F*	H	4861.344
	Fe	3099.943	b_4	Fe	5167.510
R	Ca	3181.277		Mg	5167.330
	Ca	3179.343	b_2	Mg	5172.700
Q	Fe	3286.773	b_1	Mg	5183.621
P	Ti	3361.194	E_2	Fe	5269.557
O	Fe	3441.020	D_2	Na	5889.977
N	Fe	3581.210	D_1	Na	5895.944
M	Fe	3727.636	*C*	H	6562.816
L	Fe	3820.438	*B*	O	6869.955
K	Ca	3933.684	*A*	O	7621
H	Ca	3968.494		O	7594
			Z	....	8228.5
			Y	....	8990.0

WAVE LENGTHS FOR SPECTROSCOPE CALIBRATION

(μm)

Source	Wave length	Source	Wave length
Potassium flame	0.7699	*E*, solar	0.5270
Potassium flame	0.7665	b_1, solar or magnesium flame	0.5184
Mercury I arc	0.6907		
B, solar	0.6869	b_2, solar or magnesium flame	0.5173
Lithium flame	0.6708		
C, solar or hydrogen tube	0.6563	Mercury I arc	0.4960
Mercury I arc	0.6234	Mercury I arc	0.4916
D_1, solar or sodium flame	0.5896	*F*, solar or hydrogen tube	0.4861
D_2, solar or sodium flame	0.5890	Strontium flame	0.4608
Mercury I arc	0.5791	Mercury I arc	0.4358
Mercury I arc	0.5770	*G*, solar or hydrogen tube	0.4340
Mercury I arc	0.5461	Mercury I arc	0.4047
Thallium flame	0.5351	H_1, solar	0.3969
		K, solar	0.3934

STANDARD UNITS, SYMBOLS, AND DEFINING EQUATIONS FOR FUNDAMENTAL PHOTOMETRIC AND RADIOMETRIC QUANTITIES

Submitted by Abraham Abramowitz
from Z-7.1-1967

Radiometric Quantities
(See Note at bottom of page)

Quantity*	Symbol*	Defining Equation**	Commonly Used Units	Symbol
Radiant energy	$Q, (Q_e)$		erg ‡ joule kilowatt-hour	 J kWh
Radiant density	$w, (w_e)$	$w = dQ/dV$	‡ joule per cubic meter erg per cubic centimeter	J/m^3 erg/cm^3
Radiant flux	$\Phi, (\Phi_e)$	$\Phi = dQ/dt$	erg per second †watt	erg/s W
Radiant flux density at a surface				
Radiant exitance (Radiant emittance)†	$M, (M_e)$	$M = d\Phi/dA$	watt per square centimeter	W/cm^2
Irradiance	$E, (E_e)$	$E = d\Phi/dA$	‡watt per square meter, etc.	W/m^2
Radiant intensity	$I, (I_e)$	$I = d\Phi/d\omega$ (ω = solid angle through which flux from point source is radiated)	‡watt per steradian	W/sr
Radiance	$L, (L_e)$	$L = d^2\Phi/d\omega(dA \cos\theta)$ $= dI/(dA \cos\theta)$ (θ = angle between line of sight and normal to surface considered)	watt per steradian and square centimeter ‡watt per steradian and square meter	$W \cdot sr^{-1} cm^{-2}$ $W \cdot sr^{-1} m^{-2}$
Emissivity	ε	$\varepsilon = M/M_{blackbody}$ (M and $M_{blackbody}$ are respectively the radiant exitance of the measured specimen and that of a blackbody at the same temperature as the specimen)	one (numeric)	—

Note: The symbols for photometric quantities (see following table) are the same as those for the corresponding radiometric quantities (see above). When it is necessary to differentiate them the subscripts v and e respectively should be used, e.g., Q_v and Q_e.

*Quantities may be restricted to a narrow wavelength band by adding the word spectral and indicating the wavelength. The corresponding symbols are changed by adding a subscript λ, e.g., Q_λ for a spectral concentration or a λ in parentheses, e.g., $K(\lambda)$, for a function of wavelength.

**The equations in this column are given merely for identification.

***Φ_i = incident flux
Φ_a = absorbed flux
Φ_r = reflected flux
Φ_t = transmitted flux

†to be deprecated.

‡International System (SI) unit.

PHOTOMETRIC QUANTITIES

Quantity*	Symbol*	Defining Equation**	Commonly Used Units	Symbol
Absorptance	$\alpha, (\alpha_v, \alpha_e)$	$\alpha = \Phi_a/\Phi_i$***	one (numeric)	—
Reflectance	$\rho, (\rho_v, \rho_e)$	$\rho = \Phi_r/\Phi_i$***	one (numeric)	—
Transmittance	$\tau, (\tau_v, \tau_e)$	$\tau = \Phi_t/\Phi_i$***	one (numeric)	—
Luminous energy (quantity of light)	$Q, (Q_v)$	$Q_v = \int_{380}^{760} K(\lambda)\, Q_e\lambda\, d\lambda$	lumen-hour ‡lumen-second (talbot)	$lm \cdot h$ $lm \cdot s$
Luminous density	$w, (w_v)$	$w = dQ/dV$	‡lumen-second per cubic meter	$lm \cdot s \cdot m^{-3}$
Luminous flux	$\Phi, (\Phi_v)$	$\Phi = dQ/dt$	‡lumen	lm
Luminous flux density at a surface				
Luminous exitance (Luminous emittance)†	$M, (M_v)$	$M = d\Phi/dA$	lumen per square foot	lm/ft^2
Illumination (illuminance)	$E, (E_v)$	$E = d\Phi/dA$	footcandle (lumen per square foot) ‡lux (lm/m^2) phot (lm/cm^2)	fc lx ph
Luminous intensity (candlepower)	$I, (I_v)$	$I = d\Phi/d\omega$ (ω = solid angle through which flux from point source is radiated)	‡candela (lumen per steradian)	cd
Luminance (photometric brightness)	$L, (L_v)$	$L = d^2\Phi/d\omega\,(dA\cos\theta)$ $= dI/(dA\cos\theta)$ (θ = angle between line of sight and normal to surface considered)	candela per unit area stilb (cd/cm^2) nit (†cd/m^2) footlambert ($cd/\pi ft^2$) lambert ($cd/\pi cm^2$) apostilb ($cd/\pi m^2$)	cd/in^2, etc. sb nt fL L asb
Luminous efficacy	K	$K = \Phi_v/\Phi_e$	‡lumen per watt	lm/W
Luminous efficiency	V	$V = K/K_{maximum}$ ($K_{maximum}$ = maximum value of $K(\lambda)$ function)	one (numeric)	—

ILLUMINATION CONVERSION FACTORS

1 lumen = 1/680 lightwatt
1 lumen-hour = 60 lumen-minutes
1 footcandle = 1 $lumen/ft^2$
1 watt-second = 1 joule = 10^7 ergs
1 phot = 1 $lumen/cm^2$
1 lux = 1 $lumen/m^2$

Number of → Multiplied by↘ Equals Number of ↓	Footcandles	*Lux	Phots	Milliphots
Footcandles	1	0.0929	929	0.929
*Lux	10.76	1	10,000	10
Phot	0.00108	0.0001	1	0.001
Milliphot	1.076	0.1	1,000	1

*The International Standard (SI) unit.

LINE SPECTRA OF THE ELEMENTS

Edited by Joseph Reader and Charles H. Corliss*
National Bureau of Standards

These tables were prepared under the auspices of the Committee on Line Spectra of the Elements of the National Academy of Sciences—National Research Council. They contain the outstanding spectral lines of neutral (I), singly ionized (II), doubly ionized (III), triply ionized (IV), and quadruply ionized (V) atoms. Listed are lines that appear in emission from the vacuum ultraviolet to the far infrared. For most atoms these lines were selected from much larger lists in such a way as to include the stronger observed lines in each spectral region. In a few cases prominent monoxide band heads are also given.

The data were compiled by the following contributors, whose initials are given in the headings of the tables that they prepared:

- J. G. Conway—Lawrence Berkeley Laboratory
- C. H. Corliss—National Bureau of Standards
- R. D. Cowan—Los Alamos Scientific Laboratory
- C. R. Cowley—University of Michigan
- Henry M. and Hannah Crosswhite—Argonne National Laboratory
- S. P. Davis—University of California, Berkeley
- V. Kaufman—National Bureau of Standards
- R. L. Kelly—Naval Postgraduate School
- J. F. Kielkopf—University of Louisville
- W. C. Martin—National Bureau of Standards
- T. K. McCubbin—Pennsylvania State University
- L. J. Radziemski—Los Alamos Scientific Laboratory
- J. Reader—National Bureau of Standards
- C. J. Sansonetti—National Bureau of Standards
- G. V. Shalimoff—Lawrence Berkeley Laboratory
- R. W. Stanley—Purdue University
- J. O. Stoner, Jr.—University of Arizona
- H. H. Stroke—New York University
- D. R. Wood—Wright State University
- E. F. Worden—Lawrence Livermore Laboratory
- J. J. Wynne—International Business Machines Corporation
- R. Zalubas—National Bureau of Standards

The literature references are collected at the end of the entire set of tables.

All wavelengths are given in Angstroms. Below 2000 Å the wavelengths are in vacuum; above 2000 Å the wavelengths are in air. Wavelengths given to three decimal places have an uncertainty of less than 0.001 Å and are therefore suitable for the calibration of most spectrographs. In the air region, the elements used most commonly for calibration purposes are Ne, Ar, Kr, Fe, Th, and Hg; in the vacuum region, the most common are C, N, O, Si, Cu.

A large number of the lines for neutral and singly ionized atoms were extracted from the National Bureau of Standards (NBS) Tables of Spectral-Line Intensities.[1] The intensities of these lines represent quantitative estimates of relative line strengths that take account of varying detection sensitivity at different wavelengths. They are on a linear scale. For nearly all of the other lines the intensities represent qualitative estimates of the relative strengths of lines not greatly separated in wavelength. Because different observers frequently use different scales for their intensity estimates, these intensities are useful only as a rough indication of the appearance of a spectrum. In some cases the intensity scale is not intended to be linear. In the tables of first and second spectra the intensities of the lines of the singly ionized atom relative to those of the neutral atom should be used with caution, inasmuch as the concentration of ions in a light source depends greatly on the excitation conditions.

Descriptive symbols used in the tables have the following meanings:

- c—complex
- d—line consists of two unresolved lines
- h—hazy
- l—shaded to longer wavelengths
- s—shaded to shorter wavelengths
- p—perturbed by a close line
- b—band head
- r—easily reversed
- w—wide

ACTINIUM (Ac)
Z = 89

Ac I and II
Ref. 193 — J.G.C.

Intensity		Wavelength	
		Air	
8	h	2100.00	II
20		2712.50	II
10		2726.23	II
10	h	2760.18	II
10	h	2781.56	II
20		2797.59	II
20		2806.76	II
8		2833.47	II
150	h	2847.16	II
8		2895.20	II
30		2896.82	II
30		2923.02	II
200		2994.17	II
500		3043.30	II
200		3069.36	II
100		3078.07	II
100		3086.04	II
100		3087.37	II
200		3112.83	II
100		3120.16	II
500	s	3153.09	II
600	s	3154.41	II
200	s	3164.81	II
300	s	3230.59	II
150	s	3237.70	II
500		3260.91	II
100	s	3318.01	II
200	s	3383.53	II
200		3413.84	II
500		3417.77	II
500	s	3481.16	II
200		3489.53	II
100		3529.24	II
100		3534.63	II
200	s	3554.99	II
1000	s	3565.59	II
100		3694.88	II
300	s	3756.67	II
200		3799.82	II
2000	s	3863.12	II
100		3914.47	II
400	s	4061.60	II
3000	s	4088.44	II
3000	s	4168.40	II
100		4179.98	I
20		4183.12	I
20		4194.40	I
300	s	4209.69	II
300		4359.13	II
20	l	4384.53	I
1000	l	4386.41	II
20		4396.71	I
20		4462.73	I
1000	l	4507.20	II
500		4605.45	II
10		4716.58	I
400	s	4720.16	II
300		4812.22	II
100		4945.18	II
100		4958.23	II
100		4960.87	II
150		5446.38	II
300	l	5732.05	II
400		5758.97	II
1000		5910.85	II
600	l	6164.75	II
200	l	6167.83	II
400		6242.83	II
20		6359.86	I
20	l	6691.27	I
6		7290.40	I
6	l	7866.10	I

Ac III
Ref. 193 — J.G.C.

Intensity		Wavelength	
		Air	
1000	h	2626.44	III
50	h	2682.90	III
2000	h	2952.55	III
2000	h	3392.78	III
3000		3487.59	III
2000	h	4413.09	III
3000	h	4569.87	III
8	h	5193.21	III

Ac IV
Ref. 193 — J.G.C.

Intensity		Wavelength	
		Air	
20	h	2062.00	IV
30	h	2502.12	IV
100	h	2558.08	IV
5	h	2790.83	IV
50	h	2793.90	IV
20	l	3224.7	IV

ALUMINUM (Al)
Z = 13

Al I and II
Ref. 81, 89, 144, 227, 228, 282 — E.F.W.

Intensity	Wavelength	
	Vacuum	
40	1177.43	II
50	1191.812	II
150	1350.18	II
800	1539.830	II
100	1569.385	II
125	1596.059	II
150	1625.627	II
100	1644.235	II
100	1644.809	II
1000	1670.787	II
100	1686.250	II
800	1719.440	II
500	1721.244	II
900	1721.271	II
500	1724.952	II
900	1724.984	II
350	1760.104	II
300	1761.975	II
290	1763.00	I
500	1763.869	II
700	1763.952	II
450	1765.64	I

* Charles H. Corliss is now retired.

Aluminum (Cont.)

300		1765.815	II
450		1766.38	I
400		1767.731	II
450		1769.14	I
600		1828.588	II
400		1832.837	II
250		1834.808	II
300		1855.929	II
700		1858.026	II
120		1859.980	II
1000		1862.311	II
200		1929.978	II
150		1931.048	II
200		1932.377	II
400		1934.503	II
150		1934.713	II
150		1936.907	II
220		1939.261	II
700		1990.531	II

Air

150		2016.052	II
150		2016.234	II
100		2016.368	II
200		2074.008	II
700		2094.264	II
150		2094.744	II
300		2094.791	II
100		2095.104	II
200		2095.141	II
60		2150.70	I
60		2181.00	I
400		2269.10	I
120		2269.22	I
60		2312.49	I
70		2313.53	I
90		2317.48	I
60		2319.06	I
140		2321.56	I
460		2367.05	I
110		2367.61	I
110		2368.11	I
180		2369.30	I
140		2370.22	I
70		2370.73	I
160		2372.07	I
850		2373.12	I
170		2373.35	I
110		2373.57	I
60		2378.40	I
60		2513.30	I
240		2567.98	I
480		2575.10	I
60		2575.40	I
80		2631.55	II
110		2637.70	II
150		2652.48	I
200		2660.39	I
160		2669.17	II
650		2816.19	II
90		2837.96	I
90		2840.10	I
150		3041.28	II
360		3050.07	I
60		3054.68	I
450		3057.14	I
90		3064.29	I
60		3066.14	I
150		3074.64	II
4500	rS	3082.153	I
7200	r	3092.710	I
1800	r	3092.839	I
150		3428.92	II
70		3439.35	I
150		3443.64	I
70		3444.86	I
70		3458.22	I
60		3479.81	I
60		3482.63	I
450		3586.56	II
360		3587.07	II
290		3587.45	II
220		3651.06	II
110		3651.10	II
150		3654.98	II
290		3655.00	II
450		3900.68	II
60		3932.00	I
4500	r	3944.006	I
9000	r	3961.520	I
110		3995.86	II
290		4226.81	II
150		4585.82	II
110		4588.19	II
550		4666.80	II
110		4898.76	II
110		4902.77	II
150		5280.21	II
70		5107.52	I
290		5283.77	II
150		5285.85	II
110		5312.32	II
220		5316.07	II
150		5371.84	II
180		5557.06	I
110		5557.95	I
450		5593.23	II
110		5853.62	II
220		5971.94	II
290		6001.76	II
220		6001.88	II
450		6006.42	II
150		6061.11	II
290		6068.43	II
110		6068.53	II
450		6073.23	II
110		6181.57	II
150		6181.68	II
290		6182.28	II
220		6182.45	II
450	h	6183.42	II
450		6201.52	II
360		6201.70	II
290		6226.18	II
360		6231.78	II
450		6243.36	II
450		6335.74	II
360		6696.02	I
230		6698.67	I
60		7083.97	I
70		7084.64	I
110		7361.57	I
140		7362.30	I
60		7606.16	I
90		7614.82	I
230		7835.31	I
290		7836.13	I
60		7993.05	I
90		8003.19	I
70		8065.97	I
110		8075.35	I
290		8640.70	II
360		8772.87	I
450		8773.90	I
110		8828.91	I
180		8841.28	I
90		8912.90	I
140		8923.56	I
60		9089.91	I
70		9139.95	I
150		9290.65	II
110		9290.75	II
150		10076.29	II
110		10768.36	I
140		10782.04	I
110		10872.98	I
230		10891.73	I
450		11253.19	I
570		11254.88	I
570		13123.41	I
450		13150.76	I
230		16718.96	I
300		16750.56	I
140		16763.36	I
300		21093.04	I
360		21163.75	I

Al III
Ref. 127 — E.F.W.

Intensity		Wavelength	
		Vacuum	
70		486.884	III
30		486.912	III
250		511.138	III
150		511.191	III
500		560.317	III
200		560.433	III
100		670.068	III
200		671.118	III
500		695.829	III
400		696.217	III
200		725.683	III
300		726.915	III
400		855.034	III
500		856.746	III
400		892.024	III
50		893.887	III
450		893.897	III
10		1162.59	III
5		1162.62	III
100		1352.81	III
5		1352.82	III
70		1352.86	III
600		1379.67	III
800		1384.13	III
700		1605.766	III
100		1611.814	III
800		1611.874	III
1000		1854.716	III
600		1862.790	III
300		1935.840	III
15		1935.863	III
200		1935.949	III
		Air	
110		2399.00	III
285		2762.77	III
220		2762.87	III
450		2906.93	III
360		3348.52	III
290		3350.88	III
870		3601.63	III
550		3601.93	III
750		3612.36	III
450		3702.11	III
550		3713.12	III
110	h	3980.14	III
110		4082.45	III
150		4088.61	III
110	h	4142.37	III
650		4149.92	III
650		4150.17	III
110	h	4364.64	III
650		4479.89	III
650		4479.97	III
760		4512.56	III
550		4528.94	III
870		4529.19	III
110		4701.15	III
150		4701.41	III
110	h	4904.10	III
110	h	5151.01	III
110	h	5163.89	III
1200		5696.60	III
1000		5722.73	III
110		6055.21	III
220	h	7635.37	III
150		7660.26	III
220		7681.97	III
360		7881.79	III
150		7882.52	III
290		7905.51	III
290	h	8243.59	III
360	h	8275.11	III
290		9571.52	III
360		9605.99	III

Al IV
Ref. 8, 146 — E.F.W.

Intensity		Wavelength	
		Vacuum	
400		124.03	IV
700		129.73	IV
800		160.07	IV
700		161.69	IV
500		1027.34	IV
800		1042.17	IV
700		1048.52	IV
500		1058.90	IV
500		1061.43	IV
600		1064.89	IV
500		1066.57	IV
600		1069.44	IV
400		1105.74	IV
600		1118.82	IV
500		1125.61	IV
400		1136.82	IV
400		1198.50	IV
400		1220.55	IV
900		1237.19	IV
600		1240.21	IV
700		1240.86	IV
700		1248.79	IV
900		1257.62	IV
800		1264.18	IV
1000		1272.76	IV
400		1337.90	IV
500		1376.62	IV
500		1388.79	IV
600		1431.94	IV
700		1441.82	IV
800		1447.51	IV
600		1457.96	IV
700		1486.89	IV
800		1494.79	IV
400		1519.07	IV
800		1537.54	IV
500		1550.19	IV
1000		1557.25	IV
500		1559.03	IV
700		1564.16	IV
900		1582.04	IV
800		1584.46	IV
400		1589.28	IV
400		1606.65	IV
400		1617.81	IV
600		1627.54	IV
500		1636.82	IV
800		1639.06	IV
1000		1818.56	IV
700		1881.16	IV
		Air	
400		2515.87	IV
500		3208.20	IV
500		3267.21	IV
600		3285.13	IV
400		3344.46	IV
500		3473.54	IV
900		3492.23	IV
800		3508.46	IV
500		3511.28	IV
700		3517.56	IV
400		3527.03	IV
500		3541.08	IV

Al V
Ref. 6 — E.F.W.

Intensity		Wavelength	
		Vacuum	
300		103.80	V
400		103.88	V
250		104.07	V
250		104.18	V
600		107.95	V
300		108.06	V
300		108.11	V
250		118.50	V
900		125.53	V
800		126.07	V
800		130.41	V
1000		130.85	V
900		131.00	V
900		131.44	V
500		132.63	V
1000		278.69	V
900		281.39	V
250		1068.26	V
500		1088.67	V
300		1090.14	V
300		1150.30	V
350		1165.42	V
250		1168.48	V
500		1287.70	V
400		1330.06	V
400		1350.52	V
400		1363.35	V
600		1369.20	V
300		1373.70	V
400		1445.87	V
300		1455.26	V
600		1475.64	V
300		1486.05	V
700		1508.37	V
1000		1526.14	V
500		1539.12	V
300		1577.90	V
350		1589.87	V

AMERICIUM (Am)
Z = 95

Am I and II
Ref. 92 — J.G.C.

Intensity		Wavelength	
		Air	
100	s	2706.35	II
100	s	2728.69	II
200	s	2756.55	II
100	1	2812.10	II
200	s	2812.92	II
1000	1	2815.28	II
100	s	2815.98	II
100	1	2831.24	II
5000	s	2832.26	II
100	1	2833.95	II
100	1	2861.92	II
100	s	2866.20	II
1000	1	2888.51	II
100	1	2893.29	II
200	1	2899.56	II
100	1	2909.86	II
200	1	2911.13	II
1000	s	2920.59	II
200	1	2927.53	II
200	1	2936.99	II
100	1	2939.08	II
500	1	2950.39	II
100	s	2957.05	II
100	1	2958.39	II
100	1	2963.02	II
1000	1	2966.71	II
1000	1	2969.29	II
1000	s	2987.24	II
500	1	2993.51	II
1000	s	3004.25	II

Americium (Cont.)

500	s	3027.99	II
100	1	3028.86	II
500	1	3038.36	II
200	s	3053.69	II
2000	s	3120.49	II
200	s	3161.83	II
100	s	3167.86	II
100	1	3203.26	II
500	s	3282.32	II
200	1	3286.67	II
100	1	3343.87	I
500	s	3362.55	II
200	1	3395.01	I
200	s	3419.66	II
200		3446.19	I
1000	1	3452.10	II
5000	1	3483.31	II
5000		3510.13	I
1000	1	3530.95	I
200	s	3562.68	II
5000		3569.16	I
100	s	3596.07	II
500		3603.41	I
5000		3673.12	I
100	1	3684.57	II
1000	s	3696.42	II
100	1	3707.86	II
5000	1	3777.50	II
5000	1	3926.25	II
1000	1	3952.58	II
100	s	4020.25	I
100	s	4035.81	I
500	s	4036.37	II
5000	s	4089.29	II
100	1	4089.32	II
100	s	4140.96	I
1000	s	4188.12	II
1000		4265.55	I
5000		4289.26	I
200	s	4309.65	II
2000	s	4324.57	II
2000	s	4441.36	II
5000	1	4509.45	II
5000	1	4575.59	II
1000	s	4593.31	II
100	1	4649.12	I
100	1	4653.45	I
5000	1	4662.79	I
2000	1	4681.65	I
2000	1	4699.70	II
1000	1	4706.80	I
2000	1	4872.22	II
200	1	4990.79	I
100	s	5000.21	I
1000	1	5020.96	II
200	1	5215.99	II
1000	s	5402.62	I
1000	s	5424.70	I
1000	1	5584.21	II
1000	s	5598.13	I
10000	1	6054.64	I
1000	1	6405.11	I
500	1	6544.16	I
500	s	6955.58	I

ANTIMONY (Sb)
Z = 51

Sb I and II
Ref. 167, 194 — L.J.R. and J.R.

Intensity		Wavelength	
		Vacuum	
1		691.20	II
1		764.43	II
1		814.85	II
1		849.39	II
2		855.08	II
4		876.84	II
4		921.07	II
6		983.57	II
6		1001.13	II
6		1009.43	II
6		1052.21	II
8		1056.27	II
8		1057.32	II
6		1073.81	II
6		1230.30	II
8		1274.98	II
8		1327.40	II
6		1358.04	II
8		1384.70	II
6		1407.83	II
10		1430.76	I
8		1436.49	II
10	h	1464.19	I
20	r	1486.57	I
40	h	1491.36	I
50	r	1512.57	I
120	r	1532.74	I
80	r	1535.06	I
6		1565.51	II
8		1576.11	II
7		1581.36	II
80	r	1599.96	I
10		1606.98	II
200	w	1612.8	I
100	w	1623.3	I
50	h	1651.20	I
20		1657.04	II
100	w	1662.6	I
50		1698.85	I
80	r	1716.93	I
150	r	1717.45	I
150	r	1723.43	I
100	r	1736.19	I
8		1736.43	II
50		1757.79	I
100	h	1765.76	I
100	r	1780.87	I
100	r	1788.24	I
150		1800.18	I
50	r	1810.50	I
80	r	1814.20	I
100		1829.50	I
60		1858.89	I
50	r	1868.17	I
300	r	1871.15	I
150	r	1882.56	I
70		1891.28	I
70		1899.39	I
100		1927.08	I
200	r	1950.39	I
80	h	1964.3	I
60		1986.05	I
6		1990.60	II
		Air	
50		2024.00	I
60	r	2029.49	I
70	r	2039.77	I
150	r	2049.57	I
50		2063.43	I
1000	r	2068.33	I
100		2079.56	I
50	r	2098.41	I
80	r	2118.48	I
100	r	2127.39	I
50	r	2137.05	I
100	r	2139.69	I
10		2141.80	II
50	r	2141.83	I
100	r	2144.86	I
50		2158.91	I
1500	r	2175.81	I
250	r	2179.19	I
6		2179.25	II
200	r	2201.32	I
300	r	2208.45	I
6		2208.50	II
150	r	2220.73	I
100		2221.98	I
120	r	2224.93	I
6		2225.15	II
300	r	2262.51	I
120		2288.98	I
150	r	2293.44	I
300	r	2306.46	I
2500	r	2311.47	I
150		2315.89	I
400	h	2373.67	I
300	h	2383.64	I
100		2395.22	I
150		2422.13	I
250		2426.35	I
400	r	2445.51	I
400		2478.32	I
150		2480.44	I
8		2480.46	II
100		2510.54	I
2000	r	2528.52	I
15		2528.54	II
10		2567.75	II
150		2574.06	I
1500	r	2598.05	I
500	r	2598.09	I
300	r	2612.31	I
200	r	2652.60	I
12		2656.55	II
300	r	2670.64	I
200	r	2682.76	I
120		2692.25	I
150	r	2718.90	I
400	r	2769.95	I
12		2851.09	II
100		2851.11	I
1000	r	2877.92	I
12		2966.10	II
15		2980.96	II
500	r	3029.83	I
12		3034.01	II
12		3040.67	II
600	r	3232.52	I
20		3241.28	II
700	r	3267.51	I
12		3383.09	II
100		3383.15	I
15		3498.46	II
12		3520.47	II
25		3637.80	II
250		3637.83	I
20		3722.78	II
200	r	3722.79	I
20		3850.22	II
200		4033.55	I
20		4033.56	II
20		4133.63	II
15		4140.54	II
15		4195.17	II
20		4219.07	II
20		4314.32	II
12		4344.83	II
12		4411.42	II
12		4446.48	II
12		4506.92	II
15		4514.50	II
30		4596.90	II
20		4599.09	II
15		4604.77	II
30		4647.32	II
20		4675.74	II
40		4711.26	II
12		4735.44	II
20		4757.81	II
20		4765.36	II
12		4766.91	II
30		4784.03	II
20		4802.01	II
20		4832.82	II
20		4877.24	II
15		4947.40	II
15		5044.56	II
12		5166.32	II
15		5176.55	II
20		5238.94	II
20		5354.24	II
15		5464.08	II
30	h	5490.32	I
40	h	5556.10	I
15		5568.13	II
30	1	5602.19	I
100	1	5632.02	I
30		5639.75	II
60	h	5830.34	I
15		5895.09	II
100		6005.21	II
20		6053.41	II
30		6079.80	II
50		6130.04	II
20		6154.94	II
12		6302.76	II
20		6611.49	I
30		6647.44	II
15		6688.01	II
6		6806.67	II
30	h	7648.28	I
80		7844.44	I
200		7924.65	I
40	h	7969.55	I
60		8411.69	I
150		8572.64	I
100		8619.55	I
30	h	8682.7	I
30		9132.21	I
400		9518.68	I
30		9866.78	I
400		9949.14	I
200		10078.49	I
300		10261.01	I
50		10364.33	I
50	h	10488.3	I
200		10585.60	I
1000		10677.41	I
800		10741.94	I
80		10794.11	I
600		10839.73	I
200		10868.58	I
400		10879.55	I
300		11012.79	I
40	h	11079.95	I
30	h	11084.98	I
50	h	11104.84	I
50	h	11108.52	I
30		11189.61	I
150		11266.23	I
30		11863.37	I
1		11957.7	I
5		12116.06	I
2		12276.6	I
2		12466.75	I

Sb III
Ref. 164 — L.J.R. and J.R.

Intensity	Wavelength	
	Vacuum	
10	691.18	III
10	698.69	III
15	722.86	III
8	724.81	III
15	732.33	III
15	999.62	III
40	1011.94	III
10	1056.58	III
40	1065.90	III
20	1069.93	III
20	1070.43	III
5	1073.76	III
30	1075.82	III
5	1078.10	III
20	1084.06	III
10	1098.34	III
10	1135.43	III
30	1151.49	III
40	1157.74	III
12	1166.96	III
50	1205.20	III
50	1210.64	III
20	1306.69	III
8	1379.58	III
20	1404.18	III
10	1429.57	III
15	1673.89	III
3	1710.23	III
15	1711.84	III
15	1725.33	III
12	1762.30	III
12	1839.32	III
10	1946.13	III
	Air	
3	2054.10	III
2	2091.85	III
5	2127.00	III
5	2507.71	III
15	2590.13	III
1	2614.20	III
12	2617.17	III
1	2617.63	III
20	2669.39	III
5	2785.87	III
20	2790.27	III
20	3336.61	III
50	3504.07	III
15	3519.06	III
15	3533.45	III
40	3559.18	III
40	3566.25	III
30	3738.90	III
40	4265.09	III
50	4352.16	III
30	4591.89	III
30	4692.91	III
1	5247.71	III
1	5690.8	III
1	5717.3	III
3	5845.5	III
5	6246.7	III
3	6287.6	III

Sb IV
Ref. 386 — L.J.R.

Intensity	Wavelength		
Vacuum			
861.5	SB	IV	1
873.5	SB	IV	1
888.3	SB	IV	1
891.2	SB	IV	1
1087.6	SB	IV	1
1099.3	SB	IV	1
1115.1	SB	IV	1
1120.4	SB	IV	1
1145.9	SB	IV	1
1151.5	SB	IV	1
1171.4	SB	IV	1
1192.9	SB	IV	1
1199.1	SB	IV	1
1499.2	SB	IV	1

Sb V
Ref. 406 — L.J.R.

Intensity	Wavelength	
	Vacuum	
3	699.22	V
1	746.06	V
6	831.00	V
1	898.02	V
8	1104.32	V
12	1226.00	V
12	1505.70	V
12	1524.47	V

ARGON (Ar)
Z = 18

Ar I and II
Ref. 190, 203, 204, 219, — E.F.W.

Intensity		Wavelength	
		Vacuum	
30		487.227	II
50		490.650	II
30		490.701	II
30		519.327	II
30		542.912	II
200		543.203	II
70		547.461	II
70		556.817	II
70		573.362	II
30		576.736	II
70		580.263	II
30		583.437	II
70		597.700	II
30		602.858	II
30		612.372	II
500		661.867	II
30		664.562	II
200		666.011	II
1000		670.946	II
3000		671.851	II
70		676.242	II
30		677.952	II
30		679.218	II
200		679.401	II
200		718.090	II
3000		723.361	II
500		725.548	II
70		730.930	II
200		740.269	II
200		744.925	II
70		745.322	II
20		802.859	I
100		806.471	I
60		806.869	I
30		807.218	I
40		807.653	I
50		809.927	I
120		816.232	I
70		816.464	I
80		820.124	I
120		825.346	I
120		826.365	I
150		834.392	I
100		835.002	I
100		842.805	I
180		866.800	I
150		869.754	I
180	r	876.058	I
180	r	879.947	I
150		894.310	I
1000		919.781	II
1000		932.054	II
1000	r	1048.220	I
500	r	1066.660	I
		Air	
5		2420.456	II
10		2516.789	II
10		2534.709	II
15		2562.087	II
25		2891.612	II
200		2942.893	II
100		2979.050	II
50		3033.508	II
50		3093.402	II
8		3200.37	I
20		3243.689	II
25		3293.640	II
20		3307.228	II
7		3319.34	I
25		3350.924	II
7		3373.47	I
25		3376.436	II
25		3388.531	II
7		3393.73	I
7		3461.07	I
70		3476.747	II
20		3478.232	II
50		3491.244	II
100		3491.536	II
70		3509.778	II
70		3514.388	II
70		3545.596	II
70		3545.845	II
7		3554.306	I
100		3559.508	II
100		3561.030	II
70		3576.616	II
25		3581.608	II
50		3582.355	II
70		3588.441	II
7		3606.522	I
25		3622.138	II
20		3639.833	II
35		3718.206	II
70		3729.309	II
50		3737.889	II
150		3765.270	II
50		3766.119	II
20		3770.369	I
20		3770.520	II
25		3780.840	II
25		3803.172	II
50		3809.456	II
7		3834.679	I
70		3850.581	II
35		3868.528	II
35		3925.719	II
50		3928.623	II
25		3932.547	II
70		3946.097	II
7		3947.505	I
35		3948.979	I
20		3979.356	II
35		3994.792	II
50		4013.857	II
50		4033.809	II
20		4035.460	II
150		4042.894	II
50		4044.418	I
100		4052.921	II
200		4072.005	II
70		4072.385	II
25		4076.628	II
35		4079.574	II
25		4082.387	II
150		4103.912	II
300		4131.724	II
35		4156.086	II
400		4158.590	I
50		4164.180	I
35		4179.297	II
50		4181.884	I
100		4190.713	I
50		4191.029	I
200		4198.317	I
400		4200.674	I
25		4218.665	II
25		4222.637	II
25		4226.988	II
100		4228.158	II
100		4237.220	II
25		4251.185	I
200		4259.362	I
100		4266.286	I
70		4266.527	II
150		4272.169	I
550		4277.528	II
20		4282.898	II
100		4300.101	I
25		4300.650	II
70		4309.239	II
200		4331.200	II
50		4332.030	II
100		4333.561	I
50		4335.338	I
25		4345.168	I
800		4348.064	II
50		4352.205	II
25		4362.066	II
50		4367.832	II
200		4370.753	II
70		4371.329	II
50		4375.954	II
150		4379.667	II
50		4385.057	II
70		4400.097	II
200		4400.986	II
400		4426.001	II
150		4430.189	II
50		4430.996	II
50		4433.838	II
20		4439.461	II
35		4448.879	II
100		4474.759	II
200		4481.811	II
100		4510.733	I
20		4522.323	I
20		4530.552	II
400		4545.052	II
20		4564.405	II
400		4579.350	II
400		4589.898	II
15		4596.097	I
550		4609.567	II
7		4628.441	I
35		4637.233	II
400		4657.901	II
15		4702.316	I
20		4721.591	II
550		4726.868	II
50		4732.053	II
300		4735.906	II
800		4764.865	II
550		4806.020	II
150		4847.810	II
50		4865.910	II
800		4879.864	II
70		4889.042	II
20		4904.752	II
35		4933.209	II
200		4965.080	II
50		5009.334	II
70		5017.163	II
70		5062.037	II
20		5090.495	II
100		5141.783	II
70		5145.308	II
5		5151.391	I
15		5162.285	I
25		5165.773	II
20		5187.746	I
20		5216.814	II
7		5221.271	I
5		5421.352	I
10		5451.652	I
25		5495.874	I
5		5506.113	I
25		5558.702	I
10		5572.541	I
35		5606.733	I
20		5650.704	I
10		5739.520	I
5		5834.263	I
10		5860.310	I
15		5882.624	I
25		5888.584	I
50		5912.085	I
15		5928.813	I
5		5942.669	I
7		5987.302	I
5		5998.999	I
5		6025.150	I
70		6032.127	I
35		6043.223	I
10		6052.723	I
20		6059.372	I
7		6098.803	I
10		6105.635	I
100		6114.923	II
10		6145.441	I
7		6170.174	I
150		6172.278	II
10		6173.096	I
10		6212.503	I
5		6215.938	I
25		6243.120	II
7		6296.872	I
15		6307.657	I
7		6369.575	I
20		6384.717	I
70		6416.307	I
25		6483.082	II
15		6538.112	I
15		6604.853	I
25		6638.221	II
20		6639.740	II
50		6643.698	II
5		6660.676	I
5		6664.051	I
25		6666.359	II
100		6677.282	I
35		6684.293	II
150		6752.834	I
5		6756.163	I
15		6766.612	I
20		6861.269	II
150		6871.289	I
5		6879.582	I
10		6888.174	I
50		6937.664	I
7		6951.478	I
7		6960.250	I
10000		6965.431	I
150		7030.251	I
10000		7067.218	I
100		7068.736	I
25		7107.478	I
25		7125.820	I
1000		7147.042	I
15		7158.839	I
70		7206.980	I
15		7265.172	I
7		7270.664	I
2000		7272.936	I
35		7311.716	I
25		7316.005	I
5		7350.814	I
70		7353.293	I
200		7372.118	I
20		7380.426	II
10000		7383.980	I
20		7392.980	I
15		7412.337	I
10		7425.294	I
25		7435.368	I
10		7436.297	I
20000		7503.869	I
15000		7514.652	I
25000		7635.106	I
15000		7723.761	I
10000		7724.207	I
10		7891.075	I
20000		7948.176	I
20000		8006.157	I
25000		8014.786	I
7		8053.308	I
20000		8103.693	I
35000		8115.311	I
10000		8264.522	I
20		8392.27	I
15000		8408.210	I
20000		8424.648	I
15000		8521.442	I
7		8605.776	I
4500		8667.944	I
20		8771.860	II
180		8849.91	I
20		9075.394	I
35000		9122.967	I
550		9194.638	I
15000		9224.499	I
400		9291.531	I
1600		9354.220	I
25000		9657.786	I
4500		9784.503	I
180		10052.06	I
30		10332.72	I
100		10467.177	II
1600		10470.054	I
13		10478.034	I
180		10506.50	I
200		10673.565	I
11		10681.773	I
7		10683.034	II
30		10733.87	I
30		10759.16	I
7		10812.896	II
11		11078.869	I
30		11106.46	I
12		11441.832	I
400		11488.109	I
200		11668.710	I
12		11719.488	I
200		12112.326	I
50		12139.738	I
50		12343.393	I
200		12402.827	I
200		12439.321	I
100		12456.12	I
200		12487.663	I
150		12702.281	I
30		12733.418	I
12		12746.232	I
200		12802.739	I
50		12933.195	I
500		12956.659	I
200		13008.264	I
200		13213.99	I
200		13228.107	I
100		13230.90	I
500		13272.64	I
1000		13313.210	I
1000		13367.111	I
30		13499.41	I
1000		13504.191	I
11		13573.617	I
30		13599.333	I
400		13622.659	I
200		13678.550	I
1000		13718.577	I
10		13825.715	I
10		13907.478	I
200		14093.640	I
100		15046.50	I
25		15172.69	I
10		15329.34	I
30		15989.49	I
30		16519.86	I
500		16940.58	I
12		18427.76	I
50		20616.23	I
30		20986.11	I
20		23133.20	I
20		23966.52	I

Ar III
Ref. 367, 372, 373, 375 — E.F.W.

Intensity	Wavelength	
	Vacuum	
12	769.15	III

Argon (Cont.)

10	871.10	III
9	875.53	III
12	878.73	III
8	879.62	III
9	883.18	III
10	887.40	III
7	1669.67	III
7	1673.42	III
7	1675.48	III
9	1914.40	III
7	1915.56	III

Air

10	2125.16	III
15	2133.87	III
10	2138.50	III
10	2148.73	III
15	2166.19	III
10	2168.26	III
20	2170.23	III
25	2177.22	III
8	2184.06	III
10	2188.22	III
15	2192.06	III
7	2248.73	III
10	2279.10	III
7	2281.22	III
7	2282.21	III
12	2293.03	III
10	2300.85	III
15	2302.17	III
9	2317.00	III
15	2317.47	III
12	2318.04	III
10	2319.13	III
10	2319.37	III
9	2345.17	III
7	2351.67	III
9	2360.26	III
10	2395.63	III
12	2399.15	III
10	2413.20	III
7	2415.61	III
10	2418.82	III
12	2423.52	III
12	2423.93	III
7	2443.69	III
8	2472.95	III
7	2476.10	III
12	2488.86	III
7	2631.90	III
10	2654.63	III
8	2674.02	III
9	2678.38	III
10	2724.84	III
7	2762.23	III
7	2842.88	III
8	2855.29	III
9	2884.12	III
10	3010.02	III
12	3024.05	III
12	3054.82	III
10	3064.77	III
10	3078.15	III
7	3110.41	III
7	3127.90	III
25	3285.85	III
20	3301.88	III
15	3311.25	III
7	3323.59	III
25	3336.13	III
20	3344.72	III
15	3358.49	III
7	3361.28	III
15	3391.85	III
7	3417.49	III
9	3424.25	III
8	3438.04	III
9	3471.32	III
20	3480.55	III
12	3499.67	III
15	3503.58	III
8	3511.12	III
20	3795.37	III
10	3858.32	III
7	3907.84	III
8	3960.53	III
6	4023.60	III
5	4146.70	III

Ar IV
Ref. 367, 368, 374 — E.F.W.

Intensity		Wavelength	
		Vacuum	
4		396.87	IV
4		398.55	IV
6		623.77	IV
10		683.28	IV
7		688.39	IV
12	p	689.01	IV
6		699.41	IV
8		700.28	IV
4		754.20	IV
5		761.47	IV
5		800.57	IV
10		801.09	IV
10		801.41	IV
5		801.91	IV
15		840.03	IV
20		843.77	IV
25		850.60	IV
5		900.36	IV
9		901.17	IV
		Air	
4		2299.72	IV
8		2447.71	IV
12		2513.28	IV
6		2518.40	IV
9		2525.69	IV
12		2562.17	IV
10		2568.07	IV
7		2569.53	IV
12		2599.47	IV
10		2608.06	IV
7		2608.44	IV
12		2615.68	IV
6		2619.98	IV
12		2621.36	IV
12		2624.92	IV
15		2640.34	IV
9		2682.63	IV
14		2757.92	IV
10		2776.26	IV
12		2784.47	IV
14		2788.96	IV
7		2797.11	IV
16		2809.44	IV
10		2830.25	IV
6		2874.40	IV
12		2913.00	IV
11		2926.33	IV
6		3037.98	IV
8		3077.40	IV

Ar V
Ref. 414, 421 — E.F.W.

Intensity		Wavelength	
		Vacuum	
3		336.56	V
3		337.56	V
6		338.00	V
2		338.43	V
2		339.01	V
3		339.89	V
3		350.88	V
2		436.67	V
5		446.00	V
8		446.95	V
4		447.53	V
18		449.06	V
4		449.49	V
3		458.12	V
2		458.98	V
6	p	461.23	V
3		462.42	V
7		463.94	V
3		522.09	V
5		524.19	V
6		527.69	V
2		554.50	V
5		558.48	V
3		635.12	V
3		705.35	V
5		709.20	V
4		715.60	V
3		715.65	V
2		725.11	V
4		822.16	V
5		827.05	V
3		827.35	V
4	p	834.88	V
2		836.13	V

ARSENIC (As)
Z = 33

As I and II
Ref. 168, 197 — R.L.K.

Intensity		Wavelength	
		Vacuum	
165		761.24	II
165		802.83	II
340		1021.96	II
340		1082.35	II
500		1139.40	II
615		1149.31	II
555		1181.51	II
555		1189.87	II
615		1196.38	II
615		1196.56	II
340		1207.44	II
800		1211.17	II
800		1218.10	II
340		1223.15	II
760		1241.31	II
965		1243.08	II
870		1245.67	II
800		1258.58	II
965		1263.77	II
800		1266.34	II
800		1267.59	II
715		1280.99	II
715		1287.54	II
715		1305.70	II
340		1307.74	II
760		1333.15	II
965		1341.55	II
760		1355.93	II
965		1369.77	II
800		1373.65	II
1000		1375.07	II
760		1375.78	II
800		1394.64	II
800		1400.31	II
500		1448.59	II
500		1558.88	II
500		1570.99	II
100	r	1593.60	I
500		1660.55	II
100		1758.60	I
170		1806.15	I
340		1860.34	II
1000	r	1890.42	I
500		1912.94	II
800	r	1937.59	I
585	r	1972.62	I
170	r	1990.35	I
100	r	1991.13	I
100	r	1995.43	I
		Air	
230	r	2003.34	I
100	r	2009.19	I
100		2013.32	I
100		2112.99	I
100		2144.08	I
135		2165.52	I
350	r	2288.12	I
350	r	2349.84	I
100	r	2370.77	I
135	r	2381.18	I
170	r	2456.53	I
340		2602.00	II
170	r	2780.22	I
300		2830.359	II
300		2831.164	II
100	r	2860.44	I
300		2884.406	II
615		2959.572	II
300		3003.819	II
300		3116.516	II
340		3842.60	II
715		4190.082	II
615		4197.40	II
615		4242.982	II
500		4315.657	II
500		4323.867	II
500		4336.64	II
500		4352.145	II
425		4352.864	II
375		4371.17	II
615		4427.106	II
615		4431.562	II
715		4458.469	II
340		4461.075	II
715		4466.348	II
500		4474.46	II
800		4494.230	II
850		4507.659	II
615		4539.74	II
715		4543.483	II
615		4602.427	II
340		4629.787	II
340		4707.586	II
340		4730.67	II
340		4888.557	II
100		5068.98	I
340		5105.58	II
500		5107.55	II
100		5121.34	I
100		5141.63	I
425		5231.38	II
500		5331.23	II
100		5408.13	I
135		5451.32	I
340		5497.727	II
425		5558.09	II
425		5651.32	II
425		6110.07	II
500		6170.27	II
300		6511.74	II
300		7092.27	II
300		7102.72	II
340		7990.53	II
300		8174.51	II
100		8428.91	I
100		8564.71	I
100		8654.14	I
135		8821.73	I
100		8869.66	I
135		9267.28	I
200		9300.61	I
230		9597.95	I
290		9626.70	I
230		9833.76	I
100		9886.05	I
140		9900.55	I
170		9915.71	I
290		9923.05	I
100		10010.63	I
290		10024.04	I
100		10453.09	I
100		10575.02	I
170		10614.07	I

As III
Ref. 163 — R.L.K.

Intensity	Wavelength	
	Vacuum	
185	849.9	III
185	866.3	III
510	871.7	III
325	889.0	III
325	927.5	III
325	953.6	III
325	937.2	III
325	963.8	III
120	1172.2	III
185	1209.3	III
	Air	
80	2926.3	III
185	2982.0	III
325	3922.6	III
185	4037.2	III

As IV
Ref. 244 — R.L.K.

Intensity	Wavelength	
	Air	
150	2253.1	IV
200	2263.2	IV
200	2301.0	IV
250	2417.5	IV
150	2446.1	IV
250	2454.0	IV
200	2461.4	IV
150	3108.8	IV

As V
Ref. 280 — R.L.K.

Intensity	Wavelength	
	Vacuum	
25	600.7	V
40	616.0	V
120	715.5	V
150	734.8	V
60	737.2	V
250	987.7	V
250	1029.5	V
40	1051.6	V
60	1056.6	V

ASTATINE (At)
Z = 85

At I
Ref. 188 — E.F.W.

Astatine (Cont.)

Intensity		Wavelength	
		Air	
8		2162.25	I
10		2244.01	I

BARIUM (Ba)
Z = 56

Ba I and II
Ref. 1, 252, 277, 279 — J.J.W.

Intensity		Wavelength	
		Vacuum	
200		1486.72	II
400		1504.01	II
300		1554.38	II
200		1572.73	II
		1573.92	II
		1630.40	II
100		1674.51	II
400		1694.37	II
		1697.16	II
		1761.75	II
		1771.03	II
		1786.93	II
100		1904.15	II
500		1924.70	II
		1985.60	II
300		1999.54	II
		Air	
		2009.20	II
400		2023.95	II
		2052.68	II
		2054.57	II
500		2214.7	II
800		2245.61	II
1000		2254.73	II
1400		2304.24	II
2000		2335.27	II
190		2347.58	II
60		2528.51	II
8	h	2596.64	I
100		2634.78	II
8		2702.63	I
18		2771.36	II
15		2785.28	I
100	r	3071.58	I
10	h	3108.21	I
8		3132.60	I
8	h	3135.72	I
10		3137.70	I
10		3155.34	I
10		3155.67	I
12		3158.05	I
12	h	3158.54	I
25		3165.60	I
15	h	3173.69	I
30		3183.16	I
15		3183.96	I
10		3193.91	I
25	h	3203.70	I
30		3221.63	I
40		3222.19	I
50		3261.96	I
60	r	3262.34	I
40		3281.50	I
15		3281.77	I
50		3322.80	I
80	h	3356.80	I
60	r	3377.08	I
20		3377.39	I
70	r	3420.32	I
25		3421.01	I
30	h	3421.48	I
40		3463.74	I
200	r	3501.11	I
80	h	3524.97	I
30	h	3531.35	I
80	h	3544.66	I
20	h	3547.68	I
100		3552.45	II
200		3567.73	II
100		3576.28	II
30		3577.62	I
80	h	3579.67	I
200		3596.57	II
40		3630.64	I
40	h	3636.83	I
20	h	3688.47	I
400		3735.75	II
200		3816.69	II
200		3842.80	II
100		3854.76	II
20		3889.33	I
1400	l	3891.78	II
20		3892.65	I
40		3909.91	I
500		3914.73	II
50		3935.72	I
20		3937.87	I
200		3939.67	II
500		3949.51	II
80		3993.40	I
30		3995.66	I
300		4036.26	II
200		4083.77	II
30	h	4084.86	I
1500	h	4130.66	II
20		4132.43	I
200		4166.00	II
500		4216.04	II
800		4267.95	II
100		4283.10	I
300		4287.80	II
200		4297.60	II
800		4309.32	II
20	h	4323.00	I
600		4325.73	II
200		4326.74	II
300		4329.62	II
80		4350.33	I
60		4402.54	I
400		4405.23	II
60	h	4488.98	I
40		4431.89	I
50	h	4493.64	I
40		4505.92	I
200		4509.63	II
60	h	4523.17	I
130		4524.93	II
65000		4554.03	II
40		4573.85	I
80		4579.64	I
30		4599.75	I
20	h	4619.92	I
25	h	4628.33	I
300		4644.10	II
30		4673.62	I
35		4691.62	I
20		4700.43	I
800		4708.94	II
40		4726.44	I
800		4843.46	II
300		4847.14	II
200		4850.84	II
30	h	4877.65	I
400		4899.97	II
15		4902.90	I
20000		4934.09	II
8		4947.35	I
1000		4957.15	II
300		4997.81	II
1000		5013.00	II
20	h	5159.94	I
20		5267.03	I
800		5361.35	II
1000		5391.60	II
200		5421.05	II
100		5424.55	I
200		5428.79	II
300		5480.30	II
200		5519.05	I
1000	r	5535.48	I
20	h	5620.40	I
10		5680.18	I
400		5777.62	I
800		5784.18	II
100		5800.23	I
20		5805.69	I
150		5826.28	I
2800		5853.68	II
15		5907.64	I
100		5971.70	I
800		5981.25	II
100		5997.09	I
300		5999.85	II
100		6019.47	I
200		6063.12	I
300		6110.78	I
400		6135.83	II
20000		6141.72	II
150		6341.68	I
500		6378.91	II
90		6450.85	I
150		6482.91	I
12000		6496.90	II
300		6498.76	I
150		6527.31	I
3000		6595.33	I
150		6654.10	I
1500		6675.27	I
1800		6693.84	I
1000		6769.62	II
600		6865.69	I
300	h	6867.85	I
1000		6874.09	II
6000		7059.94	I
2400	hS	7120.33	I
600		7195.24	I
600	hL	7228.84	I
3000		7280.30	I
1200		7392.41	I
300		7417.53	I
900	hL	7459.78	I
600		7488.08	I
450	hL	7636.90	I
600	hL	7642.91	I
1800		7672.09	I
1200		7780.48	I
180	h	7839.57	I
1500		7905.75	I
600		7911.34	I
900	h	8210.24	I
1800	h	8559.97	I
100		8710.74	II
100		8737.71	II
300	h	8799.76	I
300		8860.98	I
450		8914.99	I
300		9219.69	I
300		9308.08	I
300	h	9324.58	I
1500		9370.06	I
300		9455.92	I
450		9589.37	I
900		9608.88	I
300	h	9645.72	I
1500	hL	9830.37	I
900		10001.08	I
600		10032.10	I
1200	h	10233.23	I
300		10471.26	I
120	hL	10791.25	I
180	hL	11012.69	I
150	h	11114.42	I
240		11303.04	I
120	h	11697.45	I
120		13207.30	I
120		13810.50	I
120		14077.90	I
120		15000.40	I
120		20712.00	I
150		25515.70	I
150		29223.90	I

Ba III
Ref. III — J.J.W.

Intensity		Wavelength	
		Vacuum	
5		403.82	III
2		407.12	III
7		420.12	III
4		423.84	III
9		448.95	III
8		456.96	III
14		555.48	III
14		587.57	III
18		647.27	III
9		653.36	III
15		743.12	III
12		1097.41	III
15		1113.67	III
11		1116.01	III
14		1133.05	III
12		1151.76	III
12		1170.62	III
13		1207.29	III
11		1218.92	III
12		1224.55	III
12		1288.53	III
11		1299.18	III
11		1307.40	III
12		1308.87	III
12		1315.72	III
12		1334.01	III
11		1354.71	III
11		1369.53	III
11		1416.61	III
12		1478.85	III
12		1510.68	III
12		1514.22	III
12		1565.61	III
12		1566.12	III
12		1574.55	III
12		1596.80	III
12		1610.95	III
12		1615.78	III
12		1711.53	III
12		1861.74	III
12		1883.92	III
11		1974.76	III
		Air	
10		2001.30	III
15		2008.40	III
13		2022.45	III
10		2038.84	III
12		2070.43	III
12		2071.68	III
10		2076.00	III
12		2081.35	III
10		2134.87	III
16		2156.37	III
10		2160.76	III
20		2230.33	III
30		2280.68	III
35		2323.51	III
60		2331.10	III
25		2476.73	III
25		2505.07	III
40		2512.28	III
40		2523.83	III
25		2530.92	III
50		2559.54	III
25		2570.48	III
40		2681.89	III
30		2745.78	III
25		2938.95	III
25		2960.05	III
30		2962.48	III
20		3014.22	III
30		3043.42	III
40		3079.14	III
30		3103.92	III
30		3119.22	III
30		3152.70	III
25		3195.17	III
25		3235.04	III
25		3281.65	III
20		3286.79	III
50		3368.18	III
30		3369.68	III
25		3649.18	III
25		3926.85	III
25		3993.06	III
18	p	4053.71	III
15		4697.44	III
10		5049.55	III
10		5097.54	III
12	p	5102.25	III
10		5134.54	III
10		5998.00	III
13		6101.99	III
10		6377.11	III
10		6383.76	III
8		6526.17	III
8		7095.49	III
8		8308.69	III
8		9521.76	III

Ba IV
Ref. 78 — J.J.W.

Intensity		Wavelength	
		Vacuum	
40000		794.89	IV
50000		923.74	IV

Ba V

Ref. 259—J.R.

Intensity		Wavelength	
		Vacuum	
15		612.55	V
5		658.11	V
5		681.09	V
300		719.86	V
150		721.85	V
100		760.45	V
1000		766.87	V
100		783.61	V
100		816.41	V
40		875.69	V
300		877.41	V
15		892.28	V
200		946.26	V

BERKELIUM (Bk)
Z = 97

Bk I and II
Ref.53, 339—J.G.C.

Intensity		Wavelength	
		Air	
10000	s	2748.02	II
10000	s	2827.57	II
10000	l	2872.11	II
10000	s	2878.57	II
10000	l	2884.77	II
10000	s	2889.80	II
10000	s	2893.66	II
10000	l	2910.65	II
10000	s	2926.49	
10000	s	2927.91	
10000	l	2941.71	II
10000	l	2951.76	II
10000	l	2969.13	II
10000	l	2987.76	II
10000	l	3178.47	II
10000		3239.72	I
10000	s	3247.26	II
10000		3252.19	I
10000	s	3263.47	II
10000	l	3288.75	I

Berkelium (Cont.)

10000		3289.35	I
10000	s	3302.35	II
10000		3335.26	I
10000	s	3387.45	II
10000		3408.28	I
10000	l	3412.13	II
10000		3426.95	I
10000	l	3432.62	I
10000	l	3437.47	I
10000		3442.66	I
10000	s	3461.24	II
10000	l	3453.90	I
10000	s	3464.13	II
10000	s	3472.02	II
10000	s	3477.62	II
10000		3528.72	I
10000	l	3531.40	I
10000	l	3535.73	I
10000	s	3542.19	II
10000		3553.60	I
10000	s	3555.88	I
10000		3556.52	I
10000		3565.41	I
10000	l	3567.25	II
10000		3590.32	I
10000		3595.88	I
10000		3601.12	I
10000	s	3603.20	II
10000		3604.78	I
10000	l	3608.49	
10000		3609.61	
10000		3611.03	
10000	l	3611.93	
10000		3613.91	
10000		3616.62	
10000		3619.37	
10000	s	3621.81	
10000		3627.61	I
10000		3633.28	
10000	l	3637.05	I
10000		3640.26	I
10000	s	3640.93	II
10000	l	3675.59	I
10000	s	3681.22	II
10000	l	3684.43	I
10000	s	3685.21	I
10000	l	3686.74	I
10000		3692.73	I
10000		3695.37	I
10000		3703.28	I
10000	s	3704.02	I
10000		3705.26	I
10000	s	3711.14	II
10000		3712.93	I
10000		3725.39	I
10000		3739.92	I
10000		3743.05	I
10000		3745.40	I
10000		3750.08	I
10000		3751.91	I
10000		3757.35	I
10000		3757.85	I
10000	s	3771.06	II
10000		3780.72	I
10000		3781.17	I
10000		3785.38	I
10000		3788.21	I
10000		3791.42	I
10000		3796.21	I
10000		3797.12	
10000		3798.63	I
10000		3801.08	II
10000	s	3802.35	I
10000		3802.47	I
10000		3815.29	I
10000	s	3823.10	II
10000	s	3824.08	II
10000		3825.19	I
10000	s	3825.84	II
10000		3827.41	I
10000		3830.55	I
10000	l	3831.57	II
10000	s	3833.48	
10000	s	3835.97	II
10000		3842.19	I
10000	l	3846.62	I
10000		3847.63	I
10000		3855.03	I
10000	l	3859.89	II
10000	l	3877.94	II
10000		3880.11	I
10000	l	3882.60	I
10000	l	3894.55	II
10000	s	3906.09	II
10000	l	3912.16	II
10000	l	3916.37	II
10000		3921.42	I
10000	l	3928.05	II
10000	l	4147.13	II
10000	s	4189.69	II
10000	s	4197.44	II
10000	l	4329.58	I
10000		4351.50	I
10000		4363.64	I
10000		4423.01	I
10000		4466.46	I
10000		4685.70	
10000	l	4765.40	
10000	s	5056.73	I
10000	l	5118.24	I
10000	s	5135.53	II
10000		5170.61	I
10000		5197.55	I
10000		5212.53	I
10000		5271.95	I
10000		5392.03	I
10000		5394.24	I
10000	l	5404.62	I
10000		5449.63	I
10000	l	5467.47	I
10000	s	5484.58	I
10000		5512.22	II
10000	l	5537.93	I
10000	l	5556.80	I
10000	s	5557.09	I
10000		5581.21	I
10000	l	5656.54	I
10000		5659.03	I
10000		5702.24	I
10000		5910.71	I
10000	l	7040.85	I
10000		7107.85	I
10000	l	7176.22	I
10000		7249.26	
10000		7252.50	I
10000	l	7257.21	I
10000		7306.94	I
10000		7394.26	
10000	l	7511.26	I
10000		7551.12	I
10000		7579.77	I
10000	l	7729.93	I
10000	s	7903.90	I
10000	s	9319.30	I
10000	s	9429.13	I
10000	l	9801.18	I
10000	l	9862.39	II
10000	l	9879.29	I
10000	l	9892.38	I
10000		10126.20	I
10000	l	10186.58	I
10000		10292.44	I
10000	l	10527.71	I
10000	l	10570.53	I
10000	l	11293.14	I
10000	l	11500.30	I
10000	s	11575.34	I
10000	s	11793.09	I
10000	s	12159.05	I
10000		13061.13	I
10000		13498.36	I
10000	s	14196.93	I
10000		15136.10	I
10000		18352.31	
10000	l	19273.87	I
10000	s	19653.22	I
10000	s	23902.85	I
10000	s	24192.62	I

BERYLLIUM (Be)
Z = 4

Be I and II
Ref. 15, 44, 115, 134, 135, 198, 335 — J.O.S.

Intensity		Wavelength	
		Vacuum	
		82.58	II
		83.66	II
		89.16	I
		89.80	II
		90.04	II
		90.21	I
		90.67	I
		91.06	II
		91.36	II
		91.74	II
		92.19	I
		92.61	II
		93.14	II
		93.42	II
		93.93	II
		94.78	II
		95.76	II
		96.29	I
		97.24	I
		97.44	I
		97.86	I
		97.97	I
		98.12	I
		98.37	I
		98.66	I
		98.94	I
		99.19	I
		100.86	I
		101.20	I
		102.13	I
		102.49	II
		104.40	II
		104.67	I
		105.80	I
		107.26	I
		107.38	I
		714.0	II
		725.71	II
		743.58	II
8		775.37	II
20		842.06	II
		865.3	II
2		925.25	II
10		943.56	II
10		973.27	II
		981.4	II
		1020.1	II
8		1026.93	II
5		1036.32	II
15		1048.23	II
20		1143.03	II
		1155.9	II
60		1197.19	II
		1426.12	I
		1491.76	I
20		1512.30	II
60		1512.43	II
100		1661.49	I
15		1776.12	II
20		1776.34	II
		1907.	I
		1909.0	II
		1912.	I
		1919.	I
5		1929.67	I
10		1943.68	I
		1956.	I
50		1964.59	I
5		1985.13	I
		1997.95	I
		1997.98	I
60		1998.01	I
		Air	
		2033.25	I
		2033.28	I
		2033.38	I
50		2055.90	I
100		2056.01	I
10		2125.57	I
20		2125.68	I
25		2145.	I
55		2174.99	I
55		2175.10	I
		2273.5	II
		2324.6	II
		2337.0	I
950		2348.61	I
20		2350.66	I
60		2350.71	I
200		2350.83	I
2		2413.34	II
16		2413.46	II
20		2453.84	II
		2480.6	I
35		2494.54	I
35		2494.58	I
100		2494.73	I
16		2507.43	II
5		2617.99	II
20		2618.13	II
100		2650.45	I
60		2650.55	I
200		2650.62	I
60		2650.69	I
100		2650.76	I
5		2697.46	II
20		2697.58	II
20		2728.88	II
30		2738.05	II
		2764.2	II
20		2898.13	I
10		2898.19	I
20		2898.25	I
30		2986.06	I
10		2986.42	I
60		3019.33	I
30		3019.49	I
30		3019.53	I
20		3019.60	I
10		3046.52	II
30		3046.69	II
		3090.3	I
10		3110.81	I
10		3110.92	I
20		3110.99	I
		3120.	I
480		3130.42	II
320		3131.07	II
		3136.	I
		3150.	I
		3160.6	I
		3163.	I
		3168.	I
		3180.7	II
		3187.	I
20		3193.81	I
20		3197.10	II
30		3197.15	II
20		3208.60	I
		3220.	I
60		3229.63	I
2		3233.52	II
10		3241.62	II
30		3241.83	II
15		3269.02	I
100		3274.58	II
30		3274.67	II
30		3282.91	I
30		3321.01	I
30		3321.09	I
220		3321.34	I
20		3345.43	I
60		3367.63	I
		3405.6	II
5		3451.37	I
300		3455.18	I
20		3476.56	I
300		3515.54	I
10		3555.	I
100		3736.30	I
700		3813.45	I
40		3865.13	I
80		3865.42	I
1		3865.51	I
6		3865.72	I
100		3866.03	I
100		4253.05	I
60		4253.76	I
300		4360.66	II
500		4360.99	II
400		4407.94	I
		4526.6	I
		4548.	I
12		4572.66	I
700		4673.33	II
1000		4673.42	II
6		4709.37	I
200		4828.16	II
40		4849.16	I
2	h	4858.22	II
80		5087.75	I
8		5218.12	II
20		5218.33	II
3		5255.86	II
64		5270.28	II
500		5270.81	II
20		5403.04	II
20		5410.21	II
		5558.	I
10		6229.11	I
16		6279.43	II
30		6279.73	II
30		6473.54	I
60		6547.89	II
60		6558.36	II
30		6564.52	I
2	h	6636.44	II
1		6756.72	II
2		6757.13	II
30		6786.56	I
1	h	6884.22	I
6	h	6884.44	I
100		6982.75	I
6	h	7154.40	I
40	h	7154.65	I
100		7209.13	I
3		7401.20	II
2		7401.43	II
10		7551.90	I
10	h	7618.68	I
20	h	7618.88	I
60		8090.06	I
5	h	8158.99	I
10	h	8159.24	I
4		8254.07	I
10	h	8287.07	I
30		8547.36	I
60		8547.67	I
300		8801.37	I
6		8882.18	I
40		9190.45	I
20	h	9243.92	I
1	h	9343.89	II
40		9392.74	I
2		9476.43	II
16		9477.03	II
20		9847.32	I
10	h	9895.63	I
20	h	9895.96	I
80		9939.78	I
16		10095.52	II

Beryllium (Cont.)

20		10095.73	II
60		10119.92	II
80		10331.03	I
30		11066.46	I
		11173.	II
1		11173.73	II
120		11496.39	I
2	h	11625.16	II
		11659.	II
2		11660.25	II
100		12095.36	II
30		12098.18	II
100		14643.92	I
60		14644.75	I
200		16157.72	I
80		17855.38	I
120		17856.63	I
100		18143.54	I
160		31775.05	I
200		31778.70	I

Be III
Ref. 73, 102, 175 — J.O.S.

Vacuum

Intensity		Wavelength	
1	h	76.10	III
2		76.48	III
3		78.53	III
4		78.66	III
1	h	78.92	III
5		81.89	III
10		82.38	III
20		83.20	III
30		84.76	III
50		88.31	III
100		100.25	III
3		509.99	III
2		549.31	III
6		582.08	III
4		661.32	III
8		675.59	III
4		725.59	III
7		746.23	III
2		767.75	III
1		1114.69	III
2		1213.12	III
1		1214.32	III
2		1362.25	III
1		1401.52	III
10		1421.26	III
5		1422.86	III
1		1435.17	III
2		1440.77	III
2	h	1754.69	III
3		1917.03	III
60	h	1954.97	III

Air

75	h	2076.94	III
60	h	2080.38	III
25		2118.56	III
15	h	2122.27	III
15	h	2127.20	III
5		2137.25	III
5		2191.57	III
100		3720.36	III
		3720.92	III
		3722.98	III
90	h	4249.14	III
2		4485.52	III
100	h	4487.30	III
1		4495.09	III
140	h	4497.8	III
140	h	6142.01	III

Be IV
Ref. 171 — J.O.S.

Vacuum

Intensity		Wavelength	
		58.13	IV
		58.57	IV
		59.32	IV
		60.74	IV
		64.06	IV
		75.93	IV

BISMUTH (Bi)
Z = 83

Bi I and II
Ref. 1, 357-359 — C.H.C.

Vacuum

Intensity		Wavelength	
15		1058.88	II
20		1085.47	II
10		1099.20	II
8		1163.19	II
8		1167.06	II
10		1225.43	II
15		1232.78	II
10		1241.05	II
10		1265.35	II
15		1283.73	II
10		1306.18	II
20		1325.46	II
20		1329.47	II
20		1350.07	II
25		1372.61	II
15		1376.02	II
20		1393.92	II
45		1436.83	II
25		1447.94	II
50		1455.11	II
25		1462.14	II
35		1486.93	II
20		1502.50	II
40		1520.57	II
40		1533.17	II
30		1536.77	II
35		1538.06	II
20		1563.67	II
40		1573.70	II
60		1591.79	II
25		1601.58	II
40		1609.70	II
40		1611.38	II
20		1652.81	II
20		1749.29	II
80		1777.11	II
60		1787.47	II
70		1791.93	II
70		1823.80	II
100		1902.41	II
9000		1954.53	I
7000		1960.13	I
25		1989.35	II

Air

7000		2021.21	I
9000		2061.70	I
45	h	2068.9	II
4600		2110.26	I
2500		2133.63	I
15		2143.40	II
15		2143.46	II
60		2186.9	II
40	h	2214.0	II
360		2228.25	I
1700		2230.61	I
340		2276.58	I
16		2368.12	II
12		2368.25	II
190		2400.88	I
10		2501.0	II
25		2515.69	I
70		2524.49	I
20	h	2544.5	II
700		2627.91	I
12		2693.0	II
280	c	2696.76	I
20		2713.3	II
140	d	2730.50	I
360		2780.52	I
15		2803.42	II
11		2803.70	II
12		2805.3	II
140	c	2809.62	I
4000		2897.98	I
15		2936.7	II
3200		2938.30	I
20		2950.4	II
12		2963.4	II
2800		2989.03	I
700		2993.34	I
2400		3024.64	I
60		3034.87	I
9000	c	3067.72	I
140		3076.66	I
550	c	3397.21	I
10		3430.83	II
12		3431.23	II
500	c	3510.85	I
380	c	3596.11	I
12		3654.2	II
70	h	3792.5	II
12		3811.1	II
20		3815.8	II
10		3845.8	II
30		3863.9	II
40	h	4079.1	II
10		4097.2	II
140		4121.53	I
140		4121.86	I
75	h	4259.4	II
25		4272.0	II
70	h	4301.7	II
12	h	4339.8	II
25	h	4340.5	II
12	h	4379.4	II
25	h	4476.8	II
60	h	4705.3	II
600	c	4722.52	I
30		4730.3	II
20		4749.7	II
12		4908.2	II
10		4916.6	II
12		4969.7	II
20		4993.6	II
10		5091.6	II
50	h	5124.3	II
60	h	5144.3	II
20		5201.5	II
75	h	5209.2	II
40	h	5270.3	II
10		5397.8	II
10	c	5552.35	I
3		5599.41	I
20		5655.2	II
40	h	5719.2	II
6		5742.55	I
12		5818.3	II
20		5860.2	II
20		5973.0	II
15		6059.1	II
15		6128.0	II
6		6134.82	I
3		6475.73	I
3		6476.24	I
15		6497.7	II
10		6577.2	II
40	h	6600.2	II
50	h	6808.6	II
4	h	6991.12	I
12		7033.	II
2		7036.15	I
10	h	7381.	II
2		7502.33	I
10	h	7637.	II
10		7750.	II
3		7838.70	I
2		7840.33	I
20		7965.	II
12	h	8050.	II
15		8328.	II
15		8388.	II
30		8532.	II
2		8544.54	I
1		8579.74	I
25		8653.	II
2		8754.88	I
3		8761.54	I
25		8863.	II
2		8907.81	I
2000	d	9657.04	I
40		9827.78	I
20		10104.5	I
15		10138.8	I
20		10300.6	I
20		10536.19	I
50		11072.44	I
15		11551.6	I
1500	d	11710.37	I
40		11999.49	I
200		12165.08	I
10		12374.64	I
200		12690.04	I
100		12817.8	I
200		14330.5	I
50		16001.5	I
60		22551.6	I

Bi III
Ref. 359 — C.H.C.

Vacuum

Intensity		Wavelength	
1		590.73	III
5		670.76	III
4		775.16	III
6		803.65	III
7		920.93	III
6		925.48	III
25		1039.99	III
50	h	1045.76	III
30		1051.81	III
20		1139.01	III
15		1145.91	III
50		1224.64	III
40		1326.84	III
60		1346.12	III
35		1423.33	III
35		1423.52	III
60	h	1461.00	III
60	h	1606.40	III
20	h	1691.5	III
20		1834.32	III
10		1863.9	III
10		1912.12	III
10		1988.26	III

Air

20		2020.75	III
20		2021.15	III
10		2073.22	III
14		2073.37	III
15		2103.42	III
30		2213.55	III
75	h	2414.6	III
10		2437.6	III
30	h	2847.4	III
80	h	2855.6	III
35		3115.0	III
40	h	3451.0	III
40		3473.8	III
35		3485.5	III
15		3540.8	III
45		3613.4	III
50		3695.32	III
50		3695.68	III
12		4224.6	III
25		4327.8	III
30		4560.84	III
30		4561.54	III
40	h	4797.4	III
45	h	5079.3	III
12		6623.4	III
10	h	7381.	III
12		7551.	III
25		7598.	III
10	h	7637.	III
40		8008.	III
50		8070.	III
20		8100.	III
15		8671.	III
20		8934.	III

Bi IV
Ref. 360 — C.H.C.

Vacuum

Intensity		Wavelength	
6		420.7	IV
6		431.2	IV
6		790.5	IV
6		790.6	IV
8		792.5	IV
10		820.3	IV
9		822.9	IV
12		824.9	IV
15		872.6	IV
8		876.8	IV
9		916.7	IV
12		923.9	IV
15		943.3	IV
9		967.6	IV
8		968.8	IV
8		989.8	IV
24		1103.4	IV
7		1128.8	IV
6		1138.6	IV
6		1139.8	IV
7		1149.7	IV
60		1317.0	IV
30		1910.0	IV

Air

30		2093.	IV
100		2311.	IV
100		2326.	IV
100		2376.	IV
100		2629.	IV
100		2677.	IV
100		2767.	IV
100		2772.	IV
100		2786.	IV
100		2842.	IV
100		2924.	IV
100		2933.	IV
100		2936.	IV
100		3012.	IV
100		3042.	IV
100		3239.	IV
100		3643.	IV
100		3682.	IV
100		3734.	IV
100		3868.	IV
30		4342.	IV

Bismuth (Cont.)

30		5347.	IV

Bi V
Ref. 361 — C.H.C.

Intensity		Wavelength	
		Vacuum	
1		355.77	V
1		369.52	V
1		429.78	V
1		435.63	V
2		488.39	V
1		492.72	V
3		563.6[illegible]	V
2		678.87	V
6		686.88	V
1		706.54	V
5		730.71	V
10		738.17	V
6		849.86	V
5		855.68	V
15	d	864.45	V
6		880.17	V
6		929.81	V
15	d	1139.46	V

BORON (B)
Z = 5

B I and II
Ref. 66, 104, 171, 222 — R.L.K.

Intensity		Wavelength	
		Vacuum	
70		693.95	II
40		731.36	II
40		731.44	II
110		882.54	II
110		882.68	II
40		984.67	II
110		1081.88	II
110		1082.07	II
110		1230.16	II
220		1362.46	II
70		1600.46	I
120		1600.73	I
160		1623.58	II
110		1623.77	II
220		1624.02	II
70		1624.16	II
160		1624.34	II
100		1663.04	I
150		1666.87	I
200		1667.29	I
150		1817.86	I
200		1818.37	I
300		1825.91	I
300		1826.41	I
110		1842.81	II
		Air	
250		2066.38	I
250		2066.65	I
100		2066.93	I
300		2067.19	I
500		2088.91	I
500		2089.57	I
70		2220.30	II
40		2323.03	II
40		2328.67	II
40		2393.20	II
220		2395.05	II
40		2459.69	II
40		2459.90	II
1000		2496.77	I
1000		2497.73	I
160		2918.08	II
110		3032.26	II
70		3179.33	II
110		3323.18	II
110		3323.60	II
450		3451.29	II
285		4121.93	II
110		4194.79	II
110		4472.10	II
110		4472.85	II
70		4784.21	II
110		4940.38	II
110		6080.44	II
70		6285.47	II
70		7030.20	II
40		7031.90	II
70		8668.57	I
20		8667.22	I
800		11660.04	I
570		11662.47	I
125		15629.08	I
200		16240.38	I
250		16244.67	I
235		18994.33	I

B III
Ref. 69, 221 — R.L.K.

Intensity		Wavelength	
		Vacuum	
150		518.24	III
75		518.27	III
40		411.80	III
20		510.77	III
10		510.[illegible]	III
110		677.00	III
160		677.14	III
40		758.48	III
70		758.67	III
20		1953.83	III
		Air	
550		2065.78	III
450		2067.23	III
160		2077.09	III
40		2234.09	III
70		2234.59	III
40		4242.98	III
70		4243.61	III
220		4487.05	III
360		4497.73	III
110		7835.25	III
70		7841.41	III

B IV
Ref. 74 — R.L.K.

Intensity		Wavelength	
		Vacuum	
10		52.68	IV
30		60.31	IV
160		344.0	IV
450		385.0	IV
285		418.7	IV
70		1112.2	IV
450		1168.9	IV
70		1170.9	IV
		Air	
70		2524.7	IV
160		2530.3	IV
450		2821.68	IV
70		2824.57	IV
285		2825.85	IV

B V
Ref. 94 — R.L.K.

Intensity		Wavelength	
		Vacuum	
30		41.00	V
		48.59	V
		194.37	V
		262.37	V
		512.53	V
		749.74	V

BROMINE (Br)
Z = 35

Br I and II
Ref. 122, 124, 240, 248, 316
G.V.S.

Intensity		Wavelength	
		Vacuum	
300		711.68	II
250		815.48	II
350		856.19	II
1000		889.23	II
500		896.64	II
500		905.99	II
300		922.56	II
1000		948.97	II
500		984.93	II
500		1012.10	II
1000		1015.54	II
500		1037.02	II
1000		1049.00	II
450		1064.76	II
500		1071.87	II
250		1101.50	I
300		1134.59	I
250		1136.29	I
250		1177.23	I
400		1178.90	I
1000		1189.28	I
250		1189.38	I
1000		1189.50	I
500		1198.37	I
800		1209.76	I
1000		1210.73	I
750		1216.01	I
1000		1221.13	I
900		1221.87	I
1000		1223.24	I
1200		1224.41	I
1200		1226.90	I
750		1228.05	I
7500		1232.43	I
1200		1243.90	I
800		1249.59	I
1500		1251.66	I
1000		1255.80	I
1500		1259.20	I
1200		1261.66	I
1200		1266.20	I
1000		1279.48	I
1000		1286.26	I
3000		1309.91	I
3000		1316.74	I
1000		1317.37	I
2000		1317.70	I
12000		1384.60	I
3000		1449.90	I
50000		1488.45	I
30000		1531.74	I
25000		1540.65	I
30000		1574.84	I
20000		1576.39	I
25000		1582.31	I
75000		1633.40	I
		Air	
350		2285.17	II
350		2287.60	II
500	h	2317.30	II
400		2336.93	II
350		2386.45	II
500		2386.70	II
300		2388.69	II
450		2388.96	II
500		2389.69	II
350		2392.21	II
400		2392.42	II
300		2488.50	II
300	h	2495.22	II
450		2521.70	II
400		2541.48	II
400		2556.92	II
350		2690.17	II
400		2713.77	II
350	h	2746.52	II
300	h	2807.55	II
400	h	2893.40	II
400	h	2917.18	II
400	h	2967.21	II
500	h	2972.26	II
300		2981.86	II
300	h	2985.87	II
300	h	2986.53	II
300	h	3016.48	II
350		3423.82	II
300	h	3606.80	II
350		3714.30	II
1200		3815.65	I
350		3834.69	II
300		3871.21	II
400		3891.63	II
300		3901.24	II
300		3914.20	II
500		3914.38	II
350		3919.51	II
400		3924.09	II
300		3929.55	II
350		3939.69	II
350		3950.61	II
500		3980.38	II
1500		3992.36	I
300		4024.04	II
300		4135.66	II
300		4140.20	II
400		4179.63	II
300		4193.45	II
1000		4223.89	II
300		4236.89	II
300		4291.39	II
2000		4365.14	I
1000		4365.60	II
1500		4425.14	I
10000		4441.74	I
10000		4472.61	I
20000		4477.72	I
1000		4490.42	I
3000		4513.44	I
15000		4525.59	I
300		4529.60	II
500		4542.92	II
3000		4575.74	I
300		4601.36	II
2500		4614.58	I
350		4622.70	II
300		4642.02	II
300		4651.98	II
500		4678.70	II
400		4693.17	II
500		4704.85	II
250		4710.76	II
400		4720.36	II
300		4728.20	II
300		4735.41	II
400		4742.64	II
2500		4752.28	I
350		4766.00	II
400		4779.40	II
4000		4780.31	I
1600		4785.19	I
500		4785.50	II
300		4802.33	II
500		4816.70	II
300		4818.46	II
350		4844.81	II
350		4848.75	II
400		4921.12	II
400		4928.79	II
450		4930.66	II
300		4945.51	II
4000		4979.76	I
300		5038.74	II
300		5054.64	II
400		5164.38	II
300		5180.01	II
500		5182.35	II
300		5193.90	II
500		5238.23	II
300		5272.68	II
350		5304.10	II
400		5330.57	II
500		5332.05	II
1200		5395.48	I
400		5422.78	II
350		5424.99	II
300		5435.07	II
1200		5466.22	I
350		5478.47	II
300		5488.79	II
300		5495.06	II
500		5506.69	II
350		5589.94	II
300		5718.71	II
300		5830.78	II
1800		5852.08	I
1600		5940.48	I
2400		6122.14	I
40000		6148.60	I
300		6161.74	II
2000		6177.39	I
1500		6335.48	I
60000		6350.73	I
400		6352.94	II
2500		6410.32	I
1800		6483.56	I
1000		6514.62	I
20000		6544.57	I
1500		6548.09	I
50000	c	6559.80	I
1000		6571.31	I
1800		6579.14	I
20000		6582.17	I
1500		6620.47	I
50000	c	6631.62	I
20000		6682.28	I
10000		6692.13	I
8000		6728.28	I
2000		6760.06	I
2000		6779.48	I
2200		6786.74	I
6500		6790.04	I
1600	c	6791.48	I
1800		6861.15	I
10000		7005.19	I
2000		7260.45	I
10000		7348.51	I
40000		7512.96	I
1600		7591.61	I
1800		7595.07	I
2000		7616.41	I
30000		7803.02	I
1200		7827.23	I
2500	s	7881.45	I
2500		7881.57	I
2500		7925.81	I
30000	c	7938.68	I
3000		7947.94	I
3000		7950.18	I

Bromine (Cont.)

Intensity		Wavelength	
8000		7978.44	I
10000		7978.57	I
30000		7989.94	I
2000		8026.35	I
2500		8026.54	I
30000		8131.52	I
1000	c	8152.65	I
10000		8153.75	I
25000		8154.00	I
5000		8246.86	I
15000		8264.96	I
75000	c	8272.44	I
20000		8334.70	I
10000		8343.70	I
1200		8384.04	I
40000		8446.55	I
4000		8477.45	I
1500		8513.38	I
1000		8557.73	I
1000		8566.28	I
20000		8638.66	I
4000		8698.53	I
10000	c	8793.47	I
15000		8819.96	I
25000		8825.22	I
4000		8888.98	I
30000		8897.62	I
6000		8932.40	I
1800		8949.39	I
9000		8964.00	I
350		9024.42	II
30000		9166.06	I
15000		9173.63	I
20000		9178.16	I
40000		9265.42	I
15000		9320.86	I
300		9434.04	II
6000		9793.48	I
10000		9896.40	I
3000		10140.08	I
6000		10237.74	I
1000		10299.62	I
1500		10377.65	I
30000		10457.96	I
1000		10742.14	I
3000		10755.92	I
1700		13217.17	I
1800		14354.57	I
1250		14888.70	I
1800		16731.19	I
1200		18568.31	I
3500		19733.62	I
1000		20281.73	I
1000		20624.67	I
1200		21787.24	I
4000		22865.65	I
1000		23513.15	I
500		28346.50	I
500		30380.85	I
600		31630.13	I
120		32693.90	I
150		34181.87	I
150		38345.75	I
120		39964.36	I

Br III
Ref. 246, 250 — G.V.S.

Intensity		Wavelength	
		Vacuum	
450		611.1	III
300		620.4	III
500		665.54	III
500		677.19	III
300		677.8	III
450		687.68	III
400		690.2	III
350		696.99	III
300		727.0	III
300		736.4	III
250		769.63	III
250		817.79	III
350		949.0	III
400		960.4	III
450		984.9	III
250		1313.5	III
250		1402.9	III
		Air	
400	h	2293.44	III
300		2313.29	III
300		2462.39	III
300		2482.60	III
350		2499.25	III
350		2529.49	III
350	h	2551.09	III
350	h	2570.83	III
300	h	2573.17	III
400	h	2584.99	III
500		2589.14	III
300	h	2594.48	III
400	h	2595.98	III
450	h	2606.20	III
350	h	2608.15	III
500	h	2613.13	III
350	h	2616.26	III
500	h	2626.52	III
350	h	2629.23	III
350	h	2639.60	III
350	h	2671.53	III
350	h	2735.83	III
300		2770.50	III
300	h	2785.28	III
300		2804.16	III
400		2926.96	III
300		2936.22	III
350		2969.00	III
400		2994.04	III
500		3020.76	III
300		3033.63	III
350		3036.45	III
500		3074.42	III
350		3091.94	III
350		3117.29	III
300		3147.81	III
400		3174.08	III
300		3321.08	III
450		3333.07	III
500		3349.64	III
300		3385.25	III
450		3447.36	III
400		3487.58	III
300		3506.47	III
450		3517.36	III
500		3540.16	III
300		3551.08	III
500		3562.43	III
450		3600.71	III
250		3693.53	III
450		3820.26	III
200		3903.95	III
350		4506.55	III
200		4519.74	III
150		5175.87	III
100		5446.80	III
100		7192.8	III
100		7673.1	III

Br IV
Ref. 139, 142, 243, 249
G.V.S.

Intensity		Wavelength	
		Vacuum	
700		379.73	IV
700		400.37	IV
1000		545.43	IV
1000		559.76	IV
1000		569.19	IV
1000		576.59	IV
1000		585.10	IV
1000		586.71	IV
1000		597.51	IV
1000		600.09	IV
1000		601.27	IV
1000		607.03	IV
1000		617.85	IV
1000		619.87	IV
1000		630.14	IV
1000		642.23	IV
1000		661.53	IV
1000		683.51	IV
1000		697.72	IV
1000		715.39	IV
1000		731.00	IV
1000		800.12	IV
1000		813.66	IV
900		1274.82	IV
1000		1703.51	IV
		Air	
1000		2133.79	IV
1000		2145.02	IV
1000		2257.21	IV
1000		2272.73	IV
1000		2307.40	IV
1000		2408.16	IV
1000		2411.58	IV
700		2491.14	IV
1000		2581.19	IV
600		2661.40	IV
700		2820.87	IV
1000		2842.88	IV
1100	h	2907.71	IV
500		3041.18	IV
500		3380.56	IV

Br V
Ref. 42 — G.V.S.

Intensity	Wavelength	
	Vacuum	
600	468.37	V
800	482.11	V
900	531.97	V
1000	547.90	V
700	549.77	V
800	621.03	V
800	632.22	V
700	645.44	V
400	652.64	V
800	657.54	V
800	679.62	V
700	812.95	V
1000	850.81	V
150	855.27	V
600	1041.60	V
1000	1069.15	V
500	1080.54	V
900	1112.13	V
1000	1143.56	V
150	1429.75	V
400	1442.60	V
150	1470.35	V

CADMIUM (Cd)
Z = 48

Cd I and II
Ref. 44, 285, 296 — R.D.C.

Intensity		Wavelength	
		Vacuum	
100		1256.00	II
150		1296.43	II
100		1326.50	II
150		1370.91	II
200		1514.26	II
200		1571.58	II
100		1668.60	II
50		1702.47	II
50		1724.41	II
100		1785.84	II
100		1827.70	II
300		1922.23	II
100		1943.54	II
40		1965.54	II
30		1986.89	II
200		1995.43	II
		Air	
100		2007.49	II
50		2032.45	II
75		2036.23	II
150		2096.00	II
1000	r	2144.41	II
50		2155.06	II
100		2187.79	II
1000		2194.56	II
1000		2265.02	II
1500	r	2288.022	I
1000		2312.77	II
200		2321.07	II
40		2376.82	II
50		2418.69	II
50		2469.73	II
40		2487.93	II
3		2491.00	I
40		2495.58	II
10		2508.91	I
50		2509.11	II
30		2516.22	II
15	h	2518.59	I
25	h	2525.196	I
50		2544.613	I
50		2551.98	II
25		2553.465	I
3		2565.789	I
500		2572.93	II
50		2580.106	I
3		2584.87	I
30		2592.026	I
25	h	2602.048	I
50		2628.979	I
40		2632.190	I
75		2639.420	I
40		2659.23	II
50	h	2660.325	I
25		2668.20	II
50		2672.62	II
100		2677.540	I
25		2677.748	I
50		2707.00	II
75		2712.505	I
50		2733.820	I
1000		2748.54	II
100	h	2763.894	I
50	h	2764.230	I
50		2774.958	I
30		2823.19	II
200		2836.900	I
25		2856.46	II
100		2868.180	I
200	r	2880.767	I
50	r	2881.224	I
200		2914.67	II
50		2927.87	II
200		2929.27	II
1000	r	2980.620	I
200	r	2981.362	I
50		2981.845	I
50		3030.60	II
150		3080.822	I
25		3081.48	II
30		3082.593	I
100		3092.34	II
200		3133.167	I
50		3146.79	II
150		3250.33	II
300		3252.524	I
300		3261.055	I
50		3343.21	II
50		3385.49	II
30		3388.88	II
800		3403.652	I
50		3417.49	II
50		3442.42	II
100		3464.43	II
1000		3466.200	I
800		3467.655	I
25		3483.08	II
150		3495.44	II
25		3499.952	I
100		3524.11	II
100		3535.69	II
1000		3610.508	I
800		3612.873	I
60		3614.453	I
20		3649.558	I
10		3981.926	I
100		4029.12	II
200		4134.77	II
50		4141.49	II
100		4285.08	II
8		4306.672	I
100		4412.41	II
3		4412.989	I
1000		4415.63	II
30		4440.45	II
8		4662.352	I
200		4678.149	I
30		4744.69	II
300		4799.912	I
50		4881.72	II
50		5025.50	II
1000	h	5085.822	I
6		5154.660	I
100		5268.01	II
100		5271.60	II
1000		5337.48	II
1000		5378.13	II
200		5381.89	II
40		5843.30	II
50		5880.22	II
300		6099.142	I
100		6111.49	I
100		6325.166	I
30		6330.013	I
400		6354.72	II
500		6359.98	II
2000		6438.470	I
400		6464.94	II
25		6567.65	II
500		6725.78	II
100		6759.19	II
30		6778.116	I
50		7237.01	II
100		7284.38	II
1000		7345.670	I
50		8066.99	II
5		8200.309	I
20		9292	I
15		11655	I
35		14491	I
80		15712	I
55	d	19125	I
25		24378	I
35		25455	I

Cd III
Ref. 296 — R.D.C.

Intensity	Wavelength	
	Vacuum	
8	677.39	III

Cadmium (Cont.)

15	684.58	III
10	720.70	III
5	1383.60	III
15	1392.10	III
10	1396.78	III
25	1416.28	III
5	1420.29	III
8	1420.54	III
5	1432.86	III
20	1446.08	III
25	1447.55	III
30	1455.74	III
8	1466.14	III
5	1471.97	III
15	1491.81	III
5	1511.01	III
10	1511.65	III
5	1513.13	III
10	1523.55	III
15	1528.40	III
30	1529.30	III
5	1532.10	III
50	1545.17	III
25	1547.57	III
10	1550.07	III
20	1550.45	III
15	1550.89	III
5	1552.18	III
5	1556.48	III
15	1560.66	III
5	1566.03	III
15	1568.98	III
10	1582.39	III
40	1601.59	III
20	1604.87	III
20	1606.64	III
10	1607.28	III
15	1608.91	III
10	1609.61	III
15	1612.51	III
15	1625.27	III
25	1628.54	III
20	1651.87	III
25	1655.63	III
30	1678.15	III
10	1699.70	III
40	1707.16	III
10	1721.93	III
40	1722.95	III
30	1725.66	III
25	1739.00	III
5	1745.69	III
40	1747.67	III
15	1748.15	III
30	1768.82	III
40	1773.06	III
30	1789.19	III
75	1793.40	III
15	1796.10	III
5	1800.57	III
40	1823.41	III
50	1844.66	III
40	1851.13	III
20	1851.37	III
40	1855.85	III
200	1856.67	III
150	1874.08	III
15	1886.49	III
15	1903.48	III
25	1909.98	III
15	1910.57	III
10	1939.59	III
15	1988.81	III
	Air	
20	2000.60	III
15	2004.07	III
15	2016.12	III
40	2039.83	III
50	2045.61	III
10	2061.25	III
75	2087.91	III
10	2097.45	III
5	2100.47	III
50	2111.60	III
5	2188.13	III
5	2218.43	III
5	2224.43	III
7	2418.24	III
10	2426.36	III
25	2499.81	III
15	2618.81	III
5	2630.56	III
20	2766.99	III
30	2805.59	III
10	3035.72	III

Cd IV
Ref. 353, 399 — R.D.C.

Intensity	Wavelength	
	Vacuum	
50	427.01	IV
20	437.88	IV
50	447.85	IV
60	480.90	IV
10	489.49	IV
70	493.00	IV
70	495.13	IV
70	498.14	IV
70	498.53	IV
80	504.09	IV
70	504.20	IV
70	504.50	IV
80	506.31	IV
60	508.01	IV
50	508.95	IV
70	509.55	IV
25	509.81	IV
70	511.40	IV
80	513.00	IV
70	514.50	IV
60	519.42	IV
15	520.97	IV
80	524.41	IV
70	524.47	IV
50	524.77	IV
70	525.10	IV
60	525.19	IV
70	527.07	IV
50	530.79	IV
80	531.09	IV
80	531.51	IV
70	534.29	IV
70	536.77	IV
50	537.24	IV
60	540.90	IV
70	541.74	IV
80	542.60	IV
80	546.55	IV
40	548.01	IV
20	548.33	IV
15	548.90	IV
25	551.27	IV
20	552.90	IV
60	553.06	IV
80	554.05	IV
25	560.26	IV
10	564.16	IV
60	567.01	IV
20	1062.23	IV
150	1118.16	IV
30	1126.00	IV
20	1134.08	IV
15	1139.04	IV
20	1154.64	IV
10	1155.73	IV
100	1164.65	IV
20	1165.78	IV
40	1167.30	IV
20	1179.73	IV
15	1183.07	IV
100	1183.40	IV
40	1194.13	IV
20	1195.63	IV
30	1196.47	IV
20	1198.93	IV
15	1215.38	IV
20	1223.52	IV
20	1246.06	IV
15	1246.56	IV
15	1249.94	IV
30	1266.47	IV
15	1274.41	IV
20	1285.63	IV
20	1287.58	IV
20	1299.46	IV
40	1304.36	IV
30	1306.07	IV
15	1316.89	IV
15	1321.85	IV
20	1325.55	IV
30	1340.97	IV
15	1346.15	IV
30	1354.78	IV
20	1358.11	IV
30	1362.55	IV
60	1370.48	IV
30	1380.98	IV
20	1397.65	IV
30	1403.68	IV
15	1406.58	IV
60	1418.89	IV
15	1429.83	IV
20	1447.54	IV
20	1452.63	IV
15	1465.97	IV
15	1466.67	IV
15	1482.95	IV
20	1491.79	IV
20	1570.20	IV
20	1598.73	IV
20	1600.42	IV
15	1622.87	IV

CALCIUM (Ca)
Z = 20

Ca I and II
Ref. 70, 150, 270 — J.J.W. and H.H.S.

Intensity	Wavelength	
	Vacuum	
24	1341.89	II
12	1342.54	II
20	1433.75	II
12	1432.50	II
20	1553.18	II
32	1554.64	II
4	1642.80	II
20	1643.77	II
36	1644.44	II
60	1649.86	II
32	1651.99	II
12	1673.86	II
20	1680.05	II
2	1680.13	II
8	1691.78	II
16	1698.18	II
20	1807.34	II
40	1814.50	II
4	1814.65	II
60	1840.06	II
40	1838.01	II
20	1843.09	II
40	1850.69	II
	Air	
	2103.24	II
	2112.76	II
	2113.15	II
	2128.75	II
	2131.51	II
	2132.30	II
2	2150.80	I
	2197.79	II
5	2200.73	I
	2208.61	II
6	2275.46	I
8	2398.56	I
7	2721.65	I
9	2994.96	I
8	2997.31	I
8	2999.64	I
9	3000.86	I
10	3006.86	I
9	3009.21	I
2	3024.94	I
2	3034.54	I
2	3045.74	I
3	3055.32	I
2	3071.57	I
2	3076.95	I
2	3080.79	I
2	3099.30	I
10	3125.18	II
5	3136.02	I
6	3140.79	I
7	3150.75	I
170	3158.87	II
180	3179.33	II
5	3180.52	I
150	3181.28	II
7	3209.96	I
8	3215.17	I
6	3215.34	I
9	3225.90	I
6	3226.15	I
5	3274.67	I
6	3286.07	I
10	3308.02	II
20	3316.51	II
10	3344.51	I
10	3347.04	II
11	3350.21	I
9	3350.36	I
12	3361.92	I
9	3362.14	I
10	3452.66	II
20	3461.87	II
9	3468.48	I
11	3474.76	I
10	3485.61	II
13	3487.60	I
10	3495.16	II
15	3624.11	I
17	3630.75	I
14	3630.97	I
20	3644.41	I
14	3644.77	I
8	3644.99	I
5	3675.29	I
6	3678.21	I
30	3683.70	II
40	3694.11	II
10	3694.36	II
170	3706.03	II
180	3736.90	II
10	3739.38	II
6	3748.35	I
8	3750.29	I
9	3753.34	I
20	3755.67	II
30	3758.39	II
9	3870.48	I
11	3872.54	I
11	3872.56	I
12	3875.78	I
12	3875.80	I
6	3889.10	I
6	3923.48	I
230	3933.66	II
9	3935.29	I
6	3946.04	I
15	3948.90	I
17	3957.05	I
220	3968.47	II
8	3972.57	I
18	3973.71	I
50	4097.10	II
15	4098.53	I
15	4098.57	I
60	4109.82	II
30	4110.28	II
40	4206.18	II
50	4220.07	II
50	4226.73	I
15	4240.46	I
24	4283.01	I
22	4289.36	I
22	4298.99	I
25	4302.53	I
23	4307.74	I
22	4318.65	I
20	4355.08	I
25	4425.44	I
26	4434.96	I
25	4435.69	I
30	4454.78	I
28	4455.89	I
20	4456.61	I
20	4472.04	II
10	4479.23	II
20	4489.18	II
23	4526.94	I
22	4578.55	I
23	4581.40	I
23	4581.47	I
24	4585.87	I
24	4585.96	I
20	4685.27	I
30	4716.74	II
40	4721.03	II
40	4799.97	II
25	4878.13	I
70	5001.48	II
80	5019.97	II
40	5021.14	II
23	5041.62	I
25	5188.85	I
22	5261.71	I
23	5262.24	I
22	5264.24	I
24	5265.56	I
25	5270.27	I
60	5285.27	II
70	5307.22	II
50	5339.19	II
27	5349.47	I
23	5512.98	I
25	5581.97	I
27	5588.76	I
24	5590.12	I
26	5594.47	I
25	5598.49	I
24	5601.29	I
24	5602.85	I
30	5857.45	I
10	5922.72	II
10	5923.69	II
27	6102.72	I
29	6122.22	I
22	6161.29	I
30	6162.17	I
22	6163.76	I
24	6166.44	I
26	6169.06	I
28	6169.56	I
35	6439.07	I
30	6449.81	I
22	6455.60	I
80	6456.87	II
34	6462.57	I
29	6471.66	I

Calcium (Cont.)

32		6493.78	I
28		6499.65	I
23		6572.78	I
30		6717.69	I
33		7148.15	I
31		7202.19	I
		7291.47	II
		7323.89	II
33		7326.15	I
30		7575.81	II
60		7581.11	II
80		7601.30	II
20		7602.32	II
40		7820.78	II
60		7843.38	II
20		8017.50	II
20		8020.50	II
70		8133.05	II
100		8201.72	II
110		8248.80	II
70		8254.73	II
14		8256.67	I
10		8338.04	I
12		8339.12	I
10		8352.39	I
11		8357.17	I
130		8498.02	II
170		8542.09	II
10		8633.95	I
160		8662.14	II
12		8842.61	I
15		8909.18	I
100		8912.07	II
110		8927.36	II
12		8967.47	I
16		9099.10	I
13		9105.62	I
12		9108.82	I
10		9171.14	I
110		9213.90	II
90		9312.00	II
100		9319.56	II
110		9320.65	II
25		9416.97	I
10		9456.80	I
10		9534.88	I
11		9548.38	I
100		9567.97	II
110		9599.24	II
80		9601.82	II
10		9604.28	I
12		9663.65	I
10		9664.41	I
14		9676.30	I
14		9688.67	I
13		9701.94	I
80		9854.74	II
110		9890.63	II
90		9931.39	II
100		10223.04	II
20		10343.81	I
13		10838.97	I
13		10861.58	I
13		10863.87	I
14		10869.50	I
14		10879.87	I
20		11838.99	II
10		11949.72	II
25		12816.04	I
24		12823.86	I
25		12909.10	I
30		13033.57	I
21		13086.44	I
24		13134.95	I
20		16150.77	I
22		16157.36	I
21		16197.04	I
20		18925.47	I
24		18970.14	I
30		19046.14	I
48		19309.20	I
49		19452.99	I
47		19505.72	I
50		19776.79	I
35		19853.10	I
34		19862.22	I
23		19917.19	I
24		19933.70	I
		21389.00	II
		21428.90	II
20		22607.93	I
25		22624.93	I
30		22651.23	I

Ca III
Ref. 25, 16 — J.J.W. and H.H.S.

Intensity		Wavelength	
		Vacuum	
6		296.96	III
9		403.72	III
7		409.95	III
5		439.69	III
5		633.59	III
5		685.41	III
5		697.55	III
5		699.09	III
5		699.89	III
6		701.39	III
5		727.66	III
8		740.55	III
6		746.25	III
5		747.98	III
5		779.61	III
5		800.30	III
5		809.93	III
5		817.06	III
6		821.57	III
6		840.56	III
6		1020.07	III
5		1034.65	III
5		1187.30	III
8		1188.61	III
8		1188.61	III
5		1190.86	III
10		1262.65	III
11		1278.39	III
10		1281.55	III
12		1286.52	III
12		1298.04	III
11		1317.70	III
10		1328.95	III
11		1335.13	III
10		1360.01	III
11		1385.43	III
11		1397.69	III
13		1453.16	III
12		1459.79	III
11		1461.88	III
15		1463.34	III
16		1484.87	III
12		1496.88	III
11		1506.88	III
20		1545.29	III
15		1555.53	III
18		1562.47	III
13		1571.27	III
13		1586.13	III
10		1762.26	III
10		1783.93	III
10		1794.22	III
12		1800.21	III
13		1807.89	III
14		1812.15	III
11		1813.59	III
12		1830.06	III
10		1860.43	III
14		1870.26	III
14		1872.37	III
10		1894.12	III
11		1910.10	III
12		1935.72	III
10		1939.68	III
11		1943.01	III
12		1948.26	III
10		1953.55	III
10		1958.97	III
13		1964.61	III
13		1967.94	III
12		1972.82	III
10		1977.01	III
10		1978.55	III
11		1981.19	III
		Air	
12		2033.36	III
12		2041.53	III
13		2078.92	III
13		2098.49	III
15		2114.41	III
17		2123.03	III
14		2129.19	III
14		2133.96	III
13		2140.36	III
16		2152.43	III
12		2171.57	III
12		2276.52	III
15		2312.08	III
14		2497.74	III
15		2541.50	III
13		2587.15	III
12		2590.41	III
15		2620.82	III
15		2634.14	III
12		2686.72	III
16		2687.76	III
15		2704.86	III
14		2771.28	III
15		2791.59	III
16		2813.88	III
17		2866.54	III
18		2869.95	III
19		2881.78	III
21		2899.79	III
19		2924.33	III
20		2988.63	III
18		2989.27	III
15		3028.59	III
19		3119.67	III
15		3367.79	III
19		3372.67	III
18		3537.77	III
15		4081.77	III
15		4153.57	III
15		4164.31	III
15		4184.20	III
18		4207.24	III
17		4233.74	III
16		4240.74	III
15		4284.39	III
20		4302.81	III
15		4329.19	III
16		4333.57	III
15		4358.38	III
19		4399.59	III
17		4406.29	III
17		4431.30	III
19		4499.88	III
18		4516.59	III
18		4572.12	III
11		4708.83	III
11		4716.27	III
10		4859.17	III
10		5008.95	III
10		5050.07	III
10		5231.82	III
11		5247.37	III
13		5271.98	III
10		5301.32	III
11		5321.29	III
10		5328.06	III
11		5570.58	III
10		5579.06	III
13		6069.98	III
10		6173.22	III
12		6213.98	III
11		6294.89	III
11		6370.11	III
10		6387.55	III
12		6424.51	III
12		6485.35	III
10		6538.78	III
10		6542.24	III
10		7308.69	III
10		7843.06	III
12		7898.46	III
10	1	8217.20	III

Ca IV
Ref. 150 — J.J.W. and H.H.S.

Intensity		Wavelength	
		Vacuum	
150		249.41	IV
150		250.15	IV
150		251.35	IV
250		296.55	IV
200		299.32	IV
200		318.09	IV
50		318.39	IV
120		321.59	IV
250		329.12	IV
150		329.39	IV
200		331.44	IV
250		331.99	IV
235		332.53	IV
150		332.81	IV
200		338.83	IV
150		339.79	IV
200		340.29	IV
200		341.29	IV
200		341.46	IV
250	c	342.45	IV
100		343.19	IV
200		343.44	IV
250		343.93	IV
200		344.96	IV
215		345.13	IV
250		374.74	IV
600		434.57	IV
100		437.27	IV
250		437.77	IV
200		438.93	IV
750		443.82	IV
50		445.02	IV
500		450.57	IV
50		454.55	IV
250		456.98	IV
250		461.09	IV
150		565.46	IV
750		656.00	IV
500		669.70	IV

Ca V
Ref. 150 — J.J.W. and H.H.S.

Intensity		Wavelength	
		Vacuum	
200		190.36	V
250		190.46	V
250		196.97	V
300		199.55	V
250		200.51	V
265		257.98	V
165		260.45	V
400		267.77	V
300		270.31	V
200		271.14	V
250		272.27	V
200		272.98	V
400		280.99	V
300		284.98	V
450	c	286.96	V
500		322.17	V
250		322.76	V
300		323.22	V
250		324.48	V
250		325.28	V
300		330.94	V
200		333.44	V
200		333.57	V
300		334.55	V
250	c	335.34	V
200		336.55	V
200		337.54	V
250		338.06	V
200		343.64	V
450		352.92	V
250		356.25	V
250		377.18	V
200		387.08	V
750		425.00	V
500		558.60	V
400		637.93	V
300		643.12	V
400		646.57	V
250		647.88	V
250		651.55	V
300		656.76	V

CALIFORNIUM (Cf)
Z = 98

Cf I and II
Ref. 52, 331 — J.G.C.

Intensity		Wavelength	
		Air	
10000		2739.31	
10000	s	2759.10	
10000		2774.52	
10000	l	2852.03	
10000	s	2855.24	
10000		3298.14	
10000		3352.71	
10000	l	3367.79	
10000		3392.22	I
10000		3481.07	
10000	l	3513.47	
10000		3531.49	I
10000		3540.98	I
10000		3598.77	I
10000		3605.32	I
10000		3612.11	II
10000		3617.49	I
10000	s	3626.76	II
10000		3659.46	
10000		3662.70	I
10000	s	3699.49	
10000	l	3722.11	II
10000		3739.35	I
10000		3785.61	I
10000	l	3789.04	II
10000	s	3893.23	II
10000	l	3993.57	II
10000		4035.45	
10000		4099.12	I
10000		4242.38	I
10000		4329.03	I
10000		4335.22	I
10000		5173.96	I
10000		5179.08	I
10000		5219.24	I
10000	s	5279.01	
10000	s	5320.09	
10000	s	5339.13	
10000		5408.88	I
10000		5726.05	I
10000		6622.83	I
10000		6631.26	I

Californium (Cont.)

10000		6677.90	I
10000		6894.59	I
10000		6927.10	II
10000	l	7074.52	I
10000	s	7307.90	I
10000		8141.29	I
10000		8241.77	I
10000		8333.85	II
10000		8423.49	II
10000		8568.83	II
10000	l	9228.52	
10000		9337.70	
10000		9649.51	
10000	s	10308.41	
10000	l	10568.83	
10000	[illegible]	[illegible]	
10000		11300.19	
10000		11681.85	
10000		11941.33	
10000	l	12183.05	
10000	s	12352.72	
10000	s	12437.48	
10000	l	12789.41	
10000	l	13329.98	
10000	s	13362.98	
10000	l	13376.89	
10000	l	13474.44	
10000	l	14772.49	
10000	s	15281.32	
10000		15587.12	
10000		15675.92	
10000		16759.06	
10000	s	17626.25	
10000	l	18718.69	
10000	h	19068.71	
10000	l	19336.96	
10000	l	19576.84	
10000	l	20393.38	
10000	s	20869.98	

CARBON (C)
Z = 6

C I and II
Ref. 211 — R.L.K.

Intensity		Wavelength	
		Vacuum	
9		595.022	II
30		687.053	II
50		687.345	II
10		858.092	II
20		858.559	II
30		903.624	II
60		903.962	II
150		904.142	II
30		904.480	II
9		1009.86	II
10		1010.08	II
10		1010.37	II
80		1036.337	II
150		1037.018	II
150		1157.910	I
150		1158.019	I
150		1158.035	I
150		1188.992	I
150		1189.447	I
200		1189.631	I
300		1193.009	I
300		1193.031	I
300		1193.240	I
300		1193.264	I
100		1193.393	I
150		1193.649	I
150		1193.679	I
100		1194.064	I
100		1194.488	I
100		1261.552	I
250		1277.245	I
250		1277.282	I
300		1277.513	I
300		1277.550	I
200		1280.333	I
100		1311.363	I
9		1323.951	II
120		1329.578	I
120		1329.600	I
150		1334.532	II
300		1335.708	II
100		1354.288	I
150		1355.84	I
120		1364.164	I
100		1459.032	I
200		1463.336	I
120		1467.402	I
150		1481.764	I
150		1560.310	I
400		1560.683	I
400		1560.708	I
100		1561.341	I
400		1561.438	I
150		1656.266	I
120		1656.928	I
300		1657.008	I
120		1657.380	I
120		1657.907	I
150		1658.122	I
500		1751.823	I
1000		1930.905	I
		Air	
800		2478.56	I
250		2509.12	II
350		2512.06	II
250	h	2574.83	II
350	l	2741.28	II
250		2746.49	II
1000		2836.71	II
800		2837.60	II
800	h	2992.62	II
350		3876.19	II
350		3876.41	II
350		3876.66	II
570		3918.98	II
800		3920.69	II
250		4074.52	II
350	l	4075.85	II
800		4267.00	II
1000		4267.26	II
200		4771.75	I
200		4932.05	I
200		5052.17	I
350		5132.94	II
350		5133.28	II
350		5143.49	II
570		5145.16	II
400		5151.09	II
300		5380.34	I
250		5648.07	II
350		5662.47	II
570		5889.77	II
350		5891.59	II
200		6001.13	I
250		6006.03	I
110		6007.18	I
150		6010.68	I
300		6013.22	I
250		6014.84	I
800		6578.05	II
570		6582.88	II
200		6587.61	I
250		6783.90	II
250		7113.18	I
250		7115.19	I
250		7115.63	II
200		7116.99	I
350		7119.90	II
800		7231.32	II
1000		7236.42	II
200		7860.89	I
200		8058.62	I
520		8335.15	I
250		9061.43	I
200		9062.47	I
200		9078.28	I
250		9088.51	I
450		9094.83	I
300		9111.80	I
800		9405.73	I
150		9603.03	I
250		9620.80	I
300		9658.44	I
200		10683.08	I
300		10691.25	I
12		11619.29	I
23		11628.83	I
13		11658.85	I
47		11659.68	I
24		11669.63	I
85		11748.22	I
142		11753.32	I
114		11754.76	I
11		11777.54	I
17		11892.91	I
30		11895.75	I
26		12614.10	I
20		13502.27	I
38		14399.65	I
16		14403.25	I
61		14420.12	I
12		14429.03	I
13		14442.24	I
12		16559.66	I
50		16890.38	I
10		17338.56	I
11		17448.60	I
13		18139.80	I
23		19721.99	I

C III
Ref. 22, 211 — R.L.K.

Intensity		Wavelength	
		Vacuum	
250		371.69	III
250		371.75	III
150		371.78	III
500		386.203	III
200		450.734	III
400		459.46	III
500		459.52	III
570		459.63	III
250		511.522	III
250		535.288	III
300		[illegible]	[illegible]
350		538.149	III
400		538.312	III
350		574.281	III
800		977.03	III
370		1174.93	III
350		1175.26	III
330		1175.59	III
500		1175.71	III
350		1175.99	III
370		1176.37	III
		Air	
250		2162.94	III
800		2296.87	III
150		2697.75	III
110	l	2724.85	III
150	l	2725.30	III
150	l	2725.90	III
200		2982.11	III
150		4056.06	III
200		4067.94	III
250		4068.91	III
250		4070.26	III
150		4162.86	III
250	h	4186.90	III
200		4325.56	III
600		4647.42	III
520		4650.25	III
375		4651.47	III
200		4665.86	III
450		5695.92	III
150		5826.42	III
150		6744.38	III
150	h	7037.25	III
150		7612.65	III
300	h	8196.48	III
150		8332.99	III
300		8500.32	III

C IV
Ref. 66, 211 — R.L.K.

Intensity		Wavelength	
		Vacuum	
250		244.91	IV
200		289.14	IV
250		289.23	IV
570		312.42	IV
500		312.46	IV
650		384.03	IV
700		384.18	IV
400		419.52	IV
500		419.71	IV
1000		1548.202	IV
900		1550.774	IV
		Air	
200	l	2524.41	IV
300	s	2529.98	IV
200	w	4658.30	IV
250		5801.33	IV
200		5811.98	IV
90	w	7726.2	IV

C V
Ref. 211 — R.L.K.

Intensity		Wavelength	
		Vacuum	
110		34.973	V
450		40.268	V
110		227.19	V
160		248.66	V
160		248.74	V
		Air	
40		2270.91	V
5		2277.25	V
20		2277.92	V
5		4943.88	V
5		4944.56	V

CERIUM (Ce)
Z = 58

Ce I and II
Ref. 1 — C.H.C.

Intensity	Wavelength	
	Air	
130	2462.97	II
110	2518.51	II
200	2548.68	II
340	2651.01	II
120	2696.07	II
120	2706.88	II
120	2723.38	II
110	2741.96	II
100	2750.89	II
150	2761.42	II
120	2784.27	II
100	2785.35	II
100	2790.53	II
140	2791.42	II
270	2830.90	II
100	2833.31	II
250	2874.14	II
110	2908.42	II
100	2918.67	II
120	2955.94	II
110	2964.80	II
100	2972.58	II
400	2976.91	II
150	2977.46	II
120	2980.41	II
250	2990.87	II
110	2994.42	II
320	2995.64	II
400	3008.79	II
370	3017.20	II
210	3037.73	II
200	3051.98	II
350	3055.24	II
320	3056.78	II
680	3063.01	II
320	3083.67	II
250	3084.44	II
200	3090.37	II
370	3103.38	II
200	3107.47	II
320	3110.28	II
300	3111.17	II
220	3127.53	II
200	3130.33	II
240	3130.87	II
200	3144.60	II
290	3145.28	II
290	3146.41	II
290	3164.15	II
290	3169.18	II
290	3171.61	II
480	3183.52	II
240	3186.13	II
200	3190.34	II
710	3194.83	II
200	3199.28	II
990	3201.71	II
200	3218.38	II
710	3218.94	II
880	3221.17	II
330	3225.67	II
710	3227.11	II
240	3229.36	II
480	3231.24	II
710	3234.16	II
330	3234.89	II
390	3236.74	II
390	3243.37	II
200	3246.67	II
200	3260.98	II
200	3263.88	II
990	3272.25	II
330	3274.86	II
200	3279.84	II
330	3285.22	II
240	3295.28	II
200	3296.88	II
220	3300.15	II
240	3304.84	II
200	3312.22	II
240	3314.72	II
200	3317.80	II
200	3334.46	II
240	3341.87	II
330	3343.86	II
440	3344.76	II
200	3355.02	II

Cerium (Cont.)

240	3357.22	II
200	3360.54	II
240	3366.55	II
200	3371.18	II
200	3373.46	II
200	3373.73	II
480	3377.13	II
200	3383.68	II
200	3404.91	II
240	3405.98	II
290	3417.45	II
600	3422.71	II
390	3426.21	II
290	3441.21	II
480	3476.84	II
240	3482.35	II
710	3485.05	II
210	3507.94	II
600	3517.38	II
210	3520.52	II
330	3521.88	II
210	3526.68	II
600	3534.05	II
770	3539.08	II
210	3545.60	II
290	3546.19	II
240	3552.73	II
420	3555.00	II
1200	3560.80	II
210	3576.23	II
1000	3577.45	II
330	3590.60	II
390	3607.63	II
550	3609.69	II
420	3613.70	II
440	3622.15	II
380	3623.74	II
440	3623.84	II
200	3631.19	II
350	3646.97	II
260	3647.75	II
260	3647.95	II
420	3653.11	II
660	3653.67	II
310	3654.97	II
1800	3655.85	II
440	3659.23	II
350	3659.97	II
880	3660.64	II
880	3667.98	II
220	3672.18	I
350	3672.79	II
220	3679.42	II
300	3694.91	II
220	3704.98	II
1000	3709.29	II
1000	3709.93	II
1400	3716.37	II
420	3718.19	II
420	3718.38	II
210	3719.80	II
420	3725.68	II
490	3728.02	II
800	3728.42	II
320	3748.06	II
250	3751.45	II
200	3755.43	II
300	3762.98	II
680	3764.12	II
200	3765.04	II
300	3768.76	II
210	3770.76	II
300	3771.60	II
250	3776.61	II
620	3781.62	II
440	3782.52	II
200	3783.58	II
860	3786.63	II
520	3788.75	II
300	3792.32	II
2500	3801.52	II
800	3803.09	II
1000	3808.11	II
490	3809.21	II
250	3812.20	II
490	3815.85	II
470	3817.46	II
300	3819.02	II
470	3823.90	II
470	3830.55	II
490	3831.08	II
490	3834.55	II
270	3836.10	II
1100	3838.54	II
200	3843.76	II
220	3846.52	II
250	3848.10	II
860	3848.59	II
860	3853.15	II
1200	3854.18	II
1200	3854.31	II
620	3855.29	II
390	3857.02	II
370	3857.64	II
200	3862.46	II
200	3868.13	II
270	3874.68	II
620	3876.97	II
1100	3878.36	II
1500	3882.45	II
1000	3889.98	II
210	3890.75	II
210	3890.98	II
620	3895.11	II
590	3896.80	II
490	3898.27	II
270	3898.94	II
200	3903.34	II
250	3904.34	II
200	3906.92	II
770	3907.29	II
560	3908.41	II
390	3908.54	II
270	3909.31	II
230	3912.19	II
980	3912.44	II
390	3915.52	II
390	3916.14	II
230	3917.64	II
770	3918.28	II
480	3919.81	II
590	3921.73	II
560	3923.11	II
450	3924.64	II
770	3931.09	II
310	3931.37	II
230	3931.83	II
310	3933.73	II
560	3938.09	II
770	3940.34	II
310	3940.97	II
2000	3942.15	II
2700	3942.75	II
770	3943.89	II
310	3947.97	II
3100	3952.54	II
340	3953.66	II
310	3955.36	II
230	3956.06	II
980	3956.28	II
230	3958.27	II
230	3958.87	II
770	3960.91	II
390	3964.50	II
770	3967.05	II
450	3971.68	II
270	3972.07	II
270	3975.07	II
770	3978.65	II
560	3980.88	II
560	3982.89	II
310	3983.29	II
770	3984.68	II
370	3989.44	II
700	3992.39	II
370	3992.91	II
910	3993.82	II
2800	3999.24	II
230	4001.56	II
910	4003.77	II
370	4005.64	II
210	4007.59	II
2700	4012.39	II
910	4014.90	II
250	4015.88	II
200	4019.04	II
240	4022.27	II
840	4024.49	II
240	4025.15	II
840	4028.41	II
250	4030.34	II
840	4031.34	II
340	4037.67	II
2100	4040.76	II
910	4042.58	II
230	4045.21	II
620	4046.34	II
210	4051.43	II
210	4051.99	II
700	4053.51	II
450	4054.99	II
280	4062.22	II
230	4062.94	II
280	4067.28	II
420	4068.84	II
1100	4071.81	II
270	4072.92	II
1800	4073.48	II
210	4073.74	II
1500	4075.71	II
1500	4075.85	II
210	4076.24	II
420	4077.47	II
530	4078.32	II
270	4078.52	II
270	4080.44	II
670	4081.22	II
910	4083.23	II
450	4085.23	II
250	4087.36	II
230	4088.85	II
450	4101.77	II
250	4105.00	II
510	4107.42	II
200	4110.38	II
250	4111.39	II
420	4115.37	II
250	4117.01	II
200	4117.29	II
200	4117.59	II
770	4118.14	II
250	4119.02	II
310	4119.79	II
310	4119.88	II
450	4120.83	II
510	4123.24	II
510	4123.49	II
980	4123.87	II
510	4124.79	II
980	4127.37	II
250	4127.74	II
200	4128.07	II
530	4130.71	II
480	4131.10	II
2700	4133.80	II
270	4135.44	II
270	4137.47	II
2000	4137.65	II
270	4138.10	II
210	4138.35	II
770	4142.40	II
390	4144.49	II
670	4145.00	II
480	4146.23	II
280	4148.90	II
420	4149.79	II
980	4149.94	II
420	4150.91	II
1400	4151.97	II
230	4153.13	II
450	4159.03	II
310	4163.52	II
1300	4165.61	II
620	4166.88	II
250	4167.80	II
320	4169.77	II
320	4169.88	II
340	4176.70	II
340	4181.08	II
340	4185.33	II
3500	4186.60	II
530	4187.32	II
560	4193.09	II
370	4193.28	II
370	4193.87	II
630	4196.34	II
280	4198.00	II
280	4198.67	II
840	4198.72	II
240	4201.24	II
910	4202.94	II
270	4209.41	II
370	4214.04	II
310	4217.59	II
1500	4222.60	II
770	4227.75	II
390	4231.74	II
240	4234.21	II
200	4236.02	II
980	4239.92	II
390	4242.72	II
310	4245.89	II
310	4245.98	II
390	4246.72	II
1100	4248.68	II
390	4253.37	II
620	4255.79	II
200	4263.43	II
620	4270.19	II
390	4270.72	II
200	4278.86	II
280	4285.37	II
200	4288.66	II
200	4289.44	II
2000	4289.94	II
200	4296.07	II
1500	4296.67	II
420	4296.78	II
590	4299.36	II
770	4300.33	II
420	4305.14	II
770	4306.72	II
390	4309.74	II
560	4320.72	II
310	4330.45	II
310	4332.71	II
240	4336.23	II
980	4337.77	II
340	4339.31	II
700	4349.79	II
560	4352.71	II
910	4364.66	II
350	4373.82	II
530	4375.92	II
910	4382.17	II
700	4386.84	II
310	4388.01	II
1700	4391.66	II
200	4398.79	II
510	4399.20	II
350	4410.64	II
350	4410.76	II
310	4416.90	II
980	4418.78	II
200	4423.68	II
310	4427.07	II
480	4427.92	II
310	4428.44	II
650	4429.27	II
480	4444.39	II
450	4444.70	II
770	4449.34	II
620	4450.73	II
2400	4460.21	II
450	4461.14	II
420	4463.41	II
280	4467.54	II
1400	4471.24	II
450	4472.72	II
700	4479.36	II
700	4483.90	II
840	4486.91	II
250	4497.85	II
100	4506.41	I
110	4515.86	II
100	4519.59	II
770	4523.08	II
840	4527.35	II
840	4528.47	II
110	4532.49	II
110	4539.07	II
840	4539.75	II
210	4544.96	II
250	4551.30	II
650	4560.28	II
310	4560.96	II
2100	4562.36	II
420	4565.84	II
1100	4572.28	II
420	4582.50	II
130	4591.12	II
840	4593.93	II
420	4606.40	II
420	4624.90	II
1700	4628.16	II
170	4632.32	I
110	4650.51	I
130	4654.29	II
110	4669.50	II
150	4680.13	II
270	4684.61	II
200	4714.00	II
100	4714.81	II
110	4725.09	II
100	4733.52	II
310	4737.28	II
100	4739.53	II
160	4747.17	II
110	4757.84	II
100	4768.77	II
230	4773.94	II
110	4822.55	I
140	4847.77	I
180	4882.46	II
110	4943.44	I
130	4971.50	II
130	4994.63	I
210	5009.10	I
100	5011.77	II
120	5022.87	II
120	5037.78	II
120	5040.85	I
180	5044.02	II
120	5071.78	I
240	5075.35	II
470	5079.68	II
130	5112.70	I
160	5117.17	II
170	5129.57	I
110	5147.57	II
100	5149.99	I
280	5159.69	I
280	5161.48	I
190	5174.55	I
370	5187.46	II
210	5191.66	II
190	5211.92	I
260	5223.46	I
180	5229.75	I
140	5232.92	II
260	5245.92	I
130	5265.71	II

Cerium (Cont.)

340		5274.23	II
130		5296.56	I
130		5328.08	I
190		5330.54	II
450		5353.53	II
300		5393.40	II
150		5397.64	I
280		5409.23	II
110		5420.38	I
140		5449.24	I
140		5468.37	II
140		5472.29	II
260		5512.08	II
110		5556.25	I
170		5564.97	I
130		5565.97	I
100		5595.88	I
240		5601.28	I
190		5655.14	I
240		5669.96	I
120		5677.75	I
120		5692.94	I
300		5696.99	I
370		5699.23	I
240		5719.03	I
140		5773.12	I
120		5788.15	I
120		5812.92	I
230		5940.86	I
11		6000.18	I
55		6001.90	I
55		6005.86	I
15		6006.20	I
55		6006.82	I
19		6007.37	I
75		6013.42	I
23		6016.59	I
110		6024.20	I
15		6027.16	I
11		6031.26	I
23		6033.58	II
35		6034.20	II
23		6034.41	I
35		6035.49	II
110		6043.39	II
28		6045.42	I
55		6047.40	I
19		6051.80	II
23		6057.50	I
35		6058.00	I
23		6066.75	I
19		6069.46	I
35		6069.48	I
35		6072.00	I
35		6076.61	I
17		6077.16	I
17		6080.37	I
17		6081.28	I
19		6088.86	I
19		6088.96	I
35		6093.19	I
45		6098.34	II
11		6099.80	I
28		6108.74	II
15		6118.56	I
17		6118.90	I
45		6123.67	I
19		6132.00	II
19		6132.18	I
11		6135.45	I
23		6139.03	I
15		6142.92	I
35		6143.36	II
23		6146.43	I
19		6147.84	I
23		6151.72	I
19		6159.82	I
19		6162.14	I
19		6165.45	I
19		6175.28	I
35		6186.17	I
15		6187.97	I
15		6195.23	I
19		6195.53	I
19		6198.05	I
35		6208.98	I
11		6216.82	I
35		6228.94	I
19		6229.13	I
23		6232.45	II
28		6237.45	I
13		6238.71	I
11		6241.87	I
13		6242.91	I
15		6253.65	I
13		6257.99	I
15		6264.27	I
45		6272.05	II
15		6276.47	I
35		6295.58	I
28		6299.51	II
23		6300.21	I
13		6306.64	I
35		6310.01	I
15		6335.40	I
11		6337.21	I
13		6340.70	I
35		6343.95	II
35		6371.11	II
28		6386.84	I
23		6393.02	II
11		6395.16	I
11		6425.29	II
35		6430.07	I
19		6434.39	I
23		6436.40	I
19		6446.12	I
35		6458.03	I
19		6466.88	II
28		6467.39	I
35		6473.72	I
17		6490.97	I
11		6503.27	II
11		6507.16	II
23		6513.59	II
19		6517.31	I
19		6551.70	I
45		6555.65	I
23		6579.10	I
15		6606.35	I
15		6606.86	II
22		6612.06	I
10		6623.00	I
30		6628.93	I
13		6650.89	I
22		6652.72	II
10		6661.41	I
13		6665.59	I
10		6675.54	II
15		6686.60	I
26		6700.66	I
35		6704.27	I
13		6704.52	II
10		6706.04	II
15		6728.71	I
15		6729.57	I
15		6744.70	II
10		6746.90	I
30		6774.28	II
35		6775.59	I
10		6778.28	I
18		6807.81	I
10		6808.82	I
15		6010.23	I
10		6829.73	II
13		6847.25	I
12		6856.55	I
10		6893.66	I
10		6898.45	II
30		6924.81	I
10		6939.45	I
19		6973.50	II
10		6983.82	II
30		6986.02	I
12		7054.51	I
11		7058.68	II
11		7060.00	I
35		7061.75	II
11		7064.49	I
35		7086.35	II
11		7105.04	II
11		7115.08	II
10		7124.73	I
16		7141.42	I
19		7150.23	II
10		7151.67	I
16		7155.25	I
16		7156.99	II
16		7189.40	II
10		7191.72	I
11		7201.56	II
16		7201.89	I
10		7203.55	I
12		7210.67	I
19		7217.36	I
16		7235.71	II
22		7238.36	II
12		7241.73	I
25		7252.75	I
12		7262.64	I
11	h	7277.90	I
11		7296.17	I
19		7301.42	II
19		7313.45	II
25		7329.91	I
16		7334.68	II
12		7343.44	I
25		7397.77	I
11		7401.27	I
12		7417.94	II
11		7424.70	C
12		7433.08	I
11		7438.56	I
12		7444.44	I
10		7472.41	I
16		7486.57	II
11		7527.46	I
11		7527.68	I
10		7533.73	I
10		7551.25	I
12		7562.44	I
10		7562.86	I
10	h	7563.60	I
10		7603.10	I
25		7616.11	II
12		7646.08	I
10		7647.88	I
12		7682.47	I
25		7689.17	II
10		7732.33	I
16		7748.35	I
10		7797.70	I
12		7842.59	I
22		7844.94	II
16		7850.02	II
16		7851.18	II
22		7857.54	II
12		7864.49	I
10		7866.04	I
16		7898.96	II
11		7913.52	I
10		7927.30	C
10		7927.72	I
10		7934.50	II
30		8025.56	II
16		8070.71	I
10		8094.43	I
16		8120.36	I
10		8241.55	II
12		8261.09	I
16		8418.23	II
11		8495.82	I
12		8539.08	II
10		8612.64	I
10	h	8647.66	I
11	h	8702.38	II
25		8772.14	II
12		8810.84	I
30		8891.20	II

Ce III
Ref. 136, 305 — J.R.

Intensity	Wavelength	
Vacuum		
100	840.24	III
20	844.11	III
40	845.02	III
20	847.88	III
200	851.18	III
200	852.63	III
200	853.47	III
60	853.78	III
200	855.16	III
200	858.30	III
400	860.15	III
200	862.25	III
40	868.74	III
20	869.51	III
40	869.84	III
20	871.15	III
40	871.27	III
20	880.68	III
60	881.75	III
60	884.04	III
20	885.22	III
30	888.39	III
80	892.75	III
20	899.32	III
100	912.77	III
40	937.04	III
40	999.26	III
20	1025.25	III
20	1025.29	III
20	1026.28	III
40	1029.37	III
20	1034.55	III
20	1041.14	III
100	1042.74	III
50	1051.61	III
70	1057.40	III
100	1057.66	III
100	1058.46	III
50	1062.99	III
30	1063.26	III
50	1063.51	III
100	1067.76	III
100	1068.69	III
20	1070.54	III
200	1072.79	III
40	1073.69	III
30	1079.35	III
20	1080.82	III
30	1088.70	III
30	1090.03	III
20	1092.48	III
20	1099.25	III
40	1100.71	III
20	1107.09	III
20	1111.19	III
20	1116.30	III
20	1125.58	III
20	1129.73	III
20	1132.74	III
100	1142.55	III
20	1192.41	III
50	1201.87	III
20	1204.05	III
20	1719.43	III
20	1796.89	III
20	1836.66	III
20	1836.99	III
30	1862.32	III
30	1950.36	III
20	1990.54	III
Air		
100	2033.34	III
100	2057.65	III
100	2077.87	III
200	2083.32	III
100	2089.96	III
400	2109.07	III
100	2122.55	III
500	2136.95	III
1000	2151.44	III
3000	2166.88	III
2000	2169.48	III
5000	2180.64	III
1000	2183.71	III
2000	2203.15	III
3000	2218.11	III
5000	2222.01	III
5000	2225.08	III
3000	2227.84	III
3000	2228.05	III
5000	2242.29	III
2000	2249.25	III
2000	2264.85	III
2000	2268.20	III
2000	2287.82	III
2000	2298.70	III
3000	2300.65	III
4000	2302.09	III
5000	2317.34	III
10000	2318.64	III
5000	2324.31	III
2000	2337.66	III
5000	2350.10	III
2000	2362.54	III
2000	2367.77	III
10000	2372.34	III
5000	2377.07	III
5000	2377.48	III
10000	2380.12	III
3000	2382.28	III
3000	2385.06	III
5000	2395.04	III
3000	2406.15	III
4000	2408.08	III
2000	2410.26	III
5000	2415.60	III
2000	2417.01	III
2000	2423.02	III
3000	2428.64	III
5000	2430.24	III
10000	2431.45	III
15000	2439.80	III
3000	2441.55	III
2000	2444.78	III
10000	2454.32	III
10000	2469.95	III
3000	2471.66	III
5000	2477.25	III
8000	2479.44	III
3000	2479.51	III
10000	2483.82	III
10000	2497.50	III
3000	2503.56	III
2000	2504.43	III
20000	2531.99	III
3000	2539.27	III
3000	2557.49	III
4000	2577.67	III
2000	2578.30	III
2000	2584.71	III
10000	2603.59	III
2000	2607.96	III
2000	2615.79	III
2000	2649.38	III
2000	2662.81	III
3000	2719.30	III
2000	2730.04	III
3000	2743.71	III
4000	2748.90	III
4000	2754.87	III
4000	2768.28	III
3000	2849.40	III

Cerium (Cont.)

Intensity	Wavelength	Spectrum
2000	2861.39	III
4000	2907.05	III
10000	2923.81	III
5000	2925.26	III
10000	2931.54	III
2000	2948.53	III
5000	2973.72	III
10000	3022.75	III
50000	3031.58	III
95000	3055.59	III
20000	3056.56	III
40000	3057.23	III
20000	3057.58	III
40000	3085.10	III
20000	106.98	III
30000	3110.53	III
30000	3121.56	III
20000	3141.29	III
20000	3143.96	III
20000	3147.06	III
20000	3228.57	III
3000	3234.20	III
4000	3267.76	III
3000	3267.94	III
20000	3353.29	III
10000	3395.77	III
4000	3398.91	III
30000	3427.36	III
40000	3443.63	III
30000	3454.39	III
40000	3459.39	III
60000	3470.92	III
50000	3497.81	III
60000	3504.64	III
500	3514.41	III
50000	3544.07	III
3000	3784.29	III
800	3936.80	III
300	3957.10	III
500	4169.42	III
300	4191.70	III
500	4194.83	III
300	4213.26	III
300	4217.13	III
400	4284.77	III
300	4304.71	III
600	4346.35	III
400	4389.97	III
600	4448.32	III
500	4485.27	III
1000	4521.92	III
1000	4535.73	III
300	4576.90	III
500	4627.60	III
300	4766.07	III
500	4976.45	III
500	5650.97	III
1000	5664.20	III
500	5691.08	III
300	5710.59	III
500	5749.47	III
500	5949.83	III
2000	5962.22	III
500	5962.71	III
400	5979.56	III
1000	5983.40	III
3000	6002.63	III
10000	6032.54	III
10000	6060.91	III
500	6061.79	III
500	6097.35	III
500	6098.87	III
500	6135.10	III
300	6287.79	III
500	6308.16	III
300	6341.75	III
1000	6944.94	III
700	7739.04	III
300	7758.27	III
500	7826.80	III
300	7948.64	III
500	7960.31	III
500	7991.01	III
400	8030.80	III
300	8084.12	III
400	8177.33	III
300	8186.03	III
300	8222.16	III
300	9056.53	III
300	9079.58	III
400	9328.20	III
300	9367.03	III
300	9567.37	III
400	10458.37	III
400	10494.42	III
300	10534.36	III
400	10684.46	III
15	12756.96	III
12	12821.62	III
80	15847.58	III
80	15956.79	III
12	15960.59	III
87	16128.75	III
42	18579.82	III
38	19141.29	III
27	19377.15	III
26	19466.14	III
20	19498.14	III
55	19524.18	III
30	20685.63	III
12	21380.23	III

Ce IV
Ref. 166 — J.R.

Intensity	Wavelength	
	Vacuum	
2	447.58	IV
1	443.11	IV
8	558.92	IV
8	571.59	IV
40	741.79	IV
30	754.60	IV
12	755.75	IV
6	975.20	IV
5	1009.31	IV
2	1022.12	IV
9	1057.67	IV
1	1059.64	IV
50	1289.41	IV
75	1332.16	IV
75	1372.72	IV
2	1577.60	IV
1	1572.62	IV
15	1641.58	IV
20	1775.30	IV
20	1779.03	IV
35	1914.75	IV
10	1937.21	IV
	Air	
100	2000.42	IV
35	2003.11	IV
100	2009.94	IV
3	2433.50	IV
5	2445.50	IV

Ce V
Ref. 261 — J.R.

Intensity	Wavelength	
	Vacuum	
100	365.66	V
300	399.36	V
150	404.21	V
200	482.96	V
100	552.13	V

CESIUM (Cs)
Z = 55

Cs I and II

Ref. 82, 154, 155, 200, 263, 325—C.J.S.

Intensity		Wavelength	
		Vacuum	
250		591.04	II
2000		639.36	II
500		668.39	II
15000		718.14	II
15000		808.76	II
15000		813.84	II
35000		901.27	II
40000		926.66	II
3		1656.15	II
7		1689.46	II
2		1691.83	II
7		1717.64	II
7		1718.97	II
3		1727.79	II
3		1736.77	II
8		1807.83	II
8		1813.75	II
7		1815.16	II
18		1840.50	II
13		1859.16	II
7		1864.83	II
7		1876.72	II
18		1883.93	II
5		1914.61	II
29		1935.19	II
		Air	
72		2024.97	II
46		2028.32	II
180		2080.05	II
160		2087.20	II
100		2102.22	II
160		2205.52	II
190		2220.53	II
120		2245.81	II
1600		2267.65	II
750		2273.84	II
220		2285.41	II
250		2286.15	II
330		2315.69	II
1400		2332.46	II
240		2375.86	II
220		2379.27	II
1000		2392.86	II
370		2425.17	II
220		2543.93	II
630		2596.99	II
450		2609.43	II
170		2628.86	II
300		2699.18	II
180		2789.78	II
150		2793.31	II
680	w	2816.92	II
150		2829.04	II
140	c	2866.36	II
1600		2931.08	II
290		3151.19	II
520	c	3265.91	II
650	w	3267.12	II
850	c	3271.63	II
570		3368.57	II
150		3514.05	II
110	w	3615.01	II
170	w	3680.11	II
130		3732.56	II
1100	w	3785.44	II
2700		3805.12	II
520	c	3861.50	II
2100	c	3876.15	I
600	c	3888.61	I
3400		3896.99	II
4200		3959.51	II
2500		3965.20	II
2100		3974.25	II
8000		4039.85	II
2000		4068.78	II
320	w	4151.27	II
2000		4213.14	II
1900		4232.20	II
980		4234.41	II
14000		4264.70	II
18000	w	4277.13	II
5100		4288.38	II
1900		4300.65	II
7600		4363.30	II
3700		4373.04	II
960		4384.44	II
3900	w	4405.26	II
12000		4501.55	II
20000		4526.74	II
4100		4538.97	II
1000	c	4555.28	I
460	c	4593.17	I
100000		4603.79	II
4200	c	4616.17	II
2800		4646.52	II
990		4670.29	II
1500		4732.99	II
7000		4763.64	II
1900		4786.38	II
25000		4830.19	II
19000		4870.04	II
4900	c	4880.05	II
37000		4952.85	II
8200		4972.60	II
27000		5043.80	II
2800	c	5059.87	II
2900		5096.60	II
6500	c	5209.58	II
75000		5227.04	II
29000		5249.38	II
11000		5274.05	II
1300		5306.60	II
10000	c	5349.13	II
22000		5370.99	II
1200		5402.78	II
2900		5419.67	II
60	c	5465.94	I
37		5502.88	I
39000		5563.02	II
100		5635.21	I
210	c	5664.02	I
27		5745.72	I
4500		5814.16	II
24000		5831.14	II
59	c	5838.83	I
300		5845.14	I
51000		5925.63	II
1400	c	5984.39	II
640	c	6010.49	I
86		6034.09	I
760		6076.72	II
9800		6128.61	II
1000		6213.10	I
170		6217.60	I
320	c	6354.55	I
2000		6419.52	II
8300		6495.53	II
10000	w	6536.44	II
490		6586.51	I
530	c	6628.01	II
97		6628.66	I
8800		6646.57	II
3300	c	6723.28	I
9600		6724.47	II
200		6824.65	I
300		6870.45	I
37000		6955.50	II
4800		6973.30	I
16000		6979.67	II
980		6983.49	I
2300		7130.54	II
13000	w	7149.54	II
1600		7160.90	II
630		7188.37	II
1100		7206.04	II
790		7228.53	I
960		7248.88	II
130		7279.90	I
1100		7279.96	I
780		7369.36	II
550		7437.78	II
440	w	7523.39	II
2600	c	7608.90	I
910	w	7651.95	II
310	h	7746.98	II
2400		7852.52	II
3300		7943.88	I
22000		7997.44	II
2100		8012.98	II
3500		8015.73	I
8200	c	8047.13	II
1300		8078.50	II
510		8078.94	I
4500		8079.04	I
59000	cr	8521.13	I
680		8521.62	II
1700		8608.31	II
420	w	8695.60	II
15000	c	8761.41	I
840		8775.42	II
340	w	8857.39	II
61000	cr	8943.47	I
18000		9172.32	I
5200		9208.53	I
540	c	9212.36	II
3300	c	9220.75	II
310	c	9718.11	II
630	c	9932.91	II
1400		9994.79	II
19000		10024.36	I
4800		10123.41	I
26000		10123.60	I
3000	c	10176.02	II
2700		10379.66	II
1300		10480.93	II
4700		10504.51	II
1900	w	10807.88	II
610	w	11324.34	II
530	c	11496.56	II
410	c	11704.18	II
330	c	11797.93	II
710	c	11840.84	II
1100		12604.29	II
2000	c	12735.52	II
400	c	12746.34	II
850		13406.83	II
2900		13424.31	I
38000	c	13588.29	I
8400		13602.56	I
4200	c	13692.91	II
5700		13758.81	I
1400	c	13868.82	II
880	c	14482.21	II
55000	c	14694.91	I
350		14906.91	II
1900	c	15293.80	II
1600	c	15356.61	II
620		15445.47	II
940	c	15735.48	II
1100	c	16426.14	II
820		16535.63	I
1500		17012.32	I
340		18160.84	II
430		18179.35	II
710		18221.36	II
390		18222.40	II
510		18404.77	II
310		18407.23	II
590		18509.52	II
610		18509.85	II
100	c	18742.18	II
530	c	18921.76	II
150	c	18986.48	II
180		19924.85	II
390		20110.77	II
760		20138.47	I
200		20301.54	II
220	c	20443.87	II
92	w	21103.42	II
79	c	22344.98	II
78	c	22448.98	II
880		22811.86	I
1100		23037.98	I
3900		23344.47	I

Cesium (Cont.)

Intensity		Wavelength	
320	c	23408.41	II
180		24132.52	II
4400		24251.21	I
850		24374.96	I
120	c	24528.78	II
140		24810.89	II
170		25189.83	II
900		25220.37	II
240		25733.29	II
890	d	25763.51	II
500		25764.73	I
120		26448.83	II
340		26503.26	II
190		26727.33	II
17		26954.56	II
22		26978.68	II
15		27107.98	II
9		28649.38	II
680	c	29310.06	I
7		29542.08	II
2800		30103.27	I
9		30487.16	II
610	c	30953.06	I
1100		34900.13	I
190		36131.00	I
2	c	39177.28	I
2	d	39421.25	I
1		39424.11	I

Cs III
Ref. 78, 200, 201—C.J.S.

Intensity		Wavelength	
		Vacuum	
10000		614.01	III
2000		638.17	III
2500		666.25	III
5000		691.60	III
3500		703.89	III
20000		721.79	III
20000		722.20	III
5000		731.56	III
12000		740.29	III
7500		830.39	III
15000		920.35	III
25000	c	1054.79	III
17	c	1673.99	III
12	c	1705.25	III
10		1801.83	III
20	c	1822.40	III
11		1823.93	III
12		1824.70	III
12		1841.80	III
25		1915.50	III
25	c	1923.29	III
12		1961.33	III
17		1996.56	III
		Air	
710		2035.11	III
120		2056.43	III
330		2076.43	III
540		2077.30	III
410		2088.68	III
210		2101.63	III
200		2141.47	III
1000		2316.88	III
230		2325.95	III
390		2340.49	III
1600		2455.81	III
1600		2477.57	III
890		2485.45	III
410		2495.07	III
1400		2525.67	III
430		2573.05	III
16000		2596.86	III
390		2610.12	III
6200		2630.51	III
370		2700.32	III
710		2701.20	III
390		2776.44	III
270		2810.87	III
630		2845.70	III
3100		2859.32	III
200		2893.85	III
180		2921.13	III
3200		2976.86	III
210		3001.28	III
1700		3066.59	III
1100	c	3149.36	III
1400		3152.36	III
8400		3268.32	III
1300		3315.51	III
550		3340.60	III
430		3344.02	III
1200		3349.46	III
400		3463.45	III
580		3476.83	III
480		3559.82	III
7200		3597.45	III
1300		3608.31	III
2300		3618.19	III
300	c	3641.34	III
520		3651.08	III
4800		3661.40	III
640		3699.50	III
430		3837.46	III
2900		3888.37	III
2700		3925.60	III
680	c	4001.70	III
3100		4006.55	III
420		4006.78	III
520		4043.42	III
370		4403.86	III
1200		4410.22	III
940		4425.68	III
530		4471.48	III
1200		4506.72	III
590		4522.86	III
420	h	4620.61	III
210		4663.32	III
140		4851.59	III
370		5035.72	III
230		5380.79	III
140		5950.14	III
110		5979.97	III
150		6043.99	III
870		6079.86	III
330		6150.42	III
450		6242.96	III
510		6456.33	III
400		6753.12	III
1900	c	7219.60	III

Cs IV
Ref. 259 — J.R.

Intensity		Wavelength	
		Vacuum	
35		707.20	IV
60		759.57	IV
5		778.21	IV
500		824.80	IV
350		828.86	IV
400		868.18	IV
1000		874.84	IV
400		896.92	IV
400		923.02	IV
600		986.14	IV
300		995.14	IV
60		1019.13	IV
550		1068.91	IV
200		1282.66	IV

CHLORINE (Cl)
Z = 17

Cl I and II
Ref 238, 239 — L.J.R.

Intensity		Wavelength	
		Vacuum	
350		559.305	II
400		571.904	II
800		574.406	II
500		586.24	II
700		618.057	II
600		619.982	II
800		620.298	II
700		626.735	II
800		635.881	II
1000		636.626	II
1000		650.894	II
1000		659.811	II
1300		661.841	II
2000		663.074	II
1500		682.053	II
1500		687.656	II
1500		693.594	II
2000		725.271	II
2500		728.951	II
2000		777.562	II
5000		787.580	II
5000		788.740	II
5000		793.342	II
6000		839.297	II
8000		839.599	II
7000	p	841.41	II
5000		851.691	II
2000		888.026	II
2000		893.549	II
2000		961.499	II
30		969.92	I
40		978.284	I
25		998.372	I
25		998.432	I
75		1002.346	I
150		1013.664	I
90		1025.553	I
6000		1063.831	II
3000		1067.945	II
9000		1071.036	II
6000		1071.767	II
5000		1075.230	II
5000		1079.080	II
200		1084.667	I
200		1085.171	I
250		1085.304	I
400		1088.06	I
350		1090.271	I
250		1090.982	I
250		1092.437	I
400		1094.769	I
350		1095.148	I
350		1095.662	I
400		1095.797	I
250		1096.810	I
300		1097.369	I
200		1098.068	I
200		1099.523	I
500		1107.528	I
800		1139.214	II
800		1167.148	I
3000		1179.293	I
1200		1188.774	I
900		1201.353	I
3000		1335.726	I
10000		1347.240	I
5000		1351.657	I
12000		1363.447	I
2500		1373.116	I
20000		1379.528	I
25000		1389.693	I
20000		1389.957	I
12000		1396.527	I
500		1441.470	II
500		1528.569	II
500		1542.942	II
500		1558.144	II
500		1565.050	II
500		1857.488	II
450	h	1997.370	II
		Air	
450		2032.116	II
350	h	2088.583	II
350	h	2091.458	II
170		2427.79	II
360		2434.07	II
340		2498.53	II
470		2502.74	II
260		2546.96	II
500		2549.88	II
460		2564.84	II
320		2603.31	II
950		2658.72	II
750		2676.95	II
1200		2688.04	II
410		2912.05	II
950		2996.65	II
500		3006.06	II
950		3057.96	II
1300		3071.32	II
1400		3092.19	II
1200		3123.72	II
1900		3315.43	II
1200		3329.10	II
2500		3353.35	II
20		3726.54	I
1200		3749.96	II
1000		3781.17	II
1500		3798.76	II
1900		3805.18	II
1300		3809.46	II
1700		3820.20	II
2800		3827.59	II
4500		3833.35	II
2500		3843.20	II
3100		3845.37	II
3900		3845.65	II
1500		3845.80	II
10000		3850.99	II
7900		3851.37	II
1200		3851.65	II
25000		3860.83	II
4400		3860.99	II
1000		3861.37	II
1500		3913.87	II
1100		3916.63	II
20		3944.82	I
20		4104.79	I
10000	h	4132.50	II
65		4209.67	I
50		4226.42	I
60		4264.58	I
100		4363.27	I
100		4369.50	I
5000		4372.93	II
100		4379.90	I
100		4389.75	I
90		4390.40	I
90		4403.03	I
100		4438.49	I
90		4475.30	I
1500		4489.91	II
100		4526.19	I
80		4600.98	I
40		4623.938	I
50		4654.040	I
80		4661.208	I
45		4691.523	I
40		4721.255	I
45		4740.729	I
4300		4768.65	II
13000		4781.32	II
99000		4794.55	II
29000		4810.06	II
16000		4819.47	II
81000		4896.77	II
47000		4904.78	II
26000		4917.73	II
10000		4995.48	II
26000		5078.26	II
30		5099.789	I
56000		5217.94	II
23000		5221.36	II
15000		5392.12	II
99000		5423.23	II
10000		5423.51	II
19000		5443.37	II
10000		5444.21	II
5600		5457.02	II
40		5532.162	I
50	d	5796.305	I
45		5799.914	I
30		5856.742	I
100	d	5948.58	I
50		6019.812	I
35		6082.61	I
1900		6094.69	II
160		6114.43	I
200		6140.245	I
160		6194.757	I
160		6398.66	I
150		6434.833	I
150		6531.43	I
1400		6661.67	II
150		6678.43	I
1300		6686.02	II
1200		6713.41	II
150		6840.29	I
300		6932.903	I
300		6981.886	I
600		7086.814	I
7500		7256.62	I
5000		7414.11	I
550		7462.370	I
550		7489.47	I
700		7492.118	I
11000		7547.072	I
2300		7672.42	I
450		7702.828	I
7000		7717.581	I
10000		7744.97	I
2200		7769.16	I
650		7771.09	I
2200		7821.36	I
1700		7830.75	I
3000		7878.22	I
220		7893.34	I
2300		7899.31	I
1800		7915.08	I
3000		7924.645	I
2100		7933.89	I
1700		7935.012	I
650		7952.52	I
1500		7974.72	I
1300		7976.97	I
600		7980.60	I
2900		7997.85	I
2200		8015.61	I
1100		8023.33	I
400		8051.07	I
1700		8084.51	I
2200		8085.56	I
3000		8086.67	I
1300		8087.73	I
250		8094.67	I
2500		8194.42	I
2200		8199.13	I
2200		8200.21	I
800		8203.78	I
18000		8212.04	I
3000		8220.45	I
20000		8221.74	I
18000		8333.31	I
1000		8360.71	II
560		8361.84	II
99900		8375.94	I
180		8382.67	II
100		8392.02	II
400		8406.199	I
15000		8428.25	I
2200		8467.34	I
2200		8550.44	I
20000		8575.24	I
750		8578.02	I
75000		8585.97	I
450		8628.54	I
300		8641.71	I
3500		8686.26	I
2200		8912.92	I
3000		8948.06	I
2000		9038.982	I
2500		9045.43	I
1000		9069.656	I
2000		9073.17	I
7500		9121.15	I

Chlorine (Cont.)

3000	9191.731	I
500	9197.596	I
4000	9288.86	I
1500	9393.862	I
3500	9452.10	I
500	9486.964	I
1000	9584.801	I
3500	9592.22	I
250	9632.509	I
1000	9702.439	I
250	9744.426	I
200	9807.057	I
400	9875.970	I
331	10392.549	I
38	10432.83	II
10	10506.62	II
14	10509.12	II
19	10512.46	II
25	10514.17	II
9	10801.47	II
5	10885.42	II
1	10955.71	II
300	11123.05	I
231	11392.62	I
269	11409.69	I
1000	11436.33	I
180	11720.56	I
195	11866.76	I
172	12021.7	I
350	13243.8	I
310	13296.0	I
550	13346.8	I
525	13821.7	I
148	14369.7	I
294	14931.7	I
269	15108.0	I
381	15465.1	I
169	15467.6	I
1094	15520.3	I
1487	15730.1	I
193	15818.4	I
2780	15869.7	I
277	15883.3	I
342	15928.9	I
735	15960.0	I
283	15970.5	I
129	16077.6	I
259	16198.5	I
227	19370.3	I
717	19755.3	I
185	19766.8	I
227	20199.4	I
85	20370.1	I
100	24470.0	I
	39603.7	I
	39615.3	I
	39716.0	I
	39744.0	I
	39750.9	I
	39875.3	I
	39881.0	I
	39985.7	I
	40085.5	I
	40089.5	I
	40171.0	I
	40310.3	I
	40335.4	I
	40532.2	I

Cl III
Ref. 28, 30 — L.J.R.

Intensity	Wavelength	
	Vacuum	
100	406.27	III
400	411.37	III
400	411.81	III
600	556.23	III
700	556.61	III
700	557.12	III
600	2965.56	III
600	3104.46	III
800	3139.34	III
900	3191.45	III
700	3289.80	III
700	3320.57	III
800	3329.06	III
900	3340.42	III
800	3392.89	III
800	3393.45	III
900	3530.03	III
800	3560.68	III
900	3602.10	III
800	3612.85	III
700	3622.69	III
700	3656.95	III
700	3670.28	III
700	3682.05	III
600	3705.45	III
600	3707.34	III
800	3720.45	III
800	3748.81	III
500	3779.35	III
500	3925.87	III
700	3991.50	III
600	4018.50	III
600	4059.07	III
500	4104.23	III
500	4106.83	III
400	4370.91	III
500	4608.21	III
300	4703.14	III
100	4863.75	III
10	4971.64	III
700	561.53	III
700	561.68	III
700	561.74	III
500	606.35	III
400	621.28	III
300	670.38	III
300	673.13	III
100	936.28	III
500	1005.28	III
600	1008.78	III
700	1015.02	III
600	1822.50	III
500	1828.40	III
500	1901.61	III
500	1983.61	III
	Air	
400	2006.84	III
700	2253.07	III
500	2268.95	III
500	2278.34	III
700	2283.93	III
600	2323.50	III
500	2336.45	III
600	2340.64	III
600	2359.67	III
600	2370.37	III
700	2416.42	III
600	2447.14	III
600	2448.58	III
500	2486.91	III
500	2532.48	III
600	2580.67	III
500	2603.59	III
500	2632.67	III
500	2633.18	III
600	2665.54	III
700	2710.37	III

Cl IV
Ref. 11, 28, 30, 31 — L.J.R.

Intensity	Wavelength	
	Vacuum	
300	319.62	IV
200	331.84	IV
400	437.83	IV
400	464.86	IV
800	486.17	IV
800	534.73	IV
700	535.67	IV
600	536.15	IV
900	537.61	IV
600	538.12	IV
500	549.22	IV
400	550.02	IV
700	552.02	IV
600	553.30	IV
700	554.62	IV
500	601.50	IV
500	604.59	IV
400	608.90	IV
400	612.07	IV
400	653.70	IV
400	745.21	IV
400	831.43	IV
500	834.84	IV
500	834.97	IV
400	840.81	IV
600	840.93	IV
	865.3	IV
500	973.21	IV
600	977.56	IV
400	977.90	IV
700	984.95	IV
400	985.75	IV
300	1537.21	IV
200	1539.30	IV
200	1545.19	IV
200	1549.15	IV
200	1622.86	IV
	Air	
400	2701.36	IV
500	2724.03	IV
500	2751.23	IV
400	2770.64	IV
700	2782.47	IV
400	2835.4	IV
500	3063.13	IV
600	3076.68	IV
200	3167.87	IV

Cl V
Ref. 11, 28, 30, 85, 233 — L.J.R.

Intensity	Wavelength	
	Vacuum	
300	287.33	V
300	373.78	V
400	390.15	V
500	392.43	V
300	536.53	V
400	537.01	V
300	537.46	V
500	538.03	V
400	538.68	V
800	542.23	V
600	542.30	V
400	542.87	V
1000	545.11	V
600	546.33	V
1000	547.63	V
400	633.19	V
400	635.32	V
400	681.92	V
400	683.17	V
400	688.93	V
	715.55	V
	716.19	V
400	883.13	V
400	894.34	V
100	894.91	V
	914.5	V

CHROMIUM (Cr)
Z = 24

Cr I and II
Ref. 1 — C.H.C.

Intensity		Wavelength	
		Vacuum	
19000		2055.52	II
14000		2061.49	II
8900		2065.42	II
80	h	2364.71	I
130		2383.33	I
140		2408.62	I
170		2496.31	I
110		2502.53	I
190		2504.31	I
50		2508.11	I
60		2508.98	I
40		2513.62	I
110		2516.92	I
80		2518.71	I
390		2519.52	I
190		2527.12	I
40		2530.45	I
70		2534.34	II
50		2545.64	I
160		2549.54	I
40		2553.06	I
80		2557.15	I
130		2560.69	I
150		2571.74	I
100		2577.65	I
50		2588.20	I
380		2591.85	I
35		2603.57	I
35		2622.86	I
22		2625.32	I
18		2626.60	I
18		2629.82	I
35		2642.12	I
250		2653.59	II
250		2658.59	II
70		2661.73	II
320		2663.42	II
70		2663.68	II
440		2666.02	II
280		2668.71	II
350		2671.81	II
280		2672.83	II
1800		2677.16	II
35		2678.16	I
320		2678.79	II
18		2680.34	II
230		2687.09	II
60		2688.04	I
55		2688.29	II
26		2690.26	I
280		2691.04	II
35		2693.52	II
35		2697.91	II
180		2698.41	II
180		2698.69	II
18		2700.60	I
110		2701.99	I
18		2702.53	I
70		2703.48	I
		2703.55	II
35		2703.86	II
18		2705.43	I
60		2708.79	II
35		2709.31	II
140		2712.31	II
45		2716.18	I
55		2717.51	II
45		2718.43	II
170		2722.75	II
18		2724.04	I
420	h	2726.51	I
45		2727.26	II
280	h	2731.91	I
170	h	2736.47	I
70		2739.38	I
70		2740.10	II
95		2741.07	I
95		2742.03	II
95		2742.17	I
250		2743.64	II
35		2746.21	II
110	h	2748.29	I
330		2748.98	II
390		2750.73	II
45		2751.60	I
280		2751.87	II
110	h	2752.88	I
35		2754.28	II
22		2754.90	I
22		2755.27	I
22		2756.75	I
150		2757.10	I
350		2757.72	II
60		2758.98	II
80		2759.39	II
45		2759.73	II
90	h	2761.76	I
750		2762.59	II
22		2763.06	I
80	h	2764.35	I
750		2766.54	II
22		2767.54	I
250	h	2769.92	I
18		2771.45	I
45		2778.06	II
22		2779.14	I
80		2780.30	II
610		2780.70	I
70		2785.70	II
35		2787.63	II
35		2787.84	I
90		2792.16	II
55		2798.67	II
70		2800.77	II
80		2812.01	II
60		2818.36	II
45		2822.01	II
180		2822.37	II
22		2826.75	I
180		2830.47	II
70		2834.26	II
2500		2835.63	II
45		2836.48	II
55		2838.79	II
110		2840.02	II
1700		2843.25	II
22		2846.02	I
45		2849.29	I
1200		2849.84	II
120		2851.36	II
55		2853.22	II
55		2855.07	II
880		2855.68	II
90		2856.77	II
70		2857.40	II
610		2858.91	II
440		2860.93	II
790		2862.57	II
750		2865.11	II
55		2865.33	II
610		2866.74	II
90		2867.10	II
480		2867.65	II
210		2870.44	II
110		2871.63	I
160		2873.48	II
90		2873.82	II
320		2875.99	II
230		2876.24	II
180		2877.98	II
70		2878.45	II
120		2879.27	I
95		2880.87	II
30		2881.14	I
170		2887.00	I
55		2888.74	II
700		2889.29	I
55		2889.82	II
55		2891.42	I
370		2893.25	I
190		2894.17	I
55		2896.46	II

Chromium (Cont.)

Intensity		Wavelength	Spectrum
210		2896.75	I
55	d	2897.67	II
		2897.73	II
90		2898.54	II
80		2899.21	I
55		2899.48	II
26		2903.97	II
55		2904.68	I
180		2905.49	I
260		2909.05	I
260		2910.90	I
250		2911.14	I
45		2911.68	II
60		2913.73	I
22		2915.23	II
22		2915.46	II
80		[illegible]	II
60		2921.82	II
60		2927.08	II
80		2928.15	II
95		2928.30	II
26		2929.44	II
35		2930.85	II
26		2932.70	II
55		2933.97	II
90		2935.14	II
45		2940.22	II
60		2946.84	II
55		2953.36	II
45		2953.71	II
55		2961.73	II
45		2966.05	II
480		2967.64	I
480		2971.11	I
210		2971.91	II
480		2975.48	I
30		2976.72	II
190		2979.74	II
350		2980.79	I
110		2985.32	II
480		2985.85	I
1500		2986.00	I
2100		2986.47	I
660		2988.65	I
160		2989.19	II
480		2991.89	I
230		2994.07	I
300		2995.10	I
700		2996.58	I
210		2998.79	I
1100		3000.89	I
750		3005.06	I
140		3013.03	I
710		3013.71	I
710		3014.76	I
1400		3014.92	I
710		3015.19	I
2800		3017.57	I
430		3018.50	I
240		3018.82	I
430		3020.67	I
2800		3021.56	I
1100		3024.35	I
85		3026.65	II
170		3029.16	I
710		3030.24	I
140		3031.35	I
28		3032.93	II
390		3034.19	I
550		3037.04	I
80		3039.78	I
550		3040.85	I
		3040.91	II
55		3041.74	II
110		3050.14	II
710		3053.88	I
24		3059.52	II
85		3065.07	I
28		3067.16	II
85		3073.68	I
55		3077.83	I
28		3095.86	I
28		3109.34	I
28		3110.86	I
240		3118.65	II
45		3119.25	I
40		3119.71	I
430		3120.37	II
28		3122.60	II
470		3124.94	II
		3125.02	II
120		3128.70	II
590		3132.06	II
140		3136.68	II
140		3147.23	II
85		3148.44	I
100		3155.15	I
100		3163.76	I
240		3180.70	II
30		3181.43	II
65	h	3188.01	I
220		3197.08	II
24		3198.11	I
30		3208.59	II
170		3209.18	II
140		3217.40	II
30		3229.20	I
28		3234.06	II
65		3237.73	I
120		3245.54	I
130		3251.84	I
130		3257.82	I
95		3259.98	I
30		3295.43	II
24		3307.02	II
55		3324.06	II
28		3326.59	I
30		3328.35	II
30		3329.05	I
33		3330.33	II
130		3339.80	II
110		3342.59	II
30		3343.34	I
95		3346.02	I
95		3346.74	I
95		3347.84	II
65		3349.07	I
55		3349.32	I
30		3351.60	I
55		3351.97	I
55	h	3353.03	I
		3353.13	II
170		3358.50	II
160		3360.30	II
65		3361.77	II
55		3362.21	I
430		3368.05	II
30		3376.40	I
55		3378.34	II
30		3379.17	I
30		3379.37	II
95		3379.83	II
140		3382.68	II
95		3391.43	II
55		3392.99	II
70		3393.84	II
55		3394.30	II
30		3402.40	II
170		3403.32	II
360		3408.76	II
210		3421.21	II
270		3422.74	II
140		3433.31	II
270		3433.60	I
55		3434.11	I
160		3436.19	I
70		3441.12	I
140		3441.44	I
30		3443.79	I
170		3445.62	I
30		3447.02	I
170		3447.43	I
70		3447.76	I
190		3453.33	I
40		3453.74	I
130		3455.60	I
100		3460.43	I
65		3465.25	I
40		3467.02	I
70		3467.72	I
45		3469.59	I
16		3472.76	I
24		3472.91	I
40		3473.61	I
70		3481.30	I
55		3481.54	I
55		3494.97	I
40		3495.38	II
80		3510.54	I
40		3511.84	II
120		3550.64	I
80		3558.52	I
130		3566.16	I
130		3573.64	I
80		3574.04	I
330	h	3574.80	I
		3574.94	I
19000		3578.69	I
160	h	3584.33	I
130		3585.30	II
17000		3593.49	I
350		3601.67	I
40		3602.57	I
85		3603.74	I
		3603.78	II
13000		3605.33	I
40		3608.40	I
40		3609.48	I
40		3610.05	I
70		3612.61	I
85		3615.64	I
130		3632.84	I
350		3636.59	I
630		3639.80	I
85		3640.39	I
70		3641.47	I
220		3641.83	I
45		3646.16	I
85		3648.53	I
220		3649.00	I
170		3653.91	I
220		3656.26	I
45		3662.84	I
130		3663.21	I
45		3665.98	I
95		3666.64	I
55		3668.03	I
65		3676.32	I
40		3677.68	II
55		3677.89	II
40		3679.82	I
19		3681.69	I
120		3685.55	I
130		3686.80	I
130		3687.25	I
75		3687.54	I
19		3688.46	I
75		3712.95	II
40		3716.53	I
130		3730.81	I
150		3732.03	I
95		3742.97	I
480		3743.58	I
570		3743.88	I
85		3744.49	I
55		3748.61	I
340		3749.00	I
50		3757.17	I
230		3757.66	I
60		3758.04	I
24		3767.43	I
260		3768.24	I
95		3768.73	I
95		3788.86	I
95		3790.45	I
130		3791.38	I
130		3792.14	I
120		3793.29	I
130		3793.88	I
85		3794.61	I
140		3797.13	I
200		3797.72	I
530		3804.80	I
110		3806.83	I
110		3807.93	I
180		3815.43	I
70		38[illegible].40	I
180		3819.56	I
70		3823.52	I
130		3826.42	I
130		3830.03	I
380		3841.28	I
190		3848.98	I
140		3849.36	I
290		3850.04	I
140		3852.22	I
190		3854.22	I
110		3855.29	I
140		3855.57	I
260		3857.63	I
70		3874.53	I
660		3883.29	I
50		3883.66	I
570		3885.22	I
380		3886.79	I
60		3891.93	I
260		3894.04	I
40		3897.65	I
35		3902.11	I
360		3902.92	I
60		3903.16	I
960		3908.76	I
120	Hd	3911.82	I
		3912.00	I
120		3915.84	I
190		3916.24	I
35		3917.60	I
1900		3919.16	I
600		3921.02	I
30		3926.65	I
600		3928.64	I
410		3941.49	I
30		3951.10	I
40		3952.40	I
35		3953.16	I
1900		3963.69	I
120		3969.06	I
1600		3969.75	I
85		3971.26	I
1600		3976.66	I
85		3978.68	I
40		3979.80	I
85		3981.23	I
960		3983.91	I
190		3984.34	I
160		3989.99	I
960		3991.12	I
160		3991.67	I
190		3992.84	I
40		3993.97	I
160		4001.44	I
120		4012.47	II
30		4014.67	I
85		4022.26	I
70		4025.01	I
120		4026.17	I
85		4027.10	I
85		4030.68	I
190		4039.10	I
160		4048.78	I
120		4058.77	I
40		4065.72	I
85		4066.94	I
35		4074.86	I
60		4076.06	I
40		4077.09	I
40		4077.68	I
40		4104.87	I
40		4109.58	I
40		4120.61	I
40		4121.82	I
35		4122.16	I
40		4123.39	I
140		4126.52	I
35		4127.30	I
40		4127.64	I
40		4131.36	I
30		4152.78	I
120		4153.82	I
85		4161.42	I
140		4163.62	I
70		4165.52	I
40		4169.84	I
35		4170.20	I
40		4172.77	I
170		4174.80	I
30		4175.94	I
170		4179.26	I
35		4184.90	I
30		4186.36	I
35		4190.13	I
85		4191.27	I
35		4192.10	I
85		4193.66	I
70		4194.95	I
40		4197.23	I
85		4198.52	I
60		4203.59	I
40		4204.47	I
35		4208.36	I
110		4209.37	I
40		4209.76	I
40		4211.35	I
40		4216.36	I
85		4217.63	I
40		4221.57	I
40		4222.73	I
40		4238.96	I
60		4240.70	I
20000		4254.35	I
70		4255.50	I
60		4261.35	I
110		4263.14	I
30		4271.06	I
40		4272.91	I
16000		4274.80	I
85		4280.40	I
10000		4289.72	I
40		4291.96	I
85		4295.76	I
70		4297.74	I
35		4300.51	I
50		4301.18	I
30		4305.45	I
35		4319.64	I
60		4325.08	I
780		4337.57	I
1100		4339.45	I
380		4339.72	I
60		4340.13	I
1900		4344.51	I
70		4346.83	I
380		4351.05	I
2300		4351.77	I
570		4359.63	I
70		4363.13	I
530		4371.28	I
70		4373.25	I
110		4374.16	I
70		4375.33	I
50		4381.11	I
530		4384.98	I
60		4387.50	I
70		4391.75	I
60		4403.50	I
24		4410.30	I
60		4411.09	I
35		4412.25	I
50		4413.87	I
60		4424.28	I
24		4428.50	I
50		4430.49	I

Chromium (Cont.)

50		4432.18	I
110		4458.54	I
30		4459.74	I
30		4465.36	I
30		4482.88	I
40		4488.05	I
50		4489.47	I
60		4492.31	I
660		4496.86	I
50		4498.73	I
70		4500.30	I
50		4501.11	I
22		4501.79	I
24		4506.85	I
95		4511.90	I
12		4514.37	I
35		4514.53	I
24		4521.14	I
24		4526.11	I
380		4526.47	I
70	d	4527.34	I
		4527.47	I
24		4529.85	I
380		4530.74	I
50		4535.15	I
240		4535.72	I
40		4539.79	I
240		4540.50	I
240		4540.72	I
35		4541.07	I
19		4541.51	I
24		4542.62	I
140		4544.62	I
24		4545.34	I
600		4545.96	I
50		4556.17	I
22		4558.66	II
19		4564.17	I
120		4565.51	I
95		4569.64	I
120		4571.68	I
22		4575.12	I
360		4580.06	I
24		4586.14	I
360		4591.39	I
70		4595.59	I
50		4600.10	I
480		4600.75	I
50		4601.02	I
240		4613.37	I
600		4616.14	I
70		4619.55	I
85		4621.96	I
70		4622.49	I
24		4622.76	I
550		4626.19	I
24		4632.18	I
40		4637.18	I
50		4637.77	I
50	d	4639.52	I
		4639.70	I
1600		4646.17	I
24		4646.81	I
24		4648.13	I
24		4648.87	I
35		4649.46	I
570		4651.28	I
840		4652.16	I
35		4654.74	I
19		4656.19	I
40		4663.33	I
70		4663.83	I
95		4664.80	I
35		4665.90	I
22		4666.22	I
70		4666.51	I
50		4669.34	I
40		4680.54	I
19		4680.87	I
70		4689.37	I
60		4693.95	I
24		4695.15	I
60		4697.06	I
240	d	4698.46	I
		4698.62	CR I
35		4700.61	I
190		4708.04	I
240		4718.43	I
50		4723.10	I
50		4724.42	I
50		4727.15	I
24		4729.72	I
120		4730.71	I
140		4737.35	I
19		4745.31	I
70		4752.08	I
340		4756.11	I
50		4764.29	I
22		4766.63	I
30		4767.86	I
190		4789.32	I
95		4792.51	I
120		4801.03	I
110		4829.38	I
14		4836.86	I
17		4861.20	I
70		4861.84	I
140		4870.80	I
35		4885.78	I
19		4885.96	I
130		4887.01	I
19		4888.53	I
35		4903.24	I
260		4922.27	I
110		4936.33	I
70		4942.50	I
110		4954.81	I
35		4964.93	I
60		5013.32	I
17		5051.90	I
17		5065.91	I
40		5067.71	I
40		5072.92	I
30		5110.75	I
17		5113.13	I
17		5123.46	I
50		5139.65	I
14		5144.67	I
70		5166.23	I
35		5177.43	I
70		5184.59	I
70		5192.00	I
12		5193.49	I
85		5196.44	I
35		5200.19	I
5300		5204.52	I
8400		5206.04	I
11000		5208.44	I
19		5214.13	I
30		5221.75	I
85		5224.94	I
12		5226.89	I
19		5238.97	I
30		5243.40	I
290		5247.56	I
60		5254.92	I
60		5255.13	I
19		5261.75	I
530		5264.15	I
30		5265.16	I
180		5265.72	I
35		5272.01	I
30		5273.44	I
95	h	5275.17	I
35	h	5275.69	I
70	h	5276.03	I
19		5280.29	I
10		5287.19	I
340		5296.69	I
70	h	5297.36	I
660		5298.27	I
85		5300.75	I
17		5304.21	I
24		5312.88	I
24		5318.78	I
340	h	5328.34	I
70	h	5329.17	I
17	h	5329.72	I
14		5340.44	I
10		5344.76	I
780		5345.81	I
380		5348.32	I
30		5386.98	I
22		5387.57	I
10		5390.39	I
40		5400.61	I
22		5405.00	I
1400		5409.79	I
12		5442.41	I
19		5463.97	I
19		5480.50	I
24		5628.64	I
7		5642.36	I
12	h	5649.37	I
24		5664.04	I
7	h	5681.20	I
7	h	5682.48	I
24		5694.73	I
40		5698.33	I
24		5702.31	I
12		5712.64	I
24		5712.78	I
7		5719.82	I
7		5746.43	I
7		5753.69	I
12	h	5781.20	I
6	h	5781.81	I
24	h	5783.11	I
30	h	5783.93	I
24	h	5785.00	I
19	h	5785.82	I
60	h	5787.99	I
180	h	5791.00	I
35		6330.10	I
22		6362.87	I
19		6661.08	I
11		6669.26	I
5	h	6881.62	I
10	h	6882.38	I
21	h	6883.03	I
27	h	6924.13	I
17	h	6925.20	I
30	h	6978.48	I
11	h	6979.82	I
7		7185.52	I
6	h	7236.20	I
85		7355.90	I
130		7400.21	I
150		7462.31	I
11	h	7942.04	I
5	h	8163.18	I
9		8348.28	I
6		8450.26	I
3		8455.24	I
6		8548.86	I
40		8947.15	I
19		8976.83	I

Cr III
Ref. 412 — C.H.C.

Intensity	Wavelength	
	Vacuum	
20	969.26	III
40	1000.86	III
40	1001.04	III
30	1002.96	III
50	1017.14	III
50	1017.31	III
50	1017.57	III
30	1028.33	III
60	1030.47	III
30	1030.89	III
50	1033.23	III
50	1033.45	III
100	1033.69	III
50	1035.93	III
100	1036.03	III
30	1040.17	III
40	1040.53	III
40	1045.06	III
40	1045.14	III
60	1059.13	III
60	1060.15	III
60	1061.04	III
50	1062.68	III
30	1064.32	III
30	1064.43	III
50	1066.23	III
80	1068.41	III
30	1100.61	III
30	1101.43	III
30	1102.88	III
30	1117.19	III
30	1132.75	III
50	1136.67	III
50	1161.43	III
30	1187.65	III
60	1206.38	III
80	1209.13	III
80	1211.12	III
40	1221.07	III
40	1221.90	III
30	1225.65	III
30	1228.65	III
30	1231.88	III
50	1232.96	III
40	1236.20	III
40	1238.51	III
50	1252.61	III
40	1259.02	III
40	1261.86	III
30	1262.34	III
35	1263.61	III
35	1264.21	III
40	1287.05	III
30	1455.27	III
40	1584.60	III
30	1603.19	III
30	1679.25	III
30	1690.28	III
60	1692.89	III
60	1696.64	III
60	1701.48	III
80	1707.43	III
40	1707.78	III
45	1762.81	III
30	1766.92	III
30	1769.17	III
30	1827.26	III
	AIR	
60	2036.39	III
50	2039.63	III
80	2047.23	III
100	2113.73	III
100	2113.83	III
50	2114.26	III
50	2114.53	III
100	2114.87	III
100	2117.53	III
80	2123.53	III
80	2139.11	III
100	2141.15	III
80	2144.15	III
50	2147.16	III
50	2147.56	III
50	2148.65	III
50	2149.48	III
50	2152.76	III
100	2157.17	III
50	2163.86	III
60	2166.25	III
100	2170.70	III
50	2183.71	III
100	2185.01	III
50	2190.09	III
100	2190.76	III
100	2191.58	III
100	2197.89	III
100	2198.62	III
100	2203.22	III
60	2208.70	III
200	2226.72	III
100	2231.81	III
100	2233.81	III
200	2235.91	III
150	2237.59	III
150	2244.10	III
80	2251.45	III
50	2257.92	III
100	2273.30	III
80	2275.43	III
100	2276.38	III
80	2277.47	III
150	2284.44	III
50	2289.23	III
80	2290.66	III
60	2295.55	III
50	2309.99	III
80	2314.63	III
100	2319.07	III
150	2324.88	III
60	2340.51	III
50	2456.83	III
100	2472.88	III
100	2479.77	III
100	2483.06	III
60	2488.26	III
80	2506.41	III
80	2530.99	III
80	2537.73	III
80	2544.37	III
50	2545.17	III
80	2564.76	III
80	2616.50	III
100	2626.08	III
100	2640.73	III
50	2647.50	III
40	2655.28	III
40	2916.57	III

Cr IV
Ref. 379, 412 — C.H.C.

Intensity	Wavelength	
	Vacuum	
50	575.05	IV
30	576.24	IV
30	576.62	IV
30	595.09	IV
50	612.64	IV
40	613.75	IV
40	614.03	IV
40	614.90	IV
30	615.34	IV
30	615.60	IV
50	616.82	IV
40	618.23	IV
40	619.13	IV
100	620.66	IV
60	621.36	IV
40	622.09	IV
30	623.54	IV
40	625.04	IV
40	625.99	IV
100	629.26	IV
50	629.74	IV
80	630.30	IV
30	632.62	IV
30	637.34	IV
50	637.55	IV
50	638.13	IV

Chromium (Cont.)

30		638.54	IV
100		666.55	IV
75		667.30	IV
40		677.55	IV
40		687.12	IV
50		688.46	IV
100		693.92	IV
50		695.21	IV
50		705.98	IV
30		712.90	IV
80		1055.89	IV
60		1057.85	IV
30		1367.39	IV
40		1375.05	IV
70		1401.82	IV
100		1417.42	IV
30		1485.05	IV
80		1595.04	IV
90	d	1595.59	IV
100		1658.08	IV
120		1672.66	IV
90		1686.07	IV
100		1690.88	IV
80		1725.26	IV
90		1727.07	IV
100		1732.04	IV
40		1733.98	IV
80		1734.16	IV
50		1739.19	IV
70		1746.88	IV
80		1747.13	IV
110		1755.64	IV
120		1758.51	IV
100		1769.64	IV
100		1777.82	IV
40		1791.09	IV
140		1802.72	IV
130		1812.41	IV
60		1819.23	IV
30		1826.21	IV
30		1826.86	IV
100		1840.14	IV
50		1851.89	IV
100		1863.11	IV
140		1873.89	IV
35		1883.16	IV
40		1937.63	IV
30		1946.59	IV
140		1967.18	IV
120		1972.07	IV
40		1990.25	IV

Air

50	d	2042.91	IV
40		2055.73	IV
70		2299.21	IV
90		2299.59	IV
100		2316.85	IV
40		2324.06	IV
50		2360.40	IV
70		2405.15	IV
60		2423.32	IV

Cr V
Ref. 380 — C.H.C.

Intensity		Wavelength	
		Vacuum	
100		438.62	V
100		464.02	V
50		469.64	V
50		825.60	V
50		968.70	V
50		1045.04	V
60		1060.65	V
50		1112.45	V
60		1114.35	V
100		1116.48	V
80		1117.56	V
50		1118.16	V
150		1121.07	V
150		1127.63	V
100		1193.95	V
80		1196.04	V
50		1210.50	V
50		1259.99	V
100		1263.50	V
150		1465.86	V
50		1481.65	V
50		1482.76	V
50		1484.67	V
100		1489.71	V
150		1497.97	V
170		1519.03	V
220		1579.70	V
170		1591.72	V
150		1603.19	V
60		1638.50	V
50		1639.40	V
200		1837.44	V

COBALT (Co)
Z = 27

Co I and II
Ref. 1, 125, 276 — C.R.C.

Intensity		Wavelength	
		Vacuum	
20		1265.93	II
40		1271.94	II
20		1276.90	II
30		1293.97	II
25		1295.53	II
30		1295.86	II
20		1297.10	II
80		1299.58	II
30		1302.39	II
20		1306.76	II
80		1306.95	II
40		1311.12	II
40		1311.86	II
30		1315.42	II
30		1316.09	II
20		1318.19	II
30		1318.60	II
20		1319.84	II
20		1409.33	II
20		1466.21	II
20		1471.87	II
30		1472.90	II
30		1475.81	II
20		1484.26	II
30		1486.50	II
20		1509.23	II
20		1590.54	II
20		1595.77	II
20		1599.30	II
20		1693.34	II
8		1706.05	II
10		1723.01	II
15	h	1740.55	II
15		1743.39	II
8		1754.21	II
20	d	1808.01	II
10		1837.56	II
15		1839.37	II
1500		1842.34	I
1800		1847.89	I
1800		1852.71	I
2400		1855.05	I
1500		1878.28	I
10		1917.62	II
1800		1936.58	I
1500		1946.79	I
30		1950.09	II
1500		1951.90	I
1800		1954.22	I
1800		1955.17	I
30		1957.42	II
1500		1958.55	I
1500		1961.59	I
1500	h	1968.69	I
1500	h	1968.93	I
3000		1970.71	I
1800	h	1971.16	I
1800	h	1972.52	I
1500		1973.85	I
1800		1976.97	I
2400	h	1980.89	I
1500		1989.80	I
1800		1990.34	I
1500	l	1998.49	I
		Air	
20		2000.79	II
1500		2002.32	I
900		2008.04	I
50		2011.51	II
1200	h	2014.58	I
900		2016.17	I
50		2022.35	II
40		2025.76	II
50		2027.04	II
900		2031.96	I
30		2036.58	II
1500		2039.95	I
1200		2041.11	I
20		2049.17	II
40		2058.82	II
40		2063.78	II
50		2065.54	II
1500	h	2077.76	I
900		2085.67	I
900		2087.55	I
900		2089.35	I
900		2093.40	I
900		2094.86	I
900		2095.77	I
1200		2097.51	I
1500		2104.73	I
1500		2106.80	I
900		2108.98	I
30		2111.44	II
900	s	2117.68	I
20		2128.79	II
900		2137.78	I
900		2138.97	I
900		2163.03	I
20		2164.44	II
30		2173.33	II
1100		2174.60	I
20		2181.99	II
30		2187.01	II
20		2190.68	II
20		2192.50	II
200		2193.60	II
20		2200.40	II
40	p	2202.95	II
200		2256.73	II
150		2260.00	II
200		2283.52	II
1000		2286.15	II
200		2291.98	II
300	d	2293.38	II
300		2301.40	II
800	d	2307.85	II
2600		2309.02	I
500		2311.60	II
500		2314.05	II
300		2314.96	II
200	p	2317.06	II
2400		2323.14	I
300	p	2324.31	II
200	d	2326.11	II
500		2326.47	II
1400		2335.99	I
1600		2338.67	I
200		2347.39	II
1600		2352.85	I
200	d	2353.41	II
2000		2353.42	I
500		2363.80	II
400		2378.62	II
1400		2380.48	I
200		2381.76	II
300	p	2383.45	II
1400		2384.86	I
200		2386.36	II
500		2388.92	II
200		2397.38	II
1100	d	2402.06	I
200	p	2404.16	II
5300		2407.25	I
5300		2411.62	I
1600		2412.76	I
4800		2414.46	I
4800		2415.30	I
300		2417.65	II
4100		2424.93	I
3300		2432.21	I
2900		2436.66	I
2400		2439.05	I
200		2442.63	II
200	d	2446.03	II
200	p	2447.69	II
200		2450.00	II
200		2464.20	II
200		2486.44	II
200		2498.82	II
570		2504.52	I
500		2506.46	II
360		2506.88	I
200		2511.16	II
860		2517.87	I
500		2519.82	II
4300		2521.36	I
200	h	2524.65	II
300		2524.97	II
500		2528.62	II
2900		2528.97	I
200	p	2530.09	II
720		2530.13	I
860		2532.18	I
200	d	2533.82	II
2900		2535.96	I
860		2536.49	I
300		2541.94	II
1700		2544.25	I
200		2546.74	II
340		2548.34	I
310		2553.37	I
310		2555.07	I
300		2559.41	II
200		2560.03	II
960		2562.15	I
500		2564.04	II
1100		2567.35	I
960		2574.35	I
800		2580.32	II
300	d	2582.22	II
500		2587.22	II
500		2587.52	II
200		2588.91	II
100	p	2605.71	II
100		2612.50	II
100		2614.36	II
100	p	2628.77	II
100		2632.26	II
100		2636.07	II
310		2646.42	I
770		2648.64	I
100		2653.72	II
100		2663.53	II
200		2666.73	II
100		2675.85	II
100		2684.42	II
100		2702.02	II
200		2706.62	II
200		2707.35	II
190		2715.99	I
100		2727.78	II
80		2734.54	II
190		2745.10	I
100		2753.22	II
190		2764.19	I
100		2766.70	II
100		2774.97	II
100		2791.00	II
100		2793.73	II
150		2815.56	I
80		2835.63	II
80		2847.35	II
80		2871.22	II
190		2886.44	I
100		2918.38	II
100		2930.24	II
100		2954.73	II
690		2987.16	I
690		2989.59	I
20		3008.86	II
60		3022.59	II
30		3035.13	II
3100		3044.00	I
1700		3061.82	I
20		3352.79	II
80		3387.70	II
1100		3388.17	I
2200		3395.38	I
11000		3405.12	I
4500		3409.18	I
6700		3412.34	I
2200		3412.63	I
30		3415.77	II
2700		3417.16	I
50		3423.84	II
2500		3431.58	I
4500		3433.04	I
1600		3442.93	I
8800		3443.64	I
50		3446.39	II
4100		3449.17	I
2100		3449.44	I
21000		3453.50	I
1000		3455.23	I
5100		3462.80	I
5100		3465.80	I
8000		3474.02	I
1900		3483.41	I
4800		3489.40	I
2400		3495.69	I
40		3497.33	II
50		3501.72	II
9600		3502.28	I
7000		3506.32	I
50		3507.77	II
2900		3509.84	I
1400		3510.43	I
4800		3512.64	I
3800		3513.48	I
10		3514.23	II
30		3517.50	II
4800		3518.35	I
1300		3520.08	I
2700		3521.57	I
3800		3523.43	I
60		3523.51	II
6400		3526.85	I
2700		3529.03	I
7300		3529.81	I
1900		3533.36	I
40		3535.92	II
50		3545.03	II
20		3555.93	II
1100		3560.89	I
80		3561.07	II
20		3566.98	II
8800		3569.38	I
50		3574.95	II

Cobalt (Cont.)

1600		3574.96	I
60		3575.32	II
2500		3575.36	I
60		3577.96	II
1000		3585.16	I
6700		3587.19	I
1900		3594.87	I
1600		3602.08	I
100		3621.21	II
1000		3627.81	I
80		3643.61	II
10		3656.75	II
60		3681.35	II
5		3695.32	II
5		3714.73	II
1100		3745.50	I
20		3754.69	II
1400		3842.05	I
6900		3845.47	I
5500		3873.12	I
2800		3873.96	I
7900		3894.08	I
20	h	3911.40	II
1500		3935.97	I
80	h	3963.10	II
40	h	3976.74	II
10		3983.02	II
6000		3995.31	I
970		3997.91	I
350		4020.90	I
10	h	4036.14	II
20	h	4037.37	II
4		4040.02	II
370		4045.39	I
5	h	4050.23	II
10	h	4052.40	II
20	h	4062.73	II
2	h	4064.50	II
350		4066.37	I
5	h	4074.34	II
830		4092.39	I
1		4096.57	II
550		4110.54	I
2800		4118.77	I
4400		4121.32	I
3		4130.88	II
3		4145.13	II
3	d	4160.67	II
1		4181.13	II
90		4190.71	I
1		4208.61	II
30	s	4244.25	II
8		4272.33	II
20	h	4288.25	II
3		4328.86	II
2		4384.26	II
2	h	4396.94	II
3		4413.91	II
90		4469.56	I
10		4482.50	II
2	h	4489.12	II
4	h	4497.44	II
10	d	4500.54	II
0		4516.65	II
690		4530.96	I
2	h	4533.22	II
0	h	4537.95	II
90		4549.66	I
1	h	4559.29	II
140		4565.59	I
1		4569.26	II
190		4581.60	I
5		4616.30	II
120		4629.38	I
25	h	4660.66	II
85		4663.41	I
110		4792.86	I
10	d	4831.16	II
100		4840.27	I
150		4867.88	I
80	h	4964.18	II
10	h	4970.05	II
10	h	4990.47	II
20	h	4995.98	II
35		5146.74	I
50		5212.71	I
50		5230.22	I
45		5235.21	I
50		5247.93	I
26		5266.30	I
45		5266.49	I
26		5268.52	I
45	h	5280.65	I
26		5301.06	I
50		5342.71	I
26		5343.39	I
50		5352.05	I
26		5353.48	I
35		5369.58	I
45		5483.34	I
17		5530.77	I
17		5647.22	I
17		5991.88	I
17		6082.44	I
17		6282.63	I
45		6450.24	I
21		6455.00	I
15		6563.42	I
15		6632.44	I
14		6814.94	I
14		6872.40	I
21		7052.89	I
45		7084.99	I
8		7417.38	I
8		7712.68	I
7		7908.71	I
9		7987.38	I
13		8007.27	I
9		8093.96	I
9		8372.84	I
4		8575.35	I
3		8819.15	I

Co III
Ref. 291 — C.R.C.

Intensity	Wavelength	
	Vacuum	
1000	1696.01	III
800	1697.99	III
500	1702.79	III
1000	1707.35	III
500	1707.95	III
500	1723.97	III
400	1745.67	III
500	1755.98	III
5000	1760.35	III
500	1769.96	III
500	1773.22	III
5000	1773.57	III
500	1774.42	III
1000	1777.14	III
2000	1780.05	III
3000	1782.97	III
500	1784.06	III
1000	1787.08	III
1000	1789.07	III
500	1790.26	III
500	1791.28	III
500	1798.06	III
500	1805.54	III
400	1811.47	III
400	1821.26	III
400	1821.69	III
400	1821.77	III
1000	1823.08	III
400	1825.36	III
750	1825.95	III
2000	1830.09	III
2000	1831.44	III
750	1831.92	III
400	1832.20	III
5000	1835.00	III
1000	1837.63	III
500	1846.16	III
500	1852.92	III
400	1854.39	III
400	1854.76	III
1000	1861.78	III
2000	1863.83	III
400	1864.19	III
500	1871.87	III
1000	1881.70	III
500	1895.37	III
500	1919.12	III
500	1928.57	III
500	1940.15	III
500	1953.94	III
500	1959.41	III
400	1989.60	III
	Air	
100	2001.09	III
200	2011.62	III
200	2013.88	III
100	2031.81	III
200	2053.11	III
100	2056.21	III
10	2062.17	III
10	2079.74	III
15	2088.58	III
10	2090.51	III
10	2097.63	III
10	2134.15	III
10	2452.16	III
20	2811.75	III
10	2888.31	III
10	2933.27	III
10	2978.01	III
20	2991.89	III
25	3010.92	III
10	3116.68	III
2	3151.40	III
2	3180.64	III
20	3232.11	III
2	3249.24	III
20	3259.68	III
2	3269.23	III
10	3287.68	III
15	3305.38	III
10	3451.25	III
2	3526.24	III
1	3634.21	III
1	3636.31	III
2	3667.52	III
2	3677.23	III
3	3680.74	III
1	3762.50	III
15	3782.27	III

Co IV
Ref. 236 — C.R.C.

Intensity	Wavelength	
	Vacuum	
81	606.79	IV
74	607.59	IV
55	608.24	IV
66	609.16	IV
70	609.21	IV
64	609.28	IV
43	610.04	IV
37	610.25	IV
24	610.79	IV

Co V
Ref. 100, 159 — C.R.C.

Intensity	Wavelength	
	Vacuum	
20	355.52	V
18	355.88	V
12	356.06	V
4	1006.86	V
10	1007.51	V
15	1009.02	V
10	1010.94	V
10	1013.80	V
1	1017.43	V
10	1018.36	V
10	1021.14	V
1	1028.08	V
3	1226.31	V
8	1228.19	V
15	1231.73	V
2	1234.55	V
20	1236.95	V
2	1239.85	V
8	1246.91	V
6	1258.61	V
5	1263.28	V
20	1270.70	V
20	1272.23	V
2	1275.52	V
50	1277.01	V
30	1281.63	V
15	1284.00	V
28	1286.95	V
25	1295.55	V
40	1295.87	V
35	1301.12	V
50	1345.67	V
6	1351.22	V
15	1353.42	V
40	1355.20	V
30	1357.67	V
20	1361.32	V
30	1362.46	V
30	1364.17	V
30	1368.24	V
4	1369.30	V
10	1371.01	V
30	1373.09	V
30	1375.20	V
25	1378.12	V
10	1379.05	V
10	1380.21	V
32	1389.11	V
15	1459.77	V
35	1468.98	V
30	1476.65	V
25	1482.62	V
20	1482.91	V
25	1486.02	V
20	1488.73	V

COPPER (Cu)
Z = 29

Cu I and II

Ref. 273, 290—V.K.

Intensity	Wavelength	
	Vacuum	
80	685.141	II
100	709.313	II
100	718.179	II
150	724.489	II
200	735.520	II
250	736.032	II
80	779.295	II
100	797.455	II
150	810.998	II
200	813.883	II
300	826.996	II
150	848.808	II
250	851.303	II
250	858.487	II
400	861.994	II
400	865.390	II
250	869.336	II
150	873.263	II
200	876.723	II
250	877.012	II
200	877.555	II
500	878.699	II
100	884.133	II
250	885.847	II
600	886.943	II
600	890.567	II
500	892.414	II
800	893.678	II
400	894.227	II
600	896.759	II
400	896.976	II
600	901.073	II
400	906.113	II
800	914.213	II
600	922.019	II
500	924.239	II
400	935.232	II
600	935.898	II
600	943.335	II
600	945.525	II
500	945.965	II
200	954.383	II
250	956.290	II
400	958.154	II
200	960.414	II
250	968.042	II
200	974.759	II
250	977.567	II
100	987.657	II
250	992.953	II
300	1004.055	II
300	1008.569	II
300	1008.728	II
300	1010.269	II
250	1012.597	II
500	1018.707	II
500	1027.831	II
250	1028.328	II
200	1030.263	II
600	1036.470	II
600	1039.348	II
600	1039.582	II
800	1044.519	II
800	1044.744	II
500	1049.755	II
600	1054.690	II
400	1055.797	II
600	1056.955	II
400	1058.799	II
600	1059.096	II
600	1060.634	II
600	1063.005	II
200	1065.782	II
200	1066.134	II
500	1069.195	II
300	1073.745	II
200	1088.395	II
300	1094.402	II
250	1097.053	II
150	1119.947	II
200	1142.640	II
300	1144.856	II
100	1250.048	II
150	1265.506	II
300	1275.572	II
150	1282.455	II
150	1287.468	II
150	1298.395	II
300	1308.297	II
300	1314.337	II
100	1320.686	II
100	1326.395	II
150	1350.594	II
250	1351.837	II
150	1355.305	II
300	1358.773	II
200	1359.009	II

Copper (Cont.)

200		1362.600	II
250		1367.951	II
200		1371.840	II
100		1393.128	II
100		1398.642	II
150		1402.777	II
150		1407.169	II
100		1414.898	II
250		1418.426	II
250		1421.759	II
200		1427.829	II
400		1430.243	II
250		1434.904	II
150		1436.236	II
150		1442.139	II
200		1445.984	II
200		1449.058	II
250		1450.304	II
200		1452.294	II
300		1458.002	II
250		1459.412	II
200		1463.752	II
400		1463.838	II
200		1466.070	II
400		1470.697	II
200		1472.395	II
250		1473.978	II
200		1474.935	II
150		1476.059	II
200		1481.544	II
200		1485.328	II
750		1488.831	II
300		1492.834	II
250		1493.366	II
250		1495.430	II
350		1496.687	II
150		1503.368	II
250		1504.757	II
200		1505.388	II
300		1508.632	II
350		1510.506	II
200		1512.465	II
200		1513.366	II
500		1514.492	II
200		1517.631	II
500		1519.492	II
600		1519.837	II
200		1520.540	II
200		1524.860	II
150		1525.764	II
500		1531.856	II
300		1532.131	II
250		1533.986	II
250		1535.002	II
500		1537.559	II
200		1540.239	II
300		1540.389	II
300		1540.588	II
750		1541.703	II
400		1544.677	II
100		1547.958	II
300		1550.653	II
300		1551.389	II
500		1552.646	II
250		1553.896	II
400		1555.134	II
500		1555.703	II
300		1558.345	II
400		1565.924	II
400		1566.415	II
100		1569.416	II
300		1579.492	II
300		1580.626	II
400		1581.995	II
500		1583.682	II
400		1590.165	II
600		1593.556	II
400		1598.402	II
400		1602.388	II
200		1604.848	II
300		1605.281	II
400		1606.834	II
250		1608.639	II
150		1610.296	II
200		1617.915	II
600		1621.426	II
400		1622.428	II
250		1630.268	II
100		1636.605	II
250		1649.458	II
30	r	1655.32	I
200		1656.322	II
200		1660.001	II
300		1663.002	II
100		1672.776	II
30		1688.09	I
30		1691.08	I
30	r	1703.84	I
50	r	1713.36	I
150		1717.721	II
50	r	1725.66	I
100		1736.551	II
50	r	1741.57	I
150		1753.281	II
200	r	1774.82	I
100	r	1825.35	I
250		1929.751	II
250		1944.597	II
100		1946.493	II
200		1957.518	II
150		1970.495	II
150		1977.027	II
500		1979.956	II
300		1989.855	II

Air

250		1999.698	II
270		2035.854	II
250		2037.127	II
350		2043.802	II
300		2054.980	II
100		2078.663	II
110		2098.398	II
320		2104.797	II
300		2112.100	II
320		2117.310	II
350		2122.980	II
350		2126.044	II
420		2134.341	II
900		2135.981	II
400		2148.984	II
150		2161.320	II
1300	r	2165.09	I
250		2174.982	II
1600	r	2178.94	I
700		2179.410	II
1700	r	2181.72	I
700		2189.630	II
900		2192.268	II
400		2195.683	II
1700	r	2199.58	I
1300	r	2199.75	I
100		2200.509	II
200		2209.806	II
750		2210.268	II
1600	r	2214.58	I
250		2215.106	II
1000	r	2215.65	I
750		2218.108	II
2100	r	2225.70	I
150		2226.780	II
1600	r	2227.78	I
350		2228.868	II
2500	r	2230.08	I
1100	r	2238.45	I
900		2242.618	II
2300	r	2244.26	I
1000		2247.002	II
1300	r	2260.53	I
2200	r	2263.08	I
150		2263.786	II
200		2276.258	II
100		2286.645	II
2500	r	2293.84	I
170		2294.368	II
1000		2303.12	I
150		2369.890	II
2500	r	2392.63	I
120		2403.337	II
1500		2406.66	I
1000	r	2441.64	I
100		2485.792	II
2000	r	2492.15	I
150		2506.273	II
120		2526.593	II
300		2544.805	II
100		2571.756	II
150		2590.529	II
200		2600.270	II
2500	r	2618.37	I
200		2666.291	II
750		2689.300	II
700		2700.962	II
650		2703.184	II
700		2713.508	II
650		2718.778	II
300		2721.677	II
120		2737.342	II
270		2745.271	II
2500	r	2766.37	I
800		2769.669	II
200		2791.795	II
170		2799.528	II
100		2810.804	II
1250	r	2824.37	I
350		2837.368	II
100		2857.748	II
600		2877.100	II
270		2884.196	II
2500	r	2961.16	I
100		2986.335	II
2000		2997.36	I
2000		3010.84	I
2500		3036.10	I
2500		3063.41	I
1400		3073.80	I
1500		3093.99	I
1250		3099.93	I
2000		3108.60	I
1400	h	3126.11	I
1500		3194.10	I
1400		3208.23	I
1500	h	3243.16	I
10000	r	3247.54	I
10000	r	3273.96	I
1400	h	3282.72	I
400		3290.418	II
1500	h	3290.54	I
110		3300.881	II
250		3301.229	II
2500	h	3307.95	I
200		3316.276	II
1500		3337.84	I
150		3338.648	II
200		3365.648	II
450		3370.454	II
300		3374.952	II
200		3380.712	II
100		3384.945	II
1250	h	3483.76	I
1250		3524.23	I
2000		3530.38	I
1400		3599.13	I
1400		3602.03	I
1000		3686.555	II
150		3786.270	II
170		3797.849	II
100		3818.879	II
140		3826.921	II
160		3864.137	II
280		3884.131	II
150		3892.924	II
170		3903.177	II
140		3920.654	II
120		3933.268	II
120		3987.024	II
150		3993.302	II
140		4003.476	II
1250		4022.63	I
100		4032.647	II
600		4043.484	II
500		4043.751	II
2000		4062.64	I
120		4068.106	II
500		4131.363	II
200		4143.017	II
300		4153.623	II
500		4161.140	II
370		4164.284	II
400		4171.851	II
500		4179.512	II
500		4211.866	II
320		4230.449	II
200		4255.635	II
950		4275.11	I
300		4279.962	II
500		4292.470	II
400		4365.370	II
100		4444.831	II
400		4506.002	II
150		4516.049	II
150		4541.032	II
500		4555.920	II
100		4596.906	II
120		4649.271	II
2000		4651.12	I
120		4661.363	II
320		4671.702	II
300		4673.577	II
450		4681.994	II
100		4758.433	II
400		4812.948	II
120		4851.262	II
300		4854.988	II
100		4873.304	II
150		4901.427	II
1000		4909.734	II
500		4918.376	II
200		4926.424	II
900		4931.698	II
120		4943.026	II
700		4953.724	II
500		4985.506	II
400		5006.801	II
350		5009.851	II
400		5012.620	II
350		5021.279	II
200		5039.016	II
300		5047.348	II
900		5051.793	II
400		5058.910	II
500		5065.459	II
450		5067.094	II
350		5072.302	II
450		5088.277	II
420		5093.816	II
350		5100.067	II
1500		5105.54	I
250		5124.476	II
2000		5153.24	I
100		5158.093	II
100		5183.367	II
2500		5218.20	I
100		5269.991	II
100		5276.525	II
1650		5292.52	I
100		5368.383	II
1500		5700.24	I
1500		5782.13	I
150		5805.989	II
100		5833.515	II
200		5897.971	II
120		5937.577	II
400		5941.196	II
100		5993.260	II
650		6000.120	II
100		6023.264	II
250		6072.218	II
150		6080.343	II
150		6099.990	II
160		6107.412	II
300		6114.493	II
600		6150.384	II
750		6154.222	II
500		6172.037	II
550		6186.884	II
400		6188.676	II
300		6198.092	II
470		6204.261	II
450		6208.457	II
750		6216.939	II
700		6219.844	II
500		6261.848	II
1000		6273.349	II
350		6288.696	II
900		6301.009	II
550		6305.972	II
400		6312.492	II
120		6326.466	II
400		6373.268	II
750		6377.840	II
400		6403.384	II
850		6423.884	II
200		6442.965	II
750		6448.559	II
170		6466.246	II
950		6470.168	II
750		6481.437	II
400		6484.421	II
220		6517.317	II
400		6530.083	II
120		6551.286	II
200		6577.080	II
750		6624.292	II
800		6641.396	II
450		6660.962	II
100		6770.362	II
300		6806.216	II
400		6809.647	II
320		6823.202	II
250		6844.157	II
320		6868.791	II
270		6872.231	II
270		6879.404	II
220		6937.553	II
150		6952.871	II
150		6977.572	II
200		7022.860	II
300		7194.896	II
400		7326.008	II
300		7331.694	II
250		7382.277	II
1000		7404.354	II
270		7434.156	II
500		7562.015	II
700		7652.333	II
1000		7664.648	II
150		7681.788	II
450		7744.097	II
800		7778.738	II
750		7805.184	II
1500		7807.659	II
1000		7825.654	II
350		7860.577	II
300		7890.567	II
700		7902.553	II
1500		7933.13	I
1400		7944.438	II
400		7972.033	II
1200		7988.163	II
2000		8092.63	I
500		8277.560	II
800		8283.160	II
250		8503.396	II
750		8511.061	II
200		8609.134	II
500		9813.213	II
250		9827.978	II
200		9830.798	II
600		9861.280	II
600		9864.137	II
200		9883.969	II
550		9916.419	II
500		9917.954	II
550		9925.594	II

Copper (Cont.)

450		9938.998	II
500		9960.354	II
450		10006.588	II
550		10022.969	II
550		10038.093	II
650		10054.938	II
450		10080.354	II

Cu III
Ref. 295—V.K.

Intensity		Wavelength	
		Vacuum	
75		542.90	III
200		615.67	III
150		616.03	III
150		687.98	III
150		715.53	III
125		730.38	III
250		788.07	III
250		788.46	III
250		791.36	III
150		801.14	III
100		829.34	III
40		1048.88	III
50		1186.80	III
50		1200.96	III
300		1219.30	III
200		1244.38	III
100		1279.14	III
200		1312.39	III
300		1332.97	III
200		1339.48	III
150		1363.08	III
300	r	1376.79	III
200	r	1377.49	III
150		1423.48	III
300	r	1481.23	III
200		1543.46	III
500	r	1593.75	III
1000	r	1642.21	III
300		1679.14	III
400		1702.10	III
500		1722.37	III
600		1741.37	III
200		1768.86	III
200		1840.91	III
100		1971.95	III
		Air	
200		2013.22	III
150		2157.28	III
100		2299.47	III
500		2368.17	III
400		2391.74	III
800		2405.50	III
700		2412.34	III
2000		2444.44	III
500		2468.41	III
1000		2482.36	III
700		2486.46	III
500		2508.49	III
500		2522.38	III
500		2538.66	III
400		2566.37	III
400		2573.33	III
500		2609.32	III
200		2643.92	III
200		2696.38	III
20		2751.33	III
100		2812.94	III
100		2978.87	III
75		3548.87	III
100		3639.42	III
500		3702.92	III
800		3744.70	III
400		3748.27	III
600		3752.06	III
1000		3776.97	III
800		3790.80	III
600		3804.13	III
600		3809.18	III
300		3881.68	III
150		3953.81	III
100		4090.49	III
200		4283.40	III
500		4351.97	III
1000		4352.80	III
500		4355.24	III
500		4370.84	III
500		4371.40	III
500		4373.43	III
1000		4377.11	III
200		4386.42	III
150		4927.41	III
400		5094.28	III
200		5168.97	III
400		5208.34	III
600		5219.21	III
200		5268.59	III
400		5317.78	III
300		5369.79	III
350		5418.48	III
250	d	5494.94	III
50		5573.94	III
100		5609.00	III
75		5702.12	III
100		5768.56	III
100		5850.72	III
200		5965.25	III
30		6100.87	III
50		6369.27	III
20		6512.54	III
20		6644.13	III
50		6793.20	III

Cu IV
Ref. 199—V.K.

Intensity		Wavelength	
		Vacuum	
30		360.86	IV
20		374.40	IV
30		405.24	IV
80		406.45	IV
40		413.45	IV
70		443.68	IV
80		451.16	IV
80		463.72	IV
80		484.53	IV
90		497.00	IV
90		504.60	IV
70		509.38	IV
40		519.51	IV
40		540.65	IV
60		550.92	IV
20		584.85	IV
60		1056.13	IV
30		1074.72	IV
50		1091.65	IV
30		1105.50	IV
25		1119.43	IV
40	p	1152.18	IV
60		1227.44	IV
70		1228.87	IV
70		1258.69	IV
90		1274.84	IV
90		1293.46	IV
90		1309.41	IV
70		1321.17	IV
100	d	1340.08	IV
100		1350.42	IV
100		1362.05	IV
100		1372.14	IV
100		1377.82	IV
100		1388.80	IV
90		1405.49	IV
90		1415.27	IV
90		1434.34	IV
90		1449.69	IV
80		1466.18	IV
70		1482.77	IV
80		1499.81	IV
90		1515.28	IV
90		1535.12	IV
80		1551.12	IV
90		1567.35	IV
80		1583.47	IV
70		1595.12	IV
80	p	1608.14	IV
90		1639.75	IV
20		1650.16	IV
70		1704.37	IV
30		1797.99	IV
70		1817.56	IV
70		1819.23	IV
60		1837.04	IV
80		1849.62	IV
30		1867.24	IV
30		1918.71	IV
40		1966.31	IV

Cu V
Ref. 324—V.K.

Intensity		Wavelength	
		Vacuum	
9	h	258.95	V
49		271.33	V
49		283.97	V
22		293.41	V
56		299.64	V
65		305.83	V
51		312.51	V
66		321.05	V
74		326.57	V
82		333.56	V
81		339.88	V
81		346.00	V
86		355.41	V
77		363.96	V
65		370.63	V
74		377.76	V
70		387.40	V
51		396.06	V
25		406.94	V
13		1097.10	V
42		1106.24	V
77		1113.22	V
67		1121.20	V
76		1128.80	V
59		1133.86	V
63		1142.38	V
54		1149.06	V
72		1157.54	V
64		1167.35	V
77		1176.53	V
84		1183.63	V
76		1192.54	V
83		1201.22	V
77		1204.90	V
71		1214.36	V
70		1221.34	V
76		1230.11	V
80		1239.73	V
79		1246.99	V
65		1253.07	V
70		1260.24	V
77		1269.35	V
78		1278.20	V
68		1286.13	V
67		1292.08	V
73		1299.22	V
65		1309.72	V
55		1318.89	V
65		1323.28	V
64		1329.22	V

CURIUM (Cm)
Z = 96

Cm I and II
Ref. 51, 332—J.G.C.

Intensity		Wavelength	
		Air	
10000		2462.76	II
10000		2617.17	II
10000		2636.28	II
10000		2651.17	II
10000		2653.80	II
10000		2725.68	II
10000		2736.89	II
10000		2748.04	II
10000		2784.83	II
10000		2811.62	II
10000		2824.20	II
10000		2833.58	II
10000		2899.90	II
10000		2912.97	II
10000		2928.92	II
10000		2996.18	II
10000		2999.39	I
10000	b	3014.87	II
10000		3044.85	II
10000		3109.69	I
10000		3116.41	I
10000		3137.16	I
10000		3147.33	I
10000		3155.10	I
10000		3158.60	I
10000		3169.98	II
10000		3177.55	I
10000		3179.10	I
10000		3186.41	I
10000		3188.11	I
10000		3207.12	II
10000		3207.71	I
10000		3209.89	II
10000		3209.94	I
10000		3210.05	II
10000		3220.76	II
10000		3224.23	I
10000		3225.11	I
10000		3226.41	II
10000		3230.28	I
10000		3230.35	II
10000		3236.74	I
10000		3238.55	II
10000		3242.66	II
10000		3246.25	I
10000		3252.68	I
10000		3265.81	I
10000		3280.45	I
10000		3296.71	II
10000		3304.85	I
10000		3317.14	I
10000		3374.70	I
10000		3452.92	I
10000		3458.34	I
10000		3510.28	I
10000		3522.36	I
10000		3524.94	I
10000		3542.06	I
10000		3547.02	I
10000		3547.92	I
10000		3561.44	I
10000		3572.95	II
10000		3600.62	I
10000		3639.94	I
10000		3664.34	I
10000		3709.43	I
10000		3729.00	I
10000		3732.35	I
10000		3747.86	I
10000		3763.05	I
10000		3775.75	I
10000		3816.30	I
10000		3825.14	I
10000		3833.32	I
10000		3837.59	I
10000		3842.00	I
10000		3849.92	I
10000		3854.11	I
10000		3900.25	I
10000		3904.06	II
10000		3908.24	II
10000		3936.67	I
10000		3942.03	I
10000		3944.15	I
10000		3948.68	I
10000		3953.36	I
10000		3964.83	I
10000		3995.10	I
10000		4016.17	I
10000		4031.76	I
10000		4048.29	I
10000		4049.65	I
10000		4113.29	I
10000		4129.71	I
10000		4207.66	II
10000		4211.62	I
10000		4266.45	I
10000		4293.00	I
10000		4330.82	I
10000		4345.69	I
10000		4447.77	I
10000		4459.16	I
10000		4608.40	I
10000		5846.07	I
10000		5952.41	I
10000		6058.90	I
10000		6243.35	I
10000		6376.71	I
10000		6510.16	I
10000		6554.41	I
10000		6640.17	I
10000		6663.25	I
10000		6686.87	I
10000		6706.85	I
10000		6726.68	I
10000		6793.15	I
10000		7162.69	I
10000		7577.80	I
10000		7673.79	I
10000		7720.47	I
10000		8392.37	I
10000		9293.25	I
10000		9567.08	I
10000		9657.12	I
10000		10310.83	I
10000		10351.73	I
10000		10424.49	I
10000		10508.11	I
10000		10542.98	I
10000		10792.25	I
10000		10897.45	I
10000		11507.45	I
10000		11707.73	I
10000		11780.95	I
10000		11834.28	I
10000		12017.85	I
10000		12394.16	I
10000		12454.98	I
10000		12464.99	I
10000		13004.56	I
10000		13258.18	I
10000		13289.84	I
10000		13344.62	I
10000		13480.54	I
10000		13590.01	I
10000		13644.77	I
10000		13789.52	I
10000		13840.18	I
10000		13908.46	I
10000		13964.14	I
10000		14235.27	I
10000		14334.52	I
10000		14563.41	I
10000		14580.23	I
10000		15018.13	I
10000		15222.27	I
10000		15642.59	I

Curium (Cont.)

10000	15757.23	I
10000	15793.31	I
10000	16008.41	I
10000	17148.22	I
10000	17453.18	I
10000	17619.28	I
10000	18069.02	I
10000	19572.62	I
10000	19975.98	I
10000	20526.32	I
10000	20853.49	I
10000	20911.52	I
10000	20968.11	I
10000	21241.06	I
10000	21393.23	I

DYSPROSIUM (Dy)
Z = 66

Dy I and II
Ref. 1—C.H.C.

Intensity		Wavelength (Air)	
260		2356.91	II
65		2381.95	
130		2387.36	II
150		2392.15	
180		2402.29	II
240		2410.01	II
150		2422.75	II
260		2439.84	II
90		2455.15	II
110		2459.99	II
90		2471.40	II
110		2480.93	II
170		2490.61	II
90		2510.31	II
170		2513.55	II
170		2517.61	II
130		2543.81	II
90		2545.12	II
150		2552.29	II
180		2557.94	II
90		2560.21	II
90		2566.25	II
220		2585.30	I
90		2591.56	II
75		2592.54	II
120		2600.16	II
130		2600.76	II
75		2608.69	II
370		2623.69	I
440		2634.80	II
110		2642.15	I
110		2645.35	II
110		2667.94	I
55		2676.84	II
50		2677.34	II
85		2689.31	II
85		2692.83	II
55		2709.01	II
55		2727.17	II
85		2729.50	II
55		2735.79	I
40		2739.30	II
85		2740.70	II
220		2755.75	II
55		2757.08	II
70		2766.50	II
70		2772.42	II
110		2772.61	II
40		2779.58	II
55		2791.44	II
120		2800.33	II
110		2800.53	II
110		2801.41	II
300		2816.39	II
140		2825.42	II
140		2862.70	I
190		2877.88	II
110		2884.28	II
120		2885.53	I
120		2890.74	II
120		2900.82	II
110		2904.62	II
190		2906.39	II
390		2913.95	II
110		2934.31	II
250		2934.52	II
110		2941.05	II
140		2944.56	II
150		2947.06	II
150		2947.21	II
250		2948.31	II
170		2950.33	II
110		2952.12	II
140		2953.70	II
220		2964.60	I
110		2977.42	II
110		2985.97	II
220		3015.68	II
390		3026.16	II
210		3029.81	II
610		3038.28	II
280		3043.13	II
210		3047.56	II
280		3060.64	II
390		3062.62	II
220		3066.99	II
330		3071.91	II
280		3073.54	II
220		3078.68	II
280		3101.93	II
220		3103.24	II
410		3109.76	II
330		3128.41	II
830		3135.38	II
360		3140.64	II
500		3141.14	II
220		3143.83	II
250		3146.16	II
1200		3156.52	II
670		3162.83	II
1000		3169.99	II
400		3177.89	II
220		3178.37	II
200		3184.79	II
330		3186.38	II
240		3187.68	II
330		3193.30	II
240		3206.40	II
220		3207.12	II
290		3208.85	II
470		3215.19	II
830		3216.63	II
240		3221.49	II
290		3223.28	II
240		3225.08	II
330		3225.95	II
490		3235.89	II
290		3236.69	II
490		3245.12	II
200		3248.36	II
1200		3251.27	II
200		3252.19	II
290		3256.26	II
200		3266.21	II
240		3269.11	II
200		3272.73	II
890		3280.09	II
490		3282.77	II
200		3287.96	II
200		3293.88	II
200		3296.30	II
200		3305.40	II
200		3305.51	II
240		3306.19	II
440		3308.79	II
1100		3308.88	II
510		3312.72	II
780		3316.32	II
240		3317.12	II
1000		3319.88	II
270		3326.19	II
780		3341.00	II
270		3341.88	II
200		3347.83	II
270		3352.69	II
510		3353.58	II
240		3359.46	II
510		3368.11	II
5300		3385.02	II
210		3386.57	II
610		3388.85	II
210		3391.96	II
3800		3393.57	II
1300		3396.16	II
380		3407.16	II
5300		3407.80	II
420		3408.14	II
1300		3413.78	II
530		3414.82	II
780		3419.63	II
530		3425.06	II
420		3429.44	II
1900		3434.37	II
330		3438.94	II
560		3440.93	II
1300		3441.45	II
3800		3445.57	II
830		3446.99	II
440		3449.89	II
2700		3454.32	II
440		3454.51	II
1300		3456.56	II
4400		3460.97	II
720		3468.43	II
560		3471.14	II
560	d	3471.53	II
380		3473.70	II
1300		3477.07	II
4400		3494.49	II
560		3496.34	II
400		3497.81	II
830		3498.71	II
400		3501.50	II
830		3504.53	II
830		3505.45	II
1300		3506.81	II
560		3517.26	II
4400		3523.98	II
22000		3531.70	II
4400		3534.96	II
5500		3536.02	II
4400		3538.52	II
400		3539.37	II
1700		3542.33	II
400		3544.20	II
400		3544.35	II
1400		3546.83	II
330		3548.19	II
4400		3550.22	II
2200		3551.62	II
440	h	3558.23	II
440		3559.30	II
2200		3563.15	II
560		3563.69	II
780		3573.83	II
1400		3574.15	II
4400		3576.24	II
1700		3576.87	II
830		3577.98	II
440		3580.04	II
400		3584.42	II
3300		3585.06	II
1400		3585.78	II
560		3586.11	II
360		3590.07	II
1100		3591.41	II
560		3591.81	II
560		3592.11	II
1800		3595.04	II
400		3596.06	II
560		3600.38	II
360		3602.82	II
1800		3606.12	II
440		3618.51	II
560		3620.16	II
470		3624.27	II
1100		3629.42	II
4000		3630.24	II
440		3632.78	II
400		3635.27	II
360		3637.28	II
1100		3640.25	II
400		3643.92	II
11000		3645.40	II
360		3645.86	II
1000		3648.78	II
700		3664.62	II
400		3666.84	I
990		3672.30	II
420		3672.70	II
400		3673.14	II
1400		3674.08	II
2200		3676.59	II
640		3678.51	I
820		3684.85	I
1300		3685.78	I
4700		3694.81	II
370		3697.31	II
990		3698.21	II
540		3701.63	II
330		3707.40	II
440		3707.57	II
440		3708.22	II
420		3710.07	II
330		3711.66	II
1600		3724.45	II
300		3728.00	I
930		3739.34	I
1200		3747.82	II
1400		3753.51	II
1400		3753.75	II
1200		3757.05	I
4700		3757.37	II
640		3767.63	I
330		3771.11	I
640		3773.05	I
370		3774.71	I
420		3781.47	I
330		3785.41	II
3300		3786.18	II
1600		3788.44	II
700		3791.87	II
510		3804.14	II
580		3806.27	II
470		3812.27	I
470		3813.67	II
1400		3816.76	II
700		3825.68	II
2300		3836.50	II
370		3840.89	I
1400		3841.31	II
330		3842.00	II
330		3844.36	I
420		3846.34	II
420		3847.02	I
330		3849.39	II
1200		3853.03	II
420		3858.40	I
370		3866.58	II
560		3868.45	II
1600		3868.81	I
300		3869.42	II
820		3869.86	II
7000		3872.11	II
1200		3873.99	II
470		3879.11	II
300		3881.99	II
5800		3898.53	II
540		3914.87	II
540		3915.59	II
540	d	3917.29	I
320		3923.38	II
420		3927.86	I
540		3930.14	I
2100		3931.52	II
320		3932.22	II
370		3933.00	II
320		3934.21	II
420		3936.70	I
540		3942.53	II
10000		3944.68	II
420		3946.93	II
540		3950.39	II
420		3954.55	II
800		3957.79	II
370		3962.59	I
320		3967.51	I
14000		3968.39	II
2700		3978.57	II
1400		3981.92	II
1600		3983.65	II
800		3984.21	II
540		3991.32	II
1600		3996.69	II
8000		4000.45	II
420		4005.84	I
320		4006.07	I
540		4011.29	II
540		4013.82	I
540		4014.70	II
370		4023.71	I
420		4027.78	II
520	d	4028.32	II
520		4032.47	II
420		4033.65	II
420		4036.32	II
320		4041.98	II
12000		4045.97	I
1600		4050.56	II
520		4055.14	II
2500		4073.12	II
7400		4077.96	II
370		4085.34	I
390		4096.10	I
3900		4103.30	II
860		4103.87	I
1500		4111.34	II
490		4124.63	II
390		4128.24	II
350		4129.12	I
990		4129.42	II
350		4130.35	I
390		4133.85	I
470		4141.50	II
1200		4143.10	II
990		4146.06	I
5700		4167.97	I
370		4171.93	I
930		4183.72	I
12000		4186.82	I
320		4190.94	I
2200		4191.64	I
6800		4194.84	I
320		4195.19	II
800		4198.02	I
680		4201.30	I
680		4202.24	I
230		4205.06	I
370		4206.54	II
440		4211.24	I
16000		4211.72	I
1800		4213.18	I
3700		4215.16	I
4400		4218.09	I
4400		4221.11	I
540		4222.21	I
2700		4225.16	I
680		4232.02	I
680		4239.85	I
440		4245.91	I
440		4256.33	II
250		4276.69	I
370	d	4294.93	II
		4295.04	I
1000		4308.63	II
320		4325.86	I
200		4358.44	II

Dysprosium (Cont.)

320		4374.24	II
320		4374.76	II
540		4409.38	II
150		4444.58	I
740		4449.70	II
110		4455.60	II
250		4468.14	II
100	d	4527.58	I
		4527.76	II
100		4541.66	II
140		4565.09	I
420		4577.78	I
2100		4589.36	I
990		4612.26	I
50		4613.83	I
50		4614.82	I
60		4617.26	II
140		4620.03	II
50		4662.72	I
110		4664.66	II
85		4673.60	II
50		4682.03	II
50		4689.75	II
95		4698.68	II
85		4721.22	I
70		4727.13	II
170		4731.84	II
40		4745.73	II
60		4754.99	II
50		4760.04	II
60		4771.94	I
50		4774.80	I
120	h	4775.79	I
75		4786.92	II
95		4791.29	I
29		4800.64	I
50		4807.94	I
40		4810.28	I
50		4812.80	I
75		4819.04	I
85		4824.96	I
75		4828.88	I
50		4829.68	II
70		4832.38	I
35		4833.75	II
75		4841.75	I
40		4856.24	II
40		4868.05	II
40		4875.93	I
85		4880.16	I
40		4884.55	I
95		4888.08	I
40		4889.33	II
75		4890.10	II
50		4893.68	I
24		4899.24	I
55		4916.41	I
50		4922.22	II
65		4923.16	II
480		4957.34	II
24		4959.59	I
28		4973.57	I
40		4985.52	I
50		5003.87	I
55		5004.28	II
24		5010.60	I
24		5017.98	II
70		5022.12	I
30		5024.03	I
24		5024.54	I
40		5027.87	I
50		5033.00	I
160		5042.63	I
24		5047.25	I
50	h	5050.21	I
30		5053.35	I
24		5055.46	I
95		5070.68	I
120		5077.67	I
80		5090.38	II
80		5110.32	I
130	h	5120.04	I
30		5135.02	I
190		5139.60	II
40		5161.03	II
40		5164.12	II
50		5165.34	I
110		5169.69	II
20		5172.90	II
80		5185.30	I
40		5188.45	II
290		5192.86	II
95		5197.66	II
50		5246.94	II
70		5259.88	I
130		5260.56	I
55	B1	5263.3	D
65		5267.11	I
50		5272.25	II
50		5275.29	II
50		5279.70	II
55		5282.07	I
28		5284.99	II
40		5297.82	II
160		5301.58	I
40		5309.02	II
50		5324.69	I
24		5337.43	II
65		5340.30	I
30		5352.11	I
30		5368.20	II
20		5385.63	II
85		5389.58	II
40		5395.57	I
20	h	5398.26	D
24		5399.93	II
50		5404.19	I
80		5419.13	I
70		5423.32	I
30		5424.27	I
40		5426.70	II
30		5443.34	II
95		5451.11	I
30		5455.47	II
24		5469.10	II
28		5496.83	I
24		5502.79	I
28		5506.52	I
24		5515.41	II
30		5528.01	I
65		5547.27	I
40	d	5600.65	II
24		5605.53	I
30		5613.23	I
20		5627.49	I
100		5639.50	I
55	h	5645.99	I
80		5652.01	I
24		5685.58	I
28	h	5693.67	D
24	h	5694.10	D
28	Cw	5694.54	D
28		5698.72	II
24		5702.91	I
70	h	5718.46	I
28	h	5725.84	D
55	h	5728.64	D
24	h	5738.73	D
50		5740.20	I
55		5745.53	I
24		5750.48	I
24		5758.79	I
80	h	5832.01	D
55	h	5833.85	D
40	h	5834.86	D
28	h	5844.41	D
24		5845.65	D
40	h	5848.05	D
40		5855.56	D
55	h	5868.11	II
40		5915.16	II
20		5924.56	II
70		5945.80	I
50	l	5964.46	I
120		5974.49	I
24		5984.86	I
140		5988.56	I
24	h	6005.75	D
24	h	6006.54	D
24	h	6006.97	D
30		6008.94	I
65		6010.82	I
24		6017.26	I
24		6030.98	I
24	B1	6042.49	D
24		6058.18	I
30		6085.06	I
140		6088.26	I
24		6127.15	I
24		6133.64	I
24		6158.28	I
100		6168.43	I
20		6196.23	II
270		6259.09	I
30		6260.36	I
14		6343.32	I
40		6386.80	I
24		6396.60	II
50		6421.92	I
13	h	6436.55	I
8		6460.83	I
10		6468.58	II
11		6474.91	I
20		6483.59	II
28		6486.59	I
20		6558.02	I
160		6579.37	I
14		6594.14	II
15		6643.37	I
22		6658.36	I
29		6661.64	I
75		6667.86	I
10		6700.64	II
29		6747.93	I
10		6757.62	I
45		6765.89	I
12		6818.20	I
180		6835.42	I
80		6852.96	I
22		6856.46	I
22		6888.83	I
15		6897.97	II
65		6899.32	II
22		6906.53	II
15		6929.55	I
29		6950.28	II
11		6951.42	I
40		6958.08	I
13	h	6982.44	I
13		6991.30	I
45		6998.10	I
20		7017.42	I
35		7055.95	II
24		7075.14	II
17		7109.26	II
11		7120.81	II
13		7175.11	II
11		7213.27	I
17	h	7230.04	I
13		7250.01	I
17		7345.13	II
11		7370.23	II
20		7376.04	I
11		7407.59	I
24		7412.37	I
55		7426.86	II
20		7457.05	II
17		7516.61	II
55		7543.73	I
17	h	7553.00	I
27		7559.78	I
40		7562.96	II
20	h	7577.46	II
27	h	7591.30	I
13	h	7611.55	I
11	h	7617.70	I
35	h	7641.09	I
17		7645.86	I
13		7646.64	I
80		7662.36	I
11		7666.78	II
35		7715.33	I
45		7729.76	II
20		7751.62	II
35		7812.06	I
27		7909.38	I
11		7968.63	I
12		7982.85	II
13		8147.29	I
27		8198.77	II
100		8201.57	II
11		8218.62	II
20		8265.53	I
35		8326.10	I
35		8392.01	II
12		8405.85	II
20		8416.64	II
24		8438.58	II
11		8630.12	I
27		8655.94	II
17		8657.68	II
17		8678.49	II
11		8696.83	II
11	h	8715.95	II
20		8750.40	II
12		8780.83	I
45		8791.39	II
13		8833.08	II
24		8850.37	II

EINSTEINIUM (Es)
Z = 99

Es I and II
Ref. 333—J.G.C.

Intensity		Wavelength	
		Air	
300	l	2694.32	II
100	l	2703.84	
10000	s	2708.66	II
100	s	2716.02	II
1000	s	2724.57	II
1000	l	2765.76	
3000	l	2787.10	II
1000	l	2796.11	II
3000	l	2815.15	
100	s	2885.84	II
100	s	2886.44	
3000		2907.03	
100		3003.28	
1000	l	3065.40	
3000		3135.25	I
1000	l	3154.27	II
10000		3413.17	
300	l	3423.12	I
100	l	3424.28	
10000	s	3428.48	I
300		3437.31	
100		3437.34	
3000	s	3445.25	
300	l	3446.93	
3000		3452.36	
100		3453.16	
300	s	3470.77	
3000	s	3484.59	I
300		3494.30	
10000	s	3498.11	I
10000	l	3514.33	I
10000	s	3521.38	
10000	s	3523.49	I
300	l	3528.58	
10000	s	3536.01	
10000	s	3547.75	II
300		3549.97	
10000	l	3555.34	I
100	s	3555.53	
3000	s	3556.65	
3000	s	3560.92	
1000	s	3575.68	
100	s	3578.56	
300	s	3579.38	
1000	s	3582.95	
3000		3590.28	
1000		3595.47	
10000	s	3602.43	II
300		3605.58	
1000	s	3606.75	
100	l	3624.52	
3000	l	3631.09	
10000	s	3632.87	
100	l	3634.41	
1000	l	3651.94	
10000	l	3670.01	II
100	l	3672.32	
100	s	3713.56	
1000	l	3720.56	
300		3722.32	
10000	l	3728.55	II
100	h	3737.47	
300	s	3776.27	
10000	l	3792.99	
10000	l	3801.49	
300	s	3929.10	
3000		3930.77	
100		3957.19	
100		3995.35	
300		4077.71	
10000	s	4082.24	
3000	l	4107.59	
1000	l	4176.94	
100	s	4496.25	
100	l	4631.66	
300	l	4650.86	
1000		4789.93	
300	h	4802.17	
1000	h	4802.21	
3000	h	4958.29	
10000	s	5052.08	
1000	s	5102.93	
100	s	5155.82	
10000	s	5161.74	I
10000	l	5204.40	I
3000	L	5615.51	I
100	S	6539.71	ES

ERBIUM (Er)
Z = 68

Er I and II
Ref. 1—C.H.C.

Wavelength			Intensity
		Air	
110		2358.51	II
100		2386.58	II
120		2387.17	II
110		2396.38	III
140		2446.39	II
100		2537.02	II
110		2547.28	II
290		2586.73	II
110		2587.04	II
130		2592.57	II
120		2595.03	II
140		2624.18	II
490		2670.26	II
330		2672.25	II
100		2675.35	II
270		2739.31	II
310		2750.19	II
230		2755.01	II
610		2755.63	II
510		2770.02	II
230		2778.97	II
230		2802.53	II
310		2804.35	II
410		2820.19	II
270	d	2833.91	II

Erbium (Cont.)

390		2838.71	II
270		2848.37	II
250		2855.41	II
310		2859.84	II
310		2896.96	II
390		2897.52	II
1000		2904.47	II
210		2909.58	II
1500		2910.36	II
270		2915.62	II
350		2929.27	II
270		2945.28	II
230		2946.62	II
1500		2964.52	II
410		2968.76	II
210		2974.47	II
230		2975.68	II
270		2983.80	II
1200		3002.41	II
310		3002.65	II
230		3012.47	II
230		3016.84	II
290		3025.95	II
270		3028.27	II
370		3031.31	II
310		3036.22	II
210		3054.42	II
230		3066.22	II
450		3070.74	II
560		3072.53	II
610		3073.34	II
210		3078.87	II
720		3082.08	II
610		3084.02	II
370		3099.19	II
230		3106.78	II
310	d	3113.43	II
		3113.54	II
770		3122.72	II
290		3132.52	II
470		3132.77	II
410		3141.10	II
		3141.15	II
250		3144.33	II
410		3154.29	II
870		3181.92	II
410		3183.42	II
250		3185.25	II
310		3200.58	II
230		3205.15	II
270		3214.44	II
870		3220.73	II
610		3223.31	II
210		3227.16	II
2300		3230.58	II
250		3232.03	II
330		3237.98	II
330		3249.34	II
560		3259.05	II
		3259.11	II
2700		3264.78	II
430		3267.10	II
		3267.18	II
330		3269.41	II
250		3278.22	II
720		3279.33	II
720		3280.22	II
470		3286.77	II
330	d	3303.88	II
		3303.95	II
370		3305.56	II
2300		3312.42	II
560		3316.39	II
770		3323.19	II
290		3329.66	II
770		3332.70	II
370		3337.25	II
290		3337.79	II
250		3340.03	II
290		3341.84	II
1300		3346.04	II
470		3350.06	II
350		3350.26	II
1400		3364.08	II
1400	d	3368.02	II
		3368.13	I
450		3370.55	II
7700		3372.71	II
970		3374.17	II
290		3381.32	II
230		3382.06	I
1700		3385.08	II
450		3389.74	II
2300		3392.00	II
350		3396.07	II
290		3396.84	II
390		3401.83	II
350		3417.63	II
490		3428.39	II
770		3441.13	II
390		3442.68	I
490		3469.51	I
970		3471.71	II
610		3479.41	II
970		3485.85	II
350		3486.82	II
350		3496.86	II
6700		3499.10	II
610		3502.78	I
390		3508.38	II
490		3514.89	II
390		3518.18	II
610		3524.91	II
410		3539.59	I
310		3548.26	II
820		3549.84	II
310		3553.20	II
1500		3558.02	I
510		3558.71	I
1000		3559.90	II
310		3565.17	I
920		3570.75	II
310		3578.24	I
1000		3580.52	II
370		3586.60	I
610		3590.76	I
410		3595.84	I
610		3599.50	II
1000		3599.83	II
510		3604.90	II
410		3607.42	I
3100		3616.56	II
510		3617.85	II
510		3618.92	II
720		3628.04	I
310		3629.37	I
1000		3633.54	II
510		3634.67	I
1600		3638.68	I
900		3645.94	II
520		3650.41	II
360		3652.58	II
500		3652.87	II
360		3664.45	I
470		3669.02	II
500		3682.70	II
320		3684.01	I
380		3684.28	II
7900		3692.65	II
450		3696.25	II
380		3697.68	I
540		3700.72	II
520		3707.64	II
520		3712.39	II
320		3719.35	I
1300		3729.52	II
450		3731.26	II
540		3738.16	II
340		3741.10	II
900		3742.64	II
900		3747.43	I
540		3756.05	I
410		3781.01	II
1800		3786.84	II
560		3787.86	II
560		3791.83	II
500		3792.79	I
560		3797.06	II
1600		3810.33	I
3600		3830.48	II
540		3849.91	II
320		3851.60	II
680		3855.90	I
540		3858.39	II
7500		3862.85	I
1500		3880.61	II
1200		3882.89	II
400		3890.61	II
4200		3892.68	I
5200		3896.23	II
810		3902.76	II
250		3903.98	I
250		3904.56	II
1200		3905.40	I
11000		3906.31	II
280		3918.05	I
210		3918.35	II
280		3921.88	II
810		3932.25	II
3200		3937.01	I
2100		3938.63	II
3200		3944.42	I
550		3948.06	I
250		3951.48	I
320		3956.42	I
280		3966.35	I
2700		3973.04	I
3200		3973.58	I
1400		3974.72	II
280		3976.73	I
810		3977.02	I
1100		3982.33	I
280		3987.53	I
810		3987.66	I
230		3991.15	I
230		4004.05	I
14000		4007.96	I
230		4008.18	II
280		4009.16	II
1100		4012.58	I
350		4015.57	II
3000		4020.51	I
450		4021.55	I
230		4043.01	II
1000		4046.96	I
280		4048.34	II
200		4049.49	II
940		4055.47	II
550		4059.51	I
690		4059.78	II
420		4077.88	I
550		4081.24	II
3500		4087.63	I
210		4092.90	I
1100		4098.10	I
350		4100.56	II
320		4116.36	I
320		4118.55	I
600		4131.50	I
550		4142.91	II
6900		4151.11	I
280		4189.98	II
1000		4190.70	I
130		4205.32	I
1400		4218.43	I
200		4220.99	I
320		4230.20	II
140		4234.78	II
200		4251.94	II
140		4276.48	II
690		4286.56	I
320		4298.91	I
320		4301.60	II
140		4303.81	II
110		4319.94	II
130		4328.81	I
110		4331.36	I
140		4340.92	I
190		4348.34	I
110		4369.39	II
160		4382.17	I
300		4384.70	II
300		4386.40	I
100		4403.17	II
810		4409.34	I
180		4418.70	I
570		4419.61	II
110		4422.51	II
320		4424.57	I
370		4426.77	I
110		4437.66	I
100		4459.24	II
100		4473.50	II
130		4496.39	I
200		4500.75	II
130		4522.74	I
160		4563.26	II
1000		4606.61	I
160		4630.88	II
110		4640.60	II
110		4665.44	II
310		4673.16	I
570		4675.62	II
150		4679.06	II
230		4722.69	I
150		4729.05	I
130		4751.52	II
170		4759.65	II
190		4820.35	II
140		4857.44	I
150		4872.09	II
210		4900.08	II
210		4934.11	II
130		4944.36	I
180		4951.74	II
130		4976.42	I
250		5007.25	I
140		5028.33	I
120		5028.91	II
200		5035.94	I
210		5042.05	II
130		5043.86	I
130		5044.89	I
130		5077.59	II
120		5124.56	I
130		5127.41	II
120		5131.53	I
130		5133.83	II
170		5164.77	II
130		5172.78	I
160		5188.90	II
150		5206.52	I
60		5212.91	II
30		5215.13	II
30		5218.26	II
45		5229.34	II
140		5255.93	II
22		5256.47	II
27		5257.02	II
35		5264.77	II
80		5272.91	I
55		5277.71	I
27		5279.34	II
45		5302.30	II
55		5333.06	I
27		5333.33	II
27		5334.23	II
22		5343.94	II
30		5344.50	II
90		5348.06	I
45		5350.47	I
35		5368.85	I
35		5395.87	II
60		5414.63	II
18		5422.81	II
18	h	5451.30	I
35		5454.07	II
180		5456.62	I
35		5462.43	II
90		5468.32	I
18		5477.47	II
80		5485.97	II
27		5497.44	II
27		5516.02	I
80		5593.46	I
45	d	5601.14	I
		5601.32	I
45	h	5609.94	I
60		5611.82	I
70		5622.01	I
80		5626.53	II
30		5636.20	I
90		5640.36	I
22		5641.42	I
22	h	5658.63	II
70		5664.95	I
45		5665.44	II
55		5675.48	I
14		5695.53	II
27		5710.87	II
55		5717.48	I
70		5719.55	I
55		5726.97	I
22		5733.43	II
22		5736.56	I
22		5736.94	I
100		5739.19	I
35		5740.61	I
60		5748.65	I
55		5752.53	I
70		5757.63	II
290		5762.80	I
70		5769.92	I
45		5782.82	I
70		5784.66	I
22		5791.15	II
70		5800.79	I
22		5806.10	I
430		5826.79	I
45		5835.84	I
100		5850.07	I
120		5855.31	I
140		5872.35	I
120		5881.14	I
27		5886.30	II
27		5902.08	II
55		5906.06	I
45		5909.24	I
35		5933.50	I
22		5946.37	I
55		5968.68	I
27		5975.49	I
35		6006.79	II
22		6008.75	II
55		6014.83	I
35		6015.74	II
70		6022.56	I
22		6032.12	II
22		6045.63	II
22		6048.14	II
45		6054.85	I
70		6061.25	I
60		6076.45	II
35		6116.01	I
35		6125.32	I
30		6170.06	II
27		6183.21	II
360		6221.02	I
35		6230.90	I
55		6262.56	I
45		6267.93	I
60		6268.87	I
35		6274.94	I
30		6286.86	I
45		6299.42	I
130		6308.77	I
55		6326.13	I
22		6347.16	II
45		6388.19	I
22		6432.53	I
27		6485.87	I
55		6492.35	I

Erbium (Cont.)

22		6541.57	I
60		6583.48	I
70		6601.11	I
27		6721.91	I
70		6759.87	I
22		6762.92	I
27		6773.37	I
35		6790.92	I
22		6825.44	I
22		6825.98	I
70		6848.10	I
55		6865.13	I
27		6879.98	I
22		7001.40	I
12		7058.55	I
12	h	7065.04	I
11		7070.99	II
18		7101.27	I
8		7109.67	I
11		7155.40	II
5		7161.91	I
14	h	7197.00	I
7	h	7264.82	II
7	h	7283.95	I
14		7329.73	II
18		7355.37	I
11		7356.34	I
18		7428.67	I
55		7459.55	I
9		7460.42	I
120		7469.51	I
22		7532.34	I
6	h	7539.18	I
27		7556.26	I
6	h	7574.21	I
5	h	7590.51	I
11		7597.33	I
6		7607.23	I
11		7613.52	I
6		7623.48	I
16	h	7645.67	I
8		7650.63	I
22		7654.45	II
12	h	7658.05	I
22		7659.25	I
4		7665.64	I
35		7680.01	I
9		7722.14	I
8		7726.19	II
11	h	7747.44	I
22		7754.63	I
4		7762.16	I
9		7796.69	I
35		7797.47	I
9		7838.80	I
11		7844.00	I
16		7847.55	I
5		7875.36	I
5		7879.36	I
18		7899.55	I
8		7913.08	I
35		7921.85	I
30		7937.84	I
8		7952.93	I
12		7964.51	I
8		7979.03	II
8		7980.87	I
5		8023.03	I
12		8035.91	I
12		8181.85	I
35		8312.82	I
18		8328.57	II
5		8367.58	II
55		8409.90	I
11		8466.18	II
35		8472.42	I
14		8517.71	II
18		8521.37	II
22		8768.64	I
11	h	8776.63	II
9		8866.84	II

Er III
Ref. 301—J.R.

Intensity	Wavelength	
	Air	
2	2165.26	III
3	2190.77	III
10	2198.15	III
1	2223.98	III
4	2232.35	III
60	2235.28	III
8	2245.60	III
2	2255.95	III
80	2269.36	III
600	2277.65	III
100	2309.19	III
40	2358.69	III
10	2358.79	III
50	2359.33	III
200	2367.64	III
20	2375.50	III
10	2377.07	III
80	2381.25	III
20	2381.40	III
40	2381.75	III
6	2391.96	III
60	2393.08	III
5	2393.60	III
250	2396.40	III
10	2398.91	III
2	2402.75	III
80	2404.58	III
100	2410.47	III
200	2419.81	III
200	2422.47	III
40	2431.51	III
60	2464.60	III
2	2492.04	III
100	2508.59	III
2	2531.03	III
100	2532.36	III
8	2536.76	III
80	2540.91	III
3	2543.31	III
10	2545.95	III
50	2557.22	III
80	2570.74	III
40	2580.02	III
2	2589.55	III
80	2590.72	III
20	2591.56	III
200	2591.83	III
20	2598.39	III
3	2599.18	III
100	2603.62	III
40	2604.91	III
25	2614.53	III
30	2617.64	III
2	2618.40	III
2	2618.94	III
8	2625.19	III
20	2626.37	III
4	2637.52	III
200	2637.77	III
5	2651.49	III
25	2683.10	III
400	2723.29	III
100	2738.53	III
500	2739.27	III
8	2741.41	III
80	2746.03	III
6	2752.20	III
80	2756.20	III
400	2759.23	III
150	2761.92	III
60	2762.66	III
15	2767.11	III
100	2768.72	III
60	2772.07	III
60	2774.80	III
20	2775.55	III
2	2780.60	III
80	2783.11	III
500	2792.54	III
10	2804.10	III
100	2805.87	III
6	2808.44	III
50	2824.75	III
150	2830.34	III
1	2831.95	III
8	2833.03	III
8	2845.29	III
60	2846.08	III
6	2849.63	III
1	2869.52	III
8	2878.24	III
1	2955.93	III
1	2958.63	III
1000	3055.10	III
1000	3070.40	III
500	3100.40	III
1500	3166.25	III
3	3172.47	III
1	3173.45	III
50	3175.74	III
400	3214.95	III
2000	3301.23	III
8	3341.00	III
200	3480.54	III
8	3592.96	III
600	3715.67	III
200	3739.43	III
4000	3816.78	III
600	3962.87	III
40	4009.70	III
2	4088.58	III
1000	4288.18	III
40000	4290.06	III
300	4338.24	III
20000	4386.86	III
30	4612.93	III
15000	4735.56	III
2000	4783.12	III
8	4876.07	III
8000	5903.30	III

EUROPIUM (Eu)
Z = 63

Eu I and II
Ref. 1—C.H.C.

Intensity		Wavelength	
		Air	
21		2499.39	II
26		2554.78	II
26		2559.18	II
160		2564.17	II
110		2568.17	II
26		2574.76	II
230		2577.14	II
26		2581.86	II
26		2604.61	I
30		2635.50	II
1000		2638.77	II
380		2641.27	II
40		2653.61	II
640		2668.34	II
110		2673.42	II
250		2678.29	II
250		2685.66	II
550		2692.03	II
700		2701.14	II
800		2701.90	II
240		2705.28	II
180		2709.99	I
700		2716.98	II
70		2723.96	I
4200		2727.78	II
190		2729.33	II
380		2729.44	II
50		2731.37	I
40		2732.61	I
80		2735.25	I
160		2740.62	II
70		2743.28	I
120		2744.26	II
40		2745.61	I
70		2747.29	II
80		2747.83	I
90		2752.17	II
480		2781.89	II
1900		2802.84	II
220		2811.75	II
30		2813.08	II
3400		2813.94	II
550		2816.18	II
2000		2820.78	II
400	Cw	2828.72	II
120		2829.30	II
140		2833.26	II
80		2843.96	II
60		2852.05	II
260		2859.67	II
280		2862.57	II
25		2864.42	II
60		2876.06	II
100		2878.87	I
80		2887.85	II
200		2892.54	I
140		2893.03	I
360		2893.83	I
3200		2906.68	II
160		2908.99	I
30		2917.44	II
850		2925.04	II
60		2947.29	II
200	Cw	2952.68	II
30		2958.91	I
35		2959.47	II
260		2960.21	II
300		2991.33	II
35		2995.22	II
40		3006.26	II
35		3022.15	I
30		3040.77	II
320	Cw	3054.94	II
120		3058.98	I
35		3069.11	II
35		3076.07	II
220		3077.36	II
35		3089.35	II
120		3097.45	II
320		3106.18	I
950		3111.43	I
120		3130.73	II
40		3132.16	I
45		3149.88	II
85		3173.61	II
40		3185.54	I
420		3210.57	I
1000		3212.81	I
420		3213.75	I
45		3235.13	I
95		3241.40	I
45		3246.03	I
45		3247.32	II
100		3247.55	I
100		3266.39	II
150		3272.77	II
210		3277.78	II
150		3301.95	II
45		3304.50	II
140		3308.02	II
140		3313.33	II
65		3319.89	II
95		3321.86	II
85		3322.26	I
950		3334.33	I
45		3338.75	II
110		3350.40	I
40		3351.56	II
40		3354.38	II
45		3367.64	II
140		3369.06	II
65		3380.25	II
75		3390.78	II
190		3391.99	II
280		3396.58	II
45		3419.84	II
65		3423.09	II
150		3425.02	II
45		3426.44	II
45		3435.05	II
65		3435.20	II
40		3435.72	II
45		3440.82	II
150		3441.00	II
45		3445.18	II
85		3457.05	I
45		3457.56	II
130		3461.38	II
85		3467.88	I
75		3477.07	I
75	h	3505.30	II
470	Cw	3521.09	II
75		3531.15	II
45		3532.23	II
65		3538.08	II
150		3542.15	II
85		3543.85	II
45		3549.71	II
180		3552.52	II
75		3589.27	I
45		3591.31	II
150		3603.20	II
75		3611.57	II
45		3616.15	II
95		3622.54	II
95		3632.18	II
45		3673.19	II
45		3674.63	II
45		3678.26	II
6400		3688.42	II
60		3710.87	II
95		3713.45	II
95		3714.90	II
35		3716.94	II
35		3717.69	II
40		3719.16	I
20000	Cw	3724.94	II
45		3729.68	II
45		3729.74	II
21		3732.20	I
45		3738.08	II
350		3741.31	II
100		3743.56	II
260		3761.12	II
95		3765.93	II
40		3774.10	I
60		3781.40	II
40		3788.76	II
45		3791.50	II
130		3799.01	II
70		3801.36	
95		3807.54	II
120		3811.33	I
120		3815.50	II
39000	Cw	3819.67	II
120		3826.68	II
140		3844.23	II
190		3865.57	I
45		3872.72	I
70		3877.27	II
150		3884.75	I
23		3896.78	I
23		3900.18	I
70		3900.51	I
28000	Cw	3907.10	II
45		3915.24	II
45		3916.00	I
230		3917.29	I
23		3917.70	II
40		3918.52	I
100		3919.09	II
40		3928.87	II
32000	Cw	3930.48	II
55		3941.56	II

30	h	3942.21	II
60		3942.94	II
120		3943.08	II
30		3944.59	II
30		3945.67	II
30		3949.13	II
60		3949.60	I
45		3950.76	II
55		3951.33	II
60		3955.75	I
40		3957.92	II
30		3963.61	I
120		3964.90	II
150		3966.59	II
45		3967.18	I
30000	Cw	3971.96	II
60		3978.42	I
30		3979.63	II
[illegible]	h	[illegible]	I
40		3988.24	II
30		3993.93	II
55		3995.98	II
60		4003.71	II
180		4011.69	II
150		4017.58	II
120		4039.19	I
45	h	4078.24	I
120		4085.38	II
75		4096.80	II
60		4106.88	I
90	h	4112.04	II
45		4119.30	II
75		4127.28	I
33000	Cw	4129.70	II
30		4136.59	II
40		4137.07	I
30		4141.02	II
60		4141.72	II
30		4151.52	II
45		4151.64	II
30		4157.72	I
110		4172.80	II
30		4175.16	II
110		4182.22	I
40		4196.18	II
60000	Cw	4205.05	II
45		4221.08	II
40		4223.88	II
90	h	4227.40	II
75		4229.33	II
75		4232.45	II
90		4237.51	II
45		4238.69	II
45		4244.74	I
45		4247.06	II
45		4253.80	II
30		4270.24	II
150		4298.73	I
90		4329.36	I
75		4329.97	I
60		4330.61	II
40		4331.18	I
90		4337.68	I
240		4355.09	II
27		4361.57	II
55		4369.47	II
45	h	4372.20	II
75		4383.17	II
90		4387.88	I
21	h	4405.27	II
55		4407.07	II
18		4419.66	II
120		4434.81	II
14000	Cw	4435.56	II
75	h	4464.97	II
24		4485.15	II
3000		4522.57	II
45	h	4535.59	I
11000		4594.03	I
21		4602.63	I
9800		4627.22	I
8300		4661.88	I
30		4713.59	I
27		4740.50	I
45		4792.59	I
40	h	4829.30	I
60		4830.33	I
40	h	4840.47	I
60	h	4849.64	I
110		4867.62	I
40	h	4884.05	I
90		4894.68	I
60		4900.86	I
150		4907.18	I
180		4911.40	I
55		4953.52	I
55		4960.21	I
55		4962.55	I
45		4975.76	I
180		5013.17	I
170		5022.91	I
110		5029.54	I
90		5033.55	I
75		5067.95	I
75	h	5092.69	I
90	h	5096.44	I
170		5114.37	I
90		5124.77	I
170		5129.10	I
90		5130.08	I
210		5133.52	I
270		5160.07	I
210		5166.70	I
60		5193.74	I
200		5199.85	I
110		5200.96	I
120		5206.44	I
750		5215.10	I
300		5223.49	I
120		5239.24	I
200		5266.40	I
390		5271.96	I
110		5272.48	I
150		5282.82	I
55		5287.25	I
60		5289.25	I
120		5291.26	I
60		5293.68	I
120		5294.64	I
90		5303.85	I
30	h	5350.41	I
75	h	5351.69	I
40		5352.84	I
90		5355.10	I
540		5357.61	I
60		5360.83	I
120		5361.61	I
110		5376.94	I
120		5392.94	I
450		5402.77	I
45		5405.33	I
45		5411.86	I
55		5421.07	I
90		5426.94	I
40		5443.56	I
380		5451.51	I
260		5452.94	I
40		5457.62	I
90		5472.32	I
120		5488.65	I
45		5495.20	I
15		5500.83	I
120		5510.52	I
30		5526.63	I
30		5533.25	I
30		5542.54	I
200		5547.44	I
150		5570.33	I
200		5577.14	I
75		5579.63	I
120		5580.03	I
90		5586.24	I
75		5586.83	I
18		5592.25	I
18		5599.80	I
18		5605.86	I
40		5618.81	I
60		5622.44	I
75		5632.54	I
210		5645.80	I
15		5651.11	I
60		5673.85	I
27		5681.10	I
27		5684.24	I
60		5730.87	I
60		5739.00	I
330		5765.20	I
180		5783.69	I
15		5792.72	I
60		5800.27	I
170		5818.74	II
600	Cw	5830.98	I
27		5845.77	I
27		5860.97	I
15		5864.77	I
90		5872.98	II
15		5895.31	I
27		5902.97	I
12		5909.94	I
75		5915.74	I
12		5925.30	I
27		5926.52	I
45		5942.72	I
27		5953.49	I
27		5953.84	II
30		5954.28	I
90		5963.76	I
330		5966.07	II
480	Cw	5967.10	I
15	h	5968.43	I
30		5971.69	I
170		5972.75	I
15		5980.47	I
27		5983.14	I
27		5983.78	I
240		5992.83	I
60		6004.36	I
15	h	6005.61	I
60	h	6012.20	I
110		6012.56	I
60		6015.58	I
420		6018.15	I
60		6023.15	I
170		6029.00	I
60		6044.66	I
420		6049.51	II
140		6057.36	I
90		6075.58	I
30		6077.38	I
240		6083.84	I
240		6099.35	I
60		6108.15	I
120		6118.78	I
60		6124.67	I
330		6173.05	II
110		6178.76	I
260	Cw	6188.13	I
140		6195.07	I
15	h	6207.60	I
15		6230.51	I
90	h	6233.73	I
55		6250.47	I
240		6262.25	I
55		6266.95	I
15	h	6285.95	I
60		6291.34	I
170		6299.77	I
230		6303.41	II
24	h	6313.78	I
15		6318.58	I
75		6335.82	I
120	Cw	6350.04	I
60		6355.89	I
60		6369.25	I
55		6382.73	I
75		6383.86	I
120	Cw	6400.93	I
40		6406.11	I
180		6410.04	I
140		6411.32	I
55		6428.29	I
830		6437.64	II
18		6439.93	I
120		6457.96	I
12		6470.70	I
18		6483.02	I
45		6501.55	I
60		6519.59	I
15		6522.72	I
8	h	6549.12	I
75		6567.87	I
45		6593.79	I
18	h	6603.55	I
1400		6645.11	II
26		6685.21	I
95		6693.96	I
7	h	6701.06	I
12	h	6710.45	I
30		6744.88	I
30	h	6782.54	I
14	h	6787.48	I
140		6802.72	I
35		6816.06	I
11	h	6834.30	I
17		6840.93	I
17	h	6844.83	I
14	h	6847.04	I
360		6864.54	I
21		6898.21	I
60	h	6903.67	I
14	h	6910.17	I
30	h	6914.82	I
120		7040.20	I
12		7074.54	I
330		7077.10	II
100		7106.48	I
6		7164.66	I
30		7175.55	I
570		7194.81	II
570		7217.55	II
11	h	7224.68	I
15		7258.72	I
30		7262.77	I
11	h	7281.53	I
6	h	7297.56	I
540		7301.17	II
11		7310.46	I
12		7313.63	I
55	Cw	7336.18	I
4		7346.25	I
4		7356.65	I
11		7362.25	I
55	Cw	7369.60	I
720		7370.22	II
4		7387.36	I
12		7389.16	I
11	h	7404.41	I
300		7426.57	II
21	h	7436.59	I
8		7470.53	I
5	h	7491.00	I
50	Cw	7528.70	I
5	h	7533.02	I
6		7547.32	I
160		7583.91	I
60	Cw	7742.57	I
70		7746.19	I
8	h	7803.32	I
8		7818.21	I
35		7887.99	I
7		8015.47	I
24	Cw	8209.80	I
15	Cw	8226.81	I
6	h	8464.71	I
21	Cw	8642.67	I
7		8727.77	I
6		8782.46	I
12	Cw	8790.88	I
18		8870.30	I

Eu III

Ref. 312—J.R.

Intensity		Wavelength Air	
10		2073.40	III
10		2093.50	III
30		2124.69	III
10		2167.12	III
10		2173.59	III
10		2184.68	III
10		2190.59	III
10		2194.81	III
20		2211.85	III
20		2212.63	III
20		2214.66	III
10		2215.34	III
30		2217.23	III
30		2219.33	III
20		2219.42	III
10		2223.13	III
10		2235.17	III
10		2240.14	III
10		2261.88	III
20		2265.74	III
10		2269.39	III
20		2276.85	III
40		2291.62	III
20		2304.37	III
10		2311.92	III
10		2327.69	III
10		2334.56	III
10		2336.96	III
10		2339.84	III
10		2343.10	III
10		2346.83	III
10		2347.64	III
10		2350.38	III
200		2350.51	III
10		2352.28	III
10		2357.87	III
10		2359.08	III
10		2360.65	III
20		2363.76	III
20		2368.04	III
20		2374.08	III
10		2375.20	III
4000		2375.46	III
10		2376.42	III
20		2377.23	III
10		2381.81	III
10		2383.62	III
20		2387.29	III
20		2389.11	III
10		2389.98	III
20		2391.11	III
20		2391.90	III
10		2392.59	III
10		2394.66	III
20		2395.62	III
20		2398.79	III
20		2401.00	III
20		2402.34	III
20		2404.08	III
10		2406.14	III
20		2407.30	III
20		2408.32	III
10		2409.63	III
20		2410.08	III
40		2412.02	III
20		2412.96	III
20		2413.26	III
10		2413.41	III
10		2419.11	III
10		2419.25	III
10		2419.58	III
30		2422.00	III
10		2422.90	III
10		2425.33	III
50		2425.68	III
40		2427.67	III
40		2429.32	III
10	d	2429.66	III
10		2430.04	III

Europium (Cont.)

10		2431.49	III
10		2431.76	III
10		2432.55	III
10		2433.65	III
10		2434.19	III
100	d	2435.14	III
20		2436.39	III
10		2436.77	III
10		2438.83	III
10		2440.26	III
50		2440.67	III
1000		2444.38	III
4000		2445.99	III
30	h	2446.43	III
20		2448.57	III
20		2451.24	III
10		2451.73	III
30		2455.22	III
10		2461.79	III
30		2463.30	III
30		2464.47	III
40		2470.51	III
10		2474.94	III
10		2476.24	III
20		2476.45	III
10		2477.78	III
20		2480.02	III
20		2483.29	III
10		2486.92	III
10		2488.91	III
10		2490.50	III
10		2491.08	III
10		2492.48	III
10		2496.92	III
10		2499.17	III
2000		2513.76	III
20		2517.94	III
200		2522.14	III
10		2539.14	III
10		2548.30	III
20		2548.59	III
10		2554.50	III
10		2558.07	III
10		2560.36	III
10		2594.71	III
20		2594.76	III
10		2596.34	III
10		2604.44	III
30		2608.34	III
30		2610.09	III
50	c	2616.11	III
20		2616.26	III
10		2616.33	III
10		2616.35	III
10		2620.79	III
20		2623.33	III
10		2626.98	III
20	c	2628.46	III
10		2628.82	III
10		2631.98	III
30	c	2642.27	III
20		2645.22	III
20	c	2650.93	III
10		2653.19	III
10		2655.09	III
10		2662.24	III
20	c	2666.86	III
20	c	2668.21	III
40	c	2676.09	III
20	c	2683.21	III
20		2686.13	III
20	c	2687.74	III
40	c	2693.51	III
10		2694.80	III
20		2699.87	III
20	c	2700.78	III
20	c	2708.25	III
20	c	2708.84	III
10		2712.08	III
50	c	2720.67	III
20	c	2725.54	III
10		2743.94	III
10		2752.68	III
20		2755.12	III
10		2757.75	III
20	c	2760.21	III
10		2761.72	III
10	c	2766.26	III
20	c	2768.38	III
10	c	2768.54	III
10		2769.71	III
20	c	2780.48	III
20		2792.51	III
10		2808.09	III
10		2817.58	III
20	c	2839.56	III
20		2844.99	III
10	c	2848.44	III
20		2850.39	III
10	c	2892.60	III
40		2912.23	III
40	c	2912.64	III
10		2913.04	III
10		2928.91	III
20	c	2931.00	III
10	c	2950.20	III
20		2956.74	III
10		2956.90	III
10	c	2972.30	III
30	c	2982.29	III
20	c	3000.11	III
20		3006.37	III
20	c	3013.28	III
10	c	3018.43	III
20	c	3022.08	III
50	c	3022.69	III
20	c	3023.40	III
100	c	3023.93	III
10		3025.32	III
10	c	3026.09	III
200	c	3026.79	III
50	c	3029.92	III
20	c	3031.24	III
40	c	3032.84	III
20	c	3036.98	III
10		3038.64	III
10		3039.05	III
20	c	3039.98	III
10		3054.07	III
10	c	3054.97	III
20		3076.43	III
10	c	3089.09	III
10		3105.25	III
10		3109.67	III
10		3129.31	III
10		3142.54	III
50	c	3171.00	III
10		3178.08	III
20	h	3178.87	III
50	c	3183.78	III
10	h	3191.46	III
20	c	3194.34	III
10		3206.30	III
10		3208.95	III
10	h	3213.84	III
10		4837.98	III
50		6666.35	III
30		7221.84	III
20		7690.44	III
10		8079.07	III

FLUORINE (F)
Z = 9

F I and II
Ref. 169, 224—G.V.S.

Intensity		Wavelength	
Vacuum			
30		375.30	II
30		380.90	II
40		407.04	II
50		430.91	II
40		431.55	II
40		435.64	II
70		457.18	II
40		471.95	II
60		472.00	II
50		472.71	II
40		473.02	II
90		484.60	II
50		513.64	II
70		514.94	II
70		546.85	II
60		547.87	II
50		548.32	II
40		548.52	II
90		605.67	II
80		606.29	II
100		606.80	II
70		606.92	II
80		607.47	II
90		608.06	II
15		780.39	I
10		780.52	I
10		782.38	I
12		791.88	I
10		792.54	I
10		794.42	I
150		806.96	I
125		809.60	I
500		951.87	I
1000		954.83	I
750		955.55	I
500		958.52	I
20		972.40	I
350		973.90	I
100		976.22	I
40		976.51	I
100		977.75	I
40		1129.76	II
40		1327.06	II
50		1328.11	II
40		1333.59	II
50		1343.60	II
40		1344.04	II
50		1400.61	II
40		1407.14	II
60		1493.09	II
50		1493.24	II
40		1493.31	II
40		1702.13	II
40		1744.75	II
50		1745.55	II
60		1747.39	II
Air			
100		2556.11	II
100		2871.40	II
120		3059.99	II
140		3153.49	II
170		3202.76	II
140		3264.08	II
140		3414.65	II
150		3416.45	II
140		3416.80	II
160		3417.00	II
160		3472.96	II
150		3473.31	II
170		3474.78	II
190		3501.39	II
200		3501.45	II
200		3501.57	II
180		3502.84	II
200		3502.96	II
210		3503.11	II
170		3505.37	II
200		3505.52	II
220		3505.63	II
160		3522.89	II
150		3536.87	II
160		3541.77	II
160		3590.52	II
6		3594.10	I
170		3598.69	II
180		3601.39	II
190		3602.84	II
12		3668.17	I
180		3704.53	II
160		3710.35	II
160		3739.57	II
140		3805.83	II
270		3847.09	II
260		3849.99	II
250		3851.67	II
5		3898.48	I
190		3898.83	II
180		3901.93	II
170		3903.82	II
8		3930.69	I
5		3934.26	I
5		3948.56	I
150		3972.04	II
160		3972.67	II
170		3974.78	II
240		4024.73	II
220		4025.01	II
230		4025.49	II
160		4083.91	II
190		4103.07	II
170		4103.22	II
200		4103.51	II
180		4103.71	II
170		4103.87	II
170		4109.16	II
160		4116.54	II
150		4119.21	II
140		4207.15	II
170	h	4225.16	II
150	h	4244.12	II
200		4246.23	II
190		4246.39	II
180		4246.59	II
170		4246.77	II
160		4246.84	II
170	h	4275.36	II
160	h	4277.53	II
160	h	4278.93	II
200		4299.17	II
160		4446.53	II
170		4446.72	II
180		4447.19	II
140		4734.38	II
170		4859.39	II
160		4933.26	II
6		4960.65	I
140		5002.00	II
150		5173.25	II
15		5230.41	I
12		5279.01	I
18		5540.52	I
12		5552.43	I
10		5577.33	I
160		5589.27	II
20		5624.06	I
12		5626.93	I
15		5659.15	I
40		5667.53	I
90		5671.67	I
18		5689.14	I
25		5700.82	I
25		5707.31	I
12		5950.15	I
25		5959.19	I
70		5965.28	I
50		5994.43	I
150		6015.83	I
80		6038.04	I
900		6047.54	I
100		6080.11	I
800		6149.76	I
400		6210.87	I
13000		6239.65	I
140		6247.90	II
10000		6348.51	I
8000		6413.65	I
450		6569.69	I
300		6580.39	I
400		6650.41	I
1800		6690.48	I
400		6708.28	I
7000		6773.98	I
1500		6795.53	I
9000		6834.26	I
50000		6856.03	I
8000		6870.22	I
15000		6902.48	I
6000		6909.82	I
4000		6966.35	I
45000		7037.47	I
30000		7127.89	I
130		7179.90	II
15000		7202.36	I
130	h	7211.79	II
1000		7309.03	I
15000		7311.02	I
700		7314.30	I
5000		7331.96	I
10000		7398.69	I
4000		7425.65	I
2200		7482.72	I
2500		7489.16	I
900		7514.92	I
5000		7552.24	I
5000		7573.38	I
7000		7607.17	I
18000		7754.70	I
15000		7800.21	I
300		7879.18	I
500		7898.59	I
350		7936.31	I
300		7956.32	I
80		8016.01	II
1000		8040.93	I
900		8075.52	I
350		8077.52	I
350		8126.56	I
600		8129.26	I
300		8159.51	I
600		8179.34	I
300		8191.24	I
350		8208.63	I
2500		8214.73	I
3000		8230.77	I
500		8232.19	I
1500		8274.62	I
2000		8298.58	I
600		8302.40	I
900		8807.58	I
1000		8900.92	I
300		8912.78	I
350		9025.49	I
400		9042.10	I
350		9178.68	I
200		9433.67	I
25		9505.30	I
12		9662.04	I
25		9734.34	I
15		9822.11	I
12		9902.65	I
80	h	10047.98	II
15		10285.45	I
20		10862.31	I

F III
Ref. 225—G.V.S.

Intensity		Wavelength	
Vacuum			
50	h	230.12	III
50		255.72	III
60		255.77	III
70		255.86	III
70		261.71	III
60		261.75	III
80		263.81	III

Fluorine (Cont.)

Intensity		Wavelength	
70		279.69	III
80		315.22	III
70		315.54	III
60		315.75	III
100		429.51	III
110		430.15	III
80		430.22	III
90		464.29	III
100		465.11	III
120		508.39	III
120		567.69	III
110		567.75	III
80		630.14	III
90		630.20	III
120		656.12	III
130		656.87	III
140		658.33	III
80		1219.03	III
80		[illegible]	III
90		1267.71	III
70		1297.54	III
70		1359.92	III
110		1498.93	III
120		1502.01	III
110		1504.18	III
140		1504.79	III
130		1506.30	III
110		1506.77	III
100		1553.02	III
110		1557.59	III
100		1563.73	III
100		1565.54	III
100		1623.40	III
100		1650.76	III
130		1670.39	III
140		1677.40	III
100		1716.99	III
120		1770.09	III
150		1770.67	III
110		1772.93	III
140		1773.36	III
160		1791.65	III
110		1803.03	III
100		1804.70	III
170		1805.90	III
110		1839.30	III
120		1839.97	III
110		1840.14	III
80		1900.76	III
		Air	
100		2027.44	III
120		2030.32	III
120		2217.17	III
120		2452.07	III
130		2464.85	III
130		2470.29	III
120		2478.73	III
150		2484.37	III
120		2542.77	III
120		2580.04	III
130		2583.81	III
120		2593.23	III
130		2595.53	III
140		2599.28	III
130		2625.01	III
140		2629.70	III
120		2656.44	III
130		2755.55	III
160		2759.63	III
120		2788.15	III
160		2811.45	III
140		2833.99	III
150		2835.63	III
150		2860.33	III
120		2862.86	III
140		2887.58	III
150		2889.45	III
120		2905.30	III
140		2913.29	III
160		2916.34	III
140		2932.49	III
140		2994.28	III
120	h	2997.21	III
130		2997.53	III
120		2999.47	III
130		3039.25	III
120		3039.75	III
160		3042.80	III
150		3049.14	III
140		3113.62	III
160		3115.70	III
180		3121.54	III
140		3124.79	III
140		3134.23	III
140		3146.99	III
180		3174.17	III
170		3174.76	III
120		3214.00	III
140	h	4420.30	III
120	h	4427.35	III
120	h	4432.32	III
140	h	4479.99	III
150		5012.54	III
160		5110.99	III
140		5753.17	III
120		5761.20	III
150		6091.82	III
140		6125.50	III
130		6233.57	III
140		6363.05	III
120		7336.77	III
130		7354.94	III

F IV
Ref. 68, 226—G.V.S.

Intensity	Wavelength	
	Vacuum	
30	169.79	IV
30	169.84	IV
30	171.07	IV
40	176.37	IV
40	181.52	IV
40	181.57	IV
30	187.24	IV
50	196.39	IV
60	196.45	IV
50	199.76	IV
50	199.80	IV
50	199.85	IV
50	199.93	IV
50	200.00	IV
70	200.09	IV
60	201.01	IV
70	201.06	IV
60	201.10	IV
80	201.16	IV
60	201.22	IV
90	208.25	IV
70	213.85	IV
70	214.06	IV
70	220.77	IV
60	226.94	IV
50	227.10	IV
60	233.22	IV
50	233.39	IV
70	239.86	IV
70	240.02	IV
90	240.08	IV
70	240.15	IV
70	240.28	IV
70	240.37	IV
100	251.03	IV
60	270.23	IV
140	419.65	IV
150	420.05	IV
160	420.73	IV
150	430.76	IV
130	490.57	IV
160	491.00	IV
50	497.38	IV
60	497.83	IV
70	498.80	IV
140	570.64	IV
140	571.30	IV
150	571.39	IV
160	572.66	IV
140	676.12	IV
130	677.15	IV
150	677.22	IV
130	678.99	IV
160	679.21	IV
	Air	
40	2171.44	IV
50	2298.29	IV
40	2451.58	IV
50	2456.92	IV
40	2820.74	IV
50	2826.13	IV

F V
Ref. 68, 226—G.V.S.

Intensity	Wavelength	
	Vacuum	
40	134.54	V
40	147.95	V
50	148.00	V
40	152.51	V
40	158.54	V
40	162.27	V
40	163.50	V
50	163.56	V
90	165.98	V
100	166.18	V
40	174.70	V
50	178.43	V
40	178.59	V
40	182.98	V
40	186.72	V
40	186.79	V
50	186.84	V
40	186.97	V
40	187.01	V
60	190.57	V
70	190.84	V
40	191.97	V
40	205.55	V
100	464.37	V
110	465.37	V
120	465.98	V
100	466.99	V
90	506.16	V
100	508.08	V
70	513.97	V
60	514.08	V
80	524.59	V
90	525.29	V
100	526.30	V
70	647.67	V
100	647.77	V
110	647.87	V
70	647.97	V
130	654.03	V
110	657.23	V
140	657.33	V
60	757.04	V
60	1082.31	V
70	1088.39	V
	AIR	
10	2229.18	V
20	2252.72	V
20	2450.63	V
10	2461.33	V
10	2693.98	V
10	2702.30	V
10	2703.96	V
20	2707.17	V

FRANCIUM (Fr)
Z = 87

Fr I
Ref. 408 — C.H.C.

Intensity	Wavelength	
	Air	
	7177.	I

GADOLINIUM (Gd)
Z = 64

Gd I and II
Ref. 1—C.H.C.

Intensity	Wavelength	
	Air	
100	2468.22	II
55	2471.58	II
35	2485.67	II
70	2487.46	II
110	2488.72	II
55	2493.29	II
35	2496.35	II
45	2499.04	II
28	2543.68	II
28	2586.13	II
28	2661.50	II
70	2720.50	II
430	2750.22	II
460	2764.08	II
40	2768.51	II
320	2769.81	II
230	2770.17	II
21	2770.98	II
45	2778.76	I
45	2779.14	II
440	2781.40	II
70	2787.68	I
390	2791.96	II
100	2794.66	II
930	2796.93	II
60	2808.38	II
750	2809.72	II
160	2810.93	II
45	2814.01	II
300	2833.75	II
35	2836.69	II
70	2837.00	II
560	2840.23	II
140	2841.33	II
40	2853.91	II
60	2856.52	II
19	2859.78	II
120	2862.48	II
60	2865.06	II
40	2866.33	II
40	2871.75	II
460	2881.33	II
40	2882.13	II
130	2885.60	II
35	2907.44	II
170	2910.53	II
60	2913.08	II
45	2918.52	II
95	2923.32	II
35	2924.25	II
35	2928.34	II
35	2947.80	II
70	[illegible]	II
35	2952.43	I
35	2955.60	II
70	2960.93	II
130	2963.60	II
80	2965.43	II
29	2972.74	II
560	2980.15	II
35	2983.74	II
40	2991.52	II
95	2993.04	II
1200	2999.04	II
370	3002.86	II
100	3005.09	II
2100	3010.13	II
130	3012.19	II
1900	3027.60	II
120	3028.98	II
2100	3032.84	II
1600	3034.05	II
130	3043.01	I
160	3046.48	I
280	3053.57	II
100	3059.92	I
1000	3068.64	II
560	3072.56	II
640	3076.92	II
150	3077.08	II
2100	3081.99	II
140	3084.01	II
280	3089.95	II
140	3092.06	II
460	3098.64	II
190	3098.90	II
3500	3100.50	II
120	3101.18	II
230	3101.91	II
580	3102.55	II
130	3108.36	II
170	3111.19	I
160	3113.17	II
120	3118.60	II
120	3119.01	I
510	3119.94	II
100	3120.18	II
370	3123.99	II
120	3124.25	II
130	3128.56	II
130	3130.81	II
100	3133.09	II
460	3133.85	II
210	3135.03	II
190	3136.93	I
190	3137.30	I
120	3138.71	I
230	3143.13	II
930	3145.00	II
370	3145.52	II
230	3146.88	II
980	3156.53	II
200	3158.63	I
140	3160.69	II
980	3161.37	II
220	3190.28	I
220	3199.30	I
160	3199.58	I
110	3203.41	I
690	3223.74	II
	3223.78	I
110	3225.46	II
160	3226.32	II
220	3232.78	I
100	3250.19	II
110	3259.25	II
540	3266.73	I
250	3267.64	I
140	3268.34	II
110	3274.18	II
110	3279.53	II
100	3281.61	II
250	3282.25	I
	3282.30	II
430	3291.48	I
370	3292.21	II
430	3294.08	I
330	3313.73	II
200	3315.59	II

Gadolinium (Cont.)

430		3330.34	II
1400		3331.38	II
830		3332.13	II
1100		3336.18	II
590		3345.98	II
200		3350.10	II
5400		3350.47	II
220		3357.61	I
270		3358.43	II
4300		3358.62	II
780		3360.71	II
5400		3362.23	II
270		3364.24	II
200		3365.59	II
220		3374.69	II
220		3379.76	II
220		3380.52	II
1100		3392.53	II
540		3395.12	II
220	d	3397.22	I
		3397.32	I
200		3399.41	II
540		3399.99	II
540		3402.07	II
200		3406.92	I
1100	d	3407.56	II
		3407.61	II
250		3409.30	II
220		3411.02	I
220		3413.27	II
1400		3416.95	II
1400		3418.73	II
6900		3422.47	II
390		3422.75	II
1100		3423.90	I
		3423.92	II
830		3424.59	II
390		3425.93	II
220		3428.47	II
690		3432.99	II
1700		3439.21	II
830		3439.78	II
2700		3439.99	II
390		3449.62	II
1400		3450.38	II
1100		3451.23	II
540		3454.14	II
880		3454.90	II
200		3455.27	I
200		3457.05	II
220		3461.95	II
220		3463.00	II
2700		3463.98	II
330		3466.95	II
1700		3467.27	II
1700		3468.99	II
1400		3473.22	II
2200		3481.28	II
1700		3481.80	II
490		3482.60	II
220		3486.20	I
980		3491.95	II
1700		3494.40	II
1400		3505.51	II
780		3512.22	II
1100		3512.50	II
830		3513.65	I
980		3524.20	II
430		3528.54	II
540		3542.77	II
4300		3545.80	II
3900		3549.36	II
1400		3557.05	II
540		3558.19	II
430		3558.47	II
200		3564.05	II
690		3571.93	II
330		3574.74	II
390		3578.36	II
980		3581.91	II
5400		3584.96	II
540		3590.47	II
1100		3592.71	II
200		3593.44	II
540		3600.96	II
1100		3604.87	I
270		3605.26	II
250		3605.66	II
830		3608.75	II
830		3610.76	II
220		3610.91	II
540		3613.39	II
270	d	3614.21	II
		3614.42	I
430		3617.16	II
390		3620.46	II
270		3624.89	II
250		3629.51	II
330		3634.76	II
220		3639.05	II
250		3640.18	II
330		3641.39	II
870		3645.62	II
6100		3646.19	II
310		3649.44	II
450		3650.95	II
620		3652.54	II
3900		3654.62	II
3100		3656.15	II
210		3658.19	I
1400		3662.26	II
2700		3664.60	II
2000		3671.20	II
1000		3674.05	I
350		3679.21	I
2000		3684.13	I
720		3686.33	II
3100		3687.74	II
210		3694.03	II
2000		3697.73	II
1300		3699.73	II
2700		3712.70	II
2000		3713.57	I
1400		3716.36	II
2000		3717.48	I
1800	d	3719.45	II
		3719.53	II
250		3722.07	II
430		3725.47	II
1500		3730.84	II
270		3732.32	I
230		3732.45	II
230		3732.67	I
510		3733.08	II
490		3739.76	I
330		3740.02	II
4500		3743.47	II
620		3744.83	I
230		3757.74	II
1000		3757.94	I
1400		3758.31	II
820		3759.00	II
620		3760.71	II
290		3760.92	II
870		3762.20	I
210		3763.33	II
370		3764.20	II
870		3767.04	II
8700		3768.39	II
620		3769.45	II
1400		3770.69	II
250		3771.26	I
210		3773.45	I
210		3776.83	I
1000		3782.34	II
2900		3783.05	I
1100		3787.56	II
200		3790.63	I
770		3791.17	II
490		3792.39	II
5100		3796.37	II
720		3801.29	II
210		3804.39	I
210		3805.09	II
560		3805.52	II
3700		3813.97	II
430		3814.74	II
770		3816.64	II
430		3818.75	II
350		3826.05	II
230		3827.33	II
230		3829.46	II
370		3831.80	II
210		3832.97	I
330		3834.99	II
970		3836.91	II
1000		3839.64	II
1200		3842.20	II
1400		3843.28	I
1400		3844.58	II
3300		3850.69	II
5100		3850.97	II
4300		3852.45	II
470		3855.56	II
250		3863.05	II
1600		3866.99	I
250		3873.57	I
220		3875.46	II
1500		3894.70	II
450		3895.79	II
750		3902.40	II
300		3902.71	I
240		3904.29	I
450		3905.65	I
2200		3916.51	II
450		3923.25	II
1200		3934.79	I
		3934.82	II
220		3935.38	I
450		3941.80	I
590		3942.63	I
270		3943.24	I
220		3943.62	I
1400		3945.54	I
300		3952.00	II
590		3953.37	I
1200		3957.67	II
750		3959.44	II
		3959.52	GD II
220		3963.66	II
590		3966.28	I
590		3968.26	II
750		3969.00	I
270		3969.29	II
450		3971.75	II
390		3972.71	I
590		3973.98	II
300		3974.81	I
750		3979.33	I
450		3987.21	II
470		3987.84	I
320		3992.69	I
220		3993.21	II
650		3994.16	II
700		3996.32	II
320		3997.76	II
470		4001.26	II
260		4004.94	II
320		4008.33	I
300		4008.91	II
300		4013.80	II
200		4015.58	I
300		4017.25	I
430		4017.71	I
300		4019.73	I
300		4022.33	II
1100		4023.14	I
810		4023.35	I
220		4027.61	I
1100		4028.15	I
860		4030.88	I
700		4033.49	I
340		4035.40	I
260		4036.84	I
1400		4037.33	II
700		4037.90	II
410		4043.71	I
1600		4045.01	I
270		4046.84	II
270		4047.09	I
270		4049.20	I
1300		4049.43	II
2200		4049.86	II
270		4050.37	I
810		4053.29	II
2600		4053.64	I
810		4054.72	I
2600		4058.22	I
650		4059.88	I
270		4061.30	II
650		4062.59	II
1900		4063.39	II
540		4063.59	II
260		4066.04	I
520		4068.35	I
260		4068.74	I
750		4070.29	II
		4070.39	II
650		4073.20	II
300		4073.76	II
1300		4078.44	II
2800		4078.70	I
520		4083.70	I
1500		4085.56	II
260		4087.69	II
650		4090.41	I
1100		4092.71	I
260		4093.72	I
260		4094.48	II
2600		4098.61	II
520		4098.90	II
650		4100.26	I
390		4111.44	II
2200		4130.37	II
270		4131.48	II
1100		4132.28	II
750		4134.16	I
410		4137.10	II
280		4148.86	I
540		4162.73	II
280		4163.09	II
280		4167.16	II
		4167.27	I
2400		4175.54	I
2400		4184.25	II
2200		4190.78	I
750		4191.07	II
750		4191.63	I
450		4197.68	II
590		4204.86	II
1300		4212.00	II
970		4215.02	II
650		4217.20	II
320		4225.03	I
4800		4225.85	I
220		4227.14	II
220		4229.80	II
650		4238.78	II
200		4246.57	II
1700		4251.73	II
860		4253.37	II
650		4253.61	II
810		4260.12	I
1600		4262.09	I
650		4266.60	I
470		4267.00	I
300		4274.17	I
910		4280.49	II
430		4285.82	I
300		4286.12	I
540		4296.08	II
220		4297.17	II
430		4299.29	I
1100		4306.34	I
260		4309.29	I
1800		4313.84	I
520		4314.40	I
520		4316.05	II
370		4320.52	I
750		4321.11	II
		4321.20	I
2600	d	4325.57	II
		4325.69	I
1900		4327.12	I
370		4329.58	I
340		4330.61	II
240		4331.38	I
450		4341.28	II
910		4342.18	II
1000		4344.30	II
2200		4346.46	I
910		4346.62	I
220		4347.31	II
300		4369.77	II
970		4373.83	I
280		4392.06	I
1400		4401.86	I
520		4403.14	I
260		4406.67	II
260		4408.25	II
220		4409.25	I
520		4411.16	I
860		4414.16	I
700		4414.73	I
340		4419.03	II
1400		4422.41	I
1100		4430.63	I
240	d	4436.10	I
		4436.22	II
300		4464.74	I
300		4466.55	II
		4466.60	I
520		4467.08	I
700		4474.13	I
860		4476.12	I
220		4478.80	II
280		4481.06	II
220		4483.33	II
220		4484.70	I
280		4486.90	I
500		4497.13	I
220		4497.32	I
430		4506.21	I
140		4506.33	II
140		4514.50	II
1100		4519.66	I
300		4522.82	II
150		4524.12	I
910		4537.81	I
220		4540.02	II
300		4542.03	I
240		4548.00	I
120		4558.08	II
130		4573.81	I
260		4575.91	I
280		4579.59	I
410		4581.29	I
130		4582.53	II
410		4583.07	I
160		4586.99	I
220		4596.98	II
320		4597.91	II
410		4598.90	I
340		4601.05	II
240		4602.93	I
520		4614.50	I
140		4624.42	I
430		4636.64	I
110		4639.00	II
170		4640.04	I
170		4646.00	I
170		4647.64	I
170	d	4648.59	I
		4648.70	I
430		4653.54	I
140	h	4670.87	I
170		4679.18	I
260		4680.04	I
430		4683.33	I
140		4688.12	I
700		4694.33	I
170		4695.49	I

Gadolinium (Cont.)

430		4697.42	I
170		4703.13	I
200		4709.78	I
110		4721.46	I
150		4728.47	II
220		4732.60	II
260		4735.75	I
410		4743.65	I
110		4745.82	I
320		4758.70	I
110		4760.74	I
130		4763.82	I
470		4767.24	I
180		4781.92	I
300		4784.62	I
110		4786.75	I
140		4801.05	II
330		4807.15	I
320		4821.69	I
130		4835.26	I
110		4848.10	I
110		4862.59	I
170		4865.02	II
120		4871.50	I
280		4934.12	I
220		4938.61	I
110		4952.47	I
130		4958.79	I
65		5010.82	II
55		5011.74	I
750		5015.04	I
55		5023.13	II
65		5031.29	II
75		5039.09	I
65		5050.88	II
55		5073.74	I
55		5082.80	I
95		5092.25	II
65		5096.06	II
130		5098.38	II
55		5100.94	II
910		5103.45	I
180		5108.91	II
120		5125.56	II
65		5130.28	II
65		5135.59	I
75		5136.04	I
85		5140.84	II
75		5141.50	I
75		5142.68	I
860		5155.84	I
55		5156.76	II
75		5158.48	I
75		5163.70	I
55		5164.54	II
190		5176.28	II
55		5187.24	II
55		5187.88	I
55		5191.08	II
410		5197.77	I
55		5210.49	II
85		5217.48	I
280		5219.40	I
75		5220.30	II
130		5233.93	I
65		5246.87	I
320		5251.18	I
120		5252.14	II
85		5254.75	I
140		5255.80	I
65		5268.78	I
55		5272.91	I
55		5282.48	I
280		5283.08	I
280		5301.67	I
220		5302.76	I
55		5306.70	I
280		5307.30	I
130		5321.50	I
280		5321.78	I
110		5327.32	I
65		5328.30	I
170		5333.30	I
55		5337.53	I
300		5343.00	I
85		5345.13	I
75		5345.68	I
200		5348.67	I
300		5350.38	I
240		5353.26	I
55		5361.66	I
95		5365.38	I
95		5369.92	I
150		5370.63	I
85		5389.50	I
85		5413.20	I
85		5415.69	I
65		5453.46	I
55		5583.68	II
55	d	5591.85	I
190		5617.91	I
65		5629.55	I
110		5632.25	I
260		5643.24	I
55	Bl	5680.89	G
390		5696.22	I
95		5701.35	I
65		5709.42	I
120		5733.86	II
85		5746.36	I
85	d	5754.17	I
75		5776.02	I
240		5791.38	I
65	h	5796.80	I
55		5802.92	I
55	Hs	5807.72	I
55	h	5809.22	I
55		5815.85	II
65	Hs	5819.51	G
55		5840.47	II
230		5851.63	I
55		5855.24	II
280		5856.22	I
55		5860.73	II
65		5877.26	II
55		5886.46	I
55		5904.07	II
110		5904.56	I
170		5911.45	II
65		5913.55	II
55		5916.77	I
85		5930.29	I
85		5936.84	I
65		5937.71	I
55	h	5940.95	G
55	h	5942.78	G
55		5951.60	II
55		5956.48	II
85		5977.25	I
110	h	5988.02	I
85		5999.08	I
65		6000.96	G
75	h	6001.87	G
55		6004.57	II
55		6008.71	I
55		6021.13	I
55		6080.65	II
430		6114.07	I
55		6180.42	II
110	Bl	6182.68	G
110	Bl	6200.86	G
110	Bl	6211.71	G
110	Bl	6220.93	G
55	Bl	6231.62	G
75	Bl	6241.66	G
55	b	6252.12	G
55	Bl	6262.64	G
45	b	6273.00	G
85		6289.73	II
30		6292.87	I
75		6305.15	II
30		6309.11	II
27		6317.19	I
40		6331.35	I
17		6333.75	I
17		6336.34	I
27		6346.65	II
27	h	6351.72	I
17		6363.23	I
40		6380.95	II
17		6382.19	II
22		6408.55	I
22		6422.42	II
17		6424.52	I
19	h	6470.29	I
15		6480.11	II
40	h	6538.15	I
22		6549.25	I
55		6564.78	I
10		6568.00	II
10		6573.80	I
30		6591.60	I
15		6593.42	I
10		6610.04	II
50		6634.36	II
35		6640.08	I
10		6642.76	I
30		6643.98	I
10		6646.85	I
10		6653.55	I
10		6679.56	II
35		6681.23	II
10		6692.86	I
10		6704.18	II
14		6718.14	II
17		6727.83	II
85		6730.73	I
50		6752.67	II
14		6753.91	II
14		6783.39	I
26		6786.33	II
10		6787.18	I
12		6814.56	I
26		6816.49	I
17		6820.90	I
100		6828.25	I
35		6846.60	II
30		6857.13	II
15		6864.25	I
21		6887.63	II
14		6900.73	II
100		6916.57	I
21		6920.62	II
15		6924.99	II
21		6926.49	I
17		6945.98	II
15		6957.74	II
15		6959.24	II
14		6964.33	I
15		6971.66	II
12		6976.35	II
10		6978.27	II
26	h	6980.86	I
50		6985.09	II
10		6988.75	II
75		6991.92	I
21		6993.18	I
60		6996.76	II
17		7000.75	II
45		7006.16	II
10		7016.60	I
21		7037.26	II
14		7051.00	II
13		7054.62	II
10		7058.02	II
10		7068.09	II
18		7071.00	I
18		7073.63	I
14		7098.11	I
14		7098.73	I
10	h	7116.77	II
21		7118.86	II
35		7122.57	I
13		7135.73	II
18		7147.31	II
13		7158.28	I
170		7168.37	I
21		7172.26	II
28		7189.57	II
13		7197.08	II
13		7201.41	II
10		7228.02	I
25		7233.45	I
14		7252.70	II
28		7262.66	I
14		7291.35	I
21		7313.28	I
18		7324.89	II
14		7373.81	I
14		7376.41	I
13		7377.27	II
13		7380.28	I
13		7394.90	II
13		7430.19	I
35		7441.85	I
40		7464.36	I
55		7562.97	I
10		7563.19	II
10		7588.20	I
10		7611.78	I
21		7621.96	I
21		7650.32	I
25		7672.56	I
10		7676.06	I
13		7694.45	I
80		7733.50	I
35		7749.30	I
10		7755.97	I
10		7766.48	II
11	h	7844.87	I
10		7845.80	I
35		7846.35	II
35		7856.93	I
14		7869.72	I
25		7930.25	II
13		8077.59	I
18		8146.15	I
11		8218.08	I
10		8275.42	I
10		8349.73	I
11		8398.30	I
10		8445.47	I
13	h	8527.88	I
21		8668.63	I
11		8770.36	I
13		8784.85	I
10		8795.76	I
21	h	8832.06	II
14	h	8849.14	I
18	h	8867.31	I

Gd III

Ref. 46, 137, 151—J.F.K.

Intensity	Wavelength	
Vacuum		
600	1813.47	III
900	1946.26	III
1100	1974.34	III
2200	1975.24	III
Air		
900	2008.79	III
3400	2018.07	III
1800	2027.82	III
800	2046.02	III
500	2057.79	III
1500	2080.08	III
1800	2098.20	III
1300	2125.68	III
1400	2148.03	III
1700	2176.84	III
1700	2223.95	III
1700	2230.73	III
1200	2239.84	III
1500	2243.75	III
1700	2250.18	III
1300	2257.05	III
1300	2292.51	III
1000	2300.38	III
1200	2303.72	III
1500	2307.03	III
1100	2313.50	III
1700	2313.56	III
1000	2315.09	III
1400	2323.12	III
1700	2323.18	III
2200	2323.78	III
1900	2329.35	III
2100	2335.01	III
1600	2336.02	III
1900	2338.97	III
1600	2339.88	III
2100	2342.74	III
2500	2346.52	III
2800	2359.31	III
1900	2360.87	III
2300	2361.91	III
1400	2362.38	III
2100	2363.26	III
1600	2365.22	III
2000	2373.38	III
1300	2374.29	III
1300	2381.38	III
1600	2387.82	III
2000	2388.77	III
1200	2393.86	III
1200	2397.34	III
1200	2405.03	III
1300	2408.41	III
1200	2409.35	III
1500	2466.84	III
1100	2469.14	III
1300	2499.53	III
1600	2520.38	III
1600	2534.11	III
1300	2536.10	III
1600	2551.56	III
2200	2553.90	III
2100	2554.04	III
2500	2563.33	III
2100	2564.46	III
1000	2565.04	III
2400	2565.95	III
1800	2569.27	III
1800	2573.57	III
2000	2576.06	III
1300	2576.15	III
1400	2578.13	III
1600	2578.76	III
1700	2583.62	III
2000	2588.21	III
2000	2588.46	III
1300	2595.81	III
1800	2609.77	III
1200	2619.40	III
1200	2621.52	III
1400	2623.52	III
1400	2625.48	III
2000	2628.10	III
1300	2628.99	III
2400	2629.83	III
2100	2632.30	III
1800	2633.32	III
1400	2635.71	III
1600	2636.44	III
1700	2637.15	III
2100	2637.97	III
2100	2638.06	III
2200	2640.53	III
1600	2641.65	III
2100	2643.71	III
1600	2644.52	III
1800	2646.04	III
1800	2646.84	III
1600	2651.48	III
2000	2655.59	III
1900	2656.55	III
2200	2660.83	III

Gadolinium (Cont.)

1800	2675.75	III
1800	2679.44	III
1700	2680.63	III
1800	2682.52	III
1500	2683.91	III
1600	2692.78	III
1900	2692.86	III
1500	2694.43	III
2800	2697.39	III
1500	2702.91	III
2800	2703.28	III
1600	2704.53	III
1800	2717.35	III
2700	2727.89	III
450	2751.24	III
1800	2833.83	III
9000	2904.73	III
1800	2918.40	III
9500	2955.53	III
1000	2975.42	III
1000	2984.10	III
1000	3116.59	III
2500	3118.04	III
4000	3176.66	III
400	3253.53	III
400	3330.34	III
400	3371.05	III
400	3402.97	III
450	3624.90	III
250	3700.47	III
300	3831.73	III
300	3910.24	III
300	4016.91	III
600	4177.26	III
400	4279.96	III
300	4314.28	III
300	4445.91	III
600	4684.25	III
600	4715.06	III
600	4782.79	III
250	4976.72	III
5000	5091.70	III
300	5124.06	III
1800	5347.95	III
3000	5365.96	III
1100	5412.62	III
4000	5553.30	III
3000	5587.88	III
3000	5658.98	III
1800	5786.96	III
1500	5862.09	III
1500	5987.85	III
5000	14332.88	III
2000	17474.78	III
800	19996.34	III
800	21259.44	III
600	22493.33	III

Gd IV
Ref. 152—J.F.K.

Intensity	Wavelength	
	Vacuum	
1000	967.92	IV
1000	983.42	IV
1000	987.10	IV
1000	987.91	IV
1000	995.04	IV
1000	995.80	IV
1000	996.49	IV
1000	999.24	IV
1000	1000.36	IV
1000	1002.73	IV
1000	1004.46	IV
1000	1005.66	IV
1000	1006.55	IV
1200	1007.24	IV
1200	1063.84	IV
500	1228.37	IV
500	1307.23	IV
500	1313.29	IV
500	1316.71	IV
600	1321.42	IV
500	1330.79	IV
1100	1393.24	IV
1600	1476.98	IV
1500	1705.03	IV
1600	1706.01	IV
2000	1736.24	IV
1500	1815.32	IV
400	1997.89	IV
	Air	
800	2049.28	IV
800	2061.30	IV
800	2070.40	IV
800	2076.66	IV
1000	2094.29	IV
1000	2296.89	IV
1000	2352.66	IV
800	2379.17	IV
900	2385.65	IV
500	2390.07	IV
700	2392.30	IV
700	2393.29	IV
700	2395.76	IV
500	2396.22	IV
600	2396.27	IV
1400	2397.87	IV
700	2402.70	IV
500	2412.21	IV
900	2419.26	IV
500	2439.84	IV
600	2440.38	IV
600	2468.60	IV

GALLIUM (Ga)
Z = 31

Ga I and II
Ref. 19, 132, 195, 281—L.J.R.

Intensity		Wavelength	
		Vacuum	
2		829.60	II
2		958.67	II
1		960.57	II
2		969.19	II
2		998.52	II
3		1002.95	II
5		1012.38	II
3		1019.10	II
5		1023.80	II
8		1033.69	II
1		1113.87	II
3		1119.25	II
5		1130.81	II
1		1167.62	II
2		1173.78	II
3		1186.81	II
1		1227.13	II
5		1286.38	II
5		1327.81	II
20		1414.44	II
5		1449.49	II
2		1463.65	II
3		1473.73	II
3		1483.52	II
3		1485.95	II
3		1495.21	II
3		1504.41	II
3		1505.01	II
5		1514.57	II
3		1515.19	II
8		1535.40	II
5		1536.37	II
1		1536.91	II
3		1669.83	II
5		1695.85	II
5		1799.42	II
10		1813.98	II
15		1845.30	II
		Air	
20		2091.34	II
1		2218.04	I
1		2255.03	I
1		2259.23	I
2		2294.19	I
1		2297.87	I
3		2338.24	I
1		2338.60	I
3		2371.29	I
2		2377.53	II
4		2418.69	I
5		2438.88	II
6		2450.08	I
7		2500.19	I
3		2500.71	I
5		2513.55	II
3		2514.15	II
2		2551.26	II
3		2552.87	II
4		2555.28	II
5		2607.47	I
8		2624.82	I
10		2632.66	I
3		2659.87	I
10		2665.05	I
8		2691.29	I
20		2700.47	II
3		2719.66	I
15		2780.15	II
6		2874.24	I
1		2886.45	II
		2893.65 GA	II
2		2910.77	II
6		2943.64	I
6		2944.17	I
3		2969.41	II
1		2971.01	II
3		2971.60	II
5		2974.77	II
1		2992.84	II
2		3011.90	II
1		3158.18	II
4		3374.94	II
1		3375.95	II
2		3436.66	II
3		3446.46	II
2		3447.26	II
5		3470.34	II
1		3471.46	II
1		3472.52	II
2		3583.60	II
1		3693.93	II
2		3705.85	II
4		3734.85	II
9		3924.39	II
10		4032.99	I
10		4172.04	I
4		4251.11	II
15		4251.16	II
10	h	4254.04	II
4	h	4255.64	II
5		4255.70	II
10		4255.77	II
40		4262.00	II
3		5218.21	II
1		5338.3	II
2		5353.49	I
2		5360.6	II
1		5363.5	II
3		5416.8	II
1		5421.6	II
1		5425.6	II
10		6334.2	II
2000		6396.56	I
1000		6413.44	I
5		6419.4	II
3		6456.3	II
1		7000.0	II
3	h	7051.24	I
5	h	7106.82	I
1	h	7116.3	I
2	h	7172.9	I
5	h	7193.6	I
7		7198.7	II
10	h	7251.4	I
3	h	7289.6	I
5	h	7349.3	I
20	h	7403.0	I
30	h	7464.0	I
6	h	7556.6	I
10	h	7620.5	I
50	h	7734.77	I
2		7793.0	II
100	h	7800.01	I
4	h	7801.6	I
15	h	8002.55	I
20	h	8074.25	I
3	h	8167.5	I
5	h	8171.6	I
100	h	8311.86	I
200	h	8386.49	I
10	h	8389.30	I
7	h	8415.51	I
10	h	8419.91	I
20	h	8808.75	I
30	h	8813.56	I
20	h	8856.37	I
30	h	8944.33	I
200	h	9492.92	I
200	h	9493.12	I
300	h	9589.36	I
20	h	9594.25	I
60	h	10898.10	I
100	h	10905.95	I
10		10968.27	I
20		11103.51	I
400		11949.12	I
200		12109.78	I
40		12885.05	I
50		13057.50	I
50		14982.75	I
60		14996.64	I
20		17757.91	I
10		17868.96	I
60		22016.81	I
70		22568.71	I

Ga III
Ref. 141—L.J.R.

Intensity	Wavelength	
	Vacuum	
50	620.00	III
40	622.01	III
90	806.51	III
90	817.30	III
50	828.70	III
80	1085.00	III
60	1105.61	III
90	1150.27	III
90	1267.16	III
80	1293.46	III
60	1295.36	III
60	1323.15	III
70	1353.92	III
90	1495.07	III
50	1534.46	III
	Air	
90	2417.70	III
90	2423.98	III
15	2424.36	III
50	3521.77	III
80	3581.19	III
100	3589.34	III
10	3731.10	III
10	3806.60	III
100	4380.69	III
150	4381.76	III
100	4863.00	III
150	4993.78	III
10	5808.28	III
20	5848.25	III
15	5993.51	III

Ga IV
Ref. 141, 143—L.J.R.

Intensity	Wavelength	
	Vacuum	
14	294.53	IV
61	295.67	IV
41	304.99	IV
4	422.12	IV
25	423.18	IV
16	439.92	IV
67	1137.06	IV
70	1156.10	IV
70	1163.60	IV
75	1170.58	IV
48	1171.71	IV
68	1185.23	IV
40	1186.06	IV
73	1190.89	IV
73	1193.02	IV
75	1195.02	IV
69	1201.54	IV
72	1206.89	IV
63	1216.15	IV
50	1228.03	IV
60	1236.38	IV
60	1238.59	IV
45	1241.81	IV
75	1245.53	IV
83	1258.77	IV
81	1264.66	IV
82	1267.15	IV
81	1279.24	IV
80	1285.33	IV
82	1295.86	IV
83	1299.46	IV
82	1303.53	IV
80	1309.68	IV
80	1314.82	IV
85	1338.09	IV
77	1347.03	IV
76	1351.06	IV
74	1364.63	IV
60	1395.54	IV
77	1402.55	IV
70	1405.32	IV
73	1465.87	IV

Ga V
Ref. 2, 62, 140—L.J.R.

Intensity	Wavelength	
	Vacuum	
5	290.53	V
1	296.13	V
5	296.82	V
30	298.44	V
20	299.47	V
30	300.01	V
25	300.57	V
10	300.78	V
20	301.19	V
30	302.86	V
20	303.84	V
30	307.03	V
30	308.26	V
15	309.64	V
30	311.79	V
25	312.41	V
30	313.68	V

Gallium (Cont.)

15		315.95	V
20		316.48	V
40		319.41	V
12		320.53	V
40		322.31	V
50		322.99	V
30		323.10	V
40		324.25	V
40		324.95	V
40		326.14	V
30		326.77	V
30		328.65	V
5		336.61	V
20		878.17	V
40		973.21	V
10		977.89	V
15		979.60	V
20		[illegible]	V
40		989.75	V
90		1014.47	V
90		1019.71	V
20		1033.55	V
30		1038.76	V
30		1047.50	V
120		1050.48	V
80		1054.56	V
90		1058.12	V
80		1066.69	V
35		1068.59	V
30		1069.45	V
60		1069.60	V
55		1071.19	V
45		1071.41	V
80		1073.77	V
90		1078.83	V
110		1079.60	V
60		1080.99	V
250		1085.01	V
80		1087.37	V
40		1090.53	V
90		1091.71	V
100		1094.36	V
80		1095.10	V
70		1101.62	V
160		1102.83	V
140		1103.03	V
60		1104.93	V
75		1105.62	V
70		1106.17	V
40		1115.55	V
80		1118.34	V
55		1123.18	V
80		1123.66	V
120		1126.40	V
80		1127.75	V
130		1128.10	V
120		1128.53	V
100		1129.94	V
80		1131.43	V
40		1133.91	V
130		1136.07	V
65		1138.20	V
60		1144.30	V
50		1145.70	V
30		1148.42	V
45		1150.09	V
130		1150.23	V
120		1156.51	V
35		1157.74	V
25		1169.40	V
40		1178.95	V
80		1213.17	V
30		1265.45	V
30		1276.85	V
15		1283.64	V
10		1311.35	V

GERMANIUM (Ge)
Z = 32

Ge I and II
Ref. 5, 119, 293, 340—C.H.C.

Intensity		Wavelength	
		Vacuum	
1		822.97	II
3		835.08	II
10		850.50	II
10		862.234	II
15		875.493	II
15		905.977	II
20		920.554	II
50		999.101	II
100		1016.638	II
100		1075.072	II
300		1085.51	II
200		1098.71	II
500		1106.74	II
500		1120.46	II
200		1164.27	II
500		1181.19	II
500		1181.65	II
200		1188.73	II
100		1189.62	II
300		1191.26	II
50		1191.72	II
500		1237.059	II
500		1261.905	II
100		1264.710	II
100		1380.42	II
50		1392.26	II
200		1401.24	II
200		1538.091	II
500		1576.855	II
75		1581.070	II
100		1602.486	II
3	r	1615.57	I
2	r	1624.130	I
2	r	1630.173	I
3	r	1636.31	I
2		1638.96	I
4	r	1639.730	I
2		1647.531	I
200		1649.194	II
2		1651.528	I
4	r	1651.955	I
3		1661.345	I
4	r	1663.539	I
10	h	1665.275	I
4		1667.802	I
3	r	1670.608	I
100	r	1691.090	I
200	r	1716.784	I
100	h	1739.102	I
100		1742.195	I
50		1746.065	I
200		1750.043	I
100		1758.279	I
100	h	1764.185	I
100	h	1765.284	I
50	h	1766.433	I
200		1774.176	I
200		1785.046	I
100	h	1793.071	I
75	h	1801.432	I
200	h	1841.328	I
200	h	1842.410	I
100	h	1844.410	I
100	h	1845.872	I
100	h	1846.958	I
200		1853.134	I
500	r	1860.086	I
100		1865.052	I
300	r	1874.256	I
100		1895.197	I
500	r	1904.702	I
50	h	1908.434	I
30		1912.409	I
300	r	1917.592	I
100	h	1923.467	I
500	r	1929.826	I
10	h	1934.048	I
100	r	1937.483	I
500		1938.008	II
100	r	1938.300	I
500		1938.891	II
30	s	1944.116	I
200		1944.731	I
200		1955.115	I
500		1962.013	I
30	h	1963.373	I
30		1965.383	I
200		1970.880	I
200		1979.274	II
300	h	1987.849	I
300		1988.267	I
500	r	1998.887	I
		Air	
50		2007.04	II
200		2011.29	I
1700		2019.068	I
2400	r	2041.712	I
1600	r	2043.770	I
420		2054.461	I
220	h	2057.238	I
750	r	2065.215	I
2600	r	2068.656	I
420		2086.021	I
2000	r	2094.258	I
240		2105.824	I
95	h	2124.744	I
50	h	2186.451	I
100		2197.62	II
340	r	2198.714	I
100		2205.85	II
15		2220.375	I
18		2256.001	I
18		2314.201	I
24		2327.918	I
15		2359.233	I
20		2379.144	I
10		2389.472	I
15		2397.885	I
130		2417.367	I
30		2436.412	I
100		2478.66	II
90		2497.962	I
500		2500.54	II
70		2533.230	I
3		2556.298	I
28		2589.188	I
500		2592.534	I
8		2644.184	I
1200		2651.172	I
500		2651.568	I
500		2691.341	I
200		2704.03	II
850		2709.624	I
400		2729.78	II
40		2740.426	I
650		2754.588	I
50		2770.59	II
75		2772.35	II
70		2793.925	I
80		2829.008	I
1000		2831.843	II
50		2834.28	II
75		2839.68	II
1000		2845.527	II
75		2853.97	II
750		3039.067	I
600		3067.021	I
20		3124.816	I
50		3186.72	II
100		3221.64	II
110		3269.489	I
50		3312.56	II
75		3323.64	II
100		3455.72	II
300		3499.21	II
30		3845.11	II
70		4226.562	I
10		4685.829	I
75	h	4689.87	II
50	h	4690.02	II
1000		4741.806	II
1000		4814.608	II
50		4824.097	II
100		5131.752	II
200		5178.648	II
3		5194.583	I
6		5265.892	I
6		5513.263	I
6		5564.741	I
8		5607.010	I
6		5616.135	I
7		5621.426	I
8		5655.96	I
6		5664.226	I
5		5664.842	I
9		5691.954	I
6		5701.776	I
5		5717.877	I
6		5801.029	I
9		5802.093	I
1000		5893.389	II
500		6021.041	II
150		6078.39	II
50		6267.14	II
150		6268.07	II
100		6268.34	II
75		6283.452	II
100		6336.377	II
100		6484.181	II
6		6557.488	I
50		6780.51	II
50		7049.369	II
6		7130.12	I
30		7145.390	II
7		7330.38	I
5		7353.334	I
7		7384.208	I
6		7402.64	I
7		7511.57	I
5		7776.20	I
10		7833.575	I
7		7837.63	I
6		7853.77	I
7		7878.12	I
5		7962.26	I
5		7983.33	I
10		8031.039	I
6		8044.165	I
5		8095.29	I
5		8225.22	I
7		8226.09	I
10		8256.013	I
5		8264.15	I
5		8280.09	I
6		8281.04	I
8		8367.81	I
7		8391.70	I
5		8396.36	I
5		8429.42	I
10		8482.21	I
8		8506.70	I
5		8507.66	I
8		8564.89	I
6		8599.27	I
6		8652.42	I
5		8669.60	I
9		8700.60	I
5		8712.90	I
6		8734.78	I
6		8789.88	I
5		9068.785	I
5		9095.957	I
6		9398.868	I
20		9474.993	II
20		9475.645	II
4		9492.559	I
7		9625.664	I
5		10039.436	I
4		10200.952	I
10		10382.427	I
10		10404.913	I
8		10734.068	I
8		10947.416	I
10		11125.130	I
230		11252.83	I
24		11293.40	I
33		11318.13	I
55		11459.05	I
150		11483.77	I
175		11614.81	I
600		11714.76	I
10		11839.77	I
55		11917.01	I
10		12025.64	I
10		12055.49	I
30		12061.41	I
45		12065.76	I
1300		12069.20	I
30		12198.88	I
20		12207.73	I
60		12286.75	I
55		12338.76	I
1050		12391.58	I
48		12540.41	I
15		12636.80	I
150		12676.58	I
40		12681.28	I
115		12800.66	I
175		12836.38	I
12		12847.92	I
120		12955.73	I
15		13028.84	I
235		13107.61	I
20		13492.28	I
42		13534.85	I
28		13724.48	I
42		14116.70	I
42		14297.15	I
40		14569.84	I
12		14667.52	I
470		14822.38	I
16		14921.97	I
15		15001.75	I
13		15041.21	I
20		15504.34	I
14		16424.77	I
12		16626.64	I
70		16699.29	I
150		16759.79	I
135		17214.34	I
16		18428.30	I
35		18495.54	I
10		18764.11	I
70		18811.86	I
62		19279.24	I
28		20673.64	I
4		21518.30	I
9		22091.84	I
5		23921.92	I

Ge III
Ref. 341—C.H.C.

Intensity	Wavelength	
	Vacuum	
2	542.90	III
2	663.77	III
3	670.88	III
2	680.28	III
2	952.76	III
12	988.96	III
15	995.72	III
10	996.50	III
15	1011.21	III
10	1012.31	III
8	1032.62	III
12	1040.99	III
12	1058.91	III
40	1088.45	III
10	1137.92	III
12	1150.55	III

Germanium (Cont.)

Intensity		Wavelength	
8		1159.15	III
8		1159.62	III
8		1160.79	III
10		1173.78	III
8		1212.47	III
4		1323.24	III
10		1525.32	III
2		1527.15	III
9		1600.09	III
6		1883.26	III
2		1978.22	III

Air

Intensity		Wavelength	
2		2019.22	III
4		2022.25	III
3		2062.14	III
15		2100.05	III
15		2102.42	III
25		2104.45	III
3		2922.86	III
25		3197.56	III
35		3211.86	III
25		3214.95	III
40		3255.05	III
20		3259.90	III
5		3369.57	III
20		3414.27	III
40		3434.03	III
8		3464.59	III
40		3489.08	III
2		3724.51	III
15		3884.78	III
200		4178.96	III
12		4245.41	III
200		4260.85	III
150		4291.71	III
10		4674.36	III
10		5016.88	III
18		5134.75	III
5		5229.37	III
3		5256.61	III

Ge IV
Ref. 341—C.H.C.

Vacuum

Intensity		Wavelength	
1		440.11	IV
1		441.95	IV
3		847.80	IV
3		868.30	IV
8		915.00	IV
8		936.70	IV
4		938.90	IV
1		1073.44	IV
20		1188.99	IV
20		1229.81	IV
2		1494.89	IV
6		1500.61	IV
3		1648.14	IV

Air

Intensity		Wavelength	
2		2293.0	IV
2		2343.37	IV
15		2445.38	IV
15		2445.71	IV
30		2488.25	IV
20		2542.44	IV
5		2631.78	IV
3		2698.08	IV
15		2717.44	IV
30		2736.09	IV
30		2788.61	IV
5		3071.84	IV
60		3554.19	IV
50		3676.65	IV

Ge V
Ref. 342—C.H.C.

Vacuum

Intensity		Wavelength	
700		294.51	V
1000		295.64	V
200		304.98	V
20		621.52	V
35		716.26	V
50		724.21	V
35		733.54	V
35		735.35	V
35		741.52	V
60		746.88	V
40		750.26	V
35		755.84	V
60		760.05	V
60		958.51	V
300		971.35	V
150		984.92	V
200		988.13	V
300		990.66	V
300		1004.38	V
300		1016.66	V
250		1038.40	V
900		1045.71	V
400		1050.05	V
300		1054.59	V
300		1068.43	V
400		1069.13	V
700		1072.66	V
600		1086.65	V
500		1087.85	V
800		1089.49	V
300		1092.09	V
1000		1116.94	V
300		1122.01	V
700		1163.39	V
300		1165.26	V
200		1176.69	V
700		1222.30	V

GOLD (Au)
Z = 79

Au I and II
Ref. 38, 72, 234—C.H.C.

Vacuum

Intensity		Wavelength	
		925.72	II
		946.03	II
		950.39	II
20		957.78	II
3		967.94	II
3		974.47	II
2		982.24	II
		1062.67	II
8		1066.96	II
		1085.00	II
8		1090.78	II
5		1094.92	II
20		1103.31	II
3		1166.76	II
5		1210.86	II
20		1224.57	II
40	h	1305.34	I
25		1310.47	I
100	h	1328.37	I
3		1336.26	II
10	h	1338.37	I
10	h	1342.80	I
20		1350.09	I
20		1350.84	I
22		1351.74	I
25		1352.82	I
25		1354.14	I
30		1355.79	I
35		1357.86	I
40		1360.51	I
6		1362.33	II
20		1362.47	I
8		1363.15	II
50		1363.98	I
25		1364.15	I
10		1364.74	II
60		1368.62	I
35		1368.98	I
70		1374.82	I
50		1375.76	I
30		1378.87	I
8		1380.53	II
80		1382.75	I
50		1385.33	I
20		1389.14	I
60		1392.27	I
6		1393.80	II
50		1402.12	I
		1405.12	II
70		1407.38	I
100		1408.45	I
		1410.69	II
		1415.22	II
80		1429.19	I
50		1435.79	I
20		1436.61	II
10		1468.85	II
10		1469.17	II
10		1469.28	II
100		1481.76	I
25		1486.55	II
20		1532.82	I
20		1532.86	I
10		1562.04	II
200		1587.16	I
12		1593.41	II
70		1598.24	I
12		1611.11	II
		1616.65	AU II
2		1622.83	II
100		1624.34	I
2		1632.53	II
50		1639.90	I
150		1646.67	I
10		1656.99	II
100		1665.76	I
25		1673.59	II
7		1694.38	II
2		1698.65	II
200		1699.34	I
30		1700.69	II
10		1720.04	II
25		1725.75	II
45		1740.52	II
10		1749.80	II
35		1756.15	II
60		1783.22	II
35		1793.31	II
35		1800.58	II
25		1823.24	II
100		1879.83	I
20		1919.64	I
20		1921.64	II
45		1942.31	I
25		1951.93	I
30		1978.19	I

Air

Intensity		Wavelength	
25		2000.81	II
11000		2012.00	I
2600		2021.38	I
50		2044.54	II
150		2082.09	II
35		2095.13	II
20		2098.14	II
60		2110.68	II
30		2125.29	II
15		2126.63	I
20		2170.75	I
35		2188.81	II
25		2201.32	II
35		2215.63	II
45		2228.88	II
30		2231.18	II
25		2240.16	II
70		2248.56	II
80		2263.62	II
18		2263.88	II
25		2277.52	II
25		2283.30	II
25		2291.40	II
45		2304.69	II
25		2314.55	II
20		2315.75	II
25		2340.06	II
180		2352.65	I
20		2376.28	I
120		2387.75	I
2600		2427.95	I
60		2533.52	II
16		2544.19	I
45		2552.67	II
20		2589.25	I
30		2590.04	I
50		2616.40	II
20		2627.02	II
250		2641.48	I
3400		2675.95	I
20		2687.63	II
20		2688.16	II
30		2688.71	I
80		2700.89	I
1100		2748.25	I
20		2748.71	II
100		2780.82	I
30		2800.93	I
1000		2802.04	II
300		2819.79	II
100		2822.55	II
30		2823.13	II
100	h	2825.44	II
30		2833.03	II
300		2837.85	II
100		2846.92	II
100		2856.74	II
3		2872.36	I
300		2883.45	I
3		2886.96	I
10		2888.40	I
300		2891.96	I
100		2893.25	II
3		2905.74	I
30		2905.90	I
100		2907.04	II
300		2913.52	II
3		2914.82	I
300		2918.24	II
16		2932.19	
30		2940.67	I
100		2954.22	II
10		2973.33	I
100	h	2990.27	II
300		2994.80	II
10		3002.65	I
3		3005.85	I
10		3024.67	I
320		3029.20	I
30		3033.25	I
300		3065.42	I
10		3102.63	I
10	h	3117.01	I
100		3122.50	II
1600		3122.78	I
30		3126.86	II
30		3127.03	I
10		3164.88	I
10		3172.35	II
30		3191.76	I
100		3194.72	I
30		3200.37	I
30		3204.74	I
1		3221.86	I
30		3225.25	I
300		3230.63	I
10		3253.94	I
10	h	3265.10	I
30	h	3267.07	I
10		3271.63	I
10		3273.47	I
300		3308.30	I
300		3309.64	I
100		3320.12	I
100		3355.15	I
30	h	3368.44	I
10		3381.90	I
100	h	3391.31	I
100		3395.40	I
30	h	3440.36	I
100		3467.21	I
30		3471.61	I
30	h	3509.04	I
10		3510.82	I
3		3523.34	II
3		3545.61	I
30		3553.57	I
300		3557.36	I
10		3558.22	I
30		3565.97	I
30	h	3584.37	I
300		3586.73	I
30	h	3588.79	I
30		3598.06	I
100		3611.57	I
30	h	3614.00	I
30		3622.74	I
100	h	3631.31	I
10		3633.22	II
30		3634.53	I
10		3635.12	I
300		3637.90	I
3		3639.87	I
100	h	3645.02	I
10		3649.09	I
100		3650.74	I
10		3653.53	I
3		3654.69	I
10		3655.30	I
10		3656.90	I
30		3706.55	II
100		3709.62	I
10		3766.61	I
10		3770.76	I
100		3796.01	I
30		3801.92	I
30		3804.01	II
10		3821.85	I
10		3825.70	I
100		3874.73	I
30		3880.25	I
30		3889.48	I
100	h	3892.26	I
400		3897.86	I
30		3901.09	I
300		3909.38	I
100		3927.69	I
10	h	3959.10	I
30		3966.23	I
30		3976.65	I
30		3979.68	I
30		3991.37	I
3		4012.57	I
10		4016.07	II
400		4040.93	I
30		4052.79	II
700		4065.07	I
10		4076.35	II
3		4083.28	II
100		4084.10	I
30		4101.70	I
30		4128.59	I
30		4201.13	I
30	h	4227.88	I
100		4241.80	I
200		4315.11	I
30		4361.04	II
10		4420.61	II
120	h	4437.27	I

Gold (Cont.)

250		4488.25	I
900	h	4607.51	I
100	h	4620.56	I
1		4663.92	I
3		4663.97	I
10		4694.69	I
3		4760.17	II
500		4792.58	I
100		4811.60	I
10		4822.96	I
30	h	4950.82	I
30		5064.59	I
30	h	5108.84	I
100		5147.44	I
300		5230.26	I
100	h	5261.76	I
100		5655.77	I
100	h	5721.36	I
300		5837.37	I
100	h	5862.93	I
300	h	5956.96	I
30	h	5962.68	I
600		6278.17	I
100		6562.68	I
30		6652.89	I
600		7510.73	I
10		8145.06	I
10		9254.28	I

Au III
Ref. 72, 393, 395 — R.D.C.

Intensity		Wavelength	
		Vacuum	
30		779.73	III
30		788.78	III
50		799.93	III
40		811.83	III
50		817.95	III
40		820.06	III
80		833.16	III
100		843.44	III
100		845.14	III
80		855.49	III
80		859.90	III
80		863.42	III
80		901.03	III
80		910.45	III
80		924.02	III
200		945.10	III
100		1040.63	III
80		1044.49	III
80		1046.81	III
100	h	1239.96	III
100		1278.51	III
100		1314.84	III
200		1336.72	III
180		1341.68	III
100		1348.89	III
150		1350.32	III
150		1355.61	III
150		1356.13	III
80		1362.06	III
500		1365.40	III
200		1367.17	III
180		1377.73	III
150		1378.69	III
150		1379.98	III
125		1380.53	III
200		1381.36	III
300		1385.79	III
100		1389.41	III
180		1391.46	III
180		1396.00	III
100		1402.91	III
225		1409.50	III
250		1413.80	III
100		1414.27	III
80		1415.54	III
100		1417.09	III
125		1417.39	III
150		1427.42	III
300		1428.93	III
250		1430.06	III
275		1433.37	III
250		1435.81	III
80		1436.12	III
300		1439.12	III
200		1441.21	III
150		1446.37	III
80		1446.69	III
250		1448.42	III
250		1454.95	III
100		1464.72	III
150		1471.28	III
80		1473.32	III
100		1474.73	III
150		1481.10	III
300		1487.15	III
250		1487.91	III
200		1489.47	III
250		1500.37	III
200		1502.47	III
200		1503.74	III
80		1540.26	III
100		1542.00	III
80		1542.25	III
100		1548.50	III
80		1554.61	III
80		1562.33	III
80		1562.41	III
200		1567.54	III
80		1571.94	III
200		1574.85	III
200		1579.44	III
150		1584.10	III
200		1589.56	III
80		1589.68	III
150		1593.41	III
200		1600.51	III
250		1617.16	III
100		1617.78	III
500		1621.93	III
300	d	1629.13	III
250		1638.88	III
100		1644.17	III
250		1652.74	III
250		1664.77	III
100		1668.11	III
125		1673.93	III
1000		1693.94	III
150		1697.09	III
200		1698.98	III
200		1700.00	III
200		1702.25	III
100		1707.53	III
250		1710.16	III
200		1715.69	III
100		1716.71	III
300		1717.83	III
500		1727.31	III
100	d	1733.17	III
300		1738.48	III
150		1744.39	III
500		1746.10	III
500		1756.92	III
500		1761.95	III
300		1767.42	III
100		1774.42	III
800		1775.17	III
200		1776.40	III
100		1780.57	III
300		1786.11	III
150		1792.65	III
500		1793.76	III
200		1801.98	III
400		1805.24	III
100		1809.81	III
400		1821.17	III
400		1844.89	III
150		1848.83	III
80		1850.15	III
500		1861.80	III
150		1871.92	III
150		1918.28	III
100		1932.04	III
100		1935.42	III
200		1948.79	III
100		1958.47	III
400		1989.63	III
150		1996.85	III
		Air	
300		2083.09	III
80		2085.45	III
100		2159.08	III
80		2167.33	III
200		2172.20	III
100		2184.11	III
500		2188.97	III
300		2322.27	III
100		2382.40	III
150		2402.71	III
150		2405.12	III
100		3227.99	III
100		3309.86	III

HAFNIUM (Hf)
Z = 72

Hf I and II
Ref. 1—C.H.C.

Intensity	Wavelength	
	Air	
6200	2012.78	II
8500	2028.18	II
1200	2096.18	II
540	2210.82	II
320	2254.01	II
160	2255.15	II
250	2266.83	II
620	2277.16	II
230	2321.14	II
580	2322.47	II
300	2323.25	II
120	2324.50	II
300	2324.89	II
200	2332.97	II
200	2337.33	II
230	2343.32	II
320	2347.44	II
540	2351.22	II
110	2353.02	I
90	2365.98	II
250	2380.30	II
100	2381.00	II
170	2393.18	II
450	2393.36	II
670	2393.83	II
130	2400.78	II
70	2404.56	II
540	2405.42	II
130	2406.44	II
370	2410.14	II
90	2413.33	II
55	2415.96	II
320	2417.69	II
120	2425.98	II
45	2428.75	I
120	2428.99	II
130	2433.57	II
45	2434.74	II
35	2444.99	I
390	2447.25	II
140	2449.44	II
35	2452.30	II
110	2453.34	II
450	2460.49	II
70	2463.97	II
430	2464.19	II
90	2465.06	II
35	2465.67	I
140	2467.97	II
210	2469.18	II
100	2473.92	II
55	2481.44	II
55	2482.65	I
55	2487.16	I
290	2496.99	II
580	2512.69	II
580	2513.03	II
130	2515.48	II
890	2516.88	II
340	2531.19	II
200	2537.33	II
110	2548.20	II
320	2551.40	II
130	2559.19	II
250	2563.61	II
890	2571.67	II
320	2573.90	II
320	2576.82	II
300	2578.14	II
320	2582.54	II
130	2591.33	II
390	2606.37	II
450	2607.03	II
120	2608.45	I
230	2613.60	II
450	2622.74	II
160	2637.00	I
1100	2638.71	II
1100	2641.41	II
160	2642.75	I
670	2647.29	II
100	2651.16	II
160	2657.84	II
210	2661.88	II
290	2683.35	II
670	2705.61	I
110	2706.73	II
210	2712.42	II
140	2713.84	I
250	2718.59	I
120	2730.85	I
710	2738.76	II
200	2743.64	I
360	2751.81	II
450	2761.63	I
160	2766.96	I
170	2773.02	I
980	2773.36	II
180	2774.02	II
390	2779.37	I
100	2789.50	II
140	2789.73	II
230	2808.00	II
230	2813.86	II
170	2814.48	II
230	2817.68	I
140	2818.94	I
200	2819.74	I
1200	2820.22	II
490	2822.68	II
180	2833.28	I
110	2834.13	I
410	2845.83	I
270	2849.21	II
270	2850.96	I
180	2851.21	II
180	2860.56	I
760	2861.01	II
760	2861.70	II
2100	2866.37	I
130	2869.82	II
150	2876.33	II
210	2887.14	I
100	2887.54	I
800	2889.62	I
1000	2890.26	I
130	2898.71	II
1200	2904.41	I
890	2904.75	I
140	2909.91	II
2000	2916.48	I
580	2918.58	I
320	2919.59	II
180	2924.62	I
490	2929.63	II
450	2929.90	I
710	2937.80	II
2000	2940.77	I
160	2944.71	I
1200	2950.68	I
1100	2954.20	I
540	2958.02	I
120	2961.80	II
1400	2964.88	I
620	2966.93	I
140	2967.23	II
710	2968.81	II
110	2973.37	I
890	2975.88	II
150	2979.28	I
1100	2980.81	I
210	2982.72	I
170	3000.10	II
800	3005.56	I
1100	3012.90	II
540	3016.78	I
1100	3016.94	II
980	3018.31	I
1200	3020.53	I
140	3025.29	II
410	3031.16	II
110	3046.08	II
710	3050.76	I
1100	3057.02	I
130	3063.78	I
130	3064.68	II
850	3067.41	I
2100	3072.88	I
170	3074.10	I
250	3074.79	I
150	3080.66	II
430	3080.84	I
200	3096.76	I
340	3101.40	II
710	3109.12	II
130	3110.87	II
130	3119.98	I
710	3131.81	I
850	3134.72	II
130	3137.51	I
170	3139.65	II
120	3140.76	II
220	3145.32	II
220	3148.41	I
120	3151.63	I
450	3156.63	I
270	3159.82	I
710	3162.61	II
450	3168.39	I
890	3172.94	I
450	3176.86	II
220	3181.01	I
120	3181.15	I
130	3189.62	I
360	3193.53	II
670	3194.19	II
200	3196.93	I
130	3199.99	II
310	3206.11	I
180	3210.98	I
180	3217.30	II
180	3220.61	II
130	3230.06	I
130	3239.44	I
130	3243.35	I
360	3247.66	I
220	3249.53	I
890	3253.70	II
270	3255.28	II
120	3262.47	I
180	3273.66	II
270	3279.98	II

Hafnium (Cont.)

160		3291.05	I
210		3306.12	I
120		3309.19	I
340		3310.27	I
670		3312.86	I
180		3317.99	II
130		3328.21	II
890		3332.73	I
370		3352.06	II
130		3356.78	I
230		3358.91	I
180		3360.06	I
140		3366.68	I
180		3378.93	I
140		3384.14	II
230		3384.70	II
170		3386.21	I
800		3389.83	II
230		3392.81	I
230		3394.59	II
140		3394.98	II
230		3397.26	I
230		3397.60	I
2300		3399.80	II
170		3400.21	I
180		3402.51	I
140		3407.76	II
230		3410.17	II
230		3417.34	I
410		3419.18	I
140		3427.44	I
200		3428.37	II
250		3438.24	II
140		3438.43	I
100		3441.84	I
100		3452.31	I
140		3462.64	II
140		3467.60	I
710		3472.40	I
200		3478.99	II
480		3479.28	II
250		3495.75	II
250		3497.16	I
980		3497.49	I
100		3498.98	I
1200		3505.23	II
150		3513.28	I
130		3518.75	II
980		3523.02	I
100		3530.87	I
100		3531.23	I
980		3535.54	II
760		3536.62	I
180		3548.81	I
540		3552.70	II
150		3554.00	I
1300		3561.66	II
150		3564.31	I
270		3567.36	I
1100		3569.04	II
150		3579.90	II
110		3583.28	I
210		3597.42	II
540		3599.87	I
110		3615.04	I
800		3616.89	I
110		3617.68	I
110		3624.00	II
320		3630.87	II
100		3635.43	I
800		3644.36	II
320		3649.10	I
200		3651.84	I
140		3661.05	II
220		3665.35	II
100		3668.21	I
200		3672.27	I
480		3675.74	I
2200		3682.24	I
280		3696.51	I
100		3698.40	II
240		3699.72	II
340		3701.15	II
100		3704.92	I
120		3705.40	II
1000		3717.80	I
650		3719.28	II
140		3726.49	I
160		3729.10	I
460		3733.79	I
160		3737.88	II
120		3739.04	I
100		3744.98	II
400		3746.80	I
140		3753.22	I
100		3765.05	I
100		3765.56	I
170		3766.92	II
200		3768.25	I
1400		3777.64	I
1400		3785.46	I
650		3793.37	II
100		3798.66	I

850	d	3800.38	I
140		3806.07	II
320		3811.78	I
100		3817.20	II
100		3819.38	I
1300		3820.73	I
140		3829.67	I
280		3830.02	I
800		3849.18	I
140		3849.52	II
600		3858.31	I
230		3860.91	I
200		3872.55	II
160		3877.10	II
380		3880.82	II
200		3882.52	I
150		3883.77	II
200		3889.23	I
200		3889.33	I
620		3899.94	I
620		3918.09	II
200		3923.90	II
120		3926.42	I
150		3927.57	I
110		3929.54	II
320		3931.38	I
120		3935.65	II
120		3939.04	I
410		3951.83	I
160		3968.01	I
150	Bl	3970.05	H
200		3973.48	I
180		4032.27	I
100		4047.96	II
230		4062.84	I
140		4066.21	I
180		4083.35	I
540		4093.16	II
110		4104.23	I
140		4106.58	I
110		4113.53	II
110		4118.60	I
150		4127.80	II
140		4145.76	I
150		4162.36	II
110		4162.69	I
1100		4174.34	I
120		4190.95	I
160		4206.58	II
190		4209.70	I
170		4228.08	I
170		4232.44	II
120	Bl	4252.08	H
170		4260.98	I
200		4263.39	II
170		4272.85	II
320		4294.79	I
120		4318.14	I
160		4330.27	I
180		4336.66	II
150		4350.51	II
250		4356.33	I
110		4367.90	II
180		4370.97	II
120		4417.35	II
160		4417.91	I
200		4438.04	I
140		4457.34	I
140		4461.18	I
140		4540.93	I
250		4565.94	I
500	d	4598.80	I
230		4620.86	I
210		4655.19	I
120		4699.01	I
160		4782.74	I
310		4800.50	I
130		4859.24	I
120		4975.25	I
95		5018.20	I
15		5021.75	I
55		5040.82	II
95		5047.45	I
55	b	5074.74	H
30		5079.65	II
55	b	5093.88	H
15		5112.13	I
19		5128.53	II
30		5157.96	I
55		5167.42	I
75		5170.18	I
230		5181.86	I
30		5186.84	I
30		5187.75	II
110		5243.99	I
55		5247.10	II
25		5260.44	II
30		5264.95	II
55		5275.04	I
22		5286.09	I
120		5294.87	I
45		5298.06	II
30		5307.82	I
45		5309.68	I

55		5311.60	II
12		5324.26	II
9		5346.30	II
110		5354.73	I
110		5373.86	I
40		5389.34	I
19		5391.36	II
19		5404.47	I
28		5424.02	I
12		5435.78	I
40		5438.74	I
14		5444.07	II
75		5452.92	I
30		5463.38	II
15		5497.30	I
15		5510.12	I
15		5510.45	I
19		5524.35	II
45		5538.02	I
28		5538.26	I
230		5550.60	I
230		5552.12	I
55		5575.86	I
14		5600.77	I
95		5613.27	I
25		5614.01	I
8		5628.27	I
19		5650.83	I
40	Bl	5698.03	H
25		5713.28	I
160		5719.18	I
25	Bl	5720.16	H
12		5748.72	I
14		5767.18	II
12		5809.50	II
19		5817.47	I
25		5842.23	II
25		5845.87	I
19		5847.77	I
22		5883.66	I
15		5926.47	I
60		5933.69	I
75		5974.28	I
25		5974.72	I
60		5978.66	I
25		5992.96	I
45	c	6016.79	I
28	b	6021.12	H
25	b	6043.19	H
25		6054.17	I
95		6098.67	I
95		6185.13	I
55		6210.70	I
28		6216.82	I
45		6238.58	I
60		6248.95	II
22	h	6299.54	I
25	h	6311.85	I
19		6318.33	I
30		6338.10	I
19	h	6380.19	I
60		6386.23	I
19	h	6409.52	I
15	h	6556.50	I
28		6587.23	I
45		6644.60	II
19		6647.06	II
11		6659.40	I
30		6713.48	I
17		6754.61	II
11		6769.95	I
85		6789.27	I
160		6818.94	I
15		6826.56	I
13		6850.07	I
35		6858.70	I
45		6911.40	I
10		6926.19	I
19		6979.59	I
21		6980.91	II
7		7019.25	I
7		7030.33	II
7		7035.13	I
11		7061.90	I
15		7062.87	I
160		7063.83	I
11	h	7094.40	I
15		7100.54	I
55		7119.52	I
570		7131.81	I
650		7237.10	I
410		7240.87	I
6		7262.62	I
75		7320.05	I
16		7321.76	I
6		7356.10	I
6		7365.28	I
20		7390.70	I
6		7423.69	I
25		7437.56	I
13		7463.86	I
7		7484.56	I
15		7556.37	I
75		7562.93	I

15		7564.22	I
11		7576.95	I
11		7592.96	I
13		7608.59	I
360		7624.40	I
20		7645.64	I
110		7740.17	I
8		7743.57	I
5		7757.89	II
40		7790.90	I
7		7796.81	I
35		7814.55	I
310		7845.35	I
7		7846.56	I
130		7920.71	I
29		7938.06	I
250		7994.73	I
7		8010.58	I
25		8056.52	I
25		8080.32	I
16		8173.89	I
130		8204.58	I
7		8248.81	I
55		8276.95	I
13		8305.91	II
25		8344.25	I
5		8380.06	I
5		8382.98	I
35		8460.01	I
150		8546.48	I
160		8640.06	I
40		8711.24	I
65		9004.73	I

Hf III
Ref. 404 — R.L.K.

Intensity		Wavelength	
		Vacuum	
20		1449.83	III
30		1507.82	III
50		1683.95	III
50		1756.91	III
60		1843.64	III
60		1870.58	III
50		1874.81	III
150		1885.15	III
100		1991.44	III
		Air	
100		2037.76	III
300		2070.94	III
150	h	2085.33	III
200	h	2099.30	III
200	h	2110.31	III
100	h	2119.69	III
200		2155.66	III
200		2183.50	III
200		2195.44	III
100		2213.54	III
200		2234.59	III
200	h	2313.44	III
300		2336.47	III
150	h	2355.48	III
100		2373.30	III
120		2377.57	III
250		2383.540	III
400		2461.74	III
2000		2495.16	III
1000		2515.16	III
100		2534.33	III
400	h	2560.74	III
300	h	2567.46	III
200		2687.22	III
500		2753.60	III
100	h	3060.08	III
200	h	3279.67	III
100	h	3741.94	III

Hf IV
Ref. 369, 425 — R.L.K.

Intensity		Wavelength	
		Vacuum	
40		520.04	IV
50		569.19	IV
100		596.56	IV
100		600.90	IV
100		603.16	IV
200		618.27	IV
200		620.19	IV
100		633.58	IV
100		643.05	IV
200		644.54	IV
400		647.39	IV
600		665.65	IV
100		671.36	IV
200		673.49	IV

Hafnium (Cont.)

15	1305.24	IV
12	1357.40	IV
40	1390.39	IV
50	1491.67	IV
35	1528.82	IV
15	1560.18	IV
25	1572.03	IV
100	1717.21	IV
20	1718.57	IV

Air

100	2054.46	IV
7	2014.06	IV
20	7751.29	IV
10	7267.58	IV

Hf V
Ref. 410 — R.L.K.

Intensity	Wavelength	
	Vacuum	
220	545.41	V
180	600.00	V
100	816.81	V
100	830.69	V
100	836.74	V
100	846.87	V
100	856.32	V
100	861.80	V
135	865.16	V
270	867.25	V
180	875.88	V
135	877.87	V
135	880.37	V
100	880.85	V
180	885.58	V
135	885.80	V
135	894.24	V
100	894.41	V
180	896.14	V
100	896.47	V
135	899.70	V
180	901.54	V
135	901.92	V
135	904.95	V
135	909.70	V
135	913.68	V
135	918.48	V
180	919.10	V
270	921.67	V
135	928.01	V
135	931.50	V
135	947.12	V
245	951.62	V
180	960.12	V
180	964.74	V
160	971.51	V
135	974.62	V
120	984.64	V
135	991.50	V
100	1078.42	V
100	1079.92	V
160	1092.76	V
135	1097.28	V
135	1137.49	V
160	1201.76	V
135	1208.88	V
135	1224.62	V
135	1227.98	V
135	1230.21	V
270	1232.03	V
200	1233.59	V
160	1237.42	V
100	1238.85	V
160	1239.53	V
160	1244.46	V
100	1259.25	V
440	1396.66	V
270	1400.09	V
160	1401.70	V
135	1405.77	V
370	1407.17	V
370	1408.38	V
270	1412.28	V
270	1413.51	V
160	1421.96	V
220	1422.53	V
370	1433.43	V
370	1437.27	V
500	1437.73	V
370	1445.40	V
270	1457.91	V
270	1719.32	V
550	1729.08	V
750	1731.83	V
750	1733.96	V
440	1741.74	V
1000	1749.11	V
1000	1750.19	V
500	1760.89	V
370	1765.62	V
270	1774.02	V
135	1792.39	V

HELIUM (He)
Z = 2

He I and II
Ref. 16, 94, 173, 183, 317
W.C.M.

Intensity	Wavelength	
	Vacuum	
15	231.454	II
20	232.584	II
30	234.347	II
50	237.331	II
100	243.027	II
300	256.317	II
1000	303.780	II
500	303.786	II
10	320.293	I
2	505.500	I
3	505.684	I
4	505.912	I
5	506.200	I
7	506.570	I
10	507.058	I
15	507.718	I
20	508.643	I
25	509.998	I
35	512.098	I
50	515.616	I
100	522.213	I
400	537.030	I
1000	584.334	I
50	591.412	I
5	958.70	II
6	972.11	II
8	992.36	II
15	1025.27	II
30	1084.94	II
35	1215.09	II
50	1215.17	II
120	1640.34	II
180	1640.47	II
	Air	
7	2385.40	II
9	2511.20	II
50	2577.6	I
1	2723.19	I
12	2733.30	II
2	2763.80	I
10	2818.2	I
4	2829.08	I
10	2945.11	I
40	3013.7	I
20	3187.74	I
3	3202.96	II
15	3203.10	II
1	3354.55	I
2	3447.59	I
1	3587.27	I
3	3613.64	I
2	3634.23	I
3	3705.00	I
1	3732.86	I
10	3819.607	I
1	3819.76	I
500	3888.65	I
20	3964.729	I
1	4009.27	I
50	4026.191	I
5	4026.36	I
12	4120.82	I
2	4120.99	I
3	4143.76	I
10	4387.929	I
3	4437.55	I
200	4471.479	I
25	4471.68	I
6	4685.4	II
30	4685.7	II
30	4713.146	I
4	4713.38	I
20	4921.931	I
100	5015.678	I
10	5047.74	I
5	5411.52	II
500	5875.62	I
100	5875.97	I
8	6560.10	II
100	6678.15	I
3	6867.48	I
200	7065.19	I
30	7065.71	I
50	7281.35	I
1	7816.15	I
2	8361.69	I
2	9063.27	I
2	9210.34	I
10	9463.61	I
4	9516.60	I
3	9526.17	I
1	9529.27	I
1	9603.42	I
3	9702.60	I
6	10027.73	I
2	10031.16	I
15	10123.6	II
1	10138.50	I
10	10311.23	I
2	10311.54	I
3	10667.65	I
300	10829.09	I
1000	10830.25	I
2000	10830.34	I
9	10913.05	I
3	10917.10	I
4	11626.4	II
30	11969.12	I
20	12527.52	I
50	12784.99	I
20	12790.57	I
7	12845.96	I
10	12968.45	I
2	12984.89	I
12	15083.64	I
200	17002.47	I
1	18555.55	I
6	18636.8	II
500	18685.34	I
200	18697.23	I
100	19089.38	I
20	19543.08	I
1000	20581.30	I
80	21120.07	I
10	21121.43	I
20	21132.03	I
3	30908.5	II
4	40478.90	I

HOLMIUM (Ho)
Z = 67

Ho I and II
Ref. 1 — C.H.C.

Intensity		Wavelength	
		Air	
170		2502.91	II
80		2508.53	II
110		2513.55	II
95		2518.73	II
170		2533.80	I
130		2536.86	II
80		2556.84	I
80		2567.73	II
80		2586.52	I
60		2591.05	II
95		2592.99	I
190		2605.86	II
110		2610.51	II
95		2613.99	II
60		2625.20	II
80		2640.09	I
80		2640.30	II
60		2649.68	II
80		2666.24	II
70		2689.03	II
210		2713.65	II
230		2733.95	II
270		2750.35	II
110	c	2759.35	II
110		2766.85	II
270		2769.89	II
110		2772.83	II
140		2777.10	II
140		2794.41	II
100		2799.99	II
100		2806.72	II
160	c	2809.99	II
220		2811.36	II
180		2812.00	II
190		2814.74	II
300		2824.20	II
140		2826.64	II
270	c	2831.69	II
210		2834.99	II
110		2835.85	II
110		2844.18	II
100		2844.68	II
270		2849.10	II
100		2861.23	II
250		2861.49	II
150		2862.72	II
210		2871.99	II
230		2874.06	II
160		2874.43	II
360		2880.26	II
460		2880.98	II
340		2894.99	II
160		2895.62	II
170		2900.84	II
570	c	2909.41	II
170		2915.82	II
300		2919.62	II
110		2925.35	II
160		2926.09	II
300	c	2928.30	II
220		2942.05	II
300		2944.49	II
250	c	2953.11	II
390		2973.00	II
410	c	2979.63	II
180		2981.46	II
140		2985.48	II
410		2987.64	II
250		2990.27	II
110		2995.86	II
320	c	3008.10	II
220		3014.60	II
270		3038.69	II
480	c	3049.38	II
410	c	3054.00	II
500	c	3057.45	II
230		3074.30	II
500	c	3082.34	II
910		3084.36	II
430	c	3086.54	II
200		3108.31	II
200	c	3109.91	II
760		3118.50	II
300	c	3130.99	II
200	c	3134.39	II
300	c	3144.36	II
200		3156.18	II
270		3156.97	II
200	c	3159.67	II
580	c	3166.62	II
390	Dl	3171.72	II
810		3173.78	II
390		3174.84	II
270	c	3176.97	II
810	c	3181.50	II
390		3183.84	II
270	Cw	3184.48	II
200		3186.37	I
390	c	3197.83	II
390		3201.76	II
200		3206.86	II
270	c	3210.41	II
200	c	3221.42	II
320		3233.34	II
200		3236.90	II
200		3237.40	II
200	c	3257.45	II
390	c	3278.15	II
270		3279.25	II
980	c	3281.97	II
390		3288.46	II
270	c	3290.96	II
200	c	3305.16	II
200		3319.87	II
230		3320.25	II
200		3331.93	II
630	c	3337.23	II
390	c	3338.86	II
980	c	3343.58	II
200		3344.47	II
360		3350.49	II
320		3352.10	II
320	Cw	3353.55	II
320		3354.58	II
320		3357.91	II
320		3364.27	II
290		3370.87	II
230		3374.16	II
290	c	3390.75	II
320	c	3394.60	II
8100	c	3398.98	II
810	c	3410.26	II
1200		3421.63	II
3200		3453.14	II
390	c	3410.65	II
1400	c	3414.90	II
5400		3416.46	II
2000	c	3425.34	II
2000	c	3428.13	II
630	c	3429.18	II
320	c	3432.10	II
390		3449.35	I
810	c	3455.70	II
16000	c	3456.00	II
1600		3461.97	II
360	c	3467.07	II
810	c	3473.91	II
5400	c	3474.26	II
6300		3484.84	II
490		3489.58	II
580	c	3493.09	II

Holmium (Cont.)

2500	c	3494.76	II
810	c	3498.88	II
410	c	3506.95	II
320		3509.37	II
810		3510.73	I
4100	c	3515.59	II
410	c	3519.94	II
630		3540.76	II
1600		3546.05	II
1100	c	3556.78	II
410		3560.15	II
410	c	3573.24	II
630	c	3574.80	II
810		3579.12	I
410		3580.75	II
410		3581.83	II
630	c	3592.23	II
1100	Cw	3598.77	II
340		3599.48	I
540	c	3600.95	II
340		3613.31	II
410		3618.43	I
430	c	3626.69	II
490		3627.25	II
430	c	3631.76	II
430	c	3638.30	II
1600	c	3662.29	I
430		3662.99	I
720		3666.65	I
1400		3667.97	I
320		3669.05	II
450		3669.52	I
450	c	3674.77	II
720		3679.19	I
670		3679.70	I
720		3682.65	I
430		3685.16	II
580		3690.65	I
340		3691.95	I
410		3700.04	I
490	c	3702.35	II
320		3709.76	I
430		3712.88	I
450		3720.72	I
1100		3731.40	I
360		3732.09	I
810		3736.35	I
3200	Cw	3748.17	II
320	c	3753.73	II
340		3769.09	I
320		3788.08	II
8900	c	3796.75	II
8900	c	3810.73	II
490		3811.86	I
900	c	3813.25	II
300		3821.73	II
390		3829.27	I
320	Cw	3831.9	II
410	c	3835.35	II
1300	Cw	3837.51	II
410	c	3842.05	II
1100		3843.86	II
490	c	3846.73	II
300		3849.88	I
320		3852.40	II
1800	c	3854.07	II
390	Cw	3856.94	II
720		3857.72	II
2700	c	3861.68	II
540		3862.62	I
360		3872.05	II
320	c	3874.09	II
630		3874.68	II
540		3881.61	II
3000	c	3888.96	II
490		3890.42	I
13000	c	3891.02	II
540		3896.76	II
290		3902.23	II
320		3904.44	I
1300	Cw	3905.68	II
320		3911.80	I
320		3919.45	I
320	c	3936.44	II
220		3938.85	I
320	Cw	3940.53	II
220		3950.56	I
580		3955.73	I
230	c	3959.51	II
490		3959.68	I
220		3975.88	I
390	c	3976.93	I
220	Cw	3985.71	II
220		3993.73	II
380		3999.58	I
160	Cw	4002.59	II
220		4003.39	I
110		4013.50	I
320		4014.20	II
160	c	4018.09	II
160	c	4022.76	II
160	c	4023.94	II
110		4025.39	I
320		4027.21	I
270		4028.86	I
180	c	4031.80	I
220		4037.62	I
220	c	4038.87	II
2700		4040.81	I
5400	c	4045.44	II
220	c	4047.52	I
8100		4053.93	I
540		4054.48	II
270		4057.55	I
220		4060.31	I
1700		4065.09	II
170		4067.57	I
720		4068.05	I
270		4071.83	I
270		4073.13	I
290		4073.51	I
120	c	4080.23	II
230		4083.67	I
140		4085.09	I
170		4087.35	I
200		4087.59	I
140		4091.64	I
120		4094.78	I
230		4100.22	I
8900		4103.84	I
120		4105.04	I
270		4106.50	I
100		4107.36	I
2900		4108.62	I
300		4112.00	I
100		4112.72	I
270		4116.73	I
1500		4120.20	I
1300		4125.65	I
4300		4127.16	I
300		4134.54	I
1500		4136.22	I
130		4139.34	I
230		4142.19	I
290		4148.97	I
980	Cw	4152.61	II
8100		4163.03	I
160		4172.23	I
2500		4173.23	I
540		4194.35	I
100		4198.08	I
130		4203.21	I
100		4211.30	II
290		4222.29	I
290		4223.47	I
2000		4227.04	I
390		4229.52	II
130	h	4231.24	I
290		4243.78	I
1300	Cw	4254.43	I
130	c	4258.61	II
490		4264.05	I
300		4266.04	I
100		4273.63	II
200		4311.04	I
250		4330.64	II
300		4337.13	II
100	Cw	4346.84	II
1300		4350.73	I
290		4356.73	II
140		4363.93	II
170		4379.14	II
180	c	4384.83	II
150		4400.55	II
120		4401.24	II
180		4403.27	I
200		4420.56	II
130		4444.63	I
100		4473.59	II
300		4477.64	II
120		4484.57	II
140		4510.82	I
100	c	4526.14	II
170		4530.08	II
170	c	4531.28	I
130	c	4531.65	II
170		4534.58	I
200		4562.52	I
120	Cw	4609.32	II
130		4613.37	I
100		4618.84	I
100	c	4628.22	I
290		4629.10	II
200	c	4649.77	II
130	c	4661.33	II
140	c	4674.62	II
70		4701.17	II
80		4701.69	II
130		4709.84	II
65		4711.39	I
130	c	4717.52	I
35	c	4728.72	II
35		4738.00	II
290		4742.04	II
35	c	4749.09	II
35		4751.40	I
100	c	4757.01	I
35		4762.39	II
35		4763.57	II
55		4777.48	II
30		4779.42	I
70	c	4781.19	I
65		4782.92	I
55	c	4786.29	I
35		4791.48	II
35		4795.92	II
45	h	4798.87	I
27		4812.92	II
55		4832.31	II
30		4833.32	I
30		4855.54	II
45		4860.39	I
27	c	4889.67	II
30		4892.35	I
35		4896.44	II
55		4906.99	II
45		4922.73	I
55	c	4934.89	I
290		4939.01	I
27	c	4946.80	I
45		4948.18	II
65	c	4959.42	II
35		4961.03	II
55	Cw	4966.73	II
250	c	4967.21	II
220		4979.97	I
35	c	4988.96	I
90		4995.05	I
35	c	5012.42	I
55		5013.28	II
65		5026.53	I
30		5028.17	I
55		5032.95	II
65	c	5037.60	I
130		5042.37	I
35		5044.73	I
30		5051.44	II
30		5054.92	II
35	c	5060.75	I
65		5074.34	I
80		5093.07	I
140		5127.81	I
55		5129.27	II
130		5142.59	II
110		5143.22	II
160		5149.59	II
90	c	5167.88	I
130	c	5182.11	I
55		5187.85	I
90		5190.11	II
18		5195.23	I
45		5221.54	I
35	c	5244.47	I
65		5251.82	I
55		5275.48	I
90		5301.25	I
35		5319.24	I
35		5319.65	I
80		5330.11	I
90		5359.99	I
55		5381.40	I
30		5384.56	I
30		5384.97	I
18	h	5393.85	I
70		5403.17	I
100		5407.08	I
14		5413.62	II
16		5434.39	II
18		5435.87	I
30		5445.39	I
18		5449.8	II
30	h	5451.90	I
14		5454.0	II
30	c	5498.57	I
30		5504.51	I
27		5515.56	II
18		5516.45	II
30		5534.33	I
27		5553.14	I
35	c	5560.94	I
35	b	5563.6	H
70		5566.52	I
18		5573.96	II
35	Bl	5584.7	H
55	b	5591.1	H
55	Bl	5592.3	H
30	b	5607.1	H
27		5613.64	I
45	b	5626.4	H
65		5627.60	I
30		5628.24	II
55		5640.62	I
70	Bs	5655.9	H
65	b	5658.9	H
140		5659.58	I
70	c	5671.84	
65		5674.70	I
140	c	5691.47	I
70	Bs	5696.3	H
140	c	5696.57	I
27		5734.02	I
45		5736.4	H
55		5739.24	I
22		5749.58	I
30		5766.64	
27	b	5803.8	H
45	b	5819.2	H
27	h	5821.90	I
22		5839.47	I
45	b	5849.4	H
140	c	5860.28	I
27	h	5864.42	I
45		5870.85	I
27	b	5879.6	H
70	c	5882.99	I
35	c	5892.56	I
22		5904.29	I
70		5921.76	I
30	c	5933.71	I
70	Cw	5948.03	I
45		5955.98	I
70		5972.76	I
90		5973.52	I
22		5981.43	I
230	c	5982.90	I
55		6002.04	I
27		6005.33	I
35		6021.43	I
16		6038.97	I
27		6050.71	I
45		6060.31	I
120		6081.79	I
70	Cw	6133.60	I
35		6156.38	I
27		6156.58	I
55		6191.68	I
70		6208.65	I
18		6234.17	I
45	c	6255.75	I
70	c	6305.36	I
22		6306.68	I
30		6321.94	
30	c	6354.35	I
30	c	6372.59	I
14	h	6373.86	I
22	h	6413.41	I
27	c	6471.77	I
13		6479.17	I
11		6515.30	I
11	h	6538.99	I
70		6550.97	I
15		6560.08	I
35	d	6600.58	I
260		6604.94	I
55		6607.47	I
13		6628.35	I
120		6628.99	I
15		6632.24	I
9	h	6652.98	I
15		6662.52	I
19	c	6680.46	I
24	c	6681.62	I
15	h	6682.02	
55	Cw	6694.32	I
15	Cw	6722.34	I
40		6745.05	I
13		6766.74	I
28	c	6774.68	I
55	c	6785.43	I
13		6793.7	I
13	Cw	6811.04	I
15	Cw	6820.38	I
24		6821.64	I
17	c	6825.72	I
8	h	6826.62	
8	h	6852.97	
17	Cw	6865.85	I
9		6883.36	I
13		6888.50	I
15	c	6892.96	I
17		6897.95	I
15	h	6903.80	I
15	Cw	6913.47	I
9		6916.70	I
40	Cw	6939.49	I
45	Cw	6950.39	I
13	Hl	6955.3	I
19		6976.7	II
10		6985.11	
9		6994.38	
14	h	7000.71	
10		7079.07	
12		7098.58	
9		7242.08	I
9		7250.60	I
14		7308.55	
25		7341.43	
18		7389.40	
5	h	7496.20	I
10	h	7510.74	I
140		7555.09	I
18		7589.20	I
25		7591.87	I
9	h	7593.64	I
7	h	7594.35	I

Holmium (Cont.)

12		7602.31	II
16		7605.35	I
12		7617.05	I
14		7627.98	I
40	c	7628.42	I
9	c	7641.14	I
4		7648.16	I
14	c	7653.80	I
12	c	7667.30	I
20		7690.43	I
50	c	7693.15	I
40	Cw	7715.06	I
16		7719.05	I
16	h	7738.98	I
8	c	7752.01	I
60	Cw	7815.48	I
40	Cw	7823.63	I
8	h	7879.22	I
60		7894.64	I
10	h	8464.66	I
10	h	8482.67	I
50		8512.94	I
20		8545.61	II
18		8601.84	II
40		8670.19	I
8	h	8697.32	I
16	h	8805.48	II
20	c	8834.49	I
90		8915.98	II

HYDROGEN (H)
Z = 1

H I
Ref. 214 — W.C.M.

Intensity		Wavelength	
		Vacuum	
15		926.226	I
20		930.748	I
30		937.803	I
50		949.743	I
100		972.537	I
300		1025.722	I
1000		1215.668	I
500		1215.674	I
		Air	
5		3835.384	I
6		3889.049	I
8		3970.072	I
15		4101.74	I
30		4340.47	I
80		4861.33	I
120		6562.72	I
180		6562.852	I
5		9545.97	I
7		10049.4	I
12		10938.1	I
20		12818.1	I
40		18751.0	I
5		21655.3	I
8		26251.5	I
15		40511.6	I
4		46525.1	I
6		74578	I
3		123685	I

INDIUM (In)
Z = 49

In I and II
Ref. 1, 132, 348—350 — C.H.C.

Intensity		Wavelength	
		Vacuum	
2		1648.00	I
1	h	1676.16	I
5	h	1711.54	I
2	h	1741.23	I
1	h	1758.49	I
		Air	
10		2103.89	II
10		2166.88	II
2		2179.90	I
2		2182.40	I
2		2187.40	I
2		2190.84	I
15		2195.67	II
2		2197.41	I
2		2202.24	I
50		2205.28	II
3		2211.14	I
5		2230.70	I
3		2241.66	I
30		2255.79	II
10		2259.99	I
5		2278.20	I
40		2281.64	II
2		2283.75	I
2		2298.33	I
2		2298.70	I
2		2302.49	I
100	c	2306.05	II
25		2306.86	I
3		2309.32	I
2		2309.75	I
90	d	2313.21	II
2		2315.09	I
30		2323.40	II
5		2324.41	I
3		2324.92	I
70	d	2327.95	II
3		2332.76	I
80	h	2334.57	II
10		2340.19	I
8		2345.90	I
5		2346.56	I
50	d	2350.75	II
5		2358.70	I
15		2378.14	I
10		2379.00	I
110	d	2382.63	II
40		2389.54	I
40		2393.18	II
10		2399.18	I
50	h	2406.47	II
50		2408.76	II
50		2419.06	II
50		2419.20	II
70	h	2427.20	II
20		2429.86	I
10		2430.99	I
50		2432.73	II
60		2442.63	II
100		2447.90	II
60		2453.23	II
60		2460.08	I
30	h	2468.02	I
70		2486.15	II
110	d	2488.62	II
90		2488.95	II
80		2498.59	II
100		2499.60	II
90	d	2500.99	II
60		2508.16	II
110	d	2512.31	II
100		2521.37	I
10		2522.98	I
70		2553.56	II
160	d	2554.44	II
1100		2560.15	I
70		2565.13	II
70	d	2598.75	II
200		2601.76	I
50		2604.04	II
90	d	2654.70	II
100	d	2662.63	II
140	d	2668.65	II
140	d	2674.56	II
80		2683.12	II
1600		2710.26	I
300		2713.94	I
130	d	2749.75	II
700		2753.88	I
40		2775.37	I
60		2798.76	II
90	d	2818.97	II
180	c	2836.92	I
30	c	2858.14	I
80		2865.68	II
120	d	2890.18	II
1100		2932.63	I
100		2941.05	II
20	c	2957.01	I
60	d	2966.17	II
110	c	2999.40	II
8000		3039.36	I
8	d	3051.15	I
110	d	3099.80	II
180	c	3101.8	II
130	c	3138.60	II
80	c	3142.75	II
130	d	3146.70	II
150		3155.77	II
100	c	3158.40	II
90	c	3176.30	II
90	d	3198.11	II
13000		3256.09	I
3000		3258.56	I
90	c	3338.50	II
75	c	3376.59	II
100	c	3404.28	II
110	d	3438.40	II
180	c	3693.91	II
95	c	3708.13	II
380	w	3716.14	II
120	c	3718.30	II
160	c	3718.72	II
160	c	3723.40	II
170	w	3795.21	II
230	c	3799.21	II
250	c	3834.65	II
200	c	3842.18	II
100		3889.78	II
100	c	3902.07	II
60	d	3922.12	II
65	c	3934.40	II
250	w	3962.35	II
120	c	4004.66	II
140	d	4013.92	II
410	w	4056.94	II
17000		4101.76	I
140	c	4205.14	II
100	d	4213.04	II
110	d	4219.66	II
150	d	4372.87	II
150	c	4500.78	II
18000		4511.31	I
110	c	4549.01	II
140	c	4570.85	II
180	w	4578.02	II
180	w	4578.40	II
140	c	4616.08	II
170	c	4617.17	II
250	c	4620.14	II
150	w	4620.70	II
170	c	4627.30	II
140	c	4637.04	II
380	c	4638.16	II
220	c	4644.58	II
360	c	4655.62	II
320	w	4656.74	II
190	c	4681.11	II
450	w	4684.8	II
3		4878.37	I
90	d	4907.06	II
70	h	4924.93	II
150	c	4973.77	II
80	h	5109.36	II
100	w	5115.14	II
140	c	5117.40	II
270	c	5120.80	II
200	w	5121.75	II
80	d	5129.85	II
240	c	5175.42	II
140	c	5184.44	II
30		5254.32	I
12		5262.74	I
150	c	5309.45	II
80		5411.41	II
140	c	5418.45	II
220	w	5436.70	II
130	c	5497.50	II
140	c	5507.08	II
320	c	5513.00	II
250	w	5523.28	II
130	c	5536.50	II
190	w	5555.45	II
240	c	5576.90	II
200	w	5636.70	II
160	c	5708.50	II
50		5709.91	I
100	c	5721.80	II
50		5727.68	I
210	c	5853.15	II
490	w	5903.4	II
260	w	5915.4	II
120	c	5918.78	II
130	c	6062.9	II
250	c	6095.95	II
210	c	6108.66	II
180	w	6115.9	II
230	w	6128.7	II
240	w	6129.4	II
320	w	6132.1	II
150	c	6140.0	II
90		6143.23	II
140	c	6148.10	II
190	w	6149.5	II
80		6161.15	II
180	w	6162.45	II
100	c	6224.28	II
280	w	6228.3	II
140	w	6231.1	II
270	w	6304.8	II
290	w	6362.3	II
300	w	6469.0	II
210	c	6541.20	II
190	c	6751.88	II
180	c	6765.9	II
100	c	6783.72	II
8	h	6847.44	I
320	w	6891.5	II
4	h	6900.13	I
380	w	7182.9	II
180	c	7255.0	II
210	c	7276.5	II
180	c	7303.4	II
320	c	7350.6	II
100	c	7632.7	II
100	c	7682.9	II
210	c	7740.7	II
100	c	7776.96	II
180	c	7789.0	II
70	c	7806.8	II
70	c	7814.5	II
90	c	7840.9	II
20	h	8050.78	I
240	c	8227.0	II
30	h	8238.66	I
15	h	8314.92	I
50	c	8434.55	II
30		8678.95	I
20		8682.63	I
50		8700.25	I
100	w	8813.5	II
80	c	8832.6	II
40		8894.47	I
10		9170.08	I
120	c	9197.7	II
120	c	9202.0	II
220	w	9213.0	II
160	d	9241.1	II
40	h	9349.83	I
60	h	9370.27	I
20		9427.99	I
100		9977.86	I
200		10257.03	I
60	h	10717.42	I
100	h	10744.31	I
20		11334.72	I
20		11731.48	I
10		12912.59	I
9		13429.96	I
5		13824.48	I
6		14316.25	I
3		14419.20	I
6		14668.66	I
7		14719.08	I
2		16504.31	I
6		22291.06	I
7		23879.13	I

In III
Ref. 351 — C.H.C.

Intensity		Wavelength	
		Vacuum	
7		685.31	III
5		691.62	III
1		702.17	III
10		882.24	III
10		890.84	III
10		915.87	III
2		917.45	III
5		926.83	III
30		1403.08	III
30		1434.85	III
20		1487.70	III
20		1494.14	III
10		1524.78	III
20		1530.21	III
30		1532.95	III
100		1625.42	III
20		1642.28	III
20		1702.53	III
100		1748.83	III
2		1767.88	III
1		1810.71	III
30		1842.41	III
40		1850.30	III
15		1862.98	III
		Air	
30		2154.08	III
2		2154.42	III
10		2199.52	III
5		2232.18	III
20		2261.26	III
5		2266.26	III
5		2272.41	III
5		2272.84	III
10		2300.90	III
100		2527.41	III
50		2725.52	III
80		2726.15	III
100		2982.80	III
100		3008.08	III
30		3008.82	III
30		3293.55	III
8		3350.91	III
5		3562.32	III
100		3852.82	III
100		4023.77	III
150		4032.32	III
50		4062.30	III
100		4071.57	III
100		4072.93	III
100		4252.68	III
40		4509.58	III
200		5248.77	III

Iodine (Cont.)

I III
Ref. 20, 21, 161 — L.J.R.

Intensity	Wavelength	
	Vacuum	
6	666.81	III
8	705.11	III
7	784.64	III
7	784.80	III
8	795.52	III
5	865.97	III
5	920.38	III
6	961.17	III
6	1078.58	III
8	1094.20	III
4	1244.66	III
8	1252.35	III
5	1306.93	III
	Air	
1	2224.43	III
1	2238.12	III
3	2249.31	III
2	2309.38	III
3	2340.85	III
3	2350.43	III
2	2353.46	III
4	2367.74	III
2	2371.45	III
3	2372.45	III
4	2376.47	III
4	2387.12	III
3	2392.01	III
2	2403.06	III
2	2403.63	III
2	2414.85	III
2	2418.49	III
2	2418.85	III
2	2423.91	III
5	2426.12	III
3	2434.88	III
2	2462.50	III
3	2466.69	III
3	2466.99	III
6	2475.36	III
4	2489.27	III
2	2493.21	III
2	2494.27	III
3	2495.16	III
2	2496.07	III
3	2501.41	III
2	2516.82	III
6	2519.75	III
4	2521.72	III
3	2531.99	III
2	2537.56	III
7	2545.71	III
4	2640.77	III
4	2642.11	III
6	2652.25	III
2	2818.48	III
2	2839.44	III
4	2864.67	III
4	2885.15	III
3	2910.98	III
3	2917.35	III
2	2931.11	III
2	3005.68	III
3	3069.23	III
3	3153.88	III
3	3170.14	III
3	3181.66	III
3	3210.14	III
4	3213.49	III
4	3224.93	III
2	3300.47	III
2	3479.53	III
3	3546.92	III
3	3613.81	III
2	3754.40	III
2	3754.55	III
3	3963.16	III
3	4077.14	III

I IV
Ref. 21, 58 — L.J.R.

Intensity	Wavelength	
	Vacuum	
5	601.86	IV
6	612.46	IV
4	615.17	IV
4	654.22	IV
4	654.56	IV
7	919.28	IV
	Air	
5	2249.30	IV
4	2340.84	IV
7	2361.13	IV
5	2367.75	IV
6	2372.45	IV
7	2376.46	IV
4	2385.28	IV
8	2387.11	IV
6	2392.00	IV
4	2403.05	IV
2	2418.45	IV
3	2423.89	IV
9	2426.10	IV
6	2434.85	IV
3	2466.68	IV
3	2466.96	IV
8	2475.35	IV
4	2485.51	IV
5	2489.24	IV
4	2493.20	IV
2	2501.38	IV
3	2513.74	IV
8	2519.74	IV
6	2521.72	IV
4	2531.98	IV
5	2537.54	IV
8	2545.67	IV
4	2640.77	IV
5	2642.11	IV
8	2652.23	IV
3	2818.45	IV
6	2864.68	IV
4	2910.97	IV
5	2917.33	IV
4	3069.17	IV
4	3170.11	IV
4	3181.64	IV
4	3210.12	IV
6	3213.48	IV
6	3224.90	IV
4	3546.90	IV

IV
Ref. 84 — L.J.R.

Intensity	Wavelength	
	Vacuum	
30	363.78	V
36	380.74	V
45	565.53	V
50	607.57	V

IRIDIUM (Ir)
Z = 77

Ir I and II
Ref. 1 — C.H.C.

Intensity	Wavelength	
	Air	
9900	2010.65	I
8700	2022.35	I
15000	2033.57	I
6200	2052.22	I
5000	2060.64	I
3700	2083.22	I
3100	2085.74	I
17000	2088.82	I
14000	2092.63	I
2700	2112.68	I
1800	2119.54	I
2000	2125.44	I
4500	2126.81	II
2000	2127.52	I
4500	2127.94	I
3700	2148.22	I
2500	2150.54	I
3500	2152.68	II
2900	2155.81	I
7900	2158.05	I
2100	2162.88	I
5800	2169.42	II
4500	2175.24	I
2700	2178.17	I
1600	2187.43	II
1100	2190.38	II
740	2191.64	I
910	2208.09	II
1300	2220.37	I
790	2221.07	II
2500	2242.68	II
620	2245.76	II
2100	2253.38	I
	2253.49	I
2100	2255.10	I
1400	2255.81	I
350	2258.51	I
1400	2258.86	I
830	2264.61	I
1100	2266.33	I
1000	2268.90	I
660	2280.00	I
950	2281.02	II
660	2281.91	I
330	2284.60	I
330	2295.08	I
790	2298.05	I
	2298.16	I
460	2299.53	I
910	2300.50	I
2700	2304.22	I
410	2305.47	I
210	2307.27	I
910	2308.93	I
460	2315.38	I
410	2321.45	I
410	2321.58	I
210	2327.98	I
540	2333.30	I
740	2333.84	I
580	2334.50	I
1600	2343.18	I
740	2343.61	I
100	2352.62	I
580	2355.00	I
230	2357.53	II
410	2358.16	I
500	2360.73	I
2500	2363.04	I
370	2368.04	II
3500	2372.77	I
290	2375.09	II
250	2377.28	I
250	2377.98	I
500	2379.38	I
540	2381.62	I
210	2383.17	I
120	2386.58	II
1300	2386.89	I
2500	2390.62	I
2700	2391.18	I
230	2407.59	I
290	2409.37	I
290	2410.17	I
290	2410.73	I
540	2413.31	I
370	2415.86	I
620	2418.11	I
120	2424.32	I
120	2424.66	I
210	2424.89	I
370	2424.99	I
290	2425.66	I
170	2426.53	II
540	2427.61	I
540	2431.24	I
1300	2431.94	I
170	2432.36	I
100	2432.58	I
270	2435.14	I
250	2445.34	I
250	2447.76	I
190	2448.23	I
910	2452.81	I
1300	2455.61	I
230	2455.87	I
210	2457.03	I
210	2457.23	I
120	2465.09	I
870	2467.30	I
3300	2475.12	I
210	2478.11	I
2100	2481.18	I
100	2485.38	I
620	2493.08	I
210	2496.27	I
250	2502.63	I
4100	2502.98	I
170	2504.37	I
120	2505.74	I
120	2507.63	I
170	2509.71	I
170	2511.94	I
170	2512.58	II
210	2513.71	I
120	2515.36	I
40	2524.88	II
170	2525.05	I
120	2532.52	I
990	2533.13	I
1100	2534.46	I
580	2537.22	I
170	2537.68	I
100	2541.48	I
580	2542.02	I
40	2542.80	II
7900	2543.97	I
150	2545.54	I
790	2546.03	I
120	2547.20	I
120	2547.69	I
210	2551.40	I
190	2554.40	I
210	2555.35	I
170	2555.88	I
150	2563.28	I
910	2564.18	I
210	2569.88	I
100	2570.62	I
230	2572.70	I
740	2577.26	I
100	2578.71	I
35	2579.49	II
740	2592.06	I
740	2599.04	I
150	2602.04	I
190	2604.55	I
190	2607.52	I
700	2608.25	I
1800	2611.30	I
210	2614.98	I
330	2617.78	I
210	2619.88	I
70	2623.64	II
250	2625.32	I
100	2626.76	I
700	2634.17	I
170	2635.27	I
250	2639.42	I
3500	2639.71	I
210	2644.19	I
170	2653.76	I
100	2656.81	I
1800	2661.98	I
350	2662.63	I
2700	2664.79	I
140	2668.99	I
520	2669.91	I
520	2671.84	I
330	2673.61	I
120	2676.83	I
110	2684.04	I
270	2692.34	I
3000	2694.23	I
110	2704.03	I
160	2712.74	I
140	2744.00	I
330	2772.46	I
250	2775.55	I
520	2781.29	I
330	2785.22	I
540	2797.35	I
1600	2797.70	I
380	2798.18	I
410	2800.82	I
680	2823.18	I
1200	2824.45	I
110	2833.24	II
110	2835.66	I
820	2836.40	I
160	2837.33	I
1100	2839.16	I
820	2840.22	I
160	2842.28	I
3800	2849.72	I
110	2863.84	I
380	2875.60	I
380	2875.98	I
270	2877.68	I
140	2879.41	I
820	2882.64	I
650	2897.15	I
260	2901.95	I
260	2904.80	I
200	2907.24	I
440	2916.36	I
230	2918.57	I
4400	2924.79	I
1200	2934.64	I
880	2936.68	I
250	2938.47	I
190	2939.27	I
140	2940.54	I
2700	2943.15	I
230	2946.97	I
200	2949.76	I
1200	2951.22	I
150	2962.99	I
200	2974.95	I
440	2980.65	I
150	2985.80	I
190	2990.62	I
300	2996.08	I
180	2997.41	I
220	3002.25	I
600	3003.63	I
160	3011.69	I
120	3016.43	I
270	3017.31	I
140	3019.23	I
110	3025.82	I
380	3029.36	I
330	3039.26	I
35	3042.65	II
300	3047.16	I
300	3049.44	I
300	3057.28	I

Indium (Cont.)

100	5645.15	III
40	5723.17	III
100	5819.50	III
200	6197.72	III

In IV
Refs. 352, 435, 436 — J.R.

Intensity	Wavelength	
	Vacuum	
622	472.71	IV
689	479.39	IV
709	498.62	IV
61	945.74	IV
85	954.67	IV
87	973.50	IV
86	991.60	IV
89	1024.68	IV
85	1024.79	IV
88	1031.45	IV
82	1031.98	IV
80	1054.43	IV
84	1063.03	IV
83	1068.25	IV
82	1069.82	IV
86	1077.64	IV
90	1082.10	IV
83	1086.33	IV
82	1096.81	IV
84	1097.18	IV
85	1116.10	IV
80	1124.06	IV
90	1131.46	IV
85	1144.43	IV
80	1145.41	IV
89	1146.62	IV
83	1154.11	IV
84	1154.60	IV
90	1157.71	IV
90	1157.82	IV
85	1159.78	IV
88	1176.50	IV
85	1191.58	IV
83	1204.87	IV
90	1206.55	IV
88	1221.50	IV
85	1221.90	IV
85	1233.58	IV
87	1235.84	IV
90	1373.20	IV
90	1398.77	IV
81	1412.09	IV

In V
Ref. 353 — C.H.C.

Intensity	Wavelength	
	Vacuum	
6	368.67	V
6	370.10	V
10	372.82	V
10	372.94	V
2	374.95	V
6	375.84	V
6	376.07	V
10	376.79	V
17	378.61	V
3	379.24	V
9	380.27	V
11	381.56	V
9	382.14	V
11	382.76	V
10	383.05	V
17	386.21	V
10	386.70	V
3	388.66	V
14	388.91	V
11	390.03	V
11	390.92	V
9	392.29	V
9	392.46	V
1	393.60	V
25	393.89	V
11	395.74	V
3	397.73	V
10	399.79	V
9	400.05	V
25	400.57	V
25	402.39	V
3	405.33	V
9	407.28	V
3	407.36	V
9	407.95	V
9	417.43	V
2	418.45	V
2	423.16	V

IODINE (I)

Z = 53

I I and II
Ref. 124, 153, 176, 184
L.J.R.

Intensity		Wavelength	
		Vacuum	
2		655.80	II
6		659.00	II
8		663.98	II
8		664.52	II
8		665.06	II
150		665.70	II
1000		719.55	II
1000		722.08	II
1000		798.16	II
1200		834.10	II
600		847.80	II
1500		873.49	II
1000		875.94	II
2000		879.84	II
1500		881.88	II
1000		891.00	II
1000		893.17	II
1200		1000.57	II
1000		1003.35	II
4000		1018.58	II
10000		1034.66	II
1500		1054.74	II
2000		1066.34	II
3000		1075.21	II
5000		1105.00	II
2500		1111.16	II
1500		1117.22	II
3500		1125.25	II
2000		1131.50	II
1200		1139.75	II
10000		1139.80	II
1500		1154.67	II
1000		1159.87	II
10000		1160.56	II
20000		1166.48	II
1500		1167.05	II
5000		1175.84	II
10000		1178.65	II
15000		1187.34	II
10000		1190.85	II
15		1195.29	I
5000		1198.88	II
7000		1200.22	II
200		1218.41	I
20000		1220.89	II
600		1224.05	I
600		1224.08	I
500		1228.89	I
20000		1234.06	II
600		1251.34	I
2500		1259.15	I
3000		1259.51	I
800		1261.27	I
600		1267.57	I
600		1267.60	I
1500		1275.26	I
3000		1289.40	I
10000		1300.34	I
3000		1302.98	I
3000		1313.95	I
3000		1317.54	I
2000		1330.19	I
20000		1336.52	II
5000		1355.10	I
3000		1357.97	I
5000		1360.97	I
3000		1361.11	I
2500		1367.71	I
2500		1368.22	I
4000		1383.23	I
3000		1390.75	I
2000		1392.90	I
2000		1400.01	I
8000		1425.49	I
5000		1446.26	I
5000		1453.18	I
5000		1457.39	I
5000		1457.47	I
10000		1457.98	I
2500		1458.79	I
4000		1459.15	I
2500		1465.83	I
1000		1485.92	I
5000		1492.89	I
5000		1507.04	I
5000		1514.68	I
15000		1518.05	I
2500		1526.45	I
5000		1593.58	I
5000		1617.60	I
2500		1640.78	I
15000		1702.07	I
12000		1782.76	I
5000		1799.09	I
75000		1830.38	I
15000		1844.45	I
		Air	
2000		2061.63	I
100		2408.01	II
100		2419.18	II
100		2494.74	II
100		2533.60	II
200		2534.27	II
1000		2566.24	II
2000		2582.79	II
300		2593.46	II
200	c	2688.98	II
500		2730.12	II
20		2765.15	II
200		2808.59	II
1500		2878.63	II
1000		2993.87	II
5000		3078.75	II
200		3161.03	II
1000		3175.07	II
300		3355.53	II
250	c	3424.99	II
300	c	3497.41	II
500		3526.90	II
200		3742.14	II
200		4102.23	I
200		4129.21	I
100	d	4134.15	I
500	d	4321.84	I
300		4452.86	II
200		4599.77	II
300	c	4632.45	II
500	d	4666.48	II
1000		4675.53	II
250		4763.31	I
1000		4862.32	I
200		4916.94	I
1000		4986.92	II
400	c	5065.37	II
10000		5119.29	I
200		5149.73	II
3000	c	5161.20	II
300		5176.19	II
600		5216.27	II
500	d	5228.97	II
1000		5234.57	I
3000	c	5245.71	II
500		5269.36	II
400		5299.78	II
400		5322.80	II
10000		5338.22	II
5000	c	5345.15	II
1000	c	5369.86	II
800	c	5405.42	II
800	c	5407.36	II
600	c	5427.06	I
3000		5435.83	II
1000		5438.00	II
2000	c	5464.62	II
800		5491.50	II
1000	c	5496.94	II
1000		5504.72	II
600	c	5522.06	II
600	c	5598.52	II
1000		5600.32	II
1500		5612.89	II
10000		5625.69	II
1000		5678.08	II
2000	c	5690.91	II
500		5702.05	II
4000	c	5710.53	II
1000		5738.27	II
1000		5760.72	II
1000	d	5764.33	I
500	c	5774.83	II
500		5787.02	II
2000		5894.03	I
5000		5950.25	II
300		5984.86	I
2000	d	6024.08	I
500		6068.93	II
2000	c	6074.98	II
1000		6082.43	I
2000	c	6127.49	II
800		6191.88	I
1000		6204.86	II
500		6213.10	I
800		6244.48	I
900	c	6257.49	II
1000		6293.98	I
500		6313.13	I
800		6330.37	I
400		6333.50	I
2000		6337.85	I
1000		6339.44	I
500		6359.16	I
1000		6566.49	I
2000		6583.75	I
1000		6585.27	I
5000		6619.66	I
500		6661.11	I
600		6665.96	II
500	c	6697.29	I
300		6718.83	II
400		6732.03	I
4000		6812.57	II
1000		6958.78	II
500		6989.78	I
200	c	7085.21	II
500		7120.05	I
1200		7122.05	I
2000		7142.06	I
1000		7164.79	I
400	d	7191.66	I
700		7227.30	I
1000		7236.78	I
500		7237.84	I
500		7351.35	II
5000		7402.06	I
1000		7410.50	I
500		7416.48	I
5000		7468.99	I
500	c	7490.52	I
2000		7554.18	I
500	d	7556.65	I
2000	c	7700.20	I
500		7798.98	II
600		7897.98	I
500		7969.48	I
1000		8003.63	I
99000		8043.74	I
300	d	8065.70	I
1000		8090.76	I
800	c	8169.38	I
500	d	8222.57	I
4000		8240.05	I
10000	c	8393.30	I
150		8414.60	II
1000		8486.11	I
1500	c	8664.95	I
500	c	8700.80	I
250	d	8748.22	I
1000		8853.24	I
2000		8853.80	I
3000		8857.50	I
1000	d	8898.50	I
400		8964.69	I
400		8993.13	I
5000		9022.40	I
15000		9058.33	I
1000		9098.86	I
12000		9113.91	I
600		9128.03	I
30		9195.30	II
600		9227.74	I
1000		9335.05	I
4000		9426.71	I
3000		9427.15	I
10	c	9480.33	II
2000		9598.22	I
2000		9649.61	I
3000	d	9653.06	I
5000		9731.73	I
500		10003.05	I
750		10131.16	I
1000		10238.82	I
400		10375.20	I
400		10391.74	I
6		10405.49	II
5000		10466.54	I
1		11084.68	II
400		11236.56	I
350		11558.46	I
320		11778.34	I
450		11996.86	I
300		12033.69	I
150		12304.58	I
60		13149.16	I
140		13958.27	I
200		14287.02	I
100		14460.00	I
225		15032.57	I
105		15528.65	I
150		16037.33	I
15		18275.71	I
20		18348.52	I
15		18982.41	I
35		19070.17	I
110		19105.12	I
50		19370.02	I
10		20648.69	I
220		22183.03	I
150		22226.53	I
30		22309.21	I
32		24420.82	I
12		27365.42	I
9		27573.05	I
10		30361.93	I
8		30383.88	I
10		34295.73	I
9		34513.11	I
3		40228.54	I
2		41633.80	I

Iridium (Cont.)

1600		3068.89	I
190		3069.09	I
190		3069.71	I
170		3076.69	I
320		3083.22	I
240		3086.44	I
390		3088.04	I
510		3100.29	I
510		3100.45	I
340		3120.76	I
200		3121.78	I
3400		3133.32	I
190		3150.61	I
190		3154.74	I
190		3159.15	I
140		3168.18	I
490		3168.88	I
370		3177.58	I
170		3180.35	I
370		3198.92	I
610		3212.12	I
370		3219.51	I
5100		3220.78	I
100		3221.28	I
300		3229.28	I
100		3230.76	I
470		3241.52	I
200		3262.01	I
390		3266.44	I
160		3277.28	I
100		3287.59	I
160		3310.52	I
200		3322.60	I
130		3334.16	I
560		3368.48	I
660		3437.02	I
100		3437.50	I
410		3448.97	I
3200		3513.64	I
220		3515.95	I
410		3522.03	I
160		3557.17	I
320		3558.99	I
1200		3573.72	I
320		3594.39	I
220		3609.77	I
190		3617.21	I
160		3626.29	I
660		3628.67	I
220		3636.20	I
300		3661.71	I
300		3664.62	I
320		3674.98	I
200		3687.08	I
140		3725.38	I
200		3731.36	II
130		3738.53	I
530		3747.20	I
120		3793.79	I
3100		3800.12	I
230		3817.24	I
170		3865.64	I
480		3902.51	I
480		3915.38	I
400		3934.84	I
120		3946.27	I
590		3976.31	I
460		3992.12	I
180		4020.03	I
350		4033.76	I
130		4040.08	I
370		4069.92	I
150		4070.68	I
100		4092.61	I
140		4115.78	I
23		4127.92	I
27		4155.70	I
15		4166.04	I
90		4172.56	I
35		4182.47	I
15	h	4183.21	I
18		4185.66	I
23		4197.54	I
27		4217.76	I
13		4220.80	I
75		4259.11	I
27		4265.30	I
260		4268.10	I
23		4286.62	I
75		4301.60	I
55		4310.59	I
220		4311.50	I
18		4351.30	I
18		4352.56	I
18		4392.59	I
160		4399.47	I
65		4403.78	I
110		4426.27	I
15		4450.18	I
55		4478.48	I
16		4495.35	I
11	h	4496.03	I
55		4545.68	I
30		4548.48	I
13		4550.78	I
35		4568.09	I
18		4570.02	I
18		4604.48	I
75		4616.39	I
26		4656.18	I
17	h	4668.99	I
21		4708.88	I
50		4728.86	I
21		4731.86	I
26		4756.46	I
13		4757.96	I
65		4778.16	I
30		4795.67	I
10		4807.14	I
21		4809.47	I
10		4840.77	I
17		4845.38	I
50		4938.09	I
26		4970.48	I
25		4999.74	I
25		5002.74	I
17		5009.17	I
30		5014.98	I
17		5046.06	I
30		5123.66	I
20		5177.95	I
22		5238.92	I
12		5340.74	I
35		5364.32	I
75		5449.50	I
30		5454.50	I
7		5469.40	I
10		5620.04	I
45		5625.55	I
10		5828.55	I
10		5882.30	I
7		5887.36	I
35		5894.06	I
7		6026.10	I
12		6067.83	I
20		6110.67	I
12		6288.28	I
7		6334.44	I
5		6624.73	I
10		6686.08	I
5		6830.01	I
5		6929.88	I
4		7183.71	I
6		7834.32	I

IRON (Fe)
Z = 26

Fe I and II
Ref. 56, 63, 105, 138, 174, 278
— H.M.C. and H.C.

Intensity	Wavelength	
	Vacuum	
12	1055.27	II
15	1068.36	II
15	1071.60	II
15	1096.89	II
12	1099.12	II
18	1112.09	II
12	1121.99	II
12	1122.86	II
12	1128.07	II
12	1130.43	II
15	1133.41	II
12	1133.68	II
12	1138.64	II
12	1142.33	II
12	1143.23	II
18	1144.95	II
12	1147.41	II
15	1148.29	II
12	1151.16	II
12	1267.44	II
12	1272.00	II
12	1371.02	II
12	1563.79	II
12	1580.62	II
18	1608.46	II
12	1618.47	II
15	1621.68	II
15	1629.15	II
15	1631.12	II
18	1635.40	II
15	1636.32	II
15	1639.40	II
12	1641.76	II
12	1647.16	II
12	1670.74	II
12	1702.04	II
12	1761.38	II
20	1785.26	II
20	1786.74	II
18	1788.07	II
30	1934.538	I
25	1937.269	I
50	1946.988	I
25	1951.571	I
30	1952.59	I
30	1953.005	I
60	1957.823	I
60	1960.144	I
30	1961.25	I
50	1962.111	I
12	1963.11	II
	Air	
100	2084.122	I
50	2157.794	I
15	2162.02	II
40	2166.773	I
300	2178.118	I
250	2186.486	I
60	2186.892	I
120	2187.195	I
250	2191.839	I
150	2196.043	I
80	2200.390	I
80	2200.724	I
15	2208.41	II
20	2213.65	II
12	2218.26	II
20	2220.38	II
25	2245.58	II
25	2250.790	I
60	2251.874	I
25	2255.77	II
300	2259.511	I
60	2264.389	I
80	2267.085	I
80	2267.469	I
50	2270.862	I
150	2272.070	I
150	2276.026	I
80	2279.937	I
150	2284.086	I
150	2287.250	I
300	2292.524	I
80	2294.41	I
200	2297.787	I
600	2298.169	I
80	2299.220	I
300	2300.142	I
50	2301.684	I
100	2303.424	I
150	2303.581	I
120	2308.999	I
150	2313.104	I
200	2320.358	I
100	2327.40	II
15	2327.88	II
100	2331.31	II
15	2331.97	II
300	2332.80	II
200	2338.01	II
600	2343.49	II
80	2343.96	II
150	2344.28	II
25	2344.98	II
50	2345.34	II
200	2348.11	II
250	2348.30	II
50	2351.20	II
15	2351.67	II
25	2352.31	II
30	2353.47	II
15	2353.68	II
50	2354.48	II
40	2354.89	II
200	2359.12	II
15	2359.59	II
150	2360.00	II
120	2360.29	II
30	2360.51	II
40	2362.02	II
60	2363.86	II
200	2364.83	II
80	2365.76	II
25	2366.59	II
80	2368.59	II
80	2369.456	I
80	2369.95	II
25	2370.50	II
120	2371.430	I
300	2373.624	I
150	2373.74	II
120	2374.518	I
60	2375.19	II
120	2376.43	II
20	2378.13	II
80	2379.27	II
20	2379.41	II
40	2380.20	II
120	2380.76	II
150	2381.835	I
1000	2382.04	II
20	2382.90	II
20	2383.06	II
60	2383.25	II
50	2384.39	II
40	2388.37	II
300	2388.63	II
200	2389.973	I
30	2390.10	II
20	2390.77	II
15	2391.48	II
20	2392.58	II
40	2395.42	II
1000	2395.62	II
15	2396.72	II
300	2399.24	II
20	2400.05	II
15	2401.29	II
50	2404.43	II
800	2404.88	II
250	2406.66	II
80	2406.97	II
300	2410.52	II
200	2411.07	II
50	2411.81	II
150	2413.31	II
20	2416.45	II
80	2417.87	II
15	2418.44	II
60	2420.396	I
60	2422.69	II
60	2423.089	I
40	2423.21	II
150	2424.14	II
15	2424.39	II
30	2424.59	II
30	2428.29	II
120	2428.36	II
25	2428.80	II
25	2429.03	II
20	2429.39	II
30	2429.86	II
120	2430.08	II
25	2431.02	II
80	2432.26	II
60	2432.87	II
25	2434.06	II
15	2434.24	II
20	2434.65	II
50	2434.73	II
50	2434.95	II
25	2436.62	II
60	2438.182	I
150	2439.30	II
150	2439.74	I
80	2440.11	I
40	2440.42	II
30	2442.37	II
100	2442.57	II
60	2443.71	II
250	2443.872	I
100	2444.51	II
50	2445.11	II
50	2445.212	I
100	2445.57	II
40	2445.80	II
50	2446.11	II
30	2446.47	II
40	2447.20	II
25	2447.33	II
60	2447.709	I
30	2447.75	II
25	2449.96	II
25	2450.20	II
100	2453.476	I
20	2453.98	II
30	2454.58	II
15	2455.71	II
15	2455.90	II
15	2457.09	II
1500	2457.598	I
150	2458.78	II
40	2458.97	II
60	2460.44	II
80	2461.28	II
100	2461.86	II
100	2462.181	I
1500	2462.647	I
50	2463.29	II
50	2463.730	I
40	2464.01	II
40	2464.90	II
800	2465.149	I
50	2465.91	II
15	2466.50	II
60	2466.67	II
60	2466.82	II
60	2467.732	I
15	2468.29	II
600	2468.879	I
60	2469.51	II
25	2470.41	II
80	2470.67	II

Iron (Cont.)

80	2470.965	I
800	2472.336	I
40	2472.43	II
40	2472.60	II
1000	2472.895	I
200	2473.16	I
50	2473.32	II
30	2474.05	II
600	2474.814	I
50	2475.12	II
40	2475.54	II
15	2476.26	II
60	2476.657	I
25	2477.34	II
60	2478.57	II
120	2479.480	I
1200	2479.776	I
100	2480.16	II
15	2481.05	II
80	2482.12	II
25	2482.32	II
100	2482.66	II
10000	2483.271	I
300	2483.533	I
1000	2484.185	I
60	2484.24	II
30	2484.44	II
50	2485.990	I
800	2486.373	I
100	2486.691	I
100	2487.066	I
120	2487.370	I
4000	2488.143	I
100	2488.945	I
80	2489.48	II
1000	2489.750	I
50	2489.83	II
50	2489.913	I
3000	2490.644	I
100	2490.71	II
60	2490.86	II
2000	2491.155	I
100	2491.40	II
25	2492.34	II
100	2493.18	II
500	2493.26	II
60	2494.000	I
50	2494.251	I
100	2495.87	I
600	2496.533	I
50	2497.82	II
150	2498.90	I
40	2500.92	II
1000	2501.132	I
40	2501.31	II
50	2501.693	I
60	2502.39	II
40	2503.33	II
60	2503.87	II
80	2506.09	II
40	2506.80	II
500	2507.900	I
30	2508.34	II
50	2508.753	I
1000	2510.835	I
120	2511.76	II
80	2512.275	I
400	2512.365	I
50	2514.38	II
80	2516.570	I
50	2517.13	II
300	2517.661	I
800	2518.102	I
60	2519.05	II
150	2519.629	I
40	2521.09	II
30	2521.82	II
50	2522.480	I
4000	2522.849	I
200	2523.66	I
500	2524.293	I
100	2525.02	I
200	2525.39	II
25	2526.07	II
300	2526.29	II
2000	2527.435	I
30	2527.70	II
800	2529.135	I
25	2529.23	II
80	2529.31	I
250	2529.55	II
150	2529.836	I
40	2530.11	II
200	2530.687	I
120	2533.63	II
60	2533.80	I
100	2534.42	II
120	2535.49	II
400	2535.607	I
60	2536.67	II
200	2536.792	I
200	2536.80	II
50	2536.84	II
50	2537.14	II
50	2538.20	II
40	2538.50	II
100	2538.80	II
100	2538.91	II
150	2538.99	II
50	2539.357	I
200	2540.66	II
600	2540.972	I
80	2541.10	II
60	2541.84	II
300	2542.10	I
25	2542.78	II
60	2543.38	II
250	2543.92	I
150	2544.70	I
40	2544.97	II
40	2545.22	II
800	2545.978	I
40	2546.44	II
80	2546.67	II
80	2546.87	I
100	2548.74	II
80	2549.08	II
80	2549.39	II
60	2549.46	II
600	2549.613	I
40	2549.77	II
60	2550.03	II
25	2550.15	II
50	2550.68	II
40	2560.28	II
25	2562.09	II
400	2562.53	II
200	2563.48	II
60	2566.91	II
25	2570.52	II
30	2570.85	II
150	2574.36	II
50	2575.74	I
300	2576.691	I
25	2576.86	II
60	2577.92	II
50	2582.30	I
100	2582.58	II
1500	2584.54	I
650	2585.88	II
90	2588.00	I
90	2591.54	II
30	2592.78	II
60	2593.51	I
90	2593.73	II
650	2598.37	II
2000	2599.40	II
300	2599.57	I
20	2605.34	II
20	2605.42	II
60	2605.657	I
300	2606.51	II
800	2606.827	I
650	2607.09	II
20	2611.07	II
600	2611.87	II
320	2613.82	II
320	2617.62	II
250	2618.018	I
20	2619.07	II
90	2620.41	II
20	2620.69	II
40	2621.67	II
400	2623.53	I
50	2625.49	II
200	2625.67	II
150	2628.29	II
20	2630.07	II
250	2631.05	II
250	2631.32	II
50	2631.61	II
100	2632.237	I
300	2635.809	I
50	2641.646	I
200	2643.998	I
60	2664.66	II
30	2666.64	II
300	2666.812	I
60	2666.965	I
600	2679.062	I
500	2684.75	II
400	2689.212	I
60	2692.60	II
50	2696.28	I
200	2699.106	I
60	2703.99	II
80	2706.012	I
400	2706.582	I
60	2708.571	I
20	2709.05	II
200	2711.655	I
80	2714.41	II
50	2716.22	II
50	2716.257	I
50	2717.786	I
50	2717.87	II
250	2718.436	I
4000	2719.027	I
100	2719.420	I
50	2720.197	I
1500	2720.903	I
400	2723.578	I
30	2724.88	II
150	2724.953	I
80	2726.05	I
50	2726.235	I
25	2727.38	II
80	2727.54	II
200	2728.020	I
50	2728.820	I
80	2728.90	II
40	2730.73	II
1000	2733.581	I
100	2734.003	I
50	2734.268	I
500	2735.475	I
50	2735.612	I
500	2737.310	I
120	2737.83	I
400	2739.55	II
250	2742.254	I
800	2742.405	I
200	2743.20	II
150	2743.565	I
200	2744.068	I
80	2744.527	I
300	2746.48	II
100	2749.32	II
500	2749.48	II
1200	2750.140	I
20	2751.13	II
20	2752.15	II
80	2753.29	II
50	2753.69	I
150	2754.032	I
100	2754.426	I
30	2754.89	II
800	2755.73	II
250	2756.328	I
100	2757.316	I
50	2759.81	I
120	2761.780	I
150	2761.81	II
150	2762.026	I
120	2762.772	I
120	2763.109	I
20	2763.66	II
25	2765.13	II
80	2766.910	I
250	2767.522	I
50	2769.30	I
25	2769.35	II
300	2772.07	I
50	2773.23	I
20	2774.69	II
15	2776.91	II
60	2778.07	I
600	2778.220	I
40	2779.30	II
50	2783.69	II
30	2785.19	II
3000	2788.10	I
20	2793.89	II
200	2797.78	I
30	2799.29	II
400	2804.521	I
1500	2806.98	I
2500	2813.287	I
300	2823.276	I
600	2825.56	I
50	2825.687	I
120	2828.808	I
25	2831.56	II
1500	2832.436	I
120	2835.950	I
200	2838.119	I
30	2839.51	II
20	2839.80	II
15	2840.65	II
200	2843.631	I
1000	2843.977	I
100	2845.594	I
15	2848.11	II
15	2848.32	II
800	2851.797	I
30	2856.91	II
25	2858.34	II
50	2869.307	I
50	2872.334	I
80	2874.172	I
50	2894.504	I
120	2912.157	I
120	2929.007	I
1200	2936.903	I
60	2941.343	I
12	2944.40	II
1000	2947.876	I
60	2950.24	I
600	2953.940	I
250	2957.364	I
80	2959.99	I
150	2965.254	I
1500	2966.898	I
120	2969.36	I
800	2970.099	I
15	2970.52	II
1200	2973.132	I
500	2973.235	I
600	2981.445	I
1000	2983.570	I
60	2984.77	I
50	2984.82	II
13	2985.54	II
1000	2994.427	I
250	2994.502	I
500	2999.512	I
120	3000.451	I
800	3000.948	I
60	3001.655	I
15	3002.64	II
200	3007.282	I
500	3008.14	I
120	3009.569	I
60	3017.627	I
60	3018.983	I
60	3020.01	II
500	3020.491	I
1500	3020.639	I
600	3021.073	I
500	3024.032	I
150	3025.638	I
500	3025.842	I
80	3030.148	I
60	3031.214	I
60	3034.484	I
40	3036.96	II
800	3037.389	I
80	3041.637	I
800	3047.604	I
600	3057.446	I
1000	3059.086	I
250	3067.244	I
120	3075.719	I
120	3091.577	I
80	3098.189	I
100	3099.895	I
100	3099.968	I
60	3100.303	I
100	3100.665	I
12	3154.20	II
80	3175.445	I
150	3184.895	I
250	3191.659	I
500	3193.226	I
800	3193.299	I
12	3196.08	II
200	3196.928	I
80	3199.500	I
60	3200.47	I
50	3205.398	I
50	3211.67	I
100	3211.88	I
13	3213.31	II
200	3214.011	I
200	3214.396	I
60	3215.938	I
50	3217.377	I
80	3219.583	I
60	3219.766	I
300	3222.045	I
600	3225.78	I
13	3227.73	II
80	3227.796	I
20	3230.42	II
80	3233.05	I
50	3233.967	I
120	3234.613	I
300	3236.222	I
100	3239.433	I
80	3244.187	I
80	3246.005	I
60	3254.36	I
80	3265.046	I
50	3265.617	I
50	3271.000	I
50	3280.26	I
150	3286.75	I
120	3305.97	I
200	3306.343	I
400	3355.227	I
80	3355.517	I
60	3369.546	I
120	3370.783	I
50	3378.678	I
50	3380.110	I
60	3383.978	I
12	3388.13	II
50	3392.304	I
150	3392.651	I
150	3399.333	I
80	3404.353	I
500	3407.458	I

Iron (Cont.)

250	3413.131	I
60	3424.284	I
500	3427.119	I
60	3428.748	I
6000	3440.606	I
2500	3440.989	I
1000	3443.876	I
200	3445.149	I
15	3453.61	II
1200	3465.860	I
2000	3475.450	I
500	3476.702	I
2500	3490.574	I
500	3497.840	I
250	3513.817	I
300	3521.261	I
400	3526.040	I
100	3526.166	I
60	3526.237	I
60	3526.381	I
60	3526.467	I
100	3533.199	I
200	3536.556	I
300	3541.083	I
250	3542.075	I
80	3553.739	I
400	3554.925	I
200	3556.878	I
400	3558.515	I
1000	3565.379	I
1200	3570.097	I
800	3570.25	I
120	3571.996	I
100	3573.393	I
60	3573.829	I
60	3573.888	I
4000	3581.19	I
150	3582.199	I
150	3584.660	I
120	3584.929	I
300	3585.319	I
150	3585.705	I
200	3586.103	I
400	3586.984	I
100	3594.633	I
150	3603.204	I
200	3605.454	I
500	3606.680	I
1500	3608.859	I
250	3610.16	I
60	3612.068	I
150	3617.788	I
1500	3618.768	I
200	3621.462	I
150	3622.004	I
150	3623.19	I
100	3631.096	I
1200	3631.463	I
60	3632.041	I
100	3638.298	I
200	3640.389	I
80	3643.717	I
1500	3647.842	I
250	3649.506	I
80	3650.279	I
200	3651.467	I
120	3670.024	I
150	3670.089	I
100	3676.311	I
150	3677.629	I
1500	3679.913	I
200	3682.242	I
120	3683.054	I
150	3684.107	I
120	3685.998	I
500	3687.456	I
120	3689.477	I
150	3694.008	I
120	3695.051	I
150	3701.086	I
80	3704.462	I
1200	3705.566	I
60	3707.041	I
150	3707.821	I
300	3707.919	I
600	3709.246	I
120	3716.442	I
8000	3719.935	I
1500	3722.563	I
120	3724.377	I
60	3725.491	I
60	3727.093	I
500	3727.619	I
150	3732.396	I
1200	3733.317	I
5000	3734.864	I
120	3735.324	I
6000	3737.131	I
100	3738.306	I
400	3743.362	I
80	3743.47	I
6000	3745.561	I
1200	3745.899	I
3000	3748.262	I
80	3748.964	I
3000	3749.485	I
1500	3758.232	I
400	3760.05	I
1500	3763.788	I
400	3765.54	I
600	3767.191	I
60	3776.452	I
250	3785.95	I
100	3786.68	I
250	3787.880	I
250	3790.092	I
150	3794.34	I
400	3795.002	I
120	3797.518	I
250	3798.511	I
400	3799.547	I
200	3805.345	I
80	3806.696	I
600	3812.964	I
60	3813.059	I
1500	3815.840	I
2500	3820.425	I
150	3821.179	I
80	3824.306	I
2500	3824.444	I
1500	3825.880	I
1200	3827.823	I
1000	3834.222	I
120	3839.257	I
500	3840.437	I
800	3841.047	I
120	3843.256	I
80	3846.800	I
200	3849.96	I
120	3850.817	I
2500	3856.372	I
150	3859.212	I
10000	3859.911	I
150	3865.523	I
60	3867.215	I
250	3872.501	I
150	3873.761	I
250	3878.018	I
2000	3878.573	I
4000	3886.282	I
200	3887.048	I
300	3888.513	I
800	3895.656	I
1200	3899.707	I
400	3902.945	I
250	3906.479	I
80	3916.731	I
600	3920.258	I
1200	3922.911	I
1200	3927.920	I
2000	3930.296	I
60	3948.774	I
60	3949.953	I
50	3951.164	I
50	3952.601	I
60	3956.454	I
250	3956.68	I
60	3966.614	I
100	3969.257	I
80	3977.741	I
40	3981.771	I
50	3983.956	I
60	3994.114	I
200	3997.392	I
40	3998.053	I
400	4005.241	I
60	4009.713	I
80	4014.53	I
100	4021.867	I
50	4040.638	I
4000	4045.813	I
1500	4063.594	I
50	4066.975	I
50	4067.977	I
1200	4071.737	I
40	4076.629	I
40	4100.737	I
40	4107.489	I
150	4118.544	I
40	4127.608	I
400	4132.058	I
80	4134.676	I
40	4136.997	I
200	4143.415	I
800	4143.869	I
40	4153.898	I
50	4154.500	I
60	4156.799	I
50	4172.744	I
60	4174.912	I
50	4175.635	I
50	4177.593	I
120	4181.754	I
50	4184.891	I
120	4187.038	I
120	4187.795	I
80	4191.430	I
40	4195.329	I
150	4198.304	I
40	4199.095	I
300	4202.029	I
40	4203.984	I
80	4206.696	I
80	4210.343	I
400	4216.183	I
100	4219.360	I
50	4222.212	I
50	4225.956	I
200	4227.423	I
11	4233.17	II
100	4233.602	I
250	4235.936	I
50	4238.809	I
50	4247.425	I
200	4250.118	I
300	4250.787	I
40	4258.315	I
800	4260.473	I
250	4271.153	I
1200	4271.759	I
1200	4282.402	I
80	4291.462	I
250	4299.234	I
1200	4307.901	I
150	4315.084	I
1500	4325.761	I
80	4352.734	I
80	4369.771	I
800	4375.929	I
3000	4383.544	I
1200	4404.750	I
300	4415.122	I
600	4427.299	I
400	4461.652	I
120	4466.551	I
80	4476.017	I
80	4482.169	I
200	4482.252	I
50	4489.739	I
50	4528.613	I
11	4583.83	II
30	4647.433	I
30	4736.771	I
50	4859.741	I
120	4871.317	I
60	4872.136	I
30	4878.208	I
100	4890.754	I
250	4891.492	I
30	4903.309	I
150	4918.992	I
500	4920.502	I
12	4923.92	II
1500	4957.597	I
11	4990.50	II
80	5001.862	I
18	5001.91	II
11	5004.20	II
30	5005.711	I
100	5006.117	I
60	5012.067	I
30	5014.941	I
12	5018.43	II
11	5030.64	II
25	5030.77	I
12	5035.71	II
150	5041.755	I
30	5049.819	I
30	5051.634	I
25	5074.748	I
18	5100.73	II
15	5100.95	II
150	5110.357	I
40	5133.69	I
40	5139.251	I
100	5139.462	I
11	5144.36	II
12	5149.46	II
25	5151.910	I
30	5162.27	I
80	5166.281	I
2500	5167.487	I
80	5168.897	I
12	5169.03	II
500	5171.595	I
50	5191.454	I
80	5192.343	I
200	5194.941	I
30	5204.582	I
25	5215.179	I
150	5216.274	I
18	5216.85	II
60	5226.862	I
1000	5227.150	I
13	5227.49	II
250	5232.939	I
13	5247.95	II
13	5251.23	II
18	5260.26	II
11	5264.18	II
100	5266.555	I
1200	5269.537	I
800	5270.357	I
30	5281.789	I
60	5283.621	I
25	5302.299	I
11	5306.18	II
13	5316.23	II
150	5324.178	I
800	5328.038	I
300	5328.531	I
100	5332.899	I
14	5339.59	II
80	5339.928	I
500	5341.023	I
25	5364.87	I
40	5367.47	I
50	5369.96	I
400	5371.489	I
60	5383.37	I
14	5387.06	II
40	5393.167	I
12	5395.86	II
300	5397.127	I
15	5402.06	II
60	5404.12	I
250	5405.774	I
30	5410.91	I
60	5415.20	I
60	5424.07	I
30	5427.83	II
250	5429.695	I
13	5429.99	II
100	5434.523	I
200	5446.871	I
25	5455.45	I
120	5455.609	I
16	5465.93	II
20	5466.94	II
16	5482.31	II
14	5493.83	II
25	5497.516	I
20	5501.464	I
18	5506.20	II
30	5506.778	I
12	5510.78	II
12	5529.06	II
13	5544.76	II
30	5569.618	I
60	5572.841	I
120	5586.755	I
200	5615.644	I
20	5624.541	I
12	5645.40	II
50	5662.515	I
20	5762.990	I
11	5783.63	II
30	5862.353	I
13	5885.02	II
16	5902.82	II
30	5914.114	I
14	5955.70	II
30	5986.956	I
18	5961.71	II
30	5962.4	II
13	5965.63	II
40	6065.482	I
30	6102.159	I
40	6136.614	I
40	6137.694	I
30	6147.73	II
20	6149.24	II
15	6175.16	II
40	6191.558	I
30	6213.429	I
30	6219.279	I
40	6230.726	I
20	6238.37	II
20	6246.317	I
80	6247.56	II
30	6252.554	I
15	6305.32	II
12	6331.97	II
15	6383.75	II
20	6393.602	I
30	6399.999	I
20	6411.647	I
20	6416.90	II
20	6421.349	I
30	6430.844	I
20	6446.43	II
200	6456.38	II
60	6494.981	I
20	6516.05	II
20	6546.239	I
20	6592.913	I
40	6677.989	I
15	6855.18	I
15	6945.21	I
20	7067.44	II
15	7130.94	I
25	7164.443	I
80	7187.313	I

Iron (Cont.)

30		7207.381	I
12		7224.51	II
50		7307.97	II
40		7320.70	II
20		7376.46	II
30		7445.746	I
20		7462.38	II
40		7495.059	I
60		7511.045	I
15		7586.04	I
15		7711.71	II
30		7780.59	I
40		7832.22	I
80		7937.131	I
60		7945.984	I
80		7998.939	I
60		8046.047	I
50		8085.176	I
130		8220.41	I
120		8327.053	I
20		8331.908	I
120		8387.770	I
30		8468.404	I
15		8514.069	I
60		8661.898	I
150		8688.621	I
12		8793.38	I
12		8824.23	I
20		8866.96	I
15		8999.56	I
15		10216.32	I
13		10469.65	I
21		11119.80	I
14		11374.08	I
52		11422.32	I
87		11439.12	I
91		11593.59	I
255		11607.57	I
160		11638.26	I
230		11689.98	I
160		11783.26	I
580		11882.84	I
225		11884.08	I
1030		11973.05	I
15		12638.71	I
14		12879.76	I
17		13565.04	I
30		14236.25	I
24		14285.11	I
14		14292.38	I
16		14308.69	I
96		14400.56	I
30		14442.20	I
72		14512.23	I
50		14555.06	I
14		14565.95	I
40		14826.43	I
37		15051.77	I
28		15207.55	I
94		15294.58	I
16		15335.40	I
30		15621.67	I
25		15631.97	I
14		15723.59	I
41		15769.42	I
28		15813.13	I
13		16444.82	I
20		16486.69	I
105		18856.65	I
47		18987.01	I
25		19113.68	I
22		19791.88	I
14		22380.82	I
21		22619.85	I
38		26222.04	I
17		26659.22	I

Fe III
Ref. 71, 101 — J.R.

Intensity		Wavelength	
		Vacuum	
6		728.81	III
5		730.00	III
5		737.71	III
5		739.26	III
9		807.55	III
8		807.86	III
8		808.84	III
8	p	811.28	III
10		813.38	III
8		838.05	III
10		844.28	III
9		845.41	III
8	w	847.42	III
8		859.72	III
8	p	861.76	III
10	p	861.83	III
8		873.46	III
9		890.76	III
10		891.17	III
8		891.44	III
8		899.42	III
10		950.33	III
10		981.37	III
10	w	983.88	III
8		985.82	III
9		991.23	III
9		1017.25	III
8		1017.74	III
8		1018.29	III
8		1032.12	III
8		1063.87	III
9		1122.53	III
9		1124.88	III
8		1128.02	III
10	h	1505.17	III
10	h	1538.63	III
12	h	1550.20	III
10	h	1601.21	III
10		1869.83	III
12		1877.99	III
10		1882.05	III
12		1886.76	III
13		1890.67	III
11		1893.98	III
20		1895.46	III
10	s	1907.58	III
19		1914.06	III
15		1915.08	III
15		1922.79	III
10	p	1926.01	III
18		1926.30	III
15		1930.39	III
14		1931.51	III
14		1937.34	III
10	l	1938.90	III
14	s	1943.48	III
12		1945.34	III
10		1950.33	III
12		1951.01	III
11		1952.65	III
13		1953.32	III
10		1953.49	III
10	w	1954.22	III
11		1958.58	III
13		1960.32	III
15		1987.50	III
14		1991.61	III
13		1994.07	III
12		1995.56	III
12		1996.42	III
		Air	
10		2061.55	III
12		2068.24	III
14		2078.99	III
10		2084.35	III
12		2090.14	III
15		2097.48	III
12		2097.69	III
12		2103.80	III
10		2107.32	III
15		2151.78	III
12		2157.71	III
12		2158.47	III
10		2161.27	III
12		2166.95	III
12		2171.04	III
15		2174.66	III
12		2180.41	III
10	p	2208.85	III
10		2221.83	III
10		2229.27	III
10		2232.43	III
10		2232.69	III
10		2235.91	III
10		2238.16	III
12	p	2241.54	III
12		2261.59	III
10		2267.42	III
10		2293.06	III
15		2295.86	III
10	p	2317.70	III
10		2319.22	III
10	p	2321.71	III
10		2326.95	III
10	p	2336.77	III
10		2338.96	III
8		2389.53	III
8		2438.17	III
8	p	2582.37	III
8		2595.62	III
8		2617.15	III
9	p	2645.39	III
10	h	2695.13	III
9	h	2695.34	III
8	h	2700.02	III
8	h	2701.13	III
8		2773.31	III
10	p	2813.24	III
8	p	2895.08	III
9	p	2902.47	III
12		2904.43	III
8	p	2905.80	III
10		2907.50	III
12		2907.70	III
8		2923.90	III
8		2948.39	III
8		2963.23	III
12		3001.62	III
12	h	3007.28	III
15		3013.17	III
10	p	3136.43	III
10		3174.09	III
10		3175.99	III
10		3178.01	III
13		3266.88	III
11		3276.08	III
10		3288.81	III
[illegible]		[illegible]	III
9		3339.39	III
9		3499.59	III
9		3500.28	III
10		3501.76	III
10		3586.04	III
11		3600.94	III
11		3603.88	III
16		3954.33	III
11		3968.72	III
9		3969.49	III
10	w	3979.42	III
10		4035.42	III
11		4053.11	III
12		4081.00	III
10		4120.90	III
11		4122.02	III
11		4122.78	III
15		4137.76	III
13		4139.35	III
9		4140.48	III
9		4154.96	III
18		4164.73	III
9		4164.92	III
13		4166.84	III
13		4174.26	III
9		4210.67	III
11		4222.27	III
13		4235.56	III
9		4238.62	III
12		4243.75	III
12	h	4273.40	III
12		4279.72	III
14	h	4286.16	III
16	h	4296.85	III
18	h	4304.78	III
20	h	4310.36	III
9		4323.68	III
9	h	4372.04	III
9	h	4372.14	III
11	h	4372.31	III
14	h	4372.53	III
18	h	4372.81	III
9		4395.76	III
12		4419.60	III
9		4431.02	III
9		5111.07	III
9		5127.35	III
12		5156.12	III
10		5199.08	III
10		5235.66	III
18		5243.31	III
13	l	5260.34	III
9		5272.37	III
14		5272.98	III
15		5276.48	III
16		5282.30	III
12		5284.83	III
11		5298.12	III
12		5299.93	III
14	w	5302.60	III
10		5306.76	III
9		5310.88	III
10		5322.74	III
11		5346.88	III
12		5353.77	III
12		5363.76	III
10		5368.06	III
11	l	5375.47	III
11		5719.88	III
9		5744.19	III
10		5756.38	III
18		5833.93	III
9		5848.76	III
10		5854.62	III
9		5876.26	III
15		5891.91	III
9		5898.68	III
9		5918.96	III
10	p	5920.13	III
13	p	5929.69	III
10		5952.31	III
14		5953.62	III
9		5968.48	III
12		5979.32	III
9	h	5981.01	III
12	h	5989.08	III
18		5999.54	III
9		6031.02	III
16		6032.59	III
13		6036.56	III
11		6048.72	III
11		6054.18	III
9		6056.36	III
9		6149.99	III
9		6169.74	III
9		6185.26	III
7		6186.56	III
7		6194.79	III
6		6195.43	III
6		6201.37	III
5	s	6203.04	III
5		6259.81	III
6	p	[illegible]	III
5		6357.81	III
5	h	7317.63	III
6	h	7320.14	III
5	w	7921.17	III
5	w	8230.88	III
5	w	8231.79	III
9	w	8235.45	III
8	w	8236.75	III
6	w	8238.98	III
5		8563.49	III

Fe IV
Ref. 382 — J.R.

Intensity		Wavelength	
		Vacuum	
10		502.42	IV
11		506.69	IV
11		505.35	IV
17		525.69	IV
15		526.29	IV
10		526.57	IV
13		526.63	IV
10		530.91	IV
10		531.78	IV
10		535.55	IV
14		536.61	IV
10		536.74	IV
15		537.10	IV
13		537.26	IV
14		537.79	IV
13		537.94	IV
10		538.44	IV
10		544.20	IV
10		546.22	IV
10		548.80	IV
11		550.32	IV
10		551.77	IV
13		552.14	IV
11		552.74	IV
10		554.26	IV
10		555.66	IV
10		572.88	IV
10		576.76	IV
10		579.76	IV
14		607.53	IV
13		608.80	IV
10		609.65	IV
12		1425.73	IV
13		1431.43	IV
12		1473.20	IV
12		1489.53	IV
12		1495.18	IV
13		1526.60	IV
13		1530.26	IV
14		1532.63	IV
13		1532.91	IV
15		1533.86	IV
13		1533.95	IV
14		1536.58	IV
12		1538.29	IV
13		1542.16	IV
14		1542.70	IV
12		1546.40	IV
12		1552.35	IV
12		1552.71	IV
12		1562.46	IV
13		1566.26	IV
14		1568.27	IV
12		1570.18	IV
12		1570.42	IV
12		1571.24	IV
12		1577.20	IV
12		1577.76	IV
12		1590.62	IV
13		1591.51	IV
13		1592.05	IV
12		1596.67	IV
13		1598.01	IV
12		1600.50	IV
13		1600.58	IV
13		1601.67	IV

Iron (Cont.)

12	1602.08	IV
13	1603.18	IV
13	1603.73	IV
13	1604.88	IV
13	1605.68	IV
15	1605.97	IV
13	1606.98	IV
17	1609.10	IV
14	1609.83	IV
13	1610.47	IV
13	1611.20	IV
13	1613.64	IV
15	1614.02	IV
13	1614.64	IV
13	1615.00	IV
12	1615.61	IV
16	1616.68	IV
14	1617.68	IV
14	1619.02	IV
12	1620.91	IV
13	1621.16	IV
14	1621.57	IV
13	1623.38	IV
13	1623.53	IV
15	1626.47	IV
14	1626.90	IV
13	1628.54	IV
13	1630.18	IV
17	1631.08	IV
12	1632.08	IV
14	1632.40	IV
13	1634.01	IV
12	1638.07	IV
12	1638.30	IV
14	1639.40	IV
16	1640.04	IV
14	1640.16	IV
15	1641.87	IV
12	1642.88	IV
15	1647.09	IV
15	1651.58	IV
15	1652.90	IV
13	1653.41	IV
13	1656.11	IV
15	1656.65	IV
12	1657.82	IV
12	1658.43	IV
14	1660.10	IV
12	1661.57	IV
13	1662.32	IV
13	1662.52	IV
13	1663.54	IV
13	1668.09	IV
12	1669.61	IV
14	1671.04	IV
12	1672.86	IV
13	1673.68	IV
14	1675.66	IV
12	1676.78	IV
12	1677.12	IV
13	1681.36	IV
12	1681.95	IV
15	1687.69	IV
15	1698.88	IV
12	1700.40	IV
12	1704.93	IV
13	1709.81	IV
15	1711.41	IV
14	1712.76	IV
12	1717.11	IV
14	1717.90	IV
14	1718.16	IV
12	1718.42	IV
14	1719.46	IV
14	1722.71	IV
14	1724.06	IV
12	1724.26	IV
16	1725.63	IV
13	1761.08	IV
12	1764.92	IV
12	1767.36	IV
13	1792.10	IV
13	1796.93	IV
12	1805.32	IV
12	1820.42	IV
13	1827.98	IV
12	1840.24	IV
12	1860.42	IV
12	1869.64	IV
12	1874.23	IV

Fe V
Ref. 381 — J.R.

Intensity	Wavelength	
	Vacuum	
300	361.28	V
300	365.43	V
300	365.86	V
300	374.24	V
300	374.87	V
300	375.98	V
300	379.59	V
300	380.31	V
300	381.27	V
300	384.96	V
300	384.97	V
300	385.03	V
300	385.11	V
300	385.25	V
300	385.26	V
300	385.30	V
300	385.75	V
300	385.88	V
350	386.16	V
300	386.74	V
300	386.78	V
300	386.85	V
300	386.88	V
350	386.88	V
400	387.20	V
400	387.50	V
300	387.62	V
400	387.76	V
400	387.78	V
300	387.98	V
300	388.61	V
300	388.82	V
300	390.11	V
300	390.19	V
300	390.78	V
300	391.94	V
300	392.06	V
300	392.38	V
300	392.50	V
300	392.51	V
300	392.70	V
300	392.91	V
300	393.27	V
300	393.72	V
300	393.73	V
300	393.91	V
300	393.97	V
300	394.04	V
300	394.64	V
300	395.15	V
300	395.79	V
400	395.90	V
300	399.84	V
300	400.11	V
300	400.51	V
300	400.52	V
300	400.63	V
300	401.04	V
300	401.64	V
300	401.86	V
300	402.87	V
300	403.06	V
400	404.62	V
400	405.50	V
800	407.42	V
600	407.44	V
400	407.49	V
500	407.75	V
400	409.71	V
400	410.20	V
600	411.55	V
300	415.01	V
300	416.66	V
300	416.84	V
700	417.39	V
700	418.04	V
500	418.47	V
300	420.56	V
700	421.06	V
500	421.78	V
300	422.28	V
500	422.31	V
300	423.23	V
500	426.06	V
500	426.11	V
300	426.83	V
350	426.97	V
300	434.42	V
300	439.22	V
300	444.70	V
300	445.44	V
300	446.04	V
300	458.16	V
300	486.17	V
400	1317.86	V
300	1318.35	V
300	1320.41	V
300	1321.34	V
300	1321.49	V
400	1323.27	V
400	1330.40	V
300	1345.61	V
400	1359.01	V
300	1361.28	V
300	1361.45	V
600	1361.82	V
300	1363.08	V
300	1363.64	V
300	1365.57	V
700	1373.59	V
600	1373.67	V
300	1374.12	V
500	1376.34	V
300	1376.46	V
500	1378.56	V
300	1385.68	V
800	1387.94	V
400	1397.97	V
600	1400.24	V
800	1402.39	V
400	1406.67	V
500	1406.82	V
400	1407.25	V
300	1409.03	V
300	1409.22	V
600	1409.45	V
400	1415.20	V
300	1418.12	V
600	1420.46	V
800	1430.57	V
800	1440.53	V
300	1440.79	V
400	1442.22	V
800	1446.62	V
700	1448.85	V
400	1449.93	V
300	1455.56	V
700	1456.16	V
500	1459.83	V
400	1460.73	V
500	1462.63	V
700	1464.68	V
500	1465.38	V
400	1466.65	V
500	1469.00	V
300	1475.60	V
500	1479.47	V
300	1554.22	V

KRYTPON (Kr)
Z = 36

Kr I and II
Ref. 61, 121, 123, 147, 208, 232
— E.F.W.

Intensity		Wavelength	
		Vacuum	
60		729.40	II
200		761.18	II
100		763.98	II
60		766.20	II
200		771.03	II
60	p	773.69	II
200		782.10	II
100		783.72	II
60		818.15	II
60		830.38	II
100		844.06	II
60		864.82	II
60		868.87	II
200		884.14	II
1000		886.30	II
400		891.01	II
200		911.39	II
2000		917.43	II
50		945.44	I
50		946.54	I
20		951.06	I
50		953.40	I
50		963.37	I
2000		964.97	II
100		1001.06	I
100		1003.55	I
100		1030.02	I
200		1164.87	I
650		1235.84	I
		Air	
100	h	2464.77	II
60		2492.48	II
80	h	2712.40	II
100		2833.00	II
100	h	3607.88	II
200		3631.889	II
250		3653.928	II
80		3665.324	I
150		3669.01	II
100		3679.559	I
80		3686.182	II
300	h	3718.02	II
200		3718.595	II
150		3721.350	II
200		3741.638	II
150		3744.80	II
80		3754.245	II
500		3778.089	II
500		3783.095	II
150	h	3875.44	II
150		3906.177	II
200		3920.081	II
100		3994.840	II
100	h	3997.793	II
300		4057.037	II
300		4065.128	II
500		4088.337	II
250		4098.729	II
100		4109.248	II
250		4145.122	II
150		4250.580	II
1000		4273.969	I
100		4282.967	I
600		4292.923	II
200		4300.49	II
500	h	4317.81	II
400		4318.551	I
1000		4319.579	I
150	h	4322.98	II
100		4351.359	I
3000		4355.477	II
500		4362.641	I
200		4369.69	II
800		4376.121	I
300	h	4386.54	II
200		4399.965	I
100		4425.189	I
500		4431.685	II
600		4436.812	II
600		4453.917	I
800		4463.689	I
800		4475.014	II
400	h	4489.88	II
600		4502.353	I
400	h	4523.14	II
200	h	4556.61	II
800		4577.209	II
300		4582.978	II
150	h	4592.80	II
500		4615.292	II
1000		4619.166	II
800		4633.885	II
2000		4658.876	II
500		4680.406	II
100		4691.301	II
200		4694.360	II
3000		4739.002	II
300		4762.435	II
1000		4765.744	II
300		4811.76	II
300		4825.18	II
800		4832.077	II
700		4846.612	II
150		4857.20	II
300		4945.59	II
200		5022.40	II
250		5086.52	II
400	h	5125.73	II
500		5208.32	II
200		5308.66	II
500		5333.41	II
200		5468.17	II
500		5562.224	I
2000		5570.288	I
80		5580.386	I
100		5649.561	I
400		5681.89	II
200	h	5690.35	II
100		5832.855	I
3000		5870.914	I
200		5992.22	II
60		5993.849	I
60		6056.125	I
300		6420.18	II
100		6421.026	I
200		6456.288	I
150		6570.07	II
60		6699.228	I
100		6904.678	I
250		7213.13	II
100		7224.104	I
80		7287.258	I
400		7289.78	II
400		7407.02	II
60		7425.541	I
200		7435.78	II
100		7486.862	I
300		7524.46	II
1000		7587.411	I
2000		7601.544	I
150		7641.16	II
1000		7685.244	I
1200		7694.538	I
250		7735.69	II
150		7746.827	I
800		7854.821	I
200		7913.423	I
180		7928.597	I
200		7933.22	II
120		7973.62	II
100		7982.401	I
1500		8059.503	I
4000		8104.364	I

Krypton (Cont.)

6000		8112.899	I
60		8132.967	I
3000		8190.054	I
200		8202.72	II
80		8218.365	I
3000		8263.240	I
100		8272.353	I
5000		8298.107	I
1500		8281.050	I
100		8412.430	I
3000		8508.870	I
150		8764.110	I
6000		8776.748	I
2000		8928.692	I
500		9238.48	II
500	hL	9293.82	II
200	h	9320.99	II
300		9361.95	II
100		9362.082	I
200	h	9402.82	II
200	h	9470.93	II
500		9577.52	II
500	h	9605.80	II
400	h	9619.61	II
200		9663.34	II
200	h	9711.60	II
2000		9751.758	I
500		9803.14	II
500		9856.314	I
1000		10221.46	II
100		11187.108	I
200		11257.711	I
150		11259.126	I
500		11457.481	I
150		11792.425	I
1500		11819.377	I
600		11997.105	I
160		12077.224	I
100		12861.892	I
1100		13177.412	I
1000		13622.415	I
2400		13634.220	I
800		13658.394	I
200		13711.036	I
600		13738.851	I
150		13974.027	I
550		14045.657	I
140		14104.298	I
180		14402.22	I
2000		14426.793	I
100		14517.84	I
1600		14734.436	I
550		14762.672	I
450		14765.472	I
400		14961.894	I
120		15005.307	I
140		15209.526	I
1700		15239.615	I
130		15326.480	I
1500		15334.958	I
700		15372.037	I
200		15474.026	I
180		15681.02	I
120		15820.09	I
200		16726.513	I
2000		16785.128	I
1000		16853.488	I
2400		16890.441	I
1600		16896.753	I
1800		16935.806	I
600		17098.771	I
700		17367.606	I
120		17404.443	I
150		17616.854	I
650		17842.737	I
700		18002.229	I
2600		18167.315	I
100		18399.786	I
150		18580.896	I
300		18696.294	I
170		18785.460	I
200		18797.703	I
140		20209.878	I
300		20423.964	I
140		20446.971	I
600		21165.471	I
1800		21902.513	I
120		22485.775	I
180		23340.416	I
120		24260.506	I
180		24292.221	I
600		25233.820	I
180		28610.55	I
1000		28655.72	I
150		28769.71	I
140		28822.49	I
300		29236.69	I
300		30663.54	I
300		30979.16	I
500		39300.6	I
1100		39486.52	I
220		39557.25	I
100		39572.60	I
1400		39588.4	I
1100		39589.6	I
500		39954.8	I
300		39966.6	I
1300		40306.1	I
250		40685.16	I

Kr III
Ref. 208, 366, 390, 421 — E.F.W.

Intensity		Wavelength	
		Vacuum	
30		467.35	III
30		540.86	III
30		565.64	III
30		569.16	III
30		571.98	III
30		579.83	III
30		585.14	III
30		585.96	III
30		593.70	III
30		594.10	III
30		596.41	III
40		600.17	III
30		603.67	III
50		605.86	III
35		606.47	III
50		611.12	III
35		616.72	III
40		621.45	III
45		622.80	III
50		625.02	III
30		625.76	III
45		628.59	III
50		630.04	III
35		633.09	III
50		639.98	III
60		646.41	III
50		651.20	III
50		659.72	III
30		664.86	III
40		672.34	III
35		672.85	III
35		676.57	III
35		680.13	III
35		683.68	III
45		686.25	III
45		687.98	III
45		691.93	III
50		695.61	III
30		698.05	III
50		708.36	III
50		714.00	III
100	p	722.04	III
30		746.70	III
60		785.97	III
50		837.66	III
50		854.73	III
60		862.58	III
40		870.84	III
50		876.08	III
75		897.81	III
50		987.29	III
30		1158.74	III
6		1638.82	III
6		1914.09	III
		Air	
40		2393.94	III
40		2494.01	III
30		2563.25	III
60		2639.76	III
30		2680.32	III
40		2681.19	III
30		2841.00	III
30		2851.16	III
50		2870.61	III
100		2892.18	III
30		2909.17	III
50		2952.56	III
60		2992.22	III
50		3022.30	III
80		3024.45	III
50		3046.93	III
30		3056.72	III
60		3063.13	III
40		3097.16	III
60		3112.25	III
30		3120.61	III
100		3124.39	III
60		3141.35	III
100		3189.11	III
80		3191.21	III
40		3239.52	III
40		3240.44	III
300		3245.69	III
150		3264.81	III
100		3268.48	III
30		3271.65	III
30		3285.89	III
30		3304.75	III
50		3311.47	III
200		3325.75	III
60		3330.76	III
50		3342.48	III
100		3351.93	III
40		3374.96	III
100		3439.46	III
70		3474.65	III
100		3488.59	III
200		3507.42	III
100		3564.23	III
30		3641.34	III
30		3690.65	III
40	h	3868.70	III
50		4067.37	III
40		4131.33	III
40		4154.46	III
20	h	5016.45	III
10		5501.43	III
10	h	6037.17	III
10	h	6078.38	III
10		6310.22	III

Kr IV
Ref. 366, 409, 417 — E.F.W.

Intensity	Wavelength	
	Vacuum	
	793.44	IV
	794.11	IV
7	805.76	IV
18	816.82	IV
22	842.04	IV
	Air	
3	2237.34	IV
6	2291.26	IV
3	2329.3	IV
4	2336.75	IV
4	2348.27	IV
3	2358.5	IV
3	2388.05	IV
4	2416.9	IV
3	2428.04	IV
5	2442.68	IV
4	2451.7	IV
6	2459.74	IV
5	2474.06	IV
4	2517.0	IV
5	2518.02	IV
6	2519.38	IV
5	2524.5	IV
5	2546.0	IV
6	2547.0	IV
4	2558.08	IV
3	2586.9	IV
5	2606.17	IV
10	2609.5	IV
8	2615.3	IV
7	2621.11	IV
3	2730.55	IV
8	2748.18	IV
6	2774.70	IV
3	2829.60	IV
3	2836.08	IV
5	2853.0	IV
3	2859.3	IV
3	3142.01	IV
6	3224.99	IV
3	3261.70	IV
3	3809.30	IV
5	3860.58	IV
5	3934.29	IV

Kr V
Ref. 409, 421 — E.F.W.

Intensity	Wavelength	
	Vacuum	
150	472.16	V
100	484.39	V
250	496.25	V
120	500.77	V
200	507.20	V
60	548.04	V
120	637.87	V
	690.86	V
	691.75	V
600	708.85	V
	810.70	V

LANTHANUM (La)
Z = 57

La I and II
Ref. 1 — C.H.C.

Intensity		Wavelength	
		Air	
240		2187.87	II
770		2256.76	II
200		2319.44	II
400		2610.34	II
420		2808.39	II
130		2885.14	II
160		2893.07	II
110		2950.50	II
180		3104.59	II
130		3142.76	II
510		3245.13	II
260		3249.35	II
550		3265.67	II
800		3303.11	II
1500		3337.49	II
870		3344.56	II
200		3376.33	II
1500		3380.91	II
130		3452.18	II
180		3453.17	II
200		3574.43	I
320		3628.83	II
120		3637.15	II
170	d	3641.53	I
		3641.66	II
1000		3645.42	II
390		3650.18	II
170		3662.08	II
120		3704.54	I
320		3705.82	II
550		3713.54	II
140		3714.87	II
270		3715.53	II
2400		3759.08	II
120		3780.67	II
3700		3790.83	II
3900		3794.78	II
190		3835.08	II
600		3840.72	II
120		3846.00	II
1600		3849.02	II
130		3854.91	II
3400		3871.64	II
1700		3886.37	II
1300		3916.05	II
1100		3921.54	II
160		3927.56	I
2200		3929.22	II
180		3936.22	II
9000		3949.10	II
4400		3988.52	II
3600		3995.75	II
180		4015.39	I
250		4025.88	II
2800		4031.69	II
140		4037.21	I
3000		4042.91	II
320		4050.08	II
220		4060.33	I
160		4064.79	I
850		4067.39	II
110		4076.71	II
2800		4077.35	II
120		4079.18	I
5500		4086.72	II
180		4089.61	I
280		4099.54	II
110		4104.87	I
4400		4123.23	II
110		4137.04	I
550		4141.74	II
1100		4151.97	II
220		4152.78	II
100		4160.26	I
280		4187.32	I
280		4192.36	II
1500		4196.55	II
240		4204.04	II
300		4217.56	II
200		4230.95	II
1600		4238.38	II
140		4249.99	II
320		4263.59	II
480		4269.50	II
240		4275.64	II
300		4280.27	I
600		4286.97	II
600		4296.05	II
120		4300.44	II
440		4322.51	II
4600		4333.74	II
550		4354.40	II
110		4364.67	II
110	bl	4371.97	L
110	bl	4375.84	L
110		4378.10	II
280		4383.44	II
100		4385.20	II

Lanthanum (Cont.)

220	bl	4418.24	L
160	bl	4423.17	L
160		4423.90	I
260		4427.55	II
100	bl	4428.10	L
2000		4429.90	II
160	bl	4432.98	L
100	bl	4438.01	L
100		4452.15	I
100		4455.80	II
850		4522.37	II
170		4525.31	II
420		4526.12	II
400		4558.46	II
110		4559.29	II
160		4567.91	I
200		4570.02	I
400		4574.88	II
200		4580.06	II
160		4605.78	II
410		4613.39	II
410		4619.88	II
110		4645.28	II
540		4655.50	II
360		4662.51	II
230		4663.76	II
200		4668.91	II
160		4671.83	II
230		4692.50	II
140		4703.28	II
170		4716.44	II
140		4719.94	II
230		4728.42	II
500		4740.28	II
390		4743.09	II
320		4748.73	II
160		4766.89	I
160		4804.04	II
160		4809.01	II
200		4824.06	II
320		4860.91	II
850		4899.92	II
1000		4920.98	II
1000		4921.79	II
140		4934.83	II
110		4946.47	II
370		4949.77	I
340		4970.39	II
370		4986.83	II
140		4991.28	II
720		4999.47	II
140		5046.88	I
210		5050.57	I
170		5056.46	I
200		5106.23	I
470		5114.56	II
470		5122.99	II
450		5145.42	I
180		5156.74	II
180		5157.43	II
290		5158.69	I
120		5163.62	II
580		5177.31	I
850		5183.42	II
260		5188.22	II
170		5204.15	II
720		5211.86	I
520		5234.27	I
340		5253.46	I
110		5259.39	II
370		5271.19	I
140		5290.84	II
370		5301.98	II
140		5302.62	II
180		5303.55	II
110		5340.67	II
110		5357.86	I
130		5377.09	II
140		5380.99	II
500		5455.15	I
470		5501.34	I
110	bl	5602.50	L
160		5631.22	I
240		5648.25	I
130		5657.72	I
180		5740.66	I
160		5744.41	I
160		5761.84	I
160		5769.07	II
370		5769.34	I
320		5789.24	I
450		5791.34	I
220		5797.58	II
160		5805.78	II
140		5821.99	I
320		5930.62	I
720		6249.93	I
260	d	6262.30	II
180		6296.09	II
160		6320.39	II
110		6325.91	I
170		6390.48	II
450		6394.23	I
210		6410.99	I
250		6455.99	I
110		6526.99	II
130		6543.16	I
140		6578.51	I
180		6709.50	I
120		6774.26	II
13	b	7011.22	L
75		7023.67	I
26		7032.05	I
26	b	7040.84	L
110		7045.96	I
13	b	7054.80	L
160		7066.23	II
65		7068.37	I
21	bl	7070.79	L
13		7076.38	I
21	bl	7085.40	L
26	bl	7101.02	L
10	h	7116.8	II
19	bl	7131.58	L
10		7149.77	I
40	h	7158.08	I
50		7161.25	I
10		7162.60	L
21		7219.91	I
10	b	7257.16	L
26		7270.09	I
10		7270.30	I
110	cw	7282.34	II
10		7320.91	I
110	cw	7334.18	I
65		7345.34	I
50	bl	7379.71	L
85	bl	7380.08	L
35		7382.73	I
110	bl	7403.52	L
210	bl	7403.75	L
50	b	7411.34	L
65	bl	7434.28	L
110	bl	7434.36	L
30	b	7442.92	L
50	h	7463.08	I
50	bl	7465.25	L
95	bl	7465.48	L
75	cw	7483.50	II
40	bl	7496.50	L
95	bl	7496.78	L
50		7498.83	I
30	b	7506.79	L
19	bl	7528.21	L
50	bl	7528.39	L
30		7533.59	I
85		7539.23	I
35	bl	7560.09	L
35	bl	7592.26	L
19	h	7612.94	II
19	b	7624.99	L
21		7664.34	I
15	h	7841.80	I
21	b	7876.87	L
75	bl	7877.22	L
75	bl	7910.19	L
150	bl	7910.54	L
50	b	7944.61	L
110	bl	7944.95	L
40		7964.83	I
35	b	7979.34	L
75	bl	7979.70	L
35	h	8001.89	I
21	b	8014.43	L
65	bl	8014.79	L
30	b	8019.48	L
35	hc	8051.39	I
75		8086.05	I
15	b	8122.20	L
15	b	8159.02	L
7	h	8203.38	I
50		8247.44	I
13	h	8316.04	I
85		8324.69	I
95		8346.53	I
8	h	8379.80	I
8	b	8453.55	L
8	h	8467.62	I
26		8476.48	I
13	h	8507.37	I
13	h	8513.57	I
8	h	8514.65	II
17	b	8526.59	L
17	c	8543.46	I
65		8545.44	I
15	b	8563.54	L
9	h	8590.94	I
9	b	8600.81	L
7	h	8624.22	I
15		8638.47	I
19	hw	8672.11	I
40		8674.43	I
13	h	8720.41	I
35		8748.38	I
19		8818.93	I
35		8825.82	I
21		8839.63	I

La III
Ref. 220, 309 — J.R.

Intensity		Wavelength	
		Vacuum	
3		744.19	III
10		753.03	III
1		786.64	III
200		787.14	III
1		796.03	III
400		796.99	III
1		797.20	III
10		835.03	III
30		845.62	III
1		850.73	III
1		860.39	III
2		860.88	III
5		865.04	III
2000		870.40	III
30		872.43	III
1000		882.34	III
20		882.72	III
200		929.71	III
400		942.86	III
30		967.69	III
10		974.33	III
50		979.99	III
10		980.29	III
200		1058.63	III
1000		1072.59	III
5000		1076.91	III
50000		1081.61	III
95000		1099.73	III
5000		1100.70	III
30		1208.80	III
30		1212.29	III
200		1236.54	III
100		1253.99	III
2000		1255.63	III
100		1259.55	III
100		1322.42	III
5000		1330.04	III
10000		1349.18	III
5000		1459.49	III
2000		1466.44	III
10000		1523.79	III
500		1528.55	III
5000		1536.17	III
200		1923.34	III
500		1938.57	III
		Air	
60		2216.07	III
20		2238.36	III
25		2258.61	III
5		2260.30	III
250		2297.74	III
400		2379.37	III
10		2387.99	III
20		2392.49	III
100		2476.60	III
50		2478.65	III
2		2513.43	III
4		2588.87	III
2		2604.83	III
400		2651.50	III
100		2682.34	III
150		2684.76	III
110		2897.88	III
160		2904.58	III
7		2950.84	III
10		2953.77	III
40		2992.10	III
4		3006.19	III
15		3009.22	III
100		3075.17	III
4		3085.38	III
25		3093.03	III
15		3096.26	III
200		3111.97	III
50		3116.74	III
1000		3171.63	III
1500		3171.74	III
50		3172.69	III
20		3196.84	III
70		3289.11	III
15		3301.48	III
35		3327.66	III
500		3517.09	III
600		3517.22	III
3		4129.24	III
5		4137.43	III
200		4482.97	III
300		4499.05	III
5		5145.73	III
8		5158.41	III
6		5467.81	III
55		5491.90	III
2		5511.72	III
1		5518.19	III
45		5529.54	III
1		5744.09	III
200		5778.14	III
2		5813.45	III
3		5875.63	III
55		5888.62	III
3		5932.71	III
2		6017.11	III
20		6055.84	III
35		6119.25	III
120		6141.99	III
55		6220.00	III
60		6348.21	III
3		8114.42	III
2		8135.96	III
250		8252.60	III
100		8275.39	III
200		8287.75	III
250		8321.11	III
300		8583.45	III
120		9184.38	III
100		9212.63	III
80		9923.99	III
140		10284.79	III
20		10370.34	III
12		10937.90	III
		13894.47	III
		14096.18	III
		17898.09	III

La IV
Ref. 79 — J.R.

Intensity		Wavelength	
		Vacuum	
100		344.12	IV
7000		453.50	IV
10000		463.14	IV
15000		499.54	IV
40000		552.02	IV
30000		631.26	IV
25		724.92	IV
15		733.29	IV
10		797.03	IV
10	c	980.03	IV
50		1039.30	IV
60	p	1062.09	IV
75		1158.35	IV
50		1164.29	IV
400		1230.90	IV
75	p	1260.79	IV
300		1261.12	IV
150		1283.19	IV
2000		1302.31	IV
1200		1333.53	IV
3500		1334.96	IV
1000		1352.76	IV
25000		1368.04	IV
8000		1377.49	IV
3000		1394.32	IV
5000		1414.58	IV
7000		1432.55	IV
7000		1441.63	IV
7500		1462.15	IV
20000		1463.47	IV
7500		1467.54	IV
15000		1507.87	IV
5000		1527.19	IV
2500		1575.92	IV
1500		1583.61	IV
750		1585.11	IV
750		1637.42	IV
750	d	1645.21	IV
1000		1664.84	IV
750		1684.17	IV
2000	p	1767.65	IV
4000		1808.66	IV
1000		1851.81	IV
1500		1852.77	IV
750		1879.79	IV
1000		1881.57	IV
800		1889.22	IV
1000		1891.47	IV
5000		1902.97	IV
800		1907.44	IV
1500	c	1950.80	IV
1200	c	1957.57	IV
		Air	
3000		2012.42	IV
750		2037.43	IV
2000	c	2066.50	IV
3000	c	2073.18	IV
1500		2143.23	IV
4000	c	2197.45	IV
1000	w	2221.12	IV
900		2227.34	IV
3000		2244.95	IV
7500	w	2265.91	IV
2000		2315.89	IV
750		2348.36	IV
750		2355.31	IV
2000	c	2407.10	IV

Lanthanum (Cont.)

25000	w	2417.58	IV
1200	p	2443.92	IV
18000	c	2502.81	IV
15000		2515.02	IV
50000		2532.75	IV
900	d	2535.76	IV
45000		2582.05	IV
18000	c	2591.30	IV
95000	w	2597.50	IV
5000	c	2608.01	IV
70000	w	2662.75	IV
50000	w	2848.30	IV
12000	c	2863.30	IV
30000	c	2962.58	IV
70000	w	3009.51	IV
90000	c	3056.68	IV
3500		3522.28	IV
2000		3650.40	IV
2000	p	4270.76	IV
1500	w	4549.80	IV
500		4836.89	IV

La V
Ref. 78 — J.R.

Intensity		Wavelength	
		Vacuum	
2		389.03	V
400		390.72	V
1		398.53	V
30		399.34	V
350		405.10	V
50		416.13	V
3		421.55	V
50		423.07	V
400		424.78	V
1000		432.11	V
2500		435.28	V
700		436.14	V
700		436.84	V
300		437.11	V
700		437.55	V
20		444.01	V
10		444.07	V
1250		450.40	V
600		457.30	V
1000		463.85	V
150		476.67	V
5000		482.16	V
200		482.43	V
2000		483.30	V
7000		498.08	V
4000		499.03	V
10000		503.58	V
40		508.15	V
1500		525.71	V
12000		526.76	V
10000		531.07	V
15000		533.23	V
4000		540.20	V
6000		544.80	V
8000		547.44	V
3000		570.90	V
2500		593.18	V
750		597.70	V
2000		600.01	V
5000		600.24	V
700		611.70	V
500		617.60	V

LEAD (Pb)
Z = 82

Pb I and II
Ref. 64, 274, 283, 329, 330 — D.R.W.

Intensity		Wavelength	
		Vacuum	
2		846.04	II
2	h	849.88	II
3		855.57	II
3		863.00	II
6		873.71	II
2		877.96	II
8		889.68	II
3		896.30	II
5		926.44	II
2		958.76	II
2		960.21	II
3		965.36	II
10		967.23	II
9		972.56	II
8		982.17	II
10		986.71	II
10		995.89	II
6		1001.81	II
10		1016.61	II
10		1049.82	II
10		1050.77	II
10		1060.66	II
9		1065.58	II
10		1103.94	II
10		1108.43	II
10		1109.84	II
10		1119.57	II
10		1121.36	II
10		1133.14	II
4		1145.91	II
10		1203.63	II
10		1231.20	II
10		1331.65	II
10		1335.20	II
10		1348.37	II
10		1433.96	II
3		1449.33	II
10		1512.42	II
10		1671.53	II
10		1682.15	II
20		1726.75	II
2		1740.00	I
2		1766.64	I
2		1794.67	I
10		1796.670	II
5		1812.97	I
10		1822.050	II
4		1868.76	I
10		1904.77	I
7		1921.471	II
4		1972.44	I
2		1977.88	I
2		1991.60	I
2		1992.31	I
		Air	
5	r	2022.02	I
5		2050.88	I
8	r	2053.28	I
6		2111.758	I
10		2115.066	I
500	r	2170.00	I
7		2175.580	I
7		2187.888	I
8		2189.603	I
10		2203.534	II
20		2237.425	I
20		2246.86	I
25		2246.89	I
150		2332.418	I
180		2388.797	I
550	r	2393.792	I
140		2399.597	I
320	r	2401.940	I
320	r	2411.734	I
150	r	2443.829	I
160	r	2446.181	I
130	r	2476.378	I
8	c	2526.69	II
8	c	2576.60	II
80	r	2577.260	I
2	c	2608.38	II
500	r	2613.655	I
900	r	2614.175	I
160		2628.262	I
4		2634.256	II
10		2657.094	I
700		2663.154	I
10		2697.541	I
25000	r	2801.995	I
100		2822.58	I
14000	r	2823.189	I
35000	r	2833.053	I
6		2840.557	II
14000	r	2873.311	I
3	c	2887.30	II
3		2914.442	II
2	c	2947.43	II
3	c	2948.53	II
15		2966.460	I
15		2972.991	I
15		2980.157	I
4		2986.876	II
10	c	3016.39	II
150		3118.894	I
600		3220.528	I
100		3229.613	I
400		3240.186	I
200		3262.355	I
35000		3572.729	I
50000	r	3639.568	I
20000		3671.491	I
70000	r	3683.462	I
10		3713.982	II
25000		3739.935	I
15000		4019.632	I
95000		4057.807	I
14000		4062.136	I
5		4110.76	II
4		4113.35	II
10		4152.82	II
10		4157.814	I
10000		4168.033	I
9	c	4242.14	II
20	c	4244.92	II
7		4293.82	II
6		4296.65	II
200		4340.413	I
10		4352.74	II
20	c	4386.46	II
10		4579.051	II
10		4582.27	II
1000		5005.416	I
100		5006.572	I
50		5042.58	II
10		5070.58	II
10		5074.53	II
10		5076.35	I
30		5089.484	I
20		5090.01	I
10		5107.242	I
10		5111.64	II
2000		5201.437	I
10		5367.64	II
10		5372.099	II
10	c	5544.25	II
20	c	5608.85	II
40		5692.346	I
200		5895.624	I
2000		6001.862	I
9	c	6009.58	II
500		6011.667	I
8	c	6041.17	II
500		6059.356	I
40		6075.74	II
40		6081.409	II
50		6110.520	I
10		6159.89	II
100		6235.266	I
50	c	6660.20	II
10		6892.11	I
5		7128.94	I
20		7193.60	II
20000		7228.965	I
5		7304.68	I
8		7330.15	I
10		7346.676	I
10		7558.97	II
10		7632.56	II
4		7732.96	II
20		7809.259	I
5		7817.97	I
6		7829.01	I
5		7896.737	I
2	d	8156.91	II
10		8168.001	I
6		8191.886	I
5		8217.711	I
8		8255.61	I
40		8272.690	I
6		8335.54	II
10		8395.68	II
20		8409.384	I
10		8478.492	I
8		8532.17	I
7	c	8544.95	II
7		8709.90	II
5		8719.39	II
5		8722.810	I
10		8857.457	I
10		9050.82	II
10		9063.43	II
2	d	9245.28	II
8		9293.476	I
5		9384.35	I
5		9385.89	I
15		9438.05	I
15		9604.297	I
6		9608.73	I
15		9674.351	I
200		10290.458	I
5		10434.32	I
100		10498.965	I
50		10649.249	I
5		10759.41	I
7		10759.74	I
15		10886.688	I
40		10969.53	I
6		11059.22	I
3		11333.08	I
2	d	11479.49	II
2		11488.76	I
5		11627.91	II
1		12561.37	I
		13495.3	I
		13498.2	I
		13512.6	I
		14722.8	I
		14742.1	I
		14743.0	I
		15314.8	I
		15327.6	I
		15331.0	I
		15349.6	I
		38831.1	I
		38950.1	I
		38958.6	I
		39039.4	I

Pb III
Ref. 54, 256, 297 — D.R.W.

Intensity	Wavelength	
	Vacuum	
1	961.01	III
3	1030.5	III
12	1048.9	III
4	1069.2	III
3	1074.7	III
4	1118.67	III
4	1167.0	III
4	1250.6	III
1	1266.9	III
20	1553.1	III
1	1610.1	III
4	1711.23	III
	Air	
10	3043.85	III
4	3089.08	III
4	3102.74	III
10	3137.81	III
10	3176.50	III
5	3242.84	III
1	3530.17	III
7	3589.87	III
7	3689.31	III
3	3706.02	III
5	3728.69	III
12	3854.08	III
8	3951.92	III
3	4031.16	III
3	4094.54	III
2	4128.11	III
8	4272.66	III
6	4499.34	III
7	4571.21	III
1	4596.45	III
6	4761.12	III
4	4798.59	III
1	4826.86	III
2	4855.06	III
3	5065.12	III
4	5191.56	III
5	5523.97	III
3	5779.41	III
6	5857.96	III

Pb IV
Ref. 106 — D.R.W.

Intensity	Wavelength	
	Vacuum	
8	475.36	IV
7	478.35	IV
10	496.38	IV
12	499.94	IV
9	515.07	IV
14	529.78	IV
20	570.16	IV
8	573.90	IV
8	584.52	IV
10	648.50	IV
9	656.10	IV
10	761.09	IV
18	802.07	IV
12	802.82	IV
10	812.59	IV
8	822.07	IV
10	827.41	IV
12	832.60	IV
8	840.99	IV
8	842.88	IV
12	845.94	IV
18	857.64	IV
8	859.02	IV
16	862.33	IV
14	870.44	IV
12	879.96	IV
7	880.35	IV
14	884.96	IV
14	884.99	IV
16	890.72	IV
12	908.51	IV
12	917.90	IV
10	922.12	IV
12	922.49	IV
7	924.52	IV
10	927.64	IV
14	932.20	IV
8	937.00	IV
7	952.85	IV

Lead (Cont.)

Intensity	Wavelength	
8	1012.44	IV
14	1028.61	IV
20	1032.05	IV
16	1041.24	IV
18	1044.14	IV
15	1056.53	IV
12	1072.09	IV
7	1079.88	IV
18	1080.81	IV
20	1084.17	IV
6	1089.94	IV
6	1099.47	IV
6	1115.30	IV
20	1116.08	IV
18	1137.84	IV
8	1142.77	IV
14	1144.93	IV
20	1189.95	IV
8	1267.55	IV
8	1290.82	IV
10	1291.10	IV
20	1313.05	IV
8	1323.92	IV
12	1343.06	IV
16	1388.94	IV
6	1397.02	IV
18	1400.26	IV
10	1404.34	IV
7	1510.76	IV
14	1535.71	IV
8	1798.39	IV
8	1893.19	IV
12	1959.34	IV
16	1973.16	IV
	Air	
10	2042.58	IV
12	2049.34	IV
12	2079.22	IV
8	2151.96	IV
15	2154.01	IV
12	2177.46	IV
16	2359.53	IV
4	2864.24	IV
4	2864.50	IV
16	2417.61	IV
4	2978.14	IV
4	3052.56	IV
4	3221.17	IV
4	3962.48	IV
4	4049.80	IV
10	4496.15	IV
16	4534.60	IV
8	4605.40	IV
2	5914.54	IV

Pb V
Ref. 106 — D.R.W.

Intensity	Wavelength	
	Vacuum	
2	367.40	V
2	372.53	V
2	387.87	V
2	394.38	V
5	424.64	V
3	431.03	V
3	436.60	V
2	438.47	V
6	438.91	V
4	453.45	V
3	461.70	V
3	496.20	V
4	694.42	V
8	696.20	V
20	703.73	V
4	706.29	V
6	707.66	V
5	730.85	V
12	749.46	V
10	752.52	V
10	755.80	V
6	762.76	V
10	765.87	V
18	767.45	V
18	769.49	V
14	771.42	V
14	782.79	V
10	787.05	V
15	797.02	V
5	799.80	V
18	809.63	V
5	812.32	V
8	814.10	V
8	820.09	V
5	825.52	V
8	829.32	V
5	851.98	V
20	863.97	V
10	867.10	V
6	880.50	V
18	883.90	V
14	888.37	V
14	894.40	V
12	896.08	V
8	915.09	V
14	915.71	V
12	918.09	V
12	920.28	V
12	920.66	V
6	940.74	V
8	946.20	V
6	950.93	V
12	954.35	V
4	954.95	V
10	955.28	V
4	964.38	V
6	989.14	V
8	1005.42	V
10	1051.26	V
4	1059.26	V
10	1088.86	V
9	1096.52	V
6	1104.79	V
4	1121.33	V
10	1137.50	V
4	1152.36	V
12	1157.88	V
14	1185.43	V
11	1233.50	V
10	1248.47	V
8	1635.75	V
2	1802.87	V
2	1843.00	V
2	1888.67	V
2	1897.02	V
2	1914.33	V
4	1919.74	V
5	1957.96	V
2	1998.58	V
10	1998.83	V
	Air	
8	2078.45	V
10	2142.55	V
10	2167.97	V
20	2259.01	V
10	2276.66	V
8	2301.49	V
15	2424.81	V
4	4809.36	V
5	6650.99	V
4	6753.20	V

LITHIUM (Li)
Z = 3

Li I and II
Ref. 3, 15, 17, 18, 37, 44, 112, 284, 321, 335 — J.D.S.

Intensity		Wavelength	
		Vacuum	
		125.5	II
		136.5	II
		140.5	II
		167.21	II
		168.74	II
		171.58	II
		178.02	II
		199.28	II
		207.5	II
		456.	II
		483.	II
		540.	II
		729.	II
		800.	II
		820.	II
		861.	II
		905.5	II
		917.5	II
		936.	II
		945.	II
		965.	II
		972.	II
		988.	II
		1018.	II
		1032.	II
		1036.	II
		1093.	II
		1103.	II
		1109.	II
		1116.	II
		1132.1	II
		1141.	II
		1166.4	II
		1198.09	II
		1215.	II
		1238.	II
		1253.8	II
		1420.89	II
		1424.	II
3		1492.93	II
5		1492.97	II
1		1493.04	II
		1555.	II
3		1653.08	II
5		1653.13	II
1		1653.21	II
		1681.66	II
		1755.33	II
		Air	
		2009.	II
		2039.	I
		2068.	II
		2131.	II
		2164.	II
		2173.4	I
		2183.	II
		2214.	II
		2222.	II
		2237.	II
	H	2249.21	II
		2286.82	II
		2302.57	II
		2303.33	II
		2304.59	I
		2304.92	I
		2305.36	I
		2305.83	I
		2306.29	I
		2306.82	I
		2307.44	I
		2308.97	I
		2309.88	I
		2310.94	I
		2312.11	I
		2313.49	I
		2315.08	I
		2316.95	I
		2319.18	I
		2321.88	I
		2325.11	I
		2329.02	I
		2329.84	II
		2333.94	I
3		2336.88	II
5		2336.91	II
2		2337.00	II
		2340.15	I
		2348.22	I
		2358.93	I
		2373.54	I
		2381.54	II
		2383.20	II
1		2394.39	I
		2402.33	II
		2410.84	II
3		2425.43	I
		2429.81	II
		2460.2	I
10		2475.06	I
		2506.94	II
		2508.78	II
		2518.	I
		2539.49	II
24		2551.7	II
		2559	II
15		2562.31	I
		2605.08	II
		2640.	
2		2657.29	II
3		2657.30	II
		2674.46	II
0		2728.24	II
5		2728.29	II
2		2728.32	II
3		2730.47	II
1		2730.55	II
5		2741.20	I
		2766.99	II
		2790.31	II
		2801.	I
		2846.	I
		2868.	I
		2895.	I
2		2934.02	II
2		2934.07	II
5		2934.12	II
1		2934.25	II
		2968.	I
3		3029.12	II
3		3029.14	II
		3144.	I
3		3155.31	II
4		3155.33	II
1		3196.26	II
9		3196.33	II
4		3196.36	II
5		3199.33	II
2		3199.43	II
17		3232.66	I
		3249.87	II
		3306.28	II
		3393.	
		3488.	I
		3579.8	I
		3618.	I
		3662.	I
		3684.32	II
1		3714.00	II
5		3714.16	II
6	d	3714.27	II
8		3714.29	II
7	d	3714.40	II
10		3714.41	II
1		3714.51	II
0		3714.58	II
3		3718.7	I
6		3794.72	I
20		3915.30	I
20		3915.35	I
10		3985.48	I
10		3985.54	I
40		4132.56	I
40		4132.62	I
		4196.	I
20		4273.07	I
20		4273.13	I
5		4325.42	II
5		4325.47	II
1		4325.54	II
		4516.45	II
		4590.	
13		4602.83	I
13		4602.89	I
		4607.34	II
0		4671.51	II
6		4671.65	II
2		4671.70	II
3		4678.06	II
1		4678.29	II
		4760.	I
		4763.	II
		4788.36	II
		4843.0	II
4		4881.32	II
4		4881.39	II
1		4881.49	II
8		4971.66	I
8		4971.75	I
		5037.92	II
		5095.	
		5114.	
		5190.	
		5271.	I
		5315.	I
		5395.	I
		5440.	I
600	c	5483.55	II
600	c	5485.65	II
320		6103.54	I
320		6103.65	I
3600		6707.76	I
3600		6707.91	I
48		8126.23	I
48		8126.45	I
		8517.37	II
		9581.42	II
		10120.	II
		12232.	I
		12782.	I
		13566.	I
		17552.	I
		18697.	I
		19290.	I
		24467.	I
		40475.	I

Li III
Ref. 335 — J.O.S.

Intensity	Wavelength	
	Vacuum	
	102.9	III
	103.4	III
	104.1	III
	105.5	III
	108.0	III
	113.9	III
	135.0	III
	540.0	III
	729.1	III

LUTETIUM (Lu)
Z = 71

Lu I and II
Ref. 1 — C.H.C.

Intensity	Wavelength
	Air

Lutetium (Cont.)

1700	h	2195.54	II
95		2276.94	II
190		2297.41	II
1300		2392.19	II
120		2399.14	II
80		2419.21	II
55		2430.26	II
130		2459.64	II
80		2469.27	II
21	h	2481.72	II
370		2536.95	II
40		2546.87	II
20		2549.44	I
20		2549.72	I
35		2561.80	II
930		2571.23	II
1700		2578.70	II
80	h	2582.13	II
1800		2613.40	II
18000		2615.42	II
1800		2619.26	II
90		2657.05	
2700		2657.80	II
90	h	2677.25	I
570	h	2685.08	I
90	h	2685.54	I
4200		2701.71	II
90	h	2715.91	I
180	d	2719.09	I
480	h	2728.95	I
75	c	2738.17	II
3600		2754.17	II
750	h	2765.74	I
2700		2796.63	II
35		2821.23	II
270	c	2834.35	II
330	h	2845.13	I
3000		2847.51	II
570	h	2885.14	I
6300		2894.84	II
4500		2900.30	II
300		2903.05	I
9000		2911.39	II
270	h	2949.73	I
1200		2951.69	II
60		2955.78	II
4200		2963.32	II
2400		2969.82	II
1800		2989.27	I
3000		3020.54	II
120		3027.29	II
2100		3056.72	II
7500		3077.60	II
390		3080.11	I
5100	h	3081.47	I
3000		3118.43	I
2400		3171.36	I
100		3183.73	II
260		3191.80	II
1400		3198.12	II
4800		3254.31	II
3800		3278.97	I
7600		3281.74	I
6200		3312.11	I
7600		3359.56	I
6200		3376.50	I
950		3385.50	I
160	h	3391.55	I
1400		3396.82	I
4100		3397.07	II
4800		3472.48	II
8300	c	3507.39	II
1600		3508.42	I
4800		3554.43	II
4800		3567.84	I
340		3596.34	I
800		3623.99	II
680		3636.25	I
2600		3647.77	I
60		3684.32	I
60		3710.95	I
110		3756.70	I
110		3756.79	I
30	h	3786.18	I
150		3800.67	I
75	h	3829.07	I
2700		3841.18	I
75		3843.61	I
95		3853.29	I
40		3874.61	I
530		3876.65	II
29		3911.77	I
50		3918.86	I
35	h	3926.62	I
480		3968.46	I
50		3991.38	I
670		4054.45	I
75	Bl	4096.13	L
35	h	4107.44	I
95	h	4112.67	I
310		4122.49	I
3100		4124.73	I
150	c	4131.79	I
460		4154.08	I
24		4158.98	I
1600		4184.25	II
150		4277.50	I
250		4281.03	I
330	d	4295.97	I
		4296.09	I
150		4309.57	I
75		4332.72	I
29		4341.98	II
65	h	4420.96	I
190	c	4430.48	I
35		4438.79	I
190		4450.81	I
50	h	4471.55	I
60	h	4498.85	I
[illegible]		[illegible]	I
24	b	4560.95	L
24	b	4575.31	L
85	c	4605.39	I
95	h	4645.47	I
100	h	4648.21	I
95	h	4648.85	I
65	b	4654.03	L
1000		4658.02	I
85	h	4659.03	I
630	Bl	4661.75	L
310	Bl	4672.31	L
420	Bl	4684.16	L
270	Bl	4695.46	L
190	Bl	4708.00	L
30		4716.70	I
65	b	4720.86	L
65	h	4726.20	I
100	Bl	4735.00	L
75	Bl	4749.11	L
40	Bl	4764.22	L
150		4785.42	II
85		4815.05	I
50	c	4839.62	II
18		4865.36	II
460		4904.88	I
180		4942.34	I
800		4994.13	II
800		5001.14	I
55	h	5057.60	I
140		5134.05	I
2700		5135.09	I
130	Bl	5170.11	L
170		5196.61	I
90		5206.47	I
40		5304.40	I
80		5349.12	I
500		5402.57	I
140	c	5421.90	I
100		5437.88	I
35		5453.57	I
2100		5476.69	II
9		5664.89	II
14		5713.49	II
550		5736.55	I
55		5775.40	I
80		5800.59	I
40	h	5860.79	I
9		5866.30	I
690	Cw	5983.9	II
140		5997.13	I
1400		6004.52	I
35	h	6041.66	I
440		6055.03	I
11		6140.71	I
150		6159.94	II
160		6199.66	II
2100		6221.87	II
35		6228.14	II
80		6235.36	II
160		6242.34	II
16	h	6248.80	I
70	h	6345.35	I
18	h	6354.85	I
9		6365.79	I
16		6366.00	I
22		6441.14	I
11		6444.89	II
1100		6463.12	II
29		6477.67	I
55	c	6523.18	I
35	Cw	6611.28	II
		6611.58	LU II
		6611.80	LU II
		6611.95	LU II
		6612.04	LU II
11		6619.15	II
23	c	6677.14	I
9	h	6735.76	I
30	c	6793.77	I
11		6826.59	II
45		6917.31	I
8		6943.96	II
23		7031.24	I
14	c	7096.34	I
45		7125.84	II
9		7142.79	I
7		7143.10	I
8		7165.94	II
14	Ch	7237.98	I
5		7409.70	II
11	c	7441.52	I
8		7456.96	II
7	Ch	7640.08	I
7	Cw	7758.30	I
7	h	7815.9	I
9	c	8178.16	I
17		8382.08	I
35		8459.19	II
10	d	8478.50	I
29	c	8508.08	I
35	c	8610.98	I

Lu III
Ref. 148 — J.R.

Intensity		Wavelength	
		Vacuum	
1		677.34	III
7		691.05	III
30		700.25	III
50		714.89	III
100		738.76	III
200		755.03	III
3		755.16	III
500		810.73	III
100		830.53	III
2000		832.28	III
10		972.66	III
2		991.26	III
100		996.44	III
400		1001.18	III
1		1022.40	III
100		1029.83	III
200		1030.33	III
100		1031.54	III
50		1056.53	III
20		1061.99	III
3		1092.84	III
200		1187.34	III
50		1228.7	III
200		1277.53	III
30		1283.41	III
100		1331.93	III
1000		1854.57	III
		Air	
40		2050.72	III
1500		2065.35	III
1500	c	2070.56	III
100		2083.34	III
200		2099.44	III
1000		2236.14	III
2000		2236.22	III
500	c	2381.59	III
300		2563.51	III
4500	c	2603.35	III
200		2721.65	III
2000		2772.55	III
20		2781.16	III
10		2788.37	III
500		2800.90	III
20	p	2993.21	III
1000		3057.86	III
200		4251.44	III
300		4271.91	III
200		4490.00	III
400		4956.43	III
10		5046.12	III
150	c	5145.86	III
70		5419.42	III
60		5519.88	III
5		5526.80	III
5		5748.71	III
70		5786.46	III
80		5869.71	III
60		5889.76	III
300		6197.96	III
600		6198.13	III
100		7309.95	III
200		7310.25	III
50		7534.27	III
70	l	7936.45	III
3		8008.59	III

Lu IV
Ref. 310 — J.R.

Intensity		Wavelength	
		Vacuum	
400		876.80	IV
100		902.06	IV
300		1015.18	IV
20		1136.17	IV
50		1189.27	IV
20	p	1194.59	IV
60		1213.08	IV
15		1220.74	IV
20		1223.75	IV
20		1240.07	IV
20		1248.10	IV
40		1266.27	IV
100		1272.42	IV
20		1273.02	IV
40		1274.77	IV
20		1276.54	IV
50		1289.38	IV
60		1310.08	IV
50		1323.02	IV
15		1331.04	IV
[illegible]		[illegible]	IV
300		1334.94	IV
50		1338.20	IV
100		1339.49	IV
300		1342.58	IV
200		1351.68	IV
300		1353.74	IV
200		1355.85	IV
20		1359.67	IV
100		1363.24	IV
50		1363.37	IV
20		1367.34	IV
20		1373.54	IV
15		1375.36	IV
100		1376.02	IV
30		1379.56	IV
15		1383.18	IV
15		1389.85	IV
60		1390.07	IV
200		1390.30	IV
50		1390.69	IV
40		1392.38	IV
40		1397.18	IV
30		1401.32	IV
100		1401.46	IV
200		1406.64	IV
100		1407.00	IV
250		1407.04	IV
20		1420.32	IV
100		1421.59	IV
200		1429.08	IV
400		1429.38	IV
40		1430.80	IV
100		1440.62	IV
100		1448.14	IV
200		1452.33	IV
30		1462.65	IV
100		1483.79	IV
200		1493.24	IV
400		1511.26	IV
200		1521.06	IV
100		1522.21	IV
100		1537.77	IV
30		1549.35	IV
20		1551.59	IV
20		1562.06	IV
30		1592.55	IV
15		1594.92	IV
100	c	1607.72	IV
50	c	1631.65	IV
20		1684.50	IV
60	c	1693.67	IV
400	c	1721.42	IV
100		1725.14	IV
100	c	1735.79	IV
100		1736.78	IV
100		1741.74	IV
50	c	1743.84	IV
40	c	1752.60	IV
200		1759.61	IV
300		1772.08	IV
600		1772.57	IV
200		1782.45	IV
100		1797.52	IV
20	c	1901.63	IV
100		1983.92	IV
20		1990.52	IV
40		1996.18	IV
		Air	
100		2003.18	IV
100		2020.94	IV
20		2071.10	IV
100		2081.09	IV
400	c	2085.70	IV
600	c	2086.47	IV
400	c	2092.16	IV
100		2103.63	IV
1000	c	2104.41	IV
200	c	2107.85	IV
1000	c	2108.31	IV
100	c	2127.43	IV

Lu V
Ref. 401 — J.R.

Lutetium (Cont.)

Intensity		Wavelength	
		Vacuum	
40		555.44	V
100		563.72	V
50		601.54	V
60		614.23	V
40		628.79	V
50	p	637.44	V
40		637.53	V
50		663.29	V
60		850.06	V
100		861.92	V
70		866.93	V
40		870.84	V
50		875.89	V
50		876.45	V
100		880.32	V
40		884.21	V
70		886.16	V
50		886.32	V
50		886.44	V
100		891.81	V
60		895.01	V
40		895.15	V
40		898.42	V
100		914.72	V
40		918.26	V
50		920.92	V
40		921.32	V
60		921.90	V
40		922.73	V
80		925.79	V
50		927.22	V
40		947.80	V
50	w	1420.02	V
50		1432.50	V
100		1432.77	V
200		1441.76	V
40		1443.64	V
100		1448.14	V
40		1449.32	V
100		1450.36	V
100		1450.69	V
100		1452.64	V
200		1453.35	V
100		1454.38	V
100		1455.21	V
100		1460.11	V
40	c	1467.81	V
200		1468.99	V
50		1469.45	V
100		1471.20	V
400		1472.12	V
200		1473.71	V
40		1475.77	V
200		1485.58	V
50		1709.02	V
40		1728.90	V
50	c	1775.92	V
40	c	1777.68	V
40	c	1784.71	V
100	c	1786.25	V
60	c	1787.58	V
60	c	1793.85	V
60		1809.73	V
60		1814.24	V

MAGNESIUM (Mg)
Z = 12

Mg I and II
Ref. 49, 83, 103, 217, 269, 315, 335
— J.O.S.

Intensity	Wavelength	
	Vacuum	
	184.05	II
	184.31	II
	184.68	II
	184.81	II
	185.26	II
	185.59	II
	185.98	II
	186.47	II
	186.84	II
	187.19	II
	187.38	II
	188.54	II
	188.91	II
	189.01	II
	189.23	II
	189.37	II
	191.30	II
	191.56	II
	191.65	II
	192.40	II
	192.55	II
	192.84	II
	193.09	II
	193.31	II
	193.40	II
	193.64	II
	197.76	II
	199.31	II
	200.29	I
	202.00	II
	202.27	II
	202.51	II
	202.94	II
	203.15	I
	203.42	II
	203.53	II
	204.22	I
	209.09	II
	209.43	II
	209.84	I
	213.53	I
	215.12	I
	215.31	I
	215.45	I
	215.66	I
	215.79	I
	216.22	I
	216.36	I
	216.68	I
	217.21	I
	217.37	I
	218.19	I
	218.34	I
	218.42	I
	218.74	I
	219.04	I
	219.28	I
	220.03	I
	220.33	I
	222.03	I
	222.67	I
	223.45	I
	223.74	I
	225.18	I
	225.54	I
	226.26	I
	247.14	II
	248.47	II
	884.70	II
	884.72	II
	907.38	II
	907.41	II
8	946.70	II
9	946.77	II
14	1025.96	II
12	1026.11	II
25	1239.94	II
20	1240.40	II
6	1248.51	II
8	1249.93	II
8	1271.24	II
9	1271.94	II
8	1272.72	II
11	1273.43	II
11	1306.71	II
12	1307.88	II
12	1308.28	II
14	1309.44	II
14	1365.45	II
	1365.54	II
15	1367.26	II
15	1367.70	II
18	1369.42	II
20	1476.00	II
25	1478.01	II
20	1480.89	II
30	1482.90	II
	1625.22	I
	1625.50	I
	1625.81	I
	1626.16	I
	1626.36	I
	1626.56	I
	1626.79	I
	1627.02	I
	1627.27	I
	1627.53	I
	1627.82	I
	1628.12	I
	1628.46	I
	1628.80	I
	1629.21	I
	1629.59	I
	1630.52	I
	1631.62	I
	1632.93	I
	1634.52	I
	1636.48	I
	1638.90	I
	1641.97	I
1	1645.93	I
1	1651.16	I
2	1658.31	I
5	1668.43	I
10	1683.41	I
15	1707.06	I
40	1734.84	II
50	1737.62	II
20	1747.80	I
40	1750.65	II
50	1753.46	II
30	1827.93	I
	Air	
9	2025.82	I
3	2329.58	II
6	2449.57	II
1	2557.23	I
1	2560.94	I
1	2562.26	I
1	2564.94	I
1	2570.91	I
1	2572.25	I
2	2574.94	I
1	2577.89	I
1	2580.59	I
1	2584.22	I
2	2585.56	I
3	2588.28	I
1	2591.89	I
1	2593.23	I
2	2595.97	I
2	2602.50	I
4	2603.85	I
5	2606.62	I
1	2613.36	I
2	2614.73	I
3	2617.51	I
3	2628.66	I
6	2630.05	I
8	2632.87	I
2	2644.80	I
3	2646.21	I
4	2649.06	I
8	2660.76	II
8	2660.82	II
6	2668.12	I
8	2669.55	I
10	2672.46	I
3	2693.72	I
5	2695.18	I
6	2698.14	I
8	2731.99	I
10	2733.49	I
12	2736.53	I
5	2765.22	I
7	2768.34	I
38	2776.69	I
32	2778.27	I
90	2779.83	I
8	2781.29	I
32	2781.42	I
36	2782.97	I
13	2790.79	II
1000	2795.53	II
16	2798.06	II
600	2802.70	II
12	2846.75	I
14	2848.42	I
16	2851.65	I
6000	2852.13	I
3	2915.45	I
10	2936.74	I
12	2938.47	I
2	2942.00	I
3	2809.76	I
2	2811.11	I
1	2811.78	I
12	2846.72	I
14	2848.34	I
16	2851.66	I
2	2902.92	I
4	2906.36	I
3	2915.45	I
2	2928.75	II
3	2936.54	II
10	2936.74	I
12	2938.47	I
13	2942.00	I
1	2967.87	II
1	2971.70	II
20	3091.08	I
22	3092.99	I
14	3096.90	I
9	3104.71	II
8	3104.81	II
6	3168.98	II
6	3172.71	II
7	3175.78	II
2	3197.62	I
17	3329.93	I
6	3332.15	I
9	3336.68	I
7	3535.04	II
8	3538.86	II
7	3549.52	II
8	3553.37	II
140	3829.30	I
300	3832.30	I
500	3838.29	I
8	3848.24	II
1	3848.91	I
7	3850.40	II
2	3853.96	I
1	3854.96	I
2	3858.86	I
3	3878.31	I
2	3891.91	I
2	3893.30	I
3	3895.57	I
4	3903.86	I
6	3938.40	I
1	3984.21	I
8	3986.75	I
2	4054.69	I
10	4057.50	I
3	4075.06	I
2	4081.83	I
4	4165.10	I
15	4167.27	I
20	4351.91	I
6	4354.53	I
6	4380.38	I
9	4384.64	II
10	4390.59	II
8	4428.00	II
9	4433.99	II
5	4436.49	II
4	4436.60	II
14	4481.16	II
13	4481.33	II
6	4534.29	II
28	4571.10	I
3	4621.30	I
7	4702.99	I
10	4730.03	I
6	4739.59	II
5	4739.71	II
7	4851.10	II
75	5167.33	I
220	5172.68	I
400	5183.61	I
8	5264.21	II
7	5264.37	II
1	5345.98	I
9	5401.54	II
2	5509.60	I
6	5528.41	I
30	5711.09	I
5	5785.31	I
4	5785.56	I
7	5916.43	II
6	5918.16	II
10	6318.72	I
9	6319.24	I
7	6319.49	I
10	6346.74	II
9	6346.96	II
11	6545.97	II
5	6620.44	II
6	6620.57	II
2	6630.83	I
7	6781.45	II
8	6787.85	II
7	6812.86	II
8	6819.27	II
4	6894.90	I
6	6965.40	I
8	7060.41	I
10	7193.17	I
10	7291.06	I
5	7387.00	I
12	7387.69	I
4	7580.76	II
20	7657.60	I
19	7659.15	I
17	7659.90	I
8	7690.16	I
15	7691.55	I
1	7722.61	I
1	7746.34	I
1	7759.30	I
5	7786.50	II
4	7790.98	II
3	7811.14	I
12	7877.05	II
2	7881.67	I
13	7896.37	II
7	7930.81	I
3	8047.73	I
5	8049.85	I
7	8054.23	I
10	8098.72	I
9	8115.22	II
8	8120.43	II
1	8154.64	I
2	8159.13	I
10	8209.84	I
20	8213.03	I
10	8213.99	II
7	8222.92	II
7	8233.19	II
11	8234.64	II

Magnesium (Cont.)

7		8303.31	I
9		8305.60	I
10		8310.26	I
15		8346.12	I
2		8466.48	I
5		8468.84	I
7		8473.69	I
10		8710.18	I
12		8712.69	I
13		8717.83	I
10		8734.99	II
17		8736.02	I
11		8745.66	II
14		8806.76	I
10		8824.32	II
11		8835.08	II
20		8923.57	I
7		8989.03	I
9		8991.69	I
10		8997.16	I
14		9218.25	II
13		9244.27	II
12		9246.50	I
30		9255.78	I
10		9327.54	II
10		9340.54	II
25		9414.96	I
17		9429.81	I
19		9432.76	I
20		9438.78	I
8		9502.45	I
7		9503.11	I
5		9503.43	I
12		9631.89	II
11		9632.43	II
15		9953.20	I
15		9983.20	I
17		9986.47	I
18		9993.21	I
14		10092.16	II
5		10391.76	II
6		10392.23	II
35		10811.08	I
11		10914.23	II
7		10915.27	II
10		10951.78	II
25		10953.32	I
27		10957.30	I
28		10965.45	I
15		11032.10	I
14		11033.66	I
5		11255.93	II
4		11256.35	II
45		11828.18	I
30		12083.66	I
28		14877.62	I
35		15024.99	I
30		15040.24	I
25		15047.70	I
6		15740.71	I
8		15748.99	I
10		15765.84	I
30		17108.66	I
5		26392.90	I

Mg III
Ref. 4, 83, 177 — J.O.S.

Intensity		Wavelength	
		Vacuum	
		106.30	III
		106.92	III
		108.08	III
		110.16	III
		114.32	III
		126.50	III
15		170.80	III
15		171.39	III
15		182.24	III
12		182.97	III
20		186.51	III
20		187.20	III
10		188.53	III
100		231.73	III
80		234.26	III
10		1274.83	III
11	h	1280.70	III
12		1391.27	III
15		1393.39	III
10		1431.14	III
16		1572.71	III
12		1586.24	III
13		1687.09	III
13		1697.28	III
10		1722.04	III
22		1738.84	III
12		1747.56	III
18		1748.93	III
15		1772.98	III
20		1783.25	III
14		1794.58	III
15		1800.66	III
13		1858.19	III
12		1879.49	III
10		1908.50	III
12		1923.90	III
13		1930.67	III
11		1937.84	III
		Air	
15		2039.55	III
15		2055.49	III
25		2064.90	III
15		2085.90	III
20		2091.96	III
13		2097.93	III
15		2112.77	III
16		2134.06	III
20		2177.70	III
20		2395.15	III
15		2467.75	III
10		2490.54	III
10		2529.19	III
12		3299.05	III
13		3306.39	III
12		3335.90	III
11		3342.58	III
12		3361.41	III
10		3381.24	III
11		3382.90	III
11		3387.37	III
10		3706.74	III
10		4916.00	III
10		5839.82	III
15		6256.75	III

Mg IV
Ref. 7, 128, 129 — J.O.S.

Intensity		Wavelength	
		Vacuum	
40		118.16	IV
80	p	118.81	IV
70		123.59	IV
240		124.65	IV
300		129.86	IV
300		132.81	IV
400		146.95	IV
300		147.41	IV
300		147.54	IV
350		180.07	IV
400		180.62	IV
400		180.80	IV
350		181.34	IV
4000		320.99	IV
3000		323.31	IV
40		800.41	IV
150		857.29	IV
30		866.74	IV
50		919.03	IV
30		929.78	IV
40		1008.76	IV
30		1026.41	IV
250		1037.41	IV
80		1044.37	IV
60		1055.76	IV
300		1210.99	IV
300		1342.19	IV
800		1346.57	IV
300		1346.68	IV
600		1352.05	IV
900		1384.46	IV
500		1385.77	IV
800		1387.53	IV
300		1404.68	IV
1000		1409.36	IV
500		1437.53	IV
1000		1437.64	IV
300		1447.42	IV
300		1459.54	IV
400		1459.62	IV
400		1481.51	IV
350		1490.45	IV
300		1495.50	IV
300		1607.11	IV
500		1683.02	IV
400		1698.81	IV
300		1844.17	IV
		Air	
12	p	2518.40	IV
4		2534.79	IV

Mg V
Ref. 128 — J.O.S.

Intensity		Wavelength	
		Vacuum	
5		251.58	V
35		276.58	V
10		312.30	V
20		351.09	V
18		352.20	V
30		353.09	V
15		353.30	V
18		354.22	V
20		355.33	V

MANGANESE (Mn)
Z = 25

Mn I and II
Ref. 1, 126 — C.H.C.

Intensity		Wavelength	
		Vacuum	
20		1726.47	II
30		1732.70	II
50		1733.55	II
40		1734.49	II
30		1737.93	II
20		1740.16	II
20		1742.00	II
30		1853.27	II
20		1857.92	II
50		1902.95	II
20		1907.84	II
30		1911.41	II
20	d	1914.68	II
100		1915.10	II
20		1918.64	II
30		1919.64	II
80		1921.25	II
20		1923.07	II
20		1923.34	II
30		1925.52	II
50		1926.59	II
30		1931.40	II
20		1945.15	II
20		1947.93	II
20		1950.14	II
30		1953.23	II
20	d	1954.81	II
30		1959.25	II
20		1969.24	II
30		1994.23	II
9700		1996.06	I
14000		1999.51	I
		Air	
18000		2003.85	I
50		2037.31	II
40		2037.64	II
40		2039.97	II
30		2076.21	II
1500		2092.16	I
20		2097.46	II
20		2102.50	II
1700		2109.58	I
30		2113.96	II
290		2208.81	I
540		2213.85	I
770		2221.84	I
20		2373.36	II
20		2427.38	II
50		2427.72	II
30		2427.94	II
30		2437.37	II
20		2437.84	II
30		2452.49	II
50		2499.00	II
30		2507.60	II
20		2516.60	II
30		2516.74	II
20		2521.66	II
20		2530.72	II
20		2531.80	II
50		2532.78	II
75		2533.06	I
50		2533.33	II
30		2534.10	II
80		2534.22	II
100		2535.66	II
30		2535.98	II
100		2537.92	II
50		2541.11	II
80		2542.92	II
50		2543.45	II
100		2548.75	II
50		2551.85	II
30		2553.27	II
75		2556.57	II
30		2556.89	II
50		2557.54	II
95		2558.59	II
30		2559.41	II
150		2563.65	II
30		2565.22	II
580		2572.76	I
480		2575.51	I
12000		2576.10	II
550		2584.31	I
30		2588.97	II
45		2589.71	II
250		2592.94	I
6200		2593.73	II
250		2595.76	I
95		2598.90	II
40		2602.14	I
30		2602.72	II
45		2603.72	II
4300		2605.69	II
190		2610.20	II
500		2618.14	II
140		2622.90	I
150		2624.04	I
40		2624.80	II
200		2625.58	II
95		2626.64	I
30		2630.26	I
60		2630.57	I
190		2632.35	II
130		2638.17	II
80		2639.84	II
27		2650.99	II
60		2655.91	II
30		2666.77	II
45		2667.00	I
30		2667.03	II
110		2672.59	II
55		2673.37	II
55		2674.43	II
30		2676.33	I
45		2680.34	II
30		2680.68	II
30		2681.25	II
40		2681.72	I
45		2683.02	I
23		2683.75	I
55		2684.55	II
55		2685.94	II
110		2688.25	II
85		2692.66	I
27		2693.19	II
55		2695.36	II
27		2698.97	II
85		2701.00	II
50		2701.17	II
160		2701.70	II
100		2703.98	II
130		2705.74	II
80		2707.53	II
110		2708.45	II
45		2709.96	II
80		2710.33	II
110		2711.58	II
30		2716.80	II
30		2717.53	II
30		2719.01	II
50		2719.74	II
30		2722.10	II
30		2724.46	II
55		2728.61	II
30		2738.86	I
45	h	2760.93	I
30		2771.44	I
30	h	2776.23	I
30		2780.00	I
55		2789.20	I
60		2790.36	I
60		2791.08	I
6200		2794.82	I
5100		2798.27	I
220		2799.84	I
3700		2801.06	I
70		2804.10	I
60		2806.14	I
55		2808.02	I
110		2809.11	I
60		2812.84	I
70		2813.47	I
60		2815.02	II
30		2816.33	II
85		2817.97	I
40		2818.77	I
55		2821.45	I
55		2822.55	I
80		2830.79	I
27		2836.31	I
60		2870.08	II
30		2872.94	II
80		2879.49	II
40		2882.90	I
70		2886.68	II
160		2889.58	II
55		2892.39	II
50		2898.70	II
80		2900.16	II
40		2907.22	I
140	h	2914.60	I
190	h	2925.57	I

Manganese (Cont.)

27		2928.68	I
1100		2933.06	II
27		2934.02	I
1500		2939.30	II
250	h	2940.39	I
		2940.48	I
60		2941.04	I
1900		2949.20	II
40		3007.66	I
40		3011.16	I
40		3011.38	I
40		3014.67	I
60		3016.45	I
30		3019.92	II
70		3022.75	I
55		3031.06	II
30		3035.35	II
95		3040.60	I
27		3042.73	I
85		3043.36	I
330		3044.57	I
120		3045.59	I
200		3047.04	I
40		3048.86	I
30		3050.65	II
250		3054.36	I
140		3062.12	I
170		3066.02	I
170		3070.27	I
160		3073.13	I
90		3079.63	I
50		3081.33	I
23		3082.05	I
40		3097.06	I
40		3110.68	I
60	h	3148.18	I
90	h	3161.04	I
140	h	3178.50	I
220		3212.88	I
65		3216.95	I
1000		3228.09	I
300		3230.72	I
850		3236.78	I
330		3243.78	I
650		3248.52	I
100		3251.14	I
310		3252.95	I
65		3254.04	I
310		3256.14	I
220		3258.41	I
180		3260.23	I
180		3264.71	I
65		3296.88	I
65		3298.22	I
65		3320.69	I
70		3330.67	I
200		3330.78	II
100		3336.39	II
30		3365.02	II
30		3400.12	II
50		3438.97	II
720		3441.99	II
50		3460.03	II
360		3460.33	II
360	h	3474.04	II
		3474.13	II
290		3482.91	II
180		3488.68	II
140		3495.84	II
50		3496.81	II
100		3497.54	II
360		3531.85	I
		3532.00	I
1100		3532.12	I
1300		3547.80	I
1100		3548.03	I
390		3548.20	I
2200		3569.49	I
720		3569.80	I
		3570.04	I
1400		3577.88	I
720		3586.54	I
290		3595.12	I
420		3607.54	I
420		3608.49	I
360		3610.30	I
290		3619.28	I
220		3623.79	I
140		3629.74	I
100		3660.40	I
70		3670.52	I
70		3676.96	I
50		3682.09	I
280		3693.67	I
180		3696.57	I
70		3701.73	I
210		3706.08	I
130		3718.93	I
55		3728.89	I
130		3731.93	I
260		3790.22	I
55		3799.26	I
110		3800.55	I
55		3801.91	I
3200		3806.72	I
700		3809.59	I
55		3810.69	I
90		3816.75	I
2100		3823.51	I
390		3823.89	I
200		3829.68	I
480		3833.86	I
1300		3834.36	I
350		3839.78	I
670		3841.08	I
350		3843.98	I
65		3918.32	I
120		3926.47	I
65		3952.84	I
55		3975.89	I
65		3977.08	I
130		3982.58	I
150		3985.24	I
190		3986.83	I
150		3987.10	I
1500		4018.10	I
150		4026.44	I
27000		4030.76	I
19000		4033.07	I
11000		4034.49	I
1500		4035.73	I
55		4038.73	I
5600		4041.36	I
210	d	4045.13	I
		4045.21	I
1100		4048.76	I
80		4049.00	I
55		4051.73	I
65		4052.47	I
150		4055.21	I
1900		4055.54	I
210		4057.95	I
1100		4058.93	I
150		4059.39	I
730		4061.74	I
730		4063.53	I
80		4065.08	I
80		4068.00	I
290		4070.28	I
730		4079.24	I
730		4079.42	I
1100		4082.94	I
1100		4083.63	I
65		4089.94	I
55		4105.36	I
200		4110.90	I
150		4131.12	I
120		4135.04	I
80		4141.06	I
55		4147.53	I
80		4148.80	I
150		4176.60	I
120		4189.99	I
65		4201.76	I
65		4211.75	I
370		4235.14	I
510		4235.29	I
190		4239.72	I
290		4257.66	I
290		4265.92	I
270		4281.10	I
65		4284.08	I
65		4312.55	I
50		4323.63	II
45		4374.95	I
45		4381.70	I
55		4411.88	I
350		4414.88	I
55		4419.78	I
210		4436.35	I
800		4451.59	I
160		4453.00	I
130		4455.01	I
160		4455.32	I
110		4455.82	I
55		4457.04	I
210		4457.55	I
270		4458.26	I
55		4460.38	I
150		4461.08	I
510		4462.02	I
290		4464.68	I
200		4470.14	I
130		4472.79	I
40		4479.40	I
170		4490.08	I
240		4498.90	I
240		4502.22	I
80		4605.36	I
80		4626.54	I
35		4671.69	I
50		4701.16	I
160		4709.72	I
180		4727.48	I
130		4739.11	I
1000		4754.04	I
180		4761.53	I
750		4762.38	I
300		4765.86	I
500		4766.43	I
940		4783.42	I
1000		4823.52	I
25		4844.32	I
35		4965.88	I
19		5004.91	I
30		5074.79	I
60		5117.94	I
50		5150.89	I
50		5196.59	I
85		5255.32	I
160		5341.06	I
19		5349.88	I
95		5377.63	I
95		5394.67	I
50		5399.49	I
95		5407.42	I
35		5413.69	I
85		5420.36	I
35		5432.55	I
12		5457.47	I
60		5470.64	I
40		5481.40	I
30		5505.87	I
50		5516.77	I
40		5537.76	I
21		5551.98	I
8		5567.76	I
7		5573.01	I
8		5573.68	I
7		5738.29	I
7		5780.19	I
7		5816.84	I
140		6013.50	I
200		6016.64	I
290		6021.80	I
7		6384.67	I
17		6440.97	I
24		6491.71	I
14	h	6942.52	I
12		6989.96	I
14		7069.84	I
12		7184.25	I
10		7247.82	I
24	h	7283.82	I
35	h	7302.89	I
50		7326.51	I
12		7680.20	I
10		7712.42	I
10	h	7764.72	I
10	h	8670.92	I
12	h	8672.06	I
10	h	8673.97	I
12	h	8701.05	I
17	h	8703.76	I
30	h	8740.93	I

Mn III
Ref. 385 — C.H.C.

Intensity		Wavelength	
		Vacuum	
20		892.39	III
20		1108.16	III
30		1183.30	III
25	w	1183.86	III
30		1198.49	III
30		1219.80	III
100		1228.97	III
500		1283.58	III
400		1287.59	III
300		1291.62	III
1000		1360.72	III
800		1365.20	III
400		1369.43	III
300		1371.65	III
300		1596.95	III
500	h	1609.17	III
1000		1614.14	III
2000		1620.60	III
300		1623.91	III
400		1629.12	III
500		1633.80	III
250		1647.46	III
400		1653.57	III
400		1804.06	III
300		1806.47	III
300		1811.02	III
400		1877.62	III
300		1885.21	III
500		1941.28	III
250	w	1942.89	III
800		1943.21	III
500		1952.36	III
1000		1952.52	III
300		1956.61	III
250		1962.04	III
500		1978.95	III
400		1982.76	III
400		1989.59	III
		Air	
300		2022.19	III
1000	w	2027.83	III
500	w	2028.14	III
300		2044.57	III
400		2048.93	III
500		2049.68	III
300		2056.80	III
500		2066.38	III
1000		2069.02	III
900		2077.38	III
300		2078.13	III
800		2084.23	III
600		2090.05	III
300		2090.25	III
300		2094.14	III
500		2094.78	III
500		2097.93	III
500		2099.97	III
300		2123.25	III
1000		2169.78	III
700		2174.15	III
900		2176.87	III
800		2181.86	III
800	w	2184.87	III
600		2185.13	III
400		2211.95	III
600		2212.42	III
800		2215.21	III
900		2220.55	III
1000		2227.42	III
100		3287.49	III
100		3540.52	III
150		3601.72	III
100		3616.00	III
100		4246.17	III
200		5079.20	III
150		5100.03	III
100		5117.03	III
100		5252.23	III
100		5365.59	III
150		5454.07	III
200		5474.68	III
100		5671.12	III
200		5946.65	III
100	s	6213.11	III
200		6231.21	III
100		6238.64	III
100		6273.71	III

Mn IV
Ref. 433 — C.H.C.

Intensity		Wavelength	
		Vacuum	
60		579.79	IV
60		581.44	IV
60		581.65	IV
60		585.21	IV
90		1242.25	IV
90		1244.50	IV
85		1247.73	IV
95		1251.93	IV
95		1257.28	IV
90		1264.41	IV
70		1603.60	IV
70		1611.10	IV
75		1653.83	IV
70		1656.39	IV
70		1659.25	IV
75		1664.73	IV
80	b	1667.00	IV
70		1670.08	IV
75		1691.68	IV
75		1693.15	IV
80		1698.30	IV
75		1698.70	IV
70		1699.06	IV
75		1707.43	IV
65		1718.67	IV
75	b	1720.52	IV
75		1720.74	IV
75		1721.41	IV
65		1722.94	IV
75		1724.83	IV
85	b	1742.10	IV
85		1751.59	IV
75		1759.82	IV
70		1762.17	IV
75		1762.94	IV
85	d	1766.27	IV
75		1767.09	IV
65		1772.11	IV
75		1773.51	IV
75		1782.21	IV
75		1786.02	IV

Manganese (Cont.)

75	1787.04	IV
75	1787.38	IV
75	1788.64	IV
75	1790.44	IV
80	1795.65	IV
80	1795.79	IV
60	1907.03	IV
75	1910.25	IV
65	1997.54	IV

Mn V
Ref. 405 — C.H.C.

Intensity	Wavelength	
	Vacuum	
300	404.36	V
380	406.02	V
300	406.40	V
600	410.30	V
600	410.60	V
480	410.98	V
400	411.32	V
460	412.74	V
460	413.75	V
600	415.62	V
650	415.98	V
350	419.80	V
600	428.59	V
500	429.05	V
400	433.54	V
600	435.67	V
350	436.16	V
500	436.18	V
450	438.74	V
350	439.35	V
1000	441.72	V
850	442.49	V
400	467.32	V
300	474.82	V

MERCURY (198) (Hg)
Z = 80

Hg I and II (198)
Ref. 43, 50, 69, 145, 229, 242 — R.W.S.

Intensity	Wavelength	
	Vacuum	
80	1250.564	I
8	1259.242	I
100	1268.825	I
5	1307.751	I
20	1402.619	I
10	1435.503	I
1000	1849.492	I
	Air	
60	2262.210	II
20	2302.065	I
20	2345.440	I
100	2378.325	I
20	2380.004	I
40	2399.349	I
20	2399.729	I
20	2446.900	I
15	2464.064	I
40	2481.999	I
30	2482.713	I
40	2483.821	I
90	2534.769	I
15000	2536.506	I
25	2563.861	I
25	2576.290	I
250	2652.043	I
400	2653.683	I
100	2655.130	I
50	2698.831	I
80	2752.783	I
20	2759.710	I
40	2803.471	I
30	2804.438	I
750	2847.675	II
50	2856.939	I
150	2893.598	I
150	2916.227	II
60	2925.413	I
1200	2967.283	I
300	3021.500	I
120	3023.476	I
30	3025.608	I
50	3027.490	I
400	3125.670	I
320	3131.551	I
320	3131.842	I
80	3341.481	I
2800	3650.157	I
300	3654.839	I
80	3662.883	I
240	3663.281	I
30	3701.432	I
35	3704.170	I
30	3801.660	I
20	3901.867	I
60	3906.372	I
200	3983.839	II
1800	4046.572	I
150	4077.838	I
40	4108.057	I
250	4339.224	I
400	4347.496	I
4000	4358.337	I
80	4916.068	I
1100	5460.753	I
160	5675.922	I
240	5769.598	I
280	5790.663	I
20	6072.713	I
30	6234.402	I
160	6716.429	I
250	6907.461	I
240	11287.407	I

MERCURY (NATURAL) (Hg)
Z = 80

Hg I and II (nat.)
Ref. 34, 45, 90, 117, 133, 189, 235, 304, 327, 328 — R.W.S.

Intensity	Wavelength	
	Vacuum	
400	893.08	II
300	915.83	II
150	923.39	II
200	940.80	II
100	962.74	II
50	969.13	II
800	1099.26	II
80	1250.58	I
8	1259.24	I
100	1268.82	I
5	1307.75	I
300	1307.93	II
400	1321.71	II
400	1331.74	II
80	1350.07	II
200	1361.27	II
20	1402.62	I
200	1414.43	II
10	1435.51	I
15	1619.46	II
120	1623.95	II
20	1628.25	II
150	1649.94	II
50	1653.64	II
200	1672.41	II
100	1702.73	II
100	1707.40	II
120	1727.18	II
250	1732.14	II
20	1775.68	I
40	1783.70	II
30	1796.22	II
200	1796.90	II
60	1798.74	II
30	1803.89	II
40	1808.29	II
400	1820.34	II
5	1832.74	I
1000	1849.50	I
160	1869.23	II
300	1870.55	II
200	1875.54	II
20	1900.28	II
30	1927.60	II
300	1942.27	II
100	1972.94	II
200	1973.89	II
150	1987.98	II
	Air	
90	2026.97	II
90	2052.93	II
70	2148.00	II
5	2247.55	I
60	2262.23	II
20	2302.06	I
15	2323.20	I
5	2340.57	I
20	2345.43	I
20	2352.48	I
100	2378.32	I
20	2380.00	I
40	2399.38	I
20	2399.73	I
10	2400.49	I
60	2407.35	II
50	2414.13	II
5	2441.06	I
20	2446.90	I
15	2464.06	I
40	2482.00	I
30	2482.72	I
40	2483.82	I
90	2534.77	I
15000	2536.52	I
25	2563.86	I
25	2576.29	I
5	2578.91	I
15	2625.19	I
5	2639.78	I
250	2652.04	I
400	2653.69	I
100	2655.13	I
5	2674.91	I
50	2698.83	I
50	2699.38	I
80	2705.36	II
80	2752.78	I
20	2759.71	I
40	2803.46	I
30	2804.43	I
2	2805.34	I
2	2806.77	I
150	2814.93	II
750	2847.68	II
50	2856.94	I
150	2893.60	I
150	2916.27	II
60	2925.41	I
150	2935.94	II
400	2947.08	II
1200	2967.28	I
300	3021.50	I
120	3023.47	I
30	3025.61	I
50	3027.49	I
400	3125.67	I
320	3131.55	I
320	3131.84	I
400	3208.20	II
400	3264.06	II
80	3341.48	I
100	3385.25	II
400	3451.69	II
200	3549.42	II
2800	3650.15	I
300	3654.84	I
80	3662.88	I
240	3663.28	I
30	3701.44	I
35	3704.17	I
30	3801.66	I
100	3806.38	II
20	3901.87	I
60	3906.37	I
100	3918.92	II
200	3983.96	II
1800	4046.56	I
150	4077.83	I
40	4108.05	I
250	4339.22	I
400	4347.49	I
4000	4358.33	I
100	4398.62	II
90	4660.28	II
80	4855.72	II
5	4883.00	I
5	4889.91	I
80	4916.07	I
5	4970.37	I
5	4980.64	I
20	5102.70	I
40	5120.64	I
100	5128.45	II
20	5137.94	I
20	5290.74	I
5	5316.78	I
60	5354.05	I
30	5384.63	I
1100	5460.74	I
30	5549.63	I
160	5675.86	I
240	5769.60	I
100	5789.66	I
280	5790.66	I
140	5803.78	I
60	5859.25	I
60	5871.73	II
20	5871.98	I
20	6072.72	I
1000	6149.50	II
30	6234.40	I
80	6521.13	II
160	6716.43	I
250	6907.52	I
250	7081.90	I
200	7091.86	I
40	7346.37	II
100	7485.87	II
20	7728.82	I
100	7944.66	II
2000	10139.75	I
240	11287.40	I
120	13209.95	I
140	13426.57	I
60	13468.38	I
80	13505.58	I
500	13570.21	I
450	13673.51	I
200	13950.55	I
500	15295.82	I
100	16881.48	I
400	16920.16	I
300	16942.00	I
500	17072.79	I
400	17109.93	I
20	17116.75	I
20	17198.67	I
20	17213.20	I
70	17329.41	I
30	17436.18	I
50	18130.38	I
40	19700.17	I
	22493.28	I
250	23253.07	I
	32148.06	I
	36303.03	I

Hg III
Ref. 343 — C.H.C.

Intensity	Wavelength	
	Vacuum	
3	621.44	III
2	679.68	III
2	878.59	III
1	886.48	III
1	988.89	III
2	1009.29	III
5	1068.03	III
2	1161.95	III
9	1681.40	III
15	1759.75	III
1	1894.77	III
	Air	
7	2314.15	III
4	2380.55	III
8	2431.65	III
5	2480.56	III
7	2484.50	III
2	2612.92	III
4	2617.97	III
3	2670.49	III
70	2724.43	III
6	2769.22	III
3	2844.76	III
15	3090.05	III
5	3283.02	III
12	3312.28	III
8	3389.01	III
5	3450.77	III
3	3500.35	III
4	3538.88	III
5	3557.24	III
15	3803.51	III
70	4122.07	III
10	4140.34	III
100	4216.74	III
15	4470.58	III
12	4552.84	III
50	4797.01	III
10	4869.85	III
80	4973.57	III
30	5210.82	III
6	5695.71	III
25	6220.35	III
35	6418.98	III
40	6501.38	III
10	6584.26	III
6	6610.12	III
30	6709.29	III
12	7517.46	III
7	7808.10	III
25	7946.75	III
50	7984.51	III
5	8151.64	III

MOLYBDENUM (Mo)
Z = 42

Mo I and II
Ref. 1 — C.H.C.

Intensity		Wavelength	
		AIR	
19000		2015.11	II
40000		2020.30	II
21000		2038.44	II
17000		2045.98	II
4800		2081.68	II
2400		2089.52	II
2200		2092.50	II
4000		2093.11	II
2700		2100.84	II
1500		2104.29	II
1400		2108.02	II
400		2269.69	II
160		2304.25	II
160		2306.97	II
130		2325.94	I
240		2330.46	I
110		2332.12	II
190		2340.47	I
190		2341.59	II
80		2352.61	I
80		2355.22	I
80		2355.42	II
70		2364.37	I
50		2366.09	II
140		2372.27	I
100		2380.41	I
150		2383.52	I
110		2389.20	II
140		2403.61	II
80		2404.66	II
140		2405.86	I
40		2408.39	I
40		2412.84	II
120		2413.01	II
70		2415.33	I
80		2417.96	II
65		2419.01	II
80		2420.18	II
70		2424.00	II
65		2430.43	I
65		2435.96	II
65		2440.28	II
40		2461.81	II
50		2466.68	II
50		2466.97	II
50		2468.78	II
30		2470.04	II
150	h	2471.97	I
70		2477.57	II
70	h	2481.81	I
65		2482.57	II
40		2484.75	II
40	h	2485.31	I
24		2496.24	II
85		2498.28	II
40		2500.44	II
65		2502.84	II
50		2511.80	II
65		2515.08	II
70		2527.14	II
50		2530.34	II
70		2532.31	II
440		2538.46	II
50		2539.44	II
110	h	2540.45	I
330		2542.67	II
40		2543.61	II
330		2548.22	I
110	h	2550.85	I
65		2555.42	II
40		2556.75	II
80		2558.88	II
65		2562.08	II
85		2564.34	II
40		2566.26	II
250		2567.05	I
20		2571.45	II
320		2572.34	I
50		2574.42	II
40		2576.56	II
40		2578.36	II
250		2582.16	I
30		2585.95	II
65		2588.78	II
40		2591.77	II
250		2593.70	II
100		2595.40	I
40		2597.38	II
250		2602.80	II
40		2605.08	II
40		2605.93	II
250		2607.37	I
190		2611.20	I
290		2613.08	I
130		2615.39	I
400		2616.78	I
70		2619.34	II
140		2621.07	I
320		2627.55	I
160		2628.74	I
440		2629.85	I
330		2636.67	II
250		2638.30	I
720		2638.76	II
410		2640.99	I
600		2644.35	II
370		2646.49	II
640		2649.46	I
480		2653.35	II
560	h	2655.03	I
290		2658.11	I
640		2660.58	II
110		2665.10	I
55		2671.83	II
720		2672.84	II
250		2673.27	II
1000		2679.85	I
95		2681.36	II
640		2683.23	II
880		2684.14	II
560		2687.99	II
30		2692.61	II
55		2695.22	II
30		2696.83	II
55		2699.41	II
140		2701.03	I
480		2701.42	II
30		2701.87	II
30		2704.93	II
40		2710.19	II
30		2711.49	II
50		2712.35	II
190		2713.51	II
290		2717.35	II
110		2724.41	I
180		2725.15	I
85		2726.97	II
140		2729.68	II
80		2730.20	II
330		2732.88	II
250		2733.39	I
160		2736.96	II
80	h	2737.88	II
50		2738.60	II
40		2741.32	II
55		2741.62	II
240		2743.07	I
290		2746.30	II
320		2751.47	I
110		2756.07	II
65	d	2758.63	II
20		2760.53	II
190		2761.53	I
220		2763.62	II
110		2766.26	I
240		2769.76	II
160		2773.78	II
190		2774.39	II
1700		2775.40	II
130		2777.74	I
65		2777.86	II
880		2780.04	II
400		2784.99	II
180		2787.83	I
40		2791.54	II
240	d	2797.93	II
220		2801.47	I
400		2807.76	II
28		2812.58	II
24		2814.67	II
1700		2816.15	II
220		2817.44	II
50		2822.03	II
240		2826.54	I
80		2827.74	II
40		2831.44	II
30		2832.07	II
80		2834.39	II
80		2835.33	II
160		2842.15	II
24		2843.73	II
220		2844.39	I
1700		2848.23	II
160		2849.38	I
370		2853.23	II
50		2856.00	II
24		2863.20	II
370		2863.81	II
160		2864.31	I
140		2864.66	I
40		2865.62	II
220		2866.69	II
40		2868.11	II
40		2868.32	II
1700		2871.51	II
85		2872.88	II
220		2879.05	II
65		2888.15	II
95		2891.28	II
1300		2890.99	II
190		2892.81	II
950		2894.45	II
140		2897.63	II
70		2900.80	II
290		2903.07	II
160		2905.27	I
80		2907.12	II
600		2909.12	II
1100		2911.92	II
55		2913.81	II
120		2918.83	II
1300		2923.39	II
140		2924.32	II
65		2927.54	II
50		2930.06	II
1100		2930.50	II
55		2930.77	II
800		2934.30	II
65		2935.20	II
120		2937.66	I
40		2938.30	II
95		2940.10	II
110		2941.22	II
140		2944.21	I
150		2944.82	II
140		2945.66	I
		2945.95	II
190		2946.01	I
140		2946.42	I
140		2946.69	II
95		2947.28	II
95		2955.84	II
240		2956.06	II
70		2956.90	II
95		2960.24	II
140		2962.89	I
250		2963.79	II
50		2964.96	II
210		2965.27	II
70		2971.91	II
250		2972.61	II
80		2975.40	II
180		2978.28	I
120		2981.52	I
110		2987.92	I
160		2988.68	I
190		2989.80	I
95		2992.84	II
50		2993.52	II
190		3002.21	I
40		3004.46	II
130		3013.39	I
140	h	3013.76	I
250		3025.00	I
95		3027.77	II
100		3036.31	I
300		3041.70	I
150		3046.80	I
210		3047.31	I
210		3055.32	I
100		3060.78	II
160		3061.59	I
800		3064.28	I
250		3065.04	II
100		3068.00	I
250		3070.90	I
800		3074.37	I
85		3077.66	II
150		3079.88	I
210		3080.41	I
800		3085.62	I
270		3087.62	II
100		3089.12	I
100		3089.71	I
190		3092.07	II
560		3094.66	I
110		3099.93	I
110		3100.88	I
560		3101.34	I
1400		3112.12	I
290		3122.00	II
14000		3132.59	I
110		3138.72	II
220		3147.35	I
220		3152.82	II
55		3155.64	II
6000		3158.16	I
120		3164.53	I
8700		3170.35	I
95		3172.03	II
160		3172.74	II
370		3183.03	I
120		3184.57	I
370		3185.10	I
180		3185.71	I
120	d	3187.59	II
7600		3193.97	I
290		3195.96	I
120		3198.85	I
40		3201.50	II
330		3205.22	I
880		3205.88	I
3000		3208.83	I
240		3210.97	I
560		3215.07	I
350		3221.74	I
880		3228.22	I
600		3229.79	I
1100		3233.14	I
950		3237.08	I
65		3240.71	II
950		3256.21	I
300		3262.63	I
480		3264.40	I
800		3270.90	I
240		3285.02	I
320		3285.36	I
1100		3289.02	I
950		3290.82	I
190		3292.31	II
320		3305.56	I
320		3307.12	I
100		3313.62	II
190		3320.90	II
640		3323.95	I
360		3325.67	I
360		3327.30	I
240		3340.17	I
1300		3344.75	I
95		3346.40	II
320		3347.02	I
1600		3358.12	I
250		3361.37	I
950		3363.78	I
950		3379.97	I
320		3382.48	I
1900		3384.62	I
130		3395.36	II
640		3404.34	I
1300		3405.94	I
240		3418.52	I
250		3420.04	I
250		3422.31	I
380		3434.79	I
320		3435.45	I
640		3437.22	I
250		3438.87	I
250		3441.44	I
250		3443.26	I
130		3446.08	II
3200		3447.12	I
640		3449.07	I
300		3451.75	I
250		3452.60	I
950		3456.39	I
640		3460.78	I
320		3466.83	I
250		3467.85	I
320		3469.22	I
240		3485.93	I
800		3504.41	I
240		3505.32	I
560		3508.12	I
480		3521.41	I
240		3524.98	I
640		3537.28	I
320		3542.17	I
520		3558.10	I
400		3563.14	I
300		3566.05	I
240		3570.65	I
320		3573.88	I
1400		3581.89	I
200		3590.74	I
210		3598.88	I
270		3602.94	I
210		3608.37	I
200		3623.23	I
1400		3624.46	I
330		3626.18	I
28		3635.14	II
1000		3635.43	I
400		3657.35	I
540		3664.81	I
290		3666.72	I
590		3672.82	I
1300		3680.60	I
45		3684.22	II
65		3688.31	II
240		3690.59	I
180		3692.64	II
1400		3694.94	I
220		3702.03	I
220		3715.65	I
500		3727.69	I
330	d	3732.71	I
240		3742.28	I
80		3744.37	II
360		3770.45	I
220		3779.77	I

Molybdenum (Cont.)

360		3781.59	I
250		3797.30	I
29000		3798.25	I
290		3801.84	I
520		3826.70	I
940		3828.87	I
1700		3833.75	I
380		3847.25	I
29000		3864.11	I
580		3869.08	I
580		3886.82	I
380		3901.77	I
19000		3902.96	I
65		3941.48	II
230		3943.04	I
270		4056.01	I
1400		4062.08	I
2300		4069.88	I
1300		4081.44	I
940		4084.38	I
250		4102.15	I
730		4107.47	I
630		4120.10	I
2900		4143.55	I
230		4148.94	I
250		4155.28	I
200		4157.40	I
200		4178.27	I
480		4185.82	I
2500		4188.32	I
250		4194.56	I
1500		4232.59	I
270		4269.28	I
890		4276.91	I
1200		4277.24	I
1400		4288.64	I
680		4292.13	I
890		4293.21	I
360		4293.88	I
840		4326.14	I
250		4326.74	I
230		4350.34	I
230		4369.04	I
1900		4381.64	I
2500		4411.57	I
210		4423.62	I
990		4434.95	I
200		4442.20	I
340		4449.74	I
480		4457.36	I
630		4474.56	I
230		4491.28	I
120		4504.90	I
140		4512.15	I
230		4517.13	I
230		4524.34	I
120		4529.40	I
400		4536.80	I
110		4558.11	I
210		4576.50	I
170		4595.16	I
360		4609.88	I
100		4621.38	I
460		4626.47	I
100		4627.48	I
220		4662.76	I
130		4671.90	I
130		4688.22	I
640		4707.26	I
150		4708.22	I
220		4717.92	I
100		4729.14	I
700		4731.44	I
100		4750.39	I
770		4760.19	I
150		4776.34	I
100		4796.52	I
410		4819.25	I
410		4830.51	I
360		4868.00	I
110		4950.62	I
150		4957.54	I
210		4979.12	I
110		4999.91	I
20		5010.81	I
180		5014.60	I
26		5016.78	I
20		5019.85	I
80		5029.00	I
65		5030.78	I
23		5038.91	I
26		5046.52	I
100		5047.71	I
50		5055.00	I
35		5058.07	I
200		5059.88	I
35		5062.52	I
29		5064.64	I
35		5079.87	I
100		5080.02	I
35		5081.26	I
40		5090.97	I
35		5091.34	I
35		5092.16	I
40		5095.89	I
100		5096.65	I
130		5097.52	I
35		5098.03	I
130		5109.71	I
80		5114.97	I
35		5116.97	I
29		5123.83	I
150		5145.38	I
110		5147.39	I
80		5163.19	I
100		5167.76	I
160	d	5171.08	I
		5171.25	I
230	h	5172.94	I
160	h	5174.18	I
40		5191.66	I
110		5200.17	I
50		5200.74	I
26		5210.44	I
50		5211.86	I
80		5219.40	I
65		5231.06	I
26		5232.36	I
100		5234.26	I
460	h	5238.20	I
230	h	5240.88	I
110	h	5242.81	I
100		5245.51	I
150		5259.04	I
16		5260.17	I
65		5261.14	I
20		5268.95	I
35		5271.80	I
35		5276.28	I
65		5279.65	I
210		5280.86	I
20		5283.84	I
55		5292.08	I
35		5293.46	I
55		5295.47	I
20		5306.26	I
55		5313.89	I
35		5315.04	I
20		5319.89	I
20		5324.47	I
35		5327.06	I
20		5352.35	I
80		5354.88	I
35		5355.51	I
65		5356.48	I
560	H1	5360.56	I
110	H1	5364.28	I
35	H1	5367.11	I
35		5372.40	I
26		5388.69	I
65		5394.52	I
35		5397.38	I
50		5400.47	I
35		5405.79	I
35		5406.39	I
40		5417.38	I
23		5426.89	I
55		5435.68	I
65		5437.75	I
40		5450.51	I
35		5456.46	I
26		5460.53	I
23		5465.57	I
35		5475.90	I
35		5490.28	I
20		5492.17	I
26	h	5493.80	I
26		5498.49	I
50		5501.54	I
23		5501.87	I
26	h	5503.54	I
7800		5506.49	I
23		5520.04	I
26		5520.64	I
40		5526.52	I
40		5526.97	I
5200		5533.05	I
40		5539.41	I
50		5543.12	I
40		5544.49	I
55		5556.28	I
26		5556.72	I
20		5564.05	I
40		5568.62	I
26		5569.48	I
2500		5570.45	I
35		5575.19	I
20		5591.58	I
40		5602.76	I
23		5608.62	I
23		5609.23	I
100		5610.93	I
23		5613.07	I
20		5618.45	I
23		5619.38	I
330		5632.47	I
50		5634.86	I
230		5650.13	I
23		5673.63	I
55		5674.47	I
40		5677.89	I
35		5682.89	I
460		5689.14	I
23		5699.28	I
80		5705.72	I
23		5711.80	I
210		5722.74	I
23		5728.77	I
26	d	5729.45	I
		5729.59	I
620		5751.40	I
23		5774.55	I
40		5779.36	I
23	h	5783.33	I
520		5791.85	I
23	h	5795.77	I
26		5800.46	I
35		5802.67	I
23		5825.20	I
23		5835.59	I
20		5839.99	I
20	h	5848.86	I
55	h	5849.73	I
50	h	5851.52	I
520		5858.27	I
20		5861.38	I
50		5869.33	I
26		5876.59	I
820		5888.33	I
23		5892.29	I
50	h	5893.38	I
20		5898.78	I
		5898.82	MO I
40		5901.47	I
40	h	5926.36	I
160	h	5928.88	I
40		5988.17	I
35		6025.49	I
16		6027.27	I
1300		6030.66	I
20		6047.83	I
20		6054.81	I
20		6079.58	I
10		6081.27	I
40		6101.87	I
10		6130.63	I
10		6197.66	I
20		6217.09	I
10		6264.27	I
16		6265.88	I
15		6290.74	I
13		6301.75	I
11		6323.54	I
40		6357.22	I
16		6389.11	I
11		6391.12	I
35		6401.07	I
26		6409.11	I
10		6412.39	I
100		6424.37	I
20		6446.34	I
20		6471.20	I
20		6473.99	I
10		6493.13	I
23		6519.84	I
15	h	6611.20	I
230		6619.13	I
10		6624.57	I
50		6650.38	I
13		6659.68	I
18		6690.47	I
110		6733.98	I
21		6746.08	I
50		6746.27	I
35		6753.97	I
13		6763.50	I
10	h	6799.88	I
10		6802.62	I
10		6812.03	I
13		6825.63	I
18	d	6828.87	I
		6829.05	I
40		6838.88	I
16		6848.92	I
21		6886.28	I
16		6892.36	I
10		6898.01	I
10		6898.98	I
13		6908.20	I
35		6914.01	I
13		6934.10	I
10		6947.39	I
10		6960.64	I
16		6978.71	I
26		6988.94	I
12		6999.13	I
16		7001.60	I
22		7037.98	I
22		7060.21	I
13		7063.34	I
13		7081.22	I
110		7109.87	I
27		7134.08	I
150		7242.50	I
40		7245.85	I
22		7267.62	I
17		7300.19	I
13		7348.49	I
13		7361.65	I
10	h	7364.41	I
40		7391.36	I
10		7434.10	I
13		7447.34	I
13	h	7452.85	I
140		7485.74	I
13		7504.47	I
11		7572.64	I
11	h	7595.16	I
11		7601.84	I
17	h	7656.76	I
13		7679.49	I
27		7720.77	I
17		7829.65	I
15		7854.45	I
11		7923.15	I
15		7986.60	I
22	h	8245.06	I
40	h	8328.44	I
45	h	8389.32	I
45	h	8483.39	I

Mo III
Ref. 420 — C.H.C.

Intensity	Wavelength	
	Vacuum	
50	1166.07	III
100	1169.33	III
50	1173.67	III
50	1209.60	III
50	1225.46	III
50	1230.34	III
50	1234.63	III
30	1236.10	III
100	1254.93	III
100	1262.21	III
100	1263.74	III
100	1274.37	III
50	1274.94	III
100	1276.40	III
200	1277.40	III
200	1277.58	III
100	1278.06	III
200	1278.40	III
150	1281.90	III
150	1283.60	III
100	1258.52	III
100	1286.42	III
50	1288.07	III
50	1288.25	III
100	1290.49	III
100	1299.82	III
50	1305.58	III
50	1437.37	III
50	1452.38	III
50	1534.86	III
50	1751.22	III
50	1760.57	III
50	1807.60	III
100	1854.73	III
	Air	
75	2165.19	III
75	2170.57	III
50	2172.46	III
75	2179.37	III
100	2184.37	III
100	2211.02	III
50	2223.19	III
100	2253.18	III
150	2269.71	III
50	2275.47	III
50	2275.64	III
200	2294.97	III
80	2304.26	III
50	2326.75	III
150	2330.93	III
100	2359.76	III
80	2386.96	III
50	2403.61	III
70	2412.71	III
50	2422.18	III
200	2506.19	III
90	2597.13	III
50	2756.06	III
100	2807.74	III
125	2947.32	III

Molybdenum (Cont.)

75	2983.94	III
80	3254.70	III
200	3271.69	III

Mo IV
Ref. 383 — C.H.C.

Intensity	Wavelength	
	Vacuum	
10	857.75	IV
10	859.72	IV
25	865.24	IV
20	865.53	IV
20	863.63	IV
50	867.92	IV
10	878.43	IV
100	884.19	IV
10	884.82	IV
60	886.05	IV
50	891.74	IV
40	894.80	IV
15	895.41	IV
20	1819.50	IV
30	1821.59	IV
30	1850.69	IV
20	1877.88	IV
80	1926.26	IV
100	1929.24	IV
20	1949.44	IV
80	1971.06	IV
20	1991.41	IV
	Air	
70	2010.92	IV
20	2023.78	IV
25	2055.64	IV
50	2060.38	IV
15	2091.89	IV
10	2113.78	IV
5	2140.33	IV

NEODYMIUM (Nd)
Z = 60

Nd I and II
Ref. 1 — C.H.C.

Intensity		Wavelength	
		Air	
75		2702.46	
75		2704.54	
75		2764.98	I
60		2785.79	I
50		2863.95	
50		2921.26	
55		2962.88	II
65		2963.58	II
80		2993.20	II
40		2994.73	
95		3007.97	II
95		3014.19	II
95		3018.35	II
80		3026.47	II
50		3038.98	II
50		3043.29	
50		3051.11	II
80		3052.15	II
140		3056.71	II
130		3069.73	II
65	d	3071.43	II
		3071.50	II
160		3075.38	II
95		3079.38	II
95		3080.94	II
95		3092.73	
240		3092.92	II
140		3098.48	II
55		3099.52	II
130		3105.43	II
95		3106.18	II
65		3108.01	II
260		3115.18	II
190		3116.15	II
50		3119.75	II
160		3123.06	II
190		3124.58	II
290		3133.60	II
220		3134.90	II
100		3137.24	II
170		3141.46	II
170		3142.44	II
100		3144.55	II
100		3144.82	II
100		3148.51	II
100		3149.29	II
100		3149.51	II
100		3162.62	II
100		3175.99	II
50		3181.54	II
50		3188.73	II
50		3200.62	II
150		3203.47	II
85		3211.00	II
100		3217.12	II
50		3222.62	I
50		3228.04	II
60		3234.62	
40		3237.91	II
100		3254.08	II
50		3256.91	II
220		3259.24	II
100		3260.66	II
220		3265.12	II
50		3265.38	II
170		3267.25	II
100		3273.18	II
320		3275.22	II
50		3281.49	II
50		3282.78	II
290		3285.10	II
100		3286.62	II
50		3289.52	
100		3290.65	II
70		3293.84	II
70		3294.68	II
70		3298.61	
300		3300.16	II
200		3312.75	II
200		3325.90	II
410		3328.28	II
250		3331.57	II
290		3334.48	II
290		3339.07	II
320		3353.59	II
200		3355.93	II
270		3364.96	II
290		3393.63	II
120	h	3484.88	I
200		3527.53	II
290		3543.35	II
200		3555.77	II
410		3560.75	II
340		3568.87	II
470		3587.51	II
300		3592.59	II
340		3598.02	II
300		3600.91	II
320		3609.79	II
370		3615.82	II
300		3618.96	II
300		3631.02	II
340		3634.30	II
240		3637.00	II
240		3637.23	II
240		3640.24	II
240		3645.78	II
340		3648.20	II
240		3649.46	
240		3650.42	II
410		3653.15	II
240		3654.16	
470		3662.26	II
540		3665.18	II
540		3672.36	II
580		3673.54	II
240		3678.18	II
1200		3685.80	II
440		3687.30	II
410		3689.69	II
300		3694.81	II
410		3697.56	II
240		3702.84	
240		3704.95	II
200		3712.81	
470		3713.70	II
370		3714.20	II
640	d	3714.73	II
250		3715.04	II
200		3715.39	II
470		3715.68	II
410		3718.54	II
410		3721.35	II
220		3722.42	II
780		3723.50	II
410		3724.87	II
250		3726.90	II
710		3728.13	II
470		3730.58	II
270		3732.78	II
1000	d	3735.54	II
		3735.60	II
440		3737.10	II
1000		3738.06	II
270		3741.42	II
200		3749.85	II
320		3750.31	II
580		3752.49	II
370		3752.67	II
250		3754.83	II
370		3755.60	II
510		3757.82	II
930		3758.95	II
300		3759.79	II
930		3763.47	II
300		3766.59	II
510		3769.65	II
1400		3775.50	II
250		3776.34	II
710		3779.47	II
580		3780.40	II
510		3781.32	II
300		3783.78	II
2400		3784.25	II
270		3784.73	II
340		3791.50	II
340		3795.45	II
240		3799.55	II
370		3801.12	II
200		3801.38	II
340		3802.30	II
1200		3803.47	II
200		3804.10	II
2500		3805.36	II
340		3805.55	II
470		3807.23	II
540		3808.77	II
440		3809.06	II
580		3810.49	II
240		3811.06	II
270		3811.77	II
200		3812.53	II
710		3814.73	II
240		3819.70	II
410		3822.47	II
1200		3826.42	II
240		3828.00	II
540		3828.85	II
440		3829.16	II
510		3830.47	II
740		3836.54	II
340		3837.91	II
1700		3838.98	II
340		3839.51	II
410	d	3841.82	II
		3841.88	II
1700	d	3848.24	II
		3848.31	II
1500		3848.52	II
470		3850.22	II
2400	d	3851.66	II
		3851.74	II
340		3858.55	II
270		3860.94	II
300		3862.52	II
3700	d	3863.33	II
		3863.40	II
240		3866.52	II
220		3866.81	II
850		3869.07	II
240		3875.74	II
470		3875.87	II
1100		3878.58	II
1000		3879.55	II
780		3880.38	II
1200		3880.78	II
200		3881.59	
540		3887.87	II
370	h	3889.66	II
1300		3889.93	II
1300		3890.58	II
1300		3890.94	II
580		3891.51	II
470		3892.06	II
810		3894.63	II
270		3896.13	II
440		3897.63	II
2000		3900.21	II
1300		3901.84	II
1700		3905.89	II
200		3907.70	II
510		3907.84	II
2000		3911.16	II
850		3912.23	II
340		3913.69	II
440		3915.13	II
610		3915.95	II
340		3917.65	II
220		3919.92	II
1100		3920.96	II
510		3927.10	II
200		3929.26	II
610		3934.82	II
410		3936.11	II
510		3938.86	II
2000		3941.51	II
2000		3951.16	II
810		3952.20	II
320		3952.87	II
320		3953.52	II
240		3957.45	II
590		3958.00	II
510		3962.21	II
1400		3963.12	II
270		3963.90	II
1100		3973.30	II
740		3973.69	II
740		3976.85	II
740		3979.49	II
320		3982.36	II
470		3986.25	II
1400		3990.10	II
1000		3991.74	II
1100		3994.68	II
410		4000.50	II
540		4004.02	II
410		4007.43	II
3700		4012.25	II
540		4012.70	II
370		4018.81	II
1000		4020.87	II
1000		4021.34	II
1000		4021.78	II
1200		4023.00	II
340		4024.78	II
410		4030.47	II
1200		4031.82	II
270		4038.12	II
3000		4040.80	II
200		4041.06	II
410		4043.59	II
410		4048.81	II
850		4051.15	II
850		4059.96	II
4700		4061.09	II
1100		4069.28	II
710		4075.12	II
470		4075.28	II
240		4077.62	II
470		4080.23	II
240		4085.82	II
270		4096.13	II
220		4098.18	II
200		4106.59	II
1400		4109.08	II
2500		4109.46	II
510		4110.48	II
300	h	4113.83	II
410		4123.88	II
470		4133.36	II
510		4135.33	II
3000		4156.08	II
510		4156.26	II
340		4160.57	II
410		4168.00	II
810		4175.61	II
2400		4177.32	II
200		4178.64	II
640		4179.59	II
250		4184.98	II
470		4205.60	II
470		4211.29	II
290		4220.25	II
440		4227.73	II
1300		4232.38	II
250		4234.19	II
290	h	4235.24	II
290		4239.84	II
2000		4247.38	II
850		4252.44	II
290		4254.29	
410		4261.84	II
340		4266.71	II
240		4270.56	II
340		4272.79	II
340		4275.09	II
470		4282.44	II
240		4282.57	II
710		4284.52	II
270		4297.80	II
5400		4303.58	II
340		4304.45	II
200		4307.78	II
470		4314.52	II
1100		4325.76	II
510		4327.93	II
540		4338.70	II
680		4351.29	II
850		4358.17	II
240		4366.38	II
340		4368.64	II
470	d	4374.93	II
		4375.04	II
710		4385.66	II
250		4390.66	II
540		4400.83	II
510		4411.06	II
580		4446.39	II
1400		4451.57	II
200		4451.99	II
300		4456.40	II
740		4462.99	II
410		4501.82	II
200		4506.59	II
170		4513.34	II
250		4516.36	II

Neodymium (Cont.)

120		4527.25	I
340		4541.27	II
340		4542.61	II
100		4556.14	II
170		4559.67	I
340		4563.22	II
200		4578.89	II
200		4579.32	II
100		4586.62	I
200		4597.02	II
100		4603.82	I
100		4609.87	I
300		4621.94	I
100		4627.98	I
510		4634.24	I
340		4641.10	I
250		4645.77	II
200		[illegible]	I
300		4649.67	I
200		4654.73	I
130		4670.56	II
170		4680.74	II
310		4683.45	I
110		4684.04	I
110		4690.35	I
190		4696.44	I
130		4703.57	II
470		4706.54	II
140		4706.96	I
190		4709.71	II
190		4715.59	II
240		4719.02	I
190		4724.35	II
140		4731.77	I
120		4779.46	I
170		4789.41	II
120		4797.15	II
240		4811.34	II
140		4820.34	II
350		4825.48	II
130		4832.28	II
110		4849.06	II
280		4859.02	II
190		4866.74	I
350		4883.81	I
140		4889.10	II
220		4890.70	II
240		4891.07	I
280		4896.93	I
120		4901.53	I
210		4901.84	I
110		4902.03	II
190		4913.41	I
170		4914.37	II
330		4920.68	II
470		4924.53	I
260		4944.83	I
290		4954.78	I
290		4959.13	II
150		4961.39	II
250		4989.94	II
150		5033.52	II
110		5063.73	II
360		5076.59	II
150	h	5089.84	II
360		5092.80	II
180		5102.39	II
150	d	5105.21	II
		5105.35	I
360		5107.59	II
340		5123.79	II
680		5130.60	II
170		5132.33	II
170		5165.14	II
130		5181.17	II
120		5182.60	II
500		5191.45	II
630		5192.62	II
330		5200.12	II
310		5212.37	II
150		5213.23	I
130		5225.05	II
130		5228.43	II
450		5234.20	II
250		5239.79	II
720		5249.59	II
200		5250.82	II
360		5255.51	II
120		5269.48	II
590		5273.43	II
150		5276.88	II
110		5291.67	I
680		5293.17	II
160		5302.28	II
110		5306.47	II
220		5311.46	II
500		5319.82	II
180		5356.98	II
290		5361.47	II
150		5371.94	II
110		5385.90	II
160		5431.53	II
110		5451.12	II
170		5485.70	II
35		5501.47	I
45		5525.72	I
90		5533.82	I
55		5535.27	II
55		5543.24	I
55		5548.47	II
55		5561.17	I
27		5575.50	I
27		5576.70	I
27		5577.70	I
27		5587.61	I
240		5594.43	II
55		5601.43	I
45		5601.92	I
220		5620.54	I
65		5635.76	I
45		[illegible]	I
35		5653.57	I
70		5668.87	II
65		5669.77	I
140	d	5675.97	I
55		5676.33	I
220		5688.53	II
23		5689.51	I
30		5701.57	I
130		5702.24	II
80		5706.21	II
160		5708.28	II
80		5718.12	II
65		5726.83	II
100		5729.29	I
23		5734.55	
70		5740.86	II
55		5749.19	I
27		5749.66	I
23		5767.33	I
45		5776.12	I
45		5784.96	I
45		5788.22	I
45		5800.09	I
160		5804.02	II
80		5811.57	II
45		5813.89	I
27		5820.37	I
70		5825.87	II
30		5826.74	I
80		5842.39	II
30		5844.66	I
23		5845.95	I
55		5858.91	I
35		5867.08	I
30		5868.90	I
27		5871.04	I
30		5883.29	I
23		5886.24	I
30		5887.91	I
27		5921.22	I
27		5955.87	I
30		5994.76	I
27		5996.47	I
45		6007.67	I
35		6031.27	II
27		6033.29	I
45		6034.24	II
55		6066.03	I
27		6071.70	I
30		6073.97	I
23	d	6133.47	II
27		6149.28	I
27		6155.06	I
35		6157.83	II
23		6166.67	II
35		6170.49	II
45		6178.59	I
27		6183.91	II
27		6208.24	I
45		6223.39	I
27		6226.50	I
23		6238.50	II
35		6244.08	I
23		6257.49	I
27		6258.73	II
23		6277.29	II
27		6285.79	I
23		6292.84	II
23		6297.07	I
55		6310.49	I
27		6341.51	II
23		6382.07	II
65		6385.20	II
35		6485.69	I
45		6630.14	I
35		6637.96	II
45		6650.57	II
30		6655.67	I
25		6737.79	II
40		6740.11	II
25		6742.54	I
30		6790.37	II
30		6804.00	II
25		6846.72	II
40		6900.43	II
24		6941.39	II
17	h	7010.80	II
8		7018.85	II
17		7020.92	II
17		7024.58	II
10		7033.21	II
35		7037.30	II
7		7052.14	II
7		7054.74	II
7		7061.47	II
40		7066.89	II
8		7082.93	II
12	h	7089.71	II
12	h	7092.09	
12	h	7092.74	II
12	h	7092.94	
17	h	7093.98	I
20	h	7095.42	I
29		7129.35	II
12	h	7142.04	II
10		7143.72	
8		7151.03	II
6		7153.09	I
6	h	7185.01	
10		7189.09	II
24		7189.42	II
20		7192.01	II
10		7199.00	II
8	h	7227.01	I
15		7236.54	II
7	h	7261.64	II
9		7285.29	II
9		7288.56	II
6		7291.38	II
7		7298.72	II
12		7316.81	II
7	h	7321.43	I
7		7323.12	II
6		7334.54	I
6		7357.10	I
6		7374.04	II
7		7381.79	II
9		7401.31	I
10		7406.62	II
6		7411.20	II
10		7418.18	II
9		7427.41	II
9		7448.71	II
5		7481.28	II
12		7511.16	II
17		7513.73	II
7	h	7514.44	II
7		7516.02	II
9		7526.45	II
12		7528.99	II
10		7538.26	II
5		7540.97	II
7		7547.00	II
5		7577.54	II
7		7587.65	II
6		7590.75	II
6		7603.73	II
6		7605.92	II
5		7614.72	I
9		7639.79	II
8		7646.00	II
6		7663.52	II
12		7696.56	II
6		7718.20	II
4		7743.90	II
4		7748.92	II
10		7750.95	II
6		7773.06	II
7		7792.22	II
6		7796.40	II
8		7797.32	II
5		7798.32	II
10		7808.47	II
7		7818.83	II
5		7825.20	II
12		7863.04	II
5	h	7872.03	I
7		7886.60	II
4	h	7896.50	II
9		7900.40	II
5	h	7906.03	I
12		7917.01	II
10		7925.03	II
5		7947.93	II
10		7949.68	II
5	h	7955.38	II
12		7958.95	I
12		7965.73	II
15		7982.09	II
12		7982.68	II
12		8000.76	II
9		8007.70	I
4	h	8020.07	
8		8026.35	
10		8043.24	I
8		8051.33	II
5		8064.00	II
10		8099.17	I
10		8120.93	II
12		8122.07	II
12		8141.75	II
12		8143.27	II
7	h	8164.97	I
8		8172.56	II
9		8179.83	II
9		8182.41	II
4		8185.58	II
7	h	8205.38	II
10		8231.52	II
4		8248.76	II
5	h	8249.68	II
4	h	8262.80	II
7	h	8266.72	II
4		8272.79	II
6	h	8302.76	II
10		8307.72	II
6		8324.50	II
4		8332.01	II
12		8346.36	II
4		8375.16	II
4		8375.33	II
4		8394.71	II
7		8400.85	II
5	h	8456.87	II
4		8530.53	II
5		8582.03	II
5		8591.53	II
7		8594.87	II
8	c	8643.43	II
5		8667.07	II
5		8677.48	
6		8691.29	II
6		8695.07	II
6		8712.82	II
6		8715.03	I
17		8839.10	II

NEON (Ne)
Z = 10

Ne I and II
Ref. 56, 58, 118, 150, 230 — S.P.D.

Intensity		Wavelength	
		Vacuum	
90		352.956	II
60		354.962	II
90		361.433	II
60		362.455	II
150		405.854	II
120		407.138	II
200		445.040	II
300		446.256	II
250		446.590	II
180		447.815	II
150		454.654	II
200		455.274	II
10		456.275	II
120		456.348	II
90		456.896	II
1000		460.728	II
500		462.391	II
35		587.213	I
35		589.179	I
35		589.911	I
70		591.830	I
100		595.920	I
75		598.706	I
35		598.891	I
70		600.036	I
170		602.726	I
170		615.628	I
170		618.672	I
120		619.102	I
200		626.823	I
200		629.739	I
1000		735.896	I
400		743.720	I
60		993.88	II
70		1068.65	II
90		1131.72	II
100		1131.85	II
90		1229.83	II
90		1418.38	II
90		1428.58	II
90		1436.09	II
120		1681.68	II
180		1688.36	II
100		1888.11	II
100		1889.71	II
200		1907.49	II
500		1916.08	II
300		1930.03	II
200		1938.83	II
100	c	1945.46	II
		AIR	

Neon (Cont.)

80		2007.01	II
80		2025.56	II
150		2085.47	II
180		2096.11	II
120		2096.25	II
80	p	2562.12	II
90	w	2567.12	II
80		2623.11	II
80		2629.89	II
90	w	2636.07	II
80		2638.29	II
80		2644.10	II
80		2762.92	II
90		2792.02	II
80		2794.22	II
100		2809.48	II
80		2906.59	II
80		2906.82	II
80		2910.06	II
90		2910.41	II
80		2911.14	II
80		2915.12	II
80		2925.62	II
80	w	2932.10	II
80		2940.65	II
90		2946.04	II
150		2955.72	II
150		2963.24	II
150		2967.18	II
100		2973.10	II
15		2974.72	I
100		2979.46	II
12		2982.67	I
150		3001.67	II
120	p	3017.31	II
300		3027.02	II
300		3028.86	II
100		3030.79	II
120		3034.46	II
100		3035.92	II
100		3037.72	II
100		3039.59	II
100		3044.09	II
100		3045.56	II
120		3047.56	II
100		3054.34	II
100		3054.68	II
100		3059.11	II
100		3062.49	II
100		3063.30	II
100		3070.89	II
100		3071.53	II
100		3075.73	II
120		3088.17	II
100		3092.09	II
120		3092.90	II
100		3094.01	II
100		3095.10	II
100		3097.13	II
100		3117.98	II
120		3118.16	II
10		3126.199	I
300		3141.33	II
100		3143.72	II
100	p	3148.68	II
100		3164.43	II
100		3165.65	II
100		3188.74	II
120		3194.58	II
500		3198.59	II
60		3208.96	II
120		3209.36	II
120		3213.74	II
150		3214.33	II
150		3218.19	II
120		3224.82	II
120		3229.57	II
200		3230.07	II
120		3230.42	II
120		3232.02	II
150		3232.37	II
100		3243.40	II
100		3244.10	II
100		3248.34	II
100		3250.36	II
150		3297.73	II
150		3309.74	II
300		3319.72	II
1000		3323.74	II
150		3327.15	II
100		3329.16	II
200		3334.84	II
150		3344.40	II
300		3345.45	II
150		3345.83	II
200		3355.02	II
120		3357.82	II
200		3360.60	II
120		3362.16	II
100		3362.71	II
120		3367.22	II
12		3369.808	I
40		3369.908	I
100		3371.80	II
500		3378.22	II
150		3388.42	II
120		3388.94	II
300		3392.80	II
100		3404.82	II
120		3406.95	II
100		3413.15	II
120		3416.91	II
120		3417.69	II
50		3417.904	I
15		3418.006	I
120		3428.69	II
60		3447.703	I
50		3454.195	I
100		3456.61	II
100		3459.32	II
25		3460.524	I
30		3464.339	I
30		3466.579	I
60		3472.571	I
150		3479.52	II
200		3480.72	II
200		3481.93	II
25		3498.064	I
30		3501.216	I
25		3515.191	I
150		3520.472	I
120		3542.85	II
120		3557.80	II
100		3561.20	II
250		3568.50	II
100		3574.18	II
200		3574.61	II
50		3593.526	I
30		3593.640	I
15		3600.169	I
20		3633.665	I
150		3643.93	II
200		3664.07	II
20		3682.243	I
12		3685.736	I
200		3694.21	II
10		3701.225	I
150		3709.62	II
250		3713.08	II
250		3727.11	II
800		3766.26	II
1000		3777.13	II
100		3818.43	II
120		3829.75	II
150		4219.74	II
100		4233.85	II
120		4250.65	II
120		4369.86	II
70		4379.40	II
150		4379.55	II
100		4385.06	II
200		4391.99	II
150		4397.99	II
150		4409.30	II
100		4413.22	II
100		4421.39	II
100	p	4428.52	II
100	p	4428.63	II
150	p	4430.90	II
150	p	4430.94	II
120		4457.05	II
100		4522.72	II
10		4537.754	I
10		4540.380	I
100		4569.06	II
15		4704.395	I
12		4708.862	I
10		4710.067	I
10		4712.066	I
15		4715.347	I
10		4752.732	I
12		4788.927	I
10		4790.22	I
10		4827.344	I
10		4884.917	I
4		5005.159	I
10		5037.751	I
10		5144.938	I
25		5330.778	I
20		5341.094	I
8		5343.283	I
60		5400.562	I
5		5562.766	I
10		5656.659	I
5		5719.225	I
12		5748.298	I
80		5764.419	I
12		5804.450	I
40		5820.156	I
500		5852.488	I
100		5872.828	I
100		5881.895	I
60		5902.462	I
60		5906.429	I
100		5944.834	I
100		5965.471	I
100		5974.627	I
120		5975.534	I
80		5987.907	I
100		6029.997	I
100		6074.338	I
80		6096.163	I
60		6128.450	I
100		6143.063	I
120		6163.594	I
250		6182.146	I
150		6217.281	I
150		6266.495	I
60		6304.789	I
7		6328.165	I
100		6334.428	I
120		6382.992	I
200		6402.246	I
150		6506.528	I
60		6532.882	I
150		6598.953	I
70		6652.093	I
90		6678.276	I
20		6717.043	I
100		6929.467	I
90		7024.050	I
100		7032.413	I
50		7051.292	I
80		7059.107	I
100		7173.938	I
150		7213.20	II
150		7235.19	II
100		7245.167	I
150		7343.94	II
40		7472.439	I
90		7488.871	I
100		7492.10	II
150		7522.82	II
80		7535.774	I
60		7544.044	I
100		7724.628	I
120		7740.74	II
300		7839.055	I
120		7926.20	II
400		7927.118	I
700		7936.996	I
2000		7943.181	I
2000		8082.458	I
100		8084.34	II
1000		8118.549	I
600		8128.911	I
3000		8136.406	I
2500		8259.379	I
100		8264.81	II
2500		8266.077	I
800		8267.117	I
6000		8300.326	I
100		8315.00	II
1500		8365.749	I
100		8372.11	II
8000		8377.606	I
1000		8417.159	I
4000		8418.427	I
1500		8463.358	I
800		8484.444	I
5000		8495.360	I
600		8544.696	I
1000		8571.352	I
4000		8591.259	I
6000		8634.647	I
3000		8647.041	I
15000		8654.383	I
4000		8655.522	I
100		8668.26	II
5000		8679.492	I
5000		8681.921	I
2000		8704.112	I
4000		8771.656	I
12000		8780.621	I
10000		8783.753	I
500		8830.907	I
7000		8853.867	I
1000		8865.306	I
1000		8865.755	I
3000		8919.501	I
2000		8988.57	I
100		9079.46	II
6000		9148.67	I
6000		9201.76	I
4000		9220.06	I
2000		9221.58	I
2000		9226.69	I
1000		9275.52	I
200		9287.56	II
6000		9300.85	I
1500		9310.58	I
3000		9313.97	I
6000		9326.51	I
2000		9373.31	I
5000		9425.38	I
3000		9459.21	I
5000		9486.68	I
5000		9534.16	I
3000		9547.40	I
120		9577.01	II
1000		9665.42	I
100		9808.86	II
800		10295.42	I
2000		10562.41	I
1500		10798.07	I
2000		10844.48	I
3000		11143.020	I
3500		11177.528	I
1600		11390.434	I
1100		11409.134	I
3000		11522.746	I
1500		11525.020	I
950		11536.344	I
500		11601.537	I
1200		11614.081	I
300		11688.002	I
2000		11766.792	I
1500		11789.044	I
500		11789.889	I
1000		11984.912	I
3000		12066.334	I
800		12459.389	I
1000		12689.201	I
1100		12912.014	I
700		13219.241	I
800		15230.714	I
400		17161.930	I
400		18035.80	I
1000		18083.21	I
350		18221.11	I
250		18227.02	I
2500		18276.68	I
2000		18282.62	I
1200		18303.97	I
250		18359.12	I
1200		18384.85	I
2000		18389.95	I
1000		18402.84	I
1200		18422.39	I
300		18458.65	I
400		18475.79	I
900		18591.55	I
1600		18597.70	I
350		18618.96	I
550		18625.16	I
1200		21041.295	I
750		21708.145	I
300		22247.35	I
350		22428.13	I
2250		22530.40	I
400		22661.81	I
600		23100.51	I
1000		23260.30	I
1050		23373.00	I
850		23565.36	I
3500		23636.52	I
300		23701.64	I
1100		23709.2	I
1800		23951.42	I
600		23956.46	I
1000		23978.12	I
200		24098.54	I
500		24161.42	I
600		24249.64	I
1500		24365.05	I
800		24371.60	I
400		24447.85	I
700		24459.4	I
300		24776.46	I
550		24928.88	I
250		25161.69	I
650		25524.37	I
125		28386.21	I
150		30200.	I
250		33173.09	I
450		33352.35	I
1300		33901.	I
2200		33912.10	I
600		34131.31	I
100		34471.44	I
120		35834.78	I

Ne III
Ref. 365, 371, 402 — R.L.K.

Intensity	Wavelength	
	Vacuum	
20	251.14	III
20	251.56	III
20	251.73	III
40	267.06	III
40	267.52	III
20	267.71	III
40	283.18	III
160	283.21	III
110	283.69	III
40	283.89	III
220	301.12	III
220	313.05	III
220	313.68	III
40	313.95	III

Neon (Cont.)

Intensity	Wavelength	
220	379.31	III
285	488.10	III
220	488.87	III
450	489.50	III
70	489.64	III
220	490.31	III
360	491.05	III
20	1255.03	III
110	1255.68	III
160	1257.19	III

Air

200	2086.96	III
300	2089.43	III
240	2092.44	III
400	2095.54	III
200	2161.22	III
300	2163.77	III
200	2180.89	III
200	2209.35	III
200	2211.85	III
240	2213.76	III
300	2216.07	III
240	2263.21	III
200	2264.91	III
300	2412.73	III
240	2412.94	III
200	2413.78	III
200	2473.40	III
800	2590.04	III
600	2593.60	III
400	2595.68	III
300	2610.03	III
240	2613.41	III
200	2615.87	III
200	2638.70	III
200	2641.07	III
600	2677.90	III
500	2678.64	III

Ne IV
Ref. 69, 364, 388, 400, 413, 430, — R.L.K.

Vacuum

Intensity	Wavelength	
15	151.82	IV
15	152.22	IV
15	158.65	IV
15	158.82	IV
80	172.62	IV
80	177.16	IV
150	186.58	IV
100	194.28	IV
100	208.48	IV
100	208.73	IV
80	208.90	IV
150	212.56	IV
140	223.24	IV
120	223.60	IV
140	234.32	IV
120	234.70	IV
50	357.83	IV
200	358.72	IV
125	387.14	IV
100	388.22	IV
150	421.61	IV
140	469.77	IV
200	469.82	IV
180	469.87	IV
140	469.92	IV
120	521.74	IV
140	521.82	IV
80	541.13	IV
100	542.07	IV
150	543.89	IV

Air

65	2018.44	IV
110	2022.19	IV
30	2203.88	IV
10	2220.81	IV
250	2258.02	IV
175	2262.08	IV
110	2264.54	IV
550	2285.79	IV
30	2293.14	IV
250	2293.49	IV
250	2363.28	IV
110	2365.49	IV
250	2350.84	IV
450	2352.52	IV
700	2357.96	IV
250	2362.68	IV
350	2372.16	IV
65	2384.20	IV
350	2384.95	IV

Ne V
Ref. 69, 388, 389, 400, 413 — R.L.K.

Vacuum

Intensity	Wavelength	
66	119.01	V
200	122.52	V
66	125.12	V
45	131.99	V
50	132.04	V
150	140.76	V
150	140.79	V
100	142.44	V
100	142.50	V
150	142.72	V
100	143.27	V
150	143.34	V
150	147.13	V
66	151.23	V
120	151.42	V
45	154.50	V
100	164.02	V
100	164.14	V
500	173.93	V
400	357.96	V
500	358.47	V
500	359.38	V
1000	365.59	V
800	416.20	V
250	480.41	V
150	481.28	V
250	481.36	V
500	482.99	V
400	568.42	V
250	569.76	V
500	569.83	V
250	572.11	V
800	572.34	V

Air

75	2227.42	V
110	2232.41	V
65	2245.48	V
65	2259.57	V
65	2263.39	V
250	2265.71	V

NEPTUNIUM (Np)
Z = 93

Np I and II
Ref. 93 = J.G.C.

Air

Intensity		Wavelength	
300		3481.93	I
300	h	3501.50	I
300	l	3986.89	I
300	s	5044.66	I
300	l	5601.70	I
300	l	5652.75	I
300	l	5784.39	I
300	l	5878.04	I
300	s	6011.22	I
300		6056.09	I
300	s	6073.90	I
300	s	6080.05	I
300	l	6120.49	I
300		6188.59	I
300	l	6200.00	I
300	s	6215.90	I
300	s	6317.84	I
300	l	6341.38	I
300	l	6566.11	I
300	l	6720.68	I
300	s	6751.32	I
300	s	6795.21	I
300	l	6802.62	I
300	l	6805.81	I
300	s	6816.44	I
300	l	6865.45	I
300	s	6907.13	I
300	h	6912.91	I
1000	s	6930.31	I
300	l	6963.63	I
3000	s	6972.09	I
300		7014.02	I
300	l	7018.91	I
300	s	7039.14	I
300	s	7080.01	I
300	l	7174.83	I
300	l	7184.93	I
300	l	7284.28	I
300	l	7292.29	I
300	l	7332.52	I
300	s	7370.60	I
300	l	7381.03	I
300	l	7381.65	I
300	l	7402.70	I
300	s	7512.22	I
300	l	7515.15	I
300	l	7546.05	I
300	l	7624.83	I
300		7626.85	I
300	s	7681.01	I
300	s	7685.25	I
1000	l	7735.14	I
300	l	7761.61	I
1000	l	7765.75	I
300	s	7776.07	I
300		7787.46	I
1000	l	7791.38	I
300	l	[illegible]	I
300	l	7887.88	I
300	l	7901.71	I
300	l	7975.98	I
300	h	8080.32	I
300	s	8124.59	I
300		8155.11	I
300	l	8167.42	I
300	l	8183.06	I
300	l	8188.61	I
300	l	8247.82	I
300	l	8287.11	I
300	s	8287.75	I
300	l	8306.22	I
300	s	8313.66	I
1000	l	8339.12	I
300		8356.79	I
300	l	8367.11	I
3000		8372.88	I
3000		8529.96	I
1000	s	8696.23	I
1000	s	8906.02	I
1000		8942.70	I
1000	s	9004.75	I
1000	l	9006.31	I
10000	l	9016.18	I
3000	l	9141.30	I
3000	s	9379.33	I
3000	l	9468.66	I
3000	s	9679.13	I
3000	l	9930.55	I
10000	l	10091.99	I
10000	s	10817.45	I
10000	l	11695.15	I
10000	l	11776.64	I
10000	s	12148.18	I
10000	s	12377.42	I
10000	l	12407.99	I
10000	l	13834.33	I

NICKEL (Ni)
Z = 28

Ni I and II
Ref. 1,294 = C.H.C.

Vacuum

Intensity	Wavelength	
500	1317.22	II
400	1335.20	II
500	1370.14	II
1000	1741.55	II
500	1748.28	II

Air

1000	2165.55	II
2000	2169.10	II
2000	2174.67	II
1500	2175.15	II
500	2177.09	II
400	2177.36	II
400	2179.35	II
800	2180.47	II
800	2184.60	II
2500	2185.50	II
3000	2192.09	II
600	2201.41	II
5000	2205.55	II
4000	2206.72	II
6000	2216.48	II
800	2220.40	II
500	2221.06	II
900	2222.96	II
500	2242.68	II
500	2253.85	II
1000	2264.46	II
2000	2270.21	II
800	2277.28	II
400	2278.32	II
800	2278.77	II
500	2287.65	II
1600	2289.98	I
400	2296.55	II
400	2297.14	II
630	2300.78	I
1000	2303.00	II
2000	2310.96	I
1700	2312.34	I
1400	2313.66	I
1400	2313.98	I
1000	2316.04	II
1400	2317.16	I
500	2319.75	II
2600	2320.03	I
1900	2321.38	I
240	2322.68	I
1400	2325.79	I
940	2329.96	I
500	2334.58	II
400	2337.49	I
160	2337.82	I
500	2341.20	II
1200	2345.54	I
190	2346.63	I
400	2347.52	I
160	2360.63	I
200	2362.06	I
1000	2375.42	II
240	2386.58	I
1000	2394.52	II
2000	2416.13	II
240	2419.31	I
85	2421.23	I
70	2423.33	I
70	2423.66	I
70	2424.03	I
500	2437.89	II
85	2453.99	I
160	2472.06	I
85	2476.87	I
500	2510.87	II
500	2565.92	II
500	2606.26	II
500	2609.94	II
500	2615.06	II
45	2696.49	I
150	2798.65	I
250	2821.29	I
500	2864.02	II
50	2865.50	I
60	2907.46	I
25	2914.01	I
500	2943.91	I
570	2981.65	I
250	2984.13	I
500	2992.60	I
1000	2994.46	I
4000	3002.49	I
2200	3003.63	I
3700	3012.00	I
350	3019.14	I
120	3031.87	I
1700	3037.94	I
150	3045.01	I
3500	3050.82	I
1500	3054.32	I
1900	3057.64	I
500	3064.62	I
420	3080.76	I
260	3097.12	I
210	3099.12	I
2600	3101.55	I
1300	3101.88	I
220	3105.47	I
270	3114.12	I
2900	3134.11	I
55	3145.72	I
55	3181.74	I
100	3184.37	I
55	3195.57	I
150	3197.11	I
55	3202.14	I
180	3214.06	I
180	3217.83	I
100	3221.27	I
150	3221.65	I
210	3225.02	I
1100	3232.96	I
290	3234.65	I
600	3243.06	I
100	3248.46	I
120	3250.74	I
100	3271.12	I
120	3282.70	I
400	3292.87	II
500	3297.60	II
400	3305.71	II
660	3315.66	I
330	3320.26	I
310	3322.31	I
2000	3331.88	II
400	3335.64	II
500	3338.09	II
500	3348.84	II
500	3349.24	II

Nickel (Cont.)

Intensity		Wavelength	
600		3358.68	II
330		3361.56	I
500		3363.45	II
330		3365.77	I
330		3366.17	I
65		3366.81	I
65		3367.89	I
2900		3369.57	I
400		3371.99	I
260		3374.22	I
130		3374.64	I
500		3378.97	II
3300		3380.57	I
240		3380.85	I
1300		3391.05	I
3300		3392.99	I
500		3401.05	II
130		3409.58	I
330		3413.48	I
330		3413.94	I
8200		3414.76	I
1600		3423.71	I
2600		3433.56	I
990		3437.28	I
4800		3446.26	I
1300		3452.89	I
5000		3458.47	I
5000		3461.65	I
200		3467.50	I
240		3469.49	I
1600		3472.54	I
550		3483.77	I
130		3485.89	I
5500		3492.96	I
660		3500.85	I
65		3502.60	I
55		3507.69	I
2600		3510.34	I
260		3513.93	I
6600		3515.05	I
660		3519.77	I
8200		3524.54	I
110		3527.98	I
330		3548.18	I
55		3551.53	I
65		3561.75	I
5000		3566.37	I
990		3571.87	I
130		3587.93	I
1300		3597.70	I
1300		3610.46	I
530		3612.74	I
6600		3619.39	I
130		3624.73	I
200		3664.10	I
130		3669.24	I
180		3670.43	I
260		3674.15	I
160		3688.42	I
80		3693.93	I
120		3722.48	I
150		3736.81	I
60		3739.23	I
600		3775.57	I
700		3783.53	I
700		3807.14	I
110		3831.69	I
1200		3858.30	I
30		3889.67	I
35		3972.17	I
110		3973.56	I
110		4401.55	I
85		4459.04	I
18		4462.46	I
55		4470.48	I
35		4592.53	I
18		4600.37	I
65		4605.00	I
18		4606.23	I
75		4648.66	I
23		4686.22	I
110		4714.42	I
22		4715.78	I
30		4756.52	I
15		4763.95	I
45		4786.54	I
22		4807.00	I
22	h	4829.03	I
19		4831.18	I
45		4855.41	I
30		4866.27	I
17		4873.44	I
40		4904.41	I
22		4918.36	I
16		4935.83	I
45		4980.16	I
45		4984.13	I
500		4992.02	II
16	h	5000.34	I
18		5012.46	I
50		5017.59	I
100		5035.37	I
16		5048.85	I
100		5080.52	I
65		5081.11	I
26	h	5084.08	I
18		5099.32	I
26	h	5099.95	I
21		5115.40	I
18	h	5129.38	I
23		5137.08	I
23	h	5142.77	I
40	h	5146.48	I
40	h	5155.76	I
16		5168.66	I
13		5176.56	I
8		5435.87	I
180		5476.91	I
6		5510.00	I
6		5578.73	I
9		5587.86	I
13		5592.28	I
9		5614.79	I
5	h	5625.33	I
4		5649.70	I
5		5664.02	I
12		5682.20	I
8		5695.00	I
23		5709.56	I
10		5711.90	I
10		5715.09	I
16		5754.68	I
8		5760.85	I
10		5857.76	I
10		5892.88	I
10		6108.12	I
10		6176.81	I
10		6191.18	I
13		6256.36	I
10		6314.66	I
16		6643.64	I
22		6767.77	I
9		6772.32	I
10		6914.56	I
5		7110.90	I
26		7122.20	I
6		7182.00	I
5		7197.02	I
5		7261.93	I
5		7291.45	I
4		7385.24	I
16		7393.60	I
16		7409.35	I
5		7414.51	I
23		7422.28	I
13		7522.76	I
9		7525.12	I
19		7555.60	I
8		7574.05	I
23		7617.00	I
9		7619.21	I
16		7714.32	I
5	h	7715.58	I
19		7727.61	I
19		7748.89	I
10		7788.94	I
13		7797.59	I
2		7917.44	I
1000		8096.75	II
500		8114.21	II
700		8121.48	II
2		8809.42	I
9		8862.55	I
500	w	9900.92	II

Ni III
Ref. 422 — C.H.C.

Intensity		Wavelength	
		Vacuum	
100		625.68	III
500		630.71	III
200		637.54	III
200		662.37	III
150		663.57	III
500		676.94	III
200		700.17	III
300		713.33	III
300		713.38	III
500		718.48	III
200		721.26	III
300		722.09	III
250		725.20	III
500		729.82	III
250		730.11	III
400		731.70	III
300		732.16	III
200		738.26	III
300		747.99	III
200		749.68	III
300		750.05	III
200		752.02	III
300		757.80	III
250		758.73	III
250		758.27	III
400		770.22	III
200		772.04	III
500		778.81	III
200	d	785.02	III
300		788.04	III
200		788.30	III
200		805.01	III
500		811.57	III
500		826.14	III
200		826.50	III
500		842.14	III
400		845.24	III
300		847.43	III
200		857.09	III
300		860.64	III
300		862.88	III
300		863.22	III
300		867.51	III
200		869.70	III
200		870.84	III
300		973.79	III
400		979.59	III
200		1428.87	III
200		1434.31	III
200		1451.50	III
300		1604.54	III
300		1652.87	III
200		1653.12	III
250		1656.13	III
200		1661.79	III
400		1687.90	III
1000		1692.51	III
200		1707.35	III
200		1707.43	III
800		1709.90	III
650		1715.30	III
500		1719.46	III
200		1721.26	III
400		1722.28	III
250		1733.13	III
500		1738.25	III
300		1739.78	III
300		1741.96	III
550		1747.01	III
300		1752.43	III
400		1753.01	III
800		1764.69	III
500		1767.94	III
2000		1769.64	III
400		1776.07	III
200		1788.30	III
250		1790.40	III
200		1790.93	III
200		1791.64	III
200		1794.90	III
300		1807.24	III
200		1811.69	III
300		1819.28	III
800		1823.06	III
400		1830.01	III
200		1830.08	III
650		1847.28	III
800		1854.15	III
300		1858.75	III
200		1930.43	III
200		1952.54	III

Ni IV
Ref. 415 — C.H.C.

Intensity	Wavelength	
	Vacuum	
33	392.68	IV
32	393.24	IV
49	424.40	IV
57	444.21	IV
67	469.67	IV
65	471.24	IV
65	485.42	IV
66	536.28	IV
67	537.96	IV
58	1345.72	IV
69	1357.07	IV
76	1398.19	IV
74	1411.45	IV
69	1419.58	IV
74	1421.22	IV
70	1427.45	IV
67	1428.93	IV
67	1435.24	IV
70	1438.82	IV
73	1449.01	IV
76	1452.22	IV
70	1455.42	IV
69	1472.63	IV
68	1476.82	IV
73	1482.25	IV
67	1489.53	IV
72	1489.83	IV
69	1493.01	IV
74	1493.67	IV
68	1498.71	IV
71	1498.77	IV
72	1498.90	IV
67	1499.97	IV
70	1512.74	IV
70	1516.66	IV
73	1520.63	IV
75	1525.31	IV
74	1527.68	IV
74	1527.80	IV
76	1534.71	IV
73	1537.25	IV
69	1538.93	IV
75	1543.41	IV
74	1546.23	IV
68	1548.04	IV
69	1557.28	IV
67	1560.18	IV

Ni V
Ref. 416 — C.H.C.

Intensity	Wavelength	
	Vacuum	
29	304.02	V
55	315.24	V
56	315.71	V
63	336.79	V
68	343.93	V
78	347.34	V
70	347.46	V
67	347.72	V
71	348.10	V
69	350.77	V
69	353.59	V
72	354.18	V
76	354.42	V
68	354.49	V
68	355.61	V
70	355.78	V
65	357.37	V
69	358.57	V
68	358.58	V
66	359.47	V
69	365.62	V
70	370.62	V
67	371.31	V
68	371.76	V
67	373.60	V
72	377.68	V
70	393.91	V
66	394.31	V
66	395.24	V
41	400.59	V

NIOBIUM (Nb)
Z = 41

Nb I and II
Ref. I = C.H.C.

Intensity	Wavelength	
	Air	
3300	2029.32	II
3000	2032.99	II
2000	2109.42	II
1700	2125.21	II
1100	2126.54	II
1500	2131.18	II
370	2295.68	II
280	2302.08	II
170	2376.40	II
110	2387.09	II
140	2387.52	II
45	2388.27	II
160	2398.48	II
55	2405.34	II
55	2405.85	II
140	2412.46	II
160	2416.99	II
140	2418.69	II
75	2433.80	II
40	2435.95	II
35	2436.33	I
45	2437.42	II
40	2442.14	II
28	2442.68	II
65	2451.87	II
65	2453.95	II
55	2458.09	II
65	2462.89	I
35	2466.73	I
55	2469.08	I
110	2477.38	II
65	2478.29	II

Niobium (Cont.)

65		2479.94	II
35		2483.88	II
110		2504.65	I
110		2511.00	II
110		2521.40	II
390		2544.80	II
110		2551.38	II
130		2556.94	II
130		2562.41	II
130		2565.41	I
100		2569.03	I
110		2571.33	II
200		2578.74	I
390		2583.99	II
390		2590.94	II
270		2592.20	I
130		2616.48	I
130		2623.51	I
130		2627.44	I
130		2628.49	I
200		2642.24	II
320		2646.26	II
330		2647.50	I
240		2649.52	I
330		2654.45	I
310		2656.08	II
160		2657.62	I
110		2665.25	II
110		2666.59	II
110		2667.30	II
130		2668.29	I
400		2671.93	II
200		2673.57	II
200		2675.94	II
130		2687.15	I
160		2691.77	II
1000		2697.06	II
320		2698.86	II
320		2702.20	II
150		2702.52	II
470		2716.62	II
470		2721.98	II
310		2733.26	II
110		2737.09	II
200		2746.91	I
200		2748.85	I
190		2753.01	I
280		2758.61	I
240		2768.13	II
310		2773.20	I
270		2780.24	II
130		2782.36	I
110		2793.05	II
190		2827.08	II
150		2836.24	I
110		2840.94	I
250		2841.15	II
280		2842.65	II
160		2846.28	II
110		2851.45	I
240		2861.09	II
100		2864.32	I
100		2865.61	II
500		2868.52	II
800		2875.39	II
270		2876.95	II
530		2877.03	II
100		2880.72	II
570		2883.18	II
280		2888.83	II
470		2897.81	II
400		2899.24	II
470		2908.24	II
670		2910.59	II
470		2911.74	II
1100		2927.81	II
110		2931.47	II
870		2941.54	II
110	h	2945.88	II
110		2946.12	II
110		2946.90	II
1100		2950.88	II
400		2972.57	II
320		2974.10	II
210		2977.68	II
200		2982.11	II
330		2990.26	II
470		2994.73	II
140		3024.74	II
350		3028.44	II
300		3032.77	II
100		3044.76	II
150		3048.10	I
110		3053.09	I
100		3055.52	II
220		3064.53	II
110		3069.68	II
100		3070.90	II
110		3071.56	II
100		3073.24	II
400		3076.87	II
110		3080.35	II
1800		3094.18	II
140		3099.19	II
150		3111.45	I
270		3127.53	II
1500		3130.79	II
390		3145.40	II
140		3151.87	I
1200		3163.40	II
150		3175.78	II
390		3180.29	II
200		3187.49	I
300		3191.10	II
150		3191.43	II
1000		3194.98	II
120		3203.35	II
300		3206.34	II
390		3215.60	II
800		3225.48	II
140		3229.56	II
400		3236.40	II
200		3247.47	II
120		3248.94	II
160		3249.52	I
320		3254.07	II
230		3260.56	II
160		3263.37	II
160		3264.59	I
120		3270.47	I
100		3270.76	I
200		3272.07	I
160		3277.67	I
200		3283.46	II
230		3285.66	I
200		3287.59	I
160		3287.92	I
160		3292.02	II
320		3296.01	I
160		3299.61	I
120		3304.83	I
120		3308.05	I
120		3310.47	I
400		3312.60	I
200		3315.22	I
200		3318.98	I
120		3319.26	I
120		3319.58	II
240		3326.62	I
170		3329.36	I
110		3332.16	I
130		3341.60	II
1300		3341.97	I
1300		3343.71	I
130		3346.93	I
1700		3349.06	I
420		3349.52	I
340		3354.74	I
130		3357.04	I
1700		3358.42	I
130		3365.58	II
340		3366.96	I
130		3369.16	II
170		3371.33	I
350		3374.92	I
270		3380.41	I
130		3380.86	I
170		3386.24	II
350		3392.34	I
170		3395.93	I
120		3399.40	I
230		3405.41	I
130		3406.13	I
270		3408.38	I
230		3408.68	II
180		3409.19	II
230		3412.94	II
180		3415.97	I
180		3423.76	I
230		3425.42	II
130		3425.85	I
230		3426.57	II
230		3427.45	I
130		3429.04	I
180		3432.70	II
180		3440.59	II
170		3463.81	I
180		3465.86	I
130		3469.44	I
100		3471.19	I
140		3473.02	I
290		3478.69	I
200		3479.56	II
100		3484.05	II
230		3491.03	I
200		3497.81	I
500		3498.63	I
460		3507.96	I
200		3510.26	II
200		3515.42	II
200		3517.67	II
200		3520.06	I
2000		3535.30	I
1300		3537.48	I
250		3540.96	II
500		3544.02	I
250		3544.65	I
300		3550.45	I
250		3554.52	I
1000		3554.66	I
630		3563.50	I
630		3563.62	I
1500		3575.85	I
200		3577.72	I
5000		3580.27	I
500		3584.97	I
750		3589.11	I
500		3589.36	I
500		3593.97	I
500		3602.56	I
300		3619.51	II
200		3621.03	I
200		3639.33	I
630		[illegible]	I
250		3650.81	I
400		3651.19	II
200		3659.61	II
630		3660.37	I
900		3664.70	I
220		3669.01	I
270		3674.78	I
1500		3697.85	I
330		3711.34	I
3300		3713.01	I
480		3716.99	I
2700		3726.24	I
270		3738.42	I
2700		3739.80	I
670		3740.73	II
270		3741.78	I
1700		3742.39	I
250		3753.18	I
210		3755.77	I
530		3763.49	I
350		3765.08	I
250		3766.13	I
530		3771.85	I
870		3781.01	I
1700		3787.06	I
1300		3790.15	I
3500		3791.21	I
2700		3798.12	I
270		3801.30	I
2700		3802.92	I
670		3803.88	I
530		3804.74	I
670		3810.49	I
530		3811.03	I
530		3815.51	I
210		3818.86	II
210		3819.15	I
670		3824.88	I
350		3835.18	I
250		3836.45	I
210		3845.90	I
290		3858.95	I
350		3863.38	I
270		3867.92	I
530		3877.56	I
870		3878.82	I
670		3883.14	I
1100		3885.44	I
670		3885.68	I
210		3886.07	I
580		3891.30	I
210		3908.97	I
670		3914.70	I
530		3920.20	I
670		3937.44	I
520		3943.67	I
250		3965.69	I
910	d	3966.09	I
210		3971.85	I
1100		4032.52	I
250		4039.53	I
16000	c	4058.94	I
210		4059.51	I
350		4060.79	I
210		4068.26	I
12000		4079.73	I
270		4084.86	I
440		4100.40	I
6700		4100.92	I
310		4116.90	I
5300		4123.81	I
670		4129.43	I
770		4129.93	I
2300		4137.10	I
440		4139.44	I
2700		4139.71	I
350		4143.21	I
870		4150.12	I
4400		4152.58	I
870		4163.47	I
4400		4163.66	I
4000		4164.66	I
3500		4168.13	I
310		4184.44	I
1200		4190.88	I
870		4192.07	I
870		4195.09	I
1300		4195.66	I
310		4198.51	I
350		4201.52	I
870		4205.31	I
350		4214.73	I
420		4217.94	I
420		4229.15	I
250		4255.44	I
770		4262.05	I
420		4266.02	I
290		4270.69	I
400		4286.99	I
580		4299.60	I
580		4300.99	I
120		4309.56	I
300		4311.07	I
120		4312.45	I
350		4326.33	I
120		4327.38	I
390		4331.37	I
140		4342.82	I
140		4348.65	I
110		4349.03	I
290		4351.57	I
210		4368.43	I
140		4377.96	I
130		4388.36	I
160		4392.69	I
330		4410.21	I
190		4419.44	I
230	c	4437.22	I
290		4447.18	I
140		4456.80	I
140		4457.42	I
140		4469.71	I
140		4471.29	I
140		4472.53	I
150		4503.04	I
530		4523.41	I
480		4546.82	I
370		4564.53	I
720		4573.08	I
480		4581.62	I
1200		4606.77	I
170		4616.17	I
450		4630.11	I
450		4648.95	I
110		4649.27	I
450		4663.83	I
340		4666.24	I
240		4667.22	I
580		4672.09	I
530		4675.37	I
110		4678.48	I
320		4685.14	I
130	c	4706.14	I
260		4708.29	I
150		4713.50	I
110	c	4733.89	I
220	c	4749.70	I
110		4816.38	I
110	c	4848.37	I
130	c	4967.78	I
110		4973.14	I
190		4988.97	I
85		5000.95	I
65		5002.25	I
40		5013.27	I
230		5017.75	I
40		5019.51	I
150		5026.36	I
40		5030.13	I
210		5039.04	I
40		5047.96	I
170		5058.01	I
65		5059.35	I
130		5065.25	I
40		5077.40	I
750		5078.96	I
40	c	5094.41	I
420		5095.30	I
170		5100.16	I
170		5120.30	I
85		5121.80	I
85		5127.66	I
40		5133.34	I
210		5134.75	I
75		5140.58	I
75		5147.54	I
40		5150.64	I
75		5152.63	I
250		5160.33	I
250		5164.38	I
230		5180.31	I
110		5186.98	I
190		5189.20	I
170		5193.08	I
150		5195.84	I
65		5203.22	I
35		5205.13	I
85	c	5219.10	I

Niobium (Cont.)

Intensity		Wavelength	
65		5225.16	I
150		5232.81	I
85	c	5237.43	I
29		5240.39	I
150	d	5251.62	I
		5251.81	I
75		5253.03	I
85		5253.93	I
50		5269.92	I
270		5271.53	I
25		5272.48	I
130	c	5276.20	I
29	c	5279.43	I
50		5285.26	I
35		5296.34	I
50		5315.55	I
17		5317.01	I
250		5318.60	I
50		5319.49	I
75		5334.87	I
25		5336.81	I
50		5340.80	I
25		5343.58	I
460		5344.17	I
340		5350.74	I
40		5353.28	I
25		5355.31	I
40		5355.70	I
29		5359.19	I
17		5362.01	I
40		5375.27	I
40		5381.34	I
17		5388.30	I
21		5395.86	I
29		5396.33	I
29		5411.24	I
21		5416.30	I
65		5422.44	I
21		5431.26	I
110		5437.27	I
19		5448.31	I
19		5456.19	I
40		5458.04	I
19	h	5468.10	I
40		5481.00	I
13		5483.09	I
19		5483.49	I
13		5491.06	I
17		5499.53	I
40		5504.58	I
17		5509.12	I
35	c	5512.82	I
17		5517.39	I
50		5523.57	I
25		5541.47	I
85		5551.35	I
29		5563.00	I
17	c	5571.44	I
35	c	5576.16	I
35		5578.29	I
50		5586.97	I
17	c	5590.95	I
13		5594.89	I
17	c	5599.59	I
40		5603.52	I
13		5603.93	I
25		5628.26	I
65		5629.17	I
35	c	5635.42	I
170		5642.11	I
35		5645.30	I
17		5654.14	I
130		5664.71	I
170		5665.63	I
17		5666.86	I
65	Cw	5671.02	I
85		5671.91	I
25		5677.47	I
25		5693.09	I
35	d	5697.90	I
		5698.03	I
40		5706.16	I
85		5706.48	I
29		5709.33	I
17		5715.59	I
65		5716.35	I
25		5725.66	I
130		5729.19	I
21		5737.36	I
13		5738.20	I
85		5751.44	I
110		5760.34	I
65		5764.99	I
29		5771.08	I
50	c	5776.07	I
17		5780.34	I
85		5787.54	I
17		5789.79	I
50		5794.24	I
50		5804.03	I
29	h	5815.33	I
110		5819.43	I
35		5820.62	I
75		5834.90	I
25		5838.15	I
130	d	5838.64	I
50		5842.47	I
17		5846.09	I
65		5866.47	I
35		5874.70	I
17		5877.79	I
40		5893.44	I
190	Cw	5900.62	I
40	c	5903.80	I
29		5927.41	I
40	c	5934.16	I
40		5957.70	I
150		5983.22	I
65		5986.08	I
85	Cw	5997.93	I
50		6029.75	I
50		6031.84	I
50		6045.50	I
25		6048.72	I
29		6056.65	I
29		6107.71	I
40		6142.51	I
50		6148.13	I
50		6164.32	I
29		6213.06	I
75		6221.96	I
40	c	6251.76	I
21		6260.77	I
85	c	6430.46	I
50	c	6433.22	I
17		6497.84	I
65		6544.61	I
15		6574.73	I
19	Cw	6591.00	I
19		6606.16	I
19		6607.28	I
35		6614.15	I
19		6626.98	I
210	Cw	6660.84	I
150	Cw	6677.33	I
65		6701.20	I
130	c	6723.62	I
75		6739.88	I
25		6795.31	I
85		6828.11	I
25	c	6849.35	I
19		6870.92	I
40		6876.36	I
25	c	6902.89	I
35		6908.07	I
40		6918.32	I
17		6946.07	I
17		6972.49	I
25		6986.09	I
85		6990.32	I
17	c	6996.11	I
21		7023.48	I
17		7038.04	I
190	c	7046.81	I
8		7066.41	I
8		7075.23	I
40	c	7098.94	I
17	Cw	7102.01	I
19		7119.31	I
15		7122.95	I
35		7126.17	I
17		7130.06	I
130		7159.43	I
17		7191.37	I
19	c	7208.94	I
50		7252.35	I
15		7274.81	I
13		7317.03	I
17	c	7323.92	I
29	Cw	7328.38	I
65	c	7353.16	I
190	Cw	7372.50	I
13		7419.83	I
15		7436.02	I
19		7478.20	I
65		7515.93	I
29	c	7519.77	I
170	c	7574.58	I
17	c	7583.21	I
13		7639.81	I
13		7647.71	I
25		7703.33	I
75	c	7726.68	I
25		7757.31	I
6		7787.11	I
13	Cw	7873.41	I
35		7885.31	I
25		7938.89	I
8		7954.76	I
40		8135.20	I
13	Cw	8240.00	I
29	Cw	8320.93	I
29		8346.08	I
10		8350.04	I
17		8439.77	I
17	Cw	8475.98	I
25		8526.99	I
13	c	8547.25	I
17	c	8560.54	I
17		8575.87	I
21	c	8697.55	I
21		8740.96	I
21		8767.97	I
29	Cw	8815.56	I
35		8905.78	I

Nb III
Ref. 392 — C.H.C.

Intensity		Wavelength	
		Vacuum	
60		1314.56	III
50		1319.15	III
60		1431.92	III
60		1433.39	III
50		1435.26	III
80		1445.43	III
80		1445.98	III
80		1447.09	III
50		1448.50	III
60		1451.63	III
100		1456.68	III
80		1484.73	III
50		1486.79	III
100		1495.94	III
80		1498.02	III
80		1499.45	III
50		1501.53	III
100		1501.99	III
60		1505.03	III
50		1509.71	III
50		1512.34	III
50		1513.25	III
80		1513.81	III
50		1517.38	III
100		1524.91	III
60		1532.98	III
60		1537.50	III
50		1566.92	III
50		1570.19	III
50		1586.82	III
100		1590.21	III
80		1598.86	III
80		1604.72	III
80		1639.51	III
100		1682.77	III
60		1684.40	III
100		1705.44	III
100		1707.14	III
50		1739.30	III
50		1758.63	III
60		1763.72	III
50		1808.70	III
50		1863.13	III
100		1892.92	III
100		1938.84	III
60	h	1979.07	III
50		1985.15	III
50		1997.11	III
		Air	
50	h	2007.28	III
50		2032.47	III
50		2060.29	III
80	h	2130.24	III
60		2206.01	III
60		2240.31	III
60		2244.19	III
60		2265.63	III
80		2273.92	III
100		2275.23	III
80		2279.36	III
100		2281.51	III
80		2284.40	III
100		2290.36	III
60		2304.78	III
50		2309.92	III
100		2313.30	III
50		2330.22	III
100		2338.09	III
80		2344.12	III
90		2349.21	III
80		2355.54	III
100		2362.06	III
80		2362.50	III
80		2365.70	III
100		2372.73	III
100		2387.41	III
80		2388.23	III
50		2404.23	III
80		2404.89	III
100		2413.94	III
60		2414.50	III
100		2421.91	III
50		2437.74	III
60		2446.45	III
100		2456.99	III
50		2460.34	III
50		2460.45	III
50		2463.72	III
80		2468.72	III
60		2469.39	III
80		2475.87	III
50		2486.02	III
60		2488.74	III
60		2493.02	III
100		2499.73	III
60		2508.53	III
50		2511.95	III
100		2545.64	III
80		2557.94	III
50		2567.44	III
60		2593.75	III
80		2598.86	III
50		2628.67	III
80		2633.17	III
80		2657.99	III
50		2937.71	III
80		3001.84	III
80		3142.26	III
60		3266.11	III

Nb IV
Ref. 407 — C.H.C.

Intensity	Wavelength	
	Vacuum	
12	542.38	IV
12	543.09	IV
12	545.21	IV
10	559.94	IV
10	566.22	IV
18	981.27	IV
60	993.54	IV
18	996.16	IV
50	1002.76	IV
400	1005.72	IV
500	1007.05	IV
500	1010.19	IV
45	1030.27	IV
100	1116.08	IV
150	1120.02	IV
50	1447.48	IV
40	1473.43	IV
40	1487.23	IV
60	1502.30	IV
60	1524.36	IV
50	1534.06	IV
40	1635.68	IV
40	1910.70	IV
60	1922.41	IV
60	1978.22	IV
	Air	
40	2027.50	IV
65	2032.53	IV
40	2034.67	IV
40	2068.62	IV
55	2084.07	IV
35	2093.12	IV
50	2093.30	IV
20	2122.68	IV
25	2130.23	IV
45	2146.36	IV
18	2249.98	IV

Nb V
Ref. 431 — C.H.C.

Intensity	Wavelength	
	Vacuum	
80	464.55	V
80	468.32	V
60	753.01	V
80	763.77	V
80	774.02	V
40	1007.02	V
50	1044.90	V
70	1212.21	V
100	1258.87	V
40	1267.60	V
100	1758.33	V
100	1877.34	V

NITROGEN (N)
Z = 7

N I and II
Ref. 213 = R.L.K.

Intensity	Wavelength

Nitrogen (Cont.)

Vacuum

285		644.634	II
360		644.837	II
450		645.178	II
140		647.50	I
360		660.286	II
170		671.016	II
285		671.386	II
150		671.630	II
160		671.773	II
170		672.001	II
350		692.70	I
285		746.984	II
650		775.965	II
90		885.67	I
90		909.697	I
80		910.278	I
40		910.643	I
450		915.612	II
450		915.962	II
550		916.012	II
650		916.701	II
90		953.415	I
100		953.655	I
130		953.970	I
130		963.990	I
115		964.626	I
70		965.041	I
90		1067.614	I
60		1068.612	I
450		1083.990	II
600		1084.580	II
430		1085.546	II
650		1085.701	II
175		1097.237	I
115		1098.095	I
115		1098.260	I
105		1100.360	I
40		1100.465	I
90		1101.291	I
360		1134.165	I
385		1134.415	I
410		1134.980	I
105		1143.65	I
130		1163.884	I
60		1164.206	I
105		1164.325	I
270		1167.448	I
105		1168.334	I
60		1168.417	I
195		1168.536	I
230		1176.510	I
103		1176.630	I
195		1177.695	I
410		1199.550	I
385		1200.223	I
360		1200.710	I
175		1225.026	I
160		1225.37	I
130		1228.41	I
160		1228.79	I
360		1243.179	I
315		1243.306	I
290		1310.540	I
250		1310.95	I
230		1319.00	I
315		1319.68	I
115		1326.57	I
115		1327.92	I
360		1411.94	I
700		1492.625	I
490		1492.820	I
640		1494.675	I
775		1742.729	I
700		1745.252	I

Air

160		2095.53	II
70		2096.20	II
110		2096.86	II
110		2130.18	II
160		2142.78	II
160		2206.09	II
160		2286.69	II
110		2288.44	II
220		2316.49	II
160		2316.69	II
285		2317.05	II
160		2461.27	II
110		2496.83	II
70		2496.97	II
110		2520.22	II
160		2520.79	II
220		2522.23	II
110		2590.94	II
160		2709.84	II
110		2799.22	II
110		2823.64	II
160		2885.27	II
220		3006.83	II
360		3437.15	II
285		3838.37	II
360		3919.00	II
450		3955.85	II
1000		3995.00	II
360		4035.08	II
550		4041.31	II
360		4043.53	II
140		4099.94	I
185		4109.95	I
285		4176.16	II
285		4227.74	II
285		4236.91	II
220		4237.05	II
450		4241.78	II
285		4432.74	II
650		4447.03	II
360		4530.41	II
550		4601.48	II
430		4607.16	II
360		4613.87	II
450		4621.39	II
870		4630.54	II
550		4643.08	II
285		4788.13	II
450		4803.29	II
180		4847.38	I
285		4895.11	II
160		4914.94	I
210		4935.12	I
160		4950.23	I
350		4963.98	I
285		4987.37	II
450		4994.36	II
650		5001.48	II
360		5002.70	II
870		5005.15	II
550		5007.32	II
450		5010.62	II
360		5016.39	II
360		5025.66	II
550		5045.10	II
185		5281.20	I
140		5292.68	I
450		5495.67	II
285		5535.36	II
650		5666.63	II
550		5676.02	II
870		5679.56	II
450		5686.21	II
450		5710.77	II
285		5747.30	II
700		5752.50	I
240		5764.75	I
265		5829.54	I
235		5854.04	I
360		5927.81	II
550		5931.78	II
285		5940.24	II
650		5941.65	II
285		5952.39	II
160		5999.43	I
210		6008.47	I
285		6167.76	II
360		6379.62	II
185		6411.65	I
210		6420.64	I
210		6423.02	I
210		6428.32	I
185		6437.68	I
235		6440.94	I
185		6457.90	I
300		6468.44	I
750		6482.05	II
360		6482.70	I
300		6483.75	I
265		6481.71	I
325		6484.80	I
160		6491.22	I
210		6499.54	I
185		6506.31	I
750		6610.56	II
185		6622.54	I
185		6636.94	I
235		6644.96	I
185		6646.50	I
235		6653.46	I
210		6656.51	I
185		6722.62	I
210		7398.64	I
160		7406.12	I
265		7406.24	I
685		7423.64	I
785		7442.29	I
900		7468.31	I
185		7608.80	I
450		7762.24	II
400		8184.87	I
400		8188.02	I
250		8200.36	I
300		8210.72	I
570		8216.34	I
400		8223.14	I
400		8242.39	I
550		8438.74	II
500		8567.74	I
570		8594.00	I
650		8629.24	I
500		8655.89	I
220		8676.08	II
700		8680.28	I
650		8683.40	I
500		8686.15	I
110		8687.43	II
110	h	8699.00	II
500		8703.25	I
160	h	8710.54	II
570		8711.70	I
500		8718.83	I
250		8728.89	I
200		8747.36	I
500		9386.80	I
570		9392.79	I
250		9460.68	I
200		9863.33	I
160	h	9865.41	II
110	h	9868.21	II
160	h	9887.39	II
220	h	9891.09	II
160	h	9961.86	II
220	h	9969.34	II
285	h	10023.27	II
220	h	10035.45	II
220	h	10065.15	II
160	h	10070.12	II
250		10105.13	I
300		10108.89	I
350		10112.48	I
400		10114.64	I
110	h	10126.27	II
250		10539.57	I
200		12074.51	I
380		12186.82	I
225		12288.97	I
290		12328.76	I
310		12381.65	I
180		12438.40	I
510		12461.25	I
920		12469.62	I
500		13429.61	I
840		13581.33	I
180		13587.73	I
180		13602.27	I
290		13624.18	I
250		14757.07	I
100		14868.87	I
160		14966.60	I
180		15582.27	I
120	s	17516.58	I
100	l	17584.86	I
100		17878.26	I

N III
Ref. 66, 213 = R.L.K.

Vacuum

Intensity		Wavelength	
500		257.95	III
650		258.50	III
700		259.19	III
800		260.09	III
800		261.28	III
500		262.91	III
500		265.23	III
500		265.27	III
500		268.70	III
150		314.715	III
200		314.850	III
90		314.877	III
600		323.26	III
500		338.35	III
500		340.20	III
500		351.98	III
120		362.833	III
150		362.881	III
150		362.946	III
90		362.985	III
300		374.204	III
350		374.441	III
500		387.48	III
250		451.869	III
300		452.226	III
500		684.996	III
570		685.513	III
650		685.816	III
500		686.335	III
500		763.336	III
570		764.359	III
250		771.544	III
300		771.901	III
350		772.385	III
200		772.891	III
150		772.975	III
650		979.842	III
700		979.919	III
900		989.790	III
700		991.514	III
1000		991.579	III
500		1183.031	III
570		1184.550	III
150		1387.371	III
250		1729.945	III
570		1747.848	III
350		1751.218	III
650		1751.657	III
150		1804.486	III
200		1805.669	III
150		1846.42	III
350		1885.06	III
400		1885.22	III
200		1907.99	III
150		1919.55	III
150		1919.77	III
300		1920.65	III
150		1920.84	III
200		1921.30	III

Air

200		2064.01	III
250		2064.42	III
120		2068.68	III
90		2071.09	III
90		2117.59	III
90		2121.50	III
90		2147.31	III
200		2188.20	III
150		2188.38	III
250	w	2682.18	III
90		2689.20	III
120		3367.34	III
90		3754.67	III
120		3771.05	III
90		3938.52	III
150		3998.63	III
200		4003.58	III
250		4097.33	III
200		4103.43	III
120		4195.76	III
150		4200.10	III
90		4332.91	III
120		4345.68	III
300		4379.11	III
90		4510.91	III
120		4514.86	III
90		4634.14	III
120		4640.64	III
90		4858.82	III
150		4867.15	III
90		5314.35	III
200		5320.82	III
150		5327.18	III
90		6454.11	III
120		6467.02	III

N IV
Ref. 108, 212 = R.L.K.

Vacuum

Intensity		Wavelength	
400		181.75	IV
400		191.7	IV
400		192.9	IV
500		196.87	IV
500		197.23	IV
500		202.60	IV
500		205.94	IV
500		205.97	IV
500		206.03	IV
500		217.20	IV
500	d	217.90	IV
500	d	223.4	IV
800	w	225.12	IV
800		225.21	IV
600	w	234.12	IV
600	w	234.20	IV
600	w	234.25	IV
550		236.07	IV
500		237.99	IV
500	w	238.7	IV
600		238.80	IV
500	w	239.62	IV
900		247.20	IV
500	w	248.43	IV
500	w	248.46	IV
500	w	248.48	IV
600		260.45	IV
650		270.99	IV
250		283.42	IV
300		283.48	IV
350		283.58	IV
600		285.56	IV
600	w	297.7	IV
700		297.82	IV
650		300.32	IV
90		303.123	IV

Nitrogen (Cont.)

500		303.28	IV
150		315.053	IV
120		322.503	IV
150		322.570	IV
200		322.724	IV
120		323.175	IV
300		335.050	IV
500	w	351.93	IV
700		353.06	IV
500		420.77	IV
650		463.74	IV
570		765.148	IV
520		921.992	IV
500		922.519	IV
480		923.057	IV
520		921.992	IV
500		922.519	IV
520		924.283	IV
1000		955.335	IV
150	w	1036.16	IV
90		1078.71	IV
90		1188.01	IV
1000		1718.55	IV

Air

90		2080.34	IV
90	w	2318.09	IV
150		2477.69	IV
250		2645.65	IV
300		2646.18	IV
350		2646.96	IV
90		3078.25	IV
90		3463.37	IV
570		3478.71	IV
500		3482.99	IV
400		3484.96	IV
90		3747.54	IV
150		4057.76	IV
90		4606.33	IV
150		6380.77	IV

N V
Ref. 66, 107, 318 = R.L.K.

Intensity		Wavelength	
		Vacuum	
52		166.947	V
52		186.069	V
62		186.153	V
90		209.303	V
90		247.561	V
120		247.706	V
150		266.196	V
200		266.379	V
90		713.518	V
150		713.860	V
150		748.195	V
200		748.291	V
1000		1238.821	V
900		1242.804	V
90		1549.336	V
200	l	1616.33	V
350	l	1619.69	V
90	w	1860.37	V
		Air	
60	l	2859.16	V
90	l	2974.52	V
150	w	2980.78	V
250	w	2981.31	V
60	w	2998.43	V
350		4603.73	V
250		4619.98	V
200	w	4944.56	V
60	w	7618.46	V

OSMIUM (Os)
Z = 76

Os I and II
Ref. 1 = C.H.C.

Intensity		Wavelength	
		Air	
9600		2001.45	I
13000		2003.73	I
9000		2004.78	
17000		2010.15	I
29000		2018.14	I
29000		2020.26	
14000		2022.76	I
14000		2028.23	I
18000		2034.44	I
26000		2045.36	I
8600		2058.69	I
		2058.78	I
13000		2061.69	I
7800		2067.21	II
4200		2070.67	II
7200		2076.95	I
7200		2078.09	
14000		2079.97	I
2900		2082.54	I
2900		2089.03	I
2900		2089.21	I
6000		2097.60	I
5500		2100.63	I
2100		2117.66	I
4800		2117.96	I
6600		2119.79	
1900		2123.84	I
5300		2137.11	I
2400		2149.97	
2600		2154.59	I
1300		2157.84	I
1200		2158.53	I
2400		2161.00	
3100		2166.90	I
1100		2167.75	I
2100		2171.65	I
960		2184.68	I
840		2194.39	II
760		2202.49	I
600		2227.98	I
1100		2234.61	I
1300		2252.15	I
2000		2255.85	II
1400		2264.60	I
360		2268.28	I
960		2270.17	I
1400		2282.26	II
840		2283.67	I
570		2289.32	I
380		2297.31	I
660		2308.31	I
190		2313.75	II
550		2320.18	I
310		2323.98	I
660		2324.24	I
330		2326.99	I
310		2334.56	I
720		2336.80	II
430		2338.63	I
290		2340.69	I
430		2343.74	I
260		2345.75	I
430		2347.38	I
230		2350.23	II
360		2352.99	I
120		2355.28	II
240		2356.92	I
240		2357.25	I
310		2362.41	I
900		2362.77	I
500		2367.35	II
290		2369.24	I
500		2370.70	I
480		2371.18	I
95		2375.06	II
2600		2377.03	I
260		2377.61	I
900		2379.39	I
240		2384.62	I
1700		2387.29	I
330		2394.29	I
290		2395.39	I
1100		2395.88	I
220		2396.78	I
960		2401.13	I
260		2402.23	I
200		2403.54	I
330		2403.85	I
95		2405.08	II
290		2405.45	I
200		2405.96	I
360		2408.67	I
240		2410.98	I
290		2414.52	I
530		2417.99	I
530		2418.53	I
95		2420.02	II
200		2423.07	II
70		2424.02	II
500		2424.56	I
1400		2424.97	I
240		2426.81	I
70		2427.90	II
380		2431.19	I
380		2431.61	I
360		2446.02	I
900		2450.74	I
530		2451.73	I
530		2453.90	I
110		2454.91	II
530		2456.46	I
1800		2461.42	I
110		2468.90	II
290		2472.28	I
290		2474.78	I
900		2476.84	I
360		2482.43	I
530		2486.24	II
4500		2488.55	I
290		2491.02	I
290		2491.69	I
360		2492.42	I
2600		2498.41	I
330		2499.92	I
330		2502.29	I
500		2504.39	I
260		2504.51	I
35		2507.18	II
70		2509.71	II
660		2512.87	I
2400		2513.25	I
660		2515.04	I
500		2517.92	I
660		2518.44	I
200		2519.29	I
330		2519.79	I
200		2532.44	I
780		2538.00	II
240		2538.10	I
1000		2542.51	I
30		2548.83	II
310		2554.46	I
190		2563.16	II
600		2566.49	I
290		2566.88	I
480		2568.83	I
340		2571.78	I
150		2578.32	II
130		2580.03	II
360		2581.05	I
740		2581.96	I
1000		2590.76	I
200		2591.98	I
170		2596.00	II
210		2609.20	I
380		2609.56	I
400		2610.78	I
470		2612.63	I
1800		2613.06	I
800		2619.94	I
230		2620.62	I
530		2621.82	I
380		2628.48	I
27		2631.22	II
3800		2637.13	I
1900		2644.11	I
340		2646.89	I
380		2647.73	I
380		2649.34	I
490		2656.68	I
1900		2658.60	I
640		2659.83	I
380		2661.18	I
40		2664.29	II
580		2674.57	I
400		2674.88	I
2100		2689.82	I
510		2699.59	I
580		2706.70	I
3000		2714.64	I
580		2715.36	I
1300		2720.04	I
850		2721.86	I
580		2730.61	I
40		2731.36	II
580		2732.80	I
690		2761.42	I
470		2763.27	I
340		2765.04	I
960		2770.71	I
300		2776.91	I
740		2782.55	I
40		2783.88	II
640		2786.31	I
230		2793.99	I
230		2794.19	I
530		2796.73	I
320		2804.07	I
2800		2806.91	I
470		2808.94	I
420		2813.84	I
740		2814.20	I
300		2815.78	I
420		2829.27	I
230		2837.42	I
470		2838.17	I
5100		2838.63	I
740		2841.60	I
2300		2844.40	I
420		2846.39	I
420		2848.25	I
1500		2850.76	I
1500		2860.96	I
35		2863.37	II
360		2874.96	I
300		2878.40	I
35		2879.39	II
30		2880.20	II
260		2896.06	I
9600		2909.06	I
2100		2912.33	I
530		2917.26	I
2100		2919.79	I
300		2925.57	I
360		2929.51	I
510		2931.28	I
260		2934.64	I
200		2942.85	I
1100	h	2948.23	I
1400		2949.53	I
210	d	2949.81	I
300		2961.01	I
530		2962.15	I
450		2964.06	I
740		2970.97	I
450		2977.64	I
510		2982.90	I
340		2983.49	
260		2997.65	I
330		3013.07	I
570		3017.25	I
4400		3018.04	I
480		3019.38	I
1100		3030.70	I
2900		3040.90	I
210		3043.50	I
120		3042.74	II
230		3049.46	I
210		3050.39	I
8600		3058.66	I
290		3060.30	I
570		3062.19	I
210		3069.94	I
360		3074.08	I
290		3074.96	I
290		3077.44	I
1100		3077.72	I
360		3078.11	I
230		3078.38	I
230		3090.08	I
270		3093.59	I
310		3101.53	I
360		3105.99	I
310		3108.98	I
620		3109.38	I
250		3111.09	I
310		3118.33	I
480		3131.12	I
250		3152.67	I
290		3153.61	I
3100		3156.25	I
250		3156.78	I
310		3166.51	I
180		3173.93	II
420		3178.06	I
230		3181.88	I
230		3185.33	I
310		3186.98	I
310		3189.46	I
310		3194.23	I
150		3213.31	II
1900		3232.06	I
290		3238.63	I
190		3241.04	I
120		3248.00	I
190		3254.91	I
190		3256.92	I
190		3260.30	I
3100		3262.29	I
380		3262.75	I
3100		3267.94	I
620		3269.21	I
190		3272.16	I
530		3275.20	I
330		3277.97	I
190		3288.84	I
1200		3290.26	I
7600		3301.56	I
250		3306.23	I
620		3310.91	I
120		3315.42	I
250		3324.33	I
310		3327.42	I
960		3336.15	I
110		3351.74	I
120		3353.91	
230		3357.97	I
250		3361.15	I
190		3364.12	I
120		3370.20	I
960		3370.59	I
160		3372.08	I
120		3378.68	I
310		3384.00	I
190		3385.94	I
620		3387.84	I
120		3401.17	I
620		3401.86	I
250		3402.51	I
120		3408.76	I
120		3412.74	I

Osmium (Cont.)

120		3421.69	I
150		3427.67	I
250		3440.60	I
120		3444.46	I
160		3445.55	I
310		3449.20	I
120		3458.38	I
120		3465.44	I
120		3478.53	I
120		3482.11	I
120		3487.46	I
120		3490.33	
160		3498.54	I
250		3501.16	I
620		3504.66	I
440		3512.99	I
310		3518.72	I
120		3520.00	I
480		3523.64	I
110		3528.04	I
1200		3528.60	I
230		3530.06	I
230		3532.80	I
120		3533.41	I
230		3542.71	I
960		3559.79	I
1200		3560.86	I
120		3562.34	I
310		3569.78	I
120		3574.08	I
120		3587.32	I
620		3598.11	I
190		3601.83	I
95		3604.48	II
250		3616.57	I
120		3619.43	I
450		3640.33	I
230		3654.49	I
330		3656.90	I
120		3666.31	I
480		3670.89	I
120		3675.45	I
250		3689.06	I
190		3703.25	I
120		3706.56	I
120		3709.14	I
230		3713.73	I
210		3719.52	I
230		3720.13	I
180		3746.47	I
3700		3752.52	I
100		3757.12	I
130		3766.30	I
120		3768.14	I
120		3774.40	I
110		3774.62	I
120		3776.25	I
290		3776.99	I
2100		3782.20	I
620		3790.14	I
180		3790.73	I
370		3793.91	I
250		3836.06	I
150		3840.30	I
150		3841.29	I
190		3849.94	I
230		3857.09	I
230		3865.47	I
730		3876.77	I
250		3881.86	I
140		3900.39	I
190		3901.71	I
100		3930.00	I
250		3938.59	I
100		3949.78	I
200		3961.02	I
1000		3963.63	I
100		3964.96	I
150		3969.67	I
110	h	3975.44	I
730		3977.23	I
100		3988.18	I
150		4003.48	I
100		4004.02	I
150		4005.16	I
160		4018.26	I
100		4037.84	I
280		4041.92	I
160		4048.05	I
960		4066.69	I
250		4070.86	I
190		4071.56	I
230		4074.68	I
490		4091.82	I
120		4100.30	I
1200		4112.02	I
180		4124.60	I
180		4128.96	I
2500		4135.78	I
150		4137.84	I
180		4172.57	I
1200		4173.23	I
620		4175.63	I
120		4184.13	I
320		4189.91	I
180		4201.45	I
250		4202.06	I
1200		4211.86	I
120		4213.86	I
100		4215.16	I
170		4233.46	I
4900		4260.85	I
100		4264.75	I
120		4269.61	I
100		4285.90	I
560		4293.95	I
560		4311.40	I
110		4326.25	I
340		4328.68	I
100		4338.75	I
100		4351.53	I
210		4365.67	I
110		4370.88	I
520		4394.86	I
160		4397.26	I
160		4402.74	I
4900		4420.47	I
100		4432.41	I
290		4436.32	I
100		4439.64	I
230		4447.35	I
120		4484.76	I
110		4548.66	I
540		4550.41	I
140		4551.30	I
170		4616.78	I
170		4631.83	I
140		4663.82	I
670		4793.99	I
110		4865.60	I
55		5031.83	I
45		5039.12	I
35		5072.88	I
35		5074.77	I
35		5079.09	I
90		5103.50	I
55		5110.81	I
22		5122.23	I
22		5145.54	I
140		5149.74	I
28		5152.01	I
28		5168.98	I
40		5193.52	I
270		5202.63	I
35		5203.23	I
20		5250.46	
45		5255.82	I
55		5265.15	I
20		5283.89	I
20		5295.65	I
40		5298.78	I
13		5302.58	I
18		5336.23	I
11		5346.03	I
13		5352.25	I
110		5376.79	I
16		5403.43	I
13		5412.14	I
120		5416.34	I
45		5416.69	I
28		5417.51	I
16		5441.82	I
55		5443.31	I
22		5446.93	I
11		5447.76	I
20		5449.37	I
20		5453.40	I
22		5457.30	I
28		5470.00	I
13		5474.58	I
13		5475.13	I
9		5477.27	I
16		5481.85	I
22		5509.33	I
9		5516.01	I
270		5523.53	I
22		5546.82	I
9		5549.79	I
13		5552.88	I
11		5560.62	I
16		5580.66	I
80		5584.44	I
8		5600.50	I
35		5620.08	I
9		5637.41	I
22		5642.56	I
28		5645.25	I
7		5648.98	I
9		5660.21	I
7		5674.38	I
28		5680.88	I
11		5709.37	I
170		5721.93	I
8		5737.89	I
8		5739.72	I
22		5765.05	I
170		5780.82	I
40		5800.60	I
8		5842.49	I
110		5857.76	I
28		5860.64	I
11		5882.92	I
11		5903.98	I
11		5906.84	I
7		5908.95	I
7		5981.36	I
11		5983.22	I
65		5996.00	I
20		6015.79	I
7		6054.63	I
20		6144.53	I
11		6158.03	I
35		6227.70	I
7		6241.70	I
22		6269.41	I
11		6274.94	I
11		6286.83	I
9		6398.86	I
22		6403.15	I
9		6448.13	I
6		6520.85	I
7		6528.87	I
7		6533.14	I
11		6538.30	I
11		6576.83	I
8		6614.56	I
4		6615.43	I
7		6661.81	I
27		6729.56	I
18		6791.53	I
14		6806.61	I
5		6878.70	I
4		6901.58	I
11		6956.02	I
6		6984.95	I
15		7060.67	I
22		7145.54	I
10		7149.89	I
4		7184.10	I
10		7206.33	I
5		7209.96	I
9		7251.16	I
6		7253.49	I
6		7375.07	I
9		7407.95	I
26		7602.95	I
4		7701.46	I
7		7789.96	I
7		7852.17	I
6		7981.20	I
7		8041.29	I

OXYGEN (O)

Z = 8

OI and II

Ref. 66, 69, 209, 210, 215 — R.L.K.

Intensity		Wavelength	
		Vacuum	
250		537.83	II
300		538.26	II
220		539.09	II
200		539.55	II
150		539.85	II
150		644.148	II
200		672.95	II
150		673.77	II
70		685.544	I
900		718.484	II
600		718.562	II
70		744.794	I
70		770.793	I
90		771.056	I
70		775.321	I
70		791.973	I
300		796.66	II
90		804.267	I
70		804.848	I
70		805.295	I
80		805.810	I
240		832.762	II
450		833.332	II
600		834.467	II
40		877.879	I
80		922.008	I
90		935.193	I
40		948.686	I
90		971.738	I
40		976.448	I
160		988.773	I
40		990.204	I
250		1025.762	I
90		1027.431	I
160		1039.230	I
60		1040.942	I
40		1152.152	I
900		1302.168	I
600		1304.858	I
300		1306.029	I
		Air	
30	d	2283.42	II
30	d	2284.89	II
110		2293.32	II
200		2300.35	II
30	d	2313.05	II
30	d	2316.12	II
30	d	2316.79	II
50	d	2319.68	II
30	d	2322.15	II
30	d	2339.31	II
110		2411.60	II
80		2425.55	II
250		2433.56	II
80	d	2436.06	II
80		2444.26	II
300		2445.55	II
300		2733.34	II
110		2747.46	II
265		2972.29	I
160		3122.62	II
220		3129.44	II
450		3134.82	II
285		3138.44	II
220		3270.98	II
220		3273.52	II
220		3277.69	II
360		3287.59	II
160		3305.15	II
160		3306.60	II
220		3377.20	II
285		3390.25	II
220		3407.38	II
160		3409.84	II
285		3470.81	II
220		3712.75	II
285		3727.33	II
160		3739.92	II
360		3749.49	II
160		3803.14	II
120		3823.41	I
450		3911.96	II
160		3919.29	II
185		3947.29	I
160		3947.48	I
140		3947.59	I
220		3954.37	II
100		3954.61	I
450		3973.26	II
220		3982.20	II
160		4069.90	II
285		4072.16	II
450		4075.87	II
80	d	4083.91	II
50	d	4087.14	II
150	d	4089.27	II
110		4097.24	II
220		4105.00	II
285		4119.22	II
160		4132.81	II
50		4146.06	II
220		4153.30	II
285		4185.46	II
450		4189.79	II
80		4233.27	I
50	d	4253.74	II
50	d	4253.98	II
50	d	4275.47	II
50	d	4303.78	II
285		4317.14	II
160		4336.86	II
220		4345.56	II
285		4349.43	II
220		4366.90	II
100		4368.25	I
220		4395.95	II
450		4414.91	II
285		4416.98	II
160		4448.21	II
160		4452.38	II
50		4465.45	II
50	d	4466.28	II
50		4467.83	II
50		4469.41	II
360		4590.97	II
285		4596.17	II
80	d	4609.39	II
160		4638.85	II
360		4641.81	II
450		4649.14	II
160		4650.84	II
360		4661.64	II
285		4676.23	II
220		4699.21	II
285		4705.36	II
160		4924.60	II
220		4943.06	II
135		5329.10	I
160		5329.68	I
190		5330.74	I
90		5435.18	I
110		5435.78	I

Oxygen (Cont.)

135		5436.86	I
120		5577.34	I
160		5958.39	I
190		5958.58	I
80		5995.28	I
160		6046.23	I
190		6046.44	I
110		6046.49	I
100		6106.27	I
400		6155.98	I
450		6156.77	I
490		6158.18	I
80		6256.83	I
100		6261.55	I
100		6366.34	I
100		6374.32	I
320		6453.60	I
360		6454.44	I
400		6455.98	I
80		6604.91	I
100		6653.83	I
360		7001.92	I
450		7002.23	I
210		7156.70	I
400		7254.15	I
450		7254.45	I
320		7254.53	I
210		7476.44	I
100		7477.24	I
120		7479.08	I
120		7480.67	I
100		7706.75	I
870		7771.94	I
810		7774.17	I
750		7775.39	I
80		7886.27	I
100		7943.15	I
100		7947.17	I
235		7947.55	I
210		7950.80	I
185		7952.16	I
110		7981.94	I
135		7982.40	I
190		7986.98	I
135		7987.33	I
250		7995.07	I
400		8221.82	I
265		8227.65	I
265		8230.02	I
325		8233.00	I
120		8235.35	I
120		8426.16	I
810		8446.25	I
1000		8446.36	I
935		8446.76	I
325		8820.43	I
160	d	9057.01	I
120		9118.29	I
80		9134.71	I
80		9150.14	I
80		9151.48	I
235		9156.01	I
450		9260.81	I
490		9260.84	I
450		9260.94	I
400		9262.58	I
540		9262.67	I
590		9262.77	I
490		9265.94	I
640		9266.01	I
185		9399.19	I
120		9481.16	I
120	d	9482.88	I
235		9487.43	I
140		9492.71	I
265		9497.97	I
160		9499.30	I
235		9505.59	I
210		9521.96	I
120		9523.36	I
120		9523.96	I
100		9528.28	I
100		9622.13	I
120		9625.29	I
160		9677.38	I
80		9694.66	I
65		9694.91	I
235		9741.50	I
235		9760.65	I
120		9909.05	I
140		9936.98	I
120		9940.41	I
160		9995.31	I
120	d	10421.18	I
590		11286.34	I
640		11286.91	I
490		11287.02	I
490		11287.32	I
490		11295.10	I
540		11297.68	I
590		11302.38	I
265		11358.69	I
490		12464.02	I
450		12570.04	I
120		12990.77	I
160		13076.91	I
700		13163.89	I
750		13164.85	I
640		13165.11	I
160		16212.06	I
120		17966.70	I
590		18021.21	I
120		18041.48	I
120		18042.19	I
120		18046.23	I
140		18229.23	I
540		18243.63	I
140		26173.56	I

O III
Ref. 23, 66, 210 — R.L.K.

Intensity		Wavelength	
		Vacuum	
80	d	264.34	III
110		264.48	III
110		266.97	III
150		266.98	III
150		267.03	III
150		277.38	III
80		295.62	III
110		295.66	III
120		295.72	III
150		303.41	III
150		303.46	III
140		303.52	III
160		303.62	III
160		303.69	III
250		303.80	III
200		305.60	III
250		305.66	III
190		305.70	III
300		305.77	III
190		305.84	III
450		320.979	III
300		328.45	III
250		328.74	III
300		345.31	III
110		355.14	III
90		355.33	III
80		355.47	III
200		359.02	III
190		359.22	III
150		359.38	III
210		373.80	III
200		374.00	III
300		374.08	III
190		374.16	III
200		374.33	III
210		374.44	III
450		395.558	III
300		434.98	III
800		507.391	III
900		507.683	III
1000		508.182	III
1000		525.795	III
700		597.818	III
1000		599.598	III
110		609.70	III
160		610.04	III
200		610.75	III
100		610.85	III
800		702.332	III
800		702.822	III
900		702.899	III
1000		703.850	III
600		832.927	III
780		833.742	III
600		835.096	III
800		835.292	III
160		1476.89	III
285		1590.01	III
160		1591.33	III
220		1760.12	III
110		1760.42	III
220		1763.22	III
220		1764.48	III
750		1767.78	III
550		1768.24	III
360		1771.67	III
110		1773.00	III
110		1773.85	III
220		1779.16	III
160		1781.03	III
160		1784.85	III
220		1789.66	III
110		1848.26	III
110		1856.62	III
285		1872.78	III
285		1872.87	III
285		1874.94	III
160		1920.04	III
110		1920.75	III
110		1921.52	III
220		1923.49	III
110		1923.82	III
110		1926.94	III
		Air	
360		2013.27	III
160		2026.96	III
220		2045.67	III
160		2052.74	III
200	d	2390.44	III
80		2394.33	III
80		2422.84	III
80	d	2438.83	III
200		2454.99	III
200		2558.06	III
80		2687.53	III
110		2695.49	III
80		2959.68	III
250		2983.78	III
80		3017.63	III
80		3023.45	III
80		3043.02	III
200		3047.13	III
110		3059.30	III
80		3121.71	III
110		3132.86	III
80		3238.57	III
200		3260.98	III
300		3265.46	III
80		3267.31	III
80		3312.30	III
110		3340.74	III
80		3444.10	III
80		3455.12	III
80		3698.70	III
80		3702.75	III
80		3703.37	III
110		3707.24	III
110		3715.08	III
110		3744.00	III
150		3754.67	III
80		3757.21	III
250		3759.87	III
110		3791.26	III
200		3961.59	III
110		5592.37	III

O IV
Ref. 36, 66 = R.L.K.

Intensity		Wavelength	
		Vacuum	
150		195.86	IV
200		196.01	IV
110		207.18	IV
150		207.24	IV
140		233.46	IV
150		233.50	IV
110		233.52	IV
200		233.56	IV
110		233.60	IV
90		238.36	IV
180		238.57	IV
110		252.56	IV
110		252.95	IV
150		253.08	IV
300		260.39	IV
250		260.56	IV
300		279.63	IV
375		279.94	IV
110		285.71	IV
150		285.84	IV
200		306.62	IV
150		306.88	IV
700		553.330	IV
775		554.075	IV
850		554.514	IV
700		555.261	IV
580		608.398	IV
640		609.829	IV
270		616.952	IV
150		617.005	IV
200		617.036	IV
520		624.617	IV
580		625.130	IV
640		625.852	IV
200		779.734	IV
315		779.821	IV
360		779.912	IV
200		779.997	IV
640		787.711	IV
520		790.109	IV
700		790.199	IV
200		802.200	IV
160		802.255	IV
130		921.296	IV
160		921.366	IV
200		923.367	IV
130		923.433	IV
200		1338.612	IV
130		1342.992	IV
230		1343.512	IV
		Air	
200		2449.372	IV
200		2450.040	IV
200		2493.44	IV
200		2493.77	IV
200		2507.73	IV
230		2509.19	IV
200		2517.2	IV
160		2836.26	IV
160		2921.45	IV
460		3063.42	IV
410		3071.61	IV
160		3209.66	IV
230		3348.08	IV
270		3349.11	IV
160		3354.27	IV
200		3375.40	IV
130		3378.06	IV
360		3381.20	IV
360		3385.52	IV
270		3396.79	IV
360		3403.52	IV
230		3409.66	IV
410		3411.69	IV
230		3413.64	IV
200		3489.83	IV
160		3492.24	IV
230		3560.39	IV
270		3563.33	IV
315	w	3725.93	IV
360		3729.03	IV
410		3736.85	IV
230		3744.89	IV

O V
Ref. 24, 66 = R.L.K.

Intensity		Wavelength	
		Vacuum	
80		124.616	V
110		135.523	V
80		138.109	V
110		139.029	V
80		151.447	V
110		151.477	V
150		151.546	V
80		164.574	V
110		164.657	V
80		164.709	V
80		166.235	V
150		167.99	V
110		170.219	V
450		172.169	V
250		185.745	V
375		192.751	V
450		192.799	V
520		192.906	V
80		193.003	V
200		194.593	V
80		202.161	V
80		202.224	V
80		202.283	V
80		202.334	V
150		202.393	V
110		203.78	V
150		203.82	V
100		203.85	V
200		203.89	V
100		203.94	V
300		207.794	V
150		215.040	V
200		215.103	V
250		215.245	V
250		216.018	V
520		220.352	V
80		227.372	V
80		227.469	V
150		227.511	V
80		227.549	V
80		227.634	V
80		227.689	V
150		231.823	V
110		248.459	V
110		286.448	V
1000		629.730	V
230		681.272	V
700		758.678	V
640		759.441	V
580		760.228	V
775		760.445	V
640		761.128	V
700		762.003	V
520		774.518	V
640		1371.292	V
160	w	1506.72	V
315	w	1643.68	V
160		1707.996	V
		Air	
1000		2781.01	V
920		2786.99	V
775		2789.85	V
200		2941.33	V
210		2941.65	V
160		3144.66	V
100		4123.99	V
230	w	4930.27	V
130		5597.91	V
130		6500.24	V

PALLADIUM (Pd)
Z = 46

Pd I and II
Ref. 1, 287 — C.H.C.

Intensity		Wavelength	
		Air	
50		2162.27	II
50		2182.35	II
50		2212.15	II
100	r	2231.59	II
200	r	2296.53	II
50		2351.32	II
50		2362.31	II
75		2367.92	II
60		2372.16	II
50		2388.29	II
60		2416.73	II
75		2418.72	II
75		2424.49	II
100		2426.87	II
100		2430.94	II
100		2433.11	II
100		2435.32	II
150		2446.17	II
75		2446.72	II
1100		2447.91	I
80		2448.15	II
100		2457.29	II
60		2457.76	II
150		2469.29	II
80		2470.06	II
100		2471.18	II
50		2472.55	II
1700		2476.42	I
250		2486.52	II
300		2488.92	II
75		2489.61	II
200		2498.81	II
150		2505.73	II
50		2514.47	II
80		2534.57	II
50	h	2539.44	II
150		2551.84	II
150		2565.51	II
100		2569.56	II
60		2593.24	II
50		2628.24	II
70		2635.92	II
150		2658.75	II
1900		2763.09	I
150	h	2776.85	II
100	h	2787.92	II
50	h	2800.64	II
50	h	2807.59	II
200		2854.59	II
100	h	2871.37	II
100	h	2878.01	II
520		2922.49	I
50		2980.63	II
650		3002.65	I
45		3009.78	I
1500		3027.91	I
1100		3065.31	I
2600		3114.04	I
270		3142.81	I
11000		3242.70	I
2700		3251.64	I
3500		3258.78	I
460		3287.25	I
3600		3302.13	I
5000		3373.00	I
24000		3404.58	I
13000		3421.24	I
5000		3433.45	I
6400		3441.40	I
7700		3460.77	I
10000		3481.15	I
2000		3489.77	I
12000		3516.94	I
12000		3553.08	I
4500		3571.16	I
20000		3609.55	I
20000		3634.70	I
5500		3690.34	I
1400		3718.91	I
1500		3799.19	I
1500		3832.29	I
2200		3894.20	I
1500		3958.64	I
290		4087.34	I
90		4169.84	I
2500		4212.95	I
180		4473.59	I
55	h	4788.18	I
45	h	4817.51	I
35		4875.43	I
55		5110.81	I
75		5117.02	I
160		5163.84	I
55		5234.86	I
120		5295.63	I
18		5312.57	I
15		5345.10	I
35		5395.24	I
55		5542.80	I
35		5547.02	I
27		5619.44	I
15		5642.69	I
14		5655.42	I
75		5670.07	I
11		5690.14	I
55	h	5695.09	I
18		5736.61	I
23		6774.54	I
65		6784.52	I
4	h	6833.42	I
11		7016.44	I
13	h	7310.06	I
75		7368.12	I
27		7391.92	I
16		7486.90	I
120		7764.03	I
27		7786.67	I
45		7915.80	I
18		7961.08	I
55		8132.82	I
45		8300.83	I
9	h	8353.58	I
18	h	8532.74	I
16	h	8599.10	I
65		8761.35	I

Pd III
Ref. 424 — L.J.R.

Intensity		Wavelength	
		Vacuum	
10		688.74	III
20		689.46	III
50		689.54	III
50		695.91	III
200		705.49	III
150		707.80	III
100		709.89	III
100		717.90	III
100		719.47	III
200		727.72	III
150		738.79	III
100		756.85	III
100		757.41	III
500		763.06	III
500		766.42	III
200		772.11	III
200		776.51	III
2000		781.02	III
200		784.99	III
200		787.31	III
200		787.95	III
200		789.58	III
500		794.08	III
500		797.52	III
500		800.03	III
500		800.10	III
500		803.67	III
500		825.35	III
500		840.58	III
500		856.47	III
500		864.04	III
500		880.59	III
500		888.84	III
1000		889.29	III
300		947.78	III
300		965.52	III
100		1505.40	III
200		1517.18	III
200	h	1526.88	III
100		1542.63	III
200	h	1545.95	III
300		1596.89	III
200	h	1606.10	III
150		1630.84	III
50		1679.73	III
100		1704.33	III
50		1706.40	III
200		1719.86	III
500		1741.62	III
400		1758.19	III
4000		1782.55	III
400		1804.91	III
400		1843.49	III
1500		1851.59	III
2000		1852.27	III
1000		1859.21	III
1500		1874.63	III
2000		1885.83	III
1000		1887.40	III
1500		1891.34	III
4000		1914.62	III
1000		1930.33	III
2000		1941.64	III
400		1951.56	III
300		1972.29	III
300		1977.53	III
		Air	
800		2002.16	III
1000		2004.47	III
500		2055.11	III
500		2149.82	III
500		2177.55	III
500		2177.63	III
100		2291.45	III
100		2452.42	III
100		2633.22	III

PHOSPHORUS (P)
Z = 15

P I and II
Ref. 182 = R.L.K.

Intensity	Wavelength	
	Vacuum	
10	810.24	II
10	865.44	II
20	1249.82	II
20	1301.87	II
20	1304.47	II
15	1304.68	II
35	1305.48	II
25	1309.87	II
60	1310.70	II
15	1372.033	I
15	1373.500	I
10	1374.732	I
15	1377.080	I
15	1377.937	I
25	1379.429	I
25	1381.469	I
15	1381.637	I
30	1452.89	II
80	1532.51	II
120	1535.90	II
80	1536.39	II
120	1542.29	II
140	1671.070	I
100	1671.510	I
180	1671.680	I
140	1672.035	I
140	1672.474	I
600	1674.591	I
600	1679.695	I
140	1685.976	I
100	1694.028	I
100	1694.486	I
100	1706.376	I
100	1707.553	I
600	1774.951	I
500	1782.838	I
400	1787.656	I
140	1834.801	I
140	1847.165	I
100	1849.820	I
140	1851.194	I
100	1852.069	I
500	1858.886	I
400	1859.393	I
140	1864.348	I
180	1905.481	I
140	1906.403	I
280	1907.665	I
	Air	
280	2023.489	I
180	2024.516	I
400	2032.432	I
400	2033.477	I
400	2135.465	I
400	2136.182	I
400	2149.145	I
280	2152.940	I
500	2154.080	I
180	2235.732	I
100	2484.19	II
750	2533.976	I
950	2535.603	I
750	2553.262	I
500	2554.915	I
150	2606.06	II
100	2626.18	II
90	2636.76	II
150	3308.92	II
125	3419.34	II
100	3425.00	II
100	4178.48	II
200	4288.60	II
200	4385.35	II
400	4420.71	II
100	4452.46	II
150	4463.00	II
120	4467.98	II
200	4475.26	II
200	4499.24	II
120	4530.81	II
120	4554.83	II
120	4558.07	II
120	4581.71	II
500	4588.04	II
500	4589.86	II
600	4602.08	II
300	4626.70	II
300	4658.31	II
200	4864.42	II
150	4927.20	II
500	4943.53	II
300	4954.39	II
300	4969.71	II
100	5079.381	I
100	5098.221	I
100	5100.974	I
140	5109.628	I
140	5154.844	I
180	5162.290	I
150	5191.41	II
300	5253.52	II
140	5293.539	I
400	5296.13	II
250	5316.07	II
300	5344.75	II
180	5345.851	I
100	5364.631	I
250	5378.20	II
300	5386.88	II
200	5409.72	II
400	5425.91	II
100	5428.094	I
400	5450.74	II
140	5458.305	I
125	5461.20	II
180	5477.672	I
140	5477.860	I
140	5478.267	I
200	5483.55	II
200	5499.73	II
200	5507.19	II
100	5514.774	I
100	5516.997	I
200	5541.14	II
200	5583.27	II
250	5588.34	II
100	5727.71	II
500	6024.18	II
400	6034.04	II
500	6043.12	II
250	6055.50	II
100	6057.86	II
350	6087.82	II
180	6097.690	I
350	6165.59	II
500	6199.024	I
180	6210.499	I
100	6232.29	II
200	6367.27	II
140	6375.681	I
100	6388.579	I
250	6435.32	II
130	6436.31	II
600	6459.99	II
600	6503.46	II
600	6507.97	II
150	6713.28	II
100	6717.411	I
100	7102.200	I
100	7158.367	I
180	7165.465	I
180	7175.102	I
180	7176.660	I
250	7845.63	II
100	8046.801	I
140	8278.058	I
100	8367.856	I
140	8531.475	I
140	8613.835	I
180	8637.578	I
400	8741.529	I
100	8872.174	I
140	9153.34	I
180	9175.819	I
950	9193.85	I
600	9278.88	I
1250	9304.94	I
500	9323.50	I
100	9327.13	I
140	9372.09	I
950	9435.069	I
950	9441.86	I
600	9452.83	I
100	9481.84	I
100	9492.12	I
1250	9493.56	I
180	9521.78	I
1700	9525.73	I
1500	9545.18	I
280	9556.81	I
1700	9563.439	I
280	9593.50	I
750	9609.04	I
180	9625.80	I
180	9628.42	I
400	9638.939	I
140	9675.41	I
500	9676.24	I
180	9706.533	I
1500	9734.750	I
280	9736.680	I
1500	9750.77	I

Phosphorus (Cont.)

Intensity	Wavelength	
100	9760.77	I
100	9776.85	I
100	9779.11	I
600	9790.21	I
1700	9796.85	I
280	9834.80	I
400	9903.68	I
280	9976.67	I
229	10084.27	I
174	10432.66	I
132	10455.87	I
458	10511.58	I
962	10529.52	I
1235	10581.57	I
415	10596.90	I
435	10681.40	I
265	10813.13	I
134	10932.72	I
103	10967.37	I
180	11160.05	I
764	11183.23	I
402	11186.75	I
76	13438.43	I
86	13485.19	I
479	14241.64	I
150	14272.75	I
256	14307.83	I
173	14430.50	I
135	14470.62	I
98	14646.42	I
714	15711.52	I
228	15962.53	I
296	16254.77	I
203	16292.97	I
1627	16482.92	I
588	16590.07	I
225	16613.05	I
221	16738.68	I
419	16803.39	I
471	17112.48	I
104	17223.28	I
289	17286.91	I
145	17359.00	I
299	17423.67	I
95	17665.68	I
186	18007.63	I
106	18518.90	I
92	18881.16	I
125	20841.62	I
124	23038.83	I
287	23844.97	I
98	26134.44	I
118	27959.52	I
188	28049.42	I
127	28154.52	I
132	28284.16	I
98	28288.69	I
311	29097.16	I
92	31483.21	I
91	32270.90	I
146	35551.93	I
90	35582.27	I
192	35802.53	I
146	36417.43	I

P III
Ref. 180 = R.L.K.

Intensity	Wavelength	
	Vacuum	
90	471.146	III
90	484.278	III
120	498.180	III
200	569.853	III
200	581.831	III
200	844.646	III
150	845.038	III
250	845.664	III
300	847.669	III
200	848.016	III
120	848.465	III
150	848.639	III
250	852.686	III
350	855.624	III
200	859.406	III
500	859.652	III
250	859.729	III
300	913.971	III
300	917.120	III
350	918.665	III
250	921.849	III
200	997.999	III
250	1003.598	III
500	1334.808	III
650	1344.327	III
300	1344.845	III
250	1380.463	III
150	1381.089	III
350	1502.228	III
250	1504.663	III
150	1618.632	III
200	1618.907	III
	Air	
200	2611.147	III
300	2632.713	III
200	2680.133	III
250	2895.241	III
250	3186.186	III
300	3219.307	III
150	3233.536	III
400	3233.602	III
200	3556.546	III
200	3577.526	III
200	3904.812	III
250	3914.314	III
300	3957.641	III
350	3978.307	III
200	4057.440	III
400	4059.312	III
300	4080.084	III
500	4222.195	III
350	4246.720	III
200	4428.171	III
200	4463.668	III
250	4479.776	III
150	6083.409	III
150	6409.204	III
150	6484.440	III
150	6486.381	III
150	6992.690	III
150	8113.528	III

P IV
Ref. 336 – R.L.K.

Intensity	Wavelength	
	Vacuum	
90	282.301	IV
90	304.996	IV
120	359.293	IV
150	359.899	IV
120	361.514	IV
150	361.629	IV
120	371.299	IV
150	371.504	IV
200	372.001	IV
500	388.318	IV
120	414.604	IV
200	414.999	IV
250	415.805	IV
250	444.245	IV
300	445.158	IV
250	568.038	IV
350	629.008	IV
400	629.914	IV
500	631.779	IV
350	648.482	IV
300	756.510	IV
300	776.353	IV
650	823.179	IV
700	824.730	IV
800	827.932	IV
250	847.019	IV
350	849.799	IV
200	850.392	IV
700	877.476	IV
1000	950.655	IV
570	1025.563	IV
500	1028.096	IV
570	1030.517	IV
500	1033.111	IV
500	1035.517	IV
570	1118.551	IV
200	1206.422	IV
200	1335.705	IV
500	1366.695	IV
400	1372.674	IV
350	1377.282	IV
500	1484.507	IV
400	1487.788	IV
300	1489.098	IV
250	1862.762	IV
120	1862.893	IV
200	1863.580	IV
650	1888.523	IV
200	1910.183	IV
120	1985.682	IV
150	1985.851	IV
200	1986.114	IV
150	1987.022	IV
	Air	
200	2477.823	IV
150	2478.070	IV
250	2478.256	IV
250	2605.506	IV
400	2644.295	IV
300	2724.764	IV
400	2728.770	IV
200	2729.120	IV
500	2739.309	IV
250	2739.872	IV
200	2740.223	IV
200	2961.242	IV
650	3347.736	IV
570	3364.467	IV
400	3371.122	IV
200	3413.543	IV
200	3733.393	IV
300	4249.656	IV
250	4540.288	IV
250	4541.112	IV
150	4548.056	IV
200	4548.449	IV
150	5235.499	IV
150	5989.774	IV
150	6142.605	IV
150	6713.939	IV
120	6715.906	IV
200	7443.657	IV

P V
Ref. 179 = R.L.K.

Intensity	Wavelength	
	Vacuum	
80	255.59	V
50	255.67	V
110	310.58	V
150	311.34	V
300	328.47	V
250	328.78	V
150	347.23	V
200	348.20	V
110	378.56	V
250	389.50	V
300	390.70	V
150	410.03	V
375	475.60	V
110	534.63	V
80	534.99	V
520	542.57	V
600	544.92	V
450	673.90	V
450	865.45	V
600	871.39	V
250	997.62	V
150	1000.38	V
900	1117.98	V
700	1128.01	V
150	1379.62	V
250	1385.05	V
375	1447.83	V
450	1610.50	V
	Air	
200	2180.29	V
150	2186.42	V
375	2424.40	V
450	2440.93	V
200	2441.24	V
300	2961.00	V
450	2978.55	V
700	3175.09	V
520	3204.04	V
150	4083.18	V
110	4094.95	V
110	5156.72	V

PLATINUM (Pt)
Z = 78

Pt I and II
Ref. 1, 288 — C.H.C.

Intensity		Wavelength	
		Vacuum	
30		1621.66	II
30		1723.13	II
30		1751.70	II
50	r	1777.09	II
30		1781.86	II
30		1879.09	II
40		1883.05	II
50		1889.52	II
50		1911.70	II
30		1929.25	II
30		1929.68	II
30		1939.80	II
30		1949.90	II
30		1983.74	II
		Air	
40		2014.93	II
3200		2030.63	I
4400		2032.41	I
100		2036.46	II
40		2041.57	II
5500		2049.37	I
1500		2067.50	I
3000		2084.59	I
1000		2103.33	I
30		2115.57	II
950		2128.61	I
30		2130.69	II
1900		2144.23	I
100		2144.24	II
600		2165.17	I
1500		2174.67	I
30		2190.32	II
400		2202.22	I
50	h	2202.58	II
320		2222.61	I
50	h	2233.11	II
150		2249.30	I
30	h	2240.99	II
100		2245.52	II
30		2251.52	II
30	h	2251.92	II
190		2268.84	I
30	h	2271.72	II
280		2274.38	I
50	h	2287.50	II
30		2288.20	II
150		2289.27	I
150		2292.40	I
240		2308.04	I
50		2310.96	II
90		2315.50	I
220		2318.29	I
100		2326.10	I
170		2340.18	I
280		2357.10	I
180		2368.28	I
50		2377.28	II
130		2383.64	I
40		2386.81	I
120		2389.53	I
35		2396.17	I
70		2401.87	I
200		2403.09	I
100		2418.06	I
50		2424.87	II
80		2428.04	I
50		2428.20	I
25		2429.10	I
180		2436.69	I
650		2440.06	I
60		2450.97	I
440		2467.44	I
35		2471.01	I
1000		2487.17	I
25		2488.74	II
200		2490.12	I
160		2495.82	I
240		2498.50	I
50		2505.93	I
120		2508.50	I
50		2514.07	I
60		2515.03	I
240		2515.58	I
140		2524.30	I
40		2529.41	I
50		2536.49	I
160		2539.20	I
18		2549.46	I
50		2552.25	I
50		2596.00	I
70		2603.14	I
30		2616.76	II
50		2619.57	I
30		2625.34	II
1100		2628.03	I
130		2639.35	I
1000		2646.89	I
500		2650.86	I
20		2658.17	I
2800		2659.45	I
40		2674.57	I
440		2677.15	I
200		2698.43	I
2000		2702.40	I
1600		2705.89	I
60		2713.13	I
1300		2719.04	I
130		2729.92	I
1800		2733.96	I
70		2738.48	I
70		2747.61	I
80		2753.86	I
200		2754.92	I
30		2769.84	I
500		2771.67	I
40		2773.24	I
20		2774.00	I
50		2774.77	II
50		2793.27	I
100		2794.21	II
40	h	2799.98	II
140		2803.24	I
10		2808.51	I
50		2818.25	I
30	h	2822.27	II
1400		2830.30	I
70		2834.71	I
16		2853.11	I
80	h	2860.68	II
40	h	2865.05	II
40	h	2875.85	II
100	h	2877.52	II
25		2888.20	I
25		2893.22	I
600		2893.86	I
300		2897.87	I
60		2905.90	I
120		2912.26	I
120		2913.54	I
70		2919.34	I
30		2921.38	I

Platinum (Cont.)

1700		2929.79	I
30		2942.76	I
30		2944.75	I
25		2959.10	I
60		2960.75	I
1800		2997.97	I
35		3001.17	II
220		3002.27	I
30		3017.88	I
30	h	3031.22	II
130		3036.45	I
800		3042.64	I
3200		3064.71	I
30		3071.94	I
130		3100.04	I
320		3139.39	I
140		3156.56	I
120		3200.71	I
320		3204.04	I
30		3230.29	I
20		3233.42	I
20		3250.36	I
40		3251.98	I
160		3255.92	I
25		3268.42	I
25		3281.97	I
120		3290.22	I
500		3301.86	I
60		3315.05	I
35		3323.80	I
340		3408.13	I
35		3427.93	I
60		3483.43	I
160		3485.27	I
120		3628.11	I
70		3638.79	I
70		3643.17	I
50		3663.10	I
80		3671.99	I
80		3674.04	I
35		3699.91	I
18		3706.53	I
20		3801.05	
80		3818.69	I
40		3900.73	I
110		3922.96	I
35		3948.40	I
100		3966.36	I
20		3996.57	I
110		4118.69	I
80		4164.56	I
40		4192.43	I
18		4327.06	I
18		4391.83	I
80		4442.55	I
14		4445.55	I
25		4498.76	I
12		4520.90	I
35		4552.42	I
12		4879.53	I
14		5044.04	I
30		5059.48	I
35		5227.66	I
40		5301.02	I
12		5368.99	I
12		5390.79	I
14		5475.77	I
14		5478.50	I
6		5763.57	I
20		5840.12	I
8		5844.84	I
6		6026.04	I
7		6318.37	I
8		6326.58	I
9		6523.45	I
10		6710.42	I
20		6760.02	I
60		6842.60	I
20		7113.73	I
10		8224.74	I

PLUTONIUM (Pu)
Z = 94

Pu I and II
Ref. 91 — J.G.C.

Intensity		Wavelength Air	
10000		2781.40	II
10000		2784.48	II
10000		2806.11	II
10000		2815.77	II
10000		2897.97	II
10000		2898.94	II
10000		2904.25	II
10000		2904.94	II
10000		2910.40	II
10000		2918.00	II
10000		2918.80	II
10000		2926.08	II
10000		2928.25	II
10000		2929.71	II
10000		2930.98	II
10000		2932.32	II
10000		2933.30	II
10000		2938.54	II
10000		2938.95	II
10000		2941.39	II
10000		2945.26	II
10000		2946.00	II
10000		2950.06	II
10000		2951.62	
10000		2954.46	II
10000		2963.47	II
10000		2966.84	II
10000		2967.54	II
10000		2972.50	II
10000		2977.81	II
10000		2978.37	II
10000		2980.23	II
10000		2981.23	II
10000		2986.95	II
10000		2988.21	II
10000		2991.31	II
10000		2996.40	II
10000		3000.31	II
10000		3009.57	II
10000		3028.85	II
10000		3042.61	II
10000		3043.12	II
10000		3060.32	II
10000		3069.32	II
10000		3091.33	II
10000		3091.94	II
10000		3092.59	II
10000		3104.12	II
10000		3105.04	II
10000		3106.03	
10000		3123.87	II
10000		3159.21	II
10000		3161.73	II
10000		3163.18	II
10000		3174.49	II
10000		3179.41	II
10000		3185.12	II
10000		3187.60	II
10000		3189.23	II
10000		3193.54	
10000		3193.55	II
10000		3194.56	II
10000		3198.47	II
10000		3200.23	II
10000		3201.00	II
10000		3201.66	II
10000		3204.48	II
10000		3206.80	II
10000		3207.97	II
10000		3215.08	I
10000		3216.15	II
10000		3220.94	II
10000		3224.87	II
10000		3231.86	II
10000		3232.24	
10000		3232.63	II
10000		3241.39	II
10000		3242.96	II
10000		3243.40	II
10000		3244.16	I
10000		3245.25	II
10000		3245.71	
10000		3246.35	II
10000		3247.50	
10000		3247.56	II
10000		3252.08	I
10000		3260.54	II
10000		3265.17	
10000		3273.11	II
10000		3274.71	II
10000		3275.24	I
10000		3292.56	I
10000		3293.61	I
10000		3296.91	I
10000		3297.87	I
10000		3298.47	II
10000		3301.76	I
10000		3306.59	I
10000		3306.66	I
10000		3307.66	II
10000		3308.75	I
10000		3312.65	II
10000		3315.34	II
10000		3316.96	II
10000		3320.61	I
10000		3320.84	I
10000		3323.48	
10000		3327.19	I
10000		3330.11	
10000		3331.52	II
10000		3332.34	I
10000		3333.03	
10000		3337.71	II
10000		3338.40	II
10000		3338.94	I
10000		3347.87	II
10000		3349.63	I
10000		3351.82	II
10000		3356.61	II
10000		3358.41	II
10000		3358.84	II
10000		3362.26	II
10000		3365.20	I
10000		3365.66	
10000		3368.86	I
10000		3370.64	II
10000		3371.19	I
10000		3375.80	I
10000		3376.76	II
10000		3376.94	II
10000		3377.37	II
10000		3379.51	I
10000		3381.82	I
10000		3381.97	II
10000		3382.70	I
10000		3390.33	II
10000		3391.41	II
10000		3393.67	I
10000		3394.32	I
10000		3418.88	II
10000		3465.10	II
10000		3473.64	II
10000		3483.20	I
10000		3585.87	II
10000		3632.21	II
10000		3699.19	II
10000		3720.59	I
10000		3725.98	I
10000		3726.11	II
10000		3726.79	II
10000		3732.03	II
10000		3744.78	I
10000		3753.63	I
10000		3755.94	I
10000		3757.82	I
10000		3758.34	I
10000		3774.38	I
10000		3776.71	I
10000		3792.22	I
10000		3799.37	I
10000		3805.93	I
10000		3811.40	I
10000		3812.30	II
10000		3827.57	I
10000		3835.52	I
10000		3836.96	I
10000		3838.92	I
10000		3842.10	I
10000		3851.01	I
10000		3851.85	I
10000		3878.54	I
10000		3895.89	I
10000		3928.53	I
10000		3975.43	II
10000		4097.12	I
10000		4101.96	I
10000		4105.95	II
10000		4111.07	I
10000		4114.91	I
10000		4128.12	I
10000		4129.93	II
10000		4133.01	I
10000		4135.97	I
10000		4140.04	I
10000		4141.20	II
10000		4151.09	I
10000		4151.45	I
10000		4155.46	I
10000		4159.39	II
10000		4167.77	I
10000		4170.95	I
10000		4178.28	II
10000		4189.90	II
10000		4190.06	II
10000		4196.20	II
10000		4206.48	I
10000		4208.23	I
10000		4221.87	I
10000		4224.20	I
10000		4229.77	II
10000		4254.76	II
10000		4261.88	I
10000		4269.77	I
10000		4273.34	II
10000		4281.17	I
10000		4289.08	II
10000		4337.18	II
10000		4352.71	II
10000		4367.41	I
10000		4379.91	II
10000		4385.35	II
10000		4393.93	II
10000		4404.90	I
10000		4441.65	II
10000		4468.54	II
10000		4472.79	II
10000		4493.78	II
10000		4504.91	II
10000		4536.15	II
10000		4735.40	I
10000		4989.34	I
10000		5269.86	I
10000		5381.02	I
10000		5498.50	I
10000		5510.72	I
10000		5537.59	I
10000		5549.62	I
10000		5590.54	I
10000		5592.33	I
10000		5712.39	I
10000		5770.26	I
10000		5839.05	I
10000		5983.35	I
10000		6012.78	I
10000		6192.80	I
10000		6304.66	I
10000		6449.75	I
10000		6486.71	I
10000		6488.86	I
10000		6488.89	
10000		6535.27	I
10000		6544.21	I
10000		6608.95	I
10000		6672.72	I
10000		6784.66	I
10000		6880.16	I
10000		6891.38	I
10000		7059.23	I
10000		7068.90	I
10000		7092.46	I
10000		7116.88	I
10000		7141.66	I
10000		7177.14	I
10000		7231.09	I
10000		7258.06	I
10000		7322.23	I
10000		7325.97	I
10000		7331.81	I
10000		7431.18	I
10000		7447.99	I
10000		7507.80	I
10000		7526.93	I
10000		7547.45	I
10000		7564.50	I
10000		7571.87	I
10000		7572.93	I
10000		7609.77	I
10000		7689.40	I
10000		7758.20	I
10000		7798.54	I
10000		7953.17	I
10000		8102.54	I
10000		8130.86	I
10000		8309.61	I
10000		8435.47	I
10000		8476.13	I
10000		8495.75	
10000		8597.26	I
10000		8665.02	I
10000		8691.94	I
3000		8729.82	I
3000		8836.16	I
3000		9533.07	I
3000		10046.75	I
3000		11114.82	I
3000		12144.46	I
3000		12231.22	I
3000		15377.31	I
3000		16397.38	I

POLONIUM (Po)
Z = 84

Po I and II
Ref. 47, 48 — E.F.W.

Intensity		Wavelength Air	
250	w	2139.02	I
300	h	2203.80	
300		2220.67	I
200		2222.13	I
200		2284.22	
250		2344.61	I
250		2421.72	I
300		2426.09	I
1500	w	2450.08	I
700		2483.94	I
700		2490.53	I
200	h	2502.18	
300		2534.95	I
300		2557.33	I
1500	w	2558.01	I
400		2562.31	
300		2578.80	
400		2587.64	I
200		2637.01	
300		2645.36	I
700	h	2663.33	
200		2671.67	I
600		2761.92	I
400		2800.26	I
250		2824.11	I
300		2866.01	I
400		2919.31	I
600		2958.92	I
2500	w	3003.21	I
450		3069.31	I
200		3115.95	

Potassium (Cont.)

Intensity		Wavelength	
400		3189.02	I
600		3240.24	I
250		3286.38	I
600		3328.60	I
300		3489.79	
200		3493.65	
400		3588.33	
200		3671.36	
500		3861.93	I
200		4051.98	
1200		4170.52	I
250		4236.13	
200	h	4415.58	
800		4493.21	I
350		4611.44	I
200		4867.12	
400		4876.24	I
450		4946.81	
350		5323.23	I
300		5744.85	I
600		7962.62	I
300		8433.87	I
500		8618.26	I
250		9227.87	

POTASSIUM (K)
Z = 19

K I and II
Ref. 59, 76, 172, 268 — L.J.R.

Intensity		Wavelength	
		Vacuum	
5		261.20	II
25		441.81	II
5		465.08	II
		469.50	II
10		476.03	II
30		495.14	II
30		600.77	II
25		607.93	II
30		612.62	II
3		1725.0	II
		Air	
6		2190.00	II
4		2210.53	II
5		2265.04	II
6		2550.02	II
4		2743.55	II
		2992.12	I
		2992.42	I
		3034.76	I
		3034.92	I
5		3062.18	II
4		3101.79	I
3		3102.04	I
6		3105.00	II
5		3190.07	II
7		3217.16	I
6		3217.62	I
4		3220.60	II
5		3290.65	II
6		3345.32	II
6		3373.60	II
6		3380.62	II
6		3384.86	II
6		3404.24	II
7		3440.05	II
11		3446.37	I
10		3447.38	I
6		3481.11	II
7		3530.75	II
5		3608.88	II
6		3618.49	II
4		3626.42	II
3		3648.84	I
4		3648.98	I
6		3681.54	II
5		3716.60	II
5		3721.34	II
5		3739.13	II
5		3744.42	II
6		3767.36	II
6		3783.19	II
6		3800.14	II
6		3816.56	II
7		3817.50	II
5		3873.74	II
4		3878.62	II
8		3897.92	II
5		3923.00	II
5		3926.36	II
6		3942.53	II
6		3955.21	II
6		3966.72	II
6		3972.58	II
6		3995.10	II
7		4001.24	II
5		4012.10	II
6		4042.59	II
18		4044.14	I
17		4047.21	I
5		4093.69	II
6		4114.99	II
7		4134.72	II
7		4149.19	II
8		4186.24	II
7		4222.97	II
7		4225.67	II
7		4263.40	II
7		4305.00	II
7		4309.10	II
5		4340.03	II
7		4388.16	II
5		4466.65	II
6		4505.33	II
5		4595.65	II
8		4608.45	II
10		4641.88	I
11		4642.37	I
5		4659.38	II
4		4740.91	I
6		4744.35	I
5		4753.93	I
7		4757.39	I
5		4786.49	I
7		4791.05	I
6		4799.75	I
8		4804.35	I
9		4829.23	II
7		4849.86	I
8		4856.09	I
8		4863.48	I
9		4869.76	I
8		4942.02	I
6		4943.29	II
9		4950.82	I
9		4956.15	I
10		4965.03	I
8		5005.60	II
7		5056.27	II
10		5084.23	I
11		5097.17	I
11		5099.20	I
12		5112.25	I
5		5310.24	II
12		5323.28	I
13		5339.69	I
12		5342.97	I
14		5359.57	I
6		5470.13	II
5		5642.73	II
4		5772.32	II
16		5782.38	I
17		5801.75	I
15		5812.15	I
17		5831.89	I
2		5969.64	II
8		6120.27	II
6		6246.59	II
7		6307.29	II
5		6427.96	II
2		6595.00	II
19		6911.08	I
12		6936.28	I
20		6938.77	I
7		6964.18	I
12		6964.67	I
25		7664.90	I
24		7698.96	I
5		7955.37	I
4		7956.83	I
7		8078.11	I
6		8079.62	I
9		8250.18	I
8		8251.74	I
3		8390.22	I
		8391.44	I
2		8417.54	I
1		8420.00	I
11		8503.45	I
10		8505.11	I
4		8763.96	I
3		8767.05	I
13		8902.19	I
12		8904.02	I
5		8923.31	I
4		8925.44	I
7		9347.24	I
3		9349.25	I
6		9351.59	I
15		9595.70	I
14		9597.83	I
6		9949.67	I
5		9954.14	I
9		10479.63	I
5		10482.15	I
8		10487.11	I
17		11019.87	I
16		11022.67	I
17		11690.21	I
16		11769.62	I
17		11772.83	I
		12432.24	I
		12522.11	I
		13377.86	I
		13397.09	I
		15163.08	I
		15168.40	I
		40158.37	I

K III
Ref. 60, 76 — L.J.R.

Intensity		Wavelength	
		Vacuum	
2		325.28	III
5		327.60	III
25		330.68	III
30		341.92	III
15		348.00	III
30		379.12	III
25		380.48	III
30		382.23	III
15		398.63	III
20		402.10	III
30		406.48	III
40		408.96	III
50		413.79	III
30		414.87	III
30		416.00	III
30		417.54	III
30		418.62	III
75		434.72	III
50		435.68	III
75		444.34	III
75		448.60	III
75		466.79	III
100		470.09	III
75		471.57	III
45		474.92	III
40		479.18	III
10		482.11	III
10		482.41	III
75		497.10	III
10		514.94	III
50		520.61	III
25		523.79	III
40		529.80	III
15		539.71	III
15		546.12	III
20		708.84	III
20		765.31	III
30		765.64	III
35		778.53	III
20		872.31	III
10		873.86	III
15		874.04	III
		Air	
6		2550.02	III
5		2635.11	III
1		2736.96	III
5		2689.90	III
1		2898.90	III
5		2938.45	III
1		2948.94	III
5		2986.20	III
6		2992.24	III
6		3052.07	III
5		3056.84	III
6		3201.95	III
6		3209.34	III
6		3278.79	III
6		3289.06	III
6		3322.40	III
6		3364.22	III
6		3420.82	III
4		3421.83	III
6		3468.32	III
6		3481.11	III
5		3513.88	III
1		3885.50	III

K IV
Ref. 32, 76, 86, 150, 160, 314, 322 — L.J.R.

Intensity		Wavelength	
		Vacuum	
150		271.82	IV
100		273.06	IV
		279.88	IV
300		340.46	IV
150		340.74	IV
300		354.93	IV
150		356.26	IV
300		359.73	IV
200		359.91	IV
250		362.08	IV
150		362.15	IV
150		363.02	IV
300		375.96	IV
300		379.88	IV
250		380.48	IV
200		381.70	IV
300		382.23	IV
150		382.49	IV
200		382.65	IV
300		382.91	IV
250		384.10	IV
200		386.61	IV
250		388.92	IV
250		389.07	IV
250		390.42	IV
300		390.57	IV
200		391.46	IV
200		392.47	IV
500		393.14	IV
400		400.21	IV
300		402.91	IV
250		403.97	IV
150		404.41	IV
250		408.08	IV
150		417.28	IV
200		442.30	IV
300		443.57	IV
200		445.61	IV
250		446.83	IV
750		448.60	IV
400		456.33	IV
250		523.00	IV
200		526.45	IV
150		527.62	IV
750		646.19	IV
500		737.14	IV
500		741.95	IV
500		745.26	IV
400		746.35	IV
300		749.99	IV
150		754.19	IV
400		754.67	IV

K V
Ref. 32, 75, 76, 150, 322 — L.J.R.

Intensity		Wavelength	
		Vacuum	
100		214.35	V
150		282.35	V
150		293.33	V
300		294.84	V
200		296.17	V
200		297.06	V
200		300.25	V
200		300.50	V
200		311.24	V
250		312.77	V
200		315.18	V
250		327.38	V
250		389.07	V
200		349.50	V
500		372.15	V
200		372.46	V
200		372.77	V
300		375.96	V
250		377.76	V
300		379.12	V
300		387.80	V
250		390.11	V
250		395.40	V
200		398.36	V
200		398.88	V
200		399.75	V
250		415.05	V
200		415.79	V
400		422.18	V
300		425.16	V
500		425.59	V
250		438.02	V
200		449.71	V
200		452.90	V
250		455.67	V
400		456.33	V
200		482.71	V
200		483.75	V
750		580.32	V
250		585.51	V
500		586.32	V
250		602.27	V
400		603.43	V
250		638.67	V
300		687.50	V
300		720.43	V
400		724.42	V
600		731.86	V
150		770.29	V
150		771.46	V
		1035.60	V

PRASEODYMIUM (Pr)
Z = 59

Pr I and II
Ref. 1 — C.H.C.

Intensity		Wavelength	
		Air	
25		2558.58	II
25		2578.27	I
30		2579.31	I
40	h	2598.04	II
25		2608.92	II
25		2615.75	II
25		2648.48	II
30		2654.75	II
25		2666.70	II
20		2672.52	II
30		2685.19	II
45		2685.70	II

Praseodymium (Cont.)

50		2698.92	II
60		2700.38	II
30		2702.25	II
100	h	2707.37	II
20		2714.16	II
60		2720.17	II
30		2721.90	II
50		2726.50	II
12		2731.78	II
25		2733.12	II
50		2734.30	II
25		2737.90	II
40		2742.12	II
25		2744.66	II
20		2746.28	II
60		2760.35	II
50		2769.60	II
50	d	2775.94	II
		2776.03	II
40		2778.00	II
50		2783.31	II
30		2789.05	II
35		2792.51	II
50		2802.05	II
20		2823.17	II
20		2824.14	II
20		2828.29	II
20		2842.98	I
20		2844.01	II
20		2850.62	I
25		2853.99	II
30		2865.64	II
50		2881.60	I
30		2882.31	II
30		2884.89	II
30		2943.97	II
30		2967.58	II
30		2971.13	II
40	d	2971.40	II
		2971.46	II
50		2984.98	II
30		2986.18	II
30		2990.22	II
110		3082.11	II
100		3111.34	II
140		3121.58	II
140		3163.73	II
270		3168.24	II
160		3172.31	II
110		3191.42	II
200	d	3195.99	II
110		3199.04	II
100		3207.89	II
190		3219.48	II
100		3234.27	II
100		3245.48	II
140		3355.67	II
110		3394.62	II
110		3465.74	II
200		3584.21	II
130		3611.94	II
170		3630.96	II
100		3645.55	II
250		3645.66	II
250		3646.30	II
100		3648.30	II
150	c	3660.36	II
100		3661.62	II
370		3668.83	II
250		3687.03	II
150		3687.19	II
100		3689.71	II
150		3698.06	II
230		3706.75	II
170	c	3711.10	II
290		3714.05	II
120	c	3733.03	II
210	c	3734.41	II
250		3735.76	II
190		3736.49	II
410		3739.18	II
150		3740.99	II
120		3743.98	II
190		3750.98	II
140		3759.60	II
120		3760.08	II
680		3761.87	II
230		3764.77	II
230		3768.94	II
170	c	3772.82	II
170		3774.06	II
140		3777.62	II
170		3780.66	II
150		3785.46	II
150		3786.86	II
210		3792.51	II
190		3794.93	II
680		3800.30	II
290		3804.84	II
140		3809.18	II
390		3811.84	II
1300	h	3816.02	II
120		3817.66	II
680		3818.28	II
120		3819.14	II
310		3821.80	II
150	c	3823.18	II
120		3826.67	II
960		3830.72	II
140		3834.93	II
480		3840.99	II
270		3842.34	II
150	c	3844.54	II
580		3846.59	II
1200		3850.79	II
720	c	3851.55	II
960		3852.80	II
120		3858.25	II
110		3859.14	II
480	c	3865.45	II
210		3867.52	II
210		3870.72	II
480		3876.19	II
1700	c	3877.10	II
270		3879.20	II
680		3880.47	II
440	c	3885.19	II
440	c	3889.34	II
120	c	3891.71	II
190		3897.25	II
210		3898.84	II
250		3902.45	II
770	c	3908.05	II
630		3912.90	II
310		3913.55	II
210		3914.76	II
1300	c	3918.85	II
420		3919.63	II
250		3920.53	II
960		3925.47	II
480		3927.46	II
370		3929.29	II
370		3935.82	II
250		3938.30	II
730	c	3947.63	II
900	c	3949.43	II
900	c	3953.51	II
380		3956.75	II
190		3959.44	I
470		3962.45	II
560		3964.26	II
1600	c	3964.81	II
560	c	3966.57	II
500		3971.16	II
320		3971.67	II
620	c	3972.14	II
320		3974.85	II
1300	c	3989.68	II
230		3991.91	II
340		3992.16	II
1600		3994.79	II
270		3995.83	II
560	c	3997.04	II
230		3997.96	II
320		3999.12	II
620	c	4000.17	II
730		4004.70	II
1900		4008.69	II
620		4010.60	II
730		4015.39	II
620		4020.96	II
470		4022.71	II
360		4025.54	II
230		4026.83	II
230		4029.00	II
360	c	4029.72	II
730	c	4031.75	II
230		4032.47	II
960		4033.83	II
230		4034.33	II
230		4038.22	II
730		4038.45	II
470		4039.34	II
1300		4044.81	II
230		4045.70	II
230		4046.63	II
340		4047.08	II
450		4051.13	II
2200		4054.88	II
2200		4056.54	II
450		4058.80	II
230		4062.22	II
3400		4062.81	II
210		4068.80	II
500	c	4079.77	II
500	c	4080.98	II
790		4081.85	II
500		4083.34	II
200	c	4087.21	II
560		4096.82	II
380		4098.40	II
2900	c	4100.72	II
270	c	4113.89	II
1700	c	4118.46	II
250		4129.15	II
340		4130.77	II
200		4133.61	II
1500	c	4141.22	II
2700		4143.11	II
270	c	4146.50	II
270		4148.44	II
200		4156.50	II
1700	c	4164.16	II
270		4168.04	II
230		4169.45	II
620		4171.82	II
730		4172.25	II
250		4175.32	II
250		4175.62	II
200		4178.63	II
5200		4179.39	II
2500		4189.48	II
560	c	4191.60	II
290		4201.17	II
2500	c	4206.72	II
500		4208.32	II
320		4211.86	II
320		4217.81	II
3800		4222.93	II
3800		4225.35	II
320		4233.11	II
320	c	4236.15	II
270		4240.02	II
960		4241.01	II
340		4243.51	II
840	c	4247.63	II
500		4254.40	II
270	c	4263.78	II
320		4269.09	II
790	c	4272.27	II
470	c	4280.07	II
790	c	4282.42	II
450	c	4298.98	II
290		4303.61	II
1500		4305.76	II
210		4323.55	II
270		4329.41	II
1300		4333.97	II
200		4335.74	II
360		4338.70	II
620	Cw	4344.30	II
470	c	4347.49	II
340		4350.40	II
450		4354.91	II
410	c	4359.79	II
1200		4368.33	II
320		4371.62	II
270		4396.08	II
170		4403.60	II
100		4405.12	II
430		4405.83	II
1700		4408.82	II
410		4413.77	II
160		4419.04	II
190		4419.65	II
160	c	4421.22	II
160		4424.58	II
1200	c	4429.13	II
110		4432.28	II
730		4449.83	II
140		4451.90	II
140		4454.68	II
100		4465.97	II
960		4468.66	II
140	c	4477.26	II
1100		4496.46	II
790		4510.15	II
200	c	4517.58	II
340	c	4534.15	II
340		4535.92	II
200		4563.12	II
140		4612.08	II
270	c	4628.74	II
140		4632.28	I
140		4635.68	I
200		4639.55	I
110	c	4643.49	II
140		4646.05	II
200	c	4651.50	II
140		4664.65	II
270	c	4672.09	II
180		4687.80	I
290		4695.77	I
140	Cw	4708.07	II
140		4709.52	I
180		4730.67	I
250		4736.69	I
100		4744.16	I
150		4746.92	II
100		4762.72	II
110		4783.35	II
110		4906.99	I
140		4914.02	I
200		4924.60	I
140		4936.00	I
320		4939.74	I
160		4940.30	I
380		4951.37	I
110		4975.75	I
120		5018.59	I
200		5019.76	I
200		5026.96	I
100		5033.38	I
270		5034.41	II
110		5043.83	I
320		5045.52	I
160		5053.40	I
180		5087.12	I
360		5110.38	II
560		5110.76	II
410		5129.52	II
270		5133.44	I
270		5135.14	II
100		5139.81	I
100	c	5152.30	II
200		5161.74	II
620		5173.90	II
200		5191.32	II
120		5194.43	I
150		5195.11	II
200		5195.31	II
300		5200.33	II
150		5207.90	II
360		5219.05	II
560		5220.11	II
110		5227.97	I
680		5259.73	II
180		5263.88	II
340	c	5292.02	II
340		5292.62	II
230		5298.09	II
430		5322.76	II
200		5352.40	II
16		5501.50	I
40		5508.79	II
65		5509.15	II
16	c	5511.63	II
55		5513.58	II
28		5515.12	II
13		5519.38	II
20	c	5520.31	II
45	c	5522.79	II
28	c	5524.15	I
28	c	5525.91	II
16	c	5527.93	I
13		5530.21	I
45		5531.16	I
150		5535.17	II
28		5538.37	I
20		5538.78	II
55		5545.01	II
20		5548.33	II
11		5553.42	II
22		5561.46	II
45	c	5562.06	I
13		5565.52	I
13		5566.91	II
45		5571.83	II
11		5574.61	II
11		5578.81	I
13		5582.35	II
11		5584.02	II
22		5594.92	I
22		5597.29	II
13		5601.30	II
90		5605.65	II
13		5606.68	I
28		5608.93	II
55		5610.22	II
11		5620.06	II
20		5620.26	I
45	c	5621.89	II
110		5623.05	II
90		5624.45	II
11	h	5633.03	I
22		5636.46	II
55	c	5638.79	II
16		5640.37	II
16	Cw	5643.16	I
22		5645.41	II
35		5654.23	II
55		5659.84	II
35	h	5661.57	I
16		5662.19	II
65	c	5668.46	I
45		5669.55	II
35		5669.99	II
16		5674.14	II
16		5677.03	II
55		5681.89	II
13		5685.60	II
16		5686.52	I
22	h	5687.17	II
65		5688.44	II
22		5689.21	II
55	h	5690.97	II
22		5695.90	II
22		5704.38	I
65		5707.61	I
40		5711.63	II
22		5713.83	II
16		5716.08	II
45		5719.08	II
45	d	5719.63	II
		5719.80	II
11		5728.38	I

Praseodymium (Cont.)

40		5731.88	II
20		5747.13	II
11		5747.74	I
11		5747.95	II
22		5753.02	II
90		5756.17	II
16		5759.40	II
22		5760.20	I
22		5769.16	II
16		5769.79	II
45		5773.16	II
11		5775.91	II
16		5777.29	II
90		5779.28	I
65	c	5785.28	II
65		5786.17	II
16	h	5788.29	II
16		5788.92	II
16		5790.86	II
45		5791.36	II
22		5792.95	I
40		5810.58	II
16		5813.55	II
160	d	5815.17	II
		5815.33	II
55		5818.57	II
40		5820.62	II
16	h	5821.36	I
55		5822.59	II
90		5823.72	II
45		5830.94	II
40		5835.13	I
35	c	5844.65	II
40		5844.98	II
65		5847.13	II
65	c	5850.64	II
45		5852.63	II
11	c	5854.44	I
45		5856.07	II
55		5856.90	II
90		5859.68	II
80		5868.83	II
22		5873.83	II
35		5874.72	I
35		5878.10	I
35		5879.04	I
80		5879.25	II
35	c	5884.72	I
55		5892.23	II
22		5894.22	II
40		5903.11	II
45		5904.45	II
40		5908.67	II
11		5915.31	I
11		5915.97	I
40		5920.76	I
40		5930.66	II
16		5936.33	II
160		5939.90	II
65		5940.72	II
22		5941.65	I
35		5947.16	II
22	c	5949.76	I
55		5951.27	II
20		5951.76	II
90		5956.60	II
		5956.70	I
13		5959.25	I
20		5962.18	I
28		5963.00	I
110		5967.82	II
13	c	5976.95	I
13		5978.88	I
65		5981.19	II
40		5986.14	I
45	c	5987.14	I
		5987.29	II
13		5991.27	I
13	c	5994.89	I
11		5996.06	I
29		6002.44	II
90		6006.33	II
13		6008.54	I
55		6016.48	II
150		6017.80	II
28	c	6019.85	I
150		6025.72	II
35		6042.87	II
55		6046.66	II
35		6049.26	I
28		6050.04	II
11		6050.88	I
140		6055.13	I
13		6067.27	II
13		6085.81	I
28		6086.16	II
65		6087.52	II
20		6090.38	II
28		6093.09	II
18		6096.28	I
22		6106.72	II
18		6109.08	I
65		6114.38	II
22	c	6118.02	I
22	c	6122.15	I
35		6141.51	II
65		6148.23	I
		6148.24	II
22		6157.82	II
13		6159.10	II
190		6161.18	II
18		6165.38	I
270		6165.94	II
55		6182.34	II
13		6187.96	I
35		6197.45	II
35		6200.81	II
13		6205.63	II
13		6210.59	I
22		6212.73	I
18		6218.06	I
20	h	6236.80	I
20	h	6241.05	I
45		6244.35	II
35		6255.10	II
40		6262.55	II
18		6264.54	II
22	c	6274.66	II
		6274.81	II
40		6278.68	II
110		6281.28	II
18	c	6289.02	I
11	c	6298.01	I
11		6302.05	I
35		6302.35	II
16		6304.05	I
35		6305.23	II
11	h	6318.13	II
45	c	6322.36	I
22	h	6343.88	I
28		6347.11	II
18	c	6350.98	I
22	c	6357.20	I
55	c	6359.03	I
11	h	6363.62	II
16		6377.61	I
16		6378.59	I
11		6389.57	I
18	c	6391.99	I
40		6393.18	I
45		6397.96	II
10	h	6410.69	I
55		6411.23	I
40		6413.68	II
10		6415.43	I
45		6429.63	II
45		6431.84	II
7	h	6442.78	II
9	h	6443.91	II
16	c	6453.44	I
9		6454.84	II
9		6456.18	I
9	h	6460.19	I
18		6467.72	II
9	h	6475.26	II
35	Cw	6478.02	II
45		6486.55	I
9	h	6486.97	II
40	h	6491.75	I
9		6493.49	I
11		6494.89	I
22	c	6497.11	I
18		6498.94	II
22		6500.72	I
9		6504.09	I
8		6517.14	I
16		6518.79	II
8		6534.52	I
16		6540.47	I
7	h	6553.30	I
22		6564.62	II
45		6566.77	II
7		6571.03	I
6		6578.00	I
6		6584.56	II
9	h	6593.74	II
11		6595.48	I
15		6609.86	I
55		6616.67	I
11		6618.34	II
7	h	6631.00	I
13	h	6632.06	I
14		6647.12	I
75		6656.83	II
55		6673.41	II
75		6673.78	II
5	h	6687.51	II
4	h	6699.25	I
13		6736.79	I
35	c	6747.09	I
19	c	6749.19	I
7	c	6784.99	II
55	Cw	6798.60	I
11		6811.76	II
17	Cw	6812.87	II
13		6814.04	II
9		6817.61	I
35	Cw	6827.60	II
19		6830.50	II
9		6844.39	I
9	h	6845.47	II
9		6846.59	II
17	c	6850.46	II
11		6852.77	I
11	c	6870.44	I
7		6884.66	I
8		6892.71	I
8	h	6970.38	
8	c	6980.12	I
40		7021.51	II
10		7024.53	I
13		7042.40	I
8		7044.45	II
7		7051.07	I
10		7079.99	I
11	c	7095.18	I
20		7114.55	I
10	h	7116.90	I
11		7118.24	II
7		7137.33	II
10	h	7159.88	I
7		7167.77	II
7	h	7189.95	I
10	c	7208.85	II
24		7227.70	II
13		7231.53	I
7	c	7243.26	I
7	c	7259.21	I
7	c	7287.61	I
7	h	7289.19	I
7	Cw	7324.42	I
7		7328.47	I
7		7344.86	
16		7407.56	II
20	c	7451.74	II
11	h	7495.59	I
6	h	7499.42	I
14		7541.02	II
6		7574.86	I
20		7645.66	II
7		7704.98	II
16		7721.84	I
6	h	7786.16	II
6	Cw	7841.27	I
14		7871.67	I
6		7881.09	I
6	Cw	7888.56	II
6		7915.19	II
6		8031.92	I
6		8055.43	I
14		8067.44	I
10	Cw	8122.78	II
11		8141.10	I
5		8181.34	II
5	c	8211.93	I
6		8289.93	I
6		8379.84	I
6	h	8427.82	I
6	h	8605.27	II
10		8714.59	II

Pr III

Ref. 306, 308 — J.R.

Intensity		Wavelength	
		Vacuum	
25		1008.61	III
50		1021.35	III
25		1026.18	III
100		1029.03	III
50		1038.29	III
50		1042.96	III
25		1043.80	III
25		1044.03	III
25		1046.20	III
150		1047.24	III
100		1049.09	III
50		1052.63	III
25		1061.60	III
25		1066.03	III
25	p	1068.85	III
25		1069.88	III
25		1084.42	III
25		1088.66	III
100		1104.84	III
25		1108.82	III
30		1352.70	III
25		1881.22	III
		Air	
50		2031.46	III
100		2033.30	III
25		2043.12	III
100		2052.30	III
50		2052.87	III
50		2053.85	III
200		2058.59	III
50		2064.08	III
25		2075.08	III
200		2090.75	III
200		2093.49	III
25		2096.85	III
50		2096.94	III
10	w	2148.14	III
10	w	2194.24	III
10	w	2197.25	III
10	w	2205.48	III
10	w	2206.26	III
10	w	2214.45	III
10	w	2215.25	III
10	w	2217.12	III
10	w	2223.23	III
10	w	2230.35	III
10	w	2237.26	III
10		2239.06	III
10		2239.42	III
10	w	2242.15	III
10	w	2284.62	III
10		2307.59	III
10		2307.77	III
10		2308.41	III
10		2311.29	III
10		2311.44	III
10		2314.18	III
10	w	2315.46	III
10		2318.15	III
10		2318.36	III
10		2318.64	III
10		2318.82	III
10		2318.97	III
10	w	2319.40	III
10		2320.41	III
10	w	2328.56	III
10	w	2336.13	III
10		2365.52	III
10		2368.78	III
10		2369.08	III
10	w	2378.06	III
10	w	2378.97	III
10		2395.44	III
10		2399.70	III
10	w	2405.56	III
10		2408.19	III
10		2409.80	III
10		2412.40	III
10		2417.69	III
10		2418.95	III
10		2426.14	III
10		2426.85	III
10	w	2430.32	III
10	w	2434.18	III
10		2434.39	III
10		2435.91	III
10		2436.89	III
10		2438.63	III
10	w	2444.93	III
10		2445.49	III
10		2446.77	III
10		2448.16	III
10		2452.02	III
10	w	2452.81	III
10	w	2452.85	III
10		2454.60	III
10		2454.82	III
10	w	2459.77	III
10	w	2460.72	III
10		2462.18	III
10		2462.90	III
10		2468.20	III
10		2468.97	III
10	w	2473.42	III
10	w	2478.32	III
10		2479.98	III
10		2481.02	III
10		2483.30	III
10		2483.99	III
10	w	2484.60	III
10	w	2485.16	III
10	w	2488.72	III
10		2491.97	III
10	w	2494.20	III
10	w	2495.37	III
10	w	2495.51	III
10		2499.97	III
20	w	2587.71	III
40	w	2624.91	III
20	w	2644.62	III
20	w	2656.88	III
20	w	2667.51	III
70	w	2679.47	III
40	w	2710.30	III
20	w	2718.65	III
100	w	2724.03	III
20	l	2841.94	III
70	s	2910.61	III
50	s	2911.77	III
100	l	2914.49	III
50	s	2930.19	III
70	s	2942.43	III
70	s	2953.58	III
90	w	2954.40	III
90	s	2964.85	III
150	s	2968.83	III
80	s	2969.41	III
150	l	2976.86	III
150	s	2977.06	III
500	s	2980.54	III

Praseodymium (Cont.)

Intensity		Wavelength	Spectrum
100	l	2981.65	III
150	c	2982.42	III
500	s	2985.82	III
150	s	2997.12	III
70	l	2998.79	III
150	l	3000.46	III
150	s	3003.20	III
60	l	3006.47	III
150	s	3008.04	III
150	l	3010.61	III
90	s	3014.60	III
100	l	3015.13	III
90	s	3016.26	III
70	s	3021.77	III
70	s	3025.26	III
100	l	3029.38	III
100	l	3033.31	III
90	l	3034.25	III
70		3040.02	III
60	c	3040.94	III
60	l	3041.78	III
100	s	3042.35	III
100	l	3045.81	III
70	l	3046.98	III
120	l	3050.30	III
70	s	3055.30	III
150	l	3058.90	III
150	l	3066.71	III
70	s	3078.68	III
150	l	3080.20	III
50	w	3248.39	III
90	s	3280.92	III
90	l	3292.58	III
90	s	3296.10	III
100	l	3306.14	III
60	s	3333.26	III
90	s	3340.58	III
500	s	3341.43	III
70	s	3341.68	III
70	s	3345.38	III
70	l	3345.44	III
50	w	3351.07	III
50	s	3353.87	III
250	s	3354.91	III
500	l	3357.56	III
500	l	3359.41	III
100	l	3364.52	III
50		3364.88	III
50	s	3365.80	III
200	s	3367.35	III
500	s	3367.58	III
100	l	3371.92	III
100	s	3377.14	III
50	s	3379.13	III
150	s	3380.21	III
150	s	3381.26	III
300	s	3381.84	III
100	s	3391.08	III
1000	w	3394.22	III
600	d	3396.07	III
300	s	3396.62	III
300	l	3397.46	III
50	l	3402.97	III
500	c	3413.21	III
300	s	3415.15	III
150	s	3420.07	III
300	l	3422.22	III
50	c	3426.27	III
500	l	3427.02	III
300	l	3436.36	III
150	l	3440.62	III
50	l	3445.29	III
70	s	3454.05	III
180		3653.58	III
60		3817.25	III
60	l	3861.80	III
150		3980.51	III
200		4000.20	III
90		4018.36	III
180		4029.60	III
90		4142.46	III
120		4144.48	III
90		4147.85	III
90		4172.15	III
150		4179.77	III
180		4184.18	III
240		4197.01	III
120		4219.45	III
180		4231.45	III
180		4275.07	III
120		4286.32	III
90		4298.27	III
90		4301.73	III
90		4316.34	III
60		4354.28	III
90		4379.82	III
90		4381.47	III
120		4404.71	III
120		4421.10	III
120		4431.85	III
150	w	4447.93	III
180	w	4450.14	III
120	w	4451.00	III
120		4461.02	III
200		4461.81	III
300	w	4500.31	III
450	w	4612.02	III
600	w	4625.18	III
120		4654.16	III
600	w	4713.70	III
300	w	4725.55	III
270		4728.21	III
300		4747.11	III
300	w	4771.83	III
450	w	4775.30	III
600		4857.39	III
150		5208.51	III
150		5261.68	III
1000		5264.44	III
1500		5284.70	III
1500		5299.99	III
1500		5340.02	III
100		5427.70	III
100		5581.74	III
150		5646.80	III
600		5765.27	III
1500		5844.41	III
200		5947.98	III
7000	w	5956.05	III
900	w	5998.94	III
1500	w	6053.01	III
900		6071.09	III
9000	w	6090.02	III
5000		6160.24	III
1500	w	6161.22	III
100		6195.05	III
2000		6195.63	III
200		6310.36	III
100		6361.65	III
300		6429.26	III
300		6444.74	III
600		6500.04	III
300		6501.49	III
200		6578.90	III
100		6616.46	III
600		6706.70	III
100		6727.63	III
200		6827.96	III
100		6854.63	III
200		6857.30	III
1000		6866.80	III
1000		6899.06	III
500		6903.52	III
7000		6910.14	III
150	w	6934.55	III
500		6970.96	III
100		6979.03	III
5000		7030.39	III
100		7075.21	III
4500		7076.62	III
100		7083.99	III
500		7112.53	III
100	w	7165.64	III
250		7231.62	III
100	w	7238.26	III
250		7240.21	III
100	w	7262.32	III
150	w	7340.69	III
350		7343.70	III
200		7349.75	III
300	w	7350.61	III
100	w	7355.52	III
2000		7426.48	III
4000		7429.05	III
100		7463.96	III
250		7487.40	III
200		7493.20	III
100		7511.17	III
500		7529.11	III
100	w	7549.20	III
150	w	7588.64	III
300		7596.41	III
100	w	7625.63	III
100	w	7648.34	III
100	w	7670.65	III
200	w	7674.65	III
500		7742.34	III
250	w	7745.59	III
500		7754.31	III
100	w	7755.48	III
3000		7781.98	III
200	w	7814.74	III
1500	w	7866.14	III
1000		7888.12	III
400		7897.09	III
1000		7914.00	III
100		7923.16	III
1000	w	7972.75	III
250		8001.14	III
3000		8102.90	III
250		8119.54	III
400		8132.23	III
100		8138.34	III
150	w	8235.33	III
250	w	8244.89	III
100		8409.10	III
100	w	8494.99	III
200	w	8567.63	III
5000	w	8602.74	III
500		8691.58	III
500		8771.38	III
1000		8854.05	III
125		8886.17	III
100		8908.70	III
250		9099.98	III
250		9131.90	III
200		9222.32	III
250	w	9265.56	III
125		9320.54	III
250		9334.33	III
175		9377.44	III
175		9388.56	III
175		9549.77	III
100		9579.74	III
150		9802.98	III
175		9806.37	III
500	w	9991.16	III
500		10031.10	III
500		10160.33	III
500		10238.63	III
500		10301.58	III
500		10324.59	III
500		10716.58	III

Pr IV
Ref. 337, 338 — J.R.

Intensity		Wavelength	
		Vacuum	
20		718.23	IV
30		721.34	IV
60		722.41	IV
30		722.58	IV
20		726.04	IV
50		730.37	IV
30		731.77	IV
30		734.86	IV
20		735.04	IV
20		736.19	IV
50		736.32	IV
100		737.17	IV
100		741.45	IV
20		743.15	IV
20		743.89	IV
20		746.14	IV
40		763.16	IV
20		764.00	IV
300		1226.40	IV
2000		1228.59	IV
500		1230.69	IV
400		1238.19	IV
200		1249.35	IV
200		1255.64	IV
200		1261.27	IV
400		1268.32	IV
300		1270.58	IV
1000		1275.10	IV
200		1275.40	IV
1000		1278.65	IV
200		1279.34	IV
1000		1287.44	IV
300		1290.93	IV
1000		1292.30	IV
5000		1293.22	IV
5000		1295.28	IV
400		1296.50	IV
200		1298.26	IV
300		1298.54	IV
200		1304.71	IV
300		1306.86	IV
200		1308.08	IV
200		1310.71	IV
500		1314.96	IV
300		1315.28	IV
300		1316.96	IV
500		1320.10	IV
1000		1320.70	IV
5000		1321.36	IV
500		1322.51	IV
500		1326.38	IV
5000		1333.57	IV
300		1335.96	IV
500		1339.29	IV
1000		1340.74	IV
200		1341.32	IV
300		1344.23	IV
1000		1347.07	IV
1000		1352.81	IV
500		1354.35	IV
5000		1354.66	IV
2000		1360.64	IV
1000		1364.81	IV
2000		1365.77	IV
400		1368.90	IV
5000		1374.41	IV
1000		1382.62	IV
200		1384.23	IV
200		1385.91	IV
300		1394.11	IV
500		1397.11	IV
1000		1399.31	IV
1000		1400.96	IV
400		1410.90	IV
1000		1424.36	IV
500		1426.59	IV
5000		1435.56	IV
200		1459.95	IV
200		1461.76	IV
1000		1474.91	IV
200		1477.32	IV
500		1485.88	IV
500		1503.35	IV
400		1516.86	IV
200		1520.71	IV
2000		1520.98	IV
400		1525.46	IV
500		1553.62	IV
500		1559.49	IV
500		1570.13	IV
200		1572.80	IV
5000		1574.55	IV
5000		1575.10	IV
3000		1578.38	IV
500		1585.10	IV
300		1613.00	IV
400		1613.65	IV
1000		1618.03	IV
2000		1622.30	IV
300		1634.77	IV
400		1676.08	IV
200		1688.49	IV
200		1713.53	IV
500		1732.86	IV
300		1762.86	IV
1000		1766.88	IV
1000		1771.14	IV
500		1841.08	IV
10000		1884.87	IV
400		1951.23	IV
200		1954.61	IV
		Air	
200		2025.06	IV
1000		2039.15	IV
200		2047.05	IV
200		2050.73	IV
200		2058.48	IV
2000		2083.23	IV
500		2100.42	IV
1000		2154.31	IV
300		2193.37	IV
1000		2205.13	IV
300		2265.70	IV
200	c	2334.46	IV
200	c	2339.08	IV
500	c	2376.09	IV
2000	c	2378.98	IV
1000	c	2379.66	IV
500	c	2427.07	IV
500	c	2428.13	IV
500	c	2438.57	IV
500	c	2455.64	IV
500	c	2705.19	IV
200	c	2708.01	IV
200	c	2753.47	IV
200	c	2767.60	IV

Pr V
Ref. 149 — J.R.

Intensity		Wavelength	
		Vacuum	
200		843.78	V
7000		865.90	V
5000		869.17	V
80		869.66	V
1000		896.65	V
750		922.29	V
250		1234.07	V
250		1342.78	V
200		1958.09	V
400		1958.20	V
		AIR	
300		2246.06	V
300		2246.20	V

PROMETHIUM (Pm)
Z = 61

Pm I and II
Ref. 196, 260 — C.H.C.

Intensity		Wavelength	
		Air	
40	w	2502.12	II
40		2608.24	II
150		2632.00	II
70		2638.46	II
100		2671.05	II
50	w	2787.72	II
40		2808.05	II
100	h	2820.10	II
100	w	2840.82	II
150	w	2841.86	II
200	c	2857.46	II
100		3004.59	II
100		3008.85	II
300		3072.41	II
150		3086.02	II

Promethium (Cont.)

120		3090.19	II
150		3091.86	II
150		3108.11	II
100		3115.36	II
100		3117.22	II
100		3118.76	II
35		3162.23	I
35		3168.82	I
100		3172.77	II
35		3222.04	I
35		3238.55	I
60		3239.62	II
75		3296.63	I
60		3311.76	II
50		3313.38	I
50		3329.22	I
75		3331.57	I
100		3354.45	I
100		3358.14	II
80		3360.21	II
80		3364.44	II
300		3366.03	I
90		3377.68	II
100		3391.28	II
100		3408.06	II
500		3427.40	II
120		3441.15	II
400		3449.80	II
250		3460.25	II
200		3462.91	II
200		3480.61	II
150		3497.13	II
100		3514.85	II
200		3546.81	I
100		3559.43	II
200		3565.31	II
150		3580.10	II
200		3610.76	II
200		3629.84	II
300		3634.20	II
300		3659.39	II
300	r	3669.22	I
200		3674.85	I
200		3678.51	II
300	r	3679.85	I
200		3687.65	II
400		3689.79	II
300		3692.50	II
300		3697.50	II
300	r	3697.63	I
400		3702.63	II
800		3711.72	II
200		3715.75	II
200		3721.72	II
500		3726.01	I
200		3738.43	I
300		3740.68	I
300		3742.52	II
300	r	3742.97	I
500		3745.86	II
300		3747.09	II
500		3750.09	II
200		3761.68	I
300		3765.75	I
300		3775.42	I
300		3780.77	I
200		3781.43	I
400		3795.66	II
250		3806.06	II
300	r	3809.20	I
400		3810.93	I
200		3819.26	II
300		3820.53	II
300		3839.52	I
200		3842.88	II
300		3842.98	II
250		3845.38	II
300		3874.03	I
800		3877.62	II
300	r	3885.79	I
250		3890.97	I
1000		3892.15	II
300		3898.73	I
400		3899.78	II
250		3909.50	II
1000		3910.26	II
1000		3919.10	II
800		3936.48	II
300		3944.21	II
300	r	3954.76	I
1000		3957.74	II
500		3980.74	II
300		3995.05	II
1000	r	3998.96	II
500		4009.96	II
200		4012.72	II
250		4014.20	II
200		4019.34	II
250		4028.20	II
200		4045.36	II
300		4051.54	II
600	r	4055.20	II
200	r	4056.56	I
600		4075.84	II
200	r	4085.31	I
500		4086.10	II
250		4140.46	II
200		4185.74	II
300		4192.92	II
200		4194.70	II
200		4222.15	II
300	r	4264.32	I
300	r	4284.37	I
600		4297.78	II
200		4303.89	II
200	r	4305.64	I
400		4318.80	I
250		4325.92	II
200		4332.05	II
300		4336.54	II
200		4337.48	II
300		4342.12	II
200		4347.72	I
350	r	4363.92	I
300	r	4369.64	I
200		4381.88	II
400	r	4388.49	I
200		4388.76	II
400	r	4409.42	I
500	r	4412.47	I
1000		4417.96	II
400		4432.51	II
250	r	4435.86	I
300	r	4436.55	I
300	r	4438.68	I
500		4445.41	II
600		4446.90	II
800		4453.95	II
200		4459.97	II
250	r	4468.16	I
200		4471.48	II
300		4473.23	II
200		4477.46	II
350	r	4478.58	I
300	r	4481.60	I
300	r	4485.05	I
300	r	4490.50	I
250		4492.05	II
600		4500.15	II
350	r	4500.33	I
250		4506.84	I
100		4509.38	II
100		4513.56	II
200		4517.31	I
200		4523.32	I
600		4525.20	II
250	r	4526.12	I
250	r	4526.76	I
400	r	4527.70	I
800		4529.21	II
300		4540.06	I
300	r	4541.42	I
450	r	4541.75	I
500	r	4544.08	I
200		4545.17	I
400	r	4549.78	I
300	r	4554.03	I
200		4554.63	I
500	r	4555.34	I
200		4556.06	I
300		4557.03	I
300		4559.21	I
100		4564.83	II
300	r	4568.14	I
200		4570.37	I
300	r	4572.15	I
400	r	4575.27	I
300		4578.28	I
200	r	4578.41	I
300	r	4579.48	I
300		4581.14	I
300	r	4585.49	I
200		4593.82	I
400	r	4595.82	I
800		4597.55	I
500	r	4600.25	I
400	r	4602.96	I
400		4604.59	I
600	r	4605.66	I
500	r	4609.85	I
100		4615.87	II
600	r	4617.02	I
200		4618.40	I
400	r	4618.49	I
500	r	4619.75	I
500		4621.57	I
500		4623.31	I
700	r	4623.68	I
900		4624.41	I
500	r	4625.29	I
400	r	4627.60	I
200		4630.93	I
600	r	4633.45	I
400	r	4640.96	I
700	r	4643.36	I
700	r	4643.76	I
400		4645.94	I
600	r	4647.03	I
600	r	4650.42	I
500		4650.52	I
400	r	4653.41	I
400		4654.50	I
500	r	4655.05	I
300		4659.38	I
500	r	4660.79	I
300	r	4663.26	I
600	r	4663.46	I
400	r	4665.19	I
500	r	4671.23	I
400		4671.76	I
500	r	4674.42	I
200		4677.46	I
500	r	4677.92	I
400	r	4678.09	I
700	r	4682.92	I
500	r	4696.80	I
200		4699.51	I
250		4722.06	II
300		4727.06	I
900	r	4728.36	I
400	r	4728.68	I
800	r	4734.27	I
200	r	4737.99	I
100		4739.08	II
200		4739.78	I
350	r	4745.13	I
500	r	4757.73	I
800	r	4759.00	I
700	r	4762.57	I
700	r	4773.46	I
900	r	4781.29	I
250		4794.59	I
200		4795.43	I
700	r	4798.98	I
900	r	4801.36	I
700	r	4809.54	I
900	r	4811.96	I
400	r	4817.12	I
400	r	4827.72	I
800	r	4837.66	I
400	r	4838.92	I
300		4839.62	I
200		4844.01	I
350	r	4852.73	I
400	r	4860.62	I
700	r	4860.74	I
300	r	4865.30	I
500	r	4865.72	I
400	r	4869.80	I
700	r	4872.42	I
500	r	4887.02	I
700	r	4892.52	I
400	r	4900.30	I
300	r	4904.28	I
400	r	4918.28	I
600	r	4932.99	I
700	r	4959.46	I
100		4971.40	II
500	r	4997.10	I
200	r	5030.80	I
300	r	5058.31	I
100		5067.35	II
150		5080.52	II
150		5089.35	II
200		5092.42	I
400	r	5094.83	I
200		5096.18	I
150		5097.30	II
400	r	5100.77	I
250		5121.47	II
400	r	5127.34	I
200		5129.75	I
400	r	5145.13	I
500	r	5146.30	I
400		5153.86	II
300		5169.71	II
500		5171.58	II
300		5194.05	II
500		5208.09	II
150		5215.96	II
250		5225.12	II
500		5236.26	II
300		5236.66	II
400		5246.33	II
150		5262.42	II
500		5270.64	II
200		5293.92	II
100		5308.86	II
150		5318.58	II
200		5410.45	II
200		5424.54	II
180		5424.79	II
150		5429.04	II
100		5467.64	II
150		5495.45	II
100		5516.42	II
180		5534.96	II
200		5537.38	I
800		5546.08	II
120		5556.88	II
150		5558.39	II
200		5561.73	II
800		5576.02	II
200		5641.29	II
200		5730.81	I
200		5768.16	II
200		5776.99	I
500		5823.93	II
300	c	5868.79	II
200	c	5875.31	II
100	c	5878.76	II
150		5899.76	II
250		5904.71	I
100		5905.90	I
125		5914.96	I
250	c	5927.17	II
150		5939.66	I
400	c	5946.49	II
800		5956.42	I
200		5956.69	I
100	c	5960.08	II
150		5963.00	II
400		5967.89	I
200		5979.73	I
200		5984.82	I
100	c	5987.13	II
400		5997.12	I
200		6027.11	II
300		6030.06	I
400		6031.32	I
500		6043.39	I
150	c	6052.57	II
100	c	6067.00	II
500		6069.06	I
100		6076.40	II
200		6085.41	II
900		6100.21	I
400		6106.40	I
100		6114.90	II
400	h	6151.76	I
100		6159.53	II
400		6163.16	I
100		6184.52	II
200		6208.91	II
500		6229.64	I
400		6237.79	I
100		6263.25	II
400		6272.69	I
400		6286.06	I
500		6308.29	I
100		6314.20	II
700		6323.84	I
500		6390.31	I
100		6429.64	II
500	h	6431.93	I
100		6436.57	II
400		6487.61	I
400		6510.34	I
500		6517.25	I
200		6519.43	II
1000	d	6520.45	I
500		6542.20	I
100	h	6558.48	II
100		6586.39	II
100		6592.29	II
900		6598.15	I
800		6598.66	I
700		6606.37	I
800	w	6625.23	I
100	h	6625.54	II
700		6649.81	I
400		6659.05	II
100		6661.25	II
500		6661.68	I
400		6663.76	I
800	c	6667.51	I
700	h	6677.47	I
200		6680.89	II
500		6685.55	I
500		6685.68	I
150		6690.09	II
600		6700.33	I
100		6706.27	II
700		6714.67	I
500		6717.26	I
500		6720.71	I
700		6727.50	I
600		6743.71	I
900		6749.91	I
900		6750.48	I
200		6756.45	II
300		6772.29	II
400		6778.78	I
100		6783.09	II
100		6796.87	II
200		6811.68	II
800		6833.30	I
400		6848.37	I
50		6858.58	II

PROTACTINIUM (Pa)
Z = 91

Pa I and II
Ref. 96 — J.G.C.

Intensity		Wavelength (Air)	
3000	h	2466.85	
3000	h	2492.85	
3000		2599.16	II
3000		2699.22	II
3000		2822.79	II
3000	l	2832.14	
3000		2870.01	
3000	h	2871.42	II
3000	h	2891.14	II
3000	h	2906.93	
3000	l	3011.10	II
3000	s	3033.59	II
3000	l	3071.24	II
3000	h	3083.19	
3000	l	3093.23	II
3000	l	3126.23	II
3000	l	3146.28	II
3000	l	3170.89	II
3000	l	3171.54	II
3000		3204.16	
3000	l	3240.58	II
3000		3274.46	II
3000	l	3332.69	II
3000	s	3346.66	II
3000	l	3394.49	
3000	l	3452.82	II
3000		3504.97	I
3000	s	3530.65	II
3000		3570.56	I
3000		3571.82	I
3000		3618.07	I
10000		3636.52	I
3000		3702.74	I
3000		3752.67	I
3000		3873.35	I
3000		3931.83	I
3000	s	3952.62	II
10000	l	3957.85	II
3000	s	3970.07	II
3000		3981.82	I
10000		3982.23	I
3000	l	4012.96	II
3000	s	4018.21	II
3000		4030.16	II
3000	s	4046.93	II
10000	s	4056.20	II
10000	s	4070.40	II
3000		4117.62	
3000	l	4176.18	II
10000	l	4217.23	II
10000	s	4248.08	II
3000	s	4291.34	II
3000		4400.77	
3000		4436.13	
3000	s	4601.43	II
3000		4628.19	
3000		4820.34	
3000	s	4861.49	
3000	l	6035.78	I
3000		6162.56	I
3000		6216.35	
3000	l	6358.61	I
3000		6379.25	I
3000	l	6438.97	I
3000	h	6792.75	I
10000		6945.72	I
3000		6960.09	I
3000	h	6961.78	I
3000	s	6992.73	I
3000		7076.27	I
3000	h	7100.94	I
10000	s	7114.89	I
3000	h	7171.55	I
3000		7227.13	I
3000		7318.79	I
10000	l	7368.25	I
3000	h	7471.89	I
10000	h	7493.15	I
3000	h	7558.26	I
10000	h	7608.20	I
10000		7626.79	I
10000	s	7635.18	I
10000		7669.34	I
3000		7679.20	I
10000	h	7749.19	I
3000		7872.95	I
3000	l	7945.56	I
10000		8039.34	I
10000	h	8099.84	I
10000		8199.04	I
10000		8271.87	I
3000	s	8358.98	I
3000	s	8369.60	I
3000	h	8441.04	I
10000	h	8532.66	I
10000	s	8572.96	I
3000	h	8639.91	I
3000	h	8653.51	I
10000		8735.27	I
3000		10594.38	
3000		10923.32	I
3000		11646.78	
10000		11791.73	I
3000		12279.01	
3000		13234.09	
10000		13522.40	
10000		14344.76	I
3000		18478.61	I

RADIUM (Ra)
Z = 88

Ra I and II
Ref. 253, 254 — E.F.W.

Intensity		Wavelength (Air)	
8		2369.73	II
8		2460.55	II
10		2475.50	II
8		2586.61	II
10		2643.73	II
20		2708.96	II
10		2795.21	II
30		2813.76	II
10		3033.44	II
5		3101.80	I
100		3649.55	II
200		3814.42	II
8		4194.09	II
8		4244.72	II
100		4340.64	II
20		4436.27	II
30		4533.11	II
8		4641.29	I
100		4682.28	II
8		4699.28	I
100		4825.91	I
10		4856.07	I
10		4859.41	II
10		4927.53	II
10		5097.56	I
10		5205.93	I
10		5283.28	I
10		5320.29	I
10		5399.80	I
20		5400.23	I
20		5406.81	I
8		5482.13	I
10		5501.98	I
10		5553.57	I
20		5555.85	I
10		5616.66	I
50		5660.81	I
20		5813.63	II
30		6200.30	I
10	p	6336.90	I
20		6446.20	I
20		6487.32	I
10		6593.34	II
10		6719.32	II
20		6980.22	I
20		7118.50	I
50		7141.21	I
20		7225.16	I
10		7310.27	I
20		7838.12	I
50		8019.70	II
6		8177.31	I
5		8335.07	I
5		9932.21	I

RADON (Rn)
Z = 86

Rn I
Ref. 251 — E.F.W.

Intensity	Wavelength (Air)	
5	3514.60	I
10	3739.89	I
20	3753.65	I
10	3917.20	I
10	3941.72	I
10	3952.36	I
10	4226.06	I
80	4307.76	I
7	4335.78	I
100	4349.60	I
40	4435.05	I
50	4459.25	I
50	4508.48	I
50	4577.72	I
50	4609.38	I
30	4721.76	I
6	5722.58	I
10	6061.92	I
6	6200.75	I
6	6380.45	I
10	6557.49	I
10	6606.43	I
15	6627.23	I
6	6669.60	I
8	6704.28	I
20	6751.81	I
6	6806.79	I
8	6836.95	I
8	6837.57	I
10	6891.16	I
10	6998.90	I
200	7055.42	I
100	7268.11	I
20	7291.00	I
6	7320.98	I
10	7419.04	I
300	7450.00	I
8	7470.89	I
8	7483.13	I
8	7514.13	I
8	7516.92	I
6	7523.93	I
6	7597.55	I
8	7601.28	I
10	7657.48	I
10	7738.43	I
20	7746.64	I
100	7809.82	I
20	8049.00	I
100	8099.51	I
6	8173.84	I
100	8270.96	I
8	8314.51	I
6	8349.74	I
10	8381.05	I
10	8487.48	I
10	8494.89	I
20	8520.95	I
100	8600.07	I
10	8639.76	I
15	8675.83	I
10	8807.75	I
50	9327.02	I
6	9948.37	I
5	10106.13	I

RHENIUM (Re)
Z = 75

Re I and II
Ref. 1 — C.H.C.

Intensity		Wavelength (Air)	
25000		2003.53	I
16000		2017.87	I
27000		2049.08	I
4200		2074.70	I
3700		2083.92	I
10000		2085.59	I
4700		2092.41	II
9800		2097.12	I
2700		2109.22	I
3400		2139.04	II
1600		2142.74	II
		2142.97	I
3700		2156.67	I
4900		2167.94	I
3400		2176.21	I
4200	c	2214.26	II
2200		2214.58	I
1700		2226.42	I
920		2235.44	I
440		2255.73	I
860		2256.19	I
2000		2264.39	I
2100		2274.62	I
5200	c	2275.25	II
1600		2281.62	I
2900		2287.51	I
2700		2294.49	I
390		2298.09	II
390		2299.77	I
610		2302.99	I
680		2306.54	I
230		2312.97	I
220		2313.34	I
220		2319.19	I
370		2320.16	I
800		2322.49	I
300		2328.66	I
270		2334.33	I
270		2335.73	I
220		2336.10	I
270		2337.95	I
860		2344.78	I
230		2349.39	I
220	d	2350.46	I
680		2352.07	I
210	d	2353.95	I
250		2356.50	I
200		2365.32	I
1200		2365.90	I
570		2367.68	I
180		2368.53	II
520		2369.27	I
220		2370.76	II
210		2371.52	I
150		2373.48	II
320		2375.07	I
75		2378.53	II
370		2379.77	I
180		2386.90	II
340		2388.57	I
230		2393.65	I
320		2394.37	I
320		2396.79	I
200		2397.31	I
210	d	2400.72	I
		2400.89	I
210		2401.68	I
75		2403.04	II
1500		2405.06	I
740		2405.60	I
320		2406.70	I
270		2410.37	I
60		2418.20	II
1200		2419.81	I
300		2421.73	I
300		2421.88	I
60		2423.84	II
2500		2428.58	I
490		2431.54	I
420		2432.18	I
340	c	2441.47	I
230		2442.51	I
250		2444.94	I
610		2446.98	I
85		2449.03	II
85		2449.52	II
610		2449.71	I
200		2455.83	II
390		2461.20	I
800	c	2461.84	II
200		2467.57	II
120		2467.85	II
150	c	2469.36	II
120		2470.61	II
75		2471.05	II
150		2473.72	II
160		2475.17	II
75		2477.43	II
200		2479.02	I
1200		2483.92	I
390		2485.81	I
980		2487.33	I
75		2490.16	II
200		2492.84	I
370		2496.04	I
200		2498.22	I
370		2501.72	I
570		2502.35	II
230		2504.60	II
270		2505.94	I
1800	c	2508.99	I
570		2520.01	I
540		2521.50	I
150		2534.10	II
370		2534.80	I
570		2540.51	I
740	d	2544.74	I
370		2545.48	I
160		2550.09	II
300		2552.02	I
150	c	2553.59	II
370		2554.63	II
1000		2556.51	I
250		2559.08	I
340		2564.19	I
540		2568.64	II
370		2571.81	II
380		2586.79	I
290		2599.86	I
290		2603.89	I
660		2608.50	II
610	d	2611.54	I
160	c	2616.72	II
200		2622.76	I
310		2635.83	II
550		2636.64	I
190		2637.01	II
90		2641.02	II
270		2642.75	I
65		2648.46	II
270		2649.05	I
660		2651.90	I

Rhenium (Cont.)

400		2654.12	I
220		2663.63	I
940		2674.34	I
220		2688.53	I
1300		2715.47	I
200		2731.56	II
220		2732.21	I
610		2733.04	II
110	h	2753.64	II
220		2758.00	I
210		2763.79	I
200		2766.39	I
310		2767.74	I
220		2768.85	I
220		2769.32	I
350		2770.42	I
550		2783.57	I
220		2791.29	I
120		2803.28	II
220		2814.68	I
75		2819.78	II
880		2819.95	I
310		2834.08	I
200		2837.55	I
200		2840.35	I
220		2843.00	I
270		2850.98	I
240		2867.19	I
200		2875.28	I
200		2883.44	I
2900		2887.68	I
130	c	2888.06	II
490		2896.01	I
830	c	2902.48	I
210		2905.58	I
550		2909.82	I
65	h	2916.73	II
830	c	2927.42	I
270		2930.61	I
440		2943.14	I
130	h	2957.91	II
270		2962.27	I
720		2965.11	I
1500		2965.76	I
90		2968.98	II
310		2976.29	I
210		2978.15	I
220		2980.82	I
220		2982.19	I
220		2988.47	I
1800		2992.36	I
5500		2999.60	I
350		3001.14	I
220		3004.14	I
200		3006.42	I
500		3016.02	I
300		3016.49	I
380		3030.45	I
240		3047.25	I
200		3058.78	I
1600		3067.40	I
320		3069.94	I
260		3071.16	I
200		3072.96	I
550		3082.43	I
340		3088.76	I
200		3093.64	I
200		3095.06	I
700		3100.67	I
140		3103.06	II
700		3108.81	I
340		3110.86	I
340	c	3118.19	I
340		3121.36	I
420		3128.94	I
260		3134.02	I
250		3141.38	I
440		3151.64	I
330		3153.79	I
360	c	3158.31	I
220		3164.52	I
700		3168.37	I
220		3174.61	I
440		3177.71	I
260		3178.61	I
600		3182.87	I
1100		3184.76	I
1100		3185.57	I
260		3190.78	I
260		3192.36	I
200		3194.50	I
220		3198.58	I
1100	c	3204.25	I
380		3235.94	I
600		3258.85	I
600		3259.55	I
200		3261.56	I
300		3268.89	I
200		3294.83	I
280		3296.70	I
280		3296.99	I
280		3301.60	I
240		3302.23	I
320		3303.21	II
280		3303.75	I
240		3313.95	I
600		3322.48	I
200		3331.52	I
2000		3338.18	I
1600		3342.24	I
810		3344.32	I
320		3346.20	I
240	d	3356.33	I
200		3358.02	I
200		3362.74	I
240		3377.74	I
320		3379.06	II
320		3379.70	I
200		3385.76	I
240		3389.43	I
200		3390.25	I
4000		3399.30	I
650		3404.72	I
650		3405.89	I
240		3408.67	I
320		3409.83	I
320		3417.77	I
810		3419.41	I
8000		3424.62	I
400		3426.19	I
300		3427.61	I
320		3437.71	I
400		3449.37	I
16000	c	3451.88	I
240		3453.50	I
55000	c	3460.46	I
40000	c	3464.73	I
400		3467.96	I
240		3476.44	I
400		3480.38	I
320		3480.85	I
240		3482.23	I
560		3503.06	I
100	c	3512.28	I
320		3516.65	I
320		3517.33	I
120		3534.82	I
320		3537.46	I
160		3539.33	I
240		3549.89	I
160		3551.29	I
160		3553.65	I
160		3558.94	I
160		3568.23	I
240		3570.26	I
360		3579.12	I
810	c	3580.15	II
650		3580.97	I
810		3583.02	I
160		3596.39	I
160		3610.49	I
320		3617.08	I
160		3621.46	I
160		3625.91	I
140		3637.06	I
810		3637.84	I
440		3651.97	I
120		3669.78	I
320		3670.53	I
860	c	3689.50	I
1500	c	3691.48	I
100		3697.71	I
520		3703.24	I
100		3705.02	I
240		3709.93	I
360	c	3717.28	I
4000		3725.76	I
140		3731.87	I
140		3732.28	I
240	c	3735.01	I
810		3735.31	I
910		3740.10	I
140		3740.41	I
130		3742.26	II
300	Cw	3745.44	I
140		3766.48	I
120		3768.26	I
140		3777.66	I
700		3787.52	I
160		3796.59	I
160		3797.59	I
190		3807.74	I
120		3815.66	I
120		3836.30	I
240		3869.94	I
240		3875.26	I
240		3876.86	I
100		3908.21	I
130		3913.92	I
380	c	3917.27	I
550		3929.85	I
140		3936.90	I
110		3944.72	I
180		3945.91	I
280		3961.04	I
350	c	3962.48	I
100		4004.93	I
140		4022.96	I
100		4023.31	I
110	c	4029.63	I
220		4033.31	I
110		4037.49	I
200		4048.99	I
240		4081.43	I
140		4104.42	I
240	c	4110.89	I
190		4121.64	I
240	Cw	4133.42	I
1800		4136.45	I
700		4144.36	I
140		4149.96	I
160		4170.40	I
220		4182.90	I
220		4183.06	I
650		4221.08	I
3600	c	4227.46	I
150		4241.39	I
260	c	4257.60	I
120	c	4291.17	I
200		4304.40	I
200		4332.25	I
40		4357.98	II
380		4358.69	I
190		4367.58	I
140		4391.34	I
360	Cw	4394.38	I
110	Cw	4406.40	I
180		4415.82	I
150		4475.08	I
120		4478.39	I
120	c	4507.04	I
2600		4513.31	I
260		4516.64	I
500		4522.73	I
120		4523.88	I
120		4529.95	I
100		4545.17	I
120		4580.68	I
120		4605.73	I
100		4621.38	I
190	c	4791.42	I
2200	Cw	4889.14	I
220		4923.90	I
40		5058.56	I
70		5096.50	I
20		5120.32	I
25		5161.65	I
40	c	5178.89	I
20		5181.74	I
35		5234.31	I
50		5248.86	I
1300		5270.95	I
1600	Cw	5275.56	I
100		5278.24	I
30		5305.56	I
20		5317.28	I
35		5321.28	I
50		5327.46	I
20		5331.90	I
20		5332.76	I
20		5333.85	I
35		5369.48	I
50	c	5369.80	I
100	c	5377.10	I
25		5431.90	I
14		5437.03	I
14		5447.92	I
25		5460.64	I
14	h	5520.05	I
25		5521.10	I
50	c	5532.68	I
50	c	5563.24	I
25		5573.47	I
25		5584.72	I
10	h	5607.21	I
12	h	5612.27	I
100		5667.88	I
25		5711.43	I
18		5716.95	I
110	c	5752.93	I
110	Cw	5776.83	I
18		5791.60	I
10	c	5815.92	I
550		5834.31	I
10		5919.86	I
60		5943.24	I
10		5950.21	I
18	h	5969.77	I
10		5989.99	I
18	h	5995.73	I
30		6114.22	I
35	c	6145.81	I
50		6146.82	I
18		6203.24	I
25		6217.97	I
30	Cw	6229.42	I
35	Cw	6243.24	I
35	d	6260.02	I
		6260.24	I
18	c	6271.37	I
18		6278.76	I
10		6286.41	I
10		6303.42	I
200		6307.70	I
200		6321.90	I
80	d	6350.75	I
16	h	6382.94	I
14		6411.47	I
50		6511.47	I
14		6515.25	I
12		6544.91	I
35	c	6577.11	I
40	Cw	6592.52	I
100	Cw	6605.19	I
30	c	6623.91	I
10	c	6637.25	I
27	Cw	6652.39	I
15	h	6683.28	I
9	c	6711.30	I
30		6751.22	I
5	c	6761.19	I
180	c	6813.41	I
260		6829.90	I
85		6971.53	I
35	Cw	7006.63	I
65	Cw	7024.15	I
65	Cw	7246.67	I
13	Cw	7292.72	I
40	Cw	7578.73	I
13		7611.89	I
7	Cw	7620.25	I
50	Cw	7640.94	I
65	Cw	7912.94	I
35	Cw	7980.77	I
40		8417.13	I
29	Cw	8527.73	I

RHODIUM (Rh)
Z = 45

Rh I and II
Ref. 1 — C.H.C.

Intensity		Wavelength	
		Air	
150		2276.21	II
140		2288.57	I
110		2309.82	I
55		2318.36	I
95		2319.10	I
95		2321.73	I
350		2322.58	I
140		2326.47	I
80		2328.64	I
190		2334.77	II
55		2345.41	I
55		2352.47	I
55		2359.18	I
300		2361.92	I
110		2368.34	I
270		2382.89	I
230		2383.40	I
40		2384.65	I
270		2386.14	II
80		2407.88	I
27		2408.19	I
27		2410.25	I
80		2415.84	II
55		2418.64	
45		2419.75	I
45		2420.18	II
65		2420.98	II
75		2423.94	
65		2427.11	II
130		2427.68	I
230		2429.52	I
40		2431.85	II
40		2432.66	I
18		2437.08	I
110		2437.90	I
330		2440.34	I
50	h	2444.27	I
65		2448.84	I
50		2449.04	I
75		2450.56	I
30		2455.70	II
65		2458.90	II
90		2461.04	II
30		2463.61	I
75		2470.39	I
90		2471.47	I
30		2472.51	I
130		2473.09	I
15		2475.64	II
15		2477.54	II
25		2482.04	I
50		2483.33	I
150		2487.47	I
100		2490.77	II
30		2492.30	I
75	h	2494.51	
15		2499.02	I
40		2500.58	
130		2502.46	I

Rhodium (Cont.)

15		2503.84	II
300		2504.29	II
40		2505.10	II
150		2505.67	I
350		2509.70	I
50		2510.66	II
300		2511.03	II
75		2513.36	II
200		2515.75	I
130		2520.53	II
13		2525.99	I
13		2531.74	I
50		2532.66	
13		2533.59	
50		2534.07	
110		2536.71	
110		2537.04	II
30		2539.72	
40		2544.22	
350		2545.70	I
13		2548.60	
550		2555.36	I
25		2558.62	I
50		2565.79	I
45		2566.04	I
25		2566.92	II
50		2567.28	
25		2574.66	
25		2575.75	I
13		2576.23	
40		2587.29	II
30		2598.07	
30		2603.32	II
75		2606.44	II
75		2613.60	
150		2622.58	I
230		2625.88	I
100		2630.42	I
40		2634.99	I
30		2638.74	II
75		2643.00	I
110		2647.28	I
400		2652.66	I
30		2659.01	I
30		2671.06	I
65		2676.11	I
25		2680.28	I
100		2680.63	I
30		2681.78	I
30	h	2686.50	
30	h	2686.91	
50		2694.31	I
400		2703.73	I
40		2705.63	II
40		2707.23	I
75		2714.41	I
100		2715.31	II
75		2717.51	I
180		2718.54	I
65		2720.14	I
30		2720.52	I
160		2728.94	I
40		2736.76	I
75		2741.75	I
50		2767.73	I
100		2771.51	I
50		2778.06	I
75		2779.54	I
130		2783.03	I
25		2791.16	I
75		2796.63	I
150		2826.43	I
180		2826.68	I
30		2827.31	I
75		2834.12	I
45		2835.44	I
75		2836.69	I
50		2856.16	I
50	d	2860.68	I
		2860.76	I
280		2862.94	I
65		2864.40	I
50		2871.35	I
30		2873.62	I
110		2878.66	I
75		2880.76	I
140		2882.37	I
75		2885.97	I
75		2889.11	I
75		2889.84	I
65		2899.96	I
25		2904.81	I
160		2907.21	I
65		2910.17	II
75		2912.62	I
90		2915.42	I
30		2923.10	I
180		2924.02	I
130		2929.11	I
130		2931.94	I
30		2955.41	I
230		2968.66	I
25		2974.03	I
160		2977.68	I
450		2986.20	I
90		2986.99	I
50		2987.45	I
110		3004.46	I
50		3019.54	I
130		3023.91	I
50		3028.43	I
30		3045.77	I
30		3046.76	I
25		3057.89	I
65		3067.30	I
180		3083.96	I
29		3087.42	I
70		3114.91	I
140		3121.76	I
240		3123.70	I
35		3130.79	I
95		3137.71	I
45		3151.36	I
45		3152.60	I
130		3155.78	I
70		3179.73	I
80		3185.59	I
140		3189.05	I
470		3191.19	I
190		3197.13	I
70		3214.32	I
80		3237.66	I
520		3263.14	I
520		3271.61	I
2300		3280.55	I
110		3281.70	I
2300		3283.57	I
280		3289.14	I
45		3289.64	I
210		3294.28	I
45		3296.72	I
260		3300.46	I
4200		3323.09	I
60		3331.09	I
45		3331.24	I
330		3338.54	I
70		3342.90	I
80		3344.20	I
60		3359.90	I
280		3360.80	I
60		3362.18	I
420		3368.38	I
45		3369.68	I
1100		3372.25	I
110		3377.14	I
80		3377.71	I
110		3385.78	I
5600		3396.82	I
820		3399.70	I
160		3406.55	I
820		3412.27	I
60		3420.16	I
330		3421.22	I
120	d	3424.38	I
8200		3434.89	I
1400		3440.53	I
35		3442.63	I
120		3447.74	I
60		3448.58	I
120		3450.29	I
60		3451.15	I
400		3455.22	I
60		3455.42	I
180		3457.07	I
220		3457.93	I
5900		3462.04	I
180		3469.62	I
4700		3470.66	I
120		3472.25	I
4700		3474.78	I
2100		3478.91	I
95		3484.04	I
80		3491.07	I
110		3494.44	I
1200		3498.73	I
5900		3502.52	I
60		3505.41	I
2800		3507.32	I
60		3511.78	I
60		3513.10	I
60		3519.54	I
8800		3528.02	I
880	d	3538.14	I
		3538.26	I
280		3541.91	I
1200		3543.95	I
1800		3549.54	I
240		3564.13	I
1200		3570.18	I
4700		3583.10	I
120		3583.53	I
4700		3596.19	I
5900		3597.15	I
310		3605.86	I
3100		3612.47	I
240		3614.78	I
200		3620.46	I
1800		3626.59	I
95		3627.80	I
310		3639.51	I
350		3654.87	I
8200		3657.99	I
280		3661.86	I
1300		3666.22	I
180		3666.91	I
140		3674.76	I
560		3681.04	I
1900		3690.70	I
9400		3692.36	I
60		3694.95	I
940		3695.52	I
280		3698.26	I
380		3698.60	I
7600		3700.91	I
940		3713.02	I
60		3713.43	I
45		3714.83	I
16		3724.94	I
650		3735.28	I
420		3737.27	I
420		3744.17	I
1200		3748.22	I
240		3754.12	I
380		3754.27	I
490		3755.58	I
1000		3760.40	I
2300		3765.08	I
490		3769.97	I
70		3775.72	I
380		3778.13	I
1000		3788.47	I
1300		3792.18	I
3800		3793.22	I
4900		3799.31	I
760		3805.92	I
1300		3806.76	I
45		3809.50	I
95		3812.45	I
470		3815.01	I
760		3816.47	I
1300		3818.19	I
3800		3822.26	I
2300		3828.48	I
2000		3833.89	I
45		3834.75	I
5900		3856.52	I
490		3870.01	I
70		3872.39	I
380		3877.34	I
70		3888.34	I
29		3904.22	I
23		3912.83	I
120		3913.51	I
240		3922.19	I
2000		3934.23	I
45		3934.98	I
50		3935.84	I
590		3942.72	I
95		3958.24	I
3800		3958.86	I
45		3964.54	II
380		3975.31	I
240		3984.40	I
240		3995.61	I
380		3996.15	I
120		4023.14	I
60		4048.41	I
23		4049.04	I
40		4053.44	I
23		4056.34	I
70		4077.57	I
560		4082.78	I
19		4084.28	I
45		4087.79	I
60		4088.50	I
140		4097.52	I
45		4107.49	I
70		4116.33	I
120		4119.68	I
1100		4121.68	I
1500		4128.87	I
2100		4135.27	I
240		4154.37	I
330		4196.50	I
70		4206.62	I
3300		4211.14	I
29		4230.20	I
40		4244.44	I
60		4273.43	I
60		4278.60	I
820		4288.71	I
70		4296.77	I
23		4342.44	I
45		4373.04	I
4200		4374.80	I
95		4379.92	I
23		4433.32	I
35		4492.47	I
29		4503.78	I
23		4528.72	I
16		4544.27	I
35		4548.73	I
40		4551.64	I
19		4560.89	I
16		4565.19	I
130		4569.00	I
14		4571.31	I
29		4608.12	I
14		4619.91	I
23		4643.18	I
150		4675.03	I
19		4721.00	I
70		4745.11	I
12		4755.58	I
23		4810.49	I
21		4842.43	I
45		4843.99	I
60		4851.63	I
60		4963.71	I
60		4977.75	I
40		4979.18	I
14		5085.52	I
70		5090.63	I
23		5120.69	I
19		5130.76	I
60		5155.54	I
14		5157.09	I
40		5158.69	I
60		5175.97	I
12		5177.27	I
35		5184.19	I
95		5193.14	I
16		5206.95	I
16		5211.52	I
19		5212.73	I
16		5214.79	I
19		5222.66	I
19		5230.62	I
45		5237.16	I
9		5237.80	I
14		5269.27	I
11	h	5280.12	I
14		5292.14	I
14		5314.79	I
40	h	5329.74	I
14	h	5331.08	I
9		5349.31	I
130		5354.40	I
23		5356.47	I
45		5379.10	I
95		5390.44	I
23	h	5404.73	I
160	h	5424.07	I
19		5424.72	I
19	h	5425.45	I
12		5439.58	I
12	h	5441.36	I
9	h	5444.32	I
35	h	5445.23	I
23	h	5468.11	I
35	h	5470.85	I
12		5476.12	I
12		5481.42	I
16		5484.23	I
9		5504.65	I
29		5535.04	I
21	l	5544.58	I
160		5599.42	I
7		5607.71	I
16		5608.35	I
5		5632.77	I
9		5659.62	I
40		5686.38	I
9	h	5702.47	I
6		5727.30	I
29		5792.66	I
9		5795.79	I
9		5803.34	I
40		5806.91	I
6		5821.84	I
35		5831.58	I
7		5907.31	I
9		5918.54	I
7		5941.46	I
130		5983.60	I
9		5991.19	I
35		6102.72	I
6		6116.15	I
8		6128.06	I
8		6186.89	I
14		6199.99	I
16		6253.72	I
5		6276.66	I
8		6277.46	I
6		6293.38	I
29		6319.53	I
12		6414.72	I
16		6510.41	I
19		6519.70	I
9		6627.80	I
19		6630.16	I
40		6752.35	I
9		6796.65	I
13		6827.33	I
11		6857.68	I
20		6879.94	I

Rhodium (Cont.)

65		6965.67	I
8		6972.91	I
16		6979.15	I
16		7001.58	I
11		7038.76	I
18		7101.64	I
15		7104.45	I
6		7142.55	I
9		7219.06	I
18		7268.18	I
35		7270.82	I
12		7271.94	I
5		7273.03	I
9		7375.57	I
5	h	7386.64	I
9		7430.80	I
18	h	7442.39	I
7		7446.77	I
12		7475.74	I
12		7495.24	I
8		7542.02	I
11		7557.67	I
8		7577.22	I
11		7690.05	I
18		7772.90	I
29		7791.61	I
55		7824.91	I
15		7830.05	I
15		7846.50	I
21		8029.91	I
11	h	8036.09	I
29		8045.36	I
7		8063.50	I
15		8136.20	I
7	h	8193.67	I
5	h	8369.67	I
8		8425.59	I

Rh III
Ref. 396 — L.J.R.

Intensity		Wavelength	
		Vacuum	
10		746.28	III
30		759.54	III
50		813.44	III
30		826.01	III
30		843.63	III
30		849.08	III
40		852.70	III
40		854.77	III
40		859.89	III
40		861.34	III
50		863.78	III
50		865.22	III
50		870.40	III
80		882.51	III
100		925.75	III
150		937.28	III
100		976.12	III
500		991.62	III
400		992.48	III
500	d	1009.60	III
200		1012.22	III
200		1015.17	III
100		1050.00	III
100		1058.97	III
200		1073.87	III
100		1100.58	III
100		1113.79	III
100		1768.43	III
150		1784.24	III
200		1784.94	III
150		1796.50	III
200		1816.03	III
1000		1832.05	III
500		1859.85	III
100		1874.70	III
800		1880.66	III
500		1884.91	III
500		1887.36	III
700		1888.62	III
800		1901.32	III
500		1910.16	III
600		1919.37	III
500		1927.07	III
700		1931.79	III
500		1954.25	III
400		1965.16	III
500		1994.26	III
		Air	
400		2005.14	III
800		2013.71	III
500		2017.47	III
500		2028.53	III
800		2036.72	III
600		2037.61	III
1000		2040.18	III
3000		2048.67	III
2000		2064.11	III
800		2076.84	III
1000		2118.53	III
1000		2118.63	III
1000		2139.44	III
1000		2152.23	III
3000		2158.17	III
3000		2163.19	III
3000		2167.33	III
100		2207.00	III
100		2230.66	III
50		2250.84	III
30		2374.84	III
20		2470.65	III
50		3006.43	III
50		3052.44	III
50		3310.69	III
1		3852.98	III

RUBIDIUM (Rb)
Z = 37

Rb I and II
Ref. 12, 130, 241, 257, 264 — J.R.

Intensity		Wavelength	
		Vacuum	
10		474.88	II
40		481.118	II
90		497.430	II
20		508.434	II
150		513.266	II
300		530.173	II
75		533.801	II
40		542.887	II
200		555.036	II
2500		589.419	II
1500		643.878	II
3000		697.049	II
6000		711.187	II
10000		741.456	II
1000		1604.12	II
200		1644.96	II
200		1707.52	II
600		1716.85	II
5000		1760.50	II
200		1803.47	II
500		1809.68	II
500		1865.33	II
500		1889.42	II
500		1954.24	II
300		1956.54	II
200		1971.42	II
500		1983.19	II
		Air	
300		2042.23	II
300		2052.21	II
500		2052.80	II
2000		2068.92	II
1000		2071.50	II
10000		2075.95	II
1000		2090.29	II
200		2108.06	II
300		2116.50	II
1000		2125.25	II
400		2129.82	II
200		2143.10	II
30000		2143.83	II
200		2190.36	II
600		2197.99	II
600		2198.26	II
300		2207.86	II
10000		2217.08	II
200		2223.79	II
400		2237.72	II
500		2250.65	II
200		2251.43	II
800		2254.19	II
200		2254.55	II
200		2263.54	II
500		2263.94	II
500		2286.82	II
5000		2291.71	II
300		2298.80	II
250		2333.01	II
2000		2333.39	II
350		2353.11	II
300		2353.96	II
400		2356.97	II
300		2358.04	II
300		2364.27	II
200		2364.32	II
200		2365.15	II
300		2367.51	II
200		2373.21	II
2000		2385.34	II
250		2405.94	II
400		2434.17	II
800		2459.14	II
50000		2472.20	II
300		2484.56	II
700		2484.70	II
2000		2496.38	II
200		2502.67	II
250		2514.18	II
1000		2524.24	II
200		2594.56	II
400		2623.76	II
400		2645.58	II
1000		2684.10	II
1000		2711.76	II
250		2741.01	II
500		2812.15	II
350		2838.51	II
750		2873.88	II
1000		3051.36	II
2		3082.02	I
250		3088.58	II
10		3112.57	I
3		3113.06	I
5000	c	3148.90	II
25		3157.54	I
5		3158.26	I
1200		3161.00	II
50		3227.98	I
6		3229.16	I
2000		3270.99	II
1500		3321.49	II
1200		3340.55	II
60		3348.72	I
75		3350.82	I
750		3353.89	II
1200		3393.03	II
750		3415.58	II
1000		3434.18	II
1500		3461.50	II
3000		3521.39	II
3000	l	3531.55	II
1000		3541.15	II
100		3587.05	I
40		3591.57	I
5000		3600.60	II
10000		3600.64	II
600	c	3639.80	II
400	c	3646.26	II
350	c	3647.56	II
1000	c	3662.74	II
900	c	3663.81	II
350		3666.72	II
300		3675.66	II
2500	c	3699.58	II
350		3746.33	II
3500		3796.81	II
2500		3801.90	II
1000		3826.66	II
450		3860.74	II
250		3907.29	II
500		3922.20	II
2500	l	3926.44	II
25000		3940.51	II
1000	c	3978.15	II
1700		4029.49	II
2500	c	4083.88	II
2000	c	4104.28	II
1700	c	4136.11	II
3500		4193.08	II
1000		4201.80	I
500		4215.53	I
90000		4244.40	II
500		4266.58	II
250	c	4270.25	II
15000		4273.14	II
2500	c	4287.97	II
1500		4293.97	II
500	c	4306.26	II
1000		4346.96	II
2500		4377.12	II
300		4440.10	II
1000		4469.47	II
400	c	4493.92	II
700		4519.04	II
3000		4530.34	II
500	l	4533.79	II
400		4540.74	II
20000		4571.77	II
3000	c	4622.42	II
350	c	4631.89	II
10000		4648.57	II
500		4659.28	II
1000		4730.45	II
1000		4755.30	II
400	c	4757.82	II
30000		4775.95	II
5000	c	4782.83	II
300	c	4855.34	II
1500	c	4885.59	II
2		5087.987	I
2		5132.471	I
10		5150.134	I
10000		5152.08	II
300		5164.58	II
1		5165.023	I
2		5165.142	I
1		5169.65	I
15		5195.278	I
2		5233.968	I
20		5260.034	I
1		5260.228	I
200		5270.51	II
3		5322.380	I
40		5362.601	I
4		5390.568	I
75		5431.532	I
3		5431.830	I
500		5512.55	II
5000		5522.78	II
6		5578.788	I
5000	c	5635.99	II
40		5647.774	I
20		5653.750	I
3000	d	5699.15	II
60		5724.121	I
3		5724.614	I
200		5739.64	II
75		6070.755	I
200		6135.27	II
30	c	6159.626	I
1000	c	6199.08	II
75	c	6206.309	I
300		6269.40	II
120	c	6298.325	I
5		6299.224	I
10000		6458.33	II
1000		6555.62	II
5000		6560.81	II
3000	l	6775.07	II
100	l	7279.997	I
300	c	7316.52	II
150		7408.173	I
200	l	7618.933	I
300		7757.651	I
60		7759.436	I
90000	c	7800.27	I
5	l	7925.26	I
4		7925.54	I
45000	c	7947.60	I
40	l	8271.41	I
30		8271.71	I
2000		8603.96	II
40	l	8868.512	I
30		8868.852	I
300		8978.88	II
300		9021.77	II
3		9224.64	I
2		9234.25	I
500	c	9246.41	II
300		9338.87	II
200	w	9373.50	II
300		9391.36	II
1000		9479.32	II
700	l	9493.72	II
30	l	9522.65	I
5		9523.05	I
20	l	9540.18	I
300		9612.99	II
300		9671.54	II
2000	c	9689.05	II
200		9776.06	II
200		9934.76	II
35	l	10075.282	I
30	l	10075.708	I
100		13235.17	I
20		13442.81	I
30		13443.57	I
75		13665.01	I
1000		14752.41	I
800		15288.43	I
150		15289.48	I
20		22529.65	I
10		22932.47	I
4		27314.31	I
2		27905.37	I

Rb III
Ref. 258, 262 — J.R.

Intensity		Wavelength	
		Vacuum	
30		465.85	III
35	p	482.43	III
30	p	482.47	III
500		482.83	III
300		484.84	III
500		489.66	III
100		489.96	III
600		493.48	III
50		497.82	III
100		500.28	III
30		508.33	III
400		516.79	III
800		533.64	III
1200		535.86	III
1200		556.19	III
500		558.36	III

Rubidium (Cont.)

Intensity		Wavelength	
700		564.77	III
1500		566.71	III
1000		572.82	III
1500		576.65	III
2500		579.63	III
1500		581.26	III
500		582.34	III
800		586.77	III
100		591.42	III
900		593.65	III
1000		594.94	III
1300		595.88	III
1200		598.49	III
450		602.09	III
50		605.51	III
500		607.28	III
400		613.31	III
300		619.07	III
20		620.83	III
100		622.24	III
250		630.06	III
500		645.67	III
20		674.81	III
5000		769.04	III
2500		815.28	III

Air

Intensity		Wavelength	
100		2153.21	III
250		2164.59	III
100		2268.00	III
150		2300.12	III
500		2304.14	III
150		2304.45	III
250		2312.46	III
200		2337.07	III
100		2341.90	III
200		2345.37	III
100		2349.81	III
150		2380.44	III
100		2381.29	III
150		2418.46	III
300		2561.86	III
100		2573.71	III
100		2577.07	III
200		2586.83	III
1000		2631.75	III
350		2636.83	III
100		2656.68	III
100		2713.86	III
500		2798.86	III
150		2800.27	III
500		2807.58	III
100		2845.44	III
150		2869.77	III
500		2903.69	III
150		2949.62	III
100		2951.01	III
2000		2956.07	III
500	l	2967.45	III
150		2968.13	III
500		2970.74	III
250		2987.40	III
350		3023.61	III
200		3039.62	III
200		3041.48	III
250		3070.70	III
500		3086.84	III
100		3098.49	III
500		3111.36	III
250	s	3114.82	III
120		3118.92	III
100		3169.34	III
200		3222.60	III
500		3286.41	III
100		3330.16	III
200		3346.92	III
250		3439.26	III
100		3492.68	III

Rb IV
Ref. 109 — J.R.

Vacuum

Intensity	Wavelength	
10	595.18	IV
25	663.76	IV
25	716.24	IV
20	733.41	IV
50	740.85	IV
20	749.86	IV
20	753.75	IV
10	771.54	IV
25	776.89	IV
9	817.92	IV
15	850.18	IV
10	988.00	IV

RUTHENIUM (Ru)
Z = 44

Ru I and II
Ref. 1 — C.H.C.

Air

Intensity		Wavelength	
2400		2076.43	I
2600		2083.77	I
2400		2090.89	I
690		2255.52	I
290		2259.53	I
780		2272.09	I
240		2278.19	I
780		2279.57	I
170		2285.38	I
290		2302.54	I
480		2317.80	I
150		2322.01	I
120		2334.96	II
240		2340.69	I
190	h	2342.85	II
190		2349.34	I
310		2351.33	I
170		2357.91	II
140		2360.56	I
170		2370.17	I
240		2375.27	I
80		2375.63	II
160		2392.42	I
95		2396.71	II
780		2402.72	II
150		2407.92	II
55		2410.89	I
55		2414.82	II
130		2420.82	I
55		2422.92	I
45		2429.60	I
65		2432.93	I
30		2447.45	I
30		2450.58	I
65		2454.92	I
180		2455.53	II
150		2456.44	II
370		2456.57	II
65	h	2458.62	I
55		2462.94	I
85		2464.70	I
30		2474.04	I
110		2475.41	I
100		2476.88	I
280		2478.93	II
28		2481.11	II
30		2489.91	I
18		2491.78	I
65		2493.69	II
85		2494.02	I
45		2494.48	II
85		2495.69	II
65		2496.56	I
140		2498.42	II
140		2498.57	II
85		2499.78	I
260		2507.01	II
130		2508.27	I
110		2509.07	I
110		2512.81	I
110		2513.32	II
110		2517.32	II
150		2535.59	II
65		2543.25	II
280		2544.22	I
120		2546.67	I
280		2549.48	I
550		2549.58	I
130		2560.26	I
120		2560.83	I
110		2563.15	I
160		2568.77	I
100		2570.97	I
100		2578.57	I
100		2579.53	I
100		2589.57	I
170		2591.12	I
120		2592.02	I
100		2593.70	I
110		2594.85	I
370		2609.06	I
830		2612.07	I
100		2615.09	I
220		2631.30	I
220		2635.86	I
170		2636.67	I
110		2640.33	I
460		2642.96	I
110		2647.32	I
110		2651.29	I
330		2651.84	I
28		2656.25	II
400		2659.62	I
23		2661.17	II
330		2661.61	II
200		2664.76	I
30		2667.40	II
690		2678.76	II
220		2686.29	I
28		2687.50	II
		2688.16	II
330		2692.06	II
110		2701.34	I
110		2702.83	I
170		2709.20	I
200		2712.41	II
690		2719.52	I
130		2722.65	I
140		2725.47	II
310		2734.35	II
1800		2735.72	I
170		2739.22	I
130		2744.45	I
35		2747.97	II
75		2752.45	II
75		2752.77	II
260		2763.42	I
35		2765.44	II
90		2768.93	II
100		2778.38	II
110		2787.83	II
140		2802.81	I
35		2806.74	II
350		2810.03	I
1700		2810.55	I
350		2818.36	I
110		2822.03	I
200		2827.87	I
400		2829.16	I
130		2834.00	I
150		2840.54	I
35		2841.68	II
640		2854.07	I
180		2860.02	I
420		2861.41	I
550		2866.64	I
110		2868.31	I
1800		2874.98	I
220		2879.76	I
55		2882.12	II
130		2883.60	I
740		2886.54	I
180		2892.56	I
110		2901.94	I
140		2905.65	I
370		2908.88	I
1100		2916.26	I
150		2919.61	I
35		2927.54	II
180		2945.67	II
180		2946.99	I
370		2949.50	I
150		2954.49	I
18		2963.40	II
550		2965.16	I
170		2965.55	II
140		2976.59	II
550		2976.92	I
45		2977.23	II
75		2979.96	II
1400		2988.95	I
35		2991.62	II
110		2993.27	I
460		2994.96	I
440		3006.59	I
330		3017.24	I
310		3020.88	I
240		3033.45	I
200		3040.31	I
220		3042.48	I
110		3045.71	I
110		3048.78	I
150		3054.94	I
390		3064.84	I
170		3089.14	I
120		3089.80	I
330		3096.57	I
120		3097.60	I
830		3099.28	I
740		3100.84	I
120		3125.96	I
120		3153.82	I
290		3159.92	I
200		3168.52	I
60		3177.05	II
180		3186.04	I
240		3188.34	I
240		3189.98	I
180		3196.59	I
180		3223.27	I
110		3226.37	I
100		3227.88	I
220		3228.53	I
220		3238.53	I
120		3241.24	I
120		3243.50	I
280		3260.35	I
120	d	3264.55	I
120		3266.44	I
200		3268.21	I
200		3273.08	I
200		3274.71	I
100		3277.57	I
490		3294.11	I
370		3301.59	I
220		3306.17	I
290		3315.23	I
290		3316.39	I
100		3325.00	I
120		3335.69	I
930		3339.55	I
240		3341.66	I
200		3361.15	I
370		3368.45	I
100		3371.86	I
130		3374.65	I
120		3378.02	I
100		3379.60	I
130		3380.18	I
130		3385.14	I
130		3388.71	I
100		3389.50	I
370		3392.54	I
310		3401.74	I
310		3409.28	I
3100		3417.35	I
4900		3428.31	I
490		3430.77	I
310		3432.74	I
6400		3436.74	I
260		3438.37	I
220		3440.20	I
260		3473.75	I
240		3481.30	I
8300		3498.94	I
640		3514.49	I
330		3519.64	I
200		3528.68	I
240		3532.81	I
390		3537.95	I
790		3539.37	I
200		3541.63	I
690		3570.59	I
200		3574.58	I
390		3587.20	I
6400		3589.22	I
6900		3593.02	I
6400		3596.18	I
1300		3599.76	I
350		3625.20	I
370		3626.74	I
3100		3634.93	I
210		3637.47	I
200		3640.64	I
290		3650.32	I
310		3654.40	I
6200		3661.35	I
830		3663.37	I
650		3669.49	I
240		3678.32	I
260		3696.59	I
410		3717.00	I
260		3719.33	I
550		3726.10	I
8700		3726.93	I
11000		3728.03	I
7100		3730.43	I
280		3737.40	I
410		3739.46	I
3500		3742.28	I
870		3742.78	I
280		3744.22	I
410		3744.40	I
2800		3745.59	I
760		3753.54	I
310		3755.09	I
870		3755.93	I
1200		3759.84	I
370		3760.03	I
600		3761.51	I
600		3767.35	I
1500		3777.59	I
460		3781.18	I
600		3782.74	I
3900		3786.06	I
6000		3790.51	I
240		3794.92	I
760		3798.05	I
7600		3798.90	I
7600		3799.35	I
310		3800.26	I
310		3808.68	I
600		3812.72	I
760		3817.27	I
760		3819.03	I
650		3822.09	I
550		3824.93	I
760		3831.80	I
220		3835.05	I

Ruthenium (Cont.)

310		3838.07	I
930		3839.70	I
480		3846.68	I
760		3850.43	I
480		3856.46	I
1300		3857.55	I
220		3860.72	I
650		3862.69	I
1300		3867.84	I
260		3873.52	I
650		3892.21	I
760		3909.08	I
260		3920.92	I
1500		3923.47	I
3300		3925.92	I
600		3931.76	I
310		3933.55	I
760		3945.57	I
460		3950.21	I
310		3952.68	I
460		3964.90	I
600		3978.44	I
600		3979.42	I
870		3984.86	I
280		3995.98	I
1500		4022.16	I
600		4023.83	I
310		4039.21	I
1400		4051.40	I
710		4054.05	I
370		4064.46	I
200		4067.61	I
760		4068.37	I
200		4073.00	I
980		4076.73	I
6000		4080.60	I
310		4085.43	I
930		4097.79	I
350		4101.74	I
1900		4112.74	I
2000		4144.16	I
650		4145.74	I
260		4146.77	I
870		4167.51	I
550		4197.58	I
550		4198.88	I
7600		4199.90	I
1500		4206.02	I
5400		4212.06	I
760		4214.44	I
930		4217.27	I
370		4220.68	I
550		4230.31	I
760		4241.05	I
760		4243.06	I
370		4246.73	I
310		4258.99	I
760		4284.33	I
220		4293.28	I
260		4294.79	I
550		4295.93	I
3700		4297.71	I
930		4307.60	I
370		4318.43	I
550		4319.87	I
550		4342.07	I
350		4349.70	I
710		4354.13	I
870		4361.21	I
2400		4372.21	I
870		4385.39	I
1300		4385.65	I
1700		4390.44	I
1600		4410.03	I
160		4421.46	I
330		4428.46	I
460		4439.76	I
440		4449.34	I
1100		4460.04	I
190		4473.93	I
150		4480.45	I
350		4498.14	I
120		4510.10	I
220		4516.89	I
220		4517.82	I
110		4520.95	I
170		4547.33	I
110		4547.85	I
5400		4554.51	I
110		4559.98	I
1700		4584.44	I
110		4591.10	I
150		4592.52	I
330		4599.08	I
170		4635.69	I
200		4645.09	I
720		4647.61	I
290		4654.32	I
290		4681.79	I
190		4684.02	I
290		4690.11	I
1400		4709.48	I
140		4731.33	I
120		4733.52	I
500		4757.84	I
260		4815.52	I
120		4844.56	I
550		4869.15	I
160		4895.60	I
470		4903.05	I
120		4907.89	I
260		4921.07	I
180		4938.43	I
160		4968.90	I
160		4980.35	I
120		4992.74	I
160		5011.23	I
90		5014.95	I
90		5026.18	I
65		5028.16	I
35		5040.35	I
35		5040.74	I
65		5047.31	I
450		5057.33	I
21		5062.64	I
90		5072.97	I
120		5076.32	I
200		5093.83	I
80		5107.07	I
24		5123.73	I
55		5127.26	I
65		5133.89	I
530		5136.55	I
170		5142.76	I
250		5147.24	I
110		5151.07	I
55		5153.20	I
500		5155.14	I
55		5160.00	I
920		5171.03	I
180		5195.02	I
80		5199.87	I
45		5202.12	I
45		5213.43	I
65		5223.55	I
40		5242.38	I
55		5251.67	I
40		5257.07	I
40		5266.47	I
40		5266.83	I
40		5280.82	I
130		5284.08	I
40		5291.16	I
80		5304.86	I
260		5309.27	I
13		5315.33	I
40		5332.93	I
45	h	5334.70	I
110		5335.93	I
130		5361.77	I
65		5377.84	I
65		5385.88	I
110	h	5401.04	I
40		5401.39	I
40		5418.86	I
55		5427.59	I
26	l	5439.21	I
13		5452.71	I
80	h	5454.82	I
90		5456.13	I
13	h	5475.18	I
55		5479.40	I
26		5480.30	I
80		5484.32	I
18		5484.64	I
26		5496.69	I
13		5501.02	I
130		5510.71	I
20		5512.37	I
8		5517.86	I
12		5521.78	I
12		5530.99	I
24		5540.66	I
12		5556.52	I
90		5559.75	I
11		5569.03	I
21		5578.40	I
21		5603.14	I
8		5603.55	I
13		5606.73	I
11		5629.79	I
290		5636.24	I
11		5641.66	I
7		5649.56	I
7		5653.30	I
11		5665.20	I
16		5679.63	I
180		5699.05	I
13		5724.82	I
13		5725.73	I
16		5745.99	I
16		5747.47	I
11		5752.02	I
11		5756.83	I
11		5767.92	I
16		5804.39	I
65		5814.98	I
8		5828.06	I
16	h	5833.21	I
55		5919.34	I
80		5921.45	I
21		5926.87	I
26		5932.38	I
8		5936.65	I
8		5951.15	I
21	h	5973.38	I
8		5974.17	I
16		5988.67	I
35		5993.65	I
18		6116.77	I
26		6199.42	I
26		6225.20	I
9		6284.49	I
18		6295.22	I
13		6330.62	I
9		6336.12	I
9	h	6363.41	I
9		6376.45	I
16		6390.23	I
8		6417.57	I
26	h	6444.84	I
8		6496.44	I
11		6528.74	I
4		6560.45	I
4		6593.74	I
9		6618.20	I
21		6663.14	I
55		6690.00	I
11		6707.52	I
15		6718.30	I
15		6730.45	I
7		6756.54	I
21		6766.95	I
30		6775.02	I
13		6787.23	I
8		6813.51	I
15		6823.88	I
21		6824.17	I
7		6831.52	I
26		6911.48	I
110		6923.23	I
26		6982.01	I
26		7027.98	I
9		7086.06	I
12		7087.35	I
4		7141.72	I
6		7219.26	I
35		7238.92	I
7		7266.96	I
8		7323.56	I
16		7393.93	I
18		7468.91	I
12		7475.40	I
26		7485.79	I
70		7499.75	I
7		7532.07	I
26		7559.61	I
5		7612.94	I
18		7621.50	I
18		7722.87	I
5		7729.91	I
22		7791.86	I
4		7797.89	I
4		7806.82	I
3		7813.43	I
4		7829.81	I
5	h	7833.39	I
6	h	7841.90	I
30		7847.80	I
80		7881.49	I
16		7890.37	I
16		7924.43	I
5		7948.15	I
9		7967.84	I
9		8112.47	I
18		8264.96	I
11		8348.98	I
6		8352.94	I
4		8435.77	I
11		8473.64	I
11		8483.56	I
22		8710.84	I
14		8724.98	I
9		8777.36	I

Ru III
Ref. 423 — C.H.C.

Intensity		Wavelength	
		Vacuum	
250		850.09	III
200		850.30	III
50		851.22	III
50		852.49	III
150		856.32	III
50		867.48	III
250		919.74	III
50		921.78	III
250		928.08	III
150		937.16	III
500		940.09	III
250		940.68	III
50		941.85	III
50		942.63	III
50		943.06	III
150		945.68	III
100		946.05	III
100		947.14	III
100		949.83	III
50		950.35	III
100		950.45	III
50		952.59	III
50		957.06	III
50		957.18	III
50		961.58	III
250		961.68	III
100		962.56	III
500		966.54	III
250		967.09	III
150		967.85	III
150		967.92	III
150		971.83	III
250		972.40	III
100		973.54	III
150		973.78	III
750		974.14	III
250		974.46	III
250		977.51	III
100		978.18	III
900		979.43	III
500		981.35	III
250		983.81	III
250		983.91	III
250		985.55	III
900		986.84	III
200		987.87	III
250		991.67	III
250		992.75	III
900		994.56	III
250		995.30	III
200		1000.78	III
300		1001.65	III
250		1004.29	III
500		1009.13	III
900		1009.87	III
1500		1014.68	III
100		1018.72	III
100		1019.33	III
100		1020.77	III
200		1080.00	III
100		1184.37	III
800		1190.51	III
500		1200.07	III
100		1204.57	III
200		1204.88	III
500		1207.17	III
500		1209.77	III
300		1211.31	III
200		1232.57	III
100	h	1653.77	III
200		1699.84	III
100	h	1715.97	III
200		1759.49	III
200	h	1880.95	III
100	h	1883.56	III
200	h	1899.04	III
100	h	1899.42	III
100		1908.31	III
500		1941.35	III
100		1981.82	III
100		1982.10	III
200		1989.22	III
200		1993.32	III
100		1997.55	III
		Air	
200		2005.71	III
100		2006.46	III
500		2009.28	III
100		2011.17	III
50		2011.56	III
50		2011.66	III
100		2015.20	III
50		2018.58	III
100		2044.59	III

SAMARIUM (Sm)
Z = 62

Sm I and II
Ref. 1 — C.H.C.

Intensity	Wavelength
	Air
45	2610.07
90	2640.27

Samarium (Cont.)

35		2649.17	
45		2657.68	
70		2662.42	
120		2675.15	
100		2688.60	
45		2690.90	II
130		2693.34	
45		2693.74	
60		2696.08	
85		2707.96	
50		2732.42	
35		2739.87	
29		2762.28	
35		2764.18	
85		2767.85	II
60		2774.77	
85		2776.11	
85		2779.23	II
85		2786.64	
150		2789.38	II
130	h	2796.70	
85		2807.36	
150		2809.50	
120		2810.86	II
85		2817.20	II
29		2820.96	II
220		2830.94	
60		2840.30	
60		2847.49	II
60		2851.35	
120		2866.09	II
70		2868.40	II
70		2881.34	
85		2881.68	
60		2883.09	
45		2889.06	
60		2891.34	
85	d	2907.88	
		2907.99	II
130		2910.28	II
85		2937.48	II
70		2943.49	II
150		2953.19	II
85		2962.74	II
160		2969.02	II
100		2983.43	II
60		2991.57	II
100		3021.01	
150		3034.84	II
100		3039.13	II
120		3046.93	II
150		3067.54	
120		3071.79	II
100		3086.45	II
120		3096.88	II
100	h	3102.30	II
250		3106.52	II
220		3110.20	II
200		3117.72	II
270		3136.30	II
150		3139.97	II
150		3147.19	II
180		3152.10	II
410		3152.52	II
150		3162.30	II
360		3169.88	II
180		3178.12	II
720		3183.92	II
310		3187.01	II
430		3187.22	II
360		3187.79	II
360		3193.01	II
360		3196.18	II
150		3201.80	II
150		3204.90	II
360		3207.18	II
180		3208.17	II
600		3211.73	II
150		3214.12	II
270		3215.26	II
530		3216.85	II
600		3218.61	II
150		3219.43	II
270		3226.84	II
180		3228.50	II
270		3228.78	II
720		3230.56	II
360		3231.53	II
150		3231.95	II
430		3233.68	II
720		3236.64	II
150		3237.89	II
720		3239.66	II
530		3241.16	II
180		3241.59	II
180		3242.04	II
150		3244.69	II
240		3249.75	II
720		3250.37	II
360		3253.40	II
270		3253.94	II
850		3254.38	II
110		3255.63	II
360		3262.28	II
430		3264.94	II
180		3270.49	II
180		3270.68	II
430	d	3272.48	II
		3272.60	II
430		3272.81	II
430		3273.48	II
430		3276.75	II
270		3280.84	II
180		3285.66	II
430		3286.23	II
720	d	3290.28	II
		3290.39	II
180		3290.65	II
240		3293.37	II
360		3295.44	II
430		3295.81	II
720		3298.10	II
170		3300.98	II
340		3301.68	II
340		3304.52	II
340		3305.18	II
1700		3306.39	II
170		3306.61	II
850		3307.02	II
340		3309.52	II
850		3310.66	II
600		3312.42	II
410		3316.58	II
430		3320.16	II
110		3320.59	II
1200		3321.18	II
340		3323.77	II
340		3325.26	II
170		3325.48	II
340		3327.88	II
170		3333.64	II
170		3336.12	II
850		3340.58	II
240		3343.49	II
110		3343.64	II
240		3344.35	II
170		3347.30	II
240		3348.68	II
220		3350.88	II
410	d	3354.18	II
		3354.30	II
170		3354.72	II
1200		3365.86	II
150		3367.27	II
340		3368.57	II
340		3369.46	II
170		3370.59	II
340		3371.21	II
150		3376.48	II
1200		3382.40	II
510		3384.66	II
150		3384.86	
150		3387.66	II
410		3389.32	II
150		3391.11	II
410		3396.19	II
150		3397.76	II
150		3399.84	II
600		3402.46	II
210		3403.09	II
850		3408.68	II
270		3418.15	II
430		3418.51	II
170		3419.77	II
120		3424.78	II
170		3426.20	II
170		3433.68	II
150		3437.10	II
170		3438.06	II
240		3440.50	II
170		3453.56	II
170		3459.20	II
120		3459.42	II
240		3461.13	II
120		3464.07	II
170		3467.87	II
130		3473.96	II
130		3479.53	II
130		3480.26	II
170		3480.56	II
170		3487.41	II
170		3493.61	II
220		3499.84	II
340		3511.23	II
310		3530.60	II
220		3532.57	II
270		3535.65	II
240		3554.15	II
510		3559.10	II
220		3566.84	II
4200		3568.27	II
270		3577.79	II
390		3580.94	II
310		3583.39	II
4200		3592.60	II
340		3601.69	II
1700		3604.28	II
3400		3609.49	II
240		3620.58	II
1700		3621.23	II
240		3623.32	II
850		3627.01	II
850		3631.13	II
3400		3634.29	II
240		3634.93	II
410		3638.77	II
360		3645.29	II
300		3645.39	II
660		3649.53	II
340		3650.19	II
340		3656.22	II
2200		3661.36	II
220		3662.69	II
340		3667.93	II
340		3670.66	II
2200		3670.84	II
340		3677.79	II
270		3681.73	II
270		3688.42	II
270		3692.22	II
1100		3693.99	II
480		3706.75	II
480		3706.98	II
480		3708.41	II
930		3708.65	II
480		3711.54	II
350		3712.76	II
930		3718.88	II
930		3721.85	II
420		3724.90	II
1600		3728.47	II
2100		3731.26	II
1600		3735.98	II
800		3737.14	II
320		3737.48	II
2900		3739.12	II
		3739.20	II
800		3741.29	II
1200		3743.87	II
930		3745.46	I
		3745.60	II
480		3747.62	II
800		3755.28	II
800		3756.41	I
1200		3757.53	II
450		3758.45	II
660		3758.97	II
350		3760.04	II
1900		3760.69	II
660		3762.59	II
1100		3764.37	II
480		3767.36	II
480		3767.76	II
370	d	3773.33	I
		3773.42	II
1100		3778.14	II
660		3780.76	II
420		3780.93	II
320		3787.20	II
1500		3788.12	II
1600		3793.97	II
420		3797.28	II
1600		3797.73	II
500		3799.54	II
800		3800.89	II
320		3805.63	II
420		3808.46	II
320		3809.75	II
320		3809.88	II
420		3810.43	II
500		3812.07	II
480		3813.63	II
420		3814.63	II
930	d	3820.82	
530		3824.18	II
1600		3826.20	II
530		3830.29	II
1100		3831.50	II
530		3833.83	II
560		3834.48	I
560		3834.60	II
370		3835.72	II
500		3838.94	II
400		3840.45	II
1600		3843.50	II
530		3847.51	II
640		3848.78	II
420		3851.88	II
530		3853.30	I
2700		3854.21	II
480		3854.56	I
800		3855.90	II
480		3857.91	II
400		3858.74	I
660		3862.05	II
350		3862.23	II
320		3865.24	II
800		3871.78	II
400		3875.19	II
560		3875.54	II
800		3880.77	II
450		3881.38	II
450		3881.79	II
320		3882.50	II
3700		3885.29	II
660		3889.16	II
		3889.22	II
610		3890.08	II
320		3891.21	II
400		3894.05	II
1600		3896.98	II
1300		3903.42	II
620		3917.44	II
2500		3922.40	II
1900		3928.28	II
470		3935.76	II
1300		3941.87	II
620		3943.24	II
500		3946.51	II
740		3948.11	II
470		3951.89	I
370		3959.53	II
1500		3963.00	II
620		3966.04	II
470		3967.68	II
740		3970.53	II
1500		3971.40	II
620		3974.66	I
960		3976.27	II
1000		3976.43	II
960		3979.20	II
740		3983.14	II
740		3986.68	II
370		3987.43	II
1500		3990.00	II
		3990.02	I
740		3993.31	II
280		4003.46	II
470		4007.48	II
280		4019.98	II
880		4023.23	II
740		4035.11	II
590		4041.68	II
740		4042.72	II
880		4042.90	II
240		4044.11	II
560		4045.05	II
440		4046.16	II
740		4047.16	II
210		4048.62	II
590		4049.81	II
440		4058.87	II
560		4063.54	II
280		4064.33	II
1400		4064.58	II
810		4066.74	II
710		4068.33	II
810		4075.84	II
280		4076.65	II
240		4080.56	II
410		4082.60	II
280		4083.58	II
220		4084.40	II
1000		4092.27	II
290		4094.05	II
240		4104.13	II
810		4107.28	II
		4107.39	II
410		4109.40	II
280		4110.19	II
410		4113.90	II
1900		4118.55	II
410		4121.36	II
280		4122.51	II
710		4123.96	II
280		4129.23	II
250		4135.14	II
320		4147.71	II
810		4149.83	II
1200		4152.21	II
530		4153.33	II
560		4155.22	II
810		4169.48	II
410		4171.57	II
440		4178.02	II
530		4181.10	II
210		4183.33	I
530		4183.76	II
1000		4188.13	II
410		4191.93	II
270		4199.45	II
650		4202.92	II
1100		4203.05	II
660		4206.13	II
270		4206.62	II
660		4210.35	II
740		4220.66	II
1000		4225.33	II
740		4229.70	II
620		4234.57	II
1200		4236.74	II
500		4237.66	II
620		4244.70	II
210		4249.55	II
250		4251.78	II

Samarium (Cont.)

2100		4256.39	II
210		4258.58	II
1300		4262.68	II
500		4265.08	II
1200		4279.68	II
		4279.75	II
240		4279.94	II
2200		4280.79	II
710		4282.21	I
470		4282.83	I
240		4283.50	I
350		4286.64	II
350		4292.18	II
1600		4296.74	I
320		4304.94	II
880		4309.01	II
240		4312.85	I
1900		4318.94	II
470		4319.53	I
590		4323.28	II
240		4324.46	I
1800		4329.02	II
440		4330.02	I
1300		4334.15	II
880		4336.14	I
560		4345.86	II
1100		4347.80	II
560		4350.46	II
560		4352.10	II
560		4360.72	II
220		4361.07	II
810		4362.04	II
440		4362.91	I
220		4363.45	II
500		4368.03	II
210		4369.92	II
440		4373.46	II
320		4374.98	II
880		4378.24	II
530		4380.42	I
290		4384.29	II
1600		4390.86	II
210		4393.35	I
290		4397.34	I
410		4401.17	I
810	d	4403.06	II
		4403.13	I
410		4403.36	II
520		4409.33	II
290		4411.58	I
380		4417.58	II
470		4419.33	I
1500		4420.53	II
960		4421.14	II
2900		4424.34	II
470		4429.66	I
1600		4433.88	II
1800		4434.32	II
530		4441.81	I
440		4442.28	I
710		4444.26	II
710		4445.15	I
1300		4452.73	II
250		4452.95	I
1200		4454.63	II
1000		4458.52	II
250		4459.29	I
2200		4467.34	II
810		4470.89	I
470		4472.43	II
620		4473.02	II
740		4478.66	II
370		4499.11	I
370		4499.48	II
240		4503.38	I
180		4505.05	II
120		4511.33	I
560		4511.83	II
440		4515.09	II
880		4519.63	II
440		4523.04	II
		4523.18	I
650		4523.91	II
290		4533.80	I
270		4536.51	II
710		4537.95	II
150		4538.53	II
290		4540.19	II
380		4542.06	II
810		4543.95	II
100		4544.83	II
410		4552.66	II
270		4554.45	II
240		4560.43	II
470		4566.21	II
590		4577.69	II
290		4581.58	I
440		4581.73	I
560		4584.83	II
290		4591.82	II
380		4593.54	II
560		4595.29	II
240		4596.74	I
220		4604.18	II
290		4606.51	II
290		4615.44	II
470		4615.69	II
150		4630.21	II
880		4642.24	II
290		4645.40	I
290		4646.68	II
240		4648.16	II
380		4649.49	I
150		4655.13	II
290		4663.56	I
740		4669.40	II
620		4669.65	II
470	d	4670.75	I
		4670.83	I
1100		4674.60	II
680		4676.91	II
210		4681.55	I
370		4687.18	II
370		4688.73	I
130		4693.63	II
120		4699.34	II
530		4704.40	II
270		4713.06	II
130		4715.26	II
730		4716.10	I
270		4717.07	I
210		4717.72	II
190		4718.33	II
270		4719.84	II
130		4726.02	II
770		4728.42	I
470		4745.68	II
150		4750.72	I
730		4760.27	I
110		4770.20	I
110		4774.15	II
190		4777.85	II
580		4783.10	I
350		4785.86	I
160		4789.96	I
230		4791.58	II
430		4815.81	II
130		4829.57	II
970		4841.70	I
310		4844.21	II
140		4847.76	II
270		4848.32	I
120		4854.36	II
210		4883.77	I
730		4883.97	I
170		4904.97	I
630		4910.40	I
350		4913.25	II
430		4918.99	I
110		4924.04	I
120		4938.10	II
170		4948.63	II
120		4952.37	II
170		4961.94	II
170		4975.98	I
140		5028.44	II
400		5044.28	I
200		5052.76	II
170		5069.46	II
540		5071.20	I
170		5100.22	II
		5100.39	I
260		5103.09	II
140		5104.48	II
140		5116.70	II
510		5117.16	I
350		5122.14	I
360		5155.03	II
250		5172.74	I
470		5175.42	I
250		5200.59	I
260		5251.92	I
400		5271.40	I
250		5282.91	I
190		5320.60	I
110		5341.29	I
140		5368.36	I
130		5405.23	I
220		5453.00	I
140		5466.72	I
230		5493.72	I
80		5512.10	I
230		5516.09	I
50		5548.95	I
140		5550.40	I
45		5573.42	I
35		5588.20	I
50		5600.86	II
50		5621.79	I
70		5626.01	I
85		5644.10	I
140		5659.86	I
120		5696.73	I
85		5706.20	I
35		5710.93	I
50		5732.95	I
50		5743.35	II
45		5759.52	II
70		5773.77	I
60		5778.33	I
45		5779.24	I
45		5781.93	II
70	d	5786.98	II
60		5788.38	I
60		5800.52	I
65		5802.84	I
45		5814.89	I
45		5831.02	II
45		5836.37	II
35		5860.78	I
65		5867.79	I
45		5868.61	I
35		5871.06	I
50		5874.21	I
45		5897.39	II
50		5898.96	I
35		5938.90	II
65		5965.71	II
35	h	5968.82	II
35		5984.29	I
50		6045.00	I
45		6045.39	I
50		6070.06	I
45		6084.12	I
35	h	6091.40	I
45		6110.66	II
45	h	6159.56	I
45		6246.76	II
45		6256.54	I
45		6256.66	II
100		6267.28	II
50		6291.82	II
35		6307.06	II
70		6327.47	II
45		6426.64	II
45		6472.34	II
35		6484.52	II
35		6498.67	II
50		6542.76	II
140		6569.31	II
35	h	6570.67	II
40	h	6585.21	II
110		6589.72	II
40		6601.83	II
95		6604.56	II
40	h	6632.28	II
50		6671.51	I
70		6679.21	II
70		6693.55	II
40	d	6723.07	I
120	d	6731.84	II
70	d	6734.06	II
40	d	6734.81	II
55		6741.47	II
40	h	6778.61	II
60		6790.00	II
95		6794.20	II
55		6844.71	II
75		6856.03	II
120		6860.93	I
40		6862.82	II
30		6950.51	II
120		6955.29	II
90		7020.44	II
13		7036.73	II
90		7039.22	II
90		7042.24	II
13		7049.15	II
90		7051.52	II
16		7054.97	II
19		7074.67	I
90		7082.37	II
40	d	7085.52	II
26		7088.30	I
16		7091.16	I
30		7095.50	I
16		7096.33	I
30		7104.54	I
19		7106.23	I
26		7115.96	I
23		7117.51	II
26	h	7119.81	II
12		7122.40	II
23	h	7125.11	II
13		7131.80	I
10		7136.01	I
12		7139.39	II
40	d	7143.98	II
85	d	7149.60	II
10		7172.67	I
10		7189.57	II
9		7210.95	I
23		7213.82	I
26	d	7218.09	II
13		7220.07	I
13		7237.02	II
60		7240.90	II
9		7257.11	II
9	d	7261.52	II
13		7279.25	I
26		7281.47	II
8		7282.21	I
19		7283.33	II
16		7288.92	II
13		7290.23	I
26	h	7300.72	II
13		7327.08	II
13		7332.65	I
8		7338.04	I
26		7347.30	I
26		7376.69	II
13		7393.98	II
30		7444.56	I
26		7445.41	I
26	d	7453.03	II
13		7470.76	I
26		7481.99	II
23	h	7502.39	II
10	h	7517.00	II
23	h	7541.42	II
9		7544.74	I
10		7546.57	I
12	h	7560.03	II
19		7562.94	II
23		7570.95	II
23		7572.29	II
19		7578.09	II
30		7585.85	II
23		7588.31	II
10		7598.01	I
23	d	7607.48	II
		7607.74	I
12		7613.94	II
10	h	7631.77	II
23		7637.94	II
45		7645.09	II
12		7645.82	I
19		7648.02	II
10		7655.78	II
19		7667.20	II
8		7672.49	II
10	h	7678.79	II
10	h	7695.78	I
23		7712.04	II
30		7728.56	II
30		7736.26	II
30		7749.30	II
23		7755.20	II
10		7794.50	I
10		7801.54	I
8	h	7812.75	II
16		7820.15	II
10		7831.40	II
40	w	7835.08	II
26		7837.27	II
10		7844.82	II
6		7859.53	I
19		7863.65	II
10	h	7880.01	II
16		7895.96	I
26		7914.96	II
90		7928.14	II
9		7931.92	I
19		7937.09	II
16		7948.12	II
19	w	8001.61	II
19	w	8014.92	II
23		8025.12	II
23	w	8026.32	II
16		8032.03	II
40		8048.70	II
16		8065.16	I
45		8068.46	II
9	w	8117.16	II
9		8125.12	II
26		8161.82	II
19	w	8195.50	II
6		8206.30	II
26	w	8218.76	II
9		8230.33	I
16		8240.98	II
19	w	8289.26	II
10		8300.88	II
40	w	8305.79	II
10		8315.45	I
19	w	8348.68	II
19		8383.71	I
19		8387.77	II
30	w	8432.64	II
19	w	8473.54	II
45	w	8485.99	II
30	w	8510.90	II
23		8543.22	II
23	w	8617.03	II
23	w	8632.82	II
12	w	8677.81	II
13		8706.32	II
45	w	8708.43	II
30	w	8717.89	II
30	w	8758.28	II
16	w	8780.59	II
23	w	8788.83	II
26		8859.76	II
95		8913.66	II

SCANDIUM (Sc)
Z = 21

Sc I and II
Ref. 1, 88 — C.H.C.

Intensity		Wavelength Air	
65		2429.16	I
110		2438.62	I
560		2545.22	II
2900		2552.37	II
560		2555.82	II
2300		2560.25	II
1100		2563.21	II
40		2611.22	II
19		2684.23	II
120		2692.78	I
360		2706.77	I
210		2707.95	I
580		2711.35	I
30		2819.54	II
35		2822.15	II
60		2826.68	II
340		2965.86	I
1200		2974.01	I
1400		2980.75	I
340		2988.95	I
2200		3015.36	I
2700		3019.34	I
360		3030.76	I
30		3039.93	II
70		3045.72	II
85		3052.93	II
120	h	3056.31	I
130		3065.11	II
45		3139.75	II
990		3251.32	II
1500		3255.69	I
4400		3269.91	I
5500		3273.63	I
110	d	3343.28	II
270		3352.05	II
9900		3353.73	II
65	d	3357.30	II
2000		3359.68	II
1700		3361.27	II
1700		3361.94	II
4000		3368.95	II
6600		3372.15	II
90		3416.68	I
130		3418.51	I
65		3419.36	I
200		3429.21	I
200		3429.48	I
270		3431.36	I
530		3435.56	I
90		3439.41	I
65		3440.18	I
65		3448.49	I
270		3457.45	I
180		3462.19	I
130	d	3469.65	I
110		3471.13	I
200		3498.91	I
2700		3535.73	II
6600		3558.55	II
6100		3567.70	II
13000		3572.53	II
9900		3576.35	II
7700		3580.94	II
4000		3589.64	II
4000		3590.48	II
28000		3613.84	II
110		3617.43	I
20000		3630.75	II
13000		3642.79	II
6600		3645.31	II
110		3646.90	I
5300		3651.80	II
110		3664.25	II
290		3666.54	II
55		3675.26	II
40		3678.35	II
75	h	3717.10	I
270		3833.07	II
610		3843.03	II
90		3894.97	I
20000		3907.49	I
23000		3911.81	I
45		3923.51	II
4400		3933.38	I
45		3952.27	I
45		3989.06	II
5500		3996.61	I
530		4014.49	II
20000		4020.40	I
20000		4023.69	I
220		4030.67	I
140		4031.39	I
100		4034.23	I
220		4043.80	I
200		4046.48	I
2700		4047.79	I
120		4049.95	I
5500		4054.55	I
220		4056.59	I
160	h	4074.97	I
160		4078.57	I
6100		4082.40	I
200		4086.67	I
400		4087.16	I
40	h	4093.13	I
65		4094.85	I
55	h	4098.35	I
65		4100.33	I
440	h	4133.00	I
530	h	4140.30	I
65	h	4147.40	I
720		4152.36	I
55	h	4154.72	
90	Hd	4161.88	I
1100	h	4165.19	I
65	h	4171.56	I
45	h	4186.45	I
65	h	4187.62	I
75		4205.20	I
65		4212.34	I
45		4212.49	I
75	h	4216.10	I
110	h	4218.26	I
110	h	4219.73	I
40		4221.88	I
90	d	4225.59	I
180		4231.93	I
200		4233.61	I
100		4237.82	I
400		4238.05	I
90		4239.57	I
100		4246.12	I
15000		4246.83	II
55		4283.56	
290		4294.77	II
350		4305.71	II
4200		4314.09	II
3300		4320.74	II
2400		4325.01	II
28		4348.53	I
180		4354.61	II
110		4358.64	I
55		4359.08	I
28		4364.92	I
2000		4374.46	II
130		4384.81	II
45	h	4389.60	I
1100		4400.37	II
880		4415.56	II
28		4420.66	II
45		4431.36	II
65		4542.55	I
90		4544.68	I
120	h	4557.24	I
160	h	4573.99	I
65	h	4592.94	I
65	h	4598.45	I
55	h	4604.72	I
45		4609.53	I
45		4609.95	I
350		4670.40	II
40	h	4680.49	I
50		4698.29	II
120		4706.97	I
120		4709.34	I
200		4728.77	I
490		4729.23	I
40	h	4732.30	I
590		4734.10	I
60		4735.08	
690		4737.65	I
790		4741.02	I
1200		4743.81	I
200		4753.16	I
220		4779.35	I
90		4791.50	I
100		4827.28	I
100		4833.67	I
170		4839.44	I
40		4840.47	I
80		4847.68	I
80		4852.68	I
140	BLd	4857.79	S
		4858.09	SCO
80		4906.67	I
90		4909.76	I
90		4922.84	I
90		4934.25	I
45		4935.74	I
70		4941.33	I
170		4954.06	I
120		4973.66	I
150		4980.37	I
80		4983.45	I
140		4991.92	I
80		5018.39	I
70		5020.14	I
80		5021.51	I
530		5031.02	II
55		5032.74	I
250		5064.32	I
80		5068.86	I
530		5070.23	I
250		5075.81	I
2100		5081.56	I
1200		5083.72	I
1100		5085.55	I
750		5086.95	I
390		5087.14	I
270		5089.89	I
45		5092.46	I
390		5096.73	I
620		5099.23	I
370		5101.12	I
180		5109.06	I
150		5112.86	I
320		5116.69	I
70	b	5133.68	S
45	b	5171.06	S
390		5210.52	I
45		5211.28	I
280		5219.67	I
350		5239.82	II
280		5258.33	I
35		5284.97	I
210		5285.76	I
35		5301.94	I
22		5318.35	II
70		5331.77	I
14		5334.23	II
95		5339.41	I
120		5341.05	I
95		5342.96	I
350		5349.30	I
120		5349.71	I
60		5350.30	I
210		5355.75	I
530		5356.10	I
14		5357.19	II
270		5375.35	I
370		5392.08	I
45		5416.12	I
45		5425.57	I
45		5429.41	I
35		5432.94	I
55		5433.23	I
45		5438.22	I
55		5439.03	I
55	h	5442.60	I
270		5446.20	I
18		5447.39	I
120		5451.34	I
30		5455.21	I
18		5465.20	I
55		5468.40	I
60		5472.19	I
18		5474.64	I
750		5481.99	I
530		5484.62	I
570		5514.22	I
16		5515.39	I
660		5520.50	I
45		5526.06	I
660		5526.82	II
55		5541.04	I
30		5546.40	I
18		5550.40	I
5		5552.25	II
35		5553.59	I
16		5561.10	I
70		5564.86	I
18		5571.24	I
14		5579.76	I
110		5591.33	I
35	h	5593.38	I
22		5604.19	I
22		5631.02	I
80		5640.98	II
45		5646.36	I
16		5647.60	I
55		5649.56	I
250		5657.88	II
60		5658.34	II
55		5667.16	II
70		5669.04	II
1500		5671.81	I
95		5684.20	II
1200		5686.84	I
1100		5700.21	I
190		5708.61	I
880		5711.75	I
230		5717.28	I
180		5724.08	I
55	Bl	5736.85	S
55	Bl	5764.45	S
95	Bl	5772.74	S
55	Bl	5775.32	S
70	Bl	5809.84	S
70	Bl	5811.60	S
95	Bl	5847.73	S
70	Bl	5849.07	S
70	b	5887.38	S
35	Bl	5918.04	S
30		5919.11	I
60	Bl	5928.10	S
35		5961.49	I
60	Bl	5968.25	S
35		5969.19	I
90		5988.42	I
160	Bl	6017.07	S
60		6026.18	I
620	Bl	6036.17	S
490	Bl	6064.31	S
440	Bl	6072.65	S
620	Bl	6079.30	S
320	Bl	6101.87	S
370	Bl	6109.93	S
370	Bl	6115.97	S
180	b	6148.70	S
150	b	6153.93	S
150	b	6188.09	S
150	b	6192.90	S
620		6210.68	I
90		6239.41	I
320		6239.78	I
120		6245.63	II
110		6249.96	I
250		6258.96	I
60		6262.25	I
55		6276.31	I
45		6279.76	II
18		6300.70	II
750		6305.67	I
26		6309.90	II
16		6320.85	II
26		6344.83	I
60		6378.82	I
55	Bl	6408.41	S
90		6413.35	I
26	b	6437.08	S
55	Bl	6446.24	S
26	b	6457.78	S
35	b	6485.40	S
26	b	6495.90	S
55	b	6525.62	S
22	b	6535.30	S
45	b	6557.84	S
35	b	6566.88	S
18	b	6575.85	S
60		6604.60	II
26	Bl	6609.99	S
18	Bl	6617.94	S
18	Bl	6645.08	S
22	Bl	6654.42	S
26	Bl	6661.01	S
18	b	6700.48	S
18	b	6705.93	S
65		6737.87	I
35		6739.40	I
35		6817.08	I
50		6819.52	I
29		6829.54	I
50		6835.03	I
5	b	6963.12	S
5	Bl	6990.68	S
5	Bl	7025.72	S
8	b	7035.77	S
5	b	7072.37	S
5	b	7094.38	S
12	h	7138.14	I
14		7169.13	I
12		7257.57	I
8		7275.57	I
3	h	7300.62	I
12	h	7524.13	I
14	h	7553.96	I
15	h	7574.44	I
11		7617.45	I
14	h	7665.72	I
30		7697.73	I
18		7729.72	I
55	h	7741.17	I
5	h	7750.37	I
5		7752.72	I
6	h	7771.06	I
15		7785.17	I
8		7794.68	I
30		7800.44	I
11		7821.64	I
11	h	8196.98	I
15		8241.13	I
19	h	8761.40	I
11	h	8774.8	
15	h	8794.72	I
15	h	8823.8	I
30	h	8834.35	I
70		20616.32	
30		20985.81	
400		22051.86	I
150		22065.05	I

Scandium (Cont.)

Sc III
Ref. 323 — C.H.C.

Intensity		Wavelength	
		Vacuum	
10		730.60	III
15		731.65	III
15		1148.24	III
20		1154.52	III
20		1162.44	III
25		1168.61	III
10		1168.88	III
80		1598.00	III
180		1603.06	III
150		1610.19	III
40		1895.44	III
60		1912.62	III
90		1993.89	III
		Air	
160		2010.42	III
50		2012.26	III
350		2699.07	III
230		2734.05	III
10		2831.75	III
80		4061.21	III
100		4068.66	III
40		4309.47	III
10		4740.95	III
15		4780.87	III
50		4992.89	III
60		5032.09	III
80		5256.01	III
60		6307.60	III
90		7449.16	III
70		7548.15	III
70		7868.65	III
35		8814.29	III
50		8829.78	III
30		8865.89	III
15		8881.58	III

Sc IV
Ref. 298 — C.H.C.

Intensity		Wavelength	
		Vacuum	
8		220.28	IV
15		289.85	IV
15		296.31	IV
15		299.04	IV
10		371.16	IV
9		438.80	IV
8		557.50	IV
8		584.83	IV
9		617.08	IV
8		761.43	IV
8		769.70	IV
10		785.12	IV
8		789.00	IV
8		791.71	IV
8		861.24	IV
8		861.30	IV
8		890.87	IV
8		1219.40	IV
9		1228.20	IV
9		1424.66	IV
9		1444.10	IV
8		1489.64	IV
8		1514.96	IV
8		1535.76	IV
9		1543.86	IV
9		1549.55	IV
15		1550.80	IV
8		1555.72	IV
9		1563.81	IV
10		1574.92	IV
9		1583.41	IV
8		1584.64	IV
8		1592.23	IV
8		1660.71	IV
10		1665.92	IV
8		1746.23	IV
		Air	
10		2056.06	IV
8		2078.93	IV
12		2118.97	IV
9		2164.43	IV
11		2185.43	IV
11		2205.46	IV
14		2222.22	IV
11		2271.33	IV
9		2464.45	IV
8		2520.93	IV
11		2586.93	IV
9		2595.17	IV
8		2678.01	IV
8		2723.52	IV
8		2773.04	IV
8	d	4594.42	IV
8	d	4639.96	IV
8		5501.74	IV
9		5620.72	IV
10		5706.82	IV
14		5771.63	IV
9		6548.03	IV

Sc V
Ref. 150 — C.H.C.

Intensity		Wavelength	
		Vacuum	
150		179.42	V
350		180.14	V
200		180.82	V
200		180.96	V
50		181.55	V
200		182.39	V
300		228.56	V
100		230.85	V
40		243.82	V
500		243.87	V
400		246.42	V
400		250.98	V
500		252.85	V
500		253.73	V
50		255.38	V
300		255.64	V
200		257.16	V
150		258.24	V
40		258.81	V
50		260.05	V
400		281.00	V
900		283.91	V
800		284.45	V
600		288.29	V
900		289.59	V
1000	d	291.93	V
800		293.25	V
400		296.17	V
700		300.00	V
400		375.05	V
100		378.68	V
200		388.68	V
400		395.32	V
200		399.50	V
1000		573.36	V
600		587.94	V

# SELENIUM (Se)
Z = 34

Se I and II
Ref. 80, 181, 216, 275 — R.L.K.

Intensity	Wavelength	
	Vacuum	
285	828.5	II
360	832.7	II
285	906.6	II
360	912.9	II
360	1013.4	II
360	1014.0	II
450	1033.6	II
450	1049.6	II
360	1057.4	II
285	1097.8	II
360	1141.9	II
220	1156.0	II
285	1156.9	II
285	1168.5	II
450	1192.3	II
220	1205.7	II
220	1234.9	II
285	1291.0	II
285	1308.9	II
100	1405.4	I
100	1406.4	I
100	1406.6	I
120	1435.3	I
120	1435.8	I
100	1444.8	I
100	1446.8	I
100	1447.0	I
150	1449.2	I
120	1456.3	I
150	1500.9	I
250	1530.4	I
150	1531.3	I
200	1531.8	I
120	1547.1	I
120	1560.3	I
150	1575.3	I
150	1577.6	I
150	1577.9	I
150	1579.5	I
200	1580.0	I
150	1587.5	I
150	1593.2	I
250	1606.5	I
100	1610.7	I
100	1611.3	I
200	1617.4	I
150	1621.2	I
100	1622.7	I
120	1626.2	I
150	1643.4	I
250	1671.2	I
250	1675.3	I
250	1690.7	I
250	1793.3	I
300	1795.3	I
300	1855.2	I
250	1858.8	I
400	1898.6	I
350	1913.8	I
300	1919.2	I
500	1960.9	I
150	1995.1	I
	Air	
500	2039.8	I
500	2074.8	I
500	2164.2	I
150	2332.8	I
600	2413.5	I
300	2548.0	I
220	3038.7	II
220	3041.3	II
285	4070.2	II
360	4175.3	II
450	4180.9	II
120	4328.7	I
100	4330.3	I
285	4382.9	II
285	4446.0	II
220	4449.2	II
285	4467.6	II
500	4730.8	I
400	4739.0	I
300	4742.2	I
285	4840.6	II
360	4845.0	II
450	5227.5	II
360	5305.4	II
100	5365.5	I
120	5369.9	I
110	5374.1	I
285	5522.4	II
285	5566.9	II
285	5866.3	II
450	6056.0	II
200	6325.6	I
360	6444.2	II
285	6490.5	II
285	6535.0	II
150	6831.3	I
120	6990.690	I
100	6991.792	I
200	7010.809	I
150	7013.875	I
300	7062.065	I
200	7575.1	I
250	7583.4	I
150	7592.2	I
120	7606.8	I
300	8001.0	I
200	8036.4	I
120	8060.9	I
120	8065.3	I
120	8081.1	I
150	8093.2	I
150	8094.7	I
180	8149.3	I
150	8152.0	I
200	8157.7	I
180	8163.1	I
150	8182.9	I
100	8185.0	I
120	8194.6	I
150	8440.47	I
150	8450.38	I
150	8742.33	I
300	8918.86	I
100	8969.69	I
200	9001.97	I
200	9038.61	I
80	9083.14	I
120	9088.79	I
80	9140.83	I
60	9181.88	I
60	9271.12	I
100	9432.50	I
60	9825.58	I
200	10217.25	I
377	10307.45	I
900	10327.26	I
640	10386.36	I
124	10650.30	I
125	11934.56	I
275	11946.87	I
100	11947.92	I
105	11952.27	I
170	11952.64	I
100	11966.04	I
205	11972.93	I
115	11973.07	I
315	14817.93	I
410	14917.47	I
500	15151.44	I
115	15469.06	I
320	15471.00	I
265	15520.97	I
395	15618.40	I
115	15620.38	I
360	16659.44	I
505	16813.78	I
165	16817.76	I
205	16866.54	I
115	16972.71	I
235	21374.24	I
680	21442.56	I
415	21473.48	I
270	21716.36	I
240	21730.60	I
105	23133.66	I
150	23388.85	I
110	23628.17	I
265	24148.18	I
170	24159.23	I
185	24204.44	I
375	24385.99	I
160	24413.67	I
225	24471.17	I
255	25017.51	I
510	25127.43	I

Se III
Ref. 9, 247 — R.L.K.

Intensity	Wavelength	
	Vacuum	
220	709.2	III
220	709.4	III
220	720.6	III
360	724.3	III
285	726.4	III
220	737.2	III
220	741.9	III
285	777.3	III
220	790.8	III
360	843.0	III
220	879.2	III
285	953.7	III
220	954.4	III
220	954.7	III
160	974.1	III
360	974.8	III
285	1079.8	III
360	1099.1	III
450	1119.2	III
	Air	
285	2057.5	III
285	2767.2	III
220	2773.8	III
285	3379.8	III
450	3387.2	III
450	3413.9	III
285	3428.4	III
450	3457.8	III
360	3543.6	III
285	3570.2	III
450	3637.6	III
360	3711.7	III
450	3738.7	III
285	3743.0	III
450	3800.9	III
360	4046.7	III
220	4083.2	III
450	4169.1	III
220	4637.9	III
285	6303.8	III

Se IV
Ref. 245 — R.L.K.

Intensity	Wavelength	
	Vacuum	
285	636.0	IV
285	654.2	IV
360	652.7	IV
450	670.1	IV
285	671.9	IV
220	722.8	IV
285	734.6	IV
450	746.4	IV
285	759.0	IV
285	776.5	IV
285	803.8	IV

Selenium (Cont.)

360		959.6	IV
450		996.7	IV
220		1307.2	IV
285		1314.4	IV
		Air	
220		2090.0	IV
285		2136.6	IV
160		2165.2	IV
160		2166.6	IV
360		2665.5	IV
285		2724.3	IV
160		2951.6	IV

Se V
Ref. 245 — R.L.K.

Intensity		Wavelength	
		Vacuum	
285		596.0	V
285		601.0	V
220		608.7	V
360		613.0	V
285		614.3	V
450		759.1	V
285		785.8	V
285		804.3	V
360		808.7	V
220		814.8	V
220		820.7	V
360		830.3	V
450		839.5	V
360		845.8	V
360		1094.7	V
220		1151.0	V
450		1227.6	V

SILICON (Si)
Z = 14

Si I and II
Ref. 170, 237, 292 — L.J.R.

Intensity		Wavelength	
		Vacuum	
10	h	805.10	II
20	h	820.52	II
20	h	843.72	II
40	h	845.77	II
10		850.14	II
100		889.72	II
200		892.00	II
10		899.41	II
20		901.74	II
10		913.01	II
20		913.85	II
20		929.81	II
100		989.87	II
200		992.68	II
25		1020.70	II
50		1023.69	II
30		1057.05	II
15		1057.50	II
20	h	1127.44	II
40	h	1127.91	II
100		1190.42	II
200		1193.28	II
250		1194.50	II
100		1197.39	II
10	h	1216.12	II
20		1223.91	II
20		1224.25	II
10		1224.97	II
50		1226.81	II
20		1226.89	II
40		1226.99	II
100		1227.60	II
10		1228.44	II
25		1228.62	II
150		1228.75	II
200		1229.39	II
10		1235.92	II
100		1246.74	II
150		1248.43	II
100		1250.09	II
150		1250.43	II
200		1251.16	II
10		1255.28	I
40		1256.49	I
50		1258.80	I
1000		1260.42	II
2000		1264.73	II
200		1265.02	II
100		1304.37	II
50	h	1305.59	II
200		1309.27	II
20	h	1309.46	II
100		1346.87	II
100		1348.54	II
150		1350.06	II
20		1350.52	II
20		1350.66	II
100		1352.64	II
100		1353.72	II
10	h	1409.07	II
20	h	1410.22	II
10	h	1416.97	II
15	h	1474.65	II
15		1484.87	II
90	h	1485.02	II
30		1485.22	II
100	h	1485.51	II
100	h	1509.10	II
50	h	1512.07	II
30	p	1513.57	II
80	p	1518.91	II
500		1526.72	II
1000		1533.45	II
10		1562.45	II
15		1562.85	II
10		1563.77	II
50		1573.87	I
50		1574.82	I
50		1592.41	I
150		1594.55	I
50		1594.93	I
30		1597.95	I
100		1622.87	I
30		1625.71	I
300		1629.43	I
200		1629.92	I
75		1631.13	I
50		1633.98	I
30	h	1653.35	I
30		1664.52	I
50		1666.37	I
100		1667.62	I
100		1668.52	I
100		1672.59	I
200		1675.20	I
30		1682.68	I
30		1686.82	I
50	h	1689.29	I
30	h	1690.79	I
50		1693.29	I
50		1695.51	I
200		1696.20	I
200		1697.94	I
50		1700.42	I
30		1700.63	I
30		1702.86	I
50		1704.43	I
10	h	1710.83	II
20	h	1711.30	II
30	h	1743.88	I
50		1747.40	I
30	h	1753.11	I
50		1763.66	I
40		1765.03	I
30	h	1765.60	I
30		1766.06	I
30		1770.63	I
100	h	1770.92	I
100	h	1776.83	I
50	h	1783.23	I
100	h	1799.12	I
150		1808.00	II
50	h	1809.09	I
500	h	1814.07	I
200		1816.92	II
10		1817.45	II
50		1822.45	I
200		1836.51	I
30	h	1838.01	I
100	h	1841.15	I
200		1841.44	I
200		1843.77	I
300		1845.51	I
100		1846.10	I
400		1847.47	I
200		1848.14	I
100		1848.74	I
500		1850.67	I
30	h	1851.79	I
200		1852.46	I
50		1853.15	I
20		1869.32	II
15		1870.23	II
100		1873.10	I
500	h	1874.84	I
100		1875.81	I
200		1881.85	I
200		1887.70	I
200	h	1893.25	I
1000	h	1901.33	I
100	h	1902.46	II
50	h	1904.66	I
50	h	1910.62	II
50		1941.67	II
15		1944.59	II
10		1949.33	II
100		1949.56	II
100		1954.97	I
30		1984.43	I
50		1991.85	I
		Air	
30		2010.97	I
50		2054.83	I
50		2058.65	II
50		2059.01	II
40		2061.19	I
30		2065.52	I
200		2072.02	II
200		2072.70	II
30	h	2103.21	I
30		2114.63	I
100		2124.12	I
10	h	[illegible]	II
30	h	2136.40	II
50	h	2136.56	II
50	h	2147.91	I
110		2207.98	I
115		2210.89	I
110		2211.74	I
120		2216.67	I
120		2218.06	I
50		2218.91	I
35		2291.03	I
55		2303.06	I
30		2334.40	II
30		2334.61	II
10		2344.20	II
10	h	2349.54	II
20		2350.17	II
20	h	2353.09	II
100	h	2356.30	II
30	h	2357.18	II
50	h	2357.97	II
10	h	2360.20	II
30		2366.97	II
20		2374.26	II
10	h	2428.45	II
300		2435.15	I
65		2438.77	I
65		2443.36	I
70		2452.12	I
425		2506.90	I
375		2514.32	I
500		2516.113	I
350		2519.202	I
425		2524.108	I
450		2528.509	I
110		2532.381	I
30		2563.679	I
85		2568.641	I
45		2577.151	I
190		2631.282	I
10	h	2682.21	II
1000		2881.579	I
10	h	2887.51	II
300		2904.28	II
500		2905.69	II
55		2970.355	I
150		2987.645	I
50		3006.739	I
100	h	3030.00	II
75		3020.004	I
20	h	3021.55	II
20	h	3041.57	II
30	h	3042.19	II
100	h	3043.69	II
10	h	3043.85	II
10	h	3045.77	II
50	h	3048.30	II
150	h	3053.18	II
150		3188.97	II
50		3192.25	II
150		3193.09	II
50		3194.21	II
50		3194.69	II
100		3195.41	II
200		3199.51	II
20		3202.49	II
100	h	3203.87	II
200	h	3210.03	II
75		3214.66	II
15	h	3217.99	II
10		3220.44	II
20		3223.01	II
300		3333.14	II
500		3339.82	II
100	h	3853.66	II
500	h	3856.02	II
200	h	3862.60	II
300		3905.523	I
10	h	3955.74	II
10	h	3977.46	II
15	h	3991.77	II
10	h	3998.01	II
20	h	4075.45	II
15	h	4076.78	II
70		4102.936	I
300	h	4128.07	II
500	h	4130.89	II
10	h	4183.35	II
100	h	4190.72	II
50		4198.13	II
100		4621.42	II
150		4621.72	II
50		4782.991	I
35		4792.212	I
80		4792.324	I
15	h	4883.20	II
20	h	4906.99	II
20	h	4932.80	II
30		4947.607	I
40		5006.061	I
1000		5041.03	II
1000		5055.98	II
100		5181.90	II
100	h	5185.25	II
200	h	5192.86	II
500	h	5202.41	II
30	h	5295.19	II
100	h	5405.34	II
15	h	5417.24	II
15	h	5428.92	II
15	h	5432.89	II
100	h	5438.62	II
20	h	5447.26	II
15	h	5454.49	II
100	h	5456.45	II
500	h	5466.43	II
500	h	5466.87	II
100	h	5469.21	II
40		5493.23	I
200	h	5496.45	II
35		5517.535	I
100	h	5540.74	II
150	h	5576.66	II
30		5622.221	I
100	h	5632.97	II
200	h	5639.48	II
90		5645.611	I
150	h	5660.66	II
80		5665.554	I
1000	h	5669.56	II
30	h	5681.44	II
120		5684.484	I
300	h	5688.81	II
100		5690.425	I
90		5701.105	I
200	h	5701.37	II
100	h	5706.37	II
160		5708.397	I
45		5747.667	I
45		5753.625	I
45		5754.220	I
45		5762.977	I
70		5772.145	I
70		5780.384	I
30	h	5785.73	II
90		5793.071	I
30	h	5794.90	II
100		5797.859	I
150	h	5800.47	II
200		5806.74	II
30		5827.80	II
50		5846.13	II
10		5867.48	II
300	h	5868.40	II
40		5873.764	I
150		5915.22	II
200		5948.545	I
500		5957.56	II
500		5978.93	II
10	h	6067.45	II
20	h	6080.06	II
10	h	6086.67	II
90		6125.021	I
85		6131.574	I
90		6131.850	I
100		6142.487	I
100		6145.015	I
160		6155.134	I
160		6237.320	I
40		6238.287	I
125		6243.813	I
125		6244.468	I
180		6254.188	I
45		6331.954	I
1000		6347.10	II
1000		6371.36	II
45		6526.609	I
45		6527.199	I
45		6555.462	I
50	h	6660.52	II
15		6665.00	II
100		6671.88	II
20		6699.38	II
50	h	6717.04	II
100		6721.853	I
30		6741.64	I
20	h	6750.28	II
30		6818.45	II
50		6829.82	II
30		6848.568	I
80		6976.523	I

Silicon (Cont.)

Intensity		Wavelength	
180		7003.567	I
180		7005.883	I
30		7017.28	I
90		7017.646	I
250		7034.903	I
70		7164.69	I
200		7165.545	I
70		7184.89	I
65		7193.58	I
30		7193.90	I
100		7226.206	I
100		7235.326	I
60		7235.82	I
180		7250.625	I
160		7275.294	I
40		7282.81	I
400		7289.173	I
55		7290.26	I
35		7373.00	I
375		7405.774	I
200		7409.082	I
40		7415.35	I
275		7415.946	I
425		7423.497	I
85		7424.60	I
100		7680.267	I
40		7742.71	I
30		7800.008	I
400		7848.80	II
500		7849.72	II
30		7849.967	I
90		7918.386	I
120		7932.349	I
140		7944.001	I
35		7970.306	I
35		8035.619	I
70		8093.241	I
35		8230.642	I
40		8443.982	I
40		8501.547	I
60		8502.221	I
40		8536.165	I
120		8556.780	I
50		8648.462	I
40		8728.011	I
75		8742.451	I
100		8752.009	I
35		8790.389	I
100		9412.72	II
100		9413.506	I
30		10371.269	I
120		10585.141	I
120		10603.431	I
120		10660.975	I
30		10694.251	I
30		10727.408	I
60		10749.384	I
30		10784.550	I
80		10786.856	I
140		10827.091	I
60		10843.854	I
30		10868.79	I
130		10869.541	I
30		10882.802	I
30		10885.336	I
80		10979.308	I
30		10982.061	I
80		11017.965	I
13		11187.60	I
12		11289.84	I
12		11611.09	I
370		11984.19	I
220		11991.57	I
440		12031.51	I
150		12103.53	I
120		12270.68	I
11		13176.90	I
190		15888.39	I
40		15960.04	I
95		16060.03	I
20		16094.80	I
60		16163.71	I
11		16215.68	I
16		16381.55	I
29		16680.77	I
28		17327.29	I
26		18722.90	I
15		19385.94	I
48		19432.97	I
13		19493.38	I
110		19722.50	I
31		19928.88	I
12		20917.13	I
21		21354.24	I
12		22062.71	I

Si III
Ref. 320 — L.J.R.

Intensity		Wavelength	
		Vacuum	
8		566.61	III
6		652.22	III
8		653.33	III
5		673.48	III
5		800.07	III
9		823.41	III
5		883.40	III
7		939.09	III
9		967.95	III
10		993.52	III
13		994.79	III
16		997.39	III
7		1005.37	III
7		1031.16	III
8		1033.92	III
7		1037.05	III
6		1083.22	III
14		1108.37	III
16		1109.97	III
18		1113.23	III
6		1140.55	III
7		1141.58	III
6		1142.28	III
8		1144.31	III
6		1144.96	III
8		1145.11	III
7		1145.18	III
6		1155.00	III
6		1155.96	III
7		1158.10	III
6		1160.26	III
8		1161.58	III
5		1174.37	III
6		1174.43	III
8		1178.00	III
30		1206.51	III
30		1206.53	III
9		1207.52	III
10		1210.46	III
7		1235.43	III
6		1280.35	III
17		1294.54	III
14		1296.73	III
15		1298.89	III
18		1298.96	III
14		1301.15	III
16		1303.32	III
13		1312.59	III
8		1341.47	III
7		1342.39	III
6		1343.39	III
8		1361.60	III
5		1362.37	III
7		1363.47	III
8		1365.26	III
7		1367.05	III
5		1369.44	III
5		1373.03	III
5		1387.99	III
13		1417.24	III
6		1433.69	III
8		1435.77	III
7		1436.17	III
5		1441.73	III
6		1447.20	III
5		1457.25	III
12		1500.24	III
10		1501.19	III
9		1501.87	III
6		1506.06	III
7		1673.32	III
9		1842.55	III
		Air	
5		2176.89	III
6		2295.48	III
10		2296.87	III
8		2300.93	III
10		2308.19	III
11		2449.48	III
6		2483.20	III
25		2541.82	III
10		2546.09	III
14		2559.21	III
11		2640.79	III
14		2655.51	III
9		2817.11	III
7		2831.49	III
5		2839.62	III
5		2959.15	III
5		2980.52	III
5		3013.09	III
6		3034.73	III
8		3037.29	III
9		3040.93	III
7		3043.93	III
5		3045.08	III
7		3068.24	III
25		3086.24	III
6		3086.46	III
20		3093.42	III
5		3093.65	III
16		3096.83	III
6		3126.27	III
7		3147.37	III
8		3161.61	III
16		3185.13	III
13		3186.02	III
14		3196.50	III
15		3210.55	III
7		3216.25	III
12		3230.50	III
14		3233.95	III
15		3241.62	III
7		3253.40	III
5		3253.74	III
7		3254.80	III
12		3258.66	III
6		3270.46	III
10		3276.26	III
7		3279.26	III
15		3486.91	III
9		3525.94	III
8		3569.67	III
20		3590.47	III
8	h	3622.54	III
5	h	3639.45	III
6	h	3645.12	III
7	h	3681.40	III
5	h	3682.15	III
20	c	3791.41	III
25		3796.11	III
30		3806.54	III
7		3842.46	III
20		3924.47	III
6	h	3947.49	III
6		3963.84	III
5		3981.24	III
5	h	4101.86	III
8		4102.42	III
5	h	4115.50	III
9		4338.50	III
8		4341.40	III
8	h	4377.63	III
6	h	4405.90	III
8	h	4406.72	III
6		4494.05	III
30		4552.62	III
8		4554.00	III
25		4567.82	III
20		4574.76	III
7		4619.66	III
7		4638.28	III
8		4665.87	III
9		4683.02	III
7		4683.80	III
16		4716.65	III
7		4730.52	III
8		4800.43	III
15		4813.33	III
16		4819.72	III
18		4828.97	III
10	h	5091.42	III
7	h	5113.76	III
8	h	5114.12	III
5		5197.26	III
6		5451.46	III
7		5473.05	III
7		5704.60	III
8		5716.29	III
20		5739.73	III
10	h	5898.79	III
7		6314.46	III
6	h	6524.36	III
6	h	6831.56	III
7	h	6851.65	III
5	h	7461.89	III
8	h	7462.62	III
9	h	7466.32	III
12	h	7612.36	III
9	h	8102.86	III
11	h	8103.45	III
7	h	8190.43	III
6	h	8191.16	III
8	h	8191.68	III
9	h	8262.57	III
5	h	8265.64	III
8	h	8269.32	III
5	h	8271.38	III
6	h	8271.94	III

Si IV
Ref. 319 — L.J.R.

Intensity		Wavelength	
		Vacuum	
4		457.82	IV
3		458.16	IV
2		515.12	IV
3		516.35	IV
2		645.76	IV
5		749.94	IV
7		815.05	IV
8		818.13	IV
8		1066.63	IV
8		1122.49	IV
10		1128.34	IV
15		1393.76	IV
12		1402.77	IV
1		1634.61	IV
6		1722.53	IV
5		1727.38	IV
		Air	
3		2120.18	IV
4		2127.47	IV
5	h	2287.04	IV
2	h	2328.56	IV
2		2366.76	IV
3		2370.99	IV
2		2482.82	IV
1		2485.38	IV
7		2517.51	IV
1		2672.19	IV
4		2675.12	IV
4		2675.25	IV
1		2677.57	IV
3	h	2723.81	IV
3	h	2895.13	IV
2	h	2904.47	IV
1	h	2971.52	IV
7		3149.56	IV
9		3165.71	IV
1	h	3244.19	IV
8		3762.44	IV
6		3773.15	IV
1	h	4031.39	IV
2	h	4038.06	IV
10		4088.85	IV
9		4116.10	IV
7	h	4212.41	IV
3		4314.10	IV
5		4328.18	IV
2	h	4403.73	IV
1	h	4411.65	IV
1	h	4611.27	IV
3	h	4628.62	IV
9	h	4631.24	IV
10	h	4654.32	IV
3	h	4656.92	IV
1	h	4667.14	IV
2	h	4673.30	IV
1	h	4947.45	IV
3		4950.11	IV
2	h	5304.97	IV
1	h	5309.49	IV
5		6667.56	IV
7		6701.21	IV
3	h	6998.36	IV
6	h	7047.94	IV
4	h	7068.41	IV
2	h	7630.50	IV
4	h	7654.56	IV
4	h	7678.75	IV
5	h	7718.79	IV
6	h	7723.82	IV
2	h	7725.64	IV
1	h	7730.47	IV
1	h	7752.91	IV
1	h	8240.61	IV
2	h	8957.25	IV
1	h	9018.16	IV

Si V
Ref. 87 — L.J.R.

Intensity		Wavelength	
		Vacuum	
1		78.61	V
1		78.90	V
2		80.81	V
2		81.11	V
10		85.18	V
6		85.58	V
4		90.45	V
4		90.85	V
15		96.44	V
10		97.14	V
2		98.21	V
20		117.86	V
20		118.97	V

SILVER (Ag)
Z = 47

Ag I and II
Ref. 13, 99, 255, 286, 289 — C.H.C.

Intensity		Wavelength	
		Vacuum	
25		730.83	II

Silver (Cont.)

30		752.80	II
15		1005.32	II
10		1065.49	II
12		1072.23	II
250		1074.22	II
150		1107.03	II
150		1112.46	II
60		1195.83	II
50		1223.33	II
50		1240.80	II
50		1246.87	II
55		1256.81	II
55		1257.55	II
50		1266.63	II
70		1273.67	II
65		1297.51	II
85		1311.20	II
55		1313.81	II
50		1314.61	II
60		1323.84	II
60		1342.09	II
50		1342.57	II
70		1346.62	II
50		1353.54	II
150		1364.50	II
100		1396.00	II
100		1410.93	II
90		1419.72	II
95		1432.60	II
100		1464.72	II
50		1466.23	II
50	r	1507.37	I
100	r	1515.63	I
50	r	1548.58	I
100		1555.16	II
100		1644.50	II
60		1651.52	I
50		1652.10	I
120		1682.82	II
10		1708.11	I
50		1709.27	I
125		1736.44	II
10	h	1766.14	I
75		1790.37	II
20		1847.71	I
100		1967.38	II

Air

150		2015.96	II
150		2033.98	II
200		2061.17	I
100		2069.85	I
80	r	2113.82	II
60		2145.60	II
15		2170.00	I
50		2186.76	II
60		2229.53	II
100	r	2246.43	II
75	r	2248.74	II
75		2280.03	II
30	h	2309.56	I
10	h	2312.60	I
70	r	2317.05	II
80	r	2320.29	II
70	r	2324.68	II
80	r	2331.40	II
70		2357.92	II
50	h	2375.02	I
75		2411.41	II
90	r	2413.23	II
100	r	2437.81	II
80		2447.93	II
80		2473.84	II
60		2506.63	II
50	h	2575.63	I
60		2660.49	II
60		2721.77	I
75		2767.54	II
100	h	2824.39	I
10	h	2926.77	I
20	h	2938.42	I
20		3099.10	I
30	h	3130.02	I
10	h	3170.58	I
90		3180.70	II
15	h	3215.67	I
10		3225.15	I
15		3233.18	I
100		3267.35	II
55000	r	3280.68	I
10	h	3305.67	I
28000	r	3382.89	I
10	h	3403.78	I
30		3469.16	I
70		3475.82	II
80		3495.28	II
20	h	3501.92	I
20		3508.03	I
15	h	3513.38	I
10		3521.12	I
50		3542.61	I
10	h	3547.16	I
10	h	3557.01	I
20	h	3586.67	I
10	h	3623.49	I
50	h	3624.68	I
75		3682.46	II
30		3682.50	I
80		3683.34	II
50	h	3709.20	I
10	h	3727.42	I
20	h	3753.14	I
200		3810.94	I
50		3811.78	I
100	h	3840.74	I
15		3847.85	I
50	h	3907.41	I
50		3909.31	II
50	h	3914.40	I
70		3920.10	II
10	h	3928.01	I
10	h	3940.43	I
10	h	3942.97	I
60		3949.43	II
100	h	3981.58	I
70		3985.19	II
10	h	3992.15	I
100	h	4055.48	I
10	h	4083.43	I
80		4085.91	II
100		4185.48	II
90	h	4210.96	I
100		4212.82	I
50		4311.07	I
20		4396.23	I
50	h	4476.04	I
20	h	4556.0	I
30	h	4615.69	I
80		4620.04	II
50		4620.46	II
60	h	4668.48	I
30	h	4677.60	I
100		4788.40	II
20	h	4796.2	I
30	h	4847.82	I
100		4874.10	I
20		4888.21	I
10	h	4917.5	I
10		4935.75	I
20	h	4992.89	I
80		5027.35	II
15	h	5123.50	I
1000		5209.08	I
10	h	5333.62	I
1000		5465.50	I
100		5471.33	I
20		5475.38	I
20	h	5545.67	I
10	h	5559.58	I
100		5667.34	I
10	h	6083.78	I
10	h	6268.50	I
20		6621.08	I
20		7359.96	I
320		7687.78	I
25		8005.4	II
15		8254.7	II
500		8273.52	I
20		8324.4	II
15		8379.5	II
25		8403.8	II
15		8492.5	II
30	h	8645.70	I
10	h	8704.85	I
12		8747.6	II
15		9000.9	II
10		12551.0	I
60		16819.5	I
20		17416.7	I
15		18307.9	I
15		18382.3	I

Ag III
Ref. 363, 387, 398 — R.D.C.

Vacuum

Intensity	Wavelength	
200	709.80	III
200	713.85	III
100	717.73	III
200	718.53	III
300	726.96	III
350	730.04	III
150	730.28	III
150	730.94	III
200	736.57	III
100	738.13	III
200	740.98	III
200	742.29	III
200	748.30	III
150	755.73	III
100	758.27	III
200	767.19	III
250	768.33	III
150	769.61	III
350	776.38	III
200	782.91	III
150	785.76	III
200	789.08	III
250	792.35	III
200	796.54	III
250	797.91	III
400	799.41	III
300	808.88	III
150	816.12	III
180	822.39	III
200	838.11	III
120	1373.22	III
120	1374.76	III
110	1404.93	III
120	1413.90	III
120	1414.29	III
120	1428.61	III
120	1452.74	III
200	1456.41	III
100	1471.44	III
300	1489.01	III
150	1515.08	III
100	1524.23	III
120	1527.04	III
150	1541.14	III
150	1550.89	III
130	1553.04	III
130	1587.41	III
120	1589.28	III
100	1613.79	III
130	1619.14	III
100	1634.46	III
100	1652.24	III
130	1653.60	III
300	1654.43	III
700	1656.18	III
150	1657.10	III
100	1661.54	III
130	1670.75	III
150	1676.14	III
100	1681.07	III
500	1693.51	III
200	1705.06	III
150	1708.86	III
130	1717.68	III
200	1722.27	III
150	1726.76	III
250	1728.14	III
200	1747.34	III
120	1749.64	III
150	1750.89	III
750	1751.03	III
100	1760.57	III
150	1762.62	III
150	1768.70	III
100	1771.81	III
100	1783.85	III
100	1791.70	III
100	1792.69	III
150	1793.90	III
150	1802.24	III
150	1802.26	III
100	1802.77	III
300	1808.23	III
250	1816.83	III
150	1822.45	III
350	1828.83	III
250	1832.33	III
120	1832.50	III
100	1834.31	III
250	1836.10	III
150	1838.64	III
400	1840.14	III
100	1846.96	III
120	1849.93	III
150	1856.33	III
120	1858.91	III
100	1860.39	III
100	1860.64	III
350	1867.12	III
150	1868.10	III
100	1872.55	III
400	1873.45	III
250	1880.36	III
300	1889.57	III
400	1916.92	III
600	1917.08	III
200	1925.30	III
150	1946.32	III
100	1948.44	III
700	1957.62	III
120	1959.27	III
100	1960.86	III
400	1966.89	III
600	1975.92	III
500	1977.03	III
150	1981.87	III
200	1987.02	III
130	1995.16	III
600	2000.24	III

Air

Intensity	Wavelength	
300	2007.30	III
200	2011.49	III
150	2013.65	III
150	2041.33	III
150	2053.17	III
150	2053.83	III
200	2056.99	III
200	2081.04	III
300	2146.47	III
150	2149.19	III
600	2161.89	III
150	2166.21	III
150	2211.23	III
100	2238.40	III
500	2246.51	III
100	2286.50	III
700	2310.04	III
100	2386.85	III
300	2395.69	III
100	2469.62	III
100	2562.87	III

SODIUM (Na)
Z = 11

Na I and II
Ref. 268, 334 — T.K.M.

Vacuum

Intensity	Wavelength	
160	300.15	II
160	300.20	II
90	301.32	II
100	301.44	II
60	302.45	II
300	372.08	II
350	376.38	II
60	1293.97	II
50	1327.74	II
45	1347.54	II
90	1374.69	II
90	1404.68	II
45	1495.21	II
45	1497.73	II
80	1506.41	II
60	1506.91	II
70	1513.10	II
60	1519.63	II
60	1657.92	II
90	1776.57	II
60	1783.04	II
80	1787.19	II
45	1788.85	II
80	1798.41	II
45	1801.26	II
90	1807.09	II
60	1808.38	II
50	1821.70	II
45	1833.87	II
80	1835.22	II
45	1837.89	II
60	1841.82	II
70	1845.02	II
45	1850.15	II
70	1851.19	II
80	1853.17	II
45	1866.45	II
45	1873.37	II
60	1875.08	II
160	1881.91	II
50	1885.09	II
45	1885.74	II

Air

Intensity	Wavelength	
80	2228.53	II
80	2303.58	II
300	2315.65	II
130	2393.28	II
100	2401.01	II
300	2420.99	II
300	2424.73	II
200	2439.14	II
250	2441.50	II
200	2448.72	II
200	2452.18	II
1000	2493.15	II
300	2502.84	II
450	2506.30	II
600	2515.46	II
600	2531.54	II
20	2543.84	I
10	2543.87	I
550	2586.31	II
70	2593.87	I
35	2593.92	I
600	2594.96	II
850	2611.81	II
300	2627.41	II

Sodium (Cont.)

850	2661.00	II
350	2666.46	II
1000	2671.83	II
850	2678.09	II
200	2680.34	I
100	2680.43	I
650	2808.71	II
850	2809.52	II
600	2829.87	II
800	2839.56	II
1000	2841.72	II
400	2852.81	I
200	2853.01	I
650	2856.51	II
800	2859.49	II
750	2871.28	II
650	2872.95	II
900	2881.15	II
850	2886.26	II
2	2893.62	I
700	2893.95	II
900	2901.14	II
800	2904.72	II
1100	2904.92	II
1100	2917.52	II
1100	2919.05	II
1200	2919.85	II
1300	2920.95	II
1000	2923.49	II
750	2930.88	II
850	2934.08	II
950	2937.74	II
800	2945.70	II
950	2947.50	II
1200	2951.24	II
1100	2952.40	II
850	2960.12	II
500	2970.73	II
600	2974.24	II
750	2974.99	II
1000	2977.13	II
1100	2979.66	II
1100	2980.63	II
1300	2984.19	II
550	3004.15	II
750	3007.44	II
750	3009.14	II
600	3015.40	II
550	3053.67	II
550	3055.35	II
550	3056.16	II
550	3057.38	II
550	3057.95	II
550	3058.72	II
700	3060.25	II
800	3061.35	II
500	3064.38	II
500	3066.22	II
500	3066.54	II
550	3074.33	II
550	3078.32	II
550	3080.25	II
550	3087.06	II
550	3092.04	II
550	3092.73	II
650	3094.45	II
650	3095.55	II
500	3103.58	II
500	3104.40	II
500	3113.69	II
1700	3124.42	II
600	3125.21	II
600	3129.38	II
2500	3135.48	II
1700	3137.86	II
950	3145.71	II
2000	3149.28	II
2000	3163.74	II
700	3175.09	II
1000	3179.06	II
1700	3189.79	II
1600	3212.19	II
700	3234.93	II
1500	3257.96	II
650	3260.21	II
950	3274.22	II
1700	3285.60	II
1700	3301.35	II
1200	3302.37	I
600	3302.98	I
1500	3304.96	II
1000	3318.04	II
950	3327.69	II
50	3426.86	I
1500	3533.05	II
1200	3631.27	II
850	3711.07	II
300	4113.70	II
250	4123.08	II
250	4233.26	II
6	4238.99	I
250	4240.90	II
10	4242.08	I
1	4249.41	I
2	4252.52	I
15	4273.64	I
20	4276.79	I
2	4287.84	I
3	4291.01	I
250	4292.48	II
250	4292.86	II
250	4308.81	II
250	4309.04	II
250	4320.91	II
30	4321.40	I
40	4324.62	I
250	4337.29	II
3	4341.49	I
250	4344.11	II
5	4344.74	I
200	4368.60	II
200	4375.22	II
200	4387.49	II
40	4390.03	I
250	4392.81	II
60	4393.34	I
200	4405.12	II
5	4419.88	I
8	4423.25	I
200	4446.70	II
200	4447.41	II
200	4454.74	II
200	4455.23	II
200	4457.21	II
200	4474.63	II
200	4478.80	II
200	4481.67	II
200	4490.15	II
200	4490.87	II
60	4494.18	I
100	4497.66	I
200	4499.62	II
200	4506.97	II
200	4519.21	II
200	4524.98	II
200	4533.32	II
10	4541.63	I
15	4545.19	I
200	4551.53	II
160	4590.92	II
120	4664.811	I
200	4668.560	I
160	4722.23	II
160	4731.10	II
160	4741.67	II
20	4747.941	I
30	4751.822	I
160	4768.79	II
100	4788.79	II
200	4978.541	I
400	4982.813	I
40	5148.838	I
80	5153.402	I
100	5191.65	II
80	5208.55	II
70	5400.46	II
90	5414.55	II
280	5682.633	I
70	5688.193	I
560	5688.205	I
80000	5889.950	I
40000	5895.924	I
120	6154.225	I
240	6160.747	I
60	6175.25	II
70	6199.26	II
70	6234.68	II
80	6260.01	II
80	6274.74	II
70	6361.15	II
70	6366.41	II
90	6514.21	II
80	6524.68	II
130	6530.70	II
130	6544.04	II
130	6545.75	II
80	6552.43	II
20	7373.23	I
10	7373.49	I
50	7809.78	I
25	7810.24	I
4400	8183.256	I
800	8194.790	I
8800	8194.824	I
100	8649.92	I
60	8650.89	I
25	8942.96	I
40	9153.88	I
60	9465.94	I
80	9961.28	I
20	10566.00	I
60	10572.28	I
200	10746.44	I
80	10749.29	I
120	10834.87	I
35	11190.19	I
50	11197.21	I
400	11381.45	I
1000	11403.78	I
400	12679.17	I
60	14767.48	I
100	14779.73	I
60	16373.85	I
100	16388.85	I
400	18465.25	I
50	22056.44	I
25	22083.67	I
60	23348.41	I
100	23379.13	I

Na III
Ref. 178, 205, 207 — T.K.M.

Intensity		Wavelength	
		Vacuum	
5		183.95	III
5	h	189.35	III
5		193.80	III
5	h	194.04	III
5	h	194.17	III
5		194.29	III
6		194.68	III
6		195.53	III
6		202.15	III
6		202.19	III
8		202.49	III
5	d	202.71	III
7	d	202.72	III
8		202.76	III
8	p	203.06	III
8		203.28	III
8		203.33	III
10		207.30	III
10	c	215.34	III
12		215.86	III
12		216.12	III
15		229.87	III
12		230.59	III
50	c	250.52	III
30		251.37	III
25		266.90	III
70		267.65	III
50		267.87	III
50		268.63	III
20	p	272.08	III
20		272.45	III
100		378.14	III
70		380.10	III
7		1336.76	III
7		1337.36	III
8		1340.67	III
9	d	1342.39	III
10		1342.73	III
11		1355.28	III
12		1361.90	III
11		1372.34	III
10		1420.89	III
10		1444.19	III
12		1449.31	III
11		1562.87	III
10		1565.29	III
10		1598.18	III
11		1688.94	III
10		1699.29	III
10		1711.12	III
11		1728.27	III
10		1731.11	III
10		1755.48	III
15		1807.07	III
10		1810.77	III
11		1811.67	III
10		1816.81	III
10	d	1835.22	III
10		1838.94	III
11		1844.36	III
12	d	1847.53	III
10	d	1847.59	III
15		1849.56	III
12		1850.38	III
10		1855.92	III
10		1856.71	III
10		1861.21	III
10		1880.66	III
10	d	1887.39	III
20	d	1887.47	III
15	d	1890.75	III
15		1900.16	III
10		1918.45	III
11		1923.96	III
14		1926.26	III
12		1927.24	III
12		1932.74	III
13		1933.89	III
10		1943.52	III
12		1946.43	III
12		1950.91	III
14		1951.24	III
10		1977.16	III
13		1985.57	III
10		1995.68	III
		Air	
10		2004.21	III
11		2005.22	III
11		2008.47	III
15		2011.87	III
11		2014.17	III
12		2017.03	III
12		2028.56	III
12		2031.13	III
11		2035.90	III
12		2041.66	III
12		2043.29	III
10		2044.82	III
10		2045.44	III
11		2051.48	III
10		2060.36	III
15		2066.60	III
13		2082.91	III
15		2140.72	III
14		2144.54	III
15		2202.83	III
15		2225.93	III
30		2230.33	III
16		2232.19	III
20	h	2246.70	III
14		2251.47	III
15		2278.42	III
13		2285.66	III
15		2309.99	III
18		2386.99	III
17		2394.03	III
15		2406.59	III
25		2459.31	III
18		2468.85	III
20		2474.73	III
25		2497.03	III
17		2510.26	III
15		2530.25	III
14		2542.80	III

Na IV
Ref. 206 — T.K.M.

Intensity		Wavelength	
		Vacuum	
4		136.551	IV
4		136.854	IV
4		139.961	IV
7		142.232	IV
6		142.359	IV
8		146.064	IV
7		146.302	IV
9		150.298	IV
7		150.543	IV
7	c	150.64	IV
8		150.687	IV
7		151.299	IV
7		155.083	IV
7		155.240	IV
7		155.448	IV
8		155.510	IV
8		156.537	IV
12		162.448	IV
10		163.190	IV
12		168.411	IV
10		168.546	IV
10		190.445	IV
10		199.772	IV
10	c	205.49	IV
10		319.644	IV
10		360.76	IV
12		408.684	IV
10		409.614	IV
15		410.372	IV
10		411.334	IV
13		412.242	IV
10		1580.50	IV
11		1582.18	IV
10		1582.33	IV
11	d	1583.98	IV
12		1584.14	IV
10	d	1586.99	IV
12	d	1587.05	IV
10		1613.95	IV
11		1615.92	IV
12		1618.57	IV
11		1655.47	IV
15	c	1701.97	IV
10		1702.41	IV
12		1960.76	IV
11		1965.08	IV
10		1967.60	IV
		Air	
10		2018.39	IV
12	d	2106.33	IV
10		2114.53	IV

Sodium (Cont.)

Intensity		Wavelength	
10		2155.76	IV

Na V
Ref. 299 — T.K.M.

Intensity		Wavelength	
		Vacuum	
100		106.28	V
100		106.30	V
100		106.40	V
100		106.49	V
200	c	107.93	V
200		108.02	V
200	c	110.82	V
200		110.88	V
100		111.51	V
300	c	112.01	V
100	h	114.70	V
100		114.74	V
400		117.99	V
100		120.04	V
400		125.18	V
400		125.22	V
500		125.29	V
300		125.43	V
300		125.46	V
200		125.90	V
100		126.21	V
200		126.56	V
100		126.61	V
400		127.44	V
400		127.47	V
400		128.03	V
400		128.05	V
200		130.68	V
300		131.35	V
200		131.41	V
300	h	131.64	V
500		133.16	V
400		133.39	V
200		134.27	V
300		135.79	V
300		135.85	V
200		138.81	V
300		138.92	V
400		148.64	V
300		148.86	V
400		151.13	V
300		157.21	V
300		163.62	V
800		307.15	V
1000		308.26	V
800		332.55	V
900		333.91	V
800		360.32	V
800		360.37	V
1000		400.72	V
500		445.05	V
600		445.19	V
600		459.90	V
850		461.05	V
1000		463.26	V

STRONTIUM (Sr)
Z = 38

Sr I and II
Ref. 1, 218, 279, 313 — J.J.W.

Intensity		Wavelength	
		Air	
1400		2152.84	II
1400		2165.96	II
160		2428.10	I
120		2569.47	I
200		2931.83	I
300		3301.73	I
300		3329.99	I
400		3351.25	I
300		3366.33	I
650		3380.71	II
950		3464.46	II
120		3474.89	II
300	h	3940.80	I
600		3969.26	I
300		3970.04	I
1300		4030.38	I
300		4032.38	I
46000		4077.71	II
200		4161.80	II
32000		4215.52	II
340		4305.45	II
350	h	4438.04	I
		4526.10	II
		4585.91	II
65000		4607.33	I
3200		4722.28	I
2200		4741.92	I
1400		4784.32	I
4800		4811.88	I
3600		4832.08	I
500		4855.04	I
600		4868.70	I
3000		4872.49	I
600		4876.06	I
2000		4876.32	I
1000		4891.98	I
8000		4962.26	I
1300		4967.94	I
800	h	5156.07	I
1400		5222.20	I
2000		5225.11	I
2000		5229.27	I
2800		5238.55	I
4800		5256.90	I
		5303.13	II
350	h	5329.82	I
		5379.13	II
		5385.45	II
1500		5450.84	I
7000		5480.84	I
1100		5486.12	I
3500		5504.17	I
2600		5521.83	I
2000		5534.81	I
2000		5540.05	I
250	h	5543.36	I
		5622.94	II
		5650.54	II
		5723.70	II
		5819.00	II
200	h	5970.10	I
250	h	6345.75	I
250	h	6363.94	I
350	h	6369.96	I
1000		6380.75	I
900	h	6386.50	I
600	h	6388.24	I
9000		6408.47	I
250		6446.68	I
250	h	6465.79	I
		6483.17	II
5500		6504.00	I
		6509.20	II
1000		6546.79	I
1700		6550.26	I
3000		6617.26	I
800		6643.54	I
1800		6791.05	I
4800		6878.38	I
1200		6892.59	I
5500		7070.10	I
60		7153.09	I
250	h	7167.24	I
200		7232.27	I
2500		7309.41	I
500		7621.50	I
400	h	7673.06	I
50	hL	7850.00	I
30	h	7866.90	I
20	hL	7874.00	I
200	h	8422.80	I
120		8505.69	II
200		8688.91	II
30		8719.56	II
40		9170.00	I
30		9204.50	I
20		9283.90	I
100		9294.10	I
15		9306.60	I
30		9319.20	I
60		9380.45	I
40	h	9411.25	I
400	h	9448.95	I
600		9596.00	I
300		9624.70	I
100		9638.10	I
100	h	9647.70	II
300		10036.66	II
1000		10327.31	II
7		10872.70	II
200		10914.88	II
10		10984.00	I
13		11224.57	II
700		11241.25	I
100		12014.76	II
20		12236.20	I
60		12445.90	II
20		12479.60	I
40		12495.00	I
15		12652.20	I
75		12974.70	II
100		13123.80	II
15		13522.80	I
15		17140.90	I
30		17170.50	I
50		17447.40	I
4		17626.00	II
30		17743.00	I
15		19759.60	I
230		20261.40	I
120		20700.70	I
40		20764.50	I
15		20778.70	I
30		26023.60	I

Sr III
Ref. 231, 265 — J.J.W.

Intensity		Wavelength	
		Vacuum	
25		307.18	III
50		316.11	III
50		321.61	III
125		330.67	III
500		351.62	III
75		358.80	III
250		363.49	III
150		371.21	III
1000		437.24	III
1875		491.79	III
1250		507.04	III
3750		514.38	III
2500		562.75	III
20		968.37	III
20		975.78	III
25		992.98	III
50		1025.23	III
35		1044.91	III
20		1057.74	III
25		1060.20	III
20		1098.77	III
35		1125.49	III
20		1140.24	III
20		1168.27	III
20		1182.09	III
50		1236.23	III
20		1940.58	III
30	p	1958.44	III
30		1966.92	III
		Air	
25		2068.63	III
50		2099.59	III
25		2114.31	III
30		2118.48	III
50		2119.52	III
50		2133.12	III
30		2142.80	III
20		2145.74	III
30		2178.91	III
30		2180.14	III
50		2190.88	III
50		2203.86	III
50		2219.50	III
50		2220.05	III
50		2267.03	III
100		2273.71	III
50		2277.87	III
30		2310.33	III
50		2314.95	III
50		2334.79	III
100		2340.13	III
50		2404.17	III
30		2410.52	III
50		2454.03	III
100		2486.52	III
50		2503.59	III
30		2599.10	III
35		2622.69	III
30		2642.96	III
30		2648.51	III
35		2654.66	III
40		2722.47	III
50		2786.00	III
50		2821.42	III
30		2874.86	III
30		2929.34	III
30		2983.00	III
100		3002.61	III
200		3012.32	III
100		3021.73	III
30		3059.83	III
50		3061.43	III
30		3104.25	III
50		3182.61	III
100		3235.39	III
30		3302.72	III
50		3430.76	III
30		3874.26	III
30		3936.40	III
30		3936.72	III
30		3958.75	III
30		4094.03	III
30		4097.02	III
30		4105.63	III
35		4335.80	III
30		5071.09	III
30		5130.34	III
35		5158.26	III
40		5257.71	III
30		5262.21	III
30		5288.32	III
30		5391.03	III
40		5443.48	III
30		5463.90	III
30		5664.66	III
30		5689.72	III

Sr IV
Ref. 110 — J.J.W.

Intensity		Wavelength	
		Vacuum	
12		284.31	IV
12		291.09	IV
12		291.19	IV
12		293.22	IV
15		298.12	IV
15		300.12	IV
12		300.27	IV
12		301.67	IV
20		378.53	IV
75		392.44	IV
50		393.00	IV
45		394.90	IV
50		396.22	IV
40		399.92	IV
35		403.85	IV
35		406.94	IV
30		412.93	IV
40		413.07	IV
40		415.32	IV
30		419.78	IV
25		430.21	IV
30		430.65	IV
25		442.73	IV
25		471.76	IV
25		484.20	IV
25	p	508.14	IV
25		534.19	IV
200		664.43	IV
100		710.35	IV
20		1189.21	IV
30		1244.14	IV
20	p	1244.75	IV
20	p	1244.87	IV
20	p	1257.78	IV
20		1268.62	IV
20		1331.13	IV
30		1347.90	IV
20		1361.15	IV
25		1408.67	IV
20		1592.74	IV
25		1677.03	IV
20		1705.16	IV
20		1724.23	IV
25		1729.53	IV
20		1732.12	IV
20		1777.25	IV
20		1994.61	IV
		Air	
20		2104.38	IV
20		2117.90	IV
20		2217.99	IV
20		2230.41	IV
20		2240.49	IV
20		2253.38	IV
50		2346.97	IV
20		2357.34	IV
20		2438.93	IV
25		2441.41	IV
30		2482.79	IV
25		2483.57	IV
18		2500.57	IV
20		2508.02	IV
20		2534.03	IV
18		2548.02	IV
40		2555.60	IV
40		2571.04	IV
25		2571.58	IV
15		2589.34	IV
25		2620.35	IV
20		2621.16	IV
20		2642.16	IV
15		2830.53	IV
9		2934.60	IV
10		3019.29	IV
9		3266.52	IV
9		3566.43	IV
9		3741.05	IV
9		4298.57	IV
9		4685.08	IV

Sr V
Ref. 109 — J.J.W.

Intensity	Wavelength
	Vacuum

Strontium (Cont.)

10	517.28	V
6	540.51	V
25	578.01	V
30	624.93	V
25	642.23	V
50	649.21	V
20	659.15	V
25	660.94	V
9	669.93	V
35	686.23	V
6	715.79	V
12	747.82	V
9	862.32	V

SULFUR (S)
Z = 16

S I and II
Ref. 144, 209, 210, 266—R.L.K.

Intensity	Wavelength	
	Vacuum	
40	906.9	II
40	910.5	II
40	912.7	II
40	937.4	II
40	937.7	II
20	996.0	II
20	1000.5	II
20	1014.4	II
20	1019.5	II
20	1096.6	II
40	1102.3	II
20	1131.0	II
20	1131.6	II
40	1234.1	II
40	1250.5	II
110	1253.8	II
110	1259.5	II
275	1270.782	I
250	1277.216	I
280	1295.653	I
275	1302.337	I
235	1302.863	I
235	1303.110	I
245	1303.430	I
260	1305.883	I
265	1310.194	I
355	1316.542	I
290	1316.618	I
375	1323.515	I
355	1326.643	I
775	1381.552	I
710	1385.510	I
960	1388.435	I
640	1389.154	I
775	1392.588	I
1000	1396.112	I
300	1409.337	I
510	1425.030	I
425	1433.280	I
300	1436.968	I
300	1448.229	I
425	1472.972	I
550	1473.995	I
300	1474.380	I
355	1481.665	I
485	1483.039	I
300	1483.233	I
330	1485.622	I
390	1487.150	I
680	1666.688	I
640	1687.530	I
710	1807.311	I
680	1820.343	I
640	1826.245	I
710	1900.286	I
550	1914.698	I
	Air	
20	2629.1	II
40	2670.0	II
40	2847.7	II
285	3867.6	I
285	3902.0	I
360	3933.3	II
450	4120.8	I
280	4142.3	II
360	4145.1	II
450	4153.1	II
450	4162.7	II
450	4694.1	I
285	4695.4	I
160	4696.2	I
280	4716.2	II
450	4815.5	II
360	4924.1	II
450	4925.3	II
285	4993.5	I
360	5428.6	II
650	5432.8	II
1000	5453.8	II
1000	5473.6	II
1000	5509.7	II
280	5564.9	II
1000	5606.1	II
450	5640.0	II
450	5640.3	II
280	5647.0	II
650	5659.9	II
450	5664.7	II
160	5706.1	I
450	5819.2	II
450	6052.7	I
280	6286.4	II
450	6287.1	II
450	6305.5	II
450	6312.7	II
280	6384.9	II
280	6397.3	II
280	6398.0	II
360	6413.7	II
160	6743.6	I
285	6748.8	I
450	6757.2	I
450	7579.0	I
450	7629.8	I
285	7686.1	I
450	7696.7	I
1000	7924.0	I
160	7928.8	I
285	7930.3	I
450	7931.7	I
450	7967.4	I
450	7967.4	II
450	8314.7	I
450	8314.7	II
450	8585.6	I
285	8680.5	I
450	8694.7	I
360	8874.5	I
110	8882.5	I
220	8884.2	I
160	9035.9	I
450	9212.9	I
450	9228.1	I
450	9237.5	I
285	9413.5	I
285	9421.9	I
285	9437.1	I
650	9649.9	I
450	9672.3	I
450	9680.8	I
450	9693.7	I
285	9697.3	I
285	9739.7	I
110	9741.9	I
285	9932.3	I
285	9949.8	I
285	9958.9	I
285	10455.5	I
70	10456.8	I
285	10459.5	I

S III
Ref. 209, 210 — R.L.K.

Intensity	Wavelength	
	Vacuum	
70	729.5	III
110	732.42	III
70	735.2	III
70	738.5	III
70	789.0	III
70	796.7	III
70	824.9	III
70	836.3	III
285	1077.1	III
70	1194.0	III
70	1201.0	III
	Air	
110	2460.5	III
110	2489.6	III
160	2496.2	III
160	2499.1	III
220	2508.2	III
70	2636.9	III
220	2665.4	III
70	2680.5	III
110	2691.8	III
110	2702.8	III
220	2718.9	III
110	2721.4	III
220	2726.8	III
220	2731.1	III
110	2741.0	III
285	2756.9	III
110	2775.2	III
160	2785.5	III
70	2797.4	III
70	2856.0	III
110	2863.5	III
160	2904.3	III
70	2964.8	III
160	2986.0	III
70	3234.2	III
70	3324.9	III
110	3497.3	III
160	3632.0	III
70	3662.0	III
110	3709.4	III
160	3717.8	III
160	3838.3	III
160	3928.6	III
360	4253.6	III
110	4285.0	III
70	4332.7	III

S IV
Ref. 29, 202, 209 — R.L.K.

Intensity	Wavelength	
	Vacuum	
20	519.3	IV
20	520.1	IV
40	520.8	IV
20	522.0	IV
20	522.5	IV
20	551.2	IV
40	652.5	IV
40	653.0	IV
70	653.6	IV
40	654.0	IV
70	655.6	IV
20	655.9	IV
110	657.3	IV
40	660.9	IV
160	661.4	IV
40	663.7	IV
40	664.8	IV
70	666.1	IV
110	744.9	IV
110	748.4	IV
110	750.2	IV
110	753.8	IV
40	798.3	IV
70	800.5	IV
70	804.0	IV
70	809.7	IV
110	816.0	IV
160	1062.7	IV
160	1073.0	IV
70	1073.5	IV
20	1108.4	IV
20	1110.9	IV
20	1624.0	IV
20	1629.2	IV
	Air	
20	2387.0	IV
40	2398.9	IV
110	3097.5	IV
40	3117.7	IV

S V
Ref. 29 — R.L.K.

Intensity	Wavelength	
	Vacuum	
5	437.4	V
5	438.2	V
5	439.6	V
40	658.3	V
70	659.8	V
110	663.2	V
5	676.2	V
5	677.3	V
20	678.1	V
40	680.3	V
110	680.9	V
40	681.6	V
5	686.2	V
5	686.9	V
5	689.8	V
5	691.7	V
20	693.5	V
285	786.5	V
160	849.2	V
110	852.2	V
220	854.8	V
110	857.9	V
110	860.5	V
20	883.6	V
20	884.5	V
5	885.8	V
20	900.9	V
5	902.8	V
20	905.9	V

TANTALUM (Ta)
Z = 73

Ta I and II
Ref. 1 — C.H.C.

Intensity	Wavelength	
	Air	
1100	2140.13	II
1500	2146.87	II
740	2150.62	II
600	2165.01	II
740	2178.03	II
1200	2182.71	II
540	2193.20	II
1100	2193.88	II
1500	2196.03	II
1500	2199.67	II
500	2207.14	II
1400 d	2210.03	II
	2210.19	II
420	2215.60	II
1400	2239.48	II
240	2248.48	II
480	2249.79	II
1200	2250.76	II
260	2254.86	II
440	2255.77	II
360	2256.51	II
500	2258.71	II
840	2261.42	II
260	2261.62	II
990	2262.30	II
220	2269.56	II
740	2271.85	II
990	2272.59	II
200	2279.85	I
320	2282.19	II
130	2285.02	II
790	2285.25	II
600	2286.59	II
240	2287.27	II
990	2289.16	II
180	2292.54	II
160	2295.18	
160	2301.47	II
440	2302.24	II
440	2302.93	II
300	2303.49	II
100	2308.46	II
440	2312.60	II
420	2315.46	II
260	2319.16	II
100	2331.29	II
690	2331.98	II
550	2332.19	II
110	2334.13	II
180	2334.88	II
140	2335.75	II
300	2338.28	II
200	2340.94	II
200	2341.61	II
130	2343.64	II
100	2346.42	II
90	2351.99	II
170	2353.86	II
120	2355.22	II
170	2356.05	II
140	2356.90	II
250	2357.30	I
170	2359.16	II
260	2361.09	I
160	2362.78	II
130	2363.32	II
600	2364.24	II
50	2367.24	II
150	2369.32	II
300	2370.76	II
320	2371.58	I
100	2373.94	II
70	2375.91	I
150	2378.31	II
440	2381.13	II
240	2381.52	II
170	2383.72	II
240	2384.28	II
130	2385.73	I
1400	2387.06	II
80	2388.37	II
160	2389.11	II
70	2396.30	I
110	2399.15	I
50	2399.92	II
2400	2400.63	II
140	2402.13	II
100	2403.68	II
130	2406.55	I

Tantalum (Cont.)

130		2408.26	II
120		2414.32	I
240		2415.21	II
320		2416.89	II
220		2417.86	II
150		2418.77	II
140		2421.03	I
150		2421.85	II
170		2423.48	II
130		2425.91	II
360		2427.64	I
360		2429.71	II
170		2431.06	II
480		2432.70	II
130		2433.59	II
130		2436.51	II
110		2437.07	I
110		2438.64	II
[illegible]		2439.91	I
130		2442.39	I
100		2444.13	II
100		2447.17	I
100		2454.48	I
100		2458.68	I
100		2460.55	I
160		2463.82	II
130		2466.99	II
130		2467.37	II
380		2470.90	II
120		2471.38	I
120		2472.13	I
150		2473.13	I
120		2473.31	II
600		2474.62	I
120		2475.33	I
200		2476.67	II
150		2478.22	I
120		2481.86	II
100		2482.10	I
100		2482.58	II
100		2484.04	II
500		2484.95	I
120		2486.70	I
600		2488.70	II
500		2490.46	I
600		2504.45	I
600		2507.45	I
240		2512.65	I
1200	d	2526.35	I
600		2532.12	II
240		2545.49	II
240		2546.80	I
460	d	2551.07	I
460		2554.62	II
240		2555.05	I
1200		2559.43	I
460		2562.10	I
340		2571.51	II
430		2573.54	I
390		2573.79	I
600		2577.37	II
340		2577.78	I
210		2580.16	I
340		2584.03	II
430		2593.08	I
410		2593.66	II
310		2594.25	II
560		2595.26	I
310	l	2596.45	II
220		2600.14	I
600		2603.49	II
1400		2608.63	I
210		2609.00	I
310	d	2611.34	I
340		2615.46	I
310		2615.66	I
1200		2635.58	II
470		2636.67	I
860		2636.90	I
510		2646.22	I
600		2646.37	I
2400		2647.47	I
270		2651.22	II
2600		2653.27	I
1900		2656.61	I
1500		2661.34	I
220		2665.60	II
220		2668.07	I
600		2668.62	I
770		2675.90	II
270		2680.06	II
220		2680.66	II
600		2684.28	I
1500		2685.17	II
340		2691.31	I
260		2692.40	I
470		2694.52	II
240		2696.81	I
1000		2698.30	I
470		2706.69	I
310		2709.27	II
1200		2710.13	I
2600		2714.67	I
240		2717.18	I
470		2720.76	I
470		2727.44	II
410		2727.78	I
310		2736.25	II
210		2739.26	II
210		2743.59	I
510		2746.68	I
1200		2748.78	I
860		2749.83	I
410		2752.49	II
1000		2758.31	I
430		2761.68	II
770		2775.88	I
390		2787.69	I
680		2796.34	I
680		2797.76	II
380		2802.07	I
430		2806.30	I
510		2806.58	I
260		2817.10	II
260		2842.82	I
640		2844.25	I
290		2844.46	II
290	c	2845.35	I
560		2848.52	I
1500		2850.49	I
1900		2850.98	I
220		2858.44	II
360		2861.98	I
310		2868.65	I
470		2871.42	I
270		2873.36	I
260		2873.56	I
210		2874.17	I
380		2880.02	I
770		2891.84	I
260		2899.04	I
560		2902.05	I
210		2914.12	I
310		2915.49	I
410		2925.19	I
310		2932.70	I
1700		2933.55	I
470		2940.06	I
1200		2940.22	I
240		2942.14	I
510		2951.92	I
340		2953.56	I
1500		2963.32	I
770		2965.13	II
770		2965.54	I
340		2969.47	I
430		2973.36	I
210		3011.88	I
1800		3012.54	II
290	d	3027.48	I
290		3042.06	II
530		3049.56	I
530		3069.24	I
360		3077.24	I
560		3103.25	I
380		3124.97	I
380		3130.58	I
270		3132.64	I
320		3170.29	I
270		3173.59	I
200		3176.29	I
600		3180.95	I
240		3184.55	I
200		3198.67	I
200		3213.91	II
300		3223.83	I
230		3229.24	I
200		3242.05	I
200		3242.83	I
210		3274.95	II
1100		3311.16	I
210		3317.93	I
680		3318.84	I
330	d	3330.99	II
230		3358.47	I
640		3371.54	I
360		3385.05	I
230		3398.33	I
450		3406.94	I
230		3463.77	I
490		3480.52	I
380		3497.85	I
240		3503.87	I
490		3511.04	I
200		3513.61	I
750		3607.41	I
980		3626.62	I
500		3642.06	I
100		3686.18	I
100		3689.73	I
130		3731.02	I
140		3736.76	I
130		3746.36	I
110		3754.52	I
110		3777.10	I
110		3792.02	I
210		3833.74	II
100		3848.05	I
100		3885.20	I
210		3918.51	I
140		3922.78	I
140		3922.92	I
210		3970.10	I
210		3996.17	I
100		3999.28	I
190		4006.84	I
190		4026.94	I
140		4029.94	I
120		4040.87	I
410		4061.40	I
210		4064.63	I
100		4067.24	I
310		4067.91	I
120		4105.02	I
210		4129.38	I
230		4136.20	I
230		4147.89	I
210		4175.21	I
100		4177.92	I
130		4181.15	I
300		4205.88	I
120		4206.40	I
130		4245.35	I
130		4268.26	I
160	c	4302.98	I
110		4355.14	I
100		4378.82	I
150		4386.07	I
110		4398.45	I
180		4402.50	I
130		4415.74	I
360	c	4510.98	I
190		4530.85	I
130		4551.95	I
170		4565.85	I
340		4574.31	I
260		4619.51	I
130		4669.14	I
450		4681.88	I
130		4691.90	I
150		4740.16	I
220		4756.51	I
120		4768.98	I
220		4812.75	I
110		4920.11	I
100		4921.27	I
110		4926.00	I
150		4936.42	I
200		5037.37	I
100		5067.87	I
110		5115.84	I
100		5141.62	I
100		5143.69	I
330		5156.56	I
110		5212.74	I
110	d	5218.45	I
		5218.66	I
140		5341.05	I
200		5402.51	I
130		5419.19	I
18		5500.68	I
20		5505.66	I
15	c	5516.27	I
90		5518.91	I
9		5521.15	I
10		5523.98	I
13		5528.36	I
10	l	5545.20	I
20		5548.32	I
30		5584.02	I
15		5598.75	I
30		5599.52	I
9	c	5605.50	I
9	c	5617.71	I
40		5620.68	I
13		5628.20	I
20		5635.71	I
40		5640.18	I
150		5645.91	I
130		5664.90	I
30		5688.25	I
40		5699.24	I
15		5704.31	I
25		5706.28	I
30		5715.24	I
8		5716.53	I
23		5746.71	I
30		5755.81	I
15	h	5761.61	I
25		5766.56	I
30	c	5767.91	I
10		5771.93	I
130		5776.77	I
25		5780.02	I
90		5780.71	I
130		5811.10	I
25		5816.51	I
15	c	5843.94	I
13		5849.68	I
15		5866.61	I
240		5877.36	I
130		5882.30	I
90		5901.91	I
30		5916.51	I
90		5918.95	I
15		5925.90	I
15		5930.62	I
23		5931.05	I
20		5931.68	I
18		5935.54	I
130		5939.76	I
240		5944.02	I
25		5951.78	I
18		5960.13	I
190	c	5997.23	I
25	h	6009.89	I
25		6015.90	I
100		6020.72	I
250		6045.39	I
100		6047.23	I
25		6053.70	I
30		6090.82	I
18		6092.06	I
100		6101.58	I
25		6140.07	I
65		6144.56	I
30		6152.54	I
130		6154.50	I
40		6158.84	I
15		6170.46	I
15		6189.66	I
15		6193.11	I
25		6208.37	I
40		6249.79	I
150		6256.68	I
150		6268.70	I
50		6278.34	I
65		6281.33	I
15		6287.36	I
40		6287.91	I
40		6289.34	I
50		6309.06	I
150		6309.58	I
25	c	6312.22	I
75		6325.08	I
50		6332.91	I
65		6341.17	I
30		6346.02	I
75		6356.16	I
65		6360.84	I
40		6373.06	I
15		6379.07	I
90		6389.45	I
23	h	6392.21	I
65		6428.60	I
250		6430.79	I
13		6437.36	I
40		6444.61	I
30		6445.87	I
200		6450.36	I
20		6455.83	I
30		6459.92	I
380		6485.37	I
18	h	6502.43	I
65		6505.52	I
100		6514.39	I
100		6516.10	I
25	h	6561.60	I
25	Cw	6564.26	I
100		6574.84	I
10		6585.13	I
15	c	6587.16	I
110		6611.95	I
75		6621.30	I
15	Cw	6662.24	I
100		6673.73	I
180		6675.53	I
30		6684.00	I
15		6693.61	I
15		6706.46	I
25		6709.39	I
10		6714.44	I
15	h	6723.61	I
75	c	6740.73	I
40		6754.91	I
13	c	6755.85	I
13		6770.37	I
75		6771.74	I
40	Cw	6774.25	I
40	Cw	6788.99	I
13	c	6790.06	I
13		6799.27	I
40	c	6810.46	I
160	c	6813.25	I
20		6819.36	I
18	c	6824.96	I
13		6832.00	I
15		6850.83	I
15		6865.13	I
210		6866.23	I
180		6875.27	I
40		6877.49	I
15		6896.77	I
40		6900.55	I
150		6902.10	I
140		6927.38	I

Tantalum (Cont.)

140		6928.54	I
8	h	6939.33	I
20	c	6946.87	I
65		6951.26	I
45		6953.88	I
180		6966.13	I
8		6969.49	I
8	c	6971.31	I
9		6971.53	I
23		6983.52	I
110	d	6995.22	I
		6995.49	I
20		7000.21	I
40		7005.07	I
75		7006.96	I
50		7025.03	I
13		7031.51	I
40		7039.07	I
15	h	7081.30	I
20		7085.40	I
23		7093.02	I
8		7108.05	I
15	c	7117.52	I
20		7121.27	I
40		7125.72	I
150		7148.63	I
110		7172.90	I
13	c	7174.91	I
13		7191.35	I
8		7233.45	I
30	h	7250.27	I
11	h	7264.82	I
6		7272.29	I
30		7276.96	I
5		7277.54	I
9		7286.36	I
13	c	7296.32	I
140		7301.74	I
20		7319.84	I
11	c	7322.72	I
13		7325.95	I
11	Cw	7340.19	I
160		7346.41	I
140	c	7352.86	I
100		7356.96	I
90	Cw	7369.09	I
160		7407.89	I
11	c	7435.19	I
23		7440.17	I
30		7467.75	I
23		7486.01	I
30		7520.56	I
6		7569.23	I
9		7590.22	I
6		7649.62	I
11	h	7722.02	I
11		7763.11	I
9	c	7779.67	I
20	c	7842.76	I
100		7882.37	I
30		7950.19	I
5		7952.07	I
6		7998.75	I
6		8022.09	I
75		8026.50	I
5		8029.04	I
15		8039.08	I
8		8053.93	I
15		8068.98	I
5		8100.11	I
13		8128.76	I
9		8158.54	I
5	c	8180.74	I
13	c	8248.95	I
20	d	8264.85	I
75		8281.62	I
11		8389.06	I
5		8415.73	I
25	Cw	8447.62	I
11	h	8550.49	I
15		8575.92	I
10	Cw	8595.84	I

Ta IV
Ref. 411 — R.L.K.

Intensity		Wavelength	
		Vacuum	
10		763.14	IV
32		934.41	IV
67		999.34	IV
65		1063.53	IV
68		1067.17	IV
71		1074.47	IV
71		1086.39	IV
72		1094.60	IV
68		1100.13	IV
79		1116.10	IV
71		1118.83	IV
78		1136.17	IV
66		1138.26	IV
75		1149.72	IV
75		1150.42	IV
75		1151.92	IV
76		1172.51	IV
85		1175.51	IV
80		1189.28	IV
78		1192.52	IV
80		1192.67	IV
78		1211.94	IV
80		1212.68	IV
85		1213.09	IV
70		1214.66	IV
85		1215.53	IV
80		1220.73	IV
80		1220.96	IV
90		1223.73	IV
88		1238.12	IV
95		1240.06	IV
87		1258.34	IV
94		1264.91	IV
98		1272.42	IV
94		1275.48	IV
86		1275.94	IV
92		1308.51	IV
85		1311.35	IV
87		1315.58	IV
81		1325.19	IV
92		1332.38	IV
86		1343.30	IV
75		1350.46	IV
92		1365.88	IV
79		1376.62	IV
78		1388.80	IV
91		1398.78	IV
93		1413.40	IV
79		1430.11	IV
83		1441.54	IV
91		1454.32	IV
92		1464.41	IV
93		1469.82	IV
90		1495.25	IV
95		1514.19	IV
70		1525.69	IV
82		1565.97	IV
82		1584.64	IV
84		1594.91	IV
85		1607.70	IV
84		1631.65	IV
79		1639.82	IV
84		1668.76	IV
84		1676.45	IV
85		1712.16	IV
85		1716.13	IV
82		1753.90	IV
82		1759.04	IV
79		1763.03	IV
83		1865.92	IV
84		1901.63	IV
84		1907.66	IV
84		1924.75	IV
77		1940.25	IV
79		1985.68	IV
82		1989.44	IV
		Air	
85		2055.75	IV
83		2079.01	IV
75		2111.53	IV
90		2199.58	IV
90		2207.64	IV
68		2697.42	IV
10		3076.06	IV

Ta V
Ref. 426 — R.L.K.

Intensity		Wavelength	
		Vacuum	
20		478.29	V
60		493.07	V
200		841.31	V
1000		890.87	V
500		947.30	V
100		990.29	V
200		1066.64	V
200		1140.49	V
500		1213.42	V
100		1242.98	V
5000		1392.56	V
7000		1709.10	V

TECHNETIUM (Tc)
Z = 43

Tc I and II
Ref. 35 — C.H.C.

Intensity		Wavelength	
		Air	
15		2106.23	II
20		2116.44	II
15		2119.41	I
30		2156.27	I
30		2185.39	I
30		2189.06	I
40		2193.35	I
10		2266.22	II
10		2282.12	I
10		2282.71	I
50		2285.45	I
100		2298.08	I
30		2416.22	I
50		2423.23	I
20		2424.54	I
20		2435.83	I
10		2436.99	I
80	w	2463.69	I
20		2465.09	I
30		2466.87	I
20		2475.11	I
50		2480.70	I
50		2483.22	I
20		2486.50	I
25		2492.72	I
20		2493.43	I
100		2496.77	II
30		2510.17	I
80		2529.34	II
500		2543.23	II
60		2544.81	II
50		2547.92	II
50		2558.61	II
50		2567.01	II
30		2575.06	II
80		2576.28	II
40		2577.86	II
300	Hl	2578.79	I
200		2589.86	I
20	w	2590.19	I
100		2592.82	I
20		2597.19	II
500		2608.86	I
1000	c	2609.99	II
1500		2614.23	I
1000		2615.87	I
30		2618.28	I
200		2634.91	II
80		2636.36	I
30		2641.26	I
100		2642.37	I
40		2644.50	II
1000	c	2647.01	II
300	c	2649.21	I
100		2652.35	II
30		2653.57	I
100		2654.31	I
120		2660.88	I
100		2662.30	I
80		2681.19	II
60		2683.14	I
80		2683.89	I
80		2693.11	I
50		2696.64	I
70		2702.27	I
40		2702.96	II
100		2707.90	II
1000		2708.78	I
30		2715.20	I
30		2723.55	I
1000		2726.69	I
30	c	2728.47	I
500		2730.53	I
300		2732.87	I
150		2736.23	I
60	c	2736.83	II
100		2737.97	I
20	c	2738.83	II
100		2755.76	I
100		2762.13	I
200		2762.34	I
60		2765.95	I
500		2766.89	I
20		2777.31	II
150		2778.91	I
25		2781.22	I
1000		2782.05	I
500		2785.59	I
40		2788.89	I
500		2789.25	I
100		2794.53	I
80		2795.65	I
200		2795.78	II
1000		2802.81	I
150		2803.02	I
500		2808.36	I
50	c	2809.65	II
500		2811.61	II
30		2814.86	I
40		2819.46	I
100		2821.35	II
200		2828.04	I
60		2831.18	II
50		2840.38	II
60		2845.04	I
10		2846.39	II
60		2849.20	I
150		2850.96	I
500	h	2857.13	I
2000	c	2859.11	I
500		2864.49	I
100		2868.09	I
1000		2887.73	I
100		2888.46	I
30		2889.20	II
200		2893.16	I
150		2893.45	I
200		2894.32	I
1000		2896.34	I
40		2903.81	I
1000		2913.15	I
500		2921.91	I
20	c	2923.34	II
1000		2928.20	I
80		2933.89	I
200		2955.93	I
200		2973.65	I
100		2979.34	I
150		2985.36	I
100		3010.83	I
300		3017.23	I
150		3021.56	I
100		3022.66	I
200		3023.68	I
80		3025.26	I
300	w	3026.89	I
50		3033.16	I
80		3034.57	I
40		3036.88	I
20		3037.90	II
100		3038.23	I
40		3042.64	I
100	h	3051.55	I
40		3052.47	I
80		3062.11	I
200		3062.36	I
300		3064.67	I
100	c	3066.60	I
120	c	3068.34	I
80		3076.24	I
150		3089.34	I
1000		3099.10	I
200		3099.52	I
60		3108.25	I
40		3109.15	I
60		3115.98	I
80		3119.17	I
40		3119.66	I
700		3122.64	I
1500		3131.23	I
40	c	3150.26	I
300		3161.67	I
3000		3173.30	I
200		3180.30	I
2000		3182.37	I
2000		3183.11	I
800	c	3195.20	II
40	w	3197.53	I
300	c	3202.83	I
1000		3212.02	II
40		3220.74	I
60		3230.02	I
1000		3237.02	II
100		3241.84	I
500		3244.19	I
300		3252.05	I
40		3256.10	I
40		3261.94	I
100		3287.14	I
30		3298.84	II
100		3300.77	I
80		3305.89	I
200		3310.65	I
150		3313.65	I
200		3325.55	I
150	c	3327.10	I
100		3330.77	I
50		3332.47	I
60		3350.56	I
50	c	3350.83	I
400		3366.75	I
40		3386.67	I
50		3392.23	I
300		3394.18	I
60		3396.90	I
40		3397.83	I
300		3398.33	I
200		3402.10	I
200	c	3403.93	I
80		3405.33	I
80		3407.28	I
50		3408.33	I
40		3411.80	I
40	c	3418.20	I
100		3419.10	I

Technetium (Cont.)

60		3427.85	I
40		3431.75	I
200		3434.70	I
40		3435.68	I
150		3437.44	I
80		3438.73	I
200	c	3443.47	I
200		3451.05	I
200		3456.85	I
400		3457.24	I
40		3457.60	I
5000	c	3466.28	I
150		3470.51	I
80		3475.18	I
1000		3475.59	I
60		3484.62	I
1000	c	3486.23	I
100		3490.30	I
400		3493.39	I
500		3494.62	I
40		3499.14	I
1000		3500.70	I
200		3501.24	I
800	c	3502.70	I
100	c	3507.19	I
100		3508.27	I
100		3510.91	I
800		3525.83	I
300		3526.18	I
100		3529.83	I
150		3534.88	I
500		3535.51	I
300		3538.12	I
800		3538.68	I
2000	c	3541.77	I
6000	c	3549.72	I
4000	c	3550.64	I
300		3559.75	I
800		3560.32	I
100		3565.22	I
800		3568.85	I
100		3570.65	I
100		3575.42	I
1000		3580.06	I
600		3581.26	I
800		3582.08	I
2000		3582.63	I
4000		3587.94	I
200		3593.47	I
300		3594.57	I
1000	c	3595.66	I
1000	c	3607.32	I
200		3607.62	I
2000	c	3608.27	I
200		3618.94	I
1000	c	3627.36	I
200		3630.39	I
3000	c	3635.15	I
10000	c	3636.07	I
1000		3638.22	I
200		3638.85	I
900		3639.38	I
400		3640.23	I
1000	c	3648.04	I
600		3651.47	I
1000	c	3658.59	I
400	c	3661.45	I
200		3664.92	I
1000		3679.15	I
300		3680.32	I
5000		3684.74	I
300		3692.76	I
800		3703.83	I
300		3704.80	I
200		3706.70	I
200		3707.63	I
200		3708.26	I
1000		3712.26	I
300		3712.82	I
500		3715.94	I
10000		3718.86	I
1500		3723.67	I
2000		3724.40	I
5000		3726.35	I
200		3727.36	I
400	c	3729.18	I
500		3731.74	I
300		3737.42	I
400		3745.01	I
1000		3746.15	I
5000		3746.84	I
1000		3752.13	I
4000		3754.37	I
1000		3758.54	I
2000		3761.81	I
5000		3768.77	I
3000		3771.03	I
500		3777.27	I
2000		3779.37	I
3000	c	3780.68	I
500		3784.06	I
200		3786.06	I
500		3791.28	I
300		3791.73	I
200		3797.44	I
1000		3797.77	I
200		3814.67	I
300		3816.89	I
300		3824.47	I
500		3828.54	I
200		3830.35	I
200		3832.45	I
600		3832.82	I
1500		3837.56	I
800		3841.31	I
800		3845.97	I
500		3847.60	I
300		3851.22	I
500	c	3856.73	I
200		3863.07	I
400		3864.11	I
1000		3868.24	I
200		3875.66	I
500	c	3879.16	I
600	c	3880.72	I
300	w	3892.12	I
200		3893.22	I
600		3899.83	I
300		3919.38	I
300	c	3923.66	I
200		3927.57	I
200		3933.70	I
4000	c	3946.57	I
2000		3947.09	I
200		3955.73	I
300		3979.64	I
500		3980.35	I
10000	c	3984.97	I
400		3987.78	I
300		3994.04	I
2000		3994.51	I
200		3996.97	I
300		4004.69	I
500		4007.14	I
1000		4012.00	I
400		4016.68	I
600		4017.22	I
2000		4020.76	I
20000	c	4031.63	I
1000		4039.25	I
200		4041.78	I
10000	c	4049.11	I
500		4051.95	I
200		4053.18	I
200	c	4056.08	I
400		4083.54	I
10000		4088.71	I
200		4093.69	I
15000		4095.67	I
1000		4110.22	I
10000		4115.08	I
600		4119.27	I
8000		4124.22	I
1000		4128.27	I
300		4134.81	I
300		4139.12	I
800		4139.85	I
400		4141.27	I
6000		4144.95	I
3000		4145.08	I
200		4147.62	I
10000		4165.61	I
500		4167.42	I
1000		4169.68	I
4000		4170.27	I
5000		4172.53	I
1000		4176.28	I
800		4186.51	I
300		4218.61	I
10000	c	4238.19	I
20000		4262.27	I
1000		4262.69	I
800		4274.97	II
800		4278.90	I
30000		4297.06	I
400	c	4336.86	I
400		4358.49	I
200		4359.26	I
1000		4429.59	I
1000		4481.53	I
3000		4487.06	I
400		4495.03	I
1000		4515.98	I
10000		4522.84	I
2000		4539.53	I
400		4542.09	I
400		4552.20	I
800		4552.85	I
1000		4557.05	I
2000		4564.54	I
1000		4578.45	I
1000		4593.35	I
300	c	4609.16	I
1000		4616.86	I
200	c	4622.69	I
300		4624.96	I
1000		4630.57	I
200		4633.15	I
3000		4637.50	I
500		4643.28	I
2000		4648.33	I
2000	c	4660.21	I
2000		4669.30	I
400		4672.17	I
200		4678.90	I
400		4689.36	I
300		4694.28	I
1000		4706.92	I
200		4714.22	I
2000		4717.77	I
500	c	4719.02	I
4000	c	4719.28	I
200	c	4736.51	I
10000		4740.61	I
500		4749.61	I
1000		4752.72	I
200		4762.36	I
4000		4771.54	I
200		4773.89	I
200		4783.92	I
500		4785.60	I
200	c	4790.48	I
250		4791.62	I
300		4799.98	I
100		4805.69	I
100		4809.42	I
500		4816.79	I
10000		4820.74	I
300		4831.35	I
1000		4834.37	I
1000		4835.39	I
100		4841.36	I
20000		4853.59	I
100		4857.21	I
100	c	4862.19	I
10000		4866.73	I
200		4870.77	I
100		4888.70	I
150	c	4890.88	I
8000		4891.92	I
150		4892.49	I
1000		4908.51	I
2000		4909.57	I
500		4913.02	I
150	c	4914.70	I
200	c	4920.67	I
300		4923.60	I
400		4948.06	I
5000		4976.04	I
400		4995.00	I
200		5002.67	I
100		5005.74	I
200	c	5014.52	I
500		5026.24	I
300		5026.79	I
150		5027.89	I
80		5032.45	I
300		5055.27	I
60		5058.33	I
500		5060.69	I
80		5090.74	I
5000		5096.28	I
200	c	5103.24	I
500		5104.32	I
200		5109.81	I
100		5120.60	I
500		5139.26	I
500		5150.63	I
2000		5161.81	I
2000		5174.81	I
100		5206.56	I
200		5225.55	I
200	c	5260.22	I
200	c	5261.44	I
1000		5275.51	I
800		5285.07	I
100		5305.31	I
400		5314.96	I
600		5320.20	I
200		5334.79	I
500	c	5353.48	I
200		5356.63	I
300		5358.65	I
200		5360.14	I
500	h	5375.20	I
150	c	5423.05	I
200		5447.40	I
500	c	5451.90	I
100		5455.95	I
300		5471.96	I
70		5483.01	I
60		5485.37	I
80	c	5506.89	I
150		5524.11	I
100		5528.23	I
200		5541.94	I
80		5543.63	I
100		5550.53	I
3000	c	5589.02	I
200		5602.23	I
2000	c	5620.45	I
300		5629.94	I
1500		5642.13	I
800		5644.94	I
100		5656.00	I
60		5672.15	I
200		5687.30	I
200		5689.05	I
700		5725.31	I
500	c	5771.47	I
100		5794.65	I
80	c	5799.85	I
100	c	5814.24	I
200		5831.48	I
150	c	5836.33	I
150		5923.36	I
1000	c	5924.47	I
200		5926.29	I
600	c	5931.93	I
60		6032.36	I
60		6047.99	I
200		6065.09	I
800		6085.23	I
300		6099.39	I
500	c	6102.96	I
1000		6120.68	I
1000		6130.80	I
150	c	6132.23	I
100		6184.70	I
800		6192.66	I
600	c	6244.18	I
100		6312.18	I
100		6354.86	I
100		6356.73	I
80		6389.87	I
100		6408.83	I
1000		6455.90	I
600	c	6461.93	I
100		6470.27	I
200	c	6491.68	I
200		6526.82	I
150		6579.24	I
500	c	6625.57	I
300	c	6673.66	I
100		6687.10	I
80		6786.00	I
70		6798.63	I
60		6856.90	I
150		7002.37	I
100		7016.57	I
500	c	7086.18	I
60		7093.12	I
200		7141.28	I
200	c	7157.62	I
70		7256.08	I
100		7322.38	I
80		7329.14	I
100		7396.80	I
100	c	7402.61	I
200		7405.36	I
60		7427.15	I
150		7434.12	I
600		7452.49	I
60		7461.59	I
80		7534.95	I
800		7540.26	I
80		7543.39	I
200		7574.02	I
500		7579.26	I
90		7624.53	I
100		7684.45	I
500		7697.37	I
80	c	7698.19	I
800	c	7793.04	I
60		7798.28	I
60		7816.74	I
800		7817.72	I
100		7856.38	I
200		7861.44	I
400	d	7871.25	I
60	c	7874.76	I
70		7965.45	I
500		7999.73	I
200		8126.55	I
200		8170.55	I
150		8205.27	I
100		8206.49	I
150		8211.31	I
500	c	8237.08	I
200		8308.15	I
200		8309.16	I
60		8315.50	I
100		8531.06	I
100		8543.61	I
100	c	8707.21	I
100	c	8737.93	I
200	c	8829.82	I

TELLURIUM (Te)
Z = 52

Te I and II
Ref. 1, 344—347—C.H.C.

Tellurium (Cont.)

Intensity		Wavelength	
		Vacuum	
6		799.60	II
8		802.28	II
6		942.62	II
6		1003.73	II
6		1007.80	II
5		1014.27	II
5		1022.79	II
6		1057.00	II
8		1059.51	II
6		1068.86	II
8		1077.66	II
6		1090.11	II
6		1144.04	II
5		1153.10	II
10		1161.42	II
10		1174.34	II
12		1175.79	II
9		1208.54	II
5		1213.00	II
9		1220.98	II
9		1253.62	II
9		1270.52	II
7		1274.76	II
8		1306.53	II
10		1324.92	II
7		1336.42	II
7		1345.20	II
9		1363.24	II
8		1366.73	II
10		1374.80	II
6		1395.22	II
6		1439.52	II
6		1465.25	II
7		1489.56	II
8		1607.99	II
10		1608.41	II
10		1613.15	II
6		1638.91	II
5		1655.4	I
5		1688.5	I
6		1700.0	I
6		1701.58	II
5		1708.0	I
6		1751.0	I
5		1759.4	I
5		1775.0	I
6		1795.7	I
6		1796.3	I
10		1822.4	I
6		1825.5	I
6		1850.6	I
6		1852.1	I
6		1853.8	I
8		1857.2	I
6		1860.4	I
3		1962.88	II
7		1994.83	I
		Air	
6		2000.2	I
26000		2002.02	I
8		2070.9	I
6500		2081.16	I
18000		2142.81	I
3200		2147.25	I
360		2159.85	I
9		2208.74	I
10		2255.49	I
500		2259.02	I
10		2265.52	I
20		2373.06	II
1200		2383.26	I
1500		2385.78	I
20		2387.82	II
10		2401.63	II
10		2436.47	II
50		2438.69	II
120		2530.72	I
20		2567.82	II
10		2574.96	II
5		2576.10	II
7		2579.24	II
10		2591.12	II
10		2592.85	II
10		2605.72	II
5		2621.92	II
10		2624.86	II
20		2627.96	II
20		2641.89	II
20		2648.48	II
100		2649.66	II
40		2657.70	II
80		2661.10	II
110		2677.13	I
20		2711.58	II
6		2769.65	I
10		2841.17	II
10		2846.15	II
100		2858.29	II
20		2861.00	II
40		2868.82	II
150		2895.41	II
30		2919.89	II
50		2942.11	II
50		2946.68	II
70		2967.29	II
20		2973.67	II
50		2975.90	II
15		2997.04	II
15		3012.02	II
50		3017.58	II
20		3023.31	II
70		3047.00	II
20		3052.46	II
10		3063.16	II
15		3073.56	II
8		3104.44	II
10		3132.58	II
20		3160.66	II
100		3175.14	I
10		3189.83	II
5		3211.21	II
60		3256.80	II
30		3268.77	II
30		3282.63	II
40		3321.92	II
40		3323.11	II
60		3329.22	II
60		3352.10	II
60		3362.79	II
25		3374.10	II
150		3406.79	II
20		3419.63	II
50		3442.25	II
40		3455.12	II
20		3456.88	II
20		3480.32	II
40		3483.67	II
20		3486.11	II
50		3521.11	II
50		3552.19	II
100		3611.78	II
50		3617.57	II
40		3644.46	II
20		3679.26	II
30		3725.66	II
40		3797.22	II
20		3800.92	II
20		3905.67	II
20		3918.54	II
30		3931.49	II
20		3947.98	II
40		3969.22	II
25		3975.94	II
20		3981.77	II
50		4006.52	II
20		4011.69	II
30		4029.73	II
40		4047.17	II
30		4048.88	II
15		4073.48	II
30		4101.04	II
70		4127.32	II
30		4163.55	II
100		4169.77	II
30		4179.29	II
25		4211.31	II
80		4225.73	II
30		4246.47	II
20		4251.15	II
100		4261.11	II
30		4264.36	II
60		4273.43	II
80		4285.85	II
40		4320.90	II
30		4361.28	II
150		4364.00	II
30		4377.12	II
75		4385.10	II
60		4396.00	II
170		4478.63	II
80		4537.07	II
100		4557.78	II
70		4630.62	II
100		4641.12	II
180		4654.37	II
200		4686.91	II
100		4696.38	II
100		4706.53	II
100		4766.05	II
70		4771.56	II
100		4784.87	II
100		4827.14	II
150		4831.28	II
150		4842.90	II
130		4865.12	II
200		4866.24	II
80		4885.22	II
80		4904.44	II
60		4961.88	II
60		5000.82	II
8		5083.0	I
7		5148.7	I
50		5449.84	II
50		5487.95	II
150		5576.35	II
150		5649.26	II
100		5666.20	II
200		5708.12	II
7		5733.5	I
150		5755.85	II
8		5789.1	I
50		5936.15	II
100		5974.68	II
8	d	6273.5	I
8	h	6349.7	I
50		6367.13	II
8		6405.9	I
7	h	6456.7	I
8	h	6613.4	I
10		6648.58	II
8		6660.2	I
8	h	6690.0	I
10	h	6790.0	I
20	h	6837.6	I
20	h	6854.7	I
10		7016.06	II
10		7039.13	II
15	h	7191.1	I
10		7236.62	II
20	h	7263.5	I
8		7280.9	I
10		7289.26	II
10		7445.39	II
12		7460.98	II
15		7468.75	II
10		7481.26	II
10		7556.8	I
6		7688.61	II
15		7759.1	I
8		7818.79	II
8		7861.61	II
15		7921.69	II
15		7943.14	II
10		7950.34	II
20		7972.9	I
6		8056.15	II
30	h	8061.4	I
10	h	8082.5	I
10		8122.44	II
8		8130.39	II
8		8154.47	II
20		8186.44	II
6		8190.94	II
10		8251.5	I
15		8273.53	II
10		8276.6	I
10		8291.1	I
15		8355.8	I
10		8372.12	II
7		8469.8	I
8		8492.2	I
8		8500.8	I
12		8521.4	I
10		8535.68	II
12		8575.78	II
10		8604.63	II
8		8621.68	II
7	h	8632.1	I
15		8672.95	II
12		8701.09	I
10		8733.81	II
205		8758.18	I
12		8831.52	I
18		8851.15	I
6		8897.92	II
81		9004.37	I
18		9043.39	I
12		9196.80	I
15		9206.78	I
17		9207.64	I
12		9469.00	I
5660		9722.74	I
185		9785.54	I
109		9842.30	I
532		9868.92	I
118		9902.61	I
689		9956.30	I
37		9959.93	I
325		9977.13	I
136		9979.31	I
45		9985.85	I
5950		10051.41	I
4097		10091.01	I
104		10099.57	I
279		10106.05	I
381		10118.08	I
296		10151.06	I
397		10300.56	I
205		10323.05	I
745		10493.57	I
197		10509.86	I
1880		10918.34	I
298		11007.80	I
10200		11089.56	I
508		11163.74	I
6620		11487.23	I
280		11978.96	I
188		12566.24	I
389		12589.19	I
161		12805.50	I
400		13104.18	I
1580		13247.75	I
483		13316.63	I
217		14037.09	I
144		14072.53	I
434		14335.74	I
220		14417.46	I
1050		14513.51	I
129		14554.68	I
1480		15452.45	I
2430		15546.23	I
3760		16403.90	I
1960		17303.54	I
2780		18291.59	I
394		18777.30	I
269		19623.52	I
239		20147.54	I
1020		21043.73	I
464		21602.50	I
37		21799.64	I
74		22555.29	I
48		22755.66	I
27		23294.94	I
17		23978.70	I
25		24059.04	I
13		26428.62	I
38		26539.17	I
15		26553.74	I
7		27179.26	I

TERBIUM (Tb)
Z = 65

Tb I and II
Ref. 1 — C.H.C.

Intensity	Wavelength	
	Air	
29	2577.73	II
110	2584.61	II
29	2590.31	II
29	2591.42	II
24	2592.64	II
55	2597.71	II
40	2602.93	II
110	2608.57	II
40	2616.90	
130	2628.69	II
55	2655.96	II
50	2661.40	II
24	2661.64	II
55	2667.64	II
50	2668.86	II
140	2669.29	II
40	2674.13	II
40	2674.69	II
29	2678.15	II
40	2683.97	II
35	2687.82	II
50	2691.90	II
35	2693.05	II
55	2693.41	II
35	2695.46	II
50	2696.83	II
190	2704.07	II
130	2736.24	II
160	2759.47	II
270	2769.53	II
130	2784.49	II
180	2800.51	II
250	2802.75	II
250	2809.30	II
180	2812.64	II
190	2852.14	II
110	2857.68	II
230	2886.29	II
160	2894.45	II
320	2897.44	II
160	2898.86	II
110	2901.54	II
110	2910.30	II
160	2914.75	II
160	2915.30	II
190	2915.60	II
120	2916.24	II
120	2918.89	II
120	2924.16	II
120	2924.53	II
160	2932.89	II
150	2940.05	II
250	2956.21	II
170	2968.87	II
170	2977.78	II
110	2987.03	II
110	2988.57	II

Terbium (Cont.)

130		2996.00	II	460	d	3372.72	II	430		3743.09		480	Cw	4226.45	II
110		2999.03	II	520		3375.03	II	650		3745.04	I	260		4231.89	I
130		3005.52	II	320		3378.73	II	870		3747.17	II	480		4232.82	I
170		3009.30	II	520		3378.86	II	870		3747.34	II	300		4235.35	I
230		3010.59	II	320		3382.80	II	1100		3755.24	II	370		4255.24	I
230		3016.18	II	210		3390.60	II	430		3757.44	II	480		4258.23	II
130		3019.17	II	380		3391.28	II	430	d	3757.90	II	260		4263.66	I
170		3020.29	II	270		3398.35	II	650		3759.35	I	650		4266.34	I
110		3023.43	II	320		3399.10	II	350		3761.14	I	330		4269.69	I
170		3027.33	II	270		3400.53	II	1700		3765.14	I	220		4275.21	I
230		3031.60	II	210	d	3400.86	II	2100		3776.49	II	760	Cw	4278.52	II
230		3044.96	II	420		3402.33	II	330		3779.22		300		4285.13	II
190		3051.13	II	210		3410.40	II	600		3783.53	I	300		4289.70	I
130		3053.24	II	210		3410.68	II	410	d	3787.22	II	370		4298.36	I
460		3053.55	II	520		3413.76	II	410		3789.92	I	300		4299.90	I
130		3062.78	II	270		3416.24	II	390		3792.20	I	240		4302.95	I
230		3064.09	II	400	d	3420.34	II	600		3793.55		240		4307.18	I
110		3066.60	II	810		3430.01	II	330		3801.80	II	450		4310.42	I
230		3067.20	II	320		3433.26	II	760	d	3806.85	II	300		4311.56	I
270		3069.03	II	270		3439.72	II	1500		3830.26	I	370		4313.25	I
460		3070.05	II	520		3440.37	II	540		3833.42	I	2200		4318.83	I
270		3072.60	II	320		3444.58	II	920	d	3842.50	II	600		4322.23	I
670		3078.86	II	210		3446.40	II	370	d	3845.61	II	600		4325.83	II
480		3082.36	II	270		3449.46	II	3700		3848.73	II	3000		4326.43	I
120		3086.78	II	810		3454.06	II	450	d	3869.75	II	240		4328.90	I
250		3088.43	II	380		3460.38	II	3500	w	3874.17	II	600		4332.12	I
480		3089.58	II	230	d	3462.97	II	330		3883.34	I	870		4336.43	I
230		3102.54	II	620		3468.03	II	480		3888.22	I	600		4337.64	I
480		3102.96	II	270		3471.73	II	490		3894.64	I	1700		4338.41	I
290		3117.89	II	270		3472.37	II	330		3895.99	I	700		4340.62	I
290		3119.62	II	810	d	3472.79	II	330		3896.58	II	430	Cw	4342.53	I
230		3121.94	II	210		3473.00	II	330		3897.89	I	430	d	4353.20	II
230		3123.05	II	380		3480.17	II	2400		3899.20	II	280		4356.09	I
160		3124.54	II	230		3483.04	II	1600		3901.33	I	870		4356.81	I
110		3131.35	II	230		3483.69	II	480		3908.06	I	280		4360.16	I
250		3134.26	II	290	d	3489.51	II	380		3909.14	I	220		4367.30	II
440		3139.64	II	210	d	3492.00	II	330		3909.55	I	220		4372.02	I
190		3140.06	II	270		3494.21	II	650		3915.43	I	330		4382.45	I
230		3145.22	II	270		3495.36	II	480		3919.52	II	300		4388.23	I
150		3146.67	II	810		3500.84	II	300		3922.10	II	260		4390.91	I
310		3147.04	II	570		3507.45	II	480		3922.74	II	200		4416.27	II
310		3147.15	II	5700		3509.17	II	760		3925.45	II	140		4420.19	I
310		3148.71	II	380		3510.10	II	650		3935.24	II	350		4423.10	I
120		3155.62	II	320		3513.10	II	810	d	3939.52	II	110		4432.72	I
130		3162.42	II	570		3519.76	II	650		3946.89	II	240		4436.12	I
290		3162.93	II	1300		3523.66	II	350	d	3958.36	II	110		4439.38	I
190		3165.74	II	380		3525.14	II	2200	d	3976.84	II	240		4448.04	I
380		3167.52	II	440		3525.61	II	1800		3981.87	II	110		4467.69	I
140		3168.32	II	440		3536.32	II	300		3983.85	II	430		4493.07	I
230		3169.84	II	570		3537.94	II	350		3999.40	II	45	d	4509.04	II
190		3173.76	II	1100		3540.24	II	350	d	4002.19	II	150	h	4511.52	I
380		3174.66	II	810		3543.89	II	970		4002.59	II	43		4512.96	II
380		3180.54	II	310	d	3551.03		1900		4005.47	II	75		4514.31	II
140		3183.88	II	320		3551.96	II	300		4010.04	I	45		4519.72	II
480		3187.26	II	460	d	3558.77		760		4012.75	II	45		4525.01	II
290		3188.03	II	3200		3561.74	II	330		4013.26	I	45		4529.76	
190		3194.69	II	480		3562.90	II	370		4019.14	II	45	h	4531.83	II
380		3195.60	II	570		3565.74	II	540		4020.47	II	45		4534.13	I
480		3199.56	II	810		3567.35	II	220		4022.88	I	45		4537.14	I
1100		3218.93	II	4200		3568.52	II	370		4024.77	I	45		4537.23	I
1200		3219.98	II	1600		3568.98	II	520		4031.66	II	110		4549.07	I
250		3230.03	II	320		3572.07	II	870		4032.28	I	45		4549.72	II
250		3231.06	II	1100		3579.20	II	2100		4033.03	II	110		4550.45	I
210	d	3239.60	II	710		3585.03	II	350		4036.22	I	110		4556.46	I
250		3240.00	II	570		3587.44	II	210		4038.86	I	55		4562.24	II
480		3252.32	II	810		3596.38	II	300		4051.86	II	110		4563.69	II
250		3262.97	II	440		3598.06	II	300		4052.87	II	30		4564.85	II
230	d	3263.87	II	1600		3600.44	II	430		4054.12	I	55		4573.19	II
230		3264.90	II	320		3604.90	II	410		4060.37	I	210		4578.69	II
400		3266.40	II	320		3611.33	II	220		4060.87	II	65		4584.84	II
250		3274.14	II	320		3614.63	II	1300		4061.58	I	65		4591.56	II
250		3274.33	II	320	d	3615.66	II	220		4063.89	II	45		4592.38	I
210		3277.32	II	320		3616.58	II	390		4066.22	II	45	h	4604.10	II
760		3280.31	II	380		3617.86	II	260		4075.22	I	30		4611.96	I
760		3281.40	II	380		3619.73	II	390		4081.24	I	45	h	4615.92	II
520		3283.10	II	810		3625.54	II	210		4086.60	I	27		4617.49	I
1000		3285.04	II	570		3626.50	II	210		4092.19	I	30		4619.36	II
310		3287.55	II	380		3629.44	II	260		4094.37	II	75	d	4626.32	II
310		3291.56	II	670		3633.29	II	260		4094.49	I	95		4626.94	II
1500		3293.07	II	670		3638.46	II	260		4103.90	II	65		4632.07	I
210		3295.33	II	670		3641.66	II	650		4105.37	I	65	h	4636.59	I
310		3298.66	II	440		3647.06	II	300		4112.50	I	30		4636.99	II
210		3304.95	II	570		3647.75	II	260		4119.92	I	85		4641.00	II
420	d	3307.44	II	2300		3650.40	II	280		4143.51	I	210		4641.98	II
210		3308.51	II	810		3654.88	II	1100		4144.41	II	260	Cw	4645.31	II
210		3314.38	II	2000		3658.88	II	350		4158.53	I	80		4647.23	I
340	d	3321.15	II	450		3663.12	II	240		4169.09	I	60		4658.38	I
420		3322.28	II	3800		3676.35	II	240		4169.32	I	20		4658.73	
210		3323.38	II	300		3677.89	II	240		4171.05	I	80		4662.79	I
210		3323.89	II	810		3682.26	II	240		4172.60	I	50	c	4665.45	I
3800		3324.40	II	320		3688.15	II	240		4172.82	I	40		4669.40	I
520		3329.08	II	610		3691.15	II	260		4173.47	I	80		4676.90	I
210		3334.48	II	300		3692.95	II	240		4186.21	I	70	c	4681.87	I
250		3336.70	II	450		3693.58	I	300		4187.16	I	50		4682.52	I
310		3338.03	II	320		3696.85	II	390		4196.74	I	25	c	4682.79	II
250		3339.00	II	450		3700.12	I	450		4201.00	II	80		4688.63	II
210		3347.27	II	4700		3702.86	II	650		4203.74	I	80		4693.11	II
210		3348.07	II	300		3703.12	I	600		4206.49	I	30	h	4693.39	II
760		3349.42	II	2400		3703.92	II	300	Cw	4213.50	I	200		4702.41	II
320		3362.25	II	370		3709.30	II	300		4214.42	II	110		4707.94	II
760		3364.93	II	1000	d	3711.76	II	480		4215.09	I	40	w	4716.07	II
230	d	3370.61	II	300		3719.45	II	300		4217.56	I	40		4728.16	II
320		3371.50	II	650		3729.91	II	260		4219.16	I	60	Cw	4734.20	I
520		3372.36	II	430		3732.39	II	260		4224.28	I	80		4739.93	I

Terbium (Cont.)

Intensity		Wavelength	Spectrum
70		4747.80	I
410	Cw	4752.53	II
40		4758.44	II
40		4760.19	II
30		4762.37	II
25		4764.47	II
35		4778.36	II
35		4778.80	II
180		4786.78	I
40	Cw	4789.91	II
30		4801.87	II
100		4813.77	I
60		4837.59	II
25		4840.39	I
30	c	4842.69	II
30		4844.89	II
30		4854.81	I
20		4856.54	II
30		4858.87	II
80		4875.57	II
25		4876.12	II
80		4881.15	II
29		4894.33	
95		4915.90	I
35		4924.09	I
35		4926.83	I
50		4928.93	I
65		4931.79	I
29		4970.99	II
29		4971.42	I
29		4973.04	I
29		4980.16	II
29		4980.56	I
85		4993.82	II
50		4995.84	II
55		4997.95	I
29		5006.10	II
50		5022.16	I
29		5024.24	II
29		5024.65	I
50		5033.12	I
50	w	5042.06	II
55		5054.30	I
55		5065.79	I
110		5078.25	I
24		5080.05	II
24		5081.11	I
75		5089.12	II
24		5089.66	I
24		5101.09	I
24		5108.56	I
35		5118.39	I
24		5120.18	I
50	w	5131.69	I
50	w	5141.08	II
50		5147.58	I
24		5164.27	I
29		5170.13	I
24		5170.61	I
50		5176.51	I
50		5179.97	I
50		5184.59	I
85		5186.13	I
50		5188.48	I
50		5198.86	I
35	w	5202.77	I
40		5204.55	I
40		5207.97	I
40		5214.28	I
40		5221.99	I
120		5228.12	I
40		5235.11	I
75		5248.71	I
75	w	5262.11	II
24		5275.03	I
75		5281.05	I
65		5304.72	I
29		5308.19	I
29		5309.46	I
110		5319.23	I
35		5331.04	I
65	w	5337.90	I
35	d	5338.59	I
24		5347.83	II
160		5354.88	I
75		5369.72	I
75		5375.98	I
29	d	5402.06	II
29		5413.65	I
29		5416.20	I
50		5424.10	II
29	c	5426.43	I
35		5443.38	I
29		5457.00	I
55		5459.81	I
29	w	5470.34	II
24		5481.45	I
55		5509.61	I
50		5514.54	I
65		5524.12	I
24	c	5525.62	I
35		5565.93	I
29	c	5638.80	I
29	c	5685.74	II
40	c	5686.48	I
85	c	5747.58	I
24		5762.66	I
24		5785.18	II
75		5795.64	I
75		5803.13	II
65		5815.36	I
29		5842.97	I
65		5851.07	I
65		5870.62	I
35		5898.84	I
24		5902.40	I
35		5904.71	I
65	c	5920.78	I
50	c	5939.38	I
35		5940.17	I
24		5951.17	I
75		5967.34	II
29		6038.97	I
29		6039.38	I
24		6104.29	II
24	c	6292.43	I
35		6331.68	II
24		6334.91	II
24		6446.87	II
35	Cw	6518.68	I
24	c	6574.04	II
35		6581.82	I
30		6607.17	II
90		6677.94	II
40	Cw	6702.61	I
20	c	6706.79	II
30		6785.12	II
130		6794.58	II
40		6874.18	II
55		6896.37	II
45	h	6899.95	I
40		6901.98	I
9		7005.99	II
17		7082.85	II
11		7089.22	II
11		7112.69	I
10		7187.48	I
10	h	7195.89	II
65		7204.28	I
19	h	7234.98	I
40		7257.73	I
17		7311.57	I
45		7348.88	II
10		7398.27	II
15	h	7424.24	II
10	h	7429.62	II
9		7472.15	I
22		7484.54	I
9		7495.45	I
45		7496.12	I
17		7499.69	II
27		7511.40	I
9		7519.77	II
6		7557.59	II
27	h	7582.03	II
27		7587.49	I
45		7590.24	I
65		7596.44	I
17	h	7601.18	II
17		7616.01	II
22	h	7624.05	I
30		7627.81	I
9	h	7639.05	II
8		7672.72	II
8	h	7694.74	II
22	h	7706.16	II
22	h	7726.97	II
30		7737.63	I
22		7793.20	I
8		7807.33	II
16		7832.91	II
30		7855.79	II
15		7864.99	I
6	h	7885.70	I
6		7913.11	I
27		7927.90	II
13		7955.31	I
11		7998.03	I
17		8001.04	I
13	h	8010.16	II
30		8025.42	I
6		8053.80	I
19		8067.35	II
30		8085.06	II
27		8164.17	I
13		8171.70	I
65		8194.82	II
95		8212.57	I
11		8214.33	I
8		8259.08	I
40		8450.06	II
8	h	8465.80	II
13		8502.70	I
30	h	8511.80	I
45		8583.45	II
30		8603.40	I
9		8678.25	I
65		8765.74	II

Tb IV
Ref. 302 — J.R.

Intensity	Wavelength	
Vacuum		
30	1176.58	IV
30	1192.01	IV
70	1200.58	IV
50	1213.94	IV
80	1221.22	IV
500	1235.04	IV
1000	1259.40	IV
300	1301.48	IV
300	1308.30	IV
600	1311.70	IV
700	1315.12	IV
500	1325.56	IV
1000	1327.67	IV
100	1367.56	IV
400	1367.71	IV
700	1369.64	IV
1000	1373.86	IV
400	1376.46	IV
200	1378.23	IV
300	1381.00	IV
100	1382.83	IV
20	1389.92	IV
200	1516.17	IV
50	1530.10	IV
5000	1595.39	IV
2000	1633.19	IV
300	1649.38	IV
400	1654.75	IV
400	1667.58	IV
200	1672.55	IV
400	1681.98	IV
5	1684.46	IV
100	1685.37	IV
400	1691.95	IV
10	1695.23	IV
300	1698.36	IV
30	1701.60	IV
50	1704.79	IV
20	1705.05	IV
3	1943.94	IV
50	1970.90	IV
Air		
2000	2027.79	IV
200	2029.22	IV
400	2048.88	IV
200	2078.83	IV
1000	2089.98	IV
1000	2332.54	IV
100	2436.01	IV

THALLIUM (Tl)
Z = 81

Tl I and II
Ref. 1, 195, 348, 354 — C.H.C.

Intensity		Wavelength	
Vacuum			
3		650.90	II
5	r	670.87	II
4		674.10	II
15	r	696.30	II
5	r	709.23	II
10	r	817.18	II
5	r	836.34	II
8	r	1018.85	II
10	r	1049.73	II
8	r	1050.30	II
5	r	1074.97	II
10	r	1130.17	II
15	r	1162.55	II
10	r	1167.43	II
10	r	1183.41	II
12	r	1194.84	II
8		1231.81	II
5	r	1246.00	II
15	r	1307.50	II
8	r	1310.20	II
25	r	1321.71	II
8	r	1330.40	II
10	r	1373.52	II
1		1423.2	I
8	r	1489.65	I
5		1490.50	II
10	r	1499.30	II
10	r	1507.82	II
15	r	1561.58	II
10	r	1568.57	II
7	r	1593.26	II
5	h	1616.	I
1		1650.2	I
5		1685.40	I
1		1728.	I
10	r	1792.76	II
12	r	1814.85	II
3	h	1847.	I
8		1892.72	II
25	r	1908.64	II
Air			
100	r	2007.56	I
2		2209.75	I
100	r	2210.71	I
3		2287.6	I
30		2298.04	II
140		2315.98	I
900	h	2379.69	I
8		2451.83	II
6		2469.03	II
1		2508.2	I
20		2530.86	II
700		2580.14	I
60		2608.99	I
80		2665.57	I
420		2709.23	I
50	h	2710.67	I
4400	d	2767.87	I
280		2826.16	I
10		2849.80	II
2800		2918.32	I
440		2921.52	I
5		3029.01	II
20		3091.56	II
15		3185.51	II
15		3186.56	II
15		3187.74	II
1200		3229.75	I
15		3291.01	II
12		3319.91	II
12		3321.04	II
8		3322.25	II
15		3369.15	II
8		3381.00	II
6		3381.80	II
6	d	3460.48	II
20000		3519.24	I
5000		3529.43	I
8		3540.08	II
9		3560.68	II
5		3567.67	II
12000	Cw	3775.72	I
8		3793.95	II
10		3832.30	II
6		3869.15	II
10		3887.15	II
8		4223.05	II
20		4274.98	II
40		4306.80	II
2		4359.9	I
8		4490.77	II
20		4737.05	II
15		4981.35	II
25		5078.54	II
25		5152.14	II
6		5181.95	II
6		5183.10	II
18000		5350.46	I
15	d	5384.85	II
7		5409.92	II
10		5410.97	II
25		5949.48	II
10		6179.98	II
8	d	6239.03	II
10		6378.32	II
16	h	6549.84	I
6	h	6713.80	I
10		6966.5	II
3		7493.6	I
2		7678.93	I
10		7815.80	I
8		8130.0	I
20		8373.6	I
8		8445.8	II
10		8474.27	I
8		8632.9	II
10		8664.1	II
4		8850.4	I
5		8976.75	I
3		9038.4	I
20		9130.	II
20		9130.5	I
2	h	9183.1	I
4		9225.	II
2	h	9252.6	I
3		9254.	II
40		9509.4	I
10		9863.4	I
20		9930.4	I
2		9937.4	I
30		10011.9	I
40		10488.80	I
5		11101.61	I
4		11483.7	I
1000		11512.82	I

Thallium (Cont.)

5	11592.9	I
15	12491.8	I
150	12736.4	I
700	13013.2	I

Tl III
Ref 355 — C.H.C.

Intensity	Wavelength	
	Vacuum	
7	1231.57	III
10	1266.33	III
4	1332.36	III
10	1477.14	III
4	1506.37	III
[illegible]	[illegible]	III
8	1660.05	III
	Air	
6	3163.53	III
3	3300.80	III
9	3456.34	III
4	3507.41	III
6	3933.05	III
2	3946.02	III
7	4109.85	III
4	4155.75	III
6	4269.81	III
2	4380.57	III
4	5086.99	III
4	5362.40	III
2	5499.4	III
5	5927.8	III
4	8001.	III

Tl IV
Ref. 356 — C.H.C.

Intensity	Wavelength	
	Vacuum	
7	531.26	IV
10	570.49	IV
4	597.01	IV
1	868.99	IV
3	912.74	IV
8	917.31	IV
30	1028.69	IV
20	1034.73	IV
20	1036.61	IV
10	1049.48	IV
10	1057.56	IV
20	1068.04	IV
20	1070.47	IV
30	1079.68	IV
5	1079.70	IV
2	1092.90	IV
4	1094.41	IV
6	1099.60	IV
4	1125.52	IV
5	1139.30	IV
3	1144.07	IV
3	1225.45	IV
6	1273.03	IV
3	1304.55	IV
6	1323.66	IV
7	1337.10	IV
7	1358.56	IV
5	1374.62	IV
7	1377.75	IV
8	1404.60	IV
5	1412.93	IV
6	1434.72	IV
5	1449.37	IV
5	1883.2	IV
3	1974.6	IV

THORIUM (Th)
Z = 90

Th I and II
Ref. 1, 97, 98, 434 — J.G.C. and R.Z.

Intensity	Wavelength	
	Air	
100	2326.926	II
190	2377.84	II
90	2404.504	II
100	2413.409	II
30	2439.433	I
5	2532.894	I
150	2547.901	II
500	2565.593	II
270	2566.588	II
200	2576.688	II
230	2589.059	II
230	2597.047	II
230	2600.882	II
100	2609.855	II
230	2618.91	II
270	2623.448	II
270	2625.737	II
55	2628.812	II
270	2641.488	II
170	2650.583	II
150	2658.663	II
50	2680.692	I
360	2684.288	II
480	2692.415	II
100	2695.553	II
270	2703.958	II
170	2708.176	II
230	2721.691	II
170	2722.380	II
250	2729.327	II
250	2732.808	II
28	2735.834	II
520	2747.156	II
100	2749.530	II
410	2752.166	II
130	2760.391	II
100	2765.123	II
270	2768.841	II
200	2770.816	II
70	2773.951	II
50	2774.066	II
70	2778.706	II
35	2791.496	I
90	2794.255	II
70	2797.737	II
55	2799.114	II
110	2807.827	II
180	2808.998	II
70	2814.319	II
45	2816.071	II
100	2819.322	II
100	2820.336	II
100	2822.025	II
170	2826.855	II
70	2830.442	II
800	2832.315	II
1200	2837.295	II
320	2842.812	II
100	2848.084	I
270	2851.260	II
50	2854.343	I
30	2860.490	I
220	2861.42	II
35	2868.461	I
70	2869.916	II
550	2870.406	II
30	2878.657	I
40	2882.511	II
320	2884.289	II
360	2885.049	II
360	2887.817	II
40	2892.172	II
250	2899.720	II
45	2903.167	II
50	2908.506	I
200	2910.594	II
90	2911.320	II
90	2912.009	II
140	2919.840	II
250	2925.050	II
250	2928.254	II
50	2931.281	I
55	2934.135	II
100	2936.086	I
35	2940.589	II
340	2942.860	II
100	2943.729	I
150	2949.068	II
80	2950.438	II
35	2955.849	II
170	2957.580	II
28	2959.853	I
28	2963.607	I
270	2968.686	II
110	2971.481	II
220	2974.011	II
50	2976.104	I
55	2980.334	II
160	2985.243	II
360	2988.232	II
150	2991.062	II
110	2996.986	II
50	3002.686	I
30	3004.248	I
180	3008.497	II
50	3010.736	I
20	3018.644	I
50	3021.056	I
150	3026.575	II
40	3030.487	I
370	3034.065	II
170	3035.110	II
85	3038.598	II
130	3045.564	II
420	3049.092	II
30	3056.692	I
220	3061.699	II
450	3067.729	II
370	3072.114	II
670	3078.828	II
480	3080.217	II
240	3088.470	II
130	3090.093	II
50	3093.711	I
140	3097.266	II
200	3102.664	II
510	3108.296	II
50	3115.538	I
100	3116.263	I
510	[illegible]	II
510	3122.963	II
370	3124.387	II
480	3125.507	II
150	3131.070	II
100	3136.216	I
420	3139.306	II
420	3142.835	II
310	3146.044	II
150	3150.455	II
310	3154.300	II
50	3157.221	I
30	3161.364	I
140	3166.099	II
110	3169.328	II
420	3175.726	II
270	3179.048	II
1100	3180.193	II
310	3184.948	II
770	3188.233	II
85	3191.221	II
55	3192.585	I
55	3195.689	I
30	3202.520	I
170	3210.308	II
55	3214.380	I
560	3221.292	II
30	3223.168	I
560	3229.009	II
110	3230.868	II
480	3235.84	II
590	3238.116	II
240	3241.108	II
110	3244.448	I
280	3251.915	II
910	3256.274	II
180	3257.366	I
910	3262.668	II
180	3267.003	II
110	3272.027	I
30	3278.733	I
50	3281.048	I
130	3285.752	I
620	3287.789	II
910	3291.739	II
620	3292.520	II
240	3297.832	II
240	3301.650	I
480	3304.238	I
130	3309.365	I
50	3314.790	I
30	3318.390	I
510	3321.450	II
390	3324.752	II
840	3325.120	II
55	3328.255	II
250	3330.476	I
620	3334.604	II
620	3337.870	II
30	3342.073	II
180	3346.557	II
310	3348.768	I
980	3351.228	II
310	3354.179	II
620	3358.602	II
75	3361.738	II
390	3367.819	II
250	3374.974	I
390	3378.573	II
130	3380.859	I
310	3385.531	II
310	3386.501	II
110	3387.920	I
1300	3392.035	II
200	3396.727	I
250	3398.544	I
200	3405.558	I
250	3413.012	I
50	3417.497	I
390	3421.210	I
270	3423.989	I
50	3428.622	I
980	3433.998	II
770	3435.976	II
340	3438.949	II
110	3442.578	I
50	3446.547	I
130	3451.702	I
50	3457.068	I
340	3462.850	II
130	3465.924	II
390	3468.219	II
1300	3469.920	II
170	3471.218	I
250	3479.173	II
70	3480.052	I
200	3486.552	I
100	3489.184	I
270	3493.518	II
70	3496.810	I
130	3498.621	I
70	3503.786	I
50	3506.645	I
110	3511.157	I
110	3518.404	I
70	3521.059	I
70	3526.633	I
140	3531.450	I
670	3539.587	II
180	3544.018	I
170	3549.595	I
140	3551.401	I
200	3555.013	I
530	3559.451	II
110	3563.375	I
70	3569.820	I
200	3576.557	I
100	3583.101	I
170	3589.750	I
270	3592.780	I
270	3598.120	I
390	3601.034	II
170	3608.377	I
980	3609.445	II
200	3612.427	I
480	3615.133	II
400	3617.118	II
270	3623.970	II
390	3625.627	II
140	3632.831	I
270	3635.943	I
70	3638.644	I
210	3642.248	I
170	3649.735	I
50	3658.808	I
100	3659.629	I
220	3663.202	I
140	3668.140	I
280	3669.968	I
700	3675.567	II
150	3682.486	I
50	3688.658	I
100	3690.624	I
170	3692.566	I
180	3698.105	I
50	3703.229	I
340	3706.767	I
280	3711.305	II
590	3719.435	I
50	3719.836	I
770	3721.825	II
110	3727.902	I
50	3733.672	I
1300	3741.183	II
310	3747.539	I
650	3752.569	II
140	3757.694	I
110	3765.240	I
180	3770.056	I
50	3772.649	I
85	3776.271	I
50	3780.966	I
340	3785.600	II
100	3789.167	I
85	3795.386	I
50	3800.197	I
590	3803.075	I
50	3807.273	I
340	3813.068	II
50	3818.685	I
75	3825.133	I
450	3828.384	I
70	3836.584	I
840	3839.746	II
280	3841.960	II
100	3846.887	I
85	3852.135	I
390	3854.511	II
140	3859.840	II
450	3863.405	II
100	3869.663	I
210	3875.374	I
140	3879.644	I
100	3886.915	I
340	3895.419	I
50	3901.661	I
110	3903.102	I
170	3905.186	II
50	3908.750	I
85	3911.909	I
50	3916.417	I

Thorium (Cont.)

110		3919.023	I
140		3925.093	I
590		3929.669	II
200		3932.911	I
140		3937.040	II
50		3942.072	I
50		3948.030	I
200		3948.964	II
50		3952.760	I
110		3959.300	I
390		3967.392	I
200		3972.155	I
150		3980.089	I
110		3991.730	I
530		3994.549	II
50		3998.061	I
240		4003.309	II
250		4007.021	II
220		4008.210	I
220		4009.056	I
280		4012.495	I
4200		4019.129	II
210		4025.656	II
140		4027.009	I
250		4030.842	I
250		4036.047	I
240		4036.565	II
240		4041.204	II
55		4048.287	I
110		4050.887	I
140		4059.253	I
250		4063.407	I
300		4069.201	II
100		4069.461	I
55		4075.503	I
110		4081.368	I
85		4085.434	I
700		4086.520	II
70		4088.726	I
700		4094.747	II
150		4100.341	I
270		4105.330	II
840		4108.421	II
240		4112.754	I
280		4115.758	I
1100		4116.713	II
30		4123.600	I
200		4127.411	I
110		4131.002	I
340		4132.753	II
200		4134.067	I
220		4140.235	II
250		4142.701	II
200		4148.182	II
450		4149.986	II
110		4158.535	I
140		4165.766	I
620		4178.060	II
250		4179.714	II
30		4184.138	I
130		4193.017	I
620		4208.890	II
130		4210.923	I
28		4214.828	I
55		4220.065	I
55		4227.387	I
30		4230.824	I
85		4235.463	I
20		4241.094	I
30		4247.989	II
110		4253.538	I
70		4256.254	I
110		4260.333	I
28		4269.942	I
280		4273.357	II
480		4277.313	II
700		4282.042	II
28		4288.669	I
55		4297.306	I
85		4299.839	I
100		4307.176	I
200		4309.991	II
55		4315.254	I
110		4318.416	I
30		4325.274	I
28		4330.844	I
130		4337.277	I
85		4342.256	II
130		4344.326	II
55		4349.072	I
55		4354.484	I
85		4359.372	I
85		4365.930	I
85		4374.123	I
1300		4381.860	II
1100		4391.110	II
55		4392.974	I
55		4401.580	I
85		4408.882	I
210		4412.741	II
28		4416.845	I
50		4422.048	I
250		4432.963	II
140		4440.866	II
25		4452.565	I
85		4458.002	I
220		4465.341	II
30		4475.221	I
75		4482.169	I
50		4489.664	I
110		4498.940	I
55		4505.216	I
280		4510.527	II
70		4521.194	I
22		4530.319	I
40		4535.255	I
30		4545.915	II
70		4555.812	I
40		4563.660	I
65		4570.972	I
50		4588.426	I
75		4595.421	I
26		4612.554	II
30		4621.163	I
140		4631.761	II
140		4631.761	II
30		4641.254	I
140		4651.558	II
23		4663.202	I
50		4669.984	I
65		4676.056	I
50		4686.195	I
140		4694.091	II
50		4703.990	I
20		4712.841	I
90		4723.438	I
30		4729.128	I
190		4740.529	II
140		4752.414	II
13		4766.600	I
50		4778.294	I
20		4786.531	I
40		4789.387	I
45		4808.134	I
20		4819.193	I
26		4822.855	I
40		4826.700	I
45		4831.121	I
50		4840.843	I
30		4848.362	I
15		4852.868	I
40		4858.333	II
280		4863.163	II
40		4872.917	I
26		4878.733	I
45		4894.955	I
20		4907.209	I
240		4919.816	II
18		4927.780	I
40		4939.642	I
60		4947.575	II
50		4954.659	II
30		4965.731	I
35		4975.950	II
24		4985.372	I
50		5002.097	I
50		5002.097	I
50		5015.889	II
260		5017.255	II
20		5029.892	I
24		5039.230	I
50		5044.719	I
240		5049.796	II
85		5055.347	II
70		5058.562	II
110		5067.974	I
30		5081.446	I
50		5090.051	I
50		5098.043	II
40		5101.130	I
50		5110.867	II
30		5115.044	I
10		5125.950	I
20		5134.746	I
95		5143.267	II
120		5148.211	II
50		5151.612	I
50		5154.243	I
85		5158.604	I
70		5160.730	I
20		5168.922	I
50		5176.961	I
35		5183.990	II
50		5190.871	II
50	h	5195.814	I
50		5198.800	I
95		5199.164	I
50		5211.230	I
95		5216.596	II
50		5218.528	II
35		5219.110	I
110		5231.160	I
85		5233.225	II
85		5233.229	II
95		5247.654	II
10		5255.573	I
35		5258.360	I
12		5266.710	I
70		5277.501	II
15		5281.069	I
10		5294.397	I
30		5297.743	I
30		5307.466	II
35		5312.002	I
20		5317.494	I
60		5325.145	II
50		5326.976	I
20		5330.080	I
60		5343.581	I
14		5351.126	I
30		5358.707	I
20		5369.281	I
30		5378.836	I
70		5390.466	II
50		5392.572	I
20		5399.175	I
24		5417.486	I
60		5425.678	II
50		5435.893	II
40		5449.479	II
30		5462.615	II
15		5470.759	I
24		5484.147	II
10		5496.137	I
19		5504.302	I
35		5509.994	I
12		5524.584	I
50		5539.262	I
70		5539.911	II
35		5548.176	I
50		5558.342	I
60		5564.203	II
40		5573.354	I
60		5587.026	I
24		5595.064	I
50		5604.515	II
35		5615.320	I
7		5630.297	I
70		5639.746	II
7		5648.991	I
12		5657.925	I
15		5667.128	I
20		5677.053	I
10		5685.192	I
65		5700.918	II
95		5707.103	II
50		5720.183	I
30		5732.975	II
24		5742.084	II
30		5749.388	II
70		5760.551	I
15		5773.946	I
20		5789.645	I
35		5804.141	I
19		5815.422	II
10		5832.370	I
10		5845.919	I
15		5854.121	I
15		5868.373	I
10		5878.933	I
8		5891.451	I
15		5899.844	I
20		5914.387	II
19		5925.893	II
10		5937.162	I
10		5944.648	I
8		5957.587	I
30		5973.665	I
85		5989.044	II
24		5994.129	I
21		6007.072	I
30		6015.426	II
17		6021.036	I
17		6037.698	I
24		6044.431	II
10		6053.381	I
5		6061.536	I
30		6077.106	I
30		6087.262	II
24		6099.083	I
30		6104.580	I
40		6112.837	I
30		6120.557	II
10		6124.480	I
14		6151.993	I
10		6161.354	I
60		6169.822	I
50		6182.622	I
12		6191.906	I
24		6193.858	II
12		6203.493	I
12		6224.528	I
24	h	6234.856	I
8		6240.954	I
10		6257.424	I
21		6261.063	I
21		6261.418	I
50		6274.116	II
50		6274.117	II
30		6279.172	II
8		6291.192	I
10		6303.251	II
8		6317.185	I
21		6327.278	I
35		6342.860	I
50		6355.911	II
14		6369.140	I
40		6376.931	I
30		6411.899	I
24		6413.615	I
15		6437.762	I
15		6450.005	I
60		6457.283	I
50		6462.614	I
5		6466.717	I
14		6490.738	I
5		6501.992	I
20		6512.364	I
5		6522.044	I
50	h	6531.342	I
6		6554.160	I
5		6558.876	I
3		6565.070	I
5		6577.215	I
24		6583.907	I
24		6588.540	I
24		6593.940	I
24		6605.416	II
24		6619.946	II
24		6619.947	II
21		6644.650	II
6		6658.678	II
30		6662.269	I
6		6674.697	I
3		6683.367	I
8		6692.724	II
5		6697.712	I
16		6727.459	I
3		6735.126	I
5		6742.884	I
20		6756.453	I
6		6765.677	I
15		6778.313	I
15		6780.413	I
6		6791.236	I
3		6798.747	I
3		6809.511	I
5		6823.509	I
11		6829.036	I
14		6834.925	I
4		6862.873	I
5		6866.367	I
8		6874.754	I
20		6889.303	II
3		6908.988	I
24		6911.227	I
5		6916.129	I
5		6936.652	I
35		6943.611	I
5		6954.657	I
15		6965.947	I
3		6981.086	I
55		6989.656	I
24		6993.038	II
18		7000.806	I
18		7000.806	I
3		7015.319	I
10		7018.569	I
3		7026.462	I
7		7036.281	I
30		7045.795	II
15		7053.619	II
6		7060.654	I
24		7075.333	II
30		7084.171	I
24		7089.339	II
10		7100.512	II
3		7109.861	I
11		7124.562	I
3		7132.613	I
5		7148.560	I
10		7154.954	I
30		7168.896	I
15		7173.373	I
40		7191.132	II
7		7200.046	I
35		7208.006	I
11		7212.69	I
10		7217.755	II
11		7218.054	I
3		7242.355	I
5		7255.354	I
3		7270.558	I
7		7284.904	I
5		7298.143	I
11		7305.405	II
5		7315.067	I
7		7324.808	I
5		7339.606	I
8		7341.152	I
5		7361.349	I
5		7384.175	I
18		7385.501	I
3		7393.431	II
5		7402.252	I
3		7418.550	I

Thorium (Cont.)

21	7428.940	I
10	7430.254	I
3	7444.749	I
2	7462.993	I
10	7481.355	I
2	7493.427	I
50	7525.508	II
7	7549.314	I
18	7567.740	I
12	7585.69	I
12	7585.792	I
4	7598.204	I
2	7607.824	I
5	7627.176	I
3	7636.176	I
30	7647.380	I
4	7658.324	I
7	7676.219	I
21	7685.305	I
4	7710.269	I
4	7728.951	I
10	7731.72	II
4	7771.948	I
15	7787.79	II
15	7788.937	I
5	7798.360	I
4	7810.625	I
21	7817.771	I
8	7834.459	II
15	7847.540	I
4	7864.023	I
12	7865.95	I
6	7886.284	I
11	7900.31	I
4	7937.732	I
11	7941.72	I
5	7954.594	I
7	7972.598	I
24	7978.974	I
11	7987.97	I
5	7993.680	I
2	8014.502	I
5	8024.253	II
11	8032.433	I
11	8062.64	I
5	8085.220	I
5	8093.626	I
5	8129.407	I
11	8138.477	I
18	8143.139	I
12	8159.729	I
10	8163.125	II
7	8169.788	I
15	8186.914	I
12	8203.199	II
2	8231.408	I
5	8252.395	I
3	8259.512	I
18	8275.629	I
3	8292.529	I
15	8320.857	I
30	8330.451	I
4	8358.726	I
6	8387.104	II
18	8403.767	II
15	8416.729	I
12	8421.227	I
21	8446.509	I
5	8464.230	I
18	8478.360	I
5	8510.621	I
5	8516.557	I
5	8539.795	I
6	8554.946	I
11	8573.122	I
12	8591.838	I
3	8621.325	I
3	8638.363	I
10	8665.487	I
4	8668.116	I
2	8701.127	I
5	8709.236	I
8	8732.401	II
18	8748.033	I
15	8758.244	I
8	8775.573	I
4	8792.058	I
3	8804.590	I
5	8841.185	I
18	8842.073	II
15	8868.834	I
4	8875.233	I
2	8907.038	I
5	8955.848	I
15	8957.97	II
40	8967.641	I
3	8987.408	I
5	9031.819	I
25	9048.252	I
5	9063.953	I
6	9094.831	I
3	9107.225	I
2	9118.140	I
3	9170.825	I
10	9203.963	I
10	9266.208	I
10	9276.276	I
10	9289.563	I
4	9317.722	I
7	9340.706	I
15	9399.085	I
10	9431.603	I
15	9461.030	I
15	9474.882	I
15	9495.501	I
15	9497.191	I
10	9505.392	I
10	9561.24	I
8	9613.689	II
12	9632.647	I
10	9664.700	I
[illegible]	9670.100	I
15	9700.564	I
15	9746.46	I
10	9812.70	I
10	9826.45	I
20	9833.42	I
15	10039.364	I
15	10089.138	I
15	10133.56	II
15	10419.57	II
15	10556.45	I
15	10723.92	II
20	10726.93	I
20	10942.24	II
15	11051.90	I
30	11230.259	I
20	11354.719	I
15	11703.46	I
15	11864.25	I
15	11940.64	I
20	11984.67	II
15	12018.72	I
20	12127.30	I
20	12194.16	II
15	12206.89	I
20	12231.94	I
15	12338.00	I
15	12477.30	I
20	12646.54	I
15	12866.64	I
15	12940.65	II
15	12959.82	I
15	13145.90	II
15	13565.67	I
15	14090.25	I
15	14168.67	I
15	14424.54	I
15	14618.98	I
15	14654.91	I
15	14940.49	II
15	15240.24	II
15	15429.78	I
15	15831.75	I
20	17208.22	II
15	17307.66	I
15	17381.91	I
15	17481.04	I
15	17584.52	I
15	17936.43	II
15	18811.88	I
15	19145.60	II
15	19338.98	II
10	19774.30	II
10	20634.36	I
10	20692.06	II
10	22264.35	II

Th III
Ref. 157 — J.G.C.

Intensity	Wavelength	
	Vacuum	
100	1888.12	III
	Air	
50	2149.18	III
50	2162.82	III
50	2199.74	III
50	2206.62	III
50	2291.59	III
100	2301.18	III
100	2319.52	III
80	2324.68	III
150	2335.50	III
100	2340.58	III
100	2363.06	III
50	2368.91	III
100	2371.42	III
80	2381.47	III
100	2391.48	III
200	2413.50	III
50	2424.54	III
200	2427.94	III
200	2431.68	III
200	2441.24	III
100	2463.66	III
50	2473.93	III
100	2501.08	III
60	2512.69	III
50	2514.31	III
100	4555.73	III
50	4589.28	III
100	5376.13	III
50	5447.18	III
50	6242.95	III
50	6599.39	III
50	7461.59	III
50	8105.14	III

Th IV
Ref. 156, 165 — J.G.C.

Intensity	Wavelength	
	Vacuum	
4	797.53	IV
1	835.55	IV
30	846.91	IV
1	854.02	IV
30	882.39	IV
12	886.66	IV
100	1565.85	IV
70	1682.22	IV
30	1684.01	IV
150	1707.37	IV
200	1959.02	IV
	Air	
200	2002.34	IV
100	2066.70	IV
20	2143.91	IV
30	2146.81	IV
1	2242.11	IV
2	2261.26	IV
5	2296.81	IV
100	2693.99	IV
2	4937.09	IV
4	4952.52	IV
3	5420.38	IV
2	6711.87	IV
3	6740.37	IV
50	6901.16	IV
	9839.25	IV
	10875.05	IV

THULIUM (Tm)
Z = 69

Th I and II
Ref. 1 — C.H.C.

Intensity		Wavelength	
		Air	
360		2284.79	II
120		2329.77	II
70		2340.92	II
120		2363.91	II
45		2365.96	II
160		2367.11	II
150		2383.68	II
110		2388.95	II
450		2409.02	II
110		2412.44	II
120		2421.65	II
450		2426.17	II
140		2445.47	II
770		2480.13	II
150		2481.15	II
130		2487.52	II
250		2491.60	II
100		2499.54	II
130		2507.15	II
1300		2509.08	II
200		2520.87	II
250		2522.17	II
180		2524.11	II
130		2527.02	I
110		2527.42	II
120		2542.66	II
360		2552.76	I
540		2561.65	II
150		2563.86	II
430		2588.27	II
170	h	2596.49	I
110		2601.09	I
220		2606.02	II
810		2607.06	II
730		2624.33	II
210		2640.76	II
130		2646.45	II
160		2650.27	II
190		2658.48	II
250		2660.09	II
140		2668.20	II
310		2679.57	II
170		2697.50	II
540		2721.19	II
200		2744.08	II
270		2779.55	II
350		2785.07	II
680		2794.60	II
730		2797.27	II
250		2818.47	II
250		2827.02	II
580		[illegible]	II
200		2831.55	II
310		2844.67	II
200		2854.17	I
200		2860.12	II
200		2861.74	II
1600		2869.23	II
630		2890.94	II
210		2918.27	II
270		2925.65	II
680		2926.74	II
630		2935.99	II
350		2951.26	II
430		2965.86	II
490		2973.22	I
540		2981.48	II
350		2986.52	II
630		2990.54	II
200		2993.26	II
230		3013.71	II
430		3014.65	II
1500		3015.30	II
270		3017.09	II
330		3026.07	II
280		3042.35	II
340	d	3046.76	II
320		3050.73	II
340		3056.07	II
580		3073.08	II
360		3081.12	I
740		3098.60	II
7400		3131.26	II
2300		3133.89	II
230		3144.90	II
230		3146.16	II
1900		3151.04	II
1500		3157.34	II
450		3172.65	I
2300		3172.83	II
380		3173.58	II
230		3195.33	II
320		3210.56	II
320		3210.82	II
320		3212.01	II
230		3231.51	II
470		3235.44	II
1200		3236.81	II
1600		3240.23	II
2300		3241.54	II
320		3246.96	I
420		3247.46	II
1900		3258.05	II
400		3261.65	II
320		3264.10	II
1600		3266.64	II
1200		3267.40	II
790		3268.99	II
1100		3276.81	II
1200		3283.40	II
1200		3285.61	II
2300		3291.00	II
2000		3302.46	II
210		3306.01	II
210		3306.91	II
210		3308.01	II
1200		3309.80	II
640		3310.59	II
400		3316.88	II
210		3318.65	II
230		3349.99	I
230		3354.86	II
4000		3362.61	II
490		3374.50	II
420	d	3384.99	II
1700		3397.50	II
420		3399.95	II
850		3410.05	I
340		3412.59	I
340		3416.59	I
6400		3425.08	II
950		3425.63	II
340		3429.33	I
850		3429.96	II
420		3431.19	II
4900		3441.50	II
4900		3453.66	II
8500		3462.20	II
210		3467.51	I

Thulium (Cont.)

Intensity		Wavelength	Spectrum
340		3476.69	I
340		3480.98	I
340		3481.75	II
420		3487.38	I
210		3492.58	II
340		3499.95	I
250		3513.02	II
250		3517.60	I
250		3534.85	II
1700		3535.52	II
490		3536.21	II
850		3536.58	II
420		3537.91	I
210		3555.82	I
420		3557.79	II
340		3560.92	I
420		3563.88	I
490		3565.91	II
1300		3566.47	II
420		3567.36	I
280		3574.06	II
280		3586.07	I
2100		3608.77	II
250		3609.53	II
380		3638.41	I
950		3643.65	II
240		3647.72	II
600		3653.61	II
500		3665.81	II
1100		3668.09	II
410		3677.98	II
450	d	3678.85	II
410		3694.74	II
4800		3700.26	II
3800		3701.36	II
330		3704.85	II
7700		3717.91	II
890		3725.06	II
2400		3734.12	II
5000		3744.06	I
1700		3751.81	I
310		3756.86	II
6000		3761.33	II
4800		3761.91	II
260		3783.55	II
380		3795.16	II
7100		3795.75	II
770		3798.54	I
240		3798.75	II
600		3807.72	I
380		3810.72	II
550		3817.39	II
290		3826.39	I
1300		3838.20	II
290		3840.87	I
8900		3848.02	II
140		3857.84	II
6800		3883.13	I
1800		3883.44	II
5400		3887.35	I
440		3890.53	II
440		3896.62	I
680		3900.79	II
3500		3916.48	I
120		3928.66	II
570		3929.58	II
1500		3949.27	I
1500		3958.10	II
440		3995.58	II
1800		3996.52	II
220		4024.23	I
380		4044.47	I
10000		4094.19	I
9500		4105.84	I
120		4132.69	II
1100		4138.33	I
120		4149.14	I
120		4158.60	I
8800		4187.62	I
520		4199.92	II
6000		4203.73	I
220		4206.00	II
380		4222.67	I
3000		4242.15	II
270		4271.71	I
150		4298.36	I
2700		4359.93	I
1400		4386.43	I
200		4394.42	I
120		4395.96	I
140		4396.50	I
55		4437.40	II
80		4442.74	I
50		4447.58	I
120		4454.03	I
80		4459.99	I
50		4467.98	I
540		4481.26	II
80		4489.70	II
150		4519.60	I
260		4522.57	II
180		4529.38	II
80		4532.15	I
110		4548.60	I
40		4556.68	II
40		4561.86	II
80		4564.68	I
40		4567.11	II
95		4596.63	I
270		4599.02	I
35		4601.29	II
55		4603.43	II
40		4604.85	I
50		4613.97	I
40		4614.47	II
300		4615.94	II
35		4619.06	II
40		4621.72	I
80		4626.33	II
95		4626.56	II
40		4626.97	I
110		4634.26	II
40		4642.96	II
95		4643.12	I
35		4644.58	I
120		4655.09	I
35		4666.70	II
35		4671.99	II
35		4675.10	I
80		4675.31	I
40		4677.86	II
160		4681.92	I
70		4685.11	I
120		4691.11	I
110		4724.26	I
680		4733.34	I
35		4750.75	II
70		4759.90	I
27		4789.92	II
27		4807.48	I
35		4808.68	I
35		4813.50	I
27		4826.99	II
27		4828.97	I
80		4831.20	II
35		4835.75	I
27	d	4851.76	I
		4851.90	II
19		4872.28	II
27		4879.19	I
27		4891.64	I
24		4909.74	I
55		4923.83	I
140		4957.18	I
40		4970.87	II
27		4971.26	I
40		4975.12	II
50		4978.90	I
40		4980.68	II
55		4989.32	II
27		4993.79	II
19		4994.72	II
35		5001.02	I
27		5001.59	I
160		5009.77	II
35		5014.56	II
27		5017.87	II
160		5034.22	II
27	h	5041.00	II
22		5043.50	I
35		5045.41	I
27		5060.42	II
150		5060.90	I
27		5062.25	I
27		5065.88	I
80		5066.67	I
27		5072.42	I
27		5076.36	I
27		5077.18	I
35		5085.09	I
40		5107.53	I
95		5113.97	I
50		5114.55	I
22		5120.67	I
22		5140.28	II
40		5149.40	II
19		5182.68	I
40		5185.25	I
14		5204.51	II
80		5213.38	I
22		5228.23	II
14		5260.93	II
24		5267.34	II
40		5291.14	I
40		5294.32	II
35		5300.21	I
35		5302.69	I
55		5305.87	II
650		5307.12	I
16		5322.99	II
35		5338.90	I
		5339.03	I
80		5346.49	II
27		5372.98	II
14		5391.96	II
27		5400.46	II
27		5402.23	I
14		5405.98	II
14		5461.95	II
14		5464.14	I
14		5465.54	II
16		5500.30	II
14		5526.82	II
24		5528.34	I
14		5539.03	II
27		5566.00	I
22		5581.37	I
14		5586.65	II
14		5589.94	II
14		5606.64	I
270		5631.41	I
40		5642.60	I
27		5645.40	I
70		5658.30	I
520		5675.84	I
14		5683.59	I
40		5684.76	II
14		5696.42	II
35		5709.97	II
22		5715.79	I
14		5733.81	II
11	d	5737.20	II
		5737.25	II
14	h	5738.92	II
27		5758.02	I
55		5760.20	I
190		5764.29	I
5		5778.82	II
19		5782.36	II
22		5784.46	II
11		5799.97	II
14		5811.19	II
14	h	5816.46	I
35		5838.76	II
240		5895.63	I
35		5899.47	I
24		5901.57	I
8		5912.58	I
11		5931.70	I
27		5935.90	I
140		5971.26	I
27		5975.02	I
11		5984.87	I
19		6025.44	I
11		6067.78	II
16		6131.53	I
14		6175.29	I
14		6181.41	II
14		6299.46	II
27		6352.66	I
22		6401.44	I
8		6430.94	II
14		6440.54	I
200		6460.26	I
14		6490.70	I
14		6519.78	I
8		6575.54	I
95		6604.96	I
8		6627.25	I
35		6657.72	I
11		6658.64	I
11		6692.93	I
30		6721.36	I
9		6726.34	I
9		6727.94	II
18		6739.22	I
9	h	6767.48	I
9		6777.93	I
110		6779.77	I
14	h	6782.00	I
18		6788.52	I
13	h	6820.27	I
14		6826.95	I
14		6829.12	II
23		6831.09	I
120		6844.26	I
80		6845.76	I
18		6854.12	I
6		6898.56	I
6		6915.86	I
10		6937.37	I
5		6949.54	I
5	h	6976.69	II
5		7010.79	I
6	h	7014.31	II
10		7017.90	I
6	h	7029.40	I
12		7034.34	I
10		7056.43	II
5		7060.97	I
6		7079.78	II
10		7106.14	I
5	h	7231.33	I
5		7233.74	II
4		7257.72	I
17		7272.62	I
8		7284.30	I
11	h	7286.16	I
14		7310.51	I
11		7336.63	I
14		7432.18	I
5		7434.51	II
5		7439.95	II
75		7481.08	I
75		7490.20	I
10	h	7507.28	I
14		7545.78	I
140		7558.33	I
17	h	7580.61	I
20	h	7593.74	I
17		7595.07	II
5		7629.85	I
5	h	7648.76	II
17		7655.00	I
4		7660.32	I
7	h	7666.24	I
8		7676.04	II
8	h	7701.46	I
80		7731.53	I
4	h	7778.27	I
12	h	7782.35	I
8	h	7785.51	I
		7785.90	I
17		7803.93	I
4		7829.22	I
40		7856.08	I
3		7861.67	I
5		7918.10	I
55		7927.51	I
110		7930.84	I
6		7971.56	I
11	h	7985.93	I
14	h	8014.77	I
95		8017.90	I
3	h	8021.33	I
14		8194.19	I
5		8294.52	I
7		8365.75	I
7		8460.79	II
27		8472.01	II
7	h	8546.07	II
11		8565.73	II

Tm III
Ref. 307 — J.R.

Intensity	Wavelength Air	
500	2099.11	III
500	2107.10	III
200	2136.67	III
200	2156.29	III
800	2182.98	III
300	2183.91	III
5000	2185.94	III
100	2212.25	III
300	2230.86	III
400	2231.25	III
200	2243.34	III
400	2243.98	III
200	2246.68	III
200	2269.39	III
1000	2276.91	III
100	2280.08	III
100	2281.27	III
100	2282.86	III
200	2282.98	III
200	2286.57	III
400	2287.21	III
500	2294.73	III
20000	2296.21	III
200	2297.43	III
100	2304.64	III
400	2304.82	III
5000	2305.03	III
20000	2311.16	III
5000	2312.72	III
200	2314.88	III
400	2317.35	III
500	2320.96	III
200	2322.83	III
100	2323.71	III
100	2323.77	III
100	2324.43	III
500	2324.62	III
5000	2326.19	III
100	2327.02	III
300	2327.25	III
6000	2328.50	III
6000	2329.29	III
200	2330.87	III
3000	2331.80	III
400	2335.01	III
1000	2338.36	III
500	2341.74	III
300	2342.04	III
100	2344.59	III
500	2345.61	III
300	2347.43	III
400	2353.10	III
100	2355.65	III
3000	2357.05	III

Thulium (Cont.)

1000	2361.23	III
500	2363.97	III
1000	2375.32	III
700	2375.83	III
400	2389.52	III
4000	2406.63	III
500	2435.31	III
500	2457.86	III
500	2471.23	III
30000	2489.44	III
200	2496.25	III
2000	2504.71	III
3000	2519.78	III
10000	2552.46	III
500	2557.90	III
1000	2574.52	III
500	2574.98	III
100	2581.84	III
100	2585.48	III
300	2589.20	III
500	2608.96	III
300	2609.66	III
500	2617.22	III
500	2618.78	III
1000	2621.12	III
400	2621.35	III
400	2622.31	III
100	2627.09	III
100	2628.83	III
300	2634.66	III
200	2636.68	III
200	2637.30	III
100	2640.32	III
500	2643.58	III
100	2645.05	III
200	2649.27	III
100	2650.82	III
100	2654.05	III
100	2656.30	III
500	2661.51	III
1000	2663.00	III
500	2664.76	III
500	2664.88	III
200	2665.05	III
1000	2666.93	III
200	2668.59	III
100	2668.66	III
200	2669.18	III
100	2671.42	III
100	2675.30	III
1000	2676.64	III
500	2676.91	III
100	2678.28	III
100	2680.49	III
5000	2682.32	III
300	2682.64	III
300	2687.14	III
300	2695.69	III
400	2698.21	III
1000	2699.49	III
1000	2699.80	III
100	2703.63	III
200	2703.68	III
100	2704.93	III
2000	2707.03	III
300	2707.19	III
200	2707.44	III
500	2707.60	III
1000	2709.74	III
200	2710.79	III
1000	2713.38	III
200	2715.81	III
300	2717.56	III
100	2718.02	III
3000	2719.47	III
3000	2724.44	III
4000	2727.56	III
200	2728.13	III
1000	2731.38	III
300	2732.11	III
400	2737.98	III
800	2744.74	III
400	2745.99	III
500	2752.46	III
400	2753.20	III
400	2756.15	III
800	2765.98	III
700	2769.92	III
200	2772.64	III
100	2777.43	III
400	2781.12	III
2000	2806.77	III
300	2821.12	III
200	2849.52	III
700	2882.02	III
100	2899.29	III
100	2912.33	III
400	2921.08	III
200	2947.02	III
1000	2947.72	III
500	2953.18	III
100	2966.15	III
500	2966.85	III
400	2972.61	III
100	2974.85	III
1000	2998.28	III
100	3048.11	III
700	3078.87	III
200	3120.15	III
200	3277.26	III
200	3407.73	III
100	3415.40	III
100	3415.96	III
100	3436.93	III
400	3467.93	III
200	3529.29	III
200	3533.28	III
100	3537.47	III
300	3562.41	III
100	3563.42	III
200	3587.76	III
600	3617.96	III
1000	3629.09	III
100	3706.11	III
100	3799.41	III
300	3998.84	III
200	4021.92	III
200	4026.03	III
700	4032.13	III
100	4076.15	III
200	4335.47	III
500	4385.41	III

TIN (Sn)
Z = 50

Sn I and II
Ref. 187, 191 — C.H.C.

Intensity		Wavelength	
		Vacuum	
1		899.92	II
2		917.40	II
1		935.63	II
3		945.83	II
4		954.50	II
7		985.13	II
4		997.21	II
2		1016.26	II
4		1040.78	II
1		1041.32	II
3		1062.10	II
8		1108.19	II
4		1159.05	II
10		1161.43	II
3		1162.94	II
4		1180.51	II
9		1219.07	II
13		1223.70	II
11		1243.00	II
20		1290.86	II
20		1316.59	II
25		1400.52	II
20		1475.15	II
9		1489.22	II
7		16 9.47	II
10	r	1737.21	I
15	r	1751.46	I
10	h	1753.3	I
7		1758.00	II
20	r	1764.98	I
20		1773.40	I
30	r	1790.75	I
80	r	1804.60	I
15		1811.34	II
30		1813.04	I
40	r	1815.74	I
25		1819.31	I
120	r	1823.00	I
9		1831.89	II
50	r	1848.75	I
30		1852.00	I
200	r	1860.32	I
20		1861.42	I
20		1865.52	I
30		1865.96	I
15		1873.29	I
30		1882.64	I
80		1886.05	I
100		1891.40	I
20		1897.29	I
12		1899.91	II
50		1909.30	I
40		1911.61	I
20		1913.52	I
80		1925.31	I
20		1926.77	I
15		1927.95	I
40	h	1928.9	I
25		1933.17	I
20		1942.69	I
150		1952.15	I
15		1960.21	I
30		1971.46	I
50	h	1977.6	I
80		1984.20	I
15		1991.88	I
20		1994.98	I
		Air	
25		2008.05	I
30		2015.76	I
30		2025.98	I
50		2040.66	I
20		2040.90	I
50		2054.03	I
70		2058.31	I
20		2064.00	I
80		2068.58	I
100		2072.89	I
100		2073.08	I
25		2080.62	I
30		2091.58	I
40		2094.35	I
200		2096.39	I
100		2100.93	I
100	r	2113.93	I
50		2121.26	I
25		2140.73	I
20		2141.43	I
15		2148.46	I
1		2148.63	II
40	r	2148.73	I
20	r	2151.43	I
30		2151.54	II
80		2171.32	I
150	r	2194.49	I
300	r	2199.34	I
400	r	2209.65	I
4		2209.67	II
40		2211.05	I
80	r	2231.72	I
400	r	2246.05	I
6		2246.07	II
60		2251.17	I
30		2267.19	I
400	r	2268.91	I
20		2282.26	I
200	r	2286.68	I
600	r	2317.23	I
300	r	2334.80	I
1000	r	2354.84	I
20		2357.90	I
3		2360.34	II
22		2368.33	II
60		2380.72	I
4		2384.54	II
100		2408.15	I
800	r	2421.70	I
1000	r	2429.49	I
1		2433.52	II
15		2448.98	II
60		2455.24	I
20		2476.40	I
300		2483.39	I
13		2483.48	II
10		2486.99	II
200		2495.70	I
5		2522.61	II
90		2523.92	I
80	h	2531.17	I
400		2546.55	I
40	h	2558.01	I
500	r	2571.58	I
200		2594.42	I
50	h	2636.94	I
200	r	2661.24	I
2		2664.93	II
700	r	2706.51	I
2		2727.82	II
20		2761.78	I
150		2779.81	I
80		2785.03	I
60		2787.96	I
60		2812.59	I
80		2813.58	I
2		2825.52	II
1400	r	2839.99	I
1		2846.42	II
200		2850.62	I
1000	r	2863.32	I
1		2912.80	II
200		2913.54	I
6		2919.82	II
3		2991.00	II
7		2994.44	II
700	r	3009.14	I
1		3012.18	II
8		3023.94	II
200		3032.80	I
850	r	3034.12	I
12		3047.50	II
6		3094.69	II
60		3141.84	I
550	r	3175.05	I
40		3218.71	I
550	r	3262.34	I
50		3283.21	II
110		3330.62	I
60		3351.97	II
2		3407.48	II
10		3472.46	II
7		3537.57	II
11		3575.45	II
3		3582.39	II
2		3620.08	II
6		3620.54	II
40		3655.78	I
6		3715.23	II
280	r	3801.02	I
4		3841.44	II
1		4294.65	II
60		4524.74	I
1		4579.13	II
1		4580.29	II
2		4877.22	II
3		4944.31	II
20		4979.73	I
2		5071.14	II
2		5072.67	II
20		5174.54	I
10		5332.36	II
20		5561.95	II
25		5588.92	II
2		5596.20	II
500		5631.71	I
15		5753.59	I
1		5797.20	II
15		5799.18	II
50		5925.44	I
100		5970.30	I
150		6037.70	I
200		6054.86	+
250		6069.00	I
100		6073.46	I
6		6077.48	II
5		6079.70	II
400		6149.71	I
200		6154.60	I
150		6171.50	I
100		6310.78	I
40		6354.35	I
70		6453.50	II
8		6761.45	II
25		6844.05	II
20		7191.40	II
10		7387.79	II
20	h	7398.6	I
1		7408.62	II
30		7685.30	I
13		7741.80	II
100		7754.97	I
3		7904.00	II
100	h	8030.5	I
30	h	8039.3	I
200		8114.09	I
30	h	8121.0	I
30		8349.35	I
80		8357.04	I
300		8422.72	I
400		8552.60	I
50	h	8681.7	I
30	h	9018.95	I
50	h	9410.86	I
80	h	9415.37	I
150		9616.40	I
50		9741.1	I
100	h	9742.8	I
300	h	9805.38	I
500		9850.52	I
25		10456.47	I
11		10807.58	I
54		10894.00	I
70		11191.85	I
56		11277.66	I
17		11336.97	I
200		11454.59	I
200		11616.26	I
76		11670.77	I
25		11694.45	I
258		11739.78	I
96		11825.18	I
106		11835.82	I
254		11932.99	I
48		12009.50	I
111		12313.24	I
33		12335.6	I
42		12530.87	I
42		12536.5	I
37		12788.2	I
89		12888.5	I
187		12981.7	I
20		13000.3	I
187		13018.5	I
68		13081.5	I
378		13460.2	I
144		13608.2	I
13		15018.2	I
30		15464.2	I
20		17000.5	I

Tin (Cont.)

10	17807.5	I
20	20622.2	I
40	20861.7	I
8	21686.2	I
4	22131.7	I
3	22997.2	I
4	24327.2	I
4	24738.2	I

Sn III
Ref. 423 — C.H.C.

Intensity	Wavelength	
	Vacuum	
100	753.01	III
50	760.62	III
75	775.79	III
50	784.68	III
200	910.92	III
50	1010.92	III
50	1048.84	III
1000	1139.29	III
1000	1158.33	III
200	1161.09	III
100	1161.58	III
100	1180.62	III
1000	1184.25	III
200	1189.99	III
200	1204.06	III
2000	1210.52	III
100	1215.10	III
100	1218.14	III
100	1230.17	III
100	1231.38	III
500	1243.63	III
1000	1259.92	III
40	1276.31	III
1000	1305.97	III
1000	1327.34	III
200	1334.70	III
200	1346.05	III
1000	1347.65	III
200	1369.71	III
1000	1386.74	III
500	1410.61	III
200	1449.77	III
1000	1570.36	III
50	1674.29	III
500	1811.71	III
500	1941.86	III
50	1955.52	III

Sn IV
Ref. 423 — C.H.C.

Intensity	Wavelength	
	Vacuum	
50	605.23	IV
50	619.04	IV
50	628.73	IV
50	908.22	IV
500	956.25	IV
500	1019.72	IV
1000	1044.49	IV
100	1058.37	IV
50	1058.59	IV
1000	1073.41	IV
200	1087.50	IV
300	1096.92	IV
50	1103.24	IV
000	1119.34	IV
200	1120.68	IV
1000	1314.55	IV
1000	1437.52	IV
100	1532.90	IV

Sn V
Ref. 399, 423 — C.H.C.

Intensity	Wavelength	
	Vacuum	
120	355.14	V
150	361.01	V
100	372.55	V
200	1089.35	V
100	1132.79	V
200	1160.74	V
100	1176.26	V
100	1189.92	V
100	1205.72	V
2000	1251.38	V
100	1283.81	V
200	1294.36	V
100	1302.20	V

TITANIUM (Ti)
Z = 22

Ti I and II
Ref. 1 — C.H.C.

Intensity		Wavelength	
		Air	
140		2272.61	I
180		2273.28	I
130		2276.70	I
190		2279.96	I
150		2299.85	I
140		2302.73	I
190		2305.67	I
65		2380.81	I
35		2384.52	I
55		2418.36	I
75		2421.30	I
95		2424.24	I
40		2428.23	I
35		2433.22	I
19		2434.10	I
35		2440.21	II
65		2440.98	I
24		2450.44	II
24		2504.54	I
75		2517.43	II
40		2519.04	I
140		2520.54	I
75		2524.64	II
360		2525.60	II
29		2527.98	I
210		2529.85	I
190		2531.25	II
190		2534.62	II
130		2535.87	II
190		2541.92	I
65		2555.99	II
110		2571.03	II
50		2572.65	II
50		2580.82	I
35		2590.26	I
190		2593.64	I
65		2596.58	I
270		2599.92	I
340		2605.15	I
510		2611.28	I
75		2611.48	I
300		2619.94	I
170		2631.54	I
170		2632.42	I
640		2641.10	I
800		2644.26	I
950		2646.64	I
30		2649.30	I
15		2654.93	I
35		2657.19	I
85		2661.97	I
95		2669.60	I
130		2679.93	I
26		2684.80	I
30		2685.14	I
65		2688.82	I
26		2716.25	II
85		2725.07	I
75		2727.42	I
21		2731.13	I
40		2731.58	I
170		2733.26	I
55		2735.29	I
40		2735.61	I
85		2739.81	I
250		2742.32	I
40		2749.06	I
65		2757.40	I
95		2758.08	I
15		2761.29	II
250		2802.50	I
55		2805.70	I
30		2806.50	II
40		2809.17	I
75		2810.30	II
30		2812.98	I
30		2817.40	I
65		2817.84	I
		2817.87	II
65		2828.07	I
		2828.15	II
130		2832.16	II
190		2841.94	II
110		2851.10	II
40		2853.93	II
95		2862.32	II
55		2868.74	II
180		2877.44	II
280		2884.11	II
65		2888.93	II
55		2891.07	II
55		2905.66	I
30		2909.92	II
450		2912.08	I
340		2928.34	I
15		2931.03	
180		2933.55	I
26		2935.96	
150		2937.32	I
1100		2942.00	I
1300		2948.26	I
30		2954.58	
1600		2956.13	I
170		2956.80	I
30		2958.77	
26		2959.71	I
35		2959.99	I
170		2965.71	I
190		2967.22	I
26		2968.23	I
75		2970.38	I
30		2974.93	I
170		2983.31	I
35		3000.87	I
120		3017.19	II
140		3029.73	II
110		3046.68	II
130		3056.74	II
130		3057.40	II
170		3058.09	II
85		3059.74	II
1300	d	3066.22	II
		3066.35	II
70		3071.24	II
600		3072.11	II
1100		3072.97	II
1600		3075.22	II
2300		3078.64	II
3600		3088.02	II
180		3089.40	II
180		3097.19	II
180		3100.67	I
230		3103.80	II
230		3105.08	II
260		3106.23	II
70		3106.81	I
50		3110.67	II
50		3112.48	I
140		3117.67	II
720		3119.72	I
		3119.80	II
190		3123.07	I
240		3130.80	II
140		3141.54	I
95		3141.67	I
220		3143.76	II
240		3148.04	II
240		3152.25	II
240		3154.20	II
240		3155.67	II
500		3161.20	II
780		3161.77	II
1000		3162.57	II
1600		3168.52	II
2400		3186.45	I
1000		3190.87	II
3100		3191.99	I
50		3197.52	II
3800		3199.92	I
780		3202.54	II
50		3203.44	II
240		3203.83	I
50		3204.87	I
110		3213.14	II
260		3214.24	I
190		3214.75	II
1100		3217.06	II
110		3217.94	I
260		3218.27	II
110		3219.21	I
110		3221.38	I
1300		3222.84	II
220		3223.52	I
240		3224.24	II
140		3226.13	I
530		3228.60	II
780		3229.19	II
530		3229.42	II
110		3231.32	II
240		3232.28	II
6600		3234.52	II
220		3236.12	II
5200		3236.57	II
4100		3239.04	II
220		3239.66	II
2600		3241.99	II
1200		3248.60	II
950		3251.91	II
1200		3252.91	II
1200		3254.25	II
1200		3261.60	II
310		3271.65	II
310		3272.08	II
200		3278.29	II
260		3278.92	II
220		3282.33	II
530		3287.66	II
290		3292.08	I
170		3299.41	I
170		3306.88	I
220		3308.39	I
220		3308.81	II
260		3309.50	I
60		3309.73	I
110		3312.69	I
840		3314.42	I
		3314.52	I
290		3315.32	II
330		3318.02	II
550		3321.70	II
2900		3322.94	II
380		3326.76	II
2100		3329.46	II
550		3332.11	II
1800		3335.20	II
1100		3340.34	II
5700		3341.88	I
120		3342.15	I
260		3343.77	II
330		3346.73	II
4300		3349.04	II
12000		3349.41	II
120		3352.94	I
4100		3354.64	I
290		3358.28	I
290		3360.99	I
7200		3361.21	II
		3361.26	I
120		3361.84	I
1100		3370.44	I
4300		3371.45	I
140		3372.21	II
5700		3372.80	II
60		3374.35	II
2900	d	3377.48	I
		3377.58	I
290		3379.22	I
1400		3380.28	II
170		3382.31	I
5700		3383.76	II
170		3385.66	I
1400		3385.95	I
1400		3387.84	II
60		3388.76	II
140		3390.68	I
140		3392.71	I
1100		3394.58	II
60		3398.63	I
60		3402.42	II
60		3407.20	II
95		3409.81	II
60		3439.30	I
890		3444.31	II
60		3452.47	II
180		3456.39	II
600		3461.50	II
95		3467.26	I
600		3477.18	II
60		3478.92	I
240		3480.53	I
60		3485.69	I
60		3489.74	II
480		3491.05	II
60		3495.75	I
95		3499.10	I
890		3504.89	II
120		3506.64	I
600		3510.84	II
60		3520.25	II
310		3535.41	II
190		3547.03	I
120		3573.74	II
60		3574.24	I
60		3587.13	II
240		3596.05	II
190		3598.72	I
600		3610.16	I
190		3624.82	II
95		3635.20	I
4800		3635.46	I
120		3637.97	I
190		3641.33	II
6600		3642.68	I
180		3646.20	I
7200		3653.50	I
290		3654.59	I
660		3658.10	I
120		3659.76	II
380		3660.63	I
190		3662.24	II
380		3668.97	I
600		3671.67	I
3100		3685.20	II
120		3685.96	I
95		3687.35	I
600		3689.91	I

Titanium (Cont.)

140		3694.45	I
30		3698.18	I
60		3698.43	I
60		3700.08	I
120		3702.29	I
190		3704.30	I
140		3706.23	II
50		3707.53	I
290		3709.96	I
30		3715.40	I
450		3717.40	I
140		3721.64	II
330		3722.57	I
600		3724.57	I
380		3725.16	I
2900		3729.82	I
50		3735.67	I
60		3738.90	I
3300		3741.06	I
330		3741.64	II
160		3748.10	I
5200		3752.86	I
600		3753.64	I
140		3757.69	II
3300		3759.30	II
2900		3761.32	II
50		3761.89	II
60		3766.45	I
600		3771.66	I
30		3776.06	II
840		3786.04	I
120		3789.30	I
70		3795.90	I
60		3798.31	I
70		3818.22	I
60		3822.03	I
240		3828.19	I
95		3833.68	I
95		3836.78	I
60		3846.45	I
130		3853.05	I
130		3853.73	I
170		3858.14	I
240		3866.44	I
170		3868.40	I
120		3873.21	I
260		3875.26	I
170		3882.15	I
170		3882.33	I
500		3882.89	I
60	h	3888.02	I
70		3889.95	I
200	h	3895.25	I
85		3898.49	I
530		3900.54	II
180		3900.96	I
2600		3904.78	I
110	h	3911.19	I
500		3913.46	II
500		3914.34	I
24		3914.74	I
35		3919.82	I
290		3921.42	I
1100		3924.53	I
110		3926.32	I
890		3929.88	I
35		3932.02	II
70		3934.24	I
1100		3947.78	I
4500		3948.67	I
4500		3956.34	I
5200		3958.21	I
950		3962.85	I
950		3964.27	I
4800		3981.76	I
570		3982.48	I
60		3984.33	I
35		3985.25	I
60		3985.59	I
5700		3989.76	I
35		3994.70	I
7800		3998.64	I
70		3999.36	I
70		4002.49	I
70		4003.81	I
35		4005.97	I
70		4008.06	I
950		4008.93	I
190		4009.66	I
70		4012.39	II
180		4013.58	I
70		4015.38	I
35		4016.28	I
120	h	4017.77	I
140		4021.83	I
1200		4024.57	I
40		4025.14	II
190	h	4026.54	I
40		4027.48	I
40		4028.34	II
190	h	4030.51	I
40		4033.91	I
30		4034.91	I
110		4035.83	I
35	h	4040.32	I
290		4055.02	I
85		4057.62	I
85		4058.14	I
410		4060.26	I
200		4064.22	I
200		4065.10	I
840		4078.47	I
40		4079.72	I
290		4082.46	I
85		4099.17	I
220		4112.71	I
85		4122.17	I
40		4123.31	I
85		4123.57	I
130		4127.54	I
40		4129.17	I
40		4131.25	I
140		4137.29	I
85		4143.05	I
170		4150.96	I
85		4159.64	I
70		4163.65	II
35		4164.14	I
40		4166.32	I
85		4169.35	I
120		4171.03	I
40		4171.90	II
35		4183.30	I
360		4186.12	I
40		4188.69	I
70		4200.75	I
85		4203.46	I
35		4211.73	I
40		4224.79	I
40		4227.65	I
130		4237.89	I
85		4249.12	I
130		4256.04	I
70		4258.54	I
70		4261.60	I
330		4263.13	I
35		4265.71	I
40		4266.22	I
70		4270.14	I
85		4272.43	I
240		4274.58	I
120		4276.43	I
120		4278.23	I
30		4278.81	I
110		4281.38	I
220		4282.71	I
160		4284.99	I
890		4286.01	I
840		4287.40	I
30		4288.16	I
950		4289.07	I
120		4290.23	II
840		4290.94	I
120		4291.14	I
140		4294.12	II
840		4295.76	I
2000		4298.66	I
200		4299.23	I
200		4299.64	I
200		4300.05	II
2900		4300.56	I
4100		4301.09	I
85		4301.93	II
6000		4305.92	I
180		4307.90	II
35		4308.50	I
40		4311.65	I
85		4312.87	II
85		4314.35	I
1200		4314.80	I
360		4318.64	I
180		4321.66	I
190		4325.13	I
160		4326.36	I
30		4334.84	I
160		4337.92	II
24		4344.29	II
70		4346.11	I
35		4354.06	I
95		4360.49	I
24		4368.94	I
95		4369.68	I
60		4372.38	I
30		4388.08	I
170		4393.92	I
330		4395.04	II
60		4399.77	II
240		4404.28	I
60		4404.90	I
30		4405.68	I
60		4416.54	I
220		4417.28	I
60		4417.72	II
120		4421.76	I
120		4422.82	I
24		4424.39	I
30		4425.83	I
120		4426.06	I
890		4427.10	I
21		4430.02	I
85		4430.37	I
50		4431.28	I
30		4432.60	I
24		4433.58	I
170		4434.00	I
70		4436.59	I
30		4438.23	I
130		4440.35	I
50		4441.27	I
230		4443.80	II
24		4444.27	I
840		4449.15	I
30		4450.49	II
550		4450.90	I
840		4453.32	I
290		4453.71	I
950		4455.33	I
1100		4457.43	I
21		4462.09	I
70		4463.38	I
95		4463.54	I
290		4465.81	I
240		4468.50	II
240		4471.24	I
95		4474.85	I
95		4479.70	I
50		4480.59	I
530		4481.26	I
95		4482.69	I
19		4488.32	II
260		4489.09	I
24		4492.55	I
40		4495.01	I
240		4496.15	I
24		4497.73	I
200		4501.27	II
40		4503.78	I
21		4506.36	I
50		4511.17	I
780		4512.74	I
19		4515.62	I
1000		4518.03	I
95		4518.70	I
1000		4522.80	I
780		4527.31	I
6000		4533.24	I
240		4533.97	II
3600		4534.78	I
2400		4535.58	I
1200		4535.92	I
1300		4536.05	I
24		4537.23	I
24		4539.10	I
720		4544.69	I
950		4548.77	I
240		4549.63	II
950		4552.46	I
24		4555.08	I
720		4555.49	I
19		4557.86	I
19		4558.11	I
60		4559.92	I
50		4562.63	I
35		4563.43	I
110		4563.77	II
35		4570.91	I
240		4571.98	II
19		4585.84	
24		4589.95	II
60		4599.23	I
21		4609.37	I
950		4617.27	I
24		4619.52	I
480		4623.09	I
190		4629.34	I
50	d	4634.87	
60		4637.88	I
240		4639.37	I
220		4639.67	I
190		4639.95	I
140		4645.19	I
120		4650.02	I
24		4656.04	I
720		4656.47	I
840		4667.59	I
70		4675.12	I
950		4681.92	I
21		4686.92	I
24		4690.80	I
190		4691.34	I
40		4693.68	I
24		4696.94	I
190		4698.76	I
120		4710.19	I
24		4715.30	I
65		4722.62	I
65		4723.17	I
55		4731.17	I
45		4733.43	I
18		4734.68	I
22		4742.11	I
170		4742.79	I
22		4747.68	I
310		4758.12	I
310		4759.28	I
45		4766.33	I
28		4769.77	I
65		4778.26	I
45		4781.72	I
110		4792.49	I
45		4796.22	I
35		4797.98	I
110		4799.80	I
28		4805.10	II
110		4805.43	I
45		4808.53	I
22		4811.08	I
40		4812.25	I
200		4820.42	I
22		4825.46	I
40		4836.13	I
470		4840.87	I
65		4848.47	I
290		4856.01	I
35		4864.18	I
200		4868.26	I
250		4870.14	I
28		4880.91	I
45		4882.35	I
400		4885.08	I
380		4899.91	I
320		4913.62	I
55		4915.24	I
130		4919.87	I
180		4921.77	I
55		4925.41	I
30		4926.16	I
150		4928.34	I
30		4937.74	I
95		4938.29	I
30		4941.58	I
21		4948.19	I
21		4958.25	I
55		4964.75	I
21		4966.04	I
65		4968.58	I
75		4973.05	I
120		4975.35	I
65		4977.74	I
120		4978.20	I
5800		4981.73	I
150		4989.15	I
4600		4991.07	I
30		4995.08	I
140		4997.10	I
4000		4999.51	I
230		5001.01	I
3600		5007.21	I
120		5009.65	I
230		5013.30	I
3200	d	5014.19	I
		5014.24	I
580		5016.17	I
840		5020.03	I
840		5022.87	I
580		5024.84	I
300		5025.58	I
1200		5035.91	I
840		5036.47	I
740		5038.40	I
1200		5039.95	I
75		5040.62	I
85		5043.59	I
35		5044.27	I
55		5045.41	I
26		5048.21	I
110		5052.87	I
21		5054.08	I
110		5062.11	I
35		5064.07	I
1400		5064.66	I
95		5065.99	I
35	h	5068.33	I
65		5069.35	I
130		5071.48	I
40		5085.34	I
130		5087.07	I
21		5103.15	I
55		5109.44	I
190		5113.44	I
270		5120.42	I
30		5129.15	II
270		5145.47	I
230		5147.48	I
210		5152.20	I
21	Bl	5166.86	T
1100		5173.75	I
40		5186.34	I
85		5188.70	II
30		5189.58	I
1300		5192.98	I
85	h	5194.04	I
65		5201.10	I
120		5206.08	I
75		5207.87	I
65		5208.42	

Titanium (Cont.)

1400		5210.39	I
65		5212.29	I
150		5219.71	I
95		5222.69	I
85		5223.64	I
250		5224.32	I
95		5224.56	I
190		5224.95	I
65		5226.56	II
120		5238.58	I
21		5246.15	I
55		5246.57	I
75		5247.31	I
21		5250.95	I
110		5252.11	I
75		5255.83	I
55		5259.99	I
55		5263.50	I
150		5265.98	I
40		5282.39	I
140		5283.45	I
35		5284.39	I
26		5288.81	
65		5295.79	I
120		5297.26	I
65		5298.44	I
26		5336.81	II
17		5341.50	I
75		5351.08	I
26		5366.65	I
55		5369.64	I
40		5389.18	I
55		5389.99	I
17		5396.60	I
85		5397.09	I
35		5404.02	I
110		5409.61	I
40		5426.26	I
75		5429.15	I
26		5436.73	I
17		5438.32	I
40		5446.64	I
11	Bl	5448.34	T
30		5448.90	I
21		5449.16	I
35		5453.65	I
55		5460.51	I
75		5471.21	I
35		5472.70	I
40	h	5473.55	I
85		5474.23	I
30		5474.46	I
120	h	5477.71	I
110		5481.43	I
75		5481.87	I
85	h	5488.20	I
150		5490.15	I
26		5490.84	I
110		5503.90	I
40		5511.78	I
340		5512.53	I
270		5514.35	I
320		5514.54	I
26		5530.49	
110		5565.49	I
13		5579.16	
21	h	5582.98	
30	h	5585.68	
65	Bl	5597.85	T
55	Bl	5629.28	T
17		5635.84	
250		5644.14	I
75		5648.58	I
26	Bl	5661.55	T
190		5662.16	I
75		5662.91	I
21		5673.42	I
130		5675.44	I
30	h	5679.94	I
95		5689.47	I
75		5702.68	I
35		5708.23	I
65		5711.88	I
40	h	5713.92	I
95		5715.13	I
55		5716.48	I
35		5720.48	I
85		5739.51	I
40		5740.02	I
19		5741.22	I
21		5752.84	I
19		5756.86	I
40	h	5762.27	I
55	h	5766.35	I
75	h	5774.05	I
30		5780.78	I
75	h	5785.98	I
65	Hl	5804.26	I
21	Bl	5814.96	T
40		5823.71	I
21	h	5841.18	
21		5852.34	
400		5866.46	I
65		5880.31	I
21	h	5888.68	
230		5899.32	I
55		5903.33	I
120		5918.55	I
150		5922.12	I
75		5937.82	I
120		5941.76	I
300		5953.17	I
200		5965.84	I
270		5978.56	I
340		5999.04	I
65		5999.68	I
21		6012.73	
110		6064.63	I
120		6085.23	I
120		6091.17	I
40		6092.81	I
40	h	6098.67	I
35	h	6121.01	I
120		6126.22	I
19		6138.38	I
30		6146.22	I
21		6149.74	I
30	Bl	6162.23	T
35		6186.15	I
95	h	6215.28	I
75	h	6220.49	I
65	h	6221.41	I
380		6258.10	I
380		6258.70	I
300		6261.10	I
65		6303.75	I
55		6312.24	I
26		6318.03	I
30		6336.10	I
35		6366.35	I
11		6419.10	I
17		6497.69	I
19		6508.14	I
55		6546.28	I
65		6554.23	I
11	h	6554.83	
75		6556.07	I
19	h	6565.62	I
14	h	6575.18	I
35		6599.11	I
18	Bl	6651.46	T
18	h	6666.55	I
22	h	6667.74	
9		6668.39	
18		6677.18	I
22	b	6691.21	T
26		6716.68	I
16	Bl	6723.95	T
80		6743.12	I
22		6745.52	I
18		6844.64	
18		6860.39	
35		6861.47	I
9		6873.92	I
12		6913.19	I
14	h	6933.15	I
14	h	6943.70	I
23		6996.63	I
15		7004.66	I
14		7008.35	I
14		7010.94	I
14	h	7035.86	I
40		7038.80	I
14		7050.65	I
40	Bl	7054.51	T
23		7069.11	I
23		7072.05	
45	Bl	7087.89	T
30	b	7124.9	T
40	Bl	7125.61	T
26		7138.91	I
26		7167.13	
23		7171.53	
55		7189.89	I
26	b	7203.64	T
260		7209.44	I
60		7216.20	I
130		7244.86	I
130		7251.72	I
19		7263.40	
19		7266.29	I
19	b	7269.05	T
15		7315.56	I
26		7318.39	I
120		7344.72	I
11		7352.16	I
90		7357.74	I
60		7364.11	I
26		7440.60	I
9		7474.94	I
26		7489.61	I
19		7496.12	I
12		7580.55	I
9	Bl	7589.62	T
15		7614.50	I
23		7654.44	I
11	Bl	7705.21	T
30		7949.17	I
26	h	7961.58	I
60		7978.88	I
9		7979.07	I
30		7996.53	I
7	h	8003.55	
55		8024.84	I
30		8068.24	I
8		8267.62	
14	h	8306.31	I
9	h	8307.41	I
9	h	8311.76	I
8	h	8312.85	I
12		8334.37	I
14		8353.15	I
75		8364.24	I
100		8377.85	I
100		8382.54	I
55		8382.82	I
75		8396.87	I
120		8412.36	I
19		8416.98	I
15		8424.41	I
170		8426.52	I
490		8434.94	I
240		8435.70	I
40		8438.93	I
40		8450.89	I
9	h	8457.10	I
19	h	8467.15	I
45		8468.50	I
15		8496.04	I
19	h	8518.05	I
40		8518.32	I
14		8539.38	I
40		8548.12	I
9		8569.77	I
9	h	8598.18	I
90		8675.39	I
45		8682.99	I
23		8692.33	I
19		8734.69	I
23		8766.64	I
15	h	8778.71	I

Ti III
Ref. 378 — C.H.C.

Intensity		Wavelength	
		Vacuum	
6		1282.48	III
6		1286.23	III
15		1286.36	III
10		1289.30	III
10		1291.62	III
10		1293.23	III
15		1294.70	III
10		1295.88	III
20		1298.66	III
20		1298.97	III
12		1327.59	III
10		1420.04	III
10		1420.44	III
10		1421.63	III
10		1421.77	III
12		1422.40	III
10		1424.14	III
23		1455.19	III
10		1498.70	III
		Air	
10		2199.22	III
12		2237.77	III
10		2327.02	III
15		2331.35	III
15		2331.66	III
15		2334.34	III
17		2339.00	III
18		2346.79	III
18		2374.99	III
22		2413.99	III
25		2516.05	III
24		2527.84	III
23		2540.06	III
24		2563.44	III
23		2565.42	III
22		2567.56	III
15		2576.47	III
15		2580.46	III
10		2692.16	III
12		2701.96	III
22		2984.75	III
12	d	3354.71	III
12		3872.50	III
12		3881.21	III
12		3893.63	III
10		3896.33	III
15		3915.47	III
12		3921.38	III
10		3921.61	III
12		3922.95	III
10		3924.86	III
10		4060.21	III
10		4119.14	III
11		4215.52	III
11		4269.84	III
11		4296.70	III
10		4348.04	III
11		4433.91	III
10		4540.22	III
15		4549.84	III
10	d	4555.46	III
15	d	4572.20	III
10		4649.45	III
12		4652.86	III
10		4874.00	III
10		4950.10	III
10		4971.19	III
10		5083.80	III
14		5147.31	III
12	d	5226.28	III
11		5247.49	III
17		5278.12	III
10		5278.70	III
12		5298.43	III
16		5301.20	III
15		5306.88	III
10		5395.69	III
12		5533.01	III
10		5817.44	III
12		6611.38	III
18		6621.58	III
10		6629.37	III
14		6647.47	III
18		6667.99	III
15		6674.19	III
14		6707.76	III
12		6724.80	III
16		6734.10	III
15		6862.26	III
12		6874.35	III
10		6896.12	III
12		7015.38	III
10		7071.93	III
20		7072.64	III
18		7084.57	III
15		7124.13	III
11		7171.79	III
10		7175.92	III
10		7217.50	III
9		7225.55	III
12		7270.67	III
14		7316.30	III
10		7316.68	III
12		7379.96	III
10		7408.13	III
10		7457.85	III
15		7506.87	III
17		7507.68	III
10		7523.85	III
12		7544.29	III
9		7566.25	III
10	h	8172.21	III
9	h	8173.37	III
9	h	8178.00	III
10		8182.42	III
9	h	8192.68	III
9	h	8194.75	III
9	h	8263.67	III
15	h	8267.32	III
10		8338.54	III
12		8394.20	III
20		8466.87	III
5		8699.85	III
3		9017.10	III

Ti IV
Ref. 428 — C.H.C.

Intensity		Wavelength	
		Vacuum	
10		776.76	IV
18		779.07	IV
16		781.73	IV
8		1183.64	IV
10		1195.21	IV
18		1451.74	IV
20		1467.34	IV
12		1469.19	IV
		Air	
20		2067.56	IV
18		2103.16	IV
10		2359.14	IV
10		2359.50	IV

Titanium (Cont.)

8		2541.79	IV
10		2546.88	IV
5		2862.60	IV
6		2929.96	IV
14		2937.33	IV
12		2957.31	IV
15		3541.36	IV
17		3576.44	IV
10		3581.39	IV
13		4131.22	IV
14		4133.78	IV
10		4397.33	IV
9		4403.45	IV
15		4618.11	IV
20		5398.93	IV
8		5470.98	IV
18		5492.51	IV
10		5517.70	IV
14		5877.79	IV
15		5885.96	IV
7		5891.15	IV
6		6231.62	IV
17		6246.65	IV
11		6247.74	IV
15		6292.41	IV
12		6913.85	IV
15		6978.51	IV
9		7491.37	IV
8		7494.77	IV
5	h	7652.12	IV
8	h	7706.85	IV

Ti V
Ref. 427 — C.H.C.

Intensity		Wavelength	
		Vacuum	
12		225.35	V
10		228.91	V
17		252.96	V
7		323.36	V
7		461.41	V
8		474.69	V
8		483.99	V
10	d	488.58	V
15		498.26	V
14		502.08	V
7		502.71	V
12		504.66	V
7		506.47	V
8		513.37	V
7		523.05	V
12		524.58	V
13		526.57	V
8		529.32	V
10		535.84	V
10		535.89	V
8	d	540.14	V
8		541.46	V
9		541.71	V
7		543.10	V
7		543.34	V
7		1128.55	V
8		1192.35	V
9		1198.66	V
9		1222.36	V
10		1230.36	V
11		1239.96	V
10		1241.67	V
7		1246.13	V
8		1268.49	V
8		1306.11	V
8		1411.31	V
9		1675.15	V
8		1687.16	V
11		1717.40	V
8		1759.76	V
7		1771.45	V
10		1841.49	V
7		1864.45	V
7		1881.89	V
7		1920.16	V
7		1988.75	V

TUNGSTEN (W)
Z = 74

W I and II
Ref. 1 — C.H.C.

Intensity		Wavelength	
		Air	
5800		2001.71	II
13000		2008.07	II
5100		2009.98	II
4100		2010.23	II
4100		2014.23	II
7300		2026.08	II
15000		2029.98	II
2700		2035.03	II
5300		2049.63	II
2300		2065.57	II
3400		2071.21	II
2200		2075.59	II
9700		2079.11	II
3600		2088.19	II
2200		2089.14	II
1700		2090.48	I
6100		2094.75	II
2400		2098.60	II
2200		2100.67	II
1500		2101.54	I
1500		2106.18	II
1300		2110.34	II
2100		2118.87	II
2400		2121.59	II
850		2153.56	II
850		2157.80	II
1500		2166.32	II
480		2182.90	I
440		2194.52	II
1300		2204.48	II
460		2248.75	II
460		2249.80	I
180		2270.24	II
95		2271.37	I
510		2277.58	I
160		2284.91	I
320		2285.17	I
530	d	2294.49	I
		2294.54	II
270		2298.33	I
240		2303.83	II
240		2306.59	I
340		2309.02	I
440		2313.17	I
220		2314.17	I
190		2315.02	II
460		2321.63	I
290		2326.09	II
390	d	2326.56	I
		2326.70	I
75		2328.31	II
130		2333.77	II
210		2341.37	I
75		2349.26	II
120	d	2350.37	II
320		2354.61	I
60		2358.81	II
580		2360.44	I
850		2363.07	I
60		2364.22	II
510		2374.47	I
210		2382.99	I
670		2384.82	I
240		2389.08	I
120		2390.37	II
120		2392.93	II
730		2397.09	II
560		2397.73	I
560		2397.98	I
75		2404.24	II
1700	d	2405.58	I
		2405.69	I
75		2411.54	II
320		2414.04	I
610		2415.68	I
50		2419.34	II
50		2421.01	II
870		2424.21	I
190		2427.49	II
170		2429.39	II
580		2431.08	I
630		2433.98	I
60		2435.01	II
1800		2435.96	I
250		2436.62	I
580		2444.06	I
160		2446.39	II
270		2448.39	I
270		2451.35	I
780		2451.48	II
870		2452.00	I
430		2454.72	I
630		2454.98	I
780		2455.51	I
780		2456.53	I
1100		2459.30	I
270		2460.16	I
480		2462.79	I
270		2464.30	I
230		2466.52	II
1400		2466.85	I
75		2470.80	II
480		2472.51	I
1200		2474.15	I
290		2477.80	II
870		2480.13	I
390		2480.96	I
1500		2481.44	I
480	d	2482.10	I
		2482.21	I
29		2484.40	II
580		2484.74	I
390		2487.50	I
270		2488.77	II
390		2489.23	II
75		2492.93	II
630		2495.26	I
230		2496.64	II
95		2497.48	II
140		2499.69	II
40		2500.11	II
		2501.90	II
680		2504.70	I
270		2506.02	I
24		2508.00	II
250		2510.17	I
75		2510.47	II
60		2518.14	II
310		2520.46	I
780		2521.32	I
270		2522.04	II
780		2523.41	I
430		2527.76	I
780		2533.64	I
50		2534.82	II
580		2545.34	I
1200		2547.14	I
50		2549.09	II
40		2550.10	II
780		2550.38	I
2700		2551.35	I
450		2553.82	I
410		2554.86	II
580		2555.09	II
		2555.21	I
310		2556.75	I
290		2560.12	I
730		2561.97	I
230		2563.16	II
110		2563.91	II
530		2571.44	II
170	d	2572.24	II
		2572.35	II
75		2573.95	II
190		2579.26	II
290		2580.34	I
870		2580.49	I
40		2581.20	II
390		2584.39	I
390		2589.17	II
170		2591.49	II
110		2598.74	II
370		2601.96	I
75		2602.51	II
75		2603.02	II
270		2603.54	I
680		2606.39	I
320		2607.38	I
370		2608.32	I
970		2613.08	I
480		2613.82	I
230		2615.12	I
70		2615.44	II
210		2619.18	I
400		2620.25	I
400		2622.21	I
400		2625.22	I
210		2628.26	I
400		2632.48	I
400		2632.70	I
810		2633.13	I
290		2636.54	I
400	d	2638.62	I
		2638.75	I
210		2645.69	I
650		2646.18	I
400		2646.73	I
75		2647.74	II
40		2653.42	II
80		2653.57	II
1600		2656.54	I
400		2657.38	I
400	d	2658.04	II
		2658.18	I
810		2662.84	I
260		2664.97	I
75		2666.49	II
210		2669.30	I
810		2671.47	I
80		2673.59	II
650		2677.28	I
160	d	2677.79	II
		2677.91	I
400		2678.88	I
2100		2681.42	I
290		2683.35	I
210		2691.09	I
650		2695.67	I
210		2697.71	II
650		2699.59	I
400		2700.01	I
40		2701.48	II
160		2702.11	II
210		2702.52	I
400		2706.58	I
400		2708.59	I
400	d	2708.80	I
		2708.93	I
80		2709.58	II
40		2710.78	II
400		2715.50	I
80		2716.32	II
80		2718.04	II
2100		2718.91	I
320		2719.33	I
210		2719.86	I
2800		2724.35	I
210		2724.62	I
400		2725.03	I
		2725.06	I
80		2729.62	II
75		2740.79	II
650		2748.84	I
40	d	2760.74	II
80		2761.59	II
400		2762.34	I
400		2764.27	II
210		2768.98	I
400		2769.74	I
810		2770.88	I
210		2773.70	I
810		2774.00	I
810		2774.48	I
160		2776.50	II
40		2778.69	II
210		2787.98	I
340		2791.96	I
810		2792.70	I
80		2799.03	II
400		2799.93	I
160	d	2801.05	II
		2801.17	I
130		2805.92	II
40		2812.25	II
810		2818.06	I
160		2822.57	II
260		2829.82	I
1600		2831.38	I
810		2833.63	I
210		2835.64	I
400		2841.57	I
810		2848.02	I
650		2856.03	I
650		2866.06	I
230		2878.72	I
610		2879.11	I
610		2879.40	I
440		2896.01	I
1500		2896.44	I
230		2910.48	I
270		2911.00	I
360		2918.25	I
50		2918.63	II
360		2923.10	I
230		2923.54	I
230		2925.13	I
690		2935.00	I
2400		2944.40	I
2400		2946.99	I
480		2947.39	I
210		2952.29	II
440		2964.52	I
480		2977.11	I
		2977.21	I
730	d	2979.71	I
		2979.86	I
400		2993.61	I
240		2995.26	I
190		3009.09	I
360		3013.79	I
520		3016.47	I
770		3017.44	I
110		3024.50	II
210		3024.93	I
310	d	3026.67	I
		3026.79	I
160		3033.56	I
160		3034.19	I
160		3039.31	I
440	d	3041.73	I
		3041.86	I
270		3043.80	I
440		3046.44	I
110		3048.66	I
810		3049.69	I
110		3064.93	I
180		3073.28	I
110		3077.52	II
180	d	3084.83	I
		3084.91	I
370		3093.50	I
240		3107.23	I
240		3108.02	I

Tungsten (Cont.)

230		3117.57	I
260		3120.18	I
160		3133.88	I
130		3141.42	I
65		3149.85	II
290		3163.42	I
130		3164.44	I
130		3165.38	I
320		3176.60	I
130		3179.06	I
190		3181.82	I
130		3184.05	I
130		3184.42	I
65		3189.24	II
390		3191.57	I
390		3198.84	I
520		3207.25	I
140		3208.28	I
1000		3215.56	I
140		3221.21	I
140		3221.91	I
190		3232.49	I
140		3237.09	I
140		3242.03	I
140		3252.29	I
210		3254.36	I
140		3259.43	I
210		3259.66	I
210	d	3266.62	I
		3266.77	I
150		3281.94	I
150		3293.71	I
730		3300.82	I
440		3311.38	I
440		3326.20	I
440		3331.69	I
150		3354.45	I
150		3371.04	I
390		3373.75	I
150		3412.96	I
150		3413.53	I
150		3422.42	I
150		3427.71	I
230		3429.59	I
240		3443.00	I
160		3477.94	I
400		3495.24	I
160		3508.73	I
160		3510.02	I
160		3526.85	I
160		3535.54	I
160		3537.45	I
650		3545.22	I
160		3568.04	I
240		3570.65	I
80		3572.48	II
160		3575.22	I
80		3592.42	II
240		3606.06	I
80		3613.79	II
1900		3617.52	I
160		3622.34	I
130		3627.24	I
320		3631.94	I
240		3641.41	II
80		3646.52	II
80		3657.59	II
160		3675.55	I
650		3682.08	I
400		3683.30	I
		3683.39	I
160		3683.93	I
570		3688.06	I
810		3707.92	I
60		3716.08	II
100		3719.39	I
50		3736.22	II
120		3741.71	I
510		3757.92	I
680		3760.13	I
1000		3768.45	I
120		3769.21	I
120		3769.86	I
340		3773.71	I
1000		3780.77	I
170		3792.76	I
290		3809.22	I
190		3810.38	I
260		3810.79	I
1400		3817.48	I
110		3829.13	I
1100		3835.06	I
290		3838.51	I
730		3846.22	I
250		3847.49	I
27		3851.57	II
150		3855.55	I
150		3859.30	I
180		3864.34	I
1800		3867.99	I
250		3872.84	I
110		3874.41	I
730		3881.41	I
110		3892.72	I
140	h	3897.91	I
150		3935.03	I
120		3936.97	I
120		3947.98	I
120		3952.52	I
120		3952.90	I
160		3953.15	I
200		3955.30	I
160		3965.14	I
130		3968.59	I
150	h	3970.80	I
130		3979.29	I
130		3980.64	I
250		3983.29	I
8600		4008.75	I
540		4015.22	I
170	h	4016.52	I
220		4019.23	I
130		4022.12	I
180		4028.79	I
180		4036.86	I
140		4039.85	I
140	h	4044.28	I
910		4045.59	I
180		4064.79	I
150		4069.79	I
730		4069.95	I
340		4070.61	I
100		4071.93	I
5000		4074.36	I
150		4082.96	I
130		4088.33	I
100		4095.69	I
1000		4102.70	I
150		4109.75	I
100		4111.82	I
150		4118.05	I
100		4118.19	I
100		4120.85	I
100		4125.16	I
150		4126.80	I
100		4133.48	I
540		4137.46	I
150		4138.02	I
110		4142.25	I
140		4145.16	I
110		4145.95	I
160		4154.66	I
160		4170.53	I
450		4171.17	I
160		4204.40	I
220		4207.05	I
110		4215.38	I
250		4219.37	I
110		4222.04	I
150		4234.34	I
290		4241.44	I
540		4244.36	I
290		4259.35	I
200		4260.29	I
200		4263.30	I
1400		4269.38	I
110		4269.77	I
220		4274.55	I
160		4275.49	I
160		4276.74	I
110		4282.34	I
110		4286.01	I
110		4294.10	I
4100		4294.61	I
2200		4302.11	I
160		4306.87	I
110		4307.64	I
110		4332.13	I
100		4347.00	I
150		4355.17	I
100		4361.81	I
150		4364.78	I
100	d	4365.95	I
		4366.07	I
150		4372.52	I
200		4378.48	I
180		4384.85	I
100		4403.95	I
200		4408.28	I
130		4412.19	I
160		4436.90	I
140		4460.49	I
140		4466.34	I
140		4466.74	I
640		4484.19	I
160		4504.84	I
130		4512.88	I
120		4513.25	I
150		4543.54	I
150		4546.47	I
150		4551.82	I
140		4570.64	I
170		4588.73	I
140		4599.94	I
140		4609.89	I
160		4613.30	I
100		4642.53	I
130		4657.42	I
640		4659.87	I
640		4680.51	I
100		4693.72	I
140		4757.54	I
790		4843.81	I
380		4886.90	I
220		4982.59	I
330		5006.15	I
220		5015.30	I
820		5053.28	I
210		5054.60	I
210		5069.12	I
120		5071.74	I
770		5224.66	I
27		5500.49	I
27		5503.44	I
10		5508.61	I
220		5514.68	I
15		5531.38	I
15		5537.72	I
13		5568.09	I
13		5604.31	I
11		5631.27	I
27		5631.94	I
65		5648.37	I
35		5660.72	I
27		5674.39	I
13		5676.60	I
15		5676.90	I
15		5697.79	I
55		5735.09	I
13		5749.24	I
11		5756.10	I
13		5793.06	I
13		5796.49	I
45		5804.85	I
13	d	5806.05	I
		5806.24	I
13		5833.61	I
13		5838.97	I
17		5845.27	I
28		5851.58	I
11		5856.61	I
22		5864.63	I
11		5874.22	I
13		5880.21	I
13		5891.61	I
13		5901.20	I
40		5902.64	I
13		5928.58	I
55		5947.57	I
13		5953.96	I
13		5956.19	I
27		5960.83	I
55		5965.86	I
27		5972.51	I
20		5978.86	I
20		5983.82	I
13		6009.01	I
55		6012.78	I
40		6021.52	I
20		6028.32	I
20		6043.31	I
13		6049.92	I
13		6065.08	I
22		6081.44	I
13		6111.66	I
13		6115.52	I
22		6128.25	I
13		6143.94	I
20		6153.72	I
20		6154.87	I
20		6203.51	I
20		6254.28	I
27		6285.88	I
45		6292.02	I
20		6303.21	I
13		6386.47	I
35		6404.21	I
40		6445.12	I
11		6508.05	I
15		6532.39	I
13		6538.11	I
13		6563.20	I
20		6573.93	I
11		6607.13	I
11		6609.05	I
17		6611.62	I
11		6621.74	I
13		6678.42	I
15		6693.08	I
5		6746.56	I
5		6764.45	I
7		6805.31	I
9		6814.92	I
9		6820.27	I
8		6828.43	I
4		6853.74	I
4		6876.01	I
5		6908.29	I
9		6934.23	I
8		6964.12	I
13		6984.27	I
8		6993.27	I
4		6994.06	I
8		7017.88	I
3		7028.68	I
3		7098.22	I
4	h	7111.18	I
15		7140.52	I
9		7162.64	I
5	h	7191.33	I
5		7198.62	I
11		7200.16	I
5		7216.35	I
4		7226.06	I
8		7237.12	I
5		7274.47	I
10		7278.24	I
15		7285.81	I
15		7296.55	I
3		7298.25	I
7		7385.08	I
4		7451.39	I
3		7456.37	I
8		7483.35	I
7		7504.13	I
10		7509.00	I
3		7520.66	I
9		7537.45	I
9		7550.48	I
17		7569.92	I
5		7582.88	I
3		7612.18	I
17		7614.15	I
3		7631.29	I
3		7654.81	I
13		7688.97	I
4		7701.01	I
5		7761.16	I
3		7776.73	I
11		7784.15	I
7		7808.96	I
2		7823.82	I
4		7863.47	I
2		7867.04	I
4		7880.40	I
5		7886.48	I
9		7940.92	I
3		7957.06	I
22		8017.19	I
7		8054.89	I
22		8055.64	I
5		8060.38	I
13		8123.82	I
5		8143.19	I
3		8165.72	I
5		8210.22	I
4		8322.05	I
10		8338.08	I
4		8348.81	I
7		8358.72	I
3		8382.94	I
4		8402.60	I
4		8475.14	I
27		8585.11	I
10		8594.42	I
8		8613.27	I
3		8614.50	I
13		8865.53	I

URANIUM (U)
Z = 92

U I and II
Ref. 1, 303 — J.G.C.

Intensity	Wavelength	
	Air	
440	2565.41	II
340	2569.71	II
340	2591.25	II
610	2635.53	II
470	2645.47	II
340	2669.17	II
470	2683.28	II
320	2691.04	II
370	2706.95	II
370	2733.97	II
470	2754.16	II
340	2762.85	II
390	2770.04	II
410	2784.45	II
830	2793.94	II
870	2802.56	II
630	2807.05	II
440	2808.98	II
630	2817.96	II
870	2821.12	II
390	2824.37	II
680	2828.90	II
920	2832.06	II
360	2837.19	II
460	2839.89	II

Uranium (Cont.)

360		2842.09	II
360		2849.48	II
390		2860.47	II
970		2865.68	II
340		2870.97	II
490		2882.74	II
460		2887.25	II
410		2888.26	II
1200		2889.62	II
320		2894.14	II
410		2894.51	II
780		2906.80	II
780		2908.28	II
320		2914.25	II
360		2914.63	II
440	p	2921.68	II
320		2927.38	II
490		2928.60	II
580		2931.41	II
660		[illegible]	II
340		2933.86	II
530	p	2940.37	II
1300		2941.92	II
830		2943.90	II
340		2948.09	II
390		2954.77	II
580		2956.06	II
460		2965.03	II
580		2967.94	II
580		2971.06	II
410		2976.35	II
320		2982.74	II
530		2984.61	II
410		2992.72	II
360		3007.91	II
320		3021.22	II
630		3022.21	II
320		3024.51	II
320		3028.19	II
630		3031.99	II
490		3033.18	II
490		3044.16	II
580		3050.20	II
630		3057.91	II
460		3061.62	II
630		3062.54	II
580		3072.78	II
580		3093.01	II
320		3095.75	II
320		3098.01	II
580		3102.39	II
460		3104.15	II
970		3111.62	II
530		3119.35	II
680		3124.95	II
530		3139.61	II
410		3144.97	II
490		3145.56	II
680		3149.24	II
530		3153.11	II
340		3176.21	II
340		3177.33	II
340		3206.05	II
730		3229.50	II
680		3232.16	II
440		3244.22	II
340		3265.79	II
440		3270.12	II
440		3288.21	II
730		3291.33	II
1100		3305.89	II
390		3337.79	II
440		3341.66	II
390		3357.84	I
730		3390.38	I
340		3394.77	II
580		3424.56	II
580		3435.49	I
360		3453.55	II
320		3454.23	II
320		3457.05	II
320		3457.71	II
360		3459.92	I
320		3462.22	I
460		3463.55	I
630		3466.30	I
390		3472.52	II
320		3473.43	I
360		3480.36	I
680		3482.49	II
1600		3489.37	I
390		3493.33	II
340		3494.00	I
320		3494.84	II
530		3496.41	II
630		3500.08	I
320		3504.01	I
320		3505.07	II
780		3507.34	I
320		3508.84	II
1600		3514.61	I
390		3509.66	II
320		3513.67	I
390		3519.96	II
390		3531.11	II
630		3533.57	II
320		3534.33	I
530		3540.47	II
320		3542.57	I
390		3547.19	II
320		3549.20	I
1200		3550.82	II
320		3552.17	II
680		3555.32	I
320		3561.41	I
1200		3561.80	I
390		3563.66	I
2300		3566.59	I
530		3569.08	I
320		3574.76	I
360		3577.92	I
630		3578.72	II
360		3581.84	II
3200		3584.88	I
320		3590.50	II
390		3591.74	I
460		3593.52	II
460		3605.27	I
360		3606.32	II
320		3616.33	I
320		3616.76	II
320		3620.08	I
320		3622.70	I
390		3623.06	II
460		3630.73	II
840		3638.20	I
310		3640.76	II
420		3644.24	I
310		3645.03	II
660		3651.54	I
490		3652.06	I
960		3659.15	I
2800		3670.07	II
380		3678.75	II
540		3691.92	II
330		3693.70	II
540		3700.57	II
1100		3701.52	II
350		3713.55	I
300		3717.42	II
350		3718.11	II
350		3729.82	II
350		3732.62	II
350		3733.07	II
600		3738.04	II
300		3744.25	II
680		3746.42	II
350		3747.14	II
950		3748.68	II
600		3751.17	I
350		3752.66	II
350		3755.48	II
490		3758.35	I
350		3759.24	II
330		3763.26	I
490		3764.57	II
430		3766.89	II
330		3769.53	II
540		3773.43	I
300		3776.48	I
380		3780.71	II
1900		3782.84	II
430		3783.84	II
570		3793.10	II
380		3793.26	I
380		3793.57	II
380		3808.92	I
380		3809.22	II
1900		3811.99	I
380		3813.79	II
380		3814.06	II
750		3826.51	II
2000		3831.46	II
1200		3839.63	I
490	p	3848.60	II
620		3854.22	I
2400		3854.64	II
4900		3859.57	II
490		3861.17	II
1900		3865.92	II
380		3866.80	II
1500		3871.03	I
620		3874.04	II
620		3878.08	II
1000		3881.45	II
490		3882.36	II
380		3883.28	II
2200		3890.36	II
620		3892.68	II
490		3894.12	I
490		3896.77	II
620		3899.78	II
410		3902.55	II
460		3904.30	II
380		3906.45	I
330		3911.67	II
380		3915.88	II
330		3926.21	I
330		3926.72	I
430		3930.98	II
2000		3932.02	II
490		3935.38	II
330		3940.48	II
1200		3943.82	I
300		3948.44	I
300		3953.58	II
360		3954.67	II
350		3964.21	I
600		3966.52	II
1200		3985.79	II
460		3990.42	II
380		3992.53	II
350		3998.24	II
350		4004.06	II
430		4005.21	I
570		4017.72	II
300		[illegible]	II
1000		4042.75	I
520		4044.41	II
410		4047.61	I
1600		4050.04	II
540		4051.91	II
300		4054.30	II
430		4058.19	II
880		4062.54	II
520		4067.75	II
410		4071.12	II
300		4074.48	II
330		4076.69	II
330		4080.60	II
2200		4090.13	II
460		4093.03	II
380		4106.38	II
810		4116.10	II
410		4124.73	II
410		4128.34	II
460		4141.22	II
880		4153.97	I
380		4156.65	I
350		4163.68	II
1400		4171.59	II
300		4189.27	II
350		4222.37	I
1000		4241.67	II
520		4244.37	II
680		4341.69	II
430	h	4355.74	I
430		4362.05	I
330		4393.59	I
600		4472.33	II
240		4515.78	II
620		4543.63	II
300		4620.21	I
240		4627.07	II
210		4631.62	I
220		4646.60	II
140		4666.85	II
100		4671.40	II
170		4689.07	II
100		4702.51	II
160		4722.72	II
120		4731.59	II
100		4755.74	II
150		4756.81	I
100		4772.70	II
100		4860.99	II
110		5008.21	II
170		5027.38	I
70		5117.24	II
80		5160.32	II
55		5164.14	I
55		5184.57	II
45		5204.31	II
45		5247.75	II
45		5257.04	II
70		5280.38	I
55		5386.19	II
80		5475.70	II
70		5480.26	II
70		5481.20	II
45		5482.53	II
160		5492.95	II
70		5527.82	II
70		5564.17	I
45		5581.59	II
55		5620.78	I
70		5780.59	I
70		5798.53	II
45		5836.02	I
55		5837.68	II
230		5915.39	I
55		5971.50	I
100		5976.32	I
45		5997.31	I
28		6017.38	II
55		6051.74	II
45		6067.22	II
90		6077.29	I
28		6087.34	II
40		6171.86	I
35		6175.39	I
28		6280.18	II
28		6359.29	I
55		6372.46	I
28		6378.52	II
28		6392.77	I
90		6395.42	I
110		6449.16	I
35		6464.98	I
90		6826.92	I
35		6876.74	II
23		7074.79	I
27		7101.61	I
30		7128.90	I
16		7147.89	I
16		7254.45	I
23		7425.50	I
45		7533.93	I
16		7619.35	I
50		7881.94	I
18		7970.46	I
16		8174.66	I
18		8262.06	I
16		8318.35	I
16		8337.50	II
18		8381.87	I
16		8441.21	I
35		8445.39	I
18		8450.03	I
16		8570.52	I
75		8607.95	I
23		8691.28	I
18		8710.76	I
18		8753.69	I
30		8757.76	I
16		8951.96	I
16		8989.92	I
10		9093.67	I
10		9139.56	I
10		9201.51	I
10		9265.34	I
10		9276.44	I
10		9385.90	I
10		9653.26	I
10		9819.00	I
10		9819.05	I
10		9868.36	I
10		9932.76	I
10		9964.11	I
50		10157.91	I
50		10259.55	I
100		10554.93	I
50		10799.78	I
25		10823.93	I
25		11093.77	I
75		11167.84	I
50		11294.13	I
100		11384.13	I
25		11410.43	I
50		11503.38	I
25		11568.81	I
20		11784.72	II
100		11859.42	I
100		11908.83	I
100		12250.46	I
25		13088.28	I
100		13185.16	I
75		13306.23	I
100		13961.58	I
50		16906.00	I
50		17451.11	I
50		18136.65	I
50		18366.96	I
75		18634.43	I
25		19029.39	I
10		20201.13	I
10		20271.41	I
10		20374.13	I
10		20517.29	I
10		20690.64	I
10		20772.19	I
10		21008.38	I
10		21099.98	I
10		21112.14	I
20		21144.90	II
10		21674.51	I
10		21693.38	I
75		21910.22	I
10		22110.73	I
10		23156.76	I
10		23948.19	I
10		29557.07	I

VANADIUM (V)
Z = 23

V I and II
Ref. 1 — C.H.C.

Intensity		Wavelength	
		Air	
2100		2092.44	I

Vanadium (Cont.)

40		2384.00	II
40		2384.28	I
60		2386.96	I
60		2388.92	I
75		2390.87	I
75		2391.26	I
85		2392.90	I
70		2397.78	I
70		2398.27	I
70		2399.96	I
120		2406.75	I
110		2407.90	I
120		2415.33	I
120		2416.75	I
100		2420.12	I
100		2421.06	I
100		2421.98	I
110		2428.28	I
110		2435.52	I
140		2501.61	I
150		2506.90	I
240		2507.78	I
180		2511.65	I
180		2511.95	I
180		2517.14	I
240		2519.62	I
410		2526.22	I
210		2527.90	II
120		2528.47	II
150		2528.84	II
240		2530.18	I
110		2549.28	II
120		2552.65	I
210		2562.13	I
110		2564.82	I
230		2574.02	I
140		2630.67	II
130		2642.21	II
150		2645.26	I
140		2651.90	I
150		2656.22	I
180		2661.42	I
290		2672.00	II
380		2677.80	II
270		2678.57	II
380		2679.32	II
180		2682.87	II
180		2683.09	II
1100		2687.96	II
170		2688.72	II
150		2689.88	II
230		2690.24	II
240		2690.79	II
120		2696.99	I
120		2697.74	I
680		2700.94	II
380		2702.19	II
530		2706.17	II
150		2706.70	II
110		2707.86	II
170		2711.74	II
120		2714.20	II
640		2715.69	II
150		2722.56	I
240		2728.64	II
180		2731.35	I
100		2739.71	II
140		2753.40	II
140		2765.67	II
140		2777.73	II
120		2803.47	II
120		2846.57	I
110		2847.57	II
140		2852.87	I
140		2854.34	II
200		2855.22	I
180		2859.97	I
240		2864.36	I
170		2866.59	I
210		2868.10	II
140		2869.13	II
210		2870.55	I
110		2877.69	II
110		2879.16	II
350		2880.03	II
380		2882.50	II
380		2884.78	II
140		2888.25	II
380		2889.62	II
900		2891.64	II
530		2892.44	II
900		2892.66	II
1400		2893.32	II
360		2896.21	II
110		2899.60	I
360		2903.08	II
150		2906.13	I
900		2906.46	II
490		2907.47	II
2400		2908.82	II
710		2910.02	II
530		2910.39	II
560		2911.06	II
380		2914.93	I
120		2917.37	II
210		2919.99	II
380		2920.38	II
710		2923.62	I
2400		2924.02	II
1700		2924.64	II
710		2930.81	II
210		2934.40	II
110		2935.87	I
900		2941.37	II
450		2941.49	II
230	d	2942.33	I
230		2943.20	I
1100		2944.57	II
110		2946.53	I
230		2949.63	I
300		2950.35	II
640		2952.08	II
120		2954.33	I
260		2957.52	II
410		2962.77	I
600		2968.38	II
120		2972.25	II
120		2976.20	II
380		2976.52	II
240		2977.54	I
260		3001.20	II
140		3014.82	II
180		3016.78	II
270		3033.45	II
290		3033.82	II
230		3043.12	I
230		3043.56	I
230		3044.94	I
230		3048.22	II
170		3050.89	I
180		3053.39	II
450		3053.65	I
1200		3056.33	I
1400		3060.46	I
140		3063.25	II
2400		3066.38	I
200		3067.12	II
140		3069.64	I
170		3073.82	I
100		3075.27	I
150		3082.11	I
3800		3093.11	II
200		3094.20	II
180		3100.94	II
3000		3102.30	II
2600		3110.71	II
2000		3118.38	II
380		3121.14	II
150		3122.90	II
1500		3125.28	II
260		3126.22	II
530		3130.27	II
410		3133.33	II
210		3134.93	II
150		3136.51	II
150		3139.74	II
200		3142.48	II
150		3145.34	II
3200		3183.41	I
5300		3183.98	I
3800		3185.40	I
410		3187.71	II
530		3188.51	II
750		3190.68	II
530		3198.01	I
750		3202.38	I
450		3205.58	I
450		3207.41	I
410		3212.43	I
210		3217.11	II
150		3237.87	II
140		3254.77	II
140		3263.24	I
1100		3267.70	II
900		3271.12	II
750		3276.12	II
110		3279.84	II
140		3298.14	I
110		3329.86	I
110		3365.55	I
110		3377.62	I
170		3400.40	I
110		3425.07	I
110		3485.92	II
210		3504.44	II
560		3517.30	II
150		3520.02	II
110		3524.72	II
230		3529.74	I
230		3530.77	II
560		3533.68	I
110		3543.50	I
560		3545.20	II
110		3553.27	I
560		3556.80	II
110		3566.18	I
560		3589.76	II
490		3592.02	II
560		3592.53	I
270		3593.33	II
110		3606.69	I
110		3639.02	I
110		3644.71	I
250		3663.59	I
250		3667.74	I
110		3669.41	II
170		3671.20	I
280		3673.40	I
280		3675.70	I
170		3676.68	I
300		3680.11	I
570		3683.13	I
190		3686.26	I
470		3687.47	I
1300		3688.07	I
1000		3690.28	I
1500		3692.22	I
450		3695.34	I
1000		3695.86	I
3800		3703.58	I
1800		3704.70	I
570		3705.04	I
130		3708.72	I
320		3715.47	II
250		3727.34	II
280		3732.76	II
150		3734.43	I
230		3745.80	II
210		3750.87	II
210		3770.97	II
270		3778.68	I
520		3790.32	I
1100		3794.96	I
570		3799.91	I
570		3803.47	I
190		3806.80	I
300		3807.50	I
520		3808.52	I
230		3809.60	I
1000		3813.49	I
140	d	3817.84	I
1300		3818.24	I
230		3819.96	I
230		3821.49	I
570		3822.01	I
450		3822.89	I
300		3823.21	I
1700		3828.56	I
280		3834.22	I
160		3839.00	I
110		3839.38	I
570		3840.44	I
2600		3840.75	I
110		3841.89	I
380		3844.44	I
320		3847.33	I
110		3849.32	I
1200		3855.37	I
3000		3855.84	I
150		3862.22	I
130		3863.87	I
1300		3864.86	I
230		3867.60	I
170		3871.08	I
1500		3875.08	I
420		3875.90	I
570		3876.09	I
130		3878.71	II
700		3890.18	I
460		3892.86	I
280	h	3898.02	I
140		3899.13	II
140	h	3900.18	I
140	h	3901.15	I
2400		3902.25	I
100		3906.75	I
700		3909.89	I
100		3910.79	I
220		3912.21	I
140		3914.33	II
100		3916.41	II
100		3920.49	I
100		3921.90	I
230		3922.43	I
240		3924.66	I
150		3925.24	I
200		3927.93	I
260		3930.02	I
150		3931.34	I
260		3934.01	I
150		3935.14	I
100		3936.28	I
150		3943.66	I
100		3950.23	I
140		3951.97	II
100		3973.64	II
540		3990.57	I
260		3992.80	I
430		3998.73	I
170		4005.71	II
120		4023.39	II
120		4031.83	I
150		4035.63	II
120		4042.64	I
360		4050.96	I
360		4051.35	I
280		4057.07	I
130		4057.82	I
230		4063.93	I
230		4071.54	I
1100		4090.58	I
180		4092.41	I
1800		4092.69	I
120		4093.50	I
890		4095.49	I
2800		4099.80	I
590		4102.16	I
230		4104.40	I
260		4104.78	I
2800		4105.17	I
120		4108.22	I
2300		4109.79	I
8900		4111.78	I
120		4112.33	I
230		4113.52	I
4300		4115.18	I
1800		4116.47	I
180		4118.18	I
180		4118.64	I
230		4119.46	I
180		4120.54	I
180		4123.19	I
2000		4123.57	I
120		4124.07	I
3100		4128.07	I
120		4128.86	I
3100		4132.02	I
2300		4134.49	I
150		4159.69	I
100		4174.01	I
230		4179.42	I
150		4182.59	I
180		4189.84	I
180		4191.56	I
230		4209.86	I
120		4226.62	I
360		4232.46	I
180		4232.95	I
180		4234.00	I
120		4235.76	I
100		4257.37	I
120		4259.31	I
120		4262.16	I
560		4268.64	I
460		4271.55	I
460		4276.96	I
430		4284.06	I
330		4291.82	I
220		4296.11	I
170		4297.68	I
170		4298.03	I
170		4306.21	I
140		4307.18	I
170		4309.80	I
460		4330.02	I
510		4332.82	I
760		4341.01	I
1000		4352.87	I
130		4354.98	I
150		4355.94	I
150		4368.04	I
140	d	4373.23	I
100		4375.30	I
12000		4379.24	I
100		4380.55	I
7000		4384.72	I
4800		4389.97	I
3600		4395.23	I
1400		4400.58	I
2300		4406.64	I
2800		4407.64	I
3600		4408.20	I
4600		4408.51	I
140		4412.14	I
640		4416.47	I
120		4419.94	I
640		4421.57	I
460		4426.00	I
120		4427.31	I
310		4428.52	I
230		4429.80	I
430		4436.14	I
640		4437.84	I
830		4441.68	I
640		4444.21	I
610		4452.01	I
410		4457.48	I
120		4457.76	I
1000		4459.76	I
2000		4460.29	I
610		4462.36	I
120		4468.01	I
380		4469.71	I
120		4474.04	I
200		4474.71	I
380		4488.89	I

Vanadium (Cont.)

100		4496.06	I
120		4501.95	I
140		4524.22	I
360		4545.39	I
100		4549.65	I
280		4560.71	I
200		4571.78	I
510		4577.17	I
140		4578.73	I
640		4580.40	I
830		4586.36	I
170		4591.22	I
1300		4594.11	I
100		4606.15	I
30		4609.65	I
25		4611.74	I
230		4619.77	I
65		4624.41	I
50		[illegible]	I
100		4635.18	I
65		4640.07	I
65		4640.74	I
130		4646.40	I
30		4648.89	I
30		4666.14	I
160		4670.49	I
24		4684.45	I
35		4686.92	I
55		4706.16	I
80		4706.57	I
80		4710.56	I
65		4714.12	I
35		4715.89	I
55		4717.69	I
40		4721.51	I
40		4722.86	I
40		4729.53	I
27		4730.38	I
27		4742.63	I
24		4746.63	I
40		4748.52	I
45		4750.98	I
35		4751.56	I
40		4753.93	I
65		4757.48	I
55		4766.63	I
130		4776.36	I
		4776.52	I
110		4786.51	I
130		4796.92	I
19		4799.77	I
130		4807.53	I
130		4827.45	I
130		4831.64	I
120		4832.43	I
19		4833.02	I
19		4848.81	I
320		4851.48	I
35		4862.61	I
480		4864.74	I
21		4871.26	I
620		4875.48	I
55		4880.56	I
740		4881.56	I
27		4891.60	I
21		4894.21	I
55		4900.62	I
95	d	4904.29	I
		4904.34	I
85		4925.65	I
35		4932.03	I
23		4966.12	I
70		5002.33	I
85		5014.62	I
28		5051.63	I
35		5064.12	I
35		5105.14	I
110		5128.53	I
110		5138.42	I
25		5139.53	I
70		5148.72	I
40		5159.35	I
23		5169.94	I
70		5176.77	I
20		5192.01	I
110		5192.99	I
23		5193.62	I
110		5194.83	I
55		5195.36	I
20		5206.61	I
40		5216.59	I
35		5225.77	I
35		5233.75	I
110		5234.07	I
20		5240.20	I
110		5240.87	I
17		5260.98	I
40		5353.41	I
35		5383.43	I
40		5385.14	I
14		5388.30	I
11		5397.87	I
100		5401.93	I
140		5415.26	I
28		5418.09	I
50		5424.08	I
40		5434.18	I
11		5437.66	I
17		5458.12	I
13		5471.33	
25		5487.22	I
85		5487.92	I
25		5489.94	I
28		5504.87	I
70		5507.75	I
14		5511.18	I
23		5545.93	I
70		5547.07	I
35		5558.75	I
28		5561.66	I
140		5584.50	I
23		5586.00	I
100		5592.42	I
28		5601.38	I
70		5604.94	I
13		5624.20	I
200		5624.60	I
70		5624.89	I
55		5626.01	I
400		5627.64	I
13		5632.46	I
10		5633.90	I
13		5635.51	I
85		5646.11	I
110		5657.44	I
110		5668.36	I
310		5670.85	I
20		5683.22	I
1200		5698.52	I
920		5703.56	I
570		5706.98	I
11		5708.95	I
11	h	5716.21	I
70		5725.64	I
850		5727.03	I
170		5727.66	I
230		5731.25	I
40		5734.01	I
230		5737.06	I
110		5743.45	I
17		5747.70	I
40		5748.87	I
17		5752.74	I
17		5761.41	I
70		5772.42	I
35		5776.64	I
11		5782.61	I
11		5783.50	I
40	h	5784.38	I
55	h	5786.16	I
23		5788.56	I
35	h	5807.14	I
23		5817.06	I
35	h	5817.53	I
55	h	5830.72	I
85	h	5846.30	I
11		5850.32	I
40		5924.57	I
28		5978.91	I
20		5980.78	I
28		6002.31	I
55		6002.63	I
28		6016.12	I
20		6025.41	I
450		6039.73	I
100		6058.14	I
20		6067.26	I
480		6081.44	I
1300		6090.22	I
28		6106.98	I
280		6111.67	I
600		6119.52	I
20		6128.34	I
280		6135.38	I
180		6150.15	I
85		6170.36	I
23		6189.35	I
450		6199.19	I
130		6213.87	I
450		6216.37	I
28	h	6218.31	I
130		6224.50	I
430		6230.74	I
100		6233.20	I
55		6240.13	I
170		6242.81	I
710		6243.10	I
280		6251.82	I
85		6256.90	I
85		6258.57	I
55		6261.22	I
85		6266.32	I
130		6268.82	I
170		6274.65	I
17	h	6282.33	I
200		6285.16	I
200		6292.83	I
170		6296.49	I
28	h	6311.50	I
14		6324.66	I
70		6326.84	I
55		6339.09	I
50		6349.48	I
14		6355.58	I
50		6357.30	I
25		6358.82	I
35		6361.27	I
23		6379.36	I
14		6393.28	I
35		6430.47	I
23		6431.63	I
14		6433.18	I
11		6435.16	I
70		6452.34	I
11		6488.05	I
55		[illegible]	I
110		6531.43	I
28		6543.51	I
17		6558.02	I
11		6565.88	I
50		6605.97	I
15		6607.83	I
10		6623.54	I
50		6624.85	I
13		6633.26	I
13		6643.79	I
8		6693.66	I
8		6708.07	I
65	c	6753.00	I
10		6760.12	I
50	c	6766.49	I
40		6784.98	I
15		6786.32	I
26		6812.40	I
9	c	6829.94	I
15		6832.44	I
12		6839.58	I
12		6841.90	I
10	c	6870.88	I
8		6871.56	I
7		6894.00	I
12		6974.50	I
21		7026.07	I
7		7063.69	I
11	h	7092.08	I
6		7102.58	I
24		7148.15	I
7		7151.36	I
7		7182.08	I
14		7264.29	I
8		7321.44	I
40		7338.92	I
35		7356.54	I
11		7358.66	I
24		7361.39	I
12		7362.49	I
24		7363.16	I
9		7385.95	I
6	h	7393.49	
12	h	7485.90	I
12	h	7488.08	I
12	h	7492.44	
12	h	7578.75	I
9	h	7591.24	I
14	h	7596.92	I
12	h	7598.28	I
24		7624.81	I
5		7701.37	I
8		7704.81	I
8	h	7851.18	
14	Bl	7865.51	VO
12		7896.40	
14	h	7898.81	
24		7937.92	I
29	c	8027.39	I
14		8028.13	I
14	h	8035.38	
14	h	8045.71	
12		8051.89	
14		8093.48	I
8		8102.44	I
12		8108.59	I
9	h	8109.07	I
120	Cw	8116.80	I
11	h	8136.79	I
29		8144.59	I
9		8154.55	I
70	c	8161.07	I
14		8171.35	I
7		8180.21	I
35		8186.71	I
24		8187.33	I
29		8198.87	I
35		8203.07	I
24		8241.61	I
29	c	8253.51	I
29		8255.88	I
5	h	8280.39	I
19		8282.37	I
8		8324.42	I
14	h	8331.23	I
14		8342.03	I
7		8402.81	I
12		8499.52	I
6		8534.49	I
6	Bl	8624.86	VO
60	c	8919.85	I
29	c	8932.93	I
12		8971.62	I

V III
Ref. 394 — C.H.C.

Intensity		Wavelength	
		Vacuum	
25		616.09	III
50		633.94	III
40		635.41	III
100		864.27	III
75		948.84	III
500		1006.46	III
500		1149.94	III
400		1157.18	III
300		1160.77	III
500		1252.11	III
400		1254.01	III
500		1287.87	III
400		1289.42	III
300		1290.77	III
400		1313.35	III
300		1313.27	III
500		1331.99	III
500		1335.12	III
1000		1643.03	III
1000		1650.14	III
300		1668.03	III
300		1670.66	III
300		1679.19	III
1000		1694.78	III
400		1721.98	III
300		1724.63	III
500	d	1751.68	III
500		1757.73	III
1000		1760.07	III
300		1773.43	III
400		1778.02	III
500		1779.72	III
400		1784.44	III
1000		1788.26	III
500		1793.82	III
1000		1794.60	III
300		1796.77	III
500		1798.15	III
300		1802.55	III
500		1804.13	III
1000		1812.19	III
400		1831.15	III
400		1831.64	III
300		1845.07	III
300		1850.69	III
400		1852.01	III
500		1854.42	III
300		1855.06	III
500		1856.64	III
300		1864.51	III
300		1878.68	III
400		1880.41	III
300		1895.01	III
400		1899.81	III
500		1902.23	III
300		1934.00	III
		Air	
500		2232.91	III
400		2241.53	III
1000		2292.86	III
400		2314.18	III
500		2318.06	III
400		2319.00	III
500		2323.82	III
2500		2330.42	III
500		2331.75	III
400		2334.20	III
500		2343.10	III
500		2358.73	III
500		2366.31	III
2500		2371.06	III
1000		2382.46	III
500		2393.58	III
500		2404.18	III
500		2516.14	III
250		2521.16	III
250		2521.55	III
250		2548.21	III
150		2554.22	III
150		2563.32	III
250		2593.05	III
250		2595.10	III
100		3679.86	III
50	h	3705.35	III

Vanadium (Cont.)

40		4714.89	III
50		6597.20	III

V IV
Ref. 397 — C.H.C.

Intensity		Wavelength	
Vacuum			
200		677.34	IV
60		678.74	IV
50		679.65	IV
500		684.37	IV
100		684.45	IV
100		691.53	IV
50		693.13	IV
400		737.85	IV
150		750.11	IV
60		1226.52	IV
50		1308.06	IV
80		1355.13	IV
60		1395.00	IV
50		1414.41	IV
80		1419.58	IV
100		1426.65	IV
60		1520.14	IV
80		1601.92	IV
80		1611.88	IV
80		1806.18	IV
60		1809.85	IV
100		1817.68	IV
200		1825.84	IV
300		1861.56	IV
500		1939.06	IV
400		1951.43	IV
300		1963.10	IV
500		1997.72	IV
200		1999.32	IV
Air			
100		2002.48	IV
50	h	2088.74	IV
50	h	2146.83	IV
100	h	2155.34	IV
500		2268.30	IV
50	h	2421.32	IV
50	h	2433.53	IV
50	h	2446.80	IV
50	h	2450.87	IV
50	h	2556.92	IV
80	h	2570.72	IV
50	h	2624.21	IV
80	h	2645.54	IV
50	h	2655.41	IV
50	h	2656.87	IV
50		3284.56	IV
60		3334.79	IV
50		3448.41	IV
50		3496.42	IV
80	h	3514.25	IV
50	h	4985.65	IV
50	h	5130.78	IV
50	h	5262.16	IV
60	h	5352.32	IV
40	h	5940.12	IV

V V
Ref. 432 — C.H.C.

Intensity		Wavelength	
Vacuum			
18		224.91	V
20		225.46	V
17		227.88	V
16		239.41	V
19		239.48	V
20		251.66	V
18		252.44	V
18		285.98	V
20		286.84	V
17		312.39	V
35		483.01	V
25		484.51	V
20		820.86	V
30		829.48	V
15		962.03	V
15		1142.74	V
25		1157.58	V
20		1490.11	V
100		1680.20	V
50		1716.72	V
15		1724.99	V
25		1792.99	V
30		1811.42	V
Air			
20	d	2319.66	V
20		2577.90	V
15		2775.82	V
15		3617.97	V
12	d	3746.36	V
20		4200.32	V
15	w	4930.53	V
8		5356.07	V
7		6628.80	V
3		7595.51	V

XENON (Xe)
Z = 54

Xe I and II
Ref. 33, 116, 118, 120, 232— S.P.D.

Intensity		Wavelength	
Vacuum			
350		740.41	II
350		803.07	II
600		880.80	II
350		885.54	II
600		925.87	II
250		935.40	II
800		972.77	II
700		976.68	II
500		1032.44	II
700		1037.68	II
1100		1041.31	II
1000		1048.27	II
1200		1051.92	II
2000		1074.48	II
600		1083.86	II
1200		1100.43	II
600		1158.47	II
250		1169.63	II
800	p	1183.05	II
250		1192.04	I
600		1244.76	II
250		1250.20	I
1000		1295.59	I
600		1469.61	I
Air			
200		2864.73	II
150	h	2895.22	II
400		2979.32	II
100	h	3017.43	II
300		3128.87	II
200	h	3366.72	II
2		3400.07	I
2		3418.37	I
2		3420.00	I
3		3442.66	I
100	h	3461.26	II
4		3469.81	I
4		3472.36	I
5		3506.74	I
10		3549.86	I
10		3554.04	I
15		3610.32	I
8		3613.06	I
6		3633.06	I
10		3669.91	I
40		3685.90	I
40		3693.49	I
100	l	3907.91	II
100		4037.59	II
200	l	4057.46	II
100	h	4098.89	II
200	l	4158.04	II
1000	h	4180.10	II
500	h	4193.15	II
300	h	4208.48	II
100	h	4209.47	II
300	h	4213.72	II
100		4215.60	II
300	h	4223.00	II
400	h	4238.25	II
500	h	4245.38	II
100	l	4251.57	II
500	h	4296.40	II
500	h	4310.51	II
1000	l	4330.52	II
200	h	4369.20	II
100	l	4373.78	II
500	h	4393.20	II
500	l	4395.77	II
200	l	4406.88	II
150	l	4416.07	II
500	h	4448.13	II
1000	h	4462.19	II
500	l	4480.86	II
100	l	4521.86	II
600		4734.152	I
150		4792.619	I
500		4807.02	I
400		4829.71	I
300		4843.29	I
500		4916.51	I
500		4923.152	I
200	l	4971.71	II
400		4972.71	II
300		4988.77	II
100	l	4991.17	II
200		5028.280	I
200		5044.92	II
1000		5080.62	II
300		5122.42	II
100		5125.70	II
100		5178.82	II
300		5188.04	II
400		5191.37	II
100		5192.10	II
500		5260.44	II
500		5261.95	II
2000		5292.22	II
300		5309.27	II
1000		5313.87	II
2000		5339.33	II
200		5363.20	II
200		5368.07	II
500		5372.39	II
100		5392.80	I
3000		5419.15	II
800		5438.96	II
300		5445.45	II
200		5450.45	II
400		5460.39	II
1000		5472.61	II
100	l	5494.86	II
200		5525.53	II
600		5531.07	II
100		5566.62	I
300		5616.67	II
300		5659.38	II
600		5667.56	II
150		5670.91	II
100		5695.75	I
200		5699.61	II
200		5716.10	II
500		5726.91	II
500		5751.03	II
300		5758.65	II
300		5776.39	II
100		5815.96	II
300		5823.89	I
150		5824.80	I
100		5875.02	I
300		5893.29	II
100		5894.99	I
200		5905.13	II
100		5934.17	I
500		5945.53	II
300		5971.13	II
2000		5976.46	II
200		6008.92	II
1000		6036.20	II
2000		6051.15	II
600		6093.50	II
1500		6097.59	II
400		6101.43	II
100		6115.08	II
100		6146.45	II
150		6178.30	I
120		6179.66	I
300		6182.42	I
500		6194.07	II
100		6198.26	I
100		6220.02	II
500		6270.82	II
400		6277.54	II
100		6284.41	II
100		6286.01	I
250		6300.86	II
500		6318.06	I
400		6343.96	II
600		6356.35	II
200		6375.28	II
100		6397.99	II
300		6469.70	I
150		6472.84	I
120		6487.76	I
100		6498.72	I
200	h	6504.18	I
300		6512.83	II
200		6528.65	II
100		6533.16	I
1000		6595.01	II
100		6595.56	I
400		6597.25	II
100		6598.84	II
150		6668.92	I
300		6694.32	II
200		6728.01	I
150		6788.71	II
100		6790.37	II
1000		6805.74	II
200		6827.32	I
100		6872.11	I
300		6882.16	I
80		6910.22	II
100		6925.53	I
800	h	6942.11	II
100		6976.18	I
2000		6990.88	II
150		7082.15	II
500		7119.60	I
50	s	7147.50	II
200		7149.03	II
500		7164.83	II
100		7284.34	II
200		7301.80	II
200		7339.30	II
100		7386.00	I
150		7393.79	I
300		7548.45	II
200		7584.68	I
80		7618.57	II
500		7642.02	I
100		7643.91	I
200		7670.66	II
60		7787.04	II
100		7802.65	I
100		7881.32	I
300		7887.40	I
500		7967.34	I
100		8029.67	I
200		8057.26	I
150		8061.34	I
100		8101.98	I
150	h	8151.80	II
100		8171.02	I
700		8206.34	I
10000		8231.635	I
500		8266.52	I
7000		8280.116	I
2000		8346.82	I
100		8347.24	II
2000		8409.19	I
50	h	8515.19	II
200		8576.01	I
50	h	8604.23	II
250		8648.54	I
100		8692.20	I
200		8696.86	I
50	h	8716.19	II
300		8739.39	I
100		8758.20	I
5000		8819.41	I
300		8862.32	I
200		8908.73	I
200		8930.83	I
1000		8952.25	I
100		8981.05	I
200		8987.57	I
400		9045.45	I
500		9162.65	I
100		9167.52	I
100		9374.76	I
200		9513.38	I
50	h	9591.35	II
150		9685.32	I
50	l	9698.68	II
100		9718.16	I
2000		9799.70	I
3000		9923.19	I
100		10838.37	I
90		11742.01	I
375		12235.24	I
100		12257.76	I
300		12590.20	I
2500		12623.391	I
250		13544.15	I
2000		13657.055	I
1250		14142.444	I
800		14240.96	I
375		14364.99	I
140		14660.81	I
3000		14732.806	I
100		15099.72	I
2500		15418.394	I
150		15557.13	I
250		15979.54	I
100		16039.90	I
1000		16053.28	I
125		16554.49	I
1500		16728.15	I
1500		17325.77	I
350		18788.13	I
150		20187.19	I
3000		20262.242	I
250		21470.09	I
1250		23193.33	I
110		23279.54	I
1800		24824.71	I
175		25145.84	I
2000		26269.08	I
2500		26510.86	I
250		28381.54	I
750		28582.25	I
300		29384.41	I
150		29448.06	I

Xenon (Cont.)

Intensity		Wavelength	
100		29649.58	I
100		29813.62	I
600		30253.14	I
1500		30475.46	I
100		30504.12	I
500		30794.18	I
6000		31069.23	I
125		31336.01	I
550		31607.91	I
100		32293.08	I
1800		32739.26	I
3500		33666.69	I
150		34014.67	I
450		34335.27	I
170		34744.00	I
5000		35070.25	I
110		35246.92	I
250		36209.21	I
150		36231.74	I
450		36508.36	I
850		36788.83	I
140		38685.98	I
175		38737.82	I
270		38939.60	I
120		39955.14	I

Xe III
Ref. 33, 384, 391, 429 — R.L.K.

Intensity		Wavelength	
		Vacuum	
8		657.8	III
8		660.1	III
9		673.8	III
9		674.0	III
9		676.6	III
10		694.0	III
20		698.5	III
12		705.1	III
10		721.2	III
15		731.0	III
10		733.3	III
15		742.6	III
10		756.0	III
10		761.5	III
10		769.1	III
25		779.1	III
15		792.9	III
12		796.1	III
15		802.0	III
25		823.2	III
30		824.9	III
25		853.0	III
15		889.3	III
20		894.0	III
20		896.0	III
10		965.5	III
35		1003.4	III
35		1017.7	III
10		1047.8	III
12		1066.4	III
30		1130.3	III
25		1232.1	III
		Air	
80		2668.98	III
100		2717.33	III
30		2814.45	III
40		2815.91	III
30		2827.45	III
40		2847.65	III
30		2862.40	III
80	w	2871.10	III
60	w	2871.24	III
30		2871.7	III
30		2896.62	III
50		2906.6	III
40		2911.89	III
80	w	2912.36	III
40		2940.2	III
60		2945.2	III
40		2947.5	III
40		2948.1	III
80	w	2970.47	III
40		2992.87	III
30		3004.25	III
100		3023.81	III
40		3083.5	III
50		3091.1	III
30		3106.46	III
100	w	3138.3	III
80	c	3150.82	III
40		3185.2	III
100		3242.86	III
80		3268.98	III
30		3287.82	III
80	w	3301.55	III
40		3331.6	III
30		3358.0	III
80		3384.12	III
60		3444.2	III
70		3454.2	III
100	w	3458.7	III
40		3468.22	III
80		3522.83	III
50		3542.3	III
50		3552.1	III
40		3561.4	III
100		3579.7	III
80		3583.6	III
100	w	3595.4	III
100		3606.06	III
40		3607.0	III
100	w	3615.9	III
40		3623.1	III
600		3614.08	III
50		3676.67	III
40		3776.3	III
300		3781.02	III
100		3841.5	III
200		3877.8	III
60		3880.5	III
500		3922.55	III
300		3950.59	III
200		4050.07	III
60		4060.4	III
100		4109.1	III
100		4145.7	III
30		4285.9	III
50		4434.2	III
100	w	4462.1	III
100	w	4569.1	III
100	w	4570.1	III
100	w	4641.4	III
30		4673.7	III
60		4683.57	III
30		4723.60	III
100	w	4757.3	III
40		4869.5	III
60		5239.0	III
30		5367.1	III
50		5401.0	III
40		5524.4	III
60		6205.97	III
25		6221.7	III
60		6238.2	III
60		6259.05	III

YTTERBIUM (Yb)
Z = 70

Yb I and II
Ref. 1 = C.H.C.

Intensity		Wavelength	
		Air	
2500		2116.67	II
3000		2126.74	II
370		2161.60	II
850		2185.71	II
640		2224.46	II
140		2320.81	II
50		2362.89	II
170		2390.74	II
18		2398.02	II
28		2421.35	II
25		2447.26	II
28		2460.25	II
460		2464.50	I
14		2484.89	II
70		2502.02	II
28		2505.48	II
11		2508.07	II
140		2512.06	II
18		2516.35	II
50		2522.44	II
65		2537.65	II
270		2538.67	II
14		2550.06	II
70		2552.15	II
55		2552.70	II
21		2565.57	II
28		2571.36	II
13		2573.15	II
18		2596.16	II
28		2596.32	II
21		2615.26	II
100		2617.01	II
55		2634.31	II
45		2639.45	II
85		2641.89	II
110		2644.31	II
28		2646.44	II
28		2647.46	II
28		2648.80	II
50		2649.79	II
28		2650.73	II
990		2653.75	II
35		2656.12	II
21		2659.27	II
200		2665.04	II
55		2668.75	II
390		2671.96	I
390		2672.66	II
21		2680.40	II
14		2683.42	II
70		2684.75	II
25		2687.98	II
28		2695.43	II
14		2696.62	II
18		2700.80	II
21		2708.84	II
65		2710.54	II
25		2711.78	II
55		2712.66	II
170		2718.35	II
21		2722.20	II
110		2732.74	II
21		2734.09	II
55		2741.71	II
55		2747.58	II
18		2748.04	II
230		2748.66	II
1300		2750.48	II
85		2751.45	II
21		2759.00	II
65		2760.78	II
65		2761.37	II
35		2764.41	II
85		2771.32	II
170		2776.28	II
100		2784.66	II
18		2787.96	II
45		2793.28	II
25		2794.44	II
21		2795.07	II
18		2795.29	II
35		2797.80	II
100		2798.21	II
45		2799.38	II
50		2800.00	II
35		2800.06	II
14		2810.72	II
65		2814.53	II
28		2816.32	II
140		2821.15	II
100		2824.97	II
190		2830.99	II
18		2832.20	II
28		2834.97	II
14		2842.59	II
230	h	2847.18	II
100		2848.44	II
21		2849.34	II
360		2851.13	II
55		2851.86	II
21		2853.41	II
18		2853.68	II
55		2854.14	II
45		2854.49	II
45		2858.33	II
45		2858.46	II
100		2859.39	II
430		2859.80	II
55		2860.39	II
140		2861.21	II
100		2861.34	II
200		2867.06	II
25		2870.06	II
45		2873.49	I
28		2885.97	II
70		2886.26	II
200		2888.04	II
3600		2891.38	II
45		2893.62	II
28		2896.90	II
85		2899.70	II
18		2902.41	II
21		2902.92	II
21		2906.88	II
28		2908.33	II
35		2909.19	II
55		2909.48	II
85		2911.52	II
18		2912.86	II
170		2914.21	II
140		2915.28	II
18		2916.43	II
280		2919.35	II
55		2921.12	II
45		2924.24	II
25		2927.85	II
35		2934.36	I
55		2935.11	II
21		2937.19	II
45		2939.53	II
45		2940.52	II
28		2942.04	II
140		2945.91	II
45		2946.30	II
18		2946.76	II
28		2950.33	II
45		2955.32	II
18		2957.63	II
65		2962.52	II
21		2963.26	II
45		2963.46	II
130		2964.76	II
2000		2970.56	II
45		2982.49	II
21		2982.66	II
28		2983.70	II
200		2983.99	II
90		2985.08	II
35		2985.88	II
45		2990.37	II
65		2991.87	II
28		2992.9[illegible]	II
170		2994.80	II
28		2995.86	II
70		3000.46	II
25		3002.61	II
310		3005.77	II
100		3009.39	II
65		3010.62	II
55		3014.43	II
160		3017.56	II
160		3026.67	II
920		3031.11	II
55		3034.64	II
25		3037.99	II
55		3039.67	II
80		3042.65	II
21		3044.00	II
45		3046.48	II
35		3047.05	II
45		3063.12	II
21		3063.67	II
110		3065.04	II
18		3076.01	II
100		3089.10	II
70		3093.87	II
28		3100.74	I
45		3101.36	II
28		3102.07	II
55		3107.76	II
170		3107.90	II
85		3115.34	II
55		3116.70	II
190		3117.81	II
50		3136.76	II
230		3140.94	II
80		3141.73	II
80		3145.06	II
28		3145.54	II
28		3153.18	II
90		3153.88	II
50		3155.18	II
28		3162.29	I
70		3163.80	II
50		3165.21	II
120		3169.06	II
120		3180.92	II
390		3192.88	II
70		3198.65	II
240		3201.16	II
80		3217.18	II
50		3218.32	II
50		3225.88	II
45		3239.20	II
35		3239.58	I
35		3246.06	II
130		3261.51	II
18000		3289.37	II
130		3305.25	I
140		3305.73	II
50		3315.10	II
80		3319.41	I
50		3333.06	II
240		3337.17	II
280	d	3342.93	II
		3343.07	II
80		3346.50	
50		3347.54	II
35		3351.09	II
50		3351.26	
100		3352.49	II
100		3362.44	II
50		3363.64	II
240		3375.48	II
50		3376.62	II
28		3382.54	
140		3387.50	I
50		3390.25	II
28		3390.42	II
50		3391.10	II
50	h	3394.44	II
50		3401.01	II
35		3404.10	II
50		3412.45	I
140		3418.39	I
360		3426.04	I
80		3428.46	II
240		3431.11	I
45		3434.61	II

Ytterbium (Cont.)

50		3438.71	II
100		3438.85	II
35		3443.59	
35		3446.89	II
85		3452.40	I
500		3454.08	II
190	d	3458.29	II
		3458.39	I
360		3460.27	I
35		3462.34	II
2400		3464.37	I
500		3476.30	II
500		3478.84	II
50		3482.56	II
85		3485.76	II
85		3488.43	II
100	Hw	3495.90	II
85		3507.83	II
50		3517.00	I
230		3520.29	II
50		3545.72	II
100		3549.82	II
35		3559.03	I
200		3560.33	II
170		3560.70	II
50	h	3563.94	II
85		3570.57	II
50		3572.50	II
50		3574.58	II
360		3585.47	II
130		3606.48	II
50		3610.23	II
70		3611.30	II
200		3619.80	II
110		3634.52	
240		3637.76	II
70		3648.15	I
90		3655.73	I
240		3669.69	II
50		3670.69	II
140		3675.08	II
50		3690.56	II
32000		3694.19	II
70		3698.60	II
70		3700.58	I
50		3710.34	II
60		3724.21	II
180		3734.69	I
550		3770.10	I
80		3774.32	I
60	h	3791.74	I
170		3839.91	I
340		3872.85	I
340		3900.85	I
50		3904.81	II
140		3911.27	I
32000		3987.99	I
930		3990.88	I
50		4007.36	I
70		4052.28	I
85		4077.28	II
440		4089.68	I
120	h	4119.25	II
70		4135.09	II
470		4149.07	I
120		4174.56	I
340		4180.81	II
150	d	4218.56	II
		4218.69	I
120		4231.97	I
70		4277.74	I
120		4305.97	I
70		4316.95	II
60	h	4393.69	I
60	h	4430.21	I
440		4439.19	I
85	h	4482.42	I
85		4515.16	II
35		4553.58	II
85	h	4563.95	I
640		4576.21	I
200		4582.36	I
70		4589.21	I
140		4590.83	I
40		4598.36	II
35		4683.81	II
40		4684.27	I
190		4726.08	II
170	h	4781.87	I
170		4786.61	II
35		4816.43	I
40		4820.24	II
35		4836.96	II
40		4837.46	I
17		4851.15	II
40	h	4894.60	I
27		4912.36	I
710		4935.50	I
24		4937.22	II
140		4966.90	I
24		5009.52	II
17		5067.30	II
30		5067.80	I
70		5069.14	I
220		5074.34	I
50		5076.74	I
20		5135.98	II
14		5147.02	II
20		5184.15	II
60		5196.08	I
85		5211.60	I
35		5240.51	II
100		5244.11	I
40		5257.49	II
150	h	5277.04	I
35		5279.53	II
17		5300.94	II
170		5335.15	II
30	d	5345.66	II
		5345.83	II
60		5347.22	II
30	h	5351.29	I
150		5352.95	II
30		5358.64	II
30		5363.66	I
17		5389.84	II
14		5432.71	II
40		5449.27	II
14		5478.50	II
60		5481.92	I
40		5505.49	I
17		5524.54	I
85	h	5539.05	I
2400		5556.47	I
35		5562.09	I
20		5568.11	I
20		5586.36	I
40		5588.45	II
60		5651.98	II
7		5686.53	II
220		5719.99	I
10		5749.91	II
10	h	5755.89	I
27		5771.66	II
10		5803.44	I
10		5819.41	II
35		5833.99	II
35		5837.14	II
27		5854.51	I
8		5897.21	II
20		5908.36	II
17		5989.33	I
40		5991.51	II
10		6052.88	II
10		6054.57	I
60		6152.57	II
30		6246.97	II
60		6274.78	II
14		6308.15	II
35	h	6400.35	I
35	h	6417.91	I
20		6432.73	II
17	h	6463.15	II
340		6489.06	I
20		6643.55	I
180		6667.82	I
15		6678.17	I
25		6727.61	II
25		6768.70	I
690		6799.60	I
18		6934.05	II
20		6999.88	II
10		7043.78	II
9	h	7244.41	I
8	h	7305.22	I
10	h	7313.05	I
16	h	7350.04	I
25		7448.28	I
30	h	7527.46	I
750		7699.48	I
7		7895.08	I
70	h	8922.56	II

Yb III
Ref. 40, 192 — J.R.

Intensity		Wavelength	
		Vacuum	
5		968.46	III
20		973.16	III
10		994.56	III
10		1560.66	III
80		1561.42	III
30		1669.60	III
50		1670.78	III
50		1719.82	III
60		1739.18	III
70		1762.80	III
80	h	1765.21	III
65		1775.29	III
70		1779.74	III
70		1781.31	III
20		1793.70	III
60		1798.85	III
65		1810.88	III
60		1826.41	III
20		1826.77	III
30		1838.01	III
30		1847.30	III
30		1849.24	III
10		1849.42	III
75		1852.36	III
75		1852.94	III
90		1854.80	III
80		1857.16	III
100		1863.32	III
5		1864.85	III
10		1867.23	III
10		1867.63	III
10		1868.19	III
5		1868.92	III
10		1870.07	III
15		1870.83	III
10		1871.15	III
200		1872.03	III
800		1873.91	III
100		1875.41	III
75		1875.92	III
70		1880.30	III
80		1884.22	III
70		1885.07	III
70		1887.22	III
10		1890.34	III
10		1890.87	III
15		1892.42	III
5		1895.50	III
100		1896.18	III
10		1897.57	III
500		1898.25	III
7		1906.74	III
10		1908.50	III
100		1909.66	III
70		1910.86	III
20		1920.53	III
10		1926.76	III
70		1928.09	III
15		1930.63	III
55		1942.59	III
40		1950.34	III
15		1962.80	III
80		1967.13	III
20		1969.47	III
30		1969.73	III
10		1973.96	III
10		1974.18	III
25		1976.46	III
2		1981.74	III
5		1983.88	III
25		1984.62	III
7		1985.74	III
80		1986.43	III
80	h	1989.82	III
50		1991.14	III
45		1995.05	III
55		1997.28	III
55		1997.66	III
500		1998.82	III
		Air	
20		2054.80	III
10		2066.49	III
10		2073.64	III
30		2078.05	III
10		2087.37	III
50		2087.98	III
20		2091.23	III
20		2092.26	III
10		2094.77	III
80		2095.31	III
15		2096.79	III
30		2098.36	III
10		2106.71	III
50		2109.54	III
20		2119.18	III
20		2198.14	III
80		2202.27	III
300		2240.11	III
100		2244.28	III
200		2257.03	III
100		2262.26	III
200		2265.67	III
150		2282.99	III
100		2283.99	III
300		2305.32	III
100		2309.27	III
200		2314.49	III
200		2337.97	III
40		2361.08	III
200		2365.43	III
50		2367.46	III
30		2369.99	III
20		2377.22	III
50		2403.95	III
20		2410.04	III
60		2412.33	III
10		2429.18	III
20		2433.43	III
100		2438.27	III
20		2439.31	III
20		2440.43	III
10		2458.64	III
10		2464.59	III
200		2490.42	III
20		2491.69	III
40		2506.25	III
300		2516.82	III
15		2522.07	III
20		2529.14	III
40		2550.39	III
300		2555.29	III
100		2560.56	III
10		2561.66	III
100		2566.78	III
2000		2567.61	III
1000		2579.57	III
100		2588.62	III
20		2592.69	III
500		2597.23	III
800		2599.14	III
30		2609.14	III
600		2621.11	III
300		2627.07	III
30		2635.37	III
500		2638.06	III
300		2640.48	III
1000		2642.56	III
100		2643.62	III
1000		2651.74	III
700		2652.25	III
100		2659.98	III
70		2664.89	III
2000		2666.13	III
2000		2666.99	III
30		2673.33	III
500		2677.39	III
500		2691.01	III
30		2708.04	III
400		2712.32	III
500		2749.91	III
200		2755.94	III
200		2756.76	III
100		2765.50	III
300		2788.24	III
600		2795.60	III
400		2803.32	III
1000		2803.43	III
10		2807.22	III
50		2808.51	III
600		2816.92	III
1000		2818.72	III
15		2826.01	III
300		2842.96	III
400		2875.86	III
600		2898.30	III
1000		2906.31	III
300		2928.97	III
50		2977.84	III
800		2998.00	III
2000		3029.49	III
100		3031.62	III
30		3040.65	III
3000		3092.50	III
20		3102.18	III
4000		3126.01	III
1000		3138.58	III
100		3151.44	III
70		3179.34	III
800		3191.35	III
50		3216.27	III
2000		3228.58	III
2000		3325.51	III
50		3358.25	III
20		3364.30	III
2000		3384.01	III
150		3392.56	III
100		3397.66	III
80		3432.94	III
40		3456.18	III
150		3463.51	III
20		3469.98	III
300		3550.87	III
200		3613.89	III
30		3659.84	III
30		3663.74	III
200		3664.74	III
20		3675.78	III
400		3711.91	III
20		3879.98	III
10		3882.58	III
20		3887.17	III
150		3896.55	III
15		3912.75	III
20		3913.23	III
500		3931.23	III
100		3985.56	III
10		3991.74	III
10		3997.67	III
2000		4028.14	III
10		4033.03	III
20		4074.53	III
20		4090.67	III

Intensity		Wavelength	
20		4098.23	III
15		4121.06	III
10		4150.04	III
15		4153.11	III
100		4162.72	III
60		4172.95	III
30		4194.34	III
100		4194.95	III
10		4198.74	III
300		4213.64	III
10		4220.83	III
15		4231.07	III
20		4289.64	III
40		4301.14	III
20		4304.01	III
15		4350.80	III
10		4380.07	III
100		4517.58	III
40		4639.14	III
10		4834.93	III
15		5054.94	III
20		5256.85	III
20		5331.54	III
15		5740.83	III
10	d	5949.02	III
20		5973.05	III
40		6055.85	III
100		6214.22	III
200		6328.52	III
10		6365.88	III
150		6378.33	III
25		6466.33	III
20		6985.15	III
10		7037.04	III
15		7157.72	III
10		7311.02	III
10		7399.98	III
80		7410.01	III
15		7456.86	III
70		7664.41	III
80		7892.39	III
20		7893.10	III
100		7971.46	III
20		8056.02	III
10		8117.44	III
10		8326.86	III
30		8327.88	III
20		8400.01	III
30		8489.90	III
200		10110.60	III
100		10830.36	III

Yb IV
Ref. 40, 311 — J.R.

Intensity		Wavelength	
		Vacuum	
200		828.96	IV
200		870.35	IV
300		902.46	IV
300		927.01	IV
300		936.22	IV
400		943.04	IV
400		946.20	IV
200		975.21	IV
1000		1050.24	IV
1000		1054.46	IV
400		1092.51	IV
200		1109.96	IV
200		1110.55	IV
5000		1134.43	IV
300		1136.24	IV
500		1166.01	IV
600		1185.58	IV
200		1290.24	IV
600		1305.58	IV
900		1316.04	IV
200		1326.32	IV
800		1326.36	IV
200		1340.06	IV
300		1345.36	IV
900		1350.26	IV
200		1353.43	IV
400		1356.15	IV
200		1361.75	IV
300		1365.88	IV
300		1369.72	IV
400		1375.42	IV
300		1376.66	IV
200		1384.41	IV
250		1393.93	IV
350		1398.77	IV
400		1407.05	IV
300		1413.14	IV
400		1416.15	IV
400		1417.72	IV
200		1423.99	IV
300		1430.29	IV
200		1440.61	IV
400		1477.92	IV
300		1491.57	IV
400		1765.03	IV
200		1776.18	IV
200		1778.20	IV
200		1779.34	IV
300		1789.71	IV
800		1791.06	IV
200		1801.67	IV
250		1809.63	IV
600		1813.84	IV
400		1816.07	IV
250		1817.58	IV
300		1819.02	IV
200		1824.22	IV
		Air	
300		2106.48	IV
900		2116.65	IV
500		2121.29	IV
250		2122.84	IV
800		2123.32	IV
600		2125.72	IV
200		2129.65	IV
500		2135.21	IV
300		2137.58	IV
500		2138.35	IV
200		2138.53	IV
800		2139.99	IV
400		2141.04	IV
200		2142.20	IV
300		2143.42	IV
300		2143.89	IV
20000		2144.77	IV
400		2148.10	IV
300		2148.52	IV
15000		2154.18	IV
250		2165.55	IV
300		2169.12	IV
200		2172.16	IV
300		2177.53	IV
400		2183.32	IV
270	h	2186.13	IV
270		2187.17	IV
90	h	2189.90	IV
120		2193.34	IV
150		2198.27	IV
90		2224.64	IV
150		2231.28	IV
90		2233.30	IV
90		2244.20	IV
140		2331.36	IV

YTTRIUM (Y)
Z = 39

Y I and II
Ref. 1 — C.H.C.

Intensity		Wavelength	
		Air	
350		2243.06	II
50		2354.20	I
30		2373.83	
50		2385.24	
25		2413.93	II
560		2422.20	II
60		2460.61	II
25		2490.42	I
12		2540.28	
14		2547.57	
10		2550.17	
20		2681.65	I
60		2694.21	I
26		2695.39	I
95		2723.00	I
22		2730.08	I
22		2734.85	II
70		2742.53	I
140		2760.10	I
30		2785.21	II
12		2785.59	II
12		2791.20	I
30		2800.11	II
26		2813.64	I
18		2818.86	I
45		2822.56	I
22		2825.37	II
45		2826.38	II
70		2854.43	II
26		2856.30	II
11		2857.87	II
95		2886.48	I
18		2897.69	II
14		2898.82	II
160		2919.05	I
18	h	2930.03	II
390		2948.40	I
350		2964.96	I
18		2973.91	II
480		2974.59	I
30		2980.55	II
750		2984.26	I
70		2995.26	I
140		2996.94	I
70		3005.26	I
55		3018.95	I
130		3021.73	I
90		3022.28	I
26		3026.49	II
30		3036.59	II
45		3044.84	I
190		3045.37	I
22		3047.11	I
60		3055.22	II
60		3086.85	II
55	h	3091.70	I
22		3093.76	II
95		3095.88	II
45		3111.81	I
55		3112.04	II
22		3114.28	I
60		3128.77	II
80		3129.93	II
95		3135.17	II
110		3173.06	II
220		3179.41	II
70		3191.31	I
2300		3195.62	II
2200		3200.27	II
2200		3203.32	II
3900		3216.69	II
6200		3242.28	II
310		3280.91	II
19		3308.47	II
4700		3327.89	II
55		3340.38	I
160		3362.00	II
85		3388.59	I
45		3397.04	I
85		3412.47	I
200		3448.82	II
70		3450.95	I
110		3467.88	II
170		3485.73	I
1700		3496.09	II
80		3521.53	I
45		3546.01	II
3900		3549.01	II
130		3551.80	I
540		3552.69	I
170		3558.76	I
190		3571.43	I
260		3576.05	I
3300		3584.52	II
300		3587.75	I
100		3589.69	I
2800		3592.92	I
10000		3600.73	II
6200		3601.92	II
7800		3611.05	II
4300		3620.94	I
1900		3628.71	II
7800		3633.12	II
3000		3664.61	II
45		3668.49	II
170		3692.53	I
13000		3710.30	II
60		3718.12	I
60		3738.61	I
1200		3747.55	II
50		3749.89	I
10000		3774.33	II
1400		3776.56	II
50		3782.30	II
7400		3788.70	II
1300		3818.35	II
4000		3832.88	II
70		3847.87	II
80		3876.82	I
480		3878.28	II
30		3887.77	I
60	h	3904.59	I
50	h	3918.25	I
60	h	3930.11	I
240		3930.66	II
4400		3950.36	II
150		3951.60	II
60	h	3955.09	I
3600		3982.60	II
40		3987.50	I
940		4039.83	I
2400		4047.64	I
9400		4077.38	I
90	h	4081.22	I
2000		4083.71	I
9900		4102.38	I
60	h	4106.39	I
80		4110.81	I
320		4124.92	II
8900		4128.31	I
7500		4142.85	I
100	h	4157.63	I
2400		4167.52	I
2000		4174.14	I
8000		4177.54	II
120		4199.28	II
380		4204.70	II
80		4213.02	I
40		4213.54	I
160		4217.80	I
280	h	4220.63	I
80		4224.25	I
600		4235.73	II
2200		4235.94	I
300		4251.20	I
360	h	4302.30	I
2800		4309.63	II
50		4316.30	I
110		4330.78	I
30		4337.29	I
60		4344.65	I
440	h	4348.79	I
60		4352.33	I
60		4352.70	I
120		4357.73	I
800		4358.73	II
120		4366.03	I
12000		4374.94	II
150	h	4375.61	I
80		4379.33	I
30		4385.48	I
100		4387.74	I
30		4394.01	I
30		4394.67	I
1800		4398.02	II
890		4422.59	II
80		4437.34	I
100		4443.66	I
130		4446.63	I
20		4465.27	I
40		4473.89	I
170		4475.72	I
180		4476.96	I
160		4477.45	I
110		4487.28	I
300		4487.47	I
30		4491.75	I
25		4492.42	I
500		4505.95	I
50		4513.58	I
80		4514.01	I
40	h	4522.05	I
890		4527.25	I
440		4527.80	I
100		4544.32	I
100		4559.37	I
30		4564.39	I
60	h	4573.56	I
35		4581.32	I
30		4581.77	I
130		4596.55	I
95		4604.80	I
40		4613.00	I
2000		4643.70	I
200	h	4658.32	I
70		4658.89	I
85		4667.47	I
60		4670.82	I
2000		4674.84	I
60		4678.35	I
260		4682.32	II
85		4689.77	I
180		4696.81	I
35		4708.85	I
60		4725.85	I
170		4728.53	I
60	h	4732.37	I
85		4741.40	I
160		4752.79	I
410		4760.98	I
17		4780.18	I
120		4781.04	I
160		4786.58	II
170		4786.89	I
180		4799.30	I
50		4804.31	I
70		4804.81	I
85	Bl	4817.38	YO
140	Bl	4818.20	YO
140		4819.64	I
120		4822.13	I
190		4823.31	II
60		4839.15	I
770		4839.87	I
550		4845.68	I
410		4852.69	I
120		4854.25	I
890		4854.87	II
50		4856.70	I
330		4859.84	I
50		4879.65	I
1900		4883.69	II
50		4886.28	I
40		4886.65	I
95		4893.44	I
1100		4900.12	II
100		4906.11	I

Yttrium (Cont.)

Intensity		Wavelength	Spectrum
45		4909.00	I
150		4921.87	I
35		4930.93	I
45		4950.66	I
120		4974.30	I
120		4982.13	II
100		5006.97	I
75		5070.21	I
75		5072.19	I
1100		5087.42	II
30		5088.18	I
210		5119.11	II
450		5123.21	II
180		5135.20	I
120		5196.43	II
960		5200.41	II
1500		5205.72	II
180		5240.81	I
60		5289.82	II
45		5320.78	II
75		5380.62	I
220		5402.78	II
24		5417.03	I
90		5424.37	I
190		5438.24	I
710		5466.46	I
100		5468.47	I
90		5473.39	II
90		5480.74	II
60		5493.17	I
35		5495.59	I
240		5497.41	II
300		5503.45	I
250		5509.90	II
60		5513.64	I
120		5521.63	I
		5521.70	II
24		5526.76	I
740		5527.54	I
35		5541.63	I
120		5544.50	I
		5544.61	II
90		5546.02	II
75		5556.43	I
60		5567.75	I
180		5577.42	I
24		5581.08	I
620		5581.87	I
21		5590.96	I
21		5594.12	I
120		5606.33	I
15		5623.91	I
560		5630.13	I
24		5632.25	I
21		5632.89	I
120		5644.69	I
120		5648.47	I
740		5662.94	II
90		5675.27	I
18		5693.63	I
160		5706.73	I
24		5720.61	I
75		5728.89	II
150	Bl	5730.12	YO
21		5732.09	I
90		5743.85	I
18	Bl	5746.93	YO
24	Bl	5764.22	YO
75		5765.64	I
35		5773.95	I
100		5781.69	II
15	Bl	5800.00	YO
15	Bl	5818.58	YO
30		5821.87	I
21		5832.27	I
9	b	5838.07	YO
15		5858.83	YO
15		5871.83	I
24		5876.14	YO
24		5879.96	I
24	b	5893.94	YO
35		5902.96	I
24	b	5912.19	YO
24	b	5931.10	YO
90	Bl	5939.08	YO
45		5945.72	I
24		5950.02	I
75	b	5956.41	YO
1300	Bl	5972.04	YO
50		5981.86	I
1000	Bl	5987.64	YO
740	Bl	6003.60	YO
120		6004.65	
120		6009.19	I
620	Bl	6019.87	YO
120		6023.41	I
500	Bl	6036.60	YO
420	Bl	6053.81	YO
130	Bl	6072.78	YO
50		6088.00	I
210	Bl	6089.35	YO
160	Bl	6096.78	YO
130	Bl	6107.82	YO
130	Bl	6114.73	YO
75	Bl	6127.38	YO
1400	Bl	6132.06	YO
120		6135.04	I
150		6138.43	I
1100	Bl	6148.36	YO
120		6151.72	YO
820	Bl	6165.08	YO
560	Bl	6182.23	YO
1200		6191.73	I
590	Bl	6199.82	YO
450	Bl	6217.96	YO
300		6222.59	I
270	Bl	6236.72	YO
45		6251.05	I
120	Bl	6275.01	YO
60	b	6295.46	YO
24	b	6316.20	YO
24	b	6338.10	YO
15	b	6359.48	YO
15		6369.87	YO
75		6402.01	I
1000		6435.00	I
24		6437.18	I
18	h	6501.23	YO
18	h	6518.33	YO
18	h	6535.84	YO
90		6538.60	I
12	h	6553.84	YO
70		6557.39	I
12	h	6572.58	
35		6576.85	I
23		6584.87	I
95		6613.75	II
14		6622.49	I
19	h	6636.49	I
40		6650.61	I
21		6664.40	I
150		6687.58	I
14	h	6691.83	I
7		6694.75	I
16	h	6699.26	I
70		6700.71	I
35		6713.20	I
40		6735.99	I
190		6793.71	I
70	h	6795.41	II
12	h	6803.15	I
21		6815.16	I
14		6832.49	II
45		6845.24	I
14		6858.24	II
29		6887.22	I
21		6896.00	II
9		6908.26	I
14		6933.52	I
24	h	6950.31	I
10		6951.68	II
10		6958.04	I
24		6979.88	I
13	h	7008.97	I
10		7009.93	I
19	h	7035.18	I
29		7052.94	I
13	h	7054.28	I
9		7075.13	I
11		7127.92	I
35		7191.66	I
10	h	7195.93	I
35		7264.17	II
9	h	7293.08	I
9	h	7330.62	I
5		7332.96	II
50		7346.46	I
11	h	7398.77	I
29		7450.30	II
17		7494.88	I
7	h	7536.71	I
35		7563.13	I
8	h	7617.72	I
19	h	7622.94	I
7		7652.89	I
5		7689.49	I
8	h	7698.00	I
19		7719.89	I
19		7724.08	I
13		7788.42	I
13		7796.32	I
6		7802.52	I
17		7812.16	I
29		7855.52	I
110		7881.90	II
10	h	7999.33	I
9		8329.61	I
24		8344.43	I
8	h	8365.64	I
17		8450.36	I
8	h	8528.94	I
95		8800.62	I
19	h	8835.85	II

Y III
Ref. 77 — J.R.

Intensity		Wavelength	
		Vacuum	
1		643.68	III
4		646.69	III
6		653.87	III
10		656.98	III
25		668.74	III
40		671.98	III
100		691.72	III
4		693.85	III
200		695.20	III
9		727.91	III
4		728.47	III
2		728.83	III
20		729.73	III
600		730.49	III
15		732.70	III
800		734.36	III
15		770.78	III
10		771.79	III
20		804.26	III
5000		805.20	III
75		806.18	III
150		808.97	III
7000		809.92	III
100		855.64	III
60		857.82	III
25		984.23	III
15		987.96	III
15000		989.21	III
25000		996.37	III
20		999.19	III
150		1000.56	III
25		1003.35	III
1000		1006.58	III
1200		1007.86	III
120		1077.52	III
500		1081.35	III
75	p	1084.63	III
350	p	1088.39	III
250		1095.25	III
25		1095.87	III
150		1103.21	III
3000		1289.74	III
2500		1306.96	III
5000		1314.51	III
1500		1316.10	III
4000		1334.04	III
8		1549.08	III
15	p	1553.81	III
30		1635.14	III
75		1640.43	III
200		1779.80	III
600		1786.05	III
		Air	
10		2041.93	III
5		2042.07	III
1500		2060.58	III
4000		2068.98	III
10000		2127.98	III
16000		2191.16	III
8000		2200.76	III
8000		2206.03	III
150		2261.41	III
80		2261.57	III
10000		2284.34	III
3		2319.92	III
10000		2327.31	III
50000		2367.23	III
40000		2414.64	III
100	p	2710.30	III
90	h	2710.54	III
5		2780.11	III
70		2791.44	III
20		2803.27	III
100		2807.00	III
90000		2817.04	III
6000		2867.67	III
6000		2913.41	III
1500		2917.74	III
1600		2918.56	III
15		2940.53	III
99000		2946.01	III
20	p	2948.48	III
6000		2970.42	III
1400		3013.93	III
1500		3018.85	III
3		3267.10	III
25	l	3276.80	III
500		3866.96	III
3000		3900.74	III
4000		3914.58	III
3800		4039.60	III
3000		4040.11	III
120	h	4121.61	III
2000	c	4737.62	III
7500		5102.88	III
1300		5120.40	III
10000		5238.10	III
3000		5263.58	III
4000		5383.64	III
6000		5562.81	III
600		5567.27	III
4000		5572.24	III
400		5595.48	III
3000		5602.08	III
2000	h	7254.58	III
9000		7558.71	III
6000		7864.53	III
8000		7916.71	III
400		7989.41	III
10000		7991.43	III
8000		8171.41	III
4000		8645.09	III
10000		8796.21	III
8000		9116.59	III

Y IV
Ref. 265 — J.R.

Intensity	Wavelength	
	Vacuum	
3	211.80	IV
3	214.51	IV
6	215.97	IV
6	217.39	IV
12	221.71	IV
3	222.18	IV
6	222.98	IV
2	228.84	IV
20	228.94	IV
3	229.78	IV
3	235.17	IV
25	235.77	IV
1	242.12	IV
30	242.30	IV
3	244.14	IV
3	263.72	IV
150	264.64	IV
30	272.40	IV
150	273.03	IV
10	278.60	IV
900	355.86	IV
300	370.42	IV
500	386.82	IV
600	425.03	IV
300	473.10	IV

Y V
Ref. 419 — J.R.

Intensity	Wavelength	
	Vacuum	
5	289.18	V
50	299.99	V
3	312.89	V
200	313.35	V
40	320.47	V
40	321.69	V
150	325.58	V
200	326.57	V
175	328.34	V
50	330.40	V
900	333.09	V
500	333.80	V
100	335.12	V
400	335.14	V
500	336.62	V
500	339.02	V
200	340.02	V
5	340.42	V
500	344.59	V
100	349.65	V
2	349.75	V
10	351.36	V
100	353.98	V
100	355.56	V
300	372.05	V
400	379.96	V
200	397.77	V
300	403.45	V
1	408.81	V
200	409.31	V
200	415.03	V
100	418.18	V
150	418.59	V
50	419.79	V
300	420.74	V
1	427.87	V
50	430.75	V
100	437.66	V
3	441.62	V
30	442.96	V

Yttrium (Cont.)

2		451.97	V
15		455.84	V
85		457.84	V
4000		584.98	V
2000		630.97	V

ZINC (Zn)
Z = 30

Zn I and II
Ref. 39, 55, 113, 131, 185, 186 — R.D.C.

Intensity		Wavelength	
		Vacuum	
60		1193.23	II
60		1277.31	II
60	d	1366.68	II
		1404.12	I
60		1410.44	II
60		1439.09	II
60		1445.04	II
50		1456.91	II
		1457.57	I
60		1477.02	II
50		1514.76	II
90		1572.99	II
		1589.57	I
60		1617.68	II
60		1658.25	II
50		1713.25	II
60		1715.76	II
80	d	1735.61	II
60		1736.89	II
50		1737.90	II
75		1747.12	II
80	c	1762.19	II
75		1774.04	II
80		1790.76	II
100		1797.64	II
100	d	1811.05	II
80		1816.48	II
80		1831.38	II
100	d	1833.57	II
70		1836.01	II
75		1836.65	II
75		1847.56	II
100		1864.12	II
100		1866.08	II
100		1872.13	II
75		1894.26	II
60		1901.52	II
60		1914.81	II
100	d	1918.96	II
70		1920.27	II
100	d	1929.67	II
60		1945.58	II
60		1951.91	II
80		1953.00	II
75		1954.87	II
80		1964.54	II
100		1969.40	II
100		1982.11	II
70		1985.61	II
100		1986.99	II
50		1993.37	II
50		1996.92	II
		Air	
100		2011.94	II
500		2025.48	II
60		2039.31	II
500		2062.00	II
200		2064.23	II
120		2079.08	I
50		2079.93	II
60		2087.33	I
80		2096.93	I
300		2099.94	II
200		2102.18	II
150		2104.42	I
75		2122.74	II
800	r	2138.56	I
75		2147.42	II
60		2210.18	II
50		2273.15	II
1000		2501.99	II
150		2515.81	I
50		2527.96	II
1000		2557.95	II
50		2567.80	II
50		2567.98	II
100	h	2569.87	I
100		2582.44	I
300		2582.49	I
200		2608.56	I
300		2608.64	I
200		2670.53	I
300		2684.16	I
300		2712.49	I
200		2756.45	I
300		2770.86	I
300		2770.98	I
400		2800.87	I
100		2801.06	I
5		2801.17	I
100		2801.96	II
100		2902.30	II
125		3018.36	I
200		3035.78	I
200		3072.06	I
150		3075.90	I
100		3171.45	II
100		3172.23	II
300		3196.31	II
100		3197.10	II
500	r	3282.33	I
50		3299.42	II
800		3302.58	I
700	r	3302.94	I
75		3306.01	II
800		3345.02	I
500		3345.57	I
150		3345.94	I
5		3799.00	I
50		3806.34	II
100		3840.29	II
50		3883.34	I
15		3965.43	I
10		4113.21	I
25		4292.88	I
25		4298.33	I
35		4629.81	I
300		4680.14	I
400		4722.15	I
400		4810.53	I
800		4911.62	II
500		4924.03	II
7		5068.66	I
15		5069.58	I
200		5181.98	I
8		5308.65	I
7		5310.24	I
7		5311.02	I
4		5772.10	I
4		5775.50	I
10		5777.11	I
500		5894.33	II
500		6021.18	II
500		6102.49	II
100		6111.53	II
500		6214.61	II
8		6237.90	I
8		6239.17	I
1000	h	6362.34	I
10		6479.18	I
15		6928.32	I
8		6938.47	I
3		6943.20	I
200		7478.8	II
300		7588.5	II
100		7612.9	II
300		7732.5	II
200		7757.9	II
10		7799.36	I
100		11054.25	I
100		13053.63	I
100		13150.59	I
20		13196.61	I
100		14038.70	I
20		15680.29	I
20		16483.45	I
20		16491.98	I
20		16505.23	I
5		24044.16	I
10		24375.02	I

Zn III
Ref. 376, 377 — R.D.C.

Intensity	Wavelength	
	Vacuum	
1000	677.63	III
750	677.96	III
200	713.90	III
100	1432.15	III
200	1456.72	III
100	1464.20	III
100	1465.75	III
300	1473.41	III
100	1489.26	III
100	1490.96	III
200	1498.79	III
300	1499.42	III
300	1500.42	III
300	1505.92	III
300	1515.85	III
30	1533.09	III
300	1552.30	III
200	1552.94	III
80	1553.11	III
200	1560.79	III
150	1562.55	III
200	1581.53	III
100	1582.06	III
100	1598.52	III
100	1600.87	III
100	1619.61	III
150	1622.51	III
200	1629.19	III
200	1639.33	III
150	1644.82	III
100	1651.74	III
200	1673.05	III
100	1688.59	III
80	1695.40	III
80	1706.65	III
80	1749.63	III
50	1753.84	III
100	1767.69	III
80	1839.32	III

Zn IV
Ref. 370, 377 — R.D.C.

Intensity	Wavelength	
	Vacuum	
30	412.67	IV
30	423.42	IV
50	423.54	IV
200	425.90	IV
200	428.54	IV
80	428.79	IV
150	429.30	IV
200	430.59	IV
150	431.54	IV
120	431.62	IV
10	434.41	IV
150	435.02	IV
150	435.76	IV
80	436.25	IV
30	436.38	IV
50	436.82	IV
50	441.15	IV
20	441.52	IV
20	441.56	IV
100	441.70	IV
200	442.39	IV
150	444.39	IV
100	444.46	IV
100	446.58	IV
30	447.85	IV
10	449.13	IV
200	449.98	IV
200	450.99	IV
30	451.62	IV
20	452.80	IV
80	456.67	IV
150	457.32	IV
200	466.93	IV
150	468.43	IV
200	472.09	IV
200	472.66	IV
200	473.02	IV
200	473.51	IV
200	474.56	IV
50	475.78	IV
200	476.42	IV
200	478.65	IV
200	478.90	IV
200	482.10	IV
10	482.68	IV
10	485.48	IV
10	489.19	IV
10	490.96	IV
10	493.37	IV
10	496.72	IV
100	497.70	IV
15	1193.29	IV
15	1203.44	IV
8	1212.71	IV
5	1214.14	IV
25	1224.35	IV
20	1227.62	IV
40	1228.65	IV
30	1231.46	IV
30	1237.26	IV
50	1239.12	IV
3	1246.26	IV
15	1247.01	IV
50	1249.69	IV
30	1253.67	IV
30	1257.31	IV
8	1259.68	IV
500	1265.74	IV
5	1269.15	IV
100	1272.21	IV
200	1272.98	IV
30	1275.78	IV
25	1278.51	IV
100	1280.47	IV
100	1291.83	IV
100	1292.49	IV
10	1294.32	IV
100	1296.62	IV
100	1296.73	IV
20	1301.88	IV
500	1306.66	IV
200	1318.00	IV
200	1320.74	IV
200	1321.22	IV
200	1322.33	IV
200	1322.43	IV
200	1326.76	IV
200	1329.11	IV
100	1333.32	IV
150	1344.08	IV
200	1347.98	IV
200	1349.90	IV
50	1352.27	IV
100	1356.20	IV
200	1357.82	IV
100	1363.43	IV
200	1363.95	IV
40	1368.17	IV
200	1369.53	IV
50	1370.42	IV
100	1375.33	IV
50	1375.98	IV
200	1377.65	IV
50	1391.24	IV
80	1393.07	IV
50	1394.54	IV
30	1400.14	IV
15	1403.98	IV
30	1409.40	IV
5	1410.33	IV
30	1419.60	IV
50	1427.79	IV
20	1438.58	IV
80	1455.65	IV
200	1459.98	IV
100	1476.43	IV
100	1481.25	IV
100	1529.84	IV
50	1533.68	IV

ZIRCONIUM (Zr)
Z = 40

Zr I and II
Ref. 1 — C.H.C.

Intensity	Wavelength	
	Air	
60	2374.42	I
60	2384.17	I
50	2388.01	I
50	2389.21	I
45	2405.52	I
60	2419.41	II
150	2449.85	II
21	2457.44	II
75	2487.29	II
45	2496.48	II
180	2532.46	II
90	2539.65	I
220	2542.10	II
45	2550.51	I
220	2550.74	II
45	2556.43	I
60	2567.45	I
570	2567.64	II
1600	2568.87	II
2100	2571.39	II
75	2583.40	II
130	2589.07	II
22	2589.65	I
45	2609.43	I
150	2630.91	II
80	2635.42	I
210	2639.09	II
70	2643.40	II
55	2647.78	I
110	2650.38	II
70	2658.69	I
180	2667.80	II
55	2669.49	II
120	2670.96	II
1800	2678.63	II
35	2681.76	II
90	2687.75	I
90	2692.60	II
22	2692.92	I
160	2693.53	II
180	2694.06	II
70	2695.43	II

Zirconium (Cont.)

95		2699.60	II
750		2700.13	II
280		2711.51	II
140		2712.42	II
140		2714.26	II
1300		2722.61	II
140		2725.47	I
800		2726.49	II
490		2732.72	II
1400		2734.86	II
110		2740.51	II
140		2741.55	II
1100		2742.56	II
660		2745.86	II
660		2752.21	II
530		2758.81	II
200	d	2768.73	II
		2768.85	II
170	d	2774.04	I
		2774.16	II
200		2790.14	I
120		2792.04	I
160		2796.90	II
110		2799.15	II
180		2810.91	II
620		2814.90	I
390		2818.74	II
530		2825.56	II
110		2833.91	II
710		2837.23	I
120		2839.34	II
130		2843.52	II
660		2844.58	II
210		2848.19	II
350		2848.52	I
350		2851.97	II
340		2869.81	II
490		2875.98	I
120		2892.26	I
160		2905.23	II
300		2915.99	II
110		2916.64	II
270		2918.24	II
320		2926.99	II
160		2934.61	II
160		2936.31	II
320		2948.94	II
210		2951.48	II
320		2955.78	II
320		2960.87	I
320		2962.68	II
320		2968.96	II
120		2969.19	I
230		2969.63	II
130		2976.61	II
320		2978.05	II
230		2979.18	II
160		2981.02	II
820		2985.39	I
320		3003.74	II
100		3005.37	I
160		3005.50	I
820		3011.75	I
100		3013.32	II
160		3019.84	II
350		3020.47	II
500		3028.04	II
880		3029.52	I
180		3030.92	II
350	d	3036.39	II
100		3045.83	I
690		3054.84	II
100		3060.11	II
100		3064.63	II
110		3085.34	I
110		3094.80	I
250		3095.07	II
110		3095.82	I
280		3099.23	II
690		3106.58	II
110		3108.37	I
210		3110.88	II
350		3120.74	I
320		3125.92	II
500		3129.18	II
500		3129.76	II
140		3131.11	I
350		3132.07	I
110		3133.23	I
350		3133.48	II
180		3136.96	I
690		3138.68	II
140		3139.80	I
180		3148.82	I
290		3155.67	II
150		3157.00	II
320		3157.82	I
540		3164.31	II
150		3165.45	II
880		3165.97	II
150		3166.26	II
190		3178.09	II
190		3181.58	II
150		3181.92	II
880		3182.86	II
540		3191.21	I
210		3191.90	II
540		3212.01	I
760		3214.19	II
110		3222.47	II
200		3228.81	II
630		3231.69	II
630		3234.12	I
110		3236.58	II
760		3241.05	II
320		3250.39	I
200		3254.28	I
200		3260.11	I
190		3269.66	I
150		3271.13	II
540		3272.22	II
1000		3273.05	II
1300		3279.26	II
320	d	3282.73	I
880		3284.71	II
140		3285.88	II
150		3288.80	II
540		3305.15	II
880		3306.28	II
150		3313.70	II
210		3314.50	II
150		3319.02	II
380		3322.99	II
380		3326.80	II
380		3334.25	II
210		3334.62	II
190		3338.41	II
760		3340.56	II
380		3344.79	II
130		3353.66	I
180		3354.39	II
760		3356.09	II
540		3357.26	II
180		3359.96	II
150		3360.46	I
150		3363.82	II
150		3367.82	II
150		3370.59	I
180		3373.42	II
380		3374.73	II
110		3376.27	II
150		3377.46	II
570		3387.87	II
760		3388.30	II
5700		3391.98	II
570		3393.12	II
160		3396.33	II
380		3399.35	II
570		3404.83	II
760		3410.25	II
380		3414.66	I
1000		3430.53	II
380		3437.14	II
4700		3438.23	II
600		3447.36	I
200		3455.91	I
410		3457.56	II
200		3458.93	II
820		3463.02	II
600		3471.19	I
200		3478.79	I
1200		3479.39	II
1300		3481.15	II
760		3483.54	II
4100		3496.21	II
350		3505.48	II
820		3505.67	II
1000		3509.32	I
200		3510.46	II
2000		3519.60	I
440		3525.81	II
440		3533.22	I
210		3535.16	I
630		3542.62	II
1800		3547.68	I
210		3549.74	I
630		3550.46	I
1800		3551.95	II
2100		3556.60	II
1100		3566.10	I
210		3568.88	I
2100		3572.47	II
210		3573.08	II
1100		3575.79	I
1300		3576.85	II
880		3586.29	I
440		3587.98	II
3500		3601.19	I
690		3611.89	II
1100		3613.10	II
1100		3614.77	II
1100		3623.86	I
320		3634.15	I
260		3661.20	I
1100		3663.65	I
390		3671.27	II
800		3674.72	II
390		3697.46	II
960		3698.17	II
720		3709.26	II
270		3731.26	II
560		3745.98	II
880		3751.60	II
480		3764.39	I
480		3766.72	I
340		3766.82	II
720		3780.54	I
560		3791.40	I
210		3817.58	II
560		3822.41	I
2200		3835.96	I
1300		3836.76	II
550		3843.02	II
550		3847.01	I
550		3849.25	I
2900		3863.87	I
770		3864.34	I
990		3877.60	I
200		3879.05	I
1500		3885.42	I
2900		3890.32	I
2000		3891.38	I
400		3900.52	I
310		3915.94	II
610		3921.79	I
1200		3929.53	I
200		3934.12	II
200		3934.79	II
940		3958.22	II
490		3966.66	I
990		3968.26	I
660		3973.50	I
200		3975.29	I
200	h	3981.60	I
770		3991.13	II
770		3998.97	II
200		4007.60	I
200		4012.25	I
400		4023.98	I
770		4024.92	I
990		4027.20	I
240		4028.95	I
400		4029.68	II
490		4030.04	I
400		4035.89	I
240		4042.22	I
610		4043.58	I
490		4044.56	I
400		4045.61	II
610		4048.67	II
200		4050.33	II
200		4050.48	I
770		4055.03	I
600		4055.71	I
330		4061.53	I
1500		4064.16	I
2000		4072.70	I
310		4074.93	I
200		4076.53	I
240		4078.31	I
2000		4081.22	I
200		4108.40	I
400		4121.46	I
1200		4149.20	II
200		4152.64	I
290		4156.24	II
400		4161.21	II
400		4166.36	I
200		4183.32	I
660		4187.56	I
400		4194.76	I
610		4199.09	I
610		4201.46	I
610		4208.98	II
200		4211.88	II
400		4213.86	I
2000		4227.76	I
200		4236.06	I
2000		4239.31	I
770		4240.34	I
770		4241.20	I
1200		4241.69	I
310		4268.02	I
550		4282.20	I
550		4294.79	I
310		4302.89	I
550		4341.13	I
1000		4347.89	I
290		4359.74	II
310		4360.81	I
350		4366.45	I
240		4379.78	II
190		4413.04	I
240		4420.46	I
120		4427.24	I
160		4431.49	I
140		4443.00	II
110		4457.43	I
110		4466.91	I
110		4470.31	I
190		4470.56	I
200		4496.97	II
550		4507.12	I
610		4535.75	I
490		4542.22	I
200		4553.01	I
200		4555.13	I
140		4555.52	I
490		4575.52	I
100		4582.29	I
140		4590.55	I
350		4602.57	I
140		4604.42	I
210		4626.41	I
700		4633.98	I
210		4644.83	I
260		4683.42	I
2300		4687.80	I
510		4688.45	I
110		4707.79	I
1900		4710.08	I
160		4711.92	I
120		4717.62	I
210		4719.12	I
300		4732.33	I
1400		4739.48	I
190		4762.78	I
870		4772.31	I
210		4784.92	I
160		4788.67	I
260		4805.87	I
140		4809.47	I
190		4815.04	I
700		4815.63	I
280		4824.29	I
190		4828.04	I
110		4838.78	I
210		4851.36	I
160		4866.06	I
110		4881.24	I
110		4883.60	I
100		4994.76	I
30		5011.46	I
250		5046.58	I
85		5060.39	I
360		5064.91	I
110		5065.22	I
100		5070.26	I
75		5073.98	I
470		5078.25	I
85		5085.26	I
50		5112.27	II
140		5115.24	I
50		5120.42	I
85		5133.40	I
300		5155.45	I
200		5158.00	I
35		5158.67	I
75		5160.99	I
85		5165.96	I
17		5178.99	I
100		5183.70	I
30		5187.03	I
100		5191.60	II
100		5201.15	I
85		5209.30	I
85		5224.93	I
30		5243.47	I
120		5277.41	I
75		5280.05	I
60		5294.82	I
120		5296.79	I
60		5301.97	I
110		5311.40	I
25		5321.26	I
22		5330.84	I
12		5338.43	I
30		5350.09	II
30		5350.35	II
25		5350.90	I
25		5351.92	I
75		5362.56	I
12		5363.35	I
17		5369.39	I
20		5382.37	I
270		5385.14	I
30		5386.65	I
17		5391.18	I
17		5395.88	I
25		5405.13	I
85		5407.62	I
17		5413.93	I
20		5421.86	I
15		5426.36	I
25		5428.42	I
25		5437.76	I
15		5440.41	I
35		5448.57	I
10		5474.92	I
10		5477.40	I
35		5478.33	I
35		5480.83	I
10		5481.16	I
30		5486.09	I
140		5502.12	I

Zirconium (Cont.)

25		5507.87	I
30		5517.11	I
10		5518.05	I
75		5528.41	I
20		5532.30	I
45		5537.46	I
50		5545.32	I
22	B1	5551.75	Z
25	B1	5553.17	Z
12		5612.11	I
120		5620.14	I
35		5623.53	I
25	B1	5629.02	Z
25	B1	5629.58	Z
160		5664.51	I
20		5666.28	I
120		5680.90	I
15		5685.62	I
30		5708.89	I
75	B1	5718.21	Z
120		5735.70	I
35	B1	5748.17	Z
17	B1	5778.57	Z
160		5797.74	I
30		5847.32	I
50		5868.27	I
110		5869.50	I
340		5879.80	I
85		5885.62	I
50		5901.09	I
30	B1	5908.61	Z
140		5925.13	I
100		5935.20	I
110		5955.35	I
30	B1	5977.80	Z
100		5984.23	I
17		5995.37	I
50		6001.05	I
30		6025.36	I
85		6032.61	I
170		6045.85	I
100		6049.24	I
140		6062.84	I
50		6120.83	I
170		6121.91	I
85		6124.84	I
680		6127.44	I
340		6134.55	I
100		6140.46	I
440		6143.20	I
30		6155.61	I
75		6157.71	I
25		6160.20	I
35		6189.40	I
60		6192.96	I
85		6213.05	I
100		6214.69	I
170	B1	6226.51	Z
100		6257.26	I
50	b	6261.05	Z
35		6267.06	I
45	B1	6292.84	Z
120		6299.66	I
15		6304.34	I
300		6313.02	I
30		6314.71	I
50		6321.35	I
22		6340.36	I
50	B1	6345.10	Z
75		6345.22	I
75	B1	6378.56	Z
35		6407.00	I
50	b	6412.39	Z
12		6426.17	I
35		6434.33	I
60		6445.74	I
20		6451.62	I
20		6457.63	I
110		6470.21	I
60	B1	6473.79	Z
11		6484.35	I
110		6489.64	I
22		6493.10	I
50		6503.26	I
50		6506.36	I
50	B1	6508.15	Z
30	B1	6542.90	Z
35		6550.54	I
30		6569.43	I
20		6576.56	I
30	b	6578.06	Z
50		6591.99	I
10		6596.71	I
10		6598.84	I
50		6603.27	I
15		6620.56	I
11		6678.01	II
22		6688.18	I
11		6702.12	I
17		6709.61	I
27		6717.88	I
40		6752.73	I
75		6762.38	I
85		6769.16	I
27		6772.89	I
15		6787.15	II
35		6790.85	I
45		6828.78	I
45		6832.89	I
13		6845.33	I
17		6846.34	I
100		6846.97	I
27		6849.26	I
13		6852.56	I
120		6888.29	I
29		6900.59	I
20		6904.36	I
29		6907.37	I
20		6916.87	I
16		6932.38	I
20		6940.46	I
150		6953.84	I
60		6966.44	I
10		6975.91	I
150		6990.84	I
80		6994.32	I
10		7005.46	I
100		7027.40	I
25		7057.36	I
14		7057.96	I
140		7087.30	I
25		7089.43	I
35		7094.46	I
50		7095.59	I
540		7097.70	I
280		7102.91	I
170		7103.72	I
140		7111.68	I
40		7112.82	I
18		7113.52	I
12		7132.95	I
16		7140.74	I
12		7144.47	I
590		7169.09	I
50		7201.62	I
12		7258.17	I
35		7264.76	I
20		7306.21	I
25		7311.62	I
35		7313.72	I
90		7318.08	I
10		7327.82	I
50		7335.97	I
50		7343.96	I
20		7373.50	I
25		7383.63	I
14		7400.90	I
10		7411.39	I
10		7422.75	I
10		7433.10	I
110		7439.86	I
18		7467.57	I
16		7479.58	I
14	h	7515.70	I
12		7517.95	I
20	h	7540.62	I
20		7544.59	I
29		7551.46	I
40		7554.70	I
25		7558.45	I
12		7560.09	I
12		7562.12	I
80		7607.15	I
14		7612.08	I
20		7621.17	I
29		7658.60	I
18		7690.83	I
14		7704.27	I
10		7708.42	I
10		7766.55	I
12		7816.32	I
110		7819.35	I
35		7822.94	I
40		7826.72	I
90		7849.35	I
35		7869.99	I
14		7876.25	I
16		7882.18	I
10		7897.98	I
16		7908.46	I
20		7940.47	I
160		7944.61	I
80		7956.66	I
80		7959.98	I
20		7963.63	I
160		8005.27	I
25		8046.05	I
16		8053.06	I
20		8055.29	I
20		8055.76	I
60		8058.08	I
150		8063.09	I
790		8070.08	I
10		8114.28	I
20		8120.17	I
390		8132.99	I
20		8152.58	I
12		8188.77	I
40		8194.73	I
60		8201.73	I
280		8212.53	I
20		8240.37	I
40		8283.81	I
140		8305.90	I
14		8332.44	I
50		8370.23	I
120		8389.41	I
70		8414.00	I
50		8453.17	I
50		8464.65	I
40		8498.44	I
18		8584.21	I
10		8734.86	I
12		8749.48	I
10		8786.23	I
16		8804.98	I
70		8836.09	I
60		8899.52	I

Zr III
Ref. 403 — J.R.

Intensity	Wavelength	
	Vacuum	
25	687.64	III
50	690.39	III
25	819.59	III
30	820.21	III
35	823.69	III
25	829.50	III
30	850.61	III
25	859.56	III
30	864.86	III
25	868.64	III
25	868.99	III
25	919.59	III
30	1320.81	III
25	1375.13	III
30	1378.93	III
40	1403.48	III
35	1420.12	III
25	1420.87	III
25	1465.44	III
25	1563.24	III
40	1593.11	III
100	1612.38	III
50	1620.62	III
75	1631.31	III
50	1638.33	III
35	1675.06	III
35	1675.75	III
40	1703.36	III
40	1725.03	III
40	1754.38	III
35	1759.12	III
30	1764.75	III
30	1771.96	III
40	1773.90	III
100	1779.51	III
40	1783.35	III
200	1790.19	III
150	1793.56	III
125	1798.13	III
75	1800.03	III
25	1801.67	III
100	1805.26	III
35	1831.89	III
40	1850.06	III
40	1853.38	III
30	1859.12	III
25	1860.47	III
25	1861.77	III
75	1864.06	III
30	1865.45	III
35	1877.00	III
50	1892.07	III
65	1914.25	III
75	1921.96	III
75	1932.54	III
50	1934.32	III
40	1935.20	III
75	1936.48	III
75	1936.67	III
80	1937.27	III
200	1940.25	III
100	1941.08	III
25	1946.12	III
80	1946.61	III
50	1946.99	III
100	1953.95	III
50	1961.32	III
100	1962.01	III
40	1962.92	III
85	1966.22	III
25	1967.81	III
60	1974.99	III
40	1983.14	III
50	1989.83	III
30	1990.95	III
30	1994.46	III
	Air	
45	2000.23	III
100	2006.82	III
30	2013.30	III
35	2016.63	III
30	2021.52	III
60	2026.78	III
100	2035.42	III
50	2036.92	III
75	2056.13	III
40	2058.73	III
75	2060.83	III
50	2061.47	III
125	2070.43	III
50	2074.12	III
100	2077.92	III
100	2080.99	III
30	2081.81	III
25	2085.35	III
200	2086.78	III
40	2089.50	III
40	2097.03	III
40	2102.30	III
50	2103.16	III
25	2104.23	III
40	2113.98	III
35	2114.10	III
40	2125.06	III
35	2137.90	III
35	2138.45	III
25	2139.85	III
40	2159.24	III
40	2162.20	III
100	2175.80	III
100	2191.15	III
35	2192.05	III
60	2206.33	III
40	2206.97	III
30	2231.00	III
50	2245.36	III
25	2251.14	III
40	2257.83	III
30	2281.43	III
100	2301.60	III
75	2308.12	III
35	2405.81	III
40	2406.21	III
75	2420.65	III
25	2438.70	III
50	2444.58	III
100	2448.86	III
100	2593.64	III
250	2620.56	III
50	2621.28	III
60	2628.26	III
200	2643.79	III
100	2656.46	III
150	2664.26	III
100	2682.16	III
75	2686.28	III
70	2690.49	III
60	2698.31	III
50	2709.05	III
45	2715.76	III
40	2720.07	III
75	2735.76	III
25	2775.23	III
40	2836.18	III
40	3278.86	III

Zr IV
Ref. 362 — J.R.

Intensity	Wavelength	
	Vacuum	
15	478.97	IV
60	480.66	IV
60	497.23	IV
60	500.22	IV
4	500.34	IV
4	584.65	IV
15	585.42	IV
2	586.42	IV
60	588.89	IV
100	589.74	IV
600	628.66	IV
500	633.56	IV
200	633.63	IV
8	712.49	IV
90	754.39	IV
90	760.15	IV
150	846.40	IV
300	863.65	IV
500	864.59	IV
200	881.30	IV

Zirconium (Cont.)

Intensity		Wavelength	
200		882.59	IV
100		1099.76	IV
150		1100.00	IV
9000		1183.97	IV
9000		1201.77	IV
200		1212.71	IV
100		1213.01	IV
10000		1219.86	IV
500		1285.89	IV
500		1290.56	IV
70		1291.70	IV
500		1417.70	IV
500		1440.65	IV
500		1441.06	IV
1000		1469.47	IV
10000		1546.17	IV
10		1596.29	IV
10000		1598.95	IV
150		1605.26	IV
5000		1607.95	IV
90		1609.50	IV
35		1818.06	IV
500		1836.15	IV
500		1846.37	IV
200		1848.03	IV
70		1851.91	IV
		Air	
20		2045.12	IV
40		2047.15	IV
10000		2091.49	IV
10000		2092.36	IV
10000		2163.68	IV
10000		2286.67	IV
200		2473.75	IV
300		2476.71	IV
5		2572.32	IV
400		2573.66	IV
400		2583.32	IV

Zr V
Ref. 418 — J.R.

Intensity		Wavelength	
		Vacuum	
300		292.19	V
500		304.01	V
300		305.24	V
400	p	368.18	V
200		519.25	V
200		536.50	V
200	p	674.13	V
200		675.53	V
300		679.39	V
200		688.27	V
200	p	703.03	V
300		717.24	V
200		740.33	V
2000		740.61	V
500		742.74	V
300		752.48	V
300		764.58	V
500		766.20	V
200		773.80	V
200		775.58	V
200		779.21	V
400		784.50	V
400		797.23	V
10000		800.00	V
10000		806.89	V
200		809.75	V
10000		812.05	V
500		822.06	V
300	p	823.46	V
200		823.78	V
200		825.13	V
200		830.59	V
400		836.57	V
3000		841.40	V
300	p	852.87	V
700		853.68	V
200		873.11	V
300		885.68	V
300		900.48	V
500		906.66	V
500		915.30	V
400		923.10	V
400		940.41	V
200		949.70	V
200		978.06	V
200		980.70	V
200		984.18	V
300		995.59	V
400		1002.48	V
400		1038.69	V
200		1044.41	V
300		1047.77	V
400		1068.55	V
300		1072.25	V
300		1083.45	V
200		1087.05	V
200		1093.54	V
200		1108.79	V
300		1194.24	V
300		1200.76	V
200		1233.91	V
300		1238.93	V
300		1253.61	V
200		1259.70	V
400		1260.91	V
500		1265.38	V
300		1295.81	V
400		1302.80	V
500		1303.93	V
400		1306.76	V
200		1315.14	V
400		1320.74	V
500	p	1323.81	V
300		1332.06	V
300		1337.34	V
300		1355.21	V
500		1355.98	V
300		1361.39	V
500		1376.54	V
200		1396.79	V
300		1410.03	V
200		1413.40	V
300		1460.05	V
300		1486.90	V
300		1491.33	V
200		1520.47	V
200		1550.12	V
500		1633.03	V
500		1654.46	V
700		1725.02	V
500		1786.20	V
200	p	1790.81	V
500		1806.09	V
500		1860.48	V
600		1860.86	V
500		1878.33	V
400		1926.24	V
500		1927.43	V
400		1934.88	V
		Air	
500		2009.29	V
600		2028.54	V
600		2132.42	V
200		2150.18	V
200		2336.94	V

REFERENCES

1. **Meggers, W. F., Corliss, C. H., and Scribner, B. F.,** *Natl. Bur. Stand. (U.S.) Monogr.*, 145, Washington, D.C., 1975.
2. **Aksenov, V. P. and Ryabtsev, A. N.,** *Opt. Spectrosc.*, 37, 860, 1970.
3. **Andersen, N., Bickel, W. S., Carriveau, G. W., Jensen, K., and Veje, E.,** *Phys. Scr.*, 4, 113, 1971.
4. **Andersson, E. and Johannesson, G. A.,** *Phys. Scr.*, 3, 203, 1971.
5. **Andrew, K. L. and Meissner, K. W.,** *J. Opt. Soc. Am.*, 49, 146, 1959.
6. **Artru, M. C. and Brillet, W. U. L.,** *J. Opt. Soc. Am.*, 64, 1063, 1974.
7. **Artru, M. C. and Kaufman, V.,** *J. Opt. Soc. Am.*, 62, 949, 1972.
8. **Artru, M. C. and Kaufman, V.,** *J. Opt. Soc. Am.*, 65, 594, 1975.
9. **Badami, J. S. and Rao, K. R.,** *Proc. R. Soc. London*, 140(A), 387, 1933.
10. **Baird, K. M. and Smith, D. S.,** *J. Opt. Soc. Am.*, 48, 300, 1958.
11. **Bashkin, S. and Martinson, I.,** *J. Opt. Soc. Am.*, 61, 1686, 1971.
12. **Beacham, J. R.,** Ph.D. thesis, Purdue University, 1970.
13. **Benschop, H., Joshi, Y. N., and van Kleef, T. A. M.,** *Can. J. Phys.*, 53, 700, 1975.
14. **Berry, H. G., Bromander, J., and Buchta, R.,** *Phys. Scr.*, 1, 181, 1970.
15. **Berry, H. G., Bromander, J., Martinson, I., and Buchta, R.,** *Phys. Scr.*, 3, 63, 1971.
16. **Berry, H. G., Desesquelles, J., and Dufay, M.,** *Phys. Rev. Sect. A.*, 6, 600, 1972.
17. **Berry, H. G., Desesquelles, J., and Dufay, M.,** *Nucl. Instrum. Methods*, 110, 43, 1973.
18. **Berry, H. G., Pinnington, E. H., and Subtil, J. L.,** *J. Opt. Soc. Am.*, 62, 767, 1972.
19. **Bidelman, W. P. and Corliss, C. H.,** *Astrophys. J.*, 135, 968, 1962.
20. **Bloch, L. and Bloch, E.,** *Ann. Phys.* (Paris), 10(11), 141, 1929.
21. **Bloch, L., Bloch, E., and Felici, N.,** *J. Phys. Radium*, 8, 355, 1937.
22. **Bockasten, K.,** *Ark. Fys.*, 9, 457, 1955.
23. **Bockasten, K., Hallin, R., Johansson, K. B., and Tsui, P.,** *Phys. Lett.* (Netherlands), 8, 181, 1964.
24. **Bockasten, K. and Johansson, K. B.,** *Ark. Fys.*, 38, 563, 1969.
25. **Borgstrom, A.,** *Ark. Fys.*, 38, 243, 1968.
26. **Borgstrom, A.,** *Phys. Scr.*, 3, 157, 1971.
27. **Bowen, I. S.,** *Phys. Rev.*, 29, 231, 1927.
28. **Bowen, I. S.,** *Phys. Rev.*, 31, 34, 1928.
29. **Bowen, I. S.,** *Phys. Rev.*, 39, 8, 1932.
30. **Bowen, I. S.,** *Phys. Rev.*, 45, 401, 1934.
31. **Bowen, I. S.,** *Phys. Rev.*, 46, 377, 1934.
32. **Bowen, I. S.,** *Phys. Rev.*, 46, 791, 1934.
33. **Boyce, J. C.,** *Phys. Rev.*, 49, 730, 1936.
34. **Boyce, J. C. and Robinson, H. A.,** *J. Opt. Soc. Am.*, 26, 133, 1936.
35. **Bozman, W. R., Meggers, W. F., and Corliss, C. H.,** *J. Res. Natl. Bur. Stand. Sect. A*, 71, 547, 1967.
36. **Bromander, J.,** *Ark. Fys.*, 40, 257, 1969.
37. **Bromander, J. and Buchta, R.,** *Phys. Scr.*, 1, 184, 1970.
38. **Brown, C. M. and Ginter, M. L.,** *J. Opt. Soc. Am.*, 68, 243, 1978.
39. **Brown, C. M., Tilford, S. G., and Ginter, M. L.,** *J. Opt. Soc. Am.*, 65, 1404, 1975.
40. **Bryant, B. W.,** *Johns Hopkins Spectroscopic Report* No. 21, 1961.
41. **Buchet, J. P., Buchet-Poulizac, M. C., Berry, H. G., and Drake, G. W. F.,** *Phys. Rev. Sect. A*, 7, 922, 1973.
42. **Budhiraja, C. J. and Joshi, Y. N.,** *Can. J. Phys.*, 49, 391, 1971.

43. **Burns, K. and Adams, K. B.,** *J. Opt. Soc. Am.*, 42, 56, 1952.
44. **Burns, K. and Adams, K. B.,** *J. Opt. Soc. Am.*, 46, 94, 1956.
45. **Burns, K., Adams, K. B., and Longwell, J.,** *J. Opt. Soc. Am.*, 40, 339, 1950.
46. **Callahan, W. R.,** Ph.D. thesis, Johns Hopkins University, 1962.
47. **Charles, G. W.,** *J. Opt. Soc. Am.*, 56, 1292, 1966.
48. **Charles, G. W., Hunt, D. J., Pish, G., and Timma, D. L.,** *J. Opt. Soc. Am.*, 45, 869, 1955.
49. **Codling, K.,** *Proc. Phys. Soc.*, 77, 797, 1961.
50. **Comite Consulatif Pour La Definition du Metre,** *J. Phys. Chem. Ref. Data*, 3, 852, 1974.
51. **Conway, J. G., Blaise, J., and Verges, J.,** *Spectrochim. Acta Part B*, 31, 31, 1976.
52. **Conway, J. G., Worden, E. F., Blaise, J., and Verges, J.,** *Spectrochim. Acta Part B*, 32, 97, 1977.
53. **Conway, J. G., Worden, E. F., Blaise, J., Camus, P., and Verges, J.,** *Spectrochim. Acta Part B*, 32, 101, 1977.
54. **Crooker, A. M.,** *Can. J. Res. Sect. A*, 14, 115, 1936.
55. **Crooker, A. M. and Dick, K. A.,** *Can. J. Phys.*, 46, 1241, 1968.
56. **Crosswhite, H. M.,** *J. Res. Natl. Bur. Stand. Sect. A*, 79, 17, 1975.
58. **Crosswhite, H. M. and Dieke, G. H.,** *American Institute of Physics Handbook*, Section 7, 1972.
59. **de Bruin, T. L.,** *Z. Phys.*, 38, 94, 1926.
60. **de Bruin, T. L.,** *Z. Phys.*, 53, 658, 1929.
61. **de Bruin, T. L., Humphreys, C. J., and Meggers, W. F.,** *J. Res. Natl. Bur. Stand.*, 11, 409, 1933.
62. **Dick, K. A.,** *J. Opt. Soc. Am.*, 64, 702, 1973.
63. **Dobbie, J. C.,** *Ann. Solar Phys. Observ.* (Cambridge), 5, 1, 1938.
64. **Earls, L. T. and Sawyer, R. A.,** *Phys. Rev.*, 47, 115, 1935.
65. **Edlen, B.,** *Z. Phys.*, 85, 85, 1933.
66. **Edlen, B.,** *Nova Acta Reglae Soc. Sci. Ups.*, (IV) 9, No. 6, 1934.
67. **Edlen, B.,** *Z. Phys.*, 93, 726, 1935.
68. **Edlen, B.,** *Z. Phys.*, 94, 47, 1935.
69. **Edlen, B.,** *Rep. Prog. Phys.*, 26, 181, 1963.
70. **Edlen, B. and Risberg, P.,** *Ark. Fys.*, 10, 553, 1956.
71. **Edlen, B. and Swings, P.,** *Astrophys. J.*, 95, 532, 1942.
72. **Ehrhardt, J. C. and Davis, S. P.,** *J. Opt. Soc. Am.*, 61, 1342, 1971.
73. **Eidelsberg, M.,** *J. Phys. B*, 5, 1031, 1972.
74. **Eidelsberg, M.,** *J. Phys. B*, 7, 1476, 1974.
75. **Ekberg, J. O. and Svensson, L. A.,** *Phys. Scr.*, 2, 283, 1970.
76. **Ekefors, E.,** *Z. Phys.*, 71, 53, 1931.
77. **Epstein, G. L. and Reader, J.,** *J. Opt. Soc. Am.*, 65, 310, 1975.
78. **Epstein, G. L. and Reader, J.,** *J. Opt. Soc. Am.*, 66, 590, 1976.
79. **Epstein, G. L. and Reader, J.,** unpublished.
80. **Eriksson, K. B. S.,** *Phys. Lett. A.*, 41, 97, 1972.
81. **Eriksson, K. B. S. and Isberg, H. B. S.,** *Ark. Fys.*, 23, 527, 1963.
82. **Eriksson, K. B. S. and Wenaker, I.,** *Phys. Scr.*, 1, 21, 1970.
83. **Esteva, J. M. and Mehlman, G.,** *Astrophys. J.*, 193, 747, 1974.
84. **Even-Zohar, M. and Fraenkel, B. S.,** *J. Phys. B*, 5, 1596, 1972.
85. **Fawcett, B. C.,** *J. Phys. B*, 3, 1732, 1970.
86. **Fawcett, B. C.,** Culham Laboratory Report ARU-R4, 1971.
87. **Ferner, E.,** *Ark. Mat. Astron. Fys.*, 28(A), 4, 1941.
88. **Fischer, R. A., Knopf, W. C., and Kinney, F. E.,** *Astrophys. J.*, 130, 683, 1959.
89. **Fowler, A.,** *Report on Series in Line Spectra*, Fleetway Press, London, 1922.
90. **Fowles, G. R.,** *J. Opt. Soc. Am.*, 44, 760, 1954.
91. **Fred, M.,** *Argonne Natl. Lab.*, unpublished, 1977.
92. **Fred, M. and Tomkins, F. S.,** *J. Opt. Soc. Am.*, 47, 1076, 1957.
93. **Fred, M., Tomkins, F. S., Blaise, J. E., Camus, P., and Verges, J.,** Argonne National Laboratory Report No. 76-68, 1976.
94. **Garcia, J. D. and Mack, J. E.,** *J. Opt. Soc. Am.*, 55, 654, 1965.
96. **Giacchetti, A.,** *Argonne Natl. Lab.*, unpublished, 1975.
97. **Giacchetti, A., Blaise, J., Corliss, C. H., and Zalubas, R.,** *J. Res. Natl. Bur. Stand. Sect. A*, 78, 247, 1974.
98. **Giacchetti, A., Stanley, R. W., and Zalubas, R.,** *J. Opt. Soc. Am.*, 69, 474, 1970.
99. **Gilbert, W. P.,** *Phys. Rev.*, 47, 847, 1935.
100. **Gilroy, H. T.,** *Phys. Rev.*, 38, 2217, 1931.
101. **Glad, S.,** *Ark. Fys.*, 10, 291, 1956.
102. **Goldsmith, S.,** *J. Phys. B*, 2, 1075, 1969.
103. **Goorvitch, D., Mehlman-Balloffet, G., and Valero, F. P. J.,** *J. Opt. Soc. Am.*, 60, 1458, 1970.
104. **Goorvitch, D. and Valero, F. P. J.,** *Astrophys. J.*, 171, 643, 1972.
105. **Green, L. C.,** *Phys. Rev.*, 55, 1209, 1939.
106. **Gutman, F.,** *Diss. Abstr. Int. B*, 31, 363, 1970.
107. **Hallin, R.,** *Ark. Fys.*, 31, 511, 1966.
108. **Hallin, R.,** *Ark. Fys.*, 32, 201, 1966.
109. **Hansen, J. E. and Persson, W.,** *J. Opt. Soc. Am.*, 64, 696, 1974.
110. **Hansen, J. E. and Persson, W.,** *Phys. Scr.*, 13, 166, 1976.
111. **Hellintin, P.,** *Phys. Scr.*, 13, 155, 1976.
112. **Herzberg, G. and Moore, H. R.,** *Can. J. Phys.*, 37, 1293, 1959.
113. **Hetzler, C. W., Boreman, R, W., and Burns, K.,** *Phys. Rev.*, 48, 656, 1935.
114. **Holmstrom, J. E. and Johansson, L.,** *Ark. Fys.*, 40, 133, 1969.
115. **Hontzeas, S., Martinson, I., Erman, P., and Buchta, R.,** *Nucl. Instrum. Methods*, 110, 51, 1973.
116. **Humphreys, C. J.,** *J. Res. Natl. Bur. Stand.*, 22, 19, 1939.
117. **Humphreys, C. J.,** *J. Opt. Soc. Am.*, 43, 1027, 1953.
118. **Humphreys, C. J.,** *J. Phys. Chem. Ref. Data*, 2, 519, 1973.
119. **Humphreys, C. J. and Andrew, K. L.,** *J. Opt. Soc. Am.*, 54, 1134, 1964.
120. **Humphreys, C. J. and Meggers, W. F.,** *J. Res. Natl. Bur. Stand.*, 10, 139, 1933.
121. **Humphreys, C. J. and Paul, E., Jr.,** *J. Opt. Soc. Am.*, 60, 200, 1970.
122. **Humphreys, C. J. and Paul, E., Jr.,** *J. Opt. Soc. Am.*, 62, 432, 1972.
123. **Humphreys, C. J., Paul, E., Jr., Cowan, R. D., and Andrew, K. L.,** *J. Opt. Soc. Am.*, 57, 855, 1967.
124. **Humphreys, C. J., Paul, E., Jr., and Minnhagen, L.,** *J. Opt. Soc. Am.*, 61, 110, 1971.
125. **Iglesias, L.,** Inst. of Optics, Madrid, unpublished, 1977.

126. **Iglesias, L. and Velasco, R.**, *Publ. Inst. Opt. Madrid*, No. 23, 1964.
127. **Isberg, B.**, *Ark. Fys.*, 35, 551, 1967.
128. **Johannesson, G. A., Lundstrom, T., and Minnhagen, L.**, *Phys. Scr.*, 6, 129, 1972.
129. **Johannesson, G. A. and Lundstrom, T.**, *Phys. Scr.*, 8, 53, 1973.
130. **Johansson, I.**, *Ark. Fys.*, 20, 135, 1961.
131. **Johansson, I. and Contreras, R.**, *Ark. Fys.*, 37, 513, 1968.
132. **Johansson, I. and Litzen, U.**, *Ark. Fys.*, 34, 573, 1967.
133. **Johansson, I. and Svensson, K. F.**, *Ark. Fys.*, 16, 353, 1960.
134. **Johansson, L.**, *Ark. Fys.*, 20, 489, 1961.
135. **Johansson, L.**, *Ark. Fys.*, 23, 119, 1963.
136. **Johansson, S. and Litzen, U.**, *Phys. Scr.*, 6, 139, 1972.
137. **Johansson, S. and Litzen, U.**, *Phys. Scr.*, 8, 43, 1973.
138. **Johansson, S. and Litzen, U.**, *Phys. Scr.*, 10, 121, 1974.
139. **Joshi, Y. N.**, St. Francis Xavier Univ., Nova Scotia, unpublished.
140. **Joshi, Y. N., Bhatia, K. S., and Jones, W. E.**, *Sci. Light Tokyo*, 21, 113, 1972.
141. **Joshi, Y. N., Bhatia, K. S., and Jones, W. E.**, *Spectrochim. Acta Part B*, 28, 149, 1973.
142. **Joshi, Y. N. and Budhiraja, C. J.**, *Can. J. Phys.*, 49, 670, 1971.
143. **Joshi, Y. N. and van Kleef, T. A. M.**, *Can. J. Phys.*, 52, 1891, 1974.
144. **Kaufman, V.**, *Natl. Bur. Stand.*, unpublished.
145. **Kaufman, V.**, *J. Opt. Soc. Am.*, 52, 866, 1962.
146. **Kaufman, V., Artru, M. C., and Brillet, W. U. L.**, *J. Opt. Soc. Am.*, 64, 197, 1974.
147. **Kaufman, V. and Humphreys, C. J.**, *J. Opt. Soc. Am.*, 59, 1614, 1969.
148. **Kaufman, V. and Sugar, J.**, *J. Opt. Soc. Am.*, 61, 1693, 1971.
149. **Kaufman, V. and Sugar, J.**, *J. Res. Natl. Bur. Stand. Sect. A*, 71, 583, 1967.
150. **Kelly, R. L. and Palumbo, L. J.**, *Naval Research Laboratory Report 7599, Washington, D. C.*, 1973.
151. **Kielkopf, J. F.**, *Univ. of Louisville*, unpublished, 1975.
152. **Kielkopf, J. F.**, *Univ. of Louisville*, unpublished, 1976.
153. **Kiess, C. C. and Corliss, C. H.**, *J. Res. Natl. Bur. Stand. Sect. A*, 63, 1, 1959.
154. **Kleiman, H.**, *J. Opt. Soc. Am.*, 52, 441, 1962.
155. Eriksson, K. B., Johansson, I., and Norlen, G., *Ark. Fys.*, 28, 233, 1964.
156. **Klinkenberg, P. F. A.**, *Physica*, 15, 774, 1949.
157. **Klinkenberg, P. F. A.**, *Physica*, 16, 618, 1950.
158. **Krishnamurty, S. G.**, *Proc. Phys. Soc. London*, 48, 277, 1936.
159. **Kruger, P. G. and Gilroy, H. T.**, *Phys. Rev.*, 48, 720, 1935.
160. **Kruger, P. G. and Pattin, H. S.**, *Phys. Rev.*, 52, 621, 1937.
161. **Lacroute, P.**, *Ann. Phys.* (Paris), 3, 5, 1935.
162. **Lang, R. J.**, *Phys. Rev.*, 30, 762, 1927.
163. **Lang, R. J.**, *Phys. Rev.*, 32, 737, 1928.
164. **Lang, R. J.**, *Phys. Rev.*, 35, 445, 1930.
165. **Lang, R. J.**, *Can. J. Res. Sect. A*, 14, 43, 1936.
166. **Lang, R. J.**, *Can. J. Res. Sect. A*, 14, 127, 1936.
167. **Lang, R. J. and Vestine, E. H.**, *Phys. Rev.*, 42, 233, 1932.
168. **Li, H. and Andrew, K. L.**, *J. Opt. Soc. Am.*, 61, 96, 1971.
169. **Liden, K.**, *Ark. Fys.*, 1, 229, 1949.
170. **Litzen, U.**, *Ark. Fys.*, 28, 239, 1965.
171. **Litzen, U.**, *Phys. Scr.*, 1, 251, 1970.
172. **Litzen, U.**, *Phys. Scr.*, 1, 253, 1970.
173. **Litzen, U.**, *Phys. Scr.*, 2, 103, 1970.
174. **Litzen, U. and Verges, J.**, *Phys. Scr.*, 13, 240, 1976.
175. **Lofstrand, B.**, *Phys. Scr.*, 8, 57, 1973.
176. **Luc-Koenig, E., Morillon, C., and Verges, J.**, *Phys. Scr.*, 12, 199, 1975.
177. **Lundstrom, T.**, *Phys. Scr.*, 7, 62, 1973.
178. **Lundstrom, T. and Minnhagen, L.**, *Phys. Scr.*, 5, 243, 1972.
179. **Magnusson, C. E. and Zetterberg, P. O.**, *Phys. Scr.*, 10, 177, 1974.
180. Magnusson, C. E., and Zetterberg, P. O., *Phys. Scr.*, 15, 237, 1977.
181. **Martin, D. C.**, *Phys. Rev.*, 48, 938, 1935.
182. Svendenius, N., *Phys. Scr.*, 22, 240, 1980.
183. **Martin, W. C.**, *J. Res. Natl. Bur. Stand. Sect. A*, 64, 19, 1960.
184. **Martin, W. C. and Corliss, C. H.**, *J. Res. Natl. Bur. Stand. Sect. A*, 64, 443, 1960.
185. **Martin, W. C. and Kaufman, V.**, *J. Res. Natl. Bur. Stand. Sect. A*, 74, 11, 1970.
186. **Martin, W. C. and Kaufman, V.**, *J. Opt. Soc. Am.*, 60, 1096, 1970.
187. **McCormick, W. W. and Sawyer, R. A.**, *Phys. Rev.*, 54, 71, 1938.
188. **McLaughlin, R.**, *J. Opt. Soc. Am.*, 54, 965, 1964.
189. **McLennan, J. C., McLay, A. B., and Crawford, M. F.**, *Proc. R. Soc. London Ser. A*, 134, 41, 1931.
190. **Meissner, K. W.**, *Z. Phys.*, 39, 172, 1926.
191. **Meggers, W. F.**, *J. Res. Natl. Bur. Stand.*, 24, 153, 1940.
192. **Meggers, W. F. and Corliss, C. H.**, *J. Res. Natl. Bur. Stand. Sect. A*, 70, 63, 1966.
193. **Meggers, W. F., Fred, M., and Tomkins, F. S.**, *J. Res. Natl. Bur. Stand.*, 58, 297, 1957.
194. **Meggers, W. F. and Humphreys, C. J.**, *J. Res. Natl. Bur. Stand.*, 28, 463, 1942.
195. **Meggers, W. F. and Murphy, R. J.**, *J. Res. Natl. Bur. Stand.*, 48, 334, 1952.
196. **Meggers, W. F., Scribner, B. F., and Bozman, W. R.**, *J. Res. Natl. Bur. Stand.*, 46, 85, 1951.
197. **Meggers, W. F., Shenstone, A. G., and Moore, C. E.**, *J. Res. Natl. Bur. Stand.*, 45, 346, 1950.
198. **Mehlman, G. and Esteva, J. M.**, *Astrophys. J.*, 188, 191, 1974.
199. **Meinders, E.**, *Physica*, 84(C), 117, 1976.
200. Sansonetti, C. J., Dissertation, Purdue University, 1981.
201. Sansonetti, C. J., *Natl. Bur. Stand. (U.S.)*, unpublished.
202. **Millikan, R. A. and Bowen, I. S.** *Phys. Rev.*, 25, 600, 1925.
203. **Minnhagen, L.**, *J. Opt. Soc. Am.*, 61, 1257, 1925.
204. **Minnhagen, L.**, *J. Opt. Soc. Am.*, 63, 1185, 1973.
205. **Minnhagen, L.**, *Phys. Scr.*, 11, 38, 1975.
206. **Minnhagen, L.**, *J. Opt. Soc. Am.*, 66, 659, 1976.
207. **Minnhagen, L. and Nietsche, H.**, *Phys. Scr.*, 5, 237, 1972.
208. **Minnhagen, L., Strihed, H., and Petersson, B.**, *Ark. Fys.*, 39, 471, 1969.
209. **Moore, C. E.**, *Natl. Bur. Stand. (U.S.) Circ.*, 488, 1950.
210. **Moore, C. E.**, *Revised Multiplet Table*, Princeton University Observatory No. 20, 1945.

211. **Moore, C. E.**, National Standard Reference Data Series — National Bureau of Standards 3, Sect. 3, 1970.
212. **Moore, C. E.**, National Standard Reference Data Series — National Bureau of Standards 3, Sect. 4, 1971.
213. **Moore, C. E.**, National Standard Reference Data Series — National Bureau of Standards 3, Sect. 5, 1975.
214. **Moore, C. E.**, National Standard Reference Data Series — National Bureau of Standards 3, Sect. 6, 1972.
215. **Moore, C. E.**, *National Standard Reference Data Series — National Bureau of Standards 3, Sect.* 7, 1975.
216. **Morillon, C. and Verges, J.**, *Phys. Scr.*, 10, 227, 1974.
217. **Newsom, G. H.**, *Astrophys. J.*, 166, 243, 1971.
218. **Newsom, G. H., O'Connor, S., and Learner, R. C. M.**, *J. Phys. B*, 6, 2162, 1973.
219. **Norlen, G.**, *Phys. Scr.*, 8, 249, 1973.
220. **Odabasi, H.**, *J. Opt. Soc. Am.*, 57, 1459, 1967.
221. **Olme, A.**, *Ark. Fys.*, 40, 35, 1969.
222. **Olme, A.**, *Phys. Scr.*, 1, 256, 1970.
223. Johansson, S., and Litzen, U., *J. Opt. Soc. Am.*, 61, 1427, 1971.
224. **Palenius, H. P.**, *Ark. Fys.*, 39, 15, 1969.
225. **Palenius, H. P.**, *Phys. Scr.*, 1, 113, 1970.
226. **Palenius, H. P.**, *Univ. of Lund, Sweden*, unpublished.
227. **Paschen, F.**, *Ann. Phys.*, Series 5, 12, 509, 1932.
228. **Paschen, F. and Ritschl, R.**, *Ann. Phys.*, Series 5, 18, 867, 1933.
229. **Peck, E. R., Khanna, B. N., and Anderholm, N. C.** *J. Opt. Soc. Am.*, 52, 53, 1962.
230. **Persson, W.**, *Phys. Scr.*, 3, 133, 1971.
231. **Persson, W. and Valind S.**, *Phys. Scr.*, 5, 187, 1972.
232. **Petersson, B.**, *Ark. Fys.*, 27, 317, 1964.
233. **Phillips, L. W. and Parker, W. L.**, *Phys. Rev.*, 60, 301, 1941.
234. **Platt, J. R. and Sawyer, R. A.**, *Phys. Rev.*, 60, 866, 1941.
235. **Plyer, E. K., Blaine, L. R., and Tidwell, E.**, *J. Res. Natl. Bur. Stand.*, 55, 279, 1955.
236. **Poppe, R., van Kleef, T. A. M., and Raassen, A. J. J.**, *Physica*, 77, 165, 1974.
237. **Radziemski, L. J., Jr. and Andrew, K. L.**, *J. Opt. Soc. Am.*, 55, 474, 1965.
238. **Radziemski, L. J., Jr. and Kaufman, V.**, *J. Opt. Soc. Am.*, 59, 424, 1969.
239. **Radziemski, L. J., Jr. and Kaufman, V.**, *J. Opt. Soc. Am.*, 64, 366, 1974.
240. **Ramanadham, R. and Rao, K. R.**, *Indian J. Phys.*, 18, 317, 1944.
241. **Ramb, R.**, *Ann. Phys.*, 10, 311, 1931.
242. **Rank, D. H., Bennett, J. M., and Bennett, H. E.**, *J. Opt. Soc. Am.*, 40, 477, 1950.
243. **Rao, A. S. and Krishnamurty, S. G.**, *Proc. Phys. Soc. London*, 46, 531, 1943.
244. **Rao, K. R.**, *Proc. R. Soc. London, Ser. A*, 134, 604, 1932.
245. **Rao, K. R. and Badami, J. S.**, *Proc. R. Soc. London Ser. A*, 131, 154, 1931.
246. **Rao, K. R. and Krishnamurty, S. G.**, *Proc. R. Soc. London Ser. A*, 161, 38, 1937.
247. **Rao, K. R. and Murti, S. G. K.**, *Proc. R. Soc. London Ser. A*, 145, 681, 1934.
248. **Rao, Y. B.**, *Indian J. Phys.*, 32, 497, 1958.
249. **Rao, Y. B.**, *Indian J. Phys.*, 33, 546, 1959.
250. **Rao, Y. B.**, *Indian J. Phys.*, 35, 386, 1961.
251. **Rasmussen, E.**, *Z. Phys.*, 80, 726, 1933.
252. **Rasmussen, E.**, *Z. Phys.*, 83, 404, 1933.
253. **Rasmussen, E.**, *Z. Phys.*, 86, 24, 1934.
254. **Rasmussen, E.**, *Z. Phys.*, 87, 607, 1934.
255. **Rasmussen, E.**, *Phys. Rev.*, 57, 840, 1940.
256. **Rau, A. S. and Narayan, A. L.**, *Z. Phys.*, 59, 687, 1930.
257. **Reader, J.**, *J. Opt. Soc. Am.*, 65, 286, 1975.
258. **Reader, J.**, *J. Opt. Soc. Am.*, 65, 988, 1975.
259. **Reader, J.**, *J. Opt Soc. Am.*, 73, 349, 1983.
260. **Reader, J. and Davis, S.**, *J. Res. Natl. Bur. Stand. Sect. A*, 71, 587, 1967, and unpublished.
261. **Reader, J. and Ekberg, J. O.**, *J. Opt. Soc. Am.*, 62, 464, 1972.
262. **Reader, J. and Epstein, G. L.**, *J. Opt. Soc. Am.*, 62, 1467, 1972.
263. **Reader, J. and Epstein, G. L.**, *J. Opt. Soc. Am.*, 65, 638, 1975.
264. **Reader, J. and Epstein, G. L.**, *Natl. Bur. Stand.*, unpublished.
265. **Reader, J., Epstein, G. L., and Ekberg, J. O.**, *J. Opt. Soc. Am.*, 62, 273, 1972.
266. **Kaufman, V.**, *Phys. Scr.*, 26, 439, 1982.
267. **Ricard, R., Givord, M., and George, F.**, *C. R. Acad. Sci. Paris*, 205, 1229, 1937.
268. **Risberg, P.**, *Ark. Fys.*, 10, 583, 1956.
269. **Risberg, G.**, *Ark. Fys.*, 28, 381, 1965.
270. **Risberg, G.**, *Ark. Fys.*, 37, 231, 1968.
271. **Robinson, H. A.**, *Phys. Rev.*, 49, 297, 1936.
272. **Robinson, H. A.**, *Phys. Rev.*, 50, 99, 1936.
273. **Ross, C. B., Jr.**, Doctoral dissertation, Purdue University, 1969.
274. **Ross, C. B., Wood, D. R., and Scholl, P. S.**, *J. Opt. Soc. Am.*, 66, 36, 1976.
275. **Ruedy, J. E. and Gibbs, R. C.**, *Phys. Rev.*, 46, 880, 1934.
276. **Russell, H. N., King, R. B., and Moore, C. E.**, *Phys. Rev.*, 58, 407, 1940.
277. **Russell, H. N. and Moore, C. E.**, *J. Res. Natl. Bur. Stand.*, 55, 299, 1955.
278. **Russell, H. N., Moore, C. E., and Weeks, D. W.**, *Trans. Am. Philos. Soc.*, 34(2), 111, 1944.
279. **Saunders, F., Schneider, E., and Buckingham, E.**, *Proc. Natl. Acad. Sci.*, 20, 291, 1934.
280. **Sawyer, R. A. and Humphreys, C. J.**, *Phys. Rev.*, 32, 583, 1928.
281. **Sawyer, R. A. and Lang, R. J.**, *Phys. Rev.*, 34, 712, 1929.
282. **Sawyer, R. A. and Paschen, F.** *Ann. Phys.*, 84(4), 1, 1927.
283. **Scholl, P. S.**, M.S. thesis, Wright State Univ., 1975.
284. **Schurmann, D.**, *Z. Phys.*, 17, 4, 1975.
285. **Seguier, J.**, *C. R. Acad. Sci. Paris*, 256, 1703, 1963.
286. **Shenstone, A. G.**, *Phys. Rev.*, 31, 317, 1928.
287. **Shenstone, A. G.**, *Phys. Rev.*, 32, 30, 1928.
288. **Shenstone, A. G.**, *Trans. R. Soc. London*, 237(A), 57, 1938.
289. **Shenstone, A. G.**, *Phys. Rev.*, 57, 894, 1940.
290. **Shenstone, A. G.** *Philos. Trans. R. Soc. London Ser. A*, 241, 297, 1948.
291. **Shenstone, A. G.**, *Can. J. Phys.*, 38, 677, 1960.
292. **Shenstone, A. G.**, *Proc. R. Soc. London*, 261(A), 153, 1961.

293. **Shenstone, A. G.**, *Proc. R. Soc. London*, 276(A), 293, 1963.
294. **Shenstone, A. G.**, *J. Res. Natl. Bur. Stand. Sect. A*, 74, 801, 1970.
295. **Shenstone, A. G.**, *J. Res. Natl. Bur. Stand. Sect. A*, 79, 497, 1975.
296. **Shenstone, A. G. and Pittenger, J. T.**, *J. Opt. Soc. Am.*, 39, 219, 1949.
297. **Smith, S.**, *Phys. Rev.*, 36, 1, 1930.
298. **Smitt, R.**, *Phys. Scr.*, 8, 292, 1973.
299. **Soderqvist, J.**, *Ark. Mat. Astronom. Fys.*, 32(A), 1, 1946.
300. **Sommer, L. A.**, *Ann. Phys.*, 75, 163, 1924.
301. **Spector, N.**, *J. Opt. Soc. Am.*, 63, 358, 1973.
302. **Spector, N. and Sugar, J.**, *J. Opt. Soc. Am.*, 66, 436, 1976.
303. **Steinhaus, D. W., Radziemski, L. J., Jr., and Blaise, J.**, *Los Alamos Sci. Lab.*, unpublished, 1975.
304. **Subbaraya, T. S.**, *Z. Phys.*, 78, 541, 1932.
305. **Sugar, J.**, *J. Opt. Soc. Am.*, 55, 33, 1965.
306. **Sugar, J.**, *J. Res. Natl. Bur. Stand. Sect. A*, 73, 333, 1969.
307. **Sugar, J.**, *J. Opt. Soc. Am.*, 60, 454, 1970.
308. **Sugar, J.**, *J. Res. Natl. Bur. Stand. Sect. A*, 78, 555, 1974.
309. **Sugar, J. and Kaufman, V.**, *J. Opt. Soc. Am.*, 55, 1283, 1965.
310. **Sugar, J. and Kaufman, V.**, *J. Opt. Soc. Am.*, 62, 562, 1972.
311. **Sugar, J., Kaufman, V., and Spector, N.**, *J. Res. Natl. Bur. Stand.*, Sect. A, 83, 233, 1978.
312. **Sugar, J. and Spector, N.**, *J. Opt. Soc. Am.*, 64, 1484, 1974.
313. **Sullivan, F. J.** *Univ. Pittsburgh Bull.*, 35, 1, 1938.
314. **Svensson, L. A. and Ekberg, J. O.**, *Ark. Fys.*, 37, 65, 1968.
315. **Swensson, J. W. and Risberg, G.**, *Ark. Fys.*, 31, 237, 1966.
316. **Tech. J. L.**, *J. Res. Natl. Bur. Stand. Sect. A*, 67, 505, 1963.
317. **Tech, J. L. and Ward, J. F.**, *Phys. Rev. Lett.*, 27, 367, 1971.
318. **Tilford, S. G.**, *J. Opt. Soc. Am.*, 53, 1051, 1963.
319. **Toresson, Y. G.**, *Ark. Fys.*, 17, 179, 1960.
320. **Toresson, Y. G.**, *Ark. Fys.*, 18, 389, 1960.
321. **Toresson, Y. G. and Edlen, B.**, *Ark. Fys.*, 23, 117, 1963.
322. **Tsien, W. Z.**, *Chin. J. Phys.*, Peiping, 3, 117, 1939.
323. **van Deurzen, C. H. H., Conway, J., and Davis, S. P.**, *J. Opt. Soc. Am.*, 63, 158, 1973.
324. **van Kleef, T. A. M., Raassen, A. J. J., and Joshi, Y. N.**, *Physica*, 84(C), 401, 1976.
325. Sansonetti, C. J., Andrew, K. L., and Verges, J., *J. Opt. Soc. Am.*, 71, 423, 1981.
326. **Wheatley, M. A. and Sawyer, R. A.**, *Phys. Rev.*, 61, 591, 1942.
327. **Wilkinson, P. G.**, *J. Opt. Soc. Am.*, 45, 862, 1955.
328. **Wilkinson, P. G. and Andrew, K. L.**, *J. Opt. Soc. Am.*, 53, 710, 1963.
329. **Wood, D. and Andrew, K. L.**, *J. Opt. Soc. Am.*, 58, 818, 1968.
330. **Wood, D. R., Ross, C. B., Scholl, P. S., and Hoke, M.**, *J. Opt. Soc. Am.*, 64, 1159, 1974.
331. **Worden, E. F. and Conway, J. G.**, *Lawrence Livermore Lab.*, unpublished, 1977.
332. **Worden, E. F., Hulet, E. K., Gutmacher, R. G., Conway, J. G.**, *At. Data Nucl. Data Tables*, 18, 459, 1976.
333. **Worden, E. F., Lougheed, R. W., Gutmacher, R. G., and Conway, J. G.**, *J. Opt. Soc. Am.*, 64, 77, 1974.
334. **Wu, C. M.**, Ph.D. thesis, University of British Columbia, 1971.
335. **Zaidel, A. N., Prokofev, V. K., Raiskii, S. M., Slavnyi, V. A., and Schreider, E. Y.**, *Tables of Spectral Lines*, 3rd ed., Plenum, New York, 1970.
336. **Zetterberg, P. O. and Magnusson, C. E.**, *Phys. Scr.*, 15, 189, 1977.
337. **Sugar, J.**, *J. Opt. Soc. Am.*, 55, 1058, 1965.
338. **Sugar, J.**, *J. Opt. Soc. Am.*, 61, 727, 1971.
339. Worden, E. F., and Conway, J. G., *At. Data Nucl. Data Tables*, 22, 329, 1978.
340. **Kaufman, V. and Edlen, B.**, *J. Phys. Chem. Ref. Data*, 3, 825, 1974.
341. **Lang, R. J.**, *Phys. Rev.*, 34, 697, 1929.
342. **Ryabtsev, A. N.**, *Opt. Spectros.*, 39, 455, 1975.
343. **Foster, E. W.**, *Proc. R. Soc. London*, 200(A), 429, 1950.
344. **Morillon, C. and Verges, J.**, *Phys. Scr.*, 12, 129, 1975.
345. **Ruedy, J. E.**, *Phys. Rev.*, 41, 588, 1932.
346. **McLennan, J. C., McLay, A. B., and McLeod, J. H.**, *Philos. Mag.*, 4, 486, 1927.
347. **Handrup, M. B. and Mack, J. E.**, *Physica*, 30, 1245, 1964.
348. **Clearman, H. E.**, *J. Opt. Soc. Am.*, 42, 373, 1952.
349. **Paschen, F.**, *Ann. Physik*, 424, 148, 1938.
350. **Paschen, F. and Campbell, J. S.**, *Ann. Phys.*, 31(5), 29, 1938.
351. **Nodwell, R.**, *Univ. of British Columbia, Vancouver*, unpublished, 1955.
352. **Gibbs, R. C. and White, H. E.**, *Phys. Rev.*, 31, 776, 1928.
353. **Green, M.**, *Phys. Rev.*, 60, 117, 1941.
354. **Ellis, C. B. and Sawyer, R. A.**, *Phys. Rev.*, 49, 145, 1936.
355. **McLennan, J. C., McLay, A. B., and Crawford, M. F.**, *Proc. R. Soc. London Ser. A*, 125, 50, 1929.
356. **Mack, J. E. and Fromer, M.**, *Phys. Rev.*, 48, 346, 1935.
357. **Humphreys, C. J. and Paul, E.**, U.S. Nav. Ord. Lab., Navord Rep. 4589, 25, 1956.
358. **Walters, F. M.**, *Sci. Pap. Bur. Stand.*, 17, 161, 1921.
359. **Crawford, M. F. and McLay, A. B.**, *Proc. R. Soc. London Ser. A*, 143, 540, 1934.
360. **McLay, A. B. and Crawford, M. F.**, *Phys. Rev.*, 44, 986, 1933.
361. **Schoepfle, G. K.**, *Phys. Rev.*, 47, 232, 1935.
362. Acquista, N., and Reader, J., *J. Opt. Soc. Am.*, 70, 789, 1980.
363. Benschop, H., Joshi, Y. N., and Van Kleef, T. A. M., *Can. J. Phys.*, 53, 498, 1975.
364. Bockasten, K., Hallin, R., and Hughes, T. P., *Proc. Phys. Soc.*, 81, 522, 1963.
365. Boyce, J. C., *Phys. Rev.*, 46, 378, 1934.
366. Boyce, J. C., *Phys. Rev.*, 47, 718, 1935.
367. Boyce, J. C., *Phys. Rev.*, 48, 396, 1935.
368. Boyce, J. C., *Phys. Rev.*, 49, 351, 1936.
369. Corliss, C. H. and Meggers, W. F., *J. Res. Natl. Bur. Stand.*, 61, 269, 1958.
370. Crooker, A. M. and Dick, K. A., *Can. J. Phys.*, 42, 766, 1964.
371. De Bruin, T. L., *Z. Physik*, 77, 505, 1932.
372. De Bruin, T. L., *Proc. Roy. Acad. Amsterdam*, 36, 727, 1933.
373. De Bruin, T. L., *Zeeman Verhandelingen*, (The Hague), 1935, p. 415.
374. De Bruin, T. L., *Physica*, 3, 809, 1936.
375. De Bruin, T. L., *Proc. Roy. Acad. Amsterdam*, 40, 339, 1937.

376. Dick, K. A., *Can. J. Phys.*, 46, 1291, 1968.
377. Dick, K. A., unpublished, 1978.
378. Edlen, B. and Swensson, J. W., *Phys. Scr.*, 12, 21, 1975.
379. Ekberg, J. O., *Phys. Scr.*, 7, 55, 1973.
380. Ekberg, J. O., *Phys. Scr.*, 7, 59, 1973.
381. Ekberg, J. O., *Phys. Scr.*, 12, 42, 1975.
382. Ekberg, J. O. and Edlen, B., *Phys. Scr.*, 18, 107, 1978.
383. Eliason, A. Y., *Phys. Rev.*, 43, 745, 1933.
384. Gallardo, M., Massone, C. A., Tagliaferri, A. A., Garavaglia, M., and Persson, W., *Phys. Scr.*, to be published, 1979.
385. Garcia-Riquelme, O., *Optica Pura Y Aplicada*, 1, 53, 1968.
386. Gibbs, R. C., Vieweg, A. M., and Gartlein, C. W., *Phys. Rev.*, 34, 406, 1929.
387. Gilbert, W. P., *Phys. Rev.*, 48, 338, 1935.
388. Goldsmith, S. and Kaufman, A. S., *Proc. Phys. Soc.*, 81, 544, 1963.
389. Hermansdorfer, H., *J. Opt. Soc. Am.*, 62, 1149, 1972.
390. Humphreys, C. J., *Phys. Rev.*, 47, 712, 1935.
391. Humphreys, C. J., *J. Res. Natl. Bur. Stand.*, 16, 639, 1936.
392. Iglesias, L., *J. Opt. Soc. Am.*, 45, 856, 1955.
393. Iglesias, L., *J. Res. Natl. Bur. Stand.*, 64A, 481, 1960.
394. Iglesias, L., *Anales Fisica Y Quimica*, 58A, 191, 1962.
395. Iglesias, L., *J. Res. Natl. Bur. Stand.*, 70A, 465, 1966.
396. Iglesias, L., *Can. J. Phys.*, 44, 895, 1966.
397. Iglesias, L., *J. Res. Natl. Bur. Stand.*, 72A, 295, 1968.
398. Joshi, Y. N., *Can. Spectrosc.*, 15, 96, 1970.
399. Joshi, Y. N. and Van Kleef, T. A. M., *Can. J. Phys.*, 55, 714, 1977.
400. Kaufman, A. S., Hughes, T. P., and Williams, R. V., *Proc. Phys. Soc.*, 76, 17, 1960.
401. Kaufman, V. and Sugar, J., *J. Opt. Soc. Am.*, 68, 1529, 1978.
402. Keussler, V., *Z. Physik*, 85, 1, 1933.
403. Kiess, C. C., *J. Res. Natl. Bur. Stand.*, 56, 167, 1956.
404. Klinkenberg, P. F. A., Van Kleef, T. A. M., and Noorman, P. E., *Physica*, 27, 1177, 1961.
405. Kovalev, V. I., Romanos, A. A., and Ryabtsev, A. N., *Opt. Spectrosc.*, 43, 10, 1977.
406. Lang, R. J., *Proc. Natl. Acad. Sci.*, 13, 341, 1927.
407. Lang, R. J., *Zeeman Verhandelingen*, (The Hague), 44, 1935.
408. Liberman, S., et al., *C. R. Acad. Sci. (Paris)*, 286, 253, 1978.
409. Livingston, A. E., *J. Phys.*, B9, L215, 1976.
410. Meijer, F. G., *Physica*, 72, 431, 1974.
411. Meijer, F. G. and Metsch, B. C., *Physica*, 94C, 259, 1978.
412. Moore, F. L., thesis, Princeton, 1949.
413. Paul, F. W. and Polster, H. D., *Phys. Rev.*, 59, 424, 1941.
414. Phillips, L. W. and Parker, W. L., *Phys. Rev.*, 60, 301, 1941.
415. Poppe, R., *Physica*, 81C, 351, 1976.
416. Raassen, A. J. J., Van Kleef, T. A. M., and Metsch, B. C., *Physica*, 84C, 133, 1976.
417 Rao, A. B. and Krishnamurty, S. G., *Proc. Phys. Soc. (London)*, 51, 772, 1939.
418. Reader, J. and Acquista, N., *J. Opt. Soc. Am.*, 69, 239, 1979.
419. Reader, J. and Epstein, G. L., *J. Opt. Soc. Am.*, 62, 619, 1972.
420. Rico, F. R., *Anales, Real Soc. Esp. Fis. Quim.*, 61, 103, 1965.
421. Schonheit, E., *Optik*, 23, 409, 1966.
422. Shenstone, A. G., *J. Opt. Soc. Am.*, 44, 749, 1954.
423. Shenstone, A. G., unpublished, 1958.
424. Shenstone, A. G., *J. Res. Natl. Bur. Stand.*, 67A, 87, 1963.
425. Sugar, J. and Kaufman, V., *J. Opt. Soc. Am.*, 64, 1656, 1974.
426. Sugar, J. and Kaufman, V., *Phys. Rev.*, C12, 1336, 1975.
427. Svensson, L. A., *Phys. Scr.*, 13, 235, 1976.
428. Swensson, J. W. and Edlen, B., *Phys. Scr.*, 9, 335, 1974.
429. Tagliaferri, A. A., Gallego Lluesma, E., Garavaglia, M., Gallardo, M., and Massone, C. A., *Optica Pura Y Aplica*, 7, 89.
430. Tilford, S. G. and Giddings, L. E., *Astrophys. J.*, 141, 1222, 1965.
431. Trawick, M. W., *Phys. Rev.*, 46, 63, 1934.
432. Van Deurzen, C. H. H., *J. Opt. Soc. Am.*, 67, 476, 1977.
433. Yarosewick, S. L. and Moore, F. L., *J. Opt. Soc. Am.*, 57, 1381, 1967.
434. Zalubas, R., unpublished, 1979.
435. Bhatia, K. S., Jones, W. E., and Crooker, A. M., *Can. J. Phys.*, 50, 2421, 1972.
436. van Kleef, T. A. M. and Joshi, Y. N., *Phys. Scr.*, 24, 557, 1981.

ATOMIC TRANSITION PROBABILITIES

Compiled by W. L. Wiese and G. A. Martin

These tables were prepared under the auspices of the Committee on Line Spectra of the Elements of the National Academy of Sciences — National Research Council. They contain critically evaluated atomic transition probabilities for about 5000 selected lines of all elements for which reliable data are available on an absolute scale. The material is largely for neutral and singly ionized spectra, but includes a number of prominent lines of more highly charged ions.

Many of the data are obtained from comprehensive compilations of the National Bureau of Standards Data Center on Atomic Transition Probabilities. Specifically, data have been taken from critical compilations on V,[1] Cr,[1] Mn,[1] Fe,[2] Co,[2] and Ni[2] without changes. Material from earlier compilations for the elements H through Ne,[3] Na through Ca,[4] and Sc and Ti[5] was supplemented by more recent material taken directly from the original literature. For the higher ions, many data were derived from studies of the systematic behavior of transition probabilities.[6-8] The original literature is cited in a recent bibliography;[9] for lack of space, individual literature references are not cited here.

The wavelength range for the neutral species has normally been restricted to the visible or shorter wavelengths; only the very prominent near infrared lines are included. For the higher ions, most of the strong lines are located in the far UV. The tabulation is limited to electric dipole — including intercombination — lines and comprises essentially the fairly strong transitions with estimated uncertainties of 50% or less. With the exception of hydrogen, helium, and the alkalis, most transitions are between states with low principal quantum numbers.

The transition probability, A, is given in units of 10^8 s^{-1} and is listed to as many digits as is consistent with the indicated accuracy. A number in parentheses following the tabulated value of the transition probability indicates the power of 10 by which this value has to be multiplied. The estimated uncertainties of the A-values are indicated by code letters as follows: AA — for uncertainties within 1%; A — within 3%; B — within 10%; C — within 25%; D — within 50%.

Each transition is identified by the wavelength, λ, in angstroms; the energy level of the upper atomic state, E_k, in cm^{-1}; and the statistical weights, g_i and g_k, of the lower (i) and upper (k) states [the product g_kA (or g_if) is needed in many applications]. Whenever the wavelengths of individual lines within a multiplet are extremely close, an average wavelength for the multiplet is given, and is indicated by an asterisk (*) to the right of the g_k value. This has also been done when the transition probability for an entire multiplet has been taken from the literature and values for individual lines cannot be determined because of insufficient knowledge of the coupling of electrons. Wavelength and energy level data have been taken either from recent compilations or from the original literature cited in bibliographies published by the National Bureau of Standards Atomic Energy Levels Data Center.[10,11] Wavelength values are consistent with those given in Section E of this Handbook.

In the table for hydrogen, the energy level and uncertainty columns have been eliminated since the transition probabilities and energy levels are known very precisely for this element. Because of the hydrogen degeneracy, a ''transition'' is actually the sum of all transitions between the principal quantum numbers listed in the transition column, and the tabulation represents the properly weighted A values.

In addition to the transition probability A, the atomic oscillator-strength f and the line-strength S are often used in the literature. The conversion factors between these quantities are:

$$g_i f = 1.499 \times 10^{-8} \lambda^2 g_k A = 303.8 \lambda^{-1} S$$

where λ is in angstroms, A is in 10^8 s^{-1}, and S is in atomic units, which are $a_o^2 e^2 = 7.188 \times 10^{-59}$ m^{2}C^2 for electric dipole transitions.

We acknowledge the valuable preparatory work by D. Trahan, W. Croom, and F. Farley in arranging and compiling the numerical data.

Table 1a
TRANSITION PROBABILITIES FOR ALLOWED LINES OF HYDROGEN

Wavelength λ [Å]	Transition	Statistical weights g_i	Statistical weights g_k	Average transition probability A [10^8 s^{-1}]
914.039	1—20	2	800	3.928(−5)
914.286	1—19	2	722	5.077(−5)
914.576	1—18	2	648	6.654(−5)
914.919	1—17	2	578	8.858(−5)
915.329	1—16	2	512	1.200(−4)
915.824	1—15	2	450	1.657(−4)
916.429	1—14	2	392	2.341(−4)
917.181	1—13	2	338	3.393(−4)
918.129	1—12	2	288	5.066(−4)
919.351	1—11	2	242	7.834(−4)
920.963	1—10	2	200	1.263(−3)
923.150	1— 9	2	162	2.143(−3)
926.226	1— 8	2	128	3.869(−3)
930.748	1— 7	2	98	7.568(−3)
937.803	1— 6(Lε)	2	72	1.644(−2)
949.743	1—5 (Lδ)	2	50	4.125(−2)
972.537	1— 4(Lγ)	2	32	1.278(−1)
1025.72	1— 3(Lβ)	2	18	5.575(−1)
1215.67	1— 2(Lα)	2	8	4.699
3682.81	2—20	8	800	2.172(−5)
3686.83	2—19	8	722	2.809(−5)
3691.55	2—18	8	648	3.685(−5)
3697.15	2—17	8	578	4.910(−5)
3703.85	2—16	8	512	6.658(−5)
3711.97	2—15	8	450	9.210(−5)
3721.94	2—14	8	392	1.303(−4)
3734.37	2—13	8	338	1.893(−4)
3750.15	2—12	8	288	2.834(−4)
3770.63	2—11	8	242	4.397(−4)
3797.90	2—10	8	200	7.122(−4)
3835.38	2— 9	8	162	1.216(−3)
3889.05	2— 8	8	128	2.215(−3)
3970.07	2— 7(Hε)	8	98	4.389(−3)
4101.73	2— 6(Hδ)	8	72	9.732(−3)
4340.46	2— 5(Hγ)	8	50	2.530(−2)
4861.32	2— 4(Hβ)	8	32	8.419(−2)
6562.80	2— 3(Hα)	8	18	4.410(−1)
8392.40	3—20	18	800	1.517(−5)
8413.32	3—19	18	722	1.964(−5)
8437.96	3—18	18	648	2.580(−5)
8467.26	3—17	18	578	3.444(−5)
8502.49	3—16	18	512	4.680(−5)
8545.39	3—15	18	450	6.490(−5)
8598.40	3—14	18	392	9.211(−5)
8665.02	3—13	18	338	1.343(−4)
8750.48	3—12	18	288	2.021(−4)
8862.79	3—11	18	242	3.156(−4)
9014.91	3—10	18	200	5.156(−4)
9229.02	3— 9	18	162	8.905(−4)
9545.97	3— 8(Pε)	18	128	1.651(−3)
10049.4	3— 7(Pδ)	18	98	3.358(−3)
10938.1	3— 6(Pγ)	18	72	7.783(−3)
12818.1	3— 5(Pβ)	18	50	2.201(−2)
16407.2	4—12	32	288	1.620(−4)
16806.5	4—11	32	242	2.556(−4)
17362.1	4—10	32	200	4.235(−4)
18174.1	4— 9	32	162	7.459(−4)
18751.0	3— 4(Pα)	18	32	8.986(−2)
19445.6	4— 8	32	128	1.424(−3)
21655.3	4— 7	32	98	3.041(−3)
26251.5	4— 6	32	72	7.711(−3)
27575	5—12	50	288	1.402(−4)
28722	5—11	50	242	2.246(−4)
30384	5—10	50	200	3.800(−4)
32961	5— 9	50	162	6.908(−4)
37395	5— 8	50	128	1.388(−3)
40511.5	4— 5	32	50	2.699(−2)
43753	6—12	72	288	1.288(−4)
46525	5— 7	50	98	3.253(−3)
46712	6—11	72	242	2.110(−4)
51273	6—10	72	200	3.688(−4)
59066	6— 9	72	162	7.065(−4)
74578	5— 6	50	72	1.025(−2)
75004	6— 8	72	128	1.561(−3)
123680	6— 7	72	98	4.561(−3)

For hydrogen-like ions of nuclear charge Z, the following scaling laws hold:

$$A_Z = Z^4 A_{Hydrogen};\ f_Z = f_H;\ S_Z = Z^{-2} S_H$$

(For wavelengths $\lambda_Z = Z^{-2} \lambda_H$).

For very highly charged ions, relativistic effects need to be taken into account.[12]

Table 1b
TRANSITION PROBABILITIES FOR SELECTED ATOMIC AND IONIC SPECIES

Wavelength (λ[Å])	Upper energy level (E_k[cm^{-1}])	Statistical weights g_i	g_k	Transition Probability (A[10^8 s^{-1}])	Uncertainty
Aluminum					
Al I					
2263.5	44166	2	4	.66	C
2269.1	44169	4	6	.79	C
2269.2	44166	4	4	.13	C
2367.1	42234	2	4	.72	C
2373.1	42238	4	6	.86	C
2373.4	42234	4	4	.14	C
2568.0	38929	2	4	.23	C
2575.1	38934	4	6	.28	C
2575.4	38929	4	4	.044	C
2652.5	37689	2	2	.133	C
2660.4	37689	4	2	.264	C
3082.2	32435	2	4	.63	C
3092.7	32437	4	6	.74	C
3092.8	32435	4	4	.12	C
3944.0	25348	2	2	.493	C
3961.5	25348	4	2	.98	C
6696.0	40278	2	4	.0169	C
6698.7	40272	2	2	.0169	C
7835.3	45195	4	6	.057	D
7836.1	45195	6	8	.062	D
Al II					
1047.9	132823	1	3	.36	D
1048.6	132823	3	5	.48	D
1539.8	124794	3	5	8.8	D
1670.8	59852	1	3	14.6	B
1719.4	95551	1	3	6.79	B
1764.0	94269	5	5	9.8	C
1772.8	150493	1	3	9.5	D
1777.0	150544	5	7	17.	D
1819.0	150525	15	15*	5.6	D
1855.9	91275	1	3	.832	B
1858.0	91275	3	3	2.48	B
1862.3	91275	5	3	4.12	B
1931.0	111637	3	1	10.8	C
1990.5	110090	3	5	14.7	C
2816.2	95351	3	1	3.83	C
4663.1	106921	5	3	.53	C
6226.2	121484	1	3	.62	D
6231.8	121484	3	5	.84	D
6243.4	121484	5	7	1.1	D
6335.7	125869	5	3	.14	D
6823.4	120093	3	3	.34	D
6837.1	120093	5	3	.57	D
6920.3	121367	3	1	.96	D
7042.1	105471	3	5	.59	C
7056.7	105442	3	3	.58	C
7471.4	123471	5	7	.94	D
Al III					
560.36	178458	2	6*	.40	D
695.83	143714	2	4	.74	C
696.22	143633	2	2	.72	C
1352.8	189876	10	14*	4.40	C
1379.7	126164	2	2	4.59	C
1384.1	126164	4	2	9.1	C
1605.8	115959	2	4	12.2	B
1611.8	115959	4	4	2.42	B
1611.9	115956	4	6	14.5	B
1854.7	53917	2	4	5.40	B
1862.8	53683	2	2	5.33	B
1935.9	167613	10	14*	12.2	C
3601.6	143714	6	4	1.34	C
3601.9	143714	4	4	.149	C
3612.4	143633	4	2	1.5	C
Al X					
39.925	2504700	1	3	2220.	C
51.979	1923850	1	3	4800.	C
55.227	1965860	1	3	5200.	D
55.272	1966030	3	5	7200.	D
55.376	1966270	5	7	9500.	D
59.107	1992340	3	5	4600.	C
332.78	300490	1	3	56.	C
394.83	553783	3	1	83.	C
395.36	409690	3	5	12.	C
397.76	406517	1	3	17.	C
400.43	406517	3	3	13.	C
401.12	409690	5	5	36.	C
403.55	404574	3	1	49.	C
406.31	406517	5	3	19.	C
670.06	449732	3	5	9.8	C
2535.	1923850	1	3	.38	D
Al XI					
36.675	2726700	2	6*	1500.	C
39.091	2734100	2	4	2600.	C
39.180	2734500	4	6	3100.	C
39.530	2705700	2	2	180.	D
39.623	2705700	4	2	370.	D
48.298	2070520	2	4	3090.	B
48.338	2068770	2	2	3080.	B
52.299	2088100	2	4	8100.	C
52.446	2088530	4	6	9600.	C
52.458	2088100	4	4	1600.	C
54.217	2020450	2	2	480.	C
54.388	2020450	4	2	960.	C
99.083	3029700	2	6*	220.	C
103.6	3033700	2	4	420.	C
103.8	3033700	4	6	500.	C
141.6	2726700	2	6*	407.	C
150.31	2734100	2	4	850.	C
150.61	2734500	4	6	990.	C
157.0	2705700	2	2	130.	C
157.4	2705700	4	2	260.	C
205.0	3193600	2	6*	63.	C
308.6	3029700	2	6*	99.	C
341.3	3019700	6	2*	130.	C
550.05	181808	2	4	8.55	B
568.12	176019	2	2	7.73	B
1997.	2070520	2	4	1.07	C
2069.	2068770	2	2	.97	C
4761.	2726700	2	6*	.255	C
5172.	2088100	2	4	.0395	C
5551.	2088530	4	6	.0385	C
5687.	2088100	4	4	.0060	D
Argon					
Ar I					
3554.3	121271	5	5	.0029	D
3567.7	121165	5	7	.0012	D
3606.5	121470	3	1	.0081	D
3649.8	122791	3	1	.0085	D
3834.7	121470	3	1	.0080	D
3949.0	118460	5	3	.00467	C
4044.4	118469	3	5	.00346	C
4158.6	117184	5	5	.0145	C
4164.2	117151	5	3	.00295	C
4181.9	118460	1	3	.0058	C
4190.7	116999	5	5	.00254	C
4191.0	118407	1	3	.0056	C
4198.3	117563	3	1	.0276	C
4200.7	116943	5	7	.0103	C
4259.4	118871	3	1	.0415	C
4266.3	117184	3	5	.00333	C
4272.2	117151	3	3	.0084	C
4300.1	116999	3	5	.00394	C
4333.6	118469	3	5	.0060	C
4335.3	118460	3	3	.00387	C
4345.2	118407	3	3	.00313	C
4510.7	117563	3	1	.0123	C
4768.7	125066	3	5	.0090	D
4836.7	124772	3	5	.00106	C
4876.3	124604	3	5	.0081	D
4887.9	124555	3	3	.014	D
4894.7	124527	3	1	.019	D
5048.8	123903	3	5	.0048	D
5054.2	123882	3	3	.0047	D
5056.5	123873	3	1	.0059	D
5118.2	125150	5	7	.0028	D
5151.4	123509	3	1	.0249	C
5152.3	123505	3	5	.0011	D
5162.3	123468	3	3	.0198	C
5177.5	124772	7	5	.0025	D
5187.7	123373	3	5	.0138	C
5194.1	125335	3	1	.0081	D
5210.5	124650	7	7	.0011	D
5214.8	124788	5	3	.0022	D
5216.3	124783	5	3	.0014	D
5221.3	124610	7	9	.0092	D
5241.1	124692	5	5	.0014	D

Table 1b (Continued)
TRANSITION PROBABILITIES FOR SELECTED ATOMIC AND IONIC SPECIES

Wavelength (λ[Å])	Upper energy level (E_k[cm^{-1}])	Statistical weights g_i	g_k	Transition Probability (A[10^8 s^{-1}])	Uncertainty
		Ar I			
5252.8	124650	5	7	.0056	D
5373.5	124692	3	5	.0028	D
5394.0	124772	5	5	.0010	D
5410.5	124715	5	7	.0021	D
5421.4	123903	7	5	.0062	D
5440.0	122479	3	3	.0020	D
5442.2	123832	7	7	9.7 (–4)	D
5451.7	122440	3	5	.0049	D
5457.4	123936	5	3	.0037	D
5467.2	123903	5	5	7.9 (–4)	D
5473.5	123882	5	3	.0021	D
5490.1	123827	5	5	8.9 (–4)	D
5495.9	123653	7	9	.0176	C
5506.1	123774	5	7	.0037	D
5525.0	123557	7	7	.0018	D
5534.5	125353	5	3	.0028	D
5558.7	122087	3	5	.0148	C
5559.7	125113	3	5	.0023	D
5572.5	123557	5	7	.0069	C
5588.7	123505	5	5	.0016	D
5606.7	121933	3	3	.0229	C
5618.0	123882	3	3	.0022	D
5620.9	123873	3	1	.0038	D
5623.8	125066	5	5	.0015	D
5635.6	123827	3	5	.0010	D
5639.1	124783	1	3	.0022	D
5641.4	123809	3	5	9.1 (–4)	D
5648.7	123936	5	3	.0013	D
5650.7	121794	3	1	.0333	C
5659.1	123903	5	5	.0027	D
5681.9	123832	5	7	.0021	D
5683.7	123827	5	5	.0021	D
5700.9	123774	5	7	.0061	D
5739.5	123505	3	5	.0091	D
5772.1	123557	5	7	.0021	D
5774.0	124604	5	5	.0011	D
5783.5	123373	3	5	8.4 (–4)	D
5802.1	123468	5	3	.0044	D
5834.3	123373	5	5	.0052	C
5860.3	121161	3	3	.00285	C
5882.6	121097	3	1	.0128	C
5888.6	122440	7	5	.0134	C
5912.1	121012	3	3	.0105	C
5928.8	122479	5	3	.011	D
5940.9	123882	1	3	.0012	D
5942.7	122440	5	5	.0019	D
5949.3	123936	3	3	.0016	D
5968.3	123882	3	3	.0019	D
5971.6	123873	3	1	.011	D
5987.3	122160	7	7	.0013	D
5988.1	123827	3	5	6.4 (–4)	D
5999.0	122282	5	5	.0015	D
6005.7	123936	5	3	.0015	D
6013.7	122087	7	5	.0015	D
6025.2	123882	5	3	.0094	D
6032.1	122036	7	9	.0246	C
6043.2	122160	5	7	.0153	C
6052.7	120619	3	5	.0020	D
6059.4	120601	3	5	.00423	C
6064.8	123774	5	7	6.0 (–4)	D
6090.8	123468	1	3	.0031	D
6098.8	122479	3	3	.0054	D
6101.2	123882	3	3	.0034	D
6104.6	123873	3	1	.0035	D
6105.6	123505	3	5	.0126	C
6127.4	121933	5	3	.0011	D
6128.7	123809	3	5	9.0 (–4)	D
6145.4	123557	5	7	.0079	C
6155.2	122479	5	3	.0053	D
6165.1	123505	5	5	.00103	C
6170.2	122440	5	5	.0052	C
6173.1	122282	3	5	.0070	C
6212.5	122330	5	7	.0041	D
6215.9	123373	5	5	.0059	D
6248.4	122087	3	5	7.1 (–4)	D
6296.9	123373	3	5	.0094	D
6307.7	122087	5	5	.0063	D
6364.9	121794	3	1	.0058	D
6369.6	121933	5	3	.0044	D
6384.7	119760	3	3	.00439	C
6416.3	119683	3	5	.0121	C
6466.6	122514	1	3	.0016	D
6481.1	122479	1	3	9.8 (–4)	D
6538.1	120753	7	7	.0011	D
6604.0	120601	7	5	.0029	D
6660.7	121097	3	1	.0081	D
6664.1	120619	5	5	.0016	D
6677.3	108723	3	1	.00241	C
6698.9	121161	5	3	.0017	D
6719.2	121933	1	3	.0025	D
6752.8	118907	3	5	.0201	C
6754.4	121933	3	3	.0022	D
6756.2	122087	5	5	.0038	D
6766.6	121012	5	3	.0042	D
6779.9	123468	1	3	.00126	C
6827.2	121933	5	3	.0025	D
6851.9	122087	3	5	7.0 (–4)	D
6871.3	118651	3	3	.0290	C
6879.6	120619	3	5	.0019	D
6887.1	120753	5	7	.0014	D
6888.2	120601	3	5	.0026	D
6925.0	121933	3	3	.0012	D
6937.7	118512	3	1	.0321	C
6951.5	120619	5	5	.0023	D
6960.3	120601	5	5	.0025	D
6965.4	107496	5	3	.067	C
6992.2	121794	3	1	.0078	D
7030.3	119683	7	5	.0278	C
7067.2	107290	5	5	.0395	C
7068.7	119760	5	3	.021	D
7086.7	121161	1	3	.0016	D
7107.5	119683	5	5	.0047	D
7125.8	121161	3	3	.0063	D
7147.0	107132	5	3	.0065	C
7158.8	121097	3	1	.022	D
7207.0	121161	5	3	.0258	C
7229.9	119445	5	5	6.9 (–4)	D
7265.2	119848	3	3	.0018	D
7270.7	119213	7	7	.0011	D
7272.9	107496	3	3	.0200	C
7285.4	121012	5	3	.0013	D
7311.7	119760	3	3	.018	D
7316.0	121161	3	3	.010	D
7350.8	121097	3	1	.012	D
7353.3	119213	5	7	.010	D
7372.1	119024	7	9	.020	D
7384.0	107290	3	5	.087	C
7393.0	119760	5	3	.0075	D
7412.3	120619	3	5	.0041	D
7422.3	120601	3	5	6.9 (–4)	D
7425.3	120753	5	7	.0032	D
7435.4	119683	5	5	.0094	D
7436.3	118907	7	5	.0028	D
7484.3	119445	3	5	.0035	D
7503.9	108723	3	1	.472	C
7510.4	120601	5	5	.0047	D
7514.7	107054	3	1	.430	C
7618.3	120619	3	5	.0030	D
7628.9	120601	3	5	.0030	D
7635.1	106238	5	5	.274	C
7670.0	118651	5	3	.0029	D
7704.8	119213	5	7	6.6 (–4)	D
7723.8	106087	5	3	.057	C
7724.2	107496	1	3	.127	C
7798.6	118907	3	5	9.1 (–4)	D
7868.2	119760	1	3	.00365	C
7891.1	118907	5	5	.0099	C
7916.4	119760	3	3	.0013	D
7948.2	107132	1	3	.196	C
		Ar II			
3000.4	192712	4	4	1.5	D
3028.9	192712	2	4	2.3	D
3093.4	192557	4	6	4.4	D
3139.0	186891	6	6	1.0	D
3161.4	192712	2	4	1.8	D
3169.7	186891	4	6	.82	D
3181.0	186470	6	4	.63	D
3236.8	190592	2	4	.52	D
3243.7	186171	4	2	2.0	D
3249.8	186470	2	4	1.0	D
3293.6	190592	4	4	1.7	D

Table 1b (Continued)
TRANSITION PROBABILITIES FOR SELECTED ATOMIC AND IONIC SPECIES

Wavelength (λ[Å])	Upper energy level (E_k[cm^{-1}])	Statistical weights g_i	g_k	Transition Probability (A[10^8 s^{-1}])	Uncertainty
Ar II					
3307.2	189935	2	2	3.4	D
3350.9	200235	6	6	1.5	D
3376.4	200139	8	8	1.5	D
3388.5	190592	2	4	1.9	D
3454.1	183986	6	4	.45	D
3464.1	187589	6	6	.37	D
3476.7	183797	6	6	1.34	C
3491.2	183986	4	4	2.2	D
3509.8	184192	2	2	2.5	D
3514.4	183797	4	6	1.23	C
3520.0	186074	6	6	.80	D
3521.3	185625	8	8	.23	D
3535.3	183986	2	4	.82	D
3545.6	187589	4	6	3.4	D
3545.8	198595	6	8	3.9	D
3548.5	186341	4	4	1.1	D
3559.5	186816	6	8	3.9	D
3565.0	186470	2	4	1.1	D
3576.6	185625	6	8	2.77	C
3581.6	186341	2	4	1.8	D
3582.4	186074	4	6	3.72	C
3588.4	185093	8	10	3.39	C
3600.2	199525	4	4	2.2	D
3622.1	182951	4	2	.64	D
3639.8	199680	4	6	1.4	D
3655.3	187589	4	6	.23	D
3671.0	199447	4	2	.71	D
3680.1	199982	2	4	1.2	D
3718.2	200235	4	6	2.0	D
3724.5	200235	6	6	.34	D
3729.3	161049	6	4	.60	D
3737.9	200139	6	8	2.3	D
3765.3	181594	6	6	.98	D
3770.5	182222	2	4	.41	D
3780.8	183676	8	8	.94	D
3796.6	199680	4	6	.25	D
3803.2	199680	6	6	1.5	D
3809.5	181594	4	6	.44	D
3825.7	199525	6	4	.76	D
3850.6	161049	4	4	.47	D
3868.5	186891	4	6	1.9	D
3925.7	195867	6	4	1.4	D
3928.6	161049	2	4	.30	D
3932.5	186470	4	4	1.1	D
3946.1	195865	8	6	1.4	D
3952.7	186341	4	4	.35	D
3979.4	186171	4	2	1.3	D
4033.8	182951	4	2	.98	D
4042.9	173348	4	4	1.4	D
4072.0	173393	6	6	.57	C
4076.6	182951	2	2	.80	D
4076.9	183915	4	2	.99	D
4079.6	173348	6	4	.26	D
4131.7	172816	4	2	1.4	D
4156.1	182222	4	4	.39	D
4179.3	181594	6	6	.13	D
4218.7	183091	4	4	.36	D
4222.6	183915	4	2	.69	D
4227.0	195865	4	6	.41	D
4266.5	157673	6	6	.156	C
4275.2	183091	2	4	.26	D
4277.5	172214	6	4	1.0	D
4331.2	158168	4	4	.56	C
4337.1	195867	2	4	.34	D
4348.1	157234	6	8	1.24	C
4370.8	173348	4	4	.65	C
4371.3	155351	6	4	.233	C
4379.7	158428	2	2	1.04	C
4401.0	155043	8	6	.322	C
4426.0	157673	4	6	.83	C
4430.2	158168	2	4	.53	C
4448.9	195865	6	6	.65	D
4481.8	173393	6	6	.494	C
4545.1	160239	4	4	.413	B
4547.8	182222	4	4	.077	D
4589.9	170401	4	6	.82	C
4609.6	170530	6	8	.91	C
4637.2	170401	6	6	.090	D
4657.9	159707	4	2	.81	C
4726.9	159393	4	4	.50	C
4735.9	155351	6	4	.58	C
4764.9	160239	2	4	.575	B
4806.0	155043	6	6	.79	C
4847.8	155708	4	2	.85	C
4865.9	181594	4	6	.15	D
4879.9	158730	4	6	.78	C
4965.1	159393	2	4	.347	C
5009.3	155043	4	6	.147	C
5017.2	170401	4	6	.231	C
5062.0	155351	2	4	.221	C
5141.8	170530	6	8	.095	C
6638.2	158168	6	4	.129	C
6643.7	157234	10	8	.167	C
6684.3	157673	8	6	.113	C
Ar III					
769.15	144023	5	3	6.0	D
871.10	114798	5	3	1.59	C
875.53	115328	3	1	3.74	C
878.73	113801	5	5	2.79	C
879.62	114798	3	3	.92	C
883.18	114798	1	3	1.22	C
887.40	113801	3	5	.90	C
3024.1	240292	5	7	2.6	D
3027.2	240258	5	5	.64	D
3054.8	240258	3	5	1.9	D
3064.8	240151	3	3	1.0	D
3078.2	240151	1	3	1.4	D
3285.9	204797	5	7	2.0	D
3301.9	204649	5	5	2.0	D
3311.3	204564	5	3	2.0	D
3336.1	226646	7	9	2.0	D
3344.7	226503	5	7	1.8	D
3352.1	226503	7	7	.22	D
3358.5	226356	3	5	1.6	D
3361.3	226356	5	5	.30	D
3472.6	225403	5	7	.20	D
3480.6	225403	7	7	1.6	D
3499.7	225155	3	3	1.3	D
3500.6	225148	3	5	.26	D
3502.7	225155	5	3	.43	D
3503.6	225148	5	5	1.2	D
3511.7	225148	7	5	.26	D
Ar IV					
840.03	119044	4	2	2.73	C
843.77	118515	4	4	2.70	C
850.60	117564	4	6	2.63	C
Ar VI					
292.15	342286	2	2	69.	C
294.05	342286	4	2	136.	C
Ar VII					
250.41	514083	9	3*	278.	C
477.54	324151	9	15*	99.2	B
585.75	170720	1	3	78.3	B
637.30	271657	9	9*	67.	C
Ar VIII					
158.92	629237	2	4	110.	C
159.18	628240	2	2	111.	C
229.44	575910	2	2	112.	C
230.88	575910	4	2	221.	C
337.09	629237	4	4	12.	D
337.26	629237	6	4	100.	C
338.22	628240	4	2	110.	C
519.43	332576	2	4	63.	C
526.46	332727	4	6	72.	C
526.87	332576	4	4	12.	D
700.24	142776	2	4	25.5	C
713.81	140058	2	2	24.	C
Ar IX					
48.739	2051750	1	3	1690.	B
Ar XIII					
162.96	698650	5	3	340.	C
163.08	628610	9	3*	530.	C
184.90	625840	5	5	166.	C
186.38	698650	1	3	88.	C
207.89	496450	9	9*	95.	C
245.10	423420	9	15*	37.	C
Ar XIV					
180.29	554660	2	4	45.	C

Table 1b (Continued)

TRANSITION PROBABILITIES FOR SELECTED ATOMIC AND IONIC SPECIES

Wavelength (λ[Å])	Upper energy level (E_k[cm^{-1}])	Statistical weights g_i	Statistical weights g_k	Transition Probability (A[10^8 s^{-1}])	Uncertainty
		Ar XIV			
183.41	545230	2	2	169.	C
187.95	554660	4	4	197.	C
191.35	545230	4	2	75.	C
194.39	514430	2	2	46.	C
203.35	514430	4	2	78.	C
		Ar XV			
25.05	3992000	1	3	1.7 (+4)	B
221.10	452280	1	3	95.5	B
265.3	621500	9	9*	81.	C
		Ar XVI			
23.52	4251000	2	6*	1.43(+4)	B
24.96	4281000	6	10*	4.4 (+4)	B
353.88	282580	2	4	15.	B
389.11	257000	2	2	11.	B
1268.	4254000	2	4	1.9	B
1401.	4246000	2	2	1.4	B
2975.	4280000	2	4	.090	C
3514.	4281000	4	6	.065	C
		Arsenic			
		As I			
1890.4	52898	4	6	2.0	D
1937.6	51610	4	4	2.0	D
1972.6	50694	4	2	2.0	D
2288.1	54605	6	4	2.8	D
2344.0	60835	2	4	.35	D
2349.8	53136	4	2	3.1	D
2369.7	60835	4	4	.60	D
2370.8	60815	4	6	.42	D
2456.5	51610	6	4	.072	D
2492.9	50694	4	2	.12	D
2745.0	54605	2	4	.26	D
2780.2	54605	4	4	.78	D
2860.4	53136	2	2	.55	D
2898.7	53136	4	2	.099	D
		Barium			
		Ba I			
2409.2	41494	1	3	8.6 (–4)	C
2414.1	41411	1	3	.0015	C
2420.1	41308	1	3	.0023	C
2427.4	41184	1	3	.0056	C
2432.5	41097	1	3	.0072	C
2438.8	40991	1	3	.0014	C
2444.6	40893	1	3	.0045	C
2452.4	40764	1	3	8.1 (–4)	C
2473.2	40421	1	3	.0046	C
2500.2	39985	1	3	.015	D
2543.2	39309	1	3	.041	D
2596.6	38500	1	3	.12	D
2646.5	37775	1	3	.011	D
2702.6	36990	1	3	.025	D
2739.2	36496	1	3	.0091	D
2785.3	35893	1	3	.028	D
3071.6	32547	1	3	.41	C
3501.1	28554	1	3	.19	D
3889.3	25704	1	3	.0088	D
3909.9	34603	3	5	.49	D
3935.7	34617	5	7	.47	D
3937.9	34603	5	5	.11	D
3993.4	34631	7	9	.55	D
3995.7	34617	7	7	.088	D
4132.4	24192	1	3	.0071	D
4239.6	37095	5	3	.24	D
4242.6	36200	3	5	.056	D
4264.4	35709	1	3	.15	D
4283.1	34736	5	7	.64	D
4323.0	35762	3	5	.15	D
4325.2	36629	5	7	.071	D
4332.9	35709	3	3	.15	D
4350.3	35617	3	5	.60	D
4402.5	35344	3	5	.70	D
4406.8	36200	5	5	.10	D
4431.9	34823	1	3	1.2	D
4467.1	35894	5	7	.066	D
4489.0	35785	5	7	.42	D
4493.6	35762	5	5	.36	D
4505.9	34823	3	3	1.1	D
4523.2	35617	5	5	.96	D
4573.9	34494	3	1	2.9	D
		Ba I			
4579.6	35344	5	5	1.8	D
4591.8	30987	5	5	.016	D
4599.8	34371	3	1	1.0	D
4605.0	30744	3	1	.077	D
4619.9	33905	1	3	.093	D
4628.3	30816	5	3	.060	D
4673.6	30987	7	5	.065	D
4691.6	34823	5	3	1.6	D
4700.4	33905	3	3	.24	D
4726.4	32547	5	3	.46	D
5519.1	30751	3	5	.50	D
5535.5	18060	1	3	1.15	B
5777.6	30818	5	7	.64	D
5800.2	30751	5	5	.099	D
5805.7	26816	7	7	.011	D
5826.3	28554	5	3	.56	D
5907.6	25957	3	5	.036	D
5971.7	25957	5	5	.29	D
5997.1	25704	3	3	.27	D
6019.5	25642	3	1	1.4	D
6063.1	25704	5	3	.57	D
6110.8	25957	7	5	1.0	D
6341.7	24980	5	7	.19	D
6450.9	24532	3	5	.11	D
6482.9	26816	5	7	.44	D
6498.8	24980	7	7	.86	D
6527.3	24532	5	5	.59	D
6595.3	24192	3	3	.39	D
6675.3	24192	5	3	.19	D
6693.8	24532	7	5	.28	D
6865.7	25957	5	5	.078	D
7059.9	23757	7	9	.71	D
7120.3	23074	3	5	.21	D
7195.2	26160	1	3	.24	D
7280.3	22947	5	7	.53	D
7392.4	26160	3	3	.50	D
7417.5	23074	7	5	.025	D
7488.1	22947	7	7	.10	D
7672.1	22065	3	5	.31	D
7780.5	22065	5	5	.13	D
7905.8	26160	5	3	.63	D
7911.3	12637	1	3	.00298	C
		Ba II			
1413.4	76429	6	8	.017	D
1417.1	75438	4	6	.038	D
1444.9	74091	4	6	.081	D
1461.5	74109	6	8	.087	D
1487.0	72143	4	6	.14	D
1503.9	72170	6	8	.15	D
1554.4	69212	4	6	.26	D
1572.7	69260	6	8	.24	D
1573.9	69212	6	6	.016	D
1630.4	61336	2	2	.017	D
1674.5	64596	4	6	.22	D
1694.4	64697	6	8	.21	D
1697.2	64596	6	6	.017	D
1761.8	61636	4	4	.0039	D
1771.0	61336	4	2	.034	D
1786.9	61636	6	4	.044	D
1892.7	73102	2	4	.090	D
1904.2	57391	4	6	.011	D
1906.8	72705	2	2	.051	D
1924.7	57632	6	8	.031	D
1954.2	73122	4	6	.13	D
1955.1	73102	4	4	.018	D
1970.2	72705	4	2	.067	D
1985.6	70620	2	4	.25	D
1999.5	50011	2	4	.10	D
2009.2	70015	2	2	.086	D
2052.7	70652	4	6	.20	D
2054.6	70620	4	4	.029	D
2080.0	70015	4	2	.10	D
2153.9	66674	2	4	.53	D
2200.9	65683	2	2	.20	D
2232.8	66725	4	6	.29	D
2235.4	66674	4	4	.044	D
2286.0	65683	4	2	.13	D
2528.5	59800	2	4	.71	D
2634.8	59895	4	6	.76	D
2641.4	59800	4	4	.12	D

Table 1b (Continued)
TRANSITION PROBABILITIES FOR SELECTED ATOMIC AND IONIC SPECIES

Wavelength (λ[Å])	Upper energy level (E_k[cm^{-1}])	Statistical weights g_i	g_k	Transition Probability (A[10^8 s^{-1}])	Uncertainty
	Ba II				
2647.3	58025	2	2	.20	D
2771.4	58025	4	2	.40	D
3816.7	72143	4	6	.0023	D
3842.8	72170	6	8	.0022	D
3891.8	45949	2	4	1.67	B
4024.1	73102	6	4	.0053	D
4057.5	73122	8	6	.012	D
4130.7	46155	4	6	1.80	B
4166.0	45949	4	4	.37	D
4216.0	73102	2	4	.058	D
4287.8	72705	2	2	.024	D
4325.7	73122	4	6	.059	D
4329.6	73102	4	4	.0088	D
4405.2	72705	4	2	.039	D
4470.7	70620	6	4	.014	D
4509.6	70652	8	6	.012	D
4524.9	42355	2	2	.72	D
4554.0	21952	2	4	1.17	A
4708.9	70620	2	4	.097	D
4843.5	70652	4	6	.093	D
4847.1	70015	2	2	.041	D
4850.8	70620	4	4	.014	D
4900.0	42355	4	2	.775	B
4934.1	20262	2	2	.955	B
4997.8	70015	4	2	.061	D
5185.0	61636	2	4	.018	D
5361.4	64596	4	6	.048	D
5391.6	64697	6	8	.052	D
5413.6	66725	6	6	8.4 (−4)	D
5421.1	64596	6	6	.0019	D
5428.8	66674	6	4	.023	D
5480.3	66725	8	6	.018	D
5784.2	66674	2	4	.20	D
5853.7	21952	4	4	.048	B
5981.3	66725	4	6	.16	D
5999.9	66674	4	4	.026	D
6135.8	65683	2	2	.085	D
6141.7	21952	6	4	.37	B
6363.2	73102	6	4	.0029	D
6372.9	61636	4	4	6.7 (−4)	D
6378.9	65683	4	2	.099	D
6457.7	61636	6	4	.0030	D
6496.9	20262	4	2	.332	C
7556.8	70620	6	4	.0016	D
7678.2	70652	8	6	6.6 (−4)	D
8710.7	57632	6	8	.80	D
8737.7	57391	4	6	.93	D
	Beryllium				
	Be I				
1491.8	67035	1	3	.013	D
1661.5	60187	1	3	.20	D
2348.6	42565	1	3	5.56	B
2494.7	62054	9	15*	1.6	C
2650.6	59696	9	9*	4.31	C
4572.7	64428	3	5	.79	C
	Be II				
1197.1	115464	2	2	.47	D
1197.2	115464	4	2	.94	D
1512.3	98055	2	4	9.2	C
1512.4	98055	4	6	11.	C
1776.1	88232	2	2	1.4	C
1776.3	88232	4	2	2.9	C
2453.8	128972	2	6*	.142	C
3046.5	129310	2	4	.48	C
3046.7	129310	4	6	.59	C
3130.4	31935	2	4	1.14	B
3131.1	31929	2	2	1.15	B
3241.6	127335	2	2	.141	C
3241.8	127335	4	2	.28	C
3274.6	118761	2	4	.19	C
3274.7	118761	2	2	.19	C
4360.7	119421	2	4	.92	C
4361.0	119421	4	6	1.1	C
5255.9	134485	2	6*	.0256	C
5270.3	115464	2	2	.330	C
5270.8	115464	4	2	.66	C
6279.4	134681	2	4	.12	C
6279.7	134681	4	6	.143	C
6756.7	133556	2	2	.051	C
6757.1	133556	4	2	.102	C
7401.2	128972	2	4	.030	C
7401.4	128972	2	2	.030	C
	Bismuth				
	Bi I				
1954.5	51159	4	6	1.2	D
2021.2	49461	4	4	.060	D
2061.7	48490	4	6	.99	D
2110.3	47373	4	2	.91	D
2177.3	45916	4	2	.026	D
2228.3	44865	4	4	.89	D
2230.6	44817	4	6	2.6	D
2276.6	43913	4	4	.25	D
2515.7	51159	4	6	.043	D
2627.9	49461	4	4	.47	D
2696.8	48490	4	6	.064	D
2780.5	47373	4	2	.309	C
2798.7	51159	6	6	.036	D
2898.0	45916	4	2	1.53	C
2938.3	49461	6	4	1.23	C
2989.0	44865	4	4	.55	C
2993.3	44817	4	6	.16	C
3024.6	48490	6	6	.88	C
3067.7	32588	4	2	2.07	C
3076.7	43913	4	4	.035	D
3397.2	44865	6	4	.181	C
3402.9	44817	6	6	.016	D
3510.9	43913	6	4	.068	D
3596.1	49461	2	4	.198	C
3888.2	47373	2	2	.069	D
4121.5	45916	2	2	.164	C
4308.5	44865	2	4	.016	D
4493.0	43913	2	4	.015	D
4722.5	32588	4	2	.117	C
6134.8	49461	4	4	.018	D
	Boron				
	B I				
1378.6	72535	2	4	3.50	C
1378.9	72523	2	2	14.0	C
1378.9	72535	4	4	17.5	C
1379.2	72523	4	2	7.0	C
1465.5	97000	2	4	3.34	C
1465.7	97000	4	4	6.7	C
1465.8	97000	6	4	10.0	C
1825.9	54767	2	4	2.0	C
1826.4	54767	4	6	2.4	C
2088.9	47857	2	4	.28	D
2089.6	47857	4	6	.33	D
2496.8	40040	2	2	.85	C
2497.7	40040	4	2	1.69	C
	Bromine				
	Br I				
1488.5	67184	4	4	1.2	D
1540.7	64907	4	4	1.4	D
1574.8	67184	2	4	.20	D
1576.4	63436	4	6	.021	D
1633.4	64907	2	4	.081	D
4365.1	89786	2	4	.0075	D
4425.1	87499	4	2	.0042	D
4441.7	85944	6	4	.0075	D
4472.6	87259	4	4	.0093	D
4477.7	85763	6	8	.013	D
4513.4	85586	6	4	.0028	D
4525.6	85527	6	6	.0072	D
4575.7	89032	4	4	.016	D
4614.6	88848	4	6	.0054	D
4979.8	87259	4	4	.0026	D
5245.1	85944	2	4	.0031	D
5345.4	85586	2	4	.0076	D
7348.5	78512	4	6	.12	D
7513.0	76743	6	4	.12	D
7803.0	79696	2	4	.053	D
7938.7	88483	6	6	.19	D
8131.5	79178	2	4	.038	D
8343.7	78866	2	2	.22	D
8446.6	76743	4	4	.12	D
8638.7	75009	6	4	.097	D

Table 1b (Continued)
TRANSITION PROBABILITIES FOR SELECTED ATOMIC AND IONIC SPECIES

Wavelength (λ[Å])	Upper energy level (E_k[cm^{-1}])	Statistical weights g_i	g_k	Transition Probability (A[10^8 s^{-1}])	Uncertainty
	Br II				
4704.9	115176	5	7	1.1	D
4785.5	114818	5	5	.94	D
4816.7	114683	5	3	1.1	D
	Cadmium				
	Cd I				
2288.0	43692	1	3	5.3	C
2836.9	65354	1	3	.28	D
2880.8	65359	3	5	.42	D
2881.2	65353	3	3	.24	D
2980.6	65367	5	7	.59	D
2981.4	65359	5	5	.15	D
3261.1	30656	1	3	.00406	C
3403.7	59486	1	3	.77	C
3466.2	59498	3	5	1.2	D
3467.7	59486	3	3	.67	D
3610.5	59516	5	7	1.3	D
3612.9	59498	5	5	.35	D
4140.5	67838	3	5	.047	D
4662.4	65135	3	5	.055	C
4678.1	51484	1	3	.13	C
4799.9	51484	3	3	.41	C
5085.8	51484	5	3	.56	C
6438.5	59220	3	5	.59	C
	Cd II				
2144.4	46619	2	4	2.8	C
2265.0	44136	2	2	3.0	C
2572.9	82991	2	2	1.7	C
2748.5	82991	4	2	2.8	C
4415.6	69259	4	6	.014	B
	Calcium				
	Ca I				
2275.5	43933	1	3	.301	C
2995.0	48538	1	3	.367	C
2997.3	48564	3	5	.241	C
2999.6	48538	3	3	.279	C
3000.9	48524	3	1	1.58	C
3006.9	48564	5	5	.75	C
3009.2	48538	5	3	.430	C
3344.5	45049	1	3	.151	C
3350.2	45050	3	5	.178	C
3361.9	45052	5	7	.223	C
3624.1	42743	1	3	.212	C
3630.8	42745	3	5	.297	C
3631.0	42743	3	3	.153	C
3644.4	42747	5	7	.355	C
3644.8	42745	5	5	.094	C
3870.5	46165	3	5	.072	D
3957.1	40474	3	3	.098	C
3973.7	40474	5	3	.175	C
4092.6	44763	3	5	.11	D
4094.9	44763	5	7	.12	D
4098.5	44763	7	9	.13	D
4108.5	46182	5	7	.90	D
4226.7	23652	1	3	2.18	B
4283.0	38552	3	5	.434	C
4289.4	38465	1	3	.60	C
4299.0	38465	3	3	.466	C
4302.5	38552	5	5	1.36	C
4307.7	38418	3	1	1.99	C
4318.7	38465	5	3	.74	C
4355.1	44805	5	7	.19	D
4425.4	37748	1	3	.498	C
4435.0	37752	3	5	.67	C
4435.7	37748	3	3	.342	C
4454.8	37757	5	7	.87	C
4455.9	37752	5	5	.20	C
4526.9	43933	5	3	.41	D
4578.6	42170	3	5	.176	C
4581.4	42171	5	7	.209	C
4585.9	42171	7	9	.229	C
4685.3	44990	3	5	.080	D
4878.1	42344	5	7	.188	C
5041.6	41679	5	3	.33	D
5188.9	42919	3	5	.40	D
5261.7	39335	3	3	.15	D
5262.2	39333	3	1	.60	D
5264.2	39340	5	5	.091	D
5265.6	39335	5	3	.44	D

Wavelength (λ[Å])	Upper energy level (E_k[cm^{-1}])	Statistical weights g_i	g_k	Transition Probability (A[10^8 s^{-1}])	Uncertainty
	Ca I				
5270.3	39340	7	5	.50	D
5582.0	38259	5	7	.060	D
5588.8	38259	7	7	.49	D
5590.1	38219	3	5	.083	D
5594.5	38219	5	5	.38	D
5598.5	38192	3	3	.43	D
5601.3	38219	7	5	.086	D
5602.9	38192	5	3	.14	D
5857.5	40720	3	5	.66	D
6102.7	31539	1	3	.096	C
6122.2	31539	3	3	.287	C
6161.3	36575	5	5	.033	D
6162.2	31539	5	3	.477	C
6163.8	36555	3	3	.056	D
6166.4	36548	3	1	.22	D
6169.1	36555	5	3	.17	D
6169.6	36575	7	5	.19	D
6439.1	35897	7	9	.53	D
6449.8	35835	3	5	.090	D
6462.6	35819	5	7	.47	D
6471.7	35819	7	7	.059	D
6493.8	35730	3	5	.44	D
6499.7	35730	5	5	.081	D
	Ca II				
1341.9	74522	2	4	.015	D
1342.5	74485	2	2	.015	D
1649.9	60611	2	4	.0032	D
1652.0	60533	2	2	.0031	D
1673.9	84934	2	4	.224	C
1680.1	84936	4	6	.265	C
1680.1	84934	4	4	.0441	C
1807.3	80522	2	4	.354	C
1814.5	80526	4	6	.42	C
1814.7	80522	4	4	.070	C
1843.1	79448	2	2	.16	C
1850.7	79448	4	2	.308	C
2103.2	72722	2	4	.82	C
2112.8	72731	4	6	.97	C
2113.2	72722	4	4	.16	C
2197.8	70678	2	2	.31	C
2208.6	70678	4	2	.62	C
3158.9	56839	2	4	3.1	C
3179.3	56858	4	6	3.6	C
3181.3	56839	4	4	.58	C
3706.0	52167	2	2	.88	C
3736.9	52167	4	2	1.7	C
3933.7	25414	2	4	1.47	C
3968.5	25192	2	2	1.4	C
	Ca III				
357.97	279354	1	3	880.	D
439.69	227432	1	3	.19	D
490.55	203852	1	3	.016	D
	Ca V				
558.60	197845	5	3	22.	D
637.93	156760	5	3	3.9	D
643.12	157901	3	1	9.1	D
646.57	154671	5	5	6.9	D
647.88	156760	3	3	2.3	D
651.55	156760	1	3	2.9	D
656.76	154671	3	5	2.1	D
	Ca VII				
550.20	203616	5	5	18.	D
624.39	160158	1	3	3.3	D
630.54	160220	3	5	4.5	D
630.79	160158	3	3	2.2	D
639.15	160529	5	7	5.7	D
640.41	160220	5	5	1.3	D
	Ca VIII				
182.71	547322	2	2	160.	C
184.16	547322	4	2	320.	C
	Ca IX				
163.23	758974	5	3	376.	C
371.89	410514	1	3	88.	C
373.81	410627	3	5	116.	C
378.08	410841	5	7	150.	C
395.03	467631	3	5	220.	D
466.24	214482	1	3	112.	B
498.01	343908	3	5	24.9	C

Table 1b (Continued)
TRANSITION PROBABILITIES FOR SELECTED ATOMIC AND IONIC SPECIES

Wavelength (λ[Å])	Upper energy level (E_k[cm^{-1}])	Statistical weights g_i	g_k	Transition Probability (A[10^8 s^{-1}])	Uncertainty
		Ca IX			
506.18	343908	5	5	72.	C
515.57	340308	5	3	37.5	C
		Ca X			
110.96	901200	2	4	290.	C
111.20	899290	2	2	292.	C
151.84	832790	2	2	230.	C
153.02	832790	4	2	450.	C
206.57	901200	4	4	29.	D
206.75	901200	6	4	260.	C
207.39	899290	4	2	280.	C
411.70	417112	2	4	83.	C
419.75	417522	4	6	95.	C
420.47	417112	4	4	16.	D
557.76	179287	2	4	35.0	C
574.01	174213	2	2	32.	C
		Ca XI			
30.448	3284300	1	3	6200.	D
30.867	3239700	1	3	4.9 (+4)	D
35.212	2839940	1	3	2000.	D
		Ca XII			
140.05	709000	4	2	370.	C
147.27	709000	2	2	160.	C
		Ca XV			
141.69	814370	5	3	408.	C
142.23	728910	9	3*	630.	C
161.00	729720	5	5	190.	C
		Ca XVII			
19.558	5113000	1	3	3.8 (+4)	C
21.198	5236000	3	5	4.9 (+4)	C
192.82	518620	1	3	121.	C
218.82	726450	3	5	27.6	C
223.02	706680	1	3	34.4	C
228.72	706680	3	3	23.7	C
232.83	726450	5	5	65.	C
244.06	706680	5	3	32.8	C
		Ca XVIII			
18.71	5346000	2	6*	2.31(+4)	B
19.74	5383000	6	10*	7.0 (+4)	B
302.19	330920	2	4	20.	B
344.76	290060	2	2	13.	B
		Carbon			
		C			
945.19	105799	1	3	6.2	D
945.34	105799	3	3	18.	D
945.58	105799	5	3	31.	D
1260.7	79319	1	3	.40	D
1260.9	79323	3	1	1.2	D
1261.0	79319	3	3	.31	D
1261.1	79311	3	5	.30	D
1261.4	79319	5	3	.50	D
1261.6	79311	5	5	.93	D
1274.1	78530	5	7	.0068	D
1277.2	78293	1	3	.88	D
1277.3	78308	3	5	1.2	D
1277.5	78293	3	3	.65	D
1277.6	78318	5	7	1.5	D
1277.7	78308	5	5	.39	D
1278.0	78293	5	3	.042	D
1279.2	78216	5	7	.11	D
1279.9	78148	3	5	.21	D
1280.1	78117	1	3	.27	D
1280.3	78148	5	5	.62	D
1280.4	78117	3	3	.20	D
1280.6	78105	3	1	.81	D
1280.8	78117	5	3	.35	D
1328.8	75254	1	3	.49	D
1364.2	83498	5	5	.047	D
1431.6	103587	5	7	1.5	D
1432.1	103563	5	5	1.4	D
1432.5	103542	5	3	1.3	D
1459.0	78731	5	3	.37	D
1463.3	78530	5	7	2.1	D
1467.4	78340	5	3	.46	D
1468.4	78293	5	3	.019	D
1470.1	78216	5	7	.0088	D
1472.2	78117	5	3	.0051	D
1481.8	77680	5	5	.33	D

Wavelength (λ[Å])	Upper energy level (E_k[cm^{-1}])	Statistical weights g_i	g_k	Transition Probability (A[10^8 s^{-1}])	Uncertainty
		C I			
1560.3	64090	1	3	.82	D
1561.3	64091	5	5	.36	D
1561.4	64087	5	7	1.4	D
1656.3	60393	3	5	.80	D
1656.9	60353	1	3	1.1	D
1657.0	60393	5	5	2.4	D
1657.4	60353	3	3	.80	D
1657.9	60333	3	1	3.2	D
1658.1	60353	5	3	1.3	D
1751.8	78731	1	3	.57	D
1763.9	78340	1	3	.022	D
1765.4	78293	1	3	.0071	D
1930.9	61982	5	3	3.7	D
2478.6	61982	1	3	.18	D
2902.3	105799	1	3	.0066	D
2903.3	105799	3	3	.017	D
2905.0	105799	5	3	.022	D
4269.0	85400	3	5	.0032	D
4371.4	84852	3	3	.0097	D
4762.3	81326	1	3	.0052	D
4762.5	81344	3	5	.0038	D
4766.7	81326	3	3	.0039	D
4770.0	81311	3	1	.015	D
4771.8	81344	5	5	.012	D
4775.9	81326	5	3	.0062	D
4812.9	81105	1	3	9.7 (–4)	D
4817.4	81105	3	3	.0028	D
4826.8	81105	5	3	.0047	D
4932.1	82252	3	1	.046	D
5052.2	81770	3	5	.017	D
5380.3	80563	3	3	.016	D
5793.1	81344	7	5	.0033	D
5794.5	81344	5	5	5.8 (–4)	D
5800.2	81326	3	3	9.7 (–4)	D
5800.6	81326	5	3	.0029	D
5805.2	81311	3	1	.0039	D
6587.6	84032	3	3	.024	D
		C II			
687.35	145551	4	6	27.0	C
858.09	116538	2	2	.369	C
858.56	116538	4	2	1.11	C
903.62	110666	2	4	6.6	C
903.96	110624	2	2	26.3	C
904.14	110666	4	4	33.0	C
904.48	110624	4	2	13.3	C
1009.9	142027	2	4	5.8	C
1010.1	142027	4	4	11.5	C
1010.4	142027	6	4	17.3	C
1036.3	96494	2	2	8.0	C
1037.0	96494	4	2	15.9	C
1323.9	150467	4	4	4.53	C
1324.0	150462	6	6	4.71	C
1334.5	74933	2	4	2.41	C
1335.7	74930	4	6	2.89	C
2509.1	150467	2	4	.54	C
2511.7	150467	4	4	.106	C
2512.1	150462	4	6	.64	C
6578.1	131736	2	4	.36	C
6582.9	131724	2	2	.36	C
7231.3	145549	2	4	.36	C
7236.4	145551	4	6	.44	C
7237.2	145549	4	4	.072	C
		C III			
310.17	322404	1	3	18.	C
386.20	258931	1	3	32.2	B
459.46	270011	1	3	55.	C
459.52	270012	3	5	75.	C
459.63	270015	5	7	98.	C
574.28	276483	3	5	63.	C
977.03	102352	1	3	17.5	B
1174.9	137502	3	5	3.42	C
1175.3	137454	1	3	4.55	C
1175.6	137454	3	3	3.41	C
1175.7	137502	5	5	10.2	C
1176.0	137426	3	1	13.6	C
1176.4	137454	5	3	5.7	C
1247.4	182520	3	1	18.6	C
2296.9	145876	3	5	1.46	C
4647.4	259724	3	5	.73	C
4650.3	259711	3	3	.74	C

Table 1b (Continued)

TRANSITION PROBABILITIES FOR SELECTED ATOMIC AND IONIC SPECIES

Wavelength (λ[Å])	Upper energy level (E_k[cm^{-1}])	Statistical weights g_i	g_k	Transition Probability (A[10^8 s^{-1}])	Uncertainty
		C III			
4651.5	259706	3	1	.74	C
		C IV			
312.43	320071	2	6*	44.9	B
384.13	324886	6	10*	180.	C
1548.2	64592	2	4	2.66	B
1550.8	64484	2	2	2.64	B
5801.0	320082	2	4	.319	B
5812.0	320050	2	2	.316	B
		C V			
34.973	2859375	1	3	2554.	AA
40.268	2483371	1	3	8873.	AA
227.19	2851418	3	9*	136.3	AA
247.31	2859375	1	3	127.9	AA
248.70	2857310	9	15*	425.	A
260.19	2839562	9	3*	66.83	AA
267.27	2857529	3	5	396.	A
2273.9	2455225	3	9*	.5650	AA
3526.7	2483371	1	3	.1663	AA
8432.2	2851418	3	9*	.06870	AA
		Cesium			
		Cs I			
3203.5	31207	2	4	7.6 (–6)	C
3205.3	31189	2	4	7.9 (–6)	C
3207.5	31168	2	4	8.5 (–6)	C
3210.0	31144	2	4	9.4 (–6)	C
3212.8	31116	2	4	1.19(–5)	C
3216.2	31084	2	4	1.49(–5)	C
3220.1	31046	2	4	1.7 (–5)	C
3220.2	31045	2	2	1.07(–7)	C
3224.8	31001	2	4	2.0 (–5)	C
3225.0	30999	2	2	1.43(–7)	C
3230.5	30946	2	4	2.5 (–5)	C
3230.7	30944	2	2	1.97(–7)	C
3237.4	30880	2	4	2.8 (–5)	C
3237.6	30878	2	2	2.63(–7)	C
3245.9	30799	2	4	3.45(–5)	C
3246.2	30796	2	2	3.7 (–7)	C
3256.7	30698	2	4	4.25(–5)	C
3257.1	30694	2	2	7.0 (–7)	C
3270.5	30568	2	4	5.6 (–5)	C
3271.0	30563	2	2	9.8 (–7)	C
3288.6	30399	2	4	1.0 (–4)	C
3289.3	30393	2	2	2.7 (–6)	C
3313.1	30175	2	4	1.6 (–4)	C
3314.0	30166	2	2	5.2 (–6)	C
3347.5	29865	2	4	2.2 (–4)	C
3348.8	29853	2	2	1.1 (–5)	C
3397.9	29421	2	4	4.0 (–4)	C
3400.0	29404	2	2	2.4 (–5)	C
3476.8	28754	2	4	6.6 (–4)	C
3480.0	28727	2	2	6.6 (–5)	C
3611.4	27682	2	4	.0015	C
3617.3	27637	2	2	2.5 (–4)	C
3876.1	25792	2	4	.0038	C
3888.6	25709	2	2	9.7 (–4)	C
4555.3	21946	2	4	.0188	C
4593.2	21765	2	2	.0080	C
		Chlorine			
		Cl I			
1188.8	84120	4	6	2.33	C
1188.8	84122	4	4	.271	C
1201.4	84122	2	4	2.39	C
1335.7	74866	4	2	1.74	C
1347.2	74226	4	4	4.19	C
1351.7	74866	2	2	3.23	C
1363.4	74226	2	4	.75	C
4323.3	95612	4	4	.011	D
4363.3	95401	4	6	.0068	D
4379.9	95313	4	4	.014	D
4389.8	94732	6	8	.014	D
4526.2	96313	4	4	.051	C
4601.0	96594	2	2	.042	C
4661.2	96313	2	4	.012	D
7256.6	85735	6	4	.15	D
7414.1	85442	6	4	.047	D
7547.1	85735	4	4	.12	D
7717.6	85442	4	4	.030	D
7745.0	85735	2	4	.063	D

Wavelength (λ[Å])	Upper energy level (E_k[cm^{-1}])	Statistical weights g_i	g_k	Transition Probability (A[10^8 s^{-1}])	Uncertainty
		Cl I			
7769.2	95787	6	6	.060	D
7821.4	95701	6	8	.098	D
7830.8	95898	4	4	.097	D
7878.2	84648	6	6	.018	D
7899.3	95787	4	6	.051	D
7924.6	85442	2	4	.021	D
7986.0	96731	6	8	.039	D
7997.9	84988	4	4	.021	D
		Cl II			
3329.1	161798	5	7	1.5	D
3522.1	174855	7	7	1.4	D
3798.8	172652	5	7	1.6	D
3805.2	172743	7	9	1.8	D
3809.5	172575	3	5	1.5	D
3851.0	154624	5	7	1.8	D
3851.4	154621	5	5	1.6	D
3854.7	184660	3	5	2.2	D
3861.9	184657	5	7	2.4	D
3868.6	184630	7	9	2.7	D
3913.9	172743	9	9	.82	D
3990.2	174855	5	7	.84	D
4132.5	153259	5	5	1.6	D
4276.5	170577	9	7	.76	D
4768.7	158771	3	5	.77	D
4781.3	158788	5	7	1.0	D
4794.6	128731	5	7	1.04	C
4810.1	128644	5	5	.99	C
4819.5	128623	5	3	1.00	C
4904.8	147128	5	7	.81	D
4917.7	147056	3	5	.75	D
5078.3	146471	7	7	.77	D
5219.1	131765	3	9*	.86	C
5392.1	147607	5	7	1.0	D
		Cl III			
2298.5	248528	4	4	4.2	D
2340.6	248658	6	6	4.2	D
2370.4	258886	8	6	2.8	D
2531.8	248528	2	4	4.4	D
2532.5	248658	4	6	5.3	D
2577.1	243828	4	6	4.3	D
2580.7	244685	6	8	4.7	D
2601.2	239506	2	4	4.6	D
2603.6	239730	4	6	5.0	D
2609.5	240075	6	8	5.7	D
2617.0	240568	8	10	6.6	D
2661.6	241685	4	6	3.4	D
2665.5	242046	6	8	4.8	D
2691.5	243081	4	4	3.5	D
2710.4	242823	4	6	3.5	D
3340.4	204541	6	6	1.5	D
3392.9	217913	4	4	1.9	D
3393.5	217850	6	6	1.9	D
3530.0	216710	6	8	1.8	D
3560.7	216525	4	6	1.7	D
3602.1	202368	6	8	1.7	D
3612.9	201765	4	6	1.2	D
3720.5	205947	4	6	1.7	D
		Chromium			
		Cr I			
2000.0	58293	9	9	1.4	D
2383.3	50253	9	11	.41	D
2385.7	50211	9	9	.17	D
2389.2	49653	3	5	.23	D
2408.6	49812	9	7	.67	D
2408.7	49598	7	5	.29	D
2492.6	47918	3	5	.45	C
2496.3	47975	5	7	.56	C
2502.5	48043	7	9	.22	D
2504.3	48014	7	9	.45	C
2549.5	47022	3	3	.48	C
2560.7	46968	5	5	.43	D
2577.7	46879	7	7	.26	D
2591.9	46879	9	7	.65	C
2622.9	46422	9	9	.13	D
2702.0	45306	9	11	.21	C
2726.5	44259	5	7	.75	C
2731.9	44187	5	5	.78	C
2736.5	44126	5	3	.75	D
2752.9	44126	3	3	.87	D

Table 1b (Continued)
TRANSITION PROBABILITIES FOR SELECTED ATOMIC AND IONIC SPECIES

Wavelength (λ[Å])	Upper energy level (E_k[cm^{-1}])	Statistical weights g_i	g_k	Transition Probability (A[10^8 s^{-1}])	Uncertainty
		Cr I			
2757.1	44187	5	5	.68	C
2761.8	44126	5	3	.68	D
2764.4	44259	7	7	.37	D
2769.9	44187	7	5	1.1	C
2780.7	44259	9	7	1.4	C
2871.6	42909	7	9	.12	D
2879.3	42648	5	7	.21	D
2887.0	42439	3	5	.27	D
2889.3	42909	9	9	.66	C
2899.2	42293	3	3	.15	D
2909.1	42293	5	3	.68	C
2967.6	41782	7	9	.39	D
2971.1	41575	5	7	.71	C
2975.5	41409	3	5	.89	C
2980.8	41289	1	3	.85	D
2988.7	41043	5	7	.52	C
2991.9	41225	3	1	3.0	D
2995.1	40971	5	5	.43	D
2996.6	41289	5	3	2.0	C
2998.8	40930	5	3	.59	D
3000.9	41409	7	5	1.6	C
3005.1	41575	9	7	.92	C
3013.7	40983	3	5	.83	C
3020.7	40906	3	3	1.5	D
3021.6	41393	9	11	3.2	C
3024.4	40983	5	5	2.3	C
3030.2	41086	7	7	1.1	C
3034.2	41043	7	7	.35	D
3037.0	41225	9	9	.54	C
3040.9	40971	7	5	.74	D
3053.9	41043	9	7	1.2	C
3148.4	55686	9	11	.59	C
3155.2	55741	11	13	.54	C
3163.8	55799	13	15	.52	C
3237.7	54811	9	9	1.3	D
3238.1	54930	11	11	.20	D
3578.7	27935	7	9	1.48	B
3593.5	27820	7	7	1.50	B
3605.3	27729	7	5	1.62	B
3639.8	47986	13	11	1.8	D
3768.7	47047	7	5	.22	D
3804.8	50558	9	9	.69	D
3879.2	50058	3	5	.56	D
3963.7	45741	13	15	1.3	D
3969.8	45707	11	13	1.2	D
3981.2	46968	3	5	.11	D
3983.9	45615	7	9	1.05	C
4001.4	56362	9	11	.65	D
4030.7	56155	3	5	.79	D
4039.1	55799	15	15	.68	C
4048.8	55741	13	13	.65	D
4058.8	55686	11	11	.69	D
4161.4	59957	13	15	.80	D
4165.5	59884	11	13	.75	D
4211.4	48043	7	9	.085	D
4224.5	48562	9	9	.067	D
4239.0	47866	9	9	.071	D
4254.4	23499	7	9	.315	B
4274.8	23386	7	7	.306	B
4280.4	54405	13	15	.47	D
4289.7	23305	7	5	.313	B
4297.7	54317	11	13	.49	D
4344.5	31106	7	9	.11	C
4351.8	31280	9	11	.12	C
4374.2	47055	13	11	.10	C
4375.3	46905	11	9	.072	D
4381.1	44667	5	3	.10	D
4387.5	46986	13	11	.066	D
4413.9	51287	7	5	.41	D
4432.2	45719	1	3	.17	D
4458.5	46705	9	11	.13	D
4482.9	49477	3	3	.30	D
4488.1	56210	7	7	.63	D
4506.9	55945	13	11	.27	D
4511.9	47055	9	9	.13	D
4526.5	42606	13	13	.20	D
4530.7	42589	11	11	.20	D
4544.6	42515	5	5	.26	D
4595.6	55517	13	13	.47	D
4632.2	46688	7	7	.071	D
4639.5	46637	5	7	.095	D
4646.2	29825	9	7	.087	C
4689.4	46525	7	5	.23	D
4708.0	46783	11	9	.37	D
4718.4	46959	13	11	.42	D
4723.1	46000	7	7	.093	D
4724.4	46058	9	9	.063	D
4727.2	45349	13	13	.051	D
4729.7	46077	5	3	.17	D
4730.7	45966	7	5	.28	D
4737.4	46000	9	7	.24	D
4789.3	41393	13	11	.076	D
4792.5	45966	7	5	.26	D
4801.0	46000	9	7	.23	D
4870.8	45359	7	9	.35	D
4887.0	45354	9	11	.32	D
4922.3	45349	11	13	.40	D
4936.3	45359	7	9	.14	D
4954.8	45354	9	11	.12	D
5139.7	47048	7	7	.13	D
5177.4	46959	9	11	.061	D
5184.6	46783	7	9	.11	D
5192.0	46637	5	7	.14	D
5196.6	47055	11	9	.12	D
5200.2	46525	3	5	.16	D
5204.5	26802	5	3	.55	D
5206.0	26796	5	5	.53	D
5208.4	26788	5	7	.51	D
5243.4	46449	5	3	.20	D
5261.8	48825	7	9	.13	D
5272.0	46783	7	9	.11	D
5287.2	46637	5	7	.078	D
5304.2	46783	9	9	.066	D
5312.9	46637	7	7	.11	D
5318.8	46525	5	5	.13	D
5328.3	42261	9	11	.60	D
5340.4	46449	5	3	.16	D
5400.6	45734	5	5	.16	D
5409.8	26788	9	7	.062	D
		Cr II			
2653.6	49706	4	6	.35	D
2658.6	49565	2	4	.58	D
2666.0	49646	6	8	.59	D
2668.7	49493	4	2	1.4	D
2671.8	49565	6	4	1.0	D
2672.8	49706	8	6	.55	D
2693.5	67334	10	8	1.4	D
2727.3	67876	10	8	1.7	D
2740.1	48632	6	8	.11	D
2745.0	66727	4	6	.85	D
2768.6	74424	6	8	2.8	D
2774.4	75717	8	8	1.7	D
2778.1	75810	10	10	3.2	D
2782.4	69348	6	4	1.6	D
2785.7	69506	10	8	2.1	D
2787.6	66727	6	6	1.5	D
2792.2	69498	12	10	2.3	D
2800.8	69388	12	14	2.2	D
2818.4	68993	8	10	2.2	D
2822.0	68844	6	8	2.3	D
2832.5	70108	12	10	1.3	D
2838.8	73486	8	8	2.7	D
2840.0	65420	10	12	2.7	D
2843.3	47465	8	10	.64	D
2849.8	47228	6	8	.92	D
2851.4	65218	8	10	2.2	D
2856.8	54626	4	6	.43	D
2857.4	54785	6	8	.28	D
2860.9	46906	2	4	.69	D
2862.6	47228	8	8	.63	D
2866.7	46906	4	4	1.2	D
2867.1	54500	4	4	1.1	D
2867.7	46824	2	2	1.1	D
2870.4	54626	6	6	1.3	D
2873.8	54418	4	2	.88	D
2878.5	47228	10	8	.074	D
2880.9	54500	6	4	.79	D
2888.7	70880	10	12	.88	D
2898.5	65710	10	12	1.2	D
2921.8	65384	8	10	.90	D
2927.1	72717	10	10	2.8	D

Table 1b (Continued)
TRANSITION PROBABILITIES FOR SELECTED ATOMIC AND IONIC SPECIES

Cr II

Wavelength (λ[Å])	Upper energy level (E_k[cm^{-1}])	Statistical weights g_i	g_k	Transition Probability (A[10^8 s^{-1}])	Uncertainty
2930.9	64062	2	4	1.1	D
2935.1	64924	6	8	1.8	D
2953.4	63802	2	2	1.8	D
2953.7	68477	10	10	.92	D
2966.1	64924	10	8	.54	D
2971.9	64031	14	14	2.0	D
2979.7	63849	12	12	1.8	D
2985.3	63707	10	10	2.2	D
2989.2	63601	8	8	2.2	D
3040.9	67506	10	12	4.8	D
3041.7	68477	10	10	3.1	D
3050.1	67589	12	14	1.8	D
3093.5	70880	10	12	.67	D
3096.1	70852	10	8	.75	D
3107.6	70679	8	10	.62	D
3118.7	51584	2	4	1.7	D
3120.4	51670	4	6	1.5	D
3122.6	65710	12	12	.44	D
3128.7	51584	4	4	.81	D
3136.7	51670	6	6	.64	D
3152.2	67070	4	4	1.8	D
3180.7	51943	12	10	.70	D
3183.3	67012	8	6	.87	D
3209.2	51670	8	6	.68	D
3217.4	51584	6	4	.77	D
3234.1	65543	10	8	.92	D
3238.8	65680	12	10	.54	D
3295.4	64031	12	14	.32	D
3336.3	49493	2	2	.42	D
3339.8	49565	4	4	.49	D
3342.6	49706	6	6	.39	D
3347.8	49493	4	2	.52	D
3358.5	49565	6	4	1.1	D
3360.3	54785	8	8	1.3	D
3368.1	49706	8	6	1.4	D
3378.3	54626	8	6	.41	D
3379.4	54626	4	6	.48	D
3382.7	49352	6	6	.45	D
3391.4	49006	2	4	.19	D
3393.0	54500	2	4	.46	D
3393.8	54500	4	4	.66	D
3394.3	54500	6	4	.75	D
3402.4	54418	2	2	.80	D
3408.8	49352	8	6	.95	D
3421.2	48750	2	2	1.7	D
3422.7	49006	6	4	1.4	D
3433.3	48750	4	2	1.3	D
3511.8	48491	8	6	.079	D
4242.4	54785	10	8	.12	D

Cr XXI

Wavelength (λ[Å])	Upper energy level (E_k[cm^{-1}])	g_i	g_k	Transition Probability (A[10^8 s^{-1}])	Uncertainty
149.90	667110	1	3	160.	C

Cr XXII

Wavelength (λ[Å])	Upper energy level (E_k[cm^{-1}])	g_i	g_k	Transition Probability (A[10^8 s^{-1}])	Uncertainty
8.51	11800000	2	6*	1.2 (+4)	C
9.493	10534000	2	6*	2.5 (+4)	C
9.809	10553000	2	4	4.1 (+4)	B
9.865	10590000	4	6	4.9 (+4)	B
12.620	7924000	2	4	5.13(+4)	B
12.662	7898000	2	2	5.28(+4)	B
13.147	7964000	2	4	1.3 (+5)	B
13.294	7966000	4	6	1.6 (+5)	B
13.306	7964000	4	4	2.6 (+4)	C
25.2	11800000	2	6*	3750.	C
36.93	10534000	2	6*	7000.	C
37.52	10580000	6	10*	1.7 (+4)	B
223.00	448430	2	4	33.	B
279.69	357540	2	2	17.	B

Cr XXIII

Wavelength (λ[Å])	Upper energy level (E_k[cm^{-1}])	g_i	g_k	Transition Probability (A[10^8 s^{-1}])	Uncertainty
2.182	45830000	1	3	3.3 (+6)	B

Cobalt

Co I

Wavelength (λ[Å])	Upper energy level (E_k[cm^{-1}])	g_i	g_k	Transition Probability (A[10^8 s^{-1}])	Uncertainty
2987.2	33467	10	8	.050	C
2989.6	33440	10	10	.037	C
3013.6	33173	10	8	.016	C
3017.5	33946	8	6	.072	C
3042.5	33674	8	6	.020	C
3044.0	32842	10	10	.19	C
3048.9	34196	6	4	.078	C
3061.8	33467	8	8	.15	C
3064.4	33440	8	10	.0068	C
3082.6	32431	10	12	.026	C
3086.8	34196	4	4	.19	C
3089.6	33173	8	8	.024	C
3098.2	33674	6	6	.027	C
3139.9	32655	8	6	.028	C
3147.1	33173	6	8	.045	C
3149.3	33151	6	4	.031	C
3395.4	34134	6	8	.26	C
3405.1	32842	10	10	.98	C
3409.2	33467	8	8	.42	C
3412.3	33440	8	10	.64	C
3412.6	29295	10	8	.12	C
3417.2	33946	6	6	.32	C
3431.6	29949	8	6	.11	C
3433.0	34196	4	4	1.1	C
3442.9	30444	6	4	.12	C
3443.6	33173	8	8	.63	C
3449.2	33674	6	6	.73	C
3449.4	32465	10	10	.16	C
3455.2	30743	4	2	.18	C
3462.8	33946	4	6	.87	C
3465.8	28845	10	12	.097	C
3483.4	32842	8	10	.062	C
3489.4	36092	8	6	1.6	C
3491.3	30444	4	4	.053	C
3495.7	33674	4	6	.45	C
3496.7	32733	8	8	.036	C
3518.4	36875	6	4	1.7	C
3520.1	29216	8	6	.034	C
3521.6	31871	10	8	.12	C
3523.4	33449	4	2	1.2	C
3526.9	28346	10	10	.12	C
3529.0	29735	6	8	.090	C
3529.8	32465	8	10	.48	C
3533.4	30103	4	6	.091	C
3550.6	29563	6	4	.042	C
3560.9	33151	4	4	.24	C
3564.9	32733	6	8	.086	C
3569.4	35451	8	8	1.6	C
3575.0	32655	6	6	.18	C
3575.4	28777	8	8	.094	C
3585.2	32028	8	8	.076	C
3587.2	36330	6	6	1.9	C
3594.9	29216	6	6	.086	C
3602.1	29563	4	4	.10	C
3605.4	31871	8	8	.039	C
3627.8	31700	8	10	.052	C
3631.4	28346	8	10	.0065	C
3647.7	29216	4	6	.012	C
3652.5	28777	6	8	.0095	C
3704.1	35451	6	8	.18	C
3745.5	34134	8	8	.077	C
3842.1	33463	8	6	.31	C
3845.5	33440	8	10	.49	C
3873.1	29295	10	8	.12	C
3874.0	29949	8	6	.12	C
3881.9	30444	6	4	.11	C
3894.1	34134	6	8	.81	C
3895.0	30743	4	2	.11	C
3909.9	25569	10	12	.0019	C
3936.0	32842	8	10	.15	C
3995.3	32465	8	10	.36	C
3997.9	33467	6	8	.079	C
4020.9	28346	10	10	.0092	D
4092.4	31871	8	8	.14	D
4118.8	32733	6	8	.34	C
4121.3	31700	8	10	.24	C

Copper

Cu I

Wavelength (λ[Å])	Upper energy level (E_k[cm^{-1}])	g_i	g_k	Transition Probability (A[10^8 s^{-1}])	Uncertainty
2024.3	49383	2	6*	.098	C
2165.1	46173	2	4	.51	B
2178.9	45879	2	4	.913	B
2181.7	45821	2	2	1.0	C
2225.7	44916	2	2	.46	C
2244.3	44544	2	4	.0119	B
2441.6	40944	2	2	.020	C
2492.2	40114	2	4	.0311	B
2618.4	49383	6	4	.307	C
2766.4	49383	4	4	.096	C

Table 1b (Continued)
TRANSITION PROBABILITIES FOR SELECTED ATOMIC AND IONIC SPECIES

Wavelength (λ[Å])	Upper energy level (E_k[cm^{-1}])	Statistical weights g_i	g_k	Transition Probability (A[10^8 s^{-1}])	Uncertainty
		Cu I			
2824.4	46598	6	6	.078	C
2961.2	44963	6	8	.0376	C
3063.4	45879	4	4	.0155	C
3194.1	44544	4	4	.0155	C
3247.5	30784	2	4	1.39	B
3274.0	30535	2	2	1.37	B
3337.8	41153	6	8	.0038	C
4022.6	55388	2	4	.190	C
4062.6	55391	4	6	.210	C
4249.0	64472	2	2	.195	C
4275.1	62403	6	8	.345	C
4480.4	52849	2	2	.030	C
4509.4	64472	4	2	.275	C
4530.8	52849	4	2	.084	C
4539.7	63585	6	4	.212	C
4587.0	62948	8	6	.320	C
4651.1	62403	10	8	.380	C
4704.6	62403	8	8	.055	C
5105.5	30784	6	4	.020	C
5153.2	49935	2	4	.60	C
5218.2	49942	4	6	.75	C
5220.1	49935	4	4	.150	C
5292.5	62403	8	8	.109	C
5700.2	30784	4	4	.0024	C
5782.1	30535	4	2	.0165	C
		Cu II			
2489.7	66419	5	5	.015	D
2544.8	108015	9	7	1.1	D
2689.3	108015	7	7	.41	D
2701.0	110366	5	5	.67	D
2703.2	110084	3	3	1.2	D
2713.5	108336	5	5	.68	D
2769.7	108015	7	7	.61	D
		Dysprosium			
		Dy I			
2862.7	34922	17	15	.065	D
2964.6	33722	17	17	.065	D
3147.7	35894	15	17	.11	D
3263.2	34770	15	13	.14	D
3511.0	32608	15	13	.31	D
3571.4	32126	15	13	.20	D
3757.1	34175	17	19	3.0	D
3868.8	33406	17	17	3.1	D
3967.5	32763	17	19	.87	D
4046.0	24709	17	15	1.5	D
4103.9	31411	13	11	1.7	D
4186.8	23878	17	17	1.32	C
4194.8	23832	17	17	.72	D
4211.7	23737	17	19	2.08	C
4218.1	27835	15	15	1.85	C
4221.1	27818	15	17	1.52	C
4225.2	30712	13	15	4.5	D
4268.3	27556	15	15	.036	D
4276.7	30427	13	13	.73	D
4292.0	27427	15	15	.058	D
4577.8	21839	17	19	.022	D
4589.4	21783	17	15	.13	D
4612.3	21675	17	15	.082	D
5077.7	19689	17	17	.0057	D
5301.6	18857	17	15	.011	D
5547.3	18022	17	17	.0027	C
5639.5	17727	17	19	.0047	C
5974.5	16733	17	17	.0040	C
5988.6	16694	17	15	.0053	C
6010.8	20766	15	15	.026	D
6088.3	20555	15	13	.035	D
6168.4	20341	15	17	.025	D
6259.1	15972	17	19	.0085	C
6579.4	15195	17	15	.0075	D
		Erbium			
		Er I			
3862.9	25880	13	13	2.5	D
4008.0	24943	13	15	2.6	D
4151.1	24083	13	11	1.8	D
		Europium			
		Eu I			
2372.9	42131	8	6	.19	D

Wavelength (λ[Å])	Upper energy level (E_k[cm^{-1}])	Statistical weights g_i	g_k	Transition Probability (A[10^8 s^{-1}])	Uncertainty
		Eu I			
2375.3	42087	8	8	.20	D
2379.7	42010	8	10	.20	D
2619.3	38167	8	10	.0070	D
2643.8	37813	8	8	.0066	D
2659.4	37591	8	10	.012	D
2682.6	37266	8	6	.012	D
2710.0	36890	8	10	.14	D
2724.0	36700	8	8	.12	D
2731.4	36601	8	8	.031	D
2732.6	36584	8	6	.037	D
2735.3	36549	8	10	.047	D
2738.6	36505	8	10	.013	D
2743.3	36442	8	6	.11	D
2745.6	36411	8	6	.050	D
2747.8	36382	8	8	.052	D
2772.9	36053	8	6	.010	D
2878.9	34726	8	10	.028	D
2892.5	34562	8	8	.10	D
2893.0	34556	8	6	.10	D
2909.0	34366	8	10	.069	D
2958.9	33787	8	6	.016	D
3059.0	32681	8	8	.038	D
3067.0	32596	8	10	.0091	D
3106.2	32185	8	10	.055	D
3111.4	32130	8	10	.30	D
3168.3	31554	8	10	.069	D
3185.5	31383	8	10	.0058	D
3210.6	31138	8	8	.11	D
3212.8	31116	8	8	.29	D
3213.8	31107	8	6	.18	D
3235.1	30902	8	10	.010	D
3241.4	30842	8	8	.023	D
3246.0	30798	8	6	.014	D
3247.6	30784	8	8	.023	D
3322.3	30091	8	6	.035	D
3334.3	29983	8	6	.34	D
3350.4	29839	8	10	.015	D
3353.7	29809	8	8	.0058	D
3457.1	28918	8	8	.0084	D
3467.9	28828	8	8	.010	D
3589.3	27853	8	6	.0069	D
4594.0	21761	8	10	1.4	D
4627.2	21605	8	8	1.3	D
4661.9	21445	8	6	1.3	D
5645.8	17707	8	6	.0054	D
5765.2	17341	8	8	.011	D
6018.2	16612	8	10	.0085	D
6291.3	15891	8	6	.0018	D
6864.5	14564	8	10	.0058	D
7106.5	14068	8	8	.0026	D
		Fluorine			
		F I			
806.96	123921	4	6	3.3	C
809.60	123922	2	4	2.8	C
951.87	105056	4	2	2.6	C
954.83	104731	4	4	6.4	C
955.55	105056	2	2	5.1	C
958.52	104731	2	4	1.3	C
6239.7	118428	6	4	.25	D
6348.5	118428	4	4	.18	D
6413.7	118428	2	4	.11	D
6708.3	117309	6	4	.014	D
6774.0	117164	6	6	.10	D
6795.5	117392	4	2	.052	D
6834.3	117309	4	4	.21	D
6856.0	116987	6	8	.42	D
6870.2	117392	2	2	.38	D
6902.5	117164	4	6	.32	D
6909.8	117309	2	4	.22	D
6966.4	119082	4	2	.11	D
7037.5	118937	4	4	.30	D
7127.9	119082	2	2	.38	D
7309.0	137599	6	8	.47	D
7311.0	118405	4	2	.39	D
7314.3	137590	4	6	.48	D
7332.0	116041	6	4	.31	D
7398.7	115918	6	6	.31	D
7425.7	116144	4	2	.34	D
7482.7	116041	4	4	.056	D
7489.2	118405	2	2	.11	D

Table 1b (Continued)
TRANSITION PROBABILITIES FOR SELECTED ATOMIC AND IONIC SPECIES

Wavelength (λ[Å])	Upper energy level (E_k[cm^{-1}])	Statistical weights g_i	g_k	Transition Probability (A[10^8 s^{-1}])	Uncertainty
		F I			
7514.9	116144	2	2	.052	D
7552.2	115918	4	6	.078	D
7573.4	116041	2	4	.10	D
7607.2	117873	4	4	.070	D
7754.7	117623	4	6	.30	D
7800.2	117873	2	4	.01	D
		Gallium			
		Ga I			
2195.4	45537	2	2	.019	C
2199.7	46274	4	2	.033	C
2214.4	45972	4	6	.012	C
2235.9	45537	4	2	.043	C
2255.0	44332	2	2	.031	C
2259.2	45076	4	6	.031	C
2294.2	43575	2	4	.070	C
2297.9	44332	4	2	.058	C
2338.2	43581	4	6	.098	C
2371.3	42159	2	2	.057	C
2418.7	42159	4	2	.10	C
2450.1	40803	2	4	.28	C
2500.2	40811	4	6	.34	C
2659.9	37585	2	2	.12	C
2719.7	37585	4	2	.23	C
2874.2	34782	2	4	1.2	C
2943.6	34788	4	6	1.4	C
2944.2	34782	4	4	.27	C
4033.0	24789	2	2	.49	C
4172.0	24789	4	2	.92	C
		Ga II			
829.60	120540	1	3	.22	D
1414.4	70700	1	3	18.8	C
		Germanium			
		Ge I			
1944.7	51978	3	1	.70	C
1955.1	51705	3	3	.20	C
1988.3	51705	5	3	.25	C
1998.9	51438	5	5	.55	C
2041.7	48963	1	3	1.1	C
2065.2	48963	3	3	.85	C
2068.7	48882	3	5	1.2	C
2086.0	48480	3	5	.40	C
2094.3	49144	5	7	.97	C
2105.8	48882	5	5	.17	C
2256.0	51438	5	5	.032	C
2417.4	48480	5	5	.96	C
2498.0	40021	1	3	.13	C
2533.2	40021	3	3	.10	C
2589.2	40021	5	3	.051	C
2592.5	39118	3	5	.71	C
2651.2	39118	5	5	2.0	C
2651.6	37702	1	3	.85	C
2691.3	37702	3	3	.61	C
2709.6	37452	3	1	2.8	C
2754.6	37702	5	3	1.1	C
3039.1	40021	5	3	2.8	C
3124.8	39118	5	5	.031	C
3269.5	37702	5	3	.29	C
4226.6	40021	1	3	.21	C
4685.8	37702	1	3	.095	C
		Ge II			
999.10	100090	2	4	1.9	D
1016.6	100131	4	6	2.1	D
1017.1	100090	4	4	.35	D
1055.0	94784	2	2	.69	D
1075.1	94784	4	2	1.3	D
1237.1	80837	2	4	19.	D
1261.9	81013	4	6	22.	D
1264.7	80837	4	4	3.5	D
1602.5	62403	2	2	3.4	D
1649.2	62403	4	2	6.5	D
4741.8	100090	2	4	.46	D
4814.6	100131	4	6	.51	D
4824.1	100090	4	4	.086	D
5131.8	100318	4	6	1.9	D
5178.5	100318	6	6	.13	D
5178.6	100317	6	8	2.0	D
5893.4	79367	2	4	.92	D
6021.0	79007	2	2	.84	D
6336.4	94784	2	2	.44	D
6484.2	94784	4	2	.85	D
		Gold			
		Au I			
2428.0	41174	2	4	1.5	D
2676.0	37359	2	2	1.1	D
		Helium			
		He I			
510.00	196079	1	3	.462	B
512.10	195275	1	3	.717	B
515.62	193943	1	3	1.3	B
522.21	191493	1	3	2.46	A
537.03	186209	1	3	5.66	A
584.33	171135	1	3	17.99	AA
2696.1	196935	3	9*	.00550	B
2723.2	196567	3	9*	.00780	B
2763.8	196027	3	9*	.0111	B
2829.1	195193	3	9*	.017	B
2945.1	193801	3	9*	.0320	A
3187.7	191217	3	9*	.05639	AA
3354.6	196079	1	3	.0130	B
3447.6	195275	1	3	.0232	A
3554.4	197213	9	15*	.0131	A
3587.3	196955	9	15*	.0205	C
3613.6	193943	1	3	.0390	A
3634.2	196595	9	15*	.0261	A
3705.0	196070	9	15*	.0444	C
3819.6	195260	9	15*	.0636	A
3833.6	197213	3	5	.00971	B
3867.5	194936	9	3*	.025	B
3871.8	196956	3	5	.0126	C
3888.7	185565	3	9*	.09478	AA
3926.5	196596	3	5	.0195	A
3964.7	191493	1	3	.0719	A
4009.3	196070	3	5	.0279	C
4026.2	193917	9	15*	.116	A
4120.8	193347	9	3*	.0444	A
4143.8	195261	3	5	.0485	A
4387.9	193918	3	5	.0894	A
4437.6	193664	3	1	.033	B
4471.5	191445	9	15*	.246	A
4713.2	190298	9	3*	.0955	A
4921.9	191447	3	5	.198	A
5015.7	186210	1	3	.1338	AA
5047.7	190940	3	1	.0675	A
5875.7	186102	9	15*	.7053	AA
6678.2	186105	3	5	.6339	AA
7065.2	183237	9	3*	.2786	AA
7281.4	184865	3	1	.1829	AA
8361.7	195193	3	9*	.00334	A
9463.6	193801	3	9*	.00501	A
9603.4	195275	1	3	.00610	A
9702.6	195868	9	3*	.00858	B
10311.	195260	9	15*	.0201	A
10668.	194936	9	3*	.0152	A
10830.	169087	3	9*	.1022	AA
10913.	195262	15	21*	.0212	B
10917.	195262	5	7	.0212	B
10997.	195193	15	9*	.0013	B
11013.	193943	1	3	.0100	A
11045.	195261	3	5	.0185	A
11226.	195115	3	1	.0108	A
11969.	193917	9	15*	.0358	A
12528.	191217	3	9*	.00710	A
12756.	193943	5	3	.0012	B
12785.	193921	15	21*	.0462	B
12791.	193921	5	7	.0461	B
12846.	193347	9	3*	.0289	A
12968.	193918	3	5	.0343	A
12985.	193801	15	9*	.0025	B
		Indium			
		In I			
2710.3	39098	4	6	.4	D
2560.2	39048	2	4	.4	D
3256.1	32915	4	6	1.3	D
3039.4	32892	2	4	1.3	D
4101.8	24373	2	2	.56	C
4511.3	24373	4	2	1.02	C

Table 1b (Continued)
TRANSITION PROBABILITIES FOR SELECTED ATOMIC AND IONIC SPECIES

Wavelength (λ[Å])	Upper energy level (E_k[cm^{-1}])	Statistical weights g_i	g_k	Transition Probability (A[10^8 s^{-1}])	Uncertainty
		In II			
2941.1	97025	3	1	1.4	D
		Iodine			
		I I			
1782.8	56093	4	4	2.71	C
1830.4	54633	4	6	.16	D
		Iron			
		Fe I			
1934.5	51692	9	7	.25	C
1937.3	51619	9	7	.22	C
1940.7	51945	7	5	.26	C
2084.1	47967	9	7	.37	C
2102.4	47967	7	7	.088	C
2113.0	48290	1	3	.19	C
2132.0	46889	9	9	.076	C
2166.8	46137	9	7	2.7	C
2191.8	46314	5	5	1.2	C
2196.0	46410	3	3	1.2	C
2200.7	46314	3	5	.28	C
2259.5	44244	9	11	.070	C
2276.0	43923	9	7	.17	C
2277.1	51630	7	5	37.	C
2287.3	44411	5	3	.34	C
2294.4	44459	3	1	.61	C
2300.1	44166	5	7	.080	C
2309.0	44184	3	5	.15	C
2313.1	43923	5	7	.14	C
2320.4	43499	7	9	.12	C
2373.6	42533	7	7	.067	C
2374.5	43079	1	3	.29	C
2462.2	41018	7	5	.15	C
2462.6	40594	9	9	.58	C
2479.8	41018	5	5	1.8	C
2483.3	40257	9	11	4.9	C
2488.1	40594	7	9	4.7	C
2490.6	40842	5	7	3.8	C
2491.2	41018	3	5	3.0	C
2501.1	39970	9	7	.68	C
2510.8	40231	7	5	1.3	C
2518.1	40405	5	3	1.9	C
2522.8	39626	9	9	2.9	C
2524.3	40491	3	1	3.4	C
2527.4	39970	7	7	1.9	C
2529.1	40231	5	5	.98	C
2535.6	40405	1	3	.97	C
2541.0	40231	3	5	.92	C
2546.0	39970	5	7	.67	C
2549.6	39626	7	9	.36	C
2584.5	45608	11	13	.46	C
2719.0	36767	9	7	1.4	C
2720.9	37158	7	5	1.1	C
2723.6	37410	5	3	.64	C
2737.3	37410	3	3	.85	C
2742.4	37158	5	5	.63	C
2744.1	37410	1	3	.35	C
2750.1	36767	7	7	.39	C
2756.3	37158	3	5	.20	C
2788.1	42784	11	13	.63	C
2894.5	52916	5	5	.63	C
2899.4	52858	5	3	.61	C
2923.3	60549	11	11	1.7	C
2925.4	56423	7	9	.19	C
2936.9	34040	9	9	.14	C
2954.7	52213	5	7	.12	C
2966.9	33695	9	11	.272	B
2980.5	55791	7	7	.22	C
2983.6	33507	9	7	.280	B
2990.4	55430	9	11	.40	C
2996.4	52916	3	5	.19	C
2999.5	40257	11	11	.23	C
3000.9	34017	5	3	.642	B
3008.1	34122	3	1	1.07	B
3009.1	52613	13	11	.079	C
3009.6	40594	9	9	.18	C
3011.5	55446	7	9	.48	C
3019.0	40842	7	7	.15	C
3037.4	33802	3	5	.32	C
3042.7	40842	5	7	.066	C
		Fe I			
3047.6	33507	5	7	.284	B
3053.1	52297	3	5	.18	C
3057.4	39626	11	9	.45	C
3059.1	33096	7	9	.18	C
3067.2	39970	9	7	.35	C
3075.7	40231	7	5	.30	C
3083.7	40405	5	3	.35	C
3098.2	53983	11	11	.11	C
3100.7	39970	7	7	.16	C
3119.5	51668	11	9	.096	C
3120.4	51826	9	7	.10	C
3160.7	51192	9	9	.19	C
3161.9	50968	11	13	.12	C
3166.4	52213	9	7	.14	C
3175.4	50833	11	11	.13	C
3199.5	50808	9	9	.27	C
3205.4	51208	3	3	1.2	C
3215.9	50999	5	5	.81	C
3217.4	50423	11	9	.23	C
3225.8	50342	11	13	1.0	C
3231.0	50699	7	5	.39	C
3233.1	57028	13	15	.55	C
3234.0	50475	9	9	.20	C
3248.2	50534	7	7	.22	C
3253.6	56951	7	9	.18	C
3254.4	57070	11	13	.51	C
3265.6	48163	7	5	.39	C
3271.0	48290	5	3	.67	C
3280.3	57104	9	11	.55	C
3282.9	56859	3	5	.31	C
3292.0	56593	7	9	.62	C
3292.6	48290	3	3	.31	C
3306.0	47967	5	7	.48	C
3307.2	56334	13	13	.20	C
3314.7	56783	5	7	.70	C
3323.7	52916	5	5	.31	C
3328.9	56383	11	11	.27	C
3337.7	51668	11	9	.067	C
3355.2	56423	9	9	.33	C
3369.5	51668	9	9	.25	C
3370.8	51374	11	11	.34	C
3380.1	51826	7	7	.24	C
3384.0	47093	7	7	.11	C
3392.7	47017	7	7	.26	C
3399.3	47136	5	5	.39	C
3402.3	55490	13	13	.29	C
3406.4	55754	3	5	.30	C
3407.5	46889	7	9	.60	C
3410.2	56859	3	5	.48	C
3411.4	51305	9	9	.065	C
3413.1	47017	5	7	.37	C
3417.8	47177	3	3	.52	C
3418.5	47171	3	1	1.3	C
3424.3	46745	7	7	.21	C
3425.0	53763	9	7	.29	C
3427.1	46721	7	9	.56	C
3428.2	46889	5	5	.22	C
3445.1	46745	5	7	.28	C
3447.3	46727	5	5	.11	C
3450.3	46902	3	3	.24	C
3495.3	49243	9	7	.13	C
3497.1	46137	7	7	.15	C
3521.8	46314	3	5	.11	C
3524.1	49243	7	5	.091	C
3527.8	51335	9	9	.20	C
3529.8	51567	3	3	.78	C
3536.6	51461	5	7	.80	C
3540.1	51350	7	9	.12	C
3541.1	51229	9	11	.64	C
3542.1	51335	7	9	.76	C
3552.8	51331	5	5	.17	C
3553.7	56951	11	9	.83	C
3556.9	51103	9	11	.45	C
3559.5	52858	3	3	.22	C
3560.7	54301	7	9	.077	C
3565.4	35768	7	9	.39	C
3570.1	35379	9	11	.677	B
3572.0	50833	11	11	.25	C
3581.2	34844	11	13	1.02	B
3582.2	54014	13	11	.25	C
3587.0	35856	5	5	.17	C

Table 1b (Continued)

TRANSITION PROBABILITIES FOR SELECTED ATOMIC AND IONIC SPECIES

Wavelength (λ[Å])	Upper energy level (E_k[cm^{-1}])	Statistical weights g_i	g_k	Transition Probability (A[10^8 s^{-1}])	Uncertainty
	Fe I				
3594.6	50808	9	9	.28	C
3599.6	56593	11	9	.19	C
3603.2	49461	11	11	.27	C
3605.5	49727	9	9	.65	C
3606.7	49434	11	13	.84	C
3608.9	35856	3	5	.814	B
3610.2	50342	13	13	.50	C
3612.1	50523	11	13	.077	C
3617.8	51969	5	7	.66	C
3618.8	35612	5	7	.73	C
3621.5	49604	9	11	.52	C
3622.0	49851	7	7	.53	C
3623.2	46982	13	13	.076	C
3631.5	35257	7	9	.52	C
3632.0	52297	3	5	.50	C
3638.3	49727	7	9	.27	C
3640.4	49461	9	11	.39	C
3645.8	52512	1	3	.58	C
3647.8	34782	9	11	.292	B
3649.5	49109	11	9	.43	C
3651.5	49628	7	9	.64	C
3655.5	50187	5	5	.12	C
3659.5	47106	9	9	.068	C
3669.5	49243	9	7	.30	C
3670.1	51023	11	13	.078	C
3676.3	47835	9	11	.061	C
3677.6	49433	7	5	.82	C
3682.2	55754	5	5	1.7	C
3684.1	49135	9	7	.34	C
3686.0	50833	9	11	.26	C
3687.5	34040	11	9	.0801	B
3690.7	55907	11	11	.28	C
3694.0	51570	5	7	.70	C
3697.4	51219	7	7	.21	C
3701.1	51192	7	9	.49	C
3704.5	48703	11	9	.14	C
3709.2	34329	9	7	.156	B
3719.9	26875	9	11	.163	B
3724.4	45221	5	7	.13	C
3727.6	34547	7	5	.225	B
3730.4	51374	9	11	.13	C
3732.4	44512	5	5	.28	C
3734.9	33695	11	11	.902	B
3737.1	27167	7	9	.142	C
3738.3	53094	11	13	.38	C
3740.2	52954	7	7	.19	C
3742.6	50423	9	9	.11	C
3744.1	51208	5	3	.38	C
3745.6	27395	5	7	.115	B
3749.5	34040	9	9	.764	B
3753.6	44184	7	5	.11	C
3756.9	55430	11	11	.25	C
3758.2	34329	7	7	.634	B
3760.1	45978	13	15	.057	C
3763.8	34547	5	5	.544	B
3765.5	52655	13	15	.99	C
3767.2	34692	3	3	.640	B
3787.2	55754	5	5	.12	C
3787.9	34547	3	5	.129	B
3794.3	46136	9	11	.046	C
3795.0	34329	5	7	.115	B
3798.5	33695	9	11	.0323	B
3799.5	34040	7	9	.0732	B
3804.0	53155	11	9	.052	C
3805.3	52899	9	11	1.0	C
3806.2	53808	3	3	.25	C
3806.7	52613	11	11	.55	C
3807.5	44184	3	5	.097	C
3810.8	52858	5	3	.24	C
3813.0	33947	7	5	.0792	B
3813.9	55526	13	11	.091	C
3815.8	38175	9	7	1.3	C
3820.4	33096	11	9	.668	B
3821.2	52514	11	13	.70	C
3821.8	47197	5	5	.089	C
3825.9	33507	9	7	.598	B
3827.8	38678	7	5	1.1	C
3833.3	46721	9	9	.059	C
3834.2	33802	7	5	.453	B
3836.3	52683	5	5	.39	C
3839.3	50614	9	9	.29	C

Wavelength (λ[Å])	Upper energy level (E_k[cm^{-1}])	Statistical weights g_i	g_k	Transition Probability (A[10^8 s^{-1}])	Uncertainty
	Fe I				
3840.4	34017	5	3	.470	B
3841.0	38996	5	3	1.4	C
3843.3	50587	9	7	.48	C
3846.8	52213	7	7	.67	C
3850.0	34122	3	1	.606	B
3859.2	45295	13	11	.087	C
3859.9	25900	9	9	.0970	B
3865.5	34017	3	3	.155	B
3867.2	50187	5	5	.35	C
3871.8	49604	11	11	.070	C
3872.5	33802	5	5	.105	B
3873.8	45428	11	9	.082	C
3878.0	33507	7	7	.0772	B
3883.3	51969	7	7	.17	C
3884.4	47453	11	9	.042	C
3888.5	38678	5	5	.27	C
3891.9	53230	3	3	.40	C
3893.4	49461	11	11	.14	C
3900.5	51771	7	7	.086	C
3902.9	38175	7	7	.24	C
3903.9	49727	9	9	.097	C
3916.7	51630	13	11	.12	C
3919.1	49628	9	9	.045	C
3942.4	48305	3	5	.11	C
3951.2	51708	3	5	.36	C
3952.6	47008	11	11	.052	C
3953.2	49628	7	9	.043	C
3963.1	51705	3	5	.17	C
3967.4	51826	9	7	.24	C
3969.3	37163	9	7	.24	C
3971.3	46889	11	9	.068	C
3973.7	53763	5	7	.080	C
3977.7	42860	5	5	.082	C
3981.8	47106	9	9	.046	C
3984.0	47093	9	7	.089	C
3985.4	51708	5	5	.082	C
3997.0	54811	9	9	.074	C
3997.4	47008	9	11	.16	C
3998.1	46721	11	9	.075	C
4005.2	37521	7	5	.22	C
4014.5	53722	11	11	.24	C
4017.2	49461	9	11	.053	C
4021.9	47106	7	9	.10	C
4032.0	51201	3	5	.086	C
4045.8	36686	9	9	.75	C
4062.4	47556	3	3	.23	C
4063.6	37163	7	7	.69	C
4068.0	50475	9	9	.17	C
4070.8	50699	7	5	.14	C
4071.7	37521	5	5	.80	C
4073.8	50880	5	3	.19	C
4074.8	49109	9	9	.056	C
4076.6	50423	9	9	.20	C
4084.5	51350	11	9	.12	C
4085.3	50611	7	7	.12	C
4098.2	50534	7	7	.082	C
4107.5	47177	5	3	.25	C
4109.8	47272	3	3	.19	C
4113.0	58002	11	13	.15	C
4127.6	47272	1	3	.16	C
4132.9	47136	3	5	.11	C
4134.7	47017	5	7	.18	C
4137.0	51708	3	5	.23	C
4143.9	36686	7	9	.16	C
4149.4	50968	11	13	.043	C
4153.9	51462	7	9	.24	C
4154.8	51229	9	11	.15	C
4156.8	46889	5	5	.19	C
4170.9	48305	5	5	.072	C
4172.1	50187	7	5	.12	C
4175.6	46889	3	5	.17	C
4184.9	46727	5	5	.12	C
4187.0	43634	7	5	.23	C
4187.8	43435	9	7	.16	C
4196.2	51219	7	7	.11	C
4210.3	43764	3	3	.20	C
4217.6	51370	3	5	.24	C
4219.4	52514	11	13	.38	C
4222.2	43435	7	7	.063	C
4224.2	50833	9	11	.14	C
4225.5	51219	5	7	.17	C

Table 1b (Continued)
TRANSITION PROBABILITIES FOR SELECTED ATOMIC AND IONIC SPECIES

Wavelength (λ[Å])	Upper energy level (E_k[cm^{-1}])	Statistical weights g_i	g_k	Transition Probability (A[10^8 s^{-1}])	Uncertainty
		Fe I			
4233.6	43634	3	5	.20	C
4238.8	50980	7	9	.22	C
4246.1	52916	7	5	.069	C
4250.1	43435	5	7	.23	C
4282.4	40895	7	5	.13	C
4307.9	35768	7	9	.35	C
4315.1	40895	5	5	.090	C
4325.8	36079	5	7	.51	C
4327.1	51708	5	5	.094	C
4369.8	47453	9	9	.074	C
4383.5	34782	9	11	.46	C
4388.4	51837	7	7	.13	C
4401.3	51771	7	7	.069	C
4404.8	35257	7	9	.25	C
4415.1	35612	5	7	.13	C
4443.2	45552	1	3	.13	C
4466.6	45221	5	7	.13	C
4469.4	51837	5	7	.27	C
4528.6	39626	7	9	.063	C
4547.9	50587	5	7	.078	C
4736.8	47006	9	11	.050	C
4789.7	49477	5	5	.084	C
4859.7	43764	5	3	.15	C
4871.3	43634	7	5	.22	C
4872.1	43764	3	3	.24	C
4878.2	43764	1	3	.11	C
4890.8	43634	5	5	.21	C
4891.5	43435	9	7	.30	C
4903.3	43634	3	5	.054	C
4919.0	43435	7	7	.17	C
4920.5	43163	11	9	.36	C
4966.1	47006	11	11	.037	C
4973.1	52040	3	3	.12	C
4989.0	53546	7	7	.058	C
5001.9	51294	9	7	.40	C
5014.9	51740	7	5	.31	C
5022.2	52040	5	3	.27	C
5074.7	53739	9	11	.15	C
5090.8	53967	7	5	.21	C
5121.6	54067	5	5	.086	C
5137.4	53155	11	9	.11	C
5208.6	45334	7	5	.060	C
5232.9	42816	9	11	.15	C
5242.5	48383	13	11	.032	C
5263.3	45334	5	5	.061	C
5266.6	43163	7	9	.088	C
5281.8	43435	5	7	.038	C
5302.3	45334	3	5	.073	C
5324.2	44677	9	9	.15	C
5339.9	45061	5	7	.071	C
5367.5	54237	7	9	.59	C
5370.0	53874	9	11	.48	C
5383.4	53353	11	13	.59	C
5393.2	44677	7	9	.037	C
5410.9	54555	7	9	.49	C
5415.2	53841	11	13	.68	C
5463.3	54067	9	9	.33	C
5473.9	51771	7	7	.057	C
5569.6	45509	5	3	.21	C
5572.8	45334	7	5	.22	C
5586.8	45061	9	7	.19	C
5615.6	44677	11	9	.17	C
5624.5	45334	5	5	.062	C
5658.8	45061	7	7	.042	C
5753.1	51740	3	5	.072	C
5763.0	51294	5	7	.10	C
		Fe II			
2029.2	65110	10	8	.076	D
2040.7	64832	10	10	.46	D
2051.0	65110	8	8	.42	D
2296.7	65110	10	8	.037	D
2303.3	64832	12	10	.054	D
2369.2	64832	10	10	.026	D
2379.0	64832	8	10	.064	D
2388.4	62662	10	12	.14	D
2433.5	62662	10	12	.091	D
2555.0	65110	8	8	.019	D
2559.8	65110	6	8	.22	D
2561.6	64832	10	10	.0081	D
2573.2	64832	8	10	.11	D
2591.5	46967	6	6	.52	C
2592.8	71433	14	16	2.25	C
2598.0	64832	10	10	.020	D
2598.4	38859	8	6	1.3	C
2599.4	38459	10	10	2.22	C
2623.1	70987	14	14	.092	D
2625.5	70987	12	14	2.04	C
2625.7	38459	8	10	.34	C
2664.7	64832	8	10	1.50	C
2666.6	65110	6	8	1.62	C
2684.9	62662	12	12	.0043	D
2712.4	62662	10	12	.11	D
2753.3	62662	10	12	1.71	C
2879.2	65110	10	8	.029	D
2902.5	64832	10	10	.038	D
2910.8	65110	8	8	.0055	D
3002.3	65110	6	8	.018	D
3044.8	64832	8	10	.011	D
3131.7	64832	12	10	.012	D
3162.8	65110	8	8	.042	D
3186.7	45044	4	4	.039	C
3187.3	64832	10	10	.028	D
3213.3	44785	4	6	.065	C
3277.3	38459	8	10	.0023	D
3360.1	62662	12	12	.0084	D
4515.3	45080	6	6	.0018	D
4520.2	44754	10	8	.0010	D
4583.8	44447	10	8	.0063	C
4629.3	44233	10	10	.0013	D
4923.9	43621	6	4	.030	C
4954.0	65110	6	8	.0016	D
5018.4	43239	6	6	.026	C
5019.5	64832	8	10	.0015	D
		Fe XVI			
50.350	1986100	2	4	2120.	B
50.555	1978040	2	2	2100.	B
62.879	1867530	2	2	1110.	B
63.719	1867530	4	2	2140.	B
335.41	298140	2	4	77.5	B
360.80	277160	2	2	62.3	B
		Fe XVIII			
93.931	1064610	4	2	690.	D
103.95	1064610	2	2	260.	D
		Fe XXI			
98.37	1265800	5	3	700.	D
99.43	1095500	9	3*	1000.	D
113.34	1132280	5	5	470.	D
		Fe XXIII			
132.83	752840	1	3	195.	B
		Fe XXIV			
10.619	9417100	2	4	7.28(+4)	B
10.663	9378200	2	2	7.51(+4)	B
11.124	9467100	6	10*	2.18(+5)	B
192.04	520720	2	4	43.	B
255.10	392000	2	2	18.	B
		Krypton			
		Kr I			
4274.0	103363	5	5	.026	D
4351.4	108822	3	1	.032	D
4362.6	102887	5	3	.0084	D
4376.1	103762	3	1	.056	D
4400.0	108568	3	5	.020	D
4410.4	108514	3	3	.0044	D
4425.2	108438	3	3	.0097	D
4453.9	103363	3	5	.0078	D
4463.7	103314	3	3	.023	D
4502.4	103121	3	5	.0092	D
5562.2	97945	5	5	.0028	D
5570.3	97919	5	3	.021	D
5649.6	102887	1	3	.0037	D
5870.9	97945	3	5	.018	D
6904.7	105648	3	5	.013	D
7224.1	105007	3	5	.014	D
7587.4	94093	3	1	.51	D
7601.5	93123	5	5	.31	D
7685.2	98855	3	1	.49	D
7694.5	92964	5	3	.056	D

Table 1b (Continued)
TRANSITION PROBABILITIES FOR SELECTED ATOMIC AND IONIC SPECIES

Wavelength (λ[Å])	Upper energy level (E_k[cm^{-1}])	Statistical weights g_i	g_k	Transition Probability (A[10^8 s^{-1}])	Uncertainty
		Kr I			
7854.8	97919	1	3	.23	D
8059.5	97596	1	3	.19	D
8104.4	92307	5	5	.13	D
8112.9	92294	5	7	.36	D
8190.1	93123	3	5	.11	D
8263.2	97945	3	5	.35	D
8281.1	97919	3	3	.19	D
8298.1	92964	3	3	.32	D
8508.9	97596	3	3	.24	D
8776.7	92307	3	5	.27	D
8928.7	91169	5	3	.37	D
		Kr II			
4250.6	141996	4	4	.12	D
4292.9	138381	4	4	.96	D
4355.5	135783	6	8	1.0	D
4431.7	140163	2	2	1.8	D
4436.8	140137	2	4	.66	D
4577.2	149705	6	8	.96	D
4583.0	157885	6	4	.76	D
4615.3	140137	4	4	.54	D
4619.2	140119	4	6	.81	D
4633.9	149173	4	6	.71	D
4658.9	134288	6	4	.65	D
4739.0	133926	6	6	.76	D
4762.4	141996	2	4	.42	D
4765.7	136071	4	6	.67	D
4811.8	138381	2	4	.17	D
4825.2	141723	2	4	.19	D
4832.1	135783	4	2	.73	D
5208.3	134288	4	4	.14	D
5308.7	133926	4	6	.024	D
7407.0	133926	6	6	.070	D
		Lead			
		Pb I			
2022.0	49440	1	3	.052	D
2053.3	48687	1	3	.12	D
2170.0	46068	1	3	1.5	D
2401.9	49440	3	3	.19	D
2446.2	48687	3	3	.25	D
2476.4	48189	3	5	.28	D
2577.3	49440	5	3	.50	D
2613.7	46068	3	3	.27	D
2614.2	46061	3	5	1.9	D
2628.3	48687	5	3	.031	D
2657.1	45443	3	5	9.8 (–4)	D
2663.2	48189	5	5	.71	D
2802.0	46329	5	7	1.6	D
2823.2	46061	5	5	.26	D
2833.1	35287	1	3	.58	D
2873.3	45443	5	5	.37	D
3572.7	49440	5	3	.99	D
3639.6	35287	3	3	.34	D
3671.5	48687	5	3	.44	D
3683.5	34960	3	1	1.5	D
3739.9	48189	5	5	.73	D
4019.6	46329	5	7	.035	D
4057.8	35287	5	3	.89	D
4062.1	46068	5	3	.92	D
4168.0	45443	5	5	.012	D
5005.4	49440	1	3	.27	D
5201.4	48687	1	3	.19	D
7229.0	35287	5	3	.0089	D
		Lithium			
		Li I			
2741.2	36470	2	6*	.013	D
3232.7	30925	2	6*	.012	B
4602.8	36623	2	4	.197	C
4602.9	36623	4	6	.24	C
6103.5	31283	2	4	.60	C
6103.7	31283	4	6	.71	C
6103.7	31283	4	4	.12	C
6707.8	14904	2	4	.372	B
6707.9	14904	2	2	.372	B
		Lutetium			
		Lu I			
3376.5	29608	4	4	2.23	B
3567.8	28020	4	6	.59	C
		Lu I			
3620.3	29608	6	4	.011	D
3841.2	28020	6	6	.25	C
4518.6	22125	4	4	.21	B
		Magnesium			
		Mg I			
2025.8	49347	1	3	.84	D
2779.8	57854	9	9*	5.2	C
2850.0	56968	9	15*	.23	C
2852.1	35051	1	3	4.95	B
3094.9	54192	9	15*	.52	C
3329.9	51873	1	3	.033	C
3332.2	51873	3	3	.097	C
3336.7	51873	5	3	.16	C
3835.3	47957	9	15*	1.68	B
4703.0	56308	3	5	.255	C
5167.3	41197	1	3	.116	B
5172.7	41197	3	3	.346	B
5183.6	41197	5	3	.575	B
5528.4	53135	3	5	.199	C
		Mg II			
1239.9	80650	2	4	.014	C
1240.4	80620	2	2	.014	C
2660.8	109062	10	14*	.38	D
2790.8	71491	2	4	4.0	C
2795.5	35761	2	4	2.6	C
2797.9	71491	4	4	.79	D
2798.1	71490	4	6	4.8	C
2802.7	35669	2	2	2.6	C
2928.8	69805	2	2	1.2	C
2936.5	69805	4	2	2.3	C
3104.8	103690	10	14*	.81	C
3848.2	97469	6	4	.028	C
3848.3	97469	4	4	.0030	D
3850.4	97455	4	2	.030	C
4481.2	93800	10	14*	2.23	C
9218.3	80650	2	4	.36	C
9244.3	80620	2	2	.36	C
		Mg IV			
320.99	311532	4	2	120.	D
323.31	311532	2	2	59.	D
1219.0	679098	6	6	5.9	D
1375.5	677361	4	4	4.5	D
1459.6	612231	6	4	4.6	D
1495.5	679098	4	6	6.4	D
1510.7	678426	4	4	6.7	D
1683.0	603137	6	8	5.8	D
1698.8	604002	4	6	3.9	D
1893.9	596521	6	6	2.8	D
		Mg VI			
269.92	424600	10	6*	310.	D
292.53	424600	6	6*	90.	D
314.64	400600	6	2*	180.	D
349.15	340600	10	10*	61.	C
387.94	340600	6	10*	13.	D
399.29	250445	4	2	28.	C
400.68	249578	4	4	28.	C
403.32	247945	4	6	27.	C
		Mg VII			
277.01	362128	3	3	95.	C
278.41	362128	5	3	150.	C
280.74	397700	5	3	200.	D
319.02	354900	5	5	89.	C
366.42	274922	9	9*	44.	C
433.04	232934	9	15*	16.	C
1334.3	1125850	5	5	5.3	C
1410.0	1196770	5	5	2.57	C
1487.0	1192185	3	5	3.02	C
1487.9	1193061	5	7	3.66	C
		Mg VIII			
74.976	1335965	6	10*	4300.	D
315.02	320742	4	4	120.	C
342.29	524437	10	6*	63.	D
353.86	414380	4	4	38.9	C
356.00	414380	6	4	57.	C
428.52	465654	10	10*	32.4	C
434.62	232290	6	10*	16.	C
489.33	524437	6	6*	39.	D

Table 1b (Continued)
TRANSITION PROBABILITIES FOR SELECTED ATOMIC AND IONIC SPECIES

Wavelength (λ[Å])	Upper energy level (E_k[cm^{-1}])	Statistical weights g_i	g_k	Transition Probability (A[10^8 s^{-1}])	Uncertainty
Mg VIII					
686.92	465654	6	10*	9.4	D
Mg IX					
62.751	1593600	1	3	2870.	B
67.189	1631500	9	15*	6200.	C
71.965	1532700	9	3*	1220.	C
72.312	1654583	3	5	4430.	C
77.737	1558076	3	1	392.	C
368.07	271687	1	3	52.7	B
438.69	499640	3	1	79.	C
443.74	368500	9	9*	41.9	C
749.55	405100	3	5	8.2	C
1639.8	1654583	3	5	2.1	D
2814.2	1593600	1	3	.335	C
Mg X					
57.876	1727832	2	4	2090.	B
57.920	1726519	2	2	2090.	B
63.152	1743410	2	4	5600.	B
63.295	1743880	4	6	6700.	B
609.79	163976	2	4	7.53	B
624.94	159929	2	2	7.01	B
2212.5	1727832	2	4	.964	B
2278.7	1726519	2	2	.882	B
5918.7	1743410	2	4	.0320	C
6229.6	1743880	4	6	.0330	C
Mg XI					
7.310	13680600	1	3	1.15(+4)	B
7.473	13381100	1	3	2.27(+4)	B
7.850	12738400	1	3	5.50(+4)	B
9.169	10907300	1	3	1.97(+5)	B
Manganese					
Mn I					
2794.8	35770	6	8	3.7	C
2798.3	35726	6	6	3.6	C
2801.1	35690	6	4	3.7	C
3007.7	58520	6	8	.18	D
3011.4	58486	8	10	.31	D
3016.5	58427	10	12	.29	D
3043.4	58137	8	8	.59	D
3044.6	49888	10	8	.57	D
3045.6	58110	10	10	.67	D
3045.8	58110	8	10	.17	D
3046.6	67753	10	12	.13	D
3047.0	58075	12	12	.61	D
3054.4	50013	8	6	.46	D
3066.0	49888	8	8	.16	D
3073.1	50099	4	4	.37	D
3082.7	66569	14	14	.29	D
3110.7	59340	6	8	.27	D
3113.8	66356	12	10	.26	D
3122.9	66356	10	10	.19	D
3126.9	66395	8	6	.23	D
3132.3	66855	10	10	.21	D
3132.8	66334	8	8	.27	D
3160.2	66574	10	12	.14	D
3175.6	66523	8	10	.18	D
3175.7	66419	10	12	.12	D
3190.0	66454	6	8	.16	D
3201.1	66395	4	6	.22	D
3212.9	48168	10	10	.16	D
3228.1	48021	10	12	.64	D
3230.2	65887	10	12	.19	D
3230.7	48226	8	8	.35	D
3238.7	65909	8	10	.12	D
3243.8	48271	6	6	.53	D
3256.1	48271	4	6	.50	D
3258.4	48318	2	2	.97	D
3260.2	48301	2	4	.38	D
3264.7	47904	8	10	.14	D
3267.8	64732	14	14	.35	D
3268.7	64410	6	8	.33	D
3270.4	64820	12	12	.26	D
3273.0	64888	10	10	.27	D
3278.6	47775	8	8	.0091	D
3298.2	57512	6	4	.28	D
3420.8	63364	14	14	.12	D
3463.7	66600	8	8	.32	C
3470.0	66600	6	8	.24	C
3511.8	65887	12	12	.27	C
Mn I					
3535.3	65909	10	10	.17	C
3559.8	65873	6	6	.21	C
3577.9	44994	10	8	.94	C
3601.3	65769	12	10	.23	C
3607.5	44994	8	8	.23	C
3608.5	45156	6	6	.36	C
3610.3	45259	4	4	.42	C
3635.7	65617	10	8	.21	C
3660.4	64732	12	14	.91	C
3677.0	64820	10	12	.73	C
3680.2	64585	12	10	.19	C
3682.1	64888	8	10	.76	D
3684.9	64920	6	8	.26	C
3706.1	61226	12	14	1.4	C
3718.9	61226	10	12	.96	C
3729.5	61744	10	12	.066	D
3731.9	61211	8	10	1.0	C
3746.6	60934	12	12	.16	C
3756.6	60956	10	10	.14	C
3767.7	60957	8	8	.14	C
3768.2	61469	10	12	.071	C
3771.4	69561	14	14	.19	C
3773.9	69630	12	12	.25	C
3800.6	57306	6	8	.27	C
3801.9	51561	12	12	.064	C
3806.7	43314	10	12	.38	C
3809.6	43524	8	8	.20	C
3823.5	43429	8	10	.44	C
3823.9	43596	6	6	.36	C
3833.9	43644	4	4	.52	C
3834.4	43524	6	8	.52	D
3839.8	43673	2	2	.58	C
3844.0	43644	2	4	.29	C
3872.1	63449	10	12	.077	C
3873.2	63548	8	10	.11	C
3889.5	68843	12	14	.31	C
3898.4	59470	6	8	.17	C
3899.3	60102	4	6	.24	C
3911.1	56562	6	6	.13	D
3919.3	66738	8	8	.088	D
3923.3	59732	12	10	.13	C
3924.1	56602	2	4	.94	D
3926.5	56462	6	8	.54	C
3929.7	59784	10	8	.092	C
3931.5	63548	10	10	.082	C
3936.8	59818	8	6	.12	C
3952.8	59117	6	6	.41	D
3975.9	59990	2	4	.18	D
3977.1	59600	4	6	.16	D
3980.1	66149	10	8	.13	D
3982.2	59568	4	2	.35	D
3982.6	50383	6	4	.23	D
3982.9	66504	6	4	.55	D
3986.8	50341	12	10	.11	D
3987.1	50359	10	8	.10	D
4003.3	62393	12	10	.11	D
4011.9	66149	8	8	.23	D
4018.1	41933	10	8	.33	C
4026.4	50095	12	14	.089	D
4030.8	24802	6	8	.19	C
4031.8	50081	10	12	.073	D
4033.1	24788	6	6	.18	C
4034.5	24779	6	4	.18	C
4041.4	41789	10	10	1.0	C
4048.8	42144	6	4	.75	C
4052.5	59784	6	8	.38	D
4055.5	41933	8	8	.61	C
4058.9	42199	4	2	1.0	C
4061.7	49415	8	6	.19	D
4063.5	42054	6	6	.22	C
4065.1	58843	12	14	.25	D
4066.2	65617	10	8	.22	D
4079.4	42144	2	4	.38	C
4082.9	42054	4	6	.37	C
4083.6	41933	6	8	.28	C
4089.9	58867	8	10	.17	D
4092.4	59470	8	8	.14	D
4099.4	65617	8	8	.11	D
4105.4	59290	10	8	.17	D
4107.9	61485	12	10	.097	D
4114.4	59340	8	8	.15	D

Table 1b (Continued)

TRANSITION PROBABILITIES FOR SELECTED ATOMIC AND IONIC SPECIES

Wavelength (λ[Å])	Upper energy level (E_k[cm^{-1}])	Statistical weights g_i	Statistical weights g_k	Transition Probability (A[10^8 s^{-1}])	Uncertainty
	Mn I				
4116.6	62075	6	6	.12	D
4125.8	65262	10	10	.070	D
4132.3	68716	8	10	.15	D
4135.0	58427	12	12	.30	D
4141.1	58486	10	10	.26	D
4147.5	61005	6	6	.066	D
4148.8	58520	8	8	.23	D
4176.6	58075	14	12	.21	D
4182.3	61913	12	14	.092	D
4190.0	58110	12	10	.20	D
4201.8	58137	10	8	.23	D
4220.6	57512	6	4	.16	D
4239.7	47299	4	2	.39	D
4257.7	47299	2	2	.37	D
4261.3	61469	12	12	.081	D
4265.9	47155	4	4	.35	D
4278.7	61485	10	10	.068	D
4281.1	46901	6	6	.23	D
4300.2	60668	12	10	.087	D
4381.7	61485	8	10	.14	D
4411.9	60668	12	10	.26	D
4414.9	45941	8	6	.18	D
4419.8	60739	10	8	.21	D
4451.6	45754	8	8	.71	D
4452.5	59617	14	14	.059	D
4455.8	47216	4	6	.17	D
4457.0	47218	6	4	.20	D
4457.6	47216	6	6	.38	D
4458.3	47212	6	8	.28	D
4461.1	47212	8	8	.17	D
4462.0	47207	8	10	.43	D
4464.7	45941	6	6	.26	D
4479.4	63548	8	10	.34	D
4498.9	45941	4	6	.11	D
4502.2	45754	6	8	.078	D
4503.9	59617	12	14	.083	D
4605.4	59828	10	12	.36	D
4626.5	59617	12	14	.36	D
4709.7	44523	8	8	.077	D
4727.5	44696	6	6	.084	D
4754.0	39431	6	8	.38	D
4761.5	44815	2	4	.28	D
4762.4	44289	8	10	.57	D
4765.9	44696	4	6	.28	D
4766.4	44523	6	8	.45	D
4783.4	39431	8	8	.39	D
4823.5	39431	10	8	.45	D
	Mn II				
2933.1	43557	5	3	1.7	C
2939.3	43485	5	5	1.8	C
2949.2	43370	5	7	1.7	C
3439.0	38543	5	7	.0041	D
3442.0	43370	9	7	.43	C
3460.3	43485	7	5	.32	C
3474.0	43370	7	7	.079	C
3474.1	43557	5	3	.15	C
3482.9	43485	5	5	.20	C
3488.7	43557	3	3	.25	C
3495.8	43557	1	3	.11	C
3496.8	43370	5	7	.016	C
3497.5	43485	3	5	.051	C
	Mn XXIII				
12.03	8687000	2	4	1.5 (+5)	B
12.158	8708000	4	6	1.8 (+5)	B
	Mercury				
	Hg I				
2536.5	39412	1	3	.13	D
2752.8	73961	1	3	.057	D
2856.9	74405	3	1	.012	D
2893.6	73961	3	3	.16	D
2925.4	78216	5	3	.077	D
2967.3	71336	1	3	.45	D
3125.7	71396	3	5	.51	D
3341.5	73961	5	3	.27	D
4046.6	62350	1	3	.18	D
4077.8	63928	3	1	.041	D
4108.1	78404	3	1	.030	D
4339.2	77108	3	5	.080	D
4358.3	62350	3	3	.40	D
4916.1	74405	3	1	.13	D
5460.7	62350	5	3	.56	D
5769.6	71396	3	5	.61	D
6234.4	79964	1	3	.0053	D
6716.4	78813	1	3	.0043	D
6907.5	76001	3	5	.020	D
7728.8	76863	1	3	.0097	D
	Neodymium				
	Nd II				
3780.4	30247	16	18	.14	D
3805.4	28857	14	16	.69	D
3807.2	26772	10	12	.049	D
3863.3	25877	8	10	.15	D
3941.5	25877	10	10	.61	D
3951.2	26772	12	12	.60	D
3973.3	30247	18	18	.63	D
3979.5	26772	10	12	.27	D
3990.1	28857	16	16	.52	D
4012.3	30002	18	20	.55	D
4061.1	28419	16	18	.44	D
4106.6	28857	14	16	.068	D
4109.5	26913	14	16	.37	D
4133.4	26772	14	12	.15	D
4156.1	25524	12	14	.34	D
4205.6	28857	18	16	.18	D
4284.5	28419	18	18	.085	D
4303.6	23230	8	10	.47	D
4325.8	26913	16	16	.16	D
4358.2	25524	14	14	.15	D
4382.7	25877	12	10	.040	D
4400.8	23230	10	10	.068	D
4451.6	25524	12	14	.25	D
4456.4	28419	16	18	.064	D
4463.0	26913	14	16	.18	D
4958.1	23230	12	10	.012	D
5130.6	30002	22	20	.16	D
5192.6	28419	20	18	.17	D
5249.6	26913	18	16	.18	D
5276.9	25877	12	10	.12	D
5293.2	25524	16	14	.12	D
5302.3	30247	20	18	.11	D
5311.5	26772	14	12	.11	D
5319.8	23230	12	10	.16	D
5357.0	28857	18	16	.18	D
5371.9	30002	20	20	.051	D
5485.7	28419	18	18	.057	D
5594.4	26913	16	16	.070	D
5620.6	30247	18	18	.13	D
5688.5	25524	14	14	.059	D
5718.1	28857	16	16	.087	D
5726.8	25877	10	10	.056	D
5740.9	26772	12	12	.072	D
5804.0	23230	10	10	.046	D
5865.1	28419	16	18	.013	D
6051.9	25877	12	10	.011	D
	Neon				
	Ne I				
615.63	162436	1	3	.38	C
618.67	161637	1	3	.93	C
619.10	161524	1	3	.33	C
626.82	159535	1	3	.74	C
629.74	158796	1	3	.48	C
735.90	135889	1	3	6.11	B
743.72	134459	1	3	.486	B
3369.8	163709	5	5	.0010	D
3369.9	163708	5	3	.0076	D
3375.6	163657	5	3	.0022	D
3417.9	163709	3	5	.0092	D
3418.0	163708	3	3	.0022	D
3423.9	163657	3	3	.0010	D
3447.7	163038	5	5	.021	D
3450.8	163013	5	3	.0049	D
3454.2	163401	3	1	.037	D
3460.5	163708	1	3	.0070	D
3464.3	162899	5	5	.0067	D
3466.6	163657	1	3	.013	D
3472.6	162831	5	7	.017	D
3498.1	163038	3	5	.0051	D
3501.2	163013	3	3	.012	D

Table 1b (Continued)
TRANSITION PROBABILITIES FOR SELECTED ATOMIC AND IONIC SPECIES

Wavelength (λ[Å])	Upper energy level (E_k[cm^{-1}])	Statistical weights g_i	g_k	Transition Probability (A[10^8 s^{-1}])	Uncertainty
		Ne I			
3510.7	162518	5	3	.0022	D
3515.2	162899	3	5	.0069	D
3520.5	164286	3	1	.093	D
3593.5	163709	3	5	.0099	D
3593.6	163708	3	3	.0066	D
3600.2	163657	3	3	.0043	D
3633.7	163401	3	1	.011	D
3682.2	163038	3	5	.0016	D
3685.7	163013	3	3	.0039	D
3701.2	162899	3	5	.0022	D
4536.3	170296	3	3	.0050	D
4702.5	169517	3	3	.0021	D
4708.9	169488	3	3	.042	D
4955.4	170296	3	3	.0033	D
5113.7	167808	3	3	.010	D
5120.5	170296	3	3	.0056	D
5154.4	169517	3	3	.019	D
5191.3	170296	3	3	.013	D
5326.4	167027	3	3	.0068	D
5333.3	169517	3	3	.0053	D
5341.1	166975	3	3	.11	D
5400.6	152971	3	1	.0090	B
5418.6	169488	3	3	.0052	D
5433.7	166657	3	3	.00283	B
5652.6	167808	3	3	.0089	D
5662.5	165913	3	3	.0069	D
5852.5	152971	3	1	.682	B
5868.4	167808	3	3	.014	D
5881.9	151038	5	3	.115	B
5913.6	167027	3	3	.048	D
5939.3	166657	5	3	.00200	B
5944.8	150859	5	5	.113	B
5961.6	167808	3	3	.033	D
5975.5	150772	5	3	.0351	B
6030.0	151038	3	3	.0561	B
6046.1	166657	3	3	.00226	B
6074.3	150917	3	1	.603	B
6096.2	150859	3	5	.181	B
6118.0	166657	5	3	.00609	B
6128.5	150772	3	3	.0067	B
6143.1	150316	5	5	.282	B
6150.3	167027	3	3	.015	D
6163.6	151038	1	3	.146	B
6217.3	150122	5	3	.0637	B
6266.5	150772	1	3	.249	B
6273.0	166975	3	3	.0097	D
6293.7	166657	3	3	.00639	B
6304.8	150316	3	5	.0416	B
6328.2	166657	5	3	.0339	B
6330.9	165913	3	3	.023	D
6334.4	149824	5	5	.161	B
6351.9	166657	1	3	.00345	B
6383.0	150122	3	3	.321	B
6401.1	166657	3	3	.0139	B
6402.2	149657	5	7	.514	B
6506.5	149824	3	5	.300	B
6532.9	150122	1	3	.108	B
6599.0	151038	3	3	.232	B
6602.9	165913	3	3	.0059	D
6652.1	150917	3	1	.0029	B
6678.3	150859	3	5	.233	B
6717.0	150772	3	3	.217	B
6721.1	165913	3	3	4.9 (-4)	D
6929.5	150316	3	5	.174	B
7024.1	150122	3	3	.0189	B
7032.4	148258	5	3	.253	B
7051.3	162436	3	3	.030	D
7059.1	162420	3	5	.068	C
7173.9	149824	3	5	.0287	B
7245.2	148258	3	3	.0935	B
7304.8	166657	1	3	.00255	B
7438.9	148258	1	3	.0231	B
7472.4	161637	3	3	.040	D
7535.8	161524	3	3	.43	D
7937.0	162420	5	5	.0078	C
8082.5	148258	3	3	.0012	B
8118.5	162436	3	3	.049	D
8128.9	162420	3	5	.0072	C
8259.4	162420	5	5	.0203	C
8571.4	162436	3	3	.055	D
8582.9	162420	3	5	.0100	C
		Ne I			
8647.0	162420	5	5	.0391	C
8681.9	161637	3	3	.21	D
8767.5	161524	3	3	.0011	D
8771.7	162436	3	3	.16	D
8783.8	162420	3	5	.313	C
8865.3	159535	3	3	.0094	D
9201.8	161637	3	3	.091	D
9433.0	161637	3	3	.0011	D
9486.7	158796	3	3	.025	D
9534.2	161524	3	3	.063	D
10621.	159535	3	3	.0024	D
11409.	159535	3	3	.042	D
11525.	158796	3	3	.084	D
11767.	159535	3	3	.069	D
12459.	158796	3	3	.015	D
		Ne II			
357.03	280351	6	10*	38.	C
361.77	276677	6	2*	16.	C
406.28	246395	6	10*	18.	C
446.37	224291	6	6*	40.7	C
460.73	217048	4	2	47.	B
462.39	217048	2	2	23.	B
1907.5	276512	4	2	.28	D
1916.1	276277	4	4	.69	D
1930.0	276512	2	2	.57	D
1938.8	276277	2	4	.13	D
2858.0	281171	6	6	.79	D
2870.0	281026	6	6	.17	D
2873.0	280990	6	4	.38	D
2876.3	281171	4	6	.78	D
2876.5	280947	6	4	.33	D
2878.1	281333	2	2	.069	D
2888.4	281026	4	6	.070	D
2891.5	280990	4	4	.061	D
2897.0	280701	6	8	.052	D
2906.8	280990	2	4	.55	D
2910.1	280769	4	2	1.7	D
2910.4	280947	2	4	.59	D
2916.2	280474	6	4	.096	D
2925.6	280769	2	2	.56	D
2933.7	280269	6	6	.069	D
2955.7	252954	6	4	1.2	D
3001.7	252954	4	4	.87	D
3017.3	279325	6	4	.35	D
3027.0	279219	6	6	1.4	D
3028.7	279423	4	2	.85	D
3028.9	252954	2	4	.47	D
3034.5	279138	6	8	3.1	D
3037.7	279325	4	4	2.1	D
3045.6	279423	2	2	2.5	D
3047.6	279219	4	6	1.8	D
3054.7	279325	2	4	.94	D
3092.9	306687	6	6	1.3	D
3097.1	306688	8	8	1.3	D
3118.0	281171	8	6	.042	D
3134.1	306263	6	4	.26	D
3140.4	306243	8	6	.24	D
3151.1	281171	6	6	.048	D
3154.8	280797	8	6	.018	D
3164.4	280701	8	8	.16	D
3165.7	281026	6	6	.12	D
3173.6	280947	6	4	.045	D
3176.1	281171	4	6	.060	D
3187.6	251011	4	6	.014	D
3188.7	280797	6	6	.39	D
3190.9	281026	4	6	.15	D
3194.6	280990	4	4	.52	D
3198.6	280701	6	8	1.7	D
3198.9	280947	4	4	.23	D
3209.0	280262	8	8	.16	D
3209.4	280990	2	4	.60	D
3213.7	280947	2	4	1.7	D
3214.3	280797	4	6	2.2	D
3218.2	280173	8	10	3.6	D
3224.8	305365	6	8	3.5	D
3229.5	305365	8	8	.13	D
3229.6	305364	8	10	3.6	D
3230.1	277344	6	6	1.8	D
3230.4	277344	4	6	.14	D
3232.0	277326	6	4	.27	D

Table 1b (Continued)
TRANSITION PROBABILITIES FOR SELECTED ATOMIC AND IONIC SPECIES

Wavelength (λ[Å])	Upper energy level (E_k[cm^{-1}])	Statistical weights g_i	g_k	Transition Probability (A[10^8 s^{-1}])	Uncertainty
		Ne II			
3232.4	277326	4	4	1.6	D
3243.4	280269	6	6	.23	D
3244.1	280262	6	8	1.5	D
3248.1	280474	4	4	.24	D
3255.4	281720	6	4	.038	D
3263.4	280474	2	4	.39	D
3269.9	280269	4	6	.31	D
3270.8	249696	6	4	.057	D
3297.7	249446	6	6	.43	D
3309.7	254292	4	2	.31	D
3310.5	281720	4	4	.069	D
3311.3	249840	4	2	.26	D
3314.7	281171	6	6	.044	D
3319.7	276512	4	2	1.6	D
3320.2	279219	8	6	.21	D
3323.7	254165	4	4	1.6	D
3327.2	249696	4	4	.91	D
3329.2	279138	8	8	.88	D
3330.7	281026	6	6	.039	D
3334.8	249109	6	8	1.8	D
3336.1	306243	4	6	1.1	D
3344.4	249840	2	2	1.5	D
3345.5	276277	6	4	1.4	D
3345.8	276277	4	4	.22	D
3353.6	281333	4	2	.12	D
3355.0	249446	4	6	1.3	D
3356.3	280797	6	6	.20	D
3357.8	279219	6	6	.50	D
3360.3	306263	2	4	.86	D
3360.6	249696	2	4	.82	D
3362.9	279423	4	2	.35	D
3371.8	281171	4	6	.22	D
3374.1	279325	4	4	.30	D
3378.2	254292	2	2	1.7	D
3379.3	279423	2	2	.30	D
3386.2	279219	4	6	.055	D
3388.4	281026	4	6	2.2	D
3390.6	279325	2	4	.077	D
3392.8	254165	2	4	.44	D
3404.8	306687	4	6	1.9	D
3407.0	306688	6	8	2.3	D
3411.4	305582	4	2	.61	D
3413.2	305567	4	4	1.8	D
3414.9	280797	4	6	.018	D
3416.9	280269	6	6	.64	D
3417.7	280262	6	8	1.6	D
3438.9	305582	2	2	1.4	D
3440.7	305567	2	4	.35	D
3453.1	280474	4	4	.46	D
3454.8	306263	4	4	1.6	D
3456.6	281720	2	4	.96	D
3457.1	306243	4	6	.099	D
3459.3	306243	6	6	1.6	D
3475.2	281720	4	4	.012	D
3477.6	280269	4	6	.43	D
3481.9	252798	4	2	1.4	D
3503.6	281333	2	2	2.0	D
3522.7	281333	4	2	.023	D
3538.0	305582	4	2	.76	D
3539.9	305567	4	4	.036	D
3542.2	305567	6	4	.60	D
3542.9	281171	4	6	1.2	D
3546.2	280990	2	4	.063	D
3551.6	280947	2	4	.037	D
3557.8	252798	2	2	.19	D
3561.2	281026	4	6	.21	D
3565.8	280990	4	4	.62	D
3568.5	274409	6	8	1.4	D
3571.2	280947	4	4	.63	D
3574.2	274365	6	6	.10	D
3574.6	274365	4	6	1.3	D
3590.4	280797	4	6	.036	D
3594.2	280769	4	2	1.3	D
3612.3	280474	2	4	.26	D
3628.0	281720	4	4	.60	D
3632.7	280474	4	4	.13	D
3643.9	251522	4	4	.32	D
3644.9	281720	2	4	.99	D
3659.9	280269	4	6	.067	D
3664.1	246415	6	4	.70	D
3679.8	281333	4	2	.32	D
3694.2	246192	6	6	1.0	D
3697.1	281333	2	2	.28	D
3701.8	281171	4	6	.27	D
3709.6	246598	4	2	1.1	D
3713.1	251011	4	6	1.3	D
3721.8	281026	4	6	.20	D
3726.9	280990	4	4	.12	D
3727.1	251522	2	4	.98	D
3734.9	246415	4	4	.19	D
3744.6	280990	2	4	.26	D
3751.2	246598	2	2	.18	D
3753.8	280797	4	6	.45	D
3766.3	246192	4	6	.29	D
3777.1	246415	2	4	.42	D
3800.0	280474	4	4	.37	D
3818.4	280474	2	4	.61	D
3829.8	280269	4	6	.84	D
3942.3	249446	4	6	.010	D
		Ne V			
142.61	701945	9	9*	670.	C
143.32	698517	9	15*	1200.	C
147.13	709956	5	7	1500.	C
151.23	691540	5	5	338.	C
154.50	711210	1	3	700.	C
167.69	597083	9	9*	150.	C
358.93	279365	9	3*	210.	C
365.59	303812	5	3	135.	C
482.15	208161	9	9*	30.1	C
571.04	175876	9	15*	10.	C
2259.6	640868	3	5	1.9	D
2265.7	641646	5	7	2.4	D
		Ne VII			
97.502	1025600	1	3	1070.	B
115.46	978320	9	3*	480.	C
116.69	1071920	3	5	1600.	C
127.66	998280	3	1	190.	C
465.22	214952	1	3	40.9	B
558.61	290740	3	5	8.11	B
559.95	289850	1	3	10.7	B
561.38	289850	3	3	7.99	B
561.73	290740	5	5	23.9	B
562.99	289340	3	1	31.7	B
564.53	289850	5	3	13.1	B
		Ne VIII			
88.09	1135000	2	6*	840.	B
98.208	1147500	6	10*	2770.	B
770.41	129800	2	4	5.90	B
780.32	128150	2	2	5.69	B
2820.7	1135000	2	4	.720	B
2860.1	1135000	2	2	.688	B
		Nickel			
		Ni I			
1976.9	50790	7	9	1.1	C
1990.3	51125	5	7	.83	C
2007.0	50689	5	5	.17	C
2014.3	51344	3	5	.93	C
2026.6	49328	9	7	.24	C
2047.4	49033	7	7	.13	C
2055.5	50851	5	3	.33	C
2124.8	50458	5	3	.38	C
2147.8	47425	5	3	.47	C
2158.3	46523	7	5	.69	C
2190.2	46523	5	5	.30	C
2197.4	47208	3	3	.78	C
2201.6	48818	5	3	.73	C
2244.5	45419	5	5	.38	C
2253.6	44565	7	7	.19	C
2258.2	44475	7	5	.17	C
2290.0	43655	9	7	2.1	C
2300.8	43655	7	7	.75	C
2303.0	45122	3	3	.45	C
2312.3	44565	7	7	5.5	C
2317.2	44475	7	5	3.8	C
2320.0	43090	9	11	6.9	C
2325.8	44315	7	9	3.5	C
2330.0	45122	5	3	5.3	C
2345.5	42621	9	7	2.2	C
2943.9	34163	7	5	.11	C

Table 1b (Continued)
TRANSITION PROBABILITIES FOR SELECTED ATOMIC AND IONIC SPECIES

Wavelength (λ[Å])	Upper energy level (E_k[cm^{-1}])	Statistical weights g_i	g_k	Transition Probability (A[10^8 s^{-1}])	Uncertainty
		Ni I			
3012.0	36601	5	5	1.5	C
3037.9	33112	7	7	.32	C
3064.6	33501	5	7	.075	C
3080.8	34163	3	5	.093	C
3101.6	33112	5	7	.72	C
3101.9	35639	5	7	.49	C
3134.1	33611	3	5	.71	C
3225.0	34409	5	3	.11	C
3233.0	30923	9	11	.053	C
3271.1	31442	5	5	.0072	C
3315.7	31031	5	7	.053	C
3320.3	31442	7	5	.048	C
3361.6	30619	5	5	.045	C
3369.6	29669	9	7	.17	C
3380.6	32982	5	3	1.2	C
3393.0	29669	7	7	.24	C
3413.5	30619	7	5	.038	C
3414.8	29481	7	9	.55	C
3423.7	30913	3	3	.35	C
3433.6	29321	7	7	.17	C
3446.3	29889	5	5	.44	C
3452.9	29833	5	7	.098	C
3458.5	30619	3	5	.61	C
3461.7	29084	7	9	.27	C
3472.5	29669	5	7	.12	C
3483.8	30913	5	3	.14	C
3493.0	29501	5	3	.98	C
3510.3	30192	3	1	1.2	C
3515.1	29321	5	7	.44	C
3519.8	30619	5	5	.041	C
3524.5	28569	7	5	1.0	C
3566.4	31442	5	5	.56	C
3571.9	29321	7	7	.052	C
3597.7	29501	3	3	.14	C
3610.5	28569	5	5	.072	C
3619.4	31031	5	7	.73	C
3664.1	29501	5	3	.019	D
3807.1	29669	5	7	.043	C
3831.7	29501	5	3	.015	C
3858.3	29321	5	7	.069	C
4295.9	54251	9	7	.17	D
4401.6	48467	9	11	.38	D
4714.4	48467	13	11	.46	D
4786.5	48467	11	11	.18	D
4817.9	54251	7	7	.070	D
4838.7	54251	9	7	.22	D
4855.4	49159	5	5	.57	D
4980.2	49159	9	11	.19	D
5017.6	48467	11	11	.20	D
5080.5	49159	9	11	.32	D
5129.4	49159	7	5	.12	D
5371.3	54251	7	7	.16	D
5664.0	54251	5	7	.11	D
6176.8	49159	9	11	.047	D
6314.7	31442	5	5	.0057	D
6767.8	29501	1	3	.0033	D
7714.3	28569	5	5	.0014	D
		Ni II			
2053.3	57081	10	8	.041	D
2080.8	58706	4	4	.13	D
2090.1	58493	4	6	.11	D
2093.6	57081	8	8	.11	D
2125.1	56371	8	8	.10	D
2125.9	55418	10	8	.073	D
2128.6	57081	6	8	.40	D
2138.6	56075	8	6	.28	D
2158.7	56424	6	4	.57	D
2161.2	56371	6	8	.32	D
2165.6	54557	10	10	3.8	D
2169.1	55418	8	8	2.3	D
2174.7	55300	8	10	2.4	D
2175.2	56075	6	6	2.8	D
2184.6	56424	4	4	4.7	D
2188.1	55019	8	6	.090	D
2201.4	56075	4	6	2.1	D
2206.7	55418	6	8	2.5	D
2210.4	54557	8	10	.64	D
2216.5	53496	10	12	5.5	D
2220.4	68131	6	8	3.7	D
2223.0	53365	10	10	1.6	D
2224.4	58493	8	6	.51	D
2224.9	54263	8	8	2.5	D
2226.3	55019	6	6	2.0	D
2253.9	55019	4	6	3.2	D
2264.5	54263	6	8	2.4	D
2270.2	53365	8	10	2.5	D
2278.8	57420	8	6	4.5	D
2287.1	58706	6	4	4.5	D
2296.6	57081	8	8	3.2	D
2297.1	53635	6	4	4.6	D
2297.5	54176	4	2	5.3	D
2298.3	58493	6	6	4.5	D
2303.0	52738	8	6	4.7	D
2316.0	51558	10	8	4.9	D
2326.4	53635	4	4	1.0	D
2334.6	56371	8	8	1.3	D
2356.4	57420	6	6	.45	D
2367.4	51558	8	8	.13	D
2375.4	57081	6	8	1.1	D
2387.8	55418	8	8	.23	D
2394.5	55300	8	10	2.9	D
2413.0	56424	6	4	.13	D
2416.1	56371	6	8	3.3	D
2433.6	56075	6	6	.12	D
2437.9	54557	8	10	.87	D
2510.9	53365	8	10	.94	D
2545.9	54263	6	8	.26	D
		Nitrogen			
		N I			
1163.9	105144	6	6	.43	D
1164.0	105144	4	6	.032	D
1164.2	105120	6	4	.048	D
1164.3	105120	4	4	.43	D
1167.4	104881	6	8	1.1	D
1168.4	104810	6	6	.095	D
1168.5	104810	4	6	1.3	D
1169.7	104717	6	8	.030	D
1176.5	104222	6	4	.95	D
1176.6	104222	4	4	.11	D
1177.7	104145	4	2	1.3	D
1199.6	83365	4	6	5.5	D
1200.2	83318	4	4	5.3	D
1200.7	83284	4	2	5.5	D
1310.5	105144	4	6	1.3	D
1316.3	104810	4	6	.025	D
1492.6	86221	6	4	5.3	D
1492.8	86221	4	4	.58	D
1494.7	86137	4	2	5.0	D
4099.9	110521	2	4	.034	D
4110.0	110545	4	6	.040	D
4114.0	110521	4	4	.0068	D
4137.6	107446	2	4	.0039	D
4143.4	107446	4	4	.0078	D
4151.5	107446	6	4	.013	D
4214.8	107037	4	6	.022	D
4216.1	106996	2	4	.031	D
4218.9	106980	2	2	.012	D
4222.1	106996	4	4	.0098	D
4223.1	107037	6	6	.051	D
4224.9	106980	4	2	.061	D
4230.5	106996	6	4	.033	D
4385.5	120566	2	2	.0052	C
4392.4	120566	4	2	.0102	C
4914.9	106478	2	2	.00759	B
4935.1	106478	4	2	.0158	B
5169.6	107446	6	4	.00209	C
5181.4	107446	4	4	.00144	C
5186.6	107446	2	4	7.3 (−4)	C
5199.8	112808	2	2	.023	D
5201.6	112801	2	4	.023	D
5281.2	107037	6	6	.00282	C
5292.7	106996	6	4	.00167	C
5293.5	107037	4	6	.00113	C
5309.4	106980	4	2	.00273	C
5310.5	106996	2	4	.00137	C
5344.0	106814	6	6	6.2 (−4)	C
5356.6	106814	4	6	.00189	C
5367.0	106778	4	4	.00118	C
5372.6	106778	2	4	.00107	C
5378.3	106759	2	2	.00210	C

Table 1b (Continued)
TRANSITION PROBABILITIES FOR SELECTED ATOMIC AND IONIC SPECIES

N I

Wavelength (λ[Å])	Upper energy level (E_k[cm^{-1}])	Statistical weights g_i	Statistical weights g_k	Transition Probability (A[10^8 s^{-1}])	Uncertainty
5816.5	112681	4	6	.00278	C
5829.5	112681	6	6	.0064	C
5834.6	112610	2	4	.00383	C
5840.9	112610	4	4	.00122	C
5849.7	112565	2	2	.00152	C
5854.0	112610	6	4	.00409	C
5856.0	112565	4	2	.0076	C
6606.2	109927	4	6	7.9 (–4)	C
6622.5	109927	6	6	.0071	C
6627.0	109857	2	4	.00197	C
6636.9	109857	4	4	.0125	C
6645.0	109927	8	6	.0311	C
6646.5	109812	2	2	.0194	C
6653.5	109857	6	4	.0244	C
6656.5	109812	4	2	.0193	C
6926.7	109927	4	6	.0064	C
6945.2	109927	6	6	.0149	C
6951.6	109857	2	4	.0088	C
6960.5	109857	4	4	.00281	C
6973.1	109812	2	2	.00350	C
6979.2	109857	6	4	.0094	C
6982.0	109812	4	2	.0174	C
7423.6	96751	2	4	.052	C
7442.3	96751	4	4	.106	C
7468.3	96751	6	4	.161	C

N II

Wavelength (λ[Å])	Upper energy level (E_k[cm^{-1}])	Statistical weights g_i	Statistical weights g_k	Transition Probability (A[10^8 s^{-1}])	Uncertainty
474.89	210705	5	5	4.5	D
475.65	210240	1	3	14.	D
475.70	210266	3	5	20.	D
475.76	210240	3	3	11.	D
475.80	210302	5	7	26.	D
475.88	210266	5	5	6.8	D
508.70	196712	5	5	2.8	D
510.76	211104	5	7	31.	D
513.85	209926	5	5	6.8	D
529.36	188909	1	3	6.5	D
529.41	188937	3	1	20.	D
529.49	188909	3	3	4.9	D
529.64	188857	3	5	4.9	D
529.72	188909	5	3	8.1	D
529.87	188857	5	5	15.	D
533.51	187438	1	3	20.	D
533.58	187462	3	5	27.	D
533.65	187438	3	3	15.	D
533.73	187492	5	7	36.	D
533.82	187462	5	5	9.1	D
547.82	197859	5	3	5.2	D
559.76	211336	1	3	12.	D
574.65	189335	5	7	35.	D
582.16	187091	5	5	13.	D
635.20	190120	1	3	18.	D
644.63	155127	1	3	12.	C
644.84	155127	3	3	35.	C
645.18	155127	5	3	58.	C
660.29	166766	5	3	40.	C
671.02	149077	3	5	2.9	D
671.39	149077	5	5	8.9	D
671.41	148940	1	3	3.6	D
671.63	148940	3	3	2.7	D
671.77	148909	3	1	12.	D
672.00	148940	5	3	4.4	D
745.84	166766	1	3	10.	C
746.98	149188	5	3	40.	C
748.37	148940	5	3	2.0	D
775.97	144188	5	5	35.	C
915.61	109217	1	3	3.6	C
915.96	109224	3	1	11.	C
1084.0	92252	1	3	2.0	C
1085.5	92250	5	5	.90	C
1085.7	92237	5	7	3.6	C
2139.5	211336	3	3	.29	D
3593.6	196712	3	5	.24	D
3609.1	196592	3	3	.24	D
3615.9	196540	3	1	.24	D
3829.8	196712	3	5	.15	D
3838.4	196712	5	5	.45	D
3842.2	196592	1	3	.20	D
3847.4	196592	3	3	.15	D
3855.1	196540	3	1	.60	D
3856.1	196592	5	3	.25	D
3919.0	190120	3	3	1.00	C
3995.0	174212	3	5	1.3	D
4114.4	188909	3	3	.0019	D
4447.0	187091	3	5	1.30	C
4477.7	188909	5	3	.035	D
4507.6	188857	7	3	.038	D
4601.5	170666	3	5	.270	C
4607.2	170608	1	3	.340	C
4613.9	170608	3	3	.196	C
4621.4	170573	3	1	.90	C
4630.5	170666	5	5	.84	C
4643.1	170608	5	3	.466	C
4774.2	187462	3	5	.054	C
4779.7	187438	3	3	.269	C
4781.2	187492	5	7	.040	C
4788.1	187462	5	5	.248	C
4793.7	187438	5	3	.089	C
4803.3	187492	7	7	.313	C
4810.3	187462	7	5	.055	C
4987.4	188937	3	1	.63	C
4994.4	188909	3	3	.74	C
5001.1	186512	3	5	1.02	C
5001.5	186571	5	7	1.08	C
5002.7	168892	1	3	.085	C
5005.2	186652	7	9	1.22	C
5007.3	188857	3	5	.77	C
5010.6	168892	3	3	.268	C
5025.7	186571	7	7	.134	C
5040.7	186512	7	5	.0053	C
5045.1	168892	5	3	.410	C
5452.1	188909	1	3	.14	D
5454.2	188937	3	1	.41	D
5462.6	188909	3	3	.10	D
5478.1	188857	3	5	.10	D
5480.1	188909	5	3	.17	D
5495.7	188857	5	5	.30	D
5666.6	166582	3	5	.423	C
5676.0	166522	1	3	.310	C
5679.6	166679	5	7	.56	C
5686.2	166522	3	3	.231	C
5710.8	166582	5	5	.137	C
5927.8	187438	1	3	.315	C
5931.8	187462	3	5	.425	C
5940.2	187438	3	3	.235	C
5941.7	187492	5	7	.56	C
5952.4	187462	5	5	.140	C
6482.1	164611	3	3	.37	D
6610.6	189335	5	7	.59	D

N III

Wavelength (λ[Å])	Upper energy level (E_k[cm^{-1}])	Statistical weights g_i	Statistical weights g_k	Transition Probability (A[10^8 s^{-1}])	Uncertainty
374.20	267238	2	4	101.	C
451.87	221302	2	2	8.9	C
452.23	221302	4	2	17.8	C
685.00	145986	2	4	9.3	C
685.51	145876	2	2	37.1	C
685.82	145986	4	4	46.8	C
686.34	145876	4	2	19.0	C
763.34	131004	2	2	9.6	C
764.36	131004	4	2	18.7	C
771.54	186797	2	4	8.2	C
771.90	186797	4	4	16.5	C
772.39	186797	6	4	24.7	C
772.89	230409	6	4	20.3	C
772.98	230404	4	2	22.7	C
979.84	203089	4	4	8.9	C
979.92	203075	6	6	9.3	C
989.79	101031	2	4	4.15	C
991.51	101031	4	4	.82	C
991.58	101024	4	6	4.96	C
1747.8	203089	2	4	1.31	C
1751.2	203089	4	4	.258	C
1751.7	203075	4	6	1.57	C
4097.3	245701	2	4	.82	C
4103.4	245665	2	2	.82	C
4634.1	267238	2	4	.65	C
4640.6	267244	4	6	.78	C
4641.9	267238	4	4	.130	C

N IV

Wavelength (λ[Å])	Upper energy level (E_k[cm^{-1}])	Statistical weights g_i	Statistical weights g_k	Transition Probability (A[10^8 s^{-1}])	Uncertainty
247.20	404522	1	3	114.	C
283.52	420053	9	15*	290.	C
322.64	377285	9	3*	84.	C

Table 1b (Continued)
TRANSITION PROBABILITIES FOR SELECTED ATOMIC AND IONIC SPECIES

Wavelength (λ[Å])	Upper energy level (E_k[cm^{-1}])	Statistical weights g_i	g_k	Transition Probability (A[10^8 s^{-1}])	Uncertainty
		N IV			
335.05	429160	3	5	185.	C
387.35	388855	3	1	28.	D
765.15	130694	1	3	24.0	B
923.16	175669	9	9*	18.	B
955.34	235369	3	1	30.	C
1718.6	188883	3	5	2.37	C
3480.8	406005	3	9*	1.1	C
4057.8	429160	3	5	.68	C
6380.8	404522	1	3	.14	B
7116.7	420053	9	15*	.12	C
		N V			
209.29	477817	2	6*	118.	B
247.66	484418	6	10*	430.	C
1238.8	80722	2	4	3.41	B
1242.8	80463	2	2	3.38	B
4603.7	477842	2	4	.412	B
4620.0	477766	2	2	.408	B
		N VI			
24.898	4016390	1	3	5158.	AA
28.787	3473840	1	3	180.9(+2)	AA
161.22	4006180	3	9*	285.9	AA
173.29	4016390	1	3	269.7	AA
173.92	4013460	9	15*	876.	A
185.19	4013820	3	5	824.	A
1901.5	3438490	3	9*	.6777	AA
2896.4	3473840	1	3	.2080	AA
		Oxygen			
		O I			
1028.2	97488	1	3	.20	D
1152.2	102662	5	5	5.5	D
1217.6	115918	1	3	1.8	C
1302.2	76795	5	3	3.3	C
1304.9	76795	3	3	2.0	C
1306.0	76795	1	3	.66	C
5435.2	105019	3	5	.0061	C
5435.8	105019	5	5	.0102	C
5436.9	105019	7	5	.0142	C
6453.6	102117	3	5	.0142	B
6454.4	102117	5	5	.0237	B
6456.0	102117	7	5	.0331	B
6653.8	130943	3	1	.600	B
7156.7	116631	5	5	.473	B
7471.4	127292	5	3	.0114	B
7473.2	127288	5	5	.102	B
7476.4	127283	5	7	.408	B
7477.2	127292	3	3	.170	B
7479.1	127288	3	5	.306	B
7480.7	127292	1	3	.226	B
7771.9	86631	5	7	.340	B
7774.2	86628	5	5	.340	B
7775.4	86626	5	3	.340	B
7886.3	128595	3	5	.370	B
7939.5	113727	7	5	.00165	C
7943.2	113721	7	7	.0417	C
7947.2	113727	5	5	.058	C
7947.6	113714	7	9	.373	C
7950.8	113721	5	7	.331	C
7952.2	113727	3	5	.313	C
7981.9	101155	3	3	.12	D
7982.4	101155	1	3	.16	D
7987.0	101148	3	5	.21	D
7987.3	101148	5	5	.072	D
7995.1	101135	5	7	.29	D
		O II			
429.92	232603	4	2	39.	D
430.04	232536	4	4	39.	D
430.18	232463	4	6	39.	D
483.75	233544	4	2	.84	D
483.98	233430	6	4	.76	D
484.03	233430	4	4	.084	D
485.09	232959	6	8	25.	D
485.47	232796	6	6	1.6	D
485.52	232796	4	6	23.	D
3007.1	265999	8	10	.84	C
3007.7	265985	6	8	.72	C
3013.4	265639	6	8	.74	C
3032.1	265930	8	10	.85	C
3032.5	265763	6	8	.82	C
3134.8	238893	8	6	1.23	C
3273.5	259286	8	6	1.14	C
3377.2	233544	2	2	1.88	C
3390.3	233430	2	4	1.86	C
3407.4	259286	6	6	.75	C
3749.5	212162	6	4	.90	C
3882.2	232754	8	8	.493	C
3912.0	232527	6	4	1.27	C
3919.3	232480	4	2	1.40	C
3973.3	214229	4	4	1.27	C
4069.6	231296	2	4	1.39	C
4069.9	231350	4	6	1.49	C
4072.2	231428	6	8	1.70	C
4075.9	231530	8	10	1.98	C
4085.1	231350	6	6	.478	C
4087.2	255756	4	6	2.24	C
4089.3	255978	10	12	2.62	C
4095.6	255759	6	8	2.23	C
4097.2	255828	8	10	2.37	C
4104.7	232748	4	6	1.04	C
4105.0	232746	4	4	.80	C
4108.8	255759	8	8	.349	C
4119.2	232754	6	8	1.48	C
4120.3	232748	6	6	.443	C
4132.8	232536	2	4	.84	C
4153.3	232463	4	6	.77	C
4276.7	256123	6	8	1.82	C
4277.4	256084	2	4	1.49	C
4277.9	256123	8	8	.302	C
4281.4	255813	6	6	.60	C
4282.8	255913	4	4	1.06	C
4283.0	256088	4	6	1.58	C
4283.1	256088	6	6	.51	C
4283.8	256084	4	4	.59	C
4294.8	255813	4	6	1.39	C
4303.8	255691	6	8	1.97	C
4328.6	255622	4	2	1.21	C
4340.4	255829	6	8	2.23	C
4347.4	229968	4	4	.94	C
4349.4	208484	6	6	.74	C
4351.3	229947	6	6	.97	C
4396.0	234454	6	6	.398	C
4414.9	211713	4	6	1.15	C
4417.0	211522	2	4	.95	C
4443.1	251224	6	6	.57	C
4448.2	251221	8	8	.57	C
4489.5	255812	2	4	1.51	C
4491.3	255690	4	6	1.81	C
4596.2	228723	4	6	1.03	C
4602.1	256126	4	6	1.70	C
4609.4	256143	6	8	1.82	C
4641.8	206878	4	6	.79	C
4661.6	206786	4	4	.52	C
4701.2	253792	4	4	.87	C
4703.2	251224	4	6	.82	C
4705.4	232959	6	8	1.38	C
4871.6	253048	4	6	.435	C
4906.9	232536	4	4	.68	C
4924.6	232463	4	6	.67	C
4941.1	234402	2	4	.83	C
4943.1	234454	4	6	1.06	C
5206.7	233430	4	4	.391	C
6627.6	248514	4	4	.089	C
6666.9	248425	4	2	.0349	C
6678.2	248514	2	4	.0173	C
6718.1	248425	2	2	.068	C
6721.4	203942	4	2	.189	C
6810.6	246029	6	8	.00180	C
6844.1	245903	4	6	.00325	C
6847.0	246029	8	8	.0347	C
6869.7	245903	6	6	.059	C
6885.1	245816	4	4	.067	C
6895.3	246029	10	8	.298	C
6906.5	245903	8	6	.272	C
6908.1	245768	4	2	.332	C
6910.8	245816	6	4	.267	C
		O III			
262.88	380706	5	5	40.	D
263.69	379232	1	3	52.	D
263.73	379293	3	5	73.	D

Table 1b (Continued)
TRANSITION PROBABILITIES FOR SELECTED ATOMIC AND IONIC SPECIES

Wavelength (λ[Å])	Upper energy level (E_k[cm^{-1}])	Statistical weights g_i	g_k	Transition Probability (A[10^8 s^{-1}])	Uncertainty
O III					
263.77	379232	3	3	40.	D
263.82	379356	5	7	96.	D
263.86	379293	5	5	24.	D
277.38	380782	5	7	110.	D
279.79	377687	5	5	26.	D
295.94	381086	1	3	48.	D
303.41	329582	1	3	34.	D
303.46	329643	3	1	100.	D
303.52	329582	3	3	26.	D
303.62	329468	3	5	25.	D
303.69	329582	5	3	42.	D
303.80	329468	5	5	76.	D
305.60	327228	1	3	100.	D
305.66	327277	3	5	140.	D
305.70	327228	3	3	76.	D
305.77	327351	5	7	180.	D
305.84	327277	5	5	46.	D
320.98	331820	5	7	190.	D
328.45	324734	5	5	61.	D
345.31	332777	1	3	99.	D
374.08	267633	5	5	26.	D
395.56	273080	5	3	49.	D
507.39	197087	1	3	16.	C
507.68	197087	3	3	47.	C
508.18	197087	5	3	79.	C
525.80	210459	5	3	88.	C
597.82	210459	1	3	18.	C
599.60	187049	5	5	55.	C
702.33	142383	1	3	5.7	C
702.82	142397	3	1	17.	C
832.93	120059	1	3	3.2	C
835.10	120053	5	5	1.4	C
835.29	120025	5	7	5.7	C
1109.5	381086	3	3	2.8	D
1679.1	357111	3	5	.94	D
1686.9	356838	3	3	.94	D
1760.4	357111	3	5	.60	D
1764.5	357111	5	5	1.8	D
1766.3	356838	1	3	.78	D
1772.3	356732	3	1	2.3	D
1773.0	356838	5	3	.98	D
2390.4	332777	3	3	2.2	D
2959.7	324734	3	5	2.1	D
2996.5	327228	3	3	.51	D
3004.4	327277	5	5	.47	D
3017.6	327351	7	7	.59	D
3115.7	329643	3	1	1.5	D
3121.7	329582	3	3	1.5	D
3132.9	329468	3	5	1.4	D
3261.0	324658	5	7	1.8	D
3265.5	324836	7	9	2.1	D
3267.3	324462	3	5	1.7	D
3281.9	324462	5	5	.32	D
3284.6	324658	7	7	.23	D
3405.7	329582	1	3	.27	D
3408.1	329643	3	1	.81	D
3415.3	329582	3	3	.20	D
3428.7	329468	3	5	.20	D
3430.6	329582	5	3	.33	D
3444.1	329468	5	5	.59	D
3702.8	327228	1	3	.62	D
3707.2	327277	3	5	.83	D
3714.0	327228	3	3	.46	D
3715.1	327351	5	7	1.1	D
3725.3	327277	5	5	.27	D
3961.6	331820	5	7	1.3	D
5268.1	332777	1	3	.31	D
5508.1	324734	5	5	.11	D
5592.4	290957	3	3	.36	D
O IV					
238.36	419534	2	4	288.	C
238.57	419551	4	6	346.	C
238.58	419534	4	4	58.	C
279.63	357614	2	2	23.7	C
279.94	357614	4	2	47.7	C
553.33	180724	2	4	12.0	C
554.08	180481	2	2	47.3	C
554.51	180724	4	4	60.	C
555.26	180481	4	2	25.0	C
608.40	164366	2	2	12.5	C
609.83	164366	4	2	23.6	C
616.95	289024	6	4	25.5	C
617.01	289024	4	4	2.93	C
617.04	289015	4	2	28.6	C
624.62	231538	2	4	10.7	C
625.13	231538	4	4	21.5	C
625.85	231538	6	4	32.2	C
779.73	255185	6	4	1.53	C
779.82	255185	4	4	13.1	C
779.91	255156	6	6	13.7	C
780.00	255156	4	6	1.0	C
787.71	126950	2	4	5.9	C
790.11	126950	4	4	1.15	C
790.20	126936	4	6	7.1	C
921.30	289024	2	4	2.11	C
921.37	289015	2	2	9.2	C
923.37	289024	4	4	11.4	C
923.43	289015	4	2	4.48	C
1338.6	255185	2	4	2.22	C
1343.0	255185	4	4	.428	C
1343.5	255156	4	6	2.64	C
O V					
172.17	580825	1	3	296.	B
192.85	600766	9	15*	690.	B
215.17	546973	9	3*	170.	C
220.35	612616	3	5	440.	C
248.46	561276	3	1	65.	D
629.73	158798	1	3	28.0	B
758.68	213887	3	5	5.68	B
759.44	213618	1	3	7.55	B
760.23	213618	3	3	5.64	B
760.45	213887	5	5	16.9	B
761.13	213463	3	1	22.5	B
762.00	213618	5	3	9.34	B
774.52	287910	3	1	35.4	C
1371.3	231721	3	5	3.29	C
2784.0	582882	3	9*	1.6	D
3144.7	612616	3	5	.93	C
5114.1	580825	1	3	.17	B
5589.9	600766	9	15*	.15	C
O VI					
150.10	666218	2	6*	254.	B
173.03	674656	6	10*	885.	B
1031.9	96908	2	4	4.15	B
1037.6	96375	2	2	4.08	B
3811.4	666270	2	4	.513	B
3834.2	666113	2	2	.505	B
O VII					
18.627	5368550	1	3	9362.	AA
21.602	4629200	1	3	330.9(+2)	AA
120.33	5355670	3	9*	533.5	AA
128.20	5368550	1	3	505.3	AA
128.46	5364420	9	15*	1620.	A
135.82	5365470	3	5	1530.	A
1630.2	4585980	3	9*	.7935	AA
2450.0	4629200	1	3	.2514	AA
Phosphorus					
P I					
1671.7	59820	4	2	.39	D
1674.6	59716	4	4	.40	D
1679.7	59535	4	6	.39	D
1775.0	56340	4	6	2.17	C
1782.9	56090	4	4	2.14	C
1787.7	55939	4	2	2.13	C
2135.5	58174	4	4	.211	C
2136.2	58174	6	4	2.83	C
2149.1	57877	4	2	3.18	C
2152.9	65156	2	4	.485	C
2154.1	65156	4	4	.173	C
2154.1	65157	4	6	.58	C
2534.0	58174	2	4	.200	C
2535.6	58174	4	4	.95	C
2553.3	57877	2	2	.71	C
2554.9	57877	4	2	.300	C
P II					
1301.9	76813	1	3	.50	C
1304.5	76824	3	1	1.5	C
1304.7	76813	3	3	.37	C

Table 1b (Continued)
TRANSITION PROBABILITIES FOR SELECTED ATOMIC AND IONIC SPECIES

Wavelength (λ[Å])	Upper energy level (E_k[cm^{-1}])	Statistical weights g_i	g_k	Transition Probability (A[10^8 s^{-1}])	Uncertainty
	P II				
1305.5	76765	3	5	.38	C
1309.9	76813	5	3	.62	C
1310.7	76765	5	5	1.1	C
4475.3	127889	5	7	1.3	D
4499.2	130143	5	7	1.4	D
4530.8	127368	3	5	1.0	D
4554.8	127951	3	5	.96	D
4588.0	125130	5	7	1.7	D
4589.9	124948	3	5	1.6	D
4602.1	125392	7	9	1.9	D
4943.5	123892	7	5	.63	D
5253.5	107924	3	5	1.0	D
5425.9	105550	5	5	.69	D
6024.2	103340	3	5	.51	D
6043.1	103669	5	7	.68	D
	P III				
1334.8	74917	2	4	.55	D
1344.3	74946	4	6	.64	D
1344.8	74917	4	4	.11	D
4057.4	141514	4	4	.10	D
4059.3	141514	6	4	.90	D
4080.1	141377	4	2	.99	D
	Potassium				
	K I				
4044.1	24720	2	4	.0124	C
4047.2	24701	2	2	.0124	C
5084.2	32648	2	2	.00350	C
5099.2	32648	4	2	.0070	C
5323.3	31765	2	2	.0063	C
5339.7	31765	4	2	.0126	C
5343.0	31696	2	4	.0040	D
5359.6	31696	4	6	.0046	D
5782.4	30274	2	2	.0123	C
5801.8	30274	4	2	.0246	C
5812.2	30186	2	4	.0028	D
5831.9	30185	4	6	.0032	D
6911.1	27451	2	2	.0272	C
6938.8	27451	4	2	.054	C
7664.9	13043	2	4	.387	B
7699.0	12985	2	2	.382	B
	K II				
607.93	164496	1	3	.013	D
	K III				
2550.0	246626	6	4	2.0	D
2635.1	246626	4	4	1.2	D
2992.4	240830	6	8	2.5	D
3052.1	241444	4	6	1.7	D
3202.0	243947	4	4	1.8	D
3289.1	243121	4	6	2.0	D
3322.4	237512	6	6	1.3	D
3421.8	243448	2	4	1.5	D
	K XVI				
206.27	484800	1	3	94.	C
	K XVII				
22.020	4814800	2	4	4.7 (+4)	C
22.163	4818000	4	6	5.6 (+4)	C
22.18	4814800	4	4	9300.	C
22.60	4699000	2	2	2500.	D
22.76	4699000	4	2	4700.	C
	Praseodymium				
	Pr II				
3997.0	28010	15	15	.187	C
4062.8	28010	13	15	1.00	C
4100.7	28816	17	19	.84	C
4143.1	27128	15	17	.58	C
4179.4	25569	13	15	.52	C
4222.9	24116	11	13	.391	C
4241.0	28010	17	15	.230	C
4359.8	28010	15	15	.11	D
4405.8	27128	17	17	.090	D
4429.3	25569	15	15	.228	C
4449.8	24116	13	13	.124	C
4468.7	24116	11	13	.154	C
4510.2	25569	13	15	.116	C
4534.2	27128	15	17	.049	D
4734.2	24116	15	13	.025	D
	Pr II				
4879.1	25569	15	15	.018	D
4886.0	25569	15	15	.013	D
4912.6	28010	17	15	.057	D
5034.4	28816	19	19	.11	D
5110.8	28816	21	19	.278	C
5135.1	27128	17	17	.125	C
5173.9	27128	19	17	.318	C
5219.1	25569	15	15	.095	D
5220.1	25569	17	15	.235	C
5251.7	24116	15	13	.011	D
5259.7	24116	15	13	.224	C
5292.6	24116	13	13	.093	D
5810.6	28816	17	19	.023	D
5879.3	28010	15	15	.076	D
6200.8	27128	15	17	.018	D
6278.7	25569	13	15	.026	D
6398.0	24116	11	13	.019	D
	Rubidium				
	Rb I				
3022.5	33076	2	4	4.13(−5)	C
3032.0	32972	2	4	4.93(−5)	C
3044.2	32840	2	4	8.2 (−5)	C
3060.2	32668	2	4	1.05(−4)	C
3082.0	32437	2	4	1.49(−4)	C
3112.6	32119	2	4	2.5 (−4)	C
3113.1	32114	2	2	1.3 (−4)	C
3157.5	31661	2	4	3.38(−4)	C
3158.3	31654	2	2	2.0 (−4)	C
3228.0	30970	2	4	6.4 (−4)	C
3229.2	30959	2	2	3.8 (−4)	C
3348.7	29854	2	4	.00137	C
3350.8	29835	2	2	8.9 (−4)	C
3587.1	27870	2	4	.00397	C
3591.6	27835	2	2	.0029	C
4201.8	23793	2	4	.018	C
4215.5	23715	2	2	.015	C
7800.3	12817	2	4	.370	B
7947.6	12579	2	2	.340	B
	Scandium				
	Sc I				
2113.5	47315	4	6	.032	C
2263.0	44189	4	4	.058	C
2267.3	44105	4	2	.48	C
2271.6	44189	6	4	.46	C
2335.4	42819	4	2	.17	C
2346.8	42780	6	4	.13	C
2439.4	41163	6	6	.21	C
2439.9	41153	6	4	.022	C
2711.4	37040	6	6	.29	C
2739.8	36666	6	6	.0056	C
2974.0	33615	4	4	.45	C
2980.8	33707	6	6	.44	C
3015.4	33154	4	6	.66	C
3019.3	33278	6	8	.81	C
3269.9	30573	4	2	3.1	C
3273.6	30707	6	4	2.7	C
3907.5	25585	4	6	1.28	C
3911.8	25725	6	8	1.37	C
4020.4	24866	4	4	1.65	C
4023.7	25014	6	6	1.44	C
4753.2	21033	4	6	.010	C
4779.4	21086	6	8	.0084	C
4791.5	21033	6	6	.0023	C
5301.9	18856	4	4	.0013	C
5343.0	18711	4	2	.0051	C
5349.7	18856	6	4	.0040	C
6210.7	16097	4	4	.012	C
6239.8	16022	4	4	.0065	C
6305.7	16023	6	6	.015	C
6378.8	15673	4	4	.0016	C
6448.1	15673	6	4	2.6 (−4)	C
	Sc II				
2273.1	55716	1	3	7.7	D
3353.7	32350	5	7	2.0	D
3372.2	29824	7	5	1.2	D
3535.7	30816	5	3	.83	D
3572.5	28161	7	7	1.8	D
3576.4	28021	5	5	1.4	D

Table 1b (Continued)
TRANSITION PROBABILITIES FOR SELECTED ATOMIC AND IONIC SPECIES

Wavelength (λ[Å])	Upper energy level (E_k[cm^{-1}])	Statistical weights g_i	Statistical weights g_k	Transition Probability (A[10^8 s^{-1}])	Uncertainty
		Sc II			
3613.8	27841	7	9	1.9	D
3630.8	27602	5	7	1.6	D
3642.8	27444	3	5	1.5	D
4246.8	26081	5	5	1.5	D
4314.1	28161	9	7	.41	D
4374.5	27841	9	9	.14	D
4670.4	32350	5	7	.18	D
5031.0	30816	5	3	.49	D
5239.8	30816	1	3	.14	D
5526.8	32350	9	7	.42	D
5657.9	29824	5	5	.13	D
		Silicon			
		Si I			
1977.6	50566	1	3	.18	D
1979.2	50602	3	1	.51	D
1980.6	50566	3	3	.13	D
1983.2	50500	3	5	.14	D
1986.4	50566	5	3	.21	D
1989.0	50500	5	5	.41	D
2208.0	45276	1	3	.311	C
2210.9	45294	3	5	.416	C
2211.7	45276	3	3	.232	C
2216.7	45322	5	7	.55	C
2218.1	45294	5	5	.138	C
2506.9	39955	3	5	.466	C
2514.3	39760	1	3	.61	C
2516.1	39955	5	5	1.21	C
2519.2	39760	3	3	.456	C
2524.1	39683	3	1	1.81	C
2528.5	39760	5	3	.77	C
2532.4	54871	1	3	.26	D
2631.3	53387	1	3	.97	D
2881.6	40992	5	3	1.89	C
3905.5	40992	1	3	.118	C
4738.8	60857	3	3	.010	D
4783.0	60857	5	3	.017	D
4792.3	60816	5	5	.017	D
4818.1	60705	5	7	.011	D
4821.2	60496	3	5	.0080	D
4947.6	61198	3	1	.042	D
5006.1	60962	3	5	.028	D
5622.2	57542	3	3	.016	D
5690.4	57329	3	3	.012	D
5708.4	57468	5	5	.014	D
5754.2	57329	5	3	.015	D
5772.1	58312	3	1	.036	D
5948.5	57798	3	5	.022	D
7226.2	59111	3	5	.0079	D
7405.8	58775	3	5	.037	D
7409.1	58787	5	7	.023	D
7680.3	60301	3	5	.046	D
7918.4	60645	3	5	.052	D
7932.3	60705	5	7	.051	D
7944.0	60849	7	9	.058	D
7970.3	60645	5	5	.0071	D
		Si II			
989.87	101023	2	4	6.7	D
992.68	101025	4	6	8.0	D
1020.7	97972	2	2	1.3	D
1190.4	84005	2	4	6.9	D
1193.3	83802	2	2	28.	D
1194.5	84005	4	4	36.	D
1197.4	83802	4	2	14.	D
1248.4	123034	4	4	13.	D
1251.2	123034	6	4	19.	D
1260.4	79339	2	4	20.	D
1264.7	79355	4	6	23.	D
1304.4	76666	2	2	3.6	D
1309.3	76666	4	2	7.0	D
1526.7	65501	2	2	3.73	C
1533.5	65501	4	2	7.4	C
1808.0	55310	2	4	.037	D
2904.3	113761	4	6	.67	D
2905.7	113760	6	8	.71	D
3210.0	112395	4	6	.46	D
4128.1	103556	4	6	1.32	C
4130.9	103556	6	8	1.42	C
5041.0	101023	2	4	.98	D
5056.0	101025	4	6	1.2	D
5957.6	97972	2	2	.42	D
5978.9	97972	4	2	.81	D
6347.1	81252	2	4	.70	C
6371.4	81192	2	2	.69	C
7848.8	113761	4	6	.39	D
7849.7	113760	6	8	.42	D
		Si III			
883.40	235414	5	7	63.	D
994.79	153377	3	3	7.89	B
997.39	153377	5	3	13.1	B
1141.6	217440	3	5	30.	D
1144.3	217489	5	7	39.	D
1161.6	216190	5	5	16.	D
1206.5	82884	1	3	25.9	B
1206.5	165765	3	5	48.9	B
1207.5	205029	5	5	19.	D
1294.5	130101	3	5	5.42	B
1296.7	129842	1	3	7.19	B
1298.9	129842	3	3	5.36	B
1299.0	130101	5	5	16.1	B
1301.2	129708	3	1	21.3	B
1303.3	129842	5	3	8.85	B
1328.8	228700	1	3	27.	D
1417.2	153444	3	1	26.0	C
1435.8	235414	5	7	21.	D
1589.0	228700	5	3	11.	D
1778.7	199164	7	9	4.4	D
1783.1	199026	5	7	3.8	D
3241.6	206176	5	3	2.3	D
3486.9	230270	15	21*	1.8	D
3590.5	204331	3	5	3.9	D
4552.6	175336	3	5	1.26	C
4554.0	248773	5	3	.76	D
4567.8	175263	3	3	1.25	C
4683.0	240160	5	5	.95	D
4716.7	225526	5	7	2.8	D
5451.5	244866	3	5	.60	D
5473.1	245087	5	7	.79	D
5716.3	227089	9	7	.19	D
5739.7	176487	1	3	.47	D
7462.6	214995	5	3	.49	D
7466.3	214989	7	5	.54	D
7612.4	227665	3	5	1.1	D
		Si IV			
457.82	218429	2	4	3.6	D
458.16	218267	2	2	3.6	D
515.12	265418	2	2	4.1	D
516.35	265418	4	2	8.2	D
560.50	250008	6	10*	1.0	D
749.94	293719	10	14*	14.5	C
815.05	193979	2	2	12.3	C
818.13	193979	4	2	24.4	C
860.74	276554	10	6*	1.8	D
1066.6	254128	10	14*	39.1	C
1122.5	160376	2	4	20.5	C
1128.3	160376	4	4	4.03	C
1128.3	160374	4	6	24.2	C
1393.8	71749	2	4	7.73	B
1402.8	71288	2	2	7.58	B
1724.1	218375	10	6*	5.5	C
		Si V			
96.439	1036915	1	3	480.	D
97.143	1029407	1	3	2000.	D
117.86	848511	1	3	300.	D
		Si VI			
246.00	406497	4	2	170.	C
249.12	406497	2	2	85.	C
		Si VII			
217.83	506080	5	3	430.	D
272.64	366780	5	3	51.	C
274.18	368760	3	1	120.	C
275.35	363170	5	5	89.	C
275.67	366780	3	3	30.	C
276.84	366780	1	3	39.	C
278.45	363170	3	5	29.	C
		Si VIII			
214.76	534810	4	2	410.	D
216.92	530420	6	4	360.	D

Table 1b (Continued)

TRANSITION PROBABILITIES FOR SELECTED ATOMIC AND IONIC SPECIES

Wavelength (λ[Å])	Upper energy level (E_k[cm^{-1}])	Statistical weights g_i	g_k	Transition Probability (A[10^8 s^{-1}])	Uncertainty
	Si VIII				
232.86	534810	2	2	80.	D
235.56	530420	4	4	97.	D
250.45	504630	2	2	77.	D
250.79	504630	4	2	160.	D
314.31	318160	4	2	52.	D
316.20	316250	4	4	50.	D
319.83	312670	4	6	49.	D
	Si IX				
223.73	446967	1	3	42.	C
225.03	446967	3	3	120.	C
227.01	446967	5	3	200.	C
227.30	492890	5	3	230.	D
258.10	440390	5	5	104.	C
294.37	344100	9	9*	59.	C
347.36	292301	9	15*	22.	C
	Si X				
253.77	394040	2	4	29.	C
256.57	390050	2	2	110.	C
258.35	394040	4	4	140.	C
261.05	390050	4	2	54.	C
272.00	367670	2	2	30.	C
277.26	367670	4	2	57.	C
287.08	510300	2	4	26.	C
289.19	510300	4	4	50.	C
292.22	510300	6	4	73.	C
347.73	575450	10	10*	43.	D
353.09	287870	6	10*	21.	C
	Si XI				
43.763	2285040	1	3	6110.	B
49.116	2210220	9	3*	2450.	C
49.222	2361010	3	5	8900.	C
52.296	2241590	3	1	760.	C
303.30	329690	1	3	64.2	B
358.29	608790	3	1	103.	C
358.63	451000	3	5	13.8	C
361.41	446510	1	3	18.0	C
364.50	446510	3	3	13.2	C
365.42	451000	5	5	39.0	C
368.28	443690	3	1	51.	C
371.48	446510	5	3	20.7	C
604.14	495210	3	5	11.2	C
2300.8	2285040	1	3	.434	C
	Si XII				
40.924	2443500	2	6*	4420.	B
44.118	2464130	6	10*	1.4 (+4)	B
499.43	200290	2	4	9.56	B
520.72	191900	2	2	8.47	B
1862.	2444300	2	4	1.15	B
1949.	2441900	2	2	1.0	B
4620.	2463540	2	4	.046	C
4942.	2464530	4	6	.045	C
	Silver				
	Ag I				
2061.2	48501	2	4	.031	D
2069.9	48297	2	2	.015	D
3280.7	30473	2	4	1.4	B
3382.9	29552	2	2	1.3	B
5209.1	48744	2	4	.75	D
5465.5	48764	4	6	.86	D
5471.6	48744	4	4	.14	D
	Sodium				
	Na I				
3302.4	30273	2	4	.0281	C
3303.0	30267	2	2	.0281	C
4390.0	39729	2	4	.0077	D
4393.3	39729	4	4	.0016	D
4393.3	39729	4	6	.0092	D
4494.2	39201	2	4	.012	C
4497.7	39201	4	6	.014	C
4497.7	39201	4	4	.0024	D
4664.8	38387	2	4	.0233	C
4668.6	38387	4	4	.0041	D
4668.6	38387	4	6	.025	C
4747.9	38012	2	2	.0063	D
4751.8	38012	4	2	.0127	C
4978.5	37037	2	4	.041	C
4982.8	37037	4	4	.0082	D
	Na I				
4982.8	37037	4	6	.0489	C
5148.8	36373	2	2	.0117	C
5153.4	36373	4	2	.0233	C
5682.6	34549	2	4	.103	C
5688.2	34549	4	6	.12	C
5688.2	34549	4	4	.021	D
5890.0	16973	2	4	.622	A
5895.9	16956	2	2	.618	A
6154.2	33201	2	2	.026	C
6160.8	33201	4	2	.052	C
8183.3	29173	2	4	.453	C
8194.8	29173	4	6	.54	C
8194.8	29173	4	4	.090	D
11381.	25740	2	2	.089	C
11404.	25740	4	2	.176	C
	Na II				
300.15	333163	1	3	30.	D
301.44	331745	1	3	49.	D
372.08	268763	1	3	34.	D
	Na III				
378.14	264455	4	2	77.	C
380.10	264455	2	2	37.	C
1991.0	465399	4	6	8.3	D
2004.2	466788	2	4	4.6	D
2011.9	463971	6	8	8.4	D
2151.5	465018	2	4	4.4	D
2174.5	464390	4	6	5.3	D
2230.3	410977	6	8	3.7	D
2232.2	418418	4	4	3.3	D
2246.7	411536	4	6	2.4	D
2459.3	414282	4	6	3.0	D
2468.9	415172	2	4	2.4	D
2497.0	406190	6	6	1.7	D
	Na V				
307.89	372400	10	6*	200.	C
333.46	372400	6	6*	56.	C
369.01	568100	10	6*	120.	D
400.72	297100	10	10*	50.	C
445.14	297100	6	10*	7.1	C
459.90	217440	4	2	23.	D
461.05	216896	4	4	23.	D
463.26	215860	4	6	22.	D
510.10	569200	2	2	56.	D
511.19	567600	4	4	68.	D
	Na VI				
313.75	320589	5	3	130.	C
361.25	312175	5	5	77.	C
416.53	241341	9	9*	37.	C
492.80	204187	9	15*	13.	C
1550.6	873287	5	5	4.35	C
1567.8	872577	5	3	2.68	C
1608.5	934745	3	1	2.6	C
1649.4	933915	5	5	2.05	C
1741.5	929999	3	5	2.59	C
1747.5	930510	5	7	3.1	C
	Na VII				
94.409	1060651	6	10*	2700.	C
105.27	951347	6	2*	450.	C
353.29	285189	4	4	100.	C
381.30	264400	4	2	40.	C
397.49	367500	4	4	35.	C
399.18	367500	6	4	52.	C
483.28	412345	10	10*	29.	C
486.74	205448	2	4	11.	C
491.95	205412	4	6	13.	C
555.80	465111	4	4	23.	D
777.83	412311	4	6	6.8	D
	Na VIII				
83.34	1327500	9	15*	3940.	C
89.88	1240300	9	3*	809.	C
90.536	1347756	3	5	2860.	C
411.15	243223	1	3	44.2	C
1239.4	1432991	3	3	3.02	C
1802.7	1481521	3	1	2.70	C
1867.7	1347756	3	5	2.01	C
2059.1	1474598	3	5	1.80	C
2558.2	1513677	5	3	.0226	C
2772.0	1469055	3	5	.419	C

Table 1b (Continued)
TRANSITION PROBABILITIES FOR SELECTED ATOMIC AND IONIC SPECIES

Wavelength (λ[Å])	Upper energy level (E_k[cm^{-1}])	Statistical weights g_i	g_k	Transition Probability (A[10^8 s^{-1}])	Uncertainty
	Na VIII				
3021.0	1507690	5	7	.490	C
3108.9	1513677	1	3	.258	C
3182.3	1294214	1	3	.292	C
	Na IX				
70.615	1416130	2	4	1350.	B
70.653	1415368	2	2	1350.	B
77.764	1429980	2	4	3600.	B
77.911	1430204	4	6	4300.	B
681.72	146688	2	4	6.63	B
694.17	144038	2	2	6.30	B
2487.7	1416130	2	4	.832	B
2535.8	1415368	2	2	.789	B
6841.8	1429980	2	4	.0259	C
7103.4	1430204	4	6	.0278	C
	Strontium				
	Sr I				
2206.2	45312	1	3	.0066	C
2211.3	45208	1	3	.0085	C
2217.8	45075	1	3	.012	C
2226.3	44904	1	3	.016	C
2237.7	44676	1	3	.023	C
2253.3	44366	1	3	.037	C
2275.3	43937	1	3	.067	C
2307.3	43328	1	3	.12	C
2354.3	42462	1	3	.18	C
2428.1	41172	1	3	.17	C
2569.5	38907	1	3	.053	C
2931.8	34098	1	3	.019	C
4607.3	21698	1	3	2.01	B
	Sr II				
2018.7	73237	2	2	.12	D
2051.9	73237	4	2	.24	D
2282.0	67523	2	4	.83	D
2322.4	67563	4	6	.91	D
2324.5	67523	4	4	.15	D
2423.5	64964	2	2	.24	D
2471.6	64964	4	2	.48	D
3464.5	53373	4	6	3.1	D
3474.9	53286	4	4	.51	D
4077.7	24517	2	4	1.42	C
4161.8	47737	2	2	.65	D
4215.5	23715	2	2	1.27	C
4305.5	47737	4	2	1.4	D
4414.8	78702	4	6	.11	D
4417.5	78689	4	4	.018	D
4585.9	77858	4	2	.070	D
5303.1	74621	2	4	.19	D
5379.1	74643	4	6	.22	D
5385.5	74621	4	4	.037	D
5723.7	73237	2	2	.071	D
5819.0	73237	4	2	.14	D
8688.9	67563	4	6	.55	D
8719.6	67523	4	4	.097	D
	Sulfur				
	S I				
1295.7	77181	5	5	4.9	D
1296.2	77150	5	3	2.7	D
1302.3	77181	3	5	1.8	D
1302.9	77150	3	3	1.6	D
1303.1	77136	3	1	6.6	D
1303.4	76721	5	3	1.9	D
1305.9	77150	1	3	2.4	D
1401.5	71351	5	3	.91	D
1409.3	71351	3	3	.50	D
1412.9	71351	1	3	.16	D
1425.0	70174	5	7	4.5	D
1425.2	70166	5	5	1.2	D
1433.3	70166	3	5	3.3	D
1433.3	70165	3	3	1.9	D
1437.0	70165	1	3	2.4	D
1448.2	78288	5	3	7.3	D
1473.0	67890	5	7	.42	D
1474.0	67843	5	7	1.6	D
1474.4	67825	5	5	.50	D
1474.6	67816	5	3	.062	D
1481.7	67888	3	5	.17	D
1483.0	67825	3	5	1.2	D
1483.2	67816	3	3	.75	D
	S I				
1487.2	67816	1	3	.87	D
1666.7	69238	5	5	6.3	C
1687.5	81438	1	3	.94	D
1782.3	78288	1	3	1.9	D
1807.3	55331	5	3	3.8	C
1820.3	55331	3	3	2.2	C
1826.2	55331	1	3	.72	C
4694.1	73921	5	7	.0067	D
4695.4	73915	5	5	.0067	D
4696.2	73911	5	3	.0065	D
6403.6	79058	3	5	.0057	D
6408.1	79058	5	5	.0095	D
6415.5	79058	7	5	.013	D
6751.2	78271	15	25*	.079	D
7679.6	76464	3	5	.012	D
7686.1	76464	5	5	.020	D
7696.7	76464	7	5	.028	D
	S II				
1124.4	113461	2	4	1.0	D
1125.0	113461	4	4	4.6	D
1131.0	112937	2	2	3.5	D
1131.6	112937	4	2	1.4	D
1250.5	79968	4	2	.46	C
1253.8	79758	4	4	.42	C
1259.5	79395	4	6	.34	C
4463.6	150996	8	6	.53	D
4483.4	150531	6	4	.31	D
4486.7	150258	4	2	.66	D
4524.7	143623	4	4	.093	D
4525.0	143623	6	4	1.2	D
4552.4	143489	4	2	1.2	D
4656.7	131029	2	4	.09	D
4716.2	131029	4	4	.29	D
4815.5	131029	6	4	.88	D
4885.6	133400	2	4	.17	D
4917.2	133269	2	2	.66	D
4924.1	130134	4	6	.22	D
4925.3	129858	2	4	.24	D
4942.5	129788	2	2	.15	D
4991.9	129858	4	4	.15	D
5009.5	129788	4	2	.70	D
5014.0	133400	4	4	.84	D
5027.2	125485	4	2	.26	D
5032.4	130134	6	6	.81	D
5047.3	133269	4	2	.36	D
5103.3	129858	6	4	.50	D
5142.3	125485	2	2	.19	D
5201.0	140750	4	4	.75	D
5201.3	140750	6	4	.065	D
5212.6	140709	4	6	.098	D
5212.6	140709	6	6	.85	D
5320.7	140319	6	8	.92	D
5345.7	140230	4	6	.88	D
5345.7	140230	6	6	.11	D
5428.6	127976	2	4	.42	D
5432.8	128233	4	6	.68	D
5453.8	128599	6	8	.85	D
5473.6	127825	2	2	.73	D
5509.7	127976	4	4	.40	D
5526.2	128599	8	8	.081	D
5536.8	128233	4	6	.066	D
5556.0	127825	4	2	.11	D
5564.9	128233	6	6	.17	D
5578.8	128233	6	6	.11	D
5606.1	128599	10	8	.54	D
5616.6	127976	4	4	.12	D
5640.0	131187	4	6	.66	D
5645.6	127976	6	4	.018	D
5647.0	130641	2	4	.57	D
5659.9	127976	6	4	.46	D
5664.7	127825	4	2	.58	D
5819.2	130641	4	4	.085	D
6305.5	130134	8	6	.18	D
6312.7	130641	6	4	.30	D
	S III				
2496.2	210698	7	5	2.5	D
2508.2	209926	5	3	2.3	D
2636.9	210698	3	5	.45	D
2665.4	210698	5	5	1.4	D
2680.5	209926	1	3	.62	D

Table 1b (Continued)
TRANSITION PROBABILITIES FOR SELECTED ATOMIC AND IONIC SPECIES

Wavelength (λ[Å])	Upper energy level (E_k[cm^{-1}])	Statistical weights g_i	g_k	Transition Probability (A[10^8 s^{-1}])	Uncertainty
		S III			
2691.8	209926	3	3	.46	D
2702.8	209773	3	1	1.9	D
2718.9	206539	3	3	1.2	D
2721.4	209926	5	3	.77	D
2726.8	210698	3	5	.60	D
2731.1	206672	5	5	1.1	D
2756.9	206911	7	7	1.4	D
2785.5	209926	3	3	.61	D
2856.0	205071	5	7	5.1	D
2863.5	205561	7	9	5.7	D
2872.0	204579	3	5	4.7	D
2950.2	206672	3	5	3.0	D
2964.8	206911	5	7	4.0	D
3662.0	174036	3	3	.64	D
3717.8	174036	5	3	1.0	D
3778.9	173192	3	5	.44	D
3831.8	172786	1	3	.56	D
3837.8	172786	3	3	.42	D
3838.3	173192	5	5	1.3	D
3860.6	172631	3	1	1.6	D
3899.1	172786	5	3	.67	D
4253.6	170649	5	7	1.2	D
4285.0	170067	3	5	.90	D
		S IV			
551.17	181432	2	2	20.6	C
554.07	181432	4	2	40.8	C
3097.5	213717	2	4	2.6	D
3117.7	213507	2	2	2.5	D
		S V			
437.37	311700	1	3	11.2	C
438.19	311700	3	3	33.3	C
439.65	311700	5	3	55.	C
661.52	235000	9	15*	64.4	B
679.01	348100	9	15*	86.	D
690.75	345600	9	9*	50.	D
786.48	127149	1	3	52.5	B
854.85	200800	9	9*	41.8	C
		S VI			
248.99	401621	2	4	31.	C
249.27	401164	2	2	31.	C
388.94	362983	2	2	45.	C
390.86	362983	4	2	88.	C
706.48	247420	2	4	41.7	C
712.68	247452	4	6	48.5	C
712.84	247420	4	4	8.1	D
933.38	107137	2	4	17.	C
944.52	105874	2	2	16.	C
		S VII			
60.161	1662210	1	3	9460.	B
60.804	1644630	1	3	510.	B
72.029	1388330	1	3	861.	B
		S VIII			
198.55	503590	4	2	250.	C
202.61	503590	2	2	120.	C
		S XI			
189.90	535250	9	3*	430.	C
190.37	592500	5	3	280.	C
215.95	530180	5	5	140.	C
217.63	592500	1	3	72.	C
239.81	417040	1	3	26.	D
242.57	417420	3	5	19.	D
242.82	417040	3	3	19.	D
246.90	417420	5	5	54.	D
247.12	417040	5	3	30.	D
288.49	355260	9	15*	29.	C
		S XII			
212.14	471480	2	4	37.	C
215.18	464750	2	2	140.	C
218.20	471480	4	4	170.	C
221.44	464750	4	2	64.	C
227.50	439540	2	2	37.	C
234.48	439540	4	2	68.	C
		S XIII			
32.236	3102150	1	3	1.09(+4)	B
37.600	3049260	3	1	1300.	C
256.66	389660	1	3	87.	C
299.89	537520	3	5	17.8	C
303.37	529420	1	3	22.8	C
307.36	529420	3	3	16.4	C
308.91	537520	5	5	48.2	C
312.68	523880	3	1	63.	C
316.84	529420	5	3	25.0	C
500.42	3012200	3	5	14.3	C
		S XIV			
30.434	3285500	2	6*	8280.	B
32.517	3309780	6	10*	2.6 (+4)	B
417.67	239460	2	4	12.	B
445.71	224330	2	2	10.	B
1550.	3287000	2	4	1.4	B
1663.	3282640	2	2	1.2	B
3967.	3307840	2	4	.054	C
4153.	3311070	4	6	.057	C
		Thallium			
		Tl I			
2104.6	47500	2	4	.040	D
2118.9	47179	2	2	.020	D
2129.3	46950	2	4	.058	D
2151.9	46457	2	2	.031	D
2168.6	46099	2	4	.098	D
2237.8	44673	2	4	.19	D
2316.0	43166	2	2	.078	D
2379.7	42011	2	4	.44	C
2507.9	47655	4	2	.011	C
2538.2	47179	4	2	.016	C
2580.1	38746	2	2	.18	D
2609.0	46110	4	6	.10	C
2609.8	46099	4	4	.019	C
2665.6	45297	4	2	.057	C
2709.2	44693	4	6	.17	C
2710.7	44673	4	4	.037	C
2767.9	36118	2	4	1.26	C
2826.2	43166	4	2	.080	C
2918.3	42049	4	6	.42	C
2921.5	42011	4	4	.076	C
3229.8	38746	4	2	.173	C
3519.2	36200	4	6	1.24	C
3529.4	36118	4	4	.220	C
3775.7	26478	2	2	.625	B
5350.5	26478	4	2	.705	B
		Thulium			
		Tm I			
2513.8	39769	8	10	.069	D
2527.0	39560	8	8	.17	D
2596.5	38502	8	10	.16	D
2601.1	38434	8	6	.17	D
2622.5	38121	8	10	.061	D
2841.1	43958	6	6	.20	D
2854.2	35026	8	6	.27	D
2914.8	34297	8	8	.077	D
2933.0	34085	8	6	.10	D
2973.2	33624	8	8	.23	D
3046.9	32811	8	8	.18	D
3081.1	32446	8	8	.19	D
3122.5	40787	6	6	.52	D
3142.4	40585	6	6	.088	D
3172.7	31510	8	8	.18	D
3233.7	30915	8	10	.051	D
3247.0	39560	6	8	.30	D
3251.8	39515	6	4	.52	D
3380.7	38343	6	8	.20	D
3406.0	38123	6	8	.15	D
3410.1	29317	8	10	.10	D
3416.6	29261	8	8	.057	D
3418.6	38014	6	6	.11	D
3563.9	28051	8	6	.098	D
3567.4	28024	8	10	.042	D
3744.1	26701	8	8	.95	D
3751.8	26646	8	10	.19	D
3798.5	35090	6	4	1.2	D
3807.7	35026	6	6	.39	D
3883.1	25745	8	6	1.0	D
3887.4	25717	8	8	.38	D
3916.5	34297	6	8	1.5	D
3949.3	34085	6	6	1.0	D
4022.6	33624	6	8	.040	D

Table 1b (Continued)
TRANSITION PROBABILITIES FOR SELECTED ATOMIC AND IONIC SPECIES

Wavelength (λ[Å])	Upper energy level (E_k[cm^{-1}])	Statistical weights g_i	g_k	Transition Probability (A[10^8 s^{-1}])	Uncertainty
		Tm I			
4044.5	33489	6	4	.29	D
4094.2	24418	8	6	.90	C
4105.8	24349	8	10	.60	C
4138.3	32929	6	4	.70	D
4158.6	32811	6	8	.055	D
4187.6	23873	8	8	.61	C
4203.7	23782	8	10	.25	C
4222.7	32446	6	8	.15	D
4271.7	32174	6	6	.11	D
4359.9	22930	8	6	.13	D
4386.4	22791	8	8	.042	D
4394.4	31521	6	4	.11	D
4643.1	30302	6	6	.034	D
4681.9	30124	6	8	.039	D
4691.1	30082	6	6	.039	D
5307.1	18837	8	10	.023	D
5658.3	26439	6	8	.010	D
5675.8	17614	8	10	.013	D
5760.2	26127	6	6	.013	D
		Tin			
		Sn I			
2073.1	48222	1	3	.036	D
2199.3	47146	3	5	.29	D
2209.7	48670	5	5	.56	D
2246.1	44509	1	3	1.6	D
2268.9	47488	5	7	1.2	D
2286.7	47146	5	5	.31	D
2317.2	51755	5	7	2.0	D
2334.8	44509	3	3	.66	D
2354.8	44145	3	5	1.7	D
2380.7	43683	3	5	.031	D
2408.2	50126	5	3	.18	D
2421.7	49894	5	7	2.5	D
2429.5	44576	5	7	1.5	C
2433.5	44509	5	3	.0080	D
2455.2	44145	5	5	.011	D
2476.4	48982	5	3	.011	D
2483.4	43683	5	5	.21	D
2491.8	57282	1	3	.17	D
2495.7	48670	5	5	.62	D
2523.9	48222	5	3	.074	D
2546.6	39257	1	3	.21	D
2558.0	56244	1	3	.34	D
2571.6	47488	5	7	.45	D
2594.4	47146	5	5	.30	D
2636.9	55074	1	3	.11	D
2661.2	39257	3	3	.11	D
2706.5	38629	3	5	.66	D
2761.8	39626	5	5	.0037	D
2779.8	44576	5	7	.18	D
2785.0	44509	5	3	.14	D
2788.0	53021	1	3	.14	D
2812.6	52707	1	3	.23	D
2813.6	44145	5	5	.12	D
2840.0	38629	5	5	1.7	D
2850.6	43683	5	5	.33	D
2863.3	34914	1	3	.54	D
2913.5	51475	1	3	.83	D
3009.1	34914	3	3	.38	D
3032.8	50126	1	3	.62	D
3034.1	34641	3	1	2.0	D
3141.8	48982	1	3	.19	D
3175.1	34914	5	3	1.0	D
3218.7	48222	1	3	.047	D
3223.6	39626	5	5	.0012	D
3262.3	39257	5	3	2.7	D
3330.6	38629	5	5	.20	D
3655.8	44509	1	3	.041	D
3801.0	34914	5	3	.28	D
4524.7	39257	1	3	.26	D
5631.7	34914	1	3	.024	D
5970.3	55374	5	3	.096	D
6037.7	55187	5	5	.050	D
6069.0	51113	1	3	.046	D
6073.5	51375	3	1	.063	D
6171.5	51113	3	3	.049	D
		Sn II			
2368.3	46464	4	2	.0044	D
2449.0	99663	4	6	.37	D

Wavelength (λ[Å])	Upper energy level (E_k[cm^{-1}])	Statistical weights g_i	g_k	Transition Probability (A[10^8 s^{-1}])	Uncertainty
		Sn II			
2487.0	99659	6	8	.55	D
3283.2	89292	4	6	1.0	D
3352.0	89286	6	8	1.0	D
3472.5	100285	2	4	.16	D
3575.5	100339	4	6	.13	D
5332.4	90242	2	4	.86	D
5562.0	90354	4	6	1.2	D
5588.9	89292	4	6	.85	D
5596.2	90242	4	4	.15	D
5797.2	89292	6	6	.28	D
5799.2	89286	6	8	.81	D
6453.5	72377	2	4	1.2	D
6761.5	86280	2	2	.32	D
6844.1	71494.	2	2	.66	D
		Titanium			
		Ti I			
2276.7	44080	7	5	1.3	C
2299.9	43468	5	5	.69	C
2384.5	42311	9	7	.090	C
2424.2	41624	9	9	.17	C
2441.0	41342	9	11	.072	C
2520.5	39662	5	3	.38	C
2529.9	39686	7	5	.38	C
2541.9	39715	9	7	.43	C
2599.9	38451	5	5	.67	C
2605.2	38544	7	7	.64	C
2611.3	38671	9	9	.64	C
2611.5	38451	7	5	.33	C
2632.4	37977	5	5	.27	C
2641.1	37852	5	3	1.8	C
2644.3	37977	7	5	1.4	C
2646.6	38160	9	7	1.5	C
2662.0	37555	5	7	.089	C
2669.6	37618	7	9	.10	C
2679.9	37690	9	11	.13	C
2735.3	45041	3	1	4.1	D
2912.1	41585	5	7	1.3	D
2933.6	34079	5	7	.096	C
2942.0	33981	5	5	1.0	C
2948.3	34079	7	7	.93	C
2956.1	34205	9	9	.97	C
2967.2	34079	9	7	.11	C
2983.3	33680	7	7	.11	C
3000.9	33701	9	9	.12	C
3186.5	31374	5	7	.80	C
3192.0	31489	7	9	.85	C
3199.9	31629	9	11	.94	C
3203.8	31374	7	7	.072	C
3214.2	31489	9	9	.065	C
3341.9	29915	5	7	.65	C
3354.6	29971	7	9	.64	C
3370.4	29661	5	3	.76	C
3371.5	30039	9	11	.67	C
3377.6	29769	7	5	.69	C
3385.7	29915	9	7	.052	C
3635.5	27499	5	7	.72	C
3642.7	27615	7	9	.67	C
3653.5	27750	9	11	.66	C
3654.6	27355	5	3	.087	C
3671.7	27615	9	9	.048	C
3689.9	27480	9	7	.045	C
3717.4	26893	5	7	.043	C
3724.6	38959	9	9	.91	D
3729.8	26803	5	5	.40	C
3741.1	26893	7	7	.38	C
3752.9	27026	9	9	.47	C
3771.7	26893	9	7	.066	C
3786.0	33661	5	3	1.4	D
3924.5	25644	7	7	.073	C
3929.9	25439	5	5	.075	C
3948.7	25318	5	3	.53	C
3958.2	25644	9	7	.43	C
3981.8	25107	5	5	.38	C
3989.8	25227	7	7	.36	C
3998.6	25388	9	9	.39	C
4008.9	25107	7	5	.071	C
4024.6	25227	9	7	.063	C
4065.1	33085	3	1	.70	D
4186.1	36000	9	9	.28	D
4285.0	37359	5	5	.32	D

Table 1b (Continued)
TRANSITION PROBABILITIES FOR SELECTED ATOMIC AND IONIC SPECIES

Wavelength (λ[Å])	Upper energy level (E_k[cm^{-1}])	Statistical weights g_i	g_k	Transition Probability (A[10^8 s^{-1}])	Uncertainty
	Ti I				
4295.8	29829	3	1	1.3	D
4393.9	41040	9	11	.33	D
4417.3	37852	11	9	.36	D
4449.2	37690	11	11	.97	D
4455.3	34079	7	7	.48	D
4457.4	34205	9	9	.56	D
4481.3	36415	7	7	.57	D
4527.3	28639	3	5	.22	D
4544.7	28596	5	3	.33	D
4656.5	21469	5	7	.022	C
4667.6	21588	7	9	.023	C
4681.9	21740	9	11	.025	C
4981.7	26911	11	13	.59	C
4991.1	26773	9	11	.50	C
4999.5	26657	7	9	.50	C
5007.2	26564	5	7	.48	C
5014.2	26494	3	5	.68	C
5022.9	26564	7	7	.15	C
5024.8	26494	5	5	.15	C
5040.0	20006	7	5	.036	C
5064.7	20126	9	7	.043	C
5173.8	19323	5	5	.037	C
5193.0	19422	7	7	.032	C
5210.4	19574	9	9	.034	C
5866.5	25644	5	7	.046	C
5899.3	25439	3	5	.031	C
6258.1	27615	7	9	.091	C
7251.7	25318	5	3	.072	C
8024.8	27615	9	9	.0083	C
8068.2	27499	7	7	.0077	C
	Ti II				
2635.6	68768	4	4	1.9	D
2638.7	68845	6	6	1.7	D
2642.2	68950	8	8	1.8	D
2646.1	69081	10	10	2.7	D
2752.8	69081	8	10	1.1	D
2800.6	66997	10	8	1.8	D
2805.0	65185	6	8	4.6	D
2810.3	65307	8	10	5.1	D
2817.9	65446	10	12	3.8	D
2827.2	65094	8	10	1.0	D
2828.2	65589	12	14	4.4	D
2828.9	65307	10	10	.92	D
2834.1	65242	10	12	.79	D
2836.6	64978	8	8	1.2	D
2839.7	65446	12	12	.83	D
2846.1	65094	10	10	1.2	D
2856.2	65242	12	12	1.5	D
2931.3	65313	6	6	3.2	D
2936.2	64885	4	6	2.7	D
2938.7	64978	6	8	2.4	D
2942.0	65094	8	10	1.8	D
2943.1	65459	8	8	1.1	D
2945.5	65242	10	12	2.7	D
2954.8	68582	10	12	4.0	D
2959.0	68329	10	10	4.0	D
3081.6	62410	10	8	1.1	D
3088.0	32767	10	8	1.3	C
3089.4	47625	8	6	1.3	C
3103.8	47467	10	8	1.1	C
3127.9	63168	6	6	1.6	D
3128.6	63445	8	8	1.2	D
3190.9	40075	6	8	1.3	C
3202.5	39927	4	6	1.1	C
3224.2	43781	12	10	.70	C
3234.5	31301	10	10	1.3	C
3236.6	31114	8	8	1.1	D
3287.7	45674	8	10	1.4	C
3341.9	34543	6	8	.96	D
3349.0	34749	8	10	1.0	D
3349.4	30241	10	12	1.3	D
3361.2	29968	8	10	1.2	D
3372.8	29734	6	8	1.1	C
3383.8	29544	4	6	1.1	C
3483.8	63445	10	8	.97	D
3492.4	63168	8	6	.98	D
3504.9	43781	10	10	.82	D
3510.8	43741	8	8	.93	D
3759.3	31491	8	8	.96	D
3761.3	31207	6	6	1.0	D
	Ti III				
2375.0	83797	5	3	4.0	D
2414.0	83117	5	7	3.8	D
2516.1	78159	7	9	3.4	D
2527.8	77746	5	7	2.2	D
2540.1	77422	3	5	2.0	D
2563.4	77424	7	7	2.1	D
2565.4	77167	5	5	1.6	D
2567.6	77000	3	3	2.3	D
2576.5	77000	5	3	.92	D
2984.8	75198	5	5	1.9	D
	Ti XIX				
169.33	590580	1	3	129.	C
	Ti XX				
11.452	8732000	2	6*	1.7 (+4)	C
11.872	8749000	2	4	2.8 (+4)	C
11.958	8751000	4	6	3.4 (+4)	C
15.211	6574000	2	4	3.50(+4)	B
15.253	6556000	2	2	3.58(+4)	B
15.907	6612000	2	4	8.8 (+4)	C
16.049	6619000	4	6	1.1 (+5)	C
16.067	6612000	4	4	1.7 (+4)	C
	Uranium				
	U I				
3553.0	32413	13	13	.020	C
3553.0	32591	9	7	.014	C
3553.4	31935	15	13	.022	C
3554.5	28746	11	9	.0084	C
3554.9	31923	15	17	.0079	C
3555.3	28119	13	15	.027	C
3555.8	28115	13	11	.0041	C
3556.9	32382	13	11	.0075	C
3557.8	28099	13	13	.029	C
3558.0	33860	11	13	.016	C
3558.6	32546	9	7	.039	C
3559.4	31955	7	9	.015	C
3560.3	34071	9	7	.064	C
3561.4	31872	15	13	.055	C
3561.5	34062	9	9	.025	C
3561.8	28068	13	11	.057	C
3563.7	28053	13	13	.029	C
3563.8	31920	7	7	.011	C
3565.0	32318	13	11	.029	C
3566.0	32310	13	15	.017	C
3566.6	28650	11	11	.24	B
3568.8	32289	13	13	.038	C
3569.1	35656	17	15	.11	C
3569.4	32462	9	9	.015	C
3570.1	32278	13	11	.013	C
3570.2	28622	11	9	.0053	C
3570.6	35004	13	15	.027	C
3570.7	31798	15	15	.012	C
3571.2	28614	11	11	.0063	C
3571.6	38338	17	15	.13	C
3572.9	32256	13	15	.015	C
3573.9	27973	13	11	.040	C
3574.1	34977	13	15	.035	C
3574.8	27966	13	15	.019	C
3577.1	35594	17	15	.043	C
3577.5	31745	15	13	.0078	C
3577.8	28563	11	11	.0083	C
3577.9	27941	13	13	.023	C
3578.3	27938	13	11	.020	C
3580.0	32379	9	9	.012	C
3580.2	28543	11	9	.029	C
3580.4	28542	11	13	.0075	C
3580.9	32194	13	13	.021	C
3582.6	32180	13	13	.029	C
3584.6	31757	7	5	.024	C
3584.9	27887	13	15	.18	B
3585.4	28503	11	11	.019	C
3585.8	28500	11	9	.028	C
3587.8	32318	9	11	.013	C
3588.3	31729	7	9	.018	C
3589.7	28470	11	13	.021	C
3589.8	31650	15	13	.059	C
3590.7	32295	9	7	.022	C
3591.7	28454	11	9	.053	C

Table 1b (Continued)
TRANSITION PROBABILITIES FOR SELECTED ATOMIC AND IONIC SPECIES

Wavelength (λ[Å])	Upper energy level (E_k[cm^{-1}])	Statistical weights g_i	g_k	Transition Probability (A[10^8 s^{-1}])	Uncertainty
		U I			
3593.0	28445	11	11	.014	C
3593.2	32098	13	15	.042	C
3593.7	33581	11	11	.072	C
		Vanadium			
		V I			
3050.4	40074	10	8	.47	D
3053.7	32738	4	4	1.1	D
3056.3	32847	6	6	1.0	D
3060.5	32989	8	8	1.1	D
3066.4	33155	10	10	1.6	D
3088.1	42010	4	6	.43	D
3089.1	41999	4	4	.45	D
3093.8	42138	6	6	.36	D
3112.9	41752	4	2	.43	D
3183.4	31541	6	8	1.3	D
3185.4	31937	10	12	1.4	D
3198.0	31398	6	6	.31	D
3202.4	31541	8	8	.29	D
3205.6	42079	8	10	1.1	D
3212.4	42221	10	12	1.2	D
3218.9	41950	8	6	.31	D
3233.2	42021	10	8	.28	D
3273.0	41437	8	8	.24	D
3284.4	41539	10	10	.24	D
3309.2	39847	4	4	.28	D
3329.9	39847	6	4	.69	D
3356.4	39423	4	6	.27	D
3365.6	39249	2	4	.41	D
3376.1	39249	4	4	.28	D
3377.4	39237	4	2	.80	D
3377.6	39423	6	6	.53	D
3400.4	38116	8	8	.22	D
3529.7	37960	4	6	.36	D
3533.7	38116	6	8	.44	D
3533.8	37835	2	4	.32	D
3543.5	37757	2	2	.58	D
3545.3	37835	4	4	.32	D
3663.6	43649	4	6	2.7	D
3667.7	43707	6	8	2.4	D
3671.2	38124	8	10	.18	D
3672.4	44140	12	12	.79	D
3673.4	43788	8	10	2.4	D
3676.7	44327	14	14	1.1	D
3680.1	43894	10	12	1.9	D
3686.3	38221	10	12	.20	D
3687.5	44028	12	14	2.6	D
3688.1	29418	8	8	.28	D
3690.3	29203	2	4	.37	D
3692.2	29296	6	6	.46	D
3695.3	44190	14	16	2.5	D
3695.9	29203	4	4	.54	D
3703.6	29418	10	8	.79	D
3704.7	29296	8	6	.58	D
3705.0	29203	6	4	.31	D
3706.0	42079	10	10	.46	D
3708.7	42221	12	12	.39	D
3790.5	37475	10	8	.20	D
3795.0	28768	10	10	.21	D
3806.8	37362	10	10	.22	D
3818.2	26183	4	2	.56	C
3828.6	26249	6	4	.431	C
3840.1	36926	8	8	.18	D
3840.8	26353	8	6	.46	D
3855.4	25931	4	4	.28	D
3855.8	26480	10	8	.451	C
3863.9	36766	8	6	.27	D
3864.3	36763	8	6	.17	D
3864.9	26004	6	6	.208	C
3871.1	36926	10	8	.24	D
3875.1	26122	8	8	.17	D
3886.6	36823	10	8	.14	D
3902.3	26172	10	10	.217	C
3921.9	33967	4	2	.23	D
3922.4	34066	6	6	.23	D
3930.0	36539	10	10	.29	D
3934.0	34128	8	8	.56	D
3992.8	40039	12	10	1.1	D
3998.7	40064	14	12	.92	D
4051.0	41861	10	10	1.2	D
4051.4	41918	12	12	1.2	D
		V I			
4090.6	33155	8	10	.77	D
4092.7	26738	8	10	.21	D
4095.5	32989	6	8	.54	D
4099.8	26605	6	8	.39	D
4102.2	32847	4	6	.50	D
4104.0	40120	10	8	1.9	D
4105.2	26506	4	6	.42	D
4109.8	26438	2	4	.47	D
4111.8	26738	10	10	.91	D
4113.5	34128	6	8	.15	D
4115.2	26605	8	8	.59	D
4116.5	26506	6	6	.24	D
4123.6	26397	4	2	.94	D
4128.1	26438	6	4	.70	D
4132.0	26506	8	6	.52	D
4134.5	26605	10	8	.27	D
4232.5	39391	10	10	.86	D
4233.0	39342	8	8	.67	D
4268.6	38483	14	14	1.0	D
4271.6	38405	12	12	.84	D
4277.0	38324	10	10	.84	D
4284.1	38246	8	8	1.0	D
4291.8	40536	12	14	.78	D
4296.1	40452	10	12	.69	D
4297.7	40379	8	10	.61	D
4298.0	40315	6	8	.70	D
4342.8	38124	10	10	.13	D
4355.0	38221	12	12	.11	D
4379.2	25254	10	12	1.2	D
4384.7	25112	8	10	.97	D
4390.0	24993	6	8	.70	D
4395.2	24899	4	6	.48	D
4400.6	24830	2	4	.33	D
4406.6	25112	10	10	.19	D
4407.6	24993	8	8	.38	D
4408.2	24899	6	6	.51	D
4416.5	24789	4	2	.21	D
4452.0	37518	14	16	.80	D
4457.8	37530	10	12	.24	D
4460.3	24839	10	8	.26	D
4462.4	37404	12	14	.66	D
4468.0	37285	8	10	.20	D
4469.7	37316	10	12	.53	D
4474.0	38116	10	8	.41	D
4490.8	37211	10	12	.10	D
4496.1	37960	8	6	.35	D
4514.2	37835	6	4	.29	D
4524.2	37362	12	10	.26	D
4525.2	37757	4	2	.36	D
4529.6	37175	10	8	.21	D
4545.4	37765	10	12	.67	D
4560.7	37644	8	10	.62	D
4571.8	37556	6	8	.53	D
4578.7	37499	4	6	.59	D
4579.2	37556	8	8	.13	D
4706.2	36815	6	4	.21	D
4706.6	38483	12	14	.15	D
4710.6	38405	10	12	.17	D
4746.6	37423	4	4	.17	D
4751.0	37615	8	8	.15	D
4754.0	37758	10	10	.13	D
4757.5	37375	4	2	.65	D
4766.6	37423	6	4	.48	D
4776.4	37503	8	6	.43	D
4786.5	37615	10	8	.40	D
4796.9	37758	12	10	.42	D
4807.5	37931	14	12	.51	D
5193.0	37931	12	12	.35	D
5195.4	37615	8	8	.21	D
5234.1	38124	10	10	.41	D
5240.9	38221	12	12	.39	D
5415.3	37606	12	14	.27	D
5487.9	37362	12	10	.25	D
5507.8	37175	10	8	.30	D
5559.9	31786	6	2*	.14	D
5698.5	26122	6	8	.28	D
5703.6	26004	4	6	.19	D
5707.0	25931	2	4	.19	D
5725.6	36539	8	10	.18	D
5727.0	26172	8	10	.18	D
6090.2	25131	8	6	.13	D

Table 1b (Continued)
TRANSITION PROBABILITIES FOR SELECTED ATOMIC AND IONIC SPECIES

Wavelength (λ[Å])	Upper energy level (E_k[cm^{-1}])	Statistical weights g_i	g_k	Transition Probability (A[10^8 s^{-1}])	Uncertainty
V II					
2503.0	48580	5	7	.23	C
2506.2	48731	7	9	.23	C
2514.6	48853	9	11	.24	C
2672.0	37521	5	7	.18	D
2677.8	37369	3	5	.29	D
2679.3	37521	7	7	.26	D
2688.0	37531	9	9	.60	D
2690.8	37259	5	3	.43	D
2700.9	37352	9	11	.29	C
2702.2	37205	7	7	.20	D
2706.2	37151	7	9	.29	C
2728.6	36673	3	5	.20	C
2768.6	48731	11	9	.82	C
2774.3	48580	9	7	.88	C
2799.5	49202	5	5	.60	C
2802.8	49211	7	7	.46	C
2803.5	49269	9	9	.58	C
2836.5	48853	9	11	.25	C
2841.0	48731	7	9	.27	C
2877.7	49202	7	5	.67	C
2880.0	37521	7	7	.15	D
2882.5	37369	5	5	.27	D
2888.3	49269	11	9	.54	C
2889.6	37201	3	1	1.4	D
2891.6	37259	5	3	.92	D
2893.3	37521	9	7	.70	D
2903.1	37041	3	5	.21	D
2906.5	37205	7	7	.53	D
2907.5	37352	9	11	.19	C
2908.8	37531	11	9	1.1	D
2910.0	37041	5	5	.72	D
2910.4	36955	3	3	.87	D
2911.1	37151	7	9	.30	C
2924.6	37151	9	9	.91	C
2930.1	48580	7	7	.27	C
2944.6	36919	9	7	.65	C
2950.4	36489	3	3	.34	C
2952.1	36673	7	5	.58	C
2957.5	36489	5	3	.44	C
3048.9	49211	9	7	.54	C
3093.1	35483	11	13	1.8	D
3100.9	48580	7	7	1.0	C
3102.3	35193	9	11	1.6	D
3110.7	34947	7	9	1.5	D
3118.4	34746	5	7	1.5	D
3121.1	35193	11	11	.22	D
3125.3	34593	3	5	1.6	D
3126.2	34947	9	9	.41	D
3130.3	34746	7	7	.50	D
3133.3	34593	5	5	.48	D
3187.7	40002	5	5	.85	D
3188.5	40196	7	7	.80	D
3190.7	40430	9	9	.88	D
3208.4	40002	7	5	.17	D
3232.0	49202	3	5	.23	C
3267.7	39234	5	7	1.2	C
3271.1	39404	7	9	1.1	C
3276.1	39613	9	11	1.1	C
3321.5	49211	9	7	.15	C
3517.3	37521	9	7	.14	D
3530.8	36955	5	3	.19	D
3545.2	37041	7	5	.18	D
3556.8	37205	9	7	.20	D
3589.8	36489	5	3	.37	D
3592.0	36673	7	5	.23	D
3715.5	39613	13	11	.16	C
3727.3	40430	9	9	.22	C
3732.8	39404	11	9	.16	C
3745.8	39234	9	7	.17	C
3750.9	40196	7	7	.20	C
3771.0	40002	5	5	.22	C
V XX					
14.360	6964000	1	3	6.9 (+4)	C
160.0	625000	1	3	150.	C
V XXI					
8.843	11308000	2	6*	6000.	C
8.882	11675000	4	6	6400.	C
9.111	11316000	2	4	8900.	C
V XXI					
9.175	11316000	4	6	1.04(+4)	C
9.352	10693000	2	6*	1.0 (+4)	C
9.633	10721000	2	4	1.63(+4)	C
9.704	10722000	4	6	1.91(+4)	C
10.413	9603400	2	6*	2.0 (+4)	C
10.768	9627300	2	4	3.5 (+4)	B
10.853	9630600	4	6	4.2 (+4)	B
13.828	7231700	2	4	4.26(+4)	B
13.870	7209800	2	2	4.37(+4)	B
14.435	7268900	2	4	1.1 (+5)	B
14.578	7276300	4	6	1.3 (+5)	B
14.592	7268900	4	4	2.1 (+4)	C
27.95	10693000	2	6*	3000.	C
28.59	10722000	6	10*	6700.	C
40.20	9603400	2	6*	5800.	C
41.58	9629300	6	10*	1.4 (+4)	B
48.32	11673000	6	10*	1060.	C
58.39	11316000	6	10*	1660.	C
89.40	10722000	6	10*	2900.	C
240.0	416600	2	4	29.	B
293.7	340500	2	2	16.	B
V XXII					
2.382	41986020	1	3	2.9 (+6)	B
Xenon					
Xe I					
1043.8	95801	1	3	.59	D
1047.1	95499	1	3	1.3	D
1050.1	95229	1	3	.085	D
1056.1	94686	1	3	2.45	C
1061.2	94229	1	3	.19	D
1068.2	93619	1	3	3.99	C
1085.4	92129	1	3	.410	C
1099.7	90933	1	3	.434	C
1110.7	90033	1	3	1.5	C
1129.3	88550	1	3	.044	C
1170.4	85441	1	3	1.6	C
1192.0	83890	1	3	6.2	C
1250.2	79987	1	3	.14	D
1295.6	77186	1	3	2.5	C
1469.6	68046	1	3	2.8	B
4501.0	89279	5	3	.0062	D
4524.7	89163	5	5	.0021	D
4624.3	88687	5	5	.0072	D
4671.2	88470	5	7	.010	D
4807.0	88843	3	1	.024	D
7119.6	92445	7	9	.066	D
7967.3	88745	1	3	.0030	D
8409.2	78957	5	3	.010	D
Xe II					
4180.1	135708	4	4	2.2	D
4330.5	136598	6	8	1.4	D
4414.8	132208	6	6	1.0	D
4603.0	116783	4	4	.82	D
4844.3	113705	6	8	1.1	D
4876.5	130064	6	8	.63	D
5260.4	123255	2	4	.22	D
5262.0	131924	4	4	.85	D
5292.2	111959	6	6	.89	D
5372.4	113673	4	2	.71	D
5419.2	113512	4	6	.62	D
5439.0	121180	4	2	.74	D
5472.6	113705	8	8	.099	D
5531.1	113512	8	6	.088	D
5719.6	113512	4	6	.061	D
5976.5	111792	4	4	.28	D
6036.2	111959	6	6	.075	D
6051.2	111959	8	6	.17	D
6097.6	111792	6	4	.26	D
6270.8	128867	4	6	.18	D
6277.5	111959	4	6	.036	D
6805.7	113512	8	6	.061	D
6990.9	113705	10	8	.27	D
Ytterbium					
Yb I					
2464.5	40564	1	3	.91	C
2672.0	37415	1	3	.118	C
3464.4	28857	1	3	.62	C
3988.0	25068	1	3	1.76	C

Table 1b (Continued)
TRANSITION PROBABILITIES FOR SELECTED ATOMIC AND IONIC SPECIES

Wavelength (λ[Å])	Upper energy level (E_k[cm^{-1}])	Statistical weights g_i	g_k	Transition Probability (A[10^8 s^{-1}])	Uncertainty
		Yb I			
5556.5	17992	1	3	.0114	C
		Yb II			
3289.4	30392	2	4	1.8	C
3694.2	27062	2	2	1.4	C
		Zinc			
		Zn I			
748.29	133638	1	3	.060	C
765.60	130617	1	3	.076	C
792.05	126255	1	3	.057	C
793.85	125968	1	3	.18	C
809.92	123470	1	3	.26	C
1109.1	90158	1	3	.305	C
2138.6	46745	1	3	7.09	B
3075.9	32501	1	3	3.29(–4)	B
3282.3	62769	1	3	.90	B
3302.6	62772	3	5	1.2	B
3302.9	62769	3	3	.67	B
3345.0	62777	5	7	1.7	B
3345.6	62772	5	5	.40	B
3345.9	62769	5	3	.045	B
6362.3	62459	3	5	.474	C
11054.	55789	3	1	.243	C
		Zn II			
2025.5	49355	2	4	3.3	C
2064.2	96910	2	4	4.6	D
2099.9	96960	4	6	5.6	D
2102.2	96910	4	4	.93	D
4911.6	117264	4	6	1.6	D

REFERENCES

1. **Younger, S. M., Fuhr, J. R., Martin, G. A., and Wiese, W. L.,** *J. Phys. Chem. Ref. Data,* 7, 495, 1978.
2. **Fuhr, J. R., Martin, G. A., Wiese, W. L., and Younger, S. M.,** *J. Phys. Chem. Ref. Data,* to be published, 1980.
3. **Wiese, W. L., Smith, M. W., and Glennon, B. M.,** Atomic Transition Probabilities (H through Ne — A Critical Data Compilation), National Standard Reference Data Series, National Bureau of Standards 4, Vol. I, U.S. Government Printing Office, Washington, D.C., 1966.
4. **Wiese, W. L., Smith, M. W., and Miles, B. M.,** Atomic Transition Probabilities (Na through Ca — A Critical Data Compilation), National Standard Reference Data Series, National Bureau of Standards 22, Vol. II, U.S. Government Printing Office, Washington, D.C., 1969.
5. **Wiese, W. L. and Fuhr, J. R.,** *J. Phys. Chem. Ref. Data,* 4, 263, 1975.
6. **Wiese, W. L. and Weiss, A. W.,** *Phys. Rev.,* 175, 50, 1968.
7. **Smith, M. W. and Wiese, W. L.,** *Astrophys. J. Suppl. Ser.,* 23, No. 196, 103, 1971.
8. **Martin, G. A. and Wiese, W. L.,** *J. Phys. Chem. Ref. Data,* 5, 537, 1976.
9. **Fuhr, J. R., Miller, B. J., and Martin, G. A.,** Bibliography on Atomic Transition Probabilities (1914 through October 1977), National Bureau of Standards Special Publication 505, 1978; **Miller, B. J., Fuhr, J. R., and Martin, G. A.,** Bibliography on Atomic Transition Probabilities (November 1977 through February 1980), National Bureau of Standards Special Publication 505, Supplement 1, 1980.
10. **Moore, C. E.,** Bibliography on the Analyses of Optical Atomic Spectra, National Bureau of Standards Special Publication 306 — Section 1, 1968; Sections 2—4, 1969.
11. **Hagan, L. and Martin, W. C.,** Bibliography on Atomic Energy Levels and Spectra (July 1968 through June 1971), National Bureau of Standards Special Publication 363, 1972; **Hagan, L.,** Bibliography on Atomic Energy Levels and Spectra (July 1971 through June 1975), National Bureau of Standards Special Publication 363, Supplement 1, 1977; **Zalubas, R. and Albright, A.,** Bibliography on Atomic Energy Levels and Spectra (July 1975 through June 1979), National Bureau of Standards Special Publication 363, Supplement 2, 1980.
12. **Younger, S. M. and Weiss, A. W.,** *J. Res. Natl. Bur. Stand.,* 79A, 629, 1975.

NOMOGRAPH AND TABLE FOR DOPPLER LINEWIDTHS

Sidney O. Kastner

The Doppler width of a spectral line is given by the well-known relation $\Delta\lambda = (7.162 \times 10^{-7})\, \lambda T^{1/2} M^{-1/2}$, where wavelength units are in Angstroms, temperature is in degrees Kelvin and M is the atomic mass. This relation between four variables is amenable to representation by a nomograph, which does not appear to have been constructed but which would seem to be of practical value. Therefore such a nomograph is presented here. Its construction is briefly described so that the reader who has not made a plot of this type may follow the steps.

The Doppler relation is first rewritten as

$$\ell n\, \Delta\lambda = \ell n\, k\lambda + \tfrac{1}{2}\ell n\, T - \tfrac{1}{2}\ell n\, M$$

or equivalently $u_4 = u_1 + u_2 + u_3$, with $k = 7.162 \times 10^{-7}$. Putting $\xi - u_1 + u_2$, first, one has a linear relation $\xi - u_1 - u_2 = 0$ between three variables which can be represented (Menzel*) by the determinantal equation

$$\begin{vmatrix} \xi/2 & \tfrac{1}{2} & 1 \\ u_1 & 1 & 1 \\ u_2 & 0 & 1 \end{vmatrix} = 0$$

so that in the Cartesian (x,y) plane the function u_2 lies along the x axis, the function u_1 along the line $y=1$ and the function $\xi/2$ along the line $y = \tfrac{1}{2}$. These lines form three parallel scales with which to obtain a value of ξ, given the pair (u_1,u_2).

The original equation $u_4 = \xi + u_3$ provides a second linear relation $u_4 - \xi - u_3 = 0$, so that a second set of scales similarly results with u_3 along the x axis, ξ along the line $y=1$ and $u_4/2$ along the line $y = \tfrac{1}{2}$. The function $u_4 \equiv \log \Delta\lambda$ can then be obtained from the known pair (ξ,u_3).

In practice, for the nomograph scales to represent useful ranges of the physical variables, some shifting and magnification of the individual scales will be found to be necessary, after completion of the first diagram. The nomograph arrived at here is shown in Figure 1. The x axis runs horizontally through the values $x=0$, $T=10^5$, and the ordinates of the six scales from left to right are given by:

$$y_1 = 5\ell n(10^{5/2}\, k\lambda) \qquad y_2 = 10\ell n(k\lambda T^{1/2})$$

$$y_3 = \frac{5}{2}\, \ell n(k\lambda T^{1/2}) \qquad y_4 = 5\ell n\Delta\lambda + \frac{5}{2}\, \ell n 10$$

$$y_5 = 5\ell n(T^{1/2}/10^{5/2}) \qquad y_6 = 5\ell n(10/M)$$

The approximate range of temperatures covered by the figure is a slowly varying function of wavelength, being 1000°K-10^6°K for $\lambda = 10{,}000$Å; 4000°K-5×10^6°K for $\lambda = 5000$Å; 80,000°K-10^7°K for $\lambda = 1000$Å; and 300,000°K-10^7°K for $\lambda = 500$Å.

In use, the left-hand sides of the three lines are used to find the value of the intermediate variable x corresponding to given λ and T. The right-hand sides then give the Doppler width $\Delta\lambda$ in Å, for this value of x and the given M.

For example, suppose one wishes to find the Doppler width of the solar coronal forbidden line of Fe XIV (atomic mass $M=56$) at 5303Å, emitted at a temperature of about $T = 2\times10^6$°K. A straight line drawn between $\lambda = 5300$Å and $T = 2\times10^6$ intersects the x scale at about 4.2. A second line then drawn between $x=4.2$ and $M=56$ intersects the middle line at $\Delta\lambda \approx 0.73$Å.

To use the nomograph, one must know the appropriate atomic weight M for a given element of interest. An alternative and more accurate procedure, if a calculator is available, for obtaining any required Doppler width is to use Table I, which gives the value of the constant $k_Z \equiv 7.162\, M_Z^{-1/2}$ for any given atomic number Z, in the equivalent Doppler relation $\mathrm{lgD}\lambda = k_Z(\lambda T^{1/2}) \times 10^{-7}$. This table thereby avoids the necessity of looking up the atomic weight M. For the forbidden line example above, where $Z=26$, the Doppler width obtained from the table in $\Delta\lambda = (0.9584)(5303)(2{,}000{,}000)^{1/2}(10^{-7}) = 0.719$Å.

* D. H. Menzel, *Fundamental Formulas of Physics,* Dover Publications, New York, 1960, ch. 3.

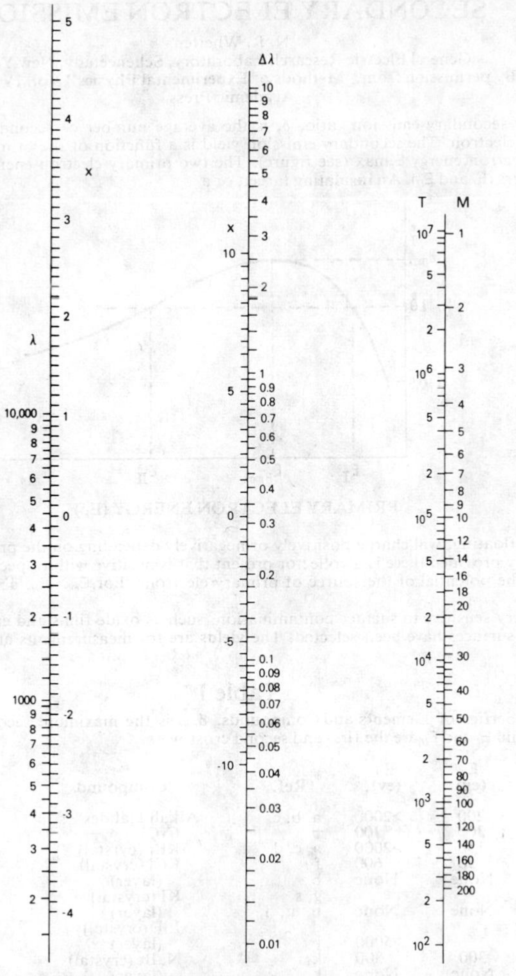

NOMOGRAPH FOR DOPPLER LINEWIDTH

λ in Å; T in °K; M atomic weight

(a) Use left-hand scales (λ, T) to find x

(b) Use right-hand scales (x, M) to find Δλ

Table 1
VALUES OF K_Z (Z ATOMIC NUMBER) FOR USE IN DOPPLER LINEWIDTH FORMULA $\Delta\lambda = K_Z(\lambda T^{1/2}) \times 10^{-7}$

Z	k_Z	Z	k_Z	Z	k_Z	Z	k_Z
1	7.1335	21	1.0682	41	0.7430	61	0.5927
2	3.5798	22	1.0348	42	0.7312	62	0.5840
3	2.7185	23	1.0035	43	0.7202	63	0.5810
4	2.3857	24	0.9932	44	0.7124	64	0.5711
5	2.1782	25	0.9663	45	0.7060	65	0.5681
6	2.0665	26	0.9584	46	0.6943	66	0.5618
7	1.9137	27	0.9329	47	0.6896	67	0.5577
8	1.7905	28	0.9347	48	0.6755	68	0.5538
9	1.6431	29	0.8984	49	0.6684	69	0.5510
10	1.5944	30	0.8858	50	0.6574	70	0.5492
11	1.4937	31	0.8577	51	0.6491	71	0.5414
12	1.4527	32	0.8406	52	0.6340	72	0.5361
13	1.3788	33	0.8274	53	0.6358	73	0.5324
14	1.3514	34	0.8060	54	0.6250	74	0.5282
15	1.2869	35	0.8012	55	0.6212	75	0.5249
16	1.2649	36	0.7824	56	0.6111	76	0.5193
17	1.2028	37	0.7747	57	0.6077	77	0.5166
18	1.1331	38	0.7651	58	0.6050	78	0.5128
19	1.1453	39	0.7596	59	0.6033	79	0.5103
20	1.1313	40	0.7499	60	0.5963	80	0.5057

SECONDARY ELECTRON EMISSION

N. R. Whetten
General Electric Research Laboratory, Schenectady, New York
By permission from "Methods of Experimental Physics" Vol. IV (1962)
Academic Press

The secondary emission yield, or secondary emission ratio, δ, is the average number of secondary electrons emitted from a bombarded material for every incident primary electron. The secondary emission yield is a function of the primary electron energy. δ_{max} is the maximum yield corresponding to a primary electron energy $E_p$max (see figure). The two primary electron energies corresponding to a yield of unity are denoted the first and second crossovers (E_I and E_{II}). An insulating target, or a

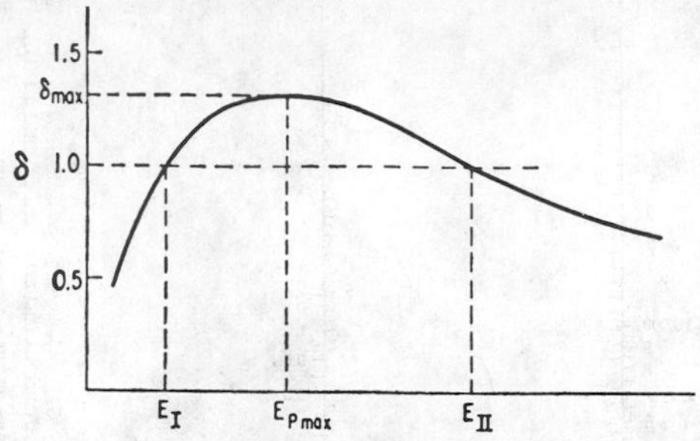

PRIMARY ELECTRON ENERGY (Ep)

conducting target that is electrically floating, will charge positively or negatively depending on the primary electron energy. For $E_I < E_p < E_{II}$, $\delta > 1$ and the surface charges positively provided there is a collector present that is positive with respect to the target. For $E_p < E_I$, $\delta < 1$, and the surface charges negatively towards the potential of the source of primary electrons. For $E_p > E_{II}$, $\delta < 1$, and the surface charges negatively to the second crossover.

The secondary emission yield is very sensitive to surface contamination, such as oxide films and carbon deposits. Whenever possible, yields believed to be most typical of clean surfaces have been selected. The yields are for measurements at room temperature and normal incidence of the primary electrons.

Table I

Secondary Electron Emission Properties of Elements and Compounds. δ_{max} is the maximum secondary emission yield. $\varepsilon_{p\ max}$ the primary electron energy for maximum yield, and E_I and E_{II} are the first and second crossovers.

Elements	δ_{max}	$E_{p\ max}$ (ev)	E_I (ev)	E_{II} (ev)	Ref.
Ag	1.5	800	200	>2000	a, b, c
Al	1.0	300	300	300	b
Au	1.4	800	150	>2000	a, c, d
B	1.2	150	50	600	f
Ba	0.8	400	None	None	b
Bi	1.2	550			g, s
Be	0.5	200	None	None	b, h, i, d
C (diamond)	2.8	750		>5000	j
(graphite)	1.0	300	300	300	k
(soot)	0.45	500	None	None	k
Cd	1.1	450	300	700	l, d
Co	1.2	600	200		m, n
Cs	0.7	400	None	None	b, o
Cu	1.3	600	200	1500	a, l, b
Fe	1.3	400	120	1400	n, c, p
Ga	1.55	500	75		q
Ge	1.15	500	150	900	f, r, s
Hg	1.3	600	350	>1200	q
K	0.7	200	None	None	t, u
Li	0.5	85	None	None	b
Mg	0.95	300	None	None	o, b
Mo	1.25	375	150	1200	a, w, c, e, p
Na	0.82	300	None	None	x
Nb	1.2	375	150	1050	a, c
Ni	1.3	550	150	>1500	a, n, m, w, p
Pb	1.1	500	250	1000	g, q
Pd	>1.3	>250	120		v
Pt	1.8	700	350	3000	c
Rb	0.9	350	None	None	t
Sb	1.3	600	250	2000	y
Si	1.1	250	125	500	f
Sn	1.35	500			g, x
Ta	1.3	600	250	>2000	a
Th	1.1	800			b
Ti	0.9	280	None	None	k
Tl	1.7	650	70	>1500	s
W	1.4	650	250	>1500	z, a, a′, c
Zr	1.1	350			k

Compounds	δ_{max}	$E_{p\ max}$ (ev)	Ref.
Alkali Halides			
CsCl	6.5		b′
KBr (cyrstal)	14	1800	c′,e′
KCl (crystal)	12	1600	p′,d′
(layer)	7.5	1200	b′,f′
KI (crystal)	10	1600	c′,d′,e′
(layer)	5.6		b′
LiF (crystal)	8.5		g′
(layer)	5.6	700	b′
NaBr (crystal)	24	1800	g′,h′,d′
(layer)	6.3		b′
NaCl (crystal)	14	1200	i′,e′,c′,d′,g′
(layer)	6.8	600	b′,j′
NaF (crystal)	14	1200	g′
(layer)	5.7		b′
NaI (crystal)	19	1300	g′
(layer)	5.5		b′
RbCl (layer)	5.8		b′
Oxides			
Ag_2O	1.0		l′
Al_2O_3 (layer)	2 to 9		k′,m′,i′,b
BaO (layer)	2.3 to 4.8	400	m′,b
BeO	3.4	2000	m′
CaO	2.2	500	m′
Cu_2O	1.2	400	b′,n′
MgO (crystal)	20 to 25	1500	t′,u′,v′
(layer)	3 to 15	400 to 1500	w,m′,r′,s′,t′
MoO_2	1.2		l′
SiO_2 (quartz)	2.1 to 4	400	k′,g′
SnO_2	3.2	640	o′
Sulfides			
MoS_2	1.1		b′
PbS	1.2	500	p′
WS_2	1.0		b′
ZnS	1.8	350	q′
Others			
BaF_2 (layer)	4.5		b′
CaF_2 (layer)	3.2		b′
$BiCs_3$	6	1000	p′
BiCs	1.9	1000	p′
GeCs	7	700	p′
Rb_3Sb	7.1	450	p′
$SbCs_3$	6	700	p′,x′
Mica	2.4	350	k′
Glasses	2 to 3	300 to 450	k′,y′

References

a R. Warnecke J. Phys. Radium, 7, 270 (1936).
b H. Bruining and J. H. deBoer, Physica, 5, 17 (1938).
c R. Kollath, Physik. Z., 38, 202 (1937).
d R. Suhrmann and W. Kundt, Z. Physik, 121, 118 (1943).
e P. L. Copeland, J. Franklin Inst., 215, 593 (1933).
f L. R. Koller and J. S. Burgess, Phys. Rev., 70, 571 (1946).
g P. M. Morozov, J. Exptl. Theoret. Phys. USSR, 11, 410, (1941).
h R. Kollath, Ann. Physik, 33, 285 (1938).
i E. G. Schneider, Phys. Rev., 54, 185 (1938).
j J. B. Johnson, Phys. Rev., 92, 843 (1953).

Table I (Continued)

References

k H. Bruining, Philips Tech. Rev., 3, 80 (1938).
l R. Suhrmann and W. Kundt, Z. Physik, 120, 363 (1943).
m D. E. Woolridge, Phys. Rev., 56, 1062 (1939).
n L. R. G. Treloar and D. H. Landon, Proc. Soc. (London), B50, 625 (1938).
o N. S. Klebnikov, Tech. Phys. USSR, 5, 593 (1938).
p R. L. Petry, Phys. Rev., 26, 346 (1925).
q J. J. Brophy, Phys. Rev., 83, 534 (1951).
r J. B. Johnson and K. G. McKay, Phys. Rev., 93, 668 (1954).
s H. Gobrecht and F. Spear, Z. Physik, 135, 602 (1953).
t A. Afanasjewa and P. W. Timofeew, Tech. Phys. USSR, 4, 953 (1937).
u M. S. Joffe and I. V. Nechlaev, J. Exptl. Theoret. Phys. USSR, 11, 93 (1941).
v H. E. Farnsworth, Phys. Rev., 25, 41 (1925).
w G. Blankenfeld, Ann. Physik. 9, 48 (1951).
x J. Woods, Proc. Phys. Soc. (London), B67, 843 (1954).
y R. Kollath, Handbuch der Physik, vol. 21 (1956), p. 232.
z R. L. Petry, Phys. Rev. 28, 362 (1926).
a′ E. A. Coomes, Phys. Rev., 55, 519 (1939).
b′ H. Bruining and J. H. deBoer, Physica, 6, 834 (1939).
c′ D. N. Dobretzov and A. S. Titkow, Doklady Acad. Nauk USSR, 100, 33 (1955).
d′ N. R. Whetten, Bull. Am. Phys. Soc. Ser. II, 5, 347 (1960).
e′ A. R. Shulman and B. P. Dementyev, J. Tech. Phys. USSR, 25, 2256 (1955).
f′ M. Knoll, O. Hachenberg, and J. Randmer, Z. Physik. 122, 137 (1944).
g′ D. N. Dobretzov and T. L. Matskevich, J. Tech. Phys. USSR, 27, 734 (1957).
h′ T. L. Matskevich, J. Tech. Phys. USSR, 26, 2399 (1956).
i′ A. R. Shulman, W. L. Makedonsky and J. D. Yaroshetsky, J. Tech. Phys. USSR 23, 1152 (1953).
j′ M. M. Vudinsky, J. Tech. Phys. USSR, 9, 271 (1939).
k′ H. Salow, Z. Tech. Phys., 21, 8 (1940).
l′ A. Afanasjewa, P. Timofeew, and A. Ignaton, Phys. Z. Sowjet, 10, 831 (1936).
m′ K. H. Geyer, Ann Phys., 42, 241 (1942).
n′ N. B. Gornij, J. Exptl. Theoret. Phys. USSR, 26, 79 (1954).
o′ H. E. Mendenhall, Phys. Rev., 72, 532 (1947).
p′ O. Hachenberg and W. Braner in Advances in Electronics and Electron Physics, Vol. XI, Academic Press, New York (1959), p. 438.
q′ N. B. Gornij, J. Exptl. Theoreto. Phys. USSR, 26, 88 (1954).
r′ P. Wargo, B. V. Haxby, and W. G. Shepherd, J. Appl. Phys. 27, 1311 (1956).
s′ P. Rappaport, J. Appl. Phys., 25, 288 (1954).
t′ N. R. Whetten and A. B. Laponsky, J. Appl. Phys., 30, 432 (1959).
n′ J. R. Johnson and K. G. McKay, Phys. Rev., 91, 582 (1953).
v′ R. G. Lye, Phys. Rev., 99, 1647 (1955).
w′ N. R. Whetten and A. B. Laponsky, J. Appl. Phys., 28, 515 (1957).
x′ N. D. Morgulis and B. I. Djatlowitskaja, J. Tech. Phys. USSR, 10, 657 (1940).
y′ C. W. Mueller, J. Appl. Phys., 16, 453 (1945).

EMISSIVITY OF TUNGSTEN

Wavelengths in μ

Temperature °K	0.25	0.30	0.35	0.40	0.50	0.60	0.70
1600	.448*	.482	.478	.481	.469	.455	.444
1800	.442*	.478*	.476	.477	.465	.452	.44
2000	.436*	.474	.473	.474	.462	.448	.436
2200	.429*	.470	.470	.471	.458	.445	.431
2400	.422	.465	.466	.468	.455	.441	.427
2600	.418	.461	.464	.464	.451	.437	.423
2800	.411	.456	.461	.461	.448	.434	.419

Temperature °K	0.80	0.90	1.0	1.1	1.2	1.3	1.4
1600	.431	.413	.39	.366	.345	.322*	.300*
1800	.425	.407	.385	.364	.344	.323*	.302*
2000	.419	.401	.381	.361	.343	.323	.305
2200	.415	.396	.378	.359	.342	.324	.306
2400	.408	.391	.372	.355	.340	.324	.309
2600	.404	.386	.369	.352	.338	.325	.310
2800	.400	.383	.367	.352	.337	.325	.313

Temperature °K	1.5	1.6	1.8	2.0	2.2	2.4	2.6
1600	.279*	.263*	.234*	.210*	.19*	.175*	.164*
1800	.282	.267*	.241*	.218*	.20*	.182*	.174*
2000	.288	.273	.247	.227	.209*	.197	.175
2200	.291	.278	.254	.235	.218	.205	.194
2400	.296	.283	.262	.244	.228	.215	.205
2600	.299	.288	.269	.251	.236	.224	.214
2800	.302	.292	.274	.259	.245	.233	.224

* Values by extrapolation.

LIQUIDS FOR INDEX BY IMMERSION METHOD

Liquid	N_D 24°C
Trimethylene chloride	1.446
Cineole	1.456
Hexahydrophenol	1.466
Decahydronaphthalene	1.477
Isoamylphthalate	1.486
Tetrachloroethane	1.492
Pentachloroethane	1.501
Trimethylene bromide	1.513
Chlorobenzene	1.523
Ethylene bromide + Chlorobenzene	1.533
α-Nitrotoluene	1.544
Xylidine	1.557
α-Toluidine	1.570
Aniline	1.584
Bromoform	1.595
Iodobenzene + Bromobenzene	1.603
Iodobenzene + Bromobenzene	1.613
Quinoline	1.622
α-Chloronaphthalene	1.633
α-Bromophthalene + α-Chloronaphthalene	1.650-1.650
α-Bromophthalene + α-Iodonaphthalene	1.660-1.690
Methylne iodide + Iodobenzene	1.700-1.730
Methylene iodide	1.738
Methylene iodide saturated with sulfur	1.78
Yellow phosphorus, sulfur and methylene iodide (8:1:1 by weight)	2.06

Can be diluted with methylene iodide to cover range 1.74-2.06. For precautions in use, cf. West, Am. Mineral, 21, p. 245-9 (1936).

HEAVY LIQUIDS FOR MINERAL SEPARATION

Liquid	Density
Tetrabromoethane (sym.)	2.964, 20°/4°
Can be diluted with carbon tetrachloride (1.595) or benzene (0.894)	3.325, 20°/4°
Methylene iodide	
Can be diluted with carbon tetrachloride or benzene.	
Thallium formate, sq.	3.5
Can be diluted with water.	
Thallium malonate-thallium formate, aq.	4.9
Can be diluted with water.	

For preparation and recovery of these liquids, cf. U. S. Bureau Mines, Rept. Inv. #2897 (1928).

INDEX OF REFRACTION

Indices of refraction for elements, inorganic, metal-organic and organic compounds and minerals will be found in the tables of physical constants for the various classes of substances in the section Properties and Physical Constants.

Values for compounds not there listed and data subsequently collected are given below.

Indices not otherwise indicated are for sodium light, λ = 589.3 mμ. Other wave lengths are indicated by the value in millimicrons or symbol in parentheses which follows the index. Wave lengths are indicated as follows: He, λ = 587.6 mμ; Li, λ = 670.8 mμ; Hg, λ = 579.1 mμ; A, λ = 759.4 mμ; C, λ = 656.3 mμ; D, λ = 589.3 mμ; F, λ = 486.1 mμ.

Temperatures are understood to be 20°C for liquids, or ordinary room temperatures in the case of solids. Other temperatures appear as superior figures with the index.

Indices for the elements and inorganic compounds will be understood to be for the solid form except as indicated by the abbreviation liq.

See also under Physical Constants of Inorganic Compounds and index of Refraction of Gases.

Elements

Name	Formula	Index
Bromine (liq.)	Br_2	1.661_{15}
Cadmium (liq.)	Cd	0.82 (579 mμ)
(sol.)		1.13
Chlorine (liq.)	Cl_2	1.385
(gas)		1.00768
Hydrogen (liq.)	H_2	$1.10974^{-252.83}$(579 mμ)
Iodine (sol.)	I_2	3.34
(gas)		1.001920
Lead	Pb	2.6 (579 mμ)
Mercury (liq.)	Hg	1.6-1.9
Nitrogen (liq.)	N_2	1.2053^{-190}
Oxygen (liq.)	O_2	1.221^{-131}
Phosphorous (yel.) (sol.)		2.1442^{25}
Selenium	Ses	3.00, 4.04
(amor.) (sol.)		2.92
Sodium (liq.)	Na	0.0045
(sol.)		4.22
Sulfur (liq.)	S_6	1.929^{110}
(amor.) (sol.)		.1998
(rhombic, α)		1.957, 2.0377, 2,2454
Tin (liq.)	Sn	2.1

Inorganic Compounds

See also under Physical Constants of Inorganic Compounds

Name	Formula	Index
Aluminum carbide	AlC_3	2.7, 2.75 (700 mμ)
chloride	$AlCl_3 \cdot 6H_2O$	1.560, 1,507
oxide	Al_2O_3	1.665-1.680, 1.63-1.65
Alums. See under appropriate element.		
Ammonium antimony tartrate	$2(NH_4 \cdot SbO \cdot C_4H_4O_6)H_2O$	β1.6229 (C)
*ortho*arsenate, di-H	NH;$4H_2AsO_4$	1.5766, 1.5217
bromide	NH_4Br	1.7108
*per*chlorate	NH_4ClO_4	1.4818, 1.4833, 1.4881
chloroplatinate	$(NH_4)_2PtCl_6$	1.8
fluoride	NH_4F	ω<1.328
acid	NH_4HF_2	1.385, 1.390, 1.394
hydrogen malate (*d*)	$NH_4C_6H_5O_5$	β1.503
nitrate	NH_4NO_3	1.413, 1.611(He), 1.63
Ammonium sulfate, acid	NH_4HSO_4	1.463, 1.473, 1.510
tartrate (*dl*)	$(NH_4)_2C_4H_4O_6 \cdot 2H_2O$	β1.564
thiocyanate	NH_4CNS	1.546, 1.685, 1.692
uranyl acetate	$NH_4C_2H_3O_2 \cdot UO_2(C_2H_3O_2)_2$	1.4808, 1.4933
Antimony bromide	$SbBr_3$	>1.74 +
iodide, tri-	SbI_3	2.78 (Li), 2.36
Barium cadmium bromide	$BaCdBr_4 \cdot 4H_2O$	β1.702
cadmium chloride	$BaCdCl_4 \cdot 4H_2O$	β1.651
calcium propionate	$BaCa_2(C_3H_5O_2)_6$	1.4442
fluochloride	$BaCl_2 \cdot BaF_2$	1.640, 1.633
fluoride	BaF_2	1.475 also 1.4741
Barium oxide	BaO	1.980
*ortho*phosphate, di-	$BaHPO_4$	1.617, 1.63±, 1.635
propionate	$Ba(C_2H_5CO_2)_2 \cdot H_2O$	β1.5175
sulfide, mono-	BaS	2.155
Cadmium ammonium chloride	$CdCl_2 \cdot 4NH_4Cl$	1.6038, 1.6042
cesium sulfate	$CdSO_4 \cdot Cs_2SO_4 \cdot 6H_2O$	1.498, 1.500, 1.506
fluoride	CdF_2	1.56
magnesium chloride	$(CdCl_2)_2 \cdot MgCl_2 \cdot 12H_2O$	1.49, 1.5331, 1.5769
oxide	CdO	2.49 (Li)
potassium chloride	$CdCl_2 \cdot 4KCl$	1.5906, 1.5907
cyanide	$Cd(CN)_2 \cdot 2KCN$	1.4213
rubidium sulfate	$CdSO_4 \cdot Rb_2SO_4 \cdot 6H_2O$	1.4798, 1.4848, 1.4948
Calcium aluminate	$Ca_3Al_2O_6$	1.710
borate	$Cao \cdot B_2O_3$	1.540, 1.656, 1.682
carbide	CaC_2	<1.75
copper acetate	$CaCu(C_2H_3O_2)_4 \cdot 6H_2O$	1.436, 1.478
cyanamide	$CaCN_2$	1.60, <1.95
dithionate	$CaS_2O_6 \cdot 4H_2O$	1.5516, 1.5414
*pyro*phosphate	$Ca_2P_2O_7$	1.585, 1.60±, 1.605
platinocyanide	$CaPt(CN)_4 \cdot 5H_2O$	1.623, 1.644, 1.767
stromtium propionate	$Ca_2Sr(C_3H_5O_2)_6$	1.4871, 1.4956
sulfide (oldhamite)	CaS	2.137
sulfite	$CaSO_3 \cdot 2H_2O$	1.590, 1.595, 1.628
thiosulfate	$CaS_2O_3 \cdot 6H_2O$	1.545, 1.560, 1.605
Carbon dioxide (liq.)	CO_2	1.195^{15}
Cerium dithionate	$Ce_2(S_2O_6)_3 \cdot 15H_2O$	β1.507
Cesium *per*chlorate	$CsClO_4$	1.4752, 1.4788, 1.4804
nitrate	$CsNO_3$	1.55, 1.56
selenate	Cs_2SeO_4	1.5989, 1.5999, 1.6003
thallium chloride	$Cs_2Tl_2Cl_9$	1.784, 1.774
Chromium cesium sulfate	$CrCs(SO_4)_2 \cdot 12H_2O$	1.4810
oxide (ic)	Cr_2O_3	2.5
potassium cyanide (ic)	$CrK_3(CN)_6$	4.5221, 1.5244, 1.5373
sulfate (ic)	$Cr_2(SO_4)_3 \cdot 18H_2O$	1.564
thallium sulfate	$CrTl(SO_4)_2 \cdot 12H_2O$	1.5228
Cobalt acetate	$Co(C_2H_3O_2)_2 \cdot 4H_2O$	β1.542
aluminate (Thenard's Blue)	$Co(AlO_2)_2$	<1.78 (red), 1.74 (blue)
ammonium selenate	$CoSeO_4 \cdot (NH_4)_2SeO_4 \cdot 6H_2O$	1.5246, 1.5311, 1.5396
cesium sulfate	$CoCs_2(SO_4)_2 \cdot 6H_2O$	1.5057, 5.5085, 1.5132
chloride (ous)	$CoCl_2 \cdot 2H_2O$	<1.624, <1.671, >1.67
potassium selenate	$C\ oSeO_4 \cdot K_2SeO_4 \cdot 6H_2O$	1.5135, 1.5195, 1.5358
rubidium sulfate	$CoSO_4 \cdot Rb_2SO_4 \cdot 6H_2O$	1.4859, 1.4916, 1.5014
selenate	$CoSeO_4 \cdot 6H_2O$	α1.5225, γ1.5227
Copper ammonium selenate	$CuSeO_4 \cdot (NH_4)_2SeO_4 \cdot 6H_2O$	1.5213, 1.5355, 1.5395
ammonium sulfate	$CuSO_4 \cdot (NH_4)_2SO_4 \cdot 6H_2O$	1.4910, 1.5007, 1.5054
cesium sulfate	$CuSO_4 \cdot Cs_2SO_4 \cdot 6H_2O$	1.5048, 1.5061, 1.5153
chloride (ic)	$CuCl_2 \cdot 2H_2O$	1.644, 1.684, 1.742
formate	$Cu(CHO_2)_24H_2O$	1.4133, 1.5423, 1.5571
Copper oxide (ous) (cuprite)	Cu_2O	2.705
potassium chloride	$CuCl_2 \cdot 2KCl \cdot 2H_2O$	1.6365, 1.6148
potassium cyanide (ous)	$CuK_3(CN)_4$	1.5215
potassium selenate	$CuSeO_4 \cdot K_2SeO_4 \cdot 6H_2O$	1.5096, 1.5235, 1.5387
potassium sulfate	$CuSO_4K_2SO_4 \cdot 6H_2O$	1.4836, 1.4864, 1.5020
strontium formate	$Cu(HCO_2)_2 \cdot 2[SrHCO_2)_2]$ $8H_2O$	1.4995, 1.5199, 1.5801
sulfate (ic)	$CuSO_4$	1.724, 1.733, 1.739
Cyanogen	C_2N_2	1.327^{18} (liq.)
Germanium bromide, tetra-	$GeBr_4$	1.6269
Gold sodium chloride	$AuNaCl_4 \cdot 2H_2O$	α1.545, γ1.75 +
Hafnium oxychloride	$HfOCl_2 \cdot 8H_2O$	1.557, 1.543
Ice	H_2O	1.3049, 1.3062 (A), 1.3001, 1.3104 (D), 1.3133, 1.3147 (F)
Iron ammonium chloride	$Fe(NH_4)_2Cl_4$	1.6439
ammonium selenate	$FeSeO_4 \cdot (NH_4)_2SeO_4 \cdot 6H_2O$	1.5201, 1.5260, 1.5356
cesium sulfate (ic)	$FeCs(SO_4)_2 \cdot 12H_2O$	1.4839
cesium sulfate (ous)	$FeSO_4 \cdot Cs_2SO_4 \cdot 6H_2O$	1.5003, 1.5035, 1.5094
rubidium sulfate	$FeRb(SO_4)_2 \cdot 12H_2O$	1.48234
sulfate (ic)	$Fe_2(SO_4)_3$	1.802, 1.814, 1.818
thallium sulfate	$FeTl(SO_4)_2 \cdot 12H_2O$	1.52365
Lanthanum sulfate	$La_2(SO_4)_3 \cdot 9H_2O$	1.564, 1.569
Lead *ortho*arsenate, di-	$PbHAsO_4$	1.8903, 1.9097, 1.9765
nitrate	$Pb(NO_3)_2$	1.782
Lithium ammonium sulfate	$LiNH_4SO_4$	β1.437 (Li)
ammonium tartrate (*d*)	$LiNH_4(C_4H_4O_6) \cdot H_2O$	β1.567, γ1.5673
ammonium tartrate (*dl*)	$LiNH_4(C_4H_4O_6) \cdot H_2O$	β1.5287
bromide	LiBr	1.784
chloride	LiCl	1.662
dithionate	$Li_2S_2O_6 \cdot H_2O$	1.5487, 1.5602, 1.5788
oxide	Li_2O	1.644
potassium sulfate	$LiKSO_4$	1.4723, 1.4717
potassium tartrate	$LiK(C_4H_4O_6) \cdot H_2O$	β1.5226 (red)
rubidium tartrate (*a*)	$LiRb(C_4H_4O_6) \cdot H_2O$	β1.552
sodium tartrate (*dl*)	$LiNa(C_4H_4O_6) \cdot 2H_2O$	β1.4904
Magnesium ammonium selenate	$MgSeO_4 \cdot (NH_4)_2SeO_4 \cdot 6H_2O$	1.5070, 1.5093, 1.5169
ammonium sulfate	$Mg(NH_4)_2 \cdot (SO_4)_2 \cdot 6H_2O$	1.4716, 1.4730, 1.4786
*ortho*borate	$3MgO \cdot B_2O_3$	1.6527, 1.6537, 1.6748
cesium sulfate	$MgCs_2(SO_4)_2 \cdot 6H_2O$	1.4857, 1.4858, 1.4916
chlorostannate	$MgSnCl_6 \cdot 6H_2O$	1.5885, 1.5970
fluosilicate	$MgSiF_6 \cdot 6H_2O$	1.3439, 1.3602
platinocyanide	$MgPt(CN)_4 \cdot 7H_2O$	1.5608, 1.91
Magnesium potassium selenate	$MgK_2(SeO_4)_2 \cdot 6H_2O$	1.4969, 1.4991, 1.5139
potassium sulfate	$MgK_2(SO_4)_2 \cdot 6H_2O$	1.407, 1.4629, 1.4755
rubidium sulfate	$MgRb_2(SO_4)_2 \cdot 6Hi2O$	1.4672, 1.4689, 1.4779
silicate	$MgSiO_3$	1.651, 1,654 (calc.), 1.660
sulfide	MgS	2.271 also 2.268
Manganese borate	$Mn_3B_4O_9$	1.617, 1.738, 1.776
cesium sulfate	$MnCs_2(SO_4)_2 \cdot 6H_2O$	1.4946, 1.4966, 1.5025
chloride	$MnCl_2 \cdot 4H_2O$	1.555, 1.575, 1.607
rubidium sulfate	$MnRb_2(SO_4)_2 \cdot 6H_2O$	1.4767, 1.4807, 1.4907
sulfate (ous)	$MnSO_4 \cdot 4H_2O$	1.508, 1.518, 1.522
	$MnSO_4 \cdot 5H_2O$	1.495, 1.508, 1.514
Mercury chloride (ic)	$HgCl_2$	1.725, 1.859, 1.965
cyanide (ic)	$Hg(CN)_2$	1.645, 1.492
iodide (ic) (red)	HgI_2	2.748, 2.455

Inorganic Compounds (Continued)

Name	Formula	Index
Nickel ammonium selenate	$Ni(NH_4)_2\cdot(SeO_4)_2\cdot 6H_2O$	1.5291, 1.5372, 1.5466
cesium sulfate	$NiCs_2(SO_4)_2\cdot 6H_2O$	1.5087, 1.5129, 1.5162
Nickel chloride	$NiCl_2\cdot 6H_2O$	α1.53, γ1.61
fluoride, acid	$NiF_2\cdot 5HF\cdot 6H_2O$	1.392, 1.408
potassium selenate	$NiK_2(SeO_4)_2\cdot 6H_2O$	1.5199, 1.5248, 1.5339
rubidium sulfate	$NiRb_2(SO_4)_2\cdot 6H_2O$	1.4895, 1.4961, 1.505
selenate	$NiSeO_4\cdot 6H_2O$	1.593, 1.5125
Platinum potassium dibromo-nitrite	$PtK_2(NO_2)_2)Br_2\cdot H_2O$	1.626, 1.6684, 1.757
Potassium carbonate	K_2CO_3	1.426, 1.531, 1.541
carbonate, acid	$KHCO_3$	1.380, 1.482, 1.598
*perch*lorate	$KC10_4$	1.4731, 1.4737, 1.4769
chloroplatinate	K_2PtCl_6	1.827 (577 mμ)
chloroplatinite	K_2PtCl_4	1.64, 1.67
*di*chromate	$K_2Cr_2O_7$	1.7202, 1.7280, 1.8107
cyanide	KCN	1.410
fluoborate	KBF_4	1.3239, 1.3245, 1.3247
fluoride	KF	1.352 (1.361)
	$KF\cdot 2H_2O$	1.345, 1.352, 1.363
fluosilicate	K_2SiF_6	1.3391
*per*iodate	$KClO_4$	1.6205, 1.6479
lithium ferrocyanide	$K_2Li_2Fe(CN)_6\cdot 3H_2O$	1.5883, 1.6007, 1.6316
*hypo*phosphate	$K_2H_2P_2O_6\cdot 2H_2O$	1.4893, 1.5314, 1.5363
	$K_2H_2P_2O_6\cdot 3H_2O$	1.4768, 1.4843, 1.4870
ruthenium cyanide	$K_4Ru(CN)_6\cdot 3H_2O$	β1.5837
silicate	K_2SiO_3	1.520, 1.521, 1.528
thiocyanate	$KCNS$	1.532, 1.660, 1.730
thionate, tetra-	$K_2S_4O_6$	1.5896, 1.6057, 1.6435
penta-	$2K_2S_5O_6\cdot 3H_2O$	1.565, 1.63, 1.655
Rhodium cesium sulfate	$RhCs(SO_4)_2\cdot 12H_2O$	1.5077
Rubidium *per*chlorate	$RbC10_4$	1.4692, 1.4701, 1.4731
chromate	Rb_2CrO_4	β1.71, γ1.72
dithionate	$Rb_2S_2O_6$	1.4574, 1.5078
fluoride	RbF	1.396
selenate	$RbSeO_4$	1.5515, 1.5537, 1.5582
Ruthenium sodium nitrate	$RuNa_2(NO_2)_5\cdot 2H_2O$	1.5889, 1.5943, 1.7163
Selenium oxide	SeO_2	>1.76
Silver cyanide	$AgCN$	1.685, 1.94
nitrate	$AgNO_3$	1.729, 1.744, 1.788
phosphate	Ag_2HPO_4	1.8036, 1.7983
potassium cyanide	$AgK(CN)_2$	1.625, 1.63
Sodium ammonium tartrate (*d*)	$NaNH_4(C_4H_4O_6)\cdot 4H_2O$	1.495, 1.498, 1.499
ammonium tartrate (*dl*)	$NaNH_4(C_4H_4O_6)\cdot H_2O$	β1.473 (red)
*ortho*arsenate	$NaH_2AsO_4\cdot H.O$	1.5382, 1.5535, 1.5607

Name	Formula	Index
	$NaH_2AsO_4\cdot 2H_2O$	1.4794, 1.5021, 1.5265
bromide	$NaBr$	1.6412
carbonate	Na_2CO_3	1.415, 1.535, 1.546
Sodium carbonate, acid	$NaHCO_3$	1.376, 1.500, 1.582
cyanide	$NaCN$	1.452
iodide	NaI	1.7745
molybdate	$3Na_2O.7MoO_3\cdot 22H_2O$	β1.627
nitrate	$NaNO_3$	1.5874, 1.3361
phosphate	$NaH_2PO_4\cdot 2H_2O$	1.4401, 1.4629, 1.4815
	$Na_2HPO_4\cdot 7H_2O$	1.4412, 1.4424, 1.4526
*hypo*phosphate	$Na_3HP_2O_6\cdot 9H_2O$	1.4653, 1.4738, 1.4804
silicate	Na_2SiO_3	1.513, 1.520, 1.528
sulfate, acid	$NaHSO_4\cdot H_2O$	1.43, 1.46, 1.47
sulfite	Na_2SO_3	1.565, 1.515
acid	$NaHSO_3$	1.474, 1.526, 1.685
tartrate, acid (d)	$NaH(C_4H_4O_6)\cdot H_2O$	β1.533
thiocyanate	$NaCNS$	1.545, 1.625, 1.695
Sodium tungstate	$Na_2WO_4\cdot 2H_2O$	1.5526, 1.5533, 1.5695
vanadate	$Na_3VO_4\cdot 10H_2O$	1.5305, ω1.5398, ε1.5475
	$Na_3VO_4\cdot 12H_2O$	1.5095, 1.5232
Strontium dichromate	$SrCr_2O_7\cdot 3H_2O$	1.7146, 1.7174, 1.812
fluoride	SrF_2	1.442 (1.438)
oxide	SrO	1.870
*ortho*phosphate, acid	$SrHPO_4$	1.608, 1.62±, 1.625
sulfide, mono-	SrS	2.107
Sulfur nitride	S_4N_4	α1.908, β2.046
Thallium chloride, mono-	$TlCl$	2.247
iodide, mono-	TlI	2.78
Tin iodide (ic)	SnI_4	2.106
Uranyl potassium sulfate	$UO_2\cdot SO_4\cdot k_2SO_4\cdot 2H_2O$	1.5144, 1.5266, 1.5705 (580 mμ)
Vanadium ammonium sulfate	$VNH_4(SO_4)_2\cdot 12H_2O$	1.475
Zinc ammonium selenate	$Zn(SeO_4)\cdot(NH_4)_2SeO_4\cdot 6H_2O$	1.5240, 1.5300, 1.5385
bromate	$Zn(BrO_3)_2\cdot 6H_2O$	1.5452
cesium sulfate	$ZnCs_2(SO_4)_2\cdot 6H_2O$	1.5022, 1.5048, 1.5093
chloride	$ZnCl_2$	1.687, 1.713
fluosilicate	$ZnSiF_6\cdot 6H_2O$	1.3824, 1.3956
potassium cyanide	$ZnK_2(CN)_4$	1.4115
selenate	$ZnK_2(SeO_4)_2\cdot 6H_2O$	1.5121, 1.5181, 1.5335
sulfate	$ZnK_2(SO_4)_2\cdot 6H_2O$	1.4775, 1.4833, 1.4969
rubidium sulfate	$ZnRb_2(SO_4)_2\cdot 6H_2O$	1.4833, 1.4884, 1.4975
silicate	$ZnSiO_3$	1.616, 1.62±, 1.623
Zirconium ammonium fluoride	$Zr(NH_4)_3F_7$	1.433

ORGANIC COMPOUNDS

See also under Physical Constants of Organic Compounds.

Name	Index
Allontoin, solid	γ1.579, λ1.660
Dimethyl thiophene (α, α'), liq	$1.51693^{13.4}$ (He)
(β, β'), liq	1.52217^{15} (He)
Ethyl carbylamine, liq	1.3659^{24}
Ethylidene cyanhydrin, liq	$1.40582^{18.4}$
Hexyl acetylene (n), liq	$1.4208^{12.5}$

Miscellaneous

Name	Index	Name	Index
Albite glass	1.4890	Magdala red	1.90
Amber	1.546	Obsidian	1.482-1.496
Anorthite glass	1.5755	Paraffin	$1.43295^{38.3}$ (C)
Asphalt	1.635	Quartz, fused	1.45640 (656 mμ)
Bell metal	1.0052		1.45843 (589 mμ)
Borax, amorphous, fused	1.4630		1.46190 (509 mμ)
Canada balsam	1.530		1.47503 (361 mμ)
Ebonite	1.66 (red)		1.49634 (275 mμ)
Fuchsin	2.70		1.53386 (214 mμ)
Gelatin, Nelson's No. 1	1.530		1.57464 (185 mμ)
Gelatin, various	1.516-1.534	Resin, aloes	1.619 (red)
Gum Arabic	1.480 (1.5,4)	colophony	1.548 (red)
	(red)	copal	1.528 (red)
Hoffman's violet	2.20	mastic	1.535 (red)
Ivory	1.539, 1.541	Peru balsam	1.593

INDEX OF REFRACTION OF ORGANIC COMPOUNDS

(See also that section of this book which contains data of Physical Constants of Organic Compounds)

The following table contains a list of organic compounds arranged in order of increasing refractive index. Measurements were made at 25°C.

Compound	n_D	Compound	n_D
Trifluoroacetic acid	1.283	Propionaldehyde	1.371
2,2,2-Trifluoroethanol	1.290	n-Hexane	1.372
Octofluoropentanol-1	1.316	2,3-Dimethylbutane	1.372
Dodecafluoroheptanol-1	1.316	3-Methylpentane	1.374
Methanol	1.326	2-Propanol	1.375
Acetonitrile	1.342	Isopropyl acetate	1.375
Ethyl ether	1.352	Propyl formate	1.375
Acetone	1.357	2-Chloropropane	1.376
Ethyl formate	1.358	2-Butanone	1.377
Ethanol	1.359	2-Chloropropane	1.377
Methyl acetate	1.360	Methylethyl ketone	1.377
Propionitrile	1.363	Butyraldehyde	1.378
2,2-Dimethylbutane	1.366	2,4-Dimethylpentane	1.379
Isopropyl ether	1.367	Propyl ether	1.379
2-methylpentane	1.369	Acetaldehyde-diethylacetal	1.379
Ethyl acetate	1.370	Butylethyl ether	1.380
Acetic acid	1.370	Nitromethane	1.380

INDEX OF REFRACTION OF ORGANIC COMPOUNDS (Continued)

Compound	n_D
Trifluoropropanol	1.381
2-Methylhexane	1.382
Butyronitrile	1.382
Propyl acetate	1.382
Ethyl propionate	1.382
2-Methyl-2-propanol	1.383
1-Propanol	1.383
Isobutyl formate	1.383
Diethyl carbonate	1.383
Heptane	1.385
2-Methyl-2-propanol	1.385
Propionic acid	1.385
3-Methylhexane	1.386
n-Propyl amine	1.386
1,1-Dimethyl-2-propanone	1.386
1-Chloropropane	1.386
2,2,3-Trimethylbutane	1.387
Methylpropyl ketone	1.387
sec-Butyl acetate	1.387
Butyl formate	1.387
β-Methylpropyl ethanoate	1.388
2,2,4-Trimethyl pentane	1.389
2,3-Dimethyl pentane	1.389
Acetic anhydride	1.389
Diisopropyl amine	1.390
2-Aminobutane	1.390
2-Pentanone	1.390
3-Pentanone	1.390
Nitroethane	1.390
Methyl-b-butyrate	1.391
Butyl acetate	1.392
2-Nitropropane	1.392
4-Methyl-2-pentnone	1.394
2-Methyl-1-propanol	1.394
Octane	1.395
1-Amino-2-methylpropane	1.395
Valeronitrile	1.395
2-Butanol	1.395
5-Hexanone	1.395
5-Methyl-3-hexanone	1.395
2-Chlorobutane	1.395
Butyric acid	1.396
2,2,2-Trimethylhexane	1.397
n-Dibutyl ether	1.397
1-Butanol	1.397
Acrolein	1.397
1-Chloro-2-methylpropane	1.397
Methacrylonitrile	1.398
3-Methyl-2-pentanone	1.398
Triethyl amine	1.399
n-Butyl amine	1.399
1,1,3,3-Tetramethyl-2-propanone	1.399
Isobutyl-n-butyrate	1.399
1-Nitropropane	1.399
n-Dodecane	1.400
Amyl acetate	1.400
1-Chlorobutane	1.400
2-Methoxy ethanol	1.400
Propionic acid anhydride	1.400
2,2,3-Trimethylpentane	1.401
1-Chlorobutane	1.401
β-Methoxypropionitrile	1.401
3-Methyl butanoic acid	1.402
n-Nonane	1.403
Dipropylamine	1.403
Isoamylacetate	1.403
Cyclopentane	1.404
2-Methyl-2-butanol	1.404
3-Methyl-1-butanol	1.404
Tetrahydrofuran	1.404
Capronitrile	1.405
2-Pentanone	1.405
2-Ethoxyethanol	1.405
2-Heptanone	1.406
Valeric acid	1.406
Diisobutylene	1.407
Methylcyclopentane	1.407
Isoamyl ether	1.407
Methylpropyl carbinol	1.407
Tributyl borate	1.407
1-Pentanol	1.408
3-Methyl-2-butanol	1.408
Diethyl oxalate	1.408
n-Decane	1.409
4-Methyl-2-pentanol	1.409
3-Isopropyl-2-pentanone	1.409
2-Methyl-1-butanol	1.409
Butyric acid anhydride	1.409
Amyl ether	1.410
Isoamyl isovalerate	1.410
1-Chloropentane	1.410
2-Propene-1-ol	1.411
2,4-Dimethyl dioxane	1.412
Ethyl lactate	1.412
Diethyl malonate	1.412
3-Chloropropene	1.413
Ethyleneglycol diacetate	1.413
2-Octanone	1.414
3-Octanone	1.414
3-Methyl-2-heptanone	1.415
Caproic acid	1.415
4-Methyldioxane	1.415
1,2-Propyleneglycol-1-monobutyl ether	1.415
Ethylcyanoacetate	1.415
Dibutylamine	1.416
2-Pentanol	1.416
1,1-Dichloroethane	1.416
Heptachlorodiethyl ether	1.416
1-Hexanol	1.416
1-Amino-3-methoxy propane	1.417
Octyl nitrile	1.418
2-Heptanol	1.418
2-Propenyl amine	1.419
1,2-Propyleneglycol carbonate	1.419
Methylpentyl carbinol	1.420
2-Ethyl-1-butanol	1.420
1-Chloro-2-mehyl-1-propene	1.420
p-Dioxane	1.420
Methylcyclohexane	1.421
4-Hydroxy-4-methyl-2-pentanone	1.421
1-Heptanol	1.422
3-Isorpopyl-2-heptanone	1.423
Cyclohexane	1.424
2-Bromopropane	1.424
3-Chloro-2-methylprop-1-ene	1.425
Caproic acid	1.426
Glycol carbonate	1.426
1-Octanol	1.427
1,1-Dimethylhexanol	1.427
N,N-Dimethylformamide	1.427
Sulfuric acid	1.427
1-Chlorooctane	1.428
Triisobutylene	1.429
N-Methylaniline nitrile	1.429
Etbylene glycol	1.429
1-Chloro-2-ethylhexane	1.430
Ethylcyclohexane	1.431
1.2-Propanediol	1.431
1-Bromopropane	1.431
2-Methyl-7-ethyl-4-nonanone	1.433
Ethyleneglycol-mono-allyl ether	1.434
Butyral lactone	1.434
2-Methyl-7-ethyl-4-undecanone	1.435
4-n-Propyl-5-ethyldioxane	1.435
1,2-Dichloro-2-methylpropane	1.435
1,2-Propyleneglycol sulfite	1.435
N-Methylmorpholine	1.436
1-Chloro-2-methyl-2-propanol	1.436
Epichlorohydrin	1.436
Triethyleneglycol-mono-butyl ether	1.437
4-Ethyl-7,7,7-trimethyl-1-heptanol	1.438
1-Methyl-3-ethyloctan-1-ol	1.438
1-Ethyl-3-ethylhexan-1-ol	1.438
Diethyl maleate	1.438
1-Butanethiol	1.440
2-Chloroethanal	1.440
Dibutyl sebacate	1.440
1-Ethyl-3-ethyloctan-1-ol	1.441
Dimethylmaleate	1.441
3-Methylpentane-2,4-diol	1.441
Ethyl sulfide	1.442
Mesityl oxide	1.442
Butyl stearate	1.442
1,2-Dichloroethane	1.444
Chloroform	1.444
trans-1,2-Dichloroethylene	1.444
Diethyleneglycol	1.445
cis-1,2-Dichloroethylene	1.445
3-(α-Butyloctyl)-oxypropyl-1-amine	1.446
2-Methylmorpholine	1.446
Dipropyleneglycol-monoethyl ether	1.446
Formamide	1.446
3-Lauryloxypropyl-1-amine	1.447
Cyclohexanone	1.448
1-Aminopropan-1-ol	1.448
Diethyleneglycol-mono-β-oxypropyl ether	1.448
1-Amino-2-methylpentan-1-ol	1.449
Tetrahydrofurfural alcohol	1.450
2-Propylcyclohexa-1-one	1.452
2-Aminoethanol	1.452
2-Butylcyclohexan-1-one	1.453
Ethylenediamine	1.454
2-(β-Methyl)-propylcyclohexan-1-one	1.454
4-Methylcyclohexanol	1.454
3-Methylcyclohexanol	1.455
bis-2-Chloroethyl ether	1.455
Cyclohexylamine	1.456
1,8-Cineol	1.456
2,2′-Dimethyl-2,2′-dipropyldieththanol amine	1.456
1,1′,2,2′-Tetramethyldiethanol amine	1.459
1-Aminopropan-3-ol	1.459
Carbon tetrachloride	1.459
3-Methyl-5-ethylheptan-2,4-diol	1.459
2-(β-Ethyl)-butylcyclohexan-1-one	1.461
2-Methylcyclohexanol	1.461
N-(n-Butyl)-diethanol amine	1.461
4,5-Chloro-1,3-dioxolane-2	1.461
2-Butylcyclohexan-1-ol	1.462

INDEX OF REFRACTION OF ORGANIC COMPOUNDS (Continued)

Compound	n_D	Compound	n_D
N-β-Oxypropyl morpholine	1.462	Propyl benzoate	1.498
2-(β-Ethyl)-hexylcyclohexanone	1.463	α-Picoline	1.499
2-Ethylcyclohexan-1-ol	1.463	1,2-Diethylbenzene	1.501
Fluorobenzene	1.463	Pentachloroethane	1.501
d-α-Pinene	1.464	1-Iodopropane	1.502
1-α-Pinene	1.465	1,2-Dimethylbenzene	1.503
Cyclohexanol	1.465	Ethyl benzoate	1.503
m-Fluorotoluene	1.465	β-Picoline	1.504
p-Fluorotoluene	1.467	Tetrachloroethylene	1.504
trans-Decahydronaphthalene	1.468	Phenetole	1.505
o-Fluorotoluene	1.468	Pyridine	1.507
3-Alloxy-2-oxypropylamine-1	1.469	Iodoethane	1.512
Ethanol-1-methylisopropanol amine	1.470	Phenylmethallyl ether	1.514
d-Limonene	1.471	Anisole	1.515
1,2,3-Trichloroisobutane	1.473	Methyl benzoate	1.515
Decahydronaphthalene	1.474	Diallylphthalate	1.517
1,2,3-Propanetriol	1.474	Benzylacetae	1.518
Trichloroethylene	1.475	2-Methyl-4-tertiarybtylphenetol	1.521
N-β-Oxyethylmorpholine	1.476	Phenylacetonitrile	1.521
Dimethylsulfoxide	1.476	Methyl salicylate	1.522
cis-Decahydronaphthalene	1.479	Chlorobenzene	1.523
N-β-Chlorallylmorpholine	1.481	Fufural	1.524
n-Dodecyl-4-tertiarybutylphenyl ether	1.482	Benzonitrile	1.526
n-Dodecylphenyl ether	1.482	Thiophene	1.526
n-Dodecyl-4-methylphenyl ether	1.483	Nonachlorodiethyl ether	1.529
2-Ethylidene cyclohexanone	1.486	Iodomethane	1.530
n-Butylbenzene	1.487	4-Phenyldioxane	1.530
p-Cymene	1.488	3-Phenylpropan-1-ol	1.532
iso-Propylbenzene	1.489	Acetophenone	1.532
Furfuralcohol	1.489	Benzyl alcohol	1.538
tert-Butylcumene	1.490	1,2-Dibromoethane	1.538
n-Propylbenzene	1.490	1,2,3,4-Tetrahydronaphthalene	1.539
sec-Butylbenzene	1.490	m-Cresol	1.542
tert-Butylbenzene	1.490	1,3-Dichlorobenzene	1.543
Dibutylphthalate	1.490	Benzaldehyde	1.544
tert-Butyltoluene	1.491	Styrene	1.545
1-Penyl-1-oxyphenylethane	1.491	Nitrobenzene	1.550
n-Hexylcumene	1.492	o-Dichlorobenzene	1.551
n-Octyltoluene	1.492	Bromobenzene	1.557
n-Octylcumene	1.492	o-Nitroanisole	1.560
p-Xylene	1.493	m-Toluidine	1.566
1,31-Diethylbenzene	1.493	Benzyl benzoate	1.568
Ethylbenzene	1.493	o-Toluidine	1.570
1,3-Dimorpholylpropan-2-ol	1.493	1-Methoxyphenyl-1-phenyl-ethane	1.571
1,12,2-Tetrachloroethane	1.493	Aniline	1.583
Toluene	1.494	o-Chloroaniline	1.586
Benzylethyl ether	1.494	Bromoform	1.587
m-Xylene	1.495	Benzenethiol	1.588
1,4-Diethylbenzene	1.496	2,4-Bis(β-phenylethyl)-phenylmethyl ether	1.590
2,3-Dichlorodioxane	1.496	Carbondisulfide	1.628
Mesitylene	1.497	1,12,2-Tetrabromomethane	1.633
2-Iodopropane	1.497	Diiodomethane	1.749
Benzene	1.498		

MOLAR REFRACTION OF ORGANIC COMPOUNDS

The molar refraction, R, is defined as:

$$R = \left(\frac{n^2 - 1}{n^2 + 2}\right)\left(\frac{M}{d}\right)$$

where n = refractive index; M = molecular weight; d = density in grams per cm³; and (M/d) is the volume occupied by 1 gram molecular weight of the compound. The units of R will then be CM^{+3}.

R is, to a first approximation, independent of temperature or physical state, and it provides an approximate measure of the actual total volume (without free space) of the molecules in one gram mole.

For a very large number of compounds R is approximately additive for the bonds present in the molecule. Using R_d based on n_D (sodium light), the following atomic, group and structural contributions to R_D are based on Vogel's extensive modern measurements published in the Journal of the Chemical Society, 1948.

INDEX OF REFRACTION OF WATER

Alcohol and Carbon Bisulfide
For sodium light, λ = .5893

Temp. °C	Water, pure relative to air	Ethyl Alcohol 99.8 relative to air	Carbon Bisulfide relative to air	Temp. °C	Water, pure relative to air	Ethyl Alcohol 99.8 relative to air	Carbon Bisulfide relative to air
14	1.33348			34	1.33136	1.35474	1.61413
15	1.33341		1.62935	36	1.33107	1.35390	1.61247
16	1.33333	1.36210	1.62858	38	1.33079	1.35306	1.61080
18	1.33317	1.36129	1.62704	40	1.33051	1.35222	1.60914
20	1.33299	1.36048	1.62546	42	1.33023	1.35138	1.60748
22	1.33281	1.35967	1.62387	44	1.32992	1.35054	1.60582
24	1.33262	1.35885	1.62226	46	1.32959	1.34969	
26	1.33241	1.35803	1.62064	48	1.32927	1.34885	
28	1.33219	1.35721	1.61902	50	1.32894	1.34800	
30	1.33192	1.35639	1.61740	52	1.32860	1.34715	
32	1.33164	1.35557	1.61577	54	1.32827	1.34629	

INDEX OF REFRACTION OF WATER (continued)

Temp. °C	Water, pure relative to air	Ethyl Alcohol 99.8 relative to air	Carbon Bisulfide relative to air	Temp. °C	Water, pure relative to air	Ethyl Alcohol 99.8 relative to air	Carbon Bisulfide relative to air
56	1.32792	1.34543		78	1.32332		
58	1.32755	1.34456		80	1.32287		
60	1.32718	1.34368		82	1.32241		
62	1.32678	1.34279		84	1.32195		
64	1.32636	1.34189		86	1.32148		
66	1.32596	1.34096		88	1.32100		
68	1.32555	1.34004		90	1.32050		
70	1.32511	1.33912		92	1.32000		
72	1.32466	1.33820		94	1.31949		
74	1.32421	1.33728		96	1.31897		
76	1.32376	1.33626		98	1.31842		
				100	1.31783		

C	2.591	Cl	5.844	=S	7.921
H	1.028	Br	8.741	C≡N	5.459
=O	2.122	I	13.954	N (primary aliphatic)	2.376
O	1.643	C_6H_5	25.463	N (secondary aliphatic	2.582
OH	2.553	$C_{10}H_7$	43.00		
F	0.81	S	7.729		

Ethylenic bond	1.575	Four membered ring	0.317
Acetylenic bond	1.977	Three membered ring	0.614
N (aromatic)	3.550		

Example: —For C_2H_5COOH: $R_{calc.} = 3(2.591) + 5(1.028) + 2.122 + 2.553 = 17.588$

$R_{obs.}$ for this compound is 17.51

For $C_6H_5NHCH_3$: $R_{calc.} = 25.463 + 3.550 + 2.591 + 4(1.028) = 35.716$

$R_{obs.}$ for this compound is 35.67

ABSOLUTE INDEX FOR PURE WATER FOR SODIUM LIGHT

Temperature	Index	Temperature	Index
15°C.	1.33377	60°C.	1.32754
20	1.33335	65	1.32652
25	1.33287	70	1.32547
30	1.33228	75	1.32434
35	1.33157	80	1.32323
40	1.33087	85	1.32208
45	1.33011	90	1.32086
50	1.32930	95	1.31959
55	1.32846	100	1.31819

INDEX OF REFRACTION OF GLASS

Relative to Air

Variety	Wave length in microns .361	.434	.486	.589 (Na)	.656	.768	1.20	2.00
Zinc crown	1.539	1.528	1.523	1.517	1.514	1.511	1.505	1.497
Higher dispersion crown	1.546	1.533	1.527	1.520	1.517	1.514	1.507	1.497
Light flint	1.614	1.594	1.585	1.575	1.571	1.567	1.559	1.549
Heavy flint	1.705	1.675	1.664	1.650	1.644	1.638	1.628	1.617
Heaviest flint		1.945	1.919	1.890	1.879	1.867	1.848	1.832

INDEX OF REFRACTION OF ROCK SALT, SYLVINE, CALCITE, FLUORITE AND QUARTZ

(Compiled from data of Martens, Paschen, and others.)

Wave length	Rock salt	Silvine, KCl	Fluorite	Calcspar, ordinary ray	Calcspar, extraordinary ray	Quartz, ordinary ray	Quartz, extraordinary ray
0.185	1.893	1.827				1.676	1690
0.198			1.496		1.578	1.651	1.664
0.340				1.701	1.506	1.567	1.577
0.589	1.544	1.490	1.434	1.658	1.486	1.544	1.553
0.760			1.431	1.650	1.483	1.539	1.548
0.884	1.534	1.481	1.430				
1.179	1.530	1.478	1.428				
1.229				1.639	1.479		
2.324					1.474	1.516	
2.357	1.526	1.475	1.421				
3.536	1.523	1.473	1.414				
5.893	1.516	1.469	1.387				
8.840	1.502	1.461	1.331				

INDEX OF REFRACTION, AQUEOUS SOLUTIONS

Substance	Density	Temp. °C	Index for λ = .5893 (Na)	Observer
Ammonium chloride	1.067	27.05	1.379	Willigen
Ammonium chloride	1.025	29.75	1.351	Willigen
Calcium chloride	1.398	25.65	1.443	Willigen
Calcium chloride	1.215	22.9	1.397	Willigen
Calcium chloride	1.143	25.8	1.374	Willigen
Hydrochloric acid	1.166	20.75	1.411	Willigen
Nitric acid	1.359	18.75	1.402	Willigen
Potash (caustic)	1.416	11.0	1.403	Frauenhofer
Potassium chloride	Nonal solution		1.343	Bender
Potassium chloride	Double normal		1.352	Bender
Potassium chloride	Triple normal		1.360	Bender
Soda (caustic)	1.376	21.6	1.413	Willigen
Sodium chloride	1.189	18.07	1.378	Schutt
Sodium chloride	1.109	18.07	1.360	Schutt
Sodium chloride	1.035	18.07	1.342	Schutt
Sodium nitrate	1.358	22.8	1.385	Willigen
Sulfuric acid	1.811	18.3	1.437	Willigen
Sulfuric acid	1.632	18.3	1.425	Willigen
Sulfuric acid	1.221	18.3	1.370	Willigen
Sulfuric acid	1.028	18.3	1.339	Willigen
Zinc chloride	1.359	26.6	1.402	Willigen
Zinc chloride	1.209	26.4	1.375	Willigen

INDEX OF REFRACTION OF FUSED QUARTZ

λ mμ, 15°C	n, 18°C	λ mμ, 15°C	n, 18°C	λ mμ, 15°C	n, 18°C	λ mμ, 15°C	n, 18°C
185.467	1.57436	434.047	1.46690	250.329	1.50745	546.072	1.46013
193.583	1.55999	435.834	1.46675	257.304	1.50379	589.29	1.45845
202.55	1.54727	467.815	1.46435	274.867	1.49617	643.847	1.45674
214.439	1.53386	479.991	1.46355	303.412	1.48594	656.278	1.45640
219.462	1.52907	486.133	1.46318	340.365	1.47867	706.520	1.45517
226.503	1.52308	508.582	1.46191	396.848	1.47061	794.763	1.45340
231.288	1.51941	533.85	1.46067	404.656	1.46968		

INDEX OF REFRACTION OF AIR (15°C, 76 cm Hg)

Corrections for reducing wavelengths and frequencies in air (15°C, 76 cm Hg) to vacuo

The indices were computed from the Cauchy formula $(n - 1)10^7 = 2726.43 + 12.288/(\lambda_2 \times 10^{-8}) + 0.3555/(\lambda^4 \times 10^{-16})$. For 0°Cand 76 cm Hg the constants of the equation become 2875.66, 13.412 and 0.3777 respectively, and for 30°C and 76 cm Hg 2589.72, 12.259 and 0.2576. Sellmeier's formula for but one absorption band closely fits the observations: $n_2 = 1 + 0.00057378\lambda^2/(\lambda^2 - 595260)$. If $n - 1$ were strictly proportional to the density, then $(n - 1)_o/(n - 1)t$ would equal $1 + \alpha t$ where α should be 0.00367. The following values of α were found to hold:

λ	0.85μ	0.75μ	0.65μ	0.55μ	0.45μ	0.35μ	0.25μ
α	0.003672	0.003674	0.003678	0.003685	0.003700	0.003738	0.003872

The indices are for dry air (0.05 ± % CO_2). Corrections to reduce to dry air the indices for moist air may be made for any wavelength by Lorenz's formula, +0.000041(m/760), where m is the vapor pressure in mm. The corresponding frequencies in waves per cm and the corrections to reduce wavelengths and frequencies in air at 15°C and 76 cm Hg pressure to vacuo are given. E.g., a light wave of 5000 angstroms in dry air at 15°C, 76 cm Hg becomes 5001.391 Å in vacuo; a frequency of 20,000 waves per cm correspondingly becomes 19994.44.

Wave-length, λ angstroms	Dry air (n — 1) × 10⁷ 15°C 76 cm Hg	Vacuo correction for λ in air (nλ — λ) add	Frequency waves per cm 1/λ in air	Vacuo correction or 1/λ in air (1/nλ — 1/λ) subtract	Wave-length, λ angstroms	Dry air (n — 1) × 10⁷ 15°C 76 cm Hg	Vacuo correction for λ in air (nλ — λ) add	Frequency waves per cm 1/λ in air	Vacuo correction for 1/λ in air (1/nλ — 1/λ) substract
2000	3256	.651	50,000	16.27	5500	2771	1.524	18,181	5.04
2100	3188	.670	47,619	15.18	5600	2769	1.551	17,857	4.94
2200	3132	.689	45,454	14.23	5700	2768	1.578	17,543	4.85
2300	3086	.710	43,478	13.41	5800	2766	1.604	17,241	4.77
2400	3047	.731	41,666	12.69	5900	2765	1.631	16,949	4.68
2500	3014	.754	40,000	12.05	6000	2763	1.658	16,666	4.60
2600	2986	.776	38,461	11.48	6100	2762	1.685	16,393	4.53
2700	2962	.800	37,037	10.97	6200	2761	1.712	16,129	4.45
2800	2941	.824	35,714	10.50	6300	2760	1.739	15,873	4.38
2900	2923	.848	34,482	10.08	6400	2759	1.766	15,625	4.31
3000	2907	.872	33,333	9.69	6500	2758	1.792	15,384	4.24
3100	2893	.897	32,258	9.33	6600	2757	1.819	15,151	4.18
3200	2880	.922	31,250	9.00	6700	2756	1.846	14,925	4.11
3300	2869	.947	30,303	8.69	6800	2755	1.873	14,705	4.05
3400	2859	.972	29,411	8.41	6900	2754	1.900	14,492	3.99
3500	2850	.998	28,571	8.14	7000	2753	1.927	14,285	3.93
3600	2842	1.023	27,777	7.89	7100	2752	1.954	14,084	3.88
3700	2835	1.049	27,027	7.66	7200	2751	1.981	13,888	3.82
3800	2829	1.075	26,315	7.44	7300	2751	2,008	13,698	3.77
3900	2823	1.101	25,641	7.24	7400	2750	2.035	13,513	3.72
4000	2817	1.127	25,000	7.04	7500	2749	2.062	13,333	3.66
4100	2812	1.153	24,390	6.86	7600	2749	2.089	13,157	3.62
4200	2808	1.179	23,809	6.68	7700	2748	2.116	12,987	3.57
4300	2803	1.205	23,255	6.52	7800	2748	2.143	12,820	3.52
4400	2799	1.232	22,727	6.36	7900	2747	2.170	12,658	3.48
4500	2796	1.258	22,222	6.21	8000	2746	2.197	12,500	3.43
4600	2792	1.284	21,739	6.07	8100	2746	2.224	12,345	3.39
4700	2789	1.311	21,276	5.93	8250	2745	2.265	12,121	3.33
4800	2786	1.338	20,833	5.80	8500	2744	2.332	11,764	3.23
4900	2784	1.364	20,406	5.68	8750	2743	2.400	11,428	3.13
5000	2781	1.391	20,000	5.56	9000	2742	2.468	11,111	3.05
5100	2779	1.417	19,607	5.45	9250	2741	2.536	10,810	2.96
5200	2777	1.444	19,230	5.34	9500	2740	2.604	10,526	2.88
5300	2775	1.471	18,867	5.23	9750	2740	2.671	10,256	2.81
5400	2773	1.497	18,518	5.13	10000	2739	2.739	10,000	2.74

INDEX OF REFRACTION, GASES

Values are relative to a vacuum and for a Temp. of 0°C, and 760 mm pressure.
(From Smithsonian Tables)

Substance	Kind of light	Indices of refraction	Observer
Acetone	D	1.001079-1.001100	
Air	D	1.0002926	Perreau
Ammonia	white	1.000381-1.000385	
Ammonia	D	1.000373-1.000379	
Argon	D	1.000281	Rayleigh
Benzene	D	1.001700-1.001823	
Bromine	D	1.001132	Mascart
Carbon dioxide	white	1.000449-1.000450	
dioxide	D	1.000448-1.000454	
disulfide	white	1.001500	Dulong
disulfide	D	1.001478-1.001485	
monoxide	White	1.000340	Dulong
monoxide	white	1.000335	Mascart
Chlorine	white	1.000772	Dulong
Chlorine	D	1.000773	Mascart
Chloroform	D	1.001436-1.001464	
Cyanogen	white	1.000834	Dulong
Cyanogen	D	1.000784-1.000825	
Ethyl alcohol	D	1.000871-1.000885	
ether	D	1.001521-1.001544	
Helium	D	1.000036	Ramsay
Hydrochloric acid	white	1.000449	Mascart
Hydrochloric acid	D	1.000447	Mascart
Hydrogen	white	1.000138-1.000143	
Hydrogen	D	1.000132	Burton
sulfide	D	1.000644	Dulong
sulfide	D	1.000623	Mascart
Methane	white	1.000443	Dulong
Methane	D	1.000444	Mascart
Methyl alcohol	D	1.000549-1.000623	
Methyl ether	D	1.000891	Marcast
Nitric oxide	white	1.000303	Dulong
Nitric oxide	D	1.000297	Mascart
Nitrogen	white	1.000295-1.000300	
Nitrogen	D	1.000296-1.000298	
Nitrous oxide	white	1.000503-1.000507	
Nitrous oxide	D	1.000516	Mascart
Oxygen	white	1.000272-1.000280	
Oxygen	D	1.000271-1.000272	
Pentane	D	1.001711	Mascart
Sulfur dioxide	white	1.000665	Dulong
Sulfur dioxide	D	1.000686	Ketteler
Water	white	1.000261	Jamin
Water	D	1.000249-1.000259	

COEFFICIENT OF TRANSPARENCY OF UVIOL GLASS FOR THE ULTRA-VIOLET

For a thickness of 1 mm.

Wave length microns	0.280	0.309	0.325	0.346	0.361	0.383	0.397
Uviol crown	0.56	0.95	0.990	0.996	0.999	1.000	1.000

INDEX OF REFRACTION OF AQUEOUS SOLUTIONS OF SUCROSE (CANE SUGAR)

The table gives the index of refraction for λ = 0.5893 of aqueous sugar solutions at 20°C from 0—85% sugar. Corrections for temperatures other than 20 are given at the end of the table

Percent sugar		.0	.1	.2	.3	.4	.5	.6	.7	.8	.9
00	1.3	330	331	333	334	336	337	338	340	341	342
1		344	345	347	348	350	351	353	355	356	357
2		359	361	362	363	365	367	368	369	371	373
3		374	375	377	378	380	381	382	384	385	387
4		388	389	391	393	394	395	397	399	400	401
5		403	405	406	407	409	411	412	413	415	417
6		418	419	421	423	424	425	427	429	430	431
7		433	435	436	437	439	441	442	443	445	447
8		448	450	451	453	454	456	458	459	461	462
9		464	465	467	469	470	471	473	475	476	477
10		479	481	482	483	485	487	488	489	491	493
11		494	496	497	499	500	502	504	505	507	508
12		510	512	513	515	516	518	520	521	523	524
13		526	527	529	531	532	533	535	537	538	539
14		541	543	544	546	547	549	551	552	554	555
15		557	559	560	562	563	565	567	568	570	571
16		573	575	576	578	580	582	583	585	587	588
17		590	592	593	595	596	598	600	601	603	604
18		606	608	609	611	612	614	616	617	619	620
19		622	624	625	627	629	631	632	634	636	637
20		639	641	652	644	645	647	649	650	652	653
21		655	657	658	660	662	663	665	667	669	670
22		672	674	675	677	679	681	682	684	686	687
23		689	691	692	694	696	698	699	701	703	704
24		706	708	709	711	713	715	716	718	720	721
25		723	725	726	728	730	731	733	735	737	738
26		740	742	744	745	747	749	751	753	754	756
27		758	760	761	763	765	767	768	770	772	773
28		775	777	779	780	782	784	786	788	789	791
29		793	795	797	798	800	802	804	806	807	809
30		811	813	815	816	818	820	822	824	825	827
31		829	831	833	834	836	838	840	842	843	845
32		847	849	851	852	854	856	858	860	861	863
33		865	867	869	870	872	874	876	878	879	881
34		883	885	887	889	891	893	894	896	898	900
35		902	904	906	907	909	911	913	915	916	918
36		920	922	924	926	928	929	931	933	935	937
37		939	941	943	945	947	949	950	952	954	956
38		958	960	962	964	966	968	970	972	974	976
39		978	980	982	984	986	987	989	991	993	995

INDEX OF REFRACTION OF AQUEOUS SOLUTIONS OF SUCROSE (CANE SUGAR) (continued)

Percent sugar		.0	.1	.2	.3	.4	.5	.6	.7	.8	.9
40		997	999	*001	*003	*005	*007	*008	*010	*012	*014
41	1.4	016	018	020	022	024	026	028	030	032	034
42		036	038	040	042	044	046	048	050	052	054
43		056	058	060	062	064	066	068	070	072	074
44		076	078	080	082	084	086	088	090	092	094
45		096	098	100	102	104	107	109	111	113	115
46		117	119	121	123	125	127	129	131	133	135
47		137	139	141	143	145	147	150	152	154	156
48		158	160	162	164	166	169	171	173	175	177
49		179	181	183	185	187	189	192	194	196	198
50		200	202	204	206	208	211	213	215	217	219
51		221	223	225	227	229	231	234	236	238	240
52		242	244	246	249	251	253	255	257	260	262
53		264	266	268	270	272	275	277	279	281	283
54		285	287	289	292	294	296	298	300	303	305
55		307	309	311	313	316	318	320	322	325	327
56		329	331	333	336	338	340	342	344	347	349
57		351	353	355	358	360	362	364	366	369	371
58		373	375	378	380	382	385	387	389	391	394
59		396	398	400	403	405	407	409	411	414	416
60		418	420	423	425	427	429	432	434	436	439
61		441	443	446	448	450	453	455	457	459	462
62		464	466	468	471	473	475	477	479	482	484
63		486	488	491	493	495	497	500	502	504	507
64		509	511	514	516	518	521	523	525	527	530
65		532	534	537	539	541	544	546	548	550	553
66		558	561	563	565	567	570	572	574	577	579
67		581	584	586	588	591	593	595	598	600	602
68		605	607	609	612	614	616	619	621	623	625
69		628	630	632	635	637	639	642	644	646	649
70		651	653	656	658	661	663	666	668	671	673
71		676	678	681	683	685	688	690	693	695	698
72		700	703	705	708	710	713	715	717	720	722
73		725	727	730	732	735	737	740	742	744	747
74		749	752	754	757	759	762	764	767	769	772
75		774	777	779	782	784	787	789	792	794	797
76		799	802	804	807	810	812	815	817	820	822
77		825	827	830	832	835	838	840	843	845	848
78		850	853	855	858	860	863	865	868	871	873
79		876	878	881	883	886	888	891	893	896	898
80		901	904	906	909	912	914	917	919	922	925
81		927	930	933	935	938	941	943	946	949	951
82		954	956	959	962	964	967	970	972	975	978
83		980	983	985	988	991	993	996	999	*001	*004
84	1.5	007	009	012	015	017	020	022	025	028	030
85		033									

Correction table for determining the percentage of sucrose by means of the refractometer when the readings are made at temperatures other than 20°C[1]

Temperature	Percentage of sucrose: 0	5	10	15	20	25	30	35	40	45	50	55	60	65	70
	Subtract from the percentage of sucrose														
°C.															
10	0.50	0.54	0.58	0.61	0.64	0.66	0.68	0.70	0.72	0.73	0.74	0.75	0.76	0.78	0.79
11	.46	.49	.53	.55	.58	.60	.62	.64	.65	.66	.67	.68	.69	.70	.71
12	.42	.45	.48	.50	.52	.54	.56	.57	.58	.59	.60	.61	.61	.63	.63
13	.37	.40	.42	.44	.46	.48	.49	.50	.51	.52	.53	.54	.54	.55	.55
14	.33	.35	.37	.39	.40	.41	.42	.43	.44	.45	.45	.46	.46	.47	.48
15	.27	.29	.31	.33	.34	.34	.35	.36	.37	.37	.38	.39	.39	.40	.40
16	.22	.24	.25	.26	.27	.28	.28	.29	.30	.30	.30	.31	.31	.32	.32
17	.17	.18	.19	.20	.21	.21	.21	.22	.22	.23	.23	.23	.23	.24	.24
18	.12	.13	.13	.14	.14	.14	.14	.15	.15	.15	.15	.16	.16	.16	.16
19	.06	.06	.06	.07	.07	.07	.07	.08	.08	.08	.08	.08	.08	.08	.08
	Add to the percentage of sucrose														
21	0.06	0.07	0.07	0.07	0.07	0.08	0.08	0.08	0.08	0.08	0.08	0.08	0.08	0.08	0.08
22	.13	.13	.14	.14	.15	.15	.15	.15	.15.	16.	16	.16	.16	.16	.16
23	.19	.20	.21	.22	.22	.23	.23	.23	.23	.24	.24	.24	.24	.24	.24
24	.26	.27	.28	.29	.30	.30	.31	.31	.31	.31	.31	.32	.32	.32	.32
25	.33	.35	.36	.37	.38	.38	.39	.40	.40	.40	.40	.40	.40	.40	.40
26	.40	.42	.43	.44	.45	.46	.47	.48	.48	.48	.48	.48	.48	.48	.48
27	.48	.50	.52	.53	.54	.55	.55	.56	.56	.56	.56	.56	.56	.56	.56
28	.56	.57	.60	.61	.62	.63	.63	.64	.64	.64	.64	.64	.64	.64	.64
29	.64	.66	.68	.69	.71	.72	.72	.73	.73	.73	.73	.73	.73	.73	.73
30	.72	.74	.77	.78	.79	.80	.80	.81	.81	.81	.81	.81	.81	.81	.81

[1] International Temperature Correction Table, 1936, adopted by the International Commission for Uniform Methods of Sugar Analysis (Int. Sugar J. **39**, 24s, 1937).

INDEX OF REFRACTION OF MALTOSE HYDRATE SOLUTIONS

McDonald: Journal of Research of the National Bureau of Standards, Vol. 46, 165, 1951.
The following table gives the refractive index of maltose solutions for λ = 5893 A at 20° and 25°C.

Percent	n_D^{20}	n_D^{25}	Δn	$\Delta n/\Delta t$	Percent	n_D^{20}	n_D^{25}	Δn	$\Delta n/\Delta t$
1	1.33438	1.33389	0.00049	0.00010	33	1.38470	1.38388	.00082	.00016
2	1.33579	1.33528	.00051	.00010	34	1.38647	1.38565	.00081	.00016
3	1.33720	1.33668	.00052	.00010	35	1.38825	1.38744	.00081	.00016
4	1.33862	1.33810	.00052	.00010	36	1.39004	1.38924	.00080	.00016
5	1.34006	1.33952	.00054	.00011	37	1.39185	1.39105	.00080	.00016
6	1.34150	1.34095	.00055	.00011	38	1.39367	1.39287	.00080	.00016
7	1.34295	1.34239	.00056	.00011	39	1.39551	1.39471	.00080	.00016
8	1.34442	1.34384	.00058	.00012	40	1.39735	1.39656	.00079	.00016
9	1.34589	1.34530	.00059	.00012	41	1.39922	1.39843	.00079	.00016
10	1.34738	1.34677	.00061	.00012	42	1.40109	1.40031	.00078	.00016
11	1.34887	1.34825	.00062	.00012	43	1.40298	1.40221	.00077	.00016
12	1.35039	1.34974	.00065	.00013	44	1.40488	1.40411	.00077	.00015
13	1.35190	1.35124	.00066	.00013	45	1.40680	1.40603	.00077	.00015
14	1.35343	1.35276	.00067	.00013	46	1.40873	1.40797	.00076	.00015
15	1.35497	1.35428	.00069	.00014	47	1.41067	1.40992	.00075	.00015
16	1.35652	1.35582	.00070	.00014	48	1.41263	1.41188	.00075	.00015
17	1.35808	1.35737	.00071	.00014	49	1.41460	1.41385	.00075	.00015
18	1.35965	1.35893	.00072	.00014	50	1.41658	1.41584	.00074	.00015
19	1.36124	1.36051	.00073	.00015	51	1.41858	1.41784	.00074	.00015
20	1.36283	1.36209	.00074	.00015	52	1.42059	1.41986	.00073	.00015
21	1.36444	1.36369	.00075	.00015	53	1.42262	1.42189	.00073	.00015
22	1.36606	1.36530	.00076	.00015	54	1.42466	1.42392	.00074	.00015
23	1.36770	1.36692	.00078	.00016	55	1.42672	1.42598	.00074	.00015
24	1.36934	1.36856	.00078	.00016	56	1.42878	1.42804	.00074	.00015
25	1.37100	1.37021	.00079	.00016	57	1.43087	1.43012	.00075	.00015
26	1.37267	1.37187	.00080	.00016	58	1.43296	1.43221	.00075	.00015
27	1.37435	1.37355	.00080	.00016	59	1.43508	1.43431	.00077	.00015
28	1.37604	1.37524	.00080	.00016	60	1.43720	1.43643	.00077	.00015
29	1.37775	1.37694	.00081	.00016	61	1.43934	1.43855	.00079	.00016
30	1.37946	1.37865	.00081	.00016	62	1.44150	1.44069	.00081	.00016
31	1.38120	1.38038	.00082	.00016	63	1.44367	1.44283	.00084	.00017
32	1.38294	1.38213	.00082	.00016	64	1.44585	1.44499	.00086	.00017
					65	1.44805	1.44716	.00089	.00018

RADIATION FROM AN IDEAL BLACK BODY

From NASA TT-F-783

Temperature dependence of the specific power radiated, Q_T, and of λ_{max} for an ideal black body according to Kirchhoff's law (σ_O = $5.68 \cdot 10^{-8}$ W/m²·deg⁴ = $4.88 \cdot 10^{-8}$ kcal/m²·hr·deg⁴)

T, °K	t, °C	Q_T, W/cm²	Q_T kcal/m²·hr	λmax μ	T, °K	t, °C	Q_T W/cm²	Q_T kcal/m²·hr	λmax μ
100	−173	$5.680 \cdot 10^{-4}$	$4.880 \cdot 10^{0}$	28.96	630	357	8.948	7.687	4.597
200	−73	$9.088 \cdot 10^{-3}$	$7.808 \cdot 10^{1}$	14.48	640	367	9.529	8.187	4.525
273	0	$3.155 \cdot 10^{-2}$	$2.711 \cdot 10^{2}$	10.608	650	377	$1.014 \cdot 10^{0}$	8.711	4.455
300	27	4.601	3.953	9.655	660	387	1.078	9.260	4.388
310	37	5.246	4.507	9.342	670	397	1.145	9.831	4.322
320	47	5.956	5.117	9.050	680	407	1.214	$1.013 \cdot 10^{4}$	4.259
330	57	6.736	5.787	8.766	690	417	1.287	1.106	4.197
340	67	7.590	6.521	8.518	700	427	1.364	1.172	4.137
350	77	8.524	7.323	8.274	710	437	1.443	1.240	4.069
360	87	9.540	8.196	8.044	720	447	1.526	1.311	4.022
370	97	1.065	9.146	7.827	730	457	$1.613 \cdot 10^{0}$	$1.386 \cdot 10^{1}$	3.967
380	107	$1.184 \cdot 10^{-1}$	$1.018 \cdot 10^{3}$	7.621	740	467	1.703	1.463	3.914
390	117	1.314	1.128	7.426	750	477	1.797	1.544	3.861
400	127	1.454	1.249	7.270	760	487	1.895	1.628	3.811
410	137	1.605	1.379	7.053	770	497	1.997	1.715	3.761
420	147	1.76	1.519	6.865	780	507	2.102	1.806	3.713
430	157	1.942	1.668	6.735	790	517	2.212	1.901	3.666
440	167	2.129	1.829	6.562	800	527	2.327	1.999	3.620
450	177	2.329	2.001	6.436	810	537	2.445	2.101	3.565
460	187	2.543	2.185	6.266	820	547	2.568	2.206	3.532
470	197	2.772	2.381	6.162	830	557	2.696	2.316	3.489
480	207	3.015	2.591	6.033	840	567	2.828	2.430	3.448
490	217	3.274	2.813	5.910	850	577	2.965	2.547	3.407
500	227	3.550	3.05	5.792	860	587	3.107	2.670	3.367
510	237	3.843	3.301	5.668	870	597	3.254	2.796	3.329
520	247	4.163	3.568	5.559	880	607	3.406	2.927	3.291
530	257	4.482	3.851	5.454	890	617	3.564	3.062	3.254
540	267	4.830	4.150	5.363	900	627	3.727	3.202	3.218
550	277	5.198		5.255	910	637	3.895	3.346	3.162
560	287	5.586	4.799	5.161	920	647	4.069	3.496	3.148
570	297	5.996	5.151	5.061	930	657	4.249	3.650	3.114
580	307	6.428	5.522	4.963	940	667	4.435	3.810	3.081
590	317	6.883	5.913	4.908	950	677	4.626	3.975	3.048
600	327	7.361	6.324	4.827	960	687	4.824	4.145	3.017
610	337	7.864	6.757	4.748	970	697	5.028	4.320	2.986
620	347	8.393	7.211	4.671	980	707	5.239	4.501	2.955

T, °K	t, °C	Q_T, W/cm²	Q_T kcal/m²·hr	λmax μ	T, °K	t, °C	Q_T W/cm²	Q_T kcal/m²·hr	λmax μ
990	717	5.456	4.688	2.925	1540	1267	3.195	2.745	1.881
1000	727	5.680	4.880	2.896	1550	1277	3.278	2.817	1.866
1010	737	5.909	5.07	2.866	1560	1287	3.363	2.890	1.856
1020	747	6.143	5.278	2.836	1570	1297	3.451	2.965	1.845
1030	757	6.394	5.494	2.812	1580	1307	3.540	3.041	1.833
1040	767	6.645	5.709	2.785	1590	1317	3.630	3.198	1.821
1050	777	6.904	5.932	2.758	1600	1327	3.722		1.810
1060	787	7.171	6.161	2.732	1610	1337	3.816	3.279	1.797
1070	797	7.445	6.397	2.707	1620	1347	3.912	3.361	1.787
1080	807	7.728	6.640	2.681	1630	1357	4.010	3.445	1.777
1090	817	8.018	6.888	2.657	1640	1367	$4.109 \cdot 10^1$	$3.530 \cdot 10^5$	1.766
1100	827	8.316	7.145	2.633	1650	1377	4.210	3.617	1.755
1110	837	8.623	7.408	2.609	1660	1387	4.313	3.706	1.745
1120	847	8.937	7.679	2.586	1670	1397	4.418	3.796	1.734
1130	857	9.261	7.956	2.563	1680	1407	4.525	3.887	1.724
1140	867	9.593	8.242	2.540	1690	1417	4.633	3.981	1.714
1150	877	9.934	8.535	2.516	1700	1427	4.744	4.076	1.704
1160	887	$1.028 \cdot 10^1$	8.836	2.497	1710	1437	4.858	4.183	1.694
1170	897	1.064	9.145	2.475	1720	1447	4.971	4.271	1.684
1180	907	1.101	9.461	2.454	1730	1457	5.088	4.371	1.674
1190	917	$1.139 \cdot 10^1$	$9.786 \cdot 10^4$	2.434	1740	1467	5.206	4.473	1.664
1200	927	1.178	$1.042 \cdot 10^5$	2.413	1750	1477	5.327	4.577	1.655
1210	937	1.218	1.046	2.393	1760	1487	5.450	4.682	1.645
1220	947	1.258	1.081	2.374	1770	1497	5.575	4.790	1.636
1230	957	1.300	1.117	2.354	1780	1507	5.703	4.900	1.627
1240	967	1.343	1.154	2.335	1790	1517	5.831	5.010	1.617
1250	977	1.387	1.191	2.317	1800	1527	5.963	5.123	1.607
1260	987	1.431	1.230	2.298	1810	1537	6.096	5.238	1.600
1270	997	1.478	1.270	2.280	1820	1547	6.232	5.354	1.591
1280	1007	1.525	1.310	2.263	1830	1557	6.370	5.473	1.583
1290	1017	1.573	1.351	2.245	1840	1567	6.511	5.594	1.574
1300	1027	1.622	1.394	2.227	1850	1577	6.653	5.717	1.565
1310	1037	1.673	1.437	2.211	1860	1587	6.798	5.841	1.557
1320	1047	1.724	1.482	2.194	1870	1597	6.946	5.967	1.549
1330	1057	1.777	1.527	2.177	1880	1607	7.095	6.096	1.540
1340	1067	1.831	1.573	2.161	1890	1617	7.248	6.227	1.532
1350	1077	1.887	1.621	2.145	1900	1627	7.402	6.360	1.524
1360	1087	1.943	1.669	2.129	1910	1637	7.559	6.495	1.516
1370	1097	2.001	1.719	2.114	1920	1647	7.720	6.632	1.509
1380	1107	2.058	1.768	2.099	1930	1657	7.881	6.771	1.501
1390	1117	2.120	1.822	2.083	1940	1667	8.046	6.912	1.493
1400	1127	2.182	1.875	2.067	1950	1677	8.213	7.056	1.485
1410	1137	2.245	1.929	2.054	1960	1687	8.382	7.202	1.478
1420	1147	2.309	1.984	2.036	1970	1697	8.555	7.350	1.470
1430	1157	2.375	2.041	2.025	1980	1707	8.730	7.500	1.463
1440	1167	2.442	2.098	2.011	1990	1717	8.907	7.653	1.455
1450	1177	2.511	2.157	1.997	2000	1727	9.088	7.808	1.448
1460	1187	2.581	2.217	1.984	2500	2227	$2.219 \cdot 10^2$	$1.906 \cdot 10^6$	1.156
1470	1197	2.652	2.279	1.970	3000	2727	4.6[illegible]1	3.953	0.965
1480	1207	2.725	2.341	1.957	3500	3227	8.524	7.323	0.826
1490	1217	2.800	2.405	1.944	4000	3727	$1.454 \cdot 10^3$	1.249	0.722
1500	1227	2.876	2.471	1.931	4500	4227	2.329	2.001	0.644
1510	1237	2.953	2.537	1.917	5000	4727	3.550	3.050	0.579
1520	1247	3.032	2.605	1.905	5500	5227	5.198	4.465	0.527
1530	1257	3.113	2.674	1.893	6000	5727	7.361	6.324	0.483

OPTICAL PROPERTIES OF METALS

J. H. Weaver

These tables list the optical properties of metals index of refraction (n), the extinction coefficient (k), and the normal-incidence reflection (R) as a function of photon energy which is expressed in electron volts (eV). To convert the energy (in eV) to wavelength (in Å) use $\lambda = 12398/E$. To compute the dielectric function, $\tilde{\varepsilon} = \varepsilon_1 + i\varepsilon_2$, from the index of refraction, $N = n + ik$, use $\varepsilon_1 = 2nk$ and $\varepsilon_2 = n^2 + k^2$.

The optical constants in these tables are abridged from a more extensive tabulation (J. H. Weaver, C. Krafka, D. W. Lynch, and E. E. Koch, "Optical Properties of Metals", Volumes I and II, *Physics Data,* Nr. 18-1 and 18-2, Fachinformationzentrum, 7514 Eggenstein-Leopoldshafen 2, Karlsruhe, West Germany). The extensive tabulation provides detailed comparison of all optical data available in the literature at the time of the compilation of the tables for *Physics Data.* In the extensive tables, metals in addition to those listed below are examined, other optical properties are given (dielectric function, loss function) and other photon energies are included. For critical applications the reader should refer to the original work. References for metals listed in this abridged table are listed at the end of the tables. Generally, tabulated values for the optical properties are accurate to 10%.

Aluminum[1]

Energy (eV)	n	k	R(ϕ = 0)
0.040	98.595	203.701	0.9923
0.050	74.997	172.199	0.9915
0.060	62.852	150.799	0.9906
0.070	53.790	135.500	0.9899
0.080	45.784	123.734	0.9895
0.090	39.651	114.102	0.9892
0.100	34.464	105.600	0.9889
0.125	24.965	89.250	0.9884
0.150	18.572	76.960	0.9882
0.175	14.274	66.930	0.9879
0.200	11.733	59.370	0.9873
0.250	8.586	48.235	0.9858
0.300	6.759	40.960	0.9844
0.350	5.438	35.599	0.9834
0.400	4.454	31.485	0.9826
0.500	3.072	25.581	0.9817
0.600	2.273	21.403	0.9806
0.700	1.770	18.328	0.9794
0.800	1.444	15.955	0.9778
0.900	1.264	14.021	0.9749
1.000	1.212	12.464	0.9697
1.100	1.201	11.181	0.9630
1.200	1.260	10.010	0.9521
1.300	1.468	8.949	0.9318
1.400	2.237	8.212	0.8852
1.500	2.745	8.309	0.8678
1.600	2.625	8.597	0.8794
1.700	2.143	8.573	0.8972
1.800	1.741	8.205	0.9069
1.900	1.488	7.821	0.9116
2.000	1.304	7.479	0.9148
2.200	1.018	6.846	0.9200
2.400	0.826	6.283	0.9228
2.600	0.695	5.800	0.9238
2.800	0.598	5.385	0.9242
3.000	0.523	5.024	0.9241
3.200	0.460	4.708	0.9243
3.400	0.407	4.426	0.9245
3.600	0.363	4.174	0.9246
3.800	0.326	3.946	0.9247
4.000	0.294	3.740	0.9248
4.200	0.267	3.552	0.9248
4.400	0.244	3.380	0.9249
4.600	0.223	3.222	0.9249
4.800	0.205	3.076	0.9249
5.000	0.190	2.942	0.9244
6.000	0.130	2.391	0.9257
6.500	0.110	2.173	0.9260
7.000	0.095	1.983	0.9262
7.500	0.082	1.814	0.9265
8.000	0.072	1.663	0.9269
8.500	0.063	1.527	0.9272
9.000	0.056	1.402	0.9277
9.500	0.049	1.286	0.9282
10.000	0.044	1.178	0.9286
10.500	0.040	1.076	0.9293
11.000	0.036	0.979	0.9298
11.500	0.033	0.883	0.9283
12.000	0.033	0.791	0.9224
12.500	0.034	0.700	0.9118
13.000	0.038	0.609	0.8960
13.500	0.041	0.517	0.8789
14.000	0.048	0.417	0.8486
14.200	0.053	0.373	0.8312
14.400	0.058	0.327	0.8102
14.600	0.067	0.273	0.7802
14.800	0.086	0.211	0.7202
15.000	0.125	0.153	0.6119
15.200	0.178	0.108	0.4903
15.400	0.234	0.184	0.3881
15.600	0.280	0.073	0.3182
15.800	0.318	0.065	0.2694
16.000	0.351	0.060	0.2326
16.200	0.380	0.055	0.2031
16.400	0.407	0.050	0.1789
16.750	0.448	0.045	0.1460
17.000	0.474	0.042	0.1278
17.250	0.498	0.040	0.1129
17.500	0.520	0.038	0.1005
17.750	0.540	0.036	0.0899
18.000	0.558	0.035	0.0809
18.500	0.591	0.032	0.0664
19.000	0.620	0.030	0.0554
19.500	0.646	0.028	0.0467
20.000	0.668	0.027	0.0398
20.500	0.689	0.025	0.0342
21.000	0.707	0.024	0.0296
21.500	0.724	0.023	0.0258
22.000	0.739	0.022	0.0226
22.500	0.753	0.021	0.0199
23.000	0.766	0.021	0.0177
23.500	0.778	0.020	0.0157
24.000	0.789	0.019	0.0140
24.500	0.799	0.018	0.0126
25.000	0.809	0.018	0.0113
25.500	0.817	0.017	0.0102
26.000	0.826	0.016	0.0092
27.000	0.840	0.015	0.0076
28.000	0.854	0.014	0.0063
29.000	0.865	0.014	0.0053
30.000	0.876	0.013	0.0044
35.000	0.915	0.010	0.0020
40.000	0.940	0.008	0.0010
45.000	0.957	0.007	0.0005
50.000	0.969	0.006	0.0003
55.000	0.979	0.005	0.0001
60.000	0.987	0.004	0.0000
65.000	0.995	0.004	0.0000
70.000	1.006	0.004	0.0000
72.500	1.025	0.004	0.0002
75.000	1.011	0.024	0.0002
77.500	1.008	0.025	0.0002
80.000	1.007	0.024	0.0002
85.000	1.007	0.028	0.0002
90.000	1.005	0.031	0.0002
95.000	0.999	0.036	0.0003
100.000	0.991	0.030	0.0002
110.000	0.994	0.025	0.0002
120.000	0.991	0.024	0.0002
130.000	0.987	0.021	0.0001
140.000	0.989	0.016	0.0001
150.000	0.990	0.015	0.0001
160.000	0.989	0.014	0.0001
170.000	0.989	0.011	0.0001
180.000	0.990	0.010	0.0000
190.000	0.990	0.009	0.000
200.000	0.991	0.007	0.0000
220.000	0.992	0.006	0.0000
240.000	0.993	0.005	0.0000
260.000	0.993	0.004	0.0000
280.000	0.994	0.003	0.0000
300.000	0.995	0.002	0.0000

Chromium[2]

Energy (eV)	n	k	R(ϕ = 0)
0.06	21.19	42.00	0.962
0.10	11.81	29.76	0.955
0.14	15.31	26.36	0.936
0.18	8.73	25.37	0.953
0.22	5.30	20.62	0.954
0.26	3.91	17.12	0.951
0.30	3.15	14.28	0.943
0.42	3.47	8.97	0.862
0.54	3.92	7.06	0.788
0.66	3.96	5.95	0.736
0.78	4.13	5.03	0.680
0.90	4.43	4.60	0.650
1.00	4.47	4.43	0.639
1.12	4.53	4.31	0.631
1.24	4.50	4.28	0.629
1.36	4.42	4.30	0.631
1.46	4.31	4.32	0.632
1.77	3.84	4.37	0.639
2.00	3.48	4.36	0.644
2.20	3.18	4.41	0.656
2.40	2.75	4.46	0.677
2.60	2.22	4.36	0.698
2.80	1.80	4.06	0.703
3.00	1.54	3.71	0.695
3.20	1.44	3.40	0.670
3.40	1.39	3.24	0.657
3.60	1.26	3.12	0.661
3.80	1.12	2.95	0.660
4.00	1.02	2.76	0.651
4.20	0.94	2.58	0.639
4.40	0.90	2.42	0.620
4.50	0.89	2.35	0.607
4.60	0.88	2.28	0.598
4.70	0.86	2.21	0.586
4.80	0.86	2.13	0.572
4.90	0.86	2.07	0.557
5.00	0.85	2.01	0.542
5.10	0.86	1.94	0.523
5.20	0.87	1.87	0.503
5.40	0.93	1.80	0.466
5.60	0.95	1.74	0.443
5.80	0.97	1.74	0.437
6.00	0.94	1.73	0.444
6.20	0.89	1.69	0.446
6.40	0.85	1.66	0.447
6.60	0.80	1.59	0.444
6.80	0.75	1.51	0.439
7.00	0.74	1.45	0.425
7.20	0.71	1.39	0.414
7.40	0.69	1.33	0.404
7.60	0.66	1.23	0.378
7.80	0.67	1.15	0.347
8.00	0.68	1.07	0.315
8.20	0.71	1.00	0.278
8.50	0.74	0.92	0.235
9.0	0.83	0.81	0.170
9.50	0.92	0.74	0.132
10.00	0.98	0.73	0.120
10.50	1.01	0.72	0.112
11.00	1.05	0.69	0.103
11.50	1.09	0.69	0.100
12.00	1.13	0.70	0.101
12.50	1.15	0.73	0.108
13.00	1.15	0.77	0.119
13.50	1.12	0.80	0.128
14.00	1.09	0.82	0.135
14.50	1.03	0.82	0.142
15.00	1.00	0.82	0.143
15.50	0.96	0.80	0.141
16.00	0.92	0.77	0.139
16.50	0.91	0.75	0.134
17.00	0.90	0.73	0.132
17.50	0.88	0.72	0.130
18.00	0.87	0.70	0.129
18.50	0.84	0.69	0.130
19.00	0.82	0.68	0.131
20.00	0.77	0.64	0.130
20.5	0.76	0.63	0.129
21.0	0.74	0.58	0.121
21.5	0.72	0.55	0.116
22.0	0.71	0.52	0.112
22.5	0.70	0.50	0.109
23.0	0.69	0.48	0.105
23.5	0.68	0.45	0.101
24.0	0.68	0.43	0.096
24.5	0.67	0.39	0.089
25.0	0.68	0.36	0.080
25.5	0.68	0.33	0.072
26.0	0.70	0.31	0.063
26.5	0.71	0.28	0.055
27.0	0.72	0.26	0.048
27.5	0.73	0.25	0.043
28.0	0.75	0.23	0.037
29.0	0.77	0.22	0.032
30.0	0.78	0.21	0.030

Cobalt (Single crystal, $\vec{E} \parallel \hat{c}$)[3]

Energy (eV)	n	k	R(ϕ = 0)
0.10	6.71	37.87	0.982

0.15	4.66	25.47	0.973
0.20	3.55	18.78	0.962
0.25	3.98	14.59	0.933
0.30	4.04	12.16	0.907
0.40	4.24	9.13	0.847
0.50	4.41	7.19	0.782
0.60	4.91	6.13	0.729
0.70	5.24	5.85	0.713
0.80	5.17	5.89	0.716
0.90	4.94	5.95	0.720
1.00	4.46	5.86	0.722
1.10	4.07	5.61	0.715
1.20	3.81	5.36	0.706
1.30	3.60	5.20	0.701
1.40	3.37	5.09	0.701
1.50	3.10	4.96	0.701
1.60	2.84	4.77	0.697
1.70	2.66	4.57	0.690
1.80	2.45	4.41	0.687
1.90	2.31	4.18	0.675
2.00	2.21	4.00	0.664
2.10	2.13	3.85	0.654
2.20	2.07	3.70	0.642
2.30	2.01	3.59	0.634
2.40	1.95	3.49	0.627
2.50	1.88	3.40	0.622
2.60	1.81	3.32	0.618
2.70	1.73	3.24	0.615
2.80	1.66	3.13	0.607
2.90	1.61	3.05	0.600
3.00	1.55	2.96	0.594
3.20	1.46	2.80	0.579
3.40	1.38	2.64	0.563
3.60	1.31	2.48	0.544
3.80	1.28	2.33	0.519
4.00	1.26	2.20	0.495
4.20	1.25	2.10	0.471
4.40	1.24	2.01	0.452
4.60	1.24	1.94	0.435
4.80	1.23	1.88	0.423
5.00	1.22	1.83	0.411
5.20	1.21	1.79	0.403
5.40	1.19	1.77	0.399
5.60	1.16	1.75	0.400
5.80	1.10	1.73	0.406
6.00	1.03	1.68	0.407
6.20	0.97	1.62	0.401
6.40	0.94	1.53	0.386
6.60	0.91	1.46	0.368
6.80	0.91	1.38	0.345
7.00	0.91	1.32	0.326
7.00	0.91	1.26	0.305
7.40	0.92	1.21	0.286
7.60	0.93	1.17	0.269
7.80	0.94	1.13	0.253
8.00	0.95	1.09	0.239

Cobalt (Single crystal, $\vec{E} \perp \hat{c}$)[3]

Energy (eV)	n	k	R(ϕ = 0)
0.10	5.83	32.36	0.979
0.15	4.24	21.37	0.965
0.20	3.87	15.53	0.042
0.30	4.34	10.01	0.865
0.40	4.66	7.39	0.785
0.50	5.17	5.75	0.709
0.60	5.77	5.17	0.682
0.70	6.15	5.20	0.685
0.80	6.08	5.61	0.702
0.90	5.57	5.93	0.715
1.00	4.83	5.94	0.721
1.10	4.31	5.60	0.711
1.20	4.02	5.34	0.701
1.30	3.78	5.16	0.694
1.40	3.55	5.05	0.692
1.50	3.26	4.93	0.692
1.60	3.03	4.74	0.687
1.70	2.83	4.60	0.684
1.80	2.61	4.45	0.683
1.90	2.41	4.27	0.677
2.00	2.25	4.09	0.670
2.10	2.13	3.89	0.659
2.20	2.04	3.72	0.646
2.30	1.99	3.56	0.632
2.40	1.95	3.44	0.620
2.50	1.90	3.34	0.611
2.60	1.86	3.26	0.605
2.70	1.79	3.19	0.602
2.80	1.72	3.11	0.596
2.90	1.66	3.03	0.591
3.00	1.60	2.94	0.586
3.20	1.50	2.78	0.571
3.40	1.42	2.62	0.553
3.60	1.36	2.47	0.533
3.80	1.33	2.33	0.511
4.00	1.31	2.21	0.488
4.20	1.28	2.12	0.471
4.40	1.27	2.03	0.452
4.60	1.26	1.95	0.435
4.80	1.25	1.90	0.423
5.00	1.24	1.84	0.411
5.20	1.22	1.80	0.403
5.40	1.21	1.78	0.399
5.60	1.17	1.76	0.400
5.80	1.11	1.74	0.406
6.00	1.04	1.69	0.407
6.20	0.98	1.62	0.401
6.40	0.94	1.54	0.386
6.60	0.92	1.46	0.368
6.80	0.91	1.38	0.345
7.00	0.91	1.32	0.326
7.20	0.91	1.26	0.305
7.40	0.92	1.21	0.285
7.60	0.93	1.17	0.269
7.80	0.94	1.13	0.253

Copper[4]

Energy (eV)	n	k	R(ϕ = 0)
0.10	29.69	71.57	0.980
0.50	1.71	17.63	0.979
1.00	0.44	8.48	0.976
1.50	0.26	5.26	0.965
1.70	0.22	4.43	0.958
1.75	0.21	4.25	0.956
1.80	0.21	4.04	0.952
1.85	0.22	3.85	0.947
1.90	0.21	3.67	0.943
2.00	0.27	3.24	0.910
2.10	0.47	2.81	0.814
2.20	0.83	2.60	0.673
2.30	1.04	2.59	0.618
2.40	1.12	2.60	0.602
2.60	1.15	2.50	0.577
2.80	1.17	2.36	0.545
3.00	1.18	2.21	0.509
3.20	1.23	2.07	0.468
3.40	1.27	1.95	0.434
3.60	1.31	1.87	0.407
3.80	1.34	1.81	0.387
4.00	1.34	1.72	0.364
4.20	1.42	1.64	0.336
4.40	1.49	1.64	0.329
4.60	1.52	1.67	0.334
4.80	1.53	1.71	0.345
5.00	1.47	1.78	0.366
5.20	1.38	1.80	0.380
5.40	1.28	1.78	0.389
5.60	1.18	1.74	0.391
5.80	1.10	1.67	0.389
6.00	1.04	1.59	0.380
6.50	0.96	1.37	0.329
7.00	0.97	1.20	0.271
7.50	1.00	1.09	0.230
8.00	1.03	1.03	0.206
8.50	1.03	0.98	0.189
9.00	1.03	0.92	0.171
9.50	1.03	0.87	0.154
10.00	1.04	0.82	0.139
11.00	1.07	0.75	0.118
12.00	1.09	0.73	0.111
13.00	1.08	0.72	0.109
14.00	1.06	0.72	0.111
14.50	1.03	0.72	0.111
15.00	1.01	0.71	0.111
15.50	0.98	0.69	0.109
16.00	0.95	0.67	0.106
17.00	0.91	0.62	0.097
18.00	0.89	0.56	0.084
19.00	0.88	0.51	0.071
20.00	0.88	0.45	0.059
21.00	0.90	0.41	0.048
22.00	0.92	0.38	0.040
23.00	0.94	0.37	0.035
24.00	0.96	0.37	0.035
25.00	0.96	0.40	0.040
26.00	0.92	0.40	0.044
27.00	0.88	0.38	0.043
28.00	0.86	0.35	0.039
29.00	0.85	0.30	0.032
30.00	0.86	0.26	0.025
31.00	0.88	0.24	0.020
32.00	0.89	0.22	0.017
33.00	0.90	0.21	0.015
34.00	0.91	0.20	0.014
35.00	0.92	0.20	0.013
36.00	0.92	0.19	0.012
37.00	0.92	0.19	0.011
38.00	0.93	0.18	0.010
39.00	0.93	0.17	0.009
40.00	0.93	0.17	0.009
41.00	0.94	0.16	0.008
42.00	0.94	0.16	0.007
43.00	0.94	0.15	0.007
44.00	0.95	0.15	0.007
45.00	0.95	0.15	0.006
46.00	0.95	0.15	0.006
47.00	0.95	0.14	0.006
48.00	0.95	0.14	0.006
49.00	0.95	0.14	0.005
50.00	0.95	0.13	0.005
51.00	0.95	0.13	0.005
52.00	0.95	0.13	0.005
53.00	0.96	0.12	0.004
54.00	0.96	0.12	0.004
55.00	0.96	0.12	0.004
56.00	0.96	0.11	0.004
57.00	0.96	0.11	0.004
58.00	0.96	0.11	0.004
59.00	0.97	0.11	0.003
60.00	0.97	0.11	0.003
61.00	0.97	0.11	0.003
62.00	0.97	0.11	0.003
63.00	0.96	0.10	0.003
64.00	0.96	0.10	0.003
65.00	0.97	0.10	0.003
66.00	0.97	0.10	0.003
67.00	0.97	0.09	0.003
68.00	0.97	0.09	0.002
69.00	0.97	0.09	0.002
70.00	0.97	0.09	0.002
75.00	0.98	0.09	0.002
80.00	0.98	0.09	0.002
85.00	0.97	0.09	0.002
90.00	0.96	0.08	0.002

Gold (Electropolished, Au [110])[5]

Energy (eV)	n	k	R(ϕ = 0)
0.10	8.17	82.83	0.995
0.20	2.13	41.73	0.995
0.30	0.99	27.82	0.995
0.40	0.59	20.83	0.995
0.50	0.39	16.61	0.994
0.60	0.28	13.78	0.994

0.70	0.22	11.75	0.994
0.80	0.18	10.21	0.993
0.90	0.15	9.01	0.993
1.00	0.13	8.03	0.992
1.20	0.10	6.54	0.991
1.40	0.08	5.44	0.989
1.60	0.08	4.56	0.986
1.80	0.09	3.82	0.979
2.00	0.13	3.16	0.953
2.10	0.18	2.84	0.925
2.20	0.24	2.54	0.880
2.30	0.32	2.22	0.807
2.40	0.50	1.86	0.647
2.50	0.82	1.59	0.438
2.60	1.24	1.54	0.331
2.70	1.43	1.72	0.356
2.80	1.46	1.77	0.368
2.90	1.50	1.79	0.368
3.00	1.54	1.80	0.369
3.10	1.54	1.81	0.371
3.20	1.54	1.80	0.368
3.30	1.55	1.78	0.362
3.40	1.56	1.76	0.356
3.50	1.58	1.73	0.349
3.60	1.62	1.73	0.346
3.70	1.64	1.75	0.351
3.80	1.63	1.79	0.360
3.90	1.59	1.81	0.366
4.00	1.55	1.81	0.369
4.10	1.51	1.79	0.368
4.20	1.48	1.78	0.367
4.30	1.45	1.77	0.368
4.40	1.41	1.76	0.370
4.50	1.35	1.74	0.370
4.60	1.30	1.69	0.364
4.70	1.27	1.64	0.354
4.80	1.25	1.59	0.344
4.90	1.23	1.54	0.332
5.00	1.22	1.49	0.319
5.20	1.21	1.40	0.295
5.40	1.21	1.33	0.275
5.60	1.21	1.27	0.256
5.80	1.21	1.20	0.236
6.00	1.22	1.14	0.218
6.20	1.24	1.09	0.203
6.40	1.25	1.05	0.190
6.60	1.27	1.01	0.177
6.80	1.30	0.97	0.167
7.00	1.34	0.95	0.162
7.20	1.36	0.95	0.161
7.40	1.38	0.96	0.164
7.60	1.38	0.98	0.169
7.80	1.35	0.99	0.171
8.00	1.31	0.96	0.165
8.20	1.30	0.92	0.155
8.40	1.30	0.89	0.147
8.60	1.31	0.88	0.144
8.80	1.31	0.86	0.140
9.00	1.30	0.83	0.133
9.20	1.31	0.81	0.126
9.40	1.33	0.78	0.122
9.60	1.36	0.78	0.121
9.80	1.37	0.79	0.124
10.00	1.37	0.80	0.126
10.20	1.36	0.80	0.127
10.40	1.35	0.80	0.125
10.60	1.34	0.79	0.123
10.80	1.34	0.77	0.120
11.00	1.34	0.76	0.116
11.20	1.34	0.74	0.113
11.40	1.35	0.73	0.111
11.60	1.36	0.72	0.109
11.80	1.38	0.71	0.108
12.00	1.39	0.71	0.109
12.40	1.44	0.73	0.115
12.80	1.45	0.79	0.127
13.20	1.42	0.84	0.137
13.60	1.37	0.86	0.140
14.00	1.33	0.86	0.140
14.40	1.29	0.86	0.139
14.80	1.26	0.84	0.135
15.20	1.24	0.83	0.132
15.60	1.22	0.81	0.127
16.00	1.21	0.79	0.123
16.40	1.20	0.78	0.119
16.80	1.19	0.76	0.116
17.20	1.19	0.75	0.114
17.60	1.19	0.74	0.111
18.00	1.19	0.74	0.109
18.40	1.19	0.73	0.109
18.80	1.20	0.74	0.110
19.20	1.21	0.76	0.116
19.60	1.21	0.80	0.125
20.00	1.18	0.83	0.133
20.40	1.14	0.85	0.141
20.80	1.10	0.87	0.149
21.20	1.05	0.88	0.156
21.60	1.00	0.88	0.162
22.00	0.94	0.86	0.164
22.40	0.89	0.83	0.163
22.80	0.85	0.79	0.157
23.20	0.82	0.75	0.149
23.60	0.80	0.70	0.138
24.00	0.80	0.66	0.125
24.40	0.80	0.62	0.113
24.80	0.80	0.58	0.101
25.20	0.82	0.56	0.090
25.60	0.83	0.54	0.084
26.00	0.84	0.52	0.079
26.40	0.85	0.51	0.074
26.80	0.85	0.50	0.071
27.20	0.86	0.49	0.068
27.60	0.86	0.49	0.065
28.00	0.87	0.48	0.063
28.40	0.88	0.48	0.062
28.80	0.88	0.48	0.062
29.20	0.88	0.48	0.062
29.60	0.87	0.48	0.064
30.00	0.86	0.48	0.064

Hafnium (Single crystal, $\vec{E} \parallel \hat{c}$)[6]

Energy (eV)	n	k	R(ϕ = 0)
0.52	1.48	4.11	0.747
0.56	1.84	3.29	0.615
0.60	2.34	2.62	0.486
0.66	3.21	2.13	0.428
0.70	3.70	2.03	0.441
0.76	4.31	2.10	0.476
0.80	4.61	2.31	0.504
0.86	4.71	2.70	0.533
0.90	4.64	2.85	0.541
0.95	4.54	2.96	0.545
1.00	4.45	3.00	0.545
1.10	4.28	3.08	0.547
1.20	4.08	3.10	0.544
1.30	3.87	3.04	0.536
1.40	3.72	2.95	0.525
1.50	3.60	2.85	0.514
1.60	3.52	2.73	0.500
1.70	3.52	2.61	0.488
1.80	3.57	2.56	0.485
1.90	3.63	2.59	0.489
2.00	3.65	2.67	0.498
2.10	3.64	2.81	0.511
2.20	3.53	2.99	0.526
2.30	3.34	3.09	0.534
2.40	3.15	3.11	0.537
2.50	2.99	3.13	0.540
2.60	2.83	3.12	0.542
2.70	2.68	3.10	0.542
2.80	2.54	3.08	0.543
2.90	2.40	3.04	0.544
3.00	2.27	3.00	0.544
3.10	2.14	2.95	0.544
3.20	2.00	2.89	0.544
3.30	1.87	2.79	0.538
3.40	1.78	2.68	0.528
3.50	1.71	2.58	0.517
3.60	1.66	2.48	0.503
3.70	1.63	2.40	0.491
3.80	1.60	2.33	0.481
3.90	1.56	2.27	0.473
4.00	1.52	2.21	0.466
4.10	1.48	2.14	0.455
4.20	1.45	2.07	0.442
4.30	1.43	2.01	0.431
4.40	1.41	1.95	0.420
4.50	1.39	1.89	0.407
4.60	1.39	1.83	0.394
4.70	1.39	1.79	0.382
4.80	1.38	1.75	0.373
4.90	1.38	1.71	0.364
5.00	1.37	1.68	0.356
5.20	1.36	1.61	0.341
5.40	1.35	1.55	0.324
5.60	1.35	1.51	0.314
5.80	1.32	1.48	0.308
6.00	1.28	1.41	0.295
6.20	1.26	1.35	0.278
6.40	1.26	1.28	0.258
6.60	1.27	1.22	0.240
6.80	1.28	1.16	0.224
7.00	1.31	1.13	0.212
7.20	1.33	1.10	0.204
7.40	1.34	1.07	0.197
7.60	1.36	1.05	0.191
7.80	1.37	1.02	0.183
8.00	1.40	1.01	0.179
8.20	1.43	1.01	0.178
8.40	1.45	1.01	0.180
8.60	1.47	1.02	0.183
8.80	1.48	1.04	0.186
9.00	1.49	1.07	0.193
9.20	1.50	1.10	0.201
9.40	1.48	1.14	0.211
9.60	1.46	1.18	0.222
9.80	1.41	1.21	0.230
10.00	1.36	1.22	0.235
10.20	1.32	1.22	0.238
10.40	1.28	1.22	0.240
10.60	1.24	1.21	0.241
10.80	1.20	1.20	0.242
11.00	1.16	1.19	0.242
11.20	1.13	1.17	0.241
11.40	1.10	1.16	0.241
11.60	1.07	1.14	0.239
11.80	1.04	1.12	0.238
12.00	1.02	1.10	0.236
12.40	0.96	1.06	0.232
12.80	0.92	1.01	0.225
13.20	0.88	0.96	0.218
13.60	0.84	0.90	0.205
14.00	0.83	0.83	0.186
14.40	0.83	0.80	0.172
14.80	0.81	0.76	0.167
15.20	0.79	0.70	0.153
15.60	0.79	0.64	0.132
16.00	0.83	0.60	0.111
16.40	0.81	0.60	0.114
16.80	0.79	0.55	0.105
17.20	0.79	0.50	0.089
17.60	0.80	0.46	0.077
18.00	0.81	0.42	0.064
18.40	0.84	0.38	0.051
18.80	0.87	0.34	0.040
19.00	0.89	0.33	0.036
19.60	0.93	0.32	0.030
20.00	0.94	0.31	0.027
20.60	0.97	0.30	0.023
21.00	0.99	0.29	0.022
21.60	1.01	0.28	0.020
22.00	1.03	0.28	0.020
22.60	1.06	0.28	0.020
23.00	1.07	0.28	0.021
23.60	1.09	0.29	0.022
24.00	1.09	0.30	0.023
24.60	1.10	0.31	0.024

Hafnium (Single crystal, $\vec{E} \perp \hat{c}$)[6]

Energy (eV)	n	k	R(ϕ = 0)
0.52	2.25	4.65	0.723
0.56	2.34	3.66	0.623
0.60	2.84	2.89	0.512
0.66	3.71	2.35	0.469
0.70	4.26	2.21	0.482
0.76	4.97	2.33	0.521
0.80	5.41	2.62	0.554
0.86	5.46	3.36	0.593
0.90	5.22	3.62	0.601
0.95	4.95	3.72	0.602
1.00	4.76	3.76	0.602
1.10	4.43	3.80	0.601
1.20	4.07	3.74	0.594
1.30	3.79	3.55	0.578
1.40	3.61	3.36	0.561
1.50	3.55	3.13	0.540
1.60	3.58	3.01	0.529
1.70	3.63	2.98	0.526
1.80	3.66	3.02	0.530
1.90	3.63	3.14	0.541
2.00	3.51	3.26	0.551
2.10	3.35	3.33	0.558
2.20	3.18	3.36	0.563
2.30	2.99	3.39	0.568
2.40	2.78	3.35	0.569
2.50	2.65	3.26	0.562
2.60	2.54	3.22	0.560
2.70	2.42	3.17	0.559
2.80	2.31	3.13	0.558
2.90	2.20	3.08	0.558
3.00	2.08	3.05	0.561
3.10	1.94	2.98	0.560
3.20	1.83	2.88	0.555
3.30	1.74	2.78	0.547
3.40	1.68	2.69	0.538
3.50	1.62	2.61	0.529
3.60	1.57	2.52	0.519
3.70	1.53	2.45	0.510
3.80	1.49	2.38	0.501
3.90	1.45	2.32	0.493
4.00	1.41	2.25	0.484
4.10	1.38	2.18	0.474
4.20	1.35	2.11	0.462
4.30	1.33	2.05	0.451
4.40	1.31	1.99	0.438
4.50	1.30	1.93	0.427
4.60	1.29	1.88	0.415
4.70	1.28	1.82	0.402
4.80	1.28	1.77	0.389
4.90	1.27	1.73	0.379
5.00	1.27	1.69	0.367
5.20	1.27	1.62	0.349
5.40	1.27	1.57	0.335
5.60	1.26	1.52	0.322
5.80	1.24	1.48	0.313
6.00	1.21	1.42	0.302
6.20	1.19	1.36	0.285
6.40	1.18	1.29	0.265
6.60	1.19	1.22	0.244
6.80	1.21	1.18	0.230
7.00	1.22	1.14	0.217
7.20	1.23	1.10	0.206
7.40	1.26	1.06	0.194
7.60	1.28	1.04	0.187
7.80	1.30	1.02	0.180
8.00	1.33	1.00	0.174
8.20	1.35	0.99	0.173
8.40	1.38	0.99	0.173
8.60	1.40	1.00	0.174
8.80	1.42	1.02	0.178
9.00	1.43	1.04	0.184
9.20	1.45	1.08	0.193
9.40	1.43	1.12	0.204
9.60	1.40	1.16	0.214
9.80	1.37	1.19	0.223
10.00	1.32	1.21	0.230
10.20	1.27	1.21	0.234
10.40	1.23	1.20	0.235
10.60	1.19	1.20	0.237
10.80	1.15	1.19	0.237
11.00	1.12	1.17	0.237
11.20	1.08	1.16	0.237
11.40	1.05	1.14	0.236
11.60	1.03	1.12	0.235
11.80	1.00	1.10	0.233
12.00	0.97	1.08	0.231
12.40	0.92	1.04	0.226
12.80	0.88	0.99	0.219
13.20	0.83	0.94	0.211
13.60	0.80	0.88	0.196
14.00	0.79	0.81	0.177
14.40	0.80	0.77	0.160
14.80	0.77	0.73	0.154
15.20	0.76	0.68	0.140
15.60	0.76	0.61	0.119
16.00	0.81	0.58	0.099
16.40	0.78	0.57	0.102
16.80	0.77	0.53	0.092
17.20	0.77	0.48	0.077
17.60	0.79	0.44	0.065
18.00	0.80	0.39	0.053
18.40	0.82	0.36	0.041
18.80	0.86	0.33	0.032
19.00	0.88	0.32	0.030
19.60	0.91	0.31	0.025
20.00	0.93	0.30	0.023
20.60	0.96	0.29	0.021
21.00	0.97	0.29	0.020
21.60	1.00	0.28	0.019
22.00	1.01	0.28	0.019
22.60	1.03	0.27	0.018
23.00	1.05	0.28	0.019
23.60	1.06	0.28	0.020
24.00	1.07	0.29	0.021
24.60	1.09	0.30	0.022

Iridium[7]

Energy (eV)	n	k	R(ϕ = 0)
0.10	28.49	60.62	0.975
0.15	15.32	45.15	0.973
0.20	9.69	35.34	0.972
0.25	6.86	28.84	0.969
0.30	5.16	24.25	0.967
0.35	4.11	20.79	0.064
0.40	3.42	18.06	0.960
0.45	3.05	15.82	0.954
0.50	2.98	14.06	0.944
0.60	2.79	11.58	0.925
0.70	2.93	9.78	0.895
0.80	3.14	8.61	0.862
0.90	3.19	7.88	0.840
1.00	3.15	7.31	0.822
1.10	3.04	6.84	0.808
1.20	2.96	6.41	0.791
1.30	2.85	6.07	0.779
1.40	2.72	5.74	0.767
1.50	2.65	5.39	0.750
1.60	2.68	5.08	0.728
1.70	2.69	4.92	0.716
1.80	2.64	4.81	0.710
1.90	2.57	4.68	0.704
2.00	2.50	4.57	0.699
2.10	2.40	4.48	0.697
2.20	2.29	4.38	0.695
2.30	2.18	4.26	0.692
2.40	2.07	4.14	0.689
2.50	1.98	4.00	0.682
2.60	1.91	3.86	0.673
2.70	1.85	3.73	0.665
2.80	1.81	3.61	0.655
2.90	1.77	3.51	0.646
3.00	1.73	3.43	0.640
3.20	1.62	3.26	0.629
3.40	1.53	3.05	0.610
3.60	1.52	2.81	0.573
3.80	1.61	2.69	0.541
4.00	1.64	2.68	0.535
4.20	1.58	2.71	0.549
4.40	1.45	2.68	0.561
4.60	1.31	2.60	0.567
4.80	1.18	2.49	0.570
5.00	1.10	2.35	0.559
5.20	1.04	2.22	0.543
5.40	1.00	2.09	0.522
5.60	0.98	1.98	0.499
5.80	0.96	1.86	0.474
6.00	0.95	1.78	0.454
6.20	0.94	1.68	0.427
6.40	0.94	1.59	0.401
6.60	0.94	1.50	0.375
6.80	0.95	1.42	0.345
7.00	0.97	1.34	0.318
7.20	0.99	1.27	0.290
7.40	1.02	1.20	0.262
7.60	1.03	1.14	0.241
7.80	1.08	1.06	0.208
8.00	1.13	1.03	0.191
8.20	1.18	1.00	0.179
8.40	1.22	0.98	0.171
8.60	1.26	0.96	0.164
8.80	1.29	0.95	0.160
9.00	1.33	0.94	0.157
9.20	1.36	0.95	0.159
9.40	1.39	0.95	0.161
9.60	1.42	0.97	0.163
9.80	1.44	0.99	0.169
10.00	1.45	1.01	0.175
10.20	1.45	1.04	0.182
10.40	1.44	1.07	0.187
10.60	1.43	1.09	0.193
10.80	1.41	1.12	0.200
11.00	1.38	1.13	0.206
11.20	1.34	1.14	0.208
11.40	1.31	1.13	0.208
11.60	1.28	1.12	0.206
11.80	1.25	1.10	0.203
12.00	1.24	1.08	0.199
12.40	1.21	1.05	0.191
12.80	1.19	1.01	0.181
13.20	1.18	0.98	0.173
13.60	1.17	0.95	0.165
14.00	1.16	0.91	0.155
14.40	1.17	0.88	0.147
14.80	1.18	0.87	0.142
15.20	1.19	0.84	0.136
15.60	1.20	0.83	0.133
16.00	1.21	0.83	0.131
16.40	1.23	0.82	0.129
16.80	1.25	0.82	0.127
17.20	1.28	0.83	0.131
17.60	1.30	0.87	0.140
18.00	1.30	0.93	0.154
18.40	1.27	0.97	0.166
18.80	1.24	1.00	0.176
19.20	1.20	1.03	0.187
19.60	1.15	1.05	0.197
20.00	1.10	1.06	0.205
20.50	1.04	1.05	0.210
21.00	0.99	1.04	0.215
21.50	0.94	1.02	0.220
22.00	0.89	1.00	0.222
22.50	0.84	0.99	0.228
23.00	0.79	0.96	0.232
23.50	0.76	0.92	0.228
24.00	0.73	0.87	0.223
24.50	0.70	0.83	0.218
25.00	0.69	0.79	0.209
25.50	0.68	0.76	0.200
26.00	0.67	0.72	0.192
26.50	0.67	0.69	0.181
27.00	0.66	0.66	0.174
27.50	0.66	0.63	0.166
28.00	0.66	0.61	0.158
28.50	0.66	0.59	0.151
29.00	0.65	0.57	0.148

29.50	0.64	0.55	0.145
30.00	0.64	0.53	0.140
32.00	0.62	0.44	0.119
34.00	0.64	0.35	0.091
36.00	0.69	0.27	0.059
38.00	0.73	0.24	0.044
40.00	0.76	0.22	0.034

Iron[3]

Energy (eV)	n	k	R(ϕ = 0)
0.10	6.41	33.07	0.978
0.15	6.26	22.82	0.956
0.20	3.68	18.23	0.958
0.26	4.98	13.68	0.911
0.30	4.87	12.05	0.892
0.36	4.68	10.44	0.867
0.40	4.42	9.75	0.858
0.50	4.14	8.02	0.817
0.60	3.93	6.95	0.783
0.70	3.78	6.17	0.752
0.80	3.65	5.60	0.725
0.90	3.52	5.16	0.700
1.00	3.43	4.79	0.678
1.10	3.33	4.52	0.660
1.20	3.24	4.26	0.641
1.30	3.16	4.07	0.626
1.40	3.12	3.87	0.609
1.50	3.05	3.77	0.601
1.60	3.00	3.60	0.585
1.70	2.98	3.52	0.577
1.80	2.92	3.46	0.573
1.90	2.89	3.37	0.563
2.00	2.85	3.36	0.563
2.10	2.80	3.34	0.562
2.20	2.74	3.33	0.563
2.30	2.65	3.34	0.567
2.40	2.56	3.31	0.567
2.50	2.46	3.31	0.570
2.60	2.34	3.30	0.576
2.70	2.23	3.25	0.575
2.80	2.12	3.23	0.580
2.90	2.01	3.17	0.580
3.00	1.88	3.12	0.583
3.10	1.78	3.04	0.580
3.20	1.70	2.96	0.576
3.30	1.62	2.87	0.572
3.40	1.55	2.79	0.565
3.50	1.50	2.70	0.556
3.60	1.47	2.63	0.548
3.70	1.43	2.56	0.542
3.83	1.38	2.49	0.534
4.00	1.30	2.39	0.527
4.17	1.26	2.27	0.510
4.33	1.23	2.18	0.494
4.50	1.20	2.10	0.482
4.67	1.16	2.02	0.470
4.83	1.14	1.93	0.451
5.00	1.14	1.87	0.435
5.17	1.12	1.81	0.425
5.33	1.11	1.75	0.408
5.50	1.09	1.17	0.401
5.67	1.09	1.65	0.383
5.83	1.10	1.61	0.373
6.00	1.09	1.59	0.366
6.17	1.08	1.57	0.365
6.33	1.04	1.55	0.365
6.50	1.02	1.51	0.358
6.67	1.00	1.47	0.351
6.83	0.97	1.43	0.346
7.00	0.96	1.39	0.333
7.17	0.94	1.35	0.327
7.33	0.94	1.30	0.311
7.50	0.94	1.26	0.298
7.67	0.94	1.23	0.288
7.83	0.94	1.21	0.279
8.00	0.94	1.18	0.272
8.17	0.94	1.16	0.265
8.33	0.94	1.14	0.258
8.50	0.94	1.12	0.251
8.67	0.94	1.10	0.246
8.83	0.93	1.08	0.240
9.00	0.93	1.07	0.236
9.17	0.92	1.06	0.233
9.33	0.91	1.04	0.231
9.50	0.90	1.02	0.226
9.67	0.90	1.00	0.221
9.83	0.89	0.99	0.218
10.00	0.88	0.97	0.213
10.17	0.87	0.94	0.203
10.33	0.87	0.91	0.196
10.50	0.87	0.89	0.189
10.67	0.88	0.87	0.179
10.83	0.89	0.85	0.170
11.00	0.91	0.83	0.162
11.17	0.92	0.83	0.159
11.33	0.93	0.84	0.159
11.50	0.93	0.84	0.160
11.67	0.93	0.84	0.162
11.83	0.92	0.84	0.163
12.00	0.91	0.84	0.163
12.17	0.90	0.84	0.165
12.33	0.89	0.83	0.164
12.50	0.98	0.83	0.165
12.67	0.87	0.82	0.166
12.83	0.86	0.81	0.166
13.00	0.85	0.80	0.162
13.17	0.84	0.79	0.161
13.33	0.84	0.78	0.160
13.50	0.83	0.77	0.159
13.67	0.82	0.76	0.157
13.83	0.81	0.75	0.154
14.00	0.81	0.73	0.151
14.17	0.80	0.72	0.149
14.33	0.80	0.71	0.146
14.50	0.79	0.70	0.144
14.67	0.79	0.69	0.141
14.83	0.78	0.67	0.138
15.00	0.78	0.66	0.135
15.17	0.78	0.65	0.131
15.33	0.78	0.64	0.128
15.50	0.77	0.63	0.126
15.67	0.77	0.62	0.123
15.83	0.77	0.61	0.119
16.00	0.77	0.60	0.116
16.17	0.78	0.58	0.112
16.33	0.78	0.58	0.110
16.50	0.78	0.57	0.107
16.67	0.77	0.56	0.106
16.83	0.78	0.55	0.103
17.00	0.78	0.55	0.102
17.17	0.78	0.54	0.100
17.33	0.78	0.54	0.098
17.50	0.77	0.53	0.097
17.67	0.77	0.52	0.095
17.83	0.78	0.51	0.092
18.00	0.78	0.51	0.091
18.17	0.78	0.51	0.090
18.33	0.78	0.50	0.089
18.50	0.77	0.50	0.089
18.67	0.77	0.50	0.088
18.83	0.77	0.49	0.087
19.00	0.77	0.49	0.087
19.17	0.76	0.49	0.088
19.33	0.76	0.48	0.087
19.50	0.75	0.47	0.086
19.67	0.75	0.47	0.085
19.83	0.75	0.46	0.084
20.00	0.74	0.45	0.083
20.17	0.74	0.44	0.081
20.33	0.74	0.44	0.081
20.50	0.74	0.43	0.080
20.67	0.73	0.43	0.079
20.83	0.73	0.42	0.078
21.00	0.73	0.41	0.077
21.17	0.72	0.40	0.076
21.33	0.72	0.39	0.074
21.50	0.72	0.38	0.073
21.67	0.72	0.38	0.071
21.83	0.72	0.37	0.070
22.00	0.72	0.36	0.068
22.17	0.71	0.35	0.067
22.33	0.72	0.34	0.064
22.50	0.72	0.34	0.063
22.67	0.72	0.33	0.062
22.83	0.72	0.32	0.059
23.00	0.72	0.31	0.058
23.17	0.72	0.30	0.056
23.33	0.72	0.29	0.054
23.50	0.73	0.28	0.050
23.67	0.73	0.28	0.049
23.83	0.74	0.27	0.047
24.00	0.74	0.27	0.045
24.17	0.74	0.26	0.044
24.33	0.74	0.26	0.043
24.50	0.74	0.25	0.042
24.67	0.75	0.25	0.040
24.83	0.75	0.24	0.039
25.00	0.75	0.24	0.038
26.00	0.76	0.21	0.031
27.00	0.78	0.18	0.026
28.00	0.79	0.16	0.021
29.00	0.81	0.14	0.017
30.00	0.82	0.13	0.014

Manganese[8]

Energy (eV)	n	k	R(ϕ = 0)
0.64	3.89	5.95	0.738
0.77	3.78	5.41	0.710
0.89	3.65	5.02	0.688
1.02	3.48	4.74	0.673
1.14	3.30	4.53	0.662
1.26	3.10	4.35	0.653
1.39	2.97	4.18	0.643
1.51	2.83	4.03	0.634
1.64	2.70	3.91	0.627
1.76	2.62	3.78	0.617
1.88	2.56	3.65	0.606
2.01	2.51	3.54	0.596
2.13	2.47	3.43	0.585
2.26	2.39	3.33	0.577
2.38	2.32	3.23	0.567
2.50	2.25	3.14	0.559
2.63	2.19	3.06	0.552
2.75	2.11	2.98	0.545
2.88	2.06	2.90	0.536
3.00	2.00	2.82	0.528
3.12	1.96	2.74	0.518
3.25	1.92	2.67	0.509
3.37	1.89	2.59	0.498
3.50	1.89	2.51	0.484
3.62	1.87	2.45	0.475
3.74	1.86	2.38	0.463
3.87	1.86	2.32	0.451
3.99	1.86	2.25	0.438
4.12	1.86	2.19	0.427
4.24	1.85	2.14	0.417
4.36	1.85	2.08	0.406
4.49	1.86	2.03	0.395
4.61	1.85	1.99	0.388
4.74	1.84	1.94	0.378
4.86	1.83	1.91	0.372
4.98	1.82	1.86	0.362
5.11	1.82	1.82	0.354
5.23	1.81	1.79	0.348
5.36	1.78	1.76	0.342
5.48	1.74	1.73	0.337
5.60	1.73	1.70	0.331
5.73	1.72	1.67	0.325
5.85	1.70	1.64	0.319
5.98	1.67	1.61	0.313
6.10	1.63	1.58	0.307
6.22	1.62	1.55	0.301
6.35	1.59	1.52	0.295
6.47	1.55	1.50	0.292
6.60	1.48	1.47	0.288

Molybdenum[9]

Energy (eV)	n	k	$R(\phi = 0)$
0.10	18.53	68.51	0.985
0.15	8.78	47.54	0.985
0.20	5.10	35.99	0.985
0.25	3.36	28.75	0.984
0.30	2.44	23.80	0.983
0.34	2.00	20.84	0.982
0.38	1.70	18.44	0.980
0.42	1.57	16.50	0.978
0.46	1.46	14.91	0.975
0.50	1.37	13.55	0.971
0.54	1.33	12.30	0.966
0.58	1.34	11.34	0.960
0.62	1.38	10.44	0.952
0.66	1.43	9.67	0.942
0.70	1.48	8.99	0.932
0.74	1.51	8.38	0.921
0.78	1.60	7.83	0.906
0.82	1.64	7.35	0.892
0.86	1.70	6.89	0.876
0.90	1.74	6.48	0.859
1.00	1.94	5.58	0.805
1.10	2.15	4.85	0.743
1.20	2.44	4.22	0.671
1.30	2.77	3.74	0.608
1.40	3.15	3.40	0.562
1.50	3.53	3.30	0.550
1.60	3.77	3.41	0.562
1.70	3.84	3.51	0.570
1.80	3.81	3.58	0.576
1.90	3.74	3.58	0.576
2.00	3.68	3.52	0.571
2.10	3.68	3.45	0.565
2.20	3.76	3.41	0.562
2.30	3.79	3.61	0.578
2.40	3.59	3.78	0.594
2.50	3.36	3.73	0.591
2.60	3.22	3.61	0.582
2.70	3.13	3.51	0.573
2.80	3.08	3.42	0.565
2.90	3.05	3.33	0.566
3.00	3.04	3.27	0.550
3.10	3.03	3.21	0.544
3.20	3.05	3.18	0.540
3.30	3.06	3.18	0.540
3.40	3.06	3.19	0.541
3.50	3.06	3.21	0.543
3.60	3.05	3.23	0.546
3.70	3.04	3.27	0.550
3.80	3.04	3.31	0.554
3.90	3.04	3.40	0.564
4.00	3.01	3.51	0.576
4.20	2.77	3.77	0.610
4.40	2.39	3.88	0.640
4.60	2.06	3.84	0.658
4.80	1.75	3.76	0.678
5.00	1.46	3.62	0.695
5.20	1.22	3.42	0.706
5.40	1.07	3.20	0.706
5.60	0.96	2.99	0.700
5.80	0.89	2.80	0.688
6.00	0.85	2.64	0.674
6.20	0.81	2.50	0.660
6.40	0.79	2.36	0.641
6.60	0.78	2.24	0.619
6.80	0.78	2.13	0.592
7.00	0.80	2.04	0.568
7.20	0.81	1.98	0.548
7.40	0.81	1.95	0.542
7.60	0.75	1.90	0.552
7.80	0.71	1.81	0.542
8.00	0.69	1.73	0.530
8.20	0.67	1.65	0.512
8.40	0.66	1.57	0.495
8.60	0.65	1.49	0.475
8.80	0.65	1.41	0.450
9.00	0.65	1.33	0.420
9.20	0.67	1.25	0.385
9.40	0.69	1.19	0.355
9.60	0.71	1.12	0.320
9.80	0.74	1.05	0.285
10.00	0.77	0.99	0.250
10.20	0.81	0.93	0.217
10.40	0.86	0.88	0.188
10.60	0.91	0.83	0.162
10.80	0.98	0.79	0.138
11.00	1.05	0.77	0.125
11.20	1.12	0.78	0.123
11.40	1.18	0.80	0.125
11.60	1.23	0.85	0.135
11.80	1.25	0.89	0.145
12.00	1.26	0.92	0.154
12.40	1.25	0.98	0.168
12.80	1.23	1.00	0.178
13.20	1.20	1.02	0.185
13.60	1.17	1.02	0.187
14.00	1.15	1.01	0.185
14.40	1.13	1.00	0.182
14.80	1.13	0.99	0.179
15.00	1.14	0.99	0.179
15.60	1.15	1.01	0.184
16.00	1.14	1.04	0.194
16.60	1.10	1.10	0.216
17.00	1.04	1.12	0.233
17.60	0.94	1.14	0.257
18.00	0.87	1.12	0.270
18.60	0.77	1.08	0.283
19.00	0.71	1.02	0.284
19.60	0.66	0.94	0.275
20.00	0.64	0.89	0.264
20.60	0.62	0.81	0.245
21.00	0.61	0.77	0.234
21.60	0.61	0.71	0.215
22.00	0.60	0.69	0.207
22.60	0.59	0.63	0.195
23.00	0.58	0.60	0.185
23.60	0.58	0.53	0.166
24.00	0.58	0.49	0.151
24.60	0.60	0.43	0.124
25.00	0.62	0.39	0.106
25.60	0.66	0.35	0.085
26.00	0.68	0.33	0.072
26.50	0.71	0.31	0.060
27.00	0.73	0.29	0.050
27.50	0.76	0.28	0.041
28.00	0.79	0.27	0.036
28.50	0.81	0.26	0.031
29.00	0.83	0.26	0.028
29.50	0.86	0.26	0.025
30.00	0.88	0.26	0.023
31.00	0.92	0.29	0.024
32.00	0.92	0.32	0.030
33.00	0.90	0.33	0.032
34.00	0.91	0.34	0.034
35.00	0.87	0.37	0.043
36.00	0.82	0.34	0.043
37.00	0.81	0.30	0.038
38.00	0.81	0.27	0.033
39.00	0.82	0.25	0.029
40.00	0.83	0.23	0.025

Nickel[10]

Energy (eV)	n	k	$(R\phi = 0)$
0.10	9.54	45.82	0.983
0.15	5.45	30.56	0.978
0.20	4.12	22.48	0.969
0.25	4.25	17.68	0.950
0.30	4.19	15.05	0.934
0.35	4.03	13.05	0.918
0.40	3.84	11.43	0.900
0.50	4.03	9.64	0.864
0.60	3.84	8.35	0.835
0.70	3.59	7.48	0.813
0.80	3.38	6.82	0.794
0.90	3.18	6.23	0.774
1.00	3.06	5.74	0.753
1.10	2.97	5.38	0.734
1.20	2.85	5.10	0.721
1.30	2.74	4.85	0.708
1.40	2.65	4.63	0.695
1.50	2.53	4.47	0.688
1.60	2.43	4.31	0.679
1.70	2.28	4.18	0.677
1.80	2.14	4.01	0.670
1.90	2.02	3.82	0.659
2.00	1.92	3.65	0.649
2.10	1.85	3.48	0.634
2.20	1.80	3.33	0.620
2.30	1.75	3.19	0.605
2.40	1.71	3.06	0.590
2.50	1.67	2.93	0.575
2.60	1.65	2.81	0.557
2.70	1.64	2.71	0.542
2.80	1.63	2.61	0.525
2.90	1.62	2.52	0.509
3.00	1.61	2.44	0.495
3.10	1.61	2.36	0.480
3.20	1.61	2.30	0.467
3.30	1.61	2.23	0.454
3.40	1.62	2.17	0.441
3.50	1.63	2.11	0.428
3.60	1.64	2.07	0.416
3.70	1.66	2.02	0.405
3.80	1.69	1.99	0.397
3.90	1.72	1.98	0.393
4.00	1.73	1.98	0.392
4.20	1.74	2.01	0.396
4.40	1.71	2.06	0.409
4.60	1.63	2.09	0.421
4.80	1.53	2.11	0.435
5.00	1.40	2.10	0.449
5.20	1.27	2.04	0.454
5.40	1.16	1.94	0.449
5.60	1.09	1.83	0.435
5.80	1.04	1.73	0.417
6.20	1.00	1.54	0.371
6.40	1.01	1.46	0.345
6.60	1.01	1.40	0.325
6.80	1.02	1.35	0.308
7.00	1.03	1.30	0.291
7.20	1.03	1.27	0.282
7.40	1.03	1.24	0.273
7.60	1.02	1.22	0.265
7.80	1.01	1.18	0.256
8.00	1.01	1.15	0.248
8.20	1.00	1.13	0.242
8.40	0.99	1.11	0.235
8.60	0.98	1.08	0.228
8.80	0.97	1.05	0.220
9.00	0.97	1.01	0.211
9.20	0.96	0.99	0.203
9.40	0.95	0.96	0.194
9.60	0.95	0.93	0.185
9.80	0.95	0.89	0.175
10.00	0.95	0.87	0.166
10.20	0.95	0.83	0.155
10.40	0.95	0.80	0.145
10.60	0.97	0.76	0.129
10.80	0.99	0.75	0.123
11.00	1.01	0.73	0.115
11.25	1.04	0.72	0.111
11.50	1.05	0.71	0.109
11.75	1.07	0.71	0.108
12.00	1.07	0.71	0.108
12.25	1.07	0.71	0.107
12.50	1.08	0.71	0.106
12.75	1.08	0.71	0.106
13.00	1.08	0.71	0.105
13.25	1.08	0.71	0.105
13.50	1.07	0.70	0.105
13.75	1.07	0.70	0.105
14.00	1.07	0.71	0.106
14.25	1.06	0.70	0.106
14.50	1.05	0.70	0.106
14.75	1.04	0.70	0.107
15.00	1.03	0.70	0.107
15.25	1.02	0.69	0.106
15.50	1.01	0.69	0.105

15.75	1.00	0.68	0.104
16.00	0.99	0.67	0.103
16.50	0.98	0.66	0.101
17.00	0.96	0.64	0.098
17.50	0.94	0.63	0.096
18.00	0.92	0.61	0.092
18.50	0.91	0.58	0.087
19.00	0.90	0.56	0.082
19.50	0.90	0.54	0.077
20.00	0.89	0.51	0.071
20.50	0.89	0.49	0.066
21.00	0.90	0.47	0.061
21.50	0.91	0.46	0.057
22.00	0.91	0.45	0.055
22.50	0.91	0.44	0.053
23.00	0.92	0.44	0.051
23.50	0.91	0.44	0.052
24.00	0.90	0.43	0.051
24.50	0.90	0.43	0.051
25.00	0.89	0.42	0.050
26.00	0.88	0.39	0.046
27.00	0.87	0.37	0.042
28.00	0.87	0.35	0.040
29.00	0.86	0.34	0.037
30.00	0.86	0.32	0.034
35.00	0.86	0.24	0.022
40.00	0.87	0.18	0.014
45.00	0.88	0.13	0.008
50.00	0.92	0.10	0.004
60.00	0.96	0.08	0.002
65.00	0.98	0.09	0.002
68.00	0.96	0.12	0.004
70.00	0.94	0.11	0.004
75.00	0.94	0.09	0.003
80.00	0.94	0.07	0.002
90.00	0.94	0.06	0.002

Niobium[11]

Energy (eV)	n	k	R(ϕ = 0)
0.12	15.99	53.20	0.979
0.20	7.25	34.14	0.976
0.24	5.47	28.88	0.975
0.28	4.26	24.95	0.974
0.35	3.11	20.03	0.970
0.45	2.28	15.58	0.964
0.55	1.83	12.67	0.956
0.65	1.57	10.59	0.947
0.75	1.41	9.00	0.935
0.85	1.35	7.74	0.918
0.95	1.35	6.70	0.893
1.05	1.44	5.86	0.857
1.15	1.55	5.18	0.814
1.25	1.65	4.63	0.768
1.35	1.76	4.13	0.715
1.45	1.95	3.68	0.650
1.55	2.15	3.37	0.595
1.65	2.36	3.13	0.552
1.75	2.54	2.99	0.527
1.85	2.69	2.89	0.510
1.95	2.82	2.86	0.505
2.05	2.89	2.87	0.505
2.15	2.92	2.87	0.505
2.25	2.93	2.87	0.505
2.35	2.92	2.88	0.506
2.45	2.89	2.90	0.509
2.55	2.83	2.92	0.512
2.65	2.74	2.90	0.511
2.75	2.66	2.86	0.507
2.85	2.58	2.80	0.500
3.00	2.51	2.68	0.485
3.10	2.48	2.60	0.475
3.20	2.45	2.53	0.465
3.30	2.44	2.45	0.453
3.40	2.46	2.38	0.442
3.50	2.48	2.33	0.435
3.60	2.52	2.29	0.428
3.70	2.56	2.27	0.426
3.80	2.59	2.28	0.427
3.90	2.62	2.29	0.429
4.00	2.64	2.33	0.434
4.20	2.64	2.42	0.447
4.40	2.53	2.56	0.467
4.60	2.39	2.56	0.470
4.80	2.32	2.52	0.465
5.00	2.26	2.57	0.475
5.20	2.16	2.62	0.487
5.40	2.00	2.68	0.505
5.60	1.81	2.67	0.518
5.80	1.63	2.60	0.522
6.00	1.49	2.49	0.520
6.20	1.38	2.38	0.512
6.40	1.31	2.25	0.496
6.60	1.26	2.14	0.480
6.80	1.24	2.04	0.460
7.00	1.23	1.96	0.441
7.20	1.22	1.91	0.430
7.40	1.20	1.88	0.427
7.60	1.14	1.85	0.430
7.80	1.07	1.78	0.428
8.00	1.02	1.69	0.412
8.20	1.00	1.60	0.390
8.40	0.99	1.51	0.365
8.60	0.99	1.43	0.340
8.70	0.99	1.39	0.328
8.80	1.00	1.36	0.315
9.00	1.01	1.29	0.290
9.20	1.04	1.22	0.265
9.40	1.07	1.18	0.245
9.60	1.10	1.13	0.227
9.80	1.13	1.09	0.209
10.00	1.18	1.05	0.194
10.20	1.23	1.04	0.187
10.40	1.27	1.04	0.185
10.60	1.30	1.06	0.190
10.80	1.32	1.08	0.195
11.00	1.32	1.10	0.200
11.20	1.31	1.12	0.204
11.40	1.30	1.13	0.207
11.60	1.28	1.13	0.209
11.80	1.27	1.13	0.210
12.00	1.25	1.12	0.209
12.40	1.24	1.10	0.204
12.80	1.24	1.09	0.200
13.20	1.24	1.09	0.201
13.60	1.23	1.12	0.208
14.00	1.20	1.13	0.216
14.40	1.16	1.15	0.225
14.80	1.11	1.16	0.234
15.00	1.08	1.16	0.238
15.60	0.99	1.14	0.247
16.00	0.92	1.11	0.250
16.60	0.85	1.04	0.245
17.00	0.80	0.99	0.240
17.20	0.79	0.96	0.236
17.40	0.77	0.93	0.230
17.80	0.75	0.87	0.217
18.00	0.74	0.85	0.209
18.60	0.73	0.77	0.185
19.00	0.72	0.72	0.170
19.60	0.72	0.66	0.150
20.00	0.72	0.62	0.137
20.60	0.71	0.55	0.119
21.00	0.72	0.50	0.100
21.60	0.75	0.43	0.075
22.00	0.78	0.40	0.063
22.60	0.82	0.35	0.045
23.00	0.85	0.33	0.038
23.60	0.88	0.30	0.029
24.00	0.91	0.29	0.025
24.60	0.94	0.28	0.022
25.00	0.96	0.27	0.020
25.60	0.99	0.26	0.018
26.00	1.00	0.26	0.017
26.60	1.03	0.25	0.016
27.00	1.04	0.25	0.015
27.60	1.06	0.25	0.015
28.00	1.08	0.24	0.015
28.60	1.11	0.24	0.016
29.00	1.13	0.25	0.017
29.60	1.16	0.26	0.020
30.00	1.18	0.28	0.023
31.00	1.18	0.31	0.026
32.00	1.20	0.34	0.031
33.00	1.21	0.38	0.038
34.00	1.20	0.42	0.044
35.20	1.17	0.47	0.051
36.00	1.15	0.50	0.056
37.50	1.07	0.53	0.064
39.50	0.95	0.50	0.063
40.50	0.92	0.47	0.059

Osmium (Polycrystalline)[5]

Energy (eV)	n	k	R(ϕ = 0)
0.10	4.08	50.23	0.994
0.15	2.90	33.60	0.990
0.20	2.44	25.11	0.985
0.25	2.35	19.99	0.977
0.30	2.23	16.54	0.969
0.35	2.33	14.06	0.955
0.40	2.45	12.32	0.940
0.45	2.43	11.02	0.927
0.50	2.41	9.97	0.913
0.55	2.33	9.12	0.901
0.60	2.21	8.37	0.890
0.65	2.11	7.68	0.877
0.70	2.02	7.04	0.862
0.75	2.00	6.46	0.842
0.80	2.00	5.95	0.820
0.85	2.01	5.51	0.796
0.90	2.03	5.10	0.769
0.95	2.05	4.74	0.742
1.00	2.09	4.41	0.712
1.10	2.15	3.84	0.651
1.20	2.16	3.35	0.592
1.30	2.25	2.77	0.506
1.40	2.49	2.23	0.419
1.50	2.84	1.80	0.369
1.60	3.36	1.62	0.379
1.70	3.70	1.75	0.411
1.80	3.78	1.83	0.423
1.90	3.81	1.75	0.418
2.00	3.98	1.60	0.418
2.10	4.26	1.54	0.432
2.20	4.58	1.62	0.457
2.30	4.84	1.76	0.479
2.40	5.10	2.01	0.506
2.50	5.28	2.38	0.532
2.60	5.36	2.82	0.557
2.70	5.30	3.29	0.580
2.80	5.07	3.78	0.603
2.90	4.65	4.18	0.624
3.00	4.05	4.40	0.639
3.20	3.29	3.96	0.614
3.40	2.93	3.79	0.607
3.60	2.75	3.45	0.577
3.80	2.73	3.32	0.562
4.00	2.71	3.34	0.565
4.20	2.53	3.44	0.584
4.40	2.24	3.44	0.599
4.60	2.01	3.31	0.598
4.80	1.88	3.19	0.592
5.00	1.74	3.12	0.596
5.20	1.58	3.00	0.597
5.40	1.46	2.88	0.593
5.60	1.36	2.77	0.589
5.80	1.27	2.65	0.582
6.00	1.20	2.54	0.575
6.20	1.13	2.44	0.571
6.40	1.06	2.33	0.562
6.60	1.01	2.21	0.548
6.80	0.97	2.11	0.532
7.00	0.95	2.00	0.514
7.20	0.92	1.91	0.497
7.40	0.91	1.81	0.476
7.60	0.90	1.72	0.451
7.80	0.90	1.63	0.426
8.00	0.91	1.55	0.400

8.20	0.91	1.48	0.375
8.40	0.94	1.40	0.344
8.60	0.96	1.34	0.319
8.80	0.98	1.29	0.296
9.00	1.01	1.24	0.274
9.20	1.04	1.19	0.255
9.40	1.08	1.16	0.238
9.60	1.10	1.14	0.229
9.80	1.13	1.11	0.217
10.00	1.16	1.10	0.209
10.20	1.19	1.08	0.203
10.30	1.20	1.08	0.201
10.40	1.22	1.08	0.200
10.50	1.23	1.09	0.201
10.60	1.24	1.10	0.203
10.80	1.25	1.11	0.206
11.00	1.24	1.13	0.213
11.20	1.23	1.14	0.217
11.40	1.19	1.15	0.223
11.60	1.17	1.12	0.216
11.80	1.16	1.10	0.211
12.00	1.15	1.08	0.205
12.40	1.14	1.03	0.191
12.80	1.15	1.01	0.183
13.20	1.16	0.98	0.174
13.60	1.17	0.97	0.170
14.00	1.17	0.96	0.169
14.40	1.16	0.94	0.165
14.80	1.16	0.91	0.156
15.20	1.17	0.89	0.148
15.60	1.20	0.86	0.140
16.00	1.25	0.87	0.140
16.40	1.28	0.90	0.147
16.80	1.28	0.94	0.157
17.20	1.27	0.97	0.167
17.60	1.26	1.01	0.178
18.00	1.23	1.04	0.189
18.40	1.19	1.08	0.200
18.80	1.14	1.10	0.210
19.20	1.10	1.10	0.219
19.60	1.05	1.11	0.227
20.00	0.96	1.10	0.239
20.40	0.93	1.09	0.240
20.80	0.89	1.05	0.240
21.20	0.86	1.02	0.237
21.60	0.83	0.99	0.235
22.00	0.80	0.96	0.230
22.40	0.78	0.93	0.226
22.80	0.77	0.90	0.220
23.20	0.75	0.88	0.217
23.60	0.75	0.86	0.211
24.00	0.73	0.84	0.209
24.40	0.72	0.82	0.207
24.80	0.70	0.80	0.205
25.20	0.69	0.77	0.202
25.60	0.67	0.75	0.199
26.00	0.66	0.72	0.195
26.40	0.65	0.69	0.189
26.80	0.63	0.66	0.183
27.20	0.65	0.62	0.165
28.00	0.64	0.59	0.156
28.40	0.64	0.57	0.148
28.80	0.65	0.55	0.140
29.20	0.65	0.53	0.134
29.60	0.65	0.51	0.128
30.00	0.65	0.49	0.121
31.00	0.65	0.45	0.111
32.00	0.66	0.41	0.095
33.00	0.68	0.37	0.079
34.00	0.70	0.34	0.068
35.00	0.72	0.31	0.057
36.00	0.74	0.29	0.048
37.00	0.77	0.27	0.040
38.00	0.79	0.26	0.035
39.00	0.81	0.26	0.031
40.00	0.84	0.26	0.026

Palladium[12]

Energy (eV)	n	k	R(ϕ = 0)
0.10	4.13	54.15	0.994
0.15	3.13	35.82	0.990
0.20	3.07	26.59	0.983
0.26	3.11	20.15	0.971
0.30	3.56	17.27	0.955
0.36	3.98	14.41	0.932
0.40	4.27	13.27	0.916
0.46	4.27	12.11	0.902
0.50	4.10	11.44	0.896
0.56	3.92	10.49	0.883
0.60	3.80	9.96	0.876
0.72	3.51	8.70	0.854
0.80	3.35	8.06	0.840
1.00	2.99	6.89	0.811
1.10	2.81	6.46	0.800
1.20	2.65	6.10	0.790
1.30	2.50	5.78	0.781
1.40	2.34	5.50	0.774
1.50	2.17	5.22	0.767
1.60	2.08	4.95	0.755
1.70	2.00	4.72	0.745
1.80	1.92	4.54	0.737
1.90	1.82	4.35	0.729
2.00	1.75	4.18	0.721
2.10	1.67	4.03	0.714
2.20	1.60	3.88	0.707
2.30	1.53	3.75	0.700
2.40	1.47	3.61	0.693
2.50	1.41	3.48	0.685
2.60	1.37	3.36	0.676
2.70	1.32	3.25	0.668
2.80	1.29	3.13	0.658
2.90	1.26	3.03	0.648
3.00	1.23	2.94	0.639
3.10	1.20	2.85	0.630
3.20	1.17	2.77	0.622
3.30	1.14	2.68	0.613
3.40	1.12	2.60	0.602
3.50	1.10	2.52	0.591
3.60	1.08	2.45	0.581
3.70	1.07	2.38	0.570
3.80	1.06	2.31	0.558
3.90	1.05	2.25	0.547
4.00	1.03	2.19	0.537
4.20	1.04	2.09	0.510
4.40	1.03	2.01	0.493
4.60	1.03	1.94	0.476
4.80	1.01	1.90	0.470
5.00	0.96	1.86	0.472
5.20	0.90	1.79	0.474
5.40	0.85	1.70	0.463
5.60	0.81	1.62	0.449
5.80	0.78	1.54	0.437
6.00	0.76	1.45	0.418
6.20	0.74	1.37	0.397
6.40	0.73	1.29	0.375
6.60	0.72	1.21	0.350
6.80	0.73	1.13	0.316
7.00	0.73	1.05	0.287
7.20	0.75	0.98	0.255
7.40	0.77	0.91	0.223
7.60	0.79	0.85	0.195
7.80	0.83	0.78	0.163
8.00	0.88	0.73	0.133
8.20	0.94	0.70	0.117
8.40	0.96	0.70	0.114
8.60	1.00	0.65	0.097
8.80	1.04	0.65	0.094
9.00	1.07	0.64	0.090
9.50	1.12	0.65	0.089
10.00	1.14	0.65	0.088
10.50	1.16	0.65	0.087
11.00	1.18	0.64	0.086
11.50	1.19	0.65	0.087
12.00	1.20	0.66	0.089
12.50	1.19	0.67	0.091
13.00	1.18	0.67	0.091
13.50	1.18	0.67	0.092
14.00	1.17	0.67	0.093
14.50	1.15	0.68	0.095
15.00	1.13	0.69	0.098
15.50	1.10	0.68	0.096
16.00	1.08	0.66	0.092
16.50	1.06	0.63	0.086
17.00	1.07	0.61	0.081
17.50	1.06	0.61	0.080
18.00	1.07	0.59	0.077
18.50	1.07	0.59	0.077
19.00	1.08	0.59	0.077
19.50	1.08	0.61	0.080
20.00	1.07	0.65	0.090
20.50	1.03	0.67	0.098
21.00	0.99	0.67	0.103
21.50	0.95	0.66	0.103
22.00	0.91	0.64	0.103
22.50	0.88	0.62	0.101
23.00	0.86	0.59	0.097
23.50	0.85	0.56	0.091
24.00	0.84	0.54	0.086
25.00	0.81	0.51	0.084
26.40	0.80	0.43	0.066
27.80	0.81	0.38	0.052
29.20	0.82	0.35	0.046

Platinum[13]

Energy (eV)	n	k	R(ϕ = 0)
0.10	13.21	44.72	0.976
0.15	8.18	31.16	0.969
0.20	5.90	23.95	0.962
0.25	4.70	19.40	0.954
0.30	3.92	16.16	0.945
0.35	3.28	13.66	0.936
0.40	2.81	11.38	0.922
0.45	3.03	9.31	0.882
0.50	3.91	7.71	0.813
0.55	4.58	7.14	0.777
0.60	5.13	6.75	0.753
0.65	5.52	6.66	0.746
0.70	5.71	6.83	0.751
0.75	5.57	7.02	0.759
0.80	5.31	7.04	0.762
0.85	5.05	6.98	0.763
0.90	4.77	6.91	0.765
0.95	4.50	6.77	0.763
1.00	4.25	6.62	0.762
1.10	3.86	6.24	0.753
1.20	3.55	5.92	0.746
1.30	3.29	5.61	0.736
1.40	3.10	5.32	0.725
1.50	2.92	5.07	0.716
1.60	2.76	4.84	0.706
1.70	2.63	4.64	0.697
1.80	2.51	4.43	0.686
1.90	2.38	4.26	0.678
2.00	2.30	4.07	0.664
2.10	2.23	3.92	0.654
2.20	2.17	3.77	0.642
2.30	2.10	3.67	0.636
2.40	2.03	3.54	0.626
2.50	1.96	3.42	0.616
2.60	1.91	3.30	0.605
2.70	1.87	3.20	0.595
2.80	1.83	3.10	0.585
2.90	1.79	3.01	0.575
3.00	1.75	2.92	0.565
3.20	1.68	2.76	0.546
3.40	1.63	2.62	0.527
3.60	1.58	2.48	0.507
3.80	1.53	2.37	0.491
4.00	1.49	2.25	0.472
4.20	1.45	2.14	0.452
4.40	1.43	2.04	0.432
4.60	1.39	1.95	0.415
4.80	1.38	1.85	0.392
5.00	1.36	1.76	0.372
5.20	1.36	1.67	0.350
5.40	1.36	1.61	0.332
5.60	1.36	1.54	0.315
5.80	1.36	1.47	0.295
6.00	1.38	1.40	0.276
6.20	1.39	1.35	0.261
6.40	1.42	1.29	0.246
6.60	1.45	1.26	0.236

6.80	1.48	1.24	0.231
7.00	1.50	1.24	0.230
7.20	1.50	1.25	0.231
7.40	1.49	1.23	0.228
7.60	1.48	1.22	0.225
7.80	1.48	1.20	0.221
8.00	1.47	1.18	0.216
8.20	1.47	1.17	0.212
8.40	1.47	1.15	0.209
8.60	1.47	1.14	0.205
8.80	1.47	1.13	0.202
9.00	1.48	1.12	0.200
9.20	1.49	1.11	0.198
9.40	1.49	1.12	0.200
9.60	1.49	1.13	0.203
9.80	1.48	1.15	0.207
10.00	1.46	1.15	0.209
10.20	1.43	1.16	0.211
10.40	1.40	1.15	0.210
10.60	1.37	1.14	0.207
10.80	1.35	1.12	0.203
11.00	1.33	1.10	0.199
11.20	1.31	1.08	0.194
11.40	1.30	1.06	0.188
11.60	1.29	1.04	0.183
11.80	1.29	1.01	0.177
12.00	1.29	1.00	0.173
12.40	1.29	0.97	0.165
12.80	1.29	0.94	0.158
13.20	1.31	0.93	0.155
13.60	1.31	0.93	0.155
14.00	1.31	0.93	0.155
14.40	1.30	0.93	0.156
14.80	1.27	0.93	0.157
15.20	1.27	0.93	0.155
15.60	1.25	0.92	0.151
16.00	1.24	0.89	0.146
16.50	1.24	0.87	0.142
17.00	1.25	0.86	0.138
17.50	1.27	0.85	0.135
18.00	1.31	0.88	0.142
18.50	1.30	0.94	0.157
19.00	1.28	0.99	0.171
19.50	1.23	1.03	0.184
20.00	1.18	1.06	0.197
20.50	1.11	1.09	0.212
21.00	1.03	1.10	0.226
21.50	0.94	1.08	0.238
22.00	0.87	1.04	0.240
22.50	0.81	0.98	0.235
23.00	0.77	0.92	0.226
23.50	0.75	0.87	0.213
24.00	0.74	0.82	0.201
24.50	0.73	0.77	0.187
25.00	0.73	0.73	0.174
25.50	0.73	0.70	0.162
26.00	0.74	0.67	0.150
26.50	0.74	0.65	0.142
27.00	0.74	0.63	0.136
27.50	0.74	0.62	0.130
28.00	0.75	0.60	0.125
28.50	0.75	0.59	0.121
29.00	0.75	0.58	0.118
29.50	0.74	0.58	0.120
30.00	0.73	0.58	0.124

Rhenium (Single crystal with $\vec{E} \parallel \hat{c}$)[5]

Energy (eV)	n	k	R(ϕ = 0)
0.10	6.06	51.03	0.991
0.15	4.66	33.96	0.984
0.20	4.16	25.36	0.975
0.25	4.03	20.10	0.962
0.30	4.37	16.69	0.943
0.35	4.50	14.53	0.925
0.40	4.53	12.96	0.909
0.45	4.53	11.78	0.893
0.50	4.53	10.88	0.878
0.55	4.50	10.26	0.867
0.60	4.29	9.75	0.861
0.65	4.07	9.35	0.856
0.70	3.80	8.94	0.853
0.75	3.48	8.55	0.850
0.80	3.21	8.10	0.846
0.85	2.96	7.68	0.841
0.90	2.73	7.24	0.835
0.95	2.56	6.79	0.826
1.00	2.45	6.36	0.813
1.10	2.38	5.61	0.778
1.20	2.35	5.02	0.742
1.30	2.39	4.54	0.702
1.40	2.44	4.13	0.662
1.50	2.50	3.79	0.624
1.60	2.59	3.49	0.587
1.70	2.70	3.27	0.557
1.80	2.82	3.10	0.535
1.90	2.90	3.00	0.520
2.00	2.97	2.91	0.510
2.10	3.03	2.86	0.504
2.20	3.06	2.84	0.501
2.30	3.07	2.82	0.499
2.40	3.06	2.81	0.498
2.50	3.02	2.80	0.497
2.60	2.96	2.77	0.493
2.70	2.89	2.68	0.482
2.80	2.89	2.57	0.468
2.90	2.99	2.47	0.457
3.00	3.11	2.57	0.470
3.20	2.90	2.68	0.482
3.40	2.83	2.50	0.459
3.60	2.93	2.48	0.457
3.80	2.86	2.56	0.467
4.00	2.81	2.51	0.460
4.20	2.86	2.55	0.466
4.40	2.81	2.74	0.489
4.60	2.56	2.83	0.504
4.80	2.41	2.71	0.493
5.00	2.39	2.68	0.488
5.20	2.34	2.75	0.500
5.40	2.20	2.81	0.515
5.60	2.02	2.84	0.530
5.80	1.83	2.80	0.538
6.00	1.65	2.71	0.541
6.20	1.54	2.59	0.532
6.40	1.45	2.50	0.526
6.80	1.32	2.31	0.508
7.00	1.26	2.23	0.500
7.20	1.20	2.15	0.493
7.40	1.16	2.06	0.480
7.60	1.12	1.99	0.470
7.80	1.08	1.89	0.454
8.00	1.05	1.80	0.435
8.20	1.05	1.71	0.411
8.40	1.05	1.62	0.386
8.60	1.06	1.55	0.360
8.80	1.09	1.48	0.336
9.00	1.11	1.43	0.317
9.20	1.13	1.39	0.301
9.40	1.16	1.34	0.281
9.60	1.18	1.32	0.274
9.80	1.20	1.29	0.264
10.00	1.23	1.26	0.252
10.20	1.25	1.25	0.246
10.40	1.28	1.25	0.242
10.60	1.29	1.25	0.242
10.80	1.30	1.26	0.244
11.00	1.30	1.27	0.247
11.20	1.29	1.28	0.249
11.40	1.28	1.28	0.252
11.60	1.26	1.28	0.252
11.80	1.24	1.26	0.249
12.00	1.23	1.24	0.244
12.40	1.22	1.21	0.237
12.80	1.21	1.18	0.230
13.20	1.22	1.16	0.222
13.60	1.22	1.13	0.215
14.00	1.24	1.12	0.209
14.40	1.27	1.11	0.204
14.80	1.29	1.15	0.213
15.20	1.29	1.19	0.225
15.60	1.26	1.22	0.236
16.00	1.23	1.25	0.248
16.40	1.19	1.27	0.259
16.80	1.14	1.29	0.269
17.00	1.12	1.30	0.275
17.40	1.07	1.30	0.286
18.00	0.99	1.30	0.300
18.40	0.93	1.29	0.311
18.80	0.87	1.28	0.321
19.20	0.81	1.25	0.330
19.60	0.77	1.21	0.332
20.00	0.73	1.18	0.333
20.40	0.70	1.14	0.332
20.80	0.67	1.11	0.332
21.20	0.64	1.08	0.334
21.60	0.61	1.04	0.335
22.00	0.58	1.01	0.340
22.40	0.55	0.97	0.341
22.80	0.53	0.93	0.338
23.20	0.51	0.89	0.334
23.60	0.50	0.85	0.329
24.00	0.48	0.80	0.319
24.40	0.48	0.76	0.307
24.80	0.47	0.72	0.296
25.20	0.47	0.68	0.282
25.60	0.47	0.65	0.270
26.00	0.47	0.61	0.255
26.40	0.48	0.57	0.240
26.80	0.48	0.54	0.225
27.20	0.49	0.51	0.208
27.60	0.50	0.48	0.193
28.00	0.51	0.45	0.176
29.00	0.54	0.39	0.145
30.00	0.57	0.33	0.114
31.00	0.62	0.29	0.086
32.00	0.66	0.26	0.065
33.00	0.68	0.24	0.054
34.00	0.72	0.21	0.041
35.00	0.76	0.20	0.031
36.00	0.79	0.20	0.025
37.00	0.82	0.19	0.021
38.00	0.85	0.20	0.018
39.00	0.89	0.21	0.016
40.00	0.88	0.26	0.022
42.00	0.88	0.26	0.022
44.00	0.89	0.29	0.026
46.00	0.85	0.32	0.035
48.00	0.82	0.30	0.036
50.00	0.80	0.30	0.038
52.00	0.78	0.30	0.044
54.00	0.72	0.30	0.055
56.00	0.66	0.24	0.061
58.00	0.65	0.16	0.055

Rhenium (Single crystal, $\vec{E} \perp \hat{c}$)[5]

Energy (eV)	n	k	R(ϕ = 0)
0.10	4.25	42.83	0.991
0.15	3.28	28.08	0.984
0.20	3.28	20.66	0.971
0.25	3.47	16.27	0.951
0.30	3.73	13.44	0.926
0.35	3.93	11.54	0.900
0.40	3.99	10.15	0.875
0.45	4.17	9.03	0.846
0.50	4.34	8.26	0.821
0.55	4.45	7.73	0.801
0.60	4.53	7.40	0.788
0.65	4.44	7.26	0.784
0.70	4.13	7.09	0.784
0.75	3.77	6.75	0.779
0.80	3.55	6.32	0.766
0.85	3.39	5.95	0.752
0.90	3.26	5.61	0.737
0.95	3.17	5.27	0.719
1.00	3.09	4.96	0.701
1.10	3.05	4.39	0.658
1.20	3.08	3.89	0.613
1.30	3.20	3.56	0.578
1.40	3.23	3.38	0.559
1.50	3.23	3.12	0.532

1.60	3.29	2.88	0.507
1.70	3.38	2.72	0.491
1.80	3.47	2.59	0.480
1.90	3.54	2.50	0.473
2.00	3.63	2.43	0.469
2.10	3.74	2.40	0.470
2.20	3.83	2.38	0.472
2.30	3.93	2.44	0.481
2.40	4.00	2.55	0.492
2.50	4.01	2.70	0.505
2.60	3.90	2.84	0.514
2.70	3.74	2.92	0.517
2.80	3.57	2.88	0.511
2.90	3.49	2.75	0.497
3.00	3.33	2.71	0.493
3.20	3.55	2.84	0.506
3.40	3.34	2.88	0.508
3.60	3.25	2.83	0.501
3.80	3.24	2.84	0.502
4.00	3.19	2.94	0.513
4.20	3.05	3.06	0.526
4.40	2.88	3.15	0.539
4.60	2.67	3.18	0.548
4.80	2.44	3.17	0.554
5.00	2.25	3.12	0.556
5.20	2.10	3.04	0.555
5.40	1.96	2.96	0.553
5.60	1.84	2.88	0.551
5.80	1.73	2.81	0.549
6.00	1.61	2.74	0.549
6.20	1.51	2.64	0.545
6.40	1.42	2.56	0.541
6.80	1.28	2.37	0.526
7.00	1.22	2.28	0.517
7.20	1.16	2.19	0.508
7.40	1.12	2.08	0.493
7.60	1.12	1.98	0.468
7.80	1.08	1.93	0.463
8.00	1.05	1.83	0.443
8.20	1.05	1.74	0.418
8.40	1.05	1.66	0.397
8.60	1.06	1.58	0.372
8.80	1.07	1.52	0.351
9.00	1.09	1.46	0.327
9.20	1.11	1.41	0.309
9.40	1.14	1.36	0.290
9.60	1.17	1.31	0.273
9.80	1.20	1.27	0.258
10.00	1.24	1.24	0.244
10.20	1.29	1.22	0.234
10.40	1.33	1.23	0.233
10.60	1.36	1.25	0.238
10.80	1.38	1.28	0.245
11.00	1.37	1.31	0.253
11.20	1.36	1.33	0.259
11.40	1.33	1.34	0.264
11.60	1.31	1.34	0.266
11.80	1.28	1.33	0.266
12.00	1.26	1.32	0.264
12.40	1.23	1.29	0.257
12.80	1.22	1.26	0.251
13.20	1.20	1.23	0.245
13.60	1.19	1.20	0.236
14.00	1.20	1.16	0.225
14.40	1.22	1.13	0.214
14.80	1.27	1.12	0.207
15.20	1.31	1.17	0.218
15.60	1.31	1.23	0.234
16.00	1.28	1.28	0.251
16.40	1.24	1.33	0.270
16.80	1.17	1.37	0.288
17.00	1.14	1.38	0.297
17.40	1.06	1.39	0.314
18.00	0.95	1.38	0.334
18.40	0.88	1.36	0.346
18.80	0.82	1.33	0.355
19.20	0.76	1.29	0.360
19.60	0.72	1.25	0.363
20.00	0.67	1.21	0.369
20.40	0.64	1.15	0.364
20.80	0.61	1.10	0.357
21.20	0.60	1.06	0.349
21.60	0.58	1.02	0.342
22.00	0.57	0.98	0.336
22.40	0.56	0.95	0.328
22.80	0.55	0.92	0.325
23.20	0.53	0.89	0.322
23.60	0.52	0.85	0.317
24.00	0.50	0.82	0.314
24.40	0.49	0.79	0.309
24.80	0.48	0.75	0.303
25.20	0.47	0.72	0.295
25.60	0.47	0.68	0.286
26.00	0.46	0.64	0.276
26.40	0.46	0.61	0.263
26.80	0.46	0.57	0.249
27.20	0.47	0.53	0.231
27.60	0.48	0.50	0.216
28.00	0.49	0.47	0.198
29.00	0.51	0.41	0.164
30.00	0.55	0.34	0.129
31.00	0.59	0.29	0.097
32.00	0.64	0.26	0.072
33.00	0.67	0.24	0.060
34.00	0.70	0.22	0.047
35.00	0.74	0.20	0.036
36.00	0.77	0.19	0.029
37.00	0.80	0.19	0.023
38.00	0.84	0.19	0.018
39.00	0.88	0.21	0.016
40.00	0.87	0.25	0.023
42.00	0.87	0.25	0.023
44.00	0.88	0.28	0.026
46.00	0.84	0.31	0.035
48.00	0.82	0.30	0.036
50.00	0.80	0.30	0.039
52.00	0.77	0.30	0.044
54.00	0.71	0.29	0.055
56.00	0.66	0.23	0.061
58.00	0.64	0.16	0.055

Rhodium[7]

Energy (eV)	n	k	R(ϕ = 0)
0.10	18.48	69.43	0.986
0.20	8.66	37.46	0.977
0.30	5.85	25.94	0.967
0.40	4.74	19.80	0.955
0.50	4.20	16.07	0.941
0.60	3.87	13.51	0.925
0.70	3.67	11.72	0.908
0.80	3.63	10.34	0.887
0.90	3.62	9.36	0.867
1.00	3.71	8.67	0.848
1.10	3.67	8.26	0.837
1.20	3.51	7.94	0.832
1.30	3.26	7.63	0.829
1.40	3.01	7.31	0.827
1.50	2.78	6.97	0.823
1.60	2.60	6.64	0.818
1.70	2.42	6.33	0.813
1.80	2.30	6.02	0.805
1.90	2.20	5.76	0.798
2.00	2.12	5.51	0.789
2.10	2.05	5.30	0.780
2.20	2.00	5.11	0.772
2.30	1.94	4.94	0.765
2.40	1.90	4.78	0.756
2.50	1.88	4.65	0.748
2.60	1.85	4.55	0.743
2.70	1.80	4.49	0.742
2.90	1.63	4.36	0.748
3.00	1.53	4.29	0.753
3.10	1.41	4.20	0.760
3.20	1.30	4.09	0.764
3.30	1.20	3.97	0.767
3.40	1.11	3.84	0.769
3.50	1.04	3.71	0.768
3.60	0.99	3.58	0.764
3.70	0.95	3.45	0.759
3.80	0.91	3.34	0.753
3.90	0.88	3.23	0.747
4.00	0.86	3.12	0.739
4.20	0.83	2.94	0.722
4.40	0.80	2.76	0.706
4.60	0.78	2.60	0.684
4.80	0.79	2.46	0.659
5.00	0.79	2.34	0.635
5.20	0.79	2.23	0.613
5.40	0.80	2.14	0.591
5.60	0.80	2.06	0.573
5.80	0.79	2.00	0.561
6.00	0.76	1.93	0.556
6.20	0.73	1.85	0.544
6.40	0.70	1.77	0.534
6.60	0.68	1.69	0.518
6.80	0.67	1.60	0.498
7.00	0.66	1.52	0.476
7.20	0.66	1.43	0.452
7.40	0.66	1.35	0.423
7.60	0.67	1.27	0.394
7.80	0.68	1.20	0.363
8.00	0.69	1.12	0.329
8.20	0.71	1.04	0.288
8.40	0.74	0.97	0.252
8.60	0.78	0.89	0.212
8.80	0.83	0.83	0.179
9.00	0.88	0.77	0.148
9.20	0.95	0.73	0.125
9.40	1.01	0.71	0.110
9.60	1.07	0.69	0.102
9.80	1.12	0.69	0.098
10.00	1.17	0.69	0.098
10.60	1.26	0.73	0.106
11.00	1.29	0.76	0.113
11.60	1.32	0.80	0.124
12.00	1.32	0.82	0.127
12.60	1.32	0.82	0.129
13.00	1.32	0.83	0.131
13.60	1.32	0.85	0.134
14.00	1.32	0.86	0.138
14.60	1.30	0.89	0.144
15.00	1.28	0.90	0.147
15.60	1.25	0.90	0.147
16.00	1.24	0.89	0.147
16.50	1.23	0.88	0.145
17.00	1.22	0.88	0.144
17.50	1.22	0.87	0.143
18.00	1.23	0.88	0.145
18.50	1.25	0.92	0.155
19.00	1.24	0.98	0.172
19.50	1.18	1.05	0.193
20.00	1.10	1.09	0.213
20.50	1.00	1.09	0.230
21.00	0.91	1.05	0.234
21.50	0.86	1.00	0.228
22.00	0.83	0.95	0.219
22.50	0.81	0.92	0.214
23.00	0.79	0.90	0.213
23.50	0.75	0.87	0.214
24.00	0.73	0.84	0.210
24.50	0.70	0.81	0.208
25.00	0.69	0.77	0.202
25.50	0.67	0.74	0.195
26.00	0.66	0.70	0.188
26.50	0.65	0.66	0.176
27.00	0.65	0.64	0.168
27.50	0.65	0.61	0.159
28.00	0.65	0.59	0.152
29.00	0.65	0.54	0.137
30.00	0.66	0.51	0.127
31.00	0.64	0.49	0.127
32.00	0.61	0.44	0.126
33.00	0.60	0.37	0.110
34.00	0.65	0.30	0.074
35.00	0.69	0.28	0.058
36.00	0.73	0.27	0.049
37.00	0.74	0.28	0.047
38.00	0.74	0.27	0.045
39.00	0.75	0.25	0.041

Ruthenium (Single crystal, $\vec{E} \parallel \hat{c}$)[5]

Energy (eV)	n	k	R(ϕ = 0)
0.10	11.50	51.38	0.984
0.20	5.93	27.14	0.970
0.30	4.33	18.50	0.953
0.40	3.60	13.97	0.933
0.50	3.18	11.04	0.909
0.60	3.28	8.89	0.865
0.70	3.62	7.73	0.822
0.80	3.42	7.02	0.801
0.90	3.25	6.12	0.766
1.00	3.39	5.33	0.715
1.10	3.66	4.83	0.675
1.20	3.84	4.57	0.654
1.30	3.94	4.38	0.638
1.40	4.02	4.19	0.624
1.50	4.16	4.07	0.614
1.60	4.33	4.08	0.615
1.70	4.42	4.21	0.624
1.80	4.40	4.38	0.636
1.90	4.29	4.61	0.651
2.00	4.04	4.81	0.667
2.10	3.69	4.90	0.679
2.20	3.35	4.82	0.683
2.30	3.09	4.70	0.681
2.40	2.89	4.55	0.677
2.50	2.74	4.40	0.671
2.60	2.64	4.25	0.663
2.70	2.58	4.14	0.656
2.80	2.54	4.05	0.650
2.90	2.48	4.03	0.650
3.00	2.38	4.03	0.656
3.10	2.26	4.00	0.661
3.20	2.13	3.96	0.666
3.30	2.00	3.91	0.671
3.40	1.87	3.83	0.673
3.50	1.76	3.74	0.674
3.60	1.66	3.65	0.675
3.70	1.57	3.55	0.673
3.80	1.49	3.45	0.672
3.90	1.42	3.35	0.668
4.00	1.37	3.24	0.661
4.20	1.29	3.08	0.649
4.40	1.22	2.93	0.639
4.60	1.16	2.79	0.628
4.80	1.11	2.67	0.617
5.00	1.06	2.56	0.607
5.20	1.01	2.46	0.600
5.40	0.95	2.35	0.593
5.60	0.92	2.23	0.576
5.80	0.90	2.14	0.559
6.00	0.88	2.05	0.545
6.20	0.87	1.98	0.531
6.40	0.84	1.91	0.521
6.60	0.82	1.84	0.510
6.80	0.79	1.77	0.500
7.00	0.76	1.69	0.489
7.20	0.75	1.61	0.472
7.40	0.73	1.54	0.455
7.60	0.73	1.46	0.433
7.80	0.73	1.39	0.411
8.00	0.72	1.33	0.391
8.20	0.72	1.26	0.366
8.40	0.73	1.20	0.342
8.60	0.74	1.14	0.318
8.80	0.74	1.08	0.295
9.00	0.75	1.02	0.267
9.20	0.77	0.97	0.243
9.40	0.79	0.91	0.217
9.60	0.82	0.86	0.190
9.80	0.85	0.81	0.167
10.00	0.88	0.76	0.144
10.20	0.92	0.72	0.125
10.40	0.96	0.69	0.110
10.60	1.01	0.67	0.100
10.80	1.05	0.66	0.094
11.00	1.09	0.65	0.090
11.20	1.12	0.65	0.088
11.40	1.15	0.65	0.087
11.60	1.18	0.65	0.088
11.80	1.21	0.66	0.090
12.00	1.23	0.67	0.092
12.40	1.26	0.69	0.098
12.80	1.27	0.72	0.104
13.20	1.28	0.74	0.108
13.60	1.28	0.75	0.111
14.00	1.28	0.76	0.114
14.40	1.27	0.76	0.114
14.80	1.27	0.76	0.114
15.00	1.27	0.76	0.114
15.60	1.28	0.77	0.115
16.00	1.30	0.78	0.118
16.50	1.32	0.80	0.123
17.00	1.34	0.85	0.136
17.50	1.32	0.93	0.155
18.00	1.26	0.99	0.173
18.50	1.18	1.02	0.185
19.00	1.11	1.02	0.192
19.50	1.05	1.02	0.199
20.00	0.99	1.02	0.208
20.50	0.92	0.99	0.212
21.00	0.86	0.94	0.209
21.50	0.83	0.90	0.203
22.00	0.81	0.86	0.193
23.00	0.77	0.79	0.182
24.00	0.74	0.74	0.171
25.00	0.71	0.69	0.163
26.00	0.68	0.63	0.154
27.00	0.67	0.57	0.140
28.00	0.66	0.51	0.124
29.00	0.67	0.46	0.107
30.00	0.67	0.43	0.097
31.00	0.67	0.37	0.084
32.00	0.69	0.33	0.070
33.00	0.71	0.30	0.058
34.00	0.73	0.27	0.048
35.00	0.75	0.25	0.039
36.00	0.77	0.24	0.035
37.00	0.79	0.23	0.030
38.00	0.80	0.22	0.027
39.00	0.82	0.22	0.024
40.00	0.83	0.22	0.022

Ruthenium (Single crystal, $\vec{E} \perp \hat{c}$)[5]

Energy (eV)	n	k	R(ϕ = 0)
0.10	11.85	50.81	0.983
0.20	6.68	27.18	0.966
0.30	4.94	18.92	0.950
0.40	3.90	14.51	0.933
0.50	3.27	11.63	0.915
0.60	2.98	9.54	0.888
0.70	2.82	7.99	0.856
0.80	2.73	6.71	0.815
0.90	2.82	5.54	0.751
1.00	3.17	4.59	0.670
1.10	3.69	3.91	0.604
1.20	4.28	3.66	0.585
1.30	4.66	3.72	0.593
1.40	4.86	3.79	0.601
1.50	4.99	3.89	0.609
1.60	5.08	4.03	0.618
1.70	5.12	4.22	0.629
1.80	5.10	4.45	0.642
1.90	4.96	4.78	0.660
2.00	4.61	5.06	0.677
2.10	4.21	5.09	0.682
2.20	3.94	5.00	0.681
2.30	3.69	4.97	0.684
2.40	3.44	4.88	0.684
2.50	3.27	4.77	0.681
2.60	3.14	4.66	0.677
2.70	3.06	4.59	0.674
2.80	2.99	4.59	0.676
2.90	2.87	4.64	0.686
3.00	2.64	4.69	0.701
3.10	2.40	4.64	0.710
3.20	2.18	4.55	0.717
3.30	2.00	4.43	0.721
3.40	1.84	4.30	0.723
3.50	1.71	4.16	0.723
3.60	1.60	4.03	0.722
3.70	1.50	3.90	0.721
3.80	1.41	3.77	0.718
3.90	1.35	3.64	0.713
4.00	1.29	3.53	0.707
4.20	1.21	3.31	0.694
4.40	1.16	3.13	0.679
4.60	1.13	2.97	0.662
4.80	1.09	2.86	0.652
5.00	1.03	2.75	0.648
5.20	0.97	2.64	0.643
5.40	0.91	2.52	0.635
5.60	0.88	2.40	0.622
5.80	0.86	2.29	0.605
6.00	0.84	2.20	0.591
6.20	0.82	2.11	0.576
6.40	0.81	2.04	0.564
6.60	0.78	1.97	0.556
6.80	0.76	1.89	0.545
7.00	0.73	1.82	0.538
7.20	0.70	1.75	0.527
7.40	0.68	1.67	0.513
7.60	0.67	1.59	0.496
7.80	0.66	1.51	0.476
8.00	0.66	1.44	0.454
8.20	0.65	1.36	0.430
8.40	0.66	1.29	0.403
8.60	0.66	1.22	0.378
8.80	0.68	1.15	0.346
9.00	0.69	1.09	0.317
9.20	0.70	1.02	0.286
9.40	0.73	0.95	0.251
9.60	0.77	0.89	0.216
9.80	0.82	0.84	0.185
10.00	0.86	0.81	0.163
10.20	0.90	0.77	0.143
10.40	0.94	0.74	0.127
10.60	0.99	0.72	0.115
10.80	1.04	0.71	0.108
11.00	1.08	0.70	0.104
11.20	1.11	0.70	0.102
11.40	1.14	0.70	0.101
11.60	1.17	0.71	0.102
11.80	1.20	0.72	0.104
12.00	1.22	0.73	0.107
12.40	1.25	0.76	0.113
12.80	1.26	0.78	0.118
13.20	1.27	0.81	0.124
13.60	1.27	0.83	0.129
14.00	1.26	0.84	0.132
14.40	1.25	0.84	0.132
14.80	1.25	0.84	0.133
15.00	1.25	0.84	0.133
15.60	1.25	0.85	0.134
16.00	1.27	0.85	0.134
16.50	1.28	0.89	0.145
17.00	1.28	0.94	0.158
17.50	1.25	1.00	0.175
18.00	1.19	1.04	0.190
18.50	1.12	1.05	0.200
19.00	1.07	1.05	0.205
19.50	1.02	1.04	0.212
20.00	0.97	1.04	0.219
20.50	0.91	1.03	0.228
21.00	0.85	1.01	0.234
21.50	0.80	0.97	0.234
22.00	0.77	0.94	0.233
23.00	0.71	0.87	0.229
24.00	0.67	0.79	0.218
25.00	0.64	0.73	0.205
26.00	0.61	0.66	0.194
27.00	0.60	0.59	0.177
28.00	0.60	0.53	0.155
29.00	0.61	0.48	0.134
30.00	0.62	0.45	0.123
31.00	0.61	0.40	0.114
32.00	0.63	0.34	0.093
33.00	0.65	0.31	0.077
34.00	0.67	0.28	0.065

35.00	0.70	0.26	0.054
36.00	0.72	0.25	0.047
37.00	0.73	0.23	0.041
38.00	0.75	0.22	0.035
39.00	0.77	0.22	0.031
40.00	0.79	0.22	0.028

Silver[4]

Energy (eV)	n	k	R(ϕ = 0)
0.10	9.91	90.27	0.995
0.20	2.84	45.70	0.995
0.30	1.41	30.51	0.994
0.40	0.91	22.89	0.993
0.50	0.67	18.32	0.992
1.00	0.28	9.03	0.987
1.50	0.27	5.79	0.969
2.00	0.27	4.18	0.944
2.50	0.24	3.09	0.914
3.00	0.23	2.27	0.864
3.25	0.23	1.86	0.816
3.50	0.21	1.42	0.756
3.60	0.23	1.13	0.671
3.70	0.30	0.77	0.475
3.77	0.53	0.40	0.154
3.80	0.73	0.30	0.053
3.90	1.30	0.36	0.040
4.00	1.61	0.60	0.103
4.10	1.73	0.85	0.153
4.20	1.75	1.06	0.194
4.30	1.73	1.13	0.208
4.50	1.69	1.28	0.238
4.75	1.61	1.34	0.252
5.00	1.55	1.36	0.257
5.50	1.45	1.34	0.257
6.00	1.34	1.28	0.246
6.50	1.25	1.18	0.225
7.00	1.18	1.06	0.196
7.50	1.14	0.91	0.157
8.00	1.16	0.75	0.114
9.00	1.33	0.56	0.074
10.00	1.46	0.56	0.082
11.00	1.52	0.56	0.088
12.00	1.61	0.59	0.100
13.00	1.66	0.64	0.112
14.00	1.72	0.78	0.141
14.50	1.64	0.88	0.152
15.00	1.56	0.92	0.156
16.00	1.42	0.91	0.151
17.00	1.33	0.86	0.139
18.00	1.28	0.80	0.124
19.00	1.27	0.75	0.111
20.00	1.29	0.71	0.103
21.00	1.35	0.75	0.112
21.50	1.37	0.80	0.124
22.00	1.34	0.87	0.141
22.50	1.26	0.93	0.157
23.00	1.17	0.94	0.163
23.50	1.10	0.93	0.165
24.00	1.04	0.90	0.165
24.50	0.99	0.87	0.160
25.00	0.95	0.83	0.154
25.50	0.91	0.78	0.144
26.00	0.90	0.74	0.133
26.50	0.89	0.69	0.121
27.00	0.89	0.65	0.109
27.50	0.89	0.62	0.099
28.00	0.90	0.59	0.090
28.50	0.91	0.57	0.084
29.00	0.92	0.56	0.079
30.00	0.93	0.54	0.074
31.00	0.93	0.53	0.072
32.00	0.92	0.53	0.072
33.00	0.90	0.51	0.071
34.00	0.88	0.49	0.067
35.00	0.86	0.45	0.061
36.00	0.89	0.44	0.055
38.00	0.89	0.39	0.043
40.00	0.90	0.37	0.039
42.00	0.90	0.35	0.036
44.00	0.90	0.33	0.033
46.00	0.90	0.32	0.031
48.00	0.89	0.31	0.030
50.00	0.88	0.29	0.027
52.00	0.89	0.28	0.024
54.00	0.88	0.27	0.024
56.00	0.87	0.26	0.024
58.00	0.87	0.24	0.021
60.00	0.87	0.22	0.018
62.00	0.88	0.21	0.016
64.00	0.88	0.21	0.016
66.00	0.88	0.21	0.016
68.00	0.87	0.21	0.017
70.00	0.83	0.20	0.021
72.00	0.85	0.18	0.016
74.00	0.85	0.17	0.014
76.00	0.85	0.16	0.013
78.00	0.85	0.15	0.013
80.00	0.85	0.14	0.012
85.00	0.85	0.11	0.011
90.00	0.85	0.08	0.009
95.00	0.86	0.06	0.007
100.00	0.87	0.04	0.005

Tantalum[9]

Energy (eV)	n	k	R (ϕ = 0)
0.10	10.14	66.39	0.984
0.15	9.45	46.41	0.983
0.20	5.77	35.46	0.982
0.26	3.67	27.53	0.981
0.30	2.87	23.90	0.980
0.38	2.03	18.87	0.978
0.50	1.37	14.26	0.974
0.58	1.15	12.19	0.970
0.70	0.96	9.92	0.962
0.78	0.89	8.77	0.956
0.90	0.84	7.38	0.942
1.00	0.89	6.47	0.992
1.10	0.93	5.75	0.899
1.20	0.98	5.14	0.872
1.30	1.00	4.62	0.842
1.40	1.04	4.15	0.805
1.50	1.09	3.73	0.762
1.60	1.15	3.33	0.707
1.70	1.24	2.95	0.640
1.80	1.35	2.60	0.560
1.90	1.57	2.24	0.460
2.00	1.83	1.99	0.388
2.10	2.10	1.84	0.354
2.20	2.36	1.81	0.351
2.30	2.56	1.86	0.365
2.40	2.68	1.92	0.378
2.50	2.75	1.98	0.388
2.60	2.80	2.02	0.395
2.70	2.84	2.08	0.405
2.80	2.85	2.14	0.412
2.90	2.84	2.20	0.420
3.00	2.81	2.24	0.425
3.20	2.73	2.31	0.432
3.40	2.61	2.33	0.435
3.60	2.49	2.30	0.430
3.80	2.40	2.22	0.418
4.00	2.36	2.14	0.406
4.20	2.35	2.06	0.392
4.40	2.39	2.01	0.384
4.60	2.45	2.00	0.384
4.80	2.53	2.06	0.394
5.00	2.58	2.20	0.416
5.20	2.52	2.44	0.450
5.40	2.31	2.61	0.480
5.60	2.06	2.67	0.501
5.80	1.83	2.63	0.510
6.00	1.63	2.56	0.515
6.20	1.48	2.45	0.512
6.40	1.37	2.33	0.504
6.60	1.29	2.22	0.492
6.80	1.23	2.11	0.478
7.00	1.18	2.01	0.462
7.20	1.15	1.91	0.445
7.40	1.13	1.82	0.425
7.60	1.12	1.75	0.406
7.80	1.11	1.68	0.390
8.00	1.11	1.61	0.370
8.20	1.12	1.55	0.350
8.40	1.13	1.50	0.332
8.60	1.14	1.45	0.317
8.80	1.17	1.41	0.301
9.00	1.19	1.40	0.294
9.20	1.21	1.38	0.289
9.40	1.21	1.38	0.287
9.60	1.21	1.38	0.285
9.80	1.21	1.37	0.285
10.00	1.20	1.37	0.286
10.20	1.19	1.37	0.286
10.40	1.18	1.37	0.287
10.60	1.16	1.36	0.288
10.80	1.15	1.36	0.289
11.00	1.13	1.35	0.290
11.20	1.11	1.35	0.292
11.40	1.09	1.34	0.293
11.60	1.07	1.33	0.294
11.80	1.05	1.32	0.295
12.00	1.02	1.31	0.296
12.20	1.00	1.29	0.295
12.40	0.98	1.28	0.294
12.60	0.96	1.26	0.292
12.80	0.94	1.24	0.289
13.00	0.93	1.22	0.286
13.60	0.91	1.16	0.272
14.00	0.90	1.15	0.272
14.60	0.85	1.15	0.285
15.00	0.80	1.13	0.293
15.60	0.72	1.08	0.301
16.00	0.68	1.04	0.304
16.60	0.63	0.97	0.301
17.00	0.60	0.92	0.296
17.60	0.60	0.92	0.296
18.00	0.55	0.79	0.274
18.60	0.53	0.71	0.254
19.00	0.53	0.65	0.236
19.60	0.53	0.57	0.207
20.00	0.54	0.52	0.185
20.60	0.55	0.44	0.153
21.00	0.57	0.39	0.127
21.60	0.64	0.34	0.089
22.00	0.64	0.32	0.081
22.60	0.69	0.27	0.058
23.00	0.73	0.24	0.043
23.60	0.80	0.26	0.033
24.00	0.80	0.26	0.034
24.60	0.82	0.25	0.029
25.00	0.83	0.25	0.026
25.60	0.86	0.24	0.022
26.00	0.88	0.25	0.022
26.60	0.87	0.26	0.023
27.00	0.87	0.25	0.022
27.60	0.89	0.23	0.019
28.00	0.90	0.23	0.017
28.60	0.91	0.22	0.015
29.00	0.92	0.22	0.014
29.60	0.94	0.22	0.014
30.00	0.95	0.22	0.014
31.00	0.97	0.23	0.014
32.00	0.98	0.24	0.015
33.00	0.98	0.25	0.015
34.00	0.99	0.25	0.016
35.00	0.99	0.26	0.017
36.00	0.99	0.27	0.018
37.00	0.99	0.28	0.019
38.00	0.98	0.28	0.021
39.00	0.97	0.29	0.022
40.00	0.95	0.29	0.023

Titanium (Polycrystalline)[14]

Energy (eV)	n	k	R (ϕ = 0)
0.10	5.03	23.38	0.965
0.15	3.00	15.72	0.954
0.20	2.12	11.34	0.939

0.25	2.05	8.10	0.890
0.30	6.39	9.94	0.833
0.35	2.74	6.21	0.792
0.40	2.49	4.68	0.708
0.45	3.35	3.25	0.545
0.50	4.43	3.22	0.555
0.60	4.71	3.77	0.597
0.70	4.38	3.89	0.603
0.80	4.04	3.82	0.596
0.90	3.80	3.65	0.582
1.00	3.62	3.52	0.570
1.10	3.47	3.40	0.560
1.20	3.35	3.30	0.550
1.30	3.28	3.25	0.546
1.40	3.17	3.28	0.549
1.50	2.98	3.32	0.557
1.60	2.74	3.30	0.559
1.70	2.54	3.23	0.557
1.80	2.36	3.11	0.550
1.90	2.22	2.99	0.540
2.00	2.11	2.88	0.530
2.10	2.01	2.77	0.520
2.20	1.92	2.67	0.509
2.30	1.86	2.56	0.495
2.40	1.81	2.47	0.483
2.50	1.78	2.39	0.471
2.60	1.75	2.34	0.462
2.70	1.71	2.29	0.456
2.80	1.68	2.25	0.451
2.90	1.63	2.21	0.447
3.00	1.59	2.17	0.444
3.10	1.55	2.15	0.442
3.20	1.50	2.12	0.442
3.30	1.44	2.09	0.442
3.40	1.37	2.06	0.443
3.50	1.30	2.01	0.443
3.60	1.24	1.96	0.441
3.70	1.17	1.90	0.436
3.80	1.11	1.83	0.430
3.85	1.08	1.78	0.423
3.90	1.06	1.73	0.413
4.00	1.04	1.62	0.389
4.20	1.05	1.45	0.333
4.40	1.13	1.33	0.284
4.60	1.17	1.29	0.265
4.80	1.21	1.23	0.244
5.00	1.24	1.21	0.236
5.20	1.27	1.20	0.228
5.40	1.17	1.16	0.228
5.60	1.24	1.21	0.234
5.80	1.21	1.22	0.241
6.00	1.15	1.21	0.244
6.20	1.11	1.18	0.240
6.40	1.08	1.14	0.232
6.60	1.04	1.06	0.212
6.80	1.05	1.02	0.198
7.00	1.06	0.97	0.182
7.20	1.07	0.95	0.175
7.40	1.11	0.94	0.167
7.60	1.09	0.92	0.165
7.80	1.11	0.93	0.165
8.00	1.10	0.94	0.169
8.20	1.10	0.95	0.171
8.40	1.08	0.95	0.175
8.60	1.04	0.96	0.181
8.80	1.02	0.95	0.181
9.00	1.00	0.94	0.182
9.20	0.97	0.93	0.182
9.40	0.95	0.91	0.181
9.60	0.94	0.90	0.179
9.80	0.91	0.88	0.179
10.00	0.89	0.88	0.180
10.20	0.86	0.85	0.178
10.40	0.85	0.83	0.175
10.60	0.81	0.79	0.167
10.80	0.80	0.76	0.162
11.00	0.79	0.72	0.152
11.20	0.81	0.69	0.139
11.40	0.81	0.69	0.139
11.60	0.79	0.68	0.139
11.80	0.78	0.67	0.137

12.00	0.77	0.65	0.132
12.80	0.76	0.55	0.106
13.20	0.76	0.52	0.097
13.60	0.76	0.48	0.087
14.00	0.77	0.45	0.077
14.40	0.77	0.42	0.069
14.80	0.79	0.38	0.058
15.20	0.79	0.36	0.052
15.60	0.79	0.32	0.045
16.00	0.83	0.31	0.037
16.40	0.84	0.28	0.030
16.80	0.87	0.27	0.025
17.20	0.90	0.25	0.020
17.60	0.93	0.25	0.017
18.00	0.94	0.24	0.165
18.40	0.94	0.23	0.017
18.80	0.95	0.24	0.016
19.20	0.96	0.25	0.016
19.60	0.97	0.25	0.017
20.00	0.98	0.27	0.018
20.40	0.98	0.27	0.019
20.60	1.00	0.29	0.020
21.20	0.99	0.31	0.023
21.60	0.99	0.31	0.024
22.00	0.98	0.32	0.025
22.40	0.98	0.33	0.027
22.80	0.97	0.33	0.028
23.20	0.96	0.34	0.030
23.60	0.95	0.35	0.031
24.00	0.92	0.35	0.033
24.5	0.91	0.34	0.032
25.0	0.91	0.33	0.032
25.5	0.89	0.33	0.032
26.0	0.89	0.33	0.032
26.5	0.88	0.32	0.032
27.0	0.86	0.31	0.032
27.5	0.85	0.30	0.033
28.0	0.84	0.29	0.033
28.5	0.82	0.26	0.029
29.0	0.83	0.25	0.027
30.0	0.84	0.22	0.022

Tungsten[15]

Energy (eV)	n	k	R (ϕ = 0)
0.10	14.06	54.71	0.983
0.20	3.87	28.30	0.981
0.25	2.56	22.44	0.980
0.30	1.83	18.32	0.979
0.34	1.71	15.71	0.973
0.38	1.86	13.88	0.963
0.42	1.92	12.63	0.954
0.46	1.69	11.59	0.952
0.50	1.40	10.52	0.952
0.54	1.23	9.45	0.948
0.58	1.17	8.44	0.938
0.62	1.28	7.52	0.917
0.66	1.45	6.78	0.888
0.70	0.59	6.13	0.856
0.74	1.83	5.52	0.810
0.78	2.12	—	0.759
0.82	2.36	4.61	0.710
0.86	2.92	4.37	0.661
0.90	3.11	4.44	0.660
0.94	3.15	4.43	0.658
0.98	3.15	4.36	0.653
1.00	3.14	1.32	0.649
1.10	3.05	4.04	0.627
1.20	3.00	3.64	0.590
1.30	3.12	3.24	0.545
1.40	3.29	2.96	0.515
1.50	3.48	2.79	0.500
1.60	3.67	2.68	0.494
1.70	3.84	2.79	0.507
1.80	3.82	2.91	0.518
1.90	3.70	2.94	0.518
2.00	3.60	2.89	0.512
2.10	3.54	2.84	0.506
2.20	3.49	2.76	0.497
2.30	3.49	2.72	0.494
2.40	3.45	2.72	0.493
2.50	3.38	2.68	0.487
2.60	3.34	2.62	0.480
2.70	3.31	2.55	0.472
2.80	3.31	2.49	0.466
2.90	3.32	2.45	0.461
3.00	3.35	2.42	0.459
3.10	3.39	2.41	0.460
3.20	3.43	2.45	0.465
3.30	3.45	2.55	0.476
3.40	3.39	2.66	0.485
3.50	3.24	2.70	0.488
3.60	3.13	2.67	0.482
3.70	3.05	2.62	0.476
3.80	2.99	2.56	0.468
3.90	2.96	2.50	0.460
4.00	2.95	2.43	0.451
4.20	3.02	2.33	0.440
4.40	3.13	2.32	0.442
4.60	3.24	2.41	0.455
4.80	3.33	2.57	0.475
5.00	3.40	2.85	0.505
5.20	3.27	3.27	0.548
5.40	2.92	3.58	0.586
5.60	2.43	3.70	0.618
5.80	2.00	3.61	0.637
6.00	1.70	3.42	0.643
6.20	1.47	3.24	0.646
6.40	1.32	3.04	0.640
6.60	1.21	2.87	0.631
6.80	1.12	2.70	0.619
7.00	1.06	2.56	0.607
7.20	1.01	2.43	0.593
7.40	0.98	2.30	0.573
7.60	0.95	2.18	0.556
7.80	0.93	2.06	0.533
8.00	0.94	1.95	0.505
8.20	0.94	1.86	0.481
8.40	0.96	1.76	0.449
8.60	0.99	1.70	0.422
8.80	1.01	1.65	0.401
9.00	1.01	1.60	0.388
9.20	1.02	1.55	0.369
9.40	1.03	1.50	0.352
9.60	1.05	1.44	0.329
9.80	1.09	1.38	0.307
10.00	1.13	1.34	0.287
10.20	1.19	1.33	0.274
10.40	1.24	1.34	0.270
10.60	1.27	1.36	0.274
10.80	1.29	1.39	0.282
11.00	1.28	1.42	0.290
11.20	1.27	1.44	0.297
11.40	1.25	1.46	0.305
11.60	1.22	1.48	0.313
11.80	1.20	1.48	0.318
12.00	1.16	1.48	0.323
12.40	1.10	1.47	0.329
12.80	1.04	1.44	0.333
13.20	0.98	1.40	0.332
13.60	0.94	1.35	0.325
14.00	0.91	1.28	0.312
14.40	0.90	1.23	0.296
14.80	0.90	1.17	0.276
15.20	0.93	1.13	0.255
15.60	0.97	1.12	0.246
16.00	0.98	1.14	0.249
16.40	0.97	1.17	0.260
16.80	0.94	1.19	0.273
17.20	0.90	1.21	0.289
17.60	0.85	1.21	0.304
18.00	0.80	1.20	0.317
18.40	0.74	1.18	0.330
18.80	0.69	1.15	0.340
19.20	0.64	1.11	0.347
19.60	0.60	1.07	0.353
20.00	0.56	1.02	0.354
20.40	0.54	0.97	0.350
20.80	0.52	0.92	0.342
21.20	0.50	0.87	0.331
21.60	0.50	0.82	0.318

22.00	0.49	0.77	0.303
22.40	0.49	0.73	0.287
22.80	0.49	0.69	0.272
23.20	0.49	0.66	0.263
23.60	0.48	0.62	0.252
24.00	0.49	0.57	0.234
24.40	0.50	0.53	0.213
24.80	0.51	0.49	0.191
25.20	0.53	0.46	0.171
25.60	0.55	0.43	0.150
26.00	0.57	0.40	0.132
26.40	0.59	0.38	0.117
26.80	0.61	0.37	0.105
27.00	0.62	0.36	0.099
27.50	0.64	0.34	0.085
28.00	0.67	0.32	0.073
28.50	0.69	0.31	0.065
29.00	0.71	0.30	0.057
29.50	0.73	0.30	0.052
30.00	0.75	0.29	0.047
31.00	0.78	0.29	0.042
32.00	0.79	0.29	0.040
33.00	0.82	0.28	0.033
34.00	0.84	0.29	0.032
35.00	0.85	0.31	0.033
36.00	0.85	0.32	0.036
37.00	0.84	0.33	0.039
38.00	0.83	0.33	0.040
39.00	0.81	0.33	0.042
40.00	0.80	0.33	0.045

Vanadium[9]

Energy (eV)	n	k	R ($\phi = 0$)
0.10	12.83	45.89	0.978
0.20	3.90	24.30	0.975
0.28	2.13	17.35	0.973
0.36	1.54	13.32	0.966
0.44	1.28	10.74	0.957
0.52	1.16	8.93	0.945
0.60	1.10	7.59	0.929
0.68	1.07	6.54	0.909
0.76	1.08	5.67	0.882
0.80	1.10	5.30	0.864
0.90	1.18	4.50	0.811
1.00	1.34	3.80	0.730
1.10	1.60	3.26	0.632
1.20	1.93	2.88	0.543
1.30	2.25	2.71	0.498
1.40	2.48	2.72	0.491
1.50	2.57	2.79	0.499
1.60	2.57	2.84	0.507
1.70	2.52	2.88	0.512
1.80	2.45	2.88	0.515
1.90	2.36	2.85	0.514
2.00	2.34	2.81	0.509
2.10	2.31	2.78	0.506
2.20	2.28	2.80	0.510
2.30	2.23	2.83	0.516
2.40	2.15	2.88	0.528
2.50	2.02	2.91	0.540
2.60	1.89	2.92	0.552
2.70	1.74	2.89	0.561
2.80	1.61	2.85	0.569
2.90	1.48	2.80	0.577
3.00	1.36	2.73	0.582
3.20	1.16	2.55	0.585
3.40	0.99	2.37	0.586
3.60	0.87	2.17	0.575
3.80	0.80	1.96	0.547
4.00	0.78	1.76	0.503
4.20	0.80	1.60	0.449
4.40	0.83	1.47	0.400
4.60	0.87	1.38	0.355
4.80	0.90	1.31	0.326
5.00	0.91	1.26	0.304
5.25	0.93	1.18	0.271
5.50	0.94	1.14	0.258
5.75	0.96	1.09	0.235
6.00	0.98	1.06	0.223
6.25	0.97	1.02	0.212
6.50	0.97	0.98	0.199
6.75	0.97	0.94	0.185
7.00	0.98	0.91	0.175
7.33	0.97	0.89	0.170
7.66	0.98	0.87	0.162
8.00	0.98	0.85	0.155
8.33	0.98	0.81	0.146
8.66	0.98	0.81	0.145
9.00	0.96	0.79	0.142
9.50	0.94	0.77	0.136
10.00	0.91	0.74	0.133
10.50	0.89	0.71	0.126
11.00	0.87	0.65	0.112
11.50	0.88	0.58	0.091
12.00	0.90	0.58	0.089
12.50	0.89	0.57	0.086
13.00	0.88	0.55	0.082
13.50	0.87	0.53	0.079
14.00	0.86	0.51	0.075
14.50	0.86	0.49	0.070
15.00	0.86	0.47	0.065
15.50	0.86	0.46	0.062
16.00	0.85	0.45	0.061
16.50	0.84	0.43	0.059
17.00	0.84	0.41	0.056
17.50	0.83	0.40	0.054
18.00	0.82	0.38	0.051
18.50	0.82	0.37	0.048
19.00	0.82	0.35	0.045
19.50	0.82	0.34	0.043
20.00	0.81	0.32	0.041
20.50	0.81	0.31	0.038
21.00	0.81	0.29	0.036
21.50	0.81	0.28	0.033
22.00	0.81	0.27	0.032
22.50	0.81	0.25	0.029
23.00	0.82	0.24	0.027
23.50	0.82	0.23	0.025
24.00	0.82	0.22	0.024
24.50	0.83	0.21	0.022
25.00	0.83	0.20	0.020
25.50	0.83	0.19	0.019
26.00	0.83	0.18	0.018
26.50	0.84	0.17	0.016
27.00	0.84	0.16	0.015
27.50	0.85	0.16	0.014
28.00	0.85	0.15	0.013
28.50	0.86	0.14	0.012
29.00	0.86	0.14	0.011
29.50	0.86	0.13	0.010
30.00	0.87	0.13	0.009
31.00	0.88	0.12	0.008
32.00	0.90	0.11	0.007
33.00	0.90	0.10	0.005
34.00	0.91	0.10	0.005
35.00	0.92	0.09	0.004
36.00	0.94	0.10	0.004
37.00	0.94	0.10	0.004
38.00	0.95	0.11	0.004
39.00	0.95	0.12	0.004
40.00	0.95	0.13	0.005

Zirconium (Polycrystalline)[a] [14]

Energy (eV)	n	k	R ($\phi = 0$)
0.10	6.18	1.76	0.300
0.15	3.37	1.30	0.123
0.20	2.34	1.08	0.058
0.26	2.24	1.06	0.052
0.30	2.59	1.14	0.073
0.36	3.17	1.26	0.110
0.40	3.09	1.24	0.105
0.46	3.36	1.30	0.123
0.50	4.13	1.44	0.175
0.56	5.01	1.58	0.231
0.60	5.18	1.61	0.242
0.70	4.54	1.51	0.202
0.80	4.03	1.42	0.168
0.90	3.74	1.37	0.149
0.96	3.69	1.36	0.145
1.00	3.66	1.35	0.143
1.10	3.65	1.35	0.142
1.20	3.53	1.33	0.134
1.30	3.25	1.27	0.116
1.40	3.10	1.25	0.106
1.50	3.02	1.23	0.100
1.60	2.88	1.20	0.091
1.70	2.68	1.16	0.078
1.80	2.49	1.12	0.067
2.00	2.14	1.03	0.047
2.10	1.00	1.00	0.040
2.20	1.87	0.97	0.034
2.30	1.78	0.94	0.030
2.40	1.71	0.92	0.027
2.50	1.62	0.90	0.024
2.60	1.54	0.88	0.022
2.70	1.46	0.86	0.019
2.80	1.40	0.84	0.018
2.90	1.34	0.82	0.016
3.00	0.30	0.81	0.016
3.10	1.26	0.80	0.015
3.30	1.19	0.77	0.014
3.40	1.16	0.76	0.013
3.50	1.13	0.75	0.013
3.60	1.10	0.74	0.013
3.70	1.07	0.73	0.013
3.80	1.04	0.72	0.012
3.90	1.01	0.71	0.012
4.00	0.98	0.70	0.012
4.20	0.94	0.68	0.013
4.40	0.89	0.67	0.013
4.60	0.85	0.65	0.014
4.80	0.81	0.64	0.014
5.00	0.78	0.63	0.015
5.20	0.77	0.62	0.016
5.40	0.77	0.62	0.016
5.60	0.80	0.63	0.014
5.80	0.87	0.66	0.013
6.00	1.00	0.71	0.012
6.20	1.11	0.75	0.013
6.40	1.23	0.78	0.014
6.60	1.33	0.81	0.016
6.80	1.42	0.84	0.018
7.00	1.49	0.86	0.020
7.20	1.54	0.88	0.022
7.40	1.58	0.89	0.023
7.60	1.61	0.90	0.024
7.80	1.63	0.90	0.025
8.00	1.66	0.91	0.026
8.20	0.67	0.91	0.026
8.40	1.68	0.92	0.026
8.60	1.68	0.92	0.026
8.80	1.66	0.91	0.026
9.00	1.65	0.91	0.025
9.20	1.63	0.90	0.025
9.40	1.60	0.89	0.024
9.60	1.57	0.89	0.023
9.80	1.52	0.87	0.021
10.00	1.47	0.86	0.020
10.20	1.42	0.84	0.018
10.40	1.35	0.82	0.016
10.50	1.32	0.81	0.016
10.60	1.28	0.80	0.015
10.80	1.23	0.78	0.014
11.00	1.19	0.77	0.014
11.20	1.16	0.76	0.013
11.40	1.13	0.75	0.013
11.60	1.11	0.74	0.013
11.80	1.09	0.74	0.013
12.00	1.08	0.73	0.013
12.40	1.05	0.72	0.012
12.80	1.01	0.71	0.012
13.20	0.98	0.70	0.012
13.60	0.95	0.69	0.013
14.00	0.92	0.68	0.013
14.40	0.89	0.67	0.013
14.80	0.90	0.67	0.013

[a] Data for single crystal is in original publication.

15.20	0.92	0.68	0.013	22.00	1.20	0.77	0.014
15.60	0.95	0.69	0.013	22.60	1.15	0.76	0.013
16.00	0.98	0.70	0.012	23.00	1.12	0.75	0.013
16.40	1.01	0.71	0.012	23.60	1.08	0.73	0.013
16.80	1.04	0.72	0.012	24.00	1.05	0.73	0.013
17.20	1.09	0.74	0.013	24.60	1.02	0.71	0.012
17.60	1.13	0.75	0.013	25.00	1.00	0.71	0.012
18.00	1.17	0.76	0.014	25.60	0.97	0.69	0.012
18.40	1.21	0.78	0.014	26.00	0.95	0.69	0.013
18.80	1.24	0.79	0.014	26.60	0.91	0.67	0.013
19.20	1.27	0.80	0.015	27.00	0.88	0.66	0.013
19.60	1.29	0.80	0.015	27.60	0.84	0.65	0.014
20.00	1.30	0.81	0.015	28.00	0.83	0.64	0.014
20.60	1.29	0.80	0.015	28.60	0.82	0.64	0.014
21.00	1.27	0.80	0.015	29.00	0.81	0.64	0.014
21.60	1.23	0.78	0.014	29.60	0.82	0.64	0.014
				30.00	0.82	0.64	0.014

REFERENCES

1. Shiles, E., Sasaki, T., Inokuti, M., and Smith, D. Y., *Phys. Rev. Sect. B,* 22, 1612, 1980.
2. Bos, L. W. and Lynch, D. W., *Phys. Rev. Sect. B,* 2, 4567, 1970; Olson, C. G. and Lynch, D. W., unpublished.
3. Weaver, J. H., Colavita, E., Lynch, D. W., and Rosei, R., *Phys. Rev. Sect. B,* 19, 3850, 1979.
4. Hagemann, H. J., Gudat, W., and Kunz, C., *J. Opt. Soc. Am.,* 65, 742, 1975.
5. Olson, C. G., Lynch, D. W., and Weaver, J. H., unpublished.
6. Lynch, D. W., Olson, C. G., and Weaver, J. H., unpublished. Supersedes those data by the same authors in *Phys. Rev. Sect. B,* 11, 3617, 1975.
7. Weaver, J. H., Olson, C. G., and Lynch, D. W., *Phys. Rev. Sect. B,* 15, 4115, 1977.
8. Johnson, P. B. and Christy, R. W., *Phys. Rev. Sect. B,* 9, 5056, 1974.
9. Weaver, J. H., Lynch, D. W., and Olson, C. G., *Phys. Rev. Sect. B,* 10, 501, 1973.
10. Lynch, D. W., Rosei, R., and Weaver, J. H., *Solid State Commun.,* 9, 2195, 1971.
11. Weaver, J. H., Lynch, D. W., and Olson, C. G., *Phys. Rev. Sect. B,* 7, 4311, 1973.
12. Weaver, J. H. and Benbow, R. L., *Phys. Rev. Sect. B,* 12, 3509, 1975.
13. Weaver, J. H., *Phys. Rev. Sect. B,* 11, 1416, 1975.
14. Lynch, D. W., Olson, C. G., and Weaver, J. H., *Phys. Rev. Sect. B,* 11, 3671, 1975.
15. Weaver, J. H., Lynch, D. W., and Olson, C. G., *Phys. Rev. Sect. B,* 12, 1293, 1975.

REFLECTION COEFFICIENTS

Coefficients of Reflection of Miscellaneous Surfaces for Monochromatic Radiation in the Visible Spectrum
(J. L. Michaelson)

Material	Wave lengths (μ) 0.400	0.500	0.600	0.700
Carbon black in oil	0.003	0.003	0.003	0.003
Clay				
Kaolin (treated)	0.82	0.81	0.82	0.82
Kaolin (untreated)	0.75	0.79	0.85	0.86
White georgia	0.94	0.92	0.93	0.94
Magnesium oxide	0.97	0.98	0.98	0.98
Paint				
Lithopone	0.95	0.98	0.98	0.98
$MgCO_3$-Vynal acetate lacquer	0.90	0.88	0.88	0.88
ZnO-Milk	0.74	0.84	0.85	0.86
Paper				
Blotting	0.64	0.72	0.79	0.79
Calendered	0.64	0.69	0.73	0.76
Crepe, green	0.23	0.49	0.19	0.48
Crepe, red	0.03	0.02	0.21	0.69
Crepe, yellow	0.17	0.44	0.75	0.79
News print stock	0.38	0.61	0.63	0.78
Peach				
Green	0.18	0.17	0.62	0.63
Ripe	0.10	0.10	0.41	0.42
Pear				
Green	0.04	0.12	0.29	0.41
Ripe	0.08	0.19	0.46	0.53
Pigment				
Chrome yellow	0.05	0.13	0.70	0.77
French ochre	0.06	0.14	0.50	0.56
Porcelain enamel				
Blue	0.44	0.10	0.05	0.23
Orange	0.09	0.09	0.59	0.69
Red	0.05	0.03	0.08	0.62
White	0.77	0.73	0.72	0.70
Yellow	0.11	0.46	0.62	0.62
Talcum, Italian	0.94	0.89	0.88	0.88
Wheat flour	0.75	0.87	0.94	0.97

REFLECTION COEFFICIENTS OF SURFACES FOR "INCANDESCENT" LIGHT

Material	Nature of Surface	Coefficient	Authority
Aluminum, "Alzak"	Diffusing	0.77—0.81	3
"Alzak"	Specular	0.79—0.83	3
On glass	First surface	0.82—0.86	4
Polished	Specular	0.69	3
Black paper	Diffusing	0.05—0.06	4
Chromium	Specular	0.62	4
Copper	Specular	0.63	4
Gold	Specular	0.75	1
Magnesium oxide	Diffusing	0.98	5
Nickel	Specular	0.62—0.64	1,3
Platinum	Specular	0.62	1
Porcelain enamel	Glossy	0.76—0.79	3
Porcelain enamel	Ground	0.81	3
Porcelain enamel	Matt.	0.72—0.76	3
Silver	Polished	0.93	1
Silvered glass	Second surface	0.88—0.93	3
Snow	Diffusing	0.93	2
Steel	Specular	0.55	1
Stellite	Specular	0.58—0.65	4

(1) Hagen and Rubena; (2) Nutting, Jones, and Elliot; (3) J. E. Bock; (4) Frank Benford; (5) J. L. Michaelson.

EMISSIVITY AND ABSORPTION

These data are the result of investigations made by the Bureau of Standards, the British National Physical Laboratory, General Electric Research Laboratories, and several eastern universities, and were collected by W. J. King of the General Electric Company.

Low Temperature Total Emissivities

Silver, highly polished	0.02	Brass, polished	0.60
Platinum, highly polished	0.05	Oxidized copper	0.60
Zinc, highly polished	0.05	Oxidized steel	0.70
Aluminum, highly polished	0.08	Bronze paint	0.80
Monel metal, polished	0.09	Black gloss paint	0.90
Nickel, polished	0.12	White lacquer	0.95
Copper, polished	0.15	White vitreous enamel	0.95
Stellite, polished	0.18	Asbestos paper	0.95
Cast iron, polished	0.25	Green paint	0.95
Monel metal, oxidized	0.43	Gray paint	0.95
Aluminum paint	0.55	Lamp black	0.95

Coefficient of Absorption of Solar Radiation

Silver, highly polished	0.07	Stellite, polished	0.30
Platinum, highly polished	0.10	Light cream paint	0.35
Nickel, highly polished	0.15	Monel metal, polished	0.40
Aluminum*	0.15	Light yellow paint	0.45
Magnesium carbonate	0.15	Light green paint	0.50
Zinc oxide	0.15	Aluminum paint	0.55
Steel*	0.20	Zinc, polished metal	0.55
Copper	0.25	Gray paint	0.75
White lead paint	0.25	Black matte	0.97
Zinc oxide paint	0.30		

* Questionable because of scant or inconsistent data.

EMISSIVITY OF TOTAL RADIATION, ε_{tot}, FOR VARIOUS MATERIALS

Material	Temperature (° C)	ε_{tot}
Alloys		
Nickel-Chromium		
20 Ni—25 Cr—55 Fe, oxidized	200	0.90
	500	0.97
60 Ni—12 Cr—28 Fe, oxidized	270	0.89
	560	0.82
80 Ni—20 Cr	100	0.87
	600	0.87
	1300	0.89
Aluminum		
Polished	50—500	0.04—0.06
Rough surface	20—50	0.06—0.07
Strongly oxidized	55—500	0.2—0.3
	25	0.022
	100	0.028
	500	0.060
Oxidized	200	0.11
	600	0.19
Asbestos board	20	0.96
Bismuth		
Unoxidized	25	0.048
Brass		
Dull tarnished	200	0.61
Oxidized at 600° C	200	0.61
	600	0.59
Unoxidized	25	0.035
	100	0.035
Polished	200	0.03
Rolled sheet	20	0.06
Bronze		
Polished	50	0.1
Carbon		
Filament	1000—1400	0.53
Graphite	0—3600	0.7—0.8
Lamp black	20—400	0.96
Soot applied to solid	50—1000	0.96
Soot with water glass	20—200	0.96
Unoxidized	100	0.81
Chromium		
Polished	50	0.1
	500—1000	0.28—0.38
Colbalt		
Unoxidized	500	0.13
	1000	0.23
Columbium		
Unoxidized	1500	0.19
Copper		
Calorized	100	0.26
Calorized, oxidized	200	0.18
	600	0.19
Commercial, scoured to a shine	20	0.07
Oxidized	50	0.6—0.7
	500	0.88
Polished	50—100	0.02
Unoxidized	100	0.02
Unoxidized, liquid	—	0.15
Fire brick	1000	0.75
Glass	20—100	0.94—0.91
	250—1000	0.87—0.72
	1100—1500	0.7—0.67
Gold		
Carefully polished	200—600	0.02—0.03
Unoxidized	100	0.02
Enamel	100	0.37
Graphite	0—3600	0.7—0.8
Gypsum	20	0.93
Iron		
Cast		
Oxidized	200	0.64
	600	0.78
Strongly oxidized	40	0.95
	250	0.95
Unoxidized	100	0.21
Unoxidized, liquid	—	0.29
Oxidized	100	0.74
	500	0.84
	1200	0.89
Rusted	25	0.65
Wrought, dull	100	0.05
Lamp black	25	0.94
	20—400	0.96
Lead		
Oxidized	200	0.05
Unoxidized	200	0.63
Mercury		
Unoxidized	25	0.10
	100	0.12
Molybdenum	600—1000	0.08—0.13
	1500—2200	0.19—0.26
Monel metal		
Oxidized	200	0.43
	600	0.43
Nichrome		
Wire		
Clean	50	0.65
	500—1000	0.71—0.79
Oxidized	50—500	0.95—0.98
Nickel		
Industrial, polished	200—400	0.07—0.09
Oxidized	200	0.37
Oxidized at 600°C	200—600	0.37—0.48
Unoxidized	25	0.045
	100	0.06
	500	0.12
	1000	0.19
Platinum		
Clean, polished	200—600	0.05—0.1
Unoxidized	25	0.037
	100	0.047
	500	0.096
	1000	0.152
	1500	0.191
Wire	50—200	0.06—0.07
	500—1000	0.1—0.16
	1400	0.18
Porcelain		
Glazed	20	0.92
Rubber		
Hard	20	0.95
Soft, gray, rough	20	0.86
Silica brick	1000	0.80
	1100	0.85
Silver		
Clean, polished	200—600	0.02—0.03
Unoxidized	100	0.02
	500	0.035
Soot applied to a solid surface	50—1000	0.94—0.91
Soot with water glass	20—200	0.96
Steel		
Alloyed (8% Ni, 18% Cr)	500	0.35
Aluminized	50—500	0.79
Dull nickel plated	20	0.11
Flat, rough surface	50	0.95—0.98
Cast, polished	750—1050	0.52—0.56
Sheet, ground	50	0.56
	950—1100	0.55—0.61
Oxidized	200—600	0.8
Calorized, oxidized	200	0.52
	600	0.57
Sheet with shiny layer of oxide	20	0.82
Strongly oxidized	50	0.88
	500	0.98
Unoxidized	100	0.08
Unoxidized, liquid	—	0.28
Tantalum		
Unoxidized	1500	0.21
	2000	0.26
Tungsten		
Unoxidized	25	0.024
	100	0.032
	500	0.071
	1000	0.15
	1500	0.23
	2000	0.28
Varnish	40—100	0.8—0.95
Dull black	40—100	0.96—0.98
Glossy black sprayed on iron	20	0.87
Zinc		
Polished	200—300	0.04—0.05
Unoxidized	300	0.05

SPECTRAL EMISSIVITY

Prepared by Roeser and Wensel, National Bureau of Standards
Spectral Emissivity of Materials, Surface Unoxidized for 0.65μ

Element	Solid	Liquid	Element	Solid	Liquid
Beryllium	0.61	0.61	Thorium	0.36	0.40
Carbon	0.80—0.93	—	Titanium	0.63	0.65
Chromium	0.34	0.39	Tungsten	0.43	—
Cobalt	0.36	0.37	Uranium	0.54	0.34
Columbium	0.37	0.40	Vanadium	0.35	0.32
Copper	0.10	0.15	Yttrium	0.35	0.35
Erbium	0.55	0.38	Zirconium	0.32	0.30
Gold	0.14	0.22	Steel	0.35	0.37
Iridium	0.30	—	Cast Iron	0.37	0.40
Iron	0.35	0.37	Constantan	0.35	—
Manganese	0.59	0.59	Monel	0.37	—
Molybdenum	0.37	0.40	Chromel P (90Ni-10Cr)	0.35	—
Nickel	0.36	0.37	80Ni-20Cr	0.35	—
Palladium	0.33	0.37	60Ni-24Fe-16Cr	0.36	—
Platinum	0.30	0.38	Alumel (95Ni; Bal. Al, Mn, Si)	0.37	—
Rhodium	0.24	0.30	90Pt-10Rh	0.27	—
Silver	0.07	0.07			
Tantalum	0.49	—			

SPECTRAL EMISSIVITY OF OXIDES

The emissivity of oxides and oxidized metals depends to a large extent upon the roughness of the surface. In general, higher values of emissivity are obtained on the rougher surfaces.

Material	Range of observed values	Probable value for oxide formed on smooth metal
Oxide		
Aluminum	0.22—0.40	0.30
Beryllim	0.07—0.37	0.35
Cerium	0.58—0.80	—
Chromium	0.60—0.80	0.70
Cobalt	—	0.75
Columbium	0.55 0.71	0.70
Copper	0.60—0.80	0.70
Iron	0.63—0.98	0.70
Magnesium	0.10—0.43	0.20
Nickel	0.85—0.96	0.90
Thorium	0.20—0.57	0.50
Tin	0.32—0.60	—
Titanium	—	0.50
Uranium	—	0.30
Vanadium	—	0.70
Yttrium	—	0.60
Zirconium	0.18—0.43	0.40
Oxidized		
Alumel	—	0.87
Cast Iron	—	0.70
Chromel P (90Ni-10Cr)	—	0.87
80Ni-20Cr	—	0.90
60Ni-24Fe-16Cr	—	0.83
55Fe-37.5Cr-7.5Al	—	0.78
70Fe-23Cr-5Al-2Co	—	0.75
Constantan (55Cu-45Ni)	—	0.84
Carbon Steel	—	0.80
Stainless Steel (18-8)	—	0.85
Porcelain	0.25—0.50	—

PROPERTIES OF TUNGSTEN

Jones and Langmuir, General Electric Review

Temp. °K	Resistivity microhm cm	Electron emission amp./cm²	Evaporation g/cm² sec	Vapor pressure dynes/cm²	Thermal expansion per cent l_0 at 293°	Atomic heat cal./g. atom./°C.
300	5.65	—	—	—	.003	6.0
400	8.06	—	—	—	.044	6.0
500	10.56	—	—	—	.086	6.1
600	13.23	—	—	—	.130	6.1
700	16.09	—	—	—	.175	6.2
800	19.00	—	—	—	.222	6.2
900	21.94	—	—	—	.270	6.3
1000	24.93	1.07×10^{-15}	5.32×10^{-34}	1.98×10^{-29}	.320	6.4
1100	27.94	1.52×10^{-13}	2.17×10^{-30}	1.22×10^{-25}	.371	6.4
1200	30.98	9.73×10^{-12}	3.21×10^{-27}	1.87×10^{-22}	.424	6.5
1300	34.08	3.21×10^{-10}	1.35×10^{-24}	8.18×10^{-20}	.479	6.7
1400	37.19	6.62×10^{-9}	2.51×10^{-22}	1.62×10^{-17}	.535	6.8
1500	40.36	9.14×10^{-8}	2.37×10^{-20}	1.54×10^{-15}	.593	7.0
1600	43.55	9.27×10^{-7}	1.25×10^{-18}	8.43×10^{-14}	.652	7.1
1700	46.78	7.08×10^{-6}	4.17×10^{-17}	2.82×10^{-12}	.713	7.2
1800	50.05	4.47×10^{-5}	8.81×10^{-16}	6.31×10^{-11}	.775	7.4
1900	53.35	2.28×10^{-4}	1.41×10^{-14}	1.01×10^{-9}	.839	7.6
2000	56.67	1.00×10^{-3}	1.76×10^{-13}	1.33×10^{-8}	.904	7.7
2100	60.06	3.93×10^{-3}	1.66×10^{-12}	1.28×10^{-7}	.971	7.8
2200	63.48	1.33×10^{-2}	1.25×10^{-11}	9.88×10^{-7}	1.039	8.0
2300	66.91	4.07×10^{-2}	8.00×10^{-11}	6.47×10^{-6}	1.109	8.2
2400	70.39	1.16×10^{-1}	4.26×10^{-10}	3.52×10^{-5}	1.180	8.3
2500	73.91	2.98×10^{-1}	2.03×10^{-9}	1.71×10^{-4}	1.253	8.4
2600	77.49	7.16×10^{-1}	8.41×10^{-9}	7.24×10^{-4}	1.328	8.6
2700	81.04	1.63	3.19×10^{-8}	2.86×10^{-3}	1.404	8.7
2800	84.70	3.54	1.10×10^{-7}	9.84×10^{-3}	1.479	8.9
2900	88.33	7.31	3.30×10^{-7}	3.00×10^{-2}	1.561	9.0
3000	92.04	1.42×10	9.95×10^{-7}	9.20×10^{-2}	1.642	9.2
3100	95.76	2.64×10	2.60×10^{-6}	2.50×10^{-1}	1.724	9.4
3200	99.54	4.78×10	6.38×10^{-6}	6.13×10^{-1}	1.808	9.5
3300	103.3	8.44×10	1.56×10^{-5}	1.51	1.893	9.6
3400	107.2	1.42×10^{2}	3.47×10^{-5}	3.41	1.980	9.8
3500	111.1	2.33×10^{2}	7.54×10^{-5}	7.52	2.068	9.9
3600	115.0	3.73×10^{2}	1.51×10^{-4}	1.53×10	2.158	10.1
3655	117.1	4.79×10^{2}	2.28×10^{-4}	2.33×10	2.209	10.2

Roeser and Wensel, National Bureau of Standards

Temp. °K	Normal brightness new candles per cm²	Spectral emissivity		Color emissivity	Total emissivity	Brightness temp. 0.65μ	Color temp.
		0.65μ	0.467μ				
300	—	0.472	0.505	—	0.032	—	—
400	—	—	—	—	.042	—	—
500	—	—	—	—	.053	—	—
600	—	—	—	—	.064	—	—
700	—	—	—	—	.076	—	—
800	—	—	—	—	.088	—	—
900	—	—	—	—	.101	—	—
1000	0.0001	.458	.486	.395	.114	966	1007
1100	0.001	.456	.484	.392	.128	1059	1108
1200	0.006	.454	.482	.390	.143	1151	1210
1300	0.029	.452	.480	.387	.158	1242	1312
1400	0.11	.450	.478	.385	.175	1332	1414
1500	0.33	.448	.476	.382	.192	1422	1516
1600	0.92	.446	.475	.380	.207	1511	1619
1700	2.3	.444	.473	.377	.222	1599	1722
1800	5.1	.442	.472	.374	.236	1687	1825
1900	10.4	.440	.470	.371	.249	1774	1928
2000	20.0	.438	.469	.368	.260	1861	2032
2100	36	.436	.467	.365	.270	1946	2136
2200	61	.434	.466	.362	.279	2031	2241
2300	101	.432	.464	.359	.288	2115	2345
2400	157	.430	.463	.356	.296	2198	2451
2500	240	.428	.462	.353	.303	2280	2556
2600	350	.426	.460	.349	.311	2362	2662
2700	500	.424	.459	.346	.318	2443	2769
2800	690	.422	.458	.343	.323	2523	2876
2900	950	.420	.456	.340	.329	2602	2984
3000	1260	.418	.455	.336	.334	2681	3092
3100	1650	.416	.454	.333	.337	2759	3200
3200	2100	.414	.452	.330	.341	2837	3310
3300	2700	.412	.451	.326	.344	2913	3420
3400	3400	.410	.450	.323	.348	2989	3530
3500	4200	.408	.449	.320	.351	3063	3642
3600	5200	.406	.447	.317	.354	3137	3754

TRANSMISSION OF CORNING COLORED FILTERS

Supplied by R. G. Saxton

If I_o is the intensity of radiation entering a layer of some medium and I the intensity reaching the opposite surface, the ratio I/I_o is called the transmittance. In practice the ratio of intensity of radiation passing through a glass sample to that incident on its surface is often measured and plotted as transmission. The transmission is the result of two factors, the transmittance of the glass and the losses by reflection. These losses amount to about 4 % for each glass-air surface; the transmission of a sample is about 92 % of its transmittance. Since the reflection losses differ slightly with different samples, the correction is often determined and applied when the transmission is measured. Values in this table have been corrected for reflection losses.

The identifying glass number, CS number, color and properties, and nominal thickness for the Corning glasses in this table are:

Glass No.	CS	Color and properties	Nominal thickness
0160	0-54	Clear; Ultraviolet transmitting	2.0
2030	2-64	Red; Sharp cut	3.0
2403	2-58	Red; Sharp cut	3.0
2404	2-59	Red; Sharp cut	3.0
2408	2-60	Red; Sharp cut	3.0
2412	2-61	Red; Sharp cut	3.0
2418	2-62	Red; Sharp cut	3.0
2424	2-63	Red; Sharp cut	3.0
2434	2-73	Red; Sharp cut	3.0
2540	7-56	Black; IR transmitting; Visible absorbing	2.5
2550	7-57	Black; IR transmitting; Visible absorbing	2.0
2600	7-69	Black; IR transmitting; Visible absorbing	2.0
3060	3-75	Straw	2.0
3304	3-76	Dark amber	3.0
3307	3-77	Dark amber	3.0
3384	3-70	Yellow	3.0
3385	3-71	Yellow	3.0
3387	3-72	Straw	3.0
3389	3-73	Straw	3.0
3391	3-74	Straw	3.0
3480	3-66	Yellow; Sharp cut	3.0
3482	3-67	Yellow; Sharp cut	3.0
3484	3-68	Yellow; Sharp cut	3.0
3486	3-69	Yellow; Sharp cut	3.0
3718	3-94	Yellow	3.0
3750	3-79	Yellow; Yellow green fluorescing	5.0
3780	3-80	Yellow	2.0
3850	0-51	Clear; UV transmitting	4.0
3961	1-56	Bluish; IR absorbing; Visible transmitting	2.5
3962	1-57	Bluish; IR absorbing; Visible transmitting	2.5
3965	1-58	Bluish; IR absorbing; Visible transmitting	2.5
3966	1-59	Bluish; IR absorbing; Visible transmitting	2.5
4010	4-64	Green	4.0
4015	4-65	Yellow green	3.0
4060	4-67	Green	2.0
4084	4-68	Green	4.5
4303	4-72	Blue green	4.0
4305	4-71	Blue green	4.0
4308	4-70	Blue green	4.0
4309	4-69	Blue green	4.0
4445	4-74	Green	2.5
4602	1-75	Bluish; IR absorbing; Visible transmitting	3.0
4784	4-94	Blue green	5.0
5030	5-57	Blue	5.0
5031	5-56	Blue	4.5
5070	7-62	Amethyst	3.9
5071	7-63	Amethyst	3.9
5073	7-64	Amethyst	3.9
5113	5-58	Blue	4.0
5120	1-60	Smoky violet; Absorbs yellow	5.2
5300	4-106	Green	3.9
5330	1-64	Blue	4.5
5433	5-59	Blue	5.0
5543	5-60	Blue	5.0
5562	5-61	Blue	5.0
5572	1-61	Blue	5.0
5840	7-60	Black; UV transmitting; Visible absorbing	4.5
5850	7-59	Purple; UV transmitting; Visible absorbing	4.0
5860	7-37	Black; UV transmitting; Visible absorbing	5.0
5874	7-39	Black; UV transmitting; Visible absorbing	5.0
5900	1.62	Blue	5.5
5970	7-51	Black; UV transmitting; Visible absorbing	5.0
7380	0-52	Clear; UV transmitting	2.0
7740	0-53	Clear; UV transmitting	2.0
7905	9-30	Clear; UV transmitting; Long Range IR transmitting	2.0
7910	9-54	Clear; UV transmitting	2.0
8364	7-98	Gray	2.0
9780	4-76	Blue green	5.0
9782	4-96	Blue green	5.0
9788	4-97	Blue green	5.0
9830	4-77	Green	3.4
9863	7-54	Black; UV transmitting; Visible absorbing	3.0

TRANSMISSION OF CORNING COLORED FILTERS (Continued)

	Transmittance											
	Corning Glass Number											
λ(nm)	0160	2030	2403	2404	2408	2412	2418	2424	2434	2540	2550	2600
.22	.000	.000	.000	.000	.000	.000	.000	.000	.000	.000	.000	.000
.24	.000	.000	.000	.000	.000	.000	.000	.000	.000	.000	.000	.000
.26	.000	.000	.000	.000	.000	.000	.000	.000	.000	.000	.000	.000
.28	.000	.000	.000	.000	.000	.000	.000	.000	.000	.000	.000	.000
.30	.005	.000	.000	.000	.000	.000	.000	.000	.000	.000	.000	.000
.32	.642	.000	.000	.000	.000	.000	.000	.000	.000	.000	.000	.000
.34	.850	.000	.000	.000	.000	.000	.000	.000	.000	.000	.000	.000
.36	.882	.000	.000	.000	.000	.000	.000	.000	.000	.000	.000	.000
.38	.890	.000	.000	.000	.000	.000	.000	.000	.000	.000	.000	.000
.40	.892	.000	.000	.000	.000	.000	.000	.000	.000	.000	.000	.000
.41	.893	.000	.000	.000	.000	.000	.000	.000	.000	.000	.000	.000
.42	.896	.000	.000	.000	.000	.000	.000	.000	.000	.000	.000	.000
.43	.896	.000	.000	.000	.000	.000	.000	.000	.000	.000	.000	.000
.44	.898	.000	.000	.000	.000	.000	.000	.000	.000	.000	.000	.000
.45	.899	.000	.000	.000	.000	.000	.000	.000	.000	.000	.000	.000
.46	.900	.000	.000	.000	.000	.000	.000	.000	.000	.000	.000	.000
.47	.900	.000	.000	.000	.000	.000	.000	.000	.000	.000	.000	.000
.48	.900	.000	.000	.000	.000	.000	.000	.000	.000	.000	.000	.000
.49	.900	.000	.000	.000	.000	.000	.000	.000	.000	.000	.000	.000
.50	.900	.000	.000	.000	.000	.000	.000	.000	.000	.000	.000	.000
.51	.900	.000	.000	.000	.000	.000	.000	.000	.000	.000	.000	.000
.52	.900	.000	.000	.000	.000	.000	.000	.000	.000	.000	.000	.000
.53	.900	.000	.000	.000	.000	.000	.000	.000	.000	.000	.000	.000
.54	.900	.000	.000	.000	.000	.000	.000	.000	.000	.000	.000	.000
.55	.900	.000	.000	.000	.000	.000	.000	.000	.000	.000	.000	.000
.56	.901	.000	.000	.000	.000	.000	.000	.000	.000	.000	.000	.000
.57	.904	.000	.000	.000	.000	.000	.000	.000	.005	.000	.000	.000
.58	.904	.000	.000	.000	.000	.000	.000	.005	.200	.000	.000	.000
.59	.908	.000	.000	.000	.000	.000	.008	.170	.615	.000	.000	.000
.60	.910	.000	.000	.000	.000	.006	.250	.575	.808	.000	.000	.000
.61	.910	.000	.000	.000	.018	.190	.660	.790	.856	.000	.000	.000
.62	.910	.000	.000	.015	.265	.625	.822	.848	.872	.000	.000	.000
.63	.910	.000	.018	.295	.670	.828	.862	.870	.881	.000	.000	.000
.64	.910	.006	.260	.660	.828	.868	.874	.880	.887	.000	.001	.000
.65	.910	.028	.675	.796	.866	.881	.881	.887	.892	.000	.003	.000
.66	.910	.110	.838	.828	.877	.885	.885	.893	.895	.000	.005	.000
.67	.910	.305	.871	.842	.883	.887	.887	.897	.897	.000	.006	.000
.68	.910	.550	.880	.847	.886	.889	.889	.900	.899	.000	.009	.000
.69	.910	.735	.885	.851	.888	.900	.900	.901	.900	.000	.012	.000
.70	.910	.820	.886	.852	.888	.900	.900	.903	.900	.000	.017	.000
.71	.910	.853	.888	.854	.888	.889	.889	.903	.900	.000	.023	.000
.72	.910	.864	.889	.853	.888	.888	.888	.903	.899	.000	.031	.040
.73	.910	.867	.900	.851	.887	.887	.887	.903	.897	.000	.041	.175
.74	.910	.867	.900	.850	.885	.886	.886	.903	.896	.000	.055	.372
.75	.910	.866	.900	.849	.884	.885	.885	.902	.895	.000	.069	.547
.80	.910	.839	.875	.827	.870	.858	.857	.881	.866	.005	.225	.770
1.00	.912	.801	.840	.772	.840	.828	.822	.857	.842	.562	.780	.350
1.20	.908	.799	.845	.786	.845	.837	.827	.859	.849	.790	.870	.000
1.40	.909	.811	.854	.809	.854	.848	.840	.862	.857	.850	.895	.000
1.60	.913	.839	.873	.837	.873	.869	.858	.880	.879	.872	.904	.000
1.80	.909	.844	.870	.829	.870	.864	.854	.877	.872	.880	.900	.000
2.00	.904	.841	.868	.827	.868	.861	.851	.874	.871	.880	.897	.000
2.20	.888	.833	.820	.773	.825	.820	.810	.837	.835	.860	.875	.005
2.40	.875	.832	.803	.757	.809	.803	.792	.818	.812	.868	.870	.049
2.60	.868	.822	.750	.695	.754	.750	.723	.772	.767	.858	.850	.058
2.80	.690	.600	.100	.100	.100	.100	.050	.050	.260	.450	.480	.030
3.00	.630	.470	.070	.020	.070	.070	.072	.122	.400	.465	.383	.022
3.20	.500	.340	.140	.074	.150	.140	.142	.180	.470	.390	.330	.020
3.40	.379	.260	.140	.078	.140	.140	.120	.150	.350	.310	.245	.018
3.60	.320	.247	.000	.000	.000	.000	.000	.006	.015	.280	.220	.021
3.80	.310	.257	.000	.000	.000	.000	.000	.000	.020	.285	.250	.035
4.00	.311	.274	.000	.000	.000	.000	.000	.000	.015	.285	.275	.068
4.20	.251	.200	.000	.000	.000	.000	.000	.000	.017	.190	.190	.065
4.40	.110	.060	.000	.000	.000	.000	.000	.000	.002	.050	.100	.020
4.60	.012	.000	.000	.000	.000	.000	.000	.000	.000	.000	.000	.000
4.80	.004	.000	.000	.000	.000	.000	.000	.000	.000	.000	.000	.000
5.00	.000	.000	.000	.000	.000	.000	.000	.000	.000	.000	.000	.000

TRANSMISSION OF CORNING COLORED FILTERS (Continued)

λ(nm)	3060	3304	3307	3384	3385	3387	3389	3391	3480	3482	3484	3486
	Transmittance											
	Corning Glass Number											
.22	.000	.000	.000	.000	.000	.000	.000	.000	.000	.000	.000	.000
.24	.000	.000	.000	.000	.000	.000	.000	.000	.000	.000	.000	.000
.26	.000	.000	.000	.000	.000	.000	.000	.000	.000	.000	.000	.000
.28	.000	.000	.000	.000	.000	.000	.000	.000	.000	.000	.000	.000
.30	.000	.000	.000	.000	.000	.000	.000	.000	.000	.000	.000	.000
.32	.000	.000	.000	.000	.000	.000	.005	.000	.000	.000	.000	.000
.34	.000	.000	.038	.000	.000	.000	.010	.000	.000	.000	.000	.000
.36	.000	.000	.050	.000	.005	.000	.015	.000	.000	.000	.000	.000
.38	.060	.000	.027	.005	.010	.010	.020	.000	.000	.000	.000	.000
.40	.410	.000	.016	.011	.016	.020	.026	.075	.000	.000	.000	.005
.41	.517	.000	.014	.011	.016	.020	.025	.425	.000	.000	.000	.005
.42	.604	.000	.014	.010	.015	.019	.105	.655	.000	.000	.000	.005
.43	.665	.000	.016	.009	.013	.017	.437	.747	.000	.000	.000	.005
.44	.710	.000	.022	.005	.011	.050	.620	.801	.000	.000	.000	.005
.45	.748	.000	.033	.003	.010	.325	.714	.838	.000	.000	.000	.005
.46	.778	.000	.049	.002	.008	.565	.780	.860	.000	.000	.000	.004
.47	.800	.000	.070	.001	.060	.690	.820	.874	.000	.000	.000	.003
.48	.819	.000	.101	.005	.410	.763	.848	.884	.000	.000	.000	.003
.49	.836	.003	.143	.088	.640	.803	.866	.890	.000	.000	.000	.002
.50	.850	.009	.193	.350	.727	.834	.878	.895	.000	.000	.000	.001
.51	.860	.019	.250	.595	.780	.854	.886	.898	.000	.000	.000	.045
.52	.870	.037	.315	.725	.817	.868	.890	.900	.000	.000	.003	.425
.53	.875	.063	.379	.789	.840	.876	.892	.901	.000	.000	.175	.710
.54	.881	.102	.447	.825	.856	.883	.894	.902	.000	.015	.600	.792
.55	.884	.146	.504	.846	.866	.887	.895	.903	.000	.230	.774	.823
.56	.886	.200	.560	.860	.873	.889	.894	.902	.020	.675	.818	.844
.57	.885	.255	.607	.869	.876	.890	.893	.901	.325	.850	.839	.859
.58	.883	.310	.648	.873	.878	.889	.892	.900	.710	.885	.854	.868
.59	.882	.360	.680	.876	.877	.887	.890	.898	.829	.894	.865	.876
.60	.882	.404	.705	.877	.877	.884	.886	.896	.858	.900	.873	.882
.61	.882	.438	.722	.877	.877	.881	.884	.893	.869	.903	.880	.886
.62	.882	.466	.735	.875	.876	.875	.880	.890	.876	.905	.885	.888
.63	.882	.488	.744	.871	.874	.871	.876	.886	.881	.906	.888	.889
.64	.883	.505	.748	.865	.872	.866	.872	.884	.884	.907	.890	.890
.65	.885	.519	.750	.860	.867	.860	.868	.881	.885	.908	.892	.890
.66	.886	.531	.750	.856	.863	.856	.865	.876	.886	.908	.893	.890
.67	.888	.543	.749	.850	.858	.851	.860	.873	.886	.908	.894	.890
.68	.890	.552	.745	.844	.853	.846	.856	.869	.886	.908	.893	.889
.69	.891	.561	.740	.837	.847	.839	.852	.865	.885	.907	.892	.887
.70	.892	.569	.734	.831	.842	.834	.847	.860	.884	.907	.891	.885
.71	.893	.574	.727	.825	.837	.827	.842	.856	.882	.906	.890	.883
.72	.893	.575	.720	.819	.831	.822	.837	.852	.880	.905	.888	.880
.73	.892	.576	.712	.813	.825	.816	.831	.848	.877	.905	.886	.877
.74	.891	.574	.702	.807	.820	.810	.826	.844	.874	.904	.885	.875
.75	.890	.570	.694	.800	.814	.805	.820	.840	.870	.903	.882	.873
.80	.871	.526	.642	.770	.865	.780	.775	.815	.837	.878	.846	.829
1.00	.830	.435	.516	.715	.830	.725	.716	.772	.801	.857	.811	.781
1.20	.860	.429	.500	.718	.858	.735	.730	.782	.807	.859	.819	.793
1.40	.901	.475	.540	.750	.900	.768	.768	.810	.828	.870	.837	.817
1.60	.917	.580	.635	.795	.918	.812	.812	.843	.852	.884	.856	.841
1.80	.916	.627	.675	.808	.915	.818	.820	.849	.847	.882	.852	.834
2.00	.908	.620	.668	.800	.909	.822	.817	.846	.847	.884	.852	.835
2.20	.900	.630	.675	.802	.900	.823	.810	.840	.805	.865	.829	.798
2.40	.885	.651	.690	.800	.885	.825	.811	.842	.787	.853	.817	.777
2.60	.860	.650	.690	.785	.858	.815	.800	.840	.725	.818	.757	.718
2.80	.550	.345	.390	.325	.670	.360	.340	.440	.050	.060	.110	.060
3.00	.379	.320	.360	.318	.348	.348	.322	.423	.080	.088	.190	.088
3.20	.315	.240	.290	.268	.332	.324	.290	.395	.150	.158	.270	.145
3.40	.250	.151	.190	.218	.289	.288	.255	.353	.120	.132	.140	.090
3.60	.231	.130	.150	.217	.266	.280	.249	.351	.000	.000	.000	.000
3.80	.258	.140	.160	.228	.270	.298	.267	.376	.000	.000	.000	.000
4.00	.283	.140	.175	.220	.280	.290	.260	.365	.000	.000	.000	.000
4.20	.200	.090	.125	.143	.210	.210	.178	.288	.000	.000	.000	.000
4.40	.100	.020	.030	.025	.080	.070	.040	.115	.000	.000	.000	.000
4.60	.008	.008	.010	.007	.002	.003	.000	.009	.000	.000	.000	.000
4.80	.000	.005	.009	.000	.000	.000	.000	.000	.000	.000	.000	.000
5.00	.000	.001	.008	.000	.000	.000	.000	.000	.000	.000	.000	.000

	Transmittance											
	Corning Glass Number											
λ(nm)	3718	3750	3780	3850	3961	3962	3965	3966	4010	4015	4060	4084
.22	.000	.000	.000	.000	.000	.000	.000	.000	.000	.000	.000	.000
.24	.000	.000	.000	.000	.000	.000	.000	.000	.000	.000	.000	.000
.26	.000	.000	.000	.000	.000	.000	.000	.000	.000	.000	.000	.000
.28	.000	.000	.000	.000	.000	.000	.000	.000	.000	.000	.000	.000
.30	.000	.000	.000	.000	.000	.000	.000	.000	.000	.000	.000	.000
.32	.004	.000	.000	.000	.000	.000	.018	.055	.000	.000	.000	.000
.34	.030	.000	.000	.000	.000	.018	.192	.375	.000	.000	.001	.018
.36	.550	.215	.000	.005	.020	.125	.430	.630	.000	.000	.021	.128
.38	.665	.327	.000	.350	.085	.270	.558	.710	.000	.000	.080	.248
.40	.480	.113	.000	.675	.185	.395	.636	.781	.000	.000	.178	.216
.41	.443	.088	.000	.749	.218	.426	.651	.788	.000	.000	.228	.180
.42	.465	.088	.000	.788	.248	.453	.666	.795	.000	.000	.281	.151
.43	.560	.135	.000	.812	.269	.474	.678	.800	.000	.000	.335	.136
.44	.675	.255	.006	.828	.290	.494	.693	.806	.000	.000	.388	.140
.45	.748	.410	.028	.841	.313	.519	.709	.816	.000	.005	.434	.163
.46	.780	.472	.058	.850	.331	.538	.724	.824	.006	.025	.473	.200
.47	.803	.570	.092	.858	.346	.556	.737	.831	.021	.073	.506	.247
.48	.800	.555	.088	.865	.361	.570	.748	.836	.050	.145	.527	.303
.49	.802	.550	.095	.870	.370	.582	.756	.840	.100	.245	.535	.370
.50	.824	.597	.152	.874	.376	.590	.762	.842	.160	.350	.528	.430
.51	.862	.720	.325	.878	.377	.594	.765	.843	.220	.455	.503	.465
.52	.894	.825	.595	.881	.373	.593	.765	.841	.252	.537	.460	.467
.53	.904	.853	.717	.883	.364	.588	.761	.838	.247	.582	.400	.438
.54	.905	.860	.763	.884	.354	.579	.764	.833	.207	.594	.325	.376
.55	.906	.864	.783	.884	.342	.569	.746	.827	.153	.572	.252	.303
.56	.907	.867	.795	.883	.331	.559	.736	.821	.096	.525	.183	.225
.57	.907	.869	.799	.882	.317	.547	.724	.813	.054	.457	.125	.157
.58	.907	.870	.804	.880	.298	.529	.706	.802	.026	.385	.083	.107
.59	.907	.870	.811	.877	.276	.508	.685	.790	.011	.311	.053	.073
.60	.908	.876	.826	.876	.251	.481	.662	.775	.004	.245	.033	.048
.61	.908	.880	.835	.875	.225	.452	.636	.757	.000	.190	.020	.034
.62	.909	.881	.842	.874	.299	.423	.610	.739	.000	.145	.012	.024
.63	.909	.884	.848	.875	.217	.392	.577	.719	.000	.115	.007	.018
.64	.910	.885	.854	.876	.147	.359	.546	.698	.000	.098	.004	.015
.65	.910	.887	.859	.877	.125	.326	.515	.675	.000	.084	.001	.012
.66	.911	.891	.864	.880	.104	.297	.482	.652	.000	.075	.000	.010
.67	.913	.896	.869	.883	.086	.265	.450	.630	.000	.075	.000	.009
.68	.914	.900	.873	.885	.063	.235	.418	.603	.000	.071	.000	.008
.69	.915	.901	.877	.887	.033	.206	.385	.576	.000	.065	.000	.007
.70	.915	.904	.880	.888	.042	.279	.352	.550	.000	.067	.000	.007
.71	.915	.905	.883	.888	.032	.155	.322	.524	.000	.070	.000	.007
.72	.915	.906	.885	.889	.025	.133	.294	.496	.000	.075	.000	.007
.73	.915	.907	.885	.888	.018	.114	.266	.472	.000	.080	.000	.007
.74	.915	.907	.882	.887	.014	.097	.242	.448	.000	.084	.000	.008
.75	.914	.906	.882	.886	.010	.084	.220	.424	.000	.086	.000	.008
.80	.902	.875	.855	.863	.000	.033	.120	.310	.000	.109	.000	.013
1.00	.899	.860	.882	.820	.000	.002	.038	.158	.000	.215	.018	.100
1.20	.898	.882	.898	.850	.000	.002	.040	.161	.007	.393	.158	.303
1.40	.880	.810	.855	.894	.002	.018	.100	.270	.058	.549	.404	.548
1.60	.882	.805	.855	.905	.007	.050	.190	.390	.162	.663	.612	.710
1.80	.900	.844	.907	.904	.008	.057	.201	.408	.299	.740	.740	.791
2.00	.897	.819	.909	.895	.011	.070	.228	.435	.422	.783	.817	.830
2.20	.888	.720	.900	.870	.021	.105	.277	.475	.518	.803	.840	.792
2.40	.865	.668	.890	.850	.037	.140	.320	.512	.597	.817	.862	.808
2.60	.840	.570	.860	.800	.050	.005	.339	.515	.634	.813	.870	.740
2.80	.460	.075	.620	.200	.022	.092	.135	.085	.270	.460	.520	.010
3.00	.282	.028	.465	.165	.033	.133	.200	.230	.260	.418	.620	.033
3.20	.252	.016	.420	.120	.054	.150	.247	.294	.204	.345	.642	.128
3.40	.175	.007	.370	.070	.048	.112	.165	.200	.131	.235	.631	.130
3.60	.150	.003	.351	.045	.003	.008	.016	.018	.121	.192	.634	.006
3.80	.168	.000	.368	.067	.007	.015	.035	.048	.125	.200	.600	.016
4.00	.182	.000	.370	.080	.006	.010	.020	.022	.123	.200	.557	.006
4.20	.120	.000	.300	.040	.005	.013	.022	.021	.070	.130	.422	.001
4.40	.020	.000	.140	.005	.001	.001	.000	.001	.002	.015	.135	.000
4.60	.002	.000	.010	.000	.000	.000	.000	.000	.000	.000	.012	.000
4.80	.000	.000	.000	.000	.000	.000	.000	.000	.000	.000	.002	.000
5.00	.000	.000	.000	.000	.000	.000	.000	.000	.000	.000	.000	.000

	Transmittance											
	Corning Glass Number											
λ(nm)	4303	4305	4308	4309	4445	4602	4784	5030	5031	5070	5071	5073
.22	.000	.000	.000	.000	.000	.000	.000	.000	.000	.000	.000	.000
.24	.000	.000	.000	.000	.000	.000	.000	.000	.000	.000	.000	.000
.26	.000	.000	.000	.000	.000	.000	.000	.000	.000	.000	.000	.000
.28	.000	.000	.000	.000	.000	.000	.000	.000	.000	.000	.000	.000
.30	.000	.000	.000	.000	.000	.001	.000	.000	.016	.000	.000	.000
.32	.000	.000	.000	.022	.001	.106	.000	.000	.145	.012	.045	.090
.34	.011	.060	.190	.394	.018	.505	.009	.038	.420	.310	.330	.540
.36	.188	.380	.580	.740	.114	.755	.200	.285	.685	.628	.340	.757
.38	.390	.590	.740	.831	.248	.827	.450	.595	.820	.745	.420	.810
.40	.545	.723	.826	.884	.400	.835	.596	.770	.884	.712	.665	.830
.41	.588	.750	.840	.887	.454	.845	.627	.799	.894	.600	.712	.786
.42	.624	.770	.850	.890	.505	.846	.648	.808	.895	.430	.716	.705
.43	.654	.786	.857	.892	.550	.851	.666	.797	.890	.290	.694	.620
.44	.680	.798	.862	.894	.593	.856	.680	.767	.872	.170	.655	.525
.45	.698	.809	.867	.897	.631	.857	.697	.738	.865	.097	.612	.436
.46	.712	.815	.869	.898	.659	.856	.717	.702	.864	.055	.568	.365
.47	.715	.814	.866	.897	.678	.861	.735	.628	.845	.035	.531	.313
.48	.705	.802	.856	.894	.689	.866	.750	.522	.805	.023	.501	.275
.49	.678	.780	.838	.885	.687	.869	.763	.406	.750	.017	.482	.252
.50	.636	.740	.810	.872	.673	.870	.768	.288	.684	.015	.469	.237
.51	.570	.685	.770	.850	.641	.869	.767	.186	.601	.013	.463	.231
.52	.480	.610	.714	.817	.586	.866	.753	.105	.495	.013	.461	.230
.53	.387	.525	.650	.781	.520	.863	.725	.053	.388	.014	.464	.235
.54	.288	.430	.576	.736	.437	.865	.676	.022	.295	.016	.473	.245
.55	.205	.340	.502	.683	.355	.869	.615	.007	.198	.018	.486	.260
.56	.132	.255	.422	.627	.275	.868	.525	.000	.113	.023	.502	.278
.57	.082	.184	.345	.565	.202	.863	.427	.000	.057	.030	.522	.300
.58	.047	.127	.277	.505	.144	.856	.328	.000	.025	.038	.540	.324
.59	.026	.087	.218	.447	.102	.848	.235	.000	.008	.048	.555	.348
.60	.013	.057	.170	.393	.068	.838	.157	.000	.000	.058	.571	.373
.61	.006	.036	.131	.341	.046	.824	.102	.000	.000	.070	.587	.395
.62	.001	.022	.100	.296	.031	.806	.058	.000	.000	.083	.600	.415
.63	.000	.013	.074	.256	.020	.787	.032	.000	.000	.096	.612	.435
.64	.000	.007	.056	.221	.013	.767	.017	.000	.000	.108	.622	.450
.65	.000	.004	.042	.191	.008	.745	.007	.000	.000	.120	.633	.466
.66	.000	.002	.033	.167	.006	.722	.002	.000	.000	.135	.644	.482
.67	.000	.001	.025	.146	.003	.695	.000	.000	.000	.148	.658	.498
.68	.000	.000	.020	.128	.001	.665	.000	.000	.000	.165	.674	.515
.69	.000	.000	.016	.116	.000	.634	.000	.000	.000	.182	.686	.531
.70	.000	.000	.013	.104	.000	.600	.000	.000	.000	.200	.700	.548
.71	.000	.000	.010	.096	.000	.565	.000	.000	.000	.220	.712	.566
.72	.000	.000	.009	.088	.000	.531	.000	.004	.024	.245	.725	.586
.73	.000	.000	.007	.083	.000	.496	.000	.047	.119	.268	.736	.606
.74	.000	.000	.006	.079	.000	.463	.000	.190	.330	.295	.749	.675
.75	.000	.000	.005	.075	.000	.430	.000	.440	.580	.323	.759	.642
.80	.000	.000	.008	.080	.003	.258	.000	.890	.917	.505	.815	.750
1.00	.000	.005	.045	.188	.056	.019	.000	.753	.868	.860	.885	.872
1.20	.013	.060	.166	.375	.269	.009	.000	.455	.720	.890	.897	.890
1.40	.080	.180	.342	.542	.527	.016	.003	.100	.285	.892	.902	.892
1.60	.210	.342	.500	.658	.701	.035	.038	.056	.162	.890	.902	.892
1.80	.350	.482	.617	.732	.792	.059	.142	.052	.140	.878	.890	.877
2.00	.475	.590	.690	.772	.839	.048	.275	.075	.175	.860	.865	.860
2.20	.560	.653	.720	.778	.850	.038	.345	.172	.295	.840	.830	.840
2.40	.635	.704	.752	.791	.870	.045	.404	.330	.483	.812	.795	.805
2.60	.663	.710	.748	.770	.861	.066	.340	.382	.530	.804	.752	.775
2.80	.260	.370	.370	.250	.280	.018	.045	.030	.030	.550	.500	.500
3.00	.249	.250	.203	.212	.395	.000	.000	.010	.001	.390	.308	.340
3.20	.202	.212	.145	.168	.451	.000	.000	.065	.026	.222	.145	.180
3.40	.135	.136	.084	.107	.449	.000	.000	.002	.006	.120	.070	.090
3.60	.124	.120	.078	.077	.470	.000	.000	.000	.000	.078	.032	.063
3.80	.132	.125	.082	.083	.470	.000	.000	.000	.000	.078	.029	.060
4.00	.132	.131	.094	.090	.448	.000	.000	.000	.000	.093	.037	.075
4.20	.078	.073	.050	.042	.320	.000	.000	.000	.000	.079	.020	.048
4.40	.008	.009	.008	.002	.100	.000	.000	.000	.000	.020	.004	.014
4.60	.000	.000	.000	.000	.007	.000	.000	.000	.000	.002	.000	.000
4.80	.000	.000	.000	.000	.000	.000	.000	.000	.000	.000	.000	.000
5.00	.000	.000	.000	.000	.000	.000	.000	.000	.000	.000	.000	.000

TRANSMISSION OF CORNING COLORED FILTERS (Continued)

λ(nm)	Transmittance Corning Glass Number 5113	5120	5300	5330	5433	5543	5562	5572	5840	5850	5860	5874
.22	.000	.000	.000	.000	.000	.000	.000	.000	.000	.000	.000	.000
.24	.000	.000	.000	.000	.000	.000	.000	.000	.000	.000	.000	.000
.26	.000	.000	.000	.000	.000	.000	.000	.000	.000	.000	.000	.000
.28	.000	.000	.000	.000	.000	.000	.000	.000	.000	.000	.000	.000
.30	.000	.000	.000	.002	.000	.000	.000	.000	.001	.039	.000	.000
.32	.000	.000	.000	.250	.000	.000	.000	.045	.242	.490	.008	.031
.34	.000	.000	.000	.622	.000	.000	.012	.325	.600	.790	.179	.228
.36	.035	.018	.000	.796	.100	.120	.205	.660	.682	.858	.340	.447
.38	.200	.610	.000	.035	.350	.380	.493	.803	.392	.830	.085	.378
.40	.371	.670	.000	.865	.585	.600	.717	.874	.000	.788	.000	.032
.41	.371	.790	.000	.850	.636	.635	.748	.873	.000	.720	.000	.004
.42	.337	.805	.000	.823	.665	.646	.761	.865	.000	.630	.000	.000
.43	.272	.560	.000	.783	.674	.635	.759	.857	.000	.522	.000	.000
.44	.198	.386	.000	.725	.665	.602	.742	.845	.000	.410	.000	.000
.45	.118	.485	.000	.650	.635	.550	.713	.832	.000	.290	.000	.000
.46	.055	.475	.000	.555	.577	.465	.662	.808	.000	.175	.000	.000
.47	.013	.370	.008	.455	.467	.335	.565	.765	.000	.125	.000	.000
.48	.000	.385	.026	.355	.327	.190	.435	.693	.000	.022	.000	.000
.49	.000	.685	.085	.270	.205	.090	.300	.605	.000	.005	.000	.000
.50	.000	.660	.125	.197	.120	.040	.197	.523	.000	.000	.000	.000
.51	.000	.390	.106	.145	.060	.013	.110	.430	.000	.000	.000	.000
.52	.000	.305	.094	.110	.024	.002	.051	.328	.000	.000	.000	.000
.53	.000	.175	.064	.085	.008	.000	.022	.247	.000	.000	.000	.000
.54	.000	.610	.149	.068	.005	.000	.012	.216	.000	.000	.000	.000
.55	.000	.817	.147	.055	.006	.000	.015	.246	.000	.000	.000	.000
.56	.000	.230	.087	.043	.007	.000	.018	.285	.000	.000	.000	.000
.57	.000	.125	.013	.032	.000	.000	.011	.258	.000	.000	.000	.000
.58	.000	.000	.000	.022	.000	.000	.002	.175	.000	.000	.000	.000
.59	.000	.006	.000	.016	.000	.000	.000	.116	.000	.000	.000	.000
.60	.000	.180	.000	.012	.000	.000	.000	.113	.000	.000	.000	.000
.61	.000	.545	.000	.009	.000	.000	.000	.123	.000	.000	.000	.000
.62	.000	.825	.000	.010	.000	.000	.000	.126	.000	.000	.000	.000
.63	.000	.838	.000	.013	.000	.000	.000	.120	.000	.000	.000	.000
.64	.000	.878	.000	.015	.000	.000	.000	.111	.000	.000	.000	.000
.65	.000	.893	.000	.015	.000	.000	.000	.120	.000	.000	.000	.000
.66	.000	.883	.000	.014	.000	.000	.000	.165	.000	.000	.000	.000
.67	.000	.820	.000	.013	.000	.000	.000	.265	.000	.000	.000	.000
.68	.000	.705	.000	.015	.000	.000	.000	.425	.000	.000	.000	.000
.69	.000	.743	.000	.022	.000	.000	.001	.615	.000	.029	.000	.000
.70	.000	.860	.000	.041	.000	.000	.004	.756	.007	.160	.000	.017
.71	.000	.876	.000	.085	.000	.001	.004	.837	.020	.385	.000	.075
.72	.000	.815	.000	.165	.000	.002	.004	.874	.037	.615	.000	.168
.73	.000	.435	.000	.285	.000	.002	.002	.889	.060	.760	.000	.257
.74	.000	.045	.000	.460	.000	.002	.002	.895	.086	.843	.000	.320
.75	.000	.055	.000	.645	.000	.001	.001	.898	.080	.878	.000	.335
.80	.000	.030	.000	.900	.005	.002	.003	.895	.009	.890	.000	.218
1.00	.000	.860	.003	.800	.010	.015	.020	.880	.000	.716	.000	.050
1.20	.000	.770	.026	.570	.030	.020	.047	.690	.000	.169	.000	.011
1.40	.000	.550	.072	.410	.060	.040	.095	.640	.004	.042	.004	.012
1.60	.000	.620	.200	.405	.116	.060	.154	.625	.002	.036	.002	.010
1.80	.000	.580	.265	.425	.172	.090	.223	.635	.000	.048	.000	.010
2.00	.010	.580	.374	.476	.343	.247	.410	.750	.000	.168	.000	.018
2,20	.071	.747	.533	.535	.475	.400	.528	.780	.000	.338	.000	.036
2.40	.190	.400	.370	.552	.575	.520	.600	.780	.000	.492	.000	.074
2.60	.203	.530	.523	.540	.580	.532	.603	.745	.000	.510	.000	.088
2.80	.100	.250	.265	.007	.162	.132	.360	.470	.000	.170	.000	.003
3.00	.080	.080	.202	.012	.195	.151	.280	.315	.000	.128	.000	.020
3.20	.068	.072	.180	.082	.131	.105	.172	.200	.000	.084	.000	.053
3.40	.030	.029	.131	.003	.079	.065	.100	.100	.000	.065	.000	.050
3.60	.021	.017	.103	.000	.061	.030	.053	.049	.002	.070	.000	.000
3.80	.023	.021	.093	.000	.068	.032	.040	.037	.001	.082	.000	.000
4.00	.040	.025	.112	.000	.071	.039	.050	.042	.005	.090	.000	.000
4.20	.019	.010	.069	.000	.020	.015	.017	.020	.002	.055	.000	.000
4.40	.001	.001	.007	.000	.000	.000	.000	.002	.000	.002	.000	.000
4.60	.000	.000	.000	.000	.000	.000	.000	.000	.000	.000	.000	.000
4.80	.000	.000	.000	.000	.000	.000	.000	.000	.000	.000	.000	.000
5.00	.000	.000	.000	.000	.000	.000	.000	.000	.000	.000	.000	.000

	Transmittance											
	Corning Glass Number											
λ(nm)	5900	5970	7380	7740	7905	7910	8364	9780	9782	9788	9830	9863
.22	.000	.000	.000	.000	.000	.012	.000	.000	.000	.000	.000	.000
.24	.000	.000	.000	.000	.360	.505	.000	.000	.000	.000	.000	.054
.26	.000	.000	.000	.000	.495	.780	.000	.000	.000	.000	.000	.482
.28	.000	.000	.000	.004	.590	.855	.000	.000	.000	.000	.000	.731
.30	.000	.000	.000	.321	.720	.877	.000	.000	.000	.000	.000	.831
.32	.000	.138	.000	.722	.825	.900	.000	.000	.000	.000	.000	.862
.34	.008	.600	.000	.851	.880	.903	.002	.015	.000	.060	.001	.854
.36	.150	.799	.440	.889	.910	.905	.083	.290	.060	.470	.059	.816
.38	.445	.742	.795	.900	.915	.906	.136	.590	.445	.770	.160	.620
.40	.678	.190	.892	.916	.920	.920	.296	.725	.747	.885	.130	.090
.41	.688	.029	.904	.915	.920	.920	.273	.705	.790	.895	.044	.018
.42	.635	.000	.910	.915	.920	.921	.232	.770	.818	.902	.004	.003
.43	.586	.000	.913	.914	.920	.923	.191	.778	.836	.905	.000	.000
.44	.522	.000	.915	.913	.920	.924	.157	.801	.847	.906	.000	.000
.45	.458	.000	.916	.913	.922	.925	.144	.814	.855	.906	.000	.000
.46	.400	.000	.917	.914	.922	.925	.140	.823	.860	.906	.000	.000
.47	.350	.000	.917	.915	.922	.925	.141	.832	.863	.906	.000	.000
.48	.306	.000	.917	.915	.923	.925	.146	.839	.863	.905	.000	.000
.49	.275	.000	.918	.915	.930	.926	.152	.843	.859	.904	.000	.000
.50	.246	.000	.918	.915	.925	.926	.166	.843	.848	.900	.014	.000
.51	.223	.000	.919	.915	.925	.927	.178	.838	.825	.893	.180	.000
.52	.196	.000	.919	.916	.925	.928	.190	.824	.784	.880	.175	.000
.53	.172	.000	.919	.916	.923	.928	.196	.798	.720	.862	.018	.000
.54	.154	.000	.919	.916	.926	.929	.198	.756	.627	.831	.000	.000
.55	.148	.000	.920	.917	.923	.929	.197	.697	.515	.787	.050	.000
.56	.151	.000	.920	.917	.925	.930	.199	.615	.380	.728	.265	.000
.57	.146	.000	.919	.918	.925	.930	.206	.518	.255	.655	.165	.000
.58	.125	.000	.918	.919	.925	.930	.217	.414	.150	.570	.035	.000
.59	.102	.000	.918	.920	.925	.930	.222	.302	.075	.475	.004	.000
.60	.093	.000	.920	.920	.925	.930	.215	.215	.032	.380	.000	.000
.61	.087	.000	.920	.920	.925	.930	.196	.135	.010	.290	.000	.000
.62	.081	.000	.920	.920	.925	.930	.175	.080	.002	.210	.000	.000
.63	.070	.000	.920	.919	.926	.930	.161	.042	.000	.145	.000	.000
.64	.061	.000	.920	.919	.927	.931	.156	.021	.000	.094	.000	.000
.65	.055	.000	.920	.918	.927	.931	.162	.008	.000	.059	.000	.000
.66	.055	.000	.920	.917	.927	.932	.176	.003	.000	.035	.000	.000
.67	.059	.000	.920	.916	.928	.932	.200	.000	.000	.020	.000	.000
.68	.065	.000	.920	.916	.927	.932	.228	.000	.000	.010	.075	.022
.69	.068	.007	.921	.915	.927	.932	.248	.000	.000	.005	.380	.106
.70	.068	.036	.921	.915	.928	.932	.251	.000	.000	.001	.642	.234
.71	.066	.085	.922	.914	.926	.933	.237	.000	.000	.000	.694	.332
.72	.064	.145	.922	.912	.926	.933	.223	.000	.000	.000	.666	.383
.73	.060	.222	.922	.910	.926	.933	.210	.000	.000	.000	.607	.384
.74	.057	.323	.921	.909	.927	.934	.197	.000	.000	.000	.531	.358
.75	.055	.385	.921	.907	.928	.934	.180	.000	.000	.000	.445	.322
.80	.050	.287	.918	.890	.930	.932	.110	.000	.000	.000	.045	.175
1.00	.085	.032	.910	.860	.930	.928	.032	.000	.000	.000	.000	.119
1.20	.180	.021	.910	.860	.925	.928	.032	.000	.000	.005	.007	.016
1.40	.295	.109	.906	.870	.925	.930	.062	.018	.000	.080	.000	.005
1.60	.405	.088	.910	.892	.931	.930	.137	.131	.011	.266	.000	.007
1.80	.495	.040	.903	.896	.931	.930	.182	.317	.085	.430	.157	.011
2.00	.590	.008	.900	.897	.934	.929	.171	.440	.216	.512	.041	.029
2.20	.628	.002	.898	.875	.934	.835	.184	.440	.278	.455	.000	.048
2.40	.640	.009	.890	.850	.930	.890	.225	.440	.325	.433	.000	.060
2.60	.630	.018	.860	.820	.920	.780	.258	.280	.212	.252	.057	.051
2.80	.470	.030	.375	.140	.908	.180	.145	.060	.040	.060	.020	.000
3.00	.248	.030	.425	.360	.880	.695	.130	.000	.000	.000	.000	.000
3.20	.121	.030	.380	.490	.861	.760	.125	.000	.000	.000	.000	.000
3.40	.032	.020	.310	.270	.670	.620	.099	.000	.000	.000	.000	.000
3.60	.010	.015	.270	.010	.111	.080	.115	.000	.000	.000	.000	.000
3.80	.010	.020	.275	.040	.270	.240	.142	.000	.000	.000	.000	.000
4.00	.010	.030	.260	.013	.170	.150	.172	.000	.000	.000	.000	.000
4.20	.006	.017	.180	.026	.250	.230	.158	.000	.000	.000	.000	.000
4.40	.001	.002	.040	.004	.085	.080	.078	.000	.000	.000	.000	.000
4.60	.000	.000	.000	.000	.050	.020	.010	.000	.000	.000	.000	.000
4.80	.000	.000	.000	.000	.000	.000	.003	.000	.000	.000	.000	.000
5.00	.000	.000	.000	.000	.000	.000	.000	.000	.000	.000	.000	.000

TRANSMISSION OF WRATTEN FILTERS

Compiled by Allie C. Peed, Jr. for The Eastman Kodak Company

Data condensed from Kodak Wratten Filters for Scientific and Technical Use published by the Eastman Kodak Company, manufacturers of the filters.

The following pages give (1) percentage luminous transmittance at wave lengths from 400 to 700μ at intervals of 10μ for the standard illuminant "C" adopted by the International Commission of Illumination, (2) dominant wavelength in millimicrons, and (3) percentage of excitation purity. Values of wave length followed by "c" indicate the complementary wave lengths of purple filters which do not have a dominant wave length.

All colorimetric specifications are based on the 1931 standard ICI colorimetric and luminosity data.

The transmittance data are given as representing standard samples of the filters. They are intended only for the information of users in choosing filters which will meet their requirements. Values taken from the tables of data should not be used by research workers as representing precisely the absorption characteristics of a particular filter. If such precise data are needed, they should be determined for the particular filter being used.

Where the spectra extend into the ultraviolet this fact is indicated by an asterisk (*) in the transmission tables immediately beneath the filter number, and quantitative data are not given. The manufacturer should be consulted for this information. Transmission in the ultraviolet of wave lengths less than 330μ will be eliminated in the case of cemented filters, as glass absorbs ultraviolet radiation of wave lengths shorter than about 330μ.

Stability ratings are given as three letter combinations following the filter description in the table below. In establishing the stability classifications each filter is exposed to a selected light source for a specific time interval. The following grading system is used to describe the result:

Class A — stable
Class B — relatively stable
Class C — somewhat unstable
Class D — unstable

The classification letters, for example, AAA, describe the stability to the following three exposure tests in this order:

1. Two weeks' exposure to daylight in a south window
2. Twenty-four hours' exposure to a "Fade-Ometer"
3. Two weeks' exposure at two feet from a 1000-watt tungsten lamp.

Filters are supplied in two forms: as lacquered gelatin film, or as a gelatin film cemented between pieces of optical glass. Filters in glass are cemented between sheets of plane-parallel glass, which is surfaced in quantities and is of sufficient accuracy for general photographic work, and for most scientific purposes.

Most Wratten Gelatin Filters are stocked in 2- or 3-inch squares. Stocks of 2- or 3-inch square filters cemented in glass are maintained only in filters usually used for general photographic work.

The booklet "Kodak Filters and Lens Attachments" gives more valuable information on this subject.

FILTER DATA

No.	Description, use, and stability
	Colorless
0	For compensating thickness of other gelatin filters in optical systems, AAA.
1	Absorbs ultraviolet below 360 mμ, DDD.
1A	Kodak Skylight Filter — Reduces excess bluishness in outdoor color photographs in open shade under a clear, blue sky, ACA.
	Yellow
2B	Absorbs ultraviolet below 410 mμ, ACA.
3	Light yellow, CCD.
3N5	No. 3 plus 0.5 neutral density, AAA.
4	Light yellow — Approximate correction on panchromatic materials for outdoor scenes, including sky, CCC.
6	K1 — Light yellow — Partial correction outdoors, BBA.
8	K2 — Yellow — Full correction outdoors on Type B panchromatic materials. Widely used for proper sky, cloud, and foliage rendering. Green separation for Fluorescence Process, AAA.
8N5	No. 8 plus 0.5 neutral density, AAA.
9	K3 — Deep yellow. Moderate contrast in outdoor photography (with black-and-white films),AAA.
11	X1 — Greenish yellow. Correction for tungsten light on Type B panchromatic materials; also for daylight correction with Type C panchromatic materials in making outdoor portraits, darkening skies, or lightening foliage, AAA.
12	Minus blue. Haze cutting in aerial photography, AAA.
13	X2 — Yellow green. Correction for Type C panchromatic materials in tungsten light, ABA.
15	G — Deep yellow. Overcorrection in landscape photography. Contrast control in copying and in aerial infrared photography, AAA.
16	Blue absorption, AAB.
18A	Transmits ultraviolet and infrared only (glass), AAA.
	Oranges and Reds
21	Blue and blue-green absorption, CBB.
22	Yellow-orange. For increasing contrast in blue preparations in microscopy. Mercury yellow, BAC.
23A	Light red. Two-color projection — contrast effects, BAB.
24	Red for two-color photography (daylight or tungsten). White-flame-arc tricolor projection, AAB.
25	A — Tricolor red for direct color separation. Contrast effects in commercial photography and in outdoor scenes. Two-color general viewing. Aerial infrared photography and haze cutting, AAA.
26	Stereo red, AAA.
29	Red color separation from transparencies and for the Kodak Fluorescence Process. Strong contrast effects. Copying blueprints. Tungsten tricolor projection, AAA.
	Magentas and Violets
30	Green absorption, BBC.
31	Green absorption, CCA.
32	Minus green, CCD.
33	Strong green absorption, CCB.
34	Violet, CDD.
34A	Blue separation — Kodak Fluorescence Process, DCC.
35	Contrast in microscopy, CDD.
36	Dark violet, CCC.
	Blues and Blue-greens
38	Red absorption, BCA.
38A	Red absorption. Increasing contrast in visual microscopy, BBB.
39	Contrast control in printing motion-picture duplicates (glass) AAA.
40	Green for two-olor photography (tungsten), CBC.
44	Minus red — Two-color general viewing, DDD.
44A	Minus red, DDD.
45	Contrast in microscopy, DDD.
45A	Highest resolving power in visual microscopy, CDC.
46	Blue projection (experimental), DDD.
47	Tricolor blue for direct color separation and from Kodak Ektacolor Film for Dye Transfer. Contrast effects in commercial photography. Tungsten and white-flame-arc tricolor projection, BBC.
47B	Tricolor blue for color separation from transparencies and from Kodak Ektacolor Film for Graphic Arts, BBB.
48	Green and red absorption, CBC.
48A	Green and red absorption, AAB.
49	Dark blue, BCB.
49B	Very dark blue, BBB.
50	Very dark blue. Mercury violet, CCC.
	Greens
52	Light green, AAB.

TRANSMISSION OF WRATTEN FILTERS (Continued)

FILTER DATA

No.	Description, use, and stability
53	Medium green, CCB
	Very dark green, AAA.
55	Stereo green, BBC.
56	Very light green, CBC.
57	Green for two-color photography (daylight), CBC.
57A	Light green, BBC.
58	Tricolor green for direct color separation. Contrast effects in commercial photography and microscopy, BBC.
59	Green for tricolor projection (white-flame-arc), BBB.
59A	Very light green, BBB.
60	Green for two-color photography (tungsten), BDC.
61	Green color separation from transparencies and Kodak Ektacolor Film. Tricolor projection (tungsten), ABC.
64	Red absorption (light), CDB.
65	Red absorption, ADB.
65A	Red absorption, CCD.
66	Contrast effects in microscopy and medical photography, DDC.
67A	Red absorption (light). Two-color projection, CDC.
	Narrow-band
70	Dark red. Infrared photography. Color separation for Kodak Ektacolor Film (with tungsten), ABC.
72B	Dark orange-yellow, CCC.
73	Dark yellow-green, ABB.
74	Dark green. Mercury green, BBC.
75	Dark blue-green, ACC.
76	Dark violet (compound filter), DDD.
77	Transmits 546 mμ mercury line (glass plus gelatin), AAA.
77A	Transmits 546 mμ mercury line (glass plus gelatin), AAA.
	Photometrics
78	Bluish. Photometric filter (visual), BAB.
78AA	Bluish. Photometric filter (visual), BAA.
78A	Bluish. Photometric filter (visual), AAA.
78B	Bluish. Photometric filter (visual), AAA.
78C	Bluish. Photometric filter (visual), BAA.
86	Yellowish. Photometric filter (visual), BBA.
86A	Yellowish. Photometric filter (visual), AAA.
†86B	Yellowish. Photometric filter (visual), BCA.
†86C	Yellowish. Photometric filter (visual), AAA.
	Light Balancing
80A	For Kodachrome Film, Daylight Type, and photographic flood lamps, ABA.
81	Yellowish. For warmer color rendering.
81A	Yellowish. For Kodak Ektachrome Film, Type B, with photographic flood lamps.
81B	Yellowish. For warmer color rendering.
81C	Yellowish. For Kodachrome Film, Type A, with flash lamps.
81D	Yellowish. For Kodachrome Film, Type A, with flash lamps.
81EF	Yellowish. For Kodak Ektachrome Film, Type B, with flash lamps.
82	Bluish. For cooler color rendering.
82A	Bluish. For Kodachrome Film, Type A, with 3200 K lamps.
82B	Bluish. For cooler color tendering.
82C	Bluish. For cooler color rendering.
83	Yellowish. For 16 mm Commercial Kodachrome Film and daylight exposure, BBB.
85	Orange. For Type A Kodak color films and daylight exposure, BAA.
85B	Orange. For Kodak Ektachrome Film, Type B, and daylight exposure, BAB.
	Miscellaneous
79	Photographic sensitometry. Corrects 2360 K to 5500 K, AAA.
87	For infrared photography. Absorbs visual.
87C	Absorbs visual, transmits infrared.
88A	For infrared photography. Absorbs visual.
89B	For infrared photography, AAA.
90	Narrow-band viewing filter for judging brightness scale of scenes, CCD.
96	Neutral filters for controlling luminance, AAB.
97	Dichroic absorption, AAA.
102	Correction filter for Barrier-layer photocell, ABA.
106	Correction filter for S-4 type photocell, AAA.

	Percent transmittance												
Wave length	No. 0 *	No. 1 *	No. 1A *	No. 2B *	No. 3	No. 3N5 *	No. 4	No. 6 *	No. 8	No. 8N5	No. 9	No. 11 *	No. 12 *
400	88.0	85.0	59.0	19.0	—	—	—	7.40	—	—	—	—	—
10	88.5	85.5	76.0	48.0	—	—	—	8.32	—	—	—	—	—
20	88.9	86.0	82.0	67.0	—	—	—	10.4	—	—	—	0.16	—
30	89.3	86.5	84.6	75.3	0.36	—	—	13.5	—	—	—	0.29	—
40	89.6	87.0	86.0	80.0	1.78	—	—	18.9	—	—	—	0.56	—
50	89.8	87.4	86.8	83.0	11.5	1.59	—	27.6	—	—	—	1.32	—
60	89.9	87.8	87.2	85.2	38.0	9.40	6.9	39.0	0.25	0.16	—	4.00	—
70	90.1	88.2	87.5	86.7	68.0	18.5	42.0	52.3	5.50	2.0	1.78	12.0	—
80	90.3	88.5	87.3	88.1	80.8	23.5	74.0	65.8	19.0	6.3	8.31	26.0	—
90	90.4	88.7	86.8	88.8	85.2	25.5	84.7	76.8	41.0	13.2	20.7	43.7	—
500	90.5	88.9	86.3	89.5	86.9	26.3	87.5	83.5	63.5	20.0	34.5	55.0	1.50
10	90.6	89.1	85.5	89.9	87.8	26.7	88.5	87.0	78.0	24.3	48.8	60.0	17.3
20	90.7	89.3	84.8	90.3	88.4	27.0	89.1	88.4	84.1	26.7	62.0	60.2	55.0
30	90.7	89.5	84.3	90.5	89.0	27.2	89.4	89.0	86.5	28.0	76.0	57.8	77.8
40	90.8	89.7	84.0	90.6	89.5	27.5	89.6	89.4	87.7	28.6	83.8	54.2	86.0
50	90.8	89.9	83.9	90.7	89.8	27.8	89.8	89.7	88.4	29.0	87.0	50.0	88.4
60	90.9	90.1	84.1	90.8	90.1	27.9	90.0	89.9	88.8	29.3	88.3	44.8	89.4
70	90.9	90.2	84.8	90.9	90.4	28.0	90.2	90.1	89.2	29.5	88.8	38.9	89.7
80	90.9	90.3	86.0	90.9	90.6	28.4	90.4	90.3	89.5	29.6	89.1	33.1	90.1
90	91.0	90.4	87.4	91.0	90.7	29.0	90.6	90.5	89.8	29.8	89.3	27.6	90.3
600	91.0	90.5	88.5	91.1	90.8	29.5	90.8	90.6	90.1	29.9	89.5	22.7	90.4
10	91.0	90.5	89.5	91.2	90.9	29.5	90.9	90.7	90.3	29.6	89.7	19.0	90.5
20	91.0	90.6	90.2	91.3	91.0	29.3	91.0	90.8	90.5	29.4	89.8	14.9	90.7
30	91.0	90.6	90.6	91.3	91.0	29.1	91.1	90.9	90.7	29.1	89.9	11.4	90.8
40	91.1	90.7	90.8	91.4	91.1	29.0	91.2	91.0	90.9	28.8	90.0	9.10	90.9
50	91.1	90.7	91.0	91.4	91.2	20.4	91.3	91.1	91.0	28.9	90.1	8.05	91.0
60	91.1	90.8	91.1	91.5	91.3	29.6	91.4	91.2	91.1	29.2	90.1	7.50	91.1
70	91.1	90.8	91.1	91.5	91.4	29.8	91.5	91.2	91.2	29.4	90.2	7.05	91.2
80	91.1	90.9	91.1	91.6	91.5	30.0	91.5	91.3	91.3	29.5	90.2	6.50	91.2
90	91.1	90.9	91.1	91.7	91.6	30.2	91.6	91.4	91.4	29.7	90.3	6.10	91.2
700	91.1	91.0	91.1	91.8	91.7	31.0	91.6	91.3	91.5	30.2	90.3	6.20	91.3
Luminous transmit.	90.8	89.9	85.9	90.5	88.3	27.4	87.8	87.5	82.7	27.0	76.6	40.2	73.8
Dominant wave lgth.	571.0	575.0	498.0	570.0	569.5	570.5	569.5	570.3	571.8	572.0	574.4	550.3	576.1
Excitation purity.	0.8	1.5	1.2	5.7	50.0	56.3	64.0	44.7	85.2	84.0	91.4	60.7	97.8

* Some transmission below 400 mμ. Consult the manufacturer.

TRANSMISSION OF WRATTEN FILTERS (Continued)

Wave length	Percent transmittance												
	No. 13 *	No. 15 *	No. 16	No. 18A *	No. 21	No. 22	No. 23A	No. 24	No. 25	No. 26	No. 29	No. 30 *	No. 31 *
400	—	—	—	—	—	—	—	—	—	—	—	48.6	13.8
10	—	—	—	—	—	—	—	—	—	—	—	47.4	14.5
20	—	—	—	—	—	—	—	—	—	—	—	48.5	16.4
30	0.18	—	—	—	—	—	—	—	—	—	—	50.1	25.5
40	0.50	—	—	—	—	—	—	—	—	—	—	49.4	42.7
50	1.35	—	—	—	—	—	—	—	—	—	—	43.0	50.2
60	4.08	—	—	—	—	—	—	—	—	—	—	26.5	40.4
70	11.0	—	—	—	—	—	—	—	—	—	—	13.8	22.6
80	23.5	—	—	—	—	—	—	—	—	—	—	5.00	8.20
90	39.0	—	—	—	—	—	—	—	—	—	—	0.63	1.85
500	50.8	—	—	—	—	—	—	—	—				0.12
10	55.2	1.00	—	—	—	—	—	—	—	—	—	—	—
20	56.5	16.0	3.00	—	—	—	—	—	—	—	—	—	—
30	55.0	52.1	22.0	—	—	—	—	—	—	—	—	—	—
40	51.0	70.7	48.0	—	2.50	—	—	—	—	—	—	—	—
50	46.0	84.3	69.5	—	29.0	0.25	—	—	—	—	—	—	—
60	39.2	87.5	79.5	—	65.0	19.0	—	—	—	—	—	0.10	—
70	32.0	88.7	84.0	—	80.6	60.0	11.0	—	—	—	—	10.0	—
80	25.1	89.3	86.3	—	85.4	81.0	47.0	4.55	—	—	—	45.0	—
90	18.2	89.7	87.8	—	87.3	87.0	69.6	37.3	12.6	2.90	—	76.0	0.63
600	13.5	90.0	89.0	—	88.1	88.5	82.7	72.3	50.0	30.0	—	87.4	26.0
10	9.60	90.1	89.6	—	88.7	89.0	85.8	82.9	75.0	63.2	10.0	89.5	67.2
20	6.40	90.2	90.0	—	89.0	89.5	87.2	86.4	82.6	78.9	45.3	90.2	84.0
30	3.66	90.3	90.2	—	89.5	89.8	87.9	87.8	85.5	84.0	71.4	90.5	88.1
40	2.20	90.4	90.3	—	89.9	90.0	88.5	88.5	86.7	86.1	82.7	90.7	89.8
50	1.58	90.5	90.4	—	90.2	90.1	89.0	89.0	87.6	87.2	86.6	90.8	90.2
60	1.74	90.6	90.5	—	90.4	90.2	89.4	89.3	88.2	88.1	88.4	90.9	90.4
70	2.62	90.6	90.6	—	90.5	90.3	89.6	89.7	88.5	88.5	89.4	91.0	90.5
80	3.55	90.7	90.7	—	90.5	90.4	89.8	89.9	89.0	88.9	90.0	91.1	90.7
90	4.48	90.7	90.8	0.25	90.6	90.5	90.0	90.2	89.3	89.2	90.3	91.1	90.8
700	5.25	90.8	90.8	1.20	90.6	90.6	90.2	90.3	89.5	89.5	90.4	91.1	91.0
Luminous transmit.	34.5	66.2	57.7	0.0014	45.6	35.8	25.0	17.8	14.0	11.7	6.3	26.6	12.9
Dominant wave lgth.	542.0	579.3	582.7	700.0	588.9	595.1	602.7	610.6	615.1	619.0	631.6	498.6_c	513.1_c
Excitation purity.	57.5	99.0	99.3	100.0	99.9	99.9	100.0	100.0	100.0	100.0	100.0	62.4	81.9

Wave length	Percent transmittance												
	No. 32 *	No. 33 *	No. 34 *	No. 34A *	No. 35 *	No. 36 *	No. 38 *	No. 38A *	No. 39 *	No. 40	No. 44 *	No. 44A *	No. 45
400	38.0	0.85	64.0	—	48.0	36.5	60.5	33.4	85.2	—	0.44	2.52	—
10	37.9	0.71	70.1	0.1	57.0	45.5	66.5	41.2	78.2	—	0.36	3.39	—
20	40.0	1.17	72.0	40.0	57.6	45.5	72.5	53.0	70.5	—	0.63	6.30	—
30	43.0	1.69	68.4	69.7	47.5	32.7	75.3	58.0	63.3	—	3.63	17.4	—
40	55.5	5.36	58.2	68.7	29.5	15.2	76.2	58.8	53.6	—	13.1	32.7	5.00
50	66.0	14.3	42.3	56.2	12.3	3.7	75.9	57.6	42.5	—	25.4	41.8	19.0
60	66.0	12.4	25.2	40.5	3.5	0.35	74.8	55.2	28.5	3.16	36.5	48.1	29.5
70	57.0	5.00	12.1	23.8	0.25	—	73.4	51.9	17.3	21.6	46.5	51.7	34.4
80	40.0	0.50	2.7	9.2	—	—	71.6	48.5	10.2	44.7	53.6	52.9	35.7
90	21.0	—	0.2	2.3	—	—	69.5	44.6	4.00	61.4	56.8	52.2	34.5
500	9.56	—	—	0.33	—	—	66.7	40.2	1.33	70.2	55.8	49.8	29.7
10	2.51	—	—	—	—	—	63.9	35.8	0.35	72.4	50.9	44.8	21.5
20	0.13	—	—	—	—	—	60.8	31.7	—	70.5	42.1	36.8	11.5
30	—	—	—	—	—	—	57.0	27.2	—	64.8	30.5	26.8	3.80
40	—	—	—	—	—	—	52.6	22.3	—	55.5	18.6	16.8	0.85
50	—	—	—	—	—	—	48.0	17.6	—	44.2	8.99	8.20	—
60	—	—	—	—	—	—	42.8	12.9	—	32.5	3.59	2.95	—
70	—	—	—	—	—	—	37.0	8.78	—	20.3	0.80	0.91	—
80	—	—	—	—	—	—	30.6	5.65	—	9.56	—	0.10	—
90	—	—	—	—	—	—	25.5	3.48	—	3.20	—	—	—
600	6.04	—	—	—	—	—	20.9	2.09	—	1.10	—	—	—
10	41.0	0.80	—	0.13	—	—	16.8	1.15	—	0.32	—	—	—
20	75.0	24.9	—	1.0	—	—	12.9	0.59	—	—	—	—	—
30	86.1	60.8	—	6.3	—	—	10.0	0.28	—	—	—	—	—
40	89.0	78.0	0.4	22.0	—	—	7.79	0.13	—	—	—	—	—
50	90.0	85.0	4.0	45.0	0.1	—	6.68	—	—	—	—	—	—
60	90.6	87.5	20.7	65.0	3.0	0.21	6.20	—	—	—	—	—	—
70	90.7	88.7	45.2	77.3	19.0	7.5	5.91	—	—	—	—	—	—
80	90.8	89.4	66.5	85.0	43.5	29.0	5.41	—	0.50	0.80	—	—	—
90	90.9	89.8	78.8	88.2	66.0	55.0	4.90	—	4.06	6.99	0.18	—	—
700	91.0	90.0	85.0	89.8	77.7	71.3	5.00	—	17.8	23.5	1.60	—	1.00
Luminous transmit.	12.5	5.2	1.3	2.9	0.45	0.25	42.5	17.3	1.2	33.6	15.6	14.4	5.2
Dominant wave lgth.	551.7_c	498.0_c	424.0	564.8_c	566.8	566.4_c	483.5	478.9	450.6	516.2	589.1	483.4	481.5
Excitation purity.	79.6	88.3	94.4	91.4	96.3	97.8	41.8	69.8	98.9	48.5	72.9	77.2	88.4

* Some transmission below 400 mμ. Consult the manufacturer.

TRANSMISSION OF WRATTEN FILTERS (Continued)

Wave length	Percent transmittance No. 45A	No. 46 *	No. 47 *	No. 47B *	No. 48 *	No. 48A *	No. 49 *	No. 49B *	No. 50 *	No. 52 *	No. 53 *	No. 54	No. 55
400	—	1.20	7.80	16.0	0.96	5.65	3.30	1.70	0.45	2.18	—	—	—
10	—	0.60	17.4	29.5	3.16	10.0	4.28	2.00	0.39	1.51	—	—	—
20	—	0.80	34.0	43.6	8.25	16.0	6.93	3.55	0.59	0.80	—	—	—
30	1.00	5.98	47.0	50.0	15.0	21.0	11.2	7.00	2.63	0.44	—	—	—
40	8.81	19.0	50.3	47.2	22.6	25.0	18.9	13.0	8.90	0.41	—	—	—
50	17.4	30.1	48.3	36.0	30.3	26.2	25.6	17.4	14.0	0.69	—	—	—
60	20.9	33.8	43.4	25.0	33.2	22.9	24.0	14.8	12.3	1.45	—	—	0.20
70	21.6	32.1	36.2	13.2	29.6	16.5	15.7	7.60	5.36	2.70	0.10	—	2.90
80	20.5	27.0	28.5	4.5	22.4	9.55	6.93	2.76	1.55	4.90	0.7	—	13.1
90	18.0	20.2	19.6	1.3	14.1	4.27	2.14	0.40	0.10	8.50	2.14	—	34.2
500	14.4	11.1	0.36	0.17	7.30	1.58	0.46	—	—	13.3	4.47	—	53.4
10	10.1	4.39	—	—	2.64	0.48	—	—	—	18.2	7.24	0.10	67.0
20	5.60	1.66	—	—	0.50	—	—	—	—	23.7	10.7	0.31	69.3
30	2.52	0.35	—	—	—	—	—	—	—	28.5	14.0	0.64	65.1
40	0.4	—	—	—	—	—	—	—	—	32.1	16.6	0.89	56.7
50	0.10	—	—	—	—	—	—	—	—	33.1	17.3	0.93	45.0
60	—	—	—	—	—	—	—	—	—	31.0	15.4	0.62	33.1
70	—	—	—	—	—	—	—	—	—	25.6	11.4	0.21	20.7
80	—	—	—	—	—	—	—	—	—	19.1	6.90	—	9.00
90	—	—	—	—	—	—	—	—	—	12.6	3.60	—	2.70
600	—	—	—	—	—	—	—	—	—	7.78	1.41	—	0.40
10	—	—	—	—	—	—	—	—	—	4.17	0.40	—	—
20	—	—	—	—	—	—	—	—	—	2.34	0.15	—	—
30	—	—	—	—	—	—	—	—	—	1.38	—	—	—
40	—	—	—	—	—	—	—	—	—	0.80	—	—	—
50	—	—	—	—	—	—	—	—	—	0.54	—	—	—
60	—	—	—	—	—	—	—	—	—	0.36	—	—	—
70	—	—	—	—	—	—	—	—	—	0.27	—	—	—
80	—	—	—	—	—	—	—	—	—	0.23	—	—	0.66
90	0.20	0.25	—	—	—	—	—	—	—	0.19	—	—	6.90
700	2.24	0.85	—	—	—	—	—	—	—	0.17	—	—	27.8
Luminous transmit.	2.8	2.4	2.8	0.78	1.86	0.88	0.69	0.36	0.26	20.1	9.0	0.032	31.4
Dominant wave lgth.	477.6	470.4	463.7	479.8	466.5	458.0	457.9	455.5	455.9	553.3	551.1	546.1	530.2
Excitation purity.	89.7	94.9	95.8	69.1	96.1	98.3	98.9	99.3	99.4	77.3	89.7	97.0	68.4

Wave length	Percent transmittance No. 56	No. 57	No. 57A	No. 58	No. 59 *	No. 59A *	No. 60	No. 61	No. 64 *	No. 65	No. 65A *	No. 66 *	No. 67A *
400	—	—	—	—	—	—	—	—	9.00	—	—	12.3	1.10
10	—	—	—	—	—	—	—	—	9.20	—	—	13.0	0.93
20	—	—	—	—	—	—	—	—	8.75	0.23	—	15.0	1.28
30	—	—	—	—	—	0.16	—	—	9.20	0.61	0.16	18.4	3.16
40	—	—	0.19	—	—	0.37	—	—	11.3	1.58	1.32	23.2	6.40
50	—	—	0.87	—	0.40	1.26	0.19	—	15.5	4.10	5.50	31.2	10.5
60	0.16	0.44	2.56	—	1.90	4.57	1.38	—	23.3	9.00	13.0	42.2	17.7
70	3.12	3.10	7.80	0.23	7.70	13.2	5.38	—	34.4	16.8	24.9	55.5	28.5
80	13.0	13.1	21.6	1.38	21.0	30.0	15.0	0.33	46.8	24.9	36.6	68.4	41.4
90	34.5	31.9	41.7	4.90	41.5	50.8	32.0	4.00	56.6	31.3	45.1	77.6	52.1
500	59.0	50.5	58.8	17.7	59.0	66.0	48.4	16.6	62.1	33.7	45.8	82.7	57.9
10	73.0	60.6	67.9	38.8	67.7	73.0	57.2	32.3	62.9	32.4	39.7	84.6	58.8
20	79.0	63.3	70.1	52.2	69.8	75.1	59.2	40.0	59.1	27.5	29.7	84.0	55.4
30	79.9	61.0	67.6	53.6	67.2	73.2	55.5	39.6	51.6	20.7	17.8	82.6	47.5
40	77.5	55.0	61.8	47.6	61.5	68.5	47.5	34.5	41.3	13.7	7.90	79.1	36.6
50	72.6	47.1	53.5	38.4	54.0	62.0	36.8	26.3	28.0	6.50	2.40	73.7	25.0
60	66.1	37.3	43.3	27.8	45.0	54.4	25.2	17.3	16.2	1.66	0.32	67.1	14.2
70	58.0	26.5	31.6	17.4	35.0	44.5	14.4	9.70	7.95	0.40	—	58.8	5.50
80	46.1	16.6	19.4	9.0	24.0	33.0	6.3	4.40	3.10	—	—	47.2	1.40
90	33.8	8.69	9.70	3.50	14.0	22.0	1.82	1.66	0.80	—	—	34.5	0.28
600	24.0	3.70	4.50	1.50	7.95	14.6	0.48	0.38	—	—	—	24.4	—
10	18.7	1.60	2.00	0.41	4.90	10.5	0.10	—	—	—	—	18.5	—
20	13.2	0.49	0.87	—	2.70	6.92	—	—	—	—	—	13.7	—
30	7.22	—	0.22	—	1.00	3.16	—	—	—	—	—	7.70	—
40	3.02	—	—	—	0.17	1.07	—	—	—	—	—	3.00	—
50	1.48	—	—	—	—	0.50	—	—	—	—	—	1.46	—
60	1.91	—	—	—	—	0.91	—	—	—	—	—	1.91	—
70	7.95	—	—	—	0.63	3.00	—	—	—	—	—	6.17	—
80	23.0	—	0.16	—	4.00	10.0	—	—	—	—	—	19.9	—
90	44.1	—	1.15	—	12.0	20.0	2.10	—	0.10	—	0.20	42.6	—
700	64.8	—	3.17	0.53	22.6	30.0	8.70	—	4.50	—	2.18	63.1	0.40
Luminous transmit.	52.8	32.5	37.2	23.7	38.7	45.8	26.1	16.8	25.0	9.6	9.8	58.3	22.4
Dominant wave lgth.	552.3	536.4	534.0	540.2	538.3	541.4	525.7	536.8	497.3	496.6	492.7	512.3	499.8
Excitation purity.	78.2	69.2	62.1	88.1	66.0	59.3	62.2	85.4	55.0	67.8	77.4	21.5	55.8

* Some transmission below 400 mμ. Consult the manufacturer.

TRANSMISSION OF WRATTEN FILTERS (Continued)

Percent transmittance

Wave length	No. 70	No. 72B	No. 73	No. 74	No. 75	No. 76 *	No. 77	No. 77A	No. 78 *	No. 78AA *	No. 78A *	No. 78B *
400	—	—	—	—	—	0.22	—	—	37.2	43.0	56.0	64.1
10	—	—	—	—	—	0.18	—	—	41.7	46.0	58.6	66.5
20	—	—	—	—	—	0.29	—	—	44.2	48.7	61.0	68.4
30	—	—	—	—	—	1.38	—	—	44.6	49.8	61.8	69.5
40	—	—	—	—	—	3.50	—	—	44.2	49.7	61.8	70.0
50	—	—	—	—	—	3.50	—	—	41.7	48.0	61.0	69.4
60	—	—	—	—	1.97	1.92	—	—	38.0	44.9	58.7	67.5
70	—	—	—	—	10.0	0.51	—	—	33.8	40.3	55.0	65.4
80	—	—	—	—	17.4	—	—	—	27.5	35.6	51.0	62.9
90	—	—	—	—	18.0	—	—	—	23.5	30.9	47.1	59.8
500	—	—	—	—	13.0	—	—	—	19.5	26.5	43.5	57.0
10	—	—	—	0.96	7.35	—	0.30	0.10	15.8	23.4	40.0	54.2
20	—	—	—	7.95	3.20	—	9.10	5.35	13.8	20.3	36.9	51.4
30	—	—	—	14.6	0.83	—	13.5	1.90	11.8	17.8	34.4	49.3
40	—	—	—	12.9	0.14	—	46.1	35.0	10.5	16.6	32.7	48.1
50	—	—	—	7.60	—	—	78.0	71.8	9.56	14.9	31.2	46.7
60	—	—	2.24	3.06	—	—	75.8	63.1	8.53	13.2	29.4	45.0
70	—	—	5.97	0.83	—	—	8.00	—	7.77	12.1	28.0	43.6
80	—	—	4.56	0.12	—	—	1.00	—	7.41	11.6	27.5	43.1
90	—	1.26	2.00	—	—	—	0.32	—	6.93	11.1	27.0	42.9
600	—	5.89	0.56	—	—	—	16.2	1.60	6.45	10.40	26.0	41.8
10	—	5.25	0.10	—	—	—	52.1	32.1	5.50	9.20	24.1	40.0
20	—	2.88	—	—	—	—	83.0	78.0	4.80	7.70	21.8	37.6
30	—	1.26	—	—	—	—	84.9	79.5	3.94	6.50	19.7	35.5
40	—	0.48	—	—	—	—	88.1	86.5	3.46	5.60	18.6	34.2
50	0.63	0.14	—	—	—	—	89.8	89.2	3.24	5.50	18.4	33.6
60	10.5	—	—	—	—	—	89.8	89.0	3.16	5.60	18.5	34.0
70	35.0	—	—	—	—	—	85.5	79.5	3.39	5.80	18.7	34.1
80	55.2	—	—	—	—	—	76.1	62.5	3.45	6.10	19.0	34.5
90	70.0	—	—	—	—	0.13	75.0	62.4	3.51	6.10	19.3	34.8
700	79.0	—	—	—	0.14	1.24	86.5	83.0	3.90	6.50	20.2	36.0
Luminous transmit.	0.31	0.74	1.3	4.0	1.9	0.046	32.3	25.5	10.7	15.8	31.6	46.7
Dominant wave lgth.	675.6	604.9	574.9	538.6	487.7	449.2	579.9	581.5	471.1	473.4	475.7	477.2
Excitation purity.	100.0	100.0	100.0	96.7	90.4	99.7	99.0	99.1	63.0	54.5	33.7	20.7

Percent transmittance

Wave length	No. 78C *	No. 79 *	No. 80A *	No. 81 *	No. 81A *	No. 81B *	No. 81C *	No. 81D *	No. 81EF *	No. 82 *	No. 82A *	No. 82B *
400	74.9	24.0	67.6	77.7	65.1	55.1	46.1	38.2	30.7	83.0	80.1	76.7
10	76.6	26.0	73.1	78.1	65.9	55.8	46.6	38.4	31.5	83.7	80.8	78.0
20	77.9	29.0	76.8	79.0	67.6	57.7	49.0	41.0	34.3	84.6	81.6	79.2
30	78.9	31.0	7.77	80.5	70.2	61.0	52.5	45.5	38.6	85.1	82.2	79.7
40	79.4	32.2	76.5	81.9	72.8	64.5	57.2	50.0	43.2	85.4	82.4	79.7
50	79.5	32.7	73.0	83.0	74.8	67.2	60.5	53.9	47.4	85.4	82.4	79.2
60	79.3	31.4	69.0	83.7	76.0	69.1	63.0	56.5	50.2	85.0	81.7	78.0
70	78.6	28.8	63.6	84.3	77.1	70.6	64.2	58.1	52.0	84.6	80.7	76.3
80	77.8	25.6	57.6	84.6	77.8	71.3	65.0	59.0	53.0	84.0	79.3	74.4
90	76.7	22.2	51.3	84.9	78.3	71.8	65.7	60.0	54.0	83.3	78.0	72.1
500	75.5	19.3	45.2	85.3	78.6	72.6	66.4	60.8	55.4	82.6	76.6	70.2
10	74.2	16.8	39.4	85.4	79.0	72.9	66.5	61.1	56.2	82.0	75.3	68.3
20	73.0	14.2	34.2	85.5	79.5	73.2	67.0	61.6	57.0	81.4	74.0	66.5
30	72.1	12.7	30.0	86.0	80.4	74.5	68.8	62.5	59.5	81.0	73.1	65.5
40	71.5	11.8	27.1	86.5	81.5	76.0	71.0	66.1	62.7	80.8	72.7	65.0
50	70.7	11.0	24.8	86.8	82.3	77.0	72.0	67.3	64.5	80.6	72.4	64.5
60	69.8	9.76	23.5	87.0	82.6	77.6	72.5	68.0	65.3	80.4	71.8	63.8
70	69.0	8.81	22.6	87.1	82.7	77.8	72.7	68.3	65.8	80.2	71.5	63.2
80	68.8	8.50	22.6	87.1	82.8	78.0	73.0	68.5	66.0	80.2	71.5	63.2
90	68.6	8.29	23.2	87.4	83.1	78.2	74.0	69.5	66.5	80.3	71.7	63.4
600	68.0	7.56	23.7	87.6	84.0	79.1	75.6	72.0	68.1	80.2	71.5	63.0
10	66.7	6.45	23.2	88.1	85.0	81.0	78.5	75.0	71.6	79.3	70.3	61.5
20	65.0	5.13	21.0	88.8	86.1	83.1	80.8	78.0	74.7	78.4	68.5	59.0
30	63.8	4.17	18.2	89.2	87.0	84.2	82.1	79.8	77.0	77.5	66.9	56.9
40	63.0	3.47	15.8	89.4	87.4	85.1	83.0	80.8	78.4	76.8	65.5	55.0
50	62.7	3.16	14.5	89.5	87.7	85.6	83.5	81.5	79.2	76.5	64.8	54.1
60	63.0	3.09	13.8	89.8	88.0	86.0	84.1	82.1	80.1	76.2	64.6	53.7
70	63.3	3.16	13.4	90.0	88.2	86.5	84.8	83.0	80.9	76.1	64.5	53.7
80	63.4	3.16	12.7	90.1	88.5	87.0	85.5	83.7	81.8	76.1	64.4	53.5
90	63.6	3.16	11.7	90.3	89.0	87.5	86.1	84.6	82.9	76.2	64.2	53.4
700	65.0	3.31	11.5	90.5	89.2	88.0	86.8	85.5	84.0	77.1	64.6	54.1
Luminous transmit.	70.4	11.3	28.4	86.8	82.0	76.9	72.0	67.4	64.0	80.7	72.5	64.6
Dominant wave lgth.	479.8	474.8	471.7	576.7	577.5	577.8	577.4	579.5	579.0	477.5	476.6	475.6
Excitation purity.	6.8	52.8	45.9	2.9	6.0	8.7	10.7	14.7	19.0	3.0	6.3	10.2

* Some transmission below 400 mμ. Consult the manufacturer.

TRANSMISSION OF WRATTEN FILTERS (Continued)

Wave length	Percent transmittance												
	No. 82C *	No. 83 *	No. 85 *	No. 85B *	No. 86 *	No. 86A *	No. 86B *	No. 86C *	No. 89B	No. 90	No. 96 *	No. 97	No. 102 *
400	73.4	13.5	6.0	1.59	0.50	8.00	20.0	44.0	—	—	4.28	—	1.12
10	75.0	13.1	18.0	9.32	0.81	12.2	26.1	55.0	—	—	4.91	—	0.96
20	76.4	13.5	28.4	15.5	1.55	16.7	31.6	62.0	—	—	5.50	—	0.89
30	77.2	14.1	33.4	19.0	2.88	21.5	37.5	66.6	—	—	6.17	—	0.96
40	77.2	15.6	36.2	20.8	5.50	27.8	44.0	70.8	—	—	6.92	—	1.23
50	76.6	17.8	38.1	22.1	9.10	34.2	50.1	74.3	—	—	7.50	—	1.86
60	75.2	21.0	40.4	24.3	13.5	40.4	55.4	76.8	—	—	7.81	—	3.23
70	73.2	25.5	43.0	27.5	17.8	45.0	59.5	78.7	—	—	8.15	0.22	6.45
80	70.7	30.2	45.3	30.9	21.3	48.7	62.5	80.2	—	—	8.47	0.43	14.0
90	68.1	35.8	47.2	34.3	24.5	51.2	64.6	81.2	—	—	8.60	0.39	21.6
500	65.7	43.5	48.9	38.3	26.8	52.8	66.0	81.7	—	—	8.73	0.15	30.7
10	63.5	46.3	49.2	40.7	27.9	53.4	66.4	81.9	—	—	8.85	—	41.4
20	61.5	47.2	48.2	40.6	28.6	53.7	66.6	82.0	—	—	8.90	—	51.3
30	59.9	48.3	48.3	40.7	30.4	55.0	67.6	82.4	—	—	9.01	—	59.4
40	59.1	49.6	49.2	41.6	32.5	56.5	69.0	83.0	—	—	9.07	—	64.2
50	58.3	51.8	51.0	43.2	35.0	58.5	70.2	83.5	—	—	9.20	—	66.7
60	57.2	56.5	55.8	47.1	41.2	63.0	73.0	84.6	—	9.00	9.30	—	66.3
70	56.2	65.0	64.5	56.0	53.0	70.9	78.1	86.8	—	30.5	9.20	—	63.0
80	56.1	75.5	75.0	68.1	67.5	79.0	84.0	88.9	—	34.3	9.19	—	58.0
90	56.0	83.0	83.0	78.1	76.5	85.2	87.5	89.9	—	25.2	9.54	—	51.9
600	55.0	87.3	87.2	85.0	85.0	88.1	89.3	90.6	—	16.1	9.64	—	45.2
10	53.0	89.3	88.9	88.0	88.1	89.8	90.3	91.0	—	11.3	9.73	—	37.8
20	50.2	90.4	90.0	89.6	89.6	90.5	90.7	91.1	—	7.40	9.56	—	30.5
30	47.4	90.8	90.5	90.3	90.4	90.8	90.9	91.2	—	2.91	9.27	—	25.0
40	45.2	91.0	90.7	90.7	90.7	91.1	91.1	91.3	—	0.76	9.10	—	20.6
50	44.1	91.1	90.9	90.9	91.0	91.2	91.2	91.4	—	0.29	9.07	—	17.5
60	43.6	91.3	91.0	91.0	91.1	91.3	91.3	91.5	—	0.41	9.00	—	15.2
70	43.5	91.5	91.0	91.2	91.2	91.4	91.4	91.6	—	2.30	9.13	—	13.7
80	43.1	91.5	91.0	91.3	91.3	91.4	91.5	91.6	0.10	9.52	9.08	0.44	12.8
90	42.8	91.5	91.0	91.3	91.3	91.5	91.6	91.6	1.58	28.5	9.21	5.02	12.1
700	43.5	91.5	91.0	91.3	91.3	91.5	91.6	91.6	11.2	51.9	9.52	18.7	12.0
Luminous transmit.	58.1	61.4	62.5	55.5	49.7	67.1	75.5	85.4	0.017	9.8	9.1	0.041	50.8
Dominant wave lgth.	477.2	581.5	587.7	585.7	585.7	581.7	579.6	577.6	700	583.1	572.4	555.0_c	564.9
Excitation purity.	14.5	55.4	30.3	48.0	69.7	37.1	24.1	9.0	100	100.0	12.1	48.0	80.0

Wave length	Percent transmittance												
	No. 106	CC-05R *	CC-10R *	CC-20R *	CC-30R *	CC-40R *	CC-50R *	CC-05B *	CC-10B *	CC-20B *	CC-30B *	CC-40B *	CC-50B *
400	—	81.0	73.0	61.5	51.6	42.5	36.4	87.0	85.5	82.2	80.2	77.0	74.1
10	—	81.0	72.4	60.0	50.0	40.0	33.9	87.5	86.4	84.0	82.5	80.3	78.4
20	0.10	81.0	72.0	58.6	48.2	38.2	31.9	87.7	87.2	85.0	84.0	82.2	80.7
30	0.20	81.1	71.6	57.7	47.0	36.8	30.5	88.0	87.5	85.3	84.3	82.5	81.1
40	0.35	81.2	71.5	57.2	46.4	36.0	29.7	88.1	87.5	85.0	83.5	81.3	79.8
50	0.58	81.4	71.6	57.2	46.4	36.1	29.6	88.1	87.2	83.9	81.9	78.7	76.6
60	0.98	81.7	72.4	58.5	47.5	37.5	31.0	87.9	86.4	82.5	79.5	75.9	72.9
70	1.5	82.3	73.7	60.6	49.9	40.0	33.6	87.5	85.3	80.3	76.2	72.0	67.9
80	2.3	82.8	74.9	62.0	52.0	42.5	35.9	87.0	84.0	77.8	72.5	67.5	62.7
90	3.5	83.2	75.8	63.5	53.9	44.8	37.9	86.2	82.4	74.2	68.3	62.3	56.6
500	5.2	83.3	76.6	64.6	55.2	46.1	39.4	85.2	80.5	71.2	63.8	56.7	50.1
10	7.7	83.0	76.1	64.0	54.5	46.0	38.5	84.4	78.6	67.7	58.7	51.0	44.5
20	10.7	82.4	74.9	61.5	51.6	42.5	35.0	83.5	77.0	64.0	54.4	46.0	38.6
30	15.1	81.6	73.5	59.4	48.5	38.5	31.5	82.6	75.2	61.5	50.7	41.6	34.1
40	20.2	81.2	72.5	57.8	46.5	36.6	29.4	82.1	73.9	59.5	48.3	39.0	31.3
50	25.7	81.1	72.4	57.1	45.6	35.6	28.7	81.5	73.0	58.0	46.6	36.9	29.5
60	31.0	81.4	72.8	58.0	46.9	36.4	29.7	81.4	72.7	57.5	45.9	35.9	28.6
70	35.6	82.5	74.5	60.6	49.2	39.2	32.7	81.4	73.0	57.9	46.3	36.1	28.7
80	43.2	83.9	77.3	65.0	54.6	45.0	38.6	81.9	73.9	59.3	47.9	37.8	30.5
90	53.8	85.7	80.7	71.0	61.8	53.5	47.6	82.7	75.1	61.6	50.3	40.8	33.4
600	65.6	87.6	84.0	77.0	70.5	64.0	58.7	83.4	76.3	63.5	53.0	43.5	36.0
10	77.0	89.0	86.5	82.0	77.8	73.0	69.8	83.6	76.7	64.5	54.6	44.7	37.8
20	82.8	90.0	88.9	86.2	83.5	80.8	78.3	83.5	76.5	64.3	54.4	44.3	37.5
30	86.0	90.6	89.9	88.1	87.2	85.1	84.2	83.2	76.6	63.1	53.2	42.5	35.6
40	87.6	91.1	90.5	89.8	89.2	88.0	87.5	82.8	74.5	61.6	51.5	40.2	33.7
50	88.7	91.2	90.8	90.5	90.3	89.5	89.2	82.5	74.0	60.6	50.3	39.0	32.4
60	89.5	91.3	91.1	90.8	90.6	90.4	90.1	82.5	73.8	60.1	49.5	38.4	31.8
70	90.0	91.4	91.3	91.0	90.9	90.8	90.7	82.3	73.3	59.6	49.0	37.7	31.0
80	90.5	91.6	91.5	91.2	91.1	91.0	91.1	82.0	72.8	58.6	48.2	36.5	30.0
90	90.8	91.7	91.7	91.4	91.4	91.3	91.1	81.9	72.5	58.1	47.2	35.4	29.0
700	91.0	91.9	91.9	91.5	91.5	91.4	91.2	82.2	73.0	58.5	47.5	35.6	29.0
Luminous transmit.	34.6	83.7	77.0	65.3	55.9	47.3	41.3	82.8	75.5	62.3	52.0	42.8	35.7
Dominant wave lgth.	589.4	605.0	597.8	604.2	605.8	605.5	608.5	459.0	462.0	460.0	461.0	463.2	462.5
Excitation purity.	95.2	2.0	4.7	8.5	12.3	17.3	21.4	2.8	6.3	13.2	20.2	27.7	34.2

* Some transmission below 400 mμ. Consult the manufacturer.

TRANSMISSION OF WRATTEN FILTERS (Continued)

Wave length	CC-05G *	CC-10G *	CC-20G *	CC-30G *	CC-40G *	CC-50G *	CC-05Y *	CC-10Y *	CC-20Y *	CC-30Y *	CC-40Y *	CC-50Y *
	Percent transmittance											
400	80.0	73.1	58.8	48.0	39.7	32.0	81.0	74.5	61.3	50.5	43.0	34.5
10	80.7	72.9	57.8	46.5	38.1	30.3	80.6	73.2	59.0	47.4	39.5	30.5
20	81.0	72.8	57.3	45.8	37.3	29.5	80.4	72.6	57.8	46.0	37.5	29.0
30	81.4	72.7	57.0	45.5	36.5	29.0	80.4	72.5	57.5	45.6	36.5	28.7
40	81.6	73.0	57.3	45.8	36.6	29.1	80.6	72.8	57.8	46.5	36.8	29.5
50	82.1	73.9	58.4	46.9	38.1	30.6	81.2	74.0	59.5	48.5	38.5	31.5
60	83.0	75.5	61.4	50.3	41.5	34.3	82.5	76.0	63.0	52.5	42.5	36.2
70	84.4	78.0	65.4	55.8	47.0	40.5	83.9	78.5	67.5	48.2	48.8	43.5
80	85.6	80.4	70.0	61.8	53.5	47.8	85.3	81.2	72.3	64.9	56.2	54.0
90	86.8	83.0	75.2	68.9	61.3	57.0	87.0	84.4	78.0	72.4	66.0	64.0
500	87.0	85.9	80.3	76.4	70.7	[illegible]	88.4	87.2	84.0	81.0	77.0	73.3
10	88.7	87.5	83.8	80.9	77.8	75.3	89.5	89.0	88.0	86.6	85.5	84.2
20	89.0	88.1	84.9	82.3	79.5	77.5	90.0	90.0	89.6	89.1	89.0	88.5
30	89.0	88.0	84.6	81.7	79.4	77.0	90.4	90.4	90.0	89.7	89.9	89.6
40	89.0	87.6	83.7	80.5	77.8	74.8	90.7	90.7	90.6	90.4	90.2	90.0
50	88.6	87.1	82.4	78.6	75.8	72.2	90.9	90.9	90.8	90.6	90.4	90.3
60	88.1	86.3	80.9	76.2	72.9	68.8	91.0	91.0	90.9	90.8	90.7	90.6
70	87.5	85.3	79.0	73.5	69.3	64.8	91.3	91.3	91.0	90.9	90.8	90.7
80	87.0	84.1	77.0	70.4	65.3	60.3	91.4	91.4	91.1	91.0	90.8	90.7
90	86.4	82.8	74.5	67.2	61.9	55.9	91.4	91.4	91.2	91.1	90.9	90.8
600	85.7	81.5	72.0	64.1	57.7	51.7	91.4	91.4	91.3	91.2	90.9	90.8
10	85.0	80.0	69.5	60.7	53.7	47.3	91.4	91.4	91.3	91.2	90.9	90.9
20	84.0	78.5	66.5	57.2	49.8	42.5	91.4	91.4	91.3	91.2	91.0	90.9
30	83.0	76.9	63.8	53.7	45.5	38.0	91.5	91.5	91.4	91.3	91.0	91.0
40	82.2	75.6	61.5	50.8	42.0	34.6	91.5	91.5	91.4	91.3	91.0	91.0
50	81.9	74.9	60.1	49.1	40.2	32.5	91.5	91.5	91.4	91.3	91.1	91.1
60	81.5	74.4	59.4	48.1	39.5	31.5	91.5	91.5	91.4	91.3	91.1	91.1
70	81.4	74.0	58.8	47.5	38.5	31.0	91.5	91.5	91.4	91.4	91.2	91.2
80	81.1	73.5	58.1	46.6	37.6	29.9	91.5	91.5	91.4	91.4	91.2	91.2
90	81.1	73.2	57.6	46.0	36.6	28.9	91.5	91.5	91.4	91.4	91.3	91.3
700	81.5	73.5	58.0	46.4	36.5	28.7	91.5	91.5	91.4	91.4	91.3	91.3
Luminous transmit.	87.2	84.5	77.8	72.2	67.7	63.3	90.4	90.1	89.1	88.2	87.4	86.9
Dominant wave lgth.	553.0	555.5	555.0	554.0	554.3	553.4	572.0	571.3	571.4	571.3	571.3	571.2
Excitation purity.	2.3	5.2	10.9	15.8	21.1	25.9	5.3	9.6	18.8	28.3	35.7	42.0

Wave length	CC-05M *	CC-10M *	CC-20M *	CC-30M *	CC-40M *	CC-50M *	CC-05C *	CC-10C *	CC-20C *	CC-30C *	CC-40C *	CC-50C *
	Percent transmittance											
400	87.6	86.6	85.6	84.2	82.3	80.9	87.3	86.0	83.9	82.3	80.4	78.8
10	88.2	87.7	86.6	85.7	84.6	83.6	88.2	87.5	85.2	84.5	83.4	82.7
20	88.6	88.0	87.0	85.9	85.2	84.4	88.7	88.1	86.5	86.0	85.3	84.8
30	88.7	88.0	86.9	85.6	84.4	83.6	89.0	88.6	87.5	87.0	86.3	85.9
40	88.7	87.9	86.0	84.7	82.5	81.4	89.3	89.0	87.7	87.3	86.6	86.1
50	88.6	87.5	84.9	82.8	80.0	78.1	89.5	89.1	87.8	87.5	86.6	86.0
60	88.4	86.5	83.1	80.0	76.1	73.7	89.6	89.1	87.7	87.3	86.4	85.7
70	87.8	85.2	80.8	76.4	71.3	68.0	89.7	89.0	87.5	87.0	85.8	85.2
80	87.0	83.6	77.9	72.1	65.8	61.7	89.7	89.0	87.2	86.5	85.3	84.3
90	86.0	81.8	74.4	67.0	60.0	55.0	89.7	89.0	87.0	86.0	84.4	83.4
500	85.0	79.7	70.5	61.7	53.7	48.1	89.6	89.0	86.5	85.2	83.5	82.3
10	83.8	77.5	66.7	56.5	47.7	41.6	89.6	88.7	86.0	84.4	82.4	80.8
20	82.7	75.3	63.4	52.0	42.8	36.3	89.5	88.5	85.2	83.5	81.1	79.2
30	81.8	73.7	60.5	48.6	39.0	31.9	89.4	88.0	84.3	82.4	79.6	77.3
40	81.3	72.5	58.6	46.6	36.7	29.8	89.2	87.5	83.4	81.0	77.7	75.0
50	81.2	72.2	58.0	46.0	36.0	29.1	88.9	87.0	82.3	79.0	75.3	72.2
60	81.5	72.8	58.3	46.5	36.7	29.7	88.5	86.1	80.5	76.7	72.7	69.0
70	82.5	74.6	60.5	49.8	40.2	32.3	88.0	85.0	78.5	74.0	69.3	65.0
80	84.0	77.3	64.9	55.6	46.2	39.0	87.5	83.8	76.1	70.9	65.4	60.5
90	85.8	80.8	70.6	63.3	54.9	48.7	87.0	82.5	73.9	67.5	61.6	55.8
600	88.0	84.5	77.1	71.6	64.9	59.9	86.4	81.0	71.2	64.1	57.6	51.3
10	89.3	87.0	82.2	79.2	74.9	70.7	85.5	79.5	68.5	60.4	53.4	36.2
20	90.2	88.9	86.1	84.1	81.4	79.2	84.5	77.9	65.5	56.7	49.2	42.3
30	90.6	90.0	88.7	87.4	86.0	84.5	83.8	76.3	62.7	53.1	45.0	38.0
40	90.8	90.5	90.0	89.3	88.7	87.6	83.3	75.1	60.8	50.4	42.0	34.9
50	91.0	90.7	90.5	90.2	90.0	89.7	82.8	74.4	59.5	48.8	40.2	32.9
60	91.1	91.0	90.8	90.8	90.4	90.4	82.5	74.0	58.5	48.0	39.4	32.0
70	91.2	91.2	91.0	91.0	90.7	90.7	82.4	73.6	57.9	47.2	38.6	31.0
80	91.3	91.3	91.2	91.1	91.0	91.0	82.0	73.0	57.5	46.0	37.5	29.9
90	91.4	91.4	91.4	91.3	91.3	91.3	82.0	72.8	57.4	45.5	36.7	29.1
700	91.5	91.5	91.5	91.5	91.5	91.5	82.5	74.0	58.5	46.4	37.3	29.8
Luminous transmit.	84.2	77.9	67.1	58.1	50.0	44.0	88.0	85.1	78.9	74.8	70.5	66.7
Dominant wave lgth.	541.0	547.5	551.2	550.0	550.3	551.2	489.2	487.5	486.5	486.2	486.1	485.5
Excitation purity.	3.5	7.4	14.4	21.5	28.3	34.0	1.6	4.1	8.9	12.8	17.5	20.2

* Some transmission below 400 mμ. Consult the manufacturer.

TRANSMISSION OF WRATTEN FILTERS (Continued)

Wave length	No. 87	No. 87C	No. 88A	No. 89B
	Percent transmittance			
700	—	—	—	11.2
10	—	—	—	32.4
20	—	—	—	57.6
30	—	—	7.4	69.1
40	0.10	—	32.8	77.6
50	2.19	—	56.3	83.1
60	7.95	—	69.2	85.0
70	17.4	—	74.2	86.1
80	31.6	—	77.6	87.0
90	43.7	—	79.7	87.7
800	53.8	0.32	81.4	88.1
10	61.7	3.20	82.6	88.4
20	69.2	8.90	83.7	88.6
30	74.1	17.8	84.7	88.8
40	77.7	28.2	85.5	89.0
50	81.4	41.0	86.1	89.2
60	84.0	53.8	86.6	89.4
70	85.4	61.6	87.2	89.6
80	86.8	69.2	87.5	89.8
90	87.8	74.1	87.8	89.9
900	88.4	78.5	88.0	90.0
10	88.8	81.5	88.2	90.1
20	89.1	83.6	88.4	90.2
30	89.1	85.1	88.6	90.3
40	89.1	86.0	88.8	90.4
50	89.1	87.0	89.0	90.5

* Some transmission below 400 mμ. Consult the manufacturer.

TRANSMISSIBILITY FOR RADIATIONS

Ratio of the transmitted light to the incident light for a definite thickness of the substance, usually 1 cm.

Glass

Glass in general is opaque to the ultra-violet and infrared. Uviol glass is transparent to the longer radiations of the ultra-violet.

Coefficient of transparency of glass for visible and ultra-violet radiations.

Normal incidence, thickness 1 cm.

Wave length microns	0.309	0.330	0.347	0.357	0.361	0.375	0.384	0.388	0.396
Crown, ordinary	—	—	—	—	—	.947	—	—	—
Crown, borosilicate	0.08	0.65	0.88	—	0.95	—	0.972	0.975	0.986
Flint, ordinary	—	—	—	0.72	—	—	—	0.904	—
Flint, heavy	—	—	0.01	—	0.16	—	0.58	—	—

Normal incidence, thickness 1 cm.

Wave length microns	0.400	0.415	0.419	0.425	0.434	0.455	0.500	0.580	0.677
Crown, ordinary	0.964	—	0.952	—	0.960	0.981	—	0.986	0.990
Crown, borosilicate	—	0.985	—	0.993	—	—	0.993	—	—
Flint, ordinary	—	0.959	—	—	—	—	1.00	—	—
Flint, heavy	—	—	—	0.905	—	—	—	—	—

Quartz

Quartz is very transparent to the ultra-violet and to the visible spectrum, but opaque for the infrared beyond 7.0μ.

(Pflüger.)

Wave length, microns	0.19	0.20	0.21	0.22
Transmission for 1 mm	.67	.84	.92	.94

Fluorite

Fluorite is very transparent to the ultra-violet, nearly to 0.10μ. Coefficient of transparency at λ = 186 is found by Pflüger to be 0.80.

For the infrared the values are given in a table below.

**Rock Salt and Sylvine and Fluorite
Transparency for the Infrared.
Thickness 1 cm.**

Wave length microns	Rock salt	Sylvine KCl	Fluorite
8.	—	—	.844
9.	0.995	1.000	.543
10.	.995	.988	.164
12.	.993	.995	.010
14.	.931	.975	.000
16.	.661	.936	—
18.	.275	.862	—
19.	.096	.758	—
20.7	.006	.585	—
23.7	.000	.155	—

COLORIMETRY

Selected from Judd, Jour, Opt. Soc. Amer. 23, 359 (1933)
Recommendations of the International Commission on Illumination

Standard Illuminants

A. Gas-filled tungsten incandescent lamp of color temperature 2848°K.
B. Noon Sunlight. Lamp as above in combination with the Davis-Gibson filter for converting color temperature 2848° to 4800°K.
The filter is to be composed of a layer one centimeter thick of each of two separate solutions B_1 and B_2, contained in a double cell of colorless optical glass.

Solution B_1	
Copper sulfate ($CuSO_4 \cdot 5H_2O$)	2.452 g
Mannite ($C_6H_8(OH)_6$)	2.452 g
Pyridine (C_5H_5N)	30.00 cc
Distilled water to make	1000 cc

Solution B_2	
Cobalt ammonium sulfate ($CoSO_4 \cdot (NH_4)_2SO_4 \cdot 6H_2O$)	21.71 g
Copper sulfate ($CuSO_4 \cdot 5H_2O$)	16.11 g
Sulfuric acid (density 1.835)	10.0 cc
Distilled water to make	1000 cc

C. Average Daylight. Lamp as in A in combination with Davis-Gibson filter for converting color temperature 2848° to 6500°K.
The filter is composed of a layer one centimeter thick of each of two separate solutions C_1 and C_2, contained in a double cell made of colorless optical glass.

Solution C_1	
Copper sulfate ($CuSO_4 \cdot 5H_2O$)	3.412 g
Mannite ($C_6H_8(OH)_6$)	3.412 g
Pyridine (C_5H_5N)	30.0 cc
Distilled water to make	1000 cc

Solution C_2	
Cobalt ammonium sulfate ($CoSO_4 \cdot (NH_4)_2SO_4 \cdot 6H_2O$)	30.580 g
Copper sulfate ($CuSO_4 \cdot 5H_2O$)	22.520 g
Sulfuric acid (density 1.835)	10.0 cc
Distilled water to make	1000 cc

See R. Davis and K. S. Gibson Bur. Stds. Misc. Pub. No. 114, Jan. 1931 or Bur. Stds. Jour. Research 7,796 (1931).

Standard Coordinate System

The tristimulus system of color specification is based on four chosen stimuli consisting of homogeneous radiant energy of wave lengths

700.0 546.1 435.8

mμ and of standard illuminant B (see above).
To establish the system of specification coordinates are assigned as follows:

Stimulus	x	y	z
700.0 mμ	0.73467	0.26533	0.00000
546.1 mμ	0.27376	0.71741	0.00883
435.8 mμ	0.16658	0.00886	0.82456
Standard illuminant B:	0.34842	0.35161	0.29997

The Standard Observer

The "standard observer" is determined below by the specification for the equal energy spectrum both in fractions, x, y, z of the total amount for each wave length interval of 5 mμ and directly $\bar{x}$, $\bar{y}$, $\bar{z}$. The fractional values are known as the trillnear coordinates or trichromatic coefficients of the spectrum; the direct values as the distribution functions or coefficients.
The sum of the trichromatic coefficients is unity, that is $x + y + z = 1$. Therefore the value of z may be and often is omitted from a specification.

Relative Visibility

The value of y given in the table is the standard visibility function or relative visibility.

Wave length mμ	Trichromatic coefficients x	y	z	Distribution coefficients for equal energy x	y (Rel. Vis.)	z
380	0.1741	0.0050	0.8209	0.0014	0.0000	0.0065
385	0.1740	0.0050	0.8210	0.0022	0.0001	0.0105
390	0.1738	0.0049	0.8213	0.0042	0.0001	0.0201
395	0.1736	0.0049	0.8215	0.0076	0.0002	0.0362
400	0.1733	0.0048	0.8219	0.0143	0.0004	0.0679
405	0.1730	0.0048	0.8222	0.0232	0.0006	0.1102
410	0.1726	0.0048	0.8226	0.0435	0.0012	0.2074
415	0.1721	0.0048	0.8231	0.0776	0.0022	0.3713
420	0.1714	0.0051	0.8235	0.1344	0.0040	0.6456
425	0.1703	0.0058	0.8239	0.2148	0.0073	1.0391
430	0.1689	0.0069	0.8242	0.2839	0.0116	1.3856
435	0.1669	0.0086	0.8245	0.3285	0.0168	1.6230
440	0.1644	0.0109	0.8247	0.3483	0.0230	1.7471
445	0.1611	0.0138	0.8251	0.3481	0.0298	1.7826
450	0.1566	0.0177	0.8257	0.3362	0.0380	1.7721
455	0.1510	0.0227	0.8263	0.3187	0.0480	1.7441
460	0.1440	0.0297	0.8263	0.2908	0.0600	1.6692
465	0.1355	0.0399	0.8246	0.2511	0.0739	1.5281
470	0.1241	0.0578	0.8181	0.1954	0.0910	1.2876
475	0.1096	0.0868	0.8036	0.1421	0.1126	1.0419
480	0.0913	0.1327	0.7760	0.0956	0.1390	0.8130
485	0.0687	0.2007	0.7306	0.0580	0.1693	0.6162
490	0.0454	0.2950	0.6596	0.0320	0.2080	0.4652
495	0.0235	0.4127	0.5638	0.0147	0.2586	0.3533
500	0.0082	0.5384	0.4534	0.0049	0.3230	0.2720
505	0.0039	0.6548	0.3413	0.0024	0.4073	0.2123
510	0.0139	0.7502	0.2359	0.0093	0.5030	0.1582
515	0.0389	0.8120	0.1491	0.0291	0.6082	0.1117
520	0.0743	0.8338	0.0919	0.0633	0.7100	0.0782
525	0.1142	0.8262	0.0596	0.1096	0.7932	0.0573
530	0.1547	0.8059	0.0394	0.1655	0.8620	0.0422
535	0.1929	0.7816	0.0255	0.2257	0.9149	0.0298
540	0.2296	0.7543	0.0161	0.2904	0.9540	0.0203
545	0.2658	0.7243	0.0099	0.3597	0.9803	0.0134
550	0.3016	0.6923	0.0061	0.4334	0.9950	0.0087
555	0.3373	0.6589	0.0038	0.5121	1.0002	0.0057
560	0.3731	0.6245	0.0024	0.5945	0.9950	0.0039
565	0.4087	0.5896	0.0017	0.6784	0.9786	0.0027
570	0.4441	0.5547	0.0012	0.7621	0.9520	0.0021
575	0.4788	0.5202	0.0010	0.8425	0.9154	0.0018
580	0.5125	0.4866	0.0009	0.9163	0.8700	0.0017
585	0.5448	0.4544	0.0008	0.9786	0.8163	0.0014
590	0.5752	0.4242	0.0006	1.0263	0.7570	0.0011
595	0.6029	0.3965	0.0006	1.0567	0.6949	0.0010
600	0.6270	0.3725	0.0005	1.0622	0.6310	0.0008
605	0.6482	0.3514	0.0004	1.0456	0.5668	0.0006

Wave length mμ	Trichromatic coefficients x	y	z	Distribution coefficients for equal energy x	y (Rel. Vis.)	z
610	0.6658	0.3340	0.0002	1.0026	0.5030	0.0003
615	0.6801	0.3197	0.0002	0.9384	0.4412	0.0002
620	0.6915	0.3083	0.0002	0.8544	0.3810	0.0002
625	0.7006	0.2993	0.0001	0.7514	0.3210	0.0001
630	0.7079	0.2920	0.0001	0.6424	0.2650	0.0000
635	0.7140	0.2859	0.0001	0.5419	0.2170	0.0000
640	0.7190	0.2809	0.0001	0.4479	0.1750	0.0000
645	0.7230	0.2770	0.0000	0.3608	0.1382	0.0000
650	0.7260	0.2740	0.0000	0.2835	0.1070	0.0000
655	0.7283	0.2717	0.0000	0.2187	0.0816	0.0000
660	0.7300	0.2700	0.0000	0.1649	0.0610	0.0000
665	0.7311	0.2689	0.0000	0.1212	0.0446	0.0000
670	0.7320	0.2680	0.0000	0.0874	0.0320	0.0000
675	0.7327	0.2673	0.0000	0.0636	0.0232	0.0000
680	0.7334	0.2666	0.0000	0.0468	0.0170	0.0000
685	0.7340	0.2660	0.0000	0.0329	0.0119	0.0000
690	0.7344	0.2656	0.0000	0.0227	0.0082	0.0000
695	0.7346	0.2654	0.0000	0.0158	0.0057	0.0000
700	0.7347	0.2653	0.0000	0.0114	0.0041	0.0000
705	0.7347	0.2653	0.0000	0.0081	0.0029	0.0000
710	0.7347	0.2653	0.0000	0.0058	0.0021	0.0000
715	0.7347	0.2653	0.0000	0.0041	0.0015	0.0000
720	0.7347	0.2653	0.0000	0.0029	0.0010	0.0000
725	0.7347	0.2653	0.0000	0.0020	0.0007	0.0000
730	0.7347	0.2653	0.0000	0.0014	0.0005	0.0000
735	0.7347	0.2653	0.0000	0.0010	0.0004	0.0000
740	0.7347	0.2653	0.0000	0.0007	0.0003	0.0000
745	0.7347	0.2653	0.0000	0.0005	0.0002	0.0000
750	0.7347	0.2653	0.0000	0.0003	0.0001	0.0000
755	0.7347	0.2653	0.0000	0.0002	0.0001	0.0000
760	0.7347	0.2653	0.0000	0.0002	0.0001	0.0000
765	0.7347	0.2653	0.0000	0.0001	0.0000	0.0000
770	0.7347	0.2653	0.0000	0.0001	0.0000	0.0000
775	0.7347	0.2653	0.0000	0.0000	0.0000	0.0000
780	0.7347	0.2653	0.0000	0.0000	0.0000	0.0000
			Totals	21.3713	21.3714	21.3715

SPECIFIC ROTATION

Specific rotation or rotatory power is given in degrees per decimeter for liquids and solutions and in degrees per millimeter for solids; + signifies right handed rotation, — left. Specific rotation varies with the wave length of light used, with temperature and, in the case of solutions, with the concentration. When sodium light is used, indicated by D in the wavelength column, a value of $\lambda = 0.5893$ may be assumed.

Optical rotatory power for a large number of organic compounds will be found in the International Critical Tables, Vol. VII; for sugars, Vol. II.

Solids

Substance	Wave length μ	Rotation deg./mm
Cinnabar (HgS)	D	+32.5
Lead hyposulfate	D	5.5
Potassium hyposulfate	D	8.4
Quartz	0.7604	12.668
	0.7184	14.304
	0.6867	15.746
	0.6562	17.318
	0.5895	21.684
	0.5889	21.727
	0.5269	27.543
	0.4861	32.773
	0.4307	42.604
	0.4101	47.481
	0.3968	51.193
	0.3933	52.155
	0.3820	55.625
	0.3726	58.894
Quartz (continued)	0.3609	+63.628
	0.3582	64.459
	0.3466	69.454
	0.3441	70.587
	0.3402	72.448
	0.3360	74.571
	0.3286	78.579
	0.3247	80.459
	0.3180	84.972
	0.2747	121.052
	0.2571	143.266
	0.2313	190.426
	0.2265	201.824
	0.2194	220.731
	0.2143	235.972
Sodium bromate	D	2.8
Sodium chlorate	D	3.13

Liquid

Liquid	Temp. °C	Wave length μ	Specific rotation deg./dm
Amyl alcohol	—	D	−5.7
Camphor	204	D	+70.33
Cedar oil	15	D	−30 to −40
Citrol oil	15	D	+62
Ethyl malate $(C_2H_5)_2C_4H_4O_5$	11	D	−10.3 to −12.4
Menthol	35.2	D	−49.7
Nicotine $C_{10}H_{14}N_2$	10—30	D	−162
	20	0.6563	−126
	20	0.5351	−207.5
	20	0.4861	−253.5
Turpentine $C_{10}H_6$	20	D	−37
	20	0.6563	−29.5
	20	0.5351	−45
	20	0.4861	−54.5

Solutions

Corrections for values of the specific rotation for concentration are given in the lat column. c indicates concentration in grams per 100 milliliters of solution; d indicates the concentration in grams per 100 grams of solution.

Substance	Solvent	Temp. °C	Wave length μ	Specific rotation deg./dm — Correction for concentration or temperature
Albumen	water	—	D	− 25 to −38
Arabinose	water	20	D	− 105.0
Camphor	alcohol	20	D	+ 54.4 −.135d for d = 45—91
	benzene	20	D	+ 56 −.166d for d = 47—90
	ether	—	D	+ 57
Dextrose *d*-glucose $C_6H_{12}O_6$	water	20	D	+ 52.5 + .025d for d = 1 18
			.5461	+ 62.03 + .04257c for c = 6—32
Galactose	water	—	D	+ 83.9 + .078d − .21t for d = 4—36 and t = 10—30°C
l-Glucose (β)	water	20	D	− 51.4
Invert sugar $C_6H_{12}O_6$	water	20	D	− 19.7 −.036c for c = 9—35
				$\alpha t = \alpha_{20} + .304(t - 20) + .00165 (t - 20)^2$ for t = 3—30°C
		25	.5461	− 21.5
Lactose	water	20	D	+ 52.4 + .072 (20° − t) for c = 5
			.5461	+ 61.9 + .085(20° − t) for c = 5
Levulose fruit sugar	water	25	D	− 88.5 − .145d for d = 2.6—18.6
		25	.5461	− 105.30
Maltose	water	20	D	+ 138.48 − .01837d for d = 5—35
		25	.5461	+ 153.75
Mannose	water	20	D	+ 14.1 c = 10.2
Nicotine	water	20	D	− 77 for d = 1—16
	benzene	20	D	− 164 for d = 8—100
Potassium tartrate	water	20	D	+ 27.14 + .0992c − .00094c² for c = 8—50
Quinine sulfate	water	17	D	− 214
Santonin	alcohol	20	D	− 161.0 c = 1.78
		20	D	+ 693 c = 4.05
	chloroform	20	D	− 202.7 + 309d for d = 75—96.5
	alcohol	20	.6867	+ 442 c = 4.05
			.5269	+ 991 c = 4.05
			.4861	+ 1323 c = 4.05
Sodium potassium tartrate (Rochelle salt)	water	20	D	+ 29.75 − .0078c
Sucrose (cane sugar) $C_{12}H_{22}O_{11}$	water	20	D	+ 66.412 + .01267d − .00376d² for d = 0—50
				$\alpha t = \alpha_{20}[1 - .00037 (t - 20)]$ for t = 14—30°C

Sucrose dissolved in water, 20°C.

μ	Spec. rot.	μ	Spec. rot.	μ	Spec. rot.
670.8(Li)	+ 50.51	510.6 (Cu)	+ 90.46	435.3(Fe)	+ 128.5
643.8(Cd)	55.04	508.6(Cd)	91.16	433.7(Fe)	129.8
636.2(Zn)	56.51	481.1(Zn)	103.07	431.5(Fe)	130.7
589.3(Na)	66.45	480.0(Cd)	103.62	428.2(Fe)	133.6
578.2(Cu)	69.10	472.2(Zn)	107.38	427.2(Fe)	134.2
578.0(Hg)	69.22	468.0(Zn)	109.49	426.1(Fe)	134.9
570.0(Cu)	71.24	467.8(Cd)	109.69	419.1(Fe)	140.0
546.1(Hg)	78.16	438.4(Fe)	126.5	414.4(Fe)	144.2
521.8(Cu)	86.21	437.6(Fe)	127.2	388.9(Fe)	166.7
515.3(Cu)	88.68	435.8(Hg)	128.49	383.3(Fe)	171.8
				382.6(Fe)	173.1

Substance	Solvent	°C	μ	Spec. rot. Correct.
Tartaric acid (ord)	water	20	D	+ 15.06 − .131c
		20	.6563	7.75
		20	D	8.86
		20	.5351	9.65 for d = 41
		20	.4861	9.37
Turpentine	alcohol	20	D	−37 − .00482d − .00013d for d = 0—90
	benzene	20	D	−37 − .0265d for d = 0—91
Xylose	water	20	D	+ 19.13 d = 2.7

OPTICAL ROTATION OF ACIDS AND BASES

Optical rotation of acids and bases commonly used in the resolution of racemic substances. Compiled by F. E. Ray.

Name	Formula	Solvent	Conc. %	α_D
Bromocamphor-sulfonic acid. K salt	$C_{10}H_{15}O_4BrS$	H_2O	—	72.1
Camphorsulfonic acid	$C_{10}H_{16}O_4S$	H_2O	—	23.9
Chlorocamphor-sulfonic acid	$C_{10}H_{15}ClO_4S$	H_2O	—	49.6
Codeinesulfonic acid	$C_{18}H_{21}NO_6S$	H_2O	3	−190.1
Hydroxybutyric acid	$C_4H_8O_3$	H_2O	3.3	−24.8
Lactic acid	$C_3H_6O_3$	H_2O	10.5	3.8
Malic acid	$C_3H_6O_5$	H_2O	—	2.4
Mandelic acid	$C_8H_8O_3$	H_2O	2.01	155.5
Methylene-camphor	$C_{11}H_{16}O$	C_2H_5OH	—	127
Phenylsuccinic acid	$C_{10}H_{10}O_4$	C_2H_5OH	1.5	148
Tartaric acid	$C_4H_6O_6$	C_2H_5OH and H_2O	—	3 to 25*
Brucine	$C_{23}H_{26}N_2O_4$	C_2H_5OH	5.4	−85
Cinchonidne	$C_{19}H_{22}N_2O$	C_2H_5OH	1.0	−111.0
Cinchonine	$C_{19}H_{22}N_2O$	$CHCl_3$	0.6	+209.6
Cocaine	$C_{17}H_{21}NO_4$	50% C_2H_5OH	1.1	−35.4
Coniine	$C_8H_{17}N$	$CHCl_3$	4	8.0
Codeine	$C_{18}H_{21}NO_3$	C_2H_5OH	5	−135.8
Hydrastine	$C_{21}H_{21}NO_6$	50% C_2H_5OH	0.2	115
Menthol	$C_{10}H_{20}O$	C_2H_5OH	9.6	−50.6
Menthylamine	$C_{10}H_{21}N$	C_2H_5OH	11.3	−31.9
Narcotine	$C_{22}H_{23}NO_7$	$CHCl_3$	2.6	±200.0
Quinidine	$C_{20}H_{23}N_2O_4$	C_2H_5OH	1.0	+233.6
Quinine	$C_{20}H_{23}N_2O_2$	C_6H_6	0.6	−136
Thebaine	$C_{19}H_{21}NO_2$	$CHCl_3$	5	−229.5
Strychnine	$C_{21}H_{22}N_2O_2$	C_2H_5OH	0.9	−128

* Varies greatly with temperature, solvent, and conc.

MAGNETIC ROTATORY POWER

The rotation of the plane of polarization of light by transparent substances subjected to a magnetic field may be applied to problems of molecular structure. This rotatory effect is known as the Faraday effect. Investigations of this effect by E. Verdet showed the angle of rotation (α) to depend on the nature of the substance and to be proportional to the length (l) of the column of the substance which the light traverses and to the strength (H) of the magnetic field. Thus

$$\alpha = \Lambda lH$$

where Λ is the Verdet constant for the experimental material.

Other symbols in the table have the following significance:

λ = Wavelength in μ

t = Degrees centigrade

ϱ_M = Molecular magnetic rotation of the substance under consideration compared to that of water determined in the same apparatus in the same magnetic field. Thus

$$\varrho_M = M'\alpha'\varrho'/M'\alpha'\varrho$$

where M is the molecular weight, α the angle of rotation and ϱ the density of the given substance, and M′, α', and ϱ' are the same quantities of water.

$[\Lambda]_M^{\lambda,\tau}$ = Molecular rotatory value of the substance at the given temperature and at the wavelength λ. Values are in (radians)(gauss^{-1})(cm^{-1}).

Values in this table are reproduced by permission from Volume 3, Tables de Constantes et Donnees Nuimeriques, "Pouvoir Rotatorie Magnetique, Effet Magneto-Optique de Kerr."

A much greater listing of values for both organic and inorganic compounds is in the above publication.

Formula	Name	λ	t	ϱ_M	$10^5[\Lambda]_M^{\lambda,t}$	$10^5\Lambda^{\lambda,t}$ Verdet
CH_4	Methane	578	0	—	—	*17.4
CCl_4	Tetrachloromethane	589	25.1	6.58	45.3	1.60
$CHCl_3$	Trichloromethane	589	20.0	5.535	37.98	1.60
$CHBr_3$	Tribromomethane	589	17.9	11.63	80.00	3.13
CH_2O_2	Formic acid	589	20.8	1.671	11.50	1.046
CH_2Cl_2	Dichloromethane	589	11.9	4.31	29.7	1.60
CH_2Br_2	Dibromomethane	589	15.9	8.11	55.8	2.74
CH_2I_2	Diiodomethane	589	15.0	10.83	129.5	1.51
CH_3Cl	Monochloromethane	589	23.0	2.99	20.5	1.37
CH_3Br	Monobromomethane	589	1.5	4.64	31.9	2.04
CH_3I	Monoiodomethane	589	19.5	9.01	63.2	3.35
CH_4O	Methylalcohol	589	18.7	1.640	11.28	0.958
CH_5N	Methylamine	578	—	—	—	*22.7
CF_2Cl_2	Difluorodichloromethane (Freon)	—	—	—	—	*32.7
CH_3O_2N	Mononitromethane	589	9.9	1.86	12.8	0.826
CH_4ON_2	Urea (carbamide) 40% aqueous solution	578	20.0	2.38	22.7	—
C_2H_2	Acetylene (ethyne)	578	—	—	—	*33.0
C_2H_4	Ethylene (ethene)	578	—	—	—	*34.5
C_2H_6	Ethane	578	—	—	—	*23.5
C_2N_2	Cyanogen	—	—	—	—	—
$C_2H_2O_4$	Oxalic acid (aqueous sol. 8.3%)	578	20	2.88	20.6	—
$C_2H_2O_4\cdot 2H_2O$	Oxalic acid (alcohol sol. 16.5%)	578	24	2.82	20.2	—
C_2H_3Br	Vinylbromide	589	7.8	6.22	42.8	2.10
C_2H_3N	Acetonitrile (ethanonitrile)	589	25.0	2.32	16.0	*21.0

* Verdet constant factor = 10^6.

MAGNETIC ROTATORY POWER (Continued)

Formula	Name	λ	t	ϱ_M	$10^5[\Lambda]_M^{\lambda,t}$	$10^5\Lambda^{\lambda,t}$ Verdet
C_2H_4O	Ethyleneoxide (1,2-epoxyethane)	589	8.0	1.935	13.3	0.92
C_2H_4O	Acetaldehyde (ethanal)	589	16.3	2.38	16.4	1.00
$C_2H_4O_2$	Acetic acid	589	21.0	2.525	17.37	1.044
$C_2H_4O_2$	Formic acid methylester (methylformate)	589	16.5	2.49	17.1	0.96
$C_2H_4Cl_2$	Ethylidenechloride (1,1-dichloroethane)	589	14.4	5.33	36.7	1.51
$C_2H_4Cl_2$	Ethylenechloride (1,2-dichloroethane)	589	14.4	5.49	37.7	1.65
$C_2H_4Br_2$	Ethylenebromide (1,2-dibromoethane)	589	15.2	9.70	66.7	2.66
C_2H_5Cl	Monochloroethane	589	5.0	4.04	27.8	1.36
C_2H_5Br	Monobromoethane	589	19.7	5.85	40.2	1.82
C_2H_5I	Monoiodoethane	589	18.1	10.07	69.3	2.95
C_2H_6O	Ethylalcohol (ethanol)	589	16.8	2.780	19.13	1.131
$C_2H_6O_2$	Glycol (1,2-ethanediol)	589	15.1	2.94	20.2	1.25
C_2H_6S	Ethylmercaptan	578	16.0	5.52	39.5	1.85
C_2H_7N	Ethylamine (aminoethane)	589	5.8	3.61	24.8	*34.5
$C_2H_2O_2Cl_2$	Dichloroacetic acid (dichloroethanoic acid)	589	13.5	5.30	3.65	1.52
$C_2H_3O_2Cl$	Chloroacetic acid (chloroethanoic acid)	589	64.5	3.89	26.7	1.33
$C_2H_3O_2Cl_3$	Chloralhydrate (2-2-2-trichloro-1,1-ethanediol)	589	54.6	7.10	48.8	1.65
$C_2H_5O_2N$	Mononitroethane	589	10.2	2.84	19.5	0.946
C_3H_3	Propane	578	—	—	—	*34.0
C_3H_4O	Acrolein (propenal)	578	20.0	4.74	34.0	1.76
$C_3H_4O_3$	Pyruvic acid (2-oxopropanoic acid)	589	14.5	3.56	24.2	1.21
$C_3H_4O_4$	Malonic acid (propandioic acid) 2 *n*-aqueous sol.	589	23.0	3.47	—	—
C_3H_6O	Allyl alcohol	589	18.3	4.68	32.2	1.60
C_3H_6O	Propyl alcohol (1-propanol)	589	13.6	3.33	22.9	1.09
C_3H_6O	Acetone (2-propanone)	589	20.0	3.472	23.89	1.1136
$C_3H_6O_2$	Propionic acid (propanoic acid)	589	20.3	3.462	23.82	1.10
$C_3H_6O_2$	Formic acid ethylester (ethylmethanoate)	589	18.8	3.56	24.5	1.05
$C_3H_6O_2$	Acetic acid methylester (methylacetate)	589	20.0	3.42	23.5	1.03
C_3H_7Cl	Propylchloride (1-chloropropane)	589	16.1	5.04	34.7	1.34
C_3H_7Cl	Isopropylchloride (2-chloropropane)	589	17.2	5.16	35.5	1.34
C_3H_7Br	Propylbromide (1-bromopropane)	589	19.2	6.88	47.8	1.79
C_3H_7Br	Isopropylbromide (2-bromopropane)	589	17.1	7.00	48.2	1.77
C_3H_7I	Propyliodide (1-iodopropane)	589	18.1	11.08	76.2	2.69
C_3H_7I	Isopropyliodide (2-iodopropane)	589	26.3	11.18	76.9	2.63
C_3H_8O	*n*-Propyl alcohol (1-propanol)	589	15.6	3.77	25.9	1.20
C_3H_8O	Isopropyl alcohol (2-propanol)	589	20.0	3.90	26.8	1.23
$C_3H_8O_3$	Glycerine (1,2,3-propanetriol)	589	16.0	4.11	28.3	1.33
C_3H_9N	*n*-Propylamine	589	9.6	4.56	31.4	1.33
$C_3H_5O_9N_3$	Nitroglycerine	589	13.5	5.405	37.2	0.900
$C_3H_7O_2N$	1-Nitropropane	589	18.9	3.82	26.3	1.018
C_4H_6	1,3-Butadiene (erythrene)	589	15.0	7.94	54.6	2.16
C_4H_8	1-Butene (α-butylene)	589	15.0	5.53	38.0	1.39
C_4H_8	cis-2-Butene (β-butylene)	589	15.0	5.27	36.3	1.38
C_4H_8	trans-2-Butene	589	15.0	5.07	34.9	1.29
C_4H_{10}	Butane	589	15.0	4.59	31.6	1.09
C_4H_{10}	Isobutane (2-methylpropane)	589	15.0	4.87	33.5	1.11
$C_4H_2O_3$	Maleic anhydride (cis-butenedioic anhydride)	589	25.0	4.5	31.0	—
C_4H_4O	Furan (furfuran)	589	20.0	5.48	37.7	1.78
$C_4H_4O_4$	Maleic acid (cis-butenedioic acid) 2 *n* aqueous solution	589	25.0	5.63	38.7	—
C_4H_4S	Thiophene (thiofuran)	589	20.0	9.40	64.7	2.83
$C_4H_6O_3$	Acetic anhydride (ethanoic anhydride)	589	20.0	4.28	29.5	1.115
$C_4H_6O_4$	Succinic acid (butanedioic acid) 5.9% aqueous sol.	578	20.0	4.68	33.5	—
$C_4H_6O_6$	Tartaric acid (47.8% aqueous sol.)	578	20.0	4.79	34.3	—
$C_4H_8O_2$	*n*-Butyric acid (butanoic acid)	589	18.8	4.47	30.8	1.15
$C_4H_8O_2$	Ethylacetate (ethyl ethanoate)	589	20.0	4.47	30.8	1.08
$C_4H_8O_2$	Propionic acid methylester (methylpropanoate)	589	20.0	4.37	30.1	1.07
$C_4H_8O_3$	Lactic acid methylester (methyl 2-hydroxypropanoate)	589	—	4.66	32.1	—
$C_4H_{10}O$	Ethyl ether (ethoxyethane)	589	20.0	4.78	32.9	1.09
$C_4H_{10}O$	*n*-Butyl alcohol (1-butanol)	589	20.0	4.60	31.6	1.23
$C_4H_{10}O$	Isobutyl alcohol (2-methyl-1-propanol)	589	17.7	4.94	34.0	1.27
$C_4H_{10}O$	sec-Butyl alcohol (methylethylcarbinol)	589	20.0	4.91	33.8	1.27
C_5H_6	Cyclopentadiene	589	15.0	7.03	48.8	2.02
C_5H_8	1,3-Pentadiene	589	15.0	8.80	60.5	2.08
C_5H_8	Isoprene (2-methyl-1,3-butadiene)	589	15.0	8.80	60.5	2.08
C_5H_8	Cyclopentene	589	15.0	5.69	39.1	1.52
C_5H_{10}	1-Pentene	589	15.0	6.45	44.4	1.39
C_5H_{10}	Isopentane (2-methyl-1-butane)	589	15.0	6.36	43.8	1.39
C_5H_{10}	Cyclopentane	589	20.0	4.89	33.6	1.23
C_5H_{12}	Pentane	589	15.0	5.60	38.5	1.15
C_5H_{12}	Isopentane (2-methylbutane)	589	15.0	5.75	39.6	1.17
$C_5H_4O_2$	Furfural (2-furancarbonal)	578	20.0	7.01	50.2	2.06
C_5H_5N	Pyridine	589	11.9	8.76	60.3	2.58
$C_5H_8O_4$	Glutaric acid (pentanedioic acid) 2 N aqueous sol.	589	16.0	5.48	37.7	—
$C_5H_{10}O_2$	Propionic acid ethylester	589	20.0	5.46	37.6	1.13
$C_5H_{10}O_2$	Acetic acid propylester	589	15.7	5.45	37.5	1.13
$C_5H_{14}N_2$	Cadaverine (1,5-pentanediamine)	589	14.7	7.49	51.6	1.53
C_6H_6	Benzene	589	15.0	11.27	77.5	3.00
C_6H_{12}	Cyclohexane	589	20.0	5.66	39.0	1.24
C_6H_{14}	Hexane	589	15.0	6.62	45.5	1.20
$C_6H_4Cl_2$	1,4-Dichlorobenzene (*p*-dichlorobenzene)	589	64.5	13.55	93.2	2.69
C_6H_5F	Fluorobenzene (phenylfluoride)	589	19.0	9.96	68.5	2.51
C_6H_5Cl	Chlorobenzene (phenylchloride)	589	15.0	12.51	86.1	2.92
C_6H_5Br	Bromobenzene (phenylbromide)	589	15.0	14.51	99.8	3.26
C_6H_5I	Iodobenzene (phenyliodide)	589	15.0	19.11	131.4	4.06
C_6H_6O	Phenol (hydroxybenzene)	589	39.0	12.07	83.5	3.21
C_6H_7N	Aniline (aminobenzene)	589	15.0	16.08	110.6	4.18
$C_6H_{11}Cl$	Chlorocyclohexane (cyclohexylchloride)	589	13.0	7.50	51.6	1.46
$C_6H_{12}O_3$	Paraldehyde (paraacetaldehyde)	589	17.3	6.66	45.8	1.19
$C_6H_{12}O_6$	Glucose 11H_2O (dextrose) (1 M aqueous sol.)	589	15.0	6.72	46.2	—
$C_6H_{12}O_6$	Galactose 10H_2O (1 M aqueous sol.)	589	15.0	6.89	47.4	—
$C_6H_{12}O_6$	Fructose 10H_2O (levulose) (1 M aqueous sol.)	589	15.0	6.73	46.3	—
$C_6H_{14}O$	2-Hexanol (butylmethylcarbinol)	589	20.0	6.89	47.4	1.31
$C_6H_{14}O$	3-Hexanol (ethylpropylcarbinol)	589	20.0	6.85	47.1	1.30
$C_6H_{14}O$	2-Methyl-3-Pentanol (ethylisopropylcarbinol)	589	20.0	6.90	47.5	1.32
$C_6H_4O_4N_2$	1,3-Dinitrobenzene (*m*-Dinitrobenzene)	589	17.1	9.65	66.4	2.17
$C_6H_5O_2N$	Nitrobenzene	589	15.0	9.36	64.4	2.17
C_7H_8	Toluene (methylbenzene)	589	15.0	12.16	83.7	2.71
C_7H_{14}	1-Heptene (α-heptylene)	589	18.0	8.48	58.3	1.43
C_7H_{16}	Heptane	589	15.0	7.61	52.7	1.23

MAGNETIC ROTATORY POWER (Continued)

Formula	Name	λ	t	ρ_M	$10^5[A]_M^{\lambda,t}$	$10^5\Lambda^{\lambda,t}$ Verdet
C_7H_5N	Benzonitrile (benzenecarbonitrile)	589	15.7	11.85	81.5	2.74
$C_7H_6O_2$	Benzoic acid (20% alcohol sol.)	578	20.0	11.8	84.7	—
C_7H_7Cl	*o*-Chlorotoluene (2-chloro-1-methylbenzene)	589	15.4	13.72	94.3	2.95
C_7H_7Cl	*p*-Chlorotoluene (4-chloro-1-methylbenzene)	589	15.2	13.25	90.8	2.65
C_7H_7Br	*o*-Bromotoluene (2-bromo-1-methylbenzene)	589	16.7	15.67	107.3	3.08
C_7H_7Br	*p*-Bromotoluene (4-bromo-1-methylbenzene)	589	39.0	15.09	103.6	2.88
C_7H_8O	*o*-Cresol (*o*-methylphenol)	589	16.0	13.38	92.1	3.07
C_7H_8O	*m*-Cresol (*m*-methylphenol)	589	17.9	12.77	87.6	2.89
C_7H_8O	*p*-Cresol (*p*-methylphenol)	589	17.0	12.86	88.5	2.91
C_7H_9N	*o*-Toluidine (*o*-methylaniline)	589	17.3	17.18	118.2	3.79
C_7H_9N	*m*-Toluidine (*m*-methylaniline)	589	15.0	16.21	111.5	3.56
C_7H_9N	*p*-Toluidine (*p*-methylaniline)	589	50.0	15.92	109.5	3.37
$C_7H_{14}O$	Enanthaldehyde (heptanal)	589	16.2	7.42	51.1	1.26
$C_7H_{16}O$	1-Heptanol (*n*-heptylalcohol)	589	12.6	7.85	54.0	1.33
$C_7H_{16}O$	2-Heptanol (amylmethylcarbinol)	589	20.0	7.94	54.6	1.32
$C_7H_{16}O$	3-Heptanol (butylethylcarbinol)	589	20.0	7.86	54.1	1.37
$C_7H_7O_2N$	*o*-Nitrotoluene	589	18.0	10.80	74.3	2.16
$C_7H_7O_2N$	*p*-Nitrotoluene	589	54.3	10.20	70.2	1.97
C_8H_{10}	Ethylbenzene (phenylethane)	589	15.0	13.41	92.3	2.80
C_8H_{10}	*o*-Xylene (1,2-dimethylbenzene)	589	15.0	13.36	91.9	2.62
C_8H_{10}	*m*-Xylene (1,3-dimethylbenzene)	589	15.0	12.82	88.2	2.47
C_8H_{10}	*p*-Xylene (1,4-dimethylbenzene)	589	15.0	12.80	88.1	2.46
C_8H_{16}	1-Octene (α-octylene)	589	15.0	9.00	65.5	1.44
C_8H_{16}	2-Octene (β-octylene)	589	15.0	9.33	64.2	1.43
C_8H_{18}	Octane	589	15.0	8.65	59.5	1.26
$C_8H_{18}O$	1-Octanol (*n*-octylalcohol)	589	20.0	8.88	61.1	1.33
$C_8H_{18}O$	2-Octanol (methylhexylcarbinol)	589	20.0	9.00	61.8	1.34
$C_8H_{18}O$	3-Octanol (ethylamylcarbinol)	589	20.0	8.90	61.2	1.33
C_9H_{12}	*o*-Ethyltoluene (1-ethyl-2-ethylbenzene)	589	15.0	14.56	100.2	2.32
C_9H_{12}	*m*-Ethyltoluene (1-ethyl-3-ethylbenzene)	589	15.0	14.18	97.6	2.91
C_9H_{12}	*p*-Ethylbenzene (1-ethyl-4-ethylbenzene)	589	15.0	13.98	96.2	2.37
C_9H_{12}	Mesitylene (1-3-5-trimethylbenzene)	589	15.0	13.36	91.9	2.28
C_9H_{20}	Nonane	589	15.0	9.70	66.7	1.28
$C_{10}H_8$	Naphthalene	589	89.5	24.98	171.8	4.47
$C_{10}H_{20}$	1-Decene (*n*-decylene)	589	21.0	11.65	80.1	1.45
$C_{10}H_{22}$	Decane	589	15.0	10.70	73.6	1.30
$C_{10}H_7Cl$	1-Chloronaphthalene (α-chloronaphthalene)	578	18.0	28.15	201.5	4.91
$C_{10}H_7Br$	1-Bromonaphthalene (α-bromonaphthalene)	578	20.0	31.05	222.0	5.19
$C_{10}H_8O$	β-Naphthol (2-hydroxynaphthalene)	578	136.0	27.1	194.0	4.80
$C_{10}H_9N$	1-Naphthylamine (α-naphthylamine)	589	32.6	37.23	256.1	6.84
$C_{10}H_{12}O_2$	Isoeugenol (4-propenylguaiacol)	589	19.3	21.44	147.5	3.55
$C_{10}H_{12}O_2$	Eugenol (4-allylguaiacol)	589	15.4	18.72	128.8	2.88
$C_{10}H_{12}O_2$	Benzoic acid propylester (n-propylbenzoate)	589	15.4	14.87	102.3	2.20
$C_{10}H_{12}O_2$	o-Toluic acid ethylester	589	15.2	15.06	103.6	2.25
$C_{10}H_{12}O_2$	p-Toluic acid ethylester	589	15.0	14.74	101.4	2.18
$C_{10}H_{12}O_2$	α-Toluic acid ethylester (ethylphenylacetate)	589	14.0	14.09	103.1	2.25
$C_{10}H_{12}O_3$	Methylsalicyclic acid ethylester	589	18.6	17.14	117.9	2.50
$C_{10}H_{15}N$	N,N-Diethylaniline (N-phenyldiethylamine)	589	15.3	25.16	173.1	3.74
$C_{10}H_{16}O$	Camphor	589	14.0	9.26	63.7	—
$C_{10}H_{18}O$	α-Terpineol	589	16.0	10.84	76.6	1.56
$C_{10}H_{18}O$	Citronellal	589	14.5	11.48	79.0	1.51
$C_{10}H_{18}O_4$	Dipropylsuccinate	589	11.4	10.36	71.3	1.22
$C_{10}H_{10}O_6$	Tartaric acid dipropylester (propyltartrate)	589	15.4	10.83	74.5	1.24
$C_{10}H_{20}O$	Menthol	589	45.2	10.51	72.3	1.40
$C_{11}H_{24}$	Undecane	589	20.5	11.65	80.1	1.31
$C_{12}H_{26}$	Dodecane	589	21.5	12.71	87.4	1.32
$C_{12}H_{22}O_{11}$	Saccharose $19H_2O$ (1 M aqueous sol.)	589	15.0	12.59	86.6	—
$C_{12}H_{22}O_{11}$	Maltose $20H_2O$ (1 M aqueous sol.)	589	15.0	12.69	87.3	—
$C_{12}H_{22}O_{11}$	Lactose $41H_2O$ (1 M aqueous sol.)	589	18.3	12.71	87.5	—
$C_{14}H_{10}$	Phenanthrene	578	100.0	39.7	284.0	5.84
$C_{16}H_{34}$	Hexadecane	589	15.0	16.8	115.6	1.35
$C_{18}H_{14}$	1,2-Diphenylbenzene	589	15.0	40.2	276.3	4.70
$C_{18}H_{14}$	1,3-Diphenylbenzene (m-phenyldiphenyl)	589	15.0	41.0	282.4	—
$C_{18}H_{22}$	1,6-Diphenylhexane	589	20.0	—	—	2.75

TRANSPARENCY TO OPTICAL DENSITY CONVERSION TABLE

Transparency of a layer of material is defined as the ratio of the intensity of the tra-smitted light to that of the incident light. Opacity is the reciprocal of the transparency. Optical density is the common logarithm of the opacity.

Thus,

$$\text{Transparency} = \frac{I_t}{I_i}$$

$$\text{Opacity} = \frac{1}{\text{Transparency}} = \frac{I_i}{I_t}$$

$$\text{Optical density} = \log_{10}\left(\frac{I_i}{I_t}\right)$$

where I_i = Intensity of incident light
I_t = Intensity of transmitted light.

Trans.	Density	Trans.	Density	Trans.	Density	Trans.	Density	Trans.	Density	Trans.	Density	Trans.	Density	Trans.	Density
0.000	—	.005	2.301	.010	2.000	.015	1.824	.020	1.699	.025	1.602	.030	1.523	.035	1.456
.001	3.000	.006	2.222	.011	1.959	.016	1.796	.021	1.678	.026	1.585	.031	1.509	.036	1.444
.002	2.699	.007	2.155	.012	1.921	.017	1.770	.022	1.658	.027	1.569	.032	1.495	.037	1.432
.003	2.523	.008	2.097	.013	1.886	.018	1.745	.023	1.638	.028	1.553	.033	1.482	.038	1.420
.004	2.398	.009	2.046	.014	1.854	.019	1.721	.024	1.620	.029	1.538	.034	1.469	.039	1.409

TRANSPARENCY TO OPTICAL DENSITY CONVERSION TABLE (Continued)

Trans.	Density	Trans.	Density	Trans.	Density	Trans.	Density	Trans.	Density	Trans.	Density	Trans.	Density	Trans.	Density
.040	1.398	.126	.8996	.212	.6737	.299	.5243	.386	.4134	.473	.3251	.500	.2518	.647	.1891
.041	1.387	.127	.8962	.213	.6716	.300	.5229	.387	.4123	.474	.3242	.561	.2510	.648	.1884
.042	1.377	.128	.8928	.214	.6696	.301	.5215	.388	.4112	.475	.3233	.562	.2503	.649	.1877
.043	1.367	.129	.8894	.215	.6676	.302	.5200	.389	.4101	.476	.3224	.564	.2487	.650	.1871
.044	1.357	.130	.8861	.216	.6655	.303	.5186	.390	.4089	.477	.3215	.565	.2479	.651	.1864
.045	1.347	.131	.8827	.217	.6635	.304	.5171	.391	.4078	.478	.3206	.566	.2472	.652	.1857
.046	1.337	.132	.8794	.218	.6615	.305	.5157	.392	.4067	.479	.3197	.567	.2464	.653	.1851
.047	1.328	.133	.8761	.219	.6596	.306	.5143	.393	.4056	.480	.3188	.568	.2457	.654	.1844
.048	1.319	.134	.8729	.220	.6576	.307	.5128	.394	.4045	.481	.3179	.569	.2449	.655	.1838
.049	1.310	.135	.8697	.221	.6556	.308	.5114	.395	.4034	.482	.3170	.570	.2441	.656	.1831
.050	1.301	.136	.8665	.222	.6536	.309	.5100	.396	.4023	.483	.3161	.571	.2434	.657	.1824
.051	1.292	.137	.8633	.223	.6517	.310	.5086	.397	.4012	.484	.3152	.572	.2426	.658	.1818
.052	1.284	.138	.8601	.224	.6498	.311	.5072	.398	.4001	.485	.3143	.573	.2418	.659	.1811
.053	1.276	.139	.8570	.225	.6478	.312	.5058	.399	.3990	.486	.3134	.574	.2411	.660	.1805
.054	1.268	.140	.8539	.226	.6459	.313	.5045	.400	.3979	.487	.3125	.575	.2403	.661	.1798
.055	1.260	.141	.8508	.227	.6440	.314	.5031	.401	.3969	.489	.3107	.576	.2396	.662	.1791
.056	1.252	.142	.8477	.228	.6421	.315	.5017	.402	.3958	.490	.3098	.577	.2388	.663	.1785
.057	1.244	.143	.8447	.229	.6402	.316	.5003	.403	.3947	.491	.3089	.578	.2381	.664	.1778
.058	1.237	.144	.8416	.230	.6383	.317	.4989	.404	.3936	.492	.3080	.579	.2373	.665	.1772
.059	1.229	.145	.8386	.231	.6364	.318	.4976	.405	.3925	.493	.3072	.580	.2366	.666	.1765
.060	1.222	.146	.8356	.232	.6345	.319	.4962	.406	.3915	.494	.3063	.581	.2358	.667	.1759
.061	1.215	.147	.8327	.233	.6326	.320	.4949	.407	.3904	.495	.3054	.582	.2351	.668	.1752
.062	1.208	.148	.8297	.234	.6308	.321	.4935	.408	.3893	.496	.3045	.583	.2343	.669	.1746
.063	1.201	.149	.8268	.235	.6289	.322	.4921	.409	.3883	.497	.3036	.584	.2336	.670	.1739
.064	1.194	.150	.8239	.236	.6271	.323	.4908	.410	.3872	.498	.3028	.585	.2328	.671	.1733
.065	1.187	.151	.8210	.237	.6253	.324	.4895	.411	.3862	.499	.3019	.586	.2321	.672	.1726
.066	1.180	.152	.8182	.238	.6234	.325	.4881	.412	.3851	.500	.3010	.587	.2314	.673	.1720
.067	1.174	.153	.8153	.239	.6216	.326	.4868	.414	.3830	.501	.3002	.588	.2306	.674	.1713
.068	1.168	.154	.8125	.240	.6198	.327	.4855	.415	.3819	.502	.2993	.589	.2299	.675	.1707
.069	1.161	.155	.8097	.241	.6180	.328	.4841	.416	.3809	.503	.2984	.590	.2291	.676	.1701
.070	1.155	.156	.8069	.242	.6162	.329	.4828	.417	.3799	.504	.2975	.591	.2284	.677	.1694
.071	1.149	.157	.8041	.243	.6144	.330	.4815	.418	.3788	.505	.2967	.592	.2277	.678	.1688
.072	1.143	.158	.8013	.244	.6126	.331	.4802	.419	.3778	.506	.2959	.593	.2269	.679	.1681
.073	1.137	.159	.7986	.245	.6108	.332	.4789	.420	.3768	.507	.2950	.594	.2262	.680	.1675
.074	1.131	.160	.7959	.246	.6091	.333	.4776	.421	.3757	.508	.2951	.595	.2255	.681	.1668
.075	1.125	.161	.7932	.247	.6073	.334	.4763	.422	.3747	.509	.2933	.596	.2248	.682	.1662
.076	1.119	.162	.7905	.248	.6056	.335	.4750	.423	.3737	.510	.2924	.597	.2240	.683	.1655
.077	1.114	.163	.7878	.249	.6038	.336	.4737	.424	.3726	.511	.2916	.598	.2233	.684	.1649
.078	1.108	.164	.7852	.250	.6021	.337	.4724	.425	.3716	.512	.2907	.599	.2226	.685	.1643
.079	1.102	.165	.7825	.251	.6003	.339	.4698	.426	.3706	.513	.2899	.600	.2219	.686	.1637
.080	1.097	.166	.7799	.252	.5986	.340	.4685	.427	.3696	.514	.2890	.601	.2211	.687	.1630
.081	1.092	.167	.7773	.253	.5969	.341	.4673	.428	.3685	.515	.2882	.602	.2204	.688	.1624
.082	1.086	.168	.7747	.254	.5952	.342	.4660	.429	.3675	.516	.2873	.603	.2197	.689	.1618
.083	1.081	.169	.7721	.255	.5935	.343	.4647	.430	.3665	.517	.2865	.604	.2190	.690	.1612
.084	1.076	.170	.7696	.256	.5918	.344	.4634	.431	.3655	.518	.2857	.605	.2182	.691	.1605
.085	1.071	.171	.7670	.257	.5901	.345	.4622	.432	.3645	.519	.2848	.606	.2175	.692	.1599
.086	1.066	.172	.7645	.258	.5884	.346	.4609	.433	.3635	.520	.2840	.607	.2168	.693	.1593
.087	1.060	.173	.7620	.259	.5867	.347	.4597	.434	.3625	.521	.2831	.608	.3161	.694	.1586
.088	1.055	.174	.7594	.260	.5850	.348	.4584	.435	.3615	.522	.2823	.609	.2154	.695	.1580
.089	1.051	.175	.7570	.261	.5834	.349	.4572	.436	.3605	.523	.2815	.610	.2147	.696	.1574
.090	1.046	.176	.7545	.263	.5800	.350	.4559	.437	.3595	.524	.2807	.611	.2140	.697	.1568
.091	1.041	.177	.7520	.264	.5784	.351	.4547	.438	.3585	.525	.2798	.612	.2132	.698	.1562
.092	1.036	.178	.7496	.265	.5768	.352	.4535	.439	.3575	.526	.2790	.613	.2125	.699	.1555
.093	1.032	.179	.7471	.266	.5751	.353	.4522	.440	.3565	.527	.2782	.614	.2118	.700	.1549
.094	1.027	.180	.7447	.267	.5735	.354	.4510	.441	.3556	.528	.2774	.615	.2111	.701	.1543
.095	1.022	.181	.7423	.268	.5719	.355	.4498	.442	.3546	.529	.2766	.616	.2104	.702	.1537
.096	1.018	.182	.7399	.269	.5702	.356	.4486	.443	.3536	.530	.2757	.617	.2097	.703	.1531
.097	1.013	.183	.7375	.270	.5686	.357	.4473	.444	.3526	.531	.2749	.618	.2090	.704	.1524
.098	1.009	.184	.7352	.271	.5670	.358	.4461	.445	.3516	.532	.2741	.619	.2083	.705	.1518
.099	1.004	.185	.7328	.272	.5654	.359	.4449	.446	.3507	.533	.2733	.620	.2076	.706	.1512
.100	1.000	.186	.7305	.273	.5638	.360	.4437	.447	.3497	.534	.2725	.621	.2069	.707	.1506
.101	.9957	.187	.7282	.274	.5622	.361	.4425	.448	.3487	.535	.2717	.622	.2062	.708	.1500
.102	.9914	.188	.7258	.275	.5607	.362	.4413	.449	.3478	.536	.2708	.623	.2055	.709	.1493
.103	.9872	.189	.7235	.276	.5591	.363	.4401	.450	.3468	.537	.2700	.624	.2048	.710	.1487
.104	.9830	.190	.7212	.277	.5575	.364	.4389	.451	.3458	.538	.2696	.625	.2041	.711	.1481
.105	.9788	.191	.7190	.278	.5560	.365	.4377	.452	.3449	.539	.2684	.626	.2034	.712	.1475
.106	.9747	.192	.7167	.279	.5544	.366	.4365	.453	.3439	.540	.2676	.627	.2027	.713	.1469
.107	.9706	.193	.7144	.280	.5528	.367	.4353	.454	.3429	.541	.2668	.628	.2020	.714	.1463
.108	.9666	.194	.7122	.281	.5513	.368	.4342	.455	.3420	.542	.2660	.629	.2013	.715	.1457
.109	.9626	.195	.7100	.282	.5498	.369	.4330	.456	.3410	.543	.2652	.630	.2007	.716	.1451
.110	.9586	.196	.7077	.283	.5482	.370	.4318	.457	.3401	.544	.2644	.631	.2000	.717	.1445
.111	.9547	.197	.7055	.284	.5467	.371	.4306	.458	.3391	.545	.2636	.632	.1993	.718	.1439
.112	.9508	.198	.7033	.285	.5452	.372	.4295	.459	.3382	.546	.2628	.633	.1986	.719	.1433
.113	.9469	.199	.7011	.286	.5436	.373	.4283	.460	.3372	.547	.2620	.634	.1979	.720	.1427
.114	.9431	.200	.6990	.287	.5421	.374	.4271	.461	.3363	.548	.2612	.635	.1972	.721	.1421
.115	.9393	.201	.6968	.288	.5406	.375	.4260	.462	.3354	.549	.2604	.636	.1965	.722	.1415
.116	.9356	.202	.6946	.289	.5391	.376	.4248	.463	.3344	.550	.2596	.637	.1959	.723	.1409
.117	.9318	.203	.6925	.290	.5376	.377	.4237	.464	.3335	.551	.2589	.638	.1952	.724	.1403
.118	.9281	.204	.6904	.291	.5361	.378	.4225	.465	.3325	.552	.2581	.639	.1945	.725	.1397
.119	.9244	.205	.6882	.292	.5346	.379	.4214	.466	.3316	.553	.2573	.640	.1938	.726	.1391
.120	.9208	.206	.0861	.293	.5331	.380	.4202	.467	.3307	.554	.2565	.641	.1932	.727	.1385
.121	.9172	.207	.6840	.294	.5317	.381	.4191	.468	.3298	.555	.2557	.642	.1925	.728	.1379
.122	.9137	.208	.6819	.295	.5302	.382	.4179	.460	.3288	.556	.2549	.643	.1918	.729	.1373
.123	.9101	.209	.6799	.296	.5287	.383	.4168	.470	.3279	.557	.2541	.644	.1911	.730	.1367
.124	.9066	.210	.6778	.297	.5272	.384	.4157	.471	.3270	.558	.2534	.645	.1904	.731	.1361
.125	.9031	.211	.6757	.298	.5258	.385	.4145	.472	.3260	.559	.2526	.646	.1898	.732	.1355

TRANSPARENCY TO OPTICAL DENSITY CONVERSION TABLE (Continued)

Trans.	Density	Trans.	Density	Trans.	Density	Trans.	Density	Trans.	Density
.733	.1349	.791	.1018	.849	.0711	.907	.0424	.965	.0155
.734	.1343	.792	.1013	.850	.0706	.908	.0419	.966	.0150
.735	.1337	.793	.1007	.851	.0701	.909	.0414	.967	.0146
.736	.1331	.794	.1002	.852	.0696	.910	.0410	.968	.0141
.737	.1325	.795	.0996	.853	.0690	.911	.0405	.969	.0137
.738	.1319	.796	.0991	.854	.0685	.912	.0400	.970	.0132
.739	.1314	.797	.0985	.855	.0680	.913	.0395	.971	.0128
.740	.1308	.798	.0980	.856	.0675	.914	.0391	.972	.0123
.741	.1302	.799	.0975	.857	.0670	.915	.0386	.973	.0119
.742	.1296	.800	.0969	.858	.0665	.916	.0381	.974	.0114
.743	.1290	.801	.0964	.859	.0660	.917	.0376	.975	.0110
.744	.1284	.802	.0958	.860	.0655	.918	.0371	.967	.0106
.745	.1278	.803	.0953	.861	.0650	.919	.0367	.977	.0101
.746	.1273	.804	.0948	.862	.0645	.920	.0362	.978	.0097
.747	.1267	.805	.0942	.863	.0640	.921	.0357	.979	.0092
.748	.1261	.806	.0937	.804	.0635	.922	.0353	.980	.0088
.749	.1255	.807	.0931	.865	.0630	.923	.0348	.981	.0083
.750	.1249	.808	.0926	.866	.0625	.924	.0343	.982	.0079
.751	.1244	.809	.0921	.867	.0620	.925	.0339	.983	.0074
.752	.1238	.810	.0915	.808	.0615	.926	.0334	.984	.0070
.753	.1232	.811	.0910	.869	.0610	.927	.0329	.985	.0066
.754	.1226	.812	.0904	.870	.0605	.928	.0325	.986	.0061
.755	.1221	.813	.0899	.871	.0600	.929	.0320	.987	.0057
.756	.1215	.814	.0894	.872	.0595	.930	.0315	.988	.0052
.757	.1209	.815	.0888	.873	.0590	.931	.0310	.989	.0048
.758	.1203	.816	.0883	.874	.0585	.932	.0306	.990	.0044
.759	.1198	.817	.0878	.975	.0580	.933	.0301	.991	.0039
.760	.1192	.818	.0872	.876	.0575	.934	.0296	.992	.0035
.761	.1186	.819	.0867	.877	.0570	.935	.0292	.993	.0030
.762	.1180	.820	.0862	.878	.0565	.936	.0287	.994	.0026
.763	.1175	.821	.0856	.879	.0560	.937	.0282	.995	.0022
.764	.1169	.822	.0851	.880	.0555	.938	.0278	.996	.0017
.765	.1163	.823	.0846	.881	.0550	.939	.0273	.997	.0013
.766	.1158	.824	.0841	.882	.0545	.940	.0269	.998	.0009
.767	.1152	.825	.0835	.883	.0540	.941	.0264	.999	.0004
.768	.1146	.826	.0830	.884	.0535	.942	.0260	1.000	.0000
.769	.1141	.827	.0825	.885	.0530	.943	.0255		
.770	.1135	.828	.0820	.886	.0526	.944	.0250		
.771	.1129	.829	.0815	.887	.0521	.945	.0246		
.772	.1124	.830	.0809	.888	.0516	.946	.0241		
.773	.1118	.831	.0804	.889	.0511	.947	.0237		
.774	.1113	.832	.0799	.890	.0506	.948	.2032		
.775	.1107	.833	.0794	.891	.0501	.949	.0227		
.776	.1102	.834	.0788	.892	.0496	.950	.0223		
.777	.1096	.835	.0783	.893	.0491	.951	.0218		
.778	.1090	.836	.0778	.894	.0487	.952	.0214		
.779	.1085	.837	.0773	.895	.0482	.953	.0209		
.780	.1079	.838	.0767	.896	.0477	.954	.0204		
.781	.1073	.839	.0762	.897	.0472	.955	.0200		
.782	.1068	.840	.0757	.898	.0467	.956	.0195		
.783	.1062	.841	.0752	.899	.0462	.957	.0191		
.784	.1057	.842	.0747	.900	.0458	.958	.0186		
.785	.1051	.843	.0742	.901	.0453	.959	.0182		
.786	.1046	.844	.0736	.902	.0448	.960	.0177		
.787	.1040	.845	.0731	.903	.0443	.961	.0173		
.788	.1035	.846	.0726	.904	.0438	.962	.0168		
.789	.1029	.847	.0721	.905	.0434	.963	.0164		
.790	.1024	.848	.0716	.906	.0429	.964	.0159		

DENSITY OF VARIOUS SOLIDS

The approximate density of various solids at ordinary atmospheric temperature.
In the case of substances with voids such as paper or leather the bulk density is indicated rather than the density of the solid portion.

(Selected principally from the Smithsonian Tables.)

Substance	Grams per cu. cm	Pounds per cu. ft.
Agate	2.5–2.7	156–168
Alabaster, carbonate	2.69–2.78	168–173
sulfate	2.26–2.32	141–145
Albite	2.62–2.65	163–165
Amber	1.06–1.11	66–69
Amphiboles	2.9–3.2	180–200
Anorthite	2.74–2.76	171–172
Asbestos	2.0–2.8	125–175
Asbestos slate	1.8	112
Asphalt	1.1–1.5	69–94
Basalt	2.4–3.1	150–190
Beeswax	0.96–0.97	60–61
Beryl	2.69–2.7	168–169
Biotitc	2.7–3.1	170–190
Bone	1.7–2.0	106–125
Brick	1.4–2.2	87–137
Butter	0.86–0.87	53–54
Calamine	4.1–4.5	255–280
Calcspar	2.6–2.8	162–175
Camphor	0.99	62
Caoutchouc	0.92–0.99	57–62
Cardboard	0.69	43
Celluloid	1.4	87
Cement, set	2.7–3.0	170.190
Chalk	1.9–2.8	118–175
Charcoal, oak	0.57	35
pine	0.28–0.44	18–28
Cinnabar	8.12	507
Clay	1.8–2.6	112–162
Coal, anthracite	1.4–1.8	87–112
bituminous	1.2–1.5	75–94
Cocoa butter	0.89–0.91	56–57
Coke	1.0–1.7	62–105
Copal	1.04–1.14	65–71
Cork	0.22–0.26	14–16
Cork linoleum	0.54	34
Corundum	3.9–4.0	245–250
Diamond	3.01–3.52	188–220
Dolomite	2.84	177
Ebonite	1.15	72
Emery	4.0	250
Epidote	3.25–3.50	203–218
Feldspar	2.55–2.75	159–172
Flint	2.63	164
Fluorite	3.18	198
Galena	7.3–7.6	460–470
Gamboge	1.2	75
Garnet	3.15–4.3	197–268
Gas carbon	1.88	117
Gelatin	1.27	79
Glass, common	2.4–2.8	150–175
flint	2.9–5.9	180–370
Glue	1.27	79
Granite	2.64–2.76	165–172
Graphite*	2.30–2.72	144–170
Gum arabic	1.3–1.4	81–87
Gypsum	2.31–2.33	144–145
Hematite	4.9–5.3	306–330
Hornblende	3.0	187
Ice	0.917	57.2
Ivory	1.83–1.92	114–120
Leather, dry	0.86	54
Lime, slaked	1.3–1.4	81–87
Limestone	2.68–2.76	167–171
Linoleum	1.18	74
Magnetite	4.9–5.2	306–324
Malachite	3.7–4.1	231–256
Marble	2.6–2.84	160–177
Meerschaum	0.99–1.28	62–80
Mica	2.6–3.2	165–200
Muscovite	2.76–3.00	172–187
Ochre	3.5	218
Opal	2.2	137
Paper	0.7–1.15	44–72
Paraffin	0.87–0.91	54–57
Peat blocks	0.84	52
Pitch	1.07	67
Porcelain	2.3–2.5	143–156
Porphyry	2.6–2.9	162–181
Pressed wood pulp board	0.19	12
Pyrite	4.95–5.1	309–318
Quartz	2.65	165
Resin	1.07	67
Rock salt	2.18	136
Rubber, hard	1.19	74
Rubber, soft commercial	1.1	69
pure gum	0.91–0.93	57–58
Sandstone	2.14–2.36	134–147
Serpentine	2.50–2.65	156–165
Silica, fused transparent	2.21	138
translucent	2.07	129
Slag	2.0–3.9	125–240
Slate	2.6–3.3	162–205
Soapstone	2.6–2.8	162–175
Spermacéti	0.95	59
Starch	1.53	95
Sugar	1.59	99
Talc	2.7–2.8	168–174
Tallow, beef	0.94	59
mutton	0.94	59
Tar	1.02	66
Topaz	3.5–3.6	219–223
Tourmaline	3.0–3.2	190–200
Wax, sealing	1.8	112
Wood (seasoned)		
alder	0.42–0.68	26–42
apple	0.66–0.84	41–52
ash	0.65–0.85	40–53
balsa	0.11–0.14	7–9
bamboo	0.31–0.40	19–25
basswood	0.32–0.59	20–37
beech	0.70–0.90	43–56
birch	0.51–0.77	32–48
blue gum	1.00	62
box	0.95–1.16	59–72
butternut	0.38	24
cedar	0.49–0.57	30–35
cherry	0.70–0.90	43–56
dogwood	0.76	47
ebony	1.11–1.33	69–83
elm	0.54–0.60	34–37
hickory	0.60–0.93	37–58
holly	0.76	47
juniper	0.56	35
larch	0.50–0.56	31–35
lignum vitae	1.17–1.33	73–83
locust	0.67–0.71	42–44
logwood	0.91	57
mahogany		
Honduras	0.66	41
Spanish	0.85	53
maple	0.62–0.75	39–47
oak	0.60–0.90	37–56
pear	0.61–0.73	38–45
pine, pitch	0.83–0.85	52–53
white	0.35–0.50	22–31
yellow	0.37–0.60	23–37
plum	0.66–0.78	41–49
poplar	0.35–0.5	22–31
satinwood	0.95	59
spruce	0.48–0.70	30–44
sycamore	0.40–0.60	24–37
teak, Indian	0.66–0.88	41–55
African	0.98	61
walnut	0.64–0.70	40–43
water gum	1.00	62
willow	0.40–0.60	24–37

* Some values reported as low as 1.6.

WEIGHT OF ONE GALLON OF WATER (U.S. GALLONS)

The weights are for dry air at the same temperature as the water up to 40°C and at a barometric pressure corrected to 760 mm and against brass weights of 8.4 density at 0°C. Above 40°C the temperature of the air is assumed to be 20°C, i.e., the water is allowed to cool to 20°C prior to the weighings being made. The volumetric computations are based upon the relations that one liter = 1 dm^3 and that 1 dm^3 = 61.023744 $in.^3$

Temperature (°C)	Weight in vacuo (g)	Weight in vacuo (lb)	Weight in air (g)	Weight in air (lb)
0	3784.856	8.34417	3780.543	8.33467
1	3785.078	8.34466	3780.781	8.33518
2	3785.233	8.34500	3780.953	8.33556
3	3785.326	8.34520	3781.060	8.33580
4	3785.355	8.34527	3781.105	8.33590
5	3785.325	8.34520	3781.090	8.33587
6	3785.235	8.34500	3781.015	8.33570
7	3785.089	8.34468	3780.884	8.33541
8	3784.887	8.34424	3780.698	8.33500
9	3784.633	8.34368	3780.358	8.33447
10	3784.326	8.34300	3780.167	8.33383
11	3783.966	8.34221	3779.821	8.33307
12	3783.557	8.34130	3779.426	8.33220
13	3783.099	8.34030	3778.983	8.33122
14	3782.597	8.33919	3778.495	8.33014
15	3782.049	8.33798	3777.962	8.32897
16	3781.458	8.33668	3777.415	8.32770
17	3780.824	8.33528	3776.764	8.32633
18	3780.148	8.33379	3776.103	8.32487
19	3779.430	8.33221	3775.398	8.32332
20	3778.672	8.33054	3774.653	8.32167
21	3777.873	8.32877	3773.868	8.31994
22	3777.035	8.32693	3773.044	8.31813
23	3776.158	8.32499	3772.180	8.31622
24	3775.243	8.32298	3771.279	8.31424
25	3774.291	8.32088	3770.340	8.31217
26	3773.320	8.31870	3769.364	8.31001
27	3772.277	8.31644	3768.352	8.30778
28	3771.218	8.31410	3767.306	8.30548
29	3770.123	8.31169	3766.224	8.30309
30	3768.995	8.30920	3765.109	8.30063
31	3768.995	8.30664	3763.961	8.29810
32	3766.641	8.30401	3762.780	8.29550
33	3765.416	8.30131	3761.568	8.29283
34	3764.160	8.29854	3760.324	8.29008
35	3762.874	8.29571	3759.050	8.28728
40	3756.018	8.28059	3752.255	8.27230
45	3748.41	8.2638	3744.42	8.2550
50	3740.19	8.2457	3736.22	8.2369
55	3731.34	8.2261	3727.37	8.2174
60	3721.91	8.2054	3717.95	8.1966
65	3711.88	8.1832	3707.93	8.1745
70	3701.35	8.1600	3697.42	8.1514
75	3690.30	8.1357	3686.38	8.1270
80	3678.72	8.1101	3674.81	8.1015
85	3666.68	8.0836	3662.78	8.0750
90	3654.15	8.0560	3650.27	8.0474
95	3641.21	8.0274	3637.34	8.0189
100	3627.81	7.9979	3623.95	7.9894

TEMPERATURE CORRECTION FOR VOLUMETRIC SOLUTIONS

This table gives the correction to various observed volumes of water, measured at the designated temperatures to give the volume at the standard temperature, 20°C. Conversely, by subtracting the corrections from the volume desired at 20°C., the volume that must be measured out at the designated temperatures in order to give the desired volume at 20°C., will be obtained. It is assumed that the volumes are measured in glass apparatus having a coefficient of cubical expansion of 0.000025 per degree centigrade. The table is applicable to dilute aqueous solutions having the same coefficient of expansion as water.

Temperature of measurement, °C.	Capacity of apparatus in milliliters at 20°C.: 2,000	1,000	500	400	300	250	150
	Correction in milliliters to give volume of water at 20°C.						
15	+1.54	+0.77	+0.38	+0.31	+0.23	+0.19	+0.12
16	+1.28	+ .64	+ .32	+ .26	+ .19	+ .16	+ .10
17	+ .99	+ .50	+ .25	+ .20	+ .15	+ .12	+ .07
18	+ .68	+ .34	+ .17	+ .14	+ .10	+ .08	+ .05
19	+ .35	+ .18	+ .09	+ .07	+ .05	+ .04	+ .03
21	− .37	− .18	− .09	− .07	− .06	− .05	− .03
22	− .77	− .38	− .19	− .15	− .12	− .10	− .06
23	−1.18	− .59	− .30	− .24	− .18	− .15	− .09
24	−1.61	− .81	− .40	− .32	− .24	− .20	− .12
25	−2.07	−1.03	− .52	− .41	− .31	− .26	− .15
26	−2.54	−1.27	− .64	− .51	− .38	− .32	− .19
27	−3.03	−1.52	− .76	− .61	− .46	− .38	− .23
28	−3.55	−1.77	− .89	− .71	− .53	− .44	− .27
29	−4.08	−2.04	−1.02	− .82	− .61	− .51	− .31
30	−4.62	−2.31	−1.16	− .92	− .69	− .58	− .35

In using the above table to correct the volume of certain standard solutions to 20°C. more accurate results will be obtained if the numerical values of the corrections are increased by the percentages given below:

Solution	Normality: N	N/2	N/10
HNO_3	50	25	6
H_2SO_4	45	25	5
NaOH	40	25	5
KOH	40	20	4

TEMPERATURE CORRECTION FOR GLASS VOLUMETRIC APPARATUS

This table gives the correction to be added to actual capacity (determined at certain temperatures) to give the capactiy at the standard temperature, 20°C. Conversely, by subtracting the corrections from the indicated capacity of an instrument standard at 20°C. the corresponding capacity at other temperatures is obtained. The table assumes for the cubical coefficient of expansion of glass 0.000025 per degree centigrade. The coefficients of expansion of glasses used for volumetric instruments vary from 0.000023 to 0.000028.

Temperature in degrees C.	2,000 ml	1,000 ml	500 ml	400 ml	300 ml	250 ml
15	+0.25	+1.12	+0.06	+0.05	+0.04	+0.031
16	+ .20	+ .10	+ .05	+0.04	+ .03	+ .025
17	+ .15	+ .08	+ .04	+ .03	+ .02	+ .019
18	+ .10	+ .05	+ .02	+ .02	+ .02	+ .012
19	+ .05	+ .02	+ .01	+ .01	+ .01	+ .006
21	− .05	− .02	− .01	− .01	− .01	− .006
22	− .10	− .05	− .02	− .02	− .02	− .012
23	− .15	− .08	− .04	− .03	− .02	− .019
24	− .20	− .10	− .05	− .04	− .03	− .025
25	− .25	− .12	− .06	− .05	− .04	− .031
26	− .30	− .15	− .08	− .06	− .04	− .038
27	− .35	− .18	− .09	− .07	− .05	− .044
28	− .40	− .20	− .10	− .08	− .06	− .050
29	− .45	− .22	− .11	− .09	− .07	− .056
30	− .50	− .25	− .12	− .10	− .08	− .062

DENSITY OF VARIOUS LIQUIDS

(Selected from Smithsonian Tables.)

Liquid	Grams per cu. cm	Pounds per cu. ft.	Temp °C
Acetone	0.792	49.4	20°
Alcohol, ethyl	0.791	49.4	20
methyl	0.810	50.5	0
Benzene	0.899	56.1	0
Carbolic acid	0.950—0.965	59.2—60.2	15
Carbon disulfide	1.293	80.7	0
tetrachloride	1.595	99.6	20
Chloroform	1.489	93.0	20
Ether	0.736	45.9	0
Gasoline	0.66—0.69	41.0—43.0	
Glycerin	1.260	78.6	0
Kerosene	0.82	51.2	
Mercury	13.6	849.0	
Milk	1.028—1.035	64.2—64.6	
Naphtha, petroleum ether	0.665	41.5	15
wood	0.848—0.810	52.9—50.5	0
Oils:			
castor	0.969	60.5	15
cocoanut	0.925	57.7	15
cotton seed	0.926	57.8	16
creosote	1.040—1.100	64.9—68.6	15
linseed, boiled	0.924	58.8	15
olive	0.918	57.3	15
Sea water	1.025	63.99	15
Turpentine (spirits)	0.87	54.3	
Water	1.00	62.43	4

DENSITY OF ALCOHOL

Density of Ethyl Alcohol in Grams Per Cubic Centimeter, Computed from Mendeleeff's Formula

(Selected from Smithsonian Tables.)

Temp. °C	0	1	2	3	4	5	6	7	8	9
0	.80625	.80541	.80457	.80374	.80290	.80207	.80123	.80039	.79956	.79872
10	.79788	.79704	.79620	.79535	.79451	.79367	.79283	.79198	.79114	.79029
20	.78945	.78860	.78775	.78691	.78606	.78522	.78437	.78352	.78267	.78182
30	.78097	.78012	.77927	.77841	.77756	.77671	.77585	.77500	.77414	.77329

HYDROMETERS AND DENSITY UNITS

Alcoholometer. — For testing alcoholic solutions; the scale shows the per cent of alcohol by volume; 0°—100° is the per cent.

Ammoniameter. — For testing ammonia solutions; scale 0°—40°; to convert to sp. gr. multiply by 3 and deduct from 1000.

Barktrometer or *Barkometer.* — For testing tanning liquor; scale 0°—80° Bk; the number to the right of the sp. gr. is the degree Bk; thus, 1.025 sp. gr. is 25° Bk.

Baumé. — There are two kinds in use; heavy Bé, for liquids heavier than water and light Bé for liquids lighter than water. In the former, 0° corresponds to a sp. gr. 1.000 (water at 4°C.) and 66° corresponds to a sp. gr. 1.842; in the lighter than water scale, 0° Bé is equivalent to the gravity of a 10% solution of sodium chloride and 60° Bé corresponds to a sp. gr. of 0.745. For Baumé degrees on the scale of densities greater than unity, the following equation gives the means of conversion:

$$\text{Sp.gr.} = \frac{m}{m - d}$$

where m = 145 (in the United States)
m = 144 (old scale used in Holland)
m = 146.78 (New scale or Gerlach scale)
d = Baumé reading

Beck's Hydrometer has 0° corresponding to sp. gr. 1.000 and 30° to sp. gr. 0.850; equal divisions on the scale are continued as far as required in both directions.

Brix Saccharometer or *Balling Saccharometer* shows directly the per cent of sugar (sucrose) by weight at the temperature indicated on the instrument, usually 17.5°C.; i.e., degrees Brix is the per cent sugar.

Cartier's Hydrometer floats in water at the 10° scale division and at 30° corresponds to 32° Be.

Oleometer. — For vegetable and sperm oils; scale 50°—0° corresponds to sp. gr. 0.870—0.970.

Soxhlet's Lactometer, for determining the density of milk, has a scale from 25° (sp. gr. 1.025) to 35° (sp. gr. 1.035) divided into suitable scale divisions.

Twaddell Hydrometers have the scale so arranged that the reading multiplied by 5 and added to 1000 gives the sp. gr. with reference to water as 1000; it is always used for densties greater than water.

HYDROMETER CONVERSION TABLES

Showing the Relation between Density (C. G. S.) and Degrees Baumé for Densities less than Unity

	Degrees Baumé									
Density	.00	.01	.02	.03	.04	.05	.06	.07	.08	.09
0.60	103.33	99.51	95.81	92.22	88.75	85.38	82.12	78.95	75.88	72.90
.70	70.00	67.18	64.44	61.78	59.19	56.67	54.21	51.82	49.49	47.22
.80	45.00	42.84	40.73	38.68	36.67	34.71	32.79	30.92	29.09	27.30
.90	25.56	23.85	22.17	20.54	18.94	17.37	15.83	14.33	12.86	11.41
1.00	10.00									

Showing the Relation between Density (C. G. S.) and Baumé and Twaddell Scales for Densities above Unity

Density	Degrees Baumé	Degrees Twaddell	Density	Degrees Baumé	Degrees Twaddell	Density	Degrees Baumé	Degrees Twaddell	Denisty	Degrees Baumé	Degrees Twaddell
1.00	0.00	0	1.20	24.17	40	1.41	42.16	82	1.61	54.94	122
1.01	1.44	2	1.21	25.16	42	1.42	42.89	84	1.62	55.49	124
1.02	2.84	4	1.22	26.15	44	1.43	43.60	86	1.63	56.04	126
1.03	4.22	6	1.23	27.11	46	1.44	44.31	88	1.64	56.58	128
1.04	5.58	8	1.24	28.06	48	1.45	45.00	90	1.65	57.12	130
1.05	6.91	10	1.25	29.00	50	1.46	45.68	92	1.66	57.65	132
1.06	8.21	12	1.26	29.92	52	1.47	46.36	94	1.67	58.17	134
1.07	9.49	14	1.27	30.83	54	1.48	47.03	96	1.68	58.69	136
1.08	10.74	16	1.28	31.72	56	1.49	47.68	98	1.69	59.20	138
1.09	11.97	18	1.29	32.60	58	1.50	48.33	100	1.70	59.71	140
1.10	13.18	20	1.30	33.46	60	1.51	48.97	102	1.71	60.20	142
1.11	14.37	22	1.31	34.31	62	1.52	49.60	104	1.72	60.70	144
1.12	15.54	24	1.32	35.15	64	1.53	50.23	106	1.73	61.18	146
1.13	16.68	26	1.33	35.98	66	1.54	50.84	108	1.74	61.67	148
1.14	17.81	28	1.34	36.79	68	1.55	51.45	110	1.75	62.14	150
1.15	18.91	30	1.35	37.59	70	1.56	52.05	112	1.76	62.61	152
1.16	20.00	32	1.36	38.38	72	1.57	52.64	114	1.77	63.08	154
1.17	21.07	34	1.37	39.16	74	1.58	53.23	116	1.78	63.54	156
1.18	22.12	36	1.38	39.93	76	1.59	53.80	118	1.79	63.99	158
1.19	23.15	38	1.39	40.68	78	1.60	54.38	120	1.80	64.44	160
			1.40	41.43	80						

DENSITY OF D_2O

G. S. Kell

t, °C.	ϱ, G./Cc.	t, °C.	ϱ, G./Cc.	t, °C.	ϱ, G./Cc.	t, °C.	ϱ, G./Cc.
0	1.10469	20	1.10534	50	1.09570	80	1.07824
3.813	1.10546	25	1.10445	55	1.09325	85	1.07475
5	1.10562	30	1.10323	60	1.09060	90	1.07112
10	1.10599	35	1.10173	65	1.08777	95	1.06736
11.185	1.10600	40	1.09996	70	1.08475	100	1.06346
15	1.10587	45	1.09794	75	1.08158	101.431	1.06232

VOLUME PROPERTIES OF WATER AT 1 atm*

	ρ, kg m^{-3}, Equation 1	10^6 α, K^{-1}, Equation 1	10^6 $\kappa T/bar^{-1}$ Equation 2		ρ, kg m^{-3}, Equation 1	10^6 α, K^{-1}, Equation 1	10^6 $\kappa T/bar^{-1}$ Equation 2
−30	983.854	−1400.0	80.79	9	999.7808	74.38	48.0560
−25	989.585	−955.9	70.94	10	999.6996	87.97	47.8086
−20	993.547	−660.6	64.25	11	999.6051	101.20	47.5726
−15	996.283	−450.3	59.44	12	999.4974	114.08	47.3474
−10	998.117	−292.4	55.83	13	999.3771	126.65	47.1327
−9	998.395	−265.3	55.22	14	999.2444	138.90	46.9280
−8	998.647	−239.5	54.64	15	999.0996	150.87	46.7331
−7	998.874	−214.8	54.08	16	998.9430	162.55	46.5475
−6	999.077	−191.2	53.56	17	998.7749	173.98	46.3708
−5	999.256	−168.6	53.06	18	998.5956	185.15	46.2029
−4	999.414	−146.9	52.58	19	998.4052	196.08	46.0433
−3	999.550	−126.0	52.12	20	998.2041	206.78	45.8918
−2	999.666	−106.0	51.69	21	997.9925	217.26	45.7482
−1	999.762	−86.7	51.28	22	997.7705	227.54	45.6122
0	999.8395	−68.05	50.8850	23	997.5385	237.62	45.4835
1	999.8985	−50.09	50.5091	24	997.2965	247.50	45.3619
2	999.9399	−32.74	50.1505	25	997.0449	257.21	45.2472
3	999.9642	−15.97	49.8081	26	996.7837	266.73	45.1392
4	999.9720	0.27	49.4812	27	996.5132	276.10	45.0378
5	999.9638	16.00	49.1692	28	996.2335	285.30	44.9427
6	999.9402	31.24	48.8712	29	995.9448	294.34	44.8537
7	999.9015	46.04	48.5868	30	995.6473	303.24	44.7707
8	999.8482	60.41	48.3152	31	995.3410	312.00	44.6935

PROPERTIES LIQUID DEUTERIUM OXIDE (D_2O)

Freezing point — 301.97K (3.82°C) at 0.1013325 MPa
Boiling point — 399.57K (101.42°C) at 0.101325 MPa
Maximum density at 0.101325 MPa — 1.10534 kg/dm^3
Temperature at maximum density — 309.335K (11.185°C)
Molar mass — 0.020027478 kg/mol
Specific gas constant (Universal gas constant divided by molar mass) — 415.150 J/kgK
Critical temperature — $(643.89 + \delta)$K = $(370.74 + \delta)$°C with $-0.2 \leqq \delta \geqq + 0.2$

Critical pressure — $(21.671 + 0.27\delta \pm 0.010)$MPa
Critical density — (356 ± 5)kg/m^3
Critical specific volume — $(0.00281 \pm 0.00004)$$m^3$/kg
Triple point temperature — (276.97 ± 0.02)K = (3.82 ± 0.02)°C
Triple point pressure — (661 ± 3)Pa
Liquid density at triple point — (1105.5 ± 0.2)kg/m^3
Vapor density at triple point — (0.00575 ± 0.00003)kg/m^3

VOLUME PROPERTIES OF WATER AT 1 atm* (continued)

	ρ, kg m^{-3}, Equation 1	$10^6\ \alpha$, K^{-1}, Equation 1	$10^6\ \kappa T$/bar^{-1} Equation 2
32	995.0262	320.63	44.6221
33	994.7030	329.12	44.5561
34	994.3715	337.48	44.4956
35	994.0319	345.73	44.4404
36	993.6842	353.86	44.3903
37	993.3287	361.88	44.3452
38	992.9653	369.79	44.3051
39	992.5943	377.59	44.2697
40	992.2158	385.30	44.2391
41	991.8298	392.91	44.2131
42	991.4364	400.43	44.1917
43	991.0358	407.85	44.1747
44	990.6280	415.19	44.1620
45	990.2132	422.45	44.1536
46	989.7914	429.63	44.1494
47	989.3628	436.73	44.1494
48	988.9273	443.75	44.1533
49	988.4851	450.71	44.1613
50	988.0363	457.59	44.1732
51	987.5809	464.40	44.189
52	987.1190	471.15	44.209
53	986.6508	477.84	44.232
54	986.1761	484.47	44.259
55	985.6952	491.04	44.290
56	985.2081	497.55	44.324
57	984.7149	504.01	44.362
58	984.2156	510.41	44.403
59	983.7102	516.76	44.448
60	983.1989	523.07	44.496
61	982.6817	529.32	44.548
62	982.1586	535.53	44.603
63	981.6297	541.70	44.662
64	981.0951	547.82	44.723
65	980.5548	553.90	44.788
66	980.0089	559.94	44.857
67	979.4573	565.95	44.930
68	978.9003	571.91	45.003
69	978.3377	577.84	45.081
70	977.7696	583.74	45.162
71	977.1962	589.60	45.246
72	976.6173	595.43	45.333
73	976.0332	601.23	45.424
74	975.4437	607.00	45.517
75	974.8990	612.75	45.614
76	974.2490	618.46	45.714
77	973.6439	624.15	45.817
78	973.0336	629.82	45.922
79	972.4183	635.46	46.031
80	971.7978	641.08	46.143
81	971.1723	646.67	46.258
82	970.5417	652.25	46.376
83	969.9062	657.81	46.497
84	969.2657	663.34	46.621
85	968.6203	668.86	46.748
86	967.9700	674.37	46.878
87	967.3148	679.85	47.011
88	966.6547	685.33	47.148
89	965.9898	690.78	47.287

	ρ, kg m^{-3},	$10^6\ \alpha$, K^{-1},	$10^6\ \kappa T$/bar^{-1}	
	Equation 1	Equation 1	Equation 2	Equation 3
90	965.3201	696.23	47.429	47.428
91	964.6457	701.66	47.574	47.574
92	963.9664	707.08	47.722	47.722
93	963.2825	712.49	47.874	47.873
94	962.5938	717.89	48.028	48.028
95	961.9004	723.28	48.185	48.185
96	961.2023	728.67	48.346	48.346
97	960.4996	734.04	48.509	48.510
98	959.7923	739.41	48.676	48.677
99	959.0803	744.78	48.846	48.847
100	958.3637	750.14	49.019	49.020
101	957.642	755.5		49.20
102	956.917	760.8		49.38
103	956.186	766.2		49.56
104	955.451	771.5		49.74
105	954.712	776.9		49.93
106	953.968	782.2		50.13
107	953.220	787.6		50.32
108	952.467	792.9		50.52
109	951.709	798.3		50.72
110	950.947	803.6		50.93
115	947.070	830.4		52.01
120	943.083	857.4		53.17
125	938.984	884.7		54.43
130	934.775	912.3		55.79
135	930.456	940.3		57.24
140	926.026	968.9		58.80
145	921.484	998.0		60.47
150	916.829	1027.8		62.25

Equations:

$$\rho/\text{kg m}^{-3} = (999.83952 + 16.945176\,t - 7.9870401 \times 10^{-3}\,t^2 - 46.170461 \times 10^{-6}\,t^3 + 105.56302 \times 10^{-9}\,t^4 - 280.54253 \times 10^{-12}\,t^5)/(1 + 16.879850 \times 10^{-3}\,t) \quad (1)$$

$$10^6\,\kappa_T/\text{bar}^{-1} = (50.88496 + 0.6163813\,t + 1.459187 \times 10^{-3}\,t^2 + 20.08438 \times 10^{-6}\,t^3 - 58.47727 \times 10^{-9}\,t^4 + 410.4110 \times 10^{-12}\,t^5)/(1 + 19.67348 \times 10^{-3}\,t) \quad (2)$$

$$10^6\,\kappa_T/\text{bar}^{-1} = (50.884917 + 0.62590623\,t + 1.3848668 \times 10^{-3}\,t^2 + 21.603427 \times 10^{-6}\,t^3 - 72.087667 \times 10^{-9}\,t^4 + 465.45054 \times 10^{-12}\,t^5)/(1 + 19.859983 \times 10^{-3}\,t) \quad (3)$$

$$\kappa_S = (\partial \ln \rho/\partial P)_S = \frac{1}{\rho U^2} \quad (4)$$

* Density ρ, thermal expansivity $\alpha = -(\partial \ln \rho/\partial T)_p$, and isothermal compressibility $\kappa T = (\partial \ln \rho/\partial p)T$. For purposes of this table, ordinary water is that with a maximum density of 999.972 kg m^{-3}. Equation 4 for the compressibility should be used for temperatures $0 < t < 100$°C, and Equation 3 for $100 < + < 150$°C. The liquid is metastable below 0°C and above 100°C. Values below 0°C were obtained by extrapolation, and no claim is made for their accuracy.

Reprinted with permission from Kell, G. S., *J. Chem. Eng. Data*, 20(1), 97, 1975. Copyright by the American Chemical Society.

DENSITY AND VOLUME OF MERCURY

Based on the Density of Mercury at 0° C. by Thiesen and Scheel

(Selected from Smithsonian Tables)

Temp. °C.	Mass in gr. per ml.	Vol. of 1 gr. in ml.	Temp. °C.	Mass in gr. per ml.	Vol. of 1 gr. in ml.	Temp. °C.	Mass in gr. per ml.	Vol. of 1gr. in ml.
−10	13.6202	0.0734205	17	5536	7813	90	13.3762	0.0747594
−9	6177	4338	18	5512	7947	100	3522	8939
−8	6152	4472	19	5487	8081	110	3283	50285
−7	6128	4606	20	13.5462	0.0738215	120	3044	1633
−6	6103	4739	21	5438	8348	130	2805	2982
−5	13.6078	0.0734873	22	5413	8482	140	13.2567	0.0754334
−4	6053	5006	23	5389	8616	150	2330	5688
−3	6029	5140	24	5364	8750	160	2093	7044
−2	6004	5273	25	13.5340	0.0738883	170	1856	8402
−1	5979	5407	26	5315	9017	180	1620	9764
0	13.5955	0.0735540	27	5291	9151	190	13.1384	0.0761128
1	5930	5674	28	5266	9285	200	1148	2495
2	5906	5808	29	5242	9419	210	0913	3865
3	5881	5941	30	13.5217	0.0739552	220	0678	5239
4	5856	6075	31	5193	9686	230	0443	6616
5	13.5832	0.0736209	32	5168	9820	240	13.0209	0.0767996
6	5807	6342	33	5144	9953	250	12.9975	9381
7	5782	6476	34	5119	40087	260	9741	70769
8	5758	6610	35	13.5095	0.0740221	270	9507	2161
9	5733	6744	36	5070	0354	280	9273	3558
10	13.5708	0.0736877	37	5046	0488	290	12.9039	0.0774958
11	5684	7011	38	5021	0622	300	8806	6364
12	5659	7145	39	4997	0756	310	8572	7774
13	5634	7278	40	13.4973	0.0740891	320	8339	9189
14	5610	7412	50	4729	2229	330	8105	80609
15	13.5585	0.0737546	60	4486	3569	340	12.7872	0.0782033
16	5561	7680	70	4244	4910	350	7638	3464
			80	4003	6252	360	7405	4900

SULFURIC ACID

SPECIFIC GRAVITY OF AQUEOUS SULFURIC ACID SOLUTIONS

AT $\frac{20°}{4°}$ C.

Be.	Sp. gr.	Per cent H_2SO_4	G. per liter	Lbs. per cu. ft.	Lbs. per gal.	Be.	Sp. gr.	Per cent H_2SO_4	G. per liter	Lbs. per cu. ft.	Lbs. per gal.
0.7	1.0051	1	10.05	0.6275	0.0839	41.8	1.4049	51	716.5	44.73	5.979
1.7	1.0118	2	20.24	1.263	0.1689	42.5	1.4148	52	735.7	45.93	6.140
2.6	1.0184	3	30.55	1.907	0.2550	43.2	1.4248	53	755.1	47.14	6.302
3.5	1.0250	4	41.00	2.560	0.3422	44.0	1.4350	54	774.9	48.37	6.467
4.5	1.0317	5	51.59	3.220	0.4305	44.7	1.4453	55	794.9	49.62	6.634
5.4	1.0385	6	62.31	3.890	0.5200	45.4	1.4557	56	815.2	50.89	6.803
6.3	1.0453	7	73.17	4.568	0.6106	46.1	1.4662	57	835.7	52.17	6.974
7.2	1.0522	8	84.18	5.255	0.7025	46.8	1.4768	58	856.5	53.47	7.148
8.1	1.0591	9	95.32	5.950	0.7955	47.5	1.4875	59	877.6	54.79	7.324
9.0	1.0661	10	106.6	6.655	0.8897	48.2	1.4983	60	899.0	56.12	7.502
9.9	1.0731	11	118.0	7.369	0.9851	48.9	1.5091	61	920.6	57.47	7.682
10.8	1.0802	12	129.6	8.092	1.082	49.6	1.5200	62	942.4	58.83	7.865
11.7	1.0874	13	141.4	8.825	1.180	50.3	1.5310	63	964.5	60.21	8.049
12.5	1.0947	14	153.3	9.567	1.279	51.0	1.5421	64	986.9	61.61	8.236
13.4	1.1020	15	165.3	10.32	1.379	51.7	1.5533	65	1010	63.03	8.426
14.3	1.1094	16	177.5	11.08	1.481	52.3	1.5646	66	1033	64.46	8.618
15.2	1.1168	17	189.9	11.85	1.584	53.0	1.5760	67	1056	65.92	8.812
16.0	1.1243	18	202.4	12.63	1.689	53.7	1.5874	68	1079	67.39	9.008
16.9	1.1318	19	215.0	13.42	1.795	54.3	1.5989	69	1103	68.87	9.207
17.7	1.1394	20	227.9	14.23	1.902	55.0	1.6105	70	1127	70.38	9.408
18.6	1.1471	21	240.9	15.04	2.010	55.6	1.6221	71	1152	71.90	9.611
19.4	1.1548	22	254.1	15.86	2.120	56.3	1.6338	72	1176	73.44	9.817
20.3	1.1626	23	267.4	16.69	2.231	56.9	1.6456	73	1201	74.99	10.02
21.1	1.1704	24	280.9	17.54	2.344	57.5	1.6574	74	1226	76.57	10.24
21.9	1.1783	25	294.6	18.39	2.458	58.1	1.6692	75	1252	78.15	10.45
22.8	1.1862	26	308.4	19.25	2.574	58.7	1.6810	76	1278	79.75	10.66
23.6	1.1942	27	322.4	20.13	2.691	59.3	1.6927	77	1303	81.37	10.88
24.4	1.2023	28	336.6	21.02	2.809	59.9	1.7043	78	1329	82.99	11.09
25.2	1.2104	29	351.0	21.91	2.929	60.5	1.7158	79	1355	84.62	11.31
26.0	1.2185	30	365.6	22.82	3.051	61.1	1.7272	80	1382	86.26	11.53
26.8	1.2267	31	380.3	23.74	3.173	61.6	1.7383	81	1408	87.90	11.75
27.6	1.2349	32	395.2	24.67	3.298	62.1	1.7491	82	1434	89.54	11.97
28.4	1.2432	33	410.3	25.61	3.424	62.6	1.7594	83	1460	91.16	12.19
29.1	1.2515	34	425.5	26.56	3.551	63.0	1.7693	84	1486	92.78	12.40
29.9	1.2599	35	441.0	27.53	3.680	63.5	1.7786	85	1512	94.38	12.62
30.7	1.2684	36	456.6	28.51	3.811	63.9	1.7872	86	1537	95.95	12.83
31.4	1.2769	37	472.5	29.49	3.943	64.2	1.7951	87	1562	97.49	13.03
32.2	1.2855	38	488.5	30.49	4.077	64.5	1.8022	88	1586	99.01	13.23
33.0	1.2941	39	504.7	31.51	4.212	64.8	1.8087	89	1610	100.5	13.42
33.7	1.3028	40	521.1	32.53	4.349	65.1	1.8144	90	1633	101.9	13.63
34.5	1.3116	41	537.8	33.57	4.488	65.3	1.8195	91	1656	103.4	13.82
35.2	1.3205	42	554.6	34.62	4.628	65.5	1.8240	92	1678	104.8	14.00
35.9	1.3294	43	571.6	35.69	4.770	65.7	1.8279	93	1700	106.1	14.19
36.7	1.3384	44	588.9	36.76	4.914	65.8	1.8312	94	1721	107.5	14.36
37.4	1.3476	45	606.4	37.86	5.061	65.9	1.8337	95	1742	108.7	14.54
38.1	1.3569	46	624.2	38.97	5.209	66.0	1.8355	96	1762	110.0	14.70
38.9	1.3663	47	642.2	40.09	5.359	66.0	1.8364	97	1781	111.2	14.87
39.6	1.3758	48	660.4	41.23	5.511	66.0	1.8361	98	1799	112.3	15.02
40.3	1.3854	49	678.8	42.38	5.665	65.9	1.8342	99	1816	113.4	15.15
41.1	1.3951	50	697.6	43.55	5.821	65.8	1.8305	100	1831	114.3	15.28

DENSITY AND COMPOSITION OF FUMING SULFURIC ACID

Actual H_2SO_4, %	Specific gravity	Equiv. H_2SO_4, %	Weight, lb./cu. ft.	Weight, lb. per U.S. gal.	Comb. H_2O, %	Free SO_3, %	Total SO_3, %	SO_3, lb./cu. ft.
100	1.839	100.00	114.70	15.33	18.37	0	81.63	93.63
99	1.845	100.22	115.07	15.38	18.19	1	81.81	94.14
98	1.851	100.45	115.33	15.41	18.00	2	82.00	94.57
97	1.855	100.67	115.70	15.46	17.82	3	82.18	95.08
96	1.858	100.89	115.88	15.49	17.64	4	82.36	95.44
95	1.862	101.13	116.13	15.52	17.45	5	82.55	95.87
94	1.865	101.35	116.35	15.55	17.27	6	82.73	96.23
93	1.869	101.58	116.57	15.58	17.08	7	82.92	96.66
92	1.873	101.80	116.82	15.61	16.90	8	83.10	97.12
91	1.877	102.02	117.07	15.64	16.72	9	83.28	97.50
90	1.880	102.25	117.26	15.67	16.57	10	83.47	97.88
89	1.884	102.47	117.51	15.70	16.35	11	83.65	98.30
88	1.887	102.71	117.69	15.73	16.17	12	83.83	98.66
87	1.891	102.92	117.94	15.76	15.98	13	84.02	99.09
86	1.895	103.15	118.19	15.79	15.80	14	84.20	99.52
85	1.899	103.38	118.44	15.82	15.61	15	84.39	99.95
84	1.902	103.60	118.63	15.86	15.43	16	84.57	100.33
83	1.905	103.82	118.81	15.89	15.25	17	84.75	100.69
82	1.909	104.05	119.06	15.92	15.06	18	84.94	101.13
81	1.911	104.28	119.28	15.95	14.88	19	85.12	101.45
80	1.915	104.50	119.50	15.98	14.70	20	85.30	101.93
79	1.920	104.73	119.75	16.01	14.51	21	85.49	102.37
78	1.923	104.95	119.94	16.04	14.33	22	85.67	102.75
77	1.927	105.18	120.19	16.07	14.14	23	85.86	103.20
76	1.931	105.40	120.44	16.10	13.96	24	86.04	103.63
75	1.934	105.62	120.62	16.12	13.78	25	86.22	104.00
74	1.939	105.85	120.94	16.16	13.59	26	86.41	104.50
73	1.943	106.08	121.18	16.19	13.41	27	86.59	104.93
72	1.946	106.29	121.37	16.22	13.28	28	86.72	105.31
71	1.949	106.53	121.56	16.25	13.04	29	86.96	105.71
70	1.952	106.75	121.75	16.28	12.86	30	87.14	106.09
69	1.955	106.97	121.93	16.30	12.68	31	87.32	106.47
68	1.958	107.20	122.12	16.33	12.49	32	87.51	106.87
67	1.961	107.42	122.31	16.35	12.31	33	87.69	107.25
66	1.965	107.65	122.56	16.38	12.12	34	87.88	107.71
65	1.968	107.87	122.74	16.40	11.94	35	88.06	108.08
64	1.972	108.10	122.99	16.43	11.76	36	88.24	108.53
63	1.976	108.33	123.24	16.46	11.57	37	88.43	108.98
62	1.979	108.55	123.43	16.50	11.39	38	88.61	109.37
61	1.981	108.77	123.55	16.52	11.21	39	88.79	109.70
60	1.983	109.00	123.74	16.54	11.02	40	88.98	110.10
59	1.985	109.22	123.80	16.55	10.84	41	89.16	110.38
58	1.987	109.45	123.93	16.56	10.65	42	89.35	110.83
57	1.989	109.68	124.05	16.58	10.47	43	89.53	111.06
56	1.991	109.90	124.18	16.60	10.29	44	89.71	111.40
55	1.993	110.13	124.30	16.62	10.10	45	89.90	111.75
50	2.001	111.25	124.80	16.68	9.18	50	90.72	113.34
40	2.102	113.50	131.10	17.53	7.35	60	92.65	121.46
30	1.982	115.75	123.62	16.50	5.51	70	94.49	116.81
20	1.949	118.00	121.56	16.25	3.67	80	96.33	117.10
10	1.911	120.25	119.19	15.92	1.84	90	98.16	117.00
0	1.857	122.50	115.83	15.50	0.00	100	100.00	115.83

* By permission from the 7th edition of Chemical Plant Control Data, Chemical Construction Corporation (1957).

COMPOSITION OF SOME INORGANIC ACIDS AND BASES

The following acids and bases are frequently supplied as concentrated aqueous solutions. This table presents certain data concerning these solutions.

	Formula weight	Molarity	Specific gravity	Weight percent	
Acetic acid	60.05	17.5	1.05	99—100	CH_3COOH
Ammonium hydroxide	35.05	14.8	0.90	28—30	NH_3
Hydriodic acid	127.91	5.5	1.5	47—47.5	HI
Hydrobromic acid	80.93	9.0	1.5	47—49	HBr
Hydrochloric acid	36.46	12.0	1.18	36.5—38	HCl
Hydrofluoric acid	20.01	28.9	1.17	48—51	HF
Phosphoric acid	98.00	14.7	1.7	85	H_3PO_4
Sulfuric acid	98.08	18.0	1.84	95—98	H_2SO_4

DENSITY OF MOIST AIR

The density of dry air may be determined by computation from the general relation $D = D_0(T_0/T)(P/P_0)$ where D_0 represents a known density at absolute temperature T_0 and pressure P_0 and D, the density at absolute temperature T and pressure P.

The density of moist air may be determined by a similar relation:

$D = 1.2929\,(273.13/T)\,[B - 0.3783\,e)/760]$ where T is the absolute temperature; B, the barometric pressure in mm, and e the vapor pressure of the moisture in the air in millimeters. The density will then be the product of two terms, each of which may be found by use of the tables which follow.

The first factor, $1.2929\,(273.13/T)$, may be found directly in Table I for various temperatures. For convenience, temperatures are given in the table in °C although the values of the factor have been computed with absolute temperatures. The tabular values actually represent the density of dry air at various temperatures and 760 mm pressure.

The second factor, $[(B - 0.3783\,e/760]$, must be obtained in two steps: First — the numerator of the expression is obtained by subtracting $0.3783\,e$ from the barometric pressure. The quantity $0.3783\,e$ may be found directly from the dew point in Table II. If the wet and dry bulb thermometer readings are known e may be found in the table Reduction of Psychrometric Observations given in the section Hygrometric and Barometric Tables. $0.3783\,e$ may then be found by calculation or read from the table. Second — the value of the whole factor for any value of $B - 0.3783\,e$ may be obtained from Table III.

The product of the above two factors will give the required density in a g/ℓ.

To facilitate obtaining approximate values of the density for ordinary pressures and temperatures, a table of products is given which may be entered with the temperatures in °C and the corrected (for moisture) value of the barometric pressure in mm to obtain density.

As an illustration of the use of the tables, let it be desired to find the density of air for a barometric pressure of 750 mm, a dew point of 10°C, and air temperature of 20°C.

From the dew point, the value of $0.3783\,e$ is found in Table II to be 3.48 mm. 750 − 3.48 = 746.52, the corrected pressure. The pressure factor for this value found in Table III by interpolation is 0.98226.

The temperature factor from Table I is 1.2047.

$$1.2047 \times 0.98224 = 1.1833 \text{ g/}\ell.$$

To obtain the value directly from Table IV, enter it for 20°C and 746.5 mm which gives by interpolation 1.183 g/ℓ.

TABLE I

$(1.2929 \times 273.13/T)$

(Besides being a necessary part of the determination of the density of moist air, the values in this table are actually the density of dry air in g/ℓ at 760 mm pressure for various temperatures.)

Temp. °C		0	1	2	3	4	5	6	7	8	9
−50	1.5	826	897	969	*042	*115	*189	*264	*339	*415	*491
−40	1.5	147	213	278	345	412	479	547	616	686	756
−30	1.4	524	584	645	706	767	829	892	955	*019	*083
−20	1.3	951	*006	*062	*118	*175	*232	*289	*347	*406	*465
−10	1.3	420	472	523	575	628	680	734	787	841	896
−0	1.2	929	977	*024	*073	*121	*170	*219	*269	*319	*370
+0	1.2	929	882	835	789	742	697	651	606	561	517
10	1.2	472	428	385	342	299	256	214	171	130	088
20	1.2	047	006	*965	*925	*885	*845	*805	*766	*727	*688
30	1.1	649	611	573	535	498	460	423	387	350	314
40	1.1	277	242	206	170	135	100	065	031	*996	*962
50	1.0	928	895	861	828	795	762	729	697	664	632
60	1.0	600	569	537	506	475	444	413	382	352	322

TABLE II

Vapor Pressure — Value of $0.3783\,e$

Dew point °C	Vapor press. e mm (ice)	0.3783 e	Dew point °C	Vapor press. e mm (water)	0.3783 e	Dew point °C	Vapor press. e mm (water)	0.3783 e	Dew point °C	Vapor press. e mm (ice)	0.3783 e	Dew point °C	Vapor press. e mm (water)	0.3783 e	Dew point °C	Vapor press. e mm (water)	0.3783 e
−50	0.029	0.01	−15	1.252	0.47	0	4.58	1.73	15	12.79	4.84	30	31.86	12.05	45	71.97	27.23
−45	0.054	0.02	−14	1.373	0.52	1	4.92	1.86	16	13.64	5.16	31	33.74	12.76	46	75.75	28.66
−40	0.096	0.04	−13	1.503	0.57	2	5.29	2.00	17	14.54	5.50	32	35.70	13.51	47	79.70	30.15
−35	0.169	0.06	−12	1.644	0.62	3	5.68	2.15	18	15.49	5.86	33	37.78	14.29	48	83.83	31.71
−30	0.288	0.11	−11	1.798	0.68	4	6.10	2.31	19	16.49	6.24	34	39.95	15.11	49	88.14	33.34
−25	0.480	0.18	−10	1.964	0.74	5	6.54	2.47	20	17.55	6.64	35	42.23	15.98	50	92.6	35.03
−24	0.530	0.20	− 9	2.144	0.81	6	7.01	2.65	21	18.66	7.06	36	44.62	16.88	51	97.3	36.81
−23	0.585	0.22	− 8	2.340	0.89	7	7.51	2.84	22	19.84	7.51	37	47.13	17.83	52	102.2	38.66
−22	0.646	0.24	− 7	2.550	0.96	8	8.04	3.04	23	21.09	7.98	38	49.76	18.82	53	107.3	40.59
−21	0.712	0.27	− 6	2.778	1.05	9	8.61	3.26	24	22.40	8.47	39	52.51	19.86	54	112.7	42.63
−20	0.783	0.30	− 5	3.025	1.14	10	9.21	3.48	25	23.78	9.00	40	55.40	20.96	55	118.2	44.72
−19	0.862	0.33	− 4	3.291	1.24	11	9.85	3.73	26	25.24	9.55	41	58.42	22.10	56	124.0	46.91
−18	0.947	0.36	− 3	3.578	1.35	12	10.52	3.98	27	26.77	10.13	42	61.58	23.30	57	130.0	49.18
−17	1.041	0.39	− 2	3.887	1.47	13	11.24	4.25	28	28.38	10.74	43	64.89	24.55	58	136.3	51.56
−16	1.142	0.43	− 1	4.220	1.60	14	11.99	4.54	29	30.08	11.38	44	68.35	25.86	59	142.8	54.02
															60	149.6	56.59

TABLE III

Pressure Factor — [$(B - 0.3783\,e)/760$]

The figures in the body of the table give values of the whole term [$(B - 0.3783\,e/760$] for various values of the numerator ($B - 0.3783\,e$) expressed at the left and top.

Press. mm corr.	0	1	2	3	4	5	6	7	8	9
80	.10526	.10658	.10789	.10921	.11053	.11184	.11316	.11447	.11579	.11711
90	.11842	.11974	.12105	.12237	.12368	.12500	.12632	.12763	.12895	.13026
100	.13158	.13289	.13421	.13553	.13684	.13816	.13947	.14079	.14211	.14342
110	.14474	.14605	.14737	.14868	.15000	.15132	.15263	.15395	.15526	.15658
120	.15789	.15921	.16053	.16184	.16316	.16447	.16579	.16711	.16842	.16974
130	.17105	.17237	.17368	.17500	.17632	.17763	.17895	.18026	.18158	.18289
140	.18421	.18553	.18684	.18816	.18947	.19079	.19211	.19342	.19474	.19605
150	.19737	.19868	.20000	.20132	.20263	.20395	.20526	.20658	.20789	.20921
160	.21053	.21184	.21316	.21447	.21579	.21711	.21842	.21974	.22105	.22237
170	.23368	.22500	.22032	.22763	.22895	.23026	.23158	.23289	.23421	.23553
180	.23684	.23816	.23947	.24079	.24211	.24342	.24474	.24605	.24737	.24868
190	.25000	.25132	.25263	.25395	.25526	.25658	.25789	.25921	.26053	.26184
200	.26316	.26447	.26579	.26711	.26842	.26974	.27105	.27237	.27368	.27500
210	.27632	.27763	.27895	.28026	.28158	.28289	.28421	.28553	.28684	.28816
220	.28947	.29079	.29211	.29342	.29474	.29605	.29737	.29868	.30000	.30132
230	.30263	.30395	.30526	.30658	.30789	.30921	.31053	.31184	.31316	.31447
240	.31579	.31711	.31842	.31974	.32105	.32237	.32368	.32500	.32632	.32763
250	.32895	.33026	.33158	.33289	.33421	.33553	.33684	.33816	.33947	.34079
260	.34211	.34342	.34474	.34605	.34737	.34868	.35000	.35132	.35263	.35395
270	.35526	.35658	.35789	.35921	.36053	.36184	.36316	.36447	.36579	.36711
280	.36842	.36974	.37105	.37237	.37368	.37500	.37632	.37763	.37895	.38026
290	.38158	.38289	.38421	.38553	.38684	.38816	.38947	.39079	.39211	.39342
300	.39474	.39605	.39737	.39868	.40000	.40132	.40263	.40395	.40526	.40658
310	.40789	.40921	.41053	.41184	.41316	.41447	.41579	.41711	.41842	.41974
320	.42105	.42237	.42368	.42500	.42632	.42763	.42895	.43026	.43158	.43289
330	.43421	.43553	.43684	.43816	.43947	.44079	.44211	.44342	.44474	.44605
340	.44737	.44808	.45000	.45132	.45263	.45395	.45526	.45658	.45789	.45921
350	.46053	.46184	.46316	.46447	.46579	.46711	.46842	.46974	.47105	.47237
360	.47368	.47500	.47632	.47763	.47895	.48026	.48158	.48289	.48421	.48553
370	.48084	.48816	.48947	.49079	.49211	.49342	.49474	.49605	.49737	.49868
380	.50000	.50132	.50263	.50395	.50526	.50658	.50789	.50921	.51053	.51184
390	.51316	.51447	.51579	.51711	.51842	.51974	.52105	.52237	.52368	.52500
400	.52632	.52763	.52895	.53026	.53158	.53289	.53421	.53553	.53684	.53816
410	.53947	.54079	.54211	.54342	.54474	.54605	.54737	.54868	.55000	.55132
420	.55263	.55395	.55526	.55658	.55789	.55921	.56053	.56184	.56316	.56447
430	.56579	.56711	.56842	.56974	.57105	.57237	.57368	.57500	.57632	.57763

Press. mm corr.	0	1	2	3	4	5	6	7	8	9
440	.57895	.58026	.58158	.58289	.58421	.58553	.58684	.58816	.58947	.59079
450	.59211	.59342	.59474	.59605	.59737	.59868	.60000	.60132	.60263	.60395
460	.60526	.60658	.60789	.60921	.61053	.61184	.61316	.61447	.61579	.61711
470	.61842	.61974	.62105	.62237	.62368	.62500	.62632	.62763	.62895	.63026
480	.63158	.63289	.63421	.63553	.63684	.63816	.63947	.64079	.64211	.64342
490	.64474	.64605	.64737	.64868	.65000	.65132	.65263	.65395	.65526	.65658
500	.65790	.65921	.66053	.66184	.66316	.66447	.66579	.66711	.66842	.66974
510	.67105	.67237	.67368	.67500	.67632	.67763	.67895	.68026	.68158	.68290
520	.68421	.68553	.68084	.68816	.68947	.69079	.69211	.69342	.69474	.69605
530	.69737	.69868	.70000	.70132	.70263	.70395	.70526	.70658	.70790	.70921
540	.71053	.71184	.71316	.71447	.71579	.71711	.71842	.71974	.72105	.72237
550	.72368	.72500	.72632	.72763	.72895	.73026	.73158	.73290	.73421	.73553
560	.73684	.73816	.73947	.74079	.74211	.74342	.74474	.74605	.74737	.74868
570	.75000	.75132	.75263	.75395	.75526	.75658	.75790	.75921	.76053	.76184
580	.76316	.76447	.76579	.76711	.76842	.76974	.77105	.77237	.77368	.77500
590	.77632	.77763	.77895	.78026	.78158	.78290	.78421	.78553	.78684	.78816
600	.78947	.79079	.79211	.79342	.79474	.79605	.79737	.79688	.80000	.80132
610	.80263	.80395	.80526	.80658	.80790	.80921	.81053	.81184	.81316	.81447
620	.81579	.81711	.81842	.81974	.82105	.82237	.82368	.82500	.82632	.82763
630	.82895	.83026	.83158	.83290	.83421	.83553	.83684	.83816	.83947	.84079
640	.84211	.84342	.84474	.84605	.84737	.84868	.85000	.85132	.85263	.85395
650	.85526	.85658	.85790	.85921	.86053	.86184	.86316	.86447	.86579	.86711
660	.86842	.86974	.87105	.87237	.87368	.87500	.87632	.87763	.87895	.88026
670	.88158	.88290	.88421	.88553	.88684	.88816	.88947	.89079	.89211	.89342
680	.89474	.89605	.89737	.89868	.90000	.90132	.90263	.90395	.90526	.90658
690	.90790	.90921	.91053	.91184	.91316	.91447	.91579	.91711	.91842	.91974
700	.92105	.92237	.92368	.92500	.92632	.92763	.92895	.93026	.93158	.93290
710	.93421	.93553	.93684	.93816	.93947	.94079	.94211	.94342	.94474	.94605
720	.94737	.94868	.95000	.95132	.95263	.95395	.95526	.95658	.95790	.95921
730	.96053	.96184	.96316	.96447	.96579	.96711	.96842	.96974	.97105	.97237
740	.97368	.97500	.97632	.97763	.97895	.98026	.98158	.98290	.98421	.98553
750	.98684	.98816	.98947	.99079	.99211	.99342	.99474	.99605	.99737	.99868
760	1.0000	1.0013	1.0026	1.0039	1.0053	1.0066	1.0079	1.0092	1.0105	1.0118
770	1.0132	1.0145	1.0158	1.0171	1.0184	1.0197	1.0211	1.0224	1.0237	1.0250
780	1.0263	1.0276	1.0289	1.0303	1.0316	1.0329	1.0342	1.0355	1.0368	1.0382
790	1.0395	1.0408	1.0421	1.0434	1.0447	1.0461	1.0474	1.0487	1.0500	1.0513

TABLE IV

Density of Moist Air

Values in the body of the table give the density of moist air in g/ℓ for a limited range of temperatures and corrected pressure values ($B - 0.3783\,e$). The latter may be obtained by use of Table II.

°C	600	610	620	630	640	650	660	670	680	690
5	1.0024	1.0191	1.0358	1.0525	1.0692	1.0859	1.1026	1.1193	1.1361	1.1528
6	.99876	1.0154	1.0321	1.0487	1.0654	1.0820	1.0986	1.1153	1.1319	1.1486
7	.99521	1.0118	1.0284	1.0450	1.0616	1.0781	1.0947	1.1113	1.1279	1.1445
8	.99165	1.0082	1.0247	1.0412	1.0578	1.0743	1.0908	1.1074	1.1239	1.1404
9	.98818	1.0047	1.0211	1.0376	1.0541	1.0705	1.0870	1.1035	1.1199	1.1364
10	.98463	1.0010	1.0175	1.0339	1.0503	1.0667	1.0831	1.0995	1.1159	1.1323
11	.98115	.99751	1.0139	1.0302	1.0466	1.0629	1.0793	1.0956	1.1120	1.1283
12	.97776	.99406	1.0104	1.0267	1.0430	1.0592	1.0755	1.0918	1.1081	1.1244
13	.97436	.99061	1.0068	1.0231	1.0393	1.0556	1.0718	1.0880	1.1043	1.1205
14	.97097	.98715	1.0033	1.0195	1.0357	1.0519	1.0681	1.0843	1.1004	1.1166
15	.96757	.98370	.99983	1.0160	1.0321	1.0482	1.0643	1.0805	1.0966	1.1127
16	.96426	.98033	.99641	1.0125	1.0286	1.0446	1.0607	1.0768	1.0928	1.1089
17	.96086	.97688	.99290	1.0089	1.0247	1.0409	1.0570	1.0730	1.0890	1.1050
18	.95763	.97359	.98955	1.0055	1.0215	1.0374	1.0534	1.0694	1.0853	1.1013
19	.95431	.97022	.98613	1.0020	1.0179	1.0338	1.0497	1.0656	1.0816	1.0975
20	.95107	.96693	.98278	.99864	1.0145	1.0303	1.0462	1.0620	1.0779	1.0937
21	.94784	.96364	.97944	.99524	1.0110	1.0294	1.0426	1.0584	1.0742	1.0900
22	.94460	.96035	.97609	.99184	1.0076	1.0233	1.0391	1.0548	1.0706	1.0863
23	.94144	.95714	.97283	.98852	1.0042	1.0199	1.0356	1.0513	1.0670	1.0827
24	.93829	.95393	.96957	.98521	1.0008	1.0165	1.0321	1.0478	1.0634	1.0790
25	.93513	.95072	.96630	.98189	.99748	1.0131	1.0286	1.0442	1.0598	1.0754
26	.93197	.94750	.96304	.97858	.99411	1.0096	1.0252	1.0407	1.0562	1.0718
27	.92889	.94437	.95986	.97534	.99083	1.0063	1.0218	1.0373	1.0528	1.0682
28	.92581	.94124	.95668	.97211	.98754	1.0030	1.0184	1.0338	1.0493	1.0647
29	.92273	.93811	.95350	.96888	.98426	.99963	1.0150	1.0304	1.0458	1.0612
30	.91965	.93498	.95031	.96564	.98097	.99629	1.0116	1.0270	1.0423	1.0576
31	.91665	.93193	.94721	.96249	.97777	.99304	.10083	1.0236	1.0389	1.0542
32	.91365	.92888	.94411	.95934	.97457	.98979	1.0050	1.0203	1.0355	1.0507
33	.91065	.92583	.94101	.95619	.97137	.98654	1.0017	1.0169	1.0321	1.0473
34	.90773	.92286	.93800	.95313	.96826	.98338	.99851	1.0136	1.0288	1.0439
35	.90473	.91981	.93490	.94998	.96506	.98013	.99521	1.0103	1.0254	1.0405

°C	700	710	720	730	740	750	760	770	780	790
5	1.1695	1.1862	1.2029	1.2196	1.2363	1.2530	1.2697	1.2864	1.3031	1.3198
6	1.1652	1.1819	1.1985	1.2152	1.2318	1.2485	1.2651	1.2817	1.2984	1.3150
7	1.1611	1.1777	1.1943	1.2108	1.2274	1.2440	1.2606	1.2772	1.2938	1.3104
8	1.1569	1.1735	1.1900	1.2065	1.2230	1.2396	1.2561	1.2726	1.2892	1.3057
9	1.1529	1.1694	1.1858	1.2023	1.2188	1.2352	1.2517	1.2682	1.2846	1.3011
10	1.1487	1.1651	1.1816	1.1980	1.2144	1.2308	1.2472	1.2636	1.2800	1.2964
11	1.1447	1.1610	1.1774	1.1937	1.2101	1.2264	1.2428	1.2592	1.2755	1.2919
12	1.1407	1.1570	1.1733	1.1896	1.2059	1.2222	1.2385	1.2548	1.2711	1.2874
13	1.1368	1.1530	1.1692	1.1855	1.2017	1.2180	1.2342	1.2504	1.2667	1.2829
14	1.1328	1.1490	1.1652	1.1814	1.1975	1.2137	1.2299	1.2461	1.2623	1.2784
15	1.1288	1.1450	1.1611	1.1772	1.1933	1.2095	1.2256	1.2417	1.2579	1.2740
16	1.1250	1.1410	1.1571	1.1732	1.1893	1.2053	1.2214	1.2375	1.2535	1.2696
17	1.1210	1.1370	1.1530	1.1691	1.1851	1.2011	1.2171	1.2331	1.2491	1.2651
18	1.1172	1.1332	1.1492	1.1651	1.1811	1.1970	1.2130	1.2290	1.2449	1.2609
19	1.1134	1.1293	1.1452	1.1611	1.1770	1.1929	1.2088	1.2247	1.2406	1.2565
20	1.1096	1.1254	1.1413	1.1572	1.1730	1.1888	1.2047	1.2206	1.2364	1.2522
21	1.1058	1.1216	1.1374	1.1532	1.1690	1.1848	1.2006	1.2164	1.2322	1.2480
22	1.1020	1.1178	1.1335	1.1493	1.1650	1.1808	1.1965	1.2122	1.2280	1.2437
23	1.0984	1.1140	1.1294	1.1454	1.1611	1.1768	1.1925	1.2082	1.2239	1.2396
24	1.0947	1.1103	1.1259	1.1416	1.1572	1.1729	1.1885	1.2041	1.2198	1.2354
25	1.0910	1.1066	1.1222	1.1377	1.1533	1.1689	1.1845	1.2001	1.2157	1.2313
26	1.0873	1.1028	1.1184	1.1339	1.1494	1.1650	1.1805	1.1960	1.2116	1.2271
27	1.0837	1.0992	1.1147	1.1302	1.1456	1.1611	1.1766	1.1921	1.2076	1.2230
28	1.0801	1.0955	1.1110	1.1264	1.1418	1.1573	1.1727	1.1881	1.2036	1.2090
29	1.0765	1.0919	1.1073	1.1227	1.1380	1.1534	1.1688	1.1842	1.1996	1.2149
30	1.0729	1.0883	1.1036	1.1189	1.1342	1.1496	1.1649	1.1802	1.1956	1.2109
31	1.0694	1.0847	1.1000	1.1153	1.1305	1.1458	1.1611	1.1764	1.1917	1.2069
32	1.0659	1.0812	1.0964	1.1116	1.1268	1.1421	1.1573	1.1725	1.1878	1.2030
33	1.0624	1.0776	1.0928	1.1080	1.1231	1.1383	1.1535	1.1687	1.1839	1.1990
34	1.0590	1.0742	1.0893	1.1044	1.1195	1.1347	1.1498	1.1649	1.1801	1.1952
35	1.0555	1.0706	1.0857	1.1008	1.1158	1.1309	1.1460	1.1611	1.1762	1.1912

DENSITY OF DRY AIR

AT THE TEMPERATURE t, AND UNDER THE PRESSURE H CM OF MERCURY THE DENSITY OF AIR

$$= \frac{0.001293}{1+0.00367\,t}\,\frac{H}{76}.$$

Units of this table are grams per milliliter
(From Miller's Laboratory Physics, Ginn & Co., publishers, by permission.)

t	Pressure H in Centimeters					
	72.0	73.0	74.0	75.0	76.0	77.0
°						
10	0.001182	0.001198	0.001215	0.001231	0.001247	0.001264
11	178	193	210	227	243	259
12	173	190	206	222	239	255
13	169	186	202	218	234	251
14	165	181	198	214	230	246
15	0.001161	0.001177	0.001193	0.001210	0.001226	0.001242
16	157	173	189	205	221	238
17	153	169	185	201	217	233
18	149	165	181	197	213	229
19	145	161	177	193	209	225
20	0.001141	0.001157	0.001173	0.001189	0.001205	0.001221
21	137	153	169	185	201	216
22	134	149	165	181	197	212
23	130	145	161	177	193	208
24	126	142	157	173	189	204
25	0.001122	0.001138	0.001153	0.001169	0.001185	0.001200
26	118	134	149	165	181	196
27	115	130	146	161	177	192
28	111	126	142	157	173	188
29	107	123	138	153	169	184
30	0.001104	0.001119	0.001134	0.001150	0.001165	0.001180

Proportional Parts

cm	17	16	15
0.1	2	2	1
0.2	3	3	3
0.3	5	5	4
0.4	7	6	6
0.5	8	8	7
0.6	10	10	9
0.7	12	11	10
0.8	14	13	12
0.9	15	14	13

Density of dry air at 20C and 760mm Hg = 1.204 mg/cm^3. (*Rev. Mod. Phys.*, 52, Part II, S33, 1980.)

DENSITY OF WATER

The temperature of maximum density for pure water, free from air = 3.98C (277.13K)

t, °C	d, gm/ml
0	0.99987
3.98	1.00000
5	0.99999
10	0.99973
15	0.99913
18	0.99862
20	0.99823
25	0.99707
30	0.99567
35	0.99406
38	0.99299
40	0.99224
45	0.99025
50	0.98807
55	0.98573
60	0.98324
65	0.98059
70	0.97781
75	0.97489
80	0.97183
85	0.96865
90	0.96534
95	0.96192
100	0.95838

THERMODYNAMIC AND TRANSPORT PROPERTIES OF AIR

From NASA Technical Note D-7488 by David J. Poferl and Roger Svehla (1973). The following three tables list the thermodynamic and transport properties of air over the temperature range of 300-2800K at pressures of 20, 30, and 40 atm. Factors for converting viscosity, specific heat at constant pressure, thermal conductivity, and enthalpy from cgs units to SI and English units are

Viscosity:

$$1\,\frac{\text{g}}{\text{(cm)(sec)}} = 0.1\,\frac{\text{(N)(sec)}}{\text{m}^2}$$

$$= 6.72\text{x}10^{-2}\,\frac{\text{lbm}}{\text{(ft)(sec)}}$$

$$= 241.9\,\frac{\text{lbm}}{\text{(ft)(hr)}}$$

$$= 2.089\text{x}10^{-3}\,\frac{\text{(lbf)(sec)}}{\text{ft}^2}$$

Thermal conductivity:

$$1\,\frac{\text{cal}}{\text{(cm)(sec)(K)}} = 418.4\,\frac{\text{W}}{\text{(m)(K)}}$$

$$= 0.8064\,\frac{\text{Btu}}{\text{(ft)}^2\text{(sec)(}^\circ\text{F/in.)}}$$

$$= 6.72\text{x}10^{-2}\,\frac{\text{Btu}}{\text{(ft}^2\text{)(sec)(}^\circ\text{F/ft)}}$$

$$= 241.9\,\frac{\text{Btu}}{\text{(ft)}^2\text{(hr)(}^\circ\text{F/ft)}}$$

Specific heat at constant pressure:

$$1\,\frac{\text{cal}}{\text{(g)(K)}} = 4.184\,\frac{\text{J}}{\text{(g)(K)}}$$

$$= 1\,\frac{\text{Btu}}{\text{(lbm)(}^\circ\text{F)}}$$

Enthalpy:

$$1\,\frac{\text{cal}}{\text{g}} = 4.184\,\frac{\text{J}}{\text{g}}$$

$$= 1.8\,\frac{\text{Btu}}{\text{lbm}}$$

PROPERTIES AT 20 ATM

Temperature, T, K	Isentropic exponent, γ	Molecular weight, m	Viscosity, μ g/(cm)(sec)	Specific heat at constant pressure, c_p, cal/(g)(K)	Thermal conductivity, k, cal/(cm)(sec)(K)	Prandtl number, Pr	Enthalpy, h, cal/g
2800	1.2309	28.890	821×10^{-6}	0.3850	473×10^{-6}	0.669	747.0
2700	1.2374	28.915	800	.3707	439	.676	709.2
2600	1.2437	28.933	779	.3588	409	.683	672.7
2500	1.2498	28.945	758	.3488	384	.688	637.4
2400	1.2556	28.953	736	.3404	362	.692	602.9
2300	1.2612	28.958	715	.3332	343	.696	569.3
2200	1.2666	28.961	694	.3270	325	.698	536.2
2100	1.2719	28.963	672	.3214	309	.699	503.8
2000	1.2772	28.964	651	.3163	294	.700	471.9
1900	1.2825	28.965	629	.3115	280	.701	440.6
1800	1.2879		607	.3070	266	.702	409.6
1700	1.2933		585	.3025	252	.702	379.2
1600	1.2989		563	.2981	239	.703	349.1
1500	1.3045		540	.2939	226	.703	319.5
1400	1.3103		517	.2897	213	.704	290.3
1300	1.3162		494	.2855	200	.704	261.6
1200	1.3224		470	.2814	188	.705	233.2
1100	1.3288		445	.2773	175	.705	205.3
1000	1.3356		419	.2730	162	.705	177.8
900	1.3439	28.964	391	.2681	148	.706	150.7
800	1.3537		362	.2626	135		124.2
700	1.3646		331	.2568	121		98.2
600	1.3759		299	.2511	106		72.8
500	1.3865		265	.2461	92		48.0
400	1.3951		227	.2422	78		23.6
300	1.4000		184	.2401	63		–.5

PROPERTIES AT 30 ATM

Temperature, T, K	Isentropic exponent, γ	Molecular weight, m	Viscosity, μ, g/(cm)(sec)	Specific heat at constant pressure, c_p, cal/(g)(K)	Thermal conductivity, k, cal/(cm)(sec)(K)	Prandtl number, Pr	Enthalpy, h cal/g
2800	1.2340	28.904	821×10^{-6}	0.3773	460×10^{-6}	0.674	745.0
2700	1.2399	28.924	800	.3652	430	.680	707.9
2600	1.2456	28.939	779	.3548	403	.686	671.9
2500	1.2512	28.949	758	.3462	380	.690	636.8
2400	1.2566	28.956	736	.3387	359	.694	602.6
2300	1.2619	28.960	715	.3321	341	.696	569.1
2200	1.2671	28.962	694	.3263	324	.698	536.1
2100	1.2722	28.964	672	.3211	309	.699	503.8
2000	1.2773	28.965	651	.3161	294	.700	471.9
1900	1.2826		629	.3115	280	.701	440.5
1800	1.2879		607	.3069	266	.702	409.6
1700	1.2933		585	.3025	252	.702	379.2
1600	1.2988		563	.2981	239	.703	349.1
1500	1.3045		540	.2939	226	.703	319.5
1400	1.3103		517	.2897	213	.704	290.3
1300	1.3162		494	.2855	200	.704	261.6
1200	1.3223		470	.2814	188	.705	233.2
1100	1.3288		445	.2773	175	.705	205.3
1000	1.3356		419	.2730	162	.705	177.8
900	1.3439	28.964	391	.2681	148	.706	150.7
800	1.3537		362	.2626	135		124.2
700	1.3646		331	.2568	121		98.2
600	1.3759		299	.2511	106		72.8
500	1.3865		265	.2461	92		48.0
400	1.3951		227	.2422	78		23.6
300	1.4000		184	.2401	63		–.5

PROPERTIES AT 40 ATM

Temperature, T, K	Isentropic exponent, γ	Molecular weight, m	Viscosity, μ, g/(cm)(sec)	Specific heat at constant pressure, c_p, cal/(g)(K)	Thermal conductivity, k, cal/(cm)(sec)(K)	Prandtl number, Pr	Enthalpy, h, cal/g
2800	1.2360	28.913	821 x 10^{-6}	0.3727	452 x 10^{-6}	0.677	743.8
2700	1.2414	28.930	800	.3619	424	.683	707.1
2600	1.2468	28.943	779	.3526	399	.688	671.4
2500	1.2521	28.951	758	.3446	377	.692	636.5
2400	1.2573	28.957	736	.3376	358	.695	602.4
2300	1.2623	28.961	715	.3315	340	.697	569.0
2200	1.2673	28.963	694	.3260	324	.699	536.1
2100	1.2724	28.964	672	.3209	308	.700	503.8
2000	1.2774	28.965	651	.3160	294	.700	471.9
1900	1.2826	↓	629	.3114	279	.701	440.5
1800	1.2879		607	.3069	266	.702	409.6
1700	1.2933		585	.3025	252	.702	379.2
1600	1.2988		563	.2981	239	.703	349.1
1500	1.3045		540	.2939	226	.703	319.5
1400	1.3103		517	.2897	213	.704	290.4
1300	1.3162		494	.2855	200	.704	261.6
1200	1.3223		470	.2814	188	.705	233.2
1100	1.3288		445	.2773	175	.705	205.3
1000	1.3356		419	.2730	162	.705	177.8
900	1.3439	↓	391	.2681	148	.706	150.7
800	1.3537	28.964	362	.2626	135	↓	124.2
700	1.3646	↓	331	.2568	121		98.2
600	1.3759		299	.2511	106		72.8
500	1.3865		265	.2461	92		48.0
400	1.3951		227	.2422	78		23.6
300	1.4000	↓	184	.2401	63	↓	–.5

ISOTHERMAL COMPRESSIBILITY OF LIQUIDS

J. C. McGowan

The figures in this table are for isothermal compressibilities in cgs units. The compiler suggests that, provided the pressure is not too high, the reciprocal of the isothermal compressibility varies linearly with the pressure. This suggestion also appears in papers by J. R. Macdonald, *Rev. Mod. Phys.*, 38, 669, 1966, and O. L. Anderson, *J. Phys. Chem. Solids*, 27, 547, 1966. The papers vary somewhat as to how far this linearity persists. 1 Nm^2 = 1 Pascal.

Liquid	Temp., °C[a]	Isothermal compressibility × 10^{10} $m^2 N^{-1}$: At 1 atm (or 1.013 × 10^8 Nm^2)	At 1000 atm (or 1.013 × 10^8 Nm^3)	Ref.
Acetic acid	15	8.75	–	1
	20	9.08	–	1
	30	9.72	–	1
	40	10.37	–	1
	50	11.11	–	1
	60	11.91	–	1
	70	12.77	–	1
	80	13.68	–	1
Acetic acid, ethyl ester	0	9.78	–	1
	10	10.36	–	1
	20	11.32	–	1
	30	12.37	–	1
	40	13.52	–	1
	50	14.78	–	1
	60	16.21	–	1
	70	17.90	–	1
Acetone	20	12.75	–	3
	20	12.29	–	10
	25	12.39	6.02	4
	30	13.34	–	10
	40	15.61	–	3
	40	14.64	–	10
	50	16.03	–	10
Aniline	0	4.08	–	1
	10	4.30	–	1
	20	4.53	–	1
	25	4.67	3.23	5
	40	5.04	–	1
	45	5.22	3.48	5

Liquid	Temp., °C[a]	Isothermal compressibility × 10^{10} $m^2 N^{-1}$: At 1 atm (or 1.013 × 10^5 Nm^2)	At 1000 atm (or 1.013 × 10^8 Nm^2)	Ref.
	50	5.33	–	1
	60	5.64	–	1
	65	5.84	3.76	5
	70	5.97	–	1
	80	6.32	–	1
	85	6.56	4.04	5
	90	6.70	–	1
Anisole	21	6.67	–	2
	30	7.04	–	2
	45	7.72	–	2
	60	8.50	–	2
	81	9.79	–	2
	100	11.25	–	2
	120	13.07	–	2
	140	15.45	–	2
Benzene	0	8.09	–	1
	10	8.73	–	1
	10	8.64	–	3
	20	9.44	–	1
	20	9.37	–	3
	20	9.54	–	10
	25	9.67	5.07	6
	25	9.7	–	15
	30	10.27	–	10
	30	10.18	–	1
	30	10.12	–	3
	35	10.43	5.28	6
	39.5	10.91	–	15
	40	11.05	–	10
	40	11.00	–	1

Liquid	Temp., °C[a]	Isothermal compressibility × 10^{10} $m^2 N^{-1}$ At 1 atm (or 1.013 × 10^5 Nm^2)	Isothermal compressibility × 10^{10} $m^2 N^{-1}$ At 1000 atm (or 1.013 × 10^8 Nm^3)	Ref.
	40	10.96	–	3
	45	11.32	5.50	6
	50	11.90	–	10
	50	11.89	–	1
	50	11.83	–	3
	50.1	11.91	–	15
	55	12.29	5.73	6
	60	12.78	–	10
	60	12.83	–	1
	60	12.96	–	15
	65	13.39	5.98	6
	70	13.72	–	10
	70	14.13	–	1
	75.9	14.95	–	15
	80	15.44	–	1
Benzene, bromo-	25	6.68	4.09	5
	45	7.52	4.39	5
	65	8.50	4.72	5
	85	9.65	5.06	5
Benezene, chloro-	0	6.61	–	1
	10	7.02	–	1
	20	7.45	–	1
	20	7.38	–	3
	25	7.51	4.39	5
	30	7.89	–	1
	30	7.84	–	3
	40	8.39	–	1
	40	8.32	–	3
	45	8.55	4.73	5
	50	8.92	–	1
	50	8.83	–	3
	60	9.50	–	1
	65	9.76	5.10	5
	70	10.13	–	1
	80	10.79	–	1
	85	11.23	5.49	5
Benezene, nitro-	0	4.41	–	1
	10	4.67	–	1
	20	4.93	–	1
	25	5.03	3.39	5
	30	5.23	–	1
	40	5.49	–	1
	45	5.59	3.64	5
	65	6.24	3.91	5
	85	6.99	4.20	5
n-Butyl alcohol	0	8.10	–	11
Carbon disulphide	0	8.04	–	1
	0	7.95	–	3
	10	8.64	–	1
	10	8.54	–	3
	20	9.13	–	10
	20	9.26	–	1
	20	9.19	–	3
	30	9.72	–	10
	30	9.92	–	1
	30	9.96	–	3
	40	10.65	–	1
	40	10.57	–	10
	40	10.89	–	3
	50	11.48	–	1
	50	11.95	–	3
Carbon tetrachloride	–22.9	7.84	–	12
	–13.1	8.31	–	12
	–3.1	8.87	–	12
	0	8.98	–	1
	0	8.85	–	3
	6.9	9.59	–	12
	10	9.70	–	1
	10	9.57	–	3
	10	9.45	–	16
	16.9	10.35	–	12
	20	10.46	–	1
	20	10.34	–	3
	20	10.40	–	10
	25	10.67	5.30	7
	25	10.77	–	15
	25	10.58	–	16
	26.9	11.15	–	12
	30	11.28	–	10
	30	11.29	–	1
	30	11.18	–	3
	35	11.95	5.52	7
	37.5	11.99	–	15
	40	12.20	–	10
	40	12.23	–	1
	40	12.16	–	3
	40	11.96	–	16
	45	12.54	5.75	7
	50	13.20	–	10
	50	13.32	–	1
	50	13.26	–	3
	50.3	13.28	–	15
	55	13.63	5.97	7
	55	13.51	–	15
	60	14.26	–	10
	60	14.52	–	1
	62.6	14.84	–	15
	65	14.87	6.22	7
	70	15.43	–	10
	70	15.77	–	1
	75	16.70	–	15
Chloroform	–33.1	7.13	–	12
	–23.1	7.61	–	12
	–13.1	8.12	–	12
	–3.1	8.66	–	12
	0	8.48	–	1
	0	8.55	–	3
	6.9	9.30	–	12
	10	9.17	–	1
	10	9.19	–	3
	16.9	10.35	–	12
	20	9.98	–	1
	20	9.94	–	3
	20	10.15	–	10
	25	9.74	5.34	4
	26.9	11.15	–	12
	30	10.86	–	1
	30	10.81	–	3
	30	10.99	–	10
	40	11.84	–	1
	40	11.79	–	3
	40	11.94	–	10
	50	12.90	–	1
	50	12.99	–	10
	60	14.06	–	1
Cyclohexane	25	11.20	–	16
	25	11.40	–	15
	35	12.19	–	16
	37.6	12.67	–	15
	40	12.56	–	16
	45	13.31	–	16
	50.1	14.15	–	15
	55	14.35	–	16
	60	15.20	–	16
	62.4	15.76	–	15
	75	17.84	–	15
n-Decane	25	10.95	–	16
	35	11.76	–	16
	45	12.65	–	16
	60	14.11	–	16
Dodecane	25	9.87	–	16
	35	10.52	–	16
	37.8	9.9	–	16
	45	11.27	–	16
	60	12.51	–	16
	79.4	12.8	–	16
	98.9	14.4	–	16
	115.0	16.1	–	16
	135.0	18.3	–	16
Ethane 1,1,2,2-tetrachloro-	25	6.17	3.88	4
Ethyl alcohol	0	9.87	–	1
	0	9.87	–	11
	0	9.63	–	3

ISOTHERMAL COMPRESSIBILITY OF LIQUIDS (continued)

Liquid	Temp., °C[a]	Isothermal compressibility × 10^{10} $m^2 N^{-1}$ At 1 atm (or 1.013 × 10^5 Nm^2)	At 1000 atm (or 1.013 × 10^8 Nm^3)	Ref.
	10	10.49	–	1
	10	10.30	–	3
	20	11.19	–	1
	20	10.98	–	3
	30	11.91	–	1
	30	11.80	–	3
	40	12.74	–	1
	40	12.61	–	3
	50	13.70	–	1
	50	13.60	–	3
	60	14.74	–	1
	70	15.93	–	1
	75	16.67	–	1
Ethyl bromide	0	10.76	–	1
	10	11.78	–	1
	20	12.94	–	1
	30	14.23	–	1
	40	15.52	–	1
Ethylene, 1,2-dichloro-(trans)	25	11.19	5.62	4
Ethylene, tetrachloro-	25	7.56	4.45	4
Ethylene, trichloro-	25	8.57	4.99	4
Ethylene, chloride	0	6.91	–	1
	10	7.42	–	1
	20	7.97	–	1
	20	7.82	–	10
	25	7.78	4.54	4
	30	8.41	–	10
	30	8.58	–	1
	40	9.09	–	10
	40	9.25	–	1
	50	9.86	–	10
	50	9.99	–	1
	60	10.66	–	10
	60	10.83	–	1
	70	11.54	–	10
	70	11.76	–	1
	80	12.79	–	1
Ethyl ether	0	15.10	–	1
	0	15.07	–	3
	10	16.81	–	1
	10	16.52	–	3
	20	18.65	–	1
	20	18.44	–	3
	30	20.90	–	1
	30	20.80	–	3
	35	24.15	–	1
Ethyl iodide	0	8.45	–	1
	10	9.12	–	1
	20	9.82	–	1
	30	10.59	–	1
	40	11.44	–	1
	50	12.38	–	1
	60	13.40	–	1
	70	14.49	–	1
Glycol	25	3.72	2.73	7
	45	4.00	2.89	7
	65	4.32	3.05	7
	85	4.70	3.24	7
	105	5.14	3.44	7
n-Hendecane	25	10.81	–	17
	35	11.50	–	17
	50	12.47	–	17
n-Heptane	25	14.40	6.18	8
	35	15.69	–	17
	45	17.12	–	8
	60	19.62	–	8
1-Heptanol	0	7.05	–	11
n-Hexadecane	25	8.67	–	17
	35	9.43	–	17
	50	10.29	–	17
n-Hexane	25	16.72	6.51	8
	35	18.44	–	8
	45	20.33	–	8
	60	23.84	–	8
1-Hexanol	0	7.47	–	11

Liquid	Temp. °C[a]	Isothermal compressibility × 10^{10} $m^2 N^{-1}$ At 1 atm (or 1.013 × $10^5 Nm^2$)	At 1,000 atm (or 1.013 × $10^8 Nm^3$)	Ref.
Mercury	0	0.40	–	14
	20	0.40	0.39	14
	40	0.41	–	14
	80	0.42	–	14
	120	0.44	–	14
	160	0.46	–	14
Methanol	0	10.62	–	1
	0	10.68	–	11
	0	10.78	–	3
	10	11.34	–	1
	10	11.45	–	3
	20	12.11	–	1
	20	12.18	–	3
	30	12.93	–	1
	30	12.98	–	3
	40	13.85	–	1
	40	13.82	–	3
	50	14.76	–	3
Methylene bromide	–33.1	4.95	–	12
	–23.1	5.11	–	12
	–13.1	5.50	–	12
	–3.1	5.81	–	12
	6.9	6.13	–	12
	16.9	6.47	–	12
	26.9	6.85	–	12
Methylene chloride	25	9.74	5.31	4
Methyl iodide	–33.1	6.86	–	12
	–23.1	7.41	–	12
	–13.1	7.97	–	12
	–3.1	8.55	–	12
	6.9	9.13	–	12
	16.9	9.71	–	12
	26.9	10.33	–	12
n-Nonane	25	11.77	–	8
	35	12.68	–	8
	45	13.66	–	8
	60	15.33	–	8
*n*Octadecane	60	9.4	5.1	13
	79.4	10.4	5.5	13
	98.9	11.6	5.8	13
	115.0	12.8	6.1	13
	135	14.4	6.4	13
n-Octane	25	12.80	–	8
	35	13.86	–	8
	45	15.04	–	8
	60	15.33	–	8
1-Octanol	0	6.82	–	11
n-Pentadecane	37.8	9.1	–	13
	60	10.2	5.2	13
	79.4	11.7	5.5	13
	98.9	13.2	5.8	13
	115	14.7	6.1	13
	135	16.8	6.4	13
Pentanol	0	7.71	–	11
Phenol	46	5.61	–	2
	60	6.05	–	2
	80	6.78	–	2
	110	8.12	–	2
	125	8.88	–	2
	150	10.30	–	2
	175	12.35	–	2
n-Propyl alcohol	0	8.43	–	11
Toluene	–59.3	5.27	–	9
	–41.1	5.93	–	9
	–19.2	6.87	–	9
	0	7.83	–	1
	0	7.97	–	3
	0	7.84	–	9
	10	8.38	–	1
	10	8.44	–	3
	20	8.96	–	1
	20	8.94	–	3
	30	9.60	–	1
	30	9.49	–	3
	40	10.33	–	1
	40	10.14	–	3
	50	11.13	–	1

ISOTHERMAL COMPRESSIBILITY OF LIQUIDS (continued)

Liquid	Temp., °C[a]	Isothermal compressibility × 10^{10} $m^2 N^{-1}$ At 1 atm (or 1.013 × 10^5 Nm^2)	At 1000 atm (or 1.013 × 10^8 Nm^3)	Ref.
	50	10.90	–	3
	60	11.99	–	1
	70	12.95	–	1
Water	0	5.01	–	1
	0	5.04	–	9
	10	4.78	–	1
	20	4.58	–	1
	25	4.57	3.48	7
	30	4.46	–	1
	34.8	4.44	–	9
	35	4.48	3.42	7
	40	4.41	–	1
	45	4.41	3.40	7
	50	4.40	–	1
	55	4.44	3.40	7
	60	4.43	–	1
	65	4.48	3.42	7
	70	4.49	–	1
	75	4.55	3.47	7
	80	4.57	–	1
	85	4.65	3.53	7
	90	4.60	–	1
	100	4.80	–	1
m-Xylene	0	7.44	–	1
	10	7.94	–	1
	20	8.46	–	1
	30	9.03	–	1
	40	9.63	–	1
	50	10.25	–	1
	60	11.01	–	1
	70	11.77	–	1
	80	12.56	–	1

REFERENCES

1. Tyrer, D., *J. Chem. Soc.*, 105, 2534, 1914.
2. Lutskii, A. E. and Solonko, V. N., *Russ. J. Phys. Chem.*, 38, 602, 1964.
3. Fryer, E. B., Hubbard, J. C., and Andrews, D. H., *J. Am. Chem. Soc.*, 51, 759, 1929.
4. Newitt, D. M. and Weale, K. E., *J. Chem. Soc.*, p. 3092, 1951.
5. Gibson, R. E. and Loeffler, O. H., *J. Am. Chem. Soc.*, 61, 2515, 1939.
6. Gibson, R. E. and Kincaid, J. F., *J. Am. Chem. Soc.*, 60, 511, 1938.
7. Gibson, R. E. and Loeffler, O. H., *J. Am. Chem. Soc.*, 63, 898, 1941.
8. Aicart, E., Tardajos, G., and Diaz Pena, M., *J. Chem. Eng. Data*, 26, 22, 1981.
9. Marshall, J. G., Staveley, L. A. K., and Hart, K. R., *Trans. Faraday Soc.*, 52, 23, 1956.
10. Staveley, L. A. K., Tupman, W. I., and Hart, K. R., *Trans. Faraday Soc.*, 51, 323, 1955.
11. McKinney, W. P., Skinner, G. F., and Staveley, L. A. K., *J. Chem. Soc.*, p. 2415, 1959.
12. Harrison, D. and Moelwyn-Hughes, E. A., *Proc. R. Soc. London Ser. A*, 239, 230, 1957.
13. Cutler, W. G., McMickle, R. H., Webb, W., and Schiessler, R. W., *J. Chem. Phys.*, 29, 727, 1958.
14. Moelwyn-Hughes, E. A., *J. Phys. Colloid Chem.*, 55, 1246, 1951.
15. Holder, G. A. and Walley, E., *Trans. Faraday Soc.*, 58, 2095, 1962.
16. Aicart, E., Tardajos, G., and Diaz Pena, M., *J. Chem. Eng. Data*, 25, 140, 1980.
17. Blinowska, A. and Brostow, W., *J. Chem. Thermodyn.*, 7, 787, 1975.

COEFFICIENT OF FRICTION

Compiled by Harold Minshall

The coefficient of friction between two surfaces is the ratio of the force required to move one over the other to the total force pressing the two together. If F is the force required to move one surface over another and W, the force pressing the surfaces together, the coefficient of friction,

$$\mu = \frac{F}{W}$$

Materials	Condition	Temperature °C	μ (Static)
A. STATIC FRICTION			
Non Metals			
Glass on glass	clean	—	0.9–1.0
,, ,, ,,	lubricated with paraffin oil	—	0.5–0.6
,, ,, ,,	,, ,, liquid fatty acids	—	0.3–0.6
,, ,, ,,	,, ,, solid hydrocarbons, alcohols or fatty acids	—	0.1
,, ,, metal	clean	—	0.5–0.7
,, ,, ,,	lubricated	—	0.2–0.3
Diamond on diamond	clean	—	0.1
,, ,, ,,	lubricated	—	0.05–0.1
,, ,, metal	clean	—	0.1–0.15
,, ,, ,,	lubricated	—	0.1
Sapphire on sapphire	clean or lubricated	—	0.2
,, ,, steel	,, ,, ,,	—	0.15
Hard carbon on carbon	clean	—	0.16
,, ,, ,, ,,	lubricated	—	0.12–0.14
Graphite on graphite	clean or lubricated	—	0.1
,, ,, ,,	outgassed	—	0.5–0.8
,, ,, steel	clean or lubricated	—	0.1
Mica on mica	freshly cleaved	—	1.0
,, ,, ,,	contaminated	—	0.2–0.4
Crystals of $NaNO_3$, KNO_3, NH_4Cl on self	clean	—	0.5
,, ,, ,, ,, ,, ,, ,,	lubricated with long chain polar compounds	—	0.12
Tungsten carbide on tungsten carbide	clean	room	0.17
Tungsten carbide on tungsten carbide	outgassed	room	0.58
,, ,, ,, ,, ,,	clean	820	0.35
,, ,, ,, ,, ,,	,,	970	0.40
,, ,, ,, ,, ,,	,,	1010	0.45
,, ,, ,, ,, ,,	,,	1160	0.5
,, ,, ,, ,, ,,	,,	1220	0.7
,, ,, ,, ,, ,,	,,	1440	1.2
,, ,, ,, ,, ,,	,,	1600	1.8
,, ,, ,, graphite	outgassed	room	0.62
,, ,, ,, ,,	clean	,,	0.15
,, ,, ,, ,,	,,	800	0.32
,, ,, ,, ,,	,,	910	0.30
,, ,, ,, ,,	,,	1000	0.25
,, ,, ,, ,,	,,	1120	0.29
,, ,, ,, ,,	,,	1220	0.26
,, ,, ,, ,,	,,	1300	0.25
,, ,, ,, ,,	,,	1410	0.25
,, ,, ,, ,,	,,	1800	0.24
,, ,, ,, ,,	,,	2030	0.25
,, ,, ,, steel	,,	—	0.4–0.6
,, ,, ,, ,,	lubricated	—	0.1–0.2
Polymethyl methacrylate on self	clean	—	0.8
,, ,, ,, steel	,,	—	0.4–0.5
Polystyrene on self	,,	—	0.5
,, ,, steel	,,	—	0.3–0.35
Polyethylene on self	,,	—	0.2
,, ,, steel	,,	—	0.2
Polytetrafluoroethylene on self	,,	—	0.04
,, ,, steel	,,	—	0.04
Nylon on nylon[1]	,,	—	0.15–0.25
Silk on silk	commercially clean	—	0.2–0.3
Cotton on cotton (thread)	,, ,,	—	0.3
,, ,, ,, (from cotton wool)	,, ,,	—	0.6
Rubber on solids	,, ,,	—	1–4
Wood on wood	,, ,, and dry	—	0.25–0.5
,, ,, ,,	,, ,, ,, wet	—	0.2
,, ,, metals	,, ,, ,, dry	—	0.2–0.6
,, ,, ,,	,, ,, ,, wet	—	0.2
,, ,, brick	,, ,,	—	0.6
,, ,, leather	,, ,,	—	0.3–0.4
Leather on metal	,, ,,	—	0.6
,, ,, ,,	,, ,, and wet	—	0.4
,, ,, ,,	greasy	—	0.2
Brake material on cast iron	commercially clean	—	0.4
,, ,, ,, ,, ,,	,, ,, and wet	—	0.2
,, ,, ,, ,, ,,	lubricated with mineral oil	—	0.1
Wool fiber on horn	clean (against scales)	—	0.8–1.0
,, ,, ,, ,,	,, (with scales)	—	0.4–0.6
,, ,, ,, ,,	greasy (against scales)	—	0.5–0.8
,, ,, ,, ,,	,, (with scales)	—	0.3–0.4
Metals			
Steel on steel	clean	20	0.58
,, ,, ,,	vegetable oil lubricant		
,, ,, ,,	(a) castor oil	20	0.095
		100	0.105
,, ,, ,,	(b) rape	20	0.105
		100	0.105
,, ,, ,,	(c) olive	20	0.105
		100	0.105
,, ,, ,,	(d) coconut	20	0.08
		100	0.08

[1] Registered trade name.

Materials	Condition	Temperature °C	μ (Static)
A. STATIC FRICTION (Cont.)			
Metals (Cont.)			
Steel on steel	Animal oil lubricant		
,, ,, ,,	(a) sperm	20	0.10
		100	0.10
,, ,, ,,	(b) pale whale	20	0.095
		100	0.095
,, ,, ,,	(c) neatsfoot	20	0.095
		100	0.095
,, ,, ,,	(d) lard	20	0.085
		100	0.085
,, ,, ,,	Mineral oil lubricant		
,, ,, ,,	(a) light machine	20	0.16
		100	0.19
,, ,, ,,	(b) thick gear	20	0.125
		100	0.15
,, ,, ,,	(c) solvent refined	20	0.15
		100	0.20
,, ,, ,,	(d) heavy motor	20	0.195
		100	0.205
,, ,, ,,	(e) extreme pressure	20	0.09–0.1
		100	0.09–0.1
,, ,, ,,	(f) graphited oil	20	0.13
		100	0.15
,, ,, ,,	(g) B.P. Paraffin	20	0.18
		100	0.22
,, ,, ,,	lubricated with trichloroethylene	20	0.33
,, ,, ,,	,, ,, benzene	20	0.48
,, ,, ,,	,, ,, glycerol	20	0.2
,, ,, ,,	,, ,, ethyl alcohol	20	0.43
,, ,, ,,	,, ,, butyl alcohol	room	0.3
,, ,, ,,	,, ,, octyl	,,	0.23
,, ,, ,,	,, ,, decyl	,,	0.16
,, ,, ,,	,, ,, cetyl	,,	0.10
,, ,, ,,	lubricated with nonane	room	0.26
,, ,, ,,	,, ,, decane	,,	0.23
,, ,, ,,	,, ,, acetic acid	,,	0.5
,, ,, ,,	,, ,, proprionic acid	,,	0.4
,, ,, ,,	,, ,, valeric acid	,,	0.17
,, ,, ,,	,, ,, caproic acid	,,	0.12
,, ,, ,,	,, ,, pelargonic acid	,,	0.11
,, ,, ,,	,, ,, capric acid	,,	0.11
,, ,, ,,	,, ,, lauric acid	,,	0.11
,, ,, ,,	,, ,, myristic acid	,,	0.11
,, ,, ,,	,, ,, oleic acid	20–100	0.08
,, ,, ,,	,, ,, palmitic acid	room	0.11
,, ,, ,,	,, ,, stearic acid	,,	0.10
,, ,, hard steel	,, ,, rape oil	—	0.14
,, ,, ,, ,,	,, ,, castor oil	—	0,12
,, ,, ,, ,,	,, ,, mineral oil	—	0.16
,, ,, ,, ,,	,, ,, long chain fatty acid	—	0.09
,, ,, cast iron	,, ,, rape oil	—	0.11
,, ,, ,, ,,	,, ,, castor oil	—	0.15
,, ,, ,, ,,	,, ,, mineral oil	—	0.21
,, ,, ,, ,,	clean	—	0.4
,, ,, gun metal	lubricated with rape oil	—	0.15
,, ,, ,, ,,	,, ,, castor oil	—	0.16
,, ,, ,, ,,	,, ,, mineral oil	—	0.21
,, ,, bronze	,, ,, rape oil	—	0.12
,, ,, ,,	,, ,, caster oil	—	0.12
,, ,, ,,	,, ,, mineral oil	—	0.16
,, ,, lead	,, ,, ,, ,,	—	0.5
,, ,, ,,	,, ,, long chain fatty acid	—	0.22
,, ,, base white metal	,, ,, mineral oil	—	0.1
,, ,, ,, ,, ,,	,, ,, long chain fatty acid	—	0.08
,, ,, ,,	clean	—	0.55
,, ,, tin	lubricated with mineral oil	—	0.6
,, ,, ,,	,, ,, long chain fatty acid	—	0.21
,, ,, white metal, tin base	,, ,, mineral oil	—	0.1
,, ,, ,, ,, ,, ,,	,, ,, long chain fatty acid	—	0.07
,, ,, ,, ,, ,, ,,	clean	—	0.8
,, ,, sintered bronze	lubricated with mineral oil	—	0.13
,, ,, brass	,, ,, ,, ,,	—	0.19
,, ,, ,,	,, ,, castor oil	—	0.11
,, ,, ,,	,, ,, long chain fatty acid	—	0.13
,, ,, ,,	clean	—	0.35
,, ,, copper–lead alloy	,,	—	0.22
,, ,, Wood's alloy	,,	—	0.7
,, ,, phosphor bronze	,,	—	0.35
,, ,, aluminum bronze	,,	—	0.45
,, ,, constantan	,,	—	0.4
[2] ,, ,, indium film deposited on steel	4 kg load, clean	—	0.08
[2] ,, ,, ,, ,, ,, ,,	8 kg ,, ,,	—	0.04
[2] ,, ,, ,, ,, ,, ,, silver	4 kg ,, ,, ,,	—	0.1
[2] ,, ,, ,, ,, ,, ,, ,,	8 kg ,, ,, ,,	—	0.07
[2] ,, ,, lead film deposited on copper	4 kg ,, ,,	—	0.18
[2] ,, ,, ,, ,, ,, ,,	8 kg ,, ,,	—	0.12
[2] ,, ,, copper film deposited on steel	4 kg ,, ,,	—	0.3
[2] ,, ,, ,, ,, ,, ,,	8 kg ,, ,,	—	0.2
[2] Al on Al	in air or O_2	—	1.9
,, ,, ,,	,, H_2O vapor	—	1.1
[3] Cu on Cu	,, H_2 or N_2	—	4.0
,, ,, ,,	,, air or O_2	—	1.6

[2] Hemispherical steel slider having 0.6 cm. diameter. The thin, 10^{-3} to 10^{-4} cm., thin metallic films were deposited on various substrates as indicated. Amonton's Law is not obeyed in this case.

[3] The metals which were spectroscopically pure were outgassed in a vacuum prior to other gases being admitted. When clean and in vacuum there is gross seizure.

Materials	Condition	Temperature °C	μ (Static)
A. STATIC FRICTION (Cont.)			
Metals (Cont.)			
[3]Au on Au	in H_2 or N_2	—	4.0
,, ,, ,,	,, air or O_2	—	2.8
,, ,, ,,	,, H_2O vapor	—	2.5
[3]Fe on Fe	,, in air or O_2	—	1.2
,, ,, ,,	,, H_2O vapor	—	1.2
[3]Mo on Mo	,, air or O_2	—	0.8
,, ,, ,,	,, H_2O vapor	—	0.8
[3]Ni on Ni	,, H_2 or N_2	—	5.0
,, ,, ,,	,, air or O_2	—	3.0
,, ,, ,,	,, H_2O vapor	—	1.6
[3]Pt on Pt	,, air or O_2	—	3.0
,, ,, ,,	,, H_2O vapor	—	3.0
[3]Ag on Ag	,, air or O_2	—	1.5
,, ,, ,,	,, H_2O vapor	—	1.5
Various Materials on Snow and Ice			
Ice on ice	clean	0	0.05–0.15
,, ,, ,,	,,	−12	0.3
,, ,, ,,	,,	−71	0.5
,, ,, ,,	,,	−82	0.5
,, ,, ,,	,,	−110	0.5
Polymethylmethylacrylate	on wet snow	0	0.5
,,	,, dry ,,	0	0.3
,,	,, ,, ,,	−10	0.34
,,	,, ,, ,,	−32	0.4
Polyester of teraphthalic acid and ethylene glycol	,, wet ,,	0	0.5
Polyester of teraphthalic acid and ethylene glycol	,, dry ,,	0	0.35
Polyester of teraphthalic acid and ethylene glycol	,, ,, ,,	−10	0.38
[1]Nylon	on wet snow	0	0.4
,,	,, dry ,,	0	0.3
,,	,, ,, ,,	−10	0.3
Polytetrafluoroethylene	,, wet ,,	0	0.05
,,	,, dry ,,	0	0.02
,,	,, ,, ,,	−10	0.08
,,	,, ,, ,,	−32	0.1
Paraffin wax	,, wet ,,	0	0.06
,, ,,	,, dry ,,	0	0.06
,, ,,	,, ,, ,,	−10	0.35
,, ,,	,, ,, ,,	−32	0.4
Swiss wax	,, wet ,,	0	0.05
,, ,,	,, dry ,,	0	0.03
,, ,,	,, ,, ,,	−10	0.2
,, ,,	,, ,, ,,	−32	0.2
Ski wax	,, wet ,,	0	0.1
,, ,,	,, dry ,,	0	0.04
,, ,,	,, ,, ,,	−10	0.2
,, ,,	,, ,, ,,	−32	0.2
,, laquer	,, wet ,,	0	0.2
,, ,,	,, dry ,,	0	0.1
,, ,,	,, ,, ,,	−10	0.4
,, ,,	,, ,, ,,	−32	0.4
Aluminum	,, wet ,,	0	0.4
,,	,, dry ,,	0	0.35
,,	,, ,, ,,	−10	0.38

Materials	Condition	Temperature °C	μ (Kinetic)
B. KINETIC FRICTION Various Materials			
Unwaxed hickory	4 m/sec on dry snow	−3	0.08
Waxed ,,	0.1 m/sec ,, wet ,,	0	0.14
,, ,,	0.1 m/sec ,, dry ,,	0	0.04
,, ,,	0.1 m/sec ,, ,, ,,	−3	0.09
,, ,,	4 m/sec ,, ,, ,,	−3	0.03
Waxed hickory	0.1 m/sec on dry snow	−10	0.18
,, ,,	0.1 m/sec ,, ,, ,,	−40	0.4
Ice on ice	4m/sec, clean	0	0.02
,, ,, ,,	,, ,,	−10	0.035
,, ,, ,,	,, ,,	−20	0.050
,, ,, ,,	,, ,,	−40	0.075
,, ,, ,,	,, ,,	−60	0.085
,, ,, ,,	,, ,,	−80	0.09
Ebonite	4m/sec on ice	0	0.02
,,	,, ,, ,,	−10	0.05
,,	,, ,, ,,	−20	0.065
,,	,, ,, ,,	−40	0.085
,,	,, ,, ,,	−60	0.10
,,	,, ,, ,,	−80	0.11
Brass	4m/sec on ice	0	0.02
,,	,, ,, ,,	−10	0.075
,,	,, ,, ,,	−20	0.085
,,	,, ,, ,,	−40	0.115
,,	,, ,, ,,	−60	0.14
,,	,, ,, ,,	−80	0.15
Natural rubber, vulcanized	100m/min on ground glass, clean	—	1.07
,, ,, ,,	100m/min ,, ,, , wetted with water	—	0.94
,, ,, ,,	100m/min on concrete, clean	—	1.02
,, ,, ,,	100m/min ,, ,, , wetted with water	—	0.97
,, ,, ,,	100m/min on bitumen, clean	—	1.07
,, ,, ,,	100m/min ,, ,, , wetted with water	—	0.95
,, ,, ,,	100m/min on rubber flooring or rubber tread vulcanisate, clean	—	1.16
,, ,, ,,	100m/min on bitumen containing rubber powder, clean	—	1.15 (Varies with quantity of powder)
,, ,, ,,	100m/min on bitumen containing rubber powder, wetted with water	—	1.03

[1]Registered trade name. [3]The metals which were spectroscopically pure were outgassed in a vacuum prior to other gases being admitted. When clean and in vacuum there is gross seizure.

LOW MELTING POINT ALLOYS

Melting point °C	Name	Composition, wt %				
−48	Binary Eutectic	Cs	K			
		77.0	23.0			
−40	Binary Eutectic	Cs	Rb			
		87.0	13.0			
−30	Binary Eutectic	Cs	Na			
		95.0	5.0			
−11	Binary Eutectic	K	Na			
		78.0	22.0			
−8	Binary Eutectic	Rb	Na			
		92.0	8.0			
10.7	Ternary Eutectic	Ga	In	Sn		
		62.5	21.5	16.0		
10.8	Ternary Eutectic	Ga	In	Sn		
		[illegible]	17.0	12.5		
17	Ternary Eutectic	Ga	Sn	Zn		
		82.0	12.0	6.0		
33	Binary Eutectic	Rb	K			
		68.0	32.0			
46.5	Quinternary	Sn	Bi	Pb	In	Cd
	Eutectic	10.65	40.63	22.11	18.1	8.2
47	Quinternary	Bi	Pb	Sn	Cd	In
	Eutectic	44.7	22.6	8.3	5.3	19.1
58.2	Quaternary	Bi	Pb	Sn	In	
	Eutectic	49.5	17.6	11.6	21.3	
60.5	Ternary Eutectic	In	Bi	Sn		
		51.0	32.5	16.5		
70	Wood's Metal	Bi	Pb	Sn	Cd	
		50.0	25.0	12.5	12.5	
70	Lipowitz's Metal	Bi	Pb	Sn	Cd	
		50.0	26.7	13.3	10.0	
70	Binary Eutectic	In	Bi			
		67.0	33.0			
91.5	Ternary Eutectic	Bi	Pb	Cd		
		51.6	40.2	8.2		
95	Ternary Eutectic	Bi	Pb	Sn		
		52.5	32.0	15.5		
97	Newton's Metal	Bi	Sn	Pb		
		50.0	18.8	31.2		
98	D'Arcet's Metal	Bi	Sn	Pb		
		50.0	25.0	25.0		
100	Onion's or Lichten-	Bi	Sn	Pb		
	berg's Metal	50.0	20.0	30.0		
102.5	Ternary Eutectic	Bi	Sn	Cd		
		54.0	26.0	20.0		
109	Rose's Metal	Bi	Pb	Sn		
		50.0	28.0	22.0		
117	Binary Eutectic	In	Sn			
		52.0	48.0			
120	Binary Eutectic	In	Cd			
		75.0	25.0			
123	Malotte's Metal	Bi	Sn	Pb		
		46.1	34.2	19.7		
124	Binary Eutectic	Bi	Pb			
		55.5	44.5			
130	Ternary Eutectic	Bi	Sn	Zn		
		56.0	40.0	4.0		
140	Binary Eutectic	Bi	Sn			
		58.0	42.0			
140	Binary Eutectic	Bi	Cd			
		60.0	40.0			
183	Eutectic solder	Sn	Pb			
		63.0	37.0			
185	Binary Eutectic	Tl	Bi			
		52.0	48.0			
192	Soft solder	Sn	Pb			
		70.0	30.0			
198	Binary Eutectic	Sn	Zn			
		91.0	9.0			
199	Tin foil	Sn	Zn			
		92.0	8.0			
199	White Metal	Sn	Sb			
		92.0	8.0			
221	Binary Eutectic	Sn	Ag			
		96.5	3.5			
226	Matrix	Bi	Pb	Sn	Sb	
		48.0	28.5	14.5	9.0	
227	Binary Eutectic	Sn	Cu			
		99.25	0.75			
240	Antimonial Tin	Sn	Sb			
	solder	95.0	5.0			
245	Tin-silver solder	Sn	Ag			
		95.0	5.0			

MOHS HARDNESS SCALE

Hardness number	Original scale	Modified scale
1	Talc	Talc
2	Gypsum	Gypsum
3	Calcite	Calcite
4	Fluorite	Fluorite
5	Apatite	Apatite
6	Orthoclase	Orthoclase
7	Quartz	Vitreous silica
8	Topaz	Quartz or Stellite
9	Corundum	Topaz
10	Diamond	Garnet
11	....	Fused Zirconia
12	....	Fused Alumina
13	....	Silicon Carbide
14	....	Boron Carbide
15	....	Diamond

HARDNESS OF MATERIALS

Agate	6–7	Indium	1.2
Alabaster	1.7	Iridium	6–6.5
Alum	2–2.5	Iridosmium	7
Aluminum	2–2.9	Iron	4–5
Alundum	9+	Kaolinite	2.0–2.5
Amber	2–2.5	Lead	1.5
Andalusite	7.5	Lithium	0.6
Anthracite	2.2	Loess (0°)	0.3
Antimony	3.0–3.3	Magnesium	2.0
Apatite	5	Magnetite	6
Aragonite	3.5	Manganese	5.0
Arsenic	3.5	Marble	3–4
Asbestos	5	Meerschaum	2–3
Asphalt	1–2	Mica	2.8
Augite	6	Opal	4–6
Barite	3.3	Orthoclase	6
Bell-metal	4	Osmium	7.0
Beryl	7.8	Palladium	4.8
Bismuth	2.5	Phosphorus	0.5
Boric acid	3	Phosphorbronze	4
Boron	9.5	Platinum	4.3
Brass	3–4	Plat-iridium	6.5
Cadmium	2.0	Potassium	0.5
Calamine	5	Pumice	6
Calcite	3	Pyrite	6.3
Calcium	1.5	Quartz	7
Carbon	10.0	Rock salt (halite)	2
Carborundum	9–10	Ross' metal	2.5–3.0
Cesium	0.2	Rubidium	0.3
Chromium	9.0	Ruthenium	6.5
Copper	2.5–3	Selenium	2.0
Corundum	9	Serpentine	3–4
Diamond	10	Silicon	7.0
Diatomaceous earth	1–1.5	Silver	2.5–4
Dolomite	3.5–4	Silver chloride	1.3
Emery	7–9	Sodium	0.4
Feldspar	6	Steel	5–8.5
Flint	7	Stibnite	2
Fluorite	4	Strontium	1.8
Galena	2.5	Sulfur	1.5–2.5
Gallium	1.5	Talc	1
Garnet	6.5–7	Tellurium	2.3
Glass	4.5–6.5	Tin	1.5–1.8
Gold	2.5–3	Topaz	8
Graphite	0.5–1	Tourmaline	7.3
Gypsum	1.6–2	Wax (0°)	0.2
Hematite	6	Wood's metal	3
Hornblende	5.5	Zinc	2.5

COMPARISON OF HARDNESS VALUES OF VARIOUS MATERIALS ON MOHS AND KNOOP SCALES*

Compiled by Laurence S. Foster

Substance	Formula	Mohs value	Knoop value
Talc	$3MgO \cdot 4SiO_2 \cdot H_2O$	1	
Gypsum	$CaSO_4 \cdot 2H_2O$	2	32
Cadmium	Cd	..	37
Silver	Ag	..	60
Zinc	Zn	..	119
Calcite	$CaCO_3$	3	135
Fluorite	CaF_2	4	163
Copper	Cu	..	163
Magnesia	MgO	..	370
Apatite	$CaF_2 \cdot 3Ca_3(PO_4)_2$	5	430
Nickel	Ni	..	557
Glass (soda lime)	....	..	530
Feldspar (orthoclase)	$K_2O \cdot Al_2O_3 \cdot 6SiO_2$	6	560
Quartz	SiO_2	7	820
Chromium	Cr	..	935
Zirconia	ZrO_2	..	1160
Beryllia	BeO	..	1250
Topaz	$(AlF)_2SiO_4$	8	1340
Garnet	$Al_2O_3 \cdot 3FeO \cdot 3SiO_2$	..	1360
Tungsten carbide alloy	WC, Co	..	1400–1800
Zirconium boride	ZrB_2	..	1550
Titanium nitride	TiN	9	1800
Tungsten carbide	WC	..	1880
Tantalum carbide	TaC	..	2000
Zirconium carbide	ZrC	..	2100
Alumina	Al_2O_3	..	2100
Beryllium carbide	Be_2C	..	2410
Titanium carbide	TiC	..	2470
Silicon carbide	SiC	..	2480
Aluminum boride	AlB	..	2500
Boron carbide	B_4C	..	2750
Diamond	C	10	7000

* Acknowledgment is made to N. W. Thibault, Norton Company, Worcester, Massachusetts, for many of Knoop hardness values. *Cf.* R. F. Geller, "A Study of Ceramics for Nuclear Reactors," Nucleonics, Vol. 7, No. 4, Table 1, pp. 8–9 (Oct. 1950). V. E. Lysaght, *Indentation Hardness Testing*, Reinhold 1949.

SURFACE TENSION OF LIQUID ELEMENTS

Gernot Lang

The following data were collected from many sources. As a result their accuracy varies. Users of data in this table are advised that:

1. As a rule, results from the "sessile drop" and "maximum bubble pressure" as well as from the "pendant drop" methods are preferable to results obtained from other methods for metals with very high melting points.
2. Values of single measurements are usually not as well supported by experiments as those of serial measurements at various temperatures.
3. Values in parentheses can be considered improbable.

Element	Purity (wt.%)	σ_{mp} (dyn/cm)	Atm.	σ_{t1}: t_1, °C	σ_{t1}: σ (dyn/cm)	σ_{t2}: t_2 °C	σ_{t2}: σ (dyn/cm)	σ_{t3}: t_3 °C	σ_{t3}: σ (dyn/cm)	Method	Ref.
Ag	99.99		H_2	1000	916					Bubble pressure	134
	–	(785)	vac.							Pendant drop	182
	99.96		H_2	1000	893	1150	862	1250	849	Bubble pressure	127
	–		vac.	1000	908					Sessile drop	51
	99.995		H_2	1000	907	1100	894	1200	876	Bubble pressure	128
	99.999	(828)	vac.							Pendant drop	65
	99.99		Ar, H_2	1000	890					Bubble pressure	169
	spect. pure	921			$\sigma = 1136-0.174\ T$ (°K) (1300–2200°K)					Bubble pressure	26
		918	–		$\sigma = 918-0.149\ (t-t_{mp})$ (t°C)					Bubble pressure	204
	99.999		Ar	980	905±10	1108	890±10			Sessile drop	Z3
Al	99.99	860±20	Ar							Bubble pressure	102
	99.72		vac.	950	840					Bubble pressure	46
	99.7	863±25	Ar		$\sigma_t = (863\pm25)-0.33\ (t-t_{mp})$ (t°C)					Bubble pressure	113
	99.99	865	vac.		$\sigma_t = 865-0.14\ (t-t_{mp})$ (t°C)					Sessile drop	137, 56
	99.99	(825)	Ar		($\sigma_t = 825-0.05\ (T-993)$) (T°K)					Bubble pressure	38
	99.99	866	He		$\sigma_t = 866-0.15\ (t-t_{mp})$ (t°C)					Sessile drop	10
	99.999	873	He	1600	725 $\sigma_t = 873-0.15\ (t-t_{mp})$ (t°C)					Sessile drop	199
	99.99	(760)	vac.		($\sigma_t = 948-0.202T$) (T°K.980–1090°K)					Sessile drop	Z4
	99.998	(915)	Ar		($\sigma_t = 915-0.51\ (t-t_{mp})$) (t°C) (660–800°C)					Sessile drop	Z6
	99.996	855±6	Ar		$\sigma_t = 855-0.104\ (t-t_{mp})$ (t°C) (660–911°C)					Bubble pressure	Z13
Au		(754)	vac.							Pendant drop	182
	99.999	1130	He	1200	1070	1300	1020			Sessile drop	86
	99.999	(731)	vac.							Pendant drop	65
	99.999		Ar	1108	1130±10					Sessile drop	Z3
B	99.8	1060±50	vac.							Sessile & pendant drop	187
Ba	–		Ar	720	224					Bubble pressure	3
	99.5	276			$\sigma = 351-0.075\ T$ (°K) (1410–1880°K)					Bubble pressure	25
Be	99.98		vac.	1500	1100					Sessile drop	58
Bi		376	vac.							Drop pressure	155
	99.99	376	vac.							Drop pressure	156
	99.90		H_2	800	343	1000	328			Bubble pressure	134
			vac.	450	(382)					Electro-capillarity	118
	99.9	380±10	Ar							Bubble pressure	103
			–	350	362					Drop pressure	76
			vac.	700	350					Sessile drop	29
	99.98	380±10	Ar							Bubble pressure	105
			–	450	380					Drop pressure	120
			vac.	300	379					Sessile drop	64
		378	vac., Ar, H_2							Drop weight	4
	99.99995	375			$\sigma = 423-0.088\ T$ (T°K) (1352–1555°K)					Bubble pressure	26
	99.999	380±3	Ar		$\sigma_t = 380-0.142\ (t-t_{mp})$ (MP–555°C) (t°C)					Bubble pressure	Z13a
Ca	–		Ar	850	337					Bubble pressure	3
	p.a.	360			$\sigma = 472-0.100\ T$ (T°K) (1445–1655°K)					Bubble pressure	25
Cd			–	450	600					Drop pressure	119
			–	400	600					Bubble pressure	13
			–	350	586					Drop pressure	69
	99.9	(550±10)	Ar							Bubble pressure	103
				390	604					Bubble pressure	1
		(525±30)	H_2							Solid state curvature	Z8
	99.9999	590±5	–				(non linear)			Sessile drop	Z12

SURFACE TENSION OF LIQUID ELEMENTS (*Continued*)

Element	Purity (wt.-%)	σ_{mp} (dyn/cm)	Atm.	σ_{t1} t_1, °C	σ_{t1} σ (dyn/cm)	σ_{t2} t_2, °C	σ_{t2} σ (dyn/cm)	σ_{t3} t_3, °C	σ_{t3} σ (dyn/cm)	Method	Ref.
Co			Ar	1550	1836					Sessile drop	108
	99.99		vac., Al_2O_3	1520	1800					Sessile drop	46
	99.99		He, Al_2O_3	1520	(1630)					Sessile drop	46
	99.99		He, BeO	1520	(1640)					Sessile drop	46
	99.99		He, MgO	1520	(1560)					Sessile drop	46
	99.99		H_2, Al_2O_3	1520	1780					Sessile drop	46
	99.99		He	1520	(1620)					Bubble pressure	46
	99.99		H_2	1520	(1590)					Bubble pressure	46
			vac.	1500	1870					Sessile drop	7
				1600	(1640)					Sessile drop	135
				1600	(1600)					Sessile drop	135
			vac.	1600	1815					Sessile drop	53
	99.99		vac., Al_2O_3	1600	1812					Sessile drop	54
		(1520)	H_2, He							Bubble pressure	61
	99.99		H_2, He	1550	1845					Bubble pressure	63
	99.9983	1880	vac.							Pendant drop	5
	99.99			1550	1780					Bubble pressure	59
Cr	–		vac.	1950	1590±50					Sessile drop	47
	99.9997	1700±50	Ar							Dynam. drop weight	6
Cs			Ar	62	68.4					Pendant drop	195
			Ar	62	67.5	146	62.9			Bubble pressure	193
	99.95		Ar	39	69.5	494	42.8	642	34.6	Bubble pressure	24
	99.99	73.74	Ar		$\sigma = 73.74 - 1.791 \cdot 10^{-2}\,(t-t_{mp}) - 9.610 \cdot 10^{-5}\,(t-t_{mp})^2 + 6.629 \cdot 10^{-8}\,(t-t_{mp})^3$ (t°C) (71–1011°C)					Bubble pressure	26
	99.995	68.6	He		$\sigma = 68.6 - 0.047\,(t-t_{mp})$ (t°C) (52–1100°C)					Bubble pressure	121
Cu			Ar	1120	1269±20					Sessile drop	11
	–	(1150)	vac.							Pendant drop	35
			Ar	1120	1285±10					Sessile drop	108
	99.98		H_2	1100	1301	1165	1295	1255	1287	Bubble pressure	134
		1270	vac.							Sessile drop	7
			vac.	1120	1285					Sessile drop	198
				1440	1298					Bubble pressure	129
		(1085)	vac.							Pendant drop	182
	99.99		He	1250	1290					Bubble pressure	62
	99.99		H_2	1250	1300					Bubble pressure	62
	99.9	(1180±40)	Ar							Bubble pressure	103
	–			1100	1220					Sessile drop	136
	–			1183	(1130)					Sessile drop	200
	–		vac.	1150	1370					Sessile drop	52
	99.997	1355	He, H_2							Sessile drop + Bubble pressure	138
	99.997	1352	vac.		$\sigma_t = 1352 - 0.17\,(t-t_{mp})$ (t°C)					Sessile drop	137
	99.997	1358	Ar		$\sigma_t = 1358 - 0.20\,(t-t_{mp})$ (t°C)					Bubble pressure	137
	–		Ar, He	1120	1285±10					Sessile drop	110
	99.99		H_2, He	1550	1265					Bubble pressure	63
	99.99999	1300	vac.							Pendant drop	5
			vac.	1130	1268±60					Sessile drop	15
	99.98		Ar	1600	1230					Bubble pressure	205
	99.999		N_2	1100	1341	1150	1338	1200	1335	Bubble pressure	128
	99.9	(1127)	vac.							Pendant drop	65
	99.99		Ar, H_2	1100	1320					Bubble pressure	169
Fe				1570	(1731)					Sessile drop	95
	–			1550	1860					Sessile drop	74
		1720	He							Sessile drop	74
				1580–1760	(880)					Bubble pressure	100
				1570	(1632)					Sessile drop	93
	99.99		He	1650	(1610)					Sessile drop	46
	99.99		He	1650	(1430)					Sessile drop	46
	99.99		H_2	1650	(1400)					Sessile drop	46
	–	(1384)	vac.							Pendant drop	182
		(1700)	vac.							Sessile drop	7
			vac., He	1550	1865					Sessile drop	209
	99.99		He	1650	(1430)					Bubble pressure	62
	99.99		H_2	1650	(1400)					Bubble pressure	62
				1650	(1640)					Sessile drop	136
		(1650)	He, H_2							Bubble pressure	61
	99.985		Ar	1550	1788					Sessile drop	110
	–	(1560)								Sessile drop	68

SURFACE TENSION OF LIQUID ELEMENTS (*Continued*)

Element	Purity (wt.–%)	σ_{mp} (dyn/cm)	Atm.	σ_{t1} t_1 °C	σ_{t1} σ (dyn/cm)	σ_{t2} t_2 °C	σ_{t2} σ (dyn/cm)	σ_{t3} t_3 °C	σ_{t3} σ (dyn/cm)	Method	Ref.
	99.94		vac., Al_2O_3	1560	(1710)					Sessile drop	207
	99.9998	1880	vac.							Pendant drop	5
	99.93	1860±40	He							Sessile drop	116
		(1510)	vac.							Drop weight	139
	99.97		vac., BeO	1550	1830±6					Sessile drop	45
	Armco		Ar, N_2	1550	1795					Bubble pressure	164
			vac.	1550	1754					Sessile drop	42
	99.987		vac.	1550	(1730)					Sessile drop	159
	99.85	(1619)	vac.							Pendant drop	65
	99.69		He, Al_2O_3	1550	(1727)					Sessile drop	158
	99.69		H_2, Al_2O_3	1550	(1734)					Sessile drop	158
	–	1760±20	He, H_2		$\sigma = 1760-0.35\ (t-t_{mp})$ (t°C)					Sessile drop	157
	99.9992	1773	He, H_2		$\sigma = 773+0.65\ t$ (t°C) (1550–1780°C)					Oscillating drop	Z2
	–		–	1550	1780					Sessile drop	Z11
Fr	–		–	100	58.4					calculated	145
Ga		704	Ar, He							Bubble pressure	197
	–	725±10	Ar							Bubble pressure	104
			vac.	350	718					Sessile drop	64
	–		He, Al_2O_3	1500	559					Sessile drop	57
	99.9998	718	vac., Al_2O_3		$\sigma = 718-0.101\ (t-t_{mp})$ (t°C)					Sessile drop	85
Ge		621.4								Drop weight	126
		650	vac.	1200	530					Sessile drop	99
			vac.	1000	650					Sessile drop	Z5
		632±5	N_2, He							Solid state curvature	Z8
Hf		(1460)	vac.							Pendant drop	151
	97.5+2.5 Zr	1630	vac.							Pendant drop	5
Hg			H_2	20	(542)					Pendant drop	165
			vac.	15	487.6 ± 1.1					Sessile drop	Z14
			air	20	(435.5)					Oscillating jet	175
			vac.	20	472					Oscillating jet	73
			vac.	20	(402)					Drop pressure	146
				25	476					Drop weight	75
			vac.	20	486.4 ± 1.1					Sessile drop	Z14
			vac.	20	(432)					Sessile drop	170
			H_2	25	476					Drop pressure	78
				25	472					Drop weight	81
				25	(464)					Sessile drop	82
			H_2	19	473					Bubble pressure	173
				20	(437)					Capillary depression	144
			vac.	20	480					Drop weight	21
				25	485.3 ± 1.1					Sessile drop	Z14
				25	(435)					Sessile drop	91
			vac.	25	473					Drop weight	31
				25	488					Sessile drop	32
				25	(498)					Sessile drop	30
			vac.	20	(420)					Sessile drop	174
				25	476					Sessile drop	90
			vac.	20	(410)					Drop pressure	178
			vac.	20	(455)					Drop pressure	39
				25	484±1.5					Sessile drop	89
			vac.	22	(468)					Drop pressure	162
			vac.	20	(465.2)	103	449.7	350	387.1	Drop pressure	162
				25	484.9±1.8					Sessile drop	Z7
			vac.	20	485.5±1.0 $\sigma_t = 489.5-0.20$ (t°C)					Drop pressure	17
			Ar	20	(454.7)					Bubble pressure	188
				21	(350.5)					Bubble pressure	23
	99.99		He, H_2	20	475					Bubble pressure	62
	99.9		Ar	20	(500±15)					Bubble pressure	103
			vac.	-10	487					Drop pressure	69
			vac.	25	483.5±1.0					Sessile drop	140
				22	(465)					Bubble pressure	194
				25	485.1					Sessile drop	142
				16.5	487.3	25	485.4±1.2			Sessile drop	171
				23–25	482.8±9.7					Contact angle	18
					$\sigma_t = 468.7-1.61 \cdot 10^{-1}\,t-1.815 \cdot 10^{-4}\,t^2$ (t°C)					–	206
			vac.	20	484.6±1.3					Pendant drop	132

SURFACE TENSION OF LIQUID ELEMENTS (*Continued*)

Element	Purity (wt.-%)	σ_{mp} (dyn/cm)	Atm.	σ_{t1} t_1 °C	σ_{t1} σ (dyn/cm)	σ_{t2} t_2 °C	σ_{t2} σ (dyn/cm)	σ_{t3} t_3 °C	σ_{t3} σ (dyn/cm)	Method	Ref.
			Ar	25	480					Bubble pressure	172
			vac.	20	482.5±3.0						
					$\sigma_t = 485.5 - 0.149t - 2.84 \cdot 10^{-4} t^2$ (t°C)					Bubble pressure	177
			Ar	21.5	484.9±0.3					Bubble pressure	Z13a
			vac.	50	480.0 ± 1.1					Sessile drop	Z14
In	99.93	559	H_2	600	515					Capillary method	133
				623	540					Capillary method	77
	99.995	556.0	Ar, He							Bubble pressure	196
			vac.	185	592					Drop pressure	69
			H_2	600	514					–	16
				300	541					Sessile drop	83
	99.999		Ar	200	556	400	535	550	527.8	Bubble pressure	123
	99.9994		vac.	350	539					Drop pressure	106
	99.9999	560±5	–		$\sigma = 568.0 - 0.04t - 7.08 \cdot 10^{-5} t^2$ (t°C)					Sessile drop	Z12
Ir	99.9980	2250	vac.							Pendant drop	5
K	99.895	101	Ar							Bubble pressure	189
		110.3±1	–							–	84
		117	vac.		$\sigma = 117 - 0.66\ (t - t_{mp})$					Drop weight	185
	–		Ar	87	112	457	80	677	64.8	Bubble pressure	24
	99.986	116.95	Ar		$\sigma = 116.95 - 6.742 \cdot 10^{-2}\ (t - t_{mp}) - 3.836 \cdot 10^{-5}\ (t - t_{mp})^2 + 3.707 \cdot 10^{-8} (t - t_{mp})^3$ (t°C) (77–983°C)					Bubble pressure	172
	99.936	(79.2)	He		($\sigma = 76.8 - 70.3 \cdot 10^{-4} (t - 400)$) (t°C) (600–1126°C)					Bubble pressure	121
		95±9.5	–							Drop weight	Z9
	99.97	111.35 ±0.64	He		$\sigma = 115.51 - 0.0653 \cdot$ (t°C) (70–713°C)					Bubble pressure	Z10
Li	99.95		Ar	180	397.5	300	380	500	351.5	Bubble pressure	189
	99.98		Ar	287	386	922	275	1077	253	Bubble pressure	24
Mg			Ar	681	563	789	532	894	502	Bubble pressure	208
	99.8		N_2	670	552	700	542	740	528	Bubble pressure	149
	99.9		Ar	700	550±15					Bubble pressure	114
	99.91	(525±10)	Ar							Bubble pressure	105
	99.5	583			$\sigma = 721 - 0.149$ T (T°K) (1125–1326°K)					Bubble pressure	25
Mn	99.9985	1100 ± 50	Ar							Dynam. drop weight	6
	99.94		vac.	1550	1030					Sessile drop	159
	–		–	1550	1010					Sessile drop	Z11
Mo	99.7	(1915)	vac.							Pendant drop	148
		2080	vac.							Drop weight	139
	99.9996	2250	vac.							Pendant drop	5
	99.98	2049	vac.							Pendant drop	65
	–	2130	vac.							Pendant drop	Z1
Na			Ar	110	205.7	263	198.2			Bubble pressure	208
	99.995	191	Ar							Bubble pressure	188
			vac.	123	198	129	198.5			Drop volume	2
				140	190					Sessile drop	28
		200.2±0.6									84
		202	vac.		$\sigma = 202 - 0.092\ (t - t_{mp})$; 100–1000C					Drop weight	185
	p.a.		Ar	617	144	764	130	855	120.4	Bubble pressure	24
	99.96	210.12	Ar		$\sigma = 210.12 - 8.105 \cdot 10^{-2}\ (t - t_{mp}) - 8.064 \cdot 10^{-5}\ (t - t_{mp})^2 + 3.380 \cdot 10^{-8} (t - t_{mp})^3$ (t°C) (141–992°C)					Bubble pressure	172
	99.982	187.4	He		$\sigma = 144 - 0.108\ (t - 500)$ (t°C) (400–1125°C)					Bubble pressure	121
Nb, Cb	99.9986	1900	vac.							Pendant drop	5
	99.99	(1827)	vac.							Pendant drop	65
	–	2020	vac.							Pendant drop	Z1
Nd		688	Ar	1186	674					Bubble pressure	124
Ni	99.7		He	1470	(1615)					Sessile drop	94
	99.7		H_2	1470	(1570)					Sessile drop	94
	99.7		vac.	1470	1735					Sessile drop	94
		1725	vac.							Sessile drop	141
			vac.	1475	1725					Sessile drop	115
	–		Ar	1550	(1934)					Sessile drop	108
	99.99		vac., Al_2O_3	1520	1740					Sessile drop	46
	99.99		He, Ar, Al_2O_3	1520	1770					Sessile drop	46
	99.99		H_2, Al_2O_3	1520	(1600)					Sessile drop	46
	99.99		He, MgO	1470	(1530)					Sessile drop	46
	99.99		He, BeO	1470	(1500)					Sessile drop	46
	99.99		H_2	1530	(1650)					Bubble pressure	46

SURFACE TENSION OF LIQUID ELEMENTS (*Continued*)

Element	Purity (wt.-%)	σ_{mp} (dyn/cm)	Atm.	σ_{t1} t_1 °C	σ_{t1} σ (dyn/cm)	σ_{t2} t_2 °C	σ_{t2} σ (dyn/cm)	σ_{t3} t_3 °C	σ_{t3} σ (dyn/cm)	Method	Ref.
	99.99		H_2	1470	(1530)					Bubble pressure	46
	99.99		He	1470	(1490)					Bubble pressure	46
		1725	vac.							Sessile drop	7
	99.99			1600	(1600)					Sessile drop	135
			vac.	1500	1720					Sessile drop	50
	99.99		vac., Al_2O_3	1550	1780					Sessile drop	54
				1550	1735					Bubble pressure	60
			H_2, He	1470	1700					Bubble pressure	61
	99.99		vac.	1640	1705					Sessile drop	56
	–		vac., Al_2O_3	1560	1810					Sessile drop	207
	99.9991	1770±13	vac.							Drop weight	5
	99.9991	1728±10	vac.							Drop weight	5
	99.9991	1822±8	vac.							Pendant drop	5
		(1670)	vac.							Drop weight	139
		1760	vac.							Sessile drop	55
		(1687)	vac.							Pendant drop	65
	–		He	1500	1745					Sessile drop	57
	–	1809±20	H_2, He Al_2O_3		$\sigma = 1770 - 0.39\,(t-1550)$ (t°C)					Sessile drop	157
	99.99975	(1977)	He		$\sigma = 1665 + 0.215t$ (t°C) (1475–1650°C)					Oscillating drop	Z2
Os	99.9998	2500	vac.							Pendant drop	5
P(white)				50	69.7	68.7	64.95			Bubble pressure	80
Pb	99.98		H_2, N_2	340	448	390	442	440	439	Bubble pressure	71
			air	360	452					Ring removal	97
		451	vac.	425	440					Drop pressure	101
		450	He							Pendant drop	87
				350–450	450					Ring removal	13
	99.998	480	H_2							Capillary method	133
				623	474					Capillary method	77
			vac.	362	455					Drop pressure	69
				700	428					Sessile drop	27
	99.9		H_2	1000	388					Bubble pressure	134
	99.9	(410±5)	Ar							Bubble pressure	104
				350	445					Drop pressure	153
	99.98		vac.	340	442	400	435			Drop pressure	70
	99.9995	470			$\sigma = 538 - 0.114\,T$ (T°K) (1440–1970°K)					Bubble pressure	26
	99.9994		vac.	450	438					Drop pressure	107
	99.999		He	1600	310					Sessile drop	199
				390	456					Bubble pressure	1
		424±10	Air							Solid state curvature	Z8
	99.999	470	Ar		$\sigma_t = 470 - 0.164\,(t-t_{mp})$ (MP–535°C) (t°C)					Bubble pressure	Z13a
Pd		1470	vac.							Sessile drop	50
	99.998	1500	vac.							Pendant drop	5
	99.998	1460	He							Sessile drop	199
Pt		1869	CO_2							Drop weight	167
	99.84	(1740±20)	vac.							Drop volume	48
	99.999		Ar	1800	(1699±20)					Sessile drop	109
	99.9980	1865	vac.							Pendant drop	5
Pu		550±55								–	186
Rb		(77±5)	vac.							Drop diffusion in quartz tube	201
	99.8		Ar	52	84	477	55	632	46.8	Bubble pressure	24
	99.92	91.17	Ar		$\sigma = 91.17 - 9.189 \cdot 10^{-2}\,(t-t_{mp}) + 7.228 \cdot 10^{-5}\,(t-t_{mp})^2 - 3.830 \cdot 10^{-8}\,(t-t_{mp})^3$ (t°C) (104–1006°C)					Bubble pressure	172
	99.997	85.7	He		$\sigma = 85.7 - 0.054\,(t-t_{mp})$ (t°C) (53–1115°C)					Bubble pressure	121
Re	99.4	2610	vac.							Pendant drop	148
	99.9999	2700	vac.							Pendant drop	5
Ru	99.9980	2250	vac.							Pendant drop	5
Rh		1940	vac.							Sessile drop	50
	99.9975	2000	vac.							Pendant drop	5
S	–	60.9	vac.	250	51.1					Pendant drop	143

Element	Purity (wt.-%)	σ_{mp} (dyn/cm)	Atm.	σ_{t1} t_1 °C	σ_{t1} (dyn/cm)	σ_{t2} t_2 °C	σ_{t2} (dyn/cm)	σ_{t3} t_3 °C	σ_{t3} (dyn/cm)	Method	Ref.
Sb			H_2	640	349	700	349	974	342	Drop weight	20
			H_2	750	368	900	361	1100	348	Bubble pressure	40
			vac.	640	367.9	762	364.9			Drop weight	131
	99.5	383	H_2, N_2	675	384	800	380			Bubble pressure	71
	99.99	395±20	Ar							Bubble pressure	104
	99.13	393±20	Ar							Bubble pressure	105
	99.999		N_2	800	359	1000	351	1100	345	Bubble pressure	128
	99.995		Ar	650	350.2	700	347.6	800	345.0	Bubble pressure	123
	99.999		He	1600	320					Sessile drop	199
Se	–		Ar	230–250	88.0±5					Bubble pressure	105
Si			He	1450	725					Pendant drop	87
			vac.	1550	720					Sessile drop	68
	99.99		vac.	1550	750					Sessile drop	41
	99.9999		Ar	1500	825					Pendant drop	43
Sn	–		N_2	275	612	500	572	800	520	Bubble pressure	149
			air	280	523	340	520			Ring removal	97
	99.99	537	vac.	500	524	600	508			Drop pressure	154
		530	He							Pendant drop	87
			H_2	489	543	572	528	692	503	Conical capillaries	9
			–	250	536					Drop pressure	179
			–	450	530					Drop pressure	119
			–	250	545					Drop pressure	112
	99.93		vac.	250	549	400	539	600	526	Drop pressure	156
	99.998	566	H_2							Capillary method	133
			–	623	559					Capillary method	77
		610	vac.							Pendant drop	182
			–	800	500					Sessile drop	7
			–	300	538					Drop pressure	190
			–	300	(527)					Drop pressure	76
			–	290	546					Sessile drop	190
	99.99		H_2, He	600	530					Bubble pressure	62
	99.9	(526±10)	Ar							Bubble pressure	104
			vac.	290	600					Sessile drop	147
	99.965		H_2	740	508	950	489.5	1115	479.5	Bubble pressure	127
	99.89	543.7								Bubble pressure	33
		562	vac.							Sessile drop	55
			vac.	300	554					Sessile drop	64
	99.999	590	vac.							Sessile drop	86
			H_2	290	(520)	290	(524)	(vac.)		Sessile drop	67
	99.9999		H_2	246	552.7					Sessile drop	203
	99.9994		vac.	350	537					Drop pressure	106
	99.999	555.8±1.9			$\sigma = 566.84-4.76 \cdot 10^{-2}\, t$ (t°C)					Bubble pressure	176
	99.96	552	vac.	1000	470					Sessile drop	Z5
	99.96	552	Ar		$\sigma_t = 552-0.167\,(t-t_{mp})$ (MP–500°C) (t°C)					Bubble pressure	Z13a
Sr			Ar	775	288	830	282	893	282	Bubble pressure	125
	99.5	303			$\sigma = 392-0.085T$ (T°K) (1152–1602°K)					Bubble pressure	25
Ta		2360	vac.							Pendant drop	88
		2030	vac.							Pendant drop	88
		1910	vac.							Drop weight	139
	99.9983	2150	vac.							Pendant drop	5
	99.9	(1884)	vac.							Pendant drop	65
Te	99.4	186±2	Ar							Bubble pressure	105
			vac.	460	178±1.5					Capillary method	184
	–		vac.	475	(162)					Electro-capillarity	117
		178			$\sigma = 178-0.024\,(t-t_{mp})$ (t°C)					Bubble pressure	204
Ti	98.7	1510	vac.							Capillary method	44
	99.92	1390	Ar							Pendant drop	151
		1460	vac.							Drop weight	139
	99.9991	1650	vac.							Pendant drop	5
	99.0		vac.	1680	1576					Drop weight	191
	99.99999		vac.	1680	1588					Drop weight	191
	99.85	(1880)	vac.							Pendant drop	65
	99.69	1402	vac.							Pendant drop	65
Tl		464.5	Ar							Bubble pressure	192
			–	450	452					Drop pressure	119
	–		vac.	450	450					Electro-capillarity	117

SURFACE TENSION OF LIQUID ELEMENTS (*Continued*)

Element	Purity (wt.-%)	σ_{mp} (dyn/cm)	Atm.	σ_{t1}: t_1 °C	σ_{t1}: σ (dyn/cm)	σ_{t2}: t_2 °C	σ_{t2}: σ (dyn/cm)	σ_{t3}: t_3 °C	σ_{t3}: σ (dyn/cm)	Method	Ref.
	99.999	467		σ = 536–0.119T (T°K) (1270–1695°K)						Bubble pressure	26
	99.999		vac.	450	450					Drop pressure	107
U		1500±75								–	186
		1550	Ar							Bubble pressure	34
	99.94	(1294)	vac.							Pendant drop	65
V	99.9977	1950	vac.							Pendant drop	5
	–	(1760)	vac.							Pendant drop	Z1
W	–	2310	vac.							Pendant drop	35
	99.9999	2500	vac.							Pendant drop	5
	99.8	2220	vac.							Pendant drop	148
	99.9	(2000)	vac.							Pendant drop	65
Zn	99.9	750 ± 20	Ar							Bubble pressure	104
	99.99	757.0 ± 5	vac.							Sessile drop	202
	99.999	761.0	vac.							Sessile drop	202
	99.9999	767.5	vac.							Sessile drop	202
Zr		1400	Ar							Drop weight	151
	99.5	1411 ± 70	vac.							Drop weight	180
	99.9998	1480	vac.							Pendant drop	5
	99.7	(1533)	vac.							Pendant drop	65

SURFACE TENSION OF LIQUID HALOGENS

John Fredrickson

Element	Purity (wt.-%)	σ_{mp} (dyn/cm)	Atm.	σ_{t1}: t_1 °C	σ_{t1}: σ (dyn/cm)	σ_{t2}: t_2 °C	σ_{t2}: σ (dyn/cm)	σ_{t3}: t_3 °C	σ_{t3}: σ (dyn/cm)	Method	Ref.
F			Vac.	–203.16	17.9	–197.86	16.2	–192.16	14.6	Capillary method	210
		22.33	Vac.		σ = –39.10 – 0.2797(t°C)					Calculated	211
Cl			Air	–61.5	31.61	–44.5	28.38	–28.7	25.23	Capillary method	212
		39.19	Air		σ = 19.736 – 0.1927(t°C)					Calculated	211
Br	99.999		Air	0	45.5	25	40.9	50	36.4	Capillary method	213
		46.8	Air		σ = 45.5 – 0.182(t°C)					Calculated	213
I	99.98	55.7	N_2	130	53.1	150	50.1	170	47.7	Bubble pressure	211
					σ = 70.74 – 0.135(t°C)					Calculated	211

References

1. Abdel-Aziz Abol Hassan, Neue Hütte, **15**, 304 (1970).
2. Addison, Addison, Kerridge and Lewis, J. Chem. Soc., 2262 (1955).
3. Addison, Coldrey and Pulham, J. Chem. Soc., 1227 (1963).
4. Addison and Raynor, J. Chem. Soc., 965 (1966).
5. Allen, Trans. AIME, **227**, 1175 (1963).
6. Allen, Trans. AIME, **230**, 1357 (1964).
7. Allen and Kingery, Trans. AIME, **215**, 30 (1959).
8. Astakhov, Penin and Dobkina, Zh. Fiz. Khim., **20**, 403 (1946).
9. Atterton and Hoar, J. Inst. Met., **81**, 541 (1953).
10. Ayushina, Levin and Geld, Zh. Fiz. Khim., **42**, 2799 (1968).
11. Baes and Kellogg, J. Metals, **15**, 643 (1953).
12. Baker and Gilbert, J. Amer. Chem. Soc., **62**, 2479 (1940).
13. Bakradse and Pines, Zh. Tekhn. Fiz., **23**, 1548 (1953).
14. Becker, Harders and Kornfeld, Arch. f. Eisenh., **20**, 363 (1949).
15. Belforti and Lepie, Trans. AIME, **227**, 80 (1963).
16. Bergh, J. Electrochem. S., **109**, 1199 (1962).
17. Bering and Ioileva, Doklady A.N., **93**, 85 (1953).
18. Biery and Oblak, Ind. Eng. Chem., Fund., **5**, 121 (1966).
19. Bircumshaw, Phil. Mag., **2**, 341 (1926).
20. Bircumshaw, Phil. Mag., **3**, 1286 (1927).
21. Bircumshaw, Phil. Mag., **6**, 510 (1928).
22. Bircumshaw, Phil. Mag., **12**, 596 (1931).
23. Bobyk, Przemysl Chem., **39**, 423 (1960).
24. Bohdanski and Schins, J. Inorg. Nucl. Chem., **29**, 2173 (1967).
25. Bohdanski and Schins, J. Inorg. Nucl. Chem., **30**, 2331 (1968).
26. Bohdanski and Schins, J. Inorg. Nucl. Chem., **30**, 3362 (1968).
27. Bradhurst and Buchanan, J. Phys. Chem., **63**, 1486 (1959).
28. Bradhurst and Buchanan, Austral. J. Chem., **14**, 397 (1961).
29. Bradhurst and Buchanan, Austral. J. Chem., **14**, 409 (1961).
30. Bradley, J. Phys. Chem., **38**, 234 (1934).
31. Brown, Phil. Mag., **13**, 578 (1932).
32. Burdon, Trans. Farad. Soc., **28**, 866 (1932).
33. Cahill and Kirshenbaum, J. Inorg. Nucl. Chem., **26**, 206 (1964).
34. Cahill and Kirshenbaum, J. Inorg. Nucl. Chem., **27**, 73 (1965).
35. Calverley, Proc. Phys. Soc., **70**, 1040 (1957).
36. Coffman and Parr, Ind. Eng. Chem., **19**, 1308 (1927).
37. Cook, Phys. Review, **34**, 513 (1929).
38. de L. Davies and West, J. Inst. Met., **92**, 208 (1964).
39. Didenko and Pokrovski, Doklady A.N., **31**, 233 (1941).
40. Drath and Sauerwald, Z. allg. anorg. Chem., **162**, 301 (1927).
41. Dshemilev, Popel and Zarevski, Fiz. Met.i Met., **18**, 83 (1964).
42. Dyson, Trans. AIME, **227**, 1098 (1963).
43. Eljutin, Kostikov and Levin, Izv. Vys. Uch. Sav., Tsvetn. Met., **2**, 131 (1970).
44. Eljutin and Maurakh, Izv. A.N., OTN, **4**, 129 (1956).
45. Eremenko, Ivashchenko and Bogatyrenko, The Role of Surface Phenomena in Metallurgy, 37 (1963).
46. Eremenko, Ivashchenko, Fessenko and Nishenko, Izv. A.N., OTN, **7**, 144 (1958).
47. Eremenko and Naidich, Izv. A.N., OTN, **2**, 111 (1959).
48. Eremenko and Naidich, Izv. A.N., OTN, **6**, 129 (1959).
49. Eremenko and Naidich, Izv. A.N., OTN, **6**, 100 (1961).
50. Eremenko and Naidich, Izv. A.N., OTN, **2**, 53 (1960).
51. Eremenko and Naidich, The Role of Surface Phenomena in Metallurgy, 65 (1963).
52. Eremenko, Naidich and Nossonovich, Zh. Fiz. Khim., **34**, 1018 (1960).
53. Eremenko and Nishenko, Ukr. Khim. Zh., **26**, 423 (1960).
54. Eremenko and Nishenko, Zh. Fiz. Khim., **35**, 1301 (1961).
55. Eremenko and Nishenko, Ukr. Khim. Zh., **30**, 125 (1964).
56. Eremenko, Nishenko and Naidich, Izv. A.N., OTN, **3**, 150 (1961).
57. Eremenko, Nishenko and Skljarenko, Izv. A. N., OTN, **2**, 188 (1966).

References (Continued)

58. Eremenko, Nishenko and Taj-Shou-Vej., Izv. A.N., OTN, **3**, 116 (1960).
59. Eremenko and Vassiliu, Ukr. Khim. Zh., **31**, 557 (1965).
60. Eremenko, Vassiliu and Fessenko, Zh. Fiz. Khim., **35**, 1750 (1961).
61. Fessenko, Zh. Fiz. Khim., **35**, 707 (1961).
62. Fessenko and Eremenko, Ukr. Khim. Zh., **26**, 198 (1960).
63. Fessenko, Vassiliu and Eremenko, Zh. Fiz. Khim., **36**, 518 (1962).
64. Flechsig, Thesis. Techn. Univ. Berlin, (1964).
65. Flint, J. Nucl. Mat., **16**, 260 (1965).
66. Gans, Pawlek and Roepenack, Z. Metallkunde, **54**, 147 (1963).
67. Gans and Parthey, Z. Metallkunde, **57**, 19 (1966).
68. Geld and Petrushevski, Izv. A.N., OTN, **3**, 160 (1961).
69. Gratzianski and Rjabov, Zh. Fiz. Khim., **33**, 487, 1253 (1959).
70. Gratzianski, Rjabov and Tobolich, Ukr. Khim. Zh., **29**, 1219 (1963).
71. Greenaway, J. Inst. Met., **74**, 133 (1947).
72. Grunmach, Ann.d. Physik, **3**, 660 (1900).
73. Hagemann, Thesis, Univ. of Freiburg/Br., (1914).
74. Halden and Kingery, J. Phys. Chem., **59**, 557 (1955).
75. Harkins and Ewing, J. Amer. Chem., Soc., **42**, 2539 (1920).
76. Herczynska, Z. Phys. Chem., **214**, 355 (1960).
77. Hoar and Melford, Trans. Farad. Soc., **53**, 315 (1957).
78. Hogness, J. Amer. Chem. Soc., **43**, 1621 (1921).
79. Humenik and Kingery, J. Amer. Ceram. Soc., **37**, 18 (1954).
80. Hutchinson, Trans. Farad. Soc., **39**, 229 (1943).
81. Iredale, Phil. Mag., **45**, 1088 (1923).
82. Iredale, Phil. Mag., **48**, 177 (1924).
83. Jacobj, Thesis, Univ. of Braunschweig, (1962).
84. Jordan and Lane, Austral. J. Chem., **18**, 1711 (1965).
85. Karasaev, Sadunikin and Kukhno, Zh. Fiz. Khim., **41**, 654 (1967).
86. Kaufman and Whalen, Acta Metallurg., **13**, 797 (1965).
87. Keck and van Horn, Phys. Review, **91**, 512 (1953).
88. Kelly and Calverley, SERL-Report, **80**, 53 (1959).
89. Kemball, Trans. Farad. Soc., **42**, 526 (1946).
90. Kernaghan, Phys. Review, **37**, 990 (1931).
91. Kernaghan, Phys. Review, **49**, 414 (1936).
92. Kingery, J. Amer. Ceram. Soc., **37**, 42 (1954).
93. Kingery, Kolloid-Z., **161**, 95 (1958).
94. Kingery and Humenik, J. Phys. Chem., **57**, 359 (1953).
95. Kingery and Norton, AEC-Progr. Rep., **NYO-6296**, (1954).
96. Klyachko, Zavodskaja Labor., **6**, 1376 (1937).
97. Klyachko and Kunin, Zavodskaja Labor., **14**, 66 (1948).
98. Klyachko and Kunin, Doklady A.N., **64**, 64 (1949).
99. Kolesnikova, Izv. Vys. Uch. Sav., Tsvetn. Met., **9** (1960).
100. Kolesnikova and Samarin, Izv. A.N., OTN, **5**, 63 (1956).
101. Konstantinov, Thesis, State Univ. Moscow, (1950).
102. Korolkov, Izv. A.N., OTN, **2**, 35 (1956).
103. Korolkov: Litein'e Svojstva Metallov i Splavov, Isdatelstvo, A.N., SSSR, 37 (1960).
104. Korolkov and Bychkova, Issled. Splav. Tsvetn. Met., **2**, 122 (1960).
105. Korolkov and Igumnova, Izv. A.N., OTN, **6**, 95 (1961).
106. Kovalchuk, Kusnezov and Kotlovanova, Zh. Fiz. Khim., **42**, 1754 (1968).
107. Kovalchuk, Kusnezov and Butuzova, Zh. Fiz. Khim., **42**, 2265 (1968).
108. Kozakevitch and Urbain, J. Iron-& Steel-Inst., **186**, 167 (1957).
109. Kozakevitch and Urbain, C.R., Paris, **253**, 2229 (1961).
110. Kozakevitch and Urbain, Mém. Sci. Rev. Met., **58**, 401, 517, 931 (1961).
111. Krause, Sauerwald and Michalke, Z. anorg. allg. Chem., **181**, 353 (1929).
112. Kristian, Thesis, State Univ. Moscow, (1954).
113. Kubichek, Izv. A.N., OTN, **2**, 96 (1959).
114. Kubichek and Malzev, Izv. A.N., OTN, **3**, 144 (1959).
115. Kurkjian and Kingery, J. Phys. Chem., **60**, 961 (1956).
116. Kurochkin, Baum and Borodulin, Fiz. Met.i Met., **15**, 461 (1963).

References (Continued)

117. Kusnezov, The Role of Surface Phenomena in Metallurgy, 72 (1963).
118. Kusnezov, Djakova and Malzeva, Zh. Fiz. Khim., **33**, 1551 (1959).
119. Kusnezov, Kochergin, Tishchenko and Posdynsheva, Doklady A.N., **92**, 1197 (1953).
120. Kusnezov, Popova and Duplina, Zh. Fiz. Khim., **36**, 880 (1962).
121. Kyrianenko and Solovev, Teplofiz. Vysok. Temp., **8**, 537 (1970).
122. Lasarev, Zh. Fiz. Khim., **36**, 405 (1962).
123. Lasarev, Zh. Fiz. Khim., **38**, 325 (1964).
124. Lasarev and Pershikov, Doklady A.N., **146**, 143 (1962).
125. Lasarev and Pershikov, Zh. Fiz. Khim., **37**, 907 (1963).
126. Lasarev and Pugachevich, Doklady A.N., **134**, 132 (1960).
127. Lauermann, Metzger and Sauerwald, Z. Phys. Chem., **216**, 42 (1961).
128. Lauermann and Sauerwald, Z. Metallkunde, **55**, 605 (1964).
129. Lucas, C.R., Paris, **248**, 2336 (1959).
130. Mack, Davis and Bartell, J. Phys. Chem., **45**, 846 (1941).
131. Matuyama, Sci. Rep. RITU, **16**, 555 (1927).
132. Melik-Gajkazan, Woronchikhina and Sakharova, Elektrokhim., **4**, 1420 (1968).
133. Melford and Hoar, J. Inst. Met., **85**, 197 (1957).
134. Metzger, Z. Phys. Chem., **211**, 1 (1959).
135. Monma and Suto, J. Jap. Inst. Met., **24**, 167 (1960).
136. Monma and Suto, J. Inst. Met., **1**, 69 (1960).
137. Naidich and Eremenko, Fiz. Met.i Met., **11**, 883 (1961).
138. Naidich, Eremenko, Fessenko, Vassiliu and Kirichenko, Zh. Fiz. Khim., **35**, 694 (1961).
139. Namba and Isobe, Sci. Pap. Inst. Phys. Chem. Res., Tokyo, **57**, 5154 (1963).
140. Nicholas, Joyner, Tessem and Olson, J. Phys. Chem., **65**, 1375 (1961).
141. Norton and Kingery, AEC-Progr. Rep., NYO 4632, (1955).
142. Olson and Johnson, J. Phys. Chem., **67**, 2529 (1963).
143. Ono and Matsushima, Sci. Rep. RITU, **9**, 309 (1957).
144. Oppenheimer, Z. anorg. allg. Chem., **171**, 98 (1928).
145. Osminin, Zh. Fiz. Khim., **43**, 2610 (1969).
146. Palacios, An. Soc. Esp. Fis., **18**, 294 (1920).
147. Parthey, Thesis, Techn. Univ. Berlin, (1961).
148. Pekarev, Izv. Vys. Uch. Sav., Tsvetn. Met., **6**, 111 (1963).
149. Pelzel, Berg-u. Hütt. Mon. Hefte, Leoben, **93**, 248 (1948).
150. Pelzel, Berg-u. Hütt. Mon. Hefte, Leoben, **94**, 10 (1949).
151. Peterson, Kedesdy, Keck and Schwarz, J. Appl. Phys., **29**, 213 (1958).
152. Poindexter, Phys. Review, **27**, 820 (1926).
153. Pokrovski, Ukr. Khim. Zh., **7**, 845 (1962).
154. Pokrovski and Galanina, Zh. Fiz. Khim., **23**, 324 (1949).
155. Pokrovski and Kristian, Zh. Fiz. Khim., **28**, (1954).
156. Pokrovski and Saidov, Fiz. Met.i Met., **2**, 546 (1956).
157. Popel, Shergin and Zarevski, Zh. Fiz. Khim., **43**, 2365 (1969).
158. Popel, Smirnov, Zarevski, Dshemilev and Pastukhov, Izv. A.N., **1**, 62 (1965).
159. Popel, Zarevski and Dshemilev, Fiz. Met.i Met., **18**, 468 (1964).
160. Popesco, C.R., Paris, **172**, 1474 (1921).
161. Portevin and Bastien, C.R., Paris, **202**, 1072 (1936).
162. Pugachevich, Zh. Fiz. Khim., **25**, 1365 (1951).
163. Pugachevich and Altynov, Doklady A.N., **86**, 117 (1952).
164. Pugachevich and Yashkichev, The Role of Surface Phenomena in Metallurgy, 46 (1963).
165. Quincke, Ann. d. Physik, **134**, 356 (1868).
166. Quincke, Ann. d. Physik, **135**, 621 (1868).
167. Quincke, Ann. d. Physik, **138**, 141 (1869).
168. Quincke, Ann. d. Physik, **61**, 267 (1897).
169. Raue, Metzger and Sauerwald, Metall, **20**, 1040 (1966).
170. Richards and Boyer, J. Amer. Chem. Soc., **43**, 290 (1921).
171. Roberts, J. Chem. Soc., 1907 (1964).
172. Roehlich jun., Tepper and Rankin, J. Chem. Eng. Data, **13**, 518 (1968).
173. Sauerwald and Drath, Z. anorg. allg. Chem., **154**, 79 (1926).
174. Sauerwald, Schmidt and Pelka, Z. anorg. allg. Chem., **223**, 84 (1935).

References (Continued)

175. Schmidt, Ann. d. Physik, **39**, 1108 (1912).
176. Schwaneke and Falke, US-Bur. Min., Inv. Rep. No. **7372**, (1970).
177. Schwaneke, Falke and Miller, US-Bur. Min., Inv. Rep. No. **7340**, (1970).
178. Semenchenko and Pokrovski, Uspekhii Khim., **6**, 945 (1937).
179. Semenchenko, Pokrovski and Lasarev, Doklady A.N., **89**, 1021 (1953).
180. Shunk and Burr, Trans. ASM, **55**, 786 (1962).
181. Siedentopf, Ann. d. Physik, **61**, 235 (1897).
182. Smirnova and Ormont, Zh. Fiz. Khim., **33**, 771 (1959).
183. Smith, J. Inst. Met., **12**, 168, 20 (1914).
184. Smith and Spitzer, J. Phys. Chem., **66**, 946 (1962).
185. Solovev and Makarova, Teplofiz. Vysok. Temp., **4**, 189 (1966).
186. Spriet, Mém. Sci. Rev. Mét., **60**, 531 (1963).
187. Tavadse, Bairamashvili, Khantadse and Zagareishvili, Doklady A.N., **150**, 544 (1963).
188. Taylor, J. Inst. Met., **83**, 143 (1954).
189. Taylor, Phil. Mag., **46**, 867 (1955).
190. Thyssen, Thesis, Humboldt-Univ. Berlin, (1960).
191. Tille and Kelly, Brit. J. Appl. Phys., **146**, 717 (1963).
192. Timofeyevicheva and Lasarev, Doklady A.N., **138**, 412 (1961).
193. Timofeyevicheva and Lasarev, Doklady A.N., 358 (1962).
194. Timofeyevicheva and Lasarev, Kolloidnyi-Zh., **24**, 227 (1962).
195. Timofeyevicheva, Lasarev and Pershikov, Doklady A.N., **143**, 618 (1962).
196. Timofeyevicheva and Pugachevich, Doklady A.N., **124**, 1093 (1959).
197. Timofeyevicheva and Pugachevich, Doklady A.N., **134**, 840 (1960).
198. Urbain and Lucas, Proc. N.P.L., Teddington, **9**, Pap. 4E (1959).
199. Watolin, Esin, Ukhov and Dubinin, Trudy Inst. Met., Sverdlovsk, **18**, 73 (1969).
200. Whalen and Humenik, Trans. AIME, **218**, 952 (1960).
201. Wegener, Z. Physik, **143**, 548 (1956).
202. White, Trans. AIME, **236**, 796 (1966).
203. White, Met. Review, **124**, 73 (1968).
204. Wobst and Rentzsch, Z. Phys. Chem., **240**, 36 (1969).
205. Yashkichevich and Lasarev, Izv. A.N., OTN, **1**, 170 (1964).
206. Yung Lee, Ind. Eng. Chem., Prod. Res. Dev., **7**, 66 (1968).
207. Zarevski and Popel, Fiz. Met.i Met., **13**, 451 (1962).
208. Zhivov, Trudy WAMI, SSSR, **14**, 99 (1937).
209. Zsin-Tan Wan, Karassev and Samarin, Izv. A.N., OTN, **1**, 30, 49 (1960).
210. Elverum and Doescher, J. Chem. Phys., **20**, 1834 (1952).
211. Fredrickson, J. Colloid Interfac. Sci., **48**, 506 (1974).
212. Johnson and McIntosh, J. Am. Chem. Soc., **31**, 1138 (1909).
213. Chao and Stenger, Talanta, **11**, 271 (1963).

Addendum

Z1. Eljutin, Kostikow and Penkow, Poroshk. Met., **9**, 46 (1970).
Z2. Fraser, Lu, Hamielec and Murarka, Met. Trans., **2**, 817 (1971).
Z3. Bernard and Lupis, Met. Trans., **2**, 555 (1971).
Z4. Rhee, J. Am. Ceram. Soc., **53**, 386 (1970).
Z5. Naidich, Perevertailo and Shuravlev, Zh. Fiz. Chim., **45**, 991 (1971).
Z6. Körber and Löhberg, Gieβereiforschung, **23**, 173 (1971).
Z7. Ziesing, Austral. J. Phys., **6**, 86 (1953).
Z8. Sangster and Carman, J. Chem. Phys., **23**, 1142 (1955).
Z9. Primak and Quarterman, J. Phys. Chem., **58**, 1051 (1954).
Z10. Cooke, HTLMHTTM, Oak Ridge, Vol. I, p. 66, (Nov. 1964).
Z11. Ofizerow, Izv. A.N., Metally, **4**, 91 (1971).
Z12. White, Met. Trans., **3**, 1933 (1972).
Z13. Lang, Aluminum (Germany), **49**, 231 (1972).
Z13a. Lang, J. Inst. Met. (to be published).
Z14. Perry and Roberts, *J. Chem. Eng. Data*, **26**, 266 (1981).

SURFACE TENSION

SURFACE TENSION OF INORGANIC SOLUTES IN WATER

% = Weight % of solute
γ = Surface tension in dynes/cm.

Solute	T°C	% / γ							
HCl	20	%	1.78	3.52	6.78	12.81	16.97	23.74	35.29
		γ	72.55	72.45	72.25	71.85	71.75	70.55	65.75
HNO_3	20	%	4.21	8.64	14.99	34.87			
		γ	72.15	71.65	70.95	68.75			
H_2SO_4	25	%	4.11	8.26	12.10	17.66	21.88	29.07	33.03
		γ	72.21	72.55	72.80	73.36	73.91	74.80	75.29
$HClO_4$	25	%	4.86	10.01	20.38	30.36	53.74	63.47	72.25
		γ	71.18	70.34	69.21	68.57	69.02	69.73	69.01
KOH	18	%	2.73	5.31	10.08	17.57			
		γ	73.95	74.85	76.55	79.75			
NaOH	18	%	2.72	5.66	16.66	30.56	35.90		
		γ	74.35	75.85	83.05	96.05	101.05		
NH_4OH	18	%	1.72	3.39	4.99	9.51	17.37	34.47	54.37
		γ	71.65	70.65	69.95	67.85	65.25	61.05	57.05
KCl	20	%	0.74	3.60	6.93	13.88	18.77	22.97	24.70
		γ	72.99	73.45	74.15	75.55	76.95	78.25	78.75
LiCl	25	%	5.46	7.37	10.17	13.95			
		γ	74.23	75.10	76.30	78.10			

Solute	T°C	% / γ							
NaCl	20	%	0.58	2.84	5.43	10.46	14.92	22.62	25.92
		γ	72.92	73.75	74.39	76.05	77.65	80.95	82.55
PbCl	25	%	21.57	28.52	37.74				
		γ	75.20	76.80	79.20				
$BaCl_2$	25	%	9.26	16.73	25.58				
		γ	73.50	74.93	76.38				
$MgCl_2$	20	%	0.94	4.55	8.69	16.00	22.30	25.44	
		γ	73.07	74.00	75.75	79.15	82.95	85.75	
NaBr	20	%	4.89	9.33	13.37	23.00			
		γ	73.45	74.05	74.75	76.55			
$Al_2(SO_4)_3$	25	%	2.54	4.06	9.40	14.60	19.32	23.54	25.50
		γ	72.32	72.92	73.51	74.71	76.06	78.30	79.73
$MgSO_4$	20	%	1.19	5.68	10.75	19.41	24.53		
		γ	73.01	73.78	74.85	77.35	79.25		
Na_2SO_4	20	%	2.76	6.63	12.44				
		γ	73.25	74.15	75.45				
Na_2CO_3	20	%	2.58	5.03	9.59	13.72			
		γ	73.45	74.05	75.45	76.75			
$NaNO_3$	20	%	0.85	4.08	7.84	14.53	29.82	37.30	47.06
		γ	72.87	73.75	73.95	75.15	78.35	80.25	87.05

SURFACE TENSION OF INORGANIC SOLUTES IN ORGANIC SOLVENTS

% = Mol %
γ = Surface tension in dynes/cm.

Solute	Solvent	T°C	% / γ						
LiCl	Ethyl alcohol	14	%	0.72	2.30	4.62			
			γ	22.90	23.17	23.26			
LiBr	Ethyl alcohol	14	%	0.95	2.60				
			γ	23.08	23.35				
LiI	Ethyl alcohol	14	%	1.43	2.87	5.08	10.21	19.47	26.92
			γ	23.11	23.56	24.39	26.03	28.87	31.95
KI	Methyl alcohol	14	%	0.81	1.52	2.68			
			γ	23.76	24.11	24.71			
$CaCl_2$	Ethyl alcohol	24	%	0.94	1.98	3.79	7.47		
			γ	22.62	22.58	23.23	23.97		
NaI	Methyl alcohol	14	%	0.76	1.48	4.33	8.55	12.53	
			γ	22.83	23.29	24.85	27.41	29.75	

Solute	Solvent	T°C	% / γ						
NaI	Ethyl alcohol	24	%	0.45	1.80	3.63	4.54	6.02	10.46
			γ	22.47	22.82	23.41	23.52	24.00	25.07
NaI	Acetone	14	%	0.93	2.08	5.07	6.53		
			γ	24.22	24.40	25.04	25.12		
ZnI_2	Methyl alcohol	22	%	0.90	2.79	5.07			
			γ	22.97	24.23	25.84			
ZnI_2	Ethyl alcohol	24	%	0.41	1.72	3.42	6.90		
			γ	22.70	22.90	23.71	25.49		
H_2SO_4	Ethyl ether	17	%	3	10	40	75	90	
			γ	17.30	19.55	32.50	46.30	46.83	
H_2SO_4	Nitrobenzene	17	%	3	10	40	75	90	
			γ	42.73	43.96	46.05	47.52	48.25	

SURFACE TENSION OF ORGANIC COMPOUNDS IN WATER

% = Weight % of solute
γ = Surface tension in dynes/cm.

Solute	T°C	% / γ							
Acetic acid	30	%	1.00	2.475	5.001	10.01	30.09	49.96	69.91
		γ	68.00	64.40	60.10	54.60	43.60	38.40	34.30
Acetone	25	%	5.00	10.00	20.00	50.00	75.00	95.00	100.00
		γ	55.50	48.90	41.10	30.40	26.80	24.20	23.00
Acetonitrile	20	%	1.13	3.35	11.77	20.20	37.58	61.33	81.22
		γ	69.02	63.03	47.61	39.06	31.84	30.02	29.02
o-Aminobenzoic acid	25	%	12.35	22.36	30.45	37.44			
		γ	71.96	73.23	74.54	75.79			
m-Aminobenzoic acid	25	%	12.35	22.36	30.45	37.44			
		γ	73.30	74.59	76.16	77.89			
p-Aminobenzoic acid	25	%	12.35	22.36	30.45	37.44			
		γ	73.38	74.79	76.32	78.20			
Aminobutyric acid	25	%	4.96	9.34	13.43				
		γ	71.91	71.67	71.40				
Ammonium lactate	29	%	30.00	50.00	60.00	70.00	80.00	90.00	
		γ	35.40	34.40	35.40	35.60	38.20	44.50	
n-Butanol	30	%	0.04	0.41	9.53	80.44	86.05	94.20	97.40
		γ	69.33	60.38	26.97	23.69	23.47	23.29	22.25
n-Butyric acid	25	%	0.14	0.31	1.05	8.60	25.00	79.00	100.00
		γ	69.00	65.00	56.00	33.00	28.00	27.00	26.00
Dioxan	26	%	0.44	2.20	4.70	11.14	20.17	35.20	55.00
		γ	69.83	65.64	62.45	56.90	51.57	45.30	39.27
Dioxan	26	%	67.68	76.45	83.02	91.90	95.60	97.77	
		γ	36.95	35.80	35.00	33.95	33.60	33.10	

Solute	T°C	% / γ							
Formic acid	30	%	1.00	5.00	10.00	25.00	50.00	75.00	100.00
		γ	70.07	66.20	62.78	56.29	49.50	43.40	36.51
Glycerol	18	%	5.00	10.00	20.00	30.00	50.00	85.00	100.00
		γ	72.90	72.90	72.40	72.00	70.00	66.00	63.00
Glycine	25	%	3.62	6.98	10.12	13.10			
		γ	72.54	73.11	73.74	74.18			
Hydrocinnamic acid	21.5	%	7.02	12.62	18.39	26.09	31.06	38.25	47.93
		γ	69.08	66.49	63.63	59.25	56.14	52.96	47.24
Methyl acetate	25	%	0.66	1.29	2.29	3.56			
		γ	66.33	62.92	58.22	55.08			
Morpholine	20	%	8.56	19.39	30.41	50.45	69.93	80.14	92.00
		γ	67.80	62.62	59.15	52.85	47.05	43.62	41.60
Potassium lactate	29	%	40.00	50.00	60.00	70.00			
		γ	66.40	66.40	65.40	63.40			
Phenol	20	%	0.024	0.047	0.118	0.417	0.941	3.76	5.62
		γ	72.60	72.20	71.30	66.50	61.10	46.00	42.30
n-Propanol	25	%	0.1	0.5	1.0	50.0	60.0	80.0	90.0
		γ	67.10	56.18	49.30	24.34	24.15	23.66	23.41
Propionic acid	25	%	1.91	5.84	9.80	21.70	49.80	73.90	100.00
		γ	60.00	49.00	44.00	36.00	32.00	30.00	26.00
Sodium lactate	29	%	1.00	10.00	30.00	40.00	50.00	60.00	70.00
		γ	70.40	69.60	68.50	64.80	45.40	56.70	60.70
Sucrose	25	%	10.00	20.00	30.00	40.00	55.00		
		γ	72.50	73.00	73.40	74.10	75.70		

SURFACE TENSION OF ORGANIC COMPOUNDS IN ORGANIC SOLVENTS

% = Weight % of solvent
γ = Surface tension dynes/cm.

Solute	Solvent	T°C	% / γ					
Acetic acid	Benzene	35	%	10.45	25.53	34.28	43.93	68.77
			γ	25.40	25.21	25.32	25.43	25.99
Acetic acid	Acetone	25	%	25.63	50.83	75.62		
			γ	27.50	26.61	24.90		
Acetone	Ethyl ether	30	%	21.83	40.98	61.29	75.69	89.03
			γ	16.75	17.49	19.15	19.80	21.00
Acetonitrile	Ethyl alcohol	20	%	10.83	32.42	49.90	69.78	80.19
			γ	22.92	23.92	24.36	25.08	26.51
Aniline	Cyclohexane	32	%	14.35	37.65	50.67	72.28	96.46
			γ	24.21	24.51	24.50	25.61	37.45
Carbontetrachloride	Benzene	50	%	30.40	51.69	62.38	78.29	88.62
			γ	24.39	24.09	23.78	23.47	23.21
Cyclohexane	Nitrobenzene	15	%	5.00	10.00			
			γ	37.59	33.03			
Naphthalene	Benzene	79.5	%	20.00	40.00	50.00	60.00	80.00
			γ	23.42	26.70	27.80	29.20	31.70
Naphthalene	*p*-Nitrophenol	121	%	10.21	29.74	45.94	58.01	67.57
			γ	29.90	31.80	33.70	34.80	41.80

SURFACE TENSION OF METHYL ALCOHOL IN WATER

% = Volume % of alcohol
γ = Surface tension dynes/cm.

T°C	%	7.5	10.00	25.0	50.0	60.0	80.0	90.0	100.0
20	γ	60.90	59.04	46.38	35.31	32.95	27.26	25.36	22.65
30	γ	59.33	57.27	45.30	34.52	32.26	26.48	24.42	21.58
50	γ	56.19	55.01	43.24	32.95	30.79	25.01	22.55	19.52

SURFACE TENSION OF ETHYL ALCOHOL IN WATER

% = Volume % of alcohol
γ = Surface tension dynes/cm.

T°C	%	5.00	10.00	24.00	34.00	48.00	60.00	72.00	80.00	96.00
20	γ	—	—	—	33.24	30.10	27.56	26.28	24.91	23.04
40	γ	54.92	48.25	35.50	31.58	28.93	26.18	24.91	23.43	21.38
50	γ	53.35	46.77	34.32	30.70	28.24	25.50	24.12	22.56	20.40

WATER AGAINST AIR

Temperature °C	Surface tension dynes/cm.	Temperature °C	Surface tension dynes/cm.
−8	77.0	25	71.97
−5	76.4	30	71.18
0	75.6	40	69.56
5	74.9	50	67.91
10	74.22	60	66.18
15	73.49	70	64.4
18	73.05	80	62.6
20	72.75	100	58.9

INTERFACIAL TENSION

Surface Tension at the Interface Between Two Liquids
(Each liquid saturated with the other)

Liquids	Temperature °C	γ
Benzene — Mercury	20	357
Ethyl ether — Mercury	20	379
Water — Benzene	20	35.00
Water — Carbon tetrachloride	20	45.
Water — Ethyl ether	20	10.7
Water — Heptylic acid	20	7.0
Water — n-Hexane	20	51.1
Water — Mercury	20	375.
Water-n-Octane	20	50.8
Water-n-Octyl alcohol	20	8.5

SURFACE TENSION OF VARIOUS LIQUIDS

Substance: Name	Substance: Formula	In contact with	Temperature °C	Surface tension dynes/cm
Acetaldehyde	C_2H_4O	- -vapor	20	21.2
Acetaldoxime	C_2H_5NO	- -vapor	35	30.1
Acetamide	C_2H_5NO	- -vapor	85	39.3
Acetanilide	C_8H_9NO	- -vapor	120	35.6
Acetic acid	$C_2H_4O_2$	- -vapor	10	28.8

SURFACE TENSION OF VARIOUS LIQUIDS (Continued)

Substance		In contact with	Temperature °C	Surface tension dynes/cm
Name	Formula			
	C_2H_4O	- -vapor	20	27.8
	C_2H_4O	- -vapor	50	24.8
Acetic anhydride	$C_4H_6O_3$	- -vapor	20	32.7
Acetone	C_3H_6O	- -air or vapor	0	26.21
	C_3H_6O	- -air or vapor	20	23.70
	C_3H_6O	- -air or vapor	40	21.16
Acetonitrile	C_2H_3N	- -vapor	20	29.30
Acetophenone	C_8H_8O	- -vapor	20	39.8
Acetyl chloride	C_2H_3ClO	- -vapor	14.8	26.7
Acetylene	C_2H_2	- -vapor	70.5	16.1
Acetylsalicylic acid (in aq. sol.)	$C_9H_8O_4$	- -vapor	25.9	60.06
Allyl alcohol	C_3H_6O	- -air or vapor	20	25.8
Allyl isothiocyanate	C_4H_5NS	- -air or vapor	20	34.5
Ammonia	NH_3	- -vapor	11.1	23.4
	NH_3	- -vapor	34.1	18.1
Aniline	C_6H_7N	- -air	10	44.10
	C_6H_7N	- -vapor	20	42.9
	C_6H_7N	- -air	50	39.4
Argon	A	- -vapor	−188	13.2
Azoxybenzene	$C_{12}H_{10}N_2O$	- -vapor	51	43.34
Benzaldehyde	C_7H_6O	- -air	20	40.04
Benzene	C_6H_6	- -air	10	30.22
	C_6H_6	- -air	20	28.85
	C_6H_6	- -saturated with vapor	20	28.89
	C_6H_6	- -air	30	27.56
Benzonitrile	C_7H_5N	- -air	20	39.05
Benzophenone	$C_{13}H_{10}O$	- -air or vapor	20	45.1
Benzylamine	C_7H_9N	- -vapor	20	39.5
Benzyl alcohol	C_7H_8O	- -air or vapor	20	39.0
Bromine	Br_2	- -air or vapor	20	41.5
Bromobenzene	C_6H_5Br	- -air	20	36.5
Bromoform	$CHBr_3$	- -vapor	20	41.53
p-Bromophenol	C_6H_5BrO	- -vapor	74.4	42.36
d-sec-Butyl alcohol	$C_4H_{10}O$	- -vapor	10	23.5
n-Butyl alcohol	$C_4H_{10}O$	- -air or vapor	0	26.2
	$C_4H_{10}O$	- -air or vapor	20	24.6
	$C_4H_{10}O$	- -air or vapor	50	22.1
tert-Butyl alcohol	$C_4H_{10}O$	- -air or vapor	20	20.7
n-Butylamine	$C_4H_{11}N$	- -nitrogen	41	19.7
n-Butyric acid	$C_4H_8O_2$	- -air	20	26.8
Carbon bisulfide	CS_2	- -vapor	20	32.33
Carbon dioxide	CO_2	- -vapor	20	1.16
	CO_2	- -vapor	−25	9.13
Carbon tetrachloride	CCl_4	- -vapor	20	26.95
	CCl_4	- -vapor	100	17.26
	CCl_4	- -vapor	200	6.53
Carbon monoxide	CO	- -vapor	−193	9.8
	CO	- -vapor	−203	12.1
Chloral	C_2HCl_3O	- -vapor	19.4	25.34
Chlorine	Cl_2	- -vapor	20	18.4
	Cl_2	- -vapor	−30	25.4
	Cl_2	- -vapor	−40	27.3
	Cl_2	- -vapor	−50	29.2
	Cl_2	- -vapor	−60	31.2
Chloroacetic acid	$C_2H_2Cl_2O_2$	- -nitrogen	25.7	35.4
Chlorobenzene	C_4H_5Cl	- -vapor	20	33.56
Chloroform	$CHCl_3$	- -air	20	27.14
o-Chlorophenol	C_6H_5ClO	- -vapor	12.7	42.25
Cyclohexane	C_6H_{12}	- -air	20	25.5
Dichloroacetic acid	$C_2H_2Cl_2O_2$	- -nitrogen	25.7	35.4
Dichloroethane	$C_2H_4Cl_2$	- -air	35.0	23.4
Diethylamine	$C_4H_{11}N$	- -air	56	16.4
Diethylaniline	$C_{10}H_{15}N$	- -vapor	20	34.2
Diethyl carbonate	$C_5H_{10}O$	- -air	20	26.31
Diethyl oxalate	$C_6H_{10}O_4$	- -vapor	20	32.0
Diethyl phthalate	$C_{12}H_{14}O_4$	- -vapor	20	37.5
Diethyl sulfate	$C_4H_{12}O_4S$	- -air	13	34.61
Dimethylamine	C_2H_7N	- -nitrogen	0	18.1
	C_2H_7N	- -nitrogen	5	17.7
Dimethylaniline	C_8H_{11}	- -air or vapor	20	36.6
1,5-Dimethyl-2-phenyl-3-pyrazolone	$C_{11}H_{12}N_2O$	- -vapor	25.9	63.63
Dimethyl sulfate	$C_2H_6O_4S$	- -air	18	40.12
Diphenylamine	$C_{12}H_{11}N$	- -air or vapor	80	37.7
Ethyl acetate	$C_4H_8O_2$	- -air	0	26.5
	$C_4H_8O_2$	- -air	20	23.9
	$C_4H_8O_2$	- -air	50	20.2
Ethyl acetoacetate	$C_6H_{10}O_3$	- -air or vapor	20	32.51
Ethyl alcohol	C_2H_6O	- -air	0	24.05
	C_2H_6O	- -vapor	10	23.61
	C_2H_6O	- -vapor	20	22.75
	C_2H_6O	- -vapor	30	21.89
Ethylamine	C_2H_7N	- -nitrogen	0	21.3
	C_2H_7N	- -nitrogen	9.9	20.4

SURFACE TENSION OF VARIOUS LIQUIDS (Continued)

Substance Name	Substance Formula	In contact with	Temperature °C	Surface tension dynes/cm
Ethylaniline	$C_8H_{11}N$	- -air or vapor	20	36.6
Ethylbenzene	C_8H_{10}	- -vapor	20	29.20
Ethylbenzoate	$C_9H_{10}O_2$	- -vapor	20	35.5
Ethyl bromide	C_2H_5Br	- -vapor	20	24.15
Ethyl chloroformate	$C_3H_5ClO_2$	- -vapor	15.1	27.5
Ethyl Cinnamate	$C_{11}H_{12}O_2$	- -air	20	38.37
Ethylene bromide	$C_2H_4Br_2$	- -vapor	20	38.37
Ethylene chloride	$C_2H_4Cl_2$	- -air	20	24.15
Ethylene oxide	C_2H_4O	- -vapor	−20	30.8
	C_2H_4O	- -vapor	0.0	27.6
	C_2H_4O	- -vapor	20	24.3
Ethyl ether	$C_4H_{10}O$	- -vapor	20	17.01
	$C_4H_{10}O$	- -vapor	50	13.47
Ethyl format	$C_3H_6O_2$	- -air or vapor	20	23.6
Ethyl iodide	C_2H_5I	- -vapor	20	29.4
Ethyl nitrate	$C_2H_5NO_3$	- -air or vapr	20	28.7
dl-Ethyl lactate	$C_5H_{10}O_3$	- -air	20	29.9
Ethyl mercaptan	C_2H_6S	- -air or vapor	20	22.5
Ethyl salicylate	$C_9H_{10}O_3$	- -vapor	20.5	38.33
Formamide	CH_3NO	- -vapor	20	58.2
Formic acid	CH_2O_2	- -air	20	37.6
Furfural	$C_5H_4O_2$	- -air or vapor	20	43.5
Gelatin solution (1%)		- -water	2.85	8.3
Glycerol	$C_3H_8O_3$	- -air	20	63.4
	$C_3H_8O_3$	- -air	90	58.6
	$C_3H_8O_3$	- -air	150	51.9
Glycol	$C_2H_6O_2$	- -air or vapor	20	47.7
Helium	He	- -vapor	−269	.12
	He	- -vapor	−270	.239
	He	- -vapor	−271.5	.353
n-Hexane	C_6H_{14}	- -air	20	18.43
Hydrazine	N_2H_4	- -vapor	25	91.5
Hydrogen	H_2	- -vapor	−255	2.31
Hydrogen cyanide	HCN	- -vapor	17	18.2
Hydrogen peroxide	H_2O_2	- -vapor	18.2	76.1
Isobutyl alcohol	$C_4H_{10}O$	- -vapor	20	23.0
Isobutylamine	$C_4H_{11}N$	- -air	68	17.6
Isobutyl chloride	C_4H_9Cl	- -air	20	21.94
Isobutyric acid	$C_4H_8O_2$	- -air or vapor	20	25.2
Isopentane	C_5H_{12}	- -air	20	13.72
Isopropyl alcohol	C_3H_8O	- -air or vapor	20	21.7
Methyl acetate	$C_3H_6O_2$	- -air or vapor	20	24.6
Methyl alcohol	CH_4O	- -air	0	24.49
	CH_4O	- -air	20	22.61
	CH_4O	- -vapor	50	20.14
Methylamine	CH_3NH_2	- -nitrogen	−12	22.2
	CH_3NH_2	- -vapor	−20	23.0
	CH_3NH_2	- -nitrogen	−70	29.2
N-Methylaniline	C_7H_9N	- -air or vapor	20	39.6
Methyl benzoate	$C_8H_8O_2$	- -air or vapor	20	37.6
Methyl chloride	CH_3Cl	- -air	20	16.2
Methyl ether	C_2H_6O	- -vapor	−10	16.4
	C_2H_6O	- -vapor	−40	21
Methylene chloride	CH_2Cl_2	- -air	20	26.52
Methylene iodide	CH_2I_2	- -air	20	50.76
Methyl ethyl ketone	C_4H_8O	- -air or vapor	20	24.6
Methyl formate	$C_2H_4O_2$	- -vapor	20	25.08
Methyl iodide	CH_3I	- -air	43.5	25.8
Methyl propionate	$C_4H_8O_2$	- -air or vapor	20	24.9
Methyl salicylate	$C_8H_8O_3$	- -nitrogen	94	31.9
Methyl sulfide	C_2H_6S	- -vapor	11.1	26.50
Naphthalene	$C_{10}H_8$	- -air or vapor	127	28.8
Neon	Ne	- -vapor	−248	5.50
Nitric acid (98.8%)	HNO_3	- -air	11.6	42.7
Nitrobenzene	$C_6H_5NO_2$	- -air or vapor	20	43.9
Nitroethane	$C_2H_5NO_2$	- -air or vapor	20	32.2
Nitrogen	N_2	- -vapor	−183	6.6
	N_2	- -vapor	−193	8.27
	N_2	- -vapor	−203	10.53
Nitrogen tetra oxide	N_2O_4	- -vapor	19.8	27.5
Nitromethane	CH_3NO_2	- -vapor	20	36.82
Nitrous oxide	N_2O	- -vapor	20	1.75
n-Octane	C_8H_{18}	- -vapor	20	21.80
n-Octyl alcohol	$C_8H_{18}O$	- -air	20	27.53
Oleic acid	$C_{18}H_{34}O_2$	- -air	20	32.50
Oxygen	O_2	- -vapor	−183	13.2
Oxygen (65%)	O_2	- -air	−190.5	12.2
	O_2	- -vapor	−193	15.7
	O_2	- -vapor	−203	18.3
Paraldehyde	$C_6H_{12}O_3$	- -air	20	25.9
Phenetole	$C_8H_{10}O$	- -vapor	20	32.74
Phenol	C_6H_6O	- -air or vapor	20	40.9
	C_6H_6O	- -air or vapor	30	39.88

SURFACE TENSION OF VARIOUS LIQUIDS (Continued)

Substance: Name	Formula	In contact with	Temperature °C	Surface tension dynes/cm
Phenylhydrazine	$C_6H_8N_2$	--vapor	20	46.1
Phosphorus tribromide	PBr_3	--air	24	45.8
Phosphorus trichloride	PCl_3	--vapor	20	29.1
Phosphorus triiodide	PI_3	--vapor	75.3	56.5
Propionic acid	$C_3H_6O_2$	--vapor	20	26.7
n-Propyl acetate	$C_5H_{10}O_2$	--air or vapor	20	24.3
n-Propyl alcohol	C_3H_8O	--vapor	20	23.78
n-Propylamine	C_3H_9N	--air	20	22.4
n-Propyl bromide	C_3H_7Br	--vapor	71	19.65
n-Propyl chloride	C_3H_7Cl	--air	47	18.2
n-Propyl formate	$C_4H_8O_2$	--vapor	20	24.5
Pyridine	C_5H_5N	--air	20	38.0
Quinoline	C_9H_7N	--air	20	45.0
Ricinoleic acid	$C_{18}H_{34}O_3$	--air	16	35.81
Selenium	Se	--air	217	92.4
Styrene	C_8H_8	--air	19	32.14
Sulfuric acid (98.5%)	H_2SO_4	--air or vapor	20	55.1
Tetrabromoethane 1,1,2,2-	$C_2H_2Br_4$	--air	20	49.67
Tetrachloroethane 1,1,2,2-	$C_2H_2Cl_4$	--air	22.5	36.03
Tetrachloroethylene	C_2Cl_4	--vapor	20	31.74
Toluene	C_7H_8	--vapor	10	27.7
	C_7H_8	--vapor	20	28.5
	C_7H_8	--vapor	30	27.4
m-Toluidine	C_7H_9N	--vapor	20	36.9
o-Toluidine	C_7H_9N	--air or vapor	20	40.0
p-Toluidine	C_7H_9N	--air	50	34.6
Trichloroacetic acid	$C_2HCl_3O_2$	--nitrogen	80.2	27.8
Trichloroethane 1,1,2-	$C_2H_3Cl_3$	--air	114	22.0
Triethyl phosphate	$C_6H_{15}O_4P$	--air	15.5	30.61
Trimethylamine	C_3H_9N	--nitrogen	−4	17.3
Triphenylcarbinol	$C_{19}H_{16}O$	--vapor	165.8	30.38
Vinyl acetate	$C_4H_6O_2$	--vapor	20	23.95
	$C_4H_6O_2$	--vapor	25	23.16
	$C_4H_6O_2$	--vapor	30	22.54
Water	H_2O	--air	18	73.05
m-Xylene	C_8H_{10}	--vapor	20	28.9
o-Xylene	C_8H_{10}	--air	20	30.10
p-Xylene	C_8H_{10}	--vapor	20	28.37

VISCOSITY

Viscosity. — All fluids possess a definite resistance to change of form and many solids show a gradual yielding to forces tending to change their form. This property, a sort of internal friction, is called viscosity; it is expressed in dyne-seconds per cm² or poises. Dimensions, —$[m\,l^{-1}\,t^{-1}]$. If the tangential force per unit area, exerted by a layer of fluid upon one adjacent is one dyne for a space rate of variation of the tangential velocity of unity, the viscosity is one poise.

Kinematic viscosity is the ratio of viscosity to density. The c.g.s. unit of kinematic viscosity is the **stoke.**

Flow of liquids through a tube; where l is the length of the tube, r its radius, p the difference of pressure at the ends, η the coefficient of viscosity, the volume escaping per second,

$$\nu = \frac{\pi p r^4}{8 l \eta} \text{ (Poiseuille).}$$

The volume will be given in cm³ per second if l and r are in cm, p in dynes per cm² and η in poises or dyne-seconds per cm².

VISCOSITY OF WATER BELOW 0°C

White-Twining 1914

Temperature	Viscosity centipoises	Temperature	Viscosity centipoises
0°C	1.798	−7.23	2.341
−2.10	1.930	−8.48	2.458
−4.70	2.121	−9.30	2.549
−6.20	2.250		

ABSOLUTE VISCOSITY OF WATER AT 20°C

Swindells, J. R. Coe, Jr., and T. B. Godfrey, Journal of Research, National Bureau of Standards **48**, 1, 1952.
The value found for the viscosity of water at 20°C was 0.010019 ± 0.000003 poise.
The value **0.01002** poise is to be used as the absolute value of the viscosity of water for calibration purposes.

VISCOSITY CONVERSION TABLE

Poise = c.g.s. unit of absolute viscosity $= \dfrac{\text{gm}}{\text{sec} \times \text{cm}}$

Stoke = c.g.s. unit of kinematic viscosity $= \dfrac{\text{gm}}{\text{sec} \times \text{cm} \times \text{density } (t^\circ\text{F})}$

Centipoise = 0.01 poise

Centistoke = 0.01 stoke

Centipoises = Centistokes × density (at given temperature)

To convert poises to $\dfrac{\text{lb}}{\text{sec} \times \text{ft}}$ or $\dfrac{\text{lb}}{\text{hr} \times \text{ft}}$ multiply by 0.0672 or 242 respectively.

Centi-stokes	Saybolt Seconds at 100°F	Saybolt Seconds at 130°F	Saybolt Seconds at 210°F	Redwood Seconds at 70°F	Redwood Seconds at 140°F	Redwood Seconds at 200°F	Engler Degrees at all temps.
2.0	32.6	32.7	32.8	30.2	31.0	31.2	1.14
3.0	36.0	36.1	36.3	32.7	33.5	33.7	1.22
4.0	39.1	39.2	39.4	35.3	36.0	36.3	1.31
5.0	42.3	42.4	42.6	37.9	38.5	38.9	1.40
6.0	45.5	45.6	45.8	40.5	41.0	41.5	1.48
7.0	48.7	48.8	49.0	43.2	43.7	44.2	1.56
8.0	52.0	52.1	52.4	46.0	46.4	46.9	1.65
9.0	55.4	55.5	55.8	48.9	49.1	49.7	1.75
10.0	58.8	58.9	59.2	51.7	52.0	52.6	1.84
11.0	62.3	62.4	62.7	54.8	55.0	55.6	1.93
12.0	65.9	66.0	66.4	57.9	58.1	58.8	2.02
14.0	73.4	73.5	73.9	64.4	64.6	65.3	2.22
16.0	81.1	81.3	81.7	71.0	71.4	72.2	2.43
18.0	89.2	89.4	89.8	77.9	78.5	79.4	2.64
20.0	97.5	97.7	98.2	85.0	85.8	86.9	2.87
22.0	106.0	106.2	106.7	92.4	93.3	94.5	3.10
24.0	114.6	114.8	115.4	99.9	100.9	102.2	3.34
26.0	123.3	123.5	124.2	107.5	108.6	110.0	3.58
28.0	132.1	132.4	133.0	115.3	116.5	118.0	3.82
30.0	140.9	141.2	141.9	123.1	124.4	126.0	4.07
32.0	149.7	150.0	150.8	131.0	132.3	134.1	4.32
34.0	158.7	159.0	159.8	138.9	140.2	142.2	4.57
36.0	167.7	168.0	168.9	146.9	148.2	150.3	4.83
38.0	176.7	177.0	177.9	155.0	156.2	158.3	5.08
40.0	185.7	186.0	187.0	163.0	164.3	166.7	5.34
42.0	194.7	195.1	196.1	171.0	172.3	175.0	5.59
44.0	203.8	204.2	205.2	179.1	180.4	183.3	5.85
46.0	213.0	213.4	214.5	187.1	188.5	191.7	6.11
48.0	222.2	222.6	223.8	195.2	196.6	200.0	6.37
50.0	231.4	231.8	233.0	203.3	204.7	208.3	6.63
60.0	277.4	277.9	279.3	243.5	245.3	250.0	7.90
70.0	323.4	324.0	325.7	283.9	286.0	291.7	9.21
80.0	369.6	370.3	372.2	323.9	326.6	333.4	10.53
90.0	415.8	416.6	418.7	364.4	367.4	375.0	11.84
*100.0	462.0	462.9	465.2	404.9	408.2	416.7	13.16

* At higher values use the same ratio as above for 100 centistokes; e.g., 110 centistokes = 110 × 4.620 Saybolt seconds at 100°F.

To obtain the Saybolt Universal viscosity equivalent to a kinematic viscosity determined at t °F, multiply the equivalent Saybolt Universal viscosity at 100°F by $1 + (t-100)0.000064$; e.g., 10 centistokes at 210°F are equivalent to 58.8 × 1.0070, or 59.2 Saybolt Universal seconds at 210°F.

VISCOSITY CONVERSION

Kinematic

To convert from	To	Multiply by
cm^2/sec (Stokes)	Centistokes	10^2
	ft^2/hr.	3.875
	ft^2/sec.	1.076×10^{-3}
	in.^{2}/sec.	1.550×10^{-1}
	m^2/hr.	3.600×10^{-1}
cm^2/sec × 10^2 (Centistokes)	cm^2/sec. (Stokes)	1×10^{-2}
	ft^2/hr.	3.875×10^{-2}
	ft^2/sec.	1.076×10^{-5}
	in.^{2}/sec.	1.550×10^{-3}
	m^2/hr.	3.600×10^{-3}
ft^2/hr	cm^2/sec (Stokes)	2.581×10^{-1}
	cm/sec × 10^2 (centistokes	2.581×10
	ft^2/sec.	2.778×10^{-4}
	in.^{2}/sec.	4.00×10^{-2}
	m^2/hr.	9.290×10^{-2}
ft^2/sec	cm^2/sec. (Stokes)	9.29×10^2
	cm^2/sec. × 10^2 (centistokes)	9.29×10^4
	ft^2/hr	3.60×10^3
	in.^{2}/sec.	1.44×10^2
	m^2/hr.	3.345×10^2
in.^{2}/sec	cm^2/sec. (Stokes)	6.452
	cm^2/sec. × 10^2 (centistokes)	6.452×10^2
	ft^2/hr.	2.50×10
	ft^2/sec.	6.944×10^{-3}
	m^2/hr.	2.323
m^2/hr	cm^2/sec. (Stokes)	2.778
	cm^2/sec. × 10^2 (centistokes)	2.778×10^2
	ft^2/hr.	1.076×10
	ft^2/sec.	2.990×10^{-3}
	in.^{2}/sec.	4.306×10^{-1}

VISCOSITY CONVERSION

Absolute

Absolute viscosity = kinematic viscosity × density; lb = mass pounds; lb_F = force pounds

To convert from	To	Multiply by
gm/(cm)(sec) [Poise]	gm/(cm)(sec)(10^2) [Centipoise]	10^2
	kg/(m)(hr)	3.6×10^2
	lb/(ft)(sec)	6.72×10^{-2}
	lb/(ft)(hr)	2.419×10^2
	lb/(in.)(sec)	5.6×10^{-3}
	(gm_F)(sec)/cm^2	1.02×10^{-3}
	(lb_F)(sec)/in.^{2} [Reyn]	1.45×10^{-5}
	(lb_F)(sec)/ft^2	2.089×10^{-3}
gm/(cm)(sec)(10^2) [Centipoise]	gm/(cm)(sec) [Poise]	10^{-2}
	kg/(m)(hr)	3.6
	lb/(ft)(sec)	6.72×10^{-4}
	lb/(ft)(hr)	2.419
	lb/(in.)(sec)	5.60×10^{-5}
	(gm_F)(sec)/cm^2	1.02×10^{-5}
	(lb_F)(sec)/in.^{2} [Reyn]	1.45×10^{-7}
	(lb_F)(sec)/ft^2	2.089×10^{-5}
kg/(m)(hr)	gm/(cm)(sec)	2.778×10^{-3}
	gm/(cm)(sec)(10^2) [Centipoise]	2.778×10^{-1}
	lb/(ft)(sec)	1.867×10^{-4}
	lb/(ft)(hr)	6.720×10^{-1}
	lb/(in.)(sec)	1.555×10^{-5}
	(gm_F)(sec)/cm^2	2.833×10^{-4}
	(lb_F)(sec)/in.^{2} [Reyn]	4.029×10^{-8}
	(lb_F)(sec)/ft^2	5.801×10^{-6}

VISCOSITY CONVERSION

Absolute (Continued)

To convert from	To	Multiply by
lb/(ft)(sec)	gm/(cm)(sec) [Poise]	1.488×10^{1}
	gm/(cm)(sec)(10^2) [Centipoise]	1.488×10^{3}
	kg/(m)(hr)	5.357×10^{3}
	lb/(ft)(hr)	3.60×10^{3}
	lb/(in.)(sec)	8.333×10^{-2}
	(gm_F)(sec)/cm^2	1.518×10^{-2}
	(lb_F)(sec)/in.^{2} [Reyn]	2.158×10^{-4}
	(lb_F)(sec)/ft^2	3.108×10^{-2}
lb/(ft)(hr)	gm/(cm)(sec) [Poise]	4.134×10^{-3}
	gm/(cm)(sec)(10^2) [Centipoise]	4.134×10^{-1}
	kg/(m)(hr)	1.488
	lb/(ft)(sec)	2.778×10^{-4}
	lb/(in.)(sec)	2.315×10^{-5}
	(gm_F)(sec)/cm^2	4.215×10^{-6}
	(lb_F)(sec)/in.^{2} [Reyn]	5.996×10^{-8}
	(lb_F)(sec)/ft^2	8.634×10^{-6}
lb/(in.)(sec)	gm/(cm)(sec) [Poise]	1.786×10^{2}
	gm/(cm)(sec)(10^2) [Centipoise]	1.786×10^{4}
	kg/(m)(hr)	6.429×10^{4}
	lb/(ft)(sec)	1.20×10
	lb/(ft)(hr)	4.32×10^{4}
	(gm_F)(sec)/cm^2	1.821×10^{-1}
	(lb_F)(sec)/in.^{2} [Reyn]	2.590×10^{-3}
	(lb_F)(sec)/ft^2	3.73×10^{-1}
(gm_F)(sec)/cm^2	gm/(cm)(sec)	9.807×10^{2}
	gm/(cm)(sec)(10^2) [Centipoise]	9.807×10^{4}
	kg/(m)(hr)	3.530×10^{5}
	lb/(ft)(sec)	6.590×10
	lb/(ft)(hr)	2.372×10^{5}
	lb/(in.)(sec)	5.492
	(lb_F)(sec)/in.^{2} [Reyn]	1.422×10^{-2}
	(lb_F)(sec)/ft^2	2.048
(lb_F)(sec)/in.^{2} [Reyn]	gm/(cm)(sec) [Poise]	6.895×10^{4}
	gm/(cm)(sec)(10^2) [Centipoise]	6.895×10^{6}
	kg/(m)(hr)	2.482×10^{7}
	lb/(ft)(sec)	4.633×10^{3}
	lb/(ft)(hr)	1.668×10^{7}
	lb/(in.)(sec)	3.861×10^{2}
	(gm_F)(sec)/cm^2	7.031×10
	(lb_F)(sec)/ft^2	1.440×10^{2}
(lb_F)(sec)/ft^2	gm/(cm)(sec) [Poise]	4.788×10^{2}
	gm/(cm)(sec)(10^2) [Centipoise]	4.788×10^{4}
	kg/(m)(hr)	1.724×10^{5}
	lb/(ft)(sec)	3.217×10
	lb/(ft)(hr)	1.158×10^{5}
	lb/(in.)(sec)	2.681
	(gm_F)(sec)/cm^2	4.882×10^{-1}
	(lb_F)(sec)/in.^{2} [Reyn]	6.944×10^{-3}

THE VISCOSITY OF WATER 0°C TO 100°C

Contribution from the National Bureau of Standards, not subject to copyright.

°C	η(cp)	°C	η(cp)	°C	η(cp)	°C	η(cp)
0	1.787	26	0.8705	52	0.5290	78	0.3638
1	1.728	27	.8513	53	.5204	79	.3592
2	1.671	28	.8327	54	.5121	80	.3547
3	1.618	29	.8148	55	.5040	81	.3503
4	1.567	30	.7975	56	.4961	82	.3460
5	1.519	31	.7808	57	.4884	83	.3418
6	1.472	32	.7647	58	.4809	84	.3377
7	1.428	33	.7491	59	.4736	85	.3337
8	1.386	34	.7340	60	.4665	86	.3297
9	1.346	35	.7194	61	.4596	87	.3259
10	1.307	36	.7052	62	.4528	88	.3221
11	1.271	37	.6915	63	.4462	89	.3184
12	1.235	38	.6783	64	.4398	90	.3147
13	1.202	39	.6654	65	.4335	91	.3111
14	1.169	40	.6529	66	.4273	92	.3076
15	1.139	41	.6408	67	.4213	93	.3042
16	1.109	42	.6291	68	.4155	94	.3008
17	1.081	43	.6178	69	.4098	95	.2975
18	1.053	44	.6067	70	.4042	96	.2942
19	1.027	45	.5960	71	.3987	97	.2911
20	1.002	46	.5856	72	.3934	98	.2879
21	0.9779	47	.5755	73	.3882	99	.2848
22	.9548	48	.5656	74	.3831	100	.2818
23	.9325	49	.5561	75	.3781		
24	.9111	50	.5468	76	.3732		
25	.8904	51	.5378	77	.3684		

The above table was calculated from the following empirical relationships derived from measurements in viscometers calibrated with water at 20°C (and one atmosphere), modified to agree with the currently accepted value for the viscosity at 20° of 1.002 cp:

$$0° \text{ to } 20°\text{C: } \log_{10} \eta_T = \frac{1301}{998.333 + 8.1855(T-20) + 0.00585(T-20)^2} - 1.30233$$

(R. C. Hardy and R. L. Cottington, J.Res.NBS *42*, 573 (1949).)

$$20° \text{ to } 100°\text{C: } \log_{10} \frac{\eta_T}{\eta_{20}} = \frac{1.3272(20-T) - 0.001053(T-20)^2}{T + 105}$$

(J. F. Swindells, NBS, unpublished results.)

VISCOSITY OF LIQUIDS

Viscosity of liquids in centipoises (cp) including elements, inorganic and organic compounds and mixtures.

Liquid	Temp. °C	Viscosity cp
Acetaldehyde	0	.2797
	10	.2557
	20	.22
Acetanilide	120	2.22
	130	1.90
Acetic acid	15	1.31
	18	1.30
	25.2	1.155
	30	1.04
	41	1.00
	59	.70
	70	.60
	100	.43
anhydride	0	1.24
	15	.971
	18	.90
	30	.783
	100	.49
Acetone	−92.5	2.148
	−80.0	1.487
	−59.6	.932
	−42.5	.695
	−30.0	.575
	−20.9	.510
	−13.0	.470
	−10.0	.450
	0	.399
	15	.337
	25	.316
	30	.295
	41	.280
Acetonitrile	0	.442
	15	.375
	25	.345
Acetophenone	11.9	2.28
	23.5	1.59
	25.0	1.617
	50.0	1.246
	80.0	.734
Air, liq	−192.3	.172
Alcohol. See *Ethyl*, *Methyl*, etc.		
Allyl alcohol	0	2.145
	15	1.49
	20	1.363
	30	1.07
	40	.914
	70	.553
Allylamine	130	.506
Allyl chloride	15	.347
	30	.300
Ammonia	−69	.475
	−50	.317
	−40	.276
	−33.5	.255
n-Amyl acetate	11	1.58
	45	.805
alcohol	15	4.65
	30	2.99
ether	15	1.188
Aniline	−6	13.8
	0	10.2
	5	8.06
	10	6.50
	15	5.31
	20	4.40
	25	3.71
	30	3.16
	35	2.71
	40	2.37
	50	1.85
	60	1.51
	70	1.27
	80	1.09
	90	.935
	100	.825
Anisol	0	1.78
	20	1.32
	40	1.12
Antimony, liq	645	1.55
	700	1.26
	800	1.08
	850	1.05
Benzaldehyde	25	1.39
Benzene	0	.912
	10	.758
	20	.652
	30	.564
	40	.503
	50	.442
	60	.392
	70	.358
	80	.329
Benzonitrile	25	1.24
Benzophenone	55	4.79
	120	1.38
Benzyl alcohol	20	5.8
Benzylamine	25	1.59
Benzylaniline	33	2.18
	130	1.20
Benzyl ether	0	10.5
	20	5.33
	40	3.21
Bismuth	285	1.61
	304	1.662
	365	1.46
	451	1.280
	600	.998
Bromine, liq	−4.3	1.31
	0	1.241
	12.6	1.07
	16	1.0
	19.5	.995
	28.9	.911
o-Bromoaniline	40	3.19
m-Bromoaniline	20	6.81
	40	3.70
	80	1.70
p-Bromoaniline	80	1.81
Bromobenzene	15	1.196
	30	.985
Bromoform	15	2.152
	25	1.89
	30	1.741
Butyl acetate	0	1.004
	20	.732
	40	.563
n-Butyl alcohol	−50.9	36.1
	−30.1	14.7
	−22.4	11.1
	−14.1	8.38
	0	5.186
	15	3.379
	20	2.948
	30	2.30
	40	1.782
	50	1.411
	70	.930
	100	.540
sec-Butyl alcohol	15	4.21
n-Butyl bromide	15	.626
n-Butyl chloride	15	.469
Butyl chloride, tertiary	15	.543
n-Butyl formate	0	.940
	20	.689
Butyric acid	0	2.286
	15	1.81
	20	1.540
	40	1.120
	50	.975
	70	.760
	100	.551
Cadmium, liq	349	1.44
	506	1.18
	603	1.10
Carbolic acid. See *Phenol*.		
Carbon dioxide, liq., pressure that of saturated vapor	0	.099
	10	.085
	20	.071
	30	.053

Liquid	Temp. °C	Viscosity cp
disulfide	−13	.514
	−10	.495
	0	.436
	5	.380
	20	.363
	40	.330
Carbon tetrachloride	0	1.329
	15	1.038
	20	.969
	30	.843
	40	.739
	50	.651
	60	.585
	70	.524
	80	.468
	90	.426
	100	.384
Cetyl alcohol	50	13.4
Chlorine, liq	−76.5	.729
	−70.5	.680
	−60.2	.616
	−52.4	.566
	−35.4	.494
	0	.385
Chlorobenzene	15	.900
	20	.799
	40	.631
	80	.431
	100	.367
Chloroform	−13	.855
	0	.700
	8.1	.643
	15	.596
	20	.58
	25	.542
	30	.514
	39	.500
o-Chlorophenol	25	4.11
	50	2.015
m-Chlorophenol	25	11.55
p-Chlorophenol	50	4.99
Copper, liq	1,085	3.36
	1,100	3.33
	1,150	3.22
	1,200	3.12
o-Cresol	40	4.49
m-Cresol	10	43.9
	20	20.8
	40	6.18
p-Cresol	40	7.00
Creosote	20	12.0
Cycloheptane	13.5	1.64
Cyclohexane	17	1.02
Cyclohexanol	20	68
Cyclohexene	13.5	.696
	20	.66
Cyclooctane	13.5	2.35
Cyclopentane	13.5	.493
*n*Decane	20	.92
Diethylamine	25	.346
	25	.367
Diethylaniline	.5	3.84
	20.0	2.18
	25.0	1.95
Diethylcarbinol	15.0	7.34
Diethylketone	15	.493
Dimethylaniline	10	1.69
	20	1.41
	25	1.285
	30	1.17
	40	1.04
	50	.91
Dimethyl-α-naphthylamine	130	.868
Dimethyl-β-naphthylamine	130	.952
Diphenyl	70	1.49
	100	.97
Diphenylamine	130	1.04
Dodecane	25	1.35
Ether (diethyl-)	−100	1.69
	−80	.958
	−60	.637
	−40	.461
	−20	.362
	0	.2842
	17	.240
	20	.2332
	25	.222
	40	.197
	60	.166
	80	.140
	100	.118
Ethyl acetate	0	.582
	8.96	.516
	10	.512
	15	.473
	20	.455
	25	.441
	30	.400
	50	.345
	75	.283
Ethyl alcohol	−98.11	44.0
	−89.8	28.4
	−71.5	13.2
	−59.42	8.41
	−52.58	6.87
	−32.01	3.84
	−17.59	2.68
	−.30	1.80
	0	1.773
	10	1.466
	20	1.200
	30	1.003
	40	.834
	50	.702
	60	.592
	70	.504
Ethyl alcohol, anh.	−148	8,470
	−146	5,990
	−130	467
Ethyl aniline	25	2.04
Ethylbenzene	17	.691
Ethyl benzoate	20	2.24
Ethyl bromide	−120	5.6
	−100	2.89
	−80	1.81
	0	.487
	10	.441
	15	.418
	20	.402
	30	.348
n-Ethyl butyrate	15	.711
Ethyl carbonate	15	.868
Ethylene bromide	0	2.438
	17	1.95
	20	1.721
	40	1.286
	67.3	.922
	70	.903
	82.2	.750
	99.0	.648
chloride	0	1.077
	15	.887
	19.4	.800
	40	.652
	50	.565
	70	.479
glycol	20	19.9
	40	9.13
	60	4.95
	80	3.02
	100	1.99
oxide	−49.8	.577
	−38.2	.488
	−21.0	.394
	0	.320
	20	.402
Ethyl formate	0	.727
iodide	15	.617
	20	.592
	40	.495
	70	.391
malate	24.7	3.016
oxalate	15	2.31

VISCOSITY OF LIQUIDS (Continued)

Liquid	Temp. °C	Viscosity cp
propionate	15	.564
Eugenol	0	29.9
	20	9.22
	40	4.22
Fluorobenzene	20	.598
	40	.478
	60	.389
	80	.329
	100	.275
Formamide	0	7.55
	25	3.30
Formic acid	7.59	2.3868
	10	2.262
	20	1.804
	30	1.465
	40	1.219
	70	.780
	100	.549
Furfural	0	2.48
	25	1.49
Glucose	22	9.1×10^{15}
	30	6.6×10^{13}
	40	2.8×10^{11}
	60	9.3×10^{7}
	80	6.6×10^{5}
	100	2.5×10^{4}
Glycerin	−42	6.71×10^{6}
	−36	2.05×10^{6}
	−25	2.62×10^{5}
	−20	1.34×10^{5}
	−15.4	6.65×10^{4}
	−10.8	3.55×10^{4}
	−4.2	1.49×10^{4}
	0	12,110
	6	6,260
	15	2,330
	20	1,490
	25	954
	30	629
Glycerin trinitrate	10	69.2
	20	36.0
	30	21.0
	40	13.6
	60	6.8
Heptane	0	.524
	17	.461
	20	.409
	25	.386
	40	.341
	70	.262
n-Heptyl alcohol	15	8.53
Hexadecane	20	3.34
Hexane	0	.401
	17	.374
	20	.326
	25	.294
	40	.271
	50	.248
Hydrazine	1	1.29
	10	1.12
	20	.97
Hydrogen, liq		.011
Iodine, liq	116	2.27
Iodobenzene	15	1.74
Iron, 2.5% carbon, liq	1,400	2.25
Isoamyl acetate	8.97	1.030
	19.91	.872
alcohol	10	6.20
amine	25	.724
Isobutyl alcohol	15	4.703
amine	25	.553
Isobutyric acid	15	1.44
	30	1.13
Isoeugenol	25	26.72
Isoheptane	0	.481
	20	.384
	40	.315
Isohexane	0	.376
	20	.306
	40	.254
Isopentane	0	.273
	20	.223
Isopropyl alcohol	15	2.86
	30	1.77
Isoquinoline	25	3.57
Isosafrol	25	3.981
Lead, liq	350	2.58
	400	2.33
	441	2.116
	500	1.84
	551	1.70
	600	1.38
	703	1.349
	844	1.185
Menthol, liq	55.6	6.29
	74.6	2.47
	99.0	1.04
Mercury	−20	1.855
	−10	1.764
	0	1.685
	10	1.615
	19.02	1.56
	20	1.554
	20.2	1.55
	30	1.499
	40	1.450
	40.8	1.45
	41.86	1.44
	50	1.407
	60	1.367
	70	1.331
	80	1.298
	90	1.268
	100	1.240
	150	1.130
	200	1.052
	250	.995
	300	.950
	340	.921
Methyl acetate	0	.484
	20	.381
	40	.320
Methyl alcohol	−98.30	13.9
(Methanol)	−84.23	6.8
	−72.55	4.36
	−44.53	1.98
	−22.29	1.22
	0	.82
	15	.623
	20	.597
	25	.547
	30	.510
	40	.456
	50	.403
Methyl amine	0	.236
aniline	25	2.02
	30	1.55
chloride	20	.1834
Methylene bromide	15	1.09
	30	0.92
chloride	15	.449
	30	.393
Methyl iodide	0	.606
	15	.518
	20	.500
	30	.460
	40	.424
Naphthalene	80	.967
	100	.776
Nitric acid	0	2.275
	10	1.770
Nitrobenzene	2.95	2.91
	5.69	2.71
	5.94	2.71
	9.92	2.48
	14.94	2.24
	20.00	2.03
Nitromethane	0	.853
	25	.620
o-Nitrotoluene	0	3.83
	20	2.37
	40	1.63
	60	1.21
m-Nitrotoluene	20	2.33

VISCOSITY OF LIQUIDS (Continued)

Liquid	Temp. °C	Viscosity cp
	40	1.60
	60	1.18
p-Nitrotoluene	60	1.20
n-Nonane	20	.711
n-Octane	0	.706
	16	.574
	20	.542
	40	.433
Octodecane	40	2.86
n-Octylalcohol	15	10.6
Oil, castor	10	2,420
	20	986
	30	451
	40	231
	100	16.9
cottonseed	20	70.4
cylinder, filtered	37.8	240.6
	100	18.7
cylinder, dark	37.8	422.4
	100	24.0
linseed	30	33.1
	50	17.6
	90	7.1
machine, light	15.6	113.8
	37.8	34.2
	100	4.9
machine, heavy	15.6	660.6
	37.8	127.4
Oil, olive	10	138.0
	20	84.0
	40	36.3
	70	12.4
rape	0	2,530
	10	385
	20	163
	30	96
soya bean	20	69.3
	30	40.6
	50	20.6
	90	7.8
sperm	15.6	42.0
	37.8	18.5
	100.0	4.6
Oleic acid	30	25.6
Pentadecane	22	2.81
Pentane	0	.289
	20	.240
o-Phenetidine	0	16.5
	20	6.08
	30	4.22
m-Phenetidine	30	12.9
p-Phenetidine	20	12.9
	30	8.3
Phenol	18.3	12.7
	50	3.49
	60	2.61
	70	2.03
	90	1.26
Phenylcyanide	.28	1.96
	20.0	1.33
Phosphorus, liq	21.5	2.34
	31.2	2.01
	43.2	1.73
	50.5	1.60
	60.2	1.45
	69.7	1.32
	79.9	1.21
Potassium bromide, liq	745	1.48
	775	1.34
	805	1.19
nitrate, liq	334	2.1
	358	1.7
	333	2.97
	418	2.00
Propionic acid	10	1.289
	15	1.18
	20	1.102
	40	.845
Propyl acetate	10	.66
	20	.59
	40	.44
n-Propyl alcohol	0	3.883
	15	2.52
	20	2.256
	30	1.72
	40	1.405
n-Propyl alcohol	50	1.130
	70	.760
Propyl aldehyde	10	.47
	20	.41
	10	[illegible]
bromide	0	.651
	20	.524
	40	.433
chloride	0	.436
	20	.352
	40	.291
n-Propyl ether	15	.448
Pyridine	20	.974
Salicylic acid	10	3.20
	20	2.71
	40	1.81
Salol	45	.746
Sodium bromide	762	1.42
	780	1.28
chloride, liq	841	1.30
	896	1.01
	924	.97
nitrate, liq	308	2.919
	348	2.439
	398	1.977
	418	1.828
Stearic acid	70	11.6
Sucrose (cane sugar)	109	2.8×10^6
	124.6	1.9×10^5
Sulfur (gas free)	123.0	10.94
	135.5	8.66
	149.5	7.09
	156.3	7.19
	158.2	7.59
	159.2	9.48
	159.5	14.45
	160.0	22.83
	160.3	77.32
	165.0	500.0
	171.0	4,500.0
	184.0	16,000.00
	190.5	19,700.0
	197.5	21,300.0
	200.0	21,500.0
	210.0	20,500.0
	217.0	19,100.0
	220.0	18,600.0
Sulfur dioxide, liq	−33.5	.5508
	−10.5	.4285
	0.1	.3936
Sulfuric acid	0	48.4
	15	32.8
	20	25.4
	30	15.7
	40	11.5
Sulfuric acid	50	8.82
	60	7.22
	70	6.09
	80	5.19
Tetrachloroethane	15	1.844
Tetradecane	20	2.18
Tin, liq	240	2.12
	280	1.678
	300	1.73
	301	1.680
	400	1.43
	450	1.270
	500	1.20
	600	1.08
	604	1.045
	750	.905
Toluene	0	.772
	17	.61
	20	.590
	30	.526
	40	.471

VISCOSITY OF LIQUIDS (Continued)

Liquid	Temp. °C	Viscosity cp	Liquid	Temp. °C	Viscosity cp
	70	.354		70	.728
o-Toluidine	20	4.39	Turpentine, Venice	17.3	1.3×10^5
m-Toluidine	20	3.81	*n*-Undecane	20	1.17
p-Toluidine	50	1.80	*o*-Xylene (xylol)	0	1.105
Triacetin	17	28.0		16	.876
Tributyrin	20	11.6		20	.810
Trichlorethane	20	1.2		40	.627
Tridecane	23.3	1.55	*m*-Xylene (xylol)	0	.806
Triethylcarbinol	20	6.75		15	.650
Tripalmitin	70	16.8		20	.620
Tristearin	75	18.5		40	.497
Turpentine	0	2.248	*p*-Xylene (xylol)	16	.696
	10	1.783		20	.648
	20	1.487		40	.513
	30	1.272	Zinc, liq	280	1.68
	40	1.071		357	1.42
				389	1.31

VISCOSITY OF GASES

Gas or vapor	Temp. °C	Viscosity micro-poises	Gas or vapor	Temp. °C	Viscosity micro-poises
Acetic acid, vap	119.1	107.0	Benzene, vap	14.2	73.8
Acetone, vap	100	93.1		131.2	103.1
	119.0	99.1		194.6	119.8
	190.4	118.6		252.5	134.3
	247.7	133.4		312.8	148.4
	306.4	148.1	Bromine, vap	12.8	151
Acetylene	0	93.5		65.7	170
Air	−194.2	55.1		99.7	188
	−183.1	62.7		139.7	208
	−104.0	113.0		179.7	227
	−69.4	133.3		220.3	248
	−31.6	153.9	Bromoform, vap	151.2	253.0
	0	170.8	Butyl alcohol, *n*, vap	116.9	143
	18	182.7	*tert*, vap	82.9	160
	40	190.4	chloride, *n*, vap	78	149.5
	54	195.8	iodide, vap	130	202
	74	210.2	*β*-Butylene	18.8	74.4
	229	263.8		100.4	94.5
	334	312.3		200	119.2
	357	317.5	Butyric acid, vap	161.7	130.0
	409	341.3	Carbon dioxide	−97.8	89.6
	466	350.1		−78.2	97.2
	481	358.3		−60.0	106.1
	537	368.6		−40.2	115.5
	565	375.0		−21	129.4
	620	391.6		−19.4	126.0
	638	401.4		0	139.0
	750	426.3		15	145.7
	810	441.9		19	149.9
	923	464.3		20	148.0
	1034	490.6		30	153
	1134	520.6		32	155
				35	156
Alcohol. See *Ethyl*, *Methyl*, etc.				40	157
Ammonia	−78.5	67.2		99.1	186.1
	0	91.8		104	188.9
	20	98.2		182.4	222.1
	50	109.2		235	241.5
	100	127.9		302.0	268.2
	132.9	139.9		490	330.0
	150	146.3		685	380.0
	200	164.6		850	435.8
	250	181.4		1052	478.6
	300	198.7	disulfide, vap	0	91.1
Argon	0	209.6		14.2	96.4
	20	221.7		114.3	130.3
	100	269.5		190.2	156.1
	200	322.3		309.8	196.6
	302	368.5	monoxide	−191.5	56.1
	401	411.5		−78.5	127
	493	448.4		0	166
	584	481.5		15	172
	714	525.7		21.7	175.3
	827	563.2		126.7	218.3
Arsenic hydride (Arsine)	0	145.8		227.0	254.8
	15	114.0		276.9	271.4
	100	198.1	tetrachloride, vap	76.7	195.0

Gas or vapor	Temp. °C	Viscosity micro-poises
Carbon tetrachloride,		
vap	127.9	133.4
	200.2	156.2
	314.9	190.2
Chlorine	12.7	120.7
	20	132.7
	50	146.9
	100	167.9
	150	187.5
	200	208.5
	250	227.6
Chloroform, vap	0	93.6
	14.2	98.9
	100	129
	121.3	135.7
	189.1	157.9
	250.0	177.6
	307.5	194.7
Cyanogen	0	92.8
	17	98.7
	100	127.1
Ethane	−78.5	63.4
	0	84.8
	17.2	90.1
	50.8	100.1
	100.4	114.3
	200.3	140.9
Ether (diethyl), vap	0	67.8
	14.2	71.6
	100	95.5
	121.8	98.3
	159.4	107.9
	189.9	115.2
	251.0	130.0
	277.8	135.8
Ethyl acetate, vap	0	68.4
	100	94.3
	128.1	101.8
	158.6	109.8
	192.9	119.5
	212.5	126
alcohol, vap	100	108
	130.2	117.3
	170.7	129.3
	191.8	135.5
	212.5	140
	251.7	151.9
	308.7	167.0
bromide, vap	38.4	186.5
butyrate, vap	119.8	160.0
chloride, vap	0	93.7
Ethylene	−75.7	69.9
	−44.1	76.9
	−38.6	78.5
	0	90.7
	13.8	95.4
	20	100.8
Ethylene	50	110.3
	100	125.7
	150	140.3
	200	154.1
	250	166.6
bromide, vap	131.6	221.0
chloride, vap	83.5	168.0
Ethyl formate, vap	99.8	92
iodide, vap	72.3	216.0
Helium	−257.4	27.0
	−252.6	35.0
	−191.6	87.1
	0	186.0
	20	194.1
	100	228.1
	200	267.2
	250	285.3
	282	299.2
	407	343.6
	486	370.6
	606	408.7
	676	430.3
	817	471.3
Hydrogen	−257.7	5.7
	−252.5	8.5
	−198.4	33.6
	−183.4	38.8
	−113.5	57.2
	−97.5	61.5
	−31.6	76.7
	0	83.5
	20.7	87.6
	28.1	89.2
	129.4	108.6
	229.1	126.0
	299	138.1
	412	155.4
	490	167.2
	601	182.9
	713	198.2
	825	213.7
bromide	18.7	181.9
	100.2	234.4
chloride	12.5	138.5
	16.5	140.7
	18	142.6
	100.3	182.2
iodide	20	165.5
	50	201.8
	100	231.6
	150	262.7
	200	292.4
	250	318.9
phosphide	0	106.1
	15	112.0
	100	143.8
sulfide	0	116.6
Hydrogen sulfide	17	124.1
	100	158.7
Iodine, vap	124.0	184
	170.0	204
	205.4	220
	247.1	240
Isobutyl acetate, vap	16.1	76.4
	116.4	155.0
alcohol, vap	108.4	144.5
bromide, vap	92.3	179.5
butyrate, vap	156.9	167.0
chloride, vap	68.5	150.0
iodide, vap	120	204.7
Isopentane, vap	25	69.5
	100	86.0
Isopropyl alcohol, vap	99.8	109
	120.3	103.1
	198.4	124.8
	293.1	148.8
bromide, vap	60	176.0
chloride, vap	37.0	148.5
iodide, vap	89.3	201.5
Krypton	0	232.7
	15	246
Mercury, vap	273	494
	313	551
	369	641
	380	654
Methane	−181.6	34.8
	−78.5	76.0
	0	102.6
	20	108.7
	100.0	133.1
	200.5	160.5
	284	181.3
	380	202.6
	499	226.4
Methyl acetate, vap	99.8	98
	100	100
	143.3	113.9
	218.5	134.8
alcohol, vap	66.8	135.0
	111.3	125.9
	217.5	162.0
	311.5	192.1
chloride	−15.3	92
	0	96.9
	15.0	104

VISCOSITY OF GASES (Continued)

Gas or vapor	Temp. °C	Viscosity micro-poises
	99.1	137
	182.4	168
	302.0	211
iodide, vap	44	232
Neon	0	297.3
	20	311.1
	100	364.6
	200	424.8
	250	453.2
Neon	285	470.8
	429	545.4
	502	580.2
	594	623.0
	686	662.6
	827	721.0
Nitric oxide (NO)	0	178
	20	187.6
	100	227.2
	200	268.2
Nitrogen	−21.5	156.3
	10.9	170.7
	27.4	178.1
	127.2	219.1
	226.7	255.9
	299	279.7
	490	337.4
	825	419.2
Nitrosyl chloride	15	113.9
	100	150.4
	200	192.0
Nitrous oxide (N_2O)	0	135
	26.9	148.8
	126.9	194.3
n-Nonane, vap	100.3	63.3
	202.1	78.1
n-Octane, vap	100.4	67.5
	202.2	84.8
Oxygen	0	189
	19.1	201.8
	127.7	256.8
	227.0	301.7
	283	323.3
	402	369.3
	496	401.3
	608	437.0
	690	461.2
	829	501.2
n-Pentane, vap	25	67.6
	100	84.1
Propane	17.9	79.5
	100.4	100.9
	199.3	125.1
n-Propyl alcohol, vap	99.9	93
	121.7	102.5
	209.7	126.7
	273.0	143.4
bromide, vap	99.8	119
Propylene	16.7	83.4
	49.9	93.5
	100.1	107.6
	199.4	133.8
Propyl iodide, vap	102	210.0
Sulfur dioxide	−75.0	85.8
	−20.0	107.8
	0	115.8
	0	117
	18	124.2
	20.5	125.4
	100.4	161.2
	199.4	203.8
	293	244.7
	490	311.5
Trimethylbutane, (2,2,3-), vap	70.3	73.4
	132.2	82.7
	262.1	104.8
Trimethylethylene, vap	25	70.1
	100	86.9
Water, vap	100	125.5
	150	144.5
	200	163.5
	250	182.7
	300	202.4
	350	221.8
	400	241.2
Xenon	0	210.1
	16.5	223.5
	20	226.0
	127	300.9
	177	335.1
	227	365.2
	277	395.4

DIFFUSIVITIES OF GASES IN LIQUIDS

Solute	Solvent	Temp., °C	Diffusivity x 10^5, cm^2/sec
H_2	*n*-Hexane	25.4	16.36
H_2	Cyclohexane	25.4	7.08
H_2	Ethylene glycol	25.4	0.75
H_2	Carbon tetrachloride	0	6.28
O_2	Cyclohexane	29.6	5.31
O_2	Carbon tetrachloride	25.4	3.71
O_2	Ethanol	29.6	2.64
N_2	Carbon tetrachloride	0	2.44
CH_4	Glycerol	25.4	0.95
C_2H_6	*n*-Hexane	30	6.00
C_2H_6	*n*-Heptane	30	5.60
C_2H_2	Water	0	1.10
H_2S	Water	16	1.77
CO_2	Amyl alcohol	25	1.91
CO_2	Isobutyl alcohol	25	2.20
SO_2	*n*-Heptane	20	2.70
SO_2	*n*-Nonane	20	2.50
SO_2	*n*-Decane	20	2.40

DIFFUSION COEFFICIENTS IN AQUEOUS SOLUTIONS AT 25°

The diffusion coefficient D may be defined by either of the equations

$$J = -D \frac{\partial c}{\partial x}$$

or

$$\frac{\partial c}{\partial t} = D \frac{\partial^2 c}{\partial x^2}$$

when diffusion occurs in the x-direction only. Here J is the diffusion-flux across unit area normal to the x-direction, $\frac{\partial c}{\partial x}$ is the concentration-gradient at a fixed time, $\frac{\partial c}{\partial t}$ is the rate of change of concentration with time at a fixed distance. If J is expressed in mole cm^{-2} sec^{-1} and c in mole cm^{-3}, x in cm, and t in sec, D will be given in units of cm^2 sec^{-1}. In general D varies somewhat with concentration. The values below are a selection from measurements by modern high-precision methods, mainly by H. S. Harned and collaborators, R. H. Stokes and collaborators, L. J. Gosting and collaborators, and L. G. Longsworth.

For strong electrolytes at infinite dilution, limiting diffusion coefficients may be calculated by the Nernst relation:

$$D = \frac{RT}{F^2}\left[\frac{(\nu_1 + \nu_2)(\lambda_1^0 \lambda_2^0)}{\nu_1|Z_1|(\lambda_1^0 + \lambda_2^0)}\right]$$

where R = gas constant, F = Faraday, T = absolute temperature, λ_1^0 and λ_2^0 are cation and anion limiting equivalent conductances, ν_1 and ν_2 are the numbers of cations and anions formed from one "molecule" of electrolyte, and Z_1 is the cation valency. Concentrations, unless expressed otherwise are as molarities and the diffusion coefficients are expressed as $10^5 D/cm^2\ sec^{-1}$ at 25°C.

DIFFUSION COEFFICIENTS OF STRONG ELECTROLYTES

Molarity

Solute	0.01	0.1	1.0
HCL	....	3.050	3.436
HBr	....	3.156	3.87
LiCl	1.312	1.269	1.302
LiBr	....	1.279	1.404
$LiNO_3$	1.276	1.240	1.293
NaCl	1.545	1.483	1.484
NaBr	....	1.517	1.596
NaI	....	1.520	1.662
KCl	1.917	1.844	1.892
KBr	....	1.874	1.975
KI	....	1.865	2.065
KNO_3	1.846	....	....
$KClO_4$	1.790	....	....
$CaCl_2$	1.188	1.110	1.203

C = 0.005M

Solute	0.01	0.1	1.0
$BaCl_2$	1.265	1.159	1.179
Na_2SO_4	1.123	....	....
$MgSO_4$	0.710	....	....
$LaCl_3$	1.105	....	....
$K_4Fe(CN)_6$	1.183	....	....

DIFFUSION COEFFICIENTS OF WEAK AND NON-ELECTROLYTES

Concentration

Solute	Concentration	Coefficient
Glucose	0.39%	0.673
Sucrose	0.38%	0.521
Raffinose	0.38%	0.434
Sucrose	Zero	0.5226
Mannitol	Zero	0.682
Penta-erythritol	Zero	0.761
Glycolamide	Zero	1.142
Glycine	Zero	1.064
α-alanine	0.32%	0.910
β-alanine	0.31%	0.933
Amino-benzoic acid ortho	0.24%	0.840
Amino-benzoic acid meta	0.24%	0.774
Amino-benzoic acid para	0.23%	0.843
Citric acid	0.1 M	0.661

DIFFUSION OF GASES INTO AIR

Gas or vapor	Temp. °C	Coefficient of diffusion, sq. cm/sec	Observer
Alcohol, vapor.......	40.4	0.137	Winkelmann
Carbon dioxide.......	0.0	0.139	Mean of various
Carbon disulfide......	19.9	0.102	Winkelmann
Ether, vapor.........	19.9	0.089	Winkelmann
Hydrogen............	0.0	0.634	Obermayer
Oxygen.............	0.0	0.178	Obermayer
Water, vapor..........	8.0	0.239	Guglielmo

RADIOACTIVE TRACER DIFFUSION DATA FOR PURE METALS

John Askill

The data in these tables are the most reliable set of radioactive tracer diffusion data for pure metals published in the literature from 1938 through December, 1970. For a complete listing of all published data on this subject up to December 1968 see "Tracer Diffusion Data for Metals, Alloys and Simple Oxides" by John Askill, published by Plenum Press, New York, 1970.

The diffusion coefficient D_T at a temperature T(°K) is given by the following relation:

$$D_T = D_0 e^{-Q/RT}$$

Abbreviations used in the tables are:

A.R.G. = Autoradiography
R.A. = Residual Activity
S.D. = Surface Decrease
S.S. = Serial Sectioning
P. = Polycrystalline
S. = Single Crystal
⊥c. = Perpendicular to c Direction
∥c. = Parallel to c Direction
99.95 = 99.95%

Solute (tracer)	Material (metal, crystalline form and purity)	Temperature range, °C	Form of analysis	Activation energy, Q, Kcal/mole	Frequency factor, D_0, cm^2/sec	Reference
Aluminum						
Ag^{110}	S 99.999	371–655	S.S.	27.83	0.118	1
Al^{27}	S	450–650	S.S.	34.0	1.71	2
Au^{198}	S 99.999	423–609	S.S.	27.0	0.077	3
Cd^{115}	S 99.999	441–631	S.S.	29.7	1.04	3
Ce^{141}	P 99.995	450–630	R.A.	26.60	1.9×10^{-6}	5
Co^{60}	S 99.999	369–655	S.S.	27.79	0.131	1
Cr^{51}	S 99.999	422–654	S.S.	41.74	464	1
Cu^{64}	S 99.999	433–652	S.S.	32.27	0.647	1
Fe^{59}	S 99.99	550–636	S.S.	46.0	135	3
Ga^{72}	S 99.999	406–652	S.S.	29.24	0.49	1
Ge^{71}	S 99.999	401–653	S.S.	28.98	0.481	1
In^{114}	P 99.99	400–600	S.S., R.A.	27.6	0.123	4
La^{140}	P 99.995	500–630	R.A.	27.0	1.4×10^{-6}	5
Mn^{54}	P 99.99	450–650	S.S.	28.8	0.22	2
Mo^{99}	P 99.995	400–630	R.A.	13.1	1.04×10^{-9}	6
Nb^{95}	P 99.95	350–480	R.A.	19.65	1.66×10^{-7}	7
Nd^{147}	P 99.995	450–630	R.A.	25.0	4.8×10^{-7}	5
Ni^{63}	P 99.99	360–630	R.A.	15.7	2.9×10^{-8}	8
Pd^{103}	P 99.995	400–630	R.A.	20.2	1.92×10^{-7}	9
Pr^{142}	P 99.995	520–630	R.A.	23.87	3.58×10^{-7}	5
Sb^{124}	P	448–620	R.A.	29.1	0.09	10
Sm^{153}	P 99.995	450–630	R.A.	22.88	3.45×10^{-7}	5
Sn^{113}	P	400–600	S.S., R.A.	28.5	0.245	4
V^{48}	P 99.995	400–630	R.A.	19.6	6.05×10^{-8}	11
Zn^{65}	S 99.999	357–653	S.S.	28.86	0.259	1
Beryllium						
Ag^{110}	S⊥c 99.75	650–900	R.A.	43.2	1.76	12
Ag^{110}	S∥c 99.75	650–900	R.A.	39.3	0.43	12
Be^{7}	S⊥c 99.75	565–1065	R.A.	37.6	0.52	13
Be^{7}	S∥c 99.75	565–1065	R.A.	39.4	0.62	13
Fe^{59}	S 99.75	700–1076	R.A.	51.6	0.67	12
Ni^{63}	P	800–1250	R.A.	58.0	0.2	14
Cadmium						
Ag^{110}	S 99.99	180–300		25.4	2.21	15
Cd^{115}	S 99.95	110–283	R.A.	19.3	0.14	16
Zn^{65}	S 99.99	180–300	—	19.0	0.0016	15
Calcium						
C^{14}	99.95	550–800	R.A.	29.8	3.2×10^{-5}	17
Ca^{45}	99.95	500–800	R.A.	38.5	8.3	17
Fe^{59}	99.95	500–800	R.A.	23.3	2.7×10^{-3}	17
Ni^{63}	99.95	550–800	—	28.9	1.0×10^{-5}	17
U^{235}	99.95	500–700	R.A.	34.8	1.1×10^{-5}	17
Carbon						
Ag^{110}	⊥c	750–1050	R.A.	64.3	9280	18
C^{14}		2000–2200	—	163	5	19
Ni^{63}	⊥c	540–920	R.A.	47.2	102	18
Ni^{63}	∥c	750–1060	R.A.	53.3	2.2	18
Th^{228}	⊥c	1400–2200	R.A.	145.4	1.33×10^{-5}	18
Th^{228}	∥c	1800–2200	R.A.	114.7	2.48	18
U^{232}	⊥c	1400–2200	R.A.	115.0	6760	18
U^{232}	∥c	1400–1820	R.A.	129.5	385	18
Chromium						
C^{14}	P	1200–1500	R.A.	26.5	9.0×10^{-3}	20
Cr^{51}	P 99.98	1030–1545	S.S.	73.7	0.2	21
Fe^{59}	P 99.8	980–1420	R.A.	79.3	0.47	22
Mo^{99}	P	1100–1420	R.A.	58.0	2.7×10^{-3}	20
Cobalt						
C^{14}	P 99.82	600–1400	R.A.	34.0	0.21	23
Co^{60}	P 99.9	1100–1405	S.S.	67.7	0.83	24
Fe^{59}	P 99.9	1104–1303	S.S.	62.7	0.21	24
Ni^{63}	P	1192–1297	R.A.	60.2	0.10	25
S^{35}	P 99.99	1150–1250	R.A.	5.4	1.3	26
Copper						
Ag^{110}	S, P	580–980	R.A.	46.5	0.61	27
As^{76}	P	810–1075	R.A.	42.13	0.20	28
Au^{198}	S, P	400–1050	S.S.	42.6	0.03	29
Cd^{115}	S 99.98	725–950	S.S.	45.7	0.935	30

Solute (tracer)	Material (metal, crystalline form and purity)	Temperature range, °C	Form of analysis	Activation energy, Q, Kcal/mole	Frequency factor, D_0, cm^2/sec	Reference
Ce^{141}	P 99.999	766–947	R.A.	27.6	2.17×10^{-8}	31
Cr^{51}	S, P	800–1070	R.A.	53.5	1.02	32
Co^{60}	S 99.998	701–1077	S.S.	54.1	1.93	33
Cu^{67}	S 99.999	698–1061	S.S.	50.5	0.78	34
Eu^{152}	P 99.999	750–970	S.S., R.A.	26.85	1.17×10^{-7}	31
Fe^{59}	S, P	460–1070	R.A.	52.0	1.36	32
Ga^{72}		—	—	45.90	0.55	35
Ge^{68}	S 99.998	653–1015	S.S.	44.76	0.397	36
Hg^{203}	P	—	—	44.0	0.35	35
Lu^{177}	P 99.999	857–1010	R.A.	26.15	4.3×10^{-9}	31
Mn^{54}	S 99.99	754–950	S.S.	91.4	10^{7}	37
Nb^{95}	P 99.999	807–906	R.A.	60.06	2.04	38
Ni^{63}	P	620–1080	R.A.	53.8	1.1	39
Pd^{102}	S 99.999	807–1056	S.S.	54.37	1.71	40
Pm^{147}	P 99.999	720–955	R.A.	27.5	3.62×10^{-8}	31
Pt^{195}	P	843–997	S.S.	37.5	4.8×10^{-4}	41
S^{35}	S 99.999	800–1000	R.A.	49.2	23	42
Sb^{124}	S 99.999	600–1000	S.S.	42.0	0.34	43
Sn^{113}	P	680–910	—	45.0	0.11	44
Tb^{160}	P 99.999	770–980	R.A.	27.45	8.96×10^{-9}	31
Tl^{204}	S 99.999	785–996	S.S.	43.3	0.71	45
Tm^{170}	P 99.999	705–950	R.A.	24.15	7.28×10^{-9}	31
Zn^{65}	P 99.999	890–1000	S.S.	47.50	0.73	46
Germanium						
Cd^{115}	S	750–950	R.A.	102.0	1.75×10^{9}	47
Fe^{59}	S	775–930	R.A.	24.8	0.13	48
Ge^{71}	S	766–928	S.S.	68.5	7.8	49
In^{114}	S	600–920	—	39.9	2.9×10^{-4}	50
Sb^{124}	S	720–900	—	50.2	0.22	51
Te^{125}	S	770–900	S.S.	56.0	2.0	52
Tl^{204}	S	800–930	S.S.	78.4	1700	53
Gold						
Ag^{110}	S 99.99	699–1007	S.S.	40.2	0.072	54
Au^{198}	S 99.97	850–1050	S.S.	42.26	0.107	224
Co^{60}	P 99.93	702–948	R.A.	41.6	0.068	55
Fe^{59}	P 99.93	701–948	R.A.	41.6	0.082	55
Hg^{203}	S 99.994	600–1027	—	37.38	0.116	56
Ni^{63}	P 99.96	880–940	S.S.	46.0	0.30	57
Pt^{195}	P, S 99.98	800–1060	S.S.	60.9	7.6	58
β-Hafnium						
Hf^{181}	P 97.9	1795–1995	S.S.	38.7	1.2×10^{-3}	59
Indium						
Ag^{110}	S⊥c 99.99	25–140	S.S.	12.8	0.52	60
Ag^{110}	S∥c 99.99	25–140	S.S.	11.5	0.11	60
Au^{198}	S 99.99	25–140	S.S.	6.7	9×10^{-3}	60
In^{114}	S⊥c 99.99	44–144	S.S.	18.7	3.7	61
In^{114}	S∥c 99.99	44–144	S.S.	18.7	2.7	61
Tl^{204}	S 99.99	49–157	S.S.	15.5	0.049	62
α-Iron						
Ag^{110}	P	748–888	S.S.	69.0	1950	63
Au^{198}	P 99.999	800–900	R.A.	62.4	31	64
C^{14}	P 99.98	616–844	R.A.	29.3	2.2	65
Co^{60}	P 99.995	638–768	R.A.	62.2	7.19	62
Cr^{51}	P 99.95	775–875	R.A.	57.5	2.53	66
Cu^{64}	P 99.9	800–1050	R.A.	57.0	0.57	67
Fe^{55}	P 99.92	809–889	—	60.3	5.4	68
K^{42}	P 99.92	500–800	R.A.	42.3	0.036	69
Mn^{54}	P 99.97	800–900	R.A.	52.5	0.35	70
Mo^{99}	P	750–875	R.A.	73.0	7800	71
Ni^{63}	P 99.97	680–800	R.A.	56.0	1.3	72
P^{32}	P	860–900	R.A.	55.0	2.9	73
Sb^{124}	P	800–900	R.A.	66.6	1100	74
V^{48}	P	755–875	R.A.	55.4	1.43	75
W^{185}	P	755–875	R.A.	55.1	0.29	75
γ-Iron						
Be^{7}	P 99.9	1100–1350	R.A.	57.6	0.1	76
C^{14}	P 99.34	800–1400	—	34.0	0.15	23
Co^{60}	P 99.98	1138–1340	S.S.	72.9	1.25	77
Cr^{51}	P 99.99	950–1400	R.A.	69.7	10.8	78
Fe^{59}	P 99.98	1171–1361	S.S.	67.86	0.49	79
Hf^{181}	P 99.99	1110–1360	R.A.	97.3	3600	78
Mn^{54}	P 99.97	920–1280	R.A.	62.5	0.16	70
Ni^{63}	P 99.97	930–2050	R.A.	67.0	0.77	72
P^{32}	P 99.99	950–1200	R.A.	43.7	0.01	80
S^{35}	P	900–1250	R.A.	53.0	1.7	81
V^{48}	P 99.99	1120–1380	R.A.	69.3	0.28	78
W^{185}	P 99.5	1050–1250	R.A.	90.0	1000	82
δ-Iron						
Co^{60}	P 99.995	1428–1521	R.A.	61.4	6.38	83
Fe^{59}	P 99.95	1428–1492	S.S.	57.5	2.01	83
P^{32}	P 99.99	1370–1460	R.A.	55.0	2.9	73
Lanthanum						
Au^{198}	P 99.97	600–800	S.S.	45.1	1.5	84
La^{140}	P 99.97	690–850	S.S.	18.1	2.2×10^{-2}	84
Lead						
Ag^{110}	P 99.9	200–310	R.A.	14.4	0.064	85
Au^{198}	S 99.999	190–320	S.S.	10.0	8.7×10^{-3}	86

Solute (tracer)	Material (metal, crystalline form and purity)	Temperature range, °C	Form of analysis	Activation energy, Q, Kcal/mole	Frequency factor, D_0, cm²/sec	Reference
Cd^{115}	S 99.999	150–320	S.S.	21.23	0.409	87
Cu^{64}	S	150–320	S.S.	14.44	0.046	88
Pb^{204}	S 99.999	150–320	S.S.	25.52	0.887	87
Tl^{205}	P 99.999	207–322	S.S.	24.33	0.511	89
Lithium						
Ag^{110}	P 92.5	65–161	S.S.	12.83	0.37	90
Au^{195}	P 92.5	47–153	S.S.	10.49	0.21	90
Bi	P 99.95	141–177	S.S.	47.3	5.3×10^{13}	91
Cd^{115}	P 92.5	80–174	S.S.	16.05	2.35	90
Cu^{64}	P 99.98	51–120	S.S.	9.22	0.47	93
Ga^{72}	P 99.98	58–173	S.S.	12.9	0.21	93
Hg^{203}	P 99.98	58–173	S.S.	14.18	1.04	93
In^{114}	P 92.5	80–175	S.S.	15.87	0.39	90
Li^{6}	P 99.98	35–178	S.S.	12.60	0.14	94
Na^{22}	P 92.5	52–176	S.S.	12.61	0.41	90
Pb^{204}	P 99.95	129–169	S.S.	25.2	160	91
Sb^{124}	P 99.95	141–176	S.S.	41.5	1.6×10^{10}	91
Sn^{113}	P 99.95	108–174	S.S.	15.0	0.62	91
Zn^{65}	P 92.5	60–175	S.S.	12.98	0.57	92
Magnesium						
Ag^{110}	P 99.9	476–621	S.S.	28.50	0.34	95
Fe^{59}	P 99.95	400–600	R.A.	21.2	4×10^{-6}	96
In^{114}	P 99.9	472–610	S.S.	28.4	5.2×10^{-2}	95
Mg^{28}	S⊥c	467–635	S.S.	32.5	1.5	97
Mg^{28}	S‖c	467–635	S.S.	32.2	1.0	97
Ni^{63}	P 99.95	400–600	R.A.	22.9	1.2×10^{-5}	96
U^{235}	P 99.95	500–620	R.A.	27.4	1.6×10^{-5}	96
Zn^{65}	P 99.9	467–620	S.S.	28.6	0.41	95
Molybdenum						
C^{14}	P 99.98	1200–1600	R.A.	41.0	2.04×10^{-2}	99
Co^{60}	P 99.98	1850–2350	S.S.	106.7	18	100
Cr^{51}	P	1000–1500	R.A.	54.0	2.5×10^{-4}	20
Cs^{134}	S 99.99	1000–1470	R.A., A.R.G.	28.0	8.7×10^{-11}	101
K^{42}	S	800–1100	R.A.	25.04	5.5×10^{-9}	102
Mo^{99}	P	1850–2350	S.S.	96.9	0.5	103
Na^{24}	S	800–1100	R.A.	21.25	2.95×10^{-9}	102
Nb^{95}	P 99.98	1850–2350	S.S.	108.1	14	100
P^{32}	P 99.97	2000–2200	S.S.	80.5	0.19	104
Re^{186}	P	1700–2100	A.R.G.	94.7	0.097	105
S^{35}	S 99.97	2220–2470	S.S.	101.0	320	106
Ta^{182}	P	1700–2150	R.A.	83.0	3.5×10^{-4}	20
U^{235}	P 99.98	1500–2000	R.A.	76.4	7.6×10^{-3}	107
W^{185}	P 99.98	1700–2260	S.S.	110	1.7	108
Nickel						
Au^{198}	S, P 99.999	700–1075	S.S.	55.0	0.02	109
Be^{7}	P 99.9	1020–1400	R.A.	46.2	0.019	76
C^{14}	P 99.86	600–1400	—	34.0	0.012	23
Co^{60}	P 99.97	1149–1390	R.A.	65.9	1.39	110
Cr^{51}	P 99.95	1100–1270	S.S.	65.1	1.1	111
Cu^{64}	P 99.95	1050–1360	S.S.	61.7	0.57	111
Fe^{59}	P	1020–1263	S.S.	58.6	0.074	112
Mo^{99}	P	900–1200	R.A.	51.0	1.6×10^{-3}	20
Ni^{63}	P 99.95	1042–1404	S.S.	68.0	1.9	111
Pu^{238}	P	1025–1125	A.R.G.	51.0	0.5	113
Sb^{124}	P 99.97	1020–1220	—	27.0	1.8×10^{-5}	114
Sn^{113}	P 99.8	700–1350	A.R.G.	58.0	0.83	115
V^{48}	P 99.99	800–1300	R.A.	66.5	0.87	11
W^{185}	P 99.95	1100–1300	S.S.	71.5	2.0	116
Niobium						
C^{14}	P	800–1250	R.A.	32.0	1.09×10^{-5}	117
Co^{60}	P 99.85	1500–2100	A.R.G.	70.5	0.74	118
Cr^{51}	S	943–1435	S.S.	83.5	0.30	119
Fe^{55}	P 99.85	1400–2100	A.R.G.	77.7	1.5	118
K^{42}	S	900–1100	R.A.	22.10	2.38×10^{-7}	102
Nb^{95}	P, S 99.99	878–2395	S.S.	96.0	1.1	120
P^{32}	P 99.0	1300–1800	S.S.	51.5	5.1×10^{-2}	104
S^{35}	S 99.9	1100–1500	R.A.	73.1	2600	121
Sn^{113}	P 99.85	1850–2400	S.S.	78.9	0.14	122
Ta^{182}	P, S 99.997	878–2395	S.S.	99.3	1.0	120
Ti^{44}	S	994–1492	S.S.	86.9	0.099	123
U^{235}	P 99.55	1500–2000	R.A.	76.8	8.9×10^{-3}	107
V^{48}	S 99.99	1000–1400	R.A.	85.0	2.21	124
W^{185}	P 99.8	1800–2200	R.A.	91.7	5×10^{-4}	125
Palladium						
Pd^{103}	S 99.999	1060–1500	S.S.	63.6	0.205	126
Phosphorus						
P^{32}	P	0–44	S.S.	9.4	1.07×10^{-3}	127
Platinum						
Co^{60}	P 99.99	900–1050	—	74.2	19.6	129
Cu^{64}	P	1098–1375	S.S.	59.5	0.074	41
Pt^{195}	P 99.99	1325–1600	S.S.	68.2	0.33	130
Potassium						
Au^{198}	P 99.95	5.6–52.5	S.S.	3.23	1.29×10^{-3}	131
K^{42}	S 99.7	−52–61	S.S.	9.36	0.16	132
Na^{22}	P 99.7	0–62	S.S.	7.45	0.058	133
Rb^{86}	P 99.95	0.1–59.9	S.S.	8.78	0.090	134

RADIOACTIVE TRACER DIFFUSION DATA FOR PURE METALS (Continued)

Solute (tracer)	Material (metal, crystalline form and purity)	Temperature range, °C	Form of analysis	Activation energy, Q, Kcal/mole	Frequency factor, D_0, cm²/sec	Reference
γ-Plutonium						
Pu^{238}	P	190–310	S.S.	16.7	2.1×10^{-5}	135
δ-Plutonium						
Pu^{238}	P	350–440	S.S.	23.8	4.5×10^{-3}	136
ε-Plutonium						
Pu^{238}	P	500–612	R.A.	18.5	2.0×10^{-2}	137
α-Praseodymium						
Ag^{110}	P 99.93	610–730	S.S.	25.4	0.14	138
Au^{195}	P 99.93	650–780	S.S.	19.7	4.3×10^{-2}	138
Co^{60}	P 99.93	660–780	S.S.	16.4	4.7×10^{-2}	138
Zn^{65}	P 99.96	766–603	S.S.	24.8	0.18	139
β-Praseodymium						
Ag^{110}	P 99.93	800–900	S.S.	21.5	3.2×10^{-2}	138
Au^{195}	P 99.93	800–910	S.S.	20.1	3.3×10^{-2}	138
Ho^{166}	P 99.96	800–930	S.S.	26.3	9.5	140
In^{114}	P 99.96	800–930	S.S.	28.9	9.6	140
La^{140}	P 99.96	800–930	S.S.	25.7	1.8	140
Pr^{142}	P 99.93	800–900	S.S.	29.4	8.7	140
Zn^{65}	P 99.96	822–921	S.S.	27.0	0.63	139
Selenium						
Fe^{59}	P	40–100	R.A.	8.88	—	141
Hg^{203}	P 99.996	25–100	R.A.	1.2	—	141
S^{35}	S⊥c	60–90	S.D.	29.9	1700	142
S^{35}	S∥c	60–90	S.D.	15.6	1100	142
Se^{75}	P	35–140	—	11.7	1.4×10^{-4}	143
Silicon						
Au^{198}	S	700–1300	S.S.	47.0	2.75×10^{-3}	145
C^{14}	P	1070–1400	R.A.	67.2	0.33	146
Cu^{64}	P	800–1100	R.A.	23.0	4×10^{-2}	147
Fe^{59}	S	1000–1200	R.A.	20.0	6.2×10^{-3}	148
Ni^{63}	P	450–800	—	97.5	1000	149
P^{32}	S	1100–1250	R.A.	41.5	—	150
Sb^{124}	S	1190–1398	R.A.	91.7	12.9	151
Si^{31}	S 99.99999	1225–1400	S.S.	110.0	1800	146
Silver						
Au^{198}	P 99.99	718–942	S.S.	48.28	0.85	54
Ag^{110}	S 99.999	640–955	S.S.	45.2	0.67	152
Cd^{115}	S 99.99	592–937	S.S.	41.69	0.44	153
Co^{60}	S 99.999	700–940	—	48.75	1.9	154
Cu^{64}	P 99.99	717–945	S.S.	46.1	1.23	155
Fe^{59}	S 99.99	720–930	S.S.	49.04	2.42	156
Ge^{77}	P	640–870	S.S.	36.5	0.084	157
Hg^{203}	P 99.99	653–948	S.S.	38.1	0.079	155
In^{114}	S 99.99	592–937	S.S.	40.80	0.41	153
Ni^{63}	S 99.99	749–950	S.S.	54.8	21.9	158
Pb^{210}	P	700–865	S.S.	38.1	0.22	159
Pd^{102}	S 99.999	736–939	S.S.	56.75	9.56	140
Ru^{103}	S 99.99	793–945	S.S.	65.8	180	160
S^{35}	S 99.999	600–900	R.A.	40.0	1.65	161
Sb^{124}	P 99.999	780–950	S.S., R.A.	39.07	0.234	162
Sn^{113}	S 99.99	592–937	S.S.	39.30	0.255	153
Te^{125}	P	770–940	R.A.	38.90	0.47	163
Tl^{204}	P	640–870	S.S.	37.9	0.15	157
Zn^{65}	S 99.99	640–925	S.S.	41.7	0.54	164
Sodium						
Au^{198}	P 99.99	1.0–77	S.S.	2.21	3.34×10^{-4}	165
K^{42}	P 99.99	0–91	S.S.	8.43	0.08	133
Na^{22}	P 99.99	0–98	S.S.	10.09	0.145	166
Rb^{86}	P 99.99	0–85	S.S.	8.49	0.15	133
Tantalum						
C^{14}	P	1450–2200	S.S.	40.3	1.2×10^{-2}	167
Fe^{59}	P	930–1240	—	71.4	0.505	168
Mo^{99}	P	1750–2220	R.A.	81.0	1.8×10^{-3}	20
Nb^{95}	P, S 99.996	921–2484	S.S.	98.7	0.23	169
S^{35}	P 99.0	1970–2110	R.A.	70.0	100	170
Ta^{182}	P, S 99.996	1250–2200	S.S.	98.7	1.24	226
Tellurium						
Hg^{203}	P	270–440	—	18.7	3.14×10^{-5}	171
Se^{75}	P	320–440	—	28.6	2.6×10^{-2}	171
Tl^{204}	P	360–430	—	41.0	320	172
Te^{127}	S⊥c 99.9999	300–400	S.S.	46.7	3.91×10^{4}	173
Te^{127}	S∥c 99.9999	300–400	S.S.	35.5	130	173
α-Thallium						
Ag^{110}	P⊥c 99.999	80–250	S.S.	11.8	3.8×10^{-2}	174
Ag^{110}	P∥c 99.999	80–250	S.S.	11.2	2.7×10^{-2}	174
Au^{198}	P⊥c 99.999	110–260	S.S.	2.8	2.0×10^{-5}	174
Au^{198}	P∥c 99.999	110–260	S.S.	5.2	5.3×10^{-4}	174
Tl^{204}	S⊥c 99.9	135–230	S.S.	22.6	0.4	175
Tl^{204}	S∥c 99.9	135–230	S.S.	22.9	0.4	175
β-Thallium						
Ag^{110}	P 99.999	230–310	S.S.	11.9	4.2×10^{-2}	174
Au^{198}	P 99.999	230–310	S.S.	6.0	5.2×10^{-4}	174
Tl^{204}	S 99.9	230–280	S.S.	20.7	0.7	175

Solute (tracer)	Material (metal, crystalline form and purity)	Temperature range, °C	Form of analysis	Activation energy, Q, Kcal/mole	Frequency factor, D_0, cm^2/sec	Reference
α-Thorium						
Pa^{231}	P 99.85	770–910	—	74.7	126	176
Th^{228}	P 99.85	720–880	—	71.6	395	176
U^{233}	P 99.85	700–880	—	79.3	2210	176
Tin						
Ag^{110}	$S\perp c$	135–225	S.S.	18.4	0.18	177
Ag^{110}	$S\parallel c$	135–225	S.S	12.3	7.1×10^{-3}	177
Au^{198}	$S\perp c$	135–225	S.S.	17.7	0.16	177
Au^{198}	$S\parallel c$	135–225	S.S.	11.0	5.8×10^{-3}	177
Co^{60}	S, P	140–217	R.A.	22.0	5.5	178
In^{114}	$S\perp c$ 99.998	181–221	S.S.	25.8	34.1	179
In^{114}	$S\parallel c$ 99.998	181–221	S.S.	25.6	12.2	179
Sn^{113}	$S\perp c$ 99.999	160–226	S.S.	25.1	10.7	180
Sn^{113}	$S\parallel c$ 99.999	160–226	S.S.	25.6	7.7	180
Tl^{204}	P 99.999	137–216	S.S.	14.7	1.2×10^{-3}	181
α-Titanium						
Ti^{44}	P 99.99	700–850	R.A.	35.9	8.6×10^{-6}	182
β-Titanium						
Ag^{110}	P 99.95	940–1570	S.S.	43.2	3×10^{-3}	183
Be^{7}	P 99.96	915–1300	R.A.	40.2	0.8	184
C^{14}	P 99.62	1100–1600	R.A.	20.0	3.02×10^{-3}	185
Cr^{51}	P 99.7	950–1600	A.R.G.	35.1	5×10^{-3}	186
				61.0	4.9	
Co^{60}	P 99.7	900–1600	S.S.	30.6	1.2×10^{-2}	186
				52.5	2.0	
Fe^{59}	P 99.7	900–1600	A.R.G.	31.6	7.8×10^{-3}	186
				55.0	2.7	
Mo^{99}	P 99.7	900–1600	S.S.	43.0	8.0×10^{-3}	186
				73.0	20	
Mn^{54}	P 99.7	900–1600	S.S.	33.7	6.1×10^{-3}	186
				58.0	20	
Nb^{95}	P 99.7	1000–1600	A.R.G.	39.3	5.0×10^{-3}	186
				73.0	20	
Ni^{63}	P 99.7	925–1600	A.R.G.	29.6	9.2×10^{-3}	186
				52.5	2.0	
P^{32}	P 99.7	950–1600	S.S.	24.1	3.62×10^{-3}	187
				56.5	5	
Sc^{46}	P 99.95	940–1590	S.S.	32.4	4.0×10^{-3}	183
Sn^{113}	P 99.7	950–1600	S.S.	31.6	3.8×10^{-4}	187
				69.2	10	
Ti^{44}	P 99.95	900–1540	S.S.	31.2	3.58×10^{-4}	188
				60.0	1.09	
U^{235}	P 99.9	900–1400	R.A.	29.3	5.1×10^{-4}	189
V^{48}	P 99.95	900–1545	S.S.	32.2	3.1×10^{-4}	190
				57.2	1.4	
W^{185}	P 99.94	900–1250	R.A.	43.9	3.6×10^{-3}	191
Zr^{95}	P 98.94	920–1500	R.A.	35.4	4.7×10^{-3}	191
Tungsten						
C^{14}	P 99.51	1200–1600	R.A.	53.5	8.91×10^{-3}	99
Fe^{59}	P	940–1240	—	66.0	1.4×10^{-2}	168
Mo^{99}	P	1700–2100	R.A.	101.0	0.3	20
Nb^{95}	P 99.99	1305–2367	S.S.	137.6	3.01	192
Re^{186}	S	2100–2400	R.A.	141.0	19.5	193
Ta^{182}	P 99.99	1305–2375	S.S.	139.9	3.05	192
W^{185}	P 99.99	1800–2403	S.S.	140.3	1.88	192
α-Uranium						
U^{234}	P	580–650	—	40.0	2×10^{-3}	194
β-Uranium						
Co^{60}	P 99.999	692–763	S.S.	27.45	1.5×10^{-2}	195
U^{235}	P	690–750	R.A.	44.2	2.8×10^{-3}	196
γ-Uranium						
Au^{195}	P 99.99	785–1007	S.S.	30.4	4.86×10^{-3}	197
Co^{60}	P 99.99	783–989	S.S.	12.57	3.51×10^{-4}	198
Cr^{51}	P 99.99	797–1037	S.S.	24.46	5.37×10^{-3}	198
Cu^{64}	P 99.99	787–1039	S.S.	24.06	1.96×10^{-3}	198
Fe^{55}	P 99.99	787–990	S.S.	12.0	2.69×10^{-4}	198
Mn^{54}	P 99.99	787–939	S.S.	13.88	1.81×10^{-4}	198
Nb^{95}	P 99.99	791–1102	S.S.	39.65	4.87×10^{-2}	198
Ni^{63}	P 99.99	787–1039	S.S.	15.66	5.36×10^{-4}	198
U^{233}	P 99.99	800–1070	S.S.	28.5	2.33×10^{-3}	227
Zr^{95}	P	800–1000	R.A.	16.5	3.9×10^{-4}	228
Vanadium						
C^{14}	P 99.7	845–1130	S.S.	27.3	4.9×10^{-3}	199
Cr^{51}	P 99.8	960–1200	R.A.	64.6	9.54×10^{-3}	200
Fe^{59}	P	960–1350	S.S.	71.0	0.373	201
P^{32}	P 99.8	1200–1450	R.A.	49.8	2.45×10^{-2}	202
S^{35}	P 99.8	1320–1520	R.A.	34.0	3.1×10^{-2}	184
V^{48}	S, P 99.99	880–1360	S.S.	73.65	0.36	203
V^{48}	S, P 99.99	1360–1830	S.S.	94.14	214.0	203
Yttrium						
Y^{90}	$S\perp c$	900–1300	R.A.	67.1	5.2	204
Y^{90}	$S\parallel c$	900–1300	R.A.	60.3	0.82	204
Zinc						
Ag^{110}	$S\perp c$ 99.999	271–413	S.S.	27.6	0.45	205
Ag^{110}	$S\parallel c$ 99.999	271–413	S.S.	26.0	0.32	205

Solute (tracer)	Material (metal, crystalline form and purity)	Temperature range, °C	Form of analysis	Activation energy, Q, Kcal/mole	Frequency factor, D_0, cm²/sec	Reference
Au^{198}	S⊥c 99.999	315–415	S.S.	29.72	0.29	206
Au^{198}	S‖c 99.999	315–415	S.S.	29.73	0.97	206
Cd^{115}	S⊥c 99.999	225–416	S.S.	20.12	0.117	206
Cd^{115}	S‖c 99.999	225–416	S.S.	20.54	0.114	206
Cu^{64}	S⊥c 99.999	338–415	S.S.	29.92	2.0	206
Cu^{64}	S‖c 99.999	338–415	S.S.	29.53	2.22	207
Ga^{72}	S⊥c	240–403	S.S.	18.15	0.018	207
Ga^{72}	S‖c	240–403	S.S.	18.4	0.016	207
Hg^{203}	S⊥c	260–413	S.S.	20.18	0.073	208
Hg^{203}	S‖c	260–413	S.S.	19.70	0.056	208
In^{114}	S⊥c	271–413	S.S.	19.60	0.11	205
In^{114}	S‖c	271–413	S.S.	19.10	0.062	205
Sn^{113}	S⊥c	298–400	S.S.	18.4	0.13	209
Sn^{113}	S‖c	298–400	S.S.	19.4	0.15	209
Zn^{65}	S⊥c 99.999	240–418	S.S.	23.0	0.18	210
Zn^{65}	S‖c 99.999	240–418	S.S.	21.9	0.13	210
α-Zirconium						
Cr^{51}	P 99.9	700–850	R.A.	18.0	1.19×10^{-8}	211
Fe^{55}	P	750–840	—	48.0	2.5×10^{-2}	212
Mo^{99}	P	600–850	R.A.	24.76	6.22×10^{-8}	213
Nb^{95}	P 99.99	740–857	R.A.	31.5	6.6×10^{-6}	182
Sn^{113}	P	300–700	A.R.G.	22.0	1.0×10^{-8}	214
Ta^{182}	P 99.6	700–800	R.A.	70.0	100	215
V^{48}	P 99.99	600–850	R.A.	22.9	1.12×10^{-8}	124
Zr^{95}	P 99.95	750–850	S.S.	45.5	5.6×10^{-4}	216
β-Zirconium						
Be^{7}	P 99.7	915–1300	R.A.	31.1	8.33×10^{-2}	184
C^{14}	P 96.6	1100–1600	R.A.	34.2	3.57×10^{-2}	217
Ce^{141}	P	880–1600	R.A.	41.4	3.16	218
				74.1	42.17	
Co^{60}	P 99.99	920–1600	S.S.	21.82	3.26×10^{-3}	219
Cr^{51}	P 99.9	700–850	R.A.	18.0	1.19×10^{-8}	211
Fe^{55}	P	750–840	—	48.0	2.5×10^{-2}	212
Mo^{99}	P	900–1635	R.A.	35.2	1.99×10^{-6}	218
				68.55	2.63	
Nb^{95}	P	1230–1635	R.A.	36.6	7.8×10^{-4}	220
P^{32}	P 99.94	950–1200	R.A.	33.3	0.33	221
Sn^{113}	P	300–700	A.R.G.	22.0	1×10^{-8}	214
Ta^{182}	P 99.6	900–1200	R.A.	27.0	5.5×10^{-5}	215
U^{235}	P	900–1065	S.S.	30.5	5.7×10^{-4}	222
V^{48}	P 99.99	870–1200	R.A.	45.8	7.59×10^{-3}	223
V^{48}	P 99.99	1200–1400	R.A.	57.7	0.32	223
W^{185}	P 99.7	900–1250	R.A.	55.8	0.41	223
Zr^{95}	P	1100–1500	S.S.	30.1	2.4×10^{-4}	225

REFERENCES

1. N. L. Peterson and S. J. Rothman, *Phys. Rev.* **B1** (8), 3264 (1970).
2. T. S. Lundy and J. F. Murdock, *J. Appl. Phys.* **33** (5), 1671 (1962).
3. W. B. Alexander and L. M. Slifkin, *Phys. Rev.* **B1** (8), 3274 (1970).
4. M. S. Anand and R. P. Agarwala *Phys. Stat. Solidi* **A1** (1), 41K (1970).
5. S. P. Murarka and R. P. Agarwala, *Indian At. Energy Comm.* **Report BARC-368** (1968).
6. A. R. Paul and R. P. Agarwala, *J. Appl. Phys.* **38** (9), 3790 (1967).
7. G. P. Tiwari and B. D. Sharma, *Trans. Indian Inst. Metals* **20,** 83 (1967).
8. K. Hirano, R. P. Agarwala, and M. Cohen, *Acta Met.* **10** (9), 857 (1962).
9. M. S. Anand and R. P. Agarwala, *Trans. AIME* **239** (11), 1848 (1967).
10. S. Badrinarayanan and H. B. Mathers, *Int. J. Appl. Radiat. Isotopes* **19** (4), 353 (1968).
11. S. P. Murarka, M. S. Anand, and R. P. Agarwala, *Acta Met.* **16** (1), 69 (1968).
12. M. C. Naik, J. M. Dupony, and Y. Adda, *Mem. Sci. Rev. Met.* **63,** 488 (1966).
13. J. M. Dupony, J. Mathie, and Y. Adda, *Mem. Sci. Rev. Met.* **63,** 481 (1966).
14. V. M. Ananyn, V. P. Gladkov, V. S. Zotov, and D. M. Skorov, *At. Energ.* **29** (3), 220 (1970).
15. W. Hirschwald and W. Schroedter, *Z. Physik. Chem. N.F.* **53,** 392 (1967).
16. W. Chomka, *Zess. Nauk Politech Gdansk Fizyka* **1,** 39 (1967).
17. L. V. Pavlinov, A. M. Gladyshev, and V. N. Bikov, *Fiz. Metal Metalloved.* **26** (5), 823 (1968).
18. J. R. Wolfe, D. R. McKenzie, and R. J. Borg, *J. Appl. Phys.* **36** (6), 1906 (1965).
19. M. A. Kanter, *Phys. Rev.* **107,** 655 (1957).
20. E. V. Borisov, P. L. Gruzin, and S. V. Zemskii, *Zashch Pokryt Metal* **2,** 104 (1968).
21. J. Askill and D. H. Tomlin, *Phil. Mag.* **11** (111), 467 (1965).
22. H. W. Paxton and R. A. Wolfe, *Trans. AIME* **230,** 1426 (1964).
23. I. I. Kovenski, *Fiz. Metal. i Metalloved.* **16,** 613 (1963).
24. H. W. Mead and C. E. Birchenall, *Trans. Met. Soc. AIME* **203** (9), 994 (1955).
25. K. Hirano, R. P. Agarwala, B. L. Averbach, and M. Cohen, *J. Appl. Phys.* **33** (10), 3049 (1962).
26. M. M. Pavlyuchenko and I. F. Kononyuk, *Dokl. Akad. Nauk Belorusskoi* **SSR 8,** 157 (1964).
27. G. Barreau, G. Brunel, G. Azeron, P. Lacombe, *C. R. Acad. Sci. Paris* **Ser C 270** (6), 514 (1970).
28. S. M. Klotsman, A. Ya Rabovskii, V. K. Talinskii, and A. N. Timofeeu, *Fiz. Met. Metalloved* **29** (4), 803 (1970).
29. A. Chatterjee and D. J. Fabian, *Acta Met.* **17** (9), 1141 (1969).
30. T. Hirone, N. Kunitomi, M. Sakamoto, and H. Yamaki, *J. Phys. Soc. Japan* **13** (8), 838 (1958).
31. S. Badrinarayanan and H. B. Mathur, *Indian J. Pure Appl. Phys.* **8** (6), 324 (1970).
32. G. Barreau, G. Brunel, and G. Cizeron, *C. R. Acad. Sci.* **Ser. C 272** (7), 618 (1971).
33. C. A. Machiet, *Phys. Rev.* **109** (6), 1964 (1958).
34. S. J. Rothman and N. L. Peterson, *Phys. Stat. Solidi* **35,** 305 (1969).
35. C. T. Tomizuka, cited by D. Lazarus, *Solid State Phys.* **Vol. 10** (1960).
36. F. D. Reinke and C. E. Dahlstrom, *Phil. Mag.* **22** (175), 57 (1970).
37. A. Ikushima, *J. Phys. Soc. Japan* **14,** 111 (1959).
38. M. C. Saxena and B. D. Sharma, *Trans. Indian Inst. Metals* **23** (3), 16 (1970).
39. G. Brunel, G. Cizeron, P. Lacombe, *C. R. Acad. Sci. Paris* **Ser. C 270** (4), 393 (1970).
40. N. L. Peterson, *Phys. Rev.* **132** (6), 2471 (1963).
41. R. D. Johnson and B. H. Faulkenberry, *ASD-TDR*-63-625 (July 1963).
42. F. Moya, G. E. Moya-Gontier and F. Cabane-Brouty, *Phys. Stat. Solidi* **35** (2), 893 (1969).
43. M. C. Inman and L. W. Barr, *Acta Met.* **8** (2), 112 (1960).
44. P. P. Kuzmenko, L. F. Ostrovskii, and V. S. Kovalchuk, *Fiz. Tverd. Tela* **4,** 490 (1962).
45. S. Komura and N. Kunitomi, *J. Phys. Soc. Japan* **18 (Suppl. 2),** 208 (1963).
46. S. M. Klotsman, Ya A. Rabovskii, V. K. Talinskii, and A. N. Timofeev, *Fiz. Metal. Metalloved.* **28** (6), 1025 (1969).
47. V. E. Kosenko, *Fiz. Tverd. Tela* **1**:1622 (1959); *Soviet Phys.—Solid State* (*English transl.*) **1,** 1481 (1959).
48. A. A. Bugai, V. E. Kosenko, and E. G. Miselynuk, *Sh. Tekhn. Fiz.* **27** (1), 207 (1957); *NP-tr*-448, **P.** 219 (1960).
49. H. Letaw, W. M. Portnoy, and L. Slifkin, *Phys. Rev.* **102,** 636 (1956).
50. A. V. Sandulova, M. I. Droniuk, and V. M. P'dak, *Fiz. Tverd. Tela* **3,** 2913 (1961).

51. B. I. Boltaks, V. P. Grabtchak, and T. D. Dzafarov, *Fiz. Tverd. Tela* **6**, 3181 (1964).
52. V. D. Ignatkov and V. E. Kosenko, *Fiz. Tverd. Tela* **4** (6), 1627 (1962): *Soviet Phys.—Solid State* (*Eng. transl.*) **4** (6), 1627 (1962).
53. V. I. Tagirov and A. A. Kuliev, *Fiz. Tverd. Tela* **4** (1), 272 (1962) *Soviet Phys.—Solid State* (*Eng. transl.*) **4** (1), 196 (1962).
54. W. C. Mallard, A. B. Gardner, R. F. Bass, and L. M. Slifkin, *Phys. Rev.* **129** (2), 617 (1963).
55. D. Duhl, K. Hirano, and M. Cohen, *Acta Met.* **11** (1), 1 (1963).
56. A. J. Mortlock and A. H. Rowe, *Phil. Mag.* **11** (114), 115 (1965).
57. J. E. Reynolds, B. L. Averbach, and M. Cohen, *Acta Met.* **5**, 29 (1957).
58. A. J. Mortlock, A. H. Rowe, and A. D. LeClaire, *Phil. Mag.* **5**, 803 (1960).
59. F. R. Winslow and T. S. Lundy, *Trans. Met. Soc. AIME* **233**, 1790 (1965).
60. T. R. Anthony and D. Turnbull, *Phys. Rev.* **151**, 495 (1966).
61. V. M. Amonenko, A. M. Blinkin, and I. G. Ivantsov, *Fiz. Metal. i Metalloved.* **17** (1), 56 (1964); *Phys. Metals Metallog.* (*USSR*) (*Eng. transl.*) **17** (1), 54 (1964).
62. D. W. James and G. M. Leak, *Phil. Mag.* **14**, 701 (1966).
63. A. Bondy and V. Levy, *C. R. Acad. Sci. Ser C* **272** (1), 19 (1971).
64. R. J. Borg and D. Y. F. Lai, *Acta Met.* **11** (8), 861 (1963).
65. C. G. Homan, *Acta Met.* **12**, 1071 (1964).
66. A. M. Huntz, M. Aucouturier, and P. Lacombe, *C. R. Acad. Sci. Paris Ser. C* **265** (10), 554 (1967).
67. M. S. Anand and R. P. Agarwala, *J. Appl. Phys.* **37**, 4248 (1966).
68. R. Angers and F. Claisse, *Can. Met. Quart.* **7** (2), 73 (1968).
69. A. V. Tomilov and G. V. Shcherbedinskii, *Fiz. Khim. Met. Mater.* **3** (3), 261 (1967).
70. K. Nohara and K. Hirano, *Proc. Intl. Conf. Sci. Technol. Iron Steel Tokyo*, Sept 1970.
71. V. T. Borisov, V. M. Golikov, and G. V. Shcherbedinskii, *Fiz. Metal. i Metalloved.* **22** (1), 159 (1966).
72. K. Hirano, M. Cohen, and B. L. Averbach, *Acta Met.* **9** (5), 440 (1961).
73. G. Seibel, *Compt. Rend.* **256** (22), 4661 (1963).
74. G. Bruggeman and J. Roberts, *J. Met.* **20** (8), 54 (1968).
75. V. D. Lyubimov, *Izv. Akad. Nauk. SSSR. Met.* **3** 201 (1969); V. D. Lyubimov, V. A. Tskhai, and G. B. Bogomolov, *Trudy Inst. Khim. Akad. Nauk. SSSR Ural Filial* **17**, 44, 48 (1970).
76. G. V. Grigorev and L. V. Pavlinov, *Fiz. Metal. i Metalloved.* **25** (5), 836 (1968).
77. T. Suzuoka, *Japan. Inst. Metals* **2**, 176 (1961).
78. A. W. Bowen and G. M. Leak, *Met. Trans.* **1** (6), 1695 (1970).
79. Th. Heumann and R. Imm, *J. Phys. Chem. Solids* **29** (9), 1613 (1968).
80. P. L. Gruzin and V. V. Mural, *Fiz. Metal. i Metalloved.* **16** (4), 551 (1963); *Phys. Metals Metallog.* (*USSR*) (*Eng. transl.*) **16** (4), 50 (1963).
81. A. Hoshino and R. Araki, *Tetsu to Hagane* **56** (2), 252 (1970).
82. K. Sato, *Trans. Japan Inst. Metals* **5**, 91 (1964).
83. D. W. James and G. M. Leak, *Phil. Mag.* **14**, 701 (1966).
84. M. P. Dariel, G. Erez, and G. M. J. Schmidt, *Phil. Mag.* **19** (161), 1053 (1969).
85. P. P. Kuzmenko, G. P. Grinevich and B. A. Danilchenko, *Fiz. Met. Metalloved* **29** (2), 318 (1970).
86. G. V. Kidson, *Phil. Mag.* **13**, 247 (1966).
87. J. W. Miller, *Phys. Rev.* **181** (3), 1095 (1969).
88. B. F. Dyson, T. Anthony, and D. Turnbull, *J. Appl. Phys.* **37**, 2370 (1966).
89. H. A. Resing and N. H. Nachtrieb, *Phys. Chem. Solids* **21** (1/2), 40 (1961).
90. A. Ott and A. Lodding, *Int. Conf. Vac. and Interstitials in metals.* **Julich 1**, 43 (1968); *Z. Naturforsch* **23A**, 1683 (1968), 2126 (1968).
91. A. Ott, A. Lodding, and D. Lazarus, *Phys. Rev.* **188** (3), 1088 (1969).
92. A. Ott, J. Appl. Phys. **40** (6), 2395 (1969); J. N. Mundy, A. Ott, L. Lowenberg, and A. Lodding, *Phys. Stat. Solidi.* **35** (1), 359 (1969).
93. A. Ott, *Z. Naturforschg* **25a** (10), 1477 (1970).
94. A. Lodding, J. N. Mundy, and A. Ott, *Phys. Stat. Solidi* **38** (2), 559 (1970).
95. K. Lal, *Commis. Energ. At. Report.* **CEA-R** 3136 (1967).
96. L. V. Pavlinov, A. M. Gladyshev, and V. N. Bikov, *Fiz. Metal Metalloved* **26** (5), 823 (1968).
97. P. G. Shewmon, *Trans. Met. Soc. AIME* **206**, 918 (1956).
98. Y. V. Borisov, P. L. Gruzin, and L. V. Pavlinov, *Met. i Metalloved. Chistykh Metal.* **1** 213 (1959), *translated in JPRS-5195.*
99. A. Y. Nakonechnikov, L. V. Pavlinov, and V. N. Bikov, *Fiz. Metal i Metalloved.* **22**, 234 (1966).
100. J. Askill, *Phys. Status Solidi* **9** (2), K113 (1965).
101. L. V. Pavlinov and A. A. Kordev, *Fiz. Metal. Metalloved.* **29** (6), 1326 (1970).
102. G. N. Dubinin, G. P. Benediktova, M. G. Karpman, and G. V. Shcherbedinskii, *Khim—Term Obrab Stali Splavov* **6**, 129 (1969).
103. J. Askill and D. H. Tomlin, *Phil. Mag.* **8** (90), 997 (1963).
104. B. A. Vandyshev and A. S. Panov, *Fiz. Met. Metalloved.* **26** (3), 517 (1968).
105. S. Z. Benediktova, G. N. Dubinin, M. G. Kapman, and G. V. Shcherbedinskii, *Metalloved. i Term. Obrabotka Metal.* **5**, 5 (1966).
106. B. A. Vandyshev and A. S. Panov, *Fiz. Metal. i Metalloved.* **25** (2), 321 (1968).
107. L. V. Pavlinov, A. Y. Nakonechnikov, and V. N. Bikov, *At. Energ.* (*USSR*) **19** 521 (1965).
108. J. Askill, *Phys. Status Solidi* **23**, K21 (1967).
109. A. Chatterjee and D. J. Fabian, *J. Inst. Metals* **96** (6), 186 (1968).
110. A. Hassner and W. Lange, *Phys. Status Solidi* **8**, 77 (1965).
111. K. Monma, H. Suto, and H. Oikawa, *J. Japan Inst. Metals* **28**, 188 (1964).
112. A. Ya. Shinyaev, *Izv. Akad. Nauk. S.S.S.R. Metl.* **4**, 182 (1969).
113. J. J. Blechet, A. VanGryeynest, and D. Calais, *J. Nucl. Mater.* **28** (2), 177 (1968).
114. P. P. Kuzmenko and G. P. Grinevich, *Fiz. Tverd. Tela.* **4** (11), 3266 (1962); *Soviet Physics—Solid State* **4** (11), 2390 (1962).
115. S. Z. Bokshtein, S. T. Kishkin, and L. M. Moroz, *Investigation of the Structure of Metals by Radioactive Isotope Methods; State Publishing House of the Ministry of Defense Industry, Moscow* (1959); *AEC-tr-4505* (1961).
116. M. S. Anand, S. P. Murarka, and R. P. Agarwala, *J. Appl. Phys.* **36** (12), 3860 (1965).
117. V. D. Lyubimov, *Izv. Akad. Nauk. S.S.S.R. Metal* **3**, 201 (1969); V. D. Lyubimov, V. A. Tskhai, and G. B. Bogmolov, *Toudy Inst. Khim. Akad. Nauk. S.S.S.R. Ural Filial* **17**, 44, 48 (1970).
118. R. F. Peart, D. Graham, and D. H. Tomlin, *Acta Met.* **10**, 519 (1962).
119. J. Pelleg, *J. Metals* **20** (8), 54 (1968); *J. Less Common Metals* **17**, 319 (1969); *Phil. Mag.* **19** (157), 25 (1969).
120. T. S. Lundy, F. E. Winslow, R. E. Pawel, and C. J. McHargue, *Trans. Met. Soc. AIME* **223**, 1533 (1965).
121. B. A. Vandyshev and A. S. Panov, *Izv. Akad. Nauk. S.S.S.R.* **1**, 206 (1968).
122. J. Askill, *Phys. Status Solidi* **9** (3), K167 (1965).
123. J. Pelleg, *Phil. Mag.* **21** (172), 735 (1970).
124. R. P. Agarwala, S. P. Murarka, and M. S. Anand, *Acta Met.* **16** (1), 61 (1968).
125. G. B. Fedorov, F. I. Zhomov, and E. A. Smirnov, *Met. Metalloved. Chist. Metal* **8**, 145 (1969).
126. N. L. Peterson, *Phys. Rev.* **136** (2A), A568 (1964).
127. N. H. Nachtrieb and G. S. Handler, *J. Chem. Phys.* **23**, 1187 (1955).
129. J. Kucera and T. Zemcik, *Can. Met. Quart.* **7** (2), 83 (1968).
130. G. V. Kidson and R. Ross, *Proc. 1st UNESCO Int. Conf. Radioisotopes in Scientific Res.*, p. 185 (1958).
131. F. A. Smith and L. W. Barr, *Phil. Mag.* **21** (171), 633 (1970).
132. J. N. Mundy, T. E. Miller, and R. J. Porte, *Phys. Rev.* **B3** (8), 2445 (1971).
133. L. W. Barr, J. N. Mundy, and F. A. Smith, *Phil. Mag.* **16**, 1139 (1967).
134. F. A. Smith and L. W. Barr, *Phil. Mag.* **20** (163), 205 (1969).
135. R. E. Tate and G. R. Edwards, *Symposium on Thermodynamics with Emphasis on Nuclear Materials and Atomic Solids*, **pp. 105–113** in **IAEA Symposium Vol. II**, International Atomic Energy Agcy., Vienna (1966).
136. R. E. Tate and E. M. Cramer, *Trans. Met. Soc. AIME* **230**, 639 (1964).
137. M. Dupuy and D. Calais, *Trans. AIME* **242**, 1679 (1968).
138. M. P. Dariel, G. Erez, and G. M. J. Schmidt, *J. Appl. Phys.* **40** (7), 2746 (1969).
139. M. P. Dariel, *Phil. Mag.* **22** (177), 653 (1970).
140. M. P. Dariel, G. Erez, and G. M. J. Schmidt, *Phil. Mag.* **18** (161), 1045 (1969).
141. A. A. Kuliev and D. N. Nasledov, *Zh. Tekhn. Fiz.* **28** (2), 259 (1958); *Soviet Phys.—Tech. Phys.* (*Eng. transl.*) **3** (2), 235 (1958).
142. P. Braetter and H. Gobrecht, *Phys. Stat. Sol* **41** (2), 631 (1970).
143. B. I. Boltaks, and B. T. Plachenov, *Soviet Phys.—Tech. Phys.* (*Eng. transl.*) **27** (10), 2071 (1957).
144. R. F. Peart, *Phys. Status Solidi* **15**, K119 (1966).
145. W. R. Wilcox and T. J. LaChapelle, *J. Appl. Phys.* **35** (1) 240 (1964).
146. R. C. Newman and J. Wakefield, *Phys. Chem. Solids* **19** (3), 230 (1961).
147. B. I. Boltaks and I. I. Sozinov, *Zh. Tekhn. Fiz.* **28** (3), 679 (1958); *Soviet Phys.—Tech. Phys.* (*Eng. transl.*) **3** (3), 636 (1958).
148. J. D. Struthers, *J. Appl. Phys.* **27** (12), 1560 (1958); errata **28** (4), 516 (1957).
149. H. P. Bonzel, *Phys. Status Solidi* **20**, 493 (1967).
150. S. Mackawa, *J. Phys. Soc. Japan* **17** (10), 1592 (1962).
151. J. J. Rahan, M. E. Pickering, and J. Kennedy, *J. Electrochem. Soc.* **106** (8), 705 (1959).
152. S. J. Rothman, N. L. Peterson, and J. T. Robinson, *Phys. Stat. Sol.* **39** (2), 635 (1970).
153. C. T. Tomizuka and L. M. Slifkin, *Phys. Rev.* **96**, 610 (1954).
154. J. Bernardini, A. Combe-Brun, and J. Cabane, *Ser. Met.* **4** (12), 985 (1970).
155. A. Sawatskii and F. E. Jaumot, *Trans. AIME* **209**, 1207 (1957).
156. J. G. Mullen, *Phys. Rev.* **121**, 1649 (1961).
157. R. E. Hoffman, *Acta Met.* **6**, 95 (1958).
158. T. Hirone, S. Miura, and T. Suzuoka, *J. Phys. Soc. Japan* **16** (12), 2456 (1961).
159. R. E. Hoffman, D. Turnbull, and E. W. Hart, *Acta Met.* **3**, 417 (1955).
160. C. B. Pierce and D. Lazarus, *Phys. Rev.* **114**, 686 (1959).
161. N. Barbouth, J. Ouder, and J. Cabane, *C. R. Acad. Sci. Paris Ser. C* **264** (12), 1029 (1967).
162. V. N. Kaigorodov, Ya. A. Rabovskii, and V. K. Talinskii, *Fiz. Metal. i Metalloved.* **24** (4), 661 (1967).
163. V. N. Kaigorodov, S. M. Klotsman, A. N. Timofeev, and I. Sh. Traktenberg, *Fiz. Metal. Metalloved.* **28** (1), 120 (1969).
164. A. Sawatskii and F. E. Jaumot, *Phys. Rev.* **100**, 1627 (1955).

165. L. W. Barr, J. N. Mundy, and F. A. Smith, *Phil. Mag.* **20** (164), 389 (1969).
166. J. N. Mundy, L. W. Barr, and F. A. Smith, *Phil. Mag.* **14,** 785 (1966).
167. P. Son, S. Ihara, M. Miyake, and T. Sano, *J. Japan Inst. Met.* **33** (1), 1 (1969).
168. V. P. Vasilev, I. F. Kamardin, V. I. Skatskill, S. G. Chernomorchenko, and G. N. Shuppe, *Trudy Sred. Gos. Uni. im V.I. Lenina* **65,** 47 (1955); *Translation AEC-tr-4272.*
169. R. E. Pawel and T. S. Lundy, *J. Phys. Chem. Solids* **26,** 937 (1965).
170. B. A. Vandyshev and A. S. Panov, *Izv. Akad. Nauk. SSSR Metal* **1,** 244 (1969).
171. Sh. Movalanov and A. A. Kuliev, *Fiz. Tverd. Tela* **4** (2), 542 (1962).
172. N. I. Ibraginov, M. G. Shachtachtinskii, and A. A. Kuliev, *Fiz. Tverd. Tela* **4,** 3321 (1962).
173. R. N. Ghoshtagore, *Phys. Rev.* **155** (3), 698 (1967).
174. T. R. Anthony, B. F. Dyson, and D. Turnbull, *J. Appl. Phys.* **39** (3), 1391 (1968).
175. G. A. Shirn, *Acta Met.* **3,** 87 (1955).
176. F. Schmitz and M. Fock, *J. Nucl. Materials* **21,** 317 (1967).
177. B. F. Dyson, *J. Appl. Phys.* **37,** 2375 (1966).
178. W. Chomka and J. Andruszkiewicz, *Nukleonika* **5** (10), 611 (1960).
179. A. Sawatskii, *J. Appl. Phys.* **29** (9), 1303 (1958).
180. D. Yolokoff, S. May, and Y. Adda, *Compt. Rend.* **251** (3), 2341 (1960).
181. L. Bartha and T. Szalay, *Int. J. Appl. Radiat. Isotopes* **20** (2), 825 (1969).
182. F. Dyment and C. M. Libanati, *J. Mager. Sci.* **3** (4), 349 (1968).
183. J. Askill, *Phys. Stat Sol.* **B43** (1), K1 (1971).
184. L. V. Pavlinov, G. V. Grigorev, and G. O. Gromyko, *Izv. Akad. Nauk. SSSR Metal* **3,** 207 (1969).
185. A. Y. Nakonechnikov, L. V. Pavlinov, and V. N. Bikov, *Fiz. Metal i Metalloved* **22,** 234 (1966).
186. G. B. Gibbs, D. Graham, and D. H. Tomlin, *Phil. Mag.* **8** (92), 1269 (1963).
187. J. Askill and G. B. Gibbs, *Phys. Status Solidi* **11,** 557 (1965).
188. J. F. Murdock, T. S. Lundy, and E. E. Stansbury, *Acta Met.* **12** (9), 1033 (1964).
189. L. V. Pavlinov, *Fiz. Metal Metalloved* **30** (4), 800 (1970).
190. J. F. Murdock, T. S. Lundy, and E. E. Stansbury, *Acta Met.* **12** (9), 1033 (1964).
191. L. V. Pavlinov, *Fiz. Metal. i Metalloved.* **24** (2), 272 (1967).
192. R. E. Pawel and T. S. Lundy, *Acta Met.* **17** (8), 979 (1969).
193. L. N. Larikov, V. M. Tyshkevich, and L. F. Chorna, *Ukr. Fiz. Zh.* **12** (6), 983 (1967).
194. Y. Adda, A. Kirianenko, and C. Mairy, *Compt. Rend.* **253,** 445 (1961); *J. Nucl. Mater.* **6** (1), 130 (1962).
195. M. P. Dariel, M. Blumenfeld, and G. Kimmel, *J. Appl. Phys.* **41** (4), 1480 (1970).
196. G. B. Federov, E. A. Smirnov, and S. S. Moiseenko, *Met. Metalloved. Chist. Metal.* **7,** 124 (1968).
197. S. J. Rothman, *J. Nucl. Mater.* **3** (1), 77 (1961).
198. N. L. Peterson and S. J. Rothman, *Phys. Rev.* **136** (3A), A842 (1964).
199. P. Son, S. Ihara, M. Miyake, and T. Sano, *J. Japan Inst. Met.* **33** (1), 1 (1969).
200. H. W. Paxton and R. A. Wolfe, *Trans. AIME* **230,** 1426 (1964).
201. M. G. Coleman, *Ph.D. Thesis Univ. of Illinois* (1967); *Univ. Microf.* **68–1725** (1968).
202. B. A. Vandychev and A. S. Panov, *Izv. Akad. Nauk SSSR Metal* **2,** 231 (1970).
203. R. F. Peart, *J. Phys. Chem. Solids* **26,** 1853 (1965).
204. D. S. Gorney and R. M. Altovskii, *Fiz. Metal Metalloved* **30** (1), 85 (1970).
205. J. H. Rosolowski, *Phys. Rev.* **124** (6), 1828 (1961).
206. P. B. Ghate, *Phys. Rev.* **130** (1), 174 (1963).
207. A. P. Batra and H. B. Huntington, *Phys. Rev.* **145,** 542 (1966).
208. A. P. Batra and H. B. Huntington, *Phys. Rev.* **154** (3), 569 (1967).
209. J. S. Warford and M. B. Huntington, *Phys. Rev.* **B1** (4), 1867 (1970).
210. N. L. Peterson and S. J. Rothman, *Phys. Rev.* **163** (3), 645 (1967).
211. R. P. Agarwala, S. P. Murarka, and M. S. Anand, *Trans. Met. Soc. AIME* **233,** 986 (1965).
212. A. M. Blinkin and V. V. Vorobiov, *Ukr. Fiz. Zh.* **9** (1), 91 (1964).
213. R. P. Agarwala and A. R. Paul, *Proc. Nucl. Rad. Chem. Symp. Poona, India* **p. 542** (1967).
214. P. L. Gruzin, V. S. Emelyanov, G. G. Ryabova, and G. B. Federov, *Proc. 2nd Intern. Conf. Peaceful Uses At. Energy,* **19:187** (1958).
215. Y. V. Borisov, Y. G. Godin, P. L. Gruzin, A. I. Evstyukhin, and V. S. Yemelyanov, *Met. i Met. Izdatel. Akad. Nauk. SSSR Moscow* 1958. *Translation NP-TR-448* **p. 196** (1960).
216. P. Flubacher, *E. I. R. Bericht* **49,** May (1963).
217. R. A. Andriyevskii, V. N. Zagraykin, and G. Ya. Meshcheryakov, *Phys. Met. Metalloved.* **19** (3), 146 (1966).
218. R. P. Agarwala and A. R. Paul, Report CONF-670335, *Int. Conf. Vacancies and Interstitials in Metals,* Julich (1968) **1,** 105 (1968), *Report BARC*-377 (1968).
219. G. V. Kidson and G. J. Young, *Phil. Mag.* **20** (167), 1047 (1969).
220. G. B. Federov, E. A. Smirnov, and S. M. Novikov, *Met. Metalloved Chist. Metal* **8,** 41 (1969).
221. B. A. Vandychev and A. S. Panov, *Izv. Akad. Nauk. SSSR Metal* **2,** 231 (1970).
222. G. B. Federov, E. A. Smirnov, and F. I. Zhomov, *Met. Metalloved. Chist. Metal.* **7,** 116 (1968).
223. L. V. Pavlinov, *Fiz. Metal i Metalloved* **24** (2), 272 (1967).
224. H. M. Gilder and D. Lazarus, *J. Phys. Chem. Solids* **26,** 2081 (1965).
225. G. V. Kidson and J. McGurn, *Can. J. Phys.* **39** (8), 1146 (1961).
226. R. E. Pawel and T. S. Lundy, *J. Phys. Chem. Solids* **26,** 937 (1965).
227. S. J. Rothman, L. T. Lloyd, and A. L. Harkness, *Trans. Met. Soc. AIME* **218** (4), 605 (1960).
228. G. B. Federov, E. A. Smirnov, and V. N. Gusev, *At. Energ.* **27** (2), 149 (1969).

VISCOSITY AND THERMAL CONDUCTIVITY OF OXYGEN AS A FUNCTION OF TEMPERATURE

Conversion Factors

T, °K	→ T, °F: multiply by (9/5) then subtract 459.67
T, °K	→ T, °C: subtract 273.15
T, °K	→ T, °R: multiply by (9/5)
P, atm	→ P, psia: multiply by 14.69595
P, atm	→ p, N/m²: multiply by 1.01325 × 10^5
η, g/cm-s	→ η, N-s/m²: multiply by 2.39006 × 10^{-3}
η, g/cm-s	→ η, lb_m/ft-s: multiply by 0.577789
λ, W/m-°K	→ λ, cal/cm-s°K: multiply by 0.1
λ, W/m-°K	→ λ, Btu ft-hr-°R: multiply by 6.72 × 10^{-2}

Temperature, °K	Viscosity, g/cm-s $10^3 \eta_0$	Thermal conductivity, W/m-K $10^3 \lambda_0$	Temperature, °K	Viscosity, g/cm-s $10^3 \eta_0$	Thermal conductivity, W/m-K $10^3 \lambda_0$
80	0.0585	6.94	280	0.1955	24.86
90	0.0663	7.95	290	0.2012	25.62
100	0.0740	8.96	300	0.2068	26.38
110	0.0818	9.96	310	0.2122	27.14
120	0.0894	10.94	320	0.2176	27.89
130	0.0970	11.92	330	0.2230	28.62
140	0.1045	12.89	340	0.2282	29.36
150	0.1118	13.85	350	0.2334	30.10
160	0.1191	14.78	360	0.2385	30.85
170	0.1261	15.70	370	0.2435	31.58
180	0.1331	16.60	380	0.2485	32.32
190	0.1399	17.48	390	0.2534	33.06
200	0.1465	18.35	400	0.2583	33.79
210	0.1530	19.21	410	0.2631	34.54
220	0.1595	20.05	420	0.2678	35.28
230	0.1658	20.88	430	0.2725	36.02
240	0.1719	21.69	450	0.2818	37.49
250	0.1780	22.49	470	0.2908	38.95
260	0.1840	23.29	490	0.2997	40.41
270	0.1898	24.08	510	0.3085	41.86

VISCOSITY AND THERMAL CONDUCTIVITY OF OXYGEN AS A FUNCTION OF TEMPERATURE (Continued)

Temperature, °K	Viscosity, g/cm-s $10^3\ \eta_o$	Thermal conductivity, W/m-K $10^3 \lambda_o$
530	0.3170	43.29
550	0.3255	44.72
570	0.3338	46.14
590	0.3420	47.56
610	0.3501	48.97
630	0.3580	50.37
650	0.3659	51.77
670	0.3736	53.16
690	0.3813	54.54
710	0.3888	55.89
730	0.3963	57.25
750	0.4037	58.59
770	0.4110	59.90
790	0.4183	61.22
820	0.4290	63.16
850	0.4396	65.09
880	0.4500	66.98
910	0.4603	68.89
940	0.4705	70.72
970	0.4805	72.52
1,000	0.4905	74.32
1,030[a]	0.5002	76.07
1,060	0.5100	77.84
1,090	0.5196	79.56
1,120	0.5291	81.27
1,150	0.5384	82.97
1,180	0.5478	84.66
1,210	0.5571	86.32
1,240	0.5662	89.97
1,270	0.5753	89.61
1,300	0.5843	91.24
1,330	0.5932	92.85
1,360	0.6021	94.45
1,390	0.6109	96.06
1,420	0.6196	97.64
1,450	0.6282	99.22
1,480	0.6368	100.78
1,510	0.6454	102.35
1,540	0.6538	103.90
1,570	0.6622	105.45
1,600	0.6706	106.98
1,630	0.6789	108.52
1,660	0.6871	110.05
1,690	0.6953	111.58
1,720	0.7034	113.09
1,750	0.7116	114.61
1,780	0.7196	116.12
1,810	0.7276	117.63
1,840	0.7355	119.12
1,870	0.7435	120.63
1,900	0.7514	122.13
1,930	0.7592	123.62
1,960	0.7670	125.11
1,980	0.7721	126.11
2,000	0.7773	127.10

[a]Data for all temperatures in excess of 1,000 K are extrapolated.

From Hanley, H. J. M., McCarty, R. D., and Sengers, J. V., *Viscosity and Thermal Conductivity Coefficients of Gaseous and Liquid Oxygen,* NASA CR-2440, National Aeronautics and Space Administration, Washington, D.C., August 1974 (available from National Technical Information Service, Springfield, Va. 22151).

THERMAL CONDUCTIVITY OF COMPRESSED OXYGEN

In this table, thermal conductivity, λ, is in the unit milliwatt/m-K.

Conversion Factors

T, °K → T, °F: multiply by (9/5) then subtract 459.67
T, °K → T, °C: subtract 273.15
T, °K → T, °R: multiply by (9/5)
P, atm → P, psia: multiply by 14.69595
P, atm → P, N/m²: multiply by 1.01325×10^5

η, g/cm-s → η, Ns/m²: multiply by 10^{-1}
η, g/cm-s → η, lb_m/ft-s: multiply by 0.0671969
λ, W/m-K → λ, cal/cm-s-K: multiply by (1/418.4)
λ, W/m-K → λ, Btu/ft-hr-°R: multiply by 0.578176

T, °K \ P, atm	1	5	10	15	20	25	30	35	40	45	50	55
80	164.7	164.9	165.2	165.4	165.7	165.9	166.2	166.4	166.7	166.9	167.1	167.4
90	151.7	152.0	152.3	152.6	152.9	153.3	153.6	153.9	154.2	154.5	154.8	155.1
100	9.5	138.4	138.8	139.2	139.6	140.0	140.4	140.8	141.1	141.5	141.9	142.3
110	10.4	11.4	124.9	125.3	125.8	126.3	126.8	127.3	127.7	128.2	128.6	129.1
120	11.4	12.2	13.5	110.9	111.5	112.1	112.8	113.3	113.9	114.5	115.1	115.6
130	12.3	13.1	14.2	15.6	96.3	97.1	98.0	98.8	99.6	100.3	101.1	101.8
140	13.2	14.0	15.0	16.1	17.5	19.3	81.6	82.9	84.1	85.2	86.3	87.3
150	14.1	14.8	15.7	16.7	17.9	19.2	20.9	23.3	27.8	68.4	70.4	72.1
160	15.8	15.7	16.5	17.4	18.4	19.5	20.7	22.3	24.2	26.7	30.5	37.3
170	15.9	16.5	17.3	18.1	19.0	19.9	20.9	22.1	23.4	24.9	26.8	29.0
180	16.8	17.4	18.1	18.8	19.6	20.4	21.3	22.3	23.3	24.4	25.7	27.1
190	17.7	18.2	18.9	19.5	20.3	21.0	21.8	22.6	23.5	24.4	25.4	26.4
200	18.5	19.0	19.6	20.3	20.9	21.6	22.3	23.0	23.8	24.6	25.4	26.3
210	19.3	19.8	20.4	21.0	21.6	22.3	22.9	23.6	24.2	24.9	25.7	26.4
220	20.2	20.6	21.2	21.7	22.3	22.9	23.5	24.1	24.7	25.4	26.0	26.7
230	21.0	21.4	21.9	22.5	23.0	23.6	24.1	24.7	25.3	25.9	26.5	27.1
240	21.7	22.2	22.7	23.2	23.7	24.2	24.8	25.3	25.8	26.4	26.9	27.5
250	22.5	22.9	23.4	23.9	24.4	24.9	25.4	25.9	26.4	26.9	27.5	28.0
260	23.3	23.7	24.1	24.6	25.1	25.6	26.0	26.5	27.0	27.5	28.0	28.5
270	24.0	24.4	24.9	25.3	25.8	26.2	26.7	27.1	27.6	28.1	28.5	29.0
280	24.8	25.1	25.6	26.0	26.4	26.9	27.3	27.8	28.2	28.6	29.1	29.5
290	25.5	25.9	26.3	26.7	27.1	27.5	28.0	28.4	28.8	29.2	29.6	30.1
300	26.2	26.6	27.0	27.4	27.8	28.2	28.6	29.0	29.4	29.8	30.2	30.6
310	27.0	27.3	27.7	28.1	28.5	28.8	29.2	29.6	30.0	30.4	30.8	31.2
320	27.7	28.0	28.4	28.7	29.1	29.5	29.9	30.2	30.6	31.0	31.4	31.7
330	28.4	28.7	29.0	29.4	29.8	30.1	30.5	30.9	31.2	31.6	31.9	32.3
340	29.1	29.3	29.7	30.1	30.4	30.8	31.1	31.5	31.8	32.2	32.5	32.9
350	29.7	30.0	30.4	30.7	31.1	31.4	31.7	32.1	32.4	32.8	33.1	33.4
360	30.4	30.7	31.0	31.4	31.7	32.0	32.4	32.7	33.0	33.3	33.7	34.0
370	31.1	31.4	31.7	32.0	32.3	32.7	33.0	33.3	33.6	33.9	34.3	34.6
380	31.8	32.0	32.3	32.7	33.0	33.3	33.6	33.9	34.2	34.5	34.8	35.1
390	32.4	32.7	33.0	33.3	33.6	33.9	34.2	34.5	34.8	35.1	35.4	35.7
400	33.1	33.3	33.6	33.9	34.2	34.5	34.8	35.1	35.4	35.7	36.0	36.3

THERMAL CONDUCTIVITY OF COMPRESSED OXYGEN (Continued)

T, °K \ P, atm	60	65	70	80	90	100	110	120	130	150	175
80	167.6	167.9	168.1	168.6	169.1	169.5	170.0	170.4	170.9	171.8	172.8
90	155.4	155.7	156.0	156.6	157.1	157.7	158.3	158.8	159.4	160.4	161.8
100	142.6	143.0	143.3	144.0	144.7	145.4	146.1	146.8	147.4	148.7	150.3
110	129.5	130.0	130.4	131.2	132.1	132.9	133.7	134.5	135.3	136.8	138.6
120	116.2	115.7	117.2	118.3	119.3	120.3	121.2	122.2	123.1	124.9	127.0
130	102.5	103.2	103.9	105.2	106.4	107.6	108.8	109.9	111.0	113.1	115.6
140	88.3	89.2	90.1	91.8	93.4	95.0	96.4	97.8	99.1	101.6	104.5
150	73.5	74.9	76.2	78.5	80.5	82.5	84.2	85.9	87.5	90.5	93.8
160	50.2	57.8	61.3	65.6	68.6	71.0	73.1	75.1	76.9	80.2	83.9
170	31.7	36.1	[illegible]	47.0	54.7	59.3	62.7	65.3	67.6	71.3	75.3
180	26.7	38.4	32.4	36.9	42.0	47.0	51.3	55.0	58.1	63.0	67.7
190	27.8	28.8	30.1	33.0	36.2	39.6	43.2	46.5	49.7	55.1	60.5
200	27.2	28.1	29.1	31.3	33.6	36.1	38.7	41.4	44.0	48.9	54.3
210	27.2	28.0	28.8	30.5	32.4	34.4	36.4	38.5	40.6	44.9	49.8
220	27.4	28.1	28.8	30.3	31.9	33.5	35.2	36.9	38.7	42.3	46.6
230	27.7	28.3	29.0	30.3	31.7	33.1	34.6	36.1	37.6	40.7	44.5
240	28.1	28.7	29.2	30.5	31.7	33.0	34.3	35.6	36.9	39.7	43.1
250	28.5	29.1	29.6	30.7	31.8	33.0	34.2	35.4	36.6	39.0	42.1
260	29.0	29.5	30.0	31.0	32.1	33.1	34.2	35.3	36.4	38.7	41.5
270	29.5	29.9	30.4	31.4	32.4	33.3	34.4	35.4	36.4	38.4	41.0
280	30.0	30.4	30.9	31.8	32.7	33.6	34.6	35.5	36.5	38.4	40.8
290	30.5	30.9	31.3	32.2	33.1	33.9	34.8	35.7	36.6	38.4	40.6
300	31.0	31.4	31.8	32.7	33.5	34.3	35.1	36.0	36.8	38.5	40.6
310	31.6	31.9	32.3	33.1	33.9	34.7	35.5	36.3	37.1	38.7	40.6
320	32.1	32.5	32.8	33.6	34.3	35.1	35.8	36.6	37.4	38.9	40.8
330	32.7	33.0	33.4	34.1	34.8	35.5	36.2	37.0	37.7	39.1	40.9
340	33.2	33.5	33.9	34.6	35.3	36.0	36.7	37.3	38.0	39.4	41.1
350	33.8	34.1	34.4	35.1	35.8	36.4	37.1	37.7	38.4	39.7	41.4
360	34.3	34.6	35.0	35.6	36.2	36.9	37.5	38.2	38.6	40.1	41.7
370	34.9	35.2	35.5	36.1	36.7	37.9	38.0	38.6	39.2	40.4	42.0
380	35.4	35.7	36.0	36.6	37.2	37.8	38.4	39.0	39.6	40.8	42.3
390	36.0	36.3	36.6	37.2	37.8	38.3	38.9	39.5	40.1	41.2	42.6
400	36.6	36.9	37.1	37.7	38.3	38.8	39.4	40.0	40.5	41.6	43.0

From Hanley, H. J. M., McCarty, R. D., and Sengers, J. V., *Viscosity and Thermal Conductivity Coefficients of Gaseous and Liquid Oxygen,* NASA CR-2440, National Aeronautics and Space Administration, Washington, D.C., August 1974 (available from National Technical Information Service, Springfield, Va. 22151).

TRANSPORT PROPERTIES OF OXYGEN AT SATURATION

Conversion Factors

T, °K → T, °F: multiply by (9/5) then subtract 459.67
T, °K → T, °C: subtract 273.15
T, °K → T, °R: multiply by (9/5)
P, atm → P, psia: multiply by 14.69595
P, atm → P, N/m^2: multiply by 1.01325×10^5
η, g/cm-s → η, Ns/m^2: multiply by 10^{-1}
η, g/cm-s → η, lb_m/ft-s: multiply by 0.0671969
λ, W/m-K → λ, cal/cm-s-K: multiply by (1/418.4)
λ, W/m-K → λ, Btu/ft-hr-°R: multiply by 0.578176

Temperature, °K	Pressure, atm	Viscosity, mg/cm-s		Thermal conductivity, mW/m-K	
		Vapor	Liquid	Vapor	Liquid
80	0.30	0.059	2.652	7.4	164.7
90	0.98	0.068	1.971	8.5	151.8
100	2.51	0.079	1.504	9.9	138.2
110	5.36	0.092	1.188	11.5	124.3
120	10.09	0.108	0.971	13.6	110.3
130	17.26	0.129	0.824	16.3	95.9
140	27.52	0.158	0.696	20.1	80.6
142	30.00	0.166	0.668	21.1	77.4
144	32.64	0.175	0.638	22.3	74.0
146	35.45	0.185	0.607	23.6	70.6
148	38.44	0.197	0.574	25.2	67.0
150	41.61	0.211	0.537	27.2	63.2
152	45.00	0.231	0.494	29.9	58.7
154	48.65	0.269	0.424	35.2	52.0

From Hanley, H. J. M., McCarty, R. D., and Sengers, J. V., *Viscosity and Thermal Conductivity Coefficients of Gaseous and Liquid Oxygen,* NASA CR-2440, National Aeronautics and Space Administration, Washington, D.C., August 1974 (available from National Technical Information Service, Springfield, Va. 22151).

PHYSICAL CONSTANTS OF OZONE AND OXYGEN

Physical Constant	Ozone (O_3)	Oxygen (O_2)
Molecular Weight	47.9982 g/g-mol	31.9988 g/g-mol
Boiling Point (760 mm)	−111.9 ± 0.3°C	−182.97°C
Melting Point	−192.7 ± 0.2°C	−218.4°C
Critical Temperature	−12.1 ± 0.1°C	−118.574°C
Critical Pressure	54.6 atm	49.77 atm
Critical Density	0.437 g/cc	0.436 g/cc
Critical Volume	147.1 cc/mol	73.37 cc/mol
Gas Density (0°C) (760 mm pressure)	2.144 g/liter	1.429 g/liter
Liquid Density		
−112°C	1.358 g/cc	
−183°C	1.571 g/cc	1.14 g/cc
−195.4°C	1.614 g/cc	1.201 g/cc
Surface Tension		
−195°C	43.8 ± 0.1 dyne/cm	
−182.7°C	38.1 ± 0.2 dyne/cm	
−183.0°C	38.4 ± 0.7 dyne/cm	13.2 dyne/cm
Heat Capacity of Liquid		
−183 to −145°C	0.45 cal/g°C	
Heat Capacity of Gas		
−173°C	7.95 cal/g mol°C	
0°C	9.10 cal/g mol°C	
25°C	9.37 cal/g mol°C	
100°C		6.979 cal/g mol°C
127°C	10.44 cal/g mol°C	
Viscosity of Liquid		
−195.6°C	4.14 ± 0.05 cP	
−183.0°C	1.57 ± 0.02 cP	0.1958 cp
Heat of Vaporization		
−112°C	75.6 cal/g	
−182.9°C		50.9 cal/g
Heat of Formation		
25°C	−34.4 kcal/mol	
Free Energy		
25°C	32.4 kcal/mol	
Van der Waals Constant (a)	3.545 atm liter^2/mol^2	1.36 atm liter^2/mol^2
Van der Waals Constant (b)	0.04903 liter/mol	0.03803 liter/mol
Magnetic Susceptibility		
gas (× 10^{-6})	0.002 cgs units	10.6.2 cgs units
liq (× 10^{-6})	0.150 cgs units	260.0 cgs units
Thermal Conductivity of Liquid		
−195.8°C	5.21 cal/sec cm°C × 10^4	
−183.0°C	5.31 cal/sec cm°C × 10^4	
−165.0°C	5.42 cal/sec cm°C × 10^4	
−128.0°C	5.52 cal/sec cm°C × 10^4	

Phase Boundaries—Ozone-Oxygen System

−183°C	29.8 and 72.4 wt % O_3
−195.4°C	9 and 90.8 wt % O_3

Consolute Temperature—Ozone-Oxygen System

−180 ± 0.5°C

Coefficient of Thermal Expansion for Liquid Ozone

Temp. °C	α
−195.6	1.62
−183.0	1.58
−148.0	1.47
−123.0	1.41
−112.0	1.35
−98.8	1.31

PHYSICAL CONSTANTS OF CLEAR FUSED QUARTZ

Based on information contained in Fused Quartz Catalogue Q-7A General Electric Company.

Property		Clear fused quartz
Density		2.2 g cc^{-1}
Hardness		4.9 (Mohs')
Tensile Strength		7,000 psi
Compressive Strength		> 160,000 psi
Bulk Modulus	(approx.)	5.3 × 10^6 psi
Rigidity Modulus		4.5 × 10^6 psi
Young's Modulus		10.4 × 10^6 psi
Poisson's Ratio		.16
Coefficient of Thermal Expansion	(av.)	5.5 × 10^{-7} cm cm^{-1} °C^{-1} {20°C, 320°C}
Thermal Conductivity		0.0033 cal cm^{-1} sec^{-1} °C^{-1} cm^{-1}
Specific Heat		0.18 cal g^{-1}
Softening Point	(approx.)	1665°C

Property		Clear fused quartz
Annealing Point	(approx.)	1140°C
Strain Point		1070°C
Electrical Resistance		9.5 $\log_{10}$ R for cm.^{3} at 350°C
Dielectric Constant		3.75 at 20°C. 1 MHz
Dielectric Loss Factor	less than	.0004 at 20°C. 1 MHz
Dissipation Factor	less than	.0001 at 20°C. 1 MHz
Index of Refraction		1.4585
Velocity of Sound—Shear Wave		3.75 × 10^5 cm sec^{-1}
Velocity of Sound—Compressional Wave		5.90 × 10^5 cm sec^{-1}
Sonic Attenuation	less than	.033 db ft^{-1} MHz^{-1}

FIXED POINT PROPERTIES OF OXYGEN

From NASA SP-3071, "ASRDI Oxygen Technology Survey, Thermophysical Properties", Volume I (1972), edited by Hans M. Roder and Lloyd A. Weber. This NASA publication contains an extensive bibliography and a discussion of basis for selection of these data for oxygen. The publication is available from the National Technical Information Service, Springfield, Virginia 22151.

		Triple Point		Normal Boiling Point			Standard Conditions	
PROPERTIES ↓ CONDITIONS →	Solid	Liquid	Vapor	Liquid	Vapor	Critical Point ††	STP (0°C)	NTP (20°C)
Temperature (K)		54.351		90.180		154.576	273.15	293.15
Pressure (mmHg)		1.138		760		37,823	760	760
Density (mole/cm³) x 10³	42.46	40.83	0.000336	35.65	0.1399	13.63	0.04466	0.04160
Specific Volume (cm³/mole) x 10^{-3}	0.02355	0.02449	2973	0.028047	7.1301	0.07337	22.392	24.038
Compressibility Factor, $Z = \frac{PV}{RT}$	–	0.0000082	0.9986	0.00379	0.9662	0.2879	0.9990	0.9992
Heats of Fusion & Vaporization (J/mole)	444.8	7761.4		6812.3		0	–	–
Specific Heat (J/mole-K) C_s, @saturation	46.07	53.313	–108.7	54.14	–53.2	(very large)	–	–
C_p, @ constant pressure	–	53.27	29.13	54.28	30.77	(very large)	29.33	29.40
C_v, @ constant volume	–	35.65	20.81	29.64	21.28	(38.7)	20.96	21.04
Specific Heat Ratio, $\gamma = C_p/C_v$	–	1.494	1.400	1.832	1.446	(large)	1.40	1.40
Enthalpy (J/mole)	–6634.4	–6189.6	1571.8	–4270.3	2542.0	1032.2	7937.8	8525.1
Internal Energy (J/mole)	–6634.4	–6189.6	1120.0	–4273.1	1817.5	662.3	5668.9	6089.5
Entropy (J/mole-K)	58.92	67.11	209.54	94.17	169.68	134.42	202.4	204.5
Velocity of Sound (m/sec)		1159	141	903	178	164	315	326
Viscosity, μ (N-sec/m²) × 10³	–	0.6194	0.003914	0.1958	0.00685	(0.031)	0.01924	0.02036
(centipoise)‡‡	–	0.6194	0.003914	0.1958	0.00685	(0.031)	0.01924	0.02036
Thermal Conductivity (mW/cm-K), k		1.929	0.04826	1.515	0.08544	(*)	0.2428	0.2575
Prandtl Number, $N_{pr} = \mu C_p/k$	–	5.344	0.7392	2.193	0.7714		0.7259	0.7265
Dielectric Constant, ϵ	(1.614)	1.5687	1.000004	1.4870	1.00166	1.17082	1.00053	1.00049
Index of Refraction, $n = \sqrt{\epsilon}$ †	(1.271)	1.2525	1.000002	1.219	1.00083	1.0820	1.00027	1.00025
Surface Tension (N/m) x 10³		22.65		13.20		0		
Equiv. Vol./Vol. Liquid at NBT	0.8397	0.8732	106,068	1	254.9	2.616	798.4	857.1

† Long Wavelengths
* Anomalously Large

Gas Constant: R = 62, 365.4 cm³-mm Hg/mole-K[1]
†† Values in parenthesis are estimates

Molecular Weight = 31.9988[3]
"mole" = gram mole
‡‡ Units for poise are: g/cm-sec

FIXED POINTS AND PHASE EQUILIBRIUM BOUNDARIES FOR PARAHYDROGEN

a. Triple point
P_t = 0.0704 bar
T_t = 13.803 K
ρ_t (liquid) = 38.21 mol/ℓ

b. Normal boiling point
P_b = 1.01325 bar
T_b = 20.268 K
ρ_b (liquid) = 35.11 mol/ℓ
ρ_b (gas) = 0.6636 mol/ℓ

c. Critical point
P_c = 12.928 bar
T_c = 32.976 K
ρ_c = 15.59 mol/ℓ

Note: Some data indicate that the true critical temperature is probably closer to 32.93 K. However, that value is pending further verfication.

d. Melting pressures: in atmospheres

$$P = P_t + (T - T_t)\,[A_1 e^{-\alpha/T} + A_2 T]$$

A_1 = 30.3312
A_2 = 0.6667
α = 5.693

e. Liquid-vapor coexistence densities
Liquid, density in mol/cm³:

$$\rho \text{ sat } \ell = \rho_c + A_1(\Delta T)^{0.380} + A_2(\Delta T) + A_3(\Delta T)^{4/3} + A_4(\Delta T)^{5/3} + A_5(\Delta T)^2$$

A_1 = 7.323 4603 × 10^{-3}
A_2 = –4.407 4261 × 10^{-4}
A_3 = 6.620 7946 × 10^{-4}
A_4 = –2.922 6363 × 10^{-4}
A_5 = 4.008 4907 × 10^{-5}
$\Delta T = T_c - T$

Vapor $T_b \leq T \leq T_c$, density in mol/cm³:

$$\rho \text{ sat } G = \rho_c + A_1(\Delta T)^{0.370} + A_2(\Delta T) + A_3(\Delta T)^{0.7} + A_4(\Delta T)^{0.8}$$

A_1 = –7.196 7724 × 10^{-3}
A_2 = 1.449 5527 × 10^{-3}
A_3 = 3.240 3120 × 10^{-3}
A_4 = –4.464 0177 × 10^{-3}

f. Vapor pressure: in atmospheres

For $T \leq 29$ K:

$$\log_{10} P_a = A_1 + \frac{A_2}{T + A_3} + A_4 T$$

A_1 = 2.000 620
A_2 = −50.09 708
A_3 = 1.0044
A_4 = 1.748 495 × 10^{-2}

For $T > 29$ K:

$$P = P_a + A_5 (T - 29)^3 + A_6 (T - 29)^5 + A_7 (T - 29)^7$$

A_5 = 1.317 × 10^{-3}
A_6 = −5.926 × 10^{-5}
A_7 = 3.913 × 10^{-6}

From Weber, L. A., Thermodynamic and Related Properties of Parahydrogen from the Triple Point to 300 K at Pressures to 1000 Bar, NASA SP-3008, NBSIR 74-374, 1975.

PHYSICAL PROPERTIES OF SODIUM, POTASSIUM AND Na-K ALLOYS

Temperature °C	Density (g/cm³)								Viscosity (centipoise)		
	Na			K	Alloys (wt % K)				Na	Alloys (wt % K)	
					43.4		78.6			43.3	66.0
	(a)	(b)	(c)		Experimental	(d) Calculated	Experimental	(d) Calculated			
100	.927	.927	.9265	.819	.887	.890	.847	.850	.705	.540	.529
200	.904	.904	.9037	.795	.862	.867	.823	.827	.450	.379	.354
300	.882	.880	.8805	.771	.838	.843	.799	.802	.345	.299	.276
400	.859	.856	.8570	.747	.814	.818	.775	.778	.284	.245	.229
500	.834	.831	.8331	.723	.789	.794	.751	.754	.234	.207	.195[e]
600	.809	.808	.8089	.701	.765	.771	.727	.732	.210	.178[e]	.168[e]
700	.783	.784	.....	.676	.740	.745	.703	.705	.186[e]	.257[e]	.146[e]
800	.757	.760	.....	....	....	....	....	....	.165[e]	......	......
900	....	....	.....	....	....	....	....	....	.150[e]	......	......

Temperature, °C	Thermal conductivity (watts/cm²-°C/cm)				Electrical resistivity (microohms)			Heat capacity[g] (cal/°C-g)		
	Na		Alloy (wt % K)		Na[f]	Alloy (wt % K)		K	Alloy (wt % K)	
	Experimental	Calculated	56.5	77.0		56.5	78.0		44.8	78.26
100	......	....	....	.238[h]	8.99	41.61	45.63	.1940	.2690	.2248
200	.815	.808	.249	.247	13.52	47.23	51.33	.1887	.2612	.2169
300	.757	.755	.262	.259	17.52	54.33	58.58	.1894	.2553	.2122
400	.712	.710	.269	.262	21.93	62.21	65.65	.126	.2512	.2097
500	.668	.672	.271	.259	26.96	69.37	73.48	.1818	.2498	.2088
600	.627[e]	.639	....	.255	32.65	78.29	82.61	.1825	.2484	.2092
700	.590[e]	.610	....	......	39.05	88.23	91.76	.1846	.2497	.2108
800	.547[e]	.583	....	......	46.15	99.68	104.51	.1883	.2529	.2133
900	......	....	....	......	.....	.....	......	......	.....	.....

Vapor Pressure, mm Hg

Temperature °C	Na	K	Alloys (wt % K)	
			56	78
127	2.23 × 10^{-6}	....	....	....
227	1.15 × 10^{-3}	2.88 × 10^{-2}	1.57 × 10^{-3}	1.81 × 10^{-3}
327	5.03 × 10^{-2}	9.27 × 10^{-1}	5.73 × 10^{-2}	6.14 × 10^{-2}
427	.881	9.26	3.53	5.06
527	7.53	52.22	23.0	31.87
627	39.98	201.25	101.35	136.50
727	148.5	588.62	328.7	431.85
827	453.7	1421.0	864.2	1099.52
927	1127.8	....	....	....
1027	2522.4	....	....	....
1127	4696.8	....	....	....

NOTES: (a) From plotted data.
(b) Epstein equation: $d_t = 0.9514 - 2.392 \times 10^{-4}\, t$°C.
(c) Thomson and Garelis: $d_t = 0.9490 - 22.3 \times 10^{-5} t$°C $- 1.75 \times 10^{-8} t^2$°C.
(d) Formula to calculate density: $V = M_K \cdot V_K + M_{Na} \cdot V_{Na}$ (where V, V_K and V_{Na} are the specific volumes (reciprocal of density) of the alloy, K and Na respectively, M_K and M_{Na} the mole fraction of the elements).
(e) Extrapolated by calculation, Epstein equation:

$$K = \frac{2.433 \times 10^{-2}(t + 273.16)}{6.8393 + 3.3873 \times 10^{-2} t + 1.7235 \times 10^{-5} t^2}$$

(f) Epstein equation: $r_t = 10.892 + 0.015272t + 3.6746 \times 10^{-5} t^2 - \frac{379.26}{t}$.
(g) Formula to calculate heat capacity: $C = W_{Na} \cdot C_{Na} + W_K \cdot C_K$ (where C, C_{Na} and C_K are the heat capacity of the alloy, Na and K respectively, W_{Na} and W_{Ka} the weight fractions of Na and K respectively in the alloy).
(h) 150°C.

MECHANICAL AND PHYSICAL PROPERTIES OF WHISKERS

From NASA SP-5055

This table lists some nominal values for physical and mechanical properties of some whiskers. The strength of whiskers is influenced by temperature, time, surface conditions, surface films or corrosion, crystallographic orientation, impurities and testing techniques.

Material	Tensile Strength (σt) lb/in.² × 10^{-6}	Young's Modulus (E) lb/in.² × 10^{-6}	Specific Gravity (S) lb/in.² × 10^{-6}	(σt)/(S) inch	(E)/(S) inch
Graphite	3.0	...	137	21,700	......
Al_2O_3	3.0	76	250	12,000	300,000
Iron	1.8	28	485	3,700	58,000
Si_3N_4	2.0	55	193	10,000	285,000
SiC	3.0	70	187	16,000	380,000
Si	1.1	26	143	7,000	182,000

ABSOLUTE VISCOSITY OF LIQUID SODIUM AND POTASSIUM

Sodium

T (°K)	η (10^{-2} poise)	ν (cm³/g)	$\eta\nu^{1/3}$ (10^3 poise cm/g$^{-1/3}$)	$1/T\nu$ [10^3g/(cm³ · °K)]
Experimental range				
371.00	0.690	1.078,75	7.0766	2.4987
473	.450	1.106,56	4.6544	1.9106
573	.340	1.135,72	3.5482	1.5366
673	.278	1.166,86	2.9268	1.2734
773	.239	1.200,34	2.54057	1.0776
873	.212	1.236,25	2.2754	0.9266
973	.193	1.274,37	2.0925	.8065
1073	.179	1.315,79	1.9615	.7083
1173	.167	1.360,54	1.8505	.6266
1203	.164	1.373,62	1.8230	.6052

Potassium

T (°K)	η (10^{-2} poise)	ν (cm³/g)	$\eta\nu^{1/3}$ (10^3 poise cm/g$^{-1/3}$)	$1/T\nu$ [10^3g/(cm³ · °K)]
Experimental range				
336.9	0.560	1.2062_7	5.961_2	2.4606
400	.384	1.22911	4.113_4	2.0340
500	.276	1.26711	2.986_6	1.5784
600	.221	1.3075_3	2.416_6	1.2747
700	.185	1.3506_2	2.045_0	1.0577
800	.162	1.3966_5	1.810_8	0.8949_7
900	.147	1.4459_2	1.6623	$.7684_6$
1000	.132	1.4988_0	1.5106	$.6672_0$
1100	.121	1.5556_9	1.4020	$.5843_5$
1200	.113	1.6170_7	1.3264	$.5153_3$
1300	.106	1.6835_0	1.2610	$.4569_1$
1400	.100	1.7556_1	1.2064	$.4068_7$

PHYSICAL PROPERTIES OF PIGMENTS

From Rutherford J. Gettens and George L. Stout. *Painting Materials*, Dover Publications, Inc., New York, 1966. Reprinted by permission of the Publisher.

OPAQUE WHITE PIGMENTS

Pigment Name and Chemical Composition[1]	Specific Gravity[2]	Particle Characteristics[3]	Refractive Index[4]
Titanium calcium white, TiO_2 (25%) + $CaSO_4$ (75%)	3.10	prism. or ragged gr.	mostly 1.8–2.0 (irr.) (bi.) [M*]
Titanium dioxide (rutile) + $CaSO_4$	3.25	...	Av. 1.98 [TPC]
Titanium barium white, TiO_2 (25%) + $BaSO_4$ (75%)	4.30	min. round. gr.	$n_{\Sigma C}$ 1.7–2.5 [M*]
White lead (basic sulfate) $PbSO_4 \cdot PbO$	6.46	...	Av. 1.93 [TPC]
Lithopone	4.3	...	Av. 1.84 [TPC]
Lithopone (regular), ZnS (28–30%), $BaSO_4$ (72–70%)	4.30	fine comp. gr.	2.3 (ZnS)–1.64 ($BaSO_4$) [M]
Zinc white (ordinary), ZnO	5.65	v. fine cryst. gr.	ϵ2.02, ω2.00 [M]
(acicular), ZnO	...	spicules, fourlets	ϵ2.02, ω2.00 [M*]
White lead (basic carbonate), $2PbCO_3 \cdot Pb(OH)_2$	6.70	v. fine cryst.	ϵ1.94, ω2.09 [M]
Antimony oxide, Sb_2O_3	5.75	v. fine cryst.	valentinite, α2.18, γ and β2.35 [LB, M*] senarmonite, 2.09 (isot.)
Zirconium oxide (baddeleyite), ZrO_2	5.69	...	χ2.13, γ2.20, β2.19 [LB], Av. 2.40 [TPC]
Titanium dioxide (anatase), TiO_2	3.9	min. round. gr.	ϵ and ω 2.5 (w. bi.) [M*]
(rutile), TiO_2	4.2	round. or prism. gr.	ϵ2.9, ω2.6 [M*]

PHYSICAL PROPERTIES OF PIGMENTS (Continued)

TRANSPARENT WHITE PIGMENTS†

Diatomaceous earth, SiO_2	2.31	min. fossil forms	n mostly 1.435, some 1.40 [M*]
Aluminum stearate, $Al(C_{18}H_{35}O_2)_3$	0.99	agg. of spher. gr.	1.49 (w. bi.) [W]
Pumice (volcanic glass), Na, K, Al, silicate	...	vesicular vitr. frag.	c 1.50 (isot.) [M*]
Aluminum hydrate, $Al(OH)_3$	2.45	v. fine amorph. part.	$n_{\Sigma}c$ 1.50–1.56 [M*]
Gypsum, $CaSO_4 \cdot 2H_2O$	2.36	fine cryst. gr.	α1.520, γ1.530, β1.523 [LB]
Silica quartz), SiO_2	2.66	cryst. frag.	ϵ1.553, ω1.544 [LB]
(chalcedony), SiO_2	2.6	crypt. agg.	ϵ, ω1.54 [LB, M*]
China clay (kaolinite), $Al_2O_3 \cdot 2SiO_2 \cdot 2H_2O$	2.60	fine, vermicular cryst.	α1.558, γ1.565, β1.564 (all ± .005) [LB, M*]
Talc, $3MgO \cdot 4SiO_2 \cdot H_2O$	2.77	platy frag.	α1.539, γ1.589, β1.589 [LB]
Mica (muscovite), $H_2KAl_3(SiO_4)_3$	2.89	platy frag.	α1.563, γ1.604, β1.599 [LB]
Anhydrite, $CaSO_4$	2.93	cryst. frag.	α1.570, γ1.614, β1.575 [LB]
Chalk (whiting), $CaCO_3$	2.70	hollow spherulites	$\epsilon_{\Sigma}c$ 1.510, $\omega_{\Sigma}c$ 1.645 [M*]
Barytes (barite, nat.), $BaSO_4$	4.45	cryst. frag.	α1.636, γ1.648, β1.637 [LB]
(blanc fixe, art.), $BaSO_4$	4.36	v. fine cryst. agg.	1.62–1.64 [M*]
Barium carbonate, $BaCO_3$ (witherite)	4.3	...	α1.529, γ1.677, β1.676 [LB]

†Also called "Extender" or "Inert" White Pigments.

IRON OXIDE PIGMENTS

Ochre, yellow (goethite), $Fe_2O_3 \cdot H_2O$, clay, etc.	2.9–4.0	irr. spherulites	n_{Σ}2.0 (isot. part); (α, β)$_{\Sigma}$ 2.05–2.31; γ_{Σ} 2.08–2.40 (bi. part) [M*]
Sienna, raw (goethite), $Fe_2O_3 \cdot H_2O$, clay, etc.	3.14	uneven spherulites	1.87–2.17 (mostly 2.06) (isot.) [M*]
Sienna, burnt, Fe_2O_3, clay, etc.	3.56	uneven, round. part.	c1.85 (var.) (isot.) [M]
Umber, raw, $Fe_2O_3 + MnO_2 + H_2O$, clay, etc.	3.20	uneven, round. gr.	mostly 1.87–2.17 [M*]
Umber, burnt, $Fe_2O_3 + MnO_2$, clay, etc.	3.64	uneven, round. gr.	mostly 2.2–2.3 [M*]
Iron oxide red (haematite), Fe_2O_3	5.2	min. cryst.	ϵ_{Li} 2.78, ω_{Li} 3.01 [M]

RED AND ORANGE PIGMENTS

Pigment Name and Chemical Composition[1]	Specific Gravity[2]	Particle Characteristics[3]	Refractive Index[4]
Red lead, Pb_3O_4 (c95%)	8.73	crypt. agg.	2.42$_{Li}$ (w. bi.; pleo.) [M]
**Realgar, As_2S_2	3.56	cryst. frag.	α_{Li} 2.46, γ_{Li} 2.61, β_{Li} 2.59 [LB]
Molybdate orange, $Pb(Mo,S,Cr,P)O_4$	...	min. round. gr.	β_{Li} 2.55 (s. bi.) [M*]
Chrome orange, $PbCrO_4 \cdot Pb(OH)_2$	6.7	tabular cryst.	α2.42, γ2.7 +, β2.7 [M*]
Cadmium red lithopone, $CdS(Se) + BaSO_4$	4.30	min. round. gr.	2.50–2.76 (for CdS(Se)part) (isot.) [M*]
Cadmium red, CdS(Se)	4.5	min. round. gr.	2.64 (bright red)–2.77 (deep red) (isot.) [M*]
Antimony vermilion, Sb_2S_3	...	v. fine red glob.	n_{Li} 2.65 (isot.) [M*]
Vermilion (art.), HgS	8.09	hexagonal gr. and prisms	ϵ_{Li} 3.14, ω_{Li} 2.81 [M]
**(nat., cinnabar), HgS	8.1	cryst. frag.	ϵ_{Li} 3.146, ω_{Li} 2.819 [LB]
Quinacridone red, $C_{20}H_{12}O_2N_2$ (gamma)	1.5	thin plates	n_{Na} Av. 2.04 [Du P]

YELLOW PIGMENTS

**Gamboge, organic resin	...	irr. amorph. part.	1.582–1.586 [W]
**Indian yellow, $C_{19}H_{18}O_{11}Mg \cdot 5H_2O$	...	prisms, plates	1.67 (w. bi.) [M*]
Cobalt yellow, $CoK_3(NO_2)_6 \cdot H_2O$	...	fine dendritic cryst.	1.72–1.76 (isot.) [W]
Zinc yellow, $4ZnO \cdot 4CrO_3 \cdot K_2O \cdot 3H_2O$	3.46	min. spher. gr.	1.84–1.9 (irr.; bi.) [M*]
Strontium yellow, $SrCrO_4$	...	small needles	α, β (or ω) 1.92, γ (or ϵ) 2.01 (‖ext.) [M*]
Barium yellow, $BaCrO_4$	4.49	v. fine cryst. gr.	1.94–1.98 (bi.) [M]
**Naples yellow, $Pb_3(SbO_4)_2$	...	round. gr.	2.01–2.28 (isot.) [M*]
Chrome yellow (med.), $PbCrO_4$	5.96	fine prism. gr.	$\alpha_{620m\mu} < 2.31$, $\gamma_{650m\mu}$ 2.49 [M]
Cadmium yellow lithopone, $CdS + BaSO_4$	4.25	fine comp. gr.	2.39–2.40 (for CdS part) [M*]
Cadmium yellow, CdS	4.35	min. round. gr.	2.35–2.48 (isot.) [M*]
Massicot (litharge), PbO	9.40	min. flakes	α_{Li} 2.51, γ_{Li} 2.71, β_{Li} 2.61 [M]
**Orpiment, As_2S_3	3.4	min. flakes	α_{Li} 2.4 ±, γ_{Li} 3.02, β_{Li} 2.81 [LB]

PHYSICAL PROPERTIES OF PIGMENTS (Continued)

GREEN PIGMENTS

Phthalocyanine green, chloro-copper phthalocyanine	2.1	laths	$n_{580m\mu}$ 1.40 [ACC]
**Verdigris (copper basic acetate), $Cu(C_2H_3O_2)_2 \cdot 2Cu(OH)_2$	...	cryst. frag.	α1.53, γ1.56 [M]
**Chrysocolla, $CuSiO_3 \cdot \eta H_2O$	2.4	crypt. agg.	α1.575, γ1.598, β1.597 [LB]
Green earth (celadonite and glauconite), Fe, Mg, Al, K, hydrosilicate	2.5–2.7	round. irr. gr.	n var. c 1.62, (porous) [M*]
Emerald green (Paris green), $Cu(C_2H_3O_2)_2 \cdot 3Cu(AsO_2)_2$	3.27	spherulites and disks	αΣ1.71, γΣ1.78 (w. pleo.) [M*]
Malachite, $CuCO_3 \cdot Cu(OH)_2$	4.0	cryst. frag.	α1.655, γ1.909, β1.875 [LB]
Cobalt green, $CoO \cdot \eta ZnO$	...	spher. gr.	1.94–2.0 (w. bi.) [M*]
Viridian (chromium oxide, transparent), $Cr_2O_3 \cdot 2H_2O$	3.32	spherul. gr.	α, βΣ 1.82, γΣ 2.12 [M*]
Chrome green (med.), $Fe_4[Fe(CN)_6]_3 + PbCrO_4$	4.06	fine green agg.	c 2.4 (cf. Prussian blue and chrome yellow)
Chromium oxide green, opaque, Cr_2O_3	5.10	fine cryst. agg.	n_{Li} 2.5 [M]

BLUE PIGMENTS

Pigment Name and Chemical Composition[1]	Specific Gravity[2]	Particle Characteristics[3]	Refractive Index[4]
Phthalocyanine blue, copper phthalocyanine	1.6	laths	Av. 1.38 [DuP]
Ultramarine blue (art.), $Na_{8-10}Al_6Si_6O_{24}S_{2-4}$	2.34	uniform small round. gr.	n 1.51 $_{green}$, 1.63 $_{red}$ (isot.) [M]
**(nat., lazurite), $3Na_2O \cdot 3Al_2O_3 \cdot 6SiO_2 \cdot 2Na_2S$	2.4	angular, broken frag.	1.50± (isot.) [LB]
**Maya blue, Fe, Mg, Ca, Al, silicate (?)	...	porous irr. agg.	βΣ1.54 (irr.; bi. and pleo.) [M*]
Smalt, K, Co(Al), silicate (glass)	...	splintery, vitr. frag.	1.49–1.52 [M*]
Prussian blue, $Fe_4[Fe(CN)_6]_3$	1.83	colloidal agg.	$1.56_{460m\mu}$ [M*]
**Egyptian blue, $CaO \cdot CuO \cdot 4SiO_2$	...	cryst. frag.	ε1.605, ω1.635 [APL]
Manganese blue, $BaMnO_4 + BaSO_4$	...	gr. and stubby prisms	c 1.65 [W]
**Blue verditer, $2CuCO_3 \cdot Cu(OH)_2$	...	fibrous agg.	αΣ1.72, γΣslightly > 1.74 [M*]
Cobalt blue, $CoO \cdot Al_2O_3$	3.83	round. gr.	n var.; max. c 1.74_{blue} (isot.) [M]
**Azurite, $2CuCO_3 \cdot Cu(OH)_2$	3.80	cryst. frag.	α1.730, γ1.838, β1.758 [LB]
Cerulean blue, $CoO \cdot \eta SnO_2$	...	round gr.	1.84 (isot.) [M*]

VIOLET PIGMENTS

Ultramarine violet	...	round. gr. (blue, rose and violet)	c 1.56 (isot.) [M*]
Cobalt violet, $Co_3(PO_4)_2$	...	round. gr.	ε1.65–1.79 (dull violet), ω1.68–1.81 (salmon) (s. bi.) [M*]
Manganese violet, $(NH_4)_2Mn_2(P_2O_7)_2$	...	fine cryst. gr.	α1.67, γ1.75, β1.72 (for violet) [M]
Quinacridone violet, $C_{20}H_{12}O_2N_2$ (beta)	1.5	thin plates	Av. 2.02 [DuP]

BROWN PIGMENTS

Sepia (organic)	...	angular frag.	(opaque) [M*]
Asphaltum (bitumen)	...	irr. amorph. part.	1.64–1.66 [M*]
Van Dyke brown (bituminous earth)	1.66	irr. amorph. part.	1.62–1.69 [M*]

BLACK PIGMENTS

Bone black, $C + Ca_3(PO_4)_2$	2.29	irr. coarse grains	1.65–1.70 (for larger translucent gr.) [M]
Lamp black, C	1.77	min. round. part.	(opaque)
Charcoal black, C	...	irr. splintery part.	(opaque)
Graphite, C	2.36	irr. plates	(opaque) [M]

[1]Abbreviations: art. = artificial; med. = medium; nat. = natural. The chemical formulas are those commonly accepted in chemical and mineralogical literature, but they may not compare exactly with structural formulas based on x-ray diffraction data or even on critical chemical analysis.

[2]The figures for specific gravity of the artificial pigments are mainly from H. A. Gardner, pp. 710–712, and those on the mineral pigments are chiefly from E. S. Larsen and H. Berman.

[3]Symmetry terms (monoclinic, orthorhombic, etc.) are omitted because pigments are so finely divided that it is rare when observations on crystal symmetry can be made. The term "spherulitic," as used here means aggregates that tend toward radial structure and spherical shape. "Amorphous" describes materials that are microscopically formless but may be truly crystalline on the basis of x-ray diffraction data. Abbreviations: agg. = aggregate(s); amorph. = amorphous; comp. = composite; crypt. = cryptocrystalline; cryst. = crystal(s); frag. = fragment(s); glob. = globule(s); gr. = grain(s); irr. = irregular; min. = minute; part. = particle(s); prism. = prismatic; round. = rounded; spher. = spheroidal; spherul. = spherulitic; var. = variable; v. = very; vitr. = vitreous.

[4]Unless otherwise indicated, all refractive index measurements are by sodium light. Σ is the symbol used by H. E. Merwin to indicate greater or less indefiniteness or irregularity in the case of aggregates, especially in respect to refractive index. Abbreviations: bi. = birefringent; c = *circa*; ext. = extinction; isot. = isotropic; ‖ = parallel; pleo. = pleochroic; s. = strongly; w. = weakly. The letters in brackets refer to the authorities for the refractive index data: M = H. E. Merwin; M* = H. E. Merwin, data by private communication, hitherto unpublished; W = C. D. West data by private communication, hitherto unpublished; LB = E. S. Larsen and H. Berman; APL = A. P. Laurie and co-authors; ACC = A. C. Cooper, "The refractive index of organic pigments. Its determination and significance," *Journal Oil & Colour Chemists Association*, Vol. 31 (1948), pp. 343–357; TPC = Titanium Pigment Corporation; Du P = E. I. Du Pont de Nemours & Co.

**Chiefly of historical interest.

CRITICAL TEMPERATURES AND CRITICAL PRESSURES OF THE ELEMENTS[a]

Critical temperatures are listed as degrees kelvin and critical pressures as megapascals. Some conversion factors which may be useful are:

Megapascals	×	9.8692	=	atmospheres (760mm Hg)
Megapascals	×	14.504	=	pounds per square inch
Megapascals	×	10^{-6}	=	newtons per square meter (pascals)
Megapascals	×	10.1972	=	kilograms per square centimeter
Megapascals	×	10	=	bars
Megapascals	×	7.501	×	10^3 = millimeters Hg at 0°C
Megapascals	×	4.014	=	inches of H_2O at 4°C

Element	Symbol	T_c, K	P_c, MPa
Argon	Ar	150.8	4.87
Arsenic	As	1673	22.3
Bromine	Br_2	588	10.3
Chlorine	Cl_2	416.9	7.977
Deuterium (equilibium)	D_2	38.2	1.65
Deuterium (normal)	D_2	38.4	1.66
Fluorine	F_2	144.3	5.215
Helium-3	He^3	3.31	0.114
Helium-4	He^4	5.19	0.227
Hydrogen (equilibium)	H_2	32.98	1.293
Hydrogen (normal)	H_2	33.2	1.297
Hydrogen deuteride	HD	36.0	1.48
Iodine	I_2	819	—
Mercury	Hg	1765	151.0
Krypton	Kr	209.4	5.50
Neon	Ne	44.40	2.76
Nitrogen	N_2	126.2	3.39
Oxygen	O_2	154.58	5.043
Ozone	O_3	261.1	5.57
Phosphorus	P	994	—
Radon	Rn	377	6.28
Rubidium	Rb	2105	—
Selenium	Se	1766	27.2
Sulfur	S	1314	20.7
Tritium	T_2	40.0	—
Xenon	Xe	289.73	5.840

[a] The data are from "Vapor-Liquid Critical Properties", March, 1978, D. Ambrose and R. Townsend, National Physical Laboratory, Teddington, Middlesex TW11 OLW, UK. These data are used with permission of the copyright owner.

PHYSICAL CONSTANTS OF LIQUID ALKALI METALS

Element	mp, °K	bp, °K	T_c, °K	P_c, MPa	Heat of fusion, KJ mol^{-1}	Heat of vaporization KJ mol^{-1}
Cesium	301.55	942.4	2051 ± 10	11.73	2.14	66
Lithium	453.69	1615	3223 ± 600	68.9 ± 20%	2.99	134.7
Potassium	336.4	1033	2220 ± 25	16.39 ± 0.02	2.33	77.08
Rubidium	312.04	959.2	2083 ± 20	14.54 ± 0.5	2.34	69
Sodium	370.96	1156	2508.7 ± 12.5	25.64 ± 0.02	2.6	89.6 ± 0.5%

CRITICAL TEMPERATURES AND PRESSURES

Compiled by Rudolf Loebel

Table I —Organic compounds

Table II—Inorganic compounds

Table I

Formula	Name	Critical temp. T_c °C	Critical press. P_c atm.
$CHClF_2$	Methane, monochlorodifluoro-	96	48.5
$CHCl_2F$	Methane, dichloromonofluoro-	178.5	51
$CHCl_3$	Methane, trichloro- (Chloroform)	263	54
CHF_3	Methane, trifluoro- (Fluoroform)	25.9	46.9
CH_2Cl_2	Methylene chloride	237	60
CH_3NO_2	Methane, nitro-	314.8	62.3
CH_3Br	Methane, monobromide-	194	83.4
CH_3Cl	Methane, monochloro-	143.8	65.9
CH_3F	Methane, monofluoro-	44.6	58
CH_3I	Methane, monoiodo-	254.8	72.7
CH_4	Methane	−82.1	45.8
CH_4O	Methanol (Methyl alcohol)	240	78.5
CH_4S	Methylmercaptan	196.8	71.4
CH_5N	Methylamine	156.9	40.2
$CBrF_3$	Methane, monobromotrifluoro-	67	50.3
$CClF_3$	Methane, monochlorotrifluoro-	28.85	38.2
CCl_2F_2	Methane, dichlorodifluoro-	111.5	39.6
CCl_3F	Methane, trichloromonofluoro-	198	43.2
CCl_4	Methane, tetrachloro- (Carbon tetrachloride)	283.1	45
CF_4	Methane, tetrafluoro-	−45.7	41.4
C_2H_2	Acetylene	35.5	61.6
C_2H_2	Ethyne (see Acetylene)	35.5	61.6
$C_2H_2F_2$	Ethylene, 1,1-difluoro-	30.1	—
C_2H_3N	Acetonitrile	274.7	47.7
$C_2H_3F_3$	Ethane, 1,1,1-trifluoro-	73.1	—
C_2H_4	Ethene	9.9	50.5
C_2H_4	Ethylene (see Ethene)	9.9	50.5
C_2H_4O	Acetaldehyde	187.8	54.7
C_2H_4O	Ethylene oxide	195.8	71
$C_2H_4O_2$	Acetic acid	321.6	57.1
$C_2H_4O_2$	Formic acid, methyl- (Methyl formate)	214	59.2
$C_2H_4Cl_2$	Ethane, 1,1-dichloro-	249.8	50
$C_2H_4F_2$	Ethane, 1,1 difluoro	386.7	
C_2H_5Br	Ethane, monobromo-	230.8	61.5
C_2H_5Cl	Ethane, monochloro-	187.2	52
C_2H_5F	Ethane, monofluoro-	102.16	49.6
C_2H_6	Ethane	32.2	48.2
C_2H_6O	Ether, dimethyl-	127	52.6
C_2H_6O	Ethanol (Ethyl alcohol)	243	63
$C_2H_6O_2$	Glycol, ethylene-	(374)*	
C_2H_6S	Dimethylsulfide	229.9	54.5
C_2H_6S	Ethylmercaptan	225.5	54.2
C_2H_7N	Dimethylamine	164.6	52.4
C_2H_7N	Ethylamine	183.2	55.5
$C_2Br_2F_4$	Ethane, dibromotetrafluoro-	214.5	—
C_2ClF_5	Ethane, chloropentafluoro-	80	—
$C_2Cl_2F_4$	Ethane, 1,2-dichlorotetrafluoro-	145.7	—
$C_2Cl_3F_3$	Ethane, trichlorotrifluoro-	214.2	33.7
$C_2Cl_4F_2$	Ethane, tetrachlorodifluoro-	278	32.9
C_2F_6	Ethane, hexafluoro-	24.3	—
C_3H_4	Propadiene	120	43.6
C_3H_4	Allene (see Propadiene)	120	43.6
C_3H_4	Acetylene, methyl-	127.8	52.8
C_3H_4	Propyne (see Acetylene, methyl-)	127.8	52.8
C_3H_5N	Ethyl cyanide	290.8	41.3
C_3H_5N	Propionitrile (see Ethyl cyanide)	290.8	41.3
C_3H_5OCl	Epichlorohydrin	(323)	—
C_3H_5Cl	Propene, 3-chloro- (allyl-chloride)	240.3	46.5
C_3H_6	Propylene	91.9	45.4
C_3H_6	Cyclopropane	124.7	—
C_3H_6O	Acetone	235.5	47
C_3H_6O	Allylalcohol	272	55.5
C_3H_6O	Propylene oxide	209	48.6
$C_3H_6O_2$	Formic acid, ethyl- (Ethyl formate)	235.3	46.3
$C_3H_6O_2$	Acetic acid, methyl- (Methyl acetate)	233.7	46.3
$C_3H_6O_2$	Propanoic acid	337.6	53
C_3H_7Cl	n-Propane, monochloro-	230	45.2
C_3H_8	Propane	96.8	42
C_3H_8O	Ether, ethyl methyl- (Methoxyethane)	164.7	43.4
C_3H_8O	Glycol, 1,2-propylene-	351	—
C_3H_8O	Isopropyl alcohol	235	47
C_3H_8O	n-Propyl alcohol	263.6	51
$C_3H_8O_3$	Glycerol	(452)	—
C_3H_9N	Isopropylamine	209.7	—
C_3H_9N	n-Propylamine	223.8	46.8
	*( ) uncertain		

Formula	Name	Critical temp. T_c °C	Critical press. P_c atm.
C_4H_4O	Furan	213.8	52.5
C_4H_4S	Thiophene	307	56.2
C_4H_6	1,2-Butadiene	171	44.4
C_4H_6	1,3-Butadiene	152	42.7
C_4H_8	n-Butene	146	39.7
C_4H_8	2-Butene, cis-	160	40.5
C_4H_8	2-Butene, trans-	155	41.5
C_4H_8O	1,2-Butylene oxide	243	—
C_4H_8O	Ketone, ethyl methyl- (2-Butanone)	262	41
$C_4H_8O_2$	Butanoic acid	355	52
$C_4H_8O_2$	p-Dioxane	314.8	51.4
$C_4H_8O_2$	Acetic acid, ethyl- (Ethyl acetate)	250.4	37.8
$C_4H_8O_2$	Propanoic acid, methyl- (Methyl propionate)	257.4	39.3
$C_4H_8O_2$	Formic acid, propyl-(Propyl formate)	264.9	40.1
C_4H_9Cl	n-Butane, monochloride-	269	—
C_4H_{10}	n-Butane	152	37.5
C_4H_{10}	Isobutane	134.7	35.9
$C_4H_{10}O$	Butanol (n-Butyl alcohol)	289.8	43.6
$C_4H_{10}O$	sec-Butyl alcohol	263	41.4
$C_4H_{10}O$	tert-Butyl alcohol	235	39.2
$C_4H_{10}O$	Ether, diethyl-	192.6	35.6
$C_4H_{10}O$	Isobutyl alcohol	277	42.4
$C_4H_{10}O_3$	Glycol, diethylene-	407	—
$C_4H_{10}S$	Diethyl sulfide	283.8	39.1
$C_4H_{11}N$	Butyl amine	287.9	—
$C_4H_{11}N$	Diethyl amine	223.3	36.6
$C_4H_{11}N$	Isobutyl amine	266.7	—
C_4F_{10}	Butane, perfluoro-	113.2	23
C_5H_5N	Pyridine	346.8	—
C_5H_8	Cyclopentene	232.94	47.2
C_5H_{10}	Cyclopentane	238.6	44.6
C_5H_{10}	1-Pentene	191	39.9
$C_5H_{10}O$	Ketone, diethyl	287.8	36.0
$C_5H_{10}O$	Propanoic acid, ethyl- (Ethyl propionate)	272.9	33
$C_5H_{10}O_2$	n-Butanoic acid, methyl- (n-Methyl butyrate)	281.3	34.3
$C_5H_{10}O_2$	Acetic acid, n-propyl (n-Propyl acetate)	276	32.9
$C_5H_{10}O_2$	n-Valeric acid	378	37.6
$C_5H_{11}N$	Piperidine	320.8	44.1
C_5H_{12}	Butane, 2-methyl- (See Isopentane)	187.8	32.9
C_5H_{12}	Isopentane	187.8	32.9
C_5H_{12}	Neopentane	160.6	31.6
C_5H_{12}	n-Pentane	196.6	33.3
C_5H_{12}	Propane, 2,2-dimethyl (see Neopentane)	160.6	31.6
$C_5H_{12}O$	Isoamyl alcohol	309.77	—
C_6H_5Br	Benzene, bromo-	397	44.6
C_6H_5Cl	Benzene, chloro-	359.2	44.6
C_6H_5F	Benzene, fluoro-	286.95	44.6
C_6H_6	Benzene	288.9	48.6
C_6H_6O	Phenol	421.1	60.5
C_6H_7N	Aniline	425.6	52.3
C_6H_7N	α-Picoline	348	—
C_6H_7N	β-Picoline	371.7	—
C_6H_7N	γ-Picoline	372.5	—
C_6H_{10}	Cyclohexene	287.3	—
C_6H_{10}	1,5-Hexadiene	234.4	32.6
C_6H_{12}	Cyclohexane	280.4	40
C_6H_{12}	Cyclopentane, methyl-	259.5	37.4
C_6H_{12}	1-Hexene	231	31.1
$C_6H_{12}O_2$	Formic acid, n-amyl (n-Amyl formate)	302.6	34.1
$C_6H_{12}O_2$	Acetic acid, n-butyl- (n-Butyl acetate)	305.9	30.7
$C_6H_{12}O_2$	Butanoic acid, ethyl- (Ethyl butyrate)	293	30.2
$C_6H_{12}O_2$	Propanoic acid, propyl- (Propyl propionate)	304.8	30.7
$C_6H_{12}O_3$	Paraldehyde	290	—
C_6H_{14}	Butane, 2,3-dimethyl-	226.8	30.9
C_6H_{14}	n-Hexane	234.2	29.9
C_6H_{14}	Pentane, 2-methyl-	224.3	30
$C_6H_{14}O$	Ether, isopropyl-	226.9	28.4
$C_6H_{14}O$	1-Hexyl alcohol	313.5	—
$C_6H_{14}O_4$	Glycol, triethylene-	(437)	—
$C_6H_{15}N$	Dipropyl amine	277	31
$C_6H_{15}N$	Hexyl amine	318.8	—
$C_6H_{15}N$	Triethyl amine	258.9	30
C_7H_5N	Benzonitrile	426.2	41.6
C_7H_8	Benzene, methyl- (see Toluene)	320.8	41.6

Table I (Continued)

Formula	Name	Critical temp. T_c °C	Critical press. P_c atm.
C_7H_8	Toluene	320.8	41.6
C_7H_8O	Anisole, (see Benzene, methoxy-)	368.5	41.3
C_7H_8O	Benzene, methoxy-	368.5	41.3
C_7H_8O	o-Cresol	424.4	49.4
C_7H_8O	m-Cresol	432	45
C_7H_8O	p-Cresol	431.4	50.8
C_7H_9N	Aniline, methyl-	428.4	51.3
C_7H_9N	2,3-Lutidine	382.3	—
C_7H_9N	2,4-Lutidine	374	—
C_7H_9N	2,6-Lutidine	350.6	—
C_7H_9N	3,4-Lutidine	410.6	—
C_7H_9N	3,5-Lutidine	394.1	—
C_7H_{14}	Cyclohexane, methyl-	299.1	34.3
C_7H_{14}	Cyclopentane, ethyl-	296.3	33.5
C_7H_{14}	1-Heptene	264.1	—
$C_7H_{14}O_2$	Acetic acid, isoamyl- (Isoamyl acetate)	326.1	28
C_7H_{16}	Butane, 2,2,3-trimethyl-	258.3	29.8
C_7H_{16}	n-Heptane	267.1	27
C_7H_{16}	Hexane, 2-methyl-	257.9	27.2
C_7H_{16}	Pentane, 3-ethyl-	267.6	28.6
C_7H_{16}	Pentane, 2,4-dimethyl-	247.1	27.4
$C_7H_{16}O$	1-Heptyl alcohol	365.3	29.4
C_7F_{16}	n-Heptane, perfluoro-	201.7	16
C_8H_8	Styrene	374.4	39.4
C_8H_{10}	Benzene, ethyl-	343.9	36.9
C_8H_{10}	o-Xylene	359	35.7
C_8H_{10}	m-Xylene	346	34.7
C_8H_{10}	p-Xylene	345	33.9
$C_8H_{10}O$	Xylenol	449.7	56.4
$C_8H_{10}O$	Phenetol (see Benzene, ethoxy-)	374	33.8
$C_8H_{10}O$	Benzene, ethoxy-	374	33.8
$C_8H_{11}N$	Aniline, N,N-dimethyl-	414.4	35.8
$C_8H_{11}N$	Aniline, N-ethyl-	425.4	—
C_8H_{16}	n-Octene	305	25.5
C_8H_{18}	Butane, 2,2,3,3-tetramethyl-	270.8	24.5
C_8H_{18}	Heptane, 4-methyl-	290	25.6
C_8H_{18}	Hexane, 2,4-dimethyl-	282	25.8
C_8H_{18}	n-Octane	296	24.8
C_8H_{18}	Pentane, 2,2,3-trimethyl-	294	28.2
$C_8H_{18}O$	1-Octyl alcohol	385.5	26.5
$C_8H_{19}N$	Dibutyl amine	322.6	—
C_9H_7N	Quinoline	508.8	—
C_9H_{12}	Benzene, n-propyl-	365	31.2
C_9H_{12}	Benzene, 1,2,3-trimethyl-	391.3	31
C_9H_{12}	Cumene (see Benzene, isopropyl-)	362.7	31.2
C_9H_{12}	Benzene, isopropyl-	362.7	31.2
C_9H_{20}	n-Nonane	321	22.5
C_9H_{21}	Tripropyl amine	304.3	—
$C_{10}H_8$	Naphthalene	474.8	40.6
$C_{10}H_{12}O$	Ether, diphenyl-	494	30.9
$C_{10}H_{14}$	Benzene, 1-isopropyl-4-methyl-	385.5	27.7
$C_{10}H_{14}$	Benzene, 1,2,3,5-tetramethyl-	402.8	28.6
$C_{10}H_{14}$	p-Cymene, (see Benzene, 1-isopropyl-4-methyl-)	—	—
$C_{10}H_{14}$	Isodurene, (see Benzene, 1,2,3,5-tetramethyl-)	—	—
$C_{10}H_{14}O$	3-p-Cymenol	425.1	33
$C_{10}H_{14}O$	Thymol, (see 3-p-Cymenol)	425.1	33
$C_{10}H_{18}$	Decalin, cis-	418	28.7
$C_{10}H_{18}$	Decalin, trans-	408	28.7
$C_{10}H_{22}$	n-Decane	344.4	20.8
$C_{11}H_{10}$	Naphthalene, 1-methyl-	498.8	32.1
$C_{12}H_{10}$	Biphenyl	495	31.8
$C_{12}H_{11}N$	Diphenylamine	615.5	—
$C_{12}H_{18}$	Benzene, hexamethyl-	494	23.5
$C_{12}H_{18}$	Mellitine, (see Benzene, hexamethyl-)	—	—
$C_{12}H_{26}$	n-Dodecane	386	17.9
$C_{12}H_{27}N$	Tributyl amine	365.2	—

Table II

Name	Formula	Critical temp. T_c °C	Critical press. P_c atm.
Ammonia	NH_3	132.5	112.5
Argon	Ar	−122.3	48
Boron tribromide	BBr_3	300	—
Boron trichloride	BCl_3	178.8	38.2
Boron trifluoride	BF_3	−12.26	49.2
Carbon dioxide	CO_2	31	72.9
Carbon disulfide	CS_2	279	78
Carbon monoxide	CO	−140	34.5
Carbonyl sulfide	COS	104.8	65
Chlorine	Cl_2	144	76.1
Cyanogen	C_2N_2	126.6	—
Deuterium	D_2	−234.8	16.4
Fluorine	F_2	−129	55
Germanium tetrachloride	$GeCl_4$	276.9	38
Helium	He	−267.9	2.26
Hydrogen	H_2	−239.9	12.8
Hydrogen bromide	HBr	90	84.5
Hydrogen chloride	HCl	51.4	82.1
Hydrogen deuteride	HD	237.3	14.6
Hydrogen cyanide	HCN	183.5	48.9
Hydrogen fluoride	HF	188	64
Hydrogen iodide	HI	150	81.9
Hydrogen sulfide	H_2S	100.4	88.9
Hydrazine	N_2H_2	380	—
Iodine	I_2	512	116
Krypton	Kr	−63.8	54.3
Neon	Ne	−228.7	26.9
Nitric oxide	NO	−93	64
Nitrogen dioxide	NO_2	157.8	100
Nitrogen	N_2	−147	33.5
Nitrous oxide	N_2O	36.5	71.7
Oxygen	O_2	−118.4	50.1
Ozone	O_3	−5.16	67
Phosphine	PH_3	51.3	64.5
Radon	Rn	104.04	62
Silane, chlorotrifluoro-	$SiClF_3$	34.5	34.2
Silane	SiH_4	−3.46	47.8
Silicon tetrachloride	$SiCl_4$	232.8	—
Silicon tetrafluoride	SiF_4	−14.06	36.7
Sulfur dioxide	SO_2	157.8	77.7
Stannic chloride	$SnCl_4$	318.7	37
Water	H_2O	374.1	218.3
Xenon	Xe	16.6	58

DISSOCIATION PRESSURE OF CALCIUM CARBONATE

Temp. °C	mm/Hg	Temp. °C	mm/Hg	Temp. °C	mm/Hg	Temp. °C	mm/Hg
550	0.41	727	44	819	235	894	716
587	1.0	736	54	830	255	898	760 atm.
605	2.3	743	60	840	311	906.5	1.151
671	13.5	748	70	852	381	937	1.770
680	15.8	749	72	857	420	1082.5	8.892
691	19.0	777	105	871	537	1157.7	18.687
701	23.0	786	134	881	603	1226.3	34.333
703	25.5	795	150	891	684	1241	39.094
711	32.7	800	183				

DEFINITIONS

AB — A prefix attached to the names of the practical electric units to indicate the corresponding unit in the cgs electromagnetic system (emu), e.g. abampere, abvolt.

Abcoulomb — The abcoulomb, the emu of charge, is defined as the charge which passes a given surface in 1 sec if a steady current of one abampere flows across the surface. Its dimensions are, therefore, $cm^{1/2}\ g^{1/2}$, which differ from the dimensions of the statcoulomb by a factor which has the dimensions of a speed. This relationship is connected with the fact that the ratio $2K_e/K_m$ must have the value of the square of the speed of light in any consistent system of units. It follows further that

$$1 \text{ abcoulomb} = 2.99793 \times 10^{10} \text{ statcoulomb,}$$

the speed of light in vacuo being $(2.99793 \pm 0.000003) \times 10^{10}$ cm/sec.

Aberration — 1. In astronomy, the apparent angular displacement of the position of a celestial body in the direction of motion of the observer, caused by the combination of the velocity of the observer and the velocity of light. See constant of aberration, planetary aberration. Compare parallax. 2. In optics, a specific deviation from perfect imagery, as, for example: spherical aberration, coma, astigmatism, curvature of field, and distortion.

Aberrations (of image) — Distortions in shape, color, focus, and density of images caused by imperfect optical elements (i.e., lens, prism, mirror, screen, etc.) Types such as coma, astigmatism, field curvature, distortion, and chromatic and spherical aberrations.

Absolute humidity — See Humidity.

Absolute pressure — See Pressure.

Absolute temperature — Temperature reckoned from the absolute zero. See Temperature.

Absolute units — A system of units based on the smallest possible number of independent units. Specifically, units of force, work, energy, and power not derived from or dependent on gravitation.

Absolute zero — The theoretical temperature at which molecular motion vanishes and a body would have no heat energy; the zero point of the Kelvin and Rankine temperature scales.

Absolute zero may be interpreted as the temperature at which the volume of a perfect gas vanishes or, more generally, as the temperature of the cold source which would render a Carnot cycle 100% efficient. The value of absoute zero is now estimated to be −273.15°C, −459.67°F, 0 K, and 0° Rankine.

Absorption — 1. Penetration of a substance into the body of another. 2. Transformation into other forms suffered by radiant energy passing through a material substances.

Absorption coefficient — 1. A measure of the amount of normally incident radiant energy absorbed through a unit distance of the absorbing medium. It is also referred to as linear attenuation coefficient and linear absorption coefficient.

The absorption coefficient (k) is frequently identified as:

$$I_{\lambda x} = I_{\lambda o}\, e^{-kx} \text{ or } k = \ln (I_{\lambda o}/I_{\lambda x})$$

where $I_{\lambda x}$ is the flux density of radiation of wavelength λ, initially of flux density $I_{\lambda o}$, after traversing a distance x in the absorbing medium.

2. In acoustics, the ratio of the sound energy absorbed by a surface of a medium (or material) exposed to a sound field or sound radiation to the sound energy incident on the surface. The stated values of this ratio are to hold for an infinite area of the surface. The conditions under which measurements of absorption coefficients are made are to be stated explicitly.

Three types of absorption coefficients associated with three methods of measurement are: chamber absorption coefficient, obtained in a certain reverberation chamber; free-wave absorption coefficient, obtained when a plane, progressive, sound wave is incident on the surface of the medium; sabine absorption coefficient, obtained when the sound is incident from all directions on the sample. See Absorption Factor.

Absorption cross sections — In radar, the ratio of the amount of power removed from a beam by absorption of radio energy by a target to the power in the beam incident upon the target. Compare scattering cross section. See Cross section.

Absorption factor — The ratio of the intensity loss by absorpotion to the total original intensity of radiation. If I_o represents the original intensity, I_r, the intensity of reflected radiation, I_t, the intensity of the transmitted radiation, the absorption factor is given by the expression

$$\frac{I_o - (I_r + I_t)}{I_o}$$

Also called coefficient of absorption.

Absorption, Lambert's law — If I_o is the original intensity, I the intensity after passing through a thickness x of a material whose absorption coefficient is k,

$$I = I_o e^{-kx}$$

The index of absorption k′ is given by the relation $k = (4\pi k'n)/\lambda$ where n is the index of refraction and λ the wave length in vacuo. The mass absorption is given by k/d when d is the density. The transmission factor is given by I/I_o.

Absorption spectrum — The array of absorption lines and absorption bands which results from the passage of radiant energy from a continuous source through a selectively absorbing medium cooler than the source. See electromagnetic spectrum.

The absorption spectrum is a characteristic of the absorbing medium, just as an emission spectrum is a characteristic of a radiator.

An absorption spectrum formed by a monatomic gas exhibits discrete dark lines, whereas that formed by a polyatomic gas exhibits ordered arrays (bands) of dark lines, which appear to overlap. This type of absorption is often referred to as line absorption. The spectrum formed by a selectively absorbing liquid or solid is typically continuous in nature (continuous absorption).

The spectrum obtained by the examination of light from a source, itself giving a continuous spectrum after this light has passed through an abosrbing medium in the gaseous state. The absorption spectrum will consist of dark lines or bands, being the reverse of the emission spectrum of the absorbing substance.

When the absorbing medium is in the solid or liquid state the spectrum of the transmitted light shows broad dark regions which are not resolvable into lines and have no sharp or distinct edges.

Absorptive index — The imaginary part of the complex index of refraction of a medium. It represents the energy loss by absorption and has a nonzero value for all media which are not dielectrics. Also called index of absorption. Compare absorption coefficient.

Absorptive power or absorptivity for any body is measured by the fraction of the radiant energy falling upon the body which is absorbed or transformed into heat. This ratio varies with the character of the surface and the wave length of the incident energy. It is the ratio of the radiation absorbed by any substance to that absorbed under the same conditions by a black body.

Abvolt — The cgs electromagnetic unit of potential difference and electromotive force. It is the potential difference that must exist between two points in order that one erg of work be done when one abcoulomb of charge is moved from one point to the other. One abvolt is 10^{-8} V.

Acceleration — The time rate of change of velocity in either speed or direction. Cgs unit, 1 cm/sec/sec. Dimensions, — $[l\,t^{-2}]$ — See Angular acceleration.

Acceleration due to gravity — The acceleration of a body freely falling in a vacuum. The International Committee on Weights and Measures has adopted as a standard or accepted value, 980.665 cm/sec^2 or 32.174 ft/sec^2.

Acceleration due to gravity at any latitude and elevation — If ϕ is the latitude and H the elevation in centimeters the acceleration in cgs units is, $g = 980.616 - 2.5928 \cos 2\phi + 0.0069 \cos^2 2\phi - 3.086 \times 10^{-6}$ H. (Helmert's equation)

Accelerators — Machines for speeding up subatomic particles to energies running into millions of electron volts. See Betatron, Cyclotron, etc.

Acceptor — In transistors, the P-type semiconductor, the electrode containing trivalent impurities (boron, gallium, or indium) to increase the number of holes which can accept electrons. Contrast with donor.

Achromatic — A term applied to lenses signifying their more or less complete correction for chromatic aberration.

Acid — For many purposes it is sufficient to say that an acid is a hydrogen-containing substance which dissociates on solution in water to produce one or more hydrogen ions. More generally, however, acids are defined according to other concepts. The Bronsted concept states that an acid is any compound which can furnish a proton. Thus NH_4^+ is an acid since it can give up a proton:

$$NH_4^+ \rightleftharpoons NH_3 + H^+$$

and NH_3 is a base since it accepts a proton.

A still more general concept is that of G. N. Lewis which defines an acid as anything which can attach itself to something with an unshared pair of electrons. Thus in the reaction

$$\begin{array}{l} \qquad\quad H \\ H^+ + :N{-}H \rightleftharpoons NH_4^+ \\ \qquad\quad H \end{array}$$

the NH_3 is a base because it possesses an unshared pair of electrons. This latter concept explains many phenomena, such as the effect of certain substances other than hydrogen ions in the changing of the color of indicators. It also explains acids and bases in nonaqueous systems as liquid NH_3 and SO_2.

Acoustic velocity (symbol a) = speed of sound.

Actinic — Pertaining to electromagneic radiation capable of initiating photochemical reactions, as in photography or the fading of pigments.

Because of the particularly strong action of ultra violet radiation on photochemical processes, the term has come to be almost synonymous with ultraviolet, as in actinic rays.

Actinide Series — Elements of atomic numbers 89 to103 analogous to the lanthanide series of the so-called rare earths.

Actinometer — The general name for any instrument used to measure the intensity of radiant energy, particularly that of the sun. See Actinometry. See also Bolometer, Dosimeter, Photometer, Radiometer.

Actinometers may be classified, according to the quantities which they measure, in the following manner: (a) pyrheliometer, which measures the intensity of direct solar radiation; (b) pyranometer, which measures global radiation (the combined intensity of direct solar radiation and diffuse sky radiation); and (c) pyrgeometer, which measures the effective terrestrial radiation.

Action is measured by the product of work by time. Cgs units of action are the erg-second and the joule-second. Dimensions, — $[m\,l_2\,t^{-1}]$. Planck's quantum or constant of action is $(6.62517 - 0.00023) \times 10^{-27}$ erg-sec.

Active mass of a substance is the number of gram molecular weights per liter in solution, or in gaseous form.

Activity coefficient — A factor which, when multiplied by the molecular concentration yields the active mass. The activity coefficient is evaluated by thermodynamic calculations, usually from data on the emf of certain cells, or the lowering of the freezing point of certain solutions. It is a correction factor which makes the thermodynamic calculations correct.

Adiabatic — A body is said to undergo an adiabatic change when its condition is altered without gain or loss of heat. The line on the pressure volume diagram representing the above change is called an adiabatic line.

Adiabatic atmosphere — A model atmosphere in which the pressure decreases with height according to:

$$p = p_o\,[1 - (-\,gz/c_{p,d}T_o)]\ c_{p,d}R_d$$

where p_o and T_o are the pressure and temperature (°K) at sea level or other datum; z is the geometric height; R_d is the gas constant for dry gas; $c_{p,d}$ is the specific heat for dry gas at constant pressure; and g is the acceleration of gra vity. Also called dry-adiabatic atmosphre, convective atmosphere, homogeneous atmosphere. See Homogeneous atmosphere, Barotropy.

Adiabatic process — A thermodynamic change of state of a system in which there is no transfer of heat or mass across the boundaries of the system. In an adiabatic process, compression always results in warming, expansion in cooling. See Diabatic process.

Adsorption — The condensation of gases, liquids, or dissolved substances on the surfaces of solids is called adsorption.

Air columns, frequency of vibration in — See Organ pipes.

Albedo — The ratio of the amoung of electromagnetic radiation reflected by a body to the amount incident upon it, often expressed as a percentage, as, the albedo of the earth is 34%.

The concept defined above is identical with reflectance. However, albedo is more commonly used in astronomy and meteorology and reflectance in physics.

Albedo is sometimes used to mean the flux of the reflected radiation as, the earth albedo is 0.64 cal/cm². This usage should be discouraged.

The albedo is to be distinguished from the spectral reflectance, which refers to one specific wavelength (monochromatic radiation).

Usage varies somewhat with regard to the esact wavelength interval implied in albedo figures; sometimes just the visible portion of the spectrum is considered, sometimes the totality of wavelengths in the solar spectrum.

Alfvén speed — The speed at which Alfvén waves are propagated along the magnetic field.

For a perfectly conducting fluid with a mass density of 1 kg/m³ in a magnetic field of 10,000 gauss, the Alfvén speed is about 1000 m/sec while the speed of sound in air is about 300 m/sec.

Alfvén wave — A transverse wave in a magneto-hydrodynamic field in which the driving force is the tension introduced by he magnetic field along the lines of force. Also called magneto-hydrodynamic wave.

The dynamics of such waves are analogous to those in a vibrating string, the phase speed C being given by

$$C^2 = \mu H^2/4\pi\rho$$

where μ is the permeability; H is the magnitude of the magnetic field; and ρ is the fluid density. Dissipative effects due to fluid viscosity and electrical resistance may also be present.

Allobar — A form of an element differing in isotopic composition from the naturally occurring form.

Allotropy — The property shown by certain elements or being capable of existence in more than one form, due to differences in the arrangement of atoms or molecules. See Monotropic and Enantiotropic.

Alpha (α)-particle, or alpha-ray — One of the particles emitted in radioactive decay. It is identical with the nucleus of the helium atom and consists, therefore, of two protons plus two neutrons bound together. A moving alpha particle is strongly ionizing and so loses energy rapidly in traversing through matter. Natural alpha particles will traverse only a few centimeters of air before coming to rest.

Alternating current, (A-C) — Current in which the charge-flow periodically reverses, as opposed to direct current, and whose average value is zero. Alternating current usually implies a sinusoidal variation of current and voltage. This behavior is represented mathematically in various ways:

$$I = I_o \cos(2\pi ft + \phi)$$
$$I = I_o < \phi$$
$$I = I_1 \epsilon^{j\omega t}$$

where fi is the frequency; $\omega \equiv 2\pi f$, the pulsatance, or radian frequency; ϕ the phase angle; I_o the amplitude; and I_1 the complex amplitude. In the complex rotation, it is understood that the actual current is the real part of I. For circuits involving also a capacitance C in farads and L in henrys, the impedance becomes,

$$\sqrt{R^2 + \left(2\pi fL - \frac{1}{2\pi fC}\right)^2}$$

Altitudes with the barometer — If b_1 and b_2 denote the corrected barometer readings at two stations in any pressure units, and t is the mean of the temperatures, t_1 and t_2, of the air at the two stations in °C, then the difference in elevation in meters is

$$\Delta H = (1 + 0.00367\,t)\,(18{,}430 \log(b_1/b_2)$$

for differences not over 6000 m.

An approximate equation, generally sufficient for differences not over 1000 m, is

$$\Delta H \simeq (1 + 0.00367)\,(16{,}000\,\frac{b_1 - b_2}{b_1 + b_2})$$

Amorphous — Without definite form, not crystallized.

Ampere (unit of electric current) is the constant current which, if maintained in two straight parallel conductors of infinite length, of negligible circular sections, and placed 1 m apart in a vacuum, will produce between these conductors a force equal to 2×10^{-2} newton per meter of length.

Ampère's law — A vector equation relating current flow and magnetic fields. The law states that if a current (I) flows through an infinitesimal distance (dl) along a line, at some point (P) a distance (r) away there is produced an infinitesimal element of a magnetic field (dH) such that

$$|dH| = I \sin\theta \, d\, l/ |r^2|.$$

where θ is the angle between the direction of current flow and the line joing P and dl. More compactly,

$$d\,H = I\,d\,l \times r/r^3$$

Ampere's rule — A positive charge moving horizontally is deflected by a force to the right if it is moving in a region where the magnetic field is vertically upward. This may be generalized to currents in wires by recallig that a current in a certain direction is equivalent to the motion of positive charges in that direction. The force felt by a negative charge is opposite to that felt by a positive charge.

Ampere-turn — A measure of magnetomotive force, especially as developed by an electric current, defined as the magnetomotive force developed by a coil of one turn through which a current of one ampere flows, that is, 1.26 Gb.

Amplitude — The maximum value of the displacement in an oscillatory motion.

Amplitude modulation — The variation of the amplitude of a wave in a way that corresponds to another wave. In general, the positive or negative envelope of the modulated wave (carrier wave) contains enough information to allow the modulating wave to be recovered, provided that the carrier wave has at least twice the frequency and twice the peak amplitude of the modulating signal. The modulated carrier C(t) is given by

$$C(t) = A_o\ [1 - k\ M(t)]\ \cos \omega_o t,$$

Where M(t) is the modulation.

Amplitude (of wave) — A measure of the maximum displacement of the wave crest from its undisturbed position (or the maximum electric or magnetic field strength of an electromagnetic wave).

AMU — The atomic mass unit (AMU), a unit of mass equal to 1/12 the mass of the carbon atom of mass number 12. On the atomic mass scale $^{12}C \equiv 12$.

$$\begin{aligned} 1\ \text{AMU} &= 931.4812(52)\ \text{MeV} \\ &= 1.6605655(86) \times 10^{-27}\ \text{(SI units)} \\ &= 1.6605655(86) \times 10^{-24}\ \text{(cgs units)} \end{aligned}$$

The numbers in parentheses are the standard deviation uncertainties in the last digits of the quoted value, computed on the basis of internal consistency.

Angle — The ratio between the arc and the radius of the arc. Units of angle, the radian, the angle subtended by an arc equal to the radius; the degree, 1/360 part of the total angle about a point. Dimensions, a numeric.

Angstrom — A unit of length, used especially in expressing the length of light waves, equal to one ten-thousandth of a micron, or one hundred-millionth of a centimeter (1×10^{-8} cm).

Angular acceleration — The time rate of change of angular velocity either in angular speed or in direction of the axis of rotation (precession). Cgs unit, 1 radian/sec/sec. Dimensions $[t^{-2}]$.

If the initial angular velocity is ω_o, and the velocity after time t is ω_t, the angular acceleration,

$$\alpha = \frac{\omega_t - \omega_o}{t}$$

The angular velocity after time t,

$$\omega_t = \omega_o + \alpha t$$

The angle swept out in time t,

$$\theta = \omega_o t + \tfrac{1}{2}\alpha t^2$$

The angular velocity after movement through the arc θ,

$$\omega = \sqrt{\omega_o{}^2 + 2\alpha\theta}$$

In the above equations, for angular displacement in radians, angular velocity will be in radians per second and angular acceleration in radians per second per second.

Angular aperture of an objective is the largest angular extent of wave surface which it can transmit.

Angular harmonic motion or harmonic motion of rotation — Periodic, oscillatory angular motion in which the restoring torque is proportional to the angular displacement. Torsional vibration.

Angular momentum or moment of momentum — Quantity of angular motion measured by the product of the angular velocity and the moment of inertia. Cgs unit, unnamed, its nature is expressed by g-cm²/sec. Dimentions, $[ml^2 t^{-1}]$.

The angular momentum of a mass whose moment of inertia is I, rotating with angular velocity w, is Iw.

Angular velocity — Time rate of angular motion about an axis. Cgs unit, radian/sec. Dimensions, $[t^{-1}]$.

If the angle described in time *t* is θ, the angular velocity,

$$\omega = \frac{\theta}{t}$$

θ in radians and *t* in seconds gives ω in radians per second.

Anhydride (of acid or base) — An oxide which when combined with water gives an acid or base.

Anion — A negatively charged ion.

Anisotropic — Exhibiting different properties when tested along aes in different directions.

Anode — The electrode at which oxidation occurs in a cell. It is also the electrode toward which anions travel due to the electrical potential. In spontaneous cells the anode is considered negative. In nonspontaneous or electrolytic cells the anode is considered positive.

Antiferromagnetic materials — Those in which the magnetic moments of atoms or ions tend to assume an ordered arrangement in zero applied field, such that the vector sum of the moments is zero, below a characteristic temperature called the Neel Point. The permeability of antiferromagnetic materials is comparable to that of paramagnetic materials. Above the Neel Point, these materials become paramagnetic.

Anti-matter — Matter consisting of antiparticles.

Anti-particle — Any particle with a charge of opposite sign to the same particle in normal matters.

Thus, the proton has a positive charge; the anti-proton, a negative charge. When a particle and its anti-particle collide, both may disappear with the creation of lighter particles; this process is called annihilation.

Aperture ratio — The ratio of the useful diameter of a lens to its focal length. It is the reciprocal of the f-number.

In application to an optical instrument, rather than to a lens, numerical aperture is more commonly used. The aperture ratio is then twice the tangent of the angle whose sine is the numerical aperture.

Apochromat — A term applied to photographic and microscope objectives indicating the highest degree of color correction.

Archimedes principle — A body wholly or partly immersed in a fluid is buoyed up by a force equal to the weight of the fluid displaced. A body of volume V cm³ immersed in a fluid of density ϱ grams per cm³ is buoyed up by a force in dynes,

$$F = \rho g V$$

where g is the accleration due to gravity.

A floating body displaces its own weight of liquid.

Area, unit of — The square centimeter. The area of a square whose sides are one centimeter in length. Other units of area are similarly derived. Dimensions, $[L^2]$.

Arrhenius theory of electrolytic dissociation states that the molecule of an electrolyte can give rise to two or more electrically charged atoms or ions.

Astigmatism is an error of spherical lenses peculiar to the formation of images by oblique pencils. The image of a point when astigmatism is present will consist of two focal lines at right angles to each other and separated by a measurable distance along the axis of the pencil. The error is not eliminated by reduction of aperture as is spherical aberration.

Astronomical unit (abbr. AU) — 1. A unit of length, usually defined as the distance from the earth to the sun, 149,599,000 km. This value for the AU was derived from radar observations of the distance of Venus. The value given in astronomical ephemerides, 149,599,000 km, was derived from observations of the minor planet Eros.

2. The unit of distance in terms of which, in the Kepler Third Law, $n^2a^3 = k^2(1 + m)$, the semimajor axis a of an elliptical orbit must be expressed in order that the numerical value of the Gaussian constant k may be exactly 0.01720209895 when the unit of time is the ephemeris day.

In astronomical units, the mean distance of the earth from the sun, calculated by the Kepler law from the observed mean motion n and adopted mass m, is 1.00000003.

Atmosphere, standard — 1 standard atmosphere = 1,013,250 dyn/cm² i.e., 101,325 N/m².

Atmospheric radiation — Infrared radiation emitted by or being propagated through the atmosphere. See Insolation.

Atmospheric radiation, lying almost entirely within the wavelength interval of from 3 to 80 μm, provides one of the most important mechanisms by which the heat balance of the earth-atmosphere system is maintained. Infrared radiation emitted by the earth's surface (terrestrial radiation) is partially absorbed by the water vapor of the atmosphere which in turn reemits it, partly upward, partly downward. This secondarily emitted radiation is then, in general, repeatedly absorbed and reemitted, as the radiant energy progresses through the atmosphere. The downward flux, or counterradiation, is of basic importance in the greenhouse effect; the upward flux is essential to the radiative balance of the planet.

Atom — The smallest particle of an element which can enter into a chemical combination. All chemical compounds are formed of atoms, the difference between compounds being attributable to the nature, number, and arrangement of their constituent atoms. See Isotopes, Nuclear atom.

Atomic bomb — An explosive that derives its energy from the fission or fusion of atomic nuclei.

Atomic energy — 1. The constitutive internal energy of the atom which was absorbed when it was formed. 2. Energy derived from the mass converted into energy in nuclear transformations. See Einstein's formula.

Atomic mass (atomic weight) — The mass of a neutral atom of a nuclide. It is usually expressed in terms of the physical scale of atomic masses, that is, in atomic mass units (AMU). See AMU.

Atomic mass unit (AMU) — A measure of atomic mass, defined as equal to 1/12 the mass of a carbon atom of mass 12.

Atomic number (symbol Z) — An interger that expres ses the positive charge of the nucleus in multiples of the electronic charge e. It is the number of electrons outside the nucleus of a neutral (un-ionized) atom and, according to widely accepted theory, the number of protons in the nucleus.

An element of atomic number Z occupies the Zth place in the periodic table of the elements. Its atom has a nucleus with a charge +Ze, which is normally surrounded by Z electrons, each of charge −e.

For example, the carbon isotope $_6C^{14}$ has an atomic number of 6 and an atomic mass of 14.

Atomic structure — According to the currently accepted view, the atom consists of a central part, called nucleus, and a number of electrons (called orbital or planetary electrons) circling about the latter, like planets about the sun. The nucleus is of a high specific weight; it contains most of the mass of the entire atom (its mass is considered equal to the atomic mass) and is composed of positively charged particles, called protons (the number of which always equals the atomic number, (Z), and particles of 0 charge, called neutrons (the number of which equals the difference between the atomic weight and the atomic number, A − Z). The diameter of the nucleus is between 10^{-13} and 10^{-12} cm, and the relatively vast distance in which the orbital electrons circle about it is illustrated by the fact that this nuclear diameter is only 10^{-4} to 10^{-5} of the entire atomic diameter. While the nucleus carries an integral number of positive charges (an integral number of protons) each of 1.6×10^{-19} coulomb, each electron carries one negative charge of 1.6×10^{-19} coulomb, and the number of orbital electrons is equal to the number of protons in the nucleus (i.e. to the atomic number, Z), so that the atom as a whole has a net charge of 0. The electrons are arranged in successive shells (q.v.) around the nucleus; the maximum number of electrons in each shell is determined by natural laws, and the extranuclear electronic structure of the atom is characteristic of the element. The electrons in the inner shells are tightly bound to the nucleus; this inner structure can be altered by high-energy particles, γ-rays of radium, of X-rays. The electrons in the outer shells are responsible for the chemical properties of the element. See Bohr's atomic theory, Heisenberg's theory, Shell, and Subshell.

Atomic theory — All elementary forms of matter are composed of very small unit quantities called atoms. The atoms of a given element all have the same size and weight. The atoms of different elements have different sizes and weights. Atoms of the same or different elements unite with each other to form very small unit quantities of compound substances called molecules.

Atomic weight — Atomic weight is the relative weight of the atom on the basis of $^{12}C \equiv 12$. For a pure isotope, the atomic weight rounded off to the nearest integer gives the total number of nucleons (neutrons and protons) making up the atomic nucleus. If these weights are expressed in grams they are called gram atomic weights. See Isotopes and Atomic mass.

Atomic weight (abbr at. wt.) — The weight of an atom according to a scale of atomic weight units, awu, valued as one-twelfth the mass of the carbon atom (C^{12} = 12.00000).

Thus expressed, the atomic weight to the nearest integer is identical with the mass number.

Attenuation coefficient (symbol α) — A measure of the space rate of attenuation of any transmitted electromagnetic radiation. The attenuation coefficient is defined by

$$dI = -\alpha I_0 dx$$

or

$$I = I_0 e^{-\alpha x}$$

where I is the flux density at the selected point in space; I_0 is the flux density at the source; and α is the attenuation coefficient.

ATTO — A prefix meaning one quintillionth or 10^{-18}. Symbol is a.

Avogadro's law — Equal volumes of different gases at the same pressure and temperature contain the same number of molecules.

Avogadro's number — The number of molecules in one mole or gram-molecular weight of a substance. A number of values of the Avogadro number, which is usually denoted by N, have been found by various methods, generally lying withing a range of 1% about the value 6.022045×10^{23}/gm.

Avogadro's principle (or theory) — The numbers of molecules present in equal volumes of gases at the same temperature and pressure are equal.

Azimuth — 1. Horizontal direction or bearing. Compare azimuth angle. 2. In navigation, the horizontal direction of a celestial point from a terrestrial point, expressed as the angular distance from a reference direction, usually measured from 0° at the reference direction clockwise through 360°.

An azimuth is often designated as true, magnetic, compass, grid, or relative as the reference direction is true, magnetic, compass, grid north, or heading, respectively. Unless otherwise specified, the term is generally understood to apply to true azimuth, which may be further defined as the arc of the horizon, or the angle at the zenith, between the north part of the celestial meridian or principal vertical circle and a vertical circle, measured from 0° at the north part of the principal vertical circle clockwise through 360°.

3. In astronomy, the direction of a celestial point from a terrestrial point measured clockwise from the north or the south point of the meridian plane. 4. In

surveying, the horizontal direction of an object measured clockwise from the south point of the meridian plane.

In surveying, an aximuth of a celestial body is called an astronomic azimuth.

Azimuth angle — 1. Azimuth measured from 0° at the north or south reference direction clockwise or counterclockwise through 90° or 130°.

Azimuth angle is labeled with the reference direction as a prefix and the direction of measurement from the reference direction as a suffix. Thus, azimuth angle S 144° W is 144° west of south, or azimuth 324°. When azimuth angle is measured through 180°, it is labeled N or S to agree with the latitude and E or W to agree with the meridian angle.

2. In surveying, an angle in triangulation or in traverse through which the computation of azimuth is carried.

Babo's law — The addition of a nonvolatile solid to a liquid in which it is soluble lowers the vapor pressure of the solvent in proportion to the amount of substance dissolved.

Balmer series of spectral lines. The wave lengths of a series of lines in the spectrum of hydrogen were given in angstroms by the equation

$$\lambda = 3646\ \frac{N^2}{N^2 - 4}$$

where N is an integer having values greater than 2.

Bar — International unit of pressure 10^6 dyn/cm². Unfortunately some writers have used this term for 1 dyn/cm². 1 bar = 0.987 atm = 1000 mbars = 29.53 in of mercury. See Torr.

Barn — Unit for measuring capture cross sections (q.v.) of elements. One barn = 10^{-24} cm²/nucleus.

Barotropy — The state of a fluid in which surfaces of constant density (or temperature) are coincident with surfaces of constant pressure; it is the state of zero baroclinity. Mathematically, the equation of batortopy states that the gradients of the density and pressure fields are proportional:

$$\Delta\rho = B\Delta p$$

where ρ is the density; p is the pressure; and B is a function of thermodynamic variables, called the coefficient of barotropy.

With the equation of state, this relation determines the spatial distribution of all state parameters once these are specified on any surface. For a homogeneous atmosphere, B = 0; for an adiabatic atmosphere,

$$B = c_v/c_p\ RT$$

where c_v and c_p are the specific heats at constant volume and pressure, respectively; R is the gas constant; and T is the Kelvin temperature; for an isothermal atmosphere, B = 1/RT.

Barye — The pressure unit of the centimeter-gram-second system of physical units; equal to one dyne per square centimeter (0.001 mbar). Unfortunately some writers have used this term for the bar which is equal to 10^6 dyn/cm².

Bases — For many purposes it is sufficient to say that a base is a substance which dissociates on solution in water to produce one or more hydroxyl ions. More generally, however, bases are defined according to other concepts. The Bronsted concept states that a base is any compound which can accept a proton. Thus NH_3 is a base since it can accept a proton to form ammonium ions.

$$NH_3 + H^+ \rightleftharpoons NH_4^+$$

A still more general concept is that of G. N. Lewis which defines a base as anything which has an unshared pair of electrons. Thus in the reaction

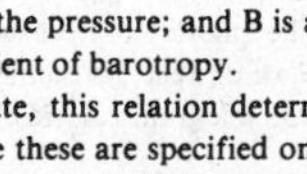

the NH_3 is a base because it possesses an unshared pair of electrons. This latter concept explains many phenomena, such as the effect of certain substances other than hydrogen ions in the changing of the color of indicators. It also explains acids and bases in nonaqueous systems as liquid NH_3 and SO_2.

Beam (of energy) — The locus of all series of wave-fronts projected from the source and directed toward given objects or positions in space.

Beam splitter — A device to produce two separate beams from one incident beam. This can be done with prisms or halfsilvered mirrors.

Beat(s) — Two vibrations of slightly different frequencies f_1 and f_2 when added together, produce in a detector sensitive to both these frequencies, a regularly varying response which rises and falls at the "beat" frequency $f_b = |f_1 - f_2|$. It is important to note that a resonator which is sharply tuned to f_b alone will not resound at all in the presence of these two beating frequencies. See Combination Frequencies.

Beat frequencies — The beat of two different frequencies of signals on a nonlinear circuit when they combine or beat together. It has a frequency equal to the difference of the two applied frequencies.

Beating — A wave phenomenon in which two or more periodic quantities of different frequencies produce a resultant having pulsations of amplitude.

This process may be contolled to produce a desired beat frequency. See Heterodyne.

Beer's law — If two solutions of the same colored compound be made in the same solvent, one of which is, say, twice the concentration of the other, the absorption due to a given thickness of the first solution should be equal to that of twice the thickness of the second.

Mathematically this may be expressed $l_1c_1 = l_2c_2$ when the intensity of light passing through the two solutions is a constant and if the intensity and wave length of light incident upon each solution are the same.

Bel — The fundamental division of a logarithmic scale for expressing the ratio of two amounts of power, the number of bels denoting such a ratio being the logarithm to the base 10 of this ratio.

With P_1 and P_2 designating two amounts of power and N the number of bels denoting their ratio, $N = \log^{10} (P_1/P_2)$ bels.

Bernoulli law or Bernoulli theorem — (After Daniel Bernoulli, 1700—1782, Swiss scientist.) 1. In aeronautics, a law or theorem stating that in a flow of incompressible fluid the sum of the static pressure and the dynamic pressure along a streamline is constant if gravity and frictional effects are disregarded.

From this law is follows that where there is a velocity increase in a fluid flow there must be a corresponding pressure decrease. Thus an airfoil, by increasing the velocity of the flow over its upper surface, derives lift from the decreased pressure.

2. As originally formulated, a statement of the conservation of energy (per unit mass) for a nonviscous fluid in steady motion. The specific energy is composed of the kinetic energy $u^2/2$, where u is the speed of the fluid; the potential energy gz, where g is the acceleration of gravity and z is the height above an arbitrary reference level; and the work done by the pressure forces of a compressible fluid $\int v\,dp$, where p is the pressure, v is the specific volume, and the integration is always with respect to values of p and v on the same parcel. Thus, the relationship

$$\frac{u_2}{2} + gz + \int v\,dp = \text{Constant along a streamline}$$

is valid for a compressible fluid in steady motion, since the streamline is also the path. If the motion is also irrotational, the same constant holds for the entire fluid.

Berthelot principle of maximum work — Of all possible chemical processes which can proceed without the aid of external energy, that process always takes place which is accompanied by the greatest evolution of heat. This law holds good for low temperatures only and does not account for endothermic reations.

Beta (β)-particle, (Beta ray) — One of the particles which can be emitted by a radioactive atomic nucleus. It has a mass about 1/1837. that of the proton. The negatively charged beta particle is identical with the ordinary electron, while the positively charged type (positron) differs from the electron in having equal but opposite electrical properties. The emission of an electron entails the change of a neutron into a proton inside the nucleus. The emission of a positron is similarly associated with the change of a proton into a neutron. Beta particles have no independent existance inside the nucleus, but are created at the instant of emission. See Neutrino.

Betatron — An accelerator used to impart high velocities to electrons (beta particles). Propellant is an electromagnetic field. A five to six Mev betatron can produce X-rays equivalent to the gamma radiation of 10 to 20 g of radium.

Bevatron — A six or more billion electron volt accelerator of protons and other atomic particles. Makes use of a Cockcroft-Walton transformer cascade accelerator and a linear (q.v.) as well as an electromagnetic field in the build-up.

Binary notation — A system of positional notation in which the digits are coefficients of powers of the base 2 in same way as the digits in the conventional decimal system are coefficients of powers of the base 10.

Binary notation employs only two digits, 1 and 0, therefore is used extensively in computers where the on and off positions of a switch or storage device can represent the two digits.

In decimal notation 111 — $(1\times10^2)+(1\times10^1)+(1\times10^0) = 100+10+1$ = one hundred and eleven.

In binary notation 111 = $(1\times2^2)+(1\times2^1)+(1\times2^0) = 4+2+1$ = seven.

Black body — If, for all values of the wave length of the incident radiant energy, all of the energy is absorbed the body is called a black body.

Bohr radius (symbol a^o) — The smallest possible radius of an electron orbit in the Bohr model of the atom, 5.29167×10^9 cm.

Bohr's atomic theory — The theory that atoms can exist for a duration solely in certain states, characterized by definite electronic orbits, i.e., by definite energy levels of their extra-nuclear electrons, and in these stationary states they do not emit radiation; the jump of an electron from an orbit to another of a smaller radius is accompanied by monochromatic radiation.

Bolometer — An instrument which measures the intensity of radiant energy by employing a thermally sensitive electrical resistor; a type of actinometer. Also called actinic balance. Compare radiometer.

Two identical, blackened, thermally sensitive electrical resistors are used in a Wheatstone bridge circuit. Radiation is allowed to fall on one of the elements, causing a change in its resistance. The change is a measure of the intensity of the radiation.

Boltzmann constant (symbol k) — The ratio of the universal gas constant to Avogadro number; equal to 1.38054×10^{-16} erg/°K. Sometimes called gas constant per molecule, Boltzmann universal conversion factor.

Bouguer law — A relationship describing the rate of decrease of flux density of a plane-parallel beam of monochromatic radiation as it penetrates a medium which both scatters and absorbs at that wavelength. This law may be expressed

$$dI_\lambda = -\alpha_\lambda I_\lambda \, dx$$

or

$$I_\lambda = I_{\lambda 0} e^{-\alpha_\lambda x}$$

where I is the flux density of the radiation; α_λ is the attenuation coefficient (or extinction coefficient) of the medium at wavelength λ; $I_{\lambda o}$ is the flux density at the source; and x is the distance from the source. Sometimes called Beer law, Lambert law of absorption. See Absorption coefficient, Scattering coefficient.

Boussinesq approximation — The assumption (frequently used in the theory of convection) that the fluid is incompressible except insofar as the thermal expansion produces a buoyancy, represented by a term $g\alpha T$, where g is the acceleration of gravity; α is the coefficient of thermal expansion; and T is the perturbation temperature.

Boyle's law for gases — At a constant temperature the volume of a given quantity of any gas varies inversely as the pressure to which the gas is subjected. For a perfect gas, changing from pressure p and volume v to pressure p′ and volume v′ without change of temperature,

$$pv = p'v'$$

Boyle-Mariotte law — The empirical generalization that for many so-called perfect gases, the product of pressure p and volume V is constnat in an isothermal process:

$$pV = F(T)$$

where the function F of the temperature T cannot be specified without reference to other laws (e.g., Charles-Gay-Lussac law). Also called Boyle law, Mariotte Law.

Brayton cycle — (After George B. Brayton, American engineer.) Same as Joule cycle.

Breakdown potential = dielectric strength.

Breeder, Reactor (Breeder pile) — A nuclear chain reactor in which transmutation produces a greater number of fissionable atoms than the number of parent atoms consumed.

Bremsstrahlung (German, braking radiation) — Electromagnetic radiation produced by the rapid change in the velocity of an electron or another fast, charged particle as it approaches an atomic nucleus and is deflected by it. See Bremsstrahlung effect.

Bremsstrahlung effect — The emission of electromagnetic radiation as a consequence of the acceleration of charged elementary particles, such as electrons, under the influence of the attractive or repulsive force fields of atomic nuclei near which the charged particle moves.

In cosmic-ray shower production, bremsstrahlung effects givve rise to emission of gamma rays as electrons encounter atmospheric nuclei. The emission of radiation in the bremsstrahlung effect is merely one instance of the general rule that electromagnetic radiation is emitted only when electric charges undergo acceleration.

Brewster window — An aperture through which light can enter into a new medium at an angle to the interface such that

$$\tan \theta_B = \frac{n_b}{n_a}$$

where θ_B is the Brewster angle. n_b and n_a are the indices of refraction of the media a and b, and light enters b from a.

Brewster's law — The tangent of the polarizing angle for a substance is equal to the index of refraction. The polarizing angle is that angle of incidence for which the reflected polarized ray is at right angles to the refracted ray. If n is the index of refraction and θ the polarizing angle, n = tan θ.

Brightness is measured by the flux emitted per unit emissive area as projected on a plane normal to the line of sight. The unit of brightness is that of a perfectly diffusing surface giving out one lumen per square centimeter of projected surface and is called the lambert. The millilambert (0.001 lambert) is a more convenient unit. Candle per square centimeter is the brightness of a surface which has, in the direction considered, a luminous intensity of 1 candle/cm².

British thermal unit — The quantity of heat required to raise the temperature of one pound of water 1°F at, or near, its point of maximum density (39.1°F). The Btu is equivalent to 0.252 kilogram-calorie or 1055 joules.

Brownian movement — A continuous agitation of particles in a colloidal solution caused by unbalanced impacts with molecules of the surrounding medium. The motion may be observed with a microscope when a strong beam of light is caused to traverse the solution across the line of sight.

Bulk modulus — The modulus of volume elasticity,

$$M_B = \frac{p_2 - p_1}{\frac{v_1 - v_2}{v_1}}$$

where p_1, p_2; v_1, v_2 are the initial and final pressure and volume respectively. It is the reciprocal of the coefficient of compressibility.

Calorie — The amount of heat necessary to raise 1 g of water at 15°C, 1°C. There are various calories depending upon the interval chosen. Sometimes the unit is written as the gram-calorie or the kilogram calorie, the meaning of which is evident. The calorie may be defined in terms of its mechanical equivalent. The National Bureau of Standards defines the calorie as 4.18400 joules. At the International Steam Table Conference held in London in 1929 the international calorie was defined at 1/860 of the international watt hour, which makes it equal to 4.1860 international joules.

With the adoption of the absolute system of electrical units, this becomes 1/859.858 watt hours or 4.18674 joules. The Btu was defined at the same time as 251.996 international calories.

Calutron — An apparatus operating on the principle of the mass spectrograph and used for separating U^{235} from U^{238}.

Candela — The candela is the luminous intensity, in the direction of the normal, of a black body surface 1/600,000 m² in area, at the temperature of solidification of platinum under a pressure of 101,325 N/m².

Candle (new unit) — 1/60 of the intensity of 1 cm^2 of a blackbody radiator at the temperature of solidification of platinum (2045K).

Capacitance is measured by the charge which must be communicated to a body to raise its potential one unit. Electrostatic unit capacitance is that which requires one electrostatic unit of charge to raise the potential one electrostatic unit. The farad = 9×10^{11} electrostatic units. A capacitance of one farad requires one coulomb of electricity to raise its potential one volt. Dimensions, $[\epsilon l]$; $[\mu^{-1} l^{-1} t^2]$.

A conductor charged with a quantity Q to a potential V has a capacitance,

$$C = \frac{Q}{V}$$

Capacitance of a spherical conductor of radius r,

$$C = Kr$$

Capacitance of two concentric spheres of radii r and r′

$$C = K \frac{rr'}{r - r'}$$

Capacitance of a parallel plate condenser, the area of whose plates is A and the distance between them d,

$$C = \frac{KA}{4\pi d}$$

Capacitances will be given in electrostatic units if the dimensions of condensers are substituted in centimeters. K is the dielectric constant of the medium.

Capillary constant or specific cohesion,

$$a^2 = \frac{2T}{(d_1 - d_2)g} = hr$$

where T is surface tension, d_1 and d_2, the densities of the two fluids, g the acceleration due to gravity, h the height of rise in a capillary tube of radius r. See Surface tension.

Carbon cycle — A sequence of atomic nuclear reactions and spontaneous radioactive decay which serves to convert matter into energy in the form of radiation and high-speed particles, and which is regarded as one of the principal sources of the energy of the sun and other similar stars.

This cycle, first suggested by Bethe in 1938, gets its name from the fact that carbon plays the role of a kind of catalyst in that it is both used by and produced by the reaction, but is not consumed itself. Four protons are, in net, converted into an alpha particle and two positrons (with accompanying neutrinos); and three gamma-ray emissions are emitted directly in addition to the two gamma emissions that ensue from annihilation of the positrons by ambient electrons. This cycle sets in at stellar interior temperatures of the order of 5 million degrees Kelvin.

An even simpler reaction, the proton-proton reaction, is also believed to occur within the sun and may be of equal or greater importance.

Carnot cycle — An idealized reversible thermodynamic cycle. The Carnot cycle consists of four stages: (a) an isothermal expansion of the gas at temperature T_1; (b) an adiabatic expansion to temperature T_2; (c) an isothermal compression at temperature T_2; (d) an adiabatic compression to the original state of the gas to complete the cycle. See Carnot engine, Thermodynamic efficiency.

In a Carnot cycle, the net work done is the difference between the heat input Q_1 at higher temperature T_1 and the heat extracted Q_2 at the lower temperature T_2.

Carnot engine — An idealized reversible heat engine working in a Carnot cycle. It is the most efficient engine that can operate between two specified temperatures; its efficiency is equivalent to the thermodynamic efficiency. The Carnot engine is capable of being run either as a conventional engine or as a refrigerator.

Cassegrain telescope — A reflecting telescope in which a small hyperboloidal mirror reflects the convergent beam from the paraboloidal primary mirror through a hole in the primary mirror to an eyepiece in back of the primary mirror. Also called Cassegrainian telescope, Cassegrain. See Newtonian Telescope.

Catalytic agent — A substance which by its mere presence alters the velocity of a reaction, and may be recovered unaltered in nature or amount at the end of the reaction.

Cathode — The electrode at which reduction occurs. It is the negative electrode in a cell through which current is being forced, but it is the positive pole of a battery. In a vacuum tube, the cathode is the electrode from which electrons are liberated. See Anode.

Cation — A positively charged ion.

Cauchy's dispersion formula —

$$n = A + \frac{B}{\lambda^2} + \frac{C}{\lambda^4} + \ldots$$

An empirical expression giving an approximate relation between the refractive index n of a medium and the wavelength λ of the light; A, B, and C being constants for a given medium.

Celsius — See Temperature, Celsius, in this section.

CENTI — A prefix meaning 1/100 or 10^{-2}. Symbol is c.

Centigrade temperature scale (abbr. C) — A temperature scale with the ice point at 0° and the boiling point of water at 100°. Now called Celsius temperature scale.

Conversion to the Fahrenheit temperature scale is according to the formula

$$°C = 5/9\ (°F - 32)$$

Centipoise — A standard unit of viscosity, equal to 0.01 poise, the c.g.s. unit of viscosity. Water at 20°C has a viscosity of 1.002 centipoise or 0.01002 poise.

Centripetal force — The force required to keep a moving mass in a circular path. Centrifugal force is the name given to the reaction against centripetal force.

Chain reaction — In general, any self-sustaining process, whether molecular or nuclear, the products of which are instrumental in, and directly contribute to the propagation of the process. Specifically, a fission chain reaction, where the energy liberated or particles produced (fission products) by the fission of an atom cause the fission of other atomic nuclei, which in turn propagate the fission reaction in the same manner.

Charles-Gay-Lussac law — An empirical generalization that in a gaseous system at constant pressure, the temperature increase and the relative volume increase stand in approximately the same proportion for all so-called perfect gases. Mathematically,

$$t - t_0 = (1/c\ [(v - v_0)/v_0]$$

where t is temperature; v is volume; and c is a coefficient of thermal expansion independent of the particular gas. If the centigrade temperature scale is used and v_o is the volume at 0°C, then the value of the constant c is approximately 1/273. Also called Charles law, Gay-Lussac law.

Charles law = Charles-Gay-Lussac law.

Chemiluminescence — Emission of light during a chemical reaction.

Christiansen effect — When finely powdered substances, such as glass or quartz, are immersed in a liquid of the same index of refraction complete transparency can only be obtained for monochromatic light. If white light is employed the transmitted color corresponds to the particular wave-length for which the two substances, solid and liquid have exactly the same index of refraction. Due to differences in dispersion the indices of refraction will match for only a narrow band of the spectrum.

Chromatic aberration — Due to the difference in the index of refraction for different wave lengths, light of various wave lengths from the same source cannot be focused at a point by a simple lens. This is called chromatic aberration.

Circularly polarized wave — An electromagnetic wave for which the electric or the magnetic field vector, or both, at a point describe a circle.

This term is usually applied to transverse waves.

Circular mil — The area of a circle with a diameter of 0.001 in, a unit used for the measurement of small circular areas, such as the cross section of a wire. One circular mil = 7.85×10^{-7} in^2.

Circular polarization — The polarization of a wave radiated by a constant electric vector rotating in a plane so as to describe a circle. See Elliptical polarization.

Circulation — 1. The flow or motion of a fluid in or through a given area or volume. 2. A precise measure of the average flow of fluid along a given closed curve. Mathema tically, circulation is the line integral.

$$\oint v \cdot dr$$

about the closed curve, where v is the fluid velocity, and dr is a vector element of the curve.

By Stokes theorem, the circulation about a plane curve is equal to the total vorticity of the fluid enclosed by the curve.

The given curve may be fixed in space or may be defined by moving fluid parcels.

Circulation integral — The line integral of an arbitrary vector taken around a closed curve. Thus,

$$\oint a \cdot dr$$

is the circulation integral of the vector a around the closed curve; dr is an infinitesimal vector element of the curve. If the vector is the velocity, this integral is called the circulation.

Clapeyron-Clausius equation — The differential equation relating pressure to temperature in a system in which two phases of a substance are in equilibrium.

$$dp/dT = L/(T\Delta V)$$

where p is pressure; T is temperature; L is the latent heat of the phase change; and ΔV is the differenc in volume of the phases. Also called Clapeyron equation, Clausius-Clapeyron equation.

Clausius-Clapeyron equation = Clapeyron-Clausius equation.

Cloud chamber — An apparatus containing moist air or other gas which on sudden expansion condenses moisture to droplets on dust particles or other nuclei. Thus charged particles or ions in the space become nuclei and their numbers and behavior, when properly illuminated, may be studied.

Coefficient of compressibility — The relative decrease of the volume of a gaseous system with increasing pressure in an isothermal process. This coefficient is

$$-(1/V)\,(\partial V/\partial p)_T$$

where V is the volume; p is the pressure; and T is the temperature. The reciprocal of this quantity is the bulk modulus. Also called compressibility. Compare coefficient of thermal expansion, coefficient of tension.

Coefficient of tension — The relative increase of pressure of a system with increasing temperature in an isochoric process. In symbols this quantity is

$$(1/p)\,(\partial p/\partial T)_V$$

where p is pressure; T is temperature; and V is volume. Compare coefficient of compressibility, coefficient of thermal expansion.

Coefficient of thermal expansion — The ratio of the change of length per unit length (linear), or change of volume per unit volume (voluminal), to the change of temperature.

Coherence length — The maximum tolerable optical path length difference between two energy beams which are forming an interference pattern. This will vary with the degree of spectral purity of the source producing the beams. For example a perfect monochromatic source would have an infinite coherence length.

Coherent (additon) — The vector addition of both the amplitude and phases of different waves of the same frequency at a given time or at a given position.

Coherent oscillator (abbr Coho) — An oscillator which provides a reference by which the radio frequency phase difference of successive received pulses may be recognized. See Coherent reference.

Coherent reference — The reference signal, usually of stable frequency, to which other signals are phase-locked to establish coherence throughout a system.

Coherent (source) — A source radiating coherent waves.

Coherent (waves) — Waves whose frequencies are equal and whose phases are related to each other at a given time or at a given place in space. Coherence can be of two types, temporal and spatial.

Cold working — Deforming metal plastically at a temperature lower than the recrystallization temperature.

Colligative property — A property numerically the same for a group of substances, independent of their chemical nature.

Colloid — A phase dispersed to such a degree that the surface forces become an important factor in determining its properties.

In general particles of colloidal dimensions are approximately 10Å to 1 μm in size. Colloidal particles are often best distinguished from ordinary molecules due to the fact that colloidal particles cannot diffuse through membranes which do allow ordinary molecules and ions to pass freely.

Colloidal system — An intimate mixture of two substances one of which, called the dispersed phase (or colloid) is uniformly distributed in a finely divided state through the second substance, called the dispersion medium (or dispersing medium). The dispersion medium may be a gas, a liquid, or a solid, and the dispersed phase may also be any of these, with the exception that one does not speak of a colloidal system of one gas in another. Also called colloidal dispersion, colloidal suspension.

A system of liquid or solid particles colloidally dispersed in a gas is called an aerosal. A system of solid substance or water-insoluble liquid colloidally dispersed in liquid water is called a hydrosol. There is no sharp line of demarcation between true solutions and colloidal systems on the one hand, or between mere suspensions and colloidal systems on the other. When the particles of the dispersed phase are smaller than about 1 millimicron in diameter, the system begins to assume the properties of a true solution; when the particles dispersed are much greater than 1 μm, separation of the dispersed phase from the dispersing medium becomes so rapid that the system is best regarded as a suspension.

Coma — An aberration of spherical lenses, occurring in the case of oblique incidence, when the bundle of rays forming the image is unsymmetrical. The image of a point is comet shaped, hence the name.

Combination frequencies — Two vibrations of arbitrary frequencies f_1 and f_2 when applied simultaneously to a nonlinear (distorting) device will excite it to a motion containing not only the original frequencies, but also members of a set of "combination" set of frequencies given by $f_* = mf_1 + nf_2$ where m and n are integers. A resonator sharply tuned to any one of these frequencies which may be produced in the nonlinear device will resound to it with an amplitude depending on the type of nonlinearity. The superheterodyne radio receiver depends on this phenomenon.

Combining volumes — Under comparable conditions of pressure and temperature the volume ratios of gases involved in chemical reactions are simple whole numbers.

Combining weight of an element or radical is its atomic weight divided by its valence.

Combining weights, law of — If the weights of elements which combine with each other be called their "combining weights," then elements always combine either in the ratio of their combining weights or of simple multiples of these weights.

Compensation Point — The temperature (below the Néel Point) at which, in some ferrimagnetic compounds, the saturation magnetization becomes zero.

Complementary color — Either of a pair of spectrum colors that when combined give a white or nearly white light.

Component substances, law of — Every material consists of one substance, or is a mixture of two or more substances, each of which exhibits a specific set of properties, independent of the other substances.

Compounds are substances containing more than one constituent element and having properties, on the whole, different from those which their constituents had as elementary substances. The composition of a given pure compound is perfectly definite, and is always the same no matter how that compound may have been formed.

Compressibility — Reciprocal of the bulk modulus.

Compton effect, (Compton recoil effect) — Elastic scattering of photons by electrons results in decrease in frequency and increase of wave length of X-rays and gamma-rays when scattered by free electrons.

Compton electron — An orbital electron of an atom which has been ejected from its orbit as a result of an impact by a high-energy quantum of radiation (X-ray or gamma-ray). Also called Compton recoil electron.

Compton wavelength (symbol λ_c) — Of a particle, the distance h/mc, where h is the Planck constant, m is the mass of the particle, and c is the velocity of light.

The Compton wavelength of the electron (symbol λ_c) is 2.4261×10^{-10} centimeter; of the proton (symbol $\lambda_{c,p}$) is 1.32140×10^{-13} centimeter.

Computer-generated holograms — A hologram made synthetically and based on computer calculations of amplitude and/or phase.

Concentration — The amount of a substance in weight, moles, or equivalents contained in unit volume.

Condensers in parallel and series — If c_1, c_2, c_3, etc. represent the capacitances of a series of condensers and C their combined capacitance,

when in parallel,

$$C = c_1 + c_2 + c_3 \ldots$$

when in series,

$$\frac{1}{C} = \frac{1}{c_1} + \frac{1}{c_2} + \frac{1}{c_3} \ldots$$

Conductance — The reciprocal of resistance, is measured by the ratio of the current flowing through a conductor to the difference of potential between its ends. The practical unit of conductance, the mho, the conductance of a body through which one ampere of current flows when the potential difference is one volt. The conductance of a body in mho is the reciprocal of the value of its resistance in ohms. Dimensions, $[\varepsilon l t^{-1}]$. $[\mu^{-1} l^{-1} t]$.

Conductivity, electrical, is measured by the quantity of electricity transferred across unit area, per unit potential gradient per unit time. Reciprocal of resistivity. Volume conductivity or specific conductance, $k = 1/\varrho$ where ϱ is the volume resistivity. Mass conductivity = k/d where d is density. Equivalent conductivity A = k/c where c is the number of equivalents per unit volume of solution. Molecular conductivity μ = k/m where m is the number of moles per unit volume of solution. Dimensions: volume conductivity, $[\varepsilon t^{-1}]$; $[\mu^{-1} l^{-2} t]$, mass conductivity, $[\varepsilon m^{-1} l^3 t^{-1}]$; $[\mu^{-1} m^{-1} lt]$.

Conductivity, thermal — Time rate of transfer of heat by conduction, through unit thickness, across unit area for unit difference of temperature. It is measured as calories per second per square centimeter for a thickness of one centimeter and a difference of temperature of 1°C.

If the two opposite faces of a rectangular solid are maintained at temperatures t_1 and t_2 the heat conducted across the solid of section a and thickness d in a time T will be,

$$Q = \frac{K(t_2 - t_1)aT}{d}$$

K is a constant depending on the nature of the substance, designated as the specific heat conductivity. K is usually given for Q in calories, t_1 and t_2 in °C, a in cem², T in sec, and d in cm. See Heat conductivity.

Conductors — A class of bodies which are incapable of supporting electric strain. A charge given to a conductor spread to all parts of the body.

Conjugate foci — Under proper conditions light divergent from a point on or near the axis of a lens or spherical mirror is focused at another point. The point of convergence and the position of the source are interchangeable and are called conjugate foci.

Conservation of energy — The principle that the total energy of an isolated system remains constant if no interconversion of mass and energy takes place.

This principle takes into account all forms of energy in the system; it therefore provides a constraint on the conversions from one form to another.

Conservation of energy, law of — Energy can neither be created nor destroyed and therefore the total amount of energy in the universe remains constant.

Conservation of mass — In all ordinary chemical changes, the total of the reactants is always equal to the total mass of the products.

Conservation of momentum, law of — For any collision, the vector sum of the momenta of the colliding bodies after collison equal the vector sum of their momenta before collision. If two bodies of masses m_1 and m_2 have, before impact velocities v_1 and v_2 and after impact velocities u_1 and u_2

$$m_1 u_1 + m_2 u_2 = m_1 v_1 + m_2 v_2$$

Constant of aberration — The maximum aberration of a star observed from the surface of the earth, 20.49 sec of arc.

The maximum occurs at the time the direction of motion of the earth in its orbit is at right angles to a line from the earth to the star.

Constitutive property — A property which depends on the constitution or structure of the molecule.

Contrast of fringes — The relative difference between the brightness or density of successive bright and dark fringes on a hologram or interferogram.

Cooling — Processing highly radioactive materials to attain lesser radioactivity for subsequent use or handling.

Coriolis acceleration — An acceleration of a particle moving in a relative coordinate system. The total acceleration of the particle, as measured in an inertial coordinate system, may be expressed as the sum of the acceleration within the relative system, the acceleration of the relative system itself, and the coriolis acceleration.

Physically, coriolis acceleration may be considered as coming from the conservation of momentum in a body moving in a direction not parallel to the axis of rotation of the relative system.

Mathematically, coriolis acceleration comes from the differentiation of terms containing the angular velocity ω in the expression for the absolute velocity of the particle.

In the case of the earth, moving with angular velocity ω, a particle moving relative to the earth with velocity v has the coriolis acceleration $2\omega \times v$. If Newton laws are to be applied in the relative system, the coriolis acceleration and the acceleration of the relative system must be treated as forces.

Cosmic rays — Highly penetrating radiations which strike the earth, assumed to originate in interstellar space. They are classed as: primary, coming from the assumed source, and secondary, those induced in upper atmospheric nuclei by collisions with primary cosmic rays.

Cosmotron — A particle accelerator capable of giving them energies to billions of electron volts.

Couette flow — The shearing flow of a fluid between two parallel surfaces in relative motion. A two-dimensional steady flow without pressure gradient in the direction of flow and caused by the tangential movement of the bounding surfaces. The only practical type is the flow between concentric rotating cylinders (as of the oil in a cylindrical bearing).

Counterradiation — The downward flux of atmospheric radiation passing through a given level surface, usually taken as the earth's surface. Also called back radiation.

This result of infrared (long-wave) absorption and reemission by the atmosphere is the principal factor in the greenhouse effect.

Coulomb (unit of quantity of electricity) — the quantity of electricity transported in 1 sec by a current of 1 A. A unit quantity of electricity. It is the quantity of electricity which must pass through a circuit to deposit 0.0011180 g of silver from a solution of silver nitrate. An ampere is 1 coulomb/sec. A coulomb is also the quantity of electricity on the positive plate of a condenser of one-farad capacity when the electromotive force is 1 v.

Couple — Two equal and oppositely directed parallel but not colinear forces acting upon a body form a couple. The moment of the couple or torque is given by the product of one of the forces by the perpendicular distance between them. Dimension, $[m\,l^2\,t^{-2}]$.

Couple acting on a magnet of magnetic moment $m\ell$ in a field of strength H. If the magnet is perpendicular to the direction of the field

$$C = Hml = HM$$

If the angle between the magnet and the field is θ

$$C = Hml \sin \theta$$

The couple will be in dyn-cm for cgs electromagnetic units of H, m and l.

Critical mass — The minimum mass the fissile material must have in order to maintain a spontaneous fission chain reaction. For pure U^{235} it is computed to be about 20 lb.

Critical point — The thermodynamic state in which liquid and gas phases of a substance coexist in equilibrium at the highest possible temperature. At higher temperatures than the critical no liquid phase can exist. For water substance the critical point is

$$P_s = 2.21 \times 10^5 \text{ mbar}$$
$$T = 647°K$$
$$v = 3.10\ g/cm^3$$

where P_s is the saturation vapor pressure of the water vapor; T is the Kelvin temperature; and v is the specific volume.

Critical temperature — 1. The temperature above which a substance cannot exist in the liquid state, regardless of the pressure. 2. As applied to materials, the temperature at which a change in phase takes place causing an appreciable change in the properties of the material.

Cross section (Nuclear cross section) — A measure of the probability of a particular process. The nuclear cross section is expressed by a/bc, where a is the number of processes occurring, b the number of incident particles, and c the number of target nuclei per cm^2. Thereare nuclear cross sections for fission, for slow neutron capture, for Compton collision, and for ionization by electron impact.

Crossed polarizer — A dual polarization filter and transducer which transforms varying orientations of polarized waves into an amplitude output.

Cryohydrate — The solid which separates when a saturated solution freezes. It contains the solvent and the solute in the same proportions as they were in the saturated solution.

Cryopumping — The process of removing gas from a system by condensing it on a surface maintained at very low temperatures.

Cryotron — A device based upon the principle that superconductivity established at temperatures near absolute zero is destroyed by the application of a magnetic field.

Crystal — The "ideal crystal" is a homogeneous portion of crystalline matter, (q.v.) whether bounded by faces or not.

Crystalline matter is matter that possesses a triperiodic structure on the atomic scale. It is characterized by discontinuous vectorial properties that give rise to "crystal planes" [(1) crystal growth (faces); (2) cohesion (cleavage planes); (3) twinning (twin planes); (4) gliding (gliding planes); (5) X-ray, electron, or neutron diffraction ("reflecting" planes); all of which are parallel to lattice planes.]

Curie — The curie is the SI unit rate of radioactive decay; the quantity of any radioactive nuclide which undergoes 3.7×10^{10} disintegrations/sec. The symbol for this unit is Ci. 1 Ci = 3.7×10^{10} Bq.

Curie's law — The intensity of magnetization,

$$I = \frac{AH}{T}$$

where H_s is the magnetic field strength, T the absolute temperature and A Curie's constant. Used for paramagnetic substances.

Curie point — All ferro-magnetic substances have a definite temperature of transition at which the phenomena of ferro-magnetism disappear and the substances become merely paramagnetic. This temperature is called the "Curie Point" and is usually lower than the melting point.

Curie-Weiss law — The Curie law was modified by Weiss to state that the susceptibility of a paramagnetic substance above the Curie point varies inversely as the excess of the temperature above that point.

This law is not valid at or below the Curie point.

Current (electric) — The rate of transfer of electricity. The transfer at the rate of one electrostatic unit of electricity in one second is the electrostatic unit of current. The electromagnetic unit of current is a current of such strength that one centimeter of the wire in which it flows is pushed sideways with a force of 1 dyn when the wire is at right angles to a magnetic field of unit intensity. The practical unit of current is the ampere, a transfer of one coulomb per second, which is one tenth the electromagnetic unit. The international ampere is the unvarying electric current which, when passed through a solution of silver nitrate in accordance with certain specifications, deposits silver at the rate of 0.00111800 g/sec. The international ampere is equivalent to 0.999835 absolute ampere. The ampere-turn is the magnetic potential produced between the two faces of a coil of one turn carrying one ampere. Dimensions, $[\epsilon^{1/2}\, m^{1/2}\, l^{1/2}\, t^{-2}]$; $[\mu^{-1/2}\, m^{1/2}\, l^{1/2}\, t^{-1}]$.

Current in a simple circuit — The current in a circuit including an external resistance R and a cell of electromotive force E and internal resistance,

$$I = \frac{E}{R + r}$$

If E is in volts and r and R in ohms the current will be in amperes.

For two cells in parallel,

$$I = \frac{E}{R + \frac{r}{2}}$$

For two cells in series,

$$I = \frac{2E}{R + 2r}$$

CW laser — Continuous wave laser — a laser that radiates its energy in an uninterpreted beam.

Cyclotron — The magnetic resonance accelerator for imparting very great velocities to heavier nuclear particles without the use of excessive voltages.

Dalton = atomic mass unit.

Dalton's law of partial pressures — The pressure exerted by a mixture of gases is equal to the sum of the separate pressures which each gas would exert if it alone occupied the whole volume. This fact is expressed in the following formula:p

$$PV = V(p_1 + p_2 + p_3, \text{etc.})$$

Day — 1. The duration of one rotation of the earth, or another celestial body, on its axis.

A day is measured by successive transits of a reference point on the celestial sphere over the meridian, and each type takes its name from the reference used. Thus, for a solar day the reference is the sun; a mean solar day if the mean sun; and an apparent solar day if the apparent sun. For a lunar day the reference is the moon for a sidereal day the vernal equinox; for a constituent day an astre fictif or fictitious star. The expression lunar day refers also to the duration of one rotation of the moon with respect to the sun. A Julian day is the consecutive number of each day, beginning with January 1, 4713 BC.

2. A period of 24 hr beginning at a specified time, as the civil day beginning at midnight, or the astronomical day beginning at noon.

DeBroglie wavelength — In quantum mechanics, a wavelength (λ) attributed to a particle by virtue of its momentum. In general

$$\lambda = \frac{h}{mv} = \frac{h}{m_0 v}\left(1 - \frac{v^2}{c^2}\right)^{1/2}$$

where m is the observed mass of the particle, m_o is its rest mass, v is its velocity, c is the velocity of light, and h is Planck's constant.

Debye-Falkenhagen effect — The increase in the conductance of an electrolytic solution produced by alternating currents of sufficiently high frequencies over that observed with low frequencies or with direct current.

Debye length — A theoretical length which describes the maximum separation at which a given electron will be influenced by the electric field of a given positive ion. Sometimes referred to as the Debye shielding distance or plasma length.

It is well known that charged particles interact through their own electric fields. In addition, Debye has shown that the attractive force between an electron and ion which would otherwise exist for very large separations is indeed cut off for a critical separation due to the presence of other positive and negative charges in between. This critical separation or Debye length decreases for increased plasma density.

Decay — Diminution of a radioactive substance due to nuclear emission of alpha or beta particles, gamma rays or positrons.

Decay constant — 1. = attenuation constant. 2. (symbol λ) A constant relating the instant rate of radioactive decay of a radioactive species to the number of atoms N present at a given time t. Thus,

$$-(\partial N/\partial t) = \lambda N$$

If N_o is the number of atoms present at time zero then

$$N = N_o e^{-\lambda t}$$

Decay product — A nuclide resulting from the radioactive disintegration of a radionuclide, being formed either directly or as the result of successive transformations in a radioactive series. Also called daughter, daughter element.

A decay product may be either radioactive or stable.

Deci — A prefix meaning one tenth or 10^{-1}. Symbol is d.

Decibel (abbr. db) — 1. A dimensionless measure of the ratio of two powers, equal to 10 times the logarithm to the base 10 of the ratio of two powers P_1/P_2. One tenth of a bel.

The power P_2 may be some reference power; in electricity, the reference power is sometimes taken as 1 milliwatt (abbr. dbm); in acoustics, the decibel is often taken as 20 times the common logarithm of the sound pressure ratio, with the reference pressure as 0.0002 dyn/cm².

Declination — 1. (symbol δ) Angular distance north or south of the celestial equator; the arc of an hour circle between the celestial equator and a point on the celestial sphere, measured northward or southward from the celestial equator through 90°, and labeled N or S to indicate the direction of measurement. 2. (symbol D) Magnetic declination. See Equatorial system.

Decomposition is the chemical separation of a substance into two or more substances, which may differ from each other and from the original substances.

Definite proportions, law of — In every sample of each compound substance the proportions by weight of the constituent elements are always the same.

Degree — Angle subtended at the center by a circular arc which is 1/360 of the circumference.

Degree of association $(1 - \alpha)$ — The degree of association of an electrolytic solution is the percentage of ions associated into nonconducting species, such as ion-pairs.

Degree of dissociation (or ionization) in general, α — The degree of dissociation (or ionization) of an electrolytic solution is the percentage of solute (or electrolyte) in the dissociated (or ionized) state in solution. Classically this degree is obtained from conductance measurements from the ratio, Λ/Λ_i where Λ_i is the equivalent conductance an electrolytic solution would have at some finite concentration if it were completely dissociated into ions at that concentration if it were completely dissociated into ions at that concentration. (See ionogens). This symbol is also used to denote the fraction of free ions in a solution when simple ions, ion pairs, and clusters higher than ion pairs are present. (See ionophores).

Degree of freedom — The number of the variables determining the state of a system (usually pressure, temperature, and concentrations of the components) to which arbitrary values can be assigned.

Deka — A prefix meaning ten or 10^1. Symbol is da.

Delayed neutrons — Neutrons emitted by excited nuclei in a radioactive process, so called because they are emitted an appreciable time after the fission. Compare prompt neutrons.

Delta ray — 1. An electron ejected by recoil when a rapidly moving alpha particle or other charged particle passes through matter. 2. By extension any secondary ionizing particle ejected by recoil when a primary particle passes through matter.

Density — Concentration of matter, measured by the mass per unit volume. Dimensions, $[m\, l^{-3}]$.

Density (of film) — The logarithm of the reciprocal of the optical transmission of the film.

Depletion layer — In a semiconductor, a region in which the mobile carrier charge density is insufficient to neutralize the net fixed charge density of donors and acceptors. Also called barrier.

Detonation wave — A shock wave in a combustible mixture, which originates as a combustion wave.

Deuterium (symbol D, d) — A heavy isotope of hydrogen having one proton and one neutron in the nucleus.

The symbol D is often used to designate deuterium in compounds, as HDO for molecules of that composition. Official chemical nomenclature uses the designation d with a number which designates the carbon atom to which the deuterium is bound; e.g., 2-d propane designates CH_3CHDCH_3.

Deuteron — Nucleus of the deuterium atom or the ion of deuterium. Its structure — one neutron and one proton.

Dewpoint — The temperature to which a given parcel of air must be cooled at constant pressure and constant water-vapor content in order for saturation to occur; the temperature at which the saturation vapor pressure of the parcel is equal to the actual vapor pressure of the contained water vapor. Any further cooling usually results in the formation of dew or frost. Also called dewpoint temperature.

When this temperatue is below 0°C, it is sometimes called the frost point.

Diabatic process — A process in a thermodynamic system in which there is a transfer of heat across the boundaries of the system.

Diabatic process is preferred to nonadiabatic process.

Diamagnetic materials — Are those within which an externally applied magnetic field is slightly reduced because of an alteration of the atomic electron orbits produced by the field. Diamagnetism is an atomic-scale consequence of the Lenz law of induction. The permeability of diagmagnetic materials is slightly less than that of empty space.

Dielectric — A material having a relatively low electrical conductivity; an insulator; a substance that contains few or no free electrons and which can support electrostatic stresses. The principal properties of a dielectric are its dielectric constant (the factor by which the electric field strength in a vacuum exceeds that in the dielectric for the same distribution of charge), its dielectric loss (the amount of energy it dissipates as heat when placed in a varying electric field), and its dielectric strength (the maximum potential gradient it can stand without breaking down).

In an electromagnetic field, the centers of the nonpolar molecules of a dielectric are displaced, and the polar molecules become oriented close to the field. The net effect is the appearance of charges at the boundaries of the dielectric. The frictional work done in orientation absorbs energy from the field which appears as heat. When the field is removed the orientation is lost by thermal agitation and so the energy is not regained. If free-charge carriers are present they too can absorb energy.

A good dielectric is one in which the absorption is a minimum. A vacuum is the only perfect dielectric. The quality of an imperfect dielectric is its dielectric strength; and the accumulation of charges within an imperfect dielectric is termed dielectric absorption.

Dielectric constant (symbol ε) — for a given substance, the ratio of the capacity of a condenser with that substance as dielectric to the capacity of the same condenser with a vacuum for dielectric. It is a measure, therefore, of the amount of electrical charge a given substance can withstand at a given electric field strength; it should not be confused with dielectric strength.

The dielectric constant ε is a function of temperature and frequency and is wirtten as a complex quantity

$$\epsilon = \epsilon' - i\epsilon''$$

where ε′ is the part that determines the displacement current and ε″ the dielectric absorption (see dielectric). For a nonabsorbing, nonmagnetic material ε′ is equal to the square of the index of refraction and the relation holds only at the particular frequency where these conditions apply.

$$F = \frac{QQ'}{\epsilon r'}$$

where F is the force of attraction between two charges Q and Q′ separated by a distance r in a uniform medium.

Dielectrics or insulators or nonconductors — A class of bodies supporting an electric strain. A charge on one part of a nonconductor is not communicated to any other part.

Dielectric strength — A measure of the resistance of a dielectric to electrical breakdown under the influence of strong electric fields; usually expressed in volts per centimeter. Sometimes called breakdown potential.

Diffraction — That phenomena produced by the spreading of waves around and past obstacles which are comparable in size to their wavelength.

Diffraction efficiency — Ratio of energy projected into the reconstructed image to the energy illuminating the hologram.

Diffraction fanning — The fanning out of a light or energy beam as it pours through a very narrow aperture (opening).

Diffraction grating — If s is the distance between the rulings, d the angle of diffraction, then the wave length where the angle of incidence is 90% is (for the nth order spectrum),

$$\lambda = \frac{s \sin d}{n}$$

If i is the angle of incidence, d the angle of diffraction, s the distance between the rulings, n the order of the spectrum, the wave length is,

$$\lambda = \frac{s}{n} (\sin i + \sin d).$$

A mask or special aperture used to break up a white light beam or composite energy beam into its various spectral components through the mechanism of diffraction.

Diffuse reflection — Scattering at all angles from the point of reflection.

Diffuse sky radiation — Solar radiation reaching the earth's surface after having been scattered from the direct solar beam by molecules or suspensoids in the atmosphere. Also called the skylight, diffuse skylight, sky radiation.

Of the total light removed from the direct solar beam by scattering in the atmosphere (approximately 25% of the incident radiation), about two-thirds ultimately reaches the earth as diffuse sky radiation.

Diffusion — If the concentration (mass of solid per unit volume of solution) at one surface of a layer of liquid is d_1 and at the other surface d_2, the thickness of the layer h and the area under consideration A, then the mass of the substance which diffuses through the cross-section A in time t is,

$$m = \Delta A \frac{(d_2 - d_1)t}{h}$$

where Δ is the coefficient of diffusion.

Diffusion Coefficient — If the concentration (mass of solid per unit volume of solution) at one surface of a layer of liquid is d_1, and at the other surface d_2, the thickness of the layer is h, the area under consideration is A, and the mass of a given substance which diffuses through the cross section A in time t is m, then the diffusion coefficient is defined as

$$D = \frac{mh}{A(d_2 - d_1)t}$$

Diffusivity — A measure of the rate of diffusion of a substance, expressed as the diffusivity coefficient K. When K is constant, the diffusion equation is

$$\frac{\partial q}{\partial t} = K\nabla^2 q$$

where q is the substance diffused; ∇^2 is the Laplacian operator; and t is time. The diffusivity has dimensions of a length times a velocity; it varies with the property diffused, and for any given property it may be considered a constant or a function of temperature, space, etc., depending on the context. Also called coefficient of diffusion. See conductivity, kinematic viscosity, exchange coefficients.

In the case of molecular diffusion the length dimension is the mean free path of the molecules. By analogy, in eddy diffusion, length becomes the mixing length. The coefficient is then called the eddy diffusivity, and is in general several orders of magnitude larger than the molecular diffusivity.

Diffusivity of heat — is given by Δ in the equation

$$\frac{dH}{dt} = -\Delta sd \frac{dT}{dx} dy\, dz$$

where dH is the quantity of heat passing through the area dy dz in the direction of x in a time dt. The rate of variation of temperature along x is given by dT/dx, s is specific heat and d, density. Dimensions, $[l^2 t^{-1}]$.

Dimensional formulae — If mass, length, and time are considered fundamental quantities, the relation of other physical quantities and their units to these three may be expressed by a formula involving the symbols l-m and t respectively, with appropriate exponents. For example; the dimensional formula for volume would be expressed, $[l^3]$; velocity, $[t^{-1}]$; force $[mlt^{-2}]$. Other fundamental quantities used in dimensional formulae may be indicated as follows: θ, temperature, ε the dielectric constant of a vacuum; μ, the magnetic permeability of a vacuum.

Diminution of pressure at the side of a moving stream — If a fluid of density d moves with a velocity v, the dimunution of pressure due to the motion is (neglecting viscosity),

$$p = \tfrac{1}{2} dv^2$$

Dip — The angle measured in a vertical plane between the direction of the earth's magnetic field and the horizontal.

Dipole — (1) A combination of two electrically or magnetically charged particles of opposite sign which are separated by a very small distance. (2) Any system of charges, such as a circulating current, which has the properties that: (a) no forces act on it in a uniform field; (b) a torque proportional to sin θ, where θ is the angle between the dipole axis and a uniform field, does act on it; (c) it produces a potential which is proportional to the inverse square of the distance from it.

Dipole moment — A mathematical entity; the product of one of the charges of a dipole unit by the distance separating the two dipolar charges. In terms of the definition of a dipole (2), the dipole moment p is related to the torque T, and the field strength E (or B) through the equation:

$$T = p \times E$$

Dipole moment, molecular — It is found from measurements of dielectric constant (i.e. by its temperature dependence, as in the Debye equation for total polarization) that certain molecules have permanent dipole moments. These moments are associated with transfer of charge within the molecule and provide valuable information as to the molecular structure.

Directly ionizing particles — are charged particles (electrons, protons, alpha particles, etc.) having sufficient kinetic energy to produce ionization by collision.

Direct solar radiation — In actinometry, that portion of the radiant energy received at the instrument direct from the sun, as distinguished from diffuse sky radiation, effective terrestrial radiation, or radiation from any other source. See global radiation.

Direct solar radiation is measured by pyrheliometers.

Dispersion — The difference between the index of refraction of any substance for any two wave lengths is a measure of the dispersion for these wave lengths, called the coefficient of dispersion.

Dispersion forces — The force of attraction between molecules possessing no permanent dipole. The interaction energy is given by

$$U_D = -\frac{3}{4} h \frac{V_0 \alpha^2}{r^6}$$

where h is Planck's constant, V_0 a characteristic frequency of the molecule, r the distance between the molecules, and α the polarizability.

Dispersive power — If n_1 and n_2 are the indices of refraction for wave lengths λ_1 and λ_2 and n the mean index or that for sodium light, the dispersive power for the specified wave length is,

$$\omega = \frac{n_2 - n_1}{n - 1}$$

Displacement is a reaction in which an elementary substance displaces and sets free a constituent element from a compound.

Displacement or elongation at any instant. The distance of a vibrating or oscillating particle from its position of equilibrium.

Dissociation-field effect — The increased dissociation (or ionization) of the molecules of weak electrolytes under the influence of high electrical fields (potential gradients).

Distribution law — A substance distributes itself between two immiscible solvents so that the ratio of its concentrations in the two solvents is approximately a constant (and equal to the ratio of the solubilities of the substance in each solvent). The above statement requires modification if more than one molecular species is formed.

Donor — In transistors, the N-type semiconductor, the electrode containing impurities which increase the number of available electrons. Contrast acceptor.

Doppler broadening — The broadening of either an emission line or an absorption line due to random motions of molecules of the gas that is emitting or absorbing the radiant energy. See pressure broadening.

In the case of an emitting gas, for example, those molecules which are approaching the observer as they emit quanta of radiant energy will, because of the Doppler effect, appear to send out a train of waves of slightly shorter wavelength than that characteristic of a stationary molecule, while receding molecules will appear to emit slightly longer wavelengths. The net effect, averaged over many molecules, is to superimpose, on the natural line width, a bell-shaped broadening that is proportional to the square root of the absolute temperature of the gas.

Doppler effect — The change in frequency with which energy reaches a receiver when the receiver and the energy source are in motion relative to each other. Also called Doppler shift.

In the case of sound, or any other wave motion where a real medium of propagation exists (excepting, therefore, light and other electromagnetic radiations) one must distinguish two principal cases: If the source is in motion with speed v relative to a medium which propagates the waves in question at speed c, then the resting observer receives waves emitted with actual frequency f as if they had a frequency f′ given by the Doppler equation

$$f' = f/[1 \pm (v/c)]$$

where the positive sign refers to the case of the source receding from the observer, and vice versa for the negative sign. If, on the other hand, the source is at rest relative to the propagating medium while the observer moves with speed v relative to the source,

$$f' = f'[1 \pm (v/c)]$$

where the positive sign now refers to the case of observer approaching the source.

For electromagnetic radition,

$$f/f' = [1 \pm (v/c)]/[1 \pm (v/c)]$$

where the top signs represent the source receding from the observer and the bottom signs, approaching the observer.

Dosimeter — 1. An instrument for measuring the ultraviolet in solar and sky radiation. Compare actinometer. 2. A device, worn by persons working around radioactive material, which indicates the dose of radiation to which they have been exposed.

Double decomposition consists of a simple exchange of the parts of two substances to form two new substances.

Double pass transmittance hologram — A hologram whose object wave was transmitted through the transparent object media to a mirror, reflected back through again, and recorded on the plate.

Dulong and Petit, law of — The specific heats of the several elements are inversely proportional to their atomic weights. The atomic heats of solid elements are constant and approximately equal to 6.3. Certain elements of low atomic weight and high melting point have, however, much lower atomic heats at ordinary temperatures.

Dynamic height — The height of a point in the atmosphere expressed in a unit proportional to the geopotential at that point. Since the geopotential at altitude z is numerically equal to the work done when a particle of unit mass is lifted from sea level up to this height, the dimensions of dynamic height are those of potential energy per unit mass. Also called geodynamic height.

The standard unit of dynamic height H_d is the dynamic meter (or geodynamic meter), defined as 10/sec²; it is related to the geopotential φ, the geometric height z in meters, and the geopotential height Z in geopotential meters by

$$d\varphi = 10dH_d = 9.8dZ = gdz$$

where g is the acceleration of gravity in meters per second squared. (Some sources prefer to give the constants 10 and 9.8 the units of meters per second squared so that the units of φ and Z would be the same as those of the geometric height.) The dynamic meter is about 2% longer than the geometric meter and the geopotential meter. One of the practical advantages of the dynamic height over the geometric height is that when the former is introduced into the hydrostatic equation the variable acceleration of gravity is eliminated. In meteorological height calculations, geopotential height is more often used than dynamic height.

Dynamic pressure (symbol q) = The pressure of a fluid resulting from its motion, equal to one-half the fluid density times the fluid velocity squared ($1/2\varrho V^2$). In incompressible flow, dynamic pressure is the difference between total pressure and static pressure. Also called kinetic pressure. Compare impact pressure.

Dynamic Viscosity — Of a fluid, the ratio of the shearing stress to the shear of the motion. It is independent of the velocity distribution, the dimensions of the system, etc. and for a gas it is independent of pressure except at very low pressures. Also called coefficient of molecular viscosity, coefficient of viscosity.

For the dynamic viscosity μ of a perfect gas, the kinetic theory of gases gives

$$\mu = 1/3\ (\rho c l)$$

where ϱ is the gas density, c is the average speed of the random heat motion of the gas molecules and is proportional to the square root of the temperature, and L is the mean free path. For dry air at 0° C, the dynamic viscosity is about 1.7×10^{-4} g/cm/sec.

Whereas the dynamic viscosity of most gases increases with increasing temperature, that of most liquids, including water, decreases rapidly with increasing temperature.

Dyne — The force necessary to give acceleration of one centimeter per second per second to one gram of mass.

Earth Current — A large-scale surge of electric charge within the earth's crust, associated with a disturbance of the ionosphere.

Current patterns of quasi-circular form and extending over areas the size of whole continents have been identified and are known to be closely related to solar-induced variations in the extreme upper atmosphere.

Eddy current — A current induced in a mass of conducting material by a varying magnetic field. Also called Foucault current.

Eddy viscosity — The turbulent transfer of momentum by eddies giving rise to an internal fluid friction, in a manner analogous to the action of molecular viscosity in laminar flow, but taking place on a much larger scale.

The value of the coefficient of eddy viscosity (an exchange coefficient) is of the order of 10^4 cm²/sec, or 100,000 times the molecular kinematic viscosity.

Effective neutron cycle time — The lifetime of an average neutron within a reactor from the time it is produced to the time it is fission captured.

This average takes into account delayed as well as prompt neutrons.

Effective radius of the earth — A fictitious value for the radius of the earth, used in place of the geometrical radius to correct for atmospheric refraction when the index of refraction in the atmosphere changes linearly with height. See modified index of refraction.

Under conditions of standard refraction the effective radius of the earth is

8.5×10^5 m, or four thirds the geometrical radius. If the effective radius is used in ray tracing diagrams, the rays may be drawn as though they were traveling in straight lines.

Effective terrestrial radiation — The amount by which outgoing infrared terrestrial radiation of the earth's surface exceeds downcoming infrared counter-radiation from the atmosphere. Also called nocturnal radiation, effective radiation. See Actinometer.

It is to be emphasized that this amount is a positive quantity, of the order of several tenths of a langley per minute, at all times of day (except under conditions of low overcast clouds). It typically attains its diurnal maximum during the midday hours when high soil temperatures create high rates of outgoing terrestrial radiation. (For this reason the synonym nocturnal radiation is apt to lead to slight confusion.) However, in daylight hours the effective terrestrial radiation is generally much smaller than the insolation, while at night it typically dominates the energy budget of the earth's surface.

Einstein theory for mass-energy equivalence — The equivalence of a quantity of mass m and a quantity of energy E by the formula $E = mc^2$. The conversion factor c^2 is the square of the velocity of light.

Elastic collision — A collision between two particles in which no change occurs in the internal energy of the particles, or in the sum of their kinetic energies. Commonly referred to as a billiard-ball collision.

Elasticity — The property by virtue of which a body resists and recovers from deformation produced by force.

Elastic limit — The smallest value of the stress producing permanent alteration.

Elastic moduli —

Young's modulus by stretching — If an elongation s is produced by the weight of the mass m, in a wire of length l, and radius r, the modulus,

$$M = \frac{mgl}{\pi r^2 s}$$

Young's modulus by bending, bar supported at both ends. If a flexure s is produced by the weight of mass m, added midway between the supports speared by a distance l for a rectangular bar with vertical dimensions of cross-section a and horizontal dimension b, the modulus is,

$$M = \frac{mgl^3}{4sa^3 b}$$

For a cylindrical bar of radius r,

$$M = \frac{mgl^3}{12\pi r^4 s}$$

For a bar supported at one end. In the case of a rectangular bar as described above,

$$M = \frac{4mgl^3}{sa^3 b}$$

For a round bar supported at one end,

$$M = \frac{4mgl^3}{3\pi r^4 s}$$

Modulus of rigidity — If a couple C (= mgx) produces a twist of θ radians in a bar of length l and radius r, the modulus is

$$M = \frac{2Cl}{\pi r^4 \theta}$$

The substitution in the above formulae for the elastic coefficients of m in grams, g in cm/sec², l, a, b, and r in cm, s in cm, and C in dyne-cm will give moduli in dyn/cm². The dimensions of elastic moduli are the same as of stress, $[ml^{-1} t^{-2}]$.

Coefficient of restitution — Two bodies moving in the same straight line, with velocities v_1 and v_2 respectively, collide and after impact move with velocities v_3 and v_4. The coefficient of restitution is

$$C = \frac{v_4 - v_3}{v_2 - v_1}$$

Electret n — A piece of dielectric material that has a permanent electric polarity; the electrostatic analog of a permanent magnet.

Electric dipole — A pair of equal and opposite charges an infinitesimal distance apart.

Electric field intensity is measured by the force exerted on unit charge. Unit field intensity is the field which exerts the force of one dyne on unit positive charge. Dimensions, $[\epsilon^{-1/2} m^{1/2} l^{-1/2} t^{-1}]$; . $[\mu^{1/2} m^{1/2} l^{1/2} t^{-2}]$.

The field intensity or force exerted on unit charge at a point distant r from a charge q in a vacuum

$$H = \frac{q}{r^2}$$

If the dielectric in the above cases is not a vacuum the dielectric constant ϵ must be introduced. The formula becomes

$$H = \frac{q}{\epsilon r^2}$$

The value of ϵ is frequently considered unity for air. If the dielectric constant of a vacuum is considered unity the value for air at 0°C and 760 mm pressure is 1.000576.

Electric potential — The work per unit charge spent in moving a charged body in an electric field from a reference point to a point of interest (P). Commonly, the reference point is chosen as infinity. The potential (V) is positive if work is done on the charge and negative if work is required of the charge to move in the existing field. Analytically, assuming an electric field of intensity E,

$$V = \int_{\infty}^{\rho} E \cdot ds$$

where ds is a vector element of the path from ∞ to P.

Electrochemical equivalent of an ion is the mass liberated by the passage of unit quantity of electricity.

Electrolysis — If a current i flows for a time t and deposits a metal whose electrochemical equivalent is e, the mass deposited is

$$m = eit$$

The value of e is usually given for mass in grams, i in amperes and t in seconds.

Electrolytic cell constant, J_e — The cell constant of an electrolytic cell is the resistance in ohms of that cell when filled with a liquid of unit resistance.

Electrolytic dissociation or ionization theory — When an acid, base or salt is dissolved in water or any other dissociating solvent, a part or all of the molecules of the dissolved substance are broken up into parts called ions, some of which are charged with positive electricity and are called cations, and an equivalent number of whch are charged with negative electricity and are called anions.

Electrolytic solution tension theory (or the Helmholtz double layer theory) — When a metal, or any other substance capable of existing in solution as ions, is placed in water or any other dissociating solvent, a part of the metal or other substances passes into solution in the form of ions, thus leaving the remainder of the metal or substances charged with an equivalent amount of electricity of opposite sign from that carried by the ions. This establishes a difference in potential between the metal and the solvent in which it is immersed.

Electromagnetic radiation — Energy, propagated through space or through material media in the form of an advancing disturbance in electric and magnetic fields existing in space or in the media. The term radiation, alone, is used commonly for this type of energy, although it actually has a broader meaning. Also called electromagnetic energy or simply radiation. See Electromagnetic spctrum.

Electromagnetic spectrum — The ordered array of known electromagnetic radiations, extending from the shortest cosmic rays, through gamma rays, X-rays, ultraviolet radiation, visible radiation, infrared radiation, and including microwave and all other wavelengths of radio energy See Ebsorption spectrum.

The division of this continuum of wavelengths (or frequencies) into a number of named subportions is rather arbitrary and, with one or two exceptions, the boundaries of the several subportions are only vaguely defined. Nevertheless, to each of the commonly identified subportions there correspond characteristic types of physical systems capable of emitting radiation of those wavelengths. Thus, gamma rays are emitted from the nuclei of atoms as they undergo any of several types of nuclear rearrangements; visible light is emitted, for the most part, by atoms whose planetary electrons are undergoing transitions to lower energy states; infrared radiations are associated with characteristic molecular vibrations and rotations; and radio waves, broadly speaking, are emitted by virtue of the accelerations of free electrons as, for example, the moving electrons in a radio antenna wire.

Electromotive force is defined as that which causes a flow of current. The electromotive force of a cell is measured by the maximum difference of potential between its plates. The electromagnetic unit of potential difference is that against which one erg of work is done in the transfer of electromagnetic unit quantity. The volt is that potential difference against which one joule of work is done in the transfer of one coulomb. One volt is equivalent to $.0^8$ electromagnetic units of potential. The international volt is the electrical potential which when steadily applied to a conductor whose resistance is one international ohm will cause a current of one international ampere to flow. The international volt = 1.00033 absolute volts. The electromotive force of a Weston standard cell is 1.0183 int. volts at 20°C. Dimensions, $[\epsilon^{-1/2}\, m^{1/2}\, t^{-1}]$, $[\mu^{1/2}\, m^{1/2}\, l^{3/2}\, t^{-2}]$.

Electromotive series is a list of the metals arranged in the decreasing order of their tendencies to pass into ionic form by losing electrons.

Electron — The electron is a small particle having a unit negative electrical charge, a small mass, and a small diameter. Its charge is $(4.80294 \pm 0.00008) \times 10^{-10}$ absolute electrostatic units, its mass. $1/1837$ that of the hydrogen nucleus, and its diameter about 10^{-12} cm. The electron has a rest mass of 0.9109534 10^{-30}kg and a magnetic moment of 9.284832×10^{-24}J T^{-1}. Every atom consists of one nucleus and one or more electrons. Cathode rays and Beta rays are electrons.

Electron-volt (eV) — Energy acquired by any charged particle carrying unit electronic charge when it falls through a potential difference of one volt. 1 electron-volt = $(1.60207 \pm 0.00007 \times 10^{-12}$ erg or 1.6020×10^{-19} joule. Multiples of this unit are also in common use: the kilo-, million-, and billion electron-volt. 1 keV = 10^3 eV; 1 meV = 10^6 eV; and 1 beV = 10^9 eV.

Electrophoretic effect — The slowing down owing to interionic attraction and repulsion, of the movement of an ion with its solvent molecules in the forward direction by ions of opposite charge with their solvent molecules moving in the reverse direction under an applied electrical field (potential gradient).

Electrostatic unit — 1. In the cgs system, the measure of electrostatic charge, defined as a charge which, if concentrated at one point in a vacuum, would repel, with a force of 1 dyn, an equal and like charge placed 1 cm away. 2. (pl.) A system of electrical units based on the electrostatic unit.

Elements are substances which cannot be decomposed by the ordinary types of chemical change, or made by chemical union.

Elliptical polarization — The polarization of a wave radiated by an electric vector rotating in a plane and simultaneously varying in amplitude so as to describe an ellipse.

Elongation — In tensile testing the elongation of a specimen is the increase in gage length, after rupture, referred to the original gage length. It is reported as percentage elongation.

Emission spectrum — The array of wavelengths and relative intensities of electromagnetic radiation emitted by a given radiator.

Each radiating substance has a unique, characteristic emission spectrum, just as every medium of transmission has its individual absorption spectrum.

Emissive power or emissivity is measured by the energy radiated from unit area of a surface in unit time for unit difference of temperature between the surface in question and surrounding bodies. For the cgs system the emissive power is given in ergs per second per square centimeter with the radiating surface at 1° absolute and the surroundings at absolute zero. See Radiation formula.

Emissivity (symbol $E\infty$) — A property of a material, measured as the emittance of a specimen of the material that is thick enough to be completely opaque and has an optically smooth surface.

Emittance (symbol E, ε) — 1. The radiant flux per unit area emitted by a body. 2. The ratio of the emitted radiant flux per unit area of a sample to that of a black body radiator at the same temperature and under the same conditions.

Spectral emittance refers to emittance measured at a specified wavelength.

Because of the two common meanings of emittance, it should be defined when used unless the context allows no misinterpretation.

Emulsion — The coating on a film or plate which is sensitive to the light illuminating it.

Enantiotropic — Crystal capable of existing in reversible equilibrium with each other.

Energy — The capability of doing work. Potential energy is energy due to position of one body with respect to another or to the relative parts of the same body. Kinetic energy is energy due to motion. Cgs units, the erg, the energy expended when a force of 1 dyn acts through a distance of 1 cm; the joule is 1 $\times 10^7$ ergs. Dimensions, $[ml^2 t^{-2}]$.

The potential energy of a mass m, raised through a distance h, where g is the acceleration due to gravity is

$$E = mgh.$$

The kinetic energy of mass m, moving with a velocity v, is

$$E = \tfrac{1}{2}mv^2$$

Energy will be given in ergs if m is in grams, g in centimeters per second square, h in centimeters and v in centimeters per second.

Energy of a charge in ergs where Q is the charge and V the potential in electrostatic units.

$$E = \tfrac{1}{2}QV$$

Energy of the electric field — If H is the electric field intensity in electrostatic units and K the specific inductive capacity, the energy of the field in ergs per cm^3 is

$$E = \frac{KH^2}{8\pi}$$

Energy of rotation — If a mass whose moment of inertia about an axis is I, rotates with angular velocity ω about this axis, the kinetic energy of rotation will be,

$$E = \tfrac{1}{2}I\omega^2$$

Energy will be given in ergs if I is in g-cm^2 and ω in radians per second.

Enthalpy, or heat content, is a thermodynamic quantity. It is equal to the sum of the internal energy of a system plus the product of the pressure-volume work done on the system. Thus

$$H = E + pv$$

where

H	=	enthalpy or heat content
E	=	internal energy of the system
p	=	pressure
v	=	volume.

Entropy — 1. A measure of the extent to which the energy of a system is unavailable. A mathematically defined thermodynamic function of state, the increase in which gives a measure of the energy of a system which has ceased to be available for work during a certain process

$$ds = (du + pdv)/T \geqslant dq/T$$

where s is specific entropy; us is specific internal energy; p is pressure; v is specific volume; T is Kelvin temperature; and q is heat per unit mass. For reversible processes,

$$ds = dq/T$$

In terms of potential temperature 0,

$$ds = c_p\,(d\theta/\theta)$$

where cp is the specific heat at constant pressure. See third law of thermodynamics.

In an adiabatic process, the entropy increases if the process is irreversible and remains unchanged if the process is reversible. Thus, since all natural processes are irreversible. and remains unchanged if the process is reversible. Thus, since all natural processes are irreversible, it is said that in an isolated system the entropy is always increasing as the system tends toward equilibrium, a statement which may be considered a form of the second law of thermodynamics.

2. In communication theory, average information content.

Ephemeris day — 86,400 ephemeris seconds. See Ephemeris time.

Ephemeris second (unit of time) — is exactly 1/31 556 925.974 7 of the tropical year of 1900, January, 0 days, and 12 hr ephemeris time.

Ephemeris time (abbr. E.T.) — The uniform measure of time defined by the laws of dynamics and determined in principle from the orbital motions of the planets, specifically the orbital motion of the earth as represented by Newcomb's Tables of the Sun. Compare universal time.

Beginning with the volume for 1960 the American Ephemeris and Nautical Almanac uses ephemeris time as the tabular argument in the fundamental ephemerides of the sun, moon, and planets.

A gravitational ephemeris expresses the position of a celestial body as a function of ephemeris time; and, at any instant, the measure of ephemeris time is the value of the argument at which the ephemeris position is the same as the actual position at the instant. The ephemeris time at any instant is obtained from observation by directly comparing observed position of the sun, moon, and planets with graviatational ephemerides of their coordinates; observations of the moon are the most effective and expeditious for this purpose. An accurate determination, however, requires observations over a more or less extended period; in practice, it takes the form of determining the time correction ΔT that must be applied to universal time (U.T.) to obtain ephemeris time:

$$\text{E.T.} = \text{U.T.} + \Delta T$$

The universal time at any instant may be obtained with little delay from observations of the diurnal motions.

The fundamental epoch from which ephemeris time is reckoned is the epoch that Newcomb designated as 1900 January 0, Greenwich mean noon, but which actually is 1900 January 0 day 12 hr E.T.; the instant to which this designation is assigned is the instant near the beginning of the calendar year A.D. 1900 when the geometric mean longitude of the Sun referred to the mean equinox of date was 279 degrees 41 minutes 48.04 seconds. Ephemeris time is the measure of time in which Newcomb's Tables of the Sun agree with observation.

The primary unit of ephemeris time is the tropical year, defined by the mean motion of the sun in longitude at the epoch 1900 January 0 day 12 hr E.T.; its length in ephemeris days is determined by the coefficient of T in Newcomb's expression for the geometric mean longitude of the sun L referred to the mean equinox of date, given among the elements of the sun.

Equatorial system — A set of celestial coordinates based on the celestial equator as the primary great circle; usually declination and hour angle or sidereal hour angle. Also called equinoctial system of coordinates, celestial equato system of coordinates.

Equilibrium, chemical — A state of affairs in which a chemical reaction and its reverse reaction are taking place at equal velocities, so that the concentrations of reacting substances remain constant.

Equilibrium constant — The product of the concentrations (or activities) of the substances produced at equilibrium in a chemical reaction divided by the product of concentrations of the reacting substances, each concentration raised to that power which is the coef-icient of the substance in the chemical equation.

Equivalent conductance of an electrolyte is defined as the conductance of a volume of solution containing one equivalent weight of dissolved substance when placed between two parallel electrodes 1 em apart, and large enough to contain between them all of the solution. Δ is never determined directly, but is calculated from the specific conductance. If C is the concentration of a solution in gram equivalents per liter, then the concentration per cubic centimeter is C/1000, and the volume containing one equivalent of the solute is, therefore, 1000/C. Since L_s is the conductance of a centimeter cube of the solution, the conductance of 1000/C cc, and hence Δ will be

$$\Delta = \frac{1000L_s}{C}$$

Equivalent temperature — 1. Isobaric equivalent temperature; the temperature that an air parcel would have if all water vapor were condensed out at constant pressure, the latent heat released being used to heat the air,

$$T_{i,e} = T[1 + (Lw/c_pT)]$$

where $T_{i,e}$ is the isobaric equivalent temperature; T is the temperature; w is the mixing ratio; L is the latent heat: and c_p is the specific heat of air at constant pressure. 2. Adiabatic equivalent temperature; The temperature that an air parcel would have after undergoing the following (physically unrealizable) process: dry-adiabatic expansion until saturated; pseudoadiabatic expansion until all moisture is precipitated out; dry-adiabatic expansion until saturated; pseudoadiabatic expansion until saturated; pseudoadiabatic expansion until all moisture is precipitated out; dry-adiabatic compression to the initial pressure. This is the equivalent temperature as read from a thermodynamic chart and is always greater than the isobaric equivalent temperature:

$$T_{a,e} = T \exp (Lw/c_pT)$$

where $T_{a,e}$ is the adiabatic equivalent temperature. Also called pseudoequivalent temperature.

Equivalent weight or combining weight of an element or ion is its atomic or formula weight divided by its valence. Elements entering into combination always do so in quantities proportional to their equivalent weights.

In oxidation-reduction reactions the equivalent weight of the reacting substances is dependent upon the change in oxidation number of the particular substance.

erg — The unit of energy or work in the centimeter-gram-second system; the work performed by a force of 1 dyne acting through a distance of 1 cm.

Escape velocity — The radial speed which a particle or larger body must attain in order to escape from the gravitational field of a planet or star. When friction is neglected, the escape velocity is

$$\sqrt{2Gm/r}$$

where G is the universal gravitational constant m is the mass of the planet or star; and r is the radial distance from the center of the planet or star. Also called escape speed.

Ettinghausen's effect (Von Ettinghausen's) — When an electric current flows across the lines of force of a magnetic field an electromotive force is observed which is at right angles to both the primary current and the magnetic field: a temperature gradient is observed which has the opposite direction to the Hall electromotive force.

Eutectic — A term applied to the mixture of two or more substances which has the lowest melting point.

Exchange coefficients — Coefficients of eddy flux (e.g., of momentum, heat, water vapor, etc) in turbulent flow, defined in analogy to those of the kinetic theory of gases (see eddy). Also called austausch coefficients, eddy coefficients, interchange coefficients.

The exchange-coefficient hypothesis states that the mean eddy flux per unit area of a conservative quantity (suitably expressed) is proportional to the gradient of the mean value of this quantity, that is,

$$\text{Mean flux per unit area} = -C_e\,(d\bar{E}/dN)$$

where C_e is the exchange coefficient; E is the mean value of the quantity; and N is the direction normal to the surface. In strict analogy to molecular properties C_e would be constant, for turbulent flow C_e turns out to depend on time and location. See eddy viscosity diffusivity.

Expansion of gases — *Charles' law or Gay-Lussac's law* — The volume of a gas at constant pressure increases proportionately to the absolute temperature. If V_1 and V_2 are volumes of the same mass of gas at absolute temperatures, T_1 and T_2,

$$\frac{V_1}{V_2} = \frac{T_1}{T_2}$$

For an original volume V_o at 0°C the volume at t° C (at constant pressure) is

$$V_t = V_o\,(1 + 0.00367t).$$

General law for gases —

$$p_t v_t = p_o v_o \left(1 + \frac{t}{273}\right)$$

where p_o, v_o, p_t, v_t represent the pressure and volume at 0° and t°C or

$$\frac{p_1 v_1}{T_1} = \frac{p_2 v_2}{T_2}$$

where p_1, v_1 and T_1 represent pressure, volume and absolute temperature in one case and p_2, v_2 and T_2 the same quantities for the same mass of gas in another.

The law may also be expressed:

$$pv = RmT$$

where m is the mass of gas at absolute temperature T. R is the gas constant which depends on the units used. Boltzmann's molecular gas constant is obtained by expressing m in terms of the number of molecules.

For volume in cm^3, pressure in dynes per cm^2 and temperature in Centigrade degrees on the absolute scale $R = 8.3136 \times 10^7$.

Reduction of a gas volume to 0°C, 760 mm pressure: — If V is the original volume at 0°C and 760 mm pressure will be,

$$V_0 = \frac{V}{(1 + \alpha t)} \frac{H}{760}$$

If d is the original density the density at 0°C and 760 mm pressure will be

$$d_0 = d(1 + \alpha t) \frac{760}{H}$$

$$\alpha = 0.00367 \text{ approximately.}$$

Extinction coefficient — The extinction coefficient ϵ is identified as

$$dI = -\epsilon I \, dx$$

or

$$I = I_o e^{-\epsilon x}$$

where I is the illuminance (luminous flux density) at the selected point in space, I_o is the illuminance at the light source, and x is the distance from the source.

When so used, the extinction coefficient equals the sum of the medium's absorption coefficient and scattering coefficient, each computed as a weighted average over all wavelengths in the visible spectrum. As long as scattering effects are primary, a in the lower atmosphere, the value of the extinction coefficient is a function of the particle size of atmospheric suspensoids. It varies in order of magnitude from 10/km with very low visibility to 0.01/km in very clear air.

Extinction cross section = scattering cross section.

Extraterrestrial radiation — In general, solar radiation received just outside the earth's atmosphere.

Fahrenheit temperature scale (abbr. F) — A temperature scale with the ice point at 32° and the boiling point of water at 212°.

Conversion with the Celsius (centrigrade) temperature scale (abbr. C) is by the formula

$$F = 9/5\,C + 32$$

Falling bodies — For bodies falling from rest conditions are as for uniformly accelerated motion except that v_o = O and g is the acceleration due to gravity. The formulae become, air resistance neglected,

$$v_t = gt, \; s = ½gt^2 \;\; v_s = \sqrt{2gs}.$$

For bodies projected vertically upward, if v is the velocity in projection, the time to reach greatest height, neglecting the resistance of the air,

$$t = \frac{v}{g}$$

Greatest height,

$$h = \frac{v^2}{2g}$$

See Projectiles,

Farad (unit of electric capacitance) — the capacitance of a capacitor between the plates of which there appears a difference of potential of 1 when it is charged by a quantity of electricity equal to 1 coulomb.

Faraday, determination of (1960) By the National Bureau of Standards which uses an electrochemical method that dissolves, rather than deposits, silver from a solution. The new value, 96,516 ± 2 coulombs (physical scale) or 96,489 ± 2 coulombs (chemical scale). NBS used its mass, time, and elecrical standards in measuring the faraday, and have found that its value agreed within 22 ppm. with the one obtained by an independent physical method using the omegratron.

Faraday constant (symbol F) — The product of the Avogadro constant N_A and the elementary charge e, $F = N_A e$ = 96,489±2 coulombs/mol.

Faraday effect — The rotation of the plane of polarization produced when plane-polarized light is passed through a substance in a magnetic field, the light traveling in a direction parallel to the lines of force. For a given substance, the rotation is proportional to the thickness traversed by the light and to the magnetic field strength.

Faraday's laws — In the process of electrolytic changes equal quantities of electricity charge or discharge equivalent quantities of ions at each electrode.

One gram equivalent weight of matter is chemically altered at each electrode for 96,489 int. coulombs, or one faraday, of electricity passed through the electrolyte.

Far field (diffraction pattern) — Diffraction pattern produced at a large range from an object which is identical to that which would be produced at an infinite range from the object. This is also called a Fraunhofer diffraction pattern

Fast neutron — A neutron of 100,000 electron-volts or greater energy.

Fast reactor — A reactor containing no moderator, so that all the fissions take place at energies on the order of 100,000 electron-volts or higher.

Femto — A prefix meaning one quadrillionth or 10^{-15}. Symbol is f.

Fermat's principle — The path followed by light (or other waves) passing through any collection of media from one specified point to another, is that-path for which the time of travel is least.

Fermi (abbr. f) — A unit of length equal to 10^{-13} cm.

Ferrimagnetic Materials — Those in which the magnetic moments of atoms or ions tend to assume an ordered but nonparallel arrangement in zero applied field, below a characteristic temperature called the Néel Point. In the usual case, within a magnetic domain, a substantial net magnetization results from the antiparallel alignment of neighboring nonequivalent sublattices. The macroscopic behavior is similar to that in ferromagnetism. Above the Neel Point, these materials become paramagnetic.

Ferrogmatic Materials — Those in which the magnetic moments of atoms or ions in a magnetic domain tend to be aligned parallel to one another in zero applied field, below a characteristic temperature called the Curie Point. Complete ordering is achieved only at the absolute zero of temperature. Within a magnetic domain, at absolute zero, the magnetization is equal to the sum of the magnetic moments of the atoms or ions per unit volume. Bulk matter, consisting of many small magnetic domains, has a net magnetization which depends upon the magnetic history of the specimen (hysteresis effect). The permeability depends on the magnetic field, and can reach values of the order of 10^6 times that of free space. Above the Currie Point, these materials become paramagnetic.

First law of thermodynamics — A statement of the conservtion of energy for thermodynamic systems (not necessarily in equilibrium). The fundamental form requires that the heat absorbed by the system serve either to raise the internal energy of the system or to do work on the environment:

$$dq = du + dw$$

where dq is the heat added per unit mass; du is the increment of specific internal energy; and dw is the specific work done by the system on the environment. Although dq and dw are not perfect differentials, their difference, dv, is always a perfect differential. Example of the application of this equation: in an adiabatic free expansion of gas into a vacuum, all three terms are zero.

For reversible processes the mechanical work is equal to the expansion against the pressure forces, i.e.,

$$dw = pdv$$

where p is the pressure and v is the specific volume. For a perfect gas, the internal energy change is proportional to the temperature change,

$$du = c_v dT$$

where c_v is the specific heat at constant volume and T is the Kelvin temperature. Therefore, the form of the first law usually used in meteorological applications is

$$dq = c_v dT + pdv$$

Use of the equation of state yields an alternative form,

$$dq = c_p dT - dp$$

where c_v is the specific heat at constant pressure.

For open systems the variation of total rather than specific quantities is important:

$$dQ = dU + pdV - hdm$$

where Q is the total heat; U is the total internal energy; V is the volume; m is the mass of the system; and h is the specific enthalpy.

If a system contains the possibility of nonmechanical work, such as work done against an electric field, this work must be included in the first law.

See Second law of thermodynamics, Third law of thermodynamics, Energy equations.

Fission — The splitting of an atomic nucleus into two more-or-less equal fragments.

Fission may occur spontaneously or may be induced by capture of bombarding particles. In addition to the fission fragments, neutrons and gamma rays are usually produced during fission.

Fleming's rule — A simple rule for relating the directions of the flux, motion, and e.m.f. in an electric machine. The forefinger, second finger and thumb, placed at right-angles to each other, represent respectively the directions of flux, e.m.f., and motion or torque. If the right hand is used the conditions are those obtaining in a generator and if the left hand is used the conditions are those obtaining in a motor.

Fluidity — The reciprocal of viscosity. The cgs unit is the rhe, the reciprocal of the poise. Dimensions, $[m^{-1} lt]$.

Fluorescence — The property of emitting radiation as the result of absorption of radiation from some other source. The emitted radiation persists only as long as the exposure is subjected to radiation which may be either electrified particles or waves. The fluorescent radiation generally has a longer wave length than that of the absorbed radiation. If the fluorescent radiation includes waves of the same length as that of the absorbed radiation it is termed resonance radiation.

f/n or f number (of optics) — The ratio of effective focal length to lens diameter.

Focal length — The distance between the optical center of a lens, or the surface of a mirror, and its focus.

Focal plane — A plane parallel to the plane of a lens or mirror and passing through the focus.

Focal point = focus, in optics.

Focus (plural focuses) — 1. That point at which parallel rays of light meet after being refracted by a lens or reflected by a mirror. Also called focal point. 2. A point having specific significance relative to a geometrical figure. See Ellipse, Hyperbola, Parabola.

Footcandle (abbr. fc) — A unit of illuminance, incident light, or illumination equal to 1 lm/ft^2. This is the illuminance provided by a light source of one candle at a distance of 1 ft, hence the name. Compare lux, phot.

Full sunlight with zenith sun produces an illuminance of the order of 10,000 footcandles on a horizontal surface at the earth's surface. Full moonlight provides an illuminance of only about 0.02 footcandle also at earth's surface. Adequate illumination for steady reading is taken to be about 10 footcandles; that for close machine work is about 30 to 40 footcandles.

Foot-lambert (abbr. ft-1) — A unit of luminance (or brightness) equal to $1/\pi$ candle per square foot, or 1 lm/ft^2.

In Great Britain this is also called the equivalent footcandle.

Force — That which changes the state of rest or motion in matter, measured by the rate of change of momentum. Absolute unit, the dyne, the force which will produce an acceleration of one centimeter per second per second in a gram mass. The gram weight or weight of a gram mass is the cgs gravitational unit. The poundal is that force which will give an acceleration of one foot per second per second to a pound mass. Dimensions, $[mlt^{-2}]$.

The force F required to produce an acceleration a in a mass m is given by

$$F = ma$$

If m is substituted in grams and a in cm per sec², F will be given in dynes.

Force between two charges, Coulomb's law — If two charges q and q′ are at a distance r in a vacuum, the force between them is,

$$F = \frac{qq'}{r^2}$$

The force will be given in dynes if q and q′ are in electrostatic units and r in centimeters.

Force between two magnetic poles — If two poles of strength m and m′ are separated by a distance r in a medium whose permeability is μ (unity for a vacuum), the force between them is,

$$F = \frac{mm'}{\mu r^2}$$

Force will be given in dynes if r is in cm and m and m′ are in cgs units of pole strength.

The strength of a magnetic field at a point distance r from an isolated pole of strength m is

$$H = \frac{m}{\mu r^2}$$

The field will be given in gauss if m and r are in cgs units.

Formula, chemical — A combination of symbols with their subscripts representing the constituents of a substance and their proportions by weight.

Fourier transform plane — Same as spatial frequency plane.

Fraunhofer's lines — When sunlight is examined through a spectroscope it is found that the spectrum is traversed by an enormous number of dark lines parallel to the length of the slit. These dark lines are known as Fraunhofer's lines. Kirchoff conceived the idea that the sun is surrounded by layers of vapors which act as filters of the white light arising from incandescent solids within and which abstract those rays which correspond in their periods of vibration to those of the components of the vapors. Thus reversed or dark lines are obtained due to the absorption by the vapor envelop, in place of the bright lines found in the emission spectrum.

Frequency — Rate of oscillation; units: 1 cycle sec^{-1} = 1 Hertz = 1 Hz. 1 megacycle = 1 megahertz = 10^6 Hz. One gigahertz = 10^9 Hz.

Frequency modulation — A form of angle modulation in which the frequency of the carrier is made to vary in accordance with the information to be transmitted, that is, given a carrier

$$C(t) = A\cos(\omega_0 t + \phi),$$

and an information function g(t), the signal transmitted is

$$C_m(t) = A\cos[(\omega_0 + g(t) - \Delta\omega)(t) + \phi].$$

When used for radio communications this system has the advantages of greater immunity from noise and other interference, at the cost of increased bandwidth.

Frequency of vibrating strings — The fundamental frequency of a stretched string is given by

$$n = \frac{1}{2l}\sqrt{\frac{T}{m}}$$

where l is the length, T, the tension and m the mass per unit length.

For a string or wire of circular section of length l, tension T, density d, and radius r, the frequency of the fundamental is

$$n = \frac{1}{2rl}\sqrt{\frac{T}{\pi d}}$$

The frequency in vibrations per second will be given if T is in dynes, r and l in cm and d in g per cm^3.

Frequency (of waves) — Number of like phase (peaks, troughs) wavefronts passing a given point in a unit of time.

Frequency standard (atomic or molecular frequency standard) — The standard is the transition between the hyperfine levels F = 4, M = 0 and F = 3, M = 0 of the ground state $^2S_{1/2}$ of the cesium-133 atom, unperturbed by external fields. This frequency of transition is assigned the value 9 192 631 770 hertz.

Fresnel — A measure of frequency, defined as equal to 10^{12} cycles/sec.

Friction, coefficient of — The coefficient of friction between two surfaces is the ratio of the force required to move one over the other to the total force pressing the two together.

If F is the force required to move one surface over another and W, the force pressing the surfaces together, the coefficient of friction,

$$k = \frac{F}{W}$$

Fringe — The locus of maximum constructive interference (light fringe) or destructive interference regions in a space where two or more coherent waves intersect. Fringes can be in two or three dimensions.

Fringe control — Methods of adjusting the position and/or characteristics of the fringe pattern of a holographic interferogram.

Fugacity (symbol f) — In thermodynamics, a measure of the tendency of a substance to escape by some chemical process from the phase in which it exists.

Fundamental units — See Mass, Length, and Time.

Fusion (atomic) — A nuclear reaction involving the combination of smaller atomic nuclei or particles into larger ones with the release of energy from mass transformation. This is also called a thermo-nuclear reaction by reason of the extremely high temperature required to initiate it.

Gal — 1 gal = cm/sec/sec. Therefore, where the value of gravity is 980 this is the same as 980 gal. The milligal is now quite commonly used since it is approximately one part in a thousand of the normal gravity of the earth.

Gamma (γ) rays (nuclear X-rays) — May be emitted from radioactive substances. They are quanta of electromagnetic wave energy similar to but of much higher energy than ordinary X-rays. The energy of a quantum is equal to hν ergs, where h is Planck's constant (6.6254×10^{-27} erg sec) and ν is the frequency of the radiation. Gamma rays are highly penetrating, an appreciable fraction being able to traverse several centimeters of lead.

Gas — A state of matter in which the molecules are practically unrestricted by cohesive forces. A gas has neither definite shape nor volume.

Gas constant (symbol R, R^*, R_o) — The constant factor in the equation of state for perfect gases. The universal gas constant is

$$R_o = 8.31441(26) \text{ joules/°K-mol}$$
$$= 1.98723 \text{ cal/°K-mol}$$

The gas constant for a particular gas, specific gas constant,

$$r = R/m$$

where m is the molecular weight of the gas. See Boltzmann constant.

Gas thermometer — where Pi0, P_s and P_x represent the total pressure with the bulb at 0°C, at the boiling-point of water and at the unknown temperature respectively, t_s the temperature of steam and t_x the unknown temperature,

$$t_x = t_s \frac{P_x - P_o}{P_s - P_o}$$

(approximately). The total pressure on the gas in the bulb is the algebraic sum of barometric pressure at the time and that measured by the manometer.

Gauss — The cgs emu of magnetic induction (flux density). It is equal to 1 maxwell per cm^2. It has such a value that magnetic field at a velocity of 1 cm, in an induction mutually perpendicular, the induced emf is one abvolt.

Gay-Lussac's law — See Charles' law.

Gay-Lussac's law of combining volumes — If gases interact and form a gaseous product, the volumes of the reacting gases and the volumes of the gaseous products are to each other in simple proportions, which can be expressed by small whole numbers.

Gee — A suffix meaning earth, as in perigee, apogee. See Perigee, note.

Geiger counter — Detector for radioactivity depending upon ionized particles that affect its mechanism. As its name indicates, it both detects and makes a count of them possible.

Geomagnetic pole — Either of two antipodal points marking the intersection of the earth's surface with the extended axis of a dipole assumed to be located at the center of the earth and approximating the source of the actual magnetic field of the earth.

That pole in the Northern Hemisphere (latitude, 78½° N; longitude, 69° W) is designated north geomagnetic pole, and that pole in the Southern Hemisphere (latitude, 78½° S, longitude, 111°E) is designated south geomagnetic pole. The great circle midway between these poles is called geomagnetic equator. The expression geomagnetic pole should not be confused with magnetic pole, which relates to the actual magnetic field of the earth.

Geometric mean — A measure of central position. The geometric mean of n quantities equals the nth root of the product of the quantities.

Geopotential — The geopotential Φ of a point at a height z above mean sea level is the work which must be accomplished against gravity in elevating a unit mass from sea level to height z.

$$\Phi = \int_o^z g\,dz$$

where g is the local acceleration of gravity at height z. For most metrological work geopotential is given in the units geopotential meter (gpm). By definition, 1 gpm = 9.8×10^4 cm^2 sec^{-2}. For most purposes the geopotential can be assumed to equal the geometric height.

Geopotential height — The height of a given point in the atmosphere in units proportional to the potential energy of unit mass (geopotential) at this height, relative to sea level.

The relation, in the cgs system, between the geopotential height H and the geometric height Z is

$$H = \frac{1}{980}\int_o^Z g\,dZ$$

where g is the acceleration of gravity, so that the two heights are numerically interchangeable for most meteorological purposes. Also, 1 geopotential meter is equal to 0.98 dynamic meter. See dynamic height.

At the present time, the geopotential height unit is used for all aerological reports, by convention of the World Meteorological Organization.

Geopotential meter — A unit of length used in measuring geopotential height; 1 geopotential meter is equal to 0.98 dynamic meter. See Dynamic height.

Gibbs free energy = Gibbs function.

Gibbs function — a mathematically defined thermodynamic function of state, which is constant during a reversible isobaric-isothermal process. Also called Gibbs free energy, thermodynamic potential. Compare Helmholtz function.

In symbols the specific Gibbs function g is

$$g = h - Ts$$

where h is specific enthalpy; T is Kelvin temperature; and s is specific entropy. By use of the first law of thermodynamics for reversible processes,

$$dg = -s\,dT + dp$$

Gibbs' phase rule — F = C + 2 − P, F, the number of degrees of freedom of a system, is the number of variable factors (temperature, pressure and concentration) of the components, which must be arbitrarily fixed in order that the condition of the system m ay be perfectly defined. C, the number of the components of the system, is chosen equal to the smallest number of independently variable constituents by means of which the composition of each phase participating in the state of equilibrium can be expressed in the form of a chemical equation; the components must be chosen from among the constituents which are present when the system is in a state of true equilibrium and which take part in that equilibrium; as components are chosen the smallest number of such constituents necessary to express the composition of each phase participating in the equilibrium, zero and negative quantities of components being permissible; in any system the number of components is definite, but may alter with changes in conditions of experiment; a qualitative but not quantitative freedom of selection of components is allowed, the choice being influenced by suitability and simplicity of application. P, the number of phases of the system, are the homogeneous, mechanically separable and physically distinct portions of a heterogeneous system; the number of phases capable of existence varies greatly in different systems; there can never be more than one gas or vapor phase since all gases are miscible in all proportions, a heterogeneous mixture of solid substances forms as many phases as there are substances present.

Giga — A prefix meaning billion or 10^9. Symbol is G.

Gilbert (abbr. Gb) — The cgs emu of magnetomotive force. 1 Gb = $10/4\pi$ ampere-turns.

Global radiation — The total of direct solar radiation and diffuse sky radiation received by a unit horizontal surface.

Graham's law — The relative rates of diffusion of gases under the same conditions are inversely proportional to the square roots of the densities of those gases.

Gram atom or gram atomic weight — The mass in grams numerically equal to the atomic weight.

Gram equivalent of a substance is the weight of a substance displacing or otherwise reacting with 1.008 g of hydrogen or combining with one-half of a gram atomic weight (8.00 g) of oxygen.

Gram molecular weight or gram molecule — A mass in grams of a substance numerically equal to its molecular weight. Gram mole.

Gram mole, gram formula weight, gram equivalent — Mass in grams numerically equal to the molecular weight, formula weight or chemical equivalent, respectively.

Gravitation — The acceleration produced by the mutual attraction of two masses, directed along the line joining their centers of masses, and of magnitude inversely proportional tothe square of the distance between the two centers of mass.

This acceleration on a unit mass has the magnitude $G(m/r^2)$, where m is the mass of the attracting body, r is the distance between the centers of mass, and G is the gravitational constant equal to $6.670 \pm 0.005 \times 10^{-11}$ N n^2 kg^{-2}.

In the case of masses in the earth's gravitational field, m is the mass of the earth, equal to 5.975×10^{27} g. However, the rotation of the earth and atmosphere modifies this field to produce the field of gravity.

Gravitational constant (symbol G) — The coefficient of proportionality in Newton law of gravitation

$$G = 6.670 \pm 0.005 \times 10^{-11}\ Nm^2\,kg^{-2}$$

Also called constant of gravitation, Newtonian universal constant of gravitation.

Gravity — 1. Viewed from a frame of reference fixed in the earth, force imparted by the earth to a mass which is at rest relative to the earth. Since the earth is rotating, the force observed as gravity is the resultant of the force of gravitation and the centrifugal force arising from this rotation and the use of an earthbound rotating frame of reference. It is directed normal to sea level and to its geopotential surfaces. See Virtual gravity, Geopotential height.

The magnitude of the force of gravity at sea level decreases from the poles, where the centrifugal force is zero, to the equator, where the centrifugal force is a maximum but directed opposite to the force of gravitation. This difference is accentuated by the shape of the earth, which is nearly that of an oblate spheroid of revolution slightly depressed at the poles. Also, because of the asymmetric distribution of the mass of the earth, the force of gravity is not directed precisely toward the earth's center.

The magnitude of the force of gravity is usually called either gravity, acceleration of gravity, or apparent gravity.

Half life — The average time required for one half the atoms in a sample of radioactive element to decay.

The half life $t_{1/2}$ is given by

$$t_{1/2} = (\ln 2)/\lambda$$

where λ is the decay constant.

Half-wave plate — An electro-optical material used to rotate the plane of polarization of a light beam.

Hall constant — In an electrical conductor, the constant of proportionality R in the relation

$$E_h = RJ \times H$$

where E_h is transverse electric field (Hall field); J is current density; and H is magnetic field.

The sign of the majority carrier can be inferred from the sign of the Hall constant.

Hall effect — When a steady current is flowing in a steady magnetic field, electromotive forces are developed which are at right angles both to the magnetic force and to the current and are proportional to the product of the intensity of the current, the magnetic force and the sine of the angle between the directions of these quantities.

Hall mobility — A measure of the flow of charged particles perpendicular to both a magnetic and an electric field.

Hardness — Property of substances determined by their ability to abrade or indent one another. An arbitrary scale of hardness is based upon ten selected minerals. For metals the diameter of the indentation made by a hardened steel sphere (Brinnell) or the height of rebound of a small drop hammer (Shore Scleroscope) serve to measure hardness.

Harmonic — A sinusoidal quantity whose frequency is an integral multiple of some fundamental frequency, that is, given a quantity, x(t), where

$$x(t) = A \cos(\omega t + \phi),$$

its harmonics, $h_n(x)$ are of the form

$$h_n(x) = A_n \cos(n\,\omega t + \phi_n),$$

where n is an integer larger than 1.

Harmonic motion — See Simple harmonic motion and Angular harmonic motion.

Heat — Energy transferred by a thermal process.

Heat can be measured in terms of the dynamical units of energy, as the erg, joule, etc., or in terms of the amount of energy required to produce a definite thermal change in some substance, as, for example, the energy required per degree to raise the temperature of a unit mass of water at some temperature (calorie, Btu).

Heat capacity — That quantity of heat required to increase the temperature of a system or substance one degree of temperature. It is usually expressed in calories per degree centigrade or joules per degree centigrade.

Molar heat capacity is the quantity of heat necessary to raise the temperature of one molecular weight of the substance one degree.

Heat effect — The heat in calories developed in a circuit by an electric current of I amperes flowing through a resistance of R ohms, with a difference of potential E volts for a time t seconds.

$$H = \frac{RI^2 t}{4.18} = \frac{Eit}{4.18}$$

Heat equivalent, or latent heat, or fusion — The quantity of heat necessary to change one gram of solid to a liquid with no temperature change. Dimensions, $[l^2 t^{-2}]$.

Heat of combustion of a substance is the amount of heat evolved by the combustion of 1 g mol wt of the substance.

Heat quantity — The cgs unit of heat is the calorie, the quantity of heat necessary to change the temperature of one gram of water from 3.5°C to 4.5°C (called a small calorie). If the temperature change involved is from 14.5 to 15.5°C, the unit is the normal calorie. The mean calorie is $^1/_{100}$ the quantity of heat necessary to raise one gram of water from 0°C to 100°C. The large calorie is equal to 1000 small calories. The British thermal unit is the heat required to raise the temperature of one pound of water at its maximum density, 1°F. It is equal to about 252 calories. Dimensions of energy, $[ml^2t^{-2}]$.

Hecto — A prefix meaning hundred or 10^2. Symbol is h.

Hehner number (value) — A number expressing the percentage (i.e. grams per hundred grams) of water-insoluble fatty acids in an oil or fat.

Heisenberg's theory of atomic structure — The currently accepted view of the structure of atom, formulated by Heisenberg in 1934, according to which the atomic nuclei are built of nucleons, which may be protons or neutrons, while the extranuclear shells consist of electrons only. The nucleons are held together by nuclear forces of attraction, with exchange forces operating between them. The number of protons is equal to the atomic number (Z) of the element, the number of neutrons is equal to the difference between the mass number and the atomic number (A — Z). The number of excess neutrons, i. e. the excess of neutrons over protons, is of paramount importance for the radioactive properties or stability of the element.

Helmholtz free energy = Helmholtz function.

Helmholtz function (symbol a) — A mathematically defined thermodynamic function of state, the decrease in which during a reversible isothermal process is equal to the work done by the system. The Helmholtz function is

$$a = u - Ts$$

where u is specific internal energy; T is Kelvin temperature; and s is specific entropy. By use of the first law of thermodynamics for reversible processes,

$$da = -s\,dT - dw$$

where dw is the work done per unit mass by the system. Also called Helmholtz free energy, work function. Compare Gibbs function.

Henry (unit of electric inductance) — The inductance of a closed circuit in which an electromotive force of 1 V is produced when the electric current in the circuit varies uniformly at a rate of 1 A/sec.

Henry's law — The mass of a slightly soluble gas that dissolves in a definite mass of a liquid at a given temperature is very nearly directly proportional to the partial pressure of that gas. This holds for gases which do not unite chemically with the solvent.

Hertz — The measure of frequency, defined as equal to 1 cycle/sec.

Hess' law of constant heat summation — The amount of heat generated by a chemical reaction is the same whether reaction takes place in one step or in several steps, or all chemical reactions which start with the same original substances and end with the same final substances liberate the same amounts of heat, irrespective of the process by which the final state is reached.

Heterodyne — To mix two radio signals of different frequencies to produce a third signal which is of lower frequency; i.e., to produce beating.

Radar receivers are of the heterodyne type (as contrasted to the superregenerative type) because the very high radio frequencies used in radar are difficult to amplify. A target signal is heterdyned with a current of lower frequency produced by a klystron oscillator and the resulting intermediate-frequency signal can then be highly amplified for subsequent presentation or analysis.

Hobbmann orbit — A minimum energy transfer orbit.

Hole — A mobile vacancy in the electronic valence of structure of a semiconductor which acts like an electron with a positive charge.

Holocamera — A device for recording or forming a hologram of an object or subject.

Hologram — A recording or picture of a three-dimensional wavefront.

Holographic — Pertaining to or using the principles of holography. For example, holographic equipment uses the principles of holography for its operation.

Holographic matched filter — A particular type of hologram which when illuminated by the type wave it is matched to will transmit a pure plane wave. This plane wave is usually focused into a correlation spot.

Holography — A recording and viewing process which allows reconstruction of three-dimensional images of diffuse objects.

Homogeneous atmosphere — 1. A hypothetical atmosphere in which the density is constant with height.

The lapse rate of temperature in such an atmosphere is known as the autoconvective lapse rate and is equal to g/R (or approximately 3.4°C/100 m) where g is the acceleration of gravity and R is the gas constant for air. A homogeneous atmosphere has a finite total thickness which is given by $R_d T_v/g$, where R_d is the gas constant for dry air and T_v is the virtual temperature (°K) at the surface. For a surface temperature of 273°K, the vertical extent of the homogeneous atmosphere on the earth is approxiamtely 8000 m. At the top of such an atmosphere both the pressure and absolute temperature vanish.

2. With respect to radio propagation, an atmosphere which has a constant index of refraction, or one in which radio waves travel in straight lines at constant speed. Free space is the ideal homogeneous atmosphere in this sense. 3. Same as adiabatic atmosphere. See Barotrophy.

Hooke's law — Within the elastic limit of any body the ratio of the stress to the strain produced is constant.

Horsepower — A measure of power, defined as equal to 33,000 foot pounds/min, or 746 watts.

Humidity, absolute — Mass of water vapor present in unit volume of the atmosphere, usually measured as grams per cubic meter. It may also be expressed in terms of the actual pressure of the water vapor present.

Huygens' principle — A very general principle applying to all forms of wave motion which states that every point on the instantaneous position of an advancing phase front (wave front) may be regarded as a source of secondary spherical wavelets. The position of the phase front a moment later is then determined as the envelope of all the secondary wavelets (ad infinitum).

Huygens' theory of light — This theory states that light is a disturbance traveling through some medium, such as the ether. Thus light is due to wave motion in ether.

Every vibrating point on the wave-front is regarded as the center of a new disturbance. These secondary disturbances traveling with equal velocity, are enveloped by a surface identical in its properties with the surface from which the secondary disturbances start and this surface forms the new wave-front.

Hydrogen equivalent of a substance is the number of replaceable hydrogen atoms in 1 mol or the number of atoms of hydrogen with which 1 mol could react.

Hydrogen ion concentration — The concentration of hydrogen ions in solution when the concentration is expressed as gram-ionic weights per liter. A convenient form of expressing hydrogen ion concentration is in terms of the negative logarithm of this concentration. The negative logarithm of the hydrogen ion concentration is called pH.

Water at 25°C has a concentration of H ion of 10^{-7} and of OH ion of 10^{-7} moles per liter. Thus the pH of water is 7 at 24°C. A greater accuracy is obtained if one substitutes the thermodynamic activity of the ion for its concentration.

Hydrolysis is a double decomposition reaction involving the splitting of water into its ions and the formation of a weak acid or base or both.

Hydrostatic equation — In numerical equations, the form assumed by the vertical component of the vector equation of motion when all coriolis, earth-curvature, frictional, and vertical-acceleration terms are considered negligible compared with those involving the vertical pressure force and the force of gravity. Thus,

$$dP = -\rho g dZ$$

where P is the atmospheric pressure; ϱ is the density; g is the acceleration of gravity; and Z is the geometric height.

Hydrostatic pressure at a distance h from the surface of a liquid of density d,

$$P = hdg$$

The total force on an area A due to hydrostatic pressure,

$$F = P\ A = Ahdg$$

Force in dynes and pressure in dynes per cm^2 will be given if h is in cm, d in g per cm^3 and g in cm per $sec.^2$

Hyperon — Any article with mass intermediate between that of the neutron and the deuteron. See Meson.

Hypersonic — 1. Pertaining to hypersonic flow. 2. Pertaining to speeds of Mach 5 or greater.

Hypersonic flow In aerodynamics, flow of a fluid over a body at speeds much greater than the speed of sound and in which the shock waves start at a finite distance from the surface of the body. Compare supersonic flow.

Hysteresis — The magnetization of a sample of iron or steel due to a magnetic field which is made to vary through a cycle of values, lags behind the field. This phenomenon is called hysteresis.

Steinmetz' equation for hysteresis gives the loss of energy in ergs per cycle per cm^3,

$$W = \eta B^{1.6}$$

where B is the maximum induction in maxwells per cm^2 and η the coefficient of hysteresis.

Ice point — The temperature at which a mixture of air-saturatated pure water and pure ice may exist in equilibrium at a pressure of one standard atmosphere.

By decision of the Tenth General Conference on Weights and Measures, Paris, October 1954 the ice point was established as 273.15°K.

Ideal gas — A gas which conforms to Boyle's law and has zero heat of free ezpansion (or also obeys Charles' law). Also called perfect gas.

Illuminance — The total luminous flux received on a unit area of a given real or imaginary surface, expressed in such units as the footcandle, lux, or phot. Illuminance is analogous to irradiance, but is to be distinguished from the latter in that illuminance refers only to light and contains the luminous efficiency weighting factor necessitated by the nonlinear wavelength-response of the human eye. Compare luminous intensity.

The only difference between illuminance and illumination is that the latter always refers to light incident upon a material surface.

A distinction should be drawn, as well, between illuminance and luminance. The latter is a measure of the light coming from a surface; thus, for a surface which is not self-luminous, luminance is entirely dependent upon the illuminance upon that surface and its reflection properties.

Illumination on any surface is measured by the luminous flux incident on unit area. The units in use are: the lux, (abbreviation lx) one lumen per square meter; the phot, (abbreviation ph) one lumen per square centimeter; the footcandle, (abbreviation fc) one lumen per square foot.

Image redundancy — Multiple storage of the same image.

Impact pressure — 1. That pressure of a moving fluid brought to rest which is in excess of the pressure the fluid has when it does not flow, i.e., total pressure less static pressure.

Impact pressure is equal to dynamic pressure in incompressible flow, but in compressible flow impact pressure includes the pressure change owing to the compressibility effect.

2. A measured quantity obtained by placing an open-ended tube, known as an impact tube or pitot tube, in a gas stream and noting the pressure in the tube on a suitable manometer.

Since the pressure is exerted at a stagnation point, the impact pressure is sometimes referred to as the stagnation pressure or total pressure.

Incandescence — Emission of light due to high temperature of the emitting material. Any other emission of light is called luminescence.

Incoherent holography — Holograms produced initially from conventional photographs or incoherent optical equipment.

Indeterminacy principle (Uncertainty principle) — The postulate that it is impossible to determine simultaneously both the exact position and the exact momentum of an electron. So this aspect of electronics can only be expressed as a probability.

Index of refraction (symbol n) — 1. A measure of the amount of refraction (a property of a dielectric substance). It is the ratio of the wavelength or phase velocity of an electromagnetic wave in a vacuum to that in the substance. Also called refractive index, absolute index of refraction, absolute refractive index, refractivity. See modified index of refraction, N-unit, potential index of refraction.

It can be a function of wavelength, temperature, and pressure. If the substance is nonabsorbing and nonmagnetic at any wavelength, then n^2 is equal to the dielectric constant at that wavelength.

The complex index of refraction is obtained when the attenuation of the wave per radian, called the absorptive index k, is paired with the index of refraction. It is written

$$n^* = n(1 - ik)$$

When the wave passes from one medium n_1 to another n_2, the angle of incidence ϕ and the angle of refraction ϕ', both measured with respect to the normal to the interface, are related by

$$\sin\phi/\sin\phi' = n_1{}^*/n_2{}^* = \text{constant}$$

which becomes, for a nonabsorbing medium, the ratios of the (noncomplex) indices of refraction. In the particular case that medium 2 is a vacuum, this ratio is the index of refraction of medium 1. This is known as Snell law, named after Willebrord Snell who discovered it about 1621.

2. A measure of the amount of refraction experienced by a ray as it passes through a refraction interface, i.e., a surface separating two media of different densities. It is the ratio of the absolute indices of refraction of the two media (see sense 1 above). Also called refractive index, relative index of refraction.

Indicators are substances which change from one color to another when the hydrogen ion concentration reaches a certain value, different for each indicator.

Indirectly ionizing particles are uncharged particles (neutrons, photons, etc.) which can liberate directly ionizing particles or can initiate nuclear transformations.

Induced electromotive force in a circuit is proportional to the rate of change of magnetic flux through the circuit.

$$E = -\frac{d\phi}{dt}$$

where $d\phi$ is the change of magnetic flux in a time dt. The induced current will be given by

$$I = -\frac{d\phi}{Rdt}$$

where R is the resistance of the circuit.

Inductance — The change in magnetic field due to the variation of a current in a conducting circuit causes an induced counter electromotive force in the circuit itself. This phenomenon is known as self-induction. If an electromotive force is induced in a neighboring circuit the term mutual induction is used. Inductance may thus be distinguished as self- or mutual and is measured by the electromotive force produced in a conductor by unit rate of variation of the current. Units of inductance are the centimeter (absolute electromagnetic) and the henry, which is equal to 10^9 centimeters of inductance. The henry is that inductance in which an induced electromotive force of one volt is produced when the inducing current is changed at the rate of one ampere per second. Dimensions, $[\epsilon^{-1}\ l^{-1}\ t^2]$; $[\mu l]$.

DEFINITIONS (Continued)

Induction — Any change in the intensity or direction of a magnetic field causes an electromotive force in any conductor in the field. The induced electromotive force generates an induced current if the conductor forms a closed circuit.

Inertia — The resistance offered by a body to a change of its state of rest or motion, a fundamental property of matter. Dimension, [m].

Information content — Containing or transmitting data involving new knowledge. When applied to waves or wavefronts, it includes both amplitude and phase of all parts of a wavefront at a given instant of time.

Infrared radiation (abbr. IR) — Electromagnetic radiation lying in the wavelength interval from about 75 μm to an indefinite upper boundary sometimes arbitrarily set at 1000 μm (0.01 cm). Also called long-wave radiation.

Inline holography — Hologram produced by single reference beam interferences with waves diffracted or scattered from a small object.

Insolation — (Contracted from incoming solar radiation.) 1. In general, solar radiation received at the earth's surface. See terrestrial radiation, extraterrestrial radiation, direct solar radiation, global radiation, effective terrestrial radiation, diffuse sky radiation, atmospheric radiation. 2. The rate at which direct solar radiation is incident upon a unit horizontal surface at any point on or above the surface of the earth. Compare solar constant.

Intensity of Illumination (properly called Illumination)—Illumination in lux of a screen by a source of illuminating power P at a distance r meters, for normal incidence,

$$I = \frac{P}{r^2}$$

If two sources of illuminating power P_1 and P_2 produce equal illumination on a screen when at distances r_1 and r_2 respectively,

$$\frac{P_1}{r_1^{\ 2}} = \frac{P_2}{r_2^{\ 2}} \quad \text{or} \quad \frac{P_1}{P_2} = \frac{r_1^{\ 2}}{r_2^{\ 2}}$$

If I_o is the illumination when the screen is normal to the incident light, then I is the illumination when the screen is at an angle θ. Thus,

$$I = I_0 \cos\theta$$

Intensity of magnetization is given by the quotient of the magnetic moment of a magnet by its volume. Unit intensity of magnetization is the intensity of a magnet which has unit magnetic moment per cubic centimeter. Dimensions, $[\epsilon^{-1/2}\ m^{-1/2}\ l^{-3/2}]$;$[\mu^{1/2}\ m^{1/2}\ l^{-1/3}\ t^{-1}]$.

Intensity of radiation is the rate of transfer of energy across unit areas by the radiation. In all forms of energy transfer by waves (radiation) the intensity I is given by I = U, where U is the energy density of the wave in the medium, and v is the velocity of propagation of the wave. The energy density U is always proportional to the square of the wave amplitude.

Intensity of sound depends upon the energy of the wave motion. The intensity is measured by the energy in ergs transmitted per second through 1 cm^3 of surface. The energy in ergs per cm^3 in a sound wave is given by

$$E = 2\pi^2 dn^2 a^2$$

where d is density in g per cm^3, n is frequency in vib. per sec and a is amplitude in cm. The energy reaching the ear in unit time will also be proportional to the velocity of propagation.

Interference hologram — A holographic interferogram produced by the superposition of two or more hologram exposures.

Interference (of waves) — The coherent addition or subtraction of two different wavefronts, which usually forms a third wave different from the first two.

Interference pattern — The pattern of light and dark fringes produced when two or more coherent waves interfere or intersect.

Interferogram — The record of an interference pattern produced by holography or by conventional optical interference techniques.

Interferometry (holographic) — The process of measuring very small movements or deformations by recording or observing interference wave patterns (either light, electronic, or acoustic).

Interionic attraction — The electrostatic attraction between ions of unlike charge (sign).

Interionic repulsion — The electrostatic repulsion between ions of like charge (sign).

Internal energy — A mathematically defined thermodynamic function of state, interpretable through statistical mechanics as a measure of the molecular activity of the system. It appears in the first law of thermodynamics as

$$du - dq - dw$$

where du is the increment of specific internal energy, dq the increment of heat, and dw the increment of work done by the system per unit mass. The differential du is a perfect differential. Its integral therefore introduces a constant of integration, the zero-point internal energy, so that care must be taken when absolute values of the internal energy are employed.

International Practical Kelvin Temperature Scale of 1960 and the **International Practical Celsius Temperature Scale of 1960** are defined by a set of interpolation equations based on the following reference temperatures:

	°K	°C
Oxygen, liquid-gas equilibrium	90.18	−182.97
Water, solid-liquid equilibrium	273.15	0.00
Water, solid-liquid-gas equilibrium	273.16	0.01
Water, liquid-gas equilibrium	373.15	100.00
Zinc, solid-liquid equilibrium	692.655	419.505
Sulphur, liquid-gas equilibrium	717.75	444.6
Silver, solid-liquid equilibrium	1233.95	960.8
Gold, solid-liquid equilibrium	1336.15	1063.0

International System of Units (abbr. SI) — The metric system of units based on the meter, kilogram, second, ampere, Kelvin degree, and candela. Also called MSKA system.

Iodine number (value) — A number expressing the percentage (i.e. grams per 100 grams) of iodine absorbed by a substance. It is a measure of the proportion of unsaturated linkages present and is usually determined in the analysis of oils and fats.

Ion — An ion is an atom or group of atoms that is not electrically neutral but instead carries a positive or negative electric charge. Positive ions are formed when neutral atoms or molecules lose valence electrons; negative ions are those which have gained electrons.

Ion atmosphere (or continuus charge distribution) — In the electrostatic effects between ions the term ion atmosphere denotes a continuous charge distribution, or charge density, ϱ(r), which is a continuous function of r, the distance from the reference ion, rather than a discrete or discontinuous charge distribution. The ion atmosphere extends from r = a to $r = O(V^{1/3}) \approx \infty$, where V is the volume of the system, and acts electrostatically somewhat like a sphere of charge −e at a distance, k^{-1}, from the reference ion of charge +e (see below for definition of k^{-1}).

Ionic equivalent conductance, λ — The ionic equivalent conductance is the equivalent conductance of an individual ion constituent of the solute (or electrolyte) of an electrolytic solution. This symbol is also used to designate the equivalent conductance of complex ions, ion pairs, ion clusters, etc., in combination with simple ions.

Ionic mobility, μ — The mobility of an ion at any finite equivalent concentration is the velocity with which the ion moves under unit potential gradient. Its unit is $cm^2\ sec^{-1}\ volt^{-1}\ equiv^{-1}$ or $cm^2\ ohm^{-1}\ F^{-1}$ where F is the Faraday expressed in coulombs (or ampere seconds) $equiv^{-1}$.

Ionization — The process by which neutral atoms or groups of atoms become electrically charged, either positively or negatively, by the loss or gain of electrons; or the state of a substance whose atoms or groups of atoms have become thus charged.

Ionization potential — The work (expressed in electron volts) required to remove a given electron from its atomic orbit and place it at rest at an infinite distance. It is customary to list values in electron volts (ev.) 1 ev. = 23,053 calories per mole.

Ionizing radiation is any radiation consisting of directly or indirectly ionizing particles or a mixture of both.

Ionogens — Substances, like acetic acid (HAc), which, although in the pure state are nonelectrolytic neutral molecules, can react with certain solvents to form products which rearrange to ion pairs which then dissociate to give conducting solutions. As an example:

$$HAc + H_2O \rightleftharpoons HAc{\cdot}H_2O \rightleftharpoons H_3O^+{\cdot}Ac^- \rightleftharpoons H_3O^+ + Ac^-$$

Ionophores — Substances, like sodium chloride, which exist only as ionic lattices in the pure crystalline form, and which when dissolved in an appropriate solvent give conductances which change according to some fractional power of the concentration. Such solutions possess no neutral molecules which can dissociate, but may contain associated ions.

Ionosphere — The atmospheric shell characterized by a high ion density. Its base is at about 70 or 80 km and it extends to an indefinite height.

Ion size or "ion-size" parameter, a (or a_i) — The ion size is formally considered to be the sum of the ionic radii of the oppositely charged ions in contact. The ion size is also called the "distance of closest approach" of the ions, or the "ion-size" parameter. Generally the ion size is greater than the sum of the crystal radii, and the "ion-size" parameter may include several factors which contribute to its numerical value.

Irradiance — A measure of the rate of energy falling on a given area.

Isentropic — Of equal or constant entropy with respect to either space or time.

Isobaric — Of equal or constant pressure, with respect to either space or time.

Isobars — For chemistry, elements of the same atomic mass but of different atomic numbers. The sum of their nucleons is the same but there are more protons in one than in the other.

Isochoric — Of equal or constant volume, usually applied to a thermodynamic process during which the volume of the system remains unchanged. Compare isosteric.

Isoclinic line — A line through points on the earth's surface having the same magnetic dip.

Isomer — 1. One of two or more nuclides having the same mass number A and atomic number Z, but existing for measurable times in different quantum states with different energies and radioactive properties.
2. One of two or more molecules having the same atomic composition and molecular weight, but differing in geometrical configuration.

Isomerism — Existence of molecules having the same number and kinds of atoms but in different configurations.

Isotherm — A line of equal or constant temperature.
A distinction is made, infrequently, between a line representing equal temperature in space, choroisotherm, and one representing constant temperature in time, chronoisotherm.

Isothermal — When a gas passes through a series of pressure and volume variations without change of temperature the changes are called isothermal. A line on a pressure-volume diagram representing these changes is called an isothermal line.

Isotope — 1. One of several nuclides having the same number of protons in their nuclei, and hence belonging to the same element, but differing in the number of neutrons and therefore in mass number A, or in energy content (isomers). For example, ${}_6C^{12}$, ${}_6C^{13}$, and ${}_6C^{14}$ are carbon isotopes. Small quantitative differences in chemical properties exist between isotopes.

Isotropic — In general, pertaining to a state in which a quantity or spatial derivatives thereof are independent of direction. Also called isotropous.

Joule constant — The ratio between heat and work units from experiments based on the first law of thermodynamics: 4.1858×10^7 ergs per 15° calorie. Also called mechanical equivalent of heat.

Joule — A unit of energy or work in the MKS system; the work done when the point of application of 1 newton is displaced a distance of 1 meter in the direction of the force.

$$1 \text{ joule} = 10^7 \text{ ergs} = 1 \text{ watt second}$$

Joule cycle — (After James Prescott Joule, 1818—1889, English physicist.) An ideal cycle for engines consisting of isentropic compression of the working substance, addition of heat at constant pressure, isentropic expansion to ambient pressure, and exhaust at constant pressure. Also called Brayton cycle.

Joule's law — A law stating that the amount of heat H produced in a conductor by the flow of a steady current I is given by

$$H = KI^2Rt.$$

where R is the resistance of the conductor, t is the time for which the current flows, and K, the constant of proportionality, has the value 0.2390 calories per joule when R is in ohms and I in amperes.

Joule-Thomson effect — The decrease in temperature which takes place when a gas expands through a throttling device as a nozzle. Also called Joule-Kelvin effect.
The rate of change of temperature T with pressure p in the Joule-Thomson effect is called the Joule-Thomson coefficient (symbol μ):

$$\mu = \left.\frac{dT}{dp}\right]_h$$

where h denotes constant enthalpy.
For the Joule-Thomson effect to take place the gas must initially be below its inversion temperature; if above the inversion temperature, the gas will gain heat on expansion. The inversion temperature of hydrogen, for example, is approximately −183°C.

Junction — In a semiconductor device, a region of transition between semiconducting regions of different electrical properties.

Kelvin — The kelvin, the unit of thermodynamic temperature, is the fraction 1/273.16 of the thermodynamic temperature of the triple point of water. The decision was made at the 13th General Conference on Weights and Measures on October 13, 1967 that the name of the unit of thermodynamic temperature would be changed from degree Kelvin (symbol: °K) to kelvin (symbol: K). The name (kelvin) and symbol (K) are to be used for expressing temperature intervals. The former convention which expressed a temperature interval in degrees Kelvin or, abbreviated, deg. K is dropped. However, the old designations are acceptable temporarily as alternatives to the new ones. One may also express temperature intervals in degrees Celsius.

Kelvin temperature scale (abbr. K)—An absolute temperature scale independent of the thermometric properties of the working substance. On this scale, the difference between two temperatures T_1 and T_2 is proportional to the heat converted into mechanical work by a Carnot engine operating between the isotherms and adiabats through T_1 and T_2. Also called absolute temperature scale, thermodynamic temperature scale.
For convenience the Kelvin degree is identified with the Celsius degree. The ice point in the Kelvin scale is 273.15 K. The triple point of water, the fundamental reference point, is 273.16 K. See Absolute zero, Rankine temperature scale.

Kepler's laws — I. The planets move about the sun in ellipses, at one focus of which the sun is situated. II. The radius vector joining each planet with the sun describes equal areas in equal times. III. The cubes of the mean distances of the planets from the sun are proportional to the squares of their times of revolution about the sun.

Kerr cell — A cell that contains electrodes immersed in nitrobenzene or other liquid. It shows double refraction in high degree and with short time lag. Used in devices in which light intensity is changed rapidly according to the voltage applied to the electrons.

Kerr effect — When plane polarized light is incident on the pole of an electromagnet, polished so as to act like a mirror, the plane of polarization of the reflected light is not the same when the magnet is "on" as when it is "off." It was found that the direction of rotation was opposite to that of the currents exciting the pole from which the light was reflected.

Kilo — A prefix meaning thousand or 10^3. Symbol is k.

DEFINITIONS (Continued)

Kilogram (unit of mass)—is the mass of a particular cylinder of platinum-iridium alloy, called the International Prototype Kilogram, which is preserved in a vault at Sèvres, France, by the International Bureau of Weights and Measures.

Kilometer (abbr. km) — A unit of distance in the metric system.
1 kilometer = 3280.8 feet = 1093.6 yards = 1000 meters = 0.62137 statute miles = 0.53996 nautical miles.

Kinematic viscosity (symbol v) — A coefficient defined as the ratio of the dynamic viscosity of a fluid to its density.

The kinematic viscosity of most gases increases with increasing temperature and decreasing pressure. For dry air at 0°C, the kinematic viscosity is about 0.13 square centimeter per second. In the theory of atmospheric turbulence the kinematic viscosity is usually replaced by the kinematic eddy viscosity to account for the increased internal friction due to turbulence.

Kinetic energy (symbol E_k) — The energy which a body possesses as a consequence of its motion, defined as one-half the product of its mass m and the square of its speed v, $\frac{1}{2}mv^2$. The kinetic energy per unit volume of a fluid parcel is thus $\frac{1}{2}\varrho v^2$, where ϱ is the density and v the speed of the parcel. See Potential energy.

For relativistic speeds the kinetic energy is given by

$$E_k = mc^2 - m_o c^2$$

where c is the velocity of light in a vacuum, m_o is the rest mass, and m is the moving mass.

Kinetic theory, expression for pressure —

$$P = \frac{1}{3} N mv^2$$

where N is the number of molecules in unit volume, m the mass of each molecule and v^2 the mean square of the velocity of the molecules.

Kinetic theory of gases — Gases are considered to be made up of minute, perfectly elastic particles which are ceaselessly moving about with high velocities, colliding with each other and with the walls of the containing vessel. The pressure exerted by a gas is due to the combined effect of the impacts of the moving molecules upon the walls of the containing vessel, the magnitude of the pressure being dependent upon the kinetic energy of the molecules and their number.

Kirchhoff law—The radiation law which states that at a given temperature the ratio of the emissivity to the absorptivity for a given wavelength is the same for all bodies and is equal to the emissivity of an ideal black body at that temperature and wavelength.

Kirchhoff's laws —
I. The algebraic sum of the currents which meet at any point is zero.
II. In any closed circuit the algebraic sum of the products of the current and the resistance in each conductor in the circuit is equal to the electromotive force in the circuit.

Kirchhoff's laws of radiation — The relation between the powers of emission and the powers of absorption for rays of the same wave-length is constant for all bodies at the same temperature. First, a substance when excited by some means or other possesses a certain power of emission; it tends to emit definite rays, whose wave-lengths depend upon the nature of the substance and upon the temperature. Second, the substance exerts a definite absorptive power, which is a maximum for the rays it tends to emit. Third, at a given temperature the ratio between the emissive and the absorptive power for a given wave-length is the same for all bodies, and is equal to the emissive power of a perfectly black body.

Knot — A nautical mile per hour, 1.1508 statute miles per hour.

Kohlrausch law of independent migration of ions — The value of the equivalent conductance, as the concentration approaches zero, is equal to the sum of the limiting ionic equivalent conductances of the ions constituting the solute of the electrolytic solution.

Kundt's law — On approaching an absorption band from the red side of the spectrum the refractive index is abnormally increased by the presence of the band, while the approach is from the blue side and the index is abnormally decreased.

Lambert (abbr. L or l) — A unit of luminance (or brightness) equal to $1/\pi$ candle per square centimeter. Physically, the lambert is the luminance of a perfectly diffusing white surface receiving an illuminance of 1 lumen/cm^2.

Lambert law — A law of physics which states that the radiant intensity (flux per unit solid angle) emitted in any direction from a unit radiating surface varies as the cosine of the angle between the normal to the surface and the direction of the radiation. The radiance (or luminance) of a radiating surface is, therefore, independent of direction. Also called Lambert cosine law.

Lambert law of absorption = Bouguer law.

Lambert's law of absorption — Each layer of equal thickness absorbs an equal fraction of the light which traverses it.

Lambert's law of illumination — The illumination of a surface on which the light falls normally from a point source is inversely proportional to the square of the distance of the surface from the source. If the normal to the surface makes an angle with the direction of the rays, the illumination is proportional to the cosine of that angle.

Landau damping — The damping of a space charge wave by electrons which move at the phase velocity of the wave and gain energy transferred from the wave.

Langley — A unit of energy per unit area, equal to 1 gram-calorie/cm^2 commonly employed in radiation theory.

Langmuir probe — A small metallic conductor or pair of conductors inserted within a plasma in order to sample the plasma current.

Lanthanide series (Lanthanides) — Rare earth elements of atomic numbers 57 through 71, which have chemical properties similar to lanthanum (#57).

Laplacian speed of sound — The phase speed of a sound wave in a compressible fluid if the expansions and compressions are assumed to be adiabatic. This speed a is given by the formula

$$a^2 = (c_p/c_v)\, RT$$

where c_p and c_v are the specific heats at constant pressure and volume, respectively; R is the gas content; and T is the Kelvin temperature. The value of this speed under standard conditions in dry air is 331 m/sec. Compare Newtonian speed of sound. See Acoustic velocity.

Laser — A device in which the excitation energy of resonant atomic or molecular systems is used to coherently amplify or generate light. Consider a system of particles with energy levels E_1, E_2 where $E_1 < E_2$. A particle passing between these levels absorbs or emits a quantum of radiation of frequency

$$v = (E_2 - E_1)/h$$

where h is Planck's constant. If N_1 is the number of particles in E_1 and N_2 the number in E_2, it follows from the laws of thermodynamics that at a positive absolute temperature $N_2 < N_1$. As a result of this, a wave of the appropriate frequency would be attenuated. If by some means N_2 can be made to exceed N_1 by a reasonably large margin, an incoming wave of frequency v will stimulate transitions from E_2 to E_1 and in the process become coherently amplified. For sustained oscillation to occur feedback must be provided; this is often done by covering the ends of the chamber in which the reaction occurs with reflective material and arranging the chamber to be resonant at the frequency of oscillation.

Latent heat of vaporization — The quantity of heat necessary to change one gram of liquid to vapor without change of temperature, measured as calories per gram. Dimensions, $[l^2t^{-2}]$.

Lattice energy — The energy required to separate the ions of a crystal to an infinite distance from each other.

LeChatelier's principle — If some stress is brought to bear upon a system in equilibrium, a change occurs, such that the equilibrium is displaced in a direction which tends to undo the effect of the stress.

Length — The name "micron", for a unit of length equal to 10^{-6} meter, and the symbol "μ" which has been used for it were dropped by action of the 13th General Conference on Weights and Measures on October 13, 1967. The symbol "μ" is to be used solely as an abbreviation for the prefix "micro-", standing for the multiplication by 10^{-6}. Thus the length previously designated as 1 micron, should be designated 1 μm.

Length, units of — Meter. 1. (Abbr. m) The basic unit of length of the metric system, defined in the October 1984 General Conference of Weights and Measures as the length of the path traveled by light in vacuum during a time interval of 1/299,792,458 of a second.

Effective 1 July 1959 in the U.S. system of measures, 1 yard = 0.9144 meter, exactly, or 1 meter = 1.094 yards = 39.37 inches. The standard inch is exactly 25.4 millimeters.

Lenses — For a single thin lens whose surfaces have radii of curvature r_1 and r_2 whose principal focus is F, the index of refraction n, and conjugate focal distances f_1 and f_2,

$$\frac{1}{F} = \frac{1}{f_1} + \frac{1}{f_2} = (n - 1)\left(\frac{1}{r_1} + \frac{1}{r_2}\right)$$

For a thick lens, of thickness t,

$$F = \frac{nr_1 r_2}{(n - 1)[n(r_1 + r_2) - t(n - 1)]}$$

Combinations of lenses—If f_1 and f_2 are the focal lengths of two thin lenses separated by a distance d the focal length of the system,

$$F = \frac{f_1 f_2}{f_1 + f - d}$$

Lenz's law — When an electromotive force is induced in a conductor by any change in the relation between the conductor and the magnetic field, the direction of the electromotive force is such as to produce a current whose magnetic field will oppose the change.

Limiting equivalent conductance, Λ_o — The limiting equivalent conductance of an electrolytic solution, Λ_o, is expressed by $\Lambda_o \equiv \lim (\sigma_{corr}/c)$ where σ_{corr} is solution conductance corrected for solvent conductance and c is the equivalent concentration. Λ_o is the value which Λ approaches as the solution is diluted so far that the effects of interionic forces become negligible (and dissociation, in the case of ionogens, is essentially complete).

Limiting ionic equivalent conductance, λ_o — The limiting ionic equivalent conductance of an individual ion constituent of the solute (or electrolyte) of an electrolytic solution is given by $\lambda_o \equiv \lim (\lambda/c)$. This symbol is also used to designate the limiting equivalent conductances of complexions, ion pairs, ion clusters, etc., in combination with simple ions.

Limiting ionic mobility, u^o — The limiting mobility of an individual ion of a solute (or electrolte) is given by $u \equiv \lim u$.

Limiting molar conductance, $\Lambda^m{}_o$ — The limiting molar conductance of an electrolytic solution, $\Lambda^m{}_o$, is expressed by $\Lambda^m{}_o \equiv \lim (\sigma_{corr}/m)$ where σ_{corr} is solution conductance corrected for solvent conductance and m is the molar concentration. $\Lambda^m{}_o$ is the value which Λ^m approaches as the solution is diluted so far that the effects of interionic forces become negligible. Seldomly used.

Line of force — A term employed in the description of an electric or magnetic field. A line such that its direction at every point is the same as the direction of the force which would act on a small positive charge (or pole) placed at that point. A line of force is defined as starting from a positive charge (or pole) and ending on a negative charge (or pole).

The line (of force) is also used as a unit of magnetic flux, equivalent to the maxwell.

Lissajous figures — The path described by a particle which is simultaneously displaced by two simple harmonic motions at right angles, when the periods of the two motions are in the ratio of two small whole numbers, shows a variety of characteristic curves called Lissajous figures.

Liquid — A state of matter in which the molecules are relatively free to change their positions with respect to each other, but restricted by cohesive forces so as to maintain a relatively fixed volume.

Logarithm — A number m related to another number p by

$$B^m = p$$

where B is an arbitrarily chosen number larger than 1. Usually m is written $\log_B p$. Logarithms have the basic properties that

$$\log pq = \log p + \log q$$

and

$$\log p^q = q \log p$$

In the systems most often used B = 10 (common logarithms) or B = e = 2.71828...(natural logarithms). The common logarithm of p is written $\log_{10} p$ or simply log p; the natural logarithm of p is written $\log_e p$ or ln p. Logarithmic adj.

Longitudinal wave energy — Waves whose amplitude displacement is in the same or opposite direction as the motion.

Loschmidt's number — The number of molecules per unit volume of an ideal gas at 0°C and normal atmospheric pressure.

$$n_o = 2.68675 \times 10^{25}\ m^{-3}$$

Loudness — The psychological response of the ear which is related to the physical quantity intensity. The loudness of a sound depends on frequency also since the ear responds more strongly to some frequency bands than to others. The loudness is roughly related to the cube root of the intensity, and for many purposes it is convenient to represent loudness as proportional to the logarithm of the intensity.

Lumen — The lumen is the unit of luminous flux. It is equal to the luminous flux through a unit-solid angle (steradian) from a uniform point source of one candle, or to the flux on a unit surface all points of which are at unit distance from a uniform point source of one candela.

Luminescence — Light emission by a process in which kinetic heat energy is not essential for the mechanism of excitation.

Electroluminescence is luminescence from electrical discharges—such as sparks or arcs. Excitation in these cases results mostly from electron or ion collision by which the kinetic energy of electrons or ions, accelerated in an electric field, is given up to the atoms or molecules of the gas present and causes light emission. Chemiluminescence results when energy, set free in a chemical reaction, is converted to light energy. The light from many chemical reactions and from many flames is of this type. Photoluminescence, or fluorescence, results from excitation by absorption of light. The term phosphorescence is usually applied to luminescence which continues after excitation by one of the above methods has ceased. Compare incandescence.

Luminous (energy) — Energy whose wavelength is such that the eye is sensitive to it.

Luminous flux — The total visible energy emitted by a source per unit time is called the total luminous flux from the source. The unit of flux, the lumen, is the flux emitted in unit solid angle (steradian) by a point source of one candela luminous intensity. A uniform point source of one candela intensity thus emits 4π lumens.

Luminous intensity or candlepower is the property of a source of emitting luminous flux and may be measured by the luminous flux emitted per unit solid angle. The SI unit of luminous intensity is the candela. The Hefner unit, which is equivalent to 0.9 international candles, is the intensity of a lamp of specified design burning amyl acetate, called the Hefner lamp.

The mean horizontal candlepower is the average intensity measured in a horizontal plane passing through the source. The mean spherical candlepower is the average candlepower measured in all directions and is equal to the total luminous flux in lumens divided by 4π.

Lux — A photometric unit of illuminance or illumination equal to 1 lumen/m². Compare footcandle, phot.

Magnetic Anisotropy — In ferro- or ferrimagnetic crystals. it is found that the magnetization prefers to lie along certain crystal directions. These are termed easy directions of magnetization. Work must be expended to turn the magnetization away from these easy directions. That work as a function of crystal direction defines the anisotropy energy surface. Directions associated with a maximum of the anisotropy energy are termed hard directions of magnetization. In general, the energy difference between easy and hard directions decreases as the temperature is increased, and vanishes at the Curie or Néel point.

Magnetic Domains — The magnetization of a ferromagnetic or a ferrimagnetic material tends to break up into regions called domains separated by thin transition regions called domain walls. Within the volume of a domain, the magnetization has its saturation value, and is directed along a single direction. The magnetizations of other domains are directed along different directions in such a way that the net magnetization of the whole sample may be zero. The application of an external magnetic field first causes some domains to grow by the motion of their walls. At higher fields, the magnetizations of the resulting domains rotate toward parallelism with the field.

Magnetic field due to a current — The intensity of the magnetic field in oersted at the center of a circular conductor of radius r in which a current I in absolute electromagnetic units is flowing,

$$H = \frac{2\pi I}{r}$$

If the circular coil has n turns the magnetic intensity at the center is,

$$H = \frac{2\pi n I}{r}$$

The magnetic field in a long solenoid of n turns per centimeter carrying a current I in absolute electromagnetic units

$$H = 4\pi n I$$

If I is given in amperes the above formulae become,

$$H = \frac{2\pi I}{10r} \qquad H = \frac{2\pi n I}{10r}, \qquad H = \frac{4\pi n I}{10}$$

Magnetic field due to a magnet — At a point on the magnetic axis prolonged, at a distance r cm from the center of the magnet of length 2l whose poles are +m and −m and magnetic moment M, the field strength in oersted is,

$$H = \frac{4mlr}{(r^2 - l^2)^2}$$

If r is large compared with l,

$$H = \frac{2M}{r^3}$$

At a point on a line bisecting the magnet at right angles, with corresponding symbols,

$$H = \frac{2ml}{(r^2 + l^2)^3}$$

For large value of r,

$$H = \frac{M}{r^3}$$

Magnetic field intensity or magnetizing force — Is measured by the force acting on unit pole. Unit field intensity, the oersted, is that field which exerts a force of one dyne on unit magnetic pole. The field intensity is also specified by the number of lines of force intersecting unit area normal to the field, equal numerically to the field strength in oersted. Magnetizing force is measured by the space rate of variation of magnetic potential and as such its unit may be the gilbert per centimeter. The gamma (γ) is equivalent to 0.00001 oersted. Dimensions, $[\epsilon^{1/2} m^{1/2} l^{1/2} t^{-2}]$; $[\mu^{-1/2} m^{1/2} l^{-1/2} t^{-1}]$.

Magnetic flux through any area perpendicular to a magnetic field is measured as the product of the area by the field strength. The units of magnetic flux, the maxwell is the flux through a square centimeter normal to a field of one gauss. The line is also a unit of flux. It is equivalent to the maxwell. Dimensions, $[\epsilon^{-1/2} m^{1/2} l^{1/2}]$; $[\mu^{1/2} m^{1/2} l^{3/2} t^{-1}]$.

Magnetic induction resulting when any substance is subjected to a magnetic field is measured as the magnetic flux per unit area taken perpendicular to the direction of the flux. The unit is the maxwell per square centimeter or its equivalent, the gauss. Dimensions, $[\epsilon^{-1/2} m^{1/2} l^{-3/2}]$; $[\mu^{1/2} m^{1/2} l^{-1/2} t^{-1}]$.

If a substance of permeability μ is placed in a magnetic field H the magnetic induction in the substance,

$$M = \mu H.$$

If I is the magnetic moment for unit volume, or intensity of magnetization,

$$M = H + 4\pi I.$$

The susceptibility,

$$k = \frac{I}{H}. \quad \mu = 1 + 4\pi k.$$

Magnetic moment of a magnet is measured by the torque experienced when it is at right angles to a uniform field of unit intensity. The value of the magnetic moment is given by the product of the magnetic pole strength by the distance between the poles. Unit magnetic moment is that possessed by a magnet formed by two poles of opposite sign and of unit strength, 1 cm apart. Dimensions, $[\mu^{1/2} m^{1/2} l^{5/2} t^{-1}]$; $[\epsilon^{-1/2} m^{1/2} l^{3/2}]$.

If the poles are separated by a distance which is great compared with the dimensions of the magnet, the magnetic moment of a magnet of length l whose poles have values of +m and −m is,

$$m = ml.$$

Magnetic permeability is a property of materials modifying the action of magnetic poles placed therein and modifying the magnetic induction resulting when the material is subjected to a magnetic field or magnetizing force. The permeability of a substance may be defined as the ratio of the magnetic induction in the substance to the magnetizing field to which it is subjected. The permeability of a vacuum is unity. Dimensions, $[\epsilon^{-1} l^{-2} t^2]$; $[\mu]$.

Magnetic pole or quantity of magnetism — Two unit quantities of magnetism concentrated at points unit distance apart in a vacuum repel each other with unit force. If the distance involved is 1 cm and the force 1 dyn, the quantity of magnetism at each point is one cgs unit of magnetism. Dimensions, $[\epsilon^{-1/2} m^{1/2} l^{1/2}]$; $[\mu^{1/2} m^{1/2} l^{3/2} t^{-1}]$.

Magnetic potential or magnetomotive force at a point measured by the work required to bring unit positive pole from an infinite distance (zero potential) to the point. The unit is the gilbert, that magnetic potential against which an erg of work is done when unit magnetic pole is transferred. Dimensions, $[\epsilon^{1/2}, m^{1/2} l^{3/2} t^{-2}]$; $[\mu^{-1/2} m^{1/2} l^{1/2} t^{-1}]$.

Magnetostriction — Change in sample dimensions as the magnitude or the direction of the magnetization in a crystal is changed.

Magnifying power of an optical instrument is the ratio of the angle subtended by the image of the object seen through the instrument to the angle subtended by the object when seen by the unaided eye. In the case of the microscope or simple magnifier the object as viewed by the unaided eye is supposed to be a distance of 25 cm (10 in.).

Maser — An abbreviation based on the English description of the function of this device, microwave amplification by stimulated emission of radiation.

Mass — Quantity of matter. Units of mass — the kilogram is one of the SI base units, defined by the international prototype kept at the BIPM. The gram is 1/1000 the quantity of matter in the International Prototype Kilogram. The U.S. and British pounds are defined as 0.45359237 kg, exactly. The mass of 1-kg secondary standards of platinum-iridium or of stainless steel is compared with the mass of the prototype by means of balances whose precision can reach 1 in 10^8 or better.

Mass — The inertial resistance of a body to acceleration, considered, in classical physics, to be a conserved quantity independent of speed. It can be shown that the parameters m_1, m_2 appearing in the gravitation equation are equivalent to the masses of the bodies in the sense given above. In relativistic physics it is shown that when the speed of a body becomes an appreciable fraction of c, the speed of light mass is given by

$$m = m_o / [1 - (v/c)^2]^{1/2}$$

where m_o is the rest mass of the body and v is its speed relative to the observer who finds its mass to be m. Further, it has been shown that mass and energy are interconvertible as given by Einstein's equation $E = mc^2$.

Mass action, law of — At a constant temperature the product of the active masses on one side of a chemical equation when divided by the product of the active masses on the other side of the chemical equation is a constant, regardless of the amounts of each substance present at the beginning of the action.

At constant temperature the rate of the reaction is proportional to the concentration of each kind of substance taking part in the reaction.

Mass and Weight — "1 The kilogram is the unit of mass; it is equal to the mass of the international prototype of the kilogram;

"2 The word weight* denotes a quantity of the same nature as a force; the weight of a body is the product of its mass and the acceleration due to gravity; in particular, the standard weight of a body is the product of its mass and the standard acceleration due to gravity;

"3 The value adopted in the international Service of Weights and Measures for the standard acceleration due to gravity is 980.665 cm/s², value already stated in the laws of some countries."

Mass by weighing on a balance with unequal arms— If W_1 is the value for one side, W_2 the value for the other, the true mass,

$$W = \sqrt{W_1 W_2}$$

Mass defect — Difference between atomic mass and mass number of a nuclide. See Packing fraction.

Mass-energy equivalence — The equivalence of a quantity of mass and a quantity of energy when the two quantities are related by the equation $E = mc^2$. The conversion factor c^2 is the square of the velocity of light. The relationship was developed from relativity theory, but has been experimentally confirmed. $1\ kg = 9 \times 10^{16}\ J = 2.5 \times 10^{10}\ kWh$.

Mass number — The whole number nearest the value of the atomic mass of an element as expressed in atomic mass units.

The mass number is assumed to represent the total number of protons and neutrons in the atomic nucleus of the element and is therefore equal to the atomic number plus the number of the neutrons. The mass number of an atom is usually written as a superscript to the element symbol, as in O^{18}, an isotope of oxygen with mass number 18.

Maxwell — The cgs emu magnetic flux is the flux through a cm² normal to a field at 1 cm from a unit magnetic pole.

Maxwell's rule — A law stating that every part of an electric circuit is acted upon by a force tending to move it in such a direction as to enclose the maximum amount of magnetic flux.

Mechanical equivalent of heat is the quantity of energy which, when transformed into heat, is equivalent to unit quantity of heat; 4.18×10^7 ergs = 1 calorie (20°C) or 4.1868 J = 1 cal_{IT}.

Mega — A prefix meaning million or 10^6. Symbol is M.

Meson — Two types of particles of mass intermediate between that of the electron and proton have been discovered in cosmic radiation and in the laboratory. The one particle with mass about 215m, is called μ-meson, the other with about 280m, π-meson. Mesons of both positive and negative charge have been found and there is now reasonably good evidence for neutral mesons. Both types of mesons decay spontaneously. Some evidence exists for a meson of mass about 1000m.

Metallic elements in general are distinguished from the non-metallic elements by their luster, malleability, conductivity and usual ability to form positive ions. Nonmetallic elements are not malleable, have low conductivity and never form positive ions.

Metamagnetic Materials — Those which are antiferromagnetic in weak fields, but which become ferromagnetically ordered in strong applied fields.

Meter — The meter is the SI base unit of length and was defined at the October 1984 General Conference of Weights and Measures as the length of the path traveled by light in vacuum during a time interval of 1/299,792,458 of a second.

MeV — = million electron volts.

Micro — A prefix meaning one millionth or 10^{-6}. Symbol is μ.

Mie scattering—That which is produced by spherical particles without special regard to comparative size of radiation wavelength ad particle diameter.

Milli — A prefix meaning 1/1000 or 10^{-3}. Symbol is m.

Minimum deviation — The deviation or change of direction of light passing through a prism is a minimum when the angle of incidence is equal to the angle of emergence. If D is the angle of minimum deviation and A the angle of the prism, the index of refraction of the prism for the wave length used is,

$$n = \frac{\sin \frac{1}{2}(A + D)}{\sin \frac{1}{2}A}$$

Minute of arc — 1/60 of a degree.

Mixtures consist of two or more substances intermingled with no constant percentage composition, and with each component retaining its essential original properties.

Moderator — A material used for slowing down neutrons in an atomic pile or reactor. Usually graphite or "heavy water" (deuterium oxide).

Modified index of refraction — An atmospheric index of refraction mathematically modified so that when its gradient is applied to energy propagation over a hypothetical flat earth it is substantially equivalent to propagation over the true curved earth with the actual index of refraction. Also called refractive modulus, modified refractive index. Compare potential index of refraction.

The modified index of refraction is usually expressed in M-units; mathematically

$$M = \left(n - 1 + \frac{h}{c}\right) 10^5 = N + \left(\frac{h}{a}\right) 10^6$$

where n is the index of refraction at a point in the atmosphere; h is the height above mean sea level of that point; a is the radius of the earth; and N is the index of refraction in N-units.

In ray tracing problems, the vertical gradient dM/dh can be used directly to obtain a ray path curvature that is relative to the curvature of the earth, i.e.,

$$\frac{dM}{dh} = \frac{dN}{dh} + \frac{10^5}{a} = \frac{10^6}{ka}$$

where k is a value by which the earth's radius is multiplied to get the radius of curvature of the ray path; ka is called the effective earth radius.

Modulus of elasticity — The stress required to produce unit strain, which may be a change of length (Young's modulus): a twist or shear (modulus of rigidity or modulus of torsion) or a change of volume (bulk modulus), expressed in dynes per square centimeter. Dimensions, the same as of stress, $[m\ l^{-1}\ t^{-2}]$.

Moire patterns — Pattern resulting from interference beats between two sets of periodic structures in an image.

Molal solution contains 1 mol/1000 g of solvent.

Molar conductance, Λ^m — The molar conductance of an electrolytic solution is the conductance of a solution containing 1 g mol of the solute (or electrolyte) when measured in a like manner to equivalent conductance. Seldomly used.

Molar solution contains 1 mol or g mol wt of the solute in 1 ℓ of solution.

Mole — the amount of substance of a system which contains as many elementary entities as there are atoms in 0.012 kg of carbon 12.

Note. When the mole is used, the elementary entities must be specified and may be atoms, molecules, ions, electrons, other particles, or specified groups of such particles.

Molecular volume — Volume occupied by 1 mol. Numerically equal to the molecular weight divided by the density.

* USA Editors' note: In the USA weight is the commonly used term for mass. Because of the dual use of the term weight as a quantity, this term should be avoided in technical practice. (American National Standard Z210.1)

Molecular weight — The sum of the atomic weights of all the atoms in a molecule.

Molecule — The smallest unit quantity of matter which can exist by itself and retain all the properties of the original substance.

Mol volume — The volume occupied by a mol or a gram molecular weight of any gas measured at standard conditions is 22.414 *l*.

Moment of force or torque — The effectiveness of a force to produce rotation about an axis, measured by the product of the force and the perpendicular distance from the line of action of the force to the axis. Cgs unit—the dyne-centimeter. Dimensions, $[m l^2 t^{-2}]$. If a force F acts to produce rotation about a center at a distance d from the line in which the force acts, the force has a torque,

$$L = Fd.$$

Moment of inertia — A measure of the effectiveness of mass in rotation. In the rotation of a rigid body not only the body's mass, but the distribution of the mass about the axis of rotation determines the change in the angular velocity resulting from the action of a given torque for a given time. Moment of inertia in rotation is analogous to mass (inertia) in simple translation. The cgs unit is g-cm². Dimensions, $[m l^2]$.

If m_1, m_2, m_3, etc. represent the masses of infinitely small particles of a body; r_1, r_2, r_3, etc. their respective distances from an axis of rotation, the moment of inertia about this axis will be

$$I = (m_1 r_1{}^2 + m_2 r_2{}^2 + m_3 r_3{}^2 + \ldots)$$

or

$$I = \Sigma\,(mr^2)$$

Momentum — Quantity of motion measured by the product of mass and velocity. Cgs unit, one gram-centimeter per second. Dimensions, $[m l t^{-1}]$.

A mass m moving with velocity v has a momentum,

$$M = mv$$

If a mass m has its velocity changed from v_1 to v_2 by the action of a force F for a time t,

$$mv_2 - mv_1 = Ft.$$

Monochromatic emissive power is the ratio of the energy of certain defined wave lengths radiated at definite temperatures to the energy of the same wave lengths radiated by a black body at the same temperature and under the same conditions.

Monochromatic (source) — All source radiation is exactly of the same wavelength. This is never achieved in practice even with a laser, so the term "quasi-monochromatic" is used to mean nearly of the same wavelength for all practical purposes.

Monotropic — Crystal forms one of which is always metastable with respect to the other.

Mosley's law — The frequencies of the characteristic X-rays of the elements show a strict linear relationship with the square of the atomic number.

Motion, laws of — See Newton's law of motion.

Multiple proportions, law of — If two elements form more than one compound, the weights of the first element which combine with a fixed weight of the second element are in the ratio of integers to each other.

Nano — A prefix meaning one billionth or 10^{-9}. Symbol is n.

Néel Point — The temperature at which ferrimagnetic and antiferromagnetic materials become paramagnetic.

Negatron — 1. A term used for electron when it is necessary to distinguish between (negative) electrons and positrons.

2. A four element vacuum tube which displays a negative resistance characteristic.

Nernst effect — When heat flows across the lines of magnetic force, there is observed an electromotive force in the mutually perpendicular direction.

Neutralization is a reaction in which the hydrogen ion of an acid and the hydroxyl ion of a base unite to form water, the other product being a salt.

Neutrino — An electrically neutral particle of very small (probably zero) rest mass and of spin quantum number ½. When the spin is oriented parallel to the linear momentum the particle is the antineutrino. When the spin is oriented antiparallel to the linear momentum the particle is the neutrino. Postulated by Pauli in explaining the beta decay process.

Whenever a beta (positron) particle is created in a radioactive decay so is an antineutrino (neutrino). The two particles and the parent nucleus share between them the available energy and momentum. Neutrinos and antineutrinos can penetrate amounts of matter measured in light years without appreciable attenuation. Detected by Reines and Cowan using antineutrinos from fission reactors and large scintillation detectors.

Neutron — A neutral elementary particle of mass number 1. It is believed to be a constituent particle of all nuclei of mass number greater than 1. It is unstable with respect to beta-decay, with a half life of about 12 min. It produces no detectable primary ionization in its passage through matter, but interacts with matter predominantly by collisions and to a lesser extent, magnetically. Some properties of the neutron are: rest mass, 1.00894 atomic mass unit; charge, 0; spin quantum number, ½; magnetic moment, −1.9125 nuclear Bohr magnetrons.

Neutron cross section — See cross section.

Newton (unit of force) — that force which gives to a mass of 1 kg an acceleration of 1 m/sec².

Newtonian speed of sound — An approximation to the speed of sound a in a perfect gas given by the relation

$$a^2 = p/\rho$$

where p is pressure and ρ is density. Compare Laplacian speed of sound. See Acoustic velocity.

Newtonian telescope — A reflecting telescope in which a small plane mirror reflects the convergent beam from the objective to an eyepiece at one side of the telescope. After the second reflection the rays travel approximately perpendicular to the longitudinal axis of the telescope. See Cassegrain telescope.

Newton laws of motion — A set of three fundamental postulates forming the basis of the mechanics of rigid bodies, formulated by Newton in 1687.

The first law is concerned with the principle of inertia and states that if a body in motion is not acted upon by an external force, its momentum remains constant (law of conservation of momentum). The second law asserts that the rate of change of momentum of a body is proportional to the force acting upon the body and is in the direction of the applied force. A familiar statement of this is the equation

$$F = ma$$

where F is vector sum of the applied forces, m is the mass, and a is the vector acceleration of the body. The third law is the principle of action and reaction, stating that for every force acting upon a body there exists a corresponding force of the same magnitude exerted by the body in the opposite direction.

Newton's law of cooling — The rate of cooling of a body under given conditions is proportional to the temperature difference between the body and its surroundings.

Nodal points — Two points on the axis of a lens such that a ray entering the lens in the direction of one, leaves as if from the other and parallel to the original direction.

Noncondensable gas — A gas whose temperature is above its critical temperature, so that it cannot be liquefied by increase of pressure alone.

Normal atmosphere — This is defined as the pressure exerted by a vertical column of 76 cm of mercury of density 13.5951 g/cm³ at a place where the gravitational acceleration is g = 980.665 cm/sec².

$$\begin{aligned} 1\text{ atm} &= 1.01325 \times 10^6\text{ dyn/cm}^2\text{ (exactly)} \\ &= 1.01325 \times 10^5\text{ Pa (exactly)} \\ &= 14.696\text{ psi} \\ &= 29.921\text{ in of Hg at }32°\text{F} \\ 1\text{mm of Hg} &= 1\text{ Torr} = 1333.22\text{ dyn/cm}^2 \end{aligned}$$

Normal salt — An ionic compound containing neither replaceable hydrogen nor hydroxyl ions.

Normal solution — contains one gram molecular weight of the dissolved substance divided by the hydrogen equivalent of the substance (that is, one gram equivalent) per liter of solution.

Nuclear atom — The atom of each element consists of a small dense nucleus which includes most of the mass of the atom. The nucleus is made up of roughly equal numbers of neutrons and protons. The positive charges of the protons enables the nucleus to surround itself with a set of negatively charged electrons which move around the nucleus in complicated orbits with well defined energies. The outermost electrons which are least tightly bound to the nucleus play the dominant part in determining the physical and chemical properties of the atom. There are as many electrons in orbits as there are protons in the nucleus.

Nuclear cross section (symbol σ) — A measure of the probability that the reaction will take place which is defined by

$$dI = In\,\sigma\,dx$$

where I is the intensity of the particle beam; n is the number of target nuclei per cubic centimeter of target; σ is the cross section for the specified process, expressed in square centimeters; and x is the target thickness in centimeters. See barn.

Nuclear fusion — See fusion.

Nuclear isomers — Isotopes of elements having the same mass number and atomic number but differing in radioactive properties such as half-life period.

Nuclear magneton symbol M_N)—A unit of magnetic moment of the proton equal to 5.0505×10^{-24} erg per gauss.

Nucleon — Any particle found in the structure of an atom's nucleus. The most plentiful ones are neutrons and protons.

Nucleus — The dense central core of the atom, in which most of the mass and all of the positive charge is concentrated. The charge on the nucleus, an integral multiple of Z of the electronic charge, is the essential factor which distinguishes one element from another. Z is called the atomic number and gives the number of protons in the nucleus, which includes a roughly equal number of neutrons. The mass number A gives the total number of neutrons plus protons. See Isotopes and Nuclear Atom.

Nuclide — A species of atom distinguished by the constitution of its nucleus. The nuclear constitution is specified by the number of protons, Z; number of neutrons, N; and energy content. (Or, by the atomic number, Z; mass number A (= N + Z) and atomic mass).

Numerical aperture is the sine of half the angular aperture, used as a measure of the optical power of an objective.

Nusselt number (symbol N_{Nu}) — (After Wilhelm Nusselt, 1882- , German engineer.) A number expressing the ratio of convective to conductive heat transfer between a solid boundary and a moving fluid, defined as hl/k where h is the heat-transfer coefficient, l is the characteristic length, and k is the thermal conductivity of the fluid.

Objective — The lens or combination of lenses which receives light rays from an object and refracts them to form an image in the focal plane of the eyepiece of an optical instrument, such as a telescope. Also called object glass.

Object wave — The scattered or reflected wave from the object which it is desired to image or "reconstruct".

Oersted — The cgs emu of magnetic intensity exists at a point where a force of 1 dyn acts upon a unit magnetic pole at that point, i.e., the intensity 1 cm from a unit magnetic pole. SI unit: 1 Oe = 79.577 A/m.

Ohm (unit of electric resistance) — The electric resistance between two points of a conductor when a constant difference of potential of 1 volt, applied between these two points, produces in this conductor a current of 1 ampere, this conductor not being the source of any electromotive force.

Ohm's law — Current in terms of electromotive force E and resistance R.

$$I = \frac{E}{R}$$

The current is given in amperes when E is in volts and R in ohms.

Optical correlation — Process of determinating the similarity of an optical signal or wave form to a reference-stored signal or wave form. The reference is usually stored as a matched filter.

Optical path length — The total phase change between the source and a given position in the energy beam as measured along the direction of travel of the beam or wavefronts.

Optical pyrometer — A device for measuring the temperature of an incandescent radiating body by comparing its brightness for a selected wavelength interval within the visible spectrum with that of a standard source; a monochromatic radiation pyrometer.

Optical thickness — Specifically, in calculations of the transfer of radiant energy, the mass of a given absorbing or emitting material lying in a vertical column of unit cross-sectional area and extending between two specific levels. Also called optical depth.

If z_1 and z_2 are the lower and upper limits, respectively, of a layer in which the variation of a density ϱ of some absorbing or emitting substance is given as a function of height z, then the quantity

$$\int_{z_1}^{z_2} \rho(z)\,dz$$

is called the optical thickness of that substance within that particular layer.

Order (of a bright fringe) — Proportional to the path difference between the two wave components producing a fringe (measured in integral numbers of wavelengths). No path difference produces zero order fringes.

Order (of a dark fringe) — Proportional to the path difference between the two wave components producing a fringe (measured as one-half the quantity of integral half wavelengths less one; i.e., path difference of 3.2 a wavelength produces a 1st order dark fringe).

Organ pipes — The frequency of vibration of a closed pipe or other air column of length l, where V is the velocity of sound in air, for the fundamental and first three overtones respectively is,

$$n_0 = \frac{V}{4l}, \qquad n_1 = \frac{3V}{4l}, \qquad n_2 = \frac{5V}{4l}, \qquad n_3 = \frac{7V}{4l}$$

For an open pipe,

$$n_0 = \frac{V}{2l}, \qquad n_1 = \frac{2V}{2l}, \qquad n_2 = \frac{3V}{2l}, \qquad n_3 = \frac{4V}{2l}$$

Osmotic-pressure effect — An enhancement in the velocity of the central ion, in the direction of the applied external field, as a result of more collisions on the central ion from ions behind the central ion than from ions in front of it.

Overall heat-transfer coefficient — The value U, in British thermal units per hour per square foot per °F in the equation

$$Q = UA(t_1 - t_2)$$

where Q is heat flow per unit time; A is area; and t is temperature.

Oxidation is any process which increases the proportion of oxygen or acid-forming or radical in a compound.

Packing fraction — Packing fraction is the difference between the actual mass of the isotope and the nearest whole number divided by the mass number. Thus the packing fraction is equal to (M − A)/A where M is the actual mass and A is the mass number. For example, one of the chlorine isotopes has a mass of 32.9860. The packing fraction for this isotope is

$$\frac{32.9860 - 33.0000}{33.0000} = -0.00042$$

Packing fractions are usually expressed as parts per 10,000 and so the packing fraction for this isotope of chlorine is written as −4.2. Since oxygen is taken as the standard, elements with positive packing fractions are less stable than oxygen, those with negative packing fractions are more stable.

It is positive for nuclides with mass number less than 16 or greater than 180, and negative for most others.

Parallax — The difference in the apparent direction or position of an object when viewed from different points expressed as an angle.

For bodies of the solar system, parallax is measured from the surface of the earth and its center and is called geocentric parallax, varying with the body's altitude and distance from the earth. The geocentric parallax when a body is in the horizon is called horizontal parallax and is the angular semidiameter of the earth as seen from the body. Parallax of the moon is called lunar parallax. For stars, parallax is measured from the earth and the sun, and is called annual, heliocentric, or stellar parallax. Compare aberration

Paramagnetic materials — Those within which an applied magnetic field is slightly increased by the alignment of electron orbits. The slight diamagnetic effect in materials having magnetic dipole moments is overshadowed by this paramagnetic alignment. As the temperature increases this paramagnetism disappears leaving only diamagnetism. The permeability of paramagnetic materials is slightly greater than that of empty space.

Parsec — A unit of length equal to the distance from the sun to a point having a heliocentric parallax of 1 second (1″), used as a measure of stellar distance.

The name parsec is derived from the words parallax second. 1 parsec = pc
= 3.086×10^{13} kilometers
= 206265 astronomical units
= 3.262 light years.

Partial pressure — The pressure exerted by a designated component or components of a gaseous mixture.

Pascal (Pa) — The unit of pressure in the SI system; 1 pascal equals 1 newton per squate meter; 100000 Pa = 1 bar.

Pascal's law — Pressure exerted at any point upon a confined liquid is transmitted undiminished in all directions.

Pauli exclusion principle — The principle that no pair of identical particles can simultaneously occupy the same quantum state. The principle applies to electrons, protons, and neutrons and accounts for the shell configurations of extranuclear electrons and for the shell structure of nuclei themselves.

Peltier effect — When a current flows across the junction of two unlike metals it gives rise to an absorption or liberation of heat. If the current flows in the same direction as the current at the hot junction in a thermoelectric circuit of the two metals, heat is absorbed; if it flows in the same direction as the current at the cold junction of the thermoelectric circuit heat is liberated.

Pendulum — For a simple pendulum of length l, for a small amplitude, the complete period,

$$T = 2\pi\sqrt{\frac{l}{g}} \qquad \text{or } g = 4\pi^2 \frac{l}{T^2}$$

T will be given in seconds if l is in cm and g in cm per sec^2. For a sphere suspended by a wire of negligible mass where d is the distance from the knife edge to the center of the sphere whose radius is r, the length of the equivalent simple pendulum,

$$l = d + \frac{2r^2}{5d}$$

If the period is P for an arc θ, the time of vibration in an infinitely small arc is approximately

$$T = \frac{P}{1 + \frac{1}{4}\sin^2\frac{\theta}{4}}$$

For a compound pendulum, if a body of mass m be suspended from a point about which its moment of inertia is I with its center of gravity a distance h below the point of suspension, the period

$$T = 2\pi\sqrt{\frac{I}{mgh}}$$

Penning effect — An increase in the effective ionization rate of a gas due to the pressure of a small number of foreign metastable atoms.

For instance, a neon atom has a metastable level at 16.6 V and if there are a few neon atoms in a gas of argon which has an ionization potential of 15.7 V, a collision between the neon metastable atom with an argon atom may lead to ionization of the argon. Thus, the energy which is stored in the metastable atom can be used to increase the ionization rate. Other gases where this effect is used are helium, with a metastable level at 19.8 V, and mercury, with an ionization level at 10.4 V.

Perfect fluid — In simplifying assumptions, a fluid chiefly characterized by lack of viscosity and, usually, by incompressibility. Also called an ideal fluid, inviscid fluid. See Perfect gas.

A perfect fluid is sometimes further characterized as homogeneous and continuous.

Perfect gas — A gas which has the following characteristics: (a) it obeys the Boyle-Mariotte law and the Charles-Gay-Lussac law, thus satisfying the equation of state for perfect gases; (b) it has internal energy as a function of temperature alone; and (c) it has specific heats with values independent of temperature. Also called ideal gas. Compare perfect fluid.

The normal volume of a perfect gas is 2.24136×10^4 cm^3/mol.

Perigee — That orbital point nearest the earth when the earth is the center of attraction.

Perihelion — That point in a solar orbit which is nearest the sun.

Period in uniform circular motion is the time of one complete revolution. In any oscillatory motion it is the time of a complete oscillation. Dimension, [t].

Periodic law — Elements when arranged in the order of their atomic weights or atomic numbers show regular variations in most of their physical and chemical properties.

Permeance, the reciprocal of reluctance. Unit permeance is the permeance of a cylinder one square centimeter crosssection and one centimeter length taken in a vacuum. Dimensions $[\epsilon^{-1} l^{-1} t^2]$; $[\mu l]$.

Phase angle — The phase difference of two periodically recurring phenomena of the same frequency, expressed in angular measure.

Phase of oscillatory motion — The fraction of a whole period which has elapsed since the moving particle last passed through its middle position in a positive direction.

Phase (of wave) — The distance between the position of an amplitude crest of a wave train and a reference position measured in units of wavelength, degrees, or radians (one wavelength equals 360°, or 2π radians).

Phon — The unit of loudness level. The loudness level of a sound in phons is equal to the sound pressure level in decibels re 0.0002 microbar of a pure tone of 1000 Hz which a group of listeners judge to be equally loud.

Phosphorescence — Emission of light which continues after the exciting mechanism has ceased. See luminescence. Compare fluorescence. An example of phosphorescence is the glowing of an oscilloscope screen after the exciting beam of electrons has moved to another part of the screen.

Phot — A photometric unit of illuminance or illumination equal to 1 lumen per square centimeter. Compare footcandle, lux.

Photoelectric effect — The emission of an electron from a surface as the surface absorbs a photon of electromagnetic radiation. Electrons so emitted are termed photoelectrons.

Photoelectron — An electron which has been ejected from its parent atom by interaction between that atom and a high-energy photon.

Photographic density — The density D of silver deposit on a photographic plate or film is defined by the relation

$$D = \log O$$

where O is the opacity. If I_o and I are the incident and transmitted intensities respectively the opacity is given by I_o/I. The transparency is the reciprocal of the opacity or I/I_o.

Photometer — An instrument for measuring the intensity of light or the relative intensity of a pair of lights. Also called illuminometer.

Photon — According to the quantum theory of radiation, the elementary quantity, or quantum, of radiant energy. It is regarded as a discrete quantity having a momentum equal to hv/c, where h is Planck constance, v is the frequency of the radiation, and c is the speed of light in a vacuum. The photon is never at rest, has no electric charge and no magnetic moment, but does have a spin moment. The energy of a photon (the unit quantum of energy) is equal to hv. Photons are generated in collisions between nuclei or electrons and in any other process in which an electrically charged particle changes its momentum. Conversely photons can be absorbed (i.e., annihilated) by any charged particle.

Pico — A prefix meaning one trillionth or 10^{-12}. Symbol is p.

Piezo-electric effect — The phenomemon exhibited by certain crystals of expansion along one axis and contraction along another when subjected to an electric field. Conversely compression of certain crystals generate an electrostatic voltage across the crystal. Piezoelectricity is only possible in crystal classes which do not possess a center of symmetry.

Pinch effect — When an electric current, either direct or alternating, passes through a liquid conductor, that conductor tends to contract in cross-section, due to electromagnetic forces.

Pitch — Psychological response of the ear, primarily dependent upon the frequency of viration of the air. The intensity of the sound also has a certain effect on the pitch. Pitch of a screw is the axial distance between adjacent turns of a single thread on the screw.

Planck's constance (h) — A universal constant of nature which relates the energy of a quantum of radiation to the frequency of the oscillator which emitted it. It has the dimensions of action (energy × time). Expressed by E = hv where E is the energy of the quantum and v is its frequency. Its numerical value is $6.626176\ (36) \times 10^{-27}$ erg sec.

Planck law — An expression for the variation of monochromatic radiant flux per unit area of source as a function of wavelength of black-body radiation at a given temperature; it is the most fundamental of the radiation laws. Mathematically, Planck law is

$$dw = [c_1 \lambda^{-5} / (e^{c_2/T\lambda} - 1)]\ d\lambda$$

where dw is the radiant flux from a black body in the wavelength interval $d\lambda$, centered around wavelength λ, per unit area of black-body surface at temperature T; c_1 and c_2 are radiation constants.

Planetary aberration — A displacement in the apparent position of a planet in the celestial sphere due to the relative movement of the observer and the planet. See Aberration.

Plane wave — A wave in which the wave fronts are everywhere parallel planes normal to the direction of propagation.

Plasma — An assembly of ions, electrons, neutral atoms and molecules in which the motion of the particles is dominated by electromagnetic interaction. The temperature of the collection of these particles is sufficiently high for the ionization to be above 5%. The plasma taken as a whole is electrically neutral. A plasma is further characterized by relatively large intermolecular distances, large amounts of energy stored in the internal energy of the particles and the presence of a plasma sheath at all boundaries of the plasma. Plasmas are sometimes referred to as the fourth state of matter.

Plutonium — A fissile element, artificially produced in the pile by neutron bombardment of U^{238}.

Poise — A unit of coefficient of viscosity, defined as the tangential force per unit area (dyn/cm²) required to maintain unit difference in velocity (1 cm/sec) between two parallel planes separated by 1 cm of fluid;

$$1 \text{ poise} = 1 \text{ dyne sec/cm}^2 = 1 \text{ g/cm sec.}$$

Poiseuille flow — The steady laminar flow of a fluid through a narrow horizontal circular cylinder according to the relation

$$u = (1/4\mu)\ (\partial p/\partial x)\ (a^2 - r^2)$$

where u is the fluid velocity along the cylinder's axis at a distance r from the cylinder's axis; μ is the dynamic viscosity of the fluid: a is the cylinder radius; and $\partial p/\partial x$ is the pressure gradient along the axis of the cylinder. The velocity profile across the cylinder is seen to be parabolic, and this relation affords a convenient experimental means of determining a fluid's viscosity. Also called Hagen-Poiseuille flow. Compare Couette flow.

Poisson constant (symbol μ) — The ratio of the gas constant to the specific heat of a gas at constant pressure.

Poisson distribution — A one-parameter discrete frequency distribution giving the probability that n points (or events) will be (or occur) in an interval (or time) x, provided that these points are individually independent and that the number occurring in a subinterval does not influence the number occurring in any other nonoverlapping subinterval. It has the form

$$f(n,x) = e^{-\sigma x}\ (\sigma x)^n/n!$$

The mean and variance are both σx, and σ is the average density (or rate) with which the events occur. When σx is large, the Poisson distribution approaches the normal distribution. The binomial distribution approaches the Poisson when the number of events n becomes large and the probability of success P becomes small in such a way that $nP \rightarrow \sigma x$.

Poisson's ratio is the ratio of the transverse contraction per unit dimension of a bar of uniform cross-section to its elongation per unit length, when subjected to a tensile stress.

Polarized light — Light which exhibits different directions at right angles to the line of propagation is said to be polarized. Specific rotation is the power of liquids to rotate the plane of polarization. It is stated in terms of specific rotation or the rotation in degrees per decimeter per unit density.

Polymorphism — The ability to exist in two or more crystalline forms.

Positron — A particle of the same mass M_e, as an ordinary electron. It has a positive electrical charge of exactly the same amount as that of an ordinary electron (which is sometimes called negatron). Positrons are created either by the radioactive decay of certain unstable nuclei or, together with a negatron, in a collision between an energetic (more than one Mev) photon and an electrically charged particle (or another photon). A positron does not deçay spontaneously but on passing through matter it sooner or later collides with an ordinary electron and in this collision the positron-negatron pair is annihilated. The rest energy of the two particles, which is given by Einstein's relation $E = mc^2$ and amounts to 1.0216 mev altogether, is converted into electromagnetic radiation in the form of one or more photons.

Potential (electric) at any point is measured by the work necessary to bring unit positive charge from an infinite distance. Difference of potential between two points is measured by the work necessary to carry unit positive charge from one to the other. If the work involved is one erg we have the electrostatic unit of potential. Dimensions, $[\varepsilon^{-1/2}\, m^{1/2}\, l^{1/2}\, t^{-1}]$, $[\mu^{1/2}\, m^{1/2}\, l^{3/2}\, t^{-1/2}]$. The potential at a point due to a charge q at a distance r in a medium whose dielectric constant is ε is,

$$V = \frac{q}{\varepsilon r}$$

Potential energy — Energy possessed by a body by virtue of its position in a gravity field in contrast with kinetic energy, that possessed by virtue of its motion.

Potential index of refraction — An atmospheric index of refraction so formulated that it would have no height variation in an adiabatic atmosphere. Also called potential refractive index. Compare modified index of refraction. The potential index of refraction is usually expressed in terms of B-units.

Pound (abbr. lb) — 1. A unit of mass equal in the U.S. to 0.45359237 kg, exactly. 2. Specifically, a unit of measurement of the thrust or force of a reaction engine representing the weight the engine can move, as an engine with 100,000 pounds of thrust. See Poundal, Pound mass. 3. The force exerted on 1 pound mass by the standard acceleration of gravity. See Gravity, sense 2.

Poundal — Force necessary to give 1 lb mass acceleration of 1 ft/sec².

Pound mass — 1. A mass equal to 0.45359237 kg. 2. A unit of measure of the inertial property equal to the mass of a body weighing 1 lb at the standard acceleration of gravity (980.665 cm/sec²).

Pound weight — A force equal to the earth's attraction for a mass of 1 lb. This force, acting on a 1 lb mass, will produce an acceleration of 32.1747 ft/sec².

Power — The time rate at which work is done. Units of power, the watt, one joule (ten million ergs) per second; the kilowatt is equal to 1000 watts; the horse-power, 33,000 foot-pounds per minute, is equal to 746 watts. Dimensions, $[m\ l^2\ t^{-3}]$. If an amount of work W is done in time t the power or rate of doing work is

$$P = \frac{W}{t}$$

Power will be obtained in watts if W is expressed in joules (10^7 ergs) and t in sec.

Power in watts for alternating current —

$$P = EI \cos \phi$$

where E and I are the effective values of the electromotive force and current in volts and amperes respectively and ϕ the phase angle between the current and the impressed electromotive force. The ratio,

$$\frac{P}{EI} = \cos \phi$$

is called the power factor.

Power developed by a direct current — The power in watts developed by an electric current flowing in a conductor, where E is the difference of potential at its terminals in volts, R its resistance in ohms, and I the current in amperes,

$$P = EI = RI^2$$

The work done in joules in a time t sec is,

$$W = EIt{:}RI^2 t.$$

Power ratios in telephone engineering are measured in decibels. The gain or loss of power expressed in decibels is ten times the logarithm of the power ratio. By reference to an arbitrarily chosen "power level" the actual power may be expressed in decibels. The numerical values thus used will not be proportional to the actual power level but roughly to the sensation on the ear produced when the electrical power is converted into sound. A difference of 1 decibel in the power supply to a telephone receiver produces approximately the smallest change in volume of sound which a normal ear can detect.

Pressure — Force applied to, or distributed, over a surface; measured as force per unit area. Cgs unit, the barye, one dyne per square centimeter. The megabarye is equal to 10^6 dynes per square centimeter. Pressure is also measured by the height of the column of mercury or water which it supports. Dimensions, $[m\ l^{-1}\ t^{-2}]$.

The pressure due to a force F distributed over an area A,

$$P = \frac{F}{A}$$

Absolute pressure — Pressure measured with respect to zero pressure.

Gauge pressure — Pressure measured with respect to that of the atmosphere.

Primary colors — Any three colors that when mixed in suitable proportions produce any color. Primary colors may be subtractive, where the primaries absorb colors from white light (e.g., magenta, cyan, yellow; red, blue, yellow) used in developing color photography; or additive, where the primaries form a color by the addition of their light (e.g., red, green, blue).

Principal focus of a lens or spherical mirror is the point of convergence of light coming from a source at an infinite distance.

Projectiles — For bodies projected with velocity v at an angle a above the horizontal, the time to highest point of flight.

$$t = \frac{v \sin a}{g}$$

Total time of flight to reach the original horizontal plane,

$$T = \frac{2v \sin a}{g}$$

Maximum height,

$$h = \frac{v^2 \sin^2 a}{2g}$$

Horizontal range,

$$R = \frac{v^2 \sin 2a}{g}$$

In the above equations the resistance of the air is neglected, g is the acceleration due to gravity.

Prompt neutrons — In nuclear fission, those neutrons released coincident with the fission process, as opposed to the neutrons subsequently released.

Proton — A positively charged subatomic particle having a mass of 1.67252 $\times 10^{-24}$ g, slightly less than that of a neutron but about 1836 times greater than that of an electron.

Proton-proton reaction — A thermonuclear reaction in which two protons collide at very high velocities and combine to form a deuteron. The resultant deuteron may capture another proton to form tritium and the latter may undergo proton capture to form helium.

Proton storm — The flux of protons sent into space by a solar flare.

Pulsed laser — A laser that radiates its energy during short bursts of times (pulses) and then is inactive until the next burst or pulse. The frequency of these pulses is called the pulse repetition frequency (PRF) of the laser.

Purkinje effect — A phenomenon associated with the human eye, making it more sensitive to blue light when the illumination is poor (less than about 0.1 lumen/ft^2) and to yellow light when the illumination is good.

Pyron — A unit of radiant intensity of electromagnetic radiation equal to 1 calorie/cm^2/min.

Quality or timbre of sound depends on the coexistence with the fundamental of other vibrations of various frequencies and amplitudes.

Quantity of electricity or charge — The electrostatic unit of charge, the quantity which when concentrated at a point and placed at unit distance from an equal and similarly concentrated quantity, is repelled with unit force. If the distance is one centimeter and force of repulsion one dyne and the surrounding medium a vacuum, we have the electrostatic unity of quantity. The electrostatic unit of quantity may be defined as that transferred by electrostatic unit current in unit time. The quantity transferred by one ampere in one second is the coulomb, the practical unit. The faraday is the electrical charge carried by one gram equivalent. The coulomb = 3×10^9 electrostatic units. Dimensions, $[\epsilon^{1/2}\ m^{1/2}\ l^{1/2}\ t^{-1}]$; $[\mu^{-1/2}\ m^{1/2}\ l^{1/2}]$.

Quantum — Unit quantity of energy postulated in the quantum theory. The photon is a quantum of the electromagnetic field, and in nuclear field theories, the meson is considered to be the quantum of the nuclear field.

Quantum theory — The theory first stated by Max Planck (before the Physical Society of Berlin on December 14, 1900) that all electromagnetic radiation is emitted and absorbed in quanta, each of magnitude hv, h being the Planck constant and v the frequency of the radiation.

Quasi-monochromatic — Nearly of the same wavelength (see monochromatic).

Rad — An ionizing radiation unit corresponding to an absorption of energy in any medium of 100 ergs/g (1 rad in tissue = 100/93 rep).

Radar — A system of radio detection and ranging which detects objects by beaming rf pulses that are reflected back by the object and measures its distance by the time elapsed between transmission and reception. The strength of the echo signal is determined by the radar equation

$$W_R = W_T \frac{G^2 \lambda^2 \sigma}{(4\pi)^3 R^4}$$

where W_R is received echo power. W_T is transmitted power, G is antenna gain, λ is radar carrier wavelength, σ is target cross section, and R is range.

Radar nautical mile — The time that a radar signal takes to hit a target one nautical mile away and return; 12.261 microseconds.

Radian — The angle subtended at the center of a circle by an arc equal in length to a radius of the circle. It is equal to $360°/2\pi$ or approximately 57 degrees 17 minutes 44.8 seconds. The radian is the SI unit of plane angle.

Radiation — The emission and propagation of energy through space or through a material medium in the form of waves.

The term may be extended to include streams of sub-atomic particles as alpha-rays, or beta-rays, and cosmic rays as well as electromagnetic radiation. Often used to designate the energy alone without reference to its character. In the case of light this energy is transmitted in bundles (photons).

Radiation formula, Planck's — The emission power of a black body at wave length λ may be written

$$E\lambda = \frac{c_1 \lambda^{-5}}{e^{c_2/\lambda T} - 1}$$

where c_1 and c_2 are constants with c_1 being 3.7403×10^{10} microwatts microns^4 per cm^2 or 3.7403×10^{-12} watt cm^2, c_2 being 14384 micron degrees and T the absolute temperature.

Radiation laws — 1. The four physical laws which, together, fundamentally describe the behavior of black-body radiation: (a) the Kirchoff law is essentially a thermodynamic relationship between emission and absorption of any given wavelength at a given temperature; (b) the Planck law describes the variation of intensity of black-body radiation at a given temperature, as a function of wavelength; (c) the Stefan-Boltzmann law relates the time rate of radiant energy emission from a black body to its absolute temperature; (d) the Wien law relates the wavelength of maximum intensity emitted by a black body to its absolute temperature. 2. All the more inclusive assemblage of empirical and theoretical laws describing all manifestations of radiative phenomena; e.g., Bouguer law and Lambert law.

Radioactive nuclides — Atoms that disintegrate by emission of corpuscular or electromagnetic radiations. The rays most commonly emitted are alpha or beta or gamma rays. The three classes are:

1. Primary, which have half-life times exceeding 10^8 years. These may be alpha-emitters or beta-emitters.
2. Secondary, which are formed in radioactive transformations starting with U^{238}, U^{235}, or Th^{232}.
3. Induced, having geologically short lifetimes and formed by induced nuclear reactions occurring in nature. All these reactions result in transmutation.

Radioactivity — 1. Spontaneous disintegration of atomic nuclei with emission of corpuscular or electromagnetic radiations.

2. The number of spontaneous disintegrations per unit mass and per unit time of a given unstable (radioactive) element, usually measured in curies.

Radiometer — An instrument for detecting and, usually, measuring radiant energy. Compare bolometer. See Actinometer, Photometer.

Radius of gyration may be defined as the distance from the axis of rotation at which the total mass of a body might be concentrated without changing its moment of inertia. The product of total mass and the square of the radius of gyration will give (the) moment of inertia.

Raman scattering — Raman scattering of light from a gas, liquid, or solid is that scattering in which a shift in wavelength from that of the usually monochromatic radiation occurs. The amount of shift is a function of the scattering particles and wavelengths.

Rankine cycle — An idealized thermodynamic cycle consisting of two constant-pressure processes and two isentropic processes.

Rankine scale of temperature — The Rankine scale, with its size of degree equal to that of the Fahrenheit scale, also has its zero at absolute zero of temperature. Thus, $T°R = t°F + 459.67$. Therefore, $0°R = -459.67°F$ and the normal boiling point of water is 671.67°R.

Raoult's law — Molar weights of non-volatile nonelectrolytes when dissolved in a definite weight of a given solvent under the same conditions lower the solvent's freezing point, elevate its boiling point and reduces its vapor pressure equally for all such solutes.

Rayleigh number — The nondimensional ratio between the product of buoyancy forces and heat advection and the product of viscous forces and heat conduction in a fluid. It is written as

$$N_{Ra} = \frac{g\,\Delta_z T\,\alpha d^3}{\nu k}$$

where g is the acceleration of gravity; $\Delta_z T$ is a characteristic vertical temperature difference in the characteristic depth d; α is the coefficient of expansion, ν is the kinematic viscosity; and k is the thermometric conductivity.

Rayleigh scattering — This is a coherent scattering in which the intensity of the light of wavelength λ, scattered in any direction making an angle θ with the incident light, is directly proportional to $1 + \cos^2\theta$ and inversely proportional to λ^4. The latter point is noteworthy in that it shows how much greater the scattering of the short wavelengths is. These relations apply when the scattering particles are much smaller than the wavelength of the radiation. Thus the sky is blue because blue light is scattered more than red. The unscattered light is, of course, complementary to blue, i.e., orange or yellow, which explains the "warm" hues of the sunset.

Real image — Image formed by converging rays which form a focused image in space.

Reduction is any process which increases the proportion of hydrogen or base-forming elements or radicals in a compound. Reduction is also the gaining of electrons by an atom, an ion, or an element thereby reducing the positive valence of that which gained the electron.

Reflecting telescope — A telescope which collects light by means of a concave mirror.

Reflection coefficient or reflectivity is the ratio of the light reflected from a surface to the total incident light. The coefficient may refer to diffuse or to specular reflection. In general it varies with the angle of incidence and with the wavelength of the light.

Reflection of light by a transparent medium in air (Fresnel's formulae) — If i is the angle of incidence, r the angle of refraction, n_1 the index of refraction for air (nearly equal to unity), n_2 index of refraction for a medium, then the ratio of the reflected light to the incident light is,

$$R = \tfrac{1}{2}\left(\frac{\sin^2(i-r)}{\sin^2(i+r)} + \frac{\tan^2(i-r)}{\tan^2(i+r)}\right)$$

If $i = 0$ (normal incidence), and $n_1 = 1$ (approximate for air),

$$R = \left(\frac{n_2 - 1}{n_2 + 1}\right)^2$$

Refracting telescope — A telescope which collects light by means of a lens or system of lenses. Also called refractor.

Refraction — See Index of refraction, Snell's law.

Refraction — The bending of a light or energy wave as it passes through material with varying wave velocities or indexes of refraction.

Refraction at a spherical surface — If u be the distance of a point source, v the distance of the point image or the intersection of the refracted ray with the axis, n_1 and n_2 the indices of refraction of the first and second medium, and r the radius of curvature of the separating surface,

$$\frac{n_2}{v} + \frac{n_1}{u} = \frac{n_2 - n_1}{r}$$

If the first medium is air the equation becomes,

$$\frac{n}{v} + \frac{1}{u} = \frac{n-1}{r}$$

Refractivity — 1. The algebraic difference between an index of refraction and unity.

For the atmosphere, refractivity may be more conveniently expressed in N-units:

$$N = (n - 1)\,10^5$$

The deviation of the refractivity at any altitude from the usual standard profile is expressed in B-units (for radio frequencies up to 20 kilomegacycles):

$$B = N + 0.012h$$

where h is altitude in feet.

The deviation of the refractivity at any altitude from the gradient at which the refraction curvature of a tangential ray will match the curvature of the earth may be expressed in M-units:

$$M = N + 0.048h$$

where 0.048 is 10^6 divided by the radius of the earth in feet.

Relative biological effectiveness (RBE) — The biological effectiveness of any type of energy of ionizing radiation in producing a specific damage (e.g., leukemia, anemia, sterility, carcinogenesis, catasacts, shortening of life span, etc.).

Relative humidity — The ratio of the quantity of water vapor present in the atmosphere to the quantity which would saturate at the existing temperature. It is also the ratio of the pressure of water vapor present to the pressure of saturated water vapor at the same temperature.

Relativistic mass equation — The equation

$$m = m_0\ [1 - (v^2/c^2)]^{-1/2} = m_0/(1 - \beta^2)^{1/2}$$

where $\beta \equiv v/c$ for the relativistic mass m of a particle or body of rest mass m_o when its velocity is v. See Relativistic velocity.

Relativistic particle — A particle with a velocity so large that its relativistic mass exceeds its rest mass by an amount which is significant for the computation or other considerations at hand. See Relativistic velocity.

Relativistic velocity — A velocity sufficiently high that some properties of a particle of this velocity have values significantly different from those obtaining when the particle is at rest. See Rest mass.

The property of most interest is the mass. For many purposes, the velocity is relativistic when it exceeds about one tenth the velocity of light.

Relaxation-field effect — The delay in the ion atmosphere in maintaining its symmetry around a central ion as the central ion moves in the forward direction under an applied electrical field (potential gradient).

Reluctance is that property of a magnetic circuit which determines the total magnetic flux in the circuit when a given magnetomotive force is applied. Unit, the reluctance of one centimeter length and one square centimeter cross-section of space taken in a vacuum. Dimensions, $[\epsilon l\ t^{-2}]$; $[\mu^{-1}\ l^{-1}]$.

Reluctivity or specific reluctance is the reciprocal of magnetic permeability. The reluctivity of empty space is taken as unity. Dimensions, $[\epsilon l^2\ t^{-2}]$; $[\mu^{-1}]$.

Rem — Abbreviation for roentgen-equivalent-man.

Resistance is a property of conductors depending on their dimensions, material and temperature which determines the current produced by a given difference of potential. The practical unit of resistance, the ohm is that resistance through which a difference of potential of one volt will produce a current of one ampere. The international ohm is the resistance offered to an unvarying current by a column of mercury at 0°C; 14,4521 in mass, of constant cross-sectional area and 106.300 cm in length, sometimes called the legal ohm. Dimensions, $[\epsilon^{-1}\ l^{-1}\ t]$; $[\mu l\ t^{-1}]$.

Resistance of a conductor at 0°C, of length l, cross-section s and specific resistance ϱ

$$R_0 = \rho\frac{l}{s}$$

The resistivity may be expressed as ohm-cm when R is in ohms, l in cm and s in cm^2.

Resistance of a conductor at a temperature t whose resistance at 0°C is R_o and whose temperature resistance coefficient is α

$$R_t = R_0(1 + \alpha t)$$

Resistance of conductors in series and parallel — The total resistance of any number of resistances joined in series is the sum of the separate resistances. The total resistance of conductors in parallel whose separate resistances are r_1, r_2, $r_3 \cdots r_n$ is given by the formula

$$\frac{1}{R} = \frac{1}{r_1} + \frac{1}{r_2} + \frac{1}{r_3} \ldots + \frac{1}{r_n}$$

Where R is the total resistance. For two terms this becomes,

$$R = \frac{r_1 r_2}{r_1 + r_2}$$

Resistance, specific (Resistivity) — A proportionality factor characteristic of different substances equal to the resistance that a centimeter cube of the substance offers to the passage of electricity, the current being perpendicular to two parallel faces. It is defined by the expression:

$$R = \rho\frac{l}{A}$$

where R is the resistance of a uniform conductor, l is its length, A is its cross sectional area, and ϱ is its resistivity. Resistivity is usually expressed in ohm-centimeters.

Resistivity (symbol ϱ) — In electricity, a characteristic proportionality factor equal to the resistance of a centimeter cube of a substance to the passage of an electric current perpendicular to two parallel faces. Also called specific resistance.

$$R = \rho(l/A)$$

where R is the resistance of a uniform conductor, l is its length, A is its cross-sectional area, and ϱ is its resistivity.

Resolving power of a telescope or microscope is indicated by the minimum separation of two objects for which they appear distinct and separate when viewed through the instrument.

Resonance — 1. The phenomenon of amplification of a free wave or oscillation of a system by a forced wave or oscillation of exactly equal period. The forced wave may arise from an impressed force upon the system or from a boundary condition. The growth of the resonant amplitude is characteristically linear in time. 2. Of a system in forced oscillation, the condition which exists when any change, however small, in the frequency of excitation causes a decrease in the response of the system.

Resonance (chemical) — The moving of electrons from one atom of a molecule or ion to another atom of that molecule or ion. It is simply the oriented movement of the bonds between atoms.

Restitution, coefficient of, for two bodies on impact. The ratio of the difference in velocity, after impact to the difference before impact.

Rest mass — According to relativistic theory, the mass which a body has when it is at absolute rest. Mass increases when the body is in motion according to

$$m = m_0/\sqrt{1 - (v^2/c^2)}$$

where m is its mass in motion, m_o is its rest mass; v is the body's speed of motion, and c is the speed of light.

Newtonian physics, in contrast with relativistic physics, makes no distinction between rest mass and mass in general.

Reverberation — 1. The persistence of sound in an enclosed space, as a result of multiple reflections after the sound source has stopped. 2. The sound that persists in an enclosed space, as a result of repeated reflection or scattering, after the source of the sound has stopped.

Reversible reaction — One which can be caused to proceed in either direction by suitable variation in the conditions of temperature, volume, pressure or of the quantities of reacting substances.

Rochon prism — A birefringent electro-optical crystal which divides incident unpolarized optical beam into two polarized components.

Roentgen (R) — That quantity of X or gamma radiation such that the associated corpuscular emission per 0.001293 g of dry air (equals 1 cc at 0°C and 760 mm Hg) produces, in air, ions carrying 1 esu of quantity of electricity of either sign.

Roentgen equivalent man (REM) — That amount of ionizing radiation of any type which produces the same damage to man as 1 roentgen of about 200-kv X radiation (1 rem = 1 rad in tissue/RBE). It should be noted that, when the physical dose is measured in rep units, the approximate definition is used: 1 rem ≈ 1 rep/RBE.

Roentgen equivalent physical (REP) — That amount of ionizing radiation of any type which results in the absorpion of energy at the point in question in soft tissue to the extent of 93 ergs/g. It is approximately equal to 1 roentgen of about 200-kv X radiation in soft tissue.

Root-mean-square error (symbol σ). In statistics, the square root of the arithmetic mean of the squares of the deviations of the various items from the arithmetic mean of the whole. Also termed standard deviation.

Rotatory power is the power of rotating the plane of polarized light, given in general by θ/l where θ is the total rotation which occurs in a distance l.

The molecular or atomic rotatory power is the product of the specific rotatory power by the molecular or atomic weight. Magnetic rotatory power is given by

$$\theta/e\ H \cos\alpha$$

where H the intensity of the magnetic field, and α the angle between the field and the direction of the light.

Rydberg formula — A formula, similar to that of Balmer, for expressing the wave-numbers (v) of the lines in a spectral series:

$$V = R\left[\frac{1}{(n+a)^2} - \frac{1}{(m+b)^2}\right]$$

where n and m are integers and m > n, a and b are constants for a particular series, and R is the Rydberg constant, 109737.3 cm^{-1} for hydrogen.

Salt — Any substance which yields ions, other than hydrogen or hydroxyl ions. A salt is obtained by displacing the hydrogen of an acid by a metal.

Scale height (symbol h, h_s) — A measure of the relationship between density and temperature at any point in an atmosphere; the thickness of a homogeneous atmosphere which would give the observed temperature:

$$h = kT/mg = R^*T/Mg$$

where k is the Boltzmann constant; T is the absolute temperature; m and M are the mean molecular mass and weight, respectively, of the layer; g is the acceleration of gravity; and R^* is the universal gas constant. Compare virtual height.

Scattered (light) — Reflection of light from a surface in all directions in a nonuniform manner.

Scattering Coefficient — A measure of the attenuation due to scattering of radiation as it traverses a medium containing scattering particles. Also called total scattering coefficient.

Scattering cross section — The hypothetical area normal to the incident radiation that would geometrically intercept the total amount of radiation actually scattered by a scattering particle. It is also defined, equivalently, as the cross-section area of an isotropic scatterer (a sphere) which would scatter the same amount of radiation as the actual amount. Also called extinction cross section, effective area.

Schlieren (photography) — A picture or image in which density gradients in a volume of flow are made visible. The image is produced by refraction and scattering from regions of changing refractive index.

Scintillation counter — The combination of phosphor, photomultiplier tube, and associated circuits for counting scintillations. Also called scintillating counter.

Scintillometer — An instrument which detects radiation by emitting flashes of light.

Second — The second is the unit of time of the International System of Units. The definition adopted at the October 13, 1967 meeting of the 13th General Conference on Weights and Measures is: "The second is the duration of 9 192 631 770 periods of the radiation corresponding to the transition between the two hyperfine levels of the ground state of the atom of cesium 133." The frequency 9 192 631 770 (Hz) which the definition assigns to the cesium radiation was carefully chosen to make it impossible, by any existing experimental evidence, to distinguish the new second from the "ephemeris second" based on the earth's motion. Therefore no changes need to be made in data stated in terms of the old standard in order to convert them to the new one. The atomic definition has two important advantages over the previous definition: (1) it can be realized (i.e., generated by a suitable clock) with sufficient precision, ± 1 part per hundred billion (10^{11}) or better, to meet the most exacting demands of modern metrology; and (2) it is available to anyone who has access to or who can build an atomic clock controlled by the specified cesium radiation. (A description of such clocks is given in "Atomic Frequency Standards," NBS Tech. News Bull. 45, 8—11 (Jan., 1961). For more recent developments and technical details, see R. E. Beehler, R. C. Mockler, and J. M. Richardson, "Cesium Beam Atomic Time and Frequency Standards," Metrologia 1, 114—131 (July, 1965)). In addition one can compare other high-precision clocks directly with such a standard in a relatively short time — an hour or so compared against years with the astronomical standard. Laboratory-type atomic clocks are complex and expensive, so that most clocks and frequency generators will continue to be calibrated against a standard such as the NBS Frequency Standard, controlled by a cesium atomic beam, at the Radio Standards Laboratory in Boulder, Colorado. In most cases the comparison will be by way of the standard-frequency and time-interval signals broadcast by NBS radio stations WWV, WWVH, WWVB, and WWVL.

Second law of thermodynamics — An inequality asserting that it is impossible to transfer heat from a colder to a warmer system without the occurrence of other simultaneous changes in the two systems or in the environment.

It follows from this law that during an adiabatic process, entropy cannot decrease. For reversible adiabetic processes entropy remains constant, and for irreversible adiabatic processes it increases.

Another equivalent formulation of the law is that it is impossible to convert the heat of a system into work without the occurrence of other simultaneous changes in the system or its environment. This version, which requires an engine to have a cold source as well as a heat source, is particularly useful in engineering applications. See first law of thermodynamics.

Second of arc — 1/60 of a minute of arc.

Seebeck effect — If a circuit consists of two metals, one junction hotter than the other, a current flows in the circuit. The direction of the flow depends on the metals and the temperature of the junctions.

Semiconductor — An electronic conductor, with resistivity in the range between metals and insulators, in which the electrical charge carrier concentration increases with increasing temperature over some temperature range. Certain semiconductors possess two types of carriers, namely, negative electrons and positive holes.

Sensitiveness of a balance — Assuming the three knife edges of a balance to lie on a straight line; if M is the weight of the beam, h the distance of the center of gravity below the knife edge, L the length of the balance arms and m a small mass added to one pan, the deflection θ produced is given by

$$\tan\theta = \frac{mL}{Mh}$$

Shear strength — "The stress, usually expressed in pounds per square inch, required to produce fracture when impressed perpendicularly upon the cross-section of a material."

Shell — According to Pauli's exclusion principle (q.v.), the extranuclear electrons do not circle around the nucleus all in orbits of the same radius, but are arranged in orbits at various distances from the nucleus. The extranuclear orbital electrons are thus assumed to be arranged in a series of concentric spheres, called shells, which are designated, in the order of increasing distance from the nucleus, as K, L, M, N, O, P, and Q shells. The number of the electrons which each of these shells can contain is limited. All electrons arranged in the same shell have the same principal quantum number. The electrons in the same shell are grouped into various subshells (q.v.), and all the electrons in the same subshell have the same orbital angular momentum. See Subshell.

Sidereal day. The duration of one rotation of the earth on its axis, with respect to the vernal equinox. It is measured by successive transits of the vernal equinox over the upper branch of a meridian.

Because of the precession of the equinoxes, the sidereal day thus defined is slightly less than the period of rotation with respect to the stars, but the difference is less than 0.01 sec. The length of the mean sidereal day is 24 hr of sidereal time or 23 hr 56 min 4.09054 sec of mean solar time.

Sidereal month — The average period of revolution of the moon with respect to the stars, a period of 27 days 7 hr 43 min 11.5 sec, or approximately 27⅓ days.

Sidereal time — Time based upon the rotation of the earth relative to the vernal equinox.

Sidereal time may be designated as local or Greenwich as the local or Greenwich meridian is used as the reference. When adjusted for nutation, to eliminate slight irregularities in the rate, it is called mean sidereal time.

Sidereal year — The period of one apparent revolution of the earth around the sun, with respect to the stars, averaging 365 days 6 hr 9 min 9.55 sec in 1955, and increasing at the rate of 0.000095 sec annually.

Because of the precession of the equinoxes this is about 20 min longer than a tropical year.

Simple harmonic motion — Periodic oscillatory motion in a straight line in which the restoring force is proportional to the displacement. If a point moves uniformly in a circle, the motion of its projection on the diameter (or any straight line in the same plane) is simple harmonic motion.

If r is the radius of the reference circle, ω the angular velocity of the point in the circle, θ the angular displacement at the time t after the particle passes the midpoint of its path, the linear displacement,

$$x = r \sin\theta = r \sin \omega t$$

The velocity at the same instant,

$$v = r\omega \cos\theta = \omega\sqrt{r^2 - x^2}$$

The acceleration,

$$a = -\omega^2 x.$$

The force for a mass m,

$$F = m\omega^2 x = -\frac{4\pi^2 mx}{T^2}$$

The period,

$$T = 2\pi\sqrt{\frac{x}{a}}$$

In the above equations the cgs system calls for x and r in centimeters, v in centimeters per second, a in centimeters per second squared, T in seconds, M in grams, F in dynes, θ in radians and ω in radians per second.

Simple machine — A contrivance for the transfer of energy and for increased convenience in the performance of work.

Mechanical advantage is the ratio of the resistance overcome to the force applied. Velocity ratio is the ratio of the distance through which force is applied to the distance through which resistance is overcome.

Efficiency is the ratio of the work done by a machine to the work done upon it.

If a force f applied to a machine through a distance S results in a force F exerted by the machine through a distance s, neglecting friction,

$$fS = Fs$$

The theoretical mechanical advantage or velocity ratio in the above case is,

$$\frac{S}{s}$$

Actually, the force obtained from the machine will have a smaller value than will satisfy the equation above. If F′ be the actual force obtained, the practical mechanical advantage will be,

$$\frac{F'}{f}$$

The efficiency of the machine,

$$E = \frac{F's}{fS}$$

Snell's law of refraction — If i is the angle of incidence, r the angle of refraction, v the velocity of light in the first medium, v′ the velocity in the second medium, the index of refraction n,

$$n = \frac{\sin i}{\sin r} = \frac{V}{V'}$$

Solar constant — The rate at which solar radiation is received outside the earth's atmosphere on a surface normal to the incident radiation and at the earth's mean distance from the sun.

Measurements of solar radiation at the earth's surface by the Smithsonian Institution for several decades give a best value for the solar constant of 1.934 calories per square centimeter per minute. Measurements from rockets of the intensity of the ultraviolet end of the spectrum have corrected this value to 2.00 calories per square centimeter per minute with a probable error of ± 2%.

Solar wind — Streams of plasma flowing approximately radially outward from the sun.

Solid — A state of matter in which the relative motion of the molecules is restricted and they tend to retain a definite fixed position relative to each other, giving rise to crystal structure. A solid may be said to have a definite shape and volume.

Solid angle — Measured by the ratio of the surface of the portion of a sphere enclosed by the conical surface forming the angle, to the square of the radius of the sphere. Unit of solid angle, the steradian, the solid angle which encloses a surface on the sphere equivalent to the square of the radius. Dimensions, unity.

Solubility of one liquid or solid in another is the mass of a substance contained in a solution which is in equilibrium with an excess of the substance. Under these conditions the solution is said to be saturated. Solubility of a gas is the ratio of concentration of gas in the solution to the concentration of gas above the solution.

Solubility product or precipitation value is the product of the concentration of the ions of a substance in a saturated solution of the substance. These concentrations are frequently expressed as moles of solute per liter of solution.

Solute — That constituent of a solution which is considered to be dissolved in the other, the solvent. The solvent is usually present in larger amount than the solute.

A solution is saturated if it contains at given temperature as much of a solute as it can retain in the presence of an excess of that solute.

A true solution is a mixture, liquid, solid or gaseous, in which the components are uniformly distributed throughout the mixture. The proportion of the constituents may be varied within certain limits.

Solvent — That constituent of a solution which is present in larger amount; or, the constituent which is liquid in the pure state, in the case of solutions or solids or gases in liquids.

Sone — A unit of loudness. A simple tone of frequency 1000 cycles/sec, 40 decibels above a listener's threshold, produces a loudness of 1 sone.

Spatial frequency plane — The focusing plane of an optical lens or system where the image represents the spatial Fourier transform of the object spatial function.

Specific conductance, σ_v, — The specific conductance, or conductivity, of a conductor of electricity is the conductance of the material between opposite sides of a cube, 1 cm in each direction. The unit of specific conductance is $ohm^{-1}\ cm^{-1}$ or $mho\ cm^{-1}$.

Specific gravity — The ratio of the mass of a body to the mass of an equal volume of water at 4°C or other specified temperature. Dimensions, unity.

Specific heat of a substance is the ratio of its thermal capacity to that of water at 15°C. Dimensions, unity.

If a quantity of heat H calories is necessary to raise the temperature of m grams of a substance from t_1 to t_2°C, the specific heat, or more properly, thermal capacity of the substance,

$$s = \frac{H}{m(t_2 - t_1)}$$

Specific heat by the method of mixtures— Where a mass m_1 of the substance is heated to a temperature t_1, then placed in a mass of water m_2 t a temperature t_2 contained in a calorimeter with stirrer (of same material) of mass m_3, specific heat of the calorimeter c, t_3 the final temperature

$$m_1 s(t_1 - t_3) = (m_3 c + m_2)(t_3 - t_2)$$

Black's ice calorimeter — If a body of mass m and temperature t melts a mass m′ of ice, its temperature being reduced to 0°C, the specific heat of the substance is,

$$s = \frac{80.1\ m'}{mt}$$

Bunsen's ice calorimeter — A body of mass m at temperature t causes a motion of the mercury column of l centimeters in a tube whose volume per unit length is v. The specific heat is

$$s = \frac{884lv}{mt}$$

Specific inductive capacity — The ratio of the capacitance of a condenser with a given substance as dielectric to the capacitance of the same condenser with air or a vacuum as dielectric is called the specific inductive capacity. The ratio of the dielectric constant of a substance to that of a vacuum.

Specific rotation — If there are n grams of active substance in v cubic centimeters of solution and the light passes through l decimeters, r being the observed rotation in degrees, the specific rotation (for 1 centimeter),

$$[\alpha] = \frac{rv}{nl}$$

Specific volume is the reciprocal of density. Dimensions, $[m^{-1} l^3]$.

Spectral series are spectral lines or groups of lines which occur in an orderly sequence.

Specular reflection — Mirror scattering at one angle from the point of reflection.

Speed — Time rate of motion measured by the distance moved over in unit time. Cgs unit, 1 cm/sec. Dimension $[l\ t^{-1}]$.

Speed of sound (symbol c_t) — The speed of propagation of sound waves. In the atmosphere

$$c_t = [\gamma\ (R^*/M_0)\ T_M]^{1/2}$$

where γ is the ratio of specific heat of air at constant pressure to that a constant volume, R^* is the universal gas constant, M_o is the mean molecular weight of air at sea level, and T_M is the molecular scale temperature.

At sea level in the standard atmosphere, the speed of sound is 340.294 meters per second (1116.45 ft/sec).

The concept of the speed of sound in the atmosphere loses its applicability at about 90 km where the mean free path of air molecules approaches the wavelengths of sound waves.

Spherical aberration — When large surfaces of spherical mirrors or lenses are used the light divergent from a point source is not exactly focused at a point. The phenomenon is known as spherical aberration. For axial pencils the error is known as axial spherical aberration; for oblique pencils, coma.

Spherical mirrors — If R is the radius of curvature, F principal focus, and f_1 and f_2 any two conjugate focal distances,

$$\frac{1}{f_1} + \frac{1}{f_2} = \frac{1}{F} = \frac{2}{R}$$

If the linear dimensions of the object and image be O and I respectively and u and v their distances from the mirror,

$$\frac{O}{I} = \frac{u}{v}$$

Spin — In nuclear physics, used to describe the angular momentum of elementary particles or of nuclei.

Spontaneous-ignition temperature — In testing fuels, the lowest temperature of a plate or other solid surface adequate to cause ignition in air of a fuel upon the surface.

Stagnation pressure — 1. The pressure at a stagnation point. 2. In compressible flow, the pressure exhibited by a moving gas or liquid brought to zero velocity by an isentropic process. 3. Equals total pressure. 4. Equals impact pressure.

Because of the lack of a standard meaning, stagnation pressure should be defined when it is used.

Standard conditions for gases — Measured volumes of gases are quite generally recalculated to 0°C temperature and 760 mm pressure, which have been arbitrarily chosen as standard conditions.

Stark effect — The splitting of a single spectrum line into multiple lines which occurs when the emitting material is placed in a strong electric field. The observed effect depends on the angle between the direction of the field and the direction of observation. The effect is due to the shifting of the energy states of certain orbits which all have the same energy in zero field.

Statcoulomb — The unit of electric charge in the metric system. 3×10^9 statcoulombs = 1 coulomb.

Static pressure (symbol p) — 1. The pressure with respect to a stationary surface tangent to the mass-flow velocity vector. 2. The pressure with respect to a surface at rest in relation to the surrounding fluid.

Stationary or standing waves are produced in a medium by the simultaneous transmission, in opposite directions of two similar wave motions. Fixed points of minimum amplitude are called nodes. A segment extends from one node to the next. An antinode or loop is the point of maximum amplitude between two nodes.

Statute mile — 5280 feet = 1.6093 kilometers = 0.869 nautical mile. Also called land mile.

Stefan-Boltzmann constant (symbol σ) — A universal constant of proportionality between the radiant emittance of a black body and the fourth power of the body's absolute temperature; 5.6697×10^{-5} erg per centimeter squared second °K^4.

Stefan-Boltzmann law — A law stating that the total radiation E from a black body is given by

$$E = \sigma T^4$$

where T is the absolute temperature of the body and σ is the Stefan-Boltzmann constant, equal to 5.672×10^{-5} erg/sec cm^2 deg^4.

Steradian — The steradian is the SI unit of solid angle.

The steradian is the solid angle which, having its vertex in the center of a sphere, cuts off an area of the surface of the sphere equal to that of a square with sides of length equal to the radius of the sphere. There are 4π steradians in a sphere.

Stereoscopic image — An image which appears as a three-dimensional object located in space.

Stoichiometric — Pertaining to weight relations in chemical reactions.

Stoke — See under Viscosity.

Stokes' law — 1. Gives the rate of fall of a small sphere in a viscous fluid. When a small sphere falls under the action of gravity through a viscous medium it ultimately acquires a constant velocity,

$$V = \frac{2ga^2(d_1 - d_2)}{9\eta}$$

where a is the radius of the sphere, d_1 and d_2 the densities of the sphere and the medium respectively, and η the coefficient of viscosity. V will be in cm per sec if g is in cm per sec^2, a in cm, d_1 and d_2 in g per cm^3 and η in dyne-sec per cm^2 or poises.

2. The empirical law stating that the wavelength of light emitted by a flu-

orescent material is longer than that of the radiation used to excite the fluorescence. In modern language the emitted photons carry off less energy than is brought in by the exciting photons; the details accord with the energy conservation principle.

Strain — The deformation resulting from a stress measured by the ratio of the change to the total value of the dimension in which the change occurred. Dimensions, unity.

Stress — The force producing or tending to produce deformation in a body measured by the force applied per unit area Cgs units, one dyne per square centimeter. Dimensions, $[m\, l^{-1}\, t^{-2}]$.

Subshell — The electrons within the same shell (energy level) of the atom are characterized by the same principal quantum number (n), and are further divided into groups according to the value of their azimuthal quantum numbers (l); the electrons which possess the same azimuthal quantum number for the same principal quantum number are considered to occupy the same subshell (or sublevel). The individual subshells are designated with the letters s, p, d, f, g, and h, as follows:

l value	designation of subshell
0	s
1	p
2	d
3	f
4	g
5	h

An electron assigned to the s-subshell is called an s-electron, one assigned to the p-subshell is referred to as a p-electron, etc. In formulae of electron structure, the value of the principal quantum number (n) is prefixed to the letter indicating the azimuthal quantum number (l) of the electron; thus, e.g., a 4f-electron is an electron which has the principal quantum number 4 (i.e., assigned to the N-shell) and the orbital angular momentum 3 (f-subshell).

Substance, amount of — See Mole

Supersonic flow — In aerodynamics, flow of a fluid over a body at speeds greater than the acoustic velocity and in which the shock waves start at the surface of the body. Compare hypersonic flow.

Surface density of electricity — Quantity of electricity per unit area. Dimensions, $[\epsilon^{1/2}\, m^{1/2}\, l^{-1/2}\, t^{-1}]$; $[\mu^{-1/2}\, m^{1/2}\, l^{-1/2}]$.

Surface density of magnetism — Quantity of magnetism per unit area. Dimensions, $[\epsilon^{-1}\, m^{1/2}\, l^{-1/2}]$, $[\mu^{1/2}\, m^{1/2}\, l^{-1/2}\, t^{-1}]$.

Surface tension — Two fluids in contact exhibit phenomena, due to molecular attractions which appear to arise from a tension in he surface of separation. It may be expressed as dynes per centimeter or as ergs per square centimeter. Dimensions, $[m\, t^{-2}]$.

The total force along a line of length l on the surface of a liquid whose surface tension is T,

$$F = lT.$$

Capillary tubes — If a liquid of density d rises a height h in a tube of internal radius r the surface tension is,

$$T = \frac{rhdg}{2}$$

The tension will be in dynes per cm if r and h are in cm, d in g per cm³ and g in cm per sec².

Drops and bubbles — Pressure in dynes per cm² due to surface tension on a drop of radius r centimeter for a liquid whose surface tension is T dynes per centimeter,

$$P = \frac{2T}{r}$$

For a bubble of mean radius r cm,

$$P = \frac{4T}{r}$$

Susceptibility (magnetic) is measured by the ratio of the intensity of magnetization produced in a substance to the magnetizing force or intensity of field to which it is subjected. The susceptibility of a substance will be unity when unit intensity of magnetization is produced by a field of one gauss. Dimensions, $[\epsilon^{-1}\, l^{-2}\, t^{2}]$, $[\mu]$.

"Synthetic" aperture" sidelooking radar — A sidelooking radar that generates very high resolution data by integrating its return signals during the time that the physical aircraft antenna or aperture is traveling through a large distance (making up the synthetic aperture).

Tangent galvanometer — A tangent galvanometer with n turns, of radius r, in the earth's field H, has a deflection θ. The current flowing is,

$$i = \frac{Hr}{2\pi n} \tan\theta$$

If $2\pi n$ = G (the galvanometer constant).

$$i = \frac{H}{G} \tan\theta$$

Tektite — Small glassy bodies containing no crystals, composed of at least 65% silicon dioxide, bearing no relation to the geological formations in which they occur, and believed to be of extraterrestrial origin.

Tektites are found in certain large areas called strewn fields. They are named, as are minerals, with the suffix ite, as australite, found in Australia, billitonite, indochinite, and rizalite, found in Southeast Asia, bediasite from Texas, and moldavite from Bohemia and Moravia.

Temperature may be defined as the condition of a body which determines the transfer of heat to or from other bodies. Particularly it is a manifestation of the average translational kinetic energy of the molecules of a substance due to heat agitation. The customary unit of temperature is the Centigrade degree, 1/100 the difference between the temperature of melting ice and that of water boiling under standard atmospheric pressure. Celsius scale — the Celsius temperature scale is a designation of the scale also known as the centigrade scale. The degree Fahrenheit is 1/180, and the degree Reaumur 1/80 the same difference of temperature.

The fundamental temperature scale is the absolute, thermodynamic or Kelvin scale in which the temperature measure is based on the average kinetic energy per molecule of a perfect gas. The zero of the Kelvin scale is −273.15°C. The temperature scale adopted by the International Bureau of Weights and Measures is that of the constant volume hydrogen gas thermometer. The magnitude of the degree in both these scales is defined as 1/100 the difference between the temperature of melting ice and that of boiling water at 760 mm pressure. Frequently the Kelvin scale is defined as degrees C + 273.15 and the Rankine scale as degrees F + 459.67°F.

The fundamental temperature scale is now defined by means of the equation

$$\theta(X) = 273.15°\,K\, \frac{X}{X_3}$$

where θ denotes the temperature; X the thermometric property (P, V, ...); the subscript 3 refers to the triple point of water; and 273.16°K is the arbitrary fixed point for the temperature associated with the triple point of water.

The ideal gas temperature θ, (numerically equal to the Kelvin temperature), in particular, is defined by either of the two equations:

$$\theta = \begin{cases} 273.15° \lim_{P_3 \to 0} \frac{P}{P_3}, \text{ const. V} \\ 273.15° \lim_{P \to 0} \frac{V}{V_3}, \text{ const. P} \end{cases}$$

Temperature resistance coefficient — The ratio of the change of resistance in a wire due to a change of temperature of 1°C to its resistance at 0°C. Dimension, $[\theta - 1]$.

Tera — A prefix meaning trillion or 10^{12}. Symbol is T.

Terrestrial radiation — 1. The total infrared radiation emitted from the earth's surface; to be carefully distinguished from effective terrestrial radiation, atmospheric radiation (which is sometimes erroneously used as a synonym for terrestrial radiation), and insolation. Also called earth radiation, eradiation.

Thermal capacity of a substance is the quantity of heat necessary to produce unit change of temperature in unit mass. It is ordinarily expressed as calories per gram per degree Centigrade. Numerically equivalent to specific heat.

Thermal capacity or water equivalent — The total quantity of heat necessary to raise any body or system unit temperature, measured as calories per degree centigrade in the cgs system. Dimension, [m].

Thermal expansion — The coefficient of linear expansion or expansivity is the ratio of the change in length per degree C to the length at 0°C. The coefficient of volume expansion (for solids) is approximately three times the linear coefficient. The coefficient of volume expansion for liquids is the ratio of the change in volume per degree to the volume at 0°C. The value of the coefficient varies with temperature. The coefficient of volume expansion for a gas under constant pressure is nearly the same for all gases and temperatures and is equal to 0.00367 for 1°C. Dimension, $[\theta^{-1}]$.

If l_o is the length at 0°C, α the coefficient of linear expansion, the length at t°C is,

$$l_t = l_0(1 + \alpha t)$$

General formula for thermal expansion — The rate of thermal expansion varies with the temperature. The genral equation giving the magnitude m, (length or volume) at a temperature t, where m_o is the magnitude at 0° C, is

$$m_t = m_0(1 + \alpha t + \beta t^2 + \gamma t^3 \ \ldots)$$

where α, β, γ, etc. are empirically determined coefficients.

Volume expansion — If V represents volume and β the coefficient of expansion,

$$V_t = V_0(1 + \beta t).$$

For solids,

$$\beta = 3\alpha \text{ (approximately).}$$

Thermal neutrons — Neutrons slowed down by a moderator to an energy of a fraction of an electron volt — about 0.025 ev. at 15°.

Thermionic emission — The emission of ions or electrons from a metal as the result of heating. The thermionic current density, J, in amperes/cm^2 is equal to $AT_2e^{-\phi/kT}$ where A is a constant, T is degrees Kelvin, k is Boltzmann's constant, and ϕ is the work function of the emitter.

Thermodynamic efficiency — In thermodynamics, the ratio of the work done by a heat engine to the total heat supplied by the heat source. Also called thermal efficiency, Carnot efficiency.

Thermodynamic probability — Under specified conditions, the number of equally likely states in which a substance may exist. The thermodynamic probability P is related to the entropy S by

$$S = k \ln P$$

where k is Boltzmann constant. See Third law of thermodynamics.

Thermodynamic temperature scale — The Kelvin thermodynamic scale is recognized as the fundamental scale to which all temperatures should ultimately be referable. The kelvin, unit of thermodynamic temperature, is the fraction 1/273.16 of the thermodynamic temperature of the triple point of water. The unit kelvin and its symbol K are used to express an interval or a difference of temperature. In addition to the thermodynamic temperature (symbol T), expressed in kelvins, use is also made of Celsius temperature (symbol t) defined by the equation

$$t = T - T_0$$

where T_o = 273.15 K by definition. The unit "degree Celsius" is equal to the unit "kelvin", but "degree Celsius" is a special name in place of "kelvin" for expressing Celsius temperature. A temperature interval or a Celsius temperature difference can be expressed in degrees Celsius as well as in kelvins.

Thermodynamics, law of —

I. When mechanical work is transformed into heat or heat into work, the amount of work is always equivalent to the quantity of heat.

II. It is impossible by any continuous self-sustaining process for heat to be transferred from a colder to a hotter body.

Thermoelectric power is measured by the electromotive force produced by a thermocouple for unit difference of temperature between the two junctions. It varies with the average temperature and is usually expressed in microvolts per degree C. It is customary to list the thermoelectric power of the various metals with respect to lead. Dimensions, $[\varepsilon^{-1/2} m^{1/2} t^{-1} \theta^{-1}]$; $[\mu^{1/2} m^{1/2} l^{1/2} t^{-1} \theta^{-1}]$.

Thickness or average radius of ion atmosphere, κ^{-1} — The average distance of the ion atmosphere from the reference ion in angstrom units. This average distance decreases in magnitude with the square root of the ionic concentration. Mathematically, K^{-1} is the distance at which the average charge, dq, in a spherical shell of volume $4\pi r^2 dr$ reaches a maximum using the continuous density, ϱ (r), approximation.

Third law of thermodynamics — The statement that every substance has a finite positive entropy, and that the entropy of a crystalline substance is zero at the temperature of absolute zero. See Thermodynamic probability.

Modern quantum theory has shown that the entropy of crystals at 0° absolute is not necessarily zero. If the crystal has any asymmetry, it may exist in more than one state; and there is, in addition, an entropy residue deriving from nuclear spin.

Thomson thermoelectric effect is the designation of the potential gradient along a conductor which accompanies a temperature gradient. The magnitude and direction of the potential varies with the substance.

The coefficient of the Thomson effect or specific heat of electricity is expressed in joules per coulomb per degree Centigrade. Dimensions, $[\varepsilon^{-1/2} m^{1/2} l^{1/2} t^{-1} \theta^{-1}]$, $[\mu^{1/2} m^{1/2} l^{1/2} t^{-2} \theta^{-1}]$.

Time, unit of — The fundamental invariable unit of time is the ephemeris second, which is defined as 1/31,556,925.9747 of the tropical year for 1900 January $0^d 12^h$ ephemeris time. The ephemeris day is 86,400 ephemeris seconds.

The former unit of time was the mean solar second, defined as 1/86,400 of the mean solar day.

Torque prouced by the action of one magnet on another — The turning moment experienced by a magnet of pole strength m′ and length 2l′ placed at a distance r from another magnet of length 2l and pole strength m, where the center of the first magnet is on the axis (extended) of the second and the axis of the first is perpendicular to the axis of the second,

$$C = 8\,\frac{mm'll'}{r^3} = \frac{2MM'}{r^3}$$

If the first magnet is deflected through an angle θ, the expression becomes,

$$C = \frac{2MM'}{r^3}\cos\theta$$

Torr — Provisional international standard term to replace the English term millimeter of mercury and its abbreviation mm of Hg (or the French mm de Hg).

The torr is defined as 1/760 of a standard atmosphere or 1,013,250/760 dynes per square centimeter. This is equivalent to defining the torr as 1333.22 microbars and differs by only one part in 7 million from the International Standard millimeter of mercury. The prefixes milli and micro are attached without hyphenation.

Torsional vibration — See Angular harmonic motion.

Total pressure — 1. Equals stagnation pressure. 2. Equals impact pressure. 3. The pressure a moving fluid would have if it were brought to rest without losses. 4. The pressure determined by all the molecular species crossing the imaginary surface.

Total reflection — When light passes from any medium to one in which the velocity is greater, refraction ceases and total reflection begins at a certain critical angle of incidence θ such that

$$\sin\theta = \frac{1}{n}$$

where n is the index of the first medium with respect to the second. If the second medium is air n has the ordinary value for the first medium. For any other second medium,

$$n = \frac{n_1}{n_2}$$

where n_1 and n_2 are the ordinary indices of refraction for the first and second medium respectively.

Tractive force of a magnet — If a magnet with induction B has a pole face of area A the force is,

$$F = \frac{B^2 A}{8\pi}$$

If B and A are in cgs units, F will be in dynes.

Transducer (acoustic) — A device to convert electrical oscillatory input energy into mechanical acoustical wave energy.

Transference (or transport) number — The transference number of each ion of a solute (or electrolyte) in an electrolytic solution is the fraction of the total current carried by that ion, and is given by the ratio of the mobility of the ion to the sum of the mobilities of the ions of the solute constituting the electrolytic solution.

Transmutation — A nuclear change producing a new element from an old one.

Transuranic elements — Elements of atomic numbers above 92. All of them are radioactive and are products of artificial nuclear changes. All are members of the actinide group.

Transverse electromagnetic energy — Wave whose electric field and magnetic field displacements are at right angles to each other and to the direction of propagation (motion) of the wave.

Triangle or polygon of forces — If three or more forces acting on the same point are in equilibrium, the vectors representing them form, when added, a closed figure.

Triple point — The thermodynamic state at which three phases of a substance exist in equilibrium.

The triple point of water occurs at a saturation vapor pressure of 6.11 millibar and at a temperature of 273.16° K.

Tritium — An isotope of hydrogen with a mass of three, structure, two neutrons, and one proton in its nucleus.

Ultrasound camera — A camera that converts a sound pressure "acoustic image" into an electrical TV-like image by means of the piezo-electric effect.

Uncertainty principle — See Indeterminancy.

Uniform circular motion — If r is the radius of a circle, v the linear speed in the arc, ω the angular velocity and T the period or time of one revolution,

$$\omega = \frac{v}{r} = \frac{2\pi}{T}$$

The acceleration toward the center is

$$a = \frac{v^2}{r} = \omega^2 r = \frac{4\pi^2 r}{T^2}$$

The centrifugal force for a mass m,

$$F = \frac{mv^2}{r} = m\omega^2 r = \frac{4\pi^2 mr}{T^2}$$

In the above equations ω will be in radians per second and a in cm per sec² if r is in cm, v in cm per sec and T in sec. F will be in dynes if mass is in grams and other units as above.

Application to the solar system — If M is the mass of the sun, G the constant of gravitation, P the period of the planet and r the distance of the planet from the sun, then the mass of the sun

$$M = \frac{4\pi^2 r^3}{GP^2}, \; (G = 6.670 \times 10^{-8} \text{ for cgs units})$$

If P is the period and r the distance of a satellite revolving around the planet, the above expression for M gives the mass of the planet. The formula is written on the assumption that the orbit of the planet or satellite is circular, which is only approximately true.

Uniformly accelerated rectilinear motion — If v_o is the initial velocity, v_t the velocity after time t, the acceleration,

$$a = \frac{v_t - v_0}{t}$$

The velocity after time t,

$$v_t = v_0 + at$$

Space passed over in time t,

$$s = v_0 t + \tfrac{1}{2}at^2$$

Velocity after passing over space s,

$$v = \sqrt{v_0^2 + 2as}$$

Space passed over in the nth second

$$s = v_0 + \tfrac{1}{2}a(2n - 1)$$

In the above and following similar equations the values of the space, velocity, and acceleration must be substituted in the same system. For space in cm, velocity will be in centimeter per second and acceleration in centimeter per second per second.

Unit — Specific magnitude of a quantity, set apart by appropriate definition, which is to serve as a basis of comparison or measurement for other quantities of the same nature.

Universal time — Time defined by the rotational motion of the earth and determined from the apparent diurnal motions which reflect this rotation; because of variations in the rate of rotation, universal time is not rigorously uniform. Also called Greenwich mean time. Compare ephemeris time.

In the years preceding 1960 the arguments of the ephemerides in the American Ephemeris and Nautical Almanac were designated as universal time.

Valence of an atom of an element is that property which is measured by the number of atoms of hydrogen (or its equivalent) one atom of that element can hold in combination if negative, or can displace in a reaction if it is positive.

Valence electrons of the atom are electrons which are gained, lost or shared in chemical reactions.

Van Allen belt, Van Allen radiation belt — The zone of high-intensity particulate radiation surrounding the earth beginning at altitudes of approximately 1000 km.

The radiation of the Van Allen belt is composed of protons and electrons temporarily trapped in the earth's magnetic field. The intensity of radiation varies with the distance from the earth.

Van der Waals' equation of state — This equation is expressed by:

$$\left(p + \frac{a}{v^2}\right)(v - b) = RT$$

It makes allowance both for the volume occupied by the molecules and for the attractive force between the molecules. b is the effective volume of molecules in one mole of gas. a is a measure of the attractive force between the molecules. For values of R, a, and b see index for table of Van der Waals' constants for gases.

Van't Hoff's principle — If the temperature of interacting substances in equilibrium is raised, the equilibrium concentrations of the reaction are changed so that the products of that reaction which absorb heat are increased in quantity, or if the temperature for such an equilibrium is lowered, the products which evolve heat in their formation are increased in amounts.

Vapor — The words vapor and gas are often used interchangeably. Vapor is more frequently used for a substance which, though present in the gaseous phase, generally exists as a liquid or solid at room temperature. Gas is more frequently used for a substance that generally exists in the gaseous phase at room temperature. Thus one would speak of iodine or carbon tetrachloride vapors and of oxygen gas.

Vapor pressure — The pressure exerted when a solid or liquid is in equilibrium with its own vapor. The vapor pressure is a function of the substance and of the temperature.

DEFINITIONS (Continued)

Vectors, composition of — If the angle between two vectors is A, and their magnitude a and b, their resultant,

$$C = \sqrt{a^2 + b^2 - 2ab \cos A.}$$

Velocity — Time rate of motion in a fixed direction. Cgs units, one centimeter per second. Dimensions, $[l\,t^{-1}]$.

If s is space passed over in time t, the velocity,

$$\bar{v} = \frac{s}{t}$$

Velocity of a compressional wave — The velocity of a compressional wave in an elastic medium, in terms of elasticity E (bulk modulus) and density d,

$$V = \sqrt{\frac{E}{d}}$$

For the velocity of sound in air, where p is the pressure and d the density,

$$V = \sqrt{\frac{1.4p}{d}}$$

Velocity of efflux of a liquid — If h is the distance from the opening to the free surface of the liquid, the velocity of efflux is

$$V = \sqrt{2gh}$$

The above is the theoretical discharge velocity disregarding friction and the shape of orifice. For water issuing through a circular opening with sharp edges of area, A, the volume discharged per second is given approximately by,

$$Q = 0.624\sqrt{2gh}$$

Velocity of sound, variation with temperature — The velocity in meters per sec at any temperature t in °C is given approximately by

$$V = V_0\sqrt{1 + \frac{t}{273}}$$

$$V = 331.5 + 0.607t$$

The variation with humidity is given by the equation

$$V_d = V_h\sqrt{1 - \frac{e}{p}\;\frac{\gamma\omega}{\gamma a} - \frac{5}{8}}$$

where V_d is the velocity in dry air, V_h that in air at barometric pressure p in which the pressure of water vapor is e. γ_ω and γ_a are the specific heat ratios for water vapor and for air respectively.

Velocity of a transverse wave is a stretched cord. If T is the tension of the cord and m the mass per unit length,

$$V = \sqrt{\frac{T}{m}}$$

Velocity of water waves — If the depth h is small compared with the wave length, the velocity,

$$V = \sqrt{gh}$$

In deep water for a wave length λ,

$$V = \sqrt{\frac{g\lambda}{2\pi}}$$

If the wavelength is very small, less than about 1.6 cm, the velocity increases as the wave length decreases and is expressed by the following,

$$V = \sqrt{\frac{2\pi T}{\lambda d} + \frac{g\lambda}{2\pi}}$$

where T is the surface tension and d the density of the liquid V will be given in cm per sec if h and λ are in cm, g in cm per sec², T in dyes per cm and d in g per cm³.

Velocity of a wave — The velocity of propagation in terms of wavelength λ and the period T or frequency n is,

$$V = \frac{\lambda}{T} = n\lambda$$

Virtual gravity — The force of gravity on an atmospheric parcel, reduced by centrifugal force due to the motion of the parcel relative to the earth. The virtual gravity g* is

$$g^* = g = V^2/a - 2\Omega_n V$$

where g is the magnitude of the acceleration of gravity; V is the parcel speed; a is the earth's radius; and Ω_n is the component of the earth's angular velocity vector normal to the motion of the parcel.

For reasonable atmospheric values, the correction terms are of the order of 0.01% of the magnitude of gravity. The identity of g* and g is implied by the assumption of hydrostatic equilibrium.

Virtual height — The apparent height of an ionized atmospheric layer determined from the time interval between the transmitted signal and the ionospheric echo at vertical incidence, assuming that the velocity of propagation is the velocity of light in a vacuum over the entire path. See Ionospheric recorder. Compare scale height.

Virtual image — An image that cannot be shown on a surface but is visible, as in a mirror.

Virtual mass — The actual mass of a body, plus its apparent additional mass.

Viscosity — All fluids possess a definite resistance to change of form and many solids show a gradual yielding to forces tending to change their form. This property, a sort of internal friction, is called viscosity; it is expressed in dyne-seconds per cm² or poises. Dimensions, $[m\,l^{-1}\,t^{-1}]$. If the tangential force per unit area, exerted by a layer of fluid upon one adjacent is one dyne for a space rate of variation of the tangential velocity of unity, the viscosity is one poise.

Kinematic viscosity is the ratio of viscosity to density. The c. g. s. unit of kinematic viscosity is the stoke.

Flow of liquids through a tube; where l is the length of the tube, r its radius, p the difference of pressure at the ends, η the coefficient of viscosity, the volume escaping per second,

$$v = \frac{\pi p r^4}{8 l \eta} \text{ (Poiseuille).}$$

The volume will be given in cm³ per second if l and r are in cm, p in dynes per cm² and η in poises or dyne-seconds per cm².

Viscosity effect — An alteration in the velocity of a given ion as a result of the contribution to the bulk viscosity owing to the ions of opposite charge. This effect applies to ions of large size.

Visibility is measured by the ratio of the luminous flux in lumens to the total radiant energy in ergs per second or in watts.

Volt (V) — The unit of electric potential difference and electromotive force, equal to the difference of electric potential between two points of a conductor carrying a constant current of 1 ampere when the power dissipated between these points equals 1 watt.

Volume, unit of — The cubic centimeter, the volume of a cube whose edges are one centimeter in length. Other units of volume are derived in a similar manner. Dimension, $[l^3]$.

Volume velocity — Volume velocity is the rate of alternating flow of the medium through a specificed surface due to a sound wave.

Walden's rule, $\Lambda_o\eta_o$ — Walden's rule states that the product of the limiting equivalent conductance of an electrolytic solution, Λ_o, and the viscosity of the solvent, η_o, in which the solute (or electrolyte) is dissolved is a constant at a particular temperature. Walden's rule is an approximation which would be valid only for ions which behave hydrodynamically like Stokes spheres in a continuum.

Watt — Rate of doing work or expending power. A watt is that power which gives rise to the production of energy at a rate of 1 joule per second or 10^7 ergs per second.

Wave — A wave is a disturbance which is propagated in a medium in such a manner that at any point in the medium the quantity serving as measure of disturbance is a function of the time while at any instant the displacement at a point is a function of the position of the point.

Wavefront — Surface whose points in an energy beam are all of equal phase or optical path length.

Wavelength — Distance between successive wavefronts of like phase (i.e., from peak to peak, or trough to trough).

Wave motion — A progressive disturbance propagated in a medium by the periodic vibration of the particles of the medium. Transverse wave motion is that in which the vibration of the particles is perpendicular to the direction of propagation. Longitudinal wave motion is that in which the vibration of the particles is parallel to the direction of propagation.

Wave number (symbol ν) — The reciprocal of wavelength; the number of waves per unit distance in the direction of propagation; or, sometimes 2π times this quantity.

In spectroscopy, wave number is usually expressed in reciprocal centimeters, as 100,000 cm^{-1} (100,000 per centimeter).

Weber (unit of magnetic flux) — The magnetic flux which, linking a circuit of one turn, produces in it an electromotive force of 1 volt as it is reduced to zero at a uniform rate in 1 second. (Symbol Wb)

Weight — The force with which a body is attracted toward the earth. Cgs unit, — the dyne. Dimensions, $[m\,l\,t^{-2}]$.

Although the weight of a body varies with its location, the weights of various standards of mass are often used as units of force as, pound weight, or pound force, gram weight, etc. The weight of mass m, where g is the acceleration due to gravity,

$$W = mg$$

The weight will be given in dynes when m is in grams and g in centimeter per second squared.

Wheatstone's bridge — If the resistances r_1, r_2, r_3, and r_4 form the arms of a Wheatstone's bridge in order as the circuit (omitting cell and galvanometer connections) is traced, when the bridge is balanced,

$$\frac{r_1}{r_2} = \frac{r_4}{r_3} \quad \text{or} \quad \frac{r_1}{r_4} = \frac{r_2}{r_3}$$

Wien distribution law — A relation, derived on purely thermodynamic reasoning by Wien, between the monochromatic emittance of an ideal black body and that body's temperature.

$$J\lambda/T^5 = f(\lambda, T)$$

where J_λ is the monochromatic emittance (emissive power) of a black body at wavelength λ and absolute temperature T, and f(λ, T) is a function which cannot be determined purely on classical thermodynamic grounds. Conpare Wien law.

Wien effect — The increase in the conductance of an electrolytic solution produced by high electrical fields (potential gradients).

Wien's displacement law — When the temperature of a radiating black body increases, the wave length corresponding to maximum energy decreases in such a way that the product of the absolute temperature and wave length is constant.

$$\lambda_{max} T = w$$

w is known as Wien's displacement constant.

Work — When a force acts against resistance to produce motion in a body the force is said to do work. Work is measured by the product of the force acting and the distance moved through against the resistance. Cgs units of work, the erg, a force of one dyne acting through a distance of one centimeter. The joule is 1×10^7 ergs. Dimensions, $[m\,l^2\,t^{-2}]$. The foot-pound is the work required to raise a mass of one pound a vertical distance of one foot where g = 32.174 ft/sec². The foot-poundal is the work done by a force of one poundal acting through a distance of one foot. The International joule, a unit of electrical energy, is the work expended per second by a current of one International ampere flowing through one International ohm. The kilowatt-hour is the total amount of energy developed in one hour by a power of one kilowatt.

If a force F act through a space s, the work done is

$$W = Fs$$

Work will be given in ergs if F is in dynes and s in centimeters.

Work done in rotation. If a torque L dyne-cm acts through an angle θ radians, the work done in ergs is

$$W = L\theta$$

Work function — The energy required for an electron to escape a solid surface. See Helmholtz function.

X-rays — A type of radiation of higher frequency than visible light but lower than gamma rays. Usually produced by high energy electrons impinging upon a metal target.

X units — X-ray wavelengths have been measured in two kinds of units. The older measurements are given in X units (XU) which are based on the effective lattice constant of rock salt being 2,814.00 XU. More recently X-ray wavelengths have been directly connected, through measurements with ruled gratings, to the wavelengths in the optical region and through them to the standard meter. It turned out that the XU which was originally intended as 10^{-11} cm was 0.202% larger than this value. It has become customary to give X-ray wavelengths in Angstrom units (A) when the absolute scale is used (1 A = 10^{-8} cm). The two are related by

$$1000\ XU = 1.00202 \pm 0.00003)\ \text{Å}$$

and wavelengths given in XU must be multiplied by 1.00202 and then divided by 1000 in order to convert them into Angstrom units.

Yield point — The stress at which a marked increase in deformation takes place without increase in the load.

Yield strength — "The stress at which a material exhibits a specified permanent set."

Zeeman effect — The splitting of a spectrum line into several symmetrically disposed components, which occurs when the source of light is placed in a strong magnetic field. The components are polarized, the directions of polarization and the appearance of the effect depending on the direction from which the source is viewed relative to the lines of force.

SI DERIVED UNITS WITH SPECIAL NAMES

Name	Quantity	Symbol	Expression in terms of other units	Expression in terms of SI base units
Becquerel	Activity (of a radionuclide)	Bq		s^{-1}
Coulomb	Quantity of electricity, electric charge	C	$A \cdot s$	$s \cdot A$
Degrees Celsius	Celsius temperature	°C		K
Gray	Absorbed dose, specific energy imparted, kerma, absorbed dose index	Gy	J/kg	$m^2 \cdot s^{-2}$
Farad	Capacitance	F	C/V	$m^{-2} \cdot kg^{-1} \cdot s^4 \cdot A^2$
Henry	Inductance	H	Wb/A	$m^2 \cdot kg \cdot s^{-2} \cdot A^{-2}$
Hertz	Frequency	Hz		s^{-1}
Joule	Energy, work, quantity of heat	J	$N \cdot m$	$m^2 \cdot kg \cdot s^{-2}$
Lumen	Luminous flux	lm		$cd \cdot sr^a$
Lux	Illuminance	lx	lm/m^2	$m^{-2} \cdot cd \cdot sr^a$
Newton	Force	N		$m \cdot kg \cdot s^{-2}$
Ohm	Electric resistance		V/A	$m^2 \cdot kg \cdot s^{-3} \cdot A^{-2}$
Pascal	Pressure, stress	Pa	N/m^2	$m^{-1} \cdot kg \cdot s^{-2}$
Siemens	Conductance	S	A/V	$m^{-2} \cdot kg^{-1} \cdot s^3 \cdot A^2$
Tesla	Magnetic flux density	T	Wb/m^2	$kg \cdot s^{-2} \cdot A^{-1}$
Volt	Electric potential, potential difference, electromotive force	V	W/A	$m^2 \cdot kg \cdot s^{-3} \cdot A^{-1}$
Watt	Power, radiant flux	W	J/s	$m^2 \cdot kg \cdot s^{-3}$
Weber	Magnetic flux	Wb	$V \cdot s$	$m^2 \cdot kg \cdot s^{-2} \cdot A^{-1}$

[a] In this expression the steradian is treated as a base unit.

PRIMARY FIXED POINTS OF INTERNATIONAL PRACTICAL TEMPERATURE SCALE-68 (IPTS-68)

Primary fixed points of IPTS-68 are given in degrees Celsius and kelvin. Also shown are the comparative values from previous international temperature scales.

Fixed point	ITS-27, °C	ITS-48, °C	IPTS-48, °C	IPTS-68 °C	IPTS-68 K	IPTS-68 Uncertainty (K)
Equilibrium between the liquid and vapor phases of water (boiling point of water)	100.000	100	100	100	373.15	0.005
Equilibrium between the solid and liquid phases of zinc (freezing point of zinc)				419.58	692.73	0.03
Equilibrium between the liquid and vapor phases of sulfur (boiling point of sulfur)	444.60	444.600	444.6			
Equilibrium between the solid and liquid phases of silver (freezing point of silver)	960.5	960.8	960.8	961.93	1,235.08	0.2
Equilibrium between the solid and liquid phases of gold (freezing point of gold)	1,063	1,063.0	1,063	1,064.43	1,337.58	0.2

From Sparks, L. L., *ASRDI Oxygen Technology Survey*, Vol. 4, Scientific and Technical Information Office, National Aeronautics and Space Administration, Washington, D. C., 1974.

STANDARD TYPES OF STAINLESS AND HEAT RESISTING STEELS

Chemical Ranges and Limits
Subject to Tolerances for Check Analyses
By permission of American Iron and Steel Institute

Type number	C	Mn Max.	P Max	S Max.	Si Max.	Cr	Ni	Mo	Zr	Se	Cb-Ta	Ta	Al	N
	Chemical composition (%)													
201[c]	0.15 Max.	5.50/ 7.50	0.060	0.030	1.00	16.00/ 18.00	3.50/ 5.50							0.25 Max.
202[c]	0.15 Max.	7.50/ 10.00	0.060	0.030	1.00	17.00/ 19.00	4.00/ 6.00							0.25 Max.
301[c]	0.15 Max.	2.00	0.045	0.030	1.00	16.00/ 18.00	6.00/ 8.00							
302[c]	0.15 Max.	2.00	0.045	0.030	1.00	17.00/ 19.00	8.00/ 10.00							
302B[c]	0.15 Max.	2.00	0.045	0.030	2.00/ 3.00	17.00/ 19.00	8.00/ 10.00							
303[c]	0.15 Max.	2.00	0.20	0.15 Min.	1.00	17.00/ 19.00	8.00/ 10.00	0.60 Max.[a]	0.60 Max.[a]					
303 Se[c]	0.15 Max.	2.00	0.20	0.06	1.00	17.00/ 19.00	8.00/ 10.00			0.15 Min.				
304[c]	0.08 Max.	2.00	0.045	0.030	1.00	18.00/ 20.00	8.00/ 12.00							
304L[c]	0.03 Max.	2.00	0.045	0.030	1.00	18.00/ 20.00	8.00/ 12.00							
305[c]	0.12 Max.	2.00	0.045	0.030	1.00	17.00/ 19.00	10.00/ 13.00							
308[c]	0.08 Max.	2.00	0.045	0.030	1.00	19.00/ 21.00	10.00/ 12.00							
309[c]	0.20 Max.	2.00	0.045	0.030	1.00	22.00/ 24.00	12.00/ 15.00							
309S[c]	0.08 Max.	2.00	0.045	0.030	1.00	22.00/ 24.00	12.00/ 15.00							
310[c]	0.25 Max.	2.00	0.045	0.030	1.50	24.00/ 26.00	19.00/ 22.00							
310S[c]	0.08 Max.	2.00	0.045	0.030	1.50	24.00/ 26.00	19.00/ 22.00							
314[c]	0.25 Max.	2.00	0.045	0.030	1.50/ 3.00	23.00/ 26.00	19.00/ 22.00							
316[c]	0.08 Max.	2.00	0.045	0.030	1.00	16.00/ 18.00	10.00/ 14.00	2.00/ 3.00						
316L[c]	0.03 Max.	2.00	0.045	0.030	1.00	16.00/ 18.00	10.00/ 14.00	2.00 3.00						
317[c]	0.08 Max.	2.00	0.045	0.030	1.00	18.00/ 20.00	11.00/ 15.00	3.00/ 4.00						
321[c]	0.08 Max.	2.00	0.045	0.030	1.00	17.00/ 19.00	9.00/ 12.00				5 × C Min.			
347[c]	0.08 Max.	2.00	0.045	0.030	1.00	17.00/ 19.00	9.00/ 13.00					10 × C Min.		
348[c]	0.08 Max.	2.00	0.045	0.030	1.00	17.00/ 19.00	9.00/ 13.00				10 × C Min.	0.10 Max.		
403[b]	0.15 Max.	1.00	0.040	0.030	0.50	11.50/ 13.00								
405[d]	0.08 Max.	1.00	0.040	0.030	1.00	11.50/ 14.50							0.10/ 0.30	
410[b]	0.15 Max.	1.00	0.040	0.030	1.00	11.50/ 13.50								
414[b]	0.15 Max.	1.00	0.040	0.030	1.00	11.50/ 13.50	1.25/ 2.50							
416[b]	0.15 Max.	1.25	0.06	0.15 Min.	1.00	12.00/ 14.00		0.60 Max.[a]	0.60 Max.[a]					
416 Se[b]	0.15 Max.	1.25	0.06	0.06	1.00	12.00/ 14.00				0.15 Min.				
420[b]	Over 0.15	1.00	0.040	0.030	1.00	12.00/ 14.00								
430[a]	0.12 Max.	1.00	0.040	0.030	1.00	14.00/ 18.00								
430F[d]	0.12 Max.	1.25	0.06	0.15 Min.	1.00	14.00/ 18.00		0.60 Max.	0.60 Max.[a]					
430F Se[d]	0.12 Max.	1.25	0.06	0.06	1.00	14.00/ 18.00				0.15 Min.				
431[b]	0.20 Max.	1.00	0.040	0.030	1.00	15.00/ 17.00	1.25/ 2.50							
440A[b]	0.60/ 0.75	1.00	0.040	0.030	1.00	16.00/ 18.00		0.75 Max.						

STANDARD TYPES OF STAINLESS AND HEAT RESISTING STEELS (continued)

Chemical composition (%)

Type number	C	Mn Max.	P Max	S Max.	Si Max.	Cr	Ni	Mo	Zr	Se	Cb-Ta	Ta	Al	N
400B[b]	0.75/ 0.95	1.00	0.040	0.030	1.00	16.00/ 18.00		0.75 Max.						
440C[b]	0.95/ 1.20	1.00	0.040	0.030	1.00	16.00/ 18.00		0.75 Max.						
446[d]	0.20 Max.	1.50	0.040	0.030	1.00	23.00/ 27.00								0.25 Max.
501[b]	Over 0.10	1.00	0.040	0.030	1.00	4.00/ 6.00		0.40/ 0.65						
502[b]	0.10 Max.	1.00	0.040	0.030	1.00	4.00/ 6.00		0.40/ 0.65						

[a] At producer's option; reported only when intentionally added.
[b] Heat treatable
[c] Not heat treatable.
[d] Essentially not heat treatable.

STANDARD TEST SIEVES (WIRE CLOTH)

Sieve Designation: Standard	Sieve Designation: Alternative	Nominal Sieve Opening in	Permissible Variation of Average Opening from the Standard Sieve Designation	Maximum Opening Size for Not More than 5 percent of Openings	Maximum Individual Opening	Nominal Wire Diameter, mm[a]
(1)	(2)	(3)	(4)	(5)	(6)	(7)
125 mm	5 in.	5	±3.7 mm	130.0 mm	130.9 mm	8.0
106 mm	4.24 in.	4.24	±3.2 mm	110.2 mm	111.1 mm	6.40
100 mm	4 in.	4	±3.0 mm	104.0 mm	104.8 mm	6.30
90 mm	3½ in.	3.5	±2.7 mm	93.6 mm	94.4 mm	6.08
75 mm	3 in.	3	±2.2 mm	78.1 mm	78.7 mm	5.80
63 mm	2½ in.	2.5	±1.9 mm	65.6 mm	66.2 mm	5.50
53 mm	2.12 in.	2.12	±1.6 mm	55.2 mm	55.7 mm	5.15
50 mm	2 in.	2	±1.5 mm	52.1 mm	52.6 mm	5.05
45 mm	1¾ in.	1.75	±1.4 mm	46.9 mm	47.4 mm	4.85
37.5 mm	1½ in.	1.5	±1.1 mm	39.1 mm	39.5 mm	4.59
31.5 mm	1¼ in.	1.25	±1.0 mm	32.9 mm	33.2 mm	4.23
26.5 mm	1.06 in.	1.06	±0.8 mm	27.7 mm	28.0 mm	3.90
25.0 mm	1 in.	1	±0.8 mm	26.1 mm	26.4 mm	3.80
22.4 mm	⅞ in.	0.875	±0.7 mm	23.4 mm	23.7 mm	3.50
19.0 mm	¾ in.	0.750	±0.6 mm	19.9 mm	20.1 mm	3.30
16.0 mm	⅝ in.	0.625	±0.5 mm	16.7 mm	17.0 mm	3.00
13.2 mm	0.530 in.	0.530	±0.41 mm	13.83 mm	14.05 mm	2.75
12.5 mm	½ in.	0.500	±0.39 mm	13.10 mm	13.31 mm	2.67
11.2 mm	7/16 in.	0.438	±0.35 mm	11.75 mm	11.94 mm	2.45
9.5 mm	⅜ in.	0.375	±0.30 mm	9.97 mm	10.16 mm	2.27
8.0 mm	5/16 in.	0.312	±0.25 mm	8.41 mm	8.58 mm	2.07
6.7 mm	0.265 in.	0.265	±0.21 mm	7.05 mm	7.20 mm	1.87
6.3 mm	¼ in.	0.250	±0.20 mm	6.64 mm	6.78 mm	1.82
5.6 mm	No. 3½	0.223	±0.18 mm	5.90 mm	6.04 mm	1.68
4.75 mm	No. 4	0.187	±0.15 mm	5.02 mm	5.14 mm	1.54
4.00 mm	No. 5	0.157	±0.13 mm	4.23 mm	4.35 mm	1.37
3.35 mm	No. 6	0.132	±0.11 mm	3.55 mm	3.66 mm	1.23
2.80 mm	No. 7	0.111	±0.095 mm	2.975 mm	3.070 mm	1.10
2.36 mm	No. 8	0.0937	±0.080 mm	2.515 mm	2.600 mm	1.00
2.00 mm	No. 10	0.0787	±0.070 mm	2.135 mm	2.215 mm	0.900
1.70 mm	No. 12	0.0661	±0.060 mm	1.820 mm	1.890 mm	0.810
1.40 mm	No. 14	0.0555	±0.050 mm	1.505 mm	1.565 mm	0.725
1.18 mm	No. 16	0.0469	±0.045 mm	1.270 mm	1.330 mm	0.650
1.00 mm	No. 18	0.0394	±0.040 mm	1.080 mm	1.135 mm	0.580
850 μm	No. 20	0.0331	±35 μm	925 μm	970 μm	0.510
710 μm	No. 25	0.0278	±30 μm	775 μm	815 μm	0.450
600 μm	No. 30	0.0234	±25 μm	660 μm	695 μm	0.390
500 μm	No. 35	0.0197	±20 μm	550 μm	585 μm	0.340
425 μm	No. 40	0.0165	±19 μm	471 μm	502 μm	0.290
355 μm	No. 45	0.0139	±16 μm	396 μm	425 μm	0.247
300 μm	No. 50	0.0117	±14 μm	337 μm	363 μm	0.215
250 μm	No. 60	0.0098	±12 μm	283 μm	306 μm	0.180
212 μm	No. 70	0.0083	±10 μm	242 μm	263 μm	0.152
180 μm	No. 80	0.0070	±9 μm	207 μm	227 μm	0.131
150 μm	No. 100	0.0059	±8 μm	174 μm	192 μm	0.110
125 μm	No. 120	0.0049	±7 μm	147 μm	163 μm	0.091
106 μm	No. 140	0.0041	±6 μm	126 μm	141 μm	0.076
90 μm	No. 170	0.0035	±5 μm	108 μm	122 μm	0.064
75 μm	No. 200	0.0029	±5 μm	91 μm	103 μm	0.053
63 μm	No. 230	0.0025	±4 μm	77 μm	89 μm	0.044
53 μm	No. 270	0.0021	±4 μm	66 μm	76 μm	0.037
45 μm	No. 325	0.0017	±3 μm	57 μm	66 μm	0.030
38 μm	No. 400	0.0015	±3 μm	48 μm	57 μm	0.025

[a] The average diameter of the warp and of the shoot wires, taken separately, of the cloth of any sieve shall not deviate from the nominal values by more than the following:

Sieves coarser than 600 μm	5 percent
Sieves 600 to 125 μm	7½ percent
Sieves finer than 125 μm	10 percent

PHYSICAL PROPERTIES OF GLASS SEALING AND LEAD WIRE MATERIALS

From General Electric Company

Glass Sealing Materials

	Copper OFC	Copper 0.02P	Dumet	42 Ni	Gas Free 42% Ni	42-6 Ni Cr.	46 Ni	52 Ni.	27 Cr.	® Kovar	W	Mo
Analysis: Carbon				0.10	0.05	0.10	0.10	0.10	0.15	0.02		
Manganese				0.50	0.50	0.50	0.50	0.50	0.60	0.30		
Silicon			*Copper Clad*	0.25	0.25	0.25	0.25	0.25	0.40	0.20		
Chromium			*42% Nickel*	—	—	5.75	—	—	28.00	—		
Nickel			*Iron*	42	42	42.5	46	51	0.50	29		
Copper	99.95	99.90	*See GE Spec.*	—	—	—	—	—	—	—		
Other		P 0.02	*DS 8311-01*	Bal.Fe	Ti0.4 Bal.Fe	Bal.Fe	Ti0.4 Bal.Fe	Bal.Fe	Bal.Fe	Co 17 Bal.Fe		
Density: grams/cc	8.94	8.94	8.26 to 8.32	8.12	8.12	8.12	8.17	8.30	7.60	8.36	19.3	10.2
lbs. per cu. in.	0.323	0.323	0.298—0.301	0.293	0.293	0.293	0.295	0.300	0.274	0.302	0.697	0.369
Thermal Conductivity 20-100°/C												
Cal/cm/sec/cm²/°C	0.948	0.8	0.2—0.3	0.025	0.025	0.029	0.028	0.032	0.054	0.04	0.31	0.34
Btu/in/hr/sq. ft/°F	2750	2320	580—870	74	74	84	81	93	158	116	900	1000
Electrical Resistivity (20°C) Microhm—cm	1.71	2.03	7.3 to 12.0	72	72	95	46	43	63	49	5.5	5.2
Ohm per cir. mil ft.	10.3	12.2	44 to 72	430	430	570	275	258	380	294	33	31
Elec. Cond. % IACS	101	85	23 to 14	2.3	2.3	1.8	3.6	3.9	2.8	3.4	31	33
Curie Temperature °C	—	—	380	380	380	295	460	530	610	435	—	—
Melting Temperature °C	1083	1083	— —	1425	1425	1425	1425	1425	1425	1450	3410	2610
°F	1981	1981	— —	2597	2597	2597	2597	2597	2597	2642	6170	4730
Specific Heat cal/gr.	0.092	0.092	0.11	0.12	0.12	0.12	0.12	0.12	0.14	0.11	0.033	0.066
Thermal Expansion in/in/°Cx10⁷												
25-100°C	168	168	Radial 60 to 80	50.1	43.4	65.5	71.0	99.5	94.6	58.6	45	51
25-200°C	172	172		47.1	44.1	70.8	73.7	101.0	100.5	52.0	46	
25-300°C	177	177	60 to 80	47.6	46.1	82.6	75.0	101.0	105.3	51.3	46	
25-350°C	178	178	65 to 85	50.5		90.4	74.4	100.0	107.0	48.9	46	
25-400°C	181	181	80 to 100	62.5	64.1	100.0	74.3	100.0	107.8	50.6	46	
25-500°C	183	183	100 to 140	83.2	85.6	115.0	86.8	102.1	111.2	61.5	46	57
25-600°C	188	188		99.0	100.1	125.8	100.2	110.0	112.6	78.0		
Mechanical Properties (Annealed)												
Ultimate Str. (1000 psi)	35	35	74	82	80	80	82	80	85	75	490	120
Yield Str. (1000 psi)			50	34	34	40	34	40	55	50	360	110
% Elong. (2")			30	30	30	30	27	35	25	30	8	30
Rockwell Hardness				B76	B76	B80	B76	B83	B85	B68	C25	B88
Elastic Modulus (10⁶ psi)	16	16		21.5	21.0	23	23	24	30	20	50	47

Metal Alloy Lead Wires

	Nickel 200	Nickel 211	AISI Type 302	AISI Type 316	AISI Type 430	400 Monel	600 Inco-nel	Ad-vance ®	CDA 752	70-30 Brass
Analysis: Carbon	0.06	0.10	0.15Max	0.08Max	0.12Max	0.12	0.04			
Manganese	0.25	4.75	2.00Max	2.00Max	1.00Max	0.90	0.20		0.50Max	
Silicon	0.05	0.05	1.00Max	1.00Max	1.00Max	0.15				
Chromium			17–19	16–18	14–18		16			
Nickel	99.5	95.0	8–10	10–14	0.50Max	66	76	43	18	
Copper	0.05			—		31.5		Bal.	65	70
Other				Mo 2–3			Fe 7.20		Bal. Zn	Bal. Zn
Density: grams/cc	8.89	8.72	7.9	7.9	7.7	8.84	8.41	8.9	8.73	8.53
lbs. per cu. in.	0.321	0.315	0.285	0.285	0.278	0.319	0.304	0.322	0.316	0.308
Thermal Conductivity 20-100°/C										
Cal/cm/sec/cm²/°C	0.15	0.12	0.04	0.04	0.05	0.062	0.036	0.051	0.08	0.29
Btu/in/hr/sq. ft/°F	435	350	116	116	145	180	104	148	232	845
Electrical Resistivity (20°C) Microhm—cm	9.5	18.3	72	74	60	48.2	98.1	49	28.7	6.16
Ohm per cir. mil ft.	57	110	433	445	361	290	590	294	173	37
Elec. Cond. % IACS	18	3	2.3	2.3	2.9	3.6	1.7	3.5	6	28
Curie Temperature °C	360	352			610	43/60	-125	—	—	—
Melting Temperature °C	1455	1427	1421	1399	1510	1349	1427	1210	1110	954
°F	2651	2600	2590	2550	2750	2460	2600	2210	2030	1750
Specific Heat cal/gr.	0.13	0.13	0.12	0.12	0.11	0.102	0.109	0.094	0.09	0.09
Thermal Expansion in/in/°Cx10⁷										
25-100°C	133	133	166	166	101	140	115	149		
25-200°C	139	139				145				
25-300°C	144	144	171	175	110	150			160	199
25-350°C	146	146								
25-400°C	148	148				155				
25-500°C	172	153	180	180	120	160		163		
25-600°C						165				
Mechanical Properties (Annealed)										
Ultimate Str. (1000 psi)	70		90	85	75	85	100	60	60	50
Yield Str. (1000 psi)	25		37	35	45	40	45		30	—
% Elong. (2")	45		55	55	30	40	40		40	60
Rockwell Hardness	B62		B82	B80	B82	B68	B74		—	—
Elastic Modulus (10⁶ psi)	30	30	29	29	29	26	31	18	18	16

Clad Materials (Also see Dumet) and Pure Metals (Also see W and Mo glass sealing materials)

	Kul-grid ®	40% CCFe	30% CCFe	Au	Ag	Al	Pt	Ta	Fe	Ti
Analysis: Nickel	27									
Copper		37.5	26							
Other	Cu Core	Fe Core	Fe Core							
Density: grams/cc	8.89	8.15	8.15	19.3	10.5	2.69	21.45	16.6	7.87	4.51
lbs. per cu. in.	0.321	0.294	0.294	0.698	0.379	0.097	0.775	0.600	0.284	0.163
Thermal Conductivity 20-100°/C										
Cal/cm/sec/cm²/°C		0.46	0.38	0.71	1.00	0.57	0.165	0.13	0.18	0.43
Btu/in/hr/sq. ft/°F		1330	1100	2050	2900	1650	480	380	520	1250
Electrical Resistivity (20°C) Microhm—cm	2.3	4.4	5.9	2.19	1.629	2.65	9.83	12.45	9.71	7.0
Ohm per cir. mil ft.	4	26.4	35.3	13	9.8	16	59	75	58	42
Elec. Cond. % IACS	70	40	30	78	105	65	16	14	18	25
Curie Temperature °C		770	770	—	—	—	—	—	770	—
Melting Temperature °C				1063	960	660	1769	3000	1536	1668
°F				1945	1760	1220	3217	5425	2797	3035
Specific Heat cal/gr.		0.10	0.10	0.031	0.056	0.225	0.031	0.034	0.11	0.124
Thermal Expansion in/in/°Cx10⁷										
25-100°C				142	196	239	91	65	122	88
25-200°C						243			129	91
25-300°C						253				
25-350°C										
25-400°C									138	94
25-500°C				152	206		96	66		
25-600°C						287			145	97
Mechanical Properties (Annealed)										
Ultimate Str. (1000 psi)				20	22	13	40	55	40	
Yield Str. (1000 psi)					8	5	12		20	
% Elong. (2")				45	48	40	30	30	45	
Rockwell Hardness										
Elastic Modulus (10⁶ psi)		24	24	11.6	11	9	21.3	27	30	16.8

WIRE TABLES COMPARISON OF WIRE GAUGES

Diameter of Wire in Inches

Gauge No.	Brown & Sharpe	Birmingham or Stubs'	Washburn & Moen	Imperial or Brit. Std.	Stubs' Steel	U.S. Std. plate
00000000						
0000000				.500		
000000				.464		.46875
00000				.432		.4375
0000	.4600	.454	.3938	.400		.40625
000	.4096	.425	.3625	.372		.375
00	.3648	.380	.3310	.348		.34375
0	.3249	.340	.3065	.324		.3125
1	.2893	.300	.2830	.300	.227	.28125
2	.2576	.284	.2625	.276	.219	.26525
3	.2294	.259	.2437	.252	.212	.25
4	.2043	.238	.2253	.232	.207	.234375
5	.1819	.220	.2070	.212	.204	.21875
6	.1620	.203	.1920	.192	.201	.203125
7	.1443	.180	.1770	.176	.199	.1875
8	.1285	.165	.1620	.160	.197	.171875
9	.1144	.148	.1483	.144	.194	.15625
10	.1019	.134	.1350	.128	.191	.140625
11	.09074	.120	.1205	.116	.188	.125
12	.08081	.109	.1055	.104	.185	.109375
13	.07196	.095	.0915	.092	.182	.09375
14	.06408	.083	.0800	.080	.180	.078125
15	.05707	.072	.0720	.072	.178	.0703125
16	.05082	.065	.0625	.064	.175	.0625
17	.04526	.058	.0540	.056	.172	.05625
18	.04030	.049	.0475	.048	.168	.05
19	.03589	.042	.0410	.040	.164	.04375
20	.03196	.035	.0348	.036	.161	.0375
21	.02846	.032	.0318	.032	.157	.034375
22	.02535	.028	.0286	.028	.155	.03125
23	.02257	.025	.0258	.024	.153	.028125
24	.02010	.022	.0230	.022	.151	.025
25	.01790	.020	.0204	.020	.148	.021875
26	.01594	.018	.0181	.018	.146	.01875
27	.01419	.016	.0173	.0164	.143	.0171875
28	.01264	.014	.0162	.0149	.139	.015625
29	.01126	.013	.0150	.0136	.134	.0140625
30	.01003	.012	.0140	.0124	.127	.0125
31	.008928	.010	.0132	.0116	.120	.0109375
32	.007950	.009	.0128	.0108	.115	.01015625
33	.007080	.008	.0118	.0100	.112	.009375
34	.006304	.007	.0104	.0092	.110	.00859375
35	.005614	.005	.0095	.0084	.108	.0078125
36	.005000	.004	.0090	.0076	.106	.00703125
37	.004453		.0085	.0068	.103	.00664062
38	.003965		.0080	.0060	.101	.00625
39	.003531		.0075	.0052	.099	
40	.003145		.0070	.0048	.097	
41			.0066	.0044	.095	
42			.0062	.0040	.092	
43			.0060	.0036	.088	
44			.0058	.0032	.085	
45			.0055	.0028	.081	
46			.0052	.0024	.079	
47			.0050	.0020	.077	
48			.0048	.0016	.075	
49			.0046	.0012	.072	
50			.0044	.0010	.069	

Diameter of Wire in Centimeters

Gauge No.	Brown & Sharpe	Birmingham or Stubs'	Washburn & Moen	Imperial or Brit. Std.	Stubs' Steel	U.S. Std. plate
00000000						
0000000			1.245	1.27		1.27
000000			1.172	1.18		1.191
00000			1.093	1.10		1.111
0000	1.168	1.15	1.000	1.02		1.032
000	1.040	1.08	0.9208	0.945		0.9525
00	0.9266	0.965	0.8407	0.884		0.8731
0	0.8252	0.864	0.7785	0.823		0.7938
1	0.7348	0.762	0.7188	0.762	0.577	0.7144
2	0.6543	0.721	0.6668	0.701	0.556	0.6747
3	0.5827	0.658	0.6190	0.640	0.538	0.6350
4	0.5189	0.605	0.5723	0.589	0.526	0.5953
5	0.4620	0.559	0.5258	0.538	0.518	0.5556
6	0.4115	0.516	0.4877	0.488	0.511	0.5159
7	0.3665	0.457	0.4496	0.447	0.505	0.4763
8	0.3264	0.419	0.4115	0.406	0.500	0.4366
9	0.2906	0.376	0.3767	0.366	0.493	0.3969
10	0.2588	0.304	0.3429	0.325	0.485	0.3572
11	0.2305	0.305	0.3061	0.295	0.478	0.3175
12	0.2053	0.277	0.2680	0.264	0.470	0.2778
13	0.1828	0.241	0.232	0.234	0.462	0.2381
14	0.1628	0.211	0.203	0.203	0.457	0.1984
15	0.1450	0.183	0.183	0.183	0.452	0.1786
16	0.1291	0.165	0.159	0.163	0.445	0.1588
17	0.1150	0.147	0.137	0.142	0.437	0.1429
18	0.1024	0.124	0.121	0.122	0.427	0.1270
19	0.09116	0.107	0.104	0.102	0.417	0.1111
20	0.08118	0.089	0.0884	0.0914	0.409	0.09525
21	0.07229	0.081	0.0808	0.0813	0.399	0.08731
22	0.06439	0.071	0.0726	0.0711	0.394	0.07938
23	0.05733	0.064	0.0655	0.0610	0.389	0.07144
24	0.05105	0.056	0.0584	0.0559	0.384	0.06350
25	0.04547	0.051	0.0518	0.0508	0.376	0.05556
26	0.04049	0.046	0.0460	0.0457	0.371	0.04763
27	0.03604	0.041	0.0439	0.0417	0.363	0.04366
28	0.03211	0.036	0.0411	0.0378	0.353	0.03969
29	0.02860	0.033	0.0381	0.0345	0.340	0.03572
30	0.02548	0.030	0.0356	0.0315	0.323	0.03175
31	0.02268	0.025	0.0335	0.0295	0.305	0.02778
32	0.02019	0.023	0.0325	0.0274	0.292	0.02580
33	0.01798	0.020	0.0300	0.0254	0.284	0.02381
34	0.01601	0.018	0.0264	0.0234	0.279	0.02183
35	0.01426	0.013	0.024	0.0213	0.274	0.01984
36	0.01270	0.010	0.023	0.0193	0.269	0.01786
37	0.01131		0.022	0.0173	0.262	0.01687
38	0.01007		0.020	0.0152	0.257	0.01588
39	0.008969		0.019	0.0132	0.251	
40	0.007988		0.018	0.0122	0.246	
41			0.017	0.0112	0.241	
42			0.016	0.0102	0.234	
43			0.015	0.0091	0.224	
44			0.015	0.0081	0.216	
45			0.014	0.0071	0.206	
46			0.013	0.0061	0.201	
47			0.013	0.0051	0.196	
48			0.012	0.0041	0.191	
49			0.012	0.0030	0.183	
50			0.011	0.0025	0.175	

TWIST DRILL AND STEEL WIRE GAUGE

Inches

No.	Size	No.	Size	No.	Size	No.	Size	No.	Size
1	0.2280	17	0.1730	33	0.1130	49	0.0730	65	0.0350
2	0.2210	18	0.1695	34	0.1110	50	0.0700	66	0.0330
3	0.2130	19	0.1660	35	0.1100	51	0.0670	67	0.0320
4	0.2090	20	0.1610	36	0.1065	52	0.0635	68	0.0310
5	0.2055	21	0.1590	37	0.1040	53	0.0595	69	0.02925
6	0.2040	22	0.1570	38	0.1015	54	0.0550	70	0.0280
7	0.2010	23	0.1540	39	0.0995	55	0.0520	71	0.0260
8	0.1990	24	0.1520	40	0.0980	56	0.0465	72	0.0250
9	0.1960	25	0.1495	41	0.0960	57	0.0430	73	0.0240
10	0.1935	26	0.1470	42	0.0935	58	0.0420	74	0.0225
11	0.1910	27	0.1440	43	0.0890	59	0.0410	75	0.0210
12	0.1890	28	0.1405	44	0.0860	60	0.0400	76	0.0200
13	0.1850	29	0.1360	45	0.0820	61	0.0390	77	0.0180
14	0.1820	30	0.1285	46	0.0810	62	0.0380	78	0.0160
15	0.1800	31	0.1200	47	0.0785	63	0.0370	79	0.0145
16	0.1770	32	0.1160	48	0.0760	64	0.0360	80	0.0135

Centimeters

No.	Size	No.	Size	No.	Size	No.	Size	No.	Size
1	0.5791	17	0.4394	33	0.2870	49	0.1854	65	0.0889
2	0.5613	18	0.4305	34	0.2819	50	0.1778	66	0.0838
3	0.5410	19	0.4216	35	0.2794	51	0.1702	67	0.0813
4	0.5309	20	0.4089	36	0.2705	52	0.1613	68	0.0787
5	0.5220	21	0.4039	37	0.2642	53	0.1511	69	0.0743
6	0.5182	22	0.3988	38	0.2578	54	0.1397	70	0.0711
7	0.5105	23	0.3912	39	0.2527	55	0.1321	71	0.0660
8	0.5055	24	0.3861	40	0.2489	56	0.1181	72	0.0635
9	0.4978	25	0.3797	41	0.2438	57	0.1092	73	0.0610
10	0.4915	26	0.3734	42	0.2375	58	0.1067	74	0.0572
11	0.4851	27	0.3658	43	0.2261	59	0.1041	75	0.0533
12	0.4801	28	0.3569	44	0.2184	60	0.1016	76	0.0508
13	0.4699	29	0.3454	45	0.2083	61	0.0991	77	0.0457
14	0.4623	30	0.3264	46	0.2057	62	0.0965	78	0.0406
15	0.4572	31	0.3048	47	0.1994	63	0.0940	79	0.0368
16	0.4496	32	0.2946	48	0.1930	64	0.0914	80	0.0343

DIMENSIONS OF WIRE

Stubs' Gauge

Giving the diameter and cross-section in English and metric system for the Birmingham or Stubs' gauge.

Gauge No.	Diameter in in.	Section in sq. in.	Diameter in cm	Section in sq. cm
0000	.454	.16188	1.1532	1.0444
000	.425	.14186	1.0795	.9152
00	.380	.11341	.9652	.7317
0	.340	.09079	.8636	.5858
1	.300	.07069	.7620	.4560
2	.284	.06335	.7214	.4087
3	.259	.05269	.6579	.3399
4	.238	.04449	.6045	.2870
5	.220	.03801	.5588	.2452
6	.203	.03237	.5156	.20881
7	.180	.02545	.4572	.16147
8	.165	.02138	.4191	.13795
9	.148	.01720	.3759	.11099
10	.134	.01410	.3404	.09098
11	.120	.011310	.3048	.07297
12	.109	.009331	.2769	.06160
13	.095	.007088	.2413	.04573
14	.083	.005411	.2108	.03491
15	.072	.004072	.1829	.02627
16	.065	.0033183	.15610	.021409
17	.058	.0026421	.14732	.017046
18	.049	.0018857	.12446	.012166
19	.042	.0013854	.10668	.008938
20	.035	.0009621	.08890	.006207
21	.032	.0008042	.08128	.005189
22	.028	.0006158	.07112	.003973
23	.025	.0004909	.06350	.003167
24	.022	.0003801	.05588	.002452
25	.020	.0003142	.05080	.002027
26	.018	.0002545	.04572	.0016417
27	.016	.0002011	.04064	.0012972
28	.014	.0001539	.03556	.0009932
29	.013	.0001327	.03302	.0008563
30	.012	.0001181	.03048	.0007297
31	.010	.00007854	.02540	.0005067
32	.009	.00006362	.02286	.0004104
33	.008	.00005027	.02032	.0003243
34	.007	.00003848	.01778	.0002483
35	.005	.00001963	.01270	.0001267
36	.004	.00001257	.01016	.0000811

British Standard Gauge

Giving the diameter and cross-section in English and metric system for the British Standard Gauge.

Gauge No.	Diameter in in.	Section in sq. in.	Diameter in cm	Section in sq. cm
0000000	.500	.1963	1.2700	1.267
000000	.464	.1691	1.1786	1.091
00000	.432	.1466	1.0973	.9456
0000	.400	.1257	1.0160	.8107
000	.372	.1087	.9449	.7012
00	.348	.0951	.8839	.6136
0	.324	.0825	.8230	.5319
1	.300	.07069	.7620	.4560
2	.276	.05983	.7010	.3858
3	.252	.04988	.6401	.3218
4	.232	.04227	.5893	.2727
5	.212	.03530	.5385	.2277
6	.192	.02895	.4877	.28679
7	.176	.02433	.4470	.15696
8	.160	.02010	.4064	.12973
9	.144	.01629	.3658	.10507
10	.128	.01287	.3251	.08302
11	.116	.010568	.2946	.06818
12	.104	.008495	.2642	.05480
13	.092	.006648	.2337	.04289
14	.080	.005027	.2032	.03243
15	.072	.004071	.1829	.02627
16	.064	.003217	.16256	.020755
17	.056	.002463	.14224	.015890
18	.048	.001810	.12192	.011675
19	.040	.001257	.10160	.008107
20	.036	.001018	.09144	.006567
21	.032	.0008042	.08128	.005189
22	.028	.0006158	.07112	.003973
23	.024	.0004524	.06096	.002922
24	.022	.0003801	.05588	.002452
25	.020	.0003142	.05080	.002027
26	.0180	.0002545	.04572	.0016417
27	.0164	.0002112	.04166	.0013628
28	.0148	.0001728	.03759	.0011099
29	.0136	.0001453	.03454	.0009363
30	.0124	.0001208	.03150	.0007791
31	.0116	.00010568	.02946	.0006818
32	.0108	.00009161	.02743	.0005910
33	.0100	.00007854	.02540	.0005067
34	.0092	.00006648	.02337	.0004289
35	.0084	.00005542	.02134	.0003575
36	.0076	.00004536	.01930	.0002927
37	.0068	.00003632	.01727	.0002343
38	.0060	.00002827	.01524	.0001824
39	.0052	.00002124	.01321	.0001370
40	.0048	.00001810	.01219	.0001167
41	.0044	.00001512	.01118	.0000982
42	.0040	.00001257	.01016	.0000811
43	.0036	.00001018	.00914	.0000656
44	.0032	.00000804	.00813	.0000519
45	.0028	.00000616	.00711	.0000397
46	.0024	.00000452	.00610	.0000212
47	.0020	.00000314	.00508	.0000203
48	.0016	.00000201	.00406	.0000129
49	.0012	.00000113	.00305	.0000073
50	.0010	.00000079	.00254	.0000051

PLATINUM WIRE

Mass in Grams per Foot

B. & S. Gauge	Diameter, inches	Mass, g per ft.
10	.1019	37.5
11	.09074	28.0
12	.08081	22.0
13	.07196	17.5
14	.06408	14.0
15	.05707	11.0
16	.05082	9.0
17	.04526	7.0
18	.04030	5.7
19	.03589	4.4
20	.03196	3.4
21	.02846	2.9
22	.02535	2.3
23	.02257	1.8
24	.02010	1.4
25	.01790	1.1
26	.01594	0.9
27	.01420	0.7
28	.01264	0.6
29	.01126	0.45
30	.01003	0.35
31	.008928	0.28
32	.007950	0.22
33	.007080	0.17
34	.006305	0.15
35	.005615	0.11

ALLOWABLE CARRYING CAPACITIES OF CONDUCTORS

(National Electrical Code)

The ratings in the following tabulation are those permitted by the National Electrical Code for flexible cords and for interior wiring of houses, hotels, office buildings, industrial plants, and other buildings.

The values are for copper wire. For aluminum wire the allowable carrying capacities shall be taken as 84% of those given in the table for the respective sizes of copper wire with the same kind of covering.

Size A.W.G.	Area Circular Mils	Diameter of Solid Wires Mils	Rubber Insulation Amperes	Varnished Cambric Insulation Amperes	Other Insulations and Bare Conductors Amperes
18	1,624.	40.3	3*		6†
16	2,583.	50.8	6*		10†
14	4,107.	64.1	15	18	20
12	6,530.	80.8	20	25	30
10	10,380.	101.9	25	30	35
8	16,510.	128.5	35	40	50
6	26,250.	162.0	50	60	70
5	33,100.	181.9	55	65	80
4	41,740.	204.3	70	85	90
3	52,630.	229.4	80	95	100
2	66,370.	257.6	90	110	125
1	83,690.	289.3	100	120	150
0	105,500.	325.0	125	150	200
00	135,100.	364.8	150	180	225
000	167,800.	409.6	175	210	275
0000	211,600.	460	225	270	325

* The allowable carrying capacities of No. 18 and 16 are 5 and 7 amperes respectively, when in flexible cords.

† The allowable carrying capacities of No. 18 and 16 are 10 and 15 amperes respectively, when in cords for portable heaters. Types AFS, AFSJ, HC, HPD, and HSJ.

WIRE TABLE, STANDARD ANNEALED COPPER

American Wire Gauge (B. & S.) English Units

Gauge No.	Diameter in mils at 20°C	Cross section at 20°C Circular mils	Sq. inches	Ohms per 1000 feet* 0°C (32°F)	20°C (68°F)	50°C (122°F)	75°C (167°F)
0000	460.0	211600	0.1662	0.04516	0.04901	0.05479	0.05961
000	409.6	167800	.1318	.05695	.06180	.06909	.07516
00	364.8	133100	.1045	.07181	.07793	.08712	.09478
0	324.9	105500	.08289	.09055	.09827	.1099	.1195
1	289.3	83690	.06573	.1142	.1239	.1385	.1507
2	257.6	66370	.05213	.1440	.1563	.1747	.1900
3	229.4	52640	.04134	.1816	.1970	.2203	.2396
4	204.3	41740	.03278	.2289	.2485	.2778	.3022
5	181.9	33100	.02600	.2887	.3133	.3502	.3810
6	162.0	26250	.02062	.3640	.3951	.4416	.4805
7	144.3	20820	.01635	.4590	.4982	.5569	.6059
8	128.5	16510	.01297	.5788	.6282	.7023	.7640
9	114.4	13090	.01028	.7299	.7921	.8855	.9633
10	101.9	10380	.008155	.9203	.9989	1.117	1.215
11	90.74	8234	.006467	1.161	1.260	1.408	1.532
12	80.81	6530	.005129	1.463	1.588	1.775	1.931
13	71.96	5178	.004067	1.845	2.003	2.239	2.436
14	64.08	4107	.003225	2.327	2.525	2.823	3.071
15	57.07	3257	.002558	2.934	3.184	3.560	3.873
16	50.82	2583	.002028	3.700	4.016	4.489	4.884
17	45.26	2048	.001609	4.666	5.064	5.660	6.158
18	40.30	1624	.001276	5.883	6.385	7.138	7.765
19	35.89	1288	.001012	7.418	8.051	9.001	9.792
20	31.96	1022	.0008023	9.355	10.15	11.35	12.35
21	28.45	810.1	.0006363	11.80	12.80	14.31	15.57
22	25.35	642.4	.0005046	14.87	16.14	18.05	19.63
23	22.57	509.5	.0004002	18.76	20.36	22.76	24.76
24	20.10	404.0	.0003173	23.65	25.67	28.70	31.22
25	17.90	320.4	.0002517	29.82	32.37	36.18	39.36
26	15.94	254.1	.0001996	37.61	40.81	45.63	49.64
27	14.20	201.5	.0001583	47.42	51.47	57.53	62.59
28	12.64	159.8	.0001255	59.80	64.90	72.55	78.93
29	11.26	126.7	.00009953	75.40	81.83	91.48	99.52
30	10.03	100.5	.00007894	95.08	103.2	115.4	125.5
31	8.928	79.70	.00006260	119.9	130.1	145.5	158.2
32	7.950	63.21	.00004964	151.2	164.1	183.4	199.5
33	7.080	50.13	.00003937	190.6	206.9	231.3	251.6
34	6.305	39.75	.00003122	240.4	260.9	291.7	317.3
35	5.615	31.52	.00002476	303.1	329.0	367.8	400.1
36	5.000	25.00	.00001964	382.2	414.8	463.7	504.5
37	4.453	19.83	.00001557	482.0	523.1	584.8	636.2
38	3.965	15.72	.00001235	607.8	659.6	737.4	802.2
39	3.531	12.47	.000009793	766.4	831.8	929.8	1012
40	3.145	9.888	.000007766	966.5	1049	1173	1276

* Resistance at the stated temperatures of a wire whose length is 1000 feet at 20°C.

WIRE TABLE, STANDARD ANNEALED COPPER

American Wire Gauge (B. & S.) English Units (Continued)

Gauge No.	Pounds per 1000 feet	Feet per pound	Feet per ohm* 0°C (32°F)	20°C (68°F)	50°C (122°F)	75°C (167°F)
0000	640.5	1.561	22140	20400	18250	16780
000	507.9	1.968	17560	16180	14470	13300
00	402.8	2.482	13930	12830	11480	10550
0	319.5	3.130	11040	10180	9103	8367
1	253.3	3.947	8758	8070	7219	6636
2	200.9	4.977	6946	6400	5725	5262
3	159.3	6.276	5508	5075	4540	4173
4	126.4	7.914	4368	4025	3600	3309
5	100.2	9.980	3464	3192	2855	2625
6	79.46	12.58	2747	2531	2264	2081
7	63.02	15.87	2179	2007	1796	1651
8	49.98	20.01	1728	1592	1424	1309
9	39.63	25.23	1370	1262	1129	1038
10	31.43	31.82	1087	1001	895.6	823.2
11	24.92	40.12	861.7	794.0	710.2	652.8
12	19.77	50.59	683.3	629.6	563.2	517.7
13	15.68	63.80	541.9	499.3	446.7	410.6
14	12.43	80.44	429.8	396.0	354.2	325.6
15	9.858	101.4	340.8	314.0	280.9	258.2
16	7.818	127.9	270.3	249.0	222.8	204.8
17	6.200	161.3	214.3	197.5	176.7	162.4
18	4.917	203.4	170.0	156.6	140.1	128.8
19	3.899	256.5	134.8	124.2	111.1	102.1
20	3.092	323.4	106.9	98.50	88.11	80.99
21	2.452	407.8	84.78	78.11	69.87	64.23
22	1.945	514.2	67.23	61.95	55.41	50.94
23	1.542	648.4	53.32	49.13	43.94	40.39
24	1.223	817.7	42.28	38.96	34.85	32.03
25	0.9699	1031	33.53	30.90	27.64	25.40
26	.7692	1300	26.59	24.50	21.92	20.15
27	.6100	1639	21.09	19.43	17.38	15.98
28	.4837	2067	16.72	15.41	13.78	12.67
29	.3836	2607	13.26	12.22	10.93	10.05
30	.3042	3287	10.52	9.691	8.669	7.968
31	.2413	4145	8.341	7.685	6.875	6.319
32	.1913	5227	6.614	6.095	5.452	5.011
33	.1517	6591	5.245	4.833	4.323	3.974
34	.1203	8310	4.160	3.833	3.429	3.152
35	.09542	10480	3.299	3.040	2.719	2.499
36	.07568	13210	2.616	2.411	2.156	1.982
37	.06001	16660	2.075	1.912	1.710	1.572
38	.04759	21010	1.645	1.516	1.356	1.247
39	.03774	26500	1.305	1.202	1.075	0.9886
40	.02993	33410	1.035	0.9534	0.8529	.7840

* Length at 20°C of a wire whose resistance is 1 ohm at the stated temperatures.

WIRE TABLE, STANDARD ANNEALED COPPER (Continued)

American Wire Gauge (B. & S.) English Units

Gauge No.	Diameter in mils at 20°C	Ohms per pound 0°C (32°F)	Ohms per pound 20°C (68°F)	Ohms per pound 50°C (122°F)	Lbs. per ohm 20°C (68°F)
0000	460.0	0.00007051	0.00007652	0.00008554	13070
000	409.6	.0001121	.0001217	.0001360	8219
00	364.8	.0001783	.0001935	.0002163	5169
0	324.9	.0002835	.0003076	.0003439	3251
1	289.3	.0004507	.0004891	.0005468	2044
2	257.6	.0007166	.0007778	.0008695	1286
3	229.4	.001140	.001237	.001383	808.6
4	204.3	.001812	.001966	.002198	508.5
5	181.9	.002881	.003127	.003495	319.8
6	162.0	.004581	.004972	.005558	201.1
7	144.3	.007284	.007905	.008839	126.5
8	128.5	.01158	.01257	.01405	79.55
9	114.4	.01842	.01999	.02234	50.03
10	101.9	.02928	.03178	.03553	31.47
11	90.74	.04656	.05053	.05649	9.79
12	80.81	.07404	.08035	.08983	12.45
13	71.96	.1177	.1278	.1428	7.827
14	64.08	.1872	.2032	.2271	4.922
15	57.07	.2976	.3230	.3611	3.096
16	50.82	.4733	.5136	.5742	1.947
17	45.26	.7525	.8167	.9130	1.224
18	40.30	1.197	1.299	1.452	0.7700
19	35.89	1.903	2.065	2.308	.4843
20	31.96	3.025	3.283	3.670	.3046
21	28.46	4.810	5.221	5.836	.1915
22	25.35	7.649	8.301	9.280	.1205
23	22.57	12.16	13.20	14.76	.07576
24	20.10	19.34	20.99	23.46	.04765
25	17.90	30.75	33.37	37.31	.02997
26	15.94	48.89	53.06	59.32	.01885
27	14.20	77.74	84.37	94.32	.01185
28	12.64	123.6	134.2	150.0	.007454
29	11.26	196.6	213.3	238.5	.004688
30	10.03	312.5	339.2	379.2	.002948
31	8.928	497.0	539.3	602.9	.001854
32	7.950	790.2	857.6	958.7	.001166
33	7.080	1256	1364	1524	.0007333
34	6.305	1998	2168	2424	.0004612
35	5.615	3177	3448	3854	.0002901
36	5.000	5051	5482	6128	.0001824
37	4.453	8032	8717	9744	.0001147
38	3.965	12770	13860	15490	.00007215
39	3.531	20310	22040	24640	.00004538
40	3.145	32290	35040	39170	.00002854

Gauge No.	Diameter in mm at 20°C	Cross section in mm² at 20°C	Ohms per kilometer* 0°C	Ohms per kilometer* 20°C	Ohms per kilometer* 50°C	Ohms per kilometer* 75°C
0000	11.68	107.2	0.1482	0.1608	0.1798	0.1956
000	10.40	85.03	.1868	.2028	.2267	.2466
00	9.266	67.43	.2356	.2557	.2858	.3110
0	8.252	53.48	.2971	.3224	.3604	.3921
1	7.348	42.41	.3746	.4066	.4545	.4944
2	6.544	33.63	.4724	.5127	.5731	.6235
3	5.827	26.67	.5956	.6465	.7227	.7862
4	5.189	21.15	.7511	.8152	.9113	.9914
5	4.621	16.77	.9471	1.028	1.149	1.250
6	4.115	13.30	1.194	1.296	1.449	1.576
7	3.665	10.55	1.506	1.634	1.827	1.988
8	3.264	8.366	1.899	2.061	2.304	2.506
9	2.906	6.634	2.395	2.599	2.905	3.161
10	2.588	5.261	3.020	3.277	3.663	3.985
11	2.305	4.172	3.807	4.132	4.619	5.025
12	2.053	3.309	4.801	5.211	5.825	6.337
13	1.828	2.624	6.054	6.571	7.345	7.991
14	1.628	2.081	7.634	8.285	9.262	10.08
15	1.450	1.650	9.627	10.45	11.68	12.71
16	1.291	1.309	12.14	13.17	14.73	16.02
17	1.150	1.038	15.31	16.61	18.57	20.20
18	1.024	.8231	19.30	20.95	23.42	25.48
19	.9116	.6527	24.34	26.42	29.53	32.12
20	.8118	.5176	30.69	33.31	37.24	40.51
21	.7230	.4105	38.70	42.00	46.95	51.08
22	.6438	.3255	48.80	52.96	59.21	64.41
23	.5733	.2582	61.54	66.79	74.66	81.22
24	.5106	.2047	77.60	84.21	94.14	102.4
25	.4547	.1624	97.85	106.2	118.7	129.1
26	.4049	.1288	123.4	133.9	149.7	162.9
27	.3606	.1021	155.6	168.9	188.8	205.4
28	.3211	.08098	196.2	212.9	238.0	258.9
29	.2859	.06422	247.4	268.5	300.1	326.5
30	.2546	.05093	311.9	338.6	378.5	411.7
31	.2268	.04039	393.4	426.9	477.2	519.2
32	.2019	.03203	496.0	538.3	601.8	654.7
33	.1798	.02540	625.5	678.8	758.8	825.5
34	.1601	.02014	788.7	856.0	956.9	1041
35	.1426	.01597	994.5	1079	1207	1313
36	.1270	.01267	1254	1361	1522	1655
37	.1131	.01005	1581	1716	1919	2087
38	.1007	.007967	1994	2164	2419	2632
39	.08969	.006318	2514	2729	3051	3319
40	.07987	.005010	3171	3441	3847	4185

* Resistance at the stated temperatures of a wire whose length is 1 kilometer at 20°C.

Gauge No.	Diameter in mm at 20°C	Kilograms per kilometer	Meters per gram	Meters per ohm* 0°C	Meters per ohm* 20°C	Meters per ohm* 50°C	Meters per ohm* 75°C
0000	11.68	953.2	0.001049	6749	6219	5563	5113
000	10.40	755.9	.001323	5352	4932	4412	4055
00	9.266	599.5	.001668	4245	3911	3499	3216
0	8.252	475.4	.002103	3366	3102	2774	2550
1	7.348	377.0	.002652	2669	2460	2200	2022
2	6.544	299.0	.003345	2117	1951	1745	1604
3	5.827	237.1	.004217	1679	1547	1384	1272
4	5.189	188.0	.005318	1331	1227	1097	1009
5	4.621	149.1	.006706	1056	972.9	870.2	799.9
6	4.115	118.2	.008457	837.3	771.5	690.1	634.4
7	3.665	93.78	.01066	664.0	611.8	547.3	503.1
8	3.264	74.37	.01345	526.6	485.2	434.0	399.0
9	2.906	58.98	.01696	417.6	384.8	344.2	316.4
10	2.588	46.77	.02138	331.2	305.1	273.0	250.9
11	2.305	37.09	.02696	262.6	242.0	216.5	199.0
12	2.053	29.42	.03400	208.3	191.9	171.7	157.8
13	1.828	23.33	.04287	165.2	152.2	136.1	125.1
14	1.628	18.50	.05406	131.0	120.7	108.0	99.24
15	1.450	14.67	.06816	103.9	95.71	85.62	78.70
16	1.291	11.63	.08595	82.38	75.90	67.90	62.41
17	1.150	9.226	.1084	65.33	60.20	53.85	49.50
18	1.024	7.317	.1367	51.81	47.74	42.70	39.25
19	0.9116	5.803	.1723	41.09	37.86	33.86	31.13
20	.8118	4.602	.2173	32.58	30.02	26.86	24.69
21	.7230	3.649	.2740	25.84	23.81	21.30	19.58
22	.6438	2.894	.3455	20.49	18.88	16.89	15.53
23	.5733	2.295	.4357	16.25	14.97	13.39	12.31
24	.5106	1.820	.5494	12.89	11.87	10.62	9.764
25	.4547	1.443	.6928	10.22	9.417	8.424	7.743
26	.4049	1.145	.8736	8.105	7.468	6.680	6.141
27	.3606	0.9078	1.102	6.428	5.922	5.298	4.870
28	.3211	.7199	1.389	5.097	4.697	4.201	3.862
29	.2859	.5709	1.752	4.042	3.725	3.332	3.063
30	.2546	.4527	2.209	3.206	2.954	2.642	2.429
31	.2268	.3590	2.785	2.542	2.342	2.095	1.926
32	.2019	.2847	3.512	2.016	1.858	1.662	1.527
33	.1798	.2258	4.429	1.599	1.473	1.318	1.211
34	.1601	.1791	5.584	1.268	1.168	1.045	0.9606
35	.1426	.1420	7.042	1.006	0.9265	0.8288	.7618
36	.1270	.1126	8.879	0.7974	.7347	.6572	.6041
37	.1131	.08931	11.20	.6324	.5827	.5212	.4791
38	.1007	.07083	14.12	.5015	.4621	.4133	.3799
39	.08969	.05617	17.80	.3977	.3664	.3278	.3013
40	.07987	.04454	22.45	.3154	.2906	.2600	.2390

* Length at 20°C of a wire whose resistance is 1 ohm at the stated temperatures.

Gauge No.	Ohms per kilogram 0°C	Ohms per kilogram 20°C	Ohms per kilogram 50°C	Grams per ohm 20°C
0000	0.0001554	0.0001687	0.0001886	5928000
000	.0002472	.0002682	.0002999	3728000
00	.0003930	.0004265	.0004768	2344000
0	.0006249	.0006782	.0007582	1474000
1	.0009936	.001078	.001206	927300
2	.001580	.001715	.001917	583200
3	.002512	.002726	.003048	366800
4	.003995	.004335	.004846	230700
5	.006352	.006893	.007706	145100
6	.01010	.01096	.01225	91230
7	.01606	.01743	.01948	57380
8	.02553	.02771	.03098	36080
9	.04060	.04406	.04926	22690
10	.06456	.07007	.07833	14270
11	.1026	.1114	.1245	8976
12	.1632	.1771	.1980	5645
13	.2595	.2817	.3149	3550
14	.4127	.4479	.5007	2233
15	.6562	.7122	.7961	1404
16	1.043	1.132	1.266	883.1
17	1.659	1.801	2.013	555.4
18	2.638	2.863	3.201	349.3
19	4.194	4.552	5.089	219.7
20	6.670	7.238	8.092	138.2
21	10.60	11.51	12.87	86.88
22	16.86	18.30	20.46	54.64
23	26.81	29.10	32.53	34.36
24	42.63	46.27	51.73	21.61
25	67.79	73.57	82.25	13.59
26	107.8	117.0	130.8	8.548
27	171.4	186.0	207.9	5.376
28	272.5	295.8	330.6	3.381
29	433.3	470.3	525.7	2.126
30	689.0	747.8	836.0	1.337
31	1096	1189	1329	0.8410
32	1742	1891	2114	.5289
33	2770	3006	3361	.3326
34	4404	4780	5344	.2092
35	7003	7601	8497	.1316
36	11140	12090	13510	.08274
37	17710	19220	21480	.05204
38	28150	305360	34160	.03273
39	44770	48590	54310	.02058
40	71180	77260	88360	.01294

ALUMINUM WIRE TABLE

Hard-Drawn Aluminum Wire at 20°C (or, 68°F) American Wire Gauge (B. & S.) English Units

Gauge No.	Diameter in mils	Cross section: Circular mils	Cross section: Square inches	Ohms per 1000 ft.	Pounds per 1000 ft.	Pounds per ohm	Feet per ohm
0000	460	212000	0.166	0.0804	195	2420	12400
000	410	168000	.132	.101	154	1520	9860
00	365	133000	.105	.128	122	957	7820
0	325	106000	.0829	.161	97.0	602	6200
1	289	83700	.0657	.203	76.9	379	4920
2	258	66400	.0521	.256	61.0	238	3900
3	229	52600	.0413	.323	48.4	150	3090
4	204	41700	.0328	.408	38.4	94.2	2450
5	182	33100	.0260	.514	30.4	59.2	1950
6	162	26300	.0206	.648	24.1	37.2	1540
7	144	20800	.0164	.817	19.1	23.4	1220
8	128	16500	.0130	1.03	15.2	14.7	970
9	114	13100	.0103	1.30	12.0	9.26	770
10	102	10400	.00815	1.64	9.55	5.83	610
11	91	8230	.00647	2.07	7.57	3.66	484
12	81	6530	.00513	2.61	6.00	2.30	384
13	72	5180	.00407	3.29	4.76	1.45	304
14	64	4110	.00323	4.14	3.78	0.911	241
15	57	3260	.00256	5.22	2.99	.573	191
16	51	2580	.00203	6.59	2.37	.360	152
17	45	2050	.00161	8.31	1.88	.227	120
18	40	1620	.00128	10.5	1.49	.143	95.5
19	36	1290	.00101	13.2	1.18	.0897	75.7
20	32	1020	.000802	16.7	0.939	.0564	60.0
21	28.5	810	.000636	21.0	.745	.0355	47.6
22	25.3	642	.000505	26.5	.591	.0223	37.8
23	22.6	509	.000400	33.4	.468	.0140	29.9
24	20.1	404	.000317	42.1	.371	.00882	23.7
25	17.9	320	.000252	53.1	.295	.00555	18.8
26	15.9	254	.000200	67.0	.234	.00349	14.9
27	14.2	202	.000158	84.4	.185	.00219	11.8
28	12.6	160	.000126	106.	.147	.00138	9.39
29	11.3	127	.0000995	134.	.117	.000868	7.45
30	10.0	101	.0000789	169.	.0924	.000546	5.91
31	8.9	79.7	.0000626	213.	.0733	.000343	4.68
32	8.0	63.2	.0000496	269.	.0581	.000216	3.72
33	7.1	50.1	.0000394	339.	.0461	.000136	2.95
34	6.3	39.8	.0000312	428.	.0365	.0000854	2.34
35	5.6	31.5	.0000248	540.	.0290	.0000537	1.85
36	5.0	25.0	.0000196	681.	.0230	.0000338	1.47
37	4.5	19.8	.0000156	858.	.0182	.0000212	1.17
38	4.0	15.7	.0000123	1080.	.0145	.0000134	0.924
39	3.5	12.5	.00000979	1360.	.0115	.00000840	.733
40	3.1	9.9	.0000077	1720.	.0091	.00000528	.581

Gauge No.	Diameter in mm	Cross section in mm²	Ohms per kilometer	Kilograms per kilometer	Grams per ohm	Meters per ohm
0000	11.7	107	.0264	289	1100000	3790
000	10.4	85.0	.333	230	690000	3010
00	9.3	67.4	.419	182	434000	2380
0	8.3	53.5	.529	144	273000	1890
1	7.3	42.4	.667	114.	172000	1500
2	6.5	33.6	.841	90.8	108000	1190
3	5.8	26.7	1.06	72.0	67900	943
4	5.2	21.2	1.34	57.1	42700	748
5	4.6	16.8	1.69	45.3	26900	593
6	4.1	13.3	2.13	35.9	16900	470
7	3.7	10.5	2.68	28.5	10600	373
8	3.3	8.37	3.38	22.6	6680	296
9	2.91	6.63	4.26	17.9	4200	235
10	2.59	5.26	5.38	14.2	2640	186
11	2.30	4.17	6.78	11.3	1660	148
12	2.05	3.31	8.55	8.93	1050	117
13	1.83	2.62	10.8	7.08	657	92.8
14	1.63	2.08	13.6	5.62	413	73.6
15	1.45	1.65	17.1	4.46	260	58.4
16	1.29	1.31	21.6	3.53	164	46.3
17	1.15	1.04	27.3	2.80	103	36.7
18	1.02	0.823	34.4	2.22	64.7	29.1
19	0.91	.653	43.3	1.76	40.7	23.1
20	.81	.518	54.6	1.40	25.6	18.3
21	.72	.411	68.9	1.11	16.1	14.5
22	.64	.326	86.9	0.879	10.1	11.5
23	.57	.258	110	.697	6.36	9.13
24	.51	.205	138	.553	4.00	7.24
25	.45	.162	174	.438	2.52	5.74
26	.40	.129	220	.348	1.58	4.55
27	.36	.102	277	.276	0.995	3.61
28	.32	.0810	349	.219	.626	2.86
29	.29	.0642	440	.173	.394	2.27
30	.25	.0509	555	.138	.248	1.80
31	.227	.0404	700	.109	.156	1.43
32	.202	.0320	883	.0865	.0979	1.13
33	.180	.0254	1110	.0686	.0616	0.899
34	.160	.0201	1400	.0544	.0387	.712
35	.143	.0160	1770	.0431	.0244	.565
36	.127	.0127	2230	.0342	.0153	.448
37	.113	.0100	2820	.0271	.00963	.355
38	.101	.0080	3550	.0215	.00606	.282
39	.090	.0063	4480	.0171	.00381	.223
40	.080	.0050	5640	.0135	.00240	.177

CROSS-SECTION AND MASS OF WIRES

U. S. Measure

Diameters are given in mils (1 mil = .001 in.), and area in square mils (1 sq. mil = .000001 sq. in.). For sections and masses for one-tenth the diameters given, divide by 100 and for sections and masses for ten times the diameter multiply by 100.

Diam. in mils	Cross-sec. in sq. mils	Pounds per foot: Copper, density 8.90	Iron, density 7.80	Brass, density 8.56	Aluminum, density 2.67
10	78.54	0.000303	0.0002656	0.0002915	0.0000909
11	95.03	0367	03214	03527	01100
12	113.10	0436	03825	04197	01309
13	132.73	0512	04488	04926	01536
14	153.94	0594	05206	05713	01782
15	176.71	0682	05976	06558	02045
16	201.06	0776	06799	07461	02327
17	226.98	0876	07675	08423	02627
18	254.47	0982	08605	09443	02946
19	283.53	1094	09588	10522	03282
20	314.16	1212	1062	1166	03636
21	346.36	1336	1171	1285	04009
22	380.13	1467	1286	1411	04400
23	415.48	1603	1405	1542	04809
24	452.39	1746	1530	1679	05237
25	490.87	1894	1660	1822	05682
26	530.93	2046	1795	1970	01647
27	572.56	2209	1936	2125	06628
28	615.75	2376	2082	2285	07127
29	660.52	2549	2234	2451	07646
30	706.86	2727	2390	2623	08182
31	754.77	2912	2552	2801	08737
32	804.25	3103	2720	2985	09309
33	855.30	3300	2892	3174	09900
34	907.92	3503	3070	3369	10509
35	962.11	3712	3253	3570	1114
36	1017.88	3927	3442	3777	1178
37	1075.21	4149	3636	3990	1245
38	1134.11	4376	3844	4218	1316
39	1194.59	4609	4040	4433	1383

Diam. in mils	Cross-sec. in sq. mils	Pounds per foot: Copper, density 8.90	Iron, density 7.80	Brass, density 8.56	Aluminum, density 2.67
40	1256.64	0.004849	0.004249	0.004664	0.001455
41	1320.25	5094	4465	4900	1528
42	1385.44	5346	4685	5141	1604
43	1452.20	5603	4911	5389	1681
44	1520.53	5867	5142	5643	1760
45	1590.43	6137	5378	5902	1841
46	1661.90	6412	5620	6167	1924
47	1734.94	6694	5867	6438	2008
48	1809.56	6982	6119	6715	2095
49	1885.74	7276	6377	6998	2183
50	1963.50	7576	6640	7287	2273
51	2042.82	7882	6908	7581	2365
52	2123.72	8194	7181	7881	2458
53	2206.18	8512	7460	8187	2554
54	2290.22	8837	7744	8499	2651
55	2375.83	9167	8034	8817	2750
56	2463.01	9504	8329	9140	2851
57	2551.76	9846	8629	9470	2954
58	2642.08	10195	8934	9805	3058
59	2733.97	10549	9245	10146	3165
60	2827.43	1091	956	1049	3273
61	2922.47	1128	988	1085	3383
62	3019.07	1165	1021	1120	3495
63	3117.25	1203	1054	1157	3608
64	3216.99	1241	1088	1194	3724
65	3318.31	1280	1122	1231	3841
66	3421.19	1320	1157	1270	3960
67	3525.65	1360	1192	1308	4081
68	3631.68	1401	1228	1348	4204
69	3739.28	1443	1264	1388	4328

CROSS-SECTION AND MASS OF WIRES (continued)

U. S. Measure

Diam. in mils	Cross-sec. in sq. mils	Pounds per foot: Copper, density 8.90	Iron, density 7.80	Brass, density 8.56	Aluminum, density 2.67
70	3848.45	0.01485	0.01302	0.01429	0.004456
71	3959.19	1528	1339	1469	4583
72	4071.50	1571	1377	1511	4713
73	4185.39	1615	1415	1553	4845
74	4300.84	1660	1454	1596	4978
75	4417.86	1705	1494	1639	5114
76	4536.46	1751	1534	1684	5251
77	4656.63	1797	1575	1728	5390
78	4778.36	1844	1616	1773	5531
79	4901.67	1892	1658	1819	5674
80	5026.55	1939	1700	1865	5818
81	5153.00	1988	1743	1912	5965
82	5281.02	2038	1786	1960	6113
83	5410.61	2088	1830	2008	6263
84	5541.77	2138	1874	2057	6415
85	5674.50	2189	1919	2106	6568

Diam. in mils	Cross-sec. in sq. mils	Pounds per foot: Copper, density 8.90	Iron, density 7.80	Brass, density 8.56	Aluminum, density 2.67
86	5808.80	0.02241	0.01964	0.02156	0.006724
87	5941.68	2294	2010	2206	6881
88	6082.12	2347	2057	2257	7040
89	6221.14	2400	2104	2309	7201
90	6361.73	2455	2151	2360	7364
91	6503.88	2509	2199	2414	7528
92	6647.61	2565	2248	2467	7695
93	6792.91	2621	2297	2521	7863
94	6939.78	2678	2347	2575	8033
95	7088.22	2735	2397	2630	8205
96	7238.23	2793	2448	2686	8378
97	7389.81	2851	2499	2742	8554
98	7542.96	2910	2551	2799	8731
99	7697.69	2970	2603	2857	8910
100	7853.98	3030	2656	2915	9091

Metric Measure

Diameters are given in thousandths of a centimeter and area of section in square thousandths of a centimeter. 1 $(cm/1000)^2$ = .000001 sq. cm. For sections and masses for diameters 1/10 or 10 times those of the table, divide or multiply by 100.

Diam. in thousandths of a cm	Cross-section in square thousandths of a cm	Grams per meter: Copper, density 8.90	Iron, density 7.80	Brass, density 8.56	Aluminum, density 2.67
10	78.54	0.06990	0.06126	0.06723	0.02097
11	95.03	.08458	.07412	.08135	.02537
12	113.10	.10065	.08822	.09681	.03020
13	132.73	.11813	.10353	.11362	.03544
14	153.94	.13701	.12008	.13177	.04110
15	176.71	.1573	.1378	.1513	.04718
16	201.06	.1789	.1568	.1721	.05368
17	226.98	.2020	.1770	.1943	.06060
18	254.47	.2265	.1985	.2178	.06794
19	283.53	.2523	.2212	.2427	.07570
20	314.16	.2796	.2450	.2689	.08388
21	346.36	.3083	.2702	.2965	.09248
22	380.13	.3383	.2965	.3254	.10149
23	415.48	.3698	.3241	.3557	.11093
24	452.39	.4026	.3529	.3872	.12079
25	490.87	.4369	.3829	.4202	.1311
26	530.93	.4725	.4141	.4545	.1418
27	572.56	.5096	.4466	.4901	.1529
28	615.75	.5480	.4803	.5271	.1644
29	660.52	.5879	.5152	.5654	.1764
30	706.86	.6291	.5514	.6051	.1887
31	754.77	.6717	.5887	.6461	.2015
32	804.25	.7158	.6273	.6884	.2147
33	855.30	.7612	.6671	.7321	.2284
34	907.92	.8081	.7082	.7772	.2424
35	962.11	.856	.7504	.8236	.2569
36	1017.88	.906	.7939	.8713	.2718
37	1075.21	.957	.8387	.9204	.2871
38	1134.11	1.012	.8866	.9730	.3035
39	1194.59	.063	.9318	1.0230	.3190
40	1256.64	1.118	.980	1.076	.3355
41	1320.25	.175	1.030	.130	.3525
42	1385.44	.233	.081	.186	.3699
43	1452.20	.292	.133	.243	.3877
44	1520.53	.353	.186	.302	.4060
45	1590.43	1.415	1.241	1.361	.4246
46	1661.90	.479	.296	.423	.4437
47	1734.94	.544	.353	.485	.4632
48	1809.56	.611	.411	.549	.4832
49	1885.74	.678	.471	.614	.5035
50	1963.50	1.748	1.532	1.681	.5243
51	2042.82	.818	.593	.753	.5454
52	2123.72	.890	.657	.818	.5670
53	2206.18	.964	.721	.888	.5891
54	2290.22	2.038	.786	.960	.6115

Diam. in thousandths of a cm	Cross-section in square thousandths of a cm	Grams per meter: Copper, density 8.90	Iron, density 7.80	Brass, density 8.56	Aluminum, density 2.67
55	235.83	2.114	1.853	2.034	0.6343
56	2463.01	.192	.921	.108	.6576
57	2551.76	.271	.990	.184	.6813
58	2642.08	.351	2.061	.262	.7054
59	2733.97	.433	.132	.340	.7300
60	2827.43	2.516	2.205	2.420	.7549
61	2922.47	.601	.280	.502	.7803
62	3019.07	.687	.355	.584	.8061
63	3117.25	.774	.431	.668	.8323
64	3216.99	.863	.509	.760	.8589
65	3318.31	2.953	2.588	2.840	.8860
66	3421.19	3.045	.669	.929	.9135
67	3525.65	.138	.750	3.018	.9413
68	3631.68	.232	.833	.109	.9697
69	3739.28	.328	.917	.201	.9984
70	3848.45	3.426	3.003	3.295	1.028
71	3959.19	.524	.088	.389	.057
72	4071.50	.624	.176	.485	.087
73	4185.39	.725	.265	.583	.117
74	4300.84	.828	.355	.682	.148
75	4417.86	3.932	3.446	3.782	1.180
76	4536.46	4.037	.538	.883	.211
77	4656.63	.144	.632	.986	.243
78	4778.36	.253	.727	4.090	.276
79	4901.67	.362	.823	.177	.309
80	5026.55	4.474	3.921	4.303	1.342
81	5153.00	.586	4.019	.411	.376
82	5281.02	.700	.119	.521	.410
83	5410.61	.815	.220	.631	.445
84	5541.77	.932	.323	.744	.480
85	5674.50	5.050	4.426	4.857	1.515
86	5808.80	.170	.531	.972	.551
87	5944.68	.291	.637	5.089	.587
88	6082.12	.413	.744	.206	.624
89	6221.14	.537	.852	.325	.661
90	6361.73	5.662	4.962	5.446	1.699
91	6503.88	.788	5.073	.567	.737
92	6647.61	.916	.185	.690	.775
93	6792.91	6.046	.298	.815	.814
94	6939.78	.176	.413	.940	.853
95	7088.22	6.309	5.529	6.068	1.893
96	7238.23	.442	.646	.196	.933
97	7389.81	.577	.764	.326	.973
98	7542.96	.713	.884	.457	2.014
99	7697.69	.851	6.004	.589	.055
100	7853.98	6.990	6.126	6.723	2.097

RESISTANCE OF WIRES

The following table gives the approximate resistance of various metallic conductors. The values have been computed from the resistivities at 20°C, except as otherwise stated, and for the dimensions of wire indicated. Owing to differences in purity in the case of elements and of composition in alloys, the values can be considered only as approximations.

The following dimensions have been adopted in the computations.

B. & S. gauge	Diameter mm	Diameter mils 1 mil = .001 in.	B. & S. gauge	Diameter mm	Diameter mils 1 mil = .001 in.
10	2.588	101.9	26	0.4049	15.94
12	2.053	80.81	27	0.3606	14.20
14	1.628	64.08	28	0.3211	12.64
16	1.291	50.82	30	0.2546	10.03
18	1.024	40.30	32	0.2019	7.950
20	0.8118	31.96	34	0.1601	6.305
22	0.6438	25.35	36	0.1270	5.000
24	0.5106	20.10	40	0.07987	3.145

***Advance (0°C) $\varrho = 48. \times 10^{-6}$ ohm cm**

B. & S. No.	Ohms per cm	Ohms per ft.
10	.000912	.0278
12	.00145	.0442
14	.00231	.0703
16	.00367	.112
18	.00583	.178
20	.00927	.283
22	.0147	.449
24	.0234	.715
26	.0373	1.14
27	.0470	1.43
28	.0593	1.81
30	.0942	2.87
32	.150	4.57
34	.238	7.26
36	.379	11.5
40	.958	29.2

Aluminum $\varrho = 2.828 \times 10^{-6}$ ohm cm

B. & S. No.	Ohms per cm	Ohms per ft.
10	.0000538	.00164
12	.0000855	.00260
14	.000136	.00414
16	.000216	.00658
18	.000344	.0105
20	.000546	.0167
22	.000869	.0265
24	.00138	.0421
26	.00220	.0669
27	.00277	.0844
28	.00349	.106
30	.00555	.169
32	.00883	.269
34	.0140	.428
36	.0223	.680
40	.0564	1.72

Eureka (0°C) $\varrho = 47. \times 10^{-6}$ ohm cm

B. & S. No.	Ohms per cm	Ohms per ft.
10	.000893	.0272
12	.00142	.0433
14	.00226	.0688
16	.00359	.109
18	.00571	.174
20	.00908	.277
22	.0144	.440
24	.0230	.700
26	.0365	1.11
27	.0460	1.40
28	.0580	1.77
30	.0923	2.81
32	.147	4.47
34	.233	7.11
36	.371	11.3
40	.938	28.6

Excello $\varrho = 92. \times 10^{-6}$ ohm cm

B. & S. No.	Ohms per cm	Ohms per ft.
10	.00175	.0533
12	.00278	.0847
14	.00442	.135
16	.00703	.214
18	.0112	.341
20	.0178	.542
22	.0283	.861
24	.0449	1.37
26	.0714	2.18
27	.0901	2.75
28	.114	3.46
30	.181	5.51
32	.287	8.75
34	.457	13.9
36	.726	22.1
40	1.84	56.0

Brass $\varrho = 7.00 \times 10^{-6}$ ohm cm

B. & S. No.	Ohms per cm	Ohms per ft.
10	.000133	.00406
12	.000212	.00645
14	.000336	.0103
16	.000535	.0163
18	.000850	.0259
20	.00135	.0412
22	.00215	.0655
24	.00342	.104
26	.00543	.166
27	.00686	.209
28	.00864	.263
30	.0137	.419
32	.0219	.666
34	.0348	1.06
36	.0552	1.68
40	.140	4.26

Climax $\varrho = 87. \times 10^{-6}$ ohm cm

B. & S. No.	Ohms per cm	Ohms per ft.
10	.00165	.0504
12	.00263	.0801
14	.00418	.127
16	.00665	.203
18	.0106	.322
20	.0168	.512
22	.0267	.815
24	.0425	1.30
26	.0675	2.06
27	.0852	2.60
28	.107	3.27
30	.171	5.21
32	.272	8.28
34	.432	13.2
36	.687	20.9
40	1.74	52.9

German silver $\varrho = 33. \times 10^{-6}$ ohm cm

B. & S. No.	Ohms per cm	Ohms per ft.
10	.000627	.0191
12	.000997	.0304
14	.00159	.0483
16	.00252	.0768
18	.00401	.122
20	.00638	.194
22	.0101	.309
24	.0161	.491
26	.0256	.781
27	.0323	.985
28	.0408	1.24
30	.0648	1.97
32	.103	3.14
34	.164	4.99
36	.260	.794
40	.659	20.1

Gold $\varrho = 2.44 \times 10^{-6}$ ohm cm

B. & S. No.	Ohms per cm	Ohms per ft.
10	.0000464	.00141
12	.0000737	.00225
14	.000117	.00357
16	.000186	.00568
18	.000296	.00904
20	.000471	.0144
22	.000750	.0228
24	.00119	.0363
26	.00189	.0577
27	.00239	.0728
28	.00301	.0918
30	.00479	.146
32	.00762	.232
34	.0121	.369
36	.0193	.587
40	.0487	1.48

Constantan (0°C) $\varrho = 44.1 \times 10^{-6}$ ohm cm

B. & S. No.	Ohms per cm	Ohms per ft.
10	.000838	.0255
12	.00133	.0406
14	.00212	.0646
16	.00337	.103
18	.00536	.163
20	.00852	.260
22	.0135	.413
24	.0215	.657
26	.0342	1.04
27	.0432	1.32
28	.0545	1.66
30	.0866	2.64
32	.138	4.20
34	.219	6.67
36	.348	10.6
40	.880	26.8

Copper, annealed $\varrho = 1.724 \times 10^{-6}$ ohm cm

B. & S. No.	Ohms per cm	Ohms per ft.
10	.0000328	.000999
12	.0000521	.00159
14	.0000828	.00253
16	.000132	.00401
18	.000209	.00638
20	.000333	.0102
22	.000530	.0161
24	.000842	.0257
26	.00134	.0408
27	.00169	.0515
28	.00213	.0649
30	.00339	.103
32	.00538	.164
34	.00856	.261
36	.0136	.415
40	.0344	1.05

Iron $\varrho = 10. \times 10^{-6}$ ohm cm

B. & S. No.	Ohms per cm	Ohms per ft.
10	.000190	.00579
12	.000302	.00921
14	.000481	.0146
16	.000764	.0233
18	.00121	.0370
20	.00193	.0589
22	.00307	.0936
24	.00489	.149
26	.00776	.237
27	.00979	.299
28	.0123	.376
30	.0196	.598
32	.0312	.952
34	.0497	1.51
36	0.789	2.41
40	.200	6.08

Lead $\varrho = 22. \times 10^{-6}$ ohm cm

B. & S. No.	Ohms per cm	Ohms per ft.
10	.000418	.0127
12	.000665	.0203
14	.00106	.0322
16	.00168	.0512
18	.00267	.0815
20	.00425	.130
22	.00676	.206
24	.0107	.328
26	.0171	.521
27	.0215	.657
28	.0272	.828
30	.0432	1.32
32	.0687	2.09
34	.109	3.33
36	.174	5.29
40	.439	13.4

* Trade mark.

RESISTANCE OF WIRES (Continued)

Magnesium $\varrho = 4.6 \times 10^{-6}$ ohm cm

B. & S. No.	Ohms per cm	Ohms per ft.
10	.0000874	.00267
12	.000139	.00424
14	.000221	.00674
16	.000351	.0107
18	.000559	.0170
20	.000889	.0271
22	.00141	.0431
24	.00225	.0685
26	.00357	.109
27	.00451	.137
28	.00568	.173
30	.00903	.275
32	.0144	.438
34	.0228	.696
36	.0363	1.11
40	.0918	2.80

Manganin $\varrho = 44. \times 10^{-6}$ ohm cm

B. & S. No.	Ohms per cm	Ohms per ft.
10	.000836	.0255
12	.00133	.0405
14	.00211	.0644
16	.00336	.102
18	.00535	.163
20	.00850	.259
22	.0135	.412
24	.0215	.655
26	.0342	1.04
27	.0431	1.31
28	.0543	1.66
30	.0864	2.63
32	.137	4.19
34	.218	6.66
36	.347	10.6
40	.878	26.8

Platinum $\varrho = 10. \times 10^{-6}$ ohm cm

B. & S. No.	Ohms per cm	Ohms per ft.
10	.000190	.00579
12	.000302	.00921
14	.000481	.0146
16	.000764	.0233
18	.00121	.0370
20	.00193	.0589
22	.00307	.0936
24	.00488	.149
26	.00776	.237
27	.00979	.299
28	.0123	.376
30	.0196	.598
32	.0312	.952
34	.0497	1.51
36	.0789	2.41
40	.200	6.08

Silver (18°C) $\varrho = 1.629 \times 10^{-6}$ ohm cm

B. & S. No.	Ohms per cm	Ohms per ft.
10	.0000310	.000944
12	.0000492	.00150
14	.0000783	.00239
16	.000124	.00379
18	.000198	.00603
20	.000315	.00959
22	.000500	.0153
24	.000796	.0243
26	.00126	.0386
27	.00160	.0486
28	.00201	.0613
30	.00320	.0975
32	.00509	.155
34	.00809	.247
36	.0129	.392
40	.0325	.991

Molybdenum $\varrho = 5.7 \times 10^{-6}$ ohm cm

B. & S. No.	Ohms per cm	Ohms per ft.
10	.000108	.00330
12	.000172	.00525
14	.000274	.00835
16	.000435	.0133
18	.000693	.0211
20	.00110	.0336
22	.00175	.0534
24	.00278	.0849
26	.00443	.135
27	.00558	.170
28	.00704	.215
30	.0112	.341
32	.0178	.542
34	.0283	.863
36	.0450	1.37
40	.114	3.47

Monel Metal $\varrho = 42. \times 10^{-6}$ ohm cm

B. & S. No.	Ohms per cm	Ohms per ft.
10	.000798	.0243
12	.00127	.0387
14	.00202	.0615
16	.00321	.0978
18	.00510	.156
20	.00811	.247
22	.0129	.393
24	.0205	.625
26	.0326	.994
27	.0411	1.25
28	.0519	1.58
30	.0825	2.51
32	.131	4.00
34	.209	6.36
36	.331	10.1
40	.838	25.6

Steel, piano wire (0°C) $\varrho = 11.8 \times 10^{-6}$ ohm cm

B. & S. No.	Ohms per cm	Ohms per ft.
10	.000224	.00684
12	.000357	.0109
14	.000567	.0173
16	.000901	.0275
18	.00143	.0437
20	.00228	.0695
22	.00363	.110
24	.00576	.176
26	.00916	.279
27	.0116	.352
28	.0146	.444
30	.0232	.706
32	.0368	1.12
34	.0586	1.79
36	.0931	2.84
40	.236	7.18

Steel, invar (35% Ni) $\varrho = 81. \times 10^{-6}$ ohm cm

B. & S. No.	Ohms per cm	Ohms per ft.
10	.00154	.0469
12	.00245	.0746
14	.00389	.119
16	.00619	.189
18	.00984	.300
20	.0156	.477
22	.0249	.758
24	.0396	1.21
26	.0629	1.92
27	.0793	2.42
28	.100	3.05
30	.159	4.85
32	.253	7.71
34	.402	12.3
36	.639	19.5
40	1.62	49.3

***Nichrome $\varrho = 150. \times 10^{-6}$ ohm cm**

B. & S. No.	Ohms per cm	Ohms per ft.
10	.0021281	.06488
12	.0033751	.1029
14	.0054054	.1648
16	.0085116	.2595
18	.0138383	.4219
20	.0216218	.6592
22	.0346040	1.055
24	.0548088	1.671
26	.0875760	2.670
28	.1394328	4.251
30	.2214000	6.750
32	.346040	10.55
34	.557600	17.00
36	.885600	27.00
38	1.383832	42.19
40	2.303872	70.24

Nickel $\varrho = 7.8 \times 10^{-6}$ ohm cm

B. & S. No.	Ohms per cm	Ohms per ft.
10	.000148	.00452
12	.000236	.00718
14	.000375	.0114
16	.000596	.0182
18	.000948	.0289
20	.00151	.0459
22	.00240	.0730
24	.00381	.116
26	.00606	.185
27	.00764	.233
28	.00963	.294
30	.0153	.467
32	.0244	.742
34	.0387	1.18
36	.0616	1.88
40	.156	4.75

Tantalum $\varrho = 15.5 \times 10^{-6}$ ohm cm

B. & S. No.	Ohms per cm	Ohms per ft.
10	.000295	.00898
12	.000468	.0143
14	.000745	.0227
16	.00118	.0361
18	.00188	.0574
20	.00299	.0913
22	.00476	.145
24	.00757	.231
26	.0120	.367
27	.0152	.463
28	.0191	.583
30	.0304	.928
32	.0484	1.47
34	.0770	2.35
36	.122	3.73
40	.309	9.43

Tin $\varrho = 11.5 \times 10^{-6}$ ohm cm

B. & S. No.	Ohms per cm	Ohms per ft.
10	.000219	.00666
12	.000348	.0106
14	.000553	.0168
16	.000879	.0268
18	.00140	.0426
20	.00222	.0677
22	.00353	.108
24	.00562	.171
26	.00893	.272
27	.0113	.343
28	.0142	.433
30	.0226	.688
32	.0359	1.09
34	.0571	1.74
36	.0908	2.77
40	.230	7.00

Tungsten $\varrho = 5.51 \times 10^{-6}$ ohm cm

B. & S. No.	Ohms per cm	Ohms per ft.
10	.000105	.00319
12	.000167	.00508
14	.000265	.00807
16	.000421	.0128
18	.000669	.0204
20	.00106	.0324
22	.00169	.0516
24	.00269	.0820
26	.00428	.130
27	.00540	.164
28	.00680	.207
30	.0108	.330
32	.0172	.524
34	.0274	.834
36	.0435	1.33
40	.110	3.35

Zinc (0°C) $\varrho = 5.75 \times 10^{-6}$ ohm cm

B. & S. No.	Ohms per cm	Ohms per ft.
10	.000109	.00333
12	.000174	.00530
14	.000276	.00842
16	.000439	.0134
18	.000699	.0213
20	.00111	.0339
22	.00177	.0538
24	.00281	.0856
26	.00446	.136
27	.00563	.172
28	.00710	.216
30	.0113	.344
32	.0180	.547
34	.0286	.870
36	.0454	1.38
40	.115	3.50

ELECTRICAL RESISTIVITY AND TEMPERATURE COEFFICIENTS OF ELEMENTS

Element	Temperature °C	Microhm-Cm	Temperature Coefficient per °C
Aluminum, 99.996%	20	2.6548	0.00429^{20} [i]
Antimony	0	39.0	
Arsenic	20	33.3	
Beryllium[a]	20	4.0	0.025^{20} [i]
Bismuth	0	106.8	
Boron	0	1.8×10^{12}	
Cadmium	0	6.83	0.0042^{0} [i]
Calcium	0	3.91	0.00416^{0} [i]
Carbon[b]	0	1375.0	
Cerium	25	75.0	$0.00087^{0\text{-}25}$
Cesium	20	20	
Chromium	0	12.9	0.003^{0} [i]
Cobalt	20	6.24	$0.00604^{0\text{-}100}$
Copper	20	1.678	$0.0068^{0\text{-}500}$ [g]
Dysprosium[c]	25	57.0	$0.00119^{0\text{-}25}$
Erbium	25	107.0	$0.00201^{0\text{-}25}$
Europium	25	90.0	
Gadolinium	25	140.5	$0.00176^{0\text{-}25}$
Gallium[d]	20	17.4	
Germanium[e]	22	46×10^{6}	
Gold	20	2.24	$0.0083^{0\text{-}100}$ [g]
Hafnium	25	35.1	0.0038^{25} [i]
Holmium	25	87.0	$0.00171^{0\text{-}25}$
Indium	20	8.37	
Iodine	20	1.3×10^{15}	
Iridium	20	5.3	$0.003925^{0\text{-}100}$
Iron, 99.99%	20	9.71	0.00651^{20} [i]
Lanthanum	25	5.70	$0.00218^{0\text{-}25}$
Lead	20	20.648	$0.00336^{20\text{-}40}$
Lithium	0	8.55	
Lutetium	25	79.0	$0.00240^{0\text{-}25}$
Magnesium[f]	20	4.45	0.0165^{20} [i]
Manganese α	23-100	185.0	
Mercury	50	98.4	
Molybdenum	0	5.2	
Neodymium	25	64.0	$0.00164^{0\text{-}25}$
Nickel	20	6.84	$0.0069^{0\text{-}100}$
Niobium (Columbium)[g]	0	12.5	
Osmium	20	9.5	$0.0042^{0\text{-}100}$
Palladium	20	10.54	$0.00374^{0\text{-}60}$ [g]
Phosphorus, white	11	1×10^{17}	
Platinum, 99.85%	20	10.6	$0.003927^{0\text{-}100}$
Plutonium	107	141.4	
Potassium	0	6.15	
Praseodymium	25	68	$0.00171^{0\text{-}25}$
Rhenium	20	19.3	$0.00395^{0\text{-}100}$
Rhodium	20	4.51	$0.0042^{0\text{-}100}$
Rubidium	20	12.5	
Ruthenium	0	7.6	
Samarium	25	88.0	$0.00184^{0\text{-}25}$
Scandium[h]	22	61.0	$0.00282^{0\text{-}25}$
Selenium[k]	0	10^{6}	
Silicon	0	$3\text{-}4 \times 10^{6}$ [j]	
Silver	20	1.586	$0.0061^{0\text{-}100}$ [g]
Sodium	0	4.2	
Strontium	20	23.0	
Sulfur, yellow	20	2×10^{23}	
Tantalum	25	12.45	$0.00383^{0\text{-}100}$
Tellurium	25	4.36×10^{6}	
Thallium	0	18.0	
Thorium	0	13.0	$0.0038^{0\text{-}100}$
Thulium	25	79.0	$0.00195^{0\text{-}25}$
Tin	0	11.0	$0.0047^{0\text{-}100}$
Titanium	20	42.0	
Tungsten	27	5.65	
Uranium		30.0	
Vanadium	20	24.8-26.0	
Ytterbium	25	29.0	$0.0013^{0\text{-}25}$
Yttrium	25	57.0	$0.0027^{0\text{-}25}$
Zinc	20	5.916	$0.00419^{0\text{-}100}$
Zirconium	20	40.0	0.0044^{20} [i]

[a] Annealed, comm. pure.
[b] Graphite.
[c] Polycrystalline.
[d] Hard Wire.
[e] Intrinsic Ge.
[f] Polycrystalline.
[g] High Purity.
[h] Zone refined bar.
[i] Data not available to indicate range over which coefficient is valid.
[j] Very sensitive to purity.
[k] Crystalline.

ELECTRICAL RESISTIVITY OF THE ALKALI METALS

Metal	Atomic no.	Atomic wt.	Density at 293.15 K	M.P. (°K)	B.P. (°K)	Resistivity (μ-ohm·cm at K)	
Li	3	6.941	0.534	453.7	1617	8.53	273.15
						9.28	293.15
						11.45	350.
						15.59(s)	453.7
						24.80(l)	453.7
Na	11	22.989	0.971	371.0	1157	4.33	273.15
						4.77	293.15
						6.23	350.
						6.86(s)	371.
						9.43(l)	371.
K	19	39.098	0.871	336.35	1032	6.49	273.15
						7.20	293.15
						9.22(s)	336.35
						13.95(l)	336.35
						14.64	350.
Rb	37	85.4678	1.53	312.64	961	11.54	273.15
						12.84	293.15
						14.21(s)	312.64
						22.52(l)	312.64
						25.42	350.
Cs	55	132.9054	1.873	301.55	944	18.75	273.15
						20.46	293.15
						21.16(s)	301.55
						36.93(l)	301.55
						42.11	350.

LOW MELTING POINT SOLDERS

(From N.B.S. Circular 492)

Nominal Composition, Weight Percent: Pb	Sn	Bi	Liquidus Temperature, °F
25	25	50	266
50	37.5	12.5	374
25	50	25	336

BRAZING FILLER METALS (SOLDERS)

AWS-ASTM Classification	Solidus, °F	Liquidus, °F	Brazing Temperature Range, °F	Ag	Al	As	Au	B	Be	Bi	C	Cd	Cr	Cu	Fe	Li	Mg	Mn	Ni	P	Pb	Sb	Si	Sn	Ti	Zn	Other: Each	Other: Total	AWS-ASTM Classification
ALUMINUM-SILICON																													
BAlSi-2	1070	1135	1110-1150		Bal.									0.25	0.8			0.10					6.8-8.2			0.20	0.05	0.15	BAlSi-2
BAlSi-3	970	1085	1060-1120		Bal.								0.15	3.3-4.7	0.8		0.15	0.15					9.3-10.7			0.20	0.05	0.15	BAlSi-3
BAlSi-4	1070	1080	1080-1120		Bal.									0.30	0.8		0.10	0.15					11.0-13.0			0.20	0.05	0.15	BAlSi-4
BAlSi-5	1070	1095	1090-1120		Bal.									0.30	0.8		0.05	0.05					9.0-11.0		0.20	0.10	0.05	0.15	BAlSi-5
COPPER-PHOSPHORUS																													
BCuP-1	1310	1650	1450-1700											Bal.						4.75-5.25								0.15	BCuP-1
BCuP-2	1310	1460	1350-1550											Bal.						7.00-7.25								0.15	BCuP-2
BCuP-3	1190	1485	1300-1500	4.75-5.25										Bal.						5.75-6.25								0.15	BCuP-3
BCuP-4	1190	1335	1300-1450	5.75-6.25										Bal.						7.00-7.50								0.15	BCuP-4
BCuP-5	1190	1475	1300-1500	14.50-15.50										Bal.						4.75-5.25								0.15	BCuP-5
SILVER																													
BAg-1	1125	1145	1145-1400	44-46								23-25		14-16												14-18		0.15	BAg-1
BAg-1a	1160	1175	1175-1400	49-51								17-19		14.5-16.5												14.5-18.5		0.15	BAg-1a
BAg-2	1125	1295	1295-1550	34-36								17-19		25-27												19-23		0.15	BAg-2
BAg-3	1170	1270	1270-1500	49-51								15-17		14.5-16.5					2.5-3.5							13.5-17.5		0.15	BAg-3
BAg-4	1240	1435	1435-1650	39-41										29-31					1.5-2.5							26-30		0.15	BAg-4
BAg-5	1250	1370	1370-1550	44-46										29-31												23-27		0.15	BAg-5
BAg-6	1270	1425	1425-1600	49-51										33-35												14-18		0.15	BAg-6
BAg-7	1145	1205	1205-1400	55-57										21-23										4.5-5.5		15-19		0.15	BAg-7
BAg-8	1435	1435	1435-1650	71-73										Bal														0.15	BAg-8
BAg-8a	1410	1410	1410-1600	71-73										Bal.		0.15-0.3												0.15	BAg-8a
BAg-13	1325	1575	1575-1775	53-55										Bal.												4.0-6.0		0.15	BAg-13
BAg-18	1115	1125	1325-1550	59-61										Bal.						0.025				9.5-10.5				0.15	BAg-18
BAg-19	1435	1635	1610-1800	92-93										Bal.		0.15-0.3												0.15	BAg-19
PRECIOUS METALS																													
BAu-1	1815	1860	1860-2000				37.0 +1 −0							Bal.														0.15	BAu-1
BAu-2	1635	1635	1635-1850				79.5 +1 −0							Bal.														0.15	BAu-2
BAu-3	1785	1885	1885-1995				34.5 +1 −0							Bal.					2.5-3.5									0.15	BAu-3
BAu-4	1740	1740	1740-1840				81.5 +1 −0												Bal.									0.15	BAu-4
COPPER AND COPPER-ZINC																													
BCu-1	1980	1980	2000-2100		0.01									99.90 min.						0.075	0.02							0.10	BCu-1
BCu-1a	1980	1980	2000-2100											99.0 min.														0.30[a]	BCu-1a
BCu-2	1980	1980	2000-2100											86.5 min.														0.50[c]	BCu-2[b]
RBCuZn-A	1630	1650	1670-1750		0.01*									57-61	*			*			0.05*		*	0.25-1.00		Bal.		0.50[d]	RBCuZn-A[e]
RBCuZn-D	1690	1715	1720-1800		0.01*									46-50					9.0-11.0	0.25	0.05*		0.04-0.25			Bal.		0.50[d]	RBCuZn-D[e]
MAGNESIUM																													
BMg-1	830	1110	1120-1160		8.3-9.7				0.0002-0.0008					0.05	0.005		Bal.	0.15 min.	0.005				0.05			1.7-2.3		0.30	BMg-1
BMg-2	770	1050	1080-1130		11.0-13.0												Bal.									4.5-5.5		0.30	BMg-2
BMg-2a	770	1050	1080-1130		11.0-13.0				0.0002-0.0008								Bal.									4.5-5.5		0.30	BMg-2a

[a] Total other elements requirement pertains only to the metallic elements for this filler metal.

[b] These chemical requirements pertain only to the copper oxide and do not include requirements for the organic vehicle in which the copper oxide is suspended.

[c] Total other elements requirement pertains only to metallic elements for this filler metal. The following limitations are placed on the nonmetallic elements:

Constituent	per cent (max)
Chlorides	0.4
Sulfates	0.1
Oxygen	remainder
Nitric acid insoluble	0.3
Acetone soluble matter	0.5

[d] Total other elements, including the elements marked with an asterisk (*), shall not exceed the value specified.

[e] This AWS-ASTM classification is intended to be identical with the same classification that appears in the Specification for Copper and Copper-Alloy Welding Rods (AWS Designation A5.7; ASTM Designation B 259).[3]

AWS-ASTM Classification	Solidus, °F	Liquidus, °F	Brazing Temperature Range, °F	Ag	Al	As	Au	B	Be	Bi	C	Cd	Cr	Cu	Fe	Li	Mg	Mn	Ni	P	Pb	Sb	Si	Sn	Ti	Zn	Other		AWS-ASTM Classification
																											Each	Total	
NICKEL																													
BNi-1	1790	1900	1950-2200					2.75-4.00			0.6-0.9		13.0-15.0		4.0-5.0				Bal.				3.0-5.0					0.50	BNi-1
BNi-2	1780	1830	1850-2150					2.75-3.5			0.15		6.0-8.0		2.0-4.0				Bal.				4.0-5.0					0.50	BNi-2
BNi-3	1800	1900	1850-2150					2.75-3.5			0.06				1.5				Bal.				4.0-5.0					0.50	BNi-3
BNi-4	1800	1950	1850-2150					1.0-2.2			0.06				1.5				Bal.				3.0-4.0					0.50	BNi-4
BNi-5	1975	2075	2100-2200								0.15		18.0-20.0						Bal.				9.75-10.5					0.50	BNi-5
BNi-6	1610	1610	1700-1875								0.15								Bal.	10.0-12.0								0.50	BNi-6
BNi-7	1630	1630	1700-1900										11.0-15.0						Bal.	9.0-11.0								0.50	BNi-7
SOLDERS—ASTM DESIGNATION B-32-60T, REVISED 1966																													
70A	361	378			0.005 max	0.03 max				0.25 max				0.08 max	0.02 max						30			70		0.005 max			70A
70B					0.005 max	0.03 max				0.25 max				0.08 max	0.02 max						30	0.2-0.5		70		0.005 max			70B
63A	361	361			0.005 max	0.03 max				0.25 max				0.08 max	0.02 max						37	0.12 max		63		0.005 max			63A
63B					0.005 max	0.03 max				0.25 max				0.08 max	0.02 max						37	0.2-0.5		63		0.005 max			63B
60A	361	374			0.005 max	0.03 max				0.25 max				0.08 max	0.02 max						40	0.12 max		60		0.005 max			60A
60B					0.005 max	0.03 max				0.25 max				0.08 max	0.02 max						40	0.2-0.5		60		0.005 max			60B
50A	361	421			0.005 max	0.03 max				0.25 max				0.08 max	0.02 max						50	0.12 max		50		0.005 max			50A
50B	360	420			0.005 max	0.03 max				0.25 max				0.08 max	0.02 max						50	0.2-0.5		50		0.005 max			50B
45A	361	441			0.005 max	0.03 max				0.25 max				0.08 max	0.02 max						55	0.12 max		45		0.005 max			45A
45B					0.005 max	0.03 max				0.25 max				0.08 max	0.02 max						55	0.2-0.5		45		0.005 max			45B
40A	361	460			0.005 max	0.03 max				0.25 max				0.08 max	0.02 max						60	0.12 max		40		0.005 max			40A
40B	360	460			0.005 max	0.03 max				0.25 max				0.08 max	0.02 max						60	0.2-0.5		40		0.005 max			40B
40C	365	448			0.005 max	0.03 max				0.25 max				0.08 max	0.02 max						58	1.8-2.4		40		0.005 max			40C
35A	361	477			0.005 max	0.03 max				0.25 max				0.08 max	0.02 max						65	0.25 max		35		0.005 max			35A
35B					0.005 max	0.03 max				0.25 max				0.08 max	0.02 max						65	0.2-0.5		35		0.005 max			35B
35C	365	470			0.005 max	0.03 max				0.25 max				0.08 max	0.02 max						63.2	1.6-2.0		35		0.005 max			35C
30A	361	491			0.005 max	0.03 max				0.25 max				0.08 max	0.02 max						70	0.25 max		30		0.005 max			30A
30B					0.005 max	0.03 max				0.25 max				0.08 max	0.02 max						70	0.2-0.5		30		0.005 max			30B
30C	364	482			0.005 max	0.03 max				0.25 max				0.08 max	0.02 max						68.4	1.4-1.8		30		0.005 max			30C
25A	361	511			0.005 max	0.03 max				0.25 max				0.08 max	0.02 max						75	0.25 max		25		0.005 max			25A
25B	360	510			0.005 max	0.03 max				0.25 max				0.08 max	0.02 max						75	0.2-0.5		25		0.005 max			25B
25C					0.005 max	0.03 max				0.25 max				0.08 max	0.02 max						73.7	1.1-1.5		25		0.005 max			25C
20B	361	531			0.005 max	0.03 max				0.25 max				0.08 max	0.02 max						80	0.2-0.5		20		0.005 max			20B
20C	363	517			0.005 max	0.03 max				0.25 max				0.08 max	0.02 max						79	0.8-1.2		20		0.005 max			20C
15B	440	550			0.005 max	0.03 max				0.25 max				0.08 max	0.02 max						85	0.2-0.5		15		0.005 max			15B
10B					0.005 max	0.03 max				0.25 max				0.08 max	0.02 max						90	0.2-0.5		10		0.005 max			10B
5A	518	594			0.005 max	0.03 max				0.25 max				0.08 max	0.02 max						95	0.12 max		5*		0.005 max			5A
5B					0.005 max	0.03 max				0.25 max				0.08 max	0.02 max						95	0.2-0.5		5*		0.005 max			5B
2A					0.005 max	0.03 max				0.25 max				0.08 max	0.02 max						98	0.12 max		2**		0.005 max			2A
2B					0.005 max	0.03 max				0.25 max				0.08 max	0.02 max						98	0.2-0.5		2**		0.005 max			2B
2.5S	579	579			0.005 max	0.03 max				0.25 max				0.08 max	0.02 max						97.5	0.4 max		0***		0.005 max			2.5S
1.5S	588	588		2.3-2.7	0.005 max	0.03 max				0.25 max				0.08 max	0.02 max						97.5	0.4 max		1****		0.005 max			1.5S
95TA	452	464		1.3-1.7	0.005 max	0.03 max				0.25 max				0.08 max	0.02 max						0.2	4.5-5.5		95		0.005 max			95TA
96.5TS				3.3-3.7	0.005 max	0.03 max				0.25 max				0.08 mas	0.02 max						0.2	0.2-0.5		96.5		0.005 max			96.5TS

* Permissible tin range, 4.5–5.5%.
** Permissible tin range, 1.5–2.5%.
*** Tin maximum, 0.25%.
**** Permissible tin range, 0.75–1.25%.

PHYSICAL AND PHOTOMETRIC DATA FOR PLANETS AND SATELLITES

Planet	Mass (10^{24} kg)	Radius (equ.) km	Angular diameter (see Note 3)	Distance from Earth (see Note 3)	Flattening (geom.)	Mean density (g/cm³)	Coefficients of potential $10^3 J_2$	$10^6 J_3$	$10^6 J_4$
Mercury	0.33022	2,439	11″.0	0.613	0	5.43		—	—
Venus	4.8690	6,052	60″.2	0.277	0	5.24	0.027	—	—
Earth	5.9742	6,378.140			0.00335281	5.515	1.08263	−2.54	−1.61
(Moon)	0.073483	1,738	31″.08	0.00257	0	3.34	0.2027	—	—
Mars	0.64191	3,393.4	17″.9	0.524	0.0051865	3.94	1.964	36	—
Jupiter	1,898.8	71,398	46″.8	4.203	0.0648088	1.33	14.75	—	−580
Saturn	568.50	60,000	19″.4	8.539	0.1076209	0.70	16.45	—	−1000
Uranus	86.625	25,400	3″.9	18.182	0.030	1.30	12	—	—
Neptune	102.78	24,300	2″.3	29.06	0.0259	1.76	4	—	—
Pluto	0.015	1,500	0″.1	38.44	0	1.1		—	—

Planet	Sidereal period of rotation (d)	Inclination of equator to orbit (°)	Geometric albedo	Visual magnitude $V(1,0)$	V_0	Color indices $B-V$	$U-B$
Mercury	58.6462	0.0	0.106	−0.42	—	0.93	0.41
Venus	−243.01	177.3	0.65	−4.40	—	0.82	0.50
Earth	0.99726968	23.45	0.367	−3.86	—	—	—
(Moon)	27.32166	6.68	0.12	+0.21	−12.74	0.92	0.46
Mars	1.02595675	25.19	0.150	−1.52	−2.01	1.36	0.58
Jupiter	0.41354 (System III)	3.12	0.52	−9.40	−2.70	0.83	0.48
Saturn	0.4375 (System III)	26.73	0.47	−8.88	+0.67	1.04	0.58
Uranus	−0.65	97.86	0.51	−7.19	+5.52	0.56	0.28
Neptune	0.768	29.56	0.41	−6.87	+7.84	0.41	0.21
Pluto	−6.3867	118?	0.3	−1.0	+15.12	0.80	0.31

NOTES:

1. The values for the masses include the atmospheres but exclude satellites.
2. The mean equatorial radii are given.
3. The angular diameters correspond to the distances from the Earth (in au) given in the adjacent column: they refer to inferior conjunction for Mercury and Venus and to mean opposition for the other planets. (1″.0 = 4.848 microradians.)
4. The flattening is the ratio of the difference of the equatorial and polar radii to the equatorial radius.
5. The notation for the coefficients of the gravitational potential is given in *Trans. IAU*, XI B, 173, 1962.
6. The period of rotation refers to the rotation at the equator with respect to a fixed frame of reference: a negative sign indicates that the rotation is retrograde with respect to the pole that lies to the north of the invariable plane of the solar system. The period is given in days of 86,400 SI seconds.
7. The data on equatorial radii, flattening, period of rotation and inclination of equator to orbit are based on the report of the IAU Working Group on Cartographic Coordinates and Rotational Elements of the Planets and Satellites, 1982.
8. The geometric albedo is the ratio of the illumination at the Earth from the planet for phase angle zero to the illumination produced by a plane, absolutely white Lambert surface of the same radius as the planet placed at the same position.
9. The quantity $V(1,0)$ is the visual magnitude of the planet reduced to a distance of 1 au from both the Sun and Earth and phase angle zero: V_0 is the mean opposition magnitude. The photometric quantities for Saturn refer to the disk only.

MINOR PLANETS
(OPPOSITION DATES MAGNITUDES AND OSCULATING ELEMENTS FOR EPOCH 1986 JUNE 19·0 TDT, ECLIPTIC AND EQUINOX J2000·0)

Name	No.	$B(1,0)$	Diameter (km)	Inclination (i) (°)	Long of asc. node (Ω) (°)	Argument of perihelion (ω) (°)	Mean distance (a)	Daily motion (n) (°)	Eccentricity (e)	Mean anomaly (M) (dg)
Ceres	1	4.5	1003	10.605	80.709	72.584	2.7672	0.21411	0.0784	28.881
Juno	3	6.5	247	12.998	170.577	246.897	2.6677	0.22620	0.2580	203.843
Vesta	4	4.3	538	7.138	104.069	150.792	2.3625	0.27143	0.0897	76.694
Hebe	6	7.0	201	14.780	139.080	238.613	2.4253	0.26095	0.2019	195.934
Iris	7	6.8	209	5.511	260.117	144.810	2.3862	0.26739	0.2293	171.246
Flora	8	7.7	151	5.887	111.151	285.032	2.2014	0.30175	0.1563	173.985
Metis	9	7.8	151	5.585	69.126	4.896	2.3869	0.26728	0.1218	310.516
Hygiea	10	6.5	450	3.842	283.855	316.533	3.1342	0.17763	0.1202	179.481
Parthenope	11	7.8	150	4.621	125.715	193.657	2.4515	0.25677	0.1003	81.899
Victoria	12	8.4	126	8.376	235.877	68.822	2.3336	0.27647	0.2204	57.724
Egeria	13	8.1	224	16.505	43.504	80.003	2.5777	0.23816	0.0870	207.976
Irene	14	7.5	158	9.111	86.887	95.091	2.5872	0.23685	0.1651	203.842
Thetis	17	9.1	109	5.586	125.694	136.017	2.4681	0.25419	0.1377	136.752
Melpomene	18	7.7	150	10.134	150.750	227.391	2.2961	0.28329	0.2179	107.403
Fortuna	19	8.4	215	1.568	211.789	181.570	2.4422	0.25825	0.1586	338.212

Name	No.	$B(1,0)$	Diameter (km)	Inclination (i) (°)	Long of asc. node (Ω) (°)	Argument of perihelion (ω) (°)	Mean distance (a)	Daily motion (n) (°)	Eccentricity (e)	Mean anomaly (M) (dg)
Massalia	20	7.7	131	0.699	207.163	254.737	2.4079	0.26378	0.1454	195.860
Kalliope	22	7.3	177	13.699	66.490	354.967	2.9100	0.19855	0.0980	335.063
Thalia	23	8.2	111	10.155	67.357	59.513	2.6286	0.23126	0.2310	79.960
Themis	24	8.3	234	0.763	36.077	111.140	3.1300	0.17798	0.1342	15.820
Phocaea	25	9.3	72	21.581	214.418	90.404	2.4015	0.26484	0.2542	231.796
Euterpe	27	8.4	108	1.584	94.828	355.936	2.3481	0.27392	0.1712	286.733
Bellona	28	8.2	126	9.404	144.725	342.672	2.7817	0.21244	0.1480	239.875
Amphitrite	29	7.1	195	6.113	356.658	62.659	2.5553	0.24129	0.0732	268.469
Euphrosyne	31	7.3	370	26.350	31.357	63.230	3.1461	0.17662	0.2276	128.642
Pomona	32	8.8	93	5.518	220.781	336.400	2.5857	0.23705	0.0844	72.769
Atalante	36	9.8	118	18.493	358.961	46.391	2.7446	0.21677	0.3050	224.239
Fides	37	8.4	95	3.078	7.896	61.834	2.6420	0.22951	0.1771	108.210
Laetitia	39	7.4	163	10.373	157.508	209.163	2.7676	0.21406	0.1137	254.136
Harmonia	40	8.3	100	4.257	94.439	269.697	2.2669	0.28877	0.0467	243.187
Daphne	41	8.2	204	15.779	178.528	45.663	2.7711	0.21366	0.2685	78.718
Isis	42	8.8	97	8.542	84.797	235.887	2.4400	0.25860	0.2258	344.086
Nysa	44	7.8	82	3.705	131.735	341.856	2.4225	0.26140	0.1513	189.490
Eugenia	45	8.3	226	6.596	148.163	85.989	2.7208	0.21961	0.0836	285.191
Hestia	46	9.6	133	2.328	181.352	175.419	2.5251	0.24563	0.1724	4.932
Nemausa	51	8.7	151	9.962	176.314	0.697	2.3654	0.27093	0.0662	87.568
Europa	52	7.6	289	7.454	129.476	335.290	3.1076	0.17992	0.1031	237.883
Alexandra	54	8.9	180	11.788	313.839	344.130	2.7122	0.22066	0.1965	248.706
Melete	56	9.5	146	8.081	193.918	103.092	2.5994	0.23518	0.2322	302.302
Mnemosyne	57	8.4	109	15.221	199.596	217.045	3.1481	0.17646	0.1177	180.390
Concordia	58	9.9	110	5.060	161.548	31.834	2.7028	0.22181	0.0433	151.384
Echo	60	10.0	51	3.592	192.154	269.247	2.3953	0.26587	0.1828	256.348
Angelina	64	8.8	56	1.313	310.134	178.286	2.6852	0.22399	0.1249	225.596
Maja	66	10.5	85	3.055	8.125	43.131	2.6457	0.22903	0.1739	151.681
Asia	67	9.7	58	6.008	203.091	105.508	2.4209	0.26167	0.1873	20.408
Leto	68	8.2	126	7.968	44.593	304.509	2.7829	0.21230	0.1852	232.488
Niobe	71	8.3	115	23.300	316.481	266.627	2.7535	0.21572	0.1751	198.102
Frigga	77	9.7	67	2.434	1.677	61.057	2.6691	0.22603	0.1327	91.229
Sappho	80	9.2	83	8.656	219.094	138.885	2.2954	0.28341	0.2003	298.083
Alkmene	82	9.5	65	2.844	25.965	110.442	2.7590	0.21507	0.2234	120.145
Beatrix	83	9.8	123	4.982	27.953	165.746	2.4314	0.25996	0.0842	354.190

SATELLITES: ORBITAL DATA

Planet		Satellite	Orbital period[a] R = Retrograde (Days)	Maximum elongation at mean opposition (° ′ ″)	Semimajor axis (× 10^3 km)	Orbital eccentricity	Orbital inclination to planetary equator (°)	Motion of node on fixed plane[d] (°/yr)
Earth		Moon	27.321661		384.400	0.054900489	18.28—28.58	19.34[f]
Mars	I	Phobos	0.31891023	25	9.378	0.015	1.0	158.8
	II	Deimos	1.2624407	1 02	23.459	0.0005	0.9—2.7	6.614
Jupiter	I	Io	1.769137786	2 18	422	0.004	0.04	48.6
	II	Europa	3.551181041	3 40	671	0.009	0.47	12.0
	III	Ganymede	7.15455296	5 51	1070	0.002	0.21	2.63
	IV	Callisto	16.6890184	10 18	1883	0.007	0.51	0.643
	V	Amalthea	0.49817905	59	181	0.003	0.40	914.6
	VI	Himalia	250.5662	1 02 46	11480	0.15798	27.63	
	VII	Elara	259.6528	1 04 10	11737	0.20719	24.77	
	VIII	Pasiphae	735 R	2 08 26	23500	0.378	145	
	IX	Sinope	758 R	2 09 31	23700	0.275	153	
	X	Lysithea	259.22	1 04 04	11720	0.107	29.02	
	XI	Carme	692 R	2 03 31	22600	0.20678	164	
	XII	Ananke	631 R	1 55 52	21200	0.16870	147	
	XIII	Leda	238.72	1 00 39	11094	0.14762	26.07	
	XIV	Thebe	0.6745	1 13	222	0.015	0.8	
	XV	Adrastea	0.29826	42	129			
	XVI	Metis	0.294780	42	128			
Saturn	I	Mimas	0.942421813	30	185.52	0.0202	1.53	365.0
	II	Enceladus	1.370217855	38	238.02	0.00452	0.00	156.2[e]
	III	Tethys	1.887802160	48	294.66	0.00000	1.86	72.25
	IV	Dione	2.736914742	1 01	377.40	0.002230	0.02	30.85[e]
	V	Rhea	4.517500436	1 25	527.04	0.00100	0.35	10.16
	VI	Titan	15.94542068	3 17	1221.83	0.029192	0.33	0.5213[e]
	VII	Hyperion	21.2766088	3 59	1481.1	0.104	0.43	

Planet		Satellite	Orbital period[a] R = Retrograde (Days)	Maximum elongation at mean opposition	Semimajor axis (× 10^3 km)	Orbital eccentricity	Orbital inclination to planetary equator (°)	Motion of node on fixed plane[d] (°/yr)
	VIII	Iapetus	79.3301825	9 35	3561.3	0.02828	14.72	
	IX	Phoebe	550.48 R	34 51	12952	0.16326	177[b]	
	X	Janus	0.6945	24	151.472	0.007	0.14	
	XI	Epimctheus	0.6942	24	151.422	0.009	0.34	
	XII	1980S6	2.7369	1 01	377.40	0.005	0.0	
	XIII	Telesto	1.8878	48	294.66			
	XIV	Calypso	1.8878	48	294.66			
	XV	Atlas	0.6019	22	137.670	0.000	0.3	
		1980S26	0.6285	23	141.700	0.004	0.0	
		1980S27	0.6130	23	139.353	0.003	0.0	
Uranus	I	Ariel	2.52037935	14	191.02	0.0034	0.3	6.8
	II	Umbriel	4.1441772	20	266.30	0.0050	0.36	3.6
	III	Titania	8.7058717	33	435.91	0.0022	0.14	2.0
	IV	Oberon	13.4632389	44	583.52	0.0008	0.10	1.4
	V	Miranda	1.41347925	10	129.39	0.0027	4.2	19.8
Neptune	I	Triton	5.8768433 R	17	354.29	<0.01	159.00	0.578
	II	Nereid	360.2	4 21	5511	0.7483	27.6[c]	
Pluto		(Charon)	6.3871	<1	19.7		94[c]	

[a] Sidereal periods except for satellites of Saturn; tropical periods for those.
[b] Relative to ecliptic plane.
[c] To equator of 1950.0.
[d] Rate of decrease (or increase) in the longitude of the ascending node.
[e] Rate of increase in the longitude of the apse.
[f] On ecliptic plane.

SATELLITES: PHYSICAL AND PHOTOMETRIC DATA

Planet		Satellite	Mass (ℓ/Planet)	Radius (km)	Sidereal period of rotation S = Synchr.[a] (Days)	Geometric albedo $(V)^c$	$V(1,0)$	V_0	$(B - V)$	$(U - B)$
Earth		Moon	0.01230002	1738	S	0.12	+0.21	−12.74	0.92	0.46
Mars	I	Phobos	1.5×10^{-8}	13.5 × 10.8 × 9.4	S	0.06	+11.8	11.3	0.6	
	II	Deimos	3×10^{-9}	7.5 × 6.1 × 5.5	S	0.07	+12.89	12.40	0.65	0.18
Jupiter	I	Io	4.68×10^{-5}	1815	S	0.61	−1.68	5.02	1.17	1.30
	II	Europa	2.52×10^{-5}	1569	S	0.64	−1.41	5.29	0.87	0.52
	III	Ganymede	7.80×10^{-5}	2631	S	0.42	−2.09	4.61	0.83	0.50
	IV	Callisto	5.66×10^{-5}	2400	S	0.20	−1.05	5.65	0.86	0.55
	V	Amalthea	38×10^{-10}	135 × 83 × 75	S	0.05	+7.4	14.1	1.50	
	VI	Himalia	50×10^{-10}	93	0.4	0.03	+8.14	14.84	0.67	0.30
	VII	Elara	4×10^{-10}	38	0.5	0.03	+10.07	16.77	0.69	0.28
	VIII	Pasiphae	1×10^{-10}	25			+10.33	17.03	0.63	0.34
	IX	Sinope	0.4×10^{-10}	18			+11.6	18.3	0.7	
	X	Lysithea	0.4×10^{-10}	18			+11.7	18.4	0.7	
	XI	Carme	0.5×10^{-10}	20			+11.3	18.0	0.7	
	XII	Ananke	0.2×10^{-10}	15			+12.2	18.9	0.7	
	XIII	Leda	0.03×10^{-10}	8			+13.5	20.2	0.7	
	XIV	Thebe	4×10^{-10}	55 × 45		0.05	+8.9	15.6		
	XV	Adrastea	0.1×10^{-10}	12.5 × 10 × 7.5		0.05	+12.4	19.1		
	XVI	Metis	0.5×10^{-10}	20	0.05		+10.8	17.5		
Saturn	I	Mimas	8.0×10^{-8}	196	S	0.5	+3.3	12.9		
	II	Enceladus	1.3×10^{-7}	250	S	1.0	+2.1	11.7	0.70	0.28
	III	Tethys	1.3×10^{-6}	530	S	0.9	+0.6	10.2	0.73	0.30
	IV	Dione	1.85×10^{-6}	560	S	0.7	+0.8	10.4	0.71	0.31
	V	Rhea	4.4×10^{-6}	765	S	0.7	+0.1	9.7	0.78	0.38
	VI	Titan	2.38×10^{-4}	2575	S	0.21	−1.28	8.28	1.28	0.75
	VII	Hyperion	3×10^{-8}	205 × 130 × 110		0.3	+4.63	14.19	0.78	0.33
	VIII	Iapetus	3.3×10^{-6}	730	S	0.2[b]	+1.5	11.1	0.72	0.30
	IX	Phoebe	7×10^{-10}	110	0.4	0.06	+6.89	16.45	0.70	0.34
	X	Janus		110 × 100 × 80	S	0.8	+4.4:	14:		
	XI	Epimetheus		70 × 60 × 50	S	0.8	+5.4:	15:		
	XII	1980S6		18 × 16 × 15		0.7	+8.4:	18:		
	XIII	Telesto		17 × 14 × 13		0.5	+8.9:	18.5:		
	XIV	Calypso		17 × 11 × 11		0.6	+9.1:	18.7:		
	XV	Atlas		20 × 10		0.9	+8.4:	18:		
		1980S26		55 × 45 × 35		0.9	+6.4:	16:		
		1980S27		70 × 50 × 40		0.6	+6.4:	16:		

Planet	Satellite		Mass (ℓ/ Planet)	Radius (km)	Sidereal period of rotation S = Synchr.[a] (Days)	Geometric albedo (V)[c]	V(1,0)	V_0	(B − V)	(U − B)
Uranus	I	Ariel	1.8×10^{-5}	665		0.2	+1.7	14.4		
	II	Umbriel	1.2×10^{-5}	555		0.1	+2.6	15.3		
	III	Titania	6.8×10^{-5}	800		0.21	+1.27	13.98	0.70	0.28
	IV	Oberon	6.9×10^{-5}	815	S	0.16	+1.52	14.23	0.68	0.20
	V	Miranda	0.2×10^{-5}	160			+3.8	16.5		
Neptune	I	Triton	1.3×10^{-3}	1900	S		−1.02	13.69	0.72	0.29
	II	Nereid	2×10^{-7}	150			+4.0	18.7		
Pluto		(Charon)	0.125(?)	1000(?)			+0.9	16.8		

[a] Rotation period same as orbital period.
[b] Bright side, 0.5; faint side, 0.05.
[c] V (Sun) = −26.8.

HOHMANN ELLIPSE TRANSFER DATA (MINIMUM ENERGY OF TRANSFER)

W. Joseph Armento

The "from" body is the body of departure and the "to" body is the body of arrival. The three lines of data for each planetary entry are:

top—from perhelion of inner planet to aphelion of outer planet
middle—from mean of inner planet to mean of outer planet
bottom—from aphelion of inner planet to perhelion of outer planet

The orbital parameters and velocities are given for the transfer orbit. The orbital velocity, change in velocity necessary to reach the new Hohmann orbital velocity, and the maximum change in velocity to reach the Hohmann orbital velocity are given for departure. For arrival data, look in the table listing the arrival planet as "from" and the departure planet as "to". The maximum Hohmann velocity change is assumed to be a single impulse to overcome the necessary energy for change in velocity and escape velocity from the surface of the departure planet.

Body	Mean orbital Semimajor axis (km × 10⁻⁶)	Eccentricity ε	Aphelion (km × 10⁻⁶)	Perhelion (km × 10⁻⁶)	Orbital Velocity (km/sec) Mean	Orbital Velocity (km/sec) Aphelion	Orbital Velocity (km/sec) Perhelion	Orbital Velocity of Departure Body (km/sec)	Time of one way trip (½ orbit) (Solar Seconds)	Departure Δ_V orbit (km/sec)	Departure Maximum Δ_V (km/sec)
From Mercury to:											
	77.440	0.4055	103.85	46.035	41.38	26.91	63.62	58.92	5.880 E6	4.699	6.284
Venus	83.030	0.3021	108.11	57.950	39.96	29.26	54.58	47.83	6.528 E6	6.748	7.934
	88.620	0.2116	107.37	69.864	38.68	31.20	47.95	38.83	7.198 E6	9.124	10.03
	99.052	0.5352	152.07	46.035	36.58	20.13	66.49	58.92	8.506 E6	7.569	8.643
Earth	103.76	0.4415	149.57	57.950	35.75	22.25	57.43	47.83	9.119 E6	9.596	10.46
	108.47	0.3559	147.07	69.864	34.96	24.10	50.72	38.83	9.747 E6	11.90	12.61
	147.58	0.6881	249.12	46.035	29.97	12.88	69.72	58.92	1.547 E7	10.80	11.58
Mars	142.89	0.5945	227.84	57.950	30.46	15.36	60.40	47.83	1.474 E7	12.57	13.24
	138.21	0.4945	206.56	69.864	30.97	18.01	53.25	38.83	1.402 E7	14.43	15.02
	430.92	0.8932	815.80	46.035	17.54	4.167	73.84	58.92	7.718 E7	14.91	15.49
Jupiter	418.04	0.8614	778.14	57.950	17.81	4.860	65.26	47.83	7.375 E7	17.43	17.92
	405.17	0.8276	740.48	69.864	18.09	5.556	58.89	38.83	7.037 E7	20.06	20.49
	775.26	0.9406	1504.5	46.035	13.08	2.287	74.76	58.92	1.862 E8	15.83	16.37
Saturn	742.47	0.9220	1427.0	57.950	13.36	2.693	66.31	47.83	1.746 E8	18.48	18.94
	709.69	0.9016	1349.5	69.864	13.67	3.110	60.07	38.83	1.631 E8	21.24	21.65
	1524.2	0.9698	3002.3	46.035	9.326	1.155	75.32	58.92	5.134 E8	16.39	16.92
Uranus	1464.1	0.9604	2870.3	57.950	9.515	1.352	66.97	47.83	4.834 E8	19.14	19.59
	1404.1	0.9502	2738.3	69.864	9.717	1.552	60.83	38.83	4.539 E8	22.01	22.40
	2291.4	0.9799	4536.8	46.035	7.606	0.7662	75.51	58.92	9.464 E8	16.59	17.10
Neptune	2278.9	0.9746	4499.9	57.950	7.627	0.8655	67.21	47.83	9.337 E8	19.38	19.83
	2266.4	0.9692	4463.0	69.864	7.648	0.9569	61.13	38.83	9.310 E8	22.30	22.69
	3710.5	0.9876	7375.0	46.035	5.977	0.4723	75.66	58.92	1.950 E9	16.73	17.24
Pluto	2983.5	0.9806	5909.0	57.950	6.666	0.6601	67.31	47.83	1.406 E9	19.48	19.93
	2256.4	0.9690	4443.0	69.864	7.665	0.9612	61.13	38.83	9.248 E8	22.30	22.69
From Venus to:											
	77.440	0.4055	108.85	46.035	41.38	26.91	63.62	34.78	5.880 E6	7.873	13.02
Mercury	83.030	0.3021	108.11	57.950	39.96	29.26	54.59	35.02	6.528 E6	5.763	11.86
	88.620	0.2116	107.37	69.864	38.68	31.20	47.95	35.26	7.198 E6	4.058	11.13
	129.72	0.1723	152.07	107.37	31.97	26.86	38.04	35.26	1.275 E7	2.787	10.73
Earth	128.84	0.1609	149.57	108.11	32.08	27.27	37.73	35.02	1.262 E7	2.712	10.71
	127.96	0.1494	147.07	108.85	32.19	27.69	37.42	34.78	1.249 E7	2.635	10.69
	178.25	0.3976	249.12	107.37	27.27	17.90	41.54	35.26	2.053 E7	6.283	12.12
Mars	167.97	0.3564	227.84	108.11	28.09	19.35	40.78	35.02	1.878 E7	5.766	11.86
	157.70	0.3098	206.56	108.85	28.99	21.05	39.94	34.78	1.709 E7	5.161	11.58
	461.59	0.7674	815.80	107.37	16.95	6.148	46.71	35.26	8.557 E7	11.46	15.45
Jupiter	443.12	0.7560	778.14	108.11	17.30	6.447	46.40	35.02	8.048 E7	11.39	15.40
	424.66	0.7437	740.48	108.85	17.67	6.774	46.09	34.78	7.551 E7	11.30	15.34
	805.93	0.8668	1504.5	107.37	12.83	3.426	48.01	35.26	1.974 E8	12.75	16.43
Saturn	767.55	0.8592	1427.0	108.11	13.14	3.617	47.75	35.02	1.835 E8	12.73	16.42
	729.18	0.8507	1349.5	108.85	13.48	3.829	47.48	34.78	1.699 E8	12.70	16.39
	1554.9	0.9309	3002.3	107.37	9.234	1.746	48.83	35.26	5.290 E8	13.57	17.08
Uranus	1489.2	0.9274	2870.3	108.11	9.435	1.831	48.62	35.02	4.958 E8	13.60	17.10
	1423.6	0.9235	2738.3	108.85	9.650	1.924	48.40	34.78	4.634 E8	13.62	17.12
	2322.6	0.9538	4536.8	107.37	7.556	1.162	49.12	35.26	9.655 E8	13.86	17.31
Neptune	2304.0	0.9531	4499.9	108.11	7.586	1.176	48.94	35.02	9.542 E8	13.92	17.36
	2285.9	0.9524	4463.0	108.85	7.616	1.189	48.77	34.78	9.430 E8	13.98	17.41
	3741.2	0.9713	7375.0	107.37	5.953	0.7183	49.34	35.26	1.974 E9	14.08	17.48
Pluto	3008.6	0.9641	5909.0	108.11	6.638	0.8979	49.08	35.02	1.424 E9	14.06	17.47
	2275.9	0.9522	4443.0	108.85	7.632	1.195	48.76	34.78	9.368 E8	13.98	17.40

HOHMANN ELLIPSE TRANSFER DATA (Continued)

Body	Mean orbital Semimajor axis (km × 10⁻⁶)	Eccentricity ε	Aphelion (km × 10⁻⁶)	Perhelion (km × 10⁻⁶)	Orbital Velocity (km/sec) Mean	Orbital Velocity (km/sec) Aphelion	Orbital Velocity (km/sec) Perhelion	Orbital Velocity of Departure Body (km/sec)	Time of one way trip (½ orbit) (Solar Seconds)	Departure Δ_v orbit (km/sec)	Departure Maximum Δ_v (km/sec)
From Earth to:											
Mercury	99.052	0.5352	152.07	46.035	36.58	20.13	66.49	29.28	8.506 E6	9.150	14.45
	103.76	0.4415	149.57	57.950	35.75	22.25	57.43	29.77	9.119 E6	7.523	13.47
	108.47	0.3559	147.07	69.864	34.96	24.10	50.72	30.27	9.747 E6	6.178	12.77
Venus	129.72	0.1723	152.07	107.37	31.97	26.86	38.04	29.28	1.275 E7	2.416	11.44
	128.84	0.1609	149.57	108.11	32.08	27.27	37.73	29.77	1.262 E7	2.500	11.46
	127.96	0.1494	147.07	108.85	32.19	27.69	37.42	30.27	1.249 E7	2.583	11.47
Mars	198.10	0.2576	249.12	147.07	25.87	19.88	33.67	30.27	2.406 E7	3.396	11.68
	188.70	0.2074	227.84	149.57	26.51	21.48	32.71	29.77	2.237 E7	2.942	11.56
	179.31	0.1519	206.56	152.07	27.19	23.33	31.69	29.28	2.072 E7	2.411	11.44
Jupiter	481.44	0.6945	815.80	147.07	16.59	7.046	39.08	30.27	9.114 E7	8.810	14.23
	463.85	0.6776	778.14	149.57	16.91	7.412	38.56	29.77	8.620 E7	8.789	14.22
	446.27	0.6592	740.48	152.07	17.24	7.811	38.03	29.28	8.134 E7	8.755	14.20
Saturn	825.78	0.8219	1504.5	147.07	12.67	3.962	40.53	30.27	2.047 E8	10.25	15.17
	788.28	0.8103	1427.0	149.57	12.97	4.199	40.06	29.77	1.910 E8	10.29	15.19
	750.79	0.7975	1349.5	152.07	13.29	4.461	39.59	29.28	1.775 E8	10.31	15.21
Uranus	1574.7	0.9066	3002.3	147.07	9.176	2.031	41.46	30.27	5.392 E8	11.18	15.81
	1509.9	0.9009	2870.3	149.57	9.370	2.139	41.05	29.77	5.062 E8	11.28	15.88
	1445.2	0.8948	2738.3	152.07	9.578	2.257	40.64	29.28	4.740 E8	11.36	15.94
Neptune	2341.9	0.9372	4536.8	147.07	7.524	1.355	41.79	30.27	9.779 E8	11.51	16.05
	2324.7	0.9357	4499.9	149.57	7.552	1.377	41.42	29.77	9.671 E8	11.65	16.15
	2307.5	0.9341	4463.0	152.07	7.580	1.399	41.06	29.28	9.564 E8	11.78	16.24
Pluto	3761.0	0.9609	7375.0	147.07	5.937	0.8384	42.04	30.27	1.990 E9	11.77	16.23
	3029.3	0.9506	5909.0	149.57	6.615	1.053	41.58	29.77	1.439 E9	11.81	16.26
	2297.5	0.9338	4443.0	152.07	7.596	1.405	41.06	29.28	9.502 E8	11.78	16.24
From Mars to:											
Mercury	147.58	0.6881	249.12	46.035	29.97	12.88	69.72	21.97	1.547 E7	9.081	10.38
	142.89	0.5945	227.84	57.950	30.46	15.36	60.40	24.12	1.474 E7	8.761	10.10
	138.21	0.4945	206.56	69.864	30.97	18.01	53.25	26.49	1.402 E7	8.479	9.858
Venus	178.25	0.3976	249.12	107.37	27.27	17.90	41.54	21.97	2.053 E7	4.060	6.463
	167.97	0.3564	227.84	108.11	28.09	19.35	40.78	24.12	1.878 E7	4.770	6.931
	157.70	0.3098	206.56	108.85	28.99	21.05	39.94	26.49	1.709 E7	5.444	7.411
Earth	198.10	0.2576	249.12	147.07	25.87	19.88	33.67	21.97	2.406 E7	2.088	5.444
	188.70	0.2074	227.84	149.57	26.51	21.48	32.71	24.12	2.237 E7	2.646	5.682
	179.31	0.1519	206.56	152.07	27.19	23.33	31.69	26.49	2.072 E7	3.161	5.939
Jupiter	511.18	0.5959	815.80	206.56	16.10	8.104	32.00	26.49	9.972 E7	5.514	7.462
	502.99	0.5470	778.14	227.84	16.23	8.785	30.00	24.12	9.733 E7	5.881	7.737
	494.80	0.4965	740.48	249.12	16.37	9.494	28.22	21.97	9.496 E7	6.256	8.026
Saturn	855.52	0.7586	1504.5	206.56	12.45	4.613	33.60	26.49	2.159 E8	7.105	8.704
	827.42	0.7246	1427.0	227.84	12.66	5.058	31.68	24.12	2.054 E8	7.556	9.076
	799.32	0.6883	1349.5	249.12	12.38	5.533	29.97	21.97	1.950 E8	8.010	9.457
Uranus	1604.4	0.8713	3002.3	206.56	9.090	2.384	34.66	26.49	5.545 E8	8.165	9.589
	1549.1	0.8529	2870.3	227.84	9.251	2.606	32.84	24.12	5.260 E8	8.713	10.06
	1493.7	0.8332	2738.3	249.12	9.421	2.842	31.23	21.97	4.981 E8	9.269	10.55
Neptune	2371.7	0.9129	4536.8	206.56	7.477	1.595	35.04	26.49	9.966 E8	8.548	9.917
	2363.9	0.9036	4499.9	227.84	7.489	1.685	33.28	24.12	9.916 E8	9.160	10.45
	2356.1	0.8943	4463.0	249.12	7.501	1.772	31.75	21.97	9.867 E8	9.785	11.00
Pluto	3790.8	0.9455	7375.0	206.56	5.914	0.9897	35.34	26.49	2.014 E9	8.846	10.17
	3068.4	0.9257	5909.0	227.84	6.573	1.291	33.47	24.12	1.467 E9	9.352	10.62
	2346.0	0.8938	4443.0	249.12	7.517	1.780	31.75	21.97	9.804 E8	9.781	11.00
From Jupiter to:											
Mercury	430.92	0.8932	815.80	46.036	17.54	4.167	73.84	12.44	7.718 E7	8.269	60.80
	418.04	0.8614	778.14	57.950	17.81	4.860	65.26	13.05	7.375 E7	8.193	60.79
	405.17	0.8276	740.48	69.865	18.09	5.556	58.89	13.70	7.037 E7	8.144	60.79
Venus	461.59	0.7674	815.80	107.37	16.95	6.148	46.71	12.44	8.557 E7	6.287	60.57
	443.12	0.7560	778.14	108.11	17.30	6.447	46.40	13.05	8.048 E7	6.606	60.60
	424.66	0.7437	740.48	108.85	17.67	6.774	46.09	13.70	7.551 E7	6.926	60.64
Earth	481.44	0.6945	815.80	147.07	16.59	7.046	39.08	12.44	9.114 E7	5.390	60.48
	463.85	0.6776	778.14	149.57	16.91	7.412	38.56	13.05	8.620 E7	5.641	60.50
	446.27	0.6592	740.48	152.07	17.24	7.811	38.03	13.70	8.134 E7	5.890	60.53
Mars	511.18	0.5959	815.80	206.56	16.10	8.104	32.00	12.44	9.972 E7	4.332	60.40
	502.99	0.5470	778.14	227.84	16.23	8.785	30.00	13.05	9.733 E7	4.268	60.39
	494.80	0.4965	740.48	249.12	16.37	9.494	28.22	13.70	9.496 E7	4.206	60.39
Saturn	1122.5	0.3403	1504.5	740.48	10.87	7.624	15.49	13.70	3.245 E8	1.790	60.27
	1102.6	0.2942	1427.0	778.14	10.97	8.097	14.85	13.05	3.159 E8	1.797	60.27
	1082.7	0.2465	1349.5	815.80	11.07	8.604	14.23	12.44	3.074 E8	1.797	60.27
Uranus	1871.4	0.6043	3007.3	740.48	8.417	4.180	16.95	13.70	6.985 E8	3.248	60.33
	1824.2	0.5743	2870.3	778.14	8.525	4.439	16.37	13.05	6.723 E8	3.320	60.33
	1777.0	0.5409	2738.3	815.80	8.637	4.715	15.82	12.44	6.463 E8	3.389	60.33
Neptune	2638.6	0.7194	4536.8	740.48	7.088	2.864	17.55	13.70	1.169 E9	3.845	60.36
	2639.0	0.7051	4499.9	778.14	7.088	2.947	17.04	13.05	1.170 E9	3.992	60.37
	2639.4	0.6909	4463.0	815.80	7.087	3.030	16.58	12.44	1.170 E9	4.141	60.38

HOHMANN ELLIPSE TRANSFER DATA (Continued)

Body	Mean orbital Semimajor axis (km × 10⁻⁶)	Eccentricity ε	Aphelion (km × 10⁻⁶)	Perhelion (km × 10⁻⁶)	Orbital Velocity (km/sec) Mean	Aphelion	Perhelion	Orbital Velocity of Departure Body (km/sec)	Time of one way trip (½ orbit) (Solar Seconds)	Departure Δv orbit (km/sec)	Maximum Δv (km/sec)
	4057.7	0.8175	7375.0	740.48	5.716	1.811	18.04	13.70	2.230 E9	4.339	60.40
Pluto	3343.6	0.7673	5909.0	778.14	6.297	2.285	17.35	13.05	1.668 E9	4.299	60.39
	2629.4	0.6897	4443.0	815.80	7.101	3.043	16.57	12.44	1.163 E9	4.135	60.38
From Saturn to:											
	775.26	0.9406	1504.5	46.036	13.08	2.287	74.76	9.129	1.862 E8	6.841	36.71
Mercury	742.47	0.9220	1427.0	57.950	13.36	2.693	66.31	9.639	1.746 E8	6.946	36.73
	709.69	0.9016	1349.5	69.860	13.67	3.110	60.07	10.18	1.631 E8	7.067	36.75
	805.93	0.8668	1504.5	107.37	12.83	3.426	45.01	9.129	1.974 E8	5.702	36.51
Venus	767.55	0.8592	1427.0	108.11	13.14	3.617	47.75	9.639	1.835 E8	6.021	36.56
	729.18	0.8507	1349.5	108.85	13.48	3.829	47.48	10.18	1.699 E8	6.348	36.62
	825.78	0.8219	1504.5	147.07	12.67	3.962	40.53	9.129	2.047 E8	5.167	36.43
Earth	788.28	0.8103	1427.0	149.57	12.97	4.199	40.06	9.639	1.910 E8	5.440	36.47
	750.79	0.7975	1349.5	152.07	13.29	4.461	39.59	10.18	1.775 E8	5.716	36.51
	855.52	0.7586	1504.5	206.56	12.45	4.613	33.60	9.129	2.159 E8	4.516	36.35
Mars	827.42	0.7246	1427.0	227.84	12.66	5.058	31.68	9.639	2.054 E8	4.581	36.35
	799.32	0.6883	1349.5	249.12	12.88	5.533	29.97	10.18	1.950 E8	4.644	36.36
	1122.5	0.3403	1504.5	740.48	10.87	7.624	15.49	9.129	3.245 E8	1.504	36.10
Jupiter	1102.6	0.2942	1427.0	778.14	10.97	8.097	14.85	9.639	3.159 E8	1.541	36.10
	1082.7	0.2465	1349.5	815.80	11.07	8.604	14.23	10.18	3.074 E8	1.573	36.10
	2175.9	0.3798	3002.3	1349.5	7.806	5.233	11.64	10.18	8.758 E8	1.465	36.09
Uranus	2148.6	0.3359	2870.3	1427.0	7.855	5.539	11.14	9.639	8.593 E8	1.502	36.10
	2121.4	0.2908	2738.3	1504.5	7.905	5.860	10.67	9.129	8.430 E8	1.536	36.10
	2943.2	0.5415	4536.8	1349.5	6.712	3.660	12.31	10.18	1.378 E9	2.129	36.13
Neptune	2963.4	0.5185	4499.9	1427.0	6.689	3.767	11.88	9.639	1.392 E9	2.239	36.13
	2983.7	0.4958	4463.0	1504.5	6.666	3.870	11.48	9.129	1.406 E9	2.352	36.14
	4362.3	0.6906	7375.0	1349.5	5.513	2.358	12.89	10.18	2.486 E9	2.710	36.17
Pluto	3668.0	0.6110	5909.0	1427.0	6.012	2.954	12.23	9.639	1.917 E9	2.595	36.16
	2973.7	0.4941	4443.0	1504.5	6.677	3.885	11.47	9.129	1.399 E9	2.345	36.14
From Uranus to:											
	1524.2	0.9698	3002.3	46.036	9.326	1.155	75.32	6.490	5.134 E8	5.336	22.82
Mercury	1464.1	0.9604	2870.3	57.950	9.516	1.352	66.97	6.796	4.834 E8	5.444	22.85
	1404.1	0.9502	2738.3	69.865	9.717	1.552	60.83	7.116	4.539 E8	5.564	22.88
	1554.9	0.9309	3002.3	107.37	9.234	1.746	48.83	6.490	5.290 E8	4.744	22.69
Venus	1489.2	0.9274	2870.3	108.11	9.435	1.831	48.62	6.796	4.958 E8	4.965	22.74
	1423.6	0.9235	2738.3	108.85	9.650	1.924	48.40	7.116	4.634 E8	5.192	22.69
	1574.7	0.9066	3002.3	147.07	9.176	2.031	41.46	6.490	5.392 E8	4.460	22.63
Earth	1509.9	0.9009	2870.3	149.57	9.370	2.139	41.05	6.796	5.062 E8	4.657	22.67
	1445.2	0.8948	2738.3	152.07	9.578	2.257	40.64	7.116	4.740 E8	4.859	22.72
	1604.4	0.8813	3002.3	206.56	9.090	2.384	34.66	6.490	5.545 E8	4.106	22.57
Mars	1549.1	0.8529	2870.3	227.84	9.251	2.606	32.84	6.796	5.260 E8	4.190	22.58
	1493.7	0.8332	2738.3	249.12	9.421	2.842	31.23	7.116	4.981 E8	4.275	22.60
	1817.4	0.6043	3002.3	740.48	8.417	4.180	16.95	6.490	6.985 E8	2.310	22.31
Jupiter	1824.2	0.5743	2870.3	778.14	8.525	4.439	16.37	6.796	6.723 E8	2.358	22.31
	1777.0	0.5409	2738.3	815.80	8.637	4.715	15.82	7.116	6.463 E8	2.402	22.32
	2175.9	0.3798	3002.3	1349.5	7.806	5.233	11.64	6.490	8.758 E8	1.257	22.23
Saturn	2148.6	0.3359	2870.3	1427.0	7.855	5.539	11.14	6.796	8.593 E8	1.258	22.23
	2121.4	0.2908	2738.3	1504.5	7.905	5.860	10.67	7.116	8.430 E8	1.257	22.23
	3637.5	0.2472	4536.8	2738.3	6.037	4.690	7.771	7.116	1.893 E9	0.6544	22.20
Neptune	3685.1	0.2211	4499.9	2870.3	5.998	4.790	7.510	6.796	1.930 E9	0.7139	22.20
	3732.7	0.1957	4463.0	3002.3	5.960	4.888	7.266	6.490	1.968 E9	0.7757	22.20
	5056.6	0.4585	7375.0	2738.3	5.120	3.120	8.403	7.116	3.102 E9	1.287	22.23
Pluto	4389.6	0.3461	5909.0	2870.3	5.496	3.830	7.885	6.796	2.509 E9	1.089	22.22
	3722.7	0.1935	4443.0	3002.3	5.968	4.906	7.260	6.490	1.960 E9	0.7691	22.20
From Neptune to:											
	2291.4	0.9799	4536.8	46.037	7.606	0.7662	75.51	5.384	9.464 E8	4.617	24.97
Mercury	2278.9	0.9746	4499.9	57.948	7.627	0.8655	67.21	5.428	9.387 E8	4.562	24.96
	2266.4	0.9692	4463.0	69.864	7.648	0.9569	61.13	5.473	9.310 E8	4.516	24.95
	2322.1	0.9538	4536.8	107.37	7.556	1.162	49.12	5.384	9.655 E8	4.221	24.90
Venus	2304.0	0.9531	4499.9	108.11	7.586	1.176	48.94	5.428	9.542 E8	4.252	24.90
	2285.9	0.9524	4463.0	108.85	7.616	1.189	48.77	5.473	9.430 E8	4.283	24.91
	2341.9	0.9372	4536.8	147.07	7.524	1.355	41.79	5.384	9.779 E8	4.029	24.86
Earth	2324.7	0.9357	4499.9	149.57	7.552	1.377	41.42	5.428	9.667 E8	4.051	24.87
	2307.5	0.9341	4463.0	152.07	7.580	1.399	41.06	5.473	9.564 E8	4.073	24.87
	2371.7	0.9129	4536.8	206.56	7.477	1.595	35.04	5.384	9.966 E8	3.788	24.83
Mars	2363.9	0.9036	4499.9	227.84	7.489	1.685	33.28	5.428	9.916 E8	3.743	24.82
	2356.1	0.8943	4463.0	249.12	7.501	1.772	31.75	5.473	9.867 E8	3.700	24.81
	2638.6	0.7194	4536.8	740.48	7.088	2.864	17.55	5.384	1.169 E9	2.520	24.66
Jupiter	2639.0	0.7051	4499.9	778.14	7.088	2.947	17.04	5.428	1.170 E9	2.480	24.66
	2639.4	0.6909	4463.0	815.80	7.088	3.030	16.58	5.473	1.170 E9	2.442	24.66

HOHMANN ELLIPSE TRANSFER DATA (Continued)

Body	Mean orbital Semimajor axis (km × 10^{-6})	Eccentricity ε	Aphelion (km × 10^{-6})	Perhelion (km × 10^{-6})	Orbital Velocity (km/sec) Mean	Aphelion	Perhelion	Orbital Velocity of Departure Body (km/sec)	Time of one way trip (½ orbit) (Solar Seconds)	Departure Δ_V orbit (km/sec)	Maximum Δ_V (km/sec)
	2943.2	0.5415	4536.8	1349.5	6.712	3.660	12.31	5.384	1.378 E9	1.723	24.60
Saturn	2963.4	0.5185	4499.9	1427.0	6.689	3.767	11.88	5.428	1.392 E9	1.661	24.59
	2983.7	0.4958	4463.0	1504.5	6.666	3.870	11.48	5.473	1.406 E9	1.602	24.59
	3637.5	0.2472	4536.8	2738.3	6.037	4.690	7.771	5.384	1.893 E9	0.6933	24.54
Uranus	3685.1	0.2211	4499.9	2870.3	5.998	4.790	7.510	5.428	1.930 E9	0.6375	24.54
	3732.7	0.1957	4463.0	3002.3	5.960	4.888	7.266	5.473	1.968 E9	0.5845	24.54
	5919.0	0.2460	7375.0	4463.0	4.733	3.682	6.084	5.473	3.929 E9	0.6112	24.54
Pluto	5204.4	0.1354	5909.0	4499.9	5.047	4.404	5.784	5.428	3.240 E9	0.3557	24.54
	4489.9	−0.1045	4443.0	4536.8	5.434	5.491	5.377	5.384	2.596 E9	−0.0061	24.54
From Pluto to:											
	3710.5	0.9876	7375.0	46.037	5.977	0.4723	75.66	3.676	1.950 E9	3.204	5.955
Mercury	2983.5	0.9806	5909.0	57.948	6.666	0.6601	67.31	4.737	1.406 E9	4.077	6.466
	2256.4	0.9690	4443.0	69.864	7.665	0.9612	67.13	6.103	9.248 E8	5.141	7.185
	3741.2	0.9713	7375.0	107.37	5.953	0.7183	49.34	3.676	1.974 E9	2.958	5.826
Venus	3008.6	0.9641	5909.0	108.11	6.638	0.8979	49.08	4.737	1.424 E9	3.839	6.319
	2275.9	0.9522	4443.0	108.85	7.632	1.195	48.76	6.103	9.368 E8	4.908	7.020
	3761.0	0.9609	7375.0	147.07	5.937	0.8384	42.04	3.676	1.990 E9	2.838	5.766
Earth	3029.3	0.9506	5909.0	149.57	6.615	1.053	41.58	4.737	1.439 E9	3.684	6.226
	2297.5	0.9338	4443.0	152.07	7.596	1.405	41.06	6.103	9.502 E8	4.697	6.874
	3790.8	0.9455	7375.0	206.56	5.914	0.9897	35.34	3.676	2.014 E9	2.687	5.693
Mars	3068.4	0.9257	5909.0	227.84	6.573	1.291	33.47	4.737	1.467 E9	3.446	6.088
	2346.0	0.8938	4443.0	249.12	7.517	1.780	31.75	6.103	9.804 E8	4.323	6.624
	4057.7	0.8175	7375.0	740.48	5.716	1.811	18.04	3.676	2.230 E9	1.865	5.355
Jupiter	3343.6	0.7673	5909.0	778.14	6.297	2.285	17.35	4.737	1.668 E9	2.452	5.586
	2629.4	0.6897	4443.0	815.80	7.101	3.043	16.57	6.103	1.163 E9	3.060	5.879
	4362.3	0.6906	7375.0	1349.5	5.513	2.358	12.89	3.676	2.486 E9	1.318	5.190
Saturn	3668.0	0.6110	5909.0	1427.0	6.012	2.954	12.23	4.737	1.917 E9	1.782	5.326
	2973.7	0.4941	4443.0	1504.5	6.677	3.885	11.47	6.103	1.399 E9	2.217	5.487
	5056.6	0.4585	7375.0	2738.3	5.120	3.120	8.403	3.676	3.102 E9	0.5564	5.050
Uranus	4389.6	0.3461	5909.0	2870.3	5.496	3.830	7.885	4.737	2.509 E9	0.9065	5.101
	3722.7	0.1935	4443.0	3002.3	5.968	4.906	7.260	6.103	1.960 E9	1.197	5.160
	5919.0	0.2460	7375.0	4463.0	4.733	3.682	6.084	3.676	3.929 E9	−5.164	5.019
Neptune	5204.4	0.1354	5909.0	4499.9	5.047	4.404	5.784	4.737	3.240 E9	0.3323	5.030
	4489.9	−0.1045	4443.0	4536.8	5.434	5.491	5.377	6.103	2.596 E9	0.6117	5.056

CONSTANTS FOR SATELLITE GEODESY

Defining Constants

1. Number of ephemeris seconds in 1 tropical year (1900) . . . . s = 31 556 925.9747
2. Gaussian gravitational constant, defining the a. u. . . . . k = 0.017 202 09895

Primary Constants

3. Velocity of light in meters per second (in vacuum) . . . . c = 2.99792458(1.2) × 10^8
4. Dynamical form-factor for Earth . . . . J_2 = 0.001 082 7
5. Sidereal mean motion of Moon in radians per second (1900) . . . . $n_☾^*$ = 2.661 699 489 × 10^{-6}
6. General precession in longitude per tropical century (1900) . . . . p = 5025″.64
7. Constant of nutation (1900) . . . . N = 9″.210

Auxiliary Constants and Factors

$k/86400$, for use when the unit of time is 1 second . . . . k' = 1.990 983 675 × 10^{-7}

Number of seconds of arc in 1 radian . . . . = 206 264.806

Factor for constant of aberration (note 10) . . . . F_1 = 1.000 142

Factor for mean distance of Moon (note 12) . . . . F_2 = 0.999 093 142

Factor for parallactic inequality (note 15) . . . . F_3 = 49853″.2

Derived Constants

8. Solar parallax . . . . $\arcsin(\alpha_\odot/A) = \pi_\odot$ = 8″.79405 (8″.794)
9. Light-time for unit distance . . . . $A/c' = \tau_A$ = 499ˢ.012 = 1ˢ/0.002 003 96
10. Constant of aberration . . . . $F_1 k' \tau_A = \kappa$ = 20″.4958 (20″.496)
11. Ratio of masses of Sun and Earth + Moon . . . . $S/E(1+\mu)$ = 328 912
12. Perturbed mean distance of Moon, in meters . . . . $F_2(GE(1+\mu)/n_☾^{*2})^{1/3} = \alpha_☾$ = 384 400 × 10^3
13. Constant of sine parallax for Moon . . . . $\alpha_\odot/\alpha_☾ = \sin \pi_☾$ = 3422″.451
14. Constant of lunar inequality . . . . $\frac{\mu}{1+\mu}\frac{\alpha_☾}{A} = L$ = 6″.43987 (6″.440)
15. Constant of parallactic inequality . . . . $F_3 \frac{1-\mu}{1+\mu}\frac{\alpha_☾}{A} = P_☾$ = 124″.986

THE EARTH: ITS MASS, DIMENSIONS AND OTHER RELATED QUANTITIES

From NASA Technical Translation NASA TT-F-533

TABLE A

Quantity	Unit of Measurement	Symbol	Numerical Value	Sources; Remarks[1]
Mass	Proportion of the mass of the sun	M	1/331950	I. D. Zhongolovich
	gram		$5.9763 \cdot 10^{27}$	
Major Orbital semi-axis	Astronomical unit	a_{orb}	1.000000	1961 data of Soviet radar determinations
	km		149,457,000	
Distance from sun at perihelion	a.u.	r_{π}	0.983298	for 1962
Distance from sun at aphelion	a.u.	r_{α}	1.016744	for 1962
Moment of perihelion passage		T_{π}	Jan. 2, $4^{hr}52^{m}$	for 1962 USSR Astron. Yearbook for 1962
Moment of aphelion passage		T_{α}	Jul 4 $5^{hr}05^{m}$	for 1962
Siderial rotation period around sun	sec	P_{orb}	$31.558 \cdot 10^{6}$	
Mean rotational velocity	km/sec	U_{orb}	29.8	
Mean equatorial radius	km	$\bar{a}$	6,378.245 6,378.077	A. A. Izotov, 1950 I. D. Zhongolovich, 1956
Mean polar compression		α	1/298.3 1/296.6 1/298.2	A. A. Izotov, 1950 I. D. Zhongolovich, 1952 D. G. King-Hele, R. Merson, 1959. On observations on the movements of artificial earth satellites.
Difference in equatorial and polar semi-axes	km	$a - c$	21.382 21.500	A. A. Izotov, 1950 I. D. Zhongolovich, 1956
Compression of meridian of major equatorial axis		α_a	1/295.2	I. D. Zhongolovich, 1952
Compression of meridian of minor equatorial axis		α_b	1/298.0	I. D. Zhongolovich, 1952
Equatorial compression		ϵ	1/30 000 1/32 000	A. A. Izotov, 1950 I. D. Zhongolovich, 1952

[1] The source does not indicate whether the value given is generally accepted or has merely been calculated by the author, G. N. Katterfel'd.

TABLE A Continued

Quantity	Unit of Measurement	Symbol	Numerical Value	Sources; Remarks[1]
Difference in equatorial semi-axes	m	$a - b$	213 199	A. A. Izotov, 1950 I. D. Zhongolovich, 1952
Meridian of longitude of minor equatorial semi-axis		λ_a	15°E - 6°W	A. A. Izotov, 1950 I. D. Zhongolovich, 1952
Meridian of longitude of minor equatorial axis		λ_b	105°E - 75° W; 84°E - 96°W	A. A. Izotov, 1950 I. D. Zhongolovich, 1952
Difference in polar semi-axes	m	$C_N - C_S$	~ 70 <100	I. D. Zhongolovich, 1952 Based on observations of artificial satellites[1]
Polar asymmetry		η	~1.10^{-5}	
Mean acceleration of gravity at equator		g_e	978,057.3	I. D. Zhongolovich, 1952
Mean acceleration of gravity at poles	milligals (mGal)	g_ρ	983,225.1	
Difference in acceleration of gravity at pole and at equator		$g_p - g_e$	+5,167.8	
Difference in acceleration of gravity at equator	mGal	$g_a - g_b$	+30.2	I. D. Zhongolovich, 1952
Difference in acceleration of gravity at poles		$g_N - g_S$	+30	I. D. Zhongolovich, 1952
Mean acceleration of gravity for entire surface of terrestrial ellipsoid	mGal	g	979,783.0	I. D. Zhongolovich, 1952
Mean radius	km	R	6,370.949	
Area of surface	km^2	S	510.0501 · 10^6	
Volume	km^3	V	1,083.1579 · 10^9	
Mean density	g/cm^3	δ	5.5170	I. D. Zhongolovich, 1952
Siderial rotational period	sec	P	86,164.09	
Angular rotational velocity	rad/sec	ω	7.292116 · 10^{-5}	
Mean equatorial rotational velocity	km/sec	ν	0.465	

TABLE A Continued

Quantity	Unit of Measurement	Symbol	Numerical Value	Sources; Remarks[1]
Ratio of centrifugal force to attractive force at equator		q	$\frac{1}{289}$	
Ratio of centrifugal force to force of gravity at equator		q_c	$0.0034677 = \frac{1}{288}$	I. D. Zhongolovich, 1952
Coefficients characterizing the radial distribution of densities within the earth		κ_1 κ	0.966 0.331	Based on observations of artificial satellites; several larger values were given earlier[1]
Radius of inertia	km; proportion of mean radius	R_i	3,674.735 0.5768	
Geocentric latitude of inertial parallel		ϕ_i	54°47′	
Moment of inertia	$gr \cdot cm^2$	I	$8.070 \cdot 10^{44}$	
Moment of rotation	$gr \cdot cm^2/sec$	L	$5.885 \cdot 10^{40}$	
Relative true secular braking of earth's rotation due to tidal friction		$\frac{\Delta\omega_e}{\omega}$	$-4.2 \cdot 10^{-8}$ per century	
Relative proper secular acceleration of earth's rotation		$\frac{\Delta\omega_i}{\omega}$	$+1.4 \cdot 10^{-8}$ per century	N. N. Pariyskiy, 1955
Relative observed secular braking of earth's rotation		$\frac{\Delta\omega}{\omega}$	$-2.8 \cdot 10^{-8}$ per century	
Mean rotational velocity of terrestrial radius due to abyssal compression	cm/century	$\frac{\Delta R}{\Delta t}$	~5 Assumed invariability of mass (M) 4.5 and distribution of masses (κ)	B. Meyermann, 1928, 1928a N. N. Pariyskiy, 1955
Secular variation in potential gravitational energy of earth accompanying reduction of terrestrial radius by 5 cm and corresponding increase in earth's kinetic energy	erg/century	ΔE	$\sim 17 \cdot 10^{30}$	Assumed uniformity of compression of entire planet. P. N. Kropotkin, 1948 and A. T. Aslanyan, 1955

[1] On the basis of materials published in the Astron. J., Vol. 64, 1272, 1959.

TABLE A Continued

Quantity	Unit of Measurement	Symbol	Numerical Value	Sources; Remarks[1]
Probable value of total energy of tectonic deformation of earth	erg/century	E_t	$\sim 1 \cdot 10^{30}$	With allowance for earthquakes, volcanic eruptions and other forms of tectonic activity P. N. Kropotkin, 1948
Secular loss of heat of earth through radiation into space	erg/century cal/century	$\Delta' E_k$	$1 \cdot 10^{30}$ $2.4 \cdot 10^{22}$	P. N. Kropotkin, 1948
Portion of earth's kinetic energy transformed into heat as a result of lunar and solar tides in the hydrosphere	erg/century cal/century	$\Delta'' E_k$	$0.11 \cdot 10^{30}$ $0.26 \cdot 10^{22}$	Heiskanen, 1922 and de Sitter, 1927
Difference in duration of days in March and August	sec	ΔP	0.0025 (March-Aug.)	N. N. Pariyskiy, 1955
Corresponding relative annual variation in earth's rotational velocity		$\frac{\Delta^* \omega}{\omega}$	$2.9 \cdot 10^{-8}$ (Aug.-March)	N. N. Pariyskiy, 1955
Presumed variation in earth's radius between August and March	cm	$\Delta^* R$	−9.2 (Aug.-March)	
Annual variation in level of world ocean	cm	Δh_0	~10 (Sept.-March)	N. N. Pariyskiy, 1955

The Earth's Lithosphere, Hydrosphere, Atmosphere and Biosphere

TABLE B

Quantity	Unit of Measurement	Symbol	Numerical Value	Source
Area of continents	km²; in % of area of surface of earth	S_C	$149 \cdot 10^6$ 29.2	E. Kossina, 1933
Area of world ocean	km²; in % of area of surface of earth	S_o	$361 \cdot 10^6$ 70.8	E. Kossina, 1933
Mean height of continents above sea level	m	h_C	875	E. Kossina, 1933
Mean depth of world ocean	m	h_o	=3794	Morskoy Atlas, Vol. II, 1953
Mean position of earth's surface with respect to sea level	m	h_m	=2430	E. Kossina, 1933

TABLE B Continued

Quantity	Unit of Measurement	Symbol	Numerical Value	Source
Mean thickness of lithosphere within the limits of the continents	km	$h_{c.l.}$	35	M. Yuing and F. Press, 1955
Mean thickness of lithosphere within the limits of the ocean	km	$h_{o.l.}$	4.7	Kh. Khess, 1955
Mean rate of thickening of continental lithosphere	$m/10^6$ yr	$\frac{\Delta h}{\Delta t}$	10 – 40	V. I. Popov, 1955
Mean rate of horizontal extension of continental lithosphere	$km/10^6$ yr	$\frac{\Delta l}{\Delta t}$	0.75 – 20	V. I. Popov, 1955
Mass of lithosphere	gr	m_l	$2.367 \cdot 10^{25}$	A. Poldervart, 1955
Amount of water released from the mantle and core in the course of geological time	gr		$3.400 \cdot 10^{24}$	Kalp, 1951
Total reserve of water in the mantle	gr		$2 \cdot 10^{26}$	A. P. Vinogradov, 1959
Present day content of free and bound water in the earth's lithosphere	gr		$2.2 - 2.6 \cdot 10^{24}$ $1.8 - 2.7 \cdot 10^{24}$	Kalp, 1951 A. Poldervart, 1955
Mass of hydrosphere	gr	m_h	$1.664 \cdot 10^{24}$	A. Poldervart, 1955
Amount of oxygen bound in the earth's crust	gr		$1.300 \cdot 10^{24}$	A. Poldervart, 1955
Amount of free oxygen	gr		$1.5 \cdot 10^{21}$	A. Poldervart, 1955
Mass of atmosphere	gr	m_a	$5.136 \cdot 10^{21}$	A. Poldervart, 1955
Mass of biosphere	gr	m_b	$1.148 \cdot 10^{19}$	A. Poldervart, 1955
Mass of living matter in the biosphere	gr		$3.6 \cdot 10^{17}$	A. Poldervart, 1955
Density of living matter on dry land	gr/cm^2		0.1	A. Poldervart, 1955
Density of living matter in ocean	gr/cm^3		$15 \cdot 10^{-8}$	A. Poldervart, 1955

ESTIMATED AGE OF EARTH

TABLE C

Object of Study	Age in 10^9 years	Source
Fossils of the most ancient organisms	2.7	L. Arens, 1955
Most ancient known terrestrial rocks:		
Mica (biotite) found within migmatites on the Kola Peninsula and at the Great Rapids of the Voron'ya River in 1958;	3.6	A. A. Polkanov and E. K. Gerling, 1961
Rock found in South Africa	4	A. L. Heils, 1960[1]
Lithosphere	~4	V. I. Baranov, 1958
Earth	~5	A. P. Vinogradov, 1959 and others

[1] See *Priroda*, No. 11, 1960, p. 113.

Approximate Scale of Geologic Time,[1] Based on the Data of Soviet Research, 1960.

TABLE D

Era	Period or Epoch	Beginning and end, in 10^6 years	Approximate Duration, in 10^6 years		
Cenozoic	Quaternian				
	Contemporary	0 – 10,000 yrs ± 2,000 yrs	8 – 12,000 years		
	Pleistocene	10,000 – 1,000,000 yrs ± 50,000 yrs	1		
	Tertiary				
	Pliocene	1 – 10	9	69	
	Miocene	10 – 25	15		
	Oligocene	25 – 40	15		
	Eocene	40 – 60	20		
	Paleocene	60 – 70	10		570
Mesozoic	Cretaceous	70 – 140	70		
	Jurassic	140 – 185	45		
	Triassic	185 – 225	40		
Paleozoic	Permian	225 – 270	45		
	Carboniferous	270 – 320	50		
	Devonian	320 – 400	80		
	Silurian	400 – 420	20		
	Ordovician	420 – 480	60		
	Cambrian	480 – 570	90		
	Pre-Cambrian IV (Riphean)[2]	570 – 1,200	630		
	Pre-Cambrian III (Proterozoic)[3]	1,200 – 1,900	700		
	Pre-Cambrian II (Archean)	1,900 – 2,700	800		
	Pre-Cambrian I (Catarchean)	2,700 – 3,500	800		
	Pregeological era	3,500 – 5,000	1,500		

[1] Based on the geochronological scale of the Commission for Determining the Absolute Age of Geological Formations, published in *Izv. AN SSSR*, Geological series, No. 10, 1960.

[2] Proterozoic II.

[3] Proterozoic I.

ACCELERATION DUE TO GRAVITY AND LENGTH OF THE SECONDS PENDULUM

For Sea Level at Various Latitudes

Based on the formula of the U. S. Coast and Geodetic Survey. The length of the simple pendulum whose period is two seconds, that is which beats seconds, is computed in each case from the corresponding value of the acceleration.

Latitude	Acceleration due to gravity		Length of seconds pendulum	
°	cm/sec.^{2}	ft./sec.^{2}	cm	in.
0	978.039	32.0878	99.0961	39.0141
5	978.078	32.0891	99.1000	39.0157
10	978.195	32.0929	99.1119	39.0204
15	978.384	32.0991	99.1310	39.0279
20	978.641	32.1076	99.1571	39.0382
25	978.960	32.1180	99.1894	39.0509
30	979.329	32.1302	99.2268	39.0656
31	979.407	32.1327		
32	979.487	32.1353		
33	979.569	32.1380		
34	979.652	32.1407		
35	979.737	32.1435	99.2681	39.0819
36	979.822	32.1463		
37	979.908	32.1491		
38	979.995	32.1520		
39	980.083	32.1549		
40	980.171	32.1578	99.3121	39.0992
41	980.261	32.1607		
42	980.350	32.1636		
43	980.440	32.1666		
44	980.531	32.1696		
45	980.621	32.1725	99.3577	39.1171
46	980.711	32.1755		
47	980.802	32.1785		
48	980.892	32.1814		
49	980.981	32.1844		
50	981.071	32.1873	99.4033	39.1351
51	981.159	32.1902		
52	981.247	32.1931		
53	981.336	32.1960		
54	981.422	32.1988		
55	981.507	32.2016	99.4475	39.1525
56	981.592	32.2044		
57	981.675	32.2071		
58	981.757	32.2098		
59	981.839	32.2125		
60	981.918	32.2151	99.4891	39.1689
65	982.288	32.2272	99.5266	39.1836
70	982.608	32.2377	99.5590	39.1964
75	982.868	32.2463	99.5854	39.2068
80	983.059	32.2525	99.6047	39.2144
85	983.178	32.2564	99.6168	39.2191
90	983.217	32.2577	99.6207	39.2207

Free Air Correction for Altitude

−0.0003086 cm/sec.^{2}/m for altitude in meters.
−0.000003086 ft./sec.^{2}/ft. for altitude in feet.

Altitude meters	Correction cm/sec.^{2}	Altitude feet	Correction ft./sec.^{2}
200	−0.0617	200	−0.000617
300	0.0926	300	0.000926
400	0.1234	400	0.001234
500	0.1543	500	0.001543
600	0.1852	600	0.001852
700	0.2160	700	0.002160
800	0.2469	800	0.002469
900	0.2777	900	0.002777

DATA IN REGARD TO THE EARTH

Quadrant of the equator, 10,019,150 meters, 6,225.60 miles.
Quadrant of the meridian, 10,002,290 meters, 6,215.12 miles.
1° latitude at the equator = 69.41 miles.
1° latitude at the pole = 68.70 miles.
Mean surface density of the continents, 2.67 g/cm^3, 166.7 lb./ft.^{3}
Land area, 148.847 × 10^6 km^2, 57.470 × 10^6 sq. mi.
Ocean area, 361.254 × 10^6 km^2, 139.480 × 10^6 sq. mi.
Highest mountain, Everest, 8,840 meters, 29,003 ft.
Greatest sea depth, 10,430 meters, 34,219 ft.
Thermal gradient of the earth, higher at increasing depths, 30°C per km, 48°C per mi. (uncertain).

ELEMENTS IN THE EARTH'S CRUST

A. Demayo

The elements in this table are listed in decreasing average concentration as they occur in the earth's crust. They are average concentrations and there will be variations in composition from point to point throughout the earth's crust.

Element	Concentration (mg/kg)	Element	Concentration (mg/kg)
Oxygen	4.64×10^5	Ytterbium	3.0×10^0
Silicon	2.82×10^5	Beryllium	2.8×10^0
Aluminum	8.32×10^4	Erbium	2.8×10^0
Iron	5.63×10^4	Uranium	2.7×10^0
Calcium	4.15×10^4	Bromine	2.5×10^0
Sodium	2.36×10^4	Tantalum	2.0×10^0
Magnesium	2.33×10^4	Tin	2.0×10^0
Potassium	2.09×10^4	Arsenic	1.8×10^0
Titanium	5.70×10^3	Molybdenum	1.5×10^0
Hydrogen	1.40×10^3	Tungsten (Wolfram)	1.5×10^0
Phosphorus	1.05×10^3	Europium	1.2×10^0
Manganese	9.50×10^2	Holmium	1.2×10^0
Fluorine	6.25×10^2	Cesium	1.0×10^0
Barium	4.25×10^2	Terbium	9×10^{-1}
Strontium	3.75×10^2	Iodine	5×10^{-1}
Sulfur	2.60×10^2	Lutetium	5×10^{-1}
Carbon	2.00×10^2	Thulium	4.8×10^{-1}
Zirconium	1.65×10^2	Thallium	4.5×10^{-1}
Vanadium	1.35×10^2	Antimony	2×10^{-1}
Chlorine	1.30×10^2	Cadmium	2×10^{-1}
Chromium	1.00×10^2	Bismuth	1.7×10^{-1}
Rubidium	9.0×10^1	Indium	1×10^{-1}
Nickel	7.5×10^1	Mercury	8×10^{-2}
Zinc	7.0×10^1	Silver	7×10^{-2}
Cerium	6.0×10^1	Selenium	5×10^{-2}
Copper	5.5×10^1	Palladium	1×10^{-2}
Yttrium	3.3×10^1	Helium	8×10^{-3}
Lanthanum	3.0×10^1	Neon	5×10^{-3}
Neodimium	2.8×10^1	Platinum	5×10^{-3}
Cobalt	2.5×10^1	Rhenium	5×10^{-3}
Scandium	2.2×10^1	Gold	4×10^{-3}
Lithium	2.0×10^1	Osmium	1.5×10^{-3}
Niobium (Columbium)	2.0×10^1	Iridium	1×10^{-3}
Nitrogen	2.0×10^1	Rhodium	1×10^{-3}
Gallium	1.5×10^1	Ruthenium	1×10^{-3}
Lead	1.25×10^1	Tellurium	1×10^{-3}
Boron	1.0×10^1	Krypton	1×10^{-4}
Thallium	9.6×10^0	Xenon	3×10^{-5}
Praeseodymium	8.2×10^0	Protactinium	1.4×10^{-6}
Samarium	6.0×10^0	Radium	9×10^{-7}
Gadolinium	5.4×10^0	Actinium	5.5×10^{-10}
Germanium	5.4×10^0	Polonium	2×10^{-10}
Argon	3.5×10^0	Radon	4×10^{-13}
Dysprosium	3.0×10^0		

CHEMICAL COMPOSITION OF ROCKS

Reprinted from "Sedimentary Rocks" (1948) with the permission of F. J. Pettijohn, author, and Harper Brothers, publishers.

Element	Average igneous rock	Average shale	Average sandstone	Average limestone	Average sediment
SiO_2	59.14	58.10	78.33	5.19	57.95
TiO_2	1.05	0.65	0.25	0.06	0.57
Al_2O_3	15.34	15.40	4.77	0.81	13.39
Fe_2O_3	3.08	4.02	1.07	0.54	3.47
FeO	3.80	2.45	0.30		2.08
MgO	3.49	2.44	1.16	7.89	2.65
CaO	5.08	3.11	5.50	42.57	5.89
Na_2O	3.84	1.30	0.45	0.05	1.13
K_2O	3.13	3.24	1.31	0.33	2.86
H_2O	1.15	5.00	1.63	0.77	3.23
P_2O_5	0.30	0.17	0.08	0.04	0.13
CO_2	0.10	2.63	5.03	41.54	5.38
SO_3		0.64	0.07	0.05	0.54
BaO	0.06	0.05	0.05		
C		0.80			0.66
	99.56	100.00	100.00	99.84	99.93

SOLAR SPECTRAL IRRADIANCE

From NASA Technical Report R-351, "The Solar Constant and the Solar Spectrum Measured from a Research Aircraft". Report edited by Matthew P. Thekaekara, Goddard Space Flight Center, Greenbelt, Maryland 20771. Discussion of previously reported values and of measurements and calculations leading to the following table are contained in the NASA Technical Report R-351.

SOLAR SPECTRAL IRRADIANCE—PROPOSED STANDARD CURVE

λ —Wavelength in microns
P_λ —Solar spectral irradiance averaged over small bandwidth centered at λ, in Watts $cm^{-2}\mu^{-1}$ *
A_λ—Area under the solar spectral irradiance curve in the wavelength range 0 to λ, mW cm^{-2}
D_λ—Percentage of the solar constant associated with wavelengths shorter than λ
Solar Constant—135.30 mW cm^{-2}, or 1.940 cal min^{-1} cm^{-2} **

λ	P_λ	A_λ	D_λ	λ	P_λ	A_λ	D_λ	λ	P_λ	A_λ	D_λ
0.120	0.000010	0.00059992	0.00044	0.475	0.2044	25.6001	18.921	2.4	0.0064	129.695	95.858
0.140	0.000003	0.00072999	0.00053	0.480	0.2074	26.6296	19.681	2.5	0.0054	130.285	96.294
0.150	0.000007	0.00077999	0.00057	0.485	0.1976	27.6421	20.430	2.6	0.0048	130.795	96.671
0.160	0.000023	0.00092999	0.00068	0.490	0.1950	28.6236	21.155	2.7	0.0043	131.250	97.007
0.170	0.000063	0.00135999	0.00100	0.495	0.1960	29.6011	21.878	2.8	0.00390	131.660	97.3103
0.180	0.000125	0.00229999	0.00169	0.500	0.1942	30.5766	22.599	2.9	0.00350	132.030	97.5838
0.190	0.000271	0.00427999	0.00316	0.505	0.1920	31.5421	23.312	3.0	0.00310	132.360	97.8277
0.200	0.00107	0.010984	0.0081	0.510	0.1882	32.4926	24.015	3.1	0.00260	132.645	98.0383
0.210	0.00229	0.027784	0.0205	0.515	0.1833	33.4214	24.701	3.2	0.00226	132.888	98.2179
0.220	0.00575	0.067984	0.0502	0.520	0.1833	34.3379	25.379	3.3	0.00192	133.097	98.3724
0.225	0.00649	0.098584	0.0728	0.525	0.1852	35.2591	26.059	3.4	0.00166	133.276	98.5047
0.230	0.00667	0.131484	0.0971	0.530	0.1842	36.1826	26.742	3.5	0.00146	133.432	98.6200
0.235	0.00593	0.162984	0.1204	0.535	0.1818	37.0976	27.418	3.6	0.00135	133.573	98.7238
0.240	0.00630	0.193559	0.1430	0.540	0.1783	37.9979	28.084	3.7	0.00123	133.702	98.8192
0.245	0.00723	0.227384	0.1680	0.545	0.1754	38.8821	28.737	3.8	0.00111	133.819	98.9056
0.250	0.00704	0.263059	0.1944	0.550	0.1725	39.7519	29.380	3.9	0.00103	133.926	98.9847
0.255	0.0104	0.306659	0.226	0.555	0.1720	40.6131	30.017	4.0	0.00095	134.025	99.0579
0.260	0.0130	0.365159	0.269	0.560	0.1695	41.4669	30.648	4.1	0.00087	134.116	99.1252
0.265	0.0185	0.443909	0.328	0.565	0.1705	42.3169	31.276	4.2	0.00078	134.198	99.1861
0.270	0.0232	0.548159	0.405	0.570	0.1712	43.1711	31.907	4.3	0.00071	134.273	99.2412
0.275	0.0204	0.657159	0.485	0.575	0.1719	44.0289	32.541	4.4	0.00065	134.341	99.2915
0.280	0.0222	0.763659	0.564	0.580	0.1715	44.8874	33.176	4.5	0.00059	134.403	99.3373
0.285	0.0315	0.897909	0.663	0.585	0.1712	45.7441	33.809	4.6	0.00053	134.459	99.3787
0.290	0.0482	1.09715	0.810	0.590	0.1700	46.5971	34.439	4.7	0.00048	134.509	99.4160
0.295	0.0584	1.36365	1.007	0.595	0.1682	47.4426	35.064	4.8	0.00045	134.556	99.4504
0.300	0.0514	1.63815	1.210	0.600	0.1666	48.2796	35.683	4.9	0.00041	134.599	99.482195
0.305	0.0603	1.91740	1.417	0.605	0.1647	49.1079	36.295	5.0	0.0003830	134.63905	99.511500
0.310	0.0689	2.24040	1.655	0.610	0.1635	49.9284	36.902	6.0	0.0001750	134.91805	99.717708
0.315	0.0764	2.60365	1.924	0.620	0.1602	51.5469	38.098	7.0	0.0000990	135.05505	99.818965
0.320	0.0830	3.00215	2.218	0.630	0.1570	53.1329	39.270	8.0	0.0000600	135.13455	99.877723
0.325	0.0975	3.45340	2.552	0.640	0.1544	54.6899	40.421	9.0	0.0000380	135.18355	99.913939
0.330	0.1059	3.96190	2.928	0.650	0.1511	56.2174	41.550	10.0	0.0000250	135.21505	99.937220
0.335	0.1081	4.49690	3.323	0.660	0.1486	57.7159	42.657	11.0	0.0000170	135.23605	99.952742
0.340	0.1074	5.03565	3.721	0.670	0.1456	59.1869	43.744	12.0	0.0000120	135.25055	99.963458
0.345	0.1069	5.57140	4.117	0.680	0.1427	60.6284	44.810	13.0	0.0000087	135.26090	99.971108
0.350	0.1093	6.11190	4.517	0.690	0.1402	62.0429	45.855	14.0	0.0000055	135.26800	99.976356
0.355	0.1083	6.65590	4.919	0.700	0.1369	63.4284	46.879	15.0	0.0000049	135.27320	99.980199
0.360	0.1068	7.19365	5.316	0.710	0.1344	64.7849	47.882	16.0	0.0000038	135.27755	99.983414
0.365	0.1132	7.74365	5.723	0.720	0.1314	66.1139	48.864	17.0	0.0000031	135.28100	99.985964
0.370	0.1181	8.32190	6.150	0.730	0.1290	67.4159	49.826	18.0	0.0000024	135.28375	99.987997
0.375	0.1157	8.90640	6.582	0.740	0.1260	68.6909	50.769	19.0	0.0000020	135.28595	99.989623
0.380	0.1120	9.47565	7.003	0.750	0.1235	69.9384	51.691	20.0	0.0000016	135.28775	99.990953
0.385	0.1098	10.0301	7.413	0.800	0.1107	75.7934	56.018	25.0	0.000000610	135.29328	99.995036
0.390	0.1098	10.5791	7.819	0.850	0.0988	81.0309	59.889	30.0	0.000000300	135.29555	99.996718
0.395	0.1189	11.1509	8.241	0.900	0.0889	85.7234	63.358	35.0	0.000000160	135.29670	99.997568
0.400	0.1429	11.8054	8.725	0.950	0.0835	90.0334	66.543	40.0	0.000000094	135.29734	99.998037
0.405	0.1644	12.5736	9.293	1.000	0.0746	93.9859	69.464	50.0	0.000000038	135.29800	99.998525
0.410	0.1751	13.4224	9.920	1.100	0.0592	100.675	74.409	60.0	0.000000019	135.29828	99.998736
0.415	0.1774	14.3036	10.571	1.200	0.0484	106.055	78.385	80.0	0.000000007	135.29854	99.998928
0.420	0.1747	15.1839	11.222	1.300	0.0396	110.455	81.637	100.0	0,000000003	135.29864	99.999002
0.425	0.1693	16.0439	11.858	1.400	0.0336	114.115	84.342	1000.0	0	135.30000	100.000000
0.430	0.1639	16.8769	12.473	1.500	0.0287	117.230	86.645				
0.435	0.1663	17.7024	13.083	1.600	0.0244	119.885	88.607				
0.440	0.1810	18.5706	13.725	1.700	0.0202	122.115	90.255				
0.445	0.1922	19.5036	14.415	1.800	0.0159	123.920	91.589				
0.450	0.2006	20.4856	15.140	1.900	0.0126	125.345	92.642				
0.455	0.2057	21.5014	15.891	2.000	0.0103	126.490	93.489				
0.460	0.2066	22.5321	16.653	2.100	0.0090	127.455	94.202				
0.465	0.2048	23.5606	17.413	2.200	0.0079	128.300	94.826				
0.470	0.2033	24.5809	18.167	2.300	0.0068	129.035	95.370				

*The variable $P_\lambda(\lambda)$, expressed in the units Watts $cm^{-2}\mu^{-1}$, is related to $P_\nu(\nu)$, expressed in the units Watts cm^{-2} Hz^{-1}, through the relation $P_\nu(\nu) = (d\lambda/d\nu)P_\lambda(\lambda) = (c/\nu^2)P_\lambda(\lambda)$ where c is the speed of light (2.998×10^{14} μ/sec) and ν is the frequency. More explicitly, $P_\nu(\nu) = (2.998 \times 10^{14})/\nu^2 \; P_\lambda(c/\nu)$.

The conversion of $P_\nu(\nu)$ or $P_\lambda(\lambda)$ to *Spectral Energy Density* is given by $\mu(\nu)$[joule cm^{-3} Hz^{-1}] = $(1/c)P_\nu(\nu)$ and $\mu(\lambda)$[joule cm^{-3} μ^{-1}] = $(1/c)P_\lambda(\lambda)$ where c now has the value, 2.998×10^{10} cm sec^{-1}.

** Different agencies have cited different values for the solar constant. For example, the Smithsonian Institution for a number of decades listed the best value as 1.934 cal/cm^2/min. Later measurements from rockets of the intensity of the ultraviolet end of the spectrum resulted in the value being revised to 2.00 cal/cm^2/min with a probable error of ±2%.

TOTAL MONTHLY SOLAR RADIATION IN A CLOUDLESS SKY

Total radiation is the sum of the direct and scattered radiation that strikes the earth's surface. It is influenced by the degree of cloudiness, atmospheric transparency, duration of sunshine, and height of elevation at which the measurements are taken. Deviations from values in this table are bound to occur; however, it is believed that the deviations do not exceed ±10% for the summer months and ±15% for the winter months. The radiation units in this table are kcal/cm².

$\varphi°$	January	February	March	April	May	June	July	August	September	October	November	December
90 N	0	0	0.1	10.0	21.9	26.0	23.8	12.9	2.4	0	0	0
85	0	0	0.7	10.2	21.8	25.8	23.4	13.1	3.0	0	0	0
80	0	0	2.4	10.8	21.4	25.2	23.0	13.4	4.3	0.5	0	0
75	0	0.5	4.0	11.7	21.0	24.5	22.2	13.8	5.8	1.3	0	0
70	0	1.6	6.0	13.1	20.5	23.6	21.2	14.6	7.5	2.7	0.5	0
65	0.7	2.8	8.0	14.5	20.1	22.8	21.0	15.6	9.5	4.3	1.4	0.2
60	1.8	4.3	9.9	16.0	20.8	22.9	21.4	16.7	11.3	6.1	2.6	1.1
55	3.1	6.2	11.7	17.3	21.4	23.4	21.9	17.9	12.9	7.8	4.0	2.3
50	4.8	8.2	13.3	18.5	22.2	23.7	22.6	19.1	14.4	9.7	5.8	3.9
45	6.7	10.3	14.8	19.5	22.6	23.9	23.2	20.1	15.8	11.5	7.8	5.9
40	8.8	12.2	16.4	20.3	23.0	24.0	23.4	20.9	17.0	13.2	9.7	7.7
35	10.7	14.0	17.6	21.0	23.0	24.0	23.6	21.6	18.1	14.7	11.4	9.7
30	12.5	15.5	18.6	21.4	23.0	23.8	23.4	21.8	19.1	16.1	13.1	11.5
25	14.1	16.8	19.5	21.6	23.0	23.4	23.1	21.8	19.8	17.4	14.6	13.1
20	15.5	17.9	20.2	21.6	22.5	22.8	22.6	21.6	20.4	18.5	16.1	14.7
15	16.9	19.0	20.8	21.4	21.9	22.0	21.9	21.2	20.9	19.3	17.4	16.1
10	18.1	19.8	21.1	21.2	21.2	21.0	21.1	20.6	21.2	20.1	18.5	17.5
5	19.3	20.4	21.4	21.0	20.2	19.9	20.0	20.0	21.3	20.6	19.5	18.8
0	20.2	20.9	21.5	20.4	19.3	18.8	19.1	19.3	21.2	21.2	20.4	19.2
5S	20.1	21.4	21.4	19.9	18.3	17.6	17.9	18.3	20.8	21.4	21.8	21.0
10	22.0	21.8	21.1	19.2	17.7	16.3	16.3	17.3	20.4	21.4	21.8	22.0
15	22.6	22.0	20.6	18.3	16.0	14.9	15.4	16.1	19.7	21.3	22.4	22.9
20	23.2	22.0	20.0	17.2	14.7	13.4	13.8	14.9	18.9	21.0	22.6	23.6
25	23.6	22.0	19.4	16.0	13.0	12.0	12.6	13.6	18.0	20.6	23.0	24.1
30	23.9	21.8	18.6	14.9	11.9	10.6	11.1	12.1	17.0	20.1	23.0	24.6
35	24.0	21.3	17.6	13.6	10.4	9.0	9.6	10.6	15.9	19.5	23.0	25.0
40	24.0	20.6	16.4	12.2	8.7	7.3	8.1	9.0	14.3	18.7	22.8	25.2
45	24.0	19.9	15.2	10.7	7.1	5.5	6.3	7.3	13.4	17.7	22.4	25.2
50	23.6	18.9	13.0	9.2	5.4	3.8	4.6	5.5	12.0	16.6	21.8	25.0
55	23.2	17.8	12.3	7.5	3.8	2.3	3.0	3.8	10.3	15.4	21.2	24.6
60	22.6	16.6	10.8	5.6	2.4	1.0	1.6	2.3	8.5	14.1	21.0	24.4
65	22.4	15.3	9.1	3.9	1.1	0.1	0.4	1.0	6.7	12.6	20.8	24.5
70	22.6	14.2	7.3	2.2	0.1	0	0	0	5.0	11.4	21.0	24.9
75	23.2	13.4	5.7	0.9	0	0	0	0	3.5	10.4	21.2	25.4
80	24.0	12.8	4.3	0	0	0	0	0	2.1	9.7	21.9	26.0
85	24.6	12.4	2.9	0	0	0	0	0	0.9	9.2	22.4	26.6
90	24.9	12.3	1.7	0	0	0	0	0	0	9.0	22.6	27.0

ELEMENTS IN SEA WATER

A. Demayo

The elements in sea water are listed in decreasing average concentration. These are average concentrations and there will be variations in composition as a function of the location where the sample was collected.

Element	Concentration (mg/ℓ)	Element	Concentration (mg/ℓ)	Element	Concentration (mg/ℓ)
Oxygen	8.57×10^{5}	Iron	1×10^{-2}	Selenium	9×10^{-5}
Hydrogen	1.08×10^{5}	Indium	$<2 \times 10^{-2}$	Germanium	7×10^{-5}
Chlorine	1.90×10^{4}	Molybdenum	1×10^{-2}	Xeon	5.2×10^{-5}
Sodium	1.05×10^{4}	Zinc	1×10^{-2}	Chromium	5×10^{-5}
Magnesium	1.35×10^{3}	Nickel	5.4×10^{-3}	Thorium	5×10^{-5}
Sulfur	8.85×10^{2}	Arsenic	3×10^{-3}	Gallium	3×10^{-5}
Calcium	4.00×10^{2}	Copper	3×10^{-3}	Mercury	3×10^{-5}
Potassium	3.80×10^{2}	Tin	3×10^{-3}	Lead	3×10^{-5}
Bromine	6.5×10^{1}	Uranium	3×10^{-3}	Zirconium	2.2×10^{-5}
Carbon	2.8×10^{1}	Krypton	2.5×10^{-3}	Bismuth	1.7×10^{-5}
Strontium	8.1×10^{0}	Manganese	2×10^{-3}	Lanthanum	1.2×10^{-5}
Boron	4.6×10^{0}	Vanadium	2×10^{-3}	Gold	1.1×10^{-5}
Silicon	3×10^{0}	Titanium	1×10^{-3}	Niobium	1×10^{-5}
Fluorine	1.3×10^{0}	Cesium	5×10^{-4}	Thallium	$<1 \times 10^{-5}$
Argon	6×10^{-1}	Cerium	4×10^{-4}	Hafnium	$<8 \times 10^{-6}$
Nitrogen	5×10^{-1}	Antimony	3.3×10^{-4}	Helium	6.9×10^{-6}
Lithium	1.8×10^{-1}	Silver	3×10^{-4}	Selenium	$<4 \times 10^{-6}$
Rubidium	1.2×10^{-1}	Yttrium	3×10^{-4}	Tantalum	$<2.5 \times 10^{-6}$
Phosphorus	7×10^{-2}	Cobalt	2.7×10^{-4}	Beryllium	6×10^{-7}
Iodine	6×10^{-2}	Neon	1.4×10^{-4}	Protoactinium	2×10^{-9}
Barium	3×10^{-2}	Cadmium	1.1×10^{-4}	Radium	6×10^{-11}
Aluminum	1×10^{-2}	Tungsten	1×10^{-4}	Radon	6×10^{-16}

THE pH OF NATURAL MEDIA AND ITS RELATION TO THE PRECIPITATION OF HYDROXIDES

Reprinted from "Principles of Geochemistry" (1952) with the permission of Brian Mason, author, and John Wiley and Sons, publishers.

pH	Precipitation of hydroxides	Natural media	pH
11 }	Magnesium		11
10 }		Alkali soils	10
9 }	Bivalent manganese		{ 9
8 }		Seawater	{ 8
7	Bivalent iron	River water	7
6	Zinc copper	Rain water	6
5	Aluminum		5
4		Peat water	4
3 }	Trivalent iron	Mine waters	3
2 }		Acid thermal springs	{ 2
1			{ 1

PROPERTIES OF THE EARTH'S ATMOSPHERE AT ELEVATIONS UP TO 160 KILOMETERS

The average atmosphere up to 160 km based on pressure and density data obtained on rocket flights above White Sands, New Mexico.
Havens, Koll, and LaGow, Journal of Geophysical Research, March, 1952.

Altitude km above sea level	Pressure mm Hg	Density gm/meters³	Temperatures °K (N_2, O_2) M = 29	Temperatures °K (N_2, O) M = 24	Velocity of Sound m. sec	Mean Free Path cm (N_2)
0	760	1220	290		345	6.5×10^{-6}
10	210	425	230		310	1.9×10^{-5}
20	42	92	210		295	8.6×10^{-5}
30	9.5	19	235		315	4.2×10^{-4}
40	2.4	4.3	260		325	1.8×10^{-3}
50	7.5×10^{-1}	1.3	270		330	6.1×10^{-3}
60	2.1×10^{-1}	3.8×10^{-1}	260		325	2.1×10^{-2}
70	5.4×10^{-2}	1.2×10^{-1}	210		295	6.6×10^{-1}
80	1.0×10^{2}	2.5×10^{-2}	190		280	3.2×10^{-1}
90	1.9×10^{-3}	4.0×10^{-3}	210		295	2.0
100	4.2×10^{-4}	8.0×10^{-4}	240		315	10.0
110	1.2×10^{-4}	2.0×10^{-4}	270	220	330	40.0
120	3.5×10^{-5}	5.0×10^{-5}	330	270	370	1.5×10^{2}
130	1.5×10^{-5}	2.0×10^{-5}	390	320	400	4.0×10^{2}
140	7×10^{-6}	7.0×10^{-6}	450	370	430	1.0×10^{3}
150	3×10^{-6}	3.0×10^{-6}	510	420	460	2.5×10^{3}
160	2×10^{-6}	1.5×10^{-6}	570	470	480	5.0×10^{3}

VELOCITY OF SEISMIC WAVES

Depth km	Longitudinal or condensational km/sec.	Transverse or distortional km/sec.
0–20	5.4 –5.6	3.2
20–45	6.25–6.75	3.5
1300	12.5	6.9
2400	13.5	7.5

ATMOSPHERIC AND METEOROLOGICAL DATA

Total mass of the atmosphere, estimated by Ekholm, 5.2×10^{21} g, 11.4×10^{18} pounds, 5.70×10^{15} tons.
Evidence of extent: twilight, 63 km, 39 mi.: meteors, 200 km, 124 mi.: aurora 44–360 km, 27–224 mi.

*Distance to Earth.

U.S. STANDARD ATMOSPHERE (1976)

From, "U.S. Standard Atmosphere, 1976", National Oceanic and Atmospheric Administration, National Aeronautics and Space Administration and the United States Air Force, 1976. The above referenced book is available from the Superintendent of Documents, U.S. Government Printing Office, Washington, D.C. 20402. The book contains considerably more extensive tables than those presented below plus the development of the equations used in the calculations of the tables as well as a discussion of the bases for selection of constants used in the equations.

The U.S. Standard Atmosphere, 1976 is an idealized, steady-state representation of the earth's atmosphere from the surface to 1000 km, as it is assumed to exist in a period of moderate solar activity. The air is assumed to be dry, and at heights sufficiently below 86 km, the atmosphere is assumed to be homogeneously mixed with a relative-volume composition leading to a mean molecular weight M. The molecular weights and assumed fractional-volume composition of sea-level dry air were

Gas species	Molecular weight M_i (kg/kmol)	Fractional volume F_i (dimensionless)
N_2	28.0134	0.78084
O_2	31.9988	0.209476
Ar	39.948	0.00934
CO_2	44.00995	0.000314
Ne	20.183	0.00001818
He	4.0026	0.00000524
Kr	83.80	0.00000114
Xe	131.30	0.000000087
CH_4	16.04303	0.000002
H_2	2.01594	0.0000005

SYMBOLS, ABBREVIATIONS, AND UNITS RELATING TO FOLLOWING TABLE

C_s	Speed of sound expressed as meters per second.
g	Acceleration due to gravity expressed as meters per second.
H	Geopotential altitude expressed in meters. The unit of measurement of geopotential is the standard geopotential meter (m′) which represents the work done by lifting a unit mass 1 geometric meter, through a region in hich the acceleration of gravity is uniformly 9.80665 m/s^2. The geopotential at any point with respect to sea level (assumed zero potential), expressed in geopotential meters, is called geopotential altitude.
H_p	Pressure scale height. The quantity, $H_pRT/(g \cdot m)$, which has the dimensions of length, is the quantity commonly associated with the concept of scale height, and is the defining form of pressure scale height used in this model.
K	Kelvin, the thermodynamic temperature.
L	Mean free path. This is the mean value of the distances traveled by each of the neutral particles in a selected volume between successive collisions with other particles in the volume.
m	Meter, the unit of length.
M	Molecular weight expressed as kg/kmol or lb/lbmol.
n	The number of neutral atmospheric gas particles per cubic meter of the atmosphere at the geometric height Z or the geopotential height H.
N	Newton, a force having the units $kg \cdot m/s^2$.
P	Pressure, expressed as millibars ($100N/m^2$).
R	The gas constant, $R = 8.31432 \times 10^3\ N \cdot m/(kmol \cdot K)$.
s	Seconds, time.
T	Temperature, degrees Kelvin.
V	Mean air particle speed, expressed as meters per second, is the arithmetic average of the speeds of all air particles in the volume element being considered.
Z	Geometric altitude expressed in kilometers.
η	Kinematic viscosity in the units m^2/s.
κ	Thermal conductivity in the units $J/m \cdot s \cdot K$.
μ	Dynamic viscosity in the units $N \cdot s/m^2$.
ν	Mean collision frequency. Average speed of air particles within a selected volume divided by the mean free path of the particles in that volume. Thus $\nu = V/L$.
ϱ	Density expressed as kg/m^2.
r	Mean radius of earth at equator, 6,370,949 m.

Some equations useful for computing certain values of the Standard Atmosphere are:

$-1{,}524 \leqslant H < 11{,}000$ m

$$T = 288.15 - 0.006500\,H$$
$$P = 1013.25(288.15/T)^{-5.255877}$$
$$H = 44{,}331.514 - 11{,}880.516\,P^{0.1902632}$$

$11{,}000 \leqslant H < 20{,}000$ m

$$T = 216.650$$
$$P = 226.32\,e^{-0.000156768832(H - 11{,}000.00)}$$
$$H = 45{,}383.967 - 6{,}341.6237\,\ln P$$

$20{,}000 \leqslant H < 32{,}000$ m

$$T = 216.650 + 0.00100(H - 19{,}999.997)$$
$$P = 54.7487\,(216.650/T)^{34.16319}$$
$$H = -196{,}650.0 + 243{,}580.85/P^{0.0292713}$$

$-1{,}524 \leqslant H < 32{,}000$ m

$$\varrho = 0.3483677\,(P/T)$$
$$H = rZ/(r + Z)$$
$$Z = rH/(r - H)$$

Altitude		Acceleration due to gravity	Pressure scale height	Number density	Particle speed	Collision frequency	Mean free path	Molecular weight	Temperature	Pressure	Density	Sound speed	Dynamic viscosity	Kinematic viscosity	Thermal conductivity
Z(m)	H(m)	g(m/s²)	H_p(m)	n(m⁻³)	V (m/s)	ν (s⁻¹)	L (m)	M (kg/kmol)	T(K)	P(mb)	ρ(kg/m³)	C_s(m/s)	μ (N·s/m²)	η (m²/s)	(J/m·s·K)
−5000	−5004	9.8221	9371.8	4.0151 (+25)	484.15	1.1506 (+10)	4.2078 (−8)	28.964	320.676	1.7776 (+3)	1.9311 (+0)	358.99	1.9422 (−5)	1.0058 (−5)	2.7882 (−5)
−4500	−4503	9.8295	9278.2	3.8445 (+25)	481.69	1.0961 (+10)	4.3945 (−8)	28.964	317.421	1.6848 (+3)	1.8491 (+0)	357.16	1.9273 (−5)	1.0423 (−5)	2.7634 (−5)
−4000	−4003	9.8190	9184.5	3.6795 (+25)	479.22	1.0437 (+10)	4.5915 (−8)	28.964	314.166	1.5959 (+3)	1.7697 (+0)	355.32	1.9123 (−5)	1.0806 (−5)	2.7384 (−5)
−3500	−3502	9.8175	9090.8	3.5201 (+25)	476.73	9.9328 (+9)	4.7995 (−8)	28.964	310.913	1.5109 (+3)	1.6930 (+0)	353.48	1.8972 (−5)	1.1206 (−5)	2.7134 (−5)
−3000	−3001	9.8159	8997.1	3.3660 (+25)	474.23	9.4481 (+9)	5.0193 (−8)	28.964	307.659	1.4297 (+3)	1.6189 (+0)	351.63	1.8820 (−5)	1.1625 (−5)	2.6884 (−5)
−2500	−2501	9.8144	8903.4	3.2171 (+25)	471.71	8.9824 (+9)	5.2515 (−8)	28.964	304.406	1.3520 (+3)	1.5473 (+0)	349.76	1.8668 (−5)	1.2065 (−5)	2.6632 (−5)
−2000	−2001	9.8128	8809.6	3.1017 (+25)	469.69	8.6231 (+9)	5.4469 (−8)	28.964	301.154	1.2778 (+3)	1.4782 (+0)	347.89	1.8515 (−5)	1.2525 (−5)	2.6380 (−5)
−1500	−1500	9.8113	8715.9	2.9346 (+25)	466.65	8.1056 (+9)	5.7571 (−8)	28.964	297.902	1.2069 (+3)	1.4114 (+0)	346.00	1.8361 (−5)	1.3009 (−5)	2.6126 (−5)
−1000	−1000	9.8097	8622.1	2.8007 (+25)	464.09	7.6934 (+9)	6.0324 (−8)	28.964	294.651	1.1393 (+3)	1.3470 (+0)	344.11	1.8206 (−5)	1.3516 (−5)	2.5872 (−5)
−500	−500	9.8082	8528.3	2.6715 (+25)	461.53	7.2980 (+9)	6.3240 (−8)	28.964	291.400	1.0747 (+3)	1.2849 (+0)	342.21	1.8050 (−5)	1.4048 (−5)	2.5618 (−5)
0	0	9.8066	8434.5	2.5470 (+25)	458.94	6.9189 (+9)	6.6332 (−8)	28.964	288.150	1.01325 (+3)	1.2250 (+0)	340.29	1.7894 (−5)	1.4607 (−5)	2.5326 (−5)
500	500	9.8051	8340.7	2.4269 (+25)	456.35	6.5555 (+9)	6.9613 (−8)	28.964	284.900	9.5461 (+2)	1.1673 (+0)	338.37	1.7737 (−5)	1.5195 (−5)	2.5106 (−5)
1000	1000	9.8036	8246.9	2.3113 (+25)	453.74	6.2075 (+9)	7.3095 (−8)	28.964	281.651	8.9876 (+2)	1.1117 (+0)	336.43	1.7579 (−5)	1.5813 (−5)	2.4849 (−5)
1500	1500	9.8020	8153.0	2.2000 (+25)	451.12	5.8743 (+9)	7.6795 (−8)	28.964	278.402	8.4559 (+2)	1.0581 (+0)	334.49	1.7420 (−5)	1.6463 (−5)	2.4591 (−5)
2000	1999	9.8005	8059.2	2.0928 (+25)	448.48	5.5554 (+9)	8.0728 (−8)	28.964	275.154	7.9501 (+2)	1.0066 (+0)	332.53	1.7260 (−5)	1.7147 (−5)	2.4333 (−5)
2500	2499	9.7989	7965.3	1.9897 (+25)	445.82	5.2504 (+9)	8.4912 (−8)	28.964	271.906	7.4691 (+2)	9.5695 (−1)	330.56	1.7099 (−5)	1.7868 (−5)	2.4073 (−5)
3000	2999	9.7974	7871.4	1.8905 (+25)	443.15	4.9588 (+9)	8.9367 (−8)	28.964	268.659	7.0121 (+2)	9.0925 (−1)	328.58	1.6938 (−5)	1.8628 (−5)	2.3813 (−5)
3500	3498	9.7959	7777.5	1.7952 (+25)	440.47	4.6802 (+9)	9.4113 (−8)	28.964	265.413	6.5780 (+2)	8.6340 (−1)	326.59	1.6775 (−5)	1.9429 (−5)	2.3552 (−5)
4000	3997	9.7943	7683.6	1.7036 (+25)	437.76	4.4141 (+9)	9.9173 (−8)	28.964	262.166	6.1660 (+2)	8.1935 (−1)	324.59	1.6612 (−5)	2.0275 (−5)	2.3290 (−5)
4500	4497	9.7928	7589.7	1.6156 (+25)	435.05	4.1602 (+9)	1.0457 (−7)	28.964	258.921	5.7752 (+2)	7.7704 (−1)	322.57	1.6448 (−5)	2.1167 (−5)	2.3028 (−5)
5000	4996	9.7912	7495.7	1.5312 (+25)	432.31	3.9180 (+9)	1.1034 (−7)	28.964	255.676	5.4048 (+2)	7.3643 (−1)	320.55	1.6282 (−5)	2.2110 (−5)	2.2765 (−5)
5500	5495	9.7897	7401.8	1.4502 (+25)	429.56	3.6871 (+9)	1.1650 (−7)	28.964	252.431	5.0539 (+2)	6.9747 (−1)	318.50	1.6116 (−5)	2.3107 (−5)	2.2500 (−5)
6000	5994	9.7882	7307.8	1.3725 (+25)	426.79	3.4671 (+9)	1.2310 (−7)	28.964	249.187	4.7217 (+2)	6.6011 (−1)	316.45	1.5949 (−5)	2.4161 (−5)	2.2236 (−5)
6500	6493	9.7866	7213.8	1.2980 (+25)	424.00	3.2577 (+9)	1.3016 (−7)	28.964	245.943	4.4075 (+2)	6.6431 (−1)	314.39	1.5781 (−5)	2.5278 (−5)	2.1970 (−5)
7000	6992	9.7851	7119.8	1.2267 (+25)	421.20	3.0584 (+9)	1.3772 (−7)	28.964	242.700	4.1105 (+2)	5.9002 (−1)	312.31	1.5612 (−5)	2.6461 (−5)	2.1703 (−5)
7500	7491	9.7836	7025.8	1.1585 (+25)	418.37	2.8689 (+9)	1.4583 (−7)	28.964	239.457	3.8299 (+2)	5.5719 (−1)	310.21	1.5442 (−5)	2.7714 (−5)	2.1436 (−5)
8000	7990	9.7820	6931.7	1.0932 (+25)	415.53	2.6888 (+9)	1.5454 (−7)	28.964	236.215	3.5651 (+2)	5.2579 (−1)	308.11	1.5271 (−5)	2.9044 (−5)	2.1168 (−5)
8500	8489	9.7805	6837.7	1.0308 (+25)	412.67	2.5178 (+9)	1.6390 (−7)	28.964	232.974	3.3154 (+2)	4.9576 (−1)	305.98	1.5099 (−5)	3.0457 (−5)	2.0899 (−5)
9000	8987	9.7789	6743.6	9.7110 (+24)	409.79	2.3555 (+9)	1.7397 (−7)	28.964	229.733	3.0800 (+2)	4.6706 (−1)	303.85	1.4926 (−5)	3.1957 (−5)	2.0630 (−5)
9500	9486	9.7774	6649.5	9.1413 (+24)	406.89	2.2016 (+9)	1.8482 (−7)	28.964	226.492	2.8584 (+2)	4.3966 (−1)	301.70	1.4752 (−5)	3.3553 (−5)	2.0359 (−5)
10000	9984	9.7759	6555.4	8.5976 (+24)	403.97	2.0558 (+9)	1.9651 (−7)	28.964	223.252	2.6499 (+2)	4.1351 (−1)	299.53	1.4577 (−5)	3.5251 (−5)	2.0088 (−5)
10500	10483	9.7743	6461.3	8.0790 (+24)	401.03	1.9177 (+9)	2.0912 (−7)	28.964	220.013	2.4540 (+2)	3.8857 (−1)	297.35	1.4400 (−5)	3.7060 (−5)	1.9816 (−5)
11000	10981	9.7728	6367.2	7.5854 (+24)	398.07	1.7871 (+9)	2.2274 (−7)	28.964	216.774	2.2699 (+2)	3.6480 (−1)	295.15	1.4223 (−5)	3.8988 (−5)	1.9543 (−5)
11500	11479	9.7713	6364.6	7.0157 (+24)	397.95	1.6525 (+9)	2.4081 (−7)	28.964	216.650	2.0984 (+2)	3.3743 (−1)	295.07	1.4216 (−5)	4.2131 (−5)	1.9533 (−5)
12000	11977	9.7697	6365.6	6.4857 (+24)	397.95	1.5277 (+9)	2.6049 (−7)	28.964	216.650	1.9399 (+2)	3.1194 (−1)	295.07	1.4216 (−5)	4.5574 (−5)	1.9533 (−5)
12500	12475	9.7682	6366.6	5.9958 (+24)	397.95	1.4123 (+9)	2.8178 (−7)	28.964	216.650	1.7934 (+2)	2.8838 (−1)	295.07	1.4216 (−5)	4.9297 (−5)	1.9533 (−5)
13000	12973	9.7667	6367.6	5.5430 (+24)	397.95	1.3056 (+9)	3.0479 (−7)	28.964	216.650	1.6579 (+2)	2.6660 (−1)	295.07	1.4216 (−5	5.3325 (−5)	1.9533 (−5)
13500	13471	9.7651	6368.6	5.1244 (+24)	397.95	1.2070 (+9)	3.2969 (−7)	28.964	216.650	1.5327 (+2)	2.4646 (−1)	295.07	1.4216 (−5)	5.7680 (−5)	1.9533 (−5)
14000	13969	9.7636	6369.6	4.7375 (+24)	397.95	1.1159 (+9)	3.5662 (−7)	28.964	216.650	1.4170 (+2)	2.2786 (−1)	295.07	1.4216 (−5)	6.2391 (−5)	1.9533 (−5)
14500	14467	9.7621	6370.6	4.3799 (+24)	397.95	1.0317 (+9)	3.8574 (−7)	28.964	216.650	1.3100 (+2)	2.1066 (−1)	295.07	1.4216 (−5)	6.7485 (−5)	1.9533 (−5)
15000	14965	9.7605	6371.6	4.0493 (+24)	397.95	9.5380 (+8)	4.1723 (−7)	28.964	216.650	1.2111 (+2)	1.9476 (−1)	295.07	1.4216 (−5)	7.2995 (−5)	1.9533 (−5)
16000	15960	9.7575	6373.6	3.4612 (+24)	397.95	8.1528 (+8)	4.8812 (−7)	28.964	216.650	1.0352 (+2)	1.6647 (−1)	295.07	1.4216 (−5)	8.5397 (−5)	1.9533 (−5)
17000	16955	9.7544	6375.6	2.9587 (+24)	397.95	6.9691 (+8)	5.7102 (−7)	28.964	216.650	8.8497 (+1)	1.4230 (−1)	295.07	1.4216 (−5)	9.9901 (−5)	1.9533 (−5)
18000	17949	9.7513	6377.6	2.5292 (+24)	397.95	5.9576 (+8)	6.6797 (−7)	28.964	216.650	7.5652 (+1)	1.2165 (−1)	295.07	1.4216 (−5)	1.1686 (−4)	1.9533 (−5)
19000	18943	9.7483	6379.6	2.1622 (+24)	397.95	5.0931 (+8)	7.8135 (−7)	28.964	216.650	6.4674 (+1)	1.0400 (−1)	295.07	1.4216 (−5)	1.3670 (−4)	1.9533 (−5)
20000	19937	9.7452	6381.6	1.8486 (+24)	397.95	4.3543 (+8)	9.1393 (−7)	28.964	216.650	5.5293 (+1)	8.8910 (−2)	295.07	1.4216 (−5)	1.5989 (−4)	1.9533 (−5)
21000	20931	9.7422	6411.0	1.5742 (+24)	398.81	3.7161 (+8)	1.0732 (−6)	28.964	217.581	4.7289 (+1)	7.5715 (−2)	295.07	1.4267 (−5)	1.8843 (−4)	1.9611 (−5)
22000	21924	9.7391	6442.3	1.3413 (+24)	399.72	3.1733 (+8)	1.2596 (−6)	28.964	218.574	4.0475 (+1)	6.4510 (−2)	296.38	1.4322 (−5)	2.2201 (−4)	1.9695 (−5)
23000	22917	9.7361	6473.6	1.1437 (+24)	400.62	2.7119 (+8)	1.4772 (−6)	28.964	219.567	3.4668 (+1)	5.5006 (−2)	297.05	1.4376 (−5)	2.6135 (−4)	1.9778 (−5)
24000	23910	9.7330	6504.9	9.7591 (+23)	401.53	2.3194 (+8)	1.7312 (−6)	28.964	220.560	2.9717 (+1)	4.6938 (−2)	297.72	1.4430 (−5)	3.0743 (−4)	1.9862 (−5)
25000	24902	9.7300	6536.2	8.3341 (+23)	402.43	1.9852 (+8)	2.0272 (−6)	28.964	221.552	2.5492 (+1)	4.0084 (−2)	298.39	1.4484 (−5)	3.6135 (−4)	1.9945 (−5)
26000	25894	9.7269	6567.5	7.1225 (+23)	403.33	1.7004 (+8)	2.3720 (−6)	28.964	222.544	2.1883 (+1)	3.4257 (−2)	299.06	1.4538 (−5)	4.2439 (−4)	2.0029 (−5)
27000	26886	9.7239	6598.9	6.0916 (+23)	404.23	1.4575 (+8)	2.7734 (−6)	28.964	223.536	1.8799 (+1)	2.9298 (−2)	299.72	1.4592 (−5)	4.9805 (−4)	2.0112 (−5)
28000	27877	9.7208	6630.2	5.2138 (+23)	405.12	1.2502 (+8)	3.2404 (−6)	28.964	224.527	1.6161 (+1)	2.5076 (−2)	300.39	1.4646 (−5)	5.8405 (−4)	2.0195 (−5)
29000	28868	9.7178	6661.6	4.4657 (+23)	406.01	1.0732 (+8)	3.7832 (−6)	28.964	225.518	1.3904 (+1)	2.1478 (−2)	301.05	1.4699 (−5)	6.8437 (−4)	2.0278 (−5)
30000	29859	9.7147	6692.0	3.8278 (+23)	406.91	9.2192 (+8)	4.4137 (−6)	28.964	226.509	1.1970 (+1)	1.8410 (−2)	301.71	1.4753 (−5)	8.0134 (−4)	2.0361 (−5)
31000	30850	9.7117	6724.3	3.2833 (+23)	407.79	7.9251 (+8)	5.1456 (−6)	28.964	227.500	1.0312 (+1)	1.5792 (−2)	302.37	1.4806 (−5)	9.3759 (−4)	2.0443 (−5)
32000	31840	9.7087	6755.7	2.8133 (+23)	408.68	6.8175 (+8)	5.9945 (−6)	28.964	228.490	8.8906 (+0)	1.3555 (−2)	303.02	1.4859 (−5)	1.0962 (−3)	2.0526 (−5)
33000	32830	9.7056	6831.2	2.4062 (+23)	410.90	5.8522 (+8)	7.0212 (−6)	28.964	230.973	7.6730 (+0)	1.1573 (−2)	304.67	1.4992 (−5)	1.2955 (−3)	2.0733 (−5)
34000	33819	9.7026	6915.4	2.0558 (+23)	413.35	5.0297 (+8)	8.2182 (−6)	28.964	233.743	6.6341 (+0)	9.8874 (−3)	306.49	1.5140 (−5)	1.5312 (−3)	2.0963 (−5)
35000	34808	9.6995	6999.5	1.7597 (+23)	415.79	4.3307 (+8)	9.6010 (−6)	28.964	236.513	5.7459 (+0)	8.4634 (−3)	308.30	1.5287 (−5)	1.8062 (−3)	2.1193 (−5)
36000	35797	9.6965	7083.7	1.5090 (+23)	418.22	3.7356 (+8)	1.1196 (−5)	28.964	239.282	4.9852 (+0)	7.2579 (−3)	310.10	1.5433 (−5)	2.1264 (−3)	2.1422 (−5)

Altitude		Acceleration due to gravity	Pressure scale height	Number density	Particle speed	Collision frequency	Mean free path	Molecular weight	Temperature	Pressure	Density	Sound speed	Dynamic viscosity	Kinematic viscosity	Thermal conductivity
Z(m)	H(m)	g(m/s²)	H_p (m)	n (m⁻³)	V (m/s)	ν (s⁻¹)	L (m)	M (kg/kmol)	T(K)	P(mb)	ρ(kg/m³)	C_s(m/s)	μ (N·s/m²)	η (m²/s)	(J/m·s·K)
38000	37774	9.6904	7252.1	1.1158 (+23)	423.03	2.7939 (+7)	1.5141 (−5)	28.964	244.818	3.7713 (+0)	5.3666 (−3)	313.67	1.5723 (−5)	2.9297 (−3)	2.1878 (−5)
40000	39750	9.6844	7420.6	8.3077 (+22)	427.78	2.1036 (+7)	2.0336 (−5)	28.964	250.350	2.8714 (+0)	3.9957 (−3)	317.19	1.6009 (−5)	4.0066 (−3)	2.2331 (−5)
42000	41724	9.6783	7589.2	6.2266 (+22)	432.48	1.5939 (+7)	2.7133 (−5)	28.964	255.878	2.1996 (+0)	2.9948 (−3)	320.67	1.6293 (−5)	5.4404 (−3)	2.2781 (−5)
44000	43698	9.6723	7757.9	4.6965 (+22)	437.13	1.2152 (+7)	3.5973 (−5)	28.964	261.403	1.6949 (+0)	2.2589 (−3)	324.12	1.6573 (−5)	7.3371 (−3)	2.3229 (−5)
46000	45669	9.6662	7926.7	3.5640 (+22)	441.72	9.3182 (+6)	4.7404 (−5)	28.964	266.925	1.3134 (+0)	1.7142 (−3)	327.52	1.6851 (−5)	9.8305 (−3)	2.3764 (−5)
48000	47640	9.6602	8042.4	2.7376 (+22)	444.79	7.2075 (+6)	6.1713 (−5)	28.964	270.650	1.0229 (+0)	1.3167 (−3)	329.80	1.7037 (−5)	1.2939 (−2)	2.3973 (−5)
50000	49610	9.6542	8047.4	2.1351 (+22)	444.79	5.6201 (+6)	7.9130 (−5)	28.964	270.650	7.9779 (−1)	1.0269 (−3)	329.80	1.7037 (−5)	1.6591 (−2)	2.3973 (−5)
52000	51578	9.6482	8004.3	1.6750 (+22)	443.46	4.3966 (+6)	1.0086 (−4)	28.964	269.031	6.2214 (−1)	8.0562 (−3)	328.81	1.6956 (−5)	2.1047 (−2)	2.3843 (−5)
54000	53545	9.6421	7845.3	1.3286 (+22)	438.90	3.4515 (+6)	1.2716 (−4)	28.964	263.524	4.8337 (−1)	6.3901 (−3)	325.43	1.6600 (−5)	2.6104 (−2)	2.3400 (−5)
56000	55511	9.6361	7686.2	1.0488 (+22)	434.29	2.6961 (+6)	1.6108 (−4)	28.964	258.019	3.7362 (−1)	5.0445 (−4)	322.01	1.6402 (−5)	3.2514 (−2)	2.2955 (−5)
58000	57476	9.6316	7527.0	8.2390 (+21)	429.63	2.0952 (+6)	2.0506 (−4)	28.964	252.518	2.8723 (−1)	3.9627 (−4)	318.56	1.6121 (−5)	4.0682 (−2)	2.2508 (−5)
60000	59439	9.6241	7367.8	6.4387 (+21)	424.93	1.6195 (+6)	2.6239 (−4)	28.964	247.021	2.1958 (−1)	3.0968 (−4)	315.07	1.5837 (−5)	5.1141 (−2)	2.2058 (−5)
65000	64342	9.6091	6969.1	3.3934 (+21)	412.95	8.2945 (+5)	4.9787 (−4)	28.964	233.292	1.0929 (−1)	1.6321 (−4)	306.19	1.5116 (−5)	9.2617 (−2)	2.0926 (−5)
70000	69238	9.5942	6569.9	1.7222 (+21)	400.64	4.0839 (+5)	9.8104 (−4)	28.964	219.585	5.2209 (−2)	8.2829 (−5)	297.06	1.4377 (−5)	1.7357 (−1)	1.9780 (−5)
75000	74125	9.5793	6244.9	8.3003 (+20)	390.30	1.9175 (+5)	2.0354 (−3)	28.964	208.399	2.3881 (−2)	3.9921 (−5)	289.40	1.3759 (−5)	3.4465 (−1)	1.8834 (−5)
80000	79006	9.5644	5961.7	3.8378 (+20)	381.05	8.6559 (+5)	4.4022 (−3)	28.964	198.639	1.0524 (−3)	1.8458 (−5)	282.54	1.3208 (−5)	7.1557 (−1)	1.8001 (−5)
85000	83878	9.5496	5678.0	1.7090 (+20)	371.59	9.8858 (+5)	9.8858 (−3)	28.964	188.893	4.4568 (−3)	8.2196 (−6)	275.52	1.2647 (−5)	1.5386 (+0)	1.7162 (−5)
85500	84365	9.5481	5649.6	1.5727 (+20)	370.63	3.4501 (+5)	1.0743 (−2)	28.964	187.920	4.0802 (−3)	7.5641 (−6)	274.81	1.2590 (−5)	1.6645 (+0)	1.7078 (−5)
86000	84852	9.5466	5621	1.447 (+20)	369.7	3.17 (+4)	1.17 (−2)	28.95	186.87	3.7338 (−3)	6.958 (−6)	—	—	—	—
90000	88744	9.5348	5636	7.116 (+19)	369.9	1.56 (+4)	2.37 (−2)	28.91	186.87	1.8359	3.416	—	—	—	—
95000	93610	9.5200	5727	2.920 (+19)	372.6	6.44 (+4)	5.79 (−2)	28.73	188.42	7.5966 (−4)	1.393	—	—	—	—
100000	98451	9.5052	6009	1.189 (+19)	381.4	2.68 (+3)	1.42 (−1)	28.40	195.08	3.2011	5.604 (−7)	—	—	—	—
110000	108129	9.4759	7723	2.144 (+18)	431.7	5.48 (+2)	7.88 (−1)	27.27	240.00	7.1042 (−5)	9.708 (−8)	—	—	—	—
120000	117777	9.4466	12091	5.107 (+17)	539.3	1.63 (+2)	3.31 (+0)	26.20	360.00	2.5382	2.222	—	—	—	—
130000	127395	9.4175	16288	1.930 (+17)	625.0	7.1 (+1)	8.8 (+0)	25.44	469.27	1.2505	8.152 (−9)	—	—	—	—
140000	136983	9.3886	20025	9.322 (+16)	691.9	3.8 (+1)	1.8 (+1)	24.75	559.63	7.2028 (−6)	3.831	—	—	—	—
150000	146542	9.3597	23380	5.186 (+16)	746.5	2.3 (+1)	3.3 (+1)	24.10	634.39	4.5422	2.076	—	—	—	—
160000	156072	9.3310	26414	3.162 (+16)	792.2	1.5 (+1)	5.3 (+1)	23.49	696.29	3.0395	1.233	—	—	—	—
170000	165572	9.3024	29175	2.055 (+16)	831.3	1.0 (+1)	8.2 (+1)	22.90	747.57	2.1210	7.815 (−10)	—	—	—	—
180000	175043	9.2740	31703	1.400 (+16)	865.3	7.2 (+0)	1.2 (+2)	22.34	790.07	1.5271	5.194	—	—	—	—
190000	184486	9.2457	34030	9.887 (+15)	895.1	5.2 (+0)	1.7 (+2)	21.81	825.16	1.1266	3.581	—	—	—	—
200000	193899	9.2175	36183	7.182 (+15)	921.6	3.9 (+0)	2.4 (+2)	21.30	854.56	8.4736 (−7)	2.541	—	—	—	—
210000	—	—	—	—	—	—	—	—	878.84	6.4756	1.846	—	—	—	—
220000	212641	9.1615	40043	4.040 (+15)	966.5	2.3 (+0)	4.2 (+2)	20.37	899.01	5.0149	1.367	—	—	—	—
240000	231268	9.1061	43405	2.420 (+15)	1003.2	1.4 (+0)	7.0 (+2)	19.56	929.73	3.1059	7.858 (−11)	—	—	—	—
260000	249784	9.0511	46346	1.515 (+15)	1033.5	9.3 (−1)	1.1 (+3)	18.85	950.99	1.9894	4.742	—	—	—	—
280000	268187	8.9966	48925	9.807 (+14)	1058.7	6.1 (−1)	1.7 (+3)	18.24	965.75	1.3076	2.971	—	—	—	—
300000	286480	8.9427	51193	6.509 (+14)	1079.7	4.2 (−1)	2.6 (+3)	17.73	976.01	8.7704 (−8)	1.916	—	—	—	—
320000	304663	8.8892	53199	4.405 (+14)	1097.4	2.9 (−1)	3.8 (+3)	17.29	983.16	5.9796	1.264	—	—	—	—
340000	322738	8.8361	54996	3.029 (+14)	1112.4	2.0 (−1)	5.6 (+3)	16.91	988.15	4.1320	8.503 (−12)	—	—	—	—
360000	340705	8.7836	56637	2.109 (+14)	1125.5	1.4 (−1)	8.0 (+3)	16.57	991.65	2.8878 (−8)	5.805 (−12)	—	—	—	—
380000	358565	8.7315	58178	1.485 (+14)	1137.4	1.0 (−1)	1.1 (+4)	16.27	994.10	2.0384 (−8)	4.013 (−12)	—	—	—	—
400000	376320	8.6799	59678	1.056 (+14)	1148.5	7.2 (−2)	1.6 (+4)	15.98	995.83	1.4518 (−8)	2.803 (−12)	—	—	—	—
450000	420250	8.5529	63644	4.678 (+13)	1177.4	3.3 (−2)	3.6 (+4)	15.25	998.22	6.4468 (−9)	1.184 (−12)	—	—	—	—
500000	463540	8.4286	68785	2.192 (+13)	1215.0	1.6 (−2)	7.7 (+4)	14.33	999.24	3.0236 (−9)	5.215 (−13)	—	—	—	—
550000	506202	8.3070	76427	1.097 (+13)	1271.5	8.4 (−3)	1.5 (+5)	13.09	999.67	1.5137 (−9)	2.384 (−13)	—	—	—	—
600000	548252	8.1880	88244	5.950 (+12)	1356.4	4.8 (−3)	2.8 (+5)	11.51	999.85	8.2130 (−10)	1.137 (−13)	—	—	—	—
650000	589701	8.0716	105992	3.540 (+12)	1476.0	3.1 (−3)	4.8 (+5)	9.72	999.93	4.8865 (−10)	5.712 (−14)	—	—	—	—
700000	630563	7.9576	130630	2.311 (+12)	1627.0	2.2 (−3)	7.3 (+5)	8.00	999.97	3.1908 (−10)	3.070 (−14)	—	—	—	—
750000	670850	7.8460	161074	1.637 (+12)	1793.9	1.7 (−3)	1.0 (+6)	6.58	999.00	2.2599 (−10)	1.788 (−14)	—	—	—	—
800000	710574	7.7368	193862	1.234 (+12)	1954.3	1.4 (−3)	1.4 (+6)	5.54	999.99	1.7036 (−10)	1.136 (−14)	—	—	—	—
850000	749747	7.6298	224737	9.717 (+11)	2089.6	1.2 (−3)	1.7 (+6)	4.85	1000.00	1.3415 (−10)	7.824 (−15)	—	—	—	—
900000	788380	7.5250	250894	7.876 (+11)	2192.6	1.0 (−3)	2.1 (+6)	4.40	1000.00	1.0873 (−10)	5.759 (−15)	—	—	—	—
950000	826484	7.4224	271754	6.505 (+11)	2266.4	8.7 (−4)	2.6 (+6)	4.12	1000.00	8.9816 (−11)	4.453 (−15)	—	—	—	—
1000000	864071	7.3218	288203	5.442 (+11)	2318.1	7.5 (−4)	3.1 (+6)	3.94	1000.00	7.5138 (−11)	3.561 (−15)	—	—	—	—

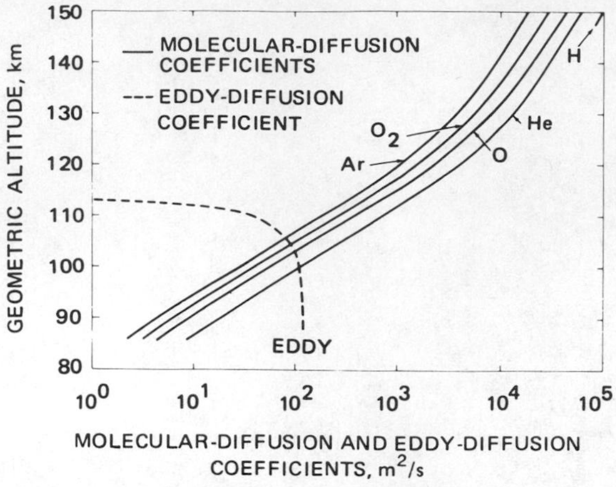

FIGURE 1. Molecular-diffusion and eddy-diffusion coefficients as a function of geometric altitude.

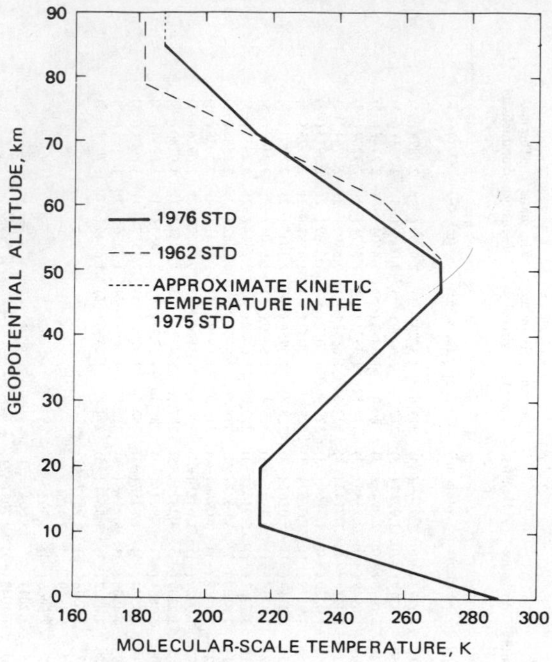

FIGURE 3. Molecular-scale temperature as a function of geopotential altitude.

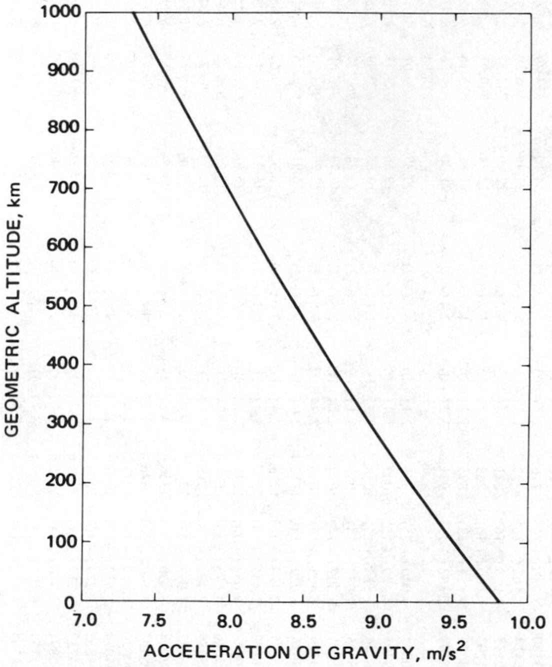

FIGURE 2. Acceleration of gravity as a function of geometric altitude.

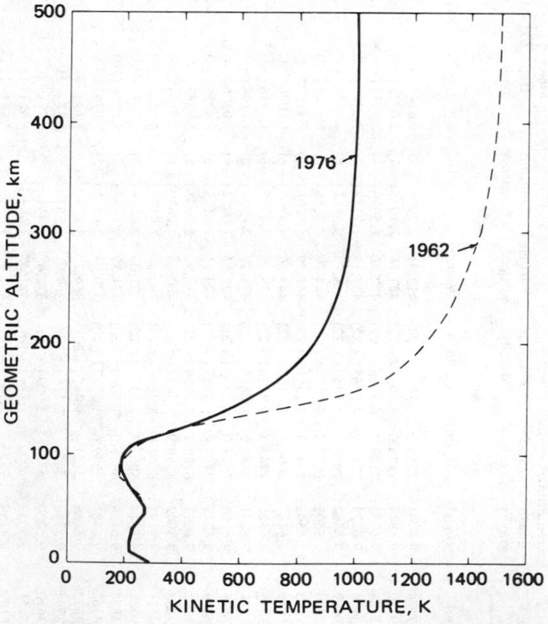

FIGURE 4. Kinetic temperature as a function of geometric altitude.

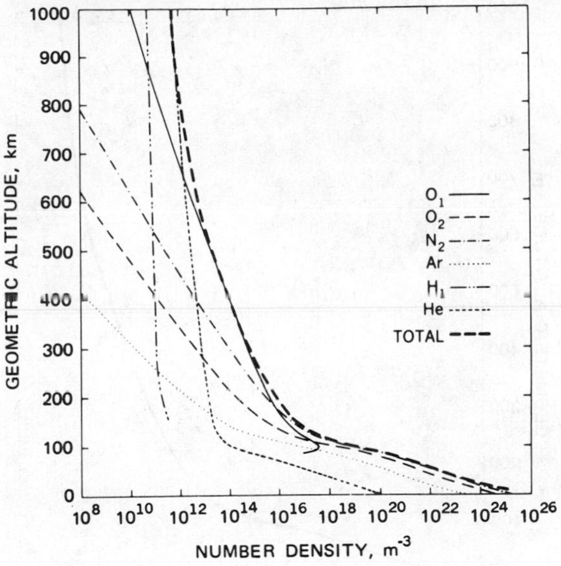

FIGURE 5. Number density of individual species and total number density as a function of geometric altitude.

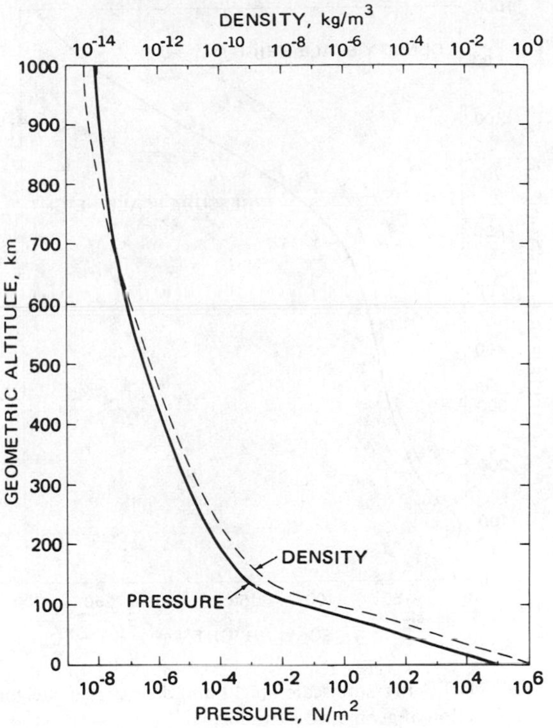

FIGURE 7. Total pressure and mass density as a function of geometric altitude.

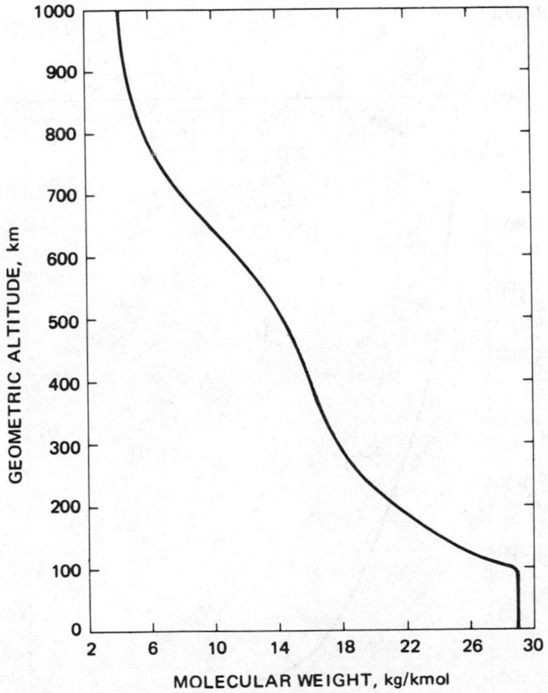

FIGURE 6. Mean molecular weight as a function of geometric altitude.

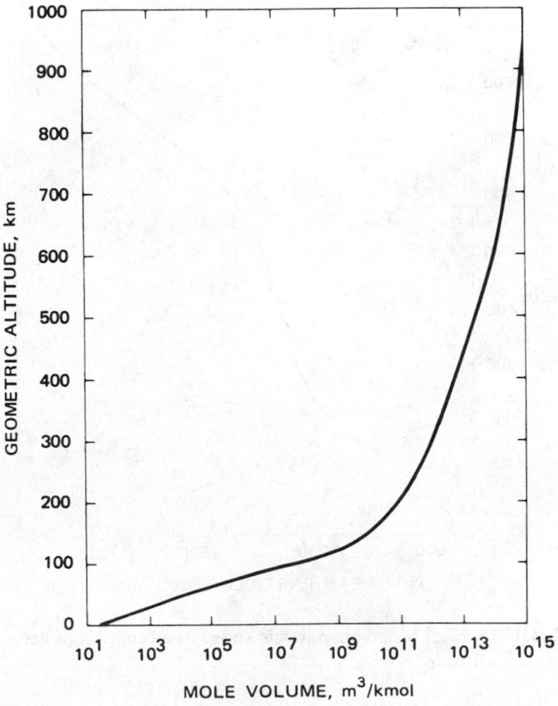

FIGURE 8. Mole volume as a function of geometric altitude.

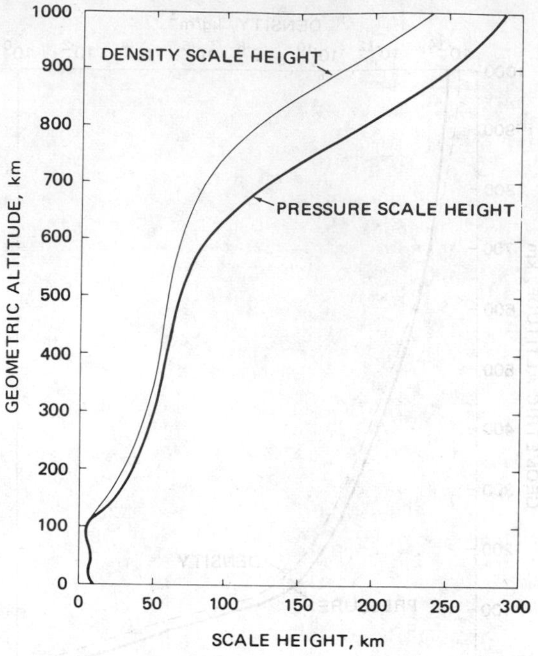

FIGURE 9. Pressure scale height and density scale height as a function of geometric altitude.

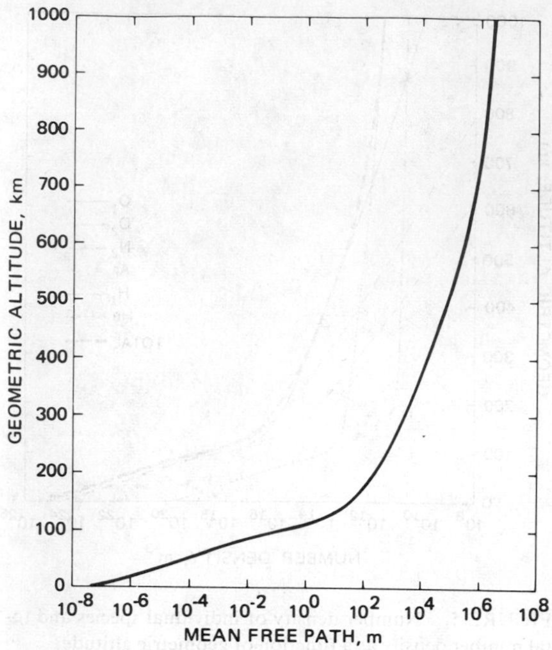

FIGURE 11. Mean free path as a function of geometric altitude.

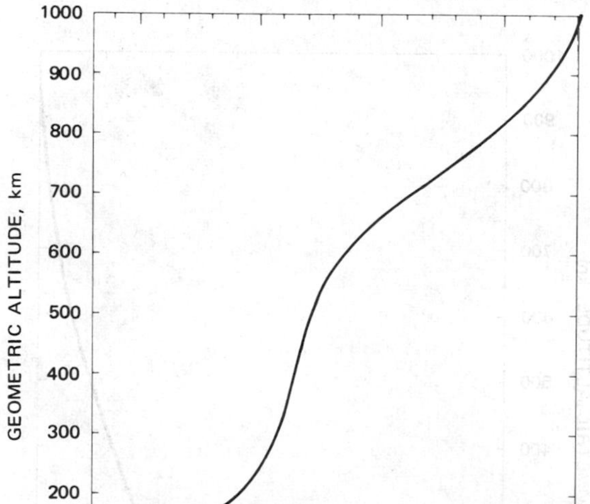

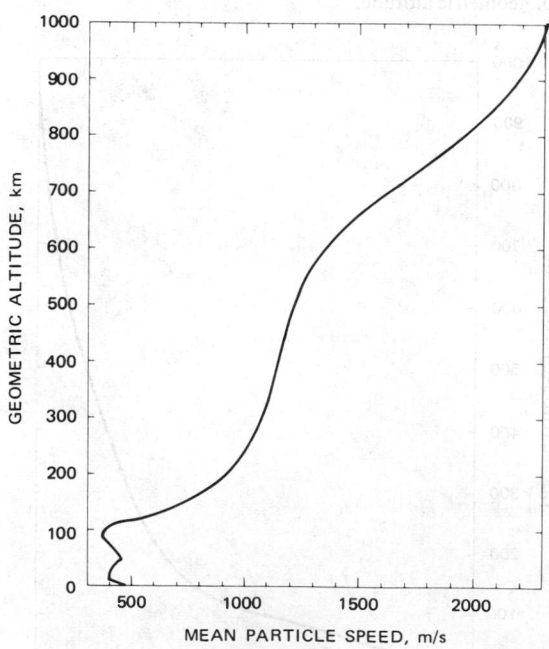

FIGURE 10. Mean air-particle speed as a function of geometric altitude.

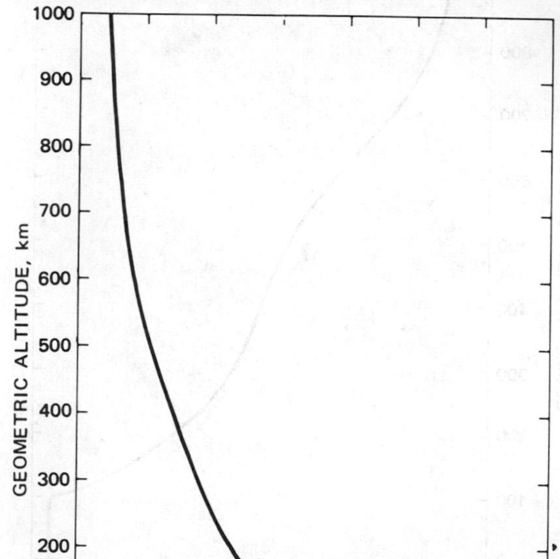

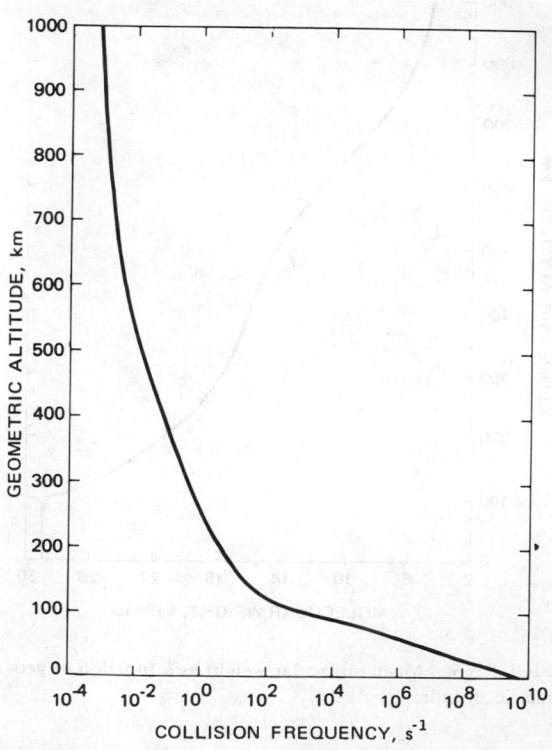

FIGURE 12. Collision frequency as a function of geometric altitude.

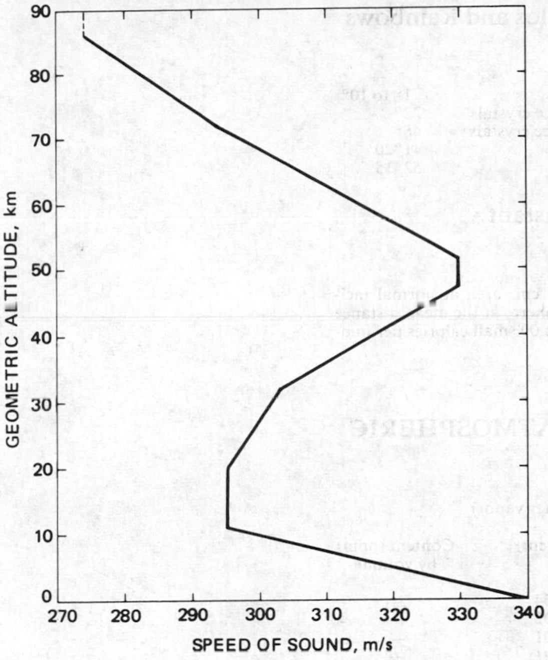

FIGURE 13. Speed of sound as a function of geometric altitude.

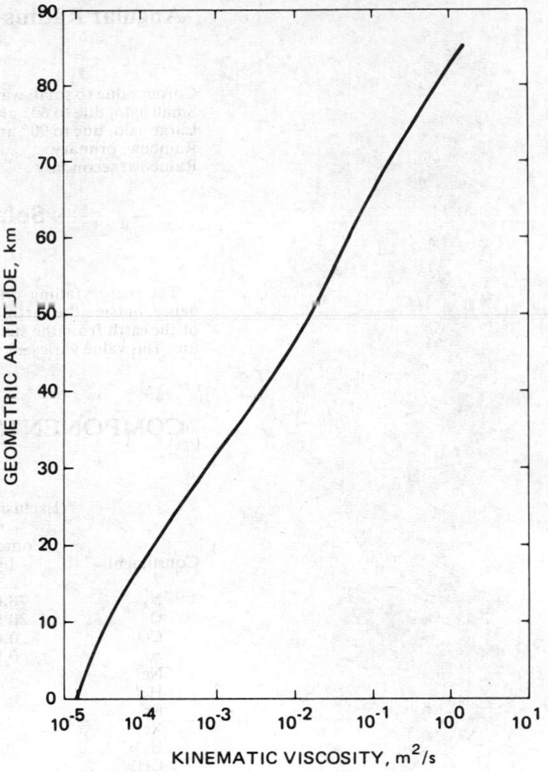

FIGURE 15. Kinematic viscosity as a function of geometric altitude.

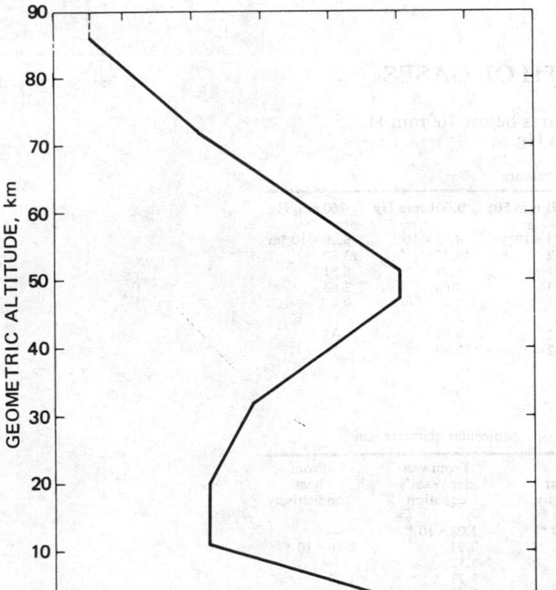

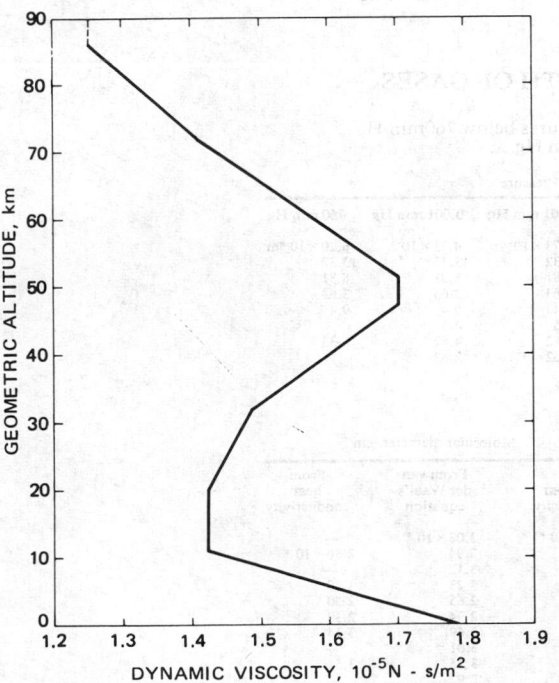

FIGURE 14. Dynamic viscosity as a function of geometric altitude.

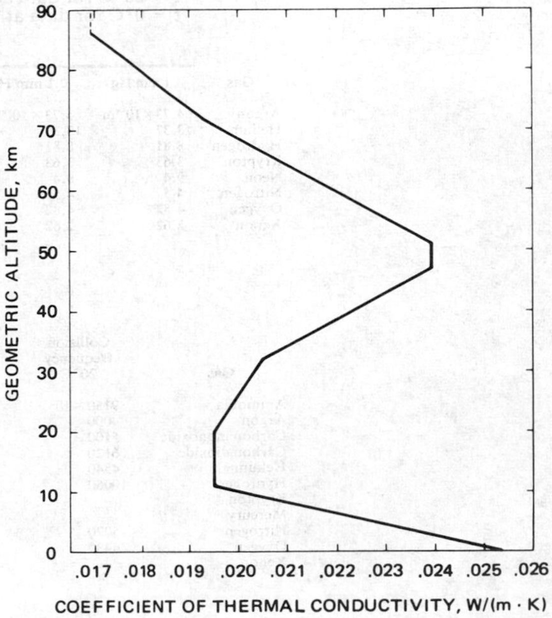

FIGURE 16. Coefficient of thermal conductivity as a function of geometric altitude.

Angular Radius of Halos and Rainbows

Coronae due to small water drops	1° to 10°
Small halo, due to 60° angles of ice crystals	22°
Large halo, due to 90° angles of ice crystals	46°
Rainbow, primary	41°20′
Rainbow, secondary	52°15′

Solar Constant

The energy falling on one sq. cm. area at normal incidence, outside the earth's atmosphere, at the mean distance of the earth from the sun equals 2.00 small calories per minute. This value varies ±2%.

COMPONENTS OF ATMOSPHERIC AIR

(Exclusive of water vapor)

Constituent	Content (per cent) by volume	Content (ppm) by volume
N_2	78.084±0.004	—
O_2	20.946±0.002	—
CO_2	0.033±0.001	—
Ar	0.934±0.001	—
Ne	—	18.18 ±0.04
He	—	5.24 ±0.004
Kr	—	1.14 ±0.01
Xe	—	0.087±0.001
H_2	—	0.5
CH_4	—	2
N_2O	—	0.5 ±0.1

MEAN FREE PATH OF GASES

t = 20°C for data at pressures below 760mm Hg
t = 0°C for data at 760mm Hg

Gas	Pressure				
	1 mm Hg	0.1 mm Hg	0.01 mm Hg	0.001 mm Hg	760 mm Hg
Argon	4.73×10^{-5}m	4.73×10^{-4}m	4.73×10^{-3}m	4.73×10^{-2}	6.20×10^{-8}m
Helium	13.32	13.32	13.32	13.32	13.32
Hydrogen	8.81	8.81	8.81	8.81	8.81
Krypton	3.63	3.63	3.63	3.63	3.63
Neon	9.4	9.4	9.4	9.4	9.4
Nitrogen	4.5	4.5	4.5	4.5	
Oxygen	4.82	4.82	4.82	4.82	6.33
Xenon	2.62	2.62	2.62	2.62	

Gas	Collision frequency 20°C	Molecular diameter, cm		
		From viscosity	From van der Waal's equation	From heat conductivity
Ammonia	9150×10^{6}	2.97×10^{-8}	3.08×10^{-8}	—
Argon	4000	2.88	2.94	2.86×10^{-6}
Carbon monoxide	5100	3.19	3.12	—
Carbon dioxide	6120	3.34	3.23	3.40
Helium	4540	1.90	2.65	2.30
Hydrogen	10060	2.40	2.34	2.32
Krypton	—	—	(3.69)	3.14
Mercury	—	—	3.01	—
Nitrogen	5070	3.15	3.15	3.53
Oxygen	4430	2.98	2.92	—
Xenon	—	—	4.02	3.42

ATMOSPHERIC ELECTRICITY

HANS DOLEZALEK

Fair-weather electricity. Electrically, a fair-weather situation is given when the activity of local generators is low. Usually, this is supposed to be the case when there are no hydrometeors (rain, snow, hail, fog, droplets) and no highly charged dust in the air, cloudiness not more than 3/10 and not concentrated overhead, wind speed below 5 m/s (no white caps on sea). Under these conditions the atmospheric electric parameters over land have the average or typical values as shown in the table.

The "Classical Picture" is a group of hypotheses to explain the electrification of the atmosphere, the main hypothesis assuming that the ionosphere is charged by the thunderstorms to a high positive potential everywhere the same but varying in time. The classical picture is probably fundamentally correct but it has never been proven and probably requires supplements and corrections. In particular, extraterrestrial influences probably have to be admitted.

ELECTRICAL PARAMETERS OF THE CLEAR (FAIR WEATHER) ATMOSPHERE, PERTINENT TO THE CLASSICAL PICTURE OF ATM. ELECTRICITY (ELECTRIC STANDARD ATMOSPHERE)

All currents and fields listed are part of the global circuit, i.e., circuits of local generators are not included. Values are subject to variations due to latitude and altitude of the point of observation above sea level, locality with respect to sources of disturbances, meteorological and climatological factors, and man-made changes.

Part of atmosphere for which the values are calculated (Elements are in free, cloudless atmosphere):	Currents, I, in A; and current densities, i, in A/m²	Potential Differences, U, in V; field strength E in V/m; U = 0 at sea level	Resistances, R, in Ω; Columnar resistances, R_c, in Ω m²; and resistivities, ρ, in Ω m	Conductances, G, in $Ω^{-1}$; Columnar conductances G_c, in $Ω^{-1} m^{-2}$; total conductivities, γ, in $Ω^{-1} m^{-1}$	Capacitances, C, in F; Columnar capacitances, C_c, in F m^{-2} and capacitivities, ε, in F m^{-1}	Time constants τ, in seconds
Volume element at about sea level, one cubic meter	$i = 3 \times 10^{-12}$*	$E_0 = 1.2 \times 10^2$*	$\rho_0 = 4 \times 10^{13}$	$\gamma_0 = 2.5 \times 10^{-14}$	$\varepsilon_0 = 8.9 \times 10^{-12}$*	$\tau_0 = 3.6 \times 10^2$
Lower column of 1 m² cross section from sea level to 2 km height	same as above	at upper end: $U_1 = 1.8 \times 10^5$	$R_{c1} = 6 \times 10^{16}$	$G_{c1} = 1.7 \times 10^{-17}$	$C_{c1} = 4.4 \times 10^{-15}$	$\tau_{c1} = 2.6 \times 10^2$
Volume element at about 2 km height, 1 m³	same as above	$E_2 = 6.6 \times 10^1$	$\rho_2 = 2.2 \times 10^{13}$(*)	$\gamma_2 = 4.5 \times 10^{-14}$	$\varepsilon_2 = 8.9 \times 10^{-12}$	$\tau_2 = 2 \times 10^2$
Center column of 1 m² cross section from 2 to 12 km	same as above	at upper end: $U_m = 3.15 \times 10^5$	$R_{cm} = 4.5 \times 10^{16}$	$G_{cm} = 5 \times 10^{-17}$	$C_{cm} = 8.8 \times 10^{-16}$	$\tau_{cm} = 1.8 \times 10^1$
Volume element at about 12 km height, 1 m³	same as above	$E_{12} = 3.3 \times 10^0$	$\rho_{12} = 1.3 \times 10^{12}$(*)	$\gamma_{12} = 7.7 \times 10^{-13}$	$\varepsilon_{12} = 8.9 \times 10^{-12}$	$\tau_{12} = 1.2 \times 10^1$
Upper column of 1 m² cross section from 12 to 65 km height	same as above	at upper end: $U_u = 3.5 \times 10^5$	$R_{cu} = 1.5 \times 10^{16}$	$G_{cu} = 2.5 \times 10^{-17}$	$C_{cu} = 1.67 \times 10^{-16}$	$\tau_{cu} = 6.7 \times 10^0$
Whole column of 1 m² cross section from 0 to 65 km height	same as above	at upper end: $U = 3.5 \times 10^5$	$R_c = 1.2 \times 10^{17}$	$G_c = 8.3 \times 10^{-18}$	$C_c = 1.36 \times 10^{-16}$	$\tau_c = 1.64 \times 10^1$
Total spherical capacitor area: 5×10^{14} m²	$I = 1.5 \times 10^3$	$U = 3.5 \times 10^5$*	$R = 2.4 \times 10^2$	$G = 4.2 \times 10^{-3}$	$C = 6.8 \times 10^{-2}$	$\tau = 1.64 \times 10^1$

Values with a star, *, are rough average values from measurement. A star in parentheses, (*), points to a typical value from one or a few measurements. All other values have been calculated from starred values, under the assumption, that at 2 km 50% and at 12 km 90% of the columnar resistance is reached. Voltage drop along one of the partial columns can be calculated by subtracting the value for the lower column from that of the upper one. Columnar resistances, conductances, and capacitances are valid for that particular part of the column which is indicated at left. Capacitances are calculated with the formula for plate capacitors, and this fact must be considered also for the time constants for columns.

According to measurements, U, the potential difference between 0 m and 65 km may vary by a factor of approx. 2 (two). The total columnar resistance, R_c, is estimated to vary up to a factor of 3 (three), the variation being due to either reduction of conductivity in the exchange layer (about lowest 2 km of this table) or to the presence of high mountains, in both cases the variation is caused in the troposphere. Smaller variations in the stratosphere and mesosphere discussed because of aerosol there. The air-earth current density in fair-weather varies by a factor of 3 to 6 accordingly. Conductivity near the ground varies by a factor of about 3 (three) but only decreasing, increase of conductivity due to extraordinary radioactivity is a singular event. The field strength near the ground varies as a consequence of variations of air-earth current density and conductivity from about 1/3 to about 10 of the value quoted in the table. Conductivity near the ground shows a diurnal and an annual variation which depends strongly on the locality; air-earth current density shows a diurnal and annual variation because the earth-ionosphere potential difference undergoes such variations, and because also the columnar resistance is supposed to have a diurnal and probably an annual variation.

Conductivities and air-earth current densities on high mountains are by factors of up to 10 greater than at sea level. Conductivity decreases when atmospheric humidity increases. Values for space charges not quoted because measurements are too few to allow calculation of average values. Values of parameters over the oceans still rather uncertain.

Ions. Fast ions. Ionization rate at ground level over land: from radioactive substances in the ground: 4.0; from radioactive substances in the air: 4.6; from cosmic radiation 1.5–1.8, total (10.1–10.4) × 10^6 ion pairs per m³ and sec. Over ocean far from shore: cosmic radiation only, but in spite of smaller ionization over the ocean conductivity is nearly equal to that over land because of longer ion life time (a consequence of smaller aerosol density).

Contribution from radioactive material over land decreases with altitude in the free air, from cosmic radiation increases (up to a certain level) and depends on latitude. Number densities of fast ions over land and ocean average at (4–5) × 10^8 m^{-3}, varying with height by only about one order of magnitude up to ionospheric C-layer. Their life time varies from about 10 s in aerosol-rich air to 300 s in pure air. All values quoted are approximate averages.

Ion classes in lower atmosphere	Designation	Mobilities in $m^2 V^{-1} s^{-1}$	Size ranges (radii in m)
Primary atmospheric ions:	Fast ions (also called small or light ions)	$k > 10^{-4}$	$r < 6.6 \times 10^{-10}$
Secondary atmospheric ions:	fast intermediate ions	$10^{-4} > k > 10^{-6}$	$6.6 \times 10^{-10} < r < 7.8 \times 10^{-9}$
slow ions (also called large or heavy ions)	Slow intermediate ions	$10^{-6} > k > 10^{-7}$	$7.8 \times 10^{-9} < r < 2.5 \times 10^{-8}$
	Langevin ions	$10^{-7} > k > 2.5 \times 10^{-8}$	$2.5 \times 10^{-8} < r < 5.7 \times 10^{-8}$
	Ultralarge ions	$k < 2.5 \times 10^{-8}$	$5.7 \times 10^{-8} < r < 10^{-7}$

Electric conduction current in atmosphere carried nearly exclusively by the fast ions because number densities of the intermediate ions generally one order of magnitude smaller.

Nature of ions in pure air ("Standard Atmosphere"): initially all molecules are ionized with equal probability according to their abundance, but within nanoseconds these ions transform into the "fast ions" with the following composition:

positive ions: $(H_3O)^+(H_2O)_n$

negative ions: $O_2^-(H_2O)_n$, or $CO_4^-(H_2O)_n$;

the n depending on atmospheric temperature and water vapor pressure, varies very often in the life time of an ion. Most probable values of n for positive ions: troposphere 6 or 7, stratosphere 4, mesosphere 3 or 2.

This is valid for the "Standard Atmosphere" which consists of nitrogen, oxygen, carbon dioxide, and water vapor (and Argon). The real atmosphere contains numerous trace gases which, in spite of their tenuous distribution of often parts per trillion or even less, can be encountered by ions participating in the Brownian movement because of the extremely high collision frequency. Many of these collisions cause a change of the chemical nature of the ion (e.g., by just transferring the charge to another molecule or by forming new compounds). It is therefore possible that an ion changes its nature many times in each second, and it is practically impossible to predict what that nature will be in a given instant. The new compounds have sometimes a quite exotic nature; some of them are possible only when electric charge is present, and disintegrate when the ion is discharged at, say, a conducting surface; others may persist, and in such a case an ion may act as a catalyst for the creation of many individuals of a species. Figures 1 and 2 depict examples for negative and positive ions. In both cases, these ions were generated artificially (present measuring methods do not yet allow to reliably measure the chemical nature of natural ions in the near-ground atmosphere, because of their small number densities). The figures give only examples, many more species have been found.

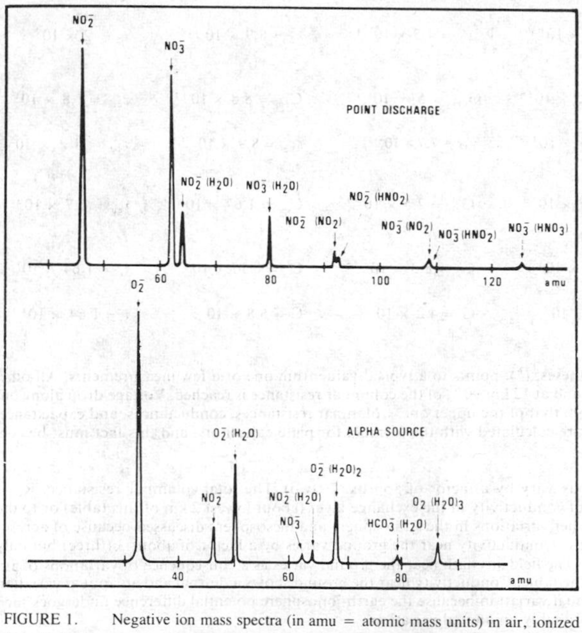

FIGURE 1. Negative ion mass spectra (in amu = atomic mass units) in air, ionized by point discharge (upper part), and an alpha-radiating source (lower part). The ordinate has arbitrary units. The measurements were made under low pressure (about 4 kPa) as required by the mass spectrometer.[1]

Slow ions are aerosol particles which have acquired electrical charge either during the process of aerosol generation or by impact of a fast ion; their nature is determined by the nature of the aerosol present in the atmosphere under consideration. Depending on the size of the aerosol particle (expressed, e.g., by its radius r in meter) and the electric field (E in volts per meter) present at its location, an aerosol particle may have one or several elementary charges (always a whole number) with the maximum charge determined by:

$$q_{max} = 12\pi r^2 \epsilon E \text{ coulombs}$$

where ϵ is the capacitivity of the air, being 8.8×10^{-12} F/m (White, 1963). An example for the distribution of slow ions in a house is given by Figure 3. The values may vary considerably, depending among other factors on the amount of radioactivity present in the walls or as radon or thoron or radioactive aerosol in the air.

References

1. **Huertas, M. L. and Fontan, J.,** Of the Nature of Tropospheric Negative Ions and on the Influence of Various Polluting Gases on the Nature of Tropospheric Positive and Negative Ions. In: Ruhnke and Latham (1983, see there); p. 30—32.
2. **Ruhnke, L. H. and Latham, J., Eds.,** Proceedings in Atmospheric Electricity, A. Deepak Publishing, Hampton, Virginia, XIII + 427 pp., 1983.
3. **Salm, J. and Reinart, R.,** Measurement of the Mobility Spectrum of Ions Over a Wide Range (in Russian). Electricheskye Yavleniya v Gazovykh i Aerosolnikh Sredakh; Acta et Commentations Universitatis Tartuensis, No. 648, p. 41—15, Tartu, Esthonia, USSR, 1983.
4. **White, H. M.,** Industrial Electrostatic Precipitation, Addison-Wesley, 1963.

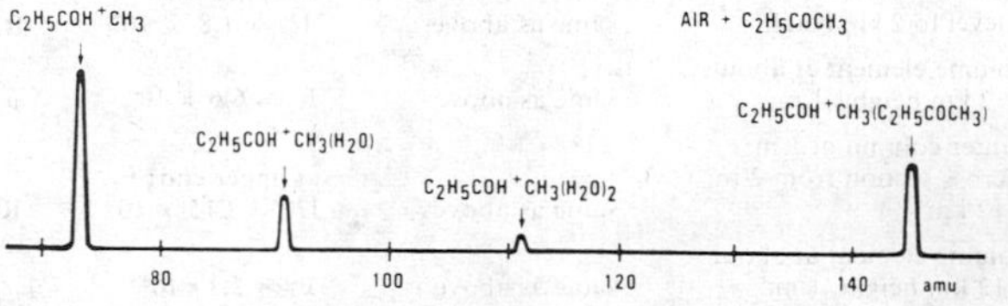

FIGURE 2. Positive ion mass spectrum (in amu) of air polluted by 2-butanone (ethyl methyl ketone). Ordinate: arbitrary units. Pressure: 4 kPa.[1]

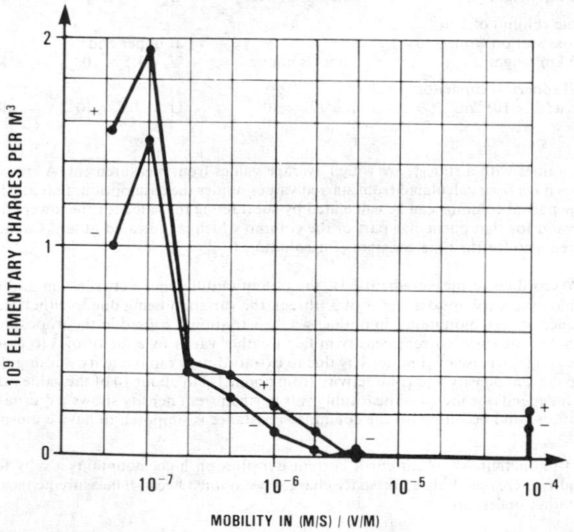

FIGURE 3. Mobility spectrum of atmospheric ions in an ordinary room, measured over a period of 6 months, during 1 hr/day. The figure shows the average derived from more than 600 mobility spectra. Fast ions (positive and negative) are depicted at the right end of the abscissa.[3]

Thunderstorm electricity. Clouds are generally more or less electrified. In thunderstorms, mechanical, thermodynamical and maybe chemical energy is transformed into electrical energy (generator) with separate poles. Most thunderstorms have during most of their life time (phases: development 10–20, mature 15–30, decay 30 min) a positive charge center above and a negative one below, probably another positive one near the base. Electrification occurs (according to processes not yet fully understood) in periods of tens of minutes and spaces of about 1 km^3, and there may be several such centers in a cloud. Thunderclouds mostly reach beyond freezing level, in moderate latitudes from about 3 to 10 km, in the tropics often up to 16–20 km, seldom more. 1000 to 2000 thunderstorms active at any time. Updrafts in storms up to 30 m/s. Water content 10^8–10^9 kg. Total energy about 10^{15} joule, electric energy 10^{12}–10^{13} J. Electric potential vs. ground at negative and positive charge centers: minus and plus 10–100 MV, resp.; field strength not yet measured reliably. Electric resistances of an atmospheric vertical column of 50 km^2 cross section: several 100 MΩ each between ground and negative center, negative and positive center, and positive center and ionosphere. Total current flowing upwards towards ionosphere derived from (still unsufficient) measurements of current densities: 0.5–2 A per thunderstorm cell. Current below cloud carried by conduction by fair-weather ions, conduction by corona ions, convection by vertical winds, precipitation, and lightning. Field strength at ground under storm: varying in strength and polarity, often 10 kV/m and more.

Lightning, during its life time of up to a second, undergoes numerous variations and all its parameters change by orders of magnitude in these. Also, there are many different forms of lightning. Even disregarding particular forms such as ball lightning, we find at least four classification criteria which, taken together, give a great number of differences: intercloud, intracloud, and cloud-to-ground discharges and combinations of these; short high-current and long low-current flashes; lightnings beginning in the cloud and moving towards ground, and lightnings moving upwards; lightnings lowering positive and lightnings lowering negative charges, and those which do first the one and then the other. The usual cloud-to-ground lightning begins with a stepped leader of low luminosity, probably a meter or so in diameter, followed by a return stroke with a diameter in the order of centimeters, temperatures of about 30,000 K, and pressures of up to a MN/m^2. After this, a dart leader may again move downwards causing a second return stroke. This may repeat several times (multiple stroke flash).

Typical voltage drop in ground or other conductors after lightning impact in the neighborhood: 10 kV/m (dangerous!).

Intracloud lightnings observed with up to 100 km length.

There are probably about 100 lightnings occurring on earth at any time. Lightning frequency over oceans only about 1/10 from that over continents. Diurnal variations on continents show maximum in late afternoon and early evening, over oceans late evening until after midnight.

Energy delivered to a stroke about 100 kJ/m.

Upward lightnings initiated mostly from high masts and towers and mountains. Over flat terrain mostly downward flashes, probably generally met by short (decameters) upwards moving darts.

Height from which a lightning points at a target probably several decameters, depending on conductivity distribution in ground.

Long-lasting low-current (hundreds of amperes) flashes more dangerous to man and more damaging to objects (forest fires) than short high current flashes.

Sferics, etc. Lightning generates a sudden variation of electric field (decreasing with r^{-3}), a quickly varying magnetic field (r^{-2}) and an electromagnetic radiation (r^{-1}) from a few hertz (in case of excitation of ionosphere-earth resonance range around the globe) through VLF (maximum intensity, range up to 10 Mm and more depending on direction and time of day) into VHF. Main frequency of thunder about 200 Hz.

Data for a normal cloud-to-ground lightning discharge bringing negative charge to earth. The values listed are intended to convey a rough feeling for the various physical parameters of lightning. No great accuracy is claimed since the results of different investigators are often not in good agreement. These values may in fact, depend on the particular environment in which the lightning discharge is generated. The choice of some of the entries in the table is arbitrary.

	Minimum*	Representative	Maximum*
Stepped leader			
Length of step, *m*	3	50	200
Time interval between steps, μsec	30	50	125
Average velocity of propagation of stepped leader, m/sec†	1.0×10^5	1.5×10^5	2.6×10^6
Charge deposited on stepped-leader channel, coulombs	3	5	20
Dart leader			
Velocity of propagation, m/sec†	1.0×10^6	2.0×10^6	2.1×10^7
Charge deposited on dart-leader channel, coulombs	0.2	1	6
Return stroke‡			
Velocity of propagation, m/sec†	2.0×10^7	5.0×10^7	1.4×10^8
Current rate of increase, kA/μsec §	<1	10	>80
Time to peak current, μsec §	<1	2	30
Peak current, kA §		10–20	110
Time to half of peak current μsec	10	40	250
Charge transferred excluding continuing current, coulombs	0.2	2.5	20
Channel length, km	2	5	14
Lightning flash			
Number of strokes per flash	1	3–4	26
Time interval between strokes in absence of continuing current, msec	3	40	100
Time duration of flash, sec	10^{-2}	0.2	2¶
Charge transferred including continuing current, coulombs	3	25	90

* The words maximum and minimum are used in the sense that most measured values fall between these limits.

† Velocities of propagation are generally determined from photographic data and thus represent "two-dimensional" velocities. Since many lightning flashes are not vertical, values stated are probably slight underestimates of actual values.

‡ First return strokes have slower average velocities of propagation, slower current rates of increase, longer times to current peak, and generally larger charge transfer than subsequent return strokes in a flash.

§ Current measurements are made at the ground.

¶ A lightning flash lasting 15 to 20 sec has been reported by Godlonton (1896).

OUR SOLAR SYSTEM

Object	Average distance from sun (astronomical units)[a]	Radius (Earth = 1)[b]	Mass (Earth = 1)[b]	Density (g/cc) (water = 1)	Atmospheric constituents	Surface (or interior) pressure (bars)	Surface (or cloud top) temperature (K)	Number of satellites
Mercury	0.38	0.38	0.05	5.4	Helium, argon, neon, hydrogen	$>10^{-12}$	600—700 (day); ∼95 (night)	0
Venus	0.72	0.96	0.82	5.2	Carbon dioxide (96%), nitrogen (∼3%), helium, sulfur dioxide, water vapor, argon, neon, carbon monoxide, oxygen, sulfuric acid, hydrogen chloride, hydrogen fluoride, hydrogen sulfide	90	∼730	0
Earth	1.00	1.00	1.00	5.5	Nitrogen (78%), oxygen (21%), argon (0.9%), water vapor, carbon dioxide, neon, methane, krypton, helium, xenon, hydrogen, nitrous oxide, carbon monoxide, nitrogen dioxide, sulfur dioxide, ozone	1.0	250—310	1
Mars	1.52	0.53	0.11	4.0	Carbon dioxide (95%), nitrogen (3%), argon (∼2%), water vapor, oxygen, neon, carbon monoxide, krypton, xenon, ozone	0.005 to 0.008	∼200—245	2
Jupiter	5.20	10.8	318	1.3	Hydrogen (89%), helium (11%), methane, ammonia, water, ethane, acetylene, phosphine, germanium tetrahydride, hydrogen cyanide?, ammonium hydrosulfide	$(<10^6)$	(∼140)	16
Saturn	9.5	9.0	95	0.7	Hydrogen (94%), helium (6%), methane, ammonia, ethane, phosphine, acetylene, propane?, methylacetylene?	$(>10^6)$	(∼100?)	17
Uranus	19.2	4.1	15	1.2	Hydrogen (<90%), methane, helium?	$(>10^6)$	(∼50—60)	5
Neptune	30.0	3.85	17	1.7	Hydrogen (<90%), methane, helium?	$(>10^6)$	(∼50—60)	2
Pluto	39.5	∼0.22 to 0.25?	∼0.002?	<2	Methane, argon?, neon?	$\sim 10^{-3}$ to 10^{-4}?	∼50—60	1
Moon	—	0.27	0.0123	3.3	Neon, argon, helium	$\sim 2 \times 10^{-14}$	370 (day); 120 (night)	—

[a] Astronomical unit defined as average distance between the centers of the sun and Earth (1.496×10^8 km)).
[b] Compared to Earth as unity: Earth radius is 6371 km; mass is 5.975×10^{24} kg.

COSMIC RADIATION

A. Gregory and R. W. Clay

THE NATURE OF COSMIC RAYS

Primary cosmic radiation, in the form of high-energy nuclear particles, electrons, and photons, from outside the solar system and from the Sun, continually bombards our atmosphere.

Secondary radiation, resulting from the interaction of primary cosmic rays with atmospheric gas, is present at sea-level and throughout the atmosphere. The secondary radiation is collimated by absorption and scattering in the atmosphere and consists of a number of components associated with different particle species. High-energy primary particles can produce a large number of secondary particles forming an extensive air shower. Thus, a number of particles may be detected in coincidence at sea level. Primary particle energies range from $\sim 10^8$ eV to $\sim 10^{20}$ eV.[1,15]

PRIMARY PARTICLE ENERGY SPECTRUM

Figure 1 shows the spectrum of primary particle energies. In differential form it is roughly a power law (with an index of −3). There appears to be a knee (a steepening) at about 10^{15} eV and an ankle (a flattening) above $\sim 10^{18}$ eV.

At energies below $\sim 10^{13}$ eV, solar system magnetic fields and plasma can modulate the primary component and Figure 2 shows the extent of this modulation between solar maximum and minimum.[10,12]

PRIMARY PARTICLE ENERGY DENSITY

If the above spectrum is corrected for solar effects the energy density above a particle energy of 10^9 eV outside the solar system is found to be $\sim 5 \times 10^5$ eV m^{-3}. As the threshold energy is increased, the energy density decreases rapidly, being 2×10^4 eV m^{-3} above 10^{12} eV and 10^2 eV m^{-3} above 10^{15} eV.[16]

PRIMARY PARTICLE ISOTROPY

This is measured as an anisotropy $(I_{max} - I_{min})/(I_{max} + I_{min}) \times 100\%$, where I, the intensity (m^{-2}s^{-1}sr^{-1}), is usually measured with an angular resolution of a few degrees.

The anisotropy is small and energy dependent. It is roughly constant at between 0.05 and 0.1% for energies between 10^{11} eV and 10^{14} eV and increases at higher energies roughly as $0.4 \times (\text{energy (eV)}/10^{16})^{1/2}\%$ up to $\sim 10^{18}$ eV.[16]

PRIMARY PARTICLE COMPOSITION

The composition of low energy cosmic rays (it is uncertain at high energies) is close to universal abundances except where propagation effects are present, e.g. Li, Be, B which are spallation products, are over-abundant by ~ 6 orders of magnitude.

Composition at 10^{11} eV/Nucleus

Charge	1	2	(3—5)	(6—8)	(10—14)	(16—24)	(26—28)	≥30
% Composition	50	25	1	12	7	4	4	0.1

(~10% uncertainty)

Cosmic ray composition at low energies is often quoted at a fixed *energy per nucleon.* When presented in this way, protons constitute roughly 90% of the flux, helium nuclei ~10% and the remainder sum to a total of ~1%.

Certain radioactive isotopic ratios show lifetime effects. The ratio of Be^{10}/Be^9 abundances is used to measure an 'age' of cosmic rays since Be^{10} is unstable (half life = 1.6×10^6 years). A ratio of 0.6 would be expected in the absence of Be^{10} decay and a ratio of ~0.2 is found experimentally.[10,13]

PRIMARY ELECTRONS

Primary electrons constitute ~1% of the cosmic ray beam. The positron to electron ratio is about 10%.[14]

ANTIMATTER IN THE PRIMARY BEAM

The ratio of antiprotons to protons in the primary cosmic ray beam (at ~100 GeV) is $\sim 5 \times 10^{-4}$.[11]

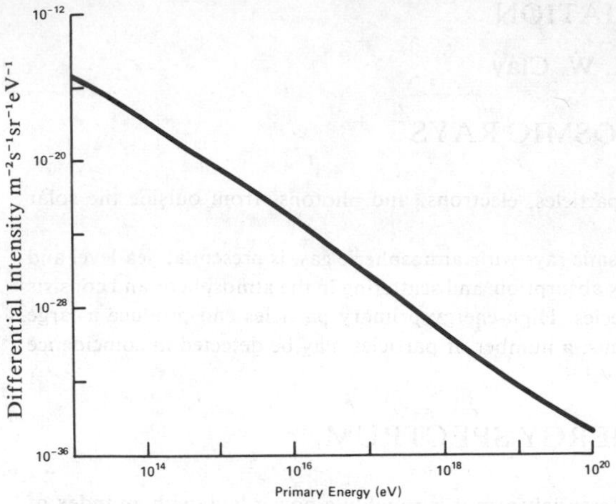

FIGURE 1. Energy Spectrum of Cosmic Ray Particles. This spectrum is of a differential form and can be converted to an integral spectrum by integrating over all energies above a required threshold (E).

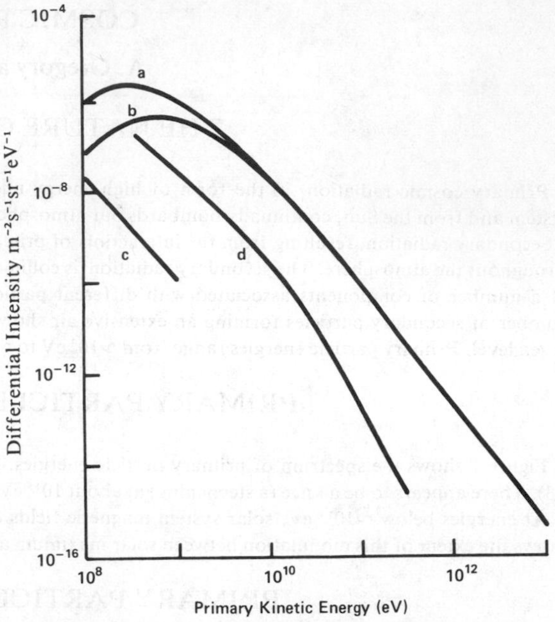

FIGURE 2. Energy Spectrum of Cosmic Ray Particles at Lower Energies. (a) Solar minimum proton energy spectrum, (b) solar maximum proton energy spectrum, (c) γ-ray energy spectrum, (d) local interstellar electron spectrum.

PRIMARY GAMMA RAYS

The flux of primary gamma rays is low at high energies. At 1 GeV the ratio of gamma rays to protons is about 10^{-6}. The arrival directions of these gamma rays are strongly concentrated in the plane of the Milky Way although there is also a diffuse near-isotropic background flux.

Since the absorption cross section for gamma rays above 100 MeV is approximately 20 mbarn/electron, less than 10% of gamma rays reach mountain altitudes.[16] Gamma rays from point sources have confidently been identified with energies up to 10^{16} eV.

SEA LEVEL COSMIC RADIATION

The sea level cosmic ray dose is 30 millirad. year^{-1} and the sea level ionisation is 2.22×10^6 ion pairs m^{-1}s^{-1}. The sea level flux has a soft component, which can be absorbed in ~10cm of lead and a more penetrating (largely muon) hard component. The sea level radiation is a secondary component produced in the atmosphere and its flux is dependent somewhat on the solar cycle and the geomagnetic latitude of the observer.

Absolute Flux of Hard Component[2]

Vertical integral intensity	(I) ~ 100 m^{-2}s^{-1}sr^{-1}
Angular dependence	$I(\theta) \sim I(O)\cos^2\theta$
Integrated intensity	~ 200 m^{-2}s^{-1}

Flux of Soft Component

In free air the soft component comprises about one third of the total cosmic ray flux.

Latitude Effect

The geomagnetic field influences the trajectories of lower energy cosmic rays approaching the Earth. As a result, the background flux is reduced by about 7% at the geomagnetic equator. The effect decreases towards the poles and is negligible at latitudes above about 40°.

Flux of Protons

The proton component is strongly attenuated by the atmosphere with an attenuation length (reduction by a factor of e) of ~ 120 g·cm^{-2}, it constitutes about 1% of the total vertical sea level flux.

Absorption

The soft component is absorbed in about 100 g·cm^{-2} of lead. The hard component is absorbed much more slowly:

Absorption in lead	6%/100 g·cm^{-2}
Absorption in rock	8.5%/100 g·cm^{-2}
Absorption in water	10%/100 g·cm^{-2}

Absorption for depths less than 100 g·cm^{-2} is given by Greisen in Reference 7.

Altitude Dependence

Cosmic ray background in the atmosphere has a maximum intensity of about 15 times that at sea level at a depth of ~ 150 g·cm^{-2} (15 km altitude). At maximum intensity the soft and hard component contribute roughly equally but the hard component is then attenuated more slowly.[8]

COSMIC RAY SHOWERS

High energy cosmic rays (< 10^{13}eV) produce measurable cascades of secondary particles in the atmosphere. The primary particle progressively loses energy which is transferred through the production of successive generations of secondary particles to a cascade of hadrons, an electromagnetic shower component (electrons and gamma rays) and muons. The secondary particles are relativistic and all travel effectively at the speed of light. As a result, they reach sea level at approximately the same time, but, due to Coulomb scattering and production angles, are spread laterally into a disk-like shower front with a typical width of ~ 100 m and thickness ~ 2 to 3 m. The number of particles at sea level is roughly proportional to the primary energy.

Number of particles at sea level ~ 10^{-10} × energy (eV). At altitudes below a few kilometres, the number of particles in a shower attenuates with an *attenuation length* ~ 200 g·cm^{-2}, i.e., $N = N_o \times \exp(-\text{depth}/200)$. The rate of observation of showers of a given size at different depths of absorber attenuates with an *absorption length* of ~ 100 g·cm^{-2}.[15]

ATMOSPHERIC BACKGROUND LIGHT FROM COSMIC RAYS

Cosmic ray particles produce Cerenkov light in the atmosphere and produce fluorescent light through the excitation of atmospheric molecules.

Cerenkov Light

High energy charged particles will emit Cerenkov light in air if their energies are above ~30 MeV (electrons). This threshold is pressure (and hence altitude) dependent. A typical Cerenkov light pulse (at sea level, 100 m from core, 10^{16} eV primary energy, in the wavelength band 430 to 530 nm) has a width of a few nanoseconds and in this time has a flux of ~10^{14} photons m^{-2}s^{-1}. For comparison, the general night sky background flux is ~6 × 10^{11} photons m^{-2}sec^{-1}sr^{-1} in the same wavelength band.

Fluorescent Light

Cosmic ray particles in the atmosphere excite atmospheric molecules which emit fluorescent light. This is weak compared to the Cerenkov component when viewed in the direction of the incident cosmic ray particle but is emitted isotropically. Typical pulse widths are expected to be longer than 50 ns and may be up to a few microseconds for distant large showers.[5,9]

CERENKOV EFFECTS IN TRANSPARENT MEDIA

Background cosmic ray particles will produce Cerenkov light in transparent material with a photon yield

$$\sim \frac{2\pi}{137} \sin^2\theta_c \int_{\lambda_1}^{\lambda_2} \frac{d\lambda}{\lambda^2} \text{ photons (unit length)}^{-1}$$

where θ_c, the Cerenkov angle, = $\cos^{-1}$ (1/refractive index). This background light is known to affect sensitive light detectors, e.g. photomultipliers, and can be a major source of background noise.[6]

EFFECTS ON ELECTRONIC COMPONENTS

If background cosmic rays pass through electronic components, they may deposit sufficient energy to affect the state of, e.g. a transistor flip-flop. This effect may be significant where reliability is of great importance and the background flux is high. For instance, it has been estimated that in communication satellite operation an error rate of ~ 2 × 10^{-3} per transistor per year may be found. Permanent damage may also result. A significant error rate may be found even at sea level in large electronic memories. This error rate is dependent on the sensitivity of the component devices to the deposition of electrons in their sensitive volumes.[17]

BIOPHYSICAL SIGNIFICANCE

When cosmic rays interact with living tissue they produce radiation damage, the amount of damage depending on the total dose of radiation. Radiation doses are commonly measured in rads (radiation absorbed dose ≡ 100 ergs g^{-1}) or rems (radiation equivalent-man ≡ Quality factor × rad). The quality factor of radiation depends on the type of particle and its energy for most cosmic ray applications it will be ~1.

At sea level the cosmic radiation dose rate is small compared with doses from other sources, but both the quantity, and quality of the radiation change rapidly with altitude. Approximate dose rates under various conditions

	Cosmic rays			
Conditions	Sea level	10 km (subsonic jets)	18 km (supersonic transport)	Mean total dose rate at sea level
Dose (mrem $year^{-1}$)	30	2000	10,000	300

Astronauts would be subject to radiation from galactic (0.05 rads d^{-1}) and solar (~ few hundred rads per solar flare) cosmic rays as well as large fluxes of low energy radiation when passing through the Van Allen belts (~0.3 rads per traverse).

Both astronauts and S.S.T. travellers would be subject to a small flux of low energy heavy nuclei stopping in the body. Such particles are capable of destroying cell nuclei and could be particularly harmful in the early stages of development of an embryo. The rates of heavy nuclei stopping in tissue in supersonic transports and spacecraft are approximately as follows (from Reference 1 and 2)

Conditions (altitude)	16 km SST	20 km SST	Spacecraft
Stopping nuclei (cm^3 tissue)$^{-1}$ hr^{-1}	5.10^{-4}	5.10^{-3}	0.15

CARBON DATING

Radiocarbon is produced in the atmosphere due to the action of cosmic ray slow neutrons. Solar cycle modulation of the low energy cosmic rays causes an anticorrelation of the atmospheric ^{14}C activity with sunspot number with a mean amplitude of ~0.5%. In the long term, modulation of cosmic rays by a varying geomagnetic field may be important.[4]

PRACTICAL USES OF COSMIC RAYS

There are few direct practical uses of cosmic rays. Their attenuation in water and snow have however enabled automatic monitors of water and snow depth to be constructed, and a search for hidden cavities in pyramids has been carried out using a muon 'telescope'.

OTHER EFFECTS

Stellar X-rays have been observed to affect the transmission times of radio signals between distant stations by altering the depth of the ionospheric reflecting layer. It has also been suggested that variations in ionization of the atmosphere due to solar modulation may have observable effects on climatic conditions.

REFERENCES

1. **Allkofer, O. C.,** *Introduction to Cosmic Radiation,* Verlag Karl Thiemig, Munchen, Germany, 1975.
2. **Allkofer, O. C.,** *J. Phys.,* Sect. G, 1, L51, 1975.
3. **Allkofer, O. C. and Heinrich, W.,** *Health Phys.,* 27, 543, 1974.
4. **Burchuladze, A. A., Pagava, S. V., Povinec, P., Togonidze, G. I., and Usacev, S.,** Proc. 16th Int. Cosmic Ray Conf. Kyoto, 3, 201, Univ. of Tokyo, 3, 201, 1979.
5. **Cassiday, G. L., et al.,** Proc. 15th Int Cosmic Ray Conf., Plovdiv, Bulgarian Academy of Sciences, 8, 258, 1977.
6. **Clay, R. W. and Gregory, A. G.,** *J. Phys.,* Sect. A, 10, 135, 1977.
7. **Greisen, K.,** *Physical Rev.,* 63, 323, 1943.
8. **Hayakawa, S.,** *Cosmic Ray Physics,* Wiley-Interscience, New York, 1969.
9. **Jelley, J. V.,** *Prog. in Elementary Particle and Cosmic Ray Physics,* 9, 41, 1967.
10. **Juliusson, E.,** Proc. 14th Int. Cos Ray Conf. Munich, 8, 2689, Max-Planck Institute for Extraterrestriche Physik, Munchen, Germany, 1975.

11. **Kiraly, P., Szabelski, J., Wdowczyk, J., and Wolfendale, A. W.,** *Nature,* 293, 120, 1981.
12. **Linsley, J.,** *Origin of Cosmic Rays,* I.A.U. Symposium 94, 53, D Reidel Publishing, Dordrecht, Holland, 1981.
13. **Meyer, P.,** *Origin of Cosmic Rays,* I.A.U. Symposium 94, 7, D. Reidel Publishing, Dordrecht, Holland, 1981.
14. **Tan, L. C., and Ng, L. K.,** *J. Phys.,* Sect. G, 7, 1135, 1981.
15. **Wilson, J. G.,** *Cosmic Rays,* Wykeham Publishing, London, 1976.
16. **Wolfendale, A. W.,** *Pramana,* 12, 631, 1979.
17. **Ziegler, J. F.,** *IEEE Trans. Electron Devices,* ED-28, 560, 1981.

CRYSTAL IONIC RADII OF THE ELEMENTS

Numerical values of the radii of the ions may vary depending on how they were measured. They may have been calculated from wavefunctions and determined from the lattice spacings or crystal structure of various salts. Different values are obtained depending on the kind of salt used or the method of calculating. Data for many of the rare-earth ions were furnished by F. H. Spedding and K. Gschneidner.

Element	Charge	Atomic number	Radius in A	Element	Charge	Atomic number	Radius in A	Element	Charge	Atomic number	Radius in A
Ac	+3	89	1.18	Hf	+4	72	0.78	Ra	+2	88	1.43
Ag	+1	47	1.26	Hg	+1	80	1.27	Rb	+1	37	1.47
	+2		0.89		+2		1.10	Re	+4	75	0.72
Al	+3	13	0.51	Ho	+3	67	0.894		+7		0.56
Am	+3	95	1.07	I	−1	53	2.20	Rh	+3	45	0.68
	+4		0.92		+5		0.62	Ru	+4	44	0.67
Ar	+1	18	1.54		+7		0.50	S	−2	16	1.84
As	−3	33	2.22	In	+3	49	0.81		+2		2.19
	+3		0.58	Ir	+4	77	0.68		+4		0.37
	+5		0.46	K	+1	19	1.33		+6		0.30
At	+7	85	0.62	La	+1	57	1.39	Sb	−3	51	2.45
Au	+1	79	1.37		+3		1.061		+3		0.76
	+3		0.85	Li	+1	3	0.68		+5		0.62
B	+1	5	0.35	Lu	+3	71	0.85	Sc	+3	21	0.732
	+3		0.23	Mg	+1	12	0.82	Se	−2	34	1.91
Ba	+1	56	1.53		+2		0.66		−1		2.32
	+2		1.34	Mn	+2	25	0.80		+1		0.66
Be	+1	4	0.44		+3		0.66		+4		0.50
	+2		0.35		+4		0.60		+6		0.42
Bi	+1	83	0.98		+7		0.46	Si	−4	14	2.71
	+3		0.96	Mo	+1	42	0.93		−1		3.84
	+5		0.74		+4		0.70		+1		0.65
Br	−1	35	1.96		+6		0.62		+4		0.42
	+5		0.47	N	−3	7	1.71	Sm	+3	62	0.964
	+7		0.39		+1		0.25	Sn	−4	50	2.94
C	−4	6	2.60		+3		0.16		−1		3.70
	+4		0.16		+5		0.13		+2		0.93
Ca	+1	20	1.18	NH_4	+1		1.43		+4		0.71
	+2		0.99	Na	+1	11	0.97	Sr	+2	38	1.12
Cd	+1	48	1.14	Nb	+1	41	1.00	Ta	+5	73	0.68
	+2		0.97		+4		0.74	Tb	+3	65	0.923
Ce	+1	58	1.27		+5		0.69		+4		0.84
	+3		1.034	Nd	+3	60	0.995	Tc	+7	43	0.979
	+4		0.92	Ne	+1	10	1.12	Te	−2	52	2.11
Cl	−1	17	1.81	Ni	+2	28	0.69		−1		2.50
	+5		0.34	Np	+3	93	1.10		+1		0.82
	+7		0.27		+4		0.95		+4		0.70
Co	+2	27	0.72		+7		0.71		+6		0.56
	+3		0.63	O	−2	8	1.32	Th	+4	90	1.02
Cr	+1	24	0.81		−1		1.76	Ti	+1	22	0.96
	+2		0.89		+1		0.22		+2		0.94
	+3		0.63		+6		0.09		+3		0.76
	+6		0.52	Os	+4	76	0.88	Ti	+4		0.68
Cs	+1	55	1.67		+6		0.69	Tl	+1	81	1.47
Cu	+1	29	0.96	P	−3	15	2.12		+3		0.95
	+2		0.72		+3		0.44	Tm	+3	69	0.87
Dy	+3	66	0.908		+5		0.35	U	+4	92	0.97
Er	+3	68	0.881	Pa	+3	91	1.13		+6		0.80
Eu	+3	63	0.950		+4		0.98	V	+2	23	0.88
	+2		1.09		+5		0.89		+3		0.74
F	−1	9	1.33	Pb	+2	82	1.20		+4		0.63
	+7		0.08		+4		0.84		+5		0.59
Fe	+2	26	0.74	Pd	+2	46	0.80	W	+4	74	0.70
	+3		0.64		+4		0.65		+6		0.62
Fr	+1	87	1.80	Pm	+3	61	0.979	Y	+3	39	0.893
Ga	+1	31	0.81	Po	+6	84	0.67	Yb	+2	70	0.93
	+3		0.62	Pr	+3	59	1.013		+3		0.858
Gd	+3	64	0.938		+4		0.90	Zn	+1	30	0.88
Ge	−4	32	2.72	Pt	+2	78	0.80		+2		0.74
	+2		0.73		+4		0.65	Zr	+1	40	1.09
	+4		0.53	Pu	+3	94	1.08		+4		0.79
H	−1	1	1.54		+4		0.93				

BOND LENGTHS BETWEEN CARBON AND OTHER ELEMENTS

Prepared by Olga Kennard.

The tables are based on bond distance determinations, by experimental methods, mainly X-ray and electron diffraction, and include values published up to January 1, 1956. In the present tables, for the sake of completeness individual values of bond distances of lower accuracy are quoted with limits of error indicated where possible. Values for tungsten and bismuth should be treated with particular caution.

According to the statistical theory of errors if an average quantity $\bar{\mu}$ and a standard deviation σ can be evaluated there is a 95% probability that the true value lies within the interval $\bar{\mu} \pm 2\sigma$. Too much reliance should, however, not be placed on σ values in bond distance determinations since the derivation of these certain sources of error may have been neglected.

Values of the bond lengths and the limits of error are each given in Ångstrom units.

Reproduced by permission from International Tables for X-ray Crystallography.

BOND LENGTHS BETWEEN CARBON AND OTHER ELEMENTS

Reference: HCP and "Tables of interatomic distances" Chem. Soc. of London, 1958

Group	Bond type	Element						
I	All types	H**						
		1.056 − 1.115						
II		Be	Hg					
		1.93	2.07 ± 0.01					
III		B	Al	In				
		1.56 ± 0.01	2.24 ± 0.04	2.16 ± 0.04				
IV	All types	C**	Ge	Si	Sn	Pb		
	Alkyls (CH_3XH_3)	1.54 − 1.20	1.98 ± 0.03	1.865 ± 0.008	2.143 ± 0.008	2.29 ± 0.05		
	Aryl ($C_6H_5XH_3$)			1.84 ± 0.01	2.18 ± 0.02			
	Neg. Subst. (CH_3XCl_3)			1.88 ± 0.01				
V	All types	N**	P	As	Sb	Bi		
	Paraffinic $(CH_3)_3X$	1.47 − 1.1	1.87 ± 0.02	1.98 ± 0.02	2.202 ± 0.016	2.30*		
VI		O**	S**	Cr	Se	Te	Mo	W
		1.43 − 1.15	1.81 − 1.55	1.92 ± 0.04	1.98 − 1.71	2.05 ± 0.14	2.08 ± 0.04	2.06 ± 0.01*
VII	Paraffinic (monosubstituted)	F	Cl	Br	I			
	(CH_3X)	1.831 ± 0.005	1.767 ± 0.002	1.937 ± 0.003	2.13_5 ± 0.01			
	Paraffinic (disubstituted) (CH_2X_2)	1.334 ± 0.004	1.767 ± 0.002	1.937 ± 0.003	2.13_5 ± 0.1			
	Olefinic (CH_2:CHX)	1.32_5 ± 0.1	1.72 ± 0.01	1.89 ± 0.01	2.092 ± 0.005			
	Aromatic (C_6H_3X)	1.30 ± 0.01	1.70 ± 0.01	1.85 ± 0.01	2.05 ± 0.01			
	Acetylenic (HC:CX)		1.635 ± 0.004	1.79_5 ± 0.01	1.99 ± 0.02			
VIII		Fe	Co	Ni	Pd			
		1.84 ± 0.02	1.83 ± 0.02	1.82 ± 0.03	2.27 ± 0.04			

* Error uncertain.

** See following individual tables.

CARBON-CARBON

Single Bond	
Paraffinic	1.541 ± 0.003
In diamond (18°C)	1.54452 ± 0.00014
Partial Double Bond	
(1) Shortening of single bond in presence of carbon carbon double bond, e.g. $(CH_3)_2C:CH_2$; or of aromatic ring e.g. C_6H_5. CH_3	1.53 ± 0.01
(2) Shortening in presence of a carbon oxygen double bond e.g. CH_3CHO	1.516 ± 0.005
(3) Shortening in presence of two carbon-oxygen double bonds, e.g. $(CO_2H)_2$	1.49 ± 0.01
(4) Shortening in presence of one carbon-carbon triple bond, e.g. CH_3.C⋮CH	1.460 ± 0.003
(5) In compounds with tendency to dipole formation, e.g. C:C.C⋮N	1.44 ± 0.01
(6) In graphite (at 15°C)	1.4210 ± 0.0001
(7) In aromatic compounds	1.395 ± 0.003
(8) In presence of two carbon carbon triple bonds, e.g. HC⋮C.C⋮CH	1.373 ± 0.004
Double Bond	
(1) Simple	1.337 ± 0.006
(2) Partial triple bond, e.g. CH_2:C:CH_2	1.309 ± 0.005
Triple Bond	
(1) Simple, e.g. C_2H_2	1.204 ± 0.002
(2) Conjugated, e.g. $CH_3.(C:C)_2$.H	1.206 ± 0.004

CARBON-HYDROGEN

(1) Paraffinic (a) in methane	1.091
(b) in monosubstitured carbon	1.101 ± 0.003
(c) in disubstituted carbon	1.073 ± 0.004
(d) in trisubstituted carbon	1.070 ± 0.007
(2) Olefinic, e.g. CH_2:CH_2	1.07 ± 0.01
(3) Aromatic in C_6H_6	1.084 ± 0.006
(4) Acetylenic, e.g. CH⋮C.X	1.056 ± 0.003
(5) Shortening in presence of a carbon triple bond, e.g. CH_3CN	1.115 ± 0.004
(6) In small rings, e.g. $(CH_2)_2S$	1.081 ± 0.007

CARBON-NITROGEN

Single Bond	
(1) Paraffinic (a) 4 co-valent nitrogen	1.479 ± 0.005
(b) 3 co-valent nitrogen	1.472 ± 0.005
(2) In C−N= e.g. CH_3NO_2	1.475 ± 0.010
(3) Aromatic in $C_6H_5NHCOCH_3$	1.426 ± 0.012
(4) Shortened (partial double bond) in heterocyclic systems, e.g. C_5H_5N	1.352 ± 0.005
(5) Shortened (partial double bond) in N−C=O e.g. $HCONH_2$	1.322 ± 0.003
Triple Bond	
(1) in R.C:N	1.158 ± 0.002

BOND LENGTHS BETWEEN CARBON AND OTHER ELEMENTS (Continued)

CARBON-OXYGEN

Single Bond	
(1) Paraffinic	1.43 ± 0.01
(2) Strained e.g. epoxides	1.47 ± 0.01
(3) Shortened (partial double bond) as in carboxylic acids or through influence of aromatic ring, e.g. salicylic acid	1.36 ± 0.01
Double Bond	
(1) In aldehydes, ketones, caboxylic acids, esters	1.23 ± 0.01
(2) In zwitterion forms, e.g. DL serine	1.26 ± 0.01
(3) Shortened (partial triple bond) as in conjugated systems	1.207 ± 0.006
(4) Partial triple bond as in acyl halides or isocyanates	1.17 ± 0.01

CARBON-SULFUR

Single Bond	
(1) Paraffinic, e.g. CH_3SH	1.81(5) ± 0.01
(2) Lengthened in presence of fluorine, e.g. $(CF_3)_2S$	1.83(5) ± 0.01
(3) Shortened (partial double bond) as in heterocyclic systems, e.g. C_4H_4S	1.73 ± 0.01
Double Bond	
(1) In ethylene thiourea	1.71 ± 0.02
(2) Shortened (partial triple bond) in presence of second caron double bond, e.g. COS	1.558 ± 0.003

BOND LENGTHS OF ELEMENTS

Element	Bond	Å
Ac	Ac–Ac	3.756
Ag (25°C)	Ag–Ag	2.8894
Al (25°C)	Al–Al	2.863
As	As–As	2.49
As_4	As–As	2.44 ± 0.03
Au (25°C)	Au–Au	2.8841
B_2	B–B	1.589
Ba (room temp.)	Ba–Ba	4.347
Be (α-form, 20°C)	Be–Be	2.2260
Bi (25°C)	Bi–Bi	3.09
Br_2	Br–Br	2.290
Ca (α-form, 18°C)	Ca–Ca	3.947 (f.c.c.)
(β-form, 500°C		3.877 (b.c.c.)
Cd (21°C)	Cd–Cd	2.9788
Cl_2	Cl–Cl	1.988
Ce	Ce–Ce	3.650
Co (18°C)	Co–Co	2.5061
Cr (α-form, 20°C)	Cr–Cr	2.4980
(β-form, >1850°C		2.61
Cs (−10°C)	Cs–Cs	5.309
Cu (20°C)	Cu–Cu	2.5560
Dy	Dy–Dy	3.503
Er	Er–Er	3.468
Eu	Eu–Eu	3.989
F_2	F–F	1.417 ± 0.001
Fe (α-form, 20°C)	Fe–Fe	2.4823 (b.c.c.)
(γ-form, 916°C)		2.578 (f.c.c.)
(δ-form, 1394°C)		2.539 (b.c.c.)
Ga (20°C)	Ga–Ga	2.442
Gd (20°C)	Gd–Gd	3.573
Ge (20°C)	Ge–Ge	2.4498
H_2	H–H in H_2	0.74611
	H–D in HD	0.74136
	D–D in D_2	0.74164
He	He–He in $[He_2]^+$	1.08_0
Hf (α-form, 24°C)	Hf–Hf	3.1273 (h.c.p.)
Hg (−46°C)	Hg–Hg	3.005
Ho	Ho–Ho	3.486
I_2	I–I	2.662
In (20°C)	In–In	3.2511
Ir (room temp.)	Ir–Ir	2.714
K (78°K)	K–K	4.544
La (α-form)	La–La	3.739 (h.c.p.)
(β-form)		3.745 (f.c.c.)
Li (20°C)	Li–Li	3.0390
Lu	Lu–Lu	3.435
Mg (25°C)	Mg–Mg	3.1971
Mn (γ-form, 1095°C)	Mn–Mn	2.7311 (f.c.c.)
(δ-form, 1134°C)		2.6679 (b.c.c.)
Mo (20°C)	Mo–Mo	2.7251
N_2	N–N	1.0975_8 ± 0.0001
Na (20°C)	Na–Na	3.7157
Nb (20°C)	Nb–Nb	2.8584
Nd	Nd–Nd	3.628
Ni (18°C)	Ni–Ni	2.4916

Element	Bond	Å
Np (α-form, 20° C)	Np–Np	2.60 (orthorhombic)
(β-form, 313°C)		2.76 (tetragon)
(γ-form, 600°C)		3.05 (b.c.c.)
O_2	O–O	1.208
O_3 angle 116.8 ± 0.5°		1.278 ± 0.003
Os (20°C)	Os–Os	2.6754
P black	P–P	2.18
P_4	P–P	2.21 ± 0.02
Pa	Pa–Pa	3.212
Pb (25°C)	Pb–Pb	3.5003
Pd (25°C)	Pd–Pd	2.7511
Po (α-form, 10°C)	Po–Po	3.345 (cubic)
(β-form, 75°C)		3.359 (rh. hedr.)
Pr (α-form)	Pr–Pr	3.640 (tetrag.)
(β-form)		3.649 (f.c.c.)
Pt (20°C)	Pt–Pt	2.746
Pu (γ-form, 235°C)	Pu–Pu	3.026 (f.c.c.)
(δ-form, 313°C)		3.279 (f.c.c.)
(ε-form, 500°C)		3.150 (b.c.c.)
Rb (20°C)	Rb–Rb	4.95
Re (room temp.)	Re–Re	2.741
Rh (20°C)	Rh–Rh	2.6901
Ru (25°C)	Ru–Ru	2.6502
S_2	S–S	1.887
S_8	S–S	2.07 ± 0.02
Sb (25°C)	Sb–Sb	2.90
Sc (room temp.)	Sc–Sc	3.212
Se (20°C)	Se–Se	2.321
Se_2	Se–Se	2.152 ± 0.003
Se_8	Se–Se	2.32 ± 0.003
Si (20°C)	Si–Si	2.3517
Sn (α-form, 20°C)	Sn–Sn diamond	2.8099
(β-form, 25°C)	type lattice	3.022 (tetrag.)
Sr (α-form, 25°C)	Sr–Sr	4.302 (f.c.c.)
(β-form, 248°C)		4.32 (h.c.p.)
(γ-form, 614°C)		4.20 (b.c.c.)
Ta (20°C)	Ta–Ta	2.86
Tb	Tb–Tb	3.525
Tc (room temp.)	Tc–Tc	2.703
Te (25°C)	Te–Te	2.864
Th (α-form, 25°C)	Th–Th	3.595 (f.c.c.)
(β-form, 1450°C)		3.56 (b.c.c.)
Ti (α-form, 25°C)	Ti–Ti	2.8956 (h.c.p.)
(β-form, 900°C)		2.8636 (b.c.c.)
Tl (α-form, 18°C)	Tl–Tl	3.4076 (h.c.p.)
(β-form, 262°C)		3.362 (b.c.c.)
Tm	Tm–Tm	3.447
U (α-form)	U–U	2.77
(β-form, 805°C)		3.058 (b.c.c.)
V (30°C)	V–V	2.6224
W (25°C)	W–W	2.7409
Y	Y–Y	3.551
Yb	Yb–Yb	3.880
Zn (25°C)	Zn–Zn	2.6694
Zr	Zr–Zr	3.179

BOND LENGTH AND ANGLE VALUES BETWEEN ELEMENTS

Elements	In	Bond length (Å)	Bond angle (°)
Boron			
B–B	B_2H_6	1.770 ± 0.013	H–B–H 121.5 ± 7.5
B–Br	BBr	1.88_7	–
	BBr_3	1.87 ± 0.02	Br–B–Br 120 ± 6
B–Cl	BCl	1.715_7	–
	BCl_3	1.72 ± 0.01	Cl–B–Cl 120 ± 3
B–F	BF	1.262	–
	BF_3	1.29_5 ± 0.01	F–B–F 120
B–H	Hydrides	1.21 ± 0.02	–
B–H Bridge	Hydrides	1.39 ± 0.02	–
B–N	$(BClNH)_3$	1.42 ± 0.01	B–N–B 121
B–O	BO	1.2049	–
	$B(OH)_3$	1.362 ± 0.005 (av.)	O–B–O 119.7
Nitrogen			
N–Cl	NO_2Cl	1.79 ± 0.02	–
N–F	NF_3	1.36 ± 0.02	F–N–F 102.5 ± 1.5
N–H	$[NH_4]^+$	1.034 ± 0.003	–
	NH	1.038	–
	ND	1.041	–
	HNCS	1.013 ± 0.005	H–N–C 130.25 ± 0.25
N–N	N_3H	1.02 ± 0.01	H–N–N′ 112.65 ± 0.5
	N_2O	1.126 ± 0.002	–
	$[N_2]^+$	1.116_2	–
N–O	NO_2Cl	1.24 ± 0.01	O–N–O 126 ± 2
	NO_2	1.188 ± 0.005	O–N–O 134.1 ± 0.25
N=O	N_2O	1.186 ± 0.002	–
	$[NO]^+$	1.0619	–
N–Si	SiN	1.572	–
Oxygen			
O–H	$[OH]^+$	1.0289	–
	OD	0.9699	–
	H_2O_2	0.960 ± 0.005	O–O–H 100 ± 2
O–O	H_2O_2	1.48 ± 0.01	–
	$[O_2]^+$	1.227	–
	$[O_2]^-$	1.26 ± 0.02	–
	$[O_2]^{--}$	1.49 ± 0.02	–
Phosphorus			
P–D	PD	1.429	–
P–H	$[PH_4]^+$	1.42 ± 0.02	–
P–N	PN	1.4910	–
P–S	$PSBr_3(Cl_3,F_3)$	1.86 ± 0.02	–
Sulfur			
S–Br	$SOBr_2$	2.27 ± 0.02	Br–S–Br 96 ± 2
S–F	SOF_2	1.585 ± 0.005	F–S–F 92.8 ± 1
S–D	SD	1.3473	–
	SD_2	1.345	–
S–O	SO_2	1.4321	O–S–O 119.54
	$SOCl_2$	1.45 ± 0.02	–
S–S	S_2Cl_2	2.04 ± 0.01	–
Silicon			
Si–Br	$SiBr_4$	2.17 ± 1.01	–
Si–Cl	$SiCl_4$	2.03 ± 1.01 (av.)	–
Si–F	SiF_4	1.561 ± 0.003 (av.)	–
SiH	SiH_4	1.480 ± 0.005	–
Si–O	$[SiO]^+$	1.504	–
Si–Si	Si_2Cl_2	2.30 ± 0.02	–

BOND LENGTHS AND ANGLES OF CHEMICAL COMPOUNDS

A. Inorganic Compounds

Compound	Formula	Bond lengths in Å		Bond angles (°)	
Ammonia	NH_3	N–H	1.008 ± 0.004	H–N–H	1.07.3 ± 0.2
Antimony tribromide	$SbBr_3$	Sb–Br	2.51 ± 0.02	Br–Sb–Br	97 ± 2
Antimony trichloride	$SbCl_3$	Sb–Cl	2.352 ± 0.005	Cl–Sb–Cl	99.5 ± 1.5
Antimony triiodide	SbI_3	Sb–I	2.67 ± 0.03	I–Sb–I	99.0 ± 1
Arsenic tribromide	$AsBr_3$	As–Br	2.33 ± 0.02	Br–As–Br	100.5 ± 1.5
Arsenic trichloride	$AsCl_3$	As–Cl	2.161 ± 0.004	Cl–As–Cl	98.4 ± 0.5
Arsenic trifluoride	AsF_3	As–F	1.712 ± 0.005	F–As–F	102.0 ± 2
Arsenic triiodide	AsI_3	As–I	2.55 ± 0.03	I–As–I	101.0 ± 1.5
Arsenic trioxide	As_4O_6	As–O	1.78 ± 0.02	O–As–O	99.0 ± 2
				As–O–As	128.0 ± 2
Arsine	AsH_3	As–H	1.5192 ± 0.002	H–As–H	91.83 ± 0.33
Bismuthum tribromide	$BiBr_3$	Bi–Br	2.63 ± 0.02	Br–Bi–Br	100.0 ± 4
Bismuthum trichloride	$BiCl_3$	Bi–Cl	2.48 ± 0.02	Cl–Bi–Cl	100.0 ± 6
Bromosilane	SiH_3Br	Si–H	1.57 ± 0.03	H–Si–H	111.3 ± 1
Chlorine dioxide	ClO_2	Cl–O	1.49	O–Cl–O	118.5
Chlorogermane	GeH_3Cl	Ge–H	1.52 ± 0.03	H–Ge–H	110.9 ± 1.5
Chlorosilane	SiH_3Cl	Si–H	1.483 ± 0.001	H–Si–H	110 ± 0.03
		Si–Cl	2.0479 ± 0.007		

BOND LENGTHS AND ANGLES OF CHEMICAL COMPOUNDS (CONTINUED)

A. Inorganic Compounds (Continued)

Compound	Formula	Bond lengths in Å		Bond angles (°)	
Chromium oxychloride	$Cr(OCl)_2$	Cr–O	1.57 ± 0.03	O–Cr–O	105 ± 4
		Cr–Cl	2.12 ± 0.02	Cl–Cr–Cl	113 ± 3
				Cl–Cr–O	109 ± 3
Cyanuric triazide	C_3N_{12}	C–N	1.38	C–N=C	113.0
Dichlorosilane	SiH_2Cl_2	Si–H	1.46		
		SiCl	2.02 ± 0.03	Cl–Si–Cl	110 ± 1
Difluorodiazine	N_2F_2	N–F	1.44 ± 0.04		
		N–N	1.25 ± 0.02	F–N=N	115 ± 5
Difluoromethylsilane	Ch_3SiHf_2			F–Si–F	106 ± 0.5
				H–Si–C	116.2 ± 1
				C–Si–F	109.8 ± 0.5
Disilicon hexachloride	Si_2Cl_6	Si–Cl	2.02 ± 0.02	Cl–Si–Cl	109.5 ± 1
(hexachlorisilane)		Si–Si	2.34 ± 0.06		
Fluorosilane	SiH_3F	Si–H	1.460 ± 0.01	H–Si–H	109.3 ± 0.3
		Si–F	1.595 ± 0.002		
Hydrogen phosphide	PH_3	P–H	1.415 ± 0.003	H–P–H	93.3 ± 0.2
Hydrogen selenide	SeH_2	Se–H	1.47	H–Se–H	91.0
Hydrogen sulfide	SH_2	S–H	1.3455	H–S–H	93.3
Hydrogen telluride	$Te–H_2$			H–Te–H	89.5 ± 1
Iodo silane	$Si–H_3I$	Si–H	1.48 ± 0.01	H–Si–H	109.9 ± 0.4
		Si–I	2.4 ± 0.09		
Methylgermane	Ch_3GeH_3			H–C–H	108.2 ± 0.5
				H–Ge–H	108.6 ± 0.5
Nitrosyl bromide	NOBr	O–N	1.15 ± 0.04	Br–N=O	117 ± 3
		N–Br	2.14 ± 0.02		
Nitrosyl chloride	NOCl	N–O	1.14 ± 0.02	Cl–N=O	113.0 ± 2
		N–Cl	1.97 ± 0.01		
Nitrosyl fluoride	NOF	N–O	1.13	F–N=O	110.0
		N–F	1.52		
cis-Nitrous acid	NO(OH)	H–O	0.98	O–N=O	114 ± 2
		O–N	1.46		
		N–O′	1.20		
trans-Nitrous acid	NO(OH)	H–O	0.98	O–N=O	118 ± 2
		O–N	1.46		
		NO′	1.20		
Oxygen chloride	OCl_2	O–Cl	1.70_1 ± 0.02	Cl–O–Cl	110.8 ± 1
Oxygen fluoride	OF_2	O–F	1.418	F–O–F	103.2
Phosphorus oxychloride	$POCl_3$	P–Cl	1.99_5 ± 0.02	Cl–P–Cl	103.5 ± 1
Phosphorus oxysulfide	$P_4O_6S_4$	P–O	1.61 ± 0.02	P–O–P	128.5 ± 1.5
				O–P–O	101.5 ± 1
		P–S	1.85 ± 0.02	O–P–S	116.5 ± 1
Phosphorus pentoxide	P_4O_{10}	P–O	1.62 ± 0.02	O–P–O	101.5 ± 1
		P–O′	1.38 ± 0.02	O–P–O′	116.5 ± 1
				P–O–P	123.5 ± 1
Phosphorus tribromide	PBr_3	P–Br	2.18 ± 0.03	Br–P–Br	101.5 ± 1.5
Phosphorus trichloride	PCl_3	P–Cl	2.043 ± 0.003	Cl–P–Cl	100.1 ± 0.3
Phosphorus trifluoride	PF_3	P–F	1.535	F–P–F	100.0
Phosphorus trioxide	P_4O_6	P–O	1.65 ± 0.02	O–P–O	99.0 ± 1
				P–O–P	127.5 ± 1
Stibine	SbH_3	Sb–H	1.7073 ± 0.0025	H–Sb–H	91.3 ± 0.33
Sulfur dichloride	SCl_2	S–Cl	1.99 ± 0.03	Cl–S–Cl	101.0 ± 4
Sulfur dioxide	SO_2	S–O	1.4321	O–S–O	119.536
Sulfur monochloride	S_2Cl_2	S–Cl	2.01 ± 0.07	Cl–S–S	104.5 ± 0.25
Sulfurylchloride	SO_2Cl_2	S–O	1.43 ± 0.02	O–S–O	119.75 ± 5
		S–Cl	1.99 ± 0.02	Cl–S–Cl	111.20 ± 2
				Cl–O–O	106.5 ± 2
Tellurium bromide	$TeBr_2$	Te–Br	2.51 ± 0.02	Br–Te–Br	98.0 ± 3
Tribromo silane	$SiHBr_3$	Si–Br	2.16 ± 0.03	Br–Si–Br	110.5 ± 1.5
Trichloro germane	$GeHCl_3$	Ge–H	1.55 ± 0.04	Cl–Ge–Cl	108.3 ± 0.2
Trichloro silane	$Si–HCl_3$	Si–H	1.47	Cl–Si–Cl	109.4 ± 0.3
Trifluorochlorosilane	$SiClF_3$	Si–F	1.560 ± 0.005	F–Si–F	108.5 ± 1
		Si–Cl	1.989 ± 0.018		
Trifluorochlorogermane	$GeClF_3$	Ge–F	1.688 ± 0.0017	F–Ge–F	107.7 ± 1.5
		Gr–Cl	2.067 ± 0.005		
Trifluorosilane	$SiHF_3$	Si–H	1.455 ± 0.01		
		Si–F	1.565 ± 0.005	F–Si–F	108.3 ± 0.5
Vanadium oxytrichloride	$VOCl_3$	V–O	1.56 ± 0.04	Cl–V–Cl	111.2 ± 2
		V–Cl	2.12 ± 0.03	Cl–V–O	108.2 ± 2
Water	H_2O	O–H	0.958_4	H–O–H	104.45

B. Organic Compounds

Compound	Formula	Bond lengths in Å		Bond angles (°)	
Acetaldehyde	CH_3COH	C–H	1.09	C–C=O	121 ± 2
		C–C	1.50 ± 0.02		
		C–O	1.22 ± 0.02		
Bromomethane	Ch_3Br	C–H	1.11 ± 0.01	H–C–H	111.2 ± 0.5
		C–Br	1.929		
Carbon tetrachloride	CCl_4	C–Cl	1.766 ± 0.003	Cl–C–Cl	109.5
		Cl–Cl	2.887 ± 0.004		
Chloromethane	CH_3Cl	C–H	1.11 ± 0.01	H–C–H	110 ± 2
		C–Cl	1.784 ± 0.003		
Dichloromethane	CH_2Cl_2	C–H	1.068 ± 0.005	H–C–H	112 ± 0.3
		C–Cl	1.7724 ± 0.0005	Cl–C–Cl	111.8

BOND LENGTHS AND ANGLES OF CHEMICAL COMPOUNDS (CONTINUED)

Compound	Formula	Bond lengths in Å		Bond angles (°)	
Difluorochloromethane	$CHCLF_2$	C–H	1.06	F–C–F	110.5 ± 1
		C–Cl	1.73 ± 0.03	Cl–C–Cl	110.5 ± 1
		C–F	1.36 ± 0.03		
1,1-Difluoroethylene	$C_2H_2F_2$	C–H	1.07 ± 0.02	F–C=C	125.2 ± 0.2
		C–F	1.321 ± 0.015	H–C–H	117 ± 7
		C–C	1.311 ± 0.035		
Difluoromethane	CH_2F_2	C–F	1.360 ± 0.005	F–C–F	108.2 ± 0.8
		C–H	1.09 ± 0.03	H–C–H	112.5 ± 6
p-Dinitrobenzene	$C_6H_4(NO_2)_2$	C–C	1.38	O–N=O	124.0
		C–N	1.48		
		N–O	1.21		
Dithio oxamide	$NH_2CSCSNH_2$	C–C	1.54	N–C=S	124.8
		C–N	1.30	S–C–C	124.87
		C–S	1.66	N–C–C	155.25
Ethane	C_2H_6	C–H	1.107	H–C–H	109.3
		C–C	1.536		
Ethylidene fluoride	CH_3CHF_2	C–F	1.345 ± 0.001	F–C–F	109.15 ± 0.001
		C–C	1.540	C–C–F	109.4
		C–H	1.100	H–C–C	110.2
Fluorochloromethane	CH_2ClF	C–H	1.078 ± 0.005	Cl–C–F	100.0 ± 0.1
		C–Cl	1.759 ± 0.003		
		C–F	1.378 ± 0.006		
Fluorotrichloromethane	$CFCl_3$	C–Cl	1.76 ± 0.02	Cl–C–Cl	113 ± 3
		C–F	1.44 ± 0.04		
Formaldehyde	CH_2O	C–H	1.060 ± 0.038	H–C–H	125.8 ± 7
		C–O	1.230 ± 0.017		
Formamide	$HCONH_2$	C–O	1.25_6	N–C=O	121.5
		C–N	1.300		
Formic acid	HCOOH	C–O′	1.245	O–C=O′	124.3
		C–O	1.312	H–C–O′	117.8
		C–H	1.085	C–O–H	107.8
		O–H	0.95		
Glycine	NH_2CH_2COOH	C–C	1.52	C–C–O	119.0
		C–O	1.27	O–C–O	122.0
		C–N	1.39	C–C–N	112.0
Hexachloroethane	C_2Cl_6	C–Cl	1.74 ± 0.01	Cl–C–Cl	109.3 ± 0.01
		C–C	1.57 ± 0.06		
Iodomethane	CH_3I	C–H	1.11 ± 0.01	H–C–H	111.4 ± 0.1
		C–I	2.139		
Methane	CH_4	C–H	1.091		
Methanethiol	CH_3SH	C–H	1.1039 ± 0.002		110.3 ± 0.2
		C–S	1.8177 ± 0.0002	H–S–C	100.3 ± 0.2
		S–H	1.329 ± 0.004		
Methanol	CH_3OH	C–H	1.096 ± 0.01	H–C–H	109.3 ± 0.75
		C–O	1.427 ± 0.007	C–O–H	108.9 ± 2
		O–H	0.956 ± 0.015		
Methylamine	CH_3NH_2	C–H	1.093	H–C–H	109.5 ± 1
		C–N	1.474 ± 0.005	H–N–H	105.8 ± 1
		N–H	1.014		
Methylether	$(CH_3)_2O$	C–O	1.43 ± 0.03	C–O–C	110.0 ± 3
Methylnitrite	CH_3NO_2	C–H	1.00	O–N=O	127 ± 4
		C–N	1.49 ± 0.02		
		N–O	1.22 ± 0.1		
Methylsulfide	$(CH_3)_2S$	C–S	1.82 ± 0.01	C–S–C	105 ± 3
		C–H	1.06	H–C–H	109.5
Oxamide	$NH_3COCONH_2$	C–C	1.54	N–C=O	125.7 ± 0.3
		C–O	1.24		
		C–N	1.32		
Phosgene	CCl_2O	C–Cl	4.745 ± 0.004	Cl–C=O	124.3 ± 0.1
		C–O	1.166 ± 0.002	Cl–C–Cl	111.3 ± 0.1
Propylene	C_3H_6			C–C=C	124.75 ± 0.3
Propynal	CHC.COH	C_1–C_2	1.204	C–C=O	123.0
		C_2–C_3	1.46	C–C–H	120.0
		C–O	1.21		
		C_1–H	1.06		
		C_3–H	1.08		
Tribromomethane	$CHBr_2$	C–H	1.068 ± 0.01	Br–C–Br	110.8 ± 0.3
		C–Br	1.930 ± 0.003		
Trichlorobromomethane	$CBrCl_3$	C–Br	1.936	Cl–C–Cl	111.2 ± 1
		C–Cl	1.764		
Trifluorochloromethane	$CClF_3$	C–Cl	1.751 ± 0.004	F–C–F	108.6 ± 0.4
		C–F	1.328± 0.02		
Trifluoromethane	CHF_3	C–H	1.098	F–C–F	108 ± 0.75
		C–F	1.332 ± 0.008		
Triiodomethane	CHI_3	C–I	2.12 ± 0.04	I–C–I	113.0 ± 1
Trimethylamine	$(CH_3)_3N$	C–N	1.47 ± 0.01	C–N–C	108.0 ± 4
		C–H	1.06	H–C–H	109.5
Trimethylarsine	$(CH_3)_3As$	C–H	1.09	C–As–C	96 ± 5
		C–As	1.98 ± 0.02		
Trimethylphosphine	$(CH_3)_3P$	C–H	1.09	C–P–C	100.0 ± 4
		C–P	1.87 ± 0.02		

ELECTRON BINDING ENERGIES OF THE ELEMENTS (UNITS ARE ELECTRON VOLTS)

Prepared from a chart containing data compiled by Gwyn P. Williams of the National Synchrotron Light Source, Brookhaven National Laboratory, Upton, New York 11973, U.S.A.

Actinium (89)

K 1s	106755	N_{IV} $4d_{3/2}$	675*
L_{I} 2s	19840	N_{V} $4d_{5/2}$	639*
L_{II} $2p_{1/2}$	19083	N_{VI} $4f_{5/2}$	319*
L_{III} $2p_{3/2}$	15871	N_{VII} $4f_{7/2}$	319*
M_{I} 3s	5002	O_{I} 5s	272*
M_{II} $3p_{1/2}$	4656	O_{II} $5p_{1/2}$	215*
M_{III} $3p_{3/2}$	3909	O_{III} $5p_{3/2}$	167*
M_{IV} $3d_{3/2}$	3370	O_{IV} $5d_{3/2}$	80*
M_{V} $3d_{5/2}$	3219	O_{V} $5d_{5/2}$	80*
N_{I} 4s	1269*	P_{I} 6s	—
N_{II} $4p_{1/2}$	1080*	P_{II} $6p_{1/2}$	—
N_{III} $4p_{3/2}$	890*	P_{III} $6p_{3/2}$	—

Aluminum (13)

K 1s	1562.3*	L_{II} $2p_{1/2}$	72.9*
L_{I} 2s	117.8*	L_{III} $2p_{3/2}$	72.5*

Antimony (51)

K 1s	30491	M_{IV} $3d_{3/2}$	537.5†
L_{I} 2s	4698	M_{V} $3d_{5/2}$	528.2†
L_{II} $2p_{1/2}$	4380	N_{I} 4s	153.2†
L_{III} $2p_{3/2}$	4132	N_{II} $4p_{1/2}$	95.6†a
M_{I} 3s	946 †	N_{III} $4p_{3/2}$	95.6†a
M_{II} $3p_{1/2}$	812.7†	N_{IV} $4d_{3/2}$	33.3†
M_{III} $3p_{3/2}$	766.4†	N_{V} $4d_{5/2}$	32.1†

Argon (18)

K 1s	3205.9*	M_{I} 3s	29.3*
L_{I} 2s	326.3*	M_{II} $3p_{1/2}$	15.9*
L_{II} $2p_{1/2}$	250.6*	M_{III} $3p_{3/2}$	15.7*
L_{III} $2p_{3/2}$	248.4*		

Arsenic (33)

K 1s	11867	M_{II} $3p_{1/2}$	146.2*
L_{I} 2s	1527.0*b	M_{III} $3p_{3/2}$	141.2*
L_{II} $2p_{1/2}$	1359.1*b	M_{IV} $3d_{3/2}$	41.7*
L_{III} $2p_{3/2}$	1323.6*b	M_{V} $3d_{5/2}$	41.7*
M_{I} 3s	204.7*		

Astatine (85)

K 1s	95730	N_{III} $4p_{3/2}$	740*
L_{I} 2s	17493	N_{IV} $4d_{3/2}$	533*
L_{II} $2p_{1/2}$	16785	N_{V} $4d_{5/2}$	507*
L_{III} $2p_{3/2}$	14214	N_{VI} $4f_{5/2}$	210*
M_{I} 3s	4317	N_{VII} $4f_{7/2}$	210*
M_{II} $3p_{1/2}$	4008	O_{I} 5s	195*
M_{III} $3p_{3/2}$	3426	O_{II} $5p_{1/2}$	148*
M_{IV} $3d_{3/2}$	2909	O_{III} $5p_{3/2}$	115*
M_{V} $3d_{5/2}$	2787	O_{IV} $5d_{3/2}$	40*
N_{I} 4s	1042*	O_{V} $5d_{5/2}$	40*
N_{II} $4p_{1/2}$	886*		

Barium (56)

K 1s	37441	N_{II} $4p_{1/2}$	192
L_{I} 2s	5989	N_{III} $4p_{3/2}$	178.6†
L_{II} $2p_{1/2}$	5624	N_{IV} $4d_{3/2}$	92.6†
L_{III} $2p_{3/2}$	5247	N_{V} $4d_{5/2}$	89.9†
M_{I} 3s	1293 *b	N_{VI} $4f_{5/2}$	—
M_{II} $3p_{1/2}$	1137 *b	N_{VII} $4f_{7/2}$	—
M_{III} $3p_{3/2}$	1063 *b	O_{I} 5s	30.3†
M_{IV} $3d_{3/2}$	795.7*	O_{II} $5p_{1/2}$	17.0†
M_{V} $3d_{5/2}$	780.5*	O_{III} $5p_{3/2}$	14.8†
N_{I} 4s	253.5†		

Beryllium (4)

K 1s	111.5*

Bismuth (83)

K 1s	90526	N_{III} $4p_{3/2}$	678.8†
L_{I} 2s	16388	N_{IV} $4d_{3/2}$	464.0†
L_{II} $2p_{1/2}$	15711	N_{V} $4d_{5/2}$	440.1†
L_{III} $2p_{3/2}$	13419	N_{VI} $4f_{5/2}$	162.3†
M_{I} 3s	3999	N_{VII} $4f_{7/2}$	157.0†
M_{II} $3p_{1/2}$	3696	O_{I} 5s	159.3*b
M_{III} $3p_{3/2}$	3177	O_{II} $5p_{1/2}$	119.0†
M_{IV} $3d_{3/2}$	2688	O_{III} $5p_{3/2}$	92.6†
M_{V} $3d_{5/2}$	2580	O_{IV} $5d_{3/2}$	26.9†
N_{I} 4s	939 †	O_{V} $5d_{5/2}$	23.8†
N_{II} $4p_{1/2}$	805.2†		

Boron (5)

K 1s	188*

Bromine (35)

K 1s	13474	M_{II} $3p_{1/2}$	189*
L_{I} 2s	1782*	M_{III} $3p_{3/2}$	182*
L_{II} $2p_{1/2}$	1596*	M_{IV} $3d_{3/2}$	70*
L_{III} $2p_{3/2}$	1550*	M_{V} $3d_{5/2}$	69*
M_{I} 3s	257*		

Cadmium (48)

K 1s	26711	M_{IV} $3d_{3/2}$	411.9†
L_{I} 2s	4018	M_{V} $3d_{5/2}$	405.2†
L_{II} $2p_{1/2}$	3727	N_{I} 4s	109.8†
L_{III} $2p_{3/2}$	3538	N_{II} $4p_{1/2}$	63.9†a
M_{I} 3s	772.0†	N_{III} $4p_{3/2}$	63.9†a
M_{II} $3p_{1/2}$	652.6†	N_{IV} $4d_{3/2}$	11.7†
M_{III} $3p_{3/2}$	618.4†	N_{V} $4d_{5/2}$	10.7†

Calcium (20)

K 1s	4038.5*	M_{I} 3s	44.3†
L_{I} 2s	438.4†	M_{II} $3p_{1/2}$	25.4†
L_{II} $2p_{1/2}$	349.7†	M_{III} $3p_{3/2}$	25.4†
L_{III} $2p_{3/2}$	346.2†		

Carbon (6)

K 1s	284.2*

Cerium (58)

K 1s	40443	N_{II} $4p_{1/2}$	223.3
L_{I} 2s	6548	N_{III} $4p_{3/2}$	206.5*
L_{II} $2p_{1/2}$	6164	N_{IV} $4d_{3/2}$	109 *
L_{III} $2p_{3/2}$	5723	N_{V} $4d_{5/2}$	—
M_{I} 3s	1436 *b	N_{VI} $4f_{5/2}$	.1
M_{II} $3p_{1/2}$	1274 *b	N_{VII} $4f_{7/2}$	.1
M_{III} $3p_{3/2}$	1187 *b	O_{I} 5s	37.8
M_{IV} $3d_{3/2}$	902.4*	O_{II} $5p_{1/2}$	19.8*
M_{V} $3d_{5/2}$	883.8*	O_{III} $5p_{3/2}$	17.0*
N_{I} 4s	291.0*		

Cesium (55)

K 1s	35985	N_{II} $4p_{1/2}$	172.4*
L_{I} 2s	5714	N_{III} $4p_{3/2}$	161.3*
L_{II} $2p_{1/2}$	5359	N_{IV} $4d_{3/2}$	79.8*
L_{III} $2p_{3/2}$	5012	N_{V} $4d_{5/2}$	77.5*
M_{I} 3s	1211 *b	N_{VI} $4f_{5/2}$	—
M_{II} $3p_{1/2}$	1071 *	N_{VII} $4f_{7/2}$	—
M_{III} $3p_{3/2}$	1003 *	O_{I} 5s	22.7
M_{IV} $3d_{3/2}$	740.5*	O_{II} $5p_{1/2}$	14.2*
M_{V} $3d_{5/2}$	726.6*	O_{III} $5p_{3/2}$	12.1*
N_{I} 4s	232.3*		

Chlorine (17)

K 1s	2823	L_{II} $2p_{1/2}$	202*
L_{I} 2s	270*	L_{III} $2p_{3/2}$	200*

Chromium (24)

K 1s	5989	M_{I} 3s	74.1†
L_{I} 2s	695.7†	M_{II} $3p_{1/2}$	42.2†
L_{II} $2p_{1/2}$	583.8†	M_{III} $3p_{3/2}$	42.2†
L_{III} $2p_{3/2}$	574.1†		

Cobalt (27)

K 1s	7709	M_{I} 3s	101.0†
L_{I} 2s	925.1†	M_{II} $3p_{1/2}$	58.9†
L_{II} $2p_{1/2}$	793.3†	M_{III} $3p_{3/2}$	58.9†
L_{III} $2p_{3/2}$	778.1†		

Copper (29)

K 1s	8979	M_{I} 3s	122.5†
L_{I} 2s	1096.7†	M_{II} $3p_{1/2}$	77.3†
L_{II} $2p_{1/2}$	952.3†	M_{III} $3p_{3/2}$	75.1†
L_{III} $2p_{3/2}$	932.5†		

Dysprosium (66)

K 1s	53789	N_{II} $4p_{1/2}$	333.5*
L_{I} 2s	9046	N_{III} $4p_{3/2}$	293.2*
L_{II} $2p_{1/2}$	8581	N_{IV} $4d_{3/2}$	153.6*
L_{III} $2p_{3/2}$	7790	N_{V} $4d_{5/2}$	153.6*
M_{I} 3s	2047	N_{VI} $4f_{5/2}$	8.0*
M_{II} $3p_{1/2}$	1842	N_{VII} $4f_{7/2}$	4.3*
M_{III} $3p_{3/2}$	1676	O_{I} 5s	49.9*
M_{IV} $3d_{3/2}$	1333	O_{II} $5p_{1/2}$	26.3
M_{V} $3d_{5/2}$	1292 *	O_{III} $5p_{3/2}$	26.3
N_{I} 4s	414.2*		

Erbium (68)

K 1s	57486	N_{II} $4p_{1/2}$	366.2
L_{I} 2s	9751	N_{III} $4p_{3/2}$	320.2*
L_{II} $2p_{1/2}$	9264	N_{IV} $4d_{3/2}$	167.6*
L_{III} $2p_{3/2}$	8358	N_{V} $4d_{5/2}$	167.6*
M_{I} 3s	2206	N_{VI} $4f_{5/2}$	—
M_{II} $3p_{1/2}$	2006	N_{VII} $4f_{7/2}$	4.7*
M_{III} $3p_{3/2}$	1812	O_{I} 5s	50.6*
M_{IV} $3d_{3/2}$	1453	O_{II} $5p_{1/2}$	31.4*
M_{V} $3d_{5/2}$	1409	O_{III} $5p_{3/2}$	24.7*
N_{I} 4s	449.8*		

Europium (63)

K 1s	48519	N_{I} 4s	360
L_{I} 2s	8052	N_{II} $4p_{1/2}$	284
L_{II} $2p_{1/2}$	7617	N_{III} $4p_{3/2}$	257
L_{III} $2p_{3/2}$	6977	N_{IV} $4d_{3/2}$	133
M_{I} 3s	1800	N_{V} $4d_{5/2}$	133
M_{II} $3p_{1/2}$	1614	N_{VI} $4f_{5/2}$	0
M_{III} $3p_{3/2}$	1481	N_{VII} $4f_{7/2}$	0
M_{IV} $3d_{3/2}$	1158.6*	O_{I} 5s	32
M_{V} $3d_{5/2}$	1127.5*	O_{II} $5p_{1/2}$	22
		O_{III} $5p_{3/2}$	22

Fluorine (9)

K 1s	696.7*

Francium (87)

K 1s	101137	N_{IV} $4d_{3/2}$	603*
L_{I} 2s	18639	N_{V} $4d_{5/2}$	577*
L_{II} $2p_{1/2}$	17907	N_{VI} $4f_{5/2}$	268*

L$_{III}$ 2p$_{3/2}$	15031	N$_{VII}$ 4f$_{7/2}$	268*
M$_{I}$ 3s	4652	O$_{I}$ 5s	234*
M$_{II}$ 3p$_{1/2}$	4327	O$_{II}$ 5p$_{1/2}$	182*
M$_{III}$ 3p$_{3/2}$	3663	O$_{III}$ 5p$_{3/2}$	140*
M$_{IV}$ 3d$_{3/2}$	3136	O$_{IV}$ 5d$_{3/2}$	58*
M$_{V}$ 3d$_{5/2}$	3000	O$_{V}$ 5d$_{5/2}$	58*
N$_{I}$ 4s	1153*	P$_{I}$ 6s	34
N$_{II}$ 4p$_{1/2}$	980*	P$_{II}$ 6p$_{1/2}$	15
N$_{III}$ 4p$_{3/2}$	810*	P$_{III}$ 6p$_{3/2}$	15

Gadolinium (64)

K 1s	50239	N$_{II}$ 4p$_{1/2}$	286
L$_{I}$ 2s	8376	N$_{III}$ 4p$_{3/2}$	271
L$_{II}$ 2p$_{1/2}$	7930	N$_{IV}$ 4d$_{3/2}$	—
L$_{III}$ 2p$_{3/2}$	7243	N$_{V}$ 4d$_{5/2}$	127.7*
M$_{I}$ 3s	1881	N$_{VI}$ 4f$_{5/2}$	8.6*
M$_{II}$ 3p$_{1/2}$	1688	N$_{VII}$ 4f$_{7/2}$	8.6*
M$_{III}$ 3p$_{3/2}$	1544	O$_{I}$ 5s	36
M$_{IV}$ 3d$_{3/2}$	1221.9*	O$_{II}$ 5p$_{1/2}$	20
M$_{V}$ 3d$_{5/2}$	1189.6*	O$_{III}$ 5p$_{3/2}$	20
N$_{I}$ 4s	378.6*		

Gallium (31)

K 1s	10367	M$_{II}$ 3p$_{1/2}$	103.5†
L$_{I}$ 2s	1299.0*b	M$_{III}$ 3p$_{3/2}$	103.5†
L$_{II}$ 2p$_{1/2}$	1143.2†	M$_{IV}$ 3d$_{3/2}$	18.7†
L$_{III}$ 2p$_{3/2}$	1116.4†	M$_{V}$ 3d$_{5/2}$	18.7†
M$_{I}$ 3s	159.5†		

Germanium (32)

K 1s	11103	M$_{II}$ 3p$_{1/2}$	124.9*
L$_{I}$ 2s	1414.6*b	M$_{III}$ 3p$_{3/2}$	120.8*
L$_{II}$ 2p$_{1/2}$	1248.1*b	M$_{IV}$ 3d$_{3/2}$	29.0*
L$_{III}$ 2p$_{3/2}$	1217.0*b	M$_{V}$ 3d$_{5/2}$	29.0*
M$_{I}$ 3s	180.1*		

Gold (79)

K 1s	80725	N$_{II}$ 4p$_{1/2}$	642.7†
L$_{I}$ 2s	14353	N$_{III}$ 4p$_{3/2}$	546.3†
L$_{II}$ 2p$_{1/2}$	13734	N$_{IV}$ 4d$_{3/2}$	353.2†
L$_{III}$ 2p$_{3/2}$	11919	N$_{V}$ 4d$_{5/2}$	335.1†
M$_{I}$ 3s	3425	N$_{VI}$ 4f$_{5/2}$	87.6†
M$_{II}$ 3p$_{1/2}$	3148	N$_{VII}$ 4f$_{7/2}$	83.9†
M$_{III}$ 3p$_{3/2}$	2743	O$_{I}$ 5s	107.2*b
M$_{IV}$ 3d$_{3/2}$	2291	O$_{II}$ 5p$_{1/2}$	74.2†
M$_{V}$ 3d$_{5/2}$	2206	O$_{III}$ 5p$_{3/2}$	57.2†
N$_{I}$ 4s	762.1†		

Hafnium (72)

K 1s	65351	N$_{II}$ 4p$_{1/2}$	438.2†
L$_{I}$ 2s	11271	N$_{III}$ 4p$_{3/2}$	380.7†
L$_{II}$ 2p$_{1/2}$	10739	N$_{IV}$ 4d$_{3/2}$	220.0†
L$_{III}$ 2p$_{3/2}$	9561	N$_{V}$ 4d$_{5/2}$	211.5†
M$_{I}$ 3s	2601	N$_{VI}$ 4f$_{5/2}$	15.9†
M$_{II}$ 3p$_{1/2}$	2365	N$_{VII}$ 4f$_{7/2}$	14.2†
M$_{III}$ 3p$_{3/2}$	2107	O$_{I}$ 5s	64.2†
M$_{IV}$ 3d$_{3/2}$	1716	O$_{II}$ 5p$_{1/2}$	38 *
M$_{V}$ 3d$_{5/2}$	1662	O$_{III}$ 5p$_{3/2}$	29.9†
N$_{I}$ 4s	538*		

Helium (2)

K 1s	24.6*

Holmium (67)

K 1s	55618	N$_{II}$ 4p$_{1/2}$	343.5
L$_{I}$ 2s	9394	N$_{III}$ 4p$_{3/2}$	308.2*
L$_{II}$ 2p$_{1/2}$	8918	N$_{IV}$ 4d$_{3/2}$	160 *
L$_{III}$ 2p$_{3/2}$	8071	N$_{V}$ 4d$_{5/2}$	160 *
M$_{I}$ 3s	2128	N$_{VI}$ 4f$_{5/2}$	8.6*
M$_{II}$ 3p$_{1/2}$	1923	N$_{VII}$ 4f$_{7/2}$	5.2*
M$_{III}$ 3p$_{3/2}$	1741	O$_{I}$ 5s	49.3*
M$_{IV}$ 3d$_{3/2}$	1392	O$_{II}$ 5p$_{1/2}$	30.8*
M$_{V}$ 3d$_{5/2}$	1351	O$_{III}$ 5p$_{3/2}$	24.1*
N$_{I}$ 4s	432.4*		

Hydrogen (1)

K 1s	16.0*

Indium (49)

K 1s	27940	M$_{IV}$ 3d$_{3/2}$	451.4†
L$_{I}$ 2s	4238	M$_{V}$ 3d$_{5/2}$	443.9†
L$_{II}$ 2p$_{1/2}$	3938	N$_{I}$ 4s	122.7†
L$_{III}$ 2p$_{3/2}$	3730	N$_{II}$ 4p$_{1/2}$	73.5†a
M$_{I}$ 3s	827.2†	N$_{III}$ 4p$_{3/2}$	73.5†a
M$_{II}$ 3p$_{1/2}$	703.2†	N$_{IV}$ 4d$_{3/2}$	17.7†
M$_{III}$ 3p$_{3/2}$	665.3†	N$_{V}$ 4d$_{5/2}$	16.4†

Iodine (53)

K 1s	33169	M$_{IV}$ 3d$_{3/2}$	631*
L$_{I}$ 2s	5188	M$_{V}$ 3d$_{5/2}$	620*
L$_{II}$ 2p$_{1/2}$	4852	N$_{I}$ 4s	186*
L$_{III}$ 2p$_{3/2}$	4557	N$_{II}$ 4p$_{1/2}$	123*
M$_{I}$ 3s	1072*	N$_{III}$ 4p$_{3/2}$	123*
M$_{II}$ 3p$_{1/2}$	931*	N$_{IV}$ 4d$_{3/2}$	50*
M$_{III}$ 3p$_{3/2}$	875*	N$_{V}$ 4d$_{5/2}$	50*

Iridium (77)

K 1s	76111	N$_{II}$ 4p$_{1/2}$	577.8†
L$_{I}$ 2s	13419	N$_{III}$ 4p$_{3/2}$	495.8†
L$_{II}$ 2p$_{1/2}$	12824	N$_{IV}$ 4d$_{3/2}$	311.9†
L$_{III}$ 2p$_{3/2}$	11215	N$_{V}$ 4d$_{5/2}$	296.3†
M$_{I}$ 3s	3174	N$_{VI}$ 4f$_{5/2}$	63.8†
M$_{II}$ 3p$_{1/2}$	2909	N$_{VII}$ 4f$_{7/2}$	60.8†
M$_{III}$ 3p$_{3/2}$	2551	O$_{I}$ 5s	95.2*b
M$_{IV}$ 3d$_{3/2}$	2116	O$_{II}$ 5p$_{1/2}$	63.0*b
M$_{V}$ 3d$_{5/2}$	2040	O$_{III}$ 5p$_{3/2}$	48.0†
N$_{I}$ 4s	691.1†		

Iron (26)

K 1s	7112	M$_{I}$ 3s	91.3†
L$_{I}$ 2s	844.6†	M$_{II}$ 3p$_{1/2}$	52.7†
L$_{II}$ 2p$_{1/2}$	719.9†	M$_{III}$ 3p$_{3/2}$	52.7†
L$_{III}$ 2p$_{3/2}$	706.8†		

Krypton (36)

K 1s	14326	M$_{III}$ 3p$_{3/2}$	214.4*
L$_{I}$ 2s	1921	M$_{IV}$ 3d$_{3/2}$	95.0*
L$_{II}$ 2p$_{1/2}$	1730.9*	M$_{V}$ 3d$_{5/2}$	93.8*
L$_{III}$ 2p$_{3/2}$	1678.4*	N$_{I}$ 4s	27.5*
M$_{I}$ 3s	292.8*	N$_{II}$ 4p$_{1/2}$	14.1*
M$_{II}$ 3p$_{1/2}$	222.2*	N$_{III}$ 4p$_{3/2}$	14.1*

Lanthanum (57)

K 1s	38925	N$_{II}$ 4p$_{1/2}$	205.8
L$_{I}$ 2s	6266	N$_{III}$ 4p$_{3/2}$	196.0*
L$_{II}$ 2p$_{1/2}$	5891	N$_{IV}$ 4d$_{3/2}$	105.3*
L$_{III}$ 2p$_{3/2}$	5483	N$_{V}$ 4d$_{5/2}$	102.5*
M$_{I}$ 3s	1362*b	N$_{VI}$ 4f$_{5/2}$	—
M$_{II}$ 3p$_{1/2}$	1209*b	N$_{VII}$ 4f$_{7/2}$	—
M$_{III}$ 3p$_{3/2}$	1128*b	O$_{I}$ 5s	34.3*
M$_{IV}$ 3d$_{3/2}$	853*	O$_{II}$ 5p$_{1/2}$	19.3*
M$_{V}$ 3d$_{5/2}$	836*	O$_{III}$ 5p$_{3/2}$	16.8*
N$_{I}$ 4s	247.7*		

Lead (82)

K 1s	88005	N$_{III}$ 4p$_{3/2}$	643.5†
L$_{I}$ 2s	15861	N$_{IV}$ 4d$_{3/2}$	434.3†
L$_{II}$ 2p$_{1/2}$	15200	N$_{V}$ 4d$_{5/2}$	412.2†
L$_{III}$ 2p$_{3/2}$	13055	N$_{VI}$ 4f$_{5/2}$	141.7†
M$_{I}$ 3s	3851	N$_{VII}$ 4f$_{7/2}$	136.9†
M$_{II}$ 3p$_{1/2}$	3554	O$_{I}$ 5s	147 *b
M$_{III}$ 3p$_{3/2}$		O$_{II}$ 5p$_{1/2}$	106.4†
M$_{IV}$ 3d$_{3/2}$	3066	O$_{III}$ 5p$_{3/2}$	83.3†
M$_{V}$ 3d$_{5/2}$	2586	O$_{IV}$ 5d$_{3/2}$	20.7†
N$_{I}$ 4s	891.8†	O$_{V}$ 5d$_{5/2}$	18.1†
N$_{II}$ 4p$_{1/2}$	761.9†		

Lithium (3)

K 1s	54.7*

Lutetium (71)

K 1s	63314	N$_{II}$ 4p$_{1/2}$	412.4*
L$_{I}$ 2s	10870	N$_{III}$ 4p$_{3/2}$	359.2*
L$_{II}$ 2p$_{1/2}$	10349	N$_{IV}$ 4d$_{3/2}$	206.1*
L$_{III}$ 2p$_{3/2}$	9244	N$_{V}$ 4d$_{5/2}$	196.3*
M$_{I}$ 3s	2491	N$_{VI}$ 4f$_{5/2}$	8.9*
M$_{II}$ 3p$_{1/2}$	2264	N$_{VII}$ 4f$_{7/2}$	7.5*
M$_{III}$ 3p$_{3/2}$	2024	O$_{I}$ 5s	57.3*
M$_{IV}$ 3d$_{3/2}$	1639	O$_{II}$ 5p$_{1/2}$	33.6*
M$_{V}$ 3d$_{5/2}$	1589	O$_{III}$ 5p$_{3/2}$	26.7*
N$_{I}$ 4s	506.8*		

Magnesium (12)

K 1s	1303.0†
L$_{I}$ 2s	88.6*
L$_{II}$ 2p$_{1/2}$	49.6†
L$_{III}$ 2p$_{3/2}$	49.2*

Manganese (25)

K 1s	6539	M$_{I}$ 3s	82.3†
L$_{I}$ 2s	769.1†	M$_{II}$ 3p$_{1/2}$	47.2†
L$_{II}$ 2p$_{1/2}$	649.9†	M$_{III}$ 3p$_{3/2}$	47.2†
L$_{III}$ 2p$_{3/2}$	638.7†		

Mercury (80)

K 1s	83102	N$_{III}$ 4p$_{3/2}$	576.6†
L$_{I}$ 2s	14839	N$_{IV}$ 4d$_{3/2}$	378.2†
L$_{II}$ 2p$_{1/2}$	14209	N$_{V}$ 4d$_{5/2}$	358.8†
L$_{III}$ 2p$_{3/2}$	12284	N$_{VI}$ 4f$_{5/2}$	104.0†
M$_{I}$ 3s	3562	N$_{VII}$ 4f$_{7/2}$	99.9†
M$_{II}$ 3p$_{1/2}$	3279	O$_{I}$ 5s	127 †
M$_{III}$ 3p$_{3/2}$	2847	O$_{II}$ 5p$_{1/2}$	83.1†
M$_{IV}$ 3d$_{3/2}$	2385	O$_{III}$ 5p$_{3/2}$	64.5†
M$_{V}$ 3d$_{5/2}$	2295	O$_{IV}$ 5d$_{3/2}$	9.6†
N$_{I}$ 4s	802.2†	O$_{V}$ 5d$_{5/2}$	7.8†
N$_{II}$ 4p$_{1/2}$	680.2†		

Molybdenum (42)

K 1s	20000	M$_{III}$ 3p$_{3/2}$	394.0†
L$_{I}$ 2s	2866	M$_{IV}$ 3d$_{3/2}$	231.1†
L$_{II}$ 2p$_{1/2}$	2625	M$_{V}$ 3d$_{5/2}$	227.9†
L$_{III}$ 2p$_{3/2}$	2520	N$_{I}$ 4s	63.2†
M$_{I}$ 3s	506.3†	N$_{II}$ 4p$_{1/2}$	37.6†
M$_{II}$ 3p$_{1/2}$	410.6†	N$_{III}$ 4p$_{3/2}$	35.5†

Neodymium (60)

K 1s	43569	N$_{II}$ 4p$_{1/2}$	243.3
L$_{I}$ 2s	7126	N$_{III}$ 4p$_{3/2}$	224.6
L$_{II}$ 2p$_{1/2}$	6722	N$_{IV}$ 4d$_{3/2}$	120.5*
L$_{III}$ 2p$_{3/2}$	6208	N$_{V}$ 4d$_{5/2}$	120.5*
M$_{I}$ 3s	1575	N$_{VI}$ 4f$_{5/2}$	1.5
M$_{II}$ 3p$_{1/2}$	1403	N$_{VII}$ 4f$_{7/2}$	1.5
M$_{III}$ 3p$_{3/2}$	1297	O$_{I}$ 5s	37.5
M$_{IV}$ 3d$_{3/2}$	1003.3*	O$_{II}$ 5p$_{1/2}$	21.1
M$_{V}$ 3d$_{5/2}$	980.4*	O$_{III}$ 5p$_{3/2}$	21.1
N$_{I}$ 4s	319.2*		

Neon (10)

K 1s	870.2*
L$_{I}$ 2s	48.5*
L$_{II}$ 2p$_{1/2}$	21.7*
L$_{III}$ 2p$_{3/2}$	21.6*

Nickel (28)

K 1s	8333	M$_{I}$ 3s	110.8†
L$_{I}$ 2s	1008.6†	M$_{II}$ 3p$_{1/2}$	68.0†
L$_{II}$ 2p$_{1/2}$	870.0†	M$_{III}$ 3p$_{3/2}$	66.2†
L$_{III}$ 2p$_{3/2}$	852.7†		

Niobium (41)

K 1s :18986	MIII 3p3/2:360.6†
LI 2s : 2698	MIV 3d3/2:205.0†
LII 2p1/2: 2465	MV 3d5/2:202.3†
LIII 2p3/2: 2371	NI 4s : 56.4†
MI 3s : 466.6†	NII 4p1/2: 32.6†
MII 3p1/2: 376.1†	NIII 4p3/2: 30.8†

Nitrogen (7)

K 1s :409.9*	
LI 2s :37.3*	

Osmium (76)

K 1s :73871	NII 4p1/2:549.1†
LI 2s :12968	NIII 4p3/2:470.7†
LII 2p1/2:12385	NIV 4d3/2:293.1†
LIII 2p3/2:10871	NV 4d5/2:278.5†
MI 3s : 3049	NVI 4f5/2 : 52.4†
MII 3p1/2: 2792	NVII 4f7/2 : 50.7†
MIII 3p3/2: 2457	OI 5s : 83 †
MIV 3d3/2: 2031	OII 5p1/2: 58 *
MV 3d5/2: 1960	OIII 5p3/2: 44.5†
NI 4s : 658.2†	

Oxygen (8)

K 1s :543.1*	
LI 2s :41.6*	

Palladium (46)

K 1s :24350	MIII 3p3/2:532.3†
LI 2s : 3604	MIV 3d3/2:340.5†
LII 2p1/2: 3330	MV 3d5/2:335.2†
LIII 2p3/2: 3173	NI 4s : 87.1*b
MI 3s : 671.6†	NII 4p1/2: 55.7†a
MII 3p1/2: 559.9†	NIII 4p3/2: 50.9†a

Phosphorus (15)

K 1s :2149	LII 2p1/2:136*
LI 2s : 189*	LIII 2p3/2:135*

Platinum (78)

K 1s :78395	NII 4p1/2:609.1†
LI 2s :13880	NIII 4p3/2:519.4†
LII 2p1/2:13273	NIV 4d3/2:331.6†
LIII 2p3/2:11564	NV 4d5/2:314.6†
MI 3s : 3296	NVI 4f5/2 : 74.5†
MII 3p1/2: 3027	NVII 4f7/2 : 71.2†
MIII 3p3/2: 2645	OI 5s :101.7*b
MIV 3d3/2: 2202	OII 5p1/2: 65.3*b
MV 3d5/2: 2122	OIII 5p3/2: 51.7†
NI 4s : 725.4†	

Polonium (84)

K 1s :93105	NIII 4p3/2:705*
LI 2s :16939	NIV 4d3/2:500*
LII 2p1/2:16244	NV 4d5/2:473*
LIII 2p3/2:13814	NVI 4f5/2:184*
MI 3s : 4149	NVII 4f7/2 184*
MII 3p1/2: 3854	OI 5s :177*
MIII 3p3/2: 3302	OII 5p1/2:132*
MIV 3d3/2: 2798	OIII 5p3/2:104*
MV 3d5/2: 2683	OIV 5d3/2: 31*
NI 4s : 995*	OV 5d5/2: 31*
NII 4p1/2: 851*	

Potassium (19)

K 1s :3608.4*	MI 3s :34.8*
LI 2s : 378.6*	MII 3p1/2:18.3*
LII 2p1/2: 297.3*	MIII 3p3/2:18.3*
LIII 2p3/2: 294.6*	

Praseodymium (59)

K 1s :41991	NII 4p1/2:236.3
LI 2s : 6835	NIII 4p3/2:217.6
LII 2p1/2: 6440	NIV 4d3/2:115.1*
LIII 2p3/2: 5964	NV 4d5/2:115.1*
MI 3s : 1511	NVI 4f5/2 : 2.0
MII 3p1/2: 1337	NVII 4f7/2 : 2.0
MIII 3p3/2: 1242	OI 5s : 37.4
MIV 3d3/2: 948.3*	OII 5p1/2: 22.3
MV 3d5/2: 928.8*	OIII 5p3/2: 22.3
NI 4s : 304.5	

Promethium (61)

K 1s :45184	MIV 3d3/2:1052
LI 2s : 7428	MV 3d5/2:1027
LII 2p1/2: 7013	NI 4s : —
LIII 2p3/2: 6459	NII 4p1/2: 242
MI 3s : —	NIII 4p3/2: 242
MII 3p1/2: 1403	NIV 4d3/2: 120
MIII 3p3/2: 1357	NV 4d5/2: 120

Protoactinium (91)

K 1s :112601	NIV 4d3/2: 743*
LI 2s : 21105	NV 4d5/2: 708*
LII 2p1/2: 20314	NVI 4f5/2: 371*
LIII 2p3/2: 16733	NVII 4f7/2 : 360*
MI 3s : 5367	OI 5s : 310*
MII 3p1/2: 5001	OII 5p1/2: 232*
MIII 3p3/2: 4174	OIII 5p3/2: 232*
MIV 3d3/2: 3611	OIV 5d3/2: 94*
MV 3d5/2: 3442	OV 5d5/2: 94*
NI 4s : 1387*	PI 6s : —
NII 4p1/2: 1224*	PII 6p1/2: —
NIII 4p3/2: 1007*	PIII 6p3/2: —

Radium (88)

K 1s :103922	NIV 4d3/2:636*
LI 2s : 19237	NV 4d5/2:603*
LII 2p1/2: 18484	NVI 4f5/2:299*
LIII 2p3/2: 15444	NVII 4f7/2:299*
MI 3s : 4822	OI 5s :254*
MII 3p1/2: 4490	OII 5p1/2:200*
MIII 3p3/2: 3792	OIII 5p3/2:153*
MIV 3d3/2: 3248	OIV 5d3/2: 68*
MV 3d5/2: 3105	OV 5d5/2: 68*
NI 4s : 1208*	PI 6s : 44
NII 4p1/2: 1958*	PII 6p1/2: 19
NIII 4p3/2: 879*	PIII 6p3/2: 19

Radon (86)

K 1s :98404	NIII 4p3/2:768*
LI 2s :18049	NIV 4d3/2:567*
LII 2p1/2:17337	NV 4d5/2:541*
LIII 2p3/2:14619	NVI 4f5/2:238*
MI 3s : 4482	NVII 4f7/2 :238*
MII 3p1/2: 4159	OI 5s :214*
MIII 3p3/2: 3538	OII 5p1/2:164*
MIV 3d3/2: 3022	OIII 5p3/2:127*
MV 3d5/2: 2892	OIV 5d3/2: 48*
NI 4s : 1097*	OV 5d5/2: 48*
NII 4p1/2: 929*	PI 6s : 26

Rhenium (75)

K 1s :71676	NII 4p1/2:518.7†
LI 2s :12527	NIII 4p3/2:446.8†
LII 2p1/2:11959	NIV 4d3/2:273.9†
LIII 2p3/2:10535	NV 4d5/2:260.5†
MI 3s : 2932	NVI 4f5/2 : 42.9*
MII 3p1/2: 2682	NVII 4f7/2 : 40.5*
MIII 3p3/2: 2367	OI 5s : 83 †
MIV 3d3/2: 1949	OII 5p1/2: 45.6†
MV 3d5/2: 1883	OIII 5p3/2: 34.6*b
NI 4s : 625.4†	

Rhodium (45)

K 1s :23220	MIII 3p3/2:496.5†
LI 2s : 3412	MIV 3d3/2:311.9†
LII 2p1/2: 3146	MV 3d5/2:307.2†
LIII 2p3/2: 3004	NI 4s : 81.4*b
MI 3s : 628.1†	NII 4p1/2: 50.5†
MII 3p1/2: 521.3†	NIII 4p3/2: 47.3†

Rubidium (37)

K 1s :15200	MIII 3p3/2:239.1*
LI 2s : 2065	MIV 3d3/2:113.0*
LII 2p1/2: 1864	MV 3d5/2:112 *
LIII 2p3/2: 1804	NI 4s : 30.5*
MI 3s : 326.7*	NII 4p1/2: 16.3*
MII 3p1/2: 248.7*	NIII 4p3/2: 15.3*

Ruthenium (44)

K 1s :22117	MIII 3p3/2:461.5†
LI 2s : 3224	MIV 3d3/2:284.2†
LII 2p1/2: 2967	MV 3d5/2:280.0†
LIII 2p3/2: 2838	NI 4s : 75.0†
MI 3s : 586.2†	NII 4p1/2: 46.5†
MII 3p1/2: 483.3†	NIII 4p3/2: 43.2†

Samarium (62)

K 1s :46834	NII 4p1/2:265.6
LI 2s : 7737	NIII 4p3/2:247.4
LII 2p1/2: 7312	NIV 4d3/2:129.0
LIII 2p3/2: 6716	NV 4d5/2:129.0
MI 3s : 1723	NVI 4f5/2 : 5.2
MII 3p1/2: 1541	NVII 4f7/2 : 5.2
MIII 3p3/2: 1419.8	OI 5s : 37.4
MIV 3d3/2: 1110.9*	OII 5p1/2: 21.3
MV 3d5/2: 1083.4*	OIII 5p3/2: 21.3
NI 4s : 347.2*	

Scandium (21)

K 1s :4492	MI 3s :51.1*
LI 2s : 498.0*	MII 3p1/2:28.3*
LII 2p1/2: 403.6*	MIII 3p3/2:28.3*
LIII 2p3/2: 398.7*	

Selenium (34)

K 1s :12658	MII 3p1/2:166.5*
LI 2s : 1652.0*b	MIII 3p3/2:160.7*
LII 2p1/2: 1474.3*b	MIV 3d3/2: 55.5*
LIII 2p3/2: 1433.9*b	MV 3d5/2: 54.6*
MI 3s : 229.6*	

Silicon (14)

K 1s :1839	LII 2p1/2:99.2*
LI 2s : 149.7*b	LIII 2p3/2:99.8*

Silver (47)

K 1s :25514	MIII 3p3/2:573.0†
LI 2s : 3806	MIV 3d3/2:374.0†
LII 2p1/2: 3524	MV 3d5/2:368.0†
LIII 2p3/2: 3351	NI 4s : 97.0†
MI 3s : 719.0†	NII 4p1/2: 63.7†
MII 3p1/2: 603.8†	NIII 4p3/2: 58.3†

Sodium (11)

K 1s :1070.8†	
LI 2s : 63.5†	
LII 2p1/2:30.4†	
LIII 2p3/2:30.5*	

Strontium (38)

K 1s :16105	MIII 3p3/2:270.0†
LI 2s : 2216	MIV 3d3/2:136.0†
LII 2p1/2: 2007	MV 3d5/2:134.2†
LIII 2p3/2: 1940	NI 4s : 38.9†

M_{I} 3s : 358.7†	N_{II} $4p_{1/2}$: 20.3†
M_{II} $3p_{1/2}$: 280.3†	N_{III} $4p_{3/2}$: 20.3†

Sulfur (16)

K 1s :2472	L_{II} $2p_{1/2}$:163.6*
L_{I} 2s : 230.9*b	L_{III} $2p_{3/2}$:162.5*

Tantalum (73)

K 1s :67416	N_{II} $4p_{1/2}$:463.4†
L_{I} 2s :11682	N_{III} $4p_{3/2}$:400.9†
L_{II} $2p_{1/2}$:11136	N_{IV} $4d_{3/2}$:237.9†
L_{III} $2p_{3/2}$: 9881	N_{V} $4d_{5/2}$:226.4†
M_{I} 3s : 2708	N_{VI} $4f_{5/2}$: 23.5†
M_{II} $3p_{1/2}$: 2469	N_{VII} $4f_{7/2}$: 21.6†
M_{III} $3p_{3/2}$: 2194	O_{I} 5s : 69.7†
M_{IV} $3d_{3/2}$: 1793	O_{II} $5p_{1/2}$: 42.2*
M_{V} $3d_{5/2}$: 1735	O_{III} $5p_{3/2}$: 32.7†
N_{I} 4s : 563.4†	

Technetium (43)

K 1s :21044	M_{III} $3p_{3/2}$:425*
L_{I} 2s : 3043	M_{IV} $3d_{3/2}$:257*
L_{II} $2p_{1/2}$: 2793	M_{V} $3d_{5/2}$:253*
L_{III} $2p_{3/2}$: 2677	N_{I} 4s : 68*
M_{I} 3s : 544*	N_{II} $4p_{1/2}$: 39*
M_{II} $3p_{1/2}$: 445*	N_{III} $4p_{3/2}$: 39*

Tellurium (52)

K 1s :31814	M_{IV} $3d_{3/2}$:583.4†
L_{I} 2s : 4939	M_{V} $3d_{5/2}$:573.0†
L_{II} $2p_{1/2}$: 4612	N_{I} 4s :169.4†
L_{III} $2p_{3/2}$: 4341	N_{II} $4p_{1/2}$:103.3†a
M_{I} 3s : 1006 †	N_{III} $4p_{3/2}$:103.3†a
M_{II} $3p_{1/2}$: 870.8†	N_{IV} $4d_{3/2}$: 41.9†
M_{III} $3p_{3/2}$: 820.0†	N_{V} $4d_{5/2}$: 40.4†

Terbium (65)

K 1s :51996	N_{II} $4p_{1/2}$:322.4*
L_{I} 2s : 8708	N_{III} $4p_{3/2}$:284.1*
L_{II} $2p_{1/2}$: 8252	N_{IV} $4d_{3/2}$:150.5*
L_{III} $2p_{3/2}$: 7514	N_{V} $4d_{5/2}$:150.5*
M_{I} 3s : 1968	N_{VI} $4f_{5/2}$: 7.7*
M_{II} $3p_{1/2}$: 1768	N_{VII} $4f_{7/2}$: 2.4*
M_{III} $3p_{3/2}$: 1611	O_{I} 5s : 45.6*
M_{IV} $3d_{3/2}$: 1276.9*	O_{II} $5p_{1/2}$: 28.7*
M_{V} $3d_{5/2}$: 1241.1*	O_{III} $5p_{3/2}$: 22.6*
N_{I} 4s : 396.0*	

Thallium (81)

K 1s :85530	N_{III} $4p_{3/2}$:609.5†
L_{I} 2s :15347	N_{IV} $4d_{3/2}$:405.7†
L_{II} $2p_{1/2}$:14698	N_{V} $4d_{5/2}$:385.0†
L_{III} $2p_{3/2}$:12658	N_{VI} $4f_{5/2}$:122.2†
M_{I} 3s : 3704	N_{VII} $4f_{7/2}$:117.8†
M_{II} $3p_{1/2}$: 3416	O_{I} 5s :136 *b
M_{III} $3p_{3/2}$: 2957	O_{II} $5p_{1/2}$: 94.6†
M_{IV} $3d_{3/2}$: 2485	O_{III} $5p_{3/2}$: 73.5†
M_{V} $3d_{5/2}$: 2389	O_{IV} $5d_{3/2}$: 14.7†
N_{I} 4s : 846.2†	O_{V} $5d_{5/2}$: 12.5†
N_{II} $4p_{1/2}$: 720.5†	

Thorium (90)

K 1s :109651	N_{IV} $4d_{3/2}$:712.1†
L_{I} 2s : 20472	N_{V} $4d_{5/2}$:675.2†
L_{II} $2p_{1/2}$: 19693	N_{VI} $4f_{5/2}$:342.4†
L_{III} $2p_{3/2}$: 16300	N_{VII} $4f_{7/2}$:333.1†
M_{I} 3s : 5182	O_{I} 5s :290 *a
M_{II} $3p_{1/2}$: 4830	O_{II} $5p_{1/2}$:229 *a
M_{III} $3p_{3/2}$: 4046	O_{III} $5p_{3/2}$:182 *a
M_{IV} $3d_{3/2}$: 3491	O_{IV} $5d_{3/2}$: 92.5†
M_{V} $3d_{5/2}$: 3332	O_{V} $5d_{5/2}$: 85.4†
N_{I} 4s : 1330 *	P_{I} 6s : 41.4†
N_{II} $4p_{1/2}$: 1168 *	P_{II} $6p_{1/2}$: 24.5†
N_{III} $4p_{3/2}$: 966.4†	P_{III} $6p_{3/2}$: 16.6†

Thulium (69)

K 1s :59390	N_{II} $4p_{1/2}$:385.9*
L_{I} 2s :10116	N_{III} $4p_{3/2}$:332.6*
L_{II} $2p_{1/2}$: 9617	N_{IV} $4d_{3/2}$:175.5*
L_{III} $2p_{3/2}$: 8648	N_{V} $4d_{5/2}$:175.5*
M_{I} 3s : 2307	N_{VI} $4f_{5/2}$: —
M_{II} $3p_{1/2}$: 2090	N_{VII} $4f_{7/2}$: 4.6
M_{III} $3p_{3/2}$: 1885	O_{I} 5s : 54.7*
M_{IV} $3d_{3/2}$: 1515	O_{II} $5p_{1/2}$: 31.8*
M_{V} $3d_{5/2}$: 1468	O_{III} $5p_{3/2}$: 25.0*
N_{I} 4s : 470.9*	

Tin (50)

K 1s :29200	M_{IV} $3d_{3/2}$:493.2†
L_{I} 2s : 4465	M_{V} $3d_{5/2}$:484.9†
L_{II} $2p_{1/2}$: 4156	N_{I} 4s :137.1†
L_{III} $2p_{3/2}$: 3929	N_{II} $4p_{1/2}$: 83.6†a
M_{I} 3s : 884.7†	N_{III} $4p_{3/2}$: 83.6†a
M_{II} $3p_{1/2}$: 756.5†	N_{IV} $4d_{3/2}$: 24.9†
M_{III} $3p_{3/2}$: 714.6†	N_{V} $4d_{5/2}$: 23.9†

Titanium (22)

K 1s :4966	M_{I} 3s :58.7†
L_{I} 2s : 560.9†	M_{II} $3p_{1/2}$:32.6†
L_{II} $2p_{1/2}$: 461.2†	M_{III} $3p_{3/2}$:32.6†
L_{III} $2p_{3/2}$: 453.8†	

Tungsten (74)

K 1s :69525	N_{II} $4p_{1/2}$:490.4†
L_{I} 2s :12100	N_{III} $4p_{3/2}$:423.6†
L_{II} $2p_{1/2}$:11544	N_{IV} $4d_{3/2}$:255.9†
L_{III} $2p_{3/2}$:10207	N_{V} $4d_{5/2}$:243.5†
M_{I} 3s : 2820	N_{VI} $4f_{5/2}$: 33.6*
M_{II} $3p_{1/2}$: 2575	N_{VII} $4f_{7/2}$: 31.4†
M_{III} $3p_{3/2}$: 2281	O_{I} 5s : 75.6†
M_{IV} $3d_{3/2}$: 1949	O_{II} $5p_{1/2}$:453 *b
M_{V} $3d_{5/2}$: 1809	O_{III} $5p_{3/2}$: 36.8†
N_{I} 4s : 594.1†	

Uranium (92)

K 1s :115606	N_{IV} $4d_{3/2}$:778.3†
L_{I} 2s : 21757	N_{V} $4d_{5/2}$:736.2†
L_{II} $2p_{1/2}$: 20948	N_{VI} $4f_{5/2}$:388.2*
L_{III} $2p_{3/2}$: 17166	N_{VII} $4f_{7/2}$:377.4†
M_{I} 3s : 5548	O_{I} 5s :321 *ab
M_{II} $3p_{1/2}$: 5182	O_{II} $5p_{1/2}$:257 *ab
M_{III} $3p_{3/2}$: 4303	O_{III} $5p_{3/2}$:192 *ab
M_{IV} $3d_{3/2}$: 3728	O_{IV} $5d_{3/2}$:102.8†
M_{V} $3d_{5/2}$: 3552	O_{V} $5d_{5/2}$: 94.2†
N_{I} 4s : 1439*b	P_{I} 6s : 43.9†
N_{II} $4p_{1/2}$: 1271*b	P_{II} $6p_{1/2}$: 26.8†
N_{III} $4p_{3/2}$: 1043†	P_{III} $6p_{3/2}$: 16.8†

Vanadium (23)

K 1s :5465	M_{I} 3s :66.3†
L_{I} 2s : 626.7†	M_{II} $3p_{1/2}$:37.2†
L_{II} $2p_{1/2}$: 519.8†	M_{III} $3p_{3/2}$:37.2†
L_{III} $2p_{3/2}$: 512.1†	

Xenon (54)

K 1s :34561	N_{II} $4p_{1/2}$:146.7
L_{I} 2s : 5453	N_{III} $4p_{3/2}$:145.5*
L_{II} $2p_{1/2}$: 5104	N_{IV} $4d_{3/2}$: 69.5*
L_{III} $2p_{3/2}$: 4782	N_{V} $4d_{5/2}$: 67.5*
M_{I} 3s : 1148.7*	N_{VI} $4f_{5/2}$: —
M_{II} $3p_{1/2}$: 1002.1*	N_{VII} $4f_{7/2}$: —
M_{III} $3p_{3/2}$: 940.6*	O_{I} 5s : 23.3*
M_{IV} $3d_{3/2}$: 689.0*	O_{II} $5p_{1/2}$: 13.4*
M_{V} $3d_{5/2}$: 676.4*	O_{III} $5p_{3/2}$: 12.1*
N_{I} 4s : 213.2*	

Ytterbium (70)

K 1s :61332	N_{II} $4p_{1/2}$:388.7*
L_{I} 2s :10486	N_{III} $4p_{3/2}$:339.7*
L_{II} $2p_{1/2}$: 9978	N_{IV} $4d_{3/2}$:191.2*
L_{III} $2p_{3/2}$: 8944	N_{V} $4d_{5/2}$:182.4*
M_{I} 3s : 2398	N_{VI} $4f_{5/2}$: 2.5*
M_{II} $3p_{1/2}$: 2173	N_{VII} $4f_{7/2}$: 1.3*
M_{III} $3p_{3/2}$: 1950	O_{I} 5s : 52.0*
M_{IV} $3d_{3/2}$: 1576	O_{II} $5p_{1/2}$: 30.3*
M_{V} $3d_{5/2}$: 1528	O_{III} $5p_{3/2}$: 24.1*
N_{I} 4s : 480.5*	

Yttrium (39)

K 1s :17038	M_{III} $3p_{3/2}$:298.8*
L_{I} 2s : 2373	M_{IV} $3d_{3/2}$:157.7†
L_{II} $2p_{1/2}$: 2156	M_{V} $3d_{5/2}$:155.8†
L_{III} $2p_{3/2}$: 2080	N_{I} 4s : 43.8*
M_{I} 3s : 392.0*b	N_{II} $4p_{1/2}$: 24.4*
M_{II} $3p_{1/2}$: 310.6*	N_{III} $4p_{3/2}$: 23.1*

Zinc (30)

K 1s :9659	M_{II} $3p_{1/2}$:91.4*
L_{I} 2s :1200.7*	M_{III} $3p_{3/2}$:88.6*
L_{II} $2p_{1/2}$:1044.9*	M_{IV} $3d_{3/2}$:10.2*
L_{III} $2p_{3/2}$:1021.8*	M_{V} $3d_{5/2}$:10.1*
M_{I} 3s : 139.8*	

Zirconium (40)

K 1s :17998	M_{III} $3p_{3/2}$:329.8†
L_{I} 2s : 2532	M_{IV} $3d_{3/2}$:181.1†
L_{II} $2p_{1/2}$: 2307	M_{V} $3d_{5/2}$:178.8†
L_{III} $2p_{3/2}$: 2223	N_{I} 4s : 50.6†
M_{I} 3s : 430.3†	N_{II} $4p_{1/2}$: 28.5†
M_{II} $3p_{1/2}$: 343.5†	N_{III} $4p_{3/2}$: 27.7†

Referred to the Fermi Level (metals), Valence Band Max (Semiconductors), Vacuum Level (Rare Gases)

† From Fuggle and Martensson, *J. Elect. Spect.*, 21, 275, 1980.

* From Cardona and Ley *Photoemission from Solids*, Springer Verlag, 1978. Rest from Bearden and Burr, *Rev. Mod. Phys.*, 39, 125, 1967.

[a] One-particle approximation not valid.

[b] Derived from Bearden and Burr.

STRENGTHS OF CHEMICAL BONDS*

J. A. Kerr

The strength of a chemical bond, D(R − X), often known as the bond dissociation energy, is defined as the heat of the reaction: RX → R + X. It is given by D(R − X) = ΔHf°(R) + ΔHf°(X) − ΔHf°(RX). Some authors list bond strengths for O K but here the values for 298 K are given because more thermodynamic data are available for this temperature. Bond strengths, or bond dissociation energies, are not equal to, and may differ considerably from, mean bond energies derived solely from thermochemical data on molecules and atoms.

BOND STRENGTHS IN DIATOMIC MOLECULES

These have usually been measured spectroscopically or by mass spectrometric analysis of hot gases effusing from a Knudsen cell. Excellent accounts of these and other methods are given in Reference 109. The errors quoted are those given in the original paper or review article. The references have been chosen primarily as a key to the literature. It should not be assumed that the author referred to was responsible for the determination quoted, as the reference may be only to a review article.

D°_0 values have been converted to D°_{298} by use of the simple relation

$$D^\circ_{280} = D^\circ_0 + (3/2)RT$$

The table has been arranged in alphabetical order with no regard to electronegativity or any other property of the atoms.

Table 1
BOND STRENGTHS IN DIATOMIC MOLECULES

Molecule	D°_{298}/kcal mol^{-1}	D°_{298}/kJ mol^{-1}	Ref.
Ag–Ag	39 ± 2	163 ± 8	66
Ag–Al	43.9 ± 2.2	183.7 ± 9.2	64
Ag–Au	48.5 ± 2.2	202.9 ± 9.2	5
Ag–Bi	46 ± 10	193 ± 42	192
Ag–Br	70 ± 7	293 ± 29	109
Ag–Cl	81.6	341.4	141
Ag–Cu	41.6 ± 2.2	174.1 ± 9.2	114
Ag–D	54.2	226.8	172
Ag–Eu	30.3 ± 3	126.8 ± 12.6	50
Ag–F	84.7 ± 3.9	354.4 ± 16.3	109
Ag–Ga	43 ± 4	180 ± 15	32
Ag–Ge	41.7 ± 5.0	174.9 ± 20.9	216
Ag–H	54 ± 2	226 ± 8	109
Ag–Ho	29.5 ± 4.0	123.4 ± 16.7	45
Ag–I	56 ± 7	234 ± 29	109
Ag–In	42 ± 4	176 ± 17	25
Ag–Li	42.4 ± 1.5	177.4 ± 6.3	218
Ag–Mn	24 ± 5	100 ± 21	192
Ag–Na	33.1 ± 2.5	138.5 ± 10.5	236
Ag–Nd	<50	<209	179
Ag–O	53.7	224.7	258
Ag–S	51.9	217.2	258
Ag–Se	48.4	202.5	258
Ag–Sn	32.5 ± 5.0	136.0 ± 20.9	2
Ag–Te	46.8	195.8	258
Al–Al	44.5 ± 2.2	186.2 ± 9.2	37,142
Al–As	42.9	179.5	230
Al–Au	77.9 ± 1.5	325.9 ± 6.3	128
Al–Br	106 ± 2	444 ± 8	66
Al–Cl	118 ± 3	494 ± 13	66
Al–Cu	51.8 ± 2.5	216 ± 10.5	228
Al–D	69.5	290.8	192
Al–F	158.6 ± 1.5	663.6 ± 6.3	66
Al–H	68.1 ± 1.5	284.9 ± 6.3	66
Al–I	88.4 ± 0.5	369.9 ± 2.1	66
Al–Li	42.0 ± 3.5	175.7 ± 14.6	142
Al–N	71 ± 23	297 ± 96	109
Al–O	121.3 ± 2.5	507.5 ± 10.5	107,225
Al–P	51.8 ± 3.0	216.7 ± 12.6	66
Al–Pd	60.8 ± 2.9	254.4 ± 12.0	48
Al–S	89.3 ± 1.9	373.6 ± 8.0	277
Al–Sb	51.7 ± 1.4	216 ± 6.0	232
Al–Se	80.7 ± 2.4	337.7 ± 10.1	277
Al–Si	60 ± 3	251 ± 12	37
Al–Te	64.0 ± 2.4	267.8 ± 10.1	277
Al–U	78 ± 7	326 ± 29	127
Ar–Ar	1.13 ± 0.01	4.73 ± 0.04	52,201
Ar–He	0.93	3.89	192
Ar–Hg	1.47	6.15	192
Ar–I	2.4	10.0	31
Ar–K	1.0	4.2	172
As–As	91.3 ± 2.5	382.0 ± 10.5	193
As–Cl	107	448	66
As–D	64.6	270.3	172
As–F	98	410	172
As–Ga	50.1 ± 0.3	209.6 ± 1.2	71
As–H	84	352	75
As–N	139 ± 30	582 ± 126	66
As–O	115 ± 2	481 ± 8	66
As–P	103.6 ± 3.0	433.5 ± 12.6	131
As–S	~114	~478	103,224
As–Sb	79.0 ± 1.3	330.5 ± 5.4	85a
As–Se	23	96	227
As–Tl	47.4 ± 3.5	198 .3 ± 14.6	238
At–At	~19	~80	80
Au–Au	53.8 ± 0.5	221.3 ± 2.1	196
Au–B	87.9 ± 2.5	367.8 ± 10.5	121
Au–Ba	38 ± 14	159 ± 59	109
Au–Be	68 ± 2	285 ± 8	109
Au–Bi	70 ± 20	293 ± 84	192
Au–Ca	46 ± 23	193 ± 96	109
Au–Ce	77.9 ± 3.5	325.9 ± 14.6	133
Au–Cl	82 ± 2.3	343 ± 9.6	109
Au–Co	51.3 ± 3.0	214.6 ± 12.6	177,258
Au–Cr	51.4 ± 1.5	215.1 ± 6.3	66
Au–Cu	56.3 ± 2.2	235.6 ± 9.2	5
Au–D	76.1	318.1	172
Au–Eu	57.6 ± 2.5	241.0 ± 10.5	50
Au–Fe	44.7 ± 4.0	187.0 ± 16.7	177
Au–Ga	70.2 ± 3.6	293.7 ± 15.1	32
Au–Ge	66.2 ± 3.5	277.0 ± 14.6	216
Au–H	75 ± 2.3	314 ± 9.6	109
Au–Ho	63.9 ± 4.0	267.4 ± 16.7	45,196
Au–La	80.4 ± 5.0	336.4 ± 20.9	133
Au–Li	68.0 ± 1.6	284.5 ± 6.5	218
Au–Lu	79.4 ± 4.0	332.2 ± 16.7	111
Au–Mg	58 ± 10	243 ± 42	192
Au–Mn	44.3 ± 3.0	185.4 ± 12.6	253
Au–Na	51.4 ± 3.0	215.1 ± 12.6	142
Au–Nd	71.5 ± 5.0	299.2 ± 20.9	133

* Revised to 30 June 1981.

Table 1 (continued)
BOND STRENGTHS IN DIATOMIC MOLECULES

Molecule	D°_{298}/kcal mol^{-1}	D°_{298}/kJ mol^{-1}	Ref.
Au–Ni	59 ± 5	247 ± 21	177
Au–O	54.6	228.5	258
Au–Pb	31 ± 10	130 ± 42	192
Au–Pd	34.2 ± 5.0	143.1 ± 20.9	6
Au–Pr	72.9 ± 5.0	305.0 ±20.9	133
Au–Rh	55.2 ± 7	230.9 ± 29	50
Au–S	100 ± 6	418 ± 25	110
Au–Sc	67.0 ± 4.0	280.3 ± 16.7	134
Au–Se	58.1	243.1	258
Au–Si	74.5 ± 2.9	311.7 ± 12.1	281
Au–Sn	58.4 ± 4.0	244.3 ± 16.7	2
Au–Sr	63 ± 10	264 ± 42	192
Au–Tb	70.1	293.3	114,196
Au–Te	75.9	317.6	258
Au–U	76 ± 7	318 ± 29	127
Au–V	57.5 ± 2.9	240.6 ± 12.1	147
Au–Y	72.0 ± 4.0	301.3 ± 16.7	134
B–B	71 ± 5	297 ± 21	66
B–Br	101 ± 5	423 ± 21	109
B–C	107 ± 7	448 ± 29	192
B–Ce	73 ± 5	305 ± 21	192
B–Cl	128	536	12
B–D	81.5 ± 1.5	341.0 ± 6.3	192
B–F	183 ± 3	766 ± 13	66
B–H	79.8	333.9	174
B–I	91 ± 5	381 ± 21	192
B–Ir	122.9 ± 4.1	514.2 ± 17.2	281
B–La	81 ± 15	339 ± 63	192
B–N	93 ± 5	389 ± 21	66
B–O	192.7 ± 1.2	806.3 ± 5.0	275
B–P	82.9 ± 4.0	346.9 ± 16.7	123
B–Pd	78.7 ± 5.0	329.3 ± 20.9	281
B–Pt	114.2 ± 4.0	477.8 ± 16.7	208
B–Rh	113.7 ± 5.0	475.7 ± 20.9	281
B–Ru	106.8 ± 5.0	446.9 ± 20.9	281
B–S	138.8 ± 2.2	580.7 ± 9.2	275
B–Sc	66 ± 15	276 ± 63	192
B–Se	110.4 ± 3.5	461.9 ± 14.6	275
B–Si	68.9	288.3	285
B–Te	84.7 ± 4.8	354.4 ± 20.1	275
B–Th	71	297	115
B–Ti	66 ± 15	276 ± 63	192
B–U	77 ± 8	322 ± 34	192
B–Y	70 ± 15	293 ± 63	192
Ba–Br	86.4 ± 2.2	361.5 ± 9.2	166,187
Ba–Cl	104.2 ± 2.0	436.0 ± 8.4	162,166
Ba–D	≤46.3	≤193.7	172
Ba–F	140.3 ± 1.6	587.0 ± 6.7	161
Ba–H	42 ± 3.5	176 ± 15	109
Ba–I	72.3 ± 1.0	302.5 ± 4.2	73,189
Ba–O	130.4 ± 1.0	545.6 ± 4	92,95,261
Ba–Pd	53.0± 1.2	221.8 ± 5.0	129
Ba–Rh	62.0 ± 6.0	259.4 ± 25.1	129
Ba–S	95.6 ± 4.5	400.0 ± 18.8	56
Be–Be	14	59	78
Be–Br	91 ± 20	381 ± 84	192
Be–Cl	92.8 ± 2.2	388.3 ± 9.2	99,171
Be–D	48.53	203.06	172
Be–F	138 ± 10	577 ± 42	63,99
Be–H	47.8 ± 0.3	200.0 ± 1.3	53
Be–O	105.4 ± 3	441.0 ± 13	261
Be–S	89 ± 14	372 ± 59	109
Bi–Bi	47.9 ± 1.8	200.4 ± 7.5	240,243
Bi–Br	63.9 ± 1.0	267.4 ± 4.2	61
Bi–Cl	72 ± 1	301 ± 8	63
Bi–D	67.8	283.7	206
Bi–F	62 ± 7	259 ± 29	109
Bi–Ga	38 ± 4	159 ± 17	235
Bi–H	≤67.7	≤283.3	206
Bi–I	52.1 ± 1.1	218.0 ± 4.6	62
Bi–O	81.9 ± 1.4	342.7 ± 5.9	276
Bi–P	67 ± 3	280 ± 13	131
Bi–Pb	33.9 ± 3.5	141.8 ± 14.6	243
Bi–S	75.4 ± 1.1	315.5 ± 4.6	276
Bi–Sb	60 ± 1	251 ± 4	190
Bi–Se	67.0 ± 1.4	280.3 ± 5.9	276
Bi–Te	55.5 ± 2.7	232.2 ± 11.3	276
Bi–Tl	29 ± 3	121 ± 13	72
Br–Br	46.333 ± 0.001	193.857 ± 0.004	14,172,202
Br–C	67 ± 5	280 ± 21	109
Br–Ca	74.3 ± 2.2	310.9 ± 9.2	165,187
Br–Cd	38 ± 23	159 ± 96	109
Br–Cl	52.3 ± 0.2	218.8 ± 0.8	109
Br–Co	79 ± 10	331 ± 42	192
Br–Cr	78.4 ± 5.8	328.0 ± 24.3	109
Br–Cs	95.0 ± 1.0	397.5 ± 4.2	192
Br–Cu	79 ± 6	331 ± 25	109
Br–D	88.61	370.75	172
Br–F	59.8 ± 0.2	250.2 ± 0.6	41,43
Br–Fe	59 ± 23	247 ± 96	109
Br–Ga	106 ± 4	444 ± 17	66
Br–Ge	61 ± 7	255 ± 29	109
Br–H	87.55	366.31	172
Br–Hg	17.4 ± 1	72.8 ± 4.2	66
Br–I	42.8 ± 0.1	179.1 ± 0.4	109
Br–In	99 ± 5	414 ± 21	66
Br–K	91.5 ± 2.0	382.8 ± 8.4	66
Br–Li	101 ± 5	423 ± 21	66
Br–Mg	≤78.2	≤327.2	172
Br–Mn	75.1 ± 2.3	314.2 ± 9.6	109
Br–N	66 ± 5	276 ± 21	109
Br–Na	88.5 ± 3.0	370.3 ± 12.6	66
Br–Ni	86 ± 3	360 ± 13	109
Br–O	56.2 ± 0.1	235.1 ± 0.4	66
Br–Pb	59 ± 9	247 ± 38	109
Br–Rb	93 ± 3	389 ± 13	109
Br–Sb	75 ± 14	314 ± 59	109
Br–Sc	106 ± 15	444 ± 63	192
Br–Se	71 ± 20	297 ± 84	192
Br–Si	82 ± 12	343 ± 50	109
Br–Sn	≥132	≥552	223a
Br–Sr	79.6 ± 2.2	333.1 ± 9.2	165
Br–Ti	105	439	192
Br–Tl	79.8 ± 0.4	333.9 ± 1.7	22
Br–V	105 ± 10	439 ± 42	192
Br–W	78.7	329.3	176a
Br–Y	116 ± 20	485 ± 84	192
Br–Zn	34 ± 7	142 ± 29	192
C–C	145 ± 5	607 ± 21	66
C–Ce	109 ± 7	456 ± 29	120
C–Cl	95 ± 7	397 ± 29	222
C–D	81.6	341.4	172
C–F	132	552	156
C–Ge	110 ± 5	460 ± 21	109
C–H	80.8	338.1	172
C–Hf	129 ± 6	540 ± 25	265
C–I	50 ± 5	209 ± 21	109
C–Ir	149.3 ± 3.0	624.7 ± 12.6	208
C–La	121 ± 15	506 ± 63	192
C–N	184 ± 1	770 ± 4	67
C–O	257.3 ± 0.1	1076.5 ± 0.4	66
C–Os	≥142	≥594	130
C–P	122.7 ± 2	513.4 ± 8	259
C–Pt	146.3 ± 1.5	612.1 ± 6.3	279
C–Rh	139.5 ± 1.5	583.7 ± 6.3	279
C–Ru	154.9 ± 3	648.1 ± 13	113
C–S	167.0 ± 2.0	698.7 ± 8.4	84,154
C–Sc	94 ± 15	393 ± 63	192
C–Se	141.1 ± 1.4	590.4 ± 5.9	254
C–Si	107.9	451.5	77,285
C–Th	116 ± 6	485 ± 25	265
C–Ti	104 ± 6	435 ± 25	265
C–U	111 ± 7	464 ± 29	120
C–V	112 ± 15	469 ± 63	120
C–Y	100 ± 15	418 ± 63	120
C–Zr	134 ± 6	561 ± 25	265
Ca–Ca	3.58 ± 0.11	14.98 ± 0.46	11
Ca–Cl	95 ± 3	398 ± 13	162
Ca–D	≤40.6	≤169.9	172
Ca–F	126 ± 5	527 ± 21	26,168

Table 1 (continued)
BOND STRENGTHS IN DIATOMIC MOLECULES

Molecule	D°_{298}/kcal mol^{-1}	D°_{298}/kJ mol^{-1}	Ref.
Ca–H	40.1	167.8	109
Ca–I	63.0 ± 2.5	263.6 ± 10.5	189
Ca–O	92 ± 2	385 ± 8	92,96,261
Ca–S	80.7 ± 4.5	337.7 ± 18.8	56,172
Cd–Cd	2.7 ± 0.2	11.3 ± 0.8	109
Cd–Cl	49.8	208.4	172
Cd–F	73 ± 5	305 ± 21	23
Cd–H	16.5 ± 0.1	69.0 ± 0.4	109
Cd–I	33 ± 5	138 ± 21	109
Cd–In	33	138	192
Cd–O	90.1	377.0	139
Cd–S	48	201	209
Cd–Se	46.8	195.8	136
Ce–Ce	58.6	245.2	133
Ce–F	139 ± 10	582 ± 42	192
Ce–Ir	140	586	126
Ce–N	124 ± 5	519 ± 21	122
Ce–O	190 ± 3	795 ± 13	4
Ce–Os	121 ± 8	507 ± 33	130
Ce–Pd	77.0	322.2	46
Ce–Pt	133	557	126
Ce–Rh	131	548	126
Ce–Ru	127 ± 6	531 ± 25	130
Ce–S	136	569	19,57
Ce–Se	118.2 ± 3.5	494.4 ± 14.6	215
Ce–Te	93 ± 10	389 ± 42	192
Cl–Cl	58.062 ± 0.001	242.933 ± 0.004	172,202
Cl–Co	93	389	199
Cl–Cr	87.5 ± 5.8	366.1 ± 24.3	109
Cl–Cs	105 ± 5	439 ± 21	66
Cl–Cu	91.5 ± 1.1	382.8 ± 21	144
Cl–D	104.32	436.47	172
Cl–Eu	~78	~ 326	105
Cl–F	61.24	256.3	172,219
Cl–Fe	~84	~352	109
Cl–Ga	115 ± 3	481 ± 13	66
Cl–Ge	~103	~431	172
Cl–H	103.13	431.49	172
Cl–Hg	24 ± 2	100 ± 8	109
Cl–I	50.5 ± 0.1	211.3 ± 0.4	109
Cl–In	105 ± 2	439 ± 8	66
Cl–K	102 ± 2	427 ± 8	66
Cl–Li	112 ± 3	469 ± 13	66
Cl–Mg	78.3 ± 0.5	327.6 ± 2.1	94,162
Cl–Mn	86.2 ± 2.3	360.7 ± 9.6	109
Cl–N	79.8 ± 2.3	333.9 ± 9.6	39
Cl–Na	98 ± 2	410 ± 8	66
Cl–Ni	89 ± 5	372 ± 21	109
Cl–O	65 ± 1	272 ± 4	66
Cl–P	69 ± 10	289 ± 42	192
Cl–Pb	72 ± 7	301 ± 29	109
Cl–Ra	82 ± 18	343 ± 75	109
Cl–Rb	107 ± 5	448 ± 21	66
Cl–Sb	86 ± 12	360 ± 50	109
Cl–Sc	79	331	284
Cl–Se	77	322	192
Cl–Si	109 ± 10	456 ± 42	192
Cl–Sm	⩾101 ± 3	⩾423 ± 13	292
Cl–Sn	99 ± 4	414 ± 17	192
Cl–Sr	97 ± 3	406 ± 13	162,166
Cl–Ti	118	494	192
Cl–Tl	89.1 ± 0.5	372.8 ± 2.1	22
Cl–V	114 ± 15	477 ± 63	192
Cl–W	101 ± 10	423 ± 42	192
Cl–Xe	1.6	6.7	172
Cl–Y	126 ± 20	527 ± 42	192
Cl–Yb	~77	~322	105
Cl–Zn	54.7 ± 4.7	228.9 ± 19.7	58
Cm–O	176	736	250
Co–Co	40 ± 6	167 ± 25	180
Co–Cu	38.7 ± 4.0	161.9 ± 16.7	183
Co–F	104 ± 15	435 ± 63	192
Co–Ge	57 ± 6	239 ± 25	182
Co–I	68 ± 5	285 ± 21	192
Co–O	88 ± 5	368 ± 21	27
Co–S	79	331	258
Co–Si	66 ± 4	276 ± 17	280
Cr–Cr	37 ± 5	155 ± 21	181
Cr–Cu	37 ± 5	155 ± 21	183
Cr–F	106.3 ± 4.7	444.8 ± 19.7	186
Cr–Ge	40.6 ± 7	169.9 ± 29	182
Cr–H	67 ± 12	280 ± 50	109
Cr–I	68.6 ± 5.8	287.0 ± 24.3	109
Cr–N	90.3 ± 4.5	377.8 ± 18.8	262
Cr–O	102 ± 7	427 ± 29	66
Cr–S	79	331	83
Cs–Cs	9.97 ± 0.22	41.75 ± 0.93	172
Cs–F	122.9 ± 2	514.2 ± 8	192
Cs–H	42.6 ± 0.9	178.2 ± 3.8	242,271
Cs–Hg	2	5	172
Cs–I	81 ± 1	339 ± 4	192
Cs–O	71 ± 6	297 ± 25	192
Cu–Cu	48.2 ± 2	201.7 ± 4	228
Cu–D	64.6	270.3	172
Cu–F	98.8 ± 3	413.4 ± 13	91
Cu–Ga	51.6 ± 3.6	215.9 ± 15.1	32
Cu–Ge	49.9 ± 5	208.8 ± 21	216
Cu–H	64	268	29,172
Cu–I	47 ± 5	197 ± 21	109
Cu–Li	46.1 ± 2.1	192.9 ± 8.8	218
Cu–Mn	37.9 ± 4	158.6 ± 17	183
Cu–Na	42.1 ± 4.0	176.2 ± 16.7	237
Cu–Ni	49.2 ± 4	205.9 ± 17	183
Cu–O	65.2	272.8	17,258
Cu–S	66	276	258
Cu–Se	60	251	258
Cu–Sn	42.3 ± 4	177.0 ± 17	2
Cu–Te	66.6	278.7	1
D–D	105.957	443.324	172
D–F	137.8	576.6	172
D–Ga	<65.2	<272.8	197
D–Ge	⩽77	⩽322	172
D–H	104.978 ± 0.001	439.228 ± 0.004	172
D–Hg	10.05	42.08	172
D–In	58.8	246.0	172
D–Li	57.4066 ± 0.0011	240.1892 ± 0.0046	173,269
D–Mg	32.3	135.1	172
D–Ni	⩽72.4	⩽302.9	172
D–Pt	⩽83.7	⩽350.2	172
D–S	84	352	172
D–Si	72.3	302.5	172
D–Sr	⩾65.9	⩾275.7	172
D–Zn	21.2	88.7	172
Dy–F	127	531	296
Dy–O	146 ± 10	611 ± 42	27
Dy–S	99 ± 10	414 ± 42	192
Dy–Se	77 ± 10	322 ± 42	192
Dy–Te	56 ± 10	234 ± 42	192
Er–F	135 ± 4	565 ± 17	296
Er–O	146 ± 3	611 ± 13	251
Er–S	100 ± 10	418 ± 42	192
Er–Se	78 ± 10	326 ± 42	192
Er–Te	57 ± 10	239 ± 42	192
Eu–Eu	8.0 ± 4.0	33.5 ± 16.7	50
Eu–F	126.1 ± 4.4	527.6 ± 18.4	74,294
Eu–O	112.2 ± 2.4	469.5 ± 10.0	8,74,214
Eu–Rh	55.9 ± 8.0	233.9 ± 33.5	50
Eu–S	86.6 ± 3.1	362.3 ± 13.0	215,256
Eu–Se	72 ± 3.5	301 ± 14.6	18,215
Eu–Te	58 ± 3.5	243 ± 14.6	18,215
F–F	37.83	158.3	68,172
F–Ga	138 ± 3.5	577 ± 14.6	212
F–Gd	141.1 ± 6.5	590.4 ± 27.2	294
F–Ge	116 ± 5	485 ± 21	90
F–H	136.2	569.9	172
F–Hg	~43	~180	172
F–Ho	129	540	296
F–I	⩽64.9	⩽271.6	7,42,60
F–In	121 ± 3.5	506 ± 14.6	212
F–K	118.9 ± 0.6	497.5 ± 2.5	13

Table 1 (continued)
BOND STRENGTHS IN DIATOMIC MOLECULES

Molecule	D°_{298}/kcal mol⁻¹	D°_{298}/kJ mol⁻¹	Ref.
F–La	143 ± 10	598 ± 42	192
F–Li	138 ± 5	577 ± 21	66
F–Lu	136 ± 10	569 ± 42	192
F–Mg	110.4 ± 1.2	461.9 ± 5.0	161
F–Mn	101.2 ± 3.5	423.4 ± 14.6	185
F–Mo	111.1	464.8	164
F–N	82	342	172
F–Na	124	519	172
F–Nd	130.3 ± 3.0	545.2 ± 12.6	293
F–Ni	104	435	192
F–O	53 ± 4	222 ± 17	44
F–P	105 ± 23	439 ± 96	109
F–Pb	85 ± 2	356 ± 8	292
F–Pm	129 ± 10	540 ± 42	192
F–Pr	139 ± 11	582 ± 46	192
F–Pu	128.7 ± 7	538.5 ± 29	184
F–Rb	118 ± 5	493.7 ± 21	66
F–S	81.9 ± 1.2	342.7 ± 5.0	167
F–Sb	105 ± 23	439 ± 96	109
F–Sc	140.8 ± 3	589.1 ± 13	295
F–Se	81 ± 10	339 ± 42	192
F–Si	132.1 ± 0.5	552.7 ± 2.1	100
F–Sm	126.9 ± 4.4	531.0 ± 18.4	74,291,294
F–Sn	111.5 ± 3	466.5 ± 13	292
F–Sr	129.5 ± 1.6	541.8 ± 6.7	161
F–Tb	134 ± 10	561 ± 42	192
F–Ti	136 ± 8	569 ± 34	297
F–Tl	106.4± 4.6	445.2 ± 19.3	22
F–Tm	136 ± 10	569 ± 42	192
F–V	141 ± 15	590 ± 63	192
F–W	131 ± 15	548 ± 63	192
F–Xe	4.0 ± 0.1	16.8 ± 0.4	273
F–Y	144.6 ± 5.0	605.0 ± 20.9	295
F–Yb	⩾124.6 ± 2.3	⩾521.3 ± 9.6	14,105,290
F–Zn	88 ± 15	368 ± 63	192
F–Zr	149 ± 15	623 ± 63	192
Fe–Fe	24 ± 5	100 ± 21	204
Fe–Ge	50.4 ± 7	210.9 ± 29	182
Fe–O	97.7 ± 3	408.8 ± 13	157
Fe–S	77	322	83
Fe–Si	71 ± 6	297 ± 25	280
Ga–Ga	33 ± 5	138 ± 21	192
Ga–H	<65.5	<274.1	197
Ga–I	81 ± 2.3	339 ± 9.6	109
Ga–Li	31.8 ± 3.5	133.1 ± 14.6	142
Ga–O	91	381	30
Ga–P	54.9 ± 3.0	229.7 ± 12.6	137
Ga–Sb	45.9 ± 3.0	208.8 ± 12.6	231
Ga–Te	60 ± 6	251 ± 25	278
Gd–O	171 ± 4	716 ± 17	251
Gd–S	125.9 ± 2.5	526.8 ± 10.5	106,251
Gd–Se	103 ± 3.5	431 ± 14.6	18
Gd–Te	82 ± 3.5	343 ± 14.6	18
Ge–Ge	65.4 ± 5	273.6 ± 21	216
Ge–H	⩽76.9	⩽ 321.8	16,189a
Ge–Ni	67 ± 3	280 ± 13	183
Ge–O	157.3	658.1	172
Ge–Pd	63.2	264.4	226
Ge–S	131.7 ± 0.6	551.0 ± 2.5	57,79,84
Ge–Se	115.7	484.1	172
Ge–Si	72 ± 5	301 ± 21	109
Ge–Te	99	414	53a
H–H	104.155	435.783	172
H–Hg	9.523	39.843	172
H–I	71.32	298.39	172
H–In	58.1	243.0	172
H–K	43.8 ± 3.5	183.3 ± 14.6	172
H–Li	56.895 ± 0.001	238.049 ± 0.004	286
H–Mg	30.2 ± 0.7	126.4 ± 5.9	9,10,89
H–Mn	56 ± 7	234 ± 29	109
H–N	⩽81	⩽339	172
H–Na	44	184	172
H–Ni	⩽72	⩽301	172
H–O	102.2	427.5	172

Molecule	D°_{298}/kcal mol⁻¹	D°_{298}/kJ mol⁻¹	Ref.
H–P	71	297	172
H–Pb	42 ± 5	176 ± 21	172
H–Pt	⩽80	⩽335	172
H–Rb	40 ± 5	167 ± 21	109
H–S	82.3 ± 2.9	344.3 ± 12.1	173
H–Sc	~43	~180	246
H–Se	73 ± 0.5	305 ± 2.1	145
H–Si	⩽71.5	⩽299.2	172
H–Sn	63 ± 4	264 ± 17	109
H–Sr	39 ± 2	163 ± 8	109
H–Te	64 ± 0.5	268 ± 2.1	145
H–Ti	~38	~159	245
H–Tl	45 ± 2	188 ± 8	109
H–Yb	38 ± 9	159 ± 38	109
H–Zn	20.5 ± 0.5	85.8 ± 2.1	109
He–He	0.9	3.8	172
He–Hg	1.58	6.61	192
Hf–C	131 ± 15	548 ± 63	192
Hf–N	128 ± 7	534 ± 29	191
Hf–O	189.8 ± 2.0	794.1 ± 8.4	4
Hg–Hg	4.1 ± 0.5	17.2 ± 2.1	109
Hg–I	9	38	289
Hg–K	1.97 ± 0.05	8.24 ± 0.21	192
Hg–Li	3.3	13.9	172
Hg–Na	2.2	9.2	172,300
Hg–Rb	2.0	8.4	172
Hg–S	51	213	209
Hg–Se	⩽40	⩽167	209
Hg–Te	⩽34	⩽142	192
Hg–Tl	1	4	151
Ho–Ho	20 ± 4	84 ± 17	45
Ho–O	148	619	251
Ho–S	102.4 ± 3.5	428.4 ± 14.6	251
Ho–Se	80 ± 4	335 ± 17	18
Ho–Te	62 ± 4	259 ± 17	18
I–I	36.456	152.532	172,202
I–In	80	335	12
I–K	79 ± 3	331 ± 13	66
I–Li	84 ± 3	352 ± 13	66
I–Mg	~68	~285	20
I–Mn	67.6 ± 2.3	282.8 ± 9.6	109
I–N	38 ± 4	159 ± 17	192
I–Na	72 ± 2	301 ± 8	66
I–Ni	70 ± 5	293 ± 21	109
I–O	43	180	172
I–Pb	47 ± 9	197 ± 38	109
I–Rb	80 ± 3	335 ± 13	66
I–Si	70	293	172
I–Sn	56 ± 10	234 ± 42	192
I–Sr	64.5 ± 1.4	269.9 ± 5.9	189
I–Te	46 ± 10	193 ± 42	192
I–Ti	74 ± 10	310 ± 42	192
I–Tl	65 ± 2	272 ± 8	20
I–Zn	33 ± 7	138 ± 29	109
I–Zr	73	305	188
In–In	24 ± 2	100 ± 8	192
In–Li	22.1 ± 3.5	92.5 ± 14.6	142
In–O	86 ± 5	360 ± 21	66
In–P	47.3 ± 2.0	197.9 ± 8.5	230
In–S	69 ± 4	289 ± 17	55
In–Sb	36.3 ± 2.5	151.9 ± 10.5	69
In–Se	59 ± 4	247 ± 17	55
In–Te	52 ± 4	218 ± 17	55
Ir–O	84 ± 5	352 ± 21	192
Ir–Si	110.6 ± 5.0	462.8 ± 20.9	281
Ir–Th	137	573	112
K–K	13.7 ± 1.0	57.3 ± 4.2	192
K–Kr	1.1	4.6	172
K–Li	19.6 ± 1.0	82.0 ± 4.2	299
K–Xe	1.2	5.1	172
K–Na	15.5 ± 1.0	64.9 ± 4.2	299
K–O	66 ± 3	276 ± 13	88
Kr–Kr	1.25	5.23	35,172
Kr–O	<2	<8	192

Table 1 (continued)
BOND STRENGTHS IN DIATOMIC MOLECULES

Molecule	D°_{298}/kcal mol^{-1}	D°_{298}/kJ mol^{-1}	Ref.
La–La	59 ± 5	247 ± 21	284
La–N	124 ± 10	519 ± 42	192
La–O	191 ± 3	799 ± 13	138,251
La–Rh	126 ± 4	527 ± 17	49
La–S	137.0 ± 0.4	573.4 ± 1.8	57,176,268
La–Se	114 ± 4	477 ± 17	18,215
La–Te	91 ± 4	381 ± 17	18
La–Y	48.3	202.1	284
Li–Li	26.31 ± 1	106.48 ± 4.2	282,290
Li–Na	21.6 ± 1.5	90.4 ± 6.3	299
Li–O	81.4 ± 1.5	340.6 ± 6.3	163,198
Lu–Lu	34 ± 8	142 ± 34	192
Lu–O	166 ± 3	695 ± 13	251
Lu–Pt	96 ± 8	402 ± 34	116
Lu–S	121.2 ± 3.5	507.1 ± 14.6	251
Lu–Se	100 ± 4	418 ± 17	18
Lu–Te	78 ± 4	326 ± 17	18
Mg–Mg	2.044 ± 0.001	8.552 ± 0.004	203
Mg–O	86.6 ± 0.5	362.3 ± 2.1	93,97
Mg–S	56	234	56
Mn–Mn	6.2	25.9	179
Mn–O	86	360	57
Mn–S	72 ± 4	301 ± 17	287
Mn–Se	57.2 ± 2.2	239.3 ± 9.2	260,288
Mo–Mo	97 ± 5	406 ± 20	145a
Mo–Nb	109 ± 6	454 ± 25	145a
Mo–O	145.1 ± 8	607.1 ± 34	38
N–N	225.94 ± 0.14	945.33 ± 0.59	172
N–O	150.71 ± 0.03	630.57 ± 0.13	172
N–P	147.5 ± 5.0	617.1 ± 20.9	125
N–Pu	113 ± 15	473 ± 63	192
N–S	111 ± 5	464 ± 21	192
N–Sb	72 ± 12	301 ± 50	109
N–Sc	112 ± 20	469 ± 84	192
N–Se	91 ± 15	381 ± 63	192
N–Si	105 ± 9	439 ± 38	109
N–Ta	146 ± 20	611 ± 84	192
N–Th	138.0 ± 0.5	577.4 ± 2.1	119
N–Ti	111	464	263
N–U	127.0 ± 0.5	531.4 ± 2.1	117
N–V	114.1 ± 2	477.4 ± 8	98
N–Xe	5.5	23.0	150
N–Y	115 ± 15	481 ± 63	192
N–Zr	135 ± 6	565 ± 25	119
Na–Na	18.4	77.0	249
Na–O	61.3 ± 4	256.5 ± 17	169
Na–Rb	14 ± 0.9	59 ± 3.8	109
Nb–O	180 ± 3	753 ± 13	192
Nd–Nd	<39	<163	192
Nd–O	168 ± 8	703 ± 34	251
Nd–S	112.7	471.5	18
Nd–Se	92 ± 4	385 ± 17	18,215
Nd–Te	73 ± 4	305 ± 17	18
Ne–Ne	0.94	3.93	272
Ni–Ni	62.6 ± 0.6	261.9 ± 2.5	244
Ni–O	93.6 ± 0.9	391.6 ± 3.8	244
Ni–S	82.3	344.3	83
Ni–Si	76 ± 4	318 ± 17	280
Np–O	172 ± 7	720 ± 29	3
O–O	119.016 ± 0.048	498.340 ± 0.200	28,51
O–Os	<142	<594	27
O–P	142.6	596.6	82
O–Pb	90.3 ± 1.0	377.8 ± 4.2	221,276
O–Pd	56 ± 7	234 ± 29	27
O–Pm	161 ± 15	674 ± 63	192
O–Pr	180 ± 4	753 ± 17	251
O–Pt	83 ± 8	347 ± 34	27
O–Pu	163 ± 15	682 ± 63	27
O–Rb	61 ± 20	255 ± 84	27
O–Rh	90 ± 15	377 ± 63	27
O–Ru	115 ± 15	481 ± 63	27
O–S	124.69 ± 0.03	521.70 ± 0.13	27
O–Sb	89 ± 20	372 ± 84	27
O–Sc	164.4 ± 2.3	687.9 ± 9.6	34
O–Se	≤102.6	≤429.3	172
O–Si	193.5 ± 2.6	809.6 ± 10.9	200
O–Sm	136.9 ± 2.0	572.8 ± 8.4	74,159,251
O–Sn	131 ± 5	548 ± 21	66
O–Sr	99.4 ± 2	415.9 ± 8	92,96,261
O–Ta	192.4 ± 3	805.0 ± 13	255
O–Tb	169 ± 3	707 ± 13	251
O–Te	93.4 ± 2.0	390.8 ± 8.4	211,298
O–Th	204 ± 3	854 ± 13	170,217
O–Ti	159.3 ± 1.5	666.5 ± 6.3	86,158,207
O–Tm	133 ± 3	557 ± 13	251
O–U	181.9 ± 4.0	761.1 ± 16.7	267
O–V	149	623	101
O–W	156 ± 6	653 ± 25	27
O–Xe	8.7	36.4	192
O–Y	170.9 ± 3.0	715.1 ± 3.0	34,267
O–Yb	99.8 ± 1.2	417.6 ± 5.0	59,291
O–Zn	67.9	284.1	140
O–Zr	181.6 ± 2	759.8 ± 8.4	4,213
P–P	117.0 ± 2.5	489.5 ± 10.5	125
P–Pt	≤99.6 ± 4	≤416.7 ± 17	257
P–Rh	84.4 ± 4	353.1 ± 17	257
P–S	106 ± 2	444 ± 8	81
P–Sb	85.3	356.9	195
P–Se	86.9 ± 2.4	363.6 ± 10	81
P–Si	86.9	363.6	252
P–Te	71.2 ± 2.4	297.9 ± 10.0	81
P–Th	90	377	115
P–Tl	50 ± 3	209 ± 13	232
P–U	71 ± 5	297 ± 21	192
P–W	73 ± 1	305 ± 4	124
Pb–Pb	19 ± 1.5	81 ± 6	132
Pb–S	82.7 ± 0.4	346.0 ± 1.7	276
Pb–Se	72.4 ± 1	302.9 ± 4	276
Pb–Te	60 ± 3	251 ± 13	276
Pd–Pd	17	75	205
Pd–Si	74.9 ± 3.3	313.4 ± 13.8	281
Pm–S	101 ± 15	423 ± 63	192
Pm–Se	81 ± 15	339 ± 63	192
Pm–Te	61 ± 15	255 ± 63	192
Po–Po	44.7	187.0	172
Pr–S	117.7 ± 1.1	492.5 ± 4.6	104
Pr–Se	106.7 ± 5.5	446.4 ± 23.0	215
Pr–Te	78 ± 10	326 ± 42	192
Pt–Si	119.8 ± 4.3	501.2 ± 18.0	281
Pt–Th	132	552	112
Rb–Rb	10.9 ± 0.5	45.6 ± 2.1	234
Rh–Rh	68.2 ± 5.0	285.4 ± 20.9	47,233
Rh–Si	94.4 ± 4.3	395.0 ± 18.0	281
Rh–Th	123 ± 5	513 ± 21	135
Rh–Ti	93.4 ± 3.5	390.8 ± 14.6	47
Rh–U	124 ± 4	519 ± 17	135
Ru–Si	94.9 ± 5.0	397.1 ± 20.9	281
Ru–Th	141.4 ± 10	591.6 ± 42	113
S–S	101.65	425.28	172
S–Sb	90.5	378.7	102
S–Sc	114.3 ± 3.0	478.2 ± 12.6	268,274
S–Se	88.7 ± 1.6	371.1 ± 6.7	85
S–Si	149	623	172
S–Sm	93	389	104
S–Sn	111 ± 0.8	464 ± 3.2	79
S–Sr	81	339	33
S–Tb	123 ± 10	515 ± 42	192
S–Te	81 ± 5	339 ± 21	79
S–Ti	101.8 ± 1.8	425.9 ± 7.5	87
S–Tm	88 ± 10	368 ± 42	192
S–U	124.9 ± 2.3	522.6 ± 9.6	267
S–V	117 ± 4	490 ± 16	223
S–Y	126.3 ± 2.5	528.4 ± 10.5	57,266
S–Yb	40	167	192
S–Zn	49 ± 3	205 ± 13	70
S–Zr	137.5 ± 4.0	575.3 ± 16.7	267
Sb–Sb	71.5 ± 1.5	299.2 ± 6.3	69,194
Sb–Te	66.3 ± 0.9	277.4 ± 3.8	239,270
Sb–Tl	26.1	109.2	232
Sc–Sc	38.9 ± 5	162.8 ± 21	114

Table 1 (continued)
BOND STRENGTHS IN DIATOMIC MOLECULES

Molecule	D°_{298}/kcal mol^{-1}	D°_{298}/kJ mol^{-1}	Ref.
Sc–Se	92 ± 4	385 ± 17	192
Sc–Te	69 ± 4	289 ± 17	192
Se–Se	79.5 ± 0.1	332.6 ± 0.4	85,276
Se–Si	131	548	172
Se–Sm	79.1 ± 3.5	331.0 ± 14.6	215
Se–Sn	95.9 ± 1.4	401.3 ± 5.9	54
Se–Sr	~68	~285	21
Se–Tb	101 ± 10	423 ± 42	192
Se–Te	71.0 ± 1.2	297.1 ± 5.0	79,85
Se–Ti	91 ± 10	381 ± 42	192
Se–Tm	66 ± 10	276 ± 42	192
Se–V	83 ± 5	347 ± 21	192
Se–Y	104 ± 3	435 ± 13	192
Se–Zn	32.6 ± 3.0	136.4 ± 12.6	70
Si–Si	78.1 ± 2.4	326.8 ± 10.0	37,172
Si–Te	108	452	172
Sm–Te	65.1 ± 3.5	272.4 ± 14.6	215
Sn–Sn	46.7 ± 4	195.4 ± 17	2
Sn–Te	86.0	359.8	172
Tb–Tb	31.4 ± 6.0	131.4 ± 25.1	196
Tb–Te	81 ± 10	339 ± 42	192
Te–Te	62.6	261.9	172,276
Te–Ti	69 ± 4	289 ± 17	192
Te–Tm	66 ± 10	276 ± 42	192
Te–Y	81 ± 3	339 ± 13	192
Te–Zn	23	96	172
Th–Th	⩽69	⩽289	115
Ti–Ti	33.8 ± 5	141.4 ± 21	178
Tl–Tl	<22	<92	172
U–U	53 ± 5	222 ± 21	192
V–V	57.9 ± 5	242.3 ± 21	178
Xe–Xe	1.56 ± 0.07	6.53 ± 0.30	36
Y–Y	38 ± 5	159 ± 21	192
Yb–Yb	4.9 ± 4	20.5 ± 17	143
Zn–Zn	7	29	249

REFERENCES

1. Abbasov, A. S., Azizov, T. Kh., Alleva, N. A., Aliev, I. Ya., Mustafaev, F. M., and Mamedov, A. N., *Zh. Fiz. Khim.*, 50, 2172, 1976.
2. Ackerman, M., Drowart, J., Stafford, F. E., and Verhaegen, G., *J. Chem. Phys.*, 36, 1557, 1962.
3. Ackermann, R. J., Faircloth, R. F., Rauh, E. G., and Thorn, R. J., *J. Inorg. Nucl. Chem.*, 28, 111, 1966.
4. Ackermann, R. J. and Rauh, E. G., *J. Chem. Phys.*, 60, 2266, 1974.
5. Ackerman, M., Stafford, F. E., and Drowart, J., *J. Chem. Phys.*, 33, 1784, 1960.
6. Ackerman, M., Stafford, F. E., and Verhaegen, G., *J. Chem. Phys.*, 36, 1560, 1962.
7. Appelman, E. H. and Clyne, M. A. A., *J. Chem. Soc. Faraday Trans. 1*, 71, 2072, 1975.
8. Balducci, G., Gigli, G., and Guido, M., *J. Chem. Phys.*, 67, 147, 1977.
9. Balfour, W. J. and Cartwright, H. M., *Astron. Astrophys. Suppl. Ser.*, 26, 389, 1976.
10. Balfour, W. J. and Lingren, B., *Can. J. Chem.*, 56, 767, 1978.
11. Balfour, W. J. and Whitlock, R. E., *J. Chem. Soc. D*, p. 1231, 1971.
12. Barrow, R. F., *Trans. Faraday Soc.*, 56, 952, 1960.
13. Barrow, R. F. and Caunt, A. D., *Proc. R. Soc. London Ser. A*, 219, 120, 1953.
14. Barrow, R. F., Clark, T. C., Coxon, J., and Yee, K. K., *J. Mol. Spectrosc.*, 51, 428, 1974.
15. Barrow, R. F. and Chojnicki, A. H., *J. Chem. Soc. Faraday Trans. 2*, 71, 728, 1975.
16. Barrow, R. F. and Deutsch, E. W., *Proc. Chem. Soc.*, p. 122, 1960.
17. Belyaev, V. N., Lebedeva, N. L., Krasnov, K. S., and Gurvich, L. V., *Uchebn. Zaved. Khim. Khim. Fekhnol.*, 21, 1698, 1978.
18. Bergman, C., Coppens, P., Drowart, J., and Smoes, S., *Trans. Faraday Soc.*, 66, 800, 1970.
19. Bergman, C. and Gingerich, K. A., *J. Phys. Chem.*, 76, 2332, 1972.
20. Berkowitz, J. and Chupka, W. A., *J. Chem. Phys.*, 45, 1287, 1966.
21. Berkowitz, J. and Chupka, W. A., *J. Chem. Phys.*, 45, 4289, 1966.
22. Berkowitz, J. and Walter, T., *J. Chem. Phys.*, 49, 1184, 1968.
23. Besenbruch, G., Kana'an, A. S., and Margrave, J. L., *J. Phys. Chem.*, 69, 3174, 1965.
24. Birks, J. W., Gabelnick, S. D., and Johnston, H. S., *J. Mol. Spectrosc.*, 57, 23, 1975.
25. Biron, M., *C. R. Acad. Ser. B*, 265, 1026, 1427, 1967.
26. Blue, G. D., Green, J. W., Bautista, R. G., and Margrave, J. L., *J. Phys. Chem.*, 67, 877, 1963,
27. Brewer, L. and Rosenblatt, G. M., *Adv. High Temp. Sci.*, 2, 1, 1969.
28. Brix, P. and Herzberg, G., *J. Chem. Phys.*, 21, 2240, 1953.
29. Bulewicz, E. M. and Sugden, T. M., *Trans. Faraday Soc.*, 52, 1475, 1956.
30. Burns, R. P., *J. Chem. Phys.*, 44, 3307, 1966.
31. Burns, G., Le Roy, L. J., Morris, D. J., and Blake, J. A., *Proc. R. Soc. London Ser. A*, 316, 81, 1970.
32. Carbonel, M., Bergman, C., and Laffite, M., *Colloq. Int. Cent. Nat. Rech. Sci.*, 210, 311, 1972.
33. Cater, E. D. and Johnson, E. W., *J. Chem. Phys.*, 47, 5353, 1967.
34. Chalek, C. L. and Gole, J. L., *Chem. Phys.*, 19, 59, 1977.
35. Chashchina, G. I. and Shreider, E. Ya., *Zh. Prikl. Spektrosk.*, 21, 696, 1974.
36. Chashchina, G. I. and Shreider, E. Ya., *Zh. Prikl. Spektrosk.*, 25, 163, 1976.
37. Chatillon, C., Allibert, M., and Pattoret, A., *C. R. Acad. Sci. Ser. C*, 280, 1505, 1975.
38. Choudary, U. V., Gingerich, K. A., and Kingcade, J. E., *J. Less Common Met.*, 42, 111, 1975.
39. Clarke, T. C. and Clyne, M. A. A., *Trans. Faraday Soc.*, 66, 877, 1970.
40. Clements, R. M. and Barrow, R. F., *Trans. Faraday Soc.*, 64, 2893, 1968.
41. Clyne, M. A. A., Curran, A. H., and Coxon, J. A., *J. Mol. Spectrosc.*, 63, 43, 1976.
42. Clyne, M. A. A. and McDermid, I. S., *J. Chem. Soc. Faraday Trans. 2*, 72, 2252, 1976.

43. Clyne, M. A. A. and McDermid, I. S., *J. Chem. Soc. Faraday Trans. 2*, 74, 644, 1978.
44. Clyne, M. A. A. and Watson, R. T., *Chem. Phys. Lett.*, 12, 344, 1971.
45. Cocke, D. L. and Gingerich, K. A., *J. Phys. Chem.*, 75, 3264, 1971.
46. Cocke, D. L. and Gingerich, K. A., *J. Phys. Chem.*, 76, 2332, 1972.
47. Cocke, D. L. and Gingerich, K. A., *J. Chem. Phys.*, 60, 1958, 1974.
48. Cocke, D. L., Gingerich, K. A., and Chang, C.-A., *J. Chem. Soc. Faraday Trans. 1*, 72, 268, 1976.
49. Cocke, D. L., Gingerich, K. A., and Kordis, J., *High Temp. Sci.*, 5, 474, 1973.
50. Cocke, D. L., Gingerich, K. A., and Kordis, J., *High Temp. Sci.*, 7, 61, 1975.
51. CODATA recommended key values for thermodynamics, 1973, *J. Chem. Thermodyn.*, 7, 1, 1975.
52. Colbourn, E. A. and Douglas, A. E., *J. Chem. Phys.*, 65, 1741, 1976.
53. Colin, R. and De Greef, D., *Can. J. Phys.*, 53, 2142, 1975.
53a. Colin, R. and Drowart, J., *J. Phys. Chem.*, 68, 428, 1964.
54. Colin, R. and Drowart, J., *Trans. Faraday Soc.*, 60, 673, 1964.
55. Colin, R. and Drowart, J., *Trans. Faraday Soc.*, 64, 2611, 1968.
56. Colin, R., Goldfinger, P., and Jeunehomme, M., *Trans. Faraday Soc.*, 60, 306, 1964.
57. Coppens, P., Smoes, S., and Drowart, J., *Trans. Faraday Soc.*, 63, 2140, 1967.
58. Corbett, J. D. and Lynde, R. A., *Inorg. Chem.*, 6, 2199, 1967.
59. Cosmovici, C. B., D'Anna, E., D'Innocenzo, A., Leggieri, G., Perrone, A., and Dirscherl, R., *Chem. Phys. Lett.*, 47, 241, 1977.
60. Coxon, J. A., *Chem. Phys. Lett.*, 33, 136, 1975.
61. Cubicciotti, D., *Inorg. Chem.*, 7, 208, 1968.
62. Cubicciotti, D., *Inorg. Chem.*, 7, 211, 1968.
63. Cubicciotti, D., *J. Phys. Chem.*, 71, 3066, 1967.
64. Cuthill, A. M., Fabian, D. J., and Shu-Shou-Shen, S., *J. Phys. Chem.*, 77, 2008, 1973.
65. Dagdigian, P. J., Cruze, H. W., and Zare, R. N., *J. Chem. Phys.*, 62, 1824, 1975.
66. Darwent, B. de B., *Bond Dissociation Energies in Simple Molecules*, NSRDS-NBS 31, National Bureau of Standards, Washington, D.C., 1970.
67. Davis, D. D. and Okabe, H., *J. Chem. Phys.*, 49, 5526, 1968.
68. De Corpo, J. J., Steiger, R. P., Franklin, J. L., and Margrave, J. L., *J. Chem. Phys.*, 53, 936, 1970.
69. De Maria, G., Drowart, J., and Inghram, M. G., *J. Chem. Phys.*, 31, 1076, 1959.
70. De Maria, G., Goldfinger, P., Malaspina, L., and Piacente, V., *Trans. Faraday Soc.*, 61, 2146, 1965.
71. De Maria, G., Malaspina, L., and Piacente, V., *J. Chem. Phys.*, 52, 1019, 1970.
72. De Maria, G., Malaspina, L., and Piacente, V., *J. Chem. Phys.*, 56, 1978, 1972.
73. Dickson, C. R., Kinney, J. B., and Zare, R. N., *Chem. Phys.*, 15, 243, 1976.
74. Dickson, C. R. and Zare, R. N., *Chem. Phys.*, 7, 361, 1975.
75. Dixon, R. N. and Lambertson, H. M., *J. Mol. Spectrosc.*, 25, 12, 1968.
76. Drowart, J., in *Phase Stability in Metals and Alloys*, Rudman, P. S., Ed., McGraw-Hill, New York, 1967, 305.
77. Drowart, J., De Maria, G., and Inghram, M. G., *J. Chem. Phys.*, 29, 1015, 1958.
78. Drowart, J. and Goldfinger, P., *Angew. Chem.*, 6, 581, 1967.
79. Drowart, J. and Goldfinger, P., *Q. Rev. (London)*, 20, 545, 1966.
80. Drowart, J. and Honig, R. E., *J. Phys. Chem.*, 61, 980, 1957.
81. Drowart, J., Myers, C. E., Szwarc, R., Vander Auwera-Mahieu, A., and Uy, O. M., *High Temp. Sci.*, 5, 482, 1973.
82. Drowart, J., Myers, C. E., Szwarc, R., Vander Auwera-Mahieu, A., and Uy, O. M., *J. Chem. Soc. Faraday Trans. 2*, 68, 1749, 1972.
83. Drowart, J., Pattoret, A., and Smoes, S., *Proc. Br. Ceramic Soc.*, No. 8, 67, 1967.
84. Drowart, J., Smets, J., Reynaert, J. C., and Coppens, P., *Adv. Mass Spectrom.*, 7A, 647, 1978.
85. Drowart, J. and Smoes, S., *J. Chem. Soc. Faraday Trans. 2*, 73, 1755, 1977.
85a. Drowart, J., Smoes, S., and Vander Auwera-Mahieu, A., *J. Chem. Thermodyn.*, 10, 453, 1978.
86. Dubois, L. H. and Gole, J. L., *J. Chem. Phys.*, 66, 779, 1977.
87. Edwards, J. G., Franklin, H. F., and Gilles, P. W., *J. Chem. Phys.*, 54, 545, 1971.
88. Ehlert, T. C., *High Temp. Sci.*, 9, 237, 1977.
89. Ehlert, T. C., Hilmer, R. M., and Beauchamp, E. A., *J. Inorg. Nucl. Chem.*, 30, 3112, 1968.
90. Ehlert, T. C. and Margrave, J. L., *J. Chem. Phys.*, 41, 1066, 1964.
91. Ehlert, T. C. and Wang, J. S., *J. Phys. Chem.*, 81, 2069, 1977.
92. Engelke, F., Sander, R. K., and Zare, R. N., *J. Chem. Phys.*, 65, 1146, 1976.
93. Evans, P. J. and Mackie, J. C., *Chem. Phys.*, 5, 277, 1974.
94. Farber, M. and Srivastava, R. D., *Chem. Phys. Lett.*, 42, 567, 1976.
95. Farber, M. and Srivastava, R. O., *High Temp. Sci.*, 7, 74, 1975.
96. Farber, M. and Srivastava, R. O., *High Temp. Sci.*, 8, 73, 1976.
97. Farber, M. and Srivastava, R. D., *High Temp. Sci.*, 8, 195, 1976.
98. Farber, M. and Srivastava, R. D., *J. Chem. Soc. Faraday Trans. 1*, 69, 390, 1973.
99. Farber, M. and Srivastava, R. D., *J. Chem. Soc. Faraday Trans. 1*, 70, 1581, 1974.
100. Farber, M. and Srivastava, R. D., *J. Chem. Soc. Faraday Trans. 1*, 74, 1089, 1978.
101. Farber, M., Uy, O. M., and Srivastava, R. D., *J. Chem. Phys.*, 56, 5312, 1972.
102. Faure, F. M., Mitchell, M. J., and Bartlett, R. W., *High Temp. Sci.*, 4, 181, 1972.

103. Faure, F. M., Mitchell, M. J., and Bartlett, R. W., *High Temp. Sci.*, 5, 128, 1973.
104. Fenochka, B. V. and Gorkienko, S. P., *Zh. Fiz, Khim.*, 47, 2445, 1973.
105. Filippenko, N. V., Motozov, E. V., Giricheva, N. I., and Krasnev, K. S., *Izv. Vyssh. Ucheb. Zaved Khim. Technol.*, 15, 1416, 1972.
106. Fries, J. A. and David, C. E., *J. Chem. Phys.*, 68, 3978, 1978.
107. Fu, C. M. and Burns, R. P., *High Temp. Sci.*, 8, 353, 1976.
108. Fujishiro, S., *Trans. Jpn. Inst. Met.*, 1, 125, 1960.
109. Gaydon, A. G., *Dissociation Energies and Spectra of Diatomic Molecules*, 3rd ed., Chapman and Hall, London, 1968.
110. Gingerich, K. A., *Chem. Commun.*, 580, 1970.
111. Gingerich, K. A., *Chem. Phys. Lett.*, 13, 262, 1972.
112. Gingerich, K. A., *Chem. Phys. Lett.*, 23, 270, 1973.
113. Gingerich, K. A., *Chem. Phys. Lett.*, 25, 523, 1974.
114. Gingerich, K. A., *Chimia*, 26, 619, 1972.
115. Gingerich, K. A., *High Temp. Sci.*, 1, 258, 1969.
116. Gingerich, K. A., *High Temp. Sci.*, 3, 415, 1971.
117. Gingerich, K. A., *J. Chem. Phys.*, 47, 2192, 1967.
118. Gingerich, K. A., *J. Chem. Phys.*, 49, 14, 1968.
119. Gingerich, K. A., *J. Chem. Phys.*, 49, 19, 1968.
120. Gingerich, K. A., *J. Chem. Phys.*, 50, 2255, 1969.
121. Gingerich, K. A., *J. Chem. Phys.*, 54, 2646, 1971.
122. Gingerich, K. A., *J. Chem. Phys.*, 54, 3720, 1971.
123. Gingerich, K. A., *J. Chem. Phys.*, 56, 4239, 1972.
124. Gingerich, K. A., *J. Phys. Chem.*, 68, 768, 1964.
125. Gingerich, K. A., *J. Phys. Chem.*, 73, 2734, 1969.
126. Gingerich, K. A., *J. Chem. Soc. Faraday Trans. 2*, 70, 471, 1974.
127. Gingerich, K. A. and Blue, G. D., *J. Chem. Phys.*, 47, 5447, 1967.
128. Gingerich, K. A. and Blue, G. D., *J. Chem. Phys.*, 59, 186, 1973.
129. Gingerich, K. A. and Choudary, U. V., *J. Chem. Phys.*, 68, 3265, 1978.
130. Gingerich, K. A. and Cocke, D. L., *Inorg. Chim. Acta*, 28, L171, 1978.
131. Gingerich, K. A., Cocke, D. L., and Kordis, J., *J. Phys. Chem.*, 78, 603, 1974.
132. Gingerich, K. A., Cocke, D. L., and Miller, F., *J. Chem. Phys.*, 64, 4027, 1976.
133. Gingerich, K. A. and Finkbeiner, H. C., *J. Chem. Phys.*, 54, 2621, 1971.
134. Gingerich, K. A. and Finkbeiner, H. C., Proc. 9th Rare Earth Res. Conf. 2, 795, 1971.
135. Gingerich, K. A. and Gupta, S. K., *J. Chem. Phys.*, 69, 505, 1978.
136. Goldfinger, P. and Jeunehomme, M., *Trans. Faraday Soc.*, 59, 2851, 1963.
137. Gingerich, K. A. and Piacente, V., *J. Chem. Phys.*, 54, 2498, 1971.
138. Gole, J. L. and Chalek, C. L., *J. Chem. Phys.*, 65, 4384, 1976.
139. Grade, M., Hirschwald, W., and Stolze, F., *Ber. Bunsenges. Phys. Chem.*, 82, 152, 1978.
140. Grade, M., Hirschwald, W., and Stolze, F., *Z. Phys. Chem. Frankfurt am Main*, 100, 165, 1976.
141. Graeber, P. and Weil, K. G., *Ber. Bunsenges. Phys. Chem.*, 76, 417, 1972.
142. Guggi, D. J., Neubert, A., and Zmbov, K. F., Conf. Int. Thermodyn. Chim. [C.R.] 4th, 3, 124, 1975.
143. Guido, M. and Balducci, G., *J. Chem. Phys.*, 57, 5611, 1972.
144. Guido, M., Gigli, G., and Balducci, G., *J. Chem. Phys.*, 57, 3731, 1972.
145. Gunn, S. R., *J. Phys. Chem.*, 68, 949, 1964.
145a. Gupta, S. K., Atkins, R. M., and Gingerich, K. A., *Inorg. Chem.*, 17, 3211, 1978.
146. Gupta, S. K. and Gingerich, K. A., *J. Chem. Phys.*, 69, 4318, 1978.
147. Gupta, S. K., Pelino, M., and Gingerich, K. A., *J. Chem. Phys.*, 70, 2044, 1979.
148. Hariharan, A. V. and Eick, H. A., *J. Chem. Thermodyn.*, 6, 373, 1974.
149. Hastie, J. W., *J. Chem. Phys.*, 57, 4556, 1972.
150. Herman, R. and Herman, L., *J. Phys. Radium*, 24, 73, 1963.
151. Herzberg, G., *Molecular Spectra and Molecular Structure. I. Spectra of Diatomic Molecules*, 2nd ed., Van Nostrand, New York, 1950.
152. Herzberg, G., *J. Mol. Spectrosc.*, 33, 147, 1970.
153. Herzberg, G. J. and Johns, J. W. G., *Astrophys. J.*, 158, 399, 1969.
154. Hildenbrand, D. L., *Chem. Phys. Lett.*, 15, 379, 1972.
155. Hildenbrand, D. L., *Chem. Phys. Lett.*, 20, 127, 1973.
156. Hildenbrand, D. L., *Chem. Phys. Lett.*, 32, 523, 1975.
157. Hildenbrand, D. L., *Chem. Phys. Lett.*, 34, 352, 1975.
158. Hildenbrand, D. L., *Chem. Phys. Lett.*, 44, 281, 1976.
159. Hildenbrand, D. L., *Chem. Phys. Lett.*, 48, 340, 1977.
160. Hildenbrand, D., *J. Chem. Phys.*, 48, 2457, 1968.
161. Hildenbrand, D. L., *J. Chem. Phys.*, 48, 3657, 1968.
162. Hildenbrand, D. L., *J. Chem. Phys.*, 52, 5751, 1970.
163. Hildenbrand, D. L., *J. Chem. Phys.*, 57, 4556, 1972.
164. Hildenbrand, D. L., *J. Chem. Phys.*, 65, 614, 1976.

165. Hildenbrand, D. L., *J. Chem. Phys.*, 66, 3526, 1977.
166. Hildenbrand, D. L., *J. Chem. Phys.*, 66, 3526, 1977.
167. Hildenbrand, D. L., *J. Chem. Phys.*, 77, 897, 1973.
168. Hildenbrand, D. L. and Murad, E., *J. Chem. Phys.*, 44, 1524, 1966.
169. Hildenbrand, D. L. and Murad, E., *J. Chem. Phys.*, 53, 3403, 1970.
170. Hildenbrand, D. L. and Murad, E., *J. Chem. Phys.*, 61, 1232, 1974.
171. Hildenbrand, D. L. and Theard, L. P., *J. Chem. Phys.*, 50, 5350, 1969.
172. Huber, K. P. and Herzberg, G., *Molecular Spectra and Molecular Structure Constants of Diatomic Molecules*, Van Nostrand, New York, 1979.
173. Ihle, H. R. and Wu, C. H., *J. Chem. Phys.*, 63, 1605, 1975.
174. Johns, J. W. C., Grimm, F. A., and Porter, R. F., *J. Mol. Spectrosc.*, 22, 435, 1967.
175. Johns, J. W. C. and Ramsey, D. A., *Can. J. Phys.*, 39, 210, 1961.
176. Jones, R. W. and Gole, J. L., *Chem. Phys.*, 20, 311, 1977.
176a. Kaposi, O., *Magy. Kem. Foly.*, 83, 356, 1977.
177. Kant, A., *J. Chem. Phys.*, 49, 5144, 1968.
178. Kant, A. and Lin, S-S., *J. Chem. Phys.*, 51, 1644, 1969.
179. Kant, A., Lin, S-S., and Strauss, B., *J. Chem. Phys.*, 49, 1983, 1968.
180. Kant, A., and Strauss, B. H., *J. Chem. Phys.*, 41, 3806, 1964.
181. Kant, A. and Strauss, B., *J. Chem. Phys.*, 45, 3161, 1966.
182. Kant, A. and Strauss, B., *J. Chem. Phys.*, 49, 3579, 1968.
183. Kant, A., Strauss, B., and Lin, S-S., *J. Chem. Phys.*, 52, 2384, 1970.
184. Kent, R. A., *J. Am. Chem. Soc.*, 90, 5657, 1968.
185. Kent, R. A., Ehlert, T. C., and Margrave, J. L., *J. Am. Chem. Soc.*, 86, 5090, 1964.
186. Kent, R. A. and Margrave, J. L., *J. Am. Chem. Soc.*, 87, 3582, 1965.
187. Khitrov, A. N., Ryabova, V. G., and Gurvich, L. V., *Teplofiz Vys. Tempo.*, 11, 1126, 1973.
188. Kleinschmidt, D., Cubicciotti, D., and Hildenbrand, D. L., *J. Electrochem. Soc.*, 125, 1543, 1978; *Proc. Electrochem. Soc.*, 78, 217, 1978.
189. Kleinschmidt, P. D. and Hildenbrand, D. L., *J. Chem. Phys.*, 68, 2819, 1978.
189a. Klynning, L. and Lindgren, B., *Arkiv. Fysik.*, 32, 575, 1966.
190. Kohl, F. J. and Carlson, K. D., *J. Am. Chem. Soc.*, 90, 4814, 1968.
191. Kohl, F. J. and Stearns, C. A., *J. Phys. Chem.*, 78, 273, 1974.
192. Kondratiev, V. N., *Bond Dissociation Energies, Ionization Potentials and Electron Affinities*, Mauka Publishing House, Moscow, 1974.
193. Kordis, J. and Gingerich, K. A., *J. Chem. Eng. Data*, 18, 135, 1973.
194. Kordis, J. and Gingerich, K. A., *J. Chem. Phys.*, 58, 5141, 1973.
195. Kordis, J. and Gingerich, K. A., *J. Phys. Chem.*, 76, 2336, 1972.
196. Kordis, J., Gingerich, K. A., and Seyse, R. J., *J. Chem. Phys.*, 61, 5114, 1974.
197. Kronekvist, M., Lagerqvist, A., and Neuhaus, H., *J. Mol. Spectrosc.*, 39, 516, 1971.
198. Kudo, H., Wu, C. H., and Ihle, H. R., *J. Nucl. Mater.*, 78, 380, 1978.
199. Kulkarni, M. P. and Dadape, V. V., *High Temp. Sci.*, 3, 277, 1971.
200. Kvande, H. and Wahlbeck, P. G., *High Temp. High Pressures*, 8, 45, 1976.
201. LeRoy, R. J., *J. Chem. Phys.*, 57, 573, 1972.
202. LeRoy, R. J. and Bernstein, R. B., *Chem. Phys. Lett.*, 5, 42, 1970.
203. Li, K. C. and Stwalley, W. C., *J. Chem. Phys.*, 59, 4423, 1973.
204. Lin, S-S., and Kant, A., *J. Phys. Chem.*, 73, 2450, 1969.
205. Lin, S-S., Strauss, B., and Kant, A., *J. Chem. Phys.*, 51, 2282, 1969.
206. Lindgren, B. and Nilsson, Ch., *J. Mol. Spectrosc.*, 55, 407, 1975.
207. Liu, M. B. and Wahlbeck, P. G., *J. Chem. Phys.*, 63, 1694, 1975.
208. McIntyre, N. S., Vander Auwera-Mahieu, A., and Drowart, J., *Trans. Faraday Soc.*, 64, 3006, 1968.
209. Marquart, J. R. and Berkowitz, J., *J. Chem. Phys.*, 39, 283, 1963.
210. Martin, E. and Barrow, R. F., *Phys. Scr.*, 17, 501, 1978.
211. Muenow, D. W., Hastie, J. W., Hauge, R., Bautista, R., and Margrave, J. L., *Trans Faraday Soc.*, 65, 3210, 1969.
212. Murad, E., Hildenbrand, D. L., and Main, R. P., *J. Chem. Phys.*, 45, 263, 1966.
213. Murad, E. and Hildenbrand, D. L., *J. Chem. Phys.*, 63, 1139, 1975.
214. Murad, E. and Hildenbrand, D. L., *J. Chem. Phys.*, 65, 3250, 1976.
215. Nagai, S., Shinmei, M., and Yokokawa, T., *J. Inorg. Nucl. Chem.*, 36, 1904, 1974.
216. Neckel, A. and Sodeck, G., *Monatsh. Chem.*, 103, 367, 1972.
217. Neubert, A. and Zmbov, K. F., *High Temp. Sci.*, 6, 303, 1974.
218. Neubert, A. and Zmbov, K. F., *J. Chem. Soc. Faraday Trans.* 1, 70, 2219, 1974.
219. Nordine, P. C., *J. Chem. Phys.*, 61, 224, 1974.
220. O'Hare, P. A. G., *J. Chem. Phys.*, 52, 2992, 1970.
221. Oldenberg, R. C., Dickson, C. R., and Zare, R. N., *J. Mol. Spectrosc.*, 58, 283, 1975.
222. Ovcharenko, I. E., Ya, Kuzyankov, Y., and Tatevaskii, V. M., *Opt. Spektrosk.*, 19, 528, 1965.
223. Owzarski, T. P. and Franzen, H. F., *J. Chem. Phys.*, 60, 1113, 1974.
223a. Parr, T. P., Behrens, R., Freedman, A., and Heron, R. R., *Chem. Phys. Lett.*, 56, 71, 1978.

224. Pashinkin, A. S., Molodyk, A. D., Belousov, V. I., Strel'chenko, S. S., and Fedorova, V. A., *Izv. Akad. Nauk. USSR Neorg. Mater.*, 10, 1600, 1974.
225. Pasternack, L. and Dagdigian, P. J., *J. Chem. Phys.*, 67, 3854, 1977.
226. Peeters, R., Vander Auwera-Mahieu, A., and Drowart, J., *Z. Naturforsch. Teil A*, 26, 327, 1971.
227. Pelevin, O. V., Mil'vidskii, M. G., Belyaev, A. I., and Khotin, B. A., *Izv. Akad. Nauk SSSR Neorg. Mater.*, 2, 924, 1966.
228. Perakis, J., Chatillon, C., and Pattoret, A., *C. R. Acad. Sci. Ser. C*, 276, 1357, 1973.
229. Petzel, T., *High Temp. Sci.*, 6, 246, 1974.
230. Piacente, V. and Balducci, G., *Dyn. Mass Spectrom.*, 4, 295, 1976.
231. Piacente, V. and Balducci, G., *High Temp. Sci.*, 6, 254, 1974.
232. Piacente, V-.V. and Balducci, G., *Adv. Mass Spectrom.*, 7A, 626, 1978.
233. Piacente, V., Balducci, G., and Bardi, G., *J. Less-Common Met.*, 37, 123, 1974.
234. Piacente, V., Bardi, G., and Malaspina, L., *J. Chem. Thermodyn.*, 5, 219, 1973.
235. Piacente, V. and Desideri, A., *J. Chem. Phys.*, 57, 2213, 1972.
236. Piacente, V. and Gingerich, K. A., *High Temp. Sci.*, 9, 189, 1977.
237. Piacente, V. and Gingerich, K. A., *Z. Naturforsch, Teil A*, 28, 316, 1973.
238. Piacente, V. and Malaspina, L., *J. Chem. Phys.*, 56, 1780, 1972.
239. Porter, R. F. and Spencer, C. W. J., *J. Chem. Phys.*, 32, 943, 1960.
240. Prasad, R., Venugopal, V., and Sood, D. D., *J. Chem. Thermodyn.*, 9, 593, 1977.
241. Ringstrom, U., *Ark, Fys.*, 27, 227, 1964.
242. Ringstrom, U., *J. Mol. Spectrosc.*, 36, 232, 1970.
243. Rovner, L., Drowart, A., and Drowart, J., *Trans. Faraday Soc.*, 63, 2910, 1967.
244. Rutner, E. and Haury, G. L., *J. Chem. Eng. Data*, 19, 19, 1974.
245. Scott, P. R. and Richards, W. G., *J. Phys. B*, 7, 500, 1974.
246. Scott, P. R. and Richards, W. G., *J. Phys. B*, 7, 1679, 1974.
247. Shardanand, A., *Phys. Rev.*, 160, 67, 1967.
248. Shenyavskaya, E. A., Mal'tsev, A. A., Kataev, D. I., and Gurvich, L. V., *Opt. Spektrosk.*, 26, 937, 1969.
249. Siegel, B., *Q. Rev. (London)*, 19, 77, 1965.
250. Smith, P. K. and Peterson, D. E., *J. Chem. Phys.*, 52, 4963, 1970.
251. Smoes, S., Coppens, P., Bergman, C., and Drowart, J., *Trans. Faraday Soc.*, 65, 682, 1969.
252. Smoes, S., Depiere, D., and Drowart, J., *Rev. Int. Hautes Temp. Refractaires Paris*, 9, 171, 1972.
253. Smoes, S. and Drowart, J., *Chem. Commun.*, p. 534, 1968.
254. Smoes, S. and Drowart, J., *J. Chem. Soc. Faraday Trans. 2*, 73, 1746, 1977.
255. Smoes, S., Drowart, J., and Myers, C. E., *J. Chem. Thermodyn.*, 8, 225, 1976.
256. Smoes, S., Drowart, J., and Welter, J. M., *J. Chem. Thermodyn.*, 9, 275, 1977; *Adv. Mass Spectrom.*, 7A, 622, 1978.
257. Smoes, S., Huguet, R., and Drowart, J., *Z. Naturforsch. Teil A*, 26, 1934, 1971.
258. Smoes, S., Mandy, F., Vander Auwera-Mahieu, A., and Drowart, J., *Bull. Soc. Chim. Belg.*, 81, 45, 1972.
259. Smoes, S., Myers, C. E., and Drowart, J., *Chem. Phys. Lett.*, 8, 10, 1971.
260. Smoes, S., Pattje, W. R., and Drowart, J., *High Temp. Sci.*, 10, 109, 1978.
261. Srivastava, R. D., *High Temp Sci.*, 8, 225, 1976.
262. Srivastava, R. D. and Farber, M., *High Temp. Sci.*, 5, 489, 1973.
263. Stearns, C. A. and Kohl, F. J., *High Temp. Sci.*, 2, 146, 1970.
264. Stearns, C. A. and Kohl, F. J., *High Temp. Sci.*, 5, 113, 1973.
265. Stearns, C. A. and Kohl, F., *High Temp. Sci.*, 6, 284, 1974.
266. Steiger, R. A. and Cater, E. D., *High Temp. Sci.*, 7, 204, 1975.
267. Steiger, R. A. and Cater, E. D., *High Temp. Sci.*, 7, 288, 1975.
268. Stwalley, W. C., *J. Chem. Phys.*, 65, 2038, 1970.
269. Stwalley, W. C., Way, K. R., and Velasco, R., *J. Chem. Phys.*, 60, 3611, 1974.
270. Sullivan, C. L., Zehe, M. J., and Carlson, K. D., *High Temp. Sci.*, 6, 80, 1974.
271. Tam, A. C. and Happer, W., *J. Chem. Phys.*, 64, 2456, 1976.
272. Tanaka, Y., Yushina, K., and Freeman, D. E., *J. Chem. Phys.*, 59, 564, 1973.
273. Tellinghuisen, J., Tisone, G. C., Hoffman, J. M., and Hays, A. K., *J. Chem. Phys.*, 64, 4796, 1976.
274. Tuenge, R. T., Laabs, F., and Franzen, H. F., *J. Chem. Phys.*, 65, 2400, 1976.
275. Uy, O. M. and Drowart, J., *High Temp. Sci.*, 2, 293, 1970.
276. Uy, O. M. and Drowart, J., *Trans. Faraday Soc.*, 65, 3221, 1969.
277. Uy, O. M. and Drowart, J., *Trans. Faraday Soc.*, 67, 1293, 1971.
278. Uy, O. M., Muenow, D. W., Ficalora, P. J., and Margrave, J. L., *Trans. Faraday Soc.*, 64, 2998, 1968.
279. Vander Auwera-Mahieu, A. and Drowart, J., *Chem. Phys. Lett.*, 1, 311, 1967.
280. Vander Auwera-Mahieu, A., McIntyre, N. S., and Drowart, J., *Chem. Phys. Lett.*, 4, 198, 1969.
281. Vander Auwera-Mahieu, A., Peeters, R., McIntyre, N. S., and Drowart, J., *Trans. Faraday Soc.*, 66, 809, 1970.
282. Velasco, R., Ottinger, C., and Zare, R. N., *J. Chem. Phys.*, 51, 5522, 1969.
283. Verhaegen, G., Ph.D. Thesis, University of Brussels, 1965.
284. Verhaegen, G., Smoes, S., and Drowart, J., *J. Chem. Phys.*, 40, 239, 1964.
285. Verhaegen, G., Stafford, F. E., and Drowart, J., *J. Chem. Phys.*, 40, 1622, 1964.

286. Way, K. R. and Stwalley, W. C., *J. Chem. Phys.*, 59, 5298, 1973.
287. Wiedemeier, H. and Gilles, P. W., *J. Chem. Phys.*, 42, 2765, 1965.
288. Wiedemeier, H. and Goyette, W. J., *J. Chem. Phys.*, 48, 2936, 1968.
289. Wieland, Von K., *Z. Elektrochem.*, 64, 761, 1960.
290. Wu, C. H., *J. Chem. Phys.*, 65, 3181, 1976; 65, 2040, 1976.
291. Yokozeki, A. and Menzinger, M., *Chem. Phys.*, 14, 427, 1976.
292. Zmbov, K. F., Hastie, J. W., and Margrave, J. L., *Trans. Faraday Soc.*, 64, 861, 1968.
293. Zmbov, K. F. and Margrave, J. L., *J. Chem. Phys.*, 45, 3167, 1966.
294. Zmbov, K. F. and Margrave, J. L., *J. Inorg. Nucl. Chem.*, 29, 59, 1976.
295. Zmbov, K. F. and Margrave, J. L., *J. Chem. Phys.*, 47, 3122, 1967.
296. Zmbov, K. F. and Margrave, J. L., *J. Phys. Chem.*, 70, 3379, 1966.
297. Zmbov, K. F. and Margrave, J. L., *J. Phys. Chem.*, 71, 2893, 1967.
298. Zmbov, K. F. and Miletic, M. M., *Glas. Hem. Drus. Beograd*, 43, 521, 1978.
299. Zmbov, K. F., Wu, C. H., and Ihle, H. R., *J. Chem. Phys.*, 67, 4603, 1977.
300. Zollweg, R. J., *Contrib. Pap. Int. Conf. Phenomena of Inoization Gases. 11th, 402, 1973.*

HEATS OF FORMATION OF GASEOUS ATOMS FROM ELEMENTS IN THEIR STANDARD STATES

For elements that are diatomic gases in their standard states these are readily obtained from the bond strength. For elements that are crystalline in their standard states they are derived from vapor pressure data.

Table 2
HEATS OF FORMATION OF GASEOUS ATOMS FROM ELEMENTS IN THEIR STANDARD STATES

Atom	$\Delta H^\circ_{f(298)}$/kcal mol^{-1}	$\Delta H^\circ_{f(298)}$/kJ mol^{-1}	Ref.	Atom	$\Delta H^\circ_{f(298)}$/kcal mol^{-1}	$\Delta H^\circ_{f(298)}$/kJ mol^{-1}	Ref.
Ag	68.1 ± 0.2	284.9 ± 0.8	1	Na	25.85 ± 0.15	108.16 ± 0.63	2
Al	78.8 ± 1.0	329.7 ± 4.0	1	Nb	172.4 ± 1	721.3 ± 4	2
As	72.3 ± 3	302.5 ± 13	2	Ni	102.8 ± 0.5	430.1 ± 2.1	2
Au	88.0 ± 0.5	368.2 ± 2.1	2	O	59.553 ± 0.024	249.17 ± 0.10	1
B	139 ± 3	560 ± 12	1	Os	188 ± 1.5	787 ± 6.3	2
Ba	42.5 ± 1	177.8 ± 4	2	P	79.4 ± 1.0	332.2 ± 4.2	2
Be	77.5 ± 1.5	324.3 ± 6.3	2	Pb	46.62 ± 0.3	195.06 ± 1.3	2
Bi	50.1 ± 0.5	209.6 ± 2.1	2	Pd	90.0 ± 0.5	376.6 ± 2.1	2
Br	26.86 ± 0.03	112.38	3	Pt	135.2 ± 0.3	565.7 ± 1.3	2
C	171.29 ± 0.11	716.67 ± 0.44	1	Pu	87.1 ± 4	364.4 ± 17	2
Ca	42.6 ± 0.4	178.2 ± 1.7	2	Rb	19.6 ± 0.1	82.0 ± 0.4	2
Cd	26.72 ± 0.15	111.80 ± 0.63	2	Re	185 ± 1.5	774 ± 6.3	2
Ce	101 ± 3	423 ± 13	2	Rh	133 ± 1	557 ± 4	2
Cl	29.031 ± 0.002	121.466 ± 0.008	3	Ru	155.5 ± 1.5	648.5 ± 6.3	2
Co	102.4 ± 1	428.4 ± 4	2	S	66.20 ± 0.06	276.98 ± 0.25	1
Cr	95 ± 1	398 ± 4	2	Sb	63.2 ± 0.6	264.4 ± 2.5	2
Cs	18.7 ± 0.1	78.2 ± 0.4	2	Sc	90.3 ± 1	377.8 ± 4	2
Cu	80.7 ± 0.3	337.6 ± 1.2	1	Se	54.3 ± 1	227.2 ± 4	2
Er	75.8 ± 1	317.1 ± 4	2	Si	108 ± 2	450 ± 8	1
F	18.92	79.14	3	Sm	49.4 ± 0.5	206.7 ± 2.1	2
Fe	99.3 ± 0.3	415.5 ± 1.3	2	Sn	72.2 ± 0.5	302.1 ± 2.1	2
Ga	65.4 ± 0.5	273.6 ± 2.1	2	Sr	39.1 ± 0.5	163.6 ± 2.1	2
Ge	89.5 ± 0.5	374.5 ± 2.1	2	Ta	186.9 ± 0.6	782.0 ± 2.5	2
H	52.077 ± 0.001	217.890 ± 0.005	3	Te	47.0 ± 0.5	196.7 ± 2.1	2
Hf	148 ± 1	619 ± 4	2	Th	137.5 ± 0.5	575.3 ± 2.1	2
Hg	14.69 ± 0.03	61.46 ± 0.13	2	Ti	112.3 ± 0.5	469.9 ± 2.1	2
I	25.688 ± 0.010	107.477 ± 0.040	3	Tl	43.55 ± 0.1	182.21 ± 0.4	2
In	58 ± 1	243 ± 4	2	U	126 ± 3	527 ± 13	2
Ir	160 ± 1	669 ± 4	2	V	122.9 ± 0.3	514.2 ± 1.3	2
K	21.42 ± 0.05	89.62 ± 0.21	2	W	203.1 ± 1	849.8 ± 4	2
Li	38.6 ± 0.4	161.5 ± 1.7	2	Y	101.5 ± 0.5	424.7 ± 2.1	2
Mg	35.0 ± 0.3	146.4 ± 1.3	2	Yb	36.35 ± 0.2	152.09 ± 0.8	2
Mn	67.7 ± 1	283.3 ± 4	2	Zn	31.17 ± 0.05	130.42 ± 0.20	1
Mo	157.3 ± 0.5	658.1 ± 2.1	2	Zr	145.5 ± 1	608.8 ± 4	2
N	112.97 ± 0.10	472.68 ± 0.40	1				

REFERENCES

1. CODATA recommended Key values for thermodynamics, 1975, *J. Chem. Thermodyn.*, 8, 603, 1976.
2. Brewer, L. and Rosenblatt, G. M., *Adv. High Temp. Chem.*, 2, 1, 1969.
3. Huber, K. P. and Herzberg, G., *Molecular Spectra and Molecular Structure Constants of Diatomic Molecules*, Van Nostrand, New York, 1979.

BOND STRENGTHS IN POLYATOMIC MOLECULES

The values below refer to a temperature of 298 K and have mostly been determined by kinetic methods (see References 18 and 71 following Table 3 for a full discussion of the methods).

Some have been calculated from the heats of formation of the species involved according to the equations:

$$D(R - X) = \Delta H_f^\circ(\dot{R}) + \Delta H_f^\circ(\dot{X}) - \Delta H_f^\circ(RX)$$

$$D(R - R) = 2\Delta H_f^\circ(\dot{R}) - \Delta H_f^\circ(RR)$$

The sources of the data on the heats of formation are given in the references following Table 3.

An attempt has been made to list all the important values obtained by methods that are considered to be valid. The references are intended to serve as a guide to the literature.

Table 3
BOND STRENGTHS IN POLYATOMIC MOLECULES

Bond	D°_{298}/kcal mol^{-1}	D°_{298}/kJ mol^{-1}	Ref.
H–CH	102 ± 2	427 ± 8	1,28
H–CH$_2$	110 ± 2	460 ± 8	1,28
H–CH$_3$	105.1 ± 0.15	439.7 ± 0.6	8
H–ethynyl	125 ± 1	523 ± 4	94
H–vinyl	≥108 ± 2	≥452 ± 8	63
H–C$_2$H$_5$	98 ± 1	410 ± 4	63
H–propargyl	90 ± 1	377 ± 4	80,127
H–allyl	86.6 ± 1.5	362.3 ± 6.3	115
H–cyclopropyl	106.3 ± 0.3	444.8 ± 1.1	9
H–*n*-C$_3$H$_7$	98 ± 1	410 ± 4	63
H–*i*-C$_3$H$_7$	95 ± 1	398 ± 4	63
H–CH$_2$C CCH$_3$	90.7	379.5	127
H–cyclobutyl	96.5 ± 1	403.8 ± 4	50,92
H–cyclopropylcarbinyl	97.4 ± 1.6	407.5 ± 6.7	91
H–CH(CH$_3$)CH$_2$	83 ± 1	347 ± 4	63
H–CH$_2$C(CH$_3$)$_2$	85.6 ± 1.1	358.2 ± 4.6	125,126
H–*s*-C$_4$H$_9$	95 ± 1	398 ± 4	63
H–*t*-C$_4$H$_9$	92 ± 1	385 ± 4	63
H–cyclopentadien-1,3-yl-5	81.2 ± 1.2	339.7 ± 5.0	61
H–pentadien-1,4-yl-3	80 ± 1	335 ± 4	63
H–cyclopentenyl-3	82.3 ± 1	344.3 ± 4	60
H–spiropentyl	98.8 ± 1	413.4 ± 4	50
H–C(CH$_3$)$_2$C CH	79	331	79,116
H–cyclopentyl	94.5 ± 1	395. ± 4	50,59
H–C(CH$_3$)$_2$CH$_2$	77.2 ± 1.5	323.0 ± 6.3	124
H–neo–C$_5$H$_{11}$	100.3 ± 1	419.7 ± 4	88
H–C$_6$H$_5$	110.2 ± 2.0	461.1 ± 8.4	25
H–cyclohexadien-1,3-yl-5	70 ± 5	293 ± 21	70
H–cyclohexyl	95.5 ± 1	400 ± 4	50
H–CH$_2$C(CH$_3$):C(CH$_3$)$_2$	78.0 ± 1.1	326.4 ± 4.6	113
H–C(CH$_3$)$_2$C(CH$_3$)$_2$	76.3 ± 1.1	319.2 ± 4.6	113
H–CH$_2$C$_6$H$_5$	87.9 ± 1.5	367.8 ± 6.3	115
H–cycloheptatrien-1,3-yl-7	73.4	307.1	131
H–norbornyl	96.7 ± 2.5	404.6 ± 10.5	97
H–cycloheptyl	92.5 ± 1	387.0 ± 4	50
H–CN	120 ± 1	502 ± 4	42
H–CH$_2$CN	~93	~389	75
H–CH(CH$_3$)CN	89.9 ± 2.3	376.1 ± 9.6	74
H–C(CH$_3$)$_2$CN	86.5 ± 2.0	361.9 ± 8.4	76
H–CH$_2$NH$_2$	94.6 ± 2.0	395.8 ± 8.4	31,116
H–CHO	87 ± 1	364 ± 4	63
H–CH$_2$OH	94 ± 2	393 ± 4	63
H–COCH$_3$	86.0 ± 0.8	359.8 ± 3.4	44
H–CH$_2$CHO	94.8 ± 2.0	396.6 ± 8.4	114
H–CH$_2$OCH$_3$	93 ± 1	389 ± 4	63
H–CH(CH$_3$)OH	93.0 ± 1.0	389.1 ± 4.2	2
H–COCH$_2$	87.1 ± 1.0	364.4 ± 4.2	3
H–CH$_2$COCH$_3$	98.3 ± 1.8	411.3 ± 7.5	77,122
H–COC$_2$H$_5$	87.4 ± 1.0	365.7 ± 4.2	136
H–CH(OH)CH$_2$	81.6 ± 1.8	341.4 ± 7.5	5
H–C(CH$_3$)$_2$OH	91 ± 1	381 ± 4	63
H–tetrahydrofuran-2-yl	92 ± 1	385 ± 4	63
H–CH(CH$_3$)COCH$_3$	92.3 ± 1.4	386.2 ± 5.9	123
H–COC$_6$H$_5$	86.9 ± 1	363.6 ± 4	121
H–COOCH$_3$	92.7 ± 1	387.9 ± 4	119
H–CH$_2$OCOC$_6$H$_5$	100.2 ± 1.3	419.2 ± 5.4	120
H–COCF$_3$	91.0 ± 2	380.7 ± 8	6
H–CH$_2$F	101 ± 2	423 ± 8	72
H–CHF$_2$	101 ± 2	423 ± 8	72

Table 3 (continued)
BOND STRENGTHS IN POLYATOMIC MOLECULES

Bond	D°_{298}/kcal mol^{-1}	D°_{298}/kJ mol^{-1}	Ref.
$H-CF_3$	106 ± 1	444 ± 4	63
$H-CF_2Cl$	104 ± 1	435 ± 4	90
$H-CH_2Cl$	100.9	422.2	58
$H-CHCl_2$	99.0	414.2	58
$H-CCl_3$	95.8 ± 1	400.8 ± 4	93
$H-CH_2Br$	102.0	426.8	58
$H-CHBr_2$	103.7	433.9	58
$H-CBr_3$	96.0 ± 1.6	401.7 ± 6.7	78
$H-CH_2I$	103 ± 2	431 ± 8	63
$H-CHI_2$	103 ± 2	431 ± 8	63
$H-CH_2CF_3$	106.7 ± 1.1	446.4 ± 4.6	139
$H-CF_2CH_3$	99.5 ± 1	416.3 ± 4.2	104
$H-C_2F_5$	103.1 ± 1.5	431.4 ± 6.3	10
$H-CCl_2CHCl_2$	94 ± 2	393 ± 8	37,57
$H-C_2Cl_5$	95 ± 2	398 ± 8	37,56
$H-n\text{-}C_3F_7$	104 ± 2	435 ± 8	71
$H-CHClCH_2$	88.6 ± 1.4	370.7 ± 5.9	4
$H-CH_2Si(CH_3)_3$	99.2 ± 1	415.1 ± 4.2	46
$H-NH_2$	110 ± 2	460 ± 8	21,64
$H-NHCH_3$	103 ± 2	431 ± 8	64
$H-N(CH_3)_2$	95 ± 2	398 ± 8	64
$H-NHC_6H_5$	86.4 ± 2	361.5 ± 8	34
$H-N(CH_3)C_6H_5$	84.9 ± 2	355.2 ± 8	34
$H-NO$	≤49.5	≤207.1	29
$H-NF_2$	75.7 ± 2.5	316.7 ± 10.5	99
$H-N_3$	85	356	66
$H-OH$	119 ± 1	498 ± 4	71
$H-OCH_3$	104.4 ± 1	436.8 ± 4	11,37
$H-OC_2H_5$	104.2 ± 1	436.0 ± 4	11,37
$H-OC(CH_3)_3$	105.1 ± 1	439.7 ± 4	11,37,110
$H-OCH_2C(CH_3)_3$	102.3 ± 1.5	428.0 ± 6.3	110
$H-OC_6H_5$	88 ± 5	368 ± 21	24
$H-O_2H$	87.2 ± 1.0	364.8 ± 4.2	1,66a
$H-O_2CCH_3$	106	443	71,143
$H-O_2CC_2H_5$	110 ± 4	460 ± 8	71
$H-O_2C\,n\text{-}C_3H_7$	103 ± 4	431 ± 8	71
$H-ONO$	78.3 ± 0.5	327.6 ± 2.1	1,19
$H-ONO_2$	101.2 ± 0.5	423.4 ± 2.1	1,19
$H-SH$	90.5 ± 1.1	378.7 ± 4.6	67
$H-SCH_3$	≥88	≥368	71
$H-SO$	41.3	172.8	137
$H-SiH_3$	90 ± 2	376 ± 8	48,111
$H-Si(CH_3)_3$	90 ± 2.6	376 ± 11	134
$H-SiF_3$	100 ± 1	419 ± 4	47
$H-SiCl_3$	91.3 ± 1.4	382 ± 6	135
$CH{\equiv}CH$	230 ± 2	962 ± 8	1,28
$CH_2{=}CH_2$	172 ± 2	720 ± 8	1,28
CH_3-CH_3	88 ± 2	368 ± 8	71
$CH_3-CH_2C\ CH$	76.0	318.0	80
$CH_3-CH_2C(CH_3)_2$	72.0 ± 0.8	301.2 ± 3.4	125
$CH_3-C(CH_3)_2C\ CH$	70.7 ± 1.5	295.8 ± 6.3	79
$CH_3-C(CH_3)_2CH_2$	69.4	290.4	20
$n\text{-}C_3H_7-CH_2CH_3$	70.7	295.8	128
$C_6H_5CH_2-C_2H_5$	69 ± 2	289 ± 8	71
$C_6H_5CH(CH_3)-CH_3$	71	297	20
$C_6H_5CH_2-n\text{-}C_3H_7$	67 ± 2	280 ± 8	71
CH_3-CN	123.9 ± 0.7	518.4 ± 2.9	94
C_2H-CN	144 ± 1	603 ± 4	94
CH_3-CH_2CN	72.7 ± 2	304.2 ± 8	20,71
$CH_3-CH(CH_3)CN$	78.8 ± 2	329.7 ± 8	74
$C_2H_5-CH_2CN$	76.9 ± 1.7	321.8 ± 7.1	75
$CH_3-C(CH_3)_2CN$	74.7 ± 1.6	312.6 ± 6.7	20,71,76
$C_6H_5C(CH_3)(CN)-CH_3$	59.9	250.6	20
$CN-CN$	128 ± 1	536 ± 4	42
$C_6H_5CH_2-CH_2NH_2$	65.7	274.9	31
$C_6H_5CH_2CO-CH_2C_6H_5$	65.4	273.6	20
$C_6H_5CO-CF_3$	73.8	308.8	20
$CH_3CO-COCH_3$	67.4 ± 2.3	282.0 ± 9.6	81
$C_6H_5CH_2-COOH$	68.1	284.9	20
$C_6H_5CH_2-O_2CCH_3$	67	280	20
$C_6H_5CO-COC_6H_5$	66.4	277.8	20
$C_6H_5CH_2-O_2CC_6H_5$	69	289	20
$(C_6H_5CH_2)_2CH-COOH$	59.4	248.5	20
CH_2F-CH_2F	88 ± 2	368 ± 8	72
CH_3-CF_3	101.2 ± 1.1	423.4 ± 4.6	112

Table 3 (continued)
BOND STRENGTHS IN POLYATOMIC MOLECULES

Bond	D°_{298}/kcal mol^{-1}	D°_{298}/kJ mol^{-1}	Ref.
$CF_2{=}CF_2$	76.3 ± 3	319.2 ± 13	142
$CF_3{-}CF_3$	96.9 ± 2	405.4 ± 8	35
$C_6F_5{-}C_6F_5$	116.6 ± 5.9	488.4 ± 24.5	109
$C_6H_5CH_2{-}NH_2$	71.9 ± 1	300.8 ± 4	64
$C_6H_5NH{-}CH_3$	67.7	283.3	20
$C_6H_5CH_2{-}NHCH_3$	68.7 ± 1	287.4 ± 4	64
$C_6H_5N(CH_3){-}CH_3$	65.2	272.8	20
$C_6H_5CH_2{-}N(CH_3)_2$	60.9 ± 1	254.8 ± 4	64
$CF_3{-}NF_2$	65 ± 2.5	272 ± 10.5	132
$CH_2{=}N_2$	≤41.7 ± 1	≤174.5 ± 4	89
$CH_3N{:}N{-}CH_3$	52.5	219.7	20
$C_2H_5N{:}N{-}C_2H_5$	50.0	209.2	20
i-$C_3H_7N{:}N{-}$*i*-C_3H_7	47.5	198.7	20
n-$C_4H_9N{:}N{-}$*n*-C_4H_9	50.0	209.2	20
i-$C_4H_9N{:}N{-}$*i*-C_4H_9	49.0	205.0	20
s-$C_4H_9N{:}N{-}$*s*-C_4H_9	46.7	195.4	20
t-$C_4H_9N{:}N{-}$*t*-C_4H_9	43.5	182.0	20
$C_6H_5CH_2N{:}N{-}CH_2C_6H_5$	37.6	157.3	20
$CF_3N{:}N{-}CF_3$	55.2	231.0	20
$C_6H_5{-}NO$	51.5 ± 1	215.5 ± 4	26
$CF_3{-}NO$	31	130	23
$C_6F_5{-}NO$	50.5 ± 1	211.3 ± 4	23,26
$CCl_3{-}NO$	32	134	23
$CN{-}NO$	28.8 ± 2.5	120.5 ± 10.5	65
t-$C_4H_9{-}NO$ *t*-C_4H_9	29	121	23
$C_2H_5{-}NO_2$	62	259	1,117
$O{=}CO$	127.2 ± 0.1	532.2 ± 0.4	40
$CH_3{-}OC(CH_3)_2$	66.3	277.4	141
$CH_3{-}OC_6H_5$	57 ± 2	239 ± 8	100
$CH_3{-}OCH_2C_6H_5$	67.0	280.3	33
$C_2H_5{-}OC_6H_5$	64.0	267.8	33
$CH_2CH_2{-}OC_6H_5$	50.6	211.7	33
$CH_3{-}O_2SCH_3$	66.8	279.5	20,71
allyl$-O_2SCH_3$	49.6	207.5	20,71
$C_6H_5CH_2{-}O_2SCH_3$	52.9	221.3	20,71
$C_6H_5S{-}CH_3$	67.5 ± 2.0	282.4 ± 8.4	31
$C_6H_5CH_2{-}SCH_3$	59.4 ± 2.0	248.5 ± 8.4	31
$F{-}CH_3$	108 ± 3	452 ± 13	45,71
$F{-}CF_2Cl$	117 ± 6	490 ± 25	55
$F{-}CFCl_2$	110 ± 6	460 ± 25	55
$Cl{-}CN$	97 ± 1	406 ± 4	42
$Cl{-}COC_6H_5$	74 ± 3	310 ± 13	71
$Cl{-}CF_3$	86.1 ± 0.8	360.2 ± 3.3	36
$Cl{-}CF_2Cl$	76 ± 2	318 ± 8	55
$Cl{-}CCl_2F$	73 ± 2	305 ± 8	54
$Cl{-}CCl_3$	70.4 ± 1	294.6 ± 4	93
$Cl{-}C_2F_5$	82.7 ± 1.7	346.0 ± 7.1	36
$Cl{-}CF_2CF_2Cl$	78 ± 2	326 ± 8	55
$Cl{-}NF_2$	≤65	≤272	30
$Cl{-}SiCl_3$	111	466	135
$Br{-}CH_3$	70.0 ± 1.2	292.9 ± 5.0	49
$Br{-}C_6H_5$	74.8 ± 1.3	313 ± 5.4	84
$Br{-}CN$	83 ± 1	347 ± 4	42
$Br{-}CH_2COCH_3$	62.5	261.5	141
$Br{-}COC_6H_5$	64.2	268.6	20
$Br{-}CHF_2$	69 ± 2	289 ± 8	95
$Br{-}CF_3$	70.6 ± 1.0	295.4 ± 4.2	51,53
$Br{-}CCl_3$	55.7 ± 1	233.1 ± 4.2	93
$Br{-}CBr_3$	56.2 ± 1.8	235.1 ± 7.5	78
$Br{-}CF_2CH_3$	68.6 ± 1.3	287.0 ± 5.4	105
$Br{-}C_2F_5$	68.7 ± 1.5	287.4 ± 6.3	36,52,53
Br$-$*n*-C_3F_7	66.5 ± 2.5	278.2 ± 10.5	36
$Br{-}NF_2$	≤53	≤222	30
$I{-}CH_3$	56.3 ± 1	235.6 ± 4	71
I$-$norbornyl	62.5 ± 2.5	261.5 ± 10.5	97
$I{-}CN$	73 ± 1	305 ± 4	42
$I{-}CF_3$	53.2 ± 1.0	222.6 ± 4.2	96
$I{-}CF_2CH_3$	52.1 ± 1.0	218.0 ± 4.2	103
$I{-}CH_2CF_3$	56.3 ± 1	235.6 ± 4	138
$I{-}C_2F_5$	51.2 ± 1.0	214.2 ± 4.2	96,140
I-*n*-C_3F_7	49.8 ± 1.0	208.4 ± 4.2	96
I$-$*i*-C_3F_7	49.8 ± 1.0	208.4 ± 4.2	96
I-*n*-C_4F_9	49.0 ± 1.0	205.0 ± 4.2	96
$I{-}C_6H_5$	63.7 ± 1.2	266.3 ± 4.8	82
$I{-}C_6F_5$	~66.2	~277	86
$I{-}SH$	49.4 ± 2	206.7 ± 8	68

Table 3 (continued)
BOND STRENGTHS IN POLYATOMIC MOLECULES

Bond	D°_{298}/kcal mol^{-1}	D°_{298}/kJ mol^{-1}	Ref.
$C_2H_5-ZnC_2H_5$	~47.5	~198.7	85
$CH_3-Ga(CH_3)_2$	59.5	249.0	20
CH_3-CdCH_3	54.4	227.6	20
$C_2H_5-Sn(C_2H_5)_3$	~57	~239	39
CH_3-HgCH_3	57.5	240.6	83
$C_2H_5-HgC_2H_5$	43.7 ± 1	182.8 ± 4	87
n-$C_3H_7-Hg\,n$-C_3H_7	47.1	197.1	20
i-$C_3H_7-Hg\,i$-C_3H_7	40.7	170.3	20
$C_6H_5-HgC_6H_5$	68	285	20
$CH_3-Tl(CH_3)_2$	36.4 ± 0.6	152.3 ± 2.5	108
$CH_3-Pb(CH_3)_3$	49.4 ± 1	206.7 ± 4	62
BH_3-BH_3	35	146	20
NH_2-NH_2	70.8 ± 2	296.2 ± 8	20
NH_2-NHCH_3	64.8	271.1	20
$NH_2-N(CH_3)_2$	62.7	262.3	20
$NH_2-NHC_6H_5$	51.1	213.8	20
$NO-NO_2$	9.5 ± 0.5	39.8 ± 2.1	71
NO_2-NO_2	12.9 ± 0.5	54.0 ± 2.1	71
NF_2-NF_2	21 ± 1	88 ± 4	40
$O-N_2$	40	167	1,19
$O-NO$	73	305	1,19
CH_3O-NO	41.8 ± 0.9	174.9 ± 3.8	11,15
C_2H_5O-NO	42.0 ± 1.3	175.7 ± 5.5	11,16
n-C_3H_7O-NO	40.1 ± 1.8	167.8 ± 7.5	11
i-C_3H_7O-NO	41.0 ± 1.3	171.5 ± 5.5	11,17
n-C_4H_9O-NO	42.5 ± 1.5	177.8 ± 6.3	11
i-C_4H_9O-NO	42.0 ± 1.5	175.7 ± 6.3	11
s-C_4H_9O-NO	41.5 ± 0.8	173.6 ± 3.4	11,13
t-C_4H_9O-NO	40.9 ± 0.8	171.1 ± 3.4	11,14,27
$HO-NCH_3$	49.7	208.0	20
$Cl-NF_2$	~32	~134	1,102
$I-NO$	17.3 ± 1	72.4 ± 4	129
$I-NO_2$	18.3 ± 1	76.6 ± 4	129
$HO-OH$	51 ± 1	213 ± 4	71
CH_3O-OCH_3	37.6 ± 0.2	157.3 ± 0.8	7,11,12
$HO-OC(CH_3)_3$	46.3 ± 1.9	193.7 ± 8.0	1,11,37
$C_2H_5O-OC_2H_5$	37.9	158.6	11
n-$C_3H_7O-O\,n$-C_3H_7	37.1	155.2	11
i-$C_3H_7O-O\,i$-C_3H_7	37.7	157.7	11
s-$C_4H_9O-O\,s$-C_4H_9	36.4 ± 1	152.3 ± 4	133
t-$C_4H_9O-O\,t$-C_4H_9	38.0	159.0	11
neo-$C_5H_{11}O-O$neo-C_5H_{11}	36.4 ± 1	152.3 ± 4	101
CF_3O-OCF_3	46.2	193.3	43
$(CF_3)_3CO-OC(CF_3)_3$	35.5 ± 1.1	148.7 ± 4.4	69
$O-O_2ClF$	58.4	244.3	20
$CH_3CO_2-O_2CCH_3$	30.4 ± 2	127.2 ± 8	20,71
$C_2H_5CO_2-O_2CC_2H_5$	30.4 ± 2	127.2 ± 8	20,71
n-$C_3H_7CO_2-O_2C\,n$-C_3H_7	30.4 ± 2	127.2 ± 8	20,71
$O-SO$	132 ± 2	552 ± 8	40
$F-OCF_3$	43.5 ± 0.5	182.0 ± 2.1	38
$HO-Cl$	60 ± 3	251 ± 13	71
$O-ClO$	59 ± 3	247 ± 13	40
$HO-Br$	56 ± 3	234 ± 13	71
$HO-I$	56 ± 3	234 ± 13	71
ClO_3-ClO_4	58.4	244.3	20
$O{=}PF_3$	130 ± 5	543.9 ± 21	71
$O{=}PCl_3$	122 ± 5	511 ± 21	71
$O{=}PBr_3$	119 ± 5	498 ± 21	71
$F-SF_5$	89.9 ± 3.4	376.1 ± 14.2	73
SiH_3-SiH_3	81 ± 4	339 ± 17	111
$(CH_3)_3Si-Si(CH_3)_3$	80.5	336.8	41
$(C_6H_5)_3Si-Si(C_6H_5)_3$	88 ± 7	368 ± 29	22

REFERENCES

1. A value calculated from one of the above thermochemical equations taking data from the references quoted.
2. Alfassi, Z. B. and Golden, D. M., *J. Phys. Chem.*, 76, 3314, 1972.
3. Alfassi, Z. B. and Golden, D. M., *J. Am. Chem. Soc.*, 95, 319, 1973.
4. Alfassi, Z. B., Golden, D. M., and Benson, S. W., *Int. J. Chem. Kinet.*, 5, 155, 1973: Trenwith, A. B., *Int. J. Chem. Kinet.*, 5, 67, 1973.
5. Alfassi, Z. B. and Golden, D. M., *Int. J. Chem. Kinet.*, 5, 295, 1973: Trenwith, A. B., *J. Chem. Soc. Faraday Trans. I*, 69, 1737, 1973.
6. Amphlett, J. C. and Whittle, E., *Trans. Faraday Soc.*, 66, 2016, 1970.

7. Barker, J. R., Benson, S. W., and Golden, D. M., *Int. J. Chem. Kinet.*, 9, 31, 1977.
8. Baghal-Vayjooee, M. H., Colussi, A. J., and Benson, S. W., *J. Am. Chem. Soc.*, 100, 3214, 1978; *Int. J. Chem. Kinet.*, 11, 147, 1979.
9. Baghal-Vayjooee, M. H. and Benson, S. W., *J. Am. Chem. Soc.*, 101, 2838, 1979.
10. Bassett, J. E. and Whittle, E., *J. Chem. Soc. Faraday Trans. 1*, 68, 492, 1972.
11. Batt, L., Christie, K., Milne, R. T., and Summers, A. J., *Int. J. Chem. Kinet.*, 6, 877, 1974.
12. Batt, L. and McCulloch, R. D., *Int. J. Chem. Kinet.*, 8, 491, 1976.
13. Batt, L. and McCulloch, R. D., *Int. J. Chem. Kinet.*, 8, 911, 1976.
14. Batt, L. and Milne, R. T., *Int. J. Chem. Kinet.*, 8, 59, 1976.
15. Batt, L., Milne, R. T., and McCulloch, R. D., *Int. J. Chem. Kinet.*, 9, 567, 1977.
16. Batt, L. and Milne, R. T., *Int. J. Chem. Kinet.*, 9, 549, 1977.
17. Batt, L. and Milne, R. T., *Int. J. Chem. Kinet.*, 9, 141, 1977.
18. Benson, S. W., *J. Chem. Educ.*, 42, 502, 1965.
19. Benson, S. W., *Thermochemical Kinetics, 2nd ed.*, John Wiley & Sons, New York, 1976.
20. Benson, S. W. and O'Neal, H. E., *Kinetic Data on Gas Phase Unimolecular Reactions,* National Bureau of Standards, NSRDS-NBS, Washington, D.C., 21, 1970.
21. Bohme, D. K., Hemsworth, R. S., and Rundle, H. W., *J. Chem. Phys.*, 59, 77, 1973.
22. Calle, L. M. and Kana'an, A. S., *J. Chem. Thermodyn.*, 6, 935, 1974.
23. Carmichael, P. J., Gowenlock, B. G., and Johnson, C. A. F., *J. Chem. Soc. Perkin Trans. 2,* 1853, 1973.
24. Carson, A. S., Fine, D. H., Gray, P., and Laye, P. G., *J. Chem. Soc. B,* p. 1611, 1971.
25. Chamberlain, G. A. and Whittle, E., *Trans. Faraday Soc.*, 67, 2077, 1971.
26. Choo, K. Y., Golden, D. M., and Benson, S. W., *Int. J. Chem. Kinet.*, 7, 713, 1975.
27. Choo, K. Y., Mendenhall, G. D., Golden, D. M., and Benson, S. W., *Int. J. Chem. Kinet.*, 6, 813, 1974.
28. Chupka, W. A. and Lifshitz, C., *J. Chem. Phys.*, 48, 1109, 1968.
29. Clement, M. J. Y. and Ramsay, D. A., *Can. J. Phys.*, 39, 205, 1961.
30. Clyne, M. A. A. and Connor, J., *J. Chem. Soc. Faraday Trans. 2,* 68, 1220, 1972.
31. Colussi, A. J. and Benson, S. W., *Int. J. Chem. Kinet.*, 9, 295, 1977.
32. Colussi, A. J. and Benson, S. W., *Int. J. Chem. Kinet.*, 9, 307, 1977.
33. Colussi, A. J., Zabel, F., and Benson, S. W., *Int. J. Chem. Kinet.*, 9, 161, 1977.
34. Colussi, A. J. and Benson, S. W., *Int. J. Chem. Kinet.*, 10, 1139, 1978.
35. Coomber, J. W. and Whittle, E., *Trans. Faraday Soc.*, 63, 1394, 1967.
36. Coomber, J. W. and Whittle, E., *Trans. Faraday Soc.*, 63, 2656, 1967.
37. Cox, J. D. and Pilcher, G., *Thermochemistry of Organic and Organometallic Compounds,* Academic Press, New York, 1970.
38. Czarnarski, J., Castellano, E., and Shumacher, H. J., *Chem. Commun.*, p. 1255, 1968.
39. Daly, M. and Price, S. J., *Can. J. Chem.*, 54, 1814, 1976.
40. Darwent, D. de B., *Bond Dissociation Energies in Simple Molecules,* National Bureau of Standards. NSRDS-NBS, Washington, D.C., 31, 1970.
41. Davidson, I. M. T. and Howard, A. B., *J. Chem. Soc. Faraday Trans. 1,* 71, 69, 1975.
42. Davis, D. D. and Okabe, H., *J. Chem. Phys.*, 49, 5526, 1968.
43. Descamps, B. and Forst, W., *J. Phys. Chem.*, 80, 933, 1976.
44. Devore, J. A. and O'Neal, H. E., *J. Phys. Chem.*, 73, 2644, 1969.
45. Dibeler, V. H. and Reese, R. M., *J. Res. Natl. Bur. Stand.*, 54, 127, 1955.
46. Doncaster, A. M. and Walsh, R., *J. Chem. Soc. Faraday Trans. 1,* 72, 2908, 1976.
47. Doncaster, A. M. and Walsh, R., *Int. J. Chem. Kinet.*, 10, 101, 1978.
48. Doncaster, A. M. and Walsh, R., *J. Chem. Soc. Chem. Commun.*, 904, 1979.
49. Ferguson, K. C., Okafo, E. N., and Whittle, E., *J. Chem. Soc. Faraday Trans. 1,* 69, 295, 1973.
50. Ferguson, K. C. and Whittle, E., *Trans. Faraday Soc.*, 67, 2618, 1971; Jones, S. H. and Whittle, E., *Int. J. Chem. Kinet.*, 2, 479, 1970.
51. Ferguson, K. C. and Whittle, E., *J. Chem. Soc. Faraday Trans. 1,* 68, 295, 1972.
52. Ferguson, K. C. and Whittle, E., *J. Chem. Soc. Faraday Trans. 1,* 68, 306, 1972.
53. Ferguson, K. C. and Whittle, E., *J. Chem. Soc. Faraday Trans. 1,* 68, 641, 1972.
54. Foon, R. and Tait, K. B., *J. Chem. Soc. Faraday Trans. 1,* 68, 104, 1972.
55. Foon, R. and Tait, K. B., *J. Chem. Soc. Faraday Trans. 1,* 68, 1121, 1972.
56. Franklin, J. A., Huybrechts, G. H., and Cillien, C., *Trans. Faraday Soc.*, 65, 2094, 1969.
57. Franklin, J. A., and Huybrechts, G. H., *Int. J. Chem. Kinet.*, 1, 1, 1969.
58. Furuyama, S., Golden, D. M., and Benson, S. W., *J. Am. Chem. Soc.*, 91, 7564, 1969.
59. Furuyama, S., Golden, D. M., and Benson, S. W., *Int. J. Chem. Kinet.*, 2, 83, 1970.
60. Furuyama, S., Golden, D. M., and Benson, S. W., *Int. J. Chem. Kinet.*, 2, 93, 1970.
61. Furuyama, S., Golden, D. M., and Benson, S. W., *Int. J. Chem. Kinet.*, 3, 237, 1971.
62. Gilroy, K. M., Price, S. J., and Webster, N. J., *Can. J. Chem.*, 50, 2639, 1972.
63. Golden, D. M., and Benson, S. W., *Chem. Rev.*, 69, 125, 1969.
64. Golden, D. M., Solly, R. K., Gac, N. A., and Benson, S. W., *J. Am. Chem. Soc.*, 94, 363, 1972.
65. Gowenlock, B. G., Johnson, C. A. F., Keary, C. M., and Pfab, J., *J. Chem. Soc. Perkin Trans. 2,* 351, 1975.
66. Gray, P., *Q. Rev. (London),* 17, 441, 1963.

66a. Howard, C. J. *J. Am. Chem. Soc.*, 102, 6937, 1980.
67. Hwang, R. J. and Benson, S. W., *Int. J. Chem. Kinet.*, 11, 579, 1979.
68. Hwang, R. J. and Benson, S. W., *J. Am. Chem. Soc.*, 101, 2615, 1979.
69. Ireton, R., Gordon, A. S., and Tardy, D. C., *Int. J. Chem. Kinet.*, 9, 769, 1977.
70. James, D. G. L. and Suart, R. D., *Trans. Faraday Soc.*, 64, 2752, 1968.
71. Kerr, J. A., *Chem. Rev.*, 66, 465, 1966.
72. Kerr, J. A. and Timlin, D. M., *Int. J. Chem. Kinet.*, 3, 427, 1971.
73. Kiang, T., Estler, R. C., and Zare, R. N., *J. Chem. Phys.*, 70, 5925, 1979.
74. King, K. D. and Goddard, R. D., *J. Am. Chem. Soc.*, 97, 4504, 1975.
75. King, K. D. and Goddard, R. D., *Int. J. Chem. Kinet.*, 7, 837, 1975.
76. King, K. D. and Goddard, R. D., *J. Phys. Chem.*, 80, 546, 1976.
77. King, K. D., Golden, D. M., and Benson, S. W., *J. Am. Chem. Soc.*, 92, 5541, 1970.
78. King, K. D., Golden, D. M., and Benson, S. W., *J. Phys. Chem.*, 75, 987, 1971.
79. King, K. D., *Int. J. Chem. Kinet.*, 9, 907, 1977.
80. King, K. D., *Int. J. Chem. Kinet.*, 10, 545, 1978.
81. Knoll, H., Scherker, K., and Geiseler, G., *Int. J. Chem. Kinet.*, 5, 271, 1973.
82. Kominar, R. J., Krech, M. J., and Price, S. J. W., *Can. J. Chem.*, 54, 2981, 1976.
83. Kominar, R. J. and Price, S. J., *Can. J. Chem.*, 47, 991, 1969.
84. Kominar, R. J., Krech, M. J., and Price, S. J. W., *Can. J. Chem.*, 56, 1589, 1978.
85. Koski, A. A., Price, S. J. W., and Trudell, B. C., *Can. J. Chem.*, 54, 482, 1976.
86. Krech, M. J., Price, S. J. W., and Yared, W. F., *Int. J. Chem. Kinet.*, 6, 257, 1974.
87. LaLonde, A. C. and Price, S. J. W., *Can. J. Chem.*, 49, 3367, 1971.
88. Larson, C. W., Hardwidge, E. A., and Rabinovitch, B. S., *J. Chem. Phys.*, 50, 2769, 1969.
89. Laufer, A. H. and Okabe, H., *J. Am. Chem. Soc.*, 93, 4137, 1971.
90. Leyland, L. M., Majer, J. R., and Robb, J. C., *Trans. Faraday Soc.*, 66, 898, 1970.
91. McMillen, D. F., Golden, D. M., and Benson, S. W., *Int. J. Chem. Kinet.*, 3, 359, 1971.
92. McMillen, D. F., Golden, D. M., and Benson, S. W., *Int. J. Chem. Kinet.*, 4, 487, 1972.
93. Mendenhall, G. D., Golden, D. M., and Benson, S. W., *J. Phys. Chem.*, 77, 2707, 1973.
94. Okabe, H. and Dibeler, V. H., *J. Chem. Phys.*, 59, 2430, 1973.
95. Okafo, E. N. and Whittle, E., *J. Chem. Soc. Faraday Trans. 1*, 70, 1366, 1974.
96. Okafo, E. N. and Whittle, E., *Int. J. Chem. Kinet.*, 7, 287, 1975.
97. O'Neal, H. E., Bagg, J. W., and Richardson, W. H., *Int. J. Chem. Kinet.*, 2, 493, 1970.
98. O'Neal, H. E. and Benson, S. W., in *Free Radicals*, Kochi, J. K., Ed., John Wiley & Sons, New York, 1973, 275.
99. Pankratov, A. V., Zercheninov, A. N., Chesnokov, V. I., and Zhdanova, N. N., *Zh. Fiz. Khim.*, 43, 394, 1969.
100. Paul, S. and Back, M. H., *Can. J. Chem.*, 53, 3330, 1975.
101. Perona, M. J. and Golden, D. M., *Int. J. Chem. Kinet.*, 5, 55, 1973.
102. Petry, R. C., *J. Am. Chem. Soc.*, 89, 4600, 1967.
103. Pickard, J. M. and Rodgers, A. S., *Int. J. Chem. Kinet.*, 8, 809, 1976.
104. Pickard, J. M. and Rodgers, A. S., *J. Am. Chem. Soc.*, 99, 695, 1977.
105. Pickard, J. M. and Rodgers, A. S., *Int. J. Chem. Kinet.*, 9, 759, 1977.
106. Piper, L. G., *J. Chem. Phys.*, 70, 3417, 1979.
107. Potzinger, P. and Lampe, F. W., *J. Phys. Chem.*, 74, 719, 1970.
108. Price, S. J., Richard, J. P., Rufeldt, R.C., and Jacko, M. G., *Can. J. Chem.*, 51, 1397, 1973.
109. Price, S. J. W. and Sapiano, H. J., *Can. J. Chem.*, 57, 1468, 1979.
110. Reed, K. J. and Brauman, J. L., *J. Am. Chem. Soc.*, 97, 1625, 1975.
111. Ring, M. A., Puentes, M. J., and O'Neal, H. E., *J. Am. Chem. Soc.*, 92, 4845, 1970; Steele, W. C. and Stone, F. G. A., *J. Am. Chem. Soc.*, 84, 3599, 1962; Steele, W. C., Nichols, L. D., and Stone, F. G. A., *J. Am. Chem. Soc.*, 84, 441, 1962.
112. Rodgers, A. S. and Ford, W. G. F., *Int. J. Chem. Kinet.*, 5, 965, 1973.
113. Rodgers, A. S. and Wu, M. C. R., *J. Am. Chem. Soc.*, 95, 6913, 1973.
114. Rossi, M. and Golden, D. M., *Int. J. Chem. Kinet.*, 11, 715, 1979.
115. Rossi, M. and Golden, D. M., *J. Am. Chem. Soc.*, 101, 1230, 1979.
116. Sen Sharma, D. K. and Franklin, J. L., *J. Am. Chem. Soc.*, 95, 6562, 1973.
117. Shaw, R., *Int. J. Chem. Kinet.*, 5, 261, 1973.
118. Simmons, J. W., Hase, W. L., Phillips, R. J., Porter, E. J., and Growcock, F. B., *Int. J. Chem. Kinet.*, 7, 879, 1975.
119. Solly, R. K. and Benson, S. W., *Int. J. Chem. Kinet.*, 1, 427, 1969.
120. Solly, R. K. and Benson, S. W., *Int. J. Chem. Kinet.*, 3, 509, 1971.
121. Solly, R. K. and Benson, S. W., *J. Am. Chem. Soc.*, 93, 1592, 1971.
122. Solly, R. K., Golden, D. M., and Benson, S. W., *Int. J. Chem. Kinet.*, 2, 11, 1970.
123. Solly, R. K., Golden, D. M., and Benson, S. W., *Int. J. Chem. Kinet.*, 2, 381, 1970.
124. Trenwith, A. B., *Trans. Faraday Soc.*, 66, 2805, 1970.
125. Trenwith, A. B. and Wrigley, S. P., *J. Chem. Soc. Faraday Trans. 1*, 73, 817, 1977.
126. Tsang, W., *Int. J. Chem. Kinet.*, 5, 929, 1973.
127. Tsang, W., *Int. J. Chem. Kinet.*, 10, 687, 1978.

128. Tsang, W., *Int. J. Chem. Kinet.*, 10, 1119, 1978.
129. van der Bergh, H. and Troe, J., *J. Chem. Phys.*, 64, 736, 1976.
130. Vanderwielen, A. J., Ring, M. A., and O'Neal, H. E., *J. Am. Chem. Soc.*, 97, 993, 1975.
131. Vincow, G., Dauben, H. J., Hunter, F. R., and Volland, W. V., *J. Am. Chem. Soc.*, 91, 2823, 1969.
132. Walker, L. C., *J. Chem. Thermodyn.*, 4, 219, 1972.
133. Walker, R. F. and Phillips, L., *J. Chem. Soc. A*, 2103, 1968.
134. Walsh, R. and Wells, J. M., *J. Chem. Soc. Faraday Trans. 1*, 72, 100, 1976.
135. Walsh, R. and Wells, J. M., *J. Chem. Soc. Faraday Trans. 1*, 72, 1212, 1976.
136. Watkins, K. W. and Thompson, W. W., *Int. J. Chem. Kinet.*, 5, 791, 1973.
137. White, J. N. and Gardiner, W. C., *Chem. Phys. Lett.*, 58, 470, 1978.
138. Wu, E-C. and Rodgers, A. S., *Int. J. Chem. Kinet.*, 5, 1001, 1973.
139. Wu, E-C. and Rodgers, A. S., *J. Phys. Chem.*, 78, 2315, 1974.
140. Wu, E-C. and Rodgers, A. S., *J. Am. Chem. Soc.*, 98, 6112, 1976.
141. Zabel, F., Benson, S. W., and Golden, D. M., *Int. J. Chem. Kinet.*, 10, 295, 1978.
142. Zmbov, K. F., Uy, O. M., and Margrave, J. L., *J. Am. Chem. Soc.*, 90, 5090, 1968.
143. Cook, K. D. and Taylor, J. W., *Int. J. Mass Spectrom. Ion Phys.*, 30, 93, 1979.

HEATS OF FORMATION OF FREE RADICALS

The heats of formation of the free radicals are related to the corresponding bond strengths by the equations

$$D(R - X) = \Delta H_f^\circ(\dot{R}) + \Delta H_f^\circ(\dot{X}) - \Delta H_f^\circ(RX)$$

or

$$D(R - R) = 2\Delta H_f^\circ(\dot{R}) - \Delta H_f^\circ(RR)$$

For an excellent review of the methods of determining the heats of formation of free radicals the reader is referred to "Thermochemistry of Free Radicals" by H. E. O'Neal and S. W. Benson in *Free Radicals,* Kochi, J. K., Ed., John Wiley & Sons, New York, 1973, 275.

The references are the same as given for Table 3.

Table 4
HEATS OF FORMATION OF FREE RADICALS

Radical	$\Delta H^\circ_{f(298)}$/kcal mol^{-1}	$\Delta H^\circ_{f(298)}$/kJ mol^{-1}	Ref.
CH	142 ± 1	594 ± 4	1,28
CH_2	92 ± 1	385 ± 4	1,28
$^1CH_2$	101 ± 3	423 ± 13	118
CH_3	35.1 ± 0.15	146.9 ± 0.63	8
Ethynyl	128 ± 1	536 ± 4	94
Vinyl	≥68	≥285	63
C_2H_5	25.9 ± 1.3	108.4 ± 5.4	63
Propargyl	82.1 ± 1.0	343.5 ± 4.2	80,127
Allyl	39.4 ± 1.5	164.9 ± 6.3	115
Cyclopropyl	66.9 ± 0.25	279.9 ± 1.1	9
n-C_3H_7	22.6 ± 1.8	94.6 ± 7.5	63
i-C_3H_7	18.2 ± 1.5	76.2 ± 6.3	63
$CH_3C{\equiv}CCH_2$	73.6	308	127
Methallyl	30.4 ± 1.3	127.2 ± 5.4	63
Isobutenyl	30.0 ± 1.0	125.5 ± 4.2	125,126
Cyclopropylcarbinyl	51.5 ± 1.6	215.5 ± 6.7	91
Cyclobutyl	51.2 ± 1.0	214.2 ± 4.2	1,37,50,92
s-C_4H_9	13.0 ± 2	54.4 ± 8	63
t-C_4H_9	7.6 ± 1.2	31.8 ± 5.0	63
Cyclopentadien-1,3-yl-5	60.9 ± 1.2	254.8 ± 5.0	61
Pentadienyl	52.9 ± 1.2	221.3 ± 5.0	63
Cyclopentenyl-3	38.4 ± 1	160.7 ± 4	60
Spiropentyl	91.0 ± 1	380.7 ± 4	1,37,50
$(CH_3)_2CC{\equiv}CH$	61.5 ± 2.0	257.3 ± 8.4	79,116
Dimethylallyl	18.3 ± 1.5	76.6 ± 6.3	124
Cyclopentyl	24.3 ± 1	101.7 ± 4	1,37,50,59
Neo-C_5H_{11}	7.6 ± 1	31.8 ± 4	1,37,88
Phenyl	77.7 ± 2.0	325.1 ± 8.4	25
Cyclohexadien-1,3-yl-5	44 ± 5	184 ± 21	70
Cyclohexyl	13.9 ± 1	58.2 ± 4	1,37,50
$(CH_3)_2C{=}C(CH_3)CH_2$	9.6 ± 1.1	40.2 ± 4.6	113
Benzyl	47.8 ± 1.5	200.0 ± 6.3	115
Cycloheptatrien-1,3-yl-7	64.8	271.1	131
Norbornyl	32.6 ± 2.5	136.4 ± 10.5	97
Cycloheptyl	12.2 ± 1	51.1 ± 4	1,37,50
$C_6H_5CHCH_3$	38	159	20,98
CN	101 ± 1	423 ± 4	42
CH_2NH_2	37.0 ± 2.0	154.8 ± 8.4	32,116
CH_3NH	45.4 ± 1	190.0 ± 4	64
CH_2CN	58.5 ± 2.2	209.6 ± 9.6	74

Table 4 (continued)
HEATS OF FORMATION OF FREE RADICALS

Radical	$\Delta H^\circ_{f(298)}$/kcal mol^{-1}	$\Delta H^\circ_{f(298)}$/kJ mol^{-1}	Ref.
$(CH_3)_2N$	38.5 ± 1	161.1 ± 4	64
CH_3CHCN	50.1 ± 2.3	209.6 ± 9.6	74
$(CH_3)_2CCN$	39.8 ± 2.0	166.5 ± 8.4	76
C_6H_5NH	55.1	230.5	34
$C_6H_5NCH_3$	53.2	222.6	34
$C_6H_5C(CH_3)CN$	54.4	227.6	20,98
CHO	7.7 ± 1.2	32.2 ± 5.0	63,98
CH_2OH	−6.2 ± 1.5	−25.9 ± 6.3	63,98
CH_3O	3.8 ± 0.2	15.9 ± 0.8	11,13
CH_3CO	−5.8 ± 0.4	−24.3 ± 1.7	63,98
CH_2CHO	3.0 ± 2.0	12.6 ± 8.4	114
CH_3CHOH	−15.2 ± 1.0	−63.6 ± 4.2	2
CH_3OCH_2	−2.8 ± 1.2	−11.7 ± 5.0	63,98
C_2H_5O	−4.1	−17.2	11
$CH_2{=}CHCO$	17.3	72.4	3
C_2H_5CO	−10.2 ± 1.0	−42.7 ± 4.2	136
CH_3COCH_2	−5.7 ± 2.6	−23.9 ± 10.9	77
$CH_2{=}CHCHOH$	0.0	0.0	5
$(CH_3)_2COH$	−26.6 ± 1.1	−111.3 ± 4.6	63,98
$CH_3COCHCH_3$	−16.8 ± 1.7	−70.3 ± 7.1	123
n-C_3H_7O	−9.9	−41.4	11
i-C_3H_7O	−12.5	−52.3	11
Tetrahydrofuran-2-yl	−4.3 ± 1.5	−18.0 ± 6.3	20,98
n-C_4H_9O	−14.7	−61.5	98
s-C_4H_9O	−16.6 ± 0.8	−69.5 ± 3.3	13
$(CH_3)_3CO$	−21.7	−90.8	11,13
C_6H_5O	11.4 ± 2.0	47.7 ± 8.4	24,33,10D
C_6H_5CO	26.1 ± 2	109.2 ± 8	98,121
COOH	−53.3	−223.0	20,121
$COOCH_3$	−40.4 ± 1	−169.0 ± 4	98,119
CH_3COO	−49.6 ± 1	−207.5 ± 4	98
C_2H_5COO	−54.6 ± 1	−228.5 ± 4	98
n C_3H_7COO	−59.6 ± 1	−249.4 ± 4	98
$C_6H_5COOCH_2$	−16.7 ± 2.0	−69.9 ± 8.4	98,120
CFO	≤−44	≤−184	98
CH_3S	34.2 ± 2.0	143.1 ± 8.4	31
C_6H_5S	56.8 ± 2.0	237.7 ± 8.4	31
CH_3SO_2	−57.2	−239.3	98
CH_2F	−7.8 ± 2	−32.6 ± 8	72
CHF_2	−59.2 ± 2	−247.7 ± 8	72
CF_2	−46.4 ± 2.2	−194.1 ± 9.2	95,142
CF_3	−112.0 ± 3.6	−468.6 ± 15.1	63,98
C_6F_5	−92.6 ± 2.9	−387.4 ± 12.0	109
CCl_2	57	239	98
$CHCl_2$	24.1	100.8	58
CF_2Cl	−64.3 ± 2	−269.0 ± 8	55,90
$CFCl_2$	−23	−96	54
CCl_3	19.0 ± 1	79.5 ± 4	93
CH_2Br	41.5	173.6	58
$CHBr_2$	54.3	227.2	58
CH_2I	55.0 ± 1.6	230.1 ± 6.7	63,98
CHI_2	79.8 ± 2.2	333.9 ± 9.2	63,98
CH_3CF_2	−72.3 ± 2	−302.5 ± 8.4	104
CF_3CH_2	−123.6 ± 1.2	−517.1 ± 5.0	139
C_2F_5	−217.3 ± 1	−909.2 ± 4	98,140
$CHCl_2CCl_2$	5.6 ± 2	23.4 ± 8	1,37,57
CF_2ClCF_2	−164 ± 4	−686 ± 17	55
C_2Cl_5	8.4 ± 2	35.2 ± 8	1,37,56
C_6F_5	−130.9 ± 2	−547.7 ± 8	26
NH	84.2 ± 2.3	352.3 ± 9.6	106
NH_2	47.2 ± 1	197.5 ± 4	21,64,98
NF_2	8 ± 1	34 ± 4	1,19,40
PH_2	33 ± 2	138 ± 8	98
HO	9.2 ± 1	38.5 ± 4	71
HO_2	2.5 ± 0.6	10.5 ± 2.5	66a
HS	33.6 ± 1.1	140.6 ± 4.6	67
SiH_2	58.6 ± 3.5	245.2 ± 14.6	130
SiH_3	49.1 ± 3	205.4 ± 13	107,111
CH_3SiH	50.9 ± 3.5	213.0 ± 14.6	130
$(CH_3)_3Si$	−2.6	−10.9	41
$C_6H_5Si(CH_3)_2$	39	163	98
$(C_6H_5)_2SiCH_3$	78	326	98
$(C_6H_5)_3Si$	116.2	486.2	22
SiH_3SiH	64.5 ± 3.5	269.9 ± 14.6	130
SiF_3	−235 ± 14	−983 ± 60	47
GeH_3	57	239	98

BOND STRENGTHS OF SOME ORGANIC MOLECULES

Bond strengths at 298 K expressed in kcal/mol for some organic molecules of the general formula R−X are presented below. Some are experimental values taken from the preceding tables; the remainder are calculated from the heats of formation of the radicals, listed above, and the heats of formation of the parent compounds from sources indicated by the references below. The table also includes the bond strengths for ammonia, water, and the hydrogen halides.

Table 5
BOND STRENGTHS OF SOME ORGANIC MOLECULES

R \ X	H	F	Cl	Br	I	OH	NH_2	CH_3O	CH_3	CH_3CO	NO	CF_3
H	104.207	135.9	103.2	87.4	71.4	119	110	104	104	86	50	106
CH_3	105	109[2]	84[1]	70[1]	56[1]	91[1]	87[1]	81[1]	88[1]	80[1]	42	101
C_2H_5	98	—	81[1]	68[1]	53[1]	91[1]	84[1]	81[1]	85[1]	77[1]	42	—
i-C_3H_7	95	106[1]	81[1]	68[1]	53[1]	92[1]	85[1]	81[1]	84[1]	75[1]	41	—
t-C_4H_9	92	—	81[1]	67[1]	50[1]	91[1]	84[1]	81[1]	82[1]	71[1]	41	—
C_6H_5	110	125[1]	95[1]	80	64[1]	110[1]	104[1]	98[1]	100[1]	93[1]	52	—
$C_6H_5CH_2$	85	—	69[1]	55[1]	45[1]	78[1]	—	—	72[1]	63[1]	—	—
CCl_3	96	102[1]	70	56	—	—	—	—	88[4]	—	32	—
CF_3	106	130[1]	86	71	53	—	—	—	99[1]	—	31	97
C_2F_5	103[6]	127[6]	85[6]	68[6]	53[6]	—	—	—	—	—	—	—
CH_3CO	86	119[1]	81[1]	67[1]	50[1]	106[1]	99[3]	97	80[1]	67	—	—
CN	120	—	—	83	73	—	—	—	—	—	29	—
C_6F_5	114	114[5]	92[5]	—	66[5]	107[5]	—	—	105[5]	—	50[5]	60[5]

REFERENCES

1. Cox, J. D. and Pilcher, G., *Thermochemistry of Organic and Organometallic Compounds,* Academic Press, London, 1970.
2. Lacher, J. R. and Skinner, H. A., *J. Chem. Soc. A,* 1034, 1968.
3. Benson, S. W. et al., *Chem. Rev.,* 69, 279, 1969.
4. Hu, A. T., Sinke, G. C., and Mintz, M. J., *J. Chem. Thermodyn.,* 4, 239, 1972.
5. Choo, K. Y., Mendenhall, G. D., Golden, D. M., and Benson, S. W., *Int. J. Chem. Kinet.,* 6, 813, 1974.
6. Wu, E.-C. and Rodgers, A. S., *J. Am. Chem. Soc.,* 98, 6112, 1976.

THE MADELUNG CONSTANT AND CRYSTAL LATTICE ENERGY

Donald F. Swinehart

If U is the crystal lattice energy and *M* is the Madelung constant, then

$$U = \frac{N M z_i z_j e^2}{r}(1 - 1/n)^a$$

Substance	Ion type	Crystal form[b]	*M*
Sodium chloride, NaCl	M^+, X^-	FCC	1.74756
Cesium chloride, CsCl	M^+, X^-	BCC	1.76267
Calcium chloride, $CaCl_2$	$M^{++}, 2X^-$	Cubic	2.365
Calcium fluoride, fluorite, CaF_2	$M^{++}, 2X^-$	Cubic	2.51939
Cadmium chloride, $CdCl_2$	$M^{++}, 2X^-$	Hexagonal	2.244[c]
Cadmium iodide, $CdI_2(\alpha)$	$M^{++}, 2X^-$	Hexagonal	2.355[c]
Magnesium fluoride, MgF_2	$M^{++}, 2X^-$	Tetragonal	2.381[c]
Cuprous oxide, cuprite, Cu_2O	$2M^+, X^=$	Cubic	2.22124
Zinc oxide, ZnO	$M^{++}, X^=$	Hexagonal	1.4985[c]
Sphalerite, zinc blende, ZnS	$M^{++}, X^=$	FCC	1.63806
Wurtzite, ZnS	$M^{++}, X^=$	Hexagonal	1.64132[c]
Titanium dioxide, anatase, TiO_2	$M^{4+}, 2X^=$	Tetragonal	2.400[c]
Titanium dioxide, rutile, TiO_2	$M^{4+}, 2X^=$	Tetragonal	(2.408) − (2.055)(0.721 − c/a)^2 [d]
β-Quartz, SiO_2	$M^{4+}, 2X^=$	Hexagonal	2.2197[c]
Corundum, Al_2O_3	$2M^{3+}, 3X^=$	Rhombohedral	4.1719

[a] N is Avogadro's number, z_i and z_j are the integral charges on the ions (i.e., in units of the electronic charge), and e is the charge on the electron in electrostatic units (e = 4.803 × 10^{-10} ESU). r is the shortest distance between cation-anion pairs in centimeters. Then U is in ergs.

[b] FCC = face centered cubic, BCC is body centered cubic.

[c] For tetragonal and hexagonal crystals the value of *M* depends on the details of the lattice parameters.

[d] For rutile-type structures, *M* is given to a good approximation by inserting the proper c/a ratio. c/a for rutile itself is 0.721 and *M* = 2.408.

Note: Several variations of the equation for U appear in the literature with corresponding variations in *M*. In general, using literature values for *M* requires caution. The Born exponent, n, may be obtained from the following table:

THE BORN EXPONENT, n

Ion type	n
He, Li^+	5
Ne, Na^+, F^-	7
Ar, K^+, Cu^+, Cl^-	9
Kr, Rb^+, Ag^+, Br^-	10
Xe, Cs^+, Au^+, I^-	12

For a crystal with a mixed-ion type, an average of the values of n in this table is to be used (6 for LiF, for example).

VALUES OF THE GAS CONSTANT, R

Coleman J. Major

The numerical value of the gas constant, R, defined by the equation PV = nRT, depends upon the units of P, V, n, and T. A large number of values of the constant may be calculated. The accompanying table gives 84 values of R in a convenient form using the most common units of pressure and volume. It also incorporates both the pound and gram mole and both Rankine and Kelvin temperature scales. Various combinations of metric and English units may, therefore, be used without the necessity of converting each variable to a common system of units. Conversion factors and constants used for computing the values of R are listed at the bottom of the table.

The following example illustrates the use of the table:

Calculate: The volume in ft^3 occupied by 2 lb. moles of a gas at 15°C at a pressure of 32.2 ft. of water, assuming the ideal gas law.

Solution: 15°C + 273.2 = 288.2°K

Enter the top of the table under the column headed "ft H_2O" and proceed downward to the value of 44.6 for R. (Note that this lines up horizontally with the desired units of ft^3, °K, and lb. moles shown on the left side of the table)

$$V = \frac{nRT}{P} = \frac{2 \times 44.6 \times 288.2}{32.2} = 798 \text{ ft}^3$$

$$\text{Values of Gas Constant, } R = \frac{PV}{nT}$$

Absolute Pressure

Volume	Temp.	moles	Atm	psia	mm Hg	cm Hg	in Hg	in H_2O	ft H_2O
ft^3	°K	g	0.00290	0.0426	2.20	0.220	0.0867	1.18	0.0982
		lb	1.31	19.31	999	99.9	39.3	535	44.6
	°R	g	0.00161	0.02366	1.22	0.122	0.0482	0.655	0.0546
		lb	0.730	10.73	555	55.5	21.8	297	24.8
cm^3	°K	g	82.05	1206	62,400	6240	2450	33,400	2780
		lb	37,200	547,000	2.83×10^7	2.83×10^6	1.11×10^6	1.51×10^7	1.26×10^6
	°R	g	45.6	670	34,600	3460	1360	18,500	1550
		lb	20,700	304,000	1.57×10^7	1.57×10^6	619,000	8.41×10^6	701,000
liters	°K	g	0.08205	1.206	62.4	6.24	2.45	33.4	2.78
		lb	37.2	547	28,300	2830	1113	15,140	1262
	°R	g	0.0456	0.670	34.6	3.46	1.36	18.5	1.55
		lb	20.7	304	15,700	1570	619	8410	701

Conversion Factors and Constants

1 lb. = 453.59 g
1 atm = 14.696 psia
1 atm = 760 mm Hg
1 atm = 76 cm Hg
1 atm = 29.921 in Hg
1 atm = 406.79 in H_2O
1 atm = 33.90 ft H_2O
1 atm = 1.01325×10^5 Pa

359.0 ft^3/lb mole
22,414 cm^3/g mole
1 inch = 2.54 cm
Std. temp. = 273.15°K or 491.67°R
28.316847 liters = 1 ft^3
R = 8.31441 (26) $\times 10^7$ erg $°K^{-1}$ mol^{-1}
= 1.9872 cal $°K^{-1}$ mol^{-1}
= 8.31441 (26) J $°K^{-1}$ mol^{-1}

RECOMMENDED CONSISTENT VALUES OF THE FUNDAMENTAL PHYSICAL CONSTANTS

The numbers in parentheses are the standard deviation uncertainties in the last digits of the quoted value, computed on the basis of internal consistency.

Quantity	Symbol	Value	Uncertainty, ppm
1. Permeability of Vacuum	μ_0	$4\pi \times 10^{-7}$ H m^{-1} = $12.5663706144 \times 10^{-7}$ H m^{-1}	
2. Speed of Light in Vacuum	c	$2.99792458(1.2) \times 10^{8}$ m s^{-1}	0.004
3. Permittivity of Vacuum	$\epsilon_0 = (\mu_0 c^2)^{-1}$	$8.85418782(7) \times 10^{-12}$ F m^{-1}	0.008
4. Fine Structure Constant, $\mu_0 ce^2/2h$	α	0.0072973506(60)	0.82
	α^{-1}	137.03604(11)	0.82
5. Elementary Charge	e	$1.6021892(46) \times 10^{-19}$ C	2.9
6. Planck Constant	h	$6.626176(36) \times 10^{-34}$ J Hz^{-1}	5.4
	$\hbar = h/2\pi$	$1.0545887(57) \times 10^{-34}$ J s	5.4
7. Avogadro Constant	N_A	$6.022045(31) \times 10^{23}$ mol^{-1}	5.1
8. Atomic Mass Unit	$1\text{u} = (10^{-3}\ \text{kg mol}^{-1})/N_A$	$1.6605655(86) \times 10^{-27}$ kg	5.1
9. Electron Rest Mass	m_e	$0.9109534(47) \times 10^{-30}$ kg	5.1
		$5.4858026(21) \times 10^{-4}$ u	0.38
10. Muon Rest Mass	m_μ	$1.883566(11) \times 10^{-28}$ kg	5.6
		0.11342920(26) u	2.3
11. Proton Rest Mass	m_p	$1.6726485(86) \times 10^{-27}$ kg	5.1
		1.007276470(11) u	0.011
12. Neutron Rest Mass	m_n	$1.6749543(86) \times 10^{-27}$ kg	5.1
		1.008665012(37) u	0.037
13. Ratio, Proton Mass to Electron Mass	m_p/m_e	1,836.15152(70)	0.38
14. Ratio, Muon Mass to Electron Mass	m_μ/m_e	206.76865(47)	2.3
15. Specific Electron Charge	e/m_e	$1.7588047(49) \times 10^{11}$ C kg^{-1}	2.8
16. Faraday Constant	$\mathcal{F} = N_A e$	$9.648456(27) \times 10^{4}$ C mol^{-1}	2.8
17. Magnetic Flux Quantum	$\Phi_0 = h/2e$	$2.0678506(54) \times 10^{-15}$ Wb	2.6
	h/e	$4.135701(11) \times 10^{-15}$ J Hz^{-1} C^{-1}	2.6
18. Josephson Frequency-Voltage Ratio	$2e/h$	483.5939(13) THz V^{-1}	2.6
19. Quantum of Circulation	$h/2m_e$	$3.6369455(60) \times 10^{-4}$ J Hz^{-1} kg^{-1}	1.6
	h/m_e	$7.273891(12) \times 10^{-4}$ J Hz^{-1} kg^{-1}	1.6
20. Rydberg Constant	R_∞	$1.097373177(83) \times 10^{7}$ m^{-1}	0.075
21. Bohr Radius	$a_0 = \alpha/4\pi R_\infty$	$0.52917706(44) \times 10^{-10}$ m	0.82
22. Electron Compton Wavelength	$\lambda_c = \alpha^2/2R_\infty$	$2.4263089(40) \times 10^{-12}$ m	1.6
	$\lambda\!\!\!^{-}_c = \lambda_c/2\pi = \alpha a_0$	$3.8615905(64) \times 10^{-13}$ m	1.6
23. Classical Electron Radius	$r_e = \mu_0 e^2/4\pi m_e = \alpha\lambda\!\!\!^{-}_c$	$2.8179380(70) \times 10^{-15}$ m	2.5
24. Electron g-Factor	$\frac{1}{2}g_e = \mu_e/\mu_B$	1.0011596567(35)	0.0035
25. Muon g-Factor	$\frac{1}{2}g_\mu$	1.00116616(31)	0.31
26. Proton Moment in Nuclear Magnetons	μ_p/μ_N	2.7928456(11)	0.38
27. Bohr Magneton	$\mu_B = e\hbar/2m_e$	$9.274078(36) \times 10^{-24}$ J T^{-1}	3.9
28. Nuclear Magneton	$\mu_N = e\hbar/2m_p$	$5.050824(20) \times 10^{-27}$ J T^{-1}	3.9
29. Electron Magnetic Moment	μ_e	$9.284832(36) \times 10^{-24}$ J T^{-1}	3.9
30. Proton Magnetic Moment	μ_p	$1.4106171(55) \times 10^{-26}$ J T^{-1}	3.9
31. Proton Magnetic Moment in Bohr Magnetons	μ_p/μ_B	$1.521032209(16) \times 10^{-3}$	0.011
32. Ratio, Electron to Proton Magnetic Moments	μ_e/μ_p	658.2106880(66)	0.010
33. Ratio, Muon Moment to Proton Moment	μ_μ/μ_p	3.1833402(72)	2.3
34. Muon Magnetic Moment	μ_μ	$4.490474(18) \times 10^{-26}$ J T^{-1}	3.9
35. Proton Gyromagnetic Ratio	γ_p	$2.6751987(75) \times 10^{8}$ s^{-1} T^{-1}	2.8
36. Diamagnetic Shielding Factor, Spherical H_2O Sample	$1 + \sigma(H_2O)$	1.000025637(67)	0.067
37. Proton Gyromagnetic Ratio (uncorrected)	γ'_p	$2.6751301(75) \times 10^{8}$ s^{-1} T^{-1}	2.8
	$\gamma'_p/2\pi$	42.57602(12) MHz T^{-1}	2.8
38. Proton Moment in Nuclear Magnetons (uncorrected)	μ'_p/μ_N	2.7927740(11)	0.38
39. Proton Compton Wavelength	$\lambda_{C,p} = h/m_p c$	$1.3214099(22) \times 10^{-15}$ m	1.7
	$\lambda\!\!\!^{-}_{C,p} = \lambda_{C,p}/2\pi$	$2.1030892(36) \times 10^{-16}$ m	1.7
40. Neutron Compton Wavelength	$\lambda_{C,n} = h/m_n c$	$1.3195909(22) \times 10^{-15}$ m	1.7
	$\lambda\!\!\!^{-}_{C,n} = \lambda_{C,n}/2\pi$	$2.1001941(35) \times 10^{-16}$ m	1.7
41. Molar Gas Constant	R	8.31441(26) J mol^{-1} K^{-1}	31
42. Molar Volume, Ideal Gas (T_0 = 273.15 K, p_0 = 1 atm)	$V_m = RT_0/p_0$	0.02241383(70) m^3 mol^{-1}	31
43. Boltzmann Constant	$k = R/N_A$	$1.380662(44) \times 10^{-23}$ J K^{-1}	32
44. Stefan-Boltzmann Constant	$\sigma = (\pi^2/60)k^4/\hbar^3 c^2$	$5.67032(71) \times 10^{-8}$ W m^{-2} K^{-4}	125
45. First Radiation Constant	$c_1 = 2\pi hc^2$	$3.741832(20) \times 10^{-16}$ W m^2	5.4
46. Second Radiation Constant	$c_2 = hc/k$	0.01438786(45) m K	31
47. Gravitational Constant	G	$6.6720(41) \times 10^{-11}$ N m^2 kg^{-2}	615

Data from CODATA Bulletin No. 11, ICSU CODATA Central Office, CODATA Secretariat: 51 Boulevard de Montmorency, 75016 Paris, France (copies of this bulletin are available at no cost from this office).

ENERGY CONVERSION FACTORS

Quantity	Value	Unit	Error (ppm)
1 kg	5.609538(24)	10^{29} MeV	4.4
1 amu	931.5016(26)	MeV	2.8
Electron mass	0.5110041(16)	MeV	3.1
Proton mass	938.2592(52)	MeV	5.5
Neutron mass	939.5527(52)	MeV	5.5
1 electron volt	1.6021917(70)	10^{-19} J	4.4
		10^{-12} erg	
	2.4179659(81)	10^{14} Hz	3.3
	8.065465(27)	10^{5} m^{-1}	3.3
		10^{3} cm^{-1}	
	1.160485(49)	10^{4} K	42
Energy–wavelength conversion	1.2398541(41)	10^{-6} eV·m	3.3
		10^{-4} eV·cm	
Rydberg constant, R_∞	2.179914(17)	10^{-18} J	7.6
		10^{-11} erg	
	13.605826(45)	eV	3.3
	3.2898423(11)	10^{15} Hz	0.35
	1.578936(67)	10^{5} K	43
Bohr magneton, μ_B	5.788381(18)	10^{-5} eV T^{-1}	3.1
	1.3996108(43)	10^{10} Hz T^{-1}	3.1
	46.68598(14)	$m^{-1}\cdot T^{-1}$	3.1
		10^{-2} $cm^{-1}\cdot T^{-1}$	
	0.671733(29)	K T^{-1}	43
Nuclear magneton, μ_n	3.152526(21)	10^{-8} eV T^{-1}	6.8
	7.622700(42)	10^{6} Hz T^{-1}	5.5
	2.542659(14)	10^{-2} $m^{-1}\cdot T^{-1}$	5.5
		10^{-4} $cm^{-1}\cdot T^{-1}$	
	3.65846(16)	10^{-4} K T^{-1}	44
Gas constant, R_0	8.20562(35)	10^{-2} m^3·atm $kmole^{-1}$ K^{-1}	42
Standard volume of ideal gas, V_0	22.4136	m^3 $kmole^{-1}$	

INFRARED CORRELATION CHART No. 1

Prepared from information supplied by Beckman Instruments

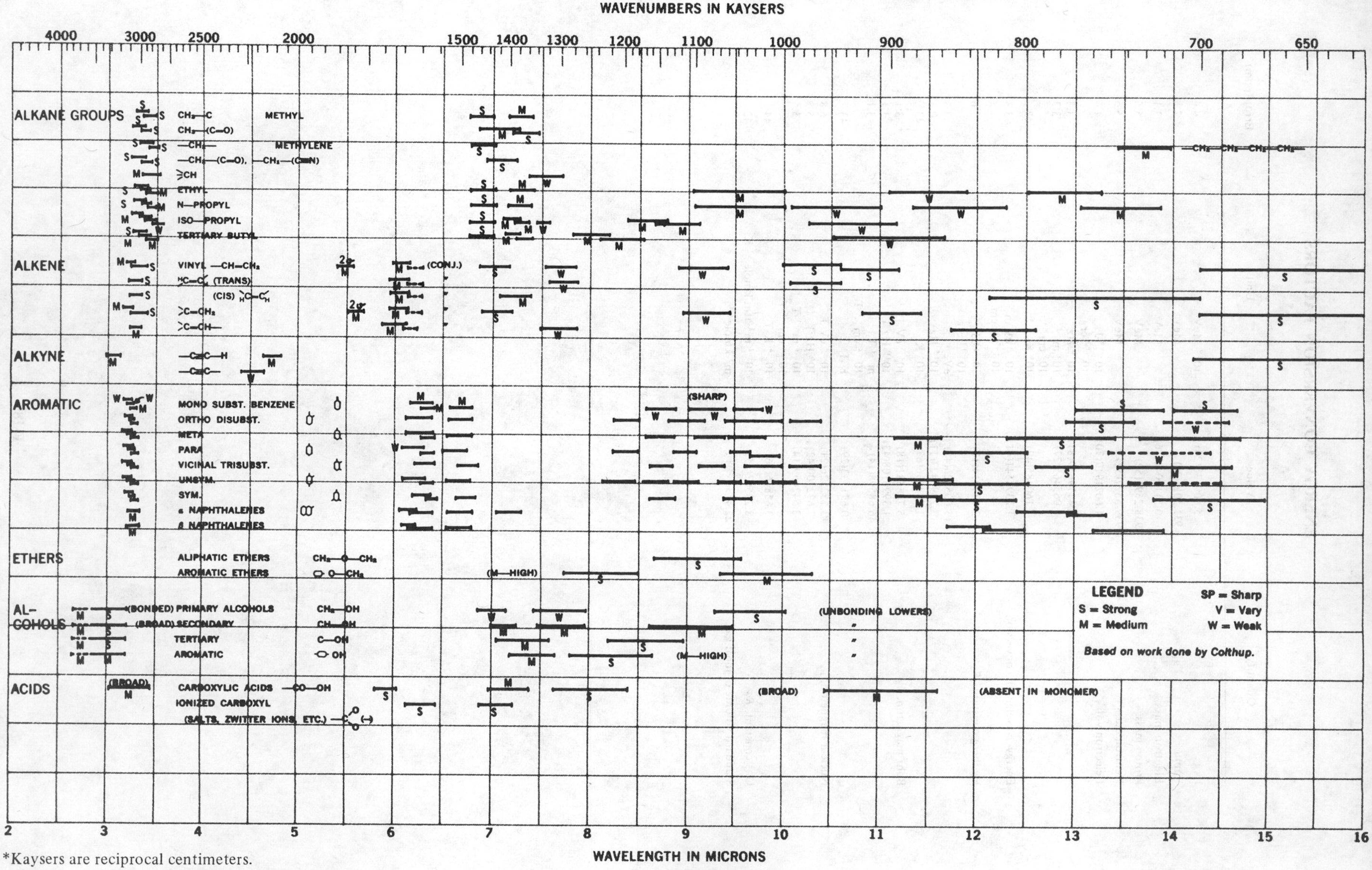

*Kaysers are reciprocal centimeters.

INFRARED CORRELATION CHART No. 1 (Con't.)

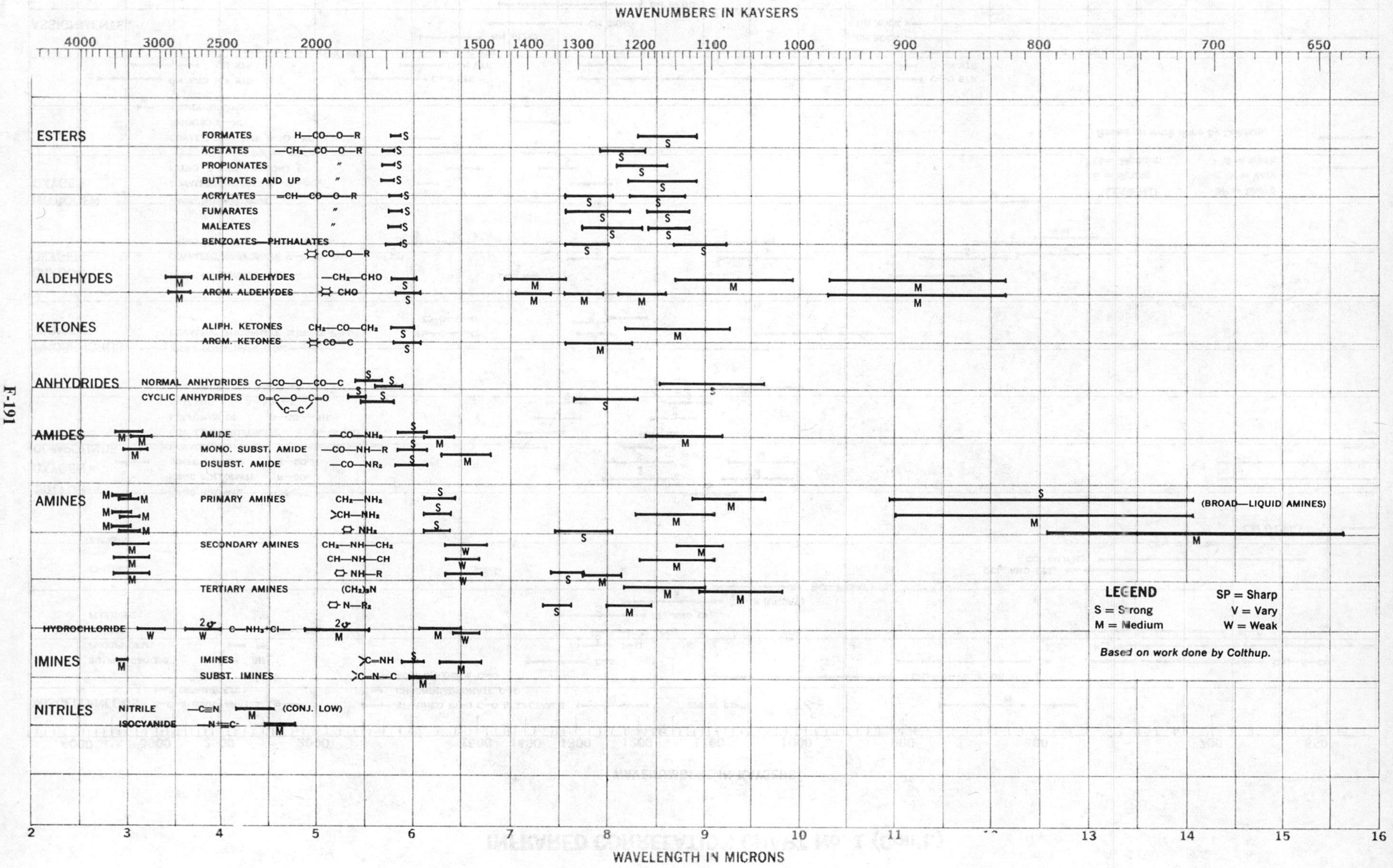

INFRARED CORRELATION CHART No. 1 (Con't.)

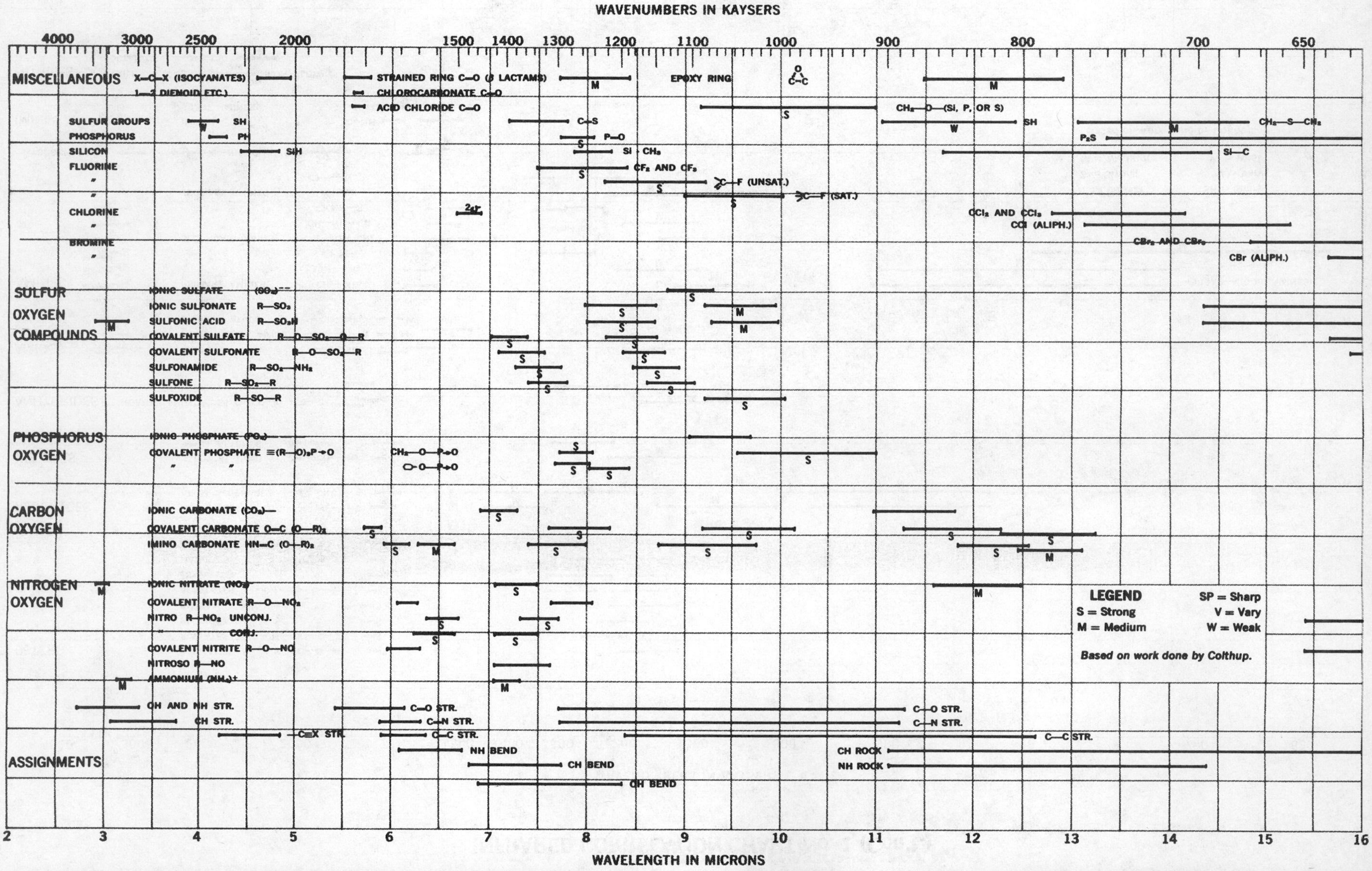

INFRARED CORRELATION CHART No. 2

Prepared from information supplied by Beckman Instruments

This chart presents some information regarding structure, double-bond vibrations, hydrogen stretching and triple-bond vibrations.

HYDROGEN STRETCHING AND TRIPLE-BOND VIBRATIONS, 3750-2000 CM.$^{-1}$ DOUBLE-BOND VIBRATIONS, ETC. 2000-1500 CM.$^{-1}$

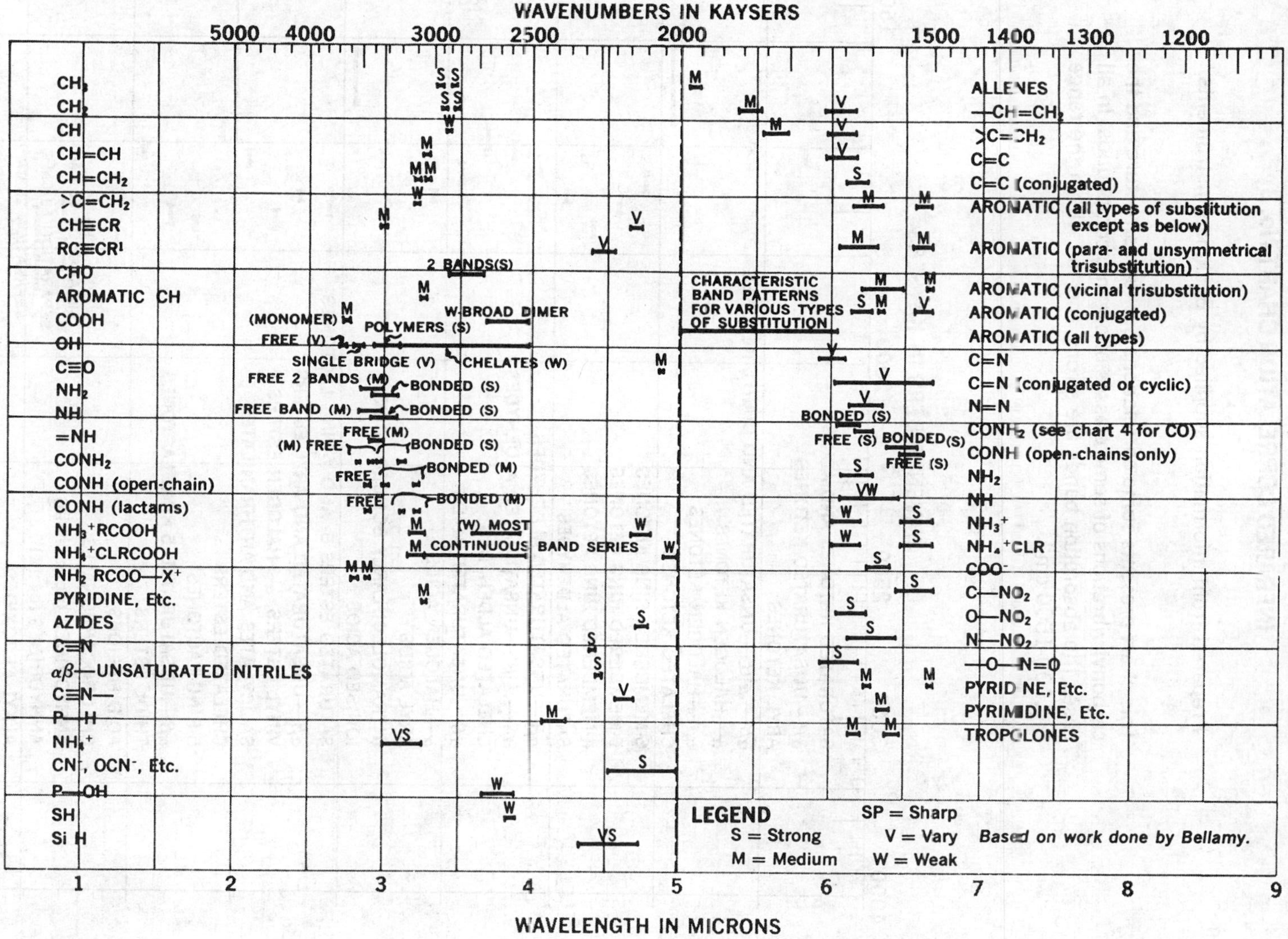

INFRARED CORRELATION CHART No. 3

Prepared from information supplied by Beckman Instruments

This chart presents some correlations between structure and the carbonyl vibrations of some classes of organic compounds. In all cases the absorption bands are strong and fall within the range of 1900-1500 cm^{-1}.

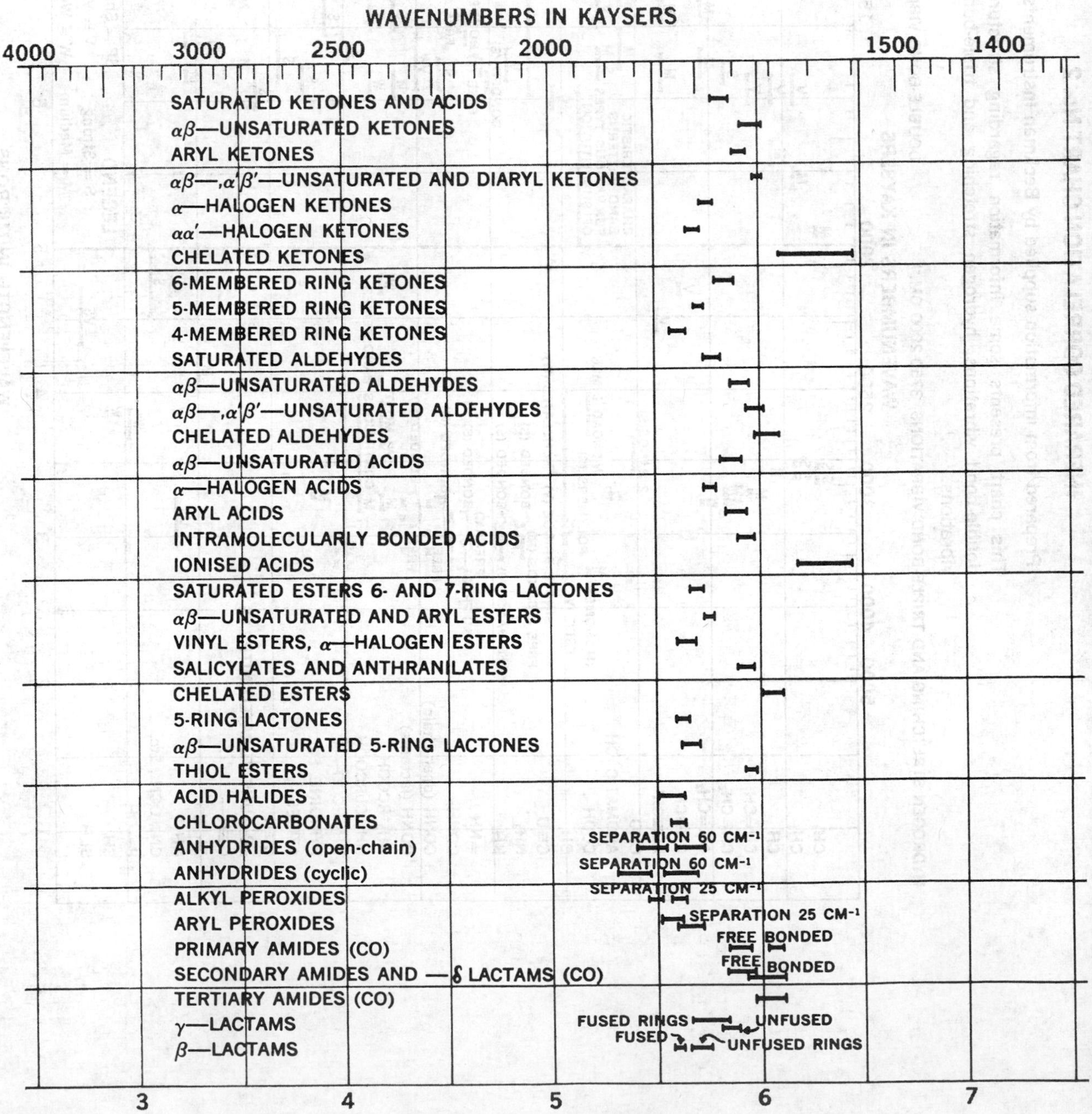

INFRARED CORRELATION CHART No. 4

Prepared from information supplied by Beckman Instruments

This chart presents some correlations between structure and single-bond vibrations for a number of classes of compounds having absorption between 1500-650 cm^{-1}

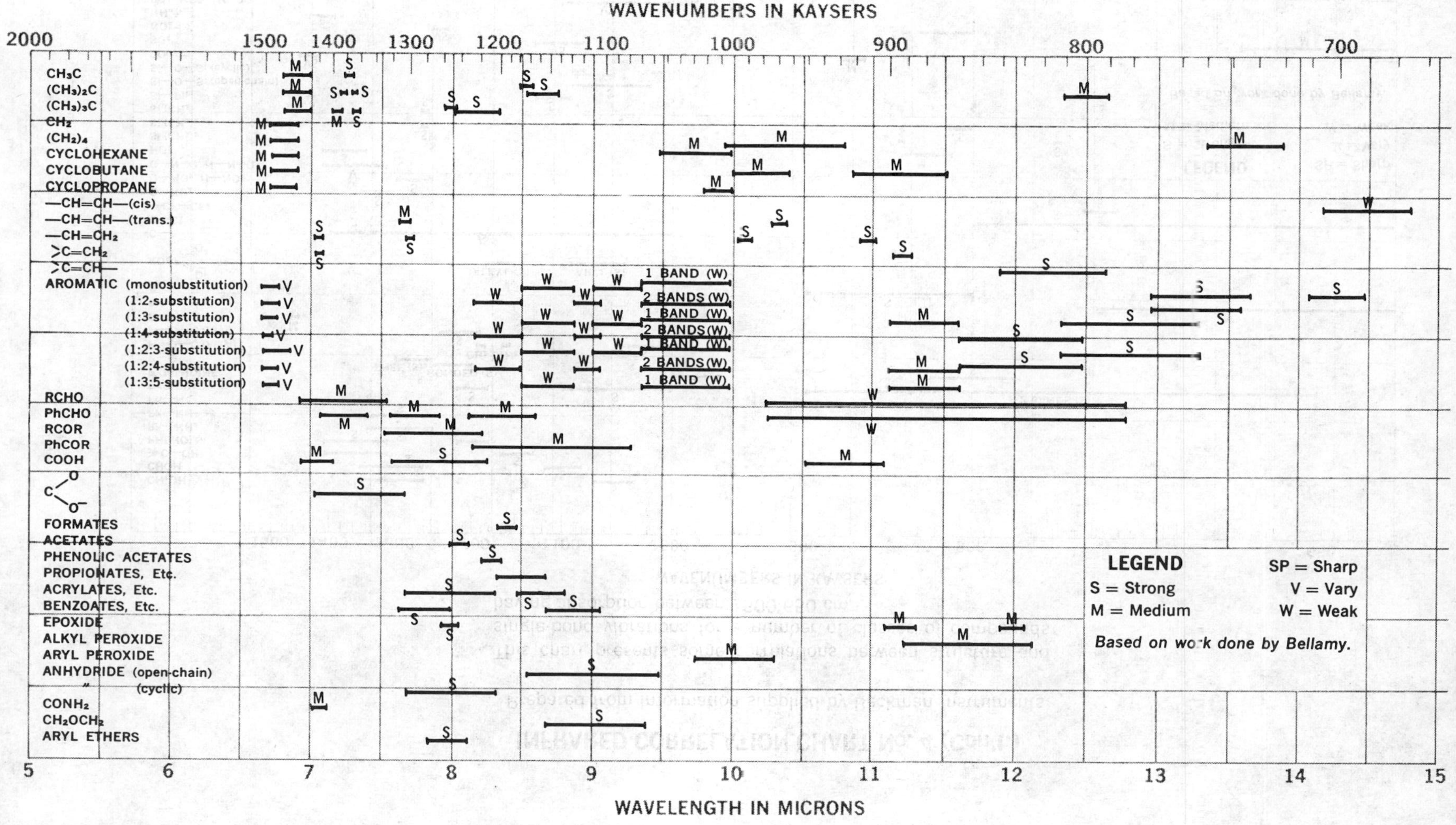

INFRARED CORRELATION CHART No. 4 (Con't.)

Prepared from information supplied by Beckman Instruments

This chart presents some correlations between structure and single-bond vibrations for a number of classes of compounds having absorption between 1500-650 cm^{-1}.

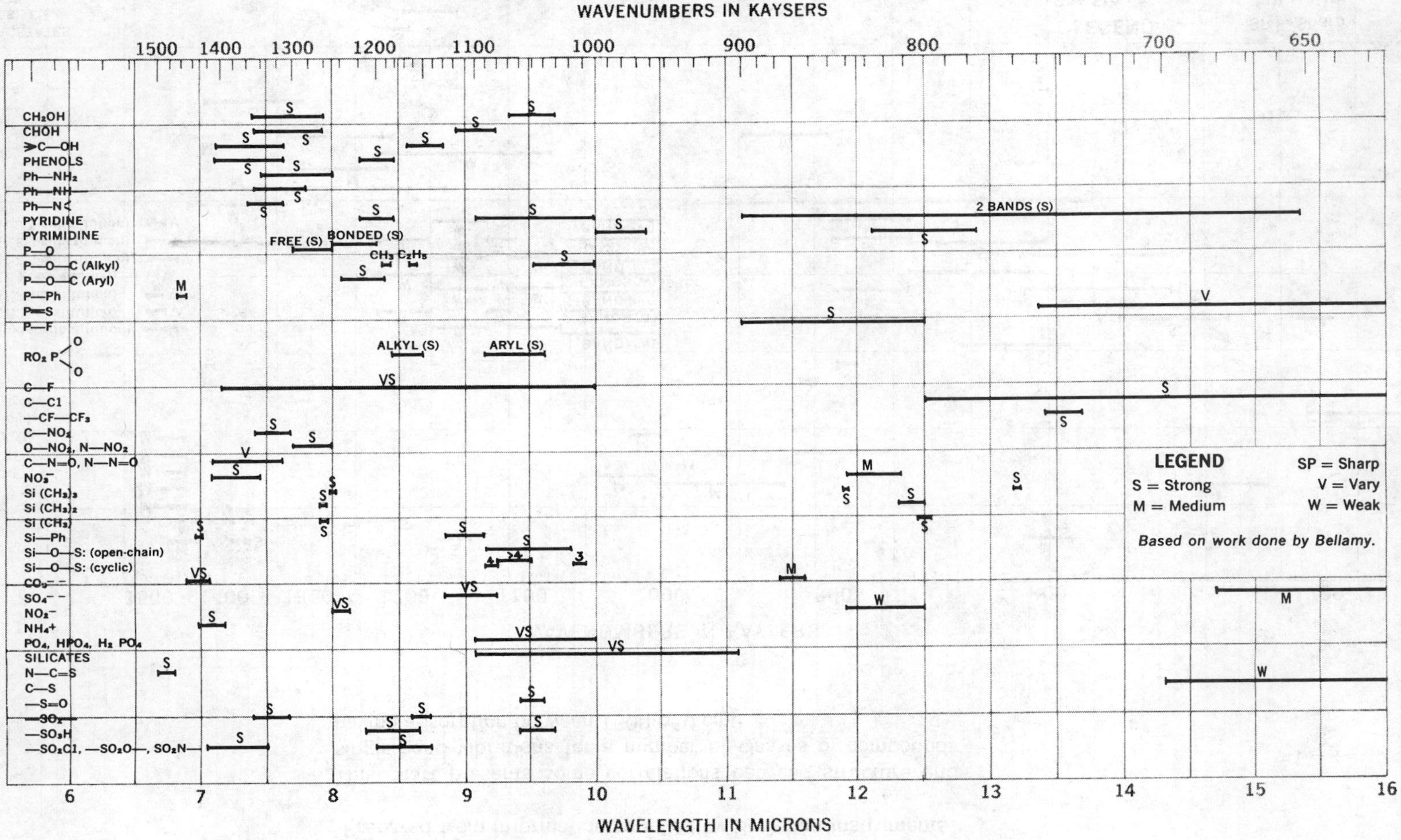

FAR INFRARED VIBRATIONAL FREQUENCY CORRELATION CHART

Based on evidence compiled by James E. Stewart of Beckman Instruments. This chart shows the vibrational frequency correlation in the far infrared region. Because research is continuing in the far infrared region, this chart is not all-inclusive.

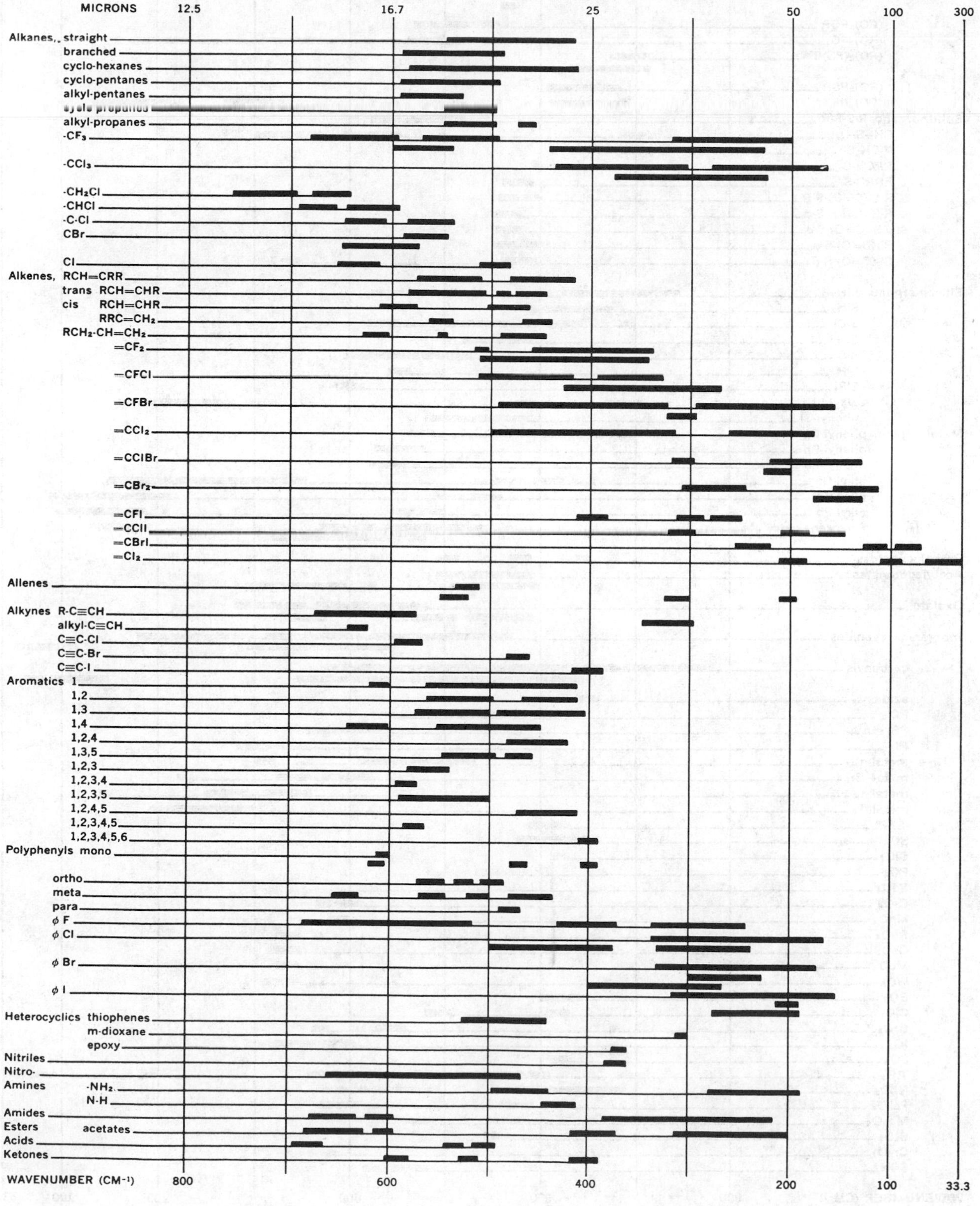

FAR INFRARED VIBRATIONAL FREQUENCY CORRELATION CHART (Con't.)

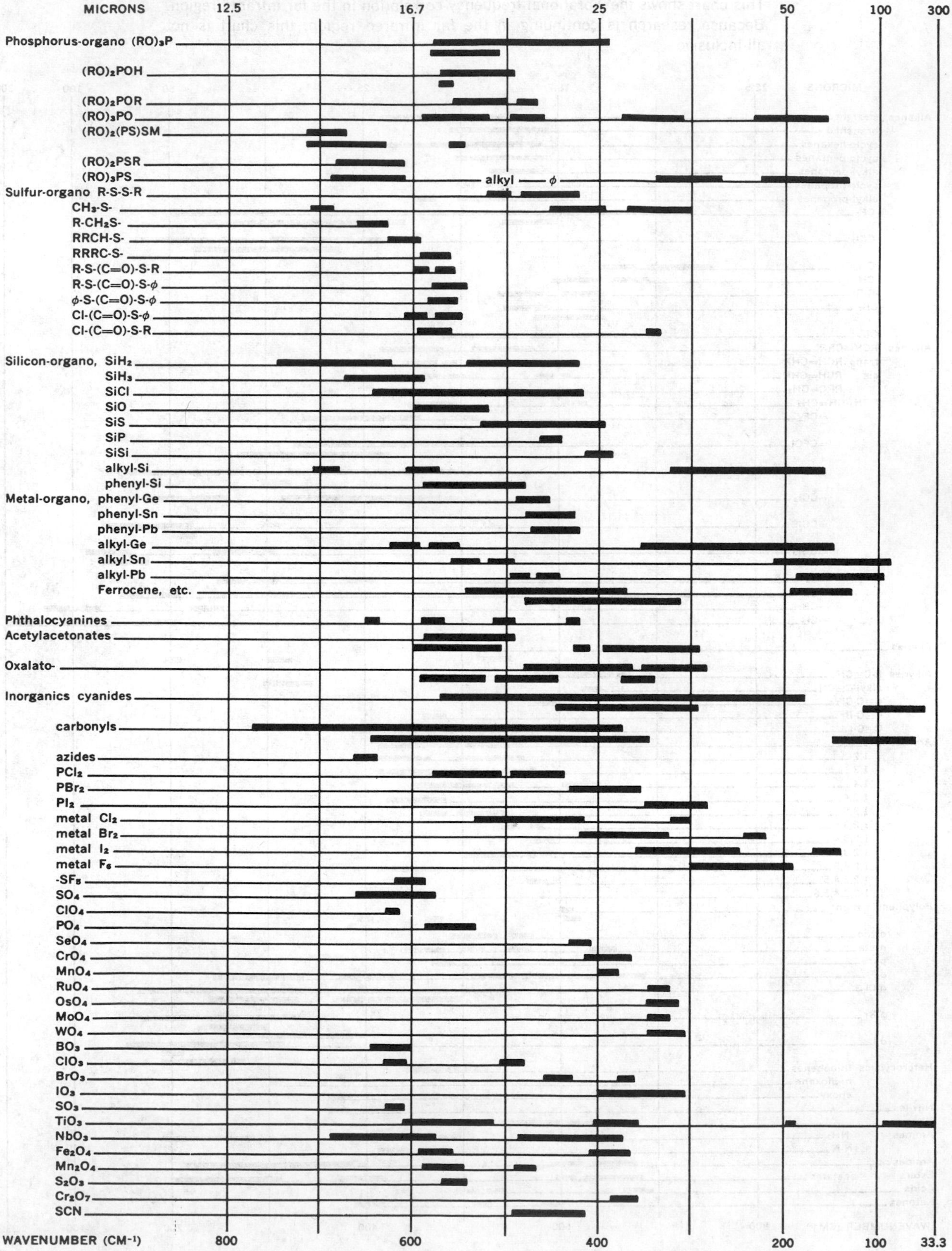

CHART OF CHARACTERISTIC FREQUENCIES BETWEEN ~ 700–300 cm^{-1}

Freeman F. Bentley, Lee D. Smithson, and Adele L. Rozek

This chart summarizes the characteristic frequencies known to occur between approximately 700–300^{-1}. Those who anticipate using this region of the spectrum should consult "Infrared Spectra and Characteristic Frequencies between ~700–300 cm^{-1}" by Interscience Publishers, a division of John Wiley and Sons, Inc. for a complete discussion of the characteristic frequencies summarized in this chart, a large collection of infrared spectra (700–300 cm^{-1}) of most of the common organic and inorganic compounds, and an extensive bibliography of references to infrared data below ~700 cm^{-1}.

In this chart the black horizontal bars indicate the range of the spectrum in which the characteristic frequencies have been observed to occur in the compounds investigated. The number of compounds investigated is given immediately to the right of the names or structures of the compounds. Obviously those characteristic frequency ranges based upon a limited number of compounds should be used with caution.

The letters above the bars indicate the relative intensities of the absorption bands. These intensities are based upon the strongest band in the spectra (700–300 cm^{-1}) of specific classes of compounds investigated, and they cannot be compared accurately with the intensities given for other classes.

When known, the specific vibration giving rise to the characteristic frequency is printed in abbreviated form immediately to the right of the bar indicating the frequency range except when lack of space prevents this. When there can be no ambiguity, this information may be printed other than to the right. In doubtful cases, arrows are used for clarification.

Naturally, the characteristic frequencies vary in their specificity and analytical value. The user is, therefore, cautioned to use this chart with some reserve. After reviewing this chart, the reader should be aware that there are many characteristic frequencies in the 700–300 cm^{-1} region. Used cautiously, this chart can be of considerable value in the elucidation of structures of unknown compounds.

It is important to emphasize that the region of the infrared between ~700–300 cm^{-1} should be used in conjunction with the more conventional 5000–700 cm^{-1} region. Much of the value of the 700–300 cm^{-1} region can only be realized after interpreting the spectrum between 5000–700 cm^{-1}.

The following symbols and abbreviations are used:

Symbol or Abbreviation	Definition
αCCC	In-plane bending of benzene ring
Antisym.	Antisymmetrical (Asymmetrical)
~	Approximately
β	In-plane bending of ring substituent bond
δ	In-plane bending
γ	Out-of-plane bending
i.p.	In-plane
m	Medium
ν	Stretching
ν_s	Symmetrical stretching
ν_{as}	Antisymmetrical stretching
o.p.	Out-of-plane
$\parallel$	Parallel
$\perp$	Perpendicular
ϕ	Phenyl
ϕCC	Out-of-plane bending of aromatic ring
r	Rocking
s	Strong
sh	Shoulder
Sym.	Symmetrical
v	Variable
w	Weak
"X" Sensitive	An aromatic vibrational mode whose frequency position is greatly dependent on the nature of the substituent.

CHART OF CHARACTERISTIC FREQUENCIES BETWEEN ~ 700–300 cm⁻¹

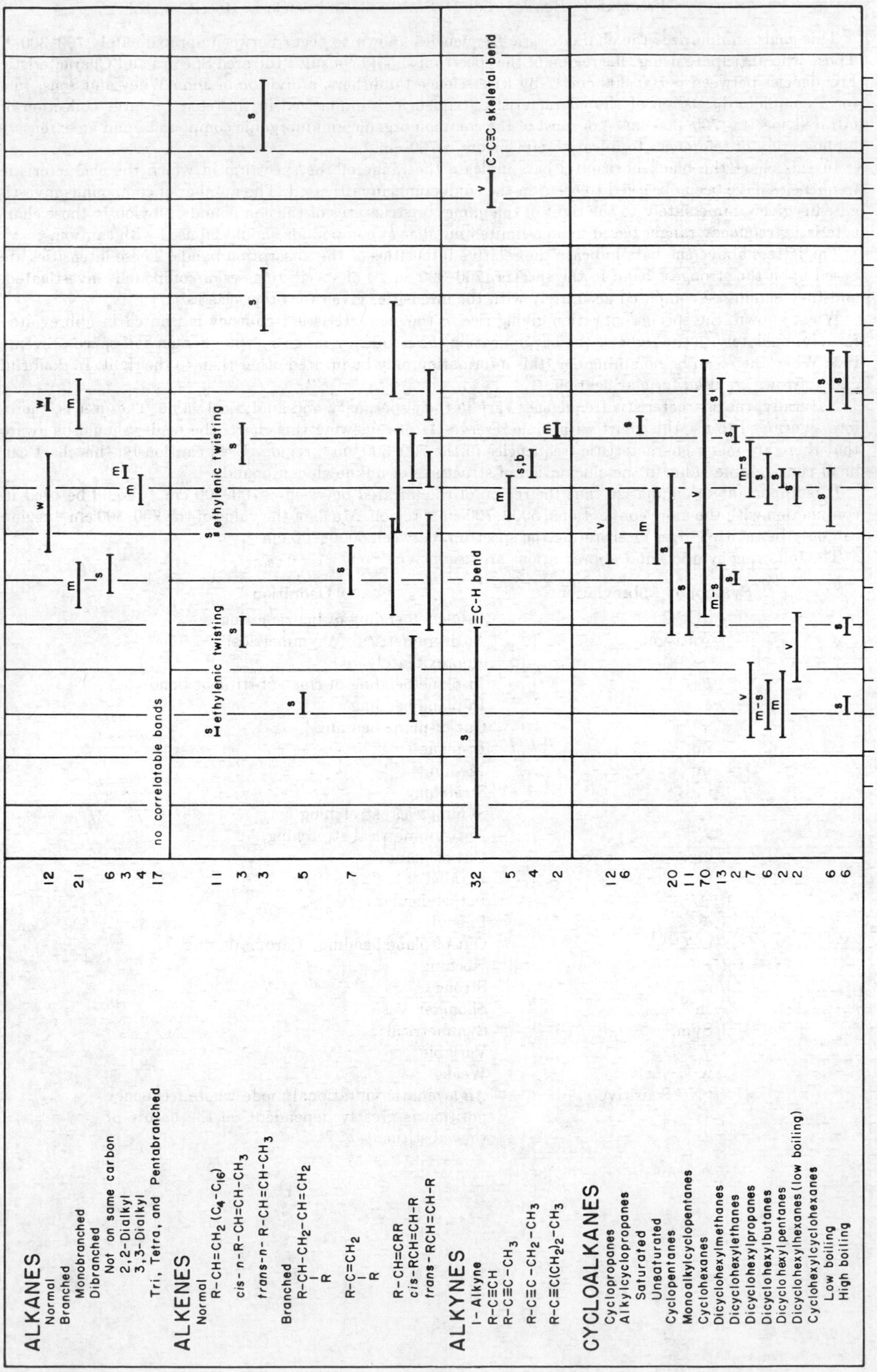

CHART OF CHARACTERISTIC FREQUENCIES BETWEEN ~ 700–300 cm⁻¹

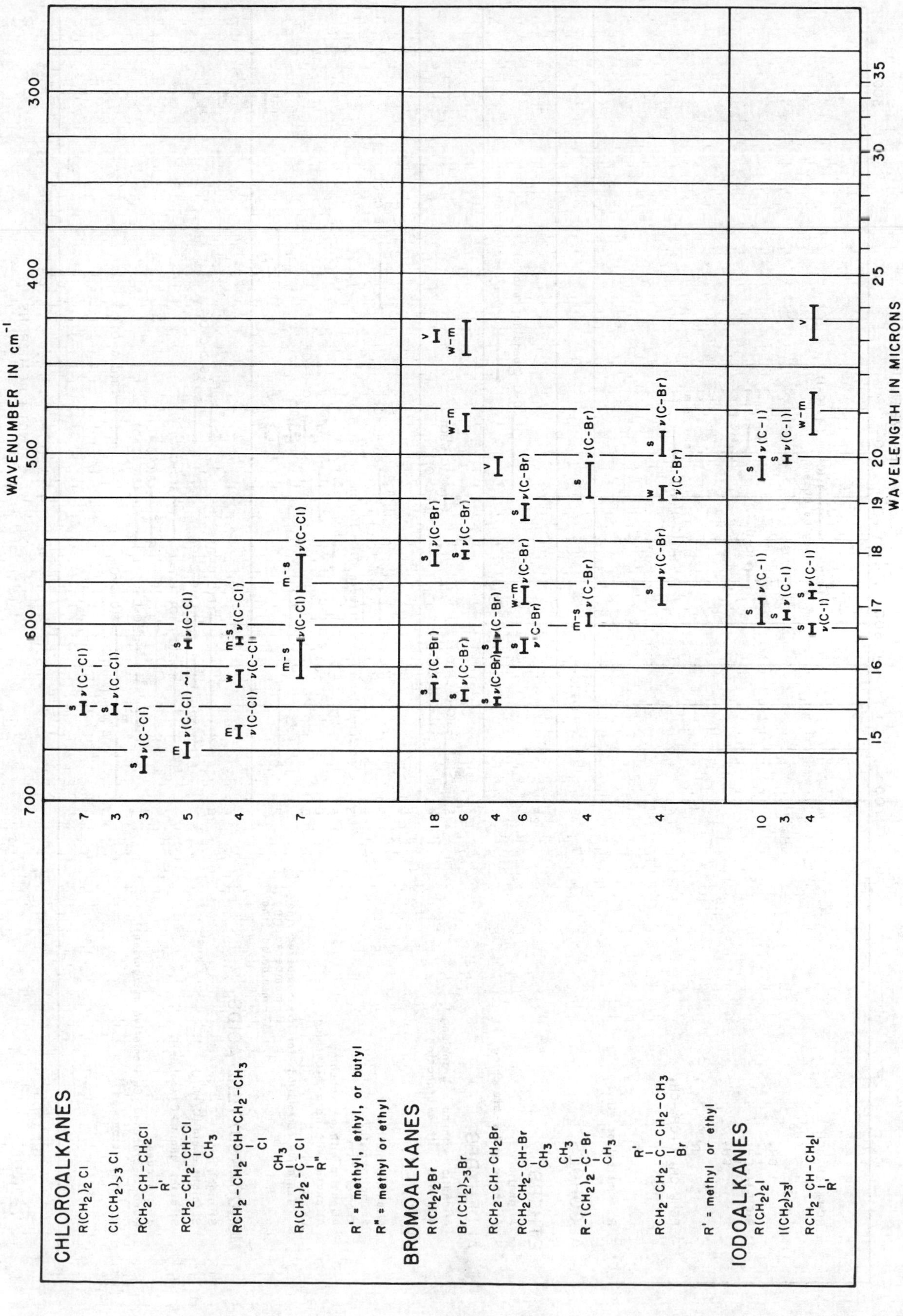

CHART OF CHARACTERISTIC FREQUENCIES BETWEEN ∽ 700–300 cm^{-1}

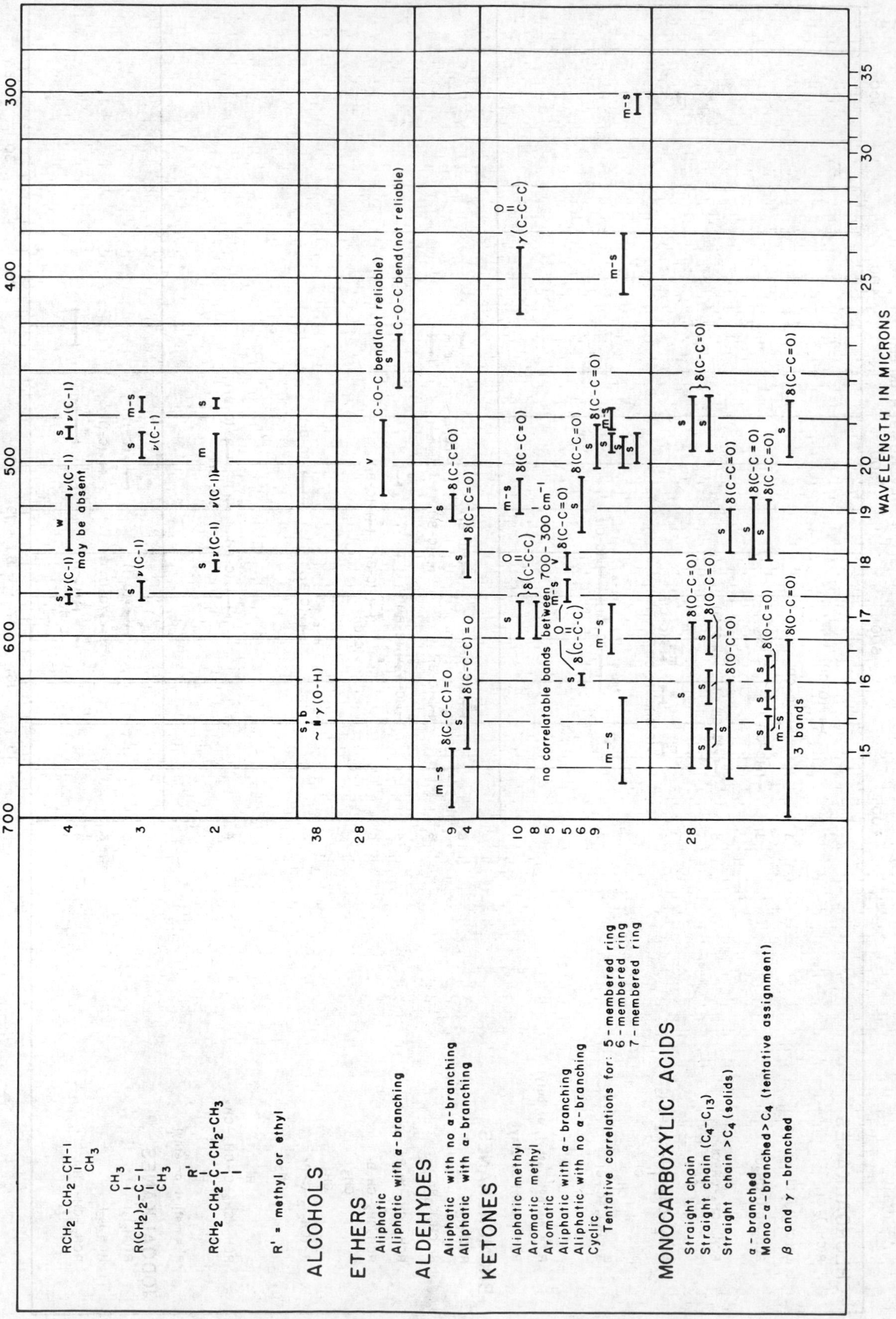

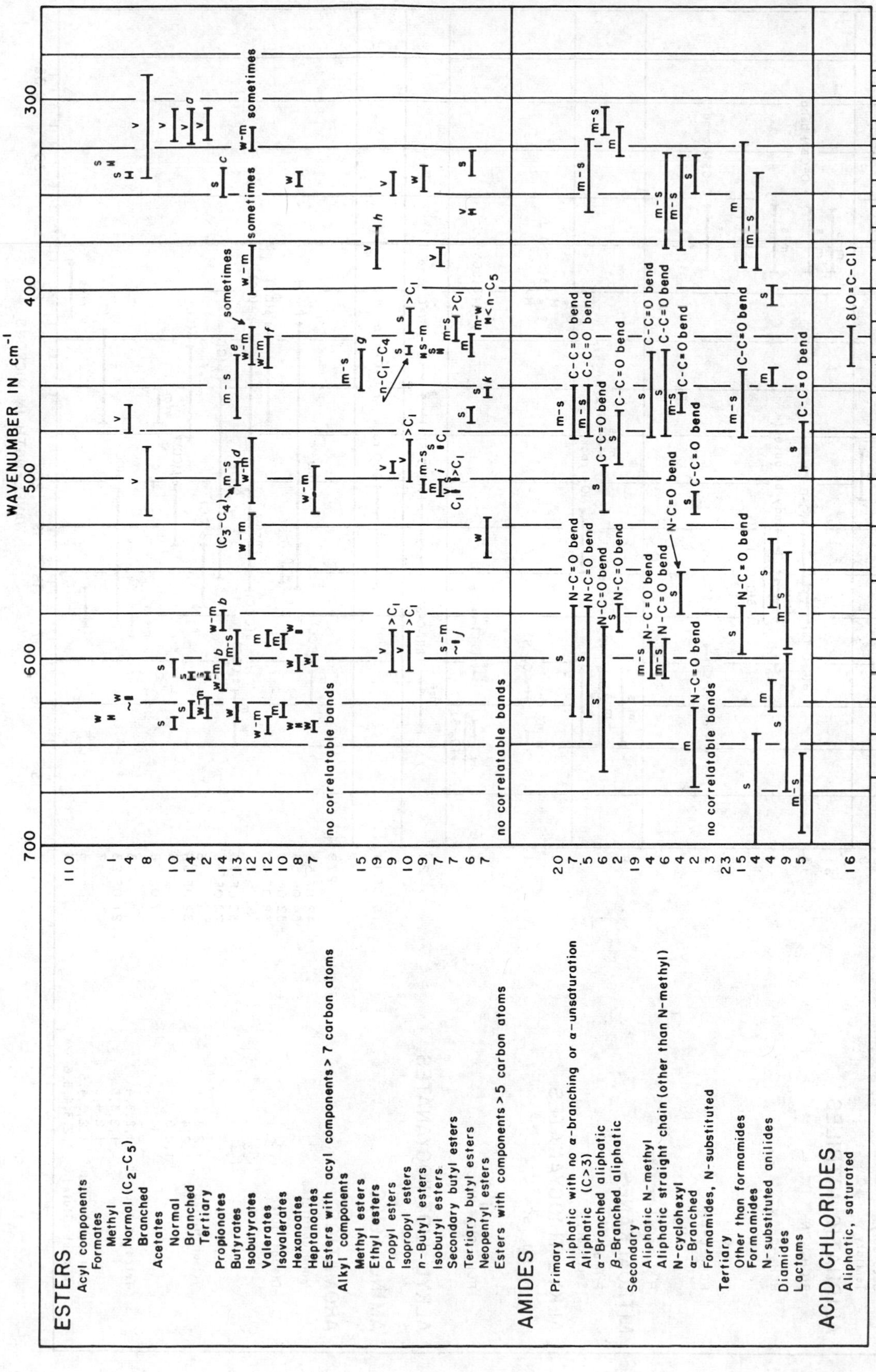

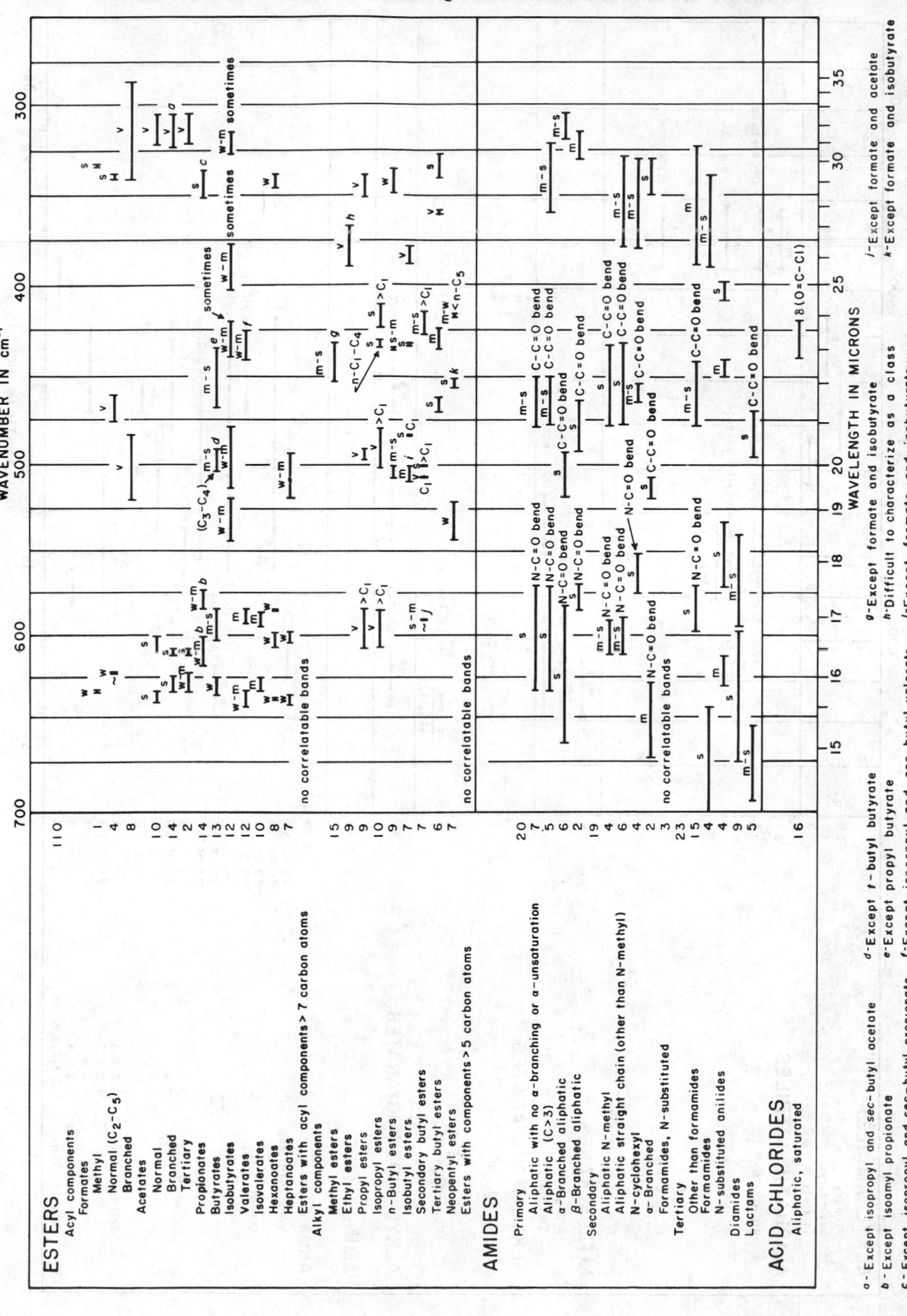

^{a}-Except isopropyl and sec-butyl acetate
^{b}-Except isoamyl propionate
^{c}-Except isopropyl and sec-butyl propionate
^{d}-Except t-butyl butyrate
^{e}-Except propyl butyrate
^{f}-Except isopropyl and sec-butyl valerate
^{g}-Except formate and isobutyrate
^{h}-Difficult to characterize as a class
^{i}-Except formate and isobutyrate
^{j}-Except formate and acetate
^{k}-Except formate and isobutyrate

CHART OF CHARACTERISTIC FREQUENCIES BETWEEN ∽ 700–300 cm^{-1}

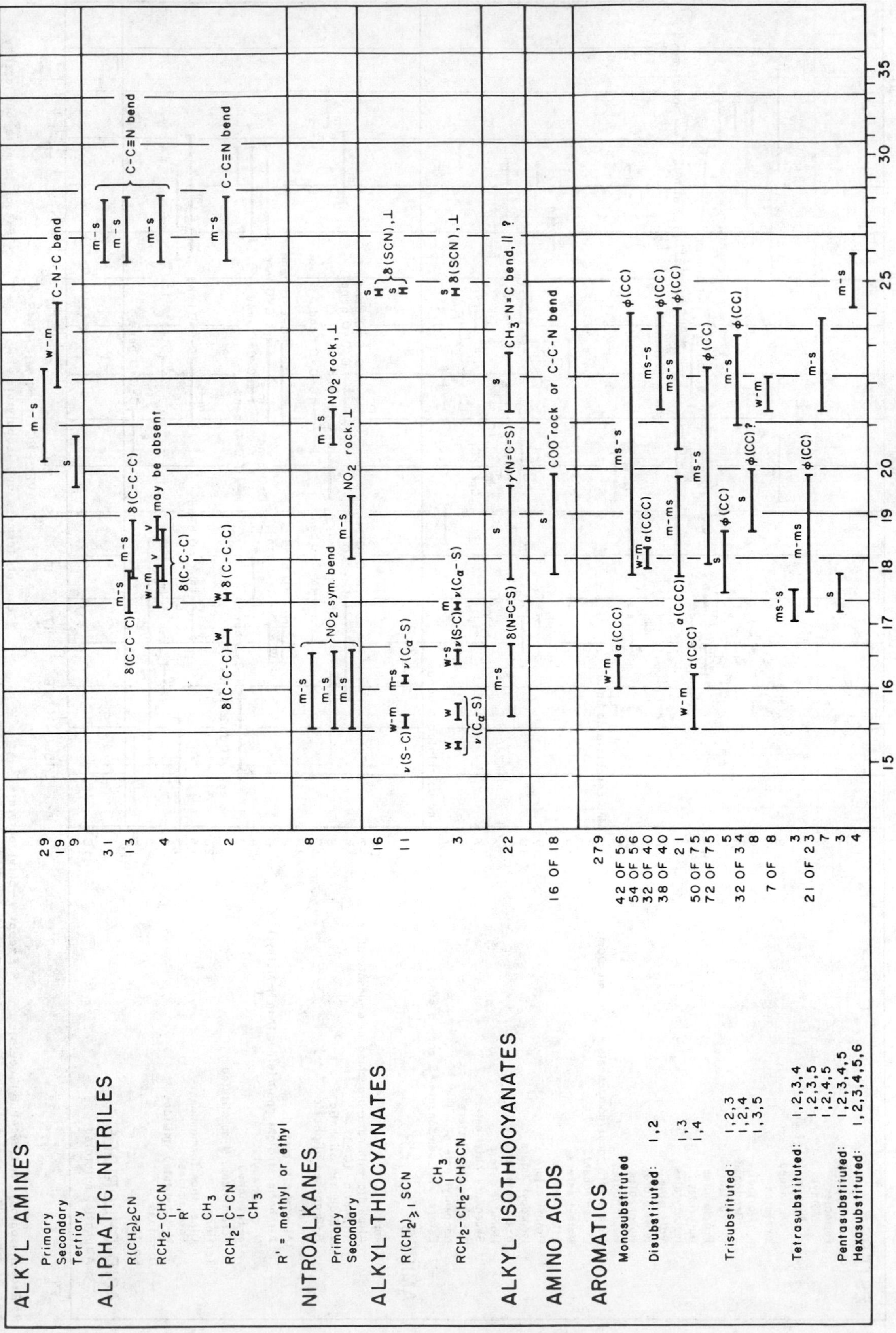

CHART OF CHARACTERISTIC FREQUENCIES BETWEEN ~ 700–300 cm^{-1}

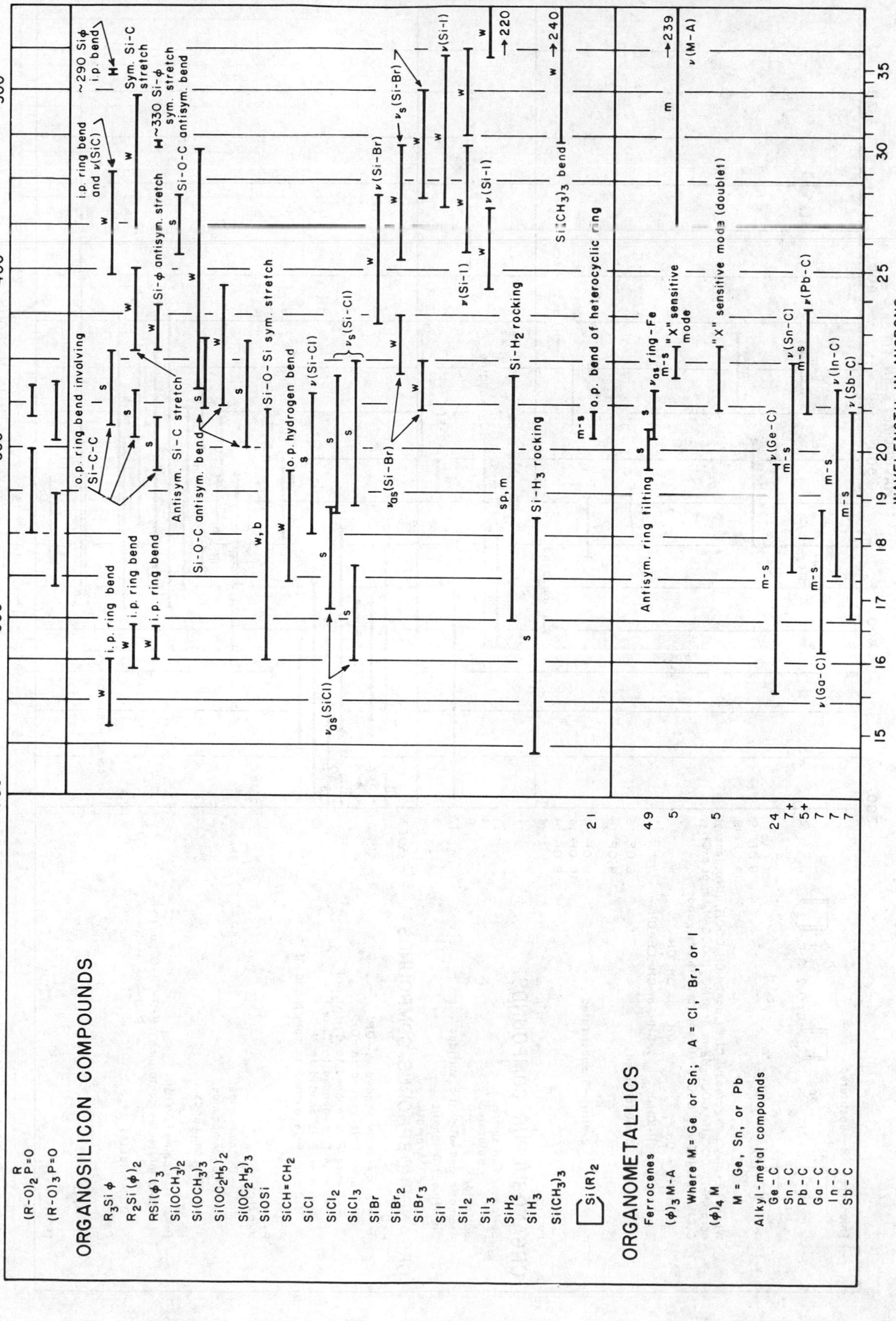

CHART OF CHARACTERISTIC FREQUENCIES BETWEEN ∽ 700–300 cm⁻¹

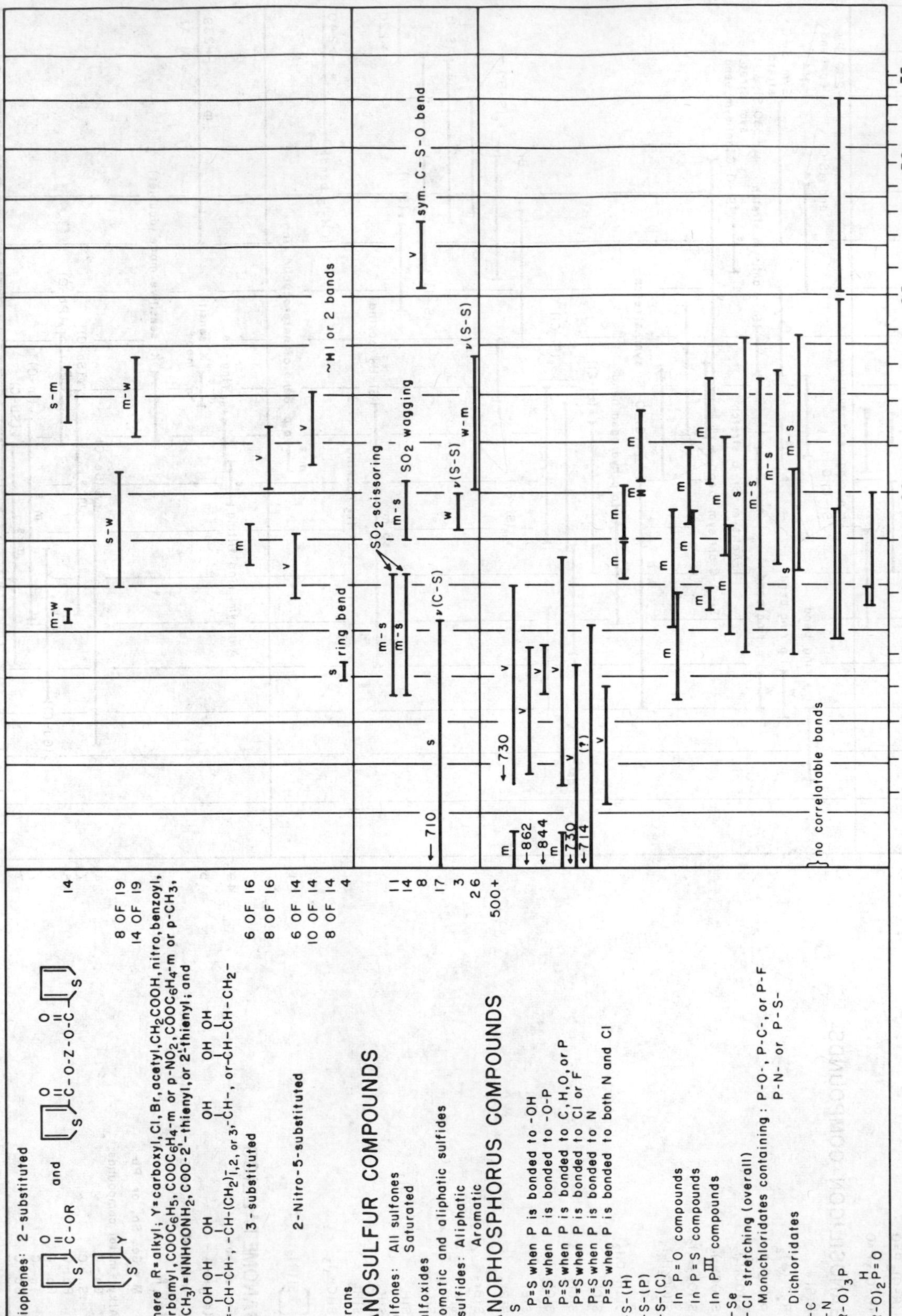

CHART OF CHARACTERISTIC FREQUENCIES BETWEEN ~ 700–300 cm^{-1}

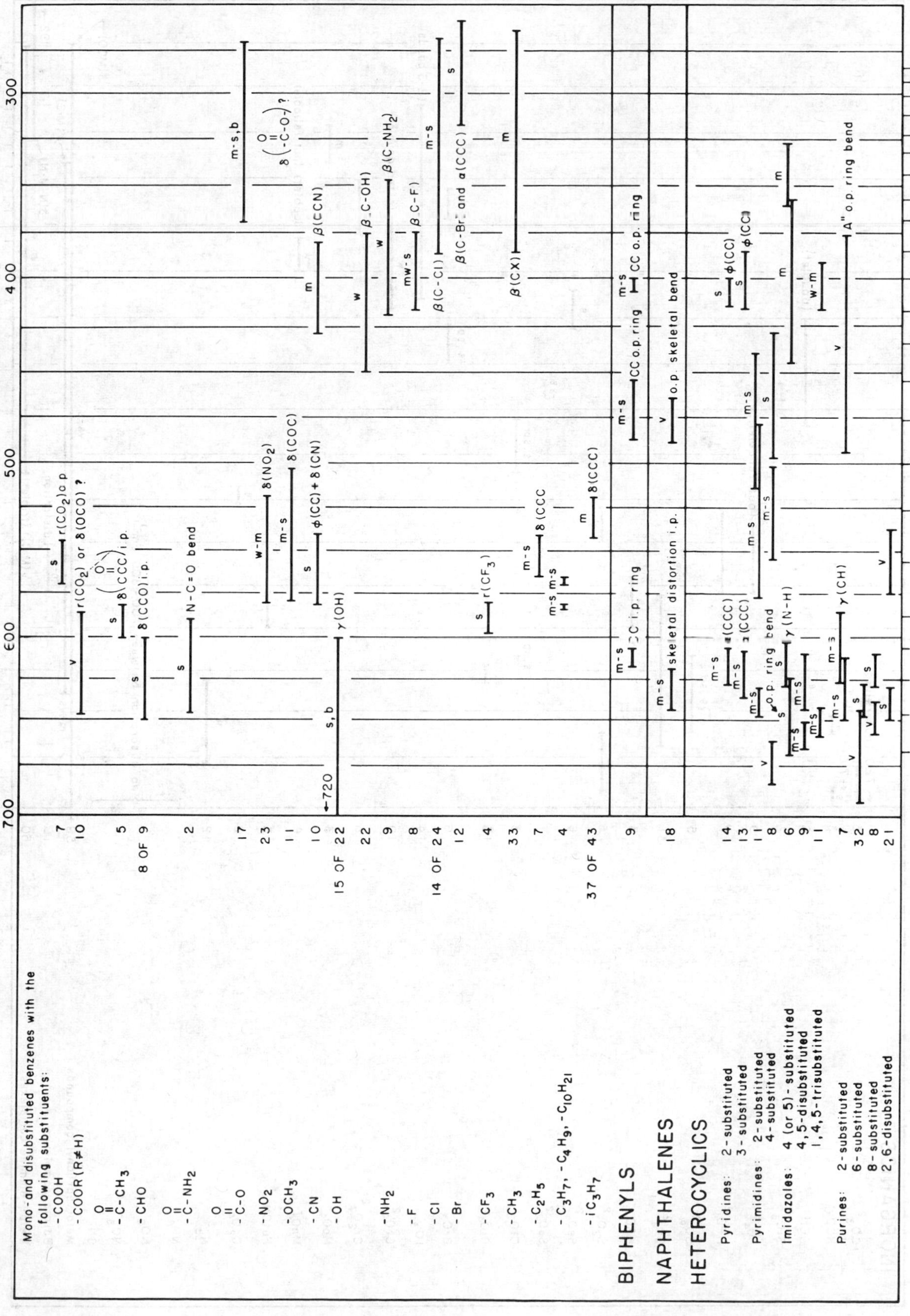

WAVENUMBER IN cm-1
WAVELENGTH IN MICRONS
Mono-and disubstituted benzenes with the following substituents:
-COOH
-COOR (R≠H)
-C-CH3
-CHO
-C-NH2
-C-O
-NO2
-OCH3
-CN
-OH
-NH2
-F
-Cl
-Br
-CF3
-CH3
-C2H5
-C3H7, -C4H9, -C10H21
-iC3H7
BIPHENYLS
NAPHTHALENES
HETEROCYCLICS
Pyridines: 2-substituted
3-substituted
Pyrimidines: 2-substituted
4-substituted
Imidazoles: 4 (or 5)-substituted
4,5-disubstituted
1,4,5-trisubstituted
Purines: 2-substituted
6-substituted
8-substituted
2,6-disubstituted
7
10
5
8 OF 9
2
17
23
11
10
15 OF 22
22
9
8
14 OF 24
12
4
33
7
4
37 OF 43
9
18
14
13
11
8
6
9
11
7
32
8
21
r(CO2) o.p.
r(CO2) or δ(OCO) ?
δ(CCC) i.p.
δ(CCO) i.p.
N-C=O bend
δ(NO2)
δ(COC)
φ(CC) + δ(CN)
γ(OH)
←720
r(CF3)
δ(CCC)
β(CCN)
β C-OH)
β(C-NH2)
β C-F)
β(C-Cl)
β(C-Br) and α(CCC)
β(CX)
CC i.p. ring
CC o.p. ring
skeletal distortion i.p.
o.p. skeletal bend
o.p. ring bend
γ(N-H)
γ(CH)
A″ o.p. ring bend

CHART OF CHARACTERISTIC FREQUENCIES BETWEEN ∽ 700–300 cm⁻¹

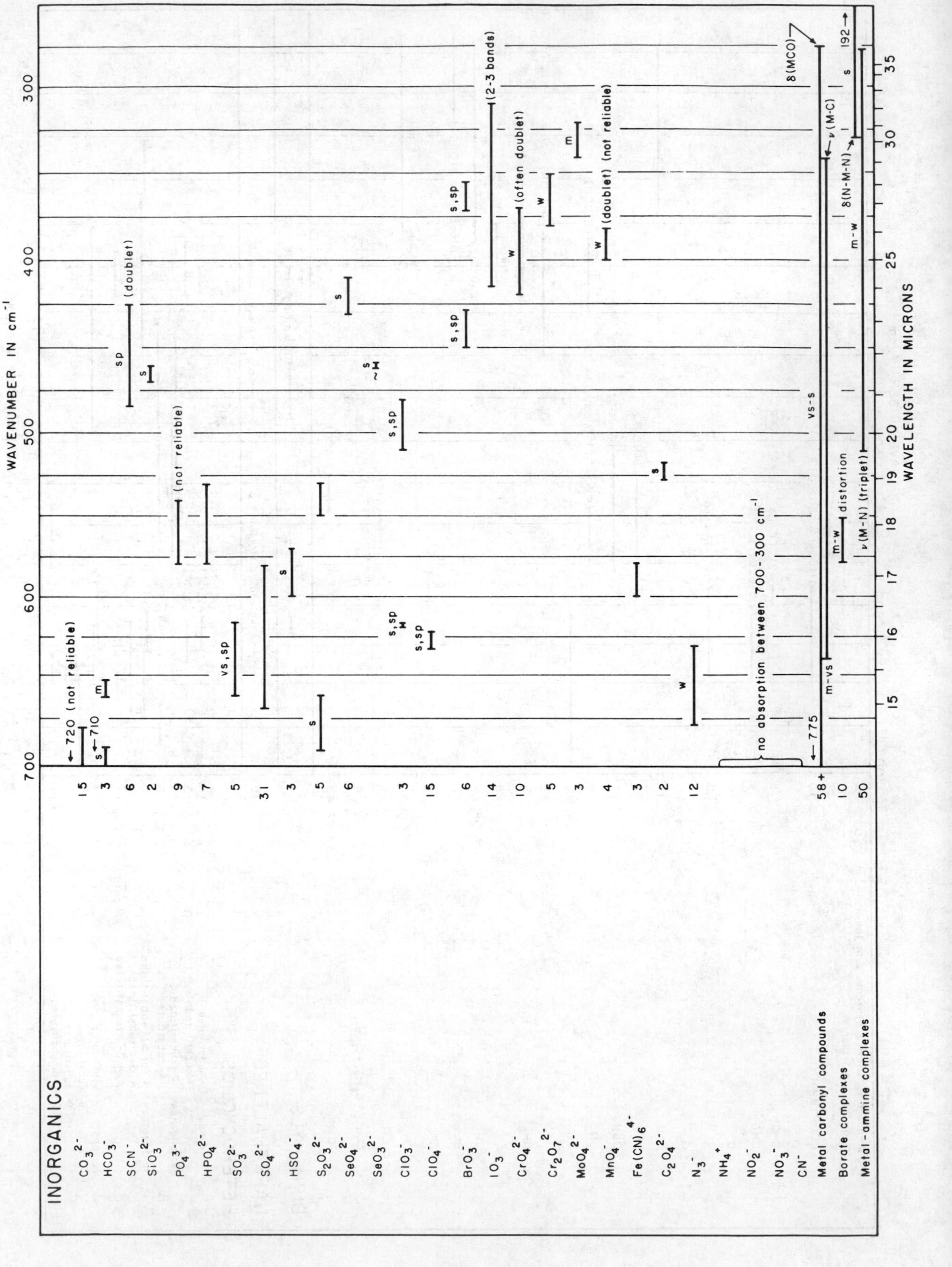

CHARACTERISTIC NMR SPECTRAL POSITIONS FOR HYDROGEN IN ORGANIC STRUCTURES

By permission from Erno Mohacsi, J. of Chemical Education, 41, 38 (1964)

This table is useful for quick qualitative determination of proton spectrum lines by providing a tabulation of line positions obtained using tetramethylsilane as an internal reference. The listing has been kept as simple as possible for this purpose. The proton spectrum lines are arranged according to the chemical shift relative to tetramethylsilane and are given in values of τ and δ. The purpose of this table is to supplement tables available in standard references and to summarize information available in the literature.

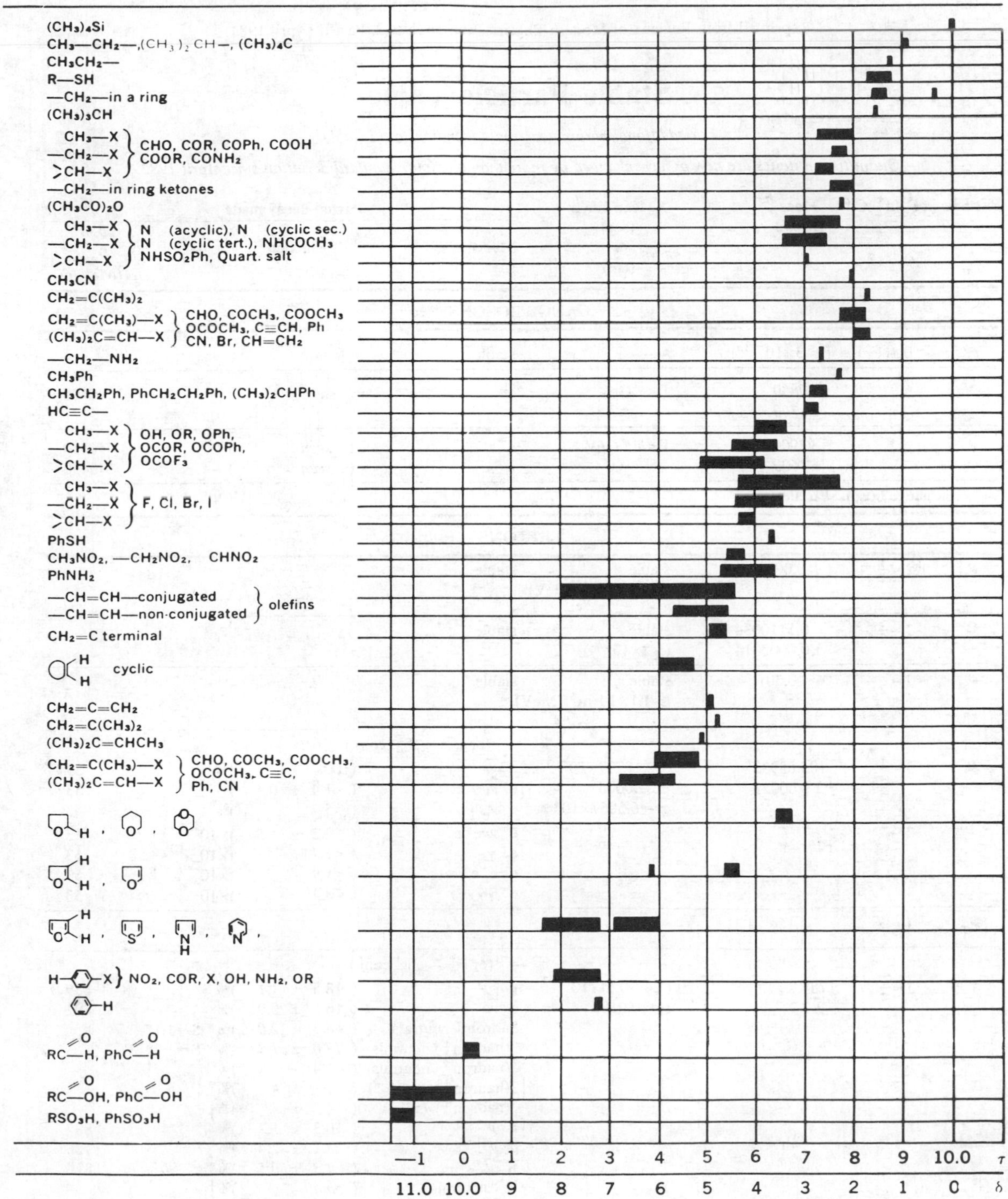

TABLES OF PARTICLE PROPERTIES

April 1984

M. Aguilar-Benitez, R.N. Cahn, R.L. Crawford, R. Frosch, G.P. Gopal, R.E. Hendrick, J.J. Hernandez, G. Höhler, M.J. Losty, L. Montanet, F.C. Porter, A. Rittenberg, M. Roos, L.D. Roper, T. Shimada, R.E. Shrock, N.A. Törnqvist, T.G. Trippe, W.P. Trower, Ch. Walck, C.G. Wohl, G.P. Yost, and B. Armstrong (Technical Associate)

(Closing date for data: Jan. 1, 1984)

Reprinted from Reviews of Modern Physics, Vol. 56, No. 2, Part II (April 1984)

Stable Particle Table

For additional parameters, see Addendum to this table.

Quantities in italics are new or have changed by more than one (old) standard deviation since April 1982.

Particle	$I^G(J^P)C$ [a]	Mass [b] (MeV)	Mean life [b] (sec) $c\tau$ (cm)	Partial decay mode: Mode	Fraction [b]	p or p_{max} [c] (MeV/c)
GAUGE BOSONS						
γ	$0,1(1^-)-$	$(<3\times10^{-33})$	-------	stable		
W		*80800*	Γ<*7* GeV	$e\nu$	(*seen*)	40400
		± *2700*				
Z		*92900*	Γ<*8.5* GeV	e^+e^-	(*seen*)	46450
		± *1600*		$\mu^+\mu^-$	(*seen*)	46450
→ weak gauge boson searches						
LEPTONS						
ν_e	$J=\frac{1}{2}$	(<0.000046) [d]	stable	stable		
			$(>3\times10^8 m_{\nu_e}$ (MeV))			
e	$J=\frac{1}{2}$	0.5110034	stable	stable		
		±0.0000014	$(>2\times10^{22}$y)			
ν_μ	$J=\frac{1}{2}$	0(<0.50)	stable	stable		
			$(>1.1\times10^5 m_{\nu_\mu}$ (MeV))			
				μ^- → (or μ^+ →chg. conj.)		
μ	$J=\frac{1}{2}$	105.65932	2.19709×10^{-6}	$e^-\bar{\nu}\nu$	(100)%	53
		±0.00029	±0.00005	†[$e^-\bar{\nu}\nu\gamma$	[e](1.4 ± 0.4)%	53
			$c\tau=6.5867\times10^4$	$e^-\nu_e\bar{\nu}_\mu$	(<5)%]	53
				$e^-\bar{\nu}\nu e^+e^-$	[e](2.2 ± 1.5)$\times10^{-5}$	53
				$e^-\gamma$	(<1.7)$\times10^{-10}$	53
				$e^-e^+e^-$	(<1.9)$\times10^{-9}$	53
				$e^-\gamma\gamma$	(<8.4)$\times10^{-9}$	53
ν_τ	$J=\frac{1}{2}$	< 164				
				τ^- → (or τ^+ →chg. conj.)		
τ	$J=\frac{1}{2}$	1784.2	$(3.4\pm0.5)\times10^{-13}$	$\mu^-\bar{\nu}\nu$	(18.5 ± 1.1)%	889
		±3.2	$c\tau=0.010$	$e^-\bar{\nu}\nu$	(16.5 ± 0.9)%	892
				hadron$^-$ neutrals	(*48.1* ± *2.0*)% S=1.1*	
				3(hadron$^\pm$) neutrals	(*17.0* ± *1.3*)% S=1.2*	
				5(hadron$^\pm$) neutrals	(<1.4)%	
				†[3(hadron$^\pm$)ν	(5 ± 4)%	
				3(hadron$^\pm$)$\nu(\geq1\gamma)$	(12 ± 4)%]	
				†[$\pi^-\nu$	(10.3 ± 1.2)%	887
				$\rho^-\nu$	(22.1 ± 2.4)%	726
				$K^-\nu$	(*1.3* ± *0.5*)%	824
				K^- neutrals	(*small*)%]	

(continued next page)

Stable Particle Table *(cont'd)*

Particle	$I^G(J^P)C$ [a]	Mass [b] (MeV)	Mean life [b] (sec) cτ (cm)	Mode	Fraction [b]		p or p_{max} [c] (MeV/c)
				τ^- → (or τ^+ → chg. conj.)			
τ (continued)				†[$K^{*-}(892)\nu$	(1.7 ± 0.7)%		669
				$K^{*-}(1430)\nu$	(<0.9)%		323
				$\pi^-\rho^0\nu$	(5.4 ± 1.7)%]		718
				e^- chgd.parts. + μ^- chgd.parts.	(<4)%		
				$\mu^-\gamma$	(<5.5)×10^{-4}		889
				$e^-\gamma$	(<6.4)×10^{-4}		892
				$\mu^-\mu^+\mu^-$	(<4.9)×10^{-4}		876
				$e^-\mu^+\mu^-$	(<3.3)×10^{-4}		886
				$\mu^-e^+e^-$	(<4.4)×10^{-4}		889
				$e^-e^+e^-$	(<4.0)×10^{-4}		892
				$\mu^-\pi^0$	(<8.2)×10^{-4}		884
				$e^-\pi^0$	(<2.1)×10^{-3}		887
				μ^-K^0	(<1.0)×10^{-3}		819
				e^-K^0	(<1.3)×10^{-3}		823
				$\mu^-\rho^0$	(<4.4)×10^{-4}		722
				$e^-\rho^0$	(<3.7)×10^{-4}		726

→ searches for massive neutrinos and lepton mixing
→ ν bounds from astrophysics and cosmology
→ heavy lepton searches

NONSTRANGE MESONS [a]

Particle	$I^G(J^P)C$	Mass (MeV)	Mean life (sec) cτ (cm)	Mode	Fraction		p or p_{max} (MeV/c)
				π^+ → (or π^- → chg. conj.)			
$\pi^{\pm}$	$1^-(0^-)$	139.5673	2.6030×10^{-8}	$\mu^+\nu$	100%		30
		±0.0007	±0.0023	$e^+\nu$	(*1.232±0.024*)×10^{-4}	S=2.0*	70
			cτ=780.4	†[$\mu^+\nu\gamma$	[e](1.24± 0.25)×10^{-4}		30
	$m_{\pi^\pm}-m_{\mu^\pm}$	=33.9080		$e^+\nu\gamma$	[e](5.6 ± 0.7)×10^{-8}]		70
		±0.0008		$e^+\nu\pi^0$	(1.033±0.034)×10^{-8}		5
				$e^+\nu e^+e^-$	(<5)×10^{-9}		70
				$\mu^+\overline{\nu}_e$	(<*1.5*)×10^{-3}		30
				$\mu^+\nu_e$	(<*8*)×10^{-3}		30
π^0	$1^-(0^-)+$	134.9630	0.83×10^{-16}	$\gamma\gamma$	(98.802 ±0.030)%		67
		±0.0038	±0.06 S=1.8*	γe^+e^-	(1.198)%		67
			cτ=2.5×10^{-6}	$\gamma\gamma\gamma$	(<3.8)×10^{-7}		67
	$m_{\pi^\pm}-m_{\pi^0}$	=4.6043		$e^+e^-e^+e^-$	[f](3.24)×10^{-5}		67
		±0.0037		$\gamma\gamma\gamma\gamma$	(<4)×10^{-6}		67
				e^+e^-	(1.8 ± 0.7)×10^{-7}		67
				$\nu\nu$	(<2.4)×10^{-5}		67
				$\mu^+e^-+\mu^-e^+$	(<*7*)×10^{-8}		26
η	$0^+(0^-)+$	548.8	Γ=(0.88±0.12)keV	$\gamma\gamma$	(39.0 ± 0.8)%		274
		±0.6	Neutral decays	$3\pi^0$	(31.8 ± 0.8)%	S=1.1*	180
		S=1.4*	(70.9±0.7)%	$\pi^0\gamma\gamma$	(0.10 ± 0.02)%		258
				$\pi^+\pi^-\pi^0$	(23.7 ± 0.5)%		175
				$\pi^+\pi^-\gamma$	(4.91± 0.13)%		236
				$e^+e^-\gamma$	(0.50± 0.12)%		274
				$\mu^+\mu^-\gamma$	(3.1 ± 0.4)×10^{-4}		253
				e^+e^-	(<3)×10^{-4}		274
				$\mu^+\mu^-$	(6.5 ± 2.1)×10^{-6}		253
			Charged decays	$\pi^+\pi^-e^+e^-$	(0.13± 0.13)%		236
			(29.1±0.7)%	$\pi^+\pi^-\gamma\gamma$	(<0.21)%		236
				$\pi^+\pi^-\pi^0\gamma$	(<6)×10^{-4}		175
				$\pi^+\pi^-$	(<0.15)%		236
				$\pi^0e^+e^-$	(<5)×10^{-5}		258
				$\pi^0\mu^+\mu^-$	(<5)×10^{-6}		211
				$\pi^0\mu^+\mu^-\gamma$	(<3)×10^{-6}		211

Stable Particle Table *(cont'd)*

Particle	$I^G(J^P)$ [a]	Mass [b] (MeV)	Mean life [b] (sec) $c\tau$ (cm)	Mode	Fraction [b]		p or p_{max} [c] (MeV/c)
				STRANGE MESONS [a]			
				$K^+ \rightarrow$ (or $K^- \rightarrow$ chg. conj.)			
$K^{\pm}$	$\frac{1}{2}(0^-)$	493.667	1.2371×10^{-8}	$\mu^+\nu$	$(63.51 \pm 0.16)\%$		236
		± 0.015	± 0.0026 S=1.9*	$\pi^+\pi^0$	$(21.17 \pm 0.15)\%$		205
			$c\tau = 370.9$	$\pi^+\pi^+\pi^-$	$(5.59 \pm 0.03)\%$	S=1.1*	125
				$\pi^+\pi^0\pi^0$	$(1.73 \pm 0.05)\%$	S=1.4*	133
				$\pi^0\mu^+\nu$	$(3.18 \pm 0.10)\%$	S=1.9*	215
		$m_{K^\pm} - m_{K^0} = -4.01$		$\pi^0 e^+\nu$	$(4.82 \pm 0.05)\%$	S=1.1*	228
		± 0.13		†[$\mu^+\nu\gamma$	[e] $(5.8 \pm 3.5) \times 10^{-3}$		236
		S=1.1*		$\pi^+\pi^0\gamma$	[g,e] $(2.75 \pm 0.16) \times 10^{-4}$		205
				$\pi^+\pi^+\pi^-\gamma$	[e] $(1.0 \pm 0.4) \times 10^{-4}$		125
				$\pi^0\mu^+\nu\gamma$	[e] $(<6) \times 10^{-5}$		215
				$\pi^0 e^+\nu\gamma$	[e] $(3.7 \pm 1.4) \times 10^{-4}$]		228
				$\pi^0\pi^0 e^+\nu$	$(1.8 \, ^{+2.4}_{-0.6}) \times 10^{-5}$		207
				$\pi^+\pi^- e^+\nu$	$(3.90 \pm 0.15) \times 10^{-5}$		203
				$\pi^+\pi^+ e^-\bar{\nu}$	$(<1.2) \times 10^{-8}$		203
				$\pi^+\pi^-\mu^+\nu$	$(1.4 \pm 0.9) \times 10^{-5}$		151
				$\pi^+\pi^+\mu^-\bar{\nu}$	$(<3.0) \times 10^{-6}$		151
				$e^+\nu$	$(1.54 \pm 0.07) \times 10^{-5}$		247
				$e^+\nu\gamma$ (SD+) [h]	$(1.52 \pm 0.23) \times 10^{-5}$		247
				$e^+\nu\gamma$ (SD−) [h]	$(<1.6) \times 10^{-4}$		247
				$\pi^+ e^+ e^-$	$(2.7 \pm 0.5) \times 10^{-7}$		227
				$\pi^- e^+ e^+$	$(<1) \times 10^{-8}$		227
				$\pi^+\mu^+\mu^-$	$(<2.4) \times 10^{-6}$		172
				$\pi^+\gamma\gamma$	[e] $(<8) \times 10^{-6}$		227
				$\pi^+\gamma\gamma\gamma$	[e] $(<1.0) \times 10^{-4}$		227
				$\pi^+\nu\bar{\nu}$	$(<1.4) \times 10^{-7}$		227
				$\pi^{\mp} e^+\mu^{\pm}$	$(<7) \times 10^{-9}$		214
				$\pi^+ e^-\mu^+$	$(<5) \times 10^{-9}$		214
				$e^+\nu\nu\bar{\nu}$	$(<6) \times 10^{-5}$		247
				$\mu^+\nu\nu\bar{\nu}$	$(<6) \times 10^{-6}$		236
				$\mu^+\nu e^+ e^-$	$(11 \pm 3) \times 10^{-7}$		236
				$\mu^-\nu e^+ e^+$	$(<2.0) \times 10^{-8}$		236
				$e^+\nu e^+ e^-$	$(2 \, ^{+2}_{-1}) \times 10^{-7}$		247
				$\mu^+\nu_e$	$(<4) \times 10^{-3}$		236
				$\mu^+\bar{\nu}_e$	$(<3.3) \times 10^{-3}$		236
				$\pi^0 e^+\bar{\nu}_e$	$(<3) \times 10^{-3}$		228
K^0 / $\bar{K}^0$	$\frac{1}{2}(0^-)$	497.67					
		± 0.13		50 % K_{Short}, 50% K_{Long}			
		S=1.1*					
K^0_S	$\frac{1}{2}(0^-)$		0.8923×10^{-10}	$\pi^+\pi^-$	$(68.61$		206
			± 0.0022	$\pi^0\pi^0$	$(31.39 \ \pm 0.24)\%$	S=1.1*	209
			$c\tau = 2.675$	†[$\pi^+\pi^-\gamma$	[e] $(1.85 \pm 0.10) \times 10^{-3}$]		206
				$\mu^+\mu^-$	$(<3.2) \times 10^{-7}$		225
				e^+e^-	$(<3.4) \times 10^{-4}$		249
				$\gamma\gamma$	$(<4) \times 10^{-4}$		249
				$\pi^+\pi^-\pi^0$	$(<8.5) \times 10^{-5}$		133
				$\pi^0\pi^0\pi^0$	$(<3.7) \times 10^{-5}$		139
K^0_L	$\frac{1}{2}(0^-)$		5.183×10^{-8}	$\pi^0\pi^0\pi^0$	$(21.5 \pm 1.0)\%$	S=1.7*	139
			± 0.040	$\pi^+\pi^-\pi^0$	$(12.39 \pm 0.20)\%$	S=1.3*	133
			$c\tau = 1554$	$\pi^{\pm}\mu^{\mp}\nu$	$(27.1 \pm 0.4)\%$	S=1.4*	216
				$\pi^{\pm}e^{\mp}\nu$	$(38.7 \pm 0.5)\%$	S=1.5*	229
		$m_{K_L} - m_{K_S} = 0.5349 \times 10^{10} \ \hbar \ sec^{-1}$		$\pi^+\pi^-$	[i] $(0.203 \pm 0.005)\%$	S=1.1*	206
		± 0.0022		$\pi^0\pi^0$	[i] $(0.094 \pm 0.018)\%$	S=1.5*	209
				†[$\pi e\nu\gamma$	[e] $(1.3 \pm 0.8)\%$		229
		$= 3.521 \times 10^{-12}$ MeV		$\pi^+\pi^-\gamma$	[e] $(4.41 \pm 0.32) \times 10^{-5}$]		206
		± 0.014		$\pi^0\gamma\gamma$	$(<2.4) \times 10^{-4}$		231
				$\gamma\gamma$	$(4.9 \pm 0.4) \times 10^{-4}$		249
				$e\mu$	[j] $(<6) \times 10^{-6}$		238
				$\mu^+\mu^-$	$(9.1 \pm 1.9) \times 10^{-9}$		225
				$\mu^+\mu^-\gamma$	$(2.8 \pm 2.8) \times 10^{-7}$		225
				$\pi^0\mu^+\mu^-$	$(<1.2) \times 10^{-6}$		177

(continued next page)

Stable Particle Table *(cont'd)*

Particle	$I^G(J^P)$ [a]	Mass [b] (MeV)	Mean life [b] (sec) $c\tau$ (cm)	Mode	Fraction [b]	p or p_{max} [c] (MeV/c)
K_L^0 (continued)				e^+e^-	[j]$(<2.0)\times10^{-7}$	249
				$e^+e^-\gamma$	$(1.7 \pm 0.9)\times10^{-5}$	249
				$\pi^0e^+e^-$	$(<2.3)\times10^{-6}$	231
				$\pi^+\pi^-e^+e^-$	$(<9)\times10^{-6}$	206
				$\pi^0\pi^\pm e^\mp\nu$	$(6.2 \pm 2.0)\times10^{-5}$	207
				$(\pi\mu$ atom$)\ \nu$	$(\mathit{1.05} \pm \mathit{0.11})\times10^{-7}$	
CHARMED NONSTRANGE MESONS [a]						
				$D^+ \rightarrow$ (or $D^- \rightarrow$ chg. conj.)		
$D^\pm$	$\frac{1}{2}(0^-)$	1869.4 ±0.6	$(9.2^{+1.7}_{-1.2})\times10^{-13}$ $c\tau=0.028$	e^+ anything	$(19\ ^{+4}_{-3})\%$	
				K^- anything	$(16 \pm 4)\%$	
				$\overline{K}^0$ any + K^0 any	$(48 \pm 15)\%$	
				K^+ anything	$(6.0 \pm 3.3)\%$	
		$m_{D^\pm} - m_{D^0} = 4.7$ ±0.3		η anything	[k]$(<13)\%$	
				$\mu^+\nu$	$(\mathit{<2})\%$	932
				†[$K^-\pi^+\pi^+$	$(4.6 \pm 1.1)\%$ S=1.3*	845
				$K^-\pi^+\pi^+\pi^0$	$(\mathit{2.6}\ ^{+\mathit{3.1}}_{-\mathit{1.0}})\%$	816
				$K^-\pi^+\pi^+\pi^+\pi^-$	$(<4)\%$	772
				$\overline{K}^0\pi^+$	$(1.8 \pm 0.5)\%$	862
				$\overline{K}^0\pi^+\pi^0$	$(13 \pm 8)\%$	845
				$\overline{K}^0\pi^+\pi^+\pi^-$	$(8.4 \pm 3.5)\%$	814
				$\overline{K}^0K^+$	$(0.45 \pm 0.30)\%$	792
				$K^+K^-\pi^+$	$(<0.6)\%$	744
				$K^+\pi^+\pi^-$	$(<0.23)\%$	845
				$\pi^+\pi^0$	$(<0.5)\%$	925
				$\pi^+\pi^+\pi^-$	$(<0.4)\%$]	908
				†[$\overline{K}^{*0}\pi^+$	$(<3.7)\%$]	714
				$D^0 \rightarrow$ (or $\overline{D}^0 \rightarrow$ chg. conj.)		
D^0 / $\overline{D}^0$	$\frac{1}{2}(0^-)$	1864.7 ±0.6	$(4.4\ ^{+0.8}_{-0.6})\times10^{-13}$ $c\tau=0.013$	e^+ anything	$(5.3\ ^{+2.9}_{-1.3})\%$	
				K^- anything	$(44 \pm 10)\%$ S=1.3*	
				$\overline{K}^0$ any + K^0 any	$(33 \pm 10)\%$	
	$\lvert m_{D_1^0} - m_{D_2^0}\rvert < \mathit{6.5\times10^{-10}}$ MeV [ℓ]			K^+ anything	$(8 \pm 3)\%$	
				η anything	[k]$(<13)\%$	
				†[$K^-\pi^+$	$(2.4 \pm 0.4)\%$	861
				$K^-\pi^+\pi^0$	$(9.3 \pm 2.8)\%$	844
	$\frac{\lvert \tau_{D_1^0} - \tau_{D_2^0}\rvert}{\text{average}} < \mathit{0.55}$ [ℓ]			$K^-\pi^+\pi^+\pi^-$	$(4.6 \pm 1.4)\%$ S=1.2*	812
				$K^-\pi^+\pi^0\pi^0$	(seen)	815
				$\overline{K}^0\pi^0$	$(2.2 \pm 1.1)\%$	860
				$\overline{K}^0\pi^+\pi^-$	$(4.2 \pm 0.8)\%$	842
	$\frac{\Gamma(D^0\rightarrow\overline{D}^0\rightarrow K^+\pi^-)}{\Gamma(D^0\rightarrow K\pi)} < 0.16$			$\pi^+\pi^-$	$(7.9 \pm 3.8)\times10^{-4}$	922
				$\pi^+\pi^+\pi^-\pi^-$	$(<1.0)\%$	880
				K^+K^-	$(2.7 \pm 0.8)\times10^{-3}$]	791
				†[$K^{*-}\pi^+$	$(3.4 \pm 1.4)\%$	711
	$\frac{\Gamma(D^0\rightarrow\overline{D}^0\rightarrow\mu^-\text{ anything})}{\Gamma(D^0\rightarrow\mu^\pm\text{ anything})} < \mathit{0.044}$			$\overline{K}^{*0}\pi^0$	$(1.4\ ^{+2.3}_{-1.4})\%$	711
				$K^-\rho^+$	$(7.2\ ^{+3.0}_{-3.1})\%$	679
				$\overline{K}^0\rho^0$	$(0.1\ ^{+0.6}_{-0.1})\%$	677
				$\overline{K}^{*0}\rho^0$	$(0.7\ ^{+0.8}_{-0.7})\%$	423
				$K^-\pi^+\rho^0$	$(3.9\ ^{+1.3}_{-1.6})\%$	613
				$\overline{K}^{*0}\pi^+\pi^-$	$(<2.3)\%$	685
				$K^-A_2^+$	$(<0.8)\%$]	198
CHARMED STRANGE MESON [a]						
				$F^+ \rightarrow$ (or $F^- \rightarrow$ chg. conj.)		
$F^\pm$	$0(0^-)$ [m]	1971 [m] ±6	$(1.9^{+1.3}_{-0.7})\times10^{-13}$ $c\tau=0.006$	$\phi\pi^+$	(*seen*)	713
				$\eta\pi^+$	(*possibly seen*)	903
				$\eta\pi^+\pi^+\pi^-$	(*possibly seen*)	857
				$\eta'\pi^+\pi^+\pi^-$	(*possibly seen*)	679
				$\phi\rho^+$	(*possibly seen*)	411

Stable Particle Table *(cont'd)*

Particle	$I^G(J^P)$ [a]	Mass [b] (MeV)	Mean life [b] (sec) $c\tau$ (cm)	Partial decay mode: Mode	Partial decay mode: Fraction [b]	p or p_{max} [c] (MeV/c)
BOTTOM MESONS [a]						
				$B^+ \to$ (or $B^- \to$ chg. conj.)		
$B^\pm$	$\frac{1}{2}(0^-)$ [n]	5270.8		$\overline{D}^0\pi^+$	(4.2 ± 4.2)%	2303
		±3.0		$D^{*-}\pi^+\pi^+$	(4.8 ± 3.0)%	2243
				$B^0 \to$ (or $\overline{B}^0 \to$ chg. conj.)		
B^0 $\overline{B}^0$	$\frac{1}{2}(0^-)$ [n]	5274.2		$\overline{D}^0\pi^+\pi^-$	(13 ± 9)%	2298
		±2.8		$D^{*-}\pi^+$	(2.6 ± 1.9)%	2253
$B^\pm$, B^0, $\overline{B}^0$			$(14\pm4)\times10^{-13}$	$e^\pm\nu$ hadrons	(13.0 ± 1.3)%	
(not separated) [p]			$c\tau$=0.042	$\mu^\pm\nu$ hadrons	(12.4 ± 3.5)%	
				D^0 anything	(80 ± 28)%	
				K anything	(seen)	
				p anything	(> 3.6)%	
				Λ anything	(> 2.2)%	
				e^+e^- anything	(< 0.8)%	
				$\mu^+\mu^-$ anything	(< 0.7)%	
NONSTRANGE BARYONS [a]						
p	$\frac{1}{2}(\frac{1}{2}^+)$	938.2796	stable ($>10^{32}$y) [q]	stable		
		±0.0027				
				$\lvert q_p\rvert - \lvert q_e\rvert < 10^{-21}\lvert q_e\rvert$ [r]		
n	$\frac{1}{2}(\frac{1}{2}^+)$	939.5731	898±16	$pe^-\overline{\nu}$	100%	1.2
		±0.0027	$c\tau$=2.7×10^13	$p\nu\overline{\nu}$ (chg.noncons.)	(<9)×10^-24	1.3
	$m_p - m_n$ = *−1.293323*					
	±0.000016			$\lvert q_n\rvert < 10^{-21}\lvert q_e\rvert$ [r]		
STRANGENESS −1 BARYONS [a]						
Λ	$0(\frac{1}{2}^+)$	1115.60	2.632×10^{-10}	$p\pi^-$	(64.2)% ± 0.5	100
		±0.05	±0.020 S=1.6*	$n\pi^0$	(35.8)% ± 0.5	104
		S=1.2*	$c\tau$ =7.89	$pe^-\nu$	(8.37 ± 0.14)×10^-4	163
				$p\mu^-\nu$	(1.57 ± 0.35)×10^-4	131
	$m_\Lambda - m_{\Sigma^0}$ = −76.86			†[$p\pi^-\gamma$	[e](8.5 ± 1.4)×10^-4]	100
	±0.08					
Σ^+	$1(\frac{1}{2}^+)$	1189.36	0.800×10^{-10}	$p\pi^0$	(51.64)% ± 0.30	189
		±0.06	±0.004	$n\pi^+$	(48.36)% ± 0.30	185
		S=1.8*	$c\tau$=2.40	$p\gamma$	(1.20 ± 0.13)×10^-3 S=1.2*	225
				†[$n\pi^+\gamma$	[e](4.5 ± 0.5)×10^-4]	185
				$\Lambda e^+\nu$	(2.0 ± 0.5)×10^-5	71
	$m_{\Sigma^+} - m_{\Sigma^-}$ = −7.97		$\frac{\Gamma(\Sigma^+\to\ell^+n\nu)}{\Gamma(\Sigma^-\to\ell^-n\nu)} < .04$	$n\mu^+\nu$	(<3.0)×10^-5	202
	±0.07			$ne^+\nu$	(<5)×10^-6	224
	S=1.3*			pe^+e^-	(<7)×10^-6	225
Σ^0	$1(\frac{1}{2}^+)$ [s]	1192.46	5.8×10^{-20}	$\Lambda\gamma$	100%	74
		±0.08	±1.3	Λe^+e^-	[f](5.45)×10^-3	74
			$c\tau$=1.7×10^-9	$\Lambda\gamma\gamma$	(<3)%	74
Σ^-	$1(\frac{1}{2}^+)$	1197.34	1.482×10^{-10}	$n\pi^-$	100%	193
		±0.05	±0.011 S=1.3*	$ne^-\nu$	(*1.022±0.034*)×10^-3	230
			$c\tau$=4.44	$n\mu^-\nu$	(4.5 ± 0.4)×10^-4	210
				$\Lambda e^-\nu$	(5.74 ± 0.27)×10^-5	79
	$m_{\Sigma^0} - m_{\Sigma^-}$ = −4.88			†[$n\pi^-\gamma$	[e](4.6 ± 0.6)×10^-4]	193
	±0.06					

Stable Particle Table *(cont'd)*

Particle	$I^G(J^P)$ [a]	Mass [b] (MeV)	Mean life [b] (sec) $c\tau$ (cm)	Partial decay mode: Mode	Fraction [b]	p or p_{max} [c] (MeV/c)
STRANGENESS −2 BARYONS [a]						
Ξ^0	$\frac{1}{2}(\frac{1}{2}^+)$ [s]	1314.9	2.90×10^{-10}	$\Lambda\pi^0$	100%	135
		±0.6	±0.10	$\Lambda\gamma$	(0.5 ± 0.5)%	184
			$c\tau$=8.69	$\Sigma^0\gamma$	(<7)%	117
				$p\pi^-$	(<3.6)$\times10^{-5}$	299
				$pe^-\nu$	(<1.3)$\times10^{-3}$	323
	$m_{\Xi^0}-m_{\Xi^-}=-6.4$			$\Sigma^+e^-\nu$	(<1.1)$\times10^{-3}$	120
	±0.6			$\Sigma^-e^+\nu$	(<0.9)$\times10^{-3}$	112
				$\Sigma^+\mu^-\nu$	(<1.1)$\times10^{-3}$	65
				$\Sigma^-\mu^+\nu$	(<0.9)$\times10^{-3}$	49
				$p\mu^-\nu$	(<1.3)$\times10^{-3}$	309
Ξ^-	$\frac{1}{2}(\frac{1}{2}^+)$ [s]	1321.32	1.641×10^{-10}	$\Lambda\pi^-$	100%	139
		±0.13	±0.016	$\Lambda e^-\nu$	(*5.5 ± 0.6*)$\times10^{-4}$ S=2.0*	190
			$c\tau$=4.92	$\Sigma^0e^-\nu$	(8.7 ± 1.7)$\times10^{-5}$	123
				$\Lambda\mu^-\nu$	(3.5 ± 3.5)$\times10^{-4}$	163
				$\Sigma^0\mu^-\nu$	(<8)$\times10^{-4}$	70
				$n\pi^-$	(<1.9)$\times10^{-5}$	303
				$ne^-\nu$	(<3.2)$\times10^{-3}$	327
				$n\mu^-\nu$	(<1.5)%	313
				$\Sigma^-\gamma$	(<1.2)$\times10^{-3}$	118
				$p\pi^-\pi^-$	(<4)$\times10^{-4}$	223
				$p\pi^-e^-\nu$	(<4)$\times10^{-4}$	304
				$p\pi^-\mu^-\nu$	(<4)$\times10^{-4}$	250
				$\Xi^0e^-\nu$	(<2.3)$\times10^{-3}$	6
STRANGENESS −3 BARYON [a]						
Ω^-	$0(\frac{3}{2}^+)$ [s]	1672.45	0.819×10^{-10}	ΛK^-	(68.6 ± 1.3)%	211
		±0.32	±0.027	$\Xi^0\pi^-$	(23.4 ± 1.3)%	294
			$c\tau$=2.46	$\Xi^-\pi^0$	(8.0 ± 0.8)%	290
				$\Xi^0e^-\nu$	(~1)%	319
				$\Xi^0(1530)\pi^-$	(~2)$\times10^{-3}$	
				$\Lambda\pi^-$	(<1.3)$\times10^{-3}$	449
				$\Xi^-\gamma$	(<3.1)$\times10^{-3}$	314
NONSTRANGE CHARMED BARYON [a]						
Λ_c^+	$0(\frac{1}{2}^+)$ [s]	2282.0	$(2.3^{+1.0}_{-0.6})\times10^{-13}$	$pK^-\pi^+$	(2.2 ± 1.0)%	820
		±3.1	$c\tau$=0.007	$p\overline{K}^0$	(1.1 ± 0.7)%	870
		S=1.8*		$p\overline{K}^0\pi^+\pi^-$	(<4, seen)%	751
				Λ anything	(33 ± 29)%	
				†[$\Lambda\pi^+$	(0.6 ± 0.5)%	861
				$\Lambda\pi^+\pi^+\pi^-$	(< 3.1, seen)%	804
				$\Sigma^0\pi^+$	(seen)]	822
				†[pK^{*0}	(0.48 ± 0.30)%	681
				$\Delta^{++}K^-$	(0.45 ± 0.27)%	706
				$pK^{*-}\pi^+$	(seen)]	575
				e^+ anything	(4.5 ± 1.7)%	
				†[pe^+ anything	(1.8 ± 0.9)%	
				Λe^+ anything	(1.1 ± 0.8)%]	

→ A^+

→ Λ_b^0

→ top hadron searches
→ free quark searches
→ magnetic monopole searches
→ axion searches
→ other stable particle searches

ADDENDUM TO

Stable Particle Table

e [t]

Magnetic Moment

$1.001\ 159\ 652\ 209 \pm .000\ 000\ 000\ 031 \ \frac{e\hbar}{2m_e c}$

μ [t]

$1.001\ 165\ 924 \pm .000\ 000\ 009 \ \frac{e\hbar}{2m_\mu c}$

μ Decay parameters [u]

$\rho = 0.752 \pm 0.003$	$\eta = -0.06 \pm 0.15$ S=1.1*	
$\xi \cdot P_\mu > 0.9959$ [v]	$\delta = 0.755 \pm 0.009$	$h = 1.01 \pm 0.06$
$\alpha' = -0.12 \pm 0.10$	$\beta' = 0.029 \pm 0.037$	$\bar{\eta} = 0.006 \pm 0.080$
$\lvert g_A/g_V \rvert = 0.91^{+0.24}_{-0.06}$	$\phi_{AV} = 180° \pm 9°$	
$\lvert g_S/g_V \rvert < 0.29$	$\lvert g_T/g_V \rvert < 0.14$	$\lvert g_P/g_V \rvert < 0.25$

τ

Michel parameter

$\rho = 0.72 \pm 0.15$

η

Mode	Left-right asymmetry	Sextant asymmetry	Quadrant asymmetry	
$\pi^+\pi^-\pi^0$	$(0.12 \pm .17)\%$	$(0.19 \pm 0.16)\%$	$(-0.17 \pm 0.17)\%$	
$\pi^+\pi^-\gamma$	$(0.88 \pm .40)\%$			$\beta = 0.047 \pm 0.062$ S=1.5*

K

Slope parameters for $K \to 3\pi$ [w]

$K^+ \to \pi^+\pi^+\pi^-$	$g = -0.215 \pm .004$	S=1.4*
$K^- \to \pi^-\pi^-\pi^+$	$g = -0.217 \pm .007$	S=2.5*
$K^\pm \to \pi^0\pi^0\pi^\pm$	$g = 0.607 \pm .030$	S=1.3*
$K^0_L \to \pi^+\pi^-\pi^0$	$g = 0.670 \pm .014$	S=1.6*

See Data Card Listings for quadratic coefficients.

$\Delta S = -\Delta Q$ in $K^0_{\ell 3}$ decay

Re x = $0.009 \pm .020$ S=1.4*

Im x = $-0.004 \pm .026$ S=1.1*

Form factors for $K_{\ell 3}$ decays [x]

K^+_{e3}: $\lambda_+ = 0.029 \pm 0.004$; $\lvert f_S/f_+ \rvert = 0.125 \pm 0.044$; $\lvert f_T/f_+ \rvert = 0.22 \pm 0.14$

K^0_{e3}: $\lambda_+ = 0.0300 \pm 0.0016$ S=1.2*; $\lvert f_S/f_+ \rvert < 0.04$; $\lvert f_T/f_+ \rvert < 0.23$

$K^+_{\mu 3}$: $\lambda_+ = 0.032 \pm 0.008$ S=2.3*; $\lambda_0 = 0.004 \pm 0.007$ S=2.3*; $\lvert f_T/f_+ \rvert = 0.02 \pm 0.12$

$K^0_{\mu 3}$: $\lambda_+ = 0.034 \pm 0.005$ S=2.3*; $\lambda_0 = 0.025 \pm 0.006$ S=2.3*; $\lvert f_T/f_+ \rvert = 0.12 \pm 0.12$

CP-violation parameters [y,i]

$\lvert \eta_{+-} \rvert = (2.274 \pm .022) \times 10^{-3}$		$\lvert \eta_{00} \rvert = (2.33 \pm .08) \times 10^{-3}$ S=1.1*
$\phi_{+-} = (44.6 \pm 1.2)°$	$\phi_{00} = (54 \pm 5)°$	Re $\epsilon = (1.621 \pm 0.088) \times 10^{-3}$
$\lvert \eta_{+-0} \rvert^2 < 0.12$	$\lvert \eta_{000} \rvert^2 < 0.1$	$\delta = (0.330 \pm .012)\%$

Baryons

	Magnetic moment ($e\hbar/2m_p c$)	Mode	Decay parameters [z] Measured α	Measured ϕ(degree)	Derived γ	Derived Δ(degree)	Coupling Constant Ratios
p [t]	2.7928456 ±.0000011						
n [t]	−1.91304184 ±.00000088	$pe^-\nu$					$g_A/g_V = -1.254 \pm 0.006$; $\phi_{AV} = (180.11 \pm 0.17)°$
Λ [t]	−0.613 ±.004	$p\pi^-$	0.642 ± 0.013	$(-6.5 \pm 3.5)°$	0.76	$(7.7 \pm 4.1)°$	
		$n\pi^0$	0.646 ± 0.044				
		$pe\nu$					$g_A/g_V = -0.694 \pm 0.025$ S=1.3*
Σ^+	2.379 ±.020	$p\pi^0$	-0.979 ± 0.016	$(36 \pm 34)°$	0.17	$(187 \pm 6)°$	
		$n\pi^+$	$+0.068 \pm 0.013$	$(167 \pm 20)°$	−0.97	$(-73^{+134}_{-10})°$	
		$p\gamma$	-0.72 ± 0.29	S=1.1*			
Σ^-	−1.10 ±.05	$n\pi^-$	-0.068 ± 0.008	$(10 \pm 15)°$	0.98	$(249^{+12}_{-116})°$	
		$ne^-\nu$					$\lvert g_A/g_V \rvert = 0.372 \pm 0.050$ S=1.9*
	S=1.5*	$\Lambda e^-\nu$					$g_V/g_A = 0.01 \pm 0.10$ S=1.5*; $g_{WM}/g_A = 2.4 \pm 1.7$
Ξ^0	−1.250 ±.014	$\Lambda\pi^0$	-0.413 ± 0.022 S=2.0*	$(21 \pm 12)°$	0.85	$(218^{+12}_{-18})°$	
Ξ^-	−1.85 ±.75	$\Lambda\pi^-$	-0.434 ± 0.015 S=1.4*	$(2 \pm 6)°$ S=1.1*	0.90	$(184 \pm 12)°$	
		$\Lambda e^-\nu$					$g_A/g_V = -0.25 \pm 0.05$
Ω^-		ΛK^-	-0.10 ± 0.38 S=1.2*				

Stable Particle Table *(cont'd)*

→ Indicates an entry in the Stable Particle Data Card Listings not entered in the Stable Particle Table.

* S = Scale factor = $\sqrt{\chi^2/(N-1)}$, where N ≈ number of experiments. S should be ≈1. If S > 1, we have enlarged the error of the mean, $\delta\bar{x}$; i.e., $\delta\bar{x} \rightarrow S\delta\bar{x}$. This convention is still inadequate, since if S >> 1 the experiments are probably inconsistent, and therefore the real uncertainty is probably even greater than $S\delta x$. See the Introduction, and ideograms in Stable Particle Data Card Listings.

† Square brackets indicate subreactions of some previous unbracketed decay mode(s). Reactions in one set of brackets may overlap with reactions in another set of brackets. A radiative mode such as $\pi \rightarrow \mu\nu\gamma$ is a subreaction of its parent mode $\pi \rightarrow \mu\nu$.

a. The strangeness *S*, charm *C*, and bottomness (beauty) *B* of the hadrons which appear in the Table are as follows:

Mesons	*S*	*C*	*B*	Mesons	*S*	*C*	*B*	Baryons	*S*	*C*	*B*
π,η	0	0	0	F^+	+1	+1	0	p,n	0	0	0
K^+,K^0	+1	0	0	F^-	−1	−1	0	Λ,Σ	−1	0	0
$K^-,\bar{K}^0$	−1	0	0	B^+, B^0	0	0	+1	Ξ	−2	0	0
D^+,D^0	0	+1	0	B^-, $\bar{B}^0$	0	0	−1	Ω^-	−3	0	0
$D^-,\bar{D}^0$	0	−1	0					Λ_c^+	0	+1	0

b. Quoted upper limits correspond to a 90% confidence level. Masses, mean lives, and partial rates evaluated assuming equality for particles and antiparticles. See Conservation Laws Section for further details.

c. In decays with more than two bodies, p_{max} is the maximum momentum that any particle can have.

d. 99% confidence level. See footnote in Stable Particle Data Card Listings.

e. See Stable Particle Data Card Listings for energy limits used in this measurement.

f. Theoretical value; see also Stable Particle Data Card Listings.

g. The direct emission branching fraction is $(1.56\pm.35)\times10^{-5}$.

h. Structure-dependent part with positive (SD+) and negative (SD−) photon helicity.

i. The $K_S^0 \rightarrow \pi\pi$ and $K_L^0 \rightarrow \pi\pi$ branching fractions are from our branching fraction and rate fits and do not include results of K_L^0–K_S^0 interference experiments. The $\pi\pi$ rate results are combined with the interference results to obtain the $|\eta_{+-}|$ and $|\eta_{00}|$ values given in the addendum.

j. The stronger limit $<2\times10^{-9}$ of Clark et al., Phys. Rev. Lett. **26**, 1667 (1971) is not listed because of possible (but unknown) systematic errors. See Stable Particle Data Card Listings.

k. This is a weighted average of $D^\pm$ (44%) and D^0 (56%) branching fractions.

ℓ. $D_1^0 - D_2^0$ limits inferred from limit on $D^0 \rightarrow \bar{D}^0 \rightarrow \mu^-$ anything.

m. F mass determined from $\phi\pi$ mode. See note on conflicting F meson results in Stable Particle Data Card Listings. Quantum numbers shown are favored but not yet established.

n. Quantum numbers not measured. Values shown are quark model predictions.

p. Except for the neutral-current decay modes ($\ell^+\ell^-$ anything), only data from $\Upsilon(10575)$ decays are used. Behrends et al. [Phys. Rev. Lett. **50**, 881 (1983)] estimate the $\Upsilon(10575) \rightarrow B^+B^-$ and $\Upsilon(10575) \rightarrow B^0\bar{B}^0$ branching fractions to be 60 ± 2 and 40 ± 2%.

q. Partial mean life for $p \rightarrow e^+\pi^0$ mode. For antiprotons the best mean life limit, inferred from observation of cosmic ray $\bar{p}$'s, is $\tau_{\bar{p}} > 10^7$ yrs, the cosmic ray storage time.

r. Limit from neutrality-of-matter experiments. Assumes $|q_n| = |q_p| - |q_e|$.

s. P for Ξ, J^P for Ω^- and Σ^0, and J for Λ_c^+ not yet measured. Values shown are quark model predictions.

t. For limits on electric dipole moment, see Conservation Laws Section. Forbidden by P and T invariance.

u. $|g_A/g_V|$ defined by $g_A^2 = |C_A|^2+|C_A'|^2$, $g_V^2 = |C_V|^2+|C_V'|^2$, and $\Sigma[\bar{e}\Gamma_i\mu][\bar{\nu}\Gamma_i(C_i+C_i'\gamma_5)\nu]$; ϕ defined by $\cos\phi = -\mathrm{Re}(C_A^*C_V'+C_A'C_V^*)/g_Ag_V$. For more details, see Data Card Listings.

v. Value assumes $\rho = \delta$. P_μ is muon longitudinal polarization from π decay. In standard V-A theory, $P_\mu = 1$ and $\rho = \delta = 3/4$.

w. The definition of the slope parameter of the Dalitz plot is as follows [see also note in Data Card Listings]:

$$|M|^2 = 1 + g\left(\frac{s_3-s_0}{m_{\pi^+}^2}\right)$$

x. For definitions of form factors f_+, f_S, and f_T, and linear t dependences λ_+ and λ_0 of $f_+(t)$ and $f_0(t)$, see note in K^+ section of Data Card Listings.

y. The definition for the CP violation parameters is as follows [see also note in Data Card Listings]:

$$\eta_{+-} = |\eta_{+-}|e^{i\phi_{+-}} = \frac{A(K_L^0\rightarrow\pi^+\pi^-)}{A(K_S^0\rightarrow\pi^+\pi^-)} \qquad \eta_{00} = |\eta_{00}|e^{i\phi_{00}} = \frac{A(K_L^0\rightarrow\pi^0\pi^0)}{A(K_S^0\rightarrow\pi^0\pi^0)}$$

$$\delta = \frac{\Gamma(K_L^0\rightarrow\ell^+)-\Gamma(K_L^0\rightarrow\ell^-)}{\Gamma(K_L^0\rightarrow\ell^+)+\Gamma(K_L^0\rightarrow\ell^-)}, \quad |\eta_{+-0}|^2 = \frac{\Gamma(K_S^0\rightarrow\pi^+\pi^-\pi^0)^{\text{CP viol.}}}{\Gamma(K_L^0\rightarrow\pi^+\pi^-\pi^0)}, \quad |\eta_{000}|^2 = \frac{\Gamma(K_S^0\rightarrow\pi^0\pi^0\pi^0)^{\text{CP viol.}}}{\Gamma(K_L^0\rightarrow\pi^0\pi^0\pi^0)}$$

z. The definition of these quantities is as follows [for more details and sign convention, see note in Data Card Listings]:

$$\alpha = \frac{2|s||p|\cos\Delta}{|s|^2+|p|^2} \qquad \beta = \sqrt{1-\alpha^2}\sin\phi \qquad g_A,\ g_V,\ g_{WM} \text{ defined by } \langle B_f|\gamma_\lambda(g_V-g_A\gamma_5)+(g_{WM}/m_{B_i})\sigma^{\lambda\nu}q_\nu|B_i\rangle$$

$$\beta = \frac{-2|s||p|\sin\Delta}{|s|^2+|p|^2} \qquad \gamma = \sqrt{1-\alpha^2}\cos\phi \qquad \phi_{AV} \text{ defined by } g_A/g_V = |g_A/g_V|e^{i\phi_{AV}}$$

Meson Table

April 1984

In addition to the entries in the Meson Table, the Meson Data Card Listings contain all substantial claims for meson resonances. See Contents of Meson Data Card Listings at end of this Table.

Quantities in italics are new or have changed by more than one (old) standard deviation since April 1982.

J^P \ G \ I		0	1	$\frac{1}{2}$
N	+	ε/f	ρ	K*,κ
	−	ω/φ	δ	
A	+	η/D	B	K,Q
	−	H	π/A	

	$I^G(J^P)C_n$ __ *estab.*	Mass M (MeV)	Full Width Γ (MeV)	Partial decay mode: Mode	Fraction(%) [Upper limits (%) are 90% CL]		p or p_{max}[b] (MeV/c)
NONSTRANGE MESONS							
$\pi^{\pm}$	$1^-(0^-)+$	139.57	0.0	See Stable Particle Table			
π^0		134.96	7.95 eV ±0.55 eV				
η	$0^+(0^-)+$	548.8 ±0.6	0.83 keV ±0.12 keV	Neutral	70.9	See Stable Particle Table	
				Charged	29.1		
ρ(770)	$1^+(1^-)-$	769‡ ±3§	154‡ ±5§	ππ	≈100		358
				πγ	0.046±0.005		372
				$\mu^+\mu^-$	0.0067±0.0012[d]		370
				e^+e^-	0.0046±0.0002[d]		384
				ηγ	seen‡		189
M and Γ from neutral mode.				For upper limits, see footnote *e*			
ω(783)	$0^-(1^-)-$	782.6 ±0.2 S=1.1*	9.9 ±0.3	$\pi^+\pi^-\pi^0$	89.9±0.5		327
				$\pi^0\gamma$	8.7±0.5		380
				$\pi^+\pi^-$	1.4±0.2		366
				$\pi^0\mu^+\mu^-$	0.010±0.002		349
				e^+e^-	0.0067±0.0004	S=1.2*	391
				ηγ	seen‡		199
				For upper limits, see footnote *f*			
η′(958)	$0^+(0^-)+$‡	957.57 ±0.25	0.29 ±0.05	ηππ	65.3±1.6		231
				$\rho^0\gamma$	30.0±1.6		170
				ωγ	2.8±0.5		159
				γγ	1.9±0.2		479
				$\mu^+\mu^-\gamma$	0.009±0.002		467
				For upper limits, see footnote *g*			
S(975) or S*	$0^+(0^+)+$	975[c] ±4 S=1.4*	33[c] ±6	ππ	78±3		467
				$K\bar{K}$	22±3		
See note on ππ and $K\bar{K}$ S wave.‡							
δ(980)‡	$1^-(0^+)+$	983[h] ±2	54[h] ±7	ηπ	seen		320
				$K\bar{K}$	seen		
φ(1020)	$0^-(1^-)-$	*1019.5* *±0.1* S=1.2*	4.22 ±0.13	K^+K^-	49.3±1.0	S=1.3*	127
				K_LK_S	34.7±1.0	S=1.3*	110
				$\pi^+\pi^-\pi^0$ (incl. ρπ)	14.8±0.7	S=1.2*	462
				ηγ	*1.2±0.2*	S=1.4*	362
				$\pi^0\gamma$	0.14±0.05		501
				e^+e^-	0.031±0.001		510
				$\mu^+\mu^-$	0.025±0.003		499
				$\pi^+\pi^-$	0.02±0.01		490
				For upper limits, see footnote *i*			
H(1190)	$0^-(1^+)-$	1190 ±60	320 ±50	ρπ	seen		327
Seen in one experiment only.							
B(1235)	$1^+(1^+)-$	1234 ±10§	*150* *±10*§	ωπ	only mode seen		350
				[D/S amplitude ratio = 0.29±0.05]			
				For upper limits, see footnote *j*			

Meson Table *(cont'd)*

J^P	G \ I	0	1	$\frac{1}{2}$
N	+	ϵ/f	ρ	K^*,κ
	−	ω/ϕ	δ	
A	+	η/D	B	K,Q
	−	H	π/A	

	$I^G(J^P)C_n$ — *estab.*	Mass M (MeV)	Full Width Γ (MeV)	Partial decay mode: Mode	Fraction(%) [Upper limits (%) are 90% CL]	p or p_{max} [b] (MeV/c)
f(1270)	$0^+(2^+)+$	1274 $\pm 5^§$	178 $\pm 20^§$	$\pi\pi$	84.3 ± 1.2	622
				$2\pi^+2\pi^-$	2.9 ± 0.4 S=1.2*	559
				$K\bar{K}$	2.9 ± 0.2	398
				$\gamma\gamma$	0.0015 ± 0.0002	637
				$\pi^+\pi^-2\pi^0$	seen	562
				For upper limits, see footnote *k*		
A(1270) or A_1	$1^-(1^+)+$	$1275^‡$ ± 30	$315^‡$ ± 45	$\rho\pi$	dominant	389
				$\pi(\pi\pi)_{S-wave}$	*< 0.7*§	599
→ D(1285)	$0^+(1^+)+$	1283 $\pm 5^§$	26 $\pm 5^§$	$K\bar{K}\pi$	11 ± 3	302
				$\eta\pi\pi$	49 ± 6	482
				†[$\delta\pi$	36 ± 7]	236
				4π (prob. $\rho\pi\pi$)‡	40 ± 7	564
ϵ(1300)	$0^+(0^+)+$	~1300	200−600	$\pi\pi$	~90	635
				$K\bar{K}$	~10	418
				$\eta\eta$	*possibly seen*	348
See note on $\pi\pi$ and $K\bar{K}$ S wave.‡						
π(1300)	$1^-(0^-)+$	$1300^§$ $\pm 100^§$	200−600	$\rho\pi$	seen	407
				$\pi(\pi\pi)_{S-wave}$	seen	612
Not a well-established resonance.						
A_2(1320)	$1^-(2^+)+$	1318 $\pm 5^§$	110 $\pm 5^§$	$\rho\pi$	70.1 ± 2.2	419
				$\eta\pi$	14.5 ± 1.2	534
				$\omega\pi\pi$	10.6 ± 2.5	361
				$K\bar{K}$	4.9 ± 0.8	434
				$\eta'\pi$	< 2 (CL=97%)	286
				$\pi\gamma$	*0.27 ± 0.06*	652
				$\gamma\gamma$	0.0007 ± 0.0002	659
E(1420)‡	$0^+(1^+)+$	1418 $\pm 10^§$	52 $\pm 10^§$	$K\bar{K}\pi$ (incl. $K^*\bar{K}+K\bar{K}^*$)	seen	423
				$\eta\pi\pi$	possibly seen	565
				†[$\delta\pi$	possibly seen]	348
ι(1440)‡	$0^+(0^-)+$	*1440*§ *± 10*§	*76*§ *± 10*§	$K\bar{K}\pi$ (incl. $K^*\bar{K}+K\bar{K}^*$)	*seen*	441
				$\eta\pi\pi$	*seen*	579
				†[$\delta\pi$	*seen*]	366
f′(1525)	$0^+(2^+)+$	1525 $\pm 5^§$	70 $\pm 10^§$	$K\bar{K}$	dominant	578
				$\pi\pi$	possibly seen	750
				$\gamma\gamma$	*0.0011 ± 0.0002*	763
→ ρ(1600) or ρ'	$1^+(1^-)-$	$1590^{‡§}$ $\pm 20^§$	$260^{‡§}$ $\pm 100^§$	4π (incl. $\rho\pi^+\pi^-$, A(1270)π)	*60 ± 7*§	733
				$\pi\pi$	*23 ± 7*§	783
				$K^*\bar{K} + \bar{K}^*K$	*9 ± 2*	377
				$\eta\pi\pi$	*7 ± 2*	669
				$K\bar{K}$	*1 ± 0.5*	623
				e^+e^-	*0.003 ± 0.001*	795
ω(1670)	$0^-(3^-)-$	1668 ± 5	166 $\pm 15^§$ S=1.1*	3π	seen	806
				†[$\rho\pi$	seen]	648
				5π	seen	740
				†[$\omega\pi\pi$ (prob. Bπ)	seen]	616
A(1680)‡ or A_3	$1^-(2^-)+$	$1680^§$ $\pm 30^§$	$250^§$ $\pm 50^§$	fπ	53 ± 5	336
				$\rho\pi$	34 ± 6	656
				$\pi(\pi\pi)_{S-wave}$	9 ± 5	813
				$K^*\bar{K} + \bar{K}^*K$	*4 ± 1.4*	459
				For upper limits, see footnote *ℓ*		
ϕ(1680) or ϕ'	$0^-(1^-)-$	1685 $\pm 10^§$	*150*§ *± 30*§	$K^*\bar{K} + \bar{K}^*K$	dominant	466
				$\omega\pi\pi$	seen	624
				$K\bar{K}$	seen	683
				e^+e^-	*seen*	842
				$\pi^+\pi^-\pi^0$	*possibly seen*	814

Meson Table *(cont'd)*

J^P \ G \ I		0	1	$\frac{1}{2}$
N	+	ϵ/f	ρ	K^*,κ
	−	ω/ϕ	δ	
A	+	η/D	B	K,Q
	−	H	π/A	

	$I^G(J^P)C_n$ — *estab.*	Mass M (MeV)	Full Width Γ (MeV)	Mode	Fraction(%) |Upper limits (%) are 90% CL|		p or p_{max}[b] (MeV/c)
g(1690)	$1^+(3^-)-$	1691 $\pm 5^{\S}$	$200^{\S}$ $\pm 20^{\S}$	2π	23.8 ± 1.3		834
				4π (incl. $\pi\pi\rho,\rho\rho,A_2\pi,\omega\pi$)	70.9 ± 1.9		787
				$K\bar{K}\pi$ (incl. $K^*\bar{K}+K\bar{K}^*$)	*3.8 ± 1.2*		625
				$K\bar{K}$	1.5 ± 0.3	S=1.3*	684
J^P, M, and Γ from the 2π and $K\bar{K}$ modes.							
$\theta(1690)$	$0^+(2^+)+$	*1690* *± 30*	*180* *± 50*	$\eta\eta$	*seen*		643
				$K\bar{K}$	*seen*		683
$\phi(1850)$	$0^-(3^-)-$	*1853* *± 10*	*96* *± 32*	$K\bar{K}$	*seen*		784
				$K^*\bar{K} + \bar{K}^*K$	*seen*		601
h(2030)	$0^+(4^+)+$	2027 ± 12	*220* *± 30*	$\pi\pi$	*17 ± 2*		1004
				$K\bar{K}$	$0.7^{+0.4}_{-0.2}$		883
$\eta_c(2980)$	$0^+(0^-)+$	2981 ± 6	< 20	$\eta\pi^+\pi^-$	seen		1426
				$2(\pi^+\pi^-)$	seen		1458
				$K^+K^-\pi^+\pi^-$	seen		1343
				$p\bar{p}$	seen		1158
$J/\psi(3100)$	$0^-(1^-)-$	3096.9 ± 0.1	0.063 ± 0.009	e^+e^-	7.4 ± 1.2		1548
				$\mu^+\mu^-$	7.4 ± 1.2		1545
				hadrons + radiative	85 ± 2		

Decay modes into stable hadrons			Decay modes into hadronic resonances			
†[$2(\pi^+\pi^-)\pi^0$	3.7 ± 0.5	1496	†[$\rho\pi$	1.22 ± 0.12		1449
$3(\pi^+\pi^-)\pi^0$	2.9 ± 0.7	1433	$\omega 2\pi^+ 2\pi^-$	0.85 ± 0.34		1392
$\pi^+\pi^-\pi^0K^+K^-$	1.2 ± 0.3	1368	ρA_2	0.84 ± 0.45		1126
$4(\pi^+\pi^-)\pi^0$	0.9 ± 0.3	1345	$\omega\pi\pi$	0.68 ± 0.19		1435
$\pi^+\pi^-K^+K^-$	0.72 ± 0.23	1407	$K^{*0}(892)\bar{K}^{*0}(1430)$+c.c.	0.67 ± 0.26		1009
$p\bar{p}\pi^+\pi^-$	0.53 ± 0.06	1107	$K^{\pm}K^{*\mp}(892)$	0.34 ± 0.05		1373
$2(\pi^+\pi^-)$	0.4 ± 0.1	1517	$B^{\pm}(1235)\pi^{\mp}$	0.29 ± 0.07		1298
$3(\pi^+\pi^-)$	0.4 ± 0.2	1466	$K^0\bar{K}^{*0}(892)$+c.c.	0.27 ± 0.06		1370
$n\bar{n}\pi^+\pi^-$	0.38 ± 0.36	1106	ωf	0.23 ± 0.08	S=1.2*	1143
$\Xi\bar{\Xi}$	0.32 ± 0.08	818	$\phi\pi^+\pi^-$	0.21 ± 0.09		1365
$2(\pi^+\pi^-)K^+K^-$	0.31 ± 0.13	1320	$\eta' p\bar{p}$	0.18 ± 0.06		596
$K_S^0K^{\pm}\pi^{\mp}$	0.26 ± 0.07	1440	$\phi K\bar{K}$	0.18 ± 0.08		1176
$\Sigma^+\bar{\Sigma}^-$	0.24 ± 0.26	988	$\omega p\bar{p}$	0.16 ± 0.03		768
$p\bar{p}\eta$	0.23 ± 0.04	948	$\omega K\bar{K}$	0.16 ± 0.10		1265
$p\bar{p}$	0.22 ± 0.02	1232	$\phi\eta$	0.10 ± 0.06		1320
$p\bar{n}\pi^-$ or $\bar{p}n\pi^+$	0.21 ± 0.02	1174	$\phi f'(1525)$	0.037 ± 0.013		871
$n\bar{n}$	0.18 ± 0.09	1231	$\phi S(975)$	*0.026 ± 0.006*		1184
$p\bar{p}\pi^+\pi^-\pi^0$	0.16 ± 0.06[m]	1033	$\pi^{\pm}A_2^{\mp}$	< 0.43		1263
$\Sigma^0\bar{\Sigma}^0$	0.13 ± 0.04	988	$K^{*0}(1430)\bar{K}^{*0}(1430)$	< 0.29		606
$\Lambda\bar{\Lambda}$	0.11 ± 0.02	1074	$K^0\bar{K}^{*0}(1430)$+c.c.	< 0.2		1158
$p\bar{p}\pi^0$	0.11 ± 0.01	1176	$K^{\pm}K^{*\mp}(1430)$	< 0.2		1159
$2(K^+K^-)$	0.07 ± 0.03	1131	$\phi 2\pi^+ 2\pi^-$	< 0.15		1318
K^+K^-	0.022 ± 0.008	1468	$\phi\eta'$	< 0.13		1192
$\pi^+\pi^-$	0.011 ± 0.005	1542	$K^{*0}(892)\bar{K}^{*0}(892)$	< 0.05		1261
$\Lambda\bar{\Sigma}$	< 0.015	1032	ϕf	< 0.037		1037
$K_S^0K_L^0$	< 0.009]	1466	$\omega f'(1525)$	< 0.016]		1003

Radiative decay modes			Radiative decay modes (cont'd)		
†[$\gamma 2(\pi^+\pi^-)$	*0.49 ± 0.17*	1517	$\gamma\eta_c(2980)$	*seen*	114
$\gamma\rho\rho$	*seen*	1344	$\gamma\theta(1690)$	*seen*	1087
$\gamma\iota(1440) \rightarrow \gamma K\bar{K}\pi$	0.42 ± 0.12[n]	1214	$\gamma\eta\pi\pi$	*seen*	1487
$\gamma\eta'$	0.36 ± 0.05	1400	$\gamma D(1285)$	< 0.6	1283
γf	0.15 ± 0.04	1286	2γ	< 0.05	1548
$\gamma\eta$	0.086 ± 0.009	1500	$\gamma f'(1525)$	< 0.03	1173
$\gamma\pi^0$	0.007 ± 0.005	1546	$\gamma p\bar{p}$	< 0.01	1232
			3γ	< 0.006]	1548

Meson Table *(cont'd)*

J^P	G \ I	0	1	$\frac{1}{2}$
N	+	ε/f	ρ	K*,κ
	−	ω/φ	δ	
A	+	η/D	B	K,Q
	−	H	π/A	

	$I^G(J^P)C_n$ — *estab.*	Mass M (MeV)	Full Width Γ (MeV)	Partial decay mode: Mode	Fraction(%) [Upper limits (%) are 90% CL]	p or p_{max} [b] (MeV/c)
χ(3415)	$0^+(0^+)+$	3415.0 ± 1.0		$2(\pi^+\pi^-)$ (incl. $\pi\pi\rho$)	4.3 ± 0.9	1679
				$\pi^+\pi^-K^+K^-$ (incl. $\pi K\bar{K}^*$)	3.4 ± 0.9	1580
				$3(\pi^+\pi^-)$	1.7 ± 0.6	1633
				$\pi^+\pi^-$	0.9 ± 0.2	1702
				$\gamma J/\psi(3100)$	0.8 ± 0.3	303
				K^+K^-	0.8 ± 0.2	1635
				$p\bar{p}\pi^+\pi^-$	0.6 ± 0.2	1320
				For upper limits, see footnote *o*		
χ(3510)	$0^+(1^+)+$	3510.0 ± 0.6		$\gamma J/\psi(3100)$	28 ± 3	389
				$3(\pi^+\pi^-)$	2.4 ± 0.9	1683
				$2(\pi^+\pi^-)$ (incl. $\pi\pi\rho$)	1.8 ± 0.5	1727
				$\pi^+\pi^-K^+K^-$ (incl. $\pi K\bar{K}^*$)	1.0 ± 0.4	1632
				$\pi^+\pi^-p\bar{p}$	0.15 ± 0.10	1381
				For upper limits, see footnote *p*		
χ(3555)	$0^+(2^+)+$	3555.8 ± 0.6		$\gamma J/\psi(3100)$	15.5 ± 1.8	429
				$2(\pi^+\pi^-)$ (incl. $\pi\pi\rho$)	2.3 ± 0.5	1750
				$\pi^+\pi^-K^+K^-$ (incl. $\pi K\bar{K}^*$)	2.0 ± 0.5	1656
				$3(\pi^+\pi^-)$	1.2 ± 0.8	1706
				$\pi^+\pi^-p\bar{p}$	0.35 ± 0.14	1410
				$\pi^+\pi^-$	0.20 ± 0.11	1772
				K^+K^-	0.16 ± 0.12	1708
				For upper limits, see footnote *q*		
→ ψ(3685)	$0^-(1^-)-$	3686.0 ± 0.1	0.215 ± 0.040	e^+e^-	0.9 ± 0.1	1843
				$\mu^+\mu^-$	0.8 ± 0.2	1840
				hadrons + radiative	98.1 ± 0.3	

$m_{\psi(3685)} - m_{\psi(3100)} = 589.06 \pm 0.13$

Radiative decay modes			Decay modes into hadrons		
†[$\gamma\chi(3415)$	8.2 ± 1.4	261	†[$J/\psi\pi^+\pi^-$	33 ± 2	477
$\gamma\chi(3510)$	8.0 ± 1.3	172	$J/\psi\pi^0\pi^0$	17 ± 2	481
$\gamma\chi(3555)$	7.4 ± 1.3	128	$J/\psi\eta$	2.8 ± 0.6§	196
$\gamma\eta_c(2980)$	0.43 ± 0.26	638	$2(\pi^+\pi^-)\pi^0$	0.35 ± 0.15	1799
$\gamma\eta_c'(3590)$	0.2 to 1.3	91	$\pi^+\pi^-K^+K^-$	0.16 ± 0.04	1726
$\gamma\pi^0$	< 0.5 (CL=95%)	1841	$J/\psi\pi^0$	0.10 ± 0.03	528
$\gamma\eta$	< 0.02	1802	$p\bar{p}\pi^+\pi^-$	0.08 ± 0.02	1491
$\gamma\eta'$	< 0.02	1719	$K^{*0}(892)K^-\pi^+$ + cc.	0.067 ± 0.025	1674
$\gamma\iota(1440)\to\gamma K\bar{K}\pi$	< 0.012[n]]	1562	$2(\pi^+\pi^-)$	0.05 ± 0.01	1817
			$\rho^0\pi^+\pi^-$	0.042 ± 0.015	1751
			$p\bar{p}$	0.019 ± 0.005	1586
			$3(\pi^+\pi^-)$	0.015 ± 0.010	1774
			K^+K^-	0.010 ± 0.007	1776
			$\pi^+\pi^-$	0.008 ± 0.005	1838
			$\rho\pi$	< 0.1	1760
			$\Lambda\bar{\Lambda}$	< 0.04]	1467

	$I^G(J^P)C_n$	Mass M (MeV)	Full Width Γ (MeV)	Mode	Fraction(%)	p or p_{max} (MeV/c)
ψ(3770)	$(1^-)-$	3770 ± 3	25 ± 3	e^+e^-	0.0011 ± 0.0002	1885
				$D\bar{D}$	dominant	242

$m_{\psi(3770)} - m_{\psi(3685)} = 83.9 \pm 2.4$ S=1.8*

	$I^G(J^P)C_n$	Mass M (MeV)	Full Width Γ (MeV)	Mode	Fraction(%)	p or p_{max} (MeV/c)
ψ(4030)	$(1^-)-$	4030§ ± 5§	52 ± 10	e^+e^-	0.0014 ± 0.0004	2015
				hadrons	dominant	
				†[$D\bar{D}$	*seen*	752
				$D\bar{D}^* + D^*\bar{D}$	*seen*	559
				$D^*\bar{D}^*$	*seen*]	177
ψ(4160)	$(1^-)-$	4159 ± 20	78 ± 20	e^+e^-	0.0010 ± 0.0004	2079
				hadrons	dominant	
ψ(4415)	$(1^-)-$	4415 ± 6	43 ± 20§	e^+e^-	0.0010 ± 0.0003 S=1.4*	2207
				hadrons	dominant	

Meson Table *(cont'd)*

J^P	G \ I	0	1	$\frac{1}{2}$
N	+	ε/f	ρ	K*,κ
N	−	ω/φ	δ	
A	+	η/D	B	K,Q
A	−	H	π/A	

	$I^G(J^P)C_n$ — *estab.*	Mass M (MeV)	Full Width Γ (MeV)	Partial decay mode: Mode	Fraction(%) [Upper limits (%) are 90% CL]	p or p_{max} [b] (MeV/c)
ϒ(9460) or ϒ(1S)	$\underline{(1^-)-}$	9460.0 ±0.3 S=1.6*	0.0443 ±0.0066	$\mu^+\mu^-$	2.9±0.5	4729
				e^+e^-	2.5±0.5	4730
				$\tau^+\tau^-$	*3.4±0.8*	4381
$\chi_b(9875)$ or $\chi_b(1\,^3P_0)^r$	( )+	*9872.9 ±5.8*		γϒ(9460)	*seen*	404
$\chi_b(9895)$ or $\chi_b(1\,^3P_1)^r$	( )+	*9894.5 ±3.5*		γϒ(9460)	*43±11*	425
$\chi_b(9915)$ or $\chi_b(1\,^3P_2)^r$	( )+	*9914.6 ±2.4*		γϒ(9460)	*20.0±4.4*	444
ϒ(10025) or ϒ(2S)	$\underline{(1^-)-}$	10023.4 ±0.3	0.0296 ±0.0047	$\mu^+\mu^-$	*1.9±1.8*	5011
				e^+e^-	1.6±0.3	5012
				ϒ(9460)ππ	*19.5±1.7*	476
$m_{\Upsilon(10025)} - m_{\Upsilon(9460)}$ = *563.3±0.4*				$\gamma\chi_b(9875)$	*3.5±1.4*	149
				$\gamma\chi_b(9895)$	*5.9±1.4*	128
				$\gamma\chi_b(9915)$	*6.1±1.4*	108
$\chi_b(10255)$ or $\chi_b(2\,^3P_1)^r$	( )+	*10253.7 ±3.4*		γϒ(9460)	*seen*	763
				γϒ(10025)	*seen*	228
$\chi_b(10270)$ or $\chi_b(2\,^3P_2)^r$	( )+	*10271.0 ±2.4*		γϒ(9460)	*seen*	779
				γϒ(10025)	*seen*	245
ϒ(10355) or ϒ(3S)	$\underline{(1^-)-}$	10355.5 ±0.5	*0.0177 ±0.0051*	e^+e^-	*2.0±0.7*	5178
				$\mu^+\mu^-$	*3.3±2.0*	5177
				$\Upsilon(9460)\pi^+\pi^-$	*5.1±1.1*	814
$m_{\Upsilon(10355)} - m_{\Upsilon(9460)}$ = *895.5±.6*				$\Upsilon(10025)\pi^+\pi^-$	*3±3*	177
				$\gamma\chi_b(10235)$	*7.6±3.5*	122
				$\gamma\chi_b(10255)$	*15.6±4.2*	101
				$\gamma\chi_b(10270)$	*12.7±4.1*	84
ϒ(10575) or ϒ(4S)	$\underline{(1^-)-}$	10573 ±4	14 ±5	e^+e^-	0.0017±0.0007	5286
$m_{\Upsilon(10575)} - m_{\Upsilon(9460)}$ = 1113±4						
STRANGE MESONS						
K^+	$\underline{1/2(0^-)}$	493.67		See Stable Particle Table		
K^0		497.67				
K*(892)	$\underline{1/2(1^-)}$	892.1 ±0.4 S=1.4*	51.3 ±1.0 S=1.1*	Kπ	≈ 100	288
				Kγ	0.10±0.01	309
				Kππ	< 0.05	216
M and Γ from charged mode; $m^0 - m^\pm = 6.7 \pm 1.2$ MeV.						
Q(1280) or Q_1	$\underline{1/2(1^+)}$	1270§ ±10§	90§ ±20§	Kρ	42±6	45
				κ(1350)π	28±4	
				K*(892)π	16±5	298
				Kω	11±2	
				Kε	3±2	
κ(1350)	$\underline{1/2(0^+)}$	~1350	~250	Kπ	seen	574
See note on Kπ S wave.‡						
Q(1400) or Q_2	$\underline{1/2(1^+)}$	1406 ±10	184 ±9	K*(892)π	94±6	403
				Kρ	3±3	299
				Kε	2±2	
				Kω	1±1	285

Meson Table *(cont'd)*

J^P	G \ I	0	1	$\frac{1}{2}$
N	+	ε/f	ρ	K*,κ
	−	ω/φ	δ	
A	+	η/D	B	K,Q
	−	H	π/A	

	$I^G(J^P)$ — *estab.*	Mass M (MeV)	Full Width Γ (MeV)	Partial decay mode: Mode	Fraction(%) [Upper limits (%) are 90% CL]		p or p_{max}[b] (MeV/c)
K*(1430)	1/2(2+)	1425§ ±5§	100§ ±10§	Kπ	44.8±2.3	S=2.7*	618
				K*(892)π	24.5±2.0	S=1.1*	417
				K*(892)ππ	13.0±2.6	S=1.1*	366
				Kρ	8.8±1.0	S=1.2*	324
				Kω	4.2±1.5		310
				Kη	5±5§		485
				Kγ	*0.24±0.05*		627
→							
→ L(1770)‡	1/2(2−)	~1770§	~200§	K*(1430)π	dominant		286
				K*(892)π	seen		651
				Kf	seen		
				Kφ	*seen*		816
See note on L(1770).‡							
K*(1780)‡ or K*	1/2(3−)	1780 ±10§	160 ±20§	Kππ	large		796
				†[Kρ	large]		620
				†[K*(892)π	large]		657
				Kπ	17±5§		815
See note on K*(1780).‡							
→ K*(2060)	1/2(4+)	*2060§ ±30§*	*210§ ±40§*	Kπ	*7±1*		966
				K*(892)ππ	*seen*		809
				ρKπ	*seen*		751
				ωKπ	*seen*		744
Not a well-established resonance.				K*(892)πππ	*seen*		775
→ → →							

CHARMED, NONSTRANGE MESONS

	$I^G(J^P)$	Mass M (MeV)	Full Width Γ (MeV)	Mode	Fraction(%)	p or p_{max} (MeV/c)
D^+	1/2(0−)	1869.4		See Stable Particle Table		
D^0		1864.7				
D^{*+}(2010)	1/2(1−)	2010.1 ±0.7	< 2.0	$D^0\pi^+$	64±11	39
				$D^+\pi^0$	28±9	38
				$D^+\gamma$	8±7	136
		$m_{D^{*+}} - m_{D^0} = 145.4 \pm 0.2$ MeV				
D^{*0}(2010)	1/2(1−)	2007.2 ±2.1	< 5	$D^0\pi^0$	55±15	44
				$D^0\gamma$	45±15	137

CHARMED, STRANGE MESON

	$I^G(J^P)$	Mass M (MeV)	
F^+	0(0−)	1971	See Stable Particle Table
→			

BOTTOM, NONSTRANGE MESON

	$I^G(J^P)$	Mass M (MeV)	
B^+	1/2(0−)	5271	See Stable Particle Table
B^0		5274	
→			

→ Indicates an entry in the Meson Data Card Listings not entered in the Meson Table. We do not regard these as established resonances. All the entries in the Listings can be found in the Table of Contents of the Meson Data Card Listings immediately following these footnotes.

‡ See Meson Data Card Listings.

* Quoted error includes scale factor $S = \sqrt{\chi^2/(N-1)}$. See footnote to Stable Particle Table.

† Square brackets indicate a subreaction of the previous (unbracketed) decay mode(s).

§ This is only an educated guess; the error given is larger than the error on the average of the published values. (See the Meson Data Card Listings for the latter.)

a. ΓM is approximately the half-width of the resonance when plotted against M^2.

b. For decay modes into ≥ 3 particles, p_{max} is the maximum momentum that any of the particles in the final state can have. The momenta have been calculated by using the averaged central mass values, without taking into account the widths of the resonances.

d. The e^+e^- branching fraction is from $e^+e^- \rightarrow \pi^+\pi^-$ experiments only. The $\omega\rho$ interference is then due to $\omega\rho$ mixing only, and is expected to be small. See note in the Meson Data Card Listings. The $\mu^+\mu^-$ branching fraction is compiled from 3 experiments, each possibly with substantial $\omega\rho$ interference. The error reflects this uncertainty; see notes in the Meson Data Card Listings. If $e\mu$ universality holds, $\Gamma(\rho^0 \rightarrow \mu^+\mu^-) = \Gamma(\rho^0 \rightarrow e^+e^-) \times 0.99785$.

e. Empirical limits on fractions for other decay modes of $\rho(770)$ are $\pi^{\pm}\eta < 0.8\%$ (CL=84%), $\pi^+\pi^+\pi^-\pi^- < 0.15\%$, $\pi^{\pm}\pi^+\pi^-\pi^0 < 0.2\%$ (CL=84%).

f. Empirical limits on fractions for other decay modes of $\omega(783)$ are $\pi^+\pi^-\gamma < 5\%$, $\pi^0\pi^0\gamma < 1\%$, η + neutral(s) $< 1.5\%$, $\mu^+\mu^- < 0.02\%$.

g. Empirical limits on fractions for other decay modes of $\eta'(958)$ are $\pi^+\pi^- < 2\%$ (CL=84%), $\pi^+\pi^-\pi^0 < 5\%$ (CL=84%), $\pi^+\pi^+\pi^-\pi^- < 1\%$ (CL=95%), $\pi^+\pi^+\pi^-\pi^-\pi^0 < 1\%$ (CL=84%), $6\pi < 1\%$, $\pi^+\pi^-e^+e^- < 0.6\%$, $\pi^0e^+e^- < 1.3\%$ (CL=84%), $\eta e^+e^- < 1.1\%$, $\pi^0\rho^0 < 4\%$, $\eta\mu^+\mu^- < 1.5 \times 10^{-5}$, $\pi^0\mu^+\mu^- < 6 \times 10^{-5}$.

h. The mass and width are from the $\eta\pi$ mode only. If the $K\overline{K}$ channel is strongly coupled, the width may be larger.

i. Empirical limits on fractions for other decay modes of $\phi(1020)$ are $\pi^+\pi^-\gamma < 0.7\%$, $\omega\gamma < 5\%$ (CL=84%), $\rho\gamma < 2\%$ (CL=84%), $2\pi^+2\pi^-\pi^0 < 1\%$ (CL=95%), $2\pi^+2\pi^- < 0.1\%$.

j. Empirical limits on fractions for other decay modes of B(1235) are $\pi\pi < 15\%$, $K\overline{K} < 2\%$ (CL=84%), $4\pi < 50\%$ (CL=84%), $\phi\pi < 1.5\%$ (CL=84%), $\eta\pi < 25\%$, $(\overline{K}K)^{\pm}\pi^0 < 8\%$, $K_SK_S\pi^{\pm} < 2\%$, $K_SK_L\pi^{\pm} < 6\%$.

k. Empirical limits (CL=95%) on fractions for other decay modes of f(1270) are $\eta\pi\pi < 1\%$, $K^0K^-\pi^+$ + c.c. $< 0.4\%$, $\eta\eta < 2\%$.

ℓ. Empirical limits on fractions for other decay modes of A(1680) are $\eta\pi < 10\%$, $5\pi < 10\%$.

m. Includes $p\overline{p}\pi^+\pi^-\gamma$ and excludes $p\overline{p}\eta$, $p\overline{p}\omega$, $p\overline{p}\eta'$.

n. See E(1420) mini-review.

o. Empirical limits on fractions for other decay modes of $\chi(3415)$ are $2\gamma < 0.17\%$, $p\overline{p} < 0.11\%$.

p. Empirical limits on fractions for other decay modes of $\chi(3510)$ are ($\pi^+\pi^-$ and K^+K^-) $< 0.2\%$, $\gamma\gamma < 0.16\%$, $p\overline{p} < 0.13\%$.

q. Empirical limits on fractions for other decay modes of $\chi(3555)$ are $2\gamma < 0.06\%$, $p\overline{p} < 0.10\%$, $J/\psi\pi^+\pi^-\pi^0 < 1.5\%$.

r. Spectroscopic labeling for these states is theoretical, pending experimental information.

Contents of Meson Data Card Listings

Non-strange (S = 0; C,B = 0)						Strange (|S| = 1; C,B = 0)	
entry	$I^G(J^P)C_n$	entry	$I^G(J^P)C_n$	entry	$I^G(J^P)C_n$	entry	$I(J^P)$
π	$1^-(0^-)+$	ω (1670)	$0^-(3^-)-$	→ e^+e^- (1100—2200)	(1^-)	K	$1/2(0^-)$
η	$0^+(0^-)+$	A (1680)	$1^-(2^-)+$	→ $\overline{N}N$ (1200—3600)		K* (892)	$1/2(1^-)$
ρ (770)	$1^+(1^-)-$	ϕ (1680)	$0^-(1^-)-$	→ X (1900—3600)		Q (1280)	$1/2(1^+)$
ω (783)	$0^-(1^-)-$	g (1690)	$1^+(3^-)-$	η_c (2980)	$0^+\ \ \ +$	κ (1350)	$1/2(0^+)$
η' (958)	$0^+(0^-)+$	θ (1690)	$0^+(\ ^+)+$	J/ψ (3100)	$0^-(1^-)-$	Q (1400)	$1/2(1^+)$
S (975)	$0^+(0^+)+$	→ η (1700)	$^+$	χ (3415)	$0^+(0^+)+$	→ K (1400)	$1/2(0^-)$
δ (980)	$1^-(0^+)+$	→ S (1730)	$0^+(0^+)+$	χ (3510)	$0^+(1^+)+$	K* (1430)	$1/2(2^+)$
ϕ (1020)	$0^-(1^-)-$	→ π (1770)	$1^-(0^-)+$	χ (3555)	$0^+(2^+)+$	→ L (1580)	$1/2(2^-)$
H (1190)	$0^-(1^+)-$	→ f (1810)	$0^+(2^+)+$	→ η_c (3590)	$+$	→ K* (1650)	$1/2(1^-)$
B (1235)	$1^+(1^+)-$	ϕ (1850)	0	ψ (3685)	$0^-(1^-)-$	L (1770)	$1/2(2^-)$
→ g_S (1240)	$0^+(0^+)+$	→ S (1935)		ψ (3770)	$(1^-)-$	K* (1780)	$1/2(3^-)$
→ ρ (1250)	$1^+(1^-)-$	h (2030)	$0^+(4^+)+$	ψ (4030)	$(1^-)-$	→ K (1830)	$1/2(0^-)$
f (1270)	$0^+(2^+)+$	→ δ (2040)	$1^-(4^+)+$	ψ (4160)	$(1^-)-$	K* (2060)	$1/2(4^+)$
A (1270)	$1^-(1^+)+$	→ A (2050)	$1^-(3^+)+$	ψ (4415)	$(1^-)-$	→ K (2250)	$1/2(2^-)$
→ η (1275)	$0^+(0^-)+$	→ A (2100)	$1^-(2^-)+$	Υ (9460)	$(1^-)-$	→ K (2320)	$1/2(3^+)$
D (1285)	$0^+(1^+)+$	→ ρ (2150)	$1^+(1^-)-$	χ_b (9875)	$(\ \ )+$	→ K (2500)	$1/2(4^-)$
ϵ (1300)	$0^+(0^+)+$	→ ϵ (2150)	$0^+(2^+)+$	χ_b (9895)	$(\ \ )+$	**Charmed (|C| = 1)**	
π (1300)	$1^-(0^-)+$	→ ξ (2220)	$0\ (\ ^+)$	χ_b (9915)	$(\ \ )+$	D	$1/2(0^-)$
A_2 (1320)	$1^-(2^+)+$	→ g_T (2240)	$0^+(2^+)+$	Υ (10025)	$(1^-)-$	D* (2010)	$1/2(1^-)$
E (1420)	$0^+(1^+)+$	→ ρ (2250)	$1^+(3^-)-$	→ χ_b (10235)	$(\ \ )+$	F	$0\ (0^-)$
ι (1440)	$0^+(0^-)+$	→ ϵ (2300)	$0^+(4^+)+$	χ_b (10255)	$(\ \ )+$	→ F* (2140)	
f′ (1525)	$0^+(2^+)+$	→ ρ (2350)	$1^+(5^-)-$	χ_b (10270)	$(\ \ )+$	**Bottom (Beauty) (|B| = 1)**	
→ D (1530)	$0^+(1^+)+$	→ δ (2450)	$1^-(6^+)$	Υ (10355)	$(1^-)-$	B	
ρ (1600)	$1^+(1^-)-$	→ r (2510)	$0^+(6^+)$	Υ (10575)	$(1^-)-$	→ Exotics	

Baryon Table

April 1984

The following short list gives the name, the nominal mass, the quantum numbers (where known), and the status of each of the Baryon States in the Data Card Listings. States with 3- or 4-star status are included in the Baryon Table below; the others are omitted because the evidence for the existence of the effect and/or for its interpretation as a resonance is open to question.

N(939) P11	****	Δ(1232) P33	****	Z0(1780) P01	*	Σ(1193) P11	****	Ξ(1318) P11	****
N(1440) P11	****	Δ(1550) P31	*	Z0(1865) D03	*	Σ(1385) P13	****	Ξ(1530) P13	****
N(1520) D13	****	Δ(1600) P33	**	Z1(1900) P13	*	Σ(1480)	*	Ξ(1630)	*
N(1535) S11	****	Δ(1620) S31	****	Z1(2150)	*	Σ(1560)	**	Ξ(1680)	**
N(1540) P13	*	Δ(1700) D33	****	Z1(2500)	*	Σ(1580) D13	**	Ξ(1820) 13	***
N(1650) S11	****	Δ(1900) S31	***			Σ(1620) S11	**	Ξ(1940)	**
N(1675) D15	****	Δ(1905) F35	****	Λ(1116) P01	****	Σ(1660) P11	***	Ξ(2030) 1	***
N(1680) F15	****	Δ(1910) P31	****	Λ(1405) S01	****	Σ(1670) D13	****	Ξ(2120)	*
N(1700) D13	***	Δ(1920) P33	***	Λ(1520) D03	****	Σ(1690)	**	Ξ(2250)	**
N(1710) P11	***	Δ(1930) D35	***	Λ(1600) P01	***	Σ(1750) S11	***	Ξ(2370) 1	**
N(1720) P13	****	Δ(1940) D33	*	Λ(1670) S01	****	Σ(1770) P11	*	Ξ(2500)	*
N(1990) F17	**	Δ(1950) F37	****	Λ(1690) D03	****	Σ(1775) D15	****		
N(2000) F15	**	Δ(2150) S31	*	Λ(1800) S01	***	Σ(1840) P13	*	Ω(1672) P03	****
N(2080) D13	**	Δ(2200) G37	*	Λ(1800) P01	***	Σ(1880) P11	**	$Λ_c$(2282)	****
N(2090) S11	*	Δ(2300) H39	**	Λ(1820) F05	****	Σ(1915) F15	****		
N(2100) P11	*	Δ(2350) D35	*	Λ(1830) D05	****	Σ(1940) D13	***	$Σ_c$(2450)	**
N(2190) G17	****	Δ(2390) F37	*	Λ(1890) P03	****	Σ(2000) S11	*		
N(2200) D15	**	Δ(2400) G39	**	Λ(2000)	*	Σ(2030) F17	****	A(2460)	*
N(2220) H19	****	Δ(2420) H311	****	Λ(2020) F07	*	Σ(2070) F15	*	$Λ_b$(5500)	*
N(2250) G19	****	Δ(2750) I313	**	Λ(2100) G07	****	Σ(2080) P13	**		
N(2600) I111	***	Δ(2950) K315	**	Λ(2110) F05	***	Σ(2100) G17	*	Dibaryons	
N(2700) K113	**	Δ(~3000)		Λ(2325) D03	*	Σ(2250)	***	NN(2170) 1D2	**
N(~3000)				Λ(2350)	***	Σ(2455)	**	NN(2250) 3F3	**
				Λ(2585)	**	Σ(2620)	**	NN(?)	*
						Σ(3000)	*	ΛN(2130) 3S1	**
						Σ(3170)	*	ΞN(?)	*

****	Good, clear, and unmistakable.
***	Good, but in need of clarification or not absolutely certain.
**	Not established; needs confirmation.
*	Evidence weak; could disappear.

Particle[a]	$I(J^P)L_{2I\cdot 2J}$[b]	P^c_{beam} (GeV/c) $\sigma = 4\pi\lambda\!\!\!^{-2}$ (mb)	Mass[d] M (MeV)	Full[e] width Γ (MeV)	Partial decay modes: Mode[f]	Fraction[g] (%)	p[h] (MeV/c)
S=0 I=1/2 NUCLEON RESONANCES (N)							
p	$1/2(1/2^+)$		938.3		See Stable Particle Table		
n			939.6				
N(1440)	$1/2(1/2^+)P'_{11}$	p = 0.61	1400 to	120 to	Nπ	50-70	397
		σ = 31.0	1480	350	Nη	8-18	†
				(200)	Nππ	~30	342
					[Δπ	12-28]*	143
					Nρ	~ 7	†
					Nε]	~ 5	†
N(1520)	$1/2(3/2^-)D'_{13}$	p = 0.74	1510 to	100 to	Nπ	50-60	456
		σ = 23.5	1530	140	Nη	~0.1	149
				(125)	Nππ	35-50	410
					[Δπ	15-25]*	228
					Nρ	15-25	†
					Nε]	< 5	†

Baryon Table *(cont'd)*

Particle[a]	$I(J^P)L_{2I\cdot 2J}$[b]	P^c_{beam} (GeV/c) $\sigma = 4\pi\lambda\!\!\!^{-2}$ (mb)	Mass[d] M (MeV)	Full[e] width Γ (MeV)	Mode[f]	Fraction[g] (%)	p^h (MeV/c)
N(1535)	$1/2(1/2^-)S'_{11}$	p = 0.76	1520 to	100 to	$N\pi$	35-50	467
		σ = 22.5	1560	250	$N\eta$	~35	182
				(150)	$N\pi\pi$	~ 5	422
					[$\Delta\pi$	~ 1]*	242
					$N\rho$	~ 3	†
					$N\epsilon$	~ 2	†
N(1650)	$1/2(1/2^-)S''_{11}$	p = 0.96	1620 to	100 to	$N\pi$	55-65	547
		σ = 16.4	1680	200	$N\eta$	~1.5	346
				(150)	ΛK	~ 8	161
					ΣK	3-10	†
					$N\pi\pi$	~30	511
					[$\Delta\pi$	4-15]*	344
					$N\rho$	~20	†
					$N\epsilon$	< 5	†
N(1675)	$1/2(5/2^-)D'_{15}$	p = 1.01	1660 to	120 to	$N\pi$	30-40	563
		σ = 15.4	1690	180	$N\eta$	~ 1	374
				(155)	ΛK	~0.1	209
					$N\pi\pi$	55-70	529
					[$\Delta\pi$	50-65]*	364
					$N\rho$	~ 5	†
N(1680)	$1/2(5/2^+)F'_{15}$	p = 1.01	1670 to	110 to	$N\pi$	55-65	567
		σ = 15.2	1690	140	$N\eta$	< 1	379
				(125)	ΛK	not seen	218
					$N\pi\pi$	~40	532
					[$\Delta\pi$	~12]*	369
					$N\rho$	~10	†
					$N\epsilon$	~20	†
N(1700)	$1/2(3/2^-)D''_{13}$	p = 1.05	1670 to	70 to	$N\pi$	8-12	580
		σ = 14.5	1730	120	$N\eta$	~ 4	400
				(100)	ΛK	~0.2	250
					$N\pi\pi$	~85	547
					[$\Delta\pi$	15-40]*	385
					$N\rho$	~ 5	†
					$N\epsilon$	<40	†
N(1710)	$1/2(1/2^+)P''_{11}$	p = 1.07	1680 to	90 to	$N\pi$	10-20	587
		σ = 14.2	1740	130	$N\eta$	~25	410
				(110)	ΛK	~15	264
					ΣK	2-10	138
					$N\pi\pi$	>50	554
					[$\Delta\pi$	10-25]*	393
					$N\rho$	25-65	48
					$N\epsilon$	15-40	†
N(1720)	$1/2(3/2^+)P''_{13}$	p = 1.09	1690 to	125 to	$N\pi$	10-20	594
		σ = 13.9	1800	250	$N\eta$	~3.5	420
				(200)	ΛK	~ 5	278
					ΣK	2-5	162
					$N\pi\pi$	~70	561
					[$\Delta\pi$	~20]*	401
					$N\rho$	45-70	104
					$N\epsilon$	~20	†
N(2190)	$1/2(7/2^-)G_{17}$	p = 2.07	2120 to	200 to	$N\pi$	~14	888
		σ = 6.21	2230	500	$N\eta$	~ 3	790
				(350)	ΛK	~0.3	712

Baryon Table *(cont'd)*

Particle[a]	$I(J^P)L_{2I\cdot 2J}$[b]	P^c_{beam} (GeV/c) $\sigma = 4\pi\lambda\!\!\!^{-2}$ (mb)	Mass[d] M (MeV)	Full[e] width Γ (MeV)	Mode[f]	Fraction[g] (%)	p[h] (MeV/c)
N(2220)	$1/2(9/2^+)H_{19}$	p = 2.14	2150 to	300 to	$N\pi$	~18	905
		σ = 5.97	2300	500	$N\eta$	~0.5	811
				(400)	ΛK	~0.2	732
N(2250)	$1/2(9/2^-)G'_{19}$	p = 2.21	2130 to	200 to	$N\pi$	~10	923
		σ = 5.74	2270	500	$N\eta$	~ 2	831
				(300)	ΛK	~0.3	754
N(2600)	$1/2(11/2^-)I_{111}$	p = 3.12	2580 to	>300	$N\pi$	~ 5	1126
		σ = 3.86	2700	(400)			
S=0 I=3/2 DELTA RESONANCES (Δ)							
Δ(1232)	$3/2(3/2^+)P'_{33}$	p = 0.30	1230 to	110 to	$N\pi$	99.4	227
		σ = 94.8	1234	120	$N\gamma$	0.6	259
				(115)			
Δ(1620)	$3/2(1/2^-)S'_{31}$	p = 0.91	1600 to	120 to	$N\pi$	25-35	526
		σ = 17.7	1650	160	$N\pi\pi$	~70	488
				(140)	[$\Delta\pi$	35-50]*	318
					$N\rho$]	<40	†
Δ(1700)	$3/2(3/2^-)D'_{33}$	p = 1.05	1630 to	190 to	$N\pi$	10-20	580
		σ = 14.5	1740	300	$N\pi\pi$	~85	547
				(250)	[$\Delta\pi$	<50]*	385
					$N\rho$]	~40	†
Δ(1900)	$3/2(1/2^-)S''_{31}$	p = 1.44	1850 to	130 to	$N\pi$	6-12	710
		σ = 9.71	2000	300	ΣK	~10	410
				(150)			
Δ(1905)	$3/2(5/2^+)F_{35}$	p = 1.45	1890 to	250 to	$N\pi$	8-15	713
		σ = 9.62	1920	400	ΣK	< 3	415
				(300)	$N\pi\pi$	~80	687
					[$\Delta\pi$	10-30]*	542
					$N\rho$]	~60	421
Δ(1910)	$3/2(1/2^+)P''_{31}$	p = 1.46	1850 to	200 to	$N\pi$	20-25	716
		σ = 9.54	1950	330	ΣK	2-20	421
				(220)	$N\pi\pi$	>40	691
					[$\Delta\pi$	small]*	545
					$N\rho$]	<40	426
Δ(1920)	$3/2(3/2^+)P'''_{33}$	p = 1.48	1860 to	190 to	$N\pi$	14-20	722
		σ = 9.38	2160	300	ΣK	~ 5	431
				(250)			
Δ(1930)	$3/2(5/2^-)D'_{35}$	p = 1.50	1890 to	150 to	$N\pi$	4-14	729
		σ = 9.21	1960	350	ΣK	<10	441
				(250)			
Δ(1950)	$3/2(7/2^+)F'_{37}$	p = 1.54	1910 to	200 to	$N\pi$	35-45	741
		σ = 8.91	1960	340	ΣK	< 1	460
				(240)	$N\pi\pi$	~60	716
					[$\Delta\pi$	~40]*	574
					$N\rho$]	~20	469
Δ(2420)	$3/2(11/2^+)H_{311}$	p = 2.64	2380 to	300 to	$N\pi$	5-15	1023
		σ = 4.68	2450	500			
				(300)			

Baryon Table *(cont'd)*

Particle[a]	$I(J^P)L_{I \cdot 2J}$[b]	P^c_{beam} (GeV/c) $\sigma = 4\pi\lambda\!\!\!^{-2}$ (mb)	Mass[d] M (MeV)	Full[e] width Γ (MeV)	Partial decay modes: Mode	Fraction[g] (%)	p^h (MeV/c)
		S=−1 I=0 LAMBDA RESONANCES (Λ)					
Λ	$0(1/2^+)$		1115.6		See Stable Particle Table		
Λ(1405)	$0(1/2^-)S'_{01}$	Below K^-p threshold	1405 $\pm 5^i$	40 ± 10^i	Σπ	100	152
Λ(1520)	$0(3/2^-)D'_{03}$	p = 0.395	1519.5	15.6 ± 1.0^i	$N\bar{K}$	45 ± 1	244
		σ = 82.3	$\pm 1.0^i$		Σπ	42 ± 1	267
					Λππ	10 ± 1	252
					Σππ	0.9 ± 0.1	152
					Λγ	0.8 ± 0.2	351
Λ(1600)	$0(1/2^+)P'_{01}$	p = 0.58	1560 to	50 to	$N\bar{K}$	15-30	343
		σ = 41.6	1700	250	Σπ	10-60	336
				(150)			
Λ(1670)	$0(1/2^-)S''_{01}$	p = 0.74	1660 to	25 to	$N\bar{K}$	15-25	414
		σ = 28.5	1680	50	Σπ	20-60	393
				(35)	Λη	15-35	64
Λ(1690)	$0(3/2^-)D''_{03}$	p = 0.78	1685 to	50 to	$N\bar{K}$	20-30	433
		σ = 26.1	1695	70	Σπ	20-40	409
				(60)	Λππ	~25	415
					Σππ	~20	350
Λ(1800)	$0(1/2^-)S'''_{01}$	p = 1.01	1720 to	200 to	$N\bar{K}$	25-40	528
		σ = 17.5	1850	400	Σπ	seen	493
				(300)	Σ(1385)π	seen	345
					$N\bar{K}$*(892)	seen	†
Λ(1800)	$0(1/2^+)P''_{01}$	p = 1.01	1750 to	50 to	$N\bar{K}$	20-50	528
		σ = 17.5	1850	250	Σπ	10-40	493
				(150)	Σ(1385)π	seen	345
					$N\bar{K}$*(892)	30-60	†
Λ(1820)	$0(5/2^+)F'_{05}$	p = 1.06	1815 to	70 to	$N\bar{K}$	55-65	545
		σ = 16.5	1825	90	Σπ	8-14	508
				(80)	Σ(1385)π	5-10	362
Λ(1830)	$0(5/2^-)D_{05}$	p = 1.08	1810 to	60 to	$N\bar{K}$	3-10	553
		σ = 16.0	1830	110	Σπ	35-75	515
				(95)	Σ(1385)π	>15	371
Λ(1890)	$0(3/2^+)P'_{03}$	p = 1.21	1850 to	60 to	$N\bar{K}$	20-35	599
		σ = 13.6	1910	200	Σπ	3-10	559
				(100)	Σ(1385)π	seen	420
					$N\bar{K}$*(892)	seen	233
Λ(2100)	$0(7/2^-)G_{07}$	p = 1.68	2090 to	100 to	$N\bar{K}$	25-35	751
		σ = 8.68	2110	250	Σπ	~ 5	704
				(200)	Λη	< 3	617
					ΞK	< 3	483
					Λω	< 8	443
					$N\bar{K}$*(892)	10-20	514
Λ(2110)	$0(5/2^+)F''_{05}$	p = 1.70	2090 to	150 to	$N\bar{K}$	5-25	757
		σ = 8.53	2140	250	Σπ	10-40	711
				(200)	Λω	seen	455
					Σ(1385)π	seen	589
					$N\bar{K}$*(892)	10-60	524
Λ(2350)	$0(9/2^+)$	p = 2.29	2340 to	100 to	$N\bar{K}$	~12	915
		σ = 5.85	2370	250	Σπ	~10	867
				(150)			

Baryon Table *(cont'd)*

Particle[a]	$I(J^P)L_{I\cdot 2J}$[b]	P^c_{beam} (GeV/c) $\sigma = 4\pi\lambda\!\!\!^{-2}$ (mb)	Mass[d] M (MeV)	Full[e] width Γ (MeV)	Partial decay modes: Mode	Fraction[g] (%)	p^h (MeV/c)
S=−1 I=1 SIGMA RESONANCES (Σ)							
Σ	$1(1/2^+)$		(+)1189.4		See Stable Particle Table		
			(0)1192.5				
			(−)1197.3				
Σ(1385)	$1(3/2^+)P_{13}^{j}$	Below	(+)1382.3 ± 0.4	35 ± 1	Λπ	88 ± 2	208
		K^-p	$S=1.6^j$	$S=1.0^j$	Σπ	12 ± 2	127
		threshold	(0)1382.0 ± 2.5	~35			
			$S=1.6^j$				
			(−)1387.4 ± 0.6	40 ± 2			
			$S=2.2^j$	$S=1.9^j$			
Σ(1660)	$1(1/2^+)P'_{11}$	p = 0.72	1630 to	40 to	$N\bar{K}$	10-30	405
		σ = 29.9	1690	200	Λπ	seen	439
				(100)	Σπ	seen	385
Σ(1670)	$1(3/2^-)D''_{13}$	p = 0.74	1665 to	40 to	$N\bar{K}$	7-13	414
		σ = 28.5	1685	80	Λπ	5-15	447
				(60)	Σπ	30-60	393
Σ(1750)	$1(1/2^-)S''_{11}$	p = 0.91	1730 to	60 to	$N\bar{K}$	10-40	486
		σ = 20.7	1800	160	Λπ	seen	507
				(90)	Σπ	< 8	455
					Ση	15-55	81
Σ(1775)	$1(5/2^-)D_{15}$	p = 0.96	1770 to	105 to	$N\bar{K}$	37-43	508
		σ = 19.0	1780	135	Λπ	14-20	525
				(120)	Σπ	2-5	474
					Σ(1385)π	8-12	324
					Λ(1520)π	17-23	198
Σ(1915)	$1(5/2^+)F'_{15}$	p = 1.26	1900 to	80 to	$N\bar{K}$	5-15	618
		σ = 12.8	1935	160	Λπ	seen	622
				(120)	Σπ	seen	577
					Σ(1385)π	< 5	440
Σ(1940)	$1(3/2^-)D'''_{13}$	p = 1.32	1900 to	150 to	$N\bar{K}$	<20	637
		σ = 12.1	1950	300	Λπ	seen	639
				(220)	Σπ	seen	594
					Σ(1385)π	seen	460
					Λ(1520)π	seen	354
					$\Delta(1232)\bar{K}$	seen	410
					$N\bar{K}$*(892)	seen	320
Σ(2030)	$1(7/2^+)F_{17}$	p = 1.52	2025 to	150 to	$N\bar{K}$	17-23	702
		σ = 9.93	2040	200	Λπ	17-23	700
				(180)	Σπ	5-10	657
					ΞK	< 2	412
					Σ(1385)π	5-15	529
					Λ(1520)π	10-20	430
					$\Delta(1232)\bar{K}$	10-20	498
					$N\bar{K}$*(892)	< 5	438
Σ(2250)	1(?)	p = 2.04	2210 to	60 to	$N\bar{K}$	<10	851
		σ = 6.76	2280	150	Λπ	seen	842
				(100)	Σπ	seen	803

Baryon Table *(cont'd)*

Particle[a]	$I(J^P)L_{2I \cdot 2J}$[b]	Mass[d] M (MeV)	Full[e] width Γ (MeV)	Mode	Fraction (%)	p[h] (MeV/c)
		S=−2 I=1/2 CASCADE RESONANCES (Ξ)				
Ξ	$1/2(1/2^+)$	(0)1314.9 (−)1321.3		See Stable Particle Table		
Ξ(1530)	$1/2(3/2^+)P_{13}$	(0)1531.8 ± 0.3 S = 1.3[j] (−)1535.0 ± 0.6	9.1 ± 0.5 10.1 ± 1.9	Ξπ	100	148
→ →						
Ξ(1820)	1/2(3/2)	1823 ±6[i]	20^{+15}_{-10}[i]	$\Lambda\overline{K}$ $\Sigma\overline{K}$ Ξπ Ξ(1530)π	~45 ~10 small ~45	396 306 413 231
→						
Ξ(2030)	1/2(?)	2024 ±6[i]	16^{+15}_{-5}[i]	$\Lambda\overline{K}$ $\Sigma\overline{K}$ Ξπ Ξ(1530)π	~20 ~80 small small	587 524 573 418
→ → → →						
		OTHER BARYONS				
Ω^-	$0(3/2^+)$	1672.4		See Stable Particle Table		
→ → → → → → → → → Λ_c^+	$0(1/2^+)$	2282		See Stable Particle Table		

→ Each arrow in the left-hand margin indicates there is an entry in the Data Card Listings for a baryon that is not well enough established (status less than 3 stars) to be included here. There is a short list of *all* the baryons in the Listings, whatever their status, at the front of this Table.

†. This mode is energetically forbidden when the nominal mass of the decaying resonance (and of any resonance in the final state) is used, but is in fact allowed due to the nonzero widths of the resonance(s).

*. The modes in brackets are subreactions of the Nππ mode.

a. The nominal mass here (in MeV) is used for identification. See column 4 for the actual mass.

b. When there is more than one baryon with the same quantum numbers, one prime is attached to the spectroscopic symbol for the first of them (e.g., S'_{11}), two primes to the second, etc.

c. The quantities here are calculated using the nominal mass of column 1.

d. Usually a conservatively large range of masses rather than a statistical average of the various determinations of the mass is given. In these cases, the mass determinations are nearly entirely from various phase-shift analyses of more or less the same data. It is thus not appropriate to treat the determinations as independent measurements or to average them together. The masses, widths, and branching fractions in this Table are Breit-Wigner parameters. The Data Card Listings also include pole parameters where they are available.

e. Usually a conservatively large range of widths rather than a statistical average of the various determinations of the width is given (see note *d* for the reason). The nominal value in parentheses is then simply a best guess.

f. For information on the Nγ decay modes, see the Note on N and Δ Resonances in the Listings.

g. Most of the inelastic branching fractions come from partial-wave analyses, and these determine $\sqrt{xx'}$, where x and x′ are the elastic and inelastic branching fractions, not x′ directly. Thus any uncertainty (and it is often considerable) in x carries over into x′. When x′ so determined is really poorly known, we here simply note that the mode is seen. The values of $\sqrt{xx'}$ are given in the Data Card Listings.

h. Foı a 2-body decay mode, this is the momentum of the decay products in the rest frame of the decaying particle. For a mode with more than two decay products, this is the maximum momentum any of the products can have in this frame. The nominal mass of column 1 is used, as is the nominal mass of any resonance in the final state.

i. The error given here is only an educated guess. It is larger than the error on the weighted average of the published values (the error on this average is given in the Listings).

j. The error given here has been scaled up by the "S factor" (see the * footnote to the Stable Particle Table for how S is defined) because the various measurements disagree more seriously than one would expect from statistics.

CHARACTERISTICS OF PARTICLES AND PARTICLE DISPERSOIDS

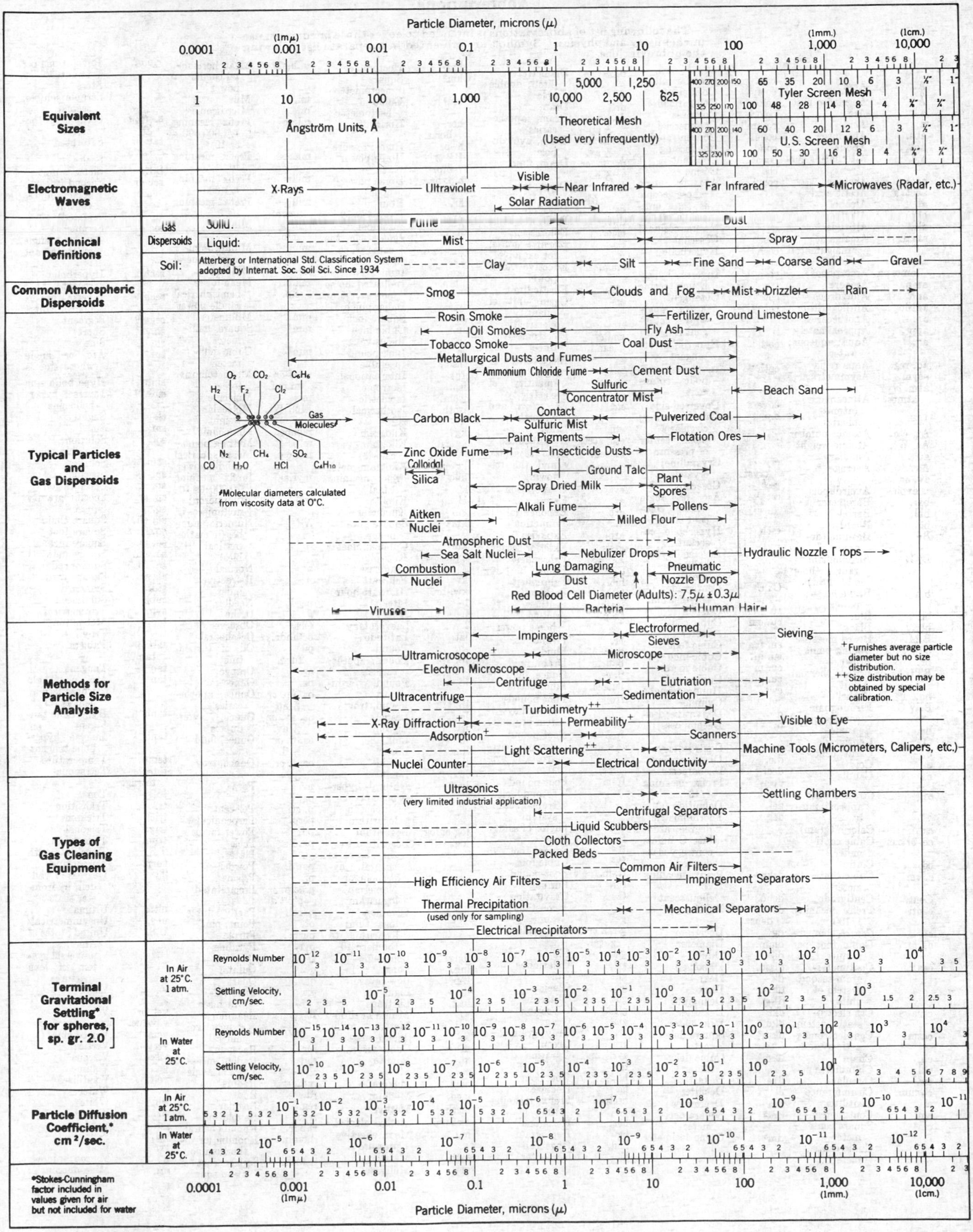

C. E. Lapple, Stanford Research Institute Journal, Vol. 5, p.95 (Third Quarter, 1961)

ABBREVIATIONS AND SYMBOLS

Abbreviations

The following list of abbreviations is intended to cover those in common use in chemistry and physics. Symbols are presented in a separate list following the abbreviations.

A.	Acre
Å	Ångström unit
a	Are
a.	Acid
abs.	Absolute
abt.	About
a.c.	Alternating current
acet.	Acetone
acet. a.	Acetic acid
al.	Alcohol
alk.	Alkali
alt.	Altitude
amal.	Amalgam; amalgamated
amor. *or* amorph.	Amorphous
amp.	Ampere
anh.	Anhydrous
antilog	Antilogarithm
ap.	Apothecaries'
appr.	Approximately
aq.	Aqua; aqueous; water
aq. reg.	Aqua regia
asym.	Asymmetrical
atm. *or* atmos.	Atmosphere (atmospheric)
At. No.	Atomic number
At. Wt.	Atomic weight
aux.	Auxiliary
Av.	Average
av. *or* avoir.	Avoirdupois
bar.	Barometer
bbl.	Barrel
bd.	Board
Bé	Beaumé (degrees)
B.G.	Birmingham gauge (hoop and sheet)
b.h.p.	Brake horse power
bl.	Blue
blk.	Black
B.M.	Board measure
b.p.	Boiling point
br.	Brown
BTU	British thermal unit
bu.	Bushel
B.W.G.	Birmingham wire gauge
bz.	Benzene
C	Centigrade
c	Carat; centi-
c.	Cold
ca	Candle
ca.	*Circa*, about; approximately
cal.	Calorie (gram)
cc. *or* c.c.	Cubic centimeter
cd.	Cord
c. cm	Cubic centimeter
Cent.	Centigrade
centi-	Prefix meaning 1/100
cf.	*Confer*, compare
c.f.m.	Cubic foot per minute
cgs	Centimeter-gram-second system of units
cgse	Cgs electrostatic system
cgsm	Cgs electromagnetic system
ch.	Chain
chl.	Chloroform
cir.	Circular
circum.	Cirumference
cl	Centiliter
cm	Centimeter
cm^2	Square centimeter
cm^3	Cubic centimeter
c.m.	Circular mil
coef.	Coefficient
colog	Cologarithm
colorl.	Colorless
comm'l	Commercial
conc.	Concentrated
cond.	Condensing
const.	Constant
cos	Cosine
cos^{-1}	Arc or angle whose cosine is . . .; anticosine of; inverse cosine of
cosec	Cosecant
cosh	Hyperbolic cosine
$cosh^{-1}$	Inverse hyperbolic cosine
cot	Cotangent
cot^{-1}	Arc or angle whose cotangent is . . .
coth	Hyperbolic cotangent
$coth^{-1}$	Inverse hyperbolic cotangent
covers	Coversed sine
c.p.	Candle power; circular pitch; center of pressure
cry. *or* cryst.	Crystalline; crystals
csc	Cosecant
csc^{-1}	Arc or angle whose cosecant is . . .
csch	Hyperbolic cosecant
$csch^{-1}$	Inverse hyperbolic cosecant
CTU	Centigrade thermal unit
cu.	Cubic
cu. cm	Cubic centimeter
cu. ft.	Cubic foot
cu. in.	Cubic inch
cu. m	Cubic meter
cu. yd.	Cubic yard
cwt.	Hundredweight
cyl.	Cylinder
d	Derivative; deci-
d.	Decomposes; day
d.	Dextrorotary
d.c.	Direct current
dec.	Decomposes
deci-	Prefix meaning 1/10
def.	Definition (s)
deg	Thermometric degree; absolute C unless contrary is indicated
deka-	Prefix meaning 10
deliq.	Deliquescent
den. *or* dens.	Density
dg	Decigram
diam.	Diameter
dil.	Dilute
dissd.	Dissolved
dk	Deka-
dk.	Dark
dkg.	Dekagram
dkl	Dekaliter
dkm	Dekameter
dkm^2	Square dekameter
dkm^3	Cubic dekameter
dks	Dekastere
dl	Deciliter
dm	Decimeter
dm^2	Square decimeter
dm^3	Cubic decimeter
d.p.	Diametral pitch; double pole
dr.	Dram
dr. ap. *or* ʒ ap.	Dram, apothecaries'
dr. av. *or* ʒ av.	Dram, avoirdupois
dr. fl. *or* ʒ fl.	Dram, fluid
dr. t. *or* ʒt.	Dram, troy
ds	Decistere
dwt.	Pennyweight
efflor.	Efflorescent
e.g.	*Exempli gratia*, for example
e.h.p.	Effective horse power
E.L.	Elastic limit
em	Cgsm unit of quantity of electricity
emf *or* e.m.f.	Electromotive force
es	Electrostatic or cgse unit of quantity of electricity
etc.	*Et. cetera*, and so forth
eth.	Ether
eth. acet.	Ethyl acetate
et. seq.	*Et sequentes*, and the following
evap.	Evaporation
ex.	Excess
exp	Exponential function
exp.	Explodes
exsec	Exterior secant
F	Fahrenheit
f.	From
fahr.	Fahrenheit
fath.	Fathom
feath.	Feathery
f.h.p.	Friction horse power
fir.	Firkin
fl.	Fluid
fl. dr.	Dram, fluid
fl. oz.	Ounce, fluid
fluores.	Fluorescent
fps	Foot-pound-second system of units
fpse	Foot-pound-second electrostatic system
fpsm	Foot-pound-second electromagnetic system
F.S.	Factor of safety
ft.	Foot
$ft.^2$	Square foot
$ft.^3$	Cubic foot
ft.-lb.	Foot-pound
fur.	Furlong
G	Gravitation constant
g	Gram
g-cal. *or* g.-cal.	Gram calorie
gal.	Gallon
gel.	Gelatinous
gi.	Gill
glac.	Glacial
glit.	Glittering
glyc.	Glycerine
gm.	Gram
gr.	Gray; grain
grn.	Green
gyr.	Gyration
h	Hecto-
h.	Hot; hour
ha	Hectare
hecto-	Prefix meaning 100
hex.	Hexagonal
hg	Hectogram
hhd.	Hogshead
hl	Hectoliter
hm	Hectometer
hm^2	Square hectometer
hm^3	Cubic hectometer
hor. *or* horiz.	Horizontal
h.-p.	High-Pressure
HP *or* h.p.	Horse power
h.p.-hr.	Horse power-hour
hr.	Hour
hyg.	Hygroscopic
i.	Insoluble
ibid.	*Ibidem*, in the same place
i.e.	*Id est*, that is
ign.	Ignites
i.h.p.	Indicated horse power
in.	Indigo; inch
$in.^2$	Square inch
$in.^3$	Cubic inch
inc.	Inclusive
in.-lb.	Inch-pound
insol.	Insoluble
Int.	International
iso.	Isotropic
isom.	Isometric
isoth.	Isothermal
k	Kilo-
kg	Kilogram
kg-cal.	Kilogram-calorie
kg-m	Kilogram-meter
kilo-	Prefix meaning 1,000
kl	Kiloliter
km	Kilometer
km^2	Square kilometer
km^3	Cubic kilometer
kva.	Kilovolt-ampere
kw.	Kilowatt
kw.-hr.	Kilowatt-hour
l	Liter
l.	Long
l	Laevorotary
lat.	Latitude
lb.	Pound
lb. ap.	Pound, apothecaries'
lb. av.	Pound, avoirdupois
lb. t.	Pound, troy
leaf.	Leaflets
lgr.	Ligroin
li.	Link
lin.	Linear
liq.	Liquid
lim.	Limit
ln	Natural hyperbolic or Napierian logarithm
log *or* log.	Logarithm
log_e	Logarithm to the base *e*; natural, hyperbolic or Napierian logarithm
log_{10}	Common logarithm; logarithm to the base 10
long.	Longitude
lng.	Long
l.-p.	Low-pressure
lt.	Light
lust.	Lustrous
♏	Minim or drop
m	Meter; milli-
m^2	Square meter
m^3	Cubic meter
m.	Minute
m.	Meta-
max.	Maximum
med.	Medium
meth.	Methyl
meth. al.	Methyl alcohol
m.e.p.	Mean effective pressure
met.	Metallic
mg	Milligram
m.h.c.p.	Mean horizontal candle power
mi.	Mile
mic.	Microscopic
micro-	Prefix meaning 1/1,000,000 or 10^{-6}
micro-micro	Prefix meaning 10^{-12}
milli-	Prefix meaning 1/1,000
milli-micro-	Prefix meaning 10^{-9}
min *or* min.	Minute
min.	Minim; minimum; mineral
ml	Milliliter
m.l.h.c.p.	Mean lower hemispherical candle power
mm	Millimeter
mm^2	Square millimeter
mm^3	Cubic millimeter
mmf *or* m.m.f.	Magnetomotive force
mol.	Molecule
Mol. Wt.	Molecular weight
monocl.	Monoclinic
m.p.	Melting point
m.s.c.p.	Mean spherical candle power
myria-	Prefix meaning 10,000 or 10
$m\mu$	Millimicron; millimicro-
N	Numeric; number (in mathematical tables)
n.	Normal
n	Refractive index
need.	Needles
o	Ortho-
Obs.	Observer
octahdr.	Octahedral
oil turp.	Oil of turpentine
or.	Orange
oz.	Ounce
oz. ap. *or* ℥ ap.	Ounce, apothecaries'
oz. av. *or* ℥ av.	Ounce, avoirdupois
oz. fl. *or* ℥ fl.	Ounce, fluid
oz. t. *or* ℥ t.	Ounce, troy
p	Para-
pa.	Pale
p. ct.	Per cent
perp.	Perpendicular
p.f.	Power factor
pk.	Peck
pl.	Plates
powd.	Powder
pr.	Prisms
precip. *or* p'p't'd	Precipitated
p. sol.	Partly soluble
pt.	Point; pint
purp.	Purple
pyr.	Pyridine
Q	Quantity
q	Quintal
qt.	Quart
q.v.	*Quod vide*, which see
R	Réaumur; radioactive mineral
rac	Racemic
rad	Radian, measure of angle
rad.	Radius
rd.	Rod
reg.	Regular
rev.	Revolution
rhbdr.	Rhombohedral
rhomb.	Rhombic or orthorhombic
R.M.S.	Square root of mean square
r.p.m.	Revolutions per minute
s	Stere
s.	Scruple; soluble; second
s. ap. *or* ℈	Scruple, apothecaries'
sat. *or* sat'd	Saturated
sc.	Scales
S.E.	Siemens unit
sec *or* sec.	Second (mean solar unless contrary is stated)
sec	Secant
sec^{-1}	Arc or angle whose secant is . . .
sech	Hyperbolic secant
$sech^{-1}$	Inverse hyperbolic secant
segm.	Segment
sh.	Short
sin	Sine
sin^{-1}	Arc or angle whose sine is . . .
sinh	Hyperbolic sine
$sinh^{-1}$	Inverse hyperbolic sine
sl.	Slightly
sm.	Small
sol.	Solution; soluble
soln.	Solution
sp.	Specific
specif.	Specification
sp. gr.	Specific gravity
sq.	Square
sq. ch.	Square chain
sq. ft.	Square foot
sq. in.	Square inch
sq. mi.	Square mile
sq. rd.	Square rod
sq. yd.	Square yard
std.	Standard
subl.	Sublimes
sym.	Symmetrical
t	Metric ton
t.	Troy
tab. *or* tabl.	Tablets
tan	Tangent
tan^{-1}	Arc or angle whose tangent is . . .
tanh	Hyperbolic tangent
$tanh^{-1}$	Inverse hyperbolic tangent
temp.	Temperature
tetr. *or* tetrag.	Tetragonal
tn.	Ton
tr.	Transition
tricl.	Triclinic
trig.	Trigonal
trim.	Trimetric
T.S.	Tensile strength
turp.	Turpentine
Tw.°	Degrees Twaddell, hydrometer scale
ult.	Ultimate
uns.	Unsymmetrical
U.S.	United States of America; universal system of lens apertures
v.	Very
v.	*Vide*, see
vel. *or* veloc.	Velocity
vers	Versed sine
vert	Vertical
visc.	Viscous
vol.	Volume
volt.	Volatilizes
w.	Water
wh.	White
wt.	Weight
yd.	Yard
yel.	Yellow
yr.	Year
μ	Micron; micro-
$\mu\mu$	Micromicron; micromicro-

SPELLING AND SYMBOLS FOR UNITS

From "Units of Weight and Measure"
L. B. Chisholm, National Bureau of Standards
Miscellaneous Publication 286 (May, 1967)

The spelling of the names of units as adopted by the National Bureau of Standards is that given in the list below. The spelling of the metric units is in accordance with that given in the law of July 28, 1866, legalizing the Metric System in the United States.

Following the name of each unit in the list below is given the symbol that the Bureau has adopted. Attention is particularly called to the following principles:

1. No period is used with symbols for units. Whenever "in" for inch might be confused with the preposition "in", "inch" should be spelled out.

2. The exponents "²" and "³" are used to signify "square" and "cubic," respectively, instead of the symbols "sq" or "cu," which are, however, frequently used in technical literature for the U. S. Customary units.

3. The same symbol is used for both singular and plural.

Some Units and Their Symbols

Unit	Symbol	Unit	Symbol	Unit	Symbol
acre	acre	fathom	fath	millimeter	mm
are	a	foot	ft	minim	minim
barrel	bbl	furlong	furlong	ounce	oz
board foot	fbm	gallon	gal	ounce, avoirdupois	oz avdp
bushel	bu	grain	grain	ounce, liquid	liq oz
carat	c	gram	g	ounce, troy	oz tr
Celsius, degree	°C	hectare	ha	peck	peck
centare	ca	hectogram	hg	pennyweight	dwt
centigram	cg	hectoliter	hl	pint, liquid	liq pt
centiliter	cl	hectometer	hm	pound	lb
centimeter	cm	hogshead	hhd	pound, avoirdupois	lb avdp
chain	ch	hundredweight	cwt	pound, troy	lb tr
cubic centimeter	cm^3	inch	in	quart, liquid	liq qt
cubic decimeter	dm^3	International Nautical Mile	INM	rod	rod
cubic dekameter	dam^3			second	s
cubic foot	ft^3	Kelvin, degree	°K	square centimeter	cm^2
cubic hectometer	hm^3	kilogram	kg	square decimeter	dm^2
cubic inch	in^3	kiloliter	kl	square dekameter	dam^2
cubic kilometer	km^3	kilometer	km	square foot	ft^2
cubic meter	m^3	link	link	square hectometer	hm^2
cubic mile	mi^3	liquid	liq	square inch	in^2
cubic millimeter	mm^3	liter	liter	square kilometer	km^2
cubic yard	yd^3	meter	m	square meter	m^2
decigram	dg	microgram	μg	square mile	mi^2
deciliter	dl	microinch	μin	square millimeter	mm^2
decimeter	dm	microliter	μl	square yard	yd^2
dekagram	dag	micron	μm	stere	stere
dekaliter	dal	mile	mi	ton, long	long ton
dekameter	dam	milligram	mg	ton, metric	t
dram, avoirdupois	dr avdp	milliliter	ml	ton, short	short ton
				yard	yd

SYMBOLS AND TERMINOLOGY FOR PHYSIOCHEMICAL QUANTITIES AND UNITS

Reproduced by permission of the Assistant Secretary, Publications, of IUPAC from the *Manual of Symbols and Terminology for Physiochemical Quantities and Units (1979),* International Union of Pure and Applied Chemistry, Division of Physical Chemistry, Commission on Symbols, Terminology and Units. The above referenced Manual is published by and is available from Pergamon Press, Inc., Maxwell House, Fairview Park, Elmsford, New York 10523.

This Handbook, the *Handbook of Chemistry and Physics,* also contains a section on symbols, terminology, etc. for Physics which immediately follows this section on symbols, terminology, etc. for Chemistry. Rules and guidance which apply to both Chemistry and Physics and which do not require separate interpretation are reproduced only in the section, "Symbols, Units and Nomenclature in Physics" and are not reproduced in the section, "Symbols and Terminology for Physiochemical Quantities and Units". Paragraphs in the following are numbered the same as those in the complete IUPAC Manual.

1. PHYSICAL QUANTITIES AND SYMBOLS FOR PHYSICAL QUANTITIES

1.2 Base Physical Quantities

Physical quantities are generally organized in a dimensional system built upon seven base quantities. These base quantities, each of which has its own dimension, and the symbols used to denote them, are as follows:

Base physical quantity	*Symbol for quantity*
Length	l
Mass	m
Time	t
Electric current	I
Thermodynamic temperature	T
Amount of substance	n
Luminous intensity	I_{v}

Luminous intensity is seldom if ever needed in physical chemistry.

One of these independent base quantities is of special importance to chemists but until recently had no generally accepted name, although units such as the mole have been used for it. The name 'amount of substance' is now reserved for this quantity.

The definition of amount of substance, as of all other physical quantities (see Section 5), has nothing to do with any choice of *unit*, and in particular has nothing to do with the particular unit of amount of substance called the mole (see Section 3.6). It is now as inconsistent to call n the 'number of moles' as it is to call m the 'number of kilograms' or l the 'number of meters', since n, m, and l are symbols for quantities not for numbers.

The amount of a substance is proportional to the number of *specified* elementary entities of that substance. The proportionality factor is the same for all substances; its reciprocal is the Avogadro constant. The specified elementary entity may be an atom, a molecule, an ion, a radical, an electron, etc., or any *specified* group of such particles.

1.3 Derived Physical Quantities

All other physical quantities are regarded as being derived from, and as having dimensions derived from, the seven independent base physical quantities by definitions involving only multiplication, division, differentiation, and/or integration. Examples of derived physical quantities are given, often with brief definitions, in Section 2.

1.4 Use of the Words 'Specific' and 'Molar' in the Names of Physical Quantities

The word 'specific' before the name of an extensive physical quantity is restricted to the meaning 'divided by mass'. For example specific volume is the volume divided by the mass. When the extensive quantity is represented by a capital letter, the corresponding specific quantity may be represented by the corresponding lower case letter. *Examples;* volume: V; specific volume: $v=V/m$; heat capacity at constant pressure: C_p; specific heat capacity at constant pressure: $c_p=C_p/m$.

The word 'molar' before the name of an extensive quantity is restricted to the meaning 'divided by amount of substance'. For example molar volume is the volume divided by the amount of substance. The subscript m attached to the symbol for the extensive quantity denotes the corresponding molar quantity. *Examples*; volume: V; Gibbs energy: G; molar volume: $V_{\mathrm{m}}=V/n$; molar Gibbs energy: $G_{\mathrm{m}}=G/n$.

The subscript m may be omitted when there is no risk of ambiguity. Lower case letters may be used to denote molar quantities when there is no risk of misinterpretation.

The symbol X_{B}, where X denotes an extensive quantity and B is the chemical symbol for a substance, denotes the partial molar quantity of the substance B defined by the relation:

$$X_B = (\partial X/\partial n_B)_{T,p,n_C,\ldots}$$

For a pure substance B the partial molar quantity X_B and the molar quantity X_m are identical. The partial molar quantity X_B of pure substance B, which is identical with the molar quantity X_m of pure substance B, may be denoted by X_B,*
where the superscript * denotes 'pure', so as to distinguish it from the partial molar quantity X_B of substance B in a mixture.

1.6 Printing of Subscripts and Superscripts

Subscripts or superscripts which are themselves symbols for physical quantities or numbers should be printed in italic (sloping) type and all others in roman (upright) type. *Examples*; C_p for heat capacity at constant pressure, but C_{B} for heat capacity of substance B.

2. RECOMMENDED NAMES AND SYMBOLS FOR QUANTITIES IN CHEMISTRY AND PHYSICS

The following list contains the recommended symbols for the most important quantities likely to be used by chemists. Whenever possible the symbol used for a physical quantity should be that recommended. In a few cases where conflicts were foreseen alternative recommendations have been made. Bold-faced italic (sloping) as well as ordinary italic (sloping) type can also sometimes be used to resolve conflicts. Further flexibility can be obtained by the use of capital letters as variants for lower-case letters, and *vice versa*, when no ambiguity is thereby introduced.

For example, d and D may be used instead of d_{i} and d_{e} for internal and external diameter in a context in which no quantity appears, such as diffusion coefficient, for which the recommended symbol is D. Again, the recommended symbol for power is P and for pressure is p or P, but P and p may be used for two powers or for two pressures; if power and pressure appear together, however, P should be used only for power and p only for pressure, and necessary distinctions between different powers or between different pressures should be made by the use of subscripts or other modifying signs.

When the above recommendations are insufficient to resolve a conflict or where a need arises for other reasons, an author is of course free to choose an *ad hoc* symbol. Any *ad hoc* symbol should be particularly carefully defined.

In the following list, where two or more symbols are indicated for a given quantity and are separated only by commas (without parentheses), they are on an equal footing; symbols within parentheses are reserve symbols.

Any description given after the name of a physical quantity is merely for identification and is not intended to be a complete definition.

Vector notation (bold-faced italic or sloping type) is used where appropriate in Section 2.6; it may be used when convenient also for appropriate quantities in other Sections.

2.1 Space, time, and related quantities

2.1.01	length	l
2.1.02	height	h
2.1.03	radius	r
2.1.04	diameter	d
2.1.05	path, length of arc	s
2.1.06	wavelength	λ
2.1.07	wavenumber: $1/\lambda$	σ[a], $\tilde{\nu}$[b]
2.1.08	plane angle	$\alpha, \beta, \gamma, \theta, \phi$
2.1.09	solid angle	ω, Ω
2.1.10	area	A, S, A_{s}[c]
2.1.11	volume	V
2.1.12	time	t
2.1.13	frequency	ν, f
2.1.14	circular frequency: $2\pi\nu$	ω
2.1.15	period: $1/\nu$	T
2.1.16	characteristic time interval, relaxation time, time constant	τ
2.1.17	velocity	v, u, w, c
2.1.18	angular velocity: $\mathrm{d}\phi/\mathrm{d}t$	ω
2.1.19	acceleration	a
2.1.20	acceleration of free fall	g

2.2 Mechanical and related quantities

2.2.01	mass	m
2.2.02	reduced mass	μ
2.2.03	specific volume (volume divided by mass)	v
2.2.04	density (mass divided by volume)	ρ
2.2.05	relative density (ratio of the density to that of a reference substance)	d
2.2.06	moment of inertia	I
2.2.07	momentum	p
2.2.08	force	F
2.2.09	weight	G, (W)
2.2.10	moment of force	M
2.2.11	angular momentum	L
2.2.12	work (force times path)	w, W
2.2.13	energy	E
2.2.14	potential energy	E_{p}, V, Φ
2.2.15	kinetic energy	E_{k}, T, K
2.2.16	Hamiltonian function	H
2.2.17	Lagrangian function	L
2.2.18	power (energy divided by time)	P
2.2.19	pressure	p, (P)
2.2.20	normal stress	σ

a In solid-state studies, wavevector $\boldsymbol{k}$ is used ($|\boldsymbol{k}| = 2\pi/\lambda$).
b For electromagnetic radiation referred to a vacuum $\tilde{\nu} = \nu/c = 1/\lambda_{\mathrm{vac}}$ is preferred.
c The symbol A_{s} may be used when necessary to avoid confusion with the symbol A for Helmholtz energy.

2.2.21	shear stress	τ
2.2.22	linear strain (relative elongation): $\Delta l/l_0$	ϵ, e
2.2.23	volume strain (bulk strain): $\Delta V/V_0$	θ
2.2.24	modulus of elasticity (normal stress divided by linear strain, Young's modulus)	E
2.2.25	shear modulus (shear stress divided by shear angle)	G
2.2.26	compressibility: $-V^{-1}(\mathrm{d}V/\mathrm{d}p)$	κ
2.2.27	compression (bulk) modulus: $-V_0(\Delta p/\Delta V)$	K
2.2.28	velocity of sound	c
2.2.29	viscosity	$\eta, (\mu)$
2.2.30	fluidity: $1/\eta$	ϕ
2.2.31	kinematic viscosity: η/ρ	ν
2.2.32	friction coefficient (frictional force divided by normal force)	$\mu, (f)$
2.2.33	surface tension	γ, σ
2.2.34	angle of contact	θ
2.2.35	diffusion coefficient	D
2.2.36	mass transfer coefficient (dimension of length divided by time)	k_d

2.3 Molecular and related quantities

2.3.01	relative atomic mass of an element formerly called 'atomic weight')[a]	A_r
2.3.02	relative molecular mass of a substance (formerly called 'molecular weight')[b]	M_r
2.3.03	molar mass (mass divided by amount of substance)	M
2.3.04	Avogadro constant	L, N_A
2.3.05	number of molecules or other entries	N
2.3.06	amount of substance[c]	$n, (\nu)$
2.3.07	mole fraction of substance B: $n_B/\Sigma_i n_i$	x_B, y_B
2.3.08	mass fraction of substance B	w_B
2.3.09	volume fraction of substance B	ϕ_B
2.3.10	molality of solute substance B (amount of B divided by mass of solvent)[d]	m_B
2.3.11	amount-of-substance concentration of substance B (amount of B divided by the volume of the solution)[e]	c_B, [B]
2.3.12	mass concentration of substance B (mass of B divided by the volume of the solution)	ρ_B
2.3.13	surface concentration, surface excess	Γ
2.3.14	collision diameter of a molecule	d, σ
2.3.15	mean free path	l, λ
2.3.16	collision number (number of collisions divided by volume and by time)	Z
2.3.17	grand partition function (system)	Ξ
2.3.18	partition function (system)	Q, Z
2.3.19	partition function (particle)	q, z
2.3.20	statistical weight	g
2.3.21	symmetry number	σ, s
2.3.22	characteristic temperature	Θ

2.4 Thermodynamic and related quantities

2.4.01	thermodynamic temperature, absolute temperature	T
2.4.02	Celsius temperature	t, θ[f]
2.4.03	(molar) gas constant	R
2.4.04	Boltzmann constant	k
2.4.05	heat	q, Q[g]
2.4.06	work	w, W[g]
2.4.07	internal energy	$U, (E)$
2.4.08	enthalpy: $U + pV$	H
2.4.09	entropy	S
2.4.10	Helmholtz energy: $U - TS$	A
2.4.11	Massieu function: $-A/T$	J
2.4.12	Gibbs energy: $H - TS$	G
2.4.13	Planck function: $-G/T$	Y
2.4.14	compression factor: pV_m/RT	Z
2.4.15	heat capacity	C
2.4.16	specific heat capacity (heat capacity divided by mass; the name 'specific heat' is not recommended)	c
2.4.17	ratio C_p/C_V	$\gamma, (\kappa)$
2.4.18	Joule–Thompson coefficient	μ
2.4.19	thermal conductivity	λ, k
2.4.20	thermal diffusivity: $\lambda/\rho c_p$	a
2.4.21	coefficient of heat transfer (density of heat flow rate divided by temperature difference)	h
2.4.22	cubic expansion coefficient: $V^{-1}(\partial V/\partial T)_p$	α
2.4.23	isothermal compressibility: $-V^{-1}(\partial V/\partial p)_T$	κ
2.4.24	pressure coefficient: $(\partial p/\partial T)_V$	β
2.4.25	chemical potential of substance B	μ_B
2.4.26	absolute activity of substance B: $\exp(\mu_B/RT)$	λ_B
2.4.27	fugacity	$f, \tilde{p}$
2.4.28	osmotic pressure	Π
2.4.29	ionic strength: ($I_m = \frac{1}{2}\Sigma_i m_i z_i^2$ or $I_c = \frac{1}{2}\Sigma_i c_i z_i^2$)	I
2.4.30	activity, relative activity of substance B	a_B
2.4.31	activity coefficient, mole fraction basis	f_B
2.4.32	activity coefficient, molality basis	γ_B
2.4.33	activity coefficient, concentration basis	y_B
2.4.34	osmotic coefficient	ϕ

2.5 Chemical reactions

[a] The ratio of the average mass per atom of an element to 1/12 of the mass of an atom of nuclide ^{12}C. *Example:* $A_r(Cl) = 35.453$.

[b] The ratio of the average mass per formula unit of a substance to 1/12 of the mass of an atom of nuclide ^{12}C. *Example:* $M_r(KCl) = 74.555$.

[c] See Section 1.2.

[d] A solution having a molality equal to 0.1 mol kg^{-1} is sometimes called a 0.1 molal solution or a 0.1 m solution.

[e] This quantity may be simply called 'concentration' when there is no risk of ambiguity. A solution with an amount-of-substance concentration of 0.1 mol dm^{-3} is often called a 0.1 molar solution or a 0.1 M solution.

[f] Where symbols are needed to represent both time and Celsius temperature, t is the preferred symbol for time and θ for Celsius temperature.

[g] It is recommended that $q > 0$ and $w > 0$ both indicate *increase* of energy of the system under discussion. Thus $\Delta U = q + w$.

2.5.01	stoichiometric coefficient of substance B (negative for reactants, positive for products)	ν_B
2.5.02	general equation for a chemical reaction	$0 = \Sigma_B \nu_B B$
2.5.03	extent of reaction: ($d\xi = dn_B/\nu_B$)	ξ
2.5.04	rate of reaction: $d\xi/dt$ (see Section 11)	$\dot{\xi}, J$
2.5.05	rate of increase of concentration of substance B: dc_B/dt	ν_B, r_B
2.5.06	rate constant	k
2.5.07	affinity of a reaction: $-\Sigma_B \nu_B \mu_B$	A, ( )
2.5.08	equilibrium constant	K
2.5.09	degree of dissociation	a

2.6 Electricity and magnetism

2.6.01	elementary charge (of a proton)	e
2.6.02	quantity of electricity	Q
2.6.03	charge density	ρ
2.6.04	surface charge density	σ
2.6.05	electric current	I
2.6.06	electric current density	j
2.6.07	electric potential	V, ϕ
2.6.08	electric potential difference: IR	$U, \Delta V, \Delta\phi$
2.6.09	electric field strength	$\boldsymbol{E}$
2.6.10	electric displacement	$\boldsymbol{D}$
2.6.11	capacitance	C
2.6.12	permittivity: ($\boldsymbol{D} = \epsilon \boldsymbol{E}$)	ϵ
2.6.13	permittivity of vacuum	ϵ_0
2.6.14	relative permittivity[a]: ϵ/ϵ_0	$\epsilon_r, (\epsilon)$
2.6.15	dielectric polarization: $\boldsymbol{D} - \epsilon_0 \boldsymbol{E}$	$\boldsymbol{P}$
2.6.16	electric susceptibility: $\epsilon_r - 1$	χ_e
2.6.17	electric dipole moment	p, p_e
2.6.18	permanent dipole moment of a molecule	$\boldsymbol{p}, \mu$
2.6.19	induced dipole moment of a molecule	p, p_i
2.6.20	electric polarizability of a molecule	a
2.6.21	magnetic flux	Φ
2.6.22	magnetic flux density, magnetic induction	$\boldsymbol{B}$
2.6.23	magnetic field strength	$\boldsymbol{H}$
2.6.24	permeability: ($\boldsymbol{B} = \mu \boldsymbol{H}$)	μ
2.6.25	permeability of vacuum	μ_0
2.6.26	relative permeability: μ/μ_0	μ_r
2.6.27	magnetization: $(\boldsymbol{B}/\mu_0) - \boldsymbol{H}$	$\boldsymbol{M}$
2.6.28	magnetic susceptibility: $\mu_r - 1$	$\chi, (\chi_m)$
2.6.29.1	Bohr magneton	μ_B
2.6.29.2	nuclear magneton	μ_N
2.6.29.3	g-factor	g
2.6.29.4	gyromagnetic ratio, magnetogyric ratio	γ
2.6.30	electromagnetic moment: ($E_p = -\boldsymbol{m} \cdot \boldsymbol{B}$)	m, μ
2.6.31	resistance	R
2.6.32	resistivity (formerly called specific resistance): ($\boldsymbol{E} = \rho \boldsymbol{j}$)	ρ
2.6.33	conductivity (formerly called specific conductance): ($\boldsymbol{j} = \kappa \boldsymbol{E}$)	$\kappa, (\sigma)$
2.6.34	self-inductance	L
2.6.35	mutual inductance	M, L_{12}
2.6.36	reactance	X
2.6.37	impedance (complex impedance): $R + iX$	Z
2.6.38	loss angle	δ
2.6.39	admittance (complex admittance): $1/Z$	Y
2.6.40	conductance: ($Y = G + iB$)	G
2.6.41	susceptance: ($Y = G + iB$)	B

2.7 Electrochemistry

2.7.01	Faraday constant	F
2.7.02	charge number of an ion B (positive for cations, negative for anions)	z_B
2.7.03	charge number of a cell reaction	$n, (z)$
2.7.04	electromotive force	E, E_{MF}
2.7.05	electrochemical potential of ionic component B: $\mu_B + z_B F \phi$	$\tilde{\mu}_B$
2.7.06	electric mobility (velocity divided by electric field strength)	u, μ
2.7.07	electrolytic conductivity (formerly called specific conductance)	$\kappa, (\sigma)$
2.7.08	molar conductivity of electrolyte or ion[b]: κ/c	Λ, λ [c]
2.7.09	transport number (transference number or migration number)	t
2.7.10	overpotential	η
2.7.11	exchange current density	i_0
2.7.12	electrochemical transfer coefficient	α
2.7.13	electrokinetic potential (zeta potential)	ζ
2.7.14	thickness of diffusion layer	δ
2.7.15	inner electric potential	ϕ
2.7.16	outer electric potential	ψ
2.7.17	surface electric potential difference: $\phi - \psi$	χ

2.8 Light and related electromagnetic radiation[d]

2.8.01	Planck constant	h
2.8.02	Planck constant divided by 2π	$\hbar$
2.8.03	radiant energy	Q [e]
2.8.04	radiant flux, radiant power	Φ [e], P
2.8.05	radiant intensity: $d\Phi/d\omega$	I [e]
2.8.06	radiance: $(dI/dS)/\cos\theta$	L [e]
2.8.07	radiant excitance: $d\Phi/dS$	M [e]
2.8.08	irradiance: $d\Phi/dS$	E [e]

a Also called dielectric constant, and sometimes denoted by D, when it is independent of E.

b The word molar, contrary to the general rule given in Section 1.4, here means 'divided by amount-of-substance concentration'.

c The formula unit whose concentration is c must be specified.
Example: $\lambda(Mg^{2+}) = 2\lambda(\frac{1}{2}Mg^{2+})$. Λ is used for an electrolyte and λ for individual ions.

d References to the symbols used in defining the quantities in 2.8 are as follows:

l	2.1.01	$\tilde{\nu}$	2.1.07	θ	2.1.08	ω	2.1.09
S	2.1.10	V	2.1.11	t	2.1.12	c_B	2.3.11
ρ_B	2.3.12	Φ	2.8.04	I	2.8.05	E	2.8.08
T	2.8.12	A	2.8.13.1	B	2.8.13.2	α	2.8.14.2
k	2.8.15	n	2.8.21.1				

e The same symbol is often used also for the corresponding luminous quantity. Subscripts e for energetic and v for visible may be added whenever confusion between these quantities might otherwise occur.

2.8.09	absorptance, absorption-factor [a] (ratio of absorbed to incident radiant or luminous flux)	α^a
2.8.10	reflectance, reflection factor[a] (ratio of reflected to incident radiant or luminous flux)	ρ^a, R
2.8.11	transmittance, transmission factor[a] (ratio of transmitted to incident radiant or luminous flux)	τ^a
2.8.12	internal transmittance[a] (transmittance of the medium itself, disregarding boundary or container influence)	τ_i^a, T
2.8.13.1	internal transmission density[a], (decadic) absorbance[b]: $\log_{10}(1/T)$	D_i^a, A
2.8.13.2	Napierian absorbance: $\ln(1/T)$	B
2.8.14.1	(linear) (decadic) absorption coefficient[a,b]: A/l	a^a, K
2.8.14.2	Napierian absorption coefficient: B/l	α
2.8.15	absorption index: $B/4\pi\tilde{\nu}l = \alpha/4\pi\tilde{\nu}$	k
2.8.16.1	specific (decadic) absorption coefficient[c]: $A/\rho_B l$	$a^{e,f}$
2.8.16.2	specific Napierian absorption coefficient[c]: $B/\rho_B l$	$\mu^{e,f}$
2.8.17.1	molar (decadic) absorption coefficient[b,d]: $A/c_B l$	$\epsilon^{e,f}$
2.8.17.2	molar Napierian absorption coefficient[d]: $B/c_B l$	$\kappa^{e,f}$
2.8.18	quantum yield	Φ
2.8.19	exposure: $\int E dt$	H
2.8.20	velocity of light *in vacuo*	c
2.8.21.1	refractive index (of a non-absorbing material)	n
2.8.21.2	complex refractive index of an absorbing material: $n + ik$	$\hat{n}$
2.8.22	molar refraction: $(n^2 - 1)V_m/(n^2 + 2)$	R_m
2.8.23	angle of optical rotation	α

2.9 Transport properties

2.9.01	Flux (of a quantity *X*)	J_x, J
2.9.02	Reynolds number: $\rho vl/\eta$	*Re*
2.9.03	Euler number: $\Delta p/\rho v^2$	*Eu*
2.9.04	Froude number: $v/(lg)^{1/2}$	*Fr*
2.9.05	Grashof number: $l^3 g\alpha\Delta\theta\rho^2/\eta^2$	*Gr*
2.9.06	Weber number: $\rho v^2 l/\gamma$	*We*
2.9.07	Mach number: v/c	*Ma*
2.9.08	Knudsen number: λ/l	*Kn*
2.9.09	Strouhal number: lf/v	*Sr*
2.9.10	Fourier number: $a\Delta t/l^2$	*Fo*
2.9.11	Peclet number: vl/a	*Pe*
2.9.12	Rayleigh number: $l^3 g\alpha\Delta\theta\rho/\eta a$	*Ra*
2.9.13	Nusselt number: hl/k	*Nu*
2.9.14	Stanton number: $h/\rho vc_p$	*St*
2.9.15	Fourier number of mass transfer: Dt/l^2	*Fo**
2.9.16	Peclet number for mass transfer: vl/D	*Pe**
2.9.17	Grashof number for mass transfer: $-l^3 g(\partial\rho/\partial x)_{T,p}\Delta x\rho/\eta^2$	*Gr**
2.9.18	Nusselt number for mass transfer[g] $k_d l/D$	*Nu**
2.9.19	Stanton number for mass transfer: k_d/v	*St**
2.9.20	Prandtl number: $\eta/\rho a$	*Pr*
2.9.21	Schmidt number: $\eta/\rho D$	*Sc*
2.9.22	Lewis number: d/D	*Le*
2.9.23	Magnetic Reynolds number: $v\mu\kappa l$	Re_m
2.9.24	Alfvén number: $v(\rho\mu)^{1/2}/B$	*Al*
2.9.25	Hartmann number: $Bl(\kappa/\eta)^{1/2}$	*Ha*
2.9.26	Cowling number: $B^2/\mu\rho v^2$	*Co*

2.10 Symbols for particular cases of physical quantities

It is much more difficult to make detailed recommendations on symbols for physical quantities in particular cases than in general cases. The reason is the incompatibility between the need for specifying numerous details and the need for keeping the printing reasonably simple. Among the most awkward things to print are superscripts to subscripts and subscripts to subscripts. Examples of symbols to be avoided are:

$$\lambda_{NO_3^-} \qquad \Delta H_{25^\circ C} \qquad (pV)^{p=0}_{0\ ^\circ C}$$

The problem is vastly reduced if it is recognized that two different kinds of notation are required for two different purposes. In the formation of general fundamental relations the most important requirement is a notation that is easy

a These names and symbols are in agreement with those adopted jointly by the International Commission of Illumination (CIE) and the International Electrotechnical Commission (IEC).

b The terms extinction (for 2.8.13.1) and extinction coefficient (for 2.8.14.1) are unsuitable because extinction is reserved for diffusion of radiation rather than absorption. Molar absorptivity (for 2.8.17.1) should be avoided because the meaning, absorptance per unit length, has been accepted internationally for the term absorptivity.

c The word specific, contrary to the general rule given in Section 1.4, here means 'divided by mass concentration'.

d The word molar, contrary to the general rule given in Section 1.4, here means 'divided by amount-of-substance concentration'.

e For measurement on solutions, $1/T$ is ordinarily replaced by T_0/T where T_0 is the internal transmittance of the solvent medium and T is the internal transmittance of the solution. If a double-beam spectrometer is used in solution spectrometry, T_0/T is given directly, provided the boundary and container influences have been equalized between the two cells; in addition to the physical matching of the sample and reference cells this requires that there be no significant difference between $n_{solvent}$ and $n_{solution}$.

f For measurements on solutions, it is tacitly assumed that the solution obeys the Beer-Lambert law unless the solute concentration is specified. The temperature should be specified.

g The name Sherwood number, symbol *Sh* has been widely used.

to understand and easy to remember. In applications to particular cases, in quoting numerical values, and in tabulation, the most important requirement is complete elimination of any possible ambiguity even at the cost of an elaborate notation.

The advantage of a dual notation is already to some extent accepted in the case of concentration. The recommended notation for the formulation of the equilibrium constant K_c for the general reaction:

$$0 = \Sigma_B \nu_B B$$

is

$$K_c = \Pi_B (c_B)^{\nu_B}$$

but when we turn to a particular example it is better to use a notation such as:

$$Br_2 + H_2O = HOBr + H^+ + Br^-$$

$$\frac{[HOBr]\,[H^+]\,[Br^-]}{[Br_2]} = K_c$$

$$K_c(25\ °C) = 6 \times 10^{-9}\ mol^2\ dm^{-6}$$

Once the principle of dual notation is accepted, its adaptability and usefulness become manifest in all fields of physical chemistry. It will here be illustrated by just a few examples.

The general relation between the molar conductivity of an electrolyte and the molar conductivities of the two ions is written most simply and most clearly as:

$$\Lambda = \lambda^+ + \lambda^-$$

but when it comes to giving values in particular cases a much more appropriate notation is:

$$\lambda\,\tfrac{1}{2}Mg^{2+} = 53\ S\ cm^2\ mol^{-1}\ \text{at}\ 25°C$$
$$\lambda(Cl^-) = 76\ S\ cm^2\ mol^{-1}\ \text{at}\ 25°C$$
$$\Lambda\,\tfrac{1}{2}MgCl_2) = 129\ S\ cm^2\ mol^{-1}\ \text{at}\ 25°C$$
$$\Lambda(MgCl_2) = 258\ S\ cm^2\ mol^{-1}\ \text{at}\ 25°C$$

The general relation between the partial molar volumes of the two components A and B of a binary mixture is written most simply:

$$n_A dV_A + n_B dV_B = 0 \qquad (T,p\ \text{const.})$$

But when it comes to specifying values, a completely different notation is called for, such as:

$$V(K_2SO_4, 0.1\ mol\ dm^{-3}\ \text{in}\ H_2O, 25\ °C) = 48\ cm^3\ mol^{-1}$$

Each kind of notation is appropriate to its purpose.

A last example will be given relating to optical rotation. The relations between the angle α of rotation of the plane of polarization and the amount n, or the number N of molecules, of the optically active substance in the path of a light beam of cross-section A can be clearly expressed in the form:

$$\alpha = n\alpha_n/A = N\alpha_N/A$$

where α_n is the molar optical rotatory power and α_N the molecular optical rotatory power. When on the other hand it is desired to record an experimental measurement, an appropriate notation would be:

$$\alpha(589.3\ nm, 20\ °C, \text{sucrose}, 10\ g\ dm^{-3}\ \text{in}\ H_2O, 10\ cm) = +66.470°$$

2.11 Recommended superscripts

The following superscripts are recommended:

$^{\circ}$ or * pure substance
$^{\infty}$ infinite dilution

^{id}ideal
$^{\circ}$ or $^{\ominus}$standard in general
‡activated complex, transition state

3. UNITS AND SYMBOLS FOR UNITS

3.5 The International System of Units

The name International System of Units has been adopted by the Conférence Générale des Poids et Mesures for the system of units based on a selected set of dimensionally independent *SI Base Units*.

The SI Base Units are the meter, kilogram, second, ampere, kelvin, candela, and mole. In the International System of Units there is one and only one *SI Unit* for each physical quantity. This is either the appropriate SI Base Unit itself (see Section 3.7) or the appropriate *SI Derived Unit* formed by multiplication and/or division of two or more SI Base Units (see Section 3.10). A few such SI Derived Units have been given special names and symbols (see Section 3.9). There are also two *SI Supplementary Units* for which it is not decided whether they are SI Base Units or SI Derived Units (see Section 3.8).

Any of the approved decimal prefixes, called *SI Prefixes*, may be used to construct decimal multiples or submultiples of SI Units (see Section 3.11).

It is recommended that only units composed of SI Units and SI Prefixes be used in science and technology.

3.6 Definitions of the SI Base Units

Meter – The meter is the SI base unit of length and was defined at the October 1984 General Conference of Weights and Measures as the length of the path traveled by light in vacuum during a time interval of $^1/299{,}792{,}458$ of a second.

Kilogram – The kilogram is the unit of mass; it is equal to the mass of the international prototype of the kilogram.

Second – The second is the duration of 9192631770 periods of the radiation corresponding to the transition between the two hyperfine levels of the ground state of the caesium-133 atom.

Ampere – The ampere is that constant current which, if maintained in two straight parallel conductors of infinite length, of negligible cross-section, and placed 1 meter apart in vacuum, would produce between these conductors a force equal to 2×10^{-7} newton per meter of length.

Kelvin – The kelvin, unit of thermodynamic temperature, is the fraction 1/273.16 of the thermodynamic temperature of the triple point of water.[a]

Candela – The candela is the luminous intensity in a given direction of a source which emits monochromatic radiation of frequency 540×10^{12} Hz and of which the radiant intensity in that direction is 1/683 W/steradian. From the 16th CGPM, *Resolution*, 3, 1979.

Mole – The mole is the amount of substance of a system which contains as many elementary entities as there are atoms in 0.012 kilogram of carbon-12. When the mole is used, the elementary entities must be specified and may be atoms, molecules, ions, electrons, other particles, or specified groups of such particles. *Some examples of the use of the mole:*

- 1 mole of HgCl has a mass of 236.04 grams
- 1 mole of Hg_2Cl_2 has a mass of 472.08 grams
- 1 mole of Hg_2^{2+} has a mass of 401.18 grams and a charge of 192.97 kilocoulombs
- 1 mole of $\frac{1}{2}Ca^{2+}$ has a mass of 20.04 grams and a charge of 96.49 kilocoulombs
- 1 mole of $Cu_{0.5}Zn_{0.5}$ has a mass of 64.46 grams
- 1 mole of $Fe_{0.91}S$ has a mass of 82.88 grams
- 1 mole of e^- has a mass of 548.60 micrograms, a charge of −96.49 kilocoulombs, and contains 6.02×10^{23} electrons
- 1 mole of a mixture containing the mole fractions $x(N_2) = 0.7809$, $x(O_2) = 0.2905$, $x(Ar) = 0.0093$, and $x(CO_2) = 0.0003$ has a mass of 28.964 grams
- 1 mole of photons whose frequency is 10^{14} Hz has energy 39.90 kilojoules

(The numerical values in these examples are approximate.)

3.7 Names and symbols for SI Base Units

Physical quantity	*Name of SI Unit*	*Symbol for SI Unit*
Length	meter	m
Mass	kilogram	kg
Time	second	s
Electric current	ampere	A
Thermodynamic temperature	kelvin	K
Amount of substance[b]	mole	mol
Luminous intensity	candela	cd

[a] In October 1967 the thirteenth Conférence Générale des Poids et Mesures recommended that the kelvin, symbol K, be used for thermodynamic temperature and for thermodynamic temperature interval, and that the unit-symbols °K and deg be abandoned.

[b] See Section 1.2.

3.8 Names and symbols for SI Supplementary Units

Physical quantity	*Name of SI Unit*	*Symbol for SI Unit*
Plane angle	radian	rad
Solid angle	steradian	sr

3.9 Special names and symbols for certain SI Derived Units

Physical quantity	*Name of SI Unit*	*Symbol for SI Unit*	*Definition of SI Unit*
Force	newton	N	m kg s^{-2}
Pressure, stress	pascal	Pa	m^{-1} kg s^{-2} (=N m^{-2})
Energy	joule	J	m^2 kg s^{-2}
Power	watt	W	m^2 kg s^{-3} (=J s^{-1})
Electric charge	coulomb	C	s A
Electric potential difference	volt	V	m^2 kg s^{-3} A^{-1} (=J A^{-1} s^{-1})
Electric resistance	ohm	Ω	m^2 kg s^{-3} A^{-2} (=V A^{-1})
Electric conductance	siemens	S	m^{-2} kg^{-1} s^3 A^2 (=A V^{-1} = Ω^{-1})
Electric capacitance	farad	F	m^{-2} kg^{-1} s^4 A^2 (=A s V^{-1})
Magnetic flux	weber	Wb	m^2 kg s^{-2} A^{-1} (=V s)
Inductance	henry	H	m^2 kg s^{-2} A^{-2} (=V A^{-1} s)
Magnetic flux density	tesla	T	kg s^{-2} A^{-1} (=V s m^{-2})
Luminous flux	lumen	lm	cd sr
Illuminance	lux	lx	m^{-2} cd sr
Frequency	hertz	Hz	s^{-1}
Activity (of radioactive source)	becquerel	Bq	s^{-1}
Absorbed dose (of radiation)	gray	Gy	m^2 s^{-2} (=J kg^{-1})

3.10 SI Derived Units and Unit-symbols for other quantities
(This list is not exhaustive)

Physical quantity	*SI Unit*	*Symbol for SI Unit*
Area	square meter	m^2
Volume	cubic meter	m^3
Density	kilogram per cubic meter	kg m^{-3}
Velocity	meter per second	m s^{-1}
Angular velocity	radian per second	rad s^{-1}
Acceleration	meter per second squared	m s^{-2}
Kinematic viscosity, diffusion coefficient	square meter per second	m^2 s^{-1}
Dynamic viscosity	newton-second per square meter	N s m^{-2}
Molar entropy, molar heat capacity	joule per kelvin mole	J K^{-1} mol^{-1}
Concentration	mole per cubic meter	mol m^{-3}
Electric field strength	volt per meter	V m^{-1}
Magnetic field strength	ampere per meter	A m^{-1}
Luminance	candela per square meter	cd m^{-2}

3.12 The degree Celsius

Physical quantity	*Name of unit*	*Symbol for unit*	*Definition of unit*
Celsius temperature	degree Celsius	°C[a]	°C = K

The Celsius temperature t, is defined by $t = T - T_0$ where T_0 = 273.15 K. This leads to t/°C = T/K − 273.15.

3.13 Decimal fractions and multiples of SI Units having special names

The following units do not belong to the International System of Units, but in view of existing practice the Comité

a The ° sign and the letter following form one symbol and there should be no space between them. *Example:* 25 °C not 25° C.

International des Poids et Mesures has considered (1969) that it was preferable to keep them for the time being (along with several other specified units not particularly relevant to chemistry) for use with those of the International System.

Physical quantity	*Name of unit*	*Symbol for unit*	*Definition of unit*
Length	ångström	Å	10^{-10} m
Cross section	barn	b	10^{-28} m^2
Volume	liter[a]	l, L	10^{-3} m^3
Mass	tonne	t	10^3 kg
Pressure	bar	bar	10^5 Pa

Other units with special names based on the c.g.s. system and the electromagnetic c.g.s. system[b] are preferably not to be used; among these are the erg (10^{-7} J), the dyne (10^{-5} N), the poise (0.1 Pa s), the stokes (10^{-4} m^2 s^{-1}), the gauss (corresponding to 10^{-4} T),[b] the oersted (corresponding to $1000/4\pi$ A m^{-1}),[b] and the maxwell (corresponding to 10^{-8} WB).[b] The name micron and symbol μ should not be used for the unit of length, 10^{-6} m, which has the SI name micrometer and symbol μm.

3.14 Some other units now exactly defined in terms of the SI units

The CIPM (1969) recognized that users of the SI will wish to employ with it certain units not part of it but which are important and are widely used. These units are given in the following table. The combination of units of this table with SI units to form compound units should, however, be authorized only in limited cases.

Units in use with the International System

Name of unit	*Symbol*	*Definition of unit*
minute	min	60 s
hour	h	3 600 s
day	d	86 400 s
degree	°	(π/180) rad
minute	′	(π/10 800) rad
second	″	(π/648 000) rad

In view of existing practice, as in the case of those units listed in Section 3.13, the CIPM (1969) has considered it preferable to retain the following units for the time being, for use with those of the SI. The definitions given in the fourth column of this table are exact.

Units to be used with the International System for a limited time

Physical quantity	*Name of unit*	*Symbol for unit*	*Definition of unit*
Radioactivity	Curie	Ci	3.7×10^{10} Bq
Exposure to X or γ radiation	röntgen	R	2.58×10^{-4} C kg^{-1}
Ionizing radiation absorbed	rad	rad[c]	10^{-2} Gy

The use of the following units is to be progressively discouraged and eventually abandoned. In the meantime it is recommended that any author who uses these units will define them in terms of SI units once in each publication in which he uses them. The definitions given here are exact. This list is not exhaustive.

Other units generally deprecated

Physical quantity	*Name of unit*	*Symbol for unit*	*Definition of unit*
Length	inch	in	2.54×10^{-2} m
Mass	pound (avoirdupois)	lb	0.453 592 37 kg
Force	kilogram-force	kgf	9.806 65 N
Pressure	standard atmosphere[d]	atm	101 325 Pa

[a] By decision of the Twelfth Conférence Générale des Poids et Mesures in October 1964, the old definition of the liter (1.000 028 dm^3) was rescinded. The word liter is now regarded as a special name for the cubic decimeter. Neither the word liter nor its symbol should be used to express results of high precision. The alternative symbol L was recommended by CIPM to CGPM in 1978.

[b] The electromagnetic c.g.s. system is a three-dimensional system of units in which the unit of electric current and units for other electric and magnetic quantities are considered to be derived from the centimeter, gram, and second as base units. The electric and magnetic units of this system cannot strictly speaking be compared to the corresponding units of the SI, which has four dimensions when only units derived from the meter, kilogram, second, and ampere are considered.

[c] Where there is a risk of confusion with the symbol for radian, rd may be used as the symbol for the unit, rad.

[d] The phrase 'standard atmosphere' remains admissable for the reference pressure 101 325 Pa.

Pressure	torr	Torr	$\frac{101\ 325}{760}$ Pa
Pressure	conventional millimeter of mercury[a]	mmHg	13.5951 × 980.665 × 10^{-2} Pa
Energy	kilowatt-hour	kW h	3.6 × 10^6 J
Energy	IT calorie	cal_{IT}	4.1868 J
Energy	thermochemical calorie	cal_{th}	4.184 J
Energy	British thermal unit	Btu	1055.055 852 62 J
Thermodynamic temperature	degree Rankine	°R	(5/9) K

3.15 Units defined in terms of the best available experimental values of certain physical constants

It is necessary to recognize outside the International System some units, useful in specialized fields, the values of which expressed in SI units can be obtained only by experiment and are therefore not known exactly. Among such units recognized by the CIPM (1969) that are relevant to chemistry are the following:

Physical quantity	*Name of unit*	*Symbol for unit*	*Conversion factor*
Energy	electronvolt	eV	1 eV ≈ 1.6021892 × 10^{-19} J
Mass	(unified) atomic mass unit	u	1 u ≈ 1.6605655 × 10^{-27} kg

3.16 'International' electrical units

These units are obsolete having been replaced by the 'absolute' (SI) units in 1948. The conversion factors which should be used with electrical measurements quoted in 'international' units depend on where and when the instruments used to make the measurements were calibrated. The following two sets of conversion factors refer respectively to the 'mean international' units estimated by the ninth Conférence Générale des Poids et Mesures in 1948, and to the 'US international' units estimated by the National Bureau of Standards (USA) as applying to published measurements made with instruments calibrated by them prior to 1948.

- 1 'mean international ohm' = 1.00049 Ω
- 1 'mean international volt' = 1.00034 V
- 1 'US international ohm' = 1.000495 Ω
- 1 'US international volt' = 1.000330 V

3.17 Electrical and magnetic units belonging to unit-systems other than the International System of Units

Definitions of units used in the obsolescent 'electrostatic CGS' and 'electromagnetic CGS' unit-systems can be found in References 13.1.05 and 13.2.

Another 'electrostatic CGS' unit used in chemistry for electric dipole moment is the debye, symbol D. 1D = $(10^{-21}/c)$ A m^2 ≈ 3.3356 × 10^{-30} C m.

5. PHYSICAL QUANTITIES, UNITS, AND NUMERICAL VALUES

The value of a *physical quantity* is equal to the product of a *numerical value* and a *unit*:

physical quantity = numerical value × unit.

Neither any physical quantity, nor the symbol used to denote it, should imply a particular choice of unit.

Operations on equations involving physical quantities, units, and numerical values, should follow the ordinary rules of algebra.

Thus the physical quantity called the critical pressure and denoted by p_c has the value for water:

$$p_c = 221.2 \text{ bar} \quad \text{or better} \quad p_c = 22.12 \text{ MPa}.$$

These equations may equally well be written in the forms:

$$p_c/\text{bar} = 221.2 \quad \text{or better} \quad p_c/\text{MPa} = 22.12,$$

which are especially useful for the headings in tables and as labels on the axes of graphs.

[a] The conventional millimeter of mercury, symbol mmHg (not mm Hg), is the pressure exerted by a column exactly 1 mm high of a fluid of density exactly 13.5951 g cm^{-3} in a place where the acceleration of free fall is exactly 980.665 cm s^{-2}. The mmHg differs from the Torr by less than 2 × 10^{-7} Torr.

6. RECOMMENDED MATHEMATICAL SYMBOLS[a]

Mathematical operators (for example d and Δ) and mathematical constants (for example e and π should always be printed in roman (upright) type. Letter symbols for numbers other than mathematical constants should be printed in italic type.

equal to	$=$
not equal to	$\neq$
identically equal to	$\equiv$
corresponds to	$\triangleq$
approximately equal to	$\approx$
approaches	$\rightarrow$
asymptotically equal to	$\simeq$
proportional to	$\propto$ $\sim$
infinity	∞
less than	$<$
greater than	$>$
less than or equal to	$\leqslant$
greater than or equal to	$\geqslant$
much less than	$\ll$
much greater than	$\gg$
plus	$+$
minus	$-$
multiplied by	$\times$ $\cdot$
a divided by *b*	$\frac{a}{b}$ a/b ab^{-1}
magnitide of *a*	$\lvert a \rvert$
a raised to the power *n*	a^n
square root of *a*	$a^{1/2}$ $a^{\frac{1}{2}}$ $\sqrt{a}$
*n*th root of *a*	$a^{1/n}$ $a^{\frac{1}{n}}$ $\sqrt[n]{a}$
mean value of *a*	$\langle a \rangle$ $\bar{a}$
natural logarithm of *a*	$\ln a$ $\log_e a$
decadic logarithm of *a*	$\lg a$ $\log_{10} a$ $\log a$
binary logarithm of *a*	$\mathrm{lb}\, a$ $\log_2 a$
exponential of *a*	$\exp a$ e^a

7. SYMBOLS FOR CHEMICAL ELEMENTS, NUCLIDES, AND PARTICLES

7.1 Definitions

A nuclide is a species of atoms of which each atom has identical atomic number (proton number) and identical mass number (nucleon number). Different nuclides having the same value of the atomic number are named isotopes or isotopic nuclides. Different nuclides having the same mass number are named isobars or isobaric nuclides.

7.2 Elements and nuclides

Symbols for chemical elements should be written in roman (upright) type. The symbol is not followed by a full stop except when it occurs at the end of a sentence in text. *Examples:* Ca C H He.

The nuclide may be specified by attaching numbers. The mass number should be placed in the left superscript position; the atomic number, if desired, may be placed as a left subscript. The number of atoms per molecule is indicated as a right subscript. Ionic charge, or state of excitation, or oxidation number[b] may be indicated in the right superscript space. *Examples;* Mass number: $^{14}N_2$, $^{35}Cl^-$; Ionic charge: Cl^-, Ca^{2+}, PO_4^{3-} or $PO_4{}^{3-}$; Excited electronic state: He*, NO*; Oxidation number: $Pb_2^{II}Pb^{IV}O_4$, $K_6M^{IV}Mo_9O_{32}$ (where M denotes a metal).

7.4 Abbreviated notation for nuclear reactions

The meaning of the symbolic expression indicating a nuclear reaction should be the following:

$$\text{initial nuclide}\left(\text{incoming particles(s) or quanta}, \text{outgoing particles(s) or quanta}\right)\text{final nuclide}$$

Examples: $^{14}N(\alpha,p)^{17}O$; $^{23}Na(\gamma,3n)^{20}Na$; $^{59}Co(n,\gamma)^{60}Co$; $^{31}P(\gamma,pn)^{29}Si$.

a Taken from Reference 13.1.11 where a more comprehensive list can be found.
b For a more detailed discussion see Reference 13.4.

8. SYMBOLS FOR SPECTROSCOPY[a]

8.1 General rules

A letter-symbol indicating the quantum state of *a system* should be printed in capital upright type. A letter-symbol indicating the quantum state of *a single particle* should be printed in lower case upright type.

9. CONVENTIONS CONCERNING THE SIGNS OF ELECTRIC POTENTIAL DIFFERENCES, ELECTROMOTIVE FORCES, AND ELECTRODE POTENTIALS [b]

9.1 The electric potential difference for a galvanic cell

The cell should be represented by a diagram, for example:

$$Zn \,|\, Zn^{2+} \,|\, Cu^{2+} \,|\, Cu$$

The electric potential difference ΔV is equal in sign and magnitude to the electric potential of a metallic conducting lead on the right minus that of an identical lead on the left.

When the reaction of the cell is written as:

$$\tfrac{1}{2}Zn + \tfrac{1}{2}Cu^{2+} \rightarrow \tfrac{1}{2}Zn^{2+} + \tfrac{1}{2}Cu$$

this implies a diagram so drawn that this reaction takes place when positive electricity flows through the cell from left to right. If this is the direction of the current when the cell is short-circuited, as it will be in the present example (unless the ratio $[Cu^{2+}]/[Zn^{2+}]$ is extremely small), the electric potential difference will be positive.

If, however, the reaction is written as:

$$\tfrac{1}{2}Cu + \tfrac{1}{2}Zn^{2+} \rightarrow \tfrac{1}{2}Cu^{2+} + \tfrac{1}{2}Zn$$

this implies the diagram:

$$Cu \,|\, Cu^{2+} \,|\, Zn^{2+} \,|\, Zn$$

and the electric potential difference of the cell so specified will be negative (unless the ratio $[Cu^{2+}]/[Zn^{2+}]$ is extremely small).

The limiting value of the electric potential difference for zero current through the cell is called the electromotive force and denoted by E_{MF} or E.

9.2 Electrode potential

The so-called electrode potential of an electrode (half-cell) is defined as the electromotive force of a cell in which the electrode on the left is a *standard hydrogen electrode* and that on the right is the electrode in question. For example, for the zinc electrode (written as $Zn^{2+} \,|\, Zn$) the cell in question is:

$$Pt \,|\, H_2 \,|\, H^+ \,|\, Zn^{2+} \,|\, Zn$$

The reaction taking place at the zinc electrode is:

$$Zn^{2+} + 2e^- \rightarrow Zn$$

The latter is to be regarded as an abbreviation for the reaction in the mentioned cell:

$$Zn^{2+} + H_2 \rightarrow Zn + 2H^+$$

In the standard state the electromotive force of this cell has a negative sign and a value of −0.763 V. The standard electrode potential of the zinc electrode is therefore −0.763 V.

The symbol $Zn \,|\, Zn^{2+}$ on the other hand implies the cell:

$$Zn \,|\, Zn^{2+} \,|\, H^+ \,|\, H_2 \,|\, Pt$$

a Taken from Reference 13.2. For further details see Reference 13.10.

b The conventions given here are in accordance with the 'Stockholm Convention' of 1953.

in which the reaction is:

$$Zn + 2H^+ \rightarrow Zn^{2+} + H_2$$

The electromotive force of this cell should *not* be called an electrode potential.

10. THE QUANTITY pH[a]

10.1 Operational definition

In all existing national standards the definition of pH is an operational one. The electromotive force E_X of the cell:

$$\text{reference electrode} \mid \text{concentrated KCl solution} \;\vdots\vdots\; \text{solution X} \mid H_2 \mid Pt$$

is measured and likewise the electromotive force E_S of the cell:

$$\text{reference electrode} \mid \text{concentrated KCl solution} \;\vdots\vdots\; \text{solution S} \mid H_2 \mid Pt$$

both cells being at the same temperature throughout and the reference electrodes and bridge solutions being identical in the two cells. The pH of the solution X, denoted by pH(X), is then related to the pH of the solution S, denoted by pH(S), by the definition:

$$\text{pH(X)} = \text{pH(S)} + \frac{(E_S - E_X)F}{RT \ln 10}$$

where R denotes the gas constant, T the thermodynamic temperature, and F the Faraday constant. Thus defined the quantity pH is a number.

To a good approximation, the hydrogen electrodes in both cells may be replaced by other hydrogen-ion-responsive electrodes, e.g. glass or quinhydrone. The two bridge solutions may be any molality not less than 3.5 mol kg^{-1}, provided they are the same (see Reference 13.5).

10.2 Standards

The difference between the pH of two solutions having been defined as above, the definition of pH can be completed by assigning a value of pH at each temperature to one or more chosen solutions designated as standards.

If the definition of pH given above is adhered to strictly, then the pH of a solution may be slightly dependent on which standard solution is used. These unavoidable deviations are caused not only by imperfections in the response of the hydrogen-ion electrodes but also by variations in the liquid junctions resulting from the different ionic compositions and mobilities of the several standards and from differences in the geometry of the liquid-liquid boundary. In fact such variations in measured pH are usually too small to be of practical significance. Moreover, the acceptance of several standards allows the use of the following alternative definition of pH.

The electromotive force E_X is measured, and likewise the electromotive forces E_1 and E_2 of two similar cells with the solution X replaced by the standard solutions S_1 and S_2 such that the E_1 and E_2 values are on either side of, and as near as possible to, E_X. The pH of solution X is then obtained by assuming linearity between pH and E, that is to say:

$$\frac{\text{pH(X)} - \text{pH}(S_1)}{\text{pH}(S_2) - \text{pH}(S_1)} = \frac{E_X - E_1}{E_2 - E_1}$$

This procedure is especially recommended when the hydrogen-ion-responsive electrode is a glass electrode.

11. DEFINITION OF RATE OF REACTION AND RELATED QUANTITIES

11.1 Rate of reaction

For the reaction

$$0 = \Sigma_B \nu_B B$$

the extent of reaction ξ is defined according to 2.5.03 by

a The symbol pH is an exception to the general rules given in Section 1.5.

$$d\xi = \nu_B^{-1}\, dn_B$$

where n_B is the amount, and ν_B is the stoichiometric number, of the substance B.

It is recommended that the *rate of reaction* be defined as the rate of increase of the extent of reaction, namely

$$\dot{\xi} = d\xi/dt = \nu_B^{-1}\, dn_B/dt$$

This definition is independent of the choice of B and is valid regardless of the conditions under which a reaction is carried out, e.g. it is valid for a reaction in which the volume varies with time, or for a reaction involving two or more phases, or for a reaction carried out in a flow reactor.

If both sides of this equation are divided by any specified volume V, not necessarily independent of time, and not necessarily that of a single phase in which the reaction is taking place, then

$$V^{-1}\, d\xi/dt = V^{-1} \nu_B^{-1}{}^{-1}\, dn_B/dt$$

If the specified volume V is independent of time, then

$$V^{-1}\, d\xi/dt = \nu_B^{-1}\, d(n_B/V)/dt$$

If this specified volume V is such that

$$n_B/V = c_B \text{ or } [B]$$

where c_B or [B] is the amount-of-substance concentration of B, then

$$V^{-1}\, d\xi/dt = \nu_B^{-1}\, dc_B/dt \text{ or } \nu_B^{-1}\, d[B]/dt$$

The quantity

$$dn_B/dt (= \nu_B d\xi/dt)$$

may be called the rate of formation of B, and the quantity

$$V^{-1} \nu_B^{-1}\, dn_B/dt (= V^{-1}\, d\xi/dt)$$

may be called the rate of reaction divided by volume, and the quantity

$$\nu_B = dc_B/dt \text{ or } d[B]/dt$$

which has often been called the rate of reaction, may be called the rate of increase of the concentration of B, but none of these three quantities should be called the rate of reaction.

11.2 Order of reaction

If it is found *experimentally* that the rate of increase of the concentration of B is given by

$$\nu_B \propto [C]^c [D]^d \ldots$$

then the reaction is described as of order c with respect to C, or order d with respect to D, . . ., and of overall order $(c + d + \ldots)$.

11.3 Labelling of elementary processes

Elementary processes should be labelled in such a manner that reverse processes are immediately recognizable. *Example:*

Elementary process	*Label*	*Rate of constant*
$Br_2 + M \rightarrow 2Br + M$	1	k_1
$Br + H_2 \rightarrow HBr + H$	2	k_2
$H + Br_2 \rightarrow HBr + Br$	3	k_3
$H + HBr \rightarrow H_2 + Br$	−2	k_{-2}
$2Br + M \rightarrow Br_2 + M$	−1	k_{-1}

11.4 Collision number

The collision number defined as the number of collisions per unit time and per unit volume and having dimensions $(\text{time})^{-1} \times (\text{volume})^{-1}$ should be denoted by Z.

The collision number divided by the product of two relevant concentrations (or by the square of the relevant concentration) and by the Avogadro constant is a second-order rate constant having dimensions $(\text{time})^{-1} \times (\text{volume}) \times (\text{amount of substance})^{-1}$ and should be denoted by z. Thus $z = Z/Lc_A c_B$.

13. REFERENCES

13.1 The ISO 31 International Standard Series will, when complete, form a comprehensive publication dealing with quantities and units in various fields of science and technology. The following parts have so far been published and can be purchased in any country belonging to ISO from the 'Member Body', usually the national standardizing organization of the country.

13.1.00 'Part 0: General principles concerning quantities, units and symbols', 2nd edition, July 1981.

13.1.01 'Part I: Quantities and units of space and time', 1st edition, March 1978.

13.1.02 'Part II: Quantities and units of periodic and related phenomena', 1st edition, March 1978.

13.1.03 'Part III: Quantities and units of mechanics', 1st edition, March 1978.

13.1.04 'Part IV: Quantities and units of heat', 1st edition, March 1978.

13.1.05 'Part 5: Quantities and units of electricity and magnetism', 2nd edition, February 1979.

13.1.06 'Part 6: Quantities and units of light and related electromagnetic radiations', 2nd edition, December 1980.

13.1.07 'Part VII: Quantities and units of acoustics', 1st edition, March 1978.

13.1.08 'Part 8: Quantities and units of physical chemistry and molecular physics', 2nd edition, December 1980.

13.1.09 'Part 9: Quantities and units of atomic and nuclear physics', 2nd edition, December 1980.

13.1.10 'Part 10: Quantities and units of nuclear reactions and ionizing radiations', 2nd edition, December 1980.

13.1.11 'Part XI: Mathematical signs and symbols for use in the physical sciences and technology', 1st edition, March 1978.

13.1.12 'Part 12: Dimensionless parameters', 2nd edition, 1981.

13.1.13 'Part 13: Quantities and units of solid state physics', 2nd edition, July 1981.

13.2 'Symbols, Units and Nomenclature in Physics', Document UIP 20 (SUN 65–3), published by IUPAP, 1978. This document supersedes Document UIP 9 (SUN 61–44) with the same title, which was published by IUPAP in 1961. Also published in *Physica*, (1978), 93A, 1.

13.3 'Manual of Physicochemical Symbols and Terminology', published for IUPAC by Butterworths Scientific Publications, London, 1959. This document was reprinted in the *J. Am. Chem. Soc.*, (1960), 82, 5517.

13.4 'Nomenclature of Inorganic Chemistry, 1970', 2nd ed., *Pure and Applied Chemistry*, (1971), 28, 1. This document has been issued for IUPAC also in book form by Butterworths, Borough Green, Sevenoaks, Kent TN158PH, UK.

13.5 *Pure and Applied Chemistry*, (1960), 1, 163.

13.6 *Pure and Applied Chemistry*, (1964), 9, 453.

13.7 *Pure and Applied Chemistry*, (1970), 21, 1.

13.8 'Le Systẽme International d'Unités (SI)', Bureau International des Poids et Mesures, 2e Ed., 1973, OFFILIB, 48, rue Gay-Lussac, F 75 Paris 5.

13.8.01 'The International System of Units (SI)', National Bureau of Standards Special Publication 330, 1977 Edition, SD Catalog No. C 13.10: 330/4, US Government Printing Office, Washington, DC 20402.

13.8.02 'SI The International System of Units', National Physical Laboratory, Third edition, 1977, Her Majesty's Stationery Office, London.

13.9.01 'Definitions, Terminology and Symbols in Colloid and Surface Chemistry', Part I, *Pure and Applied Chemistry*, (1972), 31, 577 and Part II, 'Heterogeneous Catalysis', *Pure and Applied Chemistry*, (1976), 46, 71.

13.9.02 'Electrochemical Nomenclature', *Pure and Applied Chemistry*, (1974), 37, 503.

13.10 'Report on Notation for the Spectra of Polyatomic Molecules', *J. Chem. Phys.*, (1955) 23, 1997. For notation for diatomic molecules, see F. A. Jenkins, *J. Opt. Soc. Am.*, (1953), 43, 425.

13.11 'Recommended Consistent Values of the Fundamental Physical Constants, 1973', *CODATA Bulletin No. 11*, (December 1973), Committee on Data for Science and Technology of the International Council of Scientific Unions.

13.12 'The 1973 Least-Squares Adjustment of the Fundamental Constants', E. R. Cohen and B. N. Taylor, *J. Phys. Chem. Ref. Data*, (1973), 2, 663.

APPENDIX I

DEFINITION OF ACTIVITIES AND RELATED QUANTITIES

A.I.1 *Chemical potential and absolute activity*

The chemical potential μ_B of a substance B in a mixture of substances B, C, . . . , is defined by

$$\mu_B = (\partial G/\partial n_B)_{T,p,n_C} \ldots$$

where G is the Gibbs energy of the mixture, T is the thermodynamic temperature, p is the pressure, and n_B, n_C, . . . , are the amounts of the substances B, C, . . . , in the mixture.

(In molecular theory the symbol μ_B is sometimes used for the quantity μ_B/L where L is the Avogadro constant, but this usage is not recommended.)

The absolute activity λ_B of the substance B in the mixture is a number defined by

$$\lambda_B = \exp(\mu_B/RT) \quad \text{or} \quad \mu_B = RT \ln \lambda_B$$

where R is the gas constant.

The definitions given below often take simpler, though perhaps less familiar, forms when they are expressed in terms of absolute activity rather than in terms of chemical potential. Each of the definitions given below is expressed in both of these ways.

1. Pure substances

A.I.2 *Properties of pure substances*

The superscript * attached to the symbol for a property of a substance denotes the property of the *pure* substance. It is sometimes convenient to treat a mixture of constant composition as a pure substance.

A.I.3 *Fugacity of a pure gaseous substance*

The fugacity $f_B{}^*$ of a pure gaseous substance B is a quantity with the same dimensions as pressure, defined in terms of the absolute activity $\lambda_B{}^*$ of the pure gaseous substance B by

$$f_B{}^* = \lambda_B{}^* \lim_{p\to 0}(p/\lambda_B{}^*) \qquad (T \text{ const.})$$

or in terms of the chemical potential μ_B by

$$RT \ln f_B{}^* = \mu_B{}^* + \lim_{p\to 0}(RT \ln p - \mu_B{}^*) \qquad (T \text{ const.})$$

where p is the pressure of the gas and T is its thermodynamic temperature. It follows from this definition that

$$\lim_{p\to 0}(f_B{}^*/p) = 1 \qquad (T \text{ const.})$$

and that

$$RT \ln(f_B{}^*/p) = \int_0^p (V_B{}^* - RT/p)\, dp \qquad (T \text{ const.})$$

where $V_B{}^*$ is the molar volume of the pure gaseous substance B.

A pure gaseous substance B is treated as an *ideal gas* when the approximation $f_B{}^* = p$ is used. The ratio $(f_B{}^*/p)$ may be called the fugacity coefficient.

The name activity coefficient has sometimes been used for this ratio but is not recommended.

2. Mixtures

A.I.4 *Definition of a mixture*

The word *mixture* is used to describe a gaseous or liquid or solid phase containing more than one substance, when the substances are all treated in the same way (contrast the use of the word *solution* in Section A.I.9).

A.I.5 *Partial pressure*

The partial pressure p_B of a substance B in a *gaseous* mixture is a quantity with the same dimensions as pressure defined by

$$p_B = y_B p$$

where y_B is the mole fraction of the substance B in the gaseous mixture and p is the pressure.

A.I.6 *Fugacity of a substance in a gaseous mixture*

The fugacity f_B of the substance B in a gaseous mixture containing mole fractions $y_B, y_C, \ldots$, of the substances B, C, . . . , is a quantity with the same dimensions as pressure, defined in terms of the absolute activity λ_B of the substance B in the gaseous mixture by

$$f_B = \lambda_B \lim_{p\to 0}(y_B p/\lambda_B) \qquad (T \text{ const.})$$

or in terms of the chemical potential μ_B by

$$RT \ln f_B = \mu_B + \lim_{p\to 0}\{RT \ln(y_B p) - \mu_B\} \qquad (T \text{ const.})$$

It follows from this definition that

$$\lim_{p\to 0}(f_B/y_B p) = 1 \qquad (T \text{ const.})$$

and that

$$RT \ln(f_B/y_B p) = \int_0^p (V_B - RT/p)\, dp \qquad (T \text{ const.})$$

where V_B is the partial molar volume (see Section 1.4) of the substance B in the gaseous mixture.

A gaseous mixture of B, C, . . . , is treated as an *ideal gaseous mixture* when the approximations $f_B = y_B p$, $f_C = y_C p$, . . . , are used. It follows that $pV = (n_B + n_C + \ldots) RT$ for an ideal gaseous mixture of B, C, . . .

The ratio $(f_B/y_B p)$ may be called the fugacity coefficient of the substance B. The name activity coefficient has sometimes been used for this ratio but is not recommended.

When $y_B = 1$ the definitions given in this Section for the fugacity of a substance in a gaseous mixture reduce to those given in Section A.I.3 for the fugacity of a pure gaseous substance.

A.I.7 *Activity coefficient of a substance in a liquid or solid mixture*

The activity coefficient f_B of a substance B in a liquid or solid mixture containing mole fractions $x_B, x_C, \ldots$, of the substances B, C, . . . , is a number defined in terms of the absolute activity λ_B of the substance B in the mixture by

$$f_B = \lambda_B/\lambda_B{}^* x_B$$

where $\lambda_B{}^*$ is the absolute activity of the pure substance B at the same temperature and pressure, or in terms of the chemical potential μ_B by

$$RT \ln(x_B f_B) = \mu_B - \mu_B{}^*$$

where $\mu_B{}^*$ is the chemical potential of the pure substance B at the same temperature and pressure.

It follows from this definition that

$$\lim_{x_B\to 1} f_B = 1 \qquad (T, p \text{ const.})$$

A.I.8 *Relative activity of a substance in a liquid or solid mixture*

The relative activity a_B of a substance B in a liquid or solid mixture is a number defined by

$$a_B = \lambda_B/\lambda_B{}^*$$

or by

$$RT \ln a_B = \mu_B - \mu_B{}^*$$

where the other symbols are as defined in Section A.I.7.

It follows from this definition that

$$\lim_{x_B\to 1} a_B = 1 \qquad (T, p \text{ const.})$$

A mixture of substances B, C, . . . , is treated as an *ideal mixture* when the approximations $a_B = x_B$, $a_C = x_C$, . . . , and consequently $f_B = 1$, $f_C = 1$, . . . , are used.

3. Solutions

A.I.9 *Definition of solution*

The word *solution* is used to describe a liquid or solid phase containing more than one substance, when for convenience one of the substances, which is called the *solvent* and may itself be a mixture, is treated differently from the other substances, which are called *solutes*. When, as is often but not necessarily the case, the sum of the mole fractions of the solutes is small compared with unity, the solution is called a *dilute solution*. In the following definitions the solvent substance is denoted by A and the solute substances by B, C, ...

A.I.10 *Properties of infinitely dilute solutions*

The superscript $^{\infty}$ attached to the symbol for a property of a solution denotes the property of an *infinitely dilute solution*.

For example if V_B denotes the partial molar volume (see Section 1.1) of the solute substance B in a solution containing molalities m_B, m_C, . . . , or mole fractions x_B, x_C, . . . , of solute substances B, C, . . . , in a solvent substance A, then

$$V_B^{\infty} = \lim_{\Sigma_i m_i \to 0} V_B = \lim_{\Sigma_i x_i \to 0} V_B \qquad (T, p \text{ const.})$$

where i = B, C, ...

Similarly if V_A denotes the partial molar volume of the *solvent* substance A, then

$$V_A^{\infty} = \lim_{\Sigma_i m_i \to 0} V_A = \lim_{\Sigma_i x_i \to 0} V_A = V_A{}^* \qquad (T, p \text{ const.})$$

where $V_A{}^*$ is the molar volume of the pure solvent substance A.

A.I.11 *Activity coefficient of a solute substance in a solution*

The activity coefficient γ_B of a *solute* substance B in a solution (especially in a dilute liquid solution) containing molalities m_B, m_C, . . . , of solute substances B, C, . . . , in a solvent substance A, is a number defined in terms of the absolute activity λ_B of the solute substance B in the solution by

$$\gamma_B = (\lambda_B/m_B)/(\lambda_B/m_B)^{\infty} \qquad (T, p \text{ const.})$$

or in terms of the chemical potential μ_B by

$$RT\ln(m_B\gamma_B) = \mu_B - (\mu_B - RT\ln m_B)^{\infty} \qquad (T, p \text{ const.})$$

It follows from this definition that

$$\gamma_B{}^{\infty} = 1 \qquad (T, p \text{ const.})$$

The name activity coefficient with the symbol y_B may be used for the quantity similarly defined but with amount-of-substance concentration c_B (see Section 2.3) in place of molality m_B.

Another activity coefficient, called the *rational activity coefficient* of a solute substance B and denoted by $f_{x,B}$ is sometimes used. It is defined in terms of the absolute activity λ_B by

$$f_{x,B} = (\lambda_B/x_B)/(\lambda_B/x_B)^{\infty} \qquad (T, p \text{ const.}))$$

or in terms of the chemical potential μ_B by

$$RT\ln(x_B f_{x,B}) = \mu_B - (\mu_B - RT\ln x_B)^{\infty} \qquad (T, p \text{ const.})$$

where x_B is the mole fraction of the solute substance B in the solution. The rational activity coefficient $f_{x,B}$ is related to the (practical) activity coefficient γ_B by the formula

$$f_{x,B} = \gamma_B(1 + M_A\Sigma_i m_i) = \gamma_B/(1 - \Sigma_i x_i)$$

A solution of solute substances B, C, . . . , in a solvent substance A is treated as an *ideal dilute solution* when the activity coefficients are approximated to unity, for example $\gamma_B = 1$, $\gamma_C = 1$, ...

A.I.12 *Relative activity of a solute substance in a solution*

The relative activity a_B of a *solute* substance B in a solution (especially in a dilute liquid solution) containing molalities m_B, m_C, . . . , of solute substances B, C, . . . , in a solvent substance A, is a number defined in terms of the absolute activity λ_B by

$$a_B = (\lambda_B/m^\ominus)/(\lambda_B/m_B)^\infty = m_B\gamma_B/m^\ominus \qquad (T, p \text{ const.})$$

or in terms of the chemical potential μ_B by

$$RT \ln a_B = \mu_B - RT \ln m^\ominus - (\mu_B - RT \ln m_B)^\infty$$
$$= RT \ln(m_B\gamma_B/m^\ominus)$$

where $m^\ominus$ is a standard value of molality (usually chosen to be 1 mol kg^{-1}) and where the other symbols are defined in Section A.I.11.

It follows from this definition of a_B (compare Section A.I.8) that

$$(a_B m^\ominus/m_B)^\infty = 1 \qquad (T, p \text{ const.})$$

The name activity is often used instead of the name relative activity for this quantity.

The name relative activity with the symbol $a_{c,B}$ may be used for the quantity similarly defined but with concentration c_B (see Section 2.3) in place of molality m_B, and a standard value $c^\ominus$ of concentration (usually chosen to be 1 mol dm^{-3}) in place of the standard value $m^\ominus$ of molality.

Another relative activity, called the *rational relative activity* of the solute substance B and denoted by $a_{x,B}$, is sometimes used. It is defined in terms of the absolute activity λ_B by

$$a_{x,B} = \lambda_B/(\lambda_B/x_B)^\infty = x_B f_{x,B} \qquad (T, p \text{ const.})$$

or in terms of the chemical potential μ_B by

$$RT \ln a_{x,B} = \mu_B - (\mu_B - RT \ln x_B)^\infty$$
$$= RT \ln(x_B f_{x,B}) \qquad (T, p \text{ const.})$$

where x_B is the mole fraction of the substance B in the solution. The rational relative activity $a_{x,B}$ is related to the (practical) relative activity a_B by the formula

$$a_{x,B} = a_B m^\ominus M_A$$

A.I.13 *Osmotic coefficient of the solvent substance in a solution*

The osmotic coefficient ϕ of the *solvent* substance A in a solution (especially in a dilute liquid solution) containing molalities $m_B, m_C, \ldots,$ of solute substances B, C, . . . , is a number defined in terms of the absolute activity λ_A of the solvent substance A in the solution by

$$\phi = (M_A \Sigma_i m_i)^{-1} \ln(\lambda_A{}^*/\lambda_A)$$

where $\lambda_A{}^*$ is the absolute activity of the pure solvent substance A at the same temperature and pressure, and M_A is the molar mass of the solvent substance A, or in terms of the chemical potential $\mu_A{}^*$ by

$$\phi = (\mu_A{}^* - \mu_A)/RTM_A\Sigma_i m_i$$

where $\mu_A{}^*$ is the chemical potential of the pure solvent substance A at the same temperature and pressure.

For an *ideal dilute solution* as defined in Section A.I.11 or A.I.12 it can be shown that $\phi = 1$.

Another osmotic coefficient, called the *rational osmotic coefficient* of the solvent substance A and denoted by ϕ_x, is sometimes used. It is defined in terms of the absolute activity λ_A by

$$\phi_x = \ln(\lambda_A/\lambda_A{}^*)/\ln x_A = \ln(\lambda_A/\lambda_A{}^*)/\ln(1 - \Sigma_i x_i)$$

or in terms of the chemical potential μ_A by

$$\phi_x = (\mu_A - \mu_A{}^*)/RT \ln x_A = (\mu_A - \mu_A{}^*)/RT \ln(1 - \Sigma_i x_i)$$

where x_A is the mole fraction of the solvent substance A in the solution. The rational osmotic coefficient ϕ_x is related to the (practical) osmotic coefficient ϕ by the formula

$$\phi_x = \phi M_A\Sigma_i m_i/\ln)1 + M_A\Sigma_i m_i) = -\phi M_A\Sigma_i m_i/\ln(1 - \Sigma_i x_i)$$

A.I.14 *Relative activity of the solvent substance in a solution*

The relative activity a_A of the *solvent* substance A in a solution (especially in a dilute liquid solution) containing molalities m_B, m_C, . . . , or more fractions x_B, x_C, . . . , of solute substances B, C, . . . , is a number defined in terms of the absolute activity λ_A of the solvent substance A in the solution by

$$a_A = \lambda_A/\lambda_A{}^* = \exp(-\phi M_A \Sigma_i m_i) = (1 - \Sigma_i x_i)^{\phi_x}$$

or in terms of the chemical potential μ_A by

$$RT \ln a_A = \mu_A - \mu_A{}^* = -RT\phi M_A \Sigma_i m_i = \phi_x RT \ln(1 - \Sigma_i x_i)$$

where the other quantities are as defined in Section A.I.13.

Note. The definition in this Section of the relative activity of the *solvent* in a *solution*, is identical with the definition in Section A.I.8 of the relative activity of any substance in a *mixture*. See also Section A.I.12.

SYMBOLS, UNITS AND NOMENCLATURE IN PHYSICS

From Document U.I.P. 20 (1978)
International Union of Pure and Applied Physics
Used by permission of the Secretary

1 PHYSICAL QUANTITIES–GENERAL RECOMMENDATIONS †)

Note: The German, Italian, Russian, and Spanish translations of this term are 'physikalische Größe', 'grandezza fizika', 'ФИЗИЧЕСКАЯ ВЕЛИЧИНА', and 'magnitud fisica', respectively.

1.1 Physical quantities

A physical quantity (French: 'grandeur physique') is equivalent to the product of the *numerical value*, i.e. a pure number, and a *unit*:

$$\text{physical quantity} = \text{numerical value} \times \text{unit}.$$

For a physical quantity with symbol a this relation is usually represented in the form $a = \{a\} \cdot [a]$, where $\{a\}$ stands for the numerical value and $[a]$ stands for the symbol for the unit. For dimensionless physical quantities the unit often has no name or symbol and is not explicitly indicated (see section 9.1 "unit systems").

Examples: $E = 200\ \mathrm{J}$ $n = 1{,}55$ (for quartz)
$F = 27\ \mathrm{N}$ $\nu = 3 \times 10^8\ \mathrm{Hz}$.

1.2 Symbols for physical quantities–General rules

1. *Symbols for physical quantities* should be *single letters* of the Latin or Greek alphabet with or without modifying signs: subscripts, superscripts, primes, etc.

   Remarks:
   (a) An exception to this rule is given by the two-letter symbols which are used to represent dimensionless combinations of physical quantities (see section 7.14 "dimensionless parameters"). If such a symbol, composed of two letters, appears as a factor in a product, it is recommended to separate this symbol from the other symbols by a dot or by a space or by brackets. It can be raised to a positive or negative power without using brackets.
   (b) Abbreviations, i.e. shortened forms of names or expressions, such as p.f. for partition function, should not be used in physical equations. These abbreviations in the text should be written in ordinary roman type.

2. *Symbols for physical quantities* should be printed in *italic* (or *sloping*) *type*.

   Remark: It is recommended to consider as a guiding principle for the printing of indices the criterion: only indices which are symbols for physical quantities should be printed in italic (sloping) type.

†) For further details see International Standard I.S.O. 31/0: *General principles concerning quantities, units and symbols.*

Examples:

upright indices	sloping indices
C_{g} (g = gas)	p in C_p
g_{n} (n = normal)	n in $\Sigma_n a_n \psi_n$
μ_{r} (r = relative)	x in $\Sigma_x a_x b_x$
E_{k} (k = kinetic)	i, k in g_{ik}
χ_{e} (e = electric)	x in p_x

3. *Symbols for vectors and tensors*

To avoid the usage of subscripts it is often convenient to indicate vectors and tensors of the second rank by letters of a special type. The following choice is recommended:

(*a*) Vectors should be printed in bold italic (sloping) type, e.g. ***A***, ***a***.

(*b*) Tensors of the second rank should be printed in bold sans serif italic (sloping) type, e.g. ***S***, ***T***.

Remark: When such type is not available, a vector may be indicated by an arrow and a tensor by a double arrow above the symbol; e.g. $\vec{A}$, $\vec{\vec{S}}$.

1.3 Simple mathematical operations

1. *Addition and subtraction* of two physical quantities are indicated by:

$$a + b \quad \text{and} \quad a - b.$$

2. *Multiplication* of two physical quantities may be indicated in one of the following ways:

$$ab \qquad a \cdot b \qquad a \times b.$$

3. *Division* of one quantity by another quantity may be indicated in one of the following ways:

$$\frac{a}{b} \qquad a/b \qquad ab^{-1}$$

or in any other way of writing the product of a and b^{-1}.

These procedures can be extended to cases where one of the quantities or both are themselves products, quotients, sums or differences of other quantities.

If necessary, brackets have to be used in accordance with the rules of mathematics. If the solidus is used to separate the numerator from the denominator and if there is any doubt where the numerator starts or where the denominator ends, brackets should be used.

Examples:

expressions with a horizontal bar	same expressions with a solidus
$\frac{a}{bcd}$	a/bcd
$\frac{2}{9}\sin kx$	$(2/9)\sin kx$

expressions with a horizontal bar	same expressions with a solidus
$\frac{a}{b} - c$	$a/b - c$
$\frac{a}{b - c}$	$a/(b - c)$
$\frac{a - b}{c - d}$	$(a - b)/(c - d)$
$\frac{a}{c} - \frac{b}{d}$	$a/c - b/d$

Remark: It is recommended that in expressions like:

$\sin\{2\pi(x - x_0)/\lambda\}$	$\exp\{(r - r_0)/\sigma\}$
$\exp\{-V(r)/kT\}$	$\surd(\varepsilon/c^2)$

the argument should always be placed between brackets, except when the argument is a simple product of two quantities: e.g. sin *kx*. When the horizontal bar above the square root is used no brackets are needed.

2 UNITS–GENERAL RECOMMENDATIONS

2.1 Symbols for units–General rules

1. *Symbols for units* of physical quantities should be printed in *roman* (*upright*) *type*.

2. *Symbols for units* should not contain a full stop (period) and should remain unaltered in the plural, e.g.: 7 cm and *not* 7 cms.

3. *Symbols for units* should be printed in *lower case* roman (upright) type. However, the symbol for a unit, derived from a proper name, should start with a capital roman letter, e.g.: m (metre); A (ampere); Wb (weber); Hz (hertz).

2.2 Prefixes–General rules

1. The *prefixes* which should be used to indicate decimal multiples or sub-multiples of a unit are given in table 1.

Table 1

deci; *déci*	$(= 10^{-1})$	d	deca; *déca*	$(= 10^{1})$	da
centi; *centi*	$(= 10^{-2})$	c	hecto; *hecto*	$(= 10^{2})$	h
milli; *milli*	$(= 10^{-3})$	m	kilo; *kilo*	$(= 10^{3})$	k
micro; *micro*	$(= 10^{-6})$	μ	mega; *méga*	$(= 10^{6})$	M
nano; *nano*	$(= 10^{-9})$	n	giga; *giga*	$(= 10^{9})$	G
pico; *pico*	$(= 10^{-12})$	p	tera; téra	$(= 10^{12})$	T
femto; *femto*	$(= 10^{-15})$	f	peta; *peta*	$(= 10^{15})$	P
atto; *atto*	$(= 10^{-18})$	a	exa; *exa*	$(= 10^{18})$	E

2. *Compound prefixes*, formed by juxtaposition of two or more prefixes, are not to be used.

Not: mμs,	*but:* ns	(nanosecond)
Not: kMW,	*but:* GW	(gigawatt)
Not: μμF,	*but:* pF	(picofarad)

3. When the symbol of a prefix is placed before the symbol of a unit, the *combination of the two symbols* should be considered as *one new symbol*, which can be raised to a positive or negative power without using brackets.

Examples: $\mathrm{cm^3}$ $\mathrm{mA^2}$ $\mathrm{\mu s^{-1}}$.

Remark:

$\mathrm{cm^3}$ *means always* $(0{,}01\ \mathrm{m})^3$ *but never* $0{,}01\ \mathrm{m^3}$
$\mathrm{\mu s^{-1}}$ *means always* $(10^{-6}\mathrm{s})^{-1}$ *but never* $10^{-6}\mathrm{s^{-1}}$.

2.3 Mathematical operations

1. *Multiplication* of two units may be indicated in one of the following ways:

$$\mathrm{N\,m} \qquad \mathrm{N \cdot m}.$$

2. *Division* of one unit by another unit may be indicated in one of the following ways:

$$\frac{\mathrm{m}}{\mathrm{s}} \qquad \mathrm{m/s} \qquad \mathrm{m\,s^{-1}}$$

or by any other way of writing the product of m and $\mathrm{s^{-1}}$.
Not more than one solidus should be used.

Examples:

Not: cm/s/s,	*but:* $\mathrm{cm/s^2 = cm\,s^{-2}}$
Not: J/K/mol,	*but:* $\mathrm{J/K\,mol = J\,K^{-1}\,mol^{-1}}$

3 NUMBERS

1. *Numbers* should be printed in *upright type.*

2. The *decimal* sign is a comma on the line (,). In documents in the English language a comma or a dot on the line (.) may be used.
If the magnitude of the number is less than unity, the decimal sign should be preceded by a zero.

3. The sign for *multiplication* of numbers is a cross (×) or a dot half-high (·). If a dot is used as a decimal sign, a dot should not be used as the multiplication sign.

Example: 2.3×3.4 or $2{,}3 \cdot 3{,}4$.

4. *Division* of one number by another number may be indicated in the following ways:

$$\frac{136}{273{,}15} \qquad 136/273{,}15$$

or by writing it as the product of numerator and the inverse first power of the denominator. In such cases the number under the inverse power should always be placed between brackets.

Remark: When the solidus is used and when there is any doubt where the numerator starts or the denominator ends, brackets should be used, as in the case of quantities (see section 1.3).

5. To facilitate the reading of *large numbers*, the digits may be grouped in *groups of three*, but *no* comma or point should be used except for the decimal sign.

Example: 2 573,421 736.

4 SYMBOLS FOR CHEMICAL ELEMENTS, NUCLIDES, AND PARTICLES

1. *Symbols for chemical elements* should be written in *roman (upright) type*. The symbol is not followed by a full stop.

Examples: Ca C H He.

2. The *nucleon number (mass number)* of a *nuclide* is shown as a left superscript (e.g. ^{14}N).

3. The right subscript position is used to indicate the *number of atoms* of a nuclide in a molecule (e.g. $^{14}N_2$).

4. The right superscript position should be used, if required, for indicating a *state of ionization* (e.g. Ca^{2+}, PO_4^{3-}) or an *excited state* (e.g. $^{110}Ag^m$, He*).

5. The attached numeral specifying the *spectrum of a z-fold ionized atom* is the roman number $z + 1$.

Examples: CaII, AlIII, HI (spectrum of the *neutral* hydrogen atom).

Remark: Roman numbers in right superscript position may indicate the *oxidation number* (e.g. $Pb_2^{II}Pb^{IV}O_4$; $K_6M^{IV}Mo_9O_{32}$, where M denotes a metal).

6. *Symbols for particles and quanta*

It is recommended to use the notation listed in table 2.

Table 2

nucleon	N	pion	π
proton	p	K-meson	K
neutron	n		
		electron	e
Λ-particle	Λ	muon	μ
Σ-particle	Σ	neutrino	ν
Ξ-particle	Ξ		
Ω-particle	Ω	photon	γ
deuteron	d		
triton	t		
helion ($^3He^{2+}$)	h		
α-particle	α		

The charge of a particle may be indicated by adding the superscript +, − or 0.

Examples: π^+ π^- π^0, p^+ p^-, e^+ e^-.

If in connection with the symbols p and e no charge is indicated, these symbols should refer to the positive proton and the negative electron, respectively.

The bar ¯, or sometimes the tilde ˜, above the symbol of a particle is used to indicate the corresponding anti-particle.

5 QUANTUM STATES

5.1 General rules

A letter symbol indicating the quantum state of *a system* should be printed in capital upright type.

A letter symbol indicating the quantum state of *a single particle* should be printed in lower case upright type.

5.2 Atomic spectroscopy

The letter symbols indicating atomic quantum states are:

$L, l = 0$: S, s	$L, l = 4$: G, g	$L, l = 8$: L, l
$= 1$: P, p	$= 5$: H, h	$= 9$: M, m
$= 2$: D, d	$= 6$: I, i	$= 10$: N, n
$= 3$: F, f	$= 7$: K, k	$= 11$: O, o.

A right subscript indicates the total angular momentum quantum number J or j. A left superscript indicates the spin multiplicity $2S+1$.

Example: $^2P_{3/2}$-state ($J = 3/2$, spin multiplicity 2)
$p_{3/2}$-electron ($j = 3/2$).

An atomic electron configuration is indicated symbolically by:

$$(nl)^{\kappa} \quad (n'l')^{\kappa'} \quad \ldots$$

Instead of $l = 0, 1, 2, 3 \ldots$ one uses the quantum state symbol s, p, d, f, ...

Example: the atomic electron configuration: $(1s)^2\ (2s)^2\ (2p)^3$.

5.3 Molecular spectroscopy

The letter symbols, indicating molecular electronic quantum states are in the case of *linear molecules:*

$$\Lambda, \lambda = 0: \quad \Sigma, \sigma$$
$$= 1: \quad \Pi, \pi$$
$$= 2: \quad \Delta, \delta$$

and for *non-linear molecules*

A, a ;   B, b ;   E, e ;   etc.

Remarks: A left superscript indicates the spin multiplicity. For molecules having a symmetry centre the parity symbol g or u, indicating respectively symmetric or antisymmetric behaviour on inversion, is attached as a right subscript. A + or − sign attached as a right superscript indicates the symmetry as regards reflection in any plane through the symmetry axis of the molecules.

Examples: Σ_g^+, Π_u, $^2\Sigma$, $^3\Pi$, etc.

The letter symbols indicating the vibrational angular momentum states in the case of *linear molecules* are

$$l = 0: \quad \Sigma$$
$$= 1: \quad \Pi$$
$$= 2: \quad \Delta.$$

5.4 Nuclear spectroscopy

The spin and parity assignment of a nuclear state is

$$J^\pi$$

where the parity symbol π is + for even and − for odd parity.

Examples: 3^+, 2^-, etc.

A shell model configuration is indicated symbolically by:

$$(nlj)^\kappa \quad (n'l'j')^{\kappa'}$$

where the first bracket refers to the proton shell and the second to the neutron shell. Negative values of κ or κ' indicate holes in a completed shell. Instead of $l = 0, 1, 2, 3, \ldots$ one uses the quantum state symbol s, p, d, f, ...

Example: the nuclear configuration: $(1\,d\,3/2)^3(1\,f\,7/2)^2$.

5.5 Spectroscopic transitions

1. The upper level and the lower level are indicated by ′ and ″ respectively.

   Examples: $h\nu = E' - E''$ $\sigma = T' - T''$.

2. A spectroscopic transition should be indicated by writing the upper state first and the lower state second, connected by a dash.

   Examples:

$^2P_{1/2} - {}^2S_{1/2}$	for an electronic transition
$(J', K') - (J'', K'')$	for a rotational transition
$\upsilon' - \upsilon''$	for a vibrational transition.

3. Absorption transition and emission transition may be indicated by arrows ← and → respectively.

Examples: $^2P_{1/2} \rightarrow {}^2S_{1/2}$ emission from $^2P_{1/2}$ to $^2S_{1/2}$
$(J', K') \leftarrow (J'', K'')$ absorption from (J'', K'') to (J', K').

4. The difference between two quantum numbers should be that of the upper state minus that of the lower state.

Example: $\Delta J = J' - J''$.

5. The indications of the branches of the rotation band should be as follows:

$\Delta J = J' - J'' = -2$: O-branch
$= -1$: P-branch
$= 0$: Q-branch
$= +1$: R-branch
$= +2$: S-branch.

6 NOMENCLATURE

1. *Use of the words 'specific' and 'molar'*

The word 'specific' in the English name for an extensive physical quantity should be restricted to the meaning 'divided by mass (mass of the system, if this consists of more than one component or of more than one phase)'.

Examples:

specific volume	volume/mass
specific energy	energy/mass
specific heat capacity	heat capacity/mass.

The word 'molar' in the English name for an extensive physical quantity should be restricted to the meaning 'divided by amount of substance (amount of substance of the system, if this consists of more than one component or of more than one phase)'.

Examples:

molar volume	volume/amount of substance
molar energy	energy/amount of substance
molar heat capacity	heat capacity/amount of substance.

The symbol X_B, where X denotes an extensive quantity and B is the chemical symbol for a substance, denotes the partial molar quantity of the substance B defined by the relation:

$$X_B = (\partial X / \partial n_B)_{T,p,n_C,\ldots}$$

For a pure substance B the partial molar quantity X_B and the molar quantity X_m are identical. The partial molar quantity X_B of pure substance B, which is identical with the molar quantity X_m of pure substance B, may be denoted by X_B^* where the superscript * denotes 'pure', so as to distinguish it from the partial molar quantity X_B of substance B in a mixture.

2. *Notation for covariant character of coupling*

S	Scalar coupling	A	Axial vector coupling
V	Vector coupling	P	Pseudoscalar coupling
T	Tensor coupling.		

3. *Abbreviated notation for a nuclear reaction*

The meaning of the symbolic expression indicating a nuclear reaction should be the following:

initial nuclide (incoming particle(s) or quanta, outgoing particle(s) or quanta) final nuclide.

Examples: $^{14}N(\alpha, p)^{17}O$ $\quad$ $^{59}Co(n, \gamma)^{60}Co$

$^{23}Na(\gamma, 3n)^{20}Na$ $\quad$ $^{31}P(\gamma, pn)^{29}Si$.

4. *Character of transitions*

Multipolarity of transition:

electric or magnetic monopole	E0 or M0
" " " dipole	E1 or M1
" " " quadrupole	E2 or M2
electric or magnetic octupole	E3 or M3
" " " 2^n-pole	En or Mn.

Parity change in transition:

transition *with* parity change: yes
transition *without* parity change: no.

5. *Nuclide*

A species of *atoms*, identical as regards atomic number (proton number) and mass number (nucleon number) should be indicated by the word *nuclide*, not by the word isotope.
Different nuclides having the same atomic number should be indicated as *isotopes* or *isotopic nuclides.*
Different nuclides having the same mass number should be indicated as *isobars* or *isobaric nuclides.*

6. *Sign of polarization vector* (*Basel Convention*)

In nuclear interactions the positive polarization of particles with spin $\frac{1}{2}$ is taken in the direction of the vector product

$$\boldsymbol{k}_i \times \boldsymbol{k}_o$$

where $\boldsymbol{k}_i$ and $\boldsymbol{k}_o$ are the circular wave vectors of the incoming and outgoing particles, respectively.

7. *Description of polarization effects* (*Madison Convention*)

In the symbolic expression for a nuclear reaction A(b, c)D an arrow placed over a symbol denotes a particle which is initially in a polarized state or whose state of polarization is measured.

Examples:

$A(\vec{b}, c)D$	polarized incident beam
$A(\vec{b}, \vec{c})D$	polarized incident beam; polarization of the outgoing particle c measured (polarization transfer)
$A(b, \vec{c})D$	unpolarized incident beam; polarization of the outgoing particle c measured
$\vec{A}(b, c)D$	unpolarized beam incident on a polarized target
$\vec{A}(b, \vec{c})D$	unpolarized beam incident on a polarized target; polarization of the outgoing particle c measured
$A(\vec{b}, c)\vec{D}$	polarized incident beam; measurement of the polarization of the target.

7 RECOMMENDED SYMBOLS FOR PHYSICAL QUANTITIES

Remarks:

(1) Where several symbols are given for one quantity, and no special indication is made, they are on equal footing.

(2) In general no special attention is paid to the name of the quantity.

(3) Where there is more than one form for a greek letter (ε, ϵ; ϑ, θ; $\kappa, \varkappa$; φ, ϕ) either form may be used. The form ϖ of the letter pi may be used as though it were a different letter.

7.1 Space and time

space coordinates; *coordonnées d'espace*	(x, y, z)
position vector; *vecteur de position*	$\boldsymbol{r}$
length; *longueur*	l
breadth; *largeur*	b
height; *hauteur*	h
radius; *rayon*	r
thickness; *épaisseur*	d, δ
diameter; *diamètre:* $d = 2r$	d
element of path; *élément de parcours*	$\mathrm{d}s$
area; *aire, superficie*	A, S
volume; *volume*	$V, (v)$
plane angle; *angle plan*	$\alpha, \beta, \gamma, \theta, \vartheta, \varphi$
solid angle; *angle solide*	ω, Ω
wave length; *longueur d'onde*	λ
wave number; *nombre d'onde:* $\sigma = 1/\lambda$	σ †)
wave vector; *vecteur d'onde*	$\boldsymbol{\sigma}$
circular wave number; *nombre d'onde circulaire:* $k = 2\pi/\lambda$	K
circular wave vector; *vecteur d'onde circulaire*	$\boldsymbol{K}$
attenuation coefficient; *constante d'affaiblissement*: $F(x) = \exp(-\alpha x) \cos \beta x$	α
phase coefficient; *constante de phase*	β
propagation coefficient; *constante de propagation:* $\gamma = \alpha + \mathrm{i}\beta$	γ
time; *temps*	t
period, periodic time; *période, durée d'une période*	T
frequency; *fréquence:* $\nu = 1/T$	ν, f
circular frequency, pulsatance; pulsation: $\omega = 2\pi\nu$	ω
relaxation time; *temps de relaxation:* $F(t) = \exp(-t/\tau)$	τ
damping coefficient; *coefficient d'amortissement:* $F(t) = \exp(-\delta t) \sin \omega t$	δ
logarithmic decrement; *décrément logarithmique:* $\Lambda = T\delta = T/\tau$	Λ
velocity; *vitesse:* $v = \mathrm{d}s/\mathrm{d}t$	u, v
angular velocity; *vitesse angulaire:* $\omega = \mathrm{d}\varphi/\mathrm{d}t$	ω
acceleration; *accélération:* $a = \mathrm{d}v/\mathrm{d}t$	a
angular acceleration; *accélération angulaire:* $\alpha = \mathrm{d}\omega/\mathrm{d}t$	α
acceleration of free fall; *accélération de la pesanteur*	g
standard --; -- *normale*	g_n
speed of light in empty space; *vitesse de la lumière dans le vide*	c
v/c	β
relativistic coordinates; *coordonnées relativistes:*	$(x_0 x_1 x_2 x_3)$
$x_0 = ct,\ x_1 = x,\ x_2 = y,\ x_3 = z,\ x_4 = \mathrm{i}ct$	$(x_1 x_2 x_3 x_4)$

†) In molecular spectroscopy $\bar{\nu}$ is also used.

7.2 Mechanics

mass; *masse*	m
(mass) density; *masse volumique:* $\rho = m/V$	ρ
relative density; *densité relative:* $d = \rho/\rho(H_2O)$	d
specific volume; *volume massique:* $v = V/m = 1/\rho$	v
reduced mass; *masse réduite:* $\mu = m_1m_2/(m_1 + m_2)$	μ
momentum; *quantité de mouvement:* $\mathbf{p} = m\mathbf{v}$	$\mathbf{p}$
angular momentum; *moment cinétique:* $\mathbf{L} = \mathbf{r} \times \mathbf{p}$	$\mathbf{L}$
second moment of plane area; *moment quadratique d'une aire plane:* $I_{a,y} = \int x^2 \mathrm{d}x\mathrm{d}y$	I_a
second polar moment of plane area; *moment quadratique polaire d'une aire plane:* $I_p = \int (x^2 + y^2)\,\mathrm{d}x\mathrm{d}y$	I_p
moment of inertia; *moment d'inertie:* $I_z = \int (x^2 + y^2)\,\mathrm{d}m$	I, J
force; *force*	$\mathbf{F}$
torque, moment of a couple; *torque, moment d'un couple*	$\mathbf{T}$
weight; *poids*	$G, (W, P)$
moment of force; *moment d'une force*	$\mathbf{M}$
pressure; *pression*	p
normal stress; *contrainte normale*	σ
shear stress; *contrainte tangentielle, cission*	τ
gravitational constant; *constante de gravitation:* $F(r) = G\,m_1m_2/r^2$	G
linear strain, relative elongation; *dilatation linéique relative:* $\varepsilon = \Delta l/l_0$	ε
modulus of elasticity, Young's modulus; *module d'élasticité longitudinale:* $\sigma = E\,\varepsilon$	E
shear strain; *glissement unitaire*	γ
shear modulus; *module d'élasticité de glissement:* $\tau = G\gamma$	G
volume strain, bulk strain; *dilatation volumique relative:* $\theta = \Delta V/V_0$	θ
bulk modulus; *module de compression:* $p = -K\theta$	K
Poisson ratio, *rapport de Poisson*	μ, ν
viscosity; *viscosité*	$\eta, (\mu)$
kinematic viscosity; *viscosité cinématique:* $\nu = \eta/\rho$	ν
friction coefficient; *coefficient de frottement*	$\mu, (f)$
surface tension; *tension superficielle*	γ, σ
energy; *énergie*	E, W
potential energy; *énergie potentielle*	E_p, V, Φ
kinetic energy; *énergie cinétique*	E_k, T, K
work; *travail*	W, A
power; *puissance*	P
efficiency; *rendement*	η
Hamiltonian function; *fonction de Hamilton*	H
Lagrangian function; *fonction de Lagrange*	L
principal function of Hamilton; *fonction principale de Hamilton:* $W = \int L\mathrm{d}t$	W, S_p
characteristic function of Hamilton; *fonction charactéristique de Hamilton:* $S = 2\int T\mathrm{d}t$	S
generalized coordinate; *coordonnée généralisée*	q, q_i
generalized momentum; *moment généralisé*	p, p_i
action integral; *intégrale d'action:* $J = \oint p\,\mathrm{d}q$	J

7.3 Molecular physics

number of molecules; *nombre de molécules.*	N

number density of molecules; *nombre volumique de molécules* $n = N/V$ — n

Avogadro constant; *constante d'Avogadro* — L, N_A

molecular mass; *masse moléculaire* — m

molecular velocity vector with (magnitudes of) components; *vecteur vitesse moléculaire et ses coordonnées* — $c, (c_x, c_y, c_z)$; $\boldsymbol{u}, (u_x, u_y, u_z)$

molecular position vector with coordinates; *vecteur position moléculaire et ses coordonnées* — $\boldsymbol{r}, (x, y, z)$

molecular momentum vector with (magnitudes of) components; *vecteur quantité de mouvement moléculaire et ses coordonnees* — $\boldsymbol{p}, (p_x, p_y, p_z)$

average velocity; *vitesse moyenne* — $c_0, u_0, \langle c \rangle, \langle u \rangle$

average speed; *vitesse moyenne* — $\bar{c}, \bar{u}, \langle c \rangle, \langle u \rangle$

most probable speed; *vitesse la plus probable* — $\hat{c}, \hat{v}$

mean free path; *libre parcours moyen* — l

molecular attraction energy parameter; *paramètre d'énergie d'attraction moléculaire* — ε

interaction energy between molecules i and j; *énergie d'interaction entre les molécules i et j* — φ_{ij}, V_{ij}

velocity distribution function; *fonction de distribution des vitesses:* $n = \int f \, dc_x dc_y dc_z$ — $f(c)$

Boltzmann function; *fonction de Boltzmann* — H

generalized coordinate; *coordonnée généralisée* — q

generalized momentum; *moment généralisé* — p

volume in γ phase space; *volume dans l'espace* γ — Ω

thermodynamic temperature; *température thermodynamique* — T

Boltzmann constant; *constante de Boltzmann* — k

$1/kT$ (in exponential functions; *dans les fonctions exponentielles*) — β

molar gas constant; *constante molaire des gaz* — R

partition function; *fonction de partitions* — Q, Z

symmetry number; *facteur de symétrie* — s

diffusion coefficient; *coefficient de diffusion* — D

thermal diffusion coefficient; *coefficient de thermodiffusion* — D_T

thermal diffusion ratio; *rapport de thermodiffusion* — k_T

thermal diffusion factor; *facteur de thermodiffusion* — α_T

characteristic temperature; *température caractéristique* — Θ

Debye temperature; *température de Debye:* $\Theta_D = h\nu_D/k$ — Θ_D

Einstein temperature; *température d'Einstein:* $\Theta_E = h\nu_E/k$ — Θ_E

rotational temperature; *température de rotation:* $\Theta_r = h^2/8\pi^2 Ik$ — Θ_r

vibrational temperature; *température de vibration:* $\Theta_v = h\nu/k$ — Θ_v

7.4 Thermodynamics †)

quantity of heat; *quantité de chaleur* — Q

work; *travail* — W, A

thermodynamic temperature; *température thermodynamique* — T

Celsius temperature; *température Celsius* — t, ϑ ††)

entropy; *entropie* — S

internal energy; *énergie interne* — U

Helmholtz function, *fonction de Helmholtz, énergie libre:* $F = U - TS$ — F

enthalpy; *enthalpie:* $H = U + pV$ — H

†) The index m is added in the case of molar quantities, if needed, to distinguish them from quantities referring to the whole system. For specific quantities (see section 6.1) lower case letters are used.

††) When Celsius temperature and time come together, t is reserved for time.

Gibbs function; *fonction de Gibbs, enthalpie libre:* $G = H - TS$ — G
Massieu function; *fonction de Massieu:* $J = -F/T$ — J
Planck function; *fonction de Planck:* $Y = -G/T$ — Y
pressure coefficient; *coefficient de pression:* $\beta = (\partial p/\partial T)_V$ — β
relative pressure coefficient; *coefficient de pression relative:* $\alpha_p = (1/p)(\partial p/\partial T)_V$ — α_p
compressibility; *compressibilité:* $\kappa = -(1/V)(\partial V/\partial p)_T$ — κ
linear expansion coefficient; *dilatabilité linéique* — α_l
cubic expansion coefficient; *dilatabilité volumique:* $\alpha = (1/V)(\partial V/\partial T)_p$ — α, γ
thermal conductivity; *conductivité thermique* — λ
specific heat capacity; *chaleur massique:* $c = C/m$ — c_p, c_V
heat capacity; *capacité thermique* — C_p, C_V
Joule–Thomson coefficient; *coefficient de Joule–Thomson* — μ
isentropic exponent; *exposant isentropique* *): $\kappa = -(V/p)(\partial p/\partial V)_S$ — κ
ratio of specific heat capacities; *rapport des chaleurs massiques* — $\gamma, (\kappa)$
heat flow rate; *flux thermique* — $\Phi, (q)$
heat current density; *densité de flux thermique* — $\mathbf{q}, (\varphi)$
thermal diffusivity; *diffusivité thermique:* $a = \lambda/\rho c_p$ — a

7.5 Electricity and magnetism **)

quantity of electricity; *quantité d'électricité* — Q
charge density; *charge volumique* — ρ
surface charge density; *charge surfacique* — σ
electric potential; *potentiel électrique* — V, φ
potential difference, tension; *différence de potentiel, tension* — U, V
electromotive force; *force électromotrice* — E
electric field strength; *champ électrique* — $\mathbf{E}$
electric flux, *flux électrique* — Ψ
electric displacement; *déplacement électrique* — $\mathbf{D}$
capacitance; *capacité* — C
permittivity; *permittivité:* $\mathbf{D} = \varepsilon \mathbf{E}$ — ε
permittivity of vacuum, electric constant; *permittivité du vide, constante électrique* — ε_0
relative permittivity; *permittivité relative:* $\varepsilon_r = \varepsilon/\varepsilon_0$ — ε_r
dielectric polarization; *polarisation diélectrique* : $\mathbf{D} = \varepsilon_0 \mathbf{E} + \mathbf{P}$ — $\mathbf{P}$
electric susceptibility; *susceptibilité électrique* — χ_e
polarizability; *polarisabilité* — α, γ
electric dipole moment; *moment dipolaire électrique* — $\mathbf{p}$
electric current; *courant électrique* — I
electric current density; *densité de courant électrique* — $\mathbf{j}, \mathbf{J}$
magnetic field strength; *champ magnétique* — $\mathbf{H}$
magnetic potential difference; *différence de potentiel magnétique* — U_m
magnetomotive force; *force magnétomotrice:* $F_m = \oint H_s \mathrm{d}s$ — F_m
magnetic induction, magnetic flux density; *induction magnétique, densité de flux magnétique* — $\mathbf{B}$
magnetic flux; *flux magnétique* — Φ
permeability; *perméabilité:* $\mathbf{B} = \mu \mathbf{H}$ — μ

*) For an ideal gas the isentropic exponent, κ, is equal to the ratio of specific heat capacities, γ.
**) Written according to the rationalized, 4-dimensional system of quantities, see Appendix I, section 2.

permeability of vacuum, magnetic constant; *perméabilité du vide, constante magnétique* — μ_0

relative permeability; *perméabilité relative:* $\mu_r = \mu/\mu_0$ — μ_r

magnetization; *aimantation:* $\boldsymbol{B} = \mu_0(\boldsymbol{H}+\boldsymbol{M})$ — $\boldsymbol{M}$

magnetic susceptibility; *susceptibilité magnétique* — χ_m

electromagnetic moment; *moment électromagnétique* $E_p = -\boldsymbol{m}\cdot\boldsymbol{B}$ — $\boldsymbol{\mu}, \boldsymbol{m}$

resistance; *résistance* — R

reactance; *réactance* — X

quality factor; *facteur de qualité:* $Q = |X|/R$ — Q

impedance; *impedance:* $Z = R + iX$ — Z

admittance; *admittance:* $Y = 1/Z = G + iB$ — Y

conductance; *conductance* — G

susceptance; *susceptance* — B

resistivity; *résistivité* — ρ

conductivity; *conductivité* $\gamma = 1/\rho$ — γ, σ

self inductance; *inductance propre* — L

mutual inductance; *inductance mutuelle* — M, L_{12}

coupling coefficient; *coefficient de couplage:* $k = L_{12}/(L_1L_2)^{1/2}$ — k

phase number; *nombre de phases* — m

loss angle; *angle de pertes* — δ

number of turns; *nombre de tours* — N

power; *puissance* — P

electromagnetic energy density; *énergie électromagnétique volumique* — w

Poynting vector; *vecteur de Poynting* — $\boldsymbol{S}$

magnetic vector potential; *potentiel vecteur magnétique* — $\boldsymbol{A}$

7.6 *Radiation, light* †)

radiant energy; *énergie rayonnante* — $Q, (Q_e), W$

radiant energy density; *énergie rayonnante volumique* — w

spectral concentration of radiant energy density (in terms of wavelength); *énergie rayonnante volumique spectrique (en longueur d'onde):* $w = \int w_\lambda \, d\lambda$ — w_λ

radiant flux, radiant power; *flux énergétique, puissance rayonnante:* $\Phi = \int \Phi_\lambda \, d\lambda$ — $\Phi, (\Phi_e), P$

radiant flux density; *densité de flux énergétique* — φ

radiant intensity; *intensité énergétique:* $\Phi = \int I \, d\Omega$ — $I, (I_e)$

spectral concentration of radiant intensity (in terms of frequency); *intensité énergétique spectrique (en fréquence):* $I = \int I_\nu \, d\nu$ — $I_\nu, (I_{e\nu})$

irradiance; *éclairement énergétique:* $\Phi = \int E \, dS$ — $E, (E_e)$

radiance; *luminance énergétique:* $I = \int L \cos\vartheta \, dS$ — $L, (L_e)$

radiant excitance; *excitance énergétique:* $\Phi = \int M \, dS$ — $M, (M_e)$

Stefan–Boltzmann constant; *constante de Stefan–Boltzmann:* $\sigma = 2\pi^5k^4/15h^3c^2$ — σ

first radiation constant; *première constante de rayonnement:* $c_1 = 2\pi hc^2$ — c_1

second radiation constant; *deuxième constante de rayonnement:* $c_2 = hc/k$ — c_2

†) In several cases, the same symbol is used for a pair of corresponding radiant and luminous quantities with the understanding that subscripts e for energetic and v for visible will be added, whenever confusion between these quantities might otherwise occur.

emissivity; *émissivité:* $\varepsilon = M/M_B$ — ε
(M_B: radiant excitance of a black body radiator)
luminous efficacy; *efficacité lumineuse:* $K = \Phi_v/\Phi_e$ — K
spectral luminous efficacy; *efficacité lumineuse spectrale:* $K(\lambda) = \Phi_{v\lambda}/\Phi_{e\lambda}$ — $K(\lambda)$
maximum spectral luminous efficacy; *efficacité lumineuse spectrale maximale* — K_m
luminous efficiency; *efficacité lumineuse relative:* $V = K/K_m$ — V
spectral luminous efficiency; *efficacité lumineuse relative spectrale:* $V(\lambda) = K(\lambda)/K_m$ — $V(\lambda)$
quantity of light; *quantité de lumière* — $Q, (Q_v)$
luminous flux; *flux lumineux* — $\Phi, (\Phi_v)$
luminous intensity; *intensité lumineuse:* $\Phi = \int I\, d\Omega$ — $I, (I_v)$
spectral concentration of luminous intensity (in terms of wave number); *intensité lumineuse spectrique (en nombre d'onde):* $I = \int I_\sigma\, d\sigma$ — $I_\sigma, (I_{v\sigma})$
illuminance, illumination; *éclairement lumineux:* $\Phi = \int E\, dS$ — $E, (E_v)$
luminance; *luminance:* $I = \int L \cos\vartheta\, dS$ — $L, (L_v)$
luminous excitance; *excitance lumineuse:* $\Phi = \int M\, dS$ — $M, (M_v)$
absorptance; *facteur d'absorption:* Φ_a/Φ_0 — α *)
reflectance; *facteur de réflexion:* Φ_r/Φ_0 — ρ *)
transmittance; *facteur de transmission:* Φ_{tr}/Φ_0 — τ *)
linear attenuation coefficient; *coefficient d'atténuation linéique* — μ
linear absorption coefficient; *coefficient d'absorption linéique* — a
speed of light in empty space; *vitesse de la lumière dans le vide* — c
refractive index; *indice de réfraction:* $n = c/c_n$ — n

7.7 Acoustics

velocity of sound; *vitesse du son* — c
velocity of longitudinal waves; *vitesse longitudinale* — c_l
velocity of transversal waves; *vitesse transversale* — c_t
group velocity; *vitesse de groupe* — c_g
sound energy flux; *flux d'énergie acoustique* — P, P_a
reflexion factor; *facteur de réflexion:* P_r/P_0 — ρ
acoustic absorption factor; *facteur d'absorption acoustique:* $1-\rho$ — $\alpha_a, (\alpha)$
transmission factor; *facteur de transmission:* P_{tr}/P_c — τ
dissipation factor; *facteur de dissipation:* $\alpha_a - \tau$ — δ
loudness level; *niveau d'isosonie* — L_N
sound power level; *niveau de puissance acoustique* — L_P
sound pressure level; *niveau de pression acoustique* — L_p

7.8 Quantum mechanics

complexe conjugate of Ψ; *complexe conjugué de* Ψ — Ψ^*
probability density; *densité de probabilité:* $P = \Psi^*\Psi$ — P
probability current density; *densité de courant de probabilité:* $S = (\hbar/2im)(\Psi^*\nabla\Psi - \Psi\nabla\Psi^*)$ — S
charge density of electrons; *charge volumique d'électrons:* $\rho = -eP$ — ρ
electric current density of electrons; *densité de courant électrique d'électrons:* $j = -eS$ — j

*) $\alpha(\lambda)$, $\rho(\lambda)$, and $\tau(\lambda)$ designate spectral absorptance, $\Phi_{a\lambda}/\Phi_{0\lambda}$, spectral reflectance, $\Phi_{r\lambda}/\Phi_{0\lambda}$, and spectral transmittance, $\Phi_{tr\lambda}/\Phi_{0\lambda}$, respectively.

Dirac bra vector; *vecteur bra de Dirac*	$\langle\ |$
Dirac ket vector; *vecteur ket de Dirac*	$|\ \rangle$
expectation value of A; *valeur moyenne de A*	$\langle A\rangle, \bar{A}$
commutator of A and B; *commutateur de A et B:* $[A,B] = AB - BA$	$[A,B], [A,B]_-$
anticommutator of A and B; *anticommutateur de A et B:* $[A,B]_+ = AB + BA$	$[A,B]_+$
matrix element; *élément de matrice:* $A_{ij} = \int \varphi_i^*(A\varphi_j)\,d\tau$	A_{ij}
Hermitian conjugate of operator A; *conjugué Hermitien de l'opérateur A*	$A^\dagger$
momentum operator in coordinate representation; *opérateur de quantité de mouvement*	$+(\hbar/i)\nabla$
annihilation operators; *opérateurs d'annihilation*	a, b, α, β
creation operators; *opérateurs de création*	$a^\dagger, b^\dagger, \alpha^\dagger, \beta^\dagger$

7.9 Atomic and nuclear physics

nucleon number, mass number; *nombre de nucléons, nombre de masse*	A
proton number, atomic number; *nombre de protons, nombre atomique*	Z
neutron number; *nombre de neutrons:* $N = A - Z$	N
elementary charge (equal to charge of proton); *charge élémentaire (égale à la charge du proton)*	e
electron mass; *masse de l'électron*	m, m_e
proton mass; *masse du proton*	m_p
neutron mass; *masse du neutron*	m_n
meson mass; *masse du méson*	$m_\pi, \ldots$
nuclear mass; *masse nucléaire* (of nucleus: ^{A}X)	$m_N, m_N(^{A}X)$
atomic mass; *masse atomique* (of nuclide: ^{A}X)	$m_a, m_a(^{A}X)$
(unified) atomic mass constant; *constante (unifiée) de masse atomique:* $m_u = m_a(^{12}C)/12$	m_u
relative atomic mass; *masse atomique relative:* m_a/m_u	A_r
Planck constant; *constante de Planck* ($\hbar = h/2\pi$)	h
principal quantum number (qu.n.); *nombre quantique (n.qu.) principal*	n, n_i
orbital angular momentum qu.n.; *n.qu. de moment angulaire orbital*	L, l_i
spin qu.n.; *n. qu. de spin*	S, s_i
total angular momentum qu.n.; *n.qu. de moment angulaire total*	J, j_i
magnetic qu.n.; *n.qu. magnétique*	M, m_i
nuclear spin qu.n.; *n.qu. de spin nucléaire*	I, J †)
hyperfine qu.n.; *n.qu. hyperfin*	F
rotational qu.n.; *n.qu. de rotation*	J, K
vibrational qu.n.; *n.qu. de vibration*	v
quadrupole moment; *moment quadripolaire*	Q
Rydberg constant; *constante de Rydberg* ††)	R_∞
Bohr radius; *rayon de Bohr* ††)	a_0
fine structure constant; *constante de structure fine* ††)	α
mass excess; *excès de masse:* $m_a - Am_u$	Δ
packing fraction; *packing fraction:* Δ/Am_u	f
nuclear radius; *rayon nucléaire:* $R = r_0 A^{1/3}$	R

†) I is used in atomic physics, J in nuclear physics.
††) See for definition: Appendix I, section 3.

magnetic moment of particle; *moment magnétique d'une particule* — μ
magnetic moment of proton; *moment magnétique du proton* — μ_p
magnetic moment of neutron; *moment magnétique du neutron* — μ_n
magnetic moment of electron; *moment magnétique électronique* — μ_e
Bohr magneton; *magnéton de Bohr* †) — μ_B
nuclear magneton; *magnéton nucléaire* — μ_N
g-factor; *facteur g:* e.g. $g = \mu / I\mu_N$ — g
gyromagnetic ratio, gyromagnetic coefficient; *rapport gyromagnétique, coefficient gyromagnétique* †) — γ
Larmor circular frequency; *pulsation de Larmor* †) — ω_L
level width; *largeur d'un niveau* — Γ
mean life; *vie moyenne* — τ
reaction energy; *énergie de réaction* — Q
cross section; *section efficace* — σ
macroscopic cross section; *section efficace macroscopique:* $\Sigma = n\sigma$ — Σ
impact parameter; *paramètre de collision* — b
scattering angle; *angle de diffusion* — $\vartheta, \theta, \varphi$
internal conversion coefficient; *coefficient de conversion interne* — α
disintegration energy; *énergie de désintégration* — Q
half life; *demi-vie* — $T_{1/2}$
reduced half life; *demi-vie reduite* — $fT_{1/2}$
decay constant, disintegration constant; *constante de désintégration* — λ
activity; *activité* — A
Compton wavelength; *longueur d'onde de Compton:* $\lambda_C = h/mc$ — λ_C
electron radius; *rayon de l'electron* †) — r_e
linear attenuation coefficient; *coefficient d'atténuation linéique* — μ, μ_1
atomic attenuation coefficient; *coefficient d'atténuation atomique* — μ_a
mass attenuation coefficient; *coefficient d'atténuation massique* — μ_m
linear stopping power; *pouvoir d'arrêt linéaire* — S, S_1
atomic stopping power; *pouvoir d'arrêt atomique* — S_a
linear range; *distance de pénétration linéaire* — R, R_1
recombination coefficient; *coefficient de recombinaison* — α

7.10 Solid state physics

lattice vector: a translation vector which maps the crystal lattice on itself; *vecteur du réseau: vecteur qui reproduit par translation le réseau cristallin sur lui-même* — $\boldsymbol{R}, \boldsymbol{R}_0$
fundamental translation vectors for the crystal lattice; *vecteurs de base de la maille cristalline:* $\boldsymbol{R} = n_1\boldsymbol{a}_1 + n_2\boldsymbol{a}_2 + n_3\boldsymbol{a}_3$, where n_1, n_2, n_3 are integers — $\boldsymbol{a}_1, \boldsymbol{a}_2, \boldsymbol{a}_3$; $\boldsymbol{a}, \boldsymbol{b}, \boldsymbol{c}$
(circular) reciprocal lattice vector; *vecteur du réseau réciproque:* $\boldsymbol{G} \cdot \boldsymbol{R} = m \cdot 2\pi$, where m is an integer — $\boldsymbol{G}$
(circular) fundamental translation vectors for the reciprocal lattice; *vecteur de base de la maille du réseau réciproque:* $\boldsymbol{a}_i \cdot \boldsymbol{b}_k = 2\pi\delta_{ik}$ ††), where δ_{ik} is Kronecker delta symbol — $\boldsymbol{b}_1, \boldsymbol{b}_2, \boldsymbol{b}_3$; $\boldsymbol{a}^*, \boldsymbol{b}^*, \boldsymbol{c}^*$
lattice plane spacing; *espacement entre plans réticulaires* — d

†) See for definition: Appendix I, section 3.
††) In crystallography however $\boldsymbol{a}_i \cdot \boldsymbol{b}_k = \delta_{ik}$.

Bragg angle; *angle de Bragg* ϑ
order of reflexion; *ordre de réflexion* n
short range order parameter; *paramètre d'ordre à courte distance (local)* σ
long range order parameter; *paramètre d'ordre à grande distance* s
Burgers vector; *vecteur de Burgers* $\boldsymbol{b}$
particle position vector; *vecteur de position d'une particule* *) $\boldsymbol{r}, \boldsymbol{R}$
equilibrium position vector of ion; *vecteur de position d'équilibre d'un ion* $\boldsymbol{R}_0$
displacement vector of ion; *vecteur de déplacement d'un ion* $\boldsymbol{u}$
normal coordinates; *coordonnées normales* Q_i
polarization vector; *vecteur de polarisation* $\boldsymbol{e}$
Debye–Waller factor; *facteur de Debye–Waller* D
Debye circular wave number; *nombre d'onde circulaire de Debye* q_D
Debye circular frequency; *pulsation de Debye* ω_D
Grueneisen parameter; *paramètre de Grüneisen:* $\gamma = \alpha V/\kappa C_V$ (α: cubic expansion coefficient; κ: compressibility) γ, Γ
Madelung constant; *constante de Madelung* α
mean free path of electrons; *libre parcours moyen des électrons* l, l_e
mean free path of phonons; *libre parcours moyen des phonons* Λ, l_{ph}
drift velocity; *vitesse de mouvement* v_{dr}
mobility; *mobilité* μ
one electron wave function; *fonction d'onde monoélectronique* $\psi(\boldsymbol{r})$
Bloch wave function; *fonction d'onde de Bloch:* $\psi_{\boldsymbol{k}}(\boldsymbol{r}) = u_{\boldsymbol{k}}(\boldsymbol{r}) \exp(i\boldsymbol{k} \cdot \boldsymbol{r})$ $u_{\boldsymbol{k}}(\boldsymbol{r})$
density of states; *densité (électronique) d'état:* $\mathrm{d}N(E)/\mathrm{d}E = N_E(E) = \rho(E)$ N_E, ρ
(spectral) density of vibrational modes; *densité spectrale de modes de vibration* g, N_ω
exchange integral; *intégrale d'échange* J
resistivity tensor; *tenseur de résistivité* ρ_{ik}
electric conductivity tensor; *tenseur de conductibilité électrique* σ_{ik}
thermal conductivity tensor; *tenseur de conductibilité thermique* λ_{ik}
residual resistivity; *résistivité résiduelle* ρ_R
relaxation time; *temps de relaxation* τ
Lorenz coefficient; *coefficient de Lorenz:* $L = \lambda/\sigma T$ L
Hall coefficient; *coefficient de Hall* A_H, R_H
Ettinghausen coefficient; *coefficient d'Ettinghausen* A_E
first Ettinghausen–Nernst coefficient; *premier coefficient d'Ettinghausen–Nernst* A_N
first Righi–Leduc coefficient; *premier coefficient de Righi–Leduc* A_{RL}
thermoelectromotive force between substances a and b; *force thermoélectromotrice entre deux substances* a *et* b E_{ab}
Seebeck coefficient for substances a and b; *coefficient de Seebeck pour deux substances* a *et* b: $S_{ab} = \mathrm{d}E_{ab}/\mathrm{d}T$ S_{ab}, ε_{ab}
Peltier coefficient for substances a and b; *coefficient de Peltier pour deux substances* a *et* b Π_{ab}
Thomson coefficient; *coefficient de Thomson* $\mu, (\tau)$

*) To distinguish between electron and ion position vectors, lower case and capital letters are used, respectively.

work function; *travail d'extraction* *): $\Phi = e\varphi$	φ
Richardson constant; *constante de Richardson:* $j = AT^2 \exp(-\Phi/kT)$	A
electron number density; *nombre volumique électronique (densité électronique)* **)	n, n_n, n_-
hole number density; *nombre volumique de trous (densité de trous)* **)	p, n_p, n_+
donor number density; *nombre volumique de donneurs (densité de donneurs)*	n_d
acceptor number density; *nombre volumique d'accepteurs (densité d'accepteurs)*	n_a
intrinsic number density; *nombre volumique intrinsèque (densité intrinsèque):* $n_i = (n \cdot p)^{1/2}$	n_i
energy gap: *bande d'énergie interdite (énergie gap)*	E_g
donor ionization energy; *énergie d'ionisation de donneur*	E_d
acceptor ionization energy; *énergie d'ionisation d'accepteur*	E_a
Fermi energy; *énergie de Fermi*	E_F, ε_F
circular wave vector, propagation vector (of particles); *vecteur d'onde, vecteur de propagation (de particules)*	$\boldsymbol{k}$
circular wave vector, propagation vector (of phonons); *vecteur d'onde, vecteur de propagation (de phonons)*	$\boldsymbol{q}$
Fermi circular wave vector; *vecteur de Fermi*	$\boldsymbol{k}_F$
electron annihilation operator; *opérateur d'annihilation d'électron*	a
electron creation operator; *opérateur de création d'électron*	$a^\dagger$
phonon annihilation operator; *opérateur d'annihilation de phonon*	b
phonon creation operator; *opérateur de création de phonon*	$b^\dagger$
effective mass; *masse effective* **)	m_n^*, m_p^*
mobility; *mobilité* **)	μ_n, μ_p
mobility ratio; *rapport de mobilité:* $b = \mu_n/\mu_p$	b
diffusion coefficient; *coefficient de diffusion* **)	D_n, D_p
diffusion length; *longueur de diffusion* **)	L_n, L_p
carrier life time; *durée de vie de porteur* **)	τ_n, τ_p
characteristic (Weiss) temperature; *température charactéristique (de Weiss)*	Θ, Θ_W
Curie temperature; *température de Curie*	T_C
Néel temperature; *température de Néel*	T_N
superconductor critical transition temperature; *température critique de transition supraconductrice*	T_c
superconductor (thermodynamic) critical field strength; *champ critique (thermodynamique) d'un supraconducteur*	H_c
superconductor critical field strengths (type II); *champ critique d'un supraconducteur (type II)* †)	H_{c1}, H_{c2}, H_{c3}
superconductor energy gap; *bande interdite du supraconducteur*	Δ
London penetration depth; *profondeur de pénétration de London*	λ_L
coherence length; *longueur de cohérence*	ξ
Landau–Ginzburg parameter; *paramètre de Landau–Ginzburg:* $\kappa = \lambda_L/\xi\sqrt{2}$	κ

*) The symbol W is used for the quantity $W = \Phi + \mu$, where μ is the electron chemical potential which, at $T = 0$, *is equal to the Fermi energy* E_F.

**) In general subscripts n and p or − and + may be used to denote electrons and holes, respectively.

†) H_{c1}: for magnetic flux entering the superconductor, H_{c2}: for disappearing of bulk superconductivity, H_{c3}: for disappearing of surface superconductivity.

flux (or fluxoid) quantum; *quantum de flux:* $\Phi_0 = h/2e$ **) — Φ_0

Miller indices; *indices de Miller* — h_1, h_2, h_3; h, k, l

single plane or set of parallel planes in lattice; *plan simple ou famille de plans réticulaires parallèles dans un réseau* — (h_1, h_2, h_3); (h, k, l)

full set of planes in lattice equivalent by symmetry; *famille de plans réticulaires équivalents par symétrie* — $\{h_1, h_2; h_3\}$; $\{h, k, l\}$

direction in lattice; *rangée réticulaire* — $[u, v, w]$

full set of directions in lattice equivalent by symmetry; *famille de rangées réticulaires équivalentes par symétrie* — $\langle u, v, w\rangle$

Note: If the letter symbols are replaced by numbers in the bracketed expressions, it is customary to omit the commas. A negative numerical value is commonly indicated by a bar above the number, e.g. $(\bar{1}10)$.

7.11 Molecular spectroscopy

quantum number (qu.n.) of component of electronic orbital angular momentum vector along symmetry axis; *nombre quantique (n.qu.) de la composante du moment angulaire orbital électronique suivant l'axe de symétrie* — Λ, λ_i

qu.n. of component of electronic spin along symmetry axis; *n.qu. de la composante du spin électronique suivant l'axe de symétrie* — Σ, σ_i

qu.n. of total electronic angular momentum vector along symmetry axis; *n.qu. du moment angulaire total électronique suivant l'axe de symétrie:* $\Omega = |\Lambda + \Sigma|$ — Ω, ω_i

qu.n. of electronic spin; *n.qu. du spin électronique* — S

qu.n. of nuclear spin; *n.qu. du spin nucléaire* — I

qu.n. of vibrational mode; *n.qu. d'une mode de vibration* — v

degeneracy of vibrational mode; *degré de dégénérescance d'une mode de vibration* — d

qu.n. of vibrational angular momentum; *n.qu. du moment angulaire vibrationnel* (L.M.) — l

qu.n. of total angular momentum; *n.qu. du moment angulaire total* (excluding nuclear spin) — J

qu.n. of component of $\boldsymbol{J}$ in direction of external field; *n.qu. de la composante de* $\boldsymbol{J}$ *dans la direction du champ extérieur* — M, M_J

qu.n. of component of $\boldsymbol{S}$ in direction of external field; *n.qu. de la composante de* $\boldsymbol{S}$ *dans la direction du champ extérieur* — M_S

qu.n. of total angular momentum; *n.qu. du moment angulaire total* (including nuclear spin) $\boldsymbol{F} = \boldsymbol{J} + \boldsymbol{I}$ — F

qu.n. of component of $\boldsymbol{F}$ in direction of external field; *n.qu. de la composante de* $\boldsymbol{F}$ *dans la direction du champ extérieur* — M_F

qu.n. of component of $\boldsymbol{I}$ in direction of external field; *n.qu. de la composante de* $\boldsymbol{I}$ *dans la direction du champ extérieur* — M_I

qu.n. of component of angular momentum along axis; *n.qu. de la composante du moment angulaire suivant l'axe* (L.M. and S.T.M.; excluding electron and nuclear spin; for L.M.: $(K = |\Lambda + l|)$ — K

**) $2e/h = 1/\Phi_0$ is also called characteristic constant for macroscopic coherence in superconductors.

qu.n. of total angular momentum; *n.qu. du moment angulaire total* (L.M. and S.T.M.; excluding electron and nuclear spin: $J = N + S$ *)	N

qu.n. of component of angular momentum along symmetry axis; *n.qu. de la composante du moment angulaire suivant l'axe de symétrie* (L.M. and S.T.M., excluding nuclear spin; for L.M.: $P = |K + \Sigma|$ **)	P

electronic term; *terme électronique:* $T_e = E_e/hc$ ***)	T_e

vibrational term; *terme de vibration:* $G = E_{vibr}/hc$	G

coefficients in expression for vibrational term (for D.M.); *coefficients de l'expression d'un terme de vibration:* $G = \sigma_e(v + \frac{1}{2}) - x\sigma_e(v + \frac{1}{2})^2$	$\sigma_e, x\sigma_e$

coefficients in expression for vibrational term (for P.M.); *coefficients de l'expression d'un terme de vibration:*

$G = \Sigma\sigma_j(v_j + \frac{1}{2}d_j) + \frac{1}{2}\sum_j\sum_k x_{jk}(v_j + \frac{1}{2}d_j)(v_k + \frac{1}{2}d_k)$	σ_j, x_{jk}

rotational term; *terme de rotation:* $F = E_{rot}/hc$	F

principal moments of inertia; *moments principaux d'inertie* $I_A \leq I_B \leq I_C$ †)	I_A, I_B, I_C

rotational constants; *constantes de rotation:* $A = h/8\pi^2 cI_A$, etc. †)	A, B, C

total term; *terme total:* $T = T_e + G + F$	T

Remark: L.M. = linear molecules. S.T.M. = symmetric top molecules. D.M. = diatomic molecules. P.M. = polyatomic molecules.
See for further details: Report on Notation for the Spectra of Polyatomic Molecules (Joint Commission for Spectroscopy of I.U.P.A.P. and I.A.U. 1954) J. Chem. Phys. 23 (1955) 1997.

7.12 Chemical physics

relative atomic mass; *masse atomique relative*	A_r

relative molecular mass; *masse moléculaire relative*	M_r

amount of substance; *quantité de matière*	n, ν ††)

molar mass of substance B; *masse molaire de la substance* B	M_B

concentration of substance B; *concentration de la substance* B: $c_B = n_B/V$	c_B

mole fraction of substance B; *fraction molaire de la substance* B	x_B

mass fraction of substance B; *titre en masse de la substance* B	w_B

volume fraction of substance B; *titre en volume de la substance* B	φ_B

mole ratio of solution; *rapport molaire d'une solution*	r

molality of solution; *molalité d'une solution*	m

chemical potential of substance B; *potentiel chimique de la substance* B †††)	μ_B

absolute activity of substance B (dimensionless); *activité absolue de la substance* B (*sans dimension*): $\lambda_B = \exp(\mu_B/kT)$	λ_B

relative activity; *activité relative*	a_B

*) Case of loosely coupled electron spin.
**) Case of tightly coupled electron spin.
***) All energies are taken here with respect to the ground state as reference level.
†) For diatomic molecules use I and $A = h/8\pi^2 cI$.
††) ν may be used as an alternative to n, when n is used for number density of particles.
†††) Referred to one particle.

activity coefficients; *coefficients d'activité* γ_B, f_B
osmotic pressure; *pression osmotique* Π
osmotic coefficient; *coefficient osmotique* g, φ
stoichiometric number of substance B; *nombre stœchiométrique de la substance* B ν_B
affinity; *affinité* A
extent of reaction; *état d'avancement d'une réaction* ξ
equilibrium constant; *constante d'équilibre* K
charge number of ion; *nombre de charge d'un ion, électrovalence* z
Faraday constant; *constante de Faraday* F
ionic strength; *force ionique* I
reduced activity of substance B; *activité réduite de la substance* B: $z_B = 2\pi mkT/h^2)^{3/2}\lambda_B$ z_B

7.13 *Plasma physics*

energy of particle; *énergie d'une particule* ε
dissociation energy (e.g. of molecule X); *énergie de dissociation (par ex., d'une molécule* X) E_d, $E_d(X)$
electron affinity; *affinité électronique* E_{ea}
ionization energy; *énergie d'ionisation* E_i
degree of ionization; *degré d'ionisation* x
charge number of ion (positive or negative); *charge ionique (positif ou négatif)* z
number density of ions of charge number z; *densité ionique des ions de charge* z †) n_z
degree of ionization for charge number $z \geqslant 1$; *degré d'ionisation pour un nombre de charge* $z \geqslant 1$: $x_z = n_z/(n_z + n_{z-1})$ x_z
neutral particle temperature; *température des neutres* T_n
ion temperature; *température ionique* T_i
electron temperature; *température électronique* T_e
electron number density; *densité électronique* n_e
electron plasma circular frequency; *fréquence de plasma:* $\omega_{pe}^2 = n_e e^2/\varepsilon_0 m_e$ ω_{pe}
Debye length; *longueur de Debye* λ_D
charge of particle; *charge d'une particule* q
electron cyclotron circular frequency; *fréquence cyclotron électronique:* $\omega_{ce} = (e/m_e)B$ ω_{ce}
ion cyclotron circular frequency; *fréquence cyclotron ionique:* $\omega_{ci} = (ze/m_i)B$ ω_{ci}
reduced mass; *masse réduite:* $\mu = m_1 m_2/(m_1 + m_2)$ μ, m_r
impact parameter; *paramètre d'impact* b
mean free path; *libre parcours moyen* l, λ
collision frequency; *fréquence de collision* ν_{coll}, ν_c
mean time interval between collisions; *intervalle de temps moyenne entre collisions:* $\tau_{coll} = 1/\nu_{coll}$ τ_{coll}, τ_c
cross section; *section efficace:* $\sigma = 1/ln$ σ
(electron) ionization efficiency; *efficacité d'ionisation (électronique):* $s_e = (\rho_0/\rho)\mathrm{d}N/\mathrm{d}x$ s_e
(dN: number of ion pairs formed by an ionizing electron travelling through dx in the plasma at gas density ρ; ρ_0: gas density at $p_0 = 1$ Torr, $T_0 = 273{,}15$ K)

†) When only singly charged ions need to be considered, n_{-1} and n_{+1} may be represented by n_- and n_+.

rate coefficient; *taux de réaction* — k

one-body rate coefficient; *taux de réaction unimoléculaire:* $-dn_A/dt = k_m n_A$ — k_m

relaxation time; *temps de relaxation:* (*e.g.* $\tau = 1/k_m$) — τ

two-body rate coefficient, binary rate coefficient; *taux de réaction binaire:* (e.g. $X + Y \rightarrow XY + h\nu$) $dn_{XY}/dt = k_b n_X n_Y$ — k_b

three-body rate coefficient, ternary rate coefficient; *taux de réaction ternaire:* (e.g. $X + Y + M \rightarrow XY + M^*$) $dn_{XY}/dt = k_t n_M n_X n_Y$ — k_t

Townsend (electron) ionization coefficient; *coefficient de Townsend* *) — α

Townsend (ion) ionization coefficient; *coefficient ionique de Townsend* — β

secondary electron emission coefficient; *taux d'émission secondaire* — γ

drift velocity; *vitesse de mouvement* — v_{dr}

mobility; *mobilité:* $\mu = v_{dr}/E$ — μ

positive or negative ion diffusion coefficient: *coefficient de diffusion des ions* — D_+, D_-

electron diffusion coefficient; *coefficient de diffusion des électrons* — D_e

ambipolar (ion–electron) diffusion coefficient; *coefficient de diffusion ambipolaire:* $D_a = (D_+\mu_e + D_e\mu_+)/(\mu_+ + \mu_e)$ — D_a, D_{amb}

characteristic diffusion length; *longueur caractéristique de diffusion* — L_D, Λ

ionization frequency; *fréquence d'ionisation* — ν_i

ion–ion recombination coefficient; *coefficient de recombinaison ion–ion:* $dn_-/dt = -\alpha_i n_- n_+$ — α_i

electron–ion recombination coefficient; *coefficient de recombinaison électron–ion:* $dn_e/dt = -\alpha_e n_e n_+$ — α_e

plasma pressure; *pression cinétique du plasma* — p

magnetic pressure; *pression magnétique:* $p_m = B^2/2\mu$ (μ: permeability) — p_m

magnetic pressure ratio; *coefficient* β*:* $\beta = p/p_m$ (p_m: magnetic pressure outside the plasma) — β

magnetic diffusivity; *diffusivité magnétique:* $\nu_m = 1/\mu\sigma$ (σ: electric conductivity; μ: permeability) — ν_m, η_m

Alfvén speed; *vitesse d'Alfvén:* $v_A = B/(\mu\rho)^{1/2}$ (ρ: (mass) density; μ: permeability) — v_A

7.14 Dimensionless parameters †)

1. Momentum transport

Reynolds number; *nombre de Reynolds:* $Re = vl/\nu$ — Re

Euler number; *nombre d'Euler:* $Eu = \Delta p/\rho v^2$ — Eu

Froude number; *nombre de Froude:* $Fr = v(lg)^{-1/2}$ — Fr

Grashof number; *nombre de Grashof:* $Gr = l^3 g\gamma\Delta T/\nu^2$ — Gr

Weber number; *nombre de Weber:* $We = \rho v^2 l/\sigma$ — We

Mach number; *nombre de Mach:* $Ma = v/c$ — Ma

Knudsen number; *nombre de Knudsen:* $Kn = \lambda/l$ — Kn

Strouhal number; *nombre de Strouhal:* $Sr = lf/v$ — Sr

where the symbols used in the definitions denote, respectively, l: a characteristic length; v: a characteristic velocity; ΔT: a characteristic temperature difference; Δp: pressure difference; ρ: (mass) density; η: viscosity; ν:

*) The same name is also used for $\eta = \alpha/E$, where E denotes electric field strength.

†) These symbols are those recommended in the International Standard I.S.O. 31 Part XII.

kinematic viscosity (η/ρ); σ: surface tension; g: acceleration of free fall; γ: cubic expansion coefficient ($-\rho^{-1}(\partial\rho/\partial T)_p$); λ: mean free path; f: a characteristic frequency; c: velocity of sound. $\Delta\rho/\rho$ equals $\gamma\Delta T$.

2. Transport of heat

Fourier number; *nombre de Fourier:* $Fo = a\Delta t/l^2$ — Fo

Péclet number; *nombre de Péclet:* $Pe = vl/a = Re \cdot Pr$ — Pe

Rayleigh number; *nombre de Rayleigh:*
$Ra = l^3 g\gamma\Delta T/\nu a = Gr \cdot Pr$ — Ra

Nusselt number; *nombre de Nusselt:* $Nu = hl/\lambda$ — Nu

Stanton number; *nombre de Stanton:* $St = h/\rho v c_p = Nu \cdot Pe^{-1}$ — St

where the symbols used in the definitions denote, respectively, l: a characteristic length; v: a characteristic velocity; Δt: a characteristic time interval; ΔT: a characteristic temperature difference; g: acceleration of free fall; ρ: (mass) density; η: viscosity; ν: kinematic viscosity (η/ρ); c_p: specific heat capacity at constant pressure; γ: cubic expansion coefficient ($-\rho^{-1}(\partial\rho/\partial T)_p$); λ: thermal conductivity; a: thermal diffusivity ($\lambda/\rho c_p$); h: coefficient of heat transfer (heat/(time × cross sectional area × temperature difference)).

3. Transport of matter in a binary mixture

Fourier number for mass transfer; *nombre de Fourier pour transfert de masse:* $Fo^* = D\Delta t/l^2 = Fo/Le$ — Fo^*

Péclet number for mass transfer; *nombre de Péclet pour transfert de masse:* $Pe^* \times vl/D = Re \cdot Sc = Pe \cdot Le$ — Pe^*

Grashof number for mass transfer; *nombre de Grashof pour transfert de masse:* $Gr^* = l^3 g\beta'\Delta x/\nu^2$ — Gr^*

Nusselt number for mass transfer; *nombre de Nusselt pour transfert de masse:* $Nu^* = kl/\rho D$ — Nu^*

Stanton number for mass transfer; *nombre de Stanton pour transfert de masse:* $St^* = k/\rho v = Nu^*/Pe^*$ — St^*

where the symbols used in the definitions denote, respectively, l: a characteristic length; v: a characteristic velocity; Δt: a characteristic time interval; Δx: a characteristic difference of mole fraction; g: acceleration of free fall; ρ: (mass) density; ν: kinematic viscosity (η/ρ); β': $\beta' = -\rho^{-1}(\partial\rho/\partial x)_{T,p}$; D: diffusion coefficient; k: mass transfer coefficient (mass/(time × cross sectional area × mole fraction difference)); γ: cubic expansion coefficient ($-\rho^{-1}(\partial\rho/\partial T)_p$). The quantity $-\Delta\rho/\rho$ equals $\gamma\Delta T + \beta'\Delta x$.

4. Dimensionless constants of matter

Prandtl number; *nombre de Prandtl:* $Pr = \nu/a$ — Pr

Schmidt number; *nombre de Schmidt:* $Sc = \nu/D$ — Sc

Lewis number; *nombre de Lewis:* $Le = a/D = Sc/Pr$ — Le

where the symbols used in the definitions denote, respectively, ρ: (mass) density; η: viscosity; ν: kinematic viscosity (η/ρ); D: diffusion coefficient; c_p: specific heat capacity at constant pressure; λ: thermal conductivity; a: thermal diffusivity ($\lambda/\rho c_p$).

5. Magnetohydrodynamics

Magnetic Reynolds number; *nombre de Reynolds magnétique:* $Rm = v\mu\sigma l$ — Rm

Alfvén number; *nombre d'Alfvén:* $Al = v/v_A$ — Al

Hartmann number; *nombre de Hartmann:* $Ha = Bl(\sigma/\rho\nu)^{1/2}$ — Ha

Cowling number (second Cowling number); *nombre de Cowling (deuxième nombre de Cowling):* $Co = B^2/\mu\rho v^2 = Al^{-2}$ — Co, Co_2

first Cowling number; *premier nombre de Cowling:*
$Co_1 = B^2 l\sigma/\rho v = Rm \cdot Co_2 = Ha^2/Re$ — Co_1

where the symbols used in the definitions denote, respectively, ρ: (mass) density; l: a characteristic length; v: a characteristic velocity; η: viscosity; ν: kinematic viscosity (η/ρ); μ: magnetic permeability; B: magnetic flux density; σ: electric conductivity; v_A: Alfvén speed $(B(\rho\mu)^{-1/2})$.

8 RECOMMENDED MATHEMATICAL SYMBOLS

8.1 General symbols

equal to; *égal à*	$=$
not equal to; *différent de*	$\neq$
identically equal to; *égal identiquement à*	$\equiv$
by definition equal to; *égal par définition à*	$\stackrel{\text{def}}{=}$
corresponds to; *correspond à*	$\triangleq$
approximately equal to; *égal environ à*	$\approx$
asymptotically equal to, *asymptotiquement égal à*	$\simeq$
proportional to; *proportionnel à*	$\sim, \propto$
approaches; *tend vers*	$\rightarrow$
greater than; *supérieur à*	$>$
less than; *inférieur à*	$<$
much greater than; *tres supérieur à*	$\gg$
much less than; *tres inférieur à*	$\ll$
greater than or equal to; *superiéur ou egál a*	$\geqslant, \geqq, \geq$
less than or equal to; *inférieur ou égal à*	$\leqslant, \leqq, \leq$
plus;*plus*	$+$
minus; *moins*	$-$
plus or minus; *plus ou moins*	$\pm$
a multiplied by b; *a multiplié par b*	$ab, a \cdot b, a \times b$
a divided by b; *a divisé par b*	$a/b, \frac{a}{b}, ab^{-1}$
ratio of the circumference of a circle to its diameter; *rapport de la circonférence d'un cercle à son diamètre*	π
a raised to the power n; *a puissance n*	a^n
magnitude of a; *valeur absolue de a*	$\lvert a \rvert$
square root of a; *racine carrée de a*	$\sqrt{a}, \sqrt{}a, a^{1/2}$
mean value of a; *valeur moyenne de a*	$\bar{a}, \langle a \rangle$
factorial p; *factorielle p*	$p!$
binomial coefficient; *coefficient binomial: n!/p!(n − p)!*	$\binom{n}{p}$
infinity; *infini*	∞

8.2 Letter symbols

Letter symbols and letter expressions for *mathematical operations* should be written in *roman* (i.e. *upright*) type.

exponential of x; *exponentielle de x*	$\exp x, \mathrm{e}^x$
base of natural logarithms; *base des logarithmes népériens*	e
logarithm to the base a of x; *logarithme de base a de x*	$\log_a x$
natural logarithm of x; *logarithme népérien de x*	$\ln x, \log_e x$
common logarithm of x; *logarithme décimal de x*	$\lg x, \log_{10} x$
binary logarithm of x; *logarithme binaire de x*	$\mathrm{lb}\, x, \log_2 x$
summation; *somme*	Σ
product; *produit*	Π
finite increase of x; *accroissement fini de x*	Δx †)

†) Greek capital delta, not triangle.

variation of x; *variation de x*	δx
total differential of x; *différentielle totale de x*	$\mathrm{d}x$
function of x; *fonction de x*	$f(x), \mathrm{f}(x)$
composite function of f and g; *fonction composée de f et g*: $(g \circ f)(x) = g(f, x)$	$g \circ f$
limit of $f(x)$; *limite de f(x)*	$\lim_{x \to a} f(x), \lim_{x\to a} f(x)$
derivative of f; *dérivée de f*	$\frac{\mathrm{d}f}{\mathrm{d}x}, \mathrm{d}f/\mathrm{d}x, f'$
partial derivative of f; *dérivée partielle de f*	$\frac{\partial f}{\partial x}, \partial f/\partial x, \partial_x f, f_x$
total differential of f; *différentielle totale de f*: $\mathrm{d}f(x, y) = (\partial f/\partial x)_y \,\mathrm{d}x + (\partial f/\partial y)_x \,\mathrm{d}y$	$\mathrm{d}f$
variation of f; variation de f	δf
Dirac delta function; *fonction delta de Dirac* $\delta(\boldsymbol{r}) = \delta(x)\delta(y)\delta(z)$	$\delta(x), \delta(\boldsymbol{r})$
Kronecker delta symbol; *symbole delta de Kronecker*	δ_{ij}
Unit step function; *fonction unité:* $\varepsilon(t) = 1$ for $t > 0$, $\varepsilon(t) = 0$ for $t < 0$	$\varepsilon(t)$
signum a; *signum a*: $\mathrm{sgn}\, a = a/\lvert a\rvert$ for $a \neq 0$	$\mathrm{sgn}\, a$
greatest integer $\leq a$; *le plus grand entier* $\leq a$	$\mathrm{ent}\, a$

8.3 Circular and hyperbolic functions

sine of x, *sinus x*	$\sin x$
cosine of x, *cosinus x*	$\cos x$
tangent of x, *tangente x*	$\tan x, \mathrm{tg}\, x$
cotangent of x, *cotangente x*	$\cot x, \mathrm{ctg}\, x$
secant of x, *sécante x*	$\sec x$
cosecant of x, *cosécante x*	$\mathrm{cosec}\, x$

It is recommended to use for the *inverse circular functions* the symbolic expressions for the corresponding circular function preceded by the letters: arc

Examples: $\arcsin x$, $\arccos x$, $\arctan x$, etc.

It is recommended to use for the *hyperbolic functions* the symbolic expressions for the corresponding circular function, followed by the letter: h

Examples; $\sinh x$, $\cosh x$, $\tanh x$, etc.

It is recommended to use for the *inverse hyperbolic functions* the symbolic expression for the corresponding hyperbolic function preceded by the letters: ar

Examples: $\mathrm{arsinh}\, x$, $\mathrm{arcosh}\, x$, etc.

8.4 Complex quantities

imaginary unit; *unité imaginaire* ($\mathrm{i}^2 = -1$)	i, j
real part of z; *partie réelle de z*	$\mathrm{Re}\, z, z'$
imaginary part of z; *partie imaginaire de z*	$\mathrm{Im}\, z, z''$
modulus of z; *module de z*	$\lvert z\rvert$
phase, argument of z; *phase, argument de z*: $z = \lvert z\rvert \exp \mathrm{i}\varphi$	$\arg z, \varphi$
complex conjugate of z, conjugate of z; *complexe conjugué de z, conjugué de z*	z^*

Remark: Sometimes the notation $\bar{z}$ is used for the complex conjugate of z.

8.5 Symbols for special values of periodic quantities

Symbols for special values of periodic quantities are given in table 3.

Table 3

	case A*)	case B*)
instantaneous value; *valeur instantanée*	x	x
r.m.s. value**); *valeur efficace*	$\tilde{x}$	X
maximum value†) *valeur maximale*	$\hat{x}$	$\hat{x}, \hat{X}$
mean value††) *valeur moyenne*	$\bar{x}, \langle x \rangle$	$\bar{x}, \langle x \rangle$

*) A is the case in which only a lower case or only a capital letter may be used for the quantity. B is the case in which both, lower case and capital letters, may be used for the same quantity.

**) The r.m.s. value is here defined as

$$\tilde{x} = \left(T^{-1} \int_0^T [x(t)]^2 \mathrm{d}t\right)^{1/2};$$

the symbols x_{rms} and x_{eff} are also used.

†) The minimum value of x may be indicated by x_{min} or $\check{x}$.

††) The mean value is here defined as

$$\bar{x} = T^{-1} \int_0^T x(t)\mathrm{d}t.$$

*8.6 Vector calculus**)

vector; *vecteur*	$\boldsymbol{a}, \boldsymbol{A}$
absolute value; *valeur absolue*	$\lvert\boldsymbol{A}\rvert, A$
unit vector; *vecteur unitaire:* $\boldsymbol{a}/\lvert\boldsymbol{a}\rvert$	$\boldsymbol{e}_a$
unit vectors; *vecteurs unitaires*	$\boldsymbol{e}_x, \boldsymbol{e}_y, \boldsymbol{e}_z, \boldsymbol{i}, \boldsymbol{j}, \boldsymbol{k}$ †)
scalar product of $\boldsymbol{a}$ and $\boldsymbol{b}$; *produit scalaire de* $\boldsymbol{a}$ *et* $\boldsymbol{b}$	$\boldsymbol{a} \cdot \boldsymbol{b}$
vector product of $\boldsymbol{a}$ and $\boldsymbol{b}$; *produit vectoriel de* $\boldsymbol{a}$ *et* $\boldsymbol{b}$	$\boldsymbol{a} \times \boldsymbol{b}$ $\quad \boldsymbol{a} \wedge \boldsymbol{b}$
dyadic product of $\boldsymbol{a}$ and $\boldsymbol{b}$; *produit dyadique de* $\boldsymbol{a}$ *et* $\boldsymbol{b}$	$\boldsymbol{ab}$
differential vector operator, nabla; *opérateur vectoriel, nabla*	$\partial/\partial\boldsymbol{r}, \boldsymbol{\nabla}$
gradient; *gradient*	$\mathrm{grad}\,\varphi, \boldsymbol{\nabla}\varphi$
divergence; *divergence*	$\mathrm{div}\,\boldsymbol{A}, \boldsymbol{\nabla}\cdot\boldsymbol{A}$
curl; *rotationnel*	$\mathrm{curl}\,\boldsymbol{A},\ \mathrm{rot}\,\boldsymbol{A},\ \boldsymbol{\nabla}\times\boldsymbol{A}$
Laplacian; *Laplacien*	$\triangle\varphi, \boldsymbol{\nabla}^2\varphi$
Dalembertian; *Dalembertien*	$\square\varphi$
second order tensor; *tenseur du second ordre*	$\boldsymbol{A}$

*) See also sections 1.2.3 and 1.3.2.

†) $1_x, 1_y, 1_z$ are also used.

scalar product of tensors $\boldsymbol{S}$ and $\boldsymbol{T}$; *produit scalaire des tenseurs* $\boldsymbol{S}$ *et* $\boldsymbol{T}$: $(\Sigma_{i,k} S_{ik} T_{ki})$ — $\boldsymbol{S} : \boldsymbol{T}$

tensor product of tensors $\boldsymbol{S}$ and $\boldsymbol{T}$; *produit tensoriel des tenseurs* $\boldsymbol{S}$ *et* $\boldsymbol{T}$: $(\Sigma_k S_{ik} T_{kl})$ — $\boldsymbol{S} \cdot \boldsymbol{T}$

product of tensor $\boldsymbol{S}$ and vector $\boldsymbol{A}$; *produit du tenseur* $\boldsymbol{S}$ *et vecteur* $\boldsymbol{A}$: $(\Sigma_k S_{ik} A_k)$ — $\boldsymbol{S} \cdot \boldsymbol{A}$

8.7 Matrix calculus

matrix; *matrice*	A $\begin{pmatrix} a_{11} & \dots & a_{1n} \\ \vdots & & \vdots \\ a_{m1} & \dots & a_{mn} \end{pmatrix}$
product of A and B: *produit de A et B*	AB
inverse of A: *inverse de A*	A^{-1}
unit matrix; *matrice unité*	E, I
transpose of matrix A: *matrice transposée de A*: $\tilde{A}_{ij} = A_{ji}$	$\tilde{A}$
complex conjugate of A; *matrice complexe conjugueé de A*: $A^*_{ij} = (A_{ij})^*$	A^*
Hermitian conjugate of A; *matrice conjuguée Hermitienne de A*: $A^\dagger_{ij} = A^*_{ji}$	$A^\dagger$
determinant of A; *déterminant de A*	$\det A$
trace of A; *trace de A*	$\mathrm{Tr}\, A$
Pauli matrices; *matrices de Pauli*:	$\boldsymbol{\sigma}$,
$\sigma_x = \begin{pmatrix} 0 & 1 \\ 1 & 0 \end{pmatrix}$ $\quad \sigma_y = \begin{pmatrix} 0 & -\mathrm{i} \\ \mathrm{i} & 0 \end{pmatrix}$ $\quad \sigma_z = \begin{pmatrix} 1 & 0 \\ 0 & -1 \end{pmatrix}$	$\sigma_x, \sigma_y, \sigma_z$ $\sigma_1, \sigma_2, \sigma_3$
unit matrix; *matrice unité*: $I = \begin{pmatrix} 1 & 0 \\ 0 & 1 \end{pmatrix}$	I
Dirac (4 × 4) matrices; (4 × 4) *matrices de Dirac*†):	
$\alpha_x = \begin{pmatrix} 0 & \sigma_x \\ \sigma_x & 0 \end{pmatrix}$ $\quad \alpha_y = \begin{pmatrix} 0 & \sigma_y \\ \sigma_y & 0 \end{pmatrix}$ $\quad \alpha_z = \begin{pmatrix} 0 & \sigma_z \\ \sigma_z & 0 \end{pmatrix}$	$\boldsymbol{\alpha}$ $\alpha_x, \alpha_y, \alpha_z$
$\beta = \begin{pmatrix} I & 0 \\ 0 & -I \end{pmatrix}$	β

8.8 Theory of sets

is an element of; *est un élément de*: $x \in A$	$\in$
is not an element of, *n'est pas un élément de*: $x \notin A$	$\notin$
contains as element; *contient comme élément*: $A \ni x$	$\ni$
set of elements; *ensemble des éléments*	$\{a_1, a_2, \dots\}$
the set of integers; *ensemble des nombres entiers*	Z
the set of rational numbers; *ensemble des nombres rationnels*	Q
the set of real numbers; *ensemble des nombres réels*	R
the set of complex numbers; *ensemble des nombres complexes*	C
set of elements of A for which $p(x)$ is true; *ensemble des éléments de A pour lequels p(x) est vrai*	$\{x \in A \mid p(x)\}$
is contained as subset in; *est contenu comme sous-ensemble dans*: $B \subseteq A$	$\subseteq (\subset)$
contains; *contient*: $A \supseteq B$	$\supseteq (\supset)$
is properly contained in; *est strictement contenu dans*	$\subset (\subsetneq)$

†) Sometimes a different representation is used.

contains properly; *contient strictement* $\supset$ ($\supsetneq$)

union; *réunion:* $A \cup B = \{x|x \in A \text{ or } x \in B\}$ $\cup$
intersection; *intersection:*
$A \cap B = \{x|x \in A \text{ and } x \in B\}$ $\cap$
difference; *difference:* $A \backslash B = \{x|x \in A \text{ and } x \notin B\}$ $\backslash$
complement of; *complément de:* $\complement A = \{x|x \notin A\}$ $\complement$

8.9 Symbolic logic

conjunction: $p \wedge q$ means "p and q"; *conjonction:* $p \wedge q$ *signifie* "p *et* q" $\wedge$
disjunction: $p \vee q$ means "p or q or both"; *disjonction:* $p \vee q$ *signifie* "p *ou* q *ou les deux*" $\vee$
negation; *négation* $\neg$
implication; *implication* $\Rightarrow$
equivalence, bi-implication; *équivalence, bi-implication* $\Leftrightarrow$
universal quantifier; *quantificateur universel* $\forall$
existential quantifier; *quantificateur existential* $\exists$

9. INTERNATIONAL SYMBOLS FOR UNITS

9.1 Unit systems

In a system of physical quantities and equations between them, a certain number of quantities, which *by convention* are regarded as dimensionally independent, form a set of *base quantities* for the whole system. All other quantities can be defined as *derived quantities* in terms of the base quantities and can be expressed in terms of powers of the base quantities by algebraic relations.

A *coherent system of units* is a system, based on a certain set of base units well defined in terms of actual physical phenomena, in which all *derived units* are expressed as products of powers of the base units by algebraic relations, analogous to the corresponding quantities, dropping numerical factors.

The expression of a quantity as a product of powers of the base quantities (neglecting their vectorial or tensorial character and all numerical factors including their sign) is called the *dimensional product* or the *dimension* of the quantity with respect to the chosen set of base quantities or base dimensions. The exponents of the powers, to which the various base quantities or base dimensions are raised, are called the *dimensional exponents.*

Physical quantities, which have as their dimension a product of powers of the base quantities or base dimensions with all dimensional exponents equal to zero, are called *dimensionless quantities* or *quantities of the dimension 1.*

The number of the base units of the unit system, coherent with respect to the system of quantities and equations in question, is equal to that of the corresponding set of base quantities. The base units themselves are well defined samples of the base quantities: base quantity and corresponding base unit are of the same dimension.

A derived unit, i.e. the unit for a derived quantity of the coherent unit system, is formed as a product of powers of the base units, with the same exponents as those in the product of powers of the base quantities in the algebraic expression for the derived quantity in question, and without introducing numerical factors. Derived units and their symbols are expressed algebraically in terms of base units and their symbols, respectively, by means of the mathematical signs for multiplication and division. Several of these algebraic expressions for derived units in terms of base units can be replaced by special names and their symbols which can themselves be used to form other derived units and their symbols, respectively (see sections 9.2 and 9.3).

The values of dimensionless quantities, e.g. relative density, relative permeability or refractive index, are expressed by pure numbers. The corresponding unit, which is the ratio of a unit to itself, is the *dimensionless unit* of the coherent unit system and may be expressed by the number 1.

9.2 The International System of Units (SI)

The name *International System of Units* with the international abbreviation SI has been adopted by the Conférence Générale des Poids et Mesures (CGPM) in 1960. It is based on the seven base units (CGPM 1960 and 1971) listed in table 4.

Table 4. SI base units

base quantity; *grandeur de base*	SI base unit; *unité de base SI*	
	name; *nom*	symbol; *symbole*
length; *longueur*	metre; *mètre*	m
mass; *masse*	kilogram; *kilogramme*	kg
time; *temps*	second; *seconde*	s
electric current; *courant électrique*	ampere; *ampère*	A
thermodynamic temperature; *température thermodynamique*	kelvin; *kelvin*	K
amount of substance; *quantité de matière*	mole; *mole*	mol
luminous intensity; *intensité lumineuse*	candela; *candela*	cd

The SI base units have been defined by the CGPM in the following way:

(1) metre; *mètre*

The metre is the SI base unit of length and was defined at the October 1984 General Conference of Weights and Measures as the length of the path traveled by light in vacuum during a time interval of $^1/299{,}792{,}458$ of a second.

(2) kilogram; *kilogramme*

The kilogram is the unit of mass; it is equal to the mass of the international prototype of the kilogram. (1st CGPM (1889) and 3rd CGPM (1901), Declarations.)

(3) second; *seconde*

The second is the duration of 9 192 631 770 periods of the radiation corresponding to the transition between the two hyperfine levels of the ground state of the caesium-133 atom. (13th CGPM (1967), Resolution 1.)

(4) ampere; *ampère*

The ampere is that constant current which, if maintained in two straight parallel conductors of infinite length, of negligible circular cross section, and placed 1 metre apart in vacuum, would produce between these conductors a force equal to 2×10^{-7} *newton per metre of length.* (9th CGPM (1948), Resolutions 2 and 7.)

(5) kelvin; *kelvin*

The kelvin, unit of thermodynamic temperature, is the fraction 1/273,16 of the thermodynamic temperature of the triple point of water. (13th CGPM (1967), Resolution 4.)

The 13th CGPM (1967, Resolution 3) also decided that the unit kelvin and its symbol K should be used to express an interval or a difference of temperature.
Note: In addition to the thermodynamic temperature (symbol T), expressed in kelvins, use is also made of Celsius temperature (symbol t) defined by the equation

$$t = T - T_0,$$

where $T_0 = 273{,}15$ K by definition. Celsius temperature is expressed in degree Celsius; *degré Celsius* (symbol °C). The unit "degree Celsius" is equal to the unit "kelvin" and an interval or a difference of Celsius temperature may also be expressed in degrees Celsius.

(6) mole; *mole*

The mole is the amount of substance of a system which contains as many elementary entities as there are atoms in 0.012 kilogram of carbon 12.
When the mole is used, the elementary entities must be specified and may be atoms, molecules, ions, electrons, other particles, or specified groups of such particles.
The mole is a base unit of the International System of Units. (14th CGPM (1971), Resolution 3.)

(7) candela; *candela*

The candela is the luminous intensity in a given direction of a source which emits monochromatic radiation of frequency 540×10^{12} *Hz and of which the radiant intensity in that direction is 1/683 W/steradian. From the 16th CGPM, Resolution 3, 1979.*

The coherent units of the system based on these base units are called: *SI units*. *Derived SI units having special names and symbols* are listed in table 5.

Table 5. Derived SI units with special names

quantity; *grandeur*	derived SI unit; *unité SI dérivée* name; *nom*	symbol; *symbole*	expression in terms of base units; *expression en unités de base*	expression in terms of other SI units; *expression en d'autres unités SI*
plane angle; *angle plan*	radian; *radian*	rad	$m \cdot m^{-1}$	
solid angle; *angle solide*	steradian; *stéradian*	sr	$m^2 \cdot m^{-2}$	
frequency; *fréquence*	hertz; *hertz*	Hz	s^{-1}	
force; *force*	newton; *newton*	N	$m \cdot kg \cdot s^{-2}$	J/m
pressure; *pression*	pascal; *pascal*	Pa	$m^{-1} \cdot kg \cdot s^{-2}$	N/m^2
energy, work, quantity of heat; *énergie, travail, quantité de chaleur*	joule; *joule*	J	$m^2 \cdot kg \cdot s^{-2}$	$N \cdot m$
power, radiant flux; *puissance, flux énergétique*	watt; *watt*	W	$m^2 \cdot kg \cdot s^{-3}$	J/s
quantity of electricity, electric charge; *quantité d'électricité, charge électrique*	coulomb; *coulomb*	C	$s \cdot A$	$A \cdot s$
electric potential, potential difference, electromotive force; *tension électrique, différence de potentiel, force électromotrice*	volt; *volt*	V	$m^2 \cdot kg \cdot s^{-3} \cdot A^{-1}$	W/A
capacitance; *capacité électrique*	farad; *farad*	F	$m^{-2} \cdot kg^{-1} \cdot s^4 \cdot A^2$	C/V
electric resistance; *résistance électrique*	ohm; *ohm*	Ω	$m^2 \cdot kg \cdot s^{-3} \cdot A^{-2}$	V/A
conductance; *conductance*	siemens; *siemens*	S	$m^{-2} \cdot kg^{-1} \cdot s^3 \cdot A^2$	A/V
magnetic flux; *flux d'induction magnétique*	weber; *weber*	Wb	$m^2 \cdot kg \cdot s^{-2} \cdot A^{-1}$	$V \cdot s$
magnetic flux density; *induction magnétique*	tesla; *tesla*	T	$kg \cdot s^{-2} \cdot A^{-1}$	Wb/m^2
inductance; *inductance*	henry; *henry*	H	$m^2 \cdot kg \cdot s^{-2} \cdot A^{-2}$	Wb/A
luminous flux; *flux lumineux*	lumen; *lumen*	lm		$cd \cdot sr$
illuminance; *éclairement lumineux*	lux; *lux*	lx		$m^{-2} \cdot cd \cdot sr$

Table 5 (Continued)

quantity; *grandeur*	derived SI unit; *unité SI dérivée* name; *nom*	symbol; *symbole*	expression in terms of base units; *expression en unités de base*	expression in terms of other SI units; *expression en d'autres unités SI*
activity; *activité*	becquerel; *becquerel*	Bq	s^{-1}	
absorbed dose; *dose absorbée*	gray; *gray*	Gy	$m^2 \cdot s^{-2}$	J/kg

Although it might be thought that SI units can only be base units or derived units, the 11th Conférence Générale des Poids et Mesures (1960) admitted a third class of SI units called *supplementary units*, for which it declined to state, whether they were base units or derived units. To the class of supplementary units belong only the radian as SI unit for plane angle and the steradian as SI unit for solid angle; they may be regarded either as base units or as derived units.

In physics, radian and steradian are, in general, regarded as derived units (see table 5). In some special fields, e.g. in electromagnetic radiation and light or in particle scattering problems, the steradian is sometimes *formally* treated as a base unit; then the symbol sr must *not* be replaced by the number 1.

The usage of the SI units and their decimal multiples and sub-multiples formed by using the prefixes given in table 1 (see section 2.2.1) has been especially recommended.

Remark: Among the SI base units the unit of mass is the only whose name, for historical reasons, contains a prefix. Names of decimal multiples and sub-multiples of the SI unit of mass are formed by attaching prefixes to the word "gram" (Comité International des Poids et Mesures (CIPM) 1967, Recommendation 2).

Several *sub-systems* of the SI are used in different fields of science and technology. The two most important are:

(a) *The metre-kilogram-second system (MKS system)*
The MKS system is a coherent unit system for mechanics, based on the three base units metre, kilogram and second for the three base quantities length, mass and time.

(b) *The metre-kilogram-second-ampere system (MKSA system)*
The MKSA system is a coherent unit system for mechanics, electricity and magnetism, based on the four base units metre, kilogram, second and ampere for the four base quantities length, mass, time and electric current. This system has been given the name "*Giorgi system*" by the International Electrotechnical Commission in 1935.

9.3 The centimetre-gram-second system (CGS system)

The CGS system is a coherent system of units based on the three base units listed in table 6.

Table 6. CGS base units

base quantity; *grandeur de base*	base unit; *unité de base* name; *nom*	symbol; *symbole*
length; *longueur*	centimetre; *centimètre*	cm
mass; *masse*	gram; *gramme*	g
time; *temps*	second; *seconde*	s

Derived CGS units having special names and symbols are listed in table 7.

Table 7. Derived CGS units with special name.

quantity; *grandeur*	derived CGS unit; *unité CGS dérivée*		
	name; *nom*	symbol; *symbole*	expression in terms of base units; *expression en unités de base*
force; *force*	dyne	dyn	$cm \cdot g \cdot s^{-2}$
energy; *énergie*	erg	erg	$cm^2 \cdot g \cdot s^{-2}$
viscosity; *viscosité*	poise	P	$cm^{-1} \cdot g \cdot s^{-1}$
kinematic viscosity; *viscosité cinématique*	stokes	St	$cm^2 \cdot s^{-1}$
acceleration of free fall; *accélération de la pesanteur*	gal	Gal	$cm \cdot s^{-2}$

The CGS system enlarged by the kelvin (K) as unit of thermodynamic temperature (see section 9.2) and the mole (mol) as unit of amount of substance (see section 9.2) or by the candela (cd) as unit of luminous intensity (see section 9.2) is also used in *mechanics including thermodynamics* or *photometry*, respectively.

Derived units (from cm, g, s and cd, including sr, see section 9.2), in the field of *photometry, having special names and symbols* are listed in table 8.

Table 8. CGS units in photometry with special name.

quantity; *grandeur*	derived unit; *unité dérivée*		
	name; *nom*	symbol; *symbole*	expression; *expression*
luminance; *luminance*	stilb	sb	$cm^{-2} \cdot cd$
illuminance; *éclairement lumineux*	phot	ph	$cm^{-2} \cdot cd \cdot sr$

ALPHABET TABLE

Greek letter	Greek name	English equivalent
Α α	Alpha	(ä)
Β β	Beta	(b)
Γ γ	Gamma	(g)
Δ δ	Delta	(d)
Ε ϵ	Epsilon	(e)
Ζ ζ	Zeta	(z)
Η η	Eta	(ā)
Θ θ	Theta	(th)
Ι ι	Iota	(ē)
Κ κ	Kappa	(k)
Λ λ	Lambda	(l)
Μ μ	Mu	(m)
Ν ν	Nu	(n)
Ξ ξ	Xi	(ks)
Ο ο	Omicron	(o)
Π π	Pi	(p)
Ρ ρ	Rho	(r)
Σ σ ς	Sigma	(s)
Τ τ	Tau	(t)
Υ υ	Upsilon	(ü, o͞o)
Φ φ	Phi	(f)
Χ χ	Chi	(H)
Ψ ψ	Psi	(ps)
Ω ω	Omega	(ō)

RUSSIAN letter	English equivalent
А а	(ä)
Б б	(b)
В в	(v)
Г г	(g)
Д д	(d)
Е е	(ye)
Ж ж	(zh)
З з	(z)
И и	(i, ē)
Й й	(ē) 7
К к	(k)
Л л	(l)
М м	(m)
Н н	(n)
О о	(ō, o)
П п	(p)
Р р	(r)
С с	(s)
Т т	(t)
У у	(o͞o)
Ф ф	(f)
Х х	(kh)
Ц ц	(ts)
Ч ч	(ch)
Ш ш	(sh)
Щ щ	(shch)
Ъ ъ	8
Ы ы	(ë)
Ь ь	9
Э э	(e)
Ю ю	(ū)
Я я	(yä)

AMERICAN STANDARD ABBREVIATIONS FOR SCIENTIFIC AND ENGINEERING TERMS

Reproduced by permission of The American Society of Mechanical Engineers

Introductory Notes

Scope and Purpose

1. The Executive Committee of the Sectional Committee on Scientific and Engineering Symbols and Abbreviations has made the following distinction between symbols and abbreviations: Letter symbols are letters used to represent magnitudes or physical quantities in equations and mathematical formulas. Abbreviations are shortened forms of names or expressions employed in texts and tabulations, and should not be used in equations.

Fundamental Rules

2. Abbreviations should be used sparingly in text and with due regard to the context and to the training of the reader. Terms denoting units of measurement should be abbreviated in the text only when preceded by the amounts indicated in numerals; thus "several inches," "one inch," "12 in." In tabular matter, specifications, maps, drawings, and texts for special purposes, the use of abbreviations should be governed only by the desirability of conserving space.
3. Short words such as ton, day, and mile should be spelled out.
4. Abbreviations should not be used where the meaning will not be clear. In case of doubt, spell out.
5. The same abbreviation is used for both singular and plural, as "bbl" for barrel and barrels.
6. The use of conventional signs for abbreviations in text is not recommended; thus "per," not /; "lb," not #; "in," not ". Such signs may be used sparingly in tables and similar places for conserving space.
7. The period should be omitted except in cases where the omission would result in confusion.
8. The letters of such abbreviations as ASA should not be spaced (not A S A).
9. The use in text of exponents for the abbreviations of square and cube and of the negative exponents for terms involving "per" is not recommended. The superior figures are usually not available on the keyboards of typesetting and linotype machines and composition is therefore delayed. There is also the likelihood of confusion with footnote reference numbers. These shorter forms are permissible in tables and are sometimes difficult to avoid in text.
10. A sentence should not begin with a numeral followed by an abbreviation. Abbreviations for names of units are to be used only after numerical values, such as 25 ft or 110 v.

Abbreviations*

absolute	abs
acre	spell out
acre-foot	acre-ft
air horsepower	air hp
alternating-current (as adjective)	a-c
ampere	amp
ampere-hour	amp-hr
amplitude, an elliptic function	am.
Angstrom unit	A
antilogarithm	antilog
atmosphere	atm
atomic weight	at. wt
average	avg
avoirdupois	avdp
azimuth	az or α
barometer	bar
barrel	bbl
Baumé	Bé
board feet (feet board measure)	fbm
boiler pressure	spell out
boiling point	bp
brake horsepower	bhp
brake horsepower-hour	bhp-hr
Brinell hardness number	Bhn
British thermal unit[1]	Btu or B
bushel	bu
calorie	cal
candle	c
candle-hour	c-hr
candlepower	cp
cent	c or ¢
center to center	c to c
centigram	cg
centiliter	cl
centimeter	cm
centimeter-gram-second (system)	cgs
chemical	chem
chemically pure	cp
circular	cir
circular mils	cir mils
coefficient	coef
cologarithm	colog
concentrate	conc
conductivity	cond
constant	const
continental horsepower	cont hp
cord	cd
cosecant	csc
cosine	cos
cosine of the amplitude, an elliptic function	cn
cost, insurance, and freight	cif
contangent	cot
coulomb	spell out
counter electromotive force	cemf
cubic	cu
cubic centimeter cu cm, cm^3 (liquid, meaning milliliter. ml)	
cubic foot	cu ft
cubic feet per minute	cfm
cubic feet per second	cfs
cubic inch	cu in.
cubic meter	cu m or m^3
cubic micron	cu μ or cu mu or μ^3
cubic millimeter	cu mm or mm^3
cubic yard	cu yd
current density	spell out
cycles per second	spell out or c
cylinder	cyl
day	spell out
decibel	db
degree[1]	deg or °
degree centigrade	C
degree Fahrenheit	F
degree Kelvin	K
degree Réaumur	R
delta amplitude, an elliptic function	dn
diameter	diam
direct-current (as adjective)	d-c
dollar	$
dozen	doz
dram	dr
efficiency	eff
electric	elec
electromotive force	emf
elevation	el
equation	eq

* These forms are recommended for readers whose familiarity with the terms used makes possible a maximum of abbreviations. For other classes of readers editors may wish to use less contracted combinations made up from this list. For example, the list gives the abbreviation of the term "feet per second" as 'fps". To some readers ft per sec will be more easily understood. [1]Abbreviation recommended by the A.S.M.E. Power Test Codes Committee. B = 1 Btu, kB = 1000 Btu, mB = 1,000,000 Btu. The A.S.H. & V.E. recommends the use of Mb = 1000 Btu and Mbh = 1000 Btu per hr.

[1] There are circumstances under which one or the other of these forms is preferred. In general the sign ° is used where space conditions make it necessary, as in tabular matter, and when abbreviations are cumbersome, as in some angular measurements, i.e., 59° 23' 42". In the interest of simplicity and clarity the Committee has recommended that the abbreviation for the temperature scale, F, C, K, etc., always be included in expressions for numerical temperatures, but, wherever feasible, the abbreviation for "degree" be omitted; as 69 F.

AMERICAN STANDARD ABBREVIATIONS (Continued)

external	ext
farad	spell out or f
feet board measure (board feet)	fbm
feet per minute	fpm
feet per second	fps
fluid	fl
foot	ft
foot-candle	ft-c
foot-Lambert	ft-L
foot-pound	ft-lb
foot-pound-second (system)	fps
foot-second (see cubic feet per second)	
franc	fr
free aboard ship	spell out
free alongside ship	spell out
free on board	fob
freezing point	fp
frequency	spell out
fusion point	fnp
gallon	gal
gallons per minute	gpm
gallons per second	gps
grain	spell out
gram	g
gram-calorie	g-cal
greatest common divisor	gcd
haversine	hav
hectare	ha
henry	H
high-pressure (adjective)	h-p
hogshead	hhd
horsepower	hp
horsepower-hour	hp-hr
hour	hr
hour (in astronomical tables)	h
hundred	C
hundredweight (112 lb)	cwt
hyperbolic cosine	cosh
hyperbolic sine	sinh
hyperbolic tangent	tanh
inch	in.
inch-pound	in-lb
inches per second	ips
indicated horsepower	ihp
indicated horsepower-hour	ihp-hr
inside diameter	ID
intermediate-pressure (adjective)	i-p
internal	int
joule	J
kilocalorie	kcal
kilocycles per second	kc
kilogram	kg
kilogram-calorie	kg-cal
kilogram-meter	kg-m
kilograms per cubic meter or kg/m^3	kg per cu m
kilograms per second	kgps
kiloliter	kl
kilometer	km
kilometers per second	kmps
kilovolt	kv
kilovolt-ampere	kva
kilowatt	kw
kilowatthour	kwhr
lambert	L
latitude	lat or ϕ
least common multiple	lcm
linear foot	lin ft
liquid	liq
lira	spell out
liter	l
logarithm (common)	log
logarithm (natural)	log or ln
longitude	long. or λ
low-pressure (as adjective)	l-p
lumen	l*
lumen-hour	l-hr*
lumens per watt	lpw
mass	spell out
mathematics (ical)	math
maximum	max
mean effective pressure	mep
mean horizontal candlepower	mhcp
megacycle	spell out
megohm	spell out

melting point	mp
meter	m
meter-kilogram	m-kg
mho	spell out
microampere	μa or mu a
microfarad	μf
microinch	μin.
micromicrofarad	$\mu\mu$f
micromicron	$\mu\mu$ or mu mu
micron	μ or mu
microvolt	μv
microwatt	μw or mu w
mile	spell out
miles per hour	mph
miles per hour per second	mphps
milliampere	ma
milligram	mg
millihenry	mh
millilambert	mL
milliliter	ml
millimeter	mm
millimicron	mμ or m mu
million	spell out
million gallons per day	mgd
millivolt	mv
minimum	min
minute	min
minute (angular measure)	′
minute (time) (in astronomical tables)	m
mole	spell out
molecular weight	mol. wt
month	spell out
National Electrical Code	NEC
ohm	spell out or Ω
ohm-centimeter	ohm-cm
ounce	oz
ounce-foot	oz-ft
ounce-inch	oz-in
outside diameter	OD
parts per million	ppm
peck	pk
penny (pence — New British)	p.
pennyweight	dwt
per	(see Fundamental Rules)
peso	spell out
pint	pt
potential	spell out
potential difference	spell out
pound	lb
pound-foot	lb-ft
pound-inch	lb-in.
pound sterling	£
pounds per brake horsepower-hour	lb per bhp-hr
pounds per cubic foot	lb per cu ft
pounds per square foot	psf
pounds per square inch	psi
pounds per square inch absolute	psia
power factor	spell out or pf
quart	qt
radian	spell out
reactive kilovolt-ampere	kvar
reactive volt-ampere	var
revolutions per minute	rpm
revolutions per second	rps
rod	spell out
root mean square	rms
secant	sec
second	sec
second (angular measure)	″
second-foot (see cubic feet per second)	
second (time) (in astronomical tables)	s
shaft horsepower	shp
shilling	s
sine	sin
sine of the amplitude, an elliptic function	sn
specific gravity	sp gr
specific heat	sp ht
spherical candle power	scp
square	sq
square centimeter	sq cm or cm^2
square foot	sq ft
square inch	sq in.
square kilometer	sq km or km^2
square meter	sq m or m^2
square micron	sq μ or sq mu or μ^2

* The International Commission on Illumination has changed the symbol for lumen to lm, and the symbol for lumen-hour to lm-hr. This nomenclature is used in American Standard for Illuminating Engineering Nomenclature and Photometric Standards (ASA Z7.1-1942).

AMERICAN STANDARD ABBREVIATIONS (Continued)

square millimeter	sq mm or mm²
square root of mean square	rms
standard	std
stere	s
tangent	tan
temperature	temp
tensile strength	ts
thousand	M
thousand foot-pounds	kip-ft
thousand pound	kip
ton	spell out
ton-mile	spell out
versed sine	vers
volt	v
volt-ampere	va
volt-coulomb	spell out
watt	w
watthour	whr
watts per candle	wpc
week	spell out
weight	wt
yard	yd
year	yr

ABBREVIATIONS OF COMMON UNITS OF WEIGHT AND MEASURE

From NBS Miscellaneous Publication No. 233.
Spelling and Abbreviations of Units

The spelling of the names of units as adopted by the National Bureau of Standards is that given in the list below. The spelling of the metric units is in accordance with that given in the law of July 28, 1866, legalizing the metric system in the United States.

Following the name of each unit in the list below is given the abbreviation which the Bureau has adopted. Attention is particularly called to the following principles:

1. The period is omitted after all abbreviations of units, except where the abbreviation forms an English word.
2. The exponents "²" and "³" are used to signify "square" and "cubic", respectively, instead of the abbreviations "sq" or "cu," which are, however, frequently used in technical literature for the United States Customary units. In conformity with this principle the abbreviation for cubic centimeter is "cm³" (instead of "cc" or "c cm"). The term "cubic centimeter," as used in chemical work, is, in fact, a misnomer, since the unit actually used is the "milliliter" of which "ml" is the correct abbreviation.
3. The use of the same abbreviation for both singular and plural is recommended. This practice is already established in expressing metric units and is in accordance with the spirit and chief purpose of abbreviations.
4. It is also suggested that, unless all the text is printed in capital letters, only small letters be used for abbreviations, except in such case as, A for angstrom, etc., where the use of capital letters is general.

LIST OF THE MOST COMMON UNITS OF WEIGHT AND MEASURE AND THEIR ABBREVIATIONS

Unit	Abbreviation
acre	acre
angstrom	A
are	a
avoirdupois	avdp
barrel	bbl
board foot	fbm
bushel	bu
carat	c
centare	ca
centigram	cg
centiliter	cl
centimeter	cm
chain	ch
cubic centimeter	cm³
cubic decimeter	dm³
cubic dekameter	dkm³
cubic foot	ft³
cubic hectometer	hm³
cubic inch	in.³
cubic kilometer	km³
cubic meter	m³
cubic mile	mi³
cubic millimeter	mm³
cubic yard	yd³
decigram	dg
deciliter	dl
decimeter	dm
decistere	ds
dekagram	dkg
dekaliter	dkl
dekameter	dkm
dekastere	dks
dram	dr
dram, apothecaries	dr ap or ʒ
dram, avoirdupois	dr avdp
dram, fluid	fl dr
fathom	fath
foot	ft
furlong	fur.
gallon	gal
grain	grain
gram	g
hectare	ha
hectogram	hg
hectoliter	hl
hectometer	hm
hogshead	hhd
hundredweight	cwt
inch	in.
kilogram	kg
kiloliter	kl
kilometer	km
link	li
liquid	liq
liter	liter
meter	m
metric ton	t
microgram*	μg
microinch	μin.
microliter*	μl
micron	μ
mile	mi
milligram	mg
milliliter	ml
millimeter	mm
millimicron	mμ
minim	min or ♏
ounce	oz
ounce, apothecaries	oz ap or ℥
ounce, avoirdupois	oz avdp
ounce, fluid	fl oz
ounce, troy	oz t
peck	pk
pennyweight	dwt
pint	pt
pound	lb
pound, apothecaries	lb ap
pound, avoirdupois	lb avdp
pound, troy	lb t
quart	qt
rod	rd
scruple, apothecaries	s ap or ℈
square centimeter	cm²
square chain	ch²
square decimeter	dm²
square dekameter	dkm²
square foot	ft²
square hectometer	hm²
square inch	in.²
square kilometer	km²
square link	li²
square meter	m²
square mile	mi²
square millimeter	mm²
square rod	rd²
square yard	yd²
stere	s
ton	ton
ton, metric	t
troy	t
yard	yd

* The abbreviations γ and λ for microgram and microliter, respectively have been advocated by some authorities.

CONVERSION FACTORS

L. P. Buseth

To convert from	To	Multiply by
Abampere	Ampere	*10*
Abcoulomb	Coulomb	*10*
Abfarad	Farad	1×10^{9}
Abhenry	Henry	1×10^{-9}
Abmho	Siemens (mho)	1×10^{9}
Abohm	Ohm	1×10^{-9}
Abvolt	Volt	1×10^{-8}
Acre	Hectare	0.40468564
	Square foot	*43560*
	Square kilometer	4.046856×10^{-3}
	Square meter	4046.85642
	Square mile	1.5625×10^{-3} (1/640)
	Square yard	*4840*
Acre (U.S. Survey)	Square meter	4046.872610
Acre-foot	Cubic meter	1233.482
	Cubic yard	1613.333
Acre-inch	Cubic foot	*3630*
	Cubic meter	102.7902
	Gallon (Brit.)	22610.67
	Gallon (U.S.)	27154.29
Ampere (int., mean)	Ampere	0.99985
Ampere (int., U.S.)	Ampere	0.999835
Ampere/square centimeter	Ampere/square inch	*6.4516*
Ampere/square inch	Ampere/square centimeter	0.1550003
Ampere-hour	Coulomb	*3600*
Ampere(-turn)	Gilbert	1.256637
Ångström	Nanometer	*0.1*
Apostilb	Candela/square meter	0.3183099 (1/π)
Are	Square foot	1076.391
	Square meter	*100*
Astronomical unit	Kilometer	1.4959787×10^{8}
Atmosphere	Atmosphere (tech.)	1.033227
	Bar	*1.01325*
	Foot of H_2O (conv.)	33.89854
	Inch of Hg (conv.)	29.92126
	Kilogram-force/square centimeter	1.033227
	Kilopascal	*101.325*
	Meter of H_2O (conv.)	10.33227
	Millibar	*1013.25*
	Millimeter of Hg (conv.)	760
	Newton/square centimeter	*10.1325*
	Pascal (N/square meter)	1.01325×10^{5}
	Pound-force/square foot	2116.22
	Pound-force/square inch	14.69595
	Ton-force (long)/square foot	0.944740
	Ton-force (short)/square foot	1.058108
	Ton-force (long)/square inch	6.56069×10^{-3}
	Ton-force (short)/square inch	7.34797×10^{-3}
	Torr	*760*
Atmosphere (tech.)	Atmosphere	0.967841
	Bar	*0.980665*
	Foot of H_2O (conv.)	32.8084
	Inch of Hg (conv.)	28.9590
	Kilogram-force/square centimeter	*1*
	Kilopascal	*98.0665*
	Meter of H_2O (conv.)	*10*
	Millibar	*980.665*
	Millimeter of Hg (conv.)	735.559
	Newton/square centimeter	9.80665
	Pascal (N/m^2)	98066.5
	Pound-force/square inch	14.22334
Bag (Brit.)	Gallon (Brit.)	*24*
Bar	Atmosphere	0.9869233
	Atmosphere (tech.)	1.019716
	Dyne/square centimeter	1×10^{6}
	Foot of H_2O (conv.)	33.4553
	Inch of Hg (conv.)	29.5300
	Kilogram-force/square centimeter	1.019716
	Kilopascal	*100*
	Meter of H_2O (conv.)	10.19716
	Millibar	*1000*
	Millimeter of Hg (conv.)	750.062
	Newton/square centimeter	*10*
	Pascal (N/m^2)	1×10^{5}
	Pound-force/square foot	2088.54
	Pound-force/square inch	14.50377
	Ton-force (long)/square foot	0.932385
	Ton-force (short)/square foot	1.04427
	Ton-force (long)/square inch	6.47490×10^{-3}
	Ton-force (short)/square inch	7.25189×10^{-3}
	Torr	750.062
Barleycorn (Brit.)	Inch	0.333333 (1/3)
Barn	Square meter	1×10^{-28}
Barrel (Brit., beer)	Gallon (Brit.)	*36*
	Liter	163.6592
Barrel (Brit., wine)	Gallon (Brit.)	*31.5*
	Liter	143.2018
Barrel (petroleum)	Cubic foot	5.614583
	Cubic meter	0.1589873
	Gallon (Brit.)	34.97232
	Gallon (U.S.)	*42*
	Liter	158.9873
Barrel (U.S., dry)	Bushel (U.S.)	3.281219
	Cubic foot	4.083333
	Cubic inch	*7056*
	Cubic meter	0.1156271
	Liter	115.6271
	Pint (U.S., dry)	209.998
	Quart (U.S., dry)	104.9990
Barrel (U.S., cranb.)	Cubic inch	*5826*
	Liter	95.4710
Barrel (U.S., liquid)	Cubic foot	*4.2109375*
	Cubic inch	*7276.5*
	Cubic meter	0.1192405
	Gallon (Brit.)	26.22925
	Gallon (U.S.)	*31.5*
	Liter	119.2405
Barye	Bar	1×10^{-6}
	Dyne/square centimeter	*1*
Becquerel	Curie	2.702703×10^{11}
Biot	Ampere	*10*
Board foot	Cubic foot	0.083333 (1/12)
Bolt (cloth)	Foot	*120*
Btu	Calorie	251.996
	Cubic foot-atmosphere	*0.367717*
	Foot-poundal	25036.9
	Foot-pound-force	778.169
	Horsepower-hour	3.93015×10^{-4}
	Horsepower-hour (metric)	3.98466×10^{-4}
	Joule	1055.056
	Kilocalorie	0.251996
	Kilogram-force-meter	107.586
	Kilowatt-hour	2.93071×10^{-4}
	Liter-atmosphere	10.4126
	Watt-hour	0.293071
Btu (39 °F, 4 °C)	Joule	1059.67
Btu (60 °F, 15.6 °C)	Joule	1054.68
Btu (mean)	Joule	1055.87
Btu (thermochemical)	Joule	1054.350
Btu/cubic foot	Joule/cubic meter	37258.9

To convert from	To	Multiply by
	kilocalorie/cubic meter	8.89915
Btu/°F	Calorie/°C	453.592
	Joule/°C	1899.10
Btu/hour	Btu/minute	0.0166667 (*1/60*)
	Btu/second	2.77778×10^{-4}
	Calorie/second	0.0699988
	Foot-pound-force/ second	0.216158
	Horsepower	3.93015×10^{-4}
	Watt	0.293071
Btu/(hour × square foot)	Watt/square meter	3.15459
Btu/ (hour × square foot × °F)	Calorie/ second × square meter × °C)	1.35623
	Watt/(square meter × °C)	5.67826
Btu/ (hour × square foot × °F/foot)	Watt/(meter × °C)	1.73073
Btu/ (hour × square foot × °F/inch)	Watt/(meter × °C)	0.144228
Btu/minute	Calorie/second	4.19993
	Horsepower	0.0235809
	Watt	17.5843
Btu/(minute × square foot)	Watt/square meter	189.273
Btu/pound	Calorie/gram	0.555556
	Joule/kilogram	*2326*
	Kilocalorie/kilogram	0.555556
	Watt-hour/kilogram	0.646111
Btu/(pound × °F)	Calorie/(gram × °C)	*1*
	Joule/(kilogram × °C)	*4186.8*
Btu/second	Horsepower	1.41485
	Kilowatt	1.055056
Btu/(second × square foot)	Kilowatt/square meter	11.3565
Btu/(second × square foot × °F)	Kilowatt/ (square meter × °C)	20.4417
Btu/ (second × square foot × °F/foot)	Kilowatt/(meter × °C)	6.23064
Btu/ (second × square foot × °F /inch)	Watt/(meter × °C)	519.220
Btu/square foot	Joule/square meter	11356.5
	watt-hour/square meter	3.15459
Bucket (Brit.)	Gallon (Brit.)	*4*
Bushel (Brit.)	Bushel (U.S.)	1.032057
	Gallon (Brit.)	*8*
	Liter	*36.36872*
Bushel (U.S.)	Barrel (U.S., dry)	0.3047647
	Bushel (Brit.)	0.9689390
	Cubic foot	1.244456
	Cubic inch	*2150.42*
	Gallon (Brit.)	7.751512
	Gallon (U.S., liquid)	9.309177
	Liter	35.23907
	Peck (U.S.)	*4*
	Pint (U.S., dry)	*64*
	Quart (U.S., dry)	*32*
Butt (Brit.)	Gallon (Brit.)	*108* or *126*
Cable length (int.)	Foot	607.6115
	Meter	*185.2*
	Mile (nautical)	*0.1*
Cable length (U.S.)	Foot	*720*
	Meter	*219.456*
	Mile (nautical)	0.1184968

To convert from	To	Multiply by
	Mile (statute)	0.1363636
Caliber	Inch	*0.01*
	Millimeter	*0.254*
Calorie	Btu	3.96832×10^{-3}
	Cubic foot-atmosphere	1.45922×10^{-3}
	Foot-poundal	99.3543
	Foot-pound-force	3.08803
	Horsepower-hour	1.55961×10^{-6}
	Horsepower-hour (metric)	1.58124×10^{-6}
	Joule	*4.1868*
	Kilocalorie	*0.001*
	Kilogram-force-meter	0.426935
	Kilowatt-hour	1.163×10^{-6}
	Liter-atmosphere	0.0413205
	Watt-hour	1.163×10^{-3}
Calorie (15°C)	Joule	4.1855
Calorie (20°C)	Joule	4.18190
Calorie (mean)	Joule	4.19002
Calorie (thermochem.)	Joule	*4.184*
Calorie/°C	Btu/°F	2.20462×10^{-3}
	Joule/°F	*2.326*
Calorie/gram	Btu/pound	*1.8*
	Joule/kilogram	*4186.8*
Calorie/(gram × °C)	Btu/(pound × °F)	*1*
	Joule/(kilogram × °C)	*4186.8*
Calorie/minute	Watt	*0.06978*
Calorie/(minute × square centimeter)	Watt/square meter	*697.8*
Calorie/second	Watt	*4.1868*
Calorie/ (second × square centimeter)	Kilowatt/square meter	*41.868*
Calorie/ (second × square centimeter × °C)	Kilowatt/ (square meter × °C)	*41.868*
Calorie/ (second × square centimeter × °C/ centimeter)	Watt/(meter × °C)	*418.68*
Calorie/ square centimter	Kilojoule/square meter	41.868
Candela	Hefner unit	1.11
	Lumen/steradian	*1*
Candela/ square centimeter	Candela/square foot	*929.0304*
	Candela/square inch	*6.4516*
	Lambert	3.141593 (π)
Candela/square foot	Candela/square inch	6.944444×10^{-3} (*1/144*)
	Candela/square meter	10.76391
	Foot-lambert	3.141593 (π)
	Lambert	3.381582×10^{-3}
Candela/square inch	Candela/ square centimeter	0.1550003
	Candela/square foot	*144*
	Foot-lambert	452.3893
	Lambert	0.4869478
Candela/square meter	Candela/square foot	*0.09290304*
	Lambert	3.141593×10^{-4}
Carat (metric)	Gram	*0.2*
Cental	Kilogram	*45.359237*
	Pound	*100*
°C heat unit (chu)	Btu	*1.8*
	Calorie	453.592
	Joule	1899.10
Centiliter	Cubic centimeter	*10*
	Cubic inch	0.6102374
	Drachm (Brit., fliud)	2.815606
	Dram (U.S., fluid)	2.705122

CONVERSION FACTORS (continued)

To convert from	To	Multiply by
	Ounce (Brit., fluid)	0.3519508
	Ounce (U.S., fluid)	0.3381402
Centimeter	Foot	0.03280840
	Inch	0.3937008
	Micrometer	*10000*
	Mil	393.7008
	Millimeter	*10*
	Yard	0.01093613
Centimeter of Hg (conv.)	Atmosphere	0.0131579
	Millibar	13.3322
	Millimeter of H_2O (conv.)	*135.951*
	Pascal	1333.22
	Pound-force/square inch	0.193368
Centimeter of H_2O (conv.)	Atmosphere	9.67841×10^{-4}
	Millibar	*0.980665*
	Millimeter of Hg (conv.)	0.735559
	Kilogram-force/square centimeer	*0.001*
	Pascal	*98.0665*
	Pound-force/square inch	0.0142233
Centimeter/second	Foot/minute	1.968504
	Foot/second	0.03280840
	Kilometer/hour	*0.036*
	Meter/minute	*0.6*
	Mile/hour	0.02236936
Centimeter/square second	Foot/square second	0.03280840
	Kilometer/(hour × second)	*0.036*
	Meter/square second	*0.01*
	Mile/(hour × second)	0.02236936
Centipoise	Pascal-second	*0.001*
Centistokes	Square meter/second	*1×10^{-6}*
Chain (Gunter's)	Foot	*66*
Chain (Ramsden's)	Foot	*100*
Circular inch	Circular mil	*1×10^{6}*
	Square centimeter	5.067075
	Square inch	0.7853982
Circular mil	Square inch	7.853982×10^{-7}
	Square micrometer	506.7075
	Square mil	0.7853982
Circular millimeter	Square millimeter	0.7853982
Circumference	Degree	*360*
	Gon (grade)	*400*
	Radian	6.283185 (2π)
Clo	(°C × square meter)/watt	0.2003712
Cord	Cord-foot	*8*
	Cubic foot	*128*
Cord-foot	Cord	*0.125 (1/8)*
	Cubic foot	*16*
Coulomb	Ampere-second	*1*
Cubic centimeter	Cubic foot	3.531467×10^{-5}
	Cubic inch	0.06102374
	Cubic meter	*1×10^{-6}*
	Cubic millimeter	*1000*
	Cubic yard	1.307951×10^{-6}
	Drachm (Brit., fluid)	0.2815606
	Dram (U.S., fluid)	0.2705122
	Gallon (Brit.)	2.199692×10^{-4}
	Gallon (U.S.)	2.641721×10^{-4}
	Gill (Brit.)	7.039016×10^{-3}
	Gill (U.S.)	8.453506×10^{-3}
	Liter	*0.001*
	Milliliter	*1*
	Minim (Brit.)	16.89364
	Minim (U.S.)	16.23073
	Ounce (Brit., fluid)	0.03519508
	Ounce (U.S., fluid)	0.03381402
	Pint (Brit.)	*1.759754×10^{-3}*
	Pint (U.S., dry)	1.816166×10^{-3}

To convert from	To	Multiply by
	Pint (U.S., liquid)	2.113376×10^{-3}
	Quart (Brit.)	8.798770×10^{-4}
	Quart (U.S., dry)	9.080830×10^{-4}
	Quart (U.S., liquid)	1.056688×10^{-3}
Cubic centimeter/gram	Cubic foot/pound	0.0160185
Cubic centimeter/second	Cubic foot/minute	2.118880×10^{-3}
	Liter/hour	*3.6*
Cubic centimeter-atmosphere	Joule	*0.101325*
	Watt-hour	2.814583×10^{-5}
Cubic decimeter	Cubic centimeter	*1000*
	Cubic foot	0.03531467
	Cubic inch	61.02374
	Cubic meter	*0.001*
	Liter	*1*
Cubic foot	Acre-foot	2.295684×10^{-5}
	Board foot	*12*
	Bushel (Brit.)	0.7786044
	Bushel (U.S.)	0.8035640
	Cord	*7.8125×10^{-3} (1/128)*
	Cord-foot	*0.0625 (1/16)*
	Cubic centimeter	28316.847
	Cubic inch	*1728*
	Cubic meter	0.028316847
	Cubic yard	0.03703704 *(1/27)*
	Gallon (Brit.)	6.228835
	Gallon (U.S.)	7.480519
	Liter	28.316847
	Pint (Brit.)	49.83068
	Pint (U.S., dry)	51.42809
	Pint (U.S., liquid)	59.84416
	Quart (Brit.)	24.91534
	Quart (U.S., dry)	25.71405
	Quart (U.S., liquid)	29.92208
Cubic foot/hour	Cubic centimeter/second	7.865791
	Liter/minute	0.4719474
Cubic foot/minute	Cubic centimeter/second	471.9474
	Gallon (Brit.)/second	0.1038139
	Gallon (U.S.)/second	0.1246753
Cubic foot/pound	Cubic meter/kilogram	0.06242796
Cubic foot/second	Cubic meter/hour	101.9406
	Cubic yard/minute	2.222222
	Gallon (Brit.)/minute	373.7301
	Gallon (U.S.)/minute	448.8312
	Liter/minute	1699.011
Cubic foot-atmosphere	Btu	2.71948
	Calorie	685.298
	Foot-pound-force	2116.22
	Joule	2869.205
	Kilogram-force-meter	292.577
	Liter-atmosphere	28.31685
	Watt-hour	0.7970012
Cubic foot (pound-force/square inch)	Btu	0.7970012
	Calorie	
	Joule	0.185050
	Watt-hour	46.6317
		195.238
		0.0542327
Cubic inch	Board foot	6.944444×10^{-3} *(1/144)*
	Bushel (Brit.)	4.505813×10^{-4}
	Bushel (U.S.)	4.650254×10^{-4}
	Cubic centimeter	*16.387064*
	Cubic foot	5.787037×10^{-4} *(1/1728)*
	Cubic meter	*1.6387064×10^{-5}*
	Cubic yard	2.143347×10^{-5}
	Drachm (Brit., fluid)	4.613952
	Dram (U.S., fluid)	4.432900
	Gallon (Brit.)	3.604650×10^{-3}
	Gallon (U.S.)	4.329004×10^{-3} *(1/231)*
	Liter	*0.016387064*
	Milliliter	*16.387064*
	Ounce (Brit., fluid)	0.5767440

CONVERSION FACTORS (continued)

To convert from	To	Multiply by
	Ounce (U.S., fluid)	0.5541126
	Pint (Brit.)	0.02883720
	Pint (U.S., dry)	0.02976163
	Pint (U.S., liquid)	0.03463203
	Quart (Brit.)	0.01441860
	Quart (U.S., dry)	0.01488081
	Quart (U.S., liquid)	0.01731602
Cubic inch/minute	Cubic centimeter/second	0.2731177
Cubic kilometer	Cubic mile	0.2399128
Cubic meter	Barrel (petroleum)	6.289811
	Barrel (U.S., dry)	8.648490
	Barrel (U.S., liquid)	8.386414
	Bushel (U.S.)	28.37759
	Cubic centimeter	1×10^{6}
	Cubic decimeter	*1000*
	Cubic foot	35.31467
	Cubic inch	61023.74
	Cubic yard	1.307951
	Gallon (Brit.)	219.9692
	Gallon (U.S.)	264.1721
	Liter	*1000*
	Pint (Brit.)	1759.754
	Pint (U.S., dry)	1816.166
	Pint (U.S., liquid)	2113.376
	Quart (Brit.)	879.8770
	Quart (U.S., dry)	908.0830
	Quart (U.S., liquid)	1056.688
	Register ton	0.3531467
Cubic meter/kilogram	Cubic foot/pound	16.01846
Cubic mile	Cubic kilometer	4.168182
Cubic millimeter	Cubic centimeter	*0.001*
	Cubic inch	6.102374×10^{-5}
	Minim (Brit.)	0.01689364
	Minim (U.S.)	0.01623073
Cubic yard	Bushel (Brit.)	21.02232
	Bushel (U.S.)	21.69623
	Cubic foot	*27*
	Cubic inch	*46656*
	Cubic meter	0.76455486
	Gallon (Brit.)	168.1786
	Gallon (U.S.)	201.9740
	Liter	764.5549
Cubic yard/minute	Cubic foot/second	*0.45*
	Gallon (Brit.)/second	2.802976
	Gallon (U.S.)/second	3.366234
	Liter/second	12.74258
Cubit	Inch	*18*
Cup (metric)	Milliliter	*200*
Cup (U.S.)	Milliliter	236.588
	Ounce (U.S. fluid)	*8*
Curie	Becquerel	3.7×10^{10}
Darcy	Square meter	9.869233×10^{-13}
Day (mean solar)	Hour	*24*
	Minute	*1440*
	Second	*86400*
Day (sidereal)	Second	86164.09
Decibel	Neper	0.115129255
Degree (angular)	Circumference	2.777778×10^{-3} (*1/360*)
	Gon (grade)	1.111111
	Minute (angular)	*60*
	Quadrant	0.01111111 (*1/90*)
	Radian	0.01745329
	Second (angular)	*3600*
Degree/foot	Radian/meter	0.05726146
Degree/inch	Radian/meter	0.6871375
Degree/second	revolution/minute	0.1666667 (*1/6*)
°C (temp. interval)	°Fahrenheit	*1.8*
	°Rankine	*1.8*
	Kelvin	*1*
(°C × hour)/kilocalorie	°C/watt	0.859845
(°C × hour × square meter)/kilocalorie	(°C × square meter)/watt	0.859845
°F (temp. interval)	°Celsius	0.5555556 (*5/9*)
	°Rankine	*1*
	Kelvin	0.5555556 (*5/9*)
(°F × hour)/Btu	°C/watt	1.89563
(°F × hour × square foot)/Btu	(°C × square meter)/watt	0.176110
(°F/inch × hour × square foot)/Btu	(°C × meter)/watt	6.93347
Denier	Tex	0.111111 (*1/9*)
Drachm (Brit. fluid)	Dram (U.S., fluid)	0.9607599
	Milliliter	3.551633
	Minim (Brit.)	*60*
	Ounce (Brit. fluid)	*0.125 (1/8)*
Dram (apoth. or troy)	Dram (avoirdupois)	2.1942857
	Grain	*60*
	Gram	*3.8879346*
	Ounce (apoth. or troy)	*0.125 (1/8)*
	Pennyweight	*2.5*
	Scruple	*3*
Dram (avoirdupois)	Grain	*27.34375*
	Gram	1.7718452
	Ounce (avoirdupois)	*0.0625 (1/16)*
Dram (U.S., fluid)	Cubic centimeter	3.696691
	Cubic inch	0.2255859
	Drachm (Brit. fluid)	1.040843
	Gallon (U.S.)	9.765625×10^{-4} (*1/1024*)
	Gill (U.S.)	*0.03125 (1/32)*
	Milliliter	3.696691
	Minim (U.S.)	*60*
	Ounce (U.S., fluid)	*0.125 (1/8)*
	Pint (U.S., liquid)	7.8125×10^{-3} (*1/128*)
	Quart (U.S., liquid)	3.90625×10^{-3} (*1/256*)
Dyne	Kilogram-force	1.019716×10^{-6}
	Newton	1×10^{-5}
	Poundal	7.233014×10^{-5}
	Pound-force	2.248089×10^{-6}
Dyne/centimeter	Newton/meter	*0.001*
Dyne/square centimeter	Bar	1×10^{-6}
	Kilogram-force/square centimeter	1.019716×10^{-6}
	Millimeter of Hg (conv.)	7.50062×10^{-4}
	Millimeter of H_2O (conv.)	0.01019716
	Pascal (N/square meter)	*0.1*
	Pound-force/square inch	1.450377×10^{-5}
Dyne-centimeter	Erg	*1*
	Foot-poundal	2.37304×10^{-6}
	Foot-pound-force	7.37562×10^{-8}
	Joule	1×10^{-7}
	Kilogram-force-meter	1.019716×10^{-8}
	Newton-meter	1×10^{-7}
Dyne-second/square centimeter	Poise	*1*
	Pascal-second	*0.1*
Electronvolt	Erg	1.60219×10^{-12}
	Joule	1.60219×10^{-19}
Ell	Inch	*45*
Erg	Dyne-centimeter	*1*
	Joule	1×10^{-7}
	Watt-hour	2.777778×10^{-11}
Erg/(square centimeter × second)	Watt/square meter	*0.001*
Farad (int. mean)	Farad	0.999510
Farad (int. U.S.)	Farad	0.999505
Fathom	Foot	*6*
Fermi	Meter	1×10^{-15}
Firkin (Brit.)	Gallon (Brit.)	*9*
Firkin (U.S.)	Gallon (U.S.)	*9*
Foot	Centimeter	*30.48*
	Foot (U.S. Survey)	*0.999998*
	Inch	*12*
	Meter	0.3048
	Millimeter	304.8
	Mile (nautical)	1.645788×10^{-4}
	Mile (statute)	1.893939×10^{-4}

CONVERSION FACTORS (continued)

To convert from	To	Multiply by
	Yard	0.333333 (*1/3*)
Foot (U.S. Survey)	Foot	1.000002
	Meter	0.30480060960
Foot of H_2O (conv.)	Atmosphere	0.0294998
	Bar	0.0298907
	Inch of Hg (conv.)	0.882671
	Kilogram-force/square centimeter	*0.03048*
	Millimeter of Hg (conv.)	22.4198
	Pascal (N/square meter)	2989.07
	Pound-force/square inch	0.433527
Foot/°F	Meter/°C	*0.54864*
Foot/hour	Meter/second	8.466667×10^{-5}
Foot/minute	Kilometer/hour	*0.018288*
	Knot	9.87473×10^{-3}
	Meter/second	5.08×10^{-3}
	Mile/hour	0.01136364 (*1/88*)
Foot/second	Kilometer/hour	*1.09728*
	Knot	0.5924838
	Meter/minute	*18.288*
	Meter/second	*0.3048*
	Mile/hour	0.6818182
Foot/square second	Kilometer/(hour × second)	*1.09728*
	Meter/square second	*0.3048*
	Mile/(hour × second)	0.6818182
Foot to the fourth power	Meter to the fourth power	8.630975×10^{-3}
Foot-candle	Lumen/square foot	*1*
	Lumen/square meter	10.76391
	Lux	10.76391
Foot-lambert	Candela/square centimeter	3.426259×10^{-4}
	Candela/square foot	0.3183099 ($1/\pi$)
	Candela/square meter	3.426259
	Lambert	1.076391×10^{-3}
	Meter-lambert	10.76391
Foot-poundal	Btu	3.99411×10^{-5}
	Calorie	0.0100650
	Foot-pound-force	0.0310810
	Joule	0.0421401
	Kilogram-force-meter	4.29710×10^{-3}
	Liter-atmosphere	4.15891×10^{-4}
	Watt-hour	1.17056×10^{-5}
Foot-pound-force	Btu	1.28507×10^{-3}
	Calorie	0.323832
	Cubic foot-atmosphere	4.72541×10^{-4}
	Foot-poundal	32.1740
	Horsepower-hour	5.05051×10^{-7}
	Horsepower-hour (metric)	5.12055×10^{-7}
	Joule	1.35582
	Kilogram-force-meter	0.138255
	Liter-atmosphere	0.0133809
	Newton-meter	1.35582
	Watt-hour	3.76616×10^{-4}
Foot-pound-force/hour	Watt	3.76616×10^{-4}
Foot-pound-force/minute	Horsepower	3.03030×10^{-5}
	Horsepower (metric)	3.07233×10^{-5}
	Watt	0.0225970
Foot-pound-force/second	Horsepower	1.81818×10^{-3} (*1/550*)
	Horsepower (metric)	1.84340×10^{-3}
	Watt	1.355818
Franklin	Coulomb	3.335641×10^{-10}
Furlong	Foot	*660*
	Meter	*201.168*
	Mile (statute)	*0.125 (1/8)*
	Yard	*220*
Gal	Centimeter/square second	1
	Meter/square second	*0.01*
Gallon (Brit.)	Bushel (Brit.)	*0.125 (1/8)*
	Cubic centimeter	*4546.09*
	Cubic foot	0.1605437
	Cubic inch	277.4194
	Cubic yard	5.946061×10^{-3}
	Drachm (Brit., fluid)	*1280*
	Gallon (U.S.)	1.200950
	Gill (Brit.)	*32*
	Liter	*4.54609*
	Minim (Brit.)	*76800*
	Ounce (Brit., fluid)	*160*
	Peck (Brit.)	*0.5*
	Pint (Brit.)	*8*
	Quart (Brit.)	*4*
Gallon (U.S., dry)	Bushel (U.S.)	*0.125 (1/8)*
	Cubic inch	*268.8025*
	Liter	4.404884
Gallon (U.S., liquid)	Barrel (petroleum)	0.02380952 (*1/42*)
	Cubic centimeter	3785.412
	Cubic foot	0.13368056
	Cubic inch	*231*
	Cubic yard	4.951132×10^{-3}
	Dram (U.S., fluid)	*1024*
	Gallon (Brit.)	0.8326742
	Gill (U.S.)	*32*
	Liter	3.785412
	Minim (U.S.)	*61440*
	Ounce (U.S., fluid)	*128*
	Pint (U.S., liquid)	*8*
	Quart (U.S., liquid)	*4*
Gallon (Brit.)/minute	Cubic foot/hour	9.632619
	Cubic foot/second	2.675728×10^{-3}
	Cubic meter/hour	0.2727654
	Liter/second	0.07576817
Gallon (U.S.)/minute	Cubic foot/hour	8.020834
	Cubic foot/second	2.228009×10^{-3}
	Cubic meter/hour	0.2271247
	Liter/second	0.06309020
Gamma	Tesla	1×10^{-9}
Gauss	Tesla	1×10^{-4}
	Weber/square meter	1×10^{-4}
Geepound	Slug	*1*
Gigawatt-hour	Kilowatt-hour	1×10^{6}
Gilbert	Ampere	0.7957747
Gill (Brit.)	Cubic centimeter	142.0653
	Cubic inch	8.669357
	Gallon (Brit.)	*0.03125 (1/32)*
	Gill (U.S.)	1.200950
	Milliliter	142.0653
	Ounce (Brit. fluid)	*5*
	Pint (Brit.)	*0.25 (1/4)*
	Quart (Brit.)	*0.125 (1/8)*
Gill (U.S.)	Cubic centimeter	118.2941
	Cubic inch	*7.21875*
	Gallon (U.S.)	*0.03125 (1/32)*
	Gill (Brit.)	0.8326742
	Milliliter	118.2941
	Ounce (U.S., fluid)	*4*
	Pint (U.S., liquid)	*0.25 (1/4)*
	Quart (U.S., liquid)	*0.125 (1/8)*
Gon (grade)	Circumference	*0.002 5 (1/400)*
	Degree (angular)	*0.9*
	Minute (angular)	*54*
	Radian	0.01570796
	Second (angular)	*3240*
Grain	Carat (metric)	*0.32399455*
	Dram	0.03657143
	Milligram	*64.79891*
	Ounce (avoirdupois)	2.285714×10^{-3}
	Ounce (troy)	2.083333×10^{-3} (*1/480*)
	Pennyweight	0.04166667 (*1/24*)
	Pound	1.428571×10^{-4} (*1/7000*)
	Scruple	*0.05 (1/20)*
Grain/cubic foot	Milligram/liter	2.288352

CONVERSION FACTORS (continued)

To convert from	To	Multiply by
Grain/gallon (Brit.)	Milligram/liter	14.25377
Grain/gallon (U.S.)	Milligram/liter	17.11806
	Pound/million gallons	142.8571
Gram	Carat (metric)	*5*
	Dram	0.56438339
	Grain	15.432358
	Kilogram	*0.001*
	Milligram	*1000*
	Ounce (avoirdupois)	0.035273962
	Ounce (troy)	0.032150747
	Pennyweight	0.64301493
	Pound	2.2046226×10^{-3}
	Scruple	0.77161792
	Ton (metric)	1×10^{-6}
Gram/(centimeter × second)	Poise	*1*
Gram/cubic centimeter	Kilogram/cubic decimeter	*1*
	Kilogram/cubic meter	*1000*
	Kilogram/liter	*1*
	Pound/cubic foot	62.42796
	Pound/cubic inch	0.03612729
	Pound/gallon (Brit.)	10.02241
	Pound/gallon (U.S.)	8.345404
Gram/cubic meter	Grain/cubic foot	0.4369957
Gram/liter	Grain/gallon (Brit.)	70.15689
	Grain/gallon (U.S.)	58.41783
	Gram/cubic centimeter	*0.001*
	Kilogram/cubic meter	*1*
	Pound/cubic foot	0.0624280
	Pound/gallon (Brit.)	0.0100224
	Pound/gallon (U.S.)	8.34540×10^{-3}
Gram/meter	Ounce/yard	0.03225451
Gram/milliliter	Gram/cubic centimeter	*1*
Gram/square meter	Ounce/square foot	0.3277058
	Ounce/square yard	0.02949352
Gram/ton (long)	Gram/ton (metric)	0.9842065
	Gram/ton (short)	0.8928571
	Milligram/kilogram	0.9842065
Gram/ton (metric)	Gram/ton (long)	1.016047
	Gram/ton (short)	0.9071847
	Milligram/kilogram	*1*
Gram/ton (short)	Gram/ton (long)	*1.12*
	Gram/ton (metric)	1.102311
	Milligram/kilogram	1.102311
Gram-force	Dyne	*980.665*
	Newton	9.80665×10^{-3}
Gram-force/square centimeter	Pascal	*98.0665*
Gram-force-centimeter	Erg	*980.665*
	Joule	9.80665×10^{-5}
Gray	Joule/kilogram	*1*
Hand	Inch	*4*
Hectare	Acre	2.471054
	Are	*100*
	Square foot	1.076391×10^{5}
	Square kilometer	*0.01*
	Square meter	*10000*
	Square mile	3.861022×10^{-3}
	Square yard	11959.90
Hectogram	Kilogram	*0.1*
Hectoliter	Cubic meter	*0.1*
Hefner unit	Candela	0.903
Henry (int. mean)	Henry	1.00049
Henry (int. U.S.)	Henry	1.000495
Hogshead (U.S.)	Gallon (U.S.)	*63*
Horsepower	Btu/hour	2544.43
	Btu/minute	42.4072
	Btu/second	0.706787
	Foot-pound-force/hour	1.98×10^{6}
	Foot-pound-force/minute	*33000*
	Foot-pound-force/second	*550*
	Horsepower (metric)	1.01387
	Joule/second	745.700
	Kilocalorie/hour	641.186
	Kilocalorie/minute	10.6864
	Kilocalorie/second	0.178107
	Kilogram-force-meter/second	76.0402
	Kilowatt	0.745700
Horsepower (boiler)	Kilowatt	9.80950
Horsepower (electric)	Kilowatt	*0.746*
Horsepower (metric)	Foot-pound-force/second	542.476
	Horsepower	0.986320
	Kilocalorie/hour	632.415
	Kilocalorie/minute	10.54025
	Kilocalorie/second	0.175671
	Kilogram-force-meter/second	*75*
	Kilowatt	0.735499
Horsepower (water)	Kilowatt	0.746043
Horsepower-hour	Btu	2544.43
	Foot-pound-force	1.98×10^{6}
	Horsepower-hour (metric)	1.01387
	Joule	2.68452×10^{6}
	Kilocalorie	641.186
	Kilogram-force-meter	2.73745×10^{5}
	Kilowatt-hour	0.745700
	Megajoule	2.68452
Horsepower-hour (metric)	Horsepower-hour	0.986320
	Joule	2.64780×10^{6}
	Kilocalorie	632.415
	Kilogram-force-meter	2.7×10^{5}
	Kilowatt-hour	0.735499
	Megajoule	2.64780
Hour (mean solar)	Day	0.04166667 (*1/24*)
	Minute	*60*
	Second	*3600*
	Week	5.952381×10^{-3} (*1/168*)
Hundredweight (long)	Hundredweight (short)	*1.12*
	Kilogram	*50.80234544*
	Pound	*112*
	Ton (long)	*0.05*
	Ton (metric)	0.050802345
	Ton (short)	*0.056*
Hundredweight (short)	Hundredweight (long)	0.89285714
	Kilogram	*45.359237*
	Pound	*100*
	Ton (long)	0.044642857
	Ton (metric)	*0.045359237*
	Ton (short)	*0.05*
Inch	Centimeter	*2.54*
	Foot	0.08333333 (*1/12*)
	Mil	*1000*
	Millimeter	*25.4*
	Yard	0.02777778 (*1/36*)
Inch of Hg (conv.)	Atmosphere	0.0334211
	Foot of H_2O (conv.)	*1.132925*
	Inch of H_2O (conv.)	*13.5951*
	Kilogram-force/square centimeter	0.0345316
	Millibar	33.8639
	Millimeter of H_2O (conv.)	345.316
	Pascal	3386.39
	Pound-force/square inch	0.491154
Inch of H_2O (conv.)	Inch of Hg (conv.)	0.0735559
	Kilogram-force/square centimeter	2.54×10^{-3}
	Millibar	2.49089
	Millimeter of Hg (conv.)	1.86832
	Pascal	249.089
	Pound-force/square inch	0.0361273

To convert from	To	Multiply by
Inch/°F	Millimeter/°C	*45.72*
Inch/hour	Millimeter/minute	0.4233333
	Millimeter/second	7.055556×10^{-3}
	Foot/minute	1.388889×10^{-3}
Inch/minute	Foot/hour	*5*
	Meter/hour	*1.524*
	Millimeter/second	0.4233333
Inch/second	Foot/hour	*300*
	Meter/minute	1.524
Inch to the fourth power	Meter to the fourth power	4.162314×10^{-7}
Joule	Btu	9.47817×10^{-4}
	Calorie	0.238846
	Centrigrade heat unit	5.26565
	Cubic foot-atmosphere	3.48529×10^{-4}
	Cubic foot-pound-force/square inch	5.12196×10^{-3}
	Erg	1×10^{7}
	Foot-poundal	23.7304
	Foot-pound-force	0.737562
	Horsepower-hour	3.72506×10^{-7}
	Horsepower-hour (metric)	3.77673×10^{-7}
	Kilogram-force-meter	0.101972
	Liter-atmosphere	9.86923×10^{-3}
	Newton-meter	*1*
	Watt-hour	2.777778×10^{-4} (*1/3600*)
	Watt-second	*1*
Joule/°C	Btu/°F	5.26565×10^{-4}
Joule/gram	Btu/pound	0.429923
	Kilocalorie/kilogram	0.238846
Joule/(gram × °C)	Btu/(pound × °F)	0.238846
	Kilocalorie/(kilogram × °C)	0.238846
Joule/hour	Watt	2.777778×10^{-4} (*1/3600*)
Joule/minute	Watt	0.01666667 (*1/60*)
Joule/second	Watt	*1*
Kelvin (temp. interval)	°Celsius	*1*
	°Fahrenheit	*1.8*
	°Rankine	*1.8*
Kilderkin (Brit.)	Gallon (Brit.)	*18*
Kilocalorie	Btu	3.96832
	Calorie	*1000*
	Joule	*4186.8*
Kilocalorie/cubic meter	Btu/cubic foot	0.112370
	Kilojoule/cubic meter	*4.1868*
Kilocalorie/hour	Watt	*1.163*
Kilocalorie/(hour × square meter)	Watt/square meter	*1.163*
Kilocalorie/(hour × square meter × °C)	Watt/(square meter × °C)	*1.163*
Kilocalorie/(hour × square meter × °C/centimeter)	Watt/(meter × °C)	*0.01163*
Kilocalorie/kilogram	Btu/pound	*1.8*
	Joule/gram	*4.1868*
Kilocalorie/(kilogram × °C)	Btu/(pound × °F)	*1*
	Kilojoule/(kg × °C)	*4.1868*
Kilocalorie/minute	Foot-pound-force/second	51.4671
	Horsepower	0.0935765
	Horsepower (metric)	0.0948744
	Watt	*69.78*
Kilocalorie/second	Kilowatt	*4.1868*
Kilogram	Grain	15432.358
	Gram	*1000*
	Hundredweight (long)	0.019684131
	Hundredweight (short)	0.022046226
	Ounce (avoirdupois)	35.273962
	Ounce (troy)	32.150747
	Pound	2.2046226
	Ton (long)	9.8420653×10^{-4}
	Ton (metric)	*0.001*
	Ton (short)	1.1023113×10^{-3}
Kilogram/cubic meter	Gram/cubic centimeter	*0.001*
	Gram/liter	*1*
	Pound/cubic foot	0.06242796
	Pound/cubic inch	3.612729×10^{-5}
Kilogram/meter	Gram/centimeter	*10*
	Pound/foot	0.6719690
	Pound/inch	0.05599741
Kilogram-force	Dyne	9.80665×10^{5}
	Newton	*9.80665*
	Pound-force	2.20462
	Poundal	70.9316
Kilogram-force/square centimeter	Atmosphere	0.967841
	Atmosphere (technical)	*1*
	Bar	*0.980665*
	Foot of H_2O (conv.)	32.8084
	Inch of Hg (conv.)	28.9590
	Kilogram-force/square millimeter	*0.01*
	Meter of H_2O (conv.)	*10*
	Millimeter of Hg (conv.)	735.559
	Newton/square millimeter	*0.0980665*
	Pascal (N/square meter)	*98066.5*
	Pound-force/square foot	2048.16
	Pound-force/square inch	14.22334
	Ton-force (long)/square foot	0.914358
	Ton-force (short)/square foot	1.02408
	Ton-force (long)/square inch	6.34971×10^{-3}
	Ton-force (short)/square inch	7.11167×10^{-3}
Kilogram-force/square meter	Pascal	*9.80665*
Kilogram-force/square millimeter	Newton/square millimeter	*9.80665*
	Megapascal	*9.80665*
	Pound-force/square inch	1422.334
Kilogram-force-meter	Btu	9.29491×10^{-3}
	Calorie	2.34228
	Cubic foot-atmosphere	3.41790×10^{-3}
	Erg	9.80665×10^{7}
	Foot-poundal	232.715
	Foot-pound-force	7.23301
	Horsepower-hour	3.65304×10^{-6}
	Horsepower-hour (metric)	3.70370×10^{-6}
	Joule	*9.80665*
	Liter-atmosphere	0.0967841
	Newton-meter	*9.80665*
	Watt-hour	2.72407×10^{-3}
Kilometer	Astronomical unit	6.68459×10^{-9}
	Foot	3280.840
	Light year	1.05702×10^{-13}
	Mile (nautical)	0.5399568
	Mile (statute)	0.6213712
	Yard	1093.613
Kilometer/hour	Foot/minute	54.68066
	Foot/second	0.9113444
	Inch/second	10.93613
	Knot	0.5399568
	Meter/minute	16.66667
	Meter/second	0.2777778
	Mile/hour	0.6213712
Kilometer/(hour × second)	Centimeter/square second	27.77778
	Foot/square second	0.9113444
	Meter/square second	0.2777778
	Mile/(hour × second)	0.6213712

To convert from	To	Multiply by
Kilopascal	Pound-force/square foot	20.8854
	Pound-force/square inch	0.1450377
Kilopond	Kilogram-force	*1*
	Newton	*9.80665*
Kilowatt	Btu/hour	3412.14
	Btu/minute	56.8690
	Btu/second	0.947817
	Foot-pound-force/hour	2.65522×10^6
	Foot-pound-force/minute	44253.7
	Foot-pound-force/second	737.562
	Horsepower	1.34102
	Horsepower (metric)	1.35962
	Joule/hour	3.6×10^6
	Joule/minute	*60000*
	Joule/second	*1000*
	Kilocalorie/hour	859.845
	Kilocalorie/minute	14.3308
	Kilocalorie/second	0.238846
	Kilogram-force-meter/hour	3.67098×10^5
	Kilogram-force-meter/minute	6118.30
	Kilogram-force-meter/second	101.972
Kilowatt-hour	Btu	3412.14
	Foot-pound-force	2.65522×10^6
	Horsepower-hour	1.34102
	Horsepower-hour (metric)	1.35962
	Joule	3.6×10^6
	Kilocalorie	859.845
	Kilogram-force-meter	3.67098×10^5
	Megajoule	*3.6*
Kilowatt-hour/pound	Btu/pound	3412.14
	Joule/gram	7936.641
	Kilocalorie/kilogram	1895.63
Kilowatt-hour/kilogram	Btu/pound	1547.72
Kip	Pound-force	*1000*
Kip/square inch	Newton/square millimeter	6.89476
	Megapascal	6.89476
Knot	Foot/minute	101.2686
	Foot/second	1.687810
	Kilometer/hour	*1.852*
	Meter/minute	30.86667
	Meter/second	0.5144444
	Mile (nautical)/hour	*1*
	Mile (statute)/hour	1.150779
Lambert	Candela/square centimeter	0.3183099 (*1/π*)
	Candela/square foot	295.7196
	Candela/square inch	2.053608
	Candela/square meter	3183.099
	Foot-lambert	*929.0304*
Langley	Joule/square meter	*41840*
Last (Brit.)	Gallon (Brit.)	*640*
League (nautical)	Mile (nautical)	*3*
League (statute)	Mile (statute)	*3*
Light year	Astronomical unit	63239.7
	Kilometer	9.46053×10^{12}
	Mile	5.87850×10^{12}
	Parsec	0.306595
Line	Inch	*0.1* or 0.083333 (*1/12*)
	Millimeter	*2.54* or 2.116667
Line	Weber	1×10^{-8}
Link	Chain	*0.01*
Liter	Bushel (Brit.)	0.027496156
	Bushel (U.S.)	0.02837759
	Cubic centimeter	*1000*
	Cubic decimeter	*1*
	Cubic foot	0.03531467
	Cubic inch	61.02374
	Cubic meter	*0.001*
	Cubic yard	1.307951×10^{-3}
	Drachm (Brit., fluid)	281.5606
	Dram (U.S., fluid)	270.5122
	Gallon (Brit.)	0.21996925
	Gallon (U.S.)	0.26417205
	Gill (Brit.)	7.039016
	Gill (U.S.)	8.453506
	Milliliter	1000
	Minim (Brit.)	16893.64
	Minim (U.S.)	16230.73
	Ounce (Brit., fluid)	35.19508
	Ounce (U.S., fluid)	33.81402
	Pint (Brit.)	1.759754
	Pint (U.S., dry)	1.816166
	Pint (U.S., liquid)	2.113376
	Quart (Brit.)	0.8798770
	Quart (U.S., dry)	0.9080830
	Quart (U.S., liquid)	1.056688
Liter (1901—1964)	Cubic decimeter	1.000028
Liter/minute	Cubic foot/hour	2.118880
	Cubic foot/second	5.885778×10^{-4}
	Gallon (Brit.)/hour	13.19815
	Gallon (Brit.)/second	3.666154×10^{-3}
	Gallon (U.S.)/hour	15.85032
	Gallon (U.S.)/second	4.402868×10^{-3}
Liter/second	Cubic foot/hour	127.1328
	Cubic foot/minute	2.118880
	Gallon (Brit.)/hour	791.8893
	Gallon (Brit.)/minute	13.19815
	Gallon (U.S.)/hour	951.0194
	Gallon (U.S.)/minute	15.85032
Liter-atmosphere	Btu	0.0960376
	Calorie	24.2011
	Cubic foot-atmosphere	0.0353147
	Cubic foot-pound-force/square inch	0.518983
	Foot-poundal	2404.48
	Foot-pound-force	74.7335
	Horsepower-hour	3.77442×10^{-5}
	Horsepower-hour (metric)	3.82677×10^{-5}
	Joule	*101.325*
	Kilogram-force-meter	10.3323
	Watt-hour	0.0281458
Liter-bar	Joule	*100*
Lumen/square centimeter	Lux	*10000*
	Phot	*1*
Lumen/square foot	Lux	10.76391
Lumen/square meter	Lumen/square foot	*0.09290304*
	Lux	*1*
Lux	Lumen/square meter	*1*
	Phot	1×10^{-4}
Maxwell	Weber	1×10^{-8}
Megajoule	Kilowatt-hour	0.2777778
Megapascal	Bar	*10*
	Newton/square millimeter	*1*
Megohm	Ohm	1×10^6
Meter	Ångström	1×10^{10}
	Fathom	0.5468066
	Foot	3.2808399
	Foot (U.S. Survey)	3.2808333
	Inch	39.37007874
	Micrometer	1×10^6
	Mile (nautical)	5.399568×10^{-4}
	Mile (statute)	6.213712×10^{-4}
	Nanometer	1×10^9
	Yard	1.093613298
Meter/hour	Foot/minute	0.05468066
	Foot/second	9.113444×10^{-4}
	Millimeter/minute	16.66667
	Millimeter/second	0.2777778
Meter/minute	Foot/second	0.05468066
	Kilometer/hour	*0.06*

CONVERSION FACTORS (continued)

To convert from	To	Multiply by
	Knot	0.03239741
	Mile (statute)/hour	0.03728227
	Millimeter/second	16.66667
Meter/second	Foot/minute	196.8504
	Kilometer/hour	*3.6*
	Kilometer/minute	*0.06*
	Knot	1.943844
	Mile (statute)/hour	2.236936
	Mile (statute)/minute	0.03728227
Meter/square second	Foot/square second	3.280840
	Kilometer/(hour × second)	*3.6*
	Mile/(hour × second)	2.236936
Meter-candle	Lux	*1*
Mho (ohm^{-1})	Siemens	*1*
Microfarad	Farad	1×10^{-6}
Microgram	Grain	1.5432358×10^{-5}
	Gram	1×10^{-6}
Micrometer	Ångström	*10000*
	Mil	0.03937008
	Millimeter	*0.001*
	Nanometer	*1000*
Micron	Micrometer	*1*
Mil	Inch	*0.001*
	Micrometer	*25.4*
	Millimeter	*0.0254*
Mile (nautical)	Foot	6076.1155
	Kilometer	*1.852*
	Mile (statute)	1.150779
	Yard	2025.372
Mile (statute)	Chain (Gunter's)	*80*
	Chain (Ramsden's)	*52.8*
	Foot	*5280*
	Furlong	*8*
	Inch	*63360*
	Kilometer	*1.609344*
	Light year	1.70111×10^{-13}
	Meter	*1609.344*
	Mile (nautical)	0.86897624
	Parsec	5.21552×10^{-14}
	Rod	*320*
	Yard	*1760*
Mile (U.S. Survey)	Meter	1609.3472187
Mile/gallon (Brit.)	Kilometer/liter	0.354006
Mile/gallon (U.S.)	Kilometer/liter	0.425144
Mile/hour	Foot/minute	*88*
	Foot/second	1.466667
	Kilometer/hour	*1.609344*
	Knot	0.8689762
	Meter/minute	*26.8224*
	Meter/second	*0.44704*
Mile/(hour × minute)	Centimeter/square second	0.7450667
Mile/(hour × second)	Centimeter/square second	*44.704*
Mile/minute	Foot/second	*88*
	Kilometer/hour	96.56064
	Knot	52.13857
	Meter/second	*26.8224*
Millibar	Pascal	*100*
Milligram	Carat (metric)	0.005
	Dram	5.6438339×10^{-4}
	Grain	0.015432358
	Ounce (avoirdupois)	3.5273962×10^{-5}
	Ounce (troy)	3.2150747×10^{-5}
	Pennyweight	6.4301493×10^{-4}
	Pound	2.2046226×10^{-6}
	Scruple	7.7161792×10^{-4}
Milligram/assay ton (Brit.)	Milligram/kilogram	30.612245
	Ounce (troy)/ton (long)	*1*
Milligram/assay ton (U.S.)	Milligram/kilogram	34.285714
	Ounce(troy)/ton (short)	*1*
Milligram/kilogram	Gram/ton (metric)	*1*
	Pound/ton (short)	*0.002*
Milligram/liter	Grain/gallon (Brit.)	0.07015689
	Grain/gallon (U.S.)	0.05841783
	Gram/cubic meter	*1*
	Pound/cubic foot	6.242796×10^{-5}
Milligram/cubic meter	Grain/cubic foot	4.369957×10^{-4}
Milligram-force	Dyne	*0.980665*
	Newton	9.80665×10^{-6}
Milligram-force/centimeter	Dyne/centimeter	*0.980665*
	Newton/meter	9.80665×10^{-4}
Milligram-force/inch	Dyne/centimeter	0.386089
	Newton/meter	3.86089×10^{-4}
Milliliter	Cubic centimeter	*1*
Millimeter	Ångström	1×10^{7}
	Inch	0.03937008
	Micrometer	*1000*
Millimeter of Hg (conv.)	Atmosphere	1.315789×10^{-3}
	Dyne/square centimeter	1333.224
	foot of H_2O (conv.)	0.0446033
	Gram-force/square centimeter	*1.35951*
	Millibar	1.333224
	Millimeter of H_2O (conv.)	*13.5951*
	Pascal	133.3224
	Pound-force/square foot	2.78450
	Pound-force/square inch	0.0193368
	Torr	1
Millimeter of H_2O (conv.)	Atmosphere	9.67841×10^{-3}
	Gram-force/square centimeter	*0.1*
	Millibar	*0.0980665*
	Millimeter of Hg (conv.)	0.0735559
	Pascal	*9.80665*
	Pound-force/square inch	1.42233×10^{-3}
Millimicron	Nanometer	*1*
Minim (Brit.)	Drachm (Brit., fluid)	0.01666667 (1/60)
	Milliliter	0.05919388
	Minim (U.S.)	0.9607599
	Ounce (Brit., fluid)	2.083333×10^{-3} (*1/480*)
Minim (U.S.)	Dram (U.S., fluid)	0.01666667 (*1/60*)
	Milliliter	0.06161152
	Minim (Brit.)	1.040843
	Ounce (U.S., fluid)	2.083333×10^{-3} (*1/480*)
Minute	Day	6.944444×10^{-4} (*1/1440*)
	Hour	0.01666667 (*1/60*)
	Second	*60*
	Week	9.920635×10^{-5}
Minute (angular)	Circumference	4.629630×10^{-5}
	Degree (angular)	0.01666667 (*1/60*)
	Gon (grade)	0.01851852 (*1/54*)
	Quadrant	1.851852×10^{-4}
	Radian	2.908882×10^{-4}
	Second (angular)	*60*
Month (mean of 4-year period)	Day	*30.4375*
	Hour	*730.5*
	Minute	*43830*
	Second	2.6298×10^{6}
	Week	4.348214
Nail (Brit).	Inch	*2.25*
Nanometer	Ångström	*10*
	Micrometer	*0.001*
	Mil	3.937008×10^{-5}
Neper	Decibel	8.685890
Newton	Dyne	1×10^{5}
	Kilogram-force	0.1019716
	Poundal	7.23301

To convert from	To	Multiply by
	Pound-force	0.224809
Newton/square centimeter	Newton/square millimeter	*0.01*
	Pascal	*10000*
Newton/square meter	Pascal	*1*
Newton/square millimeter	Kilogram-force/square millimeter	0.1019716
	Megapascal	*1*
	Ton-force (metric)/square meter	101.9716
Newton-meter	Foot-pound-force	0.737562
	Joule	*1*
	Kilogram-force-meter	0.1019716
	Watt-hour	2.777778×10^{-4}
	Watt-second	*1*
Nit	Candela/square meter	*1*
Noggin (Brit.)	Gill (Brit.)	*1*
Nox	Lux	*0.001*
Oersted	Ampere/meter	79.57747
Ohm (int. mean)	Ohm	1.00049
Ohm (int. U.S.)	Ohm	1.000495
Ohm/foot	Ohm/meter	3.280840
Ohm-centimeter	Ohm-meter	*0.01*
Ohm-circular mil/foot	Ohm-meter	1.662426×10^{-9}
Ohm-meter	Ohm-square millimeter/meter	*1×10^{6}*
Ohm-square millimeter/meter	Ohm-meter	*1×10^{-6}*
Ounce (avoirdupois)	Dram	*16*
	Grain	*437.5*
	Gram	28.349523
	Ounce (apot. or troy)	0.91145833
	Pennyweight	18.229167
	Pound	*0.0625 (1/16)*
	Scruple	*21.875*
Ounce (troy or ap.)	Grain	*480*
	Gram	*31.1034768*
	Ounce (avoirdupois)	1.0971429
	Penneyweight	*20*
	Pound (avoirdupois)	0.068571429
	Scruple	*24*
Ounce (Brit. fluid)	Cubic centimeter	28.41306
	Cubic inch	1.733871
	Drachm (Brit., fluid)	*8*
	Dram (U.S., fluid)	7.686079
	Gallon (Brit.)	*6.25×10^{-3} (1/160)*
	Gill (Brit.)	*0.2*
	Milliliter	28.41306
	Minim (Brit.)	*480*
	Ounce (U.S., fluid)	0.9607599
	Pint (Brit.)	*0.05*
	Quart (Brit.)	*0.025 (1/40)*
Ounce (U.S., fluid)	Cubic centimeter	29.57353
	Cubic inch	*1.8046875*
	Dram (U.S., fluid)	*8*
	Gallon (U.S.)	*7.8125×10^{-3} (1/128)*
	Gill (U.S.)	*0.25*
	Milliliter	29.57353
	Minim (U.S.)	*480*
	Ounce (Brit., fluid)	1.040843
	Pint (U.S., liquid)	*0.0625 (1/16)*
	Quart (U.S., liquid)	*0.03125 (1/32)*
Ounce (avoirdupois)/cubic foot	Kilogram/cubic meter	1.001154
Ounce (avoirdupois)/cubic inch	Kilogram/cubic meter	1729.994
Ounce (avoirdupois)/gallon (Brit.)	Kilogram/cubic meter	6.236023
Ounce (avoirdupois)/gallon (U.S.)	Kilogram/cubic meter	7.489152
Ounce (avoirdupois)/square foot	Gram/square meter	305.1517
Ounce (avoirdupois)/square yard	Gram/square meter	33.90575
Ounce (avoirdupois)/ton(long)	Gram/ton (metric)	27.90179

To convert from	To	Multiply by
	Milligram/kilogram	27.90179
Ounce (avoirdupois)/ton(short)	Gram/ton (metric)	*31.25*
	Milligram/kilogram	*31.25*
Ounce (avoirdupois)/yard	Gram/meter	31.00342
Ounce-force (avoirdupois)	Newton	0.2780139
Ounce-force (avoirdupois)/square inch	Pascal	430.922
Ounce-force (avoirdupois)-inch	Newton-meter	7.06155×10^{-3}
Pace	Foot	*2.5*
Palm	Inch	*3*
Parsec	Astronomical unit	2.06265×10^{5}
	Kilometer	3.0857×10^{13}
	Light year	3.26164
	Mile (statute)	1.91735×10^{13}
Part per million	Gram/ton (metric)	*1*
	Milligram/kilogram	*1*
	Milliliter/cubic meter	*1*
	Ounce(avoirdupois)/ton (long)	*0.03584*
	Ounce (avoirdupois)/ton(short)	*0.032*
	Ounce(troy)/ton(long)	0.03266667
	Ounce(troy)/ton(short)	0.02916667
Pascal	Atmosphere	9.869233×10^{-6}
	Bar	*1×10^{-5}*
	Dyne/square centimeter	*10*
	Foot of H_2O (conv.)	3.34552×10^{-4}
	Inch of Hg (conv.)	2.95300×10^{-4}
	Inch of H_2O (conv.)	4.01463×10^{-3}
	Kilogram-force/square centimeter	1.01972×10^{-5}
	Millibar	*0.01*
	Millimeter of Hg (conv.)	7.50062×10^{-3}
	Millimeter of H_2O (conv.)	0.101972
	Newton/square meter	*1*
	Newton/square millimeter	*1×10^{-6}*
	Poundal/square foot	0.671969
	Pound-force/square foot	0.0208854
	Pound-force/square inch	1.45038×10^{-4}
	Torr	7.50062×10^{-3}
Pascal-second	Poise	*10*
Peck (Brit.)	Gallon (Brit.)	*2*
Peck (U.S.)	Bushel (U.S.)	*0.25*
	Quart (U.S., dry)	*8*
Pennyweight	Dram	0.87771429
	Grain	*24*
	Gram	*1.55517384*
	Ounce (avoirdupois)	0.054857143
	Ounce (apoth. or troy)	*0.05*
	Pound	3.4285714×10^{-3}
Perch	Foot	*16.5*
Phot	Lux	*10000*
Pica (printer's)	Point (printer's)	*12*
Picofarad	Farad	*1×10^{-12}*
Pint (Brit.)	Cubic centimeter	*568.26125*
	Cubic inch	34.67743
	Gallon (Brit.)	*0.125 (1/8)*
	Gill (Brit.)	*4*
	Liter	*0.56826125*
	Milliliter	*568.26125*
	Ounce (Brit., fluid)	*20*
	Pint (U.S., dry)	1.032057
	Pint (U.S. liquid)	1.200950
	Quart (Brit.)	*0.5*
Pint (U.S., dry)	Bushel (U.S.)	*0.015625 (1/64)*
	Cubic centimeter	550.6105
	Cubic inch	*33.6003125*
	Liter	0.5506105
	Milliliter	550.6105

To convert from	To	Multiply by
	Peck (U.S.)	*0.0625 (1/16)*
	Pint (Brit.)	0.9689390
	Quart (U.S. dry)	*0.5*
Pint (U.S., liquid)	Cubic centimeter	473.1765
	Cubic inch	*28.875*
	Gallon (U.S.)	*0.125 (1/8)*
	Gill (U.S.)	*4*
	Liter	0.4731765
	Milliliter	473.1765
	Ounce (U.S., fluid)	*16*
	Pint (Brit.)	0.8326742
	Quart (U.S., liquid)	*0.5*
Point (printer's, Didot)	Millimeter	0.3760650
Point (printer's, U.S.)	Inch	*0.013837*
	Millimeter	*0.3514598*
Poise	Dyne-second/square centimeter	*1*
	Gram/(centimeter × second)	*1*
	Pascal-second	*0.1*
Pole (Brit.)	Foot	*16.5*
Pond	Gram-force	*1*
Pottle (Brit.)	Gallon (Brit.)	*0.5*
Pound (avoirdupois)	Dram	*256*
	Grain	*7000*
	Gram	*453.59237*
	Hundredweight (long)	8.9285714×10^{-3}
	Hundredweight (short)	*0.01*
	Kilogram	*0.45359237*
	Ounce (avoirdupois)	*16*
	Ounce (troy)	14.583333
	Pennyweight	291.66667
	Pound (troy)	1.2152778
	Scruple	*350*
	Stone (Brit.)	0.07142857 *(1/14)*
	Ton (long)	4.4642857×10^{-4}
	Ton (metric)	4.5359237×10^{-4}
	Ton (short)	5×10^{-4} *(1/2000)*
Pound (troy)	Dram (troy)	*96*
	Grain	*5760*
	Gram	*373.2417216*
	Ounce (troy)	*12*
	Pennyweight	*240*
	Pound (avoirdupois)	0.82285714
	Scruple	*288*
Pound/acre	Kilogram/hectare	1.120851
Pound/cubic foot	Gram/liter	16.01846
	Kilogram/cubic meter	16.01846
	Pound/cubic inch	5.787037×10^{-4}
Pound/cubic inch	Gram/cubic centimeter	27.679905
	Pound/cubic foot	*1728*
Pound/cubic yard	Kilogram/cubic meter	0.5932764
Pound/foot	Kilogram/meter	1.488164
Pound/(foot × hour)	Pascal-second	4.133789×10^{-4}
Pound/(foot × second)	Pascal-second	1.488164
Pound/gallon (Brit.)	Gram/cubic centimeter	0.09977637
	Gram/liter	99.77637
	Kilogram/cubic meter	99.77637
	Pound/cubic foot	6.228835
	Ton(long)/cubic yard	0.07507968
Pound/gallon (U.S.)	Gram/cubic centimeter	0.1198264
	Gram/liter	119.8264
	Kilogram/cubic meter	119.8264
	Pound/cubic foot	7.480519
	Ton(short)/cubic yard	0.1009870
Pound/hour	Gram/minute	7.559873
	Gram/second	0.1259979
	Kilogram/day	10.88622
Pound/horsepower-hour	Kilogram/megajoule	0.1689659
	Kilogram/kilowatt-hour	0.6082774
Pound/inch	Kilogram/meter	17.85797
Pound/minute	Gram/second	7.559873
	Kilogram/hour	27.21554
Pound/second	Kilogram/hour	1632.932
	Kilogram/minute	27.21554

To convert from	To	Multiply by
Pound/square foot	Kilogram/square meter	4.882428
Poundal	Gram-force	14.0981
	Newton	0.1382550
	Pound-force	0.0310810
Poundal/square foot	Pascal	1.488164
Poundal-foot	Newton-meter	0.0421401
Poundal-second/square foot	Pascal-second	1.488164
Pound-force	Kilogram-force	0.453592
	Newton	4.44822
	Poundal	32.1740
Pound-force/foot	Newton/meter	14.5939
Pound-force/inch	Newton/meter	175.127
Pound-force/square foot	Atmosphere	4.72541×10^{-4}
	Bar	4.78803×10^{-4}
	Foot of H_2O (conv.)	0.0160185
	Gram-force/square centimeter	0.488243
	Inch of Hg (conv.)	0.0141390
	Millimeter of Hg (conv.)	0.359131
	Millimeter of H_2O (conv.)	4.88243
	Pascal	47.8803
	Pound-force/square inch	6.944444×10^{-3} *(1/144)*
Pound-force/square inch	Atmosphere	0.0680460
	Bar	0.0689476
	Foot of H_2O (conv.)	2.30666
	Inch of Hg (conv.)	2.03602
	Kilogram-force/square centimeter	0.0703070
	Meter of H_2O (conv.)	0.703070
	Millibar	68.9476
	Millimeter of Hg (conv.)	51.7149
	Pascal	6894.76
	Pound-force/square foot	*144*
Pound-force-foot	Newton-meter	1.35582
Pound-force-foot/inch	Newton-meter/meter	53.3787
Pound-force-inch	Newton-meter	0.112985
Pound-force-inch/inch	Newton-meter/ meter	4.44822
Pound-force-second/square foot	Pascal-second	47.8803
Pound-force-second/square inch	Pascal-second	6894.76
Psi	Pound-force/square inch	*1*
Puncheon (Brit.)	Gallon (Brit.)	*70*
Quadrant	Degree (angular)	*90*
	Gon (grade)	*100*
	Minute (angular)	*5400*
	Radian	1.570796 (π/2)
Quart (Brit.)	Cubic centimeter	*1136.5225*
	Cubic foot	0.04013591
	Cubic inch	69.35486
	Gallon (Brit.)	*0.25 (1/4)*
	Gill (Brit.)	*8*
	Liter	*1.1365225*
	Ounce (Brit., fluid)	*40*
	Pint (Brit.)	*2*
	Quart (U.S., dry)	1.032057
	Quart (U.S., liquid)	1.200950
Quart (U.S., dry)	Bushel (U.S.)	*0.03125 (1/32)*
	Cubic centimeter	1101.221
	Cubic foot	0.03888925
	Cubic inch	*67.200625*
	Liter	1.101221
	Peck (U.S.)	*0.125 (1/8)*
	Pint (U.S., dry)	*2*
	Quart (U.S., liquid)	1.163647
Quart (U.S., liquid)	Cubic centimeter	946.35295
	Cubic foot	0.03342014
	Cubic inch	*57.75*
	Dram (U.S., fluid)	*256*
	Gallon (U.S.)	*0.25 (1/4)*

CONVERSION FACTORS (continued)

To convert from	To	Multiply by
	Gill (U.S.)	8
	Liter	0.94635295
	Ounce (U.S., fluid)	32
	Pint (U.S., liquid)	2
	Quart (Brit.)	0.8326742
	Quart (U.S., dry)	0.8593670
Quarter (Brit., cap.)	Gallon (Brit.)	64
Quarter (Brit., mass)	Pound	28
Quarter (U.S., long)	Pound	560
Quarter (U.S., short)	Pound	500
Quintal	Kilogram	100
Rad	Gray	0.01
	Joule/kilogram	0.01
Radian	Circumference	0.1591549 (1/2 π)
	Degree (angular)	57.295780
	Gon (grade)	63.66198
	Minute (angular)	3437.747
	Quadrant	0.6366198 (2/π)
	Revolution	0.1591549
	Second (angular)	2.062648×10^{5}
Radian/centimeter	Degree/millimeter	5.729578
	Degree/foot	1746.375
	Degree/inch	145.5313
Radian/second	Revolution/minute	9.549297
Radian/square second	Revolution/square minute	572.9578
Register ton	Cubic foot	100
	Cubic meter	2.831685
Rem	Sievert	0.01
Revolution	Degree (angular)	360
	Gon (Grade)	400
	Radian	6.283185 (2 π)
Revolution/minute	Degree/second	6
Reyn	Pascal-second	6894.76
Rhe	1/pascal-second	10
Right angle	Degree	90
	Gon (grade)	100
Rod	Foot	16.5
Roentgen	Coulomb/kilogram	2.58×10^{-4}
Rood (Brit.)	Acre	0.25 (1/4)
	Square meter	1011.7141
Rope (Brit.)	Foot	20
Scruple	Dram (apoth. or troy)	0.3333333 (1/3)
	Grain	20
	Gram	1.2959782
	Ounce (avoirdupois)	0.045714286
	Ounce (apoth. or troy)	0.04166667 (1/24)
	Pennyweight	0.83333333 (10/12)
	Pound	2.857143×10^{-3} (1/350)
Scruple (Brit. fluid)	Minim (Brit.)	20
Seam (Brit.)	Gallon (Brit.)	64
Second (angular)	Degree	2.777778×10^{-4} (1/3600)
	Gon (grade)	3.086420×10^{-4} (1/3240)
	Minute (angular)	0.01666667 (1/60)
	Radian	4.848137×10^{-6}
Shake	Second	1×10^{-8}
Siemens	Mho (ohm^{-1})	1
Slug	Geepound	1
	Kilogram	14.5939
	Pound	32.1740
Slug/cubic foot	Kilogram/cubic meter	515.379
Slug/(foot × second)	Pascal-second	47.8803
Span	Inch	9
Sphere	Steradian	12.56637 (4 π)
Square centimeter	Circular mil	1.973525×10^{5}
	Circular millimeter	127.3240
	Square foot	1.076391×10^{-3}
	Square inch	0.1550003
	Square meter	1×10^{-4}
	Square millimeter	100
	Square yard	1.195990×10^{-4}
Square chain(Gunter's)	Acre	0.1

To convert from	To	Multiply by
	Square foot	4356
	Square meter	404.6856
Square chain (Ramsden's)	Square foot	10000
Square chain(U.S. Survey)	Square meter	404.687261
Square degree	Steradian	3.046174×10^{-4}
Square foot	Acre	2.295684×10^{-5}
	Square centimeter	929.0304
	Square chain (Gunter's)	2.295684×10^{-4}
	Square chain (Ramsden's)	1×10^{-4}
	Square inch	144
	Square link (Gunter's)	2.295684
	Square meter	0.09290304
	Square mile	3.587006×10^{-8}
	Square rod	3.673095×10^{-3}
	Square yard	0.1111111 (1/9)
Square foot (U.S. Survey)	Square meter	0.092903412
Square foot/hour	Square meter/second	2.58064×10^{-5}
Square inch	Circular mil	1.273240×10^{6}
	Circular millimeter	821.4432
	Square centimeter	6.4516
	Square foot	6.944444×10^{-3} (1/144)
	Square millimeter	645.16
Square inch/second	Square foot/minute	0.4166667
	Square meter/hour	2.322576
Square kilometer	Acre	247.1054
	Hectare	100
	Square foot	1.076391×10^{7}
	Square meter	1×10^{6}
	Square mile	0.38610216
	Square yard	1.195990×10^{6}
Square link(Gunter's)	Square foot	0.4356
Square link(Ramsden's)	Square foot	1
Square meter	Acre	2.471054×10^{-4}
	Are	0.01
	Hectare	1×10^{-4}
	Square centimeter	10000
	Square chain (Gunter's)	2.471054×10^{-3}
	Square foot	10.76391
	Square inch	1550.003
	Square kilometer	1×10^{-6}
	Square link (Gunter's)	24.71054
	Square mile	3.861022×10^{-7}
	Square yard	1.195990
Square mil	Circular mil	1.273240
	Square inch	1×10^{-6}
	Square micrometer	645.16
	Square millimeter	6.4516×10^{-4}
Square mile	Acre	640
	Square chain (Gunter's)	6400
	Square foot	2.78784×10^{7}
	Square kilometer	2.589988110
	Square meter	2.589988×10^{6}
	Square rod	1.024×10^{5}
	Square yard	3.0976×10^{6}
	Township	0.02777778 (1/36)
Square mile (U.S. Survey)	Square kilometer	2.589998470
Square millimeter	Circular mil	1973.525
	Circular millimeter	1.273240
	Square centimeter	0.01
	Square inch	1.550003×10^{-3}
	Square mil	1550.003
Square rod	Acre	0.00625 (1/160)
	Square foot	272.25
	Square meter	25.29285
Square yard	Acre	2.066116×10^{-4}
	Square foot	9
	Square inch	1296
	Square meter	0.83612736
	Square mile	3.228306×10^{-7}

To convert from	To	Multiply by
Standard (Petrograd)	Cubic foot	*165*
Statampere	Ampere	3.335641×10^{-10}
Statcoulomb	Coulomb	3.335641×10^{-10}
Statfarad	Farad	1.112650×10^{-12}
Stathenry	Henry	8.987552×10^{11}
Statmho	Siemens	1.112650×10^{-12}
Statohm	Ohm	8.987552×10^{11}
Statvolt	Volt	299.7925
Steradian	Sphere	0.07957747 (1/4 π)
	Spherical right angle	0.6366198 (2/π)
	Square degree	3282.806
Stere	Cubic meter	*1*
Stilb	Candela/square centimeter	*1*
Stokes	Square meter/second	1×10^{-4}
Stone	Pound	*14*
Tablespoon (metric)	Milliliter	*15*
Tablespoon (U.S.)	Milliliter	14.79
Teaspoon (metric)	Milliliter	*5*
Teaspoon (U.S.)	Milliliter	4.93
Terawatt-hour	Kilowatt-hour	1×10^{9}
Tesla	Weber/square meter	*1*
Tex	Denier	9
	Gram/kilometer	*1*
Therm	Btu	1×10^{5}
Thou	Mil	*1*
Ton (assay, (Brit.)	Gram	32.66667
Ton (assay, U.S.)	Gram	29.16667
Ton (long)	Hundredweight (long)	*20*
	Hundredweight (short)	*22.4*
	Kilogram	*1016.0469088*
	Pound	*2240*
	Ton (metric)	1.016047
	Ton (short)	*1.12*
Ton (metric)	Hundredweight (long)	19.684131
	Hundredweight (short)	22.046226
	Kilogram	*1000*
	Pound	2204.6226
	Ton (long)	0.98420653
	Ton (short)	1.1023113
Ton (short)	Hundredweight (long)	17.857143
	Hundredweight (short)	*20*
	Kilogram	*907.18474*
	Pound	*2000*
	Ton (long)	0.89285714
	Ton (metric)	*0.90718474*
Ton(long)/cubic yard	Kilogram/cubic meter	1328.939
Ton (metric)/cubic meter	Gram/cubic centimeter	*1*
	Kilogram/cubic decimeter	*1*
Ton(short)/cubic yard	Kilogram/cubic meter	1186.553
Ton-force (long)	Newton	9964.02
Ton-force (metric)	Newton	*9806.65*
Ton-force (short)	Newton	8896.44
Ton-force(long)/square foot	Atmosphere	1.05849
	Bar	1.07252
	Kilogram-force/square centimeter	1.09366
	Newton/square millimeter	0.107252
	Pascal	1.07252×10^{5}
	Pound-force/square inch	15.5556
Ton-force(long)/square inch	Atmosphere	152.423
	Bar	154.443
	Kilogram-force/ square centimeter	157.488
	Newton/square millimeter	15.4443
	Pascal	1.54443×10^{7}
	Pound-force/square inch	*2240*
Ton-force(metric)/ square meter	Atmosphere	0.0967841
	Bar	*0.0980665*
	Kilogram-force/square centimeter	*0.1*
	Newton/square millimeter	9.80665×10^{-3}
	Pascal	*9806.65*
	Pound-force/square inch	1.42233
Ton-force (short)/square foot	Atmosphere	0.945083
	Bar	0.957605
	Kilogram-force/square centimeter	0.976486
	Newton/square millimeter	0.0957605
	Pascal	9.57605×10^{4}
	Pound-force/square inch	13.8889
Ton-force (short)/square inch	Atmosphere	136.092
	Bar	137.895
	Kilogram force/square centimeter	140.614
	Newton/square millimeter	13.7895
	Pascal	1.37895×10^{7}
	Pound-force/square inch	*2000*
Tonne	Kilogram	*1000*
Torr	Millibar	1.333224
	Millimeter of Hg (conv.)	1
	Pascal	133.3224
Township (U.S.)	Square kilometer	93.23957
	Square mile	*36*
Unit pole	Weber	1.256637×10^{-7}
Volt (int. mean)	Volt	1.00034
Volt (int. U.S.)	Volt	1.000330
Volt/inch	Volt/meter	39.37008
Volt-second	Weber	*1*
Watt	Btu/hour	3.41214
	Btu/minute	0.0568690
	Calorie/minute	14.3308
	Calorie/second	0.238846
	Erg/second	1×10^{7}
	Foot-pound-force/ minute	44.2537
	Foot-pound-force/ second	0.737562
	Horsepower	1.34102×10^{-3}
	Horsepower (metric)	1.35962×10^{-3}
	Joule/second	*1*
	Kilocalorie/hour	0.859845
	Kilogram-force-meter/ second	0.101972
Watt (int. mean)	Watt	1.00019
Watt (int. U.S.)	Watt	1.000165
Watt/square inch	Btu/(hour × square foot)	491.348
	Kilocalorie/(hour × square meter)	1332.76
	Watt/square meter	1550.003
Watt/square meter	Kilocalorie/(hour × square meter)	0.859845
Watt-hour	Btu	3.41214
	Calorie	859.845
	Foot-pound-force	2655.22
	Horsepower-hour	1.34102×10^{-3}
	Horsepower-hour (metric)	1.35962×10^{-3}
	Joule	*3600*
	Kilogram-force-meter	367.098
	Liter-atmosphere	35.5292
Watt-second	Erg	1×10^{7}
	Joule	*1*
	Newton-meter	*1*
Weber	Maxwell	1×10^{8}
Weber/square meter	Gauss	*10000*

To convert from	To	Multiply by
Week	Day	*7*
	Hour	*168*
	Minute	*10080*
	Month	0.2299795
	Second	*6.048 × 10⁵*
X-unit	Meter	1.00202 × 10^{-13}
Yard	Centimeter	*91.44*
	Fathom	*0.5*
	Foot	*3*
	Inch	*36*
	Meter	*0.9144*
	Mile	5.681818 × 10^{-4}
Year (calendar, mean of 4-year period)	Day	*365.25*
	Hour	*8766*
	Minute	*5.2596 × 10^5*
	Second	*3.15576 × 10^7*
	Week	52.17857
Year (leap)	Day	*366*
Year (normal calendar)	Day	*365*
	Hour	*8760*
	Minute	*5.256 × 10^5*
	Second	*3.1536 × 10^7*
	Week	52.14286
Year (sidereal)	Day	365.25636
	Second	3.155815 × 10^7
	Year (tropical)	1.0000388
Year (tropical)	Day	365.24220
	Second	3.1556926 × 10^7
	Year (sidereal)	0.9999612

DEFINED VALUES AND EQUIVALENTS

Meter	**(m)**	1 650 763.73 wave lengths in vacuo of the unperturbed transition $2p_{10} - 5d_5$ in ^{86}Kr
Kilogram	**(kg)**	mass of the international kilogram at Sèvres, France
Second	**(s)**	1/31 556 925.974 7 of the tropical year at 12^h ET, 0 January 1900
Degree Kelvin	**(°K)**	defined in the thermodynamic scale by assigning 273.16 °K to the triple point of water (freezing point, 273.15 °K = 0 °C)
Unified atomic mass unit	(u)	1/12 the mass of an atom of the ^{12}C nuclide
Mole	**(mol)**	amount of substance containing the same number of atoms as 12 g of pure ^{12}C
Standard acceleration of free fall	**(g_n)**	9.806 65 m s^{-2}, 980.665 cm s^{-2}
Normal atmospheric pressure	**(atm)**	101 325 N m^{-2}, 1 013 250 dyn cm^{-2}
Thermochemical calorie	**(cal_{th})**	4.1840 J, 4.1840 × 10^7 erg
International Steam Table calorie	**(cal_{IT})**	4.1868 J, 4.1868 × 10^7 erg
Liter	**(l)**	0.001 m^3, 1000 cm^3 (recommended by GCWM, 1964)
Inch	**(in)**	0.0254 m, 2.54 cm
Pound (avdp)	**(lb)**	0.453 592 37 kg, 453.592 37 g

FACTORS FOR THE CONVERSION OF ($LOG_{10}X$) TO (RT LOG_eX)

Units are in Calories

t°C	0	1	2	3	4	5	6	7	8	9
0	1249.4	1254.0	1258.6	1263.2	1267.7	1272.3	1276.9	1281.45	1286.0	1290.6
10	1295.2	1299.8	1304.3	1308.9	1313.5	1318.0	1322.6	1327.2	1331.8	1336.3
20	1340.9	1345.5	1350.1	1354.6	1359.2	1363.8	1368.4	1372.9	1377.5	1382.1
30	1386.7	1391.2	1395.8	1400.4	1405.0	1409.5	1414.1	1418.7	1423.2	1427.8
40	1432.4	1437.0	1441.5	1446.1	1450.7	1455.3	1459.8	1464.4	1469.0	1473.6
50	1478.1	1482.7	1487.3	1491.9	1496.4	1501.0	1505.6	1510.2	1514.7	1519.3
60	1523.9	1528.5	1533.0	1537.6	1542.2	1546.7	1551.3	1555.9	1560.5	1565.0
70	1569.6	1574.2	1578.8	1583.3	1587.9	1592.5	1597.1	1601.6	1606.2	1610.8
80	1615.4	1619.9	1624.5	1629.1	1633.7	1638.2	1642.8	1647.4	1651.9	1656.5
90	1661.1	1665.7	1670.2	1674.8	1679.4	1684.0	1688.5	1693.1	1697.7	1702.3
100	1706.8	1711.4	1716.0	1720.6	1725.1	1729.7	1734.3	1738.9	1743.4	1748.0
110	1752.6	1757.2	1761.7	1766.3	1770.9	1775.4	1780.0	1784.6	1789.2	1793.7
120	1798.3	1802.9	1807.5	1812.0	1816.6	1821.2	1825.8	1830.3	1834.9	1839.5
130	1844.1	1848.6	1853.2	1857.8	1862.4	1866.9	1871.5	1876.1	1880.6	1885.2
140	1889.8	1894.4	1898.9	1903.5	1908.1	1912.7	1917.2	1921.8	1926.4	1931.0
150	1935.5	1940.1	1944.7	1949.3	1953.8	1958.4	1963.0	1967.6	1972.1	1976.7
160	1981.3	1985.9	1990.4	1995.0	1999.6	2004.1	2008.7	2013.3	2017.9	2022.4
170	2027.0	2031.6	2036.2	2040.7	2045.3	2049.9	2054.5	2059.0	2063.6	2068.2
180	2072.8	2077.3	2081.9	2086.5	2091.1	2095.6	2100.2	2104.8	2109.4	2113.9
190	2118.5	2123.1	2127.6	2132.2	2136.8	2141.4	2145.9	2150.5	2155.1	2159.7
200	2164.2	2168.8	2173.4	2178.0	2182.5	2187.1	2191.7	2196.3	2200.8	2205.4

Differences

	Tenths	Units	Tens
1	.5	4.6	45.7
2	.9	9.1	91.5
3	1.4	13.7	137.2
4	1.8	18.3	183.0
5	2.3	22.9	228.7
6	2.7	27.4	274.4
7	3.2	32.0	320.2
8	3.7	36.6	365.9
9	4.1	41.2	411.7

Equivalent second order rate constants

A \ B	$cm^3\ mol^{-1}\ s^{-1}$	$dm^3\ mol^{-1}\ s^{-1}$	$m^3\ mol^{-1}\ s^{-1}$	$cm^3\ molecule^{-1}\ s^{-1}$	$(mm\ Hg)^{-1}\ s^{-1}$	$atm^{-1}\ s^{-1}$	$ppm^{-1}\ min^{-1}$	$m^2\ kN^{-1}\ s^{-1}$
$1\ cm^3\ mol^{-1}\ s^{-1}=$	1	10^{-3}	10^{-6}	1.66×10^{-24}	$1.604 \times 10^{-5}\ T^{-1}$	$1.219 \times 10^{-2}\ T^{-1}$	2.453×10^{-9}	$1.203 \times 10^{-4}\ T^{-1}$
$1\ dm^3\ mol^{-1}\ s^{-1}=$	10^3	1	10^{-3}	1.66×10^{-21}	$1.604 \times 10^{-2}\ T^{-1}$	$12.19\ T^{-1}$	2.453×10^{-6}	$1.203 \times 10^{-1}T^{-1}$
$1\ m^3\ mol^{-1}\ s^{-1}=$	10^6	10^3	1	1.66×10^{-18}	$16.04\ T^{-1}$	$1.219 \times 10^4\ T^{-1}$	2.453×10^{-3}	$120.3\ T^{-1}$
$1\ cm^3\ molecule^{-1}\ s^{-1}=$	6.023×10^{23}	6.023×10^{20}	6.023×10^{17}	1	$9.658 \times 10^{18}\ T^{-1}$	$7.34 \times 10^{21}\ T^{-1}$	1.478×10^{15}	$7.244 \times 10^{19}\ T^{-1}$
$1\ (mm\ Hg)^{-1}\ s^{-1}=$	$6.236 \times 10^4\ T$	$62.36\ T$	$6.236 \times 10^{-2}\ T$	$1.035 \times 10^{-19}\ T$	1	760	4.56×10^{-2}	7.500
$1\ atm^{-1}\ s^{-1}$	$82.06\ T$	$8.206 \times 10^{-2}\ T$	$8.206 \times 10^{-5}\ T$	$1.362 \times 10^{-22}\ T$	1.316×10^{-3}	1	6×10^{-5}	9.869×10^{-3}
$1\ ppm^{-1}\ min^{-1}=$ at 298 K, 1 atm total pressure	4.077×10^8	4.077×10^5	407.7	6.76×10^{-16}	21.93	1.667×10^4	1	164.5
$1\ m^2\ kN^{-1}\ s^{-1}=$	$8314\ T$	$8.314\ T$	$8.314 \times 10^{-3}\ T$	$1.38 \times 10^{-20}\ T$	0.1333	101.325	6.079×10^{-3}	1

To convert a rate constant from one set of units A to a new set B find the conversion factor for the row A under column B and multiply the old value by it, e.g. to convert $cm^3\ molecule^{-1}\ s^{-1}$ to $m^3\ mol^{-1}\ s^{-1}$ multiply by 6.023×10^{17}.
Table adapted from High Temperature Reaction Rate Data No. 5, The University, Leeds (1970).

Equivalent third order rate constants

A \ B	$cm^6\ mol^{-2}\ s^{-1}$	$dm^6\ mol^{-2}\ s^{-1}$	$m^6\ mol^{-2}\ s^{-1}$	$cm^6\ molecule^{-2}\ s^{-1}$	$(mm\ Hg)^{-2}\ s^{-1}$	$atm^{-2}\ s^{-1}$	$ppm^{-2}\ min^{-1}$	$m^4\ kN^{-2}\ s^{-1}$
$1\ cm^6\ mol^{-2}\ s^{-1}=$	1	10^{-6}	10^{-12}	2.76×10^{-48}	$2.57 \times 10^{-10}\ T^{-2}$	$1.48 \times 10^{-4}\ T^{-2}$	1.003×10^{-19}	$1.447 \times 10^{-8}\ T^{-2}$
$1\ dm^6\ mol^{-2}\ s^{-1}=$	10^6	1	10^{-6}	2.76×10^{-42}	$2.57 \times 10^{-4}\ T^{-2}$	$148\ T^{-2}$	1.003×10^{-13}	$1.447 \times 10^{-2}\ T^{-2}$
$1\ m^6\ mol^{-2}\ s^{-1}=$	10^{12}	10^6	1	2.76×10^{-36}	$257\ T^{-2}$	$1.48 \times 10^8\ T^{-2}$	1.003×10^{-7}	$1.447 \times 10^4\ T^{-2}$
$1\ cm^6 molecule^{-2}\ s^{-1}=$	3.628×10^{47}	3.628×10^{41}	3.628×10^{35}	1	$9.328 \times 10^{37}\ T^{-2}$	$5.388 \times 10^{43}\ T^{-2}$	3.64×10^{28}	$5.248 \times 10^{39}\ T^{-2}$
$1\ (mm\ Hg)^{-2}\ s^{-1}=$	$3.89 \times 10^9\ T^2$	$3.89 \times 10^3\ T^2$	$3.89 \times 10^{-3}\ T^2$	$1.07 \times 10^{-38}\ T^2$	1	5.776×10^5	3.46×10^{-5}	56.25
$1\ atm^{-2}\ s^{-1}=$	$6.733 \times 10^3\ T^2$	$6.733 \times 10^{-3}\ T^2$	$6.733 \times 10^{-9}\ T^2$	$1.86 \times 10^{-44}\ T^2$	1.73×10^{-6}	1	6×10^{-11}	9.74×10^{-5}
$1\ ppm^{-2}\ min^{-1}=$ at 298 K, 1 atm total pressure	9.97×10^{18}	9.97×10^{12}	9.97×10^6	2.75×10^{-29}	2.89×10^4	1.667×10^{10}	1	1.623×10^6
$1\ m^4\ kN^{-2}\ s^{-1}=$	$6.91 \times 10^7\ T^2$	$6.91\ T^2$	$69.1 \times 10^{-5}\ T^2$	$1.904 \times 10^{-40}\ T^2$	0.0178	1.027×10^4	6.16×10^{-7}	1

From *J. Phys. Chem. Ref. Data*, 9, 470, 1980, by permission of the authors and the copyright owner, the American Institute of Physics.

DIMENSIONLESS GROUPS

John P. Catchpole and George Fulford

 See also the Supplementary Table which follows this table.

Dimensionless groups are frequently generated in the analysis of a complex engineering problem. The more common groups thus generated are easily recognized, while the less common ones are not. Unless the less common existing groups are recognized, an already named group could unknowingly be renamed. Table A provides a tool that may be used to avoid this occurrence, by listing the groups by the variables of which they consist. These variables—i.e., length, density, diffusivity, viscosity, etc.—are further subdivided into their exponents to which they are raised in the groups in question. Thus, Reynolds number is listed under the exponent +1 for the variables, length, fluid velocity, and density, and the exponent −1 for viscosity.

To illustrate the use of the tables in the analysis of a problem, the group $(kE/\eta\sigma T^3)$ might be generated in the solution of a complex heat transfer problem. From Table A the groups containing the constituent variables are checked and the groups are listed:

Thermal conductivity	(k^{+1})	**F11**, **L7**, **R1**
Modulus of elasticity	(E^{+1})	**C1**, **E13**, **R1**
Stefan-Boltzmann coefficient	(η^{-1})	**R1**, **T6**
Surface tension	(σ^{-1})	**B9**, **C3**, **E8**, **L6**, **R1**, **W1**
Temperature	(T^{-3})	**R1**, **T6**

It is immediately apparent that the only group common to all the categories listed is the Radiation number, **R1**, which is equivalent to the previously unidentified group.

The symbol assigned to a dimensionless group is usually the first two letters of its names. Several groups, however, have nonstandard symbols, particularly in the groups which are named after persons. These symbols are listed in the nomenclature.

NOMENCLATURE

a	= annulus or clearance width, L
A	= area, L^2
A^*	= cooling area/unit volume, $1/L$
b	= bearing breadth, L
B	= groups **B6**, **B11**
c	= specific heat, $L^2/\theta^2 T$
c_A	= concentration, M/L^3
c_b	= specific vapor capacity (mass/unit mass/unit pressure change), $L\theta^2/M$
c_d, c_D	= group **D13**
c_f	= group **R9**
c_H	= group **H4**
c_m	= mass capacity, L^3/M
c_p, c_v	= specific heats at constant pressure and volume, $L^2/\theta^2 T$
c_q	= heat capacity, $L^2/\theta^2 T$
c_Q	= group **F9**
c_S	= group **S11**
C	= group **C10**, dimensional concentration, M/L^3
C_a	= groups **C11**, **C4**
d	= diameter, L
d_e	= equivalent diameter (of particles, etc.), L
d_h	= hydraulic diameter, L
D	= diffusivity (molecular, unless noted otherwise), L^2/θ
D_{AB}	= binary bulk diffusion coefficient, L^2/θ
D_{KA}	= Knudsen diffusion coefficient, L^2/θ
e	= voidage; porosity $(^-)$
e^*	= surface emissivity $(^-)$
E	= modulus of elasticity, $M/L\theta^2$
E_a	= activation energy, L^2/θ^2
E_b	= bulk modulus, $M/L\theta^2$
f	= frequency, $1/\theta$, or Group **F1**
$f(M)$	= group **F11**
F	= force, ML/θ^2
F_b	= force per unit length of bearing, M/θ^2
$F(M)$	= group **F6**
F_R	= resistance force in flow, ML/θ^2
g	= acceleration due to gravity, L/θ^2
G	= mass velocity (mass flux density; mass transfer coefficient), $M/\theta L^2$
h	= heat transfer coefficient, $M/T\theta^3$
h_c	= convective heat transfer coefficient, $M/T\theta^3$
H	= energy change per unit mass (= g × head), L^2/θ^2
H'	= fluid head, L
H_e	= field strength, $Q/L\theta$
H_0	= homochronicity number
$I(M)$	= group **F8**
j	= heat liberated per unit volume per unit time, $M/L\theta^3$
j_H, j_M	= groups **J2**, **J3**
J	= average free path/average velocity, θ, or group **L6**
k	= thermal conductivity, $ML/T\theta^3$
k_c	= mass transfer coefficient, L/θ
K	= groups **K2**, **K10**, **N5**
K_1	= group **A4**
$\overline{K}_F$	= group **E4**
K_F	= group **C2**
K_P	= group **P13**
K_Q	= group **H5**
K_r	= group **E11**
$\overline{K}_g$	= group **A1**
$\overline{K}_E$	= group **E4**
$\overline{K}_r$	= group **E12**
$\overline{K}_{rE}$	= group **E13**
$\overline{K}_s$	= group **R1**
$\overline{K}_\sigma$	= group **C1**
$\overline{K}_{\sigma g}$	= group **C2**
L	= characteristic dimension (except as noted), L
L_m	= distance from midpoint to surface, L
m_T	= group **T3**
M	= group **M10**
M_H	= group **H2**
n	= concentration, wt./wt. $(^-)$
n^*	= specific mass content, mass/mass $(^-)$
n_m	= moisture content, wt./wt. bone dry gas $(^-)$
N	= rate of rotation, $1/\theta$, and groups **M3**, **N4**
N_{B_o}	= groups **B7**, **B8**
N_c	= group **C5**
N_{cv}	= group **C10**
N_D	= groups **D7**, **D14**
N_E	= group **E1**
N_F	= group **F4**
N_H	= groups **H1**, **H9**
N_K	= group **K1**
N_{KnA}	= Knudsen number for diffusion (see Addendum)
N_1	= group **N2**
N_P	= group **P7**
N_{rf}	= group **R7**
N_{s1}	= group **S8**
N_{s2}	= group **S9**
N_T	= group **N9**
P	= pressure, $M/L\theta^2$
P	= plasticity number (see Addendum)
p_b	= bearing pressure, $M/L\theta^2$
p_s	= static pressure, $M/L\theta^2$
p_v	= vapor pressure, $M/L\theta^2$
p_σ	= capillary pressure, $M/L\theta^2$
Δp_F	= frictional pressure drop, $M/L\theta^2$
q	= heat flux (heat flow/unit time), ML^2/θ^3
q^*	= heat flux density (heat flux/unit area), M/θ^3
Q	= heat liberated/unit mass, L^2/θ^2
r	= latent heat of phase change, L^2/θ^2
r_v	= heat of vaporization, L^2/θ^2
R	= radius, L
R_H	= hydraulic radius, L
R_2'	= group **R5**
R_M	= group **M7**
R_V	= group **V3**
$\mathcal{R}$	= gas constant, $L^2/\theta^2 T$
s	= humid heat, $L^2/\theta^2 T$
$\underline{S}$	= particle area/particle volume, L^2/L^3, and group **M6**
St	= group **S14**
t	= temperature, T
T	= absolute temperature, T
Δt, ΔT	= temperature difference, T
U^+	= group **P10**
U	= reaction rate, $M/L^3\theta$
v_s	= velocity of surface (solid), L/θ
V	= fluid velocity, L/θ, and group **V1**
V_A	= velocity of Alfven magnetic waves, L/θ
V_f	= volumetric flow rate, L^3/θ
V_l	= velocity of light, L/θ
V_m	= mass flow rate, M/θ
V_s	= velocity of sound, L/θ
w	= circumferential velocity, L/θ
W	= volume of system, L^3
W^*	= gross volume, L^3
x	= entry length; distance from entrance, L
y^+	= group **P11**
Z	= group **O2**
α	= thermal diffusivity (temperature conductivity), L^2/θ

β = coefficient of bulk expansion, $1/T$, and group **D12**
β^* = Dufour coefficient, T
γ = specific gravity (¯) and group **R3**
$\dot{\gamma}$ = rate of shear, $1/\theta$
Γ = rate of change of temperature of medium, T/θ
δ = Soret or thermogradient coefficient, $1/T$, and group **D11**
Δ = difference in quantity
ε = height of roughness, L and group **A3**
ε_D = eddy mass diffusivity, L^2/θ
ζ = diffusion tortuosity (¯)
η = radiation coefficient (Stefan-Boltzmann coefficient), $M/T^4\theta^3$
θ = time, θ
θ_r = relaxation time, θ
λ = mean free path, L
μ = dynamic viscosity, $M/L\theta$
μ_e = magnetic permeability, ML/Q^2
μ_p = rigidity coefficient, $M/L\theta$
ξ = permeability, L^2
π = 3.1416. . . .
Π = power to agitator or impeller, ML^2/θ^3
ρ = density, M/L^3
ρ_n = group **P3**
σ = surface tension, M/θ^2 and group **S12**
σ_c = group **C6**
σ_e = electrical conductivity, $Q^2\theta/L^3M$
σ_t = group **T4**
τ = group **T8**
τ_w = wall shear stress, $M/L\theta^2$
τ_y = yield stress, $M/L\theta^2$
φ = group **D9**
ψ = groups **N8, P14, R10**
ω = angular velocity (of fluid, unless noted otherwise), $1/\theta$
Ω = mass transfer potential (concn.), M/L^3
— (bar over) = mean value
N.B.: $(F) = \left(\frac{ML}{\theta^2}\right)$; $(H) = \left(\frac{ML^2}{\theta^2}\right)$

TABLES FOR IDENTIFYING DIMENSIONLESS GROUPS
TABLE A

PHYSICAL PROPERTIES
General Physical Properties

Parameter	Symbol	Dimensions	Exponent	Group
Coefficient of bulk expansion	β	$1/T$	−1	**E4, 12, G2**
			+1	**G5, R5, K9, 6**
Density	ρ	M/L^3	−2	**M1**
			−1	**A1, B1, 10, C1, 2, 6, D3, 13, E3, 9, 10, 13, F1, 11, H5, J1, K4, K10, L7, M3, 6, N2, 3, 4, P7, 9, R10, 13, S4, 13, 15**
			$-\frac{2}{3}$	**C9, J3**
			$-\frac{1}{2}$	**E2, L11, O2, P13**
			$+\frac{1}{3}$	**J3, K5, 53**
			$+\frac{1}{2}$	**D10, G3, P10, 11, R4, W2**
			$+\frac{2}{3}$	**F2, N8, S2**
			+1	**A5, B1, B6, 9, C5, 7, D5, 7, 13, E4, 6, 7, 8, H6, 11, J1, 3, 4, K1, L9, P1, R5, 11, 13, S17, T1, T6, V1, W1, W3**
			+2	**C10, G1, 5, K9, R5, 6, T2**
Density gradient	$d\rho/dL$	M/L^4	+1	**R13**
Diffusivity (molecular unless noted otherwise)	D, a_m, ε_D	L^2/θ	−1	**B5, B7**[(4)]**, D2, K4**[(1)]**, K7**[(2)]**, L2, 7, 10, N7, P2, P9, S4, 13**
			$-\frac{2}{3}$	**J3**
			$-\frac{1}{2}$	**T3**
			+1	**D12, F12**[(1)]**, K7**[(3)]**, L2**[(4)]**, L9**[(1)]
Diffusivity (surface)	D_s	ML/θ^2	+1	**S18**
Diffusion tortuosity	ζ	—	−1	**K7**
Molecular weight	M	—	−1	**D14, K9, S6**
			+1	**D14**
Permeability (packed bed)	ξ	L^2	$+\frac{1}{2}$	**L_6**
Porosity (voidage)	e	—	−1	**B4**
			$-\frac{1}{2}$	**L_6**
			+1	**K7**
Specific weight	γ	—	$\pm\frac{1}{3}$	**F2**
Surface tension	σ	M/θ^2	−3	**C2**
			−2	**C1**
			−1	**B9, C3, 13, E8, L6, R1, W1, 3**
			$-\frac{1}{2}$	**D10, G3, O2, P13, R4, W2**
			+1	**M9, S17, 18**

(1) Coefficient of potential diffusion in mass transfer.
(2) Knudsen diffusion coefficient.
(3) Binary bulk diffusion coefficient.
(4) Effective diffusivity ($D + \varepsilon_D$) (molecular + eddy transfer).

PHYSICAL PROPERTIES
Electrical and Magnetic Properties

Parameter	Symbol	Dimensions	Exponent	Group
Current density	I	$Q/L^2\theta$	+1	**K10**
Electrical conductivity	σ_e	$Q^2\theta/L^3M$	−1	**E6**
			$+\frac{1}{2}$	**H2**
			+1	**L11, M3, 7**
Field strength	H_e	$Q/L\theta$	−2	**J4**
			+1	**H2, L11**
			+2	**K9, M3, 6, S6**
Magnetic permeability	μ_e	ML/Q^2	−1	**E5, J4**
			+1	**H2, M6, 7**
			$+\frac{3}{2}$	**L11**
			+2	**M3**
Voltage	E	$ML^2/Q\theta^2$	+1	**K10**

Thermal Properties

Parameter	Symbol	Dimensions	Exponent	Group
Humid heat	S	L^2/θ^2T	−1	**P15**
Latent heats of phase change	λ, r	L^2/θ^2	−1	**A3, E11, 12, 13, J1, K10, M1**
			+1	**B13, C9, K8, 11, N5**
Ratio of specific heats	γ	—	$\pm\frac{1}{2}$	**L5**
Specific heat	C, c	L^2/θ^2T	−1	**B2, 13, D3, 15, E1, F11, J2, K8, 11, L7, M10, N5, R3, S13**
			$-\frac{1}{3}$	**J2**
			$+\frac{1}{2}$	**F6, 7, 8**
			$+\frac{2}{3}$	**J2**
			+1	**A3, E4, 12, G4, J1, 4, L9, P1, 8, R3, 5, 6, 7**
Surface emissivity	e^*	—	−1	**T6**
Temperature conductivity (thermal diffusivity)	α	L^2/θ	−1	**L9, P1, 12, R5**
			+1	**C13, L7, 10**
Thermal conductivity	k or λ	$ML/T\theta^3$	−3	**M1**
			−2	**R6**
			−1	**B4, 12, C7, 9, 10, D4, G4, K3, L9, N6, P1, 5, 8, R5, S14**

Rheological and Elastic Behavior

Parameter	Symbol	Dimensions	Exponent	Group
Modulus of elasticity	E	$M/L\theta^2$	−1	**C5, E4, H11**
			+1	**C1, E13, R1**
			+3	**A1**
Rate of shear	$\dot{\gamma}$	$1/\theta$	+1	**T8**
Rigidity coefficient	μ_p	$M/L\theta$	−2	**H6**
			−1	**B3**
			−1	**E5**
Shear stress	τ	$M/L\theta^2$	$-\frac{1}{2}$	**P10**
			$+\frac{1}{2}$	**P11**
Viscosity (in all cases kinematic viscosity has been written as μ/ρ)	μ	$M/L\theta$	−2	**A1, 5, G1, 5, K1, 9, S17, T2**
			−1	**B6, C10, D5, 7, E6, 7, H8, L1, 3, 9, O1, P4, 11, R5, 6, 11, S9, 19, T1, V1**

TABLE A (Continued)

Rheological and Elastic Behavior

Parameter	Symbol	Dimensions	Exponent	Group
			$-\frac{2}{3}$	**F2, K5, N8**
			$-\frac{1}{2}$	**H2**
			$-\frac{1}{3}$	**S2, 3**
			$+\frac{1}{2}$	**E2**
			$+\frac{2}{3}$	**C9, J2, 3**
			+1	**B12, C3, 13, E3, 5, M1, O2, P8, 9, S4, 8, 15, T8**
			+2	**C1**
			+4	**C2**
Viscosity (surface)	μ_S	M/θ	+1	**S19**
Yield stress	τ_y	$M/L\theta^2$	+1	**B3, H6**

LENGTHS, AREAS AND VOLUMES

Characteristic Linear Dimensions

(In all cases kinematic viscosity has been written as μ/ρ)

Parameter	Symbol	Dimensions	Exponent	Group
General characteristic linear dimension	Various	L	−5	**P7**
			−3	**F9, L1**
			−2	**D9, 12, E3, F11, 12, H4, 5, N2, 3, 4, O1, R9, S8, 9, 15**
			−1	**B1, 10, E2, 5, 10, F1, 13, G4, H10, K6, L4, R6, 10, 15, 16, S6, 19, T5, W4**
			$-\frac{1}{2}$	**B11, D7, F14, O2**
			$+\frac{1}{2}$	**R4, T1, W2**
			+1	**B3, 4, 5, 7, 10, C7, D1, 3, 5, 11, 13, E7, 10, F1, 2, 5, G3, H2, K3, 4, 6, 10, L3, 4, 9, 11, M1, 3, 7, N6, 7, 8, O1, P1, 2, 11, R8, 10, 11, 16, S2, 7, 14, 16, 17, 18, T3, W1**
			$+\frac{3}{2}$	**D7, H7, T1**
			+2	**B9, D2, 4, E8, H6, K9, O1, P4, 5, 12, S8, 9, V1**
			+3	**A5, C10, G1, 5, K1, R5**
			+4	**T2**
			+5	**R6**
Dimension of agitator, impeller, etc.	Various	L	−5	**P7**
			−3	**F9**
			−2	**D7, H4**
			+1	**D11, F15**
			+3	**W3**
Film thickness	L_f	L	+1	**N8**
Furnace half-width	L	L	−1	**T7**
Larmor radius	L_L	L	+1	**L4**
Mean free path	λ	L	+1	**K6**
Particle dimension	d_e	L	−2	**S15**
			+1	**F2**
			+3	**A5, G1**
Pore or nozzle radius	R	L	+1	**M12, T7**
Reactor length	L	L	+1	**B7**
			−1	**C3, L9**
Thickness of liquid layer	L	L	+1	**H2, L11**
			+2	**L9**
			+3	**R5**
			+4	**T2**

Areas

Parameter	Symbol	Dimensions	Exponent	Group
Area	A	L^2	−1	**D9, F6, 7, 8**
			+1	**B2**
Area/unit volume	S, A^*	$1/L$	−1	**B6**
			+1	**M11, 12**

Volumes

Parameter	Symbol	Dimensions	Exponent	Group
Volume	—	L^3	+	**A5, H9, M11**

TIMES AND FREQUENCIES

Parameter	Symbol	Dimensions	Exponent	Group
			−1	**D8**
Time	θ	θ	+1	**D8, 12, E3, F11, 12, H1, 10, S15, T5**
Frequency	f'	$1/\theta$	+1	**H1, 10, S16, V1**

TEMPERATURES AND CONCENTRATIONS (DRIVING FORCES)

Concentrations and Related Quantities

Parameter	Symbol	Dimensions	Exponent	Group
Dimensional concentration	C, c_A	M/L^3	−1	**D1, 2, R2**
			$-\frac{1}{2}$	**T3**
			+1	**B10**
Dimensionless concentration—e.g., wt./wt. inert material, etc.	n	—	−1	**P6**
			+1	**A4, K8**
Mass capacity	C_m	L^3/M	−1	**R2**
Mole fraction	Y, Z	—	±1	**D14**
Specific mass content, mass/unit mass	n'	—	−1	**K4**
Surface concentration	Γ'	M/L^2	±1	**S18**
Vapor capacity (porous body)	C_b	$L\theta^2/M$	+1	**B13, R2**

Temperatures, Temperature Differences

Parameter	Symbol	Dimensions	Exponent	Group
Temperature, temperature difference	$T, \Delta T$	T	−3	**R1*, T6***
			−1	**A4, 6*, B12, 13, C4*, 7, 10, D3*, 4*, 15, E1, G2, 6*, K3, 8, 9, 11, L9, N5, P5*, 12*, R14*, S6**
			$-\frac{1}{2}$	**L5**
			$+\frac{1}{2}$	**F6*, 7*, 8***
			+1	**C4, G5, 6, J1, 4, L9, M1, P6, R5, 7, 14***
			+3	**S14***
Rate of temperature change	—	T/θ	+1	**P12**

* Absolute temperature; others—temperature differences.

VELOCITIES, RATES, FLUXES, TRANSFER COEFFICIENTS

Velocities

Parameter	Symbol	Dimensions	Exponent	Group
Angular velocity (rate of rotation)	N	$1/\theta$	−3	**P7**
			−2	**H4, L1**
			−1	**F9, R15**
			$-\frac{1}{2}$	**E2**
			+1	**S8, 11, 12, T1**
			+2	**F15, T2, W3**
Fluid velocity	V	L/θ	−3	**H5**
			−2	**C6, 11, D13, E9, 10, F1, M6, N2, 3, 4, R7, 8, 9, 10**
			−1	**A2, B3, C12, 14, D1, 3, F10, H7, 8, K10, L3, M3, P4, S13, 16**
			+1	**A2, B6, 7, 11, C3, 12, 14, D5, 7, E5, 7, F10, 14, H10, K2, L5, 8, M2, 4, 7, P1, 2, 3, 10, R4, 11, 15, S1, 2, 3, T5, 6, V2, W2, 4,**
			+2	**B1, 12, C5, 11, D15, E1, 11, F13, H11, W1**
			+3	**C7**
Impeller or agitator circumference	U_s	L/θ	−2	**P14**
			−1	**D9**
Light	V_l	L/θ	−1	**L8**
Sound	V_s	L/θ	−1	**M2, N1, S1**
Waves	V_w	L/θ	−1	**V1**
			−1	**P3**
Velocity gradient	dV/dL	$1/\theta$	−2	**R12**
Velocity of Alfven waves	V_A	L/θ	−1	**A2, K2, M4**
			+1	**A2, N1**
			+2	**C11**
Velocity of bearing surface	v_s	L/θ	+1	**H8, O1, S9**

Flow Rates (Mass Fluxes)

Parameter	Symbol	Dimensions	Exponent	Group
Mass flow rate (mass flux)	V_m	M/θ	−1	**B2, M11, T7**
			+1	**F6, 7, 8, G4, T7**
Mass flux density (mass flux/unit area)	G	$M/L^2\theta$	−1	**J2, 3, P15, S13**
			+1	**K4, M11**

TABLE A (Continued)

Flow Rates (Mass Fluxes)

Parameter	Symbol	Dimensions	Exponent	Group
Mass flux/unit volume (reaction rate)	U	$M/L^3\theta$	$+\frac{1}{2}$	**T3**
			$+1$	**D1, 2, 3, 4**
Reaction rate constant	K	L/θ	-1	**S5**
Volumetric flow rate	V_f	L^3/θ	-1	**H9**
			$-\frac{1}{2}$	**D11**
			$+\frac{1}{2}$	**S11, 12**
			$+1$	**D9, F9**

Heat Fluxes

Parameter	Symbol	Dimensions	Exponent	Group
Heat flux (heat flow/unit time)	q	ML^2/θ^3	$+1$	**H5**
Heat flux/unit area	q^*	M/θ^3	$+1$	**K3, R6**
Heat liberated/unit mass	Q	L^2/θ^2 (H/M)	$+1$	**D3, 4**
Rate of heat liberation/unit volume (heat source power)	j	$M/L\theta^3$	$+1$	**P5**

Transfer Coefficients

Parameter	Symbol	Dimensions	Exponent	Group
Heat transfer coefficient	h	$M/T\theta^3$	$+1$	**B2, 4, C9, J2, N6, P15, S13**
		$(H/L^2T\theta)$	$+4$	**M1**
Mass transfer coefficient	k_c	L/θ	$+1$	**B5, J3, L9, N7, S5, 7**

FORCE, HEAD, POWER, PRESSURE

Forces

Parameter	Symbol	Dimensions	Exponent	Group
Force (resistance)	F, F_R	ML/θ^2	$-\frac{1}{3}$	**S2**
			$+\frac{1}{3}$	**K5**
			$+1$	**N2, 3, 4, R9**
Force/unit length	F_b	M/θ^2	$+1$	**H8, O1, S9**

Heads, Power

Parameter	Symbol	Dimensions	Exponent	Group
Fluid head	H'	L	$-\frac{3}{4}$	**S11**
			$+1$	**H4**
Head (energy per unit mass of fluid $= gH'$)	H	L^2/θ^2	-1	**P3, T4**
			$-\frac{3}{4}$	**S12**
			$+\frac{1}{4}$	**D11**
			$+1$	**A6, P14, T4**
Power	π	ML^2/θ^3 (LF/θ)	$+1$	**L1, P7**

Pressures

Parameter	Symbol	Dimensions	Exponent	Group
Pressure	P	$M/L\theta^2$	-1	**F6, 7, 8, H9, S8, T8**
			$+1$	**B13, C6, L6, P13, R2**
Pressure drop	$\Delta P, dP$	$M/L\theta$	$+1$	**E9, 10, F1, H9, L3, R10**
Pressure gradient	$\Delta P/L$	$M/L^2\theta^2$	$+1$	**E10, F1, K1, P4, R10**

CONSTANTS AND MISCELLANEOUS QUANTITIES

Gravity Acceleration

Parameter	Symbol	Dimensions	Exponent	Group
Gravity acceleration	—	L/θ^2	-2	**A1**
			-1	**B1, F13, 15, M1, S6**
			$-\frac{3}{4}$	**S11**
			$-\frac{1}{2}$	**B11, F14, P13**
			$-\frac{1}{3}$	**C9, S3**
			$+\frac{1}{3}$	**F2, N8**
			$+1$	**A5, B9, C2, 10, D13, E8, G1, 5, H4, R5, 6, 8, 13**

Other Quantities

Parameter	Symbol	Dimensions	Exponent	Group
Avogadro's number	N	$1/M$	$+1$	**K9, S6**
Boltzmann's constant	k	—	-1	**K9, S6**
Dufour coefficient	β	T	$+1$	**A3, 4, F3**
Energy of activation	E_a	L^2/θ^2	$+1$	**A6**
Gas constant	$\mathcal{R}$	$L^2/\theta^2 T$	-1	**A6**
			$-\frac{1}{2}$	**L5**
Shape factor	ξ^*	—	$+1$	**V2**
Soret coefficient	δ	$1/T$	$+1$	**F3, P6**
Stefan constant	η	$M/T^4\theta^3$	-1	**R1, T6**
			$+1$	**S14**

TABLE B
ALPHABETICAL LIST OF NAMED GROUPS

Serial No.	Name	Symbol	Definition	Significance	Field of Use	Reference
A1	Acceleration number	K_g	$E^3/\rho g^2\mu^2 = (N_{Re}N_{Fr1})^2/(H_0)^3$	Group dependent only on physical properties	Accelerated flow	*25*
A2	Alfven number	N_{Al}	V_A/V (or V/V_A) [*cf.* Cowling No. Kármán No. (2), magnetic mach number]	Ratio of Alfven wave velocity/fluid velocity	Magneto-fluid dynamics	*5, 6*
A3	Anonymous group (1)	ε	β^*c/r [see also Fedorov No. (2)]		Transfer processes	*39*
A4	Anonymous group (2)	K_1	$\beta^*\Delta n/\Delta t$; Δt = temp. diff. [T]; Δn = conc. diff. [$^-$]		Transfer processes	*67*
A5	Archimedes number	N_{Ar}	$\frac{gL^3\rho_o}{\mu^2}(\rho-\rho_o)$; ρ_o = fluid density; ρ = particle density (*cf.* N_{Ga1})	N_{Re}, gravitational force/viscous force	Fluidization, motion of liquids due to density differences	*6, 13*
A6	Arrhenius group	—	$E_a/\mathscr{R}T$	Activation energy/potential energy of fluid	Reaction rates	*5*
B1	Bagnold number	B	$3c\alpha\rho_g V^2/4d\rho_p g$, ρ_g = gas density ρ_p = particle density	Drag force/gravitational force	Saltation studies	*12*
B2	Bansen number	N_{Ba}	$h_r A_w/V_{mc}$; h_r = radiant heat transfer coefficient; A_w = wall area of channel; (*cf.* N_{St})	Heat transferred by radiation/thermal capacity of fluid	Radiation	*1*
B3	Bingham number	N_{Bm}	$\tau_y L/\mu_p V$ (L = channel width)	Ratio of yield stress/viscous stress	Flow of Bingham plastics	*5*
B4	Biot number (heat transfer)	N_{Bih}	hL_m/k (in French literature, "Biot No." $= N_{Nu}$)	Midplane thermal internal resistance/surface film resistance	Unsteady state heat transfer	*5, 6, 13*
B5	Biot number (mass transfer)	N_{Bim}	$k_c L/D_{int}$; L = thickness of layer, D_{int} = diffusivity at interface	Mass transfer rate at interface/mass transfer rate in interior of solid wall thickness L	Mass transfer between fluid and solid	*13*
B6	Blake number	B	$V\rho/[\mu(1-e)S]$	Inertial force/viscous force	Beds of particles	*6*
B7	Bodenstein number	N_{Bo}	$VL/D_a = N_{Pe_m}$; L = reactor length, D_a = axial diffusivity (effective) (L^2/θ)		Diffusion in reactors	*6*
B8	Boltzmann number	N_{Bo}	≡ Thring radiation group			*1*
B9	Bond number	N_{Bo}	$(\rho-\rho')L^2g/\sigma = N_{We_1}/N_{Fr_1}$ if $\rho-\rho' \cong \rho$ (gas in liq.); ρ = drop or bubble density; ρ' = medium density	Gravitational force/surface tension force	Atomization, motion of bubbles and drops	*5*
B10	Bouguer number	N_{Bu}, B	$3C_D\lambda_r/4\rho DR$; C_D = wt. dust/unit bed volume (M/L^3), λ_r = mean path for radiation (L), ρ_D = dust density, $\bar{R}$ = mean particle radius. Also $N_{Bu} = kL$; L = characteristic dimension, k = absorption coefficient of medium		Radiant heat transfer to dust gas streams	*66*
B11	Boussinesq number	B	$V/(2gR_H)^{1/2}$ (*cf.* N_{Fr2})	(Inertia force/gravitational force)$^{1/2}$	Wave behavior in open channels	*6*
B12	Brinkman number	N_{Br}	$\mu V^2/k\Delta t$; Δt = temp. diff.	Heat generation/heat transferred	Viscous flow	*6*
B13	Bulygin number	N_{Bu}	$r\cdot\frac{C_b}{C_q}\cdot\frac{P}{t_m-t_o}$ t_m = temp. of medium, medium, t_o = init. temp. of body	Heat for vaporization/heat to bring liquid to boiling point	Heat transfer during evaporation	*6, 13, 14*
C1	Capillary number	$\bar{K}\sigma$	$\mu^2E/\rho\sigma^2 = (N/_{We_1})^2/H_o\,.\,(N_{Re})^2$	Depends only on physical properties	Action of surface tension in flowing media	*25*
C2	Capillarity-buoyancy number (physical properties group) (film No.)[a]	$\bar{K}\sigma_g$, K_F	$g\mu^4/\rho\sigma^3. = \sqrt{\bar{K}\sigma/Kg} = (N_{We_1})^3/(N_{Fr_1})(N_{Re})^4$	Depends only on physical properties and g	Effects of surface tension and acceleration in flowing media (two-phase flow)	*21, 25*
C3	Capillary number	Ca	$\mu V/\sigma = N_{We}/N_{Re}$	Viscous force/surface-tension force	Atomization, two-phase flow	*6*
C4	Carnot number	Ca, N_{Ca}	$(T_2-T_1)/(T_2)$; T_1, T_2 = abs. temp. of two heat sources or sinks	Theoretical efficiency of Carnot cycle operating between T_1 and T_2		*21*
C5	Cauchy number	N_c	$\rho V^2/E_b = (N_{Ma})^2$ = Hooke No.	Inertia force/compressibility force	Compressible flow	*5, 25*
C6	Cavitation number	σ_c	$[(p-P_v)/\rho]/(V^2/2)p$ = local static pressure (abs.); P_v = vapor pressure	Excess of local static head over vapor pressure head/velocity head	Cavitation	*5*
C7	Clausius number	Cl, N_{Cl}	$V^3L\rho/k\Delta T$; ΔT = temp. diff.		Heat conduction in forced flows	*25*
C8	Colburn number	—	Same as Schmidt number			*5*
C9	Condensation number (1)	N_{Co}	$(h/k)(\mu^2/\rho^2g)^{1/3}$	$N_{Co} = N_{Nu}\left[\frac{\text{(viscous force)}}{\text{(gravity force)}} \times \frac{1}{Re}\right]^{1/3}$	Condensation	*5*
C10	Condensation number (2)	N_{Cv}	$L^3\rho^2gr/k\mu\Delta t$; r = latent heat of condensation		Condensation on vertical walls	*5*
C11	Cowling number	C	$(V_A/V)^2$ ≡ (Alfven number)^2		Magneto-fluid dynamics	*6*
C12	Craya-Curtet number	C_t	$V_k/(V_d^2 - V_k^2/2)^{1/2}$; V_k = kinematic mean velocity, V_d = dynamic mean velocity		Radiant heat transfer	*3, 27*

[a] Very similar to Hu and Kintner's pH factor for drops and bubbles [*A.I.Ch.E. J.* 1, 42 (1955)].

TABLE B (Continued)

Serial No.	Name	Symbol	Definition	Significance	Field of Use	Reference
C13	Crispation group	N_{Cr}	$\mu\alpha/\sigma^* L$; σ^* = undisturbed surface tension; L = layer thickness		Convection currents	55
C14	Crocco number	N_{Cr}	$V/V_{max} = \left[1 + \frac{2}{(\gamma-1)(N_{Ma})^2}\right]^{-1/2}$ V_{max} = maximum velocity of gas expanding adiabatically	Velocity/maximum velocity	Compressible flow	5
D1	Damköhler group I	$Da\text{I}$	UL/V_{cA}	Chemical reaction rate/bulk mass flow rate	Chemical reaction, momentum, and heat transfer	5
D2	Damköhler group II	$Da\text{II}$	UL^2/D_{cA}	Chemical reaction rate/molecular diffusion rate	Chemical reaction, momentum, and heat transfer	5
D3	Damköhler group III	$Da\text{III}$	$QUL/C_p\rho Vt$	Heat liberated/bulk transport of heat	Chemical reaction, momentum, and heat transfer	5
D4	Damköhler group IV	$Da\text{IV}$	QUL^2/kt	Heat liberated/conductive heat transfer	Chemical reaction, momentum, and heat transfer	5
D5	Damköhler group V	$Da\text{V}$	$= (N_{Re})$			5
D6	Darcy number		$4f$; see Fanning friction factor			5
D7	Dean number	N_D	$(VL\rho/\mu)(L/2R)^{1/2}$; L = pipe diam.; R = radius of curvature of bend	N_{Re} (centrifugal force/inertial force)	Flow in curved channels	5, 6
D8	Deborah number	D	θ_r/θ_o; θ_o = observation time	Relaxation time/observation time	Rheology	49
D9	Delivery number	ϕ	V_f/Aw; A = impeller area = $\pi d^2/4$ ≡ [Diameter No.]$^{-3}$ [Speed No.]$^{-1}$		Flow machines	25
D10	Deryagin number	De	$L(\rho g/2\sigma)^{1/2}$; L = film thickness	Film thickness/capillary length	Coating	58
D11	Diameter group	δ	$(\pi/4)^{1/2}(2H)^{1/4}\,d/(V_f)^{1/2}$, d = impeller diam. = [pressure No.]$^{1/4}$ × [delivery No.]$^{-1/2}$		Flow machines	25
D12	Diffusion group	β	$D\theta/L_m^2$; D = diffusivity of solute through stationary solution contained in solid; *cf.* N_{Fom}		Mass transfer	37
D13	Drag coefficient	$C_d = C_D$	$(\rho-\rho')L_g/\rho V^2$; ρ = density of object; ρ' = density of medium; *cf.* f, ψ, N_e	Gravity force/inertial force	Free settling velocities, etc.	5
D14	Drew number	N_D	$\frac{Z_A(M_A-M_B)+M_B}{(Z_A-Y_{AW})(M_B-M_A)}\ln\frac{M_V}{M_W}$; M_A, M_B = mol. wt. of components A and B; M_V, M_W = mol. wt. of mixture in vapor and at wall; Y_{AW} = mole fraction of A at wall; Z_A = mole fraction of A in diffusing stream		Boundary layer mass transfer rates; velocity profile distortion; drag coefficients for binary system	22
D15	Dulong number	Du, N_{Du}	$V^2/C_p\Delta T$ = Eckert No.			25
E1	Eckert number	N_E	$V_\infty^2/C_p\Delta T$, V_∞ = velocity of fluid far from body (=2/recovery factor, *q.v.* ≡ Dulong No.)		Compressible flow	5, 24
E2	Ekman number		$(\mu/2\rho\omega L^2)^{1/2} = (N_{Ro}/N_{Re})^{1/2}$	(Viscous force/Coriolis force)$^{1/2}$	Magneto-fluid dynamics	6
E3	Elasticity number (1)	N_{El_1}	$\theta_r\mu/\rho L^2$ = pipe radius	Elastic force/inertial force	Viscoelastic flow	5
E4	Elasticity number (2)	$\overline{K}_E$	$\rho C_p/\beta E$ ≡ [Gay Lussac No.] × [Hooke No.] ÷ [Dulong No.]	Depends on physical properties only	Effect of elasticity in flow processes	25
E5	Ellis number	N_{El}	$\mu_o V/2\tau_{1/2}R$; μ_0 = zero shear viscosity, $\tau_{1/2}$ = shear stress when $\mu = \mu_o/2[M/L\theta^2]$; R = tube radius		Flow of non-Newtonian liquids	38
E6	Elsasser number	N_{El}	$\rho/\mu\sigma_e\mu_e \equiv N_{Re}$/[magnetic Reynolds No.]		Magneto-fluid dynamics	6
E7	Entry Reynolds Number	K_E	$\chi/d_h \cdot N_{Re} = \frac{\chi V\rho}{\mu}$; χ = entry length	As N_{Re}	Entry or inlet processes	25
E8	Eötvös number	N_{Eo}	$(\rho-\rho')L^2g/\sigma$ ≡ Bond No., *q.v.*			6
E9	Euler number (1)	N_{Eu_1}	$\Delta P_F/\rho V^2$; ΔP_F = pressure drop due to friction	Friction head/2 × velocity head	Fluid friction in conduits	5, 13, 25
E10	Euler number (2)	N_{Eu_2}	$d(-dp/dL)\rho V^2$; d = pipe diam., dp/dL = pressure gradient ≡ 2 × Fanning friction factor		Fluid friction in conduits	
E11	Evaporation number	K_r	V^2/r [r = heat of vaporization (L^2/θ^2)]		Evaporation processes	25
E12	Evaporation number (2)	$\overline{K}_r$	$C_p/r\beta$ (r as in **E11**) ≡ (Gay Lussac No.) × (**E11**)/(Dulong No.)		Evaporation processes	25
E13	Evaporation-elasticity number	$\overline{K}_{rE}$	$E/r\rho = K_r$/Hooke number		Evaporation processes	25
F1	Fanning friction factor	f	$d\Delta p_F/2pV^2L$, d = dimension of cross section; L = length (*cf.* resistance coeff., Ne)	Shear stress at wall expressed as number of velocity heads	Fluid friction in conduits	5
F2	Fedorov number (1)	F_{e_1}, N_{Fe_1}	$d\sqrt[3]{\frac{4g\rho^2}{3\mu^2}\left(\frac{\gamma M}{\gamma g}-1\right)}$ d_e = equiv. particle diam.; γM = sp. gr. of particles; γg = sp. gr. of gas (*cf.* N_{Ar})		Fluidized beds	6, 13

TABLE B (Continued)

Serial No.	Name	Symbol	Definition	Significance	Field of Use	Reference
F3	Fedorov number (2)	$F_{e_2} N_{Fe_2}$	$\delta\beta^* = K_1 Pn = \varepsilon \times K_o Pn$	Mass transfer analogy of Posnov number	Transport processes	*39, 67*
F4	Fenske number	N_F		Number of stages in separation process		*6*
F5	Fineness coefficient	ψ	$L/W_D^{1/3}$; W_D = volume displacement $[L^3]$		Ship modeling	*26, 43*
F6	Fliegner numbers	$F(M_a)$	Functions of ratio of specific heats and mach number			
F7	Fliegner numbers	$f(M_a)$	Functions of ratio of specific heats and mach number			
F8	Fliegner numbers	$I(M_a)$	$V_m(cT)^{1/2}/A(p_s + \rho V^2) = [\gamma M_a/(\gamma - 1)^{1/2}] \left[1 + \frac{(\gamma-1)Ma^2}{2}\right]^{1/2}$ = impulse Fliegner number; γ = ratio of specific heats, Ma = mach number, A = flow area			*43*
F9	Flow coefficient	C_Q	V_f/Nd, d = impeller diam.		Power required by fans, etc.	*32*
F10	Fluidization number		V/V_{init}, V_{init} = velocity for initial fluidization	Fluid velocity in fluidized bed/that at start of fluidization	Fluidization	*59*
F11	Fourier number (heat transfer)	N_{Foh}	$k\theta/\rho C_p L_m^2$		Unsteady state heat transfer	*5, 13, 14, 25*
F12	Fourier number (mass transfer)	N_{Fom}	$D\theta/L^2 = k_c\theta/L$ (*cf.* **D12**)		Unsteady state mass transfer	*13*
F13	Froude number (1)	N_{Fr_1}	$V^2/gL \equiv (N_{Fr_2})^2$ (*cf.* Reech No., Boussinesq No., Vedernikov No.)	Inertial force/gravitational force	Wave and surface behavior	*5, 13, 25*
F14	Froude number (2)	N_{Fr_2}	$V/\sqrt{gL} \equiv (N_{Fr_1})^{1/2}$ (*cf.* Boussinesq No.)	Velocity of open channel flow/speed of very small gravity wave	Open channel flow; free surfaces	*7, 50*
F15	Froude No. (rotating)	Fr	DN^2/g; D = impeller diam.		Agitation	*64*
G1	Galileo number	N_{Ga_1}	$L^3 g\rho^2/\mu^2$ (*cf.* N_{Ar}, Nusselt thickness group)	$N_{Ga_1} = N_{Re} \times$ gravity force/viscous force	Circulation of viscous liquid, thermal expansion	*5, 13*
G2	Gay Lussac number	Ga, N_{Ga_2}	$1/\beta\Delta T$		Thermal expansion processes	*25*
G3	Goucher number	N_{Go}	$R(\rho g/2\sigma)^{1/2}$; R = wall or wire radius	Gravitational force/surface tension force$^{1/2}$	Coating	*24*
G4	Graetz (Grätz) number	N_{Gz}	V_{mCp}/kL	Thermal capacity fluid/convective heat transfer	Streamline flow	*5, 25*
G5	Grashof number	N_{Gr}	$L^3\rho^2 g\beta\Delta t/\mu^2 \equiv N_{Ga_1}/N_{Ga_2} \equiv (N_{Re})^2/(N_{Ga_2})(N_{Fr_1})$	$N_{Gr} = N_{Re}$ (buoyancy force/viscous force)	Free convection	*5, 13, 25*
G6	Gukhman number	Gu, N_{Gu}	$(t_o - t_m)/T_o$; t_o, T_o = temp. (°C., °K.) of hot gas stream, t_m = temp. of moist surface (wet bulb temp.)	Thermodynamic criterion of evaporation under isobaric adiabatic conditions	Convective heat transfer in evaporation	*6, 23, 17, 53*
G7	Guldberg-Waage group	N_{Gw}	Given by equation relating volumes of reacting gases and reaction products		Chemical reaction in blast furnaces	*5*
H1	Hall coefficient	N_H	$f_c J$ (f_c = cyclotron frequency, J = av. free path/av. veloc.)		Magneto-fluid dynamics	*7*
H2	Hartmann number	M_H	$(\mu_e^2 H_e^2 \sigma_e L^2/\mu)^{1/2} \equiv (SR_M N_{Re})^{1/2} \equiv (N_{Re}N)^{1/2}$	Magnetically induced stress/hydrodynamic shear stress (magnetic body force/viscous force)$^{1/2}$	Magneto-fluid dynamics	*5, 6*
H3	Hatta number	β	$\gamma/\tanh\ \gamma$; $\gamma = (rCD)^{1/2}/k_c$, r = reaction rate constant $[L^3/M\theta]$ [a modified Hatta number has also been defined *35*)]		Gas absorption with chemical reaction	*41*
H4	Head coefficient	C_H	gH'/N^2d^2 (d = impeller diam.)		Flow in pumps and fans	*32*
H5	Heat transfer number	K_Q	$q/V^3L^2\rho$		Heat transfer in stream	*25*
H6	Hedstrom number	N_{He}	$\tau_y L^2\rho/\mu_p^2 \equiv (N_{Re}) \times (N_{Bm})$		Flow of Bingham plastics	*5*
H7	Helmholtz resonator group		$(d^3/W)^{1/2}/Ma$	Proportional to frequency × residence time	Pulsating combustion	*28*
H8	Hersey number		$F_b/\mu v_s$ (*cf.* truncation number)	Load force/viscous force	Lubrication	*6*
H9	Hodgson number	N_H	$Wf\Delta\rho_F/\overline{V}_{f.ps}$	Time constant of system/period of pulsation	Pulsating gas flow	*5*
H10	Homochronous number	H_{o_1}	$V\theta/L$ (θ = time for liquid to move characteristic distance L)	Duration of process/time for liquid to move through L	Choice of time scales	*13*
H11	Hooke number	H_{o_2}	$\rho V^2/E \equiv$ Vauchy No., *q.v.*		Elasticity of flowing media	*25*
J1	Jakob modulus	Ja	$C_P\rho_L\Delta t/r\rho_v$ (ρ_L, ρ_v = densities of liquid and vapor; Δt = liquid superheat temperature diff.)	Maximum bubble radius/thickness of superheated film	Boiling	*6*
J2	*J*-factor (heat transfer)	j_H	$(h/c_pG)(c_p\mu/k)^{2/3} \equiv (N_{Nu})/(N_{Re})(N_{Pr})^{1/3}$		Heat, mass and momentum transfer theory	*5*
J3	*J*-factor (mass transfer)	j_M	$(k_c\rho/G)(\mu/\rho D)^{2/3} \equiv (k_c\rho/G)(N_{Sc})^{2/3}$			*5*
J4	Joule number	J	$2\rho C_p\Delta t/\mu_e H_e^2 \equiv 2(N_{Re})(R_M)/(M_H)^2(N_E)$	Joule heating energy/magnetic field energy	Magneto-fluid dynamics	*6*
K1	Kármán number (1)	N_K	$\rho d^3(-dp/dL)/\mu^2$ (d = pipe diam., dp/dL = pressure gradient) $\equiv 2(N_{Re})^2 f^{1/2}$		Fluid friction in conduits	*5*

TABLE B (Continued)

Serial No.	Name	Symbol	Definition	Significance	Field of Use	Reference
K2	Kármán number (2)	K	V/V_A (see Alfven No.)		Magneto-fluid dynamics	6
K3	Kirpichev number for heat transfer	$K_{i_q}, N_{K_{i_q}}$	$q^* L/k\Delta t$ (*cf.* N_{Bih}, N_{Nu})	Intensity external heat transfer/internal heat transfer intensity	Heat transfer	6, *14*
K4	Kirpichev number for mass transfer	Ki_m	$GL/D\rho n^*$ (*cf.* N_{Pem}, N_{Bim})	Intensity external mass transfer/internal mass transfer intensity	Mass transfer	*13, 14*
K5	Kirpitcheff number		$(\rho F_R/\mu^2)^{1/3} = [(N_{Re})^2 cf]^{1/3}$		Flow around obstacles	*34*
K6	Knudsen number (1)	N_{Kn}	λ/L	Length of mean free path/characteristic dimension	Low pressure gas flow	*5*
K7	Knudsen number (2)	N_{KnA}	$\rho D/_{AB} D_{KA}\zeta$	Bulk diffusion/Knudsen diffusion	Gaseous diffusion in packed beds	6
K8	Kossovich number	K_o, N_{Ko}	$r_v \Delta n_m/c\Delta t$	Heat used for evaporation/heat used in raising temperature of body	Convective heat transfer during evaporation	6, *14*
K9	Kronig number	Kr	$4L^2\beta\rho^2\Delta t E_S^2 N[\alpha + 2/3(p_o^2/kT)]/u^2 M$ E_S = electric field at surface, N = Avogadro's Number, α = polarization coefficient, p_o = molecular dipole moment, k = Boltzmann's constant, M = molecular weight	(N_{Re}) (electrostatic force/viscous force	Convective heat transfer	6, *44*
K10	Kutateladze number (1)	Ku	$IEL/\rho Vu'$; I = current density $[Q/L^2\theta]$, E = voltage $[ML/Q\theta^2]$, u' = enthalpy $[L^2/\theta^2]$		Electric arcs in gas streams	*65*
K11	Kutateladze number (2)	K	$r_v/c_p(t_o - t_w)$, (t_o, t_w = stream, wall temp.)		Combined heat and mass transfer in evaporation	*17*
L1	Lagrange group (1)	La_1	$\Pi/\mu L^3 N^2$; L = characteristic dimension of agitator = $N_{Re}\, N_p$		Agitation	6
L2	Lagrange number (2)	La_2	$(D + \varepsilon_D)/D$	Combined molecular and eddy mass transfer rate/molecular mass transfer rate	Mass transfer in turbulent systems	*31*
L3	Lagrange number (3)	La_3	$\Delta PR/\mu V$		Magneto-fluid dynamics	6
L4	Larmor number	R_{Lo}	L_L/L; (L_L = Larmor radius)		Magneto-fluid dynamics	*23*
L5	Laval number	La	$V\Big/\left(\frac{2\gamma}{\gamma+1}RT\right)^{1/2}$; γ = ratio of specific heats	Linear velocity/critical velocity of sound	Compressible flow	*56*
L6	Leverett function	J	$(\xi/e)^{1/2}(P\sigma/\sigma$	Characteristic dimension of surface curvature/characteristic dimension of pores	Two-phase flow in porous media	6
L7	Lewis No.	N_{Le}	$k/\rho c_p D = \alpha/D \equiv N_{Sc}/N_{Pr}$ (N.B.: Lewis number is sometimes defined as reciprocal of this quantity)		Combined heat and mass transfer	5, *25*
L8	Lorentz number	N_{Lo}	V/V_1; (V_1 = velocity of light)	Fluid velocity/velocity of light	Magneto-fluid dynamics	6
L9	Luikov (Lȳkov) number	Lu	$k_c L/\alpha = k_c L_\rho C_p/k$	Mass diffusivity/thermal diffusivity; rate of extension of mass transfer field/rate of extension of heat transfer field	Combined heat and transfer	6, *13*
L10	Lukomskii number	Lu	α/a_m; a_m = potential conductivity of mass transfer $[L^2/\theta]$		Combined heat and mass transfer	*36*
L11	Lundquist number	N_{Lu}	$\sigma_e H_e \mu_e^{3/2} L/\rho^{1/2} = M_H(R_M/N_{Re})^{1/2}$ (L = thickness of fluid layer)		Magneto-fluid dynamics	6
L12	Lyashchenko number	Ly	$\equiv N_{Re}^3/N_{Ar}$		Fluidization	*19*
L13	Lykoudis number	N_{Ly}	$(\mu_e H_e)^2 \frac{\sigma_e}{\rho}\left[\frac{L}{g\beta\Delta t}\right]^{1/2} \equiv (M_H)^2/(N_{Gr})^{1/2}$	Magnetic body force/square root of product of the inertia and buoying force.	Magneto-fluid dynamics	6
M1	McAdams group		$h^4 L\mu\Delta t/k^3\rho^2 gr$	Constant for given surface orientation	Condensation	*5*
M2	Mach number	N_{Ma}, Ma	V/V_s; (V_s = velocity of sound in fluid) $\equiv v/\sqrt{E_b/\rho}$; (E_b = bulk modulus of fluid) (*cf.* Sarrau number)	Linear velocity/sonic velocity	Compressible flow	*5–7, 25, 50*
M3	Magnetic force parameter	N	$\mu_e^2 H_e^2 \sigma_e L/\rho V$	Magnetic body force/inertia force; resistance time of fluid in field/relaxation time of lines force	Magnetic-fluid dynamics	6
M4	Magnetic mach number	M_{Ma}	V/V_a (see Alfven number)		Magneto-fluid dynamics	6
M5	Magnetic Oseen number	k	$\frac{1}{2}(1 - N_{Al}^2)R_M$	Magnetic force/inertia force	Magneto-fluid dynamics	*2*
M6	Magnetic pressure number	S	$\mu_e H_e^2/\rho V^2$	Magnetic pressure/2 × dynamic pressure	Magneto-fluid dynamics	*5*
M7	Magnetic Reynolds number	R_M	$\sigma_e\mu_e LV$ (*cf.* velocity number)	Mass transport diffusivity/magnetic diffusivity	Magneto-fluid dynamics	*5*
M8	Maievskii number		$\equiv Ma$		Compressible flow	*12*
M9	Marangoni number	N_{Ma}	$\frac{\Delta\sigma}{\Delta t}\frac{\Delta t}{\Delta L}L^2/\mu\alpha$; L = layer thickness		Cellular convection	*46*

TABLE B (Continued)

Serial No.	Name	Symbol	Definition	Significance	Field of Use	Reference
M10	Margoulis number	M	$\equiv N_{st}$		Forced convection	6, 7
M11	Merkel number	N_{Me}	$GA^* W^*/(V_m)$ gas	Mass of water transferred in cooling per unit humidity difference/mass of dry gas	Cooling towers, liquid-gas contact	5
M12	Miniovich number	Mn	SR/e; R = pore radius		Drying	36
M13	Mondt number	N_{Mo}		Convective/conductive heat transfer	Heat transfer	6
N1	Naze number	N_a	$V_A/V_s \equiv (N_{Ma}.N_{A1})$	Velocity Alfven wave/velocity of sound	Magneto-fluid dynamics	6
N2	Newton inertial force group	N_l	$F/\rho V^2 L^2$	Imposed force/inertial group	Agitation	5
N3	Newton number	N_e	$F_R/\rho V^2 L^2$; (*cf.* f, ψ)	Resistance force/inertia force	Friction in fluid flow	25
N4	Number of velocity heads	N	$(F/\rho L^2)/(V^2/2)$	Imposed head/velocity head	Friction in conduits	6
N5	Number for similarity of phys. and chem. changes	K	$r/C_p\Delta t$	Heat flow for phase change/superheat (supercooling) of one of the phases	Changes of phase	14
N6	Nusselt number	N_{Nu}	$hL/k \equiv (N_{Re}N_{St})$ (*cf.* N_{Bi_h}, Ki_g)	Total heat transfer/conductive heat transfer	Forced convection	5, 13, 25
N7	Nusselt number for mass transfer	Nu_m, N_{Nu_m}	$k_c L/D = N_{Sh}$	Intensity of mass flux at interface/specific flux by pure molecular diffusion in layer of thickness, L	Mass transfer	13, 20
N8	Nusselt film thickness group	ψ, N_T	$L_f(\rho^2 g/\mu^2)^{1/3} = (N_{Ga})^{1/3}$; ($L_f$ = film thickness)	$= (N_{Re})^{1/3}$ (gravitational force/viscous force)$^{1/3}$	Falling films	6
O1	Ocvirk number		$(F_b/\mu V_s)(a/R)^2(D/b)^2$; ($v_s$ = shaft surface velocity; R = shaft radius; D = shaft diam.) (*cf.* N_S)	Load force/viscous force	Lubrication	6
O2	Ohnesorge number	Z	$\mu/(\rho L\sigma)^{1/2} = (N_{We_1})^{1/2}/(N_{Re})$	Viscous force/(inertia force × surface tension force)$^{1/2}$	Atomization	5
P1	Péclet number (heat)	Pe, N_{Pe_h}	$LV\rho C_p/k = LV/\alpha = (N_{Re}.N_{Pr})$	Bulk heat transfer/conductive heat transfer	Forced convection	5, 13, 25
P2	Péclet number (mass)	N_{Pe_m}	$LV/D = (N_{Re}.N_{Sc})$	Bulk mass transfer/diffusive mass transfer	Mass transfer	5
P3	Pipeline parameter	ρn	$V_W V_o/2H'_s$; (V_W = velocity water-hammer wave, V_o = initial velocity H'_s = static head × $g[L^2/\theta^2]$	Maximum pressure rise in water hammer/2 × static pressure	Water hammer	5
P4	Poiseuille number		$D^2(-dp/dL)/\mu V$(D = pipe diam., dp/dL = pressure gradient)	= 32 for laminar flow in round pipe	Laminar fluid friction	5
P5	Pomerantsev number	P_o	$jL^2/k(t_m - t_o)$ (t_m, t_o = temp. of medium, initial temp. of body) *cf.* Damköhler Group IV)		Heat transfer with heat sources in medium	6, 14, 60
P6	Posnov number	Pn	$\delta\Delta t/(\Delta n_m)$(*cf.* Fe_2)		Combined heat and mass transfer	13, 14, 67
P7	Power number	N_P	$\Pi/L^5\rho N^3$	Drag on (agitator impeller) or inertial force	Power consumption by agitators, fans, pumps, etc	5, 32
P8	Prandtl number	N_{Pr}	$C_p\mu/k$ = Da IV/Da III × Da V	Momentum diffusivity/thermal diffusivity	Forced and free convection	5, 13, 25
P9	Prandtl number (mass transfer)	Pr_m	$\mu/\rho D = N_{Sc}$, r (used in Russian, German literature)	See Schmidt number		13
P10	Prandtl velocity ratio	u^+	$V/(r_W/\rho)^{1/2}$ (V = local fluid velocity)	Inertial force/wall shear force$^{1/2}$	Turbulence studies	5
P11	Prandtl dimensionless	y^+	$L(\rho r_W)^{1/2}/\mu$ (L = distance from wall, etc.)		Turbulence studies	
P12	Predvoditelev number	Pd	$\Gamma L^2/\alpha t_o = \left(\frac{dt^*}{\alpha(N_{Fo})}\right)_{max}$ where t_o = init. temp. of body, l^* = temp. of medium relative to its initial temp.	Rate of change of temp. of medium/rate of change of temp. of body	Heat transfer	13, 14, 60
P13	Pressure number (1)	K_P	$P/\{g\sigma(\rho' - \rho'')\}^{1/2}$($\rho'$, ρ'' = density of liquid gas)	Absolute pressure in system (pressure jump on interface)		14
P14	Pressure number (2)		$H/\frac{1}{2}/_2 U_s^2$(U_s = circumferential velocity) ≡ [diameter No.]$^{-2}$ × [Speed No.]$^{-2}$		Flow machines (turbines, pumps, etc.)	25
P15	Psychrometric ratio		h_c/Gs	Heat transfer by convection/heat transfer by mass transfer	Wet and dry bulb thermometry	5
R1	Radiation number	$\bar{K}_s$	$kE/\eta\sigma T^3 = (N_{We_1})$/(Hooke No.) × (Stefan No.)		Radiant transfer	25
R2	Ramzin number	Ra	$\frac{C_b P}{C_m(\Delta\Omega)} = \frac{\text{(Bulygin No.)}}{\text{(Kosovich No.)}}$		Molar mass transfer	40
R3	Ratio of specific heats	γ	C_p/C_v (specific heats at constant pressure, volume)		Compressible flow	5
R4	Rayleigh number (1)	N_{Ra_1}	$V(\rho L/\sigma)^{1/2} = N_{We_2}$ (*q.v.*)	See N_{We}	Breakup of liquid jets	5
R5	Rayleigh number (2)	R'_2	$L^3\rho^2 g\beta c_p\Delta t/\mu k = L^3\rho g\beta\Delta t/\mu\alpha = (N_{Gr})\cdot(N_{Pr})$		Free convection	5, 25

TABLE B (Continued)

Serial No.	Name	Symbol	Definition	Significance	Field of Use	Reference
R6	Rayleigh number (3)	Ra_3	$q^* L^5 \rho^2 g \beta C_p / \mu k^2 x = (N_{Gr})(N_{Pr})(N_{Nu})(L/x)$; ($L$ = pipe diam.)		Combined free and forced convection in vertical tubes	6
R7	Recovery factor	N_{rf}	$C_p(t_{aw} - t_m)/V^2$; t_{aw} = attained adiabatic wall temp. t_m = temp. of moving medium (*cf.* Eckert No.)	Actual temp. recovery/theoretical temp. recovery	Convective heat transfer in compressible flow	5
R8	Reech number		$= 1/(N_{Fr_1})$ *q.v.*		Wave and surface behavior	6
R9	Resistance coefficient (1)	C_f	$F_R/\frac{1}{2}\rho V^2 L^2$ (*cf.* drag coeff., Newton number, Fanning factor)		Flow resistance	25
R10	Resistance coefficient (2)	ψ	$\Delta p.\ D_H/\frac{1}{2}\rho V^2 L$ (Δp = pressure drop over length, L) (*cf.* **R9**)		Fluid friction in conduits	25
R11	Reynolds number	N_{Re}	$LV\rho/\mu$	Inertia force/viscous force	Dynamic similarity	5, 25
R12	Reynolds number (rotating)	Re	$L^2 N\rho/\mu$; L = impeller diam.		Agitation	45
R13	Richardson number	N_{Ri}	$-(g/\rho)(d\rho/dL)/(dV/dL)w^2$ [L = height of liquid layer, $(dV/dL)_w$ = velocity gradient at wall]	Gravity force/inertial force	Stratified flow of multi-layer systems	54
R14	Romankov number	R'_o	T_D/T_{PROD}	Dry bulb temperature (abs.)/(product temperature (abs.)	Drying	6, 52
R15	Rossby number	N_{Ro}	$V/2\omega_e L \sin \Lambda$ (ω_e = angular velocity of earth's rotation $[1/\theta]$; Λ = angle between axis of earth's rotation and direction of fluid motion [ˉ])	Inertia force/Coriolis force	Effect of earth's rotation on flow in pipes	5
R16	Roughness factor		ε/L		Fluid friction	5
S1	Sarrau number		≡ mach number, *q.v.*		Compressible flow	6
S2	Schiller number (1)		$LV(\rho^2/\mu F_R)^{1/3}$		Flow around obstacles	34
S3	Schiller number (2)	Sch	$V\left[\frac{3}{4} \cdot \frac{\rho\gamma_m}{g\mu(\gamma_M - \gamma_m)}\right]^{1/3}$; V = velocity in fluidized bed; γ_m, γ_M = specific gravity of medium and material in bed		Fluidization	66
S4	Schmidt number	N_{Sc}	$\mu/\rho D$ (*cf.* N_{Pr_m}) (= Da II/Da I Da V)	Kinetic viscosity/molecular diffusivity	Diffusion in flowing	5, 25
S5	Semenov number	Sm	k_c/K; K = reaction rate constant $[L/\theta]$		Reaction kinetics	11
S6	Senftleben number	S_e	$NE_s^2[\alpha + 2/3(\rho_o^2/kT)] \cdot [1/4LM_g]$ Kronig number, *q.v.*		Convective heat transfer	6
S7	Sherwood number	N_{Sh}	$k_c L/D = Nu_m$ (also termed Taylor number)	Mass diffusivity/molecular diffusivity	Mass transfer	5
S8	Sommerfeld number (1)	N_{S_1}	$(\mu N/P_b)(D/a)^2$ (D = shaft diam., (*cf.*) Ocvirk number)	Viscous force/load force	Lubrication	5
S9	Sommerfeld number (2)	N_{S_2}	$(F_b/\mu V_s)(a/R)^2$ (V_s = veloc. of shaft surface; R = shaft radius) ($N_{S_2} = 4/\pi N_{S_1}$)	Viscous force/load force	Lubrication	6
S10	Spalding function	Sp	$-\left(\frac{\delta\theta}{\partial u^+}\right)_{u^+=0}$; $\theta = (T - T_\infty)/(T_w - T_\infty)$, T_w = wall temperature, T_∞ = free stream temp., u^+ = Prandtl velocity ratio	Dimensionless temp. gradient at wall	Convection	18, 33
S11	Specific speed	C_s	$N(V_f)^{1/2}/(gH')^{3/4}$ (H' = head of liquid produced by one stage) (*cf.* speed number)		Pumps and compressors	8, 32
S12	Speed number	σ	$(4\pi)^{1/2}(V_f)^{1/2} N/(2H)^{3/4}$ = (delivery number)$^{1/2}$ × pressure number)$^{-3/4}$ (*cf.* specific speed)		Flow machines	25
S13	Stanton number	N_{St}	$h/C_p\rho V = h/C_pG = (N_{Nu})/(N_{Re})(N_{Pr})$	Heat transferred/thermal capacity of fluid	Forced convection	5, 6, 13, 25
S14	Stefan number	St	$\eta LT^3/k$		Heat radiation	25
S15	Stokes number	St	$\mu\theta_v/\rho L^2$ (θ_v = vibration time) ≡ $(N_{S_1})^{-1}(N_{Re})^{-1}$		Particle dynamics	6
S16	Strouhal number	N_{S_1}, Sr	fL/V (*cf.* N_{Th})		Vortex streets; unsteady-state flow	6, 25
S17	Suratman number	Su	$\rho L\sigma/\mu^2 = (N_{Re})^2/(N_{We_1}) = (Z)^{-2}$		Particle dynamics	6
S18	Surface elasticity number	N_{El}	$-\frac{\Gamma'}{D_S} L \frac{(\partial\sigma)}{(\partial\Gamma')}$; Γ' = surface concentration of surfactant in undisturbed state, D_S = surface diffusivity, L = film thickness		Convection cells	10
S19	Surface viscosity number	N_{Vi}	$\mu_S/\mu L$; μ_S = surface viscosity, $[M/\theta]$, L = film thickness		Convection cells	10
T1	Taylor number (1)	N_{Ta_1}	$\omega_c(R_a)^{1/2}a^{3/2}\rho/\mu$; ($\omega_c$ = angular velocity of cylinder; R_a = mean radius of annulus)		Stability of flow pattern in annulus with rotating cylinder	5
T2	Taylor number (2)	N_{Ta_2}	$(2\omega L^2\rho/\mu)^2$[ω = rate of spin $(1/\theta)$; L = height of fluid layer]	α(Coriolis force/viscous force)^2	Effect of rotation on free convection	6
T3	Thiele modulus	m_T	$Q^{1/2}U^{1/2}L/k^{1/2}t^{1/2}$ = (Da IV)$^{1/2}$		Diffusion in porous catalysts	5
T4	Thoma number	σ_T	$(H_a - H_s - H_v)/H$ (H = total head; H_a = atm. pressure head; H_s = suction head; H_v = vapor pressure head)	Net positive suction head/total head	Cavitation in pumps	5
T5	Thomson number	N_{Th}	$\theta V/L$; θ = characteristic time (*cf.* N_{S_1})		Fluid flow	6
T6	Thring radiation group		$\rho C_p V/e^*\eta T^3$ (*cf.* Boltzmann number)	Bulk heat transport/heat transport by radiation	Radiation	5

TABLE B (Continued)

Serial No.	Name	Symbol	Definition	Significance	Field of Use	Reference
T7	Thring-Newby criterion	θ	$[(V_{m_1}+V_{m0})/V_{m0}](R/L)$; V_{m0}, V_{u_1} = mass flow rates of nozzle fluid and surrounding fluid $[M/\theta]$; R = equivalent nozzle radius; L = furnace half width		Combustion of fuels	*4*
T8	Truncation number	r	$\mu\gamma/P$ (*cf.* Hersey number)	Shear stress/normal stress	Viscous flow	*6*
V1	Valensi number	V	$\omega L^2\rho/\mu$; ω = circular oscillation frequency when $\mu = 0$ $[1/\theta]$		Oscillations of drops and bubble	*57*
V2	Vedernikov number	V	$\zeta^*\xi^*\overline{V}/(V_w-\overline{V}) \equiv \zeta^*\xi^*(N_{Fr_2})$ (ζ^* = exponent of hydraulic radius in formula [$^-$]; ζ^* = shape factor of channel section; V_w = absolute velocity of disturbance wave)	Generalized Froude number	Instability of open-channel flow	*6, 61*
V3	Velocity number	R_v	$\equiv$ Magnetic Reynolds number, *q.v.*			*5, 6*
W1	Weber number (1)	N_{We_1}	$V^2\rho L/\sigma = (N_{We_2})^2$	Inertia force/surface tension force	Bubble formation, etc.	*2, 25*
W2	Weber number (2)	N_{We_2}	$V(\rho L/\sigma)^{1/2} = (N_{We_1})^{1/2}$			
W3	Weber number (rotating)	W_e	$L^3N^2\rho/\sigma$; L = impeller diameter		Agitation	*64*
W4	Weissenberg number	N_{We}	$\omega_3 V/\omega_1 L$; $\omega_3 = \int_0^\infty sG(s)\,ds$, $\omega_1 = \int_0^\infty G(s)\,ds$, G = relaxation modulus of linear viscoelasticity, s = recoverable elastic strain	Viscoelastic force/viscous force	Viscoelastic flow	*63*

REFERENCES

1. Adrianov, V. N., Shorin, S. N., *AIAA J.* **1**, 1729 (1963).
2. Ahlstrom, H. G., *J. Fluid Mech.* **15**, 205 (1963).
3. Becker, H. A., Hottel, H. C., Williams, G. C., "Ninth Symposium (International) on Combustion," p. 7, Academic Press, New York, 1963.
4. Beer, J. M., Chigier, N. A., Lee, K. B., *Ibid.*, p. 892.
5. Boucher, D. F., Alves, G. E., *Chem. Eng. Progr.* **55** (9), 55 (1959).
6. *Ibid.*, **59** (8), 75 (1963).
7. British Standard 1991, "Recommendations for Letter Symbols, Signs and Abbreviations. Part 2. Chemical Engineering, Nuclear Science, and Applied Chemistry," British Standards Institution, London, 1961.
8. Brown, G. G., *et al.*, "Unit Operations," Wiley, New York, 1950.
9. Buckingham, E., *Phys. Rev.* **4**, 345 (1914).
10. Berg, J. C., Acrivos, A., *Chem. Eng. Sci.* **20**, 737 (1965).
11. Chukhanov, Z. F., *Intern. J. Heat Mass Transfer* **6**, 691 (1963).
12. Dallavalle, J. M., "Micromeritics," 2nd ed., Pitman, New York, 1948.
13. El'perin, I. T., *Inzh. Fiz. Zh. Akad. Nauk Belorussk. SSR* **4** (1), 131 (1963).
14. El'perin, I. T., *Intern. J. Heat Mass Transfer* **5**, 349 (1962).
15. Engel, F. V. A., *Z.V.D.I.* **107**, 671, 793 (1965).
16. Faller, A. J., *J. Fluid Mech.* **15**, 560 (1963).
17. Fedorov, B. I., *Inzh. Fiz. Zh. Akad. Nauk Belorussk. SSR* **7** (1), 21 (1964).
18. Gardner, G. O., Kestin, J., *Intern. J. Heat Mass Transfer* **6**, 289 (1963).
19. Gel'perin, I. T., Aĭnshteĭn, V. G., Goĭkhman, I. D., *Inzh. Fiz. Zh. Akad. Nauk Belorussk. SSR* **7** (7), 15 (1964).
20. Grassmann, P., *Chem. Ing.-Tech.* **31**, 148 (1959).
21. Grassman, P., Lemaire, L. H., *Ibid.*, **30**, 450 (1958).
22. Greene, D. F., Ph.D. Thesis, Columbia Univ., 1961 [*Dissertation Abstr.* **24** (8), 3248 (1964)].
23. Gukhman, A. A., "Introduction to the Theory of Similarity," Academic Press, New York, 1965.
24. Gutfinger, C., Tallmadge, J. A., *A.I.Ch.E. J.* **10**, 774 (1965).
25. Hahnemann, H. W., "Die Umstellung auf das internationale Einheitensystem in Mechanik und Wärmetechnik," VDI-Verlag, Düsseldorf, 1959.
26. Holt, M., "Dimensional Analysis" in "Handbook of Fluid Dynamics," V. L. Streeter, ed., McGraw-Hill, New York, 1961.
27. Hottel, H. C., Sarofim, A. F., *Intern. J. Heat Mass Transfer* **8**, 1153 (1965).
28. Hottel, H. C., Williams, G. C., Jensen, W. P., Tobey, A. C., Burrage, P. M. R., p. 923 in "Ninth Symposium (International) on Combustion," Academic Press, New York, 1963.
29. Huntley, H. E., "Dimensional Analysis," MacDonald & Co., London, 1952.
30. Johnson, S. P., "Survey of Flow Calculation Methods," p. 98, Pre-printed Papers & Program, Aeronautic & Hydraulic Divisions, A.S.M.E. Summer Meeting, June 19–21, Univ. of Calif. and Stanford Univ., 1934.
31. Kafarov, V. V., *Zh. Prikl. Khim.* **29**, 40 (1956).
32. Kay, J. M., "An Introduction to Fluid Mechanics & Heat Transfer," Cambridge Univ. Press, 1957.
33. Kestin, J., Persen, L. N., *Intern. J. Heat Mass Transfer* **5**, 143 (1962).
34. Klinkenberg, A., Mooy, H. H., *Chem. Eng. Progr.* **44**, 17 (1948).
35. Koide, K., Kubota, H., Shindo, M., *Chem. Eng.* (Japan), **28** (8), 657 (1964).
36. Lȳkov, A. V., Mikhaĭlov, Yu. A., "Theory of Energy & Mass Transfer," Prentice-Hall, Englewood Cliffs, N.J., 1961.
37. McCabe, W. L., Smith, J. C., "Unit Operations of Chemical Engineering," McGraw-Hill, New York, 1956.
38. Matsuhisa, S., Bird, R. B., *A.I.Ch.E. J.* **11**, 588 (1965).
39. Mikhaĭlov, Yu. A., Bornikova, R. M., *Inzh. Fiz. Zh. Akad. Nauk Belorussk. SSR* **6** (10), 45 (1963).
40. Mikhaĭlov, Yu. A., Romanina, I. V., *Ibid.*, **7** (1), 49 (1964).
41. Miyauchi, T., Nakano, K., Obata, K., Kimura, S., *Chem. Eng.* (Japan) **26** (9), 999 (1962).
42. Mkhitaryan, A. M., "Hydraulics & Fundamentals of Gas Dynamics," Israel Program for Scientific Translations, Jerusalem, 1964.
43. Mordell, D. L., Wu, J. H. T., *Can. Aeronaut. Space J.* **9** (4), 117 (1963).
44. Motulevich, V. P., Eroshenko, V. M., Petrov, Yu. P., in "Physics of Heat Exchange & Gas Dynamics," A. S. Predvoditelev, ed., Consultants Bureau, New York, 1963.
45. Nagata, S., *Chem. Eng.* (Japan) **27** (8), 592 (1962).
46. Nield, D. A., *J. Fluid Mech.* **19**, 341 (1964).
47. Potter, J. M. F., B.Sc. Thesis, Dept. of Chem. Engrg., Univ. of Birmingham, England, 1959.
48. Rayleigh, Lord, *Phil. Mag.* **48**, 321 (1899).
49. Reiner, M., *Phys. Today* **17** (1), 62 (1964).
50. Rouse, H. (ed.), "Engineering Hydraulics," Wiley, New York, 1950.
51. Rouse, H., Ince, S., "History of Hydraulics," Iowa Institute of Hydraulic Research, State University of Iowa, 1957.
52. Sazhin, B. S., *Inzh. Fiz. Zh. Akad. Nauk Belorussk. SSR* **5** (6), 13 (1962).
53. Sazhin, B. S., Miklin, Yu. A., *Ibid.*, **6** (10), 57 (1963).
54. Schlichting, H., "Boundary Layer Theory," 4th ed., McGraw-Hill, New York, 1960.
55. Scriven, L. E., Sternling, C. V., *J. Fluid Mech.* **19**, 321 (1964).
56. Sillem, H., *Z.V.D.I.* **106**, 398 (1964).
57. Szebehely, V. G., p. 771 in "Proc. 2nd U.S. Nat. Congress of Appl. Mech.," Ann Arbor, Mich., June 1954; A.S.M.E., New York, 1955.
58. Tallmadge, J. A., Labine, R. A., Wood, B. H., *Ind. Eng. Chem. Fundamentals* **4**, 400 (1965).
59. Tamarin, A. I., *Inzh. Fiz. Zh. Akad. Nauk Belorussk. SSR* **6** (7), 19 (1963).
60. Tartakovskii, D. F., *Ibid.*, **7** (1), 71 (1964).
61. Vedernikov, V. V., *Compt. Rend. Acad. Sci. U.R.S.S.* **48**, 239 (1945); 52, 207 (1946).
62. Weber, M., *Jahrb. Schiffbautechn. Ges.* **20**, 355 (1919).
63. White, J. L., *J. Appl. Polymer. Sci.* **8**, 2339 (1964).
64. Yamaguchi, I., Yabuta, S., Nagata, S., *Chem. Eng.* (Japan) **27** (8), 576 (1963).
65. Yas'ko, O. I., *Inzh.-Fiz. Zh. Akad. Nauk Belorussk. SSR* **7** (12), 112 (1964).
66. Zabrodskiĭ, S. S., "Flow & Heat Transfer in Fluidized Beds," to be published shortly by M.I.T. Press.
67. Zhuravleva, V. P., *Inzh.-Fiz. Zh. Akad. Nauk Belorussk. SSR* **6** (9), 73 (1963).

DIMENSIONLESS GROUPS (Supplementary Table)

George D. Fulford and John P. Catchpole

Reprinted from Industrial and Engineering Chemistry, Vol. 60, No. 3, pp. 71 to 78. Copyright 1968 by the American Chemical Society and reprinted by permission of the copyright owners and the authors.

TABLE II. ALPHABETICAL LIST OF NEW GROUPS

Serial No.	Name	Symbol	Definition	Significance	Field of Use	Reference
A0	Absorption No.	Ab	$kc_L\sqrt{\frac{xL_f}{DV'_f}}$ kc_L = liquid side mass transfer coefficient; x = length of wetted surface; L_f = film thickness; V'_f = volume flow rate per wetted perimeter $[L^2/\theta]$	Dimensionless mass transfer coefficient	Gas absorption in wetted wall column	*40*
A1a	Advance ratio	J	V/ND V = forward speed; D = propeller diameter	Special form of Strouhal No.	Propeller studies	*27*
A1b	Aeroelasticity parameter	—	≡ Cauchy No., *q.v.*	Inertia force/compressibility force	Compressible flow	*27*
A4a	Anonymous group 3	ε	Dx/V'_fL_f (symbols as in Absorption No.). $\varepsilon = (Ab)^2/(N_{Sh})^2$	Dimensionless diffusivity	Gas absorption in wetted wall column	*40*
A4b	Anonymous group 4	$1/\alpha$ $(1/\beta)$	$\tau_wR/V_\infty\mu$; R = cylinder radius; V_∞ = velocity outside boundary layer	Frictional force/viscous force dimensionless skin friction)	Laminar boundary layer flow	*12, 28, 35*
B1a	Bairstow No.	—	$V/V_{s.w}$ $V_{s\,w}$ = velocity of sound at wall (*cf.* Mach. No.)	Previously used for Mach No., now largely obsolete	—	*27*
B2a	Batchelor No.	—	$VL\sigma_e/V_l^2\varepsilon_e$ ε_e = electrical permittivity $[Q^2\theta^2/L^3M]$	—	Magnetofluid dynamics	*27*
B13a	Buoyancy parameter	—	$\frac{\Delta T}{T}\frac{gL}{V^2} = (N_{Gr})/(N_{Re})^2 = \frac{\Delta T}{T}\left(\frac{1}{N_{Fr_1}}\right)$	Buoyancy force/inertia force	Free convection	*27*
D6a	Darcy No. (2)	Da_2	VL/D'; D' = permeability coefficient of porous medium $[L^2/\theta]$	Inertia force/permeation force	Flow in porous media	*19*
D8a	Generalized Deborah No.	N_2	$\sqrt{I_e - I_w}\cdot\theta_n$ I_e = invariant of rate of strain tensor (sec.$^{-2}$); I_w = invariant of vorticity tensor (sec.$^{-2}$); θ_n = natural time (sec.)	Generalization of group **D8**	Rheology	*2*
D14a	Dufour No.	Du_2	$\mathscr{R}\Theta n'_{10}/c_p$ Θ = thermodiffusion constant = $(D_T/D)/n_{10}n_{20}$ [−]; D_T = thermal diffusion coefficient $[L^2/\theta]$; $n'_{10}, n'_{20} = n'_1/n', n'_2/n'$; n' = total No. of molecules = $n'_1 + n'_2$; n'_1, n'_2 = No. of molecules of components 1, 2, in binary mixture; also $Du_2 = (D_T/D)p/\rho c_pTn_{20}$	Heat of isothermal mass transfer/enthalpy of unit mass of mixture	Thermodiffusion	*22*
E1a	Einstein No.	—	V/V_l (V_l = speed of light) (*cf.* Lorentz No.)	Fluid velocity/velocity of light	Magnetofluid dynamics	*27*
E4a	Electric field parameter	R_E	$E/V\mu_eH_e$		Magnetohydrodynamics	*26*
E4b	Electrical characteristic No.	El	$\rho(d\chi/dT)L^2\cdot\Delta T\cdot E_1^2/\mu^2$ E_1 = electrical field strength $[ML/Q\theta^2]$; χ = dielectric susceptibility $[Q^2\theta^2/ML^3]$		Electrical effects on transfer processes	*6*
E4c	Electrical Nusselt No.	N_u	VL/D^*; $D^* = \frac{1}{2}(D^+ + D^-)$ $[L^2/\theta]$ D^+, D^- = diffusion coefficients of ions (*cf.* group *P2*)	Convection current/diffusion current	Electrochemistry	*10*
E4d	Electrical Reynolds No. (1)	—	$\varepsilon_eV/Q'Lb'$ ε_e = electrical permittivity $[Q^2\theta^2/L^3M]$; Q' = space charge density $[Q/L^3]$; b' = carrier mobility $[Q\theta/M]$		Electrical effects in flow	*27*
E4e	Electrical Reynolds No. (2)	—	Alternate name for group **E4c**, *q.v.*			*10*
E10a	Modified Euler No.	Eu'	$H_{L'}\rho_Lg/V_G^2\rho_G$ H'_L = head of liquid on tray $[L]$; ρ_L, ρ_G = densities of liquid, vapor; V_G = vapor velocity based on free area, $[L/\theta]$	Friction head/velocity head	Flow of vapor across mass transfer trays	*13*
E13a	Expansion No.	Ex	$\left(\frac{gd}{V^2}\right)\left(\frac{\rho_L - \rho_G}{\rho_L}\right)$ d = bubble diam., V = bubble veloc., ρ_L, ρ_G = densities of liquid, gas	$1/N_{Fr_1}$ × density ratio	Rise of bubbles	*13*
F12a	Fourier No. (flow)	Fo	$\mu\theta/\rho L^2$		Unsteady state flow problems	*17*
F12b	Frank-Kamenetskii No.	δ	$\frac{Q''}{k}\frac{E_a}{\mathscr{R}T_o^2}L^2k_o\exp(-E_a/\mathscr{R}T)$ Q'' = heat liberated per unit mass of material reacting/unit volume $[1/L\theta^2]$; k = thermal conductivity of reacting mixture $[ML/T\theta^3]$; k_o = preexponential constant in Arrhenius equation $[M/\theta]$	Dimensionless heat effect of reaction	Heat transfer in reacting systems	*24*
F12c	Frequency parameter	—	$\omega'L/V = 2\pi$ × Strouhal No.; $2\pi/\omega'$ = period of motion $[\theta]$	Special form of Strouhal No. (*cf.* also **T5**)	Unsteady state flow, etc.	*27*
F12d	Frequency No. (2)	N_f	ω_rL/V; L = packing element diameter $[L]$; V = interstitial fluid velocity $[L/\theta]$; ω_r = radial frequency (radians/sec.) $[1/\theta]$ (*cf.* groups **H10**, **S16**, **T5**)	Special form of group **F12c**	Flow in packed or fluidized beds	*23, 29*

TABLE II. ALPHABETICAL LIST OF NEW GROUPS (Continued)

Serial No.	Name	Symbol	Definition	Significance	Field of Use	Reference
F12e	Frössling No. (heat transfer)	Fs_h	$(N_{Nu} - 2)/(N_{Re}^{1/2} N_{Pr}^{1/3})$	Special dimensionless heat transfer coefficient	Heat transfer to spheres in turbulent streams	*8a*
F12f	Frössling No. (mass transfer)	Fs_m	$(N_{Sh} - 2)/(N_{Re}^{1/2} N_{Sc}^{1/3})$	Special dimensionless mass transfer coefficient	Mass transfer to spheres in turbulent streams	*8a*
G2a	Geometric No.	Ge	h^*/H^*; h^* = surface area of packing element/perimeter $[L]$; H = height of packing $[L]$.	Dimensionless packed bed height	Mass transfer in packed beds	*5a*
G2b	Goertler parameter	—	$\frac{VL_b\rho}{\mu}\left(\frac{L_b}{R_c}\right)^{1/2}$ L_b = boundary layer momentum thickness, $[L]$; R_c = radius of curvature $[L]$	Modified Reynolds No.	Boundary layer flow on curved surfaces	*27*
G5a	Diffusional Grashof No.	Gr_{AB}	$L^3\rho^2 g\beta'_A\Delta n'_A/\mu^2$ n'_A = mass fraction of species A, $[-]$; β'_A = coefficient of density change with n'_A, $[-]$	Buoyant forces × inertia forces/(viscous forces)^2	Interphase transfer by free convection (density changes caused by concentration differences)	*37*
H8a	Hess No.	Ge	$(KL^2/a_m)(C_o)^{n-1}$ n = order of reaction $[-]$; C_o = initial concn. $[M/L^3]$; a_m = mass transfer conductivity of reaction products $[L^2/\theta]$; K = reaction rate constant, $\left[\frac{1}{\theta}\frac{L^{3n-3}}{M^{n-1}}\right]$		Heat and mass transfer with chemical and phase changes	*22*
H9a	Homochronicity No.	Ho_3	$N\theta$ (N = mixer r.p.m., θ = mixing time)		Mixing, agitation	*31*
H11a	Hydraulic resistance group	Γc	$\Delta p_p/\rho_L gL = We_4 \times$ Laplace No. $= N_{Eu_1} \times N_{Fr_1}$ Δp_p = pressure drop across liquid on tray $[M/L\theta^2]$; ρ_L = liquid density; L = depth of liquid on tray	Characterizes development of interfacial area per unit area of tray	Pressure drop in distillation columns	*15*
I1	Ilyushin No.	I	$(Vd\rho/\mu) . 4\tau_D/3V^2\rho = 4\tau_D/3V^2\rho . N_{Re}$; τ_D = max. dynamic slip resistance $[M/L\theta^2]$		Flow of viscoplastic fluids	*20a*
K7a	Knudsen No. for diffusion	$N_{Kn_{A2}}$	$3eD_{AB}/4\zeta K_{oA}\bar{u}_A$ K_{oA} = Knudsen flow permeability constant; $\bar{u}_A$ = equilibrium mean molecular speed of species A	Differs from **K7** by numerical constant	Gaseous diffusion in packed beds	*34*
K7b	Kondrat'ev No.	Kn	ΨN_{Bih}; N_{Bih} = heat transfer Biot No.; Ψ = temp. field nonuniformity parameter = $(t_s - t_a)/(\bar{t} - t_a)$; t_a = temp. of surrounding medium; t_s = body surface temp.; $\bar{t}$ = body mean temp.		Heat transfer between fluid and body	*21*
L3a	Laplace No.	La_4	$\Delta p_p . L/\sigma = \Gamma c/We_4$ Δp_p = pressure drop across liquid on tray $[M/L\theta^2]$; L = depth of liquid on tray		Interfacial behavior on distillation trays	*32*
L5a	Lebedev No.	Le_2	$eb_T(t_a - t_o)/c_b p\rho_s$ b_T = intensity of vapor expansion in capillaries of body on heating $[M/L^3T]$; t_a = temp. of surrounding medium; t_o = initial temp.; ρ_s = density of solid $[M/L^3]$	Molar expansion flux/molar vapor transfer flux	Drying of porous materials	*22*
L5b	Leroux No.	—	≡ cavitation No. **C6**			*27*
L7a	Turbulent Lewis No.	Le_T	$c_p\rho\varepsilon_D/k_T = l_D/l_T = \varepsilon_D/\varepsilon_T$ k_T = eddy thermal conductivity $[LM/T\theta^3]$; l_D, l_T = mixing lengths for mass, heat transfer $[L]$; ε_T = eddy thermal diffusivity $[L^2/\theta]$		Combined turbulent heat and mass transfer	*11*
L7b	Lewis-Semenov No.	—	$\equiv 1/N_{Le}$; N_{Le} = group **L7**			*27*
L7c	Lock No.	—	$\rho R^4 ia'/I$ ρ = fluid density; a' = rotor lift curve slope $[L^2/M]$; i = blade chord $[L]$; R = rotor radius $[L]$; I = moment of inertia of blade about hinge $[L^4]$		Rotor blade dynamics	*27*
M3a	Magnetic Grashof No.	—	$4\pi\sigma_e\mu_e(\mu/\rho) . N_{Gr}$ N_{Gr} = group **G5**		Magnetofluid dynamics	*7*
M4a	Magnetic No.	R_m	$\mu_e H_e(\sigma_e L/\rho V)^{1/2}$ = (magnetic force parameter)$^{1/2}$	See group **M3**	Magnetohydrodynamics	*26*
M5a	Magnetic Prandtl No.	—	$\sigma_e\mu_e\frac{\mu}{\rho}$ *cf.* Magnetic Grashof No., group **M3a**	Magnetic Reynolds No./Reynolds No. (properties of fluid)	Magnetofluid dynamics	*27*
P3a	Plasticity No.	P	≡ Bingham No.			*41*
P7a	Modified Power No.	N'_P	$\frac{\Pi}{L^5\rho N^3}\left(\frac{D'_e}{L'}\right)\frac{(\Delta\omega)^{-1/2}}{(N_b N_s)^{0.67}}$ D'_e = effective agitator diameter $[L]$; L'_e = effective agitator height; $\Delta\omega$ = wall proximity factor; N_b = No. of blades on agitator; N_s = effective No. of blade edges		Agitation	*36*
P8a	Total Prandtl No.	**Pr**	$\frac{\varepsilon_M + (\mu/\rho)}{\varepsilon_T + \alpha}$ ε_M, ε_T = eddy transfer coefficient for momentum, heat $[L^2/\theta]$	Total momentum diffusivity/total thermal diffusivity	Heat transfer in combined turbulent and laminar flows	*25*
P8b	Turbulent Prandtl No.	Pr_T	$\varepsilon_M/\varepsilon_T = l/l_T$ ε_M, ε_T = eddy viscosity, eddy thermal diffusivity $[L^2/\theta]$; l, l_T = mixing lengths for momentum, heat transfer	Eddy momentum diffusivity/eddy thermal diffusivity	Heat transfer in turbulent flow	*9, 25*

TABLE II. ALPHABETICAL LIST OF NEW GROUPS (Continued)

Serial No.	Name	Symbol	Definition	Significance	Field of Use	Reference
P12a	Predvoditelev No. (mass transfer)	Pd_m	$(\Gamma_m L^2/a_m)(N_{Fo_m})$ Γ_m = rate of change of mass transfer potential of medium, (mass/unit mass)/time $[1/\theta]$; a_m = mass conductivity of material $[L^2/\theta]$; N_{Fo_m} = group **F12**	Rate of change of concn. of medium/rate of change of concn. of body	Mass transfer	*22*
P15a	Pulsation No.	N_{Pu}	$fd_e\rho/G$ d_e = equiv. diam. of channel		Transfer to pulsed fluid	*20*
R0a	Radial frequency parameter (1)	—	$\omega_r D/V^2$, $\omega_r \alpha/V^2$ D = diffusivity or dispersion coefficient of packed bed $[L^2/\theta]$, ω_r = radial frequency (radians/sec.) $[1/\theta]$		Packed and fluidized beds	*18, 33, 39*
R0b	Radial frequency parameter (2)	—	$\omega_r L^2/\alpha$ ω_r as in group **R0a**; L = tube radius $[L]$		Packed and fluidized beds	*29, 39*
R0c	Radial frequency parameter (3)	—	$\omega_r^2 DL/V^3$ (Quantities as in groups **R0a** and **R0b**)	$\dfrac{(\text{Group R0a})^2}{(\text{Group P2})}$	Packed and fluidized beds	*18, 33, 39*
R0d	Radial frequency parameter (4)	—	$L(\omega_r/2D)^{1/2} \equiv [\frac{1}{2}(\text{group R0b})^{1/2}$ (quantities as in groups **R0a** and **R0b**)	Analog of Wave No.	Packed and fluidized beds	*39*
R1a	Radiation parameter	Φ	$e^+\eta T_w^3 d_h/k$ e^+ = function of mean surface emissivity of walls, $[-]$; T_w = wall temp. (abs.) $[T]$		Effect of radiation on convective mass transfer	*14*
R6a	Reaction enthalpy No.	N_H	$(\Delta u)_A(\Delta n_A)/C_p(\Delta T)$ $(\Delta u)_A$ = enthalpy of reaction per unit mass of A produced $[L^2/\theta^2]$, n_A = mass fraction of species A $[-]$	Change in reaction energy/change in sensible energy	Interphase transfer with chemical reaction	*37*
R15a	Rossby No.	—	$V/\omega L$	More general form of group **R15**		*27*
S4a	Schmidt No. (2)	—	$\equiv$ Semenov No. (2)	No longer used		*27*
S4b	Schmidt No. (3)	$(Sc)_3$	$\mu\chi/\rho\sigma_e L^2$ χ as in group **E4b**	Diffusivity of vorticity/mass diffusivity of ions	Electrochemistry	*10*
S4c	Total Schmidt No.	**Sc**	$\dfrac{\varepsilon_M + (\mu/\rho)}{\varepsilon_D + D}$ ε_M = eddy viscosity $[L^2/\theta]$	Total momentum diffusivity/total mass diffusivity	Mass transfer in combined laminar and turbulent flows	*25*
S4d	Turbulent Schmidt No.	Sc_T	$\varepsilon_M/\varepsilon D = l/l_D$ ε_M = eddy viscosity, $[L^2/\theta]$, l, l_D = mixing lengths for momentum, mass transfer $[L]$	Eddy momentum diffusivity/eddy mass diffusivity	Mass transfer in turbulent flow	*25*
S5a	Semenov No. (2)	—	$\equiv 1/N_{Le}$ (see group **L7**)			*27*
S7a	Smoluchowski No.	—	$L/\lambda \equiv 1/N_{Kn}$ N_{Kn} = group **K6**	See group **K6**		*27*
S9a	Soret No.	S_o	$\Theta n'_{20}$ (definitions as group **D14a**)	Dimensionless thermodiffusion coefficient	Coupled heat and mass transfer	*22*
S10a	Spalding No.	B'	$c_p \Delta T/(r_v - q_r/V_m)$; q_r = radiant heat flux $[ML^2/\theta^3]$; V_m = rate of mass transfer $[M/\theta]$	Ratio (sensible heat/latent heat) for evaporated material	Droplet evaporation	*33a*
S13a	Stark No.	Sk	$\eta T^3 L/k$ L = thickness of layer $[L]$; ($\equiv$ Stefan No.)		Radiant heat transfer	*1, 30*
S14a	Stewart No.	—	$\mu_e^2 H_e^2 \sigma_e L/V\rho$	(Hartmann No.)^2/Reynolds No.	Magnetofluid dynamics	*38*
S15a	Stokes No. (2)	St_2	$1.042\ m_f g\rho(1-\rho/\rho_f)R^{*3}/\mu^2$; ρ, μ = density, viscosity of fluid; m_f, ρ_f = mass, density of float; R^* = tube radius/float radius $[-]$		Calibration of rotameters	*10a*
S18a	Surface tension No.	T_S	$\mu^2/h^*\sigma\rho$; h^* = surface area of packing element/perimeter $[L]$		Mass transfer in packed columns	*5a*
T5a	Thompson No.	N_{Th_2}	$\dfrac{-\Delta t}{\Delta L}L^2\left(\dfrac{d\sigma}{dT}\right)/\mu\alpha$ (*cf.* Marangoni No.)		Cellular convection	*3*
T7a	Thrust coefficient	T_c	$F_T/\rho V^2 d^2$ F_T = thrust force $[ML/\theta^2]$; V = forward speed $[L/\theta]$; d = tip diameter $[L]$		Propeller studies	*27*
T7b	Torque coefficient	Q_c	$F'/\rho V^2 d^3$ F' = propeller torque $[ML^2/\theta^2]$		Propeller studies	*27*
T7c	Transiency groups	K_p K_Q	K_p: $\dfrac{1}{\left(\dfrac{\partial p}{\partial L}\right)} \cdot \dfrac{\partial\left(\dfrac{\partial p}{\partial L}\right)}{\partial(Fo_f)}$ K_Q: $\dfrac{1}{(N_{Re})} \cdot \dfrac{\partial(N_{Re})}{\partial(Fo_f)}$ $\partial p/\partial L$ = pressure gradient in flow direction $[M/L^2\theta^2]$; Fo_f = group **F12a**		Transient flow behavior	*17*
W0a	Wave No.	**k**	$L(\omega_r/2\alpha)^{1/2}$ w_r = radial frequency (radians/sec.) $[1/\theta]$	Heat transfer analog group **R0d**	Cyclic heat transfer	*4*
W3a	Weber No. (3)	We_3	$\sigma/\rho_l g L^2$ ρ_L = liquid density $[M/L^3]$; L = depth of liquid on tray $[L]$	Surface tension force/gravity force	Interfacial area determination in distillation equipment	*32*
W4a	Generalized Weissenberg No.	N_1	$\sqrt{I_e}\theta_n$ (definitions as group **D8a**)	Generalization of group **W4**	Rheology	*2*

TABLE III. TABLES FOR IDENTIFYING DIMENSIONLESS GROUPS

These tables incorporate those from the earlier publication (*5*)

Basic dimensions are taken to be: *Length* [*L*]
Mass [*M*]
Electrical charge [*Q*]
Temperature [*T*]
Time [*θ*]

PHYSICAL PROPERTIES

General Physical Properties

Parameter	Symbol	Dimensions	Exponent	Groups
Coefficient of thermal bulk expansion	β	$1/T$	−1	E4, E12, G2
			+1	G5, K9, M3a, R5, R6
Coefficient of density change with concn. n_A	β_A	—	+1	G5a
Density	ρ	M/L^3	−2	M1
			−1	A1, B1, B10, C1, C2, C6, D3, D13, D14a, E3, E9, E10, E10a, E13, E13a, F1, F11, F12a, H5, H11a, I1, J1, K4, K10, L5a, L5b, L7, M3, M3a, M5a, M6, N2, N3, N4, P7, P8a, P9, R0a, R9, R10, R13, S4, S4b, S4c, S13, S14a, S15, S15a, S18a, T7a, T7b, T7c, W3a
			$-\frac{2}{3}$	C9, J3
			$-\frac{1}{2}$	E2, L11, M4a, O2, P13
			$+\frac{1}{3}$	J3, K5, S3
			$+\frac{1}{2}$	D10, G3, P10, P11, R4, W0a, W2
			$+\frac{2}{3}$	F2, N8, S2
			+1	A1b, B1, B6, B9, C5, C7, D5, D7, D13, E4, E4b, E6, E7, E8, E10a, E13a, G2b, H6, H11, I1, J1, J3, J4, K1, L7a, L7b, L7c, L9, M3a, P1, P8a, P15a, R0b, R5, R11, R13, S4a, S5a, S15a, S17, T1, T5a, T6, T7c, V1, W1, W3
			+2	A5, C10, G1, G5, G5a, K9, R5, R6, S15a, T2
Density gradient	$d\rho/dL$	M/L^4	+1	R13
Diffusivity (molecular unless noted otherwise)	D, a_m	L^2/θ	−1	B5, B7[4], D2, D14a, E4c, E4e, H8a[1], K4[1], K7[2], L2, L7, L10, N7, P2, P9, P12a[1], S4, S4c, S9a
			$-\frac{2}{3}$	J3
			$-\frac{1}{2}$	A0, R0d
			+1	A4a, F12[1], K7[3], K7a, L2[4], L7b, L9[1], R0a, R0c, S4a, S5a
Diffusivity (eddy)	E_D, ε_D	L^2/θ	−1	S4c, S4d
			+1	L2, L7a
Diffusivity (surface)	D_S	ML/θ^2	−1	S18
Diffusion tortuosity	ζ	—	−1	K7, K7a
Dispersion (permeability) coefficient	D'	L^2/θ	−1	D6a
			+1	R0a, R0c
Molecular weight	M	M	−1	D14, K9, S6
			+1	D14
Permeability (packed bed)	ξ	L^2	$+\frac{1}{2}$	L6
Porosity (voidage)	e	—	−1	B6, M12
			$-\frac{1}{2}$	L6
			+1	K7, K7a, L5a
Specific gravity	γ	—	$\pm\frac{1}{3}$	F2
Surface tension	σ	M/θ^2	−3	C2
			−2	C1
			−1	B9, C3, C13, E8, L3a, L6, R1, S18a, W1, W3
			$-\frac{1}{2}$	D10, G3, O2, P13, R4, W2
			+1	M9, S17, S18, T5a, W3a

PHYSICAL PROPERTIES

General Physical Properties

Parameter	Symbol	Dimensions	Exponent	Groups
Thermal diffusion coefficient	D_T	L^2/θ	+1	D14a, S9a

[1] Coefficient of potential diffusion in mass transfer.
[2] Knudsen diffusion coefficient.
[3] Binary bulk diffusion coefficient.
[4] Effective diffusivity (molecular and eddy) $= D + \varepsilon_D$.

Electrical and Magnetic Properties

Parameter	Symbol	Dimensions	Exponent	Groups
Carrier susceptibility	b'	$Q\theta/M$	−1	E4d
Current density	I	$Q/L^2\theta$	+1	K10
Dielectric susceptibility	χ	$Q^2\theta^2/L^3M$	+1	E4b, S4b
Electrical conductivity	σ_e	$Q^2\theta/L^3M$	−1	E6, S4b
			$+\frac{1}{2}$	H2, M4a
			+1	B2a, L11, M3, M3a, M5a, M7, S14a
Electrical permittivity	ε_e	$Q^2\theta^2/L^3\theta$	−1	B2a
			+1	E4d
Field strength	H_e	$Q/L\theta$	−2	J4
			−1	E4a
			+1	H2, L11, M4a
			+2	K9, M3, M6, S6, S14a
Magnetic permeability	μ_e	ML/Q^2	−1	E4a, E6, J4
			+1	H2, M3a, M4a, M5a, M6, M7
			$+\frac{3}{2}$	L11
			+2	M3, S14a
Space charge density	Q	Q/L^3	−1	E4d
Voltage	E	ML/θ^2Q	+1	E4a, K10
			+2	E4b

Thermal Properties

Parameter	Symbol	Dimensions	Exponent	Groups
Eddy thermal conductivity	k_T	$ML/T\theta^3$	−1	L7a
Eddy thermal diffusivity	ε_T	L^2/θ	−1	L7a, P8a, P8b
Humid heat	s	L^2/θ^2T	−1	P15
Latent heats of phase change	λ, r	L^2/θ^2	−1	A3, E11, E12, E13, J1, M1, S10a
			+1	B13, C10, K8, K11, N5
Ratio of specific heats	γ	—	$\pm\frac{1}{2}$	L5
Specific heat	C, c_p	L^2/θ^2T	−1	B2, B13, D3, D14a, D15, E1, F11, J2, K8, K11, L7, M10, N5, R0a, R3, R6a, S13
			$-\frac{1}{3}$	J2
			$+\frac{1}{2}$	F6, F7, F8, W0a
			$+\frac{2}{3}$	J2
			+1	A3, E4, E12, G4, J1, J4, L7a, L7b, L9, P1, P8, P8a, R0b, R3, R5, R6, R7, S5a, S10a, T5a
Surface emissivity	e^*	—	−1	T6
			+1	R1a
Temperature conductivity (thermal diffusivity)	$\alpha = \frac{k}{\rho c_p}$	L^2/θ	−1	L7b, L9, P1, P8a, P12, R0b, R5, S5a, T5a
			$-\frac{1}{2}$	W0a
			+1	C13, L7, L10, R0a
Thermal conductivity	k, λ	$ML/T\theta^3$	−3	M1
			−2	R6
			−1	B4, B12, C7, C9, C10, D4, F12b, G4, K3, K7b, L7b, L9, N6, P1, P5,

Thermal Properties

Parameter	Symbol	Dimensions	Exponent	Groups
				P8, R0b, R1a, R5, S4a, S5a, S13a, S14, T5a
			$-\frac{2}{3}$	J2
			$-\frac{1}{2}$	T3, W0a
			+1	F11, L7, R0a, R1

Rheological and Elastic Behavior

Parameter	Symbol	Dimensions	Exponent	Groups
Eddy viscosity	e_M	L^2/θ	+1	P8a, P8b, S4c, S4d
Invariants of rate-of-strain and vorticity tensors	I_e, I_w	θ^{-2}	$+\frac{1}{2}$	D8a, W4a
Maximum dynamic slip resistance	τ_D	$M/L\theta^2$	+1	I1
Modulus of elasticity	E	$M/L\theta^2$	−1	A1b, C5, E4, H11
			+1	C1, E13, R1
			+3	A1
Rate of shear	$\dot{\gamma}$	$1/\theta$	+1	T8
Rigidity coefficient	μ_p	$M/L\theta$	−2	H6
			−1	B3, P3a
Shear stress	τ	$M/L\theta^2$	−1	E5
			$-\frac{1}{2}$	P10
			$+\frac{1}{2}$	P11
			+1	A4b
Viscosity (in all cases, kinematic viscosity has been written as μ/ρ)	μ	$M/L\theta$	−2	A1, A5, E4b, G1, G5, G5a, K1, K9, S15a, S17, T2
			−1	A4b, B6, C10, D5, D7, E6, E7, G2b, H8, I1, L1, L3, M3a, O1, P4, P11, R5, R6, R11, S9, S19, T1, T5a, T7c, V1
			$-\frac{2}{3}$	F2, K5, N8
			$-\frac{1}{2}$	H2
			$-\frac{1}{3}$	S2, S3
			$+\frac{1}{2}$	E2
			$+\frac{2}{3}$	C9, J2, J3
			+1	B12, C3, C13, E3, E5, F12a, M1, M3a, M5a, O2, P8, P8a, P9, S4, S4b, S4c, S8, S15, T7c, T8
			+2	C1, S18a
			+4	C2
Viscosity (surface)	μ_s	M/θ	+1	S19
Yield stress	τ_y	$M/L\theta^2$	+1	B3, H6, P3a

LENGTHS, AREAS, AND VOLUMES

Characteristic Lengths

Parameter	Symbol	Dimensions	Exponent	Groups
General length dimension	Various	L	−5	P7, P7a
			−3	F9, L1, S15a, T7b
			−2	D9, D12, E3, F11, F12, F12a, H4, H5, N2, N3, N4, O1, R9, S4b, S8, S9, S15, T7a, W3a
			−1	A1a, B1, B10, E2, E4d, E5, E10, F1, F13, G2a, G4, H10, H11a, K6, L4, P7a, R6, R10, R15, R15a, R16, S6, S7a, S18a, S19, T5, T5a, T7, T7c, W4
			$-\frac{1}{2}$	B11, D7, F14, G2b, O2
			$+\frac{1}{2}$	A0, G2b, M4a, R4, T1, W2
			+1	A4a, A4b, B2a, B3, B4, B5, B7, B10, B13a, C7, D1, D3, D5, D6a, D11, D13, E4c, E4e, E7, E10, E13a, F1, F2, F5, F12c, F12d, F15, G2a, G2b, G3, H2, I1, K3, K4, K6, K7b, K10, L3, L3a, L4, L7c, L9, L11, M1, M3, M7, N6, N7, N8, P1, P2, P3a, P7a, P11, P15a, R0c, R0d, R1a, R8, R10, R11, R16, S2, S7, S7a, S13a, S14, S14a, S16, S17, S18, T3, T7c, W0a, W1
			$+\frac{3}{2}$	D7, H7, T1
			+2	B9, D2, D4, E4b, E8, F12b, H6, H8a, K9, O1, P4, P5, P12, P12a, R0b, S8, S9, T5a, T7c, V1
			+3	A5, C10, G1, G5, G5a, K1, M3a, R5, S15a, W3
			+4	L7c, T2
			+5	R6
Dimension of agitator, propeller, etc.	Various	L	−5	P7, P7a
			−3	F9, T7b
			−2	H4, T7a
			−1	A1a, P7a
			+1	D11, F15, L7c, P7a
			+3	W3
			+4	L7c
Film thickness	L_f	L	−1	A4a
			+1	N8
Furnace half-width	L	L	−1	T7
Larmor radius	L_L	L	+1	L4
Mean free path	λ	L	−1	S7a
			+1	K6
Mixing length-mass	l_D	L	−1	S4d
			+1	L7a
Mixing length-heat	l_T	L	−1	L7a, P8b
Mixing length-momentum	l	L	+1	P8b, S4d
Particle dimension	d_e	L	−2	S15
			+1	F2, F12d
			+3	A5, G1
Pore or nozzle radius	R	L	+1	B7
Thickness of liquid layer	L	L	−1	C13, M9, T5a
			+1	H2, L11
			+2	M9, T5a
			+3	R5
			+4	T2

Areas

Parameter	Symbol	Dimensions	Exponent	Groups
Area	A	L^2	−1	D9, F6, F7, F8
			+1	B2
Area/unit volume	S, A^*	L^{-1}	−1	B6
			+1	M11, M12

Volumes

Parameter	Symbol	Dimensions	Exponent	Groups
Volume	W, W^*	L^3	+1	A5, H9, M11

TIMES AND FREQUENCIES

Parameter	Symbol	Dimensions	Exponent	Groups
Time	θ	θ	−1	D8, T7c
			+1	D8, D8a, D12, E3, F11, F12, F12a, H1, H9a, H10, P12a, S15, T5, W4a
Frequency	f', ω'	$1/\theta$	+1	F12c, H1, P15a, S16, V1
Radial frequency[1] ($= 2\pi f'$)	ω, (radians/time)	$1/\theta$	$+\frac{1}{2}$	R0d, W0a
			+1	F12d, R0a, R0b
			+2	R0c

[1] Also see angular velocity.

TEMPERATURES AND CONCENTRATIONS (DRIVING FORCES)

Concentrations and Related Quantities

Parameter	Symbol	Dimensions	Exponent	Groups
Dimensional concentration	C, c_A	M/L^3	$(n-1)$	H8a (nth order reaction)
			−1	D1, D2, R2
			+1	B10
Dimensionless concentration (e.g., wt./wt., mass fraction)	n	—	−1	D14a, P6, S9a
			+1	A4, D14a, G5a, K8, R6a, S9a

TABLE III. TABLES FOR IDENTIFYING DIMENSIONLESS GROUPS (Continued)

Concentrations and Related Quantities

Parameter	Symbol	Dimensions	Exponent	Groups
Mole fraction	Y, Z	—	±1	D14
Rate of change of mass transfer potential (mass per unit mass/time)	Γ_m	$1/\theta$	+1	P12a
Specific mass content (mass/unit mass)	n^*	—	−1	K4
Surface concentration	Γ'	M/L^2	±1	S18
{ Vapor capacity }	c_b	$L\theta^2/M$	−1	L5a
{ Mass capacity (porous body) }			+1	B13, R2
Vapor expansion intensity on heating	b_T	M/L^3T	+1	L5a

Temperatures, Temperature Differences

Parameter	Symbol	Dimensions	Exponent	Groups
Dimensional temperature, temperature difference	$T, \Delta T$	T	−3	R1*, T6*
			−2	F12b*
			−1	A4, A6*, B12, B13, B13a*, C4*, C7, C10, D3*, D4*, D14a*, D15. E1, E4b*, F12b*, G2, G6*, K3, K7b, K8, K9, K11, N5, P5*, P12*, R6a, R14*, S6, T5a
			−½	L5
			+½	F6*, F7*, F8*
			+1	B13a, C4, E4b, G5, G6, J1, J4, K7b, L5a, M1, M3a, P6, R5, R7, R14*, S10a, T5a
			+3	R1a*, S13a*, S14*
Dimensionless temperature difference	Ψ	—	+1	K7b
Rate of temperature change	Γ	$1/\theta$	+1	P12

* Absolute temperature.

VELOCITIES, RATES, FLUXES, TRANSFER COEFFICIENTS

Velocities

Parameter	Symbol	Dimensions	Exponent	Groups
Angular velocity (rate of rotation) (see also frequency)	N, ω	$1/\theta$	−3	P7, P7a
			−2	H4, L1
			−1	A1a, F9, R15, R15a
			−½	E2
			+1	H9a, S8, S11, S12, T1
			+2	F15, T2, W3
Equilibrium molecular speed of A	u_A	L/θ	−1	K7a
General velocity (usually of fluid)	V	L/θ	−3	H5, R0c
			−2	B13a, C6, C11, D13, E9, E10, E13a, F1, I1, L5b, M6, N2, N3, N4, R0a, R7, R8, R9, R10, T7a, T7b
			−1	A2, A4b, B3, C12, C14, D1, D3, E4a, F10, F12c, F12d, H7, H8, I1, K10, L3, M3, P3a, P4, S13, S14a, S16, T7c, V2
			−½	M4a
			+1	A1a, A2, B1a, B2a, B6, B7, B11, C3, C12, C14, D5, D6a, D7, E1a, E4c, E4d, E4e, E5, E7, F10, F14, G2b, H7, H10, I1, K2, L5, L8, M2, M4, M7, P1, P2, P3, P10, R4, R11, R15, R15a, S1, S2, S3, T5, T6, T7c, V2, W2, W4
			+2	A1b, B1, B12, C5, C11, D15, E1, E11, F13, H11, P3, W1
Velocity of impeller or agitator periphery	U_s	L/θ	−2	P14
			−1	D9
Velocity of light	V_l	L/θ	−2	B2a
			−1	E1a, L0
Velocity of sound	V_s	L/θ	−1	B1a, M2, N1, S1
Velocity of waves	V_w	L/θ	−1	V2
			+1	P3
Velocity of Alfven waves	V_A	L/θ	−1	A2, K2, M4
			+1	A2, N1
			+2	C11
Velocity of bearing surface	v_s	L/θ	−1	H8, O1, S9
Velocity gradient	dV/dL	$1/\theta$	−2	R13

Flow Rates, Mass Fluxes, etc.

Parameter	Symbol	Dimensions	Exponent	Groups
Mass flow rate (mass flux)	V_m	M/θ	−1	B2, M11, T7
			+1	F6, F7, F8, G4, S10a, T7
Mass flow rate/unit area (mass flux density)	G	$M/L^2\theta$	−1	J2, J3, P15, P15a, S13
			+1	K4, M11
Mass flux/unit volume (reaction rate)	U	$M/L^2\theta$	+½	T3
			+1	D1, D2, D3, D4
Reaction rate constant (first order)	K	L/θ	−1	S5
Reaction rate constant (nth order	K_n	$\frac{1}{\theta}\cdot\frac{L^{3n-3}}{M^{n-1}}$	+1	H8a
Volumetric flow rate	V_f	L^3/θ	−1	H9
			−½	D11
			+½	S11, S12
			+1	D9, F9
Volumetric flow rate/wetted perimeter	$V_{f'}$	L^2/θ	−1	A4a
			−½	A0

Heat Fluxes and Related Quantities

Parameter	Symbol	Dimensions	Exponent	Groups
Heat flux (heat flow/unit time)	q	ML^2/θ^3	−1	S10a
			+1	H5
Heat flux/unit area	q^*	M/θ^2	+1	K3, R6
Heat of reaction/unit mass of material reacting	Δu	L^2/θ^2	+1	R6a
Heat liberated/unit mass	Q	L^2/θ^2	+1	D3, D4
Heat liberated per unit mass of material reacting/unit volume	Q''	$1/L\theta^2$	+1	F12b
Heat liberated/unit volume (heat source power)	j	$M/L\theta^3$	+1	P5

Transfer Coefficients

Parameter	Symbol	Dimensions	Exponent	Groups
Heat transfer coefficient	h	$M/T\theta^3$	+1	B2, B4, C9, J2, K7b, N6, P15, S13
			+4	M1
Mass transfer coefficient	k_c	L/θ	+1	A0, B5, J3, L9, N7, S5, S7

FORCE, TORQUE, POWER, PRESSURE, HEAD

Force, Torque, Power

Parameter	Symbol	Dimensions	Exponent	Groups
Force	F, F_R	ML/θ^2	−⅓	S2
			+⅓	K5
			+1	N2, N3, N4, R9, T7a
Force/unit length	F_b	M/θ^2	+1	H8, O1, S9
Torque	F'	ML^2/θ^2	+1	T7b
Power	Π	ML^2/θ^3	+1	L1, P7, P7a

TABLE III. TABLES FOR IDENTIFYING DIMENSIONLESS GROUPS (Continued)

Head, Pressure

Parameter	Symbol	Dimensions	Exponent	Groups
Fluid head (ft. fluid)	H'	L	$-\frac{3}{4}$	**S12**
			$+1$	**E10a, H4**
Head (energy/unit mass of fluid = gH')	H	L^2/θ^2	-1	**P3, T4**
			$-\frac{3}{4}$	**S11**
			$+\frac{1}{4}$	**D11**
			$+1$	**P14, T4**
Pressure	P, p	$M/L\theta^2$	-1	**F6, F7, F8, H9, L5a, S8, T8**
			$+1$	**B13, C6, D14a, L5b, L6, P13, R2**
Pressure drop	$\Delta P, dP$	$M/L\theta^2$	$+1$	**E9, E10, F1, H9, H11a, K1, L3, L3a, R10**
Pressure gradient	$\Delta P/L$	$M/L^2\theta^2$	-1	**T7c**
			$+1$	**E10, F1, H11a, K1, P4, R10, T7c**

CONSTANTS AND MISCELLANEOUS QUANTITIES

Gravity Acceleration

Parameter	Symbol	Dimensions	Exponent	Groups
Gravity acceleration	g	L/θ^2	-2	**A1**
			-1	**B1, F13, F15, H11a, M1, S6, W3a**
			$-\frac{3}{4}$	**S11**
			$-\frac{1}{2}$	**B11, F14, P13**
			$-\frac{1}{3}$	**C9, S3**
			$+\frac{1}{3}$	**F2, N8**
			$+1$	**A5, B9, B13a, C2, C10, D13, E8, E10a, E13a, G1, G5, G5a, H4, M3a, R5, R6, R8, R13, S15a**

Other Quantities

Parameter	Symbol	Dimensions	Exponent	Groups
Arrhenius pre-exponential factor	k_0	M/θ	$+1$	**F12b**
Avogadro's No.	N	$1/M$	$+1$	**K9, S6**
Boltzmann's constant	k	ML^2/θ^2T	-1	**K9, S6**
Dufour coefficient	β^*	T	$+1$	**A3, A4, F3**
Energy of activation/unit mass	E_a	L^2/θ^2	$+1$	**A6, F12b**
Gas constant/unit mass	$\mathscr{R}$	L^2/θ^2T	-1	**A6, F12b**
			$-\frac{1}{2}$	**L5**
			$+1$	**D14a**
Knudsen flow permeability	K_{oA}	L	-1	**K7a**
Mass of float	m_f	M	$+1$	**S15a**
Moment of inertia	I	L^4	-1	**L7c**
Rotor lift curve slope	a'	L^2/M	$+1$	**L7c**
Shape factor	ξ^*	—	$+1$	**V2**
Soret coefficient	δ	$1/T$	$+1$	**F3, P6**
Stefan-Boltzmann coefficient	η	$M/T^4\theta^3$	$+1$	**R1, T6**
(radiation coefficient)		–	$+1$	**R1a, S13a, S14**
Thermodiffusion constant	Θ	—	$+1$	**D14a, S9a**

REFERENCES

1. Adrianov, V. N., "Radiation-Conductive and Radiation-Convective Heat Transfer," in "Teplo- i Massoperenos," A. V. Lȳkov and B. M. Smol'skii, Eds., Vol. **II**, pp. 92–102, Nauka i Tekhnika, Minsk, 1965.

2. Astarita, G., *Ind. Eng. Chem. Fundamentals* **6**, 257 (1967).

3. Berg, J. C., Acrivos, A., Boudart, M., "Evaporative Convection," in "Advances in Chemical Engineering," T. B. Drew *et al.*, Eds., Vol. **6**, pp. 61–123, Academic Press, New York, 1966.

4. Carslaw, H. S., Jaeger, J. C., "Conduction of Heat in Solids," 2nd ed., Oxford University Press, 1959.

5. Catchpole, J. P., Fulford, G., *Ind. Eng. Chem.* **58** (3), 46 (1966).

5a. Costa Novella, E., Costa Lopez, J., *An. Real. Soc. Espan. Fis. Quim., Ser. B*, **63** (2), 145 (1967).

6. Coulson, J. M., Porter, J. E., *Trans. Inst. Chem. Engrs.* **44**, T388 (1966).

7. Ede, A. J., *Intern. J. Heat Mass Transfer* **9**, 837 (1966).

8. Gall, C. E., Hudgins, R. R., *J. Chem. Educ.* **42**, 611 (1965).

8a. Galloway, T. R., Sage, B. H., *Int. J. Heat Mass Transfer* **10**, 1195 (1967).

9. Gardner, G. O., Kestin, J., *Ibid.*, **6**, 289 (1963).

10. Gibbins, J. C., Hignett, E. T., *Electrochim. Acta* **11**, 815 (1966).

10a. Gilmont, R., Maurer, P. W., *Instr. Control Systems* **34**, 2070 (1961).

11. Ginzburg, I. P., "Heat and Mass Transfer from Bodies Interacting with Gas and Liquid Streams," in "Teplo- i Massoperenos," A. V. Lȳkov and B. M. Smol'skii, Eds., Vol. **II**, pp. 3–26, Nauka i Tekhnika, Minsk, 1965.

12. Glauert, M. B., Lighthill, M. J., *Proc. Roy. Soc.* **230A**, 188 (1955).

13. Hobler, T., "Mass Transfer and Absorbers," Pergamon Press, New York, 1966.

14. Irvine, T. F., Stein, R. P., Simon, G. A., "Effect of Radiation on Convection in a Flat Channel," in "Teplo- i Massoperenos," A. V. Lȳkov and B. M. Smol'skii, Eds., Vol. *II*, pp. 78–91, Nauka i Tekhnika, Minsk, 1965.

15. Kasatkin, A. G., Dȳtnerskii, Yu. N., Kochergin, N. V., "Mass and Heat Transfer in Tray Columns," in "Teplo- i Massoperenos," A. V. Lȳkov and B. M. Smol'skii, Eds., Vol. **IV**, pp. 12–17, Nauka i Tekhnika, Minsk, 1966.

16. Kerber, R., *Chem.-Ing.-Technik* **38**, 1133 (1966).

17. Kochenov, I. S., Kuznetsov, Yu. N., "Unsteady Flow in Pipes," in "Teplo-Massoperenos," A. V. Lȳkov and B. M. Smol'skii, Eds., Vol. **II**, pp. 306–314, Nauka i Tekhnika, Minsk, 1965.

18. Kramers, H., Alberda, G., *Chem. Eng. Sci.* **2**, 173 (1953).

19. Laslo, A., *Inzh.-Fiz. Zh.* **10** (1), 60 (1966).

20. Lemlich, R., Armour, J. C., *Chem. Engr. Progr. Symp. Ser.* **61** (57), 83 (1965).

20a. Leshchii, N. P., Mochernyuk, D. Yu, *Izv. Vysshikh Uchebn. Zavedenii, Nefti Gaz* **10** (2), 77 (1967).

21. Luikov (Lȳkov), A. V., "Heat and Mass Transfer in Capillary-Porous Bodies," Pergamon Press, New York, 1966.

22. Luikov (Lȳkov), A. V. Mikhailov, Yu. A., "Theory of Energy and Mass Transfer," revised Engl. ed., Pergamon Press, New York, 1965.

23. McHenry, K. W., Wilhelm, R. H., *A.I.Ch.E. J.* **3**, 83 (1957).

24. Merzhanov, A. G., "Problems of Heat Transfer in the Theory of Thermal Detonation," in "Teplo- i Massoperenos," A. V. Lȳkov and B. M. Smol'skii, Eds., Vol. **IV**, pp. 259–72, Nauka i Tekhnika, Minsk, 1966.

25. Opfell, J. B., Sage, B. H., "Turbulence in Thermal and Material Transport," in "Advances in Chemical Engineering," T. B. Drew *et al.*, Ed., Vol. **I**, pp. 241–88, Academic Press, New York, 1956.

26. Pai, Shih-I., "Magnetohydrodynamics of Channel Flow," in "Advances in Hydroscience," Ven Te Chow, Ed., Vol. **III**, pp. 63–110, Academic Press, New York, 1966.

27. Pankhurst, R. C., "Dimensional Analysis and Scale Factors," Chapman and Hall, London, 1964.

28. Pechoč, V., Research Institute of Chemical Fibers, Svit, Czechoslovakia, private communication, 1966.

29. Philip, J. R., *Austral. J. Phys.* **16** (4), 454 (1963).

30. Postolnik, Yu. S., *Inzh.-Fiz. Zh.* **8** (1), 64 (1965).

31. Prochazka, J., Landau, J., *Collection Czech. Chem. Commun.* **26**, 2961 (1961).

32. Rodionov, A. I., Kashnikov, A. M., Radikovskii, V. M., "Determination of Interfacial Contact Areas and Heat and Mass Transfer Coefficients on Sieve Trays," in "Teplo- i Massoperenos," A. V. Lȳkov, B. M. Smol'skii, Eds., Vol., **IV**, pp. 28–37, Nauka i Tekhnika, Minsk, 1966.

33. Rosen, J. B., Winsche, W. E., *J. Chem. Phys.* **18**, 1587 (1950).

33a. Ross, L. L., Hoffman, T. W., Proc. 3rd Internat. Heat Transfer Conference, Chicago, August 1966, **5**, 50 (A.I.Ch.E., New York, 1966).

34. Rothfield, L. B., *A.I.Ch.E. J.* **9**, 19 (1963).

35. Sakiadis, B. S., *Ibid.*, **7**, 467 (1961).

36. Skelland, A. H. P., "Non-Newtonian Flow and Heat Transfer," Wiley, New York, 1967.

37. Stewart, W. E., *Chem. Eng. Progr. Symp. Ser.* **61** (58), 16 (1965).

38. Tsinober, A. B., *Appl. Mech. Rev.* **18**, 3751 (1965).

39. Turner, G. A., Dept. of Chemical Engineering, University of Waterloo, Waterloo, Ontario, private communication, 1967.

40. Vyazovov, V. V., *Zh. Tekhn. Fiz.* **10**, 1519 (1940).

41. Wilkinson, W. L., "Non-Newtonian Fluids," Pergamon Press, London, 1960.

NATIONAL STANDARD REFERENCE DATA SYSTEM PUBLICATIONS

The National Standard Reference Data System was established in 1963 as a means of coordinating on a national scale the production and dissemination of critically evaluated reference data in the physical sciences. Under the Standard Reference Data Act (Public Law 90-396) the National Bureau of Standards of the U.S. Department of Commerce has the primary responsibility in the Federal Government for providing reliable scientific and technical reference data. The Office of Standard Reference Data of NBS coordinates a complex of data evaluation centers, located in university, industrial, and other Government laboratories as well as within the National Bureau of Standards, which are engaged in the compilation and critical evaluation of numerical data on physical and chemical properties retrieved from the world scientific literature.

The primary focus of the NSRDS is on well-defined physical and chemical properties of well-characterized materials or systems. An effort is made to assess the accuracy of data reported in the primary research literature and to prepare compilations of critically evaluated data which will serve as reliable and convenient reference sources for the scientific and technical community.

This Publication List includes NSRDS data compilations, critical reviews, and related publications which are available from various sources. Further information may be obtained from:

Office of Standard Reference Data
National Bureau of Standards
Washington, D. C. 20234
Telephone: (301) 921-2467

JOURNAL OF PHYSICAL AND CHEMICAL REFERENCE DATA (REPRINTS TO VOLUMES 1 TO 8 NO. 2)

Reprints from Volume 1 (1972)

1. *Gaseous Diffusion Coefficients,* T. R. Marrero and E. A. Mason, Vol. 1, No. 1, 1—118, 1972. $7.00.
2. *Selected Values of Critical Supersaturation for Nucleation of Liquids from the Vapor,* G. M. Pound, Vol. 1, No. 1, 119—134, 1972. $3.00.
3. *Selected Values of Evaporation and Condensation Coefficients of Simple Substances,* G. M. Pound, Vol. 1, No. 1, 135—146, 1972. $3.00.
4. *Atlas of the Observed Absorption Spectrum of Carbon Monoxide between 1060 and 1900 Å,* S. G. Tilford and J. D. Simmons, Vol. 1, No. 1, 147—188, 1972. $4.50.

*5. *Tables of Molecular Vibrational Frequencies, Part 5,* T. Shimanouchi, Vol. 1, No. 1, 189—216, 1972. $4.00.

6. *Selected Values of Heats of Combustion and Heats of Formation of Organic Compounds Containing the Elements C, H, N, O, P, and S,* Eugene S. Dolmalski, Vol. 1, No. 2, 221—278, 1972. $5.00.
7. *Thermal Conductivity of the Elements,* C. Y. Ho, R. W. Powell, and P. E. Liley, Vol. 1, No. 2, 279—422, 1972. $7.50.
8. *The Spectrum of Molecular Oxygen,* Paul H. Krupenie, Vol. 1, No. 2, 423—534, 1972. $6.50.
9. *A Critical Review of the Gas-Phase Reaction Kinetics of the Hydroxyl Radical,* W. E. Wilson, Jr., Vol. 1, No. 2, 535—574, 1972. $4.50.
10. *Molten Salts: Volume 3, Nitrates, Nitrites, and Mixtures, Electrical Conductance, Density, Viscosity, and Surface Tension Data,* G. J. Jenz, Ursula Krebs, H. F. Siegenthaler, and R. P. T. Tomkins, Vol. 1, No. 3, 581—746, 1972. $8.50.
11. *High Temperature Properties and Decomposition of Inorganic Salts—Part 3, Nitrates and Nitrites,* Kurt H. Stern, Vol. 1, No. 3, 747—772, 1972. $4.00.
12. *High-Pressure Calibration: A Critical Review,* D. L. Decker, W. A. Bassett, L. Merrill, H. T. Hall, and J. D. Barnett, Vol. 1, No. 3, 773—836, 1972. $5.00.
13. *The Surface Tension of Pure Liquid Compounds,* Joseph J. Jasper, Vol. 1, No. 4, 841—1009, 1972. $8.50.
14. *Microwave Spectra of Molecules of Astrophysical Interest, I. Formaldehyde, Formamide, and Thioformaldehyde,* Donald R. Johnson, Frank J. Lovas, and William H. Kirchhoff, Vol. 1, No. 4, 1011—1045, 1972. $4.50.
15. *Osmotic Coefficients and Mean Activity Coefficients of Uni-univalent Electrolytes in Water at 25°C,* Walter J. Hamer and Yung Chi Wu, Vol. 1, No. 4, 1047—1099, 1972. $5.00.
16. *The Viscosity and Thermal Conductivity Coefficients of Gaseous and Liquid Fluorine,* H. J. M. Hanley and R. Prydz, Vol. 1, No. 4, 1101—1113, 1972. $3.00.

Reprints from Volume 2 (1973)

17. *Microwave Spectra of Molecules of Astrophysical Interest. II. Methylenimine,* William H. Kirchhoff, Donald R. Johnson, and Frank J. Lovas, Vol. 2, No. 1, 1—10, 1973. $3.00.
18. *Analysis of Specific Heat Data in the Critical Region of Magnetic Solids,* F. J. Cook, Vol. 2, No. 1, 11—24, 1973. $3.00.
19. *Evaluated Chemical Kinetic Rate Constants for Various Gas Phase Reactions,* Keith Schofield, Vol. 2, No. 1, 25—84, 1973. $5.00.
20. *Atomic Transition Probabilities for Forbidden Lines of the Iron Group Elements. (A Critical Data Compilation for Selected Lines),* M. W. Smith and W. L. Wiese, Vol. 2, No. 1, 85—120, 1973. $4.50.

* Superseded by Reprint No. 103.

*21. *Tables of Molecular Vibrational Frequencies, Part 6,* T. Shimanouchi, Vol. 2, No. 1, 121—162, 1973. $4.50.

22. *Compilation of Energy Band Gaps in Elemental and Binary Compound Semiconductors and Insulators,* W. H. Strehlow and E. L. Cook, Vol. 2, No. 1, 163—200, 1973. $4.50.

23. *Microwave Spectra of Molecules of Astrophysical Interest. III. Methanol,* R. M. Lees, F. J. Lovas, W. H. Kirchhoff, and D. R. Johnson, Vol. 2, No. 2, 205—214, 1973. $3.00.

24. *Microwave Spectra of Molecules of Astrophysical Interest, IV. Hydrogen Sulfide,* Paul Helminger, Frank C. De Lucia, and William H. Kirchhoff, Vol. 2, No. 2, 215—224, 1973. $3.00.

*25. *Tables of Molecular Vibrational Frequencies, Part 7,* T. Shimanouchi, Vol. 2, No. 2, 225—256, 1973. $4.00.

26. *Energy Levels of Neutral Helium (^{4}He I),* W. C. Martin, Vol. 2, No. 2, 257—266, 1973. $3.00.

27. *Survey of Photochemical and Rate Data for Twenty-eight Reactions of Interest in Atmosphere Chemistry,* R. F. Hampson, Ed., W. Braun, R. L. Brown, D. Garvin, J. T. Herron, R. E. Huie, M. J. Kurylo, A. H. Laufer, J. D. McKinley, H. Okabe, M. D. Sheer, W. Tsang, and D. H. Stedman, Vol. 2, No. 2, 267—312, 1973. $4.50.

28. *Compilation of the Static Dielectric Constant of Inorganic Solids,* K. F. Young and H. P. R. Frederikse, Vol. 2, No. 2, 313—410, 1973. $6.50.

29. *Soft X-Ray Emission Spectra of Metallic Solids: Critical Review of Selected Systems,* A. J. McAlister, R. C. Dobbyn, J. R. Cuthill, and M. L. Williams, Vol. 2, No. 2, 411—426, 1973. $3.00.

30. *Ideal Gas Thermodynamic Properties of Ethane and Propane,* J. Chao, R. C. Wilhoit, and B. J. Zwolinski, Vol. 2, No. 2, 427—438, 1973. $3.00.

31. *An Analysis of Coexistence Curve Data for Several Binary Liquid Mixtures Near Their Critical Points,* A. Stein and G. F. Allen, Vol. 2, No. 3, 443—466, 1973. $4.00.

32. *Rate Constants for the Reactions of Atomic Oxygen (O^3P) with Organic Compounds in the Gas Phase,* John T. Herron and Robert E. Huie, Vol. 2, No. 3, 467—518, 1973. $5.00.

33. *First Spectra of Neon, Argon, and Xenon 136 in the 1.2—4.0 μm Region,* Curtis J. Humphreys, Vol. 2, No. 3, 519—530, 1973. $3.00.

34. *Elastic Properties of Metals and Alloys, I. Iron, Nickel, and Iron-Nickel Alloys,* H. M. Ledbetter and R. P. Reed, Vol. 2, No. 3, 531—618, 1973. $6.00.

35. *The Viscosity and Thermal Conductivity Coefficients of Dilute Argon, Krypton, and Xenon,* H. J. M. Hanley, Vol. 2, No. 3, 619—642, 1973. $4.00.

36. *Diffusion in Copper and Copper Alloys. Part I. Volume and Surface Self-Diffusion in Copper,* Daniel B. Butrymowicz, John R. Manning, and Michael E. Read, Vol. 2, No. 3, 643—656, 1973. $3.00.

37. *The 1973 Least-Squares Adjustment of the Fundamental Constants,* E. Richard Cohen and B. N. Taylor, Vol. 2, No. 4, 663—734, 1973. $5.50.

38. *The Viscosity and Thermal Conductivity Coefficients of Dilute Nitrogen and Oxygen,* H. J. M. Hanley and James F. Ely, Vol. 2, No. 4, 735—756, 1973. $4.00.

39. *Thermodynamic Properties of Nitrogen Including Liquid and Vapor Phases from 63K to 2000K with Pressures to 10,000 Bar,* Richard T. Jacobsen and Richard B. Stewart, Vol. 2, No. 4, 757—922, 1973. $8.50.

40. *Thermodynamic Properties of Helium 4 from 2 to 1500K at Pressures to 10^8Pa,* Robert D. McCarty, Vol. 2, No. 4, 923—1042, 1973. $7.00.

Reprints from Volume 3 (1974)

41. *Molten Salts: Volume 4, Part 1, Fluorides and Mixtures, Electrical Conductance, Density, Viscosity, and Surface Tension Data,* G. J. Janz, G. L. Gardner, Ursula Krebs, and R. P. T. Tomkins, Vol. 3, No. 1, 1-115, 1974. $7.00.

42. *Ideal Gas Thermodynamic Properties of Eight Chloro- and Fluoromethanes,* A. S. Rodgers, J. Chao, R. C. Wilhoit, and B. J. Zwolinski, Vol. 3, No. 1, 117—140, 1974. $4.00.

43. *Ideal Gas Thermodynamic Properties of Six Chloroethanes,* J. Chao, A. S. Rodgers, R. C. Wilhoit, and B. J. Zwolinski, Vol. 3, No. 1, 141—162, 1974. $4.00.

44. *Critical Analysis of Heat-Capacity Data and Evaluation of Thermodynamic Properties of Ruthenium, Rhodium, Palladium, Iridium, and Platinum from 0 to 300K. A Survey of the Literature Data on Osmium,* George T. Furukawa, Martin L. Reilly, and John S. Gallagher, Vol. 3, No. 1, 163—209, 1974. $4.50.

45. *Microwave Spectra of Molecules of Astrophysical Interest, V. Water Vapor,* Frank C. De Lucia, Paul Helminger, and William H. Kirchhoff, Vol. 3, No. 1, 211—219, 1974. $3.00.

46. *Microwave Spectra of Molecules of Astrophysical Interest, VI. Carbonyl Sulfide and Hydrogen Cyanide,* Arthur G. Maki, Vol. 3, No. 1, 221—244, 1974. $4.00.

47. *Microwave Spectra of Molecules of Astrophysical Interest, VII. Carbon Monoxide, Carbon Monosulfide, and Silicon Monoxide,* Frank J. Lovas and Paul H. Krupenie, Vol. 3, No. 1, 245—257, 1974. $3.00.

48. *Microwave Spectra of Molecules of Astrophysical Interest, VIII. Sulfur Monoxide,* Eberhard Tiemann, Vol. 3, No. 1, 259—268, 1974. $3.00.

*49. *Tables of Molecular Vibrational Frequencies, Part 8,* T. Shimanouchi, Vol. 3, No. 1, 269—308, 1974. $4.50.

50. *JANAF Thermochemical Tables, 1974 Supplement,* M. W. Chase, J. L. Curnutt, A. T. Hu, H. Prophet, A. N. Syverud, and L. C. Walker, Vol. 3, No. 2, 311—480, 1974. $8.50.

51. *High Temperature Properties and Decomposition of Inorganic Salts, Part 4. Oxy-Salts of the Halogens,* Kurt H. Stern, Vol. 3, No. 2, 481—526, 1974. $4.50.

52. *Diffusion in Copper and Copper Alloys, Part II. Copper-Silver and Copper-Gold Systems,* Daniel B. Butrymowicz, John R. Manning, and Michael E. Read, Vol. 3, No. 2, 527—602, 1974. $5.50.

53. *Microwave Spectral Tables, I. Diatomic Molecules,* Frank J. Lovas and Eberhard Tiemann, Vol. 3, No. 3, 609—770, 1974. $8.50.

54. *Ground Levels and Ionization Potentials for Lanthanide and Actinide Atoms and Ions,* W. C. Martin, Lucy Hagan, Joseph Reader, and Jack Sugar, Vol. 3, No. 3, 771—780, 1974. $3.00.
55. *Behavior of the Elements at High Pressures,* John Francis Cannon, Vol. 3, No. 3, 781—824, 1974. $4.50.
56. *Reference Wavelengths from Atomic Spectra in the Range 15 Å to 25000 Å,* Victor Kaufman and Bengt Edlen, Vol. 3, No. 4, 825—895, 1974. $5.50.
57. *Elastic Properties of Metals and Alloys, II. Copper,* H. M. Ledbetter and E. R. Naimon, Vol. 3, No. 4, 897—935, 1974. $4.50.
58. *A Critical Review of H-Atom Transfer in the Liquid Phase: Chlorine Atom, Alkyl Trichloromethyl, Alkoxy, and Alkylperoxy Radicals,* D. G. Hendry, T. Mill, L. Piszkiewicz, J. A. Howard, and H. K. Eigenmann, Vol. 3, No. 4, 937—978, 1974. $4.50.
59. *The Viscosity and Thermal Conductivity Coefficients for Dense Gaseous and Liquid Argon, Krypton, Xenon, Nitrogen, and Oxygen,* H. J. M. Hanley, R. D. McCarty, and W. M. Haynes, Vol. 3, No. 4, 979—1017, 1974. $4.50.

Reprints from Volume 4 (1975)

60. *JANAF Thermochemical Tables, 1975 Supplement,* M. W. Chase, J. L. Curnutt, H. Prophet, R. A. McDonald, and A. N. Syverud, Vol. 4, No. 1, 1—175, 1975. $8.50.
61. *Diffusion in Copper and Copper Alloys, Part III. Diffusion in Systems Involving Elements of the Groups IA, IIA, IIIB, IVB, VB, VIB, and VIIB,* Daniel B. Butrymowicz, John R. Manning, and Michael E. Read, Vol. 4, No. 1, 177—249, 1975. $6.00.
62. *Ideal Gas Thermodynamic Properties of Ethylene and Propylene,* Jing Chao and Bruno J. Zwolinski, Vol. 4, No. 1, 251—261, 1975. $3.00.
63. *Atomic Transition Probabilities for Scandium and Titanium (A Critical Data Compilation of Allowed Lines),* W. L. Wiese and J. R. Fuhr, Vol. 4, No. 2, 263—352, 1975. $6.00.
64. *Energy Levels of Iron, Fe I through Fe XXVI,* Joseph Reader and Jack Sugar, Vol. 4, No. 2, 353—440, 1975. $6.00.
65. *Ideal Gas Thermodynamic Properties of Six Fluoroethanes,* S. S. Chen, A. S. Rodgers, J. Chao, R. C. Wilhoit, and B. J. Zwolinski, Vol. 4, No. 2, 441—456, 1975. $3.00.
66. *Ideal Gas Thermodynamic Properties of the Eight Bromo- and Iodomethanes,* S. A. Kudchadker and A. P. Kudchadker, Vol. 4, No. 2, 457—470, 1975. $3.00.
67. *Atomic Form Factors, Incoherent Scattering Functions, and Photon Scattering Cross Sections,* J. H. Hubbell, Wm. J. Veigele, E. A. Briggs, R. T. Brown, D. T. Cromer, and R. J. Howerton, Vol. 4, No. 3, 471—538, 1975. $5.50.
68. *Binding Energies in Atomic Negative Ions,* H. Hotop and W. C. Lineberger, Vol. 4, No. 3, 539—576, 1975. $4.50.
69. *A Survey of Electron Swarm Data,* J. Dutton, Vol. 4, No. 3, 577—856, 1975. $12.00.
70. *Ideal Gas Thermodynamic Properties and Isomerization of n-Butane and Isobutane,* S. S. Chen, R. C. Wilhoit, and B. J. Zwolinski, Vol. 4, No. 4, 859—870, 1975. $3.00.
71. *Molten Salts: Volume 4, Part 2, Chlorides and Mixtures, Electrical Conductance, Density, Viscosity, and Surface Tension Data,* G. J. Janz, R. P. T. Tomkins, C. B. Allen, J. R. Downey, Jr., G. L. Gardner, U. Krebs, and S. K. Singer, Vol. 4, No. 4, 871—1178, 1975. $13.00.
72. *Property Index to Volumes 1—4 (1972—1975),* Vol. 4, No. 4, 1179—1192, 1975. $3.00.

Reprints from Volume 5 (1976)

73. *Scaled Equation of State Parameters for Gases in the Critical Region,* J. M. L. Levelt-Sengers, W. L. Greer, and J. V. Sengers, Vol. 5, No. 1, 1—51, 1976. $5.00.
74. *Microwave Spectra of Molecules of Astrophysical Interest, IX. Acetaldehyde,* A. Bauder, F. J. Lovas, and D. R. Johnson, Vol. 5, No. 1, 53—77, 1976. $4.00.
75. *Microwave Spectra of Molecules of Astrophysical Interest, X. Isocyanic Acid,* G. Winnewisser, W. H. Hocking, and M. C. L. Gerry, Vol. 5, No. 1, 79—101, 1976. $4.00.
76. *Diffusion in Copper and Copper Alloys, Part IV. Diffusion in Systems Involving Elements of Group VIII,* Daniel B. Butrymowicz, John R. Manning, and Michael E. Read, Vol. 5, No. 1, 103—200, 1976. $6.50.
77. *A Critical Review of the Stark Widths and Shifts of Spectral Lines from Non-Hydrogenic Atoms,* N. Konjevic and J. R. Roberts, Vol. 5, No. 2, 209—257, 1976. $5.00.
78. *Experimental Stark Widths and Shifts for Non-Hydrogenic Spectral Lines of Ionized Atoms,* N. Konjevic and W. L. Wiese, Vol. 5, No. 2, 259—308, 1976. $5.00.
79. *Atlas of the Absorption Spectrum of Nitric Oxide (NO) between 1420 and 1250 Å,* E. Miescner and F. Alberti, Vol. 5, No. 2, 309—317, 1976. $3.00.
80. *Ideal Gas Thermodynamic Properties of Propanone and 2-Butanone,* J. Chao and B. J. Zwolinski, Vol. 5, No. 2, 319—328, 1976. $3.00.
81. *Refractive Index of Alkali Halides and Its Wavelength and Temperature Derivatives,* H. H. Li, Vol. 5, No. 2, 329—528, 1976. $9.50.
82. *Tables of Critically Evaluated Oscillator Strengths for the Lithium Isoelectronic Sequence,* G. A. Martin and W. L. Wiese, Vol. 5, No. 3, 537—570, 1976. $4.50.
83. *Ideal Gas Thermodynamic Properties of Six Chlorofluoromethanes,* S. S. Chen, R. C. Wilhoit, and B. J. Zwolinski, Vol. 5, No. 3, 571—580, 1976. $3.00.
84. *Survey of Superconductive Materials and Critical Evaluation of Selected Properties,* B. W. Roberts, Vol. 5, No. 3, 531—821, 1976. $12.50.
85. *Nuclear Spins and Moments,* Gladys H. Fuller, Vol. 5, No. 4, 835—1092, 1976. $11.50.
86. *Nuclear Moments and Moment Ratios as Determined by Mössbauer Spectroscopy,* J. G. Stevens and B. D. Dunlop, Vol. 5, No. 4, 1093—1121, 1976. $4.00.

87. *Rate Coefficients for Ion-Molecule Reactions, I. Ions Containing C and H,* L. Wayne Sieck and Sharon G. Lias, Vol. 5, No. 4, 1123—1146, 1976. $4.00.
88. *Microwave Spectra of Molecules of Astrophysical Interest, XI. Silicon Sulfide,* Eberhard Tiemann, Vol. 5, No. 4, 1147—1156, 1976. $3.00.
89. *Property Index and Author Index to Volume 1—5 (1972—1976),* Vol. 5, No. 4, 1161—1183, 1976. $4.00.

Reprints from Volume 6 (1977)

90. *Diffusion in Copper and Copper Alloys, Part V. Diffusion in Systems Involving Elements of Group VA,* Daniel B. Butrymowicz, John R. Manning, and Michael E. Read, Vol. 6, No. 1, 1—50, 1977. $5.00.
91. *The Calculated Thermodynamic Properties of Superfluid Helium-4,* James S. Brooks and Russell J. Donnelly, Vol. 6, No. 1, 51—104, 1977. $5.00.
92. *Thermodynamic Properties of Normal and Deuterated Methanols,* S. S. Chen, R. C. Wilhoit, and B. J. Zwolinski, Vol. 6, No. 1, 105—112, 1977. $3.00.
93. *The Spectrum of Molecular Nitrogen,* Alf Lofthus and Paul H. Krupenie, Vol. 6, No. 1, 113—307, 1977. $9.50.
94. *Energy Levels of Chromium, Cr I through Cr XXIV,* Jack Sugar and Charles Corliss, Vol. 6, No. 2, 317—383, 1977. $5.50.
95. *The Activity and Osmotic Coefficients of Aqueous Calcium Chloride at 298.15K,* Bert R. Staples and Ralph L. Nuttall, Vol. 6, No. 2, 385—407, 1977. $4.00.
96. *Molten Salts: Volume 4, Part 3, Bromides and Mixtures; Iodides and Mixtures — Electrical Conductance, Density, Viscosity, and Surface Tension Data,* G. J. Janz, R. P. T. Tomkins, C. B. Allen, J. R. Downey, Jr., and S. K. Singer, Vol. 6, No. 2, 409—596, 1977. $9.00.
97. *The Viscosity and Thermal Conductivity for Dense Gaseous and Liquid Methane,* H. J. M. Hanley, W. M. Haynes, and R. D. McCarty, Vol. 6, No. 2, 597—609, 1977. $3.00.
98. *Phase Diagrams and Thermodynamic Properties of Ternary Copper-Silver Systems,* Y. Austin Chang, Daniel Goldberg, and Joachim P. Neumann, Vol. 6, No. 3, 621—673, 1977. $5.00.
99. *Crystal Data Space-Group Tables,* Alan D. Mighell, Helen M. Ondik, and Bettijoyce Breen Molino, Vol. 6, No. 3, 675—829, 1977. $8.00.
100. *Energy Levels of One-Electron Atoms,* Glen W. Erickson, Vol. 6, No. 3, 831—869, 1977. $4.50.
101. *Rate Constants for Reactions of ClO_x of Atmospheric Interest,* R. T. Watson, Vol. 6, No. 3, 871—917, 1977. $4.50.
102. *NMR Spectral Data: A Compilation of Aromatic Proton Chemical Shifts in Mono- and Di-Substituted Benzenes,* B. L. Shapiro and L. E. Mohrmann, Vol. 6, No. 3, 919—991, 1977.$5.50.
103. *Tables of Molecular Vibrational Frequencies. Consolidated Volume II,* T. Shimanouchi, Vol. 6, No. 3, 993—1102, 1977. $6.50.
104. *Effects of Isotopic Composition, Temperature, Pressure, and Dissolved Gases on the Density of Liquid Water,* George S. Kell, Vol. 6, No. 4, 1109—1132, 1977. $4.00.
105. *Viscosity of Water Substance — New International Formulation and Its Background,* A. Nagashima, Vol. 6, No. 4, 1133—1166, 1977. $4.50.
106. *A Correlation of the Existing Viscosity and Thermal Conductivity Data of Gaseous and Liquid Ethane,* H. J. M. Hanley, K. E. Gubbins, and S. Murad, Vol. 6, No. 4, 1167—1180, 1977. $3.00.
107. *Elastic Properties of Zinc: A Compilation and Review,* H. M. Ledbetter, Vol. 6, No. 4, 1181—1204, 1977. $4.00.
108. *Behavior of the AB-Type Compounds at High Pressures and High Temperatures,* Leo Merrill, Vol. 6, No. 4, 1205—1252, 1977. $4.50.
109. *Energy Levels of Manganese, Mn I through Mn XXV,* Charles Corliss and Jack Sugar, Vol. 6, No. 4, 1253—1329, 1977. $5.50.

Reprints from Volume 7 (1978)

110. *Tables of Atomic Spectral Lines for the 10,000 Å to 40,000 Å Region,* Michael Outred, Vol. 7, No. 1, 1—262, 1978. $11.50.
111. *Evaluated Activity and Osmotic Coefficients for Aqueous Solutions: The Alkaline Earth Metal Halides,* R. N. Goldberg and R. L. Nuttall, Vol. 7, No. 1, 263—310, 1978. $4.50.
112. *Microwave Spectra of Molecules of Astrophysical Interest, XII. Hydroxyl Radical,* Robert A. Beaudet and Robert L. Poynter, Vol. 7, No. 1, 311—362, 1978. $5.00.
113. *Ideal Gas Thermodynamic Properties of Methanoic and Ethanoic Acids,* Jing Chao and Bruno J. Zwolinski, Vol. 7, No. 1, 363, 1978. $3.00.
114. *Critical Review of Hydrolysis of Organic Compounds in Water Under Environmental Conditions,* W. Mabey and T. Mill, Vol. 7, No. 2, 383—415, 1978. $4.50.
115. *Ideal Gas Thermodynamic Properties of Phenol and Cresols,* S. A. Kudchadker, A. P. Kudchadker, R. C. Wilhoit, and B. J. Zwolinski, Vol. 7, No. 2, 417—423, 1978. $3.00.
116. *Densities of Liquid $CH_{4-a}X_a$ (X = Br,I) and $CH_{4-(a+b+c+d)}F_aCl_bBr_cI_d$ Halomethanes,* A. P. Kudchadker, S. A. Kudchadker, P. R. Patnaik, and P. P. Mishra, Vol. 7, No. 2, 425—439, 1978. $3.00.
117. *Microwave Spectra of Molecules of Astro-physical Interest, XIII. Cyanoacetylene,* W. J. Lafferty and F. J. Lovas, Vol. 7, No. 2, 441—493, 1978. $5.00.
118. *Atomic Transition Probabilities for Vanadium, Chromium, and Manganese (A Critical Data Compilation of Allowed Lines),* S. M. Younger, J. R. Fuhr, G. A. Martin, and W. L. Wiese, Vol. 7, No. 2, 495—629, 1978. $7.50.
119. *Thermodynamic Properties of Ammonia,* Lester Haar and John S. Gallagher, Vol. 7, No. 3, 635—792, 1978. $8.00.
120. *JANAF Thermochemical Tables, 1978 Supplement,* M. W. Chase, Jr., J. L. Curnutt, R. A. McDonald, and A. N. Syverud, Vol. 7, No. 3, 793—940, 1978. $8.00.
121. *Viscosity of Liquid Water in the Range −8 °C to 150 °C,* Joseph Kestin, Mordechai Sokolov, and William A. Wakeham, Vol. 7, No. 3, 941—948, 1978. $3.00.

122. *The Molar Volume (Density) of Solid Oxygen in Equilibrium with Vapor,* H. M. Roder, Vol. 7, No. 3, 949—957, 1978. $3.00.
123. *Thermal Conductivity of Ten Selected Binary Alloy Systems,* C. Y. Ho, M. W. Ackerman, K. Y. Wu, S. G. Oh, and T. N. Havill, Vol. 7, No. 3, 959—1177, 1978. $10.00.
124. *Semi-Empirical Extrapolation and Estimation of Rate Constants for Abstraction of H from Methane by H, O, HO, and O_2,* Robert Shaw, Vol. 7, No. 3, 1179—1190, 1978. $3.00.
125. *Energy Levels of Vanadium, V I through V XXIII,* Jack Sugar and Charles Corliss, Vol. 7, No. 3, 1191—1262, 1978. $5.50.
126. *Recommended Atomic Electron Binding Energies, 1s to $6p_{3/2}$, for the Heavy Elements, Z = 84 to 103,* F. T. Porter and M. S. Freedman, Vol. 7, No. 4, 1267—1284, 1978. $4.00.
127. *Ideal Gas Thermodynamic Properties of $CH_{4-(a+b+c+d)}F_aCl_bBr_cI_d$ Halomethanes,* Shanti A. Kudchadker and Arvind P. Kudchadker, Vol. 7, No. 4, 1285—1307, 1978. $4.00.
128. *Critical Review of Vibrational Data and Force Field Constants for Polyethylene,* John Barnes and Bruno Fanconi, Vol. 7, No. 4, 1309—1321, 1978. $3.00.
129. *Tables of Molecular Vibrational Frequencies,* Part 9, Takehiko Shimanouchi, Hiroatsu Matsuura, Yoshiki Ogawa, and Issei Harada, Vol. 7, No. 4, 1323—1443, 1978. $7.00.
130. *Microwave Spectral Tables. II. Triatomic Molecules,* Frank J. Lovas, Vol. 7, No. 4, 1445—1750, 1978. $13.00.

Reprints from Volume 8 (1979)

131. *Energy Levels of Titanium, Ti I through Ti XXII,* Charles Corliss and Jack Sugar, Vol. 8, No. 1, 1—62, 1979. $5.00.
132. *The Spectrum and Energy Levels of the Neutral Atom of Boron (B I),* G. A. Odintzova and A. R. Striganov, Vol. 8, No. 1, 63—67, 1979. $3.00.
133. *Relativistic Atomic Form Factors and Photon Coherent Scattering Cross Sections,* J. H. Hubbell and I. Øverbø, Vol. 8, No. 1, 69—105, 1979. $4.50.
134. *Microwave Spectra of Molecules of Astrophysical Interest, XIV. Vinyl Cyanide (Acrylonitrile),* M. C. L. Gerry, K. Tamada, and G. Winnewisser, Vol. 8, No. 1, 107—123, 1979. $4.00.
135. *Molten Salts: Volume 4, Part 4, Mixed Halide Melts. Electrical Conductance, Density, Viscosity, and Surface Tension Data,* G. J. Janz, R. P. T. Tomkins, and C. B. Allen, Vol. 8, No. 1, 125—302, 1979. $9.00.
136. *Atomic Radiative and Radiationless Yields for K and L Shells,* M. O. Krause, Vol. 8, No. 2, 307—327, 1979. $4.00.
137. *Natural Widths of Atomic K and L Levels, Kα X-Ray Lines and Several KLL Auger Lines,* M. O. Krause and J. H. Oliver, Vol. 8, No. 2, 329—338, 1979. $3.00.
138. *Electrical Resistivity of Alkali Elements,* T. C. Chi, Vol. 8, No. 2, 339—438, 1979. $6.50.
139. *Electrical Resistivity of Alkaline Earth Elements,* T. C. Chi, Vol. 8, No. 2, 439—497, 1979. $5.00.
140. *Vapor Pressures and Boiling Points of Selected Halomethanes,* A. P. Kudchadker, S. A. Kudchadker, P. R. Shukla, and P. R. Patnaik, Vol. 8, No. 2, 499—517, 1979. $4.00.
141. *Ideal Gas Thermodynamic Properties of Selected Bromoethanes and Iodoethanes,* S. A. Kudchadker and A. P. Kudchadker, Vol. 8, No. 2, 519—526, 1979. $3.00.
142. *Thermodynamic Properties of Normal and Deuterated Naphthalenes,* S. S. Chen, S. A. Kudchadker, and R. C. Wilhoit, Vol. 8, No. 2, 527—535, 1979. $3.00.
143. *Microwave Spectra of Molecules of Astrophysical Interest, XV. Propyne,* A. Bauer, D. Boucher, J. Burie, J. Demaison, and A. Dubrulle, Vol. 8, No. 2, 537—558, 1979. $4.00.
144. *A Correlation of the Viscosity and Thermal Conductivity Data of Gaseous and Liquid Propane,* P. M. Holland, H. J. M. Hanley, K. E. Gubbins, and J. M. Haile, Vol. 8, No. 2, 559—575, 1979. $4.00.

Supplements

Physical and Thermodynamic Properties of Aliphatic Alcohols, R. C. Wilhoit and B. J. Zwolinski, Vol. 2, Supplement 1, 420 pages, 1973. Hardcover — $33.00; Softcover — $30.00.

Thermal Conductivity of the Elements: A Comprehensive Review, C. Y. Ho, R. W. Powell, and P. E. Liley, Vol. 3, Supplement 1, 796 pages, 1974. Hardcover — $60.00; Softcover — $55.00.

Energetics of Gaseous Ions, H. M. Rosenstock, K. Draxl, B. W. Steiner, and J. T. Herron, Vol. 6, Supplement 1, 783 pages, 1977. Hardcover — $70.00; Softcover — $65.00.

Evaluated Kinetic Data for High Temperature Reactions, Volume 4. Homogeneous Gas Phase Reactions of Halogen- and Cyanide-Containing Species, D. L. Baulch, J. Duxbury, S. J. Grant, and D. C. Montague, Vol. 9, Supplement 1, 700 pages est., 1980, in press.

Journal of Physical and Chemical Reference Data Subscription Prices* for Volume 8 (1979)

	Member (ACS or AIP member or affiliated society)	Regular rate
USA, Canada, Mexico	$25.00	$100.00
Foreign (surface mail)	29.00	104.00
Air freight Europe	32.00	106.00
Air freight Asia	39.00	114.00

Subscriptions to be sent to:
American Chemical Society
Office of the Comptroller
1155 Sixteenth Street, N. W.
Washington, D. C. 20036

* Do not include supplements.

Inquiries on microfilm should be sent to:
American Institute of Physics
Department S/F
335 East 45th Street
New York, New York 10017

Special Reprint Packages

H. *Thermodynamic Properties of Industrially Important Compounds,* (14 parts) consisting of reprint nos: 6, 30, 42, 43, 62, 65, 66, 70, 80, 83, 92, 113, 115, 127. If purchased individually — $52.00; package price — $35.00.

I. *Molecular Vibrational Frequencies,* (3 parts) consisting of consolidated Volume I, NSRDS-NBS 39; reprint nos: 103, 129. If purchased individually — $18.60; package price — $12.00.

J. *Chemical Data Relevant to Air Pollution,* (7 parts) consisting of reprint nos: 1, 9, 19, 27, 32, 87, 101. If purchased individually — $34.50; package price — $23.00.

K. *Metal Properties Data,* (12 parts) consisting of reprint nos: 29, 34, 36, 44, 52, 57, 61, 76, 90, 98, 107, 123. If purchased individually — $63.00; package price — $42.00.

L. *JANAF Thermochemical Tables,* (4 parts) consisting of Second Edition of JANAF Thermochemical Tables 1971, NSRDS-NBS 37; reprint nos: 50, 60, 120. If purchased individually — $40.60; package price — $32.00.

M. *Thermophysical Properties of Fluids,* (9 parts) consisting of reprint nos: 16, 35, 38, 39, 40, 59, 97, 106, 119. If purchased individually — $45.00; package price — $28.00.

N. *Atomic Energy Levels and Spectra,* (14 parts) consisting of reprint nos: 26, 33, 54, 56, 64, 68, 77, 78, 94, 100, 109, 110, 125, 126. If purchased individually — $72.00; package price — $50.00.

O. *Molecules in Space,* (13 parts) consisting of reprint nos: 14, 17, 23, 24, 45, 46, 47, 48, 74, 75, 88, 112, 117. If purchased individually — $47.50; package price — $34.00.

P. *Molecular Spectra,* (6 parts) consisting of reprint nos: 4, 8, 53, 79, 93, 130. If purchased individually — $45.00; package price — $33.00.

Q. *Atomic Transition Probabilities,* (4 parts) consisting of reprint nos: 20, 63, 82, 118. If purchased individually — $22.50; package price — $15.00.

NSRDS-NBS SERIES

NSRDS-NBS 1, *National Standard Reference Data System Plan of Operation,* E. L. Brady and M. B. Wallenstein, 1964.

NSRDS-NBS 2, *Thermal Properties of Aqueous Uniunivalent Electrolytes,* V. B. Parker, 1965.

NSRDS-NBS 3, Sec. 1, *Selected Tables of Atomic Spectra, Atomic Energy Levels and Multiplet Tables, Si II, Si III, Si IV,* C. E. Moore, 1965.

NSRDS-NBS 3, Sec. 2, *Selected Tables of Atomic Spectra, Atomic Energy Levels and Multiplet Tables, Si I,* C. E. Moore, 1967.

NSRDS-NBS 3, Sec. 3, *Selected Tables of Atomic Spectra, Atomic Energy Levels and Multiplet Tables, CI, CII, CIII, CIV, CV, CVI,* C. E. Moore, 1970.

NSRDS-NBS 3, Sec. 4, *Selected Tables of Atomic Spectra, Atomic Energy Levels and Multiplet Tables, N IV, N V, N VI, N VII,* C. E. Moore, 1971.

NSRDS-NBS 3, Sec. 5, *Selected Tables of Atomic Spectra, Atomic Energy Levels and Multiplet Tables, N I, N II, N III,* C. E. Moore, 1975.

NSRDS-NBS 3, Sec. 6, *Selected Tables of Atomic Spectra, Atomic Energy Levels and Multiplet Tables, H I, D, T,* C. E. Moore, 1972.

NSRDS-NBS 3, Sec. 7, *Selected Tables of Atomic Spectra, Atomic Energy Levels and Multiplet Tables, O I,* C. E. Moore, 1976.

NSRDS-NBS 3, Sec. 8, *Selected Tables of Atomic Spectra, Atomic Energy Levels and Multiplet Tables, O VI, O VII, O VIII,* C. E. Moore, 1979.

NSRDS-NBS 3, Sec. 9, *Selected Tables of Atomic Spectra, Atomic Energy Levels and Multiplet Tables, O IV, O V,* C. E. Moore, 1979, in press.

NSRDS-NBS 3, Sec. 10, *Selected Tables of Atomic Spectra, Atomic Energy Levels and Multiplet Tables, O III,* C. E. Moore, 1980, in press.

NSRDS-NBS 3, Sec. 11, *Selected Tables of Atomic Spectra, Atomic Energy Levels and Multiplet Tables, O II,* C. E. Moore, 1980, in press.

NSRDS-NBS 4, *Atomic Transition Probabilities: Vol. I. Hydrogen Through Neon,* W. L. Wiese, M. W. Smith, and B. M. Glennon, 1966.

NSRDS-NBS 5, *The Band Spectrum of Carbon Monoxide,* P. H. Krupenie, 1966.

NSRDS-NBS 6, *Tables of Molecular Vibrational Frequencies,* Part 1, T. Shimanouchi, 1967; superseded by NSRDS-NBS 39.

NSRDS-NBS 7, *High Temperature Properties and Decomposition of Inorganic Salts,* Part 1, Sulfates, K. H. Stern and E. L. Weise, 1966.

NSRDS-NBS 8, *Thermal Conductivity of Selected Materials,* R. W. Powell, C. Y. Ho, and P. E. Liley, 1966.

NSRDS-NBS 9, *Tables of Bimolecular Gas Reactions,* A. F. Trotman-Dickenson and G. S. Milne, 1967.

NSRDS-NBS 10, *Selected Values of Electric Dipole Moments for Molecules in the Gas Phase,* R. D. Nelson, Jr., D. R. Lide, Jr., and A. A. Maryott, 1967.

NSRDS-NBS 11, *Tables of Molecular Vibrational Frequencies,* Part 2, T. Shimanouchi, 1967; superseded by NSRDS-NBS 39.

NSRDS-NBS 12, *Tables for the Rigid Asymmetric Rotor: Transformation Coefficients from Symmetric to Asymmetric Bases and Expectation Values of* P^2_z, P^4_z, P^6_z, R. H. Schwendeman, 1968.

NSRDS-NBS 13, *Hydrogenation of Ethylene on Metallic Catalysts,* J. Horiuti and K. Miyahara, 1968.

NSRDS-NBS 14, *X-Ray Wavelengths and X-Ray Atomic Energy Levels,* J. A. Bearden, 1967.

NSRDS-NBS 15, *Molten Salts: Vol. 1. Electrical Conductance, Density, and Viscosity Data,* G. J. Janz, F. W. Dampier, G. R. Lakshminarayanan, P. K. Lorenz, and R. P. T. Tomkins, 1968.

NSRDS-NBS 16, *Thermal Conductivity of Selected Materials,* Part 2, C. Y. Ho, R. W. Powell, and P. E. Liley, 1968.

NSRDS-NBS 17, *Tables of Molecular Vibrational Frequencies,* Part 3, T. Shimanouchi, 1968; superseded by NSRDS-NBS 39.

NSRDS-NBS 18, *Critical Analysis of the Heat-Capacity Data of the Literature and Evaluation of Thermodynamic Properties of Copper, Silver, and Gold from 0 to 300 K,* G. T. Furukawa, W. G. Saba, and M. L. Reilly, 1968.

NSRDS-NBS 19, *Thermodynamic Properties of Ammonia as an Ideal Gas,* L. Haar, 1968.

NSRDS-NBS 20, *Gas Phase Reaction Kinetics of Neutral Oxygen Species,* H. S. Johnston, 1968.

NSRDS-NBS 21, *Kinetic Data on Gas Phase Unimolecular Reactions,* S. W. Benson and H. E. O'Neil, 1970.

NSRDS-NBS 22, *Atomic Transition Probabilities, Vol. II. Sodium Through Calcium, A Critical Data Compilation,* W. L. Wiese, M. W. Smith, and B. M. Miles, 1969.

NSRDS-NBS 23, *Partial Grotrian Diagrams of Astrophysical Interest,* C. E. Moore and P. W. Merrill, 1968.

NSRDS-NBS 24, *Theoretical Mean Activity Coefficients of Strong Electrolytes in Aqueous Solutions from 0 to 100 °C,* Walter J. Hamer, 1968.

NSRDS-NBS 25, *Electron Impact Excitation of Atoms,* B. L. Moiseiwitsch and S. J. Smith, 1968.

NSRDS-NBS 26, *Ionization Potentials, Appearance Potentials and Heats of Formation of Gaseous Positive Ions,* J. L. Franklin et al., 1969; superseded by *J. Phys. Chem. Reference Data,* 6, Supplement 1, 1977.

NSRDS-NBS 27, *Thermodynamic Properties of Argon from the Triple Point to 300 K at Pressures to 1000 Atmospheres,* A. L. Gosman, R. D. McCarty, and J. G. Hust, 1969.

NSRDS-NBS 28, *Molten Salts: Vol. 2, Section 1. Electrochemistry of Molten Salts: Gibbs Free Energies and Excess Free Energies from Equilibrium-Type Cells, G. J. Janz and C. G. M. Dijkhuis; Section 2, Surface Tension Data,* G. J. Janz, G. R. Lakshminarayanan, R. P. T. Tomkins and J. Wong, 1969.

NSRDS-NBS 29, *Photon Cross Sections, Attenuation Coefficients, and Energy Absorption Coefficients from 10 keV to 100 GeV,* J. H. Hubbell, 1969.

NSRDS-NBS 30, *High Temperature Properties and Decomposition of Inorganic Salts, Part 2. Carbonates,* K. H. Stern and E. L. Weise, 1969.

NSRDS-NBS 31, *Bond Dissociation Energies in Simple Molecules,* B. deB. Darwent, 1970.

NSRDS-NBS 32, *Phase Behavior in Binary and Multicomponent Systems at Elevated Pressures: n-*Pentane and Methane-*n*-Pentane, V. M. Berry and B. H. Sage, 1970.

NSRDS-NBS 33, *Electrolytic Conductance and Conductances of the Halogen Acids in Water,* W. J. Hamer and H. J. DeWane, 1970.

NSRDS-NBS 34, *Ionization Potentials and Ionization Limits Derived from the Analyses of Optical Spectra,* C. E. Moore, 1970.

NSRDS-NBS 35, *Atomic Energy Levels as Derived from the Analyses of Optical Spectra, Vol. I.* ^{1}*H to* ^{23}*V; Vol. II.* ^{24}*Cr to* ^{41}*Nb; Vol. III.* ^{42}*Mo to* ^{57}*La,* ^{72}*Hf to* ^{89}*Ac,* C. E. Moore, 1971.

NSRDS-NBS 36, *Critical Micelle Concentrations of Aqueous Surfactant Systems,* P. Mukerjee and K. J. Mysels, 1971.

NSRDS-NBS 37, *JANAF Thermochemical Tables,* Second Edition, D. R. Stull, H. Prophet, 1971.

NSRDS-NBS 38, *Critical Review of Ultraviolet Photoabsorption Cross Sections for Molecules of Astrophysical and Aeronomic Interest,* R. D. Hudson, 1971.

NSRDS-NBS 39, *Tables of Molecular Vibrational Frequencies, Consolidated Volume I,* T. Shimanouchi, 1972.

NSRDS-NBS 40, *A Multiplet Table of Astrophysical Interest, Revised Edition, Part I. Table of Multiplets; Part II. Finding List of All Lines in the Table of Multiplets,* C. E. Moore, 1972.

NSRDS-NBS 41, *Crystal Structure Transformations in Binary Halides,* C. N. R. Rao and M. Natarajan, 1972.

NSRDS-NBS 42, *Selected Specific Rates of Reactions of the Solvated Electron in Alcohols,* E. Watson, Jr. and S. Roy, 1972.

NSRDS-NBS 43, *Selected Specific Rates of Reactions of Transients from Water in Aqueous Solution, I. Hydrated Electron,* M. Anbar, M. Bambenek, and A. B. Ross, 1972.

NSRDS-NBS 43, Supplement, *Selected Specific Rates of Reactions of Transients from Water in Aqueous Solution. Hydrated Electron, Supplemental Data,* A. B. Ross, 1975.

NSRDS-NBS 44, *The Radiation Chemistry of Gaseous Ammonia,* D. B. Peterson, 1974.

NSRDS-NBS 45, *Radiation Chemistry of Nitrous Oxide Gas. Primary Processes, Elementary Reactions, and Yields,* G. R. A. Johnson, 1973.

NSRDS-NBS 46, *Reactivity of the Hydroxyl Radical in Aqueous Solutions,* Leon M. Dorfman and Gerald E. Adams, 1973.

NSRDS-NBS 47, *Tables of Collision Integrals and Second Virial Coefficients for the (m,6,8) Intermolecular Potential Function,* Max Klein, H. J. M. Hanley, Francis J. Smith, and Paul Holland, 1974.

NSRDS-NBS 48, *Radiation Chemistry of Ethanol: A Review of Data on Yields, Reaction Rate Parameters, and Spectral Properties of Transients,* Gordon A. Freeman, 1974.

NSRDS-NBS 49, *Transition Metal Oxides, Crystal Chemistry, Phase Transition, and Related Aspects,* C. N. R. Rao and G. V. Subba Rao, 1974.

NSRDS-NBS 50, *Resonances in Electron Impact on Atoms and Diatomic Molecules,* George J. Schulz, 1973.

NSRDS-NBS 51, *Selected Specific Rates of Reactions of Transients from Water in Aqueous Solution, II. Hydrogen Atom,* Michael Anbar, Farhataziz, and A. B. Ross, 1975.

NSRDS-NBS 52, *Electronic Absorption and Internal and External Vibrational Data of Atomic and Molecular Ions Doped in Alkali Halide Crystals,* J. C. Jain, A. V. R. Warrier, and S. K. Agarwal, 1974.

NSRDS-NBS 53, *Crystal Structure Transformations in Inorganic Nitrites, Nitrates and Carbonates,* C. N. R. Rao, B. Prakash, and M. Natarajan, 1975.

NSRDS-NBS 54, *Radiolysis of Methanol: Product Yields, Rate Constants and Spectroscopic Parameters of Intermediates,* J. H. Baxendale and Peter Wardman, 1975.

NSRDS-NBS 55, *Property Index to NSRDS Data Compilations, 1964—1972,* David R. Lide, Jr., Gertrude B. Sherwood, Charles H. Douglass, Jr., and Herman M. Weisman, 1975.

NSRDS-NBS 56, *Crystal Structure Transformations in Inorganic Sulfates, Phosphates, Perchlorates, and Chromates,* B. Prakash and C. N. R. Rao, 1975.

NSRDS-NBS 57, *Yields of Free Ions Formed in Liquids by Radiation,* A. O. Allen, 1976.

NSRDS-NBS 58, *Drift Mobilities and Conduction Band Energies of Excess Electrons in Dielectric Liquids,* A. O. Allen, 1976.

NSRDS-NBS 59, *Selected Specific Rates of Reactions of Transients from Water in Aqueous Solution, III. Hydroxyl Radical and Perhydroxyl Radical and Their Radical Ions,* Farhataziz and Alberta B. Ross, 1977.

NSRDS-NBS 60, *Atomic Energy Levels — The Rare Earth Elements, The Spectra of Lanthanum Cerium, Praseodymium, Neodymium, Promethium, Samarium, Europium, Gadolinium, Terbium, Dysprosium, Holmium, Erbium, Thulium, Ytterbium, and Lutetium,* W. C. Martin, Romuald Zalubas, and Lucy Hagan, 1978.

NSRDS-NBS 61, Part I, *Physical Properties Data Compilations Relevant to Energy Storage I. Molten Salts: Eutectic Data,* George J. Janz, Carolyn B. Allen, Joseph R. Downey, Jr., and R. P. T. Tomkins, 1978.

NSRDS-NBS 61, Part II, *Physical Properties Data Compilations Relevant to Energy Storage II. Molten Salts: Data on Single and Multi-Component Systems,* G. J. Janz, C. B. Allen, N. P. Bansal, R. M. Murphy, and R. P. T. Tomkins, 1979.

NSRDS-NBS 61, Part III, *Physical Properties Data Compilations Relevant to Energy Storage, III. Batteries: Data on Sodium Beta and Beta″ Aluminas,* G. Miller, 1979.

NSRDS-NBS 61, Part IV, *Physical Properties Data Compilations Relevant to Energy Storage. IV. Flywheels: Data on Metals and Metal Alloys,* H. M. Ledbetter, 1979, in press.

NSRDS-NBS 61, Part V, *Physical Properties Data Compilations Relevant to Energy Storage. V. Flywheels: Data on Selected Non-Metallic Composite Materials,* H. E. Pebly, 1979, in press.

NSRDS-NBS 61, Part VI, *Physical Properties Data Compilations Relevant to Energy Storage. VI. Hydrogen: Data on Selected Metal Hydrides,* G. C. Carter, 1979, in press.

NSRDS-NBS 62, *Compilation of Rate Constants for the Reaction of Metal Ions in Unusual Valency States,* George V. Buxton and Robin M. Sellers, 1978.

NSRDS-NBS 63, EPA/NIH Mass Spectral Data Base, Volumes I to IV. Molecular Weights 30—1674, and 1978 Index, S. R. Heller and G. W. A. Milne, 1978.

NSRDS-NBS 64, *Rate Coefficients for Ion-Molecule Reactions, II. Organic Ions Other Than Those Containing Only C and H,* L. W. Sieck, 1979.

NSRDS-NBS 65, *Rate Constants for Reactions of Inorganic Radicals in Aqueous Solution,* A. B. Ross and P. Neta, 1979.

NSRDS DATA PUBLICATIONS FROM OTHER PUBLISHERS

Coblentz Society Evaluated Infrared Reference Spectra, edited and published by the Coblentz Society, sponsored by the Joint Committee on Atomic and Molecular Physical Data (1969—1974). Available from Sadtler Research Laboratories, Philadelphia, PA. 19104. (Microfilm available from the Coblentz Society.) Vol. 6, 1000 Spectra — $295.00; Vol. 7, 1000 Spectra — $295.00; Vol. 8, 1000 Spectra — $295.00; Vol. 9, 1000 Spectra — $295.00; Vol. 10, 1000 Spectra — $295.00; Cumulative Coblentz Indexes — $50.00.

Combustion Fundamentals for Waste Incineration, sponsored by the American Society of Mechanical Engineers Research Committee on Industrial and Municipal Wastes, 1974 (H00087). $30.00. Published by the American Society of Mechanical Engineers, New York.

Compilation of Electron Collision Cross Section Data for Modeling Gas Discharge Lasers, L. J. Kieffer, JILA Information Center Report No. 13, 1973. COM-74-11661.*

A Compilation of YN Reactions, Particle Data Group, 1970. UCRL-20000 YN.*

Compilation of Low Energy Electron Collision Cross Section Data. Part I, L. J. Kieffer, JILA Information Center Report No. 6, 1969. PB189127.*

Compilation of Low Energy Electron Collision Cross Section Data. Part II, L. J. Kieffer, JILA Information Center Report No. 7, 1969. AD696467.*

Contributions to the Data on Theoretical Metallurgy: XVI. Thermodynamic Properties of Nickel and Its Inorganic Compounds, Alla D. Mah and L. B. Pankratz, 1976. U. S. Bureau of Mines Bulletin 668.*

Crystal Data, Determinative Tables, 3rd ed., Vol. 1. Organic Compounds, J. D. H. Donnay and Helen M. Ondik, editors, 1972. Published jointly by the U. S. Department of Commerce, National Bureau of Standards and the Joint Committee on Powder Diffraction Standards. $30.00.

Crystal Data, Determinative Tables, 3rd ed., Vol. 2. Inorganic Compounds, J. D. H. Donnay and Helen M. Ondik, editors, 1973. Published jointly by the U. S. Department of Commerce, National Bureau of Standards and the Joint Committee on Powder Diffraction Standards. $50.00.

Crystal Data, Determinative Tables, 3rd ed., Vol. 3. Organic Compounds 1967—1974, Olga Kennard, David G. Watson, John R. Rogers, editors; J. D. H. Donnay, honorary editor, 1978. Published jointly by the U. S. Department of Commerce, National Bureau of Standards and the JCPDS-International Centre for Diffraction Data. $65.00.

Crystal Data, Determinative Tables, 3rd ed., Vol. 4. Inorganic Compounds 1967—1969, Helen M. Ondik and Alan D. Mighell, editors; J. D. H. Donnay, honorary editor, 1978. Published jointly by the U. S. Department of Commerce, National Bureau of Standards and the JCPDS-International Centre for Diffraction Data. $65.00.

Special price for 4-volume set of *Crystal Data.* $185.00. Order the volumes of *Crystal Data* from: JCPDS-International Centre for Diffraction Data, 1601 Park Lane, Swarthmore, PA 19081.

Derived Thermodynamic Properties of Ethylene, Roland H. Harrison and Donald R. Douslin, *Journal of Chemical and Engineering Data,* 22, 1, 24—30, January 1977.

The Desk Book of Infrared Spectra, Clara D. Craver, Editor, 1977. $90.00. The Coblentz Society, P. O. Box 9952, Kirkwood, MO 63122.

Diffusion Rate Data and Mass Transport Phenomena in Copper Systems, D. B. Butrymowicz, J. R. Manning, and M. E. Read, 1977. Published by the International Copper Research Association, Inc., New York (INCRA Monograph V). $20.00.

Dissociation in Heavy Particle Collisions, G. W. McClure and J. M. Peek, 1972. Published by Wiley-Interscience Series in Atomic and Molecular Collisional Processes. ISBN 0-471-58165-8.

Electron Impact Ionization Cross-Section Data for Atoms, Atomic Ions, and Diatomic Molecules: I. Experimental Data, L. J. Kieffer and G. H. Dunn, *Rev. Mod. Phys.,* 38, 1, 1966.

Excitation in Heavy Particle Collisions, E. W. Thomas, 1972. Published by Wiley-Interscience Series in Atomic and Molecular Collisional Processes. ISBN 0-471-85890-0. $31.00.

Infrared Spectra of Halogenated Hydrocarbons, Clara D. Craver, Editor, 1977. $40.00. The Coblentz Society, P. O. Box 9952, Kirkwood, MO 63122.

Infrared Spectra of Plasticizers and Other Additives, Clara D. Craver, Editor, 1977. $60.00. The Coblentz Society, P. O. Box 9952, Kirkwood, MO 63122.

Ion-Molecule Reactions, E. W. McDaniel, V. Cermak, A. Dalgarno, E. E. Ferguson, and L. Friedman, 1970. Published in the Wiley-Interscience Series in Atomic and Molecular Collisional Processes, Wiley-Interscience, New York. ISBN 0-471-57386-3. $29.50.

K°,N Interactions — A Compilation, Particle Data Group, 1972. LBL-55.*

LNG Materials & Fluids, A User's Manual of Property Data in Graphic Format, First Edition, Douglas Mann, General Editor, Cryogenics Division, National Bureau of Standards, 1977. Loose leaf $35.00. Make check or money order payable to "National Bureau of Standards, Department of Commerce" and send to: LNG Materials and Fluids Users' Manual, Cryogenic Division, National Bureau of Standards, Boulder, CO 80302. For additional information, contact Douglas Mann at the same address, or telephone (303) 499-1000, ext. 3652.

Low Energy Electron-Collision Cross Section Data. Part III: Total Scattering; Differential Elastic Scattering, L. J. Kieffer, *Atomic Data,* 2, 4, 293—391, 1971.

Metallic Shifts in NMR: A Review of the Theory and Comprehensive Critical Data Compilation of Metallic Materials, G. C. Carter, L. H. Bennett, and D. J. Kahan, in 4 volumes, 2350 pp., 1977. Volume 20 in *Progress in Materials Science,* Pergamon Press, Oxford, New York. ISBN 0-08-021143-7. $300.00.

NN and ND *Interactions (above 0.5 GeV/c) — A Compilation,* Particle Data Group, 1970. UCRL-20000 NN.*

NN and ND *Interactions — A Compilation,* Particle Data Group, 1972. LBL-58.*

* Available from the National Technical Information Service.

Nuclear Wallet Cards, V. S. Shirley and C. M. Lederer, editors, 1979. Copies may be obtained from the National Nuclear Data Center, Brookhaven National Laboratory.

πN Partial-Wave Amplitudes, Particle Data Group, 1970. UCRL-20030 πN.*

πN Two-Body Scattering Data I. A User's Guide to the Lovelace-Almehed Data Tape, Particle Data Group, 1973. LBL-63.*

π + p, π + n, and π + d Interactions — A Compilation: Parts I and II, Particle Data Group, 1973. LBL-53.*

Pressure, Volume, Temperature Relations of Ethylene, D. R. Douslin and R. H. Harrison, *J. Chem. Thermodyn.*, 76, 8, 301—330, Aug. 1976.

Review of Particle Properties, Particle Data Group, *Phys. Letters 75B,* No. 1, 1978.

Selected Thermodynamic Values and Phase Diagrams for Copper and Some of Its Binary Alloys, R. Hultgren and P. D. Desai, 1973. Published by the International Copper Research Association, Inc., New York; Monograph I. $20.00.

Selected Values of the Thermodynamic Properties of Binary Alloys, R. Hultgren, P. D. Desai, D. T. Hawkins, M. Gleiser, and K. K. Kelley, 1973. Published by the American Society for Metals, Metals Park, Ohio. $30.00.

Selected Values of the Thermodynamic Properties of the Elements, R. Hultgren, P. D. Desai, D. T. Hawkins, M. Gleiser, K. K. Kelley, and D. D. Wagman, 1973. Published by the American Society for Metals, Metals Park, Ohio. $20.00.

Table of Isotopes, 7th edition, C. Michael Lederer and Virginia S. Shirley, editors; Edgardo Browne, Janis M. Dairiki, and Raymond E. Doebler, 1978. John Wiley & Sons, New York. Clothbound, ISBN 0-471-04179-3 — $40.00; paperbound, ISBN 0-471-04180-7 — $26.25.

Thermodynamic Data for Waste Incineration, sponsored by the American Society of Mechanical Engineers Research Committee on Industrial and Municipal Wastes, 1979. Order from the ASME Order Department as H00141 — $30.00. Members — $15.00.

Thermodynamic Properties of Aqueous Copper Systems, Paul Duby, 1977. Published by the International Copper Research Association, Inc., New York. (Monograph IV). $20.00.

Thermodynamic Properties of Copper and Its Inorganic Compounds, E. G. King, Alla D. Mah, and L. B. Pankratz,1973. Published by the International Copper Research Association, Inc., New York. (Monograph II). $20.00.

Thermodynamic Properties of Copper-Slag Systems, Paul Duby, 1976. Published by the International Copper Research Association, Inc., New York. (Monograph III). $20.00.

Theory of Charge Exchange, R. A. Mapleton, 1971. Published by Wiley-Interscience Series in Atomic and Molecular Collisional Processes. ISBN 0-471-56781-7.

Theory of the Ionization of Atoms by Electron Impact, M. R. H. Rudge, *Rev. Mod. Phys.*, 40 (3), 564, 1968.

Total Electron-Atom Collision Cross Sections at Low Energies — A Critical Review, Benjamin Bederson and L. J. Kieffer, *Rev. Mod. Phys.*, 43 (4), 601, 1971.

OTHER NBS COMPILATIONS OF DATA

NBS Tech. Note 270-3, *Selected Values of Chemical Thermodynamic Properties, Tables for the First Thirty-Four Elements in the Standard Order of Arrangement,* D. D. Wagman, W. H. Evans, V. B. Parker, I. Halow, S. M. Bailey, and R. H. Schumm, 1968.

NBS Tech. Note 270-4, *Selected Values of Chemical Thermodynamic Properties, Tables for Elements 35 through 53 in the Standard Order of Arrangement,* D. D. Wagman, W. H. Evans, V. B. Parker, I. Halow, S. M. Bailey, and R. H. Schumm, 1969.

NBS Tech. Note 270-5, *Selected Values of Chemical Thermodynamic Properties, Tables for Elements 54 through 61 in the Standard Order of Arrangement,* D. D. Wagman, W. H. Evans, et al., 1971.

NBS Tech. Note 270-6, *Selected Values of Chemical Thermodynamic Properties, Tables for the Alkaline Earth Elements (Elements 92 through 97 in the Standard Order of Arrangement),* V. B. Parker, D. D. Wagman, and W. H. Evans, 1971.

NBS Tech. Note 270-7, *Selected Values of Chemical Thermodynamic Properties, Tables for the Lanthanide (Rare Earth) Elements (Elements 62 through 76 in the Standard Order of Arrangement),* R. H. Schumm, D. D. Wagman, S. Bailey, W. H. Evans, and V. B. Parker, 1973.

NBS Tech. Note 270-8, *Selected Values of Chemical Thermodynamic Properties, Tables for the Alkali Metals (Elements 98 through 103 in the Standard Order of Arrangement),* D. D. Wagman, et al., 1980, in press.

NBS Tech. Note 270-9, *Selected Values of Chemical Thermodynamic Properties, Tables for the Actinide Elements (Elements 77 through 90 in the Standard Order of Arrangement),* D. D. Wagman, et al., 1980, in process.

NBS Tech. Note 438, *Compendium of ab-initio Calculations of Molecular Energies and Properties,* M. Krauss, 1967.

NBS Tech. Note 474, *Critically Evaluated Transition Probabilities for Ba I and II,* B. M. Miles and W. L. Wiese, 1969.

NBS Tech. Note 484, *A Review of Rate Constants of Selected Reactions of Interest in Re-Entry Flow Fields in the Atmosphere,* M. H. Bortner, 1969.

NBS Tech. Note 968, *Total Photon Absorption Cross Section Measurements, Theoretical Analysis and Evaluation for Energies above 10 MeV,* H. A. Gimm and John H. Hubbell, 1978.

NBS Tech. Note 983, *Properties of Selected Superconductive Materials, 1978 Supplement,* B. W. Roberts, 1978. Supplements *J. Phys. Chem. Reference Data,* 5 (3), 581—821, 1976, Reprint No. 84.

NBS Spec. Publ. 513, *Reaction Rate and Photochemical Data for Atmospheric Chemistry — 1977,* Robert F. Hampson, Jr. and David Garvin, editors, 1978.

NBS Monograph 70, Vol. I. *Microwave Spectral Tables, Diatomic Molecules,* P. F. Wacker, M. Mizushima, J. D. Petersen, and J. R. Ballard, 1964.

NBS Monograph 70, Vol. II. *Microwave Spectral Tables, Line Strengths of Asymmetric Rotors,* P. F. Wacker and M. R. Pratto, 1964.

NBS Monograph 70, Vol. III. *Microwave Spectral Tables, Polyatomic Molecules with Internal Rotation,* P. F. Wacker, M. S. Cord, D. G. Burkhard, J. D. Petersen, and R. F. Kukol, 1969.

NBS Monograph 70, Vol. IV. *Microwave Spectral Tables, Polyatomic Molecules without Internal Rotation,* M. S. Cord, J. D. Petersen, M. S. Lojko, and R. H. Haas, 1968.

NBS Monograph 70, Vol. V. *Microwave Spectral Tables, Spectral Line Listing,* M. S. Cord, M. S. Lojko, and J. D. Petersen, 1968.

NBS Monograph 94, *Thermodynamic and Related Properties of Parahydrogen from the Triple Point to 100 K at Pressures to 340 Atmospheres,* H. M. Roder, L. A. Weber, and R. D. Goodwin, 1965.

NBS Monograph 115, *The Calculation of Rotational Energy Levels and Rotational Line Intensities in Diatomic Molecules,* Jon T. Hougen, 1970.

NBS Monograph 134, *Space Groups and Lattice Complexes,* Werner Fisher, Hans Burzlaff, Erwin Hellner, and J. D. H. Donnay, 1973.

NBS Monograph 153, *The First Spectrum of Hafnium (Hf I),* William F. Meggers and Charlotte E. Moore, 1976.

NBS Monograph 158, *Handbook of Rock Salt Properties Data,* L. H. Gevantman, editor, 1979, in press.

NBSIR 75-770, *Equation of State for Ammonia,* L. Haar and J. Gallagher, 1975.

NBSIR 75-968, *Selected Thermochemical Data Compatible with the CODATA Recommendations,* V. B. Parker, D. D. Wagman, and D. Garvin, 1976.

NBSIR 76-1034, *Chemical Thermodynamic Properties of Compounds of Sodium, Potassium, and Rubidium: An Interim Tabulation of Selected Material,* D. D. Wagman, W. H. Evans, V. B. Parker, and R. H. Schumm, 1976.

NBSIR 76-1061, *Annotated Bibliography on Proton Affinities,* K. Hartman, S. Lias, P. J. Ausloos, and H. M. Rosenstock, 1976.

NBSIR 77-860, *Provisional Thermodynamic Functions of Propane from 85 to 700 K at Pressures to 700 Bar,* R. D. Goodwin, 1977.

NBSIR 77-865, *Thermodynamic and Related Properties of Oxygen from the Triple Point to 300 K at Pressures to 1000 Bar,* L. A. Weber, 1977.

NBSIR 77-1300, *A Computer Assisted Evaluation of the Thermochemical Data of the Compounds of Thorium,* D. D. Wagman, R. H. Schumm, and V. B. Parker, 1977.

NBSIR 78-1479, *Thermodynamic Properties of Miscellaneous Materials,* Eugene S. Domalski, William H. Evans, and Thomas L. Jobe, Jr., 1978.

TRANSLATIONS FROM THE RUSSIAN

A number of products of the Russian standard reference data system have been translated with the support of the National Science Foundation at the request of the National Bureau of Standards. These Russian translations are available from NTIS.

Electric Conductivity of Ferroelectrics, V. M. Gurevich, 1969. Original published by the Committee on Standards, Measures and Measuring Instruments of the USSR Council of Ministers — Government Standards Service. TT70-50180, 1971, vi + 362 pp.

Handbook of Hardness Data, A. A. Ivan'ko; G. V. Samsonov, editor, 1968. Original published by the Academy of Sciences of the Ukrainian SSR, Institute of Materials Sciences. TT70-50177, 1971, iii + 66 pp.

Handbook of Phase Diagrams of Silicate Systems, Vol. I. Binary Systems, Second Revised Edition, N. A. Toropov, V. P. Barzakovskii, V. V. Lapin, and N. N. Kurtseva; N. A. Toropov, editor, 1969. Original published by the Academy of Sciences of the USSR Grebenshchikov Institute of Silicate Chemistry. TT71-50040, 1972, viii + 723 pp.

Handbook of Phase Diagrams of Silicate Systems. Vol. II. Metal-Oxygen Compounds in Silicate Systems, Second Revised Edition, N. A. Toropov, V. P. Barzakovskii, I. A. Bondar, and Yu, P. Udalov; N. A. Toropov, editor, 1969. Original published by the Academy of Sciences of the USSR Grebenshchikov Institute of Silicate Chemistry. TT71-50041, 1972, iv + 325 pp.

Handbook of Thermodynamic Data, G. B. Naumov, B. N. Ryzhenko, and I. L. Khodakovsky; A. I. Tugarinov, editor, 1971. Original published by the USSR Academy of Sciences, Atomizdat. PB226-722, xi + 328 pp.

Heavy Water, Thermophysical Properties, Ya. Z. Kazavchinskii et al.; V. A. Kirillin, editor, 1963. TT70-50094, 1971, vii + 263 pp., 2 charts.

Isochoric Heat Capacity of Water and Steam, Kh. I. Amirkhanov, G. V. Stepanov, B. G. Alibekov; M. P. Vukalovich, editor, 1969. Original published by the Dagestan Branch of the USSR Academy of Sciences. TT72-52002, xii + 203 pp.

Properties of Liquid and Solid Hydrogen, B. N. Esel'son, Yu. P. Blagoi, V. N. Grigor'ev, V. G. Manzhelii, S. A. Mikhailenko, and N. P. Neklyudov, 1969. Original published by the Committee on Standards, Measures and Measuring Instruments of the USSR Council of Ministers — Government Standards Service. TT70-50179, 1974, iii + 123 pp.

Rate Constants of Gas Phase Reactions, Reference Book, V. N. Kondratiev, 1970. Original published by the Academy of Sciences of the USSR Order-of-Lenin Institute of Chemical Physics. Translated by L. Holtschlag and R. Fristrom. COM-72-10014, 1972, vi + 428 pp.

Surface Ionization, E. Ya. Zandberg and N. L. Ionov, 1969. Original published by the Science Publishing House. Translated by E. Harnik. TT70-50148, 1971, xii + 355 pp.

Thermodynamic and Thermophysical Properties of Combustion Products, Volume I. Computation Methods, V. E. Alemasov, A. F. Dregalin, A. P. Tishin, and V. A. Khudyakov; V. P. Glushko, editor, 1971. Original published by the Academy of Sciences of the USSR. TT74-50019, 1974, xx + 433 pp.

Thermodynamic and Thermophysical Properties of Combustion Products, Volume II. Oxygen Based Propellants, V. E. Alemasov, A. F. Dregalin, V. A. Khudyakov, and V. N. Kostin; V. P. Glushko, editor, 1972. Original published by the Academy of Sciences of the USSR. TT74-50032, x + 495 pp.

Thermodynamic and Thermophysical Properties of Combustion Products, Volume III. Oxygen- and Air-Based Propellants, V. E. Alemasov, A. F. Dregalin, A. P. Tishin, V. A. Khudyakov, and V. N. Kostin; V. P. Glushko, editor, 1973. Original published by the Academy of Sciences of the USSR. TT75-50007, 1975, x + 622 pp.

Thermodynamic and Thermophysical Properties of Combustion Products, Volume IV. Nitrogen Tetroxide-Based Propellants, V. E. Alemasov, A. F. Dregalin, A. P. Tishin, V. A. Khudyakov, and V. N. Kostin; V. P. Glushko, editor, 1973. Original published by the Academy of Sciences of the USSR. TT76-50007, 1976, x + 530 pp.

Thermodynamic and Thermophysical Properties of Helium, N. V. Tsederberg, V. N. Popov, and N. A. Morozova; A. F. Alyab'ev, editor, 1969. Original published by Atomizdat. TT70-50096, 1971, v + 255 pp.

Thermodynamic and Transport Properties of Ethylene and Propylene, I. A. Neduzhii et al., 1971. Original published by the State Committee of Standards of the Soviet Ministry, USSR, State Office of Standards and Reference Data Series: Monograph No. 8. Translation published June 1972. NBSIR 75-763.

Thermophysical Properties of Air and Air Components, A. A. Vasserman, Ya. A. Kazavchinskii, and V. A. Rabinovich; A. M. Zhuravlev, editor, 1966. Original published by the Academy of Sciences of the USSR. TT70-50095, 1971, x + 383 pp., 8 charts.

Thermophysical Properties of Freon-22, A. V. Kletskii, 1970. Original published by the Committee of Standards, Measures and Measuring Instruments of the USSR Council of Ministers — Government Standards Service. TT70-50178, 1971, iv + 67 pp., 1 chart.

Thermophysical Properties of Gases and Liquids, No. 1, V. A. Rabinovich, 1968. Original published by the Committee of Standards, Measures and Measuring Instruments of the USSR Council of Ministers — Government Standards Service, Series: Physical Constants and Properties of Substances. TT69-55091, 1970, viii + 207 pp.

The translation of volumes 2, 3, and 4 of this series have the English title *Thermophysical Properties of Matter and Substances.*

Thermophysical Properties of Gaseous and Liquid Methane, V. A. Zagoruchenko and A. M. Zhuravlev, 1969. Original published by the Committee of Standards, Measures and Measuring Instruments of the USSR Council of Ministers — Government Standards Service. TT70-50097, 1970, viii + 243 pp.

Thermophysical Properties of Liquid Air and Its Components, A. A. Vasserman and V. A. Rabinovich, 1968. Original published by the Committee of Standards, Measures and Measuring Instruments of the USSR Council of Ministers — Government Standards Service. TT69-55092, 1970, viii + 235 pp., 2 charts.

Thermophysical Properties of Matter and Substances, Volume 2, V. A. Rabinovich, 1970. Original published by the Committee of Standards, Measures and Measuring Instruments of the USSR Council of Ministers — Government Standards Service, Series: Physical Constants and Properties of Matter. TT72-52001, 1974, xi + 384 pp.

Thermophysical Properties of Matter and Substances, Volume 3, V. A. Rabinovich, 1971. Original published by the Committee of Standards, Measures and Measuring Instruments of the USSR Council of Ministers — Government Standards Service, Series: Physical Constants and Properties of Matter. TT73-52009, 1974, xiii + 343 pp.

Thermophysical Properties of Matter and Substances, Volume 4, V. A. Rabinovich, 1971. Original published by the Committee of Standards, Measures and Measuring Instruments of the USSR Council of Ministers — Government Standards Service, Series: Physical Constants and Properties of Matter. TT73-52029, vii + 173 pp.

The translation of Volume 1 of this series has the English title *Thermophysical Properties of Gases and Liquids, No. 1.*

CRITICAL BIBLIOGRAPHIES AND INDEXES FROM OTHER PUBLISHERS

Bibliography of High Pressure Research, Leo Merrill, compiler, Volume XII, 1979. Bimonthly — $9.00; overseas airmail — $13.00. Subscriptions and information requests should be sent to: High Pressure Data Center, 5093 HBLL, Brigham Young University, Provo, UT 84602 USA.

Binary Fluorides, Free Molecular Structures and Force Fields, A Bibliography (1957—1975), D. T. Hawkins, L. S. Bernstein, W. E. Falconer, and W. Klemperer, IFI/Plenum, New York, 1976.

Biweekly List of Papers on Radiation Chemistry, Annual Cumulation with Keyword and Author Indexes, Vol. VII, 1974, 1975. COM-75-11475.*

Biweekly List of Papers on Radiation Chemistry, Annual Cumulation with Keyword and Author Indexes, Vol. VIII, 1975, 1976. PB 257 025.*

Biweekly List of Papers on Radiation Chemistry, Annual Cumulation with Keyword and Author Indexes, Vol. IX, 1976, 1977. PB 268 699.*

Biweekly List of Papers on Radiation Chemistry, Annual Cumulation with Keyword and Author Indexes, Vol. X, 1977, 1978. PB 291 876.*

Biweekly List of Papers on Radiation Chemistry and Photochemistry, Annual Cumulation with Keyword and Author Indexes, Vol. XI, 1978, 1979. Available from the Radiation Chemistry Data Center.

* Available from the National Technical Information Service.

Subscriptions to the *Biweekly List of Papers on Radiation Chemistry and Photochemistry (Vol. XII, 1979)* and the *Annual Cumulations* for the previous year (Vol. XI, 1978) are available on a subscription basis from the: Radiation Chemistry Data Center, Radiation Laboratory, University of Notre Dame, Notre Dame, IN 46556 USA.

Please write to the Data Center for further information.

Bulletin of Chemical Thermodynamics, IUPAC Commission on Thermodynamics and Thermochemistry, Robert D. Freeman, editor; former title: Bulletin of Thermodynamics and Thermochemistry. Volume 21, 1978 — $25.00. Subscriptions and orders for single copies and back issues should be addressed to: Bulletin of Chemical Thermodynamics, Department of Chemistry, Oklahoma State University, Stillwater, OK 74074 USA.

Checks should be made payable, in U.S. funds, to: Bulletin of Chemical Thermodynamics.

A Compilation of Volumes I—IV of Bibliography on High Pressure Research with Author and Subject Indexes, Volumes I and II, J. F. Cannon and L. Merrill, 1972. Published by and available from the High Pressure Data Center, 5093 HBLL, Brigham Young University, Provo, UT 84602 USA. In microfiche — $14.00.

Compilation of Volumes V—IX of the Bibliography on High Pressure Research with Author and Subject Indexes, L. Merrill, 1979, in press. Will be available from the High Pressure Data Center.

Equilibrium Properties of Fluid Mixtures, A Bibliography on Fluids of Cryogenic Interest, M. J. Hiza, A. J. Kidnay, and R. C. Miller, IFI/Plenum, New York, 1975. ISBN 0-306-66001-6 — $37.50.

"A Guide to Sources of Information on Materials," R. S. Marvin and G. B. Sherwood, Chapter 10, pp. 603—627, in *Handbook of Materials Science, Volume III, Nonmetallic Materials and Publications,* C. T. Lynch, editor, CRC Press, Cleveland, Ohio, 1975.

Index and Cumulative List of Papers on Radiation Chemistry, Volume IV, Numbers 27—52, July through December 1971, January 1972. COM-72-10266.*

Index and Cumulative List of Papers on Radiation Chemistry, Volume V, Numbers 1—26, January through June 1972, July 1972. COM-72-10621.*

Index and Cumulative List of Papers on Radiation Chemistry, Volume V, Numbers 27—52, July through December 1972, January 1973. COM-73-10281.*

Index and Cumulative List of Papers on Radiation Chemistry, Volume VI, Numbers 1—26, January through June 1973, July 1973. COM-73-11878/8AS.*

Index and Cumulative List of Papers on Radiation Chemistry, Volume VI, Numbers 27—52, July through December 1973, January 1974. COM-74-10899.*

An Indexed Compilation of Experimental High Energy Physics Literature, Particle Data Group, LBL90, 1978.*

Liquid Vapor Equilibria on Systems of Interest in Cryogenics, A Survey, A. J. Kidnay, M. J. Hiza, and R. C. Miller, *Cryogenics,* 13, 10, 575—599, 1973.

Mossbauer Effect Data Index, 1958—1965, A. H. Muir, K. J. Ando and H. M. Coogan, Interscience Publishers, New York, 1967.

Mossbauer Effect Data Index, Covering the 1966—1968 Literature, J. G. Stevens and V. E. Stevens, IFI/Plenum, New York, 1975. ISBN 0-306-65162-9 — $49.50.

Mossbauer Effect Data Index, Covering the 1969 Literature, 1971. ISBN 0-306-65140-8 — $37.50.

Mossbauer Effect Data Index, Covering the 1970 Literature, 1972. ISBN 0-306-65141-6 — $37.50.

Mossbauer Effect Data Index, Covering the 1971 Literature, 1973. ISBN 0-306-65142- — $37.50.

Mossbauer Effect Data Index, Covering the 1972 Literature, 1973. ISBN 0-306-65143-2 — $37.50.

Mossbauer Effect Data Index, Covering the 1973 Literature, 1975. ISBN 0-306-65144-0 — $49.50.

Mossbauer Effect Data Index, Covering the 1974 Literature, 1975. ISBN 0-306-65145-9 — $49.50.

Mossbauer Effect Data Index, Covering the 1975 Literature, 1977. ISBN 0-306-65146- — $49.50.

Mossbauer Effect Data Index, Covering the 1976 Literature, 1978. ISBN 0-306-65149-1 — $49.50.

Mossbauer Effect Reference and Data Journal, Volume 1 (1978); 10 issues per year; J. G. Stevens, V. E. Stevens, and W. L. Gettys, editors. For further information write to: Mossbauer Effect Data Center, University of North Carolina — Asheville, Asheville, NC 28804 USA. Telephone (704) 258-0200.

Thesaurus for Radiation Chemistry, Alberta B. Ross, 1977. UND-RCDC-1.*

Weekly List of Papers on Radiation Chemistry, Index and Cumulation, Volume IV, Numbers 1—26, January through June 1971, August 1971. COM-71-01103.*

CRITICAL BIBLIOGRAPHIES AND INDEXES IN NBS SERIES

NBS Misc. Publ. 281, *Bibliography on Flames Spectroscopy, Analytical Applications, 1800 to 1966,* R. Mavrodineanu, 1967.

NBS Spec. Publ. 306-1, *Bibliography on the Analyses of Optical Atomic Spectra, Section 1,* ^{1}H-^{23}V, C. E. Moore, 1968.

NBS Spec. Publ. 306-2, *Bibliography on the Analyses of Optical Atomic Spectra, Section 2,* ^{24}Cr-^{41}Nb, C. E. Moore, 1969.

NBS Spec. Publ. 306-3, *Bibliography on the Analyses of Optical Atomic Spectra, Section 3,* ^{42}Mo-^{57}La, ^{72}Hf-^{89}Ac, C. E. Moore, 1969.

NBS Spec. Publ. 306-4, *Bibliography on the Analyses of Optical Atomic Spectra, Section 4,* ^{57}La-^{71}Lu, ^{89}Ac-^{99}Es, C. E. Moore, 1969.

NBS Spec. Publ. 324, *The NBS Alloy Data Center: Permuted Materials Index,* G. C. Carter, D. J. Kahan, L. H. Bennett, J. R. Cuthill, and R. C. Dobbyn, 1971.

NBS Spec. Publ. 349, *Heavy-Atom Kinetic Isotope Effects, An Indexed Bibliography,* M. J. Stern and M. Wolfsberg, 1972.

NBS Spec. Publ. 362, *Chemical, Kinetics in the C-O-S and H-N-O-S Systems: A Bibliography — 1899 through June 1971,* F. Westley, 1972.

NBS Spec. Publ. 363, *Bibliography on Atomic Energy Levels and Spectra, July 1968 through June 1971,* L. Hagan and W. C. Martin, 1972.

NBS Spec. Publ. 363, Supplement 1, *Bibliography on Atomic Energy Levels and Spectra, July 1971 through June 1975,* Lucy Hagan, 1977.

NBS Spec. Publ. 363, Supplement 2, *Bibliography on Atomic Energy Levels and Spectra, July 1975 through December 1978,* R. Zalubas and A. Albright, 1979, in press.

NBS Spec. Publ. 366, *Bibliography on Atomic Line Shapes and Shifts (1899 through March 1972),* J. R. Fuhr, W. L. Wiese, and L. J. Roszman, 1972.

NBS Spec. Publ. 366, Supplement 1, *Bibliography on Atomic Line Shapes and Shifts (April 1972 through June 1973),* J. R. Fuhr, L. J. Roszman, and W. L. Wiese, 1974.

NBS Spec. Publ. 366, Supplement 2, *Bibliography on Atomic Line Shapes and Shifts (July 1973 through May 1975),* J. R. Fuhr, G. A. Martin, and B. Specht, 1975.

NBS Spec. Publ. 366, Supplement 3, *Bibliography on Atomic Line Shapes and Shifts (June 1975 through June 1978),* J. R. Fuhr, B. J. Miller, and G. A. Martin, 1978.

NBS Spec. Publ. 369, *Soft X-Ray Emission Spectra of Metallic Solids: Critical Review of Selected Systems and Annotated Spectral Index,* A. J. McAlister, R. C. Dobbyn, J. R. Cuthill, and M. L. Williams, 1974.

NBS Spec. Publ. 371, *A Supplementary Bibliography of Kinetic Data on Gas Phase Reactions of Nitrogen, Oxygen, and Nitrogen Oxides,* F. Westley, 1973.

NBS Spec. Publ. 371-1, *Supplementary Bibliography of Kinetic Data on Gas Phase Reactions of Nitrogen, Oxygen, and Nitrogen Oxides (1972—1973),* Francis Westley, 1975.

NBS Spec. Publ. 380, *Photonuclear Reaction Data, 1973,* E. G. Fuller, H. M. Gerstenberg, H. Vander Molen, and T. C. Dunn, 1973.

NBS Spec. Publ. 380, Supplement 1, *Photonuclear Reactions Data, July 1973 to December 1977,* E. G. Fuller and H. M. Gerstenberg, 1978.

NBS Spec. Publ. 381, *Bibliography of Ion-Molecule Reaction Rate Data (January 1950 through October 1971),* George A. Sinnott, 1973.

NBS Spec. Publ. 392, *Vibrationally Excited Hydrogen Halides: A Bibliography on Chemical Kinetics of Chemiexcitation and Energy Transfer Processes (1958 through 1973),* Francis Westley, 1974.

NBS Spec. Publ. 396-1, *Critical Surveys of Data Sources: Mechanical Properties of Metals,* R. B. Gavert, R. L. Moore, and J. H. Westbrook, 1974.

NBS Spec. Publ. 396-2, *Critical Surveys of Data Sources: Ceramics,* Dorothea M. Johnson and James F. Lynch, 1975.

NBS Spec. Publ. 396-3, *Critical Surveys of Data Sources: Corrosion of Metals,* Ronald B. Diegle and Walter K. Boyd, 1976.

NBS Spec. Publ. 396-4, *Critical Surveys of Data Sources: Electrical and Magnetic Properties of Metals,* M. J. Carr, R. B. Cavert, R. L. Moore, H. W. Wawrousek, and J. H. Westbrook, 1976.

NBS Spec. Publ. 426, *Bibliography of Low Energy Electron and Photon Cross Section Data (through December 1974),* Lee J. Kieffer, 1976.

NBS Spec. Publ. 426, Supplement 1, *Bibliography of Low Energy Electron and Photon Cross Section Data (January 1975 through December 1977),* J. W. Gallagher, J. R. Rumble, E. C. Beatty, 1979.

NBS Spec. Publ. 428, *Bibliography of Infrared Spectroscopy through 1960, Parts 1, 2, and 3,* C. N. R. Rao, S. K. Dikshit, S. A. Kudchadker, D. S. Gupta, V. A. Narayan, and J. J. Comeford, 1976.

NBS Spec. Publ. 449, *Chemical Kinetics of the Gas Phase Combustion of Fuels (A Bibliography on the Rates and Mechanisms of the Oxidation of Aliphatic C_1—C_{10} Hydrocarbons and of Their Oxygenated Derivatives),* Francis Westley, 1976.

NBS Spec. Publ. 454, *An Annotated Bibliography of Sources of Compiled Thermodynamic Data Relevant to Biochemical and Aqueous Systems (1930—1975) Equilibrium, Enthalpy, Heat Capacity, and Entropy Data,* George T. Armstrong and Robert N. Goldberg, 1976.

NBS Spec. Publ. 478, *Nitrogen Oxychlorides: A Bibliography on Data for Physical and Chemical Properties of $ClNO$, $ClNO_2$ and $ClNO_3$,* F. Westley, 1977.

NBS Spec. Publ. 485, *A Bibliography of Sources of Experimental Data Leading to Activity or Osmotic Coefficients for Polyvalent Electrolytes in Aqueous Solution,* R. N. Goldberg, B. R. Staples, R. L. Nuttail, R. Arbuckle, 1977.

NBS Spec. Publ. 505, *Bibliography on Atomic Transition Probabilities (1914 through October 1977),* J. R. Fuhr, B. J. Miller, and G. A. Martin, 1978.

NBS Spec. Publ. 537, *A Bibliography of Sources of Experimental Data Leading to Thermal Properties of Binary Aqueous Electrolyte Solutions,* D. Smith-Magowan and R. N. Goldberg, 1979.

NBS Tech. Note 291, *A Bibliography on Ion-Molecule Reactions, January 1900 to March 1966,* F. N. Harllee, H. M. Rosenstock, and J. T. Herron, 1966.

NBS Tech. Note 464, *The NBS Alloy Data Center: Function, Bibliographic System, Related Data Centers, and Reference Books,* G. C. Carter, L. H. Bennett, J. R. Cuthill, and D. J. Kahan, 1968.

NBS Tech. Note 554, *Annotated Accession List of Data Compilations of the NBS Office of Standard Reference Data,* H. M. Weisman and G. B. Sherwood, 1970.

NBS Tech. Note 848, *A Bibliography of the Russian Reference Data Holdings of the Library of the Office of Standard Reference Data,* Gertrude B. Sherwood and Howard J. White, Jr., 1974.

NBS-OSRDB-70-1, Vol. 1, *High Pressure Bibliography 1900—1968. Volume I. Section I — Bibliography, Section II. Author Index,* Leo Merrill, 1970.

NBS-OSRDB-70-1, Vol. 2, *High Pressure Bibliography 1900—1968. Volume II. Subject Index,* Leo Merrill, 1970.

NBS-OSRDB-70-2, *The NBS Alloy Data Center: Author Index,* G. C. Carter, D. J. Kahn, L. H. Bennett, J. R. Cuthill, and R. C. Dobbyn, 1970.

NBS-OSRDB-70-3, *Semiempirical and Approximate Methods for Molecular Calculations — Bibliography and KWIC Index,* George A. Henderson and Sandra Frattali, 1969.

NBS-OSRDB-70-4, *Bibliography of Photoabsorption Cross Section Data,* Robert D. Hudson and Lee J. Kieffer, 1970.

NBS-OSRDB-71-1, *Bibliography on Properties of Defect Centers in Alkali Halides,* S. C. Jain, S. A. Khan, H. K. Sehgal, V. K. Garg, and R. K. Jain, 1971.

NBS-OSRDB-71-2, *A Bibliography of Kinetic Data on Gas Phase Reactions of Nitrogen, Oxygen, and Nitrous Oxides,* Francis Westley, 1970.

The following lists the name and prices of certain publications of
Fachinformationszentrum
Energie, Physik, Mathematik Gmbh
Karlsruhe Nuclear Research Centre
Kernforschungszentrum
D-75014 Eggenstein
Leopoldshafen 2
West Germany

PHYSIK DATEN/PHYSICS DATA

		Price/plus postage
1—2 (1977):	*Survey Index of Pion-Nucleon Scattering Data,* March 1977 Edition, by K. H Augenstein, G. Hohler, E. Pietarinen and H. M. Staudenmaier, 66 pages. This issue supersedes the previous edition published as No. 1—1 (1975).	out of print
2—1 (1976):	*Stopping Cross Sections of Elements with $Z = 2$ to 87 for Li ions with Energies between 80 keV and 840 keV,* by W. Neuwirth, W. Pietsch and U. Hauser, 10 pages. A compilation of experimental and calculated stopping cross sections.	DM 3.50
3—1 (1976):	*Datensammlungen in der Physik./Data Compilations in Physics,* by H. Behrens and G. Ebel, 226 pages. A bibliography of 1457 existing tables and compilations from all fields of physics.	DM 14.—
3—2 (1977):	*Datensammlungen in der Physik./ Data Compilations in Physics,* by H. Behrens and G. Ebel. 116 pages. Supplement to No. 3—1 containing 561 further references to tables and compilations.	DM 7.—
3—3 (1978):	*Datensammlungen in der Physik./Data Compilations in Physics,* by H. Behrens and G. Ebel. 78 pages. Supplement to No. 3—1 and 3—2 containing 348 further references to tables and compilations.	DM 8.—
3—4 (1979):	*Datensammlungen in der Physik./Data Compilations in Physics,* by H. Behrens and G. Ebel. 86 pages. Supplement to No. 3—1, 3—2, and 3—3 containing 418 further references to tables and compilations.	DM 11.50
4—2 (1979):	*Compilation of Coupling Constants and Low-Energy Parameters,* 1978 Edition, reprinted from *Nuclear Physics B,* 147, 189—276, 1979, by M. M. Nagels, T. A., Rijken, J. J. de Swart, G. C. Oades, J. L. Petersen, A. C. Irving, C. Jarlskog, W. Pfeil, H. Pilkuhn, and H. P. Jakob. 92 pages. A compilation of coupling constants and other parameters which characterize the interactions of elementary particles at low energies. This issue supersedes the previous edition published as No. 4—1 (1976).	DM 7.50
5—1 (1976):	*Gases and Carbon in Metals (Thermodynamics, Kinetics, and Properties),* Part I: Alkali Metals, Alkaline Earth Metals, Light Metals (Li, Na, K, Rb, Cs; Ca, Sr, Ba; Be, Mg, Al), by E. Fromm, H. Jehn, and G. Hörz. 30 pages.	DM 9.50
5—2 (1977):	*Gases and Carbon in Metals (Thermodynamics, Kinctics, and Properties),* Part II: Group II B to V B Metals (Zn, Cd; Ga, In, Tl; Si, Ge, Sn, Pg; Bi), by H. Jehn, E. Fromm, and G. Hörz. 40 pages.	DM 9.50
5—3 (1978):	*Gases and Carbon in Metals (Thermodynamics, Kinetics, and Properties),* Part III: Group III A Metals, Rare Earth Metals (Sc, Y; La, Ce, Pr, Nd, Sm, Eu, Gd, Tb, Dy, Ho, Er, Tm, Yb, Lu), by E. Fromm, H. Jehn and G. Hörz. 32 pages.	DM 14.—
5—4 (1978):	*Gases and Carbon in Metals (Thermodynamics, Kinetics, and Properties),* Part IV: Actinides (U, Th, Pu, Pa, Np, Am, Cm, Bk, Cf), by H. Jehn, E. Fromm, and G. Hörz. 34 pages.	DM 14.50
5—5 (1979):	*Gases and Carbon in Metals (Thermodynamics, Kinetics, and Properties),* Part V: Group IV A Metals (1): Titanium (Ti), by H. Jehn, E. Fromm, and G. Hörz. 84 pages.	DM 9.—
5—6 (1979):	*Gases and Carbon in Metals (Thermodynamics, Kinetics, and Properties),* Part VI: Group IV A Metals (2); Zirconium, Hafnium (Zr, Hf), by H. Jehn, H. Speck, E. Fromm, and G. Hörz. 80 pages.	DM 12.—
5—7 (1981):	*Gases and Carbon in Metals (Thermodynamics, Kinetics, and Properties),* Part VII: Group V A Metals (1): Vanadium (V), by G. Hörz, H. Speck, E. Fromm and H. Jehn. 70 pages.	DM 9.50
5—8 (1981):	*Gases and Carbon in Metals (Thermodynamics, Kinetics, and Properties),* Part VIII: Group V A Metals (2): Niobium (Nb), by G. Hörz, H. Speck, E. Fromm, and H. Jehn, 120 pages.	DM 16.—
5—9 (1981):	*Gases and Carbon in Metals (Thermodynamics, Kinetics, and Properties),* Part IX: Group V A Metals (3): Tantalum (Ta), by G. Hörz, H. Speck, E. Fromm, and H. Jehn. 78 pages.	DM 14.—
5—10 (1980):	*Gases and Carbon in Metals (Thermodynamics, Kinetics, and Properties),* Part X: Group VI A Metals (1): Chromium, Tungsten (Cr,W), by H. Jehn, H. Speck, E. Fromm, and G. Hörz. 66 pages.	DM 9.—
5—11 (1980):	*Gases and Carbon in Metals (Thermodynamics, Kinetics, and Properties),* Part XI: Group VI A Metals (2): Molybdenum (Mo), by H. Jehn, H. Speck, E. Fromm, and G. Hörz. 56 pages.	DM 9.—
5—12 (1980):	*Gases and Carbon in Metals (Thermodynamics, Kinetics, and Properties),* Part XII: Group VII A Metals: Manganese, Technetium, Rhenium (Mn, Tc, Re), by H. Jehn, H. Speck, E. Fromm, and G. Hörz. 32 pages.	DM 7.50
5—13 (1981):	*Gases and Carbon in Metals (Thermodynamics, Kinetics, and Properties),* Part XIII: Ferrous Metals (1): Iron-Hydrogen (Fe-H), by E. Fromm, H. Speck, H. Jehn, and G. Hörz. 108 pages.	DM 16.50
5—14 (1981):	*Gases and Carbon in Metals (Thermodynamics, Kinetics, and Properties),* Part XIV: Ferrous Metals (2): Iron-Carbon (Fe-C), by E. Fromm, H. Speck, H. Jehn, and G. Hörz. 122 pages.	DM 23.50
5—15 (1982):	*Gases and Carbon in Metals (Thermodynamics, Kinetics, and Properties),* Part XV: Ferrous Metals (3) Iron-Nitrogen (Fe-N), by E. Fromm, H. Jehn, W. Hehn, and H. Speck. 94 pages.	DM 17.50
6—1 (1976):	*Shapes of Beta Spectra,* by H. Behrens and L. Szybisz. 52 pages. A compilation of all measured beta spectrum shape factors.	DM 4.50
7—1 (1977):	*Compilation of Pion Photoproduction Data,* by D. Menze, W. Pfeil, and R. Wilcke. 318 pages. Tables of differential and integrated cross sections, recoil nucleon polarizations, and polarized photon and polarized target asymmetries for photon energies up to 18 GeV.	DM 19.—
8—1 (1977):	*Optical Properties of Some Insulators in the Vacuum Ultraviolet Region,* by. R.-P. Haelbich, M. Iwan, and E. E. Koch. XII, 186 pages. Tables of available optical properties and curves of absorption spectra for alkali halides and rare gas solids in the wavelength range 2000 to 2 Å.	DM 16.—

9—1 (1977):	*Bibliography of Microwave Spectroscopy. 1945—1975,* by B. Starck, R. Mutter, C. Spreter, K. Kettemann, A. Boggs, M. Botskor, and M. Jones. X, 664 pages.	DM 18.—
10—1 (1978):	*Graphs of Neutron Capture Cross Sections of Fission Product Isotopes from FPLIB-65/ENDF/B-IV,* by M. Mattes. XXII, 276 pages.	DM 26.50
11—1 (1978):	*Nucleon-Nucleon Scattering Data,* by J. Bystricky and F. Lehar. In two parts, 782 pages.	DM 62.—
—1 (Pt. I):	Part I, 202 pages.	
—1 (Pt. II):	Part II, 580 pages.	
11—2 (1981):	*Nucleon-Nucleon Scattering Data. Summary Tables,* 1981 Edition, by J. Bystricky and F. Lehar. 232 pages. This issue supersedes the previous edition published as No. 11—1 (Pt. I) (1978).	DM 34.40
11—3 (1981):	*Nucleon-Nucleon Scattering Data. Detailed Tables,* (Supplement 1), by J. Bystricky and F. Lehar. 132 pages. Supplement to No. 11—1 (Pt. II) (1978).	DM 29.—
12—1 (1979):	*Handbook of Pion-Nucleon Scattering,* by G. Höhler, F. Kaiser, R. Koch, and E. Pietarinen. 450 pages.	DM 35.—
13—1 (1979):	*Evaluation of the Cross-Sections for the Reactions ^{24}Mg (n,p) ^{24}Na, ^{64}Zn (n,p) ^{64}Cu, ^{63}Cu (n,2n) ^{62}Cu and ^{90}Zr (n,2n) ^{89}Zr,* by S. Tagesen, H. Vonach, and B. Strohmaier. 154 pages.	DM 16.—
13—2 (1980):	*Evaluation of the Cross-Sections for the Reactions $^{19}F(n,2n)^{18}F$, $^{31}P(n,p)^{31}Si$, $^{93}Nb(n,n')^{93m}Nb$ and $^{103}Rh(n,n')^{103m}Rh$,* by B. Strohmaier, S. Tagesen, and H. Vonach. 134 pages.	DM 16.—
13—3 (1981):	*Evaluation of the Cross-Sections for the Reaction $^{27}Al(n,\alpha)^{24}Na$,* by S. Tagesen and H. Vonach. 76 pages.	DM 14.—
14—1 (1979):	*Compilation of Data from Hadronic Atoms,* by H. Poth. 138 pages.	DM 8.50
15 (1979):	*Karlsruhe Charged Particle Reaction Data Compilation,* by H. Munzel, H. Klewe-Nebenius, J. Lange, G. Pfennig, and K. Hemberle. Loosleaf collection, XLVI, 1652 pages.	DM 149.—
—1	Entry 1—35. XIV, 762 pages.	
—2	Entry 36—80. XIV, 818 pages.	
—Index	Index to Entries 1—35 and 36—80. XVIII, 72 pages.	
16—1 (1979):	*The Oxygen Framework in Garnet and its Occurrence in the Structures of $Na_3Al_2Li_3F_{12}$, $Ca_3Al_2(OH)_{12}$, $RhBi_4$ and Hg_3TeO_6,* by E. Hellner, R. Gerlich, E. Koch, and W. Fischer. 40 pages.	DM 7.50
16—2 (1981):	*The Homogeneous Frameworks of the Cubic Crystal Structures,* by E. Hellner, E. Koch, and A. Reinhardt. 78 pages.	DM 45.—
17—1 (1981):	*Compilation of Experimental Values of Internal Conversion Coefficients and Ratios for Nuclei with $Z \leqq 60$,* by H. H. Hansen. 274 pages.	DM28.—
18—1 (1981):	*Optical Properties of Metals,* Part I: The Transition Met Metals ($0.1 \leqslant hv \leqslant 500$ eV) Ti, V, Cr, Mn, Fe, Co, Ni, Zr, Nb, Mo, Ru, Rh, Pd, Hf, Ta, W, Re, Os, Ir, Pt), by J. H. Weaver, C. Krafka, D. W. Lynch, and E. E. Koch. XII, 344 pages.	DM 33.—
18—2 (1981):	*Optical Properties of Metals,* Part II: Noble Metals, Aluminium, Scandium, Yttrium, the Lanthanides and the Actinides ($0.1 \leqq hv \leqq 500$ eV) (Cu, Ag, Au; Al; Sc; Y; La, Ce, Pr, Nd, Sm, Eu, Gd, Tb, Dy, Ho, Er, Tm, Yb, Lu; Th, U, Am), by J. H. Weaver, C. Krafka, D. W. Lynch, and E. E. Koch. XIV, 320 pages.	DM 35.—
21—1 (1982):	*Bibliography of Gas Phase Electron Diffraction 1930—1979,* by J. Buck, M. Maier, R. Mutter, U. Seiter, C. Spreter, B. Starck, J. Hargittai, O. Kennard, D. G. Watson, A. Lohr, T. Pirzadeh, H. G. Schirdewahn, and Z. Maier. 714 pages.	DM 65.—
22—1 (1981):	*Phenomenological Analyses of Nucleon-Nucleon Scattering,* by P. Kroll. 280 pages.	DM 32.50
23—1 (1982):	*Temperature Dependence of Thermal Neutron Scattering Cross Sections for Hydrogen Bound in Moderators,* by Jürgen Keinert, 135 pages.	DM 28.00

INDEX

A

B

C

D

E

F

G

H

I

J

K

L

M

N

O

P

Q

R

S

T

U

V

W

X

Y

Z

ATOMIC WEIGHTS, MELTING AND BOILING POINTS OF THE ELEMENTS

Name	Symbol	Atomic number	Atomic weight	Footnotes	Melting point (°C)	Boiling point (°C)
Actinium	Ac	89	227.028	L	1050	3200 ± 300
Aluminum	Al	13	26.9815		660.37	2467
Americium	Am	95	(243)		994 ± 4	2607
Antimony (Stibium)	Sb	51	121.75		630.74	1750
Argon	Ar	18	39.948	g, r	− 189.2	− 185.7
Arsenic	As	33	74.9216		817 (28 atm)	613 (sub)
Astatine	At	85	(210)		302	337
Barium	Ba	56	137.33	g	725	1640
Berkelium	Bk	97	(247)			
Beryllium	Be	4	9.01218		1278 ± 5	2970 (5 mm)
Bismuth	Bi	83	208.980		271.3	1560 ± 5
Boron	B	5	10.81	m, r	2079	2550 (sub)
Bromine	Br	35	79.904		−7.2	58.78
Cadmium	Cd	48	112.41	g	320.9	765
Caesium (Cesium)	Cs	55	132.905		28.40 ± 0.01	669.3
Calcium	Ca	20	40.08	g	839 ± 2	1484
Californium	Cf	98	(251)			
Carbon	C	6	12.011	r, t	3652 (sub)	t
Cerium	Ce	58	140.12	g	798 ± 2	3257
Cesium (Caesium)	Cs	55	132.9054		28.40 ± 0.01	669.3
Chlorine	Cl	17	35.453		− 100.98	−34.6
Chromium	Cr	24	51.996		1857 ± 20	2672
Cobalt	Co	27	51.9332		1495	2870
Copper (Cuprum)	Cu	29	63.546	r	1083.4 ± 0.2	2567
Curium	Cm	96	(247)		1340 ± 40	
Dysprosium	Dy	66	162.50		1409	2335
Einsteinium	Es	99	(252)			
Erbium	Er	68	167.26		1522	2510
Europium	Eu	63	151.96	g	822 ± 5	1597
Fermium	Fm	100	(257)			
Fluorine	F	9	18.9984		−219.62	− 188.14
Francium	Fr	87	(223)		(27)	(677)
Gadolinium	Gd	64	157.25	g	1311 ± 1	3233
Gallium	Ga	31	69.72		29.78	2403
Germanium	Ge	32	72.59		937.4	2830
Gold (Aurum)	Au	79	196.967		1064.434	3080
Hafnium	Hf	72	178.49		2227 ± 20	4602
Helium	He	2	4.00260	g	− 272.2$^{26\ atm}$	− 268.934
Holmium	Ho	67	164.930		1470	2720
Hydrogen	H	1	1.00794	g, m, r	− 259.14	− 252.87
Indium	In	49	114.82	g	156.61	2080
Iodine	I	53	126.905		113.5	184.35
Iridium	Ir	77	192.22		2410	4130
Iron (Ferrum)	Fe	26	55.847		1535	2750
Krypton	Kr	36	83.80	g, m	− 156.6	− 152.30 ± 0.10
Lanthanum	La	57	138.906	g	920 ± 5	3454
Lawrencium	Lr	103	(260)			
Lead (Plumbum)	Pb	82	207.2	g, r	327.502	1740
Lithium	Li	3	6.941	g, m, r	180.54	1342
Lutetium	Lu	71	174.967		1656 ± 5	3315
Magnesium	Mg	12	24.305	g	648.8 ± 0.5	1090
Manganese	Mn	25	54.9380		1244 ± 3	1962
Mendelevium	Md	101	(258)			
Mercury (Hydrargyrum)	Hg	80	200.59		−38.87	356.58
Molybdenum	Mo	42	95.94	g	2617	4612
Neodymium	Nd	60	144.24	g	1010	3127
Neon	Ne	10	20.1179	g, m	− 248.67	−246.048
Neptunium	Np	93	237.048	L	640 ± 1	3902
Nickel	Ni	28	58.69		1453	2732